THE NEW VISUAL NEUROSCIENCES

THE NEW VISUAL NEUROSCIENCES

Edited by John S. Werner and Leo M. Chalupa

Associate Editors:

Marie E. Burns

Joy J. Geng

Mark S. Goldman

James Handa

Andrew T. Ishida

George R. Mangun

Kimberley McAllister

Bruno A. Olshausen

Gregg H. Recanzone

Mandyam V. Srinivasan

W. Martin Usrey

Michael A. Webster

David Whitney

THE MIT PRESS

CAMBRIDGE, MASSACHUSETTS

LONDON, ENGLAND

MIT Press books may be purchased at special quantity discounts for business or sales promotional use. For information, please email special_sales@mitpress.mit.edu or write to Special Sales Department, The MIT Press, 55 Hayward Street, Cambridge, MA 02142.

This book was set in Baskerville by Toppan Best-set Premedia Limited.
Printed and bound in the United States of America.

Library of Congress Cataloging-in-Publication Data

The new visual neurosciences / edited by John S. Werner and Leo M. Chalupa.
  pages   cm
Includes bibliographical references and index.
ISBN 978-0-262-01916-3 (hardcover : alk. paper)
1. Vision.   2. Visual pathways.   3. Visual cortex.   4. Neurosciences.   I. Werner, John Simon, editor of compilation.   II. Chalupa, Leo M., editor of compilation.
QP475.N493   2014
612.8'4—dc23
2013002297

10   9   8   7   6   5   4   3   2   1

CONTENTS

X CORTICAL MECHANISMS OF ATTENTION, COGNITION, AND
MULTIMODAL INTEGRATION

PREFACE

This book covers the extraordinary range of contemporary visual neuroscience, from molecules and cell assemblies to systems and therapy. Nowhere in the neurosciences has progress been so rapid as in the vision sciences. A book of this scope, which is even broader than the original *Visual Neurosciences* assembled 10 years ago, requires expert advice in many specialties even though the final product is intended for nonspecialists. We have been fortunate in enlisting the help of associate editors, authorities in various areas of vision science, with whom we have enjoyed a smooth working relationship through a variety of common bonds with the University of California at Davis. They reviewed chapters and solicited expertise from a large number of reviewers. We are grateful to all these colleagues for helping to bring this book to completion.

We thank Susan Garcia and Rieko Ringo for administrative support and for laboring long hours beyond their normal work day. We are also grateful to Laura Leming, Cameron Blount, and Grace Dell'Olio. In addition, the editorial staff at the MIT Press were not only cordial task masters, but reliable, supportive, and professional, especially Robert Prior, Susan Buckley, and Katherine Almeida. They provided valuable guidance at all stages. Finally, we thank the authors for adhering to tight deadlines in order to produce a book that is both up to date and of archival value, a work that we hope will inspire readers who are only beginning their quest in the visual neurosciences and those established scientists who wish to broaden the scope of their interests in *The New Visual Neurosciences*.

1 A Decade of Progress and New Directions in the Visual Neurosciences

JOHN S. WERNER AND LEO M. CHALUPA

Even Charles Darwin was initially overwhelmed by the enormous variation among species in the manner in which the seemingly effortless task of vision is accomplished. In all species, vision begins with a cascade of molecular reactions in response to the absorption of electromagnetic radiation, but so varied are the "inimitable contrivances," wrote Darwin (1859), that the thought they "could have been formed by natural selection, seems, I freely confess, absurd to the highest possible degree" (p. 186). But Darwin overcame his own skepticism, perhaps aided by the realization that, despite a long evolutionary history, the eye is not perfect and is still subject to protracted processes of natural selection.

The visual system has attracted the curiosity of many other great scientists as well. In most cases, understanding vision was not only a challenge in its own right but also a test bed for larger fundamental principles. Isaac Newton (1704), for example, after working out the laws of motion and the planetary bodies and inventing the calculus, launched the modern era of vision science with his discovery of the nature of light. His insight that "the Rays to speak properly are not coloured" (p. 124) led to the realization that the proper understanding of vision is not to be found in the nature of light but in the nature of the visual pathways. He then went on to discover the partial decussation of the human visual pathways. One hundred years later another great physicist credited with unifying our understanding of electricity and magnetism, James Clerk Maxwell (1872), concluded from his experiments on color that vision science is "essentially a mental science" (p. 261). To the list of extraordinary scientists who facilitated the visual neurosciences, we could add the names of: Thomas Young, Jan Evangelista Purkinje, Charles Wheatstone, David Ferrier, Johannes Müller, Hermann von Helmholtz, Ewald Hering, Ernst Mach, and countless others.

The rapid growth of the vision sciences and the extraordinary interdisciplinary nature of the effort made it necessary to create anthologies even in the early twentieth century. The first and most important was published in 1909 as Helmholtz's third edition of the *Handbuch der physiologischen Optik* (Helmholtz, 1909), which included not only his own writings from the first edition but also those of Gullstrand, von Kries, and Nagel. It would be many years later before edited volumes would follow, but they did so with increasing frequency as the field progressed and expanded in scope. Some that have shaped our own education include Graham's (1965) *Vision and Visual Perception* and the many volumes of *The Handbook of Sensory Physiology* (Autrum et al., 1971–1984). Other volumes integrated vision science with various areas of neuroscience such as the workshops held in Boulder, culminating in a series called, *The Neurosciences* (for example, Schmitt & Worden, 1979). The integration of visual neurophysiology and perception in a manner that could be accessible to students led to *Visual Perception: The Neurophysiological Foundations* (Spillmann & Werner, 1990). All of these volumes remain important, not just for their rich histories but for many gems lying in wait to be rediscovered and as brilliant examples of how to think about vision.

Ten years ago we assembled summaries of all areas of visual neuroscience from leaders in the field who covered the state of the art, which had by then expanded to include molecular processes, extraclassical receptive fields, and excitement about extrastriate cortex, which together with striate cortex was understood to include more than one-third of the primate brain and promised to link visual cognition to visual neurophysiology. The state of our understanding of these processes was summarized in two edited volumes called *The Visual Neurosciences* (Chalupa & Werner, 2004). It provided a comprehensive set of reviews of the field by leading international authorities in language intended for nonexperts and beginning graduate students. We could not have anticipated that, in only one decade, many of these areas would advance so fast and so far as to require much more than an update, because there is so much that is new.

The New Visual Neurosciences is a comprehensive review of the vast progress of the past decade. Most of the authors are new, and a few chapters are updated, but many cover entirely new topics. The chapters in the previous volume should be consulted along with this one, as they contain an essential body of information that is only partially repeated here.

Chapters 2–15 of the section on *Retinal Mechanisms and Processes* describe how photoreceptor signals are processed in the retina and compressed for transmission to other brain areas by some 1 million optic nerve fibers. Much of what is known about the structure of cells in the retina is owed to the Golgi stain so effectively exploited by Ramón y Cajal (1892/1972), but remarkable new techniques described in this section such as optogenetics, targeted fluorescent markers, and trans-synaptic viral labeling are revolutionizing our understanding of anatomical organization and are aided further by connectomics and multielectrode arrays that permit ensembles of cellular interactions and outputs to be discovered, quantified, and manipulated at a level that is without precedent. Topics included here that were not covered in the first edition of this book include: connectomics, gap junctions, retinal feature detection, correlated activity, intrinsically photosensitive ganglion cells, and postreceptoral adaptation.

Central mechanisms of vision encompass chapters 16–32, divided into sections on *Organization of Visual Pathways, Subcortical Processing,* and *Processing in Primary Visual Cortex.* They collectively provide an up-to-date account of our understanding of the functional organization of neural pathways that interconnect visual areas in the brain, the circuit mechanisms responsible for the generation of emergent receptive-field properties, and the influence of brain state and behavior on visual processing in early visual areas. Our understanding of these topics has grown since the discovery of what is now known as the classical receptive field (Barlow, 1953; Kuffler, 1953). The effort has benefited tremendously from the increasing use of alert monkeys for studying early visual processing as well as the addition of the genetically modifiable mouse as a major model system for understanding vision. Specific topics included here that were not covered in the first edition of this book include cortical connectomics, the role of inhibition and other identified local circuits for visual processing, the dynamic properties of neural circuits for vision, and feedback from cortex to lateral geniculate nucleus. Indeed, all of these areas are shaped by feedback from other areas, cortical and subcortical, making it possible for top-down influences to shape our visual perception by attention and prior learning.

The sections on *Brightness and Color* (chapters 33–41), *Pattern, Surface, and Shape* (chapters 42–47), *Objects and Scenes* (chapters 48–52), and *Time, Motion, and Depth* (chapters 53–58) highlight the extension of analyses of visual coding to higher stages of visual processing and to more ecologically relevant contexts where analyses of the properties of natural images continue to play an important role. This can be seen in the way subjects such as color and brightness constancy have changed to the newly added sections in this volume on topics such as face and scene perception. This myriad of visual recognition tasks is normally taken for granted but still can be accomplished to only a limited extent by artificial vision systems. More than half of the chapters in the pattern, object, scene, and motion sections are new to this work. The remaining chapters are complete revisions, focusing entirely on the last 10 years of research. The first edition provides a basic foundation that is not repeated in the new chapters. The revised and new chapters fill in crucial advances, linking low-level pattern and texture analysis to surface representation, object recognition, and, finally, scene processing. Entirely new subfields are represented in the new book, for example, on time perception. The chapters cover a wider range of techniques including psychophysics, functional imaging, and brain stimulation, and they introduce new avenues of investigation on midlevel vision and spatiotemporal perception.

A comprehensive coverage of all forms of *Eye Movements* is provided by chapters 59–68. Natural vision is an active and dynamic process that depends on eye movements to stabilize the image of objects on the retina and to direct the eyes to regions of high visual salience. These operations are carried out by complex subcortical and cortical circuits, and the dialogues between eye movements and sensory systems are essential in disambiguating signals generated by the motion of the eye itself from those present in the external world. The opening chapter of this section reviews the role of eye movements in natural behavior. Subsequent chapters describe the eye's orbital mechanics and the circuitry underlying eye movements and target selection. The final chapter reviews the neurology of eye movements and how models of the eye movement circuitry and cellular neurophysiology are being used to diagnose and treat eye movement disorders.

In this edition several chapters investigate how visual information is integrated with other sensory systems as well as the motor system in order to generate visually guided movements. How the parietal lobe and other traditionally considered multisensory cortical areas are involved in complex processing is now covered in chapters related to visual-auditory interactions and visually guided movements. These chapters are embedded with related chapters that comprise the section on *Cortical Mechanisms of Attention, Cognition, and Multimodal*

   JOHN S. WERNER AND LEO M. CHALUPA

Integration (chapters 69–78). The last decade has seen an explosion of work in the cognitive neurosciences, spurred largely by advances in methods that measure ongoing activity in the human and nonhuman primate brain during complex awake behaviors. Leaders in the field describe signature accomplishments in understanding of how extrasensory processing and multimodal integration contribute to visual perception in the earliest stages of sensory processing and the control mechanisms that support these changes.

Invertebrate Vision (chapters 79–85) was not included in the previous edition, but it has an important place in our description of neural mechanisms mediating vision. Because of the relatively small numbers of neurons in their visual pathways (compared to vertebrates) and the relative ease with which their electrophysiological responses can be recorded (by virtue of their large size, especially at the peripheral levels of the visual pathways), invertebrates have spearheaded our understanding of a number of fundamental principles of visual processing. Examples include the first investigations of phototransduction and photoreceptor physiology—in the horseshoe crab, *Limulus* (Fuortes, 1959), the first discovery of lateral inhibition and of its role in sharpening visual contours as described perceptually by Mach—again in *Limulus* (Hartline & Ratliff, 1957), and the first quantitative characterization of visual motion detection—in the beetle *Chlorophanus* (Hassenstein & Reichardt, 1956). Invertebrates continue to fascinate and challenge, with chapters from this volume elucidating, for example, the perception of polarized light (a domain of vision that lies entirely outside our own) and its use in navigation and the color vision of mantis shrimps, which possess over a dozen spectral classes of photoreceptors. This section is closely linked with the following section on *Theoretical Perspectives.*

Theoretical Perspectives (chapters 86–89) presents some of the prevailing theories regarding the evolution of the visual system and the principles that underlie its function. The opening chapter in this section describes the features of the primate visual system that distinguish it from those of other mammals, and it provides a possible account for how these features evolved and how they became diversified as an adaptation to variations in ecological niche and primate lifestyles. Two related theoretical frameworks are provided in chapters on the neural computations underlying visual perception. One attempts to account for aspects of neural coding and representation in terms of their relation to natural scene statistics, and the other attempts to show how visual systems learn and exploit these statistics for inferential computations. There is much new work on these topics since the last edition that is emphasized in both

chapters. An issue that has been a subject of much theoretical speculation and debate in visual neuroscience is the role that neural oscillations play in information processing. The past decade has seen a resurgence of interest and study on this topic, and the final chapter in this section provides an up-to-date account of the latest experimental and theoretical developments.

Chapters 90–100, *Molecular and Developmental Processes,* describe visual system development from retina to visual cortex. These chapters cover the wide range of events that constitute neural development, from prenatal events such as neurogenesis and neuronal migration through molecular specification of cell types, formation of connections, development of functional properties, and activity-dependent plasticity. Many of these topics were covered in the first edition, but the dramatically increased depth of the science included here illustrates the significant advances over the past 10 years in this field in understanding the molecular mechanisms that mediate the development of the visual system. New molecular and genetic techniques have revealed unexpected complexity in cell types within the retina and equally complex mechanisms to specify the proper connections among those cell types. The standardization of the mouse visual system for studying development and plasticity has pushed the field forward, allowing major advances in determining the learning rules that govern activity-dependent plasticity as well as in identifying central molecular pathways that mediate this plasticity. Combining the now widespread availability of knockout mice with cutting-edge genomic and imaging techniques has also provided critical insight into how projections find their targets at multiple levels within the visual system. These advances have identified many classes of new molecules involved in visual system development, including immune molecules and glial-derived factors. Finally, these accomplishments in understanding the rules that regulate visual system development have facilitated truly remarkable and groundbreaking advances in understanding and promoting optic nerve regeneration, reviewed here.

The final section of the book, *Translational Visual Neuroscience* (chapters 101–112) is closely related to the preceding section on development. The previous edition did not cover applied research, but progress in the field now requires a separate translational section. Its inclusion here is because of the amazing scientific progress in novel treatment modalities, many of which are made possible only by building on equally remarkable progress in basic science that has directed these new understandings to attack ocular disease. Disease processes begin at the molecular level and can sometimes be addressed by controlling molecular-developmental processes. A number of the molecular chapters could have been placed in the translational section and vice versa, reflecting molecular insights to specific diseases and revolutionary new treatments that will guide future research on understanding and curing major blinding diseases.

The chapters in the last two sections are not the only ones that span multiple sections. A number of basic science chapters have implications for other areas of science as well as clinical applications. Borders between research endeavors are rapidly eroding. For this reason, a bit of redundancy across chapters is inevitable and pedagogically valid when it provides an opportunity to look at phenomena from different perspectives. The scope of the overall effort to understand visual processing has intensified and broadened in the past decade, reflecting a reliance on traditional methods in combination with new technologies, ranging from nanoparticles to the functional mapping of the human brain. Much work lies ahead. We look forward to the next decade with justifiable optimism that basic and translational discoveries will resolve many more previously unanswerable questions.

REFERENCES

Autrum, H., Jung, W. R., Loewenstein, W. R., MacKay, D. M., & Teuber, H. L. (Eds.). (1971–1984). *Handbook of sensory physiology*. Berlin: Springer Verlag.

Barlow, H. B. (1953). Summation and inhibition in the frog's retina. *Journal of Physiology, 119*, 69–88.

Cajal, S. Ramón y. (1892). *La rétine des vértebres (La Cellule)*, English trans. Thorpe, S., & Glickstein, M., 1972. Springfield, IL: Charles C. Thomas.

Chalupa, L. M., & Werner, J. S. (Eds.). (2004). *The visual neurosciences*. Cambridge, MA: The MIT Press.

Darwin, C. (1859). *On the origin of species by means of natural selection, or the preservation of favoured races in the struggle for life*. London: Murray.

Fuortes, M. G. F. (1959). Initiation of impulses in visual cells of *Limulus. Journal of Physiology, 148*, 14–28.

Graham, C. H. (Ed.). (1965). *Vision and visual perception*. New York: John Wiley & Sons.

Hartline, H. K., & Ratliff, F. (1957). Inhibitory interaction of receptor units in the eye of *Limulus. Journal of General Physiology, 40*, 357–376.

Hassenstein, B., & Reichardt, W. (1956). Systemtheoretische Analyse der Zeit-, Reihenfolgen- und Vorzeichenauswertung bei der Bewegungsperzeption des Rüsselkäfers *Chlorophanus. Zeitschrift für Naturforscsung, 11b*, 513–524.

Helmholtz, H. v. (1909). *Handbuch der physiologischen Optik* (3rd ed.). Leipzig: Leopold Voss.

Kuffler, S. W. (1953). Discharge patterns and functional organization of mammalian retina. *Journal of Neurophysiology, 16*, 37–68.

Maxwell, J. C. (1872). On color vision. *Proceedings of the Royal Institution of Great Britain, 6*, 260–271.

Newton, I. (1704). *Opticks: or, a treatise of the reflections, refractions, inflexions and colours of light*. London: Smith and Walford [based on the Dover Publications, fourth edition, 1952]

Schmitt, F. O., & Worden, F. (1979). *The neurosciences: Fourth Study Program*. Cambridge, MA: The MIT Press.

Spillmann, L., & Werner, J. S. (1990). *Visual perception: The neurophysiological foundations*. New York: Academic Press.

I RETINAL MECHANISMS AND PROCESSES

2 Visual Transduction by Rod and Cone Photoreceptors

MARIE E. BURNS AND EDWARD N. PUGH, JR.

Vision in vertebrates begins in rod and cone photoreceptors, which transduce light signals into electrical responses. Exquisitely sensitive rods and nonsaturating cones govern vision over the ~9 $\log_{10}$ diurnal variation of earth's illumination (figure 2.1A). In this review we summarize our current understanding of rod and cone signaling.

RODS SIGNAL SINGLE PHOTONS BUT SATURATE, WHEREAS CONES ESCAPE SATURATION

Rods are specialized for signaling single photons. The lower limit of visual detection or "absolute thresholds" of both human and mouse rods correspond to fewer than 70 photons measured at the cornea (Hecht, Shlaer, & Pirenne, 1942; Naarendorp et al., 2010). Such threshold stimuli are estimated to yield to 10 to 25 photons absorbed by rhodopsin molecules in a retinal region (depending on the nominal image size) containing many hundreds to thousands of rods. Thus, at absolute threshold no rod captures more than 1 photon. In mice, as in humans, the absolute threshold is limited by retinal "dark light," as proposed long ago (reviewed in Stiles & Crawford, 1932). The behaviorally measured dark light of mice as measured in increment threshold experiments (figure 2.1B) corresponds precisely to the rate (0.012 R^* rod^{-1} s^{-1}, where R^* are molecules of activated rhodopsin) of spontaneous (thermal) activation of rhodopsin measured in recordings from mouse rods (Burns et al., 2002; Fu et al., 2008), confirming that such spontaneous events are the limiting source of noise in the dark adapted visual system (Barlow, 1956). Rhodopsin apparently evolved under great selective pressure to minimize this spontaneous activation, as the rate 0.012 R^* rod^{-1} s^{-1} corresponds (in a mouse rod with 7×10^7 rhodopsin molecules) to an intrinsic rate of 1.7×10^{-10} s^{-1} per molecule. In other words, an individual rhodopsin molecule would undergo spontaneous activation in ~200 years. Notably, in mice the equivalent intensity of the

"rod dark light" produced by spontaneous R^* activation is about fivefold lower than the average illuminance of the earth produced by starlight (figure 2.1A). Thus, the contrast of the light reflected from objects in starlight should be sufficient to enable discrimination of objects over the dark light produced by spontaneous R^* activation.

Visual threshold is mediated exclusively by rods over about 3–4 $\log_{10}$ units of background intensity above the dark light (figure 2.1B). At still higher intensities, both rods and cones operate in concert, with cones becoming more sensitive than rods to an incremental flash at background intensities in mice producing ~100 R^* rod^{-1} s^{-1}. Once cones become more sensitive, rods are increasingly driven toward saturation, with saturation occurring at photon capture rates exceeding 5,000 R^* s^{-1} (Aguilar & Stiles, 1954; Naarendorp et al., 2010; Nakatani, Tamura, & Yau, 1991; Sharpe, Fach, & Stockman, 1992; Umino, Solessio, & Barlow, 2008).

The functional property that most thoroughly differentiates cones from rods is that cones cannot be driven into saturation by steady light, no matter how intense (Burkhardt, 1994; Shevell, 1977). This remarkable property of cones is achieved by three mechanisms: a superior ability to adapt to steady light, negligible dark light contributed by bleached opsin photoproducts, and a faster rate of regeneration of 11-*cis*-retinal (Lamb & Pugh, 2004; Mahroo & Lamb, 2004; Wang & Kefalov, 2011). These mechanisms combine in cones to enable them to function at steady light levels that bleach large fractions of their pigment. At such levels sensitivity to small changes in contrast will be inversely proportional to the background regardless of its intensity, guaranteeing escape at all higher intensities.

The distinct functional properties of rods and cones allow the retina to function over the vast range of diurnal illumination yet are achieved with cellular structures and biochemical mechanisms that are notably similar, as now discussed.

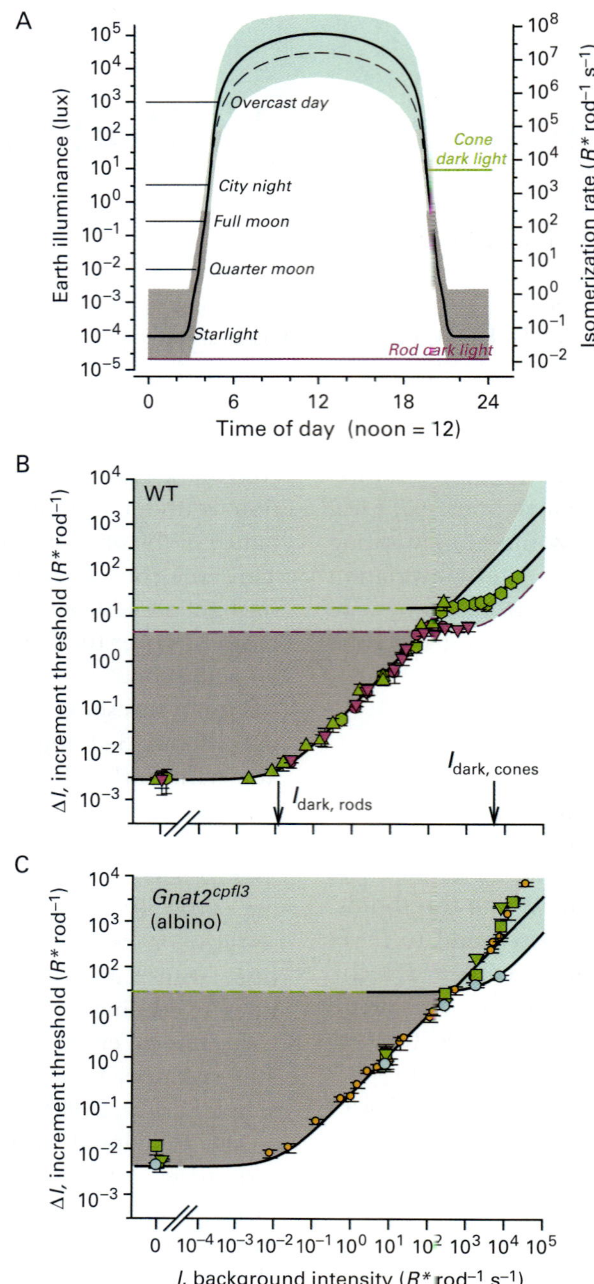

FIGURE 2.1   Earth diurnal illumination and mouse visual sensitivity. (A) Average illuminance (in photopic lux, solid curve, left ordinate) and corresponding mouse rod photoisomerization rate (dashed curve, right ordinate), which account for pupil constriction as a function of the time of day. To derive the maximal photopic illuminance, the spectral distribution of the sunlight maximum of the average latitude of the 48 contiguous United States (ASTM International; G173-03, 2008) was integrated between 280 nm and 780 nm against the Judd-Vos photopic luminosity function (V_λ), giving a maximum of 109,990 lux at noon. The diurnal variation of the illuminance was taken from a report by Newport Corporation (http://www.newport.com/ Introduction-to-Solar-Radiation/411919/1033/content.aspx) and applies between 5 h (5 AM) and 19 h (7 PM) on the graph. The named lower illuminance levels ("starlight," "quarter moon," etc.) were taken from the Wikipedia entry for "Lux" (http:// en.wikipedia.org/wiki/Lux) and used to produce a smoothed 24-h curve. Photoisomerization rates were calculated by converting the illuminance spectrum to scotopic lux, then to luminance (scotopic cd m^{-2}) assuming that the illuminance arises from a uniform sky (ganzfeld) and using the formula $L = \pi E$, where L is the luminance and E the illuminance (Wyszecki & Stiles, 1982, table I [4.4]), and finally to retinal illuminance and rod isomerization rates using mouse pupil data, as described in Naarendorp et al. (2010). The shaded regions bracketing the solid curve indicate the illuminances over which the visual system can extract contrast from roughly 20-fold below the mean to 10-fold above, with a color scheme indicating regions of pure rod function (gray), rod + cone function (gray-green), and pure cone function (blue-green). (B) Mouse incremental thresholds for brief flashes of 365 nm (purple symbols) and 500 nm (green symbols), with shaded regions illustrating the backgrounds/ increments in which rods alone govern threshold (gray), rod + cone (gray-green), and pure cone (blue-green). Note that the most intense background used in these experiments corresponds to daylight of about 6 AM and 6 PM in panel A. The rod and cone dark light levels are extracted from the increment threshold data. (C) Increment thresholds of WT albino mice (blue symbols) and albino mice lacking functional cones (green symbols), plotted along with data of a human rod monochromat (orange circles): The upward deviation of the data from the "Weber line" is the behavioral manifestation of rod saturation but also demonstrates that rods can generate detectable signals in the presence of backgrounds that produce in excess of 10^4 R^* rod^{-1} s^{-1} if the increment flash is sufficiently bright. Panels B and C adapted from Naarendorp et al. (2010).

PHOTORECEPTOR STRUCTURE IN RELATION TO FUNCTION

The ultrastructure of photoreceptors is remarkably similar across the vertebrate phylum. Photoreceptors are polarized neurons comprising four primary functional domains: (1) the outer segment, where photon capture and phototransduction occur; (2) the inner segment, which houses mitochondria, endoplasmic reticulum, and Golgi bodies, serving as the main hub of protein and membrane biosynthesis and cell metabolism; (3) the nuclear region; and (4) the synaptic terminal, which tonically releases via ribbon-based machinery the neurotransmitter glutamate in the dark onto postsynaptic bipolar and horizontal cells.

Outer Segment

The outer segments of rods are cylindrical in form, whereas those of cones of many vertebrate species are slightly conical. The outer segments are elaborated sensory cilia densely packed with disk membranes spaced at ~30 nm; the disks and their interstices house the proteins and messengers of phototransduction. In many vertebrate species, the disks of cones are continuous with the plasma membrane, allowing lateral diffusion of membrane proteins along the length of the outer segment. The situation is dramatically different in all vertebrate rods, in which the plasma membrane and disk membranes are not only physically separate but also contain distinct protein machinery. One consequence for rods is that renewal of most of the protein and lipid of the disks requires old disks to be shed from the distal tip of the outer segment and phagocytosed by pigment epithelium cells, with continual generation of new membranes at the base of the outer segment (see chapter 4 by Ruggiero and Finnemann). Renewal and shedding also occur in cones; however, as much of the entire outer segment membrane is continuous, membrane components of different ages can redistribute along the length of the cone outer segment through lateral diffusion.

In the dark, a small percentage of the cGMP-gated channels of the plasma membrane of the outer segment are maintained in the open state by cGMP, producing an inward cation current that keeps the membrane potential relatively depolarized (V_m = –40 mV). The light-triggered decrease in cGMP concentration leads to closure of the channels with submillisecond delay as a combined result of very rapid radial diffusion and extremely fast gating of channels by cGMP (Karpen et al., 1988). The concentration dependence of channel opening is cooperative, with a Hill coefficient of $n_{chan} \approx$ 3 and a dissociation constant around $K_{chan} \approx 20$ μM

cGMP. Thus, the fraction of channels in a patch of outer segment membrane exposed to a cGMP concentration cG obeys

$$\frac{N_{open}}{N_{total}} = \frac{cG^{n_{chan}}}{cG^{n_{chan}} + K_{chan}^{n_{chan}}} \qquad (2.1)$$

The value of K_{chan} decreases modestly when Ca^{2+} declines in rods and more strongly so in cones (see Calcium Feedback, below). The overall fraction of open channels in the dark-adapted rod is small: for cG ~ 4 μM, K_{chan} = 20 μM and n_{chan} = 3, ~1% of the channels are open.

The inward cGMP-gated current is balanced by an outward flow of K^+ ions through leak and voltage-gated K^+ channels of the inner segment and synaptic terminals, giving rise to the circulating, or dark current. A photon captured by an opsin in the outer segment locally closes cGMP-gated channels, suppressing a portion of the circulating current, which hyperpolarizes the membrane.

Inner Segment

Joining the outer segment to the inner segment is a narrow connecting cilium with a 9+0 arrangement of microtubules. All proteins destined for the outer segment must pass through or along this narrow connection. For discussion of the mechanisms by which the proteins are transported to the outer segment, please refer to chapter 3 by Baehr and colleagues.

The distalmost portion of the inner segment (the *ellipsoid*) contains a high density of mitochondria that supply ATP to support protein and membrane synthesis occurring in the inner segment, as well as supplying ATP via diffusion to the metabolically demanding outer segment. The high concentration of relatively dense mitochondria and other material in the ellipsoid also serves an optical purpose: This region has a higher refractive index and, because the inner segment is typically wider than the outer, helps to funnel photons into the outer segment, increasing the effective light-collecting area of the cells. This phenomenon is especially marked in cones and gives rise to the Stiles-Crawford directional sensitivity effect. Elegant investigations with adaptive optics imaging have revealed the structure of human rods and cones in vivo, in part due to the refractive boundaries between outer and inner segments (e.g., Doble et al., 2011; Dubra et al., 2011), and directly reveal their light funneling.

Synaptic Terminal

The synaptic terminals of rods and cones both utilize ribbon synapses to support the continuous release of glutamate in the dark. The overall cytoarchitecture varies considerably, with rods terminating in small spherical lobes called spherules and cones ending in large club-like endings termed pedicles (see chapter 5 by Sampath and chapter 6 by DeVries). The spherules in mammalian rods contain one or two ribbons and support release rates of up to a few dozen synaptic vesicles per second in the dark (Sheng et al., 2007), a rate limited by the replenishment of vesicles. In contrast, cone pedicles have a dozen or more ribbons that support much higher rates of release, up to several hundred vesicles per second in the dark (Jackman et al., 2009). The nerve terminals of rods and cones must modulate the rate of vesicle fusion to support synaptic transmission across a broad range of light intensities (figure 2.1) and are therefore highly dynamic (Mercer & Thoreson, 2011). Overall, a variety of mechanisms at different levels of the photoreceptors contribute to the ability of photoreceptors to signal over such a large range: several outer segment gain-control mechanisms that reduce the biochemical and electrical response per photon; hyperpolarization-activated cation channels (HCN1) expressed in the inner segment and cell body region, which tend to restore depolarization in steady light (e.g., Seeliger et al., 2011); and horizontal cell feedback at the synapse, which down-regulates intraterminal Ca^{2+} and vesicle release rates (see chapter 7 by Kramer).

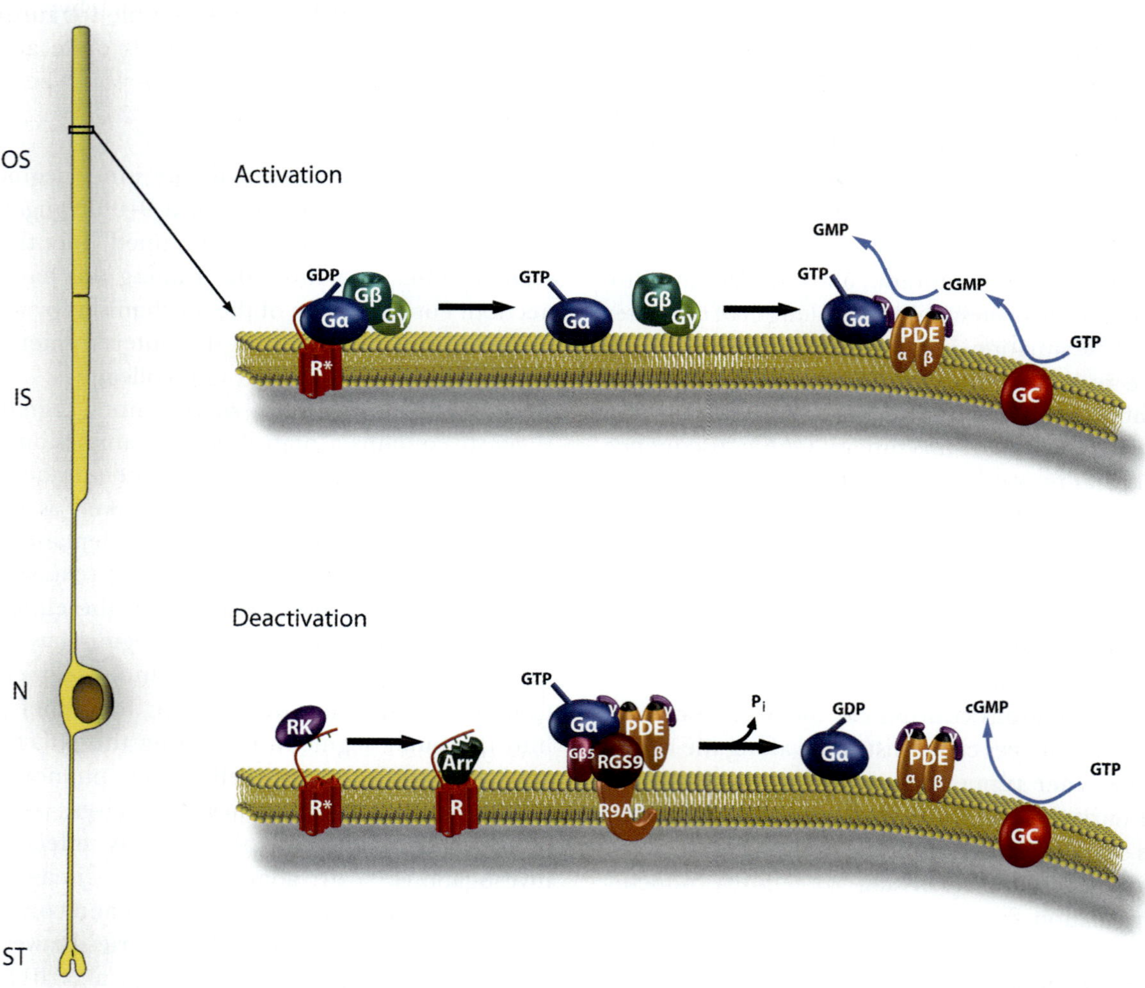

FIGURE 2.2 Rod phototransduction. (Left) Photoreceptor schematic illustrating the outer segment (OS), inner segment (IS), nuclear region (N), and synaptic terminal (ST). (Right) Phototransduction activation and deactivation reactions of the outer segment disks. Photoexcited rhodopsin (R*) activates the G-protein, and the G-protein α subunit activates PDE. cGMP produced by guanylate cyclase (GC) is hydrolyzed by PDE, causing cGMP channels to close (not shown). Deactivation of R* requires phosphorylation by rhodopsin kinase (RK) followed by arrestin (Arr) binding. The G-protein/PDE complex is deactivated by the RGS9-1·Gβ5-L·R9AP complex, which accelerates GTP hydrolysis. cGMP synthesis by GC restores cGMP to its dark level. (Reprinted with permission from Arshavsky & Burns, 2012.)

ACTIVATION OF PHOTOTRANSDUCTION

Molecular Mechanisms of Activation

The biochemical reactions that initiate vertebrate phototransduction are shown schematically in figure 2.2. After the capture of a photon by its 11-*cis*-retinal chromophore, an opsin undergoes a major conformational change into an enzymatically active state (R*) that catalyzes the activation of many copies of the G-protein transducin (to G*), and each G* in turn activates cyclic GMP phosphodiesterase (PDE) to its active form (E*). The E* hydrolyzes cGMP, causing the cytoplasmic concentration of cGMP to decrease and thereby leading to closure of cGMP-gated cation channels in the plasma membrane. Closure of these channels has the dual effect of generating the photoreceptor's electrical response (by reducing the circulating current and hyperpolarizing the membrane voltage) and of causing a reduction in cytoplasmic Ca^{2+} concentration. The molecular events of activation have been thoroughly investigated, and detailed summaries are available describing how they combine to produce the overall amplification of the early rising phase of the light response in amphibian rods (Leskov et al., 2000; Pugh & Lamb, 1993). However, such analysis of amplification requires that all deactivation of cascade components be negligible over the time period considered, and recent evidence has shown that in mammalian rods and cones deactivation and feedback begin extremely early, in less than 40 ms, and thus largely concurrently with activation (reviewed in Burns & Pugh, 2010). Thus, it is nearly impossible to experimentally separate activation and deactivation phases of the flash response in mammalian photoreceptors.

DEACTIVATION OF PHOTOTRANSDUCTION

Because the temporal resolution of the visual system is necessarily limited by that of its photoreceptors, it is important that the photoreceptor response be as brief as possible while maintaining signal amplification sufficient for reliable propagation through the retina. What then determines how long the photoreceptor's response to a brief flash of light persists?

Molecular Mechanisms of Rhodopsin Deactivation

Deactivation of R* is arguably the most important event in the recovery of the photoresponse because as long as an R* has catalytic activity, an amplified response persists that will mimic the presence of light and contribute to desensitization and/or saturation of the cell.

The first step in shutoff, phosphorylation of R*'s C-terminal residues, is performed by rhodopsin kinase, GRK1 (Kuhn & Wilden, 1987). Phosphorylation of rhodopsin reduces R*'s catalytic activity (Arshavsky et al., 1985; Wilden, Hall, & Kuhn, 1986) and is essential for normal recovery of the flash response in rods (C. K. Chen et al., 1999; J. Chen et al., 1995).

The C-terminal domain of rhodopsin has multiple serine and threonine residues serving as potential phosphorylation sites, six in mice and seven in humans. Experiments on transgenic mouse rods support the hypothesis that multiple phosphorylation is essential for rapid recovery of the single-photon response (Doan et al., 2006; Mendez et al., 2000). Mutant rhodopsins containing fewer than three of the potential phosphorylation sites display very slowed first-order deactivation kinetics, suggesting that under such conditions a single, stochastic process is responsible for turning off R*. Stochastic shutoff of this kind is quite abnormal because rhodopsin deactivation in wild-type rods is much more reproducible. Overall, these results indicate that reproducible rhodopsin deactivation involves at least three phosphorylations of R* in rods. Despite the need for multiple phosphorylation, R* deactivation is nevertheless very rapid, as the average integrated activity (effective lifetime) of R* persists for only ~40 ms (Gross & Burns, 2010; Gross, Pugh, & Burns, 2012a).

Rapid deactivation of cone R* is also likely to involve its phosphorylation. In many species, cone R* phosphorylation is probably mediated by GRK7, a cone-specific isoform of opsin kinase. GRK7 was first cloned from ground squirrel (Weiss et al., 1998), and orthologues have since been identified in many other vertebrates. Work with fish has shown that GRK7 is expressed in cones ~10-fold higher than GRK1 in rods, and that GRK7 can phosphorylate R* much more rapidly than GRK1 (Tachibanaki et al., 2005; Vogalis et al., 2011). Curiously, rodent cones express only GRK1. Both mouse M- and S-cone opsins require GRK1 for normal deactivation (Lyubarsky et al., 2000; Nikonov et al., 2005; Zhu et al., 2006), and GRK1 phosphorylates both M- and S-opsin (Zhu et al., 2003).

Following R* phosphorylation, the cytoplasmic protein arrestin (Arr1) binds with high affinity to phosphorylated rhodopsin (Wilden, Hall, & Kuhn, 1986). In mouse rods lacking Arr1 (Xu et al., 1997), the dim flash response rises to a peak amplitude comparable to that of normal rods but then recovers only half-way back to baseline. The presumably phosphorylated R* in these rods has an extremely slow deactivation since the dim flash response recovers with a final time constant of ~40 s. The failure of normal recovery in

Arr1⁻/⁻ rods shows that Arr1 binding is essential for the rapid quench of R* activity (Xu et al., 1997) and that in the absence of Arr1, phosphorylated R* is deactivated by hydrolysis of the Schiff base linkage with the retinoid (Imai et al., 2007).

Despite the clear necessity for arrestin in the timely deactivation of R*, most arrestin is sequestered in oligomeric complexes and is predominantly located in the inner segments of dark-adapted rods (see Gurevich et al., 2011, for a recent review). Following light exposure that activates at least 1% of the rhodopsin, arrestin redistributes over a time scale of minutes, increasing Arr1 levels in the outer segment and the synaptic terminal in a manner that exceeds stoichiometric proportions with R*(Strissel et al., 2006), suggesting that an amplified process triggers the translocation. The physiological significance of Arr1 movement may extend beyond its role in phototransduction, as recent evidence suggests Arr1 plays other roles such as regulating synaptic transmission (Huang, Brown, & Craft, 2010) and rod neuroprotection in bright light (Gurevich, Gurevich, & Cleghorn, 2008).

A cone-specific arrestin, Arr4, has also been identified (Craft, Whitmore, & Wiechmann, 1994). Mouse cones express both Arr1 and Arr4, and deletion of both strongly slows cone responses driven by either M- or S-opsin, revealing that binding of an arrestin to opsins in both of these anciently diverged opsin families is necessary for their normal and full deactivation (Nikonov et al., 2008). However, comparatively, cone opsin deactivation in the absence of phosphorylation (Lyubarsky et al., 2000) and arrestins (Nikonov et al., 2008;) proceeds far more quickly than that of rods in the absence of arrestins.

Deactivation of the G-Protein and PDE

Rapid R* deactivation alone is not sufficient for normal recovery of the electrical response to a flash. The activated forms of transducin and PDE must also be turned off. As in other G-protein cascades, deactivation of transducin requires hydrolysis of the GTP bound to its α-subunit, resulting in transducin dissociation from the PDE and deactivation of both enzymes. In both rods and cones the GTP hydrolysis and deactivation of the transducin-PDE complex are catalyzed by another protein complex comprising RGS9–1, Gβ5-L, and R9AP (reviewed in Anderson, Posokhova, & Martemyanov, 2009). This RGS9 complex is essential for normal recovery of the light response, both in rods (C. K. Chen et al., 2000) and in cones (Lyubarsky et al., 2001). The amount of the RGS9 complex on the disk membrane determines how rapidly the transducin-PDE complex deactivation proceeds. In normal mouse rods the rate of transducin-PDE deactivation is ~5 s⁻¹, which corresponds to the 200-ms rate-limiting time constant of recovery for dark-adapted rods. This deactivation can be sped up by overexpression of the RGS9 complex (Burns & Pugh, 2009; C. K. Chen et al., 2010; Krispel et al., 2006). Notably, the levels of the RGS9 complex are higher in cones than in rods (Cowan et al., 1998), a feature that could explain why cone response recovery appears to be rate-limited instead by deactivation of cone opsins (Matthews & Sampath, 2010).

CALCIUM FEEDBACK REGULATION OF PHOTOTRANSDUCTION

In darkness the inward flow of current into the outer segment through cGMP-activated channels is carried primarily by Na^+ ions, but a substantial proportion (~15% in rods) is carried by Ca^{2+} ions. In darkness this steady influx of Ca^{2+} is balanced by an equal and opposite efflux, driven by $Na^+/Ca^{2+},K^+$ exchangers in the plasma membrane. Measurements by different investigators have yielded estimates of the resting free cytoplasmic Ca^{2+} in dark-adapted photoreceptors in the range 250–700 nM (reviewed in Fain et al., 2001). During saturating illumination all outer segment cGMP channels close, blocking the influx of Ca^{2+} ions, with the result that the exchanger drives the cytoplasmic Ca^{2+} down to its minimum, 10–50 nM. Where the decline of Ca^{2+} has been measured in both rods and cones of the same species, it is generally much faster in cones (Korenbrot & Rebrik, 2002), in part because of the typically smaller cytoplasmic volume.

The fall in Ca^{2+} concentration during the light response orchestrates at least three distinct changes that assist restoration of the circulating current and are often referred to as *negative feedback mechanisms* because they oppose the effect of light in closing the cGMP-gated channels. In mouse rods only one of these feedback mechanisms operates *dynamically* during the dim flash response: the Ca^{2+}-dependent activation of guanylate cyclase activity, mediated by guanylate cyclase-activating proteins (GCAPs). However, during adaptation to steady light, the decline in Ca^{2+} activates two additional mechanisms: the acceleration of R* phosphorylation, mediated by the Ca^{2+}-binding protein recoverin and an increase in the sensitivity of the cGMP-activated channels to cGMP, mediated in rods by the Ca^{2+} binding protein calmodulin and in cones by the Ca^{2+} binding protein CNG-modulin. We discuss each of these processes in this section.

Calcium Regulation of Guanylate Cyclase

Photoreceptors express two guanylate cyclase isoforms (GC1 and GC2) and at least two isoforms of guanylate cyclase–activating proteins (GCAPs). Biochemical studies have shown that Ca^{2+}-free GCAPs activate GCs, whereas the Ca^{2+}-bound GCAPs inhibit GC activity. The Ca^{2+} concentration reported to half-maximally activate GCs ($K\frac{1}{2}$) varies with the Mg^{2+} concentration such that $K\frac{1}{2} \approx 200$ nM under physiological conditions. The activation by Ca^{2+} is cooperative, with a Hill coefficient of ~2. For a recent review, see Peshenko et al. (2011).

The relative expression levels of GC1/GC2 and GCAP1/GCAP2 vary in rods and cones (Stephen et al., 2008), and recent experiments suggest that regulation by each GCAP may serve a distinct function. For example, the Ca^{2+}-dependencies of GC activation are not equivalent. GCAP1 appears to respond to the fall in Ca^{2+} more quickly than GCAP2 because Ca^{2+} can dissociate from it more rapidly. As a result, GCAP1 activates GC earlier in the photoresponse than GCAP2 (Howes et al., 2002; Makino et al., 2008; Mendez et al., 2001).

Rods lacking all Ca^{2+} feedback regulation of guanylate cyclase (GCAPs$^{-/-}$ rods) generate single-photon responses that are much larger and longer lasting than those of wild-type rods (Burns et al., 2002; Okawa et al., 2010), indicating the importance of dynamic feedback during the few hundred milliseconds required for the single-photon response. In addition, single-photon responses are less reproducible from trial-to-trial in GCAPs$^{-/-}$ mouse rods, indicating that feedback also plays a role in minimizing the consequences of trial-to-trial fluctuations in R* lifetime (Gross Pugh, & Burns, 2012b). GCAPs$^{-/-}$ rods also show clear deficits in their ability to function in steady light, saturating at light intensities more than an order of magnitude dimmer than normal. Mouse cones lacking GCAPs have slowed dim flash responses and a smaller range of steady light over which they can function (Sakurai, Chen, & Kefalov, 2011). These results illustrate the importance of feedback in range extension in both rods and cones (see Light Adaptation, below).

Calcium Regulation of R Phosphorylation*

Ca^{2+}-dependent regulation of the R* lifetime arises from the dependence of the rate of rhodopsin phosphorylation on free Ca^{2+}, a process mediated by the Ca^{2+} binding protein recoverin (Kawamura, 1993). In the dark, recoverin-Ca^{2+} binds to GRK1, inhibiting rhodopsin phosphorylation (Klenchin, Calvert, & Bownds, 1995). During light exposure, when the free Ca^{2+} falls, recoverin undergoes a conformational change (Ames & Ikura, 2002) that reduces recoverin's affinity for RK and removes the inhibition of GRK1. In mouse rods the loss of recoverin causes the dim flash response to be slightly smaller and briefer but causes a pronounced decrease in overall saturation time for bright flashes delivered in darkness and a failure for responses to shorten their time in saturation times in steady light (Makino et al., 2004). Likewise, in recent experiments by Chen and colleagues, R* deactivation in intact mouse rods was speeded by the loss of recoverin (C. K. Chen et al., 2010) and modulated by background light in normal rods. Together these results demonstrate that the Ca^{2+} regulation of R* deactivation in rods is mediated by recoverin and is an important mechanism of light adaptation, effectively reducing R* lifetime in the presence of background light. A similar mechanism for shortening the lifetime cone R* in a Ca^{2+}-dependent manner has been shown to function in poikilotherm cones (Arinobu, Tachibanaki, & Kawamura, 2010).

Calcium Regulation of cGMP-Gated Ion Channels

The sensitivity of cGMP-gated channels for cGMP is sensitive to intracellular Ca^{2+} (Hsu & Molday, 1993). In rods this sensitivity is controlled by calmodulin, which binds to the β-subunit of the rod channel at high Ca^{2+}. The fall in Ca^{2+} that accompanies the response to steady prolonged light causes calmodulin to dissociate from the channel, increasing the affinity of the channel for cGMP. This mechanism extends the range of light intensities over which a photoreceptor can operate, as it allows the cGMP-gated channels to reopen at lower cGMP levels. The magnitude of this effect in rods is rather modest (Koutalos, Nakatani, & Yau, 1995; Nikonov, Lamb, & Pugh, 2000; Sagoo & Lagnado, 1996), although a similar, more powerful mechanism operates on cone channels (Rebrik & Korenbrot, 1998). The Ca^{2+} binding protein that mediates cGMP channel modulation in cones has recently been identified as CNG-modulin, which is also expressed in other neural and sensory tissues, where it may also modulate cyclic nucleotide–gated ion channels (Rebrik et al., 2012).

MECHANISMS OF PHOTORECEPTOR
LIGHT ADAPTATION

Calcium-Dependent Adaptive Mechanisms

Light adaptation underlies much of the ability of the visual system to function effectively over the ~9 $\log_{10}$ range of diurnal illumination. The term light adaptation comprises *all* the physiological changes that occur in the presence of steady illumination. Because light tends to drive cGMP-activated channels to close, the truly *adaptive* mechanisms are those that enhance the ability of photoreceptors to maintain some open cGMP-activated channels in the presence of steady light. We have already discussed three such mechanisms, all triggered by the decline in Ca^{2+} that accompanies steady illumination: (1) increased rate of cGMP synthesis mediated by GCAPs-GC; (2) increased rate of deactivation of R* mediated by disinhibition of the Ca-recoverin-GRK1; and (3) increased sensitivity of cGMP-activated channels by calmodulin in rods and CNG-modulin in cones. In addition to helping the cells to avoid saturation, during adaptation these mechanisms also speed the incremental response to a flash, enabling photoreceptors to achieve a broader signaling bandwidth than they otherwise could.

PDE Activation and Response Compression: Neither Adaptive nor Calcium Dependent

Not all the changes in photoreceptors that accompany increasing intensities are adaptive. Two particular examples are very important: desensitization caused by steady increased PDE activity and response compression. The steady PDE activity (number of G*-E*'s) in the outer segment increases with intensity, with the consequence that the steady rate of cGMP hydrolysis does also. This increase in cGMP hydrolysis reduces the average lifetime of a cGMP molecule. The increased turnover rate of cGMP is the most powerful mechanism of desensitization, and furthermore, it is Ca^{2+} independent (Nikonov, Lamb, & Pugh, 2000; Pugh, Nikonov, & Lamb, 1999).

Response compression is a direct consequence of the reduced fraction of cGMP-activated channels open in the presence of steady light. All other things being equal, if a flash produced an identical pulse of underlying PDE activity with 50% of the channels closed that would be open in darkness, and the photoreceptor had the same membrane resistance, there would be a 50% decrease in the maximal change in current and a 50% decrease in the voltage response that suppresses glutamate release. Cones, on the other hand, never saturate in intense steady light (Shevell, 1977), and the degree to which response compression attenuates light responses never exceeds a factor of 2 in lower vertebrates. Rather, cones are able to undergo voltage excursions of almost equal magnitude for intensity increments and decrements in the presence of backgrounds (Burkhardt, 1994; Normann & Perlman, 1979).

G-Protein Translocation

The activated Gα-GTP subunit of transducin (G*) is weakly water soluble and dissociates from the disk membrane on a relatively slow time scale. Thus, G* rarely dissociates from the disk under dim light conditions because it is rapidly deactivated by the RGS9 complex. However, at high intensities producing G* in excess of the total number of PDE and RGS9 complexes, the longer G* lifetime causes appreciable dissociation from the disk membrane to occur and subsequent large translocation of Gα and Gβγ from the outer segments to the inner segments in response to bright light (Lobanova et al., 2007; Martemyanov et al., 2008; Sokolov et al., 2002). The translocation of transducin to the inner segment following sustained bright light appears to be an adaptive mechanism, enabling rods exposed to the bright lights that trigger the translocation to function in brighter lights (Sokolov et al., 2002). Notably, G protein translocation during sustained bright light does not normally occur in cones because the G protein lifetime in cones is sufficiently brief even at very high light intensities to make dissociation of the subunits from the disk membrane unlikely (Lobanova et al., 2010).

Opsin Photoproducts and Bleaching

In mice complete saturation of rod sensitivity measured behaviorally occurs in response to light producing in excess of 10^4 photoisomerizations s^{-1} (figure 2.1C) (Naarendorp et al., 2010). If we assume a regeneration time constant of the order of 400 s, such an intensity corresponds to a steady-state bleach level of 4×10^5 out of $\sim 10^8$ rhodopsin molecules, or less than 0.5%. Thus, rods become completely inoperative in steady light that causes no appreciable level of bleaching. When lights that do produce substantial bleaching of rhodopsin are extinguished, the normal R* deactivation mechanisms are either too weak or are overwhelmed, and long-lived photoproducts of R* are present which activate the transduction cascade, closing all or most cGMP-activated channels for as long as 45 min. Full recovery of dark-adapted sensitivity is only reached when all the opsin has been regenerated (Lamb & Pugh, 2004).

Cones, in contrast to rods, are capable of functioning even when almost all of their opsin has been bleached (Burkhardt, 1994), and it has long been known psychophysically that the cone system does not show saturation under steady illumination (Barlow, 1972; Shevell, 1977). This suggests that cones are much less susceptible than rods to the presence of opsin bleaching products, but an additional contributory factor, no doubt, is that cone opsins exhibit much faster removal of all-*trans*-retinal through their much higher retinol dehydrogenase activity (Ala-Laurila et al., 2006). Because of the faster rates of regeneration in cones, it is unlikely that the steady-state cone opsin bleaching level experienced by human observers in daytime environments exceeds about 90%, given the normal range of light levels (figure 2.1A). Thus, bleaching desensitization likely contributes at most a factor of ~10-fold reduction in quantal catch, even though the cones are capable of functioning at even higher steady bleach levels.

SUMMARY

Photoreceptors are arguably one of the best-understood neuronal cell types in the human body. Photoreceptors have a seemingly simple job description: to accurately relay to the rest of the visual system information regarding the time-varying rate of photon absorptions and to do this across ~9 $\log_{10}$ units of intensity. To perform this task in vertebrates, nature has evolved two distinct classes of photoreceptors, which, while having homologous molecular machinery, possess a variety of distinct biochemical and electrical mechanisms that enable them to function in different light regimes: mechanisms to amplify and deactivate the response in a timely manner; mechanisms that serve to adapt the cell to light in both the short and long term; and, finally, mechanisms that maintain these functional abilities, via opsin regeneration and other homeostatic cell biological processes beyond the scope of this chapter (e.g., protein translocation; disk shedding and renewal). Our current level of understanding, though far from complete, has nonetheless revealed what astoundingly choreographed cellular and molecular processes in rods and cones seamlessly enable us to see our world throughout the diurnal cycle of illumination.

REFERENCES

Aguilar, M., & Stiles, W. S. (1954). Saturation of the rod mechanism of the retina at high levels of illumination. *Optica Acta, 1*, 59–65.

Ala-Laurila, P., Kolesnikov, A. V., Crouch, R. K., Tsina, E., Shukolyukov, S. A., Govardovskii, V. I., et al. (2006). Visual cycle: Dependence of retinol production and removal on photoproduct decay and cell morphology. *Journal of General Physiology, 128*, 153–169. doi:10.1085/jgp.200609557.

Ames, J. B., & Ikura, M. (2002). Structure and membrane-targeting mechanism of retinal Ca^{2+}-binding proteins, recoverin and GCAP-2. *Advances in Experimental Medicine and Biology, 514*, 333–348.

Anderson, G. R., Posokhova, E., & Martemyanov, K. A. (2009). The R7 RGS protein family: Multi-subunit regulators of neuronal G protein signaling. *Cell Biochemistry and Biophysics, 54*, 33–46.

Arinobu, D., Tachibanaki, S., & Kawamura, S. (2010). Larger inhibition of visual pigment kinase in cones than in rods. *Journal of Neurochemistry, 115*, 259–268.

Arshavsky, V. Y., & Burns, M. E. (2012). Photoreceptor signaling: Supporting vision across a wide range of light intensities. *Journal of Biological Chemistry, 287*, 1620–1626.

Arshavsky, V. Y., Dizhoor, A. M., Shestakova, I. K., & Philippov, P. (1985). The effect of rhodopsin phosphorylation on the light-dependent activation of phosphodiesterase from bovine rod outer segments. *FEBS Letters, 181*, 264–266.

Barlow, H. B. (1956). Retinal noise and absolute threshold. *Journal of the Optical Society of America, 46*, 634–639.

Barlow, H. B. (1972). Dark and light adaptation: Psychophysics. In D. Jameson & L. M. Hurvich (Eds.), *Handbook of sensory physiology* (Vol. VII, pp. 1–28). Berlin: Springer.

Burkhardt, D. A. (1994). Light adaptation and photopigment bleaching in cone photoreceptors in situ in the retina of the turtle. *Journal of Neuroscience, 14*(Pt 1), 1091–1105.

Burns, M. E., Mendez, A., Chen, J., & Baylor, D. A. (2002). Dynamics of cyclic GMP synthesis in retinal rods. *Neuron, 36*, 81–91.

Burns, M. E., & Pugh, E. N., Jr. (2009). RGS9 concentration matters in rod phototransduction. *Biophysical Journal, 97*, 1538–1547.

Burns, M. E., & Pugh, E. N., Jr. (2010). Lessons from photoreceptors: Turning off G-protein signaling in living cells. *Physiology (Bethesda, MD), 25*, 72–84.

Chen, C. K., Burns, M. E., He, W., Wensel, T. G., Baylor, D. A., & Simon, M. I. (2000). Slowed recovery of rod photoresponse in mice lacking the GTPase accelerating protein RGS9-1. *Nature, 403*, 557–560.

Chen, C. K., Burns, M. E., Spencer, M., Niemi, G. A., Chen, J., Hurley, J. B., et al. (1999). Abnormal photoresponses and light-induced apoptosis in rods lacking rhodopsin kinase. *Proceedings of the National Academy of Sciences of the United States of America, 96*, 3718–3722. doi:10.1073/pnas.96.7.3718.

Chen, C. K., Woodruff, M. L., Chen, F. S., Chen, D., & Fain, G. L. (2010). Background light produces a recoverin-dependent modulation of activated-rhodopsin lifetime in mouse rods. *Journal of Neuroscience, 30*, 1213–1220.

Chen, J., Makino, C. L., Peachey, N. S., Baylor, D. A., & Simon, M. I. (1995). Mechanisms of rhodopsin inactivation in vivo as revealed by a COOH-terminal truncation mutant. *Science, 267*, 374–377.

Cowan, C. W., Fariss, R. N., Sokal, I., Palczewski, K., & Wensel, T. G. (1998). High expression levels in cones of RGS9, the predominant GTPase accelerating protein of rods. *Proceedings of the National Academy of Sciences of the United States of America, 95*, 5351–5356. doi:10.1073/pnas.95.9.5351.

Craft, C. M., Whitmore, D. H., & Wiechmann, A. F. (1994). Cone arrestin identified by targeting expression of a functional family. *Journal of Biological Chemistry, 269*, 4613–4619.

Doan, T., Mendez, A., Detwiler, P. B., Chen, J., & Rieke, F. (2006). Multiple phosphorylation sites confer reproducibility of the rod's single-photon responses. *Science, 313*, 530–533.

Doble, N., Choi, S. S., Codona, J. L., Christou, J., Enoch, J. M., & Williams, D. R. (2011). In vivo imaging of the human rod photoreceptor mosaic. *Optics Letters, 36*(1), 31–33.

Dubra, A., Sulai, Y., Norris, J. L., Cooper, R. F., Dubis, A. M., Williams, D. R., et al. (2011). Noninvasive imaging of the human rod photoreceptor mosaic using a confocal adaptive optics scanning ophthalmoscope. *Biomedical Optics Express, 2*, 1864–1876.

Fain, G. L., Matthews, H. R., Cornwall, M. C., & Koutalos, Y. (2001). Adaptation in vertebrate photoreceptors. *Physiological Reviews, 81*, 117–151.

Fu, Y., Kefalov, V., Luo, D. G., Xue, T., & Yau, K. W. (2008). Quantal noise from human red cone pigment. *Nature Neuroscience, 11*, 565–571.

Gross, O. P., & Burns, M. E. (2010). Control of rhodopsin's active lifetime by arrestin-1 expression in mammalian rods. *Journal of Neuroscience, 30*, 3450–3457.

Gross, O. P., Pugh, E. N., Jr., & Burns, M. E. (2012a). Spatiotemporal cGMP dynamics in living mouse rods. *Biophysical Journal. 102*, 1775–1784.

Gross, O. P., Pugh, E. N., Jr., & Burns, M. E. (2012b). Calcium feedback to cGMP synthesis more strongly attenuates single photon responses driven by long rhodopsin lifetimes. *Neuron* (in press).

Gurevich, V. V., Gurevich, E. V., & Cleghorn, W. M. (2008). Arrestins as multi-functional signaling adaptors. *Handbook of Experimental Pharmacology, 186*, 15–37.

Gurevich, V. V., Hanson, S. M., Song, X., Vishnivetskiy, S. A., & Gurevich, E. V. (2011). The functional cycle of visual arrestins in photoreceptor cells. *Progress in Retinal and Eye Research, 30*, 405–430.

Hecht, S., Shlaer, S., & Pirenne, M. H. (1942). Energy, quanta, and vision. *Journal of General Physiology, 25*, 819–840.

Howes, K. A., Pennesi, M. E., Sokal, I., Church-Kopish, J., Schmidt, B., Margolis, D., et al. (2002). GCAP1 rescues rod photoreceptor response in GCAP1/GCAP2 knockout mice. *European Molecular Biology Organization Journal, 21*, 1545–1554. doi:10.1093/emboj/21.7.1545.

Hsu, Y. T., & Molday, R. S. (1993). Modulation of the cGMP-gated channel of rod photoreceptor cells by calmodulin. *Nature, 361*, 76–79.

Huang, S. P., Brown, B. M., & Craft, C. M. (2010). Visual arrestin 1 acts as a modulator for N-ethylmaleimide-sensitive factor in the photoreceptor synapse. *Journal of Neuroscience, 30*, 9381–9391.

Imai, H., Kefalov, V., Sakurai, K., Chisaka, O., Ueda, Y., Onishi, A., et al. (2007). Molecular properties of rhodopsin and rod function. *Journal of Biological Chemistry, 282*, 6677–6684. doi:10.1074/jbc.M610086200.

Jackman, S. L., Choi, S. Y., Thoreson, W. B., Rabl, K., Bartoletti, T. M., & Kramer, R. H. (2009). Role of the synaptic ribbon in transmitting the cone light response. *Nature Neuroscience, 12*, 303–310.

Karpen, J. W., Zimmerman, A. L., Stryer, L., & Baylor, D. A. (1988). Gating kinetics of the cyclic-GMP-activated channel of retinal rods: Flash photolysis and voltage-jump studies. *Proceedings of the National Academy of Sciences of the United States of America, 85*, 1287–1291.

Kawamura, S. (1993). Rhodopsin phosphorylation as a mechanism of cyclic GMP phosphodiesterase regulation by S-modulin. *Nature, 362*, 855–857.

Klenchin, V. A., Calvert, P. D., & Bownds, M. D. (1995). Inhibition of rhodopsin kinase by recoverin. Further evidence for a negative feedback system in phototransduction. *Journal of Biological Chemistry, 270*, 24127. doi:10.1074/jbc.270.41.24127.

Korenbrot, J. I., & Rebrik, T. I. (2002). Tuning outer segment Ca²⁺ homeostasis to phototransduction in rods and cones. *Advances in Experimental Medicine and Biology, 514*, 179–203.

Koutalos, Y., Nakatani, K., & Yau, K. W. (1995). The cGMP-phosphodiesterase and its contribution to sensitivity regulation in retinal rods. *Journal of General Physiology, 106*(5), 891–921.

Krispel, C. M., Chen, D., Melling, N., Chen, Y. J., Martemyanov, K. A., Quillinan, N., et al. (2006). RGS expression rate-limits recovery of rod photoresponses. *Neuron, 51*, 409–416. doi:10.1016/j.neuron.2006.07.010.

Kuhn, H., & Wilden, U. (1987). Deactivation of photoactivated rhodopsin by rhodopsin-kinase and arrestin. *Journal of Receptor Research, 7*, 283–298.

Lamb, T. D., & Pugh, E. N., Jr. (2004). Dark adaptation and the retinoid cycle of vision. *Progress in Retinal and Eye Research, 23*, 307–380.

Leskov, I. B., Klenchin, V. A., Handy, J. W., Whitlock, G. G., Govardovskii, V. I., Bownds, M. D., et al. (2000). The gain of rod phototransduction: Reconciliation of biochemical and electrophysiological measurements. *Neuron, 27*, 525–537. doi:10.1016/S0896-6273(00)00063-5.

Lobanova, E. S., Finkelstein, S., Song, H., Tsang, S. H., Chen, C. K., Sokolov, M., et al. (2007). Transducin translocation in rods is triggered by saturation of the GTPase-activating complex. *Journal of Neuroscience, 27*, 1151–1160.

Lobanova, E. S., Herrmann, R., Finkelstein, S., Reidel, B., Skiba, N. P., Deng, W. T., et al. (2010). Mechanistic basis for the failure of cone transducin to translocate: Why cones are never blinded by light. *Journal of Neuroscience, 30*, 6815–6824.

Lyubarsky, A. L., Chen, J., Simon, M. I., & Pugh, E. N., Jr. (2000). Mice lacking G-protein receptor kinase 1 have profoundly slowed recovery of cone-driven retinal responses. *Journal of Neuroscience, 20*, 2209–2217.

Lyubarsky, A. L., Naarendorp, F., Zhang, X., Wensel, T., Simon, M. I., & Pugh, E. N., Jr. (2001). RGS9–1 is required for normal inactivation of mouse cone phototransduction. *Molecular Vision, 7*, 71–78.

Mahroo, O. A., & Lamb, T. D. (2004). Recovery of the human photopic electroretinogram after bleaching exposures: Estimation of pigment regeneration kinetics. *Journal of Physiology, 554*(Pt 2), 417–437.

Makino, C. L., Dodd, R. L., Chen, J., Burns, M. E., Roca, A., Simon, M. I., et al. (2004). Recoverin regulates light-dependent phosphodiesterase activity in retinal rods. *Journal of General Physiology, 123*, 729–741.

Makino, C. L., Peshenko, I. V., Wen, X. H., Olshevskaya, E. V., Barrett, R., & Dizhoor, A. M. (2008). A role for GCAP2 in regulating the photoresponse. Guanylyl cyclase activation and rod electrophysiology in GUCA1B knock-out mice. *Journal of Biological Chemistry, 283*, 29135–29143. doi:10.1074/jbc.M804445200.

Martemyanov, K. A., Krispel, C. M., Lishko, P. V., Burns, M. E., & Arshavsky, V. Y. (2008). Functional comparison of RGS9 splice isoforms in a living cell. *Proceedings of the National Academy of Sciences of the United States of America, 105*, 20988–20993. doi:10.1073/pnas.0808941106.

Matthews, H. R., & Sampath, A. P. (2010). Photopigment quenching is Ca²⁺ dependent and controls response duration in salamander L-cone photoreceptors. *Journal of General Physiology, 135*, 355–366.

Mendez, A., Burns, M. E., Roca, A., Lem, J., Wu, L. W., Simon, M. I., et al. (2000). Rapid and reproducible deactivation of rhodopsin requires multiple phosphorylation sites. *Neuron, 28*, 153–164.

Mendez, A., Burns, M. E., Sokal, I., Dizhoor, A. M., Baehr, W., Palczewski, K., et al. (2001). Role of guanylate cyclase-activating proteins (GCAPs) in setting the flash sensitivity of rod photoreceptors. *Proceedings of the National Academy of Sciences of the United States of America, 98*, 9948–9953. doi:10.1073/pnas.171308998.

Mercer, A. J., & Thoreson, W. B. (2011). The dynamic architecture of photoreceptor ribbon synapses: Cytoskeletal, extracellular matrix, and intramembrane proteins. *Visual Neuroscience, 28*, 453–471.

Naarendorp, F., Esdaille, T. M., Banden, S. M., Andrews-Labenski, J., Gross, O. P., & Pugh, E. N., Jr. (2010). Dark light, rod saturation, and the absolute and incremental sensitivity of mouse cone vision. *Journal of Neuroscience, 30*, 12495–12507.

Nakatani, K., Tamura, T., & Yau, K. W. (1991). Light adaptation in retinal rods of the rabbit and two other nonprimate mammals. *Journal of General Physiology, 97*, 413–435.

Nikonov, S. S., Brown, B. M., Davis, J. A., Zuniga, F. I., Bragin, A., Pugh, E. N., Jr., et al. (2008). Mouse cones require an arrestin for normal inactivation of phototransduction. *Neuron, 59*, 462–474.

Nikonov, S. S., Daniele, L. L., Zhu, X., Craft, C. M., Swaroop, A., Pugh, E. N., et al. (2005). Photoreceptors of Nrl -/- mice coexpress functional S- and M-cone opsins having distinct inactivation mechanisms. *Journal of General Physiology, 125*, 287–304.

Nikonov, S., Lamb, T. D., & Pugh, E. N., Jr. (2000). The role of steady phosphodiesterase activity in the kinetics and sensitivity of the light-adapted salamander rod photoresponse. *Journal of General Physiology, 116*, 795–824.

Normann, R. A., & Perlman, I. (1979). The effects of background illumination on the photoresponses of red and green cones. *Journal of Physiology, 286*, 491–507.

Okawa, H., Miyagishima, K. J., Arman, A. C., Hurley, J. B., Field, G. D., & Sampath, A. P. (2010). Optimal processing of photoreceptor signals is required to maximize behavioural sensitivity. *Journal of Physiology, 588*(Pt 11), 1947–1960.

Peshenko, I. V., Olshevskaya, E. V., Savchenko, A. B., Karan, S., Palczewski, K., Baehr, W., et al. (2011). Enzymatic properties and regulation of the native isozymes of retinal membrane guanylyl cyclase (RetGC) from mouse photoreceptors. *Biochemistry, 50*, 5590–5600.

Pugh, E. N., Jr., & Lamb, T. D. (1993). Amplification and kinetics of the activation steps in phototransduction. *Biochimica et Biophysica Acta, 1141*, 111–149.

Pugh, E. N., Jr., Nikonov, S., & Lamb, T. D. (1999). Molecular mechanisms of vertebrate photoreceptor light adaptation. *Current Opinion in Neurobiology, 9*, 410–418.

Rebrik, T. I., Botchkina, I., Arshavsky, V. Y., Craft, C. M., & Korenbrot, J. I. (2012). CNG-modulin: A novel Ca-dependent modulator of ligand sensitivity in cone photoreceptor cGMP-gated ion channels. *Journal of Neuroscience, 32*, 3142–3153.

Rebrik, T. I., & Korenbrot, J. I. (1998). In intact cone photoreceptors, a Ca^{2+}-dependent, diffusible factor modulates the cGMP-gated ion channels differently than in rods. *Journal of General Physiology, 112*, 537–548.

Sagoo, M. S., & Lagnado, L. (1996). The action of cytoplasmic calcium on the cGMP-activated channel in salamander rod photoreceptors. *Journal of Physiology, 497*(Pt 2), 309–319.

Sakurai, K., Chen, J., & Kefalov, V. J. (2011). Role of guanylyl cyclase modulation in mouse cone phototransduction. *Journal of Neuroscience, 31*, 7991–8000.

Seeliger, M. W., Brombas, A., Weiler, R., Humphries, P., Knop, G., Tanimoto, N., et al. (2011). Modulation of rod photoreceptor output by HCN1 channels is essential for regular mesopic cone vision. *Nature Communication, 2*, 532.

Sharpe, L. T., Fach, C. C., & Stockman, A. (1992). The field adaptation of the human rod visual system. *Journal of Physiology, 445*, 319–343.

Sheng, Z., Choi, S. Y., Dharia, A., Li, J., Sterling, P., & Kramer, R. H. (2007). Synaptic Ca^{2+} in darkness is lower in rods than cones, causing slower tonic release of vesicles. *Journal of Neuroscience, 27*, 5033–5042.

Shevell, S. K. (1977). Saturation in human cones. *Vision Research, 17*, 427–434.

Sokolov, M., Lyubarsky, A. L., Strissel, K. J., Savchenko, A. B., Govardovskii, V. I., Pugh, E. N., Jr., et al. (2002). Massive light-driven translocation of transducing between the two major compartments of red cells: A novel mechanism of light adaptation. *Neuron, 33*, 95–106.

Stephen, R., Filipek, S., Palczewski, K., & Sousa, M. C. (2008). Ca^{2+} dependent regulation of phototransduction. *Photochemistry and Photobiology, 84*, 903–910.

Stiles, W. S., & Crawford, B. H. (1932). Equivalent adaptational levels in localized retinal areas. In *Report of a joint discussion on vision, Physical Society of London* (Mechanisms of Colour Vision, pp. 194–211). Cambridge: Cambridge University Press.

Strissel, K. J., Sokolov, M., Trieu, L. H., & Arshavsky, V. Y. (2006). Arrestin translocation is induced at a critical threshold of visual signaling and is superstoichiometric to bleached rhodopsin. *Journal of Neuroscience, 26*, 1146–1153.

Tachibanaki, S., Arinobu, D., Shimauchi-Matsukawa, Y., Tsushima, S., & Kawamura, S. (2005). Highly effective phosphorylation by G protein-coupled receptor kinase 7 of light-activated visual pigment in cones. *Proceedings of the National Academy of Sciences of the United States of America, 102*, 9329–9334. doi:10.1073/pnas.0501875102.

Umino, Y., Solessio, E., & Barlow, R. B. (2008). Speed, spatial, and temporal tuning of rod and cone vision in mouse. *Journal of Neuroscience, 28*, 189–198.

Vogalis, F., Shiraki, T., Kojima, D., Wada, Y., Nishiwaki, Y., Jarvinen, J. L., et al. (2011). Ectopic expression of cone-specific G-protein-coupled receptor kinase GRK7 in zebrafish rods leads to lower photosensitivity and altered responses. *Journal of Physiology, 589*(Pt 9), 2321–2348.

Wang, J. S., & Kefalov, V. J. (2011). The cone-specific visual cycle. *Progress in Retinal and Eye Research, 30*, 115–128.

Weiss, E. R., Raman, D., Shirakawa, S., Ducceschi, M. H., Bertram, P. T., Wong, F., et al. (1998). The cloning of GRK7, a candidate cone opsin kinase, from cone- and rod-dominant mammalian retinas. *Molecular Vision, 4*, 27.

Wilden, U., Hall, S. W., & Kuhn, H. (1986). Phosphodiesterase activation by photoexcited rhodopsin is quenched when rhodopsin is phosphorylated and binds the intrinsic 48-kDa protein of rod outer segments. *Proceedings of the National Academy of Sciences of the United States of America, 83*, 1174–1178.

Wyszecki, G., & Stiles, W. S. (1982). *Color science* (2nd ed.). New York: Wiley.

Xu, J., Dodd, R. L., Makino, C. L., Simon, M. I., Baylor, D. A., & Chen, J. (1997). Prolonged photoresponses in transgenic mouse rods lacking arrestin. *Nature, 389*, 505–509.

Zhu, X., Brown, B., Li, A., Mears, A. J., Swaroop, A., & Craft, C. M. (2003). GRK1-dependent phosphorylation of S and M opsins and their binding to cone arrestin during cone phototransduction in the mouse retina. *Journal of Neuroscience, 23*, 6152–6160.

Zhu, X., Brown, B., Rife, L., & Craft, C. M. (2006). Slowed photoresponse recovery and age-related degeneration in cones lacking G protein-coupled receptor kinase 1. *Advances in Experimental Medicine and Biology, 572*, 133–139.

3 Membrane Protein Transport in Mouse Photoreceptors: Trafficking of Visual Pigments and Transducin

WOLFGANG BAEHR, RYAN CONSTANTINE, HOUBIN ZHANG, AND JEANNE M. FREDERICK

TRAFFICKING OF ROD AND CONE VISUAL PIGMENTS

Photoreceptors are polarized neurons with very specific subcellular compartmentalization and unique requirements for protein expression capacity. Each photoreceptor contains an outer segment (OS) where phototransduction occurs, an inner segment (IS), which houses the biosynthetic machinery, and a synaptic terminal for signal transmission. Daily renewal of ~10% of the OS membrane requires continuous biosynthesis with reliable transport and targeting pathways. As modified sensory cilia with enormous biosynthetic demands, photoreceptors are an excellent model system to study synthesis, post-Golgi trafficking, and ciliary transport of proteins.

Photoreceptor Architecture

Rod and cone photoreceptors of the mammalian retina are named for the cylindrical or conical shape of their OS, respectively. The mouse retina contains approximately 6 million rods (Jeon, Strettoi, & Masland, 1998) and 200,000 cones (3% of rods), whereas the human retina contains 110 million rods and 6.4 million cones (see webvision, http://webvision.med.utah.edu/, part XIII). Exquisitely polarized, each photoreceptor consists of a photosensitive primary (nonmotile) cilium corresponding to the OS (figure 3.1), an IS, a cell body containing the nucleus, and a synaptic terminal. In mouse each rod OS contains ~800 membranous disks arranged in a vertical stack (Nickell et al., 2007). Each disk houses roughly 25,000 molecules of rhodopsin, 2,500 molecules of transducin, and 250 molecules of PDE6. Although the rod OS disks appear to float independently of the plasma membrane (PM) that surrounds them, filamentous connections between the disks and PM have been observed (Koerschen et al.,

1995; Roof & Heuser, 1982). In contrast, cone OS are formed mostly from multiple invaginations of the plasma membrane but do contain some enclosed disks that are not connected to the plasma membrane. A major structural difference between mouse photoreceptors is OS volume: the rod OS is roughly four times larger than the cone OS (Carter-Dawson & LaVail, 1979), calculated at 36 and 10 attoliters, respectively (Avasthi et al., 2009). The IS contain the metabolic machinery necessary for protein synthesis and processing, including endoplasmic reticulum (ER), the Golgi apparatus and trans-Golgi network (TGN), and mitochondria. The IS and OS are connected by a cilium (CC) through which newly synthesized proteins must pass. Photoreceptor cell bodies contain the nuclei that make up the outer nuclear layer (ONL) of the retina. Information obtained via light reception is transmitted from the photoreceptor synaptic endings (spherules in rods, pedicles in cones) to second-order neurons downstream.

Photoreceptor Outer Segment Turnover

The pioneering work of Richard Young provided evidence that rod and cone OS are renewed approximately every 10 days (Besharse & Hollyfield, 1979; LaVail, 1976; Young, 1967). Mechanisms that regulate disk membrane assembly at the proximal OS, concomitant disk shedding at the distal end, and phagocytosis of shed disk membranes by the adjacent retinal pigment epithelium (RPE) are incompletely understood (Anderson, Fisher, & Steinberg, 1978; Young & Bok, 1969) (for review, see Nachury, Seeley, & Jin, 2010; Strauss, 2005). New disks are assembled at the OS base at a rate of 80 disks/day or ~1,000 rhodopsins/min. Daily renewal of ~10% of the OS membrane requires both an extremely high rate of biosynthesis to replace OS proteins and highly reliable transport and targeting pathways

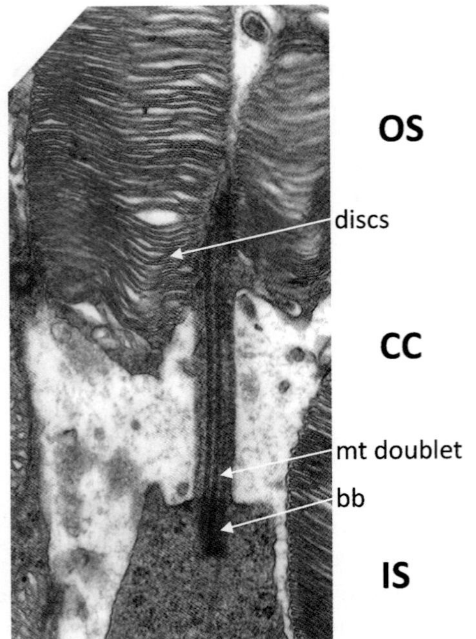

FIGURE 3.1 Electron micrograph of a mouse rod, partial view. The outer segment (OS) is connected to the inner segment (IS) by a cilium through which newly synthesized polypeptides must traffic. CC, connecting cilium; bb, basal body; mt, microtubules.

(Deretic, 1998; Deretic et al., 2005; Sung & Tai, 2000). Disk shedding at the apical end appears to be regulated by circadian processes (LaVail, 1980; LaVail & Ward, 1978), as rod disks are shed in the morning and cone disks at dusk (LaVail, 1980).

Transport of Rhodopsin

Proteins that participate in phototransduction (rhodopsin, transducin, PDE, CNG channel subunits) are synthesized in the IS and must be transported through the CC to the OS. Many details concerning rhodopsin trafficking have been studied using Southern leopard frogs, which develop very large photoreceptors (for a recent review see Deretic, 2006). Rhodopsin is a G-protein-coupled receptor (GPCR) consisting of a hep-tahelical transmembrane (TM) domain. TM proteins such as rhodopsin are synthesized by ER-associated ribosomes, glycosylated cotranslationally, and integrate into the lipid bilayer of the ER (see Lecomte, Ismail, & High, 2003, and step 1 in figure 3.2). Rhodopsin then traffics, most likely as a homodimer Milligan, 2009), to its destination in the OS following the classic secretory pathway (van Vliet et al., 2003). Briefly, rhodopsin exits the ER at specialized exit sites and transports in carrier vesicles to the Golgi (step 2) and TGN (Rodriguez-Boulan, Kreitzer, and Musch, 2005). Tam et al., using

transgenic frogs expressing GFP-rhodopsin fusion proteins, showed that the C-terminal eight amino acids of rhodopsin targeted a peripheral membrane fusion protein to the OS (Tam et al., 2000). This C-terminal sorting motif VXPX directs export from the TGN (Deretic et al., 1998) to the OS (step 3). Transport is regulated by small GTPases (e.g., Arf4) (Deretic et al., 2005), which recognize and bind the rhodopsin sorting signal for incorporation into post-Golgi rhodopsin transport vesicles. Other GTPases involved in vesicle formation and docking include Rab6 (Deretic & Papermaster, 1993), Rab8 (Deretic et al., 1995; Moritz et al., 2001), and Rab11 (Deretic, 1997; Satoh et al., 2005). Additionally, a number of guanine nucleotide exchange factors (GEFs) and GTPase-activating proteins (GAPs) have been identified (Deretic, 1998; Mazelova et al., 2009). Vesicles are likely transported along microtubules directed toward the minus end by dynein motors, terminating near the basal bodies and connecting cilium (step 4). Although the cargo composition and the nature of the cargo-dynein interaction are unknown, a dynein light chain, Tctex-1, was shown to interact directly with a dynein intermediate chain and rhodopsin (Tai et al., 1999; Yeh et al., 2006). Vesicles fuse with the cell membrane, likely by exocytosis (step 5), at the periciliary ridge complex where cargo is assembled (step 6) for anterograde intraflagellar transport (IFT). Presumably, cargo containing rhodopsin and other TM proteins (Bhowmick et al., 2009) traffics through the cilium by IFT powered by kinesin motors (step 7). Cargo is then deposited into the evaginating PM that eventually snaps off, forming independent disks. An alternative, controversial model proposes that rhodopsin-laden vesicles fuse with and enlarge nascent disks of the OS; in this model, the rhodopsin C-terminal tail may interact with the Smad anchor for receptor activation (SARA) and target these vesicles to nascent disks at the OS base (Chuang, Zhao, & Sung, 2007).

Ciliary Targeting Signals

Targeting signals are required to send newly synthesized and processed TM proteins to the cilium. Identified at rhodopsin's C-terminal region using transgenic *Xenopus* (Tam et al., 2000), the signal was shown to consist of the last five amino acids of rhodopsin (XVXPX). Transgenic mice expressing the dominant rhodopsin mutation P347S (P347 is the penultimate residue of VXPX) exhibited accumulation of extracellular vesicles near the IS/OS junction, a phenotype consistent with trafficking defects (Li et al., 1996). When transgenes of rhodopsin C-terminal truncation mutants lacking the last four amino acids (VXPX) were

    W. BAEHR, R. CONSTANTINE, H. ZHANG, AND J. M. FREDERICK

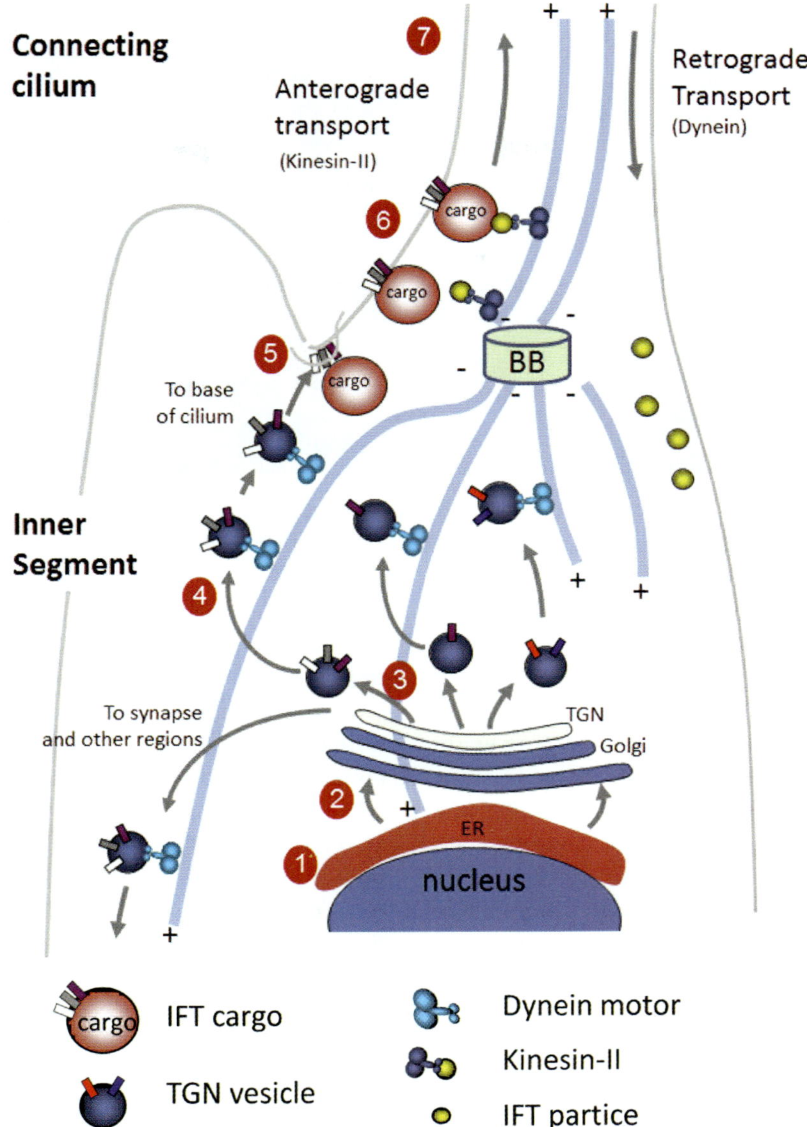

FIGURE 3.2   Schematic diagram of membrane protein trafficking in photoreceptors (adapted from Rosenbaum & Witman, 2002). Microtubules (blue lines) radiate from the basal body (BB) to all photoreceptor destinations. Plus-end-directed kinesins and minus-end-directed dyneins transport vesicles loaded with membrane proteins to their appropriate destinations, depending on the targeting signal. See text for details.

expressed on the $Rho^{-/-}$ background (Humphries et al., 1997; Lem et al., 1999), such transgenes were unable to rescue the knockout phenotype, although full-length rhodopsin did (Concepcion, Mendez, & Chen, 2002). In a different experimental approach, cone pigment has been supplied to rods. When mouse ML-opsin was knocked into the rhodopsin locus, ML-opsin was transported to the ROS, even though ML-opsin expression was low (11% of rhodopsin) (Sakurai et al., 2007). Alternatively, a transgene expressing human red cone opsin in mouse rods synthesized and trafficked enough L-opsin so that functional OS could form (Fu et al., 2008). Although cone pigments apparently carry C-terminal VXPX signals, targeting of cone pigments to the cone OS remains to be demonstrated.

*Cone Pigment Requires 11-*cis*-Retinal for Transport*

Less is known about pigment transport in cones. Presumably, newly synthesized cone pigments follow the secretory pathway and fuse with the PM at the periciliary ridge, as described for rhodopsin in rods. However, several important differences in rod versus cone pigment transport have been revealed by mouse germline knockouts such as LRAT, RPE65, and GC1. LRAT (lecithin retinol acyl transferase) and RPE65 (retinoid

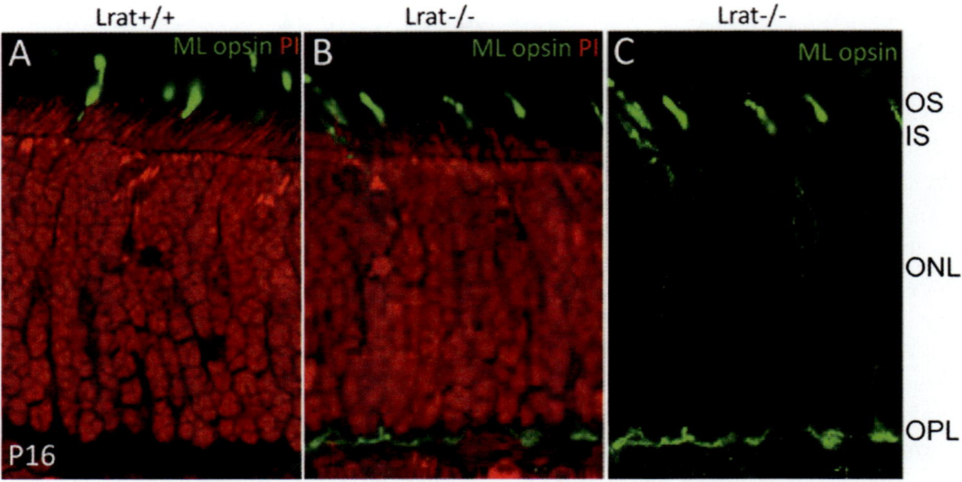

FIGURE 3.3   ML-opsin (green) mislocalization in cones of P16 Lrat$^{-/-}$ mice, contrasted with propidium iodide (PI). (A) Transverse section of wild-type retina. (B) Retina of Lrat$^{-/-}$ littermate. (C) Cones of Lrat$^{-/-}$ mouse, revealed using one channel.

isomerase) are RPE enzymes involved in recycling 11-*cis*-retinal, the chromophore for both rod and cone pigments. Both LRAT and RPE65 knockout mice lack 11-*cis*-retinal required for visual pigment regeneration (Batten et al., 2004; Redmond et al., 1998; Zhang et al., 2008). Cone opsins (S-opsin and M/L-opsin) were found to mislocalize as early as P16 in cones of *Lrat*$^{-/-}$ and *Rpe65*$^{-/-}$ mice (Rohrer et al., 2005; Zhang et al., 2008). As shown (figure 3.3C), ML-opsin can be found in the OS and IS, perinuclear regions, axons, and synaptic pedicles. Rates of cone degeneration and ML-opsin mislocalization in retinas of *Lrat*$^{-/-}$ and *RPE65*$^{-/-}$ mice are identical (Fan et al., 2008). At days P30–45, mutant OS are essentially undetectable. Early and repeated administration of exogenous 11-*cis*-retinal to *Rpe65*$^{-/-}$ *Rho*$^{-/-}$ double knockout pups prevented mislocalization, consistent with a key role of 11-*cis*-retinal bound to cone opsins for protein sorting, transport, targeting, and ultimately, cell survival. By contrast, the apoprotein opsin trafficked to the rod OS unimpeded in the absence of chromophore (Fan et al., 2008). When transgenic S-opsin was expressed in rods of *Lrat*$^{-/-}$ mice, S-opsin mislocalized and aggregated, causing rapid rod degeneration as well (Zhang et al., 2011b). Thus, 11-*cis*-retinal is key for inducing a conformation in S-opsin that enables correct trafficking regardless of the photoreceptor (rod or cone) environment.

Cone Pigments Mistraffic in GC and Kinesin-II (KIf3a) Knockouts

Severe cone pigment-mistargeting and absence of cascade components in the cone OS have been observed also in GC1 (Gucy2e) knockouts (Baehr et al., 2007) and heterotrimeric kinesin-II (KIF3A) knockouts (Avasthi et al., 2009). GC1 is a single transmembrane protein that is expressed at low levels relative to visual pigments. In contrast to opsins, mouse GC1 has no contiguous ciliary sorting signal in its cytoplasmic domain (Karan et al., 2011). Knockout of GC1 renders cones nonfunctional due to a lack of cGMP production. Cone OS form, but the disk membrane structure is affected severely. S-cone and M/L-cone pigments transport initially to the OS but become increasingly mislocalized to the IS and to vesicles that appear to have detached from the cone IS. As in both LRAT and RPE65 knockouts, proteins of the entire cone cascade were absent in *GC1*$^{-/-}$ cone OS. Because of GC1's low abundance relative to opsins and its apparent lack of a sorting signal, it is conceivable that GC1 traffics together with cone pigment along the secretory pathway. Because GC1 deletion has a severe effect on the routing of cone opsins and peripheral membrane proteins, GC1 may play a key role in assembling cargo or tethering cargo to molecular motors. Whether GC1 deficiency affects cargo at the TGN or the IFT level is unknown (Karan et al., 2008). Knockout of both rod GCs has no effect on the rhodopsin or transducin transport pathways because both are found at near normal levels prior to degeneration in GCdko ROS (Baehr et al., 2007).

The kinesin-2 subfamily consists of two anterograde motors: homodimeric and heterotrimeric kinesin. The heterotrimeric motor, kinesin-II, consists of KIF3A (kinesin family member 3a), KIF3B, and KAP3 (kinesin-associated protein 3) subunits (Yamazaki et al., 1995). Heterotrimeric kinesin-II is a molecular motor localized to the connecting cilium and axoneme of mammalian photoreceptors. In cones lacking KIF3A (the obligatory motor subunit of kinesin-II), membrane proteins involved in phototransduction (including S- and

M/L-opsins, PDE6, transducin, GRK1, and CNGA3) do not traffic to the OS, resulting in the complete absence of a photopic electroretinogram. Ultrastructurally, mutant cone OS show severe defects in membrane organization as early as P13, 1 day post–eye opening. These results indicate that kinesin-II selectively regulates trafficking of membrane proteins involved in cone phototransduction to the OS. A rod-specific knockout of kinesin-II causes rapid degeneration without affecting transport of rhodopsin or other members of the rod cascade, which again emphasizes different transport requirements.

Cone pigment mistargeting was observed in mouse models in which mutant cones lack the cone CNG channel α-subunit (CNGA3) (Michalakis et al., 2005), BBS4 (Abd-El-Barr et al., 2007), TULP1 (Grossman et al., 2011), and complement factor H (Coffey et al., 2007), suggesting that vesicular and intraflagellar transport in cones is critically dependent on several gene products/cofactors. Phenotypic similarity among several genetically altered mice in which knocking out one gene disrupts the function of seemingly unconnected gene products suggests the assembly of large cargo that is dependent on multiple factors (11-*cis*-retinal, GC1, small GTPases, IFT particles, molecular motors) for vesicle formation and anterograde intraflagellar transport.

BIOSYNTHESIS AND TRANSPORT OF TRANSDUCIN

The mechanisms underlying trafficking and ciliary targeting of heterotrimeric G proteins and GPCR-protein complexes in sensory neurons are largely unknown. Most details of this phenomenon derive from transfection of recombinant constructs in tissue culture (HEK293 cells, yeast) (Marrari et al., 2007). In the canonical model Gα subunits carrying the myristoylation consensus sequence (Farazi, Waksman, & Gordon, 2001) are acylated cotranslationally, and Gβ and Gγ subunits combine following biosynthesis and prenylation of Gγ. The field generally agrees that Gαβγ heterotrimer formation is essential for targeting (Marrari et al., 2007), but it remains unclear how the complex traffics to the PM. When Gα$_s$ and Gβγ were overexpressed in HEK cells, they targeted to the plasma membrane (PM) independent of the Golgi, most likely by diffusion. However, when a GPCR was coexpressed, PM targeting was Golgi-dependent, and both GPCR and G-protein followed the classic secretory pathway (Dupre & Hebert, 2006). Apparently, entire signaling complexes including GPCRs, G-protein, and the target enzyme are assembled following biosynthesis prior to PM targeting (Dupre et al., 2007, 2009).

Transducin subunits are synthesized on free ribosomes of the IS cytosol (figure 3.4). Tα is acylated either

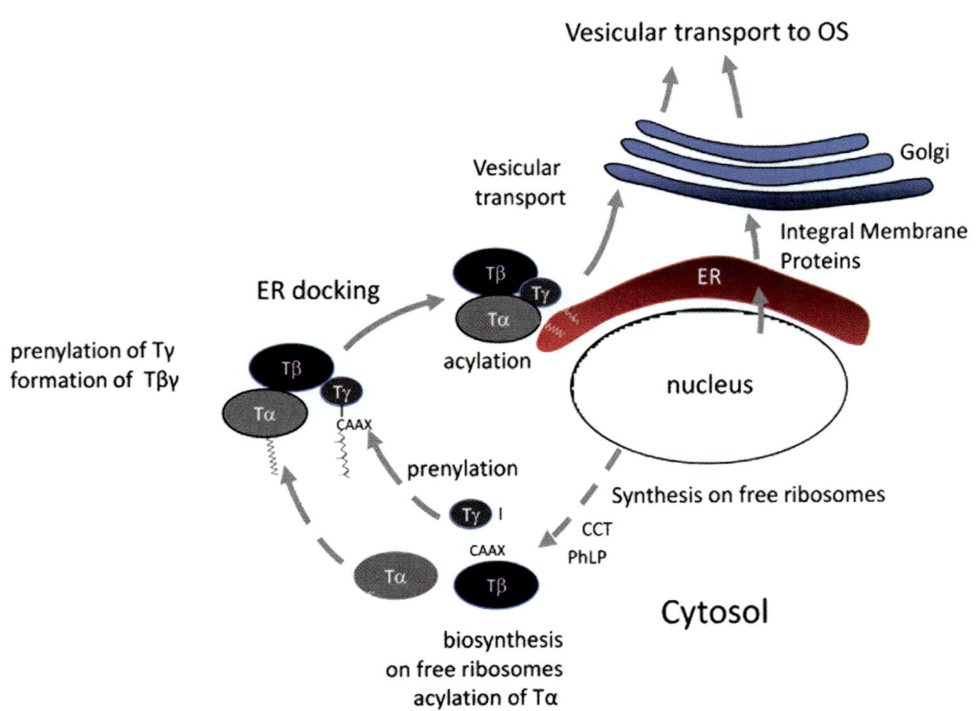

FIGURE 3.4   Transducin synthesis and processing. Tα is thought to be acylated cotranslationally, whereas Tγ is prenylated posttranslationally. Three transducin subunits combine at the ER to form a heterotrimeric G protein; CCT and PhLP are chaperones essential for correct folding of Tβ.

cotranslationally or by an unidentified resident ER acyl transferase. Acylation of Tα is essential for membrane association and correct targeting to the OS, as recombinant Tα (G2A) remains cytoplasmic and does not target to the OS (Kerov et al., 2007). Tβ biosynthesis and folding require the chaperonin CCT and the cochaperone PhLP1 (Lukov et al., 2005; Posokhova et al., 2011). Following release from CCT, Tβ combines with Tγ to form the obligate heterodimer Tβγ. Tγ is farnesylated in the cytosol by a soluble farnesyl transferase before or after heterodimer formation. The heterodimer then docks to the ER for further processing (Marrari et al., 2007). At the ER, the prenylated C-terminus of Tγ (prenyl-CAAX) is modified by removal of the three terminal amino acids (AAX) and carboxy-methylation of the C-terminal Cys to form prenyl-COOCH$_3$ (Hannoush & Sun, 2010). Genetic deletion of Tγ resulted in the degradation and mistargeting of Tα and Tβ, suggesting that heterotrimer formation is required for correct transducin localization (Lobanova et al., 2008). It is presumed that membrane-associated transducin traffics to the distal part of the IS by vesicular transport using molecular motors, most likely together with its GPCR, rhodopsin. However, trafficking of transducin subunits by diffusion to the OS has recently been proposed under specific conditions (see next paragraph). Whether by active transport, passive diffusion, or a combination of both, heterotrimeric transducin must be distributed efficiently, as it is synthesized at a rate of approximately 200,000 heterotrimers/day/rod (80 disks are replaced daily, containing about 2,500 × 80 molecules of transducin).

Light-Activated Transducin Translocation to the IS and Return to the OS

Light-activation of rhodopsin triggers GTP/GDP exchange on Tα causing Tα^{GTP} and Tβγ to dissociate. Under persistent illumination, Tα^{GTP} and Tβγ diffuse into the IS with a $t\frac{1}{2}$ of 3–5 min for Tα and a $t\frac{1}{2}$ of approximately 12 min for Tβγ (Calvert et al., 2006). Light-driven translocation results in approximately 80% of all transducin molecules moving from OS to IS (~1.6 million) in just minutes. The general consensus in the field remains that this enormous flux of molecules (1.5 million T/OS/10 min) can only occur by passive, three-dimensional diffusion. On arrival in the IS, Tα^{GTP} hydrolyzes to Tα^{GDP}, which then recombines with Tβγ to reform a heterotrimer that associates with IS membranes. The precise membrane sink where the heterotrimer docks under persistent illumination is unknown, but likely candidates include the ER, Golgi, and other intracellular membranes.

On resumption of dark adaptation, both Tα and Tβγ subunits return to the OS with a $t\frac{1}{2}$ of ~2 h. This relatively slow rate of return was suggested to occur by active transport using molecular motors (dynein). However, transducin has been shown to translocate back to the OS in eyecups depleted of ATP, without which molecular motors cannot function (Slepak & Hurley, 2008). If diffusion were to occur, the heterotrimer would have to dissociate again, perhaps forming soluble complexes with lipid-binding proteins. Tβγ is known to associate with phosducin, which reduces its affinity for both Tα and rod OS membranes (Sokolov et al., 2004). Further, it has been shown that the farnesyl group attached to Tγ may bind to PrBP/δ, as Tβγ trafficking was impeded in Pde6δ$^{-/-}$ rods (Zhang et al., 2007). Because a lipid-binding protein capable of solubilizing Tα had not yet been identified, a significant problem to be resolved was the mechanism underlying the discrepancies measured in the diffusion kinetics from OS to IS as compared to IS to OS.

The Acyl-Binding Protein UNC119

In experiments designed to address the slow rate of transducin movement from IS to OS during dark adaptation, we identified the widely expressed protein Uncoordinated 119 (UNC119A) as a chaperone that specifically enables Tα to travel from the IS to the OS (Zhang et al., 2011a; Constantine et al., 2012). UNC119 is 27-kDa polypeptide identified in the basal body proteome of various invertebrate organisms (Zhang et al., 2011a) and the mouse photoreceptor sensory cilium complex (Liu et al., 2007). UNC-119 was first discovered in *C. elegans* on the basis of a spontaneous mutation affecting locomotion, feeding behavior, and chemosensation (Maduro & Pilgrim, 1995). Independently, a protein termed Retina Gene 4 (RG4) was discovered in the retina and recognized to be a *C. elegans unc-119* orthologue (Higashide et al., 1996). UNC119 predominates in mouse photoreceptor IS and synaptic terminals and is barely detectable in the OS (Higashide, McLaren, & Inana, 1998; Ishiba et al., 2007; Swanson et al., 1998).

We have shown that UNC119 is an acyl-binding protein with specificity for G-protein α-subunits by several independent methods (Zhang et al., 2011a). First, GST-UNC119 pulled down only native Tα but not unacylated Tα(G2A); second, an acylated N-terminal Tα-mimicking peptide competed with endogenous Tα for UNC119 in pulldowns, but an unacylated peptide had no effect; third, isothermal titration calorimetry (ITC) identified a binding constant of 220 nM for the

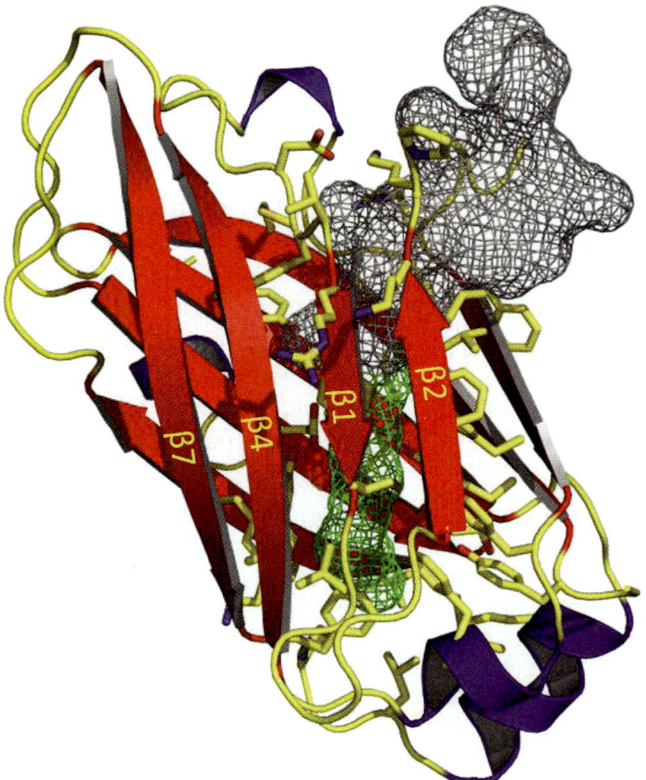

FIGURE 3.5 The UNC119 lipid-binding cavity. Acylated Tα peptide is depicted as a space-filling mesh with the lauroyl chain (green) attached to Tα residues (dark gray). UNC119 is tilted 45° to the left to orient the molecule along the vertical lipid axis. UNC119 residues (depicted as yellow sticks) are primarily hydrophobic and line the lipid-binding cavity, which is buried deeply within the UNC119 protein.

On arrival at the IS following light-induced translocation, transducin subunits recombine following GTP hydrolysis, permitting heterotrimeric transducin to dock to IS membranes, which are devoid of rhodopsin. We postulate that UNC119 plays a key role in the elution of Tα^{GTP} from the IS membranes after spontaneous GTP/GDP exchange in the absence of its GEF (rhodopsin), based on the following observations. First, in the *Unc119$^{-/-}$* mouse, the return of Tα to the OS in the dark is partially blocked. Even after 3 h of dark adaptation, a significant amount of Tα remains in the IS. Second, UNC119 elutes heterotrimeric transducin from IS membranes in the absence of GTP only very weakly. Third, UNC119 elutes Tα bound to light-adapted membranes only in the presence of GTP, suggesting that the dissociation of Tα and Tβγ is essential for UNC119/Tα complex formation. Fourth, the requirement of GTP suggests that both UNC119 and GTP/GDP exchange may be required for solubilization of Tα^{GTP} and Tβγ. In the absence of rhodopsin, GTP/GDP exchange is very slow (rate constant 10^{-4}/s) (Cowan, Wensel, & Arshavsky, 2000). The GTP requirement for Tα extraction from membranes and the slow rate constant suggest that this event may be the rate-limiting step for return to the OS. This rate constant dictates that it will take approximately 10^4 s (166 min) for transducin to return to the OS, which is consistent with return rates observed experimentally (Calvert et al., 2006; Elias et al., 2004). On solubilization of the UNC119/Tα-GTP complex, it may diffuse passively depositing Tα onto disks in the OS. After Tα^{GTP} solubilization, PrBP/δ or phosducin may extract Tβγ, thereby enabling diffusion of these subunits back to the OS (Gopalakrishna et al., 2011; Zhang et al., 2011a).

SUMMARY

Rods and cones are similar but distinct. Both cells have a light-sensing function and corresponding molecular compositions; visual pigments and transducin, major phototransduction components, are present on OS disks in high concentrations. OS are replaced every 10 days requiring highly active biosynthetic capacity and efficient transport pathways. Postbiosynthesis, rhodopsin and cone pigments are presumed to follow the conventional secretory pathway along microtubules toward the basal body and the periciliary ridge membrane complex. Vesicles carrying pigments are most likely accompanied by their heterotrimeric G-proteins and other peripheral membrane proteins (Karan et al.,

acylated peptide, whereas the unacylated peptide did not bind; and finally, a cocrystal structure of an acylated N-terminal Tα-mimicking peptide with UNC119 revealed that the acyl side chain is buried in the hydrophobic pocket formed by the β-sandwich of UNC119 (figure 3.5). The crystal structure, determined at a 2.0-Å resolution, revealed an immunoglobulin-like β-sandwich fold similar to that seen in human PrBP/δ (PDB ID:1KSH) (Hanzal-Bayer et al., 2002) and RhoGDI (PDB ID:1DOA) (Hoffman, Nassar, & Cerione, 2000). The lipid chain of the acylated Tα N-terminal peptide is buried deeply within UNC119's hydrophobic cavity. Although the β-sandwich folds of UNC119, PrBP/δ, and RhoGDI are structurally very similar, there is an important difference in lipid binding: the entrance through which the lipid can enter the hydrophobic pocket in UNC119 is located on the opposite edge of the PrBP/δ and RhoGDI pockets (Zhang et al., 2011a).

2008). An important distinction is that 11-*cis*-retinal is absolutely required for targeting of cone opsins (and the entire cone phototransduction cascade) to the OS. In contrast, rhodopsin and transducin traffic to their OS destination in the absence of chromophore. Further, absence of GC1 severely affects targeting of cone pigments, transducin, and other phototransduction components, whereas deletion of both rod GCs is known to affect PDE6 only.

Retinas of genetically engineered mice have revealed additional relationships among phototransduction components. Transducin may travel between segments by vesicular transport or diffusion. Diffusion occurs in persistent bright light when transducin is exiting the OS following rhodopsin-catalyzed GTP/GDP exchange and dissociation into Tα-GTP and Tβγ. In the IS, Tα-GTP and Tβγ components recombine and bind to membranes. Return to the OS in the dark is slow (hours), controlled by GTP/GDP exchange in the absence of a GEF and by low levels of UNC119 relative to transducin. Transducin thus appears to use two modes of trafficking: a GPCR-dependent mode in which it traffics together with rhodopsin postbiosynthetically and a GPCR-independent mode in which the subunits associate with lipid-binding proteins and diffuse.

REFERENCES

Abd-El-Barr, M. M., Sykoudis, K., Andrabi, S., Eichers, E. R., Pennesi, M. E., Tan, P. L., et al. (2007). Impaired photoreceptor protein transport and synaptic transmission in a mouse model of Bardet-Biedl syndrome. *Vision Research, 47,* 3394–3407. doi:10.1016/j.visres.2007.09.016.

Anderson, D. H., Fisher, S. K., & Steinberg, R. H. (1978). Mammalian cones: Disk shedding, phagocytosis, and renewal. *Investigative Ophthalmology & Visual Science, 17,* 117–133.

Avasthi, P., Watt, C. B., Williams, D. S., Le, Y. Z., Li, S., Chen, C. K., et al. (2009). Trafficking of membrane proteins to cone but not rod outer segments is dependent on heterotrimeric kinesin-II. *Journal of Neuroscience, 29,* 14287–14298.

Baehr, W., Karan, S., Maeda, T., Luo, D. G., Li, S., Bronson, J. D., et al. (2007). The function of guanylate cyclase 1 and guanylate cyclase 2 in rod and cone photoreceptors. *Journal of Biological Chemistry, 282,* 8837–8847. doi:10.1074/jbc. M610369200.

Batten, M. L., Imanishi, Y., Maeda, T., Tu, D. C., Moise, A. R., Bronson, D., et al. (2004). Lecithin-retinol acyltransferase is essential for accumulation of all-*trans*-retinyl esters in the eye and in the liver. *Journal of Biological Chemistry, 279,* 10422–10432. doi:10.1074/jbc.M312410200.

Besharse, J. C., & Hollyfield, J. G. (1979). Turnover of mouse photoreceptor outer segments in constant light and darkness. *Investigative Ophthalmology & Visual Science, 18,* 1019–1024.

Bhowmick, R., Li, M., Sun, J., Baker, S. A., Insinna, C., & Besharse, J. C. (2009). Photoreceptor IFT complexes containing chaperones, guanylyl cyclase 1 and rhodopsin. *Traffic (Copenhagen, Denmark), 10,* 648–663.

Calvert, P. D., Strissel, K. J., Schiesser, W. E., Pugh, E. N., Jr., & Arshavsky, V. Y. (2006). Light-driven translocation of signaling proteins in vertebrate photoreceptors. *Trends in Cell Biology, 16,* 560–568. doi:10.1016/j.tcb.2006.09.001.

Carter-Dawson, L. D., & LaVail, M. M. (1979). Rods and cones in the mouse retina. I. Structural analysis using light and electron microscopy. *Journal of Comparative Neurology, 188,* 245–262.

Chuang, J. Z., Zhao, Y., & Sung, C. H. (2007). SARA-regulated vesicular targeting underlies formation of the light-sensing organelle in mammalian rods. *Cell, 130,* 535–547.

Coffey, P. J., Gias, C., McDermott, C. J., Lundh, P., Pickering, M. C., Sethi, C., et al. (2007). Complement factor H deficiency in aged mice causes retinal abnormalities and visual dysfunction. *Proceedings of the National Academy of Sciences of the United States of America, 104,* 16651–16656.

Concepcion, F., Mendez, A., & Chen, J. (2002). The carboxyl-terminal domain is essential for rhodopsin transport in rod photoreceptors. *Vision Research, 42,* 417–426.

Constantine, R., Zhang, H., Gerstner, C. D., Frederick, J. M., & Baehr, W. (2012). Uncoordinated (UNC)119: Coordinating the trafficking of myristoylated proteins. *Vision Research,* http://dx.doi.org/10.1016/j.visres.2012.08.012.

Cowan, C. W., Wensel, T. G., & Arshavsky, V. Y. (2000). Enzymology of GTPase acceleration in phototransduction. *Methods in Enzymology, 315,* 524–538.

Deretic, D. (1997). Rab proteins and post-Golgi trafficking of rhodopsin in photoreceptor cells. *Electrophoresis, 18,* 2537–2541.

Deretic, D. (1998). Post-Golgi trafficking of rhodopsin in retinal photoreceptors. *Eye (London, England), 12*(Pt 3b), 526–530.

Deretic, D. (2006). A role for rhodopsin in a signal transduction cascade that regulates membrane trafficking and photoreceptor polarity. *Vision Research, 46,* 4427–4433.

Deretic, D., Huber, L. A., Ransom, N., Mancini, M., Simons, K., & Papermaster, D. S. (1995). rab8 in retinal photoreceptors may participate in rhodopsin transport and in rod outer segment disk morphogenesis. *Journal of Cell Science, 108,* 215–224.

Deretic, D., & Papermaster, D. S. (1993). Rab6 is associated with a compartment that transports rhodopsin from the trans-Golgi to the site of rod outer segment disk formation in frog retinal photoreceptors. *Journal of Cell Science, 106,* 803–813.

Deretic, D., Schmerl, S., Hargrave, P. A., Arendt, A., & McDowell, J. H. (1998). Regulation of sorting and post-Golgi trafficking of rhodopsin by its C- terminal sequence QVS(A)PA. *Proceedings of the National Academy of Sciences USA, 95,* 10620–10625.

Deretic, D., Williams, A. H., Ransom, N., Morel, V., Hargrave, P. A., & Arendt, A. (2005). Rhodopsin C terminus, the site of mutations causing retinal disease, regulates trafficking by binding to ADP-ribosylation factor 4 (ARF4). *Proceedings of the National Academy of Sciences of the United States of America, 102,* 3301–3306.

Dupre, D. J., Baragli, A., Rebois, R. V., Ethier, N., & Hebert, T. E. (2007). Signalling complexes associated with adenylyl cyclase II are assembled during their biosynthesis. *Cellular Signalling, 19,* 481–489.

Dupre, D. J., & Hebert, T. E. (2006). Biosynthesis and trafficking of seven transmembrane receptor signalling complexes. *Cellular Signalling, 18,* 1549–1559.

Dupre, D. J., Robitaille, M., Rebois, R. V., & Hebert, T. E. (2009). The role of Gbetagamma subunits in the organization, assembly, and function of GPCR signaling complexes. *Annual Review of Pharmacology and Toxicology, 49,* 31–56.

Elias, R. V., Sezate, S. S., Cao, W., & McGinnis, J. F. (2004). Temporal kinetics of the light/dark translocation and compartmentation of arrestin and alpha-transducin in mouse photoreceptor cells. *Molecular Vision, 10,* 672–681.

Fan, J., Rohrer, B., Frederick, J. M., Baehr, W., & Crouch, R. K. (2008). Rpe65$^{-/-}$ and Lrat$^{-/-}$ mice: Comparable models of leber congenital amaurosis. *Investigative Ophthalmology & Visual Science, 49,* 2384–2389.

Farazi, T. A., Waksman, G., & Gordon, J. I. (2001). The biology and enzymology of protein N-myristoylation. *Journal of Biological Chemistry, 276,* 39501–39504.

Fu, Y., Kefalov, V., Luo, D. G., Xue, T., & Yau, K. W. (2008). Quantal noise from human red cone pigment. *Nature Neuroscience, 11,* 565–571.

Gopalakrishna, K. N., Doddapuneni, K., Boyd, K. K., Masuho, I., Martemyanov, K. A., & Artemyev, N. O. (2011). Interaction of transducin with uncoordinated 119 protein (UNC119): Implications for the model of transducin trafficking in rod photoreceptors. *Journal of Biological Chemistry, 286,* 28954–28962.

Grossman, G. H., Watson, R. F., Pauer, G. J., Bollinger, K., & Hagstrom, S. A. (2011). Immunocytochemical evidence of Tulp1-dependent outer segment protein transport pathways in photoreceptor cells. *Experimental Eye Research, 93,* 658–668.

Hannoush, R. N., & Sun, J. (2010). The chemical toolbox for monitoring protein fatty acylation and prenylation. *Nature Chemical Biology, 6,* 498–506.

Hanzal-Bayer, M., Renault, L., Roversi, P., Wittinghofer, A., & Hillig, R. C. (2002). The complex of Arl2-GTP and PDE delta: From structure to function. *European Molecular Biology Organization Journal, 21,* 2095–2106. doi:10.1093/emboj/21.9.2095.

Higashide, T., McLaren, M. J., & Inana, G. (1998). Localization of HRG4, a photoreceptor protein homologous to Unc-119, in ribbon synapse. *Investigative Ophthalmology & Visual Science, 39,* 690–698.

Higashide, T., Murakami, A., McLaren, M. J., & Inana, G. (1996). Cloning of the cDNA for a novel photoreceptor protein. *Journal of Biological Chemistry, 271,* 1797–1804.

Hoffman, G. R., Nassar, N., & Cerione, R. A. (2000). Structure of the Rho family GTP-binding protein Cdc42 in complex with the multifunctional regulator RhoGDI. *Cell, 100,* 345–356.

Humphries, M. M., Rancourt, D., Farrar, G. J., Kenna, P., Hazel, M., Bush, R. A., et al. (1997). Retinopathy induced in mice by targeted disruption of the rhodopsin gene. *Nature Genetics, 15,* 216–219.

Ishiba, Y., Higashide, T., Mori, N., Kobayashi, A., Kubota, S., McLaren, M. J., et al. (2007). Targeted inactivation of synaptic HRG4 (UNC119) causes dysfunction in the distal photoreceptor and slow retinal degeneration, revealing a new function. *Experimental Eye Research, 84,* 473–485.

Jeon, C. J., Strettoi, E., & Masland, R. H. (1998). The major cell populations of the mouse retina. *Journal of Neuroscience, 18,* 8936–8946.

Karan, S., Tam, B. M., Moritz, O. L., & Baehr, W. (2011). Targeting of mouse guanylate cyclase 1 (*Gucy2e*) to *Xenopus laevis* rod outer segments. *Vision Research, 51,* 2304–2311.

Karan, S., Zhang, H., Li, S., Frederick, J. M., & Baehr, W. (2008). A model for transport of membrane-associated phototransduction polypeptides in rod and cone photoreceptor inner segments. *Vision Research, 48,* 442–452.

Kerov, V., Rubin, W. W., Natochin, M., Melling, N. A., Burns, M. E., & Artemyev, N. O. (2007). N-Terminal fatty acylation of transducin profoundly influences its localization and the kinetics of photoresponse in rods. *Journal of Neuroscience, 27,* 10270–10277.

Koerschen, H. G., Illing, M., Seifert, R., Sesti, F., Williams, A., Gotzes, S., et al. (1995). A 240 kDa protein represents the complete beta subunit of the cyclic nucleotide-gated channel from rod photoreceptor. *Neuron, 15,* 627–636.

LaVail, M. M. (1976). Rod outer segment disk shedding in rat retina: Relationship to cyclic lighting. *Science, 194,* 1071–1074.

LaVail, M. M. (1980). Circadian nature of rod outer segment disk shedding in the rat. *Investigative Ophthalmology & Visual Science, 19,* 407–411.

LaVail, M. M., & Ward, P. A. (1978). Studies on the hormonal control of circadian outer segment disk shedding in the rat retina. *Investigative Ophthalmology & Visual Science, 17,* 1189–1193.

Lecomte, F. J., Ismail, N., & High, S. (2003). Making membrane proteins at the mammalian endoplasmic reticulum. *Biochemical Society Transactions, 31,* 1248–1252.

Lem, J., Krasnoperova, N. V., Calvert, P. D., Kosaras, B., Cameron, D. A., Nicolo, M., et al. (1999). Morphological, physiological, and biochemical changes in rhodopsin knockout mice. *Proceedings of the National Academy of Sciences of the United States of America, 96,* 736–741.

Li, T., Snyder, W. K., Olsson, J. E., & Dryja, T. P. (1996). Transgenic mice carrying the dominant rhodopsin mutation P347S: Evidence for defective vectorial transport of rhodopsin to the outer segments. *Proceedings of the National Academy of Sciences of the United States of America, 93,* 14176–14181.

Liu, Q., Tan, G., Levenkova, N., Li, T., Pugh, E. N., Jr., Rux, J. J., et al. (2007). The proteome of the mouse photoreceptor sensory cilium complex. *Molecular & Cellular Proteomics, 6,* 1299–1317.

Lobanova, E. S., Finkelstein, S., Herrmann, R., Chen, Y. M., Kessler, C., Michaud, N. A., et al. (2008). Transducin gamma-subunit sets expression levels of alpha- and beta-subunits and is crucial for rod viability. *Journal of Neuroscience, 28,* 3510–3520.

Lukov, G. L., Hu, T., McLaughlin, J. N., Hamm, H. E., & Willardson, B. M. (2005). Phosducin-like protein acts as a molecular chaperone for G protein betagamma dimer assembly. *European Molecular Biology Organization Journal, 24,* 1965–1975.

Maduro, M., & Pilgrim, D. (1995). Identification and cloning of *unc-119,* a gene expressed in the *Caenorhabditis elegans* nervous system. *Genetics, 141,* 977–988.

Marrari, Y., Crouthamel, M., Irannejad, R., & Wedegaertner, P. B. (2007). Assembly and trafficking of heterotrimeric G proteins. *Biochemistry, 46,* 7665–7677.

Mazelova, J., Astuto-Gribble, L., Inoue, H., Tam, B. M., Schonteich, E., Prekeris, R., et al. (2009). Ciliary targeting motif VxPx directs assembly of a trafficking module through

Arf4. *European Molecular Biology Organization Journal, 28,* 183–192.

Michalakis, S., Geiger, H., Haverkamp, S., Hofmann, F., Gerstner, A., & Biel, M. (2005). Impaired opsin targeting and cone photoreceptor migration in the retina of mice lacking the cyclic nucleotide-gated channel CNGA3. *Investigative Ophthalmology & Visual Science, 46,* 1516–1524.

Milligan, G. (2009). G protein-coupled receptor hetero-dimerization: Contribution to pharmacology and function. *British Journal of Pharmacology, 158,* 5–14. doi:10.1111/j.1476-5381.2009.00169.x.

Moritz, O. L., Tam, B. M., Hurd, L. L., Peranen, J., Deretic, D., & Papermaster, D. S. (2001). Mutant rab8 impairs docking and fusion of rhodopsin-bearing post-Golgi membranes and causes cell death of transgenic *Xenopus* rods. *Molecular Biology of the Cell, 12,* 2341–2351.

Nachury, M. V., Seeley, E. S., & Jin, H. (2010). Trafficking to the ciliary membrane: How to get across the periciliary diffusion barrier? *Annual Review of Cell and Developmental Biology, 26,* 59–87.

Nickell, S., Park, P. S., Baumeister, W., & Palczewski, K. (2007). Three-dimensional architecture of murine rod outer segments determined by cryoelectron tomography. *Journal of Cell Biology, 177,* 917–925.

Posokhova, E., Song, H., Belcastro, M., Higgins, L., Bigley, L. R., Michaud, N. A., et al. (2011). Disruption of the chaperonin containing TCP-1 function affects protein networks essential for rod outer segment morphogenesis and survival. *Molecular & Cellular Proteomics, 10,* M110. doi:10.1074/mcp.M110.000570.

Redmond, T. M., Yu, S., Lee, E., Bok, D., Hamasaki, D., Chen, N., et al. (1998). Rpe65 is necessary for production of 11-*cis*-vitamin A in the retinal visual cycle. *Nature Genetics, 20,* 344–351.

Rodriguez-Boulan, E., Kreitzer, G., & Musch, A. (2005). Organization of vesicular trafficking in epithelia. *Nature Reviews. Molecular Cell Biology, 6,* 233–247.

Rohrer, B., Lohr, H. R., Humphries, P., Redmond, T. M., Seeliger, M. W., & Crouch, R. K. (2005). Cone opsin mislocalization in Rpe65$^{-/-}$ mice: A defect that can be corrected by 11-*cis* retinal. *Investigative Ophthalmology & Visual Science, 46,* 3876–3882.

Roof, D. J., & Heuser, J. E. (1982). Surfaces of rod photoreceptor disk membranes: Integral membrane components. *Journal of Cell Biology, 95,* 487–500.

Rosenbaum, J. L., & Witman, G. B. (2002). Intraflagellar transport. *Nature Reviews. Molecular Cell Biology, 3,* 813–825.

Sakurai, K., Onishi, A., Imai, H., Chisaka, O., Ueda, Y., Usukura, J., et al. (2007). Physiological properties of rod photoreceptor cells in green-sensitive cone pigment knock-in mice. *Journal of General Physiology, 130,* 21–40.

Satoh, A. K., O'Tousa, J. E., Ozaki, K., & Ready, D. F. (2005). Rab11 mediates post-Golgi trafficking of rhodopsin to the photosensitive apical membrane of *Drosophila* photoreceptors. *Development, 132,* 1487–1497.

Slepak, V. Z., & Hurley, J. B. (2008). Mechanism of light-induced translocation of arrestin and transducin in photoreceptors: Interaction-restricted diffusion. *IUBMB Life, 60,* 2–9. doi:10.1002/iub.7.

Sokolov, M., Strissel, K. J., Leskov, I. B., Michaud, N. A., Govardovskii, V. I., & Arshavsky, V. Y. (2004).

Phosducin facilitates light-driven transducin translocation in rod photoreceptors. Evidence from the phosducin knockout mouse. *Journal of Biological Chemistry, 279,* 19149–19156.

Strauss, O. (2005). The retinal pigment epithelium in visual function. *Physiological Reviews, 85,* 845–881.

Sung, C. H., & Tai, A. W. (2000). Rhodopsin trafficking and its role in retinal dystrophies. *International Review of Cytology, 195,* 215–267.

Swanson, D. A., Chang, J. T., Campochiaro, P. A., Zack, D. J., & Valle, D. (1998). Mammalian orthologs of *C. elegans* unc-119 highly expressed in photoreceptors. *Investigative Ophthalmology & Visual Science, 39,* 2085–2094.

Tai, A. W., Chuang, J. Z., Bode, C., Wolfrum, U., & Sung, C. H. (1999). Rhodopsin's carboxy-terminal cytoplasmic tail acts as a membrane receptor for cytoplasmic dynein by binding to the dynein light chain Tctex-1. *Cell, 97,* 877–887.

Tam, B. M., Moritz, O. L., Hurd, L. B., & Papermaster, D. S. (2000). Identification of an outer segment targeting signal in the COOH terminus of rhodopsin using transgenic *Xenopus laevis*. *Journal of Cell Biology, 151,* 1369–1380.

van Vliet, C., Thomas, E. C., Merino-Trigo, A., Teasdale, R. D., & Gleeson, P. A. (2003). Intracellular sorting and transport of proteins. *Progress in Biophysics and Molecular Biology, 83,* 1–45. doi:10.1016/S0079-6107(03)00019-1.

Yamazaki, H., Nakata, T., Okada, Y., & Hirokawa, N. (1995). KIF3A/B: A heterodimeric kinesin superfamily protein that works as a microtubule plus end-directed motor for membrane organelle transport. *Journal of Cell Biology, 130,* 1387–1399.

Yeh, T. Y., Peretti, D., Chuang, J. Z., Rodriguez-Boulan, E., & Sung, C. H. (2006). Regulatory dissociation of Tctex-1 light chain from dynein complex is essential for the apical delivery of rhodopsin. *Traffic (Copenhagen, Denmark), 7,* 1495–1502.

Young, R. W. (1967). The renewal of photoreceptor cell outer segments. *Journal of Cell Biology, 33,* 61–72.

Young, R. W., & Bok, D. (1969). Participation of the retinal pigment epithelium in the rod outer segment renewal process. *Journal of Cell Biology, 42,* 392–403.

Zhang, H., Constantine, R., Vorobiev, V., Chen, Y., Seetharaman, J., Huang, Y. J., et al. (2011a). UNC119 regulates G protein trafficking in sensory neurons. *Nature Neuroscience, 14,* 874–880.

Zhang, H., Fan, J., Li, S., Karan, S., Rohrer, B., Palczewski, K., et al. (2008). Trafficking of membrane-associated proteins to cone photoreceptor outer segments requires the chromophore 11-*cis*-retinal. *Journal of Neuroscience, 28,* 4008–4014.

Zhang, H., Li, S., Doan, T., Rieke, F., Detwiler, P. B., Frederick, J. M., et al. (2007). Deletion of PrBP/{delta} impedes transport of GRK1 and PDE6 catalytic subunits to photoreceptor outer segments. *Proceedings of the National Academy of Sciences of the United States of America, 104,* 8857–8862.

Zhang, T., Zhang, N., Baehr, W., & Fu, Y. (2011b). Cone opsin determines the time course of cone photoreceptor degeneration in Leber congenital amaurosis. *Proceedings of the National Academy of Sciences of the United States of America, 108,* 8879–8884.

4 Photoreceptor–RPE Interactions: Diurnal Phagocytosis

LINDA RUGGIERO AND SILVIA C. FINNEMANN

PHAGOCYTOSIS OF SHED PHOTORECEPTOR TIPS: ONE OF MANY ESSENTIAL FUNCTIONS OF THE RPE

The continuous process of rod photoreceptor outer segment renewal was first reported in 1967 (Young, 1967). In mammals formation of disks at proximal ends of outer segments is precisely balanced by shedding of distal, most aged outer segment tips and phagocytosis of the resulting outer segment fragments by the underlying retinal pigment epithelium (RPE). Shedding and phagocytosis are complex processes. Here, we first review the physiology of outer segment renewal. We then examine molecular mechanisms enabling this process, providing a snapshot of recent and ongoing research and an outlook on aspects that need further study. Finally, we briefly discuss how defects in outer segment renewal may contribute to retinal disease.

The RPE as Blood–Retinal Barrier between Choroid and Outer Retina

The RPE consists of a polarized monolayer of postmitotic epithelial cells. Early during embryonic development the RPE and neural retina are two undifferentiated layers separated by a lumen that becomes the interphotoreceptor matrix (IPM) (reviewed by Strauss, 2005). Differentiation of RPE and neural retina progress in a coordinated fashion, and maturation of either layer is dependent on the other. Once codeveloped and differentiated, mammalian RPE and photoreceptors interact for life. Basally, RPE cells adhere to Bruch's membrane, which separates the RPE from the vascularized choroid. Tight junctions physically isolate the apical aspect of the RPE and the subretinal space from the choroid, establishing the blood–retinal barrier. Apically, the RPE does not face a lumen but extends highly elongated microvillar processes that ensheath rod and cone outer segments. These microvilli carry receptors for specific outer segment or IPM components (Bonilha et al., 2004). RPE–outer segment interactions are crucial both during development and throughout adulthood.

The RPE as Support Cells Maintaining Photoreceptor Survival and Functions

RPE cells support photoreceptor homeostasis by secreting growth factors (Strauss, 2005). Selective tight junction permeability and strict polarity of transmembrane transporters and ion channels allow RPE cells to control volume and ionic composition of the subretinal space (Wimmers, Karl, & Strauss, 2007).

Reisomerization of bleached all-*trans*-retinal by RPE65 in the RPE is an essential part of the visual cycle (Redmond et al., 1998). Additionally, RPE cells protect the retina from photooxidation. Melanin pigment granules, abundant in the RPE cytoplasm, act to absorb scattered light and as sites of detoxification (Burke et al., 2011).

Functional interactions through which RPE cells support photoreceptors rely on permanent robust adhesion between RPE and outer retina. Retinal adhesion is mainly mediated by binding of RPE apical surface receptors to ligands in the IPM (Hollyfield et al., 1989) that remain poorly understood. The integrin receptor $\alpha v \beta 5$ is the only receptor to date shown to contribute to retinal adhesion in vivo (Nandrot et al., 2006). However, lack of $\alpha v \beta 5$ reduces retinal adhesion only moderately, and a ligand for this adhesive function of $\alpha v \beta 5$ remains to be identified.

Like retinal adhesion, phagocytosis of shed photoreceptor outer segment fragments (POS) depends on RPE apical surface receptors. Unlike retinal adhesion, shedding and phagocytosis of rod POS occur in a diurnal burst that is under circadian control. This complex and fascinating process will be our focus for the remainder of this chapter.

PHYSIOLOGY OF PHOTORECEPTOR OUTER SEGMENT SHEDDING AND RPE PHAGOCYTOSIS

In human, mouse, and rat retina, rod photoreceptors significantly outnumber cones, with 97.2% rods and 2.8% cones in mice (Jeon, Strettoi, & Masland, 1998). Because of their abundance, rods have been the model

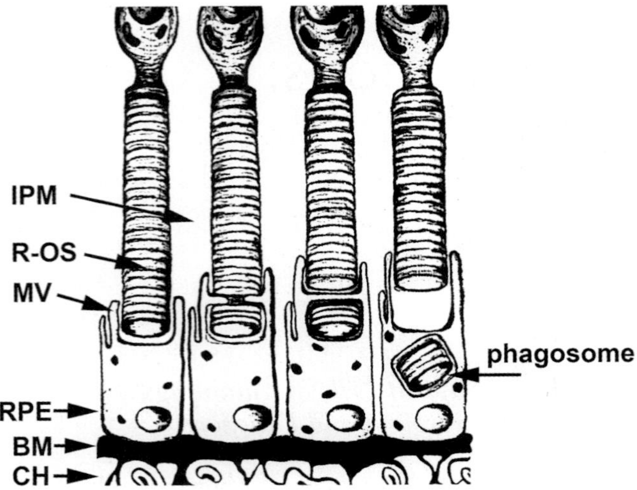

FIGURE 4.1　Anatomy of the RPE and its role in POS renewal. RPE cells extend apical microvilli (MV) into the interphotoreceptor matrix (IPM) and ensheath rod photoreceptor outer segments (R-OS). At their basolateral surface, RPE cells interact with Bruch's membrane (BM), which separates the RPE from the choroid (CH). Progression of shedding and phagocytosis of spent POS are shown from left to right and culminate in internalization of POS and formation of RPE phagosomes.

photoreceptor cell type for studies of outer segment–RPE interactions. Young (1967) reported that newly synthesized protein in rat, mouse, and frog photoreceptors was initially incorporated only into the proximal region of outer segments. In all three species radiolabeled proteins migrated as a discrete band to the rod's distal tip. Young and Bok (1969) demonstrated that rod disks shed from distal tips to yield POS of defined size that were phagocytosed by the adjacent RPE. Following ingestion, POS phagosomes matured to phagolysosomes, leading to complete digestion of engulfed material and loss of radioactive tracer. These seminal studies demonstrated that rods constantly renew their outer segments and that the RPE is responsible for engulfing and digesting rod POS (figure 4.1). It is generally assumed that photoreceptors must renew their outer segments to prevent accumulation of damaged lipids and proteins over time, for instance due to light exposure.

The Diurnal Shedding and Phagocytosis of Rod POS

POS shedding and RPE phagocytosis display a diurnal rhythm that is synchronized to external lighting conditions. In rats maintained in a constant light:dark cycle (LD), with 12 h each of light and darkness (12:12 LD), the rod POS phagosome content of the RPE is highest at light onset (LaVail, 1976). This intriguing finding led to the question of whether this rhythm was circadian in nature. In order for a rhythm to be considered circadian, it must meet specific criteria. (1) A circadian rhythm recurs daily with a period of close to 24 h. POS renewal studies on albino rats reared in cyclic light provided clear evidence for this (LaVail, 1976). (2) Circadian rhythms are endogenous. The rhythm must persist in the absence of environmental cues such as light. POS renewal meets this requirement because rats maintained in constant darkness (DD) displayed a sustained rhythm in RPE phagocytosis (Goldman, Teirstein, & O'Brien, 1980; LaVail, 1980). (3) Circadian rhythms are entrainable. Entrainment is established when the timing of a rhythm is reset or shifted by an external stimulus. In mammals the most potent entraining cue is light. Photoentrainment of POS shedding/RPE phagocytosis has been observed in rats that were shifted from one cyclic lighting regime to another. These rats adjusted the rhythm of POS renewal such that the phagocytosis peak occurred shortly after the new time of light onset (LaVail, 1980). Taken together, these studies prove that the diurnal rhythm of rod POS shedding/phagocytosis is under circadian control in the mammalian retina.

Testing effects of altering lighting regimes on the rhythm of POS shedding/phagocytosis allowed further characterization of the circadian regulation of POS renewal. The initial studies demonstrating persistence of POS renewal in DD were conducted on rats kept in constant conditions for up to 3 days. Subsequent studies examined animals maintained in DD for longer periods. After 12 days in DD, the rhythm of shedding/phagocytosis was sustained, but its period had become slightly longer, and the peak with increased phagosome number within the RPE was broadened (LaVail, 1980). In contrast, in rats placed in constant light conditions (LL), the rhythmic peak of shedding/phagocytosis was lost, suggesting that LL can inhibit POS renewal. However, transfer of the animals to darkness restored the original rhythm (Goldman, Teirstein, & O'Brien, 1980).

Synchronized rod disk shedding/phagocytosis in LD is conserved among all species studied to date, including cat (Fisher, Pfeffer, & Anderson, 1983), chicken (Young, 1978), and the frog *Xenopus laevis* (Besharse, Hollyfield, & Rayborn, 1977). In some nonmammalian species studied, rod POS renewal is not circadian but, rather, depends on light input. This is the case in the frog *Rana pipiens*, which shows decreased rod POS shedding when placed in DD and a burst of shedding stimulated by light exposure (Basinger, Hoffman, & Matthes, 1976). Similarly, in goldfish, rod POS shedding/

phagocytosis is synchronized to external lighting, with a peak 2–4 h into the light phase; however, no rhythm is observed in DD (Bassi & Powers, 1990; O'Day & Young, 1978).

Diurnal Shedding and Phagocytosis of Cone POS

Although the majority of research on outer segment renewal has focused on rods, cone outer segment renewal has been studied in several species or strains with cone-rich retinas. When Young conducted his original pulse-chase studies on photoreceptor renewal in rhesus monkeys, he did not observe the same banding of radioactive proteins in cones that he observed in rods (Young, 1971b). This led him to conclude that cones may not continuously renew their outer segments (Young, 1971a). However, when phagosomes were found within the RPE of human fovea, the cone-rich region of the retina that is virtually devoid of rods, the hypothesis that cones shed spent distal tips reemerged (Hogan, 1972). Closer examination of cones in both the cat retina and the extrafoveal region of the human retina showed that cone POS were ensheathed by long extensions of the apical RPE membrane, allowing for the necessary POS–RPE interactions (Steinberg & Wood, 1974). Electron micrographs of human retina further demonstrated phagosomes both within these apical processes and in the apical portion of RPE cells (Steinberg, Wood, & Hogan, 1977). Additional studies in tree squirrels and rhesus monkeys confirmed that mammalian cones display POS shedding/phagocytosis (Anderson & Fisher, 1975; Anderson et al., 1980). More recent research has identified the nile rat (*Arvicanthis ansorgei*), a diurnal rodent with about 10-fold more cone photoreceptors than mice, as an exciting new animal model for the study of cone POS renewal (Bobu et al., 2006). Finally, the generation of mice lacking the neural retina leucine zipper (Nrl) gene has provided an additional animal model for the study of cones. Absence of the transcription factor Nrl precludes rod differentiation. As a result, all photoreceptors in Nrl$^{-/-}$ retina display morphological and physiological characteristics of cones (Daniele et al., 2005; Mears et al., 2001). However, a high frequency of retinal tissue rosettes and progressive retinal degeneration indicate secondary abnormalities and defects in Nrl$^{-/-}$ retina that may also affect RPE–cone outer segment interactions (Mustafi et al., 2011). Therefore, studies of cone POS renewal in the Nrl$^{-/-}$ retina may need to be interpreted with caution.

Like rods, cones renew outer segments in a diurnal rhythm. However, the time of peak of cone POS phagosomes in RPE is species dependent. In some species, such as cat and tree shrew, cones shed shortly after light onset as do rods (Fisher et al., 1983; Immel & Fisher, 1985). In other species, such as lizard, chicken, goldfish, and tree squirrel, the burst of phagocytosis occurs at different time points several hours into the dark phase (O'Day & Young, 1978; Tabor, Fisher, & Anderson, 1980; Young, 1978). In LD, the nile rat displays a peak in both rod and cone POS phagosome numbers in the RPE approximately 1 h after light onset. In DD, these rhythms persist, suggesting that cone renewal, like that of rods, is under circadian control (Bobu & Hicks, 2009).

Taken together, cone POS shed and are phagocytosed by the RPE with diurnal rhythms that are species dependent and circadian in mammals. However, based on the original pulse chase assays, cone POS likely do not contain the most aged proteins and lipids of cone outer segments, as newly synthesized cone outer segment components do not persist in a common location in cones. Rather, its open disk configuration may allow protein distribution along the length of the cone outer segment.

Clock Influence on Outer Segment Renewal: Local and Central Control

In mammals circadian rhythms are generated by endogenous rhythmic activity of neurons within the suprachiasmatic nucleus (SCN) of the hypothalamus (Welsh et al., 1995). The SCN acts as the "master clock," using information from the environment, such as retinal light input, to synchronize rhythms of peripheral clocks (Green & Besharse, 2004). The mammalian retina is a peripheral clock because it displays endogenous circadian rhythms that persist when isolated from the brain. In rats with transected optic nerves, POS renewal was rhythmic despite loss of synaptic connections to the brain, suggesting that this rhythm is maintained locally, in the eye (Terman, Reme, & Terman, 1993). However, the rhythm was unable to be reset by light unless the optic nerve remained intact (Teirstein, Goldman, & O'Brien, 1980), suggesting that this aspect of circadian shedding/phagocytosis required communication with the SCN.

Application of pharmacological compounds such as excitatory amino acids to dissected amphibian retina hints at candidate molecular pathways for cellular–molecular control of POS shedding/phagocytosis (Green & Besharse, 2004; Greenberger & Besharse, 1985). Molecular components of the mammalian circadian clock regulating POS renewal and RPE phagocytosis have yet to be identified.

MOLECULAR MECHANISMS OF POS PHAGOCYTOSIS

While diurnal regulation of shedding/phagocytosis remains under investigation, both cell culture models and availability of genetically modified animals have facilitated identification of molecules that function in the process of POS clearance itself. Like phagocytosis by other cell types, POS phagocytosis by RPE cells consists of distinct recognition/binding, actin-dependent engulfment, and digestions steps, each of which requires coordination of protein activities by cell signaling pathways. Studies of RPE phagocytosis in vivo analyze POS clearance under physiological conditions and reveal functional consequences of its alterations for retinal function. However, in vivo POS shedding and the RPE phagocytic clearance process are necessarily linked, and analysis of POS turnover cannot directly distinguish between primary and secondary defects. In contrast, POS uptake assays using RPE cells in culture can experimentally isolate recognition/binding and engulfment processes (Finnemann & Rodriguez-Boulan, 1999; Mayerson & Hall, 1986). Furthermore, cell culture studies allow testing of RPE phagocytic function irrespective of other RPE–POS interactions whose changes in genetically altered experimental animals may secondarily impact phagocytosis. Taken together, complementary in vivo and cell culture assays of POS clearance are necessary to gain comprehensive knowledge of relevant molecules and mechanisms.

POS Recognition and Surface Binding by the RPE

RPE cells with active apical phagocytic machinery bind POS in a saturable, receptor-mediated recognition/tethering process. The binding step of RPE phagocytosis is governed by the integrin adhesion receptor $\alpha v\beta 5$, which is expressed solely by RPE cells in human and rodent retina (Anderson, Johnson, & Hageman, 1995; Finnemann et al., 1997; Lin & Clegg, 1998; Miceli, Newsome, & Tate, 1997). $\alpha v\beta 5$ is anchored to the actin cytoskeleton and resides in a detergent-resistant complex with the tetraspanin CD81 at the apical plasma membrane of RPE cells in situ and in culture (Chang & Finnemann, 2007; Finnemann & Rodriguez-Boulan, 1999). Blockade of either $\alpha v\beta 5$ or CD81 reduces POS binding by RPE cells, but manipulation of CD81 function is irrelevant in the absence of $\alpha v\beta 5$. Thus, CD81 regulates $\alpha v\beta 5$ binding activity for POS. Both in RPE in culture and in situ, $\alpha v\beta 5$ receptors do not bind POS directly but via the secreted integrin ligand MFG-E8. MFG-E8 is

produced by photoreceptors and RPE and localizes to the subretinal space in rodent retina (Burgess et al., 2006). POS particles must be opsonized by MFG-E8 to engage $\alpha v\beta 5$ receptors of RPE cells for POS binding (Nandrot et al., 2007).

Mechanically stable binding that withstands stringent washing protocols as in cell culture assays may largely be irrelevant to POS uptake in situ, where shedding POS and the RPE's phagocytic machinery are in continuous contact. Indeed, RPE lacking either $\alpha v\beta 5$ integrin or MFG-E8 engulf POS in vivo. However, both $\beta 5^{-/-}$ and MFG-E8$^{-/-}$ mice lack the diurnal rhythm of POS phagocytosis (Nandrot et al., 2004, 2007). Thus, POS-MFG-E8 ligation of $\alpha v\beta 5$ receptors is required for the synchronized burst of engulfment that is characteristic of the RPE.

POS Internalization by the RPE

Focal adhesion kinase (FAK) resides in a complex with unoccupied apical $\alpha v\beta 5$ integrin receptors of RPE cells, and activation of FAK is required for POS internalization (Finnemann, 2003). In response to challenge with POS opsonized with MFG-E8, outside-in signaling by $\alpha v\beta 5$ leads to phosphorylation of multiple tyrosine residues of FAK known to be essential for catalytic activity. On activation, FAK dissociates from the apical $\alpha v\beta 5$ complex but remains active. Active FAK either directly or indirectly promotes tyrosine phosphorylation of MerTK, a receptor tyrosine kinase that is essential for POS engulfment.

The Royal College of Surgeons (RCS) rat has served as experimental model for hereditary retinal degeneration for decades. Key studies in the 1970s and 1980s specified the cause of retinal dystrophy as lack of POS engulfment ability of RCS RPE (Chaitin & Hall, 1983; Mullen & LaVail, 1976). In 2000 two independent studies reported a mutation in the RCS rat's *MERTK* gene causing lack of functional MerTK (D'Cruz et al., 2000; Nandrot et al., 2000). MerTK reexpression restores engulfment function of RCS RPE (Feng et al., 2002). POS engulfment may include MerTK ligation by extracellular glycoproteins such as Gas6 or Protein S, which can activate MerTK in in vitro assays (Hall et al., 2005; Prasad et al., 2006). Gas6$^{-/-}$ retina is phenotypically normal, and in vivo relevance for POS clearance of Protein S is unknown. MerTK downstream signaling during POS phagocytosis remains poorly characterized. Myosin II redistribution in RCS RPE suggests that MerTK may be involved in regulating contractility during POS engulfment (Strick, Feng, & Vollrath, 2009). In vivo, MerTK tyrosine

 LINDA RUGGIERO AND SILVIA C. FINNEMANN

phosphorylation, likely indicative of its activity, displays a diurnal rhythm peaking shortly after light onset (Nandrot et al., 2004). The same is true for FAK activity. These rhythms are absent in mice lacking αvβ5 integrin or its ligand MFG-E8.

While ligands of RPE phagocytic receptors αvβ5 and, presumably, MerTK are secreted glycoproteins, RPE cells also express receptors for oxidized lipoproteins. Ligation of the lipid scavenger receptor CD36 affects the kinetics of POS engulfment by RPE in culture (Finnemann & Silverstein, 2001; Ryeom, Sparrow, & Silverstein, 1996). In response to light damage, rat retina generates CD36-specific ligands (Sun et al., 2006). CD36 ligation may thus increase the efficiency of RPE phagocytosis to compensate for excess POS shedding or debris degeneration in stressed retina.

The ultimate goal of phagocytic signaling is the recruitment of F-actin to tethered POS and the generation of force that pulls particles into the cell. In RPE cells, recruitment of an F-actin phagocytic cup requires activation specifically of the GTPase Rac1 (Mao & Finnemann, 2012). Like stimulation of MerTK and FAK, activation of Rac1 depends on αvβ5 integrin. Surprisingly, however, inhibition of Rac1 has no effect on tyrosine kinase signaling and vice versa during POS challenge. Thus, αvβ5 integrin engagement by MFG-E8-POS activates two distinct and independent downstream pathways toward FAK/MerTK and Rac1, both of which are essential for POS engulfment. F-actin in the phagocytic cup is likely further altered to promote engulfment by actin-binding proteins. Phosphorylation of retinal annexin-2 shortly after light onset and a small but specific delay in RPE phagocytosis in mice lacking annexin-2 support a role for this actin-binding protein in POS clearance (Law et al., 2009).

POS Digestion by the RPE

Engulfment of POS is followed by their degradation by the RPE phagolysosomal system. Surprisingly little is known about how RPE cells accomplish this essential part of the phagocytic process. Rhodopsin proteolysis involves cathepsin proteases S and D and phagolysosomal acidification (Bosch, Horwitz, & Bok, 1993; Rakoczy et al., 1994; Regan et al., 1980). A diurnal rhythm in cathepsin D activity directly correlates with RPE phagocytosis (Deguchi et al., 1994; Kim & Kwak, 1996). POS phagolysosomal degradation is altered in mice lacking the protein melanoregulin, which is also involved in pigmentation, suggesting a link between phagolysosomal and melanosomal biogenesis (Damek-Poprawa et al., 2009).

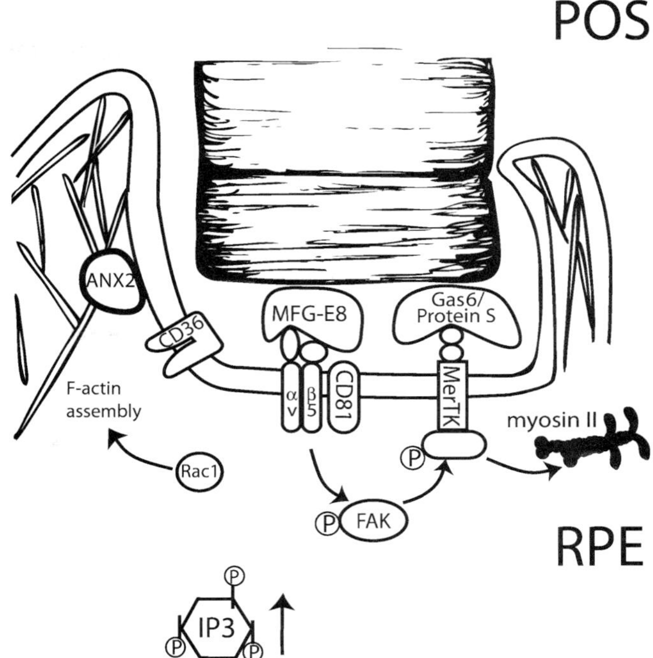

FIGURE 4.2   The molecular mechanism of RPE phagocytosis. The drawing illustrates the complex molecular mechanisms underlying receptor-mediated RPE phagocytosis of spent POS as described in detail in the text.

RPE Phagocytosis: Unique Aspect

The molecules identified as relevant for POS phagocytosis (shown in figure 4.2) are not uniquely expressed by the RPE. Rather, RPE cells utilize a molecular mechanism for POS uptake that is similar to that used for apoptotic cell phagocytosis by other cell types. Indeed, POS and apoptotic cells quantitatively compete for binding and engulfment by RPE and macrophages (Finnemann & Rodriguez-Boulan, 1999). However, RPE cells differ from other phagocytes in the strict rhythmicity and overall frequency of their phagocytic activity. Over a lifetime, postmitotic RPE cells phagocytose and degrade more material than any other cell type in the body. Each RPE cell must reliably produce a vigorous diurnal phagocytic burst and prevent inappropriate attack of intact outer segment tips at other times. RPE cells prevent excess POS interaction with αvβ5 integrin with negative feedback mechanisms that depend on activities of MerTK but not engulfment itself (Nandrot et al., 2012). Other molecular mechanisms RPE cells employ to control their phagocytic activity remain under investigation.

The circadian nature of POS renewal essentially precludes directly quantifying POS phagocytosis in tissue obtained from human donor eyes. There is therefore no direct evidence linking altered efficiency of diurnal POS clearance with a specific human retinal disease. However, insight into disease genetics and from phenotype correlations strongly suggests that defective clearance of POS by the RPE contributes to retinal degeneration in human patients. Mutations in the *MERTK* gene are associated with the development of retinitis pigmentosa (RP) (Gal et al., 2000; McHenry et al., 2004; Thompson et al., 2002). Mutant MerTK receptors in human RP patients likely retain partial activity, as they do not cause the rapid and complete photoreceptor loss of the RCS rat. Unlike most photoreceptor proteins whose mutation causes RP (rhodopsin being the most prominent), RPE proteins that form the phagocytic machinery for POS also participate in other cellular functions and in tissues other than the RPE. We speculate that dramatic dysfunction of these proteins has not been found in RP patients because such defects would cause morbidity/lethality unrelated to POS phagocytosis. However, enhanced genetic screens may lead to identification of additional mutations in human patients that cause minor protein dysfunction and changes in outer segment renewal.

Because both RPE and photoreceptors are postmitotic and generally do not turn over in the mature mammalian eye, even minor defects in diurnal POS phagocytosis by the RPE are likely to ultimately worsen with age and to impair retinal health and function. The most obvious phenotypic change with age of human RPE is the accumulation of indigestible lipofuscin in terminal storage organelles derived from POS phagolysosomes (Feeney, 1978; Sparrow & Boulton, 2005). Excess lipofuscin is associated with rare hereditary forms of human macular disease (Allikmets et al., 1997; Zhang et al., 2001) and may contribute to RPE distress in age-related macular degeneration. Low levels of lipofuscin components affect functions of RPE in culture, such as mitochondrial energy production (Vives-Bauza et al., 2008), phagolysosomal digestion (Finnemann, Leung, & Rodriguez-Boulan, 2002), and lipid trafficking (Lakkaraju, Finnemann, & Rodriguez-Boulan, 2007). Elevated levels of lipofuscin are cytotoxic, especially if cells are exposed to light (Hunter et al., 2011). Although there is agreement that lipofuscin accumulation negatively impacts RPE function, the debate continues over exactly how lipofuscin contributes to functional changes of aging human RPE and age-related macular degeneration.

Lessons from Animal Models with Aberrant POS Phagocytosis

Studies of animal models have yielded enormous insight into how specific defects in RPE phagocytosis affect RPE and retinal health. The RCS rat was the first animal model recognized both for its defect in POS engulfment by the RPE and for the rapid and complete retinal degeneration that ensues (Bok & Hall, 1971; Bourne, Campbell, & Tansley, 1938). The pathology of the RCS retina is essentially a reaction of photoreceptors to unbalanced outer segment elongation, illustrating that photoreceptors cannot tolerate a lack of phagocytosis by the RPE.

As discussed in detail earlier, $\alpha v \beta 5$ integrin and MerTK are functionally linked in the RPE. However, loss of $\alpha v \beta 5$ integrin in $\beta 5^{-/-}$ mice has very different consequences for the retina as compared to loss of MerTK. Loss of the diurnal rhythm of POS phagocytosis, which is the primary consequence of lack of $\alpha v \beta 5$ integrin, does not cause obvious buildup of shed POS, elongation of outer segments, or any obvious sign of retinal distress (Nandrot & Finnemann, 2008). This implies that in these arrhythmic mice the rates of outer segment growth and shedding/phagocytosis of POS remain precisely balanced. From 6 months of age, RPE cells fill with lipofuscin, and retinal function declines in $\beta 5^{-/-}$ mice (Nandrot et al., 2004). Dietary supplementation with natural antioxidants is sufficient to largely prevent these changes (Yu et al., 2012). Notably, antioxidant consumption only from 3 to 6 months of age or from 6 to 9 months of age is sufficient to maintain vision in $\beta 5^{-/-}$ mice with age. In contrast, antioxidants have no effect on vision if consumed only from 9 to 12 months of age. At the molecular level antioxidant consumption prevents buildup of oxidized lipids and proteins in aging $\beta 5^{-/-}$ RPE. Interestingly, actin is particularly oxidized in $\beta 5^{-/-}$ RPE with age to the extent that the F-actin cytoskeleton is significantly destabilized. How loss of RPE phagocytosis rhythm secondarily causes oxidative stress remains to be understood.

Finally, altering POS protein digestion is sufficient to cause retinal degeneration. Mice lacking cathepsin D develop postnatal retinal atrophy (Koike et al., 2003). Mice with moderately reduced cathepsin D activity demonstrate buildup of RPE lipofuscin and moderate atrophy of RPE and photoreceptors with age (Rakoczy et al., 2002; Zhang et al., 2005).

SUMMARY AND FUTURE DIRECTIONS: UNDERSTANDING THE PHOTORECEPTOR SIDE

Much progress has been made over the past decades toward understanding the physiology of outer segment renewal and toward elucidating molecular mechanisms employed by RPE cells to phagocytose spent POS. In contrast, contributions of photoreceptor cells to shedding remain obscure. The healthy retina maintains a strict balance of outer segment growth and shedding, phagocytosis, and degradation of spent distal tips, suggesting that these processes are tightly controlled. POS shedding could be an autonomous process of photoreceptors, or the RPE could contribute to shedding. If shedding were solely an intrinsic mechanism of photoreceptors, one might expect to observe detached rod POS in the subretinal space at least occasionally, but this has yet to be reported. Studies of photoreceptor shedding are still hampered by the lack of in vitro models. Examining POS renewal in situ will require development of new experimental tools.

REFERENCES

Allikmets, R., Singh, N., Sun, H., Shroyer, N. F., Hutchinson, A., Chidambaram, A., et al. (1997). A photoreceptor cell-specific ATP-binding transporter gene (ABCR) is mutated in recessive Stargardt macular dystrophy. *Nature Genetics, 15*, 236–246. doi:10.1038/ng0397-236.

Anderson, D. H., & Fisher, S. K. (1975). Disk shedding in rodlike and conelike photoreceptors of tree squirrels. *Science, 187*, 953–955. doi:10.1126/science.1145180.

Anderson, D. H., Fisher, S. K., Erickson, P. A., & Tabor, G. A. (1980). Rod and cone disk shedding in the rhesus monkey retina: A quantitative study. *Experimental Eye Research, 30*, 559–574. doi:10.1016/0014-4835(80)90040-8.

Anderson, D. H., Johnson, L. V., & Hageman, G. S. (1995). Vitronectin receptor expression and distribution at the photoreceptor–retinal pigment epithelial interface. *Journal of Comparative Neurology, 360*, 1–16. doi:10.1002/cne.903600102.

Basinger, S., Hoffman, R., & Matthes, M. (1976). Photoreceptor shedding is initiated by light in the frog retina. *Science, 194*, 1074–1076. doi:10.1126/science.1086510.

Bassi, C. J., & Powers, M. K. (1990). Shedding of rod outer segments is light-driven in goldfish. *Investigative Ophthalmology & Visual Science, 31*, 2314–2319.

Besharse, J. C., Hollyfield, J. G., & Rayborn, M. E. (1977). Turnover of rod photoreceptor outer segments. II. Membrane addition and loss in relationship to light. *Journal of Cell Biology, 75*, 507–527. doi:10.1083/jcb.75.2.507.

Bobu, C., Craft, C. M., Masson-Pevet, M., & Hicks, D. (2006). Photoreceptor organization and rhythmic phagocytosis in the nile rat *Arvicanthis ansorgei*: A novel diurnal rodent model for the study of cone pathophysiology. *Investigative Ophthalmology & Visual Science, 47*(7), 3109–3118. doi:10.1167/iovs.05-1397.

Bobu, C., & Hicks, D. (2009). Regulation of retinal photoreceptor phagocytosis in a diurnal mammal by circadian clocks and ambient lighting. *Investigative Ophthalmology & Visual Science, 50*, 3495–3502. doi:10.1167/iovs.08-3145.

Bok, D., & Hall, M. O. (1971). The role of the pigment epithelium in the etiology of inherited retinal dystrophy in the rat. *Journal of Cell Biology, 49*, 664–682. doi:10.1083/jcb.49.3.664.

Bonilha, V. L., Bhattacharya, S. K., West, K. A., Sun, J., Crabb, J. W., Rayborn, M. E., et al. (2004). Proteomic characterization of isolated retinal pigment epithelium microvilli. *Molecular & Cellular Proteomics, 3*, 1119–1127. doi:10.1074/mcp.M400106-MCP200.

Bosch, E., Horwitz, J., & Bok, D. (1993). Phagocytosis of outer segments by retinal pigment epithelium: Phagosome–lysosome interaction. *Journal of Histochemistry and Cytochemistry, 41*, 253–263. doi:10.1177/41.2.8419462.

Bourne, M. C., Campbell, D. A., & Tansley, K. (1938). Hereditary degeneration of the rat retina. *British Journal of Ophthalmology, 22*, 613–622. doi:10.1136/bjo.22.10.613.

Burgess, B. L., Abrams, T. A., Nagata, S., & Hall, M. O. (2006). MFG-E8 in the retina and retinal pigment epithelium of rat and mouse. *Molecular Vision, 12*, 1437–1447.

Burke, J. M., Kaczara, P., Skumatz, C. M., Zareba, M., Raciti, M. W., & Sarna, T. (2011). Dynamic analyses reveal cytoprotection by RPE melanosomes against non-photic stress. *Molecular Vision, 17*, 2864–2877.

Chaitin, M. H., & Hall, M. O. (1983). Defective ingestion of rod outer segments by cultured dystrophic rat pigment epithelial cells. *Investigative Ophthalmology & Visual Science, 24*, 812–820.

Chang, Y., & Finnemann, S. C. (2007). Tetraspanin CD81 is required for the $\alpha v\beta 5$ integrin-dependent particle-binding step of RPE phagocytosis. *Journal of Cell Science, 120*, 3053–3063. doi:10.1242/jcs.006361.

Damek-Poprawa, M., Diemer, T., Lopes, V. S., Lillo, C., Harper, D. C., Marks, M. S., et al. (2009). Melanoregulin (MREG) modulates lysosome function in pigment epithelial cells. *Journal of Biological Chemistry, 284*, 10877–10889. doi:10.1074/jbc.M808857200.

Daniele, L. L., Lillo, C., Lyubarsky, A. L., Nikonov, S. S., Philp, N., Mears, A. J., et al. (2005). Cone-like morphological, molecular, and electrophysiological features of the photoreceptors of the Nrl knockout mouse. *Investigative Ophthalmology & Visual Science, 46*, 2156–2167. doi:10.1167/iovs.04-1427.

D'Cruz, P. M., Yasumura, D., Weir, J., Matthes, M. T., Abderrahim, H., LaVail, M. M., et al. (2000). Mutation of the receptor tyrosine kinase gene Mertk in the retinal dystrophic RCS rat. *Human Molecular Genetics, 9*, 645–651. doi:10.1093/hmg/9.4.645.

Deguchi, J., Yamamoto, A., Yoshimori, T., Sugasawa, K., Moriyama, Y., Futai, M., et al. (1994). Acidification of phagosomes and degradation of rod outer segments in rat retinal pigment epithelium. *Investigative Ophthalmology & Visual Science, 35*, 568–579.

Feeney, L. (1978). Lipofuscin and melanin of human retinal pigment epithelium. Fluorescence, enzyme cytochemical, and ultrastructural studies. *Investigative Ophthalmology & Visual Science, 17*, 583–600.

Feng, W., Yasumura, D., Matthes, M. T., LaVail, M. M., & Vollrath, D. (2002). Mertk triggers uptake of photoreceptor outer segments during phagocytosis by cultured retinal pigment epithelial cells. *Journal of Biological Chemistry, 277*, 17016–17022. doi:10.1074/jbc.M107876200.

Finnemann, S. C. (2003). Focal adhesion kinase signaling promotes phagocytosis of integrin-bound photoreceptors. *European Molecular Biology Organization Journal, 22*, 4143–4154. doi:10.1093/emboj/cdg416.

Finnemann, S. C., Bonilha, V. L., Marmorstein, A. D., & Rodriguez-Boulan, E. (1997). Phagocytosis of rod outer segments by retinal pigment epithelial cells requires αvβ5 integrin for binding but not for internalization. *Proceedings of the National Academy of Sciences of the United States of America, 94*, 12932–12937. doi:10.1073/pnas.94.24.12932.

Finnemann, S. C., Leung, L. W., & Rodriguez-Boulan, E. (2002). The lipofuscin component A2E selectively inhibits phagolysosomal degradation of photoreceptor phospholipid by the retinal pigment epithelium. *Proceedings of the National Academy of Sciences of the United States of America, 99*, 3842–3847. doi:10.1073/pnas.052025899.

Finnemann, S. C., & Rodriguez-Boulan, E. (1999). Macrophage and retinal pigment epithelium phagocytosis: Apoptotic cells and photoreceptors compete for αvβ3 and αvβ5 integrins, and protein kinase C regulates avβ5 binding and cytoskeletal linkage. *Journal of Experimental Medicine, 190*, 861–874. doi:10.1084/jem.190.6.861.

Finnemann, S. C., & Silverstein, R. L. (2001). Differential roles of CD36 and αvβ5 integrin in photoreceptor phagocytosis by the retinal pigment epithelium. *Journal of Experimental Medicine, 194*, 1289–1298. doi:10.1084/jem.194.9.1289.

Fisher, S. K., Pfeffer, B. A., & Anderson, D. H. (1983). Both rod and cone disk shedding are related to light onset in the cat. *Investigative Ophthalmology & Visual Science, 24*, 844–856.

Gal, A., Li, Y., Thompson, D. A., Weir, J., Orth, U., Jacobson, S. G., et al. (2000). Mutations in MERTK, the human orthologue of the RCS rat retinal dystrophy gene, cause retinitis pigmentosa. *Nature Genetics, 26*, 270–271. doi:10.1038/81555.

Goldman, A. I., Teirstein, P. S., & O'Brien, P. J. (1980). The role of ambient lighting in circadian disk shedding in the rod outer segment of the rat retina. *Investigative Ophthalmology & Visual Science, 19*, 1257–1267.

Green, C. B., & Besharse, J. C. (2004). Retinal circadian clocks and control of retinal physiology. *Journal of Biological Rhythms, 19*, 91–102. doi:10.1177/0748730404263002.

Greenberger, L. M., & Besharse, J. C. (1985). Stimulation of photoreceptor disk shedding and pigment epithelial phagocytosis by glutamate, aspartate, and other amino acids. *Journal of Comparative Neurology, 239*, 361–372. doi:10.1002/cne.902390402.

Hall, M. O., Obin, M. S., Heeb, M. J., Burgess, B. L., & Abrams, T. A. (2005). Both protein S and Gas6 stimulate outer segment phagocytosis by cultured rat retinal pigment epithelial cells. *Experimental Eye Research, 81*, 581–591. doi:10.1016/j.exer.2005.03.017.

Hogan, M. J. (1972). Role of the retinal pigment epithelium in macular disease. *Transactions—American Academy of Ophthalmology and Otolaryngology, 76*, 64–80.

Hollyfield, J. G., Varner, H. H., Rayborn, M. E., & Osterfeld, A. M. (1989). Retinal attachment to the pigment epithelium. *Retina, 9*, 59–68. doi:10.1097/00006982-198909010-00008.

Hunter, J. J., Morgan, J. I., Merigan, W. H., Sliney, D. H., Sparrow, J. R., & Williams, D. R. (2011). The susceptibility of the retina to photochemical damage from visible light. *Progress in Retinal and Eye Research, 31*, 28–42. doi:10.1016/j.preteyeres.2011.11.001.

Immel, J. H., & Fisher, S. K. (1985). Cone photoreceptor shedding in the tree shrew (*Tupaia belangerii*). *Cell and Tissue Research, 239*, 667–675. doi:10.1007/BF00219247.

Jeon, C. J., Strettoi, E., & Masland, R. H. (1998). The major cell populations of the mouse retina. *Journal of Neuroscience, 18*, 8936–8946.

Kim, I. T., & Kwak, J. S. (1996). Degradation of phagosomes and diurnal changes of lysosomes in rabbit retinal pigment epithelium. *Korean Journal of Ophthalmology, 10*, 82–91.

Koike, M., Shibata, M., Ohsawa, Y., Nakanishi, H., Koga, T., Kametaka, S., et al. (2003). Involvement of two different cell death pathways in retinal atrophy of cathepsin D–deficient mice. *Molecular and Cellular Neurosciences, 22*, 146–161. doi:10.1016/S1044-7431(03)00035-6.

Lakkaraju, A., Finnemann, S. C., & Rodriguez-Boulan, E. (2007). The lipofuscin fluorophore A2E perturbs cholesterol metabolism in retinal pigment epithelial cells. *Proceedings of the National Academy of Sciences of the United States of America, 104*, 11026–11031. doi:10.1073/pnas.0702504104.

LaVail, M. M. (1976). Rod outer segment disk shedding in rat retina: Relationship to cyclic lighting. *Science, 194*, 1071–1074. doi:10.1126/science.982063.

LaVail, M. M. (1980). Circadian nature of rod outer segment disk shedding in the rat. *Investigative Ophthalmology & Visual Science, 19*, 407–411.

Law, A. L., Ling, Q., Hajjar, K. A., Futter, C. E., Greenwood, J., Adamson, P., et al. (2009). Annexin A2 regulates phagocytosis of photoreceptor outer segments in the mouse retina. *Molecular Biology of the Cell, 20*, 3896–3904. doi:10.1091/mbc.E08-12-1204.

Lin, H., & Clegg, D. O. (1998). Integrin αvβ5 participates in the binding of photoreceptor rod outer segments during phagocytosis by cultured human retinal pigment epithelium. *Investigative Ophthalmology & Visual Science, 39*, 1703–1712.

Mao, Y., & Finnemann, S. C. (2012). Essential diurnal Rac1 activation during retinal phagocytosis requires αvβ5 integrin but not tyrosine kinases FAK or MerTK. *Molecular Biology of the Cell, 23*, 1104–1114. doi:10.1091/mbc.E11-10-0840.

Mayerson, P. L., & Hall, M. O. (1986). Rat retinal pigment epithelial cells show specificity of phagocytosis in vitro. *Journal of Cell Biology, 103*, 299–308. doi:10.1083/jcb.103.1.299.

McHenry, C. L., Liu, Y., Feng, W., Nair, A. R., Feathers, K. L., Ding, X., et al. (2004). MERTK arginine-844-cysteine in a patient with severe rod-cone dystrophy: Loss of mutant protein function in transfected cells. *Investigative Ophthalmology & Visual Science, 45*, 1456–1463. doi:10.1167/iovs.03-0909.

Mears, A. J., Kondo, M., Swain, P. K., Takada, Y., Bush, R. A., Saunders, T. L., et al. (2001). Nrl is required for rod photoreceptor development. *Nature Genetics, 29*, 447–452.

Miceli, M. V., Newsome, D. A., & Tate, D. J., Jr. (1997). Vitronectin is responsible for serum-stimulated uptake of rod outer segments by cultured retinal pigment epithelial cells. *Investigative Ophthalmology & Visual Science, 38*, 1588–1597.

Mullen, R. J., & LaVail, M. M. (1976). Inherited retinal dystrophy: Primary defect in pigment epithelium determined with experimental rat chimeras. *Science, 192*, 799–801. doi:10.1126/science.1265483.

Mustafi, D., Kevany, B. M., Genoud, C., Okano, K., Cideciyan, A. V., Sumaroka, A., et al. (2011). Defective photoreceptor phagocytosis in a mouse model of enhanced S-cone syndrome causes progressive retinal degeneration. *FASEB Journal, 25*, 3157–3176. doi:10.1096/fj.11-186767.

Nandrot, E. F., Anand, M., Almeida, D., Atabai, K., Sheppard, D., & Finnemann, S. C. (2007). Essential role for MFG-E8 as ligand for αvβ5 integrin in diurnal retinal phagocytosis. *Proceedings of the National Academy of Sciences of the United States of America, 104*, 12005–12010. doi:10.1073/pnas.0704756104.

Nandrot, E. F., Anand, M., Sircar, M., & Finnemann, S. C. (2006). Novel role for αvβ5 integrin in retinal adhesion and its diurnal peak. *American Journal of Physiology. Cell Physiology, 290*, C1256–C1262. doi:10.1152/ajpcell.00480.2005.

Nandrot, E., Dufour, E. M., Provost, A. C., Pequignot, M. O., Bonnel, S., Gogat, K., et al. (2000). Homozygous deletion in the coding sequence of the c-mer gene in RCS rats unravels general mechanisms of physiological cell adhesion and apoptosis. *Neurobiology of Disease, 7*, 586–599. doi:10.1006/nbdi.2000.0328.

Nandrot, E. F., & Finnemann, S. C. (2008). Lack of αvβ5 integrin receptor or its ligand MFG-E8: Distinct effects on retinal function. *Ophthalmic Research, 40*, 120–123. doi:10.1159/000119861.

Nandrot, E. F., Kim, Y., Brodie, S. E., Huang, X., Sheppard, D., & Finnemann, S. C. (2004). Loss of synchronized retinal phagocytosis and age-related blindness in mice lacking αvβ5 integrin. *Journal of Experimental Medicine, 200*, 1539–1545. doi:10.1084/jem.20041447.

Nandrot, E. F., Silva, K. E., Scelfo, C., & Finnemann, S. C. (2012). Retinal pigment epithelial cells use a MerTK-dependent mechanism to limit the phagocytic particle binding activity of αvβ5 integrin. *Biology of the Cell, 104*, 1–16. doi:10.1111/boc.201100076.

O'Day, W. T., & Young, R. W. (1978). Rhythmic daily shedding of outer-segment membranes by visual cells in the goldfish. *Journal of Cell Biology, 76*, 593–604. doi:10.1083/jcb.76.3.593.

Prasad, D., Rothlin, C. V., Burrola, P., Burstyn-Cohen, T., Lu, Q., Garcia de Frutos, P., et al. (2006). TAM receptor function in the retinal pigment epithelium. *Molecular and Cellular Neurosciences, 33*, 96–108. doi:10.1016/j.mcn.2006.06.011.

Rakoczy, P. E., Mann, K., Cavaney, D. M., Robertson, T., Papadimitreou, J., & Constable, I. J. (1994). Detection and possible functions of a cysteine protease involved in digestion of rod outer segments by retinal pigment epithelial cells. *Investigative Ophthalmology & Visual Science, 35*, 4100–4108.

Rakoczy, P. E., Zhang, D., Robertson, T., Barnett, N. L., Papadimitriou, J., Constable, I. J., et al. (2002). Progressive age-related changes similar to age-related macular degeneration in a transgenic mouse model. *American Journal of Pathology, 161*, 1515–1524. doi:10.1016/S0002-9440(10)64427-6.

Redmond, T. M., Yu, S., Lee, E., Bok, D., Hamasaki, D., Chen, N., et al. (1998). RPE65 is necessary for production of 11-*cis*-vitamin A in the retinal visual cycle. *Nature Genetics, 20*, 344–351.

Regan, C. M., de Grip, W. J., Daemen, F. J., & Bonting, S. L. (1980). Degradation of rhodopsin by a lysosomal fraction of retinal pigment epithelium: Biochemical aspects of the visual process. XLI. *Experimental Eye Research, 30*, 183–191. doi:10.1016/0014-4835(80)90112-8.

Ryeom, S. W., Sparrow, J. R., & Silverstein, R. L. (1996). CD36 participates in the phagocytosis of rod outer segments by retinal pigment epithelium. *Journal of Cell Science, 109*, 387–395.

Sparrow, J. R., & Boulton, M. (2005). RPE lipofuscin and its role in retinal pathobiology. *Experimental Eye Research, 80*, 595–606. doi:10.1016/j.exer.2005.01.007.

Steinberg, R. H., & Wood, I. (1974). Pigment epithelial cell ensheathment of cone outer segments in the retina of the domestic cat. *Proceedings of the Royal Society of London. Series B, Biological Sciences, 187*, 461–478. doi:10.1098/rspb.1974.0088.

Steinberg, R. H., Wood, I., & Hogan, M. J. (1977). Pigment epithelial ensheathment and phagocytosis of extrafoveal cones in human retina. *Philosophical Transactions of the Royal Society of London. Series B, Biological Sciences, 277*, 459–474. doi:10.1098/rstb.1977.0028.

Strauss, O. (2005). The retinal pigment epithelium in visual function. *Physiological Reviews, 85*, 845–881. doi:10.1152/physrev.00021.2004.

Strick, D. J., Feng, W., & Vollrath, D. (2009). Mertk drives myosin II redistribution during retinal pigment epithelial phagocytosis. *Investigative Ophthalmology & Visual Science, 50*, 2427–2435. doi:10.1167/iovs.08-3058.

Sun, M., Finnemann, S. C., Febbraio, M., Shan, L., Annangudi, S. P., Podrez, E. A., et al. (2006). Light-induced oxidation of photoreceptor outer segment phospholipids generates ligands for CD36-mediated phagocytosis by retinal pigment epithelium: A potential mechanism for modulating outer segment phagocytosis under oxidant stress conditions. *Journal of Biological Chemistry, 281*, 4222–4230. doi:10.1074/jbc.M509769200.

Tabor, G. A., Fisher, S. K., & Anderson, D. H. (1980). Rod and cone disk shedding in light-entrained tree squirrels. *Experimental Eye Research, 30*, 545–557. doi:10.1016/0014-4835(80)90039-1.

Teirstein, P. S., Goldman, A. I., & O'Brien, P. J. (1980). Evidence for both local and central regulation of rat rod outer segment disk shedding. *Investigative Ophthalmology & Visual Science, 19*, 1268–1273.

Terman, J. S., Reme, C. E., & Terman, M. (1993). Rod outer segment disk shedding in rats with lesions of the suprachiasmatic nucleus. *Brain Research, 605*, 256–264. doi:10.1016/0006-8993(93)91748-H.

Thompson, D. A., McHenry, C. L., Li, Y., Richards, J. E., Othman, M. I., Schwinger, E., et al. (2002). Retinal dystrophy due to paternal isodisomy for chromosome 1 or chromosome 2, with homoallelism for mutations in RPE65 or MERTK, respectively. *American Journal of Human Genetics, 70*, 224–229. doi:10.1086/338455.

Vives-Bauza, C., Anand, M., Shirazi, A. K., Magrane, J., Gao, J., Vollmer-Snarr, H. R., et al. (2008). The age lipid A2E and mitochondrial dysfunction synergistically impair phagocytosis by retinal pigment epithelial cells. *Journal of Biological Chemistry, 283*, 24770–24780. doi:10.1074/jbc.M800706200.

Welsh, D. K., Logothetis, D. E., Meister, M., & Reppert, S. M. (1995). Individual neurons dissociated from rat suprachiasmatic nucleus express independently phased circadian firing rhythms. *Neuron, 14*, 697–706. doi:10.1016/0896-6273(95)90214-7.

Wimmers, S., Karl, M. O., & Strauss, O. (2007). Ion channels in the RPE. *Progress in Retinal and Eye Research, 26*, 263–301. doi:10.1016/j.preteyeres.2006.12.002.

Young, R. W. (1967). The renewal of photoreceptor cell outer segments. *Journal of Cell Biology, 33*, 61–72. doi:10.1083/jcb.33.1.61.

Young, R. W. (1971a). An hypothesis to account for a basic distinction between rods and cones. *Vision Research, 11*, 1–5. doi:10.1016/0042-6989(71)90201-X.

Young, R. W. (1971b). The renewal of rod and cone outer segments in the rhesus monkey. *Journal of Cell Biology, 49*, 303–318. doi:10.1083/jcb.49.2.303.

Young, R. W. (1978). The daily rhythm of shedding and degradation of rod and cone outer segment membranes in the chick retina. *Investigative Ophthalmology & Visual Science, 17*, 105–116.

Young, R. W., & Bok, D. (1969). Participation of the retinal pigment epithelium in the rod outer segment renewal process. *Journal of Cell Biology, 42*, 392–403. doi:10.1083/jcb.42.2.392.

Yu, C. C., Nandrot, E. F., Dun, Y., & Finnemann, S. C. (2012). Dietary antioxidants prevent age-related retinal pigment epithelium actin damage and blindness in mice lacking $\alpha v \beta 5$ integrin. *Free Radical Biology & Medicine, 52*, 660–670. doi:10.1016/j.freeradbiomed.2011.11.021.

Zhang, D., Brankov, M., Makhija, M. T., Robertson, T., Helmerhorst, E., Papadimitriou, J. M., et al. (2005). Correlation between inactive cathepsin D expression and retinal changes in mcd2/mcd2 transgenic mice. *Investigative Ophthalmology & Visual Science, 46*, 3031–3038. doi:10.1167/iovs.04-1510.

Zhang, K., Kniazeva, M., Han, M., Li, W., Yu, Z., Yang, Z., et al. (2001). A 5-bp deletion in ELOVL4 is associated with two related forms of autosomal dominant macular dystrophy. *Nature Genetics, 27*, 89–93. doi:10.1038/83817.

5 Information Transfer at the Rod-to-Rod Bipolar Cell Synapse

ALAPAKKAM P. SAMPATH

Our visual experience is initiated by the activation of the phototransduction cascade in our retinal rods and cones (see chapter 2 by Burns and Pugh). Phototransduction reduces the dark current of outer segments, and the subsequent membrane hyperpolarization is relayed passively to the synaptic terminal, where it influences glutamate release onto downstream retinal neurons, the bipolar and horizontal cells. Rod and cone phototransduction thus provides the fundamental input to the visual system that encodes the timing and detection of visual stimuli over a wide range of light intensities. In order for the salient features of visual space to be accurately represented downstream in the visual system, retinal bipolar cells in particular must be tuned to these features, which include the magnitude and time course of the rod hyperpolarization.

In the mammalian retina the synaptic output of rod photoreceptors is largely onto one class of ON bipolar cells, called rod bipolar cells. Our scotopic vision, or rod-mediated vision, encompasses a large range of light intensities between absolute visual threshold (~1 photoisomerization, or Rh*, per 10^4 rods; Walraven et al., 1990) up to light intensities that saturate rods (~3×10^4 Rh* rod^{-1} s^{-1}; Naarendorp et al., 2010). Over most of this range, rod bipolar cells respond robustly to light stimuli. The functional characteristics of rod bipolar light-evoked responses must therefore allow them to (1) encode single-photon responses from rod photoreceptors near absolute visual threshold, where a small minority of the rods are absorbing light, (2) remain responsive in brighter light where hundreds to thousands of Rh* per rod are generated, and (3) allow a seamless representation of light intensity as secondary rod pathways are engaged. This chapter describes the anatomical features of the rod-to-rod bipolar synapse and subsequently identifies mechanisms that allow the rod photovoltage to control the rate of glutamate release at the rod spherule and that allow the rod bipolar cells to respond to changes in glutamate release over a wide range of background light intensities. Where possible, how these mechanisms influence signal processing is emphasized.

Despite the fundamental importance of the rods to our scotopic vision, much of what we know about how rod photoresponses are relayed to ON bipolar cells is based on studies from amphibian retinas. The large cell bodies of amphibian retinal neurons and the ability to record at lower temperatures have allowed for robust recordings of rod inner segment currents and the responses of postsynaptic neurons (reviewed by Thoreson, 2007). This literature will be discussed where gaps exist in our understanding of rod-to-rod bipolar signal transfer in the mammalian retina.

THE ROD SPHERULE AND GLUTAMATE RELEASE

Unique anatomical specializations allow the synapse between rods and bipolar cells to transmit faithfully small changes in rod membrane potential—such as those generated by single-photon absorptions. Such sensitivity requires that rod bipolar cells represent glutamate release from rod photoreceptors robustly. To allow signal transmission under these circumstances, rods have developed a specialized synaptic structure, the spherule, to ensure that downstream cells sense subtle variations in glutamate release.

Structural Features of the Triad Ribbon Synapse

The "triad" synapse is shown in figure 5.1A and is defined by the contact of three cell types: a single rod synapses with two horizontal cell axons and two rod bipolar cell dendrites that make invaginating contacts into the rod spherule. The two rod bipolar cell dendrites originate in different cells (Tsukamoto et al., 2001), providing a first level of divergence of the rod photoresponses. This structure of the synapse is believed to allow multiple dendritic processes to sense the release of every vesicle and to match different cell types/glutamate receptor types with the appropriate range of glutamate concentration (see also Rao-Mirotznik, Buchsbaum, & Sterling, 1998; Rao-Mirotznik et al., 1995).

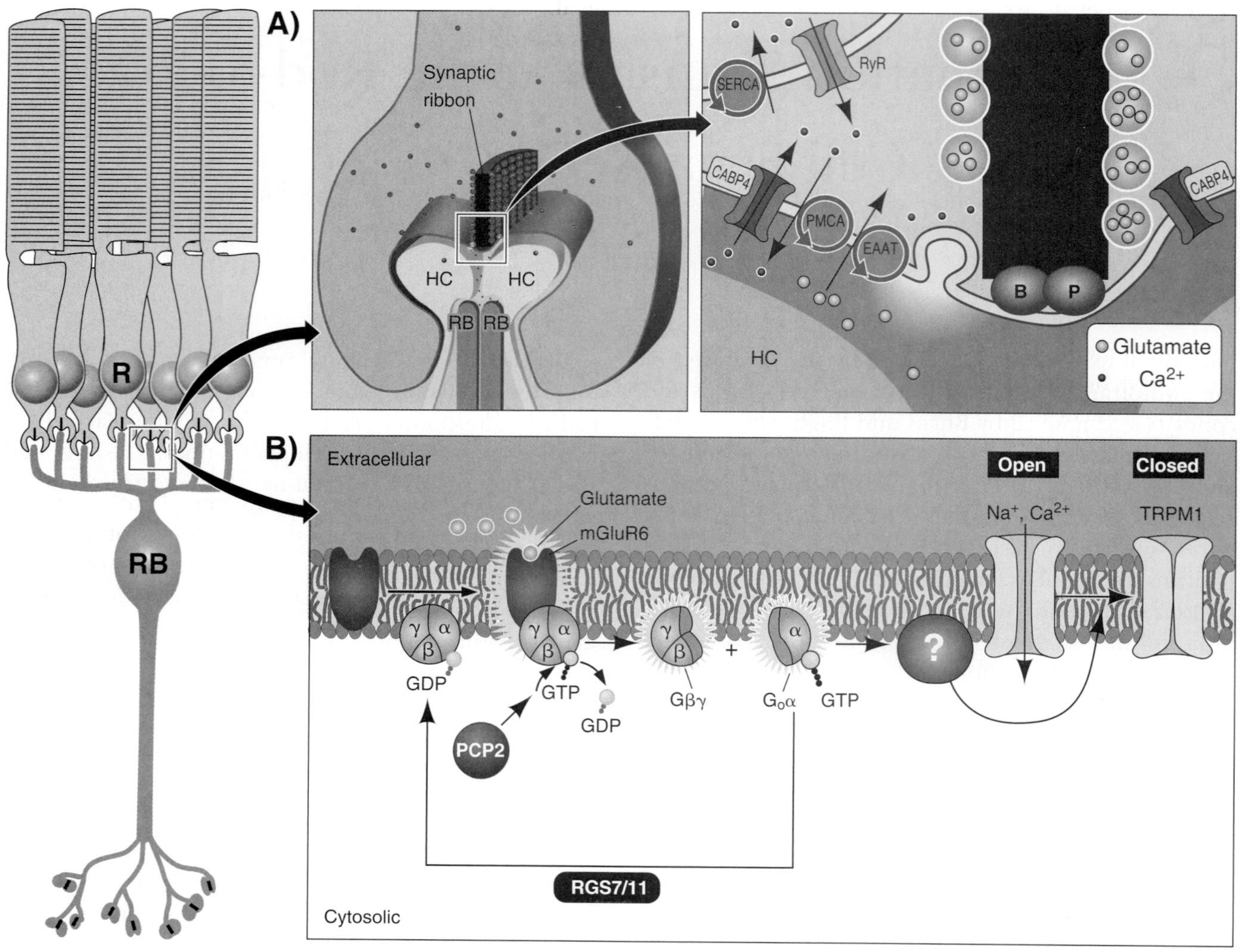

FIGURE 5.1 The rod-to-rod bipolar cell synapse. (A) The rod spherule structure (R) with the synaptic ribbon docked at the release site. The triad synapse consists of two horizontal cell axons (HC) and two rod bipolar cell dendrites (RB) protruding in the invagination. An expanded view of the spherule is shown to the right. The ribbon is held in the active zone by the proteins bassoon (B) and piccolo (P). $Ca_v1.4$ Ca^{2+} channels with CABP4 bound to tune voltage sensitivity allow the influx of Ca^{2+} and vesicle fusion. Ca^{2+} from internal stores contributes to the pool that controls glutamate release. SERCA and PMCA transporters extrude Ca^{2+} from the cytoplasm. EAAT transporters on rods take up glutamate from the synaptic cleft. (B) mGluR6 senses the release of glutamate from rods and activates a signaling cascade, which opens cationic TRPM1 channels through the action of $G\alpha_o$. Other components control the activity of $G\alpha_o$, including those that speed shutoff (RGS7/RGS11) and those that enhance nucleotide exchange (pcp2). (Modified from Pahlberg & Sampath, 2011.)

A unique feature of this synapse is its reliance on a presynaptic ribbon for the release of glutamate (figure 5.1A). Ribbon synapses are few in the nervous system, at the synaptic terminals of rod and cone photoreceptors, retinal bipolar cells, auditory and vestibular hair cells, and in auditory brainstem, and allow a high rate of glutamate release. The ribbon is a structure that consists of several proteins, the largest constituents being the protein ribeye (Schmitz, Konigstorfer, & Sudhof, 2000) and its homologue CtBP1/BARS (tom Dieck et al., 2005). In addition, other structural constituents include the proteins bassoon (tom Dieck et al.,

2005) and piccolo (Dick et al., 2001), which appear to anchor the ribbon to the active zone to allow the accumulation of synaptic vesicles (Mukherjee et al., 2010). Although the functional role of the ribbon has not been established fully, experiments from rod and cone photoreceptor and bipolar cell terminals suggest that it may be responsible for bringing glutamatergic vesicles to the plasma membrane (von Gersdorff & Matthews, 1994; Mehta et al., 2013), controlling the rate of glutamate release (Jackman et al., 2009), priming glutamatergic vesicles for release (Snellman et al., 2011), and/or allowing multivesicular release (Glowatzki & Fuchs,

2002; Singer et al., 2004). Zanazzi and Matthews (2009) and Mercer and Thoreson (2011) provide a comprehensive review of the molecular constituents of the ribbon and their functional characteristics.

The proper development of the rod-to-rod bipolar synapse also appears to be connected integrally with the expression of key signaling components with synaptic development involving two processes: anatomical synapse formation and subsequent functional development. In rod photoreceptors, expression of the synaptic Ca^{2+} channels appears to be required for normal synaptic development (Bayley & Morgans, 2007). Additionally in rod bipolar cells, the expression of functional mGluR6 (Ishii et al., 2009) and a subunit of its regulator $G\beta 5$ (Rao et al., 2007) appear to be required to form synapses anatomically. However, the postsynaptic expression of additional components is also needed for synapses to be functional. Not surprisingly, the deletion of the transduction channel TRPM1 (Bellone et al., 2008; Koike et al., 2010; Morgans et al., 2009; Shen et al., 2009) or its binding partner nyctalopin (Ball et al., 2003; Cao, Posokhova, & Martemyanov, 2011; Pearring et al., 2011) eliminates signal transmission at these synapses despite normal anatomical development. Additionally, other extracellular factors play key roles in rod-to-rod bipolar cell synapse formation including the dystroglycan ligand picachurin (Sato et al., 2008). Ultimately, the interplay of these molecules is required to set up synapses anatomically and subsequently impart function. The mechanisms by which these components aid in establishing synapses are currently under investigation.

Properties of Glutamate Release

The release of glutamate from rod photoreceptors ultimately relies on the opening of voltage-gated Ca^{2+} channels in the rod spherule. Because photoreceptor cells are continuously depolarized in darkness and release glutamate (Trifonov, 1968), these Ca^{2+} channels must remain responsive to changes in membrane potential under conditions where most voltage-gated channels become inactivated. Indeed, the expression of the L-type Ca^{2+} channel $Ca_v 1.4$ appears to be ideal for this task. $Ca_v 1.4$ channels display little voltage- or Ca^{2+}-dependent desensitization (Koschak et al., 2003; Wahl-Schott et al., 2006); thus, they are able to support continuous Ca^{2+} influx while rods remain depolarized. To allow the Ca^{2+} current to be modulated robustly over the range of physiological rod voltages (~ -40 mV to -70 mV), the voltage sensitivity of $Ca_v 1.4$ channels is shifted to more hyperpolarized potentials by their interaction with calcium-binding protein 4 (CABP4) (Haeseleer

et al., 2004). This shift increases the inward Ca^{2+} current in darkness, increasing glutamate release and extending the dynamic range of synaptic transmission. Additionally, CABP4 may serve as a component that allows the Ca^{2+} channel to anchor close to the synaptic ribbon (Haeseleer, 2008).

The Ca^{2+} concentration within the rod spherule will ultimately define the rate of glutamate release. The time course of changes in cytoplasmic Ca^{2+} concentration will depend on the balance between its influx into the spherule and its efflux and on the Ca^{2+} buffering capacity. The influx of Ca^{2+} into the cytoplasm occurs through $Ca_v 1.4$ channels (as described previously) and from intracellular stores through ryanodine receptors (Babai, Morgans, & Thoreson, 2010). Mechanisms that extrude Ca^{2+} also play a key role in setting the time course of glutamate release, as the rate of Ca^{2+} extrusion at the spherule will control the time course of the light-evoked glutamate reduction. Experiments on transgenic mice lacking the plasma membrane Ca^{2+}-ATPase 2 (PMCA2) reveal a loss of sensitivity in rod bipolar cells (Duncan et al., 2006), consistent with a strong role of PMCA2 in Ca^{2+} extrusion from rod spherules. This influence is facilitated by the co-trafficking of PMCA2 with $Ca_v 1.4$ Ca^{2+} channels (Xing, Akopian, & Krizaj, 2012).

Presynaptic Modulation of Glutamate Release

The rate of glutamate release at the rod synapse can be modulated by signaling cascades in the synaptic terminal similarly to other central synapses. For instance, metabotropic glutamate receptor group II and III agonists have been shown to inhibit glutamate release from rod spherules in the salamander retina (Higgs & Lukasiewicz, 2002) or reduce Ca^{2+} influx into rat photoreceptor spherules (Koulen et al., 1999). These experiments suggest that glutamate release may also feed back through mGluRs to control release rate (see Rabl, Bryson, & Thoreson, 2003), although the reuptake of glutamate by rod photoreceptors is fast and robust (Hasegawa et al., 2006).

More recently the proteins involved in phototransduction have been considered for roles in synaptic release. Such a role is supported by the observation that some phototransduction components move between rod outer segments and inner segments/synaptic terminals following exposure to bright light, including rod transducin ($G\alpha_t$) (Brann & Cohen, 1987; Philp, Chang, & Long, 1987; Whelan & McGinnis, 1988) and recoverin (Strissel et al., 2005), which move toward synaptic terminals, and visual arrestin, which moves toward rod outer segments (Broekhuyse et al., 1985). The movement of these components has been considered to reduce the gain of phototransduction in bright light and may be neuroprotective (Whelan & McGinnis, 1988). However, their movement toward and away from the rod spherule may also suggest a synaptic function. For instance, recoverin has been shown to play a role downstream of the rod outer segment in enhancing the sensitivity of light-evoked responses in rod bipolar cells (Sampath et al., 2005). Similar sensitizing effects for $G\alpha_t$ have been observed in transgenic mice whose $G\alpha_t$ has been replaced by one that does not translocate (Pahlberg et al., 2012). Last, visual arrestin has been shown to interact with the synaptic protein NSF, potentially to enhance glutamate release at the rod spherule (Huang, Brown, & Craft, 2010). The movement of visual arrestin to outer segments following bright light exposure may then serve to adapt the synapse by reducing the glutamate release rate. The mechanism by which any of these components influences glutamate release in rod spherules, or their relative contribution to the sensitivity of glutamate release, remains unknown.

SENSING SYNAPTIC GLUTAMATE IN ROD BIPOLAR CELL DENDRITES

The light-evoked response of rod and cone photoreceptors hyperpolarizes their membrane potential, as cGMP-gated channels close and block cation influx into the outer segment (see chapter 2 by Burns and Pugh). The hyperpolarizing response classifies photoreceptors as OFF cells. The reduction in glutamate release following membrane hyperpolarization is sensed by two classes of bipolar cells, the ON and OFF bipolar cells (see chapter 6 by DeVries). Rod bipolar cells, the predominant bipolar cells in the mammalian retina that sense the output of rod photoreceptors, depolarize in response to light exposure and are therefore classified as ON cells. The light-evoked depolarization of the membrane potential is achieved through the opening of cation channels, as opposed to the light-evoked closure of cation channels in photoreceptor cells. This sign inversion is achieved through the activity of a metabotropic signaling cascade.

mGluR6 Signaling and the ON Response

A consequence of open cGMP-gated channels in darkness and the depolarized membrane potential of rods is that Ca^{2+} influx into the rod spherule leads to continuous glutamate release (Trifonov, 1968). Glutamate is sensed postsynaptically in the rod bipolar cell dendrites by metabotropic mGluR6 receptors (Masu et al., 1995; Nakajima et al., 1993; Nomura et al., 1994), as shown in figure 5.1B. mGluR6 catalyzes the exchange of GTP for GDP on the alpha subunit of the heterotrimeric G-protein, $G\alpha_o$ (Dhingra et al., 2000, 2002; Nawy, 1999; Okawa et al., 2010b), which in turn leads to the closure of TRPM1 cationic transduction channels (Koike et al., 2010; Morgans et al., 2009; Shen et al., 2009) through mechanisms that are presently unknown. The continuous release of glutamate in darkness by rod spherules drives the activity of this signaling cascade, thereby holding TRPM1 channels closed and maintaining the rod bipolar cell membrane potential at relatively hyperpolarized levels. The light-induced reduction in glutamate release from rods reduces the activity of the mGluR6 signaling cascade, allowing TRPM1 channels to open, Na^+ and Ca^{2+} influx, and the rod bipolar cell depolarization (ON).

Modulation of Rod Bipolar Cell Sensitivity

Robust signal transfer from rods to rod bipolar cells requires the synapse to be well tuned to the properties of the rod photoresponse over a wide range of light intensities. Near visual threshold, when few rods absorb photons, the mGluR6 signaling cascade must suppress noise from rods not absorbing photons (see below) while allowing the light-induced reduction in glutamate release to open a sufficient number of TRPM1 channels. In addition, with increasing background light signals are generated by a larger fraction of the rods, and the need to suppress noise from rods not absorbing photons reduces. Under these circumstances a reduction in the efficacy of mGluR6 signaling would allow the rod-to-rod bipolar synapse to encode greater numbers of photon absorptions. Although the identities of some key components of the mGluR6 signaling cascade remain unknown, namely the effector(s), several components that modulate the mGluR6 signaling cascade have been identified (figure 5.1B), and their functional roles are just beginning to be understood (reviewed by Snellman et al., 2008).

Early studies of ON bipolar cell signal transduction suggested that cGMP played a key role in biasing TRPM1 channels to the open state, perhaps by the direct opening of cGMP-gated channels (Nawy & Jahr, 1990). In these experiments the dialysis of cGMP or GTP using patch electrodes into isolated salamander ON bipolar cells showed that the amount of glutamate-induced cation current suppression increased with time after the dialysis was initiated, suggesting to the investigators that cGMP was responsible for this current. These results were corroborated by similar work in dogfish ON bipolar cells (Shiells & Falk, 1990). The cation current in ON bipolar cells was further shown to be nonselective (de la Villa, Kurahashi, & Kaneko, 1995; Nawy & Jahr, 1991) and opened by light exposure (Nawy & Jahr, 1991). Subsequent studies did not find any direct evidence for cGMP-gated channels in ON bipolar cells (Nawy, 1999), but the earlier results clearly indicate that cGMP was interacting with the mGluR6 cascade in a manner that sensitized the signaling cascade. Such an influence has more recently been suggested in mouse rod bipolar cells to be subserved by protein kinase G (PRKG1) (Snellman & Nawy, 2004), as pharmacological agents that block PRKG1 prevent the cGMP-dependent potentiation of the cationic current. The exact role of PRKG1 in sensitizing the rod bipolar cell light-evoked responses remains unknown, as does its action as the adaptation state of the retina varies.

The desensitization of rod bipolar cell light-evoked responses is achieved through the concerted action of several mechanisms, with Ca^{2+} influx through TRPM1 channels playing a strong role (Nawy, 2000). For instance, in mouse rod bipolar cells the response to a flash is followed by a period of strong desensitization that decays with a time constant of ~375 ms (Berntson, Smith, & Taylor, 2004a). The desensitization can be prevented by stopping Ca^{2+} influx during the light-evoked response or by strongly buffering internal Ca^{2+} using the chelator BAPTA (see also Nawy, 2004, for similar work in salamander ON bipolar cells). This form of desensitization appears to be unaffected in salamander ON bipolar cells by inhibitors of calmodulin-dependent protein kinase II (CaMKII) but may involve the action of the Ca^{2+}-dependent phosphatase, calcineurin (Nawy, 2004; Snellman & Nawy, 2002). It has been suggested that these Ca^{2+}-dependent effects are directly on TRPM1 transduction channels (Nawy, 2000). Ca^{2+} may also play other roles in controlling the sensitivity of rod bipolar cell light-evoked responses. For instance, Ca^{2+} is known to contribute to the activation of protein kinase C (PKC). Recent studies on mice lacking PKCα show slightly impaired light adaptation and slowed responses of the rod bipolar cell component of the ERG (Ruether et al., 2010), although the role of Ca^{2+} in this study was not ascertained. Thus, PKCα appears to play a more modest role in the desensitization of the mGluR6 cascade and in setting the time course of responses. The target of PKCα in this signaling pathway remains unknown, but it should be noted that PKCα expression is quite robust in rod bipolar cells, and its antibody is used widely to mark these cells immunohistologically.

In addition to controlling the sensitivity of rod bipolar light-evoked responses, modulation of G-protein activity also plays a central role in setting their time course. G-protein α (Gα) subunits in vitro display a very slow rate of shutoff by GTP hydrolysis ($\tau \sim 30$ s) (Ross & Wilkie, 2000), too slow to account for the fast light-evoked responses of retinal cells, which are typically much shorter than a second. To accelerate the rate of GTPase activity, GTPase activating proteins (GAPs) interact with both the Gα and the effector to accelerate this process (Hollinger & Hepler, 2002; Ross & Wilkie, 2000). For instance, in rod phototransduction the GAP protein RGS9-1 plays a dominant role in the shutoff of $G\alpha_t$ to set the time scale of the recovery phase of rod photoresponses (C. K. Chen et al., 2000; He, Cowan, & Wensel, 1998). In addition, the faster recovery of cone photoresponses may result from the higher expression of RGS9-1 in cones (Cowan et al., 1998; Lyubarsky et al., 2001; X. Zhang, Wensel, & Kraft, 2003). Similarly, the acceleration of $G\alpha_o$ deactivation is critical to set the time course of rod bipolar light-evoked responses. However, the rate of $G\alpha_o$ deactivation sets the slope of the rod bipolar cell's rising phase. Several RGS proteins are expressed in rod bipolar cells, including RGS6, RGS7, RGS11, and ret-RGS (Dhingra et al., 2004; Morgans et al., 2007; Song et al., 2007).

Analysis of rod bipolar responses in mice where each of these components is eliminated, or where their anchoring proteins are eliminated, revealed no effect on the time course of the light-evoked response (Cao et al., 2008, 2009; F. S. Chen et al., 2010; Mojumder, Qian, & Wensel, 2009; J. Zhang et al., 2010). To determine whether there is a redundancy of these RGS proteins, many groups have further eliminated them in combination. For instance, when mice lacking RGS11 are crossed with a line carrying a hypomorphic mutation of RGS7, rod bipolar light-evoked responses are little affected (F. S. Chen et al., 2010; Mojumder et al., 2009; J. Zhang et al., 2010). The simultaneous elimination of RGS7 and RGS11, however, appears to eliminate fast rod bipolar cell light-evoked responses as assessed by ERG (Cao et al., 2012; Shim et al., 2012). Rod bipolar activity assessed using single-cell patch clamp recordings instead show that small residual responses remain (Cao et al., 2012). The rising phase of the residual response is slowed ~40-fold, and response sensitivity is reduced dramatically. It appears that the simultaneous elimination of RGS7 and RGS11 increases the level of saturation of the mGluR6 cascade in darkness for rod bipolar cells (see also below). Thus, any residual activity of RGS7 or RGS11 appears sufficient to set the time course of $G\alpha_o$ deactivation in rod bipolar cells that in turn defines the rising phase of light-evoked responses.

Metabotropic signaling cascades are known for their ability to modulate their output, as a signaling cascade provides many locations for this potential modulation. Other components of the mGluR6 pathway have also been identified that may provide finer control of TRPM1's open probability. For instance, rod bipolar cells are known to express the protein pcp2, which is also known for its robust expression in cerebellar Purkinje cells (Oberdick et al., 1990). In mice lacking pcp2, rod bipolar cells demonstrate modestly slowed light-evoked responses depending on the strength of the flash (Xu et al., 2008). This work suggested that pcp2 might act as a GTP exchange factor (GEF) to accelerate the turnover of $G\alpha_o$. Additional components of the mGluR pathway have been identified through studies of gene expression. Several groups have created expression libraries of the genes expressed selectively in ON bipolar cells (Dhingra et al., 2008; Morgans et al., 2009). Ultimately, the functional characterization of other promising candidates will require further evaluation.

SIGNALING NEAR ABSOLUTE VISUAL THRESHOLD AND THE TRANSITION TO SECONDARY ROD PATHWAYS

As do all sensory systems, the visual system desensitizes with increasing levels of the stimulus to extend the range of stimulus strengths over which the system functions. In rod-mediated vision the range of stimulus strengths spans a dynamic range of eight orders of magnitude in light intensity. At the lowest light levels the rod circuitry must encode intensities that result in the very rare absorption of single photons by rod photoreceptors, and at highest light levels rods must respond reliably in background light that results in tens of thousands of photons absorbed per rod per second. Such a wide range is achieved partially by adaptation within the photoreceptor cells (see chapter 2 by Burns and Pugh). In addition, mechanisms throughout the retina extend this range further by allowing the discrimination of signals from noise at low light levels and adapting as the light levels increase. Rod bipolar cells play a key role in this process: at low light levels they separate single-photon responses in a few rods from noise in the vast majority, and at high light levels they desensitize profoundly, allowing secondary rod pathways to take their place and recruit ON and OFF cone bipolar cells to process rod-driven signals. The roles of the mGluR6 signaling cascade in retinal processing at both these limits of light intensity are described below.

Nonlinear Threshold and Suppression of Rod Noise Near Visual Threshold

It has been more than a half-century since the seminal discovery by Hecht, Schlaer, and Pirenne (1942) that the dark-adapted visual system can resolve single-photon responses in a few rods. Hecht and his co-authors performed "frequency of seeing" experiments in which a brief flash of light was presented in the retinal periphery, and they determined based on the strength of the flash whether they could detect it or not. They interpreted their data based on a simple model in which detection threshold was set by the Poisson probability of absorbing at least some threshold number of photons, which ranged from five to seven for a spot of light that subtended ~500 rods. Many additional investigators have pushed this work further, with more flexible criteria given to subjects for identifying whether they detect a flash (Barlow, 1956; Sakitt, 1972). The upshot is that the dark-adapted visual system can sense few photons absorbed among many hundreds or thousands of rods. The functional consequences of this conclusion are that individual rods must be able to report reliably single photon absorptions (Baylor, Lamb, & Yau, 1979; Baylor, Nunn, & Schnapf, 1984) but also that the retinal ganglion cells must alter their pattern of action potential generation based on a small graded response generated in a few among the thousands of rods they sample from (Barlow, Levick, & Yoon, 1971; Mastronarde, 1983). Thus, the discriminability of single-photon responses from noise at every level of retinal processing is important in setting our absolute visual threshold (reviewed by Field, Sampath, & Rieke, 2005; Pahlberg & Sampath, 2011).

Two components of noise in darkness are present in the rod photocurrent and have an influence on the detection of single-photon responses; they are referred to as discrete and continuous noise (Baylor, Matthews, & Yau, 1980). Both of these forms of noise originate within rod phototransduction and result in fluctuations in the concentration of cGMP, which in turn influences the membrane current and potential. Discrete noise events have a size and time course that are identical to those of the single-photon response and are believed to originate in the thermal activation of rhodopsin (Baylor et al., 1980; Gozem et al., 2012). Due to the identity of the discrete noise event with the single-photon response, there is no mechanism, either linear or nonlinear, that can separate these noise events from real signals. Thus, discrete noise creates a false-positive signal that sets a fundamental limit on the absolute sensitivity of the visual system. The only recourse for minimizing this component of noise in rod photoreceptors is to reduce their frequency. In this respect rhodopsin displays a remarkable thermal stability, and discrete noise events are observed approximately every minute or so in the rods of mice (Burns et al., 2002) and monkeys (Baylor et al., 1984). Because rod outer segments in these species contain >10^8 rhodopsin molecules, a single molecule of rhodopsin undergoes thermal activation approximately every 1,000 years.

The second major form of noise in rods, and perhaps the more relevant form with respect to retinal signal processing, is continuous noise (Baylor et al., 1980). Continuous noise arises from fluctuations in the outer segment concentration of cGMP; reductions in cGMP concentration are produced by the light-independent activity of cGMP phosphodiesterase and the subsequent synthesis of cGMP by guanylyl cyclase (see chapter 2 by Burns and Pugh). Because continuous noise originates in the phototransduction cascade, its frequency composition is similar to that of light-evoked responses (Rieke & Baylor, 1996). This similarity prevents simple linear filtering from removing this noise downstream. However, unlike discrete noise events that cannot be separated from single-photon responses by any type of filtering, continuous noise fluctuations on average are smaller than the single-photon responses, although their amplitude distributions overlap significantly (Baylor et al., 1980; Rieke & Baylor, 1996). This difference in size can serve as a potential basis for separating nonlinearly single-photon responses from continuous noise (Baylor et al., 1984).

A challenge faced by rod bipolar cells is that they are the first sites of signal convergence in the primary rod pathway. As many as 20–100 rods converge on a rod bipolar cell, but as the behavioral experiments above indicate, near visual threshold perhaps one of those rods may be generating a graded single-photon response. Van Rossum and Smith (1998) proposed a model whereby signals at individual synapses between rods and rod bipolar cells needed to exceed some criterion threshold amplitude to pass. This arrangement was effective at eliminating noise from rods not absorbing photons, and they further proposed that this nonlinear threshold was produced at the level of mGluR6 receptors on rod bipolar cell dendrites. Based on recordings of the distribution of single-photon responses and noise from rod photoreceptors and recordings from rod bipolar cells, Field and Rieke (2002) showed that such a nonlinear threshold produced a substantial improvement in the signal-to-noise ratio of single-photon responses compared to linear summation of rod signals (figure 5.2; see also Berntson, Smith, & Taylor, 2004b). Their work demonstrated that the position of the nonlinear threshold was near the optimal position at the crossing point of the probability distributions of the rod continuous noise and single-photon response amplitude. What was most striking about this work is the discovery that the nonlinear threshold was set at a position that also eliminated a majority of the smaller single-photon responses. Such an arrangement seems at odds with the high sensitivity of scotopic vision, but the nonlinear threshold also eliminated nearly all the continuous noise from rod photoreceptors. Ultimately, the signal-to-noise ratio of light-evoked responses determines the ability of downstream cells to encode faithfully the signals and establish threshold; thus, such an operation will improve sensitivity.

The process of identifying the optimal position of the nonlinear threshold between rods and rod bipolar cells must also take into account an additional factor, the light level. Assuming that the probability distribution of single-photon response and continuous noise amplitude remains constant in individual rods, as light levels decline, the distribution of single-photon response amplitudes is set to lower probabilities with respect to the continuous noise (see figure 5.3). The optimal separation of single-photon responses from continuous noise under these conditions must be more stringent, or the position of the nonlinear threshold must be at higher amplitudes. The characterization of the position of the nonlinear threshold with respect to continuous noise and single-photon response probability distributions is therefore necessary to determine mechanistically how this position is set.

The release of glutamate from rod photoreceptors appears to be related linearly to the change in the rod photovoltage (see Thoreson et al., 2004). This indicates that a nonlinear threshold is set in the dendrites of the rod bipolar cells, as modeled previously (van Rossum & Smith, 1998). Ultimately, the position of the nonlinear threshold is then dependent on the properties of the mGluR6 signaling pathway. Sampath and Rieke (2004) further dissected the mechanism underlying this nonlinear threshold by distinguishing whether it was set by saturation at the mGluR6 receptors (as suggested in van Rossum & Smith, 1998) or downstream in the signaling cascade. They used the extent of nonlinearity in the relationship between the flash strength and response amplitude to estimate the position of the nonlinear threshold. Using high-affinity agonists and antagonists of the mGluR6 receptor, they were able to manipulate the extent of nonlinearity to show that the position of the nonlinear threshold was set by saturation within the signaling cascade (Sampath & Rieke, 2004). More specifically, they found that the high level of G-protein activity in rod bipolar dendrites was sufficient to hold closed nearly all the TRPM1 channels. This type of saturation in the signaling cascade makes the rod bipolar cell relatively insensitive to fluctuations in glutamate release that are caused by continuous noise in the rod outer segment or by noise in glutamate release at the

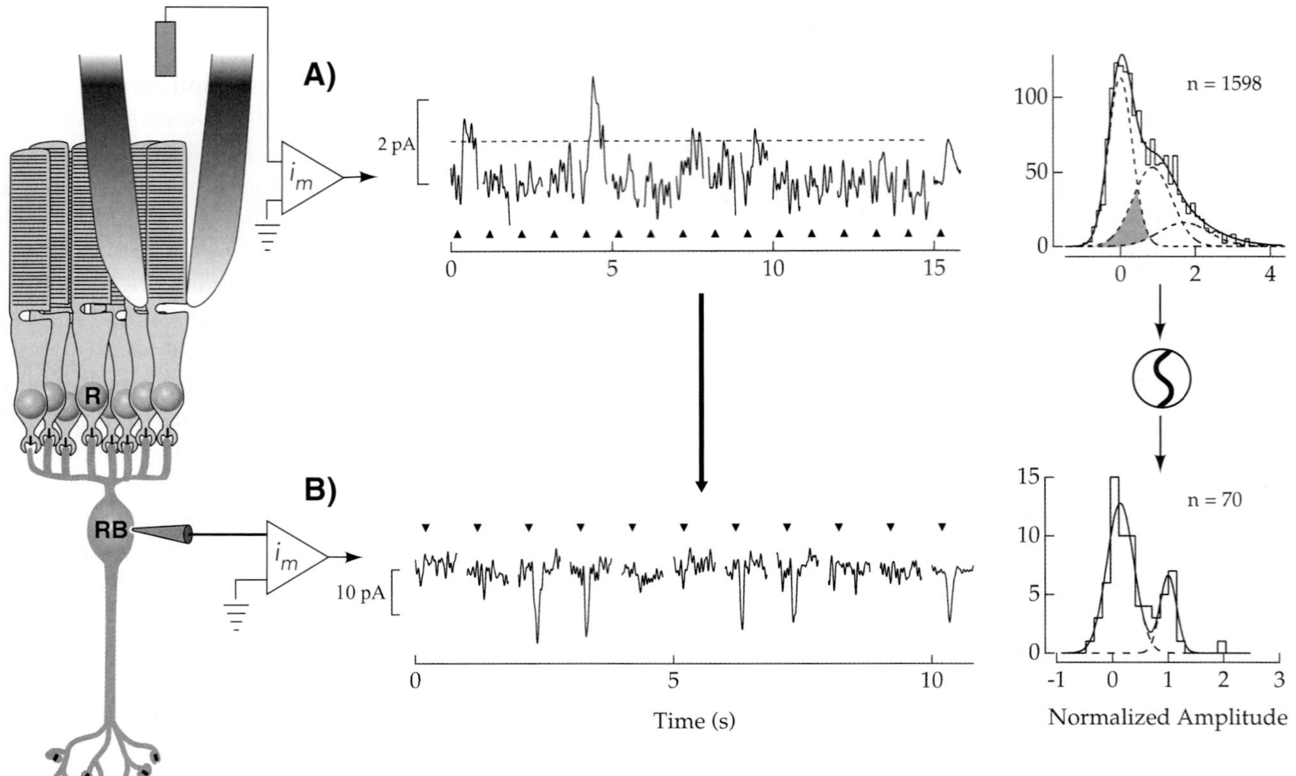

FIGURE 5.2   Nonlinear threshold between rods and rod bipolar cells eliminates noise from rods and makes single-photon response more discernable. (A) Suction electrode recordings from a mouse rod during the presentation of flashes (up arrow) that activate less than 1 Rh* per trial. The average single-photon response is on the right. Probability distributions for response amplitudes are shown at the far right, with the shaded region indicating the overlap of continuous noise and single-photon response amplitudes. (B) Despite overlapping distributions, single-photon responses, and continuous noise in rods, a nonlinear threshold in rod bipolar cells helps to separate these. Voltage-clamp recordings of a rod bipolar cell are shown for dim flashes of similar strength (up arrow). The probability distribution of continuous noise and single-photon response amplitude are shown at the far right, showing less overlap than for rods and therefore a higher signal-to-noise ratio. (Reproduced from Pahlberg & Sampath, 2011.)

rod spherule. Only larger reductions in glutamate, produced by larger single-photon responses, are able to reduce saturation in the signaling cascade and open TRPM1 channels. This mechanism can therefore eliminate largely continuous noise from rods, although at the cost of the loss of some smaller single-photon responses. Most importantly, this study provided a way to estimate the position of the nonlinear threshold with respect to the distribution of signals and noise in rods.

Setting the nonlinear threshold to the optimal position is not a trivial matter (figure 5.3). The optimal position needs to account for the probability distributions of continuous noise and single-photon response amplitude in individual rods and the light level (which sets the scaling of the probability distributions for single-photon responses and continuous noise with respect to one another). Given these factors, if the position of the nonlinear threshold is too low, then continuous noise from rods will bleed over to rod bipolar cells. If

the position of the nonlinear threshold is too high, then too many single-photon responses will be eliminated. Modulation of the position of the nonlinear threshold may be useful to provide the optimal separation of single-photon responses and noise in darkness and in dim light (Field & Rieke, 2002). How the position of the nonlinear threshold is set was studied recently in transgenic mice where the distributions of continuous noise and single-photon response amplitudes were altered: the guanylyl cyclase activating proteins 1 and 2 (GCAPs) knockout (Burns et al., 2002; Okawa et al., 2010a; see chapter 2 by Burns and Pugh). GCAPs knockout rods display approximately threefold larger single-photon responses by virtue of the lack of Ca^{2+} feedback to cGMP synthesis. They also display greater continuous noise, but the increased size of the single-photon response exceeds the increase in noise, thereby increasing the signal-to-noise ratio to the extent that discrete noise events can be counted (Burns et al.,

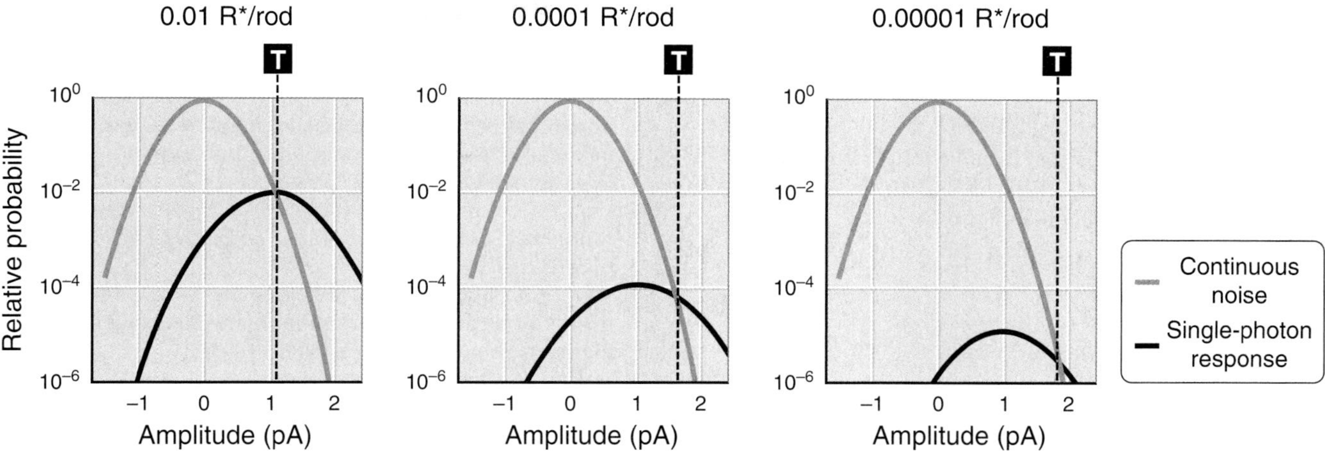

FIGURE 5.3   The optimal position of the nonlinear threshold (T) is dependent on the light intensity. The optimal position of the nonlinear threshold is the amplitude at which the probability distributions of continuous noise and single-photon response amplitude cross to separate the two. In downstream retinal cells (such as a ganglion cell), as light levels decrease, the probability of observing single-photon responses decreases compared to the probability of observing continuous noise. This results in a shift in the optimal position of the nonlinear threshold to higher current amplitudes. (Modified from Pahlberg & Sampath, 2011.)

2002). The change in the rod signal-to-noise ratio would predict a higher optimal position for the nonlinear threshold. However, the position of the nonlinear threshold in GCAPs knockout rods did not appear to be set to this new optimal position (Okawa et al., 2010a). Instead, the position appeared to remain similar to the position in normal mice, resulting in continuous noise from GCAPs knockout rods bleeding over to the response of their respective rod bipolar cells and reducing their signal-to-noise ratio. Determining the conditions under which the position of the nonlinear threshold can move, if at all, will be an important future endeavor to understand how this mechanism establishes visual threshold.

Strong Desensitization of Rod Bipolar Cells in Background Light

Behavioral experiments have indicated that absolute visual threshold occurs at light levels where the processing of the rod single-photon response is relevant. In addition other work has also shown that rod function in humans extends to much brighter light levels. Classical work by Blakemore and Rushton (1965) revealed that a rod monochromat (one lacking all cone function) could function reasonably well in bright daylight. The rod monochromat in these studies was the first author of the study, Colin B. Blakemore, a child psychiatrist trained in London who enjoyed outdoor activities and reading. Anecdotal evidence from Blakemore's experience is consistent with recent behavioral studies in mice, which show that rod saturation might occur at much higher light levels than previously appreciated ($\sim$30,000 Rh* rod^{-1} s^{-1}) (Naarendorp et al., 2010). Although considerable attention has been given to elucidating the mechanisms that allow the high sensitivity of rod vision, much less attention has been devoted to determining how rods are able to convey signals in very bright light to the retinal output.

In the range of rod-driven (scotopic) light intensities, two main retinal circuits convey rod signals to retinal ganglion cells, the rod bipolar (primary) pathway and the rod-cone and rod-OFF bipolar (secondary) pathways (reviewed by Bloomfield & Dacheux, 2001). The secondary rod pathways appear to be $\sim$10-fold less sensitive than the primary rod pathway (DeVries & Baylor, 1995; Soucy et al., 1998) and function at light levels where a majority of the rods receive photons. As light levels increase, primary rod pathways adapt, allowing a seamless transition to secondary pathways until cone photoreceptor input to the retina is initiated.

Measurements of response gain in rod bipolar cells indicate that weak background light (which does not cause adaptation in rod photoreceptors) initially increases the gain of responses by releasing the nonlinear threshold in the rod bipolar response (Dunn et al., 2006; Sampath & Rieke, 2004). However, as light intensity increases further, the gain of signaling is eventually reduced dramatically in rod bipolar cells as the gain is reduced in both phototransduction and mGluR6 transduction (Dunn et al., 2006). The profound desensitization of rod bipolar cells appears to be primarily attributable to Ca^{2+} influx during the rod bipolar response (Berntson, Smith, & Taylor, 2004a). As background light desensitizes rod bipolar cells, ON and OFF cone bipolar cells that receive rod-driven input through secondary pathways do not appear to desensitize to the same extent (Sampath et al., 2011). Thus, it appears that the profound desensitization of rod bipolar cells with increasing background light provides a mechanism that allows secondary rod pathways to dominate signal processing at high scotopic light intensities.

OVERVIEW

The sign-inverting synapse between rods and rod ON bipolar cells in the mammalian retina plays a critical role in setting our absolute visual threshold. In addition, the activity at this synapse is modulated by background light to provide a seamless transition to secondary rod pathways that provides a continuous representation of light intensity. To achieve such diverse function, the properties of glutamate release from rods and mGluR6 signal transduction in rod bipolar cells are specialized to allow signal transfer in these light regimes: (1) Near absolute visual threshold, where photons are scarce in the array of rod photoreceptors, rod single-photon responses produce a small graded membrane potential change that leads to a decrease in glutamate release and a postsynaptic depolarization in rod bipolar cells. Under these conditions, noise in the rod array is suppressed in rod bipolar cells by a nonlinear threshold that is produced by saturation in the mGluR6 signaling cascade. (2) In brighter light, the gain of rod bipolar responses first increases as the nonlinear threshold is released but then decreases dramatically as desensitization of both rod phototransduction and rod bipolar transduction is manifested. The profound desensitization of rod bipolar cells is mediated largely by the influx of Ca^{2+} through transduction channels, providing the opportunity for secondary rod pathways to transmit rod-driven signals to the retinal output. These specializations allow mammals to extend the range over which rods signal to the retinal output to subtend more than eight orders of magnitude in light intensity.

REFERENCES

Babai, N., Morgans, C. W., & Thoreson, W. B. (2010). Calcium-induced calcium release contributes to synaptic release from mouse rod photoreceptors. *Neuroscience, 165,* 1447–1456.

Ball, S. L., Pardue, M. T., McCall, M. A., Gregg, R. G., & Peachey, N. S. (2003). Immunohistochemical analysis of the outer plexiform layer in the nob mouse shows no abnormalities. *Visual Neuroscience, 20,* 267–272.

Barlow, H. B. (1956). Retinal noise and absolute threshold. *Journal of the Optical Society of America, 46,* 634–639.

Barlow, H. B., Levick, W. R., & Yoon, M. (1971). Responses to single quanta of light in retinal ganglion cells of the cat. *Vision Research* (Suppl 3), 87–101.

Bayley, P. R., & Morgans, C. W. (2007). Rod bipolar cells and horizontal cells form displaced synaptic contacts with rods in the outer nuclear layer of the nob2 retina. *Journal of Comparative Neurology, 500,* 286–298.

Baylor, D. A., Lamb, T. D., & Yau, K. W. (1979). Responses of retinal rods to single photons. *Journal of Physiology, 288,* 613–634.

Baylor, D. A., Matthews, G., & Yau, K. W. (1980). Two components of electrical dark noise in toad retinal rod outer segments. *Journal of Physiology, 309,* 591–621.

Baylor, D. A., Nunn, B. J., & Schnapf, J. L. (1984). The photocurrent, noise and spectral sensitivity of rods of the monkey *Macaca fascicularis. Journal of Physiology, 357,* 575–607.

Bellone, R. R., Brooks, S. A., Sandmeyer, L., Murphy, B. A., Forsyth, G., Archer, S., et al. (2008). Differential gene expression of TRPM1, the potential cause of congenital stationary night blindness and coat spotting patterns (LP) in the Appaloosa horse (*Equus caballus*). *Genetics, 179,* 1861–1870. doi:10.1534/genetics.108.088807.

Berntson, A., Smith, R. G., & Taylor, W. R. (2004a). Postsynaptic calcium feedback between rods and rod bipolar cells in the mouse retina. *Visual Neuroscience, 21,* 913–924.

Berntson, A., Smith, R. G., & Taylor, W. R. (2004b). Transmission of single photon signals through a binary synapse in the mammalian retina. *Visual Neuroscience, 21,* 693–702.

Blakemore, C. B., & Rushton, W. A. (1965). Dark adaptation and increment threshold in a rod monochromat. *Journal of Physiology, 181,* 612–628.

Bloomfield, S. A., & Dacheux, R. F. (2001). Rod vision: Pathways and processing in the mammalian retina. *Progress in Retinal and Eye Research, 20,* 351–384.

Brann, M. R., & Cohen, L. V. (1987). Diurnal expression of transducin mRNA and translocation of transducin in rods of rat retina. *Science, 235,* 585–587.

Broekhuyse, R. M., Tolhuizen, E. F., Janssen, A. P., & Winkens, H. J. (1985). Light induced shift and binding of S-antigen in retinal rods. *Current Eye Research, 4,* 613–618.

Burns, M. E., Mendez, A., Chen, J., & Baylor, D. A. (2002). Dynamics of cyclic GMP synthesis in retinal rods. *Neuron, 36,* 81–91.

Cao, Y., Masuho, I., Okawa, H., Xie, K., Asami, J., Kammermeier, P. J., et al. (2009). Retina-specific GTPase accelerator RGS11/G beta 5S/R9AP is a constitutive heterotrimer selectively targeted to mGluR6 in ON-bipolar neurons. *Journal of Neuroscience, 29,* 9301–9313. doi:10.1523/JNEUROSCI.1367-09.2009.

Cao, Y., Pahlberg, J., Sarria, I., Kamasawa, N., Sampath, A. P., & Martemyanov, K. A. (2012). Regulators of G protein signaling RGS7 and RGS11 determine the onset of the light response in ON bipolar neurons. *Proceedings of the National Academy of Sciences of the United States of America, 109,* 7905–7910.

Cao, Y., Posokhova, E., & Martemyanov, K. A. (2011). TRPM1 forms complexes with nyctalopin in vivo and accumulates in postsynaptic compartment of ON-bipolar neurons in mGluR6-dependent manner. *Journal of Neuroscience, 31,* 11521–11526.

Cao, Y., Song, H., Okawa, H., Sampath, A. P., Sokolov, M., & Martemyanov, K. A. (2008). Targeting of RGS7/Gbeta5 to the dendritic tips of ON-bipolar cells is independent of its association with membrane anchor R7BP. *Journal of Neuroscience, 28*, 10443–10449.

Chen, C. K., Burns, M. E., He, W., Wensel, T. G., Baylor, D. A., & Simon, M. I. (2000). Slowed recovery of rod photoresponse in mice lacking the GTPase accelerating protein RGS9–1. *Nature, 403*, 557–560.

Chen, F. S., Shim, H., Morhardt, D., Dallman, R., Krahn, E., McWhinney, L., et al. (2010). Functional redundancy of R7 RGS proteins in ON-bipolar cell dendrites. *Investigative Ophthalmology & Visual Science, 51*(2), 686–693. doi:10.1167/iovs.09-4084.

Cowan, C. W., Fariss, R. N., Sokal, I., Palczewski, K., & Wensel, T. G. (1998). High expression levels in cones of RGS9, the predominant GTPase accelerating protein of rods. *Proceedings of the National Academy of Sciences of the United States of America, 95*, 5351–5356.

de la Villa, P., Kurahashi, T., & Kaneko, A. (1995). L-Glutamate-induced responses and cGMP-activated channels in three subtypes of retinal bipolar cells dissociated from the cat. *Journal of Neuroscience, 15*(Pt 1), 3571–3582.

DeVries, S. H., & Baylor, D. A. (1995). An alternative pathway for signal flow from rod photoreceptors to ganglion cells in mammalian retina. *Proceedings of the National Academy of Sciences of the United States of America, 92*, 10658–10662. doi:10.1073/pnas.92.23.10658.

Dhingra, A., Faurobert, E., Dascal, N., Sterling, P., & Vardi, N. (2004). A retinal-specific regulator of G-protein signaling interacts with Galpha(o) and accelerates an expressed metabotropic glutamate receptor 6 cascade. *Journal of Neuroscience, 24*, 5684–5693.

Dhingra, A., Jiang, M., Wang, T. L., Lyubarsky, A., Savchenko, A., Bar-Yehuda, T., et al. (2002). Light response of retinal ON bipolar cells requires a specific splice variant of Galpha(o). *Journal of Neuroscience, 22*, 4878–4884.

Dhingra, A., Lyubarsky, A., Jiang, M., Pugh, E. N., Jr., Birnbaumer, L., Sterling, P., et al. (2000). The light response of ON bipolar neurons requires G[alpha]o. *Journal of Neuroscience, 20*, 9053–9058.

Dhingra, A., Sulaiman, P., Xu, Y., Fina, M. E., Veh, R. W., & Vardi, N. (2008). Probing neurochemical structure and function of retinal ON bipolar cells with a transgenic mouse. *Journal of Comparative Neurology, 510*, 484–496.

Dick, O., Hack, I., Altrock, W. D., Garner, C. C., Gundelfinger, E. D., & Brandstatter, J. H. (2001). Localization of the presynaptic cytomatrix protein Piccolo at ribbon and conventional synapses in the rat retina: Comparison with Bassoon. *Journal of Comparative Neurology, 439*, 224–234.

Duncan, J. L., Yang, H., Doan, T., Silverstein, R. S., Murphy, G. J., Nune, G., et al. (2006). Scotopic visual signaling in the mouse retina is modulated by high-affinity plasma membrane calcium extrusion. *Journal of Neuroscience, 26*, 7201–7211. doi:10.1523/JNEUROSCI.5230-05.2006.

Dunn, F. A., Doan, T., Sampath, A. P., & Rieke, F. (2006). Controlling the gain of rod-mediated signals in the mammalian retina. *Journal of Neuroscience, 26*, 3959–3970.

Field, G. D., & Rieke, F. (2002). Nonlinear signal transfer from mouse rods to bipolar cells and implications for visual sensitivity. *Neuron, 34*, 773–785.

Field, G. D., Sampath, A. P., & Rieke, F. (2005). Retinal processing near absolute threshold: From behavior to mechanism. *Annual Review of Physiology, 67*, 491–514.

Glowatzki, E., & Fuchs, P. A. (2002). Transmitter release at the hair cell ribbon synapse. *Nature Neuroscience, 5*, 147–154.

Gozem, S., Schapiro, I., Ferré, N., & Olivucci, M. (2012). The molecular mechanism of thermal noise in rod photoreceptors. *Science, 337*, 1225–1228. doi:10.1126/science.1220461.

Haeseleer, F. (2008). Interaction and colocalization of CaBP4 and Unc119 (MRG4) in photoreceptors. *Investigative Ophthalmology & Visual Science, 49*,2366–2375.

Haeseleer, F., Imanishi, Y., Maeda, T., Possin, D. E., Maeda, A., Lee, A., et al. (2004). Essential role of Ca^{2+}-binding protein 4, a Cav1.4 channel regulator, in photoreceptor synaptic function. *Nature Neuroscience, 7*, 1079–1087. doi:10.1038/nn1320.

Hasegawa, J., Obara, T., Tanaka, K., & Tachibana, M. (2006). High-density presynaptic transporters are required for glutamate removal from the first visual synapse. *Neuron, 50*, 63–74.

He, W., Cowan, C. W., & Wensel, T. G. (1998). RGS9, a GTPase accelerator for phototransduction. *Neuron, 20*, 95–102.

Hecht, S., Schlaer, S., & Pirenne, M. H. (1942). Energy, quanta, and vision. *Journal of General Physiology, 25*, 819–840.

Higgs, M. H., & Lukasiewicz, P. D. (2002). Activation of group II metabotropic glutamate receptors inhibits glutamate release from salamander retinal photoreceptors. *Visual Neuroscience, 19*, 275–281.

Hollinger, S., & Hepler, J. R. (2002). Cellular regulation of RGS proteins: Modulators and integrators of G protein signaling. *Pharmacological Reviews, 54*, 527–559.

Huang, S. P., Brown, B. M., & Craft, C. M. (2010). Visual Arrestin 1 acts as a modulator for N-ethylmaleimide-sensitive factor in the photoreceptor synapse. *Journal of Neuroscience, 30*, 9381–9391.

Ishii, M., Morigiwa, K., Takao, M., Nakanishi, S., Fukuda, Y., Mimura, O., et al. (2009). Ectopic synaptic ribbons in dendrites of mouse retinal ON- and OFF-bipolar cells. *Cell and Tissue Research, 338*, 355–375. doi:10.1007/s00441-009-0880-0.

Jackman, S. L., Choi, S. Y., Thoreson, W. B., Rabl, K., Bartoletti, T. M., & Kramer, R. H. (2009). Role of the synaptic ribbon in transmitting the cone light response. *Nature Neuroscience, 12*, 303–310.

Koike, C., Obara, T., Uriu, Y., Numata, T., Sanuki, R., Miyata, K., et al. (2010). TRPM1 is a component of the retinal ON bipolar cell transduction channel in the mGluR6 cascade. *Proceedings of the National Academy of Sciences of the United States of America, 107*, 332–337. doi:10.1073/pnas.0912730107.

Koschak, A., Reimer, D., Walter, D., Hoda, J. C., Heinzle, T., Grabner, M., et al. (2003). Cav1.4alpha1 subunits can form slowly inactivating dihydropyridine-sensitive L-type Ca^{2+} channels lacking Ca^{2+}-dependent inactivation. *Journal of Neuroscience, 23*, 6041–6049.

Koulen, P., Kuhn, R., Wässle, H., & Brandstatter, J. H. (1999). Modulation of the intracellular calcium concentration in photoreceptor terminals by a presynaptic metabotropic glutamate receptor. *Proceedings of the National Academy of Sciences of the United States of America, 96*, 9909–9914.

Lyubarsky, A. L., Naarendorp, F., Zhang, X., Wensel, T., Simon, M. I., & Pugh, E. N., Jr. (2001). RGS9–1 is required for normal inactivation of mouse cone phototransduction. *Molecular Vision, 7*, 71–78.

Mastronarde, D. N. (1983). Correlated firing of cat retinal ganglion cells. II. Responses of X- and Y-cells to single quantal events. *Journal of Neurophysiology, 49*, 325–349.

Masu, M., Iwakabe, H., Tagawa, Y., Miyoshi, T., Yamashita, M., Fukuda, Y., et al. (1995). Specific deficit of the ON response in visual transmission by targeted disruption of the mGluR6 gene. *Cell, 80,* 757–765. doi:10.1016/0092-8674(95)90354-2.

Mehta, B., Snellman, J., Chen, S., Li, W., & Zenisek, D. (2013). Synaptic ribbons influence the size and frequency of miniature-like evoked postsynaptic currents. *Neuron, 77,* 516–527. doi:10.1016/j.neuron.2012.11.024.

Mercer, A. J., & Thoreson, W. B. (2011). The dynamic architecture of photoreceptor ribbon synapses: Cytoskeletal, extracellular matrix, and intramembrane proteins. *Visual Neuroscience, 28,* 453–471.

Mojumder, D. K., Qian, Y., & Wensel, T. G. (2009). Two R7 regulators of G-protein signaling proteins shape retinal bipolar cell signaling. *Journal of Neuroscience, 29,* 7753–7765.

Morgans, C. W., Wensel, T. G., Brown, R. L., Perez-Leon, J. A., Bearnot, B., & Duvoisin, R. M. (2007). Gbeta5-RGS complexes co-localize with mGluR6 in retinal ON-bipolar cells. *European Journal of Neuroscience, 26,* 2899–2905.

Morgans, C. W., Zhang, J., Jeffrey, B. G., Nelson, S. M., Burke, N. S., Duvoisin, R. M., et al. (2009). TRPM1 is required for the depolarizing light response in retinal ON-bipolar cells. *Proceedings of the National Academy of Sciences of the United States of America, 106,* 19174–19178. doi:10.1073/pnas.0908711106.

Mukherjee, K., Yang, X., Gerber, S. H., Kwon, H. B., Ho, A., Castillo, P. E., et al. (2010). Piccolo and bassoon maintain synaptic vesicle clustering without directly participating in vesicle exocytosis. *Proceedings of the National Academy of Sciences of the United States of America, 107,* 6504–6509. doi:10.1073/pnas.1002307107.

Naarendorp, F., Esdaille, T. M., Banden, S. M., Andrews-Labenski, J., Gross, O. P., & Pugh, E. N., Jr. (2010). Dark light, rod saturation, and the absolute and incremental sensitivity of mouse cone vision. *Journal of Neuroscience, 30,* 12495–12507.

Nakajima, Y., Iwakabe, H., Akazawa, C., Nawa, H., Shigemoto, R., Mizuno, N., et al. (1993). Molecular characterization of a novel retinal metabotropic glutamate receptor mGluR6 with a high agonist selectivity for L-2-amino-4-phosphonobutyrate. *Journal of Biological Chemistry, 268,* 11868–11873.

Nawy, S. (1999). The metabotropic receptor mGluR6 may signal through G(o), but not phosphodiesterase, in retinal bipolar cells. *Journal of Neuroscience, 19,* 2938–2944.

Nawy, S. (2000). Regulation of the on bipolar cell mGluR6 pathway by Ca^{2+}. *Journal of Neuroscience, 20,* 4471–4479.

Nawy, S. (2004). Desensitization of the mGluR6 transduction current in tiger salamander ON bipolar cells. *Journal of Physiology, 558*(Pt 1), 137–146.

Nawy, S., & Jahr, C. E. (1990). Suppression by glutamate of cGMP-activated conductance in retinal bipolar cells. *Nature, 346,* 269–271.

Nawy, S., & Jahr, C. E. (1991). cGMP-gated conductance in retinal bipolar cells is suppressed by the photoreceptor transmitter. *Neuron, 7,* 677–683.

Nomura, A., Shigemoto, R., Nakamura, Y., Okamoto, N., Mizuno, N., & Nakanishi, S. (1994). Developmentally regulated postsynaptic localization of a metabotropic glutamate receptor in rat rod bipolar cells. *Cell, 77,* 361–369.

Oberdick, J., Smeyne, R. J., Mann, J. R., Zackson, S., & Morgan, J. I. (1990). A promoter that drives transgene expression in cerebellar Purkinje and retinal bipolar neurons. *Science, 248,* 223–226.

Okawa, H., Miyagishima, K. J., Arman, A. C., Hurley, J. B., Field, G. D., & Sampath, A. P. (2010a). Optimal processing of photoreceptor signals is required to maximize behavioural sensitivity. *Journal of Physiology, 588*(Pt 11), 1947–1960.

Okawa, H., Pahlberg, J., Rieke, F., Birnbaumer, L., & Sampath, A. P. (2010b). Coordinated control of sensitivity by two splice variants of G{alpha}o in retinal ON bipolar cells. *Journal of General Physiology, 136,* 443–454.

Pahlberg, J., Majumder, A., Boyd, K., Kerov, V., Artemyev, N., & Sampath, A. P. (2012). Transducin translocation sensitizes synaptic transmission from rods to rod bipolar cells. Paper presented at the Association for Research in Vision and Ophthalmology, Ft. Lauderdale, FL.

Pahlberg, J., & Sampath, A. P. (2011). Visual threshold is set by linear and nonlinear mechanisms in the retina that mitigate noise: How neural circuits in the retina improve the signal-to-noise ratio of the single-photon response. *BioEssays, 33,* 438–447.

Pearring, J. N., Bojang, P., Jr., Shen, Y., Koike, C., Furukawa, T., Nawy, S., et al. (2011). A role for nyctalopin, a small leucine-rich repeat protein, in localizing the TRP melastatin 1 channel to retinal depolarizing bipolar cell dendrites. *Journal of Neuroscience, 31,* 10060–10066. doi:10.1523/JNEUROSCI.1014-11.2011.

Philp, N. J., Chang, W., & Long, K. (1987). Light-stimulated protein movement in rod photoreceptor cells of the rat retina. *FEBS Letters, 225,* 127–132. doi:10.1016/0014-5793(87)81144-4.

Rabl, K., Bryson, E. J., & Thoreson, W. B. (2003). Activation of glutamate transporters in rods inhibits presynaptic calcium currents. *Visual Neuroscience, 20,* 557–566.

Rao, A., Dallman, R., Henderson, S., & Chen, C. K. (2007). Gbeta5 is required for normal light responses and morphology of retinal ON-bipolar cells. *Journal of Neuroscience, 27,* 14199–14204.

Rao-Mirotznik, R., Buchsbaum, G., & Sterling, P. (1998). Transmitter concentration at a three-dimensional synapse. *Journal of Neurophysiology, 80,* 3163–3172.

Rao-Mirotznik, R., Harkins, A. B., Buchsbaum, G., & Sterling, P. (1995). Mammalian rod terminal: Architecture of a binary synapse. *Neuron, 14,* 561–569.

Rieke, F., & Baylor, D. A. (1996). Molecular origin of continuous dark noise in rod photoreceptors. *Biophysical Journal, 71,* 2553–2572.

Ross, E. M., & Wilkie, T. M. (2000). GTPase-activating proteins for heterotrimeric G proteins: Regulators of G protein signaling (RGS) and RGS-like proteins. *Annual Review of Biochemistry, 69,* 795–827.

Ruether, K., Feigenspan, A., Pirngruber, J., Leitges, M., Baehr, W., & Strauss, O. (2010). PKC{alpha} is essential for the proper activation and termination of rod bipolar cell response. *Investigative Ophthalmology & Visual Science, 51,* 6051–6058.

Sakitt, B. (1972). Counting every quantum. *Journal of Physiology, 223,* 131–150.

Sampath, A. P., Miyagishima, K. J., Arman, A. C., Nymark, S., Pahlberg, J., & Cornwall, M. C. (2011). Bleaching and background adaptation shift rod-driven signals from primary to secondary rod pathways in the mouse retina. Poster presented at the Association for Research in Vision and Opthalmology, Ft. Lauderdale, FL.

Sampath, A. P., & Rieke, F. (2004). Selective transmission of single photon responses by saturation at the rod-to-rod bipolar synapse. *Neuron, 41*, 431–443.

Sampath, A. P., Strissel, K. J., Elias, R., Arshavsky, V. Y., McGinnis, J. F., Chen, J., et al. (2005). Recoverin improves rod-mediated vision by enhancing signal transmission in the mouse retina. *Neuron, 46*, 413–420. doi:10.1016/j.neuron.2005.04.006.

Sato, S., Omori, Y., Katoh, K., Kondo, M., Kanagawa, M., Miyata, K., et al. (2008). Pikachurin, a dystroglycan ligand, is essential for photoreceptor ribbon synapse formation. *Nature Neuroscience, 11*, 923–931. doi:10.1038/nn.2160.

Schmitz, F., Konigstorfer, A., & Sudhof, T. C. (2000). RIBEYE, a component of synaptic ribbons: A protein's journey through evolution provides insight into synaptic ribbon function. *Neuron, 28*, 857–872.

Shen, Y., Heimel, J. A., Kamermans, M., Peachey, N. S., Gregg, R. G., & Nawy, S. (2009). A transient receptor potential-like channel mediates synaptic transmission in rod bipolar cells. *Journal of Neuroscience, 29*, 6088–6093.

Shiells, R. A., & Falk, G. (1990). Glutamate receptors of rod bipolar cells are linked to a cyclic GMP cascade via a G-protein. *Proceedings. Biological Sciences, 242*, 91–94.

Shim, H., Wang, C. T., Chen, Y. L., Chau, V. Q., Fu, K. G., Yang, J., et al. (2012). Defective retinal depolarizing bipolar cells in regulators of G-protein signaling (RGS) 7 and 11 double null mice. *Journal of Biological Chemistry, 287*, 14873–14879. doi:10.1074/jbc.M112.345751.

Singer, J. H., Lassova, L., Vardi, N., & Diamond, J. S. (2004). Coordinated multivesicular release at a mammalian ribbon synapse. *Nature Neuroscience, 7*, 826–833.

Snellman, J., Kaur, T., Shen, Y., & Nawy, S. (2008). Regulation of ON bipolar cell activity. *Progress in Retinal and Eye Research, 27*, 450–463.

Snellman, J., Mehta, B., Babai, N., Bartoletti, T. M., Akmentin, W., Francis, A., et al. (2011). Acute destruction of the synaptic ribbon reveals a role for the ribbon in vesicle priming. *Nature Neuroscience, 14*, 1135–1141. doi:10.1038/nn.2870.

Snellman, J., & Nawy, S. (2002). Regulation of the retinal bipolar cell mGluR6 pathway by calcineurin. *Journal of Neurophysiology, 88*, 1088–1096.

Snellman, J., & Nawy, S. (2004). cGMP-dependent kinase regulates response sensitivity of the mouse on bipolar cell. *Journal of Neuroscience, 24*, 6621–6628.

Song, J. H., Song, H., Wensel, T. G., Sokolov, M., & Martemyanov, K. A. (2007). Localization and differential interaction of R7 RGS proteins with their membrane anchors R7BP and R9AP in neurons of vertebrate retina. *Molecular and Cellular Neurosciences, 35*, 311–319.

Soucy, E., Wang, Y., Nirenberg, S., Nathans, J., & Meister, M. (1998). A novel signaling pathway from rod photoreceptors to ganglion cells in mammalian retina. *Neuron, 21*, 481–493.

Strissel, K. J., Lishko, P. V., Trieu, L. H., Kennedy, M. J., Hurley, J. B., & Arshavsky, V. Y. (2005). Recoverin undergoes light-dependent intracellular translocation in rod photoreceptors. *Journal of Biological Chemistry, 280*, 29250–29255. doi:10.1074/jbc.M501789200.

Thoreson, W. B. (2007). Kinetics of synaptic transmission at ribbon synapses of rods and cones. *Molecular Neurobiology, 36*, 205–223.

Thoreson, W. B., Rabl, K., Townes-Anderson, E., & Heidelberger, R. (2004). A highly Ca^{2+}-sensitive pool of vesicles contributes to linearity at the rod photoreceptor ribbon synapse. *Neuron, 42*, 595–605.

tom Dieck, S., Altrock, W. D., Kessels, M. M., Qualmann, B., Regus, H., Brauner, D., et al. (2005). Molecular dissection of the photoreceptor ribbon synapse: Physical interaction of Bassoon and RIBEYE is essential for the assembly of the ribbon complex. *Journal of Cell Biology, 168*, 825–836. doi:10.1083/jcb.200408157.

Trifonov, I. U. (1968). [Study of synaptic transmission between photoreceptor and horizontal cell by electric stimulations of the retina] (Rus.). *Biofizika, 13*, 809–817.

Tsukamoto, Y., Morigiwa, K., Ueda, M., & Sterling, P. (2001). Microcircuits for night vision in mouse retina. *Journal of Neuroscience, 21*, 8616–8623.

van Rossum, M. C., & Smith, R. G. (1998). Noise removal at the rod synapse of mammalian retina. *Visual Neuroscience, 15*, 809–821.

von Gersdorff, H., & Matthews, G. (1994). Dynamics of synaptic vesicle fusion and membrane retrieval in synaptic terminals. *Nature, 367*, 735–739.

Wahl-Schott, C., Baumann, L., Cuny, H., Eckert, C., Griessmeier, K., & Biel, M. (2006). Switching off calcium-dependent inactivation in L-type calcium channels by an autoinhibitory domain. *Proceedings of the National Academy of Sciences of the United States of America, 103*, 15657–15662. doi:10.1073/pnas.0604621103.

Walraven, J., Enroth-Cugell, C., Hood, D. C., Dia, M. L., & Schnapf, J. L. (1990). *The control of visual sensitivity.* San Diego, CA: Academic Press.

Whelan, J. P., & McGinnis, J. F. (1988). Light-dependent subcellular movement of photoreceptor proteins. *Journal of Neuroscience Research, 20*, 263–270.

Xing, W., Akopian, A., & Krizaj, D. (2012). Trafficking of presynaptic PMCA signaling complexes in mouse photoreceptors requires Cav1.4 alpha(1) subunits. *Advances in Experimental Medicine and Biology, 723*, 739–744.

Xu, Y., Sulaiman, P., Feddersen, R. M., Liu, J., Smith, R. G., & Vardi, N. (2008). Retinal ON bipolar cells express a new PCP2 splice variant that accelerates the light response. *Journal of Neuroscience, 28*, 8873–8884.

Zanazzi, G., & Matthews, G. (2009). The molecular architecture of ribbon presynaptic terminals. *Molecular Neurobiology, 39*, 130–148.

Zhang, J., Jeffrey, B. G., Morgans, C. W., Burke, N. S., Haley, T. L., Duvoisin, R. M., et al. (2010). RGS7 and -11 complexes accelerate the ON-bipolar cell light response. *Investigative Ophthalmology & Visual Science, 51*, 1121–1129. doi:10.1167/iovs.09-4163.

Zhang, X., Wensel, T. G., & Kraft, T. W. (2003). GTPase regulators and photoresponses in cones of the eastern chipmunk. *Journal of Neuroscience, 23*, 1287–1297.

6 Cone Bipolar Cells: ON and OFF Pathways in the Outer Retina

STEVEN H. DEVRIES

At the cone synapse, signals diverge in parallel to 10 or more anatomically distinct bipolar cell types. The parallel pathways subserve two main functions. First, the bipolar cell types carry the cone signals to different substrata within the inner plexiform layer. In these substrata, signals are processed at synapses with distinct subsets of amacrine and ganglion cells. Second, the bipolar cell types differently transform the cone signal. Signals are transformed first during transmission at the cone synapse, then by voltage-dependent membrane currents in the bipolar cell types, and finally through inhibitory inputs at bipolar cell terminals. This review goes beyond the essential but well-described distinction between mammalian ON and OFF bipolar cells (Nelson & Kolb, 2003) and examines the evidence that each anatomical bipolar cell type carries a distinct "representation" of the visual scene (Strettoi et al., 2010). These representations have different spatial, temporal, and chromatic signaling properties. They also differ with respect to rod and cone input and potentially signal quality or fidelity. Recent reviews discuss bipolar cell processing in the context of retinal ganglion cell outputs (Masland, 2001; Wässle, 2004).

BIPOLAR CELL TYPES

There is an emerging consensus that mammalian retinas contain 11–13 cone bipolar cell types and that the classification schemes in several species are virtually complete (figure 6.1). This consensus applies to the mouse (Strettoi et al., 2010; Wässle et al., 2009), rat (Euler & Wässle, 1995), rabbit (MacNeil et al., 2004), and ground squirrel (Light et al., 2012; Puller, Ondreka, & Haverkamp, 2011) retinas, but perhaps not to the primate retina, which is thought to contain only 9–10 types (Joo et al., 2011). Our experience has been that a complete classification requires the use of several different and complementary approaches. Techniques that target all of the members of a bipolar cell type include antibody labeling and promoter-driven marker expression in transgenic mice. In rare cases antibodies label a single morphological class of bipolar cell that has a uniform dendritic and axonal tiling. Examples include recoverin in the ground squirrel (cb3b), calsenilin in the mouse (type 4), and CD15 in several species (Puller et al., 2011). More frequently, antibodies label several types (e.g., secretagogin in the mouse [Puthussery, Gayet-Primo, & Taylor, 2010]), and individual cell types are typically identified through a "combinatorial code" after labeling with two or more antibodies. Techniques that label individual bipolar cells irrespective of type include Golgi impregnation, photofilling (MacNeil et al., 2004), random genetic labeling (Badea & Nathans, 2004), and adeno-associated virus targeting (Light et al., 2012).

The utility of having a complete classification cannot be overstated. Functional and anatomical inputs and outputs differ by cell type. For example, it was not possible to appreciate that certain OFF bipolar cells use either kainate- or AMPA-type glutamate receptors to receive the cone signal without being able to distinguish among the different cone bipolar cell types (DeVries, 2000; DeVries & Schwartz, 1999). In addition, a complete anatomical classification is a precondition for functional studies that, in turn, are necessary for understanding how distinct ganglion and amacrine cell responses are created.

BIPOLAR CELL CONNECTIONS IN THE OUTER PLEXIFORM LAYER

Rod Photoreceptor–to–Cone Bipolar Cell Synapses

The segregation of rod and cone signals into separate rod and cone bipolar cell pathways is thought to be a hallmark of the mammalian retina (Dowling & Boycott, 1966), distinguishing it from fish and amphibian retinas where bipolar cells routinely contact both rods and cones (Ishida, Stell, & Lightfoot, 1980; Pang, Gao, & Wu, 2004). However, recent results in the mammalian retina challenge this idea of a strict segregation and point toward common principles of organization in vertebrates. For clarity, we define three "rod pathways." In a sensitive or classical rod pathway, signals flow from

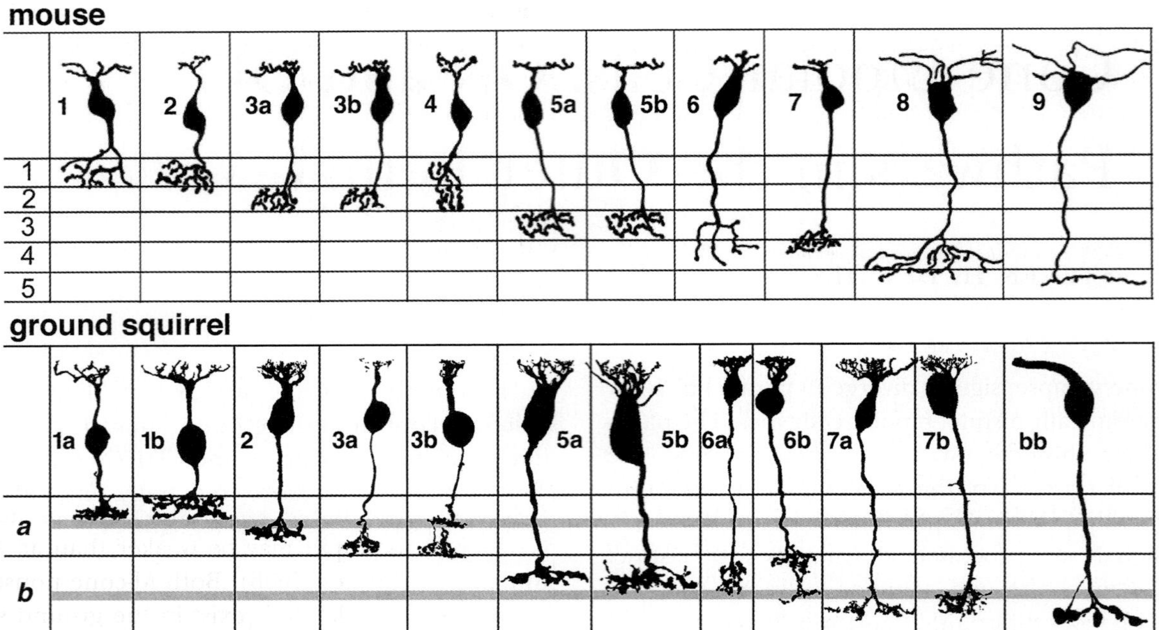

FIGURE 6.1 Comparison of cone bipolar cell types in the mouse and ground squirrel. OFF bipolar cells have axon terminals that ramify in sublamina *a* corresponding to layers 1 and 2 in the mouse. ON bipolar cells ramify in sublamina *b* (layers 2–5). Although the diagram is designed to facilitate a comparison between mouse and ground squirrel bipolar cell types, immunocytochemical and other anatomical differences preclude a direct correspondence. (Adapted from Light et al., 2012, and Wässle et al., 2009, with permission.)

rods to rod bipolar cells. In a second pathway, rod signals flow through gap junction channels to cones and then to cone bipolar cells. And finally, in a third rod pathway, signals flow from rods directly to cone ON or OFF bipolar cells.

Anatomical studies were among the first to suggest the existence of rod–to–cone bipolar cell synapses in the mammal. Ultrastructural contacts were demonstrated in tree squirrel (West, 1978), mouse (Hack, Peichl, & Brandstatter, 1999), and cat (Fyk-Kolodziej, Qin, & Pourcho, 2003) retinas. Light microscopy suggested contacts between rods and type 3a, 3b, and 4 OFF bipolar cells in the mouse (figure 6.1) (Haverkamp et al., 2008; Mataruga, Kremmer, & Muller, 2007), at least two types of OFF bipolar cells in the rabbit (Li, Keung, & Massey, 2004), and one type of OFF bipolar cell in the ground squirrel (cb2) (Li, Chen, & DeVries, 2010) retina. Notably, in the mouse, rabbit, and cat, contacts were made only with the lowest row of rod terminals, belonging to 10–20% of all rod photoreceptors. Rod–to–ON cone bipolar cell contacts were described in the mouse (type 7) (Tsukamoto et al., 2007) and ground squirrel (cb5) (Li et al., 2010) but seem to be less prevalent than the contacts with OFF cone bipolar cells.

Cone bipolar cells typically contact rods on the outside of the spherule (Hack et al., 1999),

necessitating a long (>0.5 µm) glutamate diffusion path between the ribbon release sites within the invagination and postsynaptic receptors. In spite of the unusually long diffusion path, recordings in both the ground squirrel and mouse suggest that rod–to–cone bipolar cell synapses are functional. In the ground squirrel, paired recordings demonstrate rapid and robust signaling to select OFF (cb2) and, to a lesser extent, ON (cb5) bipolar cell types (Li et al., 2010). Specific ON and OFF cone bipolar cells in the mouse also showed evidence of rod input during light responses (Pang et al., 2012). A note of caution is in order when rod-driven light responses in a bipolar cell are interpreted as resulting from rod–to–cone bipolar cell synapses: the ability of glutamate to diffuse within the outer plexiform layer (Szmajda & DeVries, 2011) raises the possibility that transmission can occur in the absence of discrete contacts.

The function of this third, direct rod–to–cone bipolar cell pathway is unclear. Nor is it evident why only some cone bipolar cell types contact rods. Experiments in the ground squirrel suggest that signaling at OFF bipolar cell AMPA receptors may be faster than that at rod bipolar cell mGluR6 receptors, thus enabling rods to signal through a fast or high temporal frequency OFF pathway (Li et al., 2010). However, other results suggest that the third pathway signals to ganglion cells that have sluggish responses (DeVries & Baylor, 1995) or produces ganglion cell responses that are temporally delayed (Soucy et al., 1998). The third pathway may also mediate signaling under mesopic light conditions when both rods and cones are actively signaling (Volgyi et al., 2004).

Blue and Green Cone-Driven Bipolar Cell Pathways

In the retinas of placental mammals, authentic short (S) and longer (M, middle, and L, long) wavelength sensitive cones are spatially distributed according to a basic plan. In this distribution a lawn of longer wavelength sensitive cones, comprising more than 90% of all cones, is studded with the minority S-cones, which are arranged in a semi-regular array (Curcio et al., 1991; Haverkamp et al., 2005; Roorda & Williams, 1999). In the excised retina the array can be visualized with either S-cone opsin-specific antibodies (Curcio et al., 1991) or antibodies that label cone terminal pre- and postsynaptic markers. When compared to M-cone terminals, S-cone terminals have the same number of ribbons (Haverkamp, Grunert, & Wässle, 2001b) but are more compact and have reduced label for certain OFF bipolar cell postsynaptic receptors (Breuninger et al., 2011; Haverkamp, Grunert, & Wässle, 2001a; Li & DeVries, 2006). Because the basic plan is conserved among dichromats and trichomats, the remaining discussion focuses on dichromats.

The primordial form of dichromatic color vision depends on comparing excitation between S- and M-cones (or M+L cones in Old World and several New World primates). To the extent that certain ganglion cell types encode chromatic signals, cone selective bipolar cells must exist to carry those signals across the retina. Indeed, many mammalian retinas contain an ON bipolar cell type that exclusively receives input from S-cones. S-ON bipolar cells have been identified by antibody labeling in the primate and ground squirrel (Kouyama & Marshak, 1992; Puller et al., 2011), electron microscopy in the primate (Calkins, Tsukamoto, & Sterling, 1998), and fluorescent protein expression in the mouse (Haverkamp et al., 2005). In the primate retina S-cone bipolar cells provide ON input to the receptive field center of a small bistratified ganglion cell (Dacey & Lee, 1994). Similar S-ON ganglion cell types are found in the guinea pig (Yin et al., 2009) and ground squirrel (Sher & DeVries, 2012). S-OFF ganglion cell responses have also been recorded in the retina, optic nerve, and lateral geniculate nucleus of primates, guinea pigs, and ground squirrels (Sher & DeVries, 2012; Wiesel & Hubel, 1966; Yin et al., 2009). Anatomical studies provide scant evidence for the existence of an S-OFF cone bipolar cell. Instead, at least one type of S-OFF ganglion cell obtains its response from an amacrine cell that inverts the sign of the S-ON bipolar cell signal (Chen & Li, 2012; Sher & DeVries, 2012). A minor exception to this pattern occurs in the central macaque retina, where OFF midget bipolar cells, which indiscriminately contact the cone types, can contact an S-cone leading to an S-OFF midget ganglion cell response (Klug et al., 2003).

While there is strong evidence for an S-cone specific bipolar cell type, there is less certainty about the prevalence of M-cone specific bipolar cell types. In part, this is because it can be difficult, even in S- and M-cone nonselective bipolar cells, to measure S-cone signals when S-cones comprise roughly 4–6% of the cone photoreceptors (Breuninger et al., 2011), and an individual S-cone synaptic input may be weaker than an M-cone input (Li & DeVries, 2006). Both M-cone nonselective and selective bipolar cells exist in the ground squirrel and mouse. In the ground squirrel at least two bipolar cell types, the cb2 and cb5, receive functional input from both S- and M-cones and can be viewed as having achromatic responses (Li & DeVries, 2006). A bipolar cell that receives input from both S- and M-cones might be suited to detect change in the visual scene (DeVries, Li, & Saszik, 2006; Saszik & DeVries, 2012). Unlike the cb2 and cb5 cells, ground squirrel cb1 and mouse type 1 OFF bipolar cells avoid S-cones (Breuninger et al., 2011; Li & DeVries, 2006; Pang et al., 2012). One idea is that this pure M-cone OFF bipolar cell could provide an input that is opposed to the pure S-cone ON input in ganglion cells such as the primate small bistratified cell. Against this idea, nonselective OFF bipolar cells have a predominant M-cone response that should still be useful for color-opponent vision. A more complete understanding of M-cone selective and nonselective bipolar signaling awaits an elucidation of the ganglion cell outputs for each bipolar cell type.

CONE TRANSMITTER RELEASE

Structure of the Cone Synapse

The design of the mammalian cone terminal is uniquely suited to distribute signals to multiple postsynaptic neurons. A terminal contains 20–40 synaptic ribbons that together can dock 200–800 vesicles at the membrane and tether severalfold more in preparation for docking. Each ribbon is located atop a 0.5- to 0.8-μm-deep invagination. Released transmitter first flows over the processes of two horizontal cells, which enter the invagination and terminate to each side of the ribbon, and then over one or more central processes, contributed mostly by ON bipolar cells that terminate beneath the ribbon at a distance of 100–200 nm (Sterling & Matthews, 2005). The invaginations empty onto a ~5-μm-diameter circular base where transmitter then flows over approximately 500 discrete basal contacts belonging mostly to OFF bipolar cells. An individual bipolar cell contributes 10–50 of these contacts (Hopkins & Boycott, 1997). Each terminal is separated from its neighbors by a Muller glial cell sheath (Burris et al., 2002) that serves to reduce but not eliminate cone–cone transmitter diffusion (Szmajda & DeVries, 2011). The Muller cell sheaths do not approach the base of the pedicle; thus, a single packet of transmitter is able to elicit responses from many different bipolar cells (DeVries et al., 2006) before it dissipates through diffusion or is bound by cone glutamate transporters (Szmajda & DeVries, 2011).

Turnover at the Cone Ribbon

The classic view of the retinal ribbon synapse is that it mediates steady transmitter release in a manner that supports graded transmission (Choi et al., 2005). Yet, paired recordings show that the cone synapse can also mediate a rapid, synchronous release of transmitter (DeVries, 2000; Rabl, Cadetti, & Thoreson, 2005). Both sustained and transient release components can convey visual information. This dual action is illustrated by examining the response of a cone OFF bipolar cell during a light step (figure 6.2). Cone terminals are depolarized in the dark, and a steady calcium influx provides for a steady rate of vesicle fusion and ribbon turnover. Each docking site turns over at a rate of 2–3 vesicles-s^{-1}; thus, steady release at the entire terminal is estimated at between 500 and 1,500 vesicles-s^{-1} (DeVries,

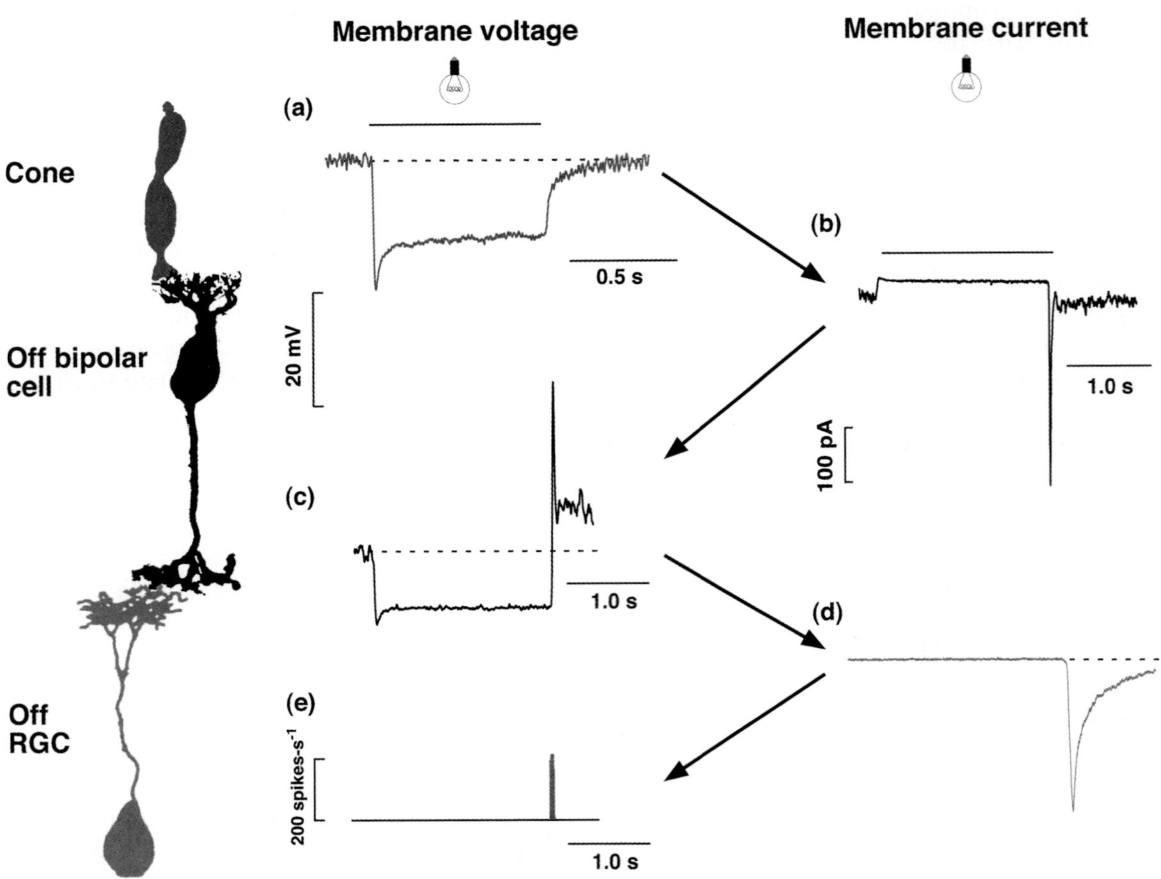

FIGURE 6.2   Transformation of the cone light response by the retinal circuitry to produce a transient burst of spikes in a ganglion cell at light-off. (a) A light pulse hyperpolarizes the cone. (b) The glutamate-gated current in a postsynaptic OFF bipolar cell responds to sustained and transient components of cone transmitter release. (c) The membrane voltage change during a light pulse measured in the same bipolar cell. (d) Depending on the amount of rectification in the release process, bipolar cell depolarization can be more effective than hyperpolarization at gating glutamate release, which leads to an inward retinal ganglion cell (RGC) membrane current at light-off. (e) The resulting membrane depolarization produces a transient burst of spikes at light-off. Traces are shaded to correspond to the cell of origin.

2000; DeVries et al., 2006). During steady release, vesicles either move slowly down the ribbon to refill empty docking sites (Jackman et al., 2009) or move rapidly into position but slowly become competent for release (Zampighi et al., 2011). In either case, the pool of membrane primed and ready–to–fuse vesicles is relatively depleted. When cones are hyperpolarized during a light step, both calcium influx and vesicle fusion cease, causing ribbon docking sites to fill with vesicles that become primed. A subsequent cone depolarization at light-off then produces a synchronous fusion of docked vesicles, resulting in a transient flood of cleft glutamate (DeVries, 2000; Jackman et al., 2009). The transient release at light-off can produce a transient depolarization in OFF bipolar cells that is several times the amplitude of the preceding hyperpolarization. This transient bipolar cell depolarization can, in turn, cause postsynaptic ganglion cells to fire a burst of spikes at light-off. The transient cone release may also affect the shape of the ON bipolar cell response, although the size of the response may be limited by receptor cascade saturation (Sampath & Rieke, 2004).

Ectopic Release

Ectopic release in either photoreceptors or bipolar cells is defined as occurring at locations that are distant from identifiable ribbons. Vesicles are nearly uniformly distributed in bipolar cell terminals, and high intracellular calcium concentrations can lead to ectopic fusion (Zenisek, 2008). Indeed, ectopic release has been implicated in bipolar cell signaling (Midorikawa et al., 2007), but a physiological role remains controversial (Snellman et al., 2011). The cone synapse is also filled with synaptic vesicles, many of which reside near the basal membrane away from ribbons. Although it is difficult to exclude ectopic fusion at the cone synapse, our results (DeVries et al., 2006) and those of Snellman et al. (2011) tend to minimize this possibility. We found that vesicles that fuse in invaginations produce a characteristic low amplitude, temporally smeared response in OFF bipolar cells that contact the cone at its base (DeVries et al., 2006), consistent with a long transmitter diffusion path. In unpublished results (Ratliff and DeVries), we have observed that these smeared events continue to dominate transmission even under conditions of sustained release approximating those that occur during a continuously varying light stimulus. Snellman et al. (2011), taking a different approach, showed that photolytically destroying cone ribbons completely and effectively abolished cone signaling. This effect would not be expected for a ribbon-independent ectopic release.

THE BIPOLAR CELL RESPONSE TO CONE TRANSMITTER RELEASE

OFF Bipolar Cells

Paired recordings at cone to OFF bipolar cell synapses in the ground squirrel first raised the possibility that each OFF bipolar cell type might receive a different signal from a common presynaptic cone. Subsequent work has shown that the OFF cone bipolar cell types differ with respect to (1) the postsynaptic glutamate receptor, AMPA or kainate (DeVries, 2000); (2) the location of dendritic contacts relative to cone transmitter release sites, invaginating or basal (DeVries et al., 2006) and (3) the number of times each cone is contacted, ranging from one to five or more (DeVries et al., 2006). As described below, each of the characteristics—postsynaptic receptor type, contact distance from release sites, and cone contact number—can lead to different postsynaptic responses under test conditions. The extent to which each characteristic contributes to signaling under more physiological conditions of transmitter ebb and flow remains an open question.

Functions of Different OFF Bipolar Cell Receptors

AMPA VERSUS KAINATE OFF bipolar cells contain ion channel glutamate receptors of the AMPA/kainate type (Gilbertson, Scobey, & Wilson, 1991). At most CNS sites AMPA receptors mediate transmission alone or in combination with kainate receptors. Rarely do kainate receptors mediate synaptic transmission on their own. Thus, it was a surprise to find that kainate receptors predominated at the ground squirrel cone–to–OFF bipolar cell synapse (DeVries & Schwartz, 1999). Subsequent work in the ground squirrel suggested a diversity of receptor types: cb1 bipolar cells use only kainate receptors, cb2 cells use only AMPA receptors, and in cb3 cells, most (70–80%) of the postsynaptic receptors are of the kainate type while the remainder are AMPA receptors (DeVries, 2000; Light et al., 2012). The AMPA and kainate receptors on OFF bipolar cells share many properties including the glutamate concentration that produces a half-maximal response (~350 μM), the kinetics of activation (20–80% rise time equal to 0.20–0.25 ms), and the percentage of receptors active (4–9%) during the steady response resulting from a saturating glutamate step (DeVries et al., 2006). However, OFF bipolar cell AMPA and kainate receptors differ with respect to one crucial property: the rate at which they recover from a desensitizing pulse of glutamate. The AMPA receptors on cb2 cells recover with an exponential time constant, τ, <20 ms, whereas the kainate receptors on cb3 cells have a recovery $\tau > 1$ s (DeVries, 2000). The kainate receptors on cb1 cells recover with an intermediate, biexponential time course. The molecular basis for the observed kainate receptor diversity is presently unclear.

The interaction between tonic glutamate release and receptors that enter long-lived desensitized states when exposed to glutamate was tested during paired recordings in which cone membrane voltage was sequentially hyperpolarized and depolarized to mimic the response to square wave of light. An initial cone depolarization produced a burst of glutamate release similar to the release at light-off. Following the burst, the slowly recovering receptors on cb3 cells entered a prolonged adapted state in which the transient component of subsequent OFF responses was abrogated (DeVries, 2000), thus producing a more sustained response. cb2 cell AMPA receptors recovered sensitivity more rapidly after a burst of transmitter release and did not display the same transient response adaptation as in cb3 cells. The case for receptor desensitization playing an adapting role during steady, as opposed to transient, release is less clear insofar as the contents of many vesicles must overlap before a population of basal (see below) kainate receptors becomes substantially desensitized.

Immunolabeling for AMPA and kainate receptor subunits suggests that the cone terminals of many mammals share the same basic organization. Antibodies to kainate receptor subunits, especially those to GluK1 (GluR5), label the base of the cone terminal in several species (Haverkamp et al., 2001a, 2003; Puller et al., 2013). In the ground squirrel the location of GluK1 labeling is consistent with the inferred basal locations of contacts from cb3 cells that use kainate receptors (DeVries, 2000; DeVries et al., 2006). Labeling for AMPA receptor subunits GluA1–4 is concentrated in and around markers for invaginations, although some labeling appears over basal regions (Haverkamp et al., 2001b). Most of the AMPA receptor subunit labeling associated with invaginations can be attributed to horizontal cells. However, in the ground squirrel, some of the labeling may come from cb2 cell terminals that express AMPA receptors and have an invaginating or semi-invaginating location (DeVries, 2000; DeVries et al., 2006).

FUNCTION OF DIFFERENT CONTACT LOCATIONS: INVAGI-NATING VERSUS BASAL Bipolar cells contact cones at two types of synapses, invaginating and basal. Basal contacts are further subdivided into non–triad associated and triad associated (or semi-invaginating). Ultrastructural studies in the primate (Hopkins & Boycott, 1997), rabbit (Merighi, Raviola, & Dacheux, 1996), and cat (Kolb, 1979) suggest that ON bipolar cell dendrites occupy the central position in invaginations, whereas OFF bipolar cell dendrites exclusively contact cones in basal positions. It is also possible to glean information about synapse structure from bipolar cell recordings. Specifically, the shape of small spontaneous and evoked excitatory postsynaptic currents (EPSCs) is determined by both receptor kinetics and the extracellular distance between a vesicle fusion site and the dendritic receptors. An assumption is that vesicles rapidly release their contents after fusing (e.g., $\tau < 0.1$ ms; Stiles et al., 1996). In the ground squirrel an analysis of small EPSC shape presents a picture at odds with the canonical ultrastructure. Modeling suggests that the fast rise (20-80% rise times ~0.2 ms) and decay ($\tau < 1$ ms) times of OFF cb2 cell events result from receptors being located within 200 nm of release sites, effectively within invaginations. The peak glutamate concentration during these events can attain 2–10 mM (DeVries et al., 2006).

The spontaneous and small evoked EPSCs in cb1 and cb3 cells are three to five times smaller and slower than those in cb2 cells. Modeling suggests that the receptors are located 0.6–0.9 μm from release sites, effectively on the basal surface of the terminal. Peak glutamate concentrations are <150 μM (DeVries et al., 2006). Two results suggest that the slow event time course in cb1 and cb3 cells is probably not due to kainate receptor kinetics. First, blocking the kainate receptors on cb3 cells reveals a smaller AMPA receptor-mediated current with a time course identical to that of the larger current (Ratliff and DeVries, unpublished observation). Second, outside-out receptor patches containing bipolar cell kainate receptors briskly respond to changes in glutamate concentration (DeVries et al., 2006).

Results in the ground squirrel point toward a major difference in signal processing at invaginating versus basal OFF bipolar cell contacts. At the invaginating contacts of cb2 cells, transmitter concentrations rapidly rise to a high level (>2 mM per fused vesicle) and then rapidly fall. The fast rise and fall facilitates the transmission of high temporal frequencies (up to 400 Hz; Ratliff and DeVries, unpublished observation), and the high glutamate concentrations attained when several vesicles fuse at once at the same ribbon suggests that synaptic receptors may saturate, a signature of high-gain transmission. Conversely, the relatively slow time course of cb1 and cb3 cell small EPSCs limits the frequencies that these bipolar cells can transmit to <200 Hz (Ratliff and DeVries, unpublished observation; the responses of cb2 cells are also shifted toward higher frequencies relative to those of cb3 cells in the physiological range). The low peak glutamate concentration attained when a single vesicle fuses suggests that basal contacts of cb1 and cb3 cells can sum many release events before saturating and thus may operate either at a lower gain or over a wider dynamic range when compared to cb2 cells.

FUNCTIONAL IMPLICATIONS OF DIFFERENCES IN CONTACT NUMBER The studies of cone terminal ultrastructure summarized by Hopkins and Boycott (1997) emphasize the differences in cone contact number between the different diffuse and midget bipolar cells in the primate. Primate ON midget bipolar cells made close to the greatest number of contacts (>25 contacts per cone), whereas diffuse ON DB6 cells made the fewest contacts (~6). Similar though less extreme differences are observed among OFF bipolar cells in the ground squirrel with light microscopy (DeVries et al., 2006). On the simple assumption that ribbons release vesicles stochastically at a steady cone voltage, it is tempting to speculate that more contacts sample more ribbons and thus provide a more accurate or less variable measure of cone voltage. In this view the signal in a cb3 cell, which makes multiple basal contacts, should be more reproducible during a repeated stimulus than the signal in a cb1a cell, which makes a single basal contact (DeVries et al., 2006). Bipolar cells that make fewer contacts with individual cones, such as the ground squirrel cb1a, could compensate for this increased variability by having greater dendrite overlap within the population. Although dendrite overlap has not been quantified in the ground squirrel, all of the bipolar cell types in the mouse share a similar degree of dendrite overlap (Wässle et al., 2009), so a compensatory mechanism based on differences in overlap seems unlikely.

ON Bipolar Cells

Mammalian ON cone and rod bipolar cells use similar transduction mechanisms. Glutamate, which is continuously released during darkness, binds to a metabotropic glutamate receptor, mGluR6 (Nakajima et al., 1993), that colocalizes with a G-protein-coupled receptor whose ligand is unknown (GPR179) (Audo et al., 2012; Peachey et al., 2012). Both are required for transduction. In the glutamate-bound state, mGluR6 activates a G-protein, G_o (Vardi, 1998), that directly or through intermediaries has been hypothesized to maintain a nonselective cation channel in the closed state. This cation channel was recently identified as TRPM1 (Koike et al., 2010; Morgans et al., 2009; Shen et al., 2009), and both activated $G_{o\alpha}$ (Koike et al., 2010) and $G_{\beta\gamma}$ (Shen et al., 2012) have been implicated in channel closure. When a bright flash of light terminates photoreceptor glutamate release, the initial slope of the depolarizing bipolar cell response depends on the rate of GTP hydrolysis, which returns $G_{o\alpha}$ to its inactive state. The rate of GTP hydrolysis is greatly facilitated by two regulators of G-protein signaling, RGS7 and RGS11 (Cao et al., 2012). In addition, the transduction current is potentiated by cGMP and reduced (outside of the regime of single-photon responses) by increases in the intracellular Ca^{2+} concentration (Snellman et al., 2008).

Although there are a plethora of mechanisms that could be used to produce different response kinetics in the ON bipolar cell types, a temporal diversity among mammalian ON bipolar cell types corresponding to that in the OFF bipolar cell types has not been described (Berntson, Smith, & Taylor, 2004). A diversity of temporal light responses has been ascribed to salamander ON bipolar cells (Awatramani & Slaughter, 2000), although a recent study suggests that the temporal differences exist within a continuum that spans anatomical types (Kaur & Nawy, 2012). Mammalian ON cone bipolar cell types can make either invaginating or basal contacts (West, 1976, 1978) or a combination of both (Hopkins & Boycott, 1997). However, the significance of contact location and number for transmission at the cone to ON bipolar cell synapse is not known.

BIPOLAR CELL MEMBRANE CURRENTS

Studies in the rat and ground squirrel provide evidence that different bipolar cell types contain distinct membrane currents that can shape their light responses. These studies examined the differential expression of Na^+ currents (Cui & Pan, 2008; Saszik & DeVries, 2012) and currents through hyperpolarization and cyclic nucleotide–activated (HCN) channels (Cui & Pan, 2008; Ivanova & Muller, 2006; Light, 2009). Most cone bipolar cell types in the rat and ground squirrel have small (<50 pA) Na^+ currents. However, two ON bipolar cell types in the ground squirrel, the cb5a and morphologically similar cb5b, could generate Na^+ currents of up to 1 nA during a voltage step from −70 to −30 mV. Type 5 ON and type 3 OFF bipolar cells in the rat have similarly large Na^+ currents. In the ground squirrel the Na^+ current in cb5b cells is large enough to produce 30- to 40-mV action potentials on depolarization from rest by either current injection or light. The Na^+ spikes provide cb5b cells with the ability to rapidly signal sharp changes in illumination at light-on (Saszik & DeVries, 2012) in an analogous fashion to the cb2 transient that occurs at light-off (see figure 6.2). HCN1, 2, and 4 channels are also variably expressed by different cone bipolar cell types. The different kinetics of these channel types can confer different dynamic behaviors in response to injected current (Mao, MacLeish, & Victor, 2003).

CONCLUSIONS/FUTURE DIRECTIONS

A central tenet of retinal processing is the division of cone signals into ON and OFF pathways at the cone–to–bipolar cell synapse. This ON/OFF dichotomy provides signals of the appropriate polarity to ganglion cell receptive field centers. The past decade has seen a coalescence around the idea that there are 11 or more anatomical types of cone bipolar cells. However, we are only beginning to appreciate that individual cell types can carry different signals from a common cone. In part, the knowledge of differential processing has been held back by a lack of labels on which to base targeted bipolar cell recordings. This barrier to progress is receding as new mouse lines with genetically labeled bipolar cells are created. Future work will need to focus on the spatial relationships of processes at the cone synapse, differences in postsynaptic receptor or cascade properties, and the characterization of voltage-dependent membrane currents in the bipolar cell types with an emphasis on how these currents shape the light response. We also lack a circuit-based framework with which to understand the selective transmission of rod and color-coded signals by some but not other types of cone bipolar cell.

The bipolar cell types provide a spatiotemporal palette from which the different types of amacrine and ganglion cells can combine signals with different properties. A thorough understanding of processing in cone bipolar cells is essential for understanding both processing in the inner retina and the retinal output.

ACKNOWLEDGMENTS

We thank Drs. Charles Ratliff and Wei Li for helpful comments on the manuscript and the National Eye Institute for support (R01 EY012141).

REFERENCES

Audo, I., Bujakowska, K., Orhan, E., Poloschek, C. M., Defoort-Dhellemmes, S., Drumare, I., et al. (2012). Whole-exome sequencing identifies mutations in GPR179 leading to autosomal-recessive complete congenital stationary night blindness. *American Journal of Human Genetics, 90*(2), 321–330. doi:10.1016/j.ajhg.2011.12.007.

Awatramani, G. B., & Slaughter, M. M. (2000). Origin of transient and sustained responses in ganglion cells of the retina. *Journal of Neuroscience, 20*, 7087–7095.

Badea, T. C., & Nathans, J. (2004). Quantitative analysis of neuronal morphologies in the mouse retina visualized by using a genetically directed reporter. *Journal of Comparative Neurology, 480*, 331–351.

Berntson, A., Smith, R. G., & Taylor, W. R. (2004). Postsynaptic calcium feedback between rods and rod bipolar cells in the mouse retina. *Visual Neuroscience, 21*, 913–924.

Breuninger, T., Puller, C., Haverkamp, S., & Euler, T. (2011). Chromatic bipolar cell pathways in the mouse retina. *Journal of Neuroscience, 31*, 6504–6517.

Burris, C., Klug, K., Ngo, I. T., Sterling, P., & Schein, S. (2002). How Muller glial cells in macaque fovea coat and isolate the synaptic terminals of cone photoreceptors. *Journal of Comparative Neurology, 453*, 100–111.

Calkins, D. J., Tsukamoto, Y., & Sterling, P. (1998). Microcircuitry and mosaic of a blue-yellow ganglion cell in the primate retina. *Journal of Neuroscience, 18*, 3373–3385.

Cao, Y., Pahlberg, J., Sarria, I., Kamasawa, N., Sampath, A. P., & Martemyanov, K. A. (2012). Regulators of G protein signaling RGS7 and RGS11 determine the onset of the light response in ON bipolar neurons. *Proceedings of the National Academy of Sciences of the United States of America, 109*, 7905–7910.

Chen, S., & Li, W. (2012). A color-coding amacrine cell may provide a blue-OFF signal in a mammalian retina. *Nature Neuroscience, 15*, 954–956. doi:10.1038/nn.3128.

Choi, S. Y., Borghuis, B. G., Rea, R., Levitan, E. S., Sterling, P., & Kramer, R. H. (2005). Encoding light intensity by the cone photoreceptor synapse. *Neuron, 48*, 555–562.

Cui, J., & Pan, Z. H. (2008). Two types of cone bipolar cells express voltage-gated Na⁺ channels in the rat retina. *Visual Neuroscience, 25*, 635–645.

Curcio, C. A., Allen, K. A., Sloan, K. R., Lerea, C. L., Hurley, J. B., Klock, I. B., et al. (1991). Distribution and morphology of human cone photoreceptors stained with anti-blue opsin. *Journal of Comparative Neurology, 312*, 610–624.

Dacey, D. M., & Lee, B. B. (1994). The "blue-on" opponent pathway in primate retina originates from a distinct bistratified ganglion cell type. *Nature, 367*, 731–735.

DeVries, S. H. (2000). Bipolar cells use kainate and AMPA receptors to filter visual information into separate channels. *Neuron, 28*, 847–856.

DeVries, S. H., & Baylor, D. A. (1995). An alternative pathway for signal flow from rod photoreceptors to ganglion cells in mammalian retina. *Proceedings of the National Academy of Sciences of the United States of America, 92*, 10658–10662. doi:10.1073/pnas.92.23.10658.

DeVries, S. H., Li, W., & Saszik, S. (2006). Parallel processing in two transmitter microenvironments at the cone photoreceptor synapse. *Neuron, 50*, 735–748.

DeVries, S. H., & Schwartz, E. A. (1999). Kainate receptors mediate synaptic transmission between cones and 'Off' bipolar cells in a mammalian retina. *Nature, 397*, 157–160.

Dowling, J. E., & Boycott, B. B. (1966). Organization of the primate retina: Electron microscopy. *Proceedings of the Royal Society of London. Series B, Biological Sciences, 166*, 80–111. doi:10.1098/rspb.1966.0086.

Euler, T., & Wässle, H. (1995). Immunocytochemical identification of cone bipolar cells in the rat retina. *Journal of Comparative Neurology, 361*, 461–478.

Fyk-Kolodziej, B., Qin, P., & Pourcho, R. G. (2003). Identification of a cone bipolar cell in cat retina which has input from both rod and cone photoreceptors. *Journal of Comparative Neurology, 464*, 104–113.

Gilbertson, T. A., Scobey, R., & Wilson, M. (1991). Permeation of calcium ions through non-NMDA glutamate channels in retinal bipolar cells. *Science, 251*, 1613–1615.

Hack, I., Peichl, L., & Brandstatter, J. H. (1999). An alternative pathway for rod signals in the rodent retina: Rod photoreceptors, cone bipolar cells, and the localization of glutamate receptors. *Proceedings of the National Academy of Sciences of the United States of America, 96,* 14130–14135. doi:10.1073/pnas.96.24.14130.

Haverkamp, S., Ghosh, K. K., Hirano, A. A., & Wässle, H. (2003). Immunocytochemical description of five bipolar cell types of the mouse retina. *Journal of Comparative Neurology, 455,* 463–476.

Haverkamp, S., Grunert, U., & Wässle, H. (2001a). Localization of kainate receptors at the cone pedicles of the primate retina. *Journal of Comparative Neurology, 436,* 471–486.

Haverkamp, S., Grunert, U., & Wässle, H. (2001b). The synaptic architecture of AMPA receptors at the cone pedicle of the primate retina. *Journal of Neuroscience, 21,* 2488–2500.

Haverkamp, S., Specht, D., Majumdar, S., Zaidi, N. F., Brandstatter, J. H., Wasco, W., et al. (2008). Type 4 OFF cone bipolar cells of the mouse retina express calsenilin and contact cones as well as rods. *Journal of Comparative Neurology, 507,* 1087–1101. doi:10.1002/cne.21612.

Haverkamp, S., Wässle, H., Duebel, J., Kuner, T., Augustine, G. J., Feng, G., et al. (2005). The primordial, blue-cone color system of the mouse retina. *Journal of Neuroscience, 25,* 5438–5445.

Hopkins, J. M., & Boycott, B. B. (1997). The cone synapses of cone bipolar cells of primate retina. *Journal of Neurocytology, 26,* 313–325.

Ishida, A. T., Stell, W. K., & Lightfoot, D. O. (1980). Rod and cone inputs to bipolar cells in goldfish retina. *Journal of Comparative Neurology, 191,* 315–335.

Ivanova, E., & Muller, F. (2006). Retinal bipolar cell types differ in their inventory of ion channels. *Visual Neuroscience, 23,* 143–154.

Jackman, S. L., Choi, S. Y., Thoreson, W. B., Rabl, K., Bartoletti, T. M., & Kramer, R. H. (2009). Role of the synaptic ribbon in transmitting the cone light response. *Nature Neuroscience, 12,* 303–310.

Joo, H. R., Peterson, B. B., Haun, T. J., & Dacey, D. M. (2011). Characterization of a novel large-field cone bipolar cell type in the primate retina: Evidence for selective cone connections. *Visual Neuroscience, 28,* 29–37.

Kaur, T., & Nawy, S. (2012). Characterization of Trpm1 desensitization in ON bipolar cells and its role in downstream signalling. *Journal of Physiology, 590,* 179–192.

Klug, K., Herr, S., Ngo, I. T., Sterling, P., & Schein, S. (2003). Macaque retina contains an S-cone OFF midget pathway. *Journal of Neuroscience, 23,* 9881–9887.

Koike, C., Obara, T., Uriu, Y., Numata, T., Sanuki, R., Miyata, K., et al. (2010). TRPM1 is a component of the retinal ON bipolar cell transduction channel in the mGluR6 cascade. *Proceedings of the National Academy of Sciences of the United States of America, 107,* 332–337. doi:10.1073/pnas.0912730107.

Kolb, H. (1979). The inner plexiform layer in the retina of the cat: Electron microscopic observations. *Journal of Neurocytology, 8,* 295–329.

Kouyama, N., & Marshak, D. W. (1992). Bipolar cells specific for blue cones in the macaque retina. *Journal of Neuroscience, 12,* 1233–1252.

Li, W., Chen, S., & DeVries, S. H. (2010). A fast rod photoreceptor signaling pathway in the mammalian retina. *Nature Neuroscience, 13,* 414–416.

Li, W., & DeVries, S. H. (2006). Bipolar cell pathways for color and luminance vision in a dichromatic mammalian retina. *Nature Neuroscience, 9,* 669–675.

Li, W., Keung, J. W., & Massey, S. C. (2004). Direct synaptic connections between rods and OFF cone bipolar cells in the rabbit retina. *Journal of Comparative Neurology, 474,* 1–12.

Light, A. C. (2009) Transient and sustained OFF bipolar cells in the ground squirrel retina. PhD thesis. Evanston: Northwestern University.

Light, A. C., Zhu, Y., Shi, J., Saszik, S., Lindstrom, S., Davidson, L., et al. (2012). Organizational motifs for ground squirrel cone bipolar cells. *Journal of Comparative Neurology, 520,* 2864–2887. doi:10.1002/cne.23068.

MacNeil, M. A., Heussy, J. K., Dacheux, R. F., Raviola, E., & Masland, R. H. (2004). The population of bipolar cells in the rabbit retina. *Journal of Comparative Neurology, 472,* 73–86.

Mao, B. Q., MacLeish, P. R., & Victor, J. D. (2003). Role of hyperpolarization-activated currents for the intrinsic dynamics of isolated retinal neurons. *Biophysical Journal, 84,* 2756–2767. doi:10.1016/S0006-3495(03)75080-2.

Masland, R. H. (2001). The fundamental plan of the retina. *Nature Neuroscience, 4,* 877–886.

Mataruga, A., Kremmer, E., & Muller, F. (2007). Type 3a and type 3b OFF cone bipolar cells provide for the alternative rod pathway in the mouse retina. *Journal of Comparative Neurology, 502,* 1123–1137.

Merighi, A., Raviola, E., & Dacheux, R. F. (1996). Connections of two types of flat cone bipolars in the rabbit retina. *Journal of Comparative Neurology, 371,* 164–178.

Midorikawa, M., Tsukamoto, Y., Berglund, K., Ishii, M., & Tachibana, M. (2007). Different roles of ribbon-associated and ribbon-free active zones in retinal bipolar cells. *Nature Neuroscience, 10,* 1268–1276.

Morgans, C. W., Zhang, J., Jeffrey, B. G., Nelson, S. M., Burke, N. S., Duvoisin, R. M., et al. (2009). TRPM1 is required for the depolarizing light response in retinal ON-bipolar cells. *Proceedings of the National Academy of Sciences of the United States of America, 106,* 19174–19178. doi:10.1073/pnas.0908711106.

Nakajima, Y., Iwakabe, H., Akazawa, C., Nawa, H., Shigemoto, R., Mizuno, N., et al. (1993). Molecular characterization of a novel retinal metabotropic glutamate receptor mGluR6 with a high agonist selectivity for L-2-amino-4-phosphonobutyrate. *Journal of Biological Chemistry, 268,* 11868–11873.

Nelson, R., & Kolb, H. (2003). ON and OFF pathways in the vertebrate retina and visual system. In L. M. Chalupa & J. S. Werner (Eds.), *The visual neurosciences* (pp. 260–278). Cambridge, MA: MIT Press.

Pang, J. J., Gao, F., Paul, D. L., & Wu, S. M. (2012). Rod, M-cone and M/S-cone inputs to hyperpolarizing bipolar cells in the mouse retina. *Journal of Physiology, 590,* 845–854.

Pang, J. J., Gao, F., & Wu, S. M. (2004). Stratum-by-stratum projection of light response attributes by retinal bipolar cells of Ambystoma. *Journal of Physiology, 558,* 249–262.

Peachey, N. S., Ray, T. A., Florijn, R., Rowe, L. B., Sjoerdsma, T., Contreras-Alcantara, S., et al. (2012). GPR179 is required for depolarizing bipolar cell function and is mutated in autosomal-recessive complete congenital stationary night blindness. *American Journal of Human Genetics, 90,* 331–339. doi:10.1016/j.ajhg.2011.12.006.

Puller, C., Ivanova, E., Euler, T., Haverkamp, S., & Schubert, T. (2013). OFF bipolar cells express distinct types of dendritic glutamate receptors in the mouse retina. *Neuroscience,* in press. doi:10.1016/j.neuroscience.2013.03.054.

Puller, C., Ondreka, K., & Haverkamp, S. (2011). Bipolar cells of the ground squirrel retina. *Journal of Comparative Neurology, 519*, 759–774.

Puthussery, T., Gayet-Primo, J., & Taylor, W. R. (2010). Localization of the calcium-binding protein secretagogin in cone bipolar cells of the mammalian retina. *Journal of Comparative Neurology, 518*, 513–525.

Rabl, K., Cadetti, L., & Thoreson, W. B. (2005). Kinetics of exocytosis is faster in cones than in rods. *Journal of Neuroscience, 25*, 4633–4640.

Roorda, A., & Williams, D. R. (1999). The arrangement of the three cone classes in the living human eye. *Nature, 397*, 520–522.

Sampath, A. P., & Rieke, F. (2004). Selective transmission of single photon responses by saturation at the rod–to–rod bipolar synapse. *Neuron, 41*, 431–443.

Saszik, S., & DeVries, S. H. (2012). A mammalian retinal bipolar cell uses both graded changes in membrane voltage and all-or-nothing Na$^+$ spikes to encode light. *Journal of Neuroscience, 32*, 297–307. doi:10.1523/JNEUROSCI.2739-08.2012.

Shen, Y., Heimel, J. A., Kamermans, M., Peachey, N. S., Gregg, R. G., & Nawy, S. (2009). A transient receptor potential-like channel mediates synaptic transmission in rod bipolar cells. *Journal of Neuroscience, 29*, 6088–6093.

Shen, Y., Rampino, M. A., Carroll, R. C., & Nawy, S. (2012). G-protein-mediated inhibition of the Trp channel TRPM1 requires the Gbetagamma dimer. *Proceedings of the National Academy of Sciences of the United States of America, 109*, 8752–8757. doi:10.1073/pnas.1117433109.

Sher, A., & DeVries, S. H. (2012). A non-canonical pathway for mammalian blue-green color vision. *Nature Neuroscience, 15*, 952–953. doi:10.1038/nn.3127.

Snellman, J., Kaur, T., Shen, Y., & Nawy, S. (2008). Regulation of ON bipolar cell activity. *Progress in Retinal and Eye Research, 27*, 450–463.

Snellman, J., Mehta, B., Babai, N., Bartoletti, T. M., Akmentin, W., Francis, A., et al. (2011). Acute destruction of the synaptic ribbon reveals a role for the ribbon in vesicle priming. *Nature Neuroscience, 14*, 1135–1141. doi:10.1038/nn.2870.

Soucy, E., Wang, Y., Nirenberg, S., Nathans, J., & Meister, M. (1998). A novel signaling pathway from rod photoreceptors to ganglion cells in mammalian retina. *Neuron, 21*, 481–493.

Sterling, P., & Matthews, G. (2005). Structure and function of ribbon synapses. *Trends in Neurosciences, 28*, 20–29. doi:10.1016/j.tins.2004.11.009.

Stiles, J. R., Van Helden, D., Bartol, T. M., Jr., Salpeter, E. E., & Salpeter, M. M. (1996). Miniature endplate current rise times less than 100 microseconds from improved dual recordings can be modeled with passive acetylcholine diffusion from a synaptic vesicle. *Proceedings of the National Academy of Sciences of the United States of America, 93*, 5747–5752. doi:10.1073/pnas.93.12.5747.

Strettoi, E., Novelli, E., Mazzoni, F., Barone, I., & Damiani, D. (2010). Complexity of retinal cone bipolar cells. *Progress in Retinal and Eye Research, 29*, 272–283.

Szmajda, B. A., & DeVries, S. H. (2011). Glutamate spillover between mammalian cone photoreceptors. *Journal of Neuroscience, 31*, 13431–13441.

Tsukamoto, Y., Morigiwa, K., Ishii, M., Takao, M., Iwatsuki, K., Nakanishi, S., et al. (2007). A novel connection between rods and ON cone bipolar cells revealed by ectopic metabotropic glutamate receptor 7 (mGluR7) in mGluR6-deficient mouse retinas. *Journal of Neuroscience, 27*, 6261–6267.

Vardi, N. (1998). Alpha subunit of Go localizes in the dendritic tips of ON bipolar cells. *Journal of Comparative Neurology, 395*, 43–52.

Volgyi, B., Deans, M. R., Paul, D. L., & Bloomfield, S. A. (2004). Convergence and segregation of the multiple rod pathways in mammalian retina. *Journal of Neuroscience, 24*, 11182–11192.

Wässle, H. (2004). Parallel processing in the mammalian retina. *Nature Reviews. Neuroscience, 5*, 747–757.

Wässle, H., Puller, C., Muller, F., & Haverkamp, S. (2009). Cone contacts, mosaics, and territories of bipolar cells in the mouse retina. *Journal of Neuroscience, 29*, 106–117.

West, R. W. (1976). Light and electron microscopy of the ground squirrel retina: Functional considerations. *Journal of Comparative Neurology, 168*, 355–377.

West, R. W. (1978). Bipolar and horizontal cells of the gray squirrel retina: Golgi morphology and receptor connections. *Vision Research, 18*, 129–136.

Wiesel, T. N., & Hubel, D. H. (1966). Spatial and chromatic interactions in the lateral geniculate body of the rhesus monkey. *Journal of Neurophysiology, 29*, 1115–1156.

Yin, L., Smith, R. G., Sterling, P., & Brainard, D. H. (2009). Physiology and morphology of color-opponent ganglion cells in a retina expressing a dual gradient of S and M opsins. *Journal of Neuroscience, 29*, 2706–2724.

Zampighi, G. A., Schietroma, C., Zampighi, L. M., Woodruff, M., Wright, E. M., & Brecha, N. C. (2011). Conical tomography of a ribbon synapse: Structural evidence for vesicle fusion. *PLoS One, 6*, e16944. doi:10.1371/journal.pone.0016944.

Zenisek, D. (2008). Vesicle association and exocytosis at ribbon and extraribbon sites in retinal bipolar cell presynaptic terminals. *Proceedings of the National Academy of Sciences of the United States of America, 105*, 4922–4927. doi:10.1073/pnas.0709067105.

7 Horizontal Cells: Lateral Interactions at the First Synapse in the Retina

RICHARD H. KRAMER

Unlike a simple camera that faithfully records patterns of light and darkness, the neural circuitry of the retina modifies the representation of images, adjusting the gain and enhancing the contrast of a visual scene. The discovery of lateral inhibition in the eye of the horseshoe crab, *Limulus,* by Hartline and Ratliff (1958) provided the first explanation for how a neural circuit could enhance visual contrast. It took an additional 13 years for Baylor, Fuortes, and O'Bryan (1971) to begin to explain the crucial role of horizontal cells (HC) in mediating lateral inhibition in the vertebrate retina. Much has been learned about the anatomy and physiology of HCs since then, but it is surprising and somewhat embarrassing that more than 40 years after these initial investigations we still have only a blurry understanding of the synaptic mechanism(s) that HCs use to enhance visual contrast.

WHAT ARE HORIZONTAL CELLS?

HCs are the first laterally projecting cells in the retina. Each HC receives and integrates synaptic signals from many photoreceptor cells, both rods and cones. In return, each HC transmits a feedback signal that alters neurotransmitter release from the photoreceptors. HC feedback establishes the antagonistic center/surround receptive field of bipolar cells, which is reflected again and again in the receptive field organization of all subsequent layers of the visual system.

At a purely circuit level, how this receptive field organization comes about is well understood. A bipolar cell receives direct synaptic input from a group of photoreceptors, forming the receptive field center (figure 7.1A). But a much larger group of photoreceptors indirectly affects the bipolar cell through lateral interactions mediated by HCs, forming the receptive field surround. The photoreceptors and HCs are connected by a reciprocal synapse (figure 7.1B), but because the photoreceptor output is excitatory and HC feedback is largely inhibitory, illumination of the center versus the surround has opposite, or antagonistic, actions on the response of the bipolar cell.

Despite their pivotal role in visual information processing, HCs are perhaps the most enigmatic cell type in the retina. HCs were originally thought to be glial cells, but it is now clear that HCs are an integral part of the synaptic circuitry of the retina. HCs possess presynaptic proteins necessary for vesicular release (Hirano et al., 2011), consistent with mechanisms found at conventional neuronal synapses. However, electron microscopy studies have shown that HCs lack some of the common hallmarks of neurons, namely clusters of small synaptic vesicles adjacent to presynaptic specializations, structures that are characteristic of vesicle fusion sites (Raviola & Gilula, 1975; Schwartz, 2002). These apparent deficiencies have fueled the idea that negative feedback from HCs to photoreceptors might be mediated by unconventional synaptic signals, for example an electrical field effect (an ephaptic signal) or a change in extracellular pH (see below).

HCs have few voltage-gated Na^+ channels, and they generate graded potentials rather than action potentials. Some HCs (named A-type) have relatively short, large-diameter dendrites, well suited for allowing spatial spread of graded signals with little electronic decay. Curiously, other HCs (named B-type) possess a long, thin, axon-like process that projects laterally for up to several millimeters. Proximal dendrites on the somata of these cells contact cones, whereas the axon terminal contacts rods. The rationale for a nonspiking cell having a long axon is not completely clear, but some very clever recent experiments on HCs in mouse retina provide a clue. These studies (Trümpler et al., 2008) show that electrical signals can pass more reliably from soma to axon terminal than vice versa, suggesting that HC-mediated feedback is asymmetrical, occurring from cones to rods but not the other way around.

Some of the uncertainties about HC function come from the fact that gap junctions electrically couple HCs into a large syncytium of cells, which makes electrophysiological analysis (especially voltage clamp) difficult or impossible. Unlike most retinal neurons that have clearly separated input and output synapses, the reciprocal synapses between HCs and photoreceptors also make

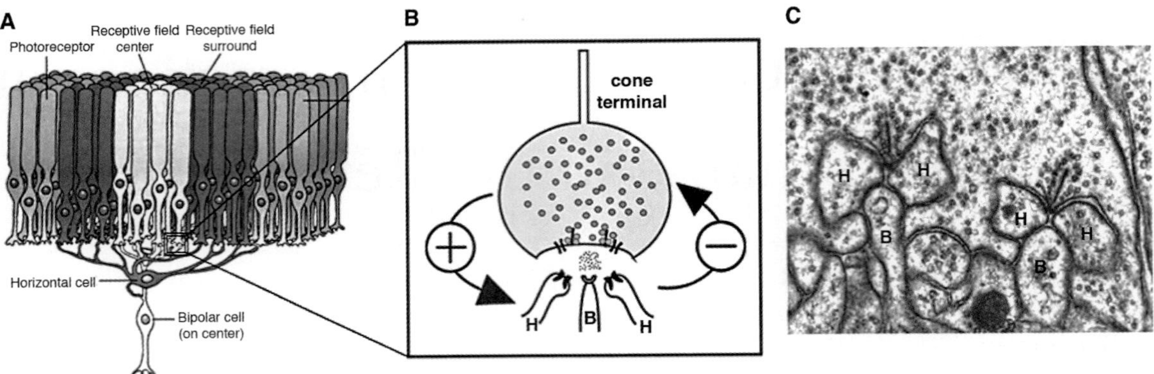

FIGURE 7.1 Synaptic organization of HCs in the retina. (A) HCs collect information from photoreceptors in the "receptive field surround" and feed back onto photoreceptors in the "receptive field center" to generate the antagonistic center/surround receptive field of bipolar cells. (B) The negative feedback synapse between cones and HCs. (C) EM picture of the invaginating cone synapse with the characterizing synaptic "triad," consisting of two lateral elements from HCs (H) and a central dendrite from a bipolar cell (B). (C is from Dowling, 1970.)

mechanistic analysis much more difficult. Other uncertainties come from the many different preparations investigators have used to study HC synapses. Lower vertebrates, including fish, reptiles, and birds, possess at least four types of HCs, whereas mammals possess only two types. HCs in lower vertebrates exhibit color-opponent responses, whereas color opponency is lacking in mammals. These and other differences make it difficult to evaluate whether conflicting experimental results are a consequence of different experimental approaches or reflect different species-specific mechanisms. But because so many other signaling mechanisms are conserved across the retinas of different vertebrate species, it seems highly likely that the synaptic mechanisms used by HCs are also conserved.

Despite these complexities, new insights are being gained both about the signal that HCs initially receive from photoreceptors and about the HC feedback signals referred back to the photoreceptors. This review focuses on the mechanisms and consequences of HC feedback. For more detailed information about the anatomy and physiology of HCs in different species, the reader is referred to other reviews (Burkhardt, 1993; Fahrenfort et al., 2005; Kamermans et al., 1999; Piccolino, 1995; Schwartz, 2002; Twig, Levy, & Perlman, 2003; Weiler et al., 2000; Wu, 1992) or Webvision (http://webvision.med.utah.edu/), an excellent online resource.

STRUCTURE OF THE HC/PHOTORECEPTOR SYNAPSE

The synaptic contact between HCs and photoreceptors occurs in an invagination that extends deep into the terminals of rods and cones (figure 7.1C). Rod terminals have only a handful of invaginations, whereas cones have as many as 25. At each invagination, the photoreceptor cell contains a cytoplasmic organelle called a synaptic ribbon.

The synaptic ribbon is a key feature of sensory neurons that continuously release neurotransmitter, namely rods, cones, and hair cells of the inner ear. Synaptic ribbons are also found in the output synapse of retinal bipolar cells. The ribbon is a proteinaceous structure composed of a handful of different proteins, but the role of these proteins in binding or transporting synaptic vesicles is unknown. The ribbon is anchored to the plasma membrane at a structure called the arciform density, which is located directly across from postsynaptic dendrites. Ten or more rows of synaptic vesicles are tethered along the flanks of the synaptic ribbon. Recent studies involving fluorophore-assisted laser inactivation have shown that the ribbon is a necessary way station for priming vesicles for fusion with the presynaptic membrane (Snellman et al., 2011), making the ribbon synapse the center of attention for understanding both synaptic output and HC feedback.

Each invagination typically has three dendritic processes: two lateral dendrites, each from a different HC, and one central dendrite from a bipolar cell. This "triad junction" (figure 7.1B, C) has been analyzed in great detail, both from ultrastructural and physiological perspectives. The deep invagination is thought to be important for retarding or preventing diffusion of chemical transmitters in or out of the synaptic cleft (Burris et al., 2002; Migdale et al., 2003), although there is recent evidence for spillover of the neurotransmitter glutamate from one cone terminal to its neighbors (Szmajda & DeVries, 2011). It has also been suggested that the invagination serves an electrical insulator that impedes extracellular current flow between different parts of the HC. The resulting tortuous path for current flow might maintain a voltage gradient between the cleft and the bulk extracellular space of the retina (Byzov & Shura-Bura, 1986; Kamermans et al., 2001), requisite for ephaptic signaling.

The invaginating HC synapse is devoid of glial cell processes (Burris et al., 2002), unlike most synapses in the brain where glial cells come into close proximity to sites of neurotransmitter release. It is interesting to speculate about the functional significance of this difference. At glutamatergic synapses in the brain it is the glial cells that contain the excitatory amino acid transporters (EAATs) that remove extracellular glutamate from the cleft to terminate synaptic transmission. Glutamate transport by EAATs is associated with a Cl^- conductance, so glutamate uptake is associated with an electrical current that hyperpolarizes the glial cells. This leaves both the presynaptic and postsynaptic neurons disengaged from this process of glutamate removal, insulating them from voltage changes associated with this process. In the invaginating HC synapse the rods and cones possess EAATs, thus making the presynaptic cell crucial for removing its own released neurotransmitter (Hasegawa et al., 2006), although Müller glial cells contribute to removing glutamate that escapes the invagination (Harada et al., 1998). One consequence of this arrangement is that photoreceptors respond to their own glutamate release with a Cl^- current that hyperpolarizes the membrane potential (Picaud et al., 1995). This constitutes a photoreceptor-autonomous form of negative feedback that may limit neurotransmitter release and help shape postsynaptic response.

THE HC LIGHT RESPONSE

How HCs receive information about light from rods and cones is well understood. Rods and cones hyperpolarize in response to light, resulting in the closure of voltage-gated Ca^{2+} channels in their synaptic terminals. This leads to a decrease in Ca^{2+} influx and a drop in the cytoplasmic Ca^{2+} concentration in the vicinity of vesicular release sites. Tonic neurotransmitter release from the photoreceptors is maintained in darkness by a relatively high intracellular Ca^{2+} concentration, so the fall in Ca^{2+} during illumination leads to a decrease in neurotransmitter release. Glutamate, the rod and cone neurotransmitter, acts on postsynaptic neurons that have different types of glutamate receptors. HCs possess AMPA receptors, which are ligand-gated ion channels that are nonselective in their permeability to Na^+ and K^+ and, to some extent, Ca^{2+} (Rivera, Blanco, & de la Villa, 2001). Also present at the synapse are dendrites from OFF-bipolar cells, which have AMPA receptors, and ON-bipolar cells, which have metabotropic glutamate receptors.

The continuous release of glutamate in darkness keeps the AMPA receptors in HCs in an activated state, maintaining the HC membrane potential at a relatively depolarized voltage (about –40 mV). Light-evoked suppression of glutamate release leads to deactivation of the AMPA receptors, allowing the HCs to hyperpolarize. The greater the light intensity, the greater the hyperpolarization, until the light response saturates at about –60 mV when all of the AMPA receptors are turned off (figure 7.2A).

PROPERTIES OF HC FEEDBACK ONTO PHOTORECEPTORS

The first indication of an HC feedback synapse onto cones was from Baylor, Fuortes, and O'Bryan (1971), who found that illumination with a large spot of light could generate a voltage response that was smaller or even opposite in polarity to the hyperpolarizing response generated by a small central spot. The antagonistic "surround" portion of the cone's receptive field extended for hundreds of micrometers from the center, consistent with the involvement of large, laterally projecting cells, that is, HCs. The kinetics of the surround response was similar to the kinetics of the HC light response, reinforcing this view. Further studies suggested that the feedback synapse functioned to increase contrast discrimination in the visual system (O'Bryan, 1973).

Early mechanistic studies on HC feedback focused on identifying the conductance change in cones that is generated in response to large spots of light or to direct manipulation of the voltage of HCs. In a classic study Samuel Wu (1991) used a micropipette to break off the outer segment of a cone in a retinal slice and then used another micropipette to obtain an intracellular recording from the remaining cell body. Removal of the outer segment eliminated the direct light response, enabling the feedback response mediated by HCs to be recorded and analyzed in isolation. In these truncated cones full-field illumination generated a very small depolarization (<5 mV), opposite in sign to the normally hyperpolarizing light response. The depolarizing response became smaller as the cone was held at more hyperpolarized potentials and apparent reversal potential near –65 mV, thought to be indicative of a Cl^- conductance.

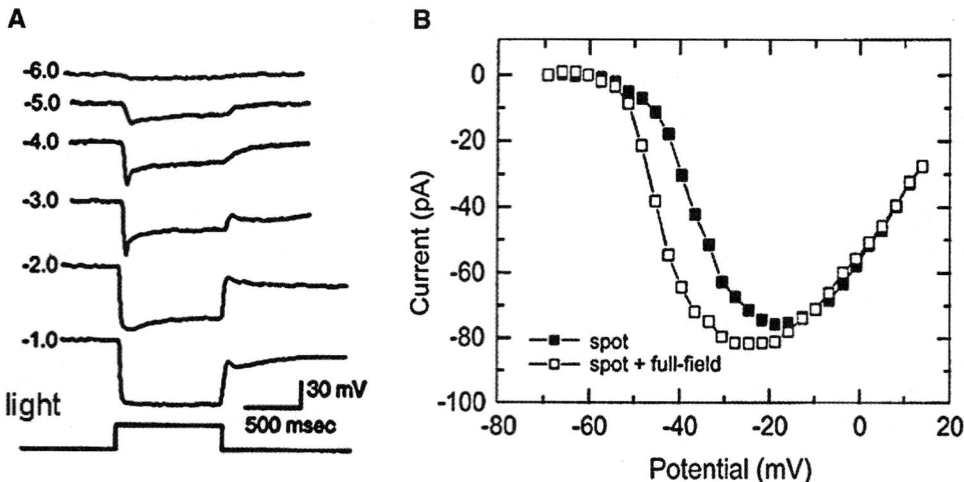

FIGURE 7.2   The light response in HCs and the negative feedback response in cones. (A) Voltage responses in HCs to increasing light intensity in rabbit retina. The "sag" in the hyperpolarizing response is a reflection of negative feedback to the photoreceptors. Stimulus intensity is given in log units relative to the most intense stimulus used. (Adapted from Bloomfield, 1992.) (B) Activation curve of the voltage-gated Ca^{2+} channels in a cone from the carp retina. Full-field illumination increases the maximal current and causes a hyperpolarizing shift in the activation curve. (Adapted from Verweij, Kamermans, & Spekreijse, 1996.)

Further studies by several groups reached a different conclusion: the primary action of negative feedback is to alter the gating of voltage-gated Ca^{2+} channels in cones, and changes in Cl^- conductance is a secondary effect, mediated by Ca^{2+}-activated Cl^- channels (Kraaij, Spekreijse, & Kamermans, 2000). Either illumination of the surround (figure 7.2B) or direct HC depolarization causes a depolarizing shift in the activation curve of L-type Ca^{2+} channels in cones (Verweij, Kamermans, & Spekreijse, 1996; Cadetti & Thoreson, 2006). This is a "sign-inverting" (or inhibitory) effect because the more the HCs are depolarized, the harder it is to open the cone Ca^{2+} channels, and less Ca^{2+}-dependent release of neurotransmitter is elicited from the cone.

Negative feedback from HCs has been studied most extensively in cones, and because the cone system is most important for high acuity vision, lateral inhibition and contrast enhancement are of particular importance for cone signaling. Negative feedback can also generate "color-opponent" responses in a particular class of HC called the "chromaticity" cell (in contrast to non–color-opponent "luminosity" HCs), which is found in lower vertebrates but not mammals. In this system one chromatic type of cone, responding to light over a particular range of wavelengths, signals through HCs to exert negative feedback onto a different type of cone, responding over a different range of wavelengths. Color opponency can be seen at the level of the cone terminal, but it is most easily observed by recording from the "chromaticity" (or C-type) HCs themselves. Depending on the exact subtype of C-type HC, the photoresponse polarity can vary in a biphasic or even triphasic manner as a function of wavelength (for review see Twig, Levy, & Perlman, 2003).

Despite the focus on cones, recent studies show that HCs also exert negative feedback onto rod terminals (Thoreson, Babai, & Bartoletti, 2008), modulating the transmission of rod light responses to downstream visual neurons. Simultaneous patch-clamp recordings from HCs and rods show that changing the HC voltage results in small changes in the activation of the rod voltage-gated Ca^{2+} current, just as it does in cones. A negative feedback response can be elicited in truncated rods lacking outer segments, just as in truncated cones. So HC feedback appears to be a general phenomenon that applies to all of the photoreceptors in the retina.

THE MECHANISM OF THE NEGATIVE FEEDBACK SYNAPSE

Since the landmark studies of Copenhagen and Jahr (1989), it has been clear that glutamate is the output neurotransmitter of photoreceptors. But the mechanism of HC feedback transmission has remained unsettled despite 40 years of study. At synapses throughout the nervous system, multiple lines of experimental evidence have converged to definitively identify specific neurotransmitters mediating specific synaptic events. But the negative feedback synapse from HCs to cones

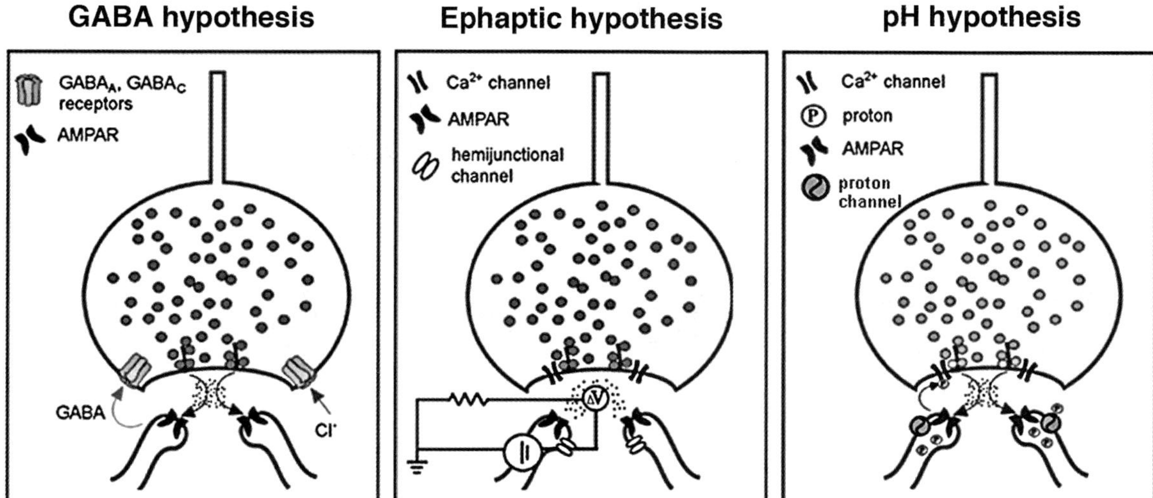

FIGURE 7.3   Three hypotheses about the putative signal mediating negative feedback from HCs to cones.

has proven to be a much tougher nut to crack, and the field has still not reached a consensus.

Identification of neurotransmitters or mediators underlying physiological events often involves three types of experiments: (1) "block it" experiments, where a specific inhibitor of the putative mediator prevents the event; (2) "move it" experiments, where exogenous application of the putative mediator can itself generate the event; and (3) "show it" experiments, where some type of indicator demonstrates that the mediator is present and can account for the event. Many "block it" and "move it" studies have been carried out over the years, but the main problem in definitively identifying the HC negative feedback transmitter is the lack of compelling "show it" studies.

Several types of chemical signals can modulate neurotransmitter release from photoreceptors but clearly do not mediate HC negative feedback. For example, nitric oxide can lead to the activation of cyclic GMP–gated cation channels, which are concentrated on cone terminals (Savchenko, Barnes, & Kramer, 1997). This leads to an elevation of intracellular Ca^{2+} in cones, which enhances neurotransmitter release. Pharmacological blockade of nitric oxide synthase reduces negative feedback. However, it is now clear that the effects of negative feedback are mediated by the voltage-gated Ca^{2+} channels of photoreceptors, not cyclic GMP–gated channels, which are voltage insensitive.

Three remaining candidate mechanisms are favored by at least some investigators (see figure 7.3). The following is a critical evaluation of the experimental evidence, both pro and con, for each of these hypothesized mechanisms.

The GABA Hypothesis

GABA, the main inhibitory neurotransmitter in the brain, was implicated early as the putative HC feedback transmitter. In fact most neuroscience textbooks still depict GABA as the transmitter mediating negative feedback and lateral inhibition. According to the GABA hypothesis (figure 7.3A) glutamate released from cones in darkness depolarizes HCs, which release GABA back onto the cone terminals. Activation of $GABA_A$ receptors leads to hyperpolarization of the membrane potential in the cone terminal. This suppresses activation of voltage-gated Ca^{2+} channels, reducing the influx of Ca^{2+} that underlies vesicular glutamate release. Hence, the process is a negative feedback loop.

At first, there was considerable evidence supporting the GABA hypothesis. HCs express the enzyme required for GABA synthesis: glutamic acid decarboxylase (GAD) (Hurd & Eldred, 1989). GABA accumulates in HCs (Lam, Lasater, & Naka, 1978). GABA is released during darkness and when the cells are depolarized by either K^+ or glutamate (Schwartz, 1982). Isolated cones have GABA receptors on their synaptic terminals, and activation of these receptors leads to an increased Cl^- conductance in cones, which hyperpolarizes the membrane potential (Kaneko & Tachibana, 1986). Surround illumination depolarizes cones in fish retina, and this effect can be eliminated by applying exogenous GABA (Murakami et al., 1982; Wu & Dowling, 1980).

The lack of large numbers of synaptic vesicles in the HCs (Raviola & Gilula, 1975; Schwartz, 2002) posed a dilemma for the GABA hypothesis. Fluorescence imaging studies using the synaptic vesicle dye FM1–43 failed to find any labeling of HCs (Rea et al., 2004), also raising doubts about feedback being mediated by vesicular release. Another challenge became apparent starting in the early 1980s when Schwartz (1982, 1987, 2002) found that GABA release from HC is independent of Ca^{2+}. Depolarization-elicited GABA release did not correlate with the voltage dependence of the small Ca^{2-} current in HCs, and it persisted even after Ca^{2+} buffer was added to both the extracellular and intracellular solution, preventing any change in Ca^{2+}. In these elegant studies, Schwartz solved the mystery of how GABA is released from HCs, at least in nonmammalian vertebrates (fish, amphibia, and reptiles). In retinas from these creatures, a voltage-sensitive Na^+-coupled GABA transporter is actually the key molecular component that couples HC depolarization with GABA efflux. In this scenario reverse membrane transport (i.e., GABA shuttled out of the cell rather than into the cell) is the mechanism of neurotransmitter release instead of vesicular fusion. However, mammalian HCs appear to lack Na^+-coupled GABA transporters (Johnson et al., 1996), limiting this mechanism to lower vertebrates.

Several other notable problems with the GABA hypothesis have arisen over the years. Ultrastructural studies show that very few GABA receptors are present in the synaptic invagination in cones, and most are located far from HC contact sites (Yazulla & Studholme, 1997). Most importantly, pharmacological studies showed that negative feedback persists after addition of a wide variety of antagonists that block all known GABA receptors, namely $GABA_A$, $GABA_B$, and $GABA_C$ receptors (Hare & Owen, 1996; Thoreson & Burkhardt, 1990). These results are particularly problematic, and no credible explanation has been offered to counter the conclusion that GABA receptors are unnecessary. So although the GABA hypothesis passes the "move it" test, it fails the "block it" test. Moreover, there is no "show it" evidence that a change in GABA concentration of appropriate magnitude and kinetics actually occurs at the synapse during negative feedback. The suppression of negative feedback by exogenous GABA could be explained by independent effects—for example, exogenous activation of $GABA_A$ receptors could decrease the input resistance of cones and HCs, shunting the current generated by the natural feedback signal.

But even if it is concluded that GABA does not mediate negative feedback, it is still possible that the transmitter could play other important roles in the outer retina, especially in lower vertebrates (Wu, 1992). The perisynaptic localization of GABA receptors suggests that, rather than mediating point-to-point rapid synaptic inhibition, GABA may function as a paracrine neurotransmitter in the outer retina. For example GABA may signal through volume transmission from HCs to cones to better inform the photoreceptors about the overall adaptation state of the retina. Even in mammals there is evidence for the existence of enzymes that synthesize and transport GABA in HC dendrites (Johnson et al., 1996; Guo et al., 2010), suggesting that GABA must play some important role. But more work is needed to elucidate the precise function of GABA.

The Ephaptic Hypothesis

The stereotypical structure of the synaptic triad and the invaginating synapse is unusual and intriguing, somewhat reminiscent of an electrical transistor. So the suggestion that HC feedback might operate through an electrical field effect (also known as ephaptic transmission) makes sense (Byzov & Shura-Bura, 1986). In this scenario an extracellular voltage change occurs when electrical current flows through ion channels in the HC dendrites and across the extracellular resistance of the invaginating synapse. The voltage-gated Ca^{2+} channels in the cone terminals sense the change in transmembrane potential, resulting in a change of their activation state, ultimately resulting in a change in Ca^{2+}-dependent neurotransmitter release. Because extracellular depolarization is equivalent to intracellular hyperpolarization, outward current from HCs in darkness should inhibit voltage-gated Ca^{2+} channels in cones, a negative feedback effect.

The ephaptic hypothesis requires that the HCs possess a sufficient density of open ion channels and that there is a sufficiently high extracellular resistance to account for the proposed change in extracellular potential. Kamermans et al. (2001) suggested that "hemichannels," comprised of the connexin-26 protein, could allow sufficient current to flow from the tips of HC dendrites, which are embedded inside the synaptic invagination. Connexins are the basis of gap junctions, aqueous pores that allow direct chemical and electrical communication between neighboring cells. Gap junctional channels are formed when connexins from two neighboring cells link up to form a transcellular pore. However, some connexins can form open hemichannnels without involving partner proteins, providing an efficient leakage path for current to flow across the membrane of a single cell.

Consistent with the ephaptic idea, immunocytochemistry in carp retina showed selective labeling of HC dendrites with an antibody against connexin-26 (Kamermans et al., 2001). Carbenoxolone, a blocker of gap-junctional channels, suppresses feedback-induced responses measured in the HCs themselves and also eliminates color opponency. More recently it was shown that genetic knockout of connexin-55.5 in zebrafish reduces HC negative feedback (Klaassen et al., 2011). The partial effect might be explained by multiple connexins contributing to the hemijunctional channels, and in zebrafish, there is evidence that both connexin-55.5 and connexin-52.6 contribute (Sun et al., 2012).

Additional studies showed that the opening of glutamate receptors found on the dendritic tips of HCs, can also suppress feedback (Kamermans et al., 2001). Hemichannels are thought to play a more important role than glutamate receptors in ephaptic signaling because they are more tightly localized within the dendritic tips, nearest the synaptic ribbons of cones (Fahrenfort et al., 2005).

The ephaptic hypothesis is attractive, but it has received serious experimental and theoretical challenges. Studies have shown that carbenoxolone has many effects aside from its potential actions on hemijunctional channels in HCs. Most importantly, carbenoxolone directly inhibits voltage-gated Ca^{2+} channels in cone terminals (Vessey et al. 2004), greatly complicating interpretation of effects on feedback. Moreover, paired HC-cone recordings in salamander retina show that carbenoxylone fails to block negative feedback regulation of cone Ca^{2+} current elicited by direct voltage changes in HCs (Cadetti & Thoreson, 2006). Furthermore, paired recordings from HCs and cones show that HC depolarization sometimes causes little or no shift in the Ca^{2+} activation curve but instead a reduction in the peak amplitude of the current (Cadetti & Thoreson, 2006; Jackman et al., 2011). Light-evoked feedback can also reduce the peak amplitude of the Ca^{2+} current (Hirasawa & Kaneko, 2003; Verweij, Kamermans, & Spekreijse, 1996). These observations are inconsistent with an ephaptic mechanism. Finally, the observation that the cone neurotransmitter glutamate (Szmajda & DeVries, 2011) and many other experimental reagents can rapidly diffuse in and out of the invaginating synapse seems inconsistent with the high resistance needed for an ephaptic mechanism. Taken together, the ephaptic hypothesis is not necessarily supported by "move it" experiments, and "block it" experiments are inconclusive. But perhaps the most critical problem is the lack of "show it" experiments to test whether an ephaptic signal is actually generated during HC feedback.

The pH Hypothesis

An equally unconventional hypothesis is that protons serve as the negative feedback signal from HCs to cones. In this scenario depolarization of the HC drives proton efflux through a channel or transporter located in the tips of HC dendrites. The resulting change in extracellular pH modulates the voltage-gated Ca^{2+} channels in the cone terminal, altering channel activation and Ca^{2+}-dependent neurotransmitter release.

The nature of the proton efflux pathway from HCs is unclear, but there have been a couple of candidates suggested by different investigators. HCs possess epithelial-type Na^+ channels (ENaC), which are highly proton permeant. Amiloride, a blocker of ENaC channels, blocks HC negative feedback (Vessey et al., 2005). However, amiloride also affects other targets, so more specific tools are needed to confirm a role for ENaC. The plasma membrane of HCs also possesses the vacuolar type H^+ pump (V-ATPase), which transports protons. Bafilomycin-A1, a selective blocker of the V-ATPase, prevents proton efflux from dissociated HCs (Jouhou et al., 2007). However, it has not been possible to test whether bafilomycin-A1 also blocks negative feedback in the retina because V-ATPase in cones is required for filling synaptic vesicles with glutamate. Hence, the drug blocks cone neurotransmitter release, precluding study of feedback. It is also possible that hemijunctional channels mediate proton efflux from HCs, but unfortunately, there are no specific inhibitors of hemijunctions, hindering "block it" analysis of this possibility.

Several observations support the pH hypothesis. First, like many ion channels, the L-type Ca^{2+} channels in cones are acutely sensitive to extracellular pH (Barnes & Bui, 1991; Barnes, Merchant, & Mahmud, 1993). This is because the extracellular side of the channel protein has many charged titratable residues, including in the S4 voltage-sensor domain, which is partly accessible to the extracellular aqueous environment (Larsson et al., 1996). Artificially acidifying or alkalinizing the extracellular pH by about 0.6 log units shifts the midpoint of Ca^{2+} channel activation by about 10 mV, enough to account for negative feedback. So "move it" evidence is consistent with the pH hypothesis.

Additional studies show that increasing the buffering capacity of the extracellular space blocks negative feedback. Supplementing bicarbonate, the natural pH buffer, with a high concentration (10–20 mM) of HEPES, a stronger and faster pH buffer, suppresses negative feedback of both the cone Ca^{2+} channels and the light responses in HCs (Hirasawa & Kaneko, 2003; Vessey et al., 2005). Further studies in primate retina show that 20 mM HEPES eliminates the antagonistic surround receptive field response in retinal ganglion cells (Davenport, Detwiler, & Dacey, 2008) one more synapse further along in the visual system. So "block it" experiments involving HEPES are consistent with the pH hypothesis.

Questions have been raised about the specificity of HEPES action in buffering extracellular pH (Fahrenfort et al., 2009). There is evidence that HEPES can alkalinize the cytoplasm of HCs and this can block hemijunctional channels. Moreover, HEPES can have a direct blocking action on connexin-55.5 hemichannels expressed in *Xenopus* oocytes. These findings raise concerns that off-target effects may be responsible for HEPES blockade of negative feedback. The aminosulfonate moiety is responsible for inhibition of hemichannels by HEPES. However, the pH buffer Tris, which has no aminosulfonate, also blocks negative feedback. Moreover, many aminosulfonates act as pH buffers, but only those few with a pK_a near 7.4 are effective in blocking negative feedback (Trenholm & Baldridge, 2010). This suggests that the pH-buffering activity, specifically, is what blocks negative feedback, not some off-target action.

The pH hypothesis has received other challenges as well. Studies on enzymatically isolated HCs from skate retina show that changing transmembrane voltage does indeed cause a change in extracellular pH, but in the wrong direction! Studies utilizing pH-sensitive dyes and a pH-selective microelectrode show that HC depolarization results in extracellular alkalinization (Jacoby et al., 2012; Kreitzer et al., 2007) instead of the acidification predicted by the pH hypothesis. However, it is possible that particular extracellular regions (e.g., outside the tips of the HC dendrites) become more acidic on HC depolarization while the rest becomes more alkaline, establishing a closed-loop proton current that persists in darkness. It is also possible that the enzyme treatment involved in cell isolation had altered the channels or transporters in the HCs that are responsible for proton efflux.

Hence, like the other hypotheses of negative feedback, the biggest problem with the pH hypothesis is the lack of "show it" experiments. The change in proton concentration resulting from HC feedback is likely to be quite small, in the 0.1 to 0.2 pH unit range, and the volume of the extracellular space in the synaptic invagination is miniscule ($\sim 10^{-18}$ L). Extracellular electrodes are invasive and have a limited spatial resolution (1–2 µm) for measuring local pH. Conventional pH-sensitive fluorescent dyes can cross the membrane of HCs, confounding measurement of extracellular pH (Jacoby et al., 2012).

Fortunately, there is a noninvasive, highly localizable, and highly sensitive tool for measuring pH on the horizon for retinal application: the pH-sensitive form of GFP, called "pHluorin." Best of all, pHluorin can be genetically expressed in particular cell types with a cell-selective promoter. We have spliced pHluorin onto a subunit of the L-type Ca^{2+} channel, and we are expressing this construct in the retina of zebrafish with a cone-specific promoter (unpublished results). This fluorescent reporter of pH should localize precisely where feedback occurs on the cone terminal. We hope that this tool will provide definitive "show it" experiments to confirm whether or not an appropriate change in pH occurs during negative feedback.

A SURPRISING TWIST IN THE STORY: A POSITIVE-FEEDBACK SYNAPSE FROM HCS TO CONES

The fundamental circuitry of the retina has been known for decades, which made it all the more surprising to discover a previously unknown type of synaptic interaction between HCs and cone photoreceptors. In 2011 we reported that a Ca^{2+}-dependent signal emanating from HCs leads to enhancement of neurotransmitter release from cones, opposite to the effect of negative feedback (Jackman et al., 2011). The initial discovery came not from electrophysiological recordings but instead from imaging studies involving the fluorescent dye FM1–43.

FM1–43 is a lipophilic dye that has gained popularity as a tool for monitoring neurotransmitter release from presynaptic terminals (Betz, Mao, & Bewick, 1992). FM1–43 partitions into the plasma membrane of cells where the hydrophobic environment enhances its fluorescence. Because FM1–43 is amphipathic (polar on one end and nonpolar on the other), it cannot "flip" around in the membrane bilayer but instead remains constrained to the outer leaflet. However, endocytosis of recycling synaptic vesicles can bring the dye inside the presynaptic terminal, trapping it in internalized vesicles. Subsequent stimulation causes these vesicles to undergo exocytosis, returning the dye to the plasma membrane, where it can escape by diffusing into the extracellular space, decreasing the fluorescence of the terminal.

The rates of loading and unloading of FM1–43 can be used as indicators for measuring the rates of endocytosis and exocytosis, respectively. Rod and cone photoreceptors continuously release neurotransmitter in the dark. So as expected, the rates of FM1–43 loading and unloading are maximal in darkness and fall off with illumination (Choi et al., 2005). We have used this information along with electron microscopy to quantify how the synaptic vesicle release rate encodes light intensity. The light used for visualizing FM1–43 can itself alter the photoreceptor release rate, but this confounding problem can be minimized with two-photon imaging, which employs infrared light.

When we began our studies on the outer retina, we fully expected that because of negative feedback, activation of glutamate receptors on HCs would lead to a decrease in the rate of FM1–43 unloading from cones. However, we were shocked to find exactly the opposite: glutamate caused the FM1–43 release rate to increase dramatically, by more than fourfold. Pharmacological studies showed that AMPA receptors in particular were responsible for this effect of glutamate on cone release. This too was a surprise because cones do not possess AMPA receptors. However, HCs have AMPA receptors, and we wondered if activation of these receptors on HCs was indirectly responsible for accelerating cone release. Support for this idea came from retinal slicing and laser ablation experiments. Treatments that disrupted or destroyed HCs eliminated the effect of AMPA on cone release, but physical or pharmacological disruption of other retinal cell types had no impact on the effect of AMPA. Taken together, these results suggested a novel sequence of events: glutamate released from cones activates AMPA receptors on HCs, and apparently, some type of positive feedback signal from the HCs to the cone terminals further accelerates cone release.

To test whether these events occur under physiological conditions, we asked whether cone release is normally accelerated by positive feedback in darkness, when glutamate release is occurring continuously. Sure enough, interrupting the system with a selective antagonist of the AMPA receptor slows FM1–43 release in darkness. Hence, AMPA receptor–mediated positive feedback is an ongoing synaptic mechanism that boosts neurotransmitter release from cones. Using the same FM1–43 strategy, we observed AMPA-elicited positive feedback in retinas from a large variety of animals, including tiger salamander, zebrafish, anole lizard, and rabbits. Hence, positive feedback is conserved among the vertebrates.

How could we reconcile this new HC-mediated positive feedback phenomenon with negative feedback that had been so extensively studied for decades? The answer lies in the intracellular signals in HCs that are responsible for triggering negative versus positive feedback. Activation of AMPA receptors leads to depolarization of the membrane potential, but because the type of AMPA receptor in HCs is Ca^{2+}-permeant (Rivera, Blanco, & de la Villa, 2001), AMPA can also lead to a rise in the intracellular Ca^{2+} concentration.

We separately investigated the effects of the HC voltage change and the HC intracellular Ca^{2+} change by carrying out paired recordings from HCs and cones. First, we found that depolarizing the HC membrane potential leads to inhibition of voltage-gated Ca^{2+} channels in cones, as many others had found previously (Cadetti & Thoreson, 2006; Verweij, Kamermans, & Spekreijse, 1996). And as reported by others, we found that inhibition of cone Ca^{2+} channels elicited by HC voltage changes can be blocked by substituting bicarbonate in the saline with a high concentration of HEPES (Hirasawa & Kaneko, 2003; Vessey et al., 2005). Hence, we confirmed that a change in HC voltage really is crucial for mediating negative feedback.

But what about positive feedback? To separately investigate the effect of intracellular Ca^{2+} in HCs we employed DM-nitrophen, a type of "caged Ca^{2+}" that can be used to generate an abrupt rise in intracellular Ca^{2+} concentration in response to UV light. We injected DM-nitrophen into individual HCs and recorded spontaneous excitatory synaptic events ("mini" EPSCs) resulting from glutamate release from rods and cones. Elevating Ca^{2+} in HCs caused a rapid and large *increase* in the frequency of these events, indicating positive feedback regulation of neurotransmitter release. Because this manipulation is specific to the single HC cell infused with the DM-nitrophen, this finding provided direct evidence that HCs are responsible for positive feedback. Moreover, these findings implicate a rise in intracellular Ca^{2+} as the underlying trigger.

We next investigated possible transmitters that might mediate positive feedback transmission from HCs to cones. Pharmacological experiments ruled out many candidates, including glutamate, GABA, glycine, dopamine, and nitric oxide. There are many other possible candidates that we did not fully evaluate, including endocannabinoids and other lipid metabolites. So at least for the time being, the identity of the positive feedback transmitter remains a mystery.

However, we do know some things about how the positive feedback leads to an increase in cone transmitter release. Unlike negative feedback, which regulates the *voltage-dependent* Ca^{2+} conductance in cones, positive feedback leads to activation of a *voltage-independent* conductance in cones (probably some type of nonselective cation channel), leading to an increase in intracellular Ca^{2+}. Blocking the voltage-dependent L-type Ca^{2+} channels in cones has no effect on the accelerated release brought about by AMPA. Hence, positive feedback turns on a Ca^{2+} influx pathway that is distinct from the voltage-gated Ca^{2+} channels studied all these years. Cone terminals possess a variety of voltage-independent, Ca^{2+}-permeant channels, including

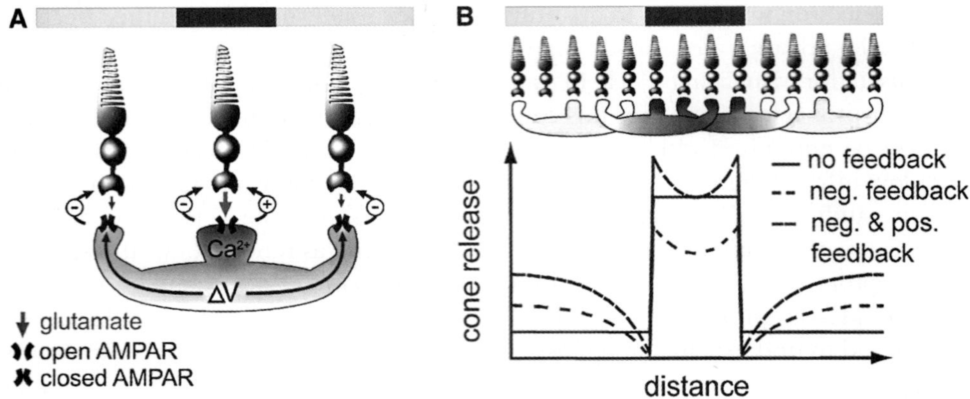

FIGURE 7.4 Functional consequences of positive and negative feedback. (A) Positive and negative feedback signals spread differently within an HC. The top bar denotes the illumination pattern. A cone depolarized in darkness releases glutamate, activating AMPA receptors (AMPARs) and causing depolarization and Ca^{2+} influx. The rise in Ca^{2+} is restricted to the dendritic branch contacting the cone, so positive feedback is localized to that cone. The depolarization spreads electrotonically through the HC, so the negative feedback signal arises from all of the dendrites. (B) Model simulations of the effects of feedback on synaptic release from cones exposed to a dark spot on a gray background. Traces show simulated cone release with no feedback, negative feedback alone, and equal amounts of negative and positive feedback. (Adapted from Jackman et al., 2011.)

several types of TRP channels, but more work is needed to determine which one mediates accelerated release from cones during positive feedback.

WHY HAVE BOTH POSITIVE AND NEGATIVE FEEDBACK?

At first glance, it might seem futile to have a feedback synapse that both enhances and suppresses neurotransmitter release. However, there is evidence that the signals that give rise to positive feedback and negative feedback spread differently through an HC. This difference could help explain why the two types of feedback do not simply cancel one another out. Previous studies had shown that the particular type of AMPA receptor found in HCs is permeant to Ca^{2+} (Rivera, Blanco, & de la Villa, 2001), and our Ca^{2+} imaging experiments in HCs confirmed this result. Furthermore, we found that local application of AMPA results in only a very local increase in Ca^{2+}. Because positive feedback from the HC onto the cone is caused by the rise in Ca^{2+}, this suggests that positive feedback will occur only in HC dendritic branches that are receiving direct excitatory input, such as those driven by cones in darkness.

In contrast, negative feedback is controlled by the HC voltage, not by Ca^{2+}. The voltage signal generated by synaptic current into individual dendrites spreads electrotonically, not only within a single HC but through the syncytium of HCs, distributing negative feedback over a relatively large area in the outer retina Thus, positive feedback appears to be more spatially constrained than negative feedback.

It is perhaps counterintuitive, but with positive and negative feedback signals spreading over different spatial domains, positive feedback can amplify, rather than suppress contrast enhancement. Figure 7.4B shows the results of a model with different spatial profiles of positive and negative feedback (Jackman et al., 2011). When a dark spot is projected on the retina, cones in the center are depolarized, maintaining glutamate release, and those in the surround are hyperpolarized, decreasing release. If there is no feedback whatsoever, the spatial profile of release mirrors the stimulus. Adding negative feedback enhances contrast by subtracting from the cone response. Thus, release in the dark spot is decreased by depolarized HCs, and release in the brighter surround is increased by hyperpolarized HCs. Cone terminals at the edge receive inputs from HCs straddling the border so the effects cancel, allowing light and dark to fully modulate cone release. Positive feedback will simply scale release in direct proportion to the local synaptic output from individual cones. As a result, positive feedback can amplify cone release without sacrificing contrast enhancement provided by negative feedback.

ACKNOWLEDGMENTS

I thank Tzu-Ming Wang and Wally Thoreson for helpful comments on the manuscript and Skyler Jackman for help with the figures. This work was supported by research grants from the NIH (R01-EY015514 and PN2-EY18241) and the Beckman Initiative for Macular Research.

REFERENCES

Barnes, S., & Bui, Q. (1991). Modulation of calcium-activated chloride current via pH-induced changes of calcium channel properties in cone photoreceptors. *Journal of Neuroscience, 11*, 4015–4023.

Barnes, S., Merchant, V., & Mahmud, F. (1993). Modulation of transmission gain by protons at the photoreceptor output synapse. *Proceedings of the National Academy of Sciences of the United States of America, 90*, 10081–10085. doi:10.1073/pnas.90.21.10081.

Baylor, D. A., Fuortes, M. G. F., & O'Bryan, P. M. (1971). Receptive fields of cones in the retina of the turtle. *Journal of Physiology, 214*, 265–294.

Betz, W. J., Mao, F., & Bewick, G. S. (1992). Activity-dependent fluorescent staining and destaining of living vertebrate motor nerve terminals. *Journal of Neuroscience, 12*, 363–375.

Bloomfield, S. A. (1992). A unique morphological subtype of horizontal cell in the rabbit retina with orientation-sensitive response properties. *Journal of Comparative Neurology, 320*, 69–85. doi:10.1002/cne.903200105.

Burkhardt, D. A. (1993). Synaptic feedback, depolarization, and color opponency in cone photoreceptors. *Visual Neuroscience, 10*, 981–989. doi:10.1017/S0952523800010087.

Burris, C., Klug, K., Ngo, I. T., Sterling, P., & Schein, S. (2002). How Müller glial cells in macaque fovea coat and isolate the synaptic terminals of cone photoreceptors. *Journal of Comparative Neurology, 453*, 100–111. doi:10.1002/cne.10397.

Byzov, A. L., & Shura-Bura, T. M. (1986). Electrical feedback mechanism in the processing of signals in the outer plexiform layer of the retina. *Vision Research, 26*, 33–44. doi:10.1016/0042-6989(86)90069-6.

Cadetti, L., & Thoreson, W. B. (2006). Feedback effects of horizontal cell membrane potential on cone calcium currents studied with simultaneous recordings. *Journal of Neurophysiology, 95*, 1992–1995. doi:10.1152/jn.01042.2005.

Choi, S. Y., Borghuis, B. G., Rea, R., Levitan, E. S., Sterling, P., & Kramer, R. H. (2005). Encoding light intensity by the cone photoreceptor synapse. *Neuron, 48*, 555–562. doi:10.1016/j.neuron.2005.09.011.

Copenhagen, D. R., & Jahr, C. E. (1989). Release of endogenous excitatory amino acids from turtle photoreceptors. *Nature, 341*, 536–539. doi:10.1038/341536a0.

Davenport, C. M., Detwiler, P. B., & Dacey, D. M. (2008). Effects of pH buffering on horizontal and ganglion cell light responses in primate retina: Evidence for the proton hypothesis of surround formation. *Journal of Neuroscience, 28*, 456–464. doi:10.1523/JNEUROSCI.2735-07.2008.

Dowling, J. (1970). *The retina: An approachable part of the brain.* Cambridge, MA: Belknap Press.

Fahrenfort, I., Klooster, J., Sjoerdsma, T., & Kamermans, M. (2005). The involvement of glutamate-gated channels in negative feedback from horizontal cells to cones. *Progress in Brain Research, 147*, 219–229. doi:10.1016/S0079-6123(04)47017-4.

Fahrenfort, I., Steijaert, M., Sjoerdsma, T., Vickers, E., Ripps, H., van Asselt, J., et al. (2009). Hemichannel-mediated and pH-based feedback from horizontal cells to cones in the vertebrate retina. *PLoS One, 4*, e6090. doi:10.1371/journal.pone.0006090.

Guo, C., Hirano, A. A., Stella, S. L., Jr., Bitzer, M., & Brecha, N. C. (2010). Guinea pig horizontal cells express GABA, the GABA-synthesizing enzyme GAD 65, and the GABA vesicular transporter. *Journal of Comparative Neurology, 518*, 1647–1669. doi:10.1002/cne.22294.

Harada, T., Harada, C., Watanabe, M., Inoue, Y., Sakagawa, T., Nakayama, N., et al. (1998). Functions of the two glutamate transporters GLAST and GLT-1 in the retina. *Proceedings of the National Academy of Sciences of the United States of America, 95*, 4663–4666. doi:10.1073/pnas.95.8.4663.

Hare, W. A., & Owen, W. G. (1996). Receptive field of the retinal bipolar cell: A pharmacological study in the tiger salamander. *Journal of Neurophysiology, 76*, 2005–2019.

Hartline, H. K., & Ratliff, F. (1958). Spatial summation of inhibitory influences in the eye of *Limulus* and the mutual interaction of receptor units. *Journal of General Physiology, 41*, 1049–1066.

Hasegawa, J., Obara, T., Tanaka, K., & Tachibana, M. (2006). High-density presynaptic transporters are required for glutamate removal from the first visual synapse. *Neuron, 50*, 63–74. doi:10.1016/j.neuron.2006.02.022.

Hirano, A. A., Brandstätter, J. H., Morgans, C. W., & Brecha, N. C. (2011). SNAP25 expression in mammalian retinal horizontal cells. *Journal of Comparative Neurology, 519*, 972–988. doi:10.1002/cne.22562.

Hirasawa, H., & Kaneko, A. (2003). pH changes in the invaginating synaptic cleft mediate feedback from horizontal cells to cone photoreceptors by modulating Ca^{2+} channels. *Journal of General Physiology, 122*, 657–671. doi:10.1085/jgp.200308863.

Hurd, L. B., II, & Eldred, W. D. (1989). Localization of GABA- and GAD-like immunoreactivity in the turtle retina. *Visual Neuroscience, 3*, 9–20. doi:10.1017/S0952523800012463.

Jackman, S. L., Babai, N., Chambers, J. J., Thoreson, W. B., & Kramer, R. H. (2011). A positive feedback synapse from retinal horizontal cells to cone photoreceptors. *PLoS Biology, 9*, e1001057. doi:10.1371/journal.pbio.1001057.

Jacoby, J., Kreitzer, M. A., Alford, S., Qian, H., Tchernookova, B. K., Naylor, E. R., et al. (2012). Extracellular pH dynamics of retinal horizontal cells examined using electrochemical and fluorometric methods. *Journal of Neurophysiology, 107*, 868–879. doi:10.1152/jn.00878.2011.

Johnson, J., Chen, T. K., Rickman, D. W., Evans, C., & Brecha, N. C. (1996). Multiple gamma-Aminobutyric acid plasma membrane transporters (GAT-1, GAT-2, GAT-3) in the rat retina. *Journal of Comparative Neurology, 375*, 212–224. doi:10.1002/(SICI)1096-9861(19961111)375:2<212::AID-CNE3>3.0.CO;2-5.s.

Jouhou, H., Yamamoto, K., Homma, A., Hara, M., Kaneko, A., & Yamada, M. (2007). Depolarization of isolated horizontal cells of fish acidifies their immediate surrounding by activating V-ATPase. *Journal of Physiology, 585*, 401–412. doi:10.1113/jphysiol.2007.142646.

Kamermans, M., Fahrenfort, I., Schultz, K., Janssen-Bienhold, U., Sjoerdsma, T., & Weiler, R. (2001). Hemichannel-mediated inhibition in the outer retina. *Science, 292*, 1178–1180. doi:10.1126/science.1060101.

Kamermans, M., & Spekreijse, H. (1999). The feedback pathway from horizontal cells to cones. A mini review with a look ahead. *Vision Research, 39*, 2449–2468. doi:10.1016/S0042-6989(99)00043-7.

Kaneko, A., & Tachibana, M. (1986). Effects of gamma-aminobutyric acid on isolated cone photoreceptors of the turtle retina. *Journal of Physiology, 373*, 443–461.

Klaassen, L. J., Sun, Z., Steijaert, M. N., Bolte, P., Fahrenfort, I., Sjoerdsma, T., et al. (2011). Synaptic transmission from horizontal cells to cones is impaired by loss of connexin hemichannels. *PLoS Biology, 9,* e1001107. doi:10.1371/journal.pbio.1001107.

Kraaij, D. A., Spekreijse, H., & Kamermans, M. (2000). The nature of surround-induced depolarizing responses in goldfish cones. *Journal of General Physiology, 115,* 3–16.

Kreitzer, M. A., Collis, L. P., Molina, A. J., Smith, P. J., & Malchow, R. P. (2007). Modulation of extracellular proton fluxes from retinal horizontal cells of the catfish by depolarization and glutamate. *Journal of General Physiology, 115,* 169–182. doi:10.1085/jgp.200709737.

Lam, D. M., Lasater, E. M., & Naka, K. I. (1978). gamma-Aminobutyric acid: A neurotransmitter candidate for cone horizontal cells of the catfish retina. *Proceedings of the National Academy of Sciences of the United States of America, 75,* 6310–6313. doi:10.1073/pnas.75.12.6310.

Larsson, H. P., Baker, O. S., Dhillon, D. S., & Isacoff, E. Y. (1996). Transmembrane movement of the shaker K^+ channel S4. *Neuron, 16,* 387–397. doi:10.1016/S0896-6273(00)80056-2.

Migdale, K., Herr, S., Klug, K., Ahmad, K., Linberg, K., Sterling, P., et al. (2003). Two ribbon synaptic units in rod photoreceptors of macaque, human, and cat. *Journal of Comparative Neurology, 455,* 100–112. doi:10.1002/cne.10501.

Murakami, M., Shimoda, Y., Nakatani, K., Miyachi, E., & Watanabe, S. (1982). GABA-mediated negative feedback from horizontal cells to cones in carp retina. *Japanese Journal of Physiology, 32,* 911–926. doi:10.2170/jjphysiol.32.911.

O'Bryan, P. M. (1973). Properties of the depolarizing synaptic potential evoked by peripheral illumination in cones of the turtle retina. *Journal of Physiology, 235,* 207–223.

Picaud, S., Larsson, H. P., Wellis, D. P., Lecar, H., & Werblin, F. (1995). Cone photoreceptors respond to their own glutamate release in the tiger salamander. *Proceedings of the National Academy of Sciences of the United States of America, 92,* 9417–9421. doi:10.1073/pnas.92.20.9417.

Piccolino, M. (1995). The feedback synapse from horizontal cells to cone photoreceptors in the vertebrate retina. *Progress in Retinal and Eye Research, 14,* 141–196. doi:10.1016/1350-9462(94)E0005-3.

Raviola, E., & Gilula, N. B. (1975). Intramembrane organization of specialized contacts in the outer plexiform layer of the retina. A freeze-fracture study in monkeys and rabbits. *Journal of Cell Biology, 65,* 192–222. doi:10.1083/jcb.65.1.192.

Rea, R., Li, J., Dharia, A., Levitan, E. S., Sterling, P., & Kramer, R. H. (2004). Streamlined synaptic vesicle cycle in cone photoreceptor terminals. *Neuron, 41,* 755–766. doi:10.1016/S0896-6273(04)00088-1.

Rivera, L., Blanco, R., & de la Villa, P. (2001). Calcium-permeable glutamate receptors in horizontal cells of the mammalian retina. *Visual Neuroscience, 18,* 995–1002.

Savchenko, A., Barnes, S., & Kramer, R. H. (1997). Cyclic-nucleotide-gated channels mediate synaptic feedback by nitric oxide. *Nature, 390,* 694–698.

Schwartz, E. A. (1982). Calcium-independent release of GABA from isolated horizontal cells of the toad retina. *Journal of Physiology, 323,* 211–227.

Schwartz, E. A. (1987). Depolarization without calcium can release gamma-aminobutyric acid from a retinal neuron. *Science, 238,* 350–355. doi:10.1126/science.2443977.

Schwartz, E. A. (2002). Transport-mediated synapses in the retina. *Physiological Reviews, 82,* 875–891.

Snellman, J., Mehta, B., Babai, N., Bartoletti, T. M., Akmentin, W., Francis, A., et al. (2011). Acute destruction of the synaptic ribbon reveals a role for the ribbon in vesicle priming. *Nature Neuroscience, 14,* 1135–1141. doi:10.1038/nn.2870.

Sun, Z., Risner, M. L., van Asselt, J. B., Zhang, D. Q., Kamermans, M., & McMahon, D. G. (2012). Physiological and molecular characterization of connexin hemichannels in zebrafish retinal horizontal cells. *Journal of Neurophysiology, 107,* 2624–2632. doi:10.1152/jn.01126.2011.

Szmajda, B. A., & DeVries, S. H. (2011). Glutamate spillover between mammalian cone photoreceptors. *Journal of Neuroscience, 31,* 13431–13441. doi:10.1523/JNEUROSCI.2105-11.2011.

Thoreson, W. B., Babai, N., & Bartoletti, T. M. (2008). Feedback from horizontal cells to rod photoreceptors in vertebrate retina. *Journal of Neuroscience, 28,* 5691–5695. doi:10.1523/JNEUROSCI.0403-08.2008.

Thoreson, W. B., & Burkhardt, D. A. (1990). Effects of synaptic blocking agents on the depolarizing responses of turtle cones evoked by surround illumination. *Visual Neuroscience, 5,* 571–583. doi:10.1017/S0952523800000730.

Trenholm, S., & Baldridge, W. H. (2010). The effect of aminosulfonate buffers on the light responses and intracellular pH of goldfish retinal horizontal cells. *Journal of Neurochemistry, 115,* 102–111. doi:10.1111/j.1471-4159.2010.06906.x.

Trümpler, J., Dedek, K., Schubert, T., de Sevilla Müller, L. P., Seeliger, M., Humphries, P., et al. (2008). Rod and cone contributions to horizontal cell light responses in the mouse retina. *Journal of Neuroscience, 28,* 6818–6825. doi:10.1523/JNEUROSCI.1564-08.2008.

Twig, G., Levy, H., & Perlman, I. (2003). Color opponency in horizontal cells of the vertebrate retina. *Progress in Retinal and Eye Research, 22,* 31–68. doi:10.1016/S1350-9462(02)00045-9.

Verweij, J., Kamermans, M., & Spekreijse, H. (1996). Horizontal cells feed back to cones by shifting the cone calcium-current activation range. *Vision Research, 36,* 3943–3953. doi:10.1016/S0042-6989(96)00142-3.

Vessey, J. P., Lalonde, M. R., Mizan, H. A., Welch, N. C., Kelly, M. E., & Barnes, S. (2004). Carbenoxolone inhibition of voltage-gated Ca channels and synaptic transmission in the retina. *Journal of Neurophysiology, 92,* 1252–1256. doi:10.1152/jn.00148.2004.

Vessey, J. P., Stratis, A. K., Daniels, B. A., Da Silva, N., Jonz, M. G., Lalonde, M. R., et al. (2005). Proton-mediated feedback inhibition of presynaptic calcium channels at the cone photoreceptor synapse. *Journal of Neuroscience, 25,* 4108–4117. doi:10.1523/JNEUROSCI.5253-04.2005.

Weiler, R., Pottek, M., He, S., & Vaney, D. I. (2000). Modulation of coupling between retinal horizontal cells by retinoic acid and endogenous dopamine. *Brain Research. Brain Research Reviews, 32,* 121–129. doi:10.1016/S0165-0173(99)00071-5.

Wu, S. M. (1991). Input-output relations of the feedback synapse between horizontal cells and cones in the tiger salamander retina. *Journal of Neurophysiology, 65,* 1197–1206. doi:10.1016/0006-8993(80)90697-6.

Wu, S. M. (1992). Feedback connections and operation of the outer plexiform layer of the retina. *Current Opinion in Neurobiology, 2,* 462–468.

Wu, S. M., & Dowling, J. E. (1980). Effects of GABA and glycine on the distal cells of the cyprinid retina. *Brain Research, 199,* 401–414. doi:10.1016/0959-4388(92)90181-J.

Yazulla, S., & Studholme, K. M. (1997). Light adaptation affects synaptic vesicle density but not the distribution of GABAA receptors in goldfish photoreceptor terminals. *Microscopy Research and Technique, 36,* 43–56. doi:10.1002/(SICI)1097-0029(19970101)36:1<43:AID-JEMT4>3.0.CO;2-#.

8 Stratification of the Inner Plexiform Layer in the Mammalian Retina

STEPHEN L. MILLS AND STEPHEN C. MASSEY

Eyes across the animal kingdom are distinguished by the collection of light by photoreceptor pigments structurally and evolutionarily related to rhodopsin. Invertebrate photoreceptors are of rhabdomeric origin and generate spikes that send information to later neural structures for further processing. Vertebrate photoreceptors, on the other hand, originate from microvillar structures and transmit their signals via graded potentials. This shift in paradigm has been accompanied by the creation of novel circuits distal to the optic nerve that immediately begin a great diversification in processing of the original camera-like visual scene, first in the outer plexiform layer (OPL) and then in the much more elaborate inner plexiform layer (IPL). This novel diversification of neural types in the retina itself may have rudimentary origins in early chordates (Lacalli, 2004; Tsuda et al., 2006) but appears very similar in broad outline across the vertebrate subphylum. A partial exception appears in one of the most ancient vertebrate lines, that of the hagfish, whose IPL is situated *below* the somata of the ganglion cells (Fritzsch & Collin, 1990). This review concentrates on mammalian retinas, which retain great similarities across the class.

Perhaps it is a consequence of the generally greater acuity of vertebrates and their large number of independent channels, which require both increased density and diversity of ganglion cells, that vertebrate eyes evolved to enable a great deal of visual "preprocessing" before the bottleneck of the optic nerve occurs. Although the first major separation of light signals into numerous parallel channels occurs at the cone pedicle, it is in the IPL that retinal preprocessing of the visual signal is brought to its fullest flower. In the process of relaying the ~10 individual bipolar cell signals to ~20 types of ganglion cells, lateral interactions of amacrine cell interneurons with bipolar cell axon terminals, ganglion cell dendrites, and with other amacrine cells are thought to provide the individual response signatures of these numerous distinct types of retinal ganglion cell. For these reasons the unique story of the vertebrate eye is the study of processing in the plexiform layer.

STRATIFICATION OF THE IPL

The vertebrate retina is a distinctly layered structure consisting of three somatic layers separated by two plexiform layers (e.g., the rabbit retina, figure 8.1). The functional stratification of the retina extends to the IPL (Dowling & Boycott, 1965), which is organized according to the polarity of bipolar cell inputs (Famiglietti, Kaneko, & Tachibana, 1977; Nelson, Famiglietti, & Kolb, 1978). Since the time of Cajal, the IPL has traditionally been divided into five strata: 1 and 2 for sublamina a and strata 3, 4, and 5 for sublamina b. Some have used finer divisions such as dividing each stratum into three, yielding 15 sublayers (Famiglietti, 2004).

In mammalian species there are 9–11 morphological types of cone bipolar cell and one rod bipolar cell, which branch at different depths in the IPL (Hartveit, 1997; MacNeil et al., 2004; McGillem & Dacheux, 2001; Wässle, 2004). The sign of a bipolar cell response is determined by the differential expression of postsynaptic glutamate receptors. The dendrites of OFF cone bipolar cells carry AMPA/kainate receptors, and they ramify in sublamina a of the IPL (DeVries, 2000). In contrast, ON cone bipolar cells and rod bipolar cells express mGluR6 receptors, and their axons descend to sublamina b (Nomura et al., 1994; Slaughter & Miller, 1981; Vardi & Morigiwa, 1997). The separation of ON and OFF pathways appears to be a fundamental principle of retinal organization that is reflected throughout the visual system (Schiller, Mandell, & Maunsell, 1986).

The story of parallel channels and IPL processing is dominated by one major variable, the level of stratification (figure 8.1). The most dramatic manifestation of this lies in the coarse division of OFF ganglion and bipolar cell processes to the outer 40% (sublamina a) of the retina and those of ON ganglion and bipolar cell processes to the inner 60% (sublamina b). Because identification of a great many of the cell types and their levels of ramification dates back to Cajal, it might surprise some readers that discovery of this fundamental division occurred within the career span of a great many researchers still active today and after the first

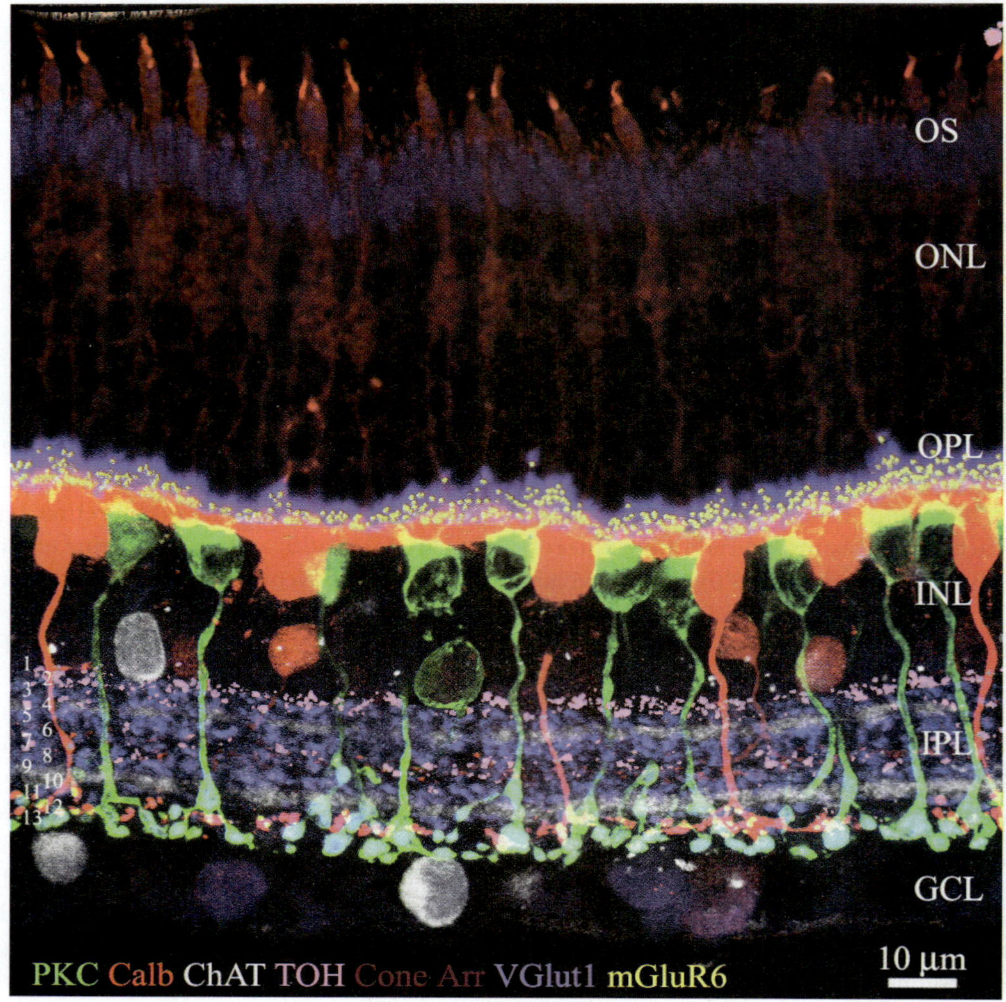

FIGURE 8.1   Multilabeled cross section of central inferior rabbit retina reveals numerous sublayers in the IPL, which are numbered according to details in the text. Vertical sections (50–100 μm) from central inferior retina were processed, and images were acquired, with a Zeiss LSM510 Meta confocal microscope. This image is a ministack of 12×0.4 μm sections. Brightness and contrast plus label-selective colors were adjusted using Photoshop (Adobe Systems Inc.).

comprehensive modern book devoted to the retina (Rodieck, 1973). Combining intracellular staining with intracellular recording, Nelson, Famiglietti, and Kaneko (Famiglietti, Kaneko, & Tachibana, 1977; Nelson, Famiglietti, & Kolb, 1978) established that all of the bipolar, ganglion, and amacrine cells from which they were able to record and stain were segregated according to their ON or OFF polarity as just described.

There is a wealth of evidence from third-order neurons to support the functional division of the IPL into ON and OFF sublaminae. For example, cholinergic amacrine cells, also known as starburst amacrine cells on account of their unique morphology, are present as mirror-image pairs (Tauchi & Masland, 1984) (figures 8.1 and 8.2C). The conventionally placed cholinergic amacrine cells have somas in the innermost layer of the INL, have OFF responses to light stimulation, and ramify in sublamina a. In contrast, the displaced cholinergic amacrine cells reside in the ganglion cell layer, produce ON responses, and ramify in sublamina b. Likewise, alpha ganglion cells are present as paramorphic pairs such that the dendritic trees of OFF alpha ganglion cells are stratified in sublamina a to receive input from OFF bipolar cells, whereas ON alpha ganglion cells ramify in sublamina b to make contact with ON bipolar cells (Peichl, Ott, & Boycott, 1987). In primate retina both midget and parasol ganglion cells are present as paramorphic pairs that conform to the stratification rules of the IPL. Finally, ON/OFF directionally selective ganglion cells (DS GC) produce both ON and OFF responses of short latency, indicating direct input, and they are bistratified with dendrites in sublaminae 2 and 4, coincident with the cholinergic bands.

    STEPHEN L. MILLS AND STEPHEN C. MASSEY

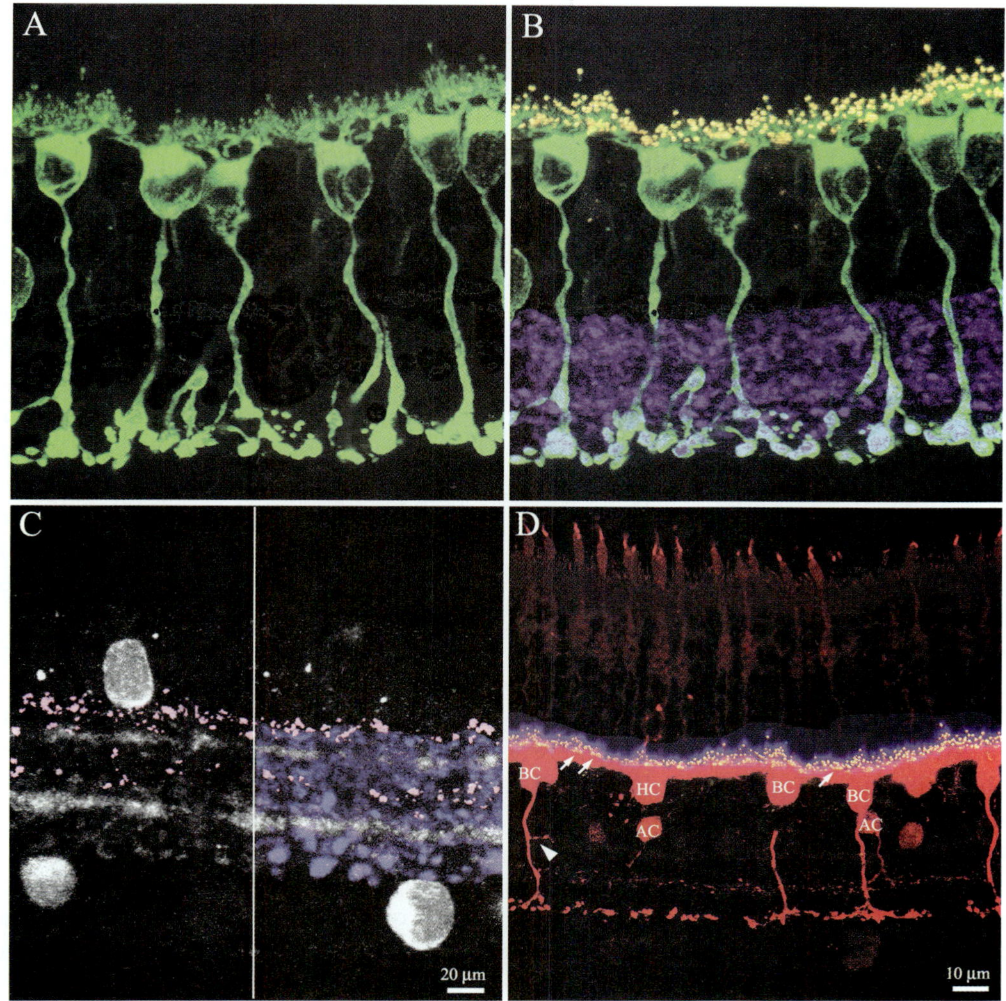

FIGURE 8.2   Multilabeled cross section of central inferior rabbit retina. All panels are subsets of immunolabels from figure 8.1 to illustrate individual patterns of organization in the IPL. (A) PKC labeling (green) shows rod bipolar cells. (B) Rod bipolar cell dendrites are decorated with mGluR6 puncta (yellow), and rod bipolar terminals appear double labeled for PKC and VGlut1 (green + blue = cyan) in stratum 5. Cone bipolar cell terminals, labeled for VGlut1 (blue), are distributed evenly throughout the IPL. VGlut1 is turned off in the OPL for clarity. (C) Labeling for choline acetyltransferase (light gray) shows both conventional and displaced amacrine cells with the classic train tracks in the IPL. Labeling for tyrosine hydroxylase (pink) is mostly confined to stratum 1, with progressively weaker bands in strata 3 and 5. The two panels in C are contiguous, but VGlut1 labeling (blue) was turned on only in the right half of the panel. (D) Cones are modestly labeled for cone arrestin (dark red) except the outer segments, which are bright. Both rod and cone terminals contain VGlut1 (blue), which has been turned off for the IPL for clarity. Calbindin labeling (red) shows horizontal cells (HC). mGluR6 labeling (yellow) shows large clusters at cone pedicles (arrows) and small puncta at rod spherules. Calbindin-labeled ON cone bipolar cells (BC) have axon terminals in stratum 4. Occasionally, these calbindin bipolar cells have fine, short branches in stratum 1 (arrowhead); these mark the sites of ectopic synapses to provide ON input to stratum 1 in sublamina a. A few calbindin-positive amacrine cells (AC) have fine processes in stratum 3. Calibration bar in C applies to A, B, and C, which are insets of figure 8.1. Panel D shows the same field as figure 8.1.

SELECTIVE LABELING REVEALS THE STRATIFICATION PATTERNS

To illustrate the layered structure of the mammalian retina, we have prepared retinal sections labeled with multiple antibodies. A combination of seven antibodies is shown in figure 8.1, and several subsets are broken out in figure 8.2 for clarity. Cones, weakly labeled for cone arrestin (dark red), have bright outer segments, somata in the top row of the ONL, and descending axons that terminate in cone pedicles at the OPL (figures 8.1 and 8.2D). The vast majority of synaptic interactions occur in the two plexiform layers, and both are well labeled with antibodies against synaptic vesicle proteins. More specifically, both photoreceptors and bipolar cells utilize glutamate as a neurotransmitter, and their synaptic terminals can be labeled with antibodies against VGlut1, the vesicular glutamate transporter. VGlut1 (blue) labels a dense band in the OPL (figures 8.1 and 8.2B, C, D) with cone pedicles appearing as large polygonal structures. Rod spherules are smaller and round, slightly higher than the cone pedicles. Each rod spherule contains a dark spot corresponding to the synaptic invagination, which has no vesicles.

Rod bipolar cells were labeled for αPKC (green, figure 8.2A). Each rod bipolar cell has a mop-head of fine dendrites, and two dendrites from different rod bipolar cells enter each rod spherule. Rod bipolar cells give ON responses to light, and each dendrite is decorated with a bright mGluR6 spot (yellow) where it enters the synaptic invagination of a rod spherule, often high in the OPL (figure 8.2B). Rod bipolar axons descend through the IPL and terminate in stratum 5. In the IPL the terminals are heavily labeled for VGlut1; hence, there is a gradient from green to cyan (figure 8.2B). Cone bipolar terminals were stained for VGlut1 (blue), and they were distributed evenly and densely throughout the IPL (figures 8.1 and 8.2B, C) except for the very top, the dopaminergic band, and the very bottom, the region devoted to rod bipolar terminals.

Horizontal cells and a subset of ON cone bipolar cells were labeled for calbindin (red). The dendritic terminals of horizontal cells contact every cone pedicle turning them pink (figure 8.2D). Each cone pedicle also contains a cluster of mGluR6 (yellow) receptors at the base (figures 8.2B, D arrows). These mGluR6 clusters are located on ON cone bipolar cell dendrites, and they are noticeably larger than the mGluR6 puncta of rod bipolar dendrites. In figure 8.2D it can be seen that the horizontal cell terminals surround the mGluR6 cluster and extend higher into the cone pedicle, reflecting the close approach of horizontal cells processes to the synaptic ribbon (arrows). In contrast, the mGluR6-labeled ON cone bipolar terminals are a little lower, further from the cone synaptic ribbons. The calbindin bipolar cells descend to sublamina 4 of the IPL, terminating just above the level of rod bipolar cells. In addition, there are occasional calbindin-labeled wide-field amacrine cells, which ramify in sublamina 3, at the midline of the IPL. Note that the descending axons of calbindin bipolar cells occasionally produce very fine side branches at the border between the INL and the IPL (arrowhead, left side). These may give rise to unusual monadic synapses with dopaminergic amacrine cells or ipRGC (see below).

The IPL is characteristically stratified. The cholinergic amacrine cells (light gray, figures 8.1 and 8.2C) form two bands in sublaminae 2 and 4 resembling train tracks along the IPL. ON/OFF DS GC are bistratified, and they run in the cholinergic bands, which are often used as reference points to determine the depth of other cell types in the IPL. The somata are located in either the amacrine cell layer or displaced to the ganglion cell layer. As expected according to the stratification rules of the IPL, the cholinergic a band has OFF responses, while cholinergic b produces ON responses to light stimulation.

Dopaminergic amacrine cells are comparatively rare, less than 0.1% of the amacrine cell population, so there are no tyrosine hydroxlase (TOH; pink)-positive somata in this section. However, the dopaminergic dendrites have a huge overlap, forming a dense band in sublamina 1 of the IPL immediately adjacent to the INL (figure 8.2C). There are no bipolar terminals at this level; the OFF cone bipolar terminals, marked by VGlut1, are slightly lower. Some dopaminergic dendrites also ramify at the midline, intermingled with the processes of calbindin amacrine cells.

ON INPUT TO THE OFF LAYER

Although the strict correspondence of response polarity to relative depth in the IPL remains the primary functional division of the IPL, a few troublesome cell types eventually appeared. Dopaminergic amacrine cells ramify high in sublamina 1, adjacent to the INL, with minor bands in stratum 3. Dopamine release is controlled by light and the daily circadian rhythm. Recordings from the mouse retina have shown that dopaminergic amacrine cells produce a variety of light responses: ON transient, ON sustained, and light independent (D. Zhang, Zhou, & McMahon, 2007). It has also been suggested that the ipRGC, intrinsically photosensitive ganglion cells, which contain melanopsin, may provide input to the dopaminergic amacrine cells because the sustained features of the response and the spectral signature match those of the melanopsin ganglion cells (D. Zhang et al., 2008). Furthermore, photoreceptor degeneration did not block all the dopaminergic amacrine cell responses, which also implicates ipRGC. However, the transient light responses of dopaminergic amacrine cells were blocked by APB, an agonist of the mGluR6 receptor, suggesting they come from an ON bipolar pathway (D. Zhang, Zhou, & McMahon, 2007). But how do these ON responses arise if most dopaminergic processes are stratified in the OFF sublamina? If dopaminergic amacrine cells are ON cells, then they break the stratification rules of the IPL.

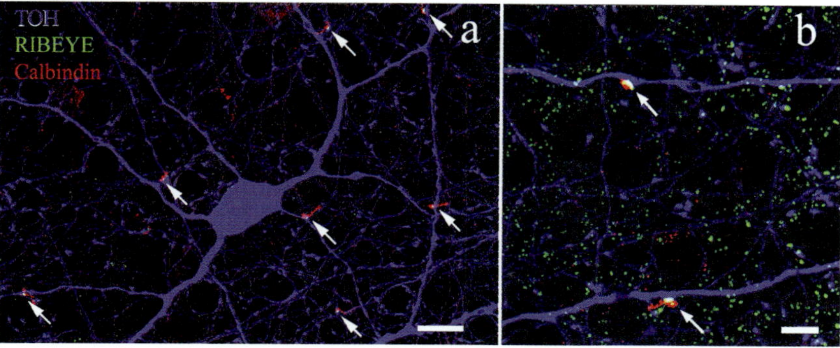

FIGURE 8.3   The sites of ectopic synapses from the calbindin+ cone bipolar cell (red) onto dopaminergic amacrine cells (blue) contain synaptic ribbons (green). (Reproduced from Hoshi et al., 2009, with permission.)

There is a similar problem with melanopsin ganglion cells or ipRGC. There are several subtypes of melanopsin ganglion cells (see chapter 14 by Berson). M1 cells ramify in sublamina 1, and M3 cells are bistratified. Both have ON driven light responses abolished by APB, which blocks ON pathways through the retina (Dacey et al., 2005; Schmidt & Kofuji, 2010). In the melanopsin knockout mouse all light responses of M3 cells were completely abolished by APB, despite the fact that much of their dendritic trees lie in sublamina a (Schmidt & Kofuji, 2010). Finally, there is a bistratified ganglion cell, known as the ON bistratified ganglion cell or the type 9 ganglion cell of Roska and Werblin (2001), that, despite the presence of dendrites in sublamina a, receives only ON input, completely abolished by APB (Hoshi et al., 2009; Roska, Molnar, & Werblin, 2006). Thus, in addition to the dopaminergic amacrine cells, there are at least two ganglion cell types that show anomalous responses to light; that is, their physiology does not match the stratification. This long-standing puzzle has recently been solved: Certain ON cone bipolar cells make axonal or ectopic synapses with dopaminergic amacrine cells (figure 8.3), ipRGC, and ON bistratified ganglion cells at the outer margins of the IPL as they descend through sublamina a (Dumitrescu et al., 2009; Hoshi et al., 2009). The vast majority of ectopic or axonal ribbons occur very high in sublamina a, in the dopaminergic band adjacent to the IPL. They are not uniformly distributed throughout sublamina a (Dumitrescu et al., 2009; Hoshi et al., 2009). These synapses are characterized by large presynaptic ribbons and postsynaptic glutamate receptors. The ON cone bipolar cell types that make axonal synapses include the calbindin bipolar cell in the rabbit (Massey & Mills, 1996) and type 6 in the mouse retina (Dumitrescu et al., 2009), although it is likely that other ON bipolar cells participate. It is important to note that the ectopic synapses do not colocalize with OFF cone bipolar cells in sublamina a, and the ON axonal ribbon synapses do not contact OFF ganglion cells such as G3 or the OFF alpha ganglion cell. In addition, in the mouse retina there are some bipolar inputs to the minor dopaminergic band in sublamina 3 that, by virtue of their depth in sublamina b, may also contribute to the ON responses of dopaminergic amacrine cells (Contini et al., 2010).

These results breach the stratification rules of the IPL and thus immediately raise further questions. What is the function of en passant or ectopic synapses? Why are the dopaminergic amacrine cells and ipRGC located in sublamina 1, and do all dopaminergic amacrine cells receive synaptic input via axonal ribbons? We can only surmise that because the release of dopamine is paracrine, reaching other retinal targets including photoreceptors by diffusion, it may be most broadly effective if positioned in the middle of the retina, high in the IPL. The axonal ribbon synapses may correspond to the bipolar inputs to dopaminergic amacrine cells described as monads (Hokoc & Mariani, 1988). Many of the axonal ribbons have input to the major dendrites of dopaminergic amacrine cells (figure 8.3), all of which made contact with descending calbindin bipolar cells in the rabbit retina (Hoshi et al., 2009). Thus, as far as we can tell, the dopaminergic amacrine cells form a uniform population. This is difficult to reconcile with the three physiological classes reported for the mouse retina (D. Zhang, Zhou, & McMahon, 2007). Either the dopaminergic amacrine cells in the rabbit are homogeneous or all variants receive ectopic inputs. There does appear to be a special relationship between dopaminergic amacrine cells and ipRGC. If there is ipRGC input to the dopaminergic amacrine cells, this may also explain the stratification of the melanopsin ganglion cells. Alternatively, ipRGC are modulated by dopamine, and they express D1 receptors (Van Hook, Wong, & Berson, 2012). Although the stratification rules of the

IPL still hold for most ganglion cell types, it is clear that an accessory ON sublayer occurs at the outer margin of the IPL, providing anomalous input to the dopamine plexus and at least two ganglion cell types (Dumitrescu et al., 2009; Hoshi et al., 2009).

LAYERS OF THE IPL

An obvious but not simple question is how many distinguishably different layers of independent processing exist, which would of course partly depend on the definition of independent. The different types of bipolar, amacrine, and ganglion cells each range from narrowly stratified (in terms of vertical width of the IPL) to broadly stratified varieties. Thus, the dendrites of a single broadly stratified bipolar or ganglion cell may span a vertical region with distinct subregions. In addition, it is clear that different cell types can occupy the same strata. An obvious example is the ON/OFF DS ganglion cell and the ON DS ganglion cell, both of which ramify in the lower cholinergic band. In fact, they may overlap only partially (Famiglietti, 1992). This may imply that the cholinergic bands themselves could be subdivided. In addition, certain bipolar cell types form pairs with partially overlapping axonal stratification (Light et al., in press).

One simple approach is to count the different layers in the multichannel image of figure 8.1, although we realize this will not provide a definitive answer. Nevertheless, it is a useful framework, based on common markers. Starting at the top of the IPL, adjacent to the INL, layer 1 is occupied by the dopaminergic amacrine cells. Bipolar terminals are scarce at this level, but, as above, the input to the dopaminergic plexus is provided by axonal ribbons from ON bipolar cells. Layer 2 contains the dendrites of ipRGC, ON bistratified cells, and uniformity detectors. Layers 1 and 2 together comprise an accessory ON layer in sublamina a.

Layer 3 contains true OFF ganglion cells such as G3 and G9 (Hoshi & Mills, 2009). G3, the JAM-B ganglion cell (Kim et al., 2008), has been called an OFF DS GC, but it also has orientation-selective responses (Venkataramani & Taylor, 2010). G9 has alpha-like morphology but is more sustained, possibly corresponding to the OFF delta cell of guinea pig retina (Manookin et al., 2008). Layer 4 is cholinergic a, the upper cholinergic band. The OFF dendrites of ON-OFF DS ganglion cells ramify here. The band between cholinergic a and the midline is wide enough to accommodate two layers. Layer 5 is just beneath cholinergic a and contains the dendrites of OFF alpha ganglion cells (J. Zhang et al., 2005). Layer 6 is empty in figure 8.1 except for bipolar cell terminals just below the OFF alpha. It may contain

the OFF dendrites of a narrowly bistratified ganglion cell, the transient ON/OFF cell that lies within the cholinergic bands (Sivyer et al., 2011).

As we move from the OFF toward the ON layers, layer 7 is the midline where certain wide field or axon-bearing amacrine cells reside, such as NOS amacrine cells, PA1, and calbindin amacrine cells (Famiglietti, 1992; Massey & Mills, 1996; Vaney & Young, 1988). There are also a few fine dopaminergic dendrites here. G1, the local edge detector (LED), has ON-OFF responses and may also be here (van Wyk, Taylor, & Vaney, 2006). There is room for two bands between the midline and cholinergic b. Layer 8 is adjacent to the midline and may hold the ON dendrites of the transient ON-OFF cell (Sivyer et al., 2011). Layer 9 is just above cholinergic b, where the transient ON DS cell ramifies (Ackert et al., 2006; Hoshi et al., 2011; Sivyer et al., 2011). Layer 10 is cholinergic b, also containing the ON dendrites of ON-OFF DS GC and the sustained ON DS GC (Buhl & Peichl, 1986; Sun et al., 2006). Layer 11 is immediately beneath cholinergic b. There are clearly cone bipolar terminals here unlabeled for calbindin, which is in the next layer (figure 8.1). The ON alpha ganglion cells and calbindin bipolar cell terminals lie in layer 12, immediately adjacent to the rod bipolar terminals. Blue cone bipolar cells and M2 ipRGC also stratify at this depth. Layer 13 is the bottom of the IPL, adjacent to the ganglion cell layer, occupied by rod bipolar terminals and the matrix of S1/S2 amacrine cells (J. Zhang et al., 2002).

Figure 8.1 illustrates the fine and complex stratification of the IPL. Because of the limitations of useful antibodies, available chromophores, and the color palette, we are able to see only a partial outline of the full panoply of sublayers. It is still unknown how many distinctly different strata exist in any retina, and to some extent this is a hypothetical exercise. Peripheral retina is noticeably thinner, and it is therefore more difficult to distinguish layers as the cell density drops. In addition, some of the sublayers may contain multiple addresses. For example, sustained ON DS and ON/OFF DS GC only partially overlap in cholinergic b (Famiglietti, 1992). This may indicate that the lower cholinergic band could be further subdivided.

BIPOLAR CELLS MATCHED TO GANGLION CELLS

Another complementary approach is to match the catalogs of bipolar and ganglion cell types. The most complete data set is from the rabbit retina. Figure 8.4, amended from MacNeil et al. (2004) and Rockhill et al. (2002), shows the level of stratification in the rabbit

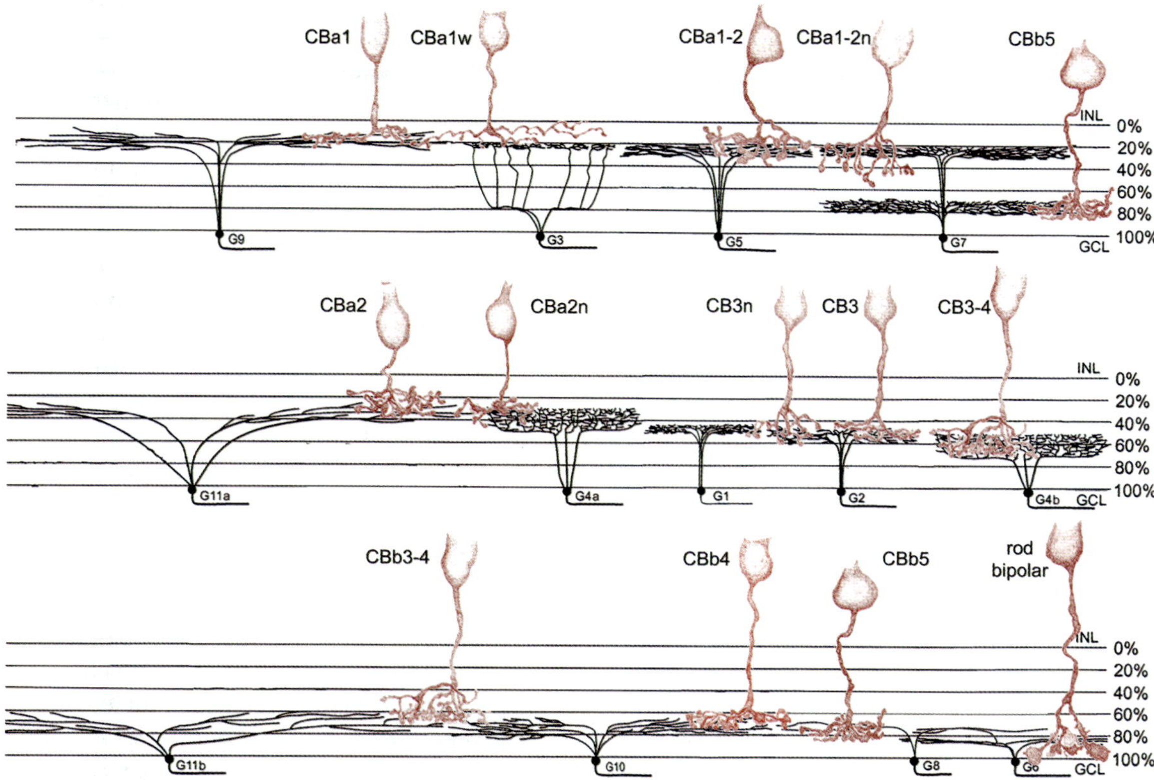

FIGURE 8.4   Likely synaptic pairs of bipolars and ganglion cells identified by Rockhill et al. (2002) and MacNeil et al. (2004).

retina for the bipolar cell and ganglion cell classes derived by the Masland lab. Two bipolar cell types are repeated for convenience. There are a number of interesting features to be derived from this figure. First, it is clear that many ganglion cell types are potentially driven by multiple bipolar cell types, as we know from the work of Sterling, Freed, and Smith (1988) and Freed (2000). Second, it is obvious that many of the ~10 types of bipolar cells must drive more than one of the ~20 types of ganglion cells and the even more diverse amacrine cells. Third, the number of distinct strata is not obvious, and, considering the different levels occupied by the more numerous amacrine cells, the number of potential strata defined by unique combinations of bipolar, ganglion, and amacrine cells is very large. Because the number of ganglion cells is about 20–25, it would seem that many of the potential strata may not be distinctly different with respect to defining individual ganglion cell types but may represent different processing zones for various ganglion cell computations. Among the bipolar cells there are pairs with overlapping stratifications also (Light et al., in press). Presumably, different ganglion cell types that overlap in depth of stratification make distinct connections that influence their physiological properties. Although there may be only ~20 distinct ganglion cell types, each gan-

glion cell almost certainly changes its information-gathering mechanisms to match the demands of the visual environment (Hosoya, Baccus, & Meister, 2005). When this occurs, some of the potential inputs may be quiescent or redundant.

PROCESSING IN THE IPL: THE PLAYERS

Synaptic processing in the IPL is dominated by bipolar, amacrine, and ganglion cells. Additionally, the dendrites of a variety of interplexiform cells whose terminals rise to the OPL have been reported in several species. Even more rarely, horizontal cells in some species send out processes that descend to the IPL. The functions of these are poorly understood and will not be addressed further here.

The degree to which the types of bipolar, amacrine, and ganglion cells are known in mammalian retina, mostly dependent on the Golgi method of staining for decades, has been supplemented more recently by a variety of techniques. A nearly complete catalog of these types can be found for the rabbit retina and ground squirrel. It is largely complete for bipolar and ganglion cell types in the mouse and primate retinas, although the full complement of amacrine cell types is unknown in these species.

Bipolar Cells

Mammalian retinas have on the order of 10–13 types of bipolar cell (see chapter 6 by DeVries). There appears to be a single type of rod bipolar cell that ramifies at the innermost margin of the IPL. Rod bipolar cells are well known *not* to synapse directly onto ganglion cells in the mammal but to synapse instead onto two types of amacrine cell. The AII amacrine cell receives rod bipolar input, which is relayed into the ON pathway via gap junctions with ON cone bipolar cells and also into the OFF pathway via glycinergic synapses onto OFF cone bipolar cells. The other amacrine cells at the rod bipolar dyad are the so-called reciprocal amacrine cells, usually called A17 after the cat nomenclature (Kolb, 1979) or S1/S2 in the rabbit (Massey, Mills, & Marc, 1992; Sandell & Masland, 1986; Vaney, 1986; J. Zhang et al., 2002).

Another bipolar cell type mostly conserved throughout the mammalian class is the ON blue cone bipolar, which collects exclusively from blue cones and ramifies in stratum 4, below cholinergic b (if we define five strata each spanning 20% of the IPL, numbered in ascending order from the INL to the GCL). The possible existence of OFF blue cone bipolar cells has often been suggested (Liu & Chiao, 2007), but the overall evidence has always been sparse compared to that of ON blue cone bipolar cells. Because most nonprimate retinas contain only two types of cones, generally a blue-sensitive cone and another cone of longer wavelength, the ON blue cone bipolar cell may provide the major chromatic input to the retinas of these species via input to blue-dominated ganglion cells whose spectral opponency probably arises from horizontal cell feedback at the level of the cone pedicles.

Specialized types called midget bipolar cells can be found in the primate retina; these are connected to a single cone and a single ganglion cell in central retina and thereby provide both superior spatial acuity and the segregation of color into private pathways. (Connectivity is less constrained with increasing eccentricity, where multiple cones contact a midget bipolar cell, which in turn contacts multiple midget ganglion cells.) Other types of bipolar cells are referred to as "diffuse," in that they contact multiple cones, usually nonselectively. Blue cone bipolars contact multiple cones but only those with blue sensitivity.

Given the specializations of midget, blue, and rod bipolar cells, there still remain many other bipolar cell types with distinct stratification patterns. What are the differences among these types that account for their diversity? One such difference is in their response kinetics, which can range from transient to sustained and is accounted for by their glutamate receptor kinetics, their intrinsic properties (Awatramani & Slaughter, 2000; DeVries, 2000), and their composition of receptors to inhibitory neurotransmitters (Lukasiewicz & Shields, 1998; see chapter 6 by DeVries for more details).

Amacrine Cells

Amacrine cells encompass the widest variety of types (>30), distinctly different morphologies, and neurotransmitter signatures of the retinal cell classes. Clear evidence of this can be seen in their differences in shape, breadth, stratification patterns, and coverage factor, which is the degree of overlap between neighbors of the same type. Although there are undoubtedly specializations across various species, the types best understood have been found to be well conserved in all studied mammals. These include the AII, dopaminergic, starburst, and A17 (reciprocal rod) amacrine cell types, which together comprise only about 18% of the amacrine cell population as a whole (MacNeil et al., 2004; Strettoi & Masland, 1996).

The large number and wide diversity of types of amacrine cells have led to several efforts to establish broad classes of types. A first useful distinction is the type of inhibitory neurotransmitter they contain, GABA or glycine. This turns out to be relatively well correlated with spatial extent in mammalian retinas, where glycinergic amacrine cells tend to have a more restricted lateral extent (less than ~145 μm in rabbit), whereas medium-field and wide-field amacrine cells typically release GABA. A few exceptions exist.

A second distinguishing feature of amacrine cell populations lies in the degree to which they overlap one another, that is, their coverage factor. This ranges from about 1 for the lobular appendages of AII (rod) amacrine cells to overlap factors of greater than 50 in, for example, starburst (cholinergic) amacrine cells and A17 (reciprocal rod) amacrine cells. It should be remembered that the term amacrine literally means "without an axon" and that these cells consequently lack clearly distinct input and output regions. Some amacrine cells do send out axons, however (Famiglietti, 1992; Wright & Vaney, 2004), and the axonal coverage factor of these cells is very large. Studies of some other large and heavily overlapping amacrine cells such as the A17 and starburst amacrine cells have led to the belief that the fundamental processing unit is a local area of each cell or group of cells rather than the cell as a whole (Grimes et al., 2010; Miller & Bloomfield, 1986). It is likely that somatic recording from these types of amacrine cells reveals little about their fundamental methods of processing.

We know very little about the remaining amacrine cell types, a situation that has changed little since the time of Cajal. The fundamental problem is a lack of markers for specific cell types. This makes it extremely difficult to differentiate among similar neurons and obstructs the ready supply needed for further studies to identify retinal circuits. The situation is starting to change as mouse lines with GFP-labeled cells become readily available (Siegert et al., 2009). In turn, this may lead to the development of molecular libraries for specific cell types (Siegert et al., 2012). It is clearly a mistake to think that the large variety of amacrine cell types is uninteresting. They account for the majority of synapses in the IPL (Koontz & Hendrickson, 1987). Every amacrine cell type that we know well has a fascinating story, from starburst amacrine cells and directional selectivity to the noise-reducing coupled network of AII amacrine cells. The remaining amacrine cell types all have unique and interesting stories as well; we just do not know what they are yet.

Ganglion Cells

The differing environments inhabited by distinct species suggest that there are likely to be specializations for specific environments. Nevertheless, the major physical demands of vision as well as the statistics of the visual scene are generally comparable, suggesting that similar types of ganglion cell are likely to be found in many related species. This is certainly true across the class *Mammalia,* all members of which seem to be drawn from a fairly similar plan. For example, all known species contain the five amacrine cell types previously listed. In addition, many ganglion cell types recur across species. These include alpha-like ganglion cells as well as small ganglion cells that set the limits of visual acuity, ganglion cells that are differentially sensitive to chromatic contrast, luminance contrast, to orientation, to motion in general, or to motion in a specific direction. Additional ganglion cell types are also differentially sensitive to stimuli in the periphery (Chiao & Masland, 2003; Olveczky, Baccus, & Meister, 2003), including responses to saccades (Roska & Werblin, 2003). Table 8.1 lists known types of ganglion cells in rabbit retina, with nomenclature from the Rockhill et al. (2002) catalog when correspondence seems likely.

This table (see also figure 8.4) defines 21 morphologically and physiologically distinct types of ganglion cell in the rabbit retina without counting subtypes of ipRGC, those based on direction of response, or some types reported in the older literature with unknown morphologies, such as dimming detectors and ON-type orientation-sensitive ganglion cells.

A persistent problem in identifying exactly the number of types and their roles in vision has been the difficulty in accurately distinguishing and locating the different types in the living retina. The use of transgenic markers holds a great deal of promise for cell-specific targeting and recording in the future (Kim et al., 2008; Rivlin-Etzion et al., 2011; Siegert et al., 2009). This will be true for other cell types as well.

SYNAPTIC INTERACTIONS IN THE IPL

Excitation in the mammalian IPL occurs via one of three modes. These include (1) release of glutamate from bipolar cells of both ON and OFF types onto kainate/AMPA/NMDA receptors on the third-order neurons, amacrine and ganglion cells, (2) release of acetylcholine from starburst amacrine cells onto amacrine, ganglion, or bipolar cells, and (3) bidirectional excitation through gap junctions between pairs of cells. There is also evidence, however, for a single type of glutamatergic amacrine cell in mouse (Johnson et al., 2004), which may play a role in interactions between the ON and OFF sublaminae (Grimes et al., 2011).

Inhibition in the IPL occurs via release of GABA or glycine from amacrine cells. The major receptors types are $GABA_A$, $GABA_C$, and glycinergic, although variation through combinations of different subunits occurs. Metabotropic $GABA_B$ receptors are also found (Rotolo & Dacheux, 2003; Shen & Slaughter, 1999; Song & Slaughter, 2010), but their location and roles have rarely been investigated.

There is a potentially wide scope for neuromodulators in the retina. The most commonly studied has been dopamine, which has D_1 and $D_{2/4}$ receptors spread throughout the IPL and which appears to be implicated at a very large number of sites and functions. Nitric oxide generators have also been found in mammalian amacrine cells, and one of the primary targets of nitric oxide, guanylyl cyclase (Haberecht et al., 1998), is nearly ubiquitous, although heavily represented in bipolar cells (Koistinaho et al., 1993). Other neuromodulators include melatonin and numerous peptides found to be contained in various amacrine cells, but whose roles are almost completely unknown. A variety of metabotropic glutamate receptors have also been localized to the IPL and presumably function as neuromodulators.

The specific response properties of the ganglion cells are mostly developed in the IPL by synaptic interactions between bipolar terminals and the dendrites of amacrine and ganglion cells. Synaptic interactions are broadly defined as being feedforward, feedback, or serial. Each of these is amply represented in the retina.

TABLE 8.1

Ganglion cell types in the rabbit retina

Name	Function	IPL Depth (figure 8.1 layers)	References
G1	Local edge detector	40–50% (7)	Rockhill et al., 2002; van Wyk, Taylor, & Vaney, 2006
G2	Unknown	50–60% (–)	Rockhill et al., 2002
G3	Vertical orientation	25% (3) 80%	Rockhill et al., 2002; Hoshi et al., 2009; Venkataramani & Taylor, 2010
G4a	OFF brisk sustained	30–50% (5,6)	Roska & Werblin, 2001; Rockhill et al., 2002; Buldyrev, Pothussery, & Taylor, 2012
G4b	ON brisk sustained	50–70% (8,9)	Roska & Werblin, 2001; Rockhill et al., 2002
G5	Unknown	20–30% (–)	Rockhill et al., 2002
G6	ipRGC M1, M2	75–85% (–)	Rockhill et al., 2002; Berson, Castrucci, & Provencio, 2010
G7	ON-OFF DS	20–30% (4) 70–80% (10)	Barlow et al., 1964; Amthor, Oyster, & Takahashi, 1984
G8	Unknown	70–80% (–)	Rockhill et al., 2002
G9	OFF delta	10–30% (3)	Roska & Werblin, 2001; Rockhill et al., 2002
G10tr	Transient ON DS	60–70% (9)	Rockhill et al., 2002; Ackert et al., 2006; Sivyer, Taylor, & Vaney, 2010; Hoshi et al., 2011
G10sust	Sustained ON DS	70–80% (10)	Buhl & Peichl, 1986; Sivyer, Taylor, & Vaney, 2010; Hoshi et al., 2011
G11a	OFF brisk transient (alpha)	30–40% (5)	Peichl et al., 1987; Roska & Werblin, 2001; Rockhill et al., 2002
G11b	ON brisk transient (alpha)	60–80% (11,12)	Peichl et al., 1987; Roska & Werblin, 2001; Rockhill et al., 2002
S-ON	Blue ON opponent	80–90% (12)	Famiglietti, 2004; Mills & Tian, 2012
Bistratified S-OFF	Blue OFF opponent APB-resistant	5–10% (2) 80–85% (12)	Mills & Tian, 2012
Inverted S-OFF	Blue OFF opponent APB-sensitive	80–90% (12)	Mills & Tian, 2012
ON bistratified	Unknown	5–10% (3) 80–85% (12)	Roska & Werblin, 2001; Hoshi et al., 2009
Unnamed	Horizontal orientation	35% (–)	Venkataramani & Taylor, 2010
Uniformity detector	Detects stimulus change	5–10% (3) 80–85% (12)	Caldwell, Daw, & Wyatt, 1978; Sivyer, Taylor, & Vaney, 2010
Transient bistratified	Unknown	30–40% (6) 60–70% (8)	Sivyer et al., 2011

These interactions are responsible for computing trigger features, orientation, motion, and directional sensitivity, and adjusting for luminance and contrast in the scene, among many other functions. Our understanding of these interactions is becoming enriched by studies examining specific patterns of interaction by GABA (Eggers & Lukasiewicz, 2011; Lukasiewicz & Shields, 1998), glycine (Molnar & Werblin, 2007), and a variety of gap junctional processing motifs (see chapters 9 by Bloomfield and Völgyi and 12 by Rieke and Chichilnisky). Gap junctions between ganglion and amacrine cells are still poorly understood, although it has been suggested that these electrical synapses contribute a slower amacrine cell component to ganglion cell synchrony and perhaps mediate long-range interactions (Olveczky et al., 2003). Exciting new findings appear certain to arise from new large-scale ultrastructural reconstructions (Anderson et al., 2011; Kleinfeld et al., 2011), including detailed cascades, novel types of connections, and new patterns of connectivity.

 STEPHEN L. MILLS AND STEPHEN C. MASSEY

The modern doctrine of retinal processing states that the point-by-point output of the photoreceptors is immediately subdivided via bipolar cells by virtue of the glutamate receptor composition of their dendrites and their intrinsic properties into about 10–12 parallel channels in mammals (more in nonmammals). In the IPL these are multiplied into ~20 channels represented by the individual ganglion cell types, although a unique 1:1 correspondence between ganglion cell types and psychophysical channels is certainly an idealization. The individual character of the bipolar cell types is further enhanced in the IPL by input from GABAergic amacrine cells. In particular, sustained inhibition is delivered via $GABA_C$ synapses, which are exclusively found on bipolar cells and, as noted, in proportions differing among bipolar cell types (Lukasiewicz & Shields, 1998).

Ultimately, each photoreceptor is thought to be sampled by at least one ganglion cell of each type, each of which projects to central structures such as the lateral geniculate of the thalamus, the superior colliculus, and other locations. Each type of ganglion cell consists of regularly spaced mosaics of that single type with relatively modest overlap and no significant unsampled space. Each portion of the mosaic is regarded as a regularly occurring single unit of a specific processing circuit or retinal hypercircuit (Werblin, 2011) that projects a unique representation of the visual scene. Each ganglion cell type tiles the retina spatially, while the various types as a group might be thought of as "tiling" the information space of the visual scene efficiently by virtue of their different representations. How these circuits and their functional representations are formed is a defining question in retinal neuroscience and occupies the remainder of this chapter.

PATTERNS OF INTERACTION IN CELL CIRCUITS

Although each ganglion cell type is thought to represent a unique processing circuit, it is likely that these circuits are comprised of recurring parts that perform similar functions within each type. For example, each ganglion cell is faced with adapting to the level of luminance and contrast in the environment. Some of this is accomplished in the outer retina, but much is not. All bipolar cell output synapses are likely to be subject to gain control, which may be a function of the reciprocal synapses found at bipolar cell dyads, as well as possibly some of the metabotropic glutamate receptors localized at these synapses. A final set of adjustments is made in the IPL.

Recent papers from the Werblin lab have enriched our understanding of some of these functional motifs by examining the pharmacology of synapses onto bipolar (Molnar & Werblin, 2007), amacrine (Hsueh, Molnar, & Werblin, 2008), and ganglion cells (Roska & Werblin, 2003) of the rabbit retina. We broadly summarize these findings in the sections below.

Inhibition in Bipolar Cells

• Amacrine cell inhibition onto bipolar cells can serve to reinforce the excitatory input (both inputs drive the membrane potential in the same direction) or to oppose (cancel) the excitatory input.

• Crossover inhibition as occurs between ON and OFF pathways is common and is glycinergic. This glycinergic inhibition serves to reinforce the excitatory light responses of the bipolar cells.

• OFF bipolar cells are not inhibited by amacrine cells driven by OFF bipolar cells.

Inhibition in Amacrine and Ganglion Cells

• As in bipolar cells, cross-laminar inhibition between ON and OFF pathways is common and is glycinergic. Again, this crossover inhibition serves to reinforce the excitatory light responses of the target cells.

• Monostratified amacrine cells tended to have no measurable inhibitory input.

• ON–OFF inhibition to amacrine cells usually arises from combined inputs of an ON glycinergic amacrine cell and an OFF GABAergic cell.

• In contrast to amacrine cells, ON–OFF inhibition to ganglion cells usually arises from wide-field GABAergic amacrine cells.

The phenomenon of crossover inhibition has only recently been established (Manookin et al., 2008; Molnar et al., 2009; Münch et al., 2009; Murphy & Rieke, 2008). This occurs when OFF or ON neurons are inhibited by glycinergic input arising from the opposite sublamina. The AII amacrine cell is known to participate in this sort of interaction at photopic levels of light, thus conferring a distinctly different function from its role at scotopic light levels as the high-gain rod pathway in the mammalian retina.

ESTABLISHING THE UNIQUE PROPERTIES OF GANGLION CELLS

A major focus in neuroscience is to identify all the classes of participating neurons. In the retina the analysis of cell types is almost complete with perhaps more progress than in other regions of the CNS. Now the goal is to identify the functions of individual ganglion cell pathways and to understand how these parallel pathways are constructed. This is a daunting task, but there have been some successes to guide us. The understanding of how alpha ganglion cells encode contrast sensitivity has progressed from early studies (Enroth-Cugell & Robson, 1966) to some new approaches that have aided in identifying the subunits (Hochstein & Shapley, 1976) with individual bipolar cells (Beaudoin, Borghuis, & Demb, 2007). Some of the unusual computations of local edge detectors have recently been established (Russell & Werblin, 2010; van Wyk, Taylor, & Vaney, 2006; Zeck, Xiao, & Masland, 2005). Recent work from the lab of Rowland Taylor has identified individual circuit elements and computations in OFF orientation-sensitive ganglion cells (Venkataramani & Taylor, 2010), a transient bistratified ganglion cell (Sivyer et al., 2011), and the rabbit OFF brisk-transient ganglion cell (Buldyrev, Puthussery, & Taylor, 2012). Most notably, nearly 50 years of research into the mechanisms of DS GC have culminated in an improved understanding of this process.

DIRECTIONALLY SELECTIVE GANGLION CELLS: A MODEL SYSTEM

DS GC were first discovered nearly 50 years ago, and they have presented a challenging puzzle in neuronal circuitry ever since (Barlow, Hill, & Levick, 1964; Lee, Kim, & Zhou, 2010; for review see Vaney, Sivyer, & Taylor, 2012). Notably, ON-OFF and ON DS GC are nearly all morphologically symmetrical, and the null/preferred axis cannot be predicted from the shape of the cell. Retrospectively, the DS GC have a unique stratification pattern in the IPL that has provided direct clues to the mechanism of directional selectivity. Namely, ON-OFF DS GC are bistratified, exactly coinciding with the two cholinergic bands (Amthor, Oyster, & Takahashi, 1984). This is completely consistent with physiological recordings that show ON/OFF DS cells to be the most sensitive to cholinergic drugs (Kittila & Massey, 1997; Masland & Ames, 1976). In other words, the stratification pattern in the IPL suggests that cholinergic amacrine cells, also known as starburst amacrine cells on account of their unique morphology, play an important role in the mechanism of DS response.

DS responses were blocked by GABA antagonists (Caldwell, Daw, & Wyatt, 1978; Kittila & Massey, 1997; Wyatt & Daw, 1976), and the dependence on GABA inhibition was explained by the initially surprising result that starburst amacrine cells contain both acetylcholine and GABA. The starburst amacrine cells were identified as the source of GABA inhibition by paired recordings.

Stimulating a nearby starburst amacrine cell on the null side generated a direct inhibitory response in ON-OFF DS GC (Fried, Münch, & Werblin, 2005; Wei et al., 2011). Two-photon calcium imaging was used to show that the individual dendrites generate directional responses as a result of their intrinsic properties and the distribution of voltage dependent channels (Euler, Detwiler, & Denk, 2002; Hausselt et al., 2007). A chloride gradient may also contribute to the directional asymmetry (Gavrikov et al., 2006). Thus, the response to centrifugal movements away from the soma was bigger than that to centripetal movements toward the soma. Finally, in a large-scale three-dimensional reconstruction, specific connections were demonstrated such that individual starburst amacrine dendrites made presumed synaptic contacts with ON-OFF DS GC with anti-parallel null axes (Briggman, Helmstaedter, & Denk, 2011). There are still some puzzling features that remain unexplained, such as the contribution of ACh, which is apparently not required for DS responses. Nevertheless, the sum of this work shows that DS responses depend on the asymmetric connections with starburst amacrine cells, which provide a GABAergic null inhibition.

Classically, two types of DS GC were recognized: the bistratified ON-OFF DS GC and the ON DS GC. ON/OFF DS GC respond to one of the four cardinal directions over a broad range of velocities. ON DS cells have three null/preferred axes and respond to much slower stimuli (Oyster & Barlow, 1967). Both cell types show the same pharmacology, and both are stratified within the cholinergic bands (Sun et al., 2006). However, lately, the DS GC types have been proliferating. This is partly because of the availability of mouse lines in which specific DS ganglion cells are genetically labeled. Also, ON DS GC have been found to consist of distinct transient and sustained types, which can be differentiated on both morphological and physiological grounds (Sivyer & Vaney, 2010; Hoshi et al., 2011; see below).

In addition, there are two more DS GC types that raise interesting questions about stratification in the IPL. The genetically labeled JAM-B GC type of the mouse has distinctly asymmetric dendrites and produces OFF DS responses. It seems to be the homologue of the rabbit G3, and it has been referred to as an OFF DS cell (Kim et al., 2008) or, in the rabbit retina, an orientation-selective GC (Venkataramani & Taylor, 2010). However, it is stratified in sublamina 3, distinctly above cholinergic a. If starburst amacrine cells are the source of asymmetric DS signals, how can the JAM-B cell be DS? It may result from the morphological asymmetry of JAM-B GC. The range of DS responses is severely limited; it produces DS responses over only a narrow range of velocities, as if a particular speed is required so that coincidental excitation and inhibition from the asymmetric dendritic field cancel. In the rabbit retina, G3 is also dye coupled to an asymmetric amacrine cell, whose dendrites all project dorsally (Hoshi & Mills, 2009). Indeed, a few aberrant JAM-B cells that were not asymmetric did not produce DS responses. The Hb9 strain of ON/OFF DS GC in the mouse retina was also asymmetric, and its DS responses were only partly reduced by GABA antagonists (Trenholm et al., 2011). Thus, there are several mechanisms that may contribute to the generation of DS responses in the retina.

Most recently, a second type of ON DS GC has also been reported. The classic ON DS GC, which ramifies in cholinergic b, has relatively sustained responses, a space-filling retroflexive dendritic tree, and is not dye coupled (Buhl & Peichl, 1986). In contrast, the transient ON DS cell has a smaller, less-branched structure, is very well dye coupled (Kanjhan & Sivyer, 2010), and has a poor response to nicotine (Hoshi et al., 2011). The transient ON DS cell is found just above cholinergic b. So again, this raises the question, how are DS responses generated if the transient ON DS cell does not receive input from starburst amacrine cells? The trigger features, orientation sensitivity in G3 (JAM-B-like) ganglion cells, and directional selectivity in the transient ON DS ganglion cells are GABA dependent, as in the traditional ON-OFF and sustained ON DS ganglion cells, but unlike these latter cells, the former cells do not ramify directly in the cholinergic bands and show little cholinergic influence, so it appears that their GABAergic input may arise from different amacrine cell types.

DEVELOPMENT OF STRATIFICATION

The specificity of stratification in the IPL, especially for different ganglion cell types, is remarkable. And, inasmuch as stratification acts as an addressing system by which appropriate contacts are established, it plays a crucial role in development. Recently, there has been progress in understanding the molecular mechanisms and developmental processes that lead to lamination. Some recent highlights follow here, but the reader is referred to the chapters on development (part XIII in this volume) for discussion in greater depth.

Retinal ganglion cells first stratify diffusely throughout the IPL but are then pruned to produce laminar specificity, probably in a way regulated by synaptic activity (Bodnarenko & Chalupa, 1993; Bodnarenko, Jeyarasasingam, & Chalupa, 1995). The use of subtype-specific markers in mouse retina revealed that different ganglion cell types employ a variety of strategies to achieve their laminar patterns. One subtype targets the correct sublamina from an early stage, one is first patterned diffusely and then remodeled, and two more develop stepwise (Kim et al., 2010). These different ganglion cell types also showed distinct central connections. In Rachel Wong's lab, in vivo imaging of the developmental sequence in live zebrafish shows that ganglion cells establish distinct laminar patterns in sequence from the inner to the outer part of the IPL. In fact, ganglion cell dendrites appeared to target pre-existing amacrine cell networks that were already laminated (Mumm et al., 2006). Developing amacrine cells apparently recognize sublamina-specific cues in the nascent IPL (Godinho et al., 2005).

Bipolar cell development has also been visualized. GFP or MYFP (monomeric yellow fluorescent protein) driven by either by the promoter for Grm6 (which encodes mGluR6) or nyx (which codes for nyctalopin, an accessory protein in the mGluR6 cascade) was expressed in ON bipolar cells. Live imaging showed that bipolar cells extend filopodial structures throughout the IPL, but subsequently, they became concentrated in specific laminae where axon terminals formed (Morgan et al., 2006; Schroeter, Wong, & Gregg, 2006). This suggests the presence of lamination cues for successful outgrowth and retraction in their absence. Three different ON bipolar cell types were followed during the time that synaptic connections with ganglion cells were established in the mouse retina. All three types targeted the appropriate sublamina and made equal numbers of close potential contacts with an ON GC type labeled biolistically. In the presence of synaptically released glutamate, one bipolar cell type converted the close contacts to multisynaptic sites (Morgan et al., 2011). Transmitter release was not required for another bipolar cell to maintain a moderate number of synaptic contacts or for the third bipolar cell type to withdraw contact forming no synapses. This last example might be expected for rod bipolar cells, which have their primary output to AII amacrine cells. Thus, the correct pattern of connections is made following lamination by the cell-type– and transmitter-dependent conversion of close appositions to synapses coupled with independent mechanisms that govern lamination and the withdrawal of inappropriate contacts (Morgan et al., 2011).

Some of the molecular cues for lamination have now been identified. A family of four closely related members of the immunoglobulin superfamily (Dscam, DscamL, Sidekick 1, and Sidekick 2) are differentially expressed by many cell types in the inner retina. Many cell types expressing the same gene target the same sublamina. In addition, reducing the expression of an adhesion molecule disrupts the stratification of certain cell types, changing them to a diffuse pattern. Finally, the expression of a different adhesion molecule changes the lamination, effectively forcing a change of address (for review see Sanes & Yamagata, 2009).

The transmembrane semaphorin Sema 6A and its receptor, Plexin A4, are also expressed in a complementary fashion in the ON and OFF layers of the IPL, respectively. Null mutants of either protein disrupt the lamination patterns of dopaminergic amacrine cells, ipRGC, and calbindin-positive amacrine cells in the IPL. All three markers, but especially the dopaminergic processes, contribute errant processes to sublamina 5 in these mutant lines. In contrast, the lamination of calretinin-positive cells, including the cholinergic amacrine cell, appears to be unchanged (Matsuoka et al., 2011b). Sema5A and Sema5B, with their respective receptors Plexin A1 and Plexin A3, play similar complementary roles in lamination (Matsuoka et al., 2011a).

In summary, this brief introduction has shown that multiple mechanisms, including molecular cues and transmitter-dependent maturation, are required to establish synaptic connectivity in the IPL. The correct stratification plays an important early role by bringing together potential synaptic partners. Thus, laminar specificity promotes synaptic specificity during the developmental sequence.

SUMMARY

The vertebrate retina with its many parallel channels is a model system for studying synaptic interactions and systems analysis. The IPL is the site of synaptic interactions among bipolar cells, amacrine cells, and ganglion cells that shape and differentiate the responses of ganglion cell output. The IPL is functionally stratified according to the polarity of bipolar cell inputs. Adjacent to the INL, there is a narrow ON layer contributed by axonal ribbons or ectopic synapses, followed by sublamina a proper, where OFF bipolar cells terminate, and sublamina b, where ON bipolar cells ramify. There are at least 13 visible layers in the IPL, primarily divided by the stratification of well-known cell types. The cholinergic bands are the primary source of directionally selective signals, but morphological asymmetry may also contribute.

The correlation of ganglion cell physiology with morphology is slowly revealing more and more ganglion cell types. A major handicap to further understanding of the IPL is the lack of markers and methods to target individual amacrine and ganglion cell types in a reliable way. This is beginning to change as new imaging methods and transgenic animals become widely available. The ability to make genetic libraries of different cell types holds great promise in the future. With the advent of new techniques such as targeted fluorescent markers, serial block-face electron microscopy, transsynaptic viral tracing (Denk, Briggman, & Helmstaedter, 2012; Viney et al., 2007), and large-scale loading of retinal cells with activity markers (Briggman & Euler, 2011), perhaps these methods will eventually tell us how many ganglion cell types contribute to the process we know as vision and how these circuits are constructed.

ACKNOWLEDGMENTS

The authors are supported by NIH grants EY 06515 to SCM, EY 10121 to SLM and Vision Core Grant EY10608, and by an unrestricted award from Research to Prevent Blindness to the Department of Ophthalmology and Visual Science, University of Texas at Houston Health Science Center. SCM is the Elizabeth Morford Professor of Ophthalmology and SLM is the John P. McGovern Distinguished Chair in Ophthalmology at UTH-HSC.

The authors regret that due to space limitations, the outstanding work of many investigators could not be included or referenced in this brief review. Also, the necessity for numerous color channels in figure 8.1 prevented its presentation in a format suitable for color-blind readers, for which we apologize.

REFERENCES

Ackert, J. M., Wu, S. H., Lee, J. C., Abrams, J., Hu, E. H., Perlman, I., et al. (2006). Light-induced changes in spike synchronization between coupled ON direction selective ganglion cells in the mammalian retina. *Journal of Neuroscience, 26,* 4206–4215.

Amthor, F. R., Oyster, C. W., & Takahashi, E. S. (1984). Morphology of on-off direction-selective ganglion cells in the rabbit retina. *Brain Research, 298,* 187–190.

Anderson, J. R., Jones, B. W., Watt, C. B., Shaw, M. V., Yang, J. H., Demill, D., et al. (2011). Exploring the retinal connectome. *Molecular Vision, 17,* 355–379.

Awatramani, G. B., & Slaughter, M. M. (2000). Origin of transient and sustained responses in ganglion cells of the retina. *Journal of Neuroscience, 20,* 7087–7095.

Barlow, H. B., Hill, R. M., & Levick, W. R. (1964). Retinal ganglion cells responding selectively to direction and speed of image motion in the rabbit. *Journal of Physiology, 173,* 377–407.

Beaudoin, D. L., Borghuis, B. G., & Demb, J. B. (2007). Cellular basis for contrast gain control over the receptive field center of mammalian retinal ganglion cells. *Journal of Neuroscience, 27,* 2636–2645.

Berson, D. M., Castrucci, A. M., & Provencio, I. (2010). Morphology and mosaics of melanopsin-expressing retinal ganglion cell types in mice. *Journal of Comparative Neurology, 518,* 405–422.

Bodnarenko, S. R., & Chalupa, L. M. (1993). Stratification of ON and OFF ganglion cell dendrites depends on glutamate-mediated afferent activity in the developing retina. *Nature, 364,* 144–146.

Bodnarenko, S. R., Jeyarasasingam, G., & Chalupa, L. M. (1995). Development and regulation of dendritic stratification in retinal ganglion cells by glutamate-mediated afferent activity. *Journal of Neuroscience, 15,* 7037–7045.

Briggman, K. L., & Euler, T. (2011). Bulk electroporation and population calcium imaging in the adult mammalian retina. *Journal of Neurophysiology, 105,* 2601–2609.

Briggman, K. L., Helmstaedter, M., & Denk, W. (2011). Wiring specificity in the direction-selectivity circuit of the retina. *Nature, 471,* 183–188.

Buhl, E. H., & Peichl, L. (1986). Morphology of rabbit retinal ganglion cells projecting to the medial terminal nucleus of the accessory optic system. *Journal of Comparative Neurology, 253,* 163–174.

Buldyrev, I., Puthussery, T., & Taylor, W. R. (2012). Synaptic pathways that shape the excitatory drive in an OFF retinal ganglion cell. *Journal of Neurophysiology, 107,* 1795–1807.

Caldwell, J. H., Daw, N. W., & Wyatt, H. J. (1978). Effects of picrotoxin and strychnine on rabbit retinal ganglion cells: Lateral interactions for cells with more complex receptive fields. *Journal of Physiology, 276,* 277–298.

Chiao, C. C., & Masland, R. H. (2003). Contextual tuning of direction-selective retinal ganglion cells. *Nature Neuroscience, 6,* 1251–1252.

Contini, M., Lin, B., Kobayashi, K., Okano, H., Masland, R. H., & Raviola, E. (2010). Synaptic input of ON-bipolar cells onto the dopaminergic neurons of the mouse retina. *Journal of Comparative Neurology, 518,* 2035–2050.

Dacey, D. M., Liao, H. W., Peterson, B. B., Robinson, F. R., Smith, V. C., Pokorny, J., et al. (2005). Melanopsin-expressing ganglion cells in primate retina signal colour and irradiance and project to the LGN. *Nature, 433,* 749–754.

Denk, W., Briggman, K. L., & Helmstaedter, M. (2012). Structural neurobiology: Missing link to a mechanistic understanding of neural computation. *Nature Reviews Neuroscience.* February 22, epub ahead of print. doi: 10.1038/nrn3169

DeVries, S. H. (2000). Bipolar cells use kainate and AMPA receptors to filter visual information into separate channels. *Neuron, 28,* 847–856.

Dowling, J. E., & Boycott, B. B. (1965). Neural connections of the retina: Fine structure of the inner plexiform layer. *Cold Spring Harbor Symposia on Quantitative Biology, 30,* 393–402.

Dumitrescu, O. N., Pucci, F. G., Wong, K. Y., & Berson, D. M. (2009). Ectopic retinal ON bipolar cell synapses in the OFF inner plexiform layer: Contacts with dopaminergic amacrine cells and melanopsin ganglion cells. *Journal of Comparative Neurology, 517,* 226–244.

Eggers, E. D., & Lukasiewicz, P. D. (2011). Multiple pathways of inhibition shape bipolar cell responses in the retina. *Visual Neuroscience, 28,* 95–108.

Enroth-Cugell, C., & Robson, J. G. (1966). The contrast sensitivity of retinal ganglion cells of the cat. *Journal of Physiology, 187,* 517–552.

Euler, T., Detwiler, P. B., & Denk, W. (2002). Directionally selective calcium signals in dendrites of starburst amacrine cells. *Nature, 418,* 845–852.

Famiglietti, E. V. (1992). Polyaxonal amacrine cells of rabbit retina: Morphology and stratification of PA1 cells. *Journal of Comparative Neurology, 316,* 391–405.

Famiglietti, E. V. (2004). Class I and class II ganglion cells of rabbit retina: A structural basis for X and Y (brisk) cells. *Journal of Comparative Neurology, 478,* 323–346.

Famiglietti, E. V., Kaneko, A., & Tachibana, M. (1977). Neuronal architecture of on and off pathways to ganglion cells in carp retina. *Science, 198,* 1267–1269.

Freed, M. A. (2000). Parallel cone bipolar pathways to a ganglion cell use different rates and amplitudes of quantal excitation. *Journal of Neuroscience, 20,* 3956–3963.

Fried, S. I., Münch, T. A., & Werblin, F. S. (2005). Directional selectivity is formed at multiple levels by laterally offset inhibition in the rabbit retina. *Neuron, 46,* 117–127.

Fritzsch, B., & Collin, S. P. (1990). Dendritic distribution of two populations of ganglion cells and the retinopetal fibers in the retina of the silver lamprey (*Ichthyomyzon unicuspis*). *Visual Neuroscience, 4,* 533–545.

Gavrikov, K. E., Nilson, J. E., Dmitriev, A. V., Zucker, C. L., & Mangel, S. C. (2006). Dendritic compartmentalization of chloride cotransporters underlies directional responses of starburst amacrine cells in retina. *Proceedings of the National Academy of Sciences of the United States of America, 103,* 18793–18798.

Godinho, L., Mumm, J. S., Williams, P. R., Schroeter, E. H., Koerber, A., Park, S. W., et al. (2005). Targeting of amacrine cell neurites to appropriate synaptic laminae in the developing zebrafish retina. *Development, 132,* 5069–5079.

Grimes, W. N., Seal, R. P., Oesch, N., Edwards, R. H., & Diamond, J. S. (2011). Genetic targeting and physiological features of VGLUT3+ amacrine cells. *Visual Neuroscience, 28,* 381–392.

Grimes, W. N., Zhang, J., Graydon, C. W., Kachar, B., & Diamond, J. S. (2010). Retinal parallel processors: More than 100 independent microcircuits operate within a single interneuron. *Neuron, 65,* 873–885.

Haberecht, M. F., Schmidt, H. H., Mills, S. L., Massey, S. C., Nakane, M., & Redburn-Johnson, D. A. (1998). Localization of nitric oxide synthase, NADPH diaphorase and soluble guanylyl cyclase in adult rabbit retina. *Visual Neuroscience, 15,* 881–890.

Hartveit, E. (1997). Functional organization of cone bipolar cells in the rat retina. *Journal of Neurophysiology, 77,* 1716–1730.

Hausselt, S. E., Euler, T., Detwiler, P. B., & Denk, W. (2007). A dendrite-autonomous mechanism for direction selectivity in retinal starburst amacrine cells. *PLoS Biology, 5,* e185. doi:10.1371/journal.pbio.0050185.

Hochstein, S., & Shapley, R. M. (1976). Linear and nonlinear spatial subunits in Y cat retinal ganglion cells. *Journal of Physiology, 262,* 265–284.

Hokoç, J. N., & Mariani, A. P. (1988). Synapses from bipolar cells onto dopaminergic amacrine cells in cat and rabbit retinas. *Brain Research, 461,* 17–26.

Hoshi, H., Liu, W. L., Massey, S. C., & Mills, S. L. (2009). ON inputs to the OFF layer: bipolar cells that break the stratification rules of the retina. *Journal of Neuroscience, 29,* 8875–8883.

Hoshi, H., & Mills, S. L. (2009). Components and properties of the G3 ganglion cell circuit in the rabbit retina. *Journal of Comparative Neurology, 513,* 69–82.

Hoshi, H., Tian, L. M., Massey, S. C., & Mills, S. L. (2011). Two distinct types of ON directionally-selective ganglion cells in the rabbit retina. *Journal of Comparative Neurology, 519*, 2509–2521.

Hosoya, T., Baccus, S. A., & Meister, M. (2005). Dynamic predictive coding by the retina. *Nature, 436*, 71–77.

Hsueh, H. A., Molnar, A., & Werblin, F. S. (2008). Amacrine-to-amacrine cell inhibition in the rabbit retina. *Journal of Neurophysiology, 100*, 2077–2088.

Johnson, J., Sherry, D. M., Liu, X., Fremeau, R. T., Jr., Seal, R. P., Edwards, R. H., et al. (2004). Vesicular glutamate transporter 3 expression identifies glutamatergic amacrine cells in the rodent retina. *Journal of Comparative Neurology, 477*, 386–398.

Kanjhan, R., & Sivyer, B. (2010). Two types of ON direction-selective ganglion cells in rabbit retina. *Neuroscience Letters, 483*, 105–109.

Kim, I. J., Zhang, Y., Meister, M., & Sanes, J. R. (2010). Laminar restriction of retinal ganglion cell dendrites and axons: Subtype-specific developmental patterns revealed with transgenic markers. *Journal of Neuroscience, 30*, 1452–1462.

Kim, I. J., Zhang, Y., Yamagata, M., Meister, M., & Sanes, J. R. (2008). Molecular identification of a retinal cell type that responds to upward motion. *Nature, 452*, 478–482.

Kittila, C. A., & Massey, S. C. (1997). Pharmacology of directionally selective ganglion cells in the rabbit retina. *Journal of Neurophysiology, 77*, 675–689.

Kleinfeld, D., Bharioke, A., Blinder, P., Bock, D. D., Briggman, K. L., Chklovskii, D. B., et al. (2011). Large-scale automated histology in the pursuit of connectomes. *Journal of Neuroscience, 31*, 16125–16138.

Koistinaho, J., Swanson, R. A., de Vente, J., & Sagar, S. M. (1993). NADPH-diaphorase (nitric oxide synthase)-reactive amacrine cells of rabbit retina: Putative target cells and stimulation by light. *Neuroscience, 57*, 587–597.

Kolb, H. (1979). The inner plexiform layer in the retina of the cat: Electron microscopic observations. *Journal of Neurocytology, 8*, 295–329.

Koontz, M. A., & Hendrickson, A. E. (1987). Stratified distribution of synapses in the inner plexiform layer of primate retina. *Journal of Comparative Neurology, 263*, 581–592.

Lacalli, T. C. (2004). Sensory systems in amphioxus: A window on the ancestral chordate condition. *Brain, Behavior and Evolution, 64*, 148–162.

Lee, S., Kim, K., & Zhou, Z. J. (2010). Role of ACh-GABA cotransmission in detecting image motion and motion direction. *Neuron, 68*, 1159–1172.

Light, A. C., Zhu, Y., Shi, J., Saszik, S., Lindstrom, S., Davidson, L., et al. (in press). Organizational motifs for ground squirrel cone bipolar cells. *Journal of Comparative Neurology*.

Liu, P. C., & Chiao, C. C. (2007). Morphologic identification of the OFF-type blue cone bipolar cell in the rabbit retina. *Investigative Ophthalmology & Visual Science, 48*, 3388–3395.

Lukasiewicz, P. D., & Shields, C. R. (1998). Different combinations of GABAA and GABAC receptors confer distinct temporal properties to retinal synaptic responses. *Journal of Neurophysiology, 79*, 3157–3167.

MacNeil, M. A., Heussy, J. K., Dacheux, R. F., Raviola, E., & Masland, R. H. (2004). The population of bipolar cells in the rabbit retina. *Journal of Comparative Neurology, 472*, 73–86.

Manookin, M. B., Beaudoin, D. L., Ernst, Z. R., Flagel, L. J., & Demb, J. B. (2008). Disinhibition combines with excitation to extend the operating range of the OFF visual pathway in daylight. *Journal of Neuroscience, 28*, 4136–4150.

Masland, R. H., & Ames, A., III. (1976). Responses to acetylcholine of ganglion cells in an isolated mammalian retina. *Journal of Neurophysiology, 39*, 1220–1235.

Massey, S. C., & Mills, S. L. (1996). A calbindin-immunoreactive bipolar cell type in the rabbit retina. *Journal of Comparative Neurology, 366*, 15–33.

Massey, S. C., Mills, S. L., & Marc, R. E. (1992). All indoleamine-accumulating cells in the rabbit retina contain GABA. *Journal of Comparative Neurology, 322*, 275–291.

Matsuoka, R. L., Chivatakarn, O., Badea, T. C., Samuels, I. S., Cahill, H., Katayama, K., et al. (2011a). Class 5 transmembrane semaphorins control selective mammalian retinal lamination and function. *Neuron, 71*, 460–473.

Matsuoka, R. L., Nguyen-Ba-Charvet, K. T., Parray, A., Badea, T. C., Chédotal, A., & Kolodkin, A. L. (2011b). Transmembrane semaphorin signalling controls laminar stratification in the mammalian retina. *Nature, 470*, 259–263.

McGillem, G. S., & Dacheux, R. F. (2001). Rabbit cone bipolar cells: Correlation of their morphologies with whole-cell recordings. *Visual Neuroscience, 18*, 675–685.

Miller, R. F., & Bloomfield, S. A. (1986). Electroanatomy of a unique amacrine cell in the rabbit retina. *Proceedings of the National Academy of Sciences of the United States of America, 80*, 3069–3073.

Mills, S. L., & Tian, M. L. (2012). The morphology and physiology of color-opponent ganglion cells in the rabbit retina. ARVO E-abstract 6915.

Molnar, A., Hsueh, H. A., Roska, B., & Werblin, F. S. (2009). Crossover inhibition in the retina: Circuitry that compensates for nonlinear rectifying synaptic transmission. *Journal of Computational Neuroscience, 27*, 569–590.

Molnar, A., & Werblin, F. (2007). Inhibitory feedback shapes bipolar cell responses in the rabbit retina. *Journal of Neurophysiology, 98*, 3423–3435.

Morgan, J. L., Dhingra, A., Vardi, N., & Wong, R. O. (2006). Axons and dendrites originate from neuroepithelial-like processes of retinal bipolar cells. *Nature Neuroscience, 9*, 85–92.

Morgan, J. L., Soto, F., Wong, R. O., & Kerschensteiner, D. (2011). Development of cell type-specific connectivity patterns of converging excitatory axons in the retina. *Neuron, 71*, 1014–1021.

Mumm, J. S., Williams, P. R., Godinho, L., Koerber, A., Pittman, A. J., Roeser, T., et al. (2006). In vivo imaging reveals dendritic targeting of laminated afferents by zebrafish retinal ganglion cells. *Neuron, 52*, 609–621.

Münch, T. A., da Silveira, R. A., Siegert, S., Viney, T. J., Awatramani, G. B., & Roska, B. (2009). Approach sensitivity in the retina processed by a multifunctional neural circuit. *Nature Neuroscience, 12*, 1308–1316.

Murphy, G. J., & Rieke, F. (2008). Signals and noise in an inhibitory interneuron diverge to control activity in nearby retinal ganglion cells. *Nature Neuroscience, 11*, 318–326.

Nelson, R., Famiglietti, E. V., & Kolb, H. (1978). Intracellular staining reveals different levels of stratification for on- and off-center ganglion cells in cat retina. *Journal of Neurophysiology, 41*, 472–483.

Nomura, A., Shigemoto, R., Nakamura, Y., Okamoto, N., Mizuno, N., & Nakanishi, S. (1994). Developmentally regulated postsynaptic localization of a metabotropic glutamate receptor in rat rod bipolar cells. *Cell, 77,* 361–369.

Olveczky, B. P., Baccus, S. A., & Meister, M. (2003). Segregation of object and background motion in the retina. *Nature, 423,* 401–408.

Oyster, C. W., & Barlow, H. B. (1967). Direction-selective units in rabbit retina: Distribution of preferred directions. *Science, 155,* 841–842.

Peichl, L., Ott, H., & Boycott, B. B. (1987). Alpha ganglion cells in mammalian retinae. *Proceedings of the Royal Society of London. Series B, Biological Sciences, 231,* 169–197.

Rivlin-Etzion, M., Zhou, K., Wei, W., Elstrott, J., Nguyen, P. L., Barres, B. A., et al. (2011). Transgenic mice reveal unexpected diversity of on-off direction-selective retinal ganglion cell subtypes and brain structures involved in motion processing. *Journal of Neuroscience, 31,* 8760–8769.

Rockhill, R. L., Daly, F. J., MacNeil, M. A., Brown, S. P., & Masland, R. H. (2002). The diversity of ganglion cells in a mammalian retina. *Journal of Neuroscience, 22,* 3831–3843.

Rodieck, R. W. (1973). *The vertebrate retina.* Sunderland, MA W. H. Freeman and Co.

Roska, B., Molnar, A., & Werblin, F. S. (2006). Parallel processing in retinal ganglion cells: How integration of space-time patterns of excitation and inhibition form the spiking output. *Journal of Neurophysiology, 95,* 3810–3822.

Roska, B., & Werblin, F. (2001). Vertical interactions across ten parallel, stacked representations in the mammalian retina. *Nature, 410,* 583–587.

Roska, B., & Werblin, F. (2003). Rapid global shifts in natural scenes block spiking in specific ganglion cell types. *Nature Neuroscience, 6,* 600–608.

Rotolo, T. C., & Dacheux, R. F. (2003). Evidence for glycine, GABAA, and GABAB receptors on rabbit OFF-alpha ganglion cells. *Visual Neuroscience, 20,* 285–296.

Russell, T. L., & Werblin, F. S. (2010). Retinal synaptic pathways underlying the response of the rabbit local edge detector. *Journal of Neurophysiology, 103,* 2757–2769.

Sandell, J. H., & Masland, R. H. (1986). A system of indoleamine-accumulating neurons in the rabbit retina. *Journal of Neuroscience, 6,* 3331–3347.

Sanes, J. R., & Yamagata, M. (2009). Many paths to synaptic specificity. *Annual Review of Cellular and Developmental Biology, 25,* 161–195.

Schiller, P. H., Sandell, J. H., & Maunsell, J. H. (1986). Functions of the ON and OFF channels of the visual system. *Nature, 322,* 824–825.

Schmidt, T. M., & Kofuji, P. (2010). Differential cone pathway influence on intrinsically photosensitive retinal ganglion cell subtypes. *Journal of Neuroscience, 30,* 16262–16271.

Schroeter, E. H., Wong, R. O., & Gregg, R. G. (2006). In vivo development of retinal ON-bipolar cell axonal terminals visualized in nyx:MYFP transgenic zebrafish. *Visual Neuroscience, 23,* 833–843.

Shen, W., & Slaughter, M. M. (1999). Metabotropic GABA receptors facilitate L-type and inhibit N-type calcium channels in single salamander retinal neurons. *Journal of Physiology, 516,* 711–718.

Siegert, S., Cabuy, E., Scherf, B. G., Kohler, H., Panda, S., Le, Y. Z., et al. (2012). Transcriptional code and disease map for adult retinal cell types. *Nature Neuroscience, 15,* 487–495.

Siegert, S., Scherf, B. G., Del Punta, K., Didkovsky, N., Heintz, N., & Roska, B. (2009). Genetic address book for retinal cell types. *Nature Neuroscience, 12,* 1197–1204.

Sivyer, B., Taylor, W. R., & Vaney, D. I. (2010). Uniformity detector retinal ganglion cells fire complex spikes and receive only light-evoked inhibition. *Proceedings of the National Academy of Sciences of the United States of America, 107,* 5628–5633.

Sivyer, B., & Vaney, D. I. (2010). Dendritic morphology and tracer-coupling pattern of physiologically identified transient uniformity detector ganglion cells in rabbit retina. *Visual Neuroscience, 27,* 159–170.

Sivyer, B., Venkataramani, S., Taylor, W. R., & Vaney, D. I. (2011). A novel type of complex ganglion cell in rabbit retina. *Journal of Comparative Neurology, 519,* 3128–3138.

Slaughter, M. M., & Miller, R. F. (1981). 2-Amino-4-phosphonobutyric acid: A new pharmacological tool for retina research. *Science, 9*(211), 182–185.

Song, Y., & Slaughter, M. M. (2010). GABA(B) receptor feedback regulation of bipolar cell transmitter release. *Journal of Physiology, 588,* 4937–4949.

Sterling, P., Freed, M. A., & Smith, R. G. (1988). Architecture of rod and cone circuits to the ON-beta ganglion cell. *Journal of Neuroscience, 8,* 623–642.

Strettoi, E., & Masland, R. H. (1996). The number of unidentified amacrine cells in the mammalian retina. *Proceedings of the National Academy of Sciences of the United States of America, 93,* 14906–14911.

Sun, W., Deng, Q., Levick, W. R., & He, S. (2006). ON direction-selective ganglion cells in the mouse retina. *Journal of Physiology, 576,* 197–202.

Tauchi, M., & Masland, R. H. (1984). The shape and arrangement of the cholinergic neurons in the rabbit retina. *Proceedings of the Royal Society of London. Series B, Biological Sciences, 223,* 101–119.

Trenholm, S., Johnson, K., Li, X., Smith, R. G., & Awatramani, G. B. (2011). Parallel mechanisms encode direction in the retina. *Neuron, 71,* 683–694.

Tsuda, M., Horie, T., Takimoto, N., Toda, Y., Kusakabe, T., Horiguchi, H., et al. (2006). Origin of the vertebrate retina: ON–type and OFF–type photoreceptors in ascidian larva. *Investigative Ophthalmology and Visual Science, 47,* E-Abstract 2834.

Vaney, D. I. (1986). Morphological identification of serotonin-accumulating neurons in the living retina. *Science, 233,* 444–446.

Vaney, D. I., Sivyer, B., & Taylor, W. R. (2012). Direction selectivity in the retina: Symmetry and asymmetry in structure and function. *Nature Reviews. Neuroscience, 13,* 194–208.

Vaney, D. I., & Young, H. M. (1988). GABA-like immunoreactivity in NADPH-diaphorase amacrine cells of the rabbit retina. *Brain Research, 474,* 380–385.

Van Hook, M. J., Wong, K. Y., & Berson, D. M. (2012). Dopaminergic modulation of ganglion-cell photoreceptors in rat. *European Journal of Neuroscience, 35,* 507–518.

van Wyk, M., Taylor, W. R., & Vaney, D. I. (2006). Local edge detectors: A substrate for fine spatial vision at low temporal frequencies in rabbit retina. *Journal of Neuroscience, 26,* 13250–13263.

Vardi, N., & Morigiwa, K. (1997). ON cone bipolar cells in rat express the metabotropic receptor mGluR6. *Visual Neuroscience, 14,* 789–794.

Venkataramani, S., & Taylor, W. R. (2010). Orientation selectivity in rabbit retinal ganglion cells is mediated by presynaptic inhibition. *Journal of Neuroscience, 30,* 15664–15676.

Viney, T. J., Balint, K., Hillier, D., Siegert, S., Boldogkoi, Z., Enquist, L. W., et al. (2007). Local retinal circuits of melanopsin-containing ganglion cells identified by transsynaptic viral tracing. *Current Biology, 17,* 981–988.

Wässle, H. (2004). Parallel processing in the mammalian retina. *Nature Reviews. Neuroscience, 5,* 747–757.

Wei, W., Hamby, A. M., Zhou, K., & Feller, M. B. (2011). Development of asymmetric inhibition underlying direction selectivity in the retina. *Nature, 469,* 402–406.

Werblin, F. S. (2011). The retinal hypercircuit: A repeating synaptic interactive motif underlying visual function. *Journal of Physiology, 589,* 3691–3702.

Wright, L. L., & Vaney, D. I. (2004). The type 1 polyaxonal amacrine cells of the rabbit retina: A tracer-coupling study. *Visual Neuroscience, 21,* 145–155.

Wyatt, H. J., & Daw, N. W. (1976). Specific effects of neurotransmitter antagonists on ganglion cells in rabbit retina. *Science, 191,* 204–205.

Zeck, G. M., Xiao, Q., & Masland, R. H. (2005). The spatial filtering properties of local edge detectors and brisk-sustained retinal ganglion cells. *European Journal of Neuroscience, 22,* 2016–2026.

Zhang, D. Q., Wong, K. Y., Sollars, P. J., Berson, D. M., Pickard, G. E., & McMahon, D. G. (2008). Intraretinal signaling by ganglion cell photoreceptors to dopaminergic amacrine neurons. *Proceedings of the National Academy of Sciences of the United States of America, 105,* 14181–14186.

Zhang, D. Q., Zhou, T. R., & McMahon, D. G. (2007). Functional heterogeneity of retinal dopaminergic neurons underlying their multiple roles in vision. *Journal of Neuroscience, 27,* 692–699.

Zhang, J., Li, W., Hoshi, H., Mills, S. L., & Massey, S. C. (2005). Stratification of alpha ganglion cells and ON/OFF directionally selective ganglion cells in the rabbit retina. *Visual Neuroscience, 22,* 535–549.

Zhang, J., Li, W., Trexler, E. B., & Massey, S. C. (2002). Confocal analysis of reciprocal feedback at rod bipolar terminals in the rabbit retina. *Journal of Neuroscience, 22,* 10871–10882.

9 Mind the Gap: The Functional Roles of Neuronal Gap Junctions in the Retina

STEWART A. BLOOMFIELD AND BÉLA VÖLGYI

Rapid communication between neurons is essential for the efficient propagation and integration of signals within the brain. As in other CNS loci, the major mode of neuronal communication in the retina is via chemically mediated synaptic transmission. However, it has long been known from serial reconstructions of electron micrographs that some neighboring retinal neurons form gap junctions between their closely apposed plasma membranes, suggesting a role for electrical transmission as well. In fact, coupling between retinal horizontal cells was described almost 50 years ago by Yamada and Ishikawa (1965), some 5 years before Goodenough and Revel (1970) coined the term "gap junction."

Over the past two decades there has been a proliferation of investigations of gap junctions, the morphological substrate of electrical synapses, as a result of new histological, electrophysiological, and genetic techniques. A seminal discovery was the connexin subunit structure of gap junctions, which today can be identified through immunocytochemical techniques to determine the distribution of gap junctions in the retina and brain (Söhl, Maxeiner, & Willecke, 2005). The use of mouse mutants in which selective gap junctions are disrupted by targeting connexin genes has become an essential tool to characterize the function of electrical synapses.

It is now clear that electrical synaptic transmission is a major mode of intercellular communication in the retina, in which each of the five neuronal types is coupled via gap junctions that express a number of different connexin proteins (Söhl, Maxeiner, & Willecke, 2005; Söhl & Willecke, 2003). In many cases the coupling strength appears regulated by ambient illumination or circadian rhythms acting through neuromodulators (Bloomfield, Xin, & Osborne, 1997; Ribelayga, Cao, & Mangel, 2008; Weiler et al., 2000; Xin & Bloomfield, 1999a,b, 2000). The broad distribution, subunit makeup, and regulatory pathways of gap junctions suggest that electrically coupled neurons provide plastic, reconfigurable circuits positioned to play key and diverse roles in the transmission and processing of visual information at every retinal level. In this chapter we review the distinct functional roles elucidated for neuronal gap junctions in the retina related to the encoding, propagation, and integration of visual signals.

STRUCTURE AND REGULATION OF GAP JUNCTIONS

Gap junctions are composed of two hemichannels or connexons that are linked across an extracellular space of 2–4 nm (figure 9.1A). They form a channel that communicates directly with the cytoplasm of neighboring cells, providing an intercellular pathway for diffusion of molecules up to a kilodalton. Hemichannels are composed of six subunit transmembrane proteins called connexins, approximately 20 different isoforms of which have been characterized in humans and mice (Willecke et al., 2002). Mammalian retinal neurons have been found to express a number of different connexin subunits, which, to date, include Cx30.2, Cx36, Cx45, Cx50, and Cx57 (Bloomfield & Völgyi, 2009; Müller et al., 2010; Willecke et al., 2002). No other CNS locus offers this defined diversity of gap junction structure and distribution (figure 9.2). The retina has thus become arguably the best model system to study the functional roles of gap junctions and electrical synaptic transmission in the brain.

Gap junctions are not simple static bridges between neighboring cells but show plasticity much like that of membrane ion channels. The conductance and trafficking of gap junctions are regulated by a number of physiological factors and agents. Most gap junction connexins are modified posttranslationally by phosphorylation mainly of serine amino acids found in their carboxyl terminus and intracellular loop regions (Lampe & Lau, 2000, 2004). Connexins are targeted by protein kinases, and the resulting phosphorylation has been implicated in a broad number of regulatory steps, including trafficking, assembly and disassembly, and conductance gating of gap junctions. These protein kinases are modulated by a number of factors, including calcium/calmodulin (DeVries & Schwartz, 1989;

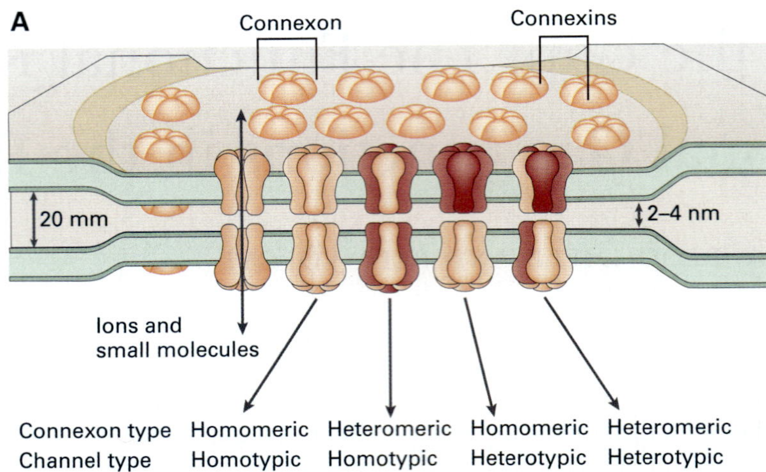

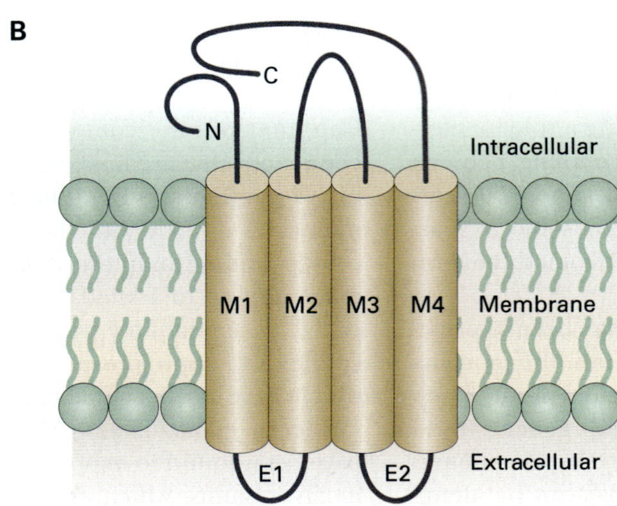

FIGURE 9.1 Structure of gap junctions. (A) Gap junctions are formed between the opposing membranes of neighboring cells. Connexons (or hemichannels) on each side dock to one another to form conductive channels between the two cells. Each connexon is comprised of six connexin protein subunits that form a central pore. Connexons can contain only one type of connexin subunit (homomeric) or a mixture of different connexins (heteromeric). Gap-junctional channels can consist of two of the same connexons (homotypic) or connexons with different subunit compositions (heterotypic). (B) Connexin subunits are proteins that have four transmembrane domains, two extracellular and one intracellular loop, and carboxy- and amino-terminal endings within the cytoplasm. Regulation of the three-dimensional connexin structure, which underlies the opening/closing of gap junction channels, is mediated at the cytoplasmic regions. (Modified from Bloomfield & Völgyi, 2009, with permission.)

Lurtz & Louis, 2007), cAMP/cGMP (Kothmann, Massey, & O'Brien, 2009; Lasater, 1987; Patel et al., 2006), pH (Chesler, 2003; González et al., 2008; Spray, Harris, & Bennett, 1981), voltage (Spray, Harris, & Bennett, 1979; Srinivas et al., 1999), and neuromodulators (Mills & Massey, 1995; Lasater & Dowling, 1985; Umino, Lee, & Dowling, 1991; Xin & Bloomfield, 2000).

In the retina, dopamine and nitric oxide are light-activated neuromodulators that are released by different amacrine cell subtypes (Koistinaho et al., 1993; Witkovsky & Dearry, 1991). Dopamine and nitric oxide activate a number of intracellular pathways involving cAMP- and cGMP-dependent protein kinases, resulting in the phosphorylation or dephosphorylation of gap junction connexins, which alter the conductance to ionic currents (DeVries & Schwartz, 1989; Kothmann, Massey, & O'Bryan, 2009; Lasater, 1987; Patel et al., 2006). This modulation varies across the population of retinal interneurons so that the conductances of some gap junctions are

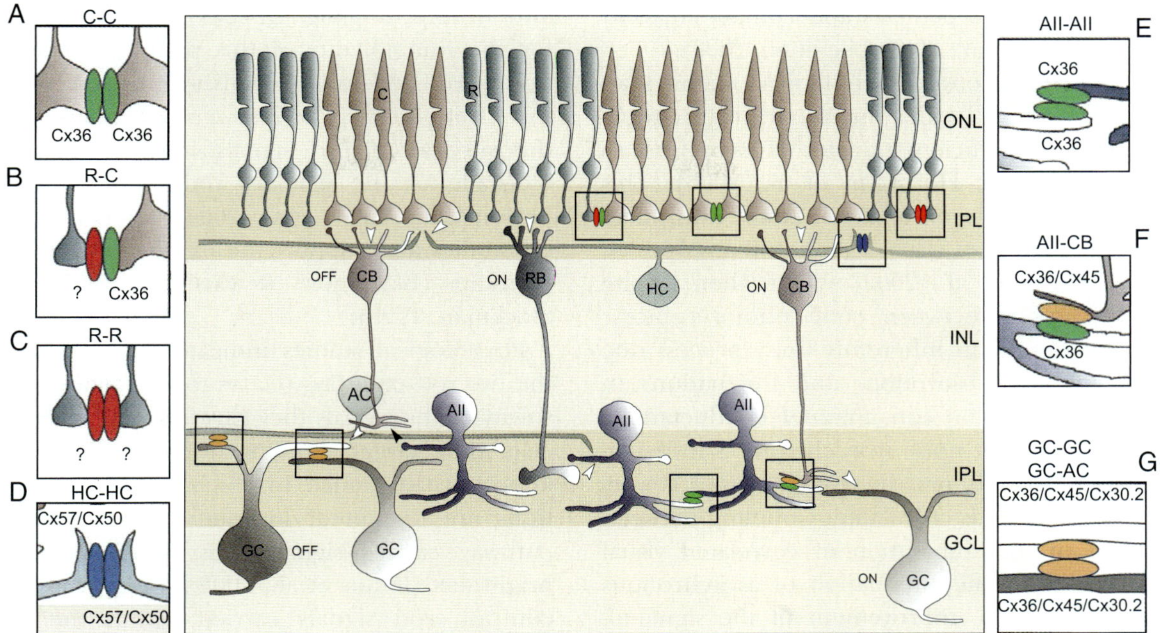

FIGURE 9.2 Gap junctions expressed by retinal neurons. This schematic shows seven examples of electrical coupling whose different functions are detailed in the text. (A) Both hemichannels of the gap junctions coupling neighboring cones express Cx36. (B) By contrast, only the hemichannel on the cone side contains Cx36, whereas the connexin expression of rods, including between rods (C), remains unknown. (D) Horizontal cell dendrites are extensively coupled. In mammals the axonless horizontal cells express Cx50, whereas the axon-bearing horizontal cells express Cx57. (E) The AII amacrine cells form two types of the gap junctions. The gap junctions between AII cells appear to be homotypic/homomeric with both hemichannels expressing Cx36. (F) By contrast, the gap junctions between AII amacrine cells and ON cone bipolar cells can be homotypic or heterotypic, with the AII cell hemichannels expressing Cx36 and the cone bipolar cell hemichannel containing either Cx36 or Cx45. (G) Ganglion cells are extensively coupled to each other and/or to neighboring amacrine cells. To date, ganglion cell gap junctions have been reported to contain Cx36, Cx45, or Cx30.2. The colored ovals depict gap junction hemichannels. The open and filled arrowheads symbolize excitatory and inhibitory chemical synapses, respectively. C, cone photoreceptor; R, rod photoreceptor; HC, horizontal cell; CB, cone bipolar cell; RB, rod bipolar cell; AII, AII amacrine cell; AC, amacrine cell; GC, ganglion cell; ONL, outer nuclear layer; OPL, outer plexiform layer; INL, inner nuclear layer; IPL, inner plexiform layer; GCL, ganglion cell layer. (Modified from Bloomfield & Völgyi, 2009, with permission.)

increased by light, whereas others are decreased (Baldridge, 2001; Hu et al., 2010; Ribelayga, Cao, & Mangel, 2008; Umino, Lee, & Dowling, 1991; Xin & Bloomfield, 1999a,b, 2000). In fact, light may increase or decrease the conductance of gap junctions within the same neuron depending on the level of brightness (Bloomfield et al., 1997; Bloomfield & Völgyi, 2004; Baldridge, Weiler, & Dowling, 1995; Xin & Bloomfield, 1999a,b). It has recently been reported that the state of phosphorylation may vary for gap junctions within a single neuron, suggesting a local regulation of electrical transmission by individual synaptic processes (Kothmann, Massey, & O'Bryan, 2009). Overall, these findings suggest that the plasticity and regulation of gap junctions are extremely complex, rivaling those shown for chemical synapses.

FUNCTIONAL ROLES OF NEURONAL GAP JUNCTIONS IN THE OUTER RETINA

Electrical Coupling between Cones Improves Signal Fidelity

Electrical coupling between cone photoreceptors was first demonstrated in the turtle retina, where a current applied extrinsically to a cone produced activity in neighboring cones within 40 μm (Baylor, Fuortes, & O'Bryan, 1971). The electrical coupling is derived from gap junctions formed between cones, a feature conserved across many species (Cohen, 1965; Kolb, 1977; Raviola & Gilula, 1973; Tsukamoto et al., 1992). Gap junctions displaying Cx36 plaques are found at the endings of fine processes called telodendria that emerge from the base of cone pedicles and contact adjacent cones (Feigenspan et al., 2004; Lee et al., 2003).

The gap junctions between neighboring mammalian cone photoreceptors have a high effective conductance, allowing for efficient transfer of visual signals (DeVries et al., 2002; Hornstein et al., 2005). This lateral interaction between adjacent cones introduces an unwanted blurring of visual signals and a reduction of acuity (DeVries et al., 2002). What, then, is the benefit of coupling between cone photoreceptors? Phototransduction is an inherently noisy process due to random photon absorptions and fluctuations in signaling molecules and ion channel conductances. Although the inherent noise in each cone is independent of that of its neighbors, the visually evoked activity of an individual cone is correlated. Coupling between cones results in the summation of correlated visual signals but an algebraic attenuation of asynchronous noise, resulting in an improvement of the signal-to-noise ratio of individual cones by nearly 80% (DeVries at al., 2002). Thus, cone coupling improves the fidelity of visual signals—but at the cost of some neural blurring of the image. However, this neural blurring appears far narrower than that produced by the optics of the eye, and so the improvement in signal-to-noise ratio of cone signals by coupling outweighs the small degradation in acuity.

Electrical Coupling between Rods and Cones Forms a Secondary Rod Pathway

In mammals rod and cone photoreceptors make chemical synapses with different bipolar cells, segregating their signals into parallel retinal streams (Boycott & Kolb, 1973; Ghosh et al., 2004). Although only a single subtype of rod bipolar cell exists, multiple pathways transmit rod signals across the retina (Bloomfield & Dacheux, 2001). In the primary rod pathway, rod photoreceptors pass signals down to rod bipolar cells, which carry them radially to the inner retina where they make synaptic contacts with AII amacrine cells. The AII cells, in turn, form sign-inverting chemical synapses with OFF cone bipolar cell axons and sign-conserving electrical synapses with ON cone bipolar cell axons. Thus, the primary rod pathway "piggybacks" onto the cone circuitry in the inner retina so that its signals can reach the output ganglion cells.

In addition, gap junctions exist between the axon terminals of rod and cone photoreceptors (Raviola & Gilula, 1975). This rod–cone coupling provides an alternative, "secondary" rod pathway for the transmission of scotopic signals. Here, rod signals are transmitted directly to cones and then to cone bipolar cells

and, in turn, to ganglion cells. Experimental evidence for the functionality of this pathway includes detection of rod-generated signals within cone photoreceptors (Nelson, 1977; Schneeweis & Schnapf, 1995) and the survival of rod signals in ganglion cells after blockade of the primary rod pathway (DeVries & Baylor, 1995; Völgyi et al., 2004). Human psychophysical studies also support the existence of multiple rod pathways (Blakemore & Rushton, 1965; Sharpe & Stockman, 1999).

Physiological studies indicate that the function of the two rod pathways relates to differences in the sensitivity of the signals they carry as well as the ganglion cells they target. Studies using retinas from Cx36 knockout (KO) mice, in which the rod-cone gap junctions are disrupted, indicate that the primary rod pathway carries signals most sensitive to stimulus brightness (Deans et al., 2002; Völgyi et al., 2004). In contrast, rod signals carried within the secondary pathway are about one log unit less sensitive. The two rod pathways thus have largely different operating ranges. Although the primary rod pathway has high sensitivity, its nonlinear synaptic transfer properties result in a narrow operating range (Dunn et al., 2006). The secondary rod pathway thus provides a route for scotopic signals to reach the ganglion cells when the primary pathway is saturated (Smith, Freed, & Sterling, 1986). Interestingly, there appear to be both segregation and convergence of signals carried by the two rod pathways, depending on the targeted ganglion cell subtype (DeVries & Baylor, 1995; Soucy et al., 1998; Völgyi et al., 2004). This indicates the complexity not only of the pathways by which rod signals are propagated within the retina but also in their transmission to central visual areas.

The conductance of the gap junctions between rod and cone photoreceptors is regulated by a circadian clock within the retina (Ribelayga, Cao, & Mangel, 2008). In this scheme the circadian clock increases dopamine release during the day, which activates $D_{2/4}$ receptors on rods and cones. This, in turn, raises intracellular cAMP and protein kinase A activity, which reduces the conductance of rod–cone gap junctions. By contrast, reduced dopamine release at night allows for robust electrical coupling between rods and cones. This circadian control ensures that the secondary rod pathway is operational during the appropriate nighttime conditions to facilitate the detection of dim objects. In addition, the reduction of rod–cone coupling during the day ensures that cone signals are not passed to a saturated and thus nonoperational network of rods, which would attenuate the signals created during bright light conditions.

A Role for Rod-to-Rod Coupling in a Third Rod Pathway

A study of mice genetically engineered to lack cones suggested a third rod pathway subserved by direct chemical synapses between rods and OFF bipolar cells (Soucy et al., 1998). Although such contacts were not thought to exist in mammals (Boycott & Dowling, 1969; Kolb, 1977), more recent studies have reported them in a number of species (Fyk-Kolodziej, Qin, & Pourcho, 2003; Hack, Peichl, & Brandstätter, 1999; Li, Keung, & Massey, 2004), and there is now strong functional evidence for this third rod circuitry (Völgyi et al., 2004). Interestingly, only one in five rods in the mouse retina forms a chemical synapse with an OFF bipolar cell, suggesting that this pathway may play a relatively limited role in scotopic vision (Tsukamoto et al., 2001). However, gap junctions exist between the axon terminals of mammalian rod photoreceptors (Hornstein et al., 2005; Tsukamoto et al., 2001). It has been proposed that rod–rod coupling pools the scotopic signals for conveyance to the ganglion cells via the third pathway. This third rod pathway may thus be useful at dusk and dawn, when relatively greater numbers of photons are available than during starlight, and the pooled signal may thereby efficiently encode faintly backlit objects (Tsukamoto et al., 2001). In support of this idea, it has been found that signals carried by the third rod pathway are less sensitive than those carried by the primary and secondary pathways (Völgyi et al., 2004). Interestingly, the ganglion cells targeted by the third rod pathways do not receive input from the other two rod pathways, consistent with its proposed unique function (Völgyi et al., 2004).

ELECTRICAL COUPLING OF HORIZONTAL CELLS UNDERLIES CONTRAST DETECTION

As mentioned, Yamada and Ishikawa (1965) first described "fused membrane structures" between neighboring horizontal cells, and there have been many subsequent morphological demonstrations of coupling between horizontal cells (Bloomfield, Xin, & Persky, 1995; Dacey, 1999; Dacheux & Raviola, 1982; Vaney, 1991). The extensive gap-junctional coupling between horizontal cells results in their characteristically large receptive fields, which can be 25 times the size of an individual cell's dendritic arbor (Bloomfield, Xin, & Persky, 1995; Naka & Rushton, 1967). Accordingly, deletion of horizontal cell gap junctions in the retina of Cx57 KO mice results in a significant reduction in their receptive field sizes (Shelley et al., 2006).

There is considerable evidence that the large receptive fields of horizontal cells provide the surround receptive field of bipolar cells and, in turn, the output ganglion cells (Mangel & Miller, 1987; Naka & Nye, 1971; Naka & Witkovsky, 1972). It remains unclear whether bipolar cell surrounds are created via direct inhibitory horizontal cell synapses onto bipolar cells or feedback inhibitory synapses of horizontal cells to photoreceptors. Although controversial, it has been advanced that the feedback inhibition is not chemical but ephaptic, whereby electrical charge moves across horizontal cell hemichannels into the extracellular space to modify the activity of cone photoreceptors (Kamermans et al., 2001). In any event, horizontal cell coupling serves to form a homogeneous electrical syncytium by which a signal of the average, ambient background illuminance is created. In essence, gap-junctional coupling between horizontal cells is believed to form the initial mechanism for contrast detection in the visual system, although this idea has recently been challenged (Dedek et al., 2008).

There is strong evidence that horizontal cell coupling is under dynamic regulation (figure 9.3). Dopamine decreases the gap-junctional conductance of horizontal cells and produces a concomitant reduction in receptive field size acting through the intracellular messenger cAMP (Lasater & Dowling, 1985; Piccolino, Neyton, & Gerschenfeld, 1984). Dopamine modifies gap junction conductances by affecting junctional densities, channel mean open times, and/or frequency of channel openings (Baldridge, Ball, & Miller, 1987; Kurz-Isler, Voigt, & Wolburg, 1992; McMahon & Brown, 1994; McMahon, Knapp, & Dowling, 1989). Nitric oxide (NO) also alters the electrical coupling of horizontal cells, acting through the intercellular messenger cGMP (DeVries & Schwartz, 1989, 1992; Lu & McMahon, 1997; Pottek, Schultz, & Weiler, 1997). Both the extracellular levels of dopamine and the production and release of NO are modulated by changes in illuminance, suggesting a light-induced modulation of horizontal cell coupling (Godley & Wurtman, 1988; Koistinaho et al., 1993; Zemel et al., 1996. Indeed, both prolonged darkness and light adaptation have been shown to uncouple horizontal cells in a number of vertebrate retinas (Mangel & Dowling, 1985; Umino, Lee, & Dowling, 1991; Xin & Bloomfield, 1999a). There is a complex triphasic relationship between the conductance of horizontal cell gap junctions and illuminance whereby coupling is poor under both dark- and light-adapted conditions and strong only under intermediate illuminance levels (Xin & Bloomfield, 1999a; Baldridge, 2001).

As expected, light-induced changes in horizontal cell coupling affect the center–surround receptive field

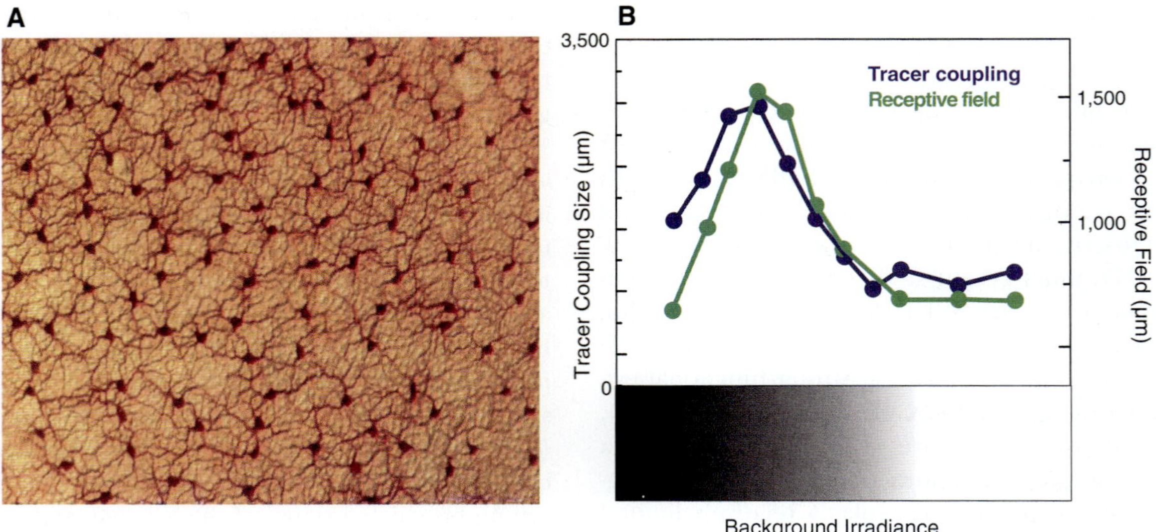

FIGURE 9.3   Horizontal cell gap junctions are modulated by light. (A) Photomicrograph of a network of coupled horizontal cells in the rabbit retina after injection of one cell with the gap junction-permeant tracer Neurobiotin. The figure shows just a portion of the coupled network, which included over 1,000 labeled cells extending over 2 mm across the retina. (B) Scatterplot showing the relationship between background illuminance and the receptive field and tracer coupling size of horizontal cells in the rabbit retina. In these experiments the retina was maintained under constant ambient illumination of different intensities, and the receptive field of horizontals cells was measured followed by intracellular injection of Neurobiotin to assay the extent of gap junctional coupling. Both the receptive field and spread of tracer were modified in a similar manner by light. These data show that the conductance of horizontal cell gap junctions is low under darkness but increases dramatically when dim light is applied. As the background light increases, the conductance of the gap junctions decreases back to values seen in darkness. This triphasic modulation of electrical coupling is thought to optimize the extraction of visual cues within an image under different ambient light conditions. (Modified from Xin & Bloomfield, 1999a, with permission.)

organization of retinal neurons and contrast signaling. Prolonged darkness attenuates the surround receptive fields of ganglion cells, which is thought to increase sensitivity at dim light levels at the expense of contrast detection (Muller & Dacheux, 1997; Peichl & Wässle, 1983; Rodieck & Stone, 1965). Likewise, reduction of horizontal cell coupling under brighter light results in smaller surround receptive fields and thus more local contrast detection consistent with higher acuity. Overall, the light-induced modulation of horizontal cell coupling optimizes the extraction of important cues within natural images including contrasts and edges (Balboa & Grzywacz, 2000a, 2000b).

ROLES OF NEURONAL GAP JUNCTIONS IN THE INNER RETINA

Gap Junctions in the Inner Retina Are Essential for Scotopic Vision

The diversity of neuronal gap junctions is greatest in the inner retina in which bipolar cells, amacrine cells, and ganglion cells display a variety of circuits generated by electrical synaptic connections (Bloomfield & Völgyi, 2009). These include homologous coupling between amacrine cells and between ganglion cells and also heterologous coupling between amacrine and ganglion cells and between amacrine cells and bipolar cells.

The function of these diverse types of coupling has perhaps best been explored for the primary rod pathway. As described earlier, rod bipolar cells do not make contacts directly with ganglion cells but instead form synapses with the intermediary AII amacrine cells. These AII cells express two types of gap junctions: neighboring AII cells form gap junctions with one another and with the axon terminals of cone bipolar cells (Famiglietti & Kolb, 1975; Strettoi, Dacheux, & Raviola, 1990). Plaques of Cx36 are found at dendritic crossings of AII cells, suggesting that AII–AII gap junctions are homotypic (Deans et al., 2002; Feigenspan et al., 2001). However, both Cx36 and Cx45 are expressed at cone bipolar cell hemichannels, indicating that at least some of the AII cell–cone bipolar cell gap junctions may be heterotypic (Dedek et al., 2006; Feigenspan et al., 2001; Han & Massey, 2005; Lin, Jakobs, & Masland, 2005; Mills et al., 2001), although the existence of Cx45/Cx36 heterotypic junctions has been questioned (Li et al., 2008). Such variations in composition could explain the different conductance and pharmacology of AII–AII and AII cell–cone bipolar cell gap junctions (Mills & Massey, 2000; Veruki & Hartveit, 2002).

Deletion of Cx36 disrupts both the AII–bipolar cell and rod–cone gap junctions resulting in the loss of signaling within the primary and secondary rod pathways, respectively (Deans et al., 2002; Güldenagel et al., 2001; Völgyi et al., 2004). As a result, all rod-driven input to ON ganglion cells is lost. These results not only show that electrical synapses play an essential part in the rod pathways but provide the first demonstration that gap junctions are obligatory elements within a defined circuit in the CNS. Based on computational models, Smith and Vardi (1995) speculated that AII–AII cell coupling serves to sum synchronous signals and subtract asynchronous noise, thereby preserving the high sensitivity of signals carried by the primary rod pathway. Consistent with this idea the intensity–response profiles of the most sensitive ganglion cells in the retina show a rightward shift when AII–AII amacrine cell electrical synapses are deleted in Cx36 KO mice (Völgyi et al., 2004). The reduced sensitivity likely results from a disruption of AII–AII cell coupling, resulting in a reduced signal-to-noise ratio and a loss of signal fidelity. AII–AII cell coupling thus underlies a unique function of the primary rod pathway: maintaining the high sensitivity of rod signals arriving in the inner retina.

The dendrites of the dopaminergic subtype of amacrine cell form a dense plexus that surrounds the AII amacrine cells, and dopamine modulates the conductance of AII–AII cell gap junctions via a cAMP-mediated PKA cascade (Kothmann et al., 2009; Mills & Massey, 2000; Voigt & Wässle, 1987). Because dopamine release is modulated by light, it is not surprising that changes in light adaptation are found to affect the coupling between AII amacrine cells in a triphasic manner as described earlier for horizontal cells (Bloomfield, Xin, & Rushton, 1997; Witkovsky & Dearry, 1991). Whereas dark-adapted AII cells are coupled in relatively small groups and show relatively small receptive fields, exposure to dim background lights brings about an approximately sevenfold increase in both parameters (Xin & Bloomfield, 1999b). Further light adaptation brings about a decrease in coupling to levels similar to those seen in dark-adapted retina. The light-induced changes in AII–AII cell coupling are believed to enable these cells, as vital elements in the rod pathway, to remain responsive throughout the entire scotopic/mesopic range. In this scheme dark adaptation is analogous to starlight conditions under which rods only sporadically absorb photons of light. Accordingly, the AII cells are relatively uncoupled in that the few correlated signals are carried only by close neighbor AII cells to sum, and

so extensive coupling would dissipate signals into a largely inactive network, thereby attenuating them. However, presentation of dim background lights, analogous to twilight conditions, brings about greater than a 10-fold increase in AII cell coupling. With the availability of more photons captured by neighboring cells, the increased coupling provides for summation of synchronous activity over a wider network of active AII cells, thereby preserving signal fidelity.

ELECTRICAL COUPLING UNDERLIES CORRELATED ACTIVITY OF NEIGHBORING GANGLION CELLS

Most ganglion cell subtypes show gap junction–mediated tracer coupling with neighboring ganglion and/or amacrine cells (Vaney, 1994; Völgyi, Chheda, & Bloomfield, 2009; Xin & Bloomfield, 1997) (figure 9.4). Interestingly, coupling between different subtypes of ganglion cells has never been reported, which suggests that ganglion cell gap junctions subserve separate electrical networks in the inner retina. At first glance the extensive coupling displayed by ganglion cells appears counterintuitive in that it suggests lateral intercellular propagation of signals across the inner plexiform layer. This would result in a reduction of visual acuity of neuronal signals just as they exit the retina. However, the receptive fields of ganglion cells approximate the extent of their dendritic arbors, irrespective of the extent of tracer coupling (Bloomfield & Xin, 1997). This is because the conductance of ganglion cell gap junctions is low, which restricts the movement of both electrical current and tracers.

Thus, ganglion cell gap junctions likely underlie local operations rather than the global processing exemplified by the extensive electrical syncytia formed by horizontal cells in the outer retina. For example, a recent study showed that electrical coupling of ganglion cells plays a critical role in establishing their absolute sensitivities (Murphy & Rieke, 2011).

Electrical coupling in the inner retina is also believed to underlie the correlated spike activity of neighboring ganglion cells. Beginning with the seminal work of Mastronarde (1983a, 1983b, 1983c), a number of studies have shown that ganglion cell coupling underlies coherent firing of neighbors, ranging from broad correlations spanning several tens of milliseconds to finely tuned spike synchrony with 1- to 3-ms latencies (Brivanlou, Warland, & Meister, 1998; DeVries, 1999; Hu & Bloomfield, 2003; Shlens, Rieke, & Chichilnisky, 2008). The homologous ganglion cell–ganglion cell and heterologous ganglion cell–amacrine cell coupling found in the inner retina are thought to produce

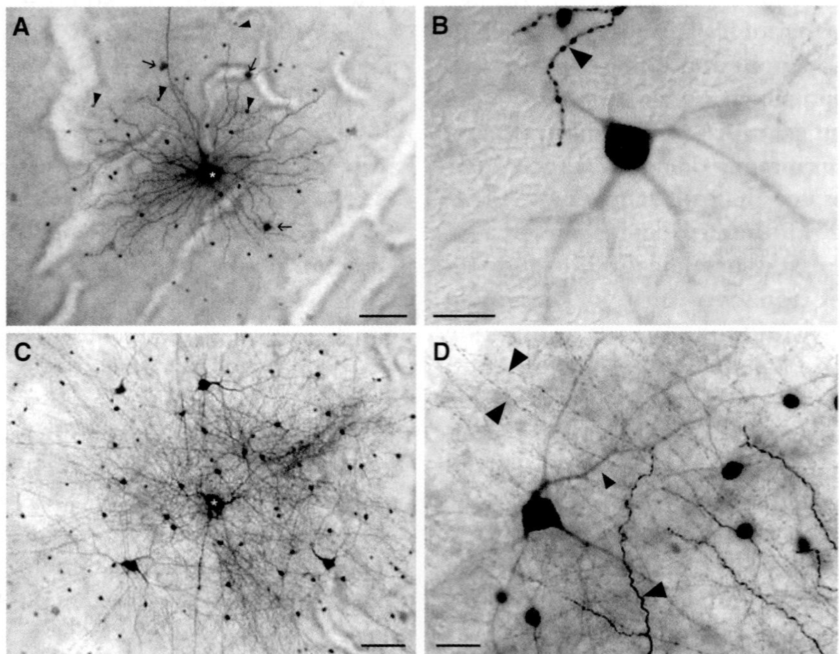

FIGURE 9.4   Light adaptation increases ganglion cell coupling to neighboring cells. (A) Typical coupling pattern of an OFF alpha-GC after injection with Neurobiotin in a dark-adapted retina. The cell is tracer coupled to neighboring alpha-GCs (arrows) as well as to an array of multiple amacrine cell subtypes (arrowheads). Note that the dendritic arbor of the injected alpha-GC is clearly labeled while those of the coupled amacrine and ganglion cells are not. Asterisk indicates soma of injected alpha-GC. Scale bar: 50 μm. (B) Tracer-coupled OFF alpha-GC in the dark-adapted retina that was particularly well labeled so that a sizable portion of its dendritic arbor could be visualized. The arbor shows the radial branching pattern typical of alpha-GCs. The arrowhead indicates the well-labeled terminal dendritic branches of the OFF alpha-GC that was injected with Neurobiotin. Scale bar: 25 μm. (C) Tracer coupling pattern seen after Neurobiotin was injected into an OFF alpha-GC in a retina light-adapted with a bright background light. Note the increased number and darker labeling of the tracer-coupled amacrine and ganglion cells compared to the pattern seen in the dark-adapted retina (panel A). Scale bar: 100 μm. (D) Detailed coupling pattern of coupled amacrine and ganglion cells in the light-adapted retina. The dendrites (open arrowhead) of the coupled alpha-GC are well labeled, as are the extensive axonal processes (gray arrowheads) of the coupled amacrine cells. The terminal dendritic ending of the injected OFF alpha-GC is indicated by the dark arrowhead. Scale bar: 25 μm. (Modified from Hu et al., 2010, with permission.)

different patterns of concerted activity in neighboring ganglion cells (Brivanlou, Warland, & Meister, 1998; DeVries, 1999; Hu & Bloomfield, 2003). Importantly, this concerted firing accounts for up to one-half of retinal spike activity, suggesting that electrical coupling plays an important part in encoding visual information (Castelo-Branco, Neuenschwander, & Singer, 1998; Schnitzer & Meister, 2003).

Correlated firing is thought to compress information for efficient transmission and thereby increase bandwidth of the optic nerve (Meister & Berry, 1999). In this scheme synchronous activity forms a separate stream of information to the brain in addition to the asynchronous signals from individual ganglion cells. Concerted spike activity is also thought to enhance the saliency of visual signals by increasing the temporal summation at central targets (Alonso, Usrey, & Reid, 1996; Stevens & Zador, 1998; Usrey & Reid, 1999). In this way concerted

ganglion cell activity may provide the temporal precision by which retinal signals are reliably transmitted to central targets (Singer, 1999).

Interestingly, a recent study showed that the synchronous spiking of neighboring ON direction-selective (DS) ganglion cells has a clearly defined and focused role, namely to encode the direction of stimulus motion (Ackert et al., 2006). Neighboring ON DS ganglion cells are coupled indirectly via gap junctions with a subtype of polyaxonal amacrine cell and show both correlated and synchronous activity. Remarkably, Ackert et al. (2006) reported that the degree of spike synchrony between neighboring ON DS cells was affected by the direction of stimulus movement whereby synchronized spiking was dramatically attenuated by stimulus movement in the null direction. It was therefore proposed that changes in coupling and spike synchrony formed a key component of how ON DS cells signal direction of movement. As mentioned, previous studies reported that changes in the state of dark/light adaptation could globally modulate the coupling between retinal neurons (Bloomfield & Xin, 1997; Hu et al., 2010; Mangel & Dowling, 1985), but this one was the first to show that specific light stimulation could effectively modulate the coupling between neurons, thereby forming a mechanism to encode specific visual information.

Ganglion cell coupling is dramatically increased under light-adapted conditions, resulting in an increase in concerted spike activity (Hu et al., 2010). Interestingly, this increase in coupling appears to be generated by a light-induced elevation of extracellular dopamine coupled with a differential activation of D_1 and $D_{2/4}$ receptors dependent on the adapational state of the retina. The light-induced regulation of ganglion cell gap junction conductance is thus opposite of that described above for horizontal cell and for AII amacrine cell coupling. It has been proposed that the increased coupling and resultant concerted activity of ganglion cells during daylight conditions strengthen the capacity and efficiency of information flow across the bottleneck of the optic nerve to higher brain centers (Hu et al., 2010).

CONCLUSIONS AND FUTURE DIRECTIONS

Work over the past decade has shown that electrical synaptic transmission via gap junctions is a common mode of intercellular communication in the retina. Not only are gap junctions and their subunit connexin proteins widely distributed within both synaptic layers but converging evidence suggests that they are expressed by most of the approximate 80 subtypes of retinal neurons. The finding that gap-junctional conductances are affected by neuromodulators acting as mediators of changes in light adaptation and circadian rhythms indicates that electrical synaptic transmission forms a complex and dynamic mode of cellular communication. Although we are just beginning to elucidate the types of connexins (and pannexins) expressed in the retina, it is already clear, as detailed in this chapter, that gap junctions have a wide variety of functions in propagating and integrating signals. Unlike studies in other parts of the CNS in which generic functions of electrical coupling are often assigned, such as increased spike synchrony, those in the retina have been able to detail the specific roles of individual gap junctions in visual processing. This makes the retina arguably the best model system to study the roles of electrical synaptic transmission in the CNS.

Ablation of gap junctions by pharmacological blockade or genetic deletion of connexins has emerged as a powerful resource to determine the functional roles of electrical synaptic transmission in the retina. Future studies using more selective genetic manipulations such as cell-specific and inducible connexin KO mouse models should elucidate the contribution of individual retinal gap junctions to visual signaling. It will also be important to link the biophysical properties of the different connexins expressed in retina, such as voltage and neuromodulator sensitivity, to the function of the gap junctions expressed by different neuronal cell types.

Beyond the task of cell signaling in the adult, it should also be noted that retinal gap junctions also play important roles in development. These include a role in the well-studied activity-dependent refinement of retinal–thalamic and intraretinal connections by modulating the correlated firing of ganglion cells associated with spontaneous waves of depolarization (Blankenship et al., 2011). Further, recent studies have now uncovered a role for gap junctions in control of cell differentiation, migration, and synaptogenesis (Cook & Becker, 2009). Studies are thus called for to determine the role(s) that connexins may play in both intra- and intercellular communication responsible for coordinating retinal development.

Finally, gap junctions have complex roles related to the neuronal loss associated with both programmed cell death and neurodegenerative conditions. In one scheme gap junctions underlie the "bystander effect" in which they act as portals for the passage of toxic molecules from dying cells to their neighbors (Decrock et al., 2011). This mechanism has been implicated in cell loss associated with a number of insults, including retinitis pigmentosa, ischemia, and excitotoxic conditions (Decrock et al., 2011; Andrade-Rozental et al., 2000; Ripps, 2002). A recent study showed that blockade of gap junctions or genetic deletion of connexins can increase cell survivability of retinal neurons by nearly 70% under a number of insult conditions (Akopian et al., 2011). These data suggest that targeting specific gap junctions and their subunit connexins will be an important new strategy for ameliorating cell loss associated with different neurodegenerative pathologies of the retina.

REFERENCES

Ackert, J. M., Wu, S. H., Lee, J. C., Abrams, J., Hu, E. H., Perlman, I., et al. (2006). Light induced changes in spike synchronization between coupled ON direction selective ganglion cells in the mammalian retina. *Journal of Neuroscience, 26,* 4206–4215.

Akopian, A., Zhang, Y., Wong, S., Pan, F., Paul, D. L., & Bloomfield, S. A. (2011). Gap junctions mediate neuronal cell death in the mammalian retina. *Investigative Ophthalmology & Visual Science, 52*(Suppl.), 1172.

Alonso, J. M., Usrey, W. M., & Reid, R. C. (1996). Precisely correlated firing in cells of the lateral geniculate nucleus. *Nature, 383*, 815–819.

Andrade-Rozental, A. F., Rozental, R., Hopperstad, M. G., Wu, J. K., Vrionis, F. D., & Spray, D. C. (2000). Gap junctions: The "kiss of death" and the "kiss of life." *Brain Research. Brain Research Reviews, 32*, 308–315.

Balboa, R. M., & Grzywacz, N. M. (2000a). The role of early retinal inhibition: More than maximizing luminance information. *Visual Neuroscience, 17*, 77–89.

Balboa, R. M., & Grzywacz, N. M. (2000b). The minimal local-asperity hypothesis of early retinal lateral inhibition. *Neural Computation, 12*, 1485–1517. doi: 10.1162/089976600300015231.

Baldridge, W. H. (2001). Triphasic adaptation of teleost horizontal cells. *Progress in Brain Research, 131*, 437–449.

Baldridge, W. H., Ball, A. K., & Miller, R. G. (1987). Dopaminergic regulation of horizontal cell gap junction particle density in goldfish retina. *Journal of Comparative Neurology, 265*, 428–436.

Baldridge, W. H., Weiler, R., & Dowling, J. E. (1995). Dark-suppression and light-sensitization of horizontal cell responses in the hybrid bass retina. *Visual Neuroscience, 12*, 611–620.

Baylor, D. A., Fuortes, M. G., & O'Bryan, P. M. (1971). Receptive fields of cones in the retina of the turtle. *Journal of Physiology, 214*, 265–294.

Blakemore, C. B., & Rushton, W. A. (1965). The rod increment threshold during dark adaptation in normal and rod monochromat. *Journal of Physiology, 181*, 629–640.

Blankenship, A. G., Hamby, A. M., Firl, A., Vyas, S., Maxeiner, S., Willecke, K., et al. (2011). The role of neuronal connexins 36 and 45 in shaping spontaneous firing patterns in the developing retina. *Journal of Neuroscience, 31*, 9998–10008.

Bloomfield, S. A., & Dacheux, R. F. (2001). Rod vision: Pathways and processing in the mammalian retina. *Progress in Retinal and Eye Research, 20*, 351–384. doi: 10.1016/S1350-9462(00)00031-8.

Bloomfield, S. A., & Völgyi, B. (2004). Function and plasticity of homologous coupling between AII amacrine cells. *Vision Research, 44*, 3297–3306. doi: 10.1016/j.visres.2004.07.012.

Bloomfield, S. A., & Völgyi, B. (2009). The diverse functional roles and regulation of neuronal gap junctions in the retina. *Nature Reviews. Neuroscience, 10*, 495–505.

Bloomfield, S. A., & Xin, D. (1997). A comparison of receptive-field and tracer-coupling size of amacrine and ganglion cells in the rabbit retina. *Visual Neuroscience, 14*, 1153–1165. doi: 10.1017/S0952523800011846.

Bloomfield, S. A., Xin, D., & Osborne, T. (1997). Light-induced modulation of coupling between AII amacrine cells in the rabbit retina. *Visual Neuroscience, 14*, 565–576.

Bloomfield, S. A., Xin, D., & Persky, S. E. (1995). A comparison of receptive field and tracer coupling size of horizontal cells in the rabbit retina. *Visual Neuroscience, 12*, 985–999.

Boycott, B. B., & Dowling, J. E. (1969). Organization of the primate retina: Light microscopy. *Philosophical Transactions of the Royal Society B (London), 255*, 109–184.

Boycott, B. B., & Kolb, H. (1973). The connections between bipolar cells and photoreceptors in the retina of the domestic cat. *Journal of Comparative Neurology, 148*, 91–114.

Brivanlou, I. H., Warland, D. K., & Meister, M. (1998). Mechanisms of concerted firing among retinal ganglion cells. *Neuron, 20*, 527–539.

Castelo-Branco, M., Neuenschwander, S., & Singer, W. (1998). Synchronization of visual responses between the cortex, lateral geniculate nucleus and retina in the anasthetized cat. *Journal of Neuroscience, 18*, 6395–6410.

Chesler, M. (2003). Regulation and modulation of pH in the brain. *Physiological Reviews, 83*, 1183–1221.

Cohen, A. I. (1965). Some electron microscopic observations on inter-receptor contacts in the human and macaque retinae. *Journal of Anatomy, 99*, 595–610.

Cook, J. E., & Becker, D. L. (2009). Gap-junction proteins in retinal development: New roles for the "nexus." *Physiology, 24*, 219–230.

Dacey, D. M. (1999). Primate retina: Cell types, circuits and color opponency. *Progress in Retinal and Eye Research, 18*, 737–763. doi: 10.1016/S1350-9462(98)00013-5.

Dacheux, R. F., & Raviola, E. (1982). Horizontal cells in the retina of the rabbit. *Journal of Neuroscience, 2*, 1486–1493.

Deans, M. R., Völgyi, B., Goodenough, D. A., Bloomfield, S. A., & Paul, D. L. (2002). Connexin36 is essential for transmission of rod-mediated visual signals in the mammalian retina. *Neuron, 36*, 703–712.

Decrock, E., Vinken, M., Bol, M., D'Herde, K., Rogiers, V., Vandenabeele, P., et al. (2011). Calcium and connexin-based intercellular communication: A deadly catch? *Cell Calcium, 50*, 310–321. doi: 10.1016/j.ceca.2011.05.007.

Dedek, K., Pandarinth, C., Alam, N.W., Wellershaus, K., Schubert, T., Willecke, K., et al. (2008). Ganglion cell adaptability: Does coupling of horizontal cells play a role? *PLoS One, 5*, e1714. doi: 10.1371/journal.pone.0001714.

Dedek, K., Schultz, K., Pieper, M., Dirks, P., Maxeiner, S., Willecke, K., et al. (2006). Localization of the heterotypic gap junctions composed of connexin45 and connexin36 in the rod pathway of the mouse retina. *European Journal of Neuroscience, 24*, 1675–1686.

DeVries, S. H. (1999). Correlated firing in rabbit retinal ganglion cells. *Journal of Neurophysiology, 81*, 908–920.

DeVries, S. H., & Baylor, D. A. (1995). An alternative pathway for signal flow from rod photoreceptors to ganglion cells in mammalian retina. *Proceedings of the National Academy of Sciences of the United States of America, 92*, 10658–10662. doi: 10.1073/pnas.92.23.10658.

DeVries, S. H., Qi, X., Smith, R., Makous, W., & Sterling, P. (2002). Electrical coupling between mammalian cones. *Current Biology, 12*, 1900–1907. doi: 10.1016/S0960-9822(02)01261-7.

DeVries, S. H., & Schwartz, E. A. (1989). Modulation of an electrical synapse between solitary pairs of catfish horizontal cells by dopamine and second messengers. *Journal of Physiology, 414*, 351–375.

DeVries, S. H., & Schwartz, E. A. (1992). Hemi-gap-junction channels in solitary horizontal cells of the catfish retina. *Journal of Physiology, 445*, 201–230.

Dunn, F. A., Doan, T., Sampath, A. P., & Rieke, F. (2006). Controlling the gain of rod-mediated signals in the mammalian retina. *Journal of Neuroscience, 26*, 3959–3970.

Famiglietti, E. V., Jr., & Kolb, H. (1975). A bistratified amacrine cell and synaptic circuitry in the inner plexiform layer of the retina. *Brain Research, 84*, 293–300.

Feigenspan, A., Janssen-Bienhold, U., Hormuzdi, S., Monyer, H., Degen, J., Söhl, G., et al. (2004). Expression of

  STEWART A. BLOOMFIELD AND BÉLA VÖLGYI

connexin36 in cone pedicles and OFF-cone bipolar cells of the mouse retina. *Journal of Neuroscience, 24,* 3325–3334.

Feigenspan, A., Teubner, B., Willecke, K., & Weiler, R. (2001). Expression of neuronal connexin36 in AII amacrine cells of the mammalian retina. *Journal of Neuroscience, 21,* 230–239.

Fyk-Kolodziej, B., Qin, P., & Pourcho, R. G. (2003). Identification of a cone bipolar cell in cat retina which has input from both rod and cone photoreceptors. *Journal of Comparative Neurology, 464,* 104–113.

Ghosh, K. K., Bujan, S., Haverkamp, S., Feigenspan, A., & Wässle, H. (2004). Types of bipolar cells in the mouse retina. *Journal of Comparative Neurology, 469,* 70–82.

Godley, B. F., & Wurtman, R. J. (1988). Release of endogenous dopamine from the superfused rabbit retina in vitro: Effect of light stimulation. *Brain Research, 452,* 393–395.

González, D., Gömez-Hernández, J. M., Larrosa, B., Gutiérrez, C., Muñoz, M. D., Fasciani, I., et al. (2008). Regulation of neuronal connexin-36 channels by pH. *Proceedings of the National Academy of Sciences of the United States of America, 105,* 17169–17174. doi: 10.1073/pnas.0804189105.

Goodenough, D. A., & Revel, J. P. (1970). A fine structural analysis of intercellular junctions in the mouse liver. *Journal of Cell Biology, 45,* 272–290.

Güldenagel, M., Ammermüller, J., Feigenspan, A., Teubner, B., Degen, J., Söhl, G., et al. (2001). Visual transmission deficits in mice with targeted disruption of the gap junction gene connexin36. *Journal of Neuroscience, 21,* 6036–6044.

Hack, I., Peichl, L., & Brandstätter, J. H. (1999). An alternative pathway for rod signals in the rodent retina: Rod photoreceptors, cone bipolar cells, and the localization of glutamate receptors. *Proceedings of the National Academy of Sciences of the United States of America, 96,* 14130–14135. doi: 10.1073/pnas.96.24.14130.

Han, Y., & Massey, S. C. (2005). Electrical synapses in retinal ON cone bipolar cells: Subtype-specific expression of connexins. *Proceedings of the National Academy of Sciences of the United States of America, 102,* 13313–13318. doi: 10.1073/pnas.0505067102.

Hornstein, E. P., Verweij, J., Li, P. H., & Schnapf, J. L. (2005). Gap-junctional coupling and absolute sensitivity of photoreceptors in macaque retina. *Journal of Neuroscience, 25,* 11201–11209.

Hu, E. H., & Bloomfield, S. A. (2003). Gap junctional coupling underlies the short-latency spike synchrony of retinal alpha ganglion cells. *Journal of Neuroscience, 23,* 6768–6777.

Hu, E. H., Pan, F., Völgyi, B., & Bloomfield, S. A. (2010). Light increases the gap junctional coupling of retinal ganglion cells. *Journal of Physiology, 588,* 4145–4163.

Kamermans, M., Fahrenfort, I., Schultz, K., Janssen-Bienhold, U., Sjoerdsma, T., & Weiler, R. (2001). Hemichannel-mediated inhibition in the outer retina. *Science, 292,* 1178–1180.

Koistinaho, J., Swanson, R. A., de Vente, J., & Sagar, S. M. (1993). NADPH-diaphorase (nitric oxide synthase)-reactive amacrine cells of rabbit retina: Putative target cells and stimulation by light. *Neuroscience, 57,* 587–597.

Kolb, H. (1977). The organization of the outer plexiform layer in the retina of the cat: Electron microscopic observations. *Journal of Neurocytology, 6,* 131–153.

Kothmann, W. W., Massey, S. C., & O'Brien, J. (2009). Dopamine-stimulated dephosphorylation of connexin 36 mediates AII amacrine cell coupling. *Journal of Neuroscience, 29,* 14903–14911.

Kurz-Isler, G., Voigt, T., & Wolburg, H. (1992). Modulation of connexon densities in gap junctions of horizontal cell perikarya and axon terminals in fish retina: Effects of light/dark cycles, interruption of the optic nerve and application of dopamine. *Cell and Tissue Research, 268,* 267–275.

Lampe, P. D., & Lau, A. F. (2000). Regulation of gap junctions by phosphorylation of connexins. *Archives of Biochemistry and Biophysics, 384,* 205–215.

Lampe, P. D., & Lau, A. F. (2004). The effects of connexin phosphorylation on gap junctional communication. *International Journal of Biochemistry & Cell Biology, 36,* 1171–1186.

Lasater, E. M. (1987). Retinal horizontal cell gap junctional conductance is modulated by dopamine through a cyclic AMP-dependent protein kinase. *Proceedings of the National Academy of Sciences of the United States of America, 84,* 7319–7323. doi: 10.1073/pnas.84.20.7319.

Lasater, E. M., & Dowling, J. E. (1985). Dopamine decreases conductance of the electrical junctions between cultured retinal horizontal cells. *Proceedings of the National Academy of Sciences of the United States of America, 82,* 3025–3029. doi: 10.1073/pnas.82.9.3025.

Lee, E. J., Han, J. W., Kim, H. J., Kim, I. B., Lee, M. Y., Oh, S. J., et al. (2003). The immunocytochemical localization of connexin 36 at rod and cone gap junctions in the guinea pig retina. *European Journal of Neuroscience, 18,* 2925–2934.

Li, W., Keung, J. W., & Massey, S. C. (2004). Direct synaptic connections between rods and OFF cone bipolar cells in the rabbit retina. *Journal of Comparative Neurology, 474,* 1–12.

Li, X., Kamasawa, N., Ciolofan, C., Olson, C. O., Lu, S., Davidson, K. G., et al. (2008). Connexin45-containing neuronal gap junctions in rodent retina also contain connexin36 in both apposing hemiplaques, forming bihomotypic gap junctions, with scaffolding contributed by zona occludens-1. *Journal of Neuroscience, 28,* 9769–9789.

Lin, B., Jakobs, T. C., & Masland, R. H. (2005). Different functional types of bipolar cells use different gap-junctional proteins. *Journal of Neuroscience, 25,* 6696–6701.

Lu, C., & McMahon, D. G. (1997). Modulation of hybrid bass retinal gap junctional channel gating by nitric oxide. *Journal of Physiology, 499,* 689–699.

Lurtz, M. M., & Louis, C. F. (2007). Intracellular calcium regulation of connexin43. *American Journal of Physiology. Cell Physiology, 293,* 1806–1813.

Mangel, S. C., & Dowling, J. E. (1985). Responsiveness and receptive field size of carp horizontal cells are reduced by prolonged darkness and dopamine. *Science, 229,* 1107–1109.

Mangel, S. C., & Miller, R. F. (1987). Horizontal cells contribute to the receptive field surround of ganglion cells in the rabbit retina. *Brain Research, 414,* 182–186.

Mastronarde, D. N. (1983a). Correlated firing of cat retinal ganglion cells. I. Spontaneously active inputs to X- and Y-cells. *Journal of Neurophysiology, 49,* 303–324.

Mastronarde, D. N. (1983b). Correlated firing of cat retinal ganglion cells. II. Responses of X- and Y-cells to single quantal events. *Journal of Neurophysiology, 49,* 325–349.

Mastronarde, D. N. (1983c). Interactions between ganglion cells in cat retina. *Journal of Neurophysiology, 49,* 350–365.

McMahon, D. G., & Brown, D. R. (1994). Modulation of gap-junction channel gating at zebrafish retinal electrical synapses. *Journal of Neurophysiology, 72,* 2257–2268.

McMahon, D. G., Knapp, A. G., & Dowling, J. E. (1989). Horizontal cell gap junctions: Single-channel conductance and modulation by dopamine. *Proceedings of the National*

Academy of Sciences of the United States of America, 86, 7639–7643. doi: 10.1073/pnas.86.19.7639.

Meister, M., & Berry, M. (1999). The neural code of the retina. *Neuron, 22,* 435–450.

Mills, S. L., & Massey, S. C. (1995). Differential properties of two gap junctional pathways made by AII amacrine cells. *Nature, 377,* 734–737.

Mills, S. L., & Massey, S. C. (2000). A series of biotinylated tracers distinguishes three types of gap junction in retina. *Journal of Neuroscience, 20,* 8629–8636.

Mills, S. L., O'Brien, J. J., Li, W., O'Brien, J., & Massey, S. C. (2001). Rod pathways in the mammalian retina use connexin36. *Journal of Comparative Neurology, 436,* 336–350.

Muller, J. F., & Dacheux, R. F. (1997). Alpha ganglion cells of the rabbit retina lose antagonistic surround responses under dark adaptation. *Visual Neuroscience, 14,* 395–401.

Müller, L. P., Dedek, K., Janssen-Bienhold, U., Meyer, A., Kreuzberg, M. M., Lorenz, S., et al. (2010). Expression and modulation of connexin 30.2, a novel gap junction protein in the mouse retina. *Visual Neuroscience, 27,* 91–101.

Murphy, G. J., & Rieke, F. (2011). Electrical synaptic input to ganglion cells underlies differences in the output and absolute sensitivity of parallel retinal circuits. *Journal of Neuroscience, 31,* 12218–12228.

Naka, K. I., & Nye, P. W. (1971). Role of horizontal cells in organization of the catfish retinal receptive field. *Journal of Neurophysiology, 34,* 785–801.

Naka, K. I., & Rushton, W. A. (1967). The generation and spread of S-potentials in fish (Cyprinidae). *Journal of Physiology, 192,* 437–461.

Naka, K. I., & Witkovsky, P. (1972). Dogfish ganglion cell discharge resulting from extrinsic polarization of the horizontal cells. *Journal of Physiology, 223,* 449–460.

Nelson, R. (1977). Cat cones have rod input: A comparison of the response properties of cones and horizontal cell bodies in the retina of the cat. *Journal of Comparative Neurology, 172,* 109–135.

Patel, L. S., Mitchell, C. K., Dubinsky, W. P., & O'Brien, J. O. (2006). Regulation of gap junction coupling through the neuronal connexin Cx35 by nitric oxide and cGMP. *Cell Communication & Adhesion, 13,* 41–54.

Peichl, L., & Wässle, H. (1983). The structural correlate of the receptive field centre of alpha cells in the cat retina. *Journal of Physiology, 341,* 309–324.

Piccolino, M., Neyton, J., & Gerschenfeld, H. M. (1984). Decrease of gap junction permeability induced by dopamine and cyclic adenosine 3':5'-monophosphate in horizontal cells of turtle retina. *Journal of Neuroscience, 4,* 2477–2488.

Pottek, M., Schultz, K., & Weiler, R. (1997). Effects of the nitric oxide on the horizontal cell network and dopamine release in the carp retina. *Vision Research, 37,* 1091–1102.

Raviola, E., & Gilula, N. B. (1973). Gap junctions between photoreceptor cells in the vertebrate retina. *Proceedings of the National Academy of Sciences of the United States of America, 70,* 1677–1681. doi: 10.1073/pnas.70.6.1677.

Raviola, E., & Gilula, N. B. (1975). Intramembrane organization of specialized contacts in the outer plexiform layer of the retina. A freeze-fracture study in monkeys and rabbits. *Journal of Cell Biology, 65,* 192–222.

Ribelayga, C., Cao, Y., & Mangel, S. C. (2008). The circadian clock in the retina controls rod-cone coupling. *Neuron, 59,* 790–801.

Ripps, H. (2002). Cell death in retinitis pigmentosa: Gap junctions and the "bystander" effect. *Experimental Eye Research, 74,* 173–178.

Rodieck, R. W., & Stone, J. (1965). Analysis of receptive fields of cat retinal ganglion cells. *Journal of Neurophysiology, 28,* 832–849.

Schneeweis, D. M., & Schnapf, J. L. (1995). Photovoltage of rods and cones in the macaque retina. *Science, 268,* 1053–1056.

Schnitzer, M. J., & Meister, M. (2003). Multineuronal firing patterns in the signal from eye to brain. *Neuron, 37,* 499–511.

Sharpe, L. T., & Stockman, A. (1999). Rod pathways: The importance of seeing nothing. *Trends in Neurosciences, 22,* 497–504. doi: 10.1016/S0166-2236(99)01458-7.

Shelley, J., Dedek, K., Schubert, T., Feigenspan, A., Schultz, K., Hombach, S., et al. (2006). Horizontal cell receptive fields are reduced in connexin57-deficient mice. *European Journal of Neuroscience, 23,* 3176–3186.

Shlens, J., Rieke, F., & Chichilnisky, E. (2008). Synchronized firing in the retina. *Current Opinion in Neurobiology, 18,* 396–402.

Singer, W. (1999). Neuronal synchrony: A versatile code for the definition of relations? *Neuron, 24,* 49–65.

Smith, R. G., Freed, M. A., & Sterling, P. (1986). Microcircuitry of the dark-adapted cat retina: Functional architecture of the rod-cone network. *Journal of Neuroscience, 6,* 3505–3517.

Smith, R. G., & Vardi, N. (1995). Simulation of the AII amacrine cell of mammalian retina: Functional consequences of electrical coupling and regenerative membrane properties. *Visual Neuroscience, 12,* 851–860.

Söhl, G., Maxeiner, S., & Willecke, K. (2005). Expression and functions of neuronal gap junctions. *Nature Reviews. Neuroscience, 6,* 191–200.

Söhl, G., & Willecke, K. (2003). An update on connexin genes and their nomenclature in mouse and man. *Cell Communication & Adhesion, 10,* 173–180.

Soucy, E., Wang, Y., Nirenberg, S., Nathans, J., & Meister, M. (1998). A novel signaling pathway from rod photoreceptors to ganglion cells in mammalian retina. *Neuron, 21,* 481–493.

Spray, D. C., Harris, A. L., & Bennett, M. V. (1979). Voltage dependence of junctional conductance in early amphibian embryos. *Science, 204,* 432–434.

Spray, D. C., Harris, A. L., & Bennett, M. V. (1981). Gap junctional conductance is a simple and sensitive function of intracellular pH. *Science, 211,* 712–715.

Srinivas, M., Costa, M., Gao, Y., Fort, A., Fishman, G. I., & Spray, D. C. (1999). Voltage dependence of macroscopic and unitary currents of gap junction channels formed by mouse connexin50 expressed in rat neuroblastoma cells. *Journal of Physiology, 517,* 673–689.

Stevens, C. F., & Zador, A. M. (1998). Input synchrony and irregular firing of cortical neurons. *Nature Neuroscience, 1,* 210–217.

Strettoi, E., Dacheux, R. F., & Raviola, E. (1990). Synaptic connections of rod bipolar cells in the inner plexiform layer of the rabbit retina. *Journal of Comparative Neurology, 295,* 449–466.

Tsukamoto, Y., Masarachia, P., Schein, S. J., & Sterling, P. (1992). Gap junctions between the pedicles of macaque foveal cones. *Vision Research, 32,* 1809–1815. doi: 10.1016/0042-6989(92)90042-H.

Tsukamoto, Y., Morigiwa, K., Ueda, M., & Sterling, P. (2001). Microcircuits for the night vision in mouse retina. *Journal of Neuroscience, 21*, 8616–8623.

Umino, O., Lee, Y., & Dowling, J. E. (1991). Effects of light stimuli on the release of dopamine from interplexiform cells in the white perch retina. *Visual Neuroscience, 7*, 451–458.

Usrey, W. M., & Reid, R. C. (1999). Synchronous activity in the visual system. *Annual Review of Physiology, 61*, 435–456.

Vaney, D. (1991). Many diverse types of retinal neurons show tracer coupling when injected with biocytin and Neurobiotin. *Neuroscience Letters, 125*, 187–190.

Vaney, D. I. (1994). Territorial organization of direction-selective ganglion cells in rabbit retina. *Journal of Neuroscience, 14*, 6301–6316.

Veruki, M. L., & Hartveit, E. (2002). Electrical synapses mediate signal transmission in the rod pathway of the mammalian retina. *Journal of Neuroscience, 22*, 10558–10566.

Voigt, T., & Wässle, H. (1987). Dopaminergic innervation of AII amacrine cells in mammalian retina. *Journal of Neuroscience, 7*, 4115–4128.

Völgyi, B., Chheda, S., & Bloomfield, S. A. (2009). Tracer coupling patterns of the ganglion cell subtypes in the mouse retina. *Journal of Comparative Neurology, 512*, 664–687.

Völgyi, B., Deans, M. R., Paul, D. L., & Bloomfield, S. A. (2004). Convergence and segregation of the multiple rod pathways in mammalian retina. *Journal of Neuroscience, 24*, 11182–11192.

Weiler, R., Pottek, M., He, S., & Vaney, D. I. (2000). Modulation of coupling between retinal horizontal cells by retinoic acid and endogenous dopamine. *Brain Research. Brain Research Reviews, 32*, 121–129.

Willecke, K., Eiberger, J., Degen, J., Eckardt, D., Romualdi, A., Güldenagel, M., et al. (2002). Structural and functional diversity of connexin genes in the mouse and human genome. *Biological Chemistry, 383*, 725–737.

Witkovsky, P., & Dearry, A. (1991). Functional roles of dopamine in the vertebrate retina. *Progress in Retinal Research, 11*, 247–292. doi: 10.1016/0278-4327(91)90031-V.

Xin, D., & Bloomfield, S. A. (1997). Tracer coupling pattern of amacrine cells in the rabbit retina. *Journal of Comparative Neurology, 383*, 512–528.

Xin, D., & Bloomfield, S. A. (1999a). Dark- and light-induced changes in coupling between horizontal cells in mammalian retina. *Journal of Comparative Neurology, 405*, 75–87.

Xin, D., & Bloomfield, S. A. (1999b). Comparison of the responses of AII amacrine cells in the dark- and light-adapted rabbit retina. *Visual Neuroscience, 16*, 653–665.

Xin, D., & Bloomfield, S. A. (2000). Effects of nitric oxide on horizontal cells in the rabbit retina. *Visual Neuroscience, 17*, 799–811.

Yamada, E., & Ishikawa, T. (1965). The fine structure of the horizontal cells in some vertebrate retinae. *Cold Spring Harbor Symposia on Quantitative Biology, 30*, 383–392.

Zemel, E., Eyal, O., Lei, B., & Perlman, I. (1996). NADPH diaphorase activity in mammalian retinas is modulated by the state of visual adaptation. *Visual Neuroscience, 13*, 865–871.

10 Retinal Connectomics: A New Era for Connectivity Analysis

ROBERT E. MARC, BRYAN W. JONES, J. SCOTT LAURITZEN, CARL B. WATT, AND JAMES R. ANDERSON

Advances in understanding retinal networks have slowed as we begin to confront the true complexity of these networks. Even with new molecular and genetic tools for visualizing and profiling neurons, definitive network architectures have not emerged. No complete subnetwork is known for any retinal neuron, and only recently has it become clear that formal network topologies cannot be discovered by optical or any other indirect means. Four developments have enabled the assembly of very large, nearly complete networks or connectomes. (1) Fast electron imaging has permitted the acquisition of archivable network data on massive scales. (2) Advances in image registration software permit assembly of navigable data volumes that would take millennia to create manually, even with fast computers. (3) Navigation and annotation software has been developed to rapidly transform imagery into network descriptors for systems analysis. (4) Data storage has become so economical that a single laboratory can afford large data servers. The initial discoveries in connectomics suggest that it will transform our understanding of retinal neurobiology. Importantly, these datasets can be made public. This chapter draws from and extends our recent review of connectomics technologies (Marc et al., in press).

THE LEGACY OF CLASSICAL ULTRASTRUCTURE

Little of the function of the mammalian retina is comprehensible without the deep understanding of connectivity provided by transmission electron microscope (TEM) imaging, notably by John Dowling, Helga Kolb, Peter Sterling, Elio Raviola, Enrica Strettoi, the late Brian Boycott, and many others. Despite this, even a description of basic photoreceptor drive in the complex retinas of nonmammalians is far from understood (Mariani & Lasansky, 1984; Normann et al., 1984; Scholes, 1975; Stell, Ishida, & Lightfoot, 1977). In addition to the uncovering of the basic topology of rod and cone divergence to separate classes of bipolar cells

(BCs) in mammals, one of the most powerful neuronal reconstructions ever achieved by TEM imaging was the mapping of the basic architecture of the AII amacrine cell (AC) by Helga Kolb and Edward V. Famiglietti Jr. (1974a,b), with subsequent refinements by Enrica Strettoi, Elio Raviola, and the late Ramon Dacheux (1992). Indeed, this cell is so complex that no physiological data exist that correctly predict its form or connectivity. Even so, our knowledge of the AII cell's complete connectivity was quite limited until recently (Anderson et al., 2011b; Marc et al., 2012). This chapter does not review historical material on retinal synaptic connectivity, as this is well covered in many book chapters and earlier versions of this series. We instead address the motivation for a new mode of anatomy, connectomics, and a new era in data curation.

THE MOTIVATION FOR CONNECTOMICS

Neither physiology nor modeling permit compete inference of networks (Diestel, 2005; Harary & Palmer, 1973). Understanding this requires delving into graph and computational complexity theory, but the effort is worth it. The importance of a shift in language when referring to retinal networks is simply that many of the key problems in understanding, tracking, and mining biological networks from imagery, as well as developing strategies for building and navigating computational connectome volumes, involve important concepts, algorithms, and analyses drawn from graph theory and imaging processing, not retinal biology. In the language of graph theory, retinal and brain networks are collections of vertices (cells or cell compartments) connected by directed (synaptic) edges, and they are classified as directed graphs with multiple edges. From biological networks is it possible to construct so many different n-vertex graphs that indirect discovery is not possible, as shown by graph enumeration computations (figure 10.1). The mammalian retina displays over 70 classes of cells (Marc, 2010). The primate brain has no fewer than

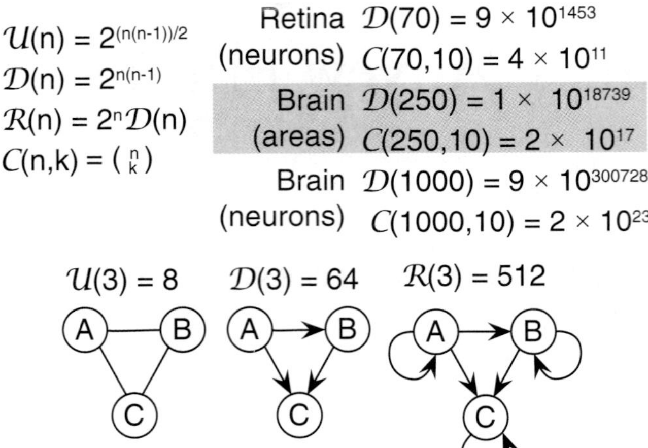

FIGURE 10.1 Graph enumeration for networks allows one to compute the number of possible networks in a system given a number of distinct neuronal classes or vertices (n). There are four basic relations for computing the number of networks as a function of vertex number and whether edges are undirected U(n), directed like synapses D(n), directed with reentrant loops like many retinal networks R(n), or combinatoric C(n,k) and limited to groups of connected cells (vertex clusters) of size k. A sample calculation is shown for a simple three-vertex model. Even a tiny network can have high complexity. Biological network complexity transcends simple analysis. Directed and combinatoric (k = 10) networks in retina (n = 70), across brain areas (n = 250) and brain neurons (n = 1,000) were calculated using the Wolfram Alpha engine (www.wolframalpha.com). (From Marc et al., 2012, by permission of the authors.)

250 regions (Van Essen et al., 2011) and likely over 1,000 neuronal classes. These calculations are independent of cell size. Adding diversity such as varied cell copy numbers and coverages (Reese, 2008) as well as varied molecular types and weights for synapses makes the number of solutions vastly larger and unmanageable by either inverse solutions using physiological methods or formal modeling. Considering the potentially greater number of synapses made and received by brain neurons only makes the problem worse. In such a large solution domain, unique mappings of transfer functions derived by physiology onto precise network topologies (i.e., bijections) do not exist (Aster, Borchers, & Thurber, 2005). From the modeling perspective, searching for specific or unique network motifs (not to mention their validation) is one of hardest mathematical problems known: the subgraph isomorphism problem (Karp, 1972). These difficulties plague genomic, proteomic, and metabolic network analyses as well (Wong et al., 2012). The correct solution lies in physically mapping networks, known in other fields as acquiring ground truth (Anderson et al., 2009). For

neuroscience this means ultrastructural anatomy, which—albeit heroic in scope—has in the past delivered only a broad view and fragments of retinal networks at best (Calkins & Sterling, 1996; 2007; Calkins, Tsukamoto, & Sterling, 1998; Klug et al., 2003; Kolb & Famiglietti, 1974b; Kolb & Nelson, 1993; Stevens et al., 1980; Strettoi, Raviola, & Dacheux, 1992; Witkovsky and Dowling, 1969). Further, none of these data have been accessioned, in contrast to genetic data. All previous ultrastructural data are collections of noncurated, low-resolution journal halftone images with no way to review either primary data or their contexts. Conflicting views on outcomes have remained unresolved, and the original data no longer exist or are completely inaccessible.

The future of anatomy in neuroscience is connectomics. Formally, a connectome is the adjacency matrix for a collection of neurons in a canonical field that maps all partners and nonpartners. An adjacency matrix is a list of all connections and nonconnections for each vertex. Indeed, finding nonpartners is critical and nontrivial. A canonical field is a volume of neural tissue that contains a defined set of elements that performs a key set of functions. For example, a field containing a minimum number of copies of the rarest cell in the neural retina should contain multiples of all other cells. Alternatively, a field could be much smaller and contain a minimum number of copies of dendrites (we call this a miniconnectome). Connectomics initiatives have included large-scale studies of regional connectivity in brain (Marcus et al., 2011; Sporns, Tononi, & Kötter, 2005; van den Heuvel & Sporns, 2011), optical mesoscale studies (Kleinfeld et al., 2011; Oberlaender et al., 2011), and synaptic connectomics in the vertebrate retina (Anderson et al., 2011b; Briggman, Helmstaedter, & Denk, 2011). Such ultrastructural connectomes are now practical because of the development of automated TEM (ATEM) imaging using software steering for high-throughput image acquisition and enterprise-scale storage. ATEM and connectomics require additional tools such as molecular markers to segment cell groups for analysis and to allow weighting of the adjacency matrix.

SAMPLE TREATMENT

Conventional TEM fixation and postfixation are optimal for connectomics. Segmentation of images by molecular or functional markers can be achieved by registering optical imagery to TEM image fields. Excitation mapping with the channel-permeant organic ion 1-amino-4-guanidobutane (AGB) (Anderson et al., 2009, 2011b) can embed small-molecule light response histories into a retinal sample. Briggmann,

    R. E. MARC, B. W. JONES, J. S. LAURITZEN, C. B. WATT, AND J. R. ANDERSON

Helmstaedter, and Denk (2011) and Bock et al. (2011) used computational alignment of optical calcium imaging data onto ultrastructural imagery to pre-identify neuron classes. TEM-based connectomes can also potentially exploit new genetic markers (Gaietta et al., 2002; Hoffmann et al., 2010; Lichtman & Smith, 2008; Shu et al., 2011). At present, molecular tags are essential for complete connectomics, and only TEM-based schemes have so far proven compliant with them for cell classification (Anderson et al., 2009; Jones et al., 2011; Jones et al., 2003; Marc & Liu, 2000; Micheva & Bruchez, 2011; Micheva et al., 2010; Micheva & Smith, 2007).

Connectome datasets are created either by ablation of a bloc surface or capturing sections. Ablation methods include in vacuo serial block-face (SBF) sectioning (Briggman & Denk, 2006; Denk & Horstmann, 2004) or ion beam milling (Knott et al., 2008), followed by SEM or scanning TEM (STEM) imaging of surface-backscattered secondary electrons. Ablation methods obligatorily have limited depth profiles and require very thin sections for tracking. In contrast, manual ultrami-crotomy (Anderson et al., 2009; Bourne & Harris, 2011) onto electron-transparent film supports followed by conventional staining and ATEM imaging (Anderson et al., 2009) generates primary electron projection images through the section thickness, optimally at 50–70 nm. Automated sectioning onto electron-opaque films has been developed for STEM imaging (Kleinfeld et al., 2011), but such platforms are currently expensive and rare. Although slicing forces the additional step of tile registration on a large scale, that is a solved problem (Tasdizen et al., 2010).

BUILDING CONNECTOMICS IMAGE SETS

Each slice of a TEM connectome volume can contain thousands of image tiles (figure 10.2), and image volumes can span 10–1,000 terabytes (Anderson et al., 2009, 2011b). The sizes of volumes are set by canonical field dimensions and the need to resolve gap junctions and synapses. This sets the minimum resolution at ~2-nm resolution with validation by reimaging at finer resolutions with goniometric tilt: techniques possible only with TEM (Anderson et al., 2009, 2011b; Bourne & Harris, 2011). SEM platforms usually acquire data at ~10 nm (Kleinfeld et al., 2011), which means that synapse identification is possible only for very large, well-oriented synapses; that small or poorly oriented synapses (the most common forms) may be missed; and that gap junctions will be invisible. True connectomics requires 2-nm resolution for sampling completeness (Anderson et al., 2009) and to enable measuring

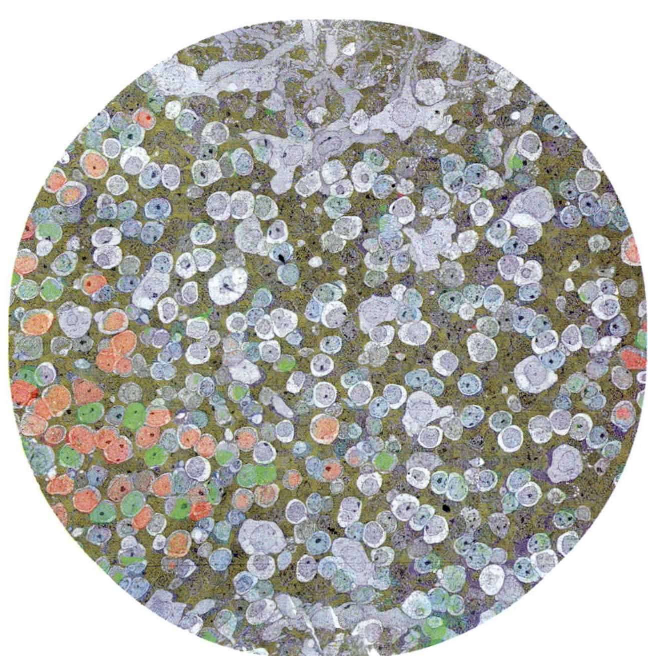

FIGURE 10.2 Connectome RC1 slice 001 composed of >1,000 high-resolution TEM tiles. The slice is augmented with a multispectral transparency mapping simultaneously displaying GABA (red), glycine (green), glutamate (blue), and a logical AND between glutamine and taurine signals as a dark gold transparency channel. GABA+ (red) neurons are amacrine cells, and glycine+ (green) neurons are either amacrine or an ON cone bipolar cell subset. Glutamate+ (blue) neurons are largely bipolar cells. Image width, 243 μm. (From Anderson et al., 2011b, by permission of the authors.)

synaptic dimensions for computing synaptic weights (Bourne & Harris, 2011). ATEM platforms are based on traditional, unmodified TEM systems augmented with advanced automation software (Mastronarde, 2005) and are economical for many existing TEM facilities. Advanced TEM camera arrays developed by Bock et al. (2011) use a modified TEM column, custom phosphor plates, and arrays of digital cameras.

There are different strategies for building navigable volumes. One approach synthesizes the entire dataset into a three-dimensional volume, requiring high-performance visualization environments. Others have opted for a paged architecture similar to Google Earth, treating each slice as a single array of two-dimensional mapped tiles, which significantly decreases computational overhead and readily enables data navigation, Web-based distribution, and open data sharing (Anderson et al., 2011a). Automated slice mosaicking and slice-to-slice registrations with subnanometer precision were developed by Tasdizen et al. (2010) and use image Fourier shift to compute displacement vectors. Resampled slice-to-slice alignments, refined by nonrigid grid alignment, are then used to automatically build 2-nm resolution volumes. As shown by Anderson et al. (2009), mechanical section differences or electron optical distortions such as magnification astigmatism inherent in TEM are invisible and inconsequential until one attempts to build a mosaic of thousands of tiles. Then they become serious problems for volume assembly. Such distortions are readily corrected by nonrigid grid methods. Further, addition of transforms from slice-to-slice aligns the volume without distortion, and remapped structures are not significantly changed from their original tile shapes or dimensions. Such approaches can also be used to align large-scale optical atlases (Berlanga et al., 2011).

How big should/can a connectome be? This depends on a mixture of resolution, the size of the canonical field desired, and the limits of acquisition speed and data storage. With resolution set at 2 nm and a canonical volume as a 0.25-mm cylinder spanning from mid–inner nuclear layer through the ganglion cell layer, the acquisition time was 5 months (now 3) and requires 16.4 terabytes of raw data. With one main data server, one mirror, and one backup, this dataset requires >64 Tb of storage. A volume 0.25 mm in diameter contains over 300 BCs, 300 Müller cells (MCs), ~40 AII ACs, over 100 neurochemically identified ACs, and 20 ganglion cells (GCs). This is more than sufficient for characterizing coarse- and fine-scale topologies of BC networks and many ACs and GCs. However, if one wanted to build complete networks of very large cells such as the largest GCs and ACs with receptive fields of ~1 mm, a more critical assessment of strategies is needed. If every major dendrite of such a large cell were representative of all others (i.e., if large neurites were functionally isotropic), a miniconnectome might suffice. Isotropism is a theory, not an established fact. It might be testable with a few miniconnectomes. But dealing with large asymmetries such as polyaxonal cells might require 1-mm-scale volumes with 2-nm resolution, similar to the requirements for cortical hypercolumns. This would require ~5 years of uninterrupted imaging time and a petabyte of storage, which is beyond any laboratory's capacity at present but might be feasible in a few years. If one's goals are merely to gauge populations of cells and use connectomics technologies for a "super-resolution" Golgi-like shape modeling with stick figures, such large fields can be acquired faster by SBF imaging with less storage because of its lower resolution.

NAVIGATION, ANNOTATION, AND ANALYSIS

Terabyte-scale imagery requires advanced software to navigate data volumes (Anderson et al., 2011a; Fiala, 2005; Jeong et al., 2010). One widely used approach uses image pyramid sets for fast image serving (Anderson et al., 2011a; Mikula et al., 2007). Anderson and colleagues developed an open-source Web-compliant Viking environment (Anderson et al., 2011a) to support multiuser visualization by converting datasets to Web-optimized tiles and delivering volume transforms to client devices via conventional Internet connections. Converting ultrastructure into three-dimensional renderings and network graphs also requires integration of annotation and database architectures. In Viking, disks placed within profiles approximate convex hulls and are linked to build three-dimensional representations (figure 10.3). Relational elements such as presynaptic complexes (ribbons, densities, vesicle accumulations), postsynaptic densities, gap junctions, and adherens junctions are located and linked to build adjacency matrices. Annotations also allow users to both enter and query data and metadata and allow building Web tours of networks.

But annotations and datasets are just the beginning of understanding and analyzing networks. These require rendering, graphing, network touring, and informatics. Vikingplot and Viz (Anderson et al., 2011a) are services using Viking databases to allow cell renderings at higher resolutions than optical methods, automated network graphs, navigation between ultrastructural data and network motifs, and automated statistical summaries (figure 10.4). Although significant efforts have been made to achieve automated tracing (Jeong et al., 2010; Jurrus et al., 2010; Luisi et al., 2011; Narayanaswamy, Wang, & Roysam, 2011), all connectomes must presently be validated by human annotation (Anderson et al., 2011a), and none are currently practical for connectomics of complex neuropil. Correcting annotation errors has proven rather straightforward so far. Completeness in network diagrams ultimately purges errors, and metadata parsing can detect early errors. Errors such as skipping between processes in tracing are flagged as forbidden switches in molecular signatures, associated synapse type, targets, or inputs. One of the best methods for error tracking and repair is parsing network graphs for wiring violations (e.g., self-synapse loops).

Connectomics datasets must be shared (Amari et al., 2002; Anderson et al., 2011a; Jeong et al., 2010), but distributing raw datasets is impractical. One solution is open-access via Web services. The Viking strategy involves open-source tools and common file formats to accommodate other widely used applications, such as Blender (www.blender.org) or Autodesk® Maya. Such approaches minimize the overhead for journals since they need not act as data repositories but also pose

VIKING TEM SURVEY VIEW

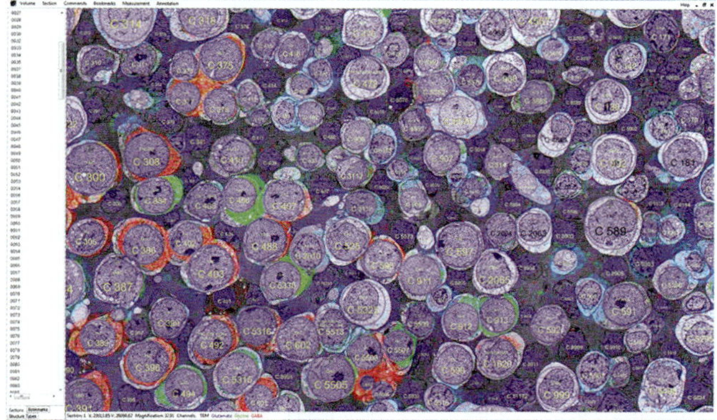

VIKING TEM SYNAPTIC VIEW

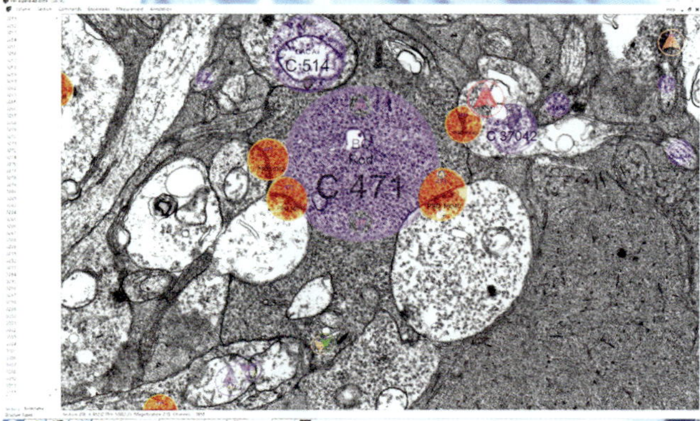

VIKINGPLOT RENDER VIEW

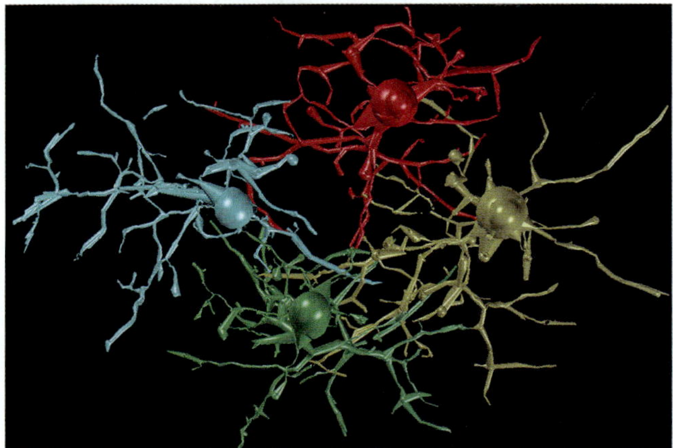

FIGURE 10.3 The Viking acquisition environment. The Viking browser allows dataset overviews (top) of populations of cells, each cell encoded with an overlay showing its ID number and linking to a list of all its connections. Zooming to synaptic resolution views (middle) then enables users to trace cells with specific annotations for cell regions and synaptic associations. Collected annotations are automatically built into three-dimensional renderings of selected cells (bottom). The cells shown here are four coupled AII ACs rendered using the Vikingplot application, calling the open-access RC1 database.

challenges to classical concepts of intellectual ownership and publication. We have opted for public sharing of our datasets and tools. The next critical stage is integrating annotated datasets and summary networks with large informatics frameworks (Akil, Martone, & Van Essen, 2011; Martone et al., 2008).

CONNECTOMICS DISCOVERY

Connectomics allows analyses not possible with previous ultrastructural techniques. Any cell can be traced multiple times, even concurrently, with full event logging. Starting from any point in a canonical volume, a nexus of synapses can be traced back to diverse sources. Simultaneously, the coordinates (locations) and logical relationships (links) for every cell (structure) or part (child structure) are captured and used to automatically build three-dimensional renderings, network graphs, and data navigation maps. As all attributes are stored as databases, all forms of data query tools are enabled, and all data can be shared. Unlike legacy anatomy, everything done in a connectomics format is fully transparent. Further, it can be shown that synaptic identification can approach 100% with connectomics, even for synapses whose section plane is parallel to the plane of the PSD (Anderson et al., 2011a). With TEM connectomics, every synapse to or from a cell can be mapped. Certainly vast numbers of synapses are missed in traditional single-section sampling (Marc & Liu, 2000) leading to underestimates of simple motifs such as serial synapses. More importantly, every potential synapse can be flagged, assigned a certainty or query status, assessed by other analysts, reviewed in its full network context, reimaged goniometrically at higher resolution if necessary, and have its identity finalized (Anderson et al., 2011a, 2011b). This was never possible with manual TEM, even with serial sections.

But has connectomics shown us anything new? In our explorations of the rabbit retinal connectome RC1 many aspects of neural organization have been significantly extended by connectomics. We here summarize several of them. The conventions we use to describe synaptic chains and their amplification properties are these: >, high-gain sign-conserving (e.g., mediated by ionotropic glutamate receptors); $>_m$, high-gain sign-inverting (mediated by mGluR6 glutamate receptors); $>_i$, low-gain sign-inverting (ionotropic glycine and GABA receptors). High-gain transfers are assigned a nominal gain of n and low-gain inhibitory transfers are assigned a separate gain of p, based on evidence that most excitatory gains are $\gg 1$ (Copenhagen, Hemilä, & Reuter, 1990; Yang & Wu, 2004) and inhibitory gains

NETWORK VIZ

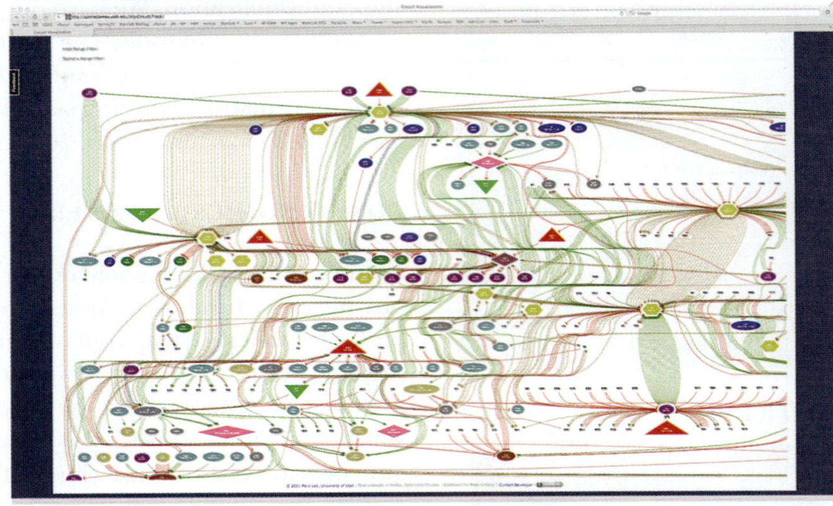

STRUCTURE VIZ

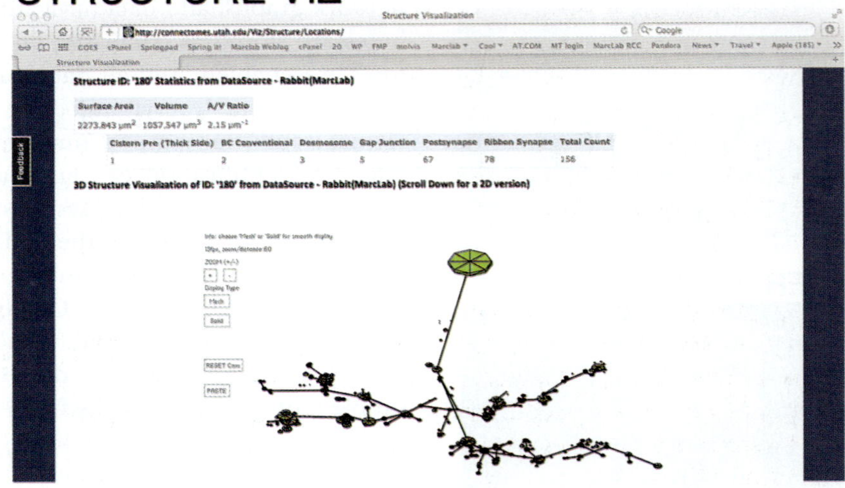

CONNECTOME QUERY

FIGURE 10.4   The Viz analysis environment. Viz is a set of browser tools that queries the RC1 database to produce network visualizations. Network Viz (top) generates complete multihop network diagrams with a number of display options including complete or reduced edges. Structure Viz (middle) generates navigable skeletons of cells so that individual synapses or features can be tracked in between the database and the Viking display. This window displays CBb cell C180. Finally, the entire database can be queried with conventional SQL strings to extract quantitative attributes of cells and connections (bottom).

are net fractional (Maltenfort, Heckman, & Rymer, 1998; Wu, 1991). The latter is not necessarily true in all cases and requires further investigation, which further justifies treating classical ionotropic inhibition as a separate parameter. Coupling is also a separate parameter, here treated as ~1 for notational simplicity, although it is certainly attenuating. Total chain gains are multiplicative. Thus, a chain of cone $>_m$ CBb > AC

   R. E. MARC, B. W. JONES, J. S. LAURITZEN, C. B. WATT, AND J. R. ANDERSON

$>_i$ GC has a total gain of n^2p and a net sign-conserving polarity (i.e., the chain copies the cone polarity into the target GC).

Refactoring the Inner Plexiform Layer

Nonmammalian bony vertebrates (reptiles, avians, amphibians, teleost fishes) all possess multistratified bipolar cells where nominal ON bipolar cells clearly have synaptic outputs in the OFF layer (e.g., Sherry & Yazulla, 1993). On occasion, various workers have noted ribbon sites in mammalian BC axons, although these were never quantified or explored. Ultimately, the idea of mixed ON/OFF signaling had never been considered for mammalians until optical evidence of ON BC axonal ribbon output in the OFF layer of the IPL was reported by several groups (Dumitrescu et al., 2009; Hattar et al., 2002; Hoshi et al., 2009). This has now been validated by ultrastructural connectomics (Anderson et al., 2011a and extended to demonstrate that the entire OFF layer contains axonal ribbon outputs (figure 10.5) from the full range of ON cone (CBb) BC classes targeting ON ganglion cells with arbors in the OFF sublayer, that is, intrinsically photo-sensitive GCs (ipGCs) and bistratified diving GCs (bsdGCs), and discrete sets of glycinergic ACs (GACs) and GABAergic ACs (γACs) for ON → OFF crossover (Lauritzen et al., 2013). But why do ipGCs and bsdGCs arborize in the OFF layer to harvest CBb cell inputs when those are also available in the ON layer? We do not understand, but connectomics shows that these cells also capture massive input from OFF γACs but not OFF cone (CBa) BCs, providing wide-field ON polarity input via OFF chains: [cone > CBa BC > γAC $>_i$ ipGC/bsdGC] with a gain of n^2p. That is, they acquire a low-gain ON signal from the OFF layer. Whether this is really a critical part of the ipGC/bsdGC profile, it appears that certain OFF ACs are preferred targets. Because most inhibitory gains p could be fractional (Wu, 1991), the direct ON chain from CBb axons should have a higher sensitivity than the OFF chain by a factor of p^{-1}. An additional refactoring of the mammalian inner plexiform layer includes the deep incursion of certain CBa cells into the ON layer, leading to a midzone of intermixed ON CBb3 and OFF CBa2 BCs, where some monostratified GCs and ACs may collect from both types to generate ON-OFF elements. An important outcome of these findings is that the level and pattern of stratification of ACs and GCs in the inner plexiform layer do not predict the response properties of a cell. Indeed there are bistratified GCs with dendrites in the OFF layer that are pure ON cells and monostratified GCs on the nominal ON layer

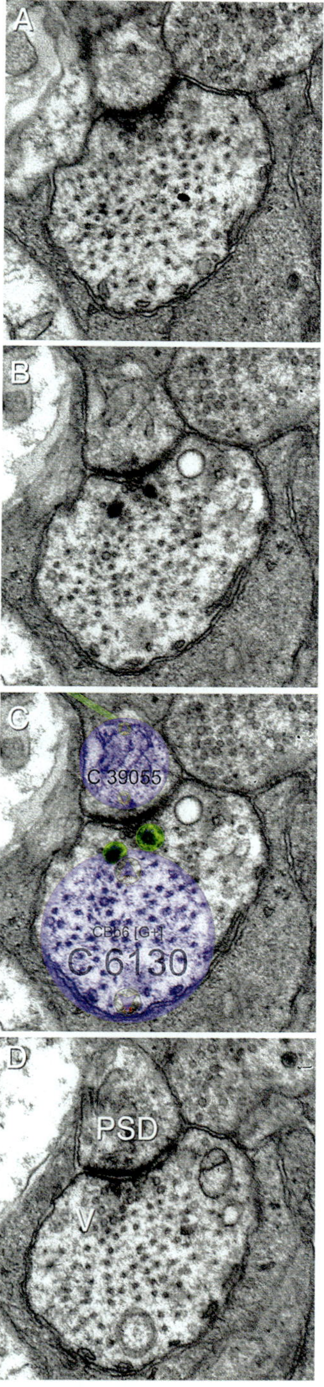

FIGURE 10.5 Axonal ribbons in ON CBb cells within the OFF layer of the inner plexiform layer. Three serial sections (A–D) in a CBb6 bipolar cell show an axonal ribbon essentially at midaxon. Each section is 70 nm thick, and the small presynaptic ribbons are oval with axes of 65 nm × 80 nm, surrounded by a small halo of vesicles (v). The target is a GC spine with a large postsynaptic density (PSD) spanning an arc of 400 nm. Section C is section B with annotation overlays. The diameters of the two green ribbon annotations are 100 nm. The PSD annotation was turned off for clarity.

that are ON-OFF. A complete ultrastructural network mapping is essential to understanding pathways.

Crossover Motifs

A major concept in retinal organization best articulated by Werblin and colleagues is that of polarity-matched ON → OFF and OFF → ON crossover pathways dominated by GAC signaling (Hsueh, Molnar, & Werblin, 2008; Werblin, 2010, 2011). Among the many possible roles of crossover are rectification compensation (Werblin, 2010) or even enhancement (Liang & Freed, 2010). The basic notion is that diffusely stratified GACs cross the ON-OFF border to mediate, for example, [CBb > GAC >$_i$ OFF GC] chains where the GAC input provides an appropriate OFF polarity via an anionic current to compensate for AMPA receptor rectification. Importantly (Werblin, 2010), refactoring of BC outputs in the inner plexiform layer shows that CBb axonal ribbons provide an ON → OFF crossover topology to monostratified GACs in the OFF layer directly. Moreover, there are also [CBb > γAC >$_i$ OFF GC] chains in the OFF layer (Lauritzen et al., 2013). Because the latter are deeply embedded in both feedback and feedforward motifs, they would be pharmacologically difficult to isolate. Another important point is that none of the CBb axonal output crossover elements involves the AII AC. Indeed, the bulk of the synaptic output of AII ACs is onto CBa cells of all classes, which means the net gain from the [AII >$_i$ CBa > OFF GC] chain is np and likely rectifying. There is extensive direct AII synaptic drive to α- and δ- OFF GCs, which are both strongly rectified GCs, but the net gain will only be p counterbalanced by even more numerous CBa2 drive with gain of np. So the AII AC is not likely a major source rectification correction for the OFF layer. Other GACs likely are.

Mapping All Rod-Cone Interaction Motifs

A major attribute of mammalian vision is rapid switching between rod- and cone-dominated operations, especially during protracted crepuscular regimes where both rods and cones operate. Psychophysical analyses, including over three decades of work by Ulf and Bjorn Stabell, reveal powerful rod–cone interactions, including mutually suppressive regimes (Brill, 1990; Buck, 2004; Frumkes & Eysteinsson, 1988; Goldberg, Frumkes, & Nygaard, 1983; Lange, Denny, & Frumkes, 1997; Stabell & Stabell, 1998, 2002; Thomas & Buck, 2006; Trezona, 1970, 1973) (also see chapter 34 by Buck), but the mechanisms have remained unknown. The notion that horizontal cells mediate such interactions is complicated by the lack of evidence for axonal signaling in mammalian horizontal cells. Connectomics reveals at least seven unique instances of high-gain mediated (C1–4) cone suppression of rod signaling (figure 10.6). Deep and wide connectomics of the rod BC cohort reveal that every rod BC gets direct inhibitory input from CBb-driven ON γACs and GACs and that ~25% of AC synapses on rod BC terminals come from ON cone ACs with a cone → rod suppression of n^2p. Further, both AI and AII ACs receive extensive cone pathway inhibition. AI ACs receive inhibition (~100 inhibitory synapses/cell) on their proximal dendrites in the OFF layer via selective CBa > γAC >$_i$ AI AC motifs, and AII ACs receive cone-driven inhibition at every level of the inner plexiform layer, including a highly selective inhibitory input from ON GACs that target rod BC, AII AC, and AI AC. Three pathways mediate rod suppression of cone signaling. Certain wide-field γACs (different from the cone → rod suppressors) collect sparse rod BC inputs and are both presynaptic and postsynaptic to CBb cells, forming a selective [RB > γAC >$_i$ CBb] suppression chain with a rod → cone suppression of n^2p. However, the most powerful motif is the [rod >$_m$ RB > AII ACs::CBb > γAC > CBb] chain with a rod → cone suppression of n^3p. Because AII cells are narrow-field elements (<100 μm) whereas the γACs are wide-field (>250 μm), each patch of rods can inhibit a vast field of surrounding cones.

Tiered Cone Bipolar Cell Coupling

Although AII::AII and AII::CBb couplings are well known, connectomics shows that cone BC coupling forms both extensive in-class sheets and cross-class tiers (Lauritzen et al., 2013). Coupling occurs between identical class pairs throughout the CBa and CBb cohorts. But coupling also occurs across neighboring pairs because their stratifications overlap vertically, and they all appear to use the similar connexins, predominantly Cx36 (Han & Massey, 2005). Thus, CBb tiered coupling occurs from CBb3::CBb4::CBb5::CBb5w::CBb6 but completely excludes any CBa or rod BC. This suggests that cone BCs may use coupling to smooth signaling transitions across BC classes with different dynamic ranges. Of course, we have, as yet, no clear data to infer the degree and modulation of BC coupling. In addition, certain cone BCs (e.g., CBb3, CBb5 cells) show covering of synaptic space (extensive terminal arbor overlap between neighboring cells) whereas others (CBb5w) show tiling of synaptic space and tip-to-tip coupling (figure 10.7). Whether CBa cells show cross-class tiering remains to be shown, but it is likely. Indeed, the presence of abundant CBa::CBa coupling is

    R. E. MARC, B. W. JONES, J. S. LAURITZEN, C. B. WATT, AND J. R. ANDERSON

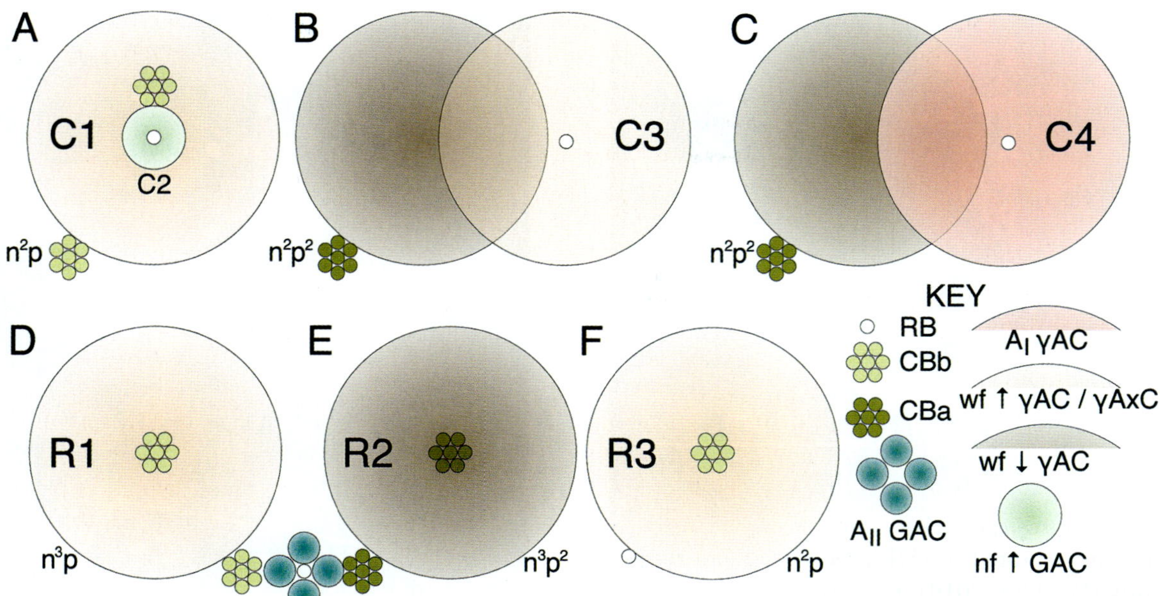

FIGURE 10.6   Seven rod-cone suppression pathways in the rabbit retina. Cone BC (hexagonal patch) suppression of rod BCs (single white circles) occurs via four discrete motifs. (A) Motifs C1/C2 are driven by patches of coupled ON CBb cells: wide-field γAC inputs (pale red C1) and narrow-field ON GAC inputs (pale green C2). (B) In motif C3, OFF CBa cells drive wide-field γAC that form a ring of extensive inhibitory synapses on A_I ACs. (C) In motif C4, a special class of γAxC (axonal cells) receives OFF layer input solely from OFF γACs and then targets rod BCs in the ON layer. Rod BC suppression of cone BCs occurs via three motifs. Single rod BCs drive patches of AII glycinergic ACs that are (D) coupled to all classes of CBb cells (motif R1) or (E) presynaptic to all classes of CBa cells (motif R2). These drive wide-field cone γACs to strongly inhibit distant cone BCs. (F) In motif C3, sparse rod BC inputs are collected by wide-field γACs that are primarily CBb-driven and CBb-feedback neurons. Each chain has a characteristic gain expressed as multiplicative excitations (n) and inhibition (p). (From Lauritzen et al., unpublished.)

consistent with the presence of many small suboptical gap junctions in the OFF layer (Kamasawa et al., 2006). How these networks are regulated is unknown, but there is evidence that neurons from the thalamic reticular nucleus can modulate gap junctions in a cell autonomous fashion (Haas, Zavala, & Landisman, 2011). Finally, some rare individual CBb cells of in-class coupled sheets are clearly identical to their neighbors but lack coupling to AII ACs. As described below, this is likely a consequence of joint distribution rules for network formation where the spatial mapping of contacts across classes cannot and need not achieve 100% fidelity.

Sparse Networks and Joint Distribution Rules

Networks are formed by the intersections of neurites in a volume (figure 10.8). But because different cell classes vary in neurite density and geometry (Reese, 2008), the probability of encounter will vary (Lauritzen et al., 2012). Most γACs cover the retina with extensive in-class overlap with coverages >>4 (i.e., simple center to center), whereas GACs, in contrast, have modest overlap closer to 4. Most (not all) GCs tile the same space with negligible overlap and coverages <4, often approaching 1. As shown above, BCs are mixed in patterning. Rod BCs and some cone BCs tile, but others cover. Because some classes are very sparse and others are dense, synaptic encounter rates will depend on joint distributions. Formally, superclasses and classes display very different Hausdorff dimensions, that is, the amount of space-filling curvature they express in a plane, which is very low for wide-field ACs and α-GCs and much higher for BCs, starburst amacrine cells, and directionally selective GCs. These geometric constraints underlie how network motifs develop and how they are sampled by other motifs. Put simply, it is impossible for every source and target to be optimized to achieve 100% contact or spatially constant contact variance. Because we do not know the sampling volumes for various cell classes, we must discover them by connectomics.

Historically, the descriptions of outflow of signals from one kind of cell to generic superclasses of targets (e.g., a given BC to ACs and GCs) has not been assessed in terms of joint distributions but rather just as percentages. But connectomics data now show that joint distributions are important (Lauritzen et al., 2013). Two examples of this are the flow of axonal ribbon signaling

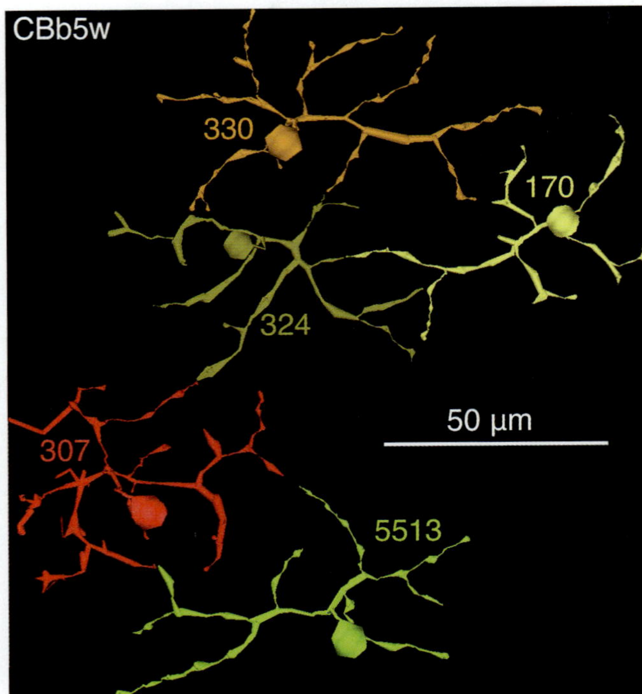

FIGURE 10.7   A coupled set of CBb5w ON BCs viewed in the XY plane (the retinal image plane). This is a new class of CBb cells that forms a precise stratum distal to the new class of CBb6 cells and are coupled in-class via tip-to-tip junctions. Each cell ID allows tracking of all features in Viking.

from CBb cell axons to targets in the OFF layer and the output of AII AC lobules on GC dendrites. In both cases there is a vast oversupply of sources relative to targets. GCs form only a small fraction of postsynaptic targets for AII AC lobules, but that is because AII ACs have much higher Hausdorff dimensions (are more space filling) than GCs and form far more lobules than are necessary for efficient GC targeting. In contrast, a single OFF α-GC dendrite traversing the neuropil in connectome RC1 receives input from every lobule it encounters and samples from every AII AC domain along its trajectory, for a sampling efficiency of 100%. Thus, the variance of output patterns of profiles across a given set has little functional meaning. A key question in any network design is: How much connectivity is required to reach an idempotent output, that is, one beyond which additional inputs are superfluous? Although we do not have formal answers to this, our analyses of cone BCs suggest that wide-field ACs need not and do not target every BC in a coupled tier to be idempotent. This means that ultrastructurally sampling one or a few BCs will never give a correct population profile for a given network's motifs. In addition, it is often stated (without real data) that neurons receive thousands of synapses, but it appears that many retinal GCs and wide-field ACs receive relatively few excitatory synapses over vast

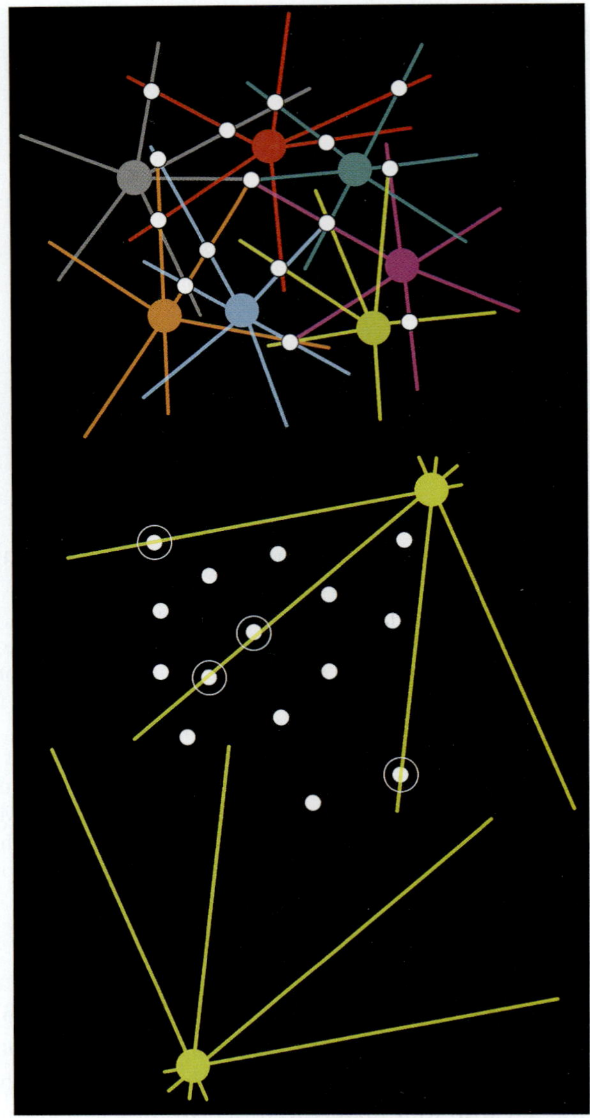

FIGURE 10.8   Sparse network topologies and joint distributions. A patch of BC axons (white) pierces the image plane of the retina. In the top field a cell class with high coverage is shown in different colors for every instance of the class. Each BC axon is contacted several times for an average contact of 2.4. In the bottom field two copies of one class of GCs (yellow) tile without overlap. Only one of them samples from the BC patch. Four circled BCs are contacted by the GC, and the GC is errorless in contacting any BC that is encountered. Cells whose dendrites do not fill a given space will sample only those inputs it encounters, and the ones it misses do not even exist as far as the GC is concerned. Thus, calculating the average outflow of BC synapses to GCs, in general, carries little useful information.

regions: instead of thousands of synapses, we may speak of tens. This has two implications: first, that noise reduction may be a major role of many feedback networks; and second, that we should not depend on snippets of TEM imagery to get a sense of a neuron's signal flow—we must map it all.

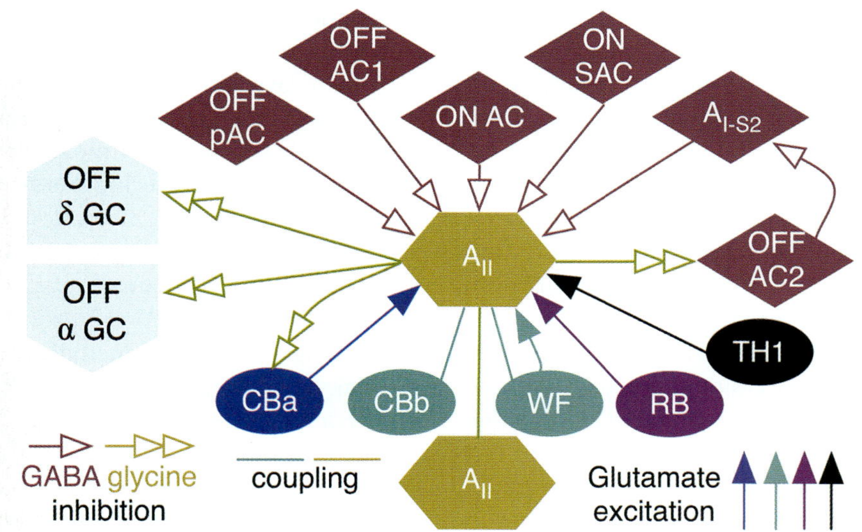

FIGURE 10.9   The connectome for class AII glycinergic ACs in the mammalian retina. The connectome shows four modes of excitation (solid arrows), three modes of coupling (lines), five modes of GABA inhibitory input (open arrows), and four glycine inhibitory output modes (double arrows). CBa, OFF cone BCs; CBb, ON cone BCs; WF, wide field ON cone BCs; RB, rod BCs; TH1, class 1 dopaminergic axonal cells; α-GCs; δ-GCs; pAC, peptidergic GABAergic AC; OFF AC1, dominant monostratified OFF cone AC population; OFF AC2, minor monostratified OFF cone AC population; ON AC, dominant monostratified ON cone AC population; ON SAC, ON starburst amacrine cell; A_{I-S2}, subclass S2 class AI rod-dominated GABAergic AC. Some of the groups can be further weighted. For example, although ON cone BCs classes (there are at least five) are coupled to AII cells via gap junctions, they differ in their gap junction areas, and one class (WF ON cone BCs) is also presynaptic via ribbon synapses. (From Marc et al. (2012), by permission of the authors.)

A Complete AII AC Profile

One of our early connectomics objectives was to completely profile AII ACs (Famiglietti & Kolb, 1975; Kolb & Famiglietti, 1974a; Strettoi, Raviola, & Dacheux, 1992). Because RC1 contains 40 AII ACs we have been able to generate an extremely quantitative view of this cell and its partnerships (Anderson et al., 2011b) and can summarize the major attributes of AII ACs (figure 10.9). The AII AC is best known for its unique intercalation in the rod signaling stream, but its engagement with photopic networks is extensive as well. Each AII AC captures input from ~10 BCs depending on the joint distribution of its field and the underlying rod BC tile. Each AII AC is nominally 8-connected (7.6 ± 1) with neighboring AII ACs via large gap junctions on the arboreal dendrites. The outflow of signals from AII AC includes coupling to all classes of CBb cells and synaptic output onto all classes of CBa cells and selective output to γAC that are feedforward elements into A_I AC in the OFF layer, as well as outputs to the sparse dendrites of OFF α- and δ-GC. AII AC also collect a range of inhibitory inputs from both wide and narrow field sources, with at least six different inputs: CBa-driven and CBb-driven wide-field γAC (of which there may be several different classes), CBa-driven wide-field peptidergic inputs, ON starburst AC inputs, sparse AI AC inputs (likely from type S2 cells), and ON narrow-field glycine AC inputs. Each AII AC also receives four to five perisomatic conventional synapses, presumably from TH1 dopaminergic axonal cells (Voigt & Wässle, 1987). Contrary to previous suppositions, we show that rabbit TH1 cells distinctively display glutamatergic and definitely not GABAergic signatures (Anderson et al., 2011b). Thus, it is likely that the ON responses of TH1 cells derived directly from CBb axonal ribbons (Dumitrescu et al., 2009; Hoshi et al., 2009) are passed to AII ACs in a sign-conserving AMPA-driven synapse. In addition, the lobules of AII ACs have long been known to receive CBa input, and we show that it derives from all classes.

The importance of mapping the full spectrum of AII AC connectivity includes developing a framework for understanding the evolutionary origin of AII ACs. If one removes all rod BC-related connections of the AII AC, over 90% of the connectivity remains, suggesting that it is an archetypal cone pathway crossover cell. In addition, AII ACs also capture sparse synaptic ribbon inputs from wide-field CBb cells, suggesting that they play some role in mediating the transition between scotopic and photopic signaling.

Classification of neurons has been based largely on shape and, more recently, molecular signatures. Connectomics provides additional quantitative and independent classification tools, even under the constraints of graph enumeration. Network mining is accelerated by molecular classifiers. The history of classification strategies in the retina and brain is a fitful progression of ad hoc attempts to discover metrics that segment cell populations (e.g., Scholl rings, fractals, diameters, number of neurites) with occasional but unpersuasive forays into multivariate and discriminant analysis. None has shown power sufficient to completely segment any cell population, and previous approaches required extensive human data pruning. In parallel, starting in the mid-1960s, a formal theory of unsupervised multidimensional classification was being developed in remote sensing (Cover & Hart, 1967; MacQueen, 1967). Based on various forms of clustering algorithms such as the K-means and isodata methods, these approaches showed that the dimensions of a true classification space must be orthogonal. Although some have suggested that there are "limits" to cluster analysis (Kong et al., 2005), this is formally incorrect. The limits are not analytic but rather are imposed by flawed data parameterizations such as improper gradient descent initialization, poor signal resolution, small sample sizes, and use of degenerate sets. The latter is a fundamental defect in many anatomical studies. With proper dimensioning these methods are very powerful (Marc, Murry, & Basinger, 1995). Importantly, the strategies of defining orthogonal dimensions in discrete space or rotating axes to generate maximal variance in continuous spaces (e.g., principal components analysis) are formally equivalent (Ding & He, 2004).

Two approaches have led to converging estimates of cell diversity in the retina: ad hoc visual categorization using Golgi impregnation or dye labeling (e.g., MacNeil et al., 1999a, 1999b; Rockhill et al., 2002), and structural or molecular profiling using cluster analysis (Badea & Nathans, 2004; Marc & Cameron, 2003; Marc & Jones, 2002; Marc, Murry, & Basinger, 1995). On balance, the number of neuronal classes in the retina is ~70 in mammalians and ~150 in nonmammalians. Golgi impregnation, although fallen from grace as a survey tool, remains the gold standard for visualizing cell diversity in many neural systems including retina (MacNeil et al., 1999b). Advanced imaging and cluster analysis have never been applied to such material, and a reevaluation is due. Dye methods (injection, diolistics, gene targeting) have also provided excellent visualizations of cell classes and, to some extent,

populations based on qualitative methods. But it is generally clear that patterns of stratification in the inner plexiform layer provide a very strong segmentation tool (Badea & Nathans, 2004). In contrast, an independent approach based on small molecule mixtures explicitly uses quantitative clustering methods and readily provides high-precision segmentation of neural classes. For example, molecular signatures based on intrinsic (GABA, glycine, glutamate, aspartate, taurine) and extrinsic markers (AGB) can provide extensive segmentation of all retinal cells, including fine segmentation of bipolar, amacrine, and ganglion cell classes, with likely near complete segmentation of the latter. The segmentation of ganglion cell classes by molecular markers closely matches that of both physiological and anatomical data but provides, in addition, important population data and information on cell precision. Because class molecular signatures are independent of other measures, they can be used to probe those metrics in ways previously untenable. For example there is a long (and failed) history of trying to classify retinal ganglion cells by diameter. Many different strategies based on tunable distributions were attempted, but all missed a key problem: the variances of cell diameters are very small, and ad hoc deconvolution could never extract them. After molecular classification the complete feature spaces of the 15 ganglion cell classes can be independently probed, showing that virtually all have unique sizes and small variances (Marc & Jones, 2002).

Although such molecular classification techniques are still in their infancy, their ability to robustly segment large, complex populations of neurons in the retina transcends all previous methods and suggests that a formal derivation of classification based on descriptive set theory should exist. Eventually, other classifier dimensions such as receptors and channels (Micheva & Bruchez, 2012; Micheva et al., 2010; Micheva & Smith, 2007) could be used as dimensions, but at present there are no efforts to do so in retina. Importantly, connectomics provides a context and need to both use and develop classifications. The need to use classification markers is based on the desire to explore the networks of neuronal trigger features. The entry of connectomics strongly changes the scenario for classification of cells. First, it provides a quantitative framework for characterizing all of the branches and nodes in a cell, and second, it allows multiple copies of cells to be assessed based on independent selection criteria such as molecular tags and network topology. By building the shapes and network maps for every cell along with its molecular feature space, three independent metrics possessing multiple dimensions converge to provide strong validations of classification. For example, the BC cohort extracted by visual inspection from Golgi and dye imaging material in rabbit retina (MacNeil et al., 2004) distinguishes 12 classes of cone bipolar cells, six OFF layer CBa cells and six ON layer CBb cells. Three-dimensional rendering of cells segmented by molecular signatures yields at least four more classes: CBa2w, dsCBab2 (an OFF BC with processes descending into the ON layer), CBb5w, and CBb6 (Lauritzen et al., 2012). In addition, each class is segmented by its coupling and tiling features, which cannot be easily extracted from single labeling methods. For example, CBb3 cells show extensive axonal overlap, whereas other CBb cells such as CBb5 show discrete tip-to-tip coupling and tiling. Whether the overlapped CBb3 cells represent subclasses is unclear, as they are, nevertheless, intimately coupled. We expect that continuing analyses will enable finer resolution of cell classes based on networks. Finally, it is important to remember that topology, network links, cell-based classification, and cell shape are related but not the same. Although there are distortions of true shape in histology (which impact interpretations of electromorphology), those distortions do not impact topology, network associations, or shape classifiers as far as we know.

CONCLUSION

Statistical completeness requires mapping all contacts and contact patterns across multiple instances of a cell class. The variance of some metric should be minimized when sampling approaches completeness, but we are still discovering those metrics, and some classical statistics may turn out to be useless or even misleading. For example, the mean rod BC ribbon synapse count onto four adjacent AII ACs in RC1 is 74 ± 5 (1 SD) with coefficient of variation (CV) of 0.066. The same cells have a mean rod BC contact count of 11.5 ± 3.7 and a CV of 0.32, a variation fivefold greater for exactly the same sample space. This suggests that neurons normalize synapse number despite varying neurite overlap geometries. On the other hand, there are spatial variances that are clearly meaningless in a physiologic context, such as the percentage of output onto various cells from AII ACs, although they may be useful as classifiers. In network flow the partitioning of output is not a usable metric, whereas the sampling of an output grid as the input to a spatially complex cell is critical. In that scenario the sampling of AII ACs by GCs is perfect (i.e., sampling from all encountered opportunities) and likely idempotent. Completeness is also gauged by edge density in network graphs where submotifs can be extracted and quantitatively compared.

Achieving completeness in connectomics involves three practical issues. Resolution sets the scale at which building connectome volumes will be feasible. A resolution of 10 nm clearly limits analysis of connectivity but admits large sampling fields. Resolution at 2 nm or better to unambiguously mark synapses and gap junctions is absolutely critical for completely mapping any network (Bourne & Harris, 2011; Kamasawa et al., 2006; Massey, 2008; O'Brien et al., 2012), and, at present, TEM is optimal for such investigations. Molecular (Anderson et al., 2011b; Shu et al., 2011) or optical (Bock et al., 2011; Briggman, Helmstaedter, & Denk, 2011) tagging to segment populations in advance of building networks is also essential both for independently verifying identities and as a strategy to speed annotation of selected networks. It is the ideal way to perform hypothesis-guided analysis in connectomics. Finally, increasing the number of ATEM platforms and sharing them are essential for synaptic connectomics (Anderson et al., 2009). It is also possible to build specialized tools, but inexpensive high-resolution commercial systems must be developed for the scientific community. The ATEM platform itself has immense power for the analysis of nonneural histoarchitectures. The next generation of tools should also facilitate comparative connectomics to explore the evolution and development of retinal networks and pathoconnectomics to explore neurodegenerations (Jones et al., 2003, 2011).

ACKNOWLEDGMENTS

Funding was provided by the National Institutes of Health (EY02576, EY015128, and EY014800), the National Science Foundation (0941717), and Research to Prevent Blindness.

DISCLOSURE STATEMENT

Robert E. Marc is a principal of Signature Immunologics, Inc., manufacturer of antibodies against small molecules used in some of the work described.

REFERENCES

Akil, H., Martone, M. E., & Van Essen, D. C. (2011). Challenges and opportunities in mining neuroscience data. *Science, 331,* 708–712.

Amari, S. I., Beltrame, F., Bjaalie, J. G., Dalkara, T., Schutter, E. D., Egan, G. F., et al. (2002). Neuroinformatics: The integration of shared databases and tools towards integrative neuroscience. *Journal of Integrative Neuroscience, 1,* 117–128.

Anderson, J. R., Grimm, B., Mohammed, S., Jones, B. W., Spaltenstein, J., Koshevoy, P., et al. (2011a). The Viking

Viewer: Scalable multiuser annotation and summarization of large connectomics datasets. *Journal of Microscopy, 241,* 13–28. doi: 10.1111/j.1365-2818.2010.03402.x.

Anderson, J. R., Jones, B. W., Watt, C. B., Shaw, M. V., Yang, J. H., DeMill, D., et al. (2011b). Exploring the retinal connectome. *Molecular Vision, 17,* 355–379.

Anderson, J. R., Jones, B. W., Yang, J. H., Shaw, M. V., Watt, C. B., Koshevoy, P., et al. (2009). A computational framework for ultrastructural mapping of neural circuitry. *PLoS Biology, 7*(3), e1000074. doi: 10.1371/journal.pbio.1000074.

Aster, R., Borchers, B., & Thurber, C. (2005). *Parameter estimation and inverse problems.* New York: Academic Press.

Badea, T. C., & Nathans, J. (2004). Quantitative analysis of neuronal morphologies in the mouse retina visualized by using a genetically directed reporter. *Journal of Comparative Neurology, 480*(4), 331–351.

Berlanga, M. L., Phan, S., Bushong, E. A., Wu, S., Kwon, O., Phung, B. S., et al. (2011). Three-dimensional reconstruction of serial mouse brain sections: Solution for flattening high-resolution large-scale mosaics. *Frontiers in Neuroanatomy, 5,* 17.

Bock, D. D., Lee, W. C., Kerlin, A. M., Andermann, M. L., Hood, G., Wetzel, A. W., et al. (2011). Network anatomy and in vivo physiology of visual cortical neurons. *Nature, 471*(7337), 177–182.

Bourne, J. N., & Harris, K. M. (2011). Nanoscale analysis of structural synaptic plasticity. *Current Opinion in Neurobiology, 22,* 1–11. doi: 10.1016/j.conb.2011.10.019.

Briggman, K. L., & Denk, W. (2006). Towards neural circuit reconstruction with volume electron microscopy techniques. *Current Opinion in Neurobiology, 16,* 562–570.

Briggman, K. L., Helmstaedter, M., & Denk, W. (2011). Wiring specificity in the direction-selectivity circuit of the retina. *Nature, 471,* 138–188.

Brill, M. H. (1990). Mesopic color matching: Some theoretical issues. *Journal of the Optical Society of America. A, Optics and Image Science, 7,* 2048–2051. doi: 10.1364/JOSAA.7.002048.

Buck, S. L. (2004). Rod-cone interactions in human vision. In L. M. Chalupa & J. Werner (Eds.), *Visual neurosciences* (pp. 863–878). Cambridge, MA: MIT Press.

Calkins, D. J., & Sterling, P. (1996). Absence of spectrally specific lateral inputs to midget ganglion cells in primate retina. *Nature, 381,* 613–615.

Calkins, D. J., & Sterling, P. (2007). Microcircuitry for two types of achromatic ganglion cell in primate fovea. *Journal of Neuroscience, 27,* 2646–2653.

Calkins, D. J., Tsukamoto, Y., & Sterling, P. (1998). Microcircuitry and mosaic of a blue-yellow ganglion cell in the primate retina. *Journal of Neuroscience, 18*(9), 3373–3385.

Copenhagen, D. R., Hemilä, S., & Reuter, T. (1990). Signal transmission through the dark-adapted retina of the toad (Bufo marinus). Gain, convergence, and signal/noise. *Journal of General Physiology, 95,* 717–732.

Cover, T., & Hart, P. (1967). Nearest neighbor pattern classification. *IEEE Transactions on Information Theory, 13,* 21–27. doi: 10.1109/TIT.1967.1053964.

Denk, W., & Horstmann, H. (2004). Serial block-face scanning electron microscopy to reconstruct three-dimensional tissue nanostructure. *PLoS Biology, 2,* e329. doi: 10.1371/journal.pbio.0020329.

Diestel, R. (2005). *Graph theory.* Heidelberg: Springer-Verlag.

Ding, C., & He, X. (2004). K-means clustering via principal component analysis. *Proceedings of the International Conference on Machine Learning, 21,* 29–36.

Dumitrescu, O. N., Pucci, F. G., Wong, K. Y., & Berson, D. M. (2009). Ectopic retinal ON bipolar cell synapses in the OFF inner plexiform layer: Contacts with dopaminergic amacrine cells and melanopsin ganglion cells. *Journal of Comparative Neurology, 517,* 226–244.

Famiglietti, E. V., Jr., & Kolb, H. (1975). A bistratified amacrine cell and synaptic circuitry in the inner plexiform layer of the retina. *Brain Research, 84,* 293–300.

Fiala, J. C. (2005). Reconstruct: A free editor for serial section microscopy. *Journal of Microscopy, 218,* 52–61. doi: 10.1111/j.1365-2818.2005.01466.x.

Frumkes, T. E., & Eysteinsson, T. (1988). The cellular basis for suppressive rod-cone interaction. *Visual Neuroscience, 1,* 263–273. doi: 10.1017/S0952523800001929.

Gaietta, G., Deerinck, T. J., Adams, S. R., Bouwer, J., Tour, O., Laird, D. W., et al. (2002). Multicolor and electron microscopic imaging of connexin trafficking. *Science, 296,* 503–507.

Goldberg, S. H., Frumkes, T. E., & Nygaard, R. W. (1983). Inhibitory influence of unstimulated rods in the human retina: Evidence provided by examining cone flicker. *Science, 221,* 180–182.

Haas, J. S., Zavala, B., & Landisman, C. E. (2011). Activity-dependent long-term depression of electrical synapses. *Science, 334,* 389–391.

Han, Y., & Massey, S. (2005). Electrical synapses in retinal ON cone bipolar cells: Subtype-specific expression of connexins. *Proceedings of the National Academy of Sciences of the United States of America, 102,* 13313–13318. doi: 10.1073/pnas.0505067102.

Harary, F., & Palmer, E. M. (1973). *Graphical enumeration.* New York: Academic Press.

Hattar, S., Liao, H. W., Takao, M., Berson, D. M., & Yau, K. W. (2002). Melanopsin-containing retinal ganglion cells: Architecture, projections, and intrinsic photosensitivity. *Science, 295,* 1065–1070.

Hoffmann, C., Gaietta, G., Zürn, A., Adams, S. R., Terrillon, S., Ellisman, M. H., et al. (2010). Fluorescent labeling of tetracysteine-tagged proteins in intact cells. *Nature Protocols, 5,* 1666–1677. doi: 10.1038/nprot.2010.129.

Hoshi, H., Liu, W. L., Massey, S. C., & Mills, S. L. (2009). ON inputs to the OFF layer: Bipolar cells that break the stratification rules of the retina. *Journal of Neuroscience, 29,* 8875–8883.

Hsueh, H. A., Molnar, A., & Werblin, F. S. (2008). Amacrine-to-amacrine cell inhibition in the rabbit retina. *Journal of Neurophysiology, 100,* 2077–2088.

Jeong, W., Beyer, J., Hadwiger, M., Blue, R., Law, C., Vazquez, A., et al. (2010). SSECRETT and NeuroTrace: Interactive visualization and analysis tools for large-scale neuroscience datasets. *IEEE Computer Graphics and Applications, 30,* 58–70. doi: 10.1109/MCG.2010.56.

Jones, B. W., Kondo, M., Terasaki, H., Watt, C. B., Rapp, K., Anderson, J., et al. (2011). Retinal degenerative disease and remodeling in a large eye model. *Journal of Comparative Neurology, 519,* 2713–2733.

Jones, B. W., Watt, C. B., Frederick, J. M., Baehr, W., Chen, C. K., Levine, E. M., et al. (2003). Retinal remodeling triggered by photoreceptor degenerations. *Journal of Comparative Neurology, 464,* 1–16.

Jurrus, E., Paiva, A. R., Watanabe, S., Jorgensen, E. M., Anderson, J., Jones, B., et al. (2010). Detection of neuron membranes in electron microscopy images using auto-context. *Medical Image Analysis, 14,* 770–783.

Kamasawa, N., Furman, C. S., Davidson, K. G., Sampson, J. A., Magnie, A. R., Gebhardt, B. R., et al. (2006). Abundance and ultrastructural diversity of neuronal gap junctions in the OFF and ON sublaminae of the inner plexiform layer of rat and mouse retina. *Neuroscience, 142,* 1093–1117.

Karp, R. M. (1972). Reducibility among combinatorial problems. In R. E. Miller & J. W. Thatcher (Eds.), *Complexity of computer computations* (pp. 85–103). New York: Plenum Press.

Kleinfeld, D., Bharioke, A., Blinder, P., Bock, D. D., Briggman, K. L., Chklovskii, D. B., et al. (2011). Large-scale automated histology in the pursuit of connectomes. *Journal of Neuroscience, 31,* 16125–16138.

Klug, K., Herr, S., Ngo, I. T., Sterling, P., & Schein, S. (2003). Macaque retina contains an S-cone OFF midget pathway. *Journal of Neuroscience, 23,* 9881–9887.

Knott, G., Marchman, H., Wall, D., & Lich, B. (2008). Serial section scanning electron microscopy of adult brain tissue using focused ion beam milling. *Journal of Neuroscience, 28,* 2959–2964.

Kolb, H., & Famiglietti, E. V. (1974a). Rod and cone pathways in the inner plexiform layer of cat retina. *Science, 186,* 47–49.

Kolb, H., & Famiglietti, E. V., Jr. (1974b). Rod and cone pathways in the retina of the cat. *Investigative Ophthalmology, 15,* 935–946.

Kolb, H., & Nelson, R. (1993). OFF-alpha and OFF-beta ganglion cells in cat retina: II. Neural circuitry as revealed by electron microscopy of HRP stains. *Journal of Comparative Neurology, 329,* 85–110.

Kong, J. H., Fish, D. R., Rockhill, R. L., & Masland, R. H. (2005). Diversity of ganglion cells in the mouse retina: Unsupervised morphological classification and its limits. *Journal of Comparative Neurology, 489,* 293–310.

Lange, G., Denny, N., & Frumkes, T. E. (1997). Suppressive rod-cone interactions: Evidence for separate retinal (temporal) and extraretinal (spatial) mechanisms in achromatic vision. *Journal of the Optical Society of America. A, Optics, Image Science, and Vision, 14,* 2487–2498. doi: 10.1364/JOSAA.14.002487.

Lauritzen, J. S., Anderson, J. R., Jones, B. W., Watt, C. B., Mohammed, S., Hoang, J. V., et al. (2013). ON cone bipolar cell axonal synapses in the OFF inner plexiform layer of the rabbit retina. *Journal of Comparative Neurology, 521,* 977–1000. doi: 10.1002/cne.23244.

Lauritzen, J. S., Jones, B. W., Watt, C. B., Mohammed, S., Anderson, J. R., & Marc, R. E. (2012). Diffusely-stratified OFF cone bipolar cell inputs to amacrine cells in the ON inner plexiform layer. *Investigative Ophthalmology and Visual Sciences, ARVO Meeting Abstracts, 53,* 3159.

Liang, Z., & Freed, M. A. (2010). The ON pathway rectifies the OFF pathway of the mammalian retina. *Journal of Neuroscience, 30,* 5533–5543.

Lichtman, J. W., & Smith, S. J. (2008). Seeing circuits assemble. *Neuron, 60,* 441–448.

Luisi, J., Narayanaswamy, A., Galbreath, Z., & Roysam, B. (2011). The FARSIGHT trace editor: An open source tool for 3-D inspection and efficient pattern analysis aided editing of automated neuronal reconstructions. *Neuroinformatics, 9,* 305–315.

MacNeil, M. A., Heussy, J. K., Dacheux, R. F., Raviola, E., & Masland, R. H. (1999a). The population of amacrine cells in a mammalian retina. *Investigative Ophthalmology & Visual Science, 40,* S437–S437.

MacNeil, M. A., Heussy, J. K., Dacheux, R. F., Raviola, E., & Masland, R. H. (1999b). The shapes and numbers of amacrine cells: Matching of photofilled with Golgi-stained cells in the rabbit retina and comparison with other mammalian species. *Journal of Comparative Neurology, 413,* 305–326.

MacNeil, M. A., Heussy, J. K., Dacheux, R. F., Raviola, E., & Masland, R. H. (2004). The population of bipolar cells in the rabbit retina. *Journal of Comparative Neurology, 472,* 73–86.

MacQueen, J. (1967). Some methods for classification and analysis of multivariate observations. Proceedings of the 5th Berkeley Symposium on Mathematical Statistics and Probability. University of California Press (pp. 281–297).

Maltenfort, M. G., Heckman, C. J., & Rymer, W. Z. (1998). Decorrelating actions of Renshaw interneurons on the firing of spinal motoneurons within a motor nucleus: A simulation study. *Journal of Neurophysiology, 80,* 309–323.

Marc, R. E. (2010). Synaptic organization of the retina. In L. A. Levin, S. F. E. Nilsson, J. Ver Hoeve, S. M. Wu, P. L. Kaufman, & A. Alm (Eds.), *Adler's physiology of the eye* (pp. 443–458). New York: Elsevier.

Marc, R. E., & Cameron, D. A. (2003). A molecular phenotype atlas of the zebrafish retina. *Journal of Neurocytology, 30,* 593–654.

Marc, R. E., & Jones, B. W. (2002). Molecular phenotyping of retinal ganglion cells. *Journal of Neuroscience, 22,* 412–427.

Marc, R. E., Jones, B. W., Lauritzen, J. S., Watt, C. B., & Anderson, J. R. (2012). Building retinal connectomes. *Current Opinion in Neurobiology, 22,* 568–574

Marc, R. E., & Liu, W. (2000). Fundamental GABAergic amacrine cell circuitries in the retina: Nested feedback, concatenated inhibition, and axosomatic synapses. *Journal of Comparative Neurology, 425,* 560–582.

Marc, R. E., Murry, R. F., & Basinger, S. F. (1995). Pattern recognition of amino acid signatures in retinal neurons. *Journal of Neuroscience, 15,* 5106–5129.

Marcus, D. S., Harwell, J., Olsen, T., Hodge, M., Glasser, M. F., Prior, F., et al. (2011). Informatics and data mining tools and strategies for the human connectome project. *Frontiers in Neuroinformatics, 5,* 4. doi: 10.3389/fninf.2011.00004.

Mariani, A. P., & Lasansky, A. (1984). Chemical synapses between turtle photoreceptors. *Brain Research, 310,* 351–354.

Martone, M. E., Tran, J., Wong, W. W., Sargis, J., Fong, L., Larson, S., et al. (2008). The cell centered database project: An update on building community resources for managing and sharing 3D imaging data. *Journal of Structural Biology, 161,* 220–231.

Massey, S. C. (2008). Circuit functions of gap junctions in the mammalian retina. In R. H. Masland & T. Albright (Eds.), *The senses* (pp. 457–472). San Diego: Academic Press.

Mastronarde, D. N. (2005). Automated electron microscope tomography using robust prediction of specimen movements. *Journal of Structural Biology, 152,* 36–51.

Micheva, K. D., & Bruchez, M. P. (2011). The gain in brain: Novel imaging techniques and multiplexed proteomic imaging of brain tissue ultrastructure. *Current Opinion in Neurobiology, 22*(1), 94. doi: 10.1016/j.conb.2011.08.004.

Micheva, K. D., Busse, B., Weiler, N. C., O'Rourke, N., & Smith, S. J. (2010). Single-synapse analysis of a diverse synapse population: Proteomic imaging methods and markers. *Neuron, 68,* 639–653.

Micheva, K. D., & Smith, S. J. (2007). Array tomography: A new tool for imaging the molecular architecture and ultrastructure of neural circuits. *Neuron, 55,* 25–36.

Mikula, S., Trotts, I., Stone, J. M., & Jones, E. G. (2007). Internet-enabled high-resolution brain mapping and virtual microscopy. *NeuroImage, 35,* 9–15. doi: 10.1016/j.neuroimage.2006.11.053.

O'Brien, J. J., Chen, X., MacLeish, P. R., O'Brien, J., & Massey, S. C. (2012). Photoreceptor coupling mediated by connexin36 in the primate retina. *Journal of Neuroscience, 32,* 4675–4687.

Narayanaswamy, A., Wang, Y., & Roysam, B. (2011). 3-D image pre-processing algorithms for improved automated tracing of neuronal arbors. *Neuroinformatics, 9,* 219–231.

Normann, R. A., Perlman, I., Kolb, H., Jones, J., & Daly, S. J. (1984). Direct excitatory interactions between cones of different spectral types in the turtle retina. *Science, 224,* 625–627.

Oberlaender, M., Boudewijns, Z. S., Kleele, T., Mansvelder, H. D., Sakmann, B., & de Kock, C. P. (2011). Three-dimensional axon morphologies of individual layer 5 neurons indicate cell type-specific intracortical pathways for whisker motion and touch. *Proceedings of the National Academy of Sciences of the United States of America, 108,* 4188–4193. doi: 10.1073/pnas.1100647108.

Reese, B. E. (2008). Mosaics, tiling and coverage by retinal neurons. In R. H. Masland & T. Albright (Eds.), *The senses* (pp. 439–456). San Diego: Academic Press.

Rockhill, R. L., Daly, F. J., MacNeil, M. A., Brown, S. P., & Masland, R. H. (2002). The diversity of ganglion cells in a mammalian retina. *Journal of Neuroscience, 22,* 3831–3843.

Scholes, J. H. (1975). Colour receptors, and their synaptic connexions, in the retina of a cyprinid fish. *Philosophical Transactions of the Royal Society of London. Series B, Biological Sciences, 270,* 61–118.

Sherry, D. M., & Yazulla, S. (1993). Goldfish bipolar cells and axon terminal patterns: A Golgi study. *Journal of Comparative Neurology, 329,* 188–200.

Shu, X., Lev-Ram, V., Deerinck, T. J., Qi, Y., Ramko, E. B., Davidson, M. W., et al. (2011). A genetically encoded tag for correlated light and electron microscopy of intact cells, tissues, and organisms. *PLoS Biology, 9,* e1001041. doi: 10.1371/journal.pbio.1001041.

Sporns, O., Tononi, G., & Kötter, R. (2005). The human connectome: A structural description of the human brain. *PLoS Computational Biology, 1,* e42. doi: 10.1371/journal.pcbi.0010042.

Stabell, B., & Stabell, U. (1998). Chromatic rod-cone interaction during dark adaptation. *Journal of the Optical Society of America. A, Optics, Image Science, and Vision, 15,* 2809–2815. doi: 10.1364/JOSAA.15.002809.

Stabell, B., & Stabell, U. (2002). Effects of rod activity on color perception with light adaptation. *Journal of the Optical Society of America. A, Optics, Image Science, and Vision, 19,* 1249–1258. doi: 10.1364/JOSAA.19.001249.

Stell, W. K., Ishida, A. T., & Lightfoot, D. O. (1977). Structural basis for on- and off-center responses in retinal bipolar cells. *Science, 198,* 1269–1271.

Stevens, J. K., Davis, T. L., Friedman, N., & Sterling, P. (1980). A systematic approach to reconstructing microcircuitry by electron microscopy of serial sections. *Brain Research, 2,* 265–293.

Strettoi, E., Raviola, E., & Dacheux, R. F. (1992). Synaptic connections of the narrow-field, bistratified rod amacrine cell (AII) in the rabbit retina. *Journal of Comparative Neurology, 325,* 152–168.

Tasdizen, T., Koshevoy, P., Grimm, B., Anderson, J. R., Jones, B. W., Whitaker, R., et al. (2010). Automatic mosaicking and volume assembly for high-throughput serial-section transmission electron microscopy. *Journal of Neuroscience Methods, 193,* 132–144.

Thomas, L. P., & Buck, S. L. (2006). Foveal and extra-foveal influences on rod hue biases. *Visual Neuroscience, 23,* 539–542.

Trezona, P. W. (1970). Rod participation in the "blue" mechanism and its effect on colour matching. *Vision Research, 10,* 317–332. doi: 10.1016/0042-6989(70)90103-3.

Trezona, P. W. (1973). The tetrachromatic colour match as a colorimetric technique. *Vision Research, 13,* 9–25. doi: 10.1016/0042-6989(73)90161-2.

van den Heuvel, M. P., & Sporns, O. (2011). Rich-Club Organization of the human connectome. *Journal of Neuroscience, 31,* 15775–15786.

Van Essen, D. C., Glasser, M. F., Dierker, D. L., Harwell, J., & Coalson, T. (2011). Parcellations and hemispheric asymmetries of human cerebral cortex analyzed on surface-based atlases. Cerebral Cortex [Epub ahead of print]. doi: 10.1093/cercor/bhr290.

Voigt, T., & Wässle, H. (1987). Dopaminergic innervation of AII amacrine cells in mammalian retina. *Journal of Neuroscience, 7*(12), 4115–4128.

Werblin, F. S. (2010). Six different roles for crossover inhibition in the retina: Correcting the nonlinearities of synaptic transmission. *Visual Neuroscience, 27,* 1–8. doi: 10.1017/S0952523810000076.

Werblin, F. S. (2011). The retinal hypercircuit: A repeating synaptic interactive motif underlying visual function. *Journal of Physiology, 589,* 3691–3702.

Witkovsky, P., & Dowling, J. E. (1969). Synaptic relationships in the plexiform layers of carp retina. *Zeitschrift fur Zellforschung und Mikroskopische Anatomie (Vienna, Austria), 100,* 60–82.

Wong, E., Baur, B., Quader, S., & Huang, C. H. (2012). Biological network motif detection: Principles and practice. *Briefings in Bioinformatics, 13,* 202–215.

Wu, S. M. (1991). Input-output relations of the feedback synapse between horizontal cells and cones in the tiger salamander retina. *Journal of Neurophysiology, 65,* 1197–1206.

Yang, X. L., & Wu, S. M. (2004). Signal transmission from cones to amacrine cells in dark- and light-adapted tiger salamander retina. *Brain Research, 1029,* 155–161.

11 Synaptic Mechanisms of Color and Luminance Coding: Rediscovering the X–Y-Cell Dichotomy in Primate Retinal Ganglion Cells

JOANNA D. CROOK, ORIN S. PACKER, JOHN B. TROY, AND DENNIS M. DACEY

BACKGROUND

A more complete and mechanistic view is rapidly emerging of the diverse retinal circuits that initiate spectral and spatial coding in the trichromatic primate retina. Synaptic currents evoked by stimuli that selectively modulate the long- (L), middle- (M), or short (S)-wavelength sensitive cone inputs to identified ganglion cell types reveal that for the two major color-sensitive pathways, the midget circuit (L vs. M cone) and the small bistratified circuit (S vs. L+M cone), opponency arises presynaptically, via bipolar cell synaptic excitation, and without recourse to amacrine cell inhibitory networks. By contrast, the achromatic (L+M cone) "parasol" pathway lacks cone opponency but displays a prominent glycinergic crossover inhibition. We revisit the long-standing but controversial hypothesis that primate midget and parasol ganglion cells are homologous to the intensively studied beta-X and alpha-Y cell types of the cat's retina, respectively. The midget cells project only to the lateral geniculate nucleus (LGN) and show the high spatial density characteristic of the beta-X cell; chromatic opponency is a variable property and appears to "piggyback" on a fundamentally X-like circuit with high achromatic contrast sensitivity. Parasol cells project to both the superior colliculus and the LGN and show nonlinear spatial summation, properties characteristic of the alpha-Y-cell pathway. As occurs in the nonprimate mammal, the small Y-cell subunits originate via summation of transient excitatory bipolar cell inputs to the parasol cell. Defining the type-specific synaptic physiology of additional diverse ganglion cell populations, including Y-like and color-opponent types, is now an ongoing and attainable goal in the primate.

Visual Pathway Origins

The visual pathway arises from diverse ganglion cell populations that display characteristic morphologies, microcircuits, and receptive field properties (see chapter 13 by Roska and Meister). As in other mammals a more complete understanding of the ganglion cells in the primate has emerged gradually (Dacey, 1994; Dacey et al., 2003, 2010; Kolb, Linberg, & Fisher, 1992; Rodieck, 1988; Rodieck, Brening, & Watanabe, 1993; Rodieck & Watanabe, 1993; Szmajda, Grunert, & Martin, 2008; Yamada, Bordt, & Marshak, 2005). At present at least 17 distinct ganglion cell types have been identified morphologically. However, estimates of the relative spatial densities of these types suggest that five populations are present at a relatively high density and comprise ~70% of the ganglion cells in the retinal periphery, and probably an even greater percentage in the macular region, where spatial resolution is critical (Dacey, 1993b). These five populations are the ON and OFF-center midget and parasol cells and the small bistratified cells, which project, respectively, to the parvo-, magno-, and koniocellular divisions of the LGN (figure 11.1). By contrast, the 12 additional cell types currently recognized account for much of the remaining 30% of the ganglion cells. Each type is low in spatial density, ranging from less than 1% to about 3% of the total ganglion cells and thus not well characterized physiologically (figure 11.1); still it is clear these cells contribute additional color-coding and achromatic pathways to the retinogeniculate projection and thus visual processing in primary visual cortex (Callaway, 2005; Crook et al., 2008a; Dacey et al., 2005; Dacey & Packer, 2003; Tailby, Solomon, & Lennie, 2008). One goal of this

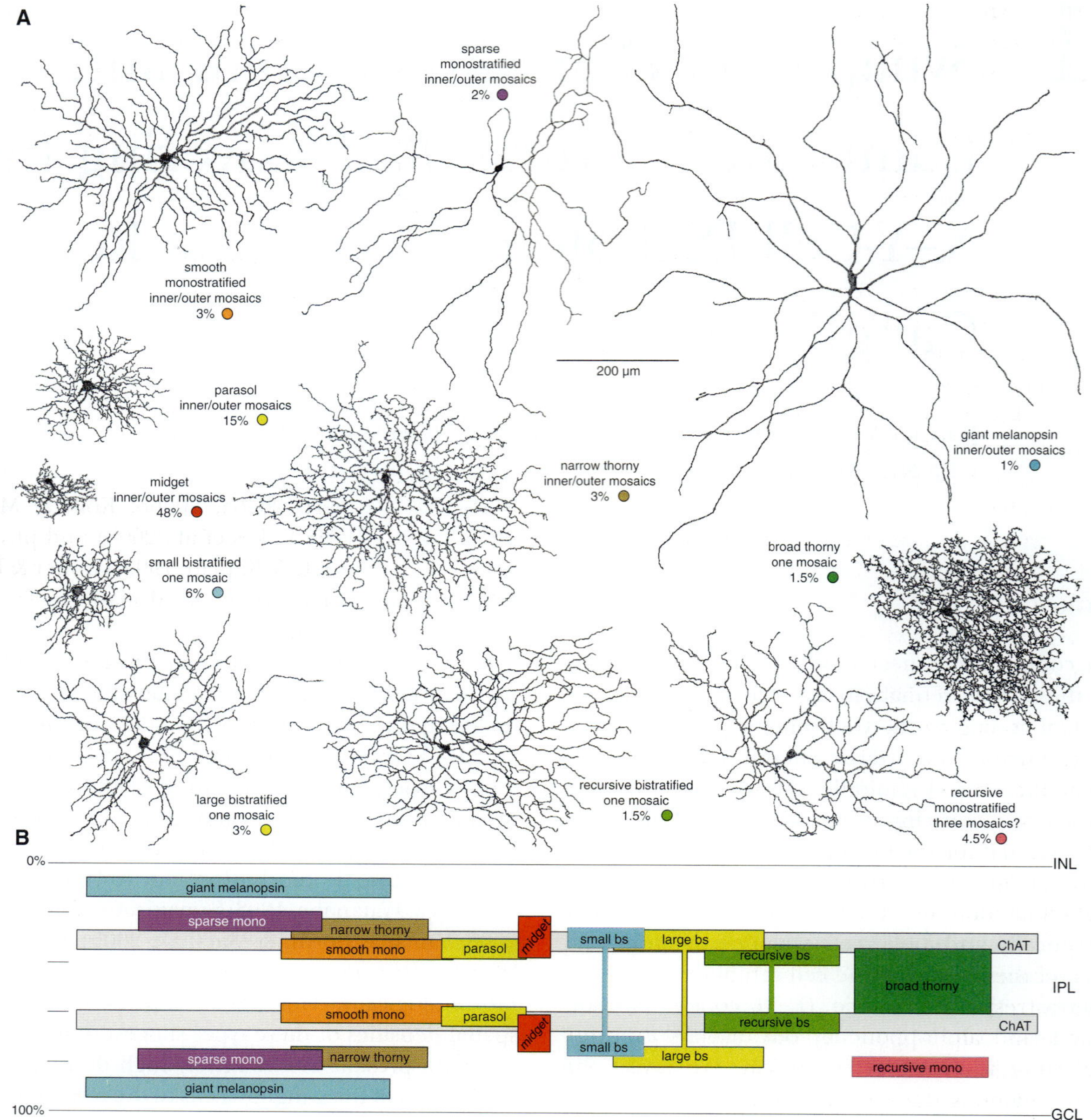

FIGURE 11.1   Dendritic morphology, relative spatial densities, and depth of dendritic stratification for diverse ganglion cell populations of the primate retina. (A) Five "high-density" types, the ON- and OFF-center midgets, ON- and OFF-center parasols, and the small bistratified cells, and 12 additional "low-density" types account for ~90% of total ganglion cells, suggesting that a few additional low-density types may yet be discovered. The relative percentage for each type was estimated by taking dendritic overlap (coverage) from the photostained mosaic of a given type (Dacey et al., 2003) and using this value to determine the percentage of total ganglion cell density at 8 mm temporal eccentricity (Wässle et al., 1989). (B) Dendritic stratification for the cells shown in A relative to the cholinergic amacrine cell processes ("ChAT bands"). Lengths of bars: relative dendritic field diameter. Vertical positioning of each bar is centered on the mean percentage stratification depth. GCL, ganglion cell layer; INL, inner nuclear layer; IPL, inner plexiform layer.

chapter is to review advances made over the last decade in understanding such extreme visual pathway diversity and begin to frame an answer to the question of why there is such a plethora of low-density ganglion cell populations.

Synaptic Mechanisms for Color and Luminance

In the last 10 years or so technical advances have made it possible to consistently make the link from single cell morphology and central projection to the underlying synaptic mechanisms that endow the unique coding properties of a given visual pathway. Perhaps the best example to date is the ON–OFF direction selective ganglion cell where the application of the voltage clamp has been used to reveal diverse excitatory and inhibitory synaptic mechanisms that give rise to direction selectivity (Briggman, Helmstaedter, & Denk, 2011; Taylor & Vaney, 2002). In the trichromatic Old World macaque monkeys, a long-standing question has been how the long- (L), middle- (M) and short- (S) wavelength sensitive cone types interact both synergistically and antagonistically to create red-green and blue-yellow spectrally opponent and nonopponent receptive fields in the high-density ganglion cell types that project to the LGN. Another goal of this chapter, therefore, is to review current understanding of the synaptic mechanisms underlying chromatic and achromatic signaling in midget, small bistratified, and parasol ganglion cells.

Comparisons across Species

Any comparison of the primate to the nonprimate mammal must begin with the visual pathways of the cat, the subject of intensive study for 50 years (Kuffler, 1953; O'Brien, Richardson, & Berson, 2003; Stone, 1983). The observation that ganglion cells, whose discharges were recorded either in the intact eye of the cat or from axons of the optic tract, show either linear or nonlinear spatial summation in their receptive fields, termed, respectively, X and Y cells (Enroth-Cugell & Robson, 1966), and the subsequent demonstration that this distinction corresponded in turn to the anatomically distinct beta and alpha ganglion cell types (Boycott & Wässle, 1974; Cleland, Levick, & Wässle, 1975; Saito, 1983), were among the early triumphs of retinal neurobiology. In primates an obvious and parsimonious correspondence, based on a number of anatomical and physiological properties, was made between the high-density beta-X and midget cells and the lower-density alpha-Y and parasol ganglion cells (Leventhal, Rodieck, & Dreher, 1981; Rodieck, Binmoeller, & Dineen, 1985; Rodieck, Brening, & Watanabe, 1993; Watanabe & Rodieck, 1989). But this midget–X-cell and parasol–Y-cell homology has been fundamentally questioned, indeed turned on its head, with the suggestion that in fact the parasol cells are a more likely homologue of the beta-X than the alpha-Y cell (Lee, 1999; Shapley & Perry, 1986). The question of the X–Y homology across species is not simply an esoteric issue of classification but fundamental to how functional studies across species acquire validity for understanding the human visual process. In addition, given the diverse origins of the visual pathway, including the retinogeniculate projection, outlined above, the question of functional and anatomical homology across species may not be as obvious as was suggested in an earlier era. A final goal of this chapter, then, is to review recent evidence from work in primates that bears on the X–Y-cell distinction and the evolutionary origin of parallel visual pathways.

MIDGET CELLS AND THE SYNAPTIC MECHANISM FOR RED–GREEN OPPONENCY

The "midget pathway" is one of the most intensively studied and clearly identified neural circuits both anatomically (e.g., Boycott & Dowling, 1969; Boycott & Wässle, 1991; Calkins et al., 1994; Calkins & Sterling, 1996; Dacey, 1993b; Dacey & Petersen, 1992; Jusuf, Martin, & Grunert, 2006a, 2006b; Kolb & Marshak, 2003; Polyak, 1941; Telkes et al., 2008) and physiologically (e.g., Benardete & Kaplan, 1999; Croner & Kaplan, 1995; Crook et al., 2011a; de Monasterio & Gouras, 1975; De Valois, Abramov, & Jacobs, 1966; Derrington, Krauskopf, & Lennie, 1984; Derrington & Lennie, 1984; Diller et al., 2004; Kaplan & Shapley, 1986; Lee, Martin, & Valberg, 1989; Lee et al., 1987, 1990; Wiesel & Hubel, 1966); yet there remains abiding and surprising disagreement about the fundamental role of the midget circuit in spatial and color vision. Indeed, it has been argued forcefully that the midget circuit principally serves to transmit either an achromatic, high-resolution spatial signal (Calkins & Sterling, 1999; Rodieck, 1991) or a pure color-related signal (Buzas et al., 2006; Cooper, Sun, & Lee, 2012; Lee, 1999; Lee et al., 2012; Martin et al., 2001; Shapley, 2006; Shapley & Perry, 1986) or play an essential and dual role in both visual channels (Ingling & Martinez-Ureigas, 1983; Lennie, Haake, & Williams, 1991; Lennie & Movshon, 2005; Martin et al., 2011; Solomon & Lennie, 2007).

Evidence for L versus M Cone Selective Wiring

One of the central questions that has prevented consensus regarding the functional architecture of the midget pathway concerns both the anatomical and synaptic basis for red–green color opponency. It is well established that the color-opponent light responses of midget cells arise by an antagonistic interaction between L and M cone signals (Buzas et al., 2006; Crook et al., 2011; Derrington, Krauskopf, & Lennie, 1984; Reid & Shapley, 1992, 2002). But there are two largely opposing views about how the midget circuit accomplishes L versus M cone opponency; each view leads to a very different picture of the development, evolution, and overall functional significance of the midget pathway.

The first suggestion, implicit in the LGN recordings of Wiesel and Hubel (1966), is that of labeled lines, whereby L and M cone type-selective connections are directed to the antagonistic receptive field center versus surround. In the central ~10° of visual angle, midget cells engage in a unique private line connection with a single cone photoreceptor (figure 11.2A–C). In the midget microcircuit the ganglion cell dendritic tree, barely 5 µm in diameter, receives virtually all excitatory synaptic input from an equally diminutive midget bipolar cell that in turn establishes a synaptic link with a single cone (Boycott & Dowling, 1969; Calkins et al., 1994; Dacey, 1993b; Jusuf, Martin, & Grunert, 2006b; Kolb & Dekorver, 1991). Thus, there is a clear anatomical basis for an excitatory receptive field center driven by input from a single L or M cone. What about the larger receptive field surround? Several recent studies argue for either complete or partial cone type selectivity to the surround as well (Buzas et al., 2006; Lee, Kremers, & Yeh, 1998; Lee et al., 2012; Martin et al., 2001; Reid & Shapley, 2002). In addition, at eccentricities beyond the macular region, midget ganglion cell dendritic trees and the corresponding receptive field centers enlarge (figure 11.2A–C) (Croner & Kaplan, 1995; Dacey & Petersen, 1992; Perry, Oehler, & Cowey, 1984; Watanabe & Rodieck, 1989). Multiple midget bipolar cells, which maintain the single cone connectivity well out into the far retinal periphery (Milam, Dacey, & Dizhoor, 1993; Wässle et al., 1994) now make convergent connections with midget ganglion cell dendrites (figure 11.2B, C) (Dacey, 1993b; Jusuf, Martin, & Grunert, 2006a). Physiological evidence has also been given that some of these midget ganglion cells in the retinal periphery show red-green opponency and a dominant or even pure input from either the L or M cone types to both center and surround (Buzas et al., 2006; Martin et al., 2001; Solomon et al., 2005).

Evidence for Indiscriminant Cone Wiring

Alternatively, it has long been recognized that, given the private-line connectivity of the midget circuit, L versus M cone opponency would be possible with no further cone selectivity to the receptive field surround (Lennie, Haake, & Williams, 1991; Paulus & Kröger-Paulus, 1983). This "mixed-surround" hypothesis has also been termed the *midget hypothesis* because it rests entirely on the evolution of the primate fovea and the restricted cone connectivity of the midget circuit (for the purpose of increased spatial resolution) and the presumption that the L and M cones, a recent adaptation in the primate lineage, are not identified as functionally distinct circuit elements. The midget hypothesis makes two simple and strong predictions. First, as the midget ganglion cell dendritic tree enlarges in the retinal periphery and the center derives from 2, 3, 4, and greater numbers of cones (figure 11.2B, C), L versus M cone opponency should become variable and decline because the center mechanism would also sum L and M cone input without selectivity. Second, the receptive field surround sums cone input indiscriminately, weighted only by the relative numbers of L and M cones located in the patch of retina that encompasses the surround.

The first prediction has been tested both anatomically and physiologically, with clear results. Anatomically the midget ganglion cells make indiscriminant contact with the terminals of every midget cone bipolar axon terminal within the span of the dendritic arbor. Indeed, the dendrites of multiple midget ganglion cells that are in proximity can share input from single midget bipolars and in the retinal periphery midget ganglion cells *do not* "skip over" individual bipolar cell terminals to create cone selective bipolar to ganglion cell wiring (Dacey, 1993b; Jusuf, Martin, & Grunert, 2006a; Telkes et al., 2008). This anatomical picture is consistent with a number of recent studies that document a large fraction of midget ganglion cells in the retinal periphery that sum L and M cone inputs to the receptive field and lack L versus M opponency (Crook et al., 2011b; Diller et al., 2004; Field et al., 2010; Martin et al., 2011).

The presence of this large and conspicuous nonopponent midget population would seem to clearly confirm the prediction that with indiscriminate wiring, opponency should decline in the periphery. However, some question does remain whether the small number of red–green opponent cells that do appear in the periphery occur via random connectivity to a patchy cone mosaic (Lennie, Haake, & Williams, 1991; Lennie & Movshon, 2005) or due to some partial or limited form of selectivity (Buzas et al., 2006; Dacey, 1993b; Field et al., 2010; Martin et al., 2001). For example, Field et al. (2010) reported a small tendency for midget ganglion cells in the far retinal periphery to sample nonrandomly from the combined L+M cone inputs to the receptive field center. Because the great majority of these midget cells lacked overt L versus M cone opponency, a question remains as to whether the small deviation from random connectivity these authors report plays any role in the limited color opponency found in the retinal periphery or simply reflects some property of the midget circuit and its development that is entirely unrelated to opponency. This critical detail will need to be addressed with a model of connectivity to center

    J. D. CROOK, O. S. PACKER, J. B. TROY, AND D. M. DACEY

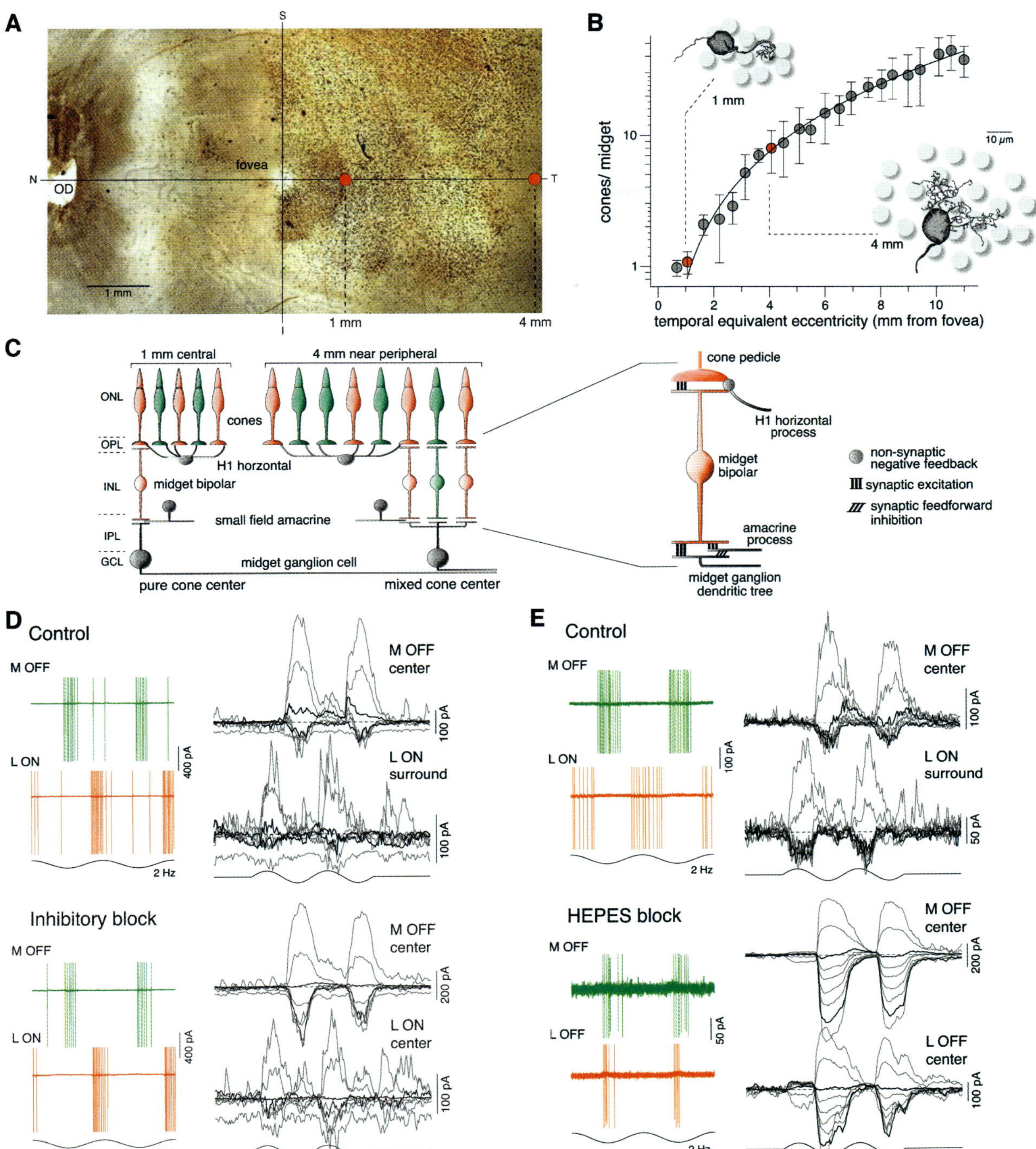

FIGURE 11.2 The midget circuit and red-green opponency. (A) Photomicrograph of macaque retina; ganglion cells retrogradely labeled in temporal quadrant after LGN tracer injection. Red dots indicate 5° (1 mm) and 20° (4 mm) eccentricity. (B) Cones per midget ganglion cell dendritic field ($n = 231$) plotted as a function of temporal equivalent eccentricity (Rodieck & Watanabe, 1993); cone density from Packer, Hendrickson, and Curcio (1989) (error bars: ±SD). Insets: "private line" (1 mm) and "peripheral" (4 mm) midget dendritic trees together with the underlying cone mosaics. (C) Key elements of the midget circuit. H1 horizontal cells contact multiple L and M cones nonselectively; midget bipolar cells contact single cones. Midget ganglion cells synapse with single midget bipolars centrally (left) and with an increasing number of bipolar cells (about two to six) in the near periphery (right). Expanded view: mixed L+M cone negative feedback from H1 horizontal cells (O) generates receptive field surround. Excitatory synapse from midget bipolar to ganglion cell (|||) transmits the center-surround field. Amacrine cells receive input from midget bipolars (|||) and make feedforward synapses (///) with ganglion cells. (D) L ON versus M OFF opponent spike discharge evoked by large-field L and M cone selective modulation (top, left) is *not* altered by block of GABAergic (GABAzine, 5 µM and TPMPA, 50 µM) and glycinergic (strychnine, 0.5 µM) transmission (bottom, left). Families of leak-subtracted synaptic currents (holding potentials range from −75 to +50 mV; thick black traces near −65 and 0 mV) recorded in response to the same stimuli show comparable synaptic excitation and feedforward inhibition for *both* L ON and M OFF opponent currents (E_{rev}, −10 vs. −6 mV; top, right); block of synaptic inhibition attenuates feedforward inhibition but does not alter L versus M opponent excitatory currents (E_{rev}, −3 vs. −5 mV; bottom, right). (E) Enhancement of retinal buffering with HEPES (20 mM; pH = 7.4) abolishes L versus M cone opponency. Reversal potentials are unchanged (L vs. M control E_{rev}, −3 vs. −3 mV, top, right; HEPES block E_{rev}, −4 vs. 0 mV, bottom right), but the surround response is lost, revealing an L cone contribution to the center.

and surround that predicts the relative numbers of opponent and nonopponent cells as a function of retinal eccentricity and that can then be tested experimentally.

The second prediction of the midget hypothesis is that the larger receptive field surround does not discriminate between L and M cones. The midget receptive field surround could be derived from H1 horizontal cell feedback to cones in the outer plexiform layer (OPL) and/or amacrine cell synaptic inhibition in the inner plexiform layer (IPL) (figure 11.2C). Evidence is strong that the H1 horizontal cells connect without selectivity to every L and M cone within a very large field created by gap-junctional coupling among horizontal cells (Boycott, Hopkins, & Sperling, 1987; Dacey et al., 1996; Wässle, Boycott, & Rohrenbeck, 1989). The result of this nonselective connectivity pattern is that the spectral sensitivity of the H1 horizontal cell simply reflects the relative numbers of L and M cones within its receptive field (Dacey et al., 2000a; Dacheux & Raviola, 1990; Deeb et al., 2000; Diller et al., 2004). Commensurate with the evidence that horizontal cell feedback to L and M cones contributes to the receptive field surround (Davenport, Detwiler, & Dacey, 2008; McMahon, Packer, & Dacey, 2004), several recent studies report mixed L and M cone input to the midget receptive field surround with no evidence of selectivity (Crook et al., 2011a, 2011b; Field et al., 2010; Martin et al., 2011). By contrast with H1 horizontal cells, little is known about functional contribution of primate amacrine cell circuits to the midget surround, although there is anatomical evidence that amacrine cells connect without selectivity to midget bipolar cells (Calkins & Sterling, 1996).

A recent study addressed the question of the horizontal versus amacrine cell contribution and cone selectivity of the midget surround directly with physiological methods (Crook et al., 2011a). Stimuli that modulated the L or M cones selectively were used to evoke synaptic currents in midget ganglion cells under voltage clamp. It was shown that L versus M cone opponency arose presynaptically and was transmitted to the midget ganglion cell dendritic tree entirely by the excitatory synaptic input (figure 11.2D, E). Moreover, pharmacological block of all GABAergic and glycinergic inhibition in the retina failed to attenuate center-surround antagonism or L versus M cone opponency in the midget cell light response (figure 11.2D). Crook et al. concluded that L versus M cone opponency in the midget circuit must therefore originate presynaptically in midget bipolar cells, which themselves display center-surround organization (Dacey et al., 2000b).

The Origin of the Midget Cell Receptive Field Surround

A striking result from Crook et al. (2011a) that is worth emphasizing is that after complete blockade of all synaptic inhibition midget cells retained center-surround organization and L versus M cone opponency. Thus, opponency must arise via a surround generating mechanism that does not involve synaptic inhibition. It is known that L and M cones acquire surrounds via negative feedback from horizontal cells (Verweij, Hornstein, & Schnapf, 2003), and there is evidence that horizontal cells provide feedback to cones by a novel, nonsynaptic, pH-dependent mechanism that acts directly on the cone calcium current (Verweij, Kamermans, & Spekreijse, 1996; Thoreson & Mangel, 2012). The precise nature of this mechanism remains controversial (Fahrenfort et al., 2009; Jacoby et al., 2012; Kreitzer et al., 2007), although a critical role for protons serving a transmitter-like function is possible (Hirasawa & Kaneko, 2003; Vessey et al., 2005; Cadetti & Thoreson, 2006). In primates, enriching the buffering capacity of the retina attenuates the surrounds of nonmidget ganglion cells consistent with a pH-dependent nonsynaptic mechanism (Crook et al., 2009a; Davenport et al., 2008). Similarly, Crook et al. (2011a) showed that buffering with HEPES completely and reversibly abolished the midget ganglion cell surround and eliminated L versus M cone opponency (figure 11.2E), in striking contrast with the lack of surround attenuation after blockade of GABAergic and glycinergic transmission (figure 11.2D). Thus, red–green opponency appears to arise exclusively via indiscriminant horizontal cell feedback to L and M cones in the outer retina without recourse to any amacrine-mediated L versus M cone selective inhibitory pathways.

Is it possible to reconcile these results with other reports of partial (Buzas et al., 2006) or complete (Lee et al., 2012) cone selectivity in the midget cell surround? Of the five recent studies that measured L and M cone contributions to the midget surround (Crook et al., 2011a, 2011b; Field et al., 2010; Lee et al., 2012; Martin et al., 2011), all but Lee et al. concluded that the surround was nonselective, consistent with known synaptic anatomy.

In sum, much of the recent work on the midget circuit appears to strongly support the original "mixed surround" hypothesis (Lennie, Haake, & Williams, 1991). First, cone opponency does not require cone type-selective synaptic inhibition but originates entirely from H1 horizontal cell feedback to L and M cones without selectivity (figure 11.2C). Second, as the midget dendritic tree enlarges and is fed by increasing numbers of cones in the retinal periphery, a preponderance of purely achromatic midget cells appears, again consistent with largely, if not completely, nonselective wiring. This view is also parsimonious; it is consistent with the evolutionary origin of the midget circuit as a pathway critical for high-resolution achromatic vision (Dacey, 1993b; Wässle et al., 1989, 1993; Williams & Coletta, 1987) and also with the recent appearance of L and M cones and trichromatic color vision in the primate lineage and the hypothesis that L versus M cone opponency arose by simply piggybacking on the unique private-line midget circuit (Martin et al., 2011; Mollon, 1989). The dramatic consequence of nonselective wiring is that the midget cells are not physiologically uniform, with many cells in the periphery showing purely achromatic responses. This returns our discussion to the question of the role that midget cells play as an achromatic pathway and the question of a "double-duty" role in both spatial and color vision. What is the nature of the achromatic response, and can this response support a key role for the midget cells as the critical pathway for spatial vision? Indeed, what is the relation of the primate midget pathway to the classical retinogeniculate beta-X pathway described first in the cat's retina (Enroth-Cugell & Robson, 1984)? These questions are addressed in a subsequent section of this chapter.

SMALL BISTRATIFIED CELLS AND THE SYNAPTIC MECHANISM FOR BLUE–YELLOW OPPONENCY

Wiesel and Hubel, recording in the LGN, were the first to identify an unusual receptive field structure, which they termed type 2, associated with certain color opponent cells (Wiesel & Hubel, 1966). In the type 2 cells color opponent fields were of comparable spatial extent, contrasting with the more numerous type 1 cells, which showed the familiar center-surround structure. Wiesel and Hubel reasoned that this atypical, spatially coextensive configuration could represent a pathway that sacrificed spatial opponency to transmit a pure color signal to the visual cortex. Subsequent work confirmed the existence of the type 2 receptive field at the retinal and LGN levels but emphasized a possible association with the blue–yellow cells, those cells in which signals from S cones were antagonistic to a combined signal from L and M cones (figure 11.3A, B) (de Monasterio, 1978c; Derrington, Krauskopf, & Lennnie, 1984; Derrington & Lennie, 1984). A breakthrough came when a morphologically distinct ganglion cell type, the small bistratified cell, was linked to blue-ON–yellow-OFF opponency and suggested a simple anatomical basis for the spatially coextensive S versus L+M receptive field (Dacey, 1993a; Dacey & Lee, 1994; Rodieck, 1991). Because this cell type was bistratified, extending dendrites in both the inner ON and outer OFF subdivisions of the inner plexiform layer (figure 11.3D left and right), it was possible that the inner dendrites received ON excitation from depolarizing cone bipolar cells that contact S cones exclusively (Kouyama & Marshak, 1992; Mariani, 1984), and the outer dendrites received OFF excitation from hyperpolarizing bipolar cells linked to L and M cones (Boycott & Wässle, 1991). This "ON–OFF pathway" hypothesis (figure 11.3D) for blue–yellow opponency suggests the origin of the type 2 receptive field as two spatially coextensive excitatory center mechanisms (Dacey, 1996), as would be present in any bistratified ON–OFF ganglion cell (e.g., Vaney, 1994b). Subsequent light and electron microscopic analysis showed directly such S ON and LM OFF bipolar contacts with the inner and outer dendritic tiers, respectively, of the small bistratified cell (Calkins, Tsukamoto, & Sterling, 1998; Ghosh & Grunert, 1999; Percival et al., 2009) supporting the ON–OFF hypothesis. However two recent physiological studies reached opposing conclusions about the fundamental spatial structure and underlying circuitry of the blue-ON small bistratified cells (Crook et al., 2009a; Field et al., 2007). Both studies utilized L-AP4, the mGluR6 receptor agonist, to block transmission in the S ON pathway and to isolate the presumptive L+M OFF pathway input to the blue-ON cells. In one study, L-AP4 completely abolished the light response, and it was concluded that the blue-ON, yellow-OFF response arose solely via the ON pathway and that the L+M signal originated presynaptically in a surround already present in the S ON bipolar input to the small bistratified cell (Field et al., 2007). This would be consistent with the recent observation of L+M cone-mediated surrounds in primate S cones (Packer et al., 2010) and the basic center surround organization of primate cone bipolar cells (Dacey et al., 2000a) but of course is inconsistent with the well-documented synaptic input from presumed OFF cone bipolar cells to the outer tier dendrites of the small bistratified cells, referenced above (figure 11.3D). By contrast, in the second study (Crook, Davenport, et al., 2009a), L-AP4 attenuated the S ON response as expected but had no effect on the L+M OFF response as predicted by the ON–OFF pathway hypothesis. In addition, this study showed that the L+M OFF response was preserved after block of synaptic inhibition and also after retinal buffering with HEPES, a condition that attenuated surrounds in parasol ganglion cells (Davenport et al., 2008) and, as noted in the previous section, midget ganglion cells (Crook et al., 2011a). This and another recent study (Solomon et al., 2005) also confirmed the earlier

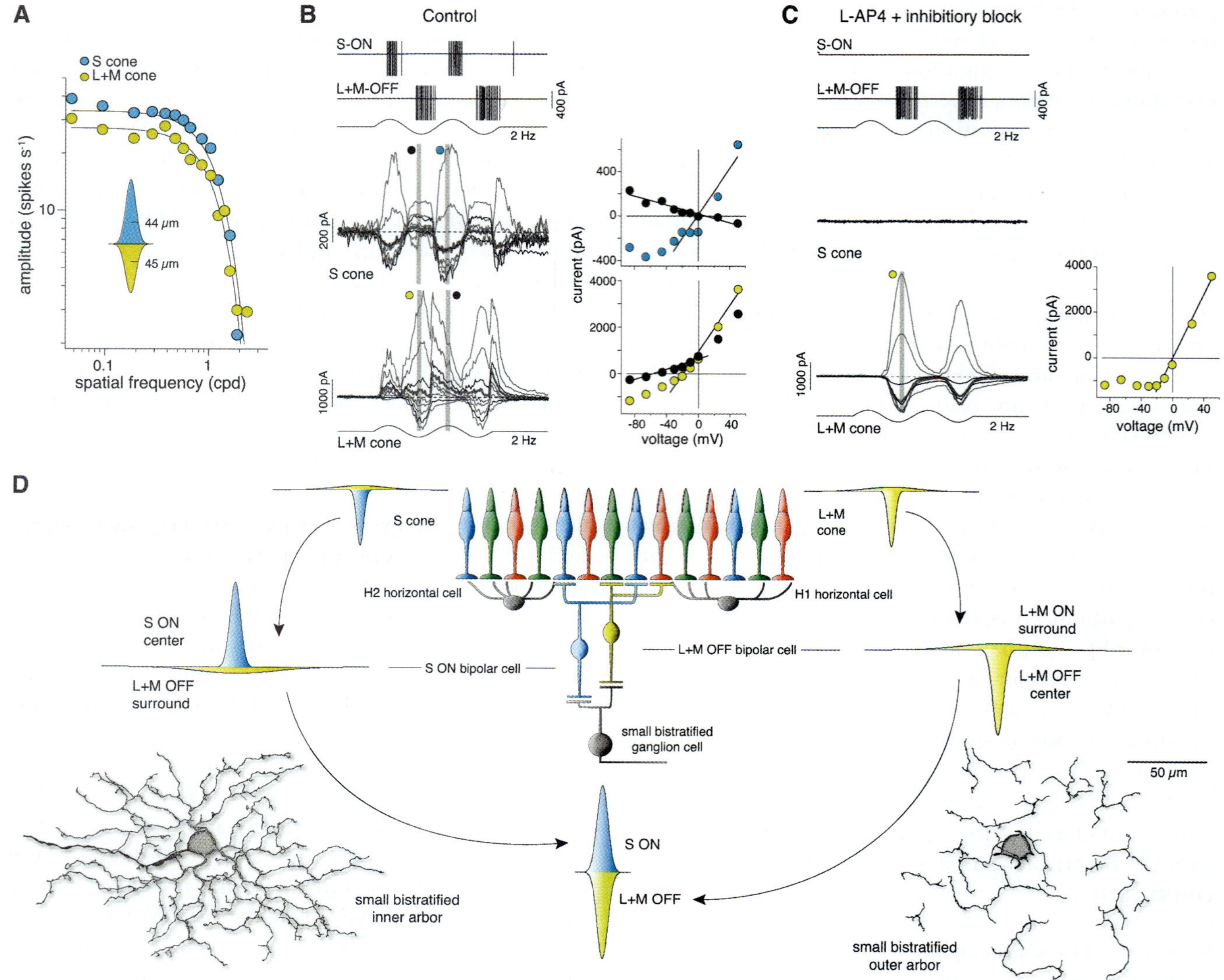

FIGURE 11.3   The small bistratified cell circuit and blue-yellow opponency. (A) Spatial frequency tuning of a small bistratified cell in response to S versus L+M cone-selective drifting sine wave gratings. Data are fit with Gaussian functions of comparable radii (inset); S ON and L+M OFF fields are thus spatially coextensive. (B) Coextensive cone-opponent receptive field structure arises postsynaptically by convergence of excitatory ON and OFF inputs. Excitatory currents underlie S versus L+M spike discharge (top). Current families, as described in figure 11.2D, to large-field S (top) and L+M (bottom) cone-isolating stimuli modulated sinusoidally at 2 Hz (80% contrast). Current–voltage plot for the S ON response (right, top) shows a pure increase in excitation at light onset (blue circles, E_{rev} 0 mV and positive slope) and decrease in excitation at light offset (black circles, E_{rev} −2 mV and negative slope). The L+M OFF response (right, bottom) shows excitation and feedforward inhibition at light offset (yellow circles, E_{rev} −18 mV and positive slope); at light onset, prominent crossover inhibition is present (black circles, E_{rev} −55 mV and positive slope). (C) After inhibitory blockade (strychnine, 0.5 μM; GABAzine, 5 μM; and TPMPA, 50 μM) and bath application of L-AP4 (40 μM), the S ON–evoked spike discharge and synaptic current were abolished while a pure excitatory L+M OFF–evoked current persisted (reversal potential, 0 mV). (D) Key elements of the small bistratified cell circuit illustrating the synaptic basis of the spatially coextensive S ON versus L+M OFF receptive field structure. The S ON bipolar's L+M OFF surround (left) arises from H2 horizontal cell feedback to the S cone (Packer et al., 2010). This receptive field sums at the small bistratified cell with input from an L+M OFF bipolar's L+M ON surround (right) created by H1 horizontal cell feedback to L and M cones. No net surround results at the ganglion cell, leaving spatially restricted S ON and L+M OFF center mechanisms as the basis for the cone-opponent receptive field (see Crook et al., 2009a, for details).

description of coextensive S ON and L+M OFF fields for this class of color opponent cells (figure 11.3A). Because cone bipolar cells typically show center-surround organization, one question raised by these results

is why a large surround is not evident in the spatial receptive field maps of the blue-ON cell. A possible answer was given by Crook et al. (2009a) when it was shown that after the S ON input had been attenuated with L-AP-4 and inhibitory block, a surround was indeed evident in spatial maps of the L+M OFF response, and this surround was subsequently attenuated by HEPES buffering, suggesting an origin in horizontal cell feedback to cones. These results suggest that an L+M OFF surround in the S ON excitatory pathway normally cancels an L+M ON surround in the L+M OFF excitatory pathway (figure 11.3D). This is consistent with such surrounds already described in S, L, and M cones (Packer et al., 2010; Verweij, Hornstein, & Schnapf, 2003) and would clearly predict the presence of such surrounds in the S cone presumed ON bipolar and the L+M cone presumed OFF bipolar cells that provide input to the inner and outer dendritic tiers of the small bistratified cell.

Finally, to more directly test the ON–OFF pathway hypothesis, a recent study used the voltage clamp to measure the excitatory and inhibitory synaptic currents evoked in the small bistratified cell by S or L+M cone selective stimuli (Crook et al., 2010; Dacey et al., 2011). Surprisingly, selective modulation of the S cone evoked a remarkably pure increase in excitation at light increment and a decrease in excitation at light decrement (figure 11.3B). Indeed, synaptic inhibition did not appear to contribute to the S-cone-evoked synaptic currents, and there was no change in reversal potential after block of both GABAergic and glycinergic transmission. The L+M OFF light-evoked current was dominated by synaptic excitation as predicted from the ON–OFF pathway hypothesis and the known anatomical wiring, but by contrast with the S-cone-driven input, the L+M OFF input was also associated with clear feedforward inhibition and an additional inhibitory input at the onset of the L+M stimulus commonly referred to as "crossover" inhibition (figure 11.3B). As found for the extracellular spike recordings, excitatory S-cone-evoked synaptic currents are sensitive to ON-pathway block, whereas the excitatory L+M OFF response was insensitive to ON-pathway and/or GABAergic and glycinergic block (figure 11.3C).

The voltage clamp results prove conclusively, therefore, that the fundamental S ON versus L+M OFF response of the small bistratified cell arises by ON versus OFF pathway excitation and is entirely consistent with the distinctive anatomy of this circuit (figure 11.3D). The voltage clamp data, however, also show clearly that synaptic inhibition is present and that there is a dramatic asymmetry in the deployment of this inhibition in which all of the inhibition, both feedforward and crossover, is associated with the L+M side of the opponent equation (figure 11.3B). The intriguing question then remains as to why the synaptic inhibition is so clearly asymmetrical. Indeed, it appears to be an exception to the standard pattern of retinal circuitry that the excitatory S cone bipolar input to the ganglion cell dendritic tree is *not* associated with feedforward amacrine cell synaptic architecture. One clue comes from the effects of blockade of all inhibition. The S-cone-mediated response is unaltered, as would be predicted from the voltage clamp reversal potentials; however, the L+M–evoked current amplitude increases. Thus, one function of the asymmetrical inhibition may be to contribute to the L+M versus S opponent balance under conditions where a weaker input from the sparse S cones opposes a very strong input from the more numerous L and M cones. Whether this unusual circuitry reflects an adaptation to the relative activation of S versus L+M cones by natural images (Lee, Wachtler, & Sejnowski, 2002; Garrigan et al., 2010) is an intriguing question.

PARASOL GANGLION CELLS AND THE SYNAPTIC ORIGIN OF THE Y-CELL SIGNATURE

Like the midget ganglion cells, the LGN magnocellular projecting parasol cells have been an accessible focus of much study, as these cells are, next to the midgets, the most numerous ganglion cells, comprising about 15% of the overall population (figure 11.1) (Dacey, 2004; Silveira & Perry, 1991; Watanabe & Rodieck, 1989). Yet here too the key functional significance of this visual pathway, as well the identity of its homologue in other mammals, has remained unresolved. Early physiological studies at the retinal (de Monasterio, 1978a, 1978b; de Monasterio & Gouras, 1975) and LGN levels (Dreher, Fukada, & Rodieck, 1976; Marrocco, McClurkin, & Young, 1982; Schiller & Malpeli, 1978; Sherman et al., 1976) recognized an obvious link to the Y-cell visual pathway that had been identified in the cat's retina. (For a broad review, including a comparison to primate, see Stone, 1983.) In brief, both cat Y cells and primate magnocellular pathway cells showed transient, achromatic light responses, high contrast and temporal sensitivity, large receptive fields and fast axonal conduction velocities, and have been identified as an input to cortical motion-processing pathways (Kaplan, Lee, & Shapley, 1990).

In the cat Y cells were identified as a distinctive anatomical type, the alpha cell (Boycott & Wässle, 1974; Wässle, Levick, & Cleland, 1975), linking a characteristic set of morphological and physiological properties (Cleland & Levick, 1974; Cleland, Levick, & Wässle, 1975; Peichl & Wässle, 1983; Wässle, Peichl, & Boycott, 1981). Parasol ganglion cells were also recognized as the likely alpha cell anatomical correlate. Like cat alpha cells, parasol cells have the largest cell bodies and thickest axons of any primate ganglion cell type and comprise two independent cell populations that stratify narrowly near the center of the IPL (Watanabe & Rodieck, 1989), are selectively labeled by a neurofibrillar stain (Peichl, Ott, & Boycott, 1987; Silveira & Perry, 1991), and show a type-specific pattern of gap-junction-mediated tracer coupling to amacrine cells (Dacey & Brace, 1992; Vaney, 1994a). Both cat alpha-Y and primate parasol cells project to LGN relay cells with the largest cell bodies (Leventhal, Rodieck, & Dreher, 1981; Perry, Oehler, & Cowey, 1984; Wässle, 1982).

Despite these connections between the parasol and alpha-Y cell, other data suggested the alternative hypothesis that the parasol cells were, in fact, more comparable to the X cells of the cat (Shapley & Perry, 1986). This suggestion is based first on the finding that many cells recorded in the magnocellular LGN layers showed linear spatial summation, a key property of X cells, and failed to demonstrate the distinctive second harmonic component in the spatial frequency response, a key feature of the alpha-Y cell (Derrington & Lennie, 1984; Enroth-Cugell & Robson, 1966; Hochstein & Shapley, 1976; Kaplan & Shapley, 1982). The same conclusion was drawn more recently from recordings of parasol cells (Petrusca et al., 2007; White et al., 2002). In addition, two tracing studies suggested that the parasol cells lacked a projection to the superior colliculus (an alpha-Y cell property) and projected exclusively to the LGN (a beta-X cell property) (Perry & Cowey, 1984; Rodieck & Watanabe, 1993) despite abundant previous results suggesting otherwise (Bunt et al., 1975; de Monasterio, 1978b; Leventhal, Rodieck, & Dreher, 1981; Schiller & Malpeli, 1977).

These two questions, whether, like cat alpha-Y cells, the parasol cells project to the superior colliculus and show nonlinear spatial summation, were reexamined in a recent study recording in the in vitro retina (Crook et al., 2008b) and utilizing a novel retrograde photostaining method that permitted ganglion cell morphology to be identified unequivocally (Dacey et al., 2003). By these methods parasols were shown clearly to project to the superior colliculus; indeed, the mosaic organization of the parasol cell population could be delineated in the retina after retrograde transport of tracer injected into the colliculus (figure 11.4A). To measure nonlinear spatial summation in the receptive field, the *Y-cell signature* (Kaplan & Shapley, 1982; Spitzer & Hochstein, 1985) was determined for a large sample of parasol cells. The Y-cell signature appears in a plot of the spatial frequency response at the stimulus temporal frequency (first harmonic, F1) to a drifting grating and the distinctive "frequency-doubled" (second harmonic, F2) response to counterphase modulation of a stationary grating (figure 11.4B). This plot clearly illustrates the

higher spatial resolution for the F2 compared to the F1 response components. And critically, the nonlinear F2 component is independent of the spatial phase (or position) of the modulated grating in the receptive field. In a typical cat alpha-Y cell a large spatial-phase-independent F2 component has a resolution three to five times that of the F1 component (Enroth-Cugell & Freeman, 1987). Crook at al. (2008b) demonstrated a Y-cell signature characterized by a large-bandpass F2 spatial frequency response to contrast reversing gratings that peaked at high spatial frequencies and was well separated from the F1 component (figure 11.4B). All parasol cells recorded also showed spatial phase invariance of the F2 response. Previous difficulties in identifying the spatial nonlinearity could have been due to measuring the F2 response component at spatial frequencies close to the peak of the F1, thus underestimating both the resolution and strength of the F2 component (Blakemore & Vital-Durand, 1986; Derrington & Lennie, 1984; Levitt et al., 2001; Petrusca et al., 2007; White et al., 2002).

The Synaptic Mechanism for Nonlinear Spatial Summation in the Parasol Y Cell

An influential early model of the circuitry underlying nonlinear spatial summation in the cat Y cell proposed that (1) linear subunits (presumed excitatory bipolar inputs) are pooled by input to a presumptive amacrine cell and (2) it is the inhibitory amacrine that is the site of the rectifying nonlinearity that elicits the F2 response component (Hochstein & Shapley, 1976; Victor & Shapley, 1979). This conception fit with the extensive inhibitory input to alpha cells (Freed & Sterling, 1988; Kolb & Nelson, 1993) and an early pharmacological study utilizing blockade of inhibition (Frishman & Linsenmeier, 1982). However, more recent recordings from alpha-Y cells in guinea pig showed that summation of partially rectifying (transient) cone bipolar input was sufficient to give rise to the F2 response (Demb et al., 2001). Similarly, blockade of inhibitory transmission has little effect on the F2 response component of primate parasol cells, again reinforcing the link between parasol and alpha-Y cells (figure 11.4C) (Crook et al., 2008b).

Most recently, voltage clamp recordings from parasol cells addressed the question of the role of synaptic inhibition directly (Crook et al., 2009b). Excitatory bipolar cell conductances combined with a small contribution of feedforward inhibition underlie ON and OFF parasol spike responses. However both ON- and OFF-center parasol cells also showed a distinctive inhibitory conductance antiphase to excitation (figure 11.4D). This

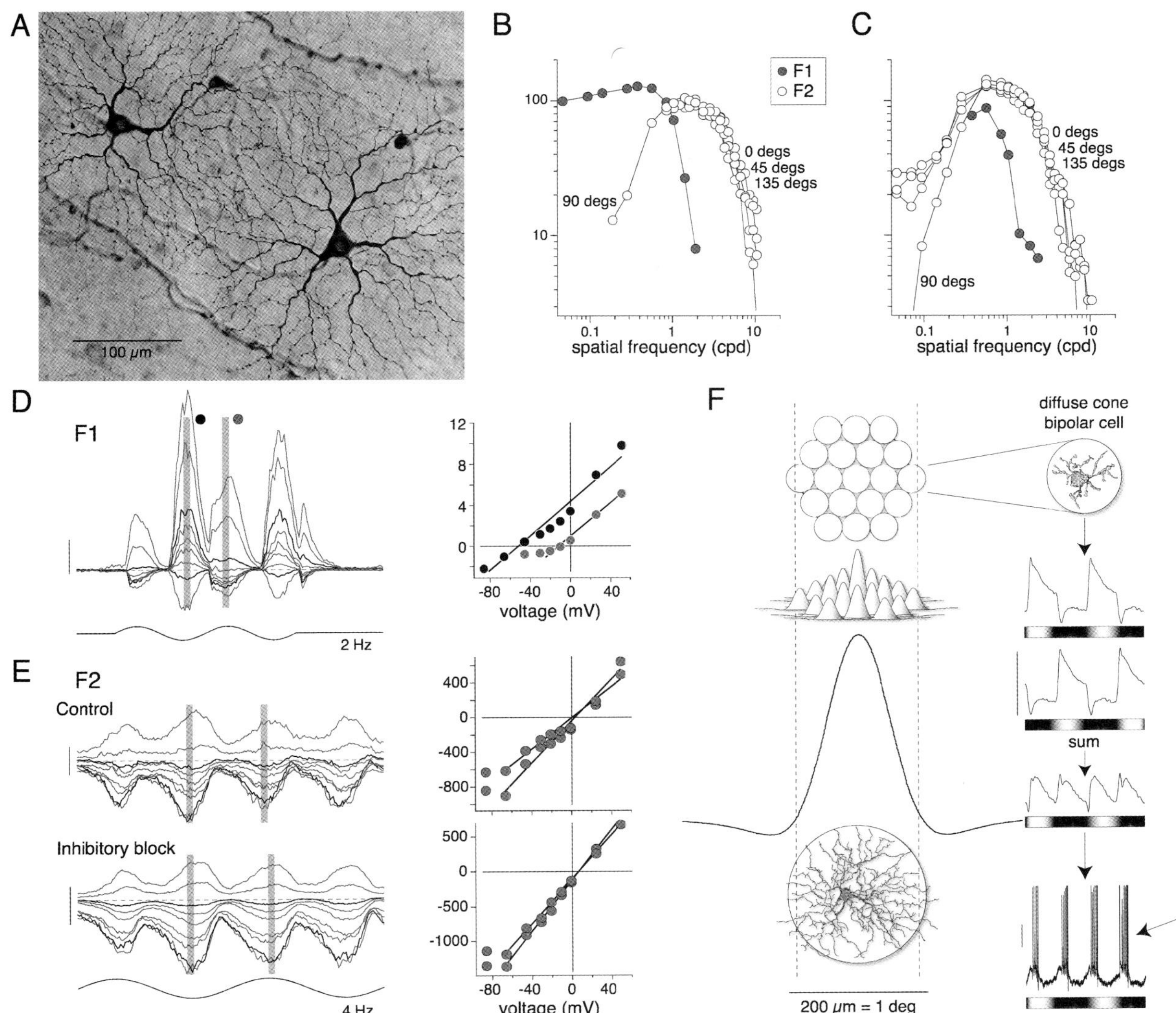

FIGURE 11.4 Parasol cell circuit and the Y-cell signature. (A) Photomicrograph of neighboring inner parasol cells retrogradely labeled from tracer injection into the superior colliculus and photostained in vitro. (B) Spatial tuning of the first (F1) and second (F2) harmonic amplitudes in response to drifting and stationary contrast-reversing gratings (4 Hz, 50% contrast), respectively. F2 response is plotted at 0°, 45°, 90°, and 135° stimulus phase. The F2 response peaks at a higher spatial frequency than the F1 and is phase independent at high spatial frequencies: the Y-cell signature. F1 (212 μm) and F2 (mean 34 μm) receptive field center diameters (3× the Gaussian radius), derived from difference-of-Gaussians fit (solid lines) to data. (C) Block of GABA (picrotoxin) or glycine (strychnine) had little effect on F2 spatial tuning, phase independence, or response amplitude. (D) Left, F1, current families (whole-cell voltage clamp, as in figure 11.2D), of an OFF parasol evoked by large-field L+M cone-selective modulation (2 Hz, 90% contrast; drifting sine wave grating). Right, Current-voltage plots, excitatory inputs (gray circles, E_{rev}, −11 mV) drive spike responses in counterphase to large crossover inhibitory input (black circles, E_{rev}, −51 mV). (E) Current families evoked by a stationary contrast-reversing grating (4 Hz, 100% contrast). Excitatory inputs (E_{rev}, 4 and 2 mV) dominate at stimulus increment and decrement and are not altered by block of inhibition as for C. (F) Summary of the origin of the F2 receptive field component in parasol cells (see Crook et al., 2008b, for details). Diffuse cone bipolar cell (top, left) excitatory input to the dendritic tree of the parasol cell (bottom, left). Intracellular recordings from a diffuse bipolar cell (tracings on right) show transient partially rectified responses. The synaptic potentials evoked by the light and dark phases of the stimulus sum at the ganglion cell to give rise to a frequency-doubled input and resultant spike discharge in the parasol cell (right column). Gaussian and difference-of-Gaussians models shown in the middle left column are taken from fits to the F2 and F1 responses in B. F1 response diameter is shown in relation to a parasol dendritic tree (center, bottom).

crossover inhibition was selectively attenuated in OFF-center parasol cells by either strychnine or L-AP4 and thus arises from the ON pathway via a bistratified, glycinergic amacrine cell. Blocking inhibition in ON-center parasols revealed an unexpected and prominent OFF excitatory conductance that was masked by crossover inhibition and isolated with the subsequent addition of L-AP4 (Crook et al., 2009b). Thus, the light-evoked synaptic currents of parasol cells are shaped in part by direct inhibitory input, a synaptic mechanism largely absent in the midget circuit and the S cone input to the small bistratified cell.

These results raise the question of whether the prominent crossover inhibition, likely arising in glycinergic amacrine cells with small receptive fields, plays any role in the Y-cell F2 response to high-spatial-frequency counterphase-modulated gratings. The question was answered by measuring the current–voltage relationship of the synaptic currents evoked by such stimuli. These currents are strongly frequency doubled, predominately excitatory (figure 11.4E top) and resistant to block of synaptic inhibition (figure 11.4E bottom). These data provide strong support for the origin of the Y-cell signature in transient, partially rectified responses of a diffuse cone bipolar cell type (Dacey et al., 2000b), which would sum at the ganglion cell to give rise to the F2 component in the ganglion cell spike discharge (figure 11.4F). It is striking that recent modeling and measurements from magnocellular LGN cells have also concluded that virtually all "M-cells" receive input from an array of rectifying subunits and are clearly Y-like (Dhruv et al., 2009).

MIDGET CELLS AS THE PRIMATE HOMOLOGUE OF THE BETA-X CELL

If the parasol cell is the homologue of the mammalian alpha-Y cell (Peichl, 1991), then what is the primate correlate of the beta-X cell? A number of clear morphological and physiological parallels between the cat beta-X cell pathway and the primate midget pathway have been recognized (Rodieck, Brening, & Watanabe, 1993). Like the midget ganglion cells (figure 11.5A), the beta-X cells are by far the most numerous ganglion cells in the cat, comprising about 50% of the total (Stein, Johnson, & Berson, 1996), and thus serve, as does the midget pathway for primates, to set the limit on the cat's spatial resolution (Wässle & Boycott, 1991). Like the midget cells, the beta-X cells project exclusively to the LGN (Humphrey et al., 1985); the light response is sustained, and the receptive field shows linear spatial summation (Enroth-Cugell & Robson, 1984; Sherman & Spear, 1982). The suggestion that midget cells are the primate X-cells also fits broadly with the midget hypothesis discussed earlier in this review, which posits that the origin of L versus M cone opponency in the midget circuit arose from the unique private-line circuit evolved to preserve the spatial resolution afforded by the high-density foveal cone array and therefore *did not* require any subsequent selective wiring to generate color opponency (figure 11.2D, E). In this view a midget–X-cell circuit, making indiscriminate connections to L and M cones, acquires a primary function in *both* spatial and color vision.

This "double-duty" role was recognized in the earliest studies of the primate retina (Gouras & Zrenner, 1979) and parvocellular pathway (DeValois et al., 1977), where it was noted that, by virtue of center-surround receptive field organization, the midget circuit could transmit an achromatic signal at high spatial and temporal frequencies (i.e., the receptive field is sensitive to achromatic stimuli that isolate the center mechanism) and a chromatic signal at low spatial and temporal frequencies (i.e., the receptive field shows L versus M cone opponency to chromatic stimuli that maximally engage center-surround antagonism). In principle it is possible for cortical circuitry to extract both the chromatic and achromatic signals from the midget cell output (e.g., Billock, 1995; Kingdom & Mullen, 1995), and there is behavioral evidence that lesions of the midget-parvocellular pathway attenuate both chromatic and achromatic contrast sensitivity as well as spatial acuity (Merigan, Katz, & Maunsell, 1991). By contrast there is also physiological evidence that midget cells and their relay cell counterparts in the parvocellular LGN layers show surprisingly poor achromatic contrast sensitivity *and* poor spatial resolution (Crook et al., 1988; Derrington & Lennie, 1984; Kaplan & Shapley, 1986). If this were true the midget would be poorly designed for double duty and a doubtful candidate for a beta-X cell homologue; rather, the midget pathway would have sacrificed achromatic sensitivity for a pure color-coding function (Cooper et al., 2012; Kaplan et al., 1990; Reid & Shapley, 1992; Shapley & Perry, 1986). A corollary of this hypothesis is that the parasol cells are X-like cells and, by virtue of high achromatic contrast sensitivity, must provide the critical substrate for spatial vision, even at the highest spatial frequencies (Lee, 1999).

Spatiochromatic Midgets: Color Piggybacks on the X Cell

The achromatic responsivity and the double-duty role of the midget pathway have been reexamined in several recent studies (Cooper, Sun, & Lee, 2012; Crook et al., 2011a, 2011b; Lee et al., 2012; Martin et al., 2011) in

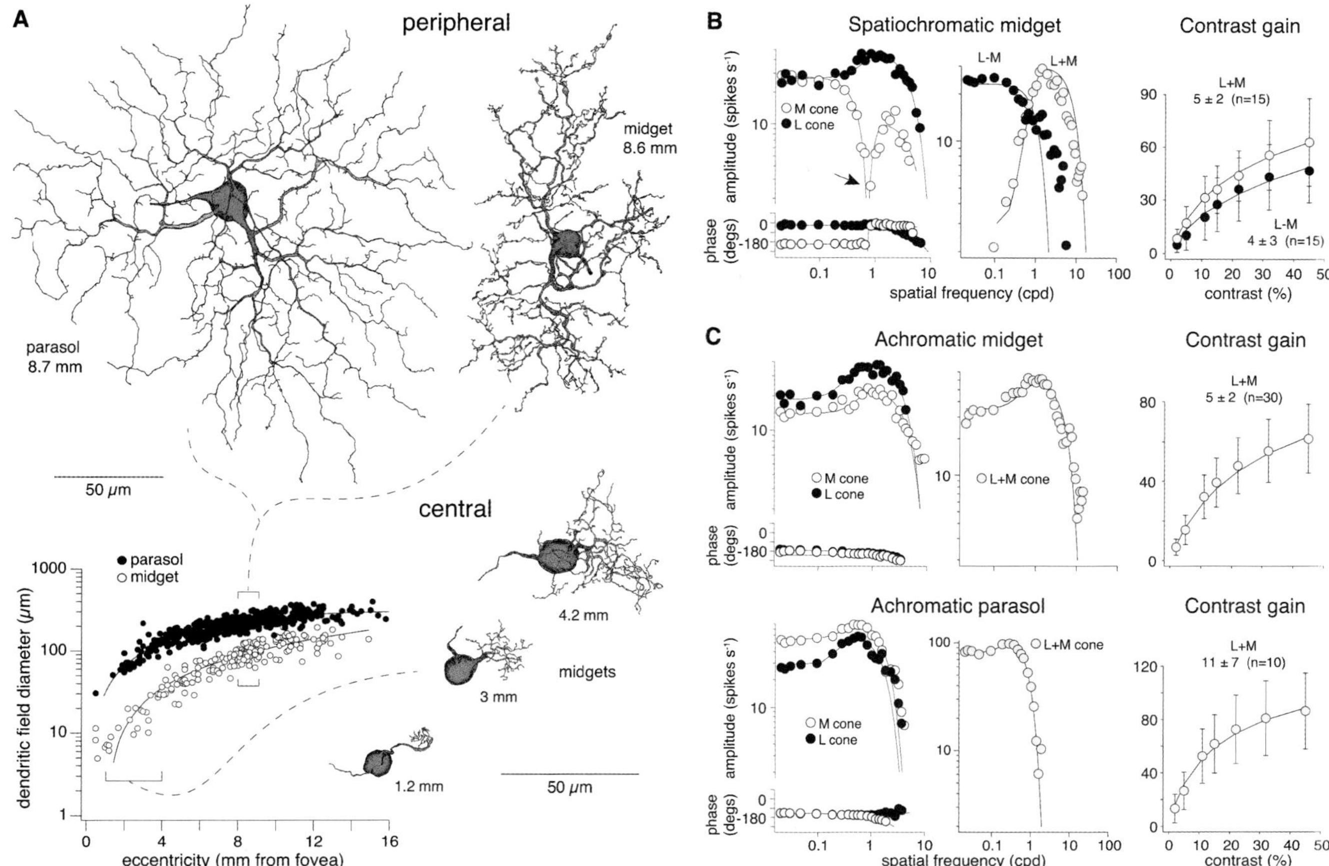

FIGURE 11.5   Spatiochromatic and achromatic midget cells: spatial organization of L and M cone inputs and contrast gain. (A) Dendritic field diameters of midget ($n = 191$) and parasol cells ($n = 437$) increase from central to peripheral retina. In the central 20° (1–4 mm) midget trees are small and gather input from one to six cones and show chromatic responses (B; also see figure 11.2); peripherally (30–50°) midget and parasol dendritic field sizes are large and gather input from many cones, and purely achromatic responses dominate (C). (B) Spatial tuning (amplitude and phase) plotted for a spatiochromatic midget cell in the near periphery in response to drifting gratings (2 Hz; 45% contrast) that specifically modulate either L or M cones or both L and M cones in phase (L+M) or antiphase (L–M). Responses are fit with a difference-of-Gaussians model. Response to L cone modulation is bandpass; L cones contribute to center and surround. Tuning to M cones shows the "spatiochromatic notch" (arrow) and follows surround phase at low frequencies (cone opponency) and center phase at high frequencies (nonopponent) (left plot). Correspondingly, tuning to chromatic stimuli (L–M) is broad and low pass; achromatic (L+M) tuning is bandpass and peaks at high spatial frequencies (middle plot). Contrast gain, measured at the peak of the spatial frequency response, is high for both chromatic (L–M) and achromatic (L+M) stimuli (right plot). Responses are fit with Naka-Rushton functions to calculate contrast gain (error bars: ±SD). (C) Plots as in B for pure achromatic midget and parasol cells in retinal periphery. Both midget and parasol types pool L and M cone inputs *in phase* at all spatial frequencies, showing typical bandpass L, M, and L+M spatial tuning functions (left and middle plots, top and bottom). Contrast gains are also similar for both midget and parasol cells in the periphery (right plot, top and bottom).

which midget pathway cells were recorded both in central and peripheral retina, where, as noted earlier, the midget cell center enlarges and collects input from multiple cones (figure 11.2B, C). Crook et al. and Martin et al. mapped the spatial tuning of L and M cone inputs to the midget receptive field at the retinal and LGN levels, respectively. Both studies were in broad agreement that L versus M cone opponency piggybacks on the high-resolution midget circuit such that red–green opponency does not come at a cost to achromatic

spatial vision; indeed, Martin et al. (2011) demonstrated that parvocellular units in both di- and trichromatic marmosets show no difference in spatial frequency tuning. The receptive field structure that underlies the midget cell response across this eccentricity range was shown in detail by Crook et al. (2011a, 2011b). When the L and M cone inputs were selectively modulated, two types of tuning functions were observed. The first was the classic bandpass response in which the cone type being modulated contributed to both the center and surround and was well fit by the standard difference-of-Gaussians receptive field model (figure 11.5B). The second was a distinctive notched tuning function (referred to as the spatiochromatic notch; figure 11.5B, arrow) in which the cone type being modulated contributed to the surround but only weakly or not at all to the center. In this case the weaker cone input to the center followed the surround phase at low spatial frequencies and the center phase at high spatial frequencies; at intermediate spatial frequencies a dip in the tuning function occurred at which center and surround almost completely canceled, and the response polarity underwent a 180° phase reversal (figure 11.5B). The consequence is that at low spatial frequencies L and M cones are in antiphase, that is, opponent, and in-phase or nonopponent at high spatial frequencies; the spatiochromatic notch separates these two response modes sharply along the spatial frequency axis. This notched tuning function is also fit by a difference-of-Gaussian receptive field model; however, it is a less familiar picture, as in this instance it is the surround that dominates at low spatial stimulus frequencies. Such a notch has been observed rarely in cat X-beta cells (Chan, Freeman, & Cleland, 1992).

Can this unique spatiochromatic tuning function be the basis for transmission of spectral and luminance information along a single channel to the cortical processing stage? When the contrast sensitivities of the chromatic and achromatic response components are measured at the peaks of the low-chromatic and high-achromatic spatial frequency bands (figure 11.5B, middle), both components show high-contrast gain (figure 11.5B, right). Indeed, the midget cells show achromatic contrast gains comparable to those of neighboring parasol cells tested with the same stimuli (figure 11.5C, bottom). Thus, midget cells, when tested at the peak of their spatial frequency response, show the high contrast sensitivity expected for the beta-X cell homologue and for the achromatic spatial channel defined psychophysically.

Purely Achromatic Midget Cells: X Cells of the Retinal Periphery

Last, as noted earlier in this review, the presence, in fact the dominance, of purely achromatic midget cells is found in the far retinal periphery (30–50° eccentricity). These midget cells show similarly weighted L and M cone inputs to both center and surround, like those found in parasol cells (figure 11.5C) and thus provide strong evidence that when the private line circuit breaks down in the periphery, the X-cell nature of the midget pathway remains, unadulterated by the chromatic response component. In the near periphery (10–20°), where the center receives input from about two to six cones (figure 11.2B, C), adjacent midget cells may show the mixed, spatiochromatic receptive fields (figure 11.5B), like that described above, or a pure achromatic response (figure 11.5C). Overall, the preponderance of achromatic midget cells in the retinal periphery is consistent with psychophysical evidence for a sharp decline in red–green relative to either blue–yellow or achromatic contrast sensitivity with increasing eccentricity (Hansen, Pracejus, & Gegenfurtner, 2009; Mullen & Kingdom, 1996, 2002; Mullen, Sakurai, & Chu, 2005).

Recent voltage clamp recordings have demonstrated that responses of both spatiochromatic and pure achromatic groups show similar synaptic currents evoked by L or M cone selective stimuli (Crook et al., 2011a, 2011b). Excitatory synaptic currents dominate combined with a small contribution of feedforward inhibition. As for the case of midgets from dichromatic versus trichromatic New World primates (Martin et al., 2011), there appears to be no fundamental difference in the circuitry of midget cells that exist as neighbors in a single spatial array, save for a small difference in the weighting of L and M cone inputs to the receptive field center that occurs when only a few cone inputs are drawn at random from a patchy L and M cone mosaic.

THE SIGNIFICANCE OF DIVERSE LOW-DENSITY GANGLION CELL POPULATIONS

Together the midget, parasol, and small bistratified cells comprise roughly 70% of the ganglion cell population in the primate retina, yet beyond these five cell types there remain on the order of a dozen additional pathways, and most of these remain to be characterized in physiological detail. The way that the retina incorporates so many additional pathways into the optic nerve output is that each of these ganglion cell types exists at a relatively low spatial density, with the extreme example being the ~3,000 giant melanopsin-expressing photoreceptive ganglion cells that form two distinct populations that tile the entire retinal landscape but sum to only about 1% of the total ganglion cells (figure 11.1A). The melanopsin-expressing cell is clearly unique, and evidence is strong that it provides an illuminance signal to the hypothalamic circadian clock and the pupillary reflex pathway (Do & Yau, 2010; Gamlin et al., 2007). The melanopsin cell projection to other targets including the LGN, and consequently a role in visual perception is also a possibility (Dacey et al., 2005; Zaidi et al., 2007; Brown et al., 2010, 2012; Estevez et al., 2012). The example of the melanopsin cells raises the question of whether the other diverse, low-density ganglion cell populations are defined by a unique pattern of central projections and function. The answer to that question appears to be a qualified no, as the majority of the remaining ganglion cell types can be retrogradely labeled from tracer injections into both the LGN and superior colliculus (Dacey, 2004; Szmajda, Grunert, & Martin, 2008; Percival et al., 2013). Although the precise distribution of the terminal fields defined by each type is unclear, they must all contribute to these two major retinal targets for visual processing. Why do so many parallel pathways contribute to the colliculus and the LGN, and how are they functionally distinct?

The start of an answer to this question is given by a recent characterization of one of these types, the smooth monostratified cell (figure 11.1). Like the parasol-Y cells the smooth cells are transient responding, achromatic cells with high temporal and contrast sensitivity; they form separate ON and OFF center populations and project to both the LGN and superior colliculus (figure 11.6A) (Crook et al., 2008a). Strikingly, the smooth cells precisely costratify with their parasol counterparts (figure 11.6A) and are thus likely to share a similar, if not identical, presynaptic circuitry. Indeed, the smooth cell receptive field demonstrates a Y-cell signature like that of the parasol-Y cell (figure 11.6B). The single major difference that distinguishes the parasol-Y from the smooth-Y cells is their spatial frequency tuning and anatomical basis. The smooth cells show a dendritic field diameter, and therefore receptive field center diameter, about twice that of the

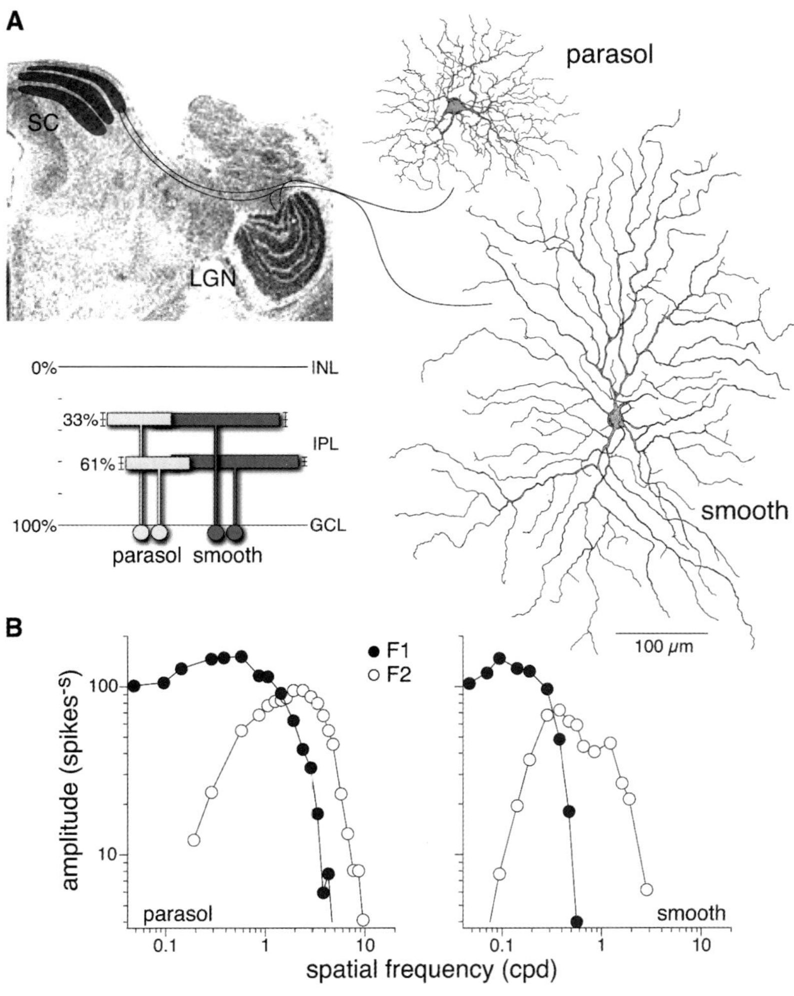

FIGURE 11.6 Smooth cells and ganglion cell type diversity. (A) Smooth cells have about twice the dendritic field diameter of parasol cells, but ON and OFF center types stratify at precisely the same depth in the IPL. Both cell types can be retrogradely labeled from tracer injections into the LGN and the superior colliculus. (B) Smooth (right) and parasol (left) cells show center surround receptive field structure (F1 component, filled circles), but smooth cells have larger receptive field center diameters. Both cell groups show high contrast sensitivity, a strong Y-cell signature, and a second harmonic response (F2 component, open circles) that extends to high spatial frequencies (see Crook et al., 2008a,b). Thus, the ON and OFF smooth and parasol populations may share a similar presynaptic circuit and overall function but create parallel spatially distinct achromatic visual channels.

parasol cells and are thus present at about a quarter of the density (figure 11.6A), comprising only about 3% of the total ganglion cell population (figure 11.1). In sum the smooth and parasol cells share central targets, key physiological properties, and overlapping or possibly identical presynaptic retinal microcircuits, differing only in receptive field size. We suggest this pattern may be analogous to the existence of multiple rabbit direction selective (DS) ganglion cell mosaics (Vaney et al., 2001). The ON–OFF DS cells form four populations that precisely costratify in the inner plexiform layer and share presynaptic circuitry, central targets, and the same basic directionally selective light response. However, each rabbit DS population independently tiles the retina with complete coverage and shows a distinctive directional preference established by subtle wiring differences (Briggman, Helmstaedter, & Denk, 2011). Thus, anatomically distinct pathways with small but critical functional differences arise from a single retinal microcircuit. Similarly the smooth and parasol cells may be components of a single microcircuit that gives rise to multiple, parallel achromatic Y-cell pathways that vary only in spatial scale. The selective pressure, then, for extreme visual pathway diversity may be the necessity to create discrete channels that vary in tuning along a single perceptual dimension. A precedent for such functional mosaicism is found in the long and rich history demonstrating multiple selectively adaptable spatial channels at the level of the visual cortex (e.g., Duong & Freeman, 2007; Movshon & Lennie, 1979).

CONCLUSIONS

Although the expectation was that cone opponency in the trichromatic primate may have involved a key role for synaptic inhibition, measurement of synaptic currents evoked by cone type-selective stimuli shows directly that amacrine cell circuits are not required for color opponency. L versus M opponency is present in the excitatory bipolar input to the midget dendritic tree and arises in the formation of center-surround receptive field organization by nonsynaptic horizontal cell feedback to cones. S versus L+M cone opponency also arises without recourse to synaptic inhibition by the convergent input of S ON and L+M OFF bipolar cells to the small bistratified cell's dendritic tree.

Recent studies of receptive field structure in parasol and midget cells support the conclusion from early work on the cat and primate retinogeniculate pathway that the fundamental beta-X and alpha-Y cell divisions of the cat are homologous respectively to the midget-parvocellular and parasol-magnocellular pathways of the primate. In support of this hypothesis it is now clear that the parasol cells project to the superior colliculus and show a clear Y-cell signature that originates by excitatory input from partially rectifying, transient bipolar cells. By contrast, the midget cells show the high spatial density and achromatic contrast sensitivity expected of the X-cell, with chromatic sensitivity piggybacking on a pathway evolved for preserving the achromatic spatial resolution afforded by the foveal cone mosaic. Accurate comparisons with other mammals remain to be established (see chapter 13 by Roska and Meister).

Finally, ongoing physiological characterization of ganglion cell types beyond the midget, parasol, and small bistratified cells suggests that extreme ganglion cell diversity may reflect the way in which a single presynaptic circuit is sampled by multiple ganglion cell dendritic trees to elaborate small changes across a single important functional dimension, such as spatial scale or direction selectivity. Continued focus on revealing the key properties and morphological relationships among the family of primate ganglion cell types should add immensely to a rich and nuanced picture of the human visual process.

ACKNOWLEDGMENTS

We thank Bill Merigan for comments on the manuscript and Beth Peterson for technical assistance. This work was supported by National Institutes of Health grants RR00166 to the Tissue Distribution Program of the National Primate Research Center at the University of Washington, EY06678 (D.M.D.), and EY01730 (Vision Research Core).

REFERENCES

Benardete, E. A., & Kaplan, E. (1999). Dynamics of primate P retinal ganglion cells: Responses to chromatic and achromatic stimuli. *Journal of Physiology, 519*(3), 775–790.

Billock, V. A. (1995). Cortical simple cells can extract achromatic information from the multiplexed chromatic and achromatic signals in the parvocellular pathway. *Vision Research, 35*, 2359–2369.

Blakemore, C., & Vital-Durand, F. (1986). Organization and post-natal development of the monkey's lateral geniculate nucleus. *Journal of Physiology, 380*, 453–491.

Boycott, B. B., & Dowling, J. E. (1969). Organization of the primate retina: Light microscopy. *Philosophical Transactions of the Royal Society of London. Series B, Biological Sciences, 255*, 109–184.

Boycott, B. B., Hopkins, J. M., & Sperling, H. G. (1987). Cone connections of the horizontal cells of the rhesus monkey's retina. *Proceedings of the Royal Society of London. Series B, Biological Sciences, 229*, 345–379.

Boycott, B. B., & Wässle, H. (1974). The morphological types of ganglion cells of the domestic cat's retina. *Journal of Physiology, 240*, 397–419.

Boycott, B. B., & Wässle, H. (1991). Morphological classification of bipolar cells of the primate retina. *European Journal of Neuroscience, 3,* 1069–1088.

Briggman, K. L., Helmstaedter, M., & Denk, W. (2011). Wiring specificity in the direction-selectivity circuit of the retina. *Nature, 471*(7337), 183–188. doi: 10.1038/nature 09818.

Brown, T. M., Gias, C., Hatori, M., Keding, S. R., Semo, M., Coffey, P. J., et al. (2010). Melanopsin contributions to irradiance coding in the thalamo-cortical visual system. *PLOS Biology, 8,* e1000558. doi:10.1371/journal.pbio. 1000558.

Brown, T. M., Tsujimura, S., Allen, A. E., Wynne, J., Bedford, R., Vickery, G., et al. (2012). Melanopsin-based brightness discrimination in mice and humans. *Current Biology, 22,* 1134–1141. doi:10.1016/j.cub.2012.04.039.

Bunt, A. H., Hendrickson, A. E., Lund, J. S., Lund, R. D., & Fuchs, A. F. (1975). Monkey retinal ganglion cells: Morphometric analysis and tracing of axonal projections with a consideration of the peroxidase technique. *Journal of Comparative Neurology, 164*(3), 265–285.

Buzas, P., Blessing, E., Szmajda, B., & Martin, P. (2006). Specificity of M and L cone inputs to receptive fields in the parvocellular pathway: Random wiring with functional bias. *Journal of Neuroscience, 26*(43), 11148–11161.

Cadetti, L., & Thoreson, W. (2006). Feedback effects of horizontal cell membrane potential on cone calcium currents studied with simultaneous readings. *Journal of Neurophysiology, 95,* 1992–1995.

Calkins, D. J., Schein, S. J., Tsukamoto, Y., & Sterling, P. (1994). M and L cones in macaque fovea connect to midget ganglion cells by different numbers of excitatory synapses. *Nature, 371,* 70–72.

Calkins, D. J., & Sterling, P. (1996). Absence of spectrally specific lateral inputs to midget ganglion cells in primate retina. *Nature, 381,* 613–615.

Calkins, D. J., & Sterling, P. (1999). Evidence that circuits for spatial and color vision segregate at the first retinal synapse. *Neuron, 24,* 313–321.

Calkins, D. J., Tsukamoto, Y., & Sterling, P. (1998). Microcircuitry and mosaic of a blue-yellow ganglion cell in the primate retina. *Journal of Neuroscience, 18,* 3373–3385.

Callaway, E. M. (2005). Structure and function of parallel pathways in the primate early visual system. *Journal of Physiology, 566*(Pt 1), 13–19.

Chan, L. H., Freeman, A. W., & Cleland, B. G. (1992). The rod-cone shift and its effect on ganglion cells in the cat's retina. *Vision Research, 32*(12), 2209–2219.

Cleland, B. G., & Levick, W. R. (1974). Brisk and sluggish concentrically organized ganglion cells in the cat's retina. *Journal of Physiology, 240,* 421–456.

Cleland, B. G., Levick, W. R., & Wässle, H. (1975). Physiological identification of a morphological class of cat retinal ganglion cells. *Journal of Physiology, 248,* 151–171.

Cooper, B., Sun, H., & Lee, B. B. (2012). Psychophysical and physiological responses to gratings with luminance and chromatic components of different spatial frequencies. *Journal of the Optical Society of America. A, Optics, Image Science, and Vision, 29*(2), A314–A323. doi: 10.1364/JOSAA. 29.00A314.

Croner, L. J., & Kaplan, E. (1995). Receptive fields of P and M ganglion cells across the primate retina. *Vision Research, 35*(1), 7–24.

Crook, J. D., Davenport, C. M., Peterson, B. B., Packer, O. S., Detwiler, P. B., & Dacey, D. M. (2009a). Parallel ON and OFF cone bipolar inputs establish spatially-coextensive receptive field structure of blue-yellow ganglion cells in primate retina. *Journal of Neuroscience, 29,* 8372–8387.

Crook, J. D., Manookin, M. B., Packer, O. S., & Dacey, D. M. (2010). Excitatory synaptic conductances mediate "blue-yellow" and "red-green" opponency in macaque monkey retinal ganglion cells. *Association for Research in Vision and Ophthalmology (ARVO),* abst. 5178, Session 503.

Crook, J. D., Manookin, M. B., Packer, O. S., & Dacey, D. M. (2011a). Horizontal cell feedback without cone type-selective inhibition mediates "red-green" color opponency in midget ganglion cells of the primate retina. *Journal of Neuroscience, 31*(5), 1762–1772.

Crook, J. D., Packer, O. S., Manookin, M. B., & Dacey, D. M. (2011b). Distinct spatio-chromatic receptive field structure mediates red-green opponency and high achromatic contrast sensitivity in primate midget ganglion cells. *Association for Research in Vision and Ophthalmology (ARVO),* abst. 3905, Session 401.

Crook, J. D., Peterson, B. B., Packer, O. S., Robinson, F. R., Gamlin, P. D., Troy, J. B., et al. (2008a). The smooth monostratified ganglion cell: Evidence for spatial diversity of the Y-cell pathway to the LGN and superior colliculus in the macaque monkey. *Journal of Neuroscience, 28*(48), 12654–12671.

Crook, J. D., Peterson, B. B., Packer, O. S., Robinson, F. R., Troy, J. B., & Dacey, D. M. (2008b). Y-cell receptive field and collicular projection of parasol ganglion cells in macaque monkey retina. *Journal of Neuroscience, 28*(44), 11277–11291.

Crook, J. D., Troy, J. B., Packer, O. S., Vrieslander, J. D., & Dacey, D. M. (2009b). Contribution of excitatory and inhibitory conductances to receptive field structure in midget and parasol ganglion cells of macaque monkey retina. *Journal of Vision, 9,* 57. doi: 10.1167/9.14.57.

Crook, J. M., Lange-Malecki, B., Lee, B. B., & Valberg, A. (1988). Visual resolution of macaque retinal ganglion cells. *Journal of Physiology, 396,* 206–224.

Dacey, D. M. (1993a). Morphology of a small-field bistratified ganglion cell type in the macaque and human retina. *Visual Neuroscience, 10,* 1081–1098.

Dacey, D. M. (1993b). The mosaic of midget ganglion cells in the human retina. *Journal of Neuroscience, 13,* 5334–5355.

Dacey, D. M. (1996). Circuitry for color coding in the primate retina. *Proceedings of the National Academy of Sciences of the United States of America, 93,* 582–588.

Dacey, D. M. (2004). Origins of perception: Retinal ganglion cell diversity and the creation of parallel visual pathways. In M. S. Gazzaniga (Ed.), *The cognitive neurosciences* (3rd ed., pp. 281–301). Cambridge, MA: MIT Press.

Dacey, D. M. (1994). Physiology, morphology and spatial densities of identified ganglion cell types in primate retina. In G. R. Bock & J. A. Goode (Eds.), *Ciba Foundation Symposium 184—Higher-order processing in the visual system.* Chichester, UK: John Wiley & Sons. doi: 10.1002/9780470514610.ch2.

Dacey, D. M., & Brace, S. (1992). A coupled network for parasol but not midget ganglion cells in the primate retina. *Visual Neuroscience, 9,* 279–290.

Dacey, D. M., Crook, J. D., Manookin, M. B., & Packer, O. S. (2011). Absence of synaptic inhibition associated with S-cone ON excitatory input to the small bistratified, blue-yellow opponent ganglion cell of the macaque monkey retina. *Association for Research in Vision and Ophthalmology (ARVO), abst. 4571/A510*, Session 455.

Dacey, D. M., Diller, L. C., Verweij, J., & Williams, D. R. (2000a). Physiology of L- and M-cone inputs to H1 horizontal cells in the primate retina. *Journal of the Optical Society of America. A, Optics, Image Science, and Vision, 17*(3), 589–596.

Dacey, D. M., Joo, H. R., Peterson, B. B., & Haun, T. J. (2010). Morphology, mosaics and targets of diverse ganglion cell populations in macaque monkey retina: Approaching a complete account. *Association for Research in Vision and Ophthalmology (ARVO), abst. 899/A182*, Session 145.

Dacey, D. M., & Lee, B. B. (1994). The blue-ON opponent pathway in primate retina originates from a distinct bistratified ganglion cell type. *Nature, 367*, 731–735.

Dacey, D. M., Lee, B. B., Stafford, D. K., Pokorny, J., & Smith, V. C. (1996). Horizontal cells of the primate retina: Cone specificity without spectral opponency. *Science, 271*, 656–659.

Dacey, D. M., Liao, H.-W., Peterson, B. B., Robinson, F. R., Smith, V. C., Pokorny, J., et al. (2005). Melanopsin-expressing ganglion cells in primate retina signal colour and irradiance and project to the LGN. *Nature, 433*(7027), 749–754. doi: 10.1038/nature03387.

Dacey, D. M., & Packer, O. S. (2003). Colour coding in the primate retina: Diverse cell types and cone-specific circuitry. *Current Opinion in Neurobiology, 13*, 421–427.

Dacey, D. M., Packer, O. S., Diller, L. C., Brainard, D. H., Peterson, B. B., & Lee, B. B. (2000b). Center surround receptive field structure of cone bipolar cells in primate retina. *Vision Research, 40*(14), 1801–1811.

Dacey, D. M., & Petersen, M. R. (1992). Dendritic field size and morphology of midget and parasol ganglion cells of the human retina. *Proceedings of the National Academy of Sciences of the United States of America, 89*, 9666–9670. doi: 10.1073/pnas.89.20.9666.

Dacey, D. M., Peterson, B. B., Robinson, F. R., & Gamlin, P. D. (2003). Fireworks in the primate retina: In vitro photodynamics reveals diverse LGN-projecting ganglion cell types. *Neuron, 37*(1), 15–27. doi: 10.1016/S0896-6273(02)01143-1.

Dacheux, R. F., & Raviola, E. (1990). Physiology of HI horizontal cells in the primate retina. *Proceedings of the Royal Society of London. Series B, Biological Sciences, 239*, 213–230.

Davenport, C. M., Detwiler, P. B., & Dacey, D. M. (2008). Effects of pH buffering on horizontal and ganglion cell light responses in primate retina: Evidence for the proton hypothesis of surround formation. *Journal of Neuroscience, 28*(2), 456–464.

Deeb, S. S., Diller, L. C., Williams, D. R., & Dacey, D. M. (2000). Interindividual and topographical variation of L:M cone ratios in monkey retinas. *Journal of the Optical Society of America. A, Optics, Image Science, and Vision, 17*(3), 538–544.

Demb, J. B., Zaghloul, K., Haarsma, L., & Sterling, P. (2001). Bipolar cells contribute to nonlinear spatial summation in the brisk-transient (Y) ganglion cell in mammalian retina. *Journal of Neuroscience, 21*(19), 7447–7454.

de Monasterio, F. M. (1978a). Center and surround mechanisms of opponent-color X and Y ganglion cells of retina of macaques. *Journal of Neurophysiology, 41*(6), 1418–1434.

de Monasterio, F. M. (1978b). Properties of concentrically organized X and Y ganglion cells of macaque retina. *Journal of Neurophysiology, 41*(6), 1394–1417.

de Monasterio, F. M. (1978c). Properties of ganglion cells with atypical receptive-field organization in retina of macaques. *Journal of Neurophysiology, 41*(6), 1435–1449.

de Monasterio, F. M., & Gouras, P. (1975). Functional properties of ganglion cells of the rhesus monkey retina. *Journal of Physiology, 251*, 167–195.

Derrington, A. M., Krauskopf, J., & Lennie, P. (1984). Chromatic mechanisms in lateral geniculate nucleus of macaque. *Journal of Physiology, 357*, 241–265.

Derrington, A. M., & Lennie, P. (1984). Spatial and temporal contrast sensitivities of neurones in lateral geniculate nucleus of macaque. *Journal of Physiology, 357*, 219–240.

De Valois, R. L., Abramov, I., & Jacobs, G. H. (1966). Analysis of response patterns of LGN cells. *Journal of the Optical Society of America, 56*(7), 966–977.

De Valois, R. L., Snodderly, D. M., Yund, E. W., & Hepler, N. K. (1977). Responses of macaque lateral geniculate cells to luminance and color figures. *Sensory Processes, 1*(3), 244–259.

Dhruv, N. T., Tailby, C., Sokol, S. H., Majaj, N. J., & Lennie, P. (2009). Nonlinear signal summation in magnocellular neurons of the macaque lateral geniculate nucleus. *Journal of Neurophysiology, 102*, 1921–1929.

Diller, L., Packer, O. S., Verweij, J., McMahon, M. J., Williams, D. R., & Dacey, D. M. (2004). L and M cone contributions to the midget and parasol ganglion cell receptive fields of macaque monkey retina. *Journal of Neuroscience, 24*(5), 1079–1088.

Do, M. T., & Yau, K. W. (2010). Intrinsically photosensitive retinal ganglion cells. *Physiological Reviews, 90*(4), 1547–1581.

Dreher, B., Fukada, Y., & Rodieck, R. W. (1976). Identification, classification and anatomical segregation of cells with X-like and Y-like properties in the lateral geniculate nucleus of old-world primates. *Journal of Physiology, 258*, 433–452.

Duong, T., & Freeman, R. (2007). Spatial frequency specific contrast adaptation originates in the primary visual cortex. *Journal of Neurophysiology, 98*, 187–195.

Enroth-Cugell, C., & Freeman, A. (1987). The receptive-field spatial structure of cat retinal Y cells. *Journal of Physiology, 384*, 49–79.

Enroth-Cugell, C., & Robson, J. G. (1966). The contrast sensitivity of retinal ganglion cells of the cat. *Journal of Physiology, 187*, 517–552.

Enroth-Cugell, C., & Robson, J. G. (1984). Functional characteristics and diversity of cat retinal ganglion cells. *Investigative Ophthalmology & Visual Science, 25*, 250–267.

Estevez, M. E., Fogerson, P. M., Ilardi, M. C., Borghuis, B. G., Chan, E., Weng, S., et al. (2012). Form and function of the M4 cell, an intrinsically photosensitive retinal ganglion cell type contributing to geniculocortical vision. *Journal of Neuroscience, 32*, 13608–13620. doi:10.1523/JNEUROSCI.1422-12.2012.

Fahrenfort, I., Steijaert, M., Sjoerdsam, T., Vickers, E., Ripps, H., van Asselt, J., et al. (2009). Hemichannel-mediated and pH-based feedback from horizontal cells to cones in the vertebrate retina. *PLoS One, 4*, 1–21. doi: 10.1371/journal.pone.0006090.

Field, G. D., Gauthier, J. L., Sher, A., Greschner, M., Machado, T. A., Jepson, L. H., et al. (2010). Functional connectivity in the retina at the resolution of photoreceptors. *Nature, 467*(7316), 673–677. doi: 10.1038/nature09424.

Field, G. D., Sher, A., Gauthier, J. L., Greschner, M., Shlens, J., Litke, A. M., et al. (2007). Spatial properties and functional organization of small bistratified ganglion cells in primate retina. *Journal of Neuroscience, 27*(48), 13261–13272.

Freed, M. A., & Sterling, P. (1988). The ON-alpha ganglion cell of the cat retina and its presynaptic cell types. *Journal of Neuroscience, 8*, 2303–2320.

Frishman, L. J., & Linsenmeier, R. A. (1982). Effects of picrotoxin and strychnine on non-linear responses of Y-type cat retinal ganglion cells. *Journal of Physiology, 324*, 347–363.

Gamlin, P. D., McDougal, D. H., Pokorny, J., Smith, V. C., Yau, K.-W., & Dacey, D. M. (2007). Human and macaque pupil responses driven by melanopsin-containing retinal ganglion cells. *Vision Research, 47*(7), 946–954.

Garrigan, P., Ratliff, C. P., Klein, J. M., Sterling, P., Brainard, D. H., & Balasuramanian, V. (2010). Design of a trichromatic cone array. *PLOS Computational Biology, 6*, e1000677. doi:10.1371/journal.pcbi.1000677.

Ghosh, K. K., & Grunert, U. (1999). Synaptic input to small bistratified (blue-ON) ganglion cells in the retina of a New World monkey, the marmoset *Callithrix jacchus. Journal of Comparative Neurology, 413*(3), 417–428.

Gouras, P., & Zrenner, E. (1979). Enhancement of luminance flicker by color-opponent mechanisms. *Science, 205*(4406), 587–589.

Hansen, T., Pracejus, L., & Gegenfurtner, K. R. (2009). Color perception in the intermediate periphery of the visual field. *Journal of Vision, 9*(4), *26*, 1–12. doi: 10.1167/9.4.26.

Hirasawa, H., & Kaneko, A. (2003). pH changes in the invaginating synaptic cleft mediate feedback from horizontal cells to cone photoreceptors by modulating Ca^{2+} channels. *Journal of General Physiology, 122*(6), 657–671.

Hochstein, S., & Shapley, R. M. (1976). Linear and nonlinear spatial subunits in Y cat retinal ganglion cells. *Journal of Physiology, 262*, 265–284.

Humphrey, A. L., Sur, M., Uhlrich, D. J., & Sherman, S. M. (1985). Projection patterns of individual X- and Y-cell axons from the lateral geniculate nucleus to cortical area 17 in the cat. *Journal of Comparative Neurology, 233*(2), 159–189.

Ingling, C., & Martinez-Ureigas, E. (1983). The relationship between spectral sensitivity and spatial sensitivity for the primate r-g X-channel. *Vision Research, 23*, 1495–1500.

Jacoby, J., Kreitzer, M. A., Alford, S., Qian, H., Tchernookova, B. K., Naylor, E. R., et al. (2012). Extracellular pH dynamics of retinal horizontal cells examined using electrochemical and fluorometric methods. *Journal of Neurophysiology, 107*(3), 868–879.

Jusuf, P., Martin, P., & Grunert, U. (2006a). Random wiring in the midget pathway of primate retina. *Journal of Neuroscience, 26*(15), 3908–3917.

Jusuf, P., Martin, P., & Grunert, U. (2006b). Synaptic connectivity in the midget-parvocellular pathway of primate central retina. *Journal of Comparative Neurology, 494*(2), 260–274.

Kaplan, E., Lee, B. B., & Shapley, R. M. (1990). New views of primate retinal function. *Progress in Retinal and Eye Research, 9*, 273–336. doi: 10.1016/0278-4327(90)90009-7.

Kaplan, E., & Shapley, R. (1982). X and Y cells in the lateral geniculate nucleus of macaque monkeys. *Journal of Physiology, 330*, 125–143.

Kaplan, E., & Shapley, R. M. (1986). The primate retina contains two types of ganglion cells, with high and low contrast

sensitivity. *Proceedings of the National Academy of Sciences of the United States of America, 83*, 2755–2757. doi: 10.1073/pnas.83.8.2755.

Kingdom, F. A. A., & Mullen, K. T. (1995). Separating colour and luminance information in the visual system. *Spatial Vision, 9*, 191–219. doi: 10.1163/156856895X00188.

Kolb, H., & Dekorver, L. (1991). Midget ganglion cells of the parafovea of the human retina: A study by electron microscopy and serial section reconstructions. *Journal of Comparative Neurology, 303*, 617–636.

Kolb, H., Linberg, K. A., & Fisher, S. K. (1992). Neurons of the human retina: A Golgi study. *Journal of Comparative Neurology, 318*, 147–187.

Kolb, H., & Marshak, D. (2003). The midget pathways of the primate retina. *Documenta Ophthalmologica, 106*(1), 67–81.

Kolb, H., & Nelson, R. (1993). OFF-Alpha and OFF-Beta ganglion cells in cat retina: II. Neural circuitry as revealed by electron microscopy of HRP stains. *Journal of Comparative Neurology, 329*, 85–110.

Kouyama, N., & Marshak, D. W. (1992). Bipolar cells specific for blue cones in the macaque retina. *Journal of Neuroscience, 12*, 1233–1252.

Kreitzer, M. A., Collis, L. P., Molina, A. J., Smith, P. J., & Malchow, R. P. (2007). Modulation of extracellular proton fluxes from retinal horizontal cells of the catfish by depolarization and glutamate. *Journal of General Physiology, 130*(2), 169–182.

Kuffler, S. W. (1953). Discharge patterns and functional organization of mammalian retina. *Journal of Neurophysiology, 16*, 37–68.

Lee, B. B. (1999). Receptor inputs to primate ganglion cells. In K. R. Gegenfurtner & L. T. Sharpe (Eds.), *Color vision: From genes to perception* (pp. 203–218). New York: Cambridge University Press.

Lee, B. B., Kremers, J., & Yeh, T. (1998). Receptive fields of primate retinal ganglion cells studied with a novel technique. *Visual Neuroscience, 15*, 161–175.

Lee, B. B., Martin, P. R., & Valberg, A. (1989). Sensitivity of macaque retinal ganglion cells to chromatic and luminance flicker. *Journal of Physiology, 414*, 223–243.

Lee, B. B., Pokorny, J., Smith, V. C., Martin, P. R., & Valberg, A. (1990). Luminance and chromatic modulation sensitivity of macaque ganglion cells and human observers. *Journal of the Optical Society of America, 7*(12), 2223–2236. doi: 10.1364/JOSAA.7.002223.

Lee, B. B., Shapley, R. M., Hawken, M. J., & Sun, H. (2012). Spatial distributions of cone inputs to cells of the parvocellular pathway investigated with cone-isolating gratings. *Journal of the Optical Society of America. A, Optics, Image Science, and Vision, 29*(2), 223–232.

Lee, B. B., Valberg, A., Tigwell, D. A., & Tryti, J. (1987). An account of responses of spectrally opponent neurons in macaque lateral geniculate nucleus to successive contrast. *Proceedings of the Royal Society of London. Series B, Biological Sciences, 230*, 293–314.

Lee, T. W., Wachtler, T., & Sejnowski, T. J. (2002). Color opponency is an efficient representation of spectral properties in natural scenes. *Vision Research, 42*(17), 2095–2103.

Lennie, P., Haake, P. W., & Williams, D. R. (1991). The design of chromatically opponent receptive fields. In M. S. Landy & J. A. Movshon (Eds.), *Computational models of visual processing* (pp. 71–82). Cambridge, MA: MIT Press.

Lennie, P., & Movshon, J. A. (2005). Coding of color and form in the geniculostriate visual pathway. *Journal of the Optical Society of America. A, Optics, Image Science, and Vision, 22*(10), 2013–2033. doi: 10.1364/JOSAA.22.002013.

Leventhal, A. G., Rodieck, R. W., & Dreher, B. (1981). Retinal ganglion cell classes in the Old World monkey: Morphology and central projections. *Science, 213*, 1139–1142.

Levitt, J. B., Schumer, R. A., Sherman, S. M., Spear, P. D., & Movshon, J. A. (2001). Visual response properties of neurons in the LGN of normally reared and visually deprived macaque monkeys. *Journal of Neurophysiology, 85*(5), 2111–2129.

Mariani, A. P. (1984). Bipolar cells in monkey retina selective for the cones likely to be blue-sensitive. *Nature, 308*, 184–186.

Marrocco, R. T., McClurkin, J. W., & Young, R. A. (1982). Spatial summation and conduction latency classification of cells of the lateral geniculate nucleus of macaques. *Journal of Neuroscience, 2*(9), 1275–1291.

Martin, P. R., Blessing, E. M., Buzas, P., Szmajda, B. A., & Forte, J. D. (2011). Transmission of colour and acuity signals by parvocellular cells in marmoset monkeys. *Journal of Physiology, 589*(Pt 11), 2795–2812.

Martin, P. R., Lee, B. B., White, A. J., Solomon, S. G., & Ruttiger, L. (2001). Chromatic sensitivity of ganglion cells in the peripheral primate retina. *Nature, 410*(6831), 933–936.

McMahon, M. J., Packer, O. S., & Dacey, D. M. (2004). The classical receptive field surround of primate parasol ganglion cells is mediated primarily by a non-GABAergic pathway. *Journal of Neuroscience, 24*(15), 3736–3745.

Merigan, W. H., Katz, L. M., & Maunsell, J. H. (1991). The effects of parvocellular lateral geniculate lesions on the acuity and contrast sensitivity of macaque monkeys. *Journal of Neuroscience, 11*(4), 994–1001.

Milam, A. H., Dacey, D. M., & Dizhoor, A. M. (1993). Recoverin immunoreactivity in mammalian cone bipolar cells. *Visual Neuroscience, 10*, 1–12. doi: 10.1017/S0952523800003175.

Mollon, J. D. (1989). The uses and origins of primate colour vision. *Journal of Experimental Biology, 146*, 21–38.

Movshon, J. A., & Lennie, P. (1979). Pattern-selective adaptation in visual cortical neurones. *Nature, 278*(5707), 850–852.

Mullen, K. T., & Kingdom, F. A. A. (1996). Losses in peripheral colour sensitivity predicted from "hit and miss" postreceptoral cone connections. *Vision Research, 36*, 1995–2000.

Mullen, K. T., & Kingdom, F. A. (2002). Differential distributions of red-green and blue-yellow cone opponency across the visual field. *Visual Neuroscience, 19*(1), 109–118.

Mullen, K. T., Sakurai, M., & Chu, W. (2005). Does L/M cone opponency disappear in human periphery? *Perception, 34*(8), 951–959. doi: 10.1068/p5374.

O'Brien, B. J., Richardson, R. C., & Berson, D. M. (2003). Inhibitory network properties shaping the light evoked responses of cat alpha retinal ganglion cells. *Visual Neuroscience, 20*(4), 351–361.

Packer, O., Hendrickson, A., & Curcio, C. A. (1989). Photoreceptor topography of the retina in the adult pigtail macaque (*Macaca nemestrina*). *Journal of Comparative Neurology, 288*(1), 165–183.

Packer, O. S., Verweij, J., Li, P. H., Schnapf, J. L., & Dacey, D. M. (2010). Blue-yellow opponency in primate S cone photoreceptors. *Journal of Neuroscience, 30*(2), 568–572.

Paulus, W., & Kröger-Paulus, A. (1983). A new concept of retinal colour coding. *Vision Research, 23*, 529–540.

Peichl, L. (1991). Alpha ganglion cells in mammalian retinae: Common properies, species differences, and some comments on other ganglion cells. *Visual Neuroscience, 7*, 155–169.

Peichl, L., Ott, H., & Boycott, B. B. (1987). Alpha ganglion cells in mammalian retinae. *Proceedings of the Royal Society of London. Series B, Biological Sciences, 231*, 169–197.

Peichl, L., & Wässle, H. (1983). The structural correlate of the receptive field centre of alpha ganglion cells in the cat retina. *Journal of Physiology, 341*, 309–324.

Percival, K. A., Jusuf, P. R., Martin, P. R., & Grunert, U. (2009). Synaptic inputs onto small bistratified (blue-ON/yellow-OFF) ganglion cells in marmoset retina. *Journal of Comparative Neurology, 517*(5), 655–669.

Percival, K. A., Martin, P. R., & Grunert, U. (2013). Organisation of koniocellular-projecting ganglion cells and diffuse bipolar cells in the primate fovea. *The European Journal of Neuroscience, 37*, 1072–1089.

Perry, V. H., & Cowey, A. (1984). Retinal ganglion cells that project to the superior colliculus and pretectum in the macaque monkey. *Neuroscience, 12*(4), 1125–1137.

Perry, V. H., Oehler, R., & Cowey, A. (1984). Retinal ganglion cells that project to the dorsal lateral geniculate nucleus in the macaque monkey. *Neuroscience, 12*(4), 1101–1123.

Petrusca, D., Grivich, M. I., Sher, A., Field, G. D., Gauthier, J. L., Greschner, M., et al. (2007). Identification and characterization of a Y-like primate retinal ganglion cell type. *Journal of Neuroscience, 27*(41), 11019–11027. doi: 10.1523/JNEUROSCI.2836-07.2007.

Polyak, S. L. (1941). *The retina*. Chicago: University of Chicago Press.

Reid, R. C., & Shapley, R. M. (1992). Spatial structure of cone inputs to receptive fields in primate lateral geniculate nucleus. *Nature, 356*, 716–718.

Reid, R. C., & Shapley, R. M. (2002). Space and time maps of cone photoreceptor signals in macaque lateral geniculate nucleus. *Journal of Neuroscience, 22*(14), 6158–6175.

Rodieck, R. W. (1988). The primate retina. In H. D. Steklis (Ed.), *Comparative Primate Biology, Vol. 4: Neurosciences* (pp. 203–278). New York: Alan R. Liss.

Rodieck, R. W. (1991). Which cells code for color? In A. Valberg & B. B. Lee (Eds.), *From pigments to perception* (pp. 83–93). New York: Plenum Press.

Rodieck, R. W., Binmoeller, K. F., & Dineen, J. (1985). Parasol and midget ganglion cells of the human retina. *Journal of Comparative Neurology, 233*, 115–132.

Rodieck, R. W., Brening, R. K., & Watanabe, M. (1993). The origin of parallel visual pathways. *Paper presented at the Proceedings of the Retina Research Foundation Symposia: Contrast Sensitivity*, Cambridge, MA.

Rodieck, R. W., & Watanabe, M. (1993). Survey of the morphology of macaque retinal ganglion cells that project to the pretectum, superior colliculus, and parvicellular laminae of the lateral geniculate nucleus. *Journal of Comparative Neurology, 338*, 289–303.

Saito, H. A. (1983). Morphology of physiologically identified X-, Y-, and W-type retinal ganglion cells of the cat. *Journal of Comparative Neurology, 221*(3), 279–288.

Schiller, P. H., & Malpeli, J. G. (1977). Properties and tectal projections of monkey retinal ganglion cells. *Journal of Neurophysiology, 40*(2), 428–445.

Schiller, P. H., & Malpeli, J. G. (1978). Functional specificity of lateral geniculate nucleus laminae of the rhesus monkey. *Journal of Neurophysiology, 41*(3), 788–796.

Shapley, R. (2006). Specificity of cone connections in the retina and color vision. Focus on "specificity of cone inputs to macaque retinal ganglion cells." *Journal of Neurophysiology, 95*(2), 587–588.

Shapley, R., & Perry, V. H. (1986). Cat and monkey retinal ganglion cells and their visual functional roles. *Trends in Neurosciences, 9*, 229–235. doi: 10.1016/0166-2236(86)90064-0.

Sherman, S. M., & Spear, P. D. (1982). Organization of visual pathways in normal and visually deprived cats. *Physiological Reviews, 62*, 738–855.

Sherman, S. M., Wilson, J. R., Kaas, J. H., & Webb, S. V. (1976). X- and Y-cells in the dorsal lateral geniculate nucleus of the owl monkey (*Aotus trivirgatus*). *Science, 192*(4238), 475–477.

Silveira, L. C. L., & Perry, V. H. (1991). The topography of magnocellular projecting ganglion cells (M-ganglion cells) in the primate retina. *Neuroscience, 40*, 217–237.

Solomon, S. G., Lee, B. B., White, A. J., Ruttiger, L., & Martin, P. R. (2005). Chromatic organization of ganglion cell receptive fields in the peripheral retina. *Journal of Neuroscience, 25*(18), 4527–4539.

Solomon, S., & Lennie, P. (2007). The machinery of colour vision. *Nature Reviews. Neuroscience, 8*(4), 276–286.

Spitzer, H., & Hochstein, S. (1985). Simple- and complex-cell response dependences on stimulation parameters. *Journal of Neurophysiology, 53*(5), 1244–1265.

Stein, J. J., Johnson, S. A., & Berson, D. M. (1996). Distribution and coverage of beta cells in the cat retina. *Journal of Comparative Neurology, 372*, 597–617.

Stone, J. (1983). *Parallel processing in the visual system: The classification of retinal ganglion cells and its impact on the neurobiology of vision.* New York: Plenum Press.

Szmajda, B. A., Grunert, U., & Martin, P. R. (2008). Retinal ganglion cell inputs to the koniocellular pathway. *Journal of Comparative Neurology, 510*(3), 251–268. doi: 10.1002/cne.21783.

Tailby, C., Solomon, S. G., & Lennie, P. (2008). Functional asymmetries in visual pathways carrying S-cone signals in macaque. *Journal of Neuroscience, 28*(15), 4078–4087.

Taylor, W. R., & Vaney, D. I. (2002). Diverse synaptic mechanisms generate direction selectivity in the rabbit retina. *Journal of Neuroscience, 22*(17), 7712–7720.

Telkes, I., Lee, S. C., Jusuf, P. R., & Grunert, U. (2008). The midget-parvocellular pathway of marmoset retina: A quantitative light microscopic study. *Journal of Comparative Neurology, 510*(5), 539–549.

Thoreson, W. B., & Mangel, S. C. (2012). Lateral interactions in the outer retina. *Progress in Retinal and Eye Research, 31*, 407–441. doi:10.1016/j.preteyeres.2012.04.003.

Vaney, D. I. (1994a). Patterns of neuronal coupling in the retina. *Progress in Retinal and Eye Research*, 13th ed. (pp. 301–355). Oxford: Pergamon Press.

Vaney, D. I. (1994b). Territorial organization of direction-selective ganglion cells in rabbit retina. *Journal of Neuroscience, 14*, 6301–6316.

Vaney, D., He, S., Taylor, W., & Levick, W. (2001). Direction-selective ganglion cells in the retina. In J. Zanker & J. Zeil (Eds.), *Motion vision: Computational, neural and ecological constraints* (pp. 13–56). Berlin: Springer Verlag.

Verweij, J., Hornstein, E. P., & Schnapf, J. L. (2003). Surround antagonism in macaque cone photoreceptors. *Journal of Neuroscience, 23*(32), 10249–10257.

Verweij, J., Kamermans, M., & Spekreijse, H. (1996). Horizontal cells feed back to cones by shifting the cone calcium-current activation range. *Vision Research, 36*, 3943–3953.

Vessey, J., Stratis, A., Daniels, B., Da Silva, N., Jonz, M., Lalonde, M., et al. (2005). Proton-mediated feedback inhibition of presynaptic calcium channels at the cone photoreceptor synapse. *Journal of Neuroscience, 25*(16), 4108–4117. doi: 10.1523/JNEUROSCI.5253-04.2005.

Victor, J. D., & Shapley, R. M. (1979). The nonlinear pathway of Y ganglion cells in the cat retina. *Journal of General Physiology, 74*, 671–687.

Wässle, H. (1982). Morphological types and central projections of ganglion cells in the cat retina. *Progress in Retinal Research, 1*, 125. doi: 10.1016/0278-4327(82)90006-2.

Wässle, H., & Boycott, B. B. (1991). Functional architecture of the mammalian retina. *Physiological Reviews, 71*, 447–480.

Wässle, H., Boycott, B. B., & Rohrenbeck, J. (1989). Horizontal cells in the monkey retina: Cone connections and dendritic network. *European Journal of Neuroscience, 1*(5), 421–435.

Wässle, H., Grünert, U., Martin, P. R., & Boycott, B. B. (1993). Color coding in the primate retina: Predictions and constraints from anatomy. In B. Albowitz, K. Albus, U. Kuhnt, H.-C. Northdurft, & P. Wahle (Eds.), *Structural and functional organization of the neocortex (Experimental Brain Research Series 24)* (pp. 94–104). Berlin: Springer-Verlag.

Wässle, H., Grünert, U., Martin, P. R., & Boycott, B. B. (1994). Immunocytochemical characterization and spatial distribution of midget bipolar cells in the macaque monkey retina. *Vision Research, 34*, 561–579.

Wässle, H., Grünert, U., Röhrenbeck, J., & Boycott, B. B. (1989). Cortical magnification factor and the ganglion cell density of the primate retina. *Nature, 341*, 643–646.

Wässle, H., Levick, W. R., & Cleland, B. G. (1975). The distribution of the alpha type of ganglion cells in the cat's retina. *Journal of Comparative Neurology, 159*(3), 419–437.

Wässle, H., Peichl, L., & Boycott, B. B. (1981). Morphology and topography of on- and off-alpha cells in the cat retina. *Proceedings of the Royal Society of London. Series B, Biological Sciences, 212*, 157–175.

Watanabe, M., & Rodieck, R. W. (1989). Parasol and midget ganglion cells of the primate retina. *Journal of Comparative Neurology, 289*, 434–454.

White, A. J., Sun, H., Swanson, W. H., & Lee, B. B. (2002). An examination of physiological mechanisms underlying the frequency-doubling illusion. *Investigative Ophthalmology & Visual Science, 43*(11), 3590–3599.

Wiesel, T. N., & Hubel, D. H. (1966). Spatial and chromatic interactions in the lateral geniculate body of the rhesus monkey. *Journal of Neurophysiology, 29*, 1115–1156.

Williams, D. R., & Coletta, N. J. (1987). Cone spacing and the visual resolution limit. *Journal of the Optical Society of America. A, Optics and Image Science, 4*(8), 1514–1523. doi: 10.1364/JOSAA.4.001514.

Yamada, E., Bordt, A., & Marshak, D. (2005). Wide-field ganglion cells in macaque retinas. *Visual Neuroscience, 22*(4), 383–393.

Zaidi, F. H., Hull, J. T., Peirson, S. N., Wulff, K., Aeschbach, D., Gooley, J. J., et al. (2007). Short-wavelength light sensitivity of circadian, pupillary, and visual awareness in humans lacking an outer retina. *Current Biology, 17*(24), 2122–2128. doi: 10.1016/j.cub.2007.11.034.

12 Correlated Activity in the Retina

FRED RIEKE AND E. J. CHICHILNISKY

A common property of signaling in neural circuits, including circuits in the visual system, is that different neurons typically do not encode information independently of one another. Instead, they exhibit correlated activity: a tendency for two or more cells to fire together more or less frequently than would be expected by chance (Averbeck, Latham, & Pouget, 2006; Shlens, Rieke, & Chichilnisky, 2008; Usrey & Reid, 1999). Correlated activity is a prevalent feature of the output signals of the retina, and understanding its structure and origins could be valuable in several ways. First, because correlated activity shapes the visual signals transmitted to the brain, understanding its functional organization is a prerequisite to a comprehensive view of neural coding by retinal output signals. Second, because correlated activity arises from interconnections in the retinal circuitry, understanding its origins may shed light on retinal connectivity and its relation to processing. Third, because correlated activity is a general feature of neural circuits, understanding its role in the comparatively well-understood circuitry and function of the retina may provide insights relevant to other structures in the brain.

Correlated activity was first documented in ganglion cells of the mammalian retina in a series of groundbreaking studies that revealed its prevalence and strength and identified several potential mechanisms (Mastronarde, 1983a, 1983b, 1983c) (figure 12.1A) (see also Arnett, 1978; Rodieck, 1967). Since that time, many fundamental aspects of retinal structure and function have been elucidated, including the importance of parallel processing and the corresponding organization of identified cell types into parallel microcircuits (Dacey & Packer, 2003; Field & Chichilnisky, 2007; Masland, 2001) (see chapter 6 by DeVries and chapter 13 by Roska and Meister). Along with these discoveries, recent technical innovations have advanced our understanding of how correlated activity shapes patterns of activity in large populations of neurons, how it originates in specific cellular mechanisms, how it propagates through the retinal circuitry, and how it influences the transmission of visual information to the brain. Collectively, these studies provide a more complete picture of correlated activity in the retinal output and suggest

principles that may apply to understanding its impact on other circuits of the central nervous system.

We begin with an overview of the different forms of correlated activity in the retina, including important candidate circuitry and mechanisms. We then delve more deeply into the experimental evidence for specific mechanisms, including quantitative analysis of key properties of correlated activity and biophysical experiments that isolate specific mechanisms. Finally, we explore the significance of correlated activity for how visual signals are represented in the activity of ganglion cell populations.

SIGNAL CORRELATIONS, NOISE CORRELATIONS, AND CANDIDATE CIRCUITRY

An important first consideration is the scope of the problem to be understood. Although correlated activity between two cells can be measured without great difficulty, characterizing the structure and function of correlated activity in a complete neural circuit has the potential to be a hopelessly complex problem. For example, in a circuit of merely 100 neurons, the number of distinct spatial patterns of correlated firing (e.g., figure 12.3A) is 2^{100}, more than the number of stars in the observable universe. Making this problem manageable requires using the connectivity and structure of the retinal circuit to identify the activity patterns that are realizable, and focusing experimental investigation on these patterns.

Fortunately, decades of work on the circuitry of the mammalian retina provide a foundation for understanding correlated activity (reviewed by Dacey & Packer, 2003; Field & Chichilnisky, 2007; Masland, 2001; Wässle, 2004; see chapter 6 by DeVries, chapter 7 by Kramer, and chapter 8 by Mills and Massey). Visual signals from rods and several spectrally distinct types of cones propagate through ~10–15 types of bipolar cells to modulate spiking activity in ~20 types of retinal ganglion cells. Two major types of horizontal cells modify signals at the junction between photoreceptor and bipolar cells, and at least 25 different types of amacrine cells modify signals in bipolar cells and ganglion cells. Photoreceptors, horizontal cells, and many bipolar and

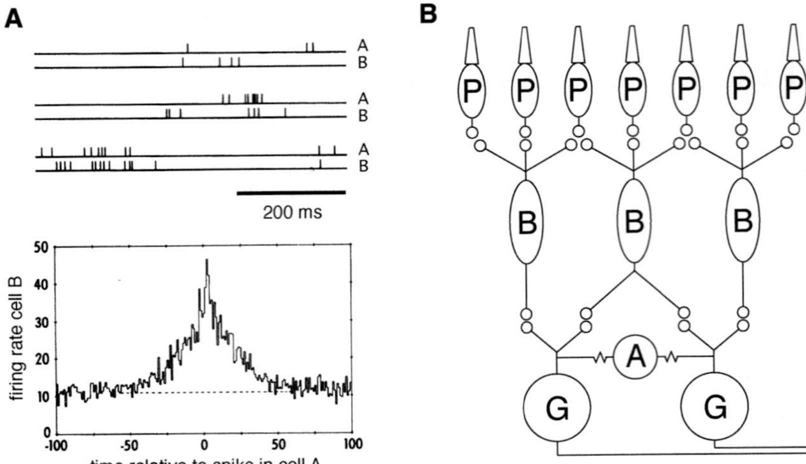

FIGURE 12.1   Correlated activity in the retina. (A) Example of correlations in spontaneous activity adapted from Mastronarde (1983b). The top traces show times of action potentials measured simultaneously from two cat ganglion cells. The cells exhibit a marked tendency to fire nearly simultaneously. This tendency is quantified in the cross-correlation function at the bottom, which shows the firing rate in cell B given that cell A generated a spike at time 0. The dashed line shows the firing rate that would be predicted if the cells fired independently. (B) Hypothetical circuit origin of correlated activity in the primate retina. Photoreceptors (P), bipolar cells (B), amacrine cells (A), and ganglion cells (G) are shown schematically. Common input to two ganglion cells from shared bipolar cells, or gap junctions between ganglion cells and amacrine cells or directly between ganglion cells (not depicted), can both contribute to correlated activity.

amacrine cells are thought not to fire spikes but instead to signal through graded changes in membrane potential. The ~60 or more cell types in the retina are organized into distinct subcircuits delineated by cell-type-specific connectivity; for example, specific types of bipolar and amacrine cells provide input only to specific types of ganglion cells. Finally, each ganglion cell type tiles the surface of the retina uniformly (Dacey, 1993; Wässle, Peichl, & Boycott, 1981), in some cases with exquisite precision (Gauthier et al., 2009), transmitting a complete representation of the visual scene to its specific central targets.

This understanding of retinal architecture suggests several considerations relevant for exploring the mechanisms and impact of correlated activity. As with all known aspects of retinal function, correlated activity is likely to be specific to the cell type(s) in question. The spatial scales of correlated activity are likely to be related to, and perhaps predictable from, the spatial extent of the cells composing each subcircuit. Finally, the exquisite precision of the retinal circuitry suggests that correlated activity itself may be a highly stereotyped aspect of retinal function. These considerations point to the importance of considering correlated activity in the context of specific species and specific retinal subcircuits.

In the primate retina, midget and parasol ganglion cells provide valuable targets for studies of correlated activity because of their well-understood input circuitry,

central projections, and roles in visual function (reviewed by Dacey & Packer, 2003; Field & Chichilnisky, 2007; see chapter 16 by Kaplan). ON (OFF) midget ganglion cells, the two most numerous ganglion cell types in primate retina, receive input from ON (OFF) midget bipolar cells. Midget ganglion cells project to the parvocellular layers of the lateral geniculate nucleus; their signals are thought to mediate sensitivity to color and fine spatial structure. The ON (OFF) parasol cells, the next most numerous ganglion cell types, each receive excitatory input from two types of ON (OFF) diffuse bipolar cells. Parasol cells project to the magnocellular layers of the LGN and are thought to mediate sensitivity to motion, to rapid changes in input, and to subtle changes in luminance. Together, the midget and parasol cell types constitute roughly 70% of all ganglion cells in primate retina (Rodieck, 1998), and the receptive fields of cells of each of these types precisely tile the visual world (Gauthier et al., 2009). The numerical dominance and precise circuitry of the parasol and midget circuits suggest that their physiological properties, including their correlated activity, may have been subjected to intense selective pressure during evolution. Several studies described below focus on these cell types.

Correlated activity could, in principle, originate at any number of sites within the retinal circuitry, with correspondingly diverse effects on visual signaling in different circuits and species. In primate midget and parasol ganglion cells, work over the past 10 years suggests that correlated activity can be largely accounted for by the simplified circuit in figure 12.1B, in which correlated activity emerges from shared components of known circuitry rather than from a novel subcircuit within the retina. In this circuit the activity of ganglion cells can covary as a consequence of two factors. First, the activity of a single photoreceptor, bipolar, or amacrine cell can diverge to produce activity simultaneously in multiple ganglion cells (Brivanlou, Warland, & Meister, 1998; DeVries, 1999; Mastronarde, 1983a, 1983b). Second, the activity of a ganglion cell can propagate laterally to another ganglion cell via gap junctions either between ganglion cells or through an intermediate amacrine cell (Brivanlou, Warland, & Meister, 1998; DeVries, 1999; Hidaka, Akahori, & Kurosawa, 2004; Hu & Bloomfield, 2003; Mastronarde, 1983c). Because these divergent and lateral sources of correlations originate in different cells and pass through different circuits, they typically have distinct impacts on the visual signal. These mechanisms of correlated activity are not exclusive, as revealed in work discussed below.

Correlated activity can involve a combination of *signal correlations* (correlated activity generated by the visual stimulus) and *noise correlations* (fluctuations in correlated activity about the average response to the stimulus) (reviewed by Averbeck, Latham, & Pouget, 2006). For example, divergent signals from photoreceptors that produce correlations in downstream ganglion cells could originate in the response to a time-varying light input (signal) or in spontaneous photopigment isomerization or other sources of cellular noise downstream of phototransduction (noise). Similarly, lateral spread of activity from one ganglion cell to another could convey the light responses of the source cell (signal) or spontaneous fluctuations in its membrane voltage or transmitter release (noise). Although signal and noise correlations may originate in the same neuron or even the same molecule, they have different effects on visual signals transmitted to the brain: signal correlations provide information relevant for behavior; noise correlations necessarily interfere with vision and behavior. Therefore, in what follows, signal and noise correlations are treated separately.

STATISTICAL ANALYSIS OF CORRELATED FIRING PATTERNS

Signal correlations can be produced by receptive field overlap or by spatially extended stimuli and in principle can be understood on the basis of sequential measurements from different cells. Noise correlations, on the other hand, emerge from fluctuations in neural responses that occur across stimulus trials and require simultaneous measurements to study. In this section we define the distinction between signal and noise correlations mathematically and describe experimental manipulations that separate them.

Consider two cells, A and B, and denote by $P(A)$ the probability of a spike in cell A, and by $P(B)$ the probability of a spike in cell B in a particular period of time. Correlated activity in the two cells means that their responses are not statistically independent, that is, $P(A,B) \neq P(A)\ P(B)$.

We define signal correlations to be those that vanish when the stimulus is held constant. Specifically, two cells exhibit only signal correlations if, conditioned on the presence of a fixed stimulus S, their signals are statistically independent: $P(A,B \mid S) = P(A \mid S)\ P(B \mid S)$. Signal correlations alone, in the absence of noise correlations, imply that the mean response of cell B to a stimulus is at least partially predictable from that of cell A but that the fluctuations in rate about the mean on a given presentation of the stimulus are independent in the two cells (figure 12.2A, left).

We define noise correlations to be those that remain when the stimulus is held constant. Specifically, two cells exhibit noise correlations if $P(A,B \mid S) \neq P(A \mid S)\ P(B \mid S)$. This nonindependence means that variation in the firing of cell B about its mean can be at least partially predicted by the variation in the firing of cell A (figure 12.2A, right). An example of noise correlations is shown by the peaked cross-correlation function (figure 12.1A, bottom) of the spike trains (figure 12.1A, top) of two cat ganglion cells recorded in the presence of steady, spatially uniform illumination. The peak indicates that the firing of the first cell was higher than average around the time of firing of the second cell, even though the stimulus was unchanging.

Statistical fluctuations in photon absorption represent a special case and deserve specific mention. A nominally constant light input will still produce statistical variations in photon absorption due to the quantal nature of light itself. These quantal fluctuations can produce noise correlations in cells with common photoreceptor input. As defined above, the signal in this case is the nominal or average light intensity or, equivalently, the rate of the Poisson process describing the times of photon absorption.

As we discuss below, signal and noise correlations in the ganglion cell output signals have different effects on the fidelity of visual signals transmitted to the brain. To understand correlated activity in the retina, then, requires methods to measure signal and noise correlations separately.

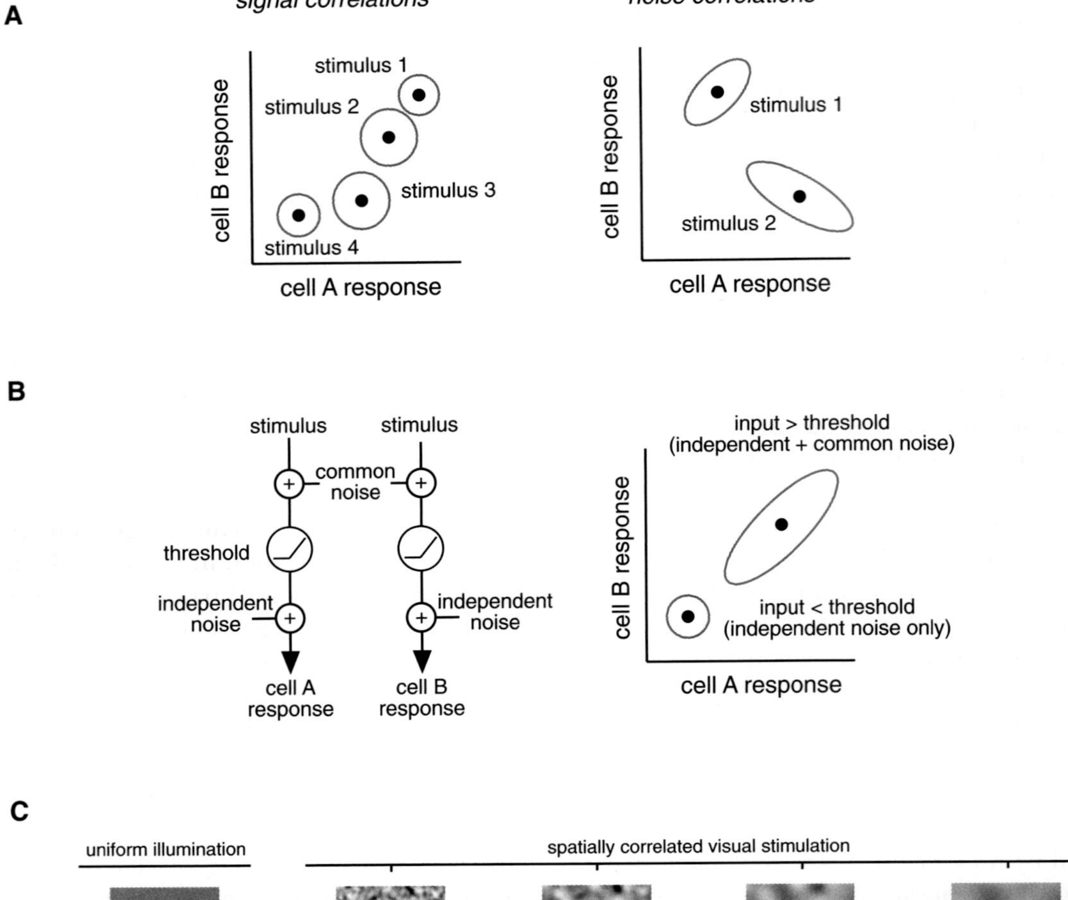

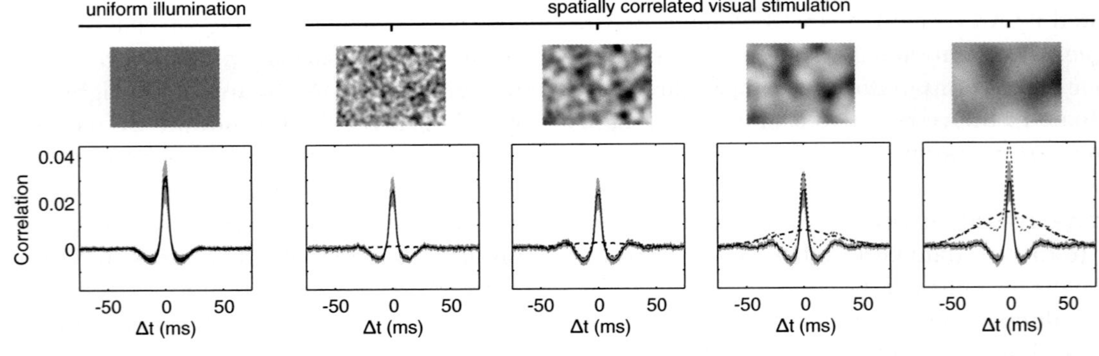

FIGURE 12.2   Signal and noise correlations. (A) Correlated activity consists of both signal and noise correlations. Signal correlations (left) cause the mean responses (points) of cells A and B to be correlated. Noise correlations (right) cause the variation in responses about the mean (ellipses) to covary. (B) A circuit in which noise correlations depend on the stimulus (left). Common noise diverges to generate correlated noise in the responses of two cells. Nonlinearities alter the strength of correlated noise by changing the relative magnitude of common and independent noise (right). (C) Separation of signal and noise correlations. (Adapted from Greschner et al., 2011.) Noise correlations can be isolated with uniform illumination (left). Signal correlations can be isolated by computing the cross-correlation of responses to different repetitions of the same stimulus (dashed lines), which in this case was spatiotemporally filtered white noise. Subtracting the signal correlations from the total correlations (dotted line) isolates noise correlations (solid lines) for several stimulus conditions.

Signal correlations can be isolated by examining the average responses of different cells to the same stimulus delivered multiple times. A typical manipulation is to measure the cross-correlation between the mean responses of the different cells to many stimulus repeats, assuming the noise has been averaged away. Alternatively, signal correlations can be measured from the cross-correlation between the responses of cells to different repetitions of the same stimulus—for example, pairing the response of cell 1 on trial n with that of cell 2 on trial $n + 1$; this procedure is often referred to as "shuffling" (figure 12.2C).

Noise correlations, on the other hand, can be isolated only by simultaneous measurements from different cells because they refer to covariation of responses about the mean. Noise correlations are most simply measured by calculating the cross-correlation in the activity of cells with no time-varying stimulus. They can also be measured in the presence of a repeated stimulus sequence in one of two (equivalent) ways: (1) by subtracting the mean response from each individual response and measuring the cross-correlation of the resulting residuals of different cells, or (2) by subtracting the signal cross-correlation (above) from the total cross-correlation (figure 12.2C).

Importantly, noise correlations measured under different stimulus conditions can differ because even in the presence of a fixed noise source, circuit nonlinearities can cause the degree of noise correlations to depend on the stimulus. The generic circuit in figure 12.2B illustrates this possibility. Noise correlations originate in this circuit from a fixed shared noise source at an early stage, whereas independent noise occurs after a threshold in the circuit. The relative strength of noise correlations in the circuit output is determined by how effectively the shared noise propagates through the circuitry. The input stimulus can increase the relative importance of common noise in the circuit output by driving the response to a level that effectively allows more shared noise through the nonlinearity (figure 12.2B, right). With a weak stimulus common noise fails to propagate, independent noise dominates, and the activity of the two cells is uncorrelated. The nonlinear element in such a scenario can originate from synapses or spike generation (de la Rocha et al., 2007).

PROPERTIES OF CORRELATED NOISE IN GANGLION CELL ACTIVITY

Noise correlations between pairs of cells such as those in figure 12.1A are a general feature of signaling in mammalian retina, shaping the response patterns of ganglion cell populations. For example, more than half of the spikes generated by a parasol ganglion cell in primate retina are correlated with a spike in another nearby parasol cell (Shlens, Rieke, & Chichilnisky, 2008). As a result of these correlations some spatial patterns of activity in the ganglion cell population occur at rates 10^3–10^6 times higher or lower than would be predicted if cells fired independently (Schneidman et al., 2006). Noise correlations alone cause some firing patterns to occur at least 100 times more frequently than would be the case with independent firing (Shlens et al., 2009). In this section we describe a set of key properties of correlated activity patterns and the suggestions these properties make about underlying mechanisms.

Pairwise Interactions Can Account for Much of Population Activity

In principle, understanding correlated activity could require specifying all interactions within large groups of cells (e.g., figure 12.3A). Because the number of possible interactions grows exponentially with the number of cells in the population, this measurement is well out of the reach of current methods. Recent work, however, indicates that patterns of activity in large populations can often be understood purely on the basis of pairwise interactions between cells, a huge simplification. Specifically, in salamander and primate retina, the degree of correlated firing in a handful of cells can be accurately predicted from their pairwise correlations and their individual firing rates (Schneidman et al., 2006; Shlens et al., 2006). This finding generalizes in a strong way to large populations: in ON and OFF parasol cells of primate retina, spatial patterns of activity generated by ~50 cells can be accurately predicted from pairwise interactions restricted to immediate neighbors in the mosaic of each cell type (Shlens et al., 2009).

The conclusion that pairwise models can capture the structure of population activity is a statistical one rather than a mechanistic one and is based on maximum-entropy analysis borrowed from statistical mechanics (Amari, 2001; Jaynes, 1957a, 1957b; Schneidman, Bialek, & Berry, 2003). The essential idea is to calculate the probability distribution of different activity patterns in a group of cells that would be predicted from pairwise interactions alone and compare this to the measured probability distribution (figure 12.3B). The prediction is obtained by calculating the distribution that exhibits the least structure (maximum entropy), subject to the measured mean firing rates and pairwise correlations between cells. The success of this approach in describing the firing patterns generated by groups of cells (figure 12.3B, C) indicates that interactions among more than two cells are not needed to account for population activity; instead, pairwise interactions, even restricted to immediate neighbors, suffice. Pairwise models have similarly proven capable of capturing much of the structure of population activity in other neural circuits (Tang et al., 2008), although in some cases pairwise models fail to completely explain population activity (Ganmor, Segev, & Schneidman, 2011; Ohiorhenuan & Victor, 2011; Schnitzer & Meister, 2003). Importantly, the success of maximum entropy models should be interpreted with caution: this approach determines the complexity of the interactions needed to empirically capture the population activity but does not make unique predictions about the underlying circuitry or mechanisms.

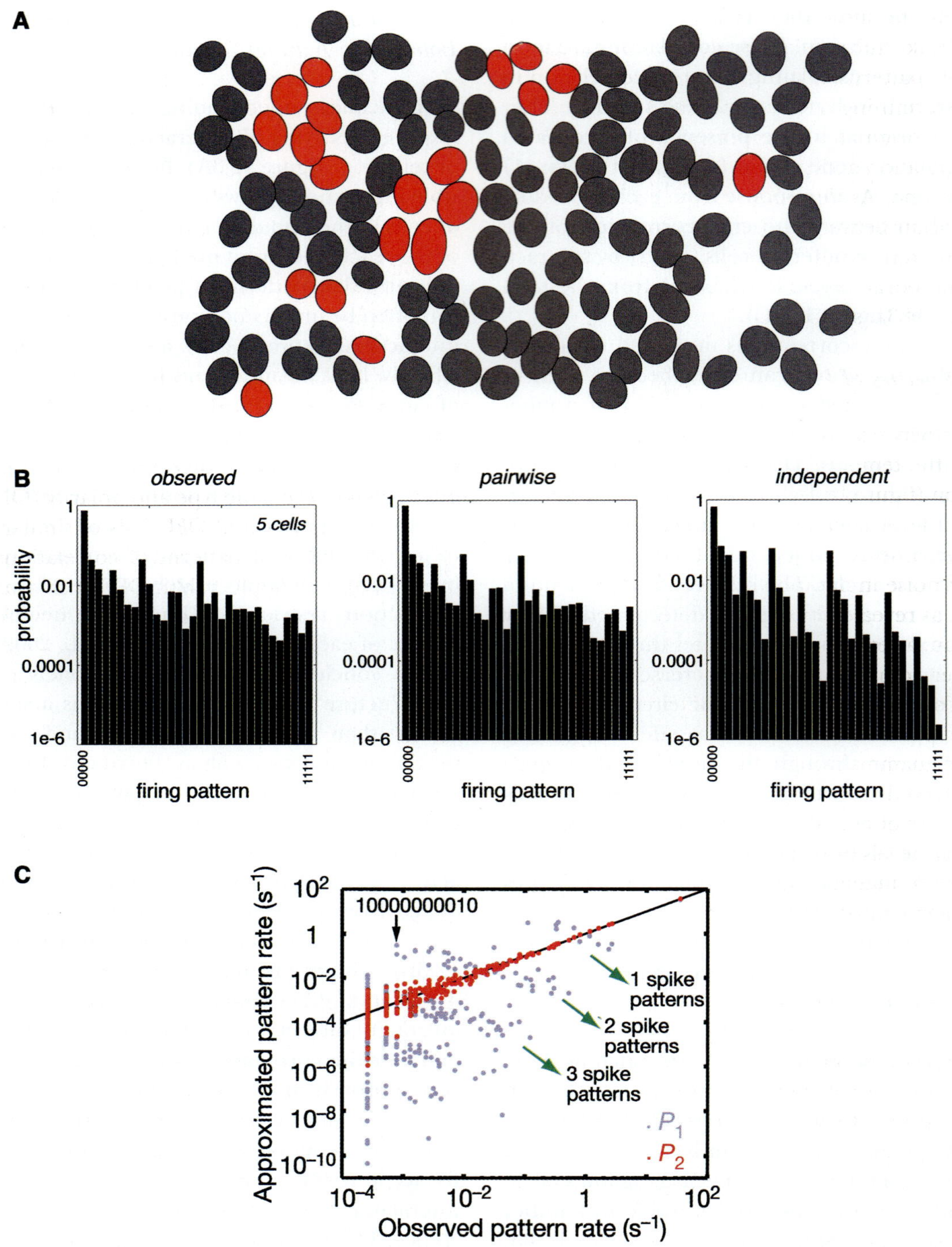

FIGURE 12.3 Pairwise interactions account for population activity. (A) Pattern of spontaneous electrical activity in a population of simultaneously recorded ON parasol ganglion cells in primate retina (Shlens et al., 2009). Ellipses indicate receptive fields of recorded cells, obtained with white noise. Red indicates cells that fired one or more spikes during a 10-ms period; gray indicates cells that did not. (B) Distribution of firing patterns and models (Shlens et al., 2006). Each panel shows a recorded or predicted frequency distribution of patterns of activity in five simultaneously recorded ON parasol cells during a 10-ms period. For example, 11000 indicates the first two cells firing and the others not. Left panel shows data recorded during spontaneous activity. Right panel shows a prediction based on assuming statistical independence of the recorded cells. Middle panel shows the prediction based on assuming only pairwise interactions between cells. (C) Comparison of empirical and predicted firing pattern probabilities. The graph shows frequencies of specific firing patterns in a group of 10 ganglion cells in salamander retina (Schneidman et al., 2006). The abscissa indicates recorded frequencies; the ordinate represents model predictions; and the line represents equality. Blue points represent predictions based on an assumption of statistical independence; red points represent predictions based on an assumption of pairwise interactions. Text insets on the right indicate the number of cells spiking in each cluster of patterns. One sample pattern with two cells firing is labeled at top.

   FRED RIEKE AND E. J. CHICHILNISKY

Nonetheless, the success of pairwise models suggests that we can make substantial progress on understanding the activity patterns of the entire ganglion cell population by determining how interactions between pairs of nearby cells originate. This reduces the problem of sampling population activity patterns to an experimentally tractable one. As described below, pairwise noise correlations occur between cells of the same type as well as cells of different types, but with a cell-type-specific spatial and temporal structure (Greschner et al., 2011; Mastronarde, 1983a; see figure 12.4).

Temporal Properties of Pairwise Correlations

Correlated activity occurs on a variety of time scales, as measured by the temporal structure of the cross-correlation function (figure 12.1A). These time scales depend on mean light level, species, and cell type and reflect a variety of mechanisms.

Correlated noise in darkness often exhibits two clear components, as revealed by the peak in the cross-correlation function: a central rapid component ~20 ms wide and a broader component ~200 ms wide. Salamander ganglion cells can exhibit different combinations of temporal components (Brivanlou, Warland, & Meister, 1998), but in mammalian retina the slow component has been observed only together with a rapid component (Greschner et al., 2011; Mastronarde, 1983b). At higher scotopic levels the slow component of correlated activity between mammalian ganglion cells recedes, while the faster component remains and sharpens in timing. Slow correlations are not apparent at photopic light levels.

In cat, slow correlations between ganglion cell responses in darkness appear to be produced by synchronous bursts of spikes (e.g., figure 12.1A, top). The rate of such bursts, their discrete nature, and their slow kinetics suggest that they originate from noise in the rod photoreceptors, specifically from the occasional spontaneous activation of rhodopsin (Mastronarde, 1983b). Weak background light increases the slow noise correlations, but the kinetics remain indistinguishable from those caused by spontaneous activation of rhodopsin; the increased noise correlations under these conditions likely result from quantal fluctuations in the stimulus.

The fast component of correlated activity, however, is more rapid than the light response of cone photoreceptors (Brivanlou, Warland, & Meister, 1998; Mastronarde, 1983a), indicating that it does not originate from quantal fluctuations in photon absorption or spontaneous activation of the cone photopigment. Instead, the rapid component could be explained by another noise source in cones that varies more rapidly than the light response or by rapid fluctuations introduced by a cell later in the circuitry. We return to this issue below.

Rapid correlated activity exhibits additional temporal structure when viewed on a fine time scale (DeVries, 1999; Hu & Bloomfield, 2003; Mastronarde, 1983c; Brivanlou, Warland, & Meister, 1998). The central peak of the cross-correlation function can be symmetrical and centered at zero, indicating that the cells tend to fire simultaneously (figure 12.5A, left). The cross-correlation function can also exhibit two peaks located symmetrically about zero (figure 12.5A, right). This structure indicates that the cells tend to fire in sequence with a fixed time lag and either cell leading, suggestive of reciprocal connectivity, for example, via an electrical synapse. Such split peak correlations have been observed only in cells of the same type and polarity (ON or OFF). In several cases ON and OFF cells of similar morphology exhibit different patterns of correlations at a fine time scale; for example, rabbit OFF brisk transient cells exhibit split peak correlations, but ON brisk transient cells do not (DeVries, 1999) (figure 12.5A, right), and the reverse is true for primate ON and OFF parasol cells (Trong & Rieke, 2008). A third possibility is that the cross-correlation function has a single peak offset from zero, indicating that one cell tends to fire before the other. This is observed in RGCs of different types (Greschner et al., 2011) and is probably attributable to their different response kinetics.

Spatial Properties of Pairwise Correlations

The precise tiling of the receptive fields of a given type of ganglion cell suggests that a similarly precise functional organization may hold for correlated activity. This possibility has been tested, within and across cell types, mainly at photopic light levels.

The sign and strength of noise correlations depends on cell type in mammalian retina (DeVries, 1999; Mastronarde, 1983a) (figure 12.4C). The activity of ON cells is generally positively correlated with that of nearby ON cells of the same or different types, and the same is true for correlations between OFF cells. The activity of ON cells is typically negatively correlated with the activity of OFF cells (Greschner et al., 2011; Mastronarde, 1983a). Thus, the polarity of noise correlations generally follows simple expectations given the polarity of the responses of the cell types involved.

In all species tested, noise correlations are spatially restricted (DeVries, 1999; Mastronarde, 1983a; Meister, Lagnado, & Baylor, 1995). Thus, the underlying mechanisms appear to operate locally, on a spatial scale not much larger than the ganglion cell receptive field.

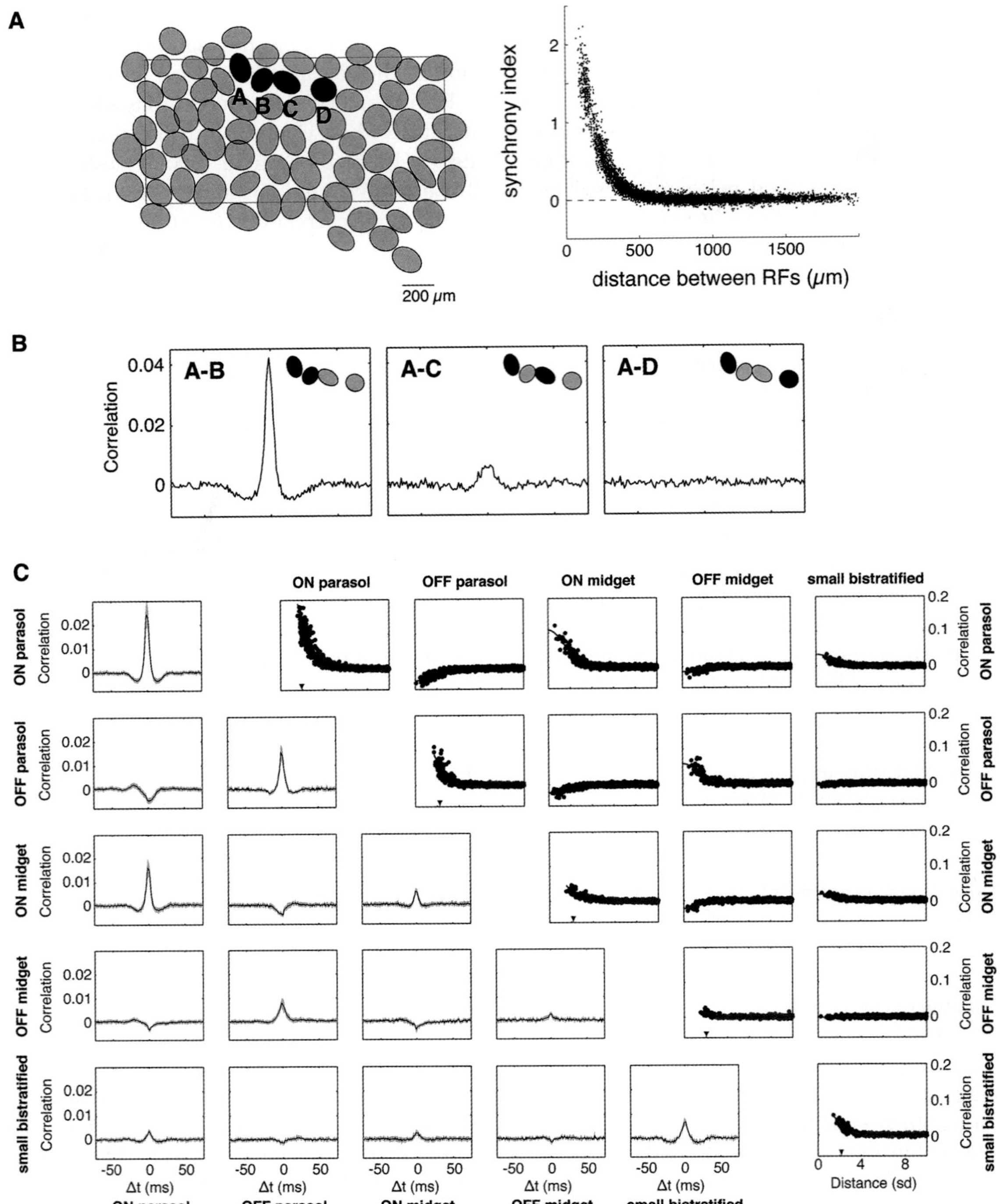

FIGURE 12.4   Noise correlations depend systematically on distance and cell type. (A) Mosaic of receptive fields of a collection of simultaneously recorded ON parasol cells (left) (Greschner et al., 2011) and typical dependence of correlation strength on distance between cells (right) (Shlens et al., 2006). (B) Example of distance dependence for collection of four highlighted cells in A (Greschner et al., 2011). (C) Noise correlations across cell types and distances (Greschner et al., 2011). The left/lower part of the grid shows average cross-correlation functions for cell pairs composed of neighboring cells of the five major types in primate retina. The right/upper part shows the dependence of correlation strength on distance, measured in units of receptive field width. For pairs composed of cells of the same type, neighboring pairs are indicated with black; other pairs are shown in gray. Curves are smoothed averages.

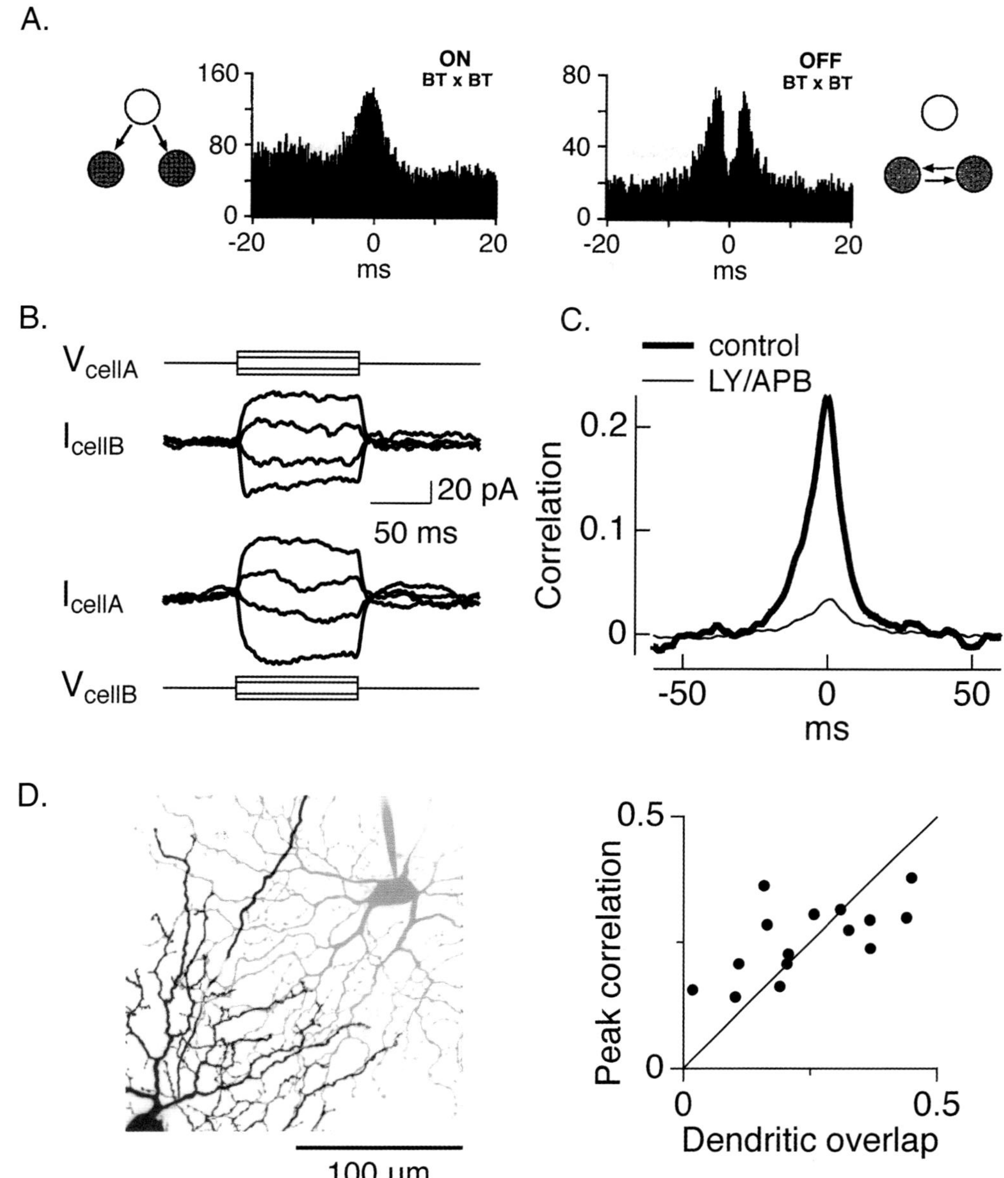

FIGURE 12.5   Combination of common input and reciprocal connections shapes correlated activity. (A) Single-peak cross-correlation functions suggest common divergent input; split-peak cross-correlation functions suggest a contribution of reciprocal connections (DeVries, 1999). (B) ON parasol cells exhibit reciprocal connections (Trong & Rieke, 2008). Stepping the voltage of one cell produced a change in current in the other cell. (C) Input correlations are largely generated by cone photoreceptors (Ala-Laurila et al., 2011). Cross-correlation functions for excitatory input to an ON parasol/ON midget ganglion cell pair. Correlations are suppressed when cone input to cone bipolar cells is blocked pharmacologically (LY/APB). (D) Dendritic overlap (left for two ON parasol cells) predicts correlation strength (Trong & Rieke, 2008).

Consistent with these observations, in primate retina the dependence of noise correlations on distance between cells (and on their receptive field overlap) for cells of the same cell type is strikingly precise (figure 12.4A,B) (Greschner et al., 2011; Shlens et al., 2006). This precision is reminiscent of the coordinated mosaics of receptive fields of each cell type (Gauthier et al., 2009). A stereotyped dependence of noise correlations on receptive field overlap is also observed for pairs consisting of cells of two different types, for example, ON parasol and ON midget (Greschner et al., 2011) (figure 12.4C).

Nonlinearities in retinal circuitry, such as spike generation in ganglion cells, could alter the balance of different mechanisms generating noise correlations (figure 12.2B). Therefore, a clearer picture of the underlying mechanisms can be obtained by measurements of correlations at different stages of the circuitry.

In primate retina correlated noise in the excitatory synaptic inputs to pairs of ganglion cells exhibits a systematic dependence on distance and cell type that resembles the dependence of spike correlations on receptive field overlap. Specifically, the strength of correlations in the excitatory input to pairs of neighboring ON parasol ganglion cells, and to ON parasol and ON midget pairs, can be predicted given the overlap of the dendrites and the expected spatial spread of cone signals as they are conveyed via bipolar cells to ganglion cells (figure 12.5D) (Ala-Laurila et al., 2011; Trong & Rieke, 2008). Consistent with spike responses, excitatory inputs to ON and OFF parasol cells are negatively correlated.

Given that different ganglion cell types share little circuitry other than photoreceptors, and that the ON and OFF division occurs at the photoreceptor synapse, widespread noise correlations that respect the sign of light responses of the cells involved strongly suggest an origin in photoreceptor noise. However, positive correlations between different cell types in some cases could arise from shared bipolar inputs, and both positive and negative correlations could arise from crosstalk between circuits via amacrine cells. These considerations motivate more direct tests of the origin of correlated noise.

DIRECT EVIDENCE FOR MECHANISMS OF CORRELATED ACTIVITY

Reciprocal Connections

Split-peak cross-correlation functions such as those in figure 12.5A indicate reciprocal interactions between the two cells. The likely mechanism for such reciprocal connections is gap junction coupling between ganglion cells and/or between ganglion cells and common amacrine cells. Indeed, both tracer injections into single ganglion cells and electron microscopy reveal gap junctions between some ganglion cell types and amacrine cells (Dacey & Brace, 1992; Jacoby et al., 1996; Vaney, 1991; see chapter 9 by Bloomfield and Völgyi), although some ganglion cell types do not exhibit tracer coupling (Dacey & Brace, 1992).

Gap junctions would be expected to cause changes in the voltage of one ganglion cell to propagate to the other cell. Such a mechanism could explain split-peak correlations because the voltage change in the originating cell would precede the voltage change in the other cell. The strength of gap junction coupling has been measured from paired recordings by stepping the voltage of one cell while measuring the current change in the other cell (figure 12.5B). The coupling strength measured in this way does not always correlate well with inferences from tracer coupling. In particular, some

ganglion cell types that exhibit tracer coupling show little or no electrical coupling (Dacey & Brace, 1992; Trong & Rieke, 2008). Furthermore, gap junctions do not appear to be required for correlated activity, as some populations of ganglion cells that exhibit robust correlated activity do not show tracer coupling or electrical coupling (DeVries, 1999; Trong & Rieke, 2008).

Gap junctions alone appear capable of supporting some of the correlated activity in salamander retina. Specifically, rapid split-peak correlations (as well as slower single-peaked correlations) survive chemical blockade of synaptic transmission (Brivanlou, Warland, & Meister, 1998). This suggests that gap junctions can account for both reciprocal coupling and common input via an amacrine cell. The situation in primate retina is quite different (Trong & Rieke, 2008): correlation strength is reduced to near zero when chemical synaptic transmission is suppressed. Correspondingly, electrical coupling measured between mammalian ganglion cells is weak: an action potential in one cell typically generates only a small (< 1 mV) change in voltage in a nearby cell.

The relatively weak reciprocal connections between mammalian ganglion cells suggest that such connections alone are not likely to generate correlated action potentials. More probably, reciprocal connections act together with common input noise (see below) to correlate action potential timing (Trong & Rieke, 2008). In this scheme common input noise correlates voltage changes in two cells, but independent noise causes the exact voltage trajectories to differ. Generation of an action potential by the cell that reaches spike threshold first can then cause, via reciprocal connections, a delayed action potential in the second cell because common noise has brought its voltage close to spike threshold. Thus, reciprocal coupling could create a split-peak cross-correlation function by shortening and regularizing the typical time interval between correlated spikes generated by common input.

Common Input

In addition to correlations in spike output, primate ganglion cells show robust correlated noise in their chemical synaptic inputs (Ala-Laurila et al., 2011; Trong & Rieke, 2008). Given the known circuitry of the retina, these synaptic inputs must come from either bipolar or amacrine cells. Thus, correlations could arise from noise generated in a bipolar or amacrine cell that makes synapses onto two or more ganglion cells. Alternatively, correlations could arise from noise in upstream photoreceptors, whose signals diverge via bipolar and amacrine cells to the ganglion cells.

In primate midget and parasol ganglion cells the evidence favors an origin in photoreceptor noise. Noise in the total excitatory synaptic input to individual ganglion cells exhibits kinetics very similar to that of correlated noise between cells (Trong & Rieke, 2008). This suggests that the independent noise in each ganglion cell and the correlated noise in pairs of cells share a common origin. However, the rapid time scale of correlated noise at photopic light levels, relative to the light response, raises a potential problem. If correlated noise originates in cones, then cone noise must vary more rapidly than the light response, and these rapid variations must be able to propagate through the retina and produce rapid variations in synaptic inputs to ganglion cells. Both of these requirements are supported by direct measurements (Ala-Laurila et al., 2011; Schneeweis & Schnapf, 1999), indicating that the kinetics of common noise can be accounted for by noise produced in the cones and retinal circuits known to convey cone inputs to ganglion cells. Further supporting the importance of cone noise, pharmacological blockade of cone inputs to ON bipolar cells suppresses correlated noise in the excitatory inputs to pairs of ON ganglion cells (figure 12.5C) (Ala-Laurila et al., 2011). Importantly, this blockade was designed to preserve mean synaptic activity and thereby avoid confounds with circuit nonlinearities (figure 12.2B).

These results, together with the observations described above about the dependence of correlations on cell type and receptive field overlap, suggest a simple picture of how correlated activity is produced: cone noise, traversing the retina through diverse pathways, produces correlations in the synaptic inputs to any ganglion cells that share cone inputs. These correlations in synaptic inputs produce correlations in spiking activity. A corollary with important implications for the fidelity of visual signals is that a significant fraction of the noise in the output of the retina is generated in the cones.

Small bistratified cells provide a notable exception to the picture above: these cells show positive noise correlations with ON parasol (and other ON ganglion) cells (Greschner et al., 2011) but receive OFF input from long- and middle-wavelength cones. Therefore, although cone noise appears to explain much of the correlated activity in the dominant circuits of the primate retina, a complete account is likely to involve additional mechanisms.

IMPLICATIONS FOR VISUAL SIGNALING BY THE RETINA

How do correlations affect the way the retina transmits visual information to the brain? Below we first describe a set of general considerations that lay the groundwork for approaching this problem. We then relate these considerations to several recently developed approaches to understanding the fidelity of coding by the ganglion cell population.

General Considerations Governing the Impact of Signal and Noise Correlations

Noise correlations (by definition) emerge from fluctuations in neural activity about the average response to the visual stimulus. Therefore, noise correlations cannot by themselves transmit information useful for visual perception or visually guided behavior. Indeed, the presence of noise in neural responses inevitably reduces visual fidelity by masking changes in neural activity that are reliably associated with the stimulus. In principle, noise can play a useful role by raising neural responses above a threshold, permitting transmission of signals that otherwise would fail to propagate (reviewed by McDonnell & Ward, 2011; Ward, Neiman, & Moss, 2002). However, this function could be implemented more effectively by raising the neural response in an unvarying manner, without incurring the cost in information transmission produced by noise. Therefore, relative to appropriate noiseless signals, noise correlations are not helpful for neural signaling. Instead, noise correlations represent two immutable facts. First, basic physical laws mean that neural signals will fluctuate as a result of photon counts that vary for nominally identical stimuli and spontaneous thermally-driven movements and changes in molecules (e.g., channels and enzymes). Second, the functional organization of the retinal circuitry causes many neurons to share input circuitry, and propagation of noise through shared circuitry inevitably produces noise correlations.

Although noise correlations are unavoidable, they are not without consequence. One way to determine their impact is to compare retinal signaling with noise correlations to what would occur in a hypothetical retinal circuit with equal-magnitude but independent noise (note that such a circuit may not be physically realizable). Relative to this hypothetical circuit, correlated noise can either improve or diminish the fidelity of visual signals, depending on how the signals themselves are correlated. Figure 12.6A shows a case in which noise correlations decrease the ability to discriminate between two stimuli. The figure depicts the joint distributions of responses of two cells, A and B, to two stimuli, 1 and 2. Solid lines show distributions with noise correlations, and dashed lines show distributions with equal-variance independent noise. Discriminating between the stimuli requires determining whether a

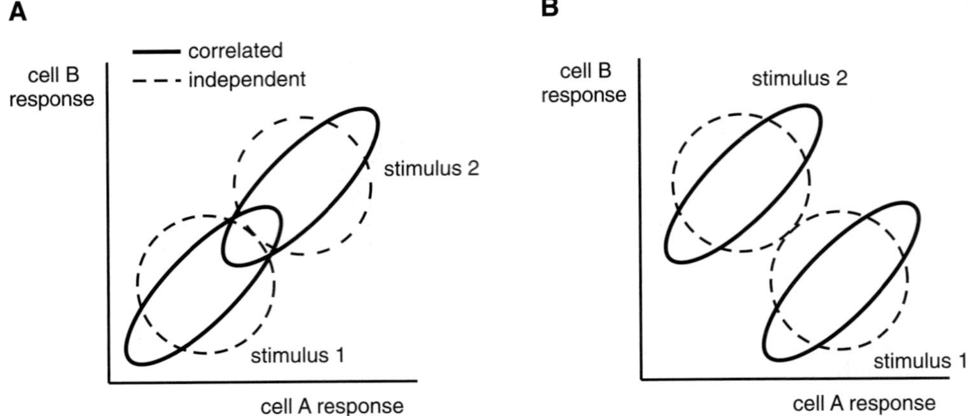

FIGURE 12.6 Impact of signal and noise correlations on discrimination. (A) Schematic of distributions of joint responses of two cells with positive signal and noise correlations. Noise correlations increase overlap and hence decrease discriminability of the two stimuli when compared to the situation with equal variance but uncorrelated noise (dashed lines). (B) As in A, but with negative signal correlations. In this case noise correlations increase discriminability compared to the situation with uncorrelated noise.

particular pair of responses from cells A and B was elicited by stimulus 1 or 2. The region of overlap of the ellipses represents the joint responses that could be elicited by either stimulus and that therefore do not contain information useful for discrimination: the smaller this overlap region, the better discrimination performance can be. In this example both stimuli increase the responses of both cells, producing positive signal correlations that are obscured by positive noise correlations. Figure 12.6B shows an example of the opposite (e.g., see Gollisch & Meister, 2008). Here, one stimulus causes a larger response in one cell and smaller response in the other; thus, signal correlations are negative. In this case positive noise correlations decrease overlap and increase the discriminability of the stimuli.

In general, when signal and noise correlations have the same sign, noise correlations will obscure the signal and decrease discriminability more than would occur with equal-variance independent noise (reviewed by Averbeck, Latham, & Pouget, 2006). When signal and noise correlations have opposite signs, correlated noise obscures the signal less than would occur with independent noise. These effects are particularly important for interpreting data from populations of cells: the common assumption that noise is independent across cells can lead to substantial errors in estimates of the fidelity of the population signal (Britten et al., 1992).

These general ideas can be illustrated with specific predictions in the primate retina. Two neighboring ON parasol cells invariably exhibit positive noise correlations (Shlens et al., 2006), and these noise correlations are roughly additive with the light response (Greschner et al., 2011), indicating that they are not

greatly affected by the stimulus (figure 12.2C). Thus, variations in the contrast of a large incremental stimulus covering the receptive fields of both cells should be less discriminable than they would be if ON parasol cells exhibited equal but independent noise. On the other hand, variations in contrast of an edge positioned to increase the firing of one cell and decrease the firing of the other should be more easily discriminated with correlated rather than independent noise. The reverse applies to pairs of nearby cells consisting of one ON parasol cell and one OFF parasol cell, which always exhibit negative noise correlations (Greschner et al., 2011). Variations in the contrast of a large uniform stimulus will increase the firing of one cell but decrease the firing of the other. These negative signal correlations will be masked by negative noise correlations. However, variations in contrast of an edge positioned to produce positive signal correlations will be more discriminable with correlated than with independent noise. In summary, the opposite sign of the signal and noise correlations in the responses of neighboring cells to edges serves to emphasize spatial contrast, compared to what would occur if the cells exhibited equal variance independent noise.

From the point of view of the organism using retinal responses to make inferences about the world and to guide behaviors, interpretation of the signals transmitted from retina to brain should take noise correlations into account. For example, responses from adjacent parasol cells of opposite sign, which are indicative of spatial contrast, should be assigned greater confidence than responses of the same sign. Given the strength of noise correlations (Greschner et al., 2011; Shlens, Rieke, & Chichilnisky, 2008), such nonuniform

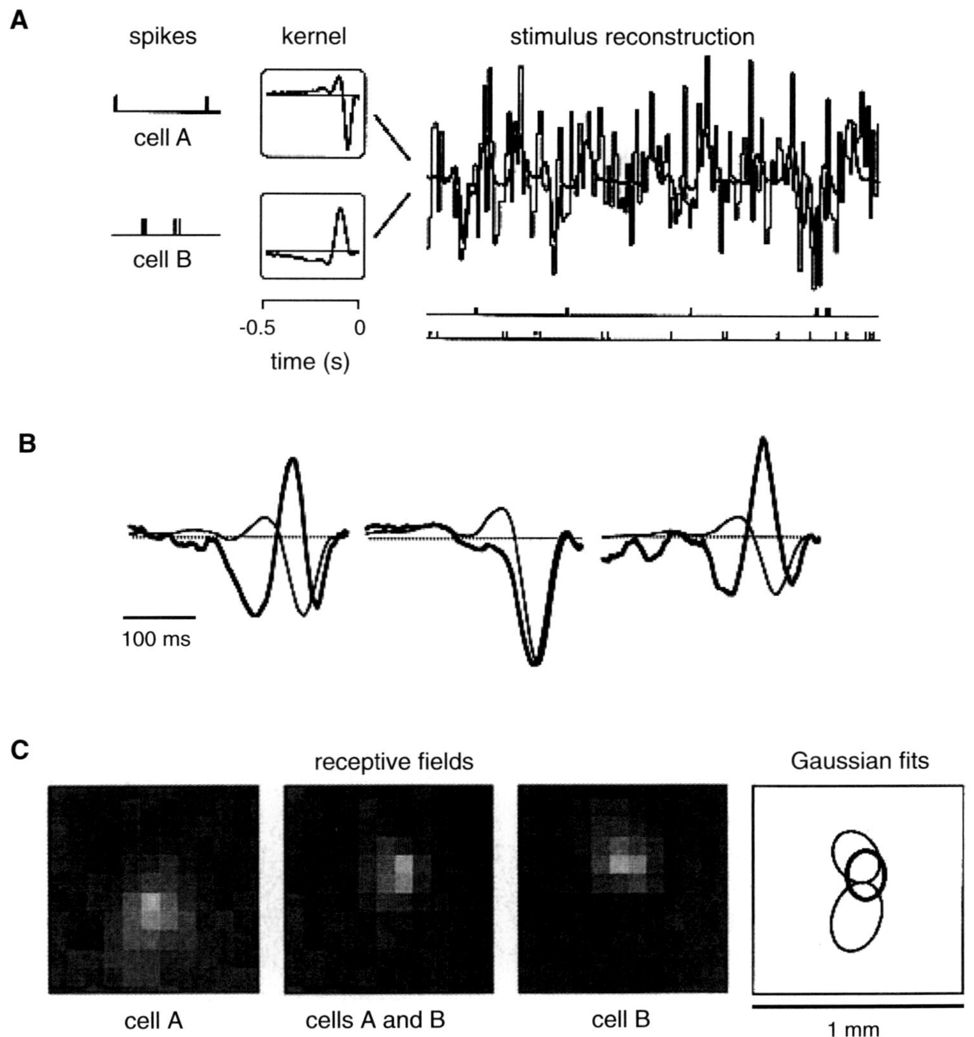

FIGURE 12.7   Impact of correlations on stimulus estimation. (A) Approach to linear stimulus estimation using spike responses of two ganglion cells (Warland, Reinagel, & Meister, 1997). Examples of linear reconstruction in which spike trains from two cells are convolved with optimal linear kernels and summed to produce an estimate of the output. (B) Examples of optimal kernels for three cell pairs calculated for individual cells when stimulus reconstruction depends on single cells only (thin traces) or pairs of cells (thick traces). The optimal kernel for one cell depends on whether another cell is used in stimulus estimation. (C) Visual signals associated with correlated spikes (Meister, Lagnado & Baylor, 1995). Receptive fields for all spikes from cell A or B and for correlated spikes, obtained by spike-triggered averaging. Gaussian receptive field fits at right reveal that the receptive field for correlated spikes is smaller than that for each cell.

weighting in interpreting the retinal output could have a sizable effect on visual performance.

Impact of Correlations on Stimulus Decoding

The impact of noise correlations on stimulus representation can be formalized quantitatively. Consider the problem of using the responses of a ganglion cell to construct an estimate of the visual stimulus. Arguably, the organism performs a task similar to this in many forms of visually guided behavior. One approach to this problem is *linear reconstruction* (Rieke et al., 1997), in which an estimate of the stimulus is created by assigning an elementary visual feature (or kernel) to each spike in each neuron and summing these over space and time. In the primate retina this approach reveals that noise correlations between nearby cells limit the degree to which averaging across cells improves coding accuracy (Ala-Laurila et al., 2011); this effect, like the magnitude of noise correlations themselves, scales predictably with receptive field overlap. Further, the reconstruction approach provides an estimate of the optimal kernel to associate with the spikes from each cell (Rieke et al., 1997). In salamander retina the optimal reconstruction kernel for spikes from one cell changes substantially when the activity of a nearby cell is taken into account (Warland, Reinagel, & Meister, 1997) (figure 12.7A, B). In other words the optimal interpretation of spiking in one cell depends on whether the other cell is active.

The importance of this effect in an entire population can be quantified, and potentially assessed more accurately, using stimulus reconstruction that exploits an accurate nonlinear model of neural response and correlations, for which the optimal reconstruction can be derived mathematically. Such an approach, applied to primate parasol cells, reveals that reconstruction of a visual stimulus exploiting knowledge of the noise correlations in a complete population of cells can result in a 20% improvement over reconstruction that incorrectly assumes no noise correlations (Pillow et al., 2008). This does not mean that noise correlations carry information; rather, it implies that accounting for their structure is important for accurately interpreting retinal output signals.

Information-Theoretic Approaches

Although the approaches described above provide insight into the problem of using retinal signals to make inferences about the stimulus, they are potentially limited by relying on particular approaches to reconstruction. In principle the amount of information contained in neural responses about the visual stimulus can be computed using methods from classical information theory (Cover & Thomas, 1991; Rieke et al., 1997; Schneidman, Bialek, & Berry, 2003) without reference to a particular reconstruction scheme. Specifically, the *information* about a stimulus contained in the response of a cell quantifies the degree to which knowing the response reduces uncertainty about the stimulus. For cells that fire independently, the joint information carried by two cells about the stimulus is the sum of the information carried by each cell individually. For cells exhibiting noise correlations, a positive (negative) difference between the joint and summed information is defined as *synergy (redundancy)*, which indicates that two cells jointly convey more (less) information than would occur if they fired independently.

In general, responses of salamander ganglion cell populations exhibit substantial redundancy, and almost never any synergy, when presented with a variety of natural scenes as well as flickering checkerboard patterns (Puchalla et al., 2005). Redundancy of this kind might be expected from noise in the retinal circuitry diverging to produce correlated firing, and indeed the degree of redundancy exhibits a dependence on receptive field overlap similar to the degree of correlated firing. However, these experiments did not separate signal correlations from noise correlations or isolate the effects of specific patterns of correlated activity (but see below). Other work, based on modifying information-theoretic approaches to analyze the accuracy of stimulus reconstruction, suggests that correlations may have limited impact (Nirenberg et al., 2001) (see also Schneidman, Bialek, & Berry, 2003).

Newer information-theoretic methods permit the analysis of synergy or redundancy contributed by individual symbols in the neural code, that is, specific patterns of activity such as one cell firing and another cell not. Analysis of data from salamander and guinea pig retina indicate that symbols consisting of cells firing simultaneously are usually redundant, whereas symbols consisting of one cell firing and others not are usually synergistic in conveying information about natural scenes (Schneidman et al., 2011). In other words observation of simultaneous spikes in two cells provides less information about the stimulus than the sum of the information transmitted by observation of a spike in each cell, whereas observation of a spike in one cell and none in the other is more informative than the sum of the information provided by each observation alone. Because synergy and redundancy of this kind can arise either from signal or noise correlations (Schneidman et al., 2011), the interpretation of this kind of combinatorial coding differs from that of average synergy and redundancy described above. However, the results provide additional evidence that the interpretation of spikes in one cell can depend on the activity of another.

Alternative Approaches

The above considerations do not account for the metabolic cost of transmitting visual information to the brain using neurons with various degrees of signal and noise correlations. This issue is significant because natural scenes exhibit a high degree of correlation across space and time and therefore have the potential to produce large signal correlations in the retina. This would effectively mean that each cell expends considerable energy transmitting visual information already being conveyed by other cells. Given the high metabolic cost of brain function (Schreiber et al., 2002), such a scheme may not be advantageous for the organism.

These considerations suggest that visual system circuitry may be organized to decorrelate the input, that is, to reduce or eliminate signal correlations in response to visual stimuli encountered by the animal in its natural environment (Barlow, 1961). Theoretical work based on simplifying assumptions about retinal circuitry

   FRED RIEKE AND E. J. CHICHILNISKY

predicts that a center-surround receptive field structure, qualitatively resembling the receptive fields of retinal ganglion cells, could lead to substantial decorrelation in the retinal output (Atick, 1990). Despite its simplicity this theory makes successful predictions about the band-pass nature of human contrast sensitivity and its dependence on light level, although its precise relationship to receptive field structure in the retina has not been tested directly. Also, recent work suggests that the nonlinear response properties of ganglion cells may contribute more to decorrelating retinal spike trains than the spatial structure of receptive fields (Pitkow & Meister, 2012).

As summarized above, diverging noise from photoreceptors appears to be a major source of correlated activity in the primate retina. Work in salamander retina, however, suggests that correlated activity could additionally reflect common input from a retinal interneuron, for example, a spiking amacrine cell that drives spikes in two or more ganglion cells. This could allow the interneuron to transmit information about the specific visual features that drive its response via those correlated spikes (signal correlations). Put another way, correlated activity could act to multiplex different visual features in the output of the retina (Meister, 1996). Consistent with this idea, correlated spikes in salamander ganglion cells exhibit a smaller receptive field than the spikes of individual cells (Meister, Lagnado, & Baylor, 1995; Schnitzer & Meister, 2003) (figure 12.7C), and correlations persist during blockade of chemical transmission (Brivanlou, Warland, & Meister, 1998). The situation appears to be different in primate retina, where correlated activity is almost completely dependent on chemical transmission (Trong & Rieke, 2008).

Some of the approaches described above to understanding the effects of correlated activity on visual signals are distinguished by comparison to different benchmarks, as is illustrated by several examples. Compared to a hypothetical noiseless retina, real noise correlations interfere with visual function rather than serving a purpose. Compared to a hypothetical retina with noise but no correlations, noise correlations increase the discriminability of some stimuli and decrease the discriminability of others. Compared to a hypothetical visual system that ignores correlated activity, more information is decoded from the retina that takes correlations into account. Although useful, these comparisons may not reflect a realistic set of possibilities on which natural selection has acted. For example, a noiseless retina is physically unrealizable. Future work may benefit from comparing what is known about correlated activity in the retina to realizable alternatives, using computational models that account for how both signal and noise traverse divergent circuitry to form the output of the retina (Trong & Rieke, 2008; Vidne et al., 2012).

SUMMARY AND CONCLUSIONS

Correlated activity is a dominant feature of information processing in the retina and in other neural circuits. Because of the experimental accessibility and well-understood circuitry and function of the retina, the origins and impact of correlated activity therein have recently come into clear focus. In addition to being widespread, correlated activity is stereotyped and cell-type specific, in a manner reminiscent of the precise, lattice-like arrangement of visual signals in different cell types. Local noise correlations, in particular, are created by a combination of lateral spread of electrical signals via gap junctions and a dominant effect of shared photoreceptor noise, particularly in the primate retina. The correlations in the activity of nearby pairs of ganglion cells provide an accurate statistical explanation for patterns of activity observed in larger populations. Although noise correlations inevitably reduce the fidelity of retinal signals relative to a noiseless retina, they can also be seen to alter the retinal output by increasing the salience of certain visual signals at the expense of others relative to what would occur with independent noise. Together, these findings provide an overview of the major sources and signaling impact of correlated activity in the retina and raise fundamental questions about how the brain efficiently decodes visual information from patterns of retinal activity.

REFERENCES

Ala-Laurila, P., Greschner, M., Chichilnisky, E. J., & Rieke, F. (2011). Cone photoreceptor contributions to noise and correlations in the retinal output. *Nature Neuroscience, 14,* 1309–1316.

Amari, S. (2001). Information geometry on hierarchical decomposition of stochastic interactions. *IEEE Transactions on Information Theory, 47*(5), 1701–1711. doi: 10.1109/18.930911.

Arnett, D. (1978). Statistical dependence between neighboring retinal ganglion cells in goldfish. *Experiments in Brain Research, 32,* 49–53.

Atick, J. J. (1990). Towards a theory of early visual processing. *Neural Computation, 2,* 308–320.

Averbeck, B. B., Latham, P. E., & Pouget, A. (2006). Neural correlations, population coding and computation. *Nature Reviews. Neuroscience, 7,* 358–366.

Barlow, H. B. (1961). Possible principles underlying the transformations of sensory messages. In W. A. Rosenblith (Ed.), *Sensory communication* (pp. 217–234). Cambridge, MA: MIT Press.

Britten, K. H., Shadlen, M. N., Newsome, W. T., & Movshon, J. A. (1992). The analysis of visual motion: A comparison of neuronal and psychophysical performance. *Journal of Neuroscience, 12*, 4745–4765.

Brivanlou, I. H., Warland, D. K., & Meister, M. (1998). Mechanisms of concerted firing among retinal ganglion cells. *Neuron, 20*, 527–539.

Cover, T. M., & Thomas, J. A. (1991). *Elements of information theory*. Wiley Online Library.

Dacey, D. M. (1993). The mosaic of midget ganglion cells in the human retina. *Journal of Neuroscience, 13*, 5334–5355.

Dacey, D. M., & Brace, S. (1992). A coupled network for parasol but not midget ganglion cells in the primate retina. *Visual Neuroscience, 9*, 279–290.

Dacey, D. M., & Packer, O. S. (2003). Colour coding in the primate retina: Diverse cell types and cone-specific circuitry. *Current Opinion in Neurobiology, 13*, 421–427.

de la Rocha, J., Doiron, B., Shea-Brown, E., Josic, K., & Reyes, A. (2007). Correlation between neural spike trains increases with firing rate. *Nature, 448*, 802–806.

DeVries, S. H. (1999). Correlated firing in rabbit retinal ganglion cells. *Journal of Neurophysiology, 81*, 908–920.

Field, G. D., & Chichilnisky, E. J. (2007). Information processing in the primate retina: Circuitry and coding. *Annual Review of Neuroscience, 30*, 1–30.

Ganmor, E., Segev, R., & Schneidman, E. (2011). Sparse low-order interaction network underlies a highly correlated and learnable neural population code. *Proceedings of the National Academy of Sciences of the United States of America, 108*, 9679–9684.

Gauthier, J. L., Field, G. D., Sher, A., Greschner, M., Shlens, J., Litke, A. M., et al. (2009). Receptive fields in primate retina are coordinated to sample visual space more uniformly. *PLoS Biology, 7*, e1000063.

Gollisch, T., & Meister, M. (2008). Rapid neural coding in the retina with relative spike latencies. *Science, 319*, 1108–1111. doi:10.1126/science.1149639.

Greschner, M., Shlens, J., Bakolitsa, C., Field, G. D., Gauthier, J. L., Jepson, L. H., et al. (2011). Correlated firing among major ganglion cell types in primate retina. *Journal of Physiology, 589*, 75–86.

Hidaka, S., Akahori, Y., & Kurosawa, Y. (2004). Dendrodendritic electrical synapses between mammalian retinal ganglion cells. *Journal of Neuroscience, 24*, 10553–10567.

Hu, E. H., & Bloomfield, S. A. (2003). Gap junctional coupling underlies the short-latency spike synchrony of retinal alpha ganglion cells. *Journal of Neuroscience, 23*, 6768–6777.

Jacoby, R., Stafford, D., Kouyama, N., & Marshak, D. (1996). Synaptic inputs to ON parasol ganglion cells in the primate retina. *Journal of Neuroscience, 16*, 8041–8056.

Jaynes, E. T. (1957a). Information theory and statistical mechanics. I. *Physical Review, 106*, 620–630.

Jaynes, E. T. (1957b). Information theory and statistical mechanics. II. *Physical Review, 108*, 171–190.

Masland, R. H. (2001). The fundamental plan of the retina. *Nature Neuroscience, 4*, 877–886.

Mastronarde, D. N. (1983a). Correlated firing of cat retinal ganglion cells. I. Spontaneously active inputs to X- and Y-cells. *Journal of Neurophysiology, 49*, 303–324.

Mastronarde, D. N. (1983b). Correlated firing of cat retinal ganglion cells. II. Responses of X- and Y-cells to single quantal events. *Journal of Neurophysiology, 49*, 325–349.

Mastronarde, D. N. (1983c). Interactions between ganglion cells in cat retina. *Journal of Neurophysiology, 49*, 350–365.

McDonnell, M. D., & Ward, L. M. (2011). The benefits of noise in neural systems: Bridging theory and experiment. *Nature Reviews. Neuroscience, 12*, 415–426.

Meister, M. (1996). Multineuronal codes in retinal signaling. *Proceedings of the National Academy of Sciences of the United States of America, 93*, 609–614.

Meister, M., Lagnado, L., & Baylor, D. A. (1995). Concerted signaling by retinal ganglion cells. *Science, 270*, 1207–1210.

Nirenberg, S., Carcieri, S. M., Jacobs, A. L., & Latham, P. E. (2001). Retinal ganglion cells act largely as independent encoders. *Nature, 411*, 698–701.

Ohiorhenuan, I. E., & Victor, J. D. (2011). Information-geometric measure of 3-neuron firing patterns characterizes scale-dependence in cortical networks. *Journal of Computational Neuroscience, 30*, 125–141.

Pillow, J. W., Shlens, J., Paninski, L., Sher, A., Litke, A. M., Chichilnisky, E. J., et al. (2008). Spatio-temporal corelations and visual signalling in a complete neuronal population. *Nature, 454*, 995–999. doi: 10.1038/nature07140.

Pitkow, X., & Meister, M. (2012). Decorrelation and efficient coding by retinal ganglion cells. *Nature Neuroscience, 15*, 628–635.

Puchalla, J. L., Schneidman, E., Harris, R. A., & Berry, M. J. (2005). Redundancy in the population code of the retina. *Neuron, 46*, 493–504.

Rieke, F., Warland, D., de Ruyter Van Steveninck, R., & Bialek, W. (1997). *Spikes: Exploring the neural code*. Cambridge, MA: MIT Press.

Rodieck, R. W. (1967). Maintained activity of cat retinal ganglion cells. *Journal of Neurophysiology, 30*, 1043–1071.

Rodieck, R. W. (1998). *The first steps in seeing*. Sunderland, MA: Sinauer Associates.

Schneeweis, D. M., & Schnapf, J. L. (1999). The photovoltage of macaque cone photoreceptors: Adaptation, noise, and kinetics. *Journal of Neuroscience, 19*, 1203–1216.

Schneidman, E., Berry, M. J. II, Segev, R., & Bialek, W. (2006). Weak pairwise correlations imply strongly correlated network states in a neural population. *Nature, 440*, 1007–1012.

Schneidman, E., Bialek, W., & Berry, M. J. II. (2003). Synergy, redundancy, and independence in population codes. *Journal of Neuroscience, 23*, 11539–11553.

Schneidman, E., Puchalla, J. L., Segev, R., Harris, R. A., Bialek, W., & Berry, M. J. II. (2011). Synergy from silence in a combinatorial neural code. *Journal of Neuroscience, 31*, 15732–15741.

Schnitzer, M. J., & Meister, M. (2003). Multineuronal firing patterns in the signal from eye to brain. *Neuron, 37*, 499–511.

Schreiber, S., Machens, C. K., Herz, A. V., & Laughlin, S. B. (2002). Energy-efficient coding with discrete stochastic events. *Neural Computation, 14*, 1323–1346.

Shlens, J., Field, G. D., Gauthier, J. L., Greschner, M., Sher, A., Litke, A. M., et al. (2009). The structure of large-scale synchronized firing in primate retina. *Journal of Neuroscience, 29*, 5022–5031.

Shlens, J., Field, G. D., Gauthier, J. L., Grivich, M. I., Petrusca, D., Sher, A., et al. (2006). The structure of multi-neuron firing patterns in primate retina. *Journal of Neuroscience, 26*, 8254–8266.

Shlens, J., Rieke, F., & Chichilnisky, E. (2008). Synchronized firing in the retina. *Current Opinion in Neurobiology, 18*, 396–402.

Tang, A., Jackson, D., Hobbs, J., Chen, W., Smith, J. L., Patel, H., et al. (2008). A maximum entropy model applied to spatial and temporal correlations from cortical networks in vitro. *Journal of Neuroscience, 28*, 505–518.

Trong, P. K., & Rieke, F. (2008). Origin of correlated activity between parasol retinal ganglion cells. *Nature Neuroscience, 11*, 1343–1351.

Usrey, W. M., & Reid, R. C. (1999). Synchronous activity in the visual system. *Annual Review of Physiology, 61*, 435–456.

Vaney, D. I. (1991). Many diverse types of retinal neurons show tracer coupling when injected with biocytin or Neurobiotin. *Neuroscience Letters, 125*, 187–190.

Vidne, M., Ahmadian, Y., Shlens, J., Pillow, J.W., Kulkarni, J., Litke, A.M., et al. (2012). Modeling the impact of common noise inputs on the network activity of retinal ganglion cells. *Journal of Computational Neuroscience, 33*, 97–121. doi: 10.1007/s10827-011-0376-2.

Ward, L. M., Neiman, A., & Moss, F. (2002). Stochastic resonance in psychophysics and in animal behavior. *Biological Cybernetics, 87*, 91–101.

Warland, D. K., Reinagel, P., & Meister, M. (1997). Decoding visual information from a population of retinal ganglion cells. *Journal of Neurophysiology, 78*, 2336–2350.

Wässle, H. (2004). Parallel processing in the mammalian retina. *Nature Reviews. Neuroscience, 5*, 747–757.

Wässle, H., Peichl, L., & Boycott, B. B. (1981). Dendritic territories of cat retinal ganglion cells. *Nature, 292*, 344–345.

13 The Retina Dissects the Visual Scene into Distinct Features

BOTOND ROSKA AND MARKUS MEISTER

TWENTY RETINAL PATHWAYS CONVEY VISUAL INFORMATION FROM THE EYE TO THE BRAIN

Classical studies of the functional architecture of the retina have found that the image projected onto the retina and captured by the photoreceptors is processed locally by multiple parallel circuits (Masland, 2001; Wässle, 2004). This parallel processing results in several different dynamic activity patterns at the retinal output (Roska & Werblin, 2001) that are simultaneously transmitted to the brain by ganglion cells, the output neurons of the retina (figure 13.1).

There is a growing consensus that the ganglion cell population comprises ~20 different types. Each type forms a subpopulation that covers the entire retina, usually in a regularly spaced arrangement called a "mosaic." Thus, the unit of cellular infrastructure that underlies parallel processing in the retina is a mosaic of ganglion cells with similar morphology and response properties, together with an associated mosaic of local circuits (figure 13.2). The retina embodies 20 such mosaics that independently extract different features from the visual world, although the underlying circuits share many of the interneurons. It is as if our eye comprised multiple different TV crews pointing their cameras at the same event but each broadcasting to their audience (the relevant brain region) a subjectively cut and processed version of the captured image flow (figure 13.3). Some workers are shared among all of the different crews, others specialize for jobs with few crews, and some are only participating in one crew.

Each ganglion cell in a given mosaic has a local circuit with different circuit elements. These elements can be ranked according to how many synapses separate them from the sensory receptors. Rods and cones are first-order neurons; bipolar and horizontal cells are second-order; amacrine and ganglion cells are third-order. The ganglion cells, as the sole output element of the retina, are positioned clearly at the top of the hierarchy in these circuits. Bipolar, amacrine, and horizontal cells each come in a variety of different morphological and physiological variants. Again, neurons with the same shapes and response properties are arranged in a mosaic (Wässle, 2004). Thus, the retina is built from multiple mosaics of cells, which we will call "cell types" for short.

Because each retinal patch contains 20 different ganglion cell circuits with more than 60 different circuit elements, it is not surprising that there are strict organizational rules for the spatial arrangement of the various circuit components. The first rule is that the cell bodies of different circuits are packed in three different cell body layers. Connections between circuit elements occur between these three layers in two "plexiform" layers of synapses. The second rule is that the dendrites of the various ganglion cell types and the axon terminals of the different bipolar cell types are stacked vertically above each other in the inner plexiform layer, forming ~10 narrow strata (Siegert et al., 2009) (figure 13.4). If the axon terminals of a bipolar cell type and the dendrites of a given ganglion cell type are in different strata, there can be no direct communication between the two. Some bipolar cells have axon terminals in more than one stratum and therefore give input to these strata. There are more ganglion cell types than bipolar cell types, and, therefore, the axon terminals of a bipolar type typically excite more than one ganglion cell type. Consistent with this, each stratum of the IPL tends to contain dendrites of more than one ganglion cell type.

Comparing the stratification of the three main cell classes of the inner retina, the bipolar, ganglion, and amacrine cells, we find striking differences. The processes of ganglion and bipolar cells are generally confined to one or two IPL strata. By contrast certain amacrine cells, such as the AII amacrines, are arranged vertically. With only a narrow horizontal extent, they receive input or provide output in several strata. Other amacrines, such as the starburst cells and polyaxonal cells, are thinly stratified with a wide horizontal extent within the stratum (Masland, 2012). It appears that the "tall and narrow" amacrines and their "flat and broad" classmates have entirely different computational roles.

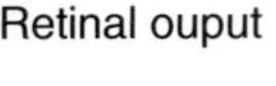

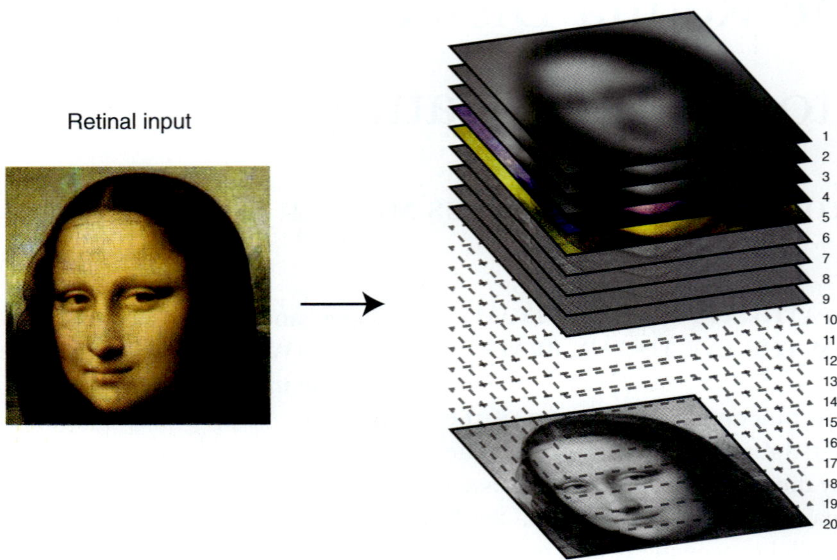

FIGURE 13.1 The retina creates 20 neural representations of the "movie" that enters the eye.

In evolution, the 20 retinal circuits could have been organized into 20 separate pairs of eyes with overlapping fields of view. Although this would have been simpler with regard to the wiring and positioning of the circuit elements, some cell types, such as photoreceptors, are needed for all circuits. Photoreceptors are indeed numerous—they account for >80% of all retinal neurons (Jeon, Strettoi, & Masland, 1998)—and it is economical to share bulk common resources across circuits. The layered structure presented above, which allows an efficient use of common resources, gives rise to a hierarchical organization of cell types. Cells at the bottom of the hierarchy, such as photoreceptors, provide input to many ganglion cell types, whereas a specialized amacrine cell higher in the hierarchy influences few ganglion cell types. Shared resources and cell type hierarchy have important consequences for understanding both retinal processing and visual disorders. Common operations needed for all circuits, such as the gain control required for light adaptation (Fain, 2011), are more likely to be carried out at the front by common elements. Similarly, one expects that cells whose dysfunction gives rise to noticeable visual defects are low in the hierarchy. On the other hand circuit elements responsible for specialized ganglion cell computations are higher in the hierarchy.

The structure of the retina appears to be tailor-made to extract many different features from the visual scene. Under daylight conditions the image is captured by cone photoreceptors. The first processing stage, an interaction with the inhibitory horizontal cells, contributes a step of lateral inhibition that affects all downstream circuits (Kamermans & Fahrenfort, 2004; Wu, 1992). In dim light visual transduction is accomplished by the rods, and their signals are subsequently fed into the cone system by several elaborate pathways (Bloomfield & Dacheux, 2001). From then on the rod-derived signals are largely processed as though they came from cones. For the purposes of this chapter, we therefore focus on circuits downstream of the cone bipolar cells.

Each cone is connected to ~10 types of bipolar cell (Wässle et al., 2009). Some of these bipolar types are distinguished by their neurotransmitter receptors with different kinetics (DeVries, 2000), and in turn they terminate at different levels of the inner plexiform layer. Therefore, the signals in different strata of the inner retina already parse the visual input according to different temporal features. Because there are more ganglion cell types than strata, the activity carried from the outer to the inner retina by each bipolar cell type is further diversified. Different features can emerge within a stratum because ganglion cell types have different spatial extents and different receptors, and their circuits may include different amacrine cell types (Taylor & Smith, 2011). Notably, features carried by bipolar cells can also be recombined locally by the action of vertical amacrine cells.

The elaboration of visual features as discussed above is a columnar operation: restricted in space and organized across or within strata. Retinal processing also occurs laterally across space as a result of the lateral

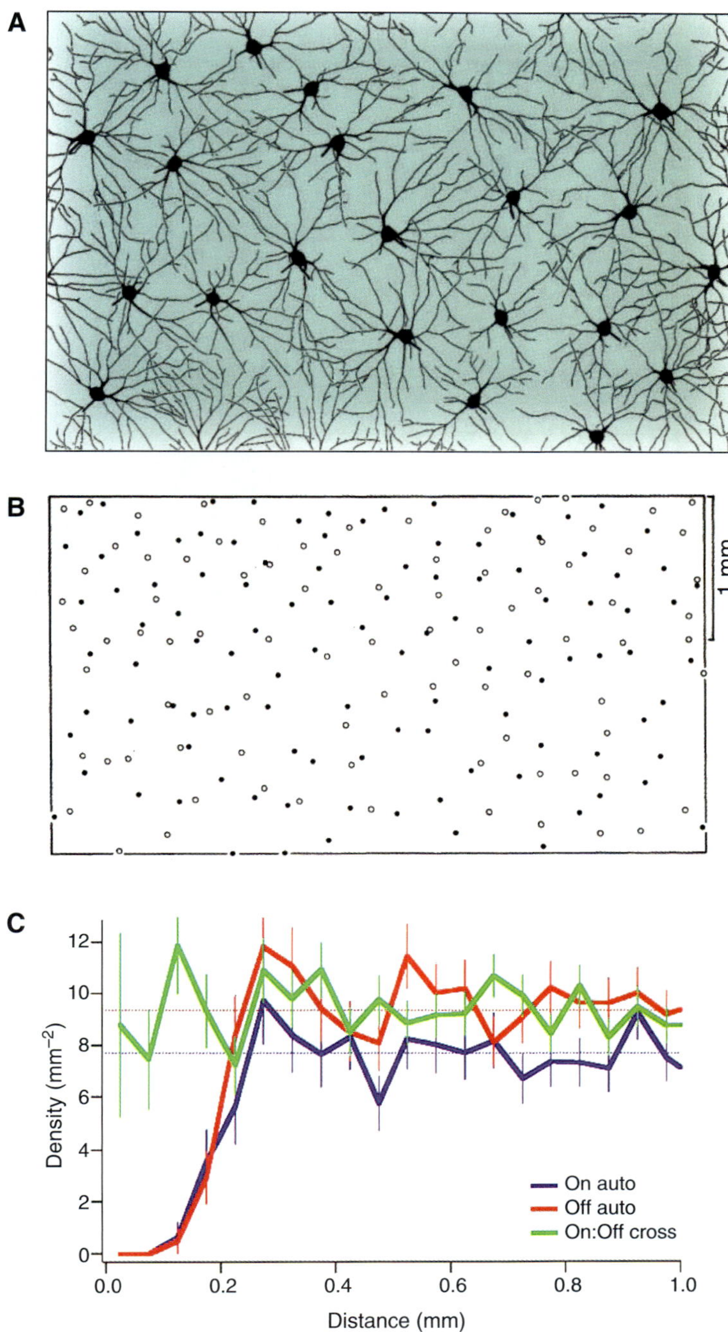

FIGURE 13.2   Ganglion cells of one type cover the retina with a regular mosaic. (A) Cell bodies and dendrites of ON alpha ganglion cells in a wholemount view of the cat retina. Note that the dendrites cover space uniformly, and the cell bodies are placed at regular distances (Wässle, 2004). (B) Cell body locations of ON alpha (open circles) and OFF alpha (closed circles) ganglion cells in a patch of cat retina (Wässle, Peichl, & Boycott, 1981). (C) Each of the two cell types forms a regular mosaic independent of the other. Spatial autocorrelation of the ON (solid blue line) and OFF (red) cell locations, showing the probability per unit area of finding a cell at a given distance from another cell of the same type. Note the prominent hole for distances <0.2 mm. Cross-correlation (green) shows the probability of finding an OFF cell at a given distance from an ON cell. Dotted lines are the average densities of ON (blue) and OFF (red) cells in this patch.

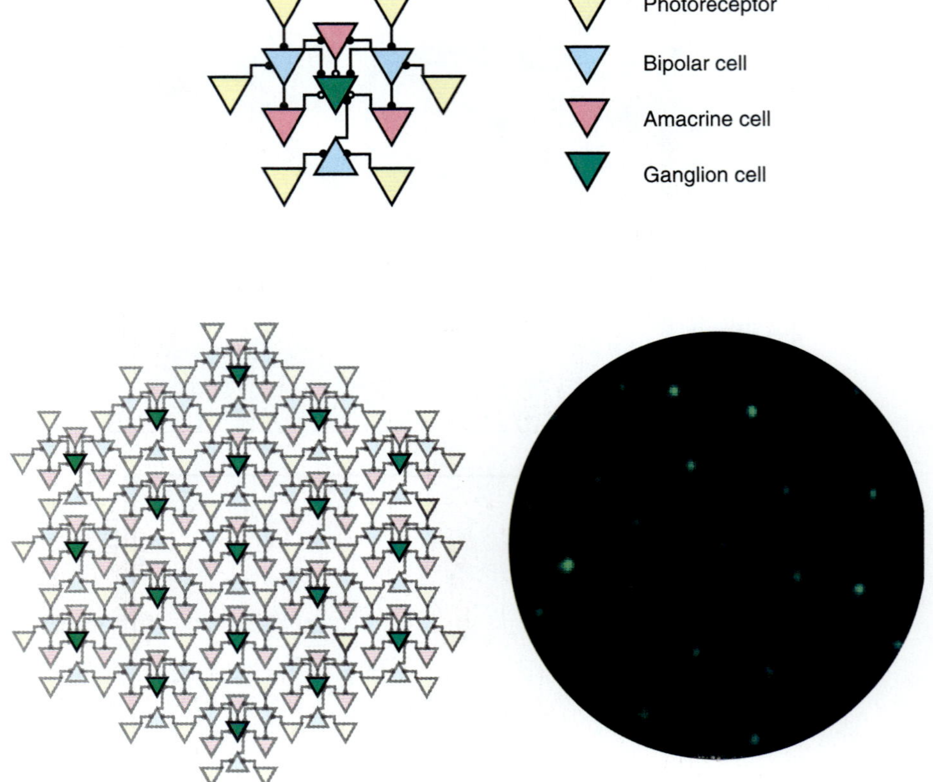

FIGURE 13.3   The unit of retinal infrastructure is a ganglion cell circuit mosaic. (Top) A single ganglion cell surrounded by first- and second-order circuit elements. (Bottom, left) The same circuit is repeated across the retina forming a mosaic. (Bottom, right) Actual mosaic locations of retinal ganglion cells of a specific type.

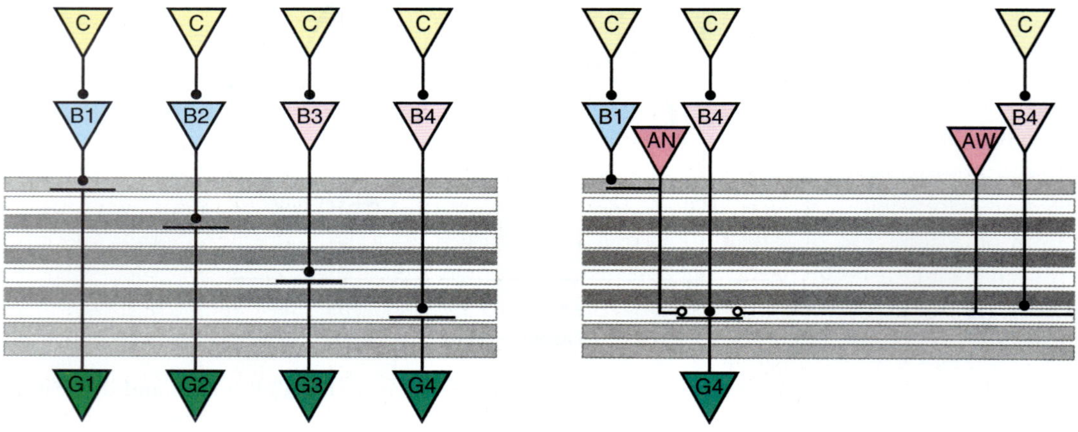

FIGURE 13.4   Retinal features are stacked in the inner retina. (Left) Bipolar cell terminals and ganglion cell dendrites are laid down in different strata of the IPL. (Right) Some amacrine cells (AN) are narrow and tall; their inputs and outputs are in different strata. Other amacrine cells (AW) are wide and flat, with long processes in one stratum; these cells carry information across the local circuits of the same mosaic.

connections by horizontal cells, large amacrine cells, and electric coupling between cells of the same and different types.

Several aspects of the functional organization of the retina are evolutionarily conserved. The layered arrangement of cell bodies and cellular processes, the major cell classes, and their general connectivity are common to all vertebrates. In comparing across mammals from mouse to human, one finds even greater similarities that extend to the level of cell types and

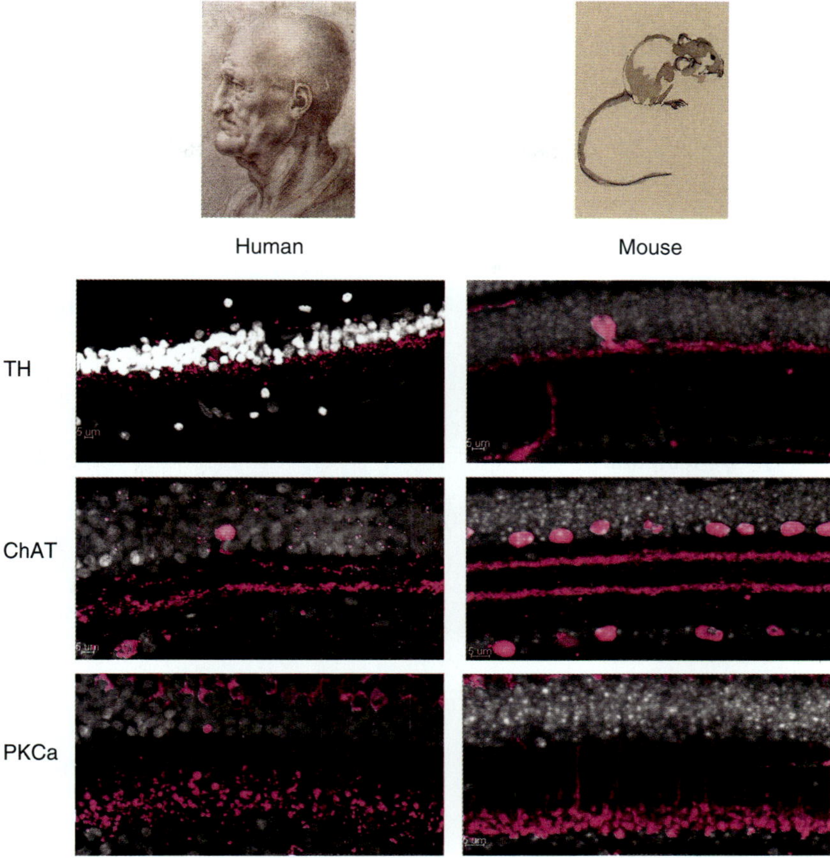

FIGURE 13.5   Comparing the retinas of humans and mice. Vertical sections of human (left) and mouse (right) retinas. Staining with three antibodies against tyrosine hydroxylase (TH), choline acetyl transferase (ChAT), and protein kinase C alpha (PKCa) identifies strata with similar positions in the two species.

their lamination (figure 13.5). Several antibody markers label the same strata across these species, and a number of cell types are conserved. For example, both the mouse (Puller & Haverkamp, 2011) and the macaque (Dacey & Packer, 2003) have a bipolar cell specialized for signals from blue cones. In table 13.1 we compile a catalog of retinal ganglion cell types across the major species in which the topic has been studied. This illustrates a number of "canonical" cell types found in many species (Berson, 2008). For other cell types the correspondence is more difficult to identify, although this may improve as we learn more about their visual responses.

There are also distinct differences among mammals. For example, in the mouse retina the spacing of cells in a given mosaic is almost uniform across the retina; at the other extreme, in the primate retina the cell density rises sharply toward a small patch of retina in the center called the fovea. The fovea has, therefore, high spatial resolution and is used for encoding details in the visual scene. Different mammals have different degrees of nonuniformity in the spatial density of ganglion cell mosaics, resulting in specialized retinal regions such as the area centralis in cats or the visual streak in rabbits.

A second difference is in the circuits processing color. Most mammals have two cone types, one expressing a short-wavelength pigment and the other medium wavelength. Some primates also have cones with a long-wavelength pigment. The circuitry connected to short-wavelength cones has common circuit motifs across mammals, such as the specialized blue cone cell, but the differential handling of color information for medium and long wavelengths is unique to a group of primates. Some mammals such as mice and rats express more than one pigment in many of their cones, and the ratio of these pigments varies in a dorsoventral gradient. Because of this gradient the part of the eye that looks at the blue sky is more sensitive at short wavelengths, and the part that looks at the ground is more sensitive at longer wavelengths.

The anatomical evidence that the retina contains 20 ganglion cell mosaics along with their associated circuits has emerged gradually over the last 50 years.

A catalog of retinal ganglion cell types in the mammalian retina[a]

Icon	Mouse	Rabbit	Cat	Macaque	Properties
	M1[1,2]			Outer melanopsin[3]	Large sparse dendrites. ON sluggish synaptic response.
	M2[1,2]			Inner melanopsin[3]	Large complex dendrites. ON sluggish synaptic response.
	ON DS temporal[4]	ON DS temporal[5]			ON DS. Preferred direction temporal.
	ON DS ventral,[4] Spig-1 EGFP[6,7]	ON DS ventral[5]			ON DS. Preferred direction ventral.
	ON, DS dorsal[4,6,7]	ON DS dorsal[5]			ON DS. Preferred direction dorsal.
	ON–OFF DS temporal[8]	ON–OFF DS temporal[9]	Theta?[10]	Recursive bistratified?[11]	ON-OFF DS. Preferred direction temporal.
	ON-OFF DS dorsal[8]	ON-OFF DS dorsal[9]	Theta?[10]	Recursive bistratified?[11]	ON-OFF DS. Preferred direction dorsal.
	Drd4-EGFP,[12] W9[8]	ON–OFF, DS nasal[9]	Theta?[10]	Recursive bistratified?[11]	ON-OFF DS. Preferred direction nasal.
	BD-CreER,[8] Hb9-EGFP[13]	ON–OFF, DS ventral[9]	Theta?[10]	Recursive bistratified?[11]	ON-OFF DS. Preferred direction ventral. Asymmetric dendrites in mouse.
	JAM-B[14]	OFF coupled[15], G3[16]			OFF DS. Preferred direction ventral. Highly asymmetric dendrites point ventral.

   BOTOND ROSKA AND MARKUS MEISTER

Icon	Mouse	Rabbit	Cat	Macaque	Properties
	ON alpha,[17] PV-Cre-1[18]	ON alpha[19]	ON alpha[20,21]		Large dendritic field. ON response.
	PV-Cre-3[18]	ON parasol[15]		ON parasol[22]	Medium dendritic field. ON response.
				ON smooth[23]	
		ON beta[15]	ON beta[24]		Small dendritic field. ON response.
				ON midget[22]	Small dendritic field. ON response.
		OFF beta[15]	OFF beta[24]		Small dendritic field. OFF response.
				OFF midget[22]	Small dendritic field. OFF response.
	PV-Cre-4[18]	OFF parasol[15]	Eta?[25]	OFF parasol[22]	Medium dendritic field. OFF response.
	OFF alpha transient[17], PV-Cre-5[18]	OFF alpha[19]	OFF alpha[24]	OFF smooth[23]	Large dendritic field. OFF response.
	OFF alpha Sustained,[17] PV-Cre-6[18]	OFF delta[15]	OFF delta[10]		Large dendritic field. OFF sustained response.

continued

A catalog of retinal ganglion cell types in the mammalian retina[a] (Continued)

Icon	Mouse	Rabbit	Cat	Macaque	Properties
		ON-bistratified?[15]		Small-bistratified[11]	ON excitation, OFF inhibition, blue-yellow opponent in macaque.
				Large-bistratified[11]	Blue-yellow opponent.
	W3[26]	Local edge Detector[5,27]	Zeta[28]	Broad thorny[11]	ON–OFF, strong surround, fast ON–OFF inhibition.
			Epsilon?[29]	Recursive monostratified[11]	
				ON narrow thorny[11]	
				OFF narrow thorny[11]	
		Uniformity Detector[30]	Uniformity Detector[31]		Transiently suppressed by visual stimuli. ON-OFF response. Dendrites just outside the ChAT bands.

[a]Each graphic icon illustrates stratification of the dendritic tree in the IPL, divided into 10 laminae (Siegert et al., 2009). For each type we list the defining morphological and physiological features and identify its plausible correspondences in four species, as supported by the cited literature. For further detail on cross-species comparisons, see Berson (2008). Note that many of these ganglion cell types have only sparse and partial entries, emphasizing the need for future work to round out the catalog of retinal output signals.

References: [1]Hattar et al. (2006). [2]Schmidt et al. (2011b). [3]Dacey et al. (2005). [4]Sun et al. (2006). [5]Barlow, Hill, & Levick (1964), but see Kanjhan & Sivyer (2010) and Hoshi et al. (2011) for finer divisions. [6]Yonehara et al. (2009). [7]Yonehara et al. (2008). [8]Kay et al. (2011). [9]Oyster & Barlow (1967). [10]Isayama, Berson, & Pu (2000). [11]Dacey (2004). [12]Huberman et al. (2009). [13]Trenholm et al. (2011) erroneously identified the preferred direction as temporal. [14]Kim et al. (2008). [15]Roska, Molnar, & Werblin (2006). [16]Hoshi et al. (2011). [17]Pang, Gao, & Wu (2003). [18]Münch et al. (2009). [19]Zhang et al. (2005). [20]Cleland, Levick, & Wässle (1975). [21]Wässle, Peichl, & Boycott (1981). [22]Dacey & Packer (2003). [23]Crook et al. (2008). [24]Wässle, Boycott, & Illing (1981). [25]Berson, Isayama, & Pu (1999). [26]Kim et al. (2010). [27]van Wyk, Taylor, & Vaney (2006). [28]Berson, Pu, & Famiglietti (1998). [29]Pu, Berson, & Pan (1994). [30]Sivyer & Vaney (2010). [31]Cleland & Levick (1974).

However, with exception of a few ganglion cell types, the functional distinctions among all these visual pathways have been more difficult to understand. Recent technical advances have greatly accelerated this research program, in particular the ability to genetically mark and manipulate cell types (Azeredo da Silveira & Roska, 2011; Huberman et al., 2009; Kay et al., 2011; Kim et al., 2008; Yonehara et al., 2008). The fundamental new insight is that the gene expression patterns of distinct cell types are quite different. With advanced molecular, genetic, and viral tools one can hijack the cellular machinery that controls these expression patterns to manipulate neurons selectively. Now it is possible to target specific cell types for physiological recording, to modify them, to observe the effects on network function, and—importantly—to communicate the scientific results without ambiguity about the identity of retinal neurons under study.

In the following two sections we discuss what these various ganglion mosaics extract from the visual scene and how the associated circuitry performs the necessary computations.

PIXEL SENSORS VERSUS FEATURE DETECTORS

What is the role of the retina for the overall function of the visual system? In the conventional view—still dominant in textbooks and held by many vision researchers today—the retina's primary task is to get the visual image transmitted to the brain, where the cortex and other heavy-duty circuits can get on with the challenges of processing the information. For that purpose, the retina must first format the image signals a bit to deal with the vicissitudes of the physical environment. Because the illumination conditions can change so dramatically, the retina applies a gain control through the cellular processes of light adaptation. And because natural images tend to be highly redundant in their pixel patterns, the retina performs some image compression through the circuits that implement lateral inhibition. In this view the defining characteristics of retinal ganglion cell function are the center-surround receptive field and gain control.

In an alternate view the retina shapes the visual representation much more dramatically. Rather than simply recoding the image for more efficient transmission through the optic nerve, the retina extracts from the scene only a few specific features and transmits those very selectively to the brain through several specific image channels. In this picture much of the raw information in the visual scene is discarded. The ganglion cells transmit signals that result from very nonlinear computations, for example, the speed of image motion in a specific direction, that relate only distantly to the raw data of image intensity.

Both of these rival conceptions of retinal processing date back to the earliest days of retinal neurophysiology (Barlow, 1953; Kuffler, 1953; Lettvin et al., 1959). Today, with a complete catalog of ganglion cell types within reach, one can envision an end to this debate. As usual, the resolution will likely be a compromise. It appears that a few ganglion cell types match the notion of "pixel sensors," whereas many others are better described as "feature detectors." Here we give these two concepts explicit meaning and assess how they apply to the different retinal pathways.

Pixel Sensor

In its idealized form a pixel sensor ganglion cell would simply measure the light intensity at a particular point on the retina and convey that value directly to the brain. A technological example of this is a single pixel sensor of a digital camera. In practice, of course, ganglion cells do not observe light at a single point but over a receptive field. Furthermore, they cannot signal instantaneously but integrate the light over the retina's response time. Finally, they cannot put out a continuous signal but only spikes. With these realistic constraints we can define a pixel sensor retinal ganglion cell as one that performs these image operations: compute a weighted average of the light intensity over the receptive field and the integration time and then use the result to modulate the firing rate accordingly (Meister & Berry, 1999). Such neural responses are generally characterized as "linear" because they derive from a linear summation of light intensity across time and space in the receptive field (see appendix 13.1).

Remarkably, there are in fact retinal ganglion cells that approach this ideal. For example, the midget P cells in the primate fovea (including our own) receive excitation mostly from a single bipolar cell, which in turn gets input from a single cone (Kolb & Marshak, 2003). Their response is truly dominated by a single pixel in the photoreceptor array, and thus, there is no opportunity for sophisticated nonlinear image computations. Actual response measurements from ganglion cells in the fovea are rare (McMahon et al., 2000), but midget cells at greater eccentricity seem to integrate light in a mostly linear manner (Benardete & Kaplan, 1997a, 1997b).

Similarly, the X cells of the cat retina respond quite linearly to light (Enroth-Cugell et al., 1983). They have a substantial maintained firing rate when the stimulus remains constant. When the light increases they fire more; when it decreases they fire less (or vice versa, depending on polarity of the ganglion cell). A brightening in one part of the receptive field can be counterbalanced by a dimming in another part to completely cancel the response, which illustrates that the circuit sums light over space (Enroth-Cugell & Robson, 1966). If the visual stimulus varies in time like a sine-wave function, the firing rate is modulated like a sine wave of the same frequency; this is a common indicator of linear processing.

Both macaque P cells and cat X cells are the smallest ganglion cells in the respective retinas. A tempting suggestion is that in any given species the ganglion cells with the finest receptive fields are pixel sensors that convey a high-resolution version of the scene to the brain. However, the mouse violates this simple notion: The smallest ganglion cells in the mouse retina (called local edge detectors or "W3" cells) do not participate at all in the signaling of routine visual scenes and respond only very sparsely to specific events (Zhang et al., 2012).

Feature Detector

A prototype of feature detection, again using a man-made example, would be the face-detection circuit used in current point-and-shoot cameras. Recognizing a face in the visual scene requires an interesting combination of selectivity and invariance: selectivity so the detector remains silent in the many image regions that do not contain a face; invariance so it responds to many different faces under different views and illuminations. Obviously, this kind of performance requires computations that are very different from mere linear filtering of the image.

Again, there exist retinal ganglion cell types whose performance approaches this ideal. One example would be the "bug perceivers" described early on in the frog retina (Lettvin et al., 1959): These cells fire if a small fly moves against a patterned background but not if the same fly and the background move together. Among the new ganglion cell types that were identified or better explored in recent years, most have the characteristics of feature detectors: highly nonlinear behavior, selectivity for a certain visual feature, and invariance to many other aspects of the scene. Often one can understand the feature selectivity from ecological and ethological considerations: the particular images produced by the natural environment, the needs of the visual system, and the observer's own behavior during active vision. In the following section we illustrate some of these cases.

MANY RETINAL GANGLION CELLS ARE FEATURE DETECTORS

For each of the sample ganglion cell types we begin by discussing what it computes, namely what aspects of the visual scene define its selectivity and its invariance. In many cases we also understand how this stimulus selectivity emerges from the interaction of retinal neurons, and we present these explanations in the form of a circuit diagram that summarizes the relevant connectivity and signal flow. These circuits are not intended to be complete and exhaustive, so some cautionary comments are in order.

First, for simplicity all the circuits begin with bipolar cells. The outer retina circuits of photoreceptors and horizontal cells perform some low-level formatting of the visual signal, including light adaptation and lateral inhibition. As a result of this processing, bipolar cells have simple center-surround receptive fields. They produce essentially linear light responses under the same conditions where ganglion cells act as nonlinear feature detectors (Baccus et al., 2008). Therefore, not much computation has occurred by the bipolar cell level. Interesting selectivity emerges largely in the inner retina through the interaction of bipolars, amacrines, and ganglion cells. Second, the circuit diagram is intended as schematic, not accurate in detail. For example, the diagram does not spell out the correct number of elements: a single component marked "A" may stand for an entire population of amacrine cells of that type. Third, the diagram is not exhaustive: It spells out the minimal circuit that has been confirmed and is essential to producing the function in question, but the full circuit likely includes other components.

Y Cells

These ganglion cells drew attention in early studies of cat retina because they clearly violate the notion of linear summation (Enroth-Cugell & Robson, 1966). If the receptive field center is divided into a dark half and a light half, and then the two regions are switched, the X cells described above will remain silent because the total light on the receptive field remains unchanged. By contrast, the Y cells fire a strong burst on each of these transitions. Y cells come in both polarities: An ON Y cell is excited transiently by an ON transition in any small region of the receptive field, even if it coincides with a dimming elsewhere. These neurons are exquisitely sensitive to a moving pattern because any such motion will produce a brightening somewhere in the receptive field (figure 13.6A). The same applies to OFF Y cells, which are excited by local OFF transitions. Thus, the Y cell shows a form of invariance: It responds well to a fine stimulus independent of where it occurs within the receptive field or of the direction of motion. However, the Y cells are not usually described as selective. They do have an antagonistic surround that also includes some nonlinear pooling over space similar to that in the center (Crook et al., 2008; Enroth-Cugell & Freeman, 1987), but it has not been associated with isolating any specific visual feature.

The unique response properties of the Y cell suggest that its circuit pools excitatory inputs from small subregions in the receptive field whose signals are individually rectified (Enroth-Cugell & Freeman, 1987) (figure

    BOTOND ROSKA AND MARKUS MEISTER

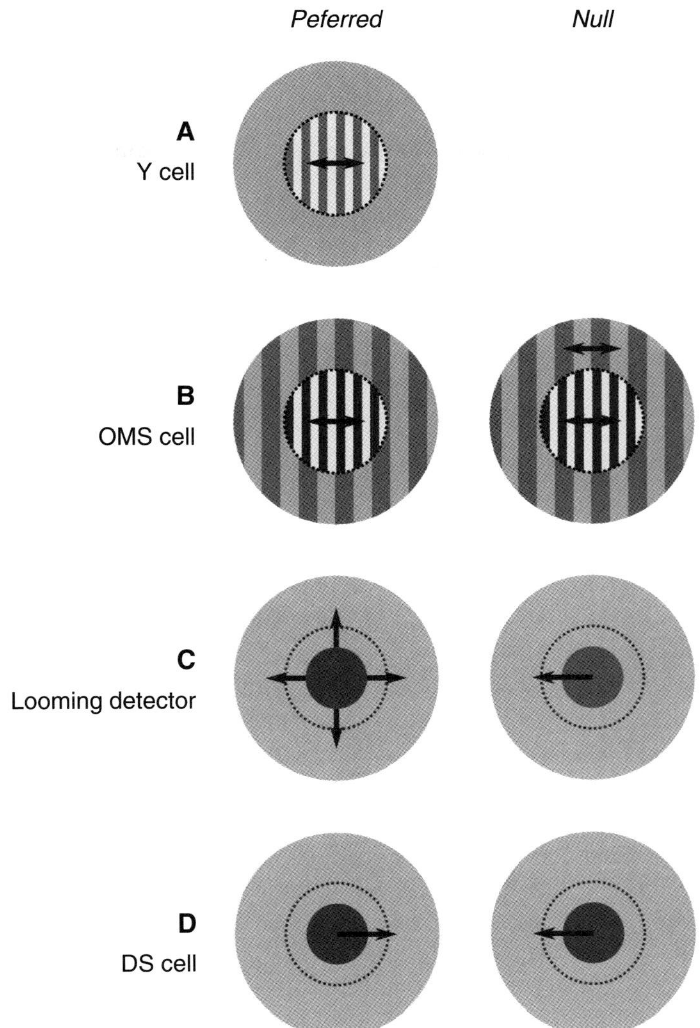

FIGURE 13.6   Visual features extracted by retinal ganglion cells. For each of the four types of ganglion cells highlighted in the text, this illustrates visual stimuli that excite the neuron (preferred) or suppress it (null). Arrows indicate movement. Dotted line marks the receptive field center. The examples are taken from conditions that occur during natural vision: (A) excitation by pattern motion on the receptive field center (Y cell and OMS cell); (B) suppression by a simultaneous pattern motion in the surround (OMS cell); (C) excitation by expanding motion but not translating motion (looming detector); and (D) movement in one direction but not in the opposite direction (DS cell).

13.7A). There is good evidence now that the subfields correspond to individual bipolar cells: These interneurons match the size of the subfields (Crook et al., 2008; Demb et al., 2001), and their synaptic output can indeed show strong rectification (Baccus et al., 2008; Demb et al., 2001). At a rectifying bipolar cell synapse, only the depolarizations of the bipolar cell are transmitted to the ganglion cell as excitation, whereas hyperpolarizations have no postsynaptic effect. This rectification arises when the basal transmitter release rate at the bipolar cell synapse is low because the resting potential lies below the activation voltage of synaptic calcium channels (Matsui, Hosoi, & Tachibana, 1998; Palmer, 2010). The Y-cell circuit (figure 13.7A) explains

qualitatively how the neuron responds to moving textures regardless of the direction or the spatial pattern. Small features of the texture activate different bipolar cells as they move around. Bipolar cells often have biphasic impulse responses (Awatramani & Slaughter, 2000; Baccus et al., 2008; DeVries, 2000), which make them sensitive to rapid changes but not to static patterns. The rectification at the bipolar cell synapse then allows accumulation of these transient signals from the activated bipolars while it prevents cancellation from other bipolars that experience opposite stimulus changes. A time-varying velocity of the image pattern leads to a time-varying firing rate, and the simple Y-cell circuit model (figure 13.7A) can indeed predict this

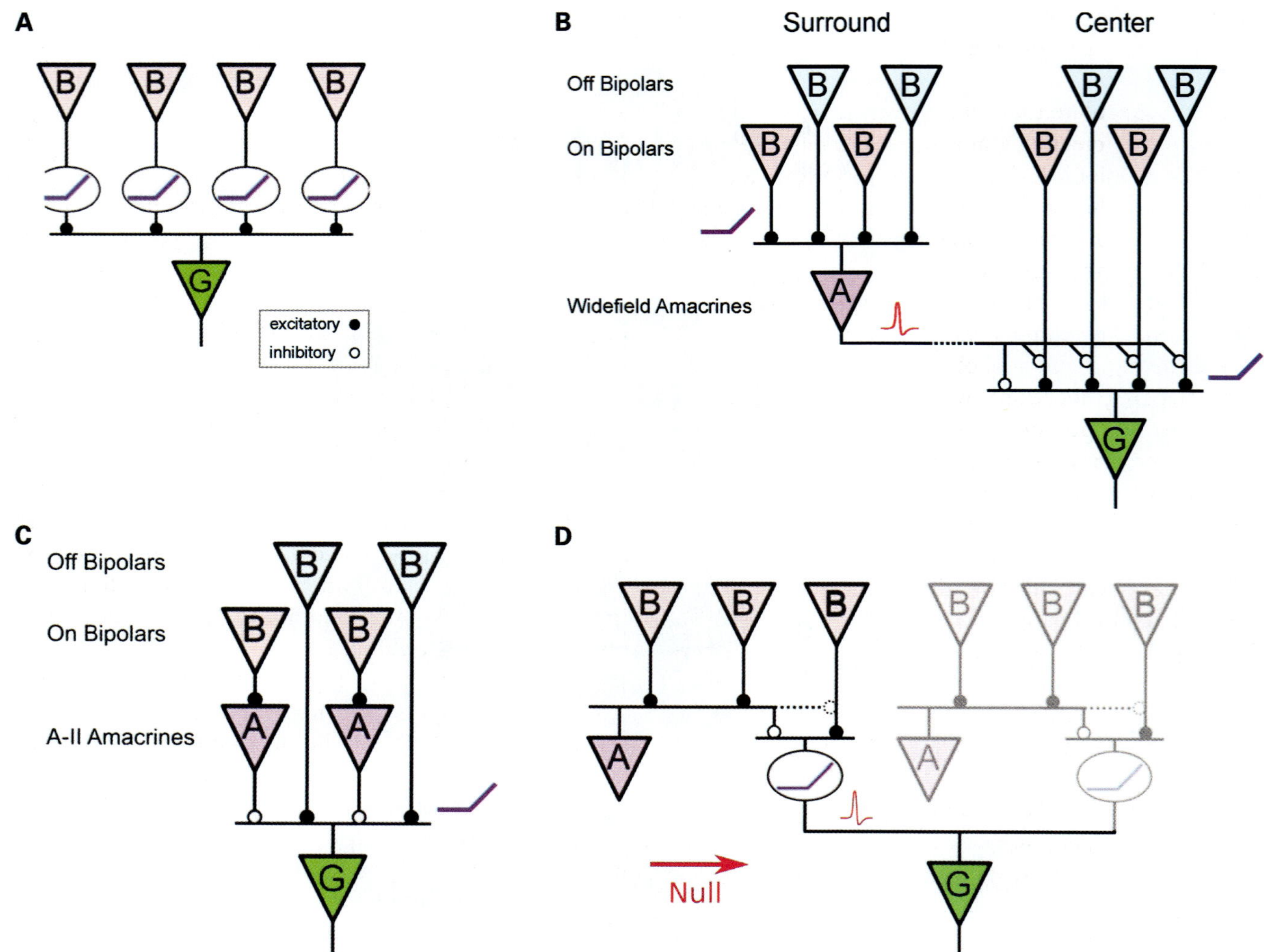

FIGURE 13.7 Retinal circuits leading to different feature detector ganglion cells. (A) The Y-type ganglion cell. This ganglion cell collects excitation from many bipolar cells. The bipolar cell synapses are rectifying: At baseline the release rate of transmitter is low, so depolarization increases transmitter release, but hyperpolarization has little or no effect. In subsequent panels, this rectifying quality is assumed for all bipolar cell synapses. (B) The object-motion-sensitive cell. Note that the ganglion cell pools over both ON and OFF bipolars, but this process is gated by the action of a wide-field amacrine cell. (C) The looming detector. Again there is pooling over ON and OFF channels, but with opposite sign because of an interposed narrow amacrine cell. (D) The direction-selective ganglion cell. The asymmetric interaction that defines the null direction occurs between the dendrite of a starburst amacrine cell and local bipolar cells. An additional threshold nonlinearity arises from spike generation within the dendritic tree of the ganglion cell.

output quantitatively (Baccus et al., 2008; Enroth-Cugell & Freeman, 1987; Victor & Shapley, 1979).

The defining Y-cell characteristic of nonlinear summation over space has now been encountered in many types of ganglion cell, but these differ strongly in other response features that confer certain selectivities, as seen in the following examples.

Object-Motion-Sensitive Cells

Ganglion cells of the "bug perceiver" type (Lettvin et al., 1959) have now been identified in several species,

and they likely represent one of the canonical types. They have been called "OMS" cells in the salamander (Ölveczky, Baccus, & Meister, 2003), "W3" cells in the mouse retina (Zhang et al., 2012), and "local edge detector" cells in the rabbit retina (Levick, 1967). They produce transient responses to both ON and OFF events in the receptive field center; thus, they process the stimulus in a very nonlinear fashion, beyond that of the Y cells. They are highly sensitive to moving patterns within the receptive field center, for the most part independent of the precise content of the pattern. But if the receptive field surround experiences pattern motion, and the motion is synchronized with that of the center, the ganglion cell remains silent (figure 13.6B).

One can speculate that such ganglion cells would be very useful for detecting a moving object within a visual scene. The ethological challenge here is that the eye of the observer is almost always in motion, be it from small fixational jitter or from the observer's own locomotion through the environment (Kowler, 1990; Martinez-Conde, Macknik, & Hubel, 2004). Thus, the default condition on the retina is one of incessant image flow, and identifying a moving object requires more than simply flagging locations where the image moves. Instead, the computation performed by the OMS cells identifies image regions that move with a trajectory different from the surroundings. These are likely small objects that move relative to the larger background, such as bugs among leaves.

We now understand how this computation is performed. The OMS ganglion cell pools excitation from many bipolar cells with rectifying synapses (figure 13.7B). Moreover, unlike the typical Y cell, it receives excitation from both ON- and OFF-type bipolars (Levick, 1967; Zhang et al., 2012). Thus, movement of a small object anywhere within the receptive field will depolarize some bipolars and hyperpolarize others, and in each case the ganglion cell receives a short pulse of excitation. This will happen regardless of the exact shape or pattern of the object or its direction of motion. We see that the nonlinear summation over space and over bipolars of opposite polarity already introduces a great deal of invariance.

What about selectivity? The same ganglion cell receives strong inhibition from amacrine cells in the receptive field surround, up to large distances from the center. The amacrine cell synapses act both directly on the ganglion cell and at the bipolar cell terminal, where they suppress transmission presynaptically (Baccus et al., 2008; Zhang et al., 2012). The amacrine cells are themselves driven by the same kind of nonlinear pooling mechanism that excites the ganglion cell (Baccus et al., 2008; Russell & Werblin, 2010; van Wyk, Taylor, & Vaney, 2006; Zhang et al., 2012). So if the visual pattern in the surround moves at the same time as the pattern in the center, excitation and inhibition cancel each other in the ganglion cell, and it remains silent. But if a spurt of motion in the center occurs independently of that in the surround, the ganglion cell fires (Ölveczky, Baccus, & Meister, 2003). This selectivity can be exquisite; for example, the mouse W3 ganglion cell remains completely silent during natural stimuli that result from the animal's own locomotion because they contain a great deal of global optic flow. These neurons are induced to fire only in special conditions where a small target moves against a static background (Zhang et al., 2012).

Looming Detectors

These ganglion cells, called PV5 in the mouse retina, fire strongly when a dark spot expands within the receptive field, as would occur when an object approaches the observer (figure 13.6C). Again, the receptive field center is sensitive to both ON and OFF events in the stimulus but now with opposite sign (Münch et al., 2009). A local dimming produces a transient excitation, whereas a local brightening produces transient inhibition. When a dark spot moves through the receptive field laterally, the leading edge produces excitation, and the trailing edge inhibition; the two effects cancel, and the ganglion cell remains silent. However, if a dark spot expands within the receptive field, there is no ON edge to contribute inhibition, and the ganglion cell fires strongly. A symmetric looming detector for bright objects has not been found.

The circuit that achieves the approach-specific responses is based on the pooling of excitation from the OFF pathway and inhibition from the ON pathway (figure 13.7C). The PV5 ganglion cell is excited by OFF bipolar cells and inhibited by AII amacrine cells (Münch et al., 2009). The AII cell is a local interneuron that in turn is excited by ON bipolar cells. Again, these synaptic inputs to the ganglion cell are rectified. As an edge travels across the receptive field, it stimulates in turn each of the small bipolar-size subunits, triggering a transient pulse of excitation or inhibition. When a dark object expands over the receptive field, the excitatory pulses are unopposed by any inhibition, and the ganglion cell fires throughout the period of expansion. If the object moves laterally, on the other hand, excitation from its leading edge is balanced by inhibition from the trailing edge, and the ganglion cell remains silent.

Inhibition thus serves to suppress responses to the nonpreferred motion signal, similar to the strategy of the OMS cell circuit. In contrast to the OMS cells, however, it is essential that the inhibition act postsynaptically rather than presynaptically at bipolar terminals because signals from different parts of the object must be combined. Again, rectification at the bipolar synapse constitutes an essential element in the circuit, but here we encounter an additional twist. This ganglion cell combines rectified excitation and inhibition from pathways of opposite polarities. It has been suggested that such "crossover inhibition" serves to make the overall response of ganglion cells more linear (Werblin, 2011), with the ON pathway implementing responses to brightening and the OFF pathway those to dimming. This is not the case for PV5 cells. A neuron that pools the light stimulus linearly over its receptive field can at best make a dimming sensor but will not be selective for looming objects. By contrast, the looming detector is excited by an expanding dark edge, even if other parts of the receptive field experience a gradual brightening. This can be understood if the inhibitory pathway has a high threshold such that a gradual brightening is ignored but the sudden brightening at a traveling ON edge gets transmitted (Münch et al., 2009).

Note that the AII amacrine cell in this circuit also serves an entirely different function during scotopic vision, namely to feed rod signals into the cone bipolar cells (Bloomfield & Dacheux, 2001; Demb & Singer, 2012). This is an interesting example of a single cell type that serves quite different roles, even signaling in opposite directions (Manookin et al., 2008).

Direction-Selective Cells

Again these ganglion cells are very sensitive to movement within the receptive field. However, they respond preferentially to motion in one direction and remain silent to motion in the opposite direction (Vaney, Sivyer, & Taylor, 2012) (figure 13.6D). In some cases this direction selectivity applies even for tiny spots moving as little as 1/10 of the RF diameter; thus, the computation is performed on a very local scale, and the overall result is pooled over the receptive field (Barlow & Levick, 1965). Such a direction-selective (DS) cell is invariant to the precise pattern or shape that moves within its receptive field but selective for the direction in which it moves.

Three classes of DS ganglion cells have been identified, distinguished by the polarity of the response in the receptive field center. ON–OFF DS cells are excited transiently by both ON and OFF steps of light (Barlow & Levick, 1965; Weng, Sun, & He, 2005), ON DS cells by ON steps only (Oyster, 1968; Sun et al., 2006), and OFF DS cells by OFF steps (Kim et al., 2008). These three classes encompass multiple distinct cell types. The four types of ON–OFF DS cells in the mammalian retina have different preferred directions of motion in the receptive field center, aligned with the cardinal directions on the eye: dorsal, ventral, nasal, and temporal (Elstrott et al., 2008; Kay et al., 2011; Oyster, 1968). Motion in the surround exerts a powerful suppression (Barlow & Levick, 1965; Wyatt & Daw, 1975). As for OMS cells, this suppression is particularly strong when surround motion matches the center motion in speed and direction (Chiao & Masland, 2003; Ölveczky, Baccus, & Meister, 2003). Thus, the ON–OFF DS ganglion cells appear tuned to the local motion of objects within the scene. Their axons project to both the thalamus and the superior colliculus (Huberman et al., 2009; Kay et al., 2011; Stewart, Chow, & Masland, 1971; Vaney, Sivyer, & Taylor, 2012) and thus make this information available to the two major streams for higher visual processing. By contrast, the ON DS cells include three types, with preferred directions on the retina roughly dorsal, ventral, and temporal (Oyster, 1968; Yonehara et al., 2009). They are not suppressed by surround motion and respond very well to moving patterns that extend over the whole retina. Thus, they can serve to encode the overall optic flow in the scene, as produced by slip of the image on the retina when the animal or the eye moves relative to the scene. Interestingly these neurons do not project to the major visual pathways but exclusively into the accessory optic system (Buhl & Peichl, 1986; Oyster et al., 1980; Yonehara et al., 2008, 2009) whose role is to sense self-motion for the regulation of eye movements (Simpson, 1984; Giolli, Blanks, & Lui, 2006). Finally, a single type of OFF DS cell has been described that prefers motion in the ventral direction (Kim et al., 2008). Again, these neurons project to both superior colliculus and thalamus, but their role in downstream processing remains unclear. This list represents the consensus types of DS ganglion cells (DSGC), but there are recent indications that the population may yet split into finer types whose distinctions and downstream projections remain to be established (Hoshi et al., 2011; Kanjhan & Sivyer, 2010; Rivlin-Etzion et al., 2011).

The retinal circuitry underlying the ON–OFF DSGC has been studied intensely, and we now have a great wealth of physiological, anatomical, and computational results available. As may be expected, there is some discordance in this large set of reports. As a result it has become difficult to integrate all the observations into a coherent model of neuronal circuitry. We present here one subcircuit that almost certainly contributes to the observed direction selectivity, although it leaves some aspects unexplained (figure 13.7D). In this we largely follow a recent review (Vaney, Sivyer, & Taylor, 2012), which is recommended for an overview of this retinal subcircuit.

The discoverers of retinal direction selectivity proposed a simple model of how it might be achieved through the interaction of excitatory and inhibitory synaptic inputs to the ganglion cell (Barlow & Levick, 1965). This model has four required ingredients: spatial asymmetry—inhibition should be laterally offset from excitation; temporal asymmetry—inhibition should be delayed relative to excitation; nonlinear pooling—the ganglion cell responds only if its pooled synaptic input exceeds a threshold; and small subunits—this pooling should occur independently within many small subunits of the receptive field, to explain the selectivity for even small motions. There now exists compelling evidence that assigns these various functions to specific cellular elements in the inner retina (Vaney, Sivyer, & Taylor, 2012).

In this circuit (figure 13.7D), the independent subunits correspond to individual electrotonically distinct dendritic compartments of the DSGC. Each such compartment pools excitation and inhibition and generates a spike if the net depolarization exceeds a threshold. These dendritic spikes travel to the soma reliably and cause a spike in the axon. Each compartment receives excitation from bipolar cells and inhibition from starburst amacrine cells (SACs). Whereas the receptive field of a bipolar cell directly overlies its terminal, the receptive field of the starburst cell is displaced laterally toward the null side (the side from which null stimuli arrive). This spatial asymmetry results from a peculiar rule of connectivity between the two cell types. First, although the SAC receives bipolar cell inputs all along the dendrite, its inhibitory terminals are at the dendritic tips. Second, the DSGC connects preferentially to those SAC dendrites that course in the null direction, by a factor of 10 to 1 (Briggman, Holmstaedter, & Denk, 2011). As a result the receptive field of the contributing starburst dendrite is displaced toward the null side of the DSGC (figure 13.7D). Another cellular mechanism contributes to asymmetry: The depolarization at the SAC dendritic tip is itself direction-selective, favoring outward motion over inward motion (Euler, Detwiler, & Denk, 2002). Thus, the DSGC receives stronger inhibition for null than for preferred motion. Finally, the temporal delay and extended duration of inhibition result from the additional synapse in the SAC pathway as well as the prolonged time course of GABA release.

Given this circuit, one can understand the direction-selective processing of moving stimuli (figure 13.7D). A small spot moving in the null direction first excites the SAC and then the bipolar cell. Because the SAC input is delayed and more sustained, inhibition and excitation arrive at the GC dendrite at the same time, the resulting signal remains below threshold, and no spikes are produced. The same sequence recurs in each of the other compartments. With motion in the preferred direction, excitation from the bipolar cell is triggered before the inhibition can quench it, and this launches a dendritic spike followed by somatic firing. The same circuit is found in both the inner and the outer starburst stratum of the IPL, allowing the DSGC to process ON and OFF edges independently.

As mentioned above, this should be considered a minimal circuit. It leaves a number of observations unexplained, for example, that excitatory inputs to the DSGC are already direction selective (Borg-Graham, 2001; Taylor & Vaney, 2002). This might occur if SACs also inhibit the bipolar cell terminals (figure 13.7D). Starburst amacrines also release acetylcholine, which excites the DSGC; the function of these synapses is unclear. The minimal circuit also does not account for the suppressive effects of motion in the surround, which may involve input from another type of amacrine cell.

The circuits for the other DS ganglion cells are less well understood. Some ON DS cells seem to interact with starburst amacrines (Yonehara et al., 2011) much as the ON–OFF DS cells do. However, the newly described ON DS types that ramify outside the starburst stratum suggest there must be other mechanisms to achieve the same effects (Hoshi et al., 2011). The same holds for the OFF DS cells. This cell type has a strongly asymmetic dendritic tree that points in the preferred direction of motion (Kim et al., 2008). In that case the key asymmetry may well be provided by the morphology of the ganglion cell itself, although the details of its function remain to be explored.

Diverse Circuits Using Common Components

Although the various feature detectors discussed above seem to select very different visual features, their underlying circuits share much in common. In fact, all these circuits make use of the same kinds of simple elements: small-field bipolar cells of two polarities, rectification at a bipolar cell synapse, spatial pooling, narrow-field amacrines for sign inversion, wide-field amacrines for lateral inhibition. The only differences lie in the sequence and combination of the elements. As in the man-made field of electronics, varying the arrangement and combination of simple elements results in dramatically different functions. Still the above account falls short of filling out the catalog of all 20 morphological ganglion cells, which suggests that other retinal feature detectors and their associated circuit computations remain to be identified (table 13.1).

OPEN QUESTIONS

Visual Features and Ecology

The above examples motivate a deeper consideration of feature selectivity. Are these neurons truly selective for just one type of stimulus, and if not, can one justify associating them with a specific feature? The answer to the first question is clearly negative: For example, every known ganglion cell will respond to a small spot flashing in the receptive field center. This includes the object-motion cells and the looming detectors. However, under conditions of natural vision, flashing spots simply do not happen very often. Except right after an eye blink, objects rarely appear on the retina out of thin air; instead, they move into a neuron's receptive field from a neighboring region, or they move within the receptive field. Within the rather constrained set of stimuli that occur commonly in natural vision, the looming detector is selective primarily for objects that expand, and the OMS cell for those that move differently from their background.

An interesting theme is that most of the feature detectors identified to date seem to process some form of image motion: wide-field, local, or differential. This has a simple ethological interpretation: Moving objects in the visual scene tend to be interesting points, either as threats or opportunities. Similarly the global image flow on the retina is a useful indicator of self-motion through the environment. It is perhaps not surprising that specific circuits have evolved to extract and separate these important cues from the image rapidly and efficiently. However, these qualitative arguments will need to be tested more seriously. One approach is to study retinal signaling under conditions that truly reflect vision in the natural environment, including the ever-present observer and eye motion. Such stimuli can be gathered now thanks to ultralight video cameras that can travel on the head of a rodent moving freely in the natural environment (Zhang et al., 2012). It will be important to test for each ganglion cell type how selective it is under these conditions and whether the trigger features are indeed those identified using the more conventional synthetic stimuli.

Downstream Processing

Where in the brain are all these different parallel representations sent? A simple suggestion, consistent with the concept of a retina with many independent image processors, would be that each ganglion cell type provides input to a different retinorecipient region. But that is not the case. There are three patterns of ganglion cell projections (figure 13.8). A few ganglion cell types project to a single target region, such as the three types of ON DS cells (Vaney, Sivyer, & Taylor, 2012). A few others project to multiple regions such as some of the melanopsin-expressing ganglion cells (Schmidt et al., 2011a). However, most ganglion cell types project to two main visual centers, the lateral geniculate nucleus (LGN) and the superior colliculus (SC). In fact, the axons of most individual ganglion cells branch and innervate both target areas.

There are two important points to note. First, many visual features are copied to both the LGN and the SC, and it is, therefore, intriguing to ask whether these copies will ever be compared or, alternatively, will live independent lives to drive behavior or perception. Second, in most species studied, including primates, many of the specialized visual features are sent to the LGN. On the other hand, most cortical researchers are convinced that only a few pathways—perhaps three—arrive at the primary cortex from the LGN. One

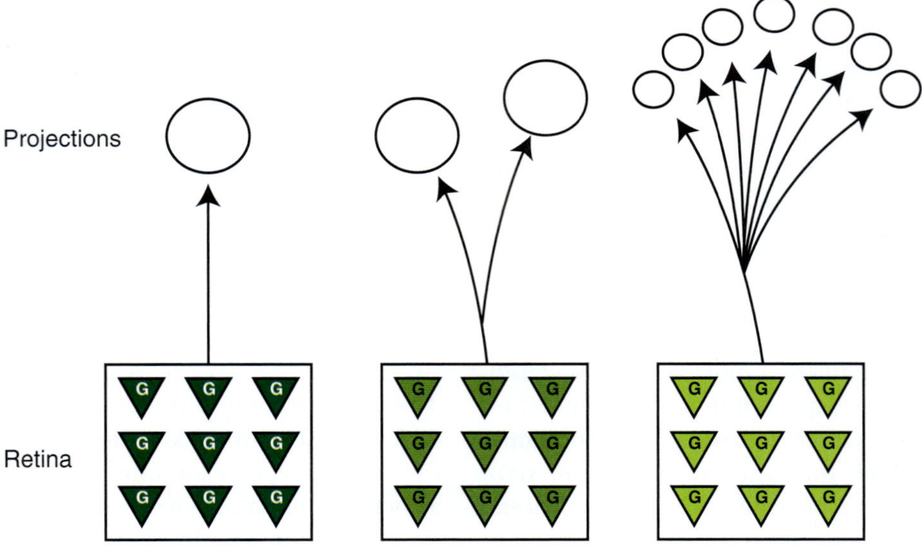

FIGURE 13.8   Three types of retinal projections. (Left) Ganglion cell mosaic projecting to a single nucleus. (Center) Ganglion cell mosaic projecting to two nuclei. (Right) Ganglion cell mosaic projecting to multiple nuclei.

possibility is that multiple retinal features combine immediately within the LGN into few visual channels. Alternatively we may still have an incomplete understanding of the pathways that drive the visual cortex. Fortunately a new set of tools is coming available that includes the means to trace the downstream pathways from genetically identified cells, to activate or silence specific types of ganglion cells, and to record activity from hundreds of neurons in the cortex. Therefore, it is likely that this controversy will be resolved soon. More broadly, it seems possible now to map out the relation between the distinct ganglion cell pathways and specific aspects of the animal's visual behavior.

So far the silencing of types of ganglion cells has led to controversial conclusions. Targeted elimination of a few or single types of melanopsin-containing ganglion cells caused well-defined behavioral deficits in mice (Chen, Badea, & Hattar, 2011; Guler et al., 2008; Hatori et al., 2008). Similar marked deficits in mouse behavior were found when the starburst amacrine cells were eliminated (Yoshida et al., 2001), and likely the directional selectivity of at least seven types of directional selective ganglion cells was abolished. In contrast, the acute silencing of all types of ON ganglion cells led to minor changes in primate visual behavior (Schiller, Sandell, & Maunsell, 1986). A mutation that is predicted to result in a similar silencing of all ON cells in humans does not lead to any major visual defects at light intensities where cones are active (Dryja et al., 2005; Zeitz et al., 2005). Many ganglion cell types come in pairs, including an ON and an OFF version. It appears that for a significant part of our visual perception and function one version of a type is enough.

Clinical Tests of Feature Processing

The conservation across species of retinal structure and function also provides new opportunities to diagnose retinal diseases. When we visit the ophthalmologist, our vision is tested on a chart from which we read small and large letters. This test mostly diagnoses the optics of the eye, based on the performance of 2 of the 20 types of ganglion cells, the ON and OFF midget cells. Although the retina incudes a massive infrastructure to analyze and dissect different categories of motion, there is, remarkably, not a single quantitative or even qualitative test used regularly by ophthalmologists that would evaluate how well we can perceive motion. If, for example, a mutation had produced a defect in the development of amacrine cell networks, the patient may be unable to see motion or, conversely, might see motion all the time. Such a patient would likely end up in the office of a psychiatrist, even though the defect originates in the sensory periphery. By understanding one by one the computations that different ganglion cells perform and by understanding the behavioral phenotypes that result from silencing identified ganglion cell mosaics, it may be possible to discover abnormalities in human visual perception that arise within the retina.

APPENDIX 13.1

Linear Visual Responses

Mathematically, a linear light response is derived by convolving the stimulus with a filter function:

$$r(t) = r_0 + \int\int s(x,t')\,F(x,t-t')\,dt'dx$$

In this expression, $s(x,t)$ is the stimulus intensity as a function of space and time, and r_0 is the firing rate in absence of any stimulus. The weighting function $F(x,t)$ specifies the weight applied to the intensity at location x and time t in the past and is commonly called the spatiotemporal receptive field of the neuron.

Clearly the range of a neuron's firing rate is restricted, namely to zero firing at the bottom, and some maximal rate determined by cellular biophysics at the top. Therefore, this linear relationship cannot persist when the light intensity varies over too large a range. In fact one generally finds distortions in the response to strong stimuli. A more general version of a pixel sensor allows for such distortions in the relation between stimulus and response:

$$r(t) = \mathcal{N}\left(\int\int s(x,t')\,F(x,t-t')\,dt'dx\right)$$

where the function $\mathcal{N}(\ )$ is the distortion function that relates the linear-weighted stimulus to the firing rate, generally with a sigmoid shape. A response function of this kind is often called an LN model: a linear filter followed by a nonlinearity (Chichilnisky, 2001). Note that the nonlinearity $\mathcal{N}(\ )$ does not fundamentally alter *what* the ganglion cell reports about the visual scene, only *how* it is reported. The visual meaning of the message is fully defined by the spatiotemporal receptive field $F(x,t)$. In summary then, we can consider a retinal ganglion cell a pixel sensor if its stimulus–response function follows the LN model under most visual stimuli. The spatial and temporal extent of the image pixel that this ganglion cell reports is embodied by the spatiotemporal receptive field $F(x,t)$.

REFERENCES

Awatramani, G. B., & Slaughter, M. M. (2000). Origin of transient and sustained responses in ganglion cells of the retina. *Journal of Neuroscience, 20,* 7087–7095.

Azeredo da Silveira, R., & Roska, B. (2011). Cell types, circuits, computation. *Current Opinion in Neurobiology, 21,* 664–671.

Baccus, S. A., Ölveczky, B. P., Manu, M., & Meister, M. (2008). A retinal circuit that computes object motion. *Journal of Neuroscience, 28,* 6807–6817.

Barlow, H. B. (1953). Summation and inhibition in the frog's retina. *Journal of Physiology, 119,* 69–88.

Barlow, H. B., Hill, R. M., & Levick, W. R. (1964). Retinal ganglion cells responding selectively to direction and speed of image motion in the rabbit. *Journal of Physiology, 173,* 377–407.

Barlow, H. B., & Levick, W. R. (1965). The mechanism of directionally selective units in rabbit's retina. *Journal of Physiology, 178,* 477–504.

Benardete, E. A., & Kaplan, E. (1997a). The receptive field of the primate P retinal ganglion cell, II: Nonlinear dynamics. *Visual Neuroscience, 14,* 187–205.

Benardete, E. A., & Kaplan, E. (1997b). The receptive field of the primate P retinal ganglion cell, I: Linear dynamics. *Visual Neuroscience, 14,* 169–185.

Berson, D. M., Isayama, T., & Pu, M. (1999). The eta ganglion cell type of cat retina. *Journal of Comparative Neurology, 408,* 204–219.

Berson, D. M., Pu, M., & Famiglietti, E. V. (1998). The zeta cell: A new ganglion cell type in cat retina. *Journal of Comparative Neurology, 399,* 269–288.

Bloomfield, S. A., & Dacheux, R. F. (2001). Rod vision: Pathways and processing in the mammalian retina. *Progress in Retinal and Eye Research, 20,* 351–384.

Borg-Graham, L. J. (2001). The computation of directional selectivity in the retina occurs presynaptic to the ganglion cell. *Nature Neuroscience, 4,* 176–183.

Briggman, K. L., Helmstaedter, M., & Denk, W. (2011). Wiring specificity in the direction-selectivity circuit of the retina. *Nature, 471,* 183–188.

Buhl, E. H., & Peichl, L. (1986). Morphology of rabbit retinal ganglion cells projecting to the medial terminal nucleus of the accessory optic system. *Journal of Comparative Neurology, 253,* 163–174.

Chen, S. K., Badea, T. C., & Hattar, S. (2011). Photoentrainment and pupillary light reflex are mediated by distinct populations of ipRGCs. *Nature, 476,* 92–95.

Chiao, C. C., & Masland, R. H. (2003). Contextual tuning of direction-selective retinal ganglion cells. *Nature Neuroscience, 6,* 1251–1252.

Chichilnisky, E. J. (2001). A simple white noise analysis of neuronal light responses. *Network, 12,* 199–213.

Cleland, B. G., Levick, W. R., & Wässle, H. (1975). Physiological identification of a morphological class of cat retinal ganglion cells. *Journal of Physiology, 248,* 151–171.

Crook, J. D., Peterson, B. B., Packer, O. S., Robinson, F. R., Troy, J. B., & Dacey, D. M. (2008). Y-cell receptive field and collicular projection of parasol ganglion cells in macaque monkey retina. *Journal of Neuroscience, 28,* 11277–11291.

Dacey, D. M. (2004). Origins of perception: Retinal ganglion cell diversity and the creation of parallel visual pathways. In M. S. Gazzaniga (Ed.), *The cognitive neurosciences* (pp. 281–301). Cambridge, MA: MIT Press.

Dacey, D. M., Liao, H. W., Peterson, B. B., Robinson, F. R., Smith, V. C., Pokorny, J., et al. (2005). Melanopsin-expressing ganglion cells in primate retina signal colour and irradiance and project to the LGN. *Nature, 433,* 749–754. doi: 10.1038/nature03387.

Dacey, D. M., & Packer, O. S. (2003). Colour coding in the primate retina: Diverse cell types and cone-specific circuitry. *Current Opinion in Neurobiology, 13,* 421–427.

Demb, J. B., & Singer, J. H. (2012). Intrinsic properties and functional circuitry of the AII amacrine cell. *Visual Neuroscience, 29,* 51–60.

Demb, J. B., Zaghloul, K., Haarsma, L., & Sterling, P. (2001). Bipolar cells contribute to nonlinear spatial summation in the brisk-transient (Y) ganglion cell in mammalian retina. *Journal of Neuroscience, 21,* 7447–7454.

DeVries, S. H. (2000). Bipolar cells use kainate and AMPA receptors to filter visual information into separate channels. *Neuron, 28,* 847–856.

Dryja, T. P., McGee, T. L., Berson, E. L., Fishman, G. A., Sandberg, M. A., Alexander, K. R., et al. (2005). Night blindness and abnormal cone electroretinogram ON responses in patients with mutations in the GRM6 gene encoding mGluR6. *Proceedings of the National Academy of Sciences of the United States of America, 102,* 4884–4889. doi: 10.1073/pnas.0501233102.

Elstrott, J., Anishchenko, A., Greschner, M., Sher, A., Litke, A. M., Chichilnisky, E. J., et al. (2008). Direction selectivity in the retina is established independent of visual experience and cholinergic retinal waves. *Neuron, 58,* 499–506.

Enroth-Cugell, C., & Freeman, A. W. (1987). The receptive-field spatial structure of cat retinal Y cells. *Journal of Physiology, 384,* 49–79.

Enroth-Cugell, C., & Robson, J. G. (1966). The contrast sensitivity of retinal ganglion cells of the cat. *Journal of Physiology, 187,* 517–552.

Enroth-Cugell, C., Robson, J. G., Schweitzer-Tong, D. E., & Watson, A. B. (1983). Spatio-temporal interactions in cat retinal ganglion cells showing linear spatial summation. *Journal of Physiology, 341,* 279–307.

Euler, T., Detwiler, P. B., & Denk, W. (2002). Directionally selective calcium signals in dendrites of starburst amacrine cells. *Nature, 418,* 845–852.

Fain, G. L. (2011). Adaptation of mammalian photoreceptors to background light: Putative role for direct modulation of phosphodiesterase. *Molecular Neurobiology, 44,* 374–382.

Giolli, R. A., Blanks, R. H., & Lui, F. (2006). The accessory optic system: Basic organization with an update on connectivity, neurochemistry, and function. *Progress in Brain Research, 151,* 407–440.

Guler, A. D., Ecker, J. L., Lall, G. S., Haq, S., Altimus, C. M., Liao, H. W., et al. (2008). Melanopsin cells are the principal conduits for rod-cone input to non-image-forming vision. *Nature, 453,* 102–105.

Hatori, M., Le, H., Vollmers, C., Keding, S. R., Tanaka, N., Buch, T., et al. (2008). Inducible ablation of melanopsin-expressing retinal ganglion cells reveals their central role in non-image forming visual responses. *PLoS One, 3,* e2451. doi: 10.1371/journal.pone.0002451.

Hattar, S., Kumar, M., Park, A., Tong, P., Tung, J., Yau, K. W., et al. (2006). Central projections of melanopsin-expressing retinal ganglion cells in the mouse. *Journal of Comparative Neurology, 497,* 326–349. doi: 10.1002/cne.20970.

Hoshi, H., Tian, L. M., Massey, S. C., & Mills, S. L. (2011). Two distinct types of ON directionally selective ganglion cells in the rabbit retina. *Journal of Comparative Neurology, 519,* 2509–2521.

Huberman, A. D., Wei, W., Elstrott, J., Stafford, B. K., Feller, M. B., & Barres, B. A. (2009). Genetic identification of an

On-Off direction-selective retinal ganglion cell subtype reveals a layer-specific subcortical map of posterior motion. *Neuron, 62,* 327–334. doi: 10.1016/j.neuron.2009.04.014.

Isayama, T., Berson, D. M., & Pu, M. (2000). Theta ganglion cell type of cat retina. *Journal of Comparative Neurology, 417,* 32–48.

Jeon, C. J., Strettoi, E., & Masland, R. H. (1998). The major cell populations of the mouse retina. *Journal of Neuroscience, 18,* 8936–8946.

Kamermans, M., & Fahrenfort, I. (2004). Ephaptic interactions within a chemical synapse: Hemichannel-mediated ephaptic inhibition in the retina. *Current Opinions in Neurobiology, 14,* 531–541. doi:10.1016/j.conb.2004.08.016.

Kanjhan, R., & Sivyer, B. (2010). Two types of ON direction-selective ganglion cells in rabbit retina. *Neuroscience Letters, 483,* 105–109. doi:10.1016/j.neulet.2010.07.071.

Kay, J. N., De la Huerta, I., Kim, I. J., Zhang, Y., Yamagata, M., Chu, M. W., et al. (2011). Retinal ganglion cells with distinct directional preferences differ in molecular identity, structure, and central projections. *Journal of Neuroscience, 31,* 7753–7762. doi: 10.1523/JNEUROSCI.0907-11.2011.

Kim, I. J., Zhang, Y., Yamagata, M., Meister, M., & Sanes, J. R. (2008). Molecular identification of a retinal cell type that responds to upward motion. *Nature, 452,* 478–482.

Kolb, H., & Marshak, D. (2003). The midget pathways of the primate retina. *Documenta Ophthalmologica, 106,* 67–81.

Kowler, E. (1990). *Eye movements and their role in visual and cognitive processes.* New York: Elsevier Science.

Kuffler, S. W. (1953). Discharge patterns and functional organization of mammalian retina. *Journal of Neurophysiology, 16,* 37–68.

Lettvin, J. Y., Maturana, H. R., McCulloch, W. S., & Pitts, W. H. (1959). What the frog's eye tells the frog's brain. *Proceedings of the Institute of Radio Engineers, 47,* 1940–1951. doi: 10.1109/JRPROC.1959.287207.

Levick, W. R. (1967). Receptive fields and trigger features of ganglion cells in the visual streak of the rabbits retina. *Journal of Physiology, 188,* 285–307.

Manookin, M. B., Beaudoin, D. L., Ernst, Z. R., Flagel, L. J., & Demb, J. B. (2008). Disinhibition combines with excitation to extend the operating range of the OFF visual pathway in daylight. *Journal of Neuroscience, 28,* 4136–4150.

Martinez-Conde, S., Macknik, S. L., & Hubel, D. H. (2004). The role of fixational eye movements in visual perception. *Nature Reviews. Neuroscience, 5,* 229–240.

Masland, R. H. (2001). The fundamental plan of the retina. *Nature Neuroscience, 4,* 877–886.

Masland, R. H. (2012). The tasks of amacrine cells. *Visual Neuroscience, 29,* 3–9.

Matsui, K., Hosoi, N., & Tachibana, M. (1998). Excitatory synaptic transmission in the inner retina: Paired recordings of bipolar cells and neurons of the ganglion cell layer. *Journal of Neuroscience, 18,* 4500–4510.

McMahon, M. J., Lankheet, M. J., Lennie, P., & Williams, D. R. (2000). Fine structure of parvocellular receptive fields in the primate fovea revealed by laser interferometry. *Journal of Neuroscience, 20,* 2043–2053.

Meister, M., & Berry, M. J. (1999). The neural code of the retina. *Neuron, 22,* 435–450.

Münch, T. A., Azeredo da Silveira, R., Siegert, S., Viney, T. J., Awatramani, G. B., & Roska, B. (2009). Approach sensitivity in the retina processed by a multifunctional neural circuit. *Nature Neuroscience, 12,* 1308–1316. doi: 10.1038/nn.2389.

Ölveczky, B. P., Baccus, S. A., & Meister, M. (2003). Segregation of object and background motion in the retina. *Nature, 423,* 401–408.

Oyster, C. W. (1968). The analysis of image motion by the rabbit retina. *Journal of Physiology, 199,* 613–635.

Oyster, C. W., & Barlow, H. B. (1967). Direction-selective units in rabbit retina: Distribution of preferred directions. *Science, 155,* 841–842.

Oyster, C. W., Simpson, J. I., Takahashi, E. S., & Soodak, R. E. (1980). Retinal ganglion cells projecting to the rabbit accessory optic system. *Journal of Comparative Neurology, 190,* 49–61.

Palmer, M. J. (2010). Characterisation of bipolar cell synaptic transmission in goldfish retina using paired recordings. *Journal of Physiology, 588,* 1489–1498.

Pang, J. J., Gao, F., & Wu, S. M. (2003). Light-evoked excitatory and inhibitory synaptic inputs to ON and OFF alpha ganglion cells in the mouse retina. *Journal of Neuroscience, 23,* 6063–6073.

Pu, M., Berson, D. M., & Pan, T. (1994). Structure and function of retinal ganglion cells innervating the cat's geniculate wing: An in vitro study. *Journal of Neuroscience, 14,* 4338–4358.

Puller, C., & Haverkamp, S. (2011). Bipolar cell pathways for color vision in non-primate dichromats. *Visual Neuroscience, 28,* 51–60.

Rivlin-Etzion, M., Zhou, K., Wei, W., Elstrott, J., Nguyen, P. L., Barres, B. A., et al. (2011). Transgenic mice reveal unexpected diversity of on-off direction-selective retinal ganglion cell subtypes and brain structures involved in motion processing. *Journal of Neuroscience, 31,* 8760–8769.

Roska, B., Molnar, A., & Werblin, F. S. (2006). Parallel processing in retinal ganglion cells: How integration of space-time patterns of excitation and inhibition form the spiking output. *Journal of Neurophysiology, 95,* 3810–3822.

Roska, B., & Werblin, F. (2001). Vertical interactions across ten parallel, stacked representations in the mammalian retina. *Nature, 410,* 583–587.

Russell, T. L., & Werblin, F. S. (2010). Retinal synaptic pathways underlying the response of the rabbit local edge detector. *Journal of Neurophysiology, 103,* 2757–2769.

Schiller, P. H., Sandell, J. H., & Maunsell, J. H. (1986). Functions of the ON and OFF channels of the visual system. *Nature, 322,* 824–825.

Schmidt, T. M., Chen, S. K., & Hattar, S. (2011a). Intrinsically photosensitive retinal ganglion cells: Many subtypes, diverse functions. *Trends in Neurosciences, 34,* 572–580. doi: 10.1016/j.tins.2011.07.001.

Schmidt, T. M., Do, M. T. H., Dacey, D., Lucas, R., Hattar, S., & Matynia, A. (2011b). Melanopsin-positive intrinsically photosensitive retinal ganglion cells: From form to function. *Journal of Neuroscience, 31,* 16094–16101. doi: 10.1523/JNEUROSCI.4132-11.2011.

Siegert, S., Scherf, B. G., Del Punta, K., Didkovsky, N., Heintz, N., & Roska, B. (2009). Genetic address book for retinal cell types. *Nature Neuroscience, 12,* 1197–1204.

Simpson, J. I. (1984). The accessory optic system. *Annual Review of Neuroscience, 7,* 13–41. doi: 10.1146/annurev.neuro.7.1.13.

Stewart, D. L., Chow, K. L., & Masland, R. H. (1971). Receptive-field characteristics of lateral geniculate neurons in the rabbit. *Journal of Neurophysiology, 34,* 139–147.

Sun, W., Deng, Q., Levick, W. R., & He, S. (2006). ON direction-selective ganglion cells in the mouse retina. *Journal of Physiology, 576*, 197–202.

Taylor, W. R., & Smith, R. G. (2011). Trigger features and excitation in the retina. *Current Opinion in Neurobiology, 21*, 672–678.

Taylor, W. R., & Vaney, D. I. (2002). Diverse synaptic mechanisms generate direction selectivity in the rabbit retina. *Journal of Neuroscience, 22*, 7712–7720.

Trenholm, S., Johnson, K., Li, X., Smith, R. G., & Awatramani, G. B. (2011). Parallel mechanisms encode direction in the retina. *Neuron, 71*, 683–694.

Vaney, D. I., Sivyer, B., & Taylor, W. R. (2012). Direction selectivity in the retina: Symmetry and asymmetry in structure and function. *Nature Reviews. Neuroscience, 13*, 194–208.

van Wyk, M., Taylor, W. R., & Vaney, D. I. (2006). Local edge detectors: A substrate for fine spatial vision at low temporal frequencies in rabbit retina. *Journal of Neuroscience, 26*, 13250–13263.

Victor, J. D., & Shapley, R. M. (1979). The nonlinear pathway of Y ganglion cells in the cat retina. *Journal of General Physiology, 74*, 671–689.

Wässle, H. (2004). Parallel processing in the mammalian retina. *Nature Reviews. Neuroscience, 5*, 747–757.

Wässle, H., Boycott, B. B., & Illing, R. B. (1981). Morphology and mosaic of on- and off-beta cells in the cat retina and some functional considerations. *Proceedings of the Royal Society of London. Series B, Biological Sciences, 212*, 177–195.

Wässle, H., Peichl, L., & Boycott, B. B. (1981). Morphology and topography of on- and off-alpha cells in the cat retina. *Proceedings of the Royal Society of London. Series B, Biological Sciences, 212*, 157–175.

Wässle, H., Puller, C., Muller, F., & Haverkamp, S. (2009). Cone contacts, mosaics, and territories of bipolar cells in the mouse retina. *Journal of Neuroscience, 29*, 106–117.

Weng, S., Sun, W., & He, S. (2005). Identification of ON-OFF direction-selective ganglion cells in the mouse retina. *Journal of Physiology, 562*, 915–923.

Werblin, F. S. (2011). The retinal hypercircuit: A repeating synaptic interactive motif underlying visual function. *Journal of Physiology, 589*, 3691–3702.

Wu, S. M. (1992). Feedback connections and operation of the outer plexiform layer of the retina. *Current Opinions in Neurobiology, 2*, 462–468. doi:10.1016/0959-4388(92)90181-J.

Wyatt, H. J., & Daw, N. W. (1975). Directionally sensitive ganglion cells in the rabbit retina: Specificity for stimulus direction, size, and speed. *Journal of Neurophysiology, 38*, 613–626.

Yonehara, K., Balint, K., Noda, M., Nagel, G., Bamberg, E., & Roska, B. (2011). Spatially asymmetric reorganization of inhibition establishes a motion-sensitive circuit. *Nature, 469*, 407–410.

Yonehara, K., Ishikane, H., Sakuta, H., Shintani, T., Nakamura-Yonehara, K., Kamiji, N. L., et al. (2009). Identification of retinal ganglion cells and their projections involved in central transmission of information about upward and downward image motion. *PLoS One, 4*, e4320. doi: 10.1371/journal.pone.0004320.

Yonehara, K., Shintani, T., Suzuki, R., Sakuta, H., Takeuchi, Y., Nakamura-Yonehara, K., et al. (2008). Expression of SPIG1 reveals development of a retinal ganglion cell subtype projecting to the medial terminal nucleus in the mouse. *PLoS One, 3*, e1533. doi: 10.1371/journal.pone.0001533.

Yoshida, K., Watanabe, D., Ishikane, H., Tachibana, M., Pastan, I., & Nakanishi, S. (2001). A key role of starburst amacrine cells in originating retinal directional selectivity and optokinetic eye movement. *Neuron, 30*, 771–780.

Zeitz, C., van Genderen, M., Neidhardt, J., Luhmann, U. F., Hoeben, F., Forster, U., et al. (2005). Mutations in GRM6 cause autosomal recessive congenital stationary night blindness with a distinctive scotopic 15-Hz flicker electroretinogram. *Investigative Ophthalmology & Visual Science, 46*, 4328–4335. doi: 10.1167/iovs.05-0526.

Zhang, Y., Kim, I.-J., Sanes, J. R., & Meister, M. (2012). The high-resolution ganglion cells of the mouse retina are feature detectors. *Proceedings of the National Academy of Sciences of the United States of America, 109*, E2391-8.

Zhang, J., Li, W., Hoshi, H., Mills, S. L., & Massey, S. C. (2005). Stratification of alpha ganglion cells and ON/OFF directionally selective ganglion cells in the rabbit retina. *Visual Neuroscience, 22*, 535–549.

14 Intrinsically Photosensitive Retinal Ganglion Cells

DAVID M. BERSON

Best known for their pivotal role in photic modulation of circadian rhythms, intrinsically photosensitive retinal ganglion cells (ipRGCs) are linked also to other reflexive, subconscious responses to bright environmental illumination, responses mediated by "non-image-forming" (NIF) neural circuits. Recent advances have enriched this picture by showing that ipRGCs blend excitatory influences from rods and cones with their own melanopsin-based photosensitivity to expand their dynamic range in the intensity and temporal domains. They serve as the primary or exclusive source of retinal influence on the circadian and pupillary systems, among other NIF functions. Although ipRGCs were originally considered to comprise a single rare and relatively homogeneous type, new findings have revealed at least five types of ipRGCs, each with distinct physiology and projections (Baver et al., 2008; Dacey et al., 2005; Ecker et al., 2010; Hattar et al., 2006; Schmidt et al., 2008, 2009, 2010, 2011; Tu et al., 2005). Together, ipRGCs are now known to contribute to a dizzying array of functional roles, including avoidance of bright light, regulation of mood, visual-system development, modulation of intraretinal signaling, and even crude pattern vision.

DISCOVERY AND EARLY CHARACTERIZATION

The early clues to possible inner-retinal photoreception and the discovery of ipRGCs have been extensively reviewed elsewhere (Do & Yau, 2010; Wong & Berson, 2011). To briefly summarize, there had been growing evidence for the surprising persistence of circadian entrainment and pupillary and melatonin responses to light in animals (including humans) with severe degeneration of rods and cones. Foster and colleagues elevated the standard of evidence to build a compelling case for the presence of non-rod, non-cone ocular photoreceptors (Freedman et al., 1999; Lucas et al., 1999). Provencio discovered the novel opsin melanopsin, localized it to a small minority of RGCs, and surmised that these could be the mysterious photoreceptors implied by Foster's work (Provencio et al., 1998, 2000). Several groups provided evidence that melanopsin was expressed in ganglion cells innervating the suprachiasmatic nucleus (SCN) of the hypothalamus (Gooley et al., 2001; Hannibal et al., 2002a; Hattar et al., 2002), providing a crucial link between these possible novel photoreceptors and the circadian system. Berson, Dunn, and Takao (2002) provided direct electrophysiological evidence that RGCs innervating the SCN were intrinsically photosensitive, Hattar et al. (2002) demonstrated that these cells expressed melanopsin, and Lucas et al. (2003) confirmed that deletion of melanopsin silenced ipRGC phototransduction. Such deletion also disrupted normal photic influence on circadian rhythms and pupil response (Lucas et al., 2003; Panda et al., 2002; Ruby et al., 2002), and melanopsin deletion in mice lacking functional rods and cones effectively eliminated all visually driven behaviors (Hattar et al., 2003; Panda et al., 2003).

UBIQUITY AMONG VERTEBRATES

In addition to mice, a wide variety of mammalian species have been shown to express melanopsin, or exhibit intrinsic photosensitivity, in a minority of RGCs. These include rat (Gooley et al., 2001; Hannibal et al., 2002a; Hattar et al., 2002), primates (Dacey et al., 2005; Hannibal et al., 2004; Jusuf et al., 2007; Neumann, Haverkamp, & Auferkorte, 2011), hamster (Morin, Blanchard, & Provencio, 2003), rabbit (Hoshi et al., 2009), cat (Semo et al., 2005; see also Pu, 1999), blind mole rat (Hannibal et al., 2002b), and dunnart (a marsupial; Pires et al., 2007). Several nonmammalian vertebrates have been shown to possess such RGCs, including chicken (Bailey & Cassone, 2005; Chaurasia et al., 2005; Contin, Verra, & Guido, 2006), salamander (Rajaraman, 2012), and zebrafish (Davies et al., 2011), although melanopsin is also expressed in several non-RGC retinal cell types in nonmammalian vertebrates.

DIVERSITY

There is both anatomical and physiological evidence for diversity among mammalian ipRGCs. This has been

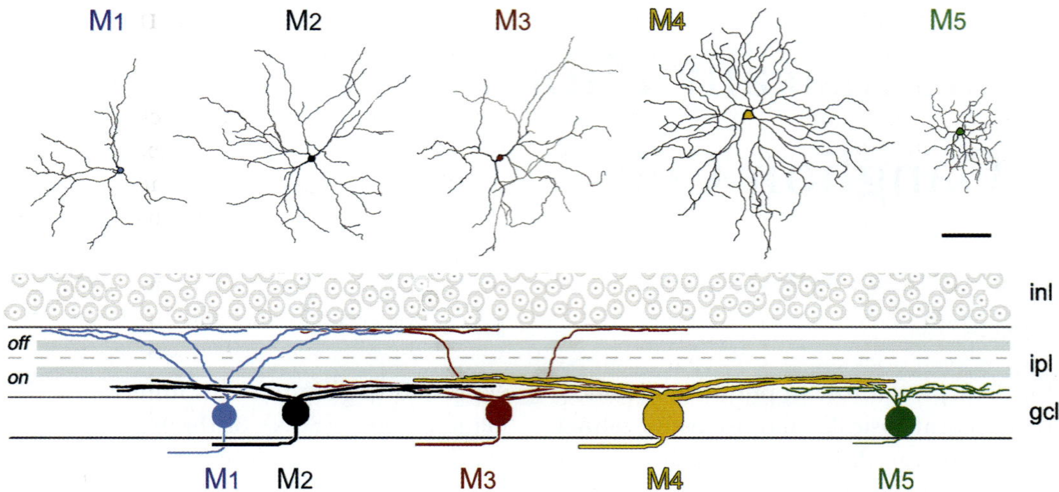

FIGURE 14.1   Morphology of five types of intrinsically photosensitive retinal ganglion cells (ipRGCs). (Top) En face view (scale: 100 μm). (Bottom) Dendritic stratification as viewed in schematic radial section. Pale blue bands are the ON and OFF cholinergic bands. There are two bands of melanopsin dendrites, both outside the cholinergic bands. One lies at the margin of the inner nuclear layer (INL), and the second, broader band sits close to the ganglion cell layer. The outer band contains processes of M1 and M3 cells, the inner one, the processes of M2, M3, M4, and M5 cells. There are subtle differences in stratification among the inner stratifying population.

most fully studied in mice, to which the following nomenclature and synopsis refers (figure 14.1).

Anatomical Diversity

M1 CELLS   These ipRGCs were the first to be identified (Berson, Dunn, & Takao, 2002; Hattar et al., 2002; Provencio, Rollag, & Castrucci, 2002). They have the highest levels of melanopsin expression among ipRGCs. Their sparse, tortuous dendrites stratify in the outermost level of the inner plexiform layer (IPL), abutting the inner nuclear layer (INL) and within the OFF sublamina. Some have cell bodies displaced to the INL. A further subdivision of this type has been proposed based on expression of the Brn3b transcription factor and patterns of central projection (Chen, Badea, & Hattar, 2011; Jain, Ravindran, & Dhingra, 2012). Dendritic overlap is substantial among M1 cells (Berson, Castrucci, & Provencio, 2010), so the type could be subdivided and still permit full retinal tiling by each subtype's dendritic fields. In mice, M1 cells comprise approximately 2% of all RGCs (~900 M1 out of 45,000 total RGCs) (Berson, Castrucci, & Provencio, 2010; Jeon, Strettoi, & Masland, 1998). Probable homologues of murine M1 cells have been described in all mammals in which melanopsin-expressing ganglion cells have been observed, including human and nonhuman primates (Dacey et al., 2005; Jusuf et al., 2007).

M2 CELLS   These cells stratify in the inner third of the IPL, within the ON sublayer. The first hint of their existence was the second, inner band of melanopsin-like immunoreactivity in the mouse retina (Provencio, Rollag, & Castrucci, 2002). Immunohistochemistry and recording in primate retina soon revealed a pair of ipRGC types in primates, one closely resembling the M1 cell in rodents and a second with dendrites restricted to the ON sublayer of the IPL (Dacey et al., 2005). Similar cells, termed M2 or type 2 cells, were soon identified in rodents as well (Schmidt & Kofuji, 2009; Viney et al., 2007). In mice these cells are intrinisically photosensitive, but their photocurrents are smaller, and sensitivity is lower, than those of M1 cells (Ecker et al., 2010; Schmidt & Kofuji, 2009; Viney et al., 2007). This is presumably because M2 cells express lower levels of melanopsin, judging both from immunohistochemical staining (Baver, Pickard, & Sollars, 2008; Berson, Castrucci, & Provencio, 2010) and from the intensity of GFP fluorescence driven by the melanopsin promoter (Schmidt & Kofuji, 2009). M2 cells make up ~2% of all mouse RGCs (~800 cells) (Berson, Castrucci, & Provencio, 2010). Cells possibly homologous to murine M2 cells have also been described in both primates (Dacey et al., 2005; Jusuf et al., 2007) and rabbits (Hoshi et al., 2009).

M3 CELLS   These cells are bistratified, with some dendritic branches costratifying with those of M1 cells in the outer IPL and others costratifying with M2-cell dendrites in the inner IPL (Berson, Castrucci, & Provencio, 2010; Schmidt & Kofuji, 2009, 2011; Viney et al., 2007; Warren et al., 2003). M3 cells appear to be relatively rare (~0.3% of all RGCs) (Berson, Castrucci, & Provencio, 2010); there may be too few to tile the retina (Berson, Castrucci, & Provencio, 2010). They closely resemble M2 cells in the sensitivity of their intrinsic light responses and in their level of melanopsin expression (Schmidt & Kofuji, 2011). M3-like cells have been described in rabbits (Hoshi et al., 2009) and possibly in cats (Pu, 1999).

M4 CELLS   M4 cells have very weak melanopsin immunoreactivity (Berson, Castrucci, & Provencio, 2010; Ecker et al., 2010; Estevez et al., 2012) but are marked by a sensitive cre-lox–based melanopsin reporter and exhibit intrinsic light responses of low amplitude and sensitivity (Ecker et al., 2010; Estevez et al., 2012). They have very large cell bodies and huge dendritic arbors stratifying near, but probably just sclerad to, the level of M2 dendrites (Estevez et al., 2012). They closely resemble, and are probably equivalent to, ON-alpha or RGA1 cells of mouse retina. Their abundance has not been established but is probably <2% of all RGCs (Berson, Castrucci, & Provencio, 2010).

M5 CELLS   Like M4 cells, M5 cells have very low melanopsin expression levels and weak intrinsic photoresponses and were detected only with the same sensitive cre-lox reporter system (Ecker et al., 2010). They have the smallest dendritic profiles of all ipRGC types. Their highly branched arbors stratify in the inner (ON) IPL. Their incidence among all RGCs is unknown, although their smaller dendritic fields compared to those of other ipRGC types could imply somewhat greater abundance.

Physiological Diversity

The existence of multiple functional subtypes of ipRGCs was first suggested based on extracellular recordings of adult mouse retina; two types were identified, differing in kinetics and sensitivity of their intrinsic response (Tu et al., 2005). Patch recordings paired with dye injection have now shown that M2, M3, M4, and M5 cells all have smaller, less sensitive photocurrents than M1 cells, but these are nonetheless capable of driving sodium spikes (Ecker et al., 2010; Estevez et al., 2012; Schmidt & Kofuji, 2010, 2011); thus, one or more of these types presumably correspond to the less sensitive type recorded by Tu et al. (2005). The strength of synaptic input is complementary to that of the direct light response, being relatively weak in M1 cells and substantially stronger in M2, M3, and M4 cells (Estevez et al., 2012; Schmidt & Kofuji, 2010, 2011).

PHYSIOLOGICAL PROPERTIES OF THE DIRECT PHOTORESPONSE

The basic features of the direct light response are consistent among ipRGC types except for sensitivity and amplitude. Light triggers a nonspecific cationic conductance that, under physiological conditions, carries an inward current of up to several hundred pA (in M1 cells). This evokes a membrane depolarization of up to 20 mV or more, which, in turn, can trigger tetrodotoxin-sensitive sodium spikes (peak current-evoked spike frequencies ~80 Hz for M1 and ~240 Hz for M2 cells) (Schmidt & Kofuji, 2009).

Kinetics

The onset latency of the intrinsic photoresponse is generally much longer than in rod and cone photoreceptors. However, it is steeply and inversely related to stimulus intensity, ranging from several hundred milliseconds for saturating stimuli to as long as 20 minutes near threshold (Berson, Dunn, & Takao, 2002; Wong, 2012). The slowness of the onset kinetics apparently stems not from sluggishness in the single-photon response but rather from the gradual summation single-photon events with low amplitude and frequency but unusually long integration time (Do et al., 2009). Poststimulus decay of the response is also remarkably slow, lasting up to several minutes. These properties may function as a low-pass temporal filter, smoothing transients in illumination and offering the brain a steadier representation of average environmental illumination.

Sensitivity, Irradiance Encoding, Adaptation, and Circadian Modulation

The threshold of the intrinsic response is high, probably roughly 6 log units above rod threshold and perhaps a log unit above cone threshold (Berson, Dunn, & Takao, 2002; Dacey et al., 2005; Do et al., 2009; Schmidt & Kofuji, 2009; Schmidt, Taniguchi, & Kofuji, 2008; Wong, 2012). The intrinsic photoresponse is subject to light adaptation (Wong, Dunn, & Berson, 2005); the mechanisms responsible have been little studied, although light-dependent phosphorylation or reductions in abundance of melanopsin may contribute (Blasic, Lane Brown, & Robinson, 2011; Hannibal et al., 2005). In any case, such adaptation is incomplete, and melanopsin can drive a remarkably sustained photocurrent and elevated spiking lasting at least ten hours (Berson, Dunn, & Takao, 2002; Mrosovsky & Hattar, 2003; Wong, 2012). The magnitude of maintained depolarization is monotonically related to stimulus intensity (Berson, Dunn, & Takao, 2002; Dacey et al., 2005; Wong, 2012), a property not exhibited by conventional ganglion cells driven only by rod and cone input. Melanopsin expression and the sensitivity of intrinsic photoresponses exhibit some circadian modulation (Gonzalez-Menendez et al., 2009; Hannibal, 2006; Hannibal et al., 2005; Owens et al., 2012; Sakamoto et al., 2005; Weng, Wong, & Berson, 2009; Zele et al., 2011). Dopaminergic amacrine cells may be one source of this modulatory influence (Van Hook, Wong, & Berson, 2012; Viney et al., 2007; Vugler et al., 2007).

Spectral Sensitivity

The intrinsic light response exhibits its peak sensitivity near 480 nm as measured in a variety of species and probed electrophysiologically or behaviorally (Berson, Dunn, & Takao, 2002; Dacey et al., 2005; Gamlin et al., 2007; Lucas, Douglas, & Foster, 2001; Yoshimura & Ebihara, 1996). The action spectrum (figure 14.2) corresponds closely to the form expected for photopigment consisting of a retinaldehyde chromophore bound to an opsin, presumably melanopsin. This inference is supported by the spectral absorption of melanopsin measured spectroscopically (Koyanagi et al., 2005; Walker et al., 2008) or in expression systems (Panda et al., 2005; Qiu et al., 2005).

PHOTOTRANSDUCTION MECHANISM

Key components of the phototransduction cascade in ipRGCs have now been identified, although many details remain sketchy, especially concerning channel gating, response termination, retinoid cycle, and adaptation. See other reviews on this topic by Berson (2007) and Do and Yau (2010).

Melanopsin Photopigment

There is compelling evidence that melanopsin is the sensory photopigment of ipRGCs. When expressed in any of various light-insensitive cell types, melanopsin induces photosensitivity, typically with spectral sensitivity closely matching that of ipRGCs (Lin et al., 2008; Melyan et al., 2005; Newman et al., 2003; Panda et al., 2005; Qiu et al., 2005). Melanopsin is expressed not only in the cell body of ipRGCs but throughout the dendrites as well, at least in M1–M3 cells (Berson, Castrucci, & Provencio, 2010; Provencio, Rollag, & Castrucci, 2002), and can thus account for dendritic photosensitivity (Berson, Dunn, & Takao, 2002). In melanopsin-null mice, the direct photoresponse of ipRGCs is abolished (Lucas et al., 2003). In mice, there are two splice isoforms of melanopsin, both of which form photosensory pigments. Both are present in M1 cells, but only the long form is present in M2 cells (Pires et al., 2009).

Melanopsin is localized almost entirely to the plasma membrane of ipRGCs (Belenky et al., 2003). These cells lack obvious microvilli or intracellular membranous organelles that might increase the surface area of melanopsin-laden membrane. Overall, the density of the photopigment in ipRGCs appears very low relative to rods and cones, and this is probably the main explanation for the low sensitivity of ipRGC phototransduction (Do et al., 2009), as assessed by current or voltage recording.

Under some conditions melanopsin apparently forms a bistable photopigment, retaining its chromophore after light isomerizes it from 11-*cis* to all-*trans* retinaldehyde (Koyanagi et al., 2005; Mure et al., 2007, 2009; Rollag, 2008; Walker et al., 2008). In this respect it resembles opsins in invertebrate photoreceptors (Hillman, Hochstein, & Minke, 1983) rather than those

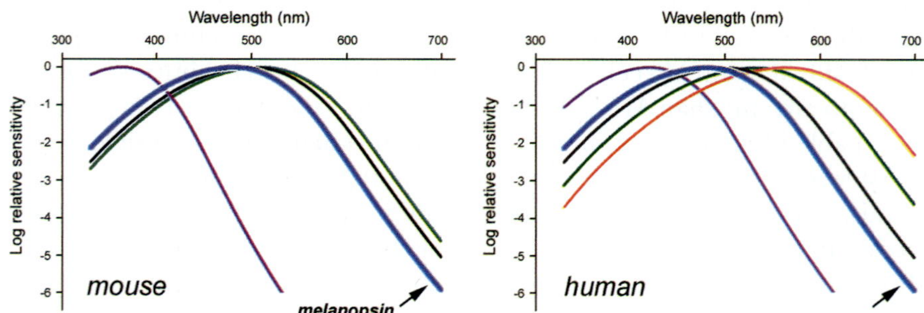

FIGURE 14.2   Spectral tuning of melanopsin (bold blue curves indicated by arrow), the photopigment of ipRGCs, in relation to those of other opsin photopigments in mouse (left) and human (right) retina. Rhodopsin, the rod photopigment, is shown in black. Remaining curves indicate the cone opsins, of which there are two in rodents and three in humans.

in rods and cones. In bistable pigments, the activated state is presumed to be thermally stable and to be restored to its ground state by absorption of a second photon. In melanopsin the optimal wavelength for this photoreversal reaction (~530–600 nm; green to yellow) (Koyanagi et al., 2005; Mure et al., 2007, 2009; Walker et al., 2008) is apparently longer than that for photoactivation (~480 nm; blue). Melanopsin bistability could help to explain both the poststimulus persistence of ipRGC photocurrents (perhaps maintained by a long-lived pool of activated pigment) and the ability of ipRGCs to regenerate their photopigment despite their physical separation from the retinal pigment epithelium and even in the face of retinoid depletion (Sexton, Buhr, & Van Gelder, 2012). However, in its native cellular context, melanopsin exhibits certain behaviors not expected of a classical bistable photopigment, and the physiological significance of melanopsin bistability remains controversial (Enezi et al., 2011; Mawad & Van Gelder, 2008; Rollag, 2008; Walker et al., 2008).

Intermediate Signaling Components

In many native and expression systems as well as biochemically, melanopsin has been shown to signal through a phosphoinositide cascade, involving G-proteins of the G_q family and activation of the effector enzyme phospholipase C (PLC) (e.g., Angueyra et al., 2012; Gomez, Angueyra, & Nasi, 2009; Isoldi et al., 2005; Koyanagi et al., 2005; Panda et al., 2005; Qiu et al., 2005; Terakita et al., 2008; Xue et al., 2011). This is now well established for ipRGCs specifically. Pharmacological blockade of G_q-type G-proteins or of the effector enzyme PLC abolishes the light response (Contin et al., 2010; Graham et al., 2008), whereas inhibitors of the cyclic nucleotide cascade that mediates rod and cone phototransduction are without effect (Warren et al., 2006). The identity of the G-alpha subunit responsible has not been established, although transcripts of all four members of the G_q family (G_q, G_{11}, G_{14}, and G_{15}) have been detected in at least some ipRGCs (Graham et al., 2008). The essential PLC isozyme seems to have been identified. G_q-family G-proteins generally signal through PLCβ (Hubbard & Hepler, 2006), of which there are four variants. All have been shown to be expressed in ipRGCs by single-cell RT-PCR, but PLCβ4 transcripts are the only ones invariably present, and PLCβ4 protein has been detected in ipRGCs immuno-histochemically (Graham et al., 2008). Most tellingly, intrinsic photoresponses of ipRGCs are abolished in *Plcb4*$^{-/-}$ mice (Xue et al., 2011).

The light-gated channel in ipRGCs appears to belong to the transient receptor potential channel (TRPC) family. The photocurrent is carried by a nonspecific cationic conductance reversing near 0 mV and exhibiting both inward and outward rectification (Warren et al., 2003), which is typical of TRPC channels. Functional studies in expression systems indicate that activation of melanopsin can trigger current through TRPC channels (Panda et al., 2005; Qiu et al., 2005). Various pharmacological agents known to block TRPC channels suppress the photocurrent (Hartwick et al., 2007; Sekaran et al., 2007; Warren et al., 2003). Both TRPC6 and TRPC7 have been detected in ipRGCs (Hartwick et al., 2007; Warren et al., 2006). Photocurrents persist in ipRGCs in knockout mice lacking single TRPC-family genes (Perez-Leighton et al., 2011; Xue et al., 2011), although maximal current may be reduced in *Trpc6*$^{-/-}$ mice. The most compelling evidence on this point, however, is that ipRGC photocurrents are abolished in TRPC6/TRPC7 double knockouts (Xue et al., 2011). These findings suggest that TRPC6 and TRPC7 subunits form light-gated channels in ipRGCs, perhaps together as heteromeric TRPC6/7 channels, or as a mixture of homomeric channels containing only TRPC6 or only TRPC7 subunits. One remaining mystery is why application of the diacylglycerol (DAG) analogue 1-oleoyl-2-acetyl-*sn*-glycerol (OAG), which invariably opens TRPC6 and TRPC7 channels in expression systems, fails to elicit a transmembrane current in ipRGCs (Graham et al., 2008; Warren et al., 2006).

As is true for TRPC channels generally, the mechanism by which activation of PLCβ4 ultimately opens the light-gated channels in ipRGCs remains obscure. Cleavage by PLC of its substrate, phosphatidylinositol-4,5-bisphosphate (PIP_2), generates DAG, which remains in the membrane, and IP_3, which enters the cytosol where it acts to mobilize Ca^{2+} from intracellular stores. Pharmacological data indicate that neither IP_3, nor IP_3 receptors, nor mobilization of intracellular Ca^{2+} are essential for phototransduction. This inference is supported by the persistence of photocurrents in inside-out patches of ipRGC plasma membrane (Graham et al., 2008). Thus, it appears that all essential signaling components lie within, or are tightly associated with, the plasma membrane. DAG itself is unlikely to gate the channel, because application of its analogue, OAG, fails to induce a current. Metabolites of DAG such as polyunsaturated fatty acids or monoacylglycerol may be involved, or the channel might be released into an open state by the reduction of membrane PIP_2.

The phototransduction cascade in ipRGCs bears many similarities to that in the microvillar or rhabdomeric photoreceptors of invertebrates, especially those of *Drosophila*, encouraging the view that ipRGCs share an evolutionary lineage with these photoreceptors (Fain, Hardie, & Laughlin, 2010; Yau & Hardie, 2009).

Other Components

Many other important facets of, and potential players in, the phototransduction process in ipRGCs have received only preliminary investigation. An area of particular ignorance concerns response termination. The slow shutoff of the photocurrent in ipRGCs is probably attributable in part to the thermal stability of activated melanopsin, as noted above, but there are likely to be unusual features in the interaction between this pigment and other regulatory macromolecules, including G-protein receptor kinases (GRKs) and arrestins. There is evidence for light-dependent phosphorylation of melanopsin, perhaps by GRK2, and consequent regulation of the lifetime of the photoactivated pigment (Blasic, Lane Brown, & Robinson, 2011). Such phosphorylation is probably a prelude to the binding of an arrestin. However, although interactions between several arrestins and melanopsin have been detected in an expression system (Panda et al., 2005), the identity of arrestins expressed in ipRGCs and their interactions with melanopsin in situ remain unknown. As there is no evidence for the presence of rod or cone arrestins in ipRGCs, the beta-arrestins are more likely candidates, and this would extend the parallels with fly photoreceptors, but this speculation awaits experimental testing. Nor is it known whether regulators of G-protein signaling (RGS) proteins play a role in regulating phototransduction kinetics in ipRGCs as they do in classical photoreceptors.

Light evokes increases in intracellular Ca^{2+} in ipRGCs (Sekaran et al., 2003), mainly by transmembrane flux through voltage-gated Ca^{2+} channels (Hartwick et al., 2007), although influx through the light-gated TRPC channels and IP_3-mediated Ca^{2+} mobilization from intracellular stores also must occur to some extent. The elevated Ca^{2+} levels are likely to contribute to adaptation and other modulatory effects on the phototransduction process, but these have not been explored. A variety of approaches have implicated protein kinase C and protein kinase A as playing modulatory, or even central, roles in the phototransduction process (Peirson et al., 2007; Van Hook, Wong, & Berson, 2012; Warren et al., 2006), but key details are lacking. The retinoid cycle in ipRGCs is also very poorly understood. Under physiological conditions in situ, does light cause true bleaching of melanopsin in ipRGCs, stripping the opsin of its chromophore? If so, how is the retinoid retained, processed, and reisomerized to regenerate the functional pigment? Does this all occur within the ipRGC, or are other cell types involved (e.g., Müller

cells)? If bleaching does not occur, what is the fate of photoactivated melanopsin? In the absence of light-driven photoreversal, does it revert to the ground state through thermal decay or interaction with other proteins, or does it persist indefinitely? Is it internalized, as can occur with invertebrate opsins and other activated G-protein-coupled receptors? Such fundamental questions are crying for concerted experimental attention.

SYNAPTIC INFLUENCES ON IPRGCS

Although ipRGCs can respond to light in the absence of any synaptic input, they are nonetheless strongly influenced by such input under physiological conditions (Dacey et al., 2005; Perez-Leon et al., 2006; Pickard et al., 2009; Wong et al., 2007). Their dendrites receive direct synaptic contacts within the inner plexiform layer from both bipolar and amacrine cells (Belenky et al., 2003; Dumitrescu et al., 2009; Hoshi et al., 2009; Jusuf et al., 2007; Neumann, Haverkamp, & Auferkorte, 2011). The excitatory inputs derive almost exclusively from ON bipolar cells, although weak input from OFF bipolar cells can be detected under special conditions (Wong et al., 2007). When activated by steps of light, the ON bipolar inputs are unusually sustained (Wong et al., 2007; Wong, 2012) with amplitudes monotonically related to stimulus irradiance (Dacey et al., 2005; Wong, 2012). Both of these features parallel the sustained, irradiance-encoding nature of the intrinsic light response. However, the influences derived from rods and cones are both more sensitive and have much faster kinetics than the intrinsic, melanopsin-mediated response (Dacey et al., 2005; Wong et al., 2007; Wong, 2012). The combined influences of classical photoreceptors and melanopsin evidently allow ipRGCs to encode irradiance over a much wider dynamic range of intensity and temporal frequency than would be possible on the basis of either influence alone.

For M2–M5 cells, the predominance of ON channel input is expected from their dendritic stratification, partly (M3) or entirely (M2, M4, M5) within the ON sublayer of the IPL. By contrast, the ON input to M1 cells was initially baffling because these cells have monostratified dendritic arbors in the OFF sublayer. It is now understood that the distal dendrites of M1 cells receive synaptic input from the axons of specific subtypes of ON bipolar cells as they pass through the OFF sublayer (Dumitrescu et al., 2009; Grunert et al., 2011; Hoshi et al., 2009). Based on light microscopic evidence, direct contacts between rod bipolar cells and ipRGC dendrites and somata have been reported in rat (Østergaard, Hannibal, & Fahrenkrug, 2007) but ruled out in primate

(Grunert et al., 2011). An unusual form of chromatic opponency has been reported in the synaptic drive to ipRGCs in primates (blue OFF, yellow ON) (Dacey et al., 2005), but murine ipRGCs appear to be spectrally broadband (Schmidt & Kofuji, 2010).

Ionotropic GABAergic and glycinergic inputs from amacrine cells strongly modulate ipRGC excitability (Belenky et al., 2003; Neumann, Haverkamp, & Auferkorte, 2011; Perez-Leon et al., 2006; Wong et al., 2007). Some of the GABAergic input may derive from dopaminergic amacrine cells, which coexpress GABA and have dendrites that costratify and make apparent contact with M1 ipRGCs (Dumitrescu et al., 2009; Viney et al., 2007; Vugler et al., 2007). Dopamine released from these amacrine cells apparently acts at D1 receptors to attenuate ipRGC phototransduction (Van Hook, Wong, & Berson, 2012). Gap junctions also couple ipRGCs to widefield amacrine cells (Müller et al., 2010), providing a basis for bidirectional electrical communication with these cells.

PROJECTIONS AND FUNCTIONS

Like other retinal ganglion cells, ipRGCs are glutamatergic. For example, the direct retinal influence on the suprachiasmatic nucleus (SCN) of the hypothalamus, which derives exclusively from ipRGCs (Baver, Pickard, & Sollars, 2008; Güler et al., 2008), is mediated by glutamatergic transmission (e.g., Colwell & Menaker, 1992; Hannibal, 2002). Furthermore, ipRGCs express the vesicular glutamate transporter VGlut2 (Engelund et al., 2010; Johnson et al., 2007). However, ipRGCs also synthesize pituitary adenylate cyclase-activating polypeptide (PACAP) and likely release it in at least one of their central targets (Hannibal, 2002).

A large number of brainstem and diencephalic nuclei receive direct input from ipRGCs. Early studies suggested that these were almost exclusively "non-image-forming" centers (Gooley et al., 2003; Hannibal & Fahrenkrug, 2004a; Hattar et al., 2002, 2006; but see Dacey et al., 2005). However, more sensitive reporters now reveal that key centers for the perception and localization of objects are also involved (Brown et al., 2010; Ecker et al., 2010). For many non-image-forming visual centers and functions, ipRGCs seem to constitute the dominant, if not exclusive, source of retinal influence. Because ipRGCs blend the convergent influences of rods, cones, and intrinsic melanopsin phototransduction, photic influence persists with either genetic disruption of rod and cone function or deletion of melanopsin, but is dramatically impaired or abolished by targeted ablation of ipRGCs (Göz et al., 2008; Güler et al., 2008; Hatori et al., 2008).

Suprachiasmatic Nucleus and Modulation of Circadian Rhythms and Melatonin Synthesis

The SCN was one of the first output targets of ipRGCs to be established, and that projection figured prominently in their initial discovery (Berson, Dunn, & Takao, 2002; Gooley et al., 2001, 2003; Hannibal et al., 2002a; Hattar et al., 2002; Panda et al., 2002; Ruby et al., 2002). Retinal input to the SCN is believed to come almost entirely from ipRGCs, especially from M1 cells (Brn3b-negative subtype), with a smaller contribution from M2 cells (Baver, Pickard, & Sollars, 2008; Chen, Badea, & Hattar, 2011). These inputs carry signals derived from rods, cones, and melanopsin (Altimus et al., 2010; Drouyer et al., 2007; Lall et al., 2010). The role of this pathway in circadian entrainment to the solar cycle is covered in detail elsewhere in this volume (chapter 20 by Meijer, Hattar, and Takahashi). ipRGCs also appear to play a crucial role in the acute photic suppression of melatonin release by the pineal gland (Brainard et al., 2001; Czeisler et al., 1995; Lucas et al., 1999; Zaidi et al., 2007), ostensibly through a polysynaptic circuit passing through the SCN, paraventricular nucleus of the hypothalamus and sympathetic nervous system. The intergeniculate leaflet (IGL) of the lateral geniculate complex of the thalamus is another major target of ipRGC input with functional links to the circadian system. Adjacent to this lies another recipient of ipRGC signals, the ventral division of the lateral geniculate nucleus (vLGN), a poorly understood visual center that lacks projections to the cortex. These ipRGC inputs to IGL and vLGN certainly originate in part from M1 cells (Hattar et al., 2002, 2006) but may also come from M2 or other ipRGC types (Ecker et al., 2010; Osterhout et al., 2011).

Olivary Pretectal Nucleus and Pupillary Control

The ipRGCs are the dominant source of the irradiance signal that drives a brainstem circuit mediating the pupillary light reflex. The reflex survives massive loss of rod and cone photoreceptors, and the residual responses precisely parallel melanopsin-mediated ipRGC photoresponses in their spectral behavior (Lucas, Douglas, & Foster, 2001). The olivary pretectal nucleus (OPN), an essential brainstem relay in the pupillary control circuit, receives direct innervation from ipRGCs (Baver, Pickard, & Sollars, 2008; Hannibal & Fahrenkrug, 2004a; Hattar et al., 2002; Viney et al., 2007). The "shell" region of the nucleus may be the critical pupillary relay (Baver, Pickard, & Sollars, 2008), and this is heavily innervated by M1 ipRGCs (Brn3b-positive subtype) (Chen, Badea, & Hattar,

2011; Hattar et al., 2002, 2006), although some input from other types of ipRGCs is also possible (Viney et al., 2007). The core of the OPN, whose function remains obscure, also gets ipRGC afferents, but not from M1 cells (Ecker et al., 2010; Hattar et al., 2006). Different spectral and temporal components of the pupillary response can be traced to melanopsin, rods, and cones (Allen, Brown, & Lucas, 2011; Gamlin et al., 2007; Kardon et al., 2009; McDougal & Gamlin, 2010). The influences are channeled mainly through ipRGCs (Chen, Badea, & Hattar, 2011; Güler et al., 2008; Hatori et al., 2008), although it remains uncertain whether ipRGC subtypes beyond M1 contribute and whether conventional RGCs play any role. Intrinsic photosensitivity of the mammalian iris muscle, long a subject of dispute, has now been definitively demonstrated in many mammals (Xue et al., 2011). This is mediated by iridial melanopsin expression and operates at relatively high irradiances.

Sleep and Negative Masking

In mice and other nocturnal rodents light exposure during the active (dark) phase of the light cycle acutely suppresses locomotor activity. Termed "negative masking" in the circadian field, this suppression of activity appears attributable in large part to the photic induction or promotion of sleep. The soporific effect of light can be mediated by classical opsins or melanopsin, but both influences appear to be routed through ipRGCs (Altimus et al., 2008; Lupi et al., 2008; Mrosovsky & Hattar, 2003; Mrosovsky, Lucas, & Foster, 2001). Direct projections from M1 cells to the ventrolateral preoptic area have been proposed as a key link between ipRGCs and sleep control circuits (Gooley et al., 2003; Hattar et al., 2006), but the projection is relatively weak, and other ipRGC types and hypothalamic circuits may be involved.

Light Avoidance and Photic Pain

Even before eye opening, neonatal mice avoid light, and this negative phototaxis is melanopsin dependent (Johnson et al., 2010). Melanopsin also mediates a preference for dark environments in adult mice (Semo et al., 2010). The circuit responsible is unknown but may be related to a projection from ipRGCs to the somatosensory thalamus that has been proposed as a possible substrate for the photophobia of migraine, which persists in cases of blindness due to outer retinal degeneration (Noseda et al., 2010). There are also projections from M1 ipRGCs to the amygdala and other limbic targets (Hattar et al., 2006) that may trigger fear or aversion to light. For further review see Schmidt et al. (2011) and Do and Yau (2010).

Lateral Geniculate and Conscious Vision

One of the most striking features of ipRGC projections in early studies was the virtual absence of projections to two key centers for spatial vision. The first of these, the dorsal lateral geniculate nucleus (dLGN), relays retinal signals to primary visual cortex; the other, the superior colliculus (SC), is a precisely retinotopic visual gaze center. The apparent lack of ipRGC projections to these centers was congruent with functional specializations of ipRGCs sacrificing spatiotemporal precision in favor of pronounced integration over space and time. Such specializations include large dendritic and receptive fields, sluggish kinetics, small target nuclei with little if any retinotopy, and mediation of visual functions requiring a simple spatiotemporally integrated irradiance signal, like that afforded by a light meter in an automatic camera. This picture began to change as at least a few axons of M1 cells were found to innervate dLGN and SC (Dacey et al., 2005; Hattar et al., 2006), but especially with the emergence of additional ipRGC types (see above). Cre-based melanopsin reporters uncovered two new cell types (M4 and M5) and labeled the axons of all five ipRGC types, revealing robust input to novel targets, including a subregion within the dLGN and the deepest purely visual layer in the SC, the stratum opticum (Brown et al., 2010; Ecker et al., 2010). These studies further revealed that melanopsin phototransduction alone could drive light responses in the dLGN and visual cortex (Brown et al., 2010) and could support crude pattern discrimination (Ecker et al., 2010). Input from ipRGCs to the dLGN may be responsible for the conscious appreciation of light in human patients with virtually total loss of conventional photoreceptors (Zaidi et al., 2007).

Of the two new ipRGC types revealed by these reporters, one—the M4 cell—resembles the ON alpha cell in other mammals, a type known to innervate the dLGN (Estevez et al., 2012). The other type, M5, has a small bushy dendritic arbor suggesting relatively high spatial precision in keeping with a possible role in pattern vision. Retinogeniculate input from ipRGCs, which is now known to derive at least in part from M4 cells (Estevez et al., 2012), terminates selectively in the ventromedial third of the nucleus. Interestingly, the same sector is also innervated by an OFF alpha cell (Huberman et al., 2008) and by the rare M1 ipRGC axons that reach the dLGN (Hattar et al., 2006). Functional subdivisions of the murine dLGN are not well understood, but geniculocortical anatomy in better studied mammals implies that signals from ipRGCs and those from conventional RGCs are probably routed to at least partly different cortical layers and/or cells.

Intraretinal Influences of ipRGCs

Convergent evidence indicates that ipRGCs are unusual among ganglion cells in their ability to influence the processing of visual signals by other retinal neurons. Such influences include modulation of the electroretinogram (Barnard et al., 2006; Hankins & Lucas, 2002), excitation of dopaminergic amacrine cells (Zhang et al., 2008), and alteration of the spontaneous excitatory events sweeping across the early postnatal retina known as "retinal waves" (Renna, Weng, & Berson, 2011). Through their actions on dopaminergic amacrine cells, ipRGCs are in a position to participate in mechanisms of light adaptation and circadian modulation that affect all retinal neurons. Their effect on retinal waves may indicate an influence over cholinergic "starburst" amacrine cells, which are the pattern generators for these events. The anatomical substrate for the centrifugal influence of ipRGCs on lower-order retinal neurons is unknown but could involve gap junctional coupling (Müller et al., 2010; Sekaran et al., 2005), dendrodendritic outputs, or axon collaterals linking ipRGCs to amacrine cells and other ganglion cells. For further review, see Wong and Berson (2011).

DEVELOPMENT

In the developing retina ipRGCs are the first fully functional photoreceptors. In mice ipRGC neurogenesis begins as early as embryonic day 11, and melanopsin expression begins almost immediately thereafter (Gonzalez-Menendez et al., 2010; McNeill et al., 2011; Tarttelin et al., 2003; but see Fahrenkrug, Nielsen, & Hannibal, 2004). Functional responses in ipRGCs are detectable at birth (Hannibal & Fahrenkrug, 2004b; Schmidt, Taniguchi, & Kofuji, 2008; Sekaran et al., 2005; Tu et al., 2005), and within 6 days, melanopsin phototransduction drives negative phototaxis (Johnson et al., 2010) and modulation of the retinal waves (Renna, Weng, & Berson, 2011). The waves are thought to play a key role in activity-dependent refinement of ocular segregation and retinotopy in retinofugal projections; light apparently modulates this process in a melanopsin-dependent manner (Renna, Weng, & Berson, 2011).

CONCLUSIONS

In the decade since their discovery ipRGCs have been implicated in a remarkable diversity of visual functions. The outlines of their phototransduction process and their projections to the brain have been sketched. Many key challenges remain, including a fuller accounting of the distinctive functional properties of ipRGC subtypes and their contributions to specific perceptual processes, reflexive visual behaviors, and developmental phenomena. Much remains to be learned about the retinoid cycle of ipRGCs and about their phototransduction cascade and its modulation by both cell-autonomous and synaptic mechanisms. We need a more comprehensive understanding of the influences of ipRGCs on the retina itself, including the synaptic circuits that support them. If what is past is prologue, there are many more surprises in store.

REFERENCES

Allen, A. E., Brown, T. M., & Lucas, R. J. (2011). A distinct contribution of short-wavelength-sensitive cones to light-evoked activity in the mouse pretectal olivary nucleus. *Journal of Neuroscience, 31*, 16833–16843.

Altimus, C. M., Güler, A. D., Alam, N. M., Arman, A. C., Prusky, G. T., Sampath, A. P., et al. (2010). Rod photoreceptors drive circadian photoentrainment across a wide range of light intensities. *Nature Neuroscience, 13*, 1107–1112.

Altimus, C. M., Güler, A. D., Villa, K. L., McNeill, D. S., Legates, T. A., & Hattar, S. (2008). Rods-cones and melanopsin detect light and dark to modulate sleep independent of image formation. *Proceedings of the National Academy of Sciences of the United States of America, 105*, 19998–20003.

Angueyra, J. M., Pulido, C., Malagon, G., Nasi, E., & Gomez, M. del P. (2012). Melanopsin-expressing amphioxus photoreceptors transduce light via a phospholipase C signaling cascade. *PLoS One, 7*, e29813. doi:10.1371/journal.pone.0029813.

Bailey, M. J., & Cassone, V. M. (2005). Melanopsin expression in the chick retina and pineal gland. *Brain Research. Molecular Brain Research, 134*, 345–348.

Barnard, A. R., Hattar, S., Hankins, M. W., & Lucas, R. J. (2006). Melanopsin regulates visual processing in the mouse retina. *Current Biology, 16*, 389–395.

Baver, S. B., Pickard, G. E., & Sollars, P. J. (2008). Two types of melanopsin retinal ganglion cell differentially innervate the hypothalamic suprachiasmatic nucleus and the olivary pretectal nucleus. *European Journal of Neuroscience, 27*, 1763–1770.

Belenky, M. A., Smeraski, C. A., Provencio, I., Sollars, P. J., & Pickard, G. E. (2003). Melanopsin retinal ganglion cells receive bipolar and amacrine cell synapses. *Journal of Comparative Neurology, 460*, 380–393.

Berson, D. M. (2007). Phototransduction in ganglion-cell photoreceptors. *Pflugers Archiv, 454*, 849–855.

Berson, D. M., Castrucci, A. M., & Provencio, I. (2010). Morphology and mosaics of melanopsin-expressing retinal ganglion cell types in mice. *Journal of Comparative Neurology, 518*, 2405–2422.

Berson, D. M., Dunn, F. A., & Takao, M. (2002). Phototransduction by retinal ganglion cells that set the circadian clock. *Science, 295*, 1070–1073.

Blasic, J. R., Jr., Lane Brown, R., & Robinson, P. R. (2011). Light-dependent phosphorylation of the carboxy tail of mouse melanopsin. *Cellular and Molecular Life Sciences, 69,* 1551–1562.

Brainard, G. C., Hanifin, J. P., Greeson, J. M., Byrne, B., Glickman, G., Gerner, E., et al. (2001). Action spectrum for melatonin regulation in humans: Evidence for a novel circadian photoreceptor. *Journal of Neuroscience, 21,* 6405–6412.

Brown, T. M., Gias, C., Hatori, M., Keding, S. R., Semo, M., Coffey, P. J., et al. (2010). Melanopsin contributions to irradiance coding in the thalamo-cortical visual system. *PLoS Biology, 8,* e1000558. doi:10.1371/journal.pbio.1000558.

Chaurasia, S. S., Rollag, M. D., Jiang, G., Hayes, W. P., Haque, R., Natesan, A., et al. (2005). Molecular cloning, localization and circadian expression of chicken melanopsin (Opn4): Differential regulation of expression in pineal and retinal cell types. *Journal of Neurochemistry, 92,* 158–170.

Chen, S. K., Badea, T. C., & Hattar, S. (2011). Photoentrainment and pupillary light reflex are mediated by distinct populations of ipRGCs. *Nature, 476,* 92–95.

Colwell, C. S., & Menaker, M. (1992). NMDA as well as non-NMDA receptor antagonists can prevent the phase-shifting effects of light on the circadian system of the golden hamster. *Journal of Biological Rhythms, 7,* 125–136.

Contin, M. A., Verra, D. M., & Guido, M. E. (2006). An invertebrate-like phototransduction cascade mediates light detection in the chicken retinal ganglion cells. *FASEB Journal, 20,* 2648–2650.

Contin, M. A., Verra, D. M., Salvador, G., Ilincheta, M., Giusto, N. M., & Guido, M. E. (2010). Light activation of the phosphoinositide cycle in intrinsically photosensitive chicken retinal ganglion cells. *Investigative Ophthalmology & Visual Science, 51,* 5491–5498.

Czeisler, C. A., Shanahan, T. L., Klerman, E. B., Martens, H., Brotman, D. J., Emens, J. S., et al. (1995). Suppression of melatonin secretion in some blind patients by exposure to bright light. *New England Journal of Medicine, 332,* 6–11. doi:10.1056/NEJM199501053320102.

Dacey, D. M., Liao, H. W., Peterson, B. B., Robinson, F. R., Smith, V. C., Pokorny, J., et al. (2005). Melanopsin-expressing ganglion cells in primate retina signal colour and irradiance and project to the LGN. *Nature, 433,* 749–754.

Davies, W. I., Zheng, L., Hughes, S., Tamai, T. K., Turton, M., Halford, S., et al. (2011). Functional diversity of melanopsins and their global expression in the teleost retina. *Cellular and Molecular Life Sciences, 68,* 4115–4132.

Do, M. T., Kang, S. H., Xue, T., Zhong, H., Liao, H. W., Bergles, D. E., et al. (2009). Photon capture and signalling by melanopsin retinal ganglion cells. *Nature, 457,* 281–287.

Do, M. T., & Yau, K. W. (2010). Intrinsically photosensitive retinal ganglion cells. *Physiological Reviews, 90,* 1547–1581.

Drouyer, E., Rieux, C., Hut, R. A., & Cooper, H. M. (2007). Responses of suprachiasmatic nucleus neurons to light and dark adaptation: Relative contributions of melanopsin and rod-cone inputs. *Journal of Neuroscience, 27,* 9623–9631.

Dumitrescu, O. N., Pucci, F. G., Wong, K. Y., & Berson, D. M. (2009). Ectopic retinal ON bipolar cell synapses in the OFF inner plexiform layer: Contacts with dopaminergic amacrine cells and melanopsin ganglion cells. *Journal of Comparative Neurology, 517,* 226–244.

Ecker, J. L., Dumitrescu, O. N., Wong, K. Y., Alam, N. M., Chen, S. K., LeGates, T., et al. (2010). Melanopsin-expressing retinal ganglion-cell photoreceptors: Cellular diversity and role in pattern vision. *Neuron, 67,* 49–60.

Enezi, J., Revell, V., Brown, T., Wynne, J., Schlangen, L., & Lucas, R. (2011). A "melanopic" spectral efficiency function predicts the sensitivity of melanopsin photoreceptors to polychromatic lights. *Journal of Biological Rhythms, 26,* 314–323.

Engelund, A., Fahrenkrug, J., Harrison, A., & Hannibal, J. (2010). Vesicular glutamate transporter 2 (VGLUT2) is co-stored with PACAP in projections from the rat melanopsin-containing retinal ganglion cells. *Cell and Tissue Research, 340,* 243–255.

Estevez, M. E., Fogerson, P. M., Ilardi, M. C., Borghuis, B. G., Chan, E., Weng, S., Auferkorte, O. N., Demb, J. B., & Berson, D. M. (2012). Form and function of the M4 cell, an intrinsically photosensitive retinal ganglion cell type contributing to geniculocortical vision. *Journal of Neuroscience, 32,* 13608–13620.

Fahrenkrug, J., Nielsen, H. S., & Hannibal, J. (2004). Expression of melanopsin during development of the rat retina. *Neuroreport, 15,* 781–784.

Fain, G. L., Hardie, R., & Laughlin, S. B. (2010). Phototransduction and the evolution of photoreceptors. *Current Biology, 20,* R114–R124.

Freedman, M. S., Lucas, R. J., Soni, B., von Schantz, M., Munoz, M., David-Gray, Z., et al. (1999). Regulation of mammalian circadian behavior by non-rod, non-cone, ocular photoreceptors. *Science, 284,* 502–504.

Gamlin, P. D., McDougal, D. H., Pokorny, J., Smith, V. C., Yau, K. W., & Dacey, D. M. (2007). Human and macaque pupil responses driven by melanopsin-containing retinal ganglion cells. *Vision Research, 47,* 946–954.

Gomez, M. del P., Angueyra, J. M., & Nasi, E. (2009). Light-transduction in melanopsin-expressing photoreceptors of amphioxus. *Proceedings of the National Academy of Sciences of the United States of America, 106,* 9081–9086.

Gonzalez-Menendez, I., Contreras, F., Cernuda-Cernuda, R., & Garcia-Fernandez, J. M. (2009). Daily rhythm of melanopsin-expressing cells in the mouse retina. *Frontiers in Cellular Neuroscience, 3,* 3. doi:10.3389/neuro.03.003.2009.

Gonzalez-Menendez, I., Contreras, F., Cernuda-Cernuda, R., Provencio, I., & Garcia-Fernandez, J. M. (2010). Postnatal development and functional adaptations of the melanopsin photoreceptive system in the albino mouse retina. *Investigative Ophthalmology & Visual Science, 51,* 4840–4847.

Gooley, J. J., Lu, J., Chou, T. C., Scammell, T. E., & Saper, C. B. (2001). Melanopsin in cells of origin of the retinohypothalamic tract. *Nature Neuroscience, 4,* 1165.

Gooley, J. J., Lu, J., Fischer, D., & Saper, C. B. (2003). A broad role for melanopsin in nonvisual photoreception. *Journal of Neuroscience, 23,* 7093–7106.

Göz, D., Studholme, K., Lappi, D. A., Rollag, M. D., Provencio, I., & Morin, L. P. (2008). Targeted destruction of photosensitive retinal ganglion cells with a saporin conjugate alters the effects of light on mouse circadian rhythms. *PLoS One, 3,* e3153. doi:10.1371/journal.pone.0003153.

Graham, D. M., Wong, K. Y., Shapiro, P., Frederick, C., Pattabiraman, K., & Berson, D. M. (2008). Melanopsin ganglion cells use a membrane-associated rhabdomeric phototransduction cascade. *Journal of Neurophysiology, 99,* 2522–2532.

Grunert, U., Jusuf, P. R., Lee, S. C., & Nguyen, D. T. (2011). Bipolar input to melanopsin containing ganglion cells in primate retina. *Visual Neuroscience, 28*, 39–50.

Güler, A. D., Ecker, J. L., Lall, G. S., Haq, S., Altimus, C. M., Liao, H. W., et al. (2008). Melanopsin cells are the principal conduits for rod/cone input to non-image forming vision. *Nature, 453*, 102–105.

Hankins, M. W., & Lucas, R. J. (2002). The primary visual pathway in humans is regulated according to long-term light exposure through the action of a nonclassical photopigment. *Current Biology, 12*, 191–198.

Hannibal, J. (2002). Neurotransmitters of the retino-hypothalamic tract. *Cell and Tissue Research, 309*, 73–88.

Hannibal, J. (2006). Regulation of melanopsin expression. *Chronobiology International, 23*, 159–166.

Hannibal, J., & Fahrenkrug, J. (2004a). Target areas innervated by PACAP-immunoreactive retinal ganglion cells. *Cell and Tissue Research, 316*, 99–113. doi:10.1007/s00441-004-0858-x.

Hannibal, J., & Fahrenkrug, J. (2004b). Melanopsin containing retinal ganglion cells are light responsive from birth. *Neuroreport, 15*, 2317–2320.

Hannibal, J., Georg, B., Hindersson, P., & Fahrenkrug, J. (2005). Light and darkness regulate melanopsin in the retinal ganglion cells of the albino Wistar rat. *Journal of Molecular Neuroscience, 27*, 147–155.

Hannibal, J., Hindersson, P., Knudsen, S. M., Georg, B., & Fahrenkrug, J. (2002a). The photopigment melanopsin is exclusively present in pituitary adenylate cyclase-activating polypeptide-containing retinal ganglion cells of the retinohypothalamic tract. *Journal of Neuroscience, 22*, RC191.

Hannibal, J., Hindersson, P., Nevo, E., & Fahrenkrug, J. (2002b). The circadian photopigment melanopsin is expressed in the blind subterranean mole rat, *Spalax*. *Neuroreport, 13*, 1411–1414.

Hannibal, J., Hindersson, P., Østergaard, J., Georg, B., Heegaard, S., Larsen, P. J., et al. (2004). Melanopsin is expressed in PACAP-containing retinal ganglion cells of the human retinohypothalamic tract. *Investigative Ophthalmology & Visual Science, 45*, 4202–4209.

Hartwick, A. T., Bramley, J. R., Yu, J., Stevens, K. T., Allen, C. N., Baldridge, W. H., et al. (2007). Light-evoked calcium responses of isolated melanopsin-expressing retinal ganglion cells. *Journal of Neuroscience, 27*, 13468–13480.

Hatori, M., Le, H., Vollmers, C., Keding, S. R., Tanaka, N., Buch, T., et al. (2008). Inducible ablation of melanopsin-expressing retinal ganglion cells reveals their central role in non-image forming visual responses. *PLoS One, 3*,e2451. doi:10.1371/journal.pone.0002451.

Hattar, S., Kumar, M., Park, A., Tong, P., Tung, J., Yau, K. W., et al. (2006). Central projections of melanopsin-expressing retinal ganglion cells in the mouse. *Journal of Comparative Neurology, 497*, 326–349.

Hattar, S., Liao, H. W., Takao, M., Berson, D. M., & Yau, K. W. (2002). Melanopsin-containing retinal ganglion cells: Architecture, projections, and intrinsic photosensitivity. *Science, 295*, 1065–1070.

Hattar, S., Lucas, R. J., Mrosovsky, N., Thompson, S., Douglas, R. H., Hankins, M. W., et al. (2003). Melanopsin and rod-cone photoreceptive systems account for all major accessory visual functions in mice. *Nature, 424*, 76–81.

Hillman, P., Hochstein, S., & Minke, B. (1983). Transduction in invertebrate photoreceptors: Role of pigment bistability. *Physiological Reviews, 63*, 668–772.

Hoshi, H., Liu, W. L., Massey, S. C., & Mills, S. L. (2009). ON inputs to the OFF layer: Bipolar cells that break the stratification rules of the retina. *Journal of Neuroscience, 29*, 8875–8883.

Hubbard, K. B., & Hepler, J. R. (2006). Cell signalling diversity of the Gqalpha family of heterotrimeric G proteins. *Cellular Signalling, 18*, 135–150.

Huberman, A. D., Manu, M., Koch, S. M., Susman, M. W., Lutz, A. B., Ullian, E. M., et al. (2008). Architecture and activity-mediated refinement of axonal projections from a mosaic of genetically identified retinal ganglion cells. *Neuron, 59*, 425–438.

Isoldi, M. C., Rollag, M. D., Castrucci, A. M., & Provencio, I. (2005). Rhabdomeric phototransduction initiated by the vertebrate photopigment melanopsin. *Proceedings of the National Academy of Sciences of the United States of America, 102*, 1217–1221.

Jain, V., Ravindran, E., & Dhingra, N. K. (2012). Differential expression of Brn3 transcription factors in intrinsically photosensitive retinal ganglion cells in mouse. *Journal of Comparative Neurology, 520*, 742–755.

Jeon, C. J., Strettoi, E., & Masland, R. H. (1998). The major cell populations of the mouse retina. *Journal of Neuroscience, 18*, 8936–8946.

Johnson, J., Fremeau, R. T., Jr., Duncan, J. L., Renteria, R. C., Yang, H., Hua, Z., et al. (2007). Vesicular glutamate transporter 1 is required for photoreceptor synaptic signaling but not for intrinsic visual functions. *Journal of Neuroscience, 27*, 7245–7255.

Johnson, J., Wu, V., Donovan, M., Majumdar, S., Renteria, R. C., Porco, T., et al. (2010). Melanopsin-dependent light avoidance in neonatal mice. *Proceedings of the National Academy of Sciences of the United States of America, 107*, 17374–17378.

Jusuf, P. R., Lee, S. C., Hannibal, J., & Grunert, U. (2007). Characterization and synaptic connectivity of melanopsin-containing ganglion cells in the primate retina. *European Journal of Neuroscience, 26*, 2906–2921.

Kardon, R., Anderson, S. C., Damarjian, T. G., Grace, E. M., Stone, E., & Kawasaki, A. (2009). Chromatic pupil responses: preferential activation of the melanopsin-mediated versus outer photoreceptor-mediated pupil light reflex. *Ophthalmology, 116*, 1564–1573.

Koyanagi, M., Kubokawa, K., Tsukamoto, H., Shichida, Y., & Terakita, A. (2005). Cephalochordate melanopsin: Evolutionary linkage between invertebrate visual cells and vertebrate photosensitive retinal ganglion cells. *Current Biology, 15*, 1065–1069.

Lall, G. S., Revell, V. L., Momiji, H., Al Enezi, J., Altimus, C. M., Güler, A. D., et al. (2010). Distinct contributions of rod, cone, and melanopsin photoreceptors to encoding irradiance. *Neuron, 66*, 417–428.

Lin, B., Koizumi, A., Tanaka, N., Panda, S., & Masland, R. H. (2008). Restoration of visual function in retinal degeneration mice by ectopic expression of melanopsin. *Proceedings of the National Academy of Sciences of the United States of America, 105*, 16009–16014.

Lucas, R. J., Douglas, R. H., & Foster, R. G. (2001). Characterization of an ocular photopigment capable of driving pupillary constriction in mice. *Nature Neuroscience, 4*, 621–626.

Lucas, R. J., Freedman, M. S., Munoz, M., Garcia-Fernandez, J. M., & Foster, R. G. (1999). Regulation of the mammalian pineal by non-rod, non-cone, ocular photoreceptors. *Science, 284*, 505–507.

Lucas, R. J., Hattar, S., Takao, M., Berson, D. M., Foster, R. G., & Yau, K. W. (2003). Diminished pupillary light reflex at high irradiances in melanopsin-knockout mice. *Science, 299*, 245–247.

Lupi, D., Oster, H., Thompson, S., & Foster, R. G. (2008). The acute light-induction of sleep is mediated by OPN4-based photoreception. *Nature Neuroscience, 11*, 1068–1073. doi:10.1038/nn.2179.

Mawad, K., & Van Gelder, R. N. (2008). Absence of long-wavelength photic potentiation of murine intrinsically photosensitive retinal ganglion cell firing in vitro. *Journal of Biological Rhythms, 23*, 387–391.

McDougal, D. H., & Gamlin, P. D. (2010). The influence of intrinsically-photosensitive retinal ganglion cells on the spectral sensitivity and response dynamics of the human pupillary light reflex. *Vision Research, 50*, 72–87.

McNeill, D. S., Sheely, C. J., Ecker, J. L., Badea, T. C., Morhardt, D., Guido, W., et al. (2011). Development of melanopsin-based irradiance detecting circuitry. *Neural Development, 6*, 8. doi:10.1186/1749-8104-6-8.

Melyan, Z., Tarttelin, E. E., Bellingham, J., Lucas, R. J., & Hankins, M. W. (2005). Addition of human melanopsin renders mammalian cells photoresponsive. *Nature, 433*, 741–745.

Morin, L. P., Blanchard, J. H., & Provencio, I. (2003). Retinal ganglion cell projections to the hamster suprachiasmatic nucleus, intergeniculate leaflet, and visual midbrain: Bifurcation and melanopsin immunoreactivity. *Journal of Comparative Neurology, 465*, 401–416.

Mrosovsky, N., & Hattar, S. (2003). Impaired masking responses to light in melanopsin-knockout mice. *Chronobiology International, 20*, 989–999.

Mrosovsky, N., Lucas, R. J., & Foster, R. G. (2001). Persistence of masking responses to light in mice lacking rods and cones. *Journal of Biological Rhythms, 16*, 585–588.

Müller, L. P., Do, M. T., Yau, K. W., He, S., & Baldridge, W. H. (2010). Tracer coupling of intrinsically photosensitive retinal ganglion cells to amacrine cells in the mouse retina. *Journal of Comparative Neurology, 518*,4813–4824.

Mure, L. S., Cornut, P. L., Rieux, C., Drouyer, E., Denis, P., Gronfier, C., et al. (2009). Melanopsin bistability: A fly's eye technology in the human retina. *PLoS One, 4*, e5991. doi:10.1371/journal.pone.0005991.

Mure, L. S., Rieux, C., Hattar, S., & Cooper, H. M. (2007). Melanopsin-dependent nonvisual responses: Evidence for photopigment bistability in vivo. *Journal of Biological Rhythms, 22*,411–424.

Neumann, S., Haverkamp, S., & Auferkorte, O. N. (2011). Intrinsically photosensitive ganglion cells of the primate retina express distinct combinations of inhibitory neurotransmitter receptors. *Neuroscience, 199*, 24–31.

Newman, L. A., Walker, M. T., Brown, R. L., Cronin, T. W., & Robinson, P. R. (2003). Melanopsin forms a functional short-wavelength photopigment. *Biochemistry, 42*, 12734–12738.

Noseda, R., Kainz, V., Jakubowski, M., Gooley, J. J., Saper, C. B., Digre, K., et al. (2010). A neural mechanism for exacerbation of headache by light. *Nature Neuroscience, 13*, 239–245.

Østergaard, J., Hannibal, J., & Fahrenkrug, J. (2007). Synaptic contact between melanopsin-containing retinal ganglion cells and rod bipolar cells. *Investigative Ophthalmology & Visual Science, 48*, 3812–3820.

Osterhout, J. A., Josten, N., Yamada, J., Pan, F., Wu, S. W., Nguyen, P. L., et al. (2011). Cadherin-6 mediates axon-target matching in a non-image-forming visual circuit. *Neuron, 71*, 632–639.

Owens, L., Buhr, E., Tu, D. C., Lamprecht, T. L., Lee, J., & Van Gelder, R. N. (2012). Effect of circadian clock gene mutations on nonvisual photoreception in the mouse. *Investigative Ophthalmology & Visual Science, 53*, 454–460.

Panda, S., Nayak, S. K., Campo, B., Walker, J. R., Hogenesch, J. B., & Jegla, T. (2005). Illumination of the melanopsin signaling pathway. *Science, 307*, 600–604.

Panda, S., Provencio, I., Tu, D. C., Pires, S. S., Rollag, M. D., Castrucci, A. M., et al. (2003). Melanopsin is required for non-image-forming photic responses in blind mice. *Science, 301*, 525–527.

Panda, S., Sato, T. K., Castrucci, A. M., Rollag, M. D., DeGrip, W. J., Hogenesch, J. B., et al. (2002). Melanopsin (Opn4) requirement for normal light-induced circadian phase shifting. *Science, 298*, 2213–2216.

Peirson, S. N., Oster, H., Jones, S. L., Leitges, M., Hankins, M. W., & Foster, R. G. (2007). Microarray analysis and functional genomics identify novel components of melanopsin signaling. *Current Biology, 17*, 1363–1372.

Perez-Leighton, C. E., Schmidt, T. M., Abramowitz, J., Birnbaumer, L., & Kofuji, P. (2011). Intrinsic phototransduction persists in melanopsin-expressing ganglion cells lacking diacylglycerol-sensitive TRPC subunits. *European Journal of Neuroscience, 33*, 856–867.

Perez-Leon, J. A., Warren, E. J., Allen, C. N., Robinson, D. W., & Brown, R. L. (2006). Synaptic inputs to retinal ganglion cells that set the circadian clock. *European Journal of Neuroscience, 24*, 1117–1123.

Pickard, G. E., Baver, S. B., Ogilvie, M. D., & Sollars, P. J. (2009). Light-induced fos expression in intrinsically photosensitive retinal ganglion cells in melanopsin knockout (opn4) mice. *PLoS One, 4*, e4984. doi:10.1371/journal.pone.0004984.

Pires, S. S., Hughes, S., Turton, M., Melyan, Z., Peirson, S. N., Zheng, L., et al. (2009). Differential expression of two distinct functional isoforms of melanopsin (Opn4) in the mammalian retina. *Journal of Neuroscience, 29*, 12332–12342.

Pires, S. S., Shand, J., Bellingham, J., Arrese, C., Turton, M., Peirson, S., et al. (2007). Isolation and characterization of melanopsin (Opn4) from the Australian marsupial *Sminthopsis crassicaudata* (fat-tailed dunnart). *Proceedings. Biological Sciences, 274*, 2791–2799.

Provencio, I., Jiang, G., De Grip, W. J., Hayes, W. P., & Rollag, M. D. (1998). Melanopsin: An opsin in melanophores, brain, and eye. *Proceedings of the National Academy of Sciences of the United States of America, 95*, 340–345.

Provencio, I., Rodriguez, I. R., Jiang, G., Hayes, W. P., Moreira, E. F., & Rollag, M. D. (2000). A novel human opsin in the inner retina. *Journal of Neuroscience, 20*, 600–605.

Provencio, I., Rollag, M. D., & Castrucci, A. M. (2002). Anatomy: Photoreceptive net in the mammalian retina. *Nature, 415*, 493–494.

Pu, M. (1999). Dendritic morphology of cat retinal ganglion cells projecting to suprachiasmatic nucleus. *Journal of Comparative Neurology, 414*, 267–274.

Qiu, X., Kumbalasiri, T., Carlson, S. M., Wong, K. Y., Krishna, V., Provencio, I., et al. (2005). Induction of photosensitivity by heterologous expression of melanopsin. *Nature, 433,* 745–749.

Rajaraman, K. (2012). ON ganglion cells are intrinsically photosensitive in the tiger salamander retina. *Journal of Comparative Neurology, 520,* 200–210.

Renna, J. M., Weng, S., & Berson, D. M. (2011). Light acts through melanopsin to alter retinal waves and segregation of retinogeniculate afferents. *Nature Neuroscience, 14,* 827–829.

Rollag, M. D. (2008). Does melanopsin bistability have physiological consequences? *Journal of Biological Rhythms, 23,* 396–399.

Ruby, N. F., Brennan, T. J., Xie, X., Cao, V., Franken, P., Heller, H. C., et al. (2002). Role of melanopsin in circadian responses to light. *Science, 298,* 2211–2213.

Sakamoto, K., Liu, C., Kasamatsu, M., Pozdeyev, N. V., Iuvone, P. M., & Tosini, G. (2005). Dopamine regulates melanopsin mRNA expression in intrinsically photosensitive retinal ganglion cells. *European Journal of Neuroscience, 22,* 3129–3136.

Schmidt, T. M., Do, M. T., Dacey, D., Lucas, R., Hattar, S., & Matynia, A. (2011). Melanopsin-positive intrinsically photosensitive retinal ganglion cells: From form to function. *Journal of Neuroscience, 31,* 16094–16101.

Schmidt, T. M., & Kofuji, P. (2009). Functional and morphological differences among intrinsically photosensitive retinal ganglion cells. *Journal of Neuroscience, 29,* 476–482.

Schmidt, T. M., & Kofuji, P. (2010). Differential cone pathway influence on intrinsically photosensitive retinal ganglion cell subtypes. *Journal of Neuroscience, 30,* 16262–16271.

Schmidt, T. M., & Kofuji, P. (2011). Structure and function of bistratified intrinsically photosensitive retinal ganglion cells in the mouse. *Journal of Comparative Neurology, 519,* 1492–1504.

Schmidt, T. M., Taniguchi, K., & Kofuji, P. (2008). Intrinsic and extrinsic light responses in melanopsin-expressing ganglion cells during mouse development. *Journal of Neurophysiology, 100,* 371–384.

Sekaran, S., Foster, R. G., Lucas, R. J., & Hankins, M. W. (2003). Calcium imaging reveals a network of intrinsically light-sensitive inner-retinal neurons. *Current Biology, 13,* 1290–1298. doi:10.1016/S0960-9822(03)00510-4.

Sekaran, S., Lall, G. S., Ralphs, K. L., Wolstenholme, A. J., Lucas, R. J., Foster, R. G., et al. (2007). 2-Aminoethoxydiphenylborane is an acute inhibitor of directly photosensitive retinal ganglion cell activity in vitro and in vivo. *Journal of Neuroscience, 27,* 3981–3986.

Sekaran, S., Lupi, D., Jones, S. L., Sheely, C. J., Hattar, S., Yau, K. W., et al. (2005). Melanopsin-dependent photoreception provides earliest light detection in the mammalian retina. *Current Biology, 15,* 1099–1107.

Semo, M., Gias, C., Ahmado, A., Sugano, E., Allen, A. E., Lawrence, J. M., et al. (2010). Dissecting a role for melanopsin in behavioural light aversion reveals a response independent of conventional photoreception. *PLoS One, 5,* e15009. doi:10.1371/journal.pone.0015009.

Semo, M., Munoz Llamosas, M., Foster, R. G., & Jeffery, G. (2005). Melanopsin (Opn4) positive cells in the cat retina are randomly distributed across the ganglion cell layer. *Visual Neuroscience, 22,* 111–116.

Sexton, T., Buhr, E., & Van Gelder, R. N. (2012). Melanopsin and mechanisms of non-visual ocular photoreception. *Journal of Biological Chemistry, 287,* 1649–1656.

Tarttelin, E. E., Bellingham, J., Bibb, L. C., Foster, R. G., Hankins, M. W., Gregory-Evans, K., et al. (2003). Expression of opsin genes early in ocular development of humans and mice. *Experimental Eye Research, 76,* 393–396.

Terakita, A., Tsukamoto, H., Koyanagi, M., Sugahara, M., Yamashita, T., & Shichida, Y. (2008). Expression and comparative characterization of Gq-coupled invertebrate visual pigments and melanopsin. *Journal of Neurochemistry, 105,* 883–890.

Tu, D. C., Zhang, D., Demas, J., Slutsky, E. B., Provencio, I., Holy, T. E., et al. (2005). Physiologic diversity and development of intrinsically photosensitive retinal ganglion cells. *Neuron, 48,* 987–999.

Van Hook, M. J., Wong, K. Y., & Berson, D. M. (2012). Dopaminergic modulation of ganglion-cell photoreceptors in rat. *European Journal of Neuroscience, 35,* 507–518.

Viney, T. J., Balint, K., Hillier, D., Siegert, S., Boldogkoi, Z., Enquist, L. W., et al. (2007). Local retinal circuits of melanopsin-containing ganglion cells identified by transsynaptic viral tracing. *Current Biology, 17,* 981–988. doi:10.1016/j.cub.2007.04.058.

Vugler, A. A., Redgrave, P., Semo, M., Lawrence, J., Greenwood, J., & Coffey, P. J. (2007). Dopamine neurones form a discrete plexus with melanopsin cells in normal and degenerating retina. *Experimental Neurology, 205,* 26–35.

Walker, M. T., Brown, R. L., Cronin, T. W., & Robinson, P. R. (2008). Photochemistry of retinal chromophore in mouse melanopsin. *Proceedings of the National Academy of Sciences of the United States of America, 105,* 8861–8865.

Warren, E. J., Allen, C. N., Brown, R. L., & Robinson, D. W. (2003). Intrinsic light responses of retinal ganglion cells projecting to the circadian system. *European Journal of Neuroscience, 17,* 1727–1735. doi:10.1046/j.1460-9568.2003.02594.x.

Warren, E. J., Allen, C. N., Brown, R. L., & Robinson, D. W. (2006). The light-activated signaling pathway in SCN-projecting rat retinal ganglion cells. *European Journal of Neuroscience, 23,* 2477–2487.

Weng, S., Wong, K. Y., & Berson, D. M. (2009). Circadian modulation of melanopsin-driven light response in rat ganglion-cell photoreceptors. *Journal of Biological Rhythms, 24,* 391–402.

Wong, K. Y. (2012). A retinal ganglion cell that can signal irradiance continuously for 10 hours. *Journal of Neuroscience, 32,* 11478–11485.

Wong, K. Y., & Berson, D. M. (2011). Ganglion-cell photoreceptors and non-image-forming vision. In P. L. Kaufman, A. Alm, L. A. Levin, S. F. E. Nilsson, J. N. Ver Hoeve, & S. M. Wu (Eds.), *Adler's physiology of the eye* (11th ed.). London: Elsevier. doi:10.1016/B978-0-323-05714-1.00026-1

Wong, K. Y., Dunn, F. A., & Berson, D. M. (2005). Photoreceptor adaptation in intrinsically photosensitive retinal ganglion cells. *Neuron, 48,* 1001–1010.

Wong, K. Y., Dunn, F. A., Graham, D. M., & Berson, D. M. (2007). Synaptic influences on rat ganglion-cell photoreceptors. *Journal of Physiology, 582,* 279–296.

Xue, T., Do, M. T., Riccio, A., Jiang, Z., Hsieh, J., Wang, H. C., et al. (2011). Melanopsin signalling in mammalian iris and retina. *Nature, 479,* 67–73.

Yau, K. W., & Hardie, R. C. (2009). Phototransduction motifs and variations. *Cell, 139*, 246–264.

Yoshimura, T., & Ebihara, S. (1996). Spectral sensitivity of photoreceptors mediating phase-shifts of circadian rhythms in retinally degenerate CBA/J (rd/rd) and normal CBA/N (+/+) mice. *Journal of Comparative Physiology. A, Neuroethology, Sensory, Neural, and Behavioral Physiology, 178*, 797–802.

Zaidi, F. H., Hull, J. T., Peirson, S. N., Wulff, K., Aeschbach, D., Gooley, J. J., et al. (2007). Short-wavelength light sensitivity of circadian, pupillary, and visual awareness in humans lacking an outer retina. *Current Biology, 17*, 2122–2128.

Zele, A. J., Feigl, B., Smith, S. S., & Markwell, E. L. (2011). The circadian response of intrinsically photosensitive retinal ganglion cells. *PLoS One, 6*, e17860. doi:10.1371/journal.pone.0017860.

Zhang, D. Q., Wong, K. Y., Sollars, P. J., Berson, D. M., Pickard, G. E., & McMahon, D. G. (2008). Intraretinal signaling by ganglion cell photoreceptors to dopaminergic amacrine neurons. *Proceedings of the National Academy of Sciences of the United States of America, 105*, 14181–14186.

15 Postreceptoral Mechanisms for Adaptation in the Retina

YANBIN V. WANG AND JONATHAN B. DEMB

Humans and animals can function across a remarkably wide range of lighting conditions. For example, a person can navigate across a field on a starlit night, when individual rod photoreceptors capture a photon only once per minute, and the same person can navigate across the beach on a cloudless day at noon, when individual cone photoreceptors capture thousands of photons per second. Between these two settings, the mean light intensity could differ by 100 millionfold. However, the output neurons of the retina—the ganglion cells—can fire at most ~20 spikes within the ~100-ms integration time of a postsynaptic neuron (O'Brien et al., 2002). Thus, there is a substantial mismatch between the retina's input (10^{10} levels) and output (10^{1}–10^{2} levels) (Shapley & Enroth-Cugell, 1984). To solve this problem the retina adapts by changing its sensitivity—or *gain*—to allow the ganglion cell's dynamic range to be matched to the statistics of the immediate environment. There are certainly limitations to the adaptation process. For example, one could only read this book or play baseball on the beach setting, whereas behavior in the starlit field would be limited to basic navigation and identification of large, slowly moving objects. Nevertheless, it is remarkable that a single organism can navigate and perform at least simple behaviors under such extreme changes in lighting conditions.

Here, we describe the mechanisms of adaptation within the retina. The focus is on two forms of adaptation that have been studied extensively: adaptation to the mean light intensity and adaptation to variations around the mean, also known as *contrast*. Figure 15.1 illustrates the basic concept of each type of adaptation as well as the impact of eye movements when viewing a natural scene. As the eyes move around a scene, there are changes in the mean and contrast over local regions corresponding to a cell's receptive field. Following a change in the mean or standard deviation (i.e., contrast) of light inputs, the stimulus–response relationship adjusts to match the range of inputs (Laughlin, 1981). This example illustrates a complete adaptation, generating a perfect match between stimulus and response, whereas real neurons typically show incomplete adaptation.

Our description of adaptation focuses on the postreceptor mechanisms—that is, the mechanisms that are implemented beyond the stage of the photoreceptors within the retinal circuitry. These mechanisms include intrinsic membrane properties of individual cells and modification of synapses between cells. Previously, there have been excellent reviews of the impact of adaptation on ganglion cell spike responses measured with extracellular recording (Shapley & Enroth-Cugell, 1984; Walraven et al., 1990); here, we focus instead on more recent studies that have measured intracellular responses at multiple stages in retinal circuitry to reveal sites and mechanisms of adaptation. We also emphasize findings in the mammalian retina, where the cell types and circuitry have been worked out most extensively (figure 15.2). We understand now that a typical mammalian retina includes about 60–80 cell types (Field & Chichilnisky, 2007; Masland, 2001, 2011; Sterling & Demb, 2004; Wässle, 2004). There are three or four types of photoreceptor, including one type of rod and two or three types of cone; a dozen types of bipolar cell, including both ON and OFF subtypes, which convey photoreceptor signals to the inner retina; one or two types of horizontal cell, and ~30–40 types of amacrine cell, which collectively refine the photoreceptor and bipolar cell signals and generate multiple computations, including lateral inhibition. Specific bipolar and amacrine cell types converge onto each of the ~20 types of ganglion cell. Each ganglion cell type codes a unique property of the visual input (e.g., direction, color, motion). One ultimately wants to understand how adaptation is implemented within each of the specific circuits that converge onto the different ganglion cell types.

GANGLION CELL RECEPTIVE FIELDS AND ADAPTATION

It is worth considering how the concept of a ganglion cell's receptive field relates to the concept of

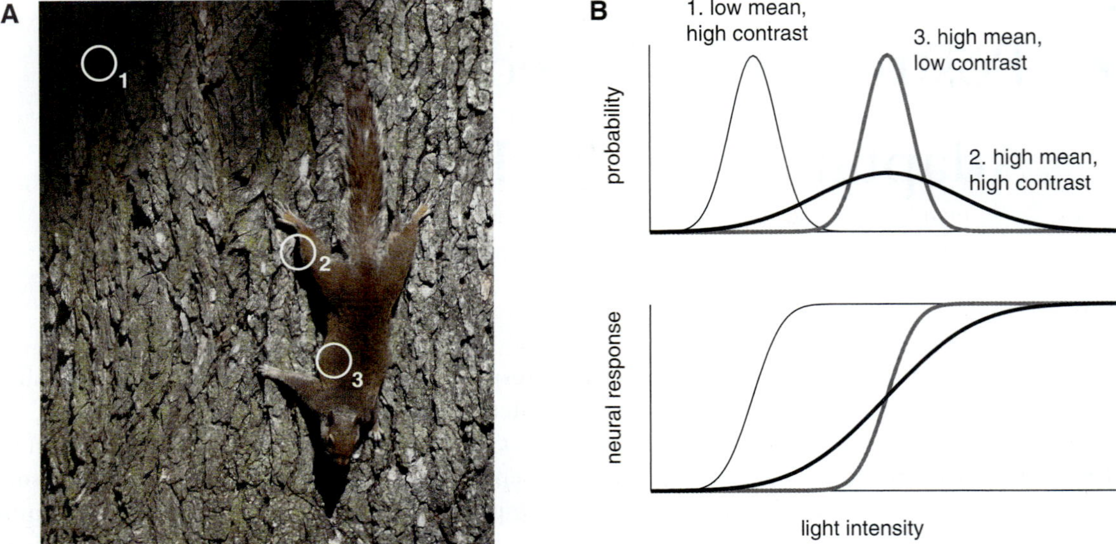

FIGURE 15.1 Challenges to viewing natural scenes and the impact of adaptation. (A) Within a scene there are regions with low mean intensity (1) and higher mean intensity (2 and 3). Given a relatively constant mean, the standard deviation of intensity—or contrast—can differ between high (2) and low (3). Because of eye movements (and squirrel movements), a ganglion cell's receptive field will experience different mean and contrast levels every few hundred milliseconds. (B) The top plot shows intensity distributions with different values of mean and contrast. Bottom plot shows how the neural input–output response relationship would adapt to match the different input distributions.

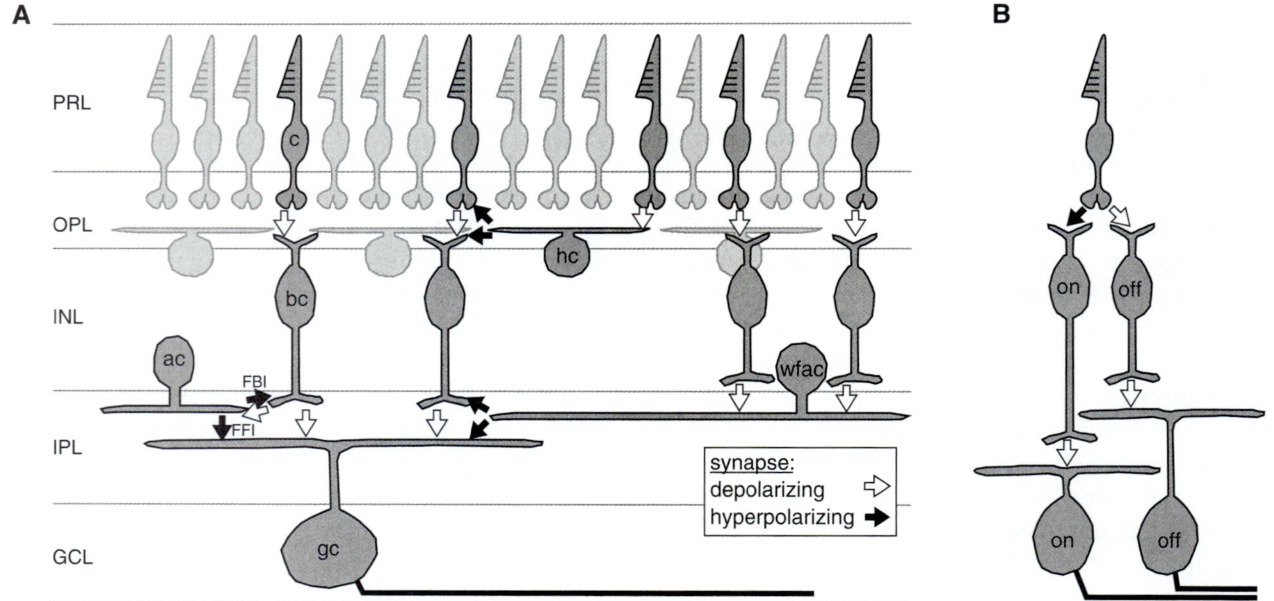

FIGURE 15.2 Basic circuitry of the retina. (A) Basic circuitry of the cone pathway. Cones (c) signal bipolar cells (bc), which in turn signal a ganglion cell (gc); this example shows the circuit for an OFF-type ganglion cell. A horizontal cell (hc) generates lateral inhibition in the outer retina, and a wide-field amacrine cell (wfac) does the same in the inner retina. A local amacrine cell (ac) can generate feedback inhibition (fbi) onto the bipolar cell terminal and feedforward inhibition (ffi) onto the ganglion cell. Abbreviations: PRL, photoreceptor layer; OPL, outer plexiform layer; INL, inner nuclear layer; IPL, inner plexiform layer; GCL, ganglion cell layer. (B) The cone signal diverges into two major pathways by stimulating either ON or OFF bipolar cells. ON bipolar cells express a metabotropic glutamate receptor (mGluR6), whereas the OFF bipolar cells express ionotropic glutamate receptors; this is why cone glutamate release has opposite effects, either hyperpolarizing (ON) or depolarizing (OFF) the bipolar cell. The axon terminals of the two bipolar cell classes terminate in different halves of the IPL, where they can excite either ON or OFF ganglion cells.

 YANBIN V. WANG AND JONATHAN B. DEMB

adaptation. The receptive field refers to the area of the retina over which changes in light intensity influence the ganglion cell's response. The standard model of a receptive field includes center and surround regions, which are antagonistic (Enroth-Cugell & Robson, 1966; Kuffler, 1953; Rodieck, 1965). The sign of the center is either ON or OFF, meaning that spiking is driven by either increases or decreases in light intensity. Stimulation of the surround has the opposite sign and counteracts the effect of center stimulation. For example, stimulating an ON cell's receptive field surround with light will attenuate the effect of stimulating the center. The ganglion cell center-surround receptive field can be described as the difference of two Gaussians (figure 15.3). The combination of positive and negative subregions of the spatial receptive field generates bandpass filtering, resulting in optimal tuning for certain spatial frequencies (figure 15.4). We should point out that the ON or OFF nature of the receptive field center is mediated by the ON or OFF types of bipolar cell, which depolarize to light increments or decrements, respectively (Nelson, Famiglietti, & Kolb, 1978; Werblin & Dowling, 1969). Surround stimulation evokes lateral inhibition mediated by a combination of horizontal and amacrine cells (Flores-Herr, Protti, & Wässle, 2001;

McMahon, Packer, & Dacey, 2004; Roska & Werblin, 2001; Zaghloul et al., 2007).

The ganglion cell's receptive field also has a temporal component. For example, under light-adapted conditions the center is typically biphasic (Chander & Chichilnisky, 2001). For an ON-center cell the filter would be positive, and then there would be a negative undershoot. The temporal component of the surround would have roughly the opposite shape: a negative region followed by a positive overshoot (figure 15.3C). The biphasic temporal receptive field generates bandpass filtering in time, resulting in optimal tuning for certain temporal frequencies (figure 15.4). Combining the spatial and temporal components of the receptive field generates a three-dimensional movie, which represents the cell's spatiotemporal receptive field (figure 15.3A).

The spatiotemporal receptive field describes a filter over space and time that can be compared to the ongoing "movie" of visual input to the retina. Typically, the relevant time window for comparing the stimulus to the filter is ~100–500 ms, and the relevant spatial region is ~0.2–2 mm in diameter (depending on the width of the surround). The spatial dimensions on the retina correspond to varying degrees of visual angle depending on the size of the eye. For example, 0.2 mm

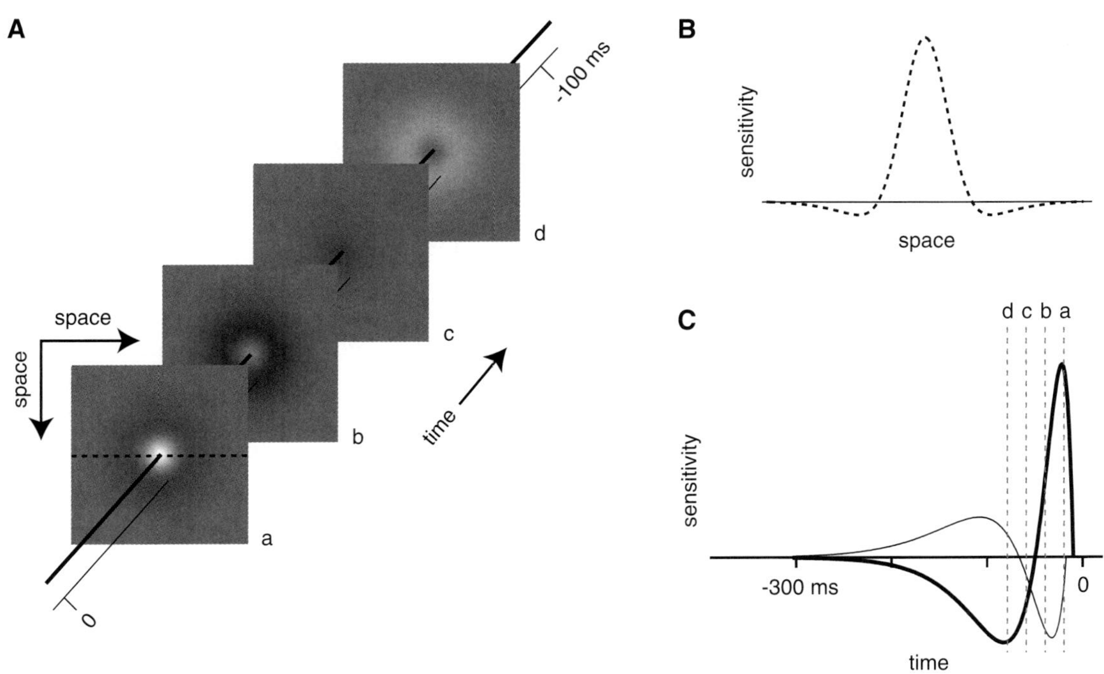

FIGURE 15.3   The ganglion cell's spatiotemporal receptive field. (A) A center-surround receptive field has components in both space and time, forming a movie. Four frames of this movie are shown, extending backward in time. The first frame (a) shows a bright center, indicative of an ON-type ganglion cell. (B) The spatial component of the receptive field (from the dashed line in frame a of part A) showing the difference-of-Gaussians profile. (C) The temporal component of the receptive field center (thick line) and surround (thin line). Both components are biphasic, and the surround component is delayed relative to the center. The times of the four frames from part A are indicated.

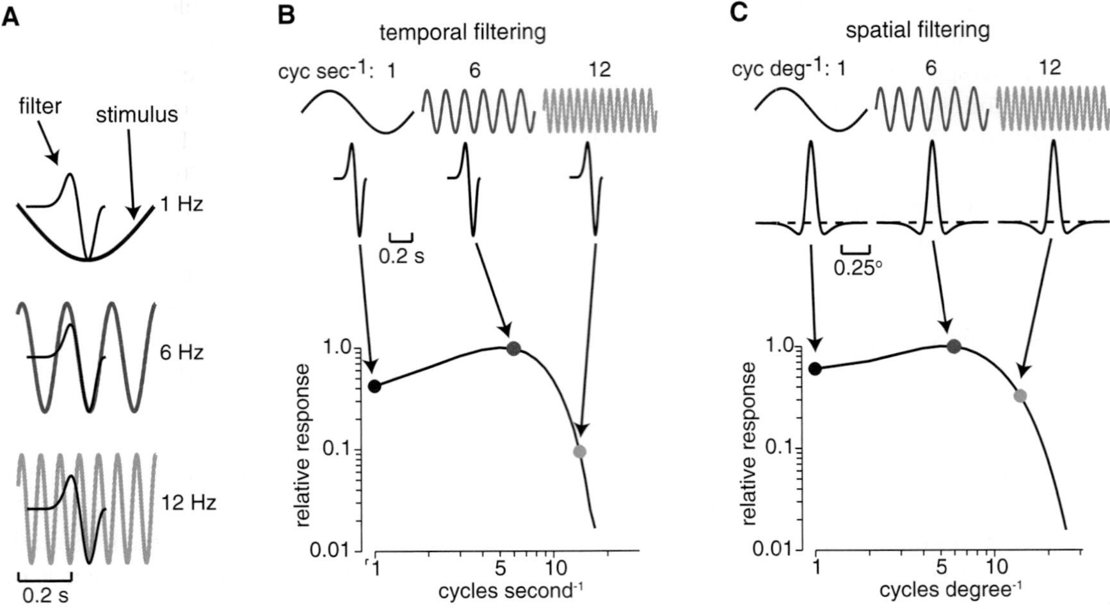

FIGURE 15.4   Spatial and temporal filtering by the receptive field. (A) A temporal filter is shown with stimuli of three frequencies. The 6-Hz stimulus provides the best match to the filter. (B) Temporal contrast sensitivity function shows the filter's normalized response as a function of temporal frequency, with peak tuning at 6 cycles/s (Hz). (C) Spatial contrast sensitivity function shows a spatial filter's normalized response as a function of spatial frequency, with peak tuning at 6 cycles/degree.

corresponds to ~1° in primates but ~7° in mouse (Dacey & Petersen, 1992; Remtulla & Hallett, 1985; Stone & Pinto, 1993). To the extent that the stimulus history matches the filter, the cell will generate a strong response. To the extent that there is a poor match between the stimulus and the filter, there will be a weak response. If a neuron could be described as a linear system, then the spatiotemporal receptive field would suffice to predict the response to an arbitrary stimulus. However, visual neurons are only approximately linear under specific conditions, at best, which necessitates additional characterization of the nonlinearities.

Nonlinearities can be divided into at least two groups. The first group includes static nonlinearities associated with thresholds and saturation of neural responses. These include the firing rates of ganglion cells and the synaptic vesicle release rates of photoreceptors and interneurons, both of which cannot be negative and also cannot exceed a maximum rate. However, the maximum rate is not a fixed property of the neuron but rather depends on the neuron's physiological state. For example, when a ganglion cell rests at a depolarized potential, there are fewer available sodium channels for spike generation, resulting in a lower maximum firing rate (Kim & Rieke, 2003). Furthermore, a bipolar cell with a depolarized resting potential would experience vesicle depletion, resulting in a lower maximum release rate (Jarsky et al., 2011).

The second group includes the adaptive nonlinearities, which are the main subject of this chapter. Adaptation represents a change in the sensitivity of the spatiotemporal profile of the receptive field. Adaptation, therefore, changes the stimulus–response relationship within a neuron's operating range, above the threshold and below the point of saturation. Thus, the receptive field represents an ongoing filtering operation, whereas adaptation represents a continuous updating of the filter itself. At any given point in time, adaptation to stimulus history has shaped the filter, and the filter can be compared to the recent history of the input to generate an output (figure 15.5). The output of this filtering operation would then be further shaped by the static nonlinearities that translate the filtered stimulus to a neural response (i.e., a firing rate) (figure 15.5). This description represents a linear–nonlinear (LN) model of a ganglion cell: an adapting linear filter followed by a static nonlinear stage.

The adapting LN model provides a good description of neural responses for certain cell types under certain stimulus conditions. For example, the model provides a useful description of ganglion cells with relatively linear receptive fields (e.g., X/beta cells in cat) and bipolar cells (Baccus & Meister, 2002; Rieke, 2001; Victor, 1987). However, the model has several limitations that are worth mentioning.

   YANBIN V. WANG AND JONATHAN B. DEMB

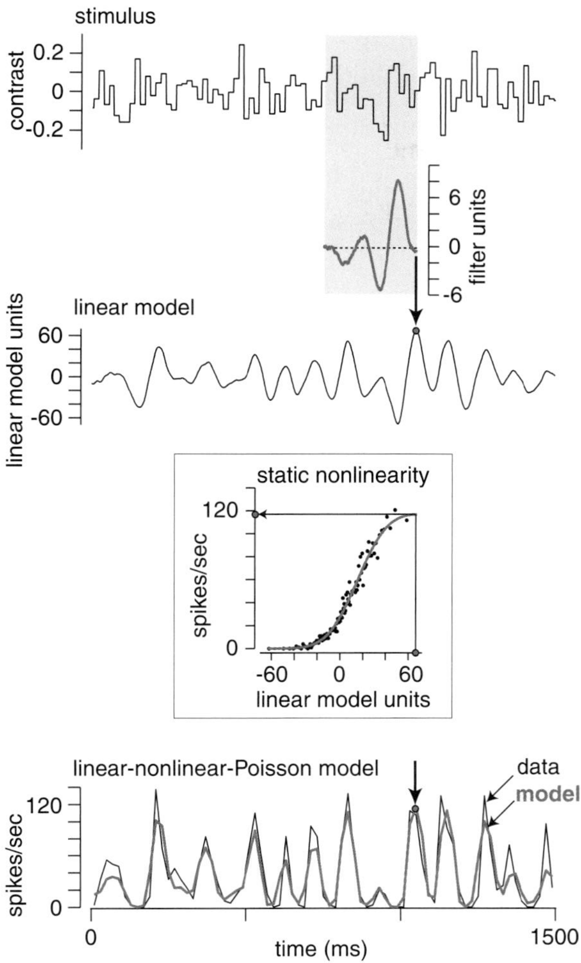

FIGURE 15.5 Linear–nonlinear model of a retinal ganglion cell's spike response. A model of an ON Y-type ganglion cell (guinea pig retina) generated from 100 s of response to a white-noise spot stimulus presented over the receptive field center (see Zaghloul, Boahen, & Demb, 2005). The model generates a linear filter (temporal receptive field) and a static nonlinearity (Chichilnisky, 2001). To predict the response to a novel dataset, the stimulus is weighted by the filter (i.e., convolution) to generate the linear model of the response. The filter is shown at a time when it closely matched the stimulus (gray box), and so the linear model response is large (+63 in arbitrary linear model units, gray circle). The linear model is translated to a spike rate using the nonlinearity, which works as a "lookup table" (box). The +63 linear model value is translated to a spike rate of 117 spikes/s (arrow). The bottom trace shows the spike rate (black line) to 1.5 s of a novel test stimulus. The test stimulus was repeated 20 times, and data were averaged and binned (bin, 20 ms). The gray line shows the output of the model. Because the model predicts spike rate but not spike times, this is a linear–nonlinear–Poisson model of the firing rate.

One major limitation is caused by the nonlinear properties of bipolar and amacrine cell synapses (figure 15.6). The receptive field center of most ganglion cell types is driven by a group of bipolar cells with nonoverlapping receptive fields. Bipolar cells have a nonlinearity at their output that depends on several factors. First, the basal release at most bipolar synapses is relatively low, and thus release can increase above the basal level more than it can decrease; this is a form of rectification (Demb et al., 2001; Hochstein & Shapley, 1976; Werblin, 2010). Second, even in cases where basal release is relatively high, the temporal properties of increased release differ from the properties of decreased release: increased release is relatively more transient (Manookin et al., 2008; Roska & Werblin, 2001). Thus, when a ganglion cell is presented with a high-spatial-frequency grating stimulus that reverses contrast, the excitation of some bipolar cells (e.g., ON bipolar cells viewing light bars) will not be canceled by the inhibition of other bipolar cells (e.g., ON bipolar cells viewing dark bars). The net effect is a burst of excitation at each reversal of grating contrast, generating a response at twice the reversal frequency (Demb et al., 2001; Enroth-Cugell & Robson, 1966; Hochstein & Shapley, 1976). There is a similar rectification at the output of amacrine cells (Demb et al., 1999; Werblin, 2010). Thus, to characterize the full spatiotemporal receptive field of a ganglion cell requires characterizing the nonlinearities at the outputs of the presynaptic neurons (figure 15.6), which is difficult. In practice, there are some useful shortcuts for designing experiments. For example, it is possible to study the temporal receptive field of a ganglion cell that integrates the nonlinear outputs of multiple bipolar cells (e.g., Y/alpha cells) by stimulating these bipolar cells synchronously with a modulated spot over the ganglion cell's receptive field center (e.g., Zaghloul, Boahen, & Demb, 2005).

The adapting LN model can describe the firing rate well on a coarse temporal scale, for example, the firing rate measured in 20-ms time bins (Beaudoin, Borghuis, & Demb, 2007; Zaghloul, Boahen, & Demb, 2003). However, the model cannot capture the properties of ganglion cell spiking on a fine scale such as the firing rate measured in 1-ms time bins (Butts et al., 2011; Keat et al., 2001; Pillow et al., 2005). In particular, the basic model generates a firing rate but does not determine spike timing. This represents a Poisson model of the firing rate—the rate is determined, but the spike times are otherwise random and independent of one another. In an actual neuron, there are absolute and relative refractory periods following a spike that regulate the timing of the following spike. This property of spike generation has been modeled explicitly in extensions of the LN model by including a feedback component driven by the model's spiking output, resembling a refractory state (Butts et al., 2011; Gaudry & Reinagel, 2007a; Keat et al., 2001; Pillow et al., 2005).

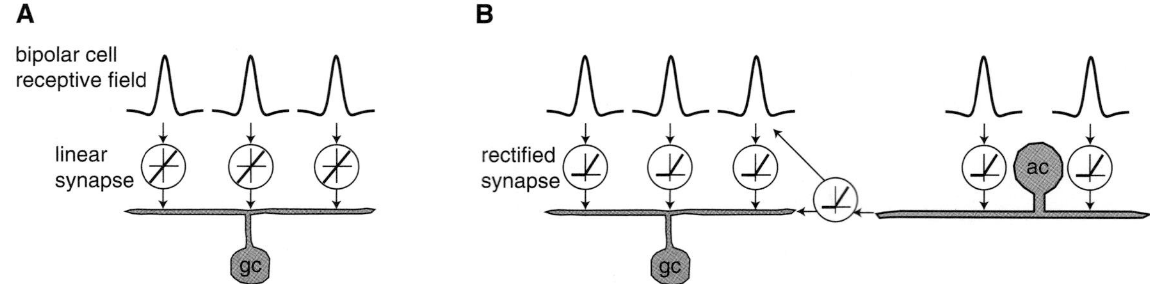

FIGURE 15.6   Linear and nonlinear spatial integration by ganglion cell receptive fields. (A) A ganglion cell (gc) integrates bipolar cell inputs linearly. Each bipolar cell's receptive field is indicated by a spatial filter (difference-of-Gaussians). The ganglion's spatial receptive field would have a center-surround profile generated by the weighted sum of the bipolar cell receptive fields. (B) A ganglion cell integrates bipolar cell inputs nonlinearly. Each bipolar cell has a rectification at its output. The ganglion cell's spatial receptive field would be composed of multiple nonlinear subunits driven by the bipolar cells. Amacrine cells (ac) also have their own receptive fields composed of multiple bipolar cell subunits. The amacrine cell's output to both ganglion cells and bipolar terminals can also be rectified.

In the following sections we consider studies of neurons that can be described by the adapting LN model. Thus, we limit our discussion to interneurons and ganglion cells that are driven by either the ON or OFF bipolar cell pathways and do not discuss cells with extreme nonlinearities (e.g., ON–OFF ganglion or amacrine cells). We also do not discuss melanopsin-expressing, intrinsically photosensitive ganglion cells, although these cells show adaptation within their phototransduction cascade that ultimately influences their spiking response (Wong, Dunn, & Berson, 2005).

ADAPTATION TO MEAN LIGHT INTENSITY

The ability to behave across different points in the diurnal cycle represents the most extreme requirement for adaptation. In part, the solution to this problem occurred on an evolutionary time scale: the coexistence of rod and cone photoreceptors within the same retina. Rods can signal the capture of a single photon (i.e., photoisomerization, R*) but saturate when the mean intensity evokes ~10^3 R*/rod/s, whereas cones operate when the mean intensity evokes ~10^2–10^5 R*/cone/s (see Sterling & Demb, 2004). Under natural conditions, the switch between rod- and cone-mediated vision and back again occurs slowly across the diurnal cycle from dawn to daylight to dusk. However, within each scene there are rapid changes in the mean light level over a ganglion cell's receptive field that occur because of motion of objects in the environment as well as the observer's body, head, and eye movements (Frazor & Geisler, 2006; Mante et al., 2005; Ölveczky, Baccus, & Meister, 2003). These sources of motion occur on a more rapid scale compared to the diurnal cycle, requiring adaptation on the time scale of ~100 ms. At this time

scale, adaptation occurs within the photoreceptors and their downstream circuits (Dunn, Lankheet, & Rieke, 2007; Dunn & Rieke, 2008; Enroth-Cugell & Shapley, 1973; Lee et al., 2003; Yeh, Lee, & Kremers, 1996).

Rods and cones converge onto the same ganglion cells using a combination of shared and distinct circuits (figure 15.7). In the dimmest lighting conditions rods signal through a specialized pathway: rod → rod bipolar cell → AII amacrine cell. The AII cell signals some types of OFF ganglion cell directly, via glycinergic synapses, and some types of ON and OFF ganglion cells indirectly, via either electrical or glycinergic synapses with the presynaptic cone bipolar terminal (Bloomfield & Dacheux, 2001; Demb & Singer, 2012). In slightly brighter conditions rods signal through electrical synapses with cones; there are also some direct chemical synapses between rods and certain types of cone bipolar cell. In bright conditions, when rods saturate, the cones signal through the cone bipolar cells.

The rod circuitry is specialized for encoding the rod's binary response (i.e., zero or one photon absorbed), which covers much of the rod's operating range. The rod bipolar cells, which are all ON-type cells, use a metabotropic glutamate receptor, mGluR6, for encoding glutamate release from rods; the same receptor is used by ON-type cone bipolar cells (Nakajima et al., 1993; Nomura et al., 1994). The mGluR6 receptor is coupled with a G-protein cascade, which closes a cation channel (TRPM1) (Koike et al., 2010; Morgans et al., 2009). Thus, glutamate causes channel closure and net hyperpolarization of the rod bipolar cell. In the dark, the rod releases glutamate continuously, saturating the rod bipolar cell's mGluR6 receptors (Sampath & Rieke, 2004). When a rod absorbs a photon, the phototransduction cascade leads to a transient decrease in

   YANBIN V. WANG AND JONATHAN B. DEMB

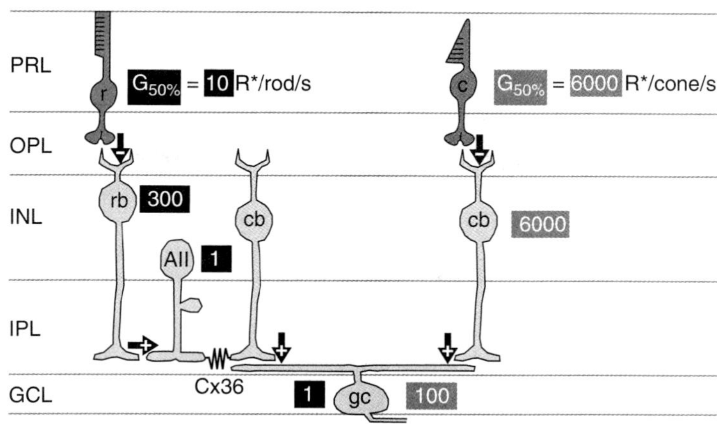

FIGURE 15.7   Gain changes at multiple stages in the rod and cone bipolar cell pathways. In the rod bipolar pathway, rods signal rod bipolar (rb) cells, which excite AII amacrine cells (AII). The AII is electrically coupled (via connexin-36 gap junctions, Cx36) to ON cone bipolar cells (cb), which excite ON ganglion cells (gc). The background light level where gain reduced by 50%, relative to the gain in darkness ($G_{50\%}$), is shown at several points in the rod bipolar circuit. For a dim background (1 R*/rod/s), gain was reduced at two points beyond the rod bipolar synapse. With brighter backgrounds, gain was reduced at earlier sites (Dunn et al., 2006). In the cone (c) bipolar pathway, a dim background reduced gain in a parasol ganglion cell (100 R*/cone/s). With a brighter background, gain was reduced at earlier sites (Dunn, Lankheet, & Rieke, 2007). Hyperpolarizing (−) and depolarizing (+) synapses are indicated by the arrows. For both rod and cone bipolar pathways, the site of adaptation switches from the bipolar synapse (dim background) to the photoreceptor (brighter background).

glutamate release, which reduces mGluR6 stimulation, causing a small depolarization in the rod bipolar cell. However, the mGluR6 receptors have a threshold, and only the relatively large single-photon responses can generate a depolarization of the rod bipolar cell; this includes only ~30% of the single-photon responses (Berntson, Smith, & Taylor, 2004; Field & Rieke, 2002; van Rossum & Smith, 1998). The other 70% are discarded, along with much of the noise at the synapse, never being transmitted beyond the rod.

In the mouse retina adaptation was studied at multiple points within the rod bipolar pathway (figure 15.7; Dunn et al., 2006). Gain was quantified by a cell's response to a flash in either darkness or on variable background levels (up to 300 R*/rod/s). The flash response is roughly the same as the linear filter described above, assuming that the flash response is in the linear range and away from the point of response saturation. In general, the response gain decreases as background level increases, but the gain reduction at a given background differs between cell types.

At a background that evokes 1 R*/rod/s, both AII amacrine cells and ON-type ganglion cells decrease their gain by ~50% (figure 15.7). However, at this background level the R* rate in individual rods is too low to evoke adaptation within the rods themselves; for a rod to experience adaptation, there must be more than one R* within the rods' ~100–200 ms integration time (rather than 1/s). Likewise, rod bipolar cells do not reduce their gain at this background. Thus, it appears that adaptation occurs at the synapse between the rod bipolar cell and the AII amacrine cell. Multiple rods converge on the rod bipolar cell, generating an adapting signal that is robust and clearly distinct from noise (Dunn & Rieke, 2006). Both AII and ganglion cells are most susceptible to response saturation (Dunn et al., 2006), so it makes sense that adaptation would occur prior to the AII cell stage but not necessarily beyond this stage.

At a background that evokes 10 R*/rod/s, rods reduce their gain by ~50% (figure 15.7). However, at this background, rod bipolar cells counteract the adaptation in rods and actually increase their gain. The relative increase in gain can be explained by the relief of the nonlinearity associated with the mGluR6 cascade mentioned above. At a background of 100 R*/rod/s, all cells within the rod pathway reduce their gain. Thus, in this range the adaptation in the rod pathway can be explained primarily by adaptation in the rods—which would then transfer their adaptation to all subsequent stages of processing. At a background that evokes 300 R*/rod/s, the rod bipolar cell reduces its gain by ~50%, and the downstream AII amacrine and ON-type ganglion cells reduce their gain by >90%. Thus, the site of adaptation changes with the background level from the rod bipolar synapse in dim light to the rods in brighter light.

The adaptation at the rod bipolar → AII amacrine cell synapse can also be measured using paired-flashes protocols. In the case of two flashes spaced 100 ms apart, the response to the second flash is weaker than the response to the first flash (Dunn & Rieke, 2008). Remarkably, flashes that cause a rod bipolar cell to respond to only one R* event are sufficient to produce paired-flash suppression. This suppression persisted after blocking of inhibitory inputs from amacrine cells. Thus, the mechanism apparently depends on a depression at the rod bipolar synapse, which can be explained by vesicle depletion. For example, the rod bipolar cell's synaptic release during the first flash would partially deplete the pool of vesicles available for the second flash response. Furthermore, increasing the background light level above ~1 R*/rod/s depolarizes the rod bipolar cell's membrane potential, which steadily increases the basal release rate. This increased basal release decreases the number of vesicles available for evoked release, to the flashes, and this explains the reduced gain in the AII amacrine cell at backgrounds >1 R*/rod/s (Jarsky et al., 2011; Oesch & Diamond, 2011; Singer & Diamond, 2006).

Similar to the rod pathway, the cone pathway shows a switch in the site of adaptation at different background levels. Adaptation was studied at several points in the cone bipolar pathway of primate retina under conditions where the rods were suppressed (figure 15.7) (Dunn, Lankheet, & Rieke, 2007). Gain was measured by the response to a brief flash presented on either darkness or backgrounds of variable intensities. Two types of ganglion cell—midget and parasol—show an ~50% reduction in gain between darkness and backgrounds that evoke 100 to 1,000 R*/cone/s (Dunn, Lankheet, & Rieke, 2007; see also Lee et al., 1990; Purpura et al., 1990). However, at these background levels there is no reduction in gain within the cones or the cone bipolar cells. When the background is increased to ~6,000 R*/cone/s, gain is reduced similarly in the cones and the downstream bipolar and ganglion cells (figure 15.7) (see also Schneeweis & Schnapf, 1999). Thus, the site of adaptation changes with the background level from the cone bipolar synapses, at dim backgrounds, to the cones, at brighter backgrounds. Adaptation is restricted within individual cones and does not spread to neighboring cones (Lee et al., 1999). In addition to reductions in gain, increases in background light cause a speeding of response kinetics and an associated shift to higher peak temporal frequency sensitivity (Dunn, Lankheet, & Rieke, 2007; Enroth-Cugell & Shapley, 1973; Smith et al., 2001; Wang, Weick, & Demb, 2011) (figure 15.8A).

In addition to adjusting the gain and temporal kinetics, changes in background level also alter the spatial receptive field. In dim light the center expands, and the surround weakens and in some cases becomes undetectable (Barlow, Fitzhugh, & Kuffler, 1957; Enroth-Cugell & Robson, 1966; Troy, Bohnsack, & Diller, 1999). Center expansion may be explained by the change in electrical coupling within the network of AII amacrine cells. For example, under twilight conditions coupling within the AII network increases about sevenfold relative to daylight conditions, which could expand the center size of ganglion cells (Xin & Bloomfield, 1999a). Dim light also increases the electrical coupling of horizontal cells, which may explain the weakening of the surround (Xin & Bloomfield, 1999b). In both cases changes in coupling depend on neuromodulators, including dopamine (see Bloomfield & Völgyi, 2009).

ADAPTATION TO CONTRAST

In addition to adapting to the mean light intensity, ganglion cells also adapt to the contrast (Shapley & Victor, 1978; Smirnakis et al., 1997; Victor, 1987). Similar to mean light adaptation, contrast adaptation should be able to adjust gain on a rapid time scale (milliseconds) to keep pace with eye movements. In a given contrast condition a cell's receptive field is probed with a stimulus that includes a broad range of temporal and/ or spatial frequencies. The most commonly used stimulus is white noise, where multiple stimuli are presented in a checkerboard pattern, and each check displays a random luminance value on each frame. The white noise can also be displayed as a single temporal sequence presented within a spot over the receptive field center. The cell's linear filter can then be determined by cross-correlating the stimulus and the response (Chichilnisky, 2001). This procedure can be performed at two or more contrast levels to determine how the response gain and kinetics change. Furthermore, the change in the cell's linear filter can be separated from the static nonlinearities mentioned above (e.g., the spike threshold and response saturation) by including a single contrast-independent nonlinear stage (Chander & Chichilnisky, 2001; Kim & Rieke, 2001; Zaghloul, Boahen, & Demb, 2005).

Increasing contrast has two effects on ganglion cell firing rates. First, there is a reduction in gain, as indicated by a reduction in the height of the linear filter (figure 15.8B). Second, there is a speeding of response kinetics, as indicated by narrowing of the filter width (Baccus & Meister, 2002; Chander & Chichilnisky, 2001; Kim & Rieke, 2001; Shapley & Victor, 1978; Smirnakis et al., 1997; Zaghloul, Boahen, & Demb, 2005). These changes in the time domain correspond to a selective reduction in the gain at low temporal frequencies (Shapley & Victor, 1978; Zaghloul, Boahen, & Demb,

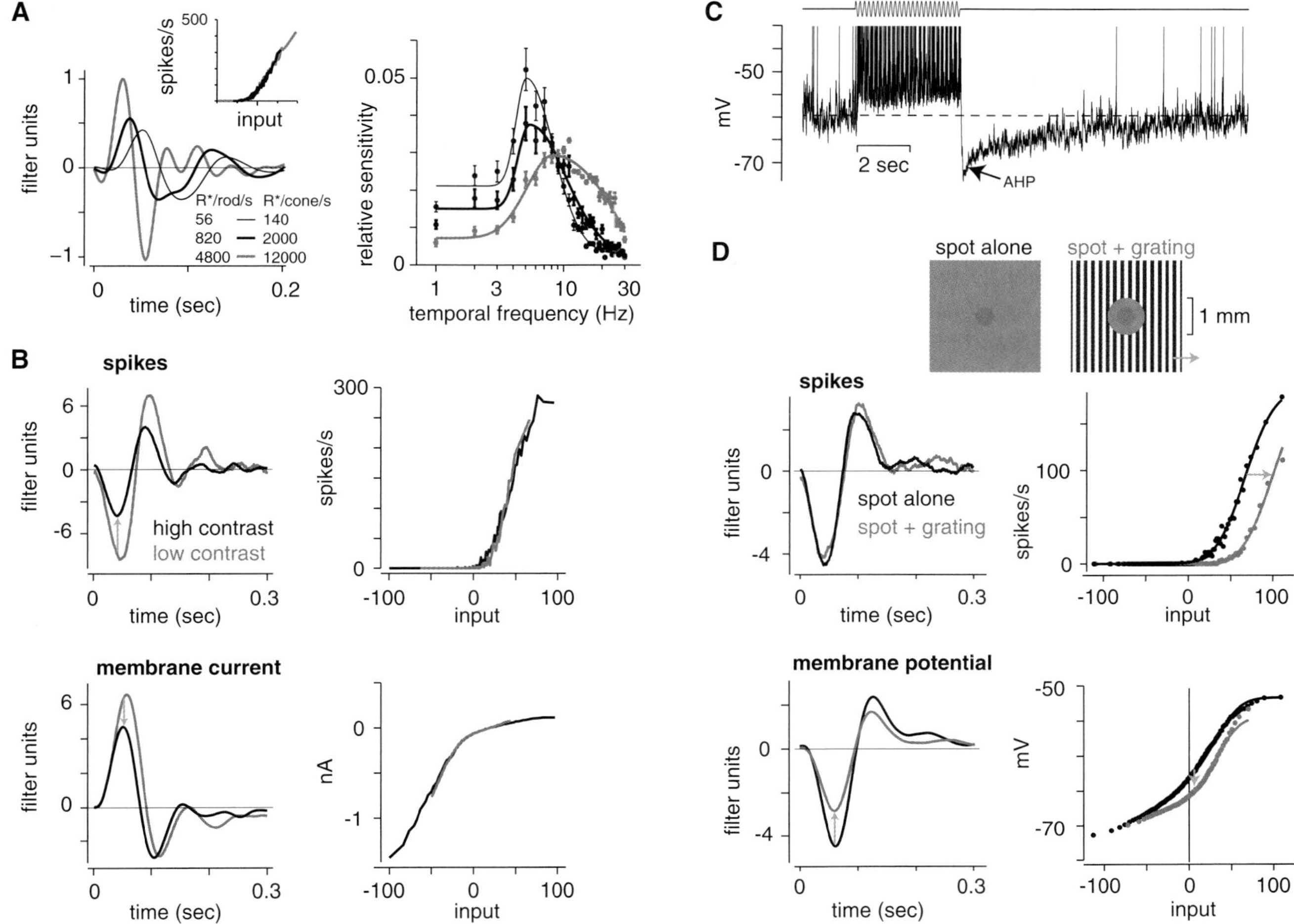

FIGURE 15.8 Different forms of ganglion cell adaptation to mean intensity and contrast. (A, left) Linear filters describing an ON ganglion cell's response at three levels of mean intensity (which correspond to the R* rates indicated). The filter changes with mean intensity, becoming faster at higher means, whereas the nonlinearities overlap (see inset: x-axis shows linear model input values in arbitrary units). The linear–nonlinear model was calculated based on white-noise stimulation using UV light in the ventral mouse retina (see Wang, Weick, & Demb, 2011). (A, right) Temporal sensitivity of mouse ganglion cells at three levels of mean intensity (n = 6 cells). At higher mean levels the peak temporal frequency shifts to higher values. The data are derived from the Fourier transform of the linear filters measured as in the plot at left. (B) Linear–nonlinear models show adaptation to contrast in a guinea pig OFF ganglion cell. For the spike response (top row), the high-contrast filter has reduced gain (arrow) and faster kinetics. For the excitatory membrane currents the high-contrast filter also showed reduced gain and faster kinetics, although the gain change is less than that measured in the spikes. In both cases the high- and low-contrast nonlinearities overlap (see Beaudoin, Borghuis, & Demb, 2007). (C) Contrast adaptation can generate a slow change in membrane potential. A guinea pig OFF ganglion cell was stimulated with a high-contrast drifting grating. At stimulus offset there was an afterhyperpolarization (AHP) that recovered over several seconds (see Manookin & Demb, 2006). (D) Contrast over a ganglion cell's receptive field surround suppresses the response to center stimulation. White-noise modulation of a spot over the center was presented alone or in the presence of a drifting grating in the surround. A linear–nonlinear model of the spike response showed that the grating increased the threshold for firing (i.e., shifted the nonlinear function rightward). For the same cell the response in the membrane potential showed that the grating suppressed the gain of the linear filter and caused tonic hyperpolarization of the membrane potential (i.e., shifted the nonlinearity downward; see Zaghloul et al., 2007).

2005). Thus, at high contrast the cell becomes relatively more sensitive to higher temporal frequencies. Contrast adaptation is typically studied using laboratory stimuli, such as white noise, but it can also be observed during more natural stimuli (Lesica et al., 2007; Mante, Bonin, & Carandini, 2008).

The mechanism of contrast adaptation depends on a combination of presynaptic and intrinsic mechanisms of the ganglion cell. To understand the presynaptic mechanisms, adaptation was measured in cells throughout the circuit. Photoreceptors and horizontal cells did not show adaptation (Baccus & Meister, 2002;

Beaudoin, Borghuis, & Demb, 2007; Rieke, 2001). Thus, contrast adaptation must originate at a site beyond the point of photoreceptor synaptic release. Bipolar cells did show adaptation, although the effects were variable (Baccus & Meister, 2002; Rieke, 2001). In one case both ON and OFF bipolar cells from salamander showed reduced gain at high contrast, but only OFF cells showed a substantial speeding of response kinetics (Rieke, 2001). In a second case some bipolar cells from salamander showed a different form of adaptation at high contrast: a slow membrane hyperpolarization following the onset of high contrast that likewise recovered slowly following the onset of low contrast (Baccus & Meister, 2002). There was additional diversity at the level of amacrine and ganglion cells, with variable combinations of the rapid changes in response gain and kinetics and the slower changes in membrane potential (Baccus & Meister, 2002). Mammalian ganglion cells show similarly diverse contrast adaptation at multiple time scales (figure 15.8C) (Chander & Chichilnisky, 2001; Manookin & Demb, 2006; Shapley & Victor, 1978; Solomon et al., 2004; Wark, Fairhall, & Rieke, 2009; Zaghloul, Boahen, & Demb, 2005).

The first stage of contrast adaptation occurs at the level of bipolar cells. Adaptation in bipolar cells appears not to require inhibitory inputs from amacrine cells because adaptation measured in bipolar cells or in the output of bipolar cells (i.e., excitatory postsynaptic currents in ganglion cells) persists after blocking of inhibitory receptors (Beaudoin, Borghuis, & Demb, 2007; Manookin & Demb, 2006; Rieke, 2001). Adaptation in the bipolar cell can be explained by activity-dependent processes either at the dendrites or at the axon terminal. In salamander bipolar cells adaptation in OFF-type cells depends on an intrinsic mechanism sensitive to intracellular Ca^{2+} level in the dendrites (Rieke, 2001). However, there could also be mechanisms for adaptation at the bipolar cell axon terminal. In mouse rod bipolar cells the synaptic release onto AII amacrine cells was studied by paired voltage-clamp recordings (Jarsky et al., 2011). To mimic the effect of changing the contrast of a light stimulus, the rod bipolar cell's membrane potential was modulated around a constant mean with low and high variance. The gain of the synapse was reduced when the variance increased. The reduced gain could be explained by vesicle depletion, the same mechanism for mean light adaptation following a steady depolarization of the rod bipolar membrane potential. Thus, a common synaptic mechanism, vesicle depletion, can apparently drive both forms of adaptation.

Although contrast adaptation clearly occurs at the level of bipolar cells, the magnitude of this adaptation cannot explain the magnitude in ganglion cells (Beaudoin, Borghuis, & Demb, 2007; Brown & Masland, 2001; Kim & Rieke, 2001; Zaghloul, Boahen, & Demb, 2005). For example, the excitatory currents from guinea pig alpha cells reduce gain by ~20% following a tripling of the contrast, whereas the firing rate of the same cells reduce gain by ~40% (figure 15.8B) (Beaudoin, Borghuis, & Demb, 2007; Beaudoin, Manookin, & Demb, 2008; Zaghloul, Boahen, & Demb, 2005). Thus, there is apparently additional adaptation that originates within the ganglion cell itself. A modeling study suggested that contrast adaptation could be generated through the process of spike generation. In particular, in the presence of relatively high firing rates at high contrast, the refractory period following each spike serves as an inhibitory feedback that accumulates over spikes and attenuates the firing rate (Gaudry & Reinagel, 2007a). Consistent with this model, isolated salamander ganglion cells showed a slow recovery from Na^+ channel inactivation. The cells were stimulated by injecting fluctuating currents, around a constant mean, with either low or high variance to mimic low or high contrast. In the presence of high-variance stimulation, the gain was reduced by Na^+ channel inactivation (Kim & Rieke, 2003). Furthermore, intact mammalian ganglion cells show evidence that Na^+ channel inactivation accounts for reduced firing rates during periods of strong stimulation that mimic high contrast (Weick & Demb, 2011).

There is a second intrinsic property of ganglion cells that can contribute to contrast adaptation. High contrast generates both extreme depolarizations and hyperpolarizations from the resting potential. The depolarizations drive spiking, which induces the Na^+ channel inactivation described above. The hyperpolarizations can remove inactivation of K^+ channels. During subsequent depolarization there are more K^+ channels available to suppress firing, and indeed, depolarizing a ganglion cell following a hyperpolarization yields fewer spikes than if the depolarization occurred in isolation (Weick & Demb, 2011). Thus, both Na^+- and K^+-channel mechanisms contribute to intrinsic mechanisms for contrast adaptation in ganglion cells.

In the experiments described above the response gain of the ganglion cell's center was reduced by increasing contrast over the center. However, the center response can also be suppressed by contrast presented over the surround (Shapley & Victor, 1979). This response suppression can be driven by stimuli with high spatial frequencies that drive amacrine cells but not horizontal cells (Baccus, Olveczky, & Meister, 2008; Enroth-Cugell & Jakiela, 1980; Solomon, Lee, & Sun, 2006; Zaghloul et al., 2007). This amacrine-mediated surround effect suppresses the gain of the subthreshold

    YANBIN V. WANG AND JONATHAN B. DEMB

response in the ganglion cell's membrane potential and also causes a tonic membrane hyperpolarization; these effects on the membrane potential cause an increased threshold for firing spikes (figure 15.8D) (Zaghloul et al., 2007). It was proposed that a common mechanism for contrast adaptation might extend broadly across the ganglion cell's center and surround and explain the suppression associated with increasing contrast in either region (Enroth-Cugell & Jakiela, 1980; Shapley & Victor, 1979). However, we can now appreciate that the cellular basis for suppression associated with center and surround stimulation differs: Increasing contrast over the center suppresses bipolar release through a mechanism that does not require amacrine cells, whereas increasing contrast over the surround drives amacrine cell–mediated inhibition (Beaudoin, Borghuis, & Demb, 2007; Zaghloul et al., 2007). The suppression from the surround may be better conceptualized as a nonlinear component of the receptive field rather than a form of adaptation (figure 15.4B), but this distinction is perhaps subtle.

FUTURE DIRECTIONS

Local regions within natural images viewed in daylight conditions show independent mean intensity and contrast values. Therefore, a ganglion cell's receptive field would experience independent changes in mean intensity and contrast across eye fixations following saccadic eye movements (Mante et al., 2005). In cells of the lateral geniculate nucleus (LGN) of the thalamus, which are postsynaptic to ganglion cells, mechanisms for mean light and contrast adaptation could be modeled as distinct and separable processes, each of which reduces response gain. However, it is not clear that the cellular mechanisms for the two processes could be distinct and separable. For example, both mean intensity and contrast adaptation depend in part on mechanisms involving bipolar cells (Dunn et al., 2006; Dunn, Lankheet, & Rieke, 2007; Dunn & Rieke, 2008; Jarsky et al., 2011). Future research will be required to further elucidate the independence of mechanisms for mean light and contrast adaptation.

Here we focused on the two most commonly studied forms of adaptation in the retina. However, ganglion cells (or postsynaptic LGN cells) also adapt to more complex features of the light input. For example, some ganglion cells show adaptation to stimulus orientation: Following exposure to a vertical grating, the cell becomes more sensitive to horizontal gratings (Hosoya, Baccus, & Meister, 2005). In one study LGN cells did not show adaptation to changes in higher statistical moments of the input distribution (i.e., skewness or

kurtosis), suggesting that adaptation depended only on the mean and contrast (standard deviation) (Bonin, Mante, & Carandini, 2006). However, another study suggested that changes in higher statistical moments affect the time course of adaptation (Wark, Fairhall, & Rieke, 2009). Furthermore, there are some cells that show the opposite of contrast adaptation, becoming temporarily more sensitive following a switch from high to low contrast (Kastner & Baccus, 2011). Working out the mechanisms for these more complex forms of adaptation will be an important area for future study.

The LN model described above has been useful for quantifying adaptation and distinguishing gain changes from other fixed nonlinear processes (e.g., response saturation). However, this modeling yields a filter for each contrast or mean light level but does not include an adaptive mechanism, whereas it would be better to have a single model that could predict how adaptation affects the response during ongoing changes in mean luminance and contrast (Victor, 1987). One such model combined a LN cascade with a third kinetic stage resembling the state diagram used to describe the behavior of an ion channel (Ozuysal & Baccus, 2012). This model generated an accurate characterization of subthreshold membrane potential responses in bipolar, amacrine, and ganglion cells. Furthermore, the apparently distinct fast contrast-dependent gain changes and slower effects on membrane potential could be predicted by the same model. This approach seems promising and may help to link the computational description of adaptation to the biophysical mechanisms (Jarsky et al., 2011).

It is surprising, given the diversity of amacrine cell types, that there are not more well-defined roles for these cells in adaptation. One type that we have not discussed is the dopamine amacrine cell, which increases release during the day and decreases release at night. Dopamine acts in a paracrine fashion stimulating receptors throughout the retina. Although there are multiple descriptions of dopamine's effect on retinal neurons, from photoreceptors to ganglion cells, a unifying theory for dopamine's effect on retinal circuitry is still needed (Bloomfield & Völgyi, 2009; Witkovsky, 2004). A recent study suggested a novel role for dopamine that would cause horizontal cells to release GABA onto rod bipolar cells (Herrmann et al., 2011); this inhibition of the rod bipolar cell might allow the cell's synaptic output to avoid saturation in the light-adapted state (Jarsky et al., 2011). There are other slow-acting neuromodulators in the retina, and it will be important to understand their possible role in light adaptation as well (Daw, Brunken, & Parkinson, 1989; Karten & Brecha, 1983). Furthermore, local GABAergic or glycinergic amacrine circuits that provide feedback inhibition onto bipolar terminals would seem ideal for generating contrast adaptation on a fast time scale (Eggers & Lukasiewicz, 2011). However, the role of feedback inhibition in contrast adaptation is not clear.

Finally, it will be important to understand further the impact of adaptation on visual behavior. There have been parallel studies of adaptation using psychophysical measurements of vision, primarily in humans, and physiological measurements of cellular activity, exclusively in experimental animals (see Shapley & Enroth-Cugell, 1984; Walraven et al., 1990). However, the link between the physiology and behavior is not always straightforward (Rieke & Rudd, 2009). Furthermore, contrast adaptation has been shown to improve information processing in physiological responses (Gaudry & Reinagel, 2007b), but a similar impact on behavior is less clear. An exciting avenue for future research would be to manipulate adaptation within specific circuits using genetic techniques in an animal where it also possible to measure behavior quantitatively (Okawa et al., 2010). For genetic manipulation, the mouse is the obvious choice. And we now understand how to study cone vision in this animal. The temporal properties of mouse ganglion cell receptive fields are quite fast and similar to those of primate cells as long as the cones are sufficiently stimulated (Wang, Weick, & Demb, 2011); in much of the mouse retina, strong cone stimulation requires an ultraviolet light stimulus. Genetic manipulations of retinal circuitry combined with quantitative estimates of photoreceptor stimulation and behavioral analysis (Busse et al., 2011; Histed, Carvalho, & Maunsell, 2012) could yield fundamental insights into the impact of adaptation on visual behavior.

REFERENCES

Baccus, S. A., & Meister, M. (2002). Fast and slow contrast adaptation in retinal circuitry. *Neuron, 36,* 909–919.

Baccus, S. A., Olveczky, B. P., & Meister, M. (2008). A retinal circuit that computes object motion. *Journal of Neuroscience, 28,* 6807–6817.

Barlow, H. B., Fitzhugh, R., & Kuffler, S. W. (1957). Change of organization in the receptive fields of the cat's retina during dark adaptation. *Journal of Physiology (London), 137,* 338–354.

Beaudoin, D. L., Borghuis, B. G., & Demb, J. B. (2007). Cellular basis for contrast gain control over the receptive field center of mammalian retinal ganglion cells. *Journal of Neuroscience, 27,* 2636–2645.

Beaudoin, D. L., Manookin, M. B., & Demb, J. B. (2008). Distinct expressions of contrast gain control in parallel synaptic pathways converging on a retinal ganglion cell. *Journal of Physiology, 586,* 5487–5502.

Berntson, A., Smith, R. G., & Taylor, W. R. (2004). Transmission of single photon signals through a binary synapse in the mammalian retina. *Visual Neuroscience, 21,* 693–702.

Bloomfield, S. A., & Dacheux, R. F. (2001). Rod vision: Pathways and processing in the mammalian retina. *Progress in Retinal and Eye Research, 20,* 351–384.

Bloomfield, S. A., & Völgyi, B. (2009). The diverse functional roles and regulation of neuronal gap junctions in the retina. *Nature Reviews. Neuroscience, 10,* 495–506.

Bonin, V., Mante, V., & Carandini, M. (2006). The statistical computation underlying contrast gain control. *Journal of Neuroscience, 26,* 6346–6353.

Brown, S. P., & Masland, R. H. (2001). Spatial scale and cellular substrate of contrast adaptation by retinal ganglion cells. *Nature Neuroscience, 4,* 44–51.

Busse, L., Ayaz, A., Dhruv, S. N., Katzner, T., Saleem, A. B., Schölvinck, M. L., et al. (2011). The detection of visual contrast in the behaving mouse. *Journal of Neuroscience, 31,* 11351–11361. doi:10.1523/JNEUROSCI.6689-10.2011.

Butts, D. A., Weng, C., Jin, J., Alonso, J. M., & Paninski, L. (2011). Temporal precision in the visual pathway through the interplay of excitation and stimulus-driven suppression. *Journal of Neuroscience, 31,* 11313–11327.

Chander, D., & Chichilnisky, E. J. (2001). Adaptation to temporal contrast in primate and salamander retina. *Journal of Neuroscience, 21,* 9904–9916.

Chichilnisky, E. J. (2001). A simple white noise analysis of neuronal light responses. *Network, 12,* 199–213.

Dacey, D. M., & Petersen, M. R. (1992). Dendritic field size and morphology of midget and parasol ganglion cells of the human retina. *Proceedings of the National Academy of Sciences of the United States of America, 89,* 9666–9670. doi:10.1073/pnas.89.20.9666.

Daw, N. W., Brunken, W. J., & Parkinson, D. (1989). The function of synaptic transmitters in the retina. *Annual Review of Neuroscience, 12,* 205–225.

Demb, J. B., Haarsma, L., Freed, M. A., & Sterling, P. (1999). Functional circuitry of the retinal ganglion cell's nonlinear receptive field. *Journal of Neuroscience, 19,* 9756–9767.

Demb, J. B., & Singer, J. H. (2012). Intrinsic properties and functional circuitry of the AII amacrine cell. *Visual Neuroscience, 29,* 51–60.

Demb, J. B., Zaghloul, K., Haarsma, L., & Sterling, P. (2001). Bipolar cells contribute to nonlinear spatial summation in the brisk-transient (Y) ganglion cell in mammalian retina. *Journal of Neuroscience, 21,* 7447–7454.

Dunn, F. A., Doan, T., Sampath, A. P., & Rieke, F. (2006). Controlling the gain of rod-mediated signals in the mammalian retina. *Journal of Neuroscience, 26,* 3959–3970.

Dunn, F. A., Lankheet, M. J., & Rieke, F. (2007). Light adaptation in cone vision involves switching between receptor and post-receptor sites. *Nature, 449,* 603–606.

Dunn, F. A., & Rieke, F. (2006). The impact of photoreceptor noise on retinal gain controls. *Current Opinion in Neurobiology, 16,* 363–370.

Dunn, F. A., & Rieke, F. (2008). Single-photon absorptions evoke synaptic depression in the retina to extend the operational range of rod vision. *Neuron, 57,* 894–904.

Eggers, E. D., & Lukasiewicz, P. D. (2011). Multiple pathways of inhibition shape bipolar cell responses in the retina. *Visual Neuroscience, 28,* 95–108.

Enroth-Cugell, C., & Jakiela, H. G. (1980). Suppression of cat retinal ganglion cell responses by moving patterns. *Journal of Physiology, 302,* 49–72.

 YANBIN V. WANG AND JONATHAN B. DEMB

Enroth-Cugell, C., & Robson, J. G. (1966). The contrast sensitivity of retinal ganglion cells of the cat. *Journal of Physiology, 187*, 517–552.

Enroth-Cugell, C., & Shapley, R. M. (1973). Adaptation and dynamics of cat retinal ganglion cells. *Journal of Physiology, 233*, 311–326.

Field, G. D., & Chichilnisky, E. J. (2007). Information processing in the primate retina: Circuitry and coding. *Annual Review of Neuroscience, 30*, 1–30.

Field, G. D., & Rieke, F. (2002). Nonlinear signal transfer from mouse rods to bipolar cells and implications for visual sensitivity. *Neuron, 34*, 773–785.

Flores-Herr, N., Protti, D. A., & Wässle, H. (2001). Synaptic currents generating the inhibitory surround of ganglion cells in the mammalian retina. *Journal of Neuroscience, 21*, 4852–4863.

Frazor, R. A., & Geisler, W. S. (2006). Local luminance and contrast in natural images. *Vision Research, 46*, 1585–1598.

Gaudry, K. S., & Reinagel, P. (2007a). Contrast adaptation in a nonadapting LGN model. *Journal of Neurophysiology, 98*, 1287–1296.

Gaudry, K. S., & Reinagel, P. (2007b). Benefits of contrast normalization demonstrated in neurons and model cells. *Journal of Neuroscience, 27*, 8071–8079.

Herrmann, R., Heflin, S. J., Hammond, T., Lee, B., Wang, J., Gainetdinov, R. R., et al. (2011). Rod vision is controlled by dopamine-dependent sensitization of rod bipolar cells by GABA. *Neuron, 72*, 101–110.

Histed, M. H., Carvalho, L. A., & Maunsell, J. H. (2012). Psychophysical measurement of contrast sensitivity in the behaving mouse. *Journal of Neurophysiology, 107*, 758–765.

Hochstein, S., & Shapley, R. M. (1976). Linear and nonlinear spatial subunits in Y cat retinal ganglion cells. *Journal of Physiology, 262*, 265–284.

Hosoya, T., Baccus, S. A., & Meister, M. (2005). Dynamic predictive coding by the retina. *Nature, 436*, 71–77.

Jarsky, T., Cembrowski, M., Logan, S. M., Kath, W. L., Riecke, H., Demb, J. B., et al. (2011). A synaptic mechanism for retinal adaptation to luminance and contrast. *Journal of Neuroscience, 31*, 11003–11015.

Karten, H. J., & Brecha, N. (1983). Localization of neuroactive substances in the vertebrate retina: Evidence for lamination in the inner plexiform layer. *Vision Research, 10*, 1197–1205.

Kastner, D. B., & Baccus, S. A. (2011). Coordinated dynamic encoding in the retina using opposing forms of plasticity. *Nature Neuroscience, 14*, 1317–1322.

Keat, J., Reinagel, P., Reid, R. C., & Meister, M. (2001). Predicting every spike: A model for the responses of visual neurons. *Neuron, 30*, 803–817.

Kim, K. J., & Rieke, F. (2001). Temporal contrast adaptation in the input and output signals of salamander retinal ganglion cells. *Journal of Neuroscience, 21*, 287–299.

Kim, K. J., & Rieke, F. (2003). Slow Na$^+$ inactivation and variance adaptation in salamander retinal ganglion cells. *Journal of Neuroscience, 23*, 1506–1516.

Koike, C., Obara, T., Uriu, Y., Numata, T., Sanuki, R., Miyata, K., et al. (2010). TRPM1 is a component of the retinal ON bipolar cell transduction channel in the mGluR6 cascade. *Proceedings of the National Academy of Sciences of the United States of America, 107*(1), 332–337. doi:10.1073/pnas.0912730107.

Kuffler, S. W. (1953). Discharge patterns and functional organization of mammalian retina. *Journal of Neurophysiology, 16*, 37–68.

Laughlin, S. (1981). A simple coding procedure enhances a neuron's information capacity. *Zeitschrift für Naturforschung C, 36*, 910–912.

Lee, B. B., Dacey, D. M., Smith, V. C., & Pokorny, J. (1999). Horizontal cells reveal cone type-specific adaptation in primate retina. *Proceedings of the National Academy of Sciences of the United States of America, 96*, 14611–14666.

Lee, B. B., Dacey, D. M., Smith, V. C., & Pokorny, J. (2003). Dynamics and sensitivity regulation in primate outer retina: The horizontal cell network. *Journal of Vision, 3*(7), 5, 513–526. doi:10.1167/3.7.5.

Lee, B. B., Pokorny, J., Smith, V. C., Martin, P. R., & Valberg, A. (1990). Luminance and chromatic modulation sensitivity of macaque ganglion cells and human observers. *Journal of the Optical Society of America. A, Optics and Image Science, 7*, 2223–2236. doi:10.1364/JOSAA.7.002223.

Lesica, N. A., Jin, J., Weng, C., Yeh, C. I., Butts, D. A., Stanley, G. B., et al. (2007). Adaptation to stimulus contrast and correlations during natural visual stimulation. *Neuron, 55*, 479–491.

Manookin, M. B., Beaudoin, D. L., Ernst, Z. R., Flagel, L. J., & Demb, J. B. (2008). Disinhibition combines with excitation to extend the operating range of the OFF visual pathway in daylight. *Journal of Neuroscience, 28*, 4136–4150.

Manookin, M. B., & Demb, J. B. (2006). Presynaptic mechanism for slow contrast adaptation in mammalian retinal ganglion cells. *Neuron, 50*, 453–464.

Mante, V., Bonin, V., & Carandini, M. (2008). Functional mechanisms shaping lateral geniculate responses to artificial and natural stimuli. *Neuron, 58*, 625–638.

Mante, V., Frazor, R. A., Bonin, V., Geisler, W. S., & Carandini, M. (2005). Independence of luminance and contrast in natural scenes and in the early visual system. *Nature Neuroscience, 8*, 1690–1697.

Masland, R. H. (2001). The fundamental plan of the retina. *Nature Neuroscience, 4*, 877–886.

Masland, R. H. (2011). Cell populations of the retina: The Proctor lecture. *Investigative Ophthalmology & Visual Science, 52*, 4581–4591. doi:10.1167/iovs.10-7083.

McMahon, M. J., Packer, O. S., & Dacey, D. M. (2004). The classical receptive field surround of primate parasol ganglion cells is mediated primarily by a non-GABAergic pathway. *Journal of Neuroscience, 24*, 3736–3745.

Morgans, C. W., Zhang, J., Jeffrey, B. G., Nelson, S. M., Burke, N. S., Duvoisin, R. M., et al. (2009). TRPM1 is required for the depolarizing light response in retinal ON-bipolar cells. *Proceedings of the National Academy of Sciences of the United States of America, 106*, 19174–19178. doi:10.1073/pnas.0908711106.

Nakajima, Y., Iwakabe, H., Akazawa, C., Nawa, H., Shigemoto, R., Mizuno, N., et al. (1993). Molecular characterization of a novel retinal metabotropic glutamate receptor mGluR6 with a high agonist selectivity for L-2-amino-4-phosphono-butyrate. *Journal of Biological Chemistry, 268*, 11868–11873.

Nelson, R., Famiglietti, E. V., Jr., & Kolb, H. (1978). Intracellular staining reveals different levels of stratification for on- and off-center ganglion cells in cat retina. *Journal of Neurophysiology, 41*, 472–483.

Nomura, A., Shigemoto, R., Nakamura, Y., Okamoto, N., Mizuno, N., & Nakanishi, S. (1994). Developmentally regulated postsynaptic localization of a metabotropic glutamate receptor in rat rod bipolar cells. *Cell, 77*, 361–369.

O'Brien, B. J., Isayama, T., Richardson, R., & Berson, D. M. (2002). Intrinsic physiological properties of cat retinal ganglion cells. *Journal of Physiology, 538*, 787–802.

Oesch, N. W., & Diamond, J. S. (2011). Ribbon synapses compute temporal contrast and encode luminance in retinal rod bipolar cells. *Nature Neuroscience, 14*, 1555–1561.

Okawa, H., Miyagishima, K. J., Arman, A. C., Hurley, J. B., Field, G. D., & Sampath, A. P. (2010). Optimal processing of photoreceptor signals is required to maximize behavioural sensitivity. *Journal of Physiology, 588*, 1947–1960.

Ölveczky, B. P., Baccus, S. A., & Meister, M. (2003). Segregation of object and background motion in the retina. *Nature, 423*, 401–408.

Ozuysal, Y., & Baccus, S. A. (2012). Linking the computational structure of variance adaptation to biophysical mechanisms. *Neuron, 73*, 1002–1015.

Pillow, J. W., Paninski, L., Uzzell, V. J., Simoncelli, E. P., & Chichilnisky, E. J. (2005). Prediction and decoding of retinal ganglion cell responses with a probabilistic spiking model. *Journal of Neuroscience, 25*, 11003–11013.

Purpura, K., Tranchina, D., Kaplan, E., & Shapley, R. M. (1990). Light adaptation in primate retina: Analysis of changes in gain and dynamics of monkey retinal ganglion cells. *Visual Neuroscience, 4a*, 75–93. doi:10.1017/S0952523800002789.

Remtulla, S., & Hallett, P. E. (1985). A schematic eye for the mouse, and comparisons with the rat. *Vision Research, 25*, 21–31.

Rieke, F. (2001). Temporal contrast adaptation in salamander bipolar cells. *Journal of Neuroscience, 21*, 9445–9454.

Rieke, F., & Rudd, M. E. (2009). The challenges natural images pose for visual adaptation. *Neuron, 64*, 605–616.

Rodieck, R. W. (1965). Quantitative analysis of cat retinal ganglion cell response to visual stimuli. *Vision Research, 5*, 583–601.

Roska, B., & Werblin, F. (2001). Vertical interactions across ten parallel, stacked representations in the mammalian retina. *Nature, 410*, 583–587.

Sampath, A. P., & Rieke, F. (2004). Selective transmission of single photon responses by saturation at the rod-to-rod bipolar synapse. *Neuron, 41*, 431–443.

Schneeweis, D. M., & Schnapf, J. L. (1999). The photovoltage of macaque cone photoreceptors: Adaptation, noise, and kinetics. *Journal of Neuroscience, 19*, 1203–1216.

Shapley, R. M., & Enroth-Cugell, C. (1984). Visual adaptation and retinal gain controls. *Progress in Retinal Research, 3*, 263–346. doi:10.1016/0278-4327(84)90011-7.

Shapley, R. M., & Victor, J. D. (1978). The effect of contrast on the transfer properties of cat retinal ganglion cells. *Journal of Physiology, 285*, 275–298.

Shapley, R. M., & Victor, J. D. (1979). Nonlinear spatial summation and the contrast gain control of cat retinal ganglion cells. *Journal of Physiology, 290*, 141–161.

Singer, J. H., & Diamond, J. S. (2006). Vesicle depletion and synaptic depression at a mammalian ribbon synapse. *Journal of Neurophysiology, 95*, 3191–3198.

Smirnakis, S. M., Berry, M. J., Warland, D. K., Bialek, W., & Meister, M. (1997). Adaptation of retinal processing to image contrast and spatial scale. *Nature, 386*, 69–73.

Smith, V. C., Pokorny, J., Lee, B. B., & Dacey, D. M. (2001). Primate horizontal cell dynamics: An analysis of sensitivity regulation in the outer retina. *Journal of Neurophysiology, 85*, 545–558.

Solomon, S. G., Lee, B. B., & Sun, H. (2006). Suppressive surrounds and contrast gain in magnocellular-pathway retinal ganglion cells of macaque. *Journal of Neuroscience, 26*, 8715–8726.

Solomon, S. G., Peirce, J. W., Dhruv, N. T., & Lennie, P. (2004). Profound contrast adaptation early in the visual pathway. *Neuron, 42*, 155–162.

Sterling, P., & Demb, J. B. (2004). Retina. In G. Shephard (Ed.), *Synaptic organization of the brain* (5th ed.) (pp. 217–269). New York: Oxford University Press.

Stone, C., & Pinto, L. H. (1993). Response properties of ganglion cells in the isolated mouse retina. *Visual Neuroscience, 10*, 31–39.

Troy, J. B., Bohnsack, D. L., & Diller, L. C. (1999). Spatial properties of the cat X-cell receptive field as a function of mean light level. *Visual Neuroscience, 16*, 1089–1104.

van Rossum, M. C., & Smith, R. G. (1998). Noise removal at the rod synapse of mammalian retina. *Visual Neuroscience, 15*, 809–821.

Victor, J. D. (1987). The dynamics of the cat retinal X cell centre. *Journal of Physiology, 386*, 219–246.

Walraven, J., Enroth-Cugell, C., Hood, D. C., MacLeod, D. I. A., & Schnapf, J. L. (1990). The control of visual sensitivity: Receptoral and postreceptoral processes. In L. Spillman & J. Werner (Eds.), *The neurophysiological foundations of visual perception* (pp. 53–101). New York: Academic Press.

Wang, Y. V., Weick, M., & Demb, J. B. (2011). Spectral and temporal sensitivity of cone-mediated responses in mouse retinal ganglion cells. *Journal of Neuroscience, 31*, 7670–7681.

Wark, B., Fairhall, A., & Rieke, F. (2009). Timescales of inference in visual adaptation. *Neuron, 61*, 750–761.

Wässle, H. (2004). Parallel processing in the mammalian retina. *Nature Reviews. Neuroscience, 5*, 747–757.

Weick, M., & Demb, J. B. (2011). Delayed-rectifier K channels contribute to contrast adaptation in mammalian retinal ganglion cells. *Neuron, 71*, 166–179.

Werblin, F. S. (2010). Six different roles for crossover inhibition in the retina: Correcting the nonlinearities of synaptic transmission. *Visual Neuroscience, 27*, 1–8. doi:10.1017/S0952523810000076.

Werblin, F. S., & Dowling, J. E. (1969). Organization of the retina of the mudpuppy, *Necturus maculosus*. II. Intracellular recording. *Journal of Neurophysiology, 32*, 339–355.

Witkovsky, P. (2004). Dopamine and retinal function. *Documenta Ophthalmologica, 108*, 17–40. doi:10.1023/B:DOOP.0000019487.88486.0a.

Wong, K. Y., Dunn, F. A., & Berson, D. M. (2005). Photoreceptor adaptation in intrinsically photosensitive retinal ganglion cells. *Neuron, 48*, 1001–1010.

Xin, D., & Bloomfield, S. A. (1999a). Comparison of the responses of AII amacrine cells in the dark- and light-adapted rabbit retina. *Visual Neuroscience, 16*, 653–665.

Xin, D., & Bloomfield, S. A. (1999b). Dark- and light-induced changes in coupling between horizontal cells in mammalian retina. *Journal of Comparative Neurology, 405*, 75–87.

Yeh, T., Lee, B. B., & Kremers, J. (1996). The time course of adaptation in macaque retinal ganglion cells. *Vision Research, 36*, 913–931.

Zaghloul, K. A., Boahen, K., & Demb, J. B. (2003). Different circuits for ON and OFF retinal ganglion cells cause different contrast sensitivities. *Journal of Neuroscience, 23*, 2645–2654.

Zaghloul, K. A., Boahen, K., & Demb, J. B. (2005). Contrast adaptation in subthreshold and spiking responses of mammalian Y-type retinal ganglion cells. *Journal of Neuroscience, 25*, 860–868.

Zaghloul, K. A., Manookin, M. B., Borghuis, B. G., Boahen, K., & Demb, J. B. (2007). Functional circuitry for peripheral suppression in mammalian Y-type retinal ganglion cells. *Journal of Neurophysiology, 97*, 4327–4340.

II ORGANIZATION OF VISUAL PATHWAYS

16 The M, P, and K Pathways of the Primate Visual System Revisited

EHUD KAPLAN

OVERVIEW

In this chapter I update the state of our knowledge about the major parts of the early primate visual system: the retinal ganglion cells (RGCs), the lateral geniculate nucleus (LGN), and the early parts of the visual cortex. I generally refrain from repeating what appeared in the previous edition (Kaplan, 2003) and in a recent review (Kaplan, 2008), where the interested reader can find additional details. Following this snapshot of our expanding database on this subject, I present some recent applications of this knowledge to basic and clinical vision research. Finally, I outline a critique of the prevailing view as it is commonly presented and suggest possible alternatives.

THE BIRTH OF THE M, P, K HYPOTHESIS

The outside world is represented on our retina only by the spatiotemporal distribution of light. Our brain transforms this physical distribution into a rich dynamical perceptual kaleidoscope, populated by visible objects. These objects are each characterized by several properties: color, brightness, location, size, motion, texture, and so on. These exist in our minds only, as Galileo (Galilei, 1623/1960) has put it: "I think that tastes, odors, colors, and so on are no more than mere names so far as the object in which we locate them are concerned, and that they reside in consciousness. Hence, if the living creature were removed, all these qualities would be wiped away and annihilated." This distinction between the external *primary qualities* and the inner *secondary qualities* was made explicit by John Locke in his influential *Essay Concerning Human Understanding* (Locke, 1690/1832). Because neuroanatomy, neurophysiology, and neurochemistry report cells that differ in size, morphology, connectivity, response type, and chemistry, and in some (but not all) cases cells of similar properties congregate together in layers, columns, or stripes, it is tempting to yield to the seductive machine metaphor, which assigns a distinct perceptual function (representing or analyzing a secondary quality) to each structure or cell population. We use the physiological properties of the neurons in these structures to guess the possible role that they might play in the analysis of visual information. This approach gave rise to the concept of functional architecture (Hubel & Wiesel, 1962, 1977), which eventually gave birth to the *visual streams hypothesis*, which states that in the early visual system of primates the visual input is handled by three main streams: the M (magnocellular), P (parvocellular), and K (koniocellular), each with its distinct cluster of properties, and each occupying a distinct niche in the multidimensional parameter space that makes up our visual world. Recent advances in neuroanatomy, functional imaging, and electrophysiological recording from populations of neurons are now providing a much more complete picture of the neuronal machinery that supports vision and bring us closer to a more realistic evaluation of this view.

AN UPDATED SUMMARY OF THE *THREE STREAMS* VIEW

Anatomy

MORPHOLOGICAL CELL TYPES   The input layer of the retina consists of only two types of photoreceptors: rods and cones (of three different wavelength selectivities). The output layer, however, includes (at the latest count) at least 20 types of RGCs, distinguished by their size, shape, connectivity, or neurochemistry. Figure 16.1 shows some of the major types of RGCs that project to the LGN in the macaque monkey (Crook et al., 2008a). Taken together, these 20 types account for ~90% of the fibers in the optic nerve—it is not yet established which unknown types make up the rest. The major (most numerous) types are midget, parasol, bistratified, and large monostratified, which together make up ~60% of the population. A significant recent distinction is between parasol cells and smooth cells, which project to the LGN and to the superior colliculus, and are distinguished by the smooth appearance of their dendrites (see figure 16.1, from Crook et al., 2008a). Their cell

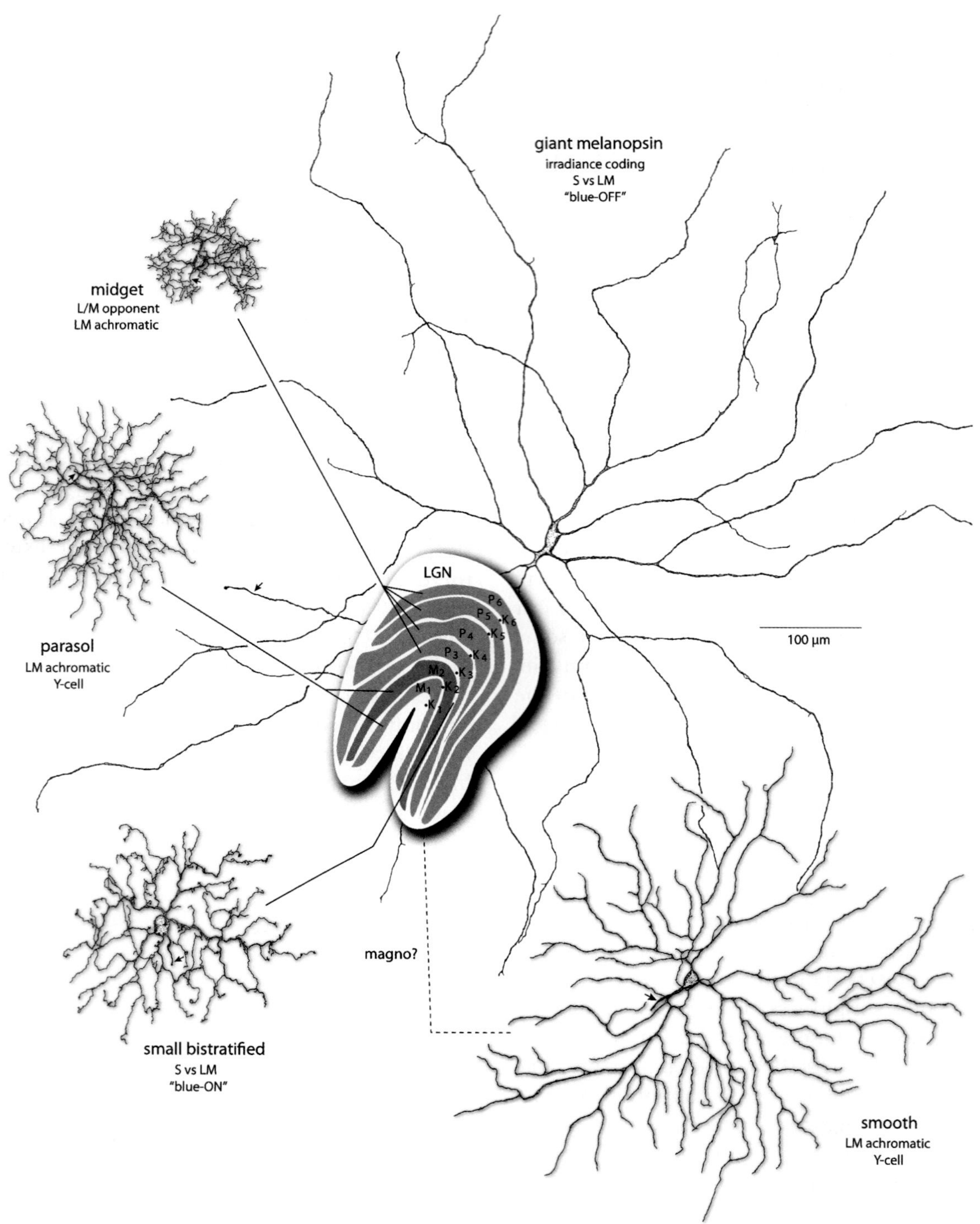

FIGURE 16.1   Retinal ganglion cells that project to the macaque LGN. Midget and several other types of RGCs project to the top four (parvocellular) layers, whereas parasol and smooth cells project to the bottom two (magnocellular) layers. Small bistratified, melanopsin-containing, and other types of RGCs project mostly to the intercalated layers between the major LGN layers and together make up the koniocellular population. (With permission from Crook et al., 2008a.)

bodies are smaller than those of the parasol RGCs, but their dendritic fields are larger, and they tile the retina in a mosaic that is less dense than that of the parasols. Their physiological properties suggest strongly that in the LGN they terminate in the magnocellular layers. A few RGCs contain the pigment melanopsin, which renders them light sensitive and allows them to report the ambient light level to the suprachiasmatic nucleus, which controls the circadian rhythm (Dacey et al., 2005). Most RGC types have ON and OFF varieties, although the small bistratified cells produce only ON responses to blue and OFF responses to yellow (Dacey & Lee, 1994; Field et al., 2007). Several types of RGCs project to the superior colliculus and are involved in the control of eye movements.

Together with this new information about the morphology of individual RGCs, we now have accurate information about the way that the mosaics of several types of RGCs sample visual space (Dacey, 1993; Field et al., 2007). This information comes from both morphological studies and the more recent progress in recording from complete populations of monkey RGCs with multielectrode arrays (for example, Gauthier et al., 2009a, 2009b; Pillow & Latham, 2008; Shlens et al., 2006). Such information is crucial if one is to ascribe a particular visual capability to a cell type. The spatial resolution of a single cell is important, but without knowing the sampling density it is impossible to say what the population could do (Kaplan, 2003). A similar argument holds for the temporal domain.

LGN   It has been known for some time that, anatomically, the RGCs that project to the parvocellular layers of the LGN are a heterogeneous population (Rodieck & Watanabe, 1993; Watanabe & Rodieck, 1989). The recent results of Dacey et al. (2003, 2009) provide a fuller account of this diversity and approach an almost complete description of the relative representation of the various types in the population as well as quantitative details of their morphology and connectivity pattern within the retina. However, aside from the melanopsin-containing cells, which are involved primarily in circadian rhythm rather than in image analysis, it is unclear at this point what role the various new groups of cells might play in the operation of the visual machinery. Whether the smooth and upsilon RGCs target specific subpopulations of LGN neurons is still an open question.

A persistent mystery is the issue of LGN lamination: Why are there four parvocellular layers? The laminar segregation might make it easier for the massive descending feedback pathway (Briggs & Usrey, 2009; Ichida & Casagrande, 2002) from V1 to find its LGN targets, but this is only a speculation at this point.

VISUAL CORTEX   Experiments in which the primary visual cortex (V1) was silenced with muscimol while recordings were made from various layers of V1 have provided new details about the cortical destinations of cells from the various LGN layers (Chatterjee & Callaway, 2003). It is now accepted that cells from the parvocellular layers terminate in layer 4Cβ of V1, whereas those originating in the magnocellular layers terminate in 4Cα. Cells from the koniocellular LGN population

project to the upper layers of V1, mostly to the cytochrome oxidase (CO) blobs (summarized in Callaway, 2005). Figure 16.2 shows the pattern of the major LGN projections into V1.

Tracer injections into area MT revealed a population of LGN neurons that project directly to MT, bypassing V1. Because many (but not all) of them stained positive for CaMK2, and many were in the intercalated layers, they are considered to be koniocellular neurons, although some were found within the parvocellular or magnocellular layers (Sincich et al., 2004). They thus join other (small) populations of LGN cells that project directly to V2 (Bullier & Kennedy, 1983) or V4 (Yukie & Iwai, 1981). These projections are thought to support "blindsight," the residual visual function that persists after damage to V1 (Rodman et al., 2001; Vakalopoulos, 2005).

The anatomical segregation of the M, P, and K LGN projections to V1 is maintained to some extent at the level of the extrastriate cortex, V2, where the various stripes that are rich or poor in CO receive projections from V1 cells that cluster together to form the CO blobs or interblob regions (Wong-Riley, 1979). This anatomical segregation in the cortex is at the root of the

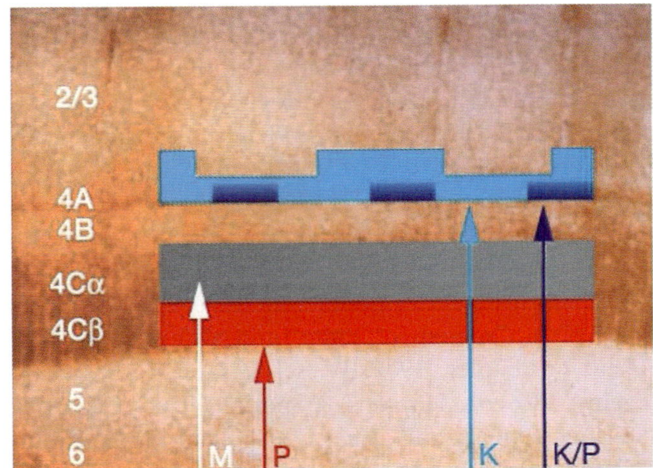

FIGURE 16.2   The projections of LGN cells to the macaque primary visual cortex. The cortex was silenced with muscimol, and recordings were obtained from the terminals of LGN fibers. Afferents recorded in layer 4Cβ of V1 had red–green color opponency and arise from LGN parvocellular cells. Afferents recorded in layer 4Cα had no chromatic opponency and arise from LGN magnocellular cells. Afferents recorded in more superficial layers of V1 had blue–yellow color opponency. Blue-OFF afferents were encountered only in layer 4A and might arise from blue-OFF midget ganglion cells via LGN P cells, but they might have some other origin. Blue-ON afferents were encountered in layers 3 and 4A and therefore arise (at least in part) from αCAM kinase/calbindin-expressing LGN koniocellular cells. (With permission from Chatterjee & Callaway, 2003.)

suggestion (Ungerleider & Mishkin, 1982) that the visual system performs its tasks with two main streams, a dorsal (M dominated) and a ventral (P dominated, with some K input); for a recent review, see Nassi and Callaway (2009). This view enjoys wide acceptance despite anatomical, physiological, and psychophysical challenges. We shall return to this topic later.

The Physiological Properties of the M, P, and K Cell Types

CHROMATIC OPPONENCY   It was known for a long time that, physiologically, RGCs were of two major types, ON and OFF (Hartline, 1940). A finer subdivision was introduced when the temporal and chromatic properties of the RGCs were investigated, and the distinction between tonic (sustained) and phasic (transient) cells was recognized in the primate retina (Gouras, 1968). Gouras also observed that the tonic cells were chromatically opponent, but the phasic cells were not. These observations suggested that the color-opponent cells that Wiesel and Hubel (1966) recorded in the parvocellular layers of the macaque LGN were driven by the tonic, color-opponent RGCs, while the nonopponent cells recorded in the magnocellular layers were driven by the phasic, nonopponent cells. This suggestion was later supported by anatomical methods (Bunt et al., 1975; Leventhal, Rodieck, & Dreher, 1981). However, the view of the M RGCs as color-blind was recently challenged by Lee and Sun (2009), who reported evidence of chromatically antagonistic input to these cells.

CONE INPUT TO THE RECEPTIVE FIELD SURROUND: RANDOM OR SELECTIVE?   The cellular mechanisms that generate chromatic opponency in the midget (P) RGCs have been debated for some time, with evidence for (Reid & Shapley, 1992; Wiesel & Hubel, 1966) and against (Boycott, Hopkins, & Sperling, 1987; Calkins & Sterling, 1996; Lennie, Haake, & Williams, 1991; Paulus & Kröger-Paulus, 1983) the notion of selective innervation of the receptive field surround mechanism by only one of the cone types. Recently, Crook et al. (2011) have reported that the origin of the red/green chromatic opponency is presynaptic to the midget cell, does not depend on GABA or glycine inhibition, and disappears (with the midget cell's receptive field surround) if the feedback from horizontal cells to cones is blocked. Recall that the horizontal cells, which provide (at least some of) the antagonistic surround to the RGCs, sum the cones within their territorial grasp promiscuously (Boycott, Hopkins, & Sperling, 1987). The results of Crook et al. (2011) are consistent with those of Buzás et al. (2006) and Field et al. (2010). However, a recent

analysis of physiological recordings from P RGCs and parvocellular LGN cells (Lee et al., 2012) finds considerable cone selectivity in many P and parvocellular receptive fields. The discrepancy could be due to the effects of retinal eccentricity: Most of the physiological evidence against a cone-selective surround (Crook et al., 2011; Field et al., 2010) comes from (in vitro) recordings of rather peripheral cells, typically beyond 10° from the fovea. The mosaic of L and M cones appears random, which implies a patchy arrangement of several L or M cones being grouped together (Roorda & Williams, 1999). Because receptive fields become smaller near the fovea, one would expect to find more purely selective surrounds there simply by chance. Lee et al. (2012) also point out that the receptive field of (at least some) P cells might be more complex than previously thought, in agreement with previous suggestions (Kaplan & Benardete, 2001).

LINEARITY OF SPATIAL SUMMATION   When the linearity of spatial summation across the receptive fields of monkey RGCs was investigated, most of the cells showed largely linear spatial summation. They were thus "X-like" as defined originally by Enroth-Cugell and Robson (1966) and refined by Hochstein and Shapley (1976) (but see Benardete & Kaplan, 1997). Approximately 25% of the M cells showed the frequency-doubled response signature of "Y-like" cells (Kaplan & Shapley, 1982). The smooth RGC reported by Crook et al. (2008a), and the upsilon cell reported by Petrusca et al. (2007) have Y-like physiological properties—high firing rate, high sensitivity to luminance contrast, and transient responses—which makes them responsive to high temporal frequencies and suitable for conveying motion information. This topic, however, is still controversial: Crook et al. (2008b) reported that *all* the parasol cells, as well as the smooth cells mentioned above (Crook et al., 2008a), had Y-like physiological characteristics in their in vitro experiments, in agreement with the in vivo results of Dhruv et al. (2009). Another distinctive dynamical characteristic of M cells that P cells lack is the change in their response phase with increasing contrast (Benardete, Kaplan, & Knight, 1992; Movshon et al., 2005).

"EXTRACLASSICAL" SUPPRESSION   Alitto and Usrey (2008) reported that LGN magnocellular cells in macaques were suppressed by stimuli beyond their classical receptive field and that the suppression was greater than what they found in parvocellular neurons, in agreement with earlier results of Solomon, White, and Martin (2002) in marmosets. The suppression acted too quickly to be due to cortical feedback and, because it was similar to what was found in the retina, was thought to originate there. An earlier anatomical study showed that the magnocellular layers had many more GABAergic cells than the parvocellular layers (Hámori, Pasik, & Pasik, 1983), and it is possible that these cells mediate the extraclassical suppression effects.

LUMINANCE CONTRAST GAIN   A prominent distinction between M and P cells is their luminance contrast gain (the slope of the response versus contrast function), which is high in M at low contrasts but low in P RGCs throughout their dynamic range (Alitto & Usrey, 2008; Kaplan & Shapley, 1986; Movshon et al., 2005). This greater gain is due primarily to the larger size of the M receptive field (Croner & Kaplan, 1995), and therefore, the upsilon and smooth cells also have similarly high contrast gain (Crook et al., 2008a; Petrusca et al., 2007).

INTERACTIONS AMONG CELLS   Recordings from large populations of RGCs from the isolated macaque retina in vitro, carried out with multielectrode arrays, have provided new information about the retinal output beyond the properties of individual cells. In agreement with the previous work of Mastronarde (1983) on the cat retina, Greschner et al. (2011) found that cells of a given type tended to have correlated firing. The correlations declined with the retinal distance between the cells. Most of the correlations could be attributed to common noise in the photoreceptors that are shared by the correlated RGCs. The correlations were relatively modest, but it was shown by Pillow et al. (2008) that such modest correlations increase significantly the amount of information that the population transmits about the visual scene.

KONIOCELLULAR CELLS   Cells in the K stream are a diverse population in terms of their chromatic selectivity, luminance contrast gain, spatial and temporal resolutions, as well as their projection targets from the LGN (Hendry & Reid, 2000; Sceniak, Chatterjee, & Callaway, 2006; White, Solomon, & Martin, 2001; Xu et al., 2001). Even the definition of what constitutes a koniocellular neuron is not straightforward, as demonstrated by the results of Sincich et al. (2004), which showed that neither the specific staining nor the layer location are definitive in fingering a cell as a koniocellular neuron.

THE USE OF STIMULI THAT FAVOR EITHER THE M OR P CELL TYPE

The apparent clustering of functional properties, especially in the LGN, where the M, P, and K populations are largely segregated in layers, has inspired attempts to create stimuli that would excite one population but not the others, in order to assess the functioning of the targeted population. This has been done using psychophysical methods, visual evoked potentials (VEPs), or fMRI, and the research aimed at either basic vision science issues or clinical questions. These studies typically took the M population to be chromatically nonopponent, highly sensitive to low spatial frequencies and low luminance contrast, to respond transiently, and to be involved in the analysis of motion and luminance. The P population was assumed to respond tonically and less well to luminance contrast and to convey chromatic and form information.

A Word of Caution

We should point out that one cannot easily deduce from single-cell data how to construct such stream-favoring stimuli. Pokorny and Smith (1997; Smith & Pokorny, 2003; Pokorny, 2011), who were at the forefront of such efforts, recognized the difficulty: "Thus, although the psychophysical data are consistent with activation of a given inferred retinal pathway, there is considerable reorganization of the retinal outputs in determining the psychophysical data" (Leonova, Pokorny, & Smith, 2003). There are other complications in the interpretation of studies that use such biasing stimuli. (1) M cells do not really saturate at moderate or high contrasts. In fact, their response versus contrast function above 30% has a contrast gain that is virtually identical to that of P cells (figure 16.3), so there is no reason to believe that they stop responding to contrast modulation at high contrasts. (2) P cells outnumber M cells by a factor of about eight and could compensate for their lower contrast gain by converging on cortical neurons. (3) Other cells (smooth monostratified, some K cells) have response properties that are similar to either the M or P cell types. Thus, at this point, the deduction of either a type-specific loss or a function for one of the cell types through the use of such stimuli should be viewed as mere speculation.

Basic Studies

Much of the basic research in this area focused on the role of the M or P streams in directing attention (for example, Brown, 2009; Cheng, Eysel, & Vidyasagar, 2004; Denison & Silver, 2012; Ries & Hopfinger, 2011; Tapia & Breitmeyer, 2011), on assessing their influence on reaction time (Burr & Corsale, 2001; Murray & Plainis, 2003), or on the relative contributions of the two streams to VEP or fMRI signals (Foxe et al., 2008;

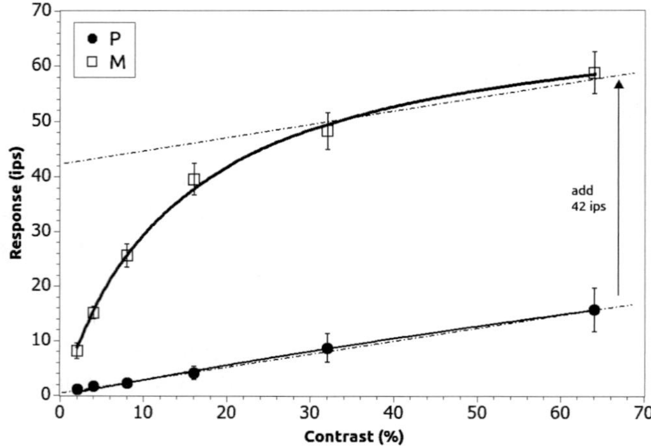

FIGURE 16.3   The contrast gain of M cells is higher than that of P cells for low-contrast stimuli. Shown are average responses of 28 P and 8 M RGCs from one rhesus monkey as a function of luminance contrast. The stimulus was a drifting black–white sine-wave grating that modulated the screen luminance at 4 Hz and had the optimal spatial frequency for each cell. The smooth solid curves are the Michaelis–Menten function, $R = R_{max} \times \dfrac{C}{C + C_{50}}$, where R is the response and C is the contrast; error bars: ± 1 SEM. C_{50}, the half-saturation contrast of the M (magnocellular projecting) ganglion cells, was 14%. Note the steeper slope (higher gain) of the M cells at low contrasts. The dashed regression line that fits the P cells' response was shifted upward by ~40 ips to emphasize the fact that M cells *continue to increase their response with increasing contrast*, even at moderate- to high-contrast stimuli. In fact, their contrast gain at the higher contrasts is similar to that of the P population. (Modified from Kaplan & Shapley, 1986.)

Grose-Fifer, Zemon, & Gordon, 1991; Lalor & Foxe, 2009; Victor et al., 1992).

Clinical Studies

The magnocellular and parvocellular populations, which are typically viewed as more homogeneous than the koniocellular cell type, have been implicated in several major neurological or psychiatric pathologies such as autism (Sutherland & Crewther, 2010), Alzheimer disease (Curcio & Drucker, 1993; Davies et al., 1995; Price et al., 1991), Parkinson disease (Silva et al., 2005), retinitis pigmentosa (Alexander et al., 2001, 2004), schizophrenia (Bedwell, Brown, & Miller, 2003, 2004; Martinez et al., 2008, 2012; Schechter et al., 2003), dyslexia (Chase et al., 2003; Chouake et al., 2012; Stein, 2001; Stein & Walsh, 1997), amblyopia (Demirci et al., 2002; Zele, Wood, & Girgenti, 2010), optic neuritis (Al-Hashmi, Kramer, & Mullen, 2011; Grigsby et al., 1991), and aging (Elliott & Werner, 2010). A recent review of how the current knowledge of the properties of the M, P, and K cell types is being used in ophthalmology, neurology, and psychiatry can be found in Yoonessi and Yoonessi (2011). These applications are not without controversy and often mirror the still-evolving exploration of the physiological properties of these cell populations in the retina and LGN (e.g., Maddess, 2011; Swanson et al., 2011) or the criticism of the claims that M cells are involved in dyslexia (Skottun, 2005; Skottun & Parke, 1999) or schizophrenia (Skottun & Skoyles, 2008, 2011).

THE PARALLEL-STREAMS HYPOTHESIS: A REASSESSMENT

In a recent review (Kaplan, 2008) I summarized the widely held view that maintains that three major cell types (M, P, and K) that are largely segregated anatomically in the LGN are each dedicated to a distinct visual function. I presented there several criteria that such hypothetical visual streams must meet: homogeneity, independence (both anatomical and perceptual), and compatibility of each stream's properties with its presumed function. I argued that various lines of evidence (anatomical, physiological, and psychophysical) show that the three streams do not meet these criteria. Rather than discuss my argument here in detail, I shall resketch it briefly, address the additional new information that bears on this hypothesis, and suggest alternatives.

The Simplistic Three-Streams Hypothesis

In its simplest form the three-streams hypothesis sees the visual system as a machine in which each cell type extracts from the visual environment the kind of information it is best suited to analyze: size, color, motion, and so on. Furthermore, cells of a feather flock together and create anatomical compartments (layers, blobs, stripes, dorsal or ventral streams) that contain cells of distinct types, and each one (or a grouping) of these brain structures thus plays a distinct role in vision. According to this view, the streams are homogeneous, independent, and have appropriate and nonoverlapping visual functions. This is an example of the view that information is processed in the brain in parallel by dedicated groups (types) of neurons (Livingstone & Hubel, 1988; Stone, 1983; Ungerleider & Mishkin, 1982).

Structure and Function: The Concept of a Cell Type

The view sketched above represents an example of one of neuroscience's central goals, linking structure with function, and it depends critically on the concept of a *cell type* (Rodieck & Brening, 1983). Clearly, the RGCs that form the optic nerve are not all alike, and this diversity of shape and response patterns suggests that if neurons were characterized along many dimensions (such as size, shape, connectivity, neurochemistry, response properties, and stimulus selectivity), an objective classifier would place each cell in a distinct, functionally homogeneous cluster. A partial categorization of this sort has been attempted for the early parts of the primate visual system (for example, table 30.1 in Kaplan, 2003). Neuroscientists then attempt to deduce the function of each cell type from the collection of properties of each cluster. So far, such a comprehensive analysis has not been carried out exhaustively in any part of the brain, although there are promising efforts in that direction (e.g., Bota & Swanson, 2007; Farrow & Masland, 2011; Masland, 2011). This task is challenging both technically and conceptually, but it is essential if we are to reach a deeper understanding of the design principles of the visual system, and indeed—the brain. For recent reviews of the challenge of linking structure and function in the brain, see Lichtman and Denk (2011) and Callaway (chapter 25, this volume).

Challenges for the Separate-Streams Hypothesis

As new techniques are brought to bear on our subject, new cell types, response patterns, connectivity schemes, and stimulus selectivities emerge. We have already mentioned the several new types of RGCs that have been described recently (Crook et al., 2008a; Dacey et al., 2003; Petrusca et al., 2007) and the new projection pathway from the koniocellular cells in the LGN to MT (Sincich et al., 2004). In addition, a disynaptic pathway was described between the parvocellular layers of the LGN and the "motion area" MT (Nassi, Lyon, & Callaway, 2006). These findings and many previous results (Economides et al., 2011; Ferrera, Nealey, & Maunsell, 1992; Lee & Sun, 2009; Leventhal et al., 1995; Nealey & Maunsell, 1994; Sincich & Horton, 2005; Vidyasagar et al., 2002) challenge the notion of segregation and independence of the streams. Likewise, additional evidence (Cropper & Wuerger, 2005; Dobkins & Albright, 1993, 1995; Gegenfurtner & Hawken, 1996; Martinovic et al., 2009; Poom, 2011; Ruppertsberg, Wuerger, & Bertamini, 2007; Wuerger et al., 2011) shows closer integration of color and motion perception than one might expect based on the parallel-stream paradigm, where motion and color analysis are often cited as prime examples of the exclusive domains of the M and P streams, respectively. It appears that motion might be perceived by several mechanisms, some color-selective and others color-blind (Gorea & Papathomas, 1989; Papathomas, Gorea, & Julesz, 1991).

We should also consider the possibility that some of the cell types, which the neuroscientist assumes must fulfill an important function simply because they exist, might be like the spandrels of San Marco and serve no important biological function (Gould & Lewontin, 1979). Finally, it is possible that the concept of *functional architecture*, which is central to so much of current brain research (Bihan, 2012; Frackowiak, 1998; Hubel & Wiesel, 1962), has lulled us into automatic acceptance of the notion that if we have a word for color, motion, soul, or consciousness, there must be some identifiable brain part that is creating it. Rather than look for brain structures with specific, distinct functions, perhaps we should be looking for functional network motifs (Sporns & Kötter, 2004).

A Refined Multistream Hypothesis

The current three-stream view might be replaced by a refined multistream hypothesis that replaces the M, P, and K streams with a larger number of streams (anatomical/functional groups), preserving the notion of parallel processing by function-specific cell types. For example, the M stream might include the well-known parasol RGCs and the newly discovered smooth cells (Crook et al., 2008a), which might serve a different function from the parasols. The P streams should be split into the midget RGCs and some of the other groups that project to the parvocellular layers (Dacey et al., 2003; Rodieck & Watanabe, 1988). As more data are gathered about the various RGCs that terminate on koniocellular neurons, that stream will surely be replaced by several more homogeneous substreams. Some of that has begun already (e.g., Sincich et al., 2004). This multistream view is close to the one advocated by Masland and Martin (2007) and Martin and Solomon (2011). Because a complete structural–functional taxonomy of the primate RGCs and LGN cells is still missing, it is impossible to predict how many streams will be needed to ensure that each cell type will be relatively homogeneous, but the number will surely be >10.

A Paradigm Shift?

Several authors have recently voiced criticism of the prevailing three-stream paradigm. They include Sincich and Horton (2005), Kaplan (2008), Schenk and McIntosh (2010), and de Haan and Cowey (2011), among others. Even one of the cornerstones of neuroscience, the belief in a tight link between structure and function, has been called into question (Wallisch & Movshon, 2008) in view of some recent neurophysiological and behavioral data (Chowdhury & DeAngelis, 2008). The criticism is typically based on the well-documented anatomical and functional crosstalk among the presumed independent streams, on the heterogeneity of the cell types that are grouped together in streams, on the fact that some of the evidence supporting the allocation and segregation of function is based on unreliable lesion experiments or observations, and on the conceptual difficulties of assigning specific distinct functions to elements within an interconnected, nonlinear dynamical network.

A Network Hypothesis

An alternative view, in which we relax the tight link between structure and function that we inherited from the machine metaphor, is that each cell type contributes something to the analysis of the visual scene, and all (or at least many) of them participate in this analysis. The different response and selectivity properties of the various cells thus provide not distinct and separate channels of information about stimulus dimensions but instead are providing a distribution of tuning functions (selectivities) along each dimension of the stimulus. This view is supported by the observation made by almost everyone who has studied the selectivity of cortical neurons: With rare exceptions, most neurons are tuned (to various degrees) along many stimulus dimensions. The details of the computation that leads from this population activity to perception are being investigated and debated and are probably statistical in nature, perhaps along the lines explored by Pouget, Dayan, and Zemel (2000, 2003) and Jazayeri and Movshon (2006). Perception is thus accomplished not by some discrete entity that weighs the contribution of each stream to reach a perceptual decision but rather is the psychological corollary or an emergent property of the coordinated activity of the entire network. Given the fact that the brain is a *highly interconnected, nonlinear dynamical system*, the assignment of distinct functions to cell types, regardless of how they are defined, might be a futile exercise.

CONCLUSIONS

The three major cell types that send axons from the retina to the LGN are widely believed to subserve distinct visual functions: M cells relay motion and luminance information, P cells report about color and form, and some of the more diverse K cells produce blue-ON responses. We saw that this three-way parcellation of the

RGC population is far too simplistic and should be replaced either by a refined, multistream paradigm or perhaps by a network-based view, in which most cell types contribute to the perception of most visual attributes. Just as the 92 elements, each with its own unique properties, interact to make up our multifaceted world, so could 20 or so cell types interact to weave the dynamic tapestry of our visual world.

ACKNOWLEDGMENTS

The author was supported by NIH grants EY16224, NIGMS 1P50GM071558, and R21MH093868–02.

REFERENCES

Alexander, K. R., Barnes, C. S., Fishman, G. A., Pokorny, J., & Smith, V. C. (2004). Contrast sensitivity deficits in inferred magnocellular and parvocellular pathways in retinitis pigmentosa. *Investigative Ophthalmology & Visual Science, 45,* 4510–4519. doi:10.1167/iovs.04-0188.

Alexander, K. R., Pokorny, J., Smith, V. C., Fishman, G. A., & Barnes, C. S. (2001). Contrast discrimination deficits in retinitis pigmentosa are greater for stimuli that favor the magnocellular pathway. *Vision Research, 41,* 671–683.

Al-Hashmi, A. M., Kramer, D. J., & Mullen, K. T. (2011). Human vision with a lesion of the parvocellular pathway: An optic neuritis model for selective contrast sensitivity deficits with severe loss of midget ganglion cell function. *Experimental Brain Research, 215,* 293–305.

Alitto, H. J., & Usrey, W. M. (2008). Origin and dynamics of extraclassical suppression in the lateral geniculate nucleus of the macaque monkey. *Neuron, 57,* 135–146.

Bedwell, J. S., Brown, J. M., & Miller, L. S. (2003). The magnocellular visual system and schizophrenia: What can the color red tell us? *Schizophrenia Research, 63,* 273–284.

Bedwell, J. S., Miller, L. S., Brown, J. M., McDowell, J. E., & Yanasak, N. E. (2004). Functional magnetic resonance imaging examination of the magnocellular visual pathway in nonpsychotic relatives of persons with schizophrenia. *Schizophrenia Research, 71,* 509–510.

Benardete, E. A., & Kaplan, E. (1997). The receptive field of the primate P retinal ganglion cell, II: Nonlinear dynamics. *Visual Neuroscience, 14,* 187–205.

Benardete, E. A., Kaplan, E., & Knight, B. W. (1992). Contrast gain control in the primate retina: P cells are not X-like, some M cells are. *Visual Neuroscience, 8,* 483–486.

Bihan, D. L. (2012). Diffusion, confusion and functional MRI. *NeuroImage, 62,* 1131–6

Bota, M., & Swanson, L. W. (2007). The neuron classification problem. *Brain Research. Brain Research Reviews, 56,* 79–88.

Boycott, B. B., Hopkins, J. M., & Sperling, H. G. (1987). Cone connections of the horizontal cells of the rhesus monkeys retina. *Proceedings of the Royal Society of London. Series B, Biological Sciences, 229,* 345–379.

Briggs, F., & Usrey, W. M. (2009). Parallel processing in the corticogeniculate pathway of the macaque monkey. *Neuron, 62,* 135–146.

Brown, J. M. (2009). Visual streams and shifting attention. *Progress in Brain Research, 176,* 47–63.

Bullier, J., & Kennedy, H. (1983). Projection of the lateral geniculate nucleus onto cortical area V2 in the macaque monkey. *Experimental Brain Research, 53,* 168–172.

Bunt, A. H., Hendrickson, A. E., Lund, J. S., Lund, R. D., & Fuchs, A. F. (1975). Monkey retinal ganglion cells: Morphometric analysis and tracing of axonal projections, with a consideration of the peroxidase technique. *Journal of Comparative Neurology, 164,* 265–286.

Burr, D. C., & Corsale, B. (2001). Dependency of reaction times to motion onset on luminance and chromatic contrast. *Vision Research, 41,* 1039–1048. doi:10.1016/S0042-6989(01)00019-0.

Buzás, P., Blessing, E. M., Szmajda, B. A., & Martin, P. R. (2006). Specificity of m and l cone inputs to receptive fields in the parvocellular pathway: Random wiring with functional bias. *Journal of Neuroscience, 26,* 11148–11161.

Calkins, D. J., & Sterling, P. (1996). Absence of spectrally specific lateral inputs to midget ganglion cells in primate retina. *Nature, 381,* 613–615.

Callaway, E. M. (2005). Structure and function of parallel pathways in the primate early visual system. *Journal of Physiology, 566,* 13–19.

Chase, C., Ashourzadeh, A., Kelly, C., Monfette, S., & Kinsey, K. (2003). Can the magnocellular pathway read? Evidence from studies of color. *Vision Research, 43,* 1211–1222. doi:10.1016/S0042-6989(03)00085-3.

Chatterjee, S., & Callaway, E. M. (2003). Parallel color-opponent pathways to primary visual cortex. *Nature, 426,* 668–671.

Cheng, A., Eysel, U. T., & Vidyasagar, T. R. (2004). The role of the magnocellular pathway in serial deployment of visual attention. *European Journal of Neuroscience, 20,* 2188–2192.

Chouake, T., Levy, T., Javitt, D. C., & Lavidor, M. (2012). Magnocellular training improves visual word recognition. *Frontiers in Human Neuroscience, 6,* 14.

Chowdhury, S. A., & DeAngelis, G. C. (2008). Fine discrimination training alters the causal contribution of macaque area MT to depth perception. *Neuron, 60,* 367–377.

Croner, L. J., & Kaplan, E. (1995). Receptive fields of P and M ganglion cells across the primate retina. *Vision Research, 35,* 7–24. doi:10.1016/0042-6989(94)E0066-T.

Crook, J. D., Manookin, M. B., Packer, O. S., & Dacey, D. M. (2011). Horizontal cell feedback without cone type-selective inhibition mediates "red-green" color opponency in midget ganglion cells of the primate retina. *Journal of Neuroscience, 31,* 1762–1772.

Crook, J. D., Peterson, B. B., Packer, O. S., Robinson, F. R., Gamlin, P. D., Troy, J. B., et al. (2008a). The smooth monostratified ganglion cell: Evidence for spatial diversity in the Y-cell pathway to the lateral geniculate nucleus and superior colliculus in the macaque monkey. *Journal of Neuroscience, 28,* 12654–12671.

Crook, J. D., Peterson, B. B., Packer, O. S., Robinson, F. R., Troy, J. B., & Dacey, D. M. (2008b). Y-cell receptive field and collicular projection of parasol ganglion cells in macaque monkey retina. *Journal of Neuroscience, 28,* 11277–11291.

Cropper, S. J., & Wuerger, S. M. (2005). The perception of motion in chromatic stimuli. *Behavioral and Cognitive Neuroscience Reviews, 4,* 192–217.

Curcio, C. A., & Drucker, D. N. (1993). Retinal ganglion cells in Alzheimer's disease and aging. *Annals of Neurology, 33,* 248–257.

Dacey, D. M. (1993). The mosaic of midget ganglion cells in the human retina. *Journal of Neuroscience, 13,* 5334–5355.

Dacey, D., Joo, H., Peterson, B., & Haun, T. (2009). Morphology, mosaics and central projections of diverse ganglion cell populations in macaque retina: Approaching a complete account. [ARVO abstract.]. *Journal of Vision, 9,* 35.

Dacey, D. M., & Lee, B. B. (1994). The "blue-ON" opponent pathway in primate retina originates from a distinct bistratified ganglion cell type. *Nature, 367,* 731–735.

Dacey, D. M., Liao, H. W., Peterson, B. B., Robinson, F. R., Smith, V. C., Pokorny, J., et al. (2005). Melanopsin-expressing ganglion cells in primate retina signal colour and irradiance and project to the LGN. *Nature, 433,* 749–754.

Dacey, D. M., Peterson, B. B., Robinson, F. R., & Gamlin, P. D. (2003). Fireworks in the primate retina: In vitro photodynamics reveals diverse LGN-projecting ganglion cell types. *Neuron, 37,* 15–27.

Davies, D. C., McCoubrie, P., McDonald, B., & Jobst, K. A. (1995). Myelinated axon number in the optic nerve is unaffected by Alzheimer's disease. *British Journal of Ophthalmology, 79,* 596–600.

de Haan, E. H. F., & Cowey, A. (2011). On the usefulness of "what" and "where" pathways in vision. *Trends in Cognitive Sciences, 15,* 460–466. doi:10.1016/j.tics.2011.08.005.

Demirci, H., Gezer, A., Sezen, F., Ovali, T., Demiralp, T., & Isoglu-Alkoc, U. (2002). Evaluation of the functions of the parvocellular and magnocellular pathways in strabismic amblyopia. *Journal of Pediatric Ophthalmology and Strabismus, 39,* 215–221.

Denison, R. N., & Silver, M. A. (2012). Distinct contributions of the magnocellular and parvo-cellular visual streams to perceptual selection. *Journal of Cognitive Neuroscience, 24,* 246–259.

Dhruv, N. T., Tailby, C., Sokol, S. H., Majaj, N. J., & Lennie, P. (2009). Nonlinear signal summation in magnocellular neurons of the macaque lateral geniculate nucleus. *Journal of Neurophysiology, 102,* 1921–1929.

Dobkins, K. R., & Albright, T. D. (1993). What happens if it changes color when it moves? Psychophysical experiments on the nature of chromatic input to motion detectors. *Vision Research, 33,* 1019–1036. doi:10.1016/0042-6989(93)90238-R.

Dobkins, K. R., & Albright, T. D. (1995). Behavioral and neural effects of chromatic isoluminance in the primate visual motion system. *Visual Neuroscience, 12,* 321–332.

Economides, J. R., Sincich, L. C., Adams, D. L., & Horton, J. C. (2011). Orientation tuning of cytochrome oxidase patches in macaque primary visual cortex. *Nature Neuroscience, 14,* 1574–1580.

Elliott, S. L., & Werner, J. S. (2010). Age-related changes in contrast gain related to the M and P pathways. *Journal of Vision, 10,* 1–15. doi:10.1167/10.4.4.

Enroth-Cugell, C., & Robson, J. G. (1966). The contrast sensitivity of retinal ganglion cells of the cat. *Journal of Physiology, 187,* 517–552.

Farrow, K., & Masland, R. H. (2011). Physiological clustering of visual channels in the mouse retina. *Journal of Neurophysiology, 105,* 1516–1530.

Ferrera, V. P., Nealey, T. A., & Maunsell, J. H. (1992). Mixed parvocellular and magnocellular geniculate signals in visual area V4. *Nature, 358,* 756–761.

Field, G. D., Gauthier, J. L., Sher, A., Greschner, M., Machado, T. A., Jepson, L. H., et al. (2010). Functional connectivity in the retina at the resolution of photoreceptors. *Nature, 467*, 673–677.

Field, G. D., Sher, A., Gauthier, J. L., Greschner, M., Shlens, J., Litke, A. M., et al. (2007). Spatial properties and functional organization of small bistratified ganglion cells in primate retina. *Journal of Neuroscience, 27*, 13261–13272.

Foxe, J. J., Strugstad, E. C., Sehatpour, P., Molholm, S., Pasieka, W., Schroeder, C. E., et al. (2008). Parvocellular and magnocellular contributions to the initial generators of the visual evoked potential: High-density electrical mapping of the "C1" component. *Brain Topography, 21*, 11–21.

Frackowiak, R. S. (1998). The functional architecture of the brain. *Daedalus, 127*(2), 105–130.

Galilei, G. (1960). The Assayer (trans. S. Drake). In S. Drake (Ed.), *The controversy on the comets of 1618*. Philadelphia: University of Pennsylvania Press. (Original work published 1623.)

Gauthier, J. L., Field, G. D., Sher, A., Greschner, M., Shlens, J., Litke, A. M., et al. (2009a). Receptive fields in primate retina are coordinated to sample visual space more uniformly. *PLoS Biology, 7*(4), e1000063. doi:10.1371/journal.pbio.1000063.

Gauthier, J. L., Field, G. D., Sher, A., Shlens, J., Greschner, M., Litke, A. M., et al. (2009b). Uniform signal redundancy of parasol and midget ganglion cells in primate retina. *Journal of Neuroscience, 29*(14), 4675–4680.

Gegenfurtner, K., & Hawken, M. (1996). Interactions between color and motion in the visual pathways. *Trends in Neurosciences, 19*, 394–401. doi:10.1098/rspb.1977.0085.

Gorea, A., & Papathomas, T. V. (1989). Motion processing by chromatic and achromatic visual pathways. *Journal of the Optical Society of America. A, Optics and Image Science, 6*, 590–602.

Gould, S. J., & Lewontin, R. C. (1979). The spandrels of San Marco and the Panglossian paradigm: A critique of the adaptationist programme. *Proceedings of the Royal Society of London. Series B, Biological Sciences, 205*, 581–598.

Gouras, P. (1968). Identification of cone mechanisms in monkey ganglion cells. *Journal of Physiology, 199*, 533–547.

Greschner, M., Shlens, J., Bakolitsa, C., Field, G. D., Gauthier, J. L., Jepson, L. H., et al. (2011). Correlated firing among major ganglion cell types in primate retina. *Journal of Physiology, 589*, 75–86.

Grigsby, S. S., Vingrys, A. J., Benes, S. C., & King-Smith, P. E. (1991). Correlation of chromatic, spatial, and temporal sensitivity in optic nerve disease. *Investigative Ophthalmology & Visual Science, 32*, 3252–3262.

Grose-Fifer, J., Zemon, V., & Gordon, J. (1991). The development of magno and parvo pathways in human infants investigated using the sweep VEP. *Investigative Ophthalmology & Visual Science (Suppl), 32*, 1045.

Hámori, J., Pasik, P., & Pasik, T. (1983). Differential frequency of P-cells and I-cells in magnocellular and parvocellular laminae of monkey lateral geniculate nucleus. An ultrastructural study. *Experimental Brain Research, 52*, 57–66.

Hartline, H. K. (1940). The receptive fields of optic nerve fibers. *American Journal of Physiology, 130*, 690–699.

Hendry, S. H. C., & Reid, C. R. (2000). The koniocellular pathway in primate vision. *Annual Review of Neuroscience, 23*, 127–153.

Hochstein, S., & Shapley, R. M. (1976). Quantitative analysis of retinal ganglion cell classifications. *Journal of Physiology, 262*, 237–264.

Hubel, D. H., & Wiesel, T. N. (1962). Receptive fields, binocular interaction and functional architecture in the cat's visual cortex. *Journal of Physiology, 160*, 106–154.

Hubel, D. H., & Wiesel, T. N. (1977). Functional architecture of macaque monkey visual cortex. *Proceedings of the Royal Society of London. Series B, Biological Sciences, 198*, 1–59. doi:10.1098/rspb.1977.0085.

Ichida, J. M., & Casagrande, V. A. (2002). Organization of the feedback pathway from striate cortex (V1) to the lateral geniculate nucleus (LGN) in the owl monkey (*Aotus trivirgatus*). *Journal of Comparative Neurology, 454*, 272–283.

Jazayeri, M., & Movshon, J. A. (2006). Optimal representation of sensory information by neural populations. *Nature Neuroscience, 9*, 690–696.

Kaplan, E. (2003). The M, P and K pathways in the primate visual system. In L. Chalupa & J. Werner (Eds.), *The visual neurosciences* (Vol. I, pp. 481–494). Cambridge, MA: MIT Press.

Kaplan, E. (2008). The M, K, and P streams in the primate visual system: What do they do for vision? In M. C. Bushnell et al. (Eds.), *The senses* (pp. 369–382). London: Elsevier.

Kaplan, E., & Benardete, E. (2001). The dynamics of primate retinal ganglion cells. *Progress in Brain Research, 134*, 17–34.

Kaplan, E., & Shapley, R. M. (1982). X and Y cells in the lateral geniculate nucleus of macaque monkeys. *Journal of Physiology, 330*, 125–143.

Kaplan, E., & Shapley, R. M. (1986). The primate retina contains two types of ganglion cells, with high and low contrast sensitivity. *Proceedings of the National Academy of Sciences of the United States of America, 83*, 2755–2757. doi:10.1073/pnas.83.8.2755.

Lalor, E. C., & Foxe, J. J. (2009). Visual evoked spread spectrum analysis (VESPA) responses to stimuli biased towards magnocellular and parvocellular pathways. *Vision Research, 49*, 127–133.

Lee, B. B., Shapley, R. M., Hawken, M. J., & Sun, H. (2012). Spatial distributions of cone inputs to cells of the parvocellular pathway investigated with cone-isolating gratings. *Journal of the Optical Society of America. A, Optics, Image Science, and Vision, 29*, A223–A232. doi:10.1364/JOSAA.29.00A223.

Lee, B. B., & Sun, H. (2009). The chromatic input to cells of the magnocellular pathway of primates. *Journal of Vision, 9*, 15, 1–18. doi:10.1167/9.2.15.

Lennie, P., Haake, P., & Williams, D. (1991). The design of chromatically opponent receptive ϕields. In J. Movshon (Ed.), *Computational models of visual processing* (pp. 71–82). Cambridge, MA: MIT Press.

Leonova, A., Pokorny, J., & Smith, V. C. (2003). Spatial frequency processing in inferred PC and MC-pathways. *Vision Research, 43*, 2133–2139. doi:10.1016/S0042-6989(03)00333-X.

Leventhal, A. G., Rodieck, R. W., & Dreher, B. (1981). Retinal ganglion cell classes in the Old World monkey: Morphology and central projections. *Science, 213*, 1139–1142.

Leventhal, A. G., Thompson, K. G., Liu, D., Zhou, Y., & Ault, S. J. (1995). Concomitant sensitivity to orientation, direction, and color of cells in layers 2, 3, and 4 of monkey striate cortex. *Journal of Neuroscience, 15*, 1808–1818.

Lichtman, J. W., & Denk, W. (2011). The big and the small: Challenges of imaging the brain's circuits. *Science, 334,* 618–623.

Livingstone, M., & Hubel, D. (1988). Segregation of form, color, movement, and depth: Anatomy, physiology, and perception. *Science, 240,* 740–749.

Locke, J. (1832). *Essay concerning human understanding.* Oxford: Oxford University Press. (Original work published 1690.)

Maddess, T. (2011). Frequency-doubling technology and parasol cells. *Investigative Ophthalmology & Visual Science, 52,* 3759, author reply 3759–3760.

Martin, P. R., & Solomon, S. G. (2011). Information processing in the primate visual system. *Journal of Physiology, 589,* 29–31.

Martinez, A., Hillyard, S. A., Bickel, S., Dias, E. C., Butler, P. D., & Javitt, D. C. (2012). Consequences of magnocellular dysfunction on processing attended information in schizophrenia. *Cerebral Cortex, 22,* 1282–1293. doi:10.1093/cercor/bhv195.

Martinez, A., Hillyard, S. A., Dias, E. C., Hagler, D. J., Butler, P. D., Guilfoyle, D. N., et al. (2008). Magnocellular pathway impairment in schizophrenia: Evidence from functional magnetic resonance imaging. *Journal of Neuroscience, 28,* 7492–7500.

Martinovic, J., Meyer, G., Müller, M. M., & Wuerger, S. M. (2009). S-cone signals invisible to the motion system can improve motion extraction via grouping by color. *Visual Neuroscience, 26,* 237–248.

Masland, R. H. (2011). Cell populations of the retina: The Proctor Lecture. *Investigative Ophthalmology & Visual Science, 52,* 4581–4591. doi:10.1167/iovs.10-7083.

Masland, R. H., & Martin, P. R. (2007). The unsolved mystery of vision. *Current Biology, 17,* R577–R582. doi:10.1016/j.cub.2007.05.040.

Mastronarde, D. N. (1983). Interactions between ganglion cells in cat retina. *Journal of Neurophysiology, 49,* 350–365.

Movshon, J. A., Kiorpes, L., Hawken, M. J., & Cavanaugh, J. R. (2005). Functional maturation of the macaque's lateral geniculate nucleus. *Journal of Neuroscience, 25,* 2712–2722.

Murray, I., & Plainis, S. (2003). Contrast coding and magno/parvo segregation revealed in reaction time studies. *Vision Research, 43,* 2707–2719. doi:10.1016/S0042-6989(03)00408-5.

Nassi, J. J., & Callaway, E. M. (2009). Parallel processing strategies of the primate visual system. *Nature Reviews. Neuroscience, 10,* 360–372.

Nassi, J. J., Lyon, D. C., & Callaway, E. M. (2006). The parvocellular LGN provides a robust disynaptic input to the visual motion area MT. *Neuron, 50,* 319–327.

Nealey, T. A., & Maunsell, J. H. (1994). Magnocellular and parvocellular contributions to the responses of neurons in macaque striate cortex. *Journal of Neuroscience, 14,* 2069–2079.

Papathomas, T. V., Gorea, A., & Julesz, B. (1991). Two carriers for motion perception: Color and luminance. *Vision Research, 31,* 1883–1891. doi:10.1016/0042-6989(91)90183-6.

Paulus, W., & Kröger-Paulus, A. (1983). A new concept of retinal colour coding. *Vision Research, 23,* 529–540.

Petrusca, D., Grivich, M. I., Sher, A., Field, G. D., Gauthier, J. L., Greschner, M., et al. (2007). Identification and characterization of a Y-like primate retinal ganglion cell type. *Journal of Neuroscience, 27,* 11019–11027.

Pillow, J., & Latham, P. (2008). Neural characterization in partially observed populations of spiking neurons. In J. C. Platt, D. Koller, Y. Singer, & S. Roweis (Eds.), *Advances in neural information processing systems, 20* (pp. 1161–1168). Cambridge, MA: MIT Press.

Pillow, J. W., Shlens, J., Paninski, L., Sher, A., Litke, A. M., Chichilnisky, E. J., et al. (2008). Spatio-temporal correlations and visual signalling in a complete neuronal population. *Nature, 454,* 995–999.

Pokorny, J. (2011). Review: Steady and pulsed pedestals, the how and why of post-receptoral pathway separation. *Journal of Vision, 11,* 1–23. doi:10.1167/11.5.7.

Pokorny, J., & Smith, V. C. (1997). Psychophysical signatures associated with magnocellular and parvocellular pathway contrast gain. *Journal of the Optical Society of America. A, Optics, Image Science, and Vision, 14,* 2477–2486. doi:10.1364/JOSAA.14.002477.

Poom, L. (2011). Motion and color generate coactivation at postgrouping identification stages. *Attention, Perception & Psychophysics, 73,* 1833–1842. doi:10.3758/s13414-011-0132-8.

Pouget, A., Dayan, P., & Zemel, R. (2000). Information processing with population codes. *Nature Reviews. Neuroscience, 1*(2), 125–132.

Pouget, A., Dayan, P., & Zemel, R. S. (2003). Inference and computation with population codes. *Annual Review of Neuroscience, 26,* 381–410.

Price, J. L., Davis, P. B., Morris, J. C., & White, D. L. (1991). The distribution of tangles, plaques and related immunohistochemical markers in healthy aging and Alzheimer's disease. *Neurobiology of Aging, 12,* 295–312.

Reid, R. C., & Shapley, R. M. (1992). Spatial structure of cone inputs to receptive fields in primate lateral geniculate nucleus. *Nature, 356,* 716–718.

Ries, A. J., & Hopfinger, J. B. (2011). Magnocellular and parvocellular influences on reflexive attention. *Vision Research, 51,* 1820–1828. doi:10.1016/j.visres.2011.06.012.

Rodieck, R. W., & Brening, R. K. (1983). Retinal ganglion cells: Properties, types, genera, pathways and trans-species comparisons. *Brain, Behavior and Evolution, 23,* 121–164.

Rodieck, R. W., & Watanabe, M. (1988). Morphology of ganglion cell types that project to the parvocellular laminae of the lateral geniculate nucleus, pretectum, and superior colliculus of primates. *Society for Neuroscience Abstract, 14,* 1120.

Rodieck, R. W., & Watanabe, M. (1993). Survey of the morphology of macaque retinal ganglion cells that project to the pretectum, superior colliculus, and parvicellular laminae of the lateral geniculate nucleus. *Journal of Comparative Neurology, 338,* 289–303.

Rodman, H. R., Sorenson, K. M., Shim, A. J., & Hexter, D. P. (2001). Calbindin immunoreactivity in the geniculo-extrastriate system of the macaque: Implications for heterogeneity in the koniocellular pathway and recovery from cortical damage. *Journal of Comparative Neurology, 431,* 168–181.

Roorda, A., & Williams, D. R. (1999). The arrangement of the three cone classes in the living human eye. *Nature, 397,* 520–522.

Ruppertsberg, A. I., Wuerger, S. M., & Bertamini, M. (2007). When S-cones contribute to chromatic global motion processing. *Visual Neuroscience, 24,* 1–8. doi:10.1017/S0952523807230081.

Sceniak, M. P., Chatterjee, S., & Callaway, E. M. (2006). Visual spatial summation in macaque geniculocortical afferents. *Journal of Neurophysiology, 96,* 3474–3484.

Schechter, I., Butler, P. D., Silipo, G., Zemon, V., & Javitt, D. C. (2003). Magnocellular and parvocellular contributions to backward masking dysfunction in schizophrenia. *Schizophrenia Research, 64,* 91–101.

Schenk, T., & McIntosh, R. D. (2010). Do we have independent visual streams for perception and action? *Cognitive Neuroscience, 1,* 52–78.

Shlens, J., Field, G. D., Gauthier, J. L., Grivich, M. I., Petrusca, D., Sher, A., et al. (2006). The structure of multi-neuron firing patterns in primate retina. *Journal of Neuroscience, 26,* 8254–8266.

Silva, M. F., Faria, P., Regateiro, F. S., Forjaz, V., Januário, C., Freire, A., et al. (2005). Independent patterns of damage within magno-, parvo- and koniocellular pathways in Parkinson's disease. *Brain, 128,* 2260–2271. doi:10.1093/brain/awh581.

Sincich, L. C., & Horton, J. C. (2005). The circuitry of V1 and V2: Integration of color, form, and motion. *Annual Review of Neuroscience, 28,* 303–326.

Sincich, L. C., Park, K. F., Wohlgemuth, M. J., & Horton, J. C. (2004). Bypassing V1: A direct geniculate input to area MT. *Nature Neuroscience, 7,* 1123–1128.

Skottun, B. C. (2005). Magnocellular reading and dyslexia. *Vision Research, 45,* 133–134, author reply 135–136.

Skottun, B. C., & Parke, L. A. (1999). The possible relationship between visual deficits and dyslexia: Examination of a critical assumption. *Journal of Learning Disabilities, 32,* 2–5. doi:10.1177/002221949903200101.

Skottun, B. C., & Skoyles, J. R. (2008). A few remarks on attention and magnocellular deficits in schizophrenia. *Neuroscience and Biobehavioral Reviews, 32,* 118–122.

Skottun, B. C., & Skoyles, J. R. (2011). On identifying magnocellular and parvocellular responses on the basis of contrast-response functions. *Schizophrenia Bulletin, 37,* 23–26.

Smith, V. C., & Pokorny, J. (2003). Psychophysical correlates of parvo- and magnocellular function. In J. Mollon, K. Knoblauch, & J. Pokorny (Eds.), *Normal and defective colour vision* (pp. 91–107). Oxford: Oxford University Press.

Solomon, S. G., White, A. J. R., & Martin, P. R. (2002). Extra-classical receptive field properties of parvocellular, magnocellular, and koniocellular cells in the primate lateral geniculate nucleus. *Journal of Neuroscience, 22,* 338–349.

Sporns, O., & Kötter, R. (2004). Motifs in brain networks. *PLoS Biology, 2,* e369. doi:10.1371/journal.pbio.0020369.

Stein, J. (2001). The magnocellular theory of developmental dyslexia. *Dyslexia (Chichester, England), 7,* 12–36.

Stein, J., & Walsh, V. (1997). To see but not to read; the magnocellular theory of dyslexia. *Trends in Neurosciences, 20,* 147–152. doi:10.1016/S0166-2236(96)01005-3.

Stone, J. (1983). *Parallel processing in the visual system.* New York: Plenum Press.

Sutherland, A., & Crewther, D. P. (2010). Magnocellular visual evoked potential delay with high autism spectrum quotient yields a neural mechanism for altered perception. *Brain, 133*(Pt 7), 2089–2097.

Swanson, W. H., Sun, H., Lee, B. B., & Cao, D. (2011). Responses of primate retinal ganglion cells to perimetric stimuli. *Investigative Ophthalmology & Visual Science, 52,* 764–771.

Tapia, E., & Breitmeyer, B. G. (2011). Visual consciousness revisited: Magnocellular and parvo-cellular contributions to conscious and nonconscious vision. *Psychological Science, 22,* 934–942.

Ungerleider, L. G., & Mishkin, M. (1982). Two cortical visual systems. In D. J. Ingle, M. A. Goodale, & R. J. W. Mansfield (Eds.), *Analysis of visual behavior* (pp. 549–586). Cambridge, MA: MIT Press.

Vakalopoulos, C. (2005). A theory of blindsight—the anatomy of the unconscious: A proposal for the koniocellular projections and intralaminar thalamus. *Medical Hypotheses, 65,* 1183–1190. doi:10.1016/j.mehy.2005.05.039.

Victor, J., Conte, M., Burton, L., & Nass, R. D. (1992). Lack of VEP evidence for magnocellular dysfunction in dyslexia. *Society for Neuroscience Abstracts, 18,* 1395.

Vidyasagar, T. R., Kulikowski, J. J., Lipnicki, D. M., & Dreher, B. (2002). Convergence of parvocellular and magnocellular information channels in the primary visual cortex of the macaque. *European Journal of Neuroscience, 16,* 945–956.

Wallisch, P., & Movshon, J. A. (2008). Structure and function come unglued in the visual cortex. *Neuron, 60,* 195–197.

Watanabe, M., & Rodieck, R. W. (1989). Parasol and midget cells of the primate retina. *Journal of Comparative Neurology, 299,* 434–454.

White, A. J., Solomon, S. G., & Martin, P. R. (2001). Spatial properties of koniocellular cells in the lateral geniculate nucleus of the marmoset *Callithrix jacchus. Journal of Physiology, 533*(Pt 2), 519–535.

Wiesel, T. N., & Hubel, D. H. (1966). Spatial and chromatic interactions in the lateral geniculate body of the rhesus monkey. *Journal of Neurophysiology, 29,* 1115–1156.

Wong-Riley, M. (1979). Changes in the visual system of monocularly sutured or enucleated cats demonstrable with cytochrome oxidase histochemistry. *Brain Research, 171,* 11–28.

Wuerger, S. M., Ruppertsberg, A., Malek, S., Bertamini, M., & Martinovic, J. (2011). The integration of local chromatic motion signals is sensitive to contrast polarity. *Visual Neuroscience, 28,* 239–246.

Xu, X., Ichida, J. M., Allison, J. D., Boyd, J. D., Bonds, A. B., & Casagrande, V. A. (2001). A comparison of koniocellular, magnocellular and parvocellular receptive field properties in the lateral geniculate nucleus of the owl monkey (*Aotus trivirgatus*). *Journal of Physiology, 531*(Pt 1), 203–218.

Yoonessi, A., & Yoonessi, A. (2011). Functional assessment of magno, parvo and konio-cellular pathways; Current state and future clinical applications. *Journal of Ophthalmic & Vision Research, 6,*119–126.

Yukie, M., & Iwai, E. (1981). Direct projection from the dorsal lateral geniculate nucleus to the prestriate cortex in macaque monkeys. *Journal of Comparative Neurology, 201,* 81–97.

Zele, A. J., Wood, J. M., & Girgenti, C. C. (2010). Magnocellular and parvocellular pathway mediated luminance contrast discrimination in amblyopia. *Vision Research, 50,* 969–976.

17 Ventral and Dorsal Cortical Processing Streams

ANDREW H. BELL, TATIANA PASTERNAK, AND LESLIE G. UNGERLEIDER

Much of our knowledge concerning visual cortical organization derives from anatomical and physiological studies of the brains of macaque monkeys, in which over 50 separate visual areas have now been described. Similar visual areas and processing stages are now known to exist in both New World monkeys (e.g., Rosa & Tweedale, 2005) and humans (e.g., Wandell, Dumoulin, & Brewer, 2007) and thus likely represent a common primate plan. These visual areas are organized into two separate processing streams, both of which originate in the primary visual cortex (striate cortex), V1, and both of which consist of multiple extrastriate areas beyond V1. The ventral stream is directed into the temporal lobe and is crucial for the visual recognition of objects, whereas the dorsal stream is directed into the parietal lobe and is crucial for appreciating the spatial relationships among objects and visual guidance toward them (figure 17.1A) (see Kravitz et al., 2011; Ungerleider & Bell, 2011, for reviews). A simple way to conceptualize the two streams is "what" versus "where."

The original evidence for separate processing streams for "what" versus "where" was based on the contrasting effects of inferior temporal and posterior parietal cortex lesions in monkeys (see Ungerleider & Mishkin, 1982). Lesions of inferior temporal cortex cause severe deficits on a wide range of visual discrimination tasks (e.g., discriminating objects, colors, patterns, and shapes), but these lesions do not affect animals' performance on visuospatial tasks (e.g., visually guided reaching and judging which of two objects lies closer to a visual landmark). Conversely, posterior parietal lesions do not affect visual discrimination ability but instead cause severe deficits on visuospatial performance (figure 17.1B).

Functional neuroimaging studies have since demonstrated separate processing streams in humans (e.g., Haxby et al., 1991), and patient studies have shown a similar dissociation in behavioral deficits, with lesions to occipitotemporal cortex resulting in visual object agnosia and achromatopsia, or cortical color blindness, and lesions to occipitoparietal cortex resulting in optic ataxia (misreaching), hemispatial neglect, constructional apraxia, gaze apraxia, and akinetopsia (an inability to perceive movement) (e.g., Newcombe, Ratcliff, & Damasio, 1987; Zihl, von Cramon, & Mai, 1983).

In the first part of this chapter, we summarize the anatomical arrangement of components of both the ventral and dorsal streams in monkey cortex. We next consider the neuronal properties and functional organization of each of the streams. We also describe the behavioral effects of selective lesions of the two streams. Finally, we describe interactions between the two processing streams as well as the differential projections from them to regions within prefrontal cortex and the role these areas play in vision.

ANATOMICAL ORGANIZATION OF THE VENTRAL AND DORSAL PROCESSING STREAMS

Regions that make up the ventral stream lie directly anterior to V1 in the occipital lobe and in progressively more anterior and ventral portions of the temporal lobe, whereas regions that make up the dorsal stream also include these early occipital-lobe areas but then occupy sites within the posterior superior temporal sulcus (STS) and include progressively more anterior sites within this sulcus and more dorsal sites within the parietal lobe. Figure 17.1C illustrates the location of these areas within the cortex, and figure 17.2 diagrams their anatomical connections.

The cortical analysis of objects begins in V1, where information about contour orientation, color, luminance, and direction of motion is represented in subsets of neurons responsible for each point in the visual field (see Sincich & Horton, 2005, for review). Information from V1 is sent forward to interdigitating thin, thick, and interstripe regions within V2. From the thin and interstripe regions in V2, representing mainly color and form information, respectively, neural signals proceed forward to area V4 on the lateral and ventromedial surfaces of the hemisphere and to a posterior inferior temporal area just in front of V4, area TEO. From both V4 and TEO, signals related to object form, color, and

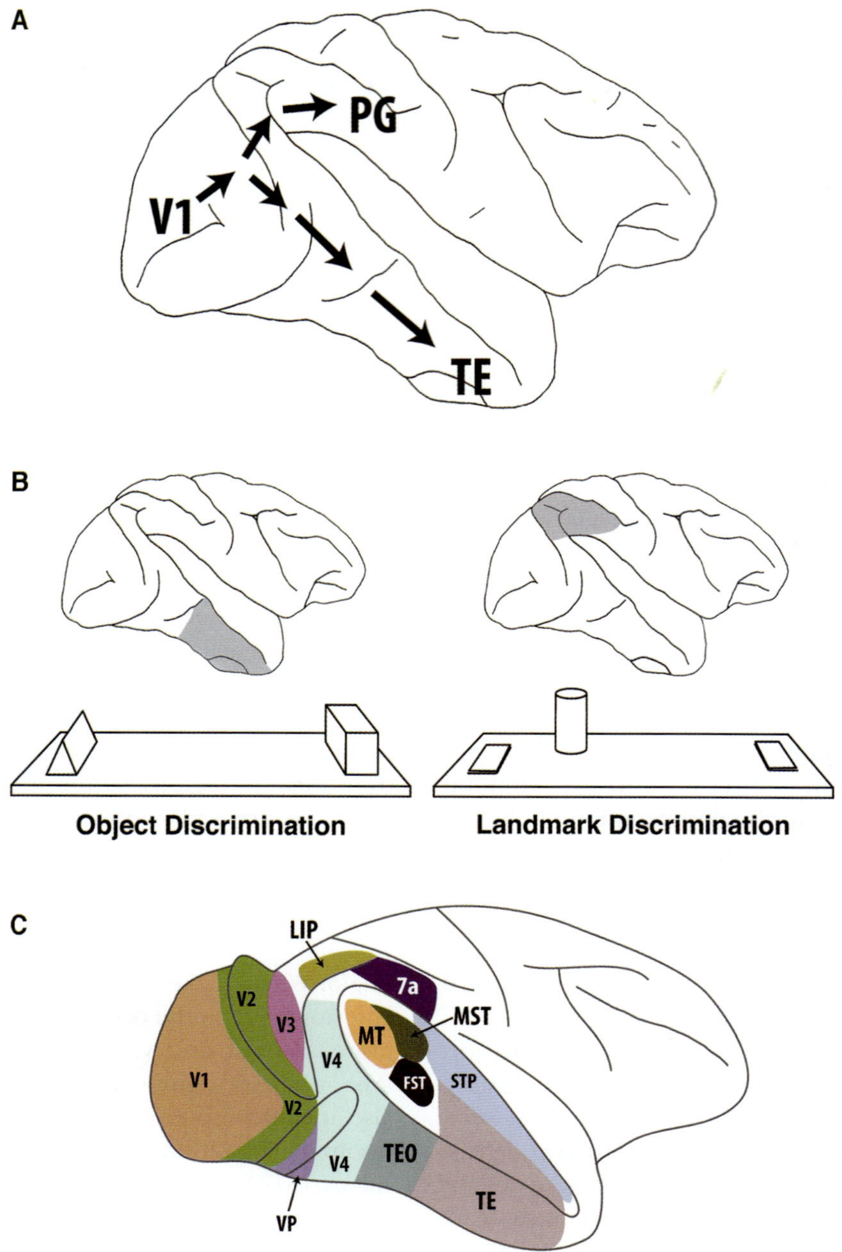

FIGURE 17.1   Two visual processing streams in monkey cortex. (A) According to the model originally proposed by Ungerleider and Mishkin (1982), both streams originate in primary visual cortex (V1), and both consist of multiple visual areas beyond V1. The ventral stream is directed into the temporal lobe, area TE, and is crucial for the visual recognition of objects, whereas the dorsal stream is directed into the parietal lobe, area PG, and is crucial for appreciating the spatial relationships among objects and for visual guidance toward them. (B) The original evidence for separate processing streams was based largely on the contrasting effects of inferior temporal and posterior parietal lesions in monkeys. Lesions of inferior temporal cortex (left, in black) cause impairments on visual discrimination tasks such as recognizing an object seen a few seconds previously, whereas posterior parietal lesions (right, in black) cause impairments on visuospatial tasks such as judging which of two identical plaques is located closer to a visual landmark—the cylinder. (C) Lateral view of the monkey brain with the sulci partially opened to illustrate the location of the multiplicity of functionally distinct visual areas that comprise the ventral and dorsal processing streams.

texture proceed forward to area TE—the last exclusively visual area within the ventral stream for object recognition. Together, areas TEO and TE comprise inferior temporal (IT) cortex.

The cortical analysis of spatial perception also begins in V1, deriving largely from the motion-sensitive neurons in layer 4B. These neurons project to the thick stripes within V2 and to V3. Together, these three early

   ANDREW H. BELL, TATIANA PASTERNAK, AND LESLIE G. UNGERLEIDER

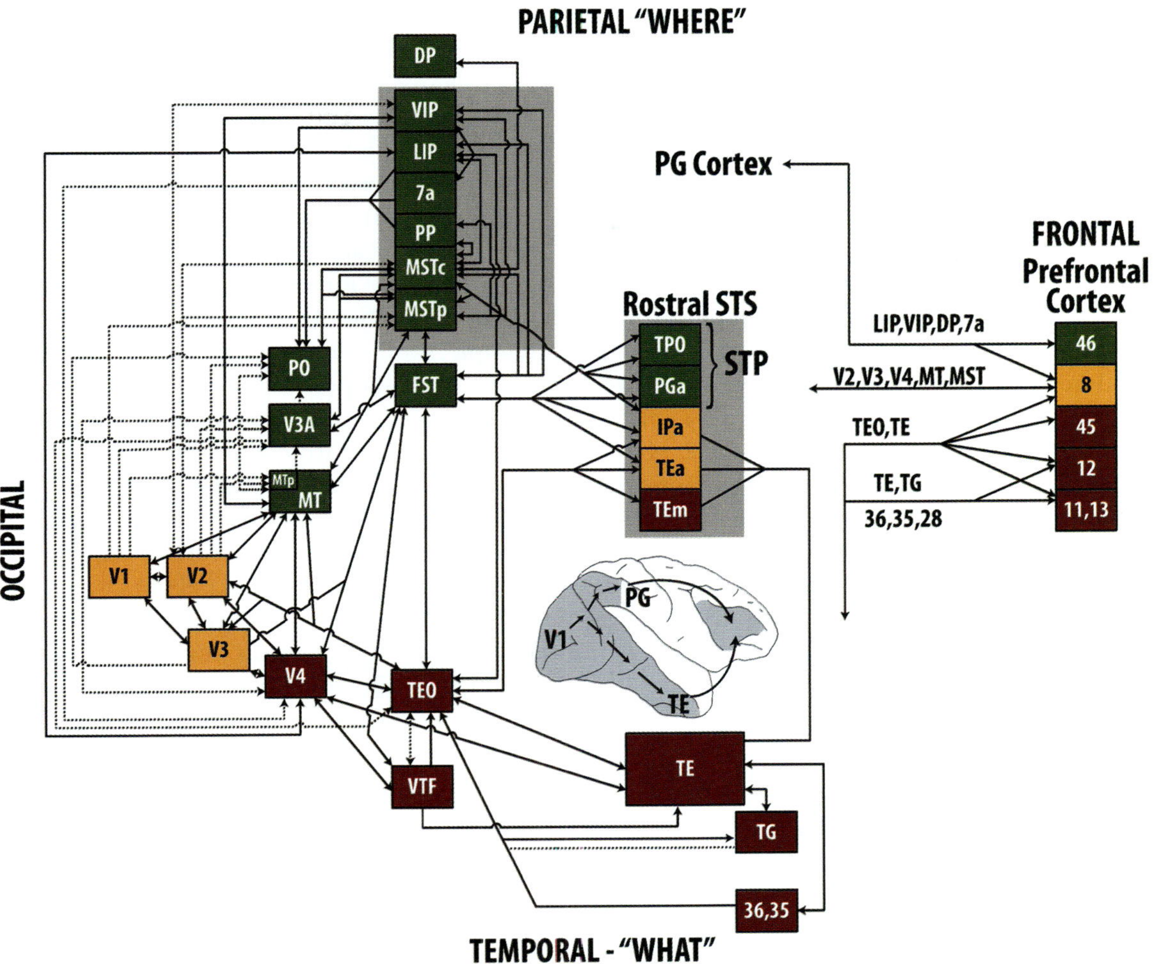

FIGURE 17.2 Anatomical connections of visual areas in the ventral and dorsal streams. Solid lines indicate connections arising from both central and peripheral visual field representations, whereas dotted lines indicate connections restricted to peripheral field representations. Red boxes indicate ventral stream areas associated primarily with object vision; green boxes indicate dorsal stream areas associated primarily with spatial vision; yellow boxes indicate areas not clearly allied with either stream. Shaded region on the lateral view of the monkey brain indicates the extent of the cortex included in the diagram.

visual areas provide input to the middle temporal visual area, MT, also known as the motion-sensitive area of the STS. From MT, information is sent forward within the STS to several additional motion-sensitive areas, including the medial superior temporal area (MST) and the fundus of the superior temporal area (FST), both of which project to the superior temporal polysensory area (STP) located farther forward on the upper bank of the STS. Portions of STP also receive ventral stream inputs, raising the possibility that these regions may serve as anatomical sites for the integration of visual information about form and motion.

MT is also the source of information for the ventral intraparietal area (VIP), which in turn projects to several additional parietal areas including the lateral intraparietal area (LIP) and area 7a, located in the inferior parietal lobule. Many of these parietal areas also receive direct inputs from the peripheral field representations of V1 and V2, bypassing MT, and these inputs may provide rapid activation of regions mediating spatial attention.

Connections at virtually all stages of both the ventral and dorsal streams are reciprocal, with each node providing feedforward and feedback projections to other nodes in the respective processing streams. Whereas feedforward projections provide bottom-up sensory-driven inputs to subsequent visual areas, the precise functions of the reciprocal feedback projections are still unknown. In addition to feedforward and feedback projections, there appear to be "intermediate-type" projections that link areas at corresponding levels of the visual hierarchy, most notably between areas of the ventral and dorsal streams.

All areas within the ventral and dorsal streams also have heavy connections with subcortical structures, including the pulvinar, claustrum, and basal ganglia.

Each stream receives subcortical modulatory inputs from ascending cholinergic projections from the basal forebrain and ascending noradrenergic projections from the locus coeruleus. Visual information is sent from the last stations of both streams to the most ventral and anterior reaches of the temporal lobe, notably peri-rhinal cortex and parahippocampal areas TF and TH. These regions in turn project, via entorhinal cortex, to medial temporal lobe structures such as the hippocam-pus, which contributes to forming long-term memories of visual objects and their contexts.

Information is also sent from both streams to pre-frontal cortex, which plays an important role in visual working memory (see Hussar & Pasternak, 2012; Miller, 2000). Finally, there are direct projections from the ventral stream to the amygdala, which is important for attaching emotional valence to a stimulus, and from the dorsal stream to the pons, which, in conjunction with the cerebellum, likely contributes to the control of eye and head velocity during tracking eye movements.

PROPERTIES OF THE VENTRAL STREAM

Neurons in the Ventral Stream Are Sensitive to Increasingly More Complex Object Features

Consistent with a role in object recognition, neurons in all areas of the ventral stream share a number of physi-ological characteristics, including sensitivity to the shape, color, or texture of visual stimuli. As one pro-gresses forward along the pathway, there is a general trend toward selectivity for increasingly complex stimu-lus features. For example, although many V1 neurons function as local spatial filters (e.g., signaling the pres-ence of contours at particular positions in the visual field), some V2 neurons respond to illusory contours (von der Heydt, Peterhans, & Baumgartner, 1984). Neurons in V2 have also been shown to be sensitive to color, binocular disparity, and/or simple shapes.

Farther along the ventral stream, the trend continues with neurons tuned for multiple stimulus dimensions, such as length, width, disparity, and/or color. Neurons in V4 and posterior portions of IT cortex are responsive to simple contours as well as particular combinations of contours, suggesting an intermediate stage of process-ing. Moving forward successively into area TEO and then into posterior and anterior portions of area TE, there is a further increase in the complexity of the fea-tures needed to activate neurons to the point where many neurons in area TE, the final stage in the ventral stream, require moderately complex features, three-dimensional configurations, or even intact objects to elicit a response (see Kourtzi & Connor, 2011, for review).

Receptive fields (RFs) become larger as one moves along the ventral stream. In V1 they can be less than a single degree of visual angle, whereas in IT cortex they can be much larger, bilateral, and can encompass the entire visual field. Further, neurons in anterior TE appear organized according to functional, not spatial, relationships. Tanaka and colleagues demonstrated that these neurons are organized into columns based on their selectivity for critical features that make up the intact objects (Ito et al., 1995; Tsunoda et al., 2001). For example, some neurons in a column responsive to faces might respond optimally to different face components, and others might be sensitive to the overall configura-tion of a face regardless of viewpoint, and still others might respond well to faces specifically in profile. Recent evidence from functional magnetic resonance imaging in awake monkeys further supports this orga-nization (Op de Beeck et al., 2008).

Thus, a given object feature is not represented by the activity of a single neuron but by the activity of many neurons within a column. Given this clustering of neurons responsive to related features, an object in its entirety would presumably be represented by a pattern of activity distributed across multiple columns, each of which may participate in representing multiple objects. This so-called combination coding (or population coding) provides the ventral stream with the robustness and flexibility necessary to encode a given object despite changes in size, orientation, etc.

Neurons in the Ventral Stream Exhibit Perceptual Constancy

The visual information reaching the retina is a two-dimensional projection of a three-dimensional world, a projection that varies dramatically as a function of changes in an object's position, distance, illumination, and orientation relative to the viewer. Thus, two of the functions of the ventral stream are to identify and encode those invariant features of objects that are useful for recognition across myriad circumstances. For example, the selectivity of neurons in TE for particular features or objects is conserved regardless of where those features or objects appear in the visual field (Ito et al., 1995). These neurons are, therefore, said to exhibit *position invariance*. In addition to invariance over position, many neurons in TE show maintained selectiv-ity for their preferred stimuli over more complex kinds of transformations such as size, distance from the observer in depth, or degree of ambient illumination so that they show invariance over a number of the changes that normally take place as a moving object is

   ANDREW H. BELL, TATIANA PASTERNAK, AND LESLIE G. UNGERLEIDER

viewed under naturalistic conditions (or as an observer moves relative to a stationary object).

Recent work by Li and DiCarlo (2008, 2010) has shown how position and size invariance might emerge through an associative learning process based on repeated exposures to the same stimulus under different viewing conditions. When an object moves through the visual field, the location of the object on the retina moves in a predictable way. The visual system can use this knowledge to associate different retinal images as the same object in different positions, thus generating a *position-invariant* representation. Li and DiCarlo (2008) tested this hypothesis by recording from neurons in IT cortex while presenting monkeys with an object at a fixed position from where the monkeys were currently looking. When the monkeys would move their eyes, the object would change to a new object. They hypothesized that if linking different retinal images according to changes in eye gaze generated invariant representations, this manipulation should result in these two stimuli being confused as the same object—which was indeed the case. The ability of downstream neurons to discriminate between the two stimuli had been abolished.

Neurons in the Ventral Stream Are Sensitive to Experience, Familiarity, and Expectation

Neurons in the ventral stream exhibit a number of other properties that further highlight their importance to visual perception. Miller and Desimone (1994) demonstrated how short-term changes in the response properties of neurons in IT cortex encode aspects of both stimulus recency and short-term stimulus significance. They trained monkeys to perform a delayed match-to-sample task in which a sample stimulus was followed by a series of potential match stimuli, one after the other. It was found that when the same stimulus was presented more than once (and became increasingly familiar to the animal) the responses to that stimulus decreased. This kind of response decrement, termed *repetition suppression*, has been found both under anesthesia and for stimuli that an awake animal is trained to ignore. By contrast, if a monkey is trained to do a matching task in which it must hold a particular visual pattern in memory over a short delay, some IT neurons increase their firing rates, specifically when the to-be-remembered stimulus is recognized. This phenomenon, termed *mnemonic enhancement*, can be thought of as a mechanism in which bottom-up processing of the stimulus is affected by top-down cognitive factors, namely the animal's expectation of the relevant stimulus.

Using paired-associate tasks, in which monkeys are trained to associate arbitrary stimulus patterns with one another for reward, Sakai and Miyashita (1991) have found that TE neurons respond best to stimuli that had been paired together during training. This pair coding by neurons is decreased by surgically interrupting the feedback projections to them from the perirhinal and entorhinal cortices (Miyashita et al., 1998). Furthermore, just as neurons in TE can signal learned stimulus associations, as demonstrated with paired-associate tasks, so too can they signal when this association is violated. Meyer and Olson (2011) recently reported that when monkeys are presented with a stimulus that violates a previously established paired association, neurons in IT cortex respond to the unexpected stimulus with increased magnitude and selectivity compared to an expected stimulus.

Some Neurons in the Ventral Stream Are Selective for Faces

One of the more striking characteristics of primate temporal cortex is the presence of neurons with responses selective for face stimuli, first described by Gross, Bender, and Rocha-Miranda (1969). Although initially only found in small number, recent studies combining functional neuroimaging with targeted neuronal recordings in nonhuman primates show that face-selective neurons are concentrated into a few "patches," located primarily along the lower bank of the STS and extending onto the adjacent IT gyrus (Bell et al., 2011; Tsao et al., 2006).

Face-selective neurons vary in the degree and nature of their preference for face stimuli. Many respond well to both real faces and face pictures but give little or no response to any other stimuli tested, including other complex objects, texture patterns, and images in which the features making up the face are rearranged or scrambled. Other neurons in TE and STP respond to specific face components, such as the presence of eyes, the distance between the eyes, or the extent of the forehead. Finally, many face-selective neurons in the ventral stream are also sensitive to different facial expressions and gaze directions, both potent social cues among primates.

Freiwald, Tsao, and Livingstone (2009) found that face-selective neurons in the so-called middle face patch (located along the inferior bank of the STS) are sensitive to features or groups of features (e.g., nose width, eye separation) that make up a face. This evidence provides a basis for how face neurons in the temporal cortex might distinguish not only between faces and nonface objects but also between individual faces.

Farther along the ventral stream, near the anterior medial temporal sulcus, Leopold, Bondar, and Giese (2006) reported finding face-selective neurons that represented facial identity according to how much they deviated from a "prototypical" face (i.e., "norm-based encoding"). How these two complementary processes interact is still unknown, but it does highlight how face processing might progress from more feature-based representations to more holistic representations within the ventral stream.

Functional Organization of the Ventral Stream

The widespread use of functional neuroimaging in neuroscience has made possible the discovery of specific subregions within ventral temporal cortex that appear to be selective for categories of visual stimuli (figure 17.3). Puce and colleagues (1995) identified a region along the fusiform gyrus that was selectively activated by faces, later labeled the *fusiform face area* (FFA) (Kanwisher, McDermott, & Chun, 1997). Other face-selective regions have since been identified in human ventral occipital cortex (the *occipital face area*, OFA), the STS, and the anterior portion of the temporal lobe (Kriegeskorte et al., 2007; Rajimehr, Young, & Tootell, 2009).

Adjacent to the fusiform gyrus is a region within the parahippocampal cortex that is more active when subjects view complex scenes (such as rooms, landscapes, landmarks, and city streets) compared to when they view objects, faces, or other kinds of visual stimuli (the *parahippocampal place area*, PPA) (Epstein & Kanwisher, 1998). Selectivity for places has also been observed in retrosplenial cortex (Epstein, 2008). Along the lateral occipitotemporal cortex (near the middle occipital gyrus) is the *extrastriate body area* (EBA), an area selectively activated by images of human body parts (such as an isolated foot or arm) (Downing et al., 2001). An additional body-part–selective region has been identified in the fusiform gyrus (the *fusiform body area,* FBA) (Schwarzlose, Baker, & Kanwisher, 2005). Finally, the so-called *visual word form area* (VWFA), located within/adjacent to the fusiform gyrus, is selective for letters, nonsense letter strings, and words (Cohen et al., 2002). Interestingly, the VWFA is found almost exclusively in the left hemisphere—the same hemisphere to contain other language-related areas such as Broca's and Wernicke's areas. Similar regions have been found in IT cortex of nonhuman primates (Bell et al., 2009; Pinsk et al., 2005; Tsao et al., 2003).

Behavioral Effects of Ventral Stream Lesions

FORM AND COLOR IMPAIRMENTS   Lesions of V4 and IT cortex increase contrast thresholds for relatively simple orientation discriminations as well as more complex shape discriminations. Although V4 has been identified as one of the areas actively involved in the processing of color, V4 lesions result in transitory and relatively modest disruption of simple color discriminations. The effects of V4 lesions on color thresholds and chromatic

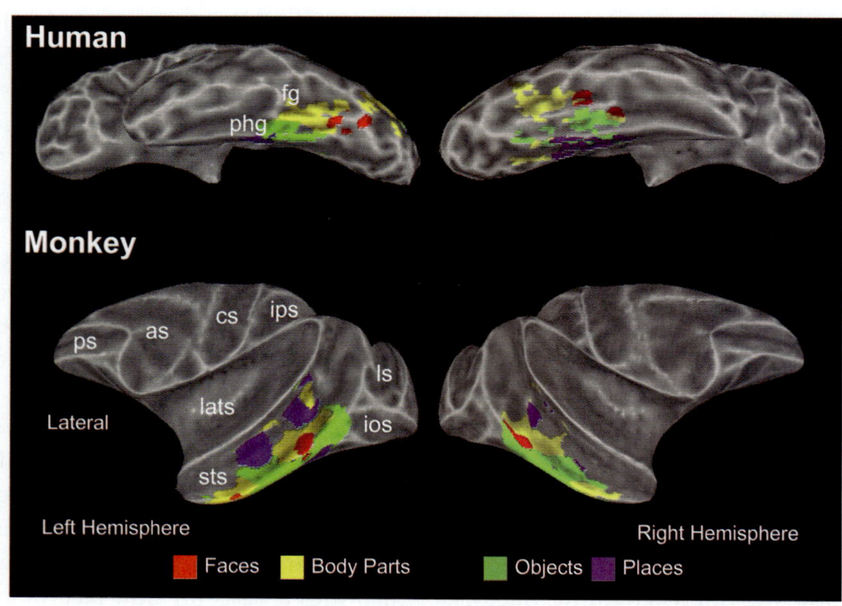

FIGURE 17.3   Category-selective regions in the human and monkey as revealed by fMRI. Voxels are colored according to their preference for faces, body parts, objects, and places. (Adapted from Bell et al., 2009.)

contrast sensitivity are also relatively mild (Dean, 1979; Merigan, 1996). This preservation of various forms of color perception is consistent with reports that some portions of the dorsal stream of both macaques and humans have access to chromatic information (Gegenfurtner et al., 1994; Seidemann et al., 1999). Color sensitivity found in the dorsal stream may help account for the failure to find large, permanent color vision losses after V4 lesions. It cannot, however, account for the more dramatic and often permanent color deficits found after IT lesions (Huxlin et al., 2000).

IMPAIRMENTS IN PERCEPTUAL CONSTANCIES FOLLOWING IT LESIONS One of the critical features of neurons within IT cortex is their tolerance to changes in simple visual features, such as size, position, and ambient illumination. Several studies have demonstrated that such perceptual constancies are lost in monkeys with IT lesions. For example, monkeys with IT lesions that have been trained to discriminate between two disks of different absolute sizes are unable to perform the discrimination when those disks are presented at variable distances (and hence with variable retinal image sizes), responding instead on the basis of either retinal size or distance. Similarly, whereas normal monkeys will easily generalize a visual discrimination learned in one hemifield to the other, monkeys with IT lesions do not, indicating that they have lost neural representations for objects that are invariant across visual field locations. Finally, whereas normal monkeys are unaffected by changes in the patterns of shadow and light falling on object surfaces, variations in illumination prevent monkeys with IT lesions from seeing the equivalence of objects.

CATEGORY-SELECTIVE AGNOSIAS AND MEMORY IMPAIRMENTS In addition to the loss of perceptual constancy described above, monkeys with IT lesions are impaired in learning to discriminate visual stimuli, including those involving shapes and objects. These deficits can be general in nature, affecting a wide range of stimulus categories, or restricted to specific categories. For instance, damage to any of the category-selective regions described above can result in difficulties in recognizing stimuli from those categories. In humans, damage that includes the FFA or OFA, for example, results in face-processing deficits, such as prosopagnosia (the inability to recognize familiar faces) (Dricot et al., 2008; Rossion et al., 2003). Similarly, damage to the parahippocampal gyrus produces difficulties in navigating familiar locations (*topographical disorientation*) (Epstein et al., 2001), and damage to the VFWA is associated with alexia (Cohen & Dehaene, 2004).

PROPERTIES OF THE DORSAL STREAM

Neurons in the Dorsal Stream Are Sensitive to Visual Motion

One of the more prominent features of neurons in the dorsal stream is selectivity for the direction of visual motion. Directionally selective neurons respond vigorously to one direction of motion and respond less or not at all when the same stimulus moves in the opposite direction. This property first appears in V1, where directionally selective neurons are found primarily in layers 4Cα and 4B, the target layers of the magnocellular neurons of the lateral geniculate nucleus (Hawken, Parker, & Lund, 1988). In area V2, directionally selective neurons are less numerous and are concentrated in the thick stripes, suggesting magnocellular influences (Burkhalter & Van Essen, 1986). In area V3, these neurons are more numerous and prefer coarse, relatively fast moving stimuli, indicating a role in processing of motion information. Indeed, neurons in V3 show the ability to integrate local motion signals (Gegenfurtner, Kiper, & Levitt, 1997), a feature indicative of higher-level motion analysis.

Processing of visual motion is even more prevalent in area MT, where the selectivity for the direction and speed of motion is found in the majority of neurons. MT neurons display directional selectivity to the motion of random dots, bars, gratings, and naturalistic stimuli (Maunsell & Van Essen, 1983; Nishimoto & Gallant, 2011), and neurons with similar directional preferences are clustered in columns. When presented with a plaid pattern consisting of two component gratings moving in orthogonal directions, MT neurons, unlike V1 neurons, respond to the direction of the plaid rather than to its components. The RFs of MT neurons have strong antagonistic surrounds that show inhibition when stimulated by the same direction and/or speed as the excitatory RF center. These properties allow these neurons to code local motion signals according to the context in which this motion appears, suggesting a role in the detection of relative motion and in figure/ground segregation.

Visual motion is also processed by neurons in the medially adjacent area MST, which receives projections from area MT. Neurons in the dorsal portion (MSTd) have very large RFs, preferring motion of full-field stimuli, a property that suggests a role in integrating visual motion signals generated during the observer's movement through the environment with eye-movement and vestibular signals. By contrast, neurons in the lateral portion (MSTl) appear to be more involved in the analysis of object motion in the environment and in the maintenance of pursuit eye movements associated with this motion. RFs of neurons in this region, similar to neurons in MT, have antagonistic surrounds and respond very strongly to object motion when the motion in the surround is in the opposite direction, suggesting a role in segmenting moving objects from backgrounds.

A number of areas within parietal cortex also respond to visual motion (e.g., areas LIP and VIP), and these responses are largely reminiscent of responses at earlier stages of processing within the dorsal stream (e.g., Colby, Duhamel, & Goldberg, 1993; Fanini & Assad, 2009).

In addition to representing visual information, there is accumulating evidence to suggest that these neurons may serve a more direct role in working memory and perceptual decisions involving visual motion (Krug, 2004; Pasternak & Greenlee, 2005). Several studies have demonstrated that neurons in MT and MST reflect the perceived direction of motion within coherently moving dot patterns (as indicated by an eye movement or other behavioral response on the part of the monkey) (e.g., Britten et al., 1996; Dodd et al., 2001; Newsome, Britten, & Movshon, 1989). Similarly, a number of studies have shown that electrical microstimulation of these areas can alter the perception of motion direction (e.g., Bisley, Zaksas, & Pasternak, 2001; DeAngelis, Cumming, & Newsome, 1998; Salzman, Britten, & Newsome, 1990) and heading (Britten & van Wezel, 1998). Microstimulation of area MT can also interfere with direction discrimination when delivered during the memory delay (figure 17.4A–C) (Bisley, Zaksas, & Pasternak, 2001), suggesting its role in transient storage of visual motion. More recently, Lui and Pasternak (2011) have shown that during the memory-guided direction comparison task, MT responses to the current motion direction can be modulated by the preceding direction (figure 17.4D), suggesting their active participation in comparison circuits underlying perceptual decisions (Hussar & Pasternak, 2012).

Neurons in the Dorsal Stream Are Sensitive to Binocular Disparity

Sensitivity to binocular disparity, a property believed to underlie depth perception, is also characteristic of many neurons in the dorsal stream. Many V1 neurons respond best when both eyes are stimulated and are sensitive to absolute retinal disparity, an early stage of processing that leads to stereoscopic depth perception. Sensitivity to absolute retinal disparity continues in V2, yet some neurons begin to show sensitivity to *relative disparity* between different locations in the visual field.

This property appears to be segregated primarily to neurons in the thick stripes of V2 (Roe & Ts'o, 1995), which project to MT. Disparity sensitivity, also present in some neurons in area V3, becomes even more prevalent in MT (Krug & Parker, 2011), where the neurons are found clustered according to preferred disparity. These neurons appear to contribute to stereoscopic depth perception because microstimulation of similarly tuned cells can bias the monkey's perceptual judgment of depth toward the preferred disparity (DeAngelis, Cumming, & Newsome, 1998).

Neurons in the Dorsal Stream Are Involved in Processing Visual Space in Active Observers

Many properties of neurons within the dorsal stream suggest a role in the processing of location information in active observers. For example, neurons in MSTd are capable of integrating visual information extracted during movement of the observer with signals related to eye and head movements (Andersen et al., 1999). Further, there is evidence that LIP neurons play a role in planning saccades to a remembered location and in storing information in both eye-centered and body-centered coordinates. Another intriguing feature of LIP neurons is that the spatial representation of a remembered stimulus is dynamic and shifts to the corresponding retinal location around the time of a saccade (Duhamel, Colby, & Goldberg, 1992). This "remapping," which is important for maintaining a continuous representation of the visual world during eye movements, is not unique to LIP, as it has also been observed in areas V2, V3, and V3A (Nakamura & Colby, 2002).

Area VIP has also been implicated in the encoding of spatial representations. It has prominent connections with MT, MST, and FST but, unlike LIP, receives little if any inputs from ventral stream areas. Neuronal responses in VIP show selectivity for local motion as well as for the optic flow, are often modulated by eye position, and appear to reflect an encoding in head-centered coordinates (Zhang, Heuer, & Britten, 2004). Some VIP neurons seem to prefer stimuli that are close to the animal (Colby, Duhamel, & Goldberg, 1993) and respond to tactile stimuli and vestibular cues (Chen, DeAngelis, & Angelaki, 2011). This observation has led to the suggestion that VIP neurons are involved in a construction of a multisensory head-centered representation of near-personal space (Avillac et al., 2005).

Area 7a represents the final stage in the dorsal stream hierarchy. It is connected with a wide range of cortical and subcortical regions, including a number of areas in parietal, prefrontal, and IT cortex, as well as the basal

    ANDREW H. BELL, TATIANA PASTERNAK, AND LESLIE G. UNGERLEIDER

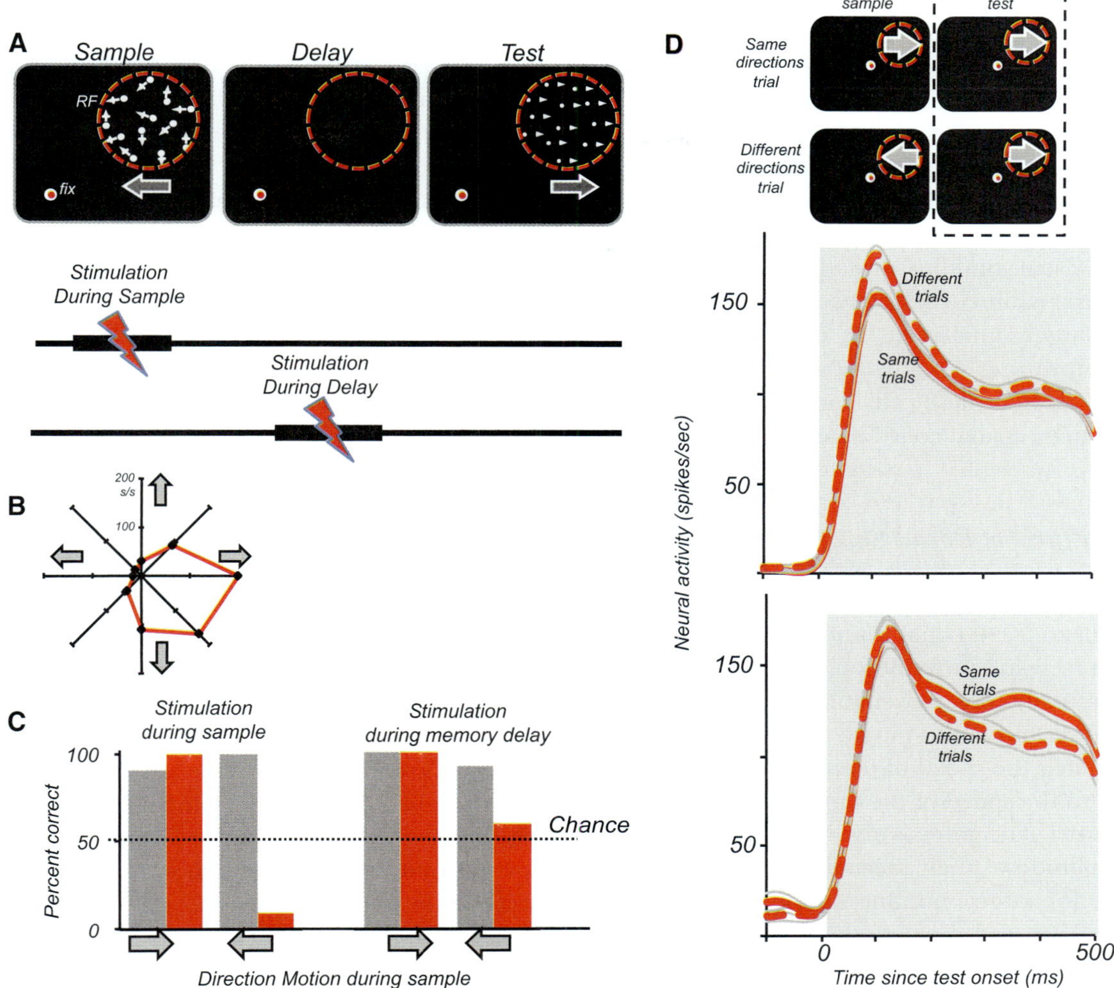

FIGURE 17.4   Participation of MT neurons in memory-guided comparisons of motion directions. Effect of microstimulation of MT applied during different trial components of a direction discrimination task. (A) Behavioral task. On each trial, the monkeys compared directions of two sequentially presented 500-ms random-dot stimuli separated by a 1,500-ms delay. Sample and test moved either in the same or opposite directions, and the monkeys pressed one of the two response buttons depending on whether the two directions were the same or different. Stimulation was applied on 25% of the trials either during the sample or during the delay, as shown in the diagram. Stimulation and nonstimulation trials were randomly interleaved. (B) Direction selectivity profile of an example stimulation site showing its multiunit activity. The preferred direction for that site was rightward. (C) Behavioral performance on trials when stimulation was applied during the stimulus (left columns). The data show that when the sample moved to the left (direction opposite to the preferred direction of the stimulated site) the monkey consistently equated it with rightward motion, and its performance was below chance (~10% correct), showing that activity generated by stimulation was interpreted by the monkey as rightward motion. However, when the same stimulation was applied during the period of storage (right columns), it no longer produced signals interpreted as unambiguous directional information, and the monkey performed at chance. These results show that MT neurons not only participate in encoding of visual motion but also in its storage by either maintaining an active connection with the circuitry involved in storage or being an integral component of that circuitry. (Adapted from Bisley, Zaksas, & Pasternak, 2001.) (D) Comparison effects in MT during direction discrimination task. Responses to the comparison test are often modulated by the preceding sample direction. This can be seen by comparing responses to the same test direction on trials when sample direction is the same as the test and on trials when sample direction is opposite (as shown in the diagram). (Top) Example neuron with stronger response on "different" trials. This type of modulation was encountered in over 30% of MT neurons. (Bottom) Example neuron with a stronger test response on "same" direction trials. This type of modulation was encountered in about 20% of MT cells. These data demonstrate that MT responses during the comparison phase of the direction discrimination task reflect similarities and differences between the two stimuli, suggesting their participation in sensory comparisons. (Adapted from Lui & Pasternak, 2011.)

ganglia, providing visual and visuomotor signals important for the execution of visually guided behavior. Neurons in area 7a have large, often bilateral, RFs and possess properties similar to those encountered at preceding stages in the dorsal stream. For example, many neurons are sensitive to complex visual motion and optic flow (Read & Siegel, 1997). Most notably, responses of 7a neurons are modulated by the behavioral relevance of the visual stimuli (Constantinidis & Steinmetz, 2001) and task parameters. Because the activity of these neurons is largely unaffected by changes in body or head position during saccades to specific retinal locations, it is likely that these cells encode information in world-referenced spatial coordinates (see Snyder et al., 1998).

Behavioral Effects of Dorsal Stream Lesions

MT LESIONS AFFECT SMOOTH PURSUIT EYE MOVEMENTS AND SPEED PROCESSING   MT and MST lesions have been shown to result in retinotopically specific deficits in smooth pursuit eye movements (Newsome et al., 1985). Monkeys trained to pursue moving targets were unable to match the speed of their smooth pursuit eye movements to the speed of the target following lesions to MT and MST. The animals also had problems adjusting the amplitude of their saccadic eye movements to compensate for target motion, although they had no problems making saccades to stationary targets. Because these effects reflected difficulties in matching the velocity of visual targets without evidence of motor abnormalities, they were interpreted as impaired perception of stimulus velocity. Subsequently, a number of studies examined the effects of MT/MST lesions on speed discriminations more directly by measuring speed difference thresholds. Substantial and largely permanent losses in the accuracy of speed judgments were found (Pasternak & Merigan, 1994; Rudolph & Pasternak, 1999).

LESIONS OF MT/MST LEAD TO DEFICITS IN COMPLEX MOTION PERCEPTION   MT and MST lesions have also been shown to affect the perception of motion (i.e., akinetopsia). Newsome and Pare (1988) found severe deficits in motion coherence thresholds when stimuli were placed in the portion of the visual field corresponding to the site of discrete MT lesions. A similar inability to extract coherent motion in the presence of noise was later reported after MT/MST lesions by Pasternak and colleagues (Pasternak & Merigan, 1994; Rudolph & Pasternak, 1999). They found that MT/MST lesions produced deficits not only in coherence thresholds measured with random dots but also

in signal-to-noise thresholds measured with drifting gratings masked by noise. This increased susceptibility to noise was specific to the domain of motion perception, as the same monkeys showed no deficit in discriminating the orientation of gratings masked by two-dimensional noise.

In addition to the increased susceptibility to noise, MT/MST lesions result in permanent deficits in the integration of local motion vectors (Pasternak & Merigan, 1994; Rudolph & Pasternak, 1999). Furthermore, during memory-guided direction discrimination tasks, such lesions affect the retention of stimuli requiring motion integration as well as affect the accuracy of direction comparisons measured with coherently moving stimuli (Bisley & Pasternak, 2000).

LESIONS OF POSTERIOR PARIETAL CORTEX AFFECT ENCODING OF VISUAL SPACE   Neuronal properties in parietal cortex suggest a role in the integration of sensory signals and in constructing a representation of extrapersonal space in preparation for motor action, which is largely supported by lesion studies performed in nonhuman primates. Latto (1986) tested monkeys after bilateral lesions of area 7a and found deficits in the performance of a spatial landmark test. Quintana and Fuster (1993), who applied cooling to areas 5 and 7 of the parietal cortex, found that monkeys show decreased speed and accuracy in both reaching and eye movements during the performance of tasks requiring the processing and retention of spatial information. Li, Mazzoni, and Andersen (1999) reported deficits in memory-guided saccadic eye movements after reversible inactivation of LIP. They concluded that LIP neurons play a direct role in processing incoming sensory information to program saccadic eye movements. Inaccuracy in reaching toward visual targets after lesions of areas 7a, 7b, and LIP was observed by Rushworth, Nixon, and Passingham (1997). Thus, lesions of parietal cortex appear to disrupt the ability of a subject to coordinate movements relative to him-/herself or relative to other objects located in extrapersonal space.

Functional Subdivisions of the Dorsal Stream

Recent evidence has suggested that the dorsal stream might be further subdivided into three anatomically and functionally distinct pathways, each serving different functions related to visuospatial processing (figure 17.5; see Kravitz et al., 2011, for review). The first of the three pathways, believed to be concerned with mediating spatial working memory, follows connections from regions within the parietal cortex (LIP and VIP) and posterior STS (MT/MST) to regions within the

    ANDREW H. BELL, TATIANA PASTERNAK, AND LESLIE G. UNGERLEIDER

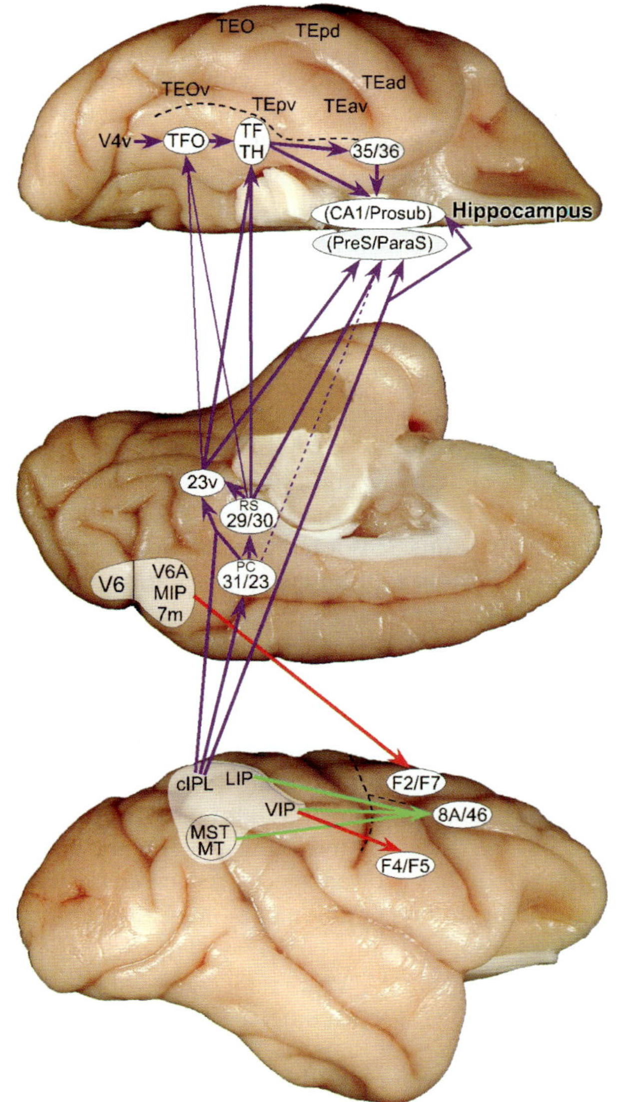

Parietal-prefrontal (Pathway 1)

Parietal-premotor (Pathway 2)

Parietal-medial temporal (Pathway 3)

1993). Therefore, such parietoparietal–prefrontal connections appear to allow parietal structures to feed spatial information to prefrontal areas. In turn, these prefrontal areas can exert influence on parietal areas to provide top-down control of spatial signals present in dorsal stream structures.

The second of the three pathways, which is associated with the control of visually guided actions, relies on direct connections between parietal and premotor cortices. Some parietal neurons receive inputs from the cerebellum and presumably translate the locations of visual objects from retinotopic (i.e., eye-centered) into the appropriate reference frames, including head-, arm-, and body-centered reference frames (Batista et al., 1999; Pertzov, Avidan, & Zohary, 2011). This information is fed to dorsal and ventral premotor cortices, which are directly connected with regions responsible for initiating eye, head, and limb movements and have been shown to encode various movement-related parameters such as target location and arm kinematics (Kurata, 2010; Yamagata et al., 2009).

The final of the three pathways links the parietal cortex with the medial temporal cortex to control visuospatial navigation. This pathway links parietal cortex, both directly and indirectly (via the retrosplenial and posterior cingulate cortices) to the hippocampus and parahippocampal cortex. Many parietal neurons exhibit world- and object-centered (allocentric) reference frames and sensitivity to optic flow, properties valuable for both encoding and interpreting movement through one's environment. Activity in inferior parietal cortex correlates with path direction during mental

dorsolateral prefrontal cortex. The dorsolateral prefrontal cortex has been shown to play active roles in tasks requiring spatial working memory. For example, neurons in the frontal eye fields (FEF) have been strongly implicated in the planning and generation of goal-oriented eye movements and spatial attention (Ekstrom et al., 2008; Moore & Fallah, 2001). Dorsolateral prefrontal neurons are active during memory-guided saccade tasks, including at the time of target presentation, during the delay period, and at the time of saccade initiation (Funahashi, Bruce, & Goldman-Rakic, 1991; Funahashi, Chafee, & Goldman-Rakic, maze navigation, and lesions to this region can result in egocentric disorientation—the inability to represent the spatial location of objects relative to self, accompanied by profound navigation impairments in novel and familiar environments. The hippocampus and parahippocampal cortex have been associated with spatial navigation, as have the retrosplenial and posterior cingulate cortices. Damage to any of these structures, for example, leads to impairments in spatial navigation (Epstein et al., 2001; Maguire, 2001).

INTEGRATION OF THE VENTRAL AND DORSAL STREAMS

Processing of information about color/form and motion/space appears to be segregated and carried largely by separate pathways, and yet there is evidence of intermixing of properties at various stages of cortical processing. For example, although there is functional segregation of neurons into thin and thick stripe and interstripe regions in area V2 with respect to color, retinal disparity, and form (see above), this segregation is not complete (Ts'o, Roe, & Gilbert, 2001). Furthermore, it is likely that the extensive spread of projections supplied by V1 neurons as well as the horizontal connections between the stripes in V2 may provide an anatomical substrate for interactions between segregated channels. Thus, even at the level of V2, there may be intermixing of signals proceeding toward the dorsal and the ventral streams.

Similarly, although the properties of area V3 point to an association with the dorsal stream (see Lyon & Connolly, 2011), it is also heavily interconnected with V4, a major component of the ventral stream. The high incidence of directional selectivity of V3 neurons and their preference for lower spatial and higher temporal frequencies suggest a role in processing motion information. Yet, the majority of V3 neurons are also orientation selective, and nearly half of all neurons in this area show selectivity for color. It is interesting that many of the neurons responding to color also show directional selectivity, and a substantial number show directional selectivity to isoluminant gratings. This interaction between color and motion suggests that area V3 may represent a stage in the analysis of the visual scene where interactions between the two streams take place.

Interactions between the processing of color and motion have also been observed in area MT, an area strongly influenced by the magnocellular pathway. Many MT neurons maintain significant responses to motion of isoluminant stimuli, and the presence of chromatic information has been shown to increase neuronal direction discrimination (Croner & Albright,

1999; Thiele, Dobkins, & Albright, 2001). Likewise, neurons in V4 have been shown to acquire sensitivity to motion after repeated exposures to moving stimuli (Tolias et al., 2005).

Although often associated with spatial processing, the posterior parietal cortex is heavily connected with areas TE and TEO in IT cortex and exhibits several properties more consistent with the ventral stream. LIP neurons in nonhuman primates have been shown to encode nonspatial stimulus features such as shape and color irrespective of location (Janssen et al., 2008; Sereno & Maunsell, 1998). Similar selectivity for shape has been demonstrated in human posterior parietal cortex (Konen & Kastner, 2008). Yet another region of convergence and integration of form and motion information is STP, which also receives inputs from both processing streams. It is a multimodal area containing neurons that respond to somatosensory and auditory as well as visual stimuli (Bruce, Desimone, & Gross, 1981). STP neurons have large gaze-centered RFs and show selectivity to visual motion similar to that observed in areas MT and MST. Some neurons also respond particularly well to biological motion, that is, motion patterns consistent with what one sees when one observes the gait of another animal (Oram & Perrett, 1996). These observations have led to the idea that the primate temporal lobe has evolved specialized mechanisms for the encoding and recognition of biologically significant stimuli.

BEYOND VISUAL PROCESSING STREAMS

Here, we have reviewed evidence for a functional separation between the dorsal and ventral projections originating from primary visual cortex in the primate brain. The respective functions of the two streams to visual processing are still actively debated. Although originally based on anatomical tracing and lesion work in macaques, the labels "what" and "where" used to summarize the function of the ventral and dorsal streams, respectively, have since been reinterpreted by some on the basis of more recent patient data. For example, Goodale and Milner (1992) reported on a patient (DF) who, following diffuse damage to various parts of extrastriate visual cortex secondary to carbon monoxide poisoning, suffered from visual agnosia but was nonetheless able to perform various visuomotor tasks that appeared to require knowledge of the target object's shape, form, and orientation—functions normally ascribed to the ventral stream. They subsequently proposed an alternate interpretation of the ventral and dorsal streams, that of *action* versus *perception,* with the dorsal stream concerned with the properties of visual stimuli necessary to perform visually guided actions, and the ventral stream concerned with the information necessary to recognize and interpret visual stimuli. There continues to be discussion regarding the precise nature of this and other patients' deficits (see Schenk, 2006) and how they impact our understanding of the separation of function within the visual domain. What is clear is that there is substantial interaction between the two streams.

Finally, in addition to the cortical pathways subserving visual processing described in this chapter, V1 has a number of connections with subcortical structures that provide yet another route for visual information to reach anterior structures in the frontal, parietal, and temporal cortices. For example, V1 projects directly to the superior colliculus, located in the midbrain, which itself projects to a number of cortical (and subcortical) structures, such as the dorsolateral prefrontal cortex (including FEF), LIP, and several others. Such cortical/subcortical pathways offer potential bypasses for early dorsal and ventral stream structures and may partially explain some of the residual visual abilities of patients with extensive cortical lesions (i.e., blindsight; see Schmid et al., 2010). A greater understanding of the role of such corticosubcortical pathways to visual perception and visually guided action is therefore essential and may provide further insights as to the respective roles of the dorsal and ventral streams.

REFERENCES

Andersen, R. A., Shenoy, K. V., Snyder, L. H., Bradley, D. C., & Crowell, J. A. (1999). The contributions of vestibular signals to the representations of space in the posterior parietal cortex. *Annals of the New York Academy of Sciences, 871*, 282–292. doi:10.1111/j.1749-6632.1999.tb09192.x.

Avillac, M., Deneve, S., Olivier, E., Pouget, A., & Duhamel, J. R. (2005). Reference frames for representing visual and tactile locations in parietal cortex. *Nature Neuroscience, 8*, 941–949.

Batista, A. P., Buneo, C. A., Snyder, L. H., & Andersen, R. A. (1999). Reach plans in eye-centered coordinates. *Science, 285*, 257–260. doi:10.1126/science.285.5425.257.

Bell, A. H., Hadj-Bouziane, F., Frihauf, J. B., Tootell, R. B., & Ungerleider, L. G. (2009). Object representations in the temporal cortex of monkeys and humans as revealed by functional magnetic resonance imaging. *Journal of Neurophysiology, 101*, 688–700.

Bell, A. H., Malecek, N. J., Morin, E. L., Hadj-Bouziane, F., Tootell, R. B. H., & Ungerleider, L. G. (2011). Relationship between fMRI-identified regions and neuronal category-selectivity. *Journal of Neuroscience, 31*, 12229–12240.

Bisley, J. W., & Pasternak, T. (2000). The multiple roles of visual cortical areas MT/MST in remembering the direction of visual motion. *Cerebral Cortex, 10*, 1053–1065. doi:10.1093/cercor/10.11.1053.

Bisley, J. W., Zaksas, D., & Pasternak, T. (2001). Microstimulation of cortical area MT affects performance on a visual working memory task. *Journal of Neurophysiology, 85*, 187–196.

Britten, K. H., Newsome, W. T., Shadlen, M. N., Celebrini, S., & Movshon, J. A. (1996). A relationship between behavioral choice and the visual responses of neurons in macaque MT. *Visual Neuroscience, 13*, 87–100. doi:10.1017/S095252380000715X.

Britten, K. H., & van Wezel, R. J. (1998). Electrical microstimulation of cortical area MST biases heading perception in monkeys. *Nature Neuroscience, 1*, 59–63. doi:10.1038/259.

Bruce, C., Desimone, R., & Gross, C. G. (1981). Visual properties of neurons in a polysensory area in superior temporal sulcus of the macaque. *Journal of Neurophysiology, 46*, 369–384.

Burkhalter, A., & Van Essen, D. C. (1986). Processing of color, form and disparity information in visual areas VP and V2 of ventral extrastriate cortex in the macaque monkey. *Journal of Neuroscience, 6*, 2327–2351.

Chen, A., DeAngelis, G. C., & Angelaki, D. E. (2011). Representation of vestibular and visual cues to self-motion in ventral intraparietal cortex. *Journal of Neuroscience, 31*, 12036–12052.

Cohen, L., & Dehaene, S. (2004). Specialization within the ventral stream: The case for the visual word form area. *NeuroImage, 22*, 466–476. doi:10.1016/j.neuroimage.2003.12.049.

Cohen, L., Lehericy, S., Chochon, F., Lemer, C., Rivaud, S., & Dehaene, S. (2002). Language-specific tuning of visual cortex? Functional properties of the visual word form area. *Brain, 125*, 1054–1069. doi:10.1093/brain/awf094.

Colby, C. L., Duhamel, J. R., & Goldberg, M. E. (1993). Ventral intraparietal area of the macaque: Anatomic location and visual response properties. *Journal of Neurophysiology, 69*, 902–914.

Constantinidis, C., & Steinmetz, M. A. (2001). Neuronal responses in area 7a to multiple-stimulus displays: I. Neurons encode the location of the salient stimulus. *Cerebral Cortex, 11*, 581–591. doi:10.1093/cercor/11.7.581.

Croner, L. J., & Albright, T. D. (1999). Segmentation by color influences responses of motion-sensitive neurons in the cortical middle temporal visual area. *Journal of Neuroscience, 19*, 3935–3951.

Dean, P. (1979). Visual cortex ablation and thresholds for successively presented stimuli in rhesus monkeys: II. Hue. *Experimental Brain Research, 35*, 69–83. doi:10.1007/BF00236785.

DeAngelis, G. C., Cumming, B. G., & Newsome, W. T. (1998). Cortical area MT and the perception of stereoscopic depth. *Nature, 394*, 677–680. doi:10.1038/29299.

Dodd, J. V., Krug, K., Cumming, B. G., & Parker, A. J. (2001). Perceptually bistable three-dimensional figures evoke high choice probabilities in cortical area MT. *Journal of Neuroscience, 21*, 4809–4821.

Downing, P. E., Jiang, Y., Shuman, M., & Kanwisher, N. (2001). A cortical area selective for visual processing of the human body. *Science, 293*, 2470–2473.

Dricot, L., Sorger, B., Schiltz, C., Goebel, R., & Rossion, B. (2008). The roles of "face" and "non-face" areas during individual face perception: Evidence by fMRI adaptation in a brain-damaged prosopagnosic patient. *NeuroImage, 40*, 318–332. doi:10.1016/j.neuroimage.2007.11.012.

Duhamel, J. R., Colby, C. L., & Goldberg, M. E. (1992). The updating of the representation of visual space in parietal cortex by intended eye movements. *Science, 255*, 90–92.

Ekstrom, L. B., Roelfsema, P. R., Arsenault, J. T., Bonmassar, G., & Vanduffel, W. (2008). Bottom-up dependent gating of frontal signals in early visual cortex. *Science, 321*, 414–417. doi:10.1126/science.1153276.

Epstein, R. A. (2008). Parahippocampal and retrosplenial contributions to human spatial navigation. *Trends in Cognitive Sciences, 12*, 388–396. doi:10.1016/j.tics.2008.07.004.

Epstein, R., DeYoe, E. A., Press, D., & Kanwisher, N. (2001). Neuropsychological evidence for a topographical learning mechanism in parahippocampal cortex. *Cognitive Psychology, 18*, 481–508.

Epstein, R., & Kanwisher, N. (1998). A cortical representation of the local visual environment. *Nature, 392*, 598–601. doi:1038/33402.

Fanini, A., & Assad, J. A. (2009). Direction selectivity of neurons in the macaque lateral intraparietal area. *Journal of Neurophysiology, 101*, 289–305.

Freiwald, W. A., Tsao, D. Y., & Livingstone, M. S. (2009). A face feature space in the macaque temporal lobe. *Nature Neuroscience, 12*, 1187–1196. doi:10.1038/nn.2363.

Funahashi, S., Bruce, C. J., & Goldman-Rakic, P. S. (1991). Neuronal activity related to saccadic eye movements in the monkey's dorsolateral prefrontal cortex. *Journal of Neurophysiology, 65*, 1464–1483.

Funahashi, S., Chafee, M. V., & Goldman-Rakic, P. S. (1993). Prefrontal neuronal activity in rhesus monkeys performing a delayed anti-saccade task. *Nature, 365*, 753–756. doi:10.1038/365753a0.

Gegenfurtner, K. R., Kiper, D. C., Beusmans, J. M., Carandini, M., Zaidi, Q., & Movshon, J. A. (1994). Chromatic properties of neurons in macaque MT. *Visual Neuroscience, 11*, 455–466. doi:10.1017/S095252380000239X.

Gegenfurtner, K. R., Kiper, D. C., & Levitt, J. B. (1997). Functional properties of neurons in macaque area V3. *Journal of Neurophysiology, 77*, 1906–1923.

Goodale, M. A., & Milner, A. D. (1992). Separate visual pathways for perception and action. *Trends in Neurosciences, 15*, 20–25. doi:10.1016/0166-2236(92)90344-8.

Gross, C. G., Bender, D. B., & Rocha-Miranda, C. E. (1969). Visual receptive fields of neurons in inferotemporal cortex of the monkey. *Science, 166*, 1303–1306. doi:10.1126/science.166.3910.1303.

Hawken, M. J., Parker, A. J., & Lund, J. S. (1988). Laminar organization and contrast sensitivity of direction-selective cells in the striate cortex of the Old World monkey. *Journal of Neuroscience, 8*, 3541–3548.

Haxby, J. V., Grady, C. L., Horwitz, B., Ungerleider, L. G., Mishkin, M., Carson, R. E., et al. (1991). Dissociation of object and spatial visual processing pathways in human extrastriate cortex. *Proceedings of the National Academy of Sciences of the United States of America, 88*, 1621–1625. doi:10.1073/pnas.88.5.1621.

Hussar, C. R., & Pasternak, T. (2012). Memory-guided sensory comparisons in the prefrontal cortex: Contribution of putative pyramidal cells and interneurons. *Journal of Neuroscience, 32*, 2747–2761.

Huxlin, K. R., Saunders, R. C., Marchionini, D., Pham, H. A., & Merigan, W. H. (2000). Perceptual deficits after lesions of inferotemporal cortex in macaques. *Cerebral Cortex, 10*, 671–683. doi:10.1093/cercor/10.7.671.

Ito, M., Tamura, H., Fujita, I., & Tanaka, K. (1995). Size and position invariance of neuronal responses in monkey inferotemporal cortex. *Journal of Neurophysiology, 73*, 218–226.

Janssen, P., Srivastava, S., Ombelet, S., & Orban, G. A. (2008). Coding of shape and position in macaque lateral intraparietal area. *Journal of Neuroscience, 28*, 6679–6690.

Kanwisher, N., McDermott, J., & Chun, M. M. (1997). The fusiform face area: A module in human extrastriate cortex specialized for face perception. *Journal of Neuroscience, 17*, 4302–4311.

Konen, C. S., & Kastner, S. (2008). Two hierarchically organized neural systems for object information in human visual cortex. *Nature Neuroscience, 11*, 224–231. doi:10.1038/nn2036.

Kourtzi, Z., & Connor, C. E. (2011). Neural representations for object perception: Structure, category, and adaptive coding. *Annual Review of Neuroscience, 34*, 45–67. doi:10.1146/annurev-neuro-060909-153218.

Kravitz, D. J., Saleem, K. S., Baker, C. I., & Mishkin, M. (2011). A new neural framework for visuospatial processing. *Nature Reviews. Neuroscience, 12*, 217–230. doi:10.1038/nrn3008.

Kriegeskorte, N., Formisano, E., Sorger, B., & Goebel, R. (2007). Individual faces elicit distinct response patterns in human anterior temporal cortex. *Proceedings of the National Academy of Sciences of the United States of America, 104*, 20600–20605. doi:10.1073/pnas.0705654104.

Krug, K. (2004). A common neuronal code for perceptual processes in visual cortex? Comparing choice and attentional correlates in V5/MT. *Philosophical Transactions of the Royal Society of London. Series B, Biological Sciences, 359*, 929–941. doi:10.1098/rstb.2003.1415.

Krug, K., & Parker, A. J. (2011). Neurons in dorsal visual area V5/MT signal relative disparity. *Journal of Neuroscience, 31*, 17892–17904.

Kurata, K. (2010). Conditional selection of contra- and ipsilateral forelimb movements by the dorsal premotor cortex in monkeys. *Journal of Neurophysiology, 103*, 262–277.

Latto, R. (1986). The role of inferior parietal cortex and the frontal eye-fields in visuospatial discriminations in the macaque monkey. *Behavioural Brain Research, 22*, 41–52. doi:10.1016/0166-4328(86)90079-3.

Leopold, D. A., Bondar, I. V., & Giese, M. A. (2006). Norm-based face encoding by single neurons in the monkey inferotemporal cortex. *Nature, 442*, 572–575. doi:10.1038/nature04951.

Li, C. S., Mazzoni, P., & Andersen, R. A. (1999). Effect of reversible inactivation of macaque lateral intraparietal area on visual and memory saccades. *Journal of Neurophysiology, 81*, 1827–1838.

Li, N., & DiCarlo, J. J. (2008). Unsupervised natural experience rapidly alters invariant object representation in visual cortex. *Science, 321*, 1502–1507. doi:10.1126/science.1160028.

Li, N., & DiCarlo, J. J. (2010). Unsupervised natural visual experience rapidly reshapes size-invariant object representation in inferior temporal cortex. *Neuron, 67*, 1062–1075. doi:10.1016/j.neuron.2010.08.029.

Lui, L. L., & Pasternak, T. (2011). Representation of comparison signals in cortical area MT during a delayed direction discrimination task. *Journal of Neurophysiology, 106*, 1260–1273.

Lyon, D. C., & Connolly, J. D. (2011). The case for primate V3. *Proceedings of the Royal Society B: Biological Sciences, 279*(1729), 625–633. doi:10.1098/rspb.2011.2048

Maguire, E. A. (2001). The retrosplenial contribution to human navigation: A review of lesion and neuroimaging findings. *Scandinavian Journal of Psychology, 42*, 225–238. doi:10.1111/1467-9450.00233.

Maunsell, J. H., & Van Essen, D. C. (1983). Functional properties of neurons in middle temporal visual area of the macaque monkey. I. Selectivity for stimulus direction, speed, and orientation. *Journal of Neurophysiology, 49*, 1127–1147.

Merigan, W. H. (1996). Basic visual capacities and shape discrimination after lesions of extrastriate area V4 in macaques. *Visual Neuroscience, 13*, 51–60. doi:10.1017/S0952523800007124.

Meyer, T., & Olson, C. R. (2011). Statistical learning of visual transitions in monkey inferotemporal cortex. *Proceedings of the National Academy of Sciences of the United States of America, 108*, 19401–19406. doi:10.1073/pnas.1112895108.

Miller, E. K. (2000). The prefrontal cortex and cognitive control. *Nature Reviews. Neuroscience, 1*(1), 59–65. doi:10.1038/35036228.

Miller, E. K., & Desimone, R. (1994). Parallel neuronal mechanisms for short-term memory. *Science, 263*, 520–522. doi:10.1126/science.8290960.

Miyashita, Y., Kameyama, M., Hasegawa, I., & Fukushima, T. (1998). Consolidation of visual associative long-term memory in the temporal cortex of primates. *Neurobiology of Learning and Memory, 70*, 197–211. doi:10.1006/nlme.1998.3848.

Moore, T., & Fallah, M. (2001). Control of eye movements and spatial attention. *Proceedings of the National Academy of Sciences of the United States of America, 98*, 1273–1276. doi:10.1073/pnas.021549498.

Nakamura, K., & Colby, C. L. (2002). Updating of the visual representation in monkey striate and extrastriate cortex during saccades. *Proceedings of the National Academy of Sciences of the United States of America, 99*, 4026–4031. doi:10.1073/pnas.052379899.

Newcombe, F., Ratcliff, G., & Damasio, H. (1987). Dissociable visual and spatial impairments following right posterior cerebral lesions: clinical, neuropsychological and anatomical evidence. *Neuropsychologia, 25*, 149–161. doi:10.1016/0028-3932(87)90127-8.

Newsome, W. T., Britten, K. H., & Movshon, J. A. (1989). Neuronal correlates of a perceptual decision. *Nature, 341*, 52–54.

Newsome, W. T., & Pare, E. B. (1988). A selective impairment of motion perception following lesions of the middle temporal visual area (MT). *Journal of Neuroscience, 8*, 2201–2211.

Newsome, W. T., Wurtz, R. H., Dursteler, M. R., & Mikami, A. (1985). Deficits in visual motion processing following ibotenic acid lesions of the middle temporal visual area of the macaque monkey. *Journal of Neuroscience, 5*, 825–840.

Nishimoto, S., & Gallant, J. L. (2011). A three-dimensional spatiotemporal receptive field model explains responses of area MT neurons to naturalistic movies. *Journal of Neuroscience, 31*, 14551–14564.

Op de Beeck, H. P., Deutsch, J. A., Vanduffel, W., Kanwisher, N. G., & DiCarlo, J. J. (2008). A stable topography of selectivity for unfamiliar shape classes in monkey inferior temporal cortex. *Cerebral Cortex, 18*, 1676–1694. doi:10.1093/cercor/bhm196.

Oram, M. W., & Perrett, D. I. (1996). Integration of form and motion in the anterior superior temporal polysensory area (STPa) of the macaque monkey. *Journal of Neurophysiology, 76*, 109–129.

Pasternak, T., & Greenlee, M. W. (2005). Working memory in primate sensory systems. *Nature Reviews. Neuroscience, 6*, 97–107. doi:10.1038/nrn1603.

Pasternak, T., & Merigan, W. H. (1994). Motion perception following lesions of the superior temporal sulcus in the monkey. *Cerebral Cortex, 4*, 247–259. doi:10.1093/cercor/4.3.247.

Pertzov, Y., Avidan, G., & Zohary, E. (2011). Multiple reference frames for saccadic planning in the human parietal cortex. *Journal of Neuroscience, 31*, 1059–1068.

Pinsk, M. A., DeSimone, K., Moore, T., Gross, C. G., & Kastner, S. (2005). Representations of faces and body parts in macaque temporal cortex: A functional MRI study. *Proceedings of the National Academy of Sciences of the United States of America, 102*, 6996–7001. doi:10.1073/pnas.0502605102.

Puce, A., Allison, T., Gore, J. C., & McCarthy, G. (1995). Face-sensitive regions in human extrastriate cortex studied by functional MRI. *Journal of Neurophysiology, 74*, 1192–1199.

Quintana, J., & Fuster, J. M. (1993). Spatial and temporal factors in the role of prefrontal and parietal cortex in visuomotor integration. *Cerebral Cortex, 3*, 122–132. doi:10.1093/cercor/3.2.122.

Rajimehr, R., Young, J. C., & Tootell, R. B. (2009). An anterior temporal face patch in human cortex, predicted by macaque maps. *Proceedings of the National Academy of Sciences of the United States of America, 106*, 1995–2000. doi:10.1073/pnas.0807304106.

Read, H. L., & Siegel, R. M. (1997). Modulation of responses to optic flow in area 7a by retinotopic and oculomotor cues in monkey. *Cerebral Cortex, 7*, 647–661. doi:10.1093/cercor/7.7.647.

Roe, A. W., & Ts'o, D. Y. (1995). Visual topography in primate V2: Multiple representation across functional stripes. *Journal of Neuroscience, 15*, 3689–3715.

Rosa, M. G., & Tweedale, R. (2005). Brain maps, great and small: Lessons from comparative studies of primate visual cortical organization. *Philosophical Transactions of the Royal Society B. Biological Sciences, 360*, 665–691. doi:10.1098/rstb.2005.1626.

Rossion, B., Caldara, R., Seghier, M., Schuller, A. M., Lazeyras, F., & Mayer, E. (2003). A network of occipito-temporal face-sensitive areas besides the right middle fusiform gyrus is necessary for normal face processing. *Brain, 126*, 2381–2395. doi:10.1093/brain/awg241.

Rudolph, K., & Pasternak, T. (1999). Transient and permanent deficits in motion perception after lesions of cortical areas MT and MST in the macaque monkey. *Cerebral Cortex, 9*, 90–100. doi:10.1093/cercor/9.1.90.

Rushworth, M. F., Nixon, P. D., & Passingham, R. E. (1997). Parietal cortex and movement. I. Movement selection and reaching. *Experimental Brain Research, 117*, 292–310. doi:10.1007/s002210050224.

Sakai, K., & Miyashita, Y. (1991). Neural organization for the long-term memory of paired associates. *Nature, 354*, 152–155. doi:10.1038/354152a0.

Salzman, C. D., Britten, K. H., & Newsome, W. T. (1990). Cortical microstimulation influences perceptual judgements of motion direction. *Nature, 346,* 174–177. doi:10.1038/346174a0.

Schenk, T. (2006). An allocentric rather than perceptual deficit in patient D.F. *Nature Neuroscience, 9,* 1369–1370. doi:10.1038/nn1784.

Schmid, M. C., Mrowka, S. W., Turchi, J., Saunders, R. C., Wilke, M., Peters, A. J., et al. (2010). Blindsight depends on the lateral geniculate nucleus. *Nature, 466,* 373–377. doi:10.1038/nature09179.

Schwarzlose, R. F., Baker, C. I., & Kanwisher, N. (2005). Separate face and body selectivity on the fusiform gyrus. *Journal of Neuroscience, 25,* 11055–11059.

Seidemann, E., Poirson, A. B., Wandell, B. A., & Newsome, W. T. (1999). Color signals in area MT of the macaque monkey. *Neuron, 24,* 911–917. doi:10.1016/S0896-6273(00)81038-7.

Sereno, A. B., & Maunsell, J. H. (1998). Shape selectivity in primate lateral intraparietal cortex. *Nature, 395,* 500–503. doi:10.1038/26752.

Sincich, L. C., & Horton, J. C. (2005). The circuitry of V1 and V2: Integration of color, form, and motion. *Annual Review of Neuroscience, 28,* 303–326. doi:10.1146/annurev.neuro.28.061604.135731.

Snyder, L. H., Grieve, K. L., Brotchie, P., & Andersen, R. A. (1998). Separate body- and world-referenced representations of visual space in parietal cortex. *Nature, 394,* 887–891. doi:10.1038/29777.

Thiele, A., Dobkins, K. R., & Albright, T. D. (2001). Neural correlates of chromatic motion perception. *Neuron, 32,* 351–358. doi:10.1016/S0896-6273(01)00463-9.

Tolias, A. S., Keliris, G. A., Smirnakis, S. M., & Logothetis, N. K. (2005). Neurons in macaque area V4 acquire directional tuning after adaptation to motion stimuli. *Nature Neuroscience, 8,* 591–593. doi:10.1038/nn1446.

Tsao, D. Y., Freiwald, W. A., Knutsen, T. A., Mandeville, J. B., & Tootell, R. B. (2003). Faces and objects in macaque cerebral cortex. *Nature Neuroscience, 6,* 989–995. doi:10.1038/nn1111.

Tsao, D. Y., Freiwald, W. A., Tootell, R. B., & Livingstone, M. S. (2006). A cortical region consisting entirely of face-selective cells. *Science, 311,* 670–674. doi:10.1126/science.1119983.

Ts'o, D. Y., Roe, A. W., & Gilbert, C. D. (2001). A hierarchy of the functional organization for color, form and disparity in primate visual area V2. *Vision Research, 41,* 1333–1349. doi:10.1016/S0042-6989(01)00076-1.

Tsunoda, K., Yamane, Y., Nishizaki, M., & Tanifuji, M. (2001). Complex objects are represented in macaque inferotemporal cortex by the combination of feature columns. *Nature Neuroscience, 4,* 832–838. doi:10.1038/90547.

Ungerleider, L. G., & Bell, A. H. (2011). Uncovering the visual "alphabet": Advances in our understanding of object perception. *Vision Research, 51,* 782–799. doi:10.1016/j.visres.2010.10.002.

Ungerleider, L. G., & Mishkin, M. (1982). Two cortical visual systems. In D. G. Ingle (Ed.), *Analysis of visual behavior* (pp. 549–586). Cambridge, MA: MIT Press.

von der Heydt, R., Peterhans, E., & Baumgartner, G. (1984). Illusory contours and cortical neuron responses. *Science, 224,* 1260–1262. doi:10.1126/science.6539501.

Wandell, B. A., Dumoulin, S. O., & Brewer, A. A. (2007). Visual field maps in human cortex. *Neuron, 56,* 366–383. doi:10.1016/j.neuron.2007.10.012.

Yamagata, T., Nakayama, Y., Tanji, J., & Hoshi, E. (2009). Processing of visual signals for direct specification of motor targets and for conceptual representation of action targets in the dorsal and ventral premotor cortex. *Journal of Neurophysiology, 102,* 3280–3294.

Zhang, T., Heuer, H. W., & Britten, K. H. (2004). Parietal area VIP neuronal responses to heading stimuli are encoded in head-centered coordinates. *Neuron, 42,* 993–1001. doi:10.1016/j.neuron.2004.06.008.

Zihl, J., von Cramon, D., & Mai, N. (1983). Selective disturbance of movement vision after bilateral brain damage. *Brain, 106,* 313–340. doi:10.1093/brain/106.2.313.

18 Network of Mouse Visual Cortex

ANDREAS BURKHALTER, OLAF SPORNS, ENQUAN GAO, AND
QUANXIN WANG

Vision is central to our lives. We use it for recognizing and identifying objects as well as for guiding eye movements, reaching, and navigating though the environment. Early studies indicated that different brain regions were associated with these tasks. The optic tectum was thought to be uniquely important for visually guided actions, whereas the cerebral cortex was specialized for object recognition (Schneider, 1967; Trevarthen, 1968). Later, the view was extended, showing that cortical lesions in monkey not only impair visual discrimination but also interfere with ability to locate objects in space (Ungerleider & Mishkin, 1982; see chapter 17 by Bell, Pasternak, and Ungerleider). Importantly, Ungerleider and Mishkin's work showed that the deficits were dependent on several distinct cortical areas and that these areas were located in different regions of the forebrain. Based on these findings it was proposed that object recognition was performed in areas of the temporal cortex, whereas the parietal cortex was necessary for locating objects in space and taking action to reach for them (Goodale & Milner, 1992; Ungerleider & Mishkin, 1982). Over the last two decades much has been learned about the anatomy and physiology of ventral and dorsal processing streams (Nassi & Callaway, 2009) including that the networks are vulnerable to genetic and developmental disorders (Atkinson & Braddick, 2011). To understand the cellular and molecular underpinnings of these diseases, the mouse visual system offers a tractable model. This raises the question whether similar functional networks exist in mouse visual cortex.

STRIATE AND EXTRASTRIATE VISUAL CORTEX

Compared to the human cortex the mouse cerebral cortex is tiny. In fact, the surface area is 1,000-fold smaller, but the cortical thickness differs barely by a factor of two (Rakic, 1995, 2009). Thus, what changes during evolution is mainly the size of the cortical sheet, which involves a 10,000-fold increase of cortical neurons (Herculano-Houzel, 2011; Kaas, 2012; Schüz & Palm, 1989). Although to the untrained eye the cortical sheet looks uniform, the classic studies of Brodmann have

shown that human cortex contains 52 structurally distinct areas (Zilles & Amunt, 2010). Over the years Brodmann's map has been revised so that today we are able to distinguish as many as 150–200 areas per hemisphere (Van Essen et al., 2012a). Slightly fewer (130–140) areas are contained in the eightfold smaller macaque monkey cortex (Van Essen et al., 2012b). Estimates of areas contained in the fingernail-size mouse cortex range between 25 and 55 (Rose, 1929; Caviness, 1975; Wang, Sporns, & Burkhalter, 2012). Evidently, the number of areas per square centimeter decreases across the mammalian radiation from 7 to 16 in mice to ~1 in monkey and ~0.2 in humans. In parallel, individual areas have grown in size. This suggests that in primates more neurons participate in a particular function represented in an area and/or an area represents a larger number of different functions. For example, in the visual system of cat and monkey each retinal ganglion cell (RGC) projects to about 100 neurons in layer 4 of V1 (primary visual cortex), whereas in mouse the RGC:V1 ratio in layers 3 and 4 is about 1:1 (Airey et al., 2005; Salinas-Navarro et al., 2009; Schüz & Palm, 1989). As a result, mouse V1 neurons whose response properties are assembled by converging inputs lose their fine-scale topography (Smith & Häusser, 2010).

Mouse V1 cortex was identified by Dräger (1974), who labeled the afferent pathway from the eye by transneuronal transport of tritiated proline. Shortly thereafter, Alan Pearlman's group showed, by retrograde tracing with wheat germ agglutinin-HRP, that inputs to V1 were relayed through the lateral geniculate nucleus (LGN) (Simmons, Lemmon, & Pearlman, 1982). These studies also showed that visual input was not confined to V1. Using single-unit recording Dräger (1975) found that receptive fields were topographically mapped in V1 and that visually responsive neurons also populated the surrounding areas 18a and 18b (Caviness, 1975). Recordings later showed that area 18a lateral to V1 contained not a single but two visuotopic maps, V2 and V3 (Wagor, Mangini, & Pearlman, 1980). The same study further revealed two additional maps in area 18b on the medial side of V1 (Wagor, Mangini, & Pearlman, 1980). The existence of multiple visuotopic maps in mouse visual cortex was later confirmed by optical imaging of the intrinsic hemodynamic response as well as endogenous flavoprotein fluorescence (Kalasky & Stryker, 2003; Schuett, Bonhoeffer, & Huebener, 2002; Tohmi et al., 2009). Despite the progress there was no consensus on how many distinct areas were contained within extrastriate cortex, and it remained an open question how these areas were related to structural borders of mouse visual cortex. Even the extent of visually responsive cortex was unknown. Although visually responsive cortex was mapped by the expression of activity-related genes, the outline of the visual territory remained approximate (Tagawa et al., 2005; Van Brussel, Gerits, & Arckens, 2009). To address this question, we used in situ hybridization of the immediate early gene *Arc* in flatmounted cortex. The gene product of *Arc* is the activity-regulated cytoskeleton protein Arc/Arg3.1, which is rapidly induced by elevation of neuronal activity and plays a role in synaptic scaling by internalizing AMPA glutamate receptors from the postsynaptic membrane (Shepherd et al., 2006). We induced the expression of the *Arc* transcript in 3- to 6-month-old C57BL/6 mice in which we additionally traced the callosal connections by injecting bisbenzimide into the right occipital cortex (Wang & Burkhalter, 2007). Three days later the mice were allowed to explore a richly furnished cage for 45 min. The first group of mice explored the cage with both eyes open. In the second group of mice both eyes were sutured closed 24 h before the exploration. At the end of the exposure we perfused the mice, flattened the cortex, cut tangential sections, and performed in situ hybridization of *Arc* mRNA with a full-length digoxygenin-labeled mouse *Arc* cDNA probe (gift from P. Worley, Johns Hopkins University). In normal mice V1 and surrounding cortex were densely labeled (figure 18.1A). Labeling of the surrounding cortex extended from the lateral border of the retrosplenial cortex into the gap between V1 and barrel cortex (S1) and continued behind and around the auditory belt alongside the rhinal fissure (figure 18.1A). Importantly, the labeling spread across all of the callosally connected as well as acallosal regions within V1 and surrounding cortex (figure 18.1B). In contrast, in lid-sutured mice labeling in V1 and surrounding medial and lateral cortex was very sparse (figure 18.1C, D). The only notable exception was the labeling in a part of the posterior parietal cortex near the anterior tip of V1 (figure 18.1C). The difference in the distribution of neuronal activation between sighted and lid-sutured mice indicates that a large part of the cortex surrounding V1 is visually responsive and can therefore be considered extrastriate visual cortex.

EXTRASTRIATE VISUAL AREAS

The notion that cortex surrounding V1 is composed of multiple anatomically distinct areas goes back to Jaime Olavarria and Vincente Montero. They traced the outputs of mouse V1 and observed multiple clusters of axon terminals in surrounding cortex, which were closely associated with acallosal regions encircled by callosal connections (Olavarria & Montero, 1989). Based on similar findings in rat, Olavarria and Montero (1984) suggested that mouse extrastriate visual cortex is subdivided into multiple areas that are common across all rodent species. The proposal was met with skepticism. How could extrastriate visual cortex on the lateral side of V1 be composed of multiple areas when V1 in cats and primates is adjoined by a single area, V2 (Rosa & Krubitzer, 1999)? Jon Kaas argued that the patchy outputs labeled by Olavarria and Montero represent distinct modules within a single area, V2 (Kaas, Krubitzer, & Johanson, 1989). Vincente Montero countered with a multitracer study, showing that in rat each patch in extrastriate cortex belongs to a separate topographic map representing the entire visual field (Montero, 1993). But Montero's proposals of multiple small areas surrounding V1 remained challenged because his areas were incomplete, not firmly anchored to the underlying architectonic structure of cortex, and not confirmed by mapping of receptive fields (Rosa & Krubitzer, 1999).

Attracted by the controversy, we traced the inputs from mouse V1 to extrastriate cortex by injecting three different points of the map with red (R), green (G), and yellow (Y) fluorescent anterograde tracers (Wang & Burkhalter, 2007). We found nine projection fields, each containing multicolored axon terminal clusters, resembling the stripes in the flag of the Republic of Lithuania. Importantly, the orientation of the "flags," the interflag spacing, and the location of the flags relative to callosal/acallosal zones were dependent on the topographic location of the injection sites, suggesting that the flags represented distinct maps. Indeed, we were able to show that many of the labeled maps represented parts of the entire visual field, which strongly argued that they correspond to distinct extrastriate visual areas. Because many of the areas were located at similar sites as the projections of V1 described by Olavarria and Montero (1989), we adopted their nomenclature: PM (posteromedial), AM (anteromedial), A (anterior), RL (rostrolateral), AL (anterolateral), LM (lateromedial), LI (laterointermediate), P (posterior), POR (postrhinal; Burwell & Amaral, 1998), and 36p (posterior area 36) (figure 18.2A). All of these areas lie in V2L and V2ML (Franklin & Paxinos, 2007) except

     ANDREAS BURKHALTER, OLAF SPORNS, ENQUAN GAO, AND QUANXIN WANG

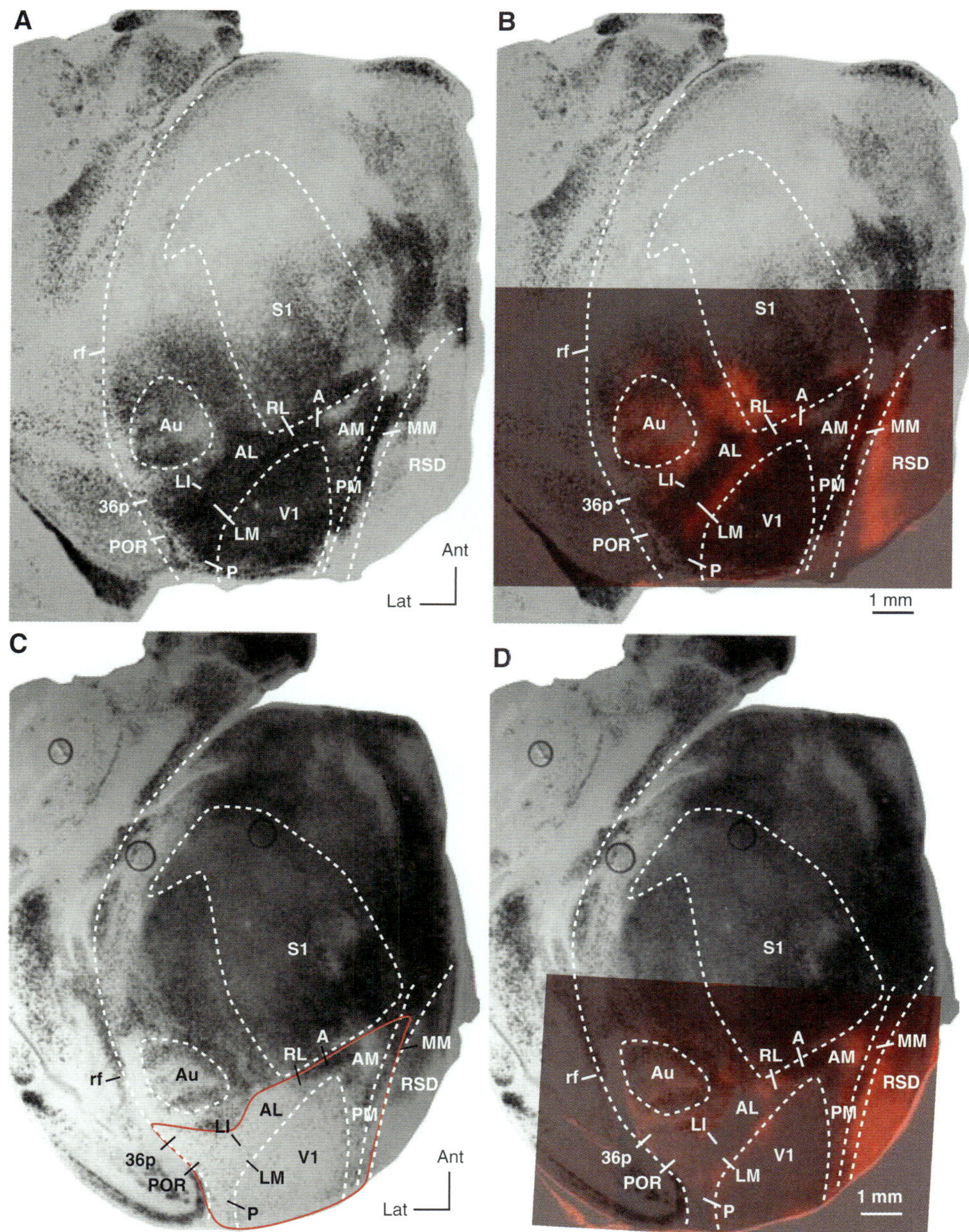

FIGURE 18.1 Extent of visually responsive cortex. In situ hybridization of the transcript of the activity-regulated immediate early gene *Arc* in tangential sections through flattened mouse cerebral cortex. (A, B) *Arc* expression in normal mice after exploring a new environment with both eyes open. Intense staining involves V1 and surrounding cortex. The red pattern (B) represents false-colored bisbenzimide-labeled callosal connections traced from the opposite hemisphere. Callosal connections provide landmarks for identifying areas 36p, POR, P, LM, LI, AL, RL, A, AM, and PM within extrastriate visual cortex (Wang & Burkhalter, 2007). V1 was identified by its distinctive myeloarchitecture. (C, D) *Arc* expression in mice that explored the new environment with both eyes closed. Notice that staining of V1 and surrounding cortex (C, red outline) is extremely sparse. (D) Same section as in C with superimposed callosal pattern (orange).

for POR and 36a. In rat the part between the tip of V1 and S1 is considered posterior parietal cortex and is coextensive with area 7 (Krieg, 1946) of Fabri and Burton's (1991) parietal medial area. Others subdivide the territory into V2ML and posterior parietal cortex (Reep & Corwin, 2009) or assign it entirely to V2ML (Franklin

& Paxinos, 2007). Our mapping studies in mouse suggest that posterior parietal cortex contains areas RL, A, and AM (Wang & Burkhalter, 2007).

Many of the maps revealed by tracing the outputs of V1 were recently confirmed by intrinsic signal imaging of neuronal activity (Andermann et al., 2011; Marshel

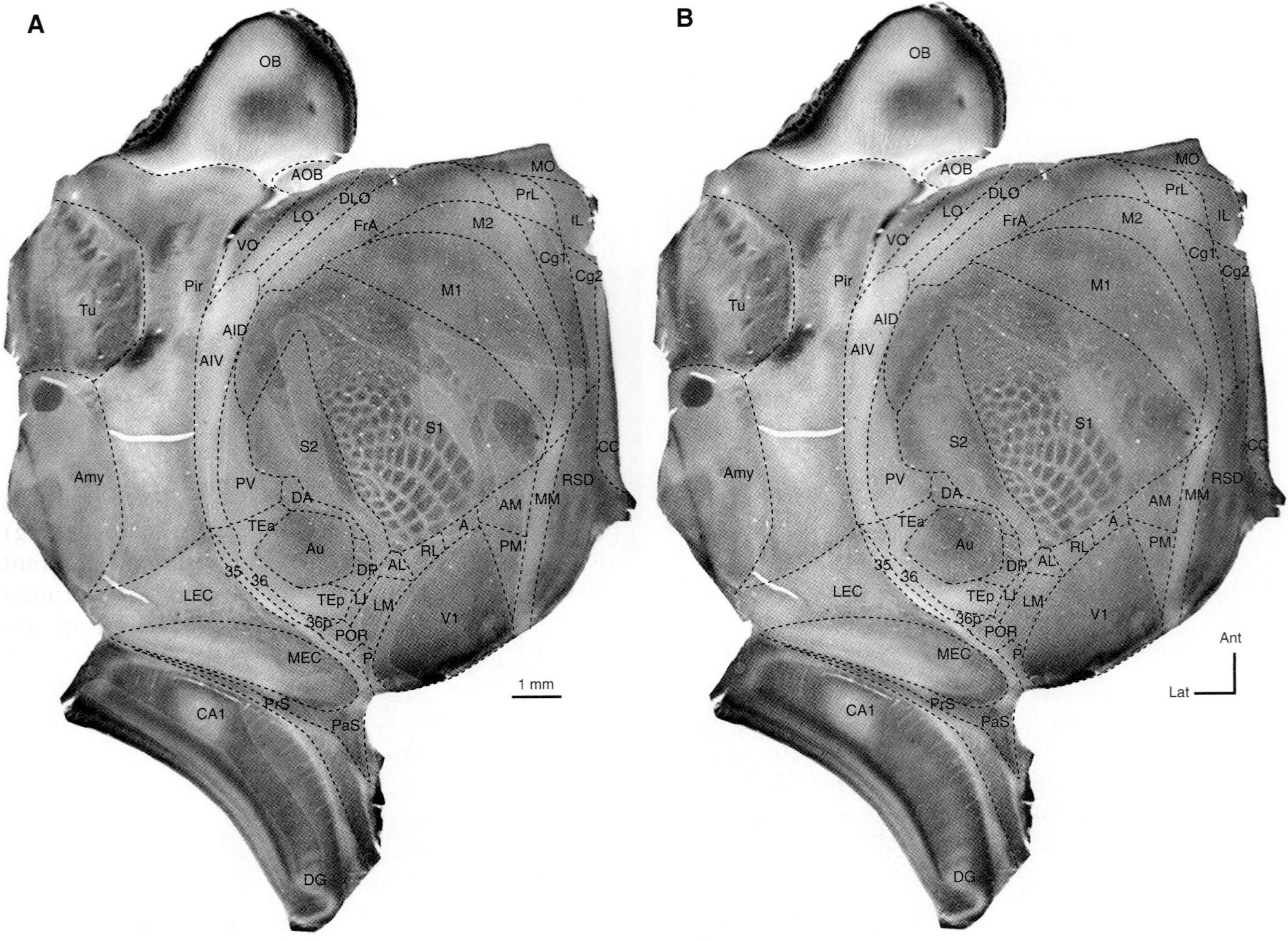

FIGURE 18.2   Areas of mouse cerebral cortex. Cytochrome oxidase reactivity in tangential sections of flattened mouse cerebral cortex. (A) Overlay of cytochrome oxidase expression in layer 4 (gray) with retrogradely bisbenzimide-labeled callosal connections (blue). (B) Same section as shown in A.

et al., 2011). This strongly supports the conclusion that unlike in primates, mouse V1 is adjoined by multiple small extrastriate areas. Each of these maps is isomorphic with the visual field, whose perimeter is represented along the areal borders. The areas are stitched together at the borders so that topographically matching points are aligned across the seams. This tiling format dictates that the vertical meridian of V1 can only be shared by a single area, LM. Thus, LM may be analogous to primate V2 (Allman & Kaas, 1974). The most surprising feature, however, is the rich parcellation of visual cortex in animals whose vision is clearly inferior to that of primates (Huberman & Niell, 2011; Prusky & Douglas, 2004). Although, cortical arealization is sparser than in the more highly visual cats and primates (Sereno & Allman, 1991), the inevitable conclusion is that mouse visual cortex is highly functionally specialized.

NETWORK OF VISUAL CORTICAL AREAS

The first attempt to determine the intracortical network of rodent visual cortex goes back to the work of Miller and Vogt (1984) in the rat. Using anterograde and retrograde tracing of connections, they showed that V1 provided input to 19 cortical areas, which were identified by their distinctive architectonics (Caviness, 1975; Krieg, 1946; Zilles, Zilles, & Schliecher, 1980). It is important to note that these criteria lumped together the subdivisions that Olavarria and Montero (1984) found in areas 18a and 18b. Although Miller and Vogt (1984) noted that projections within 18a and 18b were nonuniform, they saw no compelling reasons for interpreting patches as areas. We have since learned that rat areas 18a and 18b contain as many as 12 areas (Coogan & Burkhalter, 1993; Montero, 1993; Sereno & Allman, 1991) and at least 9 in mice (Wang & Burkhalter, 2007).

    ANDREAS BURKHALTER, OLAF SPORNS, ENQUAN GAO, AND QUANXIN WANG

This bit of history is a sobering reminder of the problem we face in determining the network supplied by afferent visual information. Although the identification of cortical targets remains a challenge, we have made progress in subdividing mouse cortex into parcels by using a combination of architectonic (Nissl, myelin), connectional (corpus callosum), histochemical, and immunological markers in tangential sections (Wang, Gao, & Burkhalter, 2011; Wang, Sporns, & Burkhalter, 2012). This raises the question of how the emerging parcellation scheme registers with existing cytoarchitectonic atlases (Dong, 2008; Franklin & Paxinos, 2007). Faced with the difficult task of assigning areal borders based on cytoarchitectonics, we found it helpful to first identify areas with immunological (e.g., type 2 muscarinic acetylcholine receptor, m2AChR) and callosal markers and then find associated cytoarchitectonic features (Wang, Gao, & Burkhalter, 2011). There is no a priori reason to assume that any marker is area specific. However, if two markers, such as m2AChR and RALD3H (retinoic acid dehydrogenase 3), show complementary patterns, it is suggestive that the common borders outline distinct subdivisions (Wang, Gao, & Burkhalter, 2011). Based on these staining patterns we were able to distinguish areas MM, TEp, 36, and TEa (Wang, Gao, & Burkhalter, 2011). It is reassuring that these subdivisions can also be seen as cytochrome oxidase (CO)-negative areas (figure 18.2B). In other regions, including V1 and surrounding cortex, CO reactivity is significantly higher (figure 18.2B). Although CO expression in cortex surrounding V1 is nonuniform, it is difficult to subdivide the surrounding belt into distinct areas without labeling of callosal connections (figure 18.2A). Even in the presence of callosal connections the PM/AM and A/AM borders are not evident (figure 18.2B). Only the LM/AL border is visible as a transition from dark to light CO expression (figure 18.2B). A similar transition at the corresponding LM/AL border can be seen in the expression of m2AChR (Wang, Gao, & Burkhalter, 2011). Evidently, multiple complementary approaches are necessary to determine areal borders. Thus, the confidence by which we are presently able to identify areas varies across cortex. This strongly suggests that the partitioning scheme shown in figure 18.2A, B is not final but at present offers a framework for referencing corticocortical connections.

Navigating by this map and classifying projections categorically as present or absent, we found that mouse V1 sends output to 25 cortical targets (Wang, Sporns, & Burkhalter, 2012). The downside of such a binary description of V1's connectivity is that it disregards the weight of inputs and provides a potentially misleading map of the flow of information across the cortical network. Although the tiny contribution of the LGN input to the overall number of synapses in cat V1 (Binzegger, Douglas, & Martin, 2004) is a reminder that the relationship between anatomical and physiological strength is complex, we determined the weight of inputs from 10 areas of mouse visual cortex to 39 cortical projection targets. We labeled the inputs by anterograde tracing with biotinylated dextran amine (BDA) and determined the strength of axonal projections by optical densitometry, which tightly correlates with bouton density (Wang, Gao, & Burkhalter, 2011). We then summed the densities of all projections of a source area and expressed the strength of individual projections as percentages of the total.

Our findings are shown in figure 18.3 in which the projection targets are ordered on the x-axis according to their ventral (red shading) and dorsal (blue shading) location in the brain. The projections of different source areas are shown on the arms of a bifurcating path leading from V1 (figure 18.3a) to ventral (figure 18.3b–e) and dorsal (figure 18.3f–k) cortex, respectively. In addition, we found that the distribution of projection weights of each source area is different. In fact, if one orders the projection targets by the weight of their inputs, each source area shows a different order of projection targets (Wang, Sporns, & Burkhalter, 2012), suggesting that long-range connections of each source have distinct connectivity profiles (Markov et al., 2011). Within each profile the weights vary over two to three orders of magnitude, which is less than half the range found in the 200× larger macaque monkey cortex (Markov et al., 2011; Wang, Sporns, & Burkhalter, 2012). As in the monkey, the weight distributions are best fit by a lognormal function (Markov et al., 2011, 2012; Wang, Sporns, & Burkhalter, 2012), suggesting that the scaling of projection strength is evolutionarily conserved. Most interestingly, figure 18.3a–e shows that the projection weights progressively increase along the ventral path involving LM, LI, P, and POR, whereas the projection weights in dorsal areas decrease. The opposite pattern was found in the projections of areas on the dorsal path, AL, PM, RL, AM, and A (figure 18.3f–k). This suggests that the cortical outputs of V1 flow through ventral and dorsal streams.

Strong support for ventral and dorsal networks derives from multidimensional scaling, using the vector angle between projections as a distance measure and principal component analysis of the correlation matrix of projection patterns. Both of these analyses showed significant differences in the projections from ventral and dorsal areas (Wang, Sporns, & Burkhalter, 2012). The community structure of the matrix of visual areas is revealed with the Kamada–Kawai (1989) energy

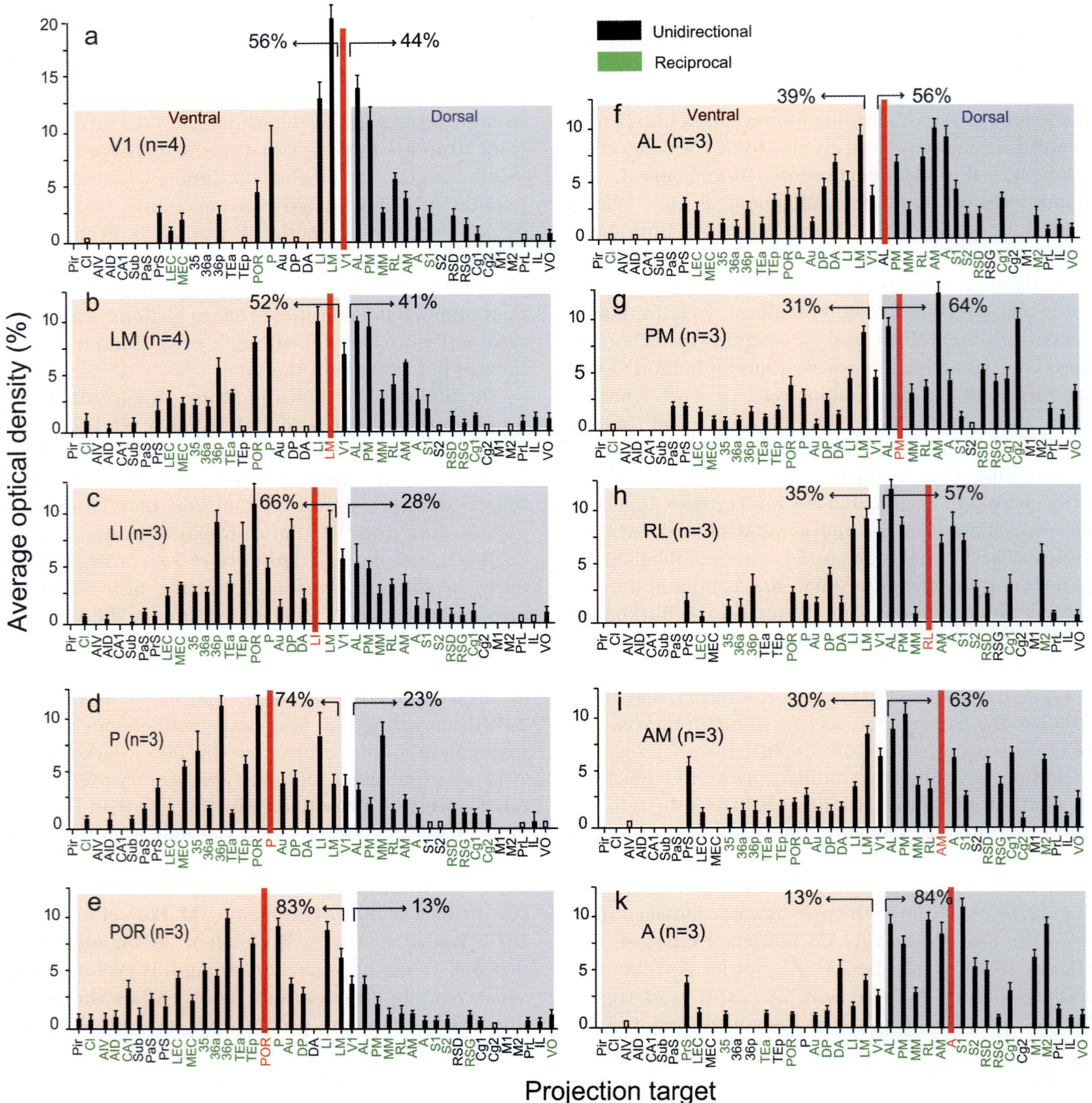

FIGURE 18.3   Relative strength of connections labeled by injections of BDA into areas V1 (a), LM (b), LI (c), P (d), POR (e), AL (f), PM (g), RL (h), AM (i) and A (k) of mouse visual cortex. Average ± SEM optical density (*y*-axis) of a specific projection target (*x*-axis) scaled by the summed density of all projections labeled by an injection of a specific area (red bar). Target regions indicated in black have unidirectional connections with the source region. Green labels indicate reciprocal connections. The percentages indicated by arrows represent the strengths of connections (excluding V1) in the ventral (red shading) and dorsal (blue shading) areas. (From Wang, Sporns, & Burkhalter, 2012.)

minimization layout algorithm. It shows that the optimal modularity score partitions the areas into ventral (LM, LI, P, POR) and dorsal (AL, RL, A, AM, PM) modules, with 61% within-module and 39% between-module connections (figure 18.4A). The outflow from the ventral module is mainly destined for higher auditory, insular, parahippocampal, and hippocampal cortex, whereas outputs of the dorsal module preferentially flow to retrosplenial, somatosensory, motor, and prefrontal cortices (figure 18.4B). The asymmetric organization of

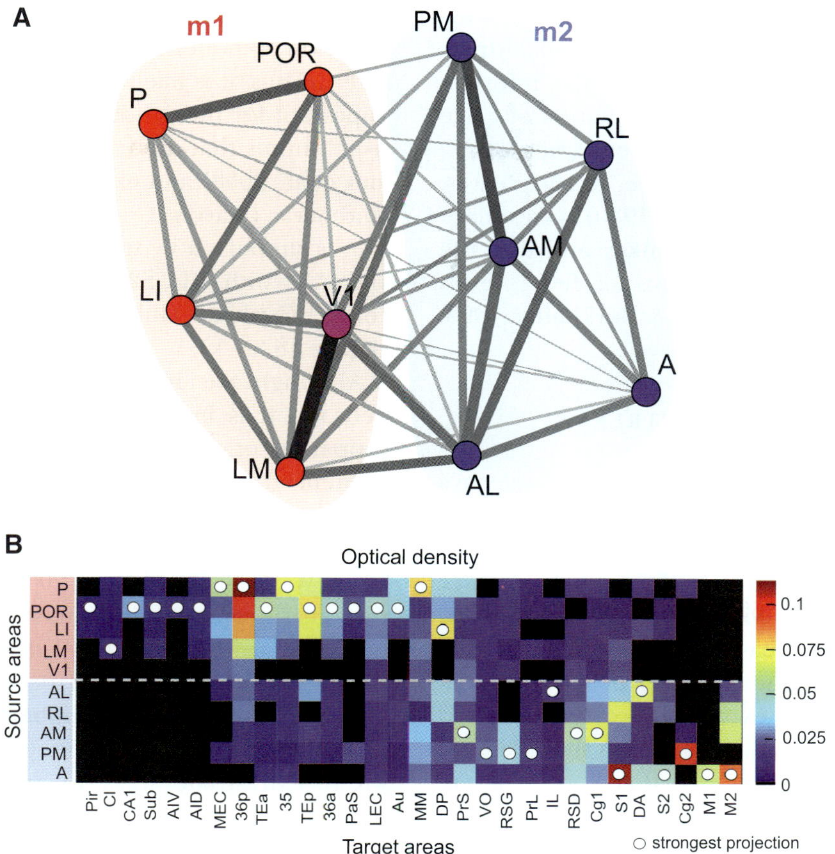

FIGURE 18.4   Modularity structure of connections in mouse cerebral cortex. (A) Graphical layout of the connectivity of the submatrix consisting of areas V1, LM, LI, P, POR, AL, RL, AM, and PM after optimizing the positions of the nodes (areas) using the Kamada–Kawai force-based energy minimization algorithm. Red areas belong to the ventral module m1. Blue areas belong to the dorsal module m2. V1 is more strongly affiliated with the ventral module and is colored purple. Connections between pairs of areas are shown as the sum of their reciprocal projection densities (darker gray/thicker lines = stronger projections). (B) Projection density for 10 source areas to 30 projection targets, arranged by module origin (*y*-axis). White dots highlight the strongest projection each area receives. (From Wang, Sporns & Burkhalter, 2012.)

the cortical network resembles the flow of visual information through ventral and dorsal streams in cat and monkey (Bell, Pasternak, & Ungerleider, 2012; Felleman & Van Essen, 1991; Hilgetag et al., 2000; Kravitz et al., 2011).

The cortical network in mice appears more complex than that in rat (Burwell & Amaral, 1998; McDonald & Mascagni, 1996; Miller & Vogt, 1984; Reep, Goodwin, & Corwin, 1990), showing twice as many connections of single areas. Because studies in rats and mice used different cortical parcellation schemes, it is difficult to know to what extent this is true. However, our finding of 99% connectivity among 10 visual areas suggests that the cortical graph in mice is at least 30% denser than that in monkey (Felleman & Van Essen, 1991; Markov et al., 2011; 2012). The almost fully interconnected matrix further suggests that most connections are reciprocal. However, figure 18.3 shows a surprisingly large number of unidirectional connections (figure 18.3a–k). We think that this might be misleading because BDA is not an effective retrograde tracer (Jiang, Johnson, & Burkhalter, 1993).

Although the existence of ventral and dorsal processing streams indicates fundamental similarities in the visual systems of mice and primates, it is important to note that they are not the same. In both species V1 projects to many "visual" areas in occipital, temporal, parietal, and frontal cortex, including projections to parahippocampal cortex, different parts of the auditory cortex, and layers of the superior colliculus (Markov et al., 2011; Wang, Sporns, & Burkhalter, 2012; Wang & Burkhalter, 2013). However, projections to somatosensory, retrosplenial, cingulate, orbitofrontal, perirhinal, entorhinal, and subicular cortex have only been found in rats and mice (Miller & Vogt, 1984; Reep, Corwin, & King, 1996; Reep, Goodwin, & Corwin, 1990; Vogt &

Miller, 1983; Wang, Sporns, & Burkhalter, 2012). This may indicate that in rodents integration of visual inputs with somatosensory, auditory, and vestibular information, including synthesis for the representation of space and memory (Brecht et al., 2004; Brett-Green et al., 2003; Chen et al., 1994; Knierim, Lee, & Hargreaves, 2006; Wagor, Mangini, & Pearlman, 1980), takes place lower in the hierarchy than in monkey and that the hierarchy in rats and mice consists of fewer levels (Burkhalter & Wang, 2008; Coogan & Burkhalter, 1993; Felleman & Van Essen, 1991).

INTRACORTICAL PROCESSING STREAMS

Based on the behavioral impairments by selective lesions in rat visual and posterior parietal cortex, Kolb (1990) proposed that rodent visual system has "two distinct visual processing routes: one for spatial guidance and one for the kind of visual analysis required for object recognition." Kolb based his proposal on studies that originated with Hughes's (1977) observation that lesions in V1 that involved extrastriate cortex produced greater deficits in pattern discrimination than damage to V1 alone. Later studies showed that a more severe learning deficit in the discrimination of stationary contrast patterns resulted from lesions in V2L (McDaniel, Coleman, & Lindsay, 1982), whereas lesions in posterior parietal cortex and V2M had little effect (McDaniel & Terrell Wall, 1988). Interestingly, the posterior part of V2L (putative LM and P) was particularly important for the discrimination of stationary patterns, whereas the anterior part of V2L (putative AL) was more important for detecting transient moving stimuli (Dean, 1990). Moreover, lesions in anterior V2L and V2M severely affected navigation in different types of mazes but left pattern discrimination intact (Kolb & Walkey, 1987). The basic conclusion of these early findings, that ventral and dorsomedial extrastriate cortical areas were specialized for different functions, was recently confirmed. Specifically, it was shown that targeted lesions in the ventral area, POR, interfere with shape learning (Zhang et al., 2010). Lesions of the more ventral perirhinal cortex (putative area 36) impaired object recognition but had no effect on the formation of associations between stimulus identity and object location (Aggleton et al., 2010). In contrast, lesions of area AM in the medial part of posterior parietal cortex impaired the acquisition of visual target location (Sànchez et al., 1997) and the association of visual with somatosensory inputs (Pinto-Hamuy, Montero, & Torrealba, 2004). Numerous studies have proposed that posterior parietal cortex plays a role in spatial navigation. For example, it was suggested that posterior parietal cortex integrates optic flow patterns that inform about the speed and direction of motion, with motor output (Save & Poucet, 2009). Thus, it is likely that posterior parietal cortex is important for the processing of self-motion–generated visual inputs that are used for path integration (Whitlock et al., 2008) and navigation based on remembered landmarks (Harvey, Coen, & Tank, 2012). This conclusion is supported by the strong connections of RL, A, and AM with S1, S2 (including putative vestibular cortex) (Nishiike, Guldin, & Bäurle, 2000), M1, and M2 (figure 18.3h–k).

Although we have emphasized the intracortical network, it is important to note that the functional properties of neurons in different areas may also be influenced by subcortical inputs from the LGN, the lateral posterior nucleus, as well as other thalamic and nonthalamic nuclei (Simmons, Lemmon, & Pearlman, 1982; Wang & Burkhalter, 2009). Thus, the view that response properties in extrastriate cortex are either inherited from those in V1 (Dräger, 1975; Gao, De Angelis, & Burkhalter, 2010; Mangini & Pearlman, 1980; Métin, Godement, & Imbert, 1988; Niell & Stryker, 2008, 2010; Van den Bergh et al., 2010), generated within extrastriate areas, or are the result of cortical network interactions may be an oversimplification. With these qualifications in mind, it is reassuring to see that the tuning properties of visual responses fit reasonably well with the proposed modularity of the network. For example, by use of two-photon calcium imaging of layer 2/3 neurons in alert (Andermann et al., 2011) and anesthetized (Marshel et al., 2011) mice, it was shown that tuning to high spatial frequency is more common in LI than in AL, RL, and AM, which are more selective for high temporal frequency and the direction of motion.

These results suggest that areas of the ventral stream encode image detail, whereas areas of the dorsal stream are specialized for visual motion (Van Essen & Gallant, 1994). Surprisingly, however, neurons of the ventral-stream area, LM, had relatively low spatial acuity and were tuned to high temporal frequency (Marshel et al., 2011). At present, we have no good explanation for this result. It suggests, however, that LM may not be *the* gateway to the ventral stream (Wang, Gao, & Burkhalter, 2011). Rather, LM may contain ventral and dorsal stream compartments similar to monkey V2 (Nassi & Callaway, 2009).

Another surprise was that the dorsal-stream area, PM, favored ventral-stream properties (Van Essen & Gallant, 1994) such as high spatial frequency sensitivity and preferred low temporal frequencies and slow speeds (Andermann et al., 2011; Marshel et al., 2011). That PM belongs to the dorsal stream is supported by

 ANDREAS BURKHALTER, OLAF SPORNS, ENQUAN GAO, AND QUANXIN WANG

the lack of connections with the amygdala (Wang & Burkhalter, 2011). A similar argument was used previously for assigning V2M to the dorsal stream in rat (McDonald & Mascagni, 1996). Because PM is strongly connected to V1 (figure 18.3g), it is possible that this input provides the information about the shape of slow-moving objects. Importantly, however, the ventral-stream properties of PM suggest that, similar to primates, the dorsal stream may have multiple branches specialized for either the processing of visual information from self-motion (e.g., RL, A, AM) and moving objects viewed from fixed locations (e.g., PM) (Kravitz et al., 2011).

Area PM stands out with strong projections to eye movement and attention centers in Cg2 and to the head direction cell–containing retrosplenial cortex (Brecht et al., 2004; Muir, Everitt, & Robbins, 1996; Taube, 2007; Wang, Sporns, & Burkhalter, 2012). Head direction cells were also found in putative rat PM, where it was shown that the responses were strongly influenced by visual input (Chen et al., 1994). This suggests that responses encoded in a head-centered reference frame may be remapped into an external reference frame used for guiding head movements. Areas of the posterior parietal cortex (RL, A, AM) are strongly connected to S1, S2, cingulate, motor, and putative vestibular cortex within S2 (Nishiike, Guldin, & Bäurle, 2000). Each of these areas is highly sensitive to coarsely topographic transient visual information (Andermann et al., 2011: Marshel et al., 2011) presumably generated by optic flow patterns during self-motion. The areal connectivity pattern suggests that these visual inputs are combined with somatosensory, proprioceptive, vestibular, and motor efferent copy signals, which are used for path integration. The network between the branches of the dorsal stream may link the internal path integration information with inputs from external landmarks and enable goal-directed navigation (Harvey, Coen, & Tank, 2012).

ABBREVIATIONS

Area 35, 36, 36a, 36p; A (anterior); AID (anterior dorsal insula); AIV (anterior ventral insula); AL (anterolateral); AM (anteromedial); Amy (amygdala); AOB (accessory olfactory bulb); Au (auditory); CA1 (hippocampus); CC (corpus callosum); Cg1 (cingulate 1); Cg2 (cingulate 2); CO (cytochrome oxidase); DA (dorsal anterior); DG (dentate gyrus); DLO (dorsal lateral orbital); DP (dorsal posterior); FrA (frontal association); IL (infra limbic); LEC (lateral entorhinal); LI (lateral intermediate); LM (lateral medial); LO (lateral orbital); m2AChR (type 2 acetylcholine receptor); M1 (motor 1); M2 (motor 2); MEC (medial entorhinal); MM (medio medial); MO (medial orbital); OB (olfactory bulb); P (posterior); PaS (parasubiculum); Pir (piriform); PM (posterior medial); POR (postrhinal); PrL (prelimbic); PrS (presubiculum); PV (parietal ventral); RALDH3 (retinoic acid dehydrogenase 3); rf (rhinal fissure); RL (rostrolateral); RSD (retrosplenial dysgranular); RSG (retrosplenial granular); S1 (somatosensory 1); S2 (somatosensory 2); TEa (temporal anterior); TEp (temporal posterior); Tu (olfactory tubercle); V1 (primary visual); V2L (lateral visual area 2), V2LM (lateromedial visual area 2), V2M (medial visual area 2); VO (ventral orbital).

ACKNOWLEDGMENTS

This work was supported National Eye Institute Grants R01EY-05935, R01EY-022090, the McDonnell Center for Systems Neuroscience, and the Human Frontier Science Program 2000B. We thank Katia Valkova for excellent technical assistance.

REFERENCES

Aggleton, J. P., Albasser, M. M., Aggleton, D. J., Poirier, G., & Pearce, J. M. (2010). Lesions of the rat perirhinal cortex spare the acquisition of a complex configural visual discrimination yet impair object recognition. *Behavioral Neuroscience, 124,* 55–68.

Airey, D. C., Robbins, A. I., Enzinger, K. M., Wu, F., & Collins, C. E. (2005). Variation in the cortical area map of C57BL/6J and DBA/2J inbred mice predicts strain identity. *BMC Neuroscience, 6,* 1–8.

Allman, J. M., & Kaas, J. H. (1974). The organization of the second area (VII) in the owl monkey: A second order transformation of the visual hemifield. *Brain Research, 76,* 247–265.

Andermann, M. L., Kerlin, A. M., Roumis, D. K., Glickfield, L., & Reid, R. C. (2011). Functional specialization of mouse higher visual cortical areas. *Neuron, 72,* 1025–1039.

Atkinson, J., & Braddick, O. (2011). From genes to brain development to phenotypic behavior: "Dorsal-stream vulnerability" in relation to spatial cognition, attention, and planning of actions in Williams syndrome (WS) and other developmental disorders. *Progress in Brain Research, 189,* 261–283.

Bell, A. H., Pasternak, T., & Ungerleider, L. G. (2012). Ventral and dorsal cortical processing streams. In preparation.

Binzegger, T., Douglas, R. J., & Martin, K. A. (2004). A quantitative map of the circuit of cat primary visual cortex. *Journal of Neuroscience, 24,* 8441–8453.

Brecht, M., Krauss, A., Muhammad, S., Sinai-Esfahani, L., Bellanca, S., & Margerie, T. W. (2004). Organization of rat vibrissa motor cortex and adjacent areas according to cytoarchitectonics, microstimulation, and intracellular stimulation of identified cells. *Journal of Comparative Neurology, 479,* 360–373.

Brett-Green, B., Fifkova, A., Laurie, D. T., Winder, J. A., & Barth, D. S. (2003). A multisensory zone in rat parietotemporal cortex: Intra- and extracellular physiology and thalamocortical connections. *Journal of Comparative Neurology, 460*, 223–237.

Burkhalter, A., & Wang, Q. (2008). Interconnections of visual cortical areas in mouse. In L. M. Chalupa & R. W. Williams (Eds.), *Eye, retina, and visual system of the mouse* (pp. 245–254). Cambridge, MA: MIT Press.

Burwell, R. C., & Amaral, D. G. (1998). Cortical afferents of the perirhinal, postrhinal and entorhinal cortices. *Journal of Comparative Neurology, 398*, 179–205.

Caviness, V. S. (1975). Architectonic map of the normal mouse. *Journal of Comparative Neurology, 164*, 247–263.

Chen, L. L., Lin, L.-H., Barnes, C. A., & McNaughton, B. L. (1994). Head-direction cells in the rat posterior cortex. II. Contributions of visual and ideothetic information to the directional firing. *Experimental Brain Research, 101*, 24–34.

Coogan, T. A., & Burkhalter, A. (1993). Hierarchical organization of area in rat visual cortex. *Journal of Neuroscience, 13*, 3749–3772.

Dean, P. (1990). Sensory cortex: Visual perceptual functions. In B. Kolb & R. C. Tees (Eds.), *The cerebral cortex of the rat* (pp. 275–307). Cambridge, MA: MIT Press.

Dong, H. W. (2008). *Allen reference atlas. A digital color brain atlas for the C57BL/6J male mouse.* Seattle, WA: Allen Institute for Brain Science.

Dräger, U. C. (1974). Autoradiography of tritiated proline and fucose transported transneuronally from the eye to the visual cortex in pigmented and albino mice. *Brain Research, 82*, 284–292.

Dräger, U. C. (1975). Receptive fields of single cells and topography in mouse visual cortex. *Journal of Comparative Neurology, 160*, 269–290.

Fabri, M., & Burton, H. (1991). Ipsilateral cortical connection of primary somatic sensory cortex in rats. *Journal of Comparative Neurology, 311*, 405–424.

Felleman, D. J., & Van Essen, D. C. (1991). Distributed hierarchical processing in the primate cerebral cortex. *Cerebral Cortex, 1*, 1–42.

Franklin, K. B. J., & Paxinos, G. (2007). *The mouse brain in stereotaxic coordinates.* Amsterdam: Elsevier.

Gao, E., De Angelis, G. C., & Burkhalter, A. (2010). Parallel input channels to mouse primary visual cortex. *Journal of Neuroscience, 30*, 5912–5926.

Goodale, M. A., & Milner, A. D. (1992). Separate visual pathways for perception and action. *Trends in Neurosciences, 15*, 20–25. doi:10.1016/0166-2236(92)90344-8.

Harvey, C. D., Coen, P., & Tank, D. W. (2012). Choice-specific sequences in parietal cortex during a virtual-navigation decision task. *Nature, 484*, 62–68.

Herculano-Houzel, S. (2011). Not all brains are made the same: New views on brain scaling in evolution. *Brain, Behavior and Evolution, 78*, 22–36.

Hilgetag, C.-C., Burn, G. A. P. C., O'Neill, M. A., Scannel, J. W., & Young, M. P. (2000). Anatomical connectivity defines the organization of clusters of cortical areas in the macaque money and cat. *Philosophical Transactions of the Royal Society of London. Series B, Biological Sciences, 355*, 91–110.

Huberman, A. D., & Niell, C. M. (2011). What can mice tell us about how vision works? *Trends in Neurosciences, 34*, 464–473.

Hughes, H. C. (1977). Anatomical and neurobehavioral investigations concerning the thalamo-cortical organization of the rat's visual system. *Journal of Comparative Neurology, 175*, 311–336.

Jiang, X., Johnson, R. R., & Burkhalter, A. (1993). Visualization of dendritic morphology of cortical projection neurons by retrograde labeling. *Journal of Neuroscience Methods, 50*(1), 45–60.

Kaas, J. H. (2012). The evolution of neocortex in primates. *Progress in Brain Research, 195*, 91–102.

Kaas, J. H., Krubitzer, L. A., & Johanson, K. L. (1989). Cortical connections of areas 17 (V-I) and 18 (V-II) of squirrels. *Journal of Comparative Neurology, 281*, 426–446.

Kalasky, V. A., & Stryker, M. P. (2003). New paradigm for optical imaging: Temporally encoded maps of intrinsic signal. *Neuron, 38*, 529–545.

Kamada, T., & Kawai, S. (1989). An algorithm for drawing general undirected graphs. *Information Processing Letters, 31*, 7–15.

Knierim, J. J., Lee, I., & Hargreaves, E. L. (2006). Hippocampal place cells: Parallel input streams, subregional processing, and implications for episodic memory. *Hippocampus, 16*, 755–764.

Kolb, B., & Walkey, J. (1987). Behavioral and anatomical studies of the posterior parietal cortex in the rat. *Behavioural Brain Research, 23*, 127–145.

Kolb, B. (1990). Posterior parietal and temporal association cortex. In B. Kolb & R. C. Tees (Eds.), *The cerebral cortex of the rat* (pp. 459–471). Cambridge, MA: MIT Press.

Kravitz, D. J., Saleem, K. S., Baker, C. I., & Mishkin, M. (2011). A new framework for visuospatial processing. *Nature Reviews Neuroscience, 12*, 217–230.

Krieg, W. J. S. (1946). Connections of the cerebral cortex. I. The albino rat. A. Topography of the cortical areas. *Journal of Comparative Neurology, 84*, 221–275.

Mangini, N. J., & Pearlman, A. L. (1980). Laminar distribution of receptive field properties in the primary visual cortex of the mouse. *Journal of Comparative Neurology, 193*, 203–222.

Markov, N. T., Ercsey-Ravasz, M. M., Ribiero Gomes, A. R., Lamy, C., Magrou, L., Vezoli, J., Misery, P., Falchier, A., Quilodran, R., Gariel, M. A., Sallet, J., Gamanut, R., Huissoud, C., Clavagnier, S., Giroud, P., Sappey-Marinier, D., Barone, P., Dehay, C., Toroczkai, Z., Knoblauch, K., Van Essen, D. C., & Kennedy, H. (2012). A weighted directed interareal connectivity matrix for macaque cerebral cortex. *Cerebral Cortex*, September 25, Epub ahead of print.

Markov, N. T., Misery, P., Falchier, A., Lamy, C., Vexoli, J., Quilodran, R., et al. (2011). Weight consistency specifies regularities of macaque cortical networks. *Cerebral Cortex, 21*, 1254–1274.

Marshel, J. H., Garrett, M. E., Nauhaus, I., & Callaway, E. M. (2011). Functional specialization of seven mouse visual cortical areas. *Neuron, 72*, 1042–1054.

McDaniel, W. F., Coleman, J., & Lindsay, J. F., Jr. (1982). A comparison of lateral peristriate and striate cortical ablations in the rat. *Behavioural Brain Research, 6*, 249–272.

McDaniel, W. F., & Terrell Wall, T. (1988). Visuospatial functions in the rat following injuries to striate, peristriate, and parietal neocortical areas. *Psychobiology, 16*, 251–260.

McDonald, A. J., & Mascagni, F. (1996). Cortico-cortical and geniculo-amygdaloid projections of the rat occipital cortex: A *Phaseolus vulgaris* leucoagglutinin study. *Neuroscience, 71*, 37–54.

Métin, C., Godement, P., & Imbert, M. (1988). The primary visual cortex in the mouse: Receptive field properties and functional organization. *Experimental Brain Research, 69,* 594–612.

Miller, M. W., & Vogt, B. A. (1984). Direct connections of rat visual cortex with sensory, motor, and association cortices. *Journal of Comparative Neurology, 226,* 184–202.

Montero, V. M. (1993). Retinotopy of cortical connections between the striate cortex and extrastriate visual areas in the rat. *Experimental Brain Research, 94,* 1–15. doi:10.1007/BF00230466.

Muir, J. L., Everitt, B. J., & Robbins, T. W. (1996). The cerebral cortex of the rat and visual attention function: Dissociable effects of mediofrontal, cingulate, anterior dorsolateral, and parietal cortex lesions on a five-choice serial reaction time task. *Cerebral Cortex, 6,* 470–481.

Nassi, J. J., & Callaway, E. M. (2009). Parallel processing strategies of the primate visual system. *Nature Reviews Neuroscience, 10,* 360–372.

Niell, C. M., & Stryker, M. P. (2008). Highly selective receptive fields in mouse visual cortex. *Journal of Neuroscience, 28,* 7520–7536.

Niell, C. M., & Stryker, M. P. (2010). Modulation of visual responses by behavioral state in mouse visual cortex. *Neuron, 65,* 472–479.

Nishiike, S., Guldin, W. O., & Bäurle, L. (2000). Corticofugal connections between cerebral cortex and the vestibular nuclei in the rat. *Journal of Comparative Neurology, 420,* 363–372.

Olavarria, J., & Montero, V. M. (1984). Relation of callosal and striate-extrastriate cortical connections in the rat: Morphological definition of extrastriate visual areas. *Experimental Brain Research, 54,* 240–252.

Olavarria, J., & Montero, V. M. (1989). Organization of visual cortex in the mouse revealed by correlating callosal and striate-extrastriate connections. *Visual Neuroscience, 3,* 59–69.

Pinto-Hamuy, T., Montero, V. M., & Torrealba, F. (2004). Neurotoxic lesion of anteromedial/posterior parietal cortex disrupts spatial maze memory in blind rats. *Behavioural Brain Research, 153,* 465–470.

Prusky, G. T., & Douglas, R. M. (2004). Characterization of mouse cortical spatial vision. *Vision Research, 44,* 3411–3418. doi:10.1016/j.visres.2004.09.001.

Rakic, P. (1995). A small step for the cell, a giant leap for mankind: A hypothesis of neocortical expansion during evolution. *Trends in Neurosciences, 18,* 383–388. doi:10.1016/0166-2236(95)93934-P.

Rakic, P. (2009). Evolution of the neocortex: A perspective from developmental biology. *Nature Reviews Neuroscience, 10*(10), 724–735.

Reep, R. L., & Corwin, J. V. (2009). Posterior parietal cortex as part of a network for directed attention in rats. *Neurobiology of Learning and Memory, 91,* 104–113.

Reep, R. L., Corwin, J. V., & King, V. (1996). Neuronal connections of orbital cortex in rats: Topography of cortical and thalamic afferents. *Experimental Brain Research, 111,* 215–232.

Reep, R. L., Goodwin, G. S., & Corwin, J. V. (1990). Topographic organization in the corticocortical connections of the medial agranular cortex in rats. *Journal of Comparative Neurology, 294,* 262–280.

Rosa, M. P. G., & Krubitzer, L. A. (1999). The evolution of visual cortex: Where is V2? *Trends in Neurosciences, 22,* 242–248. doi:10.1016/S0166-2236(99)01398-3.

Rose, M. (1929). Cytoarchitektonischer Atlas der Grosshirnrinde der Maus. *Journal für Psychologische Neurologie, 40,* 1–32.

Salinas-Navarro, M., Jiménez-López, M., Valientine-Soriano, F. J., Alarcón-Martínez, L., Avilés-Trigueros, M., Major, S., et al. (2009). Retinal ganglion cell population in adult albino and pigmented mice: A computerized analysis of the entire population and its spatial distribution. *Vision Research, 49,* 637–647.

Sànchez, R. F., Montero, V. M.., Espinoza, S. G., Diaz, E., Canitrot, M., & Pinto-Hamuy, T. (1997). Visuospatial discrimination deficit in rats after ibotenic lesions in anteromedial visual cortex. *Physiology and Behavior, 62,* 989–994.

Save, E., & Poucet, B. (2009). Role of the parietal cortex in long-term representation of spatial information in the rat. *Neurobiology of Learning and Memory, 91,* 172–178.

Schneider, G. E. (1967). Two visual systems. *Science, 163,* 895–902.

Schuett, S., Bonhoeffer, T., & Huebener, M. (2002). Mapping retinotopic structure in mouse visual cortex with optical imaging. *Journal of Neuroscience, 22,* 6549–6559.

Schüz, A., & Palm, G. (1989). Density of neurons and synapses in the cerebral cortex of the mouse. *Journal of Comparative Neurology, 286,* 442–455.

Sereno, M. I., & Allman, J. M. (1991). Cortical visual areal in mammals. In E. G. Leventhal (Ed.), *The neuronal basis of visual function* (pp. 160–172). London: Macmillan.

Shepherd, J. D., Rumbaugh, G. J., Chowdhury, S., Path, N., Kuhl, D., Huganir, R. L., et al. (2006). Arc/Arg3.1 mediates homeostatic synaptic scaling of AMPA receptors. *Neuron, 52,* 475–484.

Simmons, P. A., Lemmon, V., & Pearlman, A. L. (1982). Afferent and efferent connections of striate and extrastriate visual cortex of the normal and *reeler* mouse. *Journal of Comparative Neurology, 211,* 295–308.

Smith, S. L., & Häusser, M. (2010). Parallel processing of visual space by neighboring neurons in mouse visual cortex. *Nature Neuroscience, 13,* 1144–1149.

Tagawa, Y., Kanold, P. P., Majdan, M., & Shatz, C. J. (2005). Multiple periods of functional ocular dominance plasticity in mouse visual cortex. *Nature Neuroscience, 8,* 380–388.

Taube, J. S. (2007). The head direction signal: Origins and sensory-motor integration. *Annual Review of Neuroscience, 30,* 181–207.

Tohmi, M., Takahashi, K., Kubota, Y., Hishida, R., & Shibuki, K. (2009). Transcranial flavoprotein fluorescence imaging of mouse cortical activity and plasticity. *Journal of Neurochemistry, 109,* 3–9.

Trevarthen, C. B. (1968). Two mechanisms of vision in primates. *Psychologische Forschung, 31,* 299–348.

Ungerleider, L. G., & Mishkin, M. (1982). Two cortical systems. In D. J. Ingle, M. A. Goodale, & R. J. W. Mansfield (Eds.), *Analysis of visual behaviors* (pp. 549–586). Cambridge, MA: MIT Press.

Van Brussel, L., Gerits, A., & Arckens, L. (2009). Idenitification and localization of functional subdivisions in the visual cortex of the adult mouse. *Journal of Comparative Neurology, 514,* 107–116.

Van den Bergh, G., Zhang, B., Arckens, L., & Chino, Y. M. (2010). Receptive field properties of V1 and V2 neurons in mice and macaque monkeys. *Journal of Comparative Neurology, 518,* 2051–2070.

Van Essen, D. C., & Gallant, J. L. (1994). Neural mechanisms of form and motion processing in the primate visual system. *Neuron, 13,* 1–10.

Van Essen, D. C., Glasser, M. F., Dierker, D. L., & Harwell, J. (2012a). Cortical Parcellations of the macaque monkey analyzed on surface-based atlases. *Cerebral Cortex, 22,* 2227–2240.

Van Essen, D. C., Glasser, M. F., Dierker, D. L., Harwell, J., & Coalson, T. (2012b). Parcellations and hemispheric asymmetries of human cerebral cortex analyzed on surface-based atlases. *Cerebral Cortex, 22,* 2241–2262.

Vogt, B. A., & Miller, M. W. (1983). Cortical connections between rat cingulate and visual, motor, and postsubicular cortices. *Journal of Comparative Neurology, 216,* 192–210.

Wagor, E., Mangini, N. J., & Pearlman, A. L. (1980). Retinotopic organization of striate and extrastriate visual cortex in the mouse. *Journal of Comparative Neurology, 193,* 187–202.

Wang, Q., & Burkhalter, A. (2007). Area map of mouse visual cortex. *Journal of Comparative Neurology, 502,* 339–357.

Wang, Q., & Burkhalter, A. (2009). Stream-specific corticothalamic feedback projections in mouse visual cortex. *Society for Neuroscience Abstracts,* 453.19/Y14.

Wang, Q., & Burkhalter, A. (2011). Stream-specific connections of mouse visual cortex with the amygdala. *Society for Neuroscience Abstracts,* 378.02/005.

Wang, Q., & Burkhalter, A. (2013). Stream-related preferences of inputs to the superior colliculus from areas of dorsal and ventral streams of mouse visual cortex. *Journal of Neuroscience, 33,* 1696–1705. doi:10.1523/JNEUROSCI.3067-12.2013.

Wang, Q., Gao, E., & Burkhalter, A. (2011). Gateways of ventral and dorsal streams in mouse visual cortex. *Journal of Neuroscience, 31,* 1905–1918.

Wang, Q., Sporns, O., & Burkhalter, A. (2012). Network analysis of corticocortical connections reveals ventral and dorsal processing streams in mouse visual cortex. *Journal of Neuroscience, 32*(13), 4386–4399.

Whitlock, J. R., Sutherland, R. J., Witter, M. P., Moser, M.-B., & Moser, E. I. (2008). Navigating from hippocampus to parietal cortex. *Proceedings of the National Academy of Sciences of the United States of America, 105,* 14755–14762. doi:10.1073/pnas.0804216105.

Zhang, G.-R., Cao, H., Kong, L., O'Brien, J., Baughns, A., Jan, M., et al. (2010). Identified circuit in rat postrhinal cortex encodes essential information for performing specific visual shape discriminations. *Proceedings of the National Academy of Sciences of the United States of America, 107,* 14478–14483. doi:10.1073/pnas.0912950107.

Zilles, K., & Amunt, K. (2010). Centenary of Brodmann's map conception and fate. *Nature Reviews Neuroscience, 11,* 139–145.

Zilles, K., Zilles, B., & Schliecher, A. (1980). A quantitative approach to cytoarchitectonics. VI. The areal pattern of the cortex of the albino rat. *Anatomy and Embryology, 159,* 335–360.

III SUBCORTICAL PROCESSING

19 The Lateral Geniculate Nucleus and Pulvinar

S. MURRAY SHERMAN AND R. W. GUILLERY

The lateral geniculate nucleus[1] is the thalamic relay of retinal input to the visual cortex. It is the best understood of thalamic relays, and, because there is an overall structure shared by all thalamic nuclei, it can serve as a general model for the thalamus. We first consider the lateral geniculate nucleus and then, using this as a prototype, look at other thalamic relays on the visual pathways, the lateral posterior nucleus, and the pulvinar. For convenience, both are referred to jointly as the pulvinar. We look at features that are shared by the lateral geniculate nucleus and the pulvinar and explore the extent to which the organizational principles that are well defined for the lateral geniculate nucleus can help us to understand aspects of pulvinar organization. The lateral geniculate nucleus will be treated as a "first order" relay (Sherman & Guillery, 2006, 2011), receiving ascending visual messages from the retina and sending them to primary receiving cortex, and the pulvinar as a "higher order" relay, relaying messages from one cortical area through the thalamic relay to another cortical area and thus playing a potentially crucial role in corticocortical communication. First and higher order relays are defined more fully below, but it suffices to note here that the former represent the initial relay of a particular sort of information (e.g., visual or auditory) to cortex, and the latter represent further relay of such information via a corticothalamocortical loop.

When the distinctive receptive field properties of cells in the retina and the visual cortex were being defined (reviewed in Hubel & Wiesel, 1977), the lateral geniculate nucleus was treated as a simple machine-like relay. This reflected the great success of the receptive field approach to vision. Initially, in anesthetized animals this approach showed that receptive fields become increasingly elaborated along the synaptic hierarchies through retina and cortex, with the one glaring exception being the retinogeniculate synapse: the center/surround receptive fields of geniculate relay cells are essentially the same as those of their retinal afferents. From this arose the misleading conclusion that nothing much of interest was happening in the geniculate relay (Hubel & Wiesel, 1977; Zeki, 1993).

Indeed, this raises several questions: Why have a geniculate relay at all? Why not have retinal axons project directly to visual cortex? Or, more generally, why bother with thalamic relays? One of the main purposes of this chapter is to provide a partial answer to these questions. A second purpose is to indicate that currently we are far from having complete answers. It is highly probable that much of what the thalamus does still remains to be defined.

As we shall see, even on the evidence available today, there is much of interest happening in the geniculate relay and also throughout the thalamus. The lack of receptive field elaboration seen in the geniculate relay is not evidence of nothing happening there but, rather, evidence that something completely different but important occurs. That is, this is a crucial synapse in the visual pathways doing something other than receptive field elaboration: geniculate circuitry is involved in affecting, in a dynamic fashion, the amount and nature of information relayed to cortex. One reason this was missed in earlier studies is that these functions largely depend on the behavioral state of the animal and are suppressed in anesthetized animals (e.g., attentional mechanisms) (McAlonan, Cavanaugh, & Wurtz, 2008; Schneider & Kastner, 2009; see also chapter 32 by Swadlow and Alonso).

FUNCTIONAL ORGANIZATION OF THE LATERAL GENICULATE NUCLEUS

For those who still have in mind the lateral geniculate nucleus as a simple relay, the complexity of its organization will no doubt come as a surprise. The nucleus is organized into a number of layers with a detailed topographic map of visual space that cuts across layers, and its circuitry involves several cell types, many distinctive groups of afferents, and complex synaptic relationships.

Maps

There is a precise map of the contralateral visual hemifield in the lateral geniculate nucleus of all species so far studied, and in this the visual system is like other sensory systems whose receptive surfaces are mapped in their thalamic relays. The map in the lateral geniculate nucleus is laid out in fairly simple Cartesian coordinates in all species (see figure 19.1, which shows the relationships in a cat and where the visual field and its retinal and geniculate representations are shown as tapered arrows). Each layer maps the contralateral hemifield through one or the other eye, and all of these maps are aligned across the various layers that characterize the lateral geniculate nucleus of most mammals (see below and figure 19.1). Thus, a point in visual space is represented by a line, called a line of projection, that runs perpendicularly through all the layers. The precise alignment of these maps, which matches inputs from the nasal retina of one eye across layers with inputs from the temporal retina of the other eye is seen in all mammals and is surprising when one thinks about the necessary developmental mechanisms needed to produce such a match because the match is formed before the eyes open and before the two visual images can be matched. There are non-retinal afferent axons that innervate the lateral geniculate nucleus and that distribute terminals along the lines of projection. In this way these afferents can have a well-localized action on just one part of the visual input, even though this comes from different eyes and distributes to distinct sets of geniculate layers (see Afferents below).

Layering

The lateral geniculate nuclei of all mammalian species so far studied show some form of layering, although there is considerable difference among species as to what the layers represent functionally, how easy they are to identify, and how the sequence of the layers is arranged. For all mammals each layer receives input from only one eye, but the distribution of distinct functional types of retinal afferents to the layers differs greatly from one species to another. This is described in chapter 16 by Kaplan and chapter 86 by Kaas. Figure 19.2 shows the layering of the lateral geniculate nucleus in the macaque monkey[2] and the cat, the two best-studied species. The figure illustrates the variation in layering seen across species and also serves to introduce the several parallel pathways that are relayed through the lateral geniculate nucleus to the cortex.

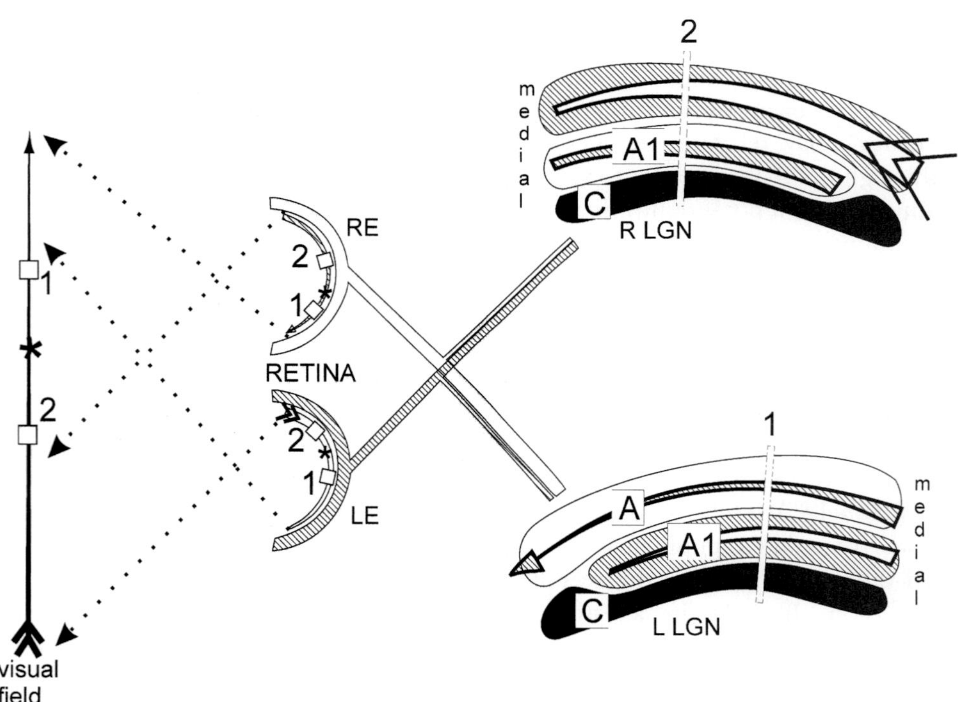

FIGURE 19.1   Schematic view of the representation of the retina and visual field in the laminae of the lateral geniculate nucleus of a cat. The visual field is represented by a straight arrow, and the projection of a part of this arrow onto each retina is shown. Small white areas of the visual field and corresponding parts of the retina are labeled "1" and "2." The representation of this part of the visual field in the lateral geniculate nucleus is shown as a corresponding white column going through all of the geniculate layers "like a toothpick through a club sandwich" (Walls, 1953). Each such column is bounded by the lines of projection, which also pass through all of the laminae. A, A1, and C identify the major geniculate layers; L LGN and R LGN, left and right lateral geniculate nuclei; LE and RE, left eye and right eye; *, central point of fixation.

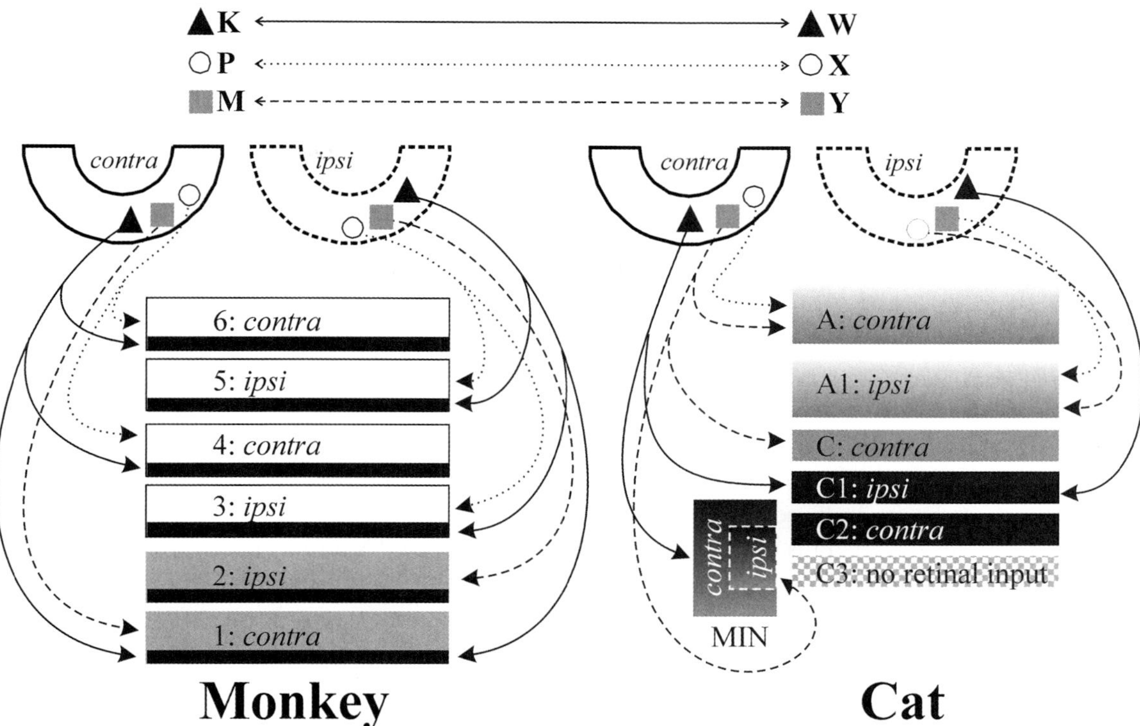

FIGURE 19.2 Comparison of layering in lateral geniculate nucleus of cat and monkey. See text for details; MIN is the medial interlaminar nucleus.

Both species have three main retinal ganglion cell classes that project to the lateral geniculate nucleus. For the monkey (Casagrande & Xu, 2004), these are the P (for parvocellular, meaning small celled), M (for magnocellular; large celled), and K (for koniocellular; tiny or dust-like cells), and these are comparable respectively to the X, Y, and W cells in the cat (Sherman, 1985; Stone, 1983). The terminology for the monkey relates to the geniculate layers to which they project. P and M cells project to parvocellular and magnocellular layers, respectively. In the monkey the K cells project to the ventral regions of all the layers, where very small cells lie scattered, and the projections overlap with those of M and P cells. However, in *Galago*, a prosimian primate, each of the homologous retinal cell types projects to a separate set of layers (not shown): koniocellular, parvocellular, and magnocellular, and it was in this species that the koniocellular pathway was first clearly recognized (Casagrande & Xu, 2004; Conley, Birecree, & Casagrande, 1985; Itoh, Conley, & Diamond, 1982). In the cat X and Y cells have overlapping projections to the A layers, Y cells also project to layer C, and W cells project to layers C1 and C2 (Sherman, 1985; Stone, 1983); there is no retinal input to layer C3, which is therefore shown in black in figure 19.2 (Hickey & Guillery, 1974). Strictly speaking, layer C3 should perhaps not be included in the lateral geniculate nucleus, which can be defined as the thalamic relay of retinal inputs. The cat and other carnivores also have a part of the lateral geniculate nucleus known as the medial interlaminar nucleus (see figure 19.2), which also has separate zones innervated by ipsilateral and contralateral inputs from W and Y axons (Sherman, 1985). What is common to cat and monkey is that each layer is innervated by only one eye. What is different is the partial segregation of parallel pathways through each layer: In the monkey the P and M pathways use separate layers, but the K pathway overlaps with each; in the cat the W pathway uses separate layers, and the Y pathway has exclusive use of layer C, but the X and Y pathways are mingled in the A layers.

The separation of the two ocular inputs to one or another set of layers is constant across species, although in most rodents there is a functional and topographic separation of the inputs from the two eyes but no histologically distinctive identification of the lamination. The separation of different functional types into distinct layers, however, is highly inconsistent, so that the number of layers and their sequence from superficial to deep show great variability across species. For instance, even in closely related carnivores, the cat, ferret, and mink, cells with ON center receptive fields comingle with those with OFF center fields in the A layers of the cat (Sherman, 1985; Stone, 1983) but occupy separate sublaminae of the A layers in the ferret (Stryker & Zahs, 1983) and lie in completely separate layers in the mink (LeVay & McConnell, 1982). Despite the overlap within layers of some of the parallel pathways, there is no functional interaction at the cellular level; within the A layers of the cat, retinal X and Y axons and ON and OFF center axons innervate their own classes of relay cell, and a similar pattern exists for the monkey, with each of the K, P, and M, or ON and OFF center retinal axons targeting separate classes of geniculate relay cell, whose axons, in turn have distinct distributions in the visual cortex (see chapters 16 by Kaplan and 25 by Callaway).

Most of the information we have for cell and circuit properties of the lateral geniculate nucleus specifically and for the thalamus more generally comes from the A layers in the cat, and thus most details described below are from these layers. However, this focus on the A layers should not obscure two important facts. One is that in terms of the general arrangements of synaptic circuitry, there is a common thalamic plan that applies to all parts of the lateral geniculate nucleus, to the pulvinar, and to most other parts of the thalamus; the other is that structural details vary significantly among different layers and species. That is, there are details of

functional organization that remain largely unexplored but almost certainly will ultimately have to be added to our account of the visual relays in the thalamus. (For some details of geniculate relays beyond the A layers in the cat and seen in other species, see Casagrande & Xu, 2004, and Jones, 2007.)

Thalamic Cell Types

There are three basic cell types in the A layers of the cat's lateral geniculate nucleus (see figure 19.3). These include the two relay cell types, X and Y, and interneurons. The relay cells use glutamate as a neurotransmitter and are excitatory, whereas interneurons use GABA and are inhibitory.

RELAY CELLS   X and Y cells represent geniculate relays of two parallel and independent geniculocortical pathways each innervated by its own retinal X or Y axons. Retinogeniculate Y axons are thicker and conduct more rapidly than do X axons (reviewed in Sherman, 1985, and Stone, 1983). The geniculate X and Y cells differ from one another with respect to their functional and morphological properties. Their receptive field differences are already present in the retinal afferents:

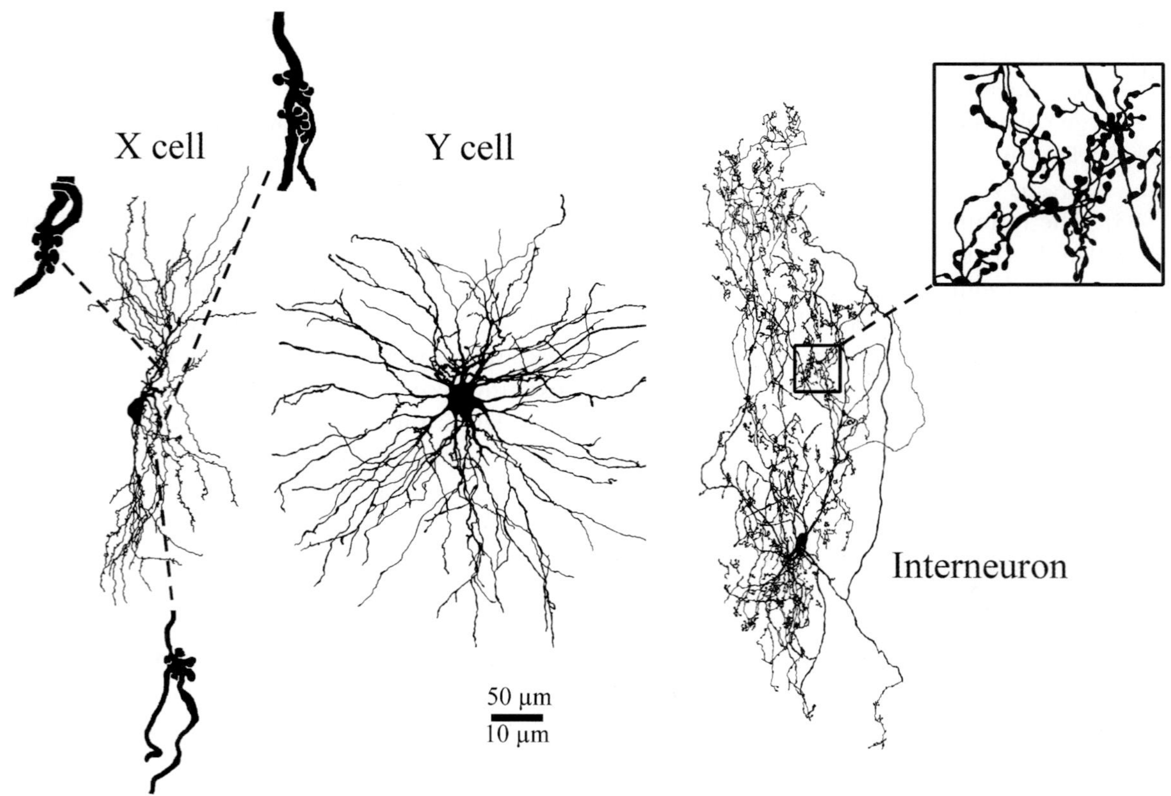

FIGURE 19.3   Reconstruction of X cell, Y cell, and interneuron from A layers of the cat's lateral geniculate nucleus. The larger scales are for the insets for the X cell and interneuron.

compared to Y cells, X cells have smaller receptive fields, more linear summation properties, and respond better to higher spatial frequencies but more poorly to higher temporal frequencies.

Morphologically, there are some differences between these relay cell types (Friedlander et al., 1981; Guillery, 1966; LeVay & Ferster, 1977; Wilson, Friedlander, & Sherman, 1984). At the light microscopic level (see figure 19.3), Y cells have larger cell bodies and smooth dendrites that have cruciate branches in a relatively spherical arbor, with peripheral segments of dendrites often crossing from one layer to another. X cells have arbors that tend to be bipolar and oriented perpendicular to the borders of the layers. Their dendrites also have numerous clusters of grape-like appendages located mostly near primary branch points (figure 19.3). The functional significance of these morphological differences remains to be defined, although the clustered appendages, which are more characteristic of X than Y cells (Wilson, Friedlander, & Sherman, 1984; Hamos et al., 1987; Datskovskaia, Carden, & Bickford, 2001) and are particularly prominent in the lateral geniculate nucleus of the cat, represent the postsynaptic site of retinal inputs, where complex synaptic relationships known as triads (see below) are formed.

INTERNEURONS Interneurons have the smallest cell bodies in the A layers and long, sinuous dendrites whose arbors are oriented perpendicular to the layers, often spanning an entire layer (see figure 19.3). The dendrites have the appearance of terminal axonal arbors and for that reason have been described as axoniform (Guillery, 1966). Terminals of these dendritic arbors contain synaptic vesicles and are both presynaptic to local dendrites and also postsynaptic to other axons, generally those coming from either the retina or from brainstem (Erişir et al., 1997; Famiglietti & Peters, 1972; Hamos et al., 1985; Ralston, 1971). In addition, most, if not all, interneurons have conventional axons that terminate in the general vicinity of the dendritic arbor. The receptive fields of the few identified interneurons that have been studied are like those of X relay cells and unlike those of Y cells, which suggests that the retinal inputs responsible for the firing of the interneurons are X axons (Friedlander et al., 1981; Sherman & Friedlander, 1988).

Afferents

The major sources of inputs to the A layers, besides the retina, include the thalamic reticular nucleus, layer 6 of cortex, and the parabrachial region[3] of the midbrain. These are summarized in figure 19.4. Other afferent sources not shown in figure 19.4 include the nucleus of the optic tract (midbrain), the dorsal raphe nucleus (midbrain and pons), and the tuberomamillary nucleus of the hypothalamus (reviewed in Sherman & Guillery, 1996, 2006).

RETINAL AFFERENTS Retinal afferents to the A layers are glutamatergic (see figure 19.4). They are relatively thick axons and have a distinct terminal structure involving richly branched, dense terminal arbors with boutons densely distributed mostly in flowery terminal

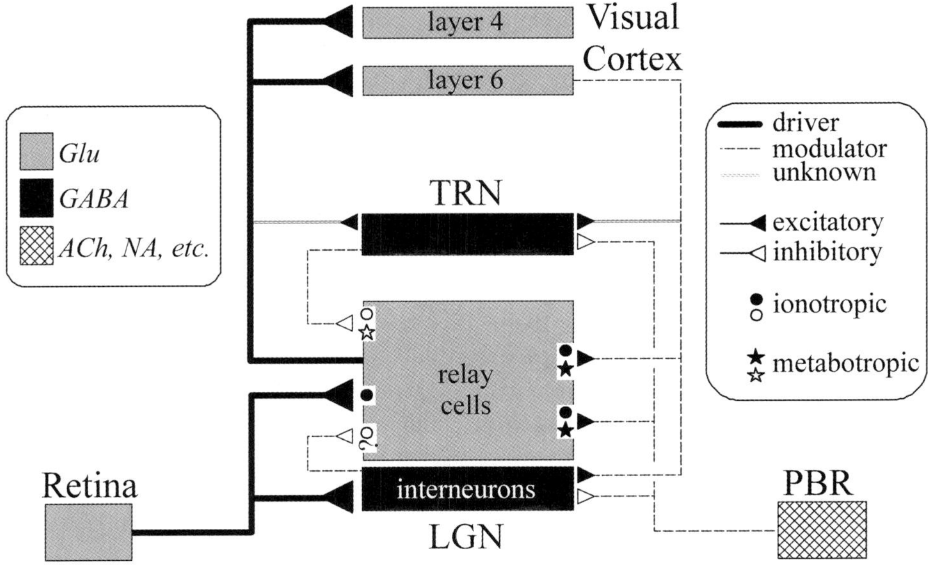

FIGURE 19.4 Neuronal circuitry related to A layers of the cat's lateral geniculate nucleus. Shown are the various inputs, the neurotransmitters associated with them, and the type of receptor, ionotropic or metabotropic, each activates. Also, driver versus modulator inputs are shown (see text for details).

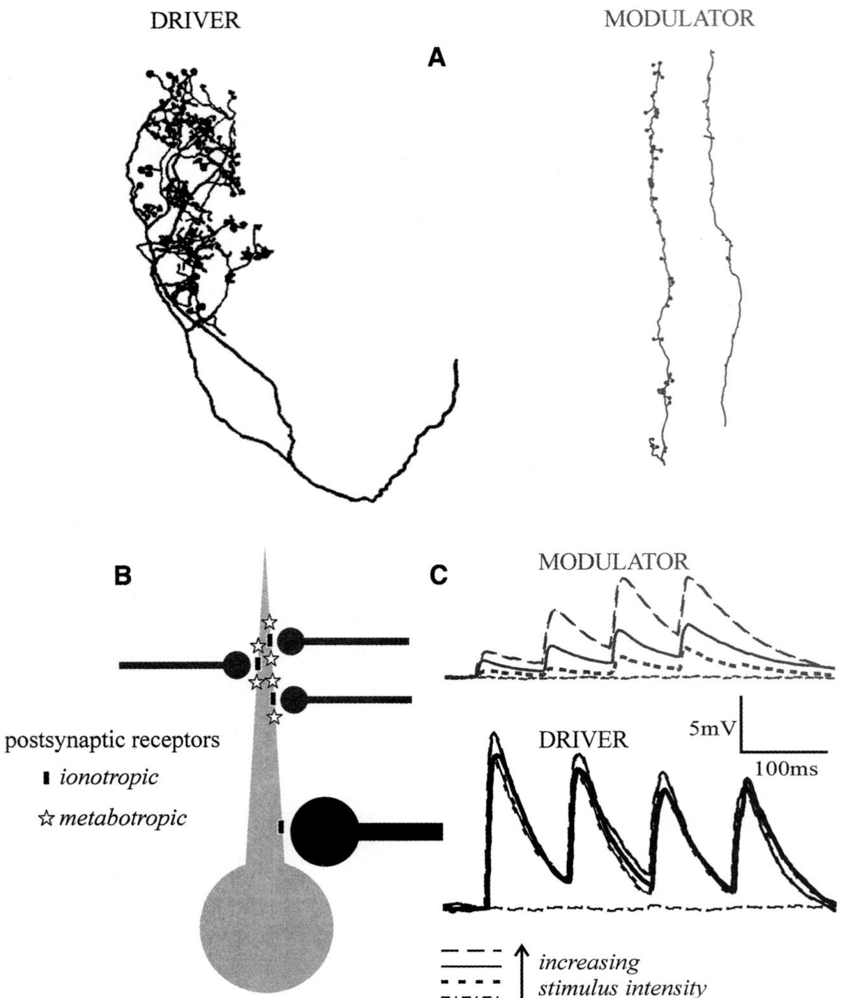

FIGURE 19.5   Distinguishing glutamatergic driver from modulator inputs to thalamus. (A) Light microscopic tracings of a driver afferent (a retinogeniculate axon from the cat) and a modulator afferent (a corticogeniculate axon from layer 6 of the cat). (B) Modulators (gray) shown contacting more peripheral dendrites than do drivers (black). Also, drivers activate only ionotropic glutamate receptors, whereas modulators also activate metabotropic glutamate receptors. (C) Effects of repetitive stimulation on EPSP amplitude: for modulators it produces paired pulse facilitation (increasing EPSP amplitudes during the stimulus train), whereas for drivers it produces paired pulse depression (decreasing EPSP amplitudes during the stimulus train). Also, increasing stimulus intensity for modulators (shown as different line styles) produces increasing EPSP amplitudes overall, whereas for drivers it does not; this indicates more convergence of modulator inputs compared to driver inputs. (Redrawn from Sherman & Guillery, 2011.)

clusters (Guillery, 1966). In contrast, most nonretinal inputs described below are thinner and have an equally distinct structure with smaller terminals en passant or on short side branches (see figure 19.5A). The retinal axons innervate both relay cells and interneurons in the A layers. The terminal arbors of retinal Y axons are much larger than those of X axons and give rise to many more synaptic terminals (Bowling & Michael, 1984; Sur et al., 1987). All retinal X and Y axons innervating the lateral geniculate nucleus branch to innervate the midbrain as well (a point that is discussed further below; see also Guillery & Sherman, 2002a,b,

2011; Sherman & Guillery, 2006), but they do not innervate the thalamic reticular nucleus.

AFFERENTS FROM THE THALAMIC RETICULAR NUCLEUS
The thalamic reticular nucleus is a thin shell of GABAergic neurons that closely abuts the entire thalamus laterally, extending somewhat dorsally and ventrally. It derives from the ventral thalamus, together with the ventral lateral geniculate nucleus (see note 1) and is divided into sectors, each related to thalamic relay nuclei concerned with a particular modality or function (e.g., visual, auditory, somatosensory, and motor), each

mapped in a distinct sector of the nucleus (Crabtree, 1992a, 1992b, 1996; Crabtree & Killackey, 1989; Guillery, Feig, & Lozsádi, 1998; Montero, Guillery, & Woolsey, 1977). There are strong reciprocal connections between relay cells and reticular cells linking corresponding parts of the reticular and the geniculate maps (Gentet & Ulrich, 2003; Pinault, Bourassa, & Deschênes, 1995; Pinault & Deschênes, 1998; Uhlrich et al., 1991), and the cortical afferents from layer 6 (next section) are mapped along the same coordinates (Murphy & Sillito, 1996). Thus, the portion of the thalamic reticular nucleus innervating the lateral geniculate nucleus[4] is mapped in retinotopic coordinates.

In addition to this, the visual sector of the reticular nucleus is also reciprocally linked to the pulvinar of the bush baby (Conley & Diamond, 1990) and rat (Pinault, Bourassa, & Deschênes, 1995) with the pulvinar representation external to the geniculate representation. These cells of the thalamic reticular nucleus in the cat, which lie just external to layer A, have moderate to large cell bodies and dendrites oriented mostly parallel to layer A (Uhlrich et al., 1991). Their axons descend into the A layers, generally along the lines of projection, with terminal arbors that are moderately branched and contain numerous boutons, mostly en passant. These terminals innervate mainly geniculate relay cells with only a very sparse innervation to interneurons (Cucchiaro, Uhlrich, & Sherman, 1991; Wang et al., 2001). Thus, the thalamic reticular nucleus provides a potent inhibitory GABAergic input to relay cells (see figure 19.4). Their receptive fields tend to be larger than those of relay cells and are often binocular (So & Shapley, 1981; Uhlrich et al., 1991).

AFFERENTS FROM LAYER 6 OF THE CORTEX Cortical afferents from layer 6 of visual cortex (see also chapter 22 by Briggs and Usrey), which are glutamatergic, have thin axons in the lateral geniculate nucleus with most boutons located at the ends of short side branches (figure 19.5A) (see Murphy & Sillito, 1996). They are topographically organized, with each axon having terminal arbors passing roughly along lines of projection and across more than one layer. These axons enter the A layers after traveling through the thalamic reticular nucleus, where they also give off branches to innervate cells there, and this corticoreticular projection, too, is topographic.

AFFERENTS FROM THE PARABRACHIAL REGION Most of the input from the brainstem to the A layers derives from the parabrachial region (Bickford et al., 1993; de Lima & Singer, 1987). Most of these axons are cholinergic, but some are noradrenergic. Light

microscopically, they resemble the cortical afferents more than the retinal afferents, but their terminal arbors are more widespread, and most appear to terminate in a nontopographic fashion. These axons contact both relay cells and interneurons in the A layers and also branch to innervate cells in the thalamic reticular nucleus.

OTHER AFFERENTS Some other afferents to the A layers not shown in figure 19.4 have been described, but they are small in number, not well documented, and are mentioned only briefly here (for further details, see Sherman & Guillery, 1996, 2006). There is a limited serotonergic input from the dorsal raphe nucleus in the midbrain and pons. GABAergic cells of the nucleus of the optic tract in the midbrain also provide a limited input. The projection from the superior colliculus to the lateral geniculate nucleus targets the C layers nearly exclusively. Finally, the tuberomamillary nucleus of the hypothalamus provides a small histaminergic input.

POSTSYNAPTIC RECEPTORS In addition to showing the inputs and their transmitters onto relay cells, figure 19.4 also shows the associated postsynaptic receptors. Note that both ionotropic and metabotropic receptors are postsynaptic in relay cells. There are a number of differences between these two receptor types, but only a few concern us here (for details, see Conn & Pin, 1997; Mott & Lewis, 1994; Nicoll, Malenka, & Kauer, 1990; Pin & Duvoisin, 1995; Recasens & Vignes, 1995).

Ionotropic receptors (iGluRs) include AMPA and NMDA receptors for glutamate, $GABA_A$ for GABA, and nicotinic receptors for acetylcholine. These receptors are complex proteins found in the postsynaptic membrane, and when the transmitter contacts the receptor, it leads to a rapid conformational change that opens an ion channel, leading to transmembrane flow of ions and a change in the postsynaptic potential. Ionotropic responses are rapid and brief, typically with a latency for postsynaptic potentials of <1 ms and a duration of 1 ms up to a few tens of milliseconds. Metabotropic receptors include various glutamate receptors (mGluRs), $GABA_B$, and various muscarinic receptors for acetylcholine. These are not directly linked to ion channels but instead involve a series of complex biochemical reactions after transmitter contact, and these reactions ultimately lead to the opening or closing of an ion channel, among other postsynaptic events. For thalamic cells this is primarily a K^+ channel that, when opened, produces an IPSP as K^+ flows out of the cell and, when closed, produces an EPSP as leakage of K^+ is reduced. However, these postsynaptic responses are slow and prolonged, with a latency of >10 ms and a duration of hundreds of milliseconds to several seconds. Also, in general, metabotropic receptors require higher firing rates from inputs to be activated; this is thought to be related to the observation in electron micrographs that these receptors are located perisynaptically, slightly further from the synaptic site than are ionotropic receptors, so that more transmitter must be released to reach them (Lujan et al., 1996).

As shown in figure 19.4, retinal inputs activate only ionotropic receptors, whereas all nonretinal inputs activate metabotropic receptors, and some of these also activate ionotropic receptors (reviewed in Sherman & Guillery, 1996, 2006).

The prolonged metabotropic responses have a number of important effects. Because retinal inputs activate only ionotropic receptors, their EPSPs are relatively fast and brief. This has the virtue that the rate of firing in the retinal afferents can reach relatively high levels before temporal summation of the EPSPs occurs, and thus each retinal action potential has a unique postsynaptic response associated with it. The prolonged metabotropic responses act as a low-pass temporal filter, and thus higher temporal frequencies are not faithfully transmitted. So the fact that retinal inputs activate only ionotropic receptors serves to produce a more faithful transfer of higher temporal frequencies. However, it is not clear whether any individual nonretinal axon can

activate both ionotropic receptors and metabotropic ones. Nonetheless, the activation of metabotropic receptors means that these inputs can create sustained changes in the baseline membrane potential, which, among other things, means that these inputs can have sustained effects on overall responsiveness of relay cells. Other consequences of these sustained postsynaptic responses are considered below.

SYNAPTIC STRUCTURES   Over 95% of all synaptic terminals in the A layers can be placed into one of four categories (reviewed in Sherman & Guillery, 1996, 2006): (1) RL (round vesicle, large profile) terminals, which are the retinal terminals, are the largest terminals in the A layers. They form asymmetric[5] contacts consistent with their identity as excitatory inputs and each terminal forms many contacts. (2) RS terminals (round vesicle, small profile) are smaller than RL terminals but also form asymmetric contacts, rarely more than one. The vast majority of these come from either layer 6 of cortex or from the parabrachial region. (3) F1 terminals (flattened vesicles) form symmetric contacts consistent with their origin from the GABAergic axons of reticular cells or interneurons. (4) F2 terminals represent the dendritic outputs of interneurons, and they also have flattened vesicles and form symmetric contacts. Unlike all of the other terminals that are strictly

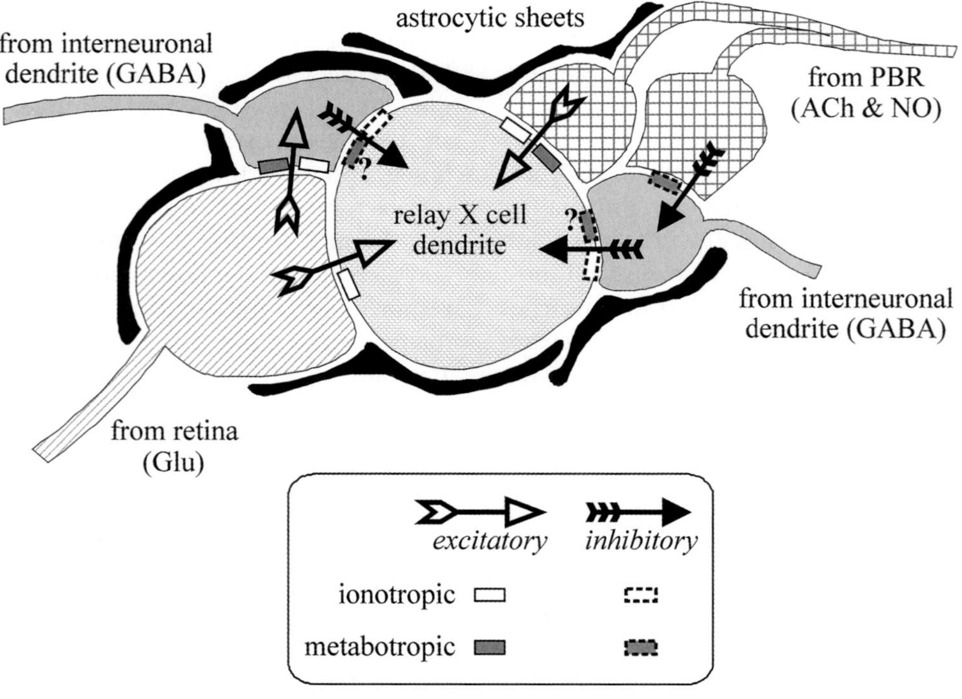

FIGURE 19.6   Schematic view of triadic circuits in a glomerulus of the lateral geniculate nucleus in the cat. The arrows indicate presynaptic-to-postsynaptic directions. The question marks postsynaptic to the dendritic terminals of interneurons indicate that it is not clear whether or not metabotropic (GABA$_B$) receptors exist there.

presynaptic, these are both presynaptic and postsynaptic, with inputs either from retinal or parabrachial terminals.

Triadic synaptic arrangements involving F2 terminals are common in the A layers (see figure 19.6). In most triads, an RL terminal contacts both an F2 terminal and the dendrite of a relay cell, and the F2 terminal contacts the same relay cell dendrite. A slightly different kind of triad can be formed by a parabrachial terminal contacting an F2 terminal and a different parabrachial terminal from a preterminal branch of the same axon contacting a relay cell dendrite, again with the F2 terminal contacting the same relay cell dendrite (see figure 19.6). Nearly all F2 terminals are involved in one or the other form of triad. Curiously, these triads are quite common for relay X cells and rare for Y cells, the latter thus receiving very few inputs from F2 terminals. Triads are typically found in complex synaptic zones that lack astrocytic processes but that are surrounded by sheets of astrocytic cytoplasm; these are called *glomeruli*. It is not at all clear how the triads function, but several suggestions have recently been offered (Guillery & Sherman, 2002b; Sherman, 2004). The arrangement of the synaptic inputs for the drivers that transmit the message to cortex for other modalities are essentially the same as those described here for the retinal inputs. We treat these as characteristic for all thalamus, including pulvinar, of inputs that carry messages for transfer to cortex.

DISTRIBUTION OF INPUTS TO RELAY CELLS   The dendritic arbors of relay cells can be divided into two distinct sectors with little or no overlap (Wilson, Friedlander, & Sherman, 1984; Erişir et al., 1997): a proximal region (up to about 100 μm from the cell body or generally close to the first branch point) and a distal region (further than about 100 μm from the cell body). Retinal terminals contact the former region, whereas cortical terminals contact the latter. F2 and parabrachial terminals also contact relay cells in the proximal zone. Axonal inputs from interneurons mostly contact the proximal zone, whereas those from reticular axons mostly contact the distal zone.

Only a small minority of synaptic inputs onto geniculate relay cells derive from the retina. In the A layers of the cat's lateral geniculate nucleus, for instance, about 5–10% of the synaptic input to relay cells comes from the retinal axons; roughly 30% comes from local GABAergic cells (interneurons plus reticular cells), 30% from the cortical input, and 30% from the parabrachial region (Erişir, Van Horn, & Sherman, 1997; Van Horn, Erişir, & Sherman, 2000; Wilson, Friedlander, & Sherman, 1984). If one had only the anatomical data, and for many other thalamic relays that is all we have, one might well conclude that the numerically small retinal input plays only a minor role in geniculate functioning. Because we have functional data, mainly in the form of receptive field comparisons, we know that it is the retinal input that provides the main information relayed to cortex, so we accept that the small number of retinal afferents serve as the inputs carrying the message that is relayed to cortex, and we refer to these inputs as the *drivers* (see Drivers and Modulators below).

Intrinsic Properties of Thalamic Cells in the A Layers

There are three factors that largely control retinogeniculate transmission, and they are considered below. First are the intrinsic membrane properties of relay cells, including their passive and active membrane properties, because these determine the effects of retinal EPSPs at the cell body or the region of action potential generation. Second is the geniculate circuitry that, by affecting many of the intrinsic membrane properties, also controls how retinal EPSPs lead to relay cell firing. Third, the nature of the postsynaptic receptors largely determines the postsynaptic response of relay (and other) cells to their active inputs; this feature is considered below.

Generally, all thalamic cells show a wide range of intrinsic membrane properties that are found generally in neurons of the brain (reviewed in Sherman & Guillery, 1996, 2006). These include passive cable properties, voltage-sensitive and -insensitive conductances, and conductances sensitive to other factors such as Ca^{2+} concentration. The conductances underlie transmembrane currents, including a leak K^+ current ($I_{K[leak]}$) that helps control resting membrane potential, various voltage- and Ca^{2+}-gated K^+ currents (I_A, $I_{K[Ca^{2+}]}$, etc.) and a voltage-gated cation current (I_h). Because these are properties found widely in the brain, they are not considered further here. (Additional details of these as they apply to thalamic neurons can be found in Sherman & Guillery, 1996, 2006.)

One feature that is of particular interest is the presence in all thalamic relay cells of a voltage-gated Ca^{2+} conductance based on T-type (for transient) Ca^{2+} channels that, when activated, leads to a current (I_T) large enough to produce an all-or-none Ca^{2+} spike (reviewed in Sherman & Guillery, 1996, 2006; see also chapter 32 by Swadlow and Alonso). This spike is large enough to induce a high-frequency burst of 2–10 action potentials riding its crest, and when this occurs, the cell is said to respond in *burst mode*; when I_T is inactivated, the cell responds with single action potentials, and this is known as *tonic mode*. The voltage and time requirements of the two modes are as follows (see figure 19.7). I_T becomes inactivated after the cell has been depolarized above about –65 mV for roughly 100 ms,[6] and the cell then responds to a suprathreshold depolarizing input (e.g., an EPSP) with action potentials that appear throughout the period of depolarization (figure 19.7A). If, however, the cell is suitably hyperpolarized for 100 ms or so, I_T is deinactivated and primed to fire a Ca^{2+} spike, so the next sufficient depolarization evokes the Ca^{2+} spike, leading to a brief burst of action potentials (figure 19.7B). Note that the same depolarizing pulse in figure 19.7A, B leads to very different patterns of response. It is also important to emphasize the point that only the conventional action potentials are transmitted to cortex: the Ca^{2+} spike propagates along the membranes of the dendritic tree and soma, but not up the axon, because the requisite T-type Ca^{2+} channels are available in the dendritic membranes but not in those of the axons. Thus, the means by which the Ca^{2+} spike affects thalamocortical transmission is through its activation of conventional action potentials.[7]

There are at least three clear consequences of these firing modes for thalamic relays. The first, as shown in figure 19.7C, is that the tonic mode is much more linear, meaning that the postsynaptic response rises monotonically with the size of the input (Zhan et al., 1999). This occurs because the greater the input depolarization or EPSP, the more action potentials are produced. However, with burst firing, the action potentials are not evoked directly from the input EPSP but rather from the Ca^{2+} spike, and because this spike is all-or-none, once the EPSP is large enough to reach threshold for this spike, larger EPSPs do not evoke a larger Ca^{2+} spike, and so the input–output relationship is more like a step function than a linear relationship (figure 19.7C). Second, because burst firing occurs after a period of hyperpolarization, the burst of action potentials occurs against a background of low spontaneous firing

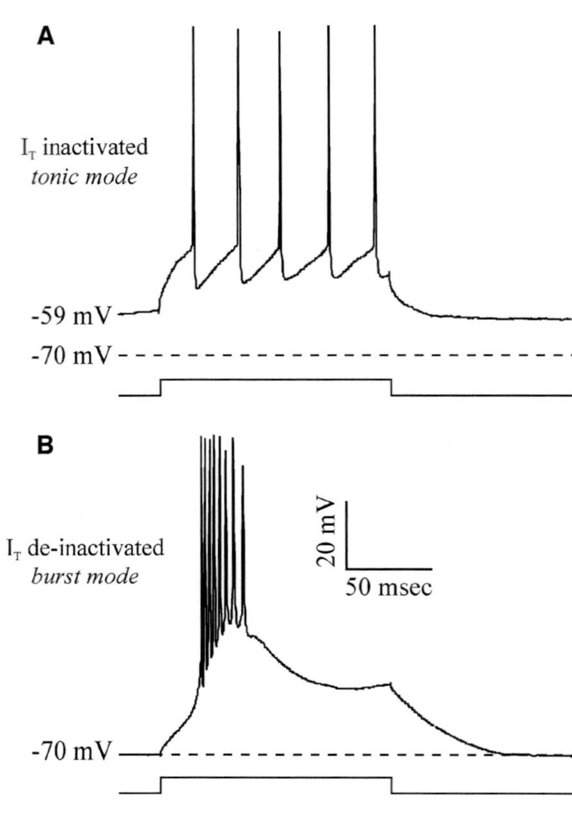

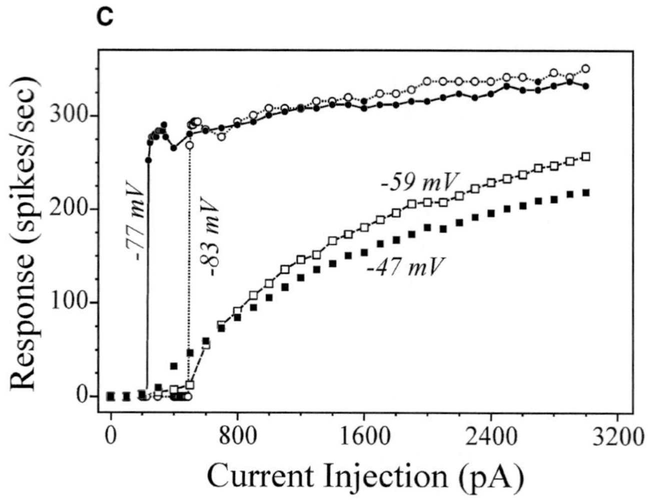

FIGURE 19.7 Properties of burst and tonic firing. (A, B) Voltage dependency of the low-threshold Ca^{2+} spike for a geniculate relay cell recorded intracellularly in vitro. Responses are shown to the same depolarizing current pulse administered intracellularly but from two different initial holding potentials. With relative depolarization (A), I_T is inactivated, and the response is a barrage of unitary action potentials lasting for the duration of the suprathreshold stimulus. This is the *tonic mode* of firing. With relative hyperpolarization (B), I_T is deinactivated, and the response is a low-threshold spike with eight action potentials riding its crest. This is the *burst mode* of firing. (C) Input–output relationship for another geniculate relay cell recorded intracellularly in vitro. The input variable is the amplitude of the depolarizing current pulse, and the output is the evoked firing frequency determined by the first six action potentials of the response, because this cell usually exhibited six action potentials per burst in this experiment. The initial holding potentials are shown: –47 mV and –59 mV reflect tonic mode, whereas –77 mV and –83 mV reflect burst mode. (Redrawn from Sherman & Guillery, 2002.)

compared to tonic mode. Spontaneous firing can be regarded as noise, and as such the signal-to-noise ratio of burst firing is considerably greater than that of tonic firing, so that the thalamic response and thus the signal that is passed to cortex, is more detectable (Sherman, 1996). Third, because a burst can occur only after I_T has been deinactivated, and this requires a 100-ms or so period of sustained hyperpolarization, it follows that the cell cannot have fired action potentials in the period before a burst.

Geniculocortical synapses show the property of paired-pulse depression (explained more fully below in Drivers and Modulators), meaning that an action potential produces a smaller EPSP if it follows another within about 100 ms or so. During tonic firing, when rates usually exceed 10 action potentials/s, the geniculocortical synapse will usually be depressed; however, a burst of action potentials would arrive at the thalamocortical synapse when it has been relieved of depression, and thus the postsynaptic response evoked would be greater. This, in turn, predicts that the first action potentials in a burst should evoke a greater response in cortex than a typical tonic action potential, and this indeed occurs (Swadlow & Gusev, 2001; Swadlow, Gusev, & Bezdudnaya, 2002; see also chapter 32 by Swadlow and Alonso).

Overall, these differences suggest that burst firing produces a larger signal that is more readily detected in cortex compared to tonic firing. However, the more linear responses of tonic firing suggest that this represents a more faithful relay mode for information transfer. Together, these differences have led to the hypothesis that burst firing can provide a "wake-up call" to cortex to strongly signal that a novel stimulus has occurred after a quiescent period; once this signal has been detected, circuitry can then be brought to bear to depolarize the relay cell to switch to tonic mode so that further presence of the novel stimulus can be faithfully relayed (Sherman, 1996).

CLASSIFICATION OF GLUTAMATERGIC INPUTS

In the section on Afferents above, we described basic features of afferent input to the lateral geniculate nucleus, but here we wish to continue this discussion with a different emphasis, namely to consider the properties of these various inputs and the classification these properties lead to. It follows from a consideration of geniculate circuitry that not all thalamic afferents are equal in their action. Inputs to geniculate relay cells include various transmitters, such as GABA, ACh, NA, and 5-HT, which are commonly considered as modulatory, but as regards a classification of inputs, that

involving an identification of transmitters involved is only a first step to a useful classification.

Drivers and Modulators

We normally regard glutamatergic inputs as the main carrier of information, but it is clear that the two main glutamatergic inputs to geniculate relay cells, from retina and layer 6 of cortex, are not functionally equivalent. A consideration of the receptive field properties of geniculate relay cells helps to clarify this point: their receptive fields, and thus the information they relay to cortex, closely match those of their retinal inputs and not, for instance, those of cortical layer 6 inputs. The layer 6 feedback evokes weak responses (Granseth & Lindström, 2003) that serve to modulate retinogeniculate transmission without major *qualitative* changes in receptive field properties (Sherman, 2007; Sherman & Guillery, 2006). Removal of the layer 6 input causes only minor effects on geniculate receptive fields (e.g., Baker & Malpeli, 1977; Geisert, Langsetmo, & Spear, 1981; Kalil & Chase, 1970; McClurkin & Marrocco, 1984), effects that can be recognized as modulation (see below). We have thus distinguished two different glutamatergic types of thalamic input (Sherman & Guillery, 1998, 2006): *drivers* and *modulators*.[8] The drivers are the information-bearing input that is to be relayed to cortex, and in the lateral geniculate nucleus this is the retinal input. All other inputs, including the cortical layer 6 input, are modulators. Details of how drivers are distinguished from modulators for glutamatergic inputs are provided in detail elsewhere (Sherman & Guillery, 1998, 2006) and summarized briefly here (see figure 19.5 and table 19.1).[9]

• Drivers activate only ionotropic receptors, mainly AMPA but some NMDA, whereas modulators in addition activate metabotropic receptors.

• Drivers produce larger initial EPSPs that show paired-pulse depression, indicating a high probability of transmitter release, whereas modulators produce smaller initial EPSPs that show paired-pulse facilitation, indicating a low probability of transmitter release (Dobrunz & Stevens, 1997).

• The available counts of afferent synapses from anatomical studies combined with a comparison of the all-or-none activation of driver inputs with the graded activation of modulators indicate that, although the driver synapses form a minority of the inputs to the target neurons, they dominate the action of the target neurons. For instance, corticogeniculate modulator inputs produce 5–10 times as many synapses as do retinal driver inputs to geniculate relay cells, and yet

Differences between drivers and modulators for glutamatergic inputs

Driver (e.g., retinal)	Modulator (e.g., layer 6)
Large EPSPs	Small EPSPs
Synapses show paired-pulse depression	Synapses show paired-pulse facilitation
Less convergence onto target	More convergence onto target
Dense terminal arbors (type 2)	Sparse terminal arbors (type 1)
Thick axons	Thin axons
Large terminals	Small terminals
Contacts target cell proximally	Contacts target cell peripherally
Activates only iGluRs	Activates iGluRs & mGluRs

driver inputs are functionally dominant (reviewed in Sherman & Guillery, 2006). Thus, assessing the relative strength of inputs based solely on anatomical numbers can be very misleading.

• Drivers have thicker axons with larger terminals that contact proximal dendrites and are distributed in denser, more tightly localized terminal arbors, and their terminals in thalamus have a characteristic light and electron microscopic appearance and synaptic relationships often involving triadic relationships in glomeruli (see above).

Some Effects of Modulation

The function of the driver input to thalamic relay cells seems obvious and straightforward; its role is to provide the main information to be relayed, that is, to carry a message about events in the brain, the body, or the world for relay to cortex. Modulatory functions are more varied and complex. That for traditional modulatory inputs, such as GABA and ACh, is often related to overall effects on excitability via inhibitory or excitatory actions, but by activating metabotropic receptors, these effects can be prolonged. The same thus applies to glutamatergic modulatory inputs via the activation of metabotropic glutamate receptors.

Because activation of metabotropic receptors produces membrane potential changes that typically last >100 ms (see above), such activation provides the necessary time and voltage shift needed to control various voltage- and time-gated ion channels. These have been described above, and the properties of the T-type Ca^{2+} channels that underlie the burst/tonic response modes help to show the importance of the prolonged changes characteristic of the metabotropic receptors. To

inactivate the channel requires a sustained (>100 ms) depolarization. Activation of ionotropic receptors such as AMPA (glutamate) or nicotinic (ACh) receptors produces brief EPSPs with little effect on the inactivation state of this channel. However, activation of appropriate muscarinic (ACh) or metabotropic glutamate receptors produces a sufficiently long EPSP to inactivate the channel. Likewise, activation of the $GABA_A$ receptor (ionotropic) produces a brief IPSP that will not lead to much deinactivation of the channel, but activation of the $GABA_B$ receptor (metabotropic) produces a sustained IPSP that will do so.

There is evidence from cortex that activation of metabotropic glutamate receptors also affects the size of evoked EPSPs from driver inputs (Mateo & Porter, 2007; DePasquale & Sherman, 2013). Data from the lateral geniculate nucleus suggest that this may is the case for the retinogeniculate synapse (Govindaiah, et al., 2012; Lam & Sherman, 2013). Here activation of metabotropic glutamate receptors located on the retinal terminals, via glutamate released either from the retinal terminals themselves or from the layer 6 modulator feedback input (Lam & Sherman, 2013), reduces the retinogeniculate EPSP amplitude.

THE FUNCTIONAL ORGANIZATION OF THE PULVINAR

The Pulvinar as a Visual Relay

MAPS IN THE PULVINAR   The pulvinar, like the lateral geniculate nucleus, receives topographically organized representations of the contralateral visual hemifield, with the region receiving from area 17 forming a mirror reversal of the geniculate map and also receiving inputs from other visual areas, including areas 18 and 19 in the cat (Berson & Graybiel, 1978; Guillery, Feig, & Van Lieshout, 2001). Lines of projection are identifiable, as in the lateral geniculate nucleus of the cat, on the basis of the driver inputs and of the thalamocortical connections, and it appears that each line of projection receives from more than one cortical area suggesting that in the pulvinar these lines represent the arrangement of distinct cortical functions, and these all relate to the same small area in visual space, but their functions are not identified at present.

PULVINAR CIRCUITRY   Although much less is known about the pulvinar than about the lateral geniculate nucleus, it provides an important pathway to many, possibly to all, higher visual cortical areas. Many or all of the cells in this complex have visual receptive fields (Bender, 1982; Casanova & Savard, 1996; Chalupa,

1991; Hutchins & Updyke, 1989), and their links with extrastriate visual cortical areas have long been recognized (Jones, 2007), but the functional nature of this link has become clear only more recently. In order to appreciate the nature of this link, it is important to recall that the lateral geniculate nucleus receives modulatory afferents from layer 6 of cortex, and driving afferents, providing the visual inputs, from the retina. These two types of afferents have been described above, and it was shown that they are clearly distinguishable in terms of their properties (see table 19.1) and their functions as driver or modulator.

Experiments using retrograde tracers have shown that the pulvinar receives afferents from layers 5 and 6 of visual cortex (Abramson & Chalupa, 1985), whereas the lateral geniculate nucleus receives afferents only from layer 6 of cortex but not from layer 5 (Gilbert & Kelly, 1975). The question of whether any part of the pulvinar receives direct retinal input is unclear.[10] Injections of anterograde tracers that labeled axons of individual cells in layer 6 of area 17 show that the terminals of these axons in the pulvinar have the structural characteristics of the modulatory corticogeniculate afferents described above for the lateral geniculate nucleus (see figure 19.5). Limited data from slice preparations of rodents indicate that the synaptic properties of the layer 6 input to the pulvinar are the same as those to the lateral geniculate nucleus: In the rat evoked synaptic properties from activating corticothalamic axons include the property of paired-pulse facilitation (see table 19.1), although the layer of origin for this input could not be determined (Li, Guido, & Bickford, 2003); in the mouse, layer 6 input to the posterior medial nucleus, which can be considered as the somatosensory thalamic equivalent of the pulvinar, shows the properties of a modulator (Reichova & Sherman, 2004).

In contrast to the layer 6 cells, which relate to pulvinar much as they do to the lateral geniculate nucleus, individual layer 5 cells in area 17 have been shown to send no labeled axons to the lateral geniculate nucleus (Bourassa & Deschênes, 1995; Rockland, 1996) but do send axons that terminate in the pulvinar, where they have the characteristics of the retinogeniculate driver afferents described above, both in terms of their light microscopical appearance (Bourassa & Deschênes, 1995; Ojima, Murakami, & Kishi, 1996) and synaptic arrangements seen with the electron microscope (Mathers, 1972; Robson & Hall, 1977b; Ogren & Hendrickson, 1979; Feig & Harting, 1998). Evidence from slice preparations of rodents indicates that the synaptic properties of these layer 5 corticothalamic afferents are very much like those of retinogeniculate inputs (Li, Guido, & Bickford, 2003; Reichova & Sherman, 2004;

Theyel et al., 2010). We thus regard them as the drivers of the pulvinar cells, and it is because these drivers come from areas of cortex classifiable as "visual" that we see visual receptive fields in the pulvinar. Further evidence that these layer 5 afferents function as drivers is provided by the fact that silencing the cortical areas that send layer 5 afferents to the pulvinar relay abolishes (Bender, 1983) or greatly diminishes (Chalupa, 1991) the visual responses of the pulvinar cells and that the receptive field properties of many pulvinar cells are not unlike the receptive field properties of cells in cortical layer 5 (Chalupa, 1991).

Besides the distinction between layer 5 and 6 inputs to pulvinar as driver or modulator there is a difference in their patterns of termination in the thalamus. The layer 6 projection is organized as a feedback, innervating the same regions of pulvinar from which it receives its thalamic input, whereas the layer 5 projection is organized in a feedforward pattern (Van Horn & Sherman, 2004). Also, whereas the layer 6 afferents send a rich innervation to the thalamic reticular nucleus, the layer 5 afferents do not but instead send branches to lower extrathalamic centers.

The Pulvinar as a Higher Order Thalamic Relay

The fact that the pulvinar receives driving afferents from layer 5 of visual cortex shows that the pulvinar serves as a relay in the visual pathways for messages that have already been through cortical processing at least once. For this reason the pulvinar has been called a higher order visual relay, in contrast to the first order relay in the lateral geniculate nucleus, which transfers ascending messages directly and for the first time to cortex (Guillery, 1995; Sherman & Guillery, 1996, 2002, 2006). We define a first order thalamic relay as one that receives its driver input from a subcortical source (e.g., retina) with no driver inputs from cortex and a higher order one as receiving a significant part of its driver input from layer 5 of cortex. This distinction between first and higher order relays is found not only for the visual pathways but for other relays to cortex as well (reviewed in Sherman & Guillery, 2006, 2011). However, only the visual relays concern us here. One important point about the higher order visual relays is that they involve a much greater volume of thalamus and also a much greater area of cortex than does the first order visual relay in the thalamus. That is, the pulvinar is far larger than is the lateral geniculate nucleus, and the areas of cortex in receipt of inputs from the pulvinar are, in total, far greater than area 17.

We have described the pulvinar as providing higher order relays to extrastriate cortical areas. However, the possibility that there may also be first order pulvinar relays of ascending afferents has not been excluded. The small direct input to the pulvinar from the retina was mentioned earlier, but, as pointed out in note 10, we regard this as a part of the lateral geniculate nucleus. The input from the superior colliculus and pretectum to a part of the pulvinar raises another issue. Are these driving or modulatory inputs?

There were strong arguments in the past (Diamond, 1973; Schneider, 1969; Sprague, 1966, 1972; Sprague, Berlucchi, & Di Berardino, 1970) for the view that there are two parallel visual pathways going to the cortex, one from retina through the lateral geniculate nucleus to cortex and the other from retina via the superior colliculus to the pulvinar and then to cortex. This was based on behavioral studies primarily in cat, hamster, and tree shrew and on anterograde tract-tracing studies that demonstrated axonal pathways from the region of the colliculus to the pulvinar (Altman & Carpenter, 1961). One logical flaw with this notion is that the superior colliculus in mammals receives massive input from both retina and visual cortex, and the latter seems to provide an input, perhaps driver, that is required for normal collicular responses to visual stimuli (Wickelgren & Sterling, 1969), and so this raises doubts about independent pathways from retina through the superior colliculus to cortex. Further, the collicular receptive fields in a normal cat resemble those of cortical cells and not of retinal cells, but after cortical silencing they resemble those of retinal cells (Wickelgren & Sterling, 1969), suggesting that if the colliculus does send information to the pulvinar for relay to cortex, it is information about cortex rather than about colliculus.

More recently, a pathway in the monkey from the superior colliculus through the pulvinar to the MT (medial temporal) area of cortex was defined (Berman & Wurtz, 2010). Two problems need to be resolved with such studies. One is the question as to whether this input to pulvinar is actually a driver or modulator. The other is technical: when stimulating the tectothalamic pathway or labeling it for anatomical study from the superior colliculus, it may be that, instead of activating or labeling these axons, the layer 5 axons that branch to innervate both the superior colliculus and pulvinar are activated or labeled (Bourassa & Deschênes, 1995).

The available evidence as to whether the tectothalamic pathway is driver or modulator (or other) is incomplete. Recordings from cells in the pulvinar of cats and monkeys have shown that the receptive field properties of cells there are dependent on cortical inputs, not on collicular inputs (Bender, 1983; Chalupa, 1991; Chalupa, Anchel, & Lindsley, 1972), suggesting that the collicular inputs are not drivers innervating a pulvinar first order relay (see also Smith & Spear, 1979). The morphological evidence on the structure of the tectopulvinar connections is conflicting (Mathers, 1971; Partlow, Colonnier, & Szabo, 1977; Robson & Hall, 1977a,b), with some reports showing terminals like those of the layer 6 modulators and others showing terminals like those of the layer 5 drivers and the retinal terminals (Kelly et al., 2003; Mathers, 1971; Partlow, Colonnier, & Szabo, 1977; Robson & Hall, 1977a).[11] This issue needs to be resolved. Because there are several distinct subdivisions of the pulvinar that are not easily compared across species, it is possible that there are some tectal drivers innervating some first order relays in some regions of the pulvinar, with other tectal inputs acting as modulators. It is possible that there are significant species differences in the extent to which such tectopulvinar driver afferents may or may not play a significant role in the transfer of visual information to higher cortical areas (Rodman, Gross, & Albright, 1989, 1990), but at present the nature of any information transmitted from colliculus to cortex through the pulvinar remains undefined.

The possibility that parts of the pulvinar may be in receipt of drivers from the tectum and also from layer 5 of cortex raises a different issue regarding the possible integration in thalamus of different driver inputs. Do they interact on single relay cells, or do they, like X cells and Y cells in the A layers of the cat's lateral geniculate nucleus, or ON and OFF center cells, form two essentially independent parallel pathways?

Driver Afferents to Pulvinar Are Branching Axons

Most or all driver inputs to both first and higher order thalamic relays, including the lateral geniculate nucleus and pulvinar, arrive via branching axons, with one or more extrathalamic branches that innervate brainstem or spinal motor centers (reviewed in Sherman & Guillery, 2006, 2011). Thus, most or all retinal axons that innervate the lateral geniculate nucleus branch to innervate midbrain regions involved in head and eye movements, accommodation, pupillary control, and so on. Details of experimental evidence for these retinofugal branching patterns can be found elsewhere (Guillery, 2003; Guillery & Sherman, 2002b, 2011). Furthermore, layer 5 corticothalamic axons from visual cortex to pulvinar also branch to innervate many of the same brainstem motor regions (Bourassa & Deschênes, 1995; Bourassa, Pinault, & Deschênes, 1995; Guillery, Feig, & Van Lieshout, 2001).

This branching pattern has suggested a novel and unexpected function for driver afferents to thalamus (Guillery & Sherman, 2011): the extrathalamic branches carry messages to motor centers, and the thalamic branches necessarily carry the same message.[12] This means that driver inputs to thalamus carry a copy of motor instructions currently on their way to subcortical centers. In this regard driver inputs to thalamus serve as efference copies[13] in the sense used by others (Sommer & Wurtz, 2004; Sperry, 1950; von Holst & Mittelstaedt, 1950) and provide information about forthcoming movements. An intriguing observation that appears to be related involves the "forward receptive fields" described for several cortical areas (Duhamel, Colby, & Goldberg, 1992; Sommer & Wurtz, 2006; Umeno & Goldberg, 1997), because these represent an anticipation of a movement, and such receptive fields may well depend on driver inputs to the appropriate thalamic relays carrying information in the form of efference copies. Further details of this hypothesis and its ramifications can be found in Guillery and Sherman (2011).

The Pulvinar Region as a Key Relay in Corticocortical Visual Communication

The distinction between first and higher order relays outlined above is based on the pathways from layer 5 cells, which provide the pulvinar with visual inputs from the cortex, and it is based on the evidence, summarized above, that these cortical inputs are drivers of pulvinar cells. That is, the pulvinar serves as a relay from one visual cortical area to other cortical areas, providing information about outputs to lower centers that are currently being sent by the first cortical area. Further, the functional parallel between retinal inputs to the lateral geniculate nucleus and cortical layer 5 inputs to the pulvinar provides a useful key for comparing these two visual relays, which we explore in subsequent sections. Before looking at these comparisons more closely, however, it is important to stress that the pulvinar, as a higher order relay, takes on a vitally important role as a key participant in corticocortical communication.

The different perspective introduced by these corticothalamic drivers is highlighted by figure 19.8, which contrasts the conventional view (figure 19.8A) with the alternative view offered here (figure 19.8B). In the conventional view, information enters striate cortex from the lateral geniculate nucleus and is then processed entirely within cortex, from primary sensory cortex through sensorimotor areas until finally reaching motor cortex, where needed motor commands are initiated. Analyses of perceptual and motor control mechanisms are commonly presented as going from the thalamus through a hierarchy of cortical areas for perceptual processing before they are passed to motor areas of

cortex (e.g., Andreas et al., 2001; Galletti et al., 2001). The processing on this interpretation strictly involves direct corticocortical connections among many discrete areas (over 30 in the monkey and probably fewer in the cat) organized into several (five or six in the monkey) hierarchical levels, with feedback as well as feedforward connections. Higher order thalamic relays such as the pulvinar are generally ignored in this conventional view except for a suggestion, without direct supporting evidence, that the pulvinar provides a modulatory influence on visual cortex in the service of attentional requirements (Olshausen, Anderson, & Van Essen, 1993; Van Essen, Anderson, & Felleman, 1992). As noted above (Driver Afferents to Pulvinar Are Branching Axons), the axons that go from cortical layer 5 represent one of several striking features that play no role in the current, conventional approaches schematized in figure 19.8A. One is that driver connections are seen in the transthalamic pathway from the primary visual cortex to the higher order thalamic relay (i.e., the pulvinar) and the target cortical areas in occipital, parietal, and temporal lobes. That is, the transthalamic pathway offers a novel route for the transmission of information from one cortical area to another (see also Theyel, Llano, & Sherman, 2010), and this must be seen as one essential part of the function of the pulvinar, which represents a far larger part of the thalamus than does the lateral geniculate nucleus. A further point is that the information that is passed from any one visual cortical area to another through a higher order thalamic relay is also a copy of descending messages that will act rapidly on motor pathways without the prior complex route through a hierarchical scheme of corticocortical connections to areas of motor cortex (Guillery & Sherman, 2011).

The important point for understanding corticocortical communication among visual cortical areas is that there are potentially important but largely unstudied and unrecognized transthalamic pathways that go from layer 5 in one cortical area through a pulvinar relay to another cortical area in the occipital, parietal, or temporal lobe. These transthalamic pathways, shown in figure 19.8B, are likely to differ in their functional properties from the more widely cited direct corticocortical routes (Van Essen, Anderson, & Felleman, 1992). Specifically, the thalamic relay has, as we have seen, properties that can modify or block transmission in accord with different functional needs, and because the direct corticocortical pathways lack such a thalamic relay, these properties are unavailable to the direct pathways.

There are a number of reasons why the thalamic input to higher cortical areas has received much less attention than has the direct corticocortical input. One

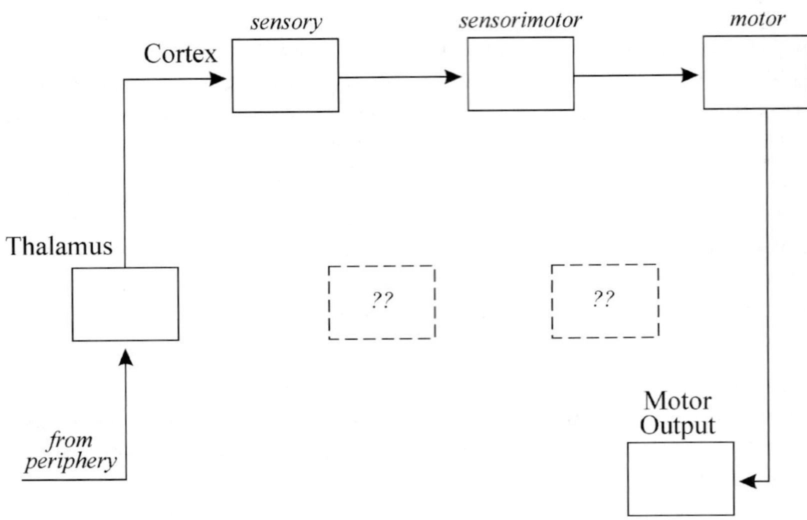

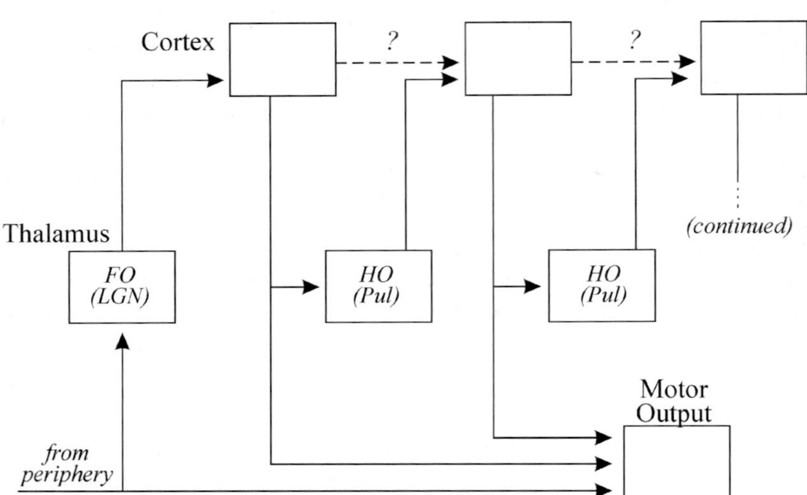

FIGURE 19.8   Comparison of conventional view (A) with the alternative view proposed here (B). FO, first order; HO, higher order; LGN, lateral geniculate nucleus; Pul, pulvinar. (Redrawn from Sherman, 2005.)

is that for many years it was simply not recognized as a possibility because the layer 5 and layer 6 afferents could not be distinguished from each other, and there was no reason to consider the layer 5 input to thalamus as a driver. A second reason concerns the numbers of axons involved. The thalamocortical afferents represent a relatively small group of afferents to cortex, and thus attention was directed at the apparently much more massive direct corticocortical connections. However, this consideration has to be viewed in relation to what we know about the first order visual relay, where the afferents from retina represent only about 5–10% of the synapses in the lateral geniculate nucleus (Van Horn, Erişir, & Sherman, 2000), and the geniculocortical afferents in area 17 also represent only about 5–10% of the synapses in cortical layer 4 (Ahmed et al., 1994; Latawiec, Martin, & Meskenaite, 2000). The modulators, in fact, far outnumber the drivers in these pathways, and, as noted above, a strategy that considered only the size of an input would not lead one to see the

   S. MURRAY SHERMAN AND R. W. GUILLERY

retinal input as the source of drivers to the lateral geniculate nucleus. The large number of synapses arising from modulators probably reflects the delicate adjustments that the modulators are capable of and may also indicate that there are modulatory functions that still remain to be explored; the numbers cannot be taken as a good indication of which pathway carries the information that the pathway is processing (e.g., reflected in the receptive fields in the visual pathways). Insofar as it is reasonable to expect some common organizational pattern to characterize all thalamocortical pathways, one should expect that a major information-bearing driver input to higher cortical areas will come from the thalamus, as it does for all first order cortical areas.

Another important and practical reason why the transthalamic corticocortical pathways have received much less attention than have the direct corticocortical pathways is that it is generally easier to explore the cortical surface than the depths of the thalamus, particularly when it comes to tracing the pathways. However, looking for evidence about the nature of corticocortical processing by studying the direct corticocortical pathways, which are readily accessible on the surface, and ignoring the deeper transthalamic pathways, which are likely to prove more difficult, can at best be justified by arguments such as those used by the proverbial drunk, searching for lost keys under the lamppost, where it was light.

So far as we know, all cortical areas receive thalamic afferents, and for almost all higher cortical areas, the functional contribution made by the thalamic afferents remains essentially unexplored. This in itself suggests that schemes tracing connections to primary cortical receiving areas, and from these through corticocortical pathways progressively to higher and higher cortical areas, for perceptual processing and eventually for motor outputs (e.g., Kandel, Schwartz, & Jessell, 2000; Van Essen, Anderson, & Felleman, 1992) or outputs to memory storage represent a false view about the nature of the cortical processes that relate to visual perception and its relation to movement control.

Not only, as indicated, do all cortical areas receive thalamic inputs, but most, probably all, have descending outputs from layer 5. Some of these layer 5 outputs have branches to thalamus, and some lack such branches, but those that innervate thalamus all have long descending extrathalamic branches (reviewed in Guillery & Sherman, 2011; Sherman & Guillery, 2006). Although the final destination of these descending branches is often undefined, many innervate regions with clear motor functions. The significance of knowing the functions of the descending branches for understanding the relevant transthalamic pathway is explored in the next section. Here it is important to look at these multiple output pathways from many cortical areas as a good reason for seeing cortical processing as being continually in touch with lower motor centers (figure 19.8B).

For example, area 17 sends axons to the superior colliculus, which is concerned with the control of head and eye movements (e.g., Tehovnik, Slocum, & Schiller, 2003), and there are comparable outputs from many other visual cortical areas that have connections with the pulvinar. The fact that area 17, like other primary sensory areas, sends a layer 5 output to motor centers starts to blur the distinction between sensory and motor cortex.

The Transpulvinar Corticocortical Pathway as a Monitor of Motor Outputs

We have seen that the messages received by higher order thalamic relay cells from layer 5 of cortex are also being sent to other centers where, directly or indirectly, they will have motor actions. That is, the relay cells of the pulvinar can be regarded as sending to cortex copies of motor instructions that are being sent out by visual cortex, not only by area 17 but also by many other visual areas that send layer 5 axons to the thalamus with branches to lower centers. This pattern of connectivity may seem surprising because it seems to turn the thalamic relay into a monitor of motor commands instead of just a sensory relay on the way to the cortex, which is how it has long been seen. The relationship is not special to the pulvinar. It can be seen in most or all thalamic relays, first order and higher order. That is, a detailed survey shows that most thalamic relays receive either afferents that are branches of axons that innervate motor centers or afferents that come from cells innervated by axons that have such branches (for a fuller account, see Guillery & Sherman, 2011; Sherman & Guillery, 2006). For the visual system, these connectivity patterns raise an important issue about the way in which activity in thalamocortical pathways is interpreted. Where one records activity that seems to have a close relationship to perceptual processing, one is likely, at the same time, also to be looking at activity that relates equally closely to motor control patterns, particularly in relation to eye movements, pupillary control, or accommodation.

Connectional and Cellular Properties in the Pulvinar Region

There is a basic similarity in the cell types seen in the pulvinar and the lateral geniculate nucleus. Relay cells and interneurons are distinguishable on the basis of the same criteria, and the general appearance of the synaptic zones is closely comparable (Feig & Harting, 1998; Mathers, 1972; Ogren & Hendrickson, 1979; Rockland, 1996, 1998). The afferents, too, are readily comparable to the afferents that innervate the lateral geniculate nucleus provided one recognizes that the drivers come from different sources: retina for the lateral geniculate nucleus and from layer 5 (and/or superior colliculus)[14] for the pulvinar. Both cell groups, in addition to their driving afferents, receive modulatory afferents from cortical layer 6, from the thalamic reticular nucleus, and from the brainstem.

There are some subtle differences, however, between the lateral geniculate nucleus and pulvinar as regards cell and circuit properties. Many of these differences are seen generally as common to first versus higher order relays:

• Electron microscopic studies indicate that the relative percentage of driver synapses is lower in pulvinar, being only 2% compared to about 7% for the lateral geniculate nucleus (Van Horn, Erişir, & Sherman, 2000; Van Horn & Sherman, 2007; Wang, Eisenback, & Bickford, 2002).
• Certain inputs relatively selectively target higher order relays, including the pulvinar. This includes a GABAergic input from the zona incerta (Lavallée et al., 2005; Power, Kolmac, & Mitrofanis, 1999) and a dopaminergic input from as yet unspecified sources (Garcia-Cabezas et al., 2007; Sanchez-Gonzalez et al., 2005).
• Modulatory serotonergic and cholinergic inputs that depolarize all first order relay cells, including those of the lateral geniculate nucleus, hyperpolarize a significant minority (about one-fifth) of higher order relay cells, including those of the pulvinar (Varela & Sherman, 2007, 2008).
• Higher order, including pulvinar, relay cells have a greater tendency to burst firing (Ramcharan, Gnadt, & Sherman, 2005), and this might be related to the above point that modulatory inputs hyperpolarize many higher order relay cells, which is a prerequisite to burst firing or to the recent observation that pulvinar relay cells have a higher density of T-type Ca^{2+} channels (Wei et al., 2011).

Properties of Pulvinar Connections to Cortex

A vital piece of information needed for understanding the function of the pulvinar is the pattern of projections from the relay cells to the various cortical areas that receive afferents from the pulvinar. This is not merely a question of enumerating the cortical areas that receive afferents from the pulvinar, although this information is, of course, essential but, unfortunately, very difficult to summarize from the literature at present. Studies of retrograde cell degeneration in the thalamus, of retrograde cell labeling, or of anterogradely labeled axonal pathways (Hackett, Stepniewska, & Kaas, 1998; Rockland et al., 1999; Walker, 1938; Wong et al., 2009; Wong-Riley, 1977) all indicate that there are widespread axonal projections from the pulvinar to the cortex. To some extent these studies indicate pathways from particular subdivisions of the pulvinar, but, in general, the information that allows one to relate each pulvinar subdivision to particular groups of cortical areas is not available, nor, where we have such information, do we know which are drivers and which are modulators.

We generally assume that pulvinar projections to cortex, like thalamocortical projections more generally, are feedforward and driver in nature. Indirect evidence from slices of mouse brain, in which analogous somatosensory and auditory higher order thalamic inputs to cortex were functionally tested, revealed these to be driver pathways (Lee & Sherman, 2008; Theyel, Llano, & Sherman, 2010. This suggests, by extrapolation, that this will prove true for pulvinar inputs to higher areas of visual cortex. However, this property of pulvinar projections to extrastriate cortical areas has yet to be explicitly tested in any species, and this remains a key unanswered question. It may be relevant in this context to note that, in the mouse, the projection from the posterior medial nucleus, a higher order somatosensory thalamic relay organized much like the pulvinar is for vision, has a projection to primary somatosensory cortex that is entirely modulatory, unlike its driver inputs to the second somatosensory area (Viaene, Petrof, & Sherman, 2011a). Recent evidence in monkeys indicates a modulatory function for the pulvinar projection to striate cortex (Purushothaman et al., 2012), suggesting the possibility to be tested that higher order thalamic relays such as pulvinar provide a driving input in feedforward connections but are modulatory in feedback.

Role of Pulvinar in Corticocortical Communication

Until recently, ideas of corticocortical communication among visual areas have been dominated by the notion that, once information reaches striate cortex from the lateral geniculate nucleus, it remains in cortex, being processed strictly via direct corticocortical pathways (Bond, 2004; Hilgetag & Kaiser, 2004; Lamme, 2003; Moore & Armstrong, 2003; Salin & Bullier, 1995; Van Essen, Anderson, & Felleman, 1992; Wise et al., 1997; Womelsdorf et al., 2006) with no role for subcortical structures such as the pulvinar. We now propose a very specific role for the pulvinar as a thalamic link in indirect corticothalamocortical circuits critical for cortical functioning. We have generally suggested such a role for all higher order thalamic relays, such as the pulvinar, and examples for the somatosensory and auditory systems are the dorsal medial geniculate nucleus and posterior medial nucleus, respectively (Sherman & Guillery, 2011).

Another important feature of the transthalamic cortical pathways in general is that often, perhaps always, a direct connection between two cortical areas is paralleled by an indirect one through thalamus. This suggests, for instance, that the direct pathway between striate cortex and the medial temporal cortex (known as MT) is paralleled by one relayed through pulvinar. This raises a series of questions related to differences between the direct and transthalamic cortical pathways, considered below, for which we have only partial answers.

Differences between Direct and Transpulvinar Visual Cortical Pathways

The basic question, from which others spring, is: Why is some information transmitted from one cortical area to another sent directly and other information sent through a thalamic relay?

WHAT IS DIFFERENT ABOUT THE SYNAPTIC PROPERTIES OF THE DIRECT VERSUS TRANSPULVINAR CIRCUITS? It is remarkable that so much thinking about cortical functioning is based on ideas about direct corticocortical interconnections when, as noted above, we know rather little about the actions of these connections. That is, schemes such as that of Van Essen and colleagues (Felleman & Van Essen, 1991; Van Essen, Anderson, & Felleman, 1992), which dominate thinking about corticocortical interactions, are based virtually entirely on neuroanatomical studies of connectivity. Only recently has there been any synaptic study of such direct corticocortical connections, based on recordings from slices of mouse brain, and this shows, for connections between areas of both visual and auditory cortices, that both driver and modulator connections exist with complex laminar relationships (Covic & Sherman, 2011; DePasquale & Sherman, 2011).

Several interesting and surprising conclusions can be drawn from these mouse data. Both presumed feedforward (primary visual or auditory cortex to secondary cortex) and feedback (secondary visual or auditory cortex to primary cortex) directions were studied, and whereas one prediction would be for a relative dominance of drivers in the feedforward direction compared to the feedback, this was not seen. Indeed, no clear differences were seen between directions in the laminar pattern of driver and modulator inputs in the corticocortical pathways, which markedly differs from the pattern suggested for the monkey (Felleman & Van Essen, 1991).

Although clearly more examples are needed from other mammalian species, it seems reasonable to conclude at least that direct corticocortical connections, including those between visual areas, contain drivers among the inputs to the target area. If so, then both the direct and transthalamic pathways include driver components and thus routes of information transfer. Nonetheless, important details about these circuits remain to be determined.

One telling difference in the circuits, besides the obvious one of whether or not a thalamic relay is present, is the relationship of the circuits to subcortical processes. As noted above, the layer 5 axons that innervate pulvinar are the first link in the transthalamic pathway branch, with the extrathalamic branch targeting various brainstem sites. The direct pathways, with rare exceptions,[15] are comprised of axons with no subcortically directed branches (Petrof, Viaene, & Sherman, 2012). Thus, the messages sent via the transthalamic pathways relate to various subcortical centers, whereas the direct pathways carry messages that are strictly cortical.

WHAT IS DIFFERENT IN THE INFORMATION CONTENT OF THE DIRECT VERSUS TRANSPULVINAR CIRCUITS? As noted above, the transthalamic route involves branching axons from layer 5 that also target subcortical sites, so the message in the transthalamic pathway may be regarded as a copy of an upcoming motor command, that is, an efference copy of the command sent to the lower motor centers via the branching axon. If so, this means that the nature of the transthalamic information is at least in part to inform higher cortical areas about motor commands sent out from lower areas. On this basis we suggest that the most parsimonious explanation for the direct connections is that they may be involved primarily in basic information processing about the environment.

The following example may help to clarify this distinction. The layer 5 projection from a visual area, say V1, involves one or more branches that innervate subcortical motor areas, such as the superior colliculus, with a message to create some form of movement, perhaps an eye movement (Tehovnik, Slocum, & Schiller, 2003), and this same message is sent through another branch to the pulvinar, where it is relayed to a higher cortical area. This provides information to each visual cortical area about prospective motor commands initiated by cortical areas lower in the hierarchy, and this is precisely what is needed to disambiguate the sensory consequences of eye and head movements from movements in the outside world. It may well be that the direct connections are primarily concerned with an analysis of the visual scene, whereas the transthalamic connections play their major role in relation to the movements, although the connectivity patterns suggest that the two functions are never fully separated.

why is the transthalamic pathway transthalamic? If the object is to send an efference copy from one cortical area to another via a branching axon, a thalamic relay for this is not an absolute requirement. That is, the same layer 5 cells that innervate pulvinar and send a branch to other subcortical centers, instead of innervating pulvinar, could project that branch directly to the target cortical area. The Meynert cell may be a rare example of this (see note 15). However, such a connection would lack the modulatory and gating effects that characterize the thalamic relay.

As noted above, certain inputs specifically target higher order relays such as the pulvinar. An interesting one is the GABAergic input from zona incerta of the rat (Lavallée et al., 2005; Power, Kolmac, & Motrofanis, 1999). Although there is no direct evidence to illustrate what effect this input may have on the pulvinar, evidence from the analogous higher order thalamic relay for the somatosensory system, the posterior medial nucleus, suggests what significance this might have for vision. Under most conditions during which a rat is not alert or actively whisking, the zona incerta is active, and its GABAergic input to the posterior medial nucleus serves to shut down relay cells there, effectively closing the thalamic gate (Barthó, Freund, & Acsády, 2002; Bokor et al., 2005; Lavallée et al., 2005; Masri et al., 2006; Trageser & Keller, 2004; Trageser et al., 2006). Two processes bring about inhibition of zona incerta cells, thus disinhibiting relay cells of the posterior medial nucleus and opening the thalamic gate. One is the level of arousal, because greater activity in parabrachial cholinergic inputs inhibits zona incerta cells (Trageser et al., 2006). The other, which is more interesting, occurs when a rat moves from passive to active whisking (i.e., to an exploratory mode); the now-active motor cortex inhibits zona incerta cells, the silencing of the zona incerta then opens the gate of the posterior medial nucleus (Urbain & Deschênes, 2007), and this then allows the transthalamic message from primary sensory cortex (S1) to be passed to higher cortical areas (S2). In other words, during active whisking, which is associated with high levels of activity in motor cortex, messages that are processed in somatosensory cortex lead to motor output signals from layer 5, and copies of these are successfully relayed through the posterior medial nucleus up the cortical hierarchy.

One can imagine a similar process applying to gating of pulvinar by the zona incerta. During active vision, messages sent by lower visual cortical areas about motor commands are transmitted through the thalamic relay to affect the higher, target cortical area. During periods when vision is not active, as happens, for instance, during drowsiness or when another sensory system captures attentional mechanisms, these messages are blocked by closing the pulvinar gate.

This sort of process for whisking or vision makes sense if the presumed motor messages sent out by cortical areas do not always result in motor actions, and a consideration of the likely evolutionary history of thalamus and cortex provides a reasonable scenario for this. As thalamocortical circuits evolved, there was no parallel evolution of separate subcortical motor circuits, although existing ones did continue to evolve, and so cortex ends up sharing these earlier evolved brainstem and spinal motor circuits to affect behavior.

If an animal is actively guiding behavior through visual stimuli, and the auditory system is passive (e.g., taking in stimuli but not guiding behavior), then messages from layer 5 cells, including copies of messages to move the eyes, would be passed up the cortical hierarchy, and the higher order gates through pulvinar would be held open. Those in the auditory system, however, would be gated shut because any messages created in an auditory cortical area and leading to firing of the layer 5 output cells are unlikely to significantly affect behavior, and these messages then would not get through the thalamic links in the transthalamic auditory pathways.

Another consequence of a transthalamic route involves the ability to modulate the message in ways not available to the direct route because, as described above, thalamic circuitry offers a number of ways to modulate messages in transfer to cortex. We consider one example of this involving the switching of relay cells between tonic and burst firing mode. The details of this, which depend on the inactivation state of T-type Ca^{2+} channels, have been described more fully above (Intrinsic Properties of Thalamic Cells in the A Layers). We have suggested that tonic mode is used for normal processing of information, whereas burst mode is used as a "wake-up call" to cortex that information is once again being relayed after a period of no relay.

This scenario can also be applied to the pulvinar and its zona incerta input as follows. When the zona incerta is active, it not only shuts down the pulvinar relay, but it also hyperpolarizes the relay cells, switching them to burst mode by deinactivating the T-type Ca²⁺ channels. Silencing of these zona incerta inputs, presumably when vision becomes active, would then lead to layer 5 inputs evoking bursts in pulvinar relay cells, which in turn would strongly signal the target cortical area that a significant change has occurred.

Consider again the scenario just suggested for the zona incerta input to the pulvinar. During passive vision the zona incerta powerfully inhibits pulvinar relay cells, placing them in burst mode, and activity in layer 5 corticothalamic cells from layer 5 of a visual cortical area is not relayed to higher areas. With a switch to active vision, new outputs from layer 5 of visual cortex inhibit zona incerta cells, removing their inhibition of cells in the pulvinar, and the next output from cortical layer 5 evokes a burst in these relay cells. That burst strongly activates circuits in target visual cortical areas. Persistent activity in layer 5 inputs, in the absence of further inhibition from zona incerta, continues to depolarize pulvinar relay cells and thereby switches their firing to tonic mode, ensuring a more faithful relay of the trans-thalamic signal.

CONCLUSIONS

Clearly, the thalamus can no longer be viewed as a passive, machine-like relay of information to cortex. We have outlined a number of important functional properties of a dynamic nature that occur during thalamic relay functions and that relate to behavioral states such as attention and alertness. This is probably the tip of the iceberg, with many additional functions likely to emerge as our understanding of thalamic properties expands.

It is important to note in this context that the thalamus offers a last "bottleneck" for general behavioral states to have an effect on information processing. Thalamic relays represent a relatively small number of neurons and synapses compared to their target cortical areas. Thus, if there is a need to increase or decrease the saliency of a particular message, say, a visual stimulus at the expense of an auditory one, it requires orders of magnitude less synaptic processing to modulate at the thalamic level than at the cortical level. For visual processing in mammals, there is no opportunity for the rest of the brain to affect processing in the retina except for possible autonomic effects on accommodation and pupillary control. The lateral geniculate nucleus is not only a convenient last bottleneck of information flow, it is the most peripheral site at which such processing can be modulated. In other sensory systems it is possible

to modulate processing more peripherally than the thalamus, but for all pathways going to cortex, the thalamic level remains the last convenient stage at which modulation can efficiently affect information flow before it is passed to cortex.

When one looks at the visual relays in the thalamus, it is necessary to recognize that there is both a relatively well-studied first order relay in the lateral geniculate nucleus and a series of more elusive higher order relays in the pulvinar. The lateral geniculate nucleus relays several functionally distinct, largely independent, topographically organized, parallel visual pathways from the retina to the cortex. It serves, among other possible but currently undefined functions, to modify transmission to visual cortex in accord with attentional needs, acting either in *tonic mode,* where accurate, linear transfer of information from the periphery to the cortex is required, or in *burst mode,* where the need is for identification of novel signals that merit attention. Messages from the retina are carried to the lateral geniculate nucleus by the retinogeniculate *drivers.* These represent only about 5–10% of the afferent synapses to geniculate relay cells and have characteristic structural features, synaptic connectivity patterns, and functional relationships in terms of transmitters and receptors. The rest of the afferent synapses are formed by *modulators*, which can serve to switch transmission from burst to tonic or from tonic to burst mode and come from several sources, including the glutamatergic feedback from layer 6 of visual cortex.

The higher order relays in the pulvinar serve to transmit information from one cortical area to other cortical areas through the thalamus. The *drivers* come from pyramidal cells in layer 5 of cortex and represent branches of axons that are going to lower (motor) centers. That is, the pulvinar serves to send copies of motor outputs, and thus efference copies, from one cortical area to another. We stress the likely importance of these transthalamic pathways in corticocortical communication, a role that has not been recognized in the past.

Finally, whereas it is now clear that direct and transthalamic corticocortical circuits exist, often or perhaps always organized in parallel, it is not entirely clear what purpose this remarkable neuronal organization serves. Part of the answer may relate to the suggestion that part of the message sent to higher cortical areas is an efference copy, and this is the message that is transthalamic, although it also carries information about the environment (Guillery & Sherman, 2011). In this regard the organization of the modulatory afferents to the pulvinar also merits closer study than it has received to date. We propose that in general terms the functions of these are comparable to the functions of the modulators in the lateral geniculate nucleus. That is, they serve to switch the relay between the burst and the tonic modes. Special GABAergic inputs to the pulvinar, chiefly from the zona incerta and pretectum, may provide an additional function to gate the pulvinar relay, and we have speculated that this may occur to prevent relays through pulvinar of efference copies that are not likely to be related to eventual motor actions when vision is not active.

These ideas provide entirely novel functional possibilities for the pulvinar relay. Clearly, much more work is needed to clarify the validity of this challenging proposal.

NOTES

1. By "lateral geniculate nucleus" throughout this chapter, we mean the *dorsal* lateral geniculate nucleus. Like all thalamic nuclei providing a relay to neocortex, the lateral geniculate nucleus is developmentally a part of the dorsal thalamus. The *ventral* lateral geniculate nucleus, which is part of the ventral thalamus, does not send axons to cortex and is not considered further.
2. Unless otherwise specified, we shall refer to this as the "monkey" in what follows, noting however, that the macaque can be regarded as representative of Old World monkeys and that the lateral geniculate nucleus in New World monkeys has a slightly different structure.
3. Another term frequently used for this area is "pedunculopontine tegmental nucleus." We prefer "parabrachial region," because the scattered cells that innervate thalamus from this area do not have a clear nuclear boundary, and they are found scattered around the brachium conjunctivum.
4. In the cat the major portion of the thalamic reticular nucleus innervating the lateral geniculate nucleus is called the "perigeniculate nucleus."
5. One ultrastructural characteristic of synaptic contacts is a thickening of the postsynaptic membrane, which includes the postsynaptic receptors (Sheng, 2001). When the thickening is especially prominent, the postsynaptic membranes are seen as thicker than the presynaptic ones, and this characterizes an asymmetric synapse. When the thickening is less prominent, there is a less pronounced difference in thickness between presynaptic and postsynaptic membranes, and this characterizes a symmetric synapse (Gray, 1959). Typically, asymmetric synapses are associated with excitatory synapses, and symmetric, with inhibitory ones.
6. The voltage and time dependencies vary somewhat among cells and also are interdependent because a greater depolarization requires less time to inactivate, and, likewise, greater hyperpolarization is associated with faster deinactivation (Sherman & Guillery, 1996, 2006).
7. Sometimes, "spike" is used to refer to a conventional action potential, but here we use it to refer to an all-or-none Ca^{2+} event. To avoid confusion, when we mean a Na^+/K^+ action potential, we use "action potential" and not "spike."
8. It is worth noting that glutamatergic inputs seen in cortical circuitry, including thalamocortical and corticocortical inputs, show the same two basic types, described here as "drivers" and "modulators." However, a new terminology has been applied—"class 1" for "driver" and "class 2" for "modulator"—for the cortical circuitry because the terms "driver" and "modulator" have implications that seem clear for thalamus but not yet for cortex (Covic & Sherman, 2011; DePasquale & Sherman, 2011; Viaene, Petrof, & Sherman, 2011a, 2011b, 2011c). Nonetheless, we stick with these original terms in this chapter because we feel that the "driver/modulator" terminology does indeed capture the presumed function of each of these inputs.
9. It should be noted that, although these bullet points were written specifically for thalamic circuitry, they apply as well to cortical circuitry (Covic & Sherman, 2011; DePasquale & Sherman, 2011, 2012; Lee & Sherman, 2008, 2009; Viaene, Petrof, & Sherman, 2011a, 2011b, 2011c).
10. A retinal projection to the pulvinar has been described in several species, but this is generally in a region close to or adjoining the lateral geniculate nucleus. One definition of the lateral geniculate nucleus is that it is the collection of neurons that receive a retinal input and whose relay cells project to cortex. If so, then any retinal projection to thalamus can be regarded as innervating a part of the lateral geniculate nucleus, and in this context, in the cat, a thalamic region innervated by retina that some refer to as the "retinorecipient zone of the pulvinar" has instead been called a part of the "geniculate wing" (Guillery et al., 1980). Thus, the issue of whether the pulvinar receives a retinal input is a matter of semantics, but in any case the issue of whether this region, pulvinar or lateral geniculate nucleus, receives other driver inputs from cortex or midbrain and how these might interact has yet to be explored.
11. The variety of tectopulvinar terminals seen suggests that both driver and modulator types of input may be represented in this pathway.
12. The message carried by any axon is coded by the pattern of action potentials that travel down it. Under normal conditions, every action potential traveling down the parent branch will lead to one and only one action potential in each of the daughter branches, thus ensuring that the same message is carried by the parent axon and each branch. Failure to generate an action potential across a branch may sometimes occur in the nonphysiological, antidromic direction, but exceedingly rarely under normal, orthodromic conduction.
13. "Corollary discharge" is often used instead of "efference copy" to describe this feature. We use the latter to stress that it is a copy of a motor output, not of a sensory input.
14. See note 10 for an explanation as to why we do not consider retinal inputs here.
15. The one known exception is the Meynert cell, which in monkeys projects via collaterals from area 17 to both MT and the superior colliculus, but a collateral to thalamus from these cells has not yet been described (Fries, Keizer, & Kuypers, 1985; Rockland & Knutson, 2001).

REFERENCES

Abramson, B. P., & Chalupa, L. M. (1985). The laminar distribution of cortical connections with the tecto- and cortico-recipient zones in the cat's lateral posterior nucleus. *Neuroscience, 15,* 81–95.

Ahmed, B., Anderson, J. C., Douglas, R. J., Martin, K. A. C., & Nelson, J. C. (1994). Polyneuronal innervation of spiny stellate neurons in cat visual cortex. *Journal of Comparative Neurology, 341,* 39–49.

Altman, J., & Carpenter, M. B. (1961). Fiber projections of the superior colliculus in the cat. *Journal of Comparative Neurology, 116,* 157–177.

Andreas, S. T., Stelios, M., Smirnakis, A., Augath, M., Trinath, T., & Logothetis, N. K. (2001). Motion processing in the macaque: Revisited with functional magnetic resonance imaging. *Journal of Neuroscience, 21,* 8594–8601.

Baker, F. H., & Malpeli, J. G. (1977). Effects of cryogenic blockade of visual cortex on the responses of lateral geniculate neurons in the monkey. *Experimental Brain Research, 29,* 433–444.

Barthó, P., Freund, T. F., & Acsády, L. (2002). Selective GABAergic innervation of thalamic nuclei from zona incerta. *European Journal of Neuroscience, 16,* 999–1014.

Bender, D. B. (1982). Receptive-field properties of neurons in the macaque inferior pulvinar. *Journal of Neurophysiology, 48,* 1–17.

Bender, D. B. (1983). Visual activation of neurons in the primate pulvinar depends on cortex but not colliculus. *Brain Research, 279,* 258–261.

Berman, R. A., & Wurtz, R. H. (2010). Functional identification of a pulvinar path from superior colliculus to cortical area MT. *Journal of Neuroscience, 30,* 6342–6354.

Berson, D. M., & Graybiel, A. M. (1978). Parallel thalamic zones in the LP-pulvinar complex of the cat identified by their afferent and efferent connections. *Brain Research, 147,* 139–148.

Bickford, M. E., Günlük, A. E., Guido, W., & Sherman, S. M. (1993). Evidence that cholinergic axons from the parabrachial region of the brainstem are the exclusive source of nitric oxide in the lateral geniculate nucleus of the cat. *Journal of Comparative Neurology, 334,* 410–430.

Bokor, H., Frere, S. G. A., Eyre, M. D., Slezia, A., Ulbert, I., Luthi, A., et al. (2005). Selective GABAergic control of higher order thalamic relays. *Neuron, 45,* 929–940.

Bond, A. H. (2004). An information-processing analysis of the functional architecture of the primate neocortex. *Journal of Theoretical Biology, 227,* 51–79.

Bourassa, J., & Deschênes, M. (1995). Corticothalamic projections from the primary visual cortex in rats: A single fiber study using biocytin as an anterograde tracer. *Neuroscience, 66,* 253–263.

Bourassa, J., Pinault, D., & Deschênes, M. (1995). Corticothalamic projections from the cortical barrel field to the somatosensory thalamus in rats: A single-fibre study using biocytin as an anterograde tracer. *European Journal of Neuroscience, 7,* 19–30.

Bowling, D. B., & Michael, C. R. (1984). Terminal patterns of single, physiologically characterized optic tract fibers in the cat's lateral geniculate nucleus. *Journal of Neuroscience, 4,* 198–216.

Casagrande, V. A., & Xu, X. (2004). Parallel visual pathways: A comparative perspective. In L. M. Chalupa & J. S. Werner (Eds.), *The visual neurosciences* (pp. 494–506). Cambridge, MA: MIT Press.

Casanova, C., & Savard, T. (1996). Responses to moving texture patterns of cells in the striate-recipient zone of the cat's lateral posterior-pulvinar complex. *Neuroscience, 70,* 439–447.

Chalupa, L. M. (1991). Visual function of the pulvinar. In A. G. Leventhal (Ed.), *The neural basis of visual function* (pp. 140–159). New York: Macmillan Press.

Chalupa, L. M., Anchel, H., & Lindsley, D. B. (1972). Visual input to the pulvinar via lateral geniculate, superior colliculus and visual cortex in the cat. *Experimental Neurology, 36,* 449–462.

Conley, M., Birecree, E., & Casagrande, V. A. (1985). Neuronal classes and their relation to functional and laminar organization to functional and laminar organization of the lateral geniculate nucleus: A Golgi study of the prosimian primate, *Galago crassicaudatus. Journal of Comparative Neurology, 242,* 561–583.

Conley, M., & Diamond, I. T. (1990). Organization of the visual sector of the thalamic reticular nucleus in *Galago. European Journal of Neuroscience, 2,* 211–226.

Conn, P. J., & Pin, J. P. (1997). Pharmacology and functions of metabotropic glutamate receptors. *Annual Review of Pharmacology and Toxicology, 37,* 205–237.

Covic, E. N., & Sherman, S. M. (2011). Synaptic properties of connections between the primary and secondary auditory cortices in mice. *Cerebral Cortex, 21,* 2425–2441.

Crabtree, J. W. (1992a). The somatotopic organization within the cat's thalamic reticular nucleus. *European Journal of Neuroscience, 4,* 1352–1361.

Crabtree, J. W. (1992b). The somatotopic organization within the rabbit's thalamic reticular nucleus. *European Journal of Neuroscience, 4,* 1343–1351.

Crabtree, J. W. (1996). Organization in the somatosensory sector of the cat's thalamic reticular nucleus. *Journal of Comparative Neurology, 366,* 207–222.

Crabtree, J. W., & Killackey, H. P. (1989). The topographical organization of the axis of projection within the visual sector of the rabbit's thalamic reticular nucleus. *European Journal of Neuroscience, 1,* 94–109.

Cucchiaro, J. B., Uhlrich, D. J., & Sherman, S. M. (1991). Electron-microscopic analysis of synaptic input from the perigeniculate nucleus to the A-laminae of the lateral geniculate nucelus in cats. *Journal of Comparative Neurology, 310,* 316–336.

Datskovskaia, A., Carden, W. B., & Bickford, M. E. (2001). Y retinal terminals contact interneurons in the cat dorsal lateral geniculate nucleus. *Journal of Comparative Neurology, 430,* 85–100.

de Lima, A. D., & Singer, W. (1987). The brainstem projection to the lateral geniculate nucleus in the cat: Identification of cholinergic and monoaminergic elements. *Journal of Comparative Neurology, 259,* 92–121.

DePasquale, R., & Sherman, S. M. (2011). Synaptic properties of corticocortical connections between the primary and secondary visual cortical areas in the mouse. *Journal of Neuroscience, 31,* 16494–16506.

DePasquale, R., & Sherman, S. M. (2012). Modulatory effects of metabotropic glutamate receptors on local cortical circuits. *Journal of Neuroscience, 32*(21), 7364–7372.

DePasquale, R., & Sherman, S. M. (2013). A modulatory effect of the feedback from higher visual areas to V1 in the mouse. *Journal of Neurophysiology,* in press.

Diamond, I. T. (1973). The evolution of the tectal-pulvinar system in mammals: Structural and behavioral studies of the visual system. *Symposia of the Zoological Society of London, 33*, 205–233.

Dobrunz, L. E., & Stevens, C. F. (1997). Heterogeneity of release probability, facilitation, and depletion at central synapses. *Neuron, 18*, 995–1008.

Duhamel, J.-R., Colby, C. L., & Goldberg, M. E. (1992). The updating of the representation of visual space in parietal cortex by intended eye movements. *Science, 255*, 90–92.

Erişir, A., Van Horn, S. C., Bickford, M. E., & Sherman, S. M. (1997). Immunocytochemistry and distribution of parabrachial terminals in the lateral geniculate nucleus of the cat: A comparison with corticogeniculate terminals. *Journal of Comparative Neurology, 377*, 535–549.

Erişir, A., Van Horn, S. C., & Sherman, S. M. (1997). Relative numbers of cortical and brainstem inputs to the lateral geniculate nucleus. *Proceedings of the National Academy of Sciences of the United States of America, 94*, 1517–1520.

Famiglietti, E. V., & Peters, A. (1972). The synaptic glomerulus and the intrinsic neuron in the dorsal lateral geniculate nucleus of the cat. *Journal of Comparative Neurology, 144*, 285–334.

Feig, S., & Harting, J. K. (1998). Corticocortical communication via the thalamus: Ultrastructural studies of corticothalamic projections from area 17 to the lateral posterior nucleus of the cat and inferior pulvinar nucleus of the owl monkey. *Journal of Comparative Neurology, 395*, 281–295.

Felleman, D. J., & Van Essen, D. C. (1991). Distributed hierarchical processing in the primate cerebral cortex. *Cerebral Cortex, 1*(1), 1–47.

Friedlander, M. J., Lin, C.-S., Stanford, L. R., & Sherman, S. M. (1981). Morphology of functionally identified neurons in lateral geniculate nucleus of the cat. *Journal of Neurophysiology, 46*, 80–129.

Fries, W., Keizer, K., & Kuypers, H. G. (1985). Large layer VI cells in macaque striate cortex (Meynert cells) project to both superior colliculus and prestriate visual area V5. *Experimental Brain Research, 58*, 613–616.

Galletti, C., Gamberini, M., Kutz, D. F., Fattori, P., Luppino, G., & Matelli, M. (2001). The cortical connections of area V6: An occipito-parietal network processing visual information. *European Journal of Neuroscience, 13*, 1572–1588.

Garcia-Cabezas, M. A., Rico, B., Sanchez-Gonzalez, M. A., & Cavada, C. (2007). Distribution of the dopamine innervation in the macaque and human thalamus. *NeuroImage, 34*, 965–984.

Geisert, E. E., Langsetmo, A., & Spear, P. D. (1981). Influence of the cortico-geniculate pathway on reponse properties of cat lateral geniculate neurons. *Brain Research, 208*, 409–415.

Gentet, L. J., & Ulrich, D. (2003). Strong, reliable and precise synaptic connections between thalamic relay cells and neurones of the nucleus reticularis in juvenile rats. *Journal of Physiology, 546*, 801–811.

Gilbert, C. D., & Kelly, J. P. (1975). The projections of cells in different layers of the cat's visual cortex. *Journal of Physiology, 163*, 81–106.

Govindaiah, G., Wang, T., Gillette, M. U., & Cox, C. L. (2012). Activity-dependent regulation of retinogeniculate signaling by metabotropic glutamate receptors. *Journal of Neuroscience, 32*, 12820–12831.

Granseth, B., & Lindström, S. (2003). Unitary EPSCs of corticogeniculate fibers in the rat dorsal lateral geniculate nucleus in vitro. *Journal of Neurophysiology, 89*, 2952–2960.

Gray, E. G. (1959). Axo-somatic and axo-dendritic synapses of the cerebral cortex. An electron microscopic study. *Journal of Anatomy (London), 93*, 420–433.

Guillery, R. W. (1966). A study of Golgi preparations from the dorsal lateral geniculate nucleus of the adult cat. *Journal of Comparative Neurology, 128*, 21–50.

Guillery, R. W. (1995). Anatomical evidence concerning the role of the thalamus in corticocortical communication: A brief review. *Journal of Anatomy, 187*, 583–592.

Guillery, R. W. (2003). Branching thalamic afferents link action and perception. *Journal of Neurophysiology, 90*, 539–548.

Guillery, R. W., Feig, S. L., & Lozsádi, D. A. (1998). Paying attention to the thalamic reticular nucleus. *Trends in Neurosciences, 21*, 28–32.

Guillery, R. W., Feig, S. L., & Van Lieshout, D. P. (2001). Connections of higher order visual relays in the thalamus: A study of corticothalamic pathways in cats. *Journal of Comparative Neurology, 438*, 66–85.

Guillery, R. W., Geisert, E. E., Polley, E. H., & Mason, C. A. (1980). An analysis of the retinal afferents to the cat's medial interlaminar nucleus and to its rostral thalamic extension, the "geniculate wing." *Journal of Comparative Neurology, 194*, 117–142.

Guillery, R. W., & Sherman, S. M. (2002a). Thalamic relay functions and their role in corticocortical communication: Generalizations from the visual system. *Neuron, 33*, 163–175.

Guillery, R. W., & Sherman, S. M. (2002b). The thalamus as a monitor of motor outputs. *Philosophical Transactions of the Royal Society of London. B, 357*, 1809–1821.

Guillery, R. W., & Sherman, S. M. (2011). Branched thalamic afferents: What are the messages that they relay to cortex? *Brain Research. Brain Research Reviews, 66*, 205–219.

Hackett, T. A., Stepniewska, I., & Kaas, J. H. (1998). Thalamocortical connections of the parabelt auditory cortex in macaque monkeys. *Journal of Comparative Neurology, 400*, 271–286.

Hamos, J. E., Van Horn, S. C., Raczkowski, D., & Sherman, S. M. (1987). Synaptic circuits involving an individual retinogeniculate axon in the cat. *Journal of Comparative Neurology, 259*, 165–192.

Hamos, J. E., Van Horn, S. C., Raczkowski, D., Uhlrich, D. J., & Sherman, S. M. (1985). Synaptic connectivity of a local circuit neurone in lateral geniculate nucleus of the cat. *Nature, 317*, 618–621.

Hickey, T. L., & Guillery, R. W. (1974). An autoradiographic study of retinogeniculate pathways in the cat and the fox. *Journal of Comparative Neurology, 156*, 239–254.

Hilgetag, C. C., & Kaiser, M. (2004). Clustered organization of cortical connectivity. *Neuroinformatics, 2*, 353–360.

Hubel, D. H., & Wiesel, T. N. (1977). Functional architecture of macaque monkey visual cortex. *Proceedings of the Royal Society of London. Series B, Containing Papers of a Biological Character, 198*, 1–59.

Hutchins, B., & Updyke, B. V. (1989). Retinotopic organization within the lateral posterior complex of the cat. *Journal of Comparative Neurology, 285*, 350–398.

Itoh, K., Conley, M., & Diamond, I. T. (1982). Retinal ganglion cell projections to individual layers of the lateral geniculate body in *Galago crassicaudatus*. *Journal of Comparative Neurology, 205*, 282–290.

Jones, E. G. (2007). *The thalamus* (2nd ed.). Cambridge: Cambridge University Press.

Kalil, R. E., & Chase, R. (1970). Corticofugal influence on activity of lateral geniculate neurons in the cat. *Journal of Neurophysiology, 33*, 459–474.

Kandel, E. R., Schwartz, J. H., & Jessell, T. M. (2000). *Principles of neural science.* New York: McGraw-Hill.

Kelly, L. R., Li, J., Carden, W. B., & Bickford, M. E. (2003). Ultrastructure and synaptic targets of tectothalamic terminals in the cat lateral posterior nucleus. *Journal of Comparative Neurology, 464*, 472–486.

Lam, Y. W., & Sherman, S. M. (2013). Activation of both group I and group II metabotropic glutamatergic receptors suppress retinogeniculate transmission. *Neuroscience*, in press.

Lamme, V. A. (2003). Recurrent corticocortical interactions in neural disease. *Archives of Neurology, 60*, 178–184.

Latawiec, D., Martin, K. A. C., & Meskenaite, V. (2000). Termination of the geniculocortical projection in the striate cortex of macaque monkey: A quantitative immunoelectron microscopic study. *Journal of Comparative Neurology, 419*, 306–319.

Lavallée, P., Urbain, N., Dufresne, C., Bokor, H., Acsády, L., & Deschênes, M. (2005). Feedforward inhibitory control of sensory information in higher order thalamic nuclei. *Journal of Neuroscience, 25*, 7489–7498.

Lee, C. C., & Sherman, S. M. (2008). Synaptic properties of thalamic and intracortical inputs to layer 4 of the first and higher order cortical areas in the auditory and somatosensory systems. *Journal of Neurophysiology, 100*, 317–326.

Lee, C. C., & Sherman, S. M. (2009). Modulator property of the intrinsic cortical projection from layer 6 to layer 4. *Frontiers in Systems Neuroscience, 3*, 1–5. doi:10.3389/neuro.06.003.2009.

LeVay, S., & Ferster, D. (1977). Relay cell classes in the lateral geniculate nucleus of the cat and the effects of visual deprivation. *Journal of Comparative Neurology, 172*, 563–584.

LeVay, S., & McConnell, S. K. (1982). ON and OFF layers in the lateral geniculate nucleus of the mink. *Nature, 300*, 350–351.

Li, J., Guido, W., & Bickford, M. E. (2003). Two distinct types of corticothalamic EPSPs and their contribution to short-term synaptic plasticity. *Journal of Neurophysiology, 90*, 3429–3440.

Lujan, R., Nusser, Z., Roberts, J. D., Shigemoto, R., & Somogyi, P. (1996). Perisynaptic location of metabotropic glutamate receptors mGluR1 and mGluR5 on dendrites and dendritic spines in the rat hippocampus. *European Journal of Neuroscience, 8*, 1488–1500.

Masri, R., Trageser, J. C., Bezdudnaya, T., Li, Y., & Keller, A. (2006). Cholinergic regulation of the posterior medial thalamic nucleus. *Journal of Neurophysiology, 96*, 2265–2273.

Mateo, Z., & Porter, J. T. (2007). Group II metabotropic glutamate receptors inhibit glutamate release at thalamocortical synapses in the developing somatosensory cortex. *Neuroscience, 146*, 1062–1072.

Mathers, L. H. (1971). Tectal projection to posterior thalamus of the squirrel monkey. *Brain Research, 35*, 357–380.

Mathers, L. H. (1972). The synaptic organization of the cortical projection to the pulvinar of the squirrel monkey. *Journal of Comparative Neurology, 146*, 43–60.

McAlonan, K., Cavanaugh, J., & Wurtz, R. H. (2008). Guarding the gateway to cortex with attention in visual thalamus. *Nature, 456*(7220), 391. doi:10.1038/nature07382.

McClurkin, J. W., & Marrocco, R. T. (1984). Visual cortical input alters spatial tuning in monkey lateral geniculate nucleus cells. *Journal of Physiology, 348*, 135–152.

Montero, V. M., Guillery, R. W., & Woolsey, C. N. (1977). Retinotopic organization within the thalamic reticular nucleus demonstrated by a double label autoradiographic technique. *Brain Research, 138*, 407–421.

Moore, T., & Armstrong, K. M. (2003). Selective gating of visual signals by microstimulation of frontal cortex. *Nature, 421*, 370–373.

Mott, D. D., & Lewis, D. V. (1994). The pharmacology and function of central GABAB receptors. *International Review of Neurobiology, 36*, 97–223.

Murphy, P. C., & Sillito, A. M. (1996). Functional morphology of the feedback pathway from area 17 of the cat visual cortex to the lateral geniculate nucleus. *Journal of Neuroscience, 16*, 1180–1192.

Nicoll, R. A., Malenka, R. C., & Kauer, J. A. (1990). Functional comparison of neurotransmitter receptor subtypes in mammalian central nervous system. *Physiological Reviews, 70*, 513–565.

Ogren, M. P., & Hendrickson, A. E. (1979). The morphology and distribution of striate cortex terminals in the inferior and lateral subdivisions of the *Macaca* monkey pulvinar. *Journal of Comparative Neurology, 188*, 179–199. doi:10.1002/cne.901880113.

Ojima, H., Murakami, K., & Kishi, K. (1996). Dual termination modes of corticothalamic fibers originating from pyramids of layers 5 and 6 in cat visual cortical area 17. *Neuroscience Letters, 208*, 57–60.

Olshausen, B. A., Anderson, C. H., & Van Essen, D. C. (1993). A neurobiological model of visual attention and invariant pattern recognition based on dynamic routing of information. *Journal of Neuroscience, 13*, 4700–4719.

Partlow, G. D., Colonnier, M., & Szabo, J. (1977). Thalamic projections of the superior colliculus in the rhesus monkey, *Macaca mulatta.* A light and electron microscopic study. *Journal of Comparative Neurology, 171*, 285–318.

Petrof, I., Viaene, A. N., & Sherman, S. M. (2012). Two populations of corticothalamic and interareal corticocortical cells in the subgranular layers of the mouse primary sensory cortices. *Journal of Comparative Neurology, 520*, 1678–1686.

Pin, J. P., & Duvoisin, R. (1995). The metabotropic glutamate receptors: Structure and functions. *Neuropharmacology, 34*, 1–26.

Pinault, D., Bourassa, J., & Deschênes, M. (1995). Thalamic reticular input to the rat visual thalamus: A single fiber study using biocytin as an anterograde tracer. *Brain Research, 670*, 147–152.

Pinault, D., & Deschênes, M. (1998). Projection and innervation patterns of individual thalamic reticular axons in the thalamus of the adult rat: A three-dimensional, graphic, and morphometric analysis. *Journal of Comparative Neurology, 391*, 180–203.

Power, B. D., Kolmac, C. I., & Mitrofanis, J. (1999). Evidence for a large projection from the zona incerta to the dorsal thalamus. *Journal of Comparative Neurology, 404*, 554–565.

Purushothaman, G., Marion, R., Li, K., & Casagrande, V. A. (2012). Gating and control of primary visual cortex by pulvinar. *Nature Neuroscience, 15*, 905–912.

Ralston, H. J. (1971). Evidence for presynaptic dendrites and a proposal for their mechanism of action. *Nature, 230*, 585–587.

Ramcharan, E. J., Gnadt, J. W., & Sherman, S. M. (2005). Higher order thalamic relays burst more than first order relays. *Proceedings of the National Academy of Sciences of the United States of America, 102,* 12236–12241.

Recasens, M., & Vignes, M. (1995). Excitatory amino acid metabotropic receptor subtypes and calcium regulation. *Annals of the New York Academy of Sciences, 757,* 418–429.

Reichova, I., & Sherman, S. M. (2004). Somatosensory corticothalamic projections: Distinguishing drivers from modulators. *Journal of Neurophysiology, 92,* 2185–2197.

Robson, J. A., & Hall, W. C. (1977a). The organization of the pulvinar in the grey squirrel (*Sciurus carolinensis*) I. Cytoarchitecture and connections. *Journal of Comparative Neurology, 173,* 355–388.

Robson, J. A., & Hall, W. C. (1977b). The organization of the pulvinar in the grey squirrel (*Sciurus carolinensis*) II. Synaptic organization and comparisons with the dorsal lateral geniculate nucleus. *Journal of Comparative Neurology, 173,* 389–416.

Rockland, K. S. (1996). Two types of corticopulvinar terminations: Round (type 2) and elongate (type 1). *Journal of Comparative Neurology, 368,* 57–87.

Rockland, K. S., Andresen, J., Cowie, R. J., & Robinson, D. L. (1999). Single axon analysis of pulvinocortical connections to several visual areas in the macaque. *Journal of Comparative Neurology, 406,* 221–250.

Rockland, K. S., & Knutson, T. (2001). Axon collaterals of Meynert cells diverge over large portions of area V1 in the macaque monkey. *Journal of Comparative Neurology, 441,* 134–147.

Rodman, H. R., Gross, C. G., & Albright, T. D. (1989). Afferent basis of visual response properties in area MT of the macaque. I. Effects of striate cortex removal. *Journal of Neuroscience, 9,* 2033–2050.

Rodman, H. R., Gross, C. G., & Albright, T. D. (1990). Afferent basis of visual response properties in area MT of the macaque. II. Effects of superior colliculus removal. *Journal of Neuroscience, 10,* 1154–1164.

Salin, P. A., & Bullier, J. (1995). Corticocortical connections in the visual system: Structure and function. *Physiological Reviews, 75,* 107–154.

Sanchez-Gonzalez, M. A., Garcia-Cabezas, M. A., Rico, B., & Cavada, C. (2005). The primate thalamus is a key target for brain dopamine. *Journal of Neuroscience, 25,* 6076–6083.

Schneider, G. E. (1969). Two visual systems. *Science, 163,* 895–902.

Schneider, K. A., & Kastner, S. (2009). Effects of sustained spatial attention in the human lateral geniculate nucleus and superior colliculus. *Journal of Neuroscience, 29,* 1784–1795.

Sheng, M. H. T. (2001). The postsynaptic specialization. In W. M. Cowan, T. C. Südhof, & C. F. Stevens (Eds.), *Synapses* (pp. 315–355). Baltimore, MD: Johns Hopkins University Press.

Sherman, S. M. (1985). Functional organization of the W-, X-, and Y-cell pathways in the cat: A review and hypothesis. In J. M. Sprague & A. N. Epstein (Eds.), *Progress in psychobiology and physiological psychology* (Vol. 11, pp. 233–314). Orlando, FL: Academic Press.

Sherman, S. M. (1996). Dual response modes in lateral geniculate neurons: Mechanisms and functions. *Visual Neuroscience, 13,* 205–213.

Sherman, S. M. (2004). Interneurons and triadic circuitry of the thalamus. *Trends in Neurosciences, 27,* 670–675.

Sherman, S. M. (2005). Thalamic relays and cortical functioning. *Progress in Brain Research, 149,* 107–126.

Sherman, S. M. (2007). The thalamus is more than just a relay. *Current Opinion in Neurobiology, 17,* 1–6.

Sherman, S. M., & Friedlander, M. J. (1988). Identification of X versus Y properties for interneurons in the A-laminae of the cat's lateral geniculate nucleus. *Experimental Brain Research, 73,* 384–392.

Sherman, S. M., & Guillery, R. W. (1996). The functional organization of thalamocortical relays. *Journal of Neurophysiology, 76,* 1367–1395.

Sherman, S. M., & Guillery, R. W. (1998). On the actions that one nerve cell can have on another: Distinguishing "drivers" from "modulators." *Proceedings of the National Academy of Sciences of the United States of America, 95,* 7121–7126.

Sherman, S. M., & Guillery, R. W. (2002). The role of thalamus in the flow of information to cortex. *Philosophical Transactions of the Royal Society of London. B, 357,* 1695–1708.

Sherman, S. M., & Guillery, R. W. (2006). *Exploring the thalamus and its role in cortical function.* Cambridge, MA: MIT Press.

Sherman, S. M., & Guillery, R. W. (2011). Distinct functions for direct and transthalamic corticocortical connections. *Journal of Neurophysiology, 106,* 1068–1077.

Smith, D. C., & Spear, P. D. (1979). Effects of superior colliculus removal on receptive-field properties of neurons in lateral suprasylvian visual area of the cat. *Journal of Neurophysiology, 42,* 57–75.

So, Y.-T., & Shapley, R. (1981). Spatial tuning of cells in and around lateral geniculate nucleus of the cat: X and Y relay cells and perigeniculate interneurons. *Journal of Neurophysiology, 45,* 107–120.

Sommer, M. A., & Wurtz, R. H. (2004). What the brain stem tells the frontal cortex. II. Role of the SC-MD-FEF pathway in corollary discharge. *Journal of Neurophysiology, 91,* 1403–1423.

Sommer, M. A., & Wurtz, R. H. (2006). Influence of the thalamus on spatial visual processing in frontal cortex. *Nature, 444,* 374–377.

Sperry, R. W. (1950). Neural basis of the spontaneous optokinetic response produced by visual inversion. *Journal of Comparative Neurology, 43,* 482–489.

Sprague, J. M. (1966). Interaction of cortex and superior colliculus in mediation of visually guided behavior in the cat. *Science, 153,* 1544–1547.

Sprague, J. M. (1972). The superior colliculus and pretectum in visual behavior. *Investigative Ophthalmology, 11,* 473–482.

Sprague, J. M., Berlucchi, G., & Di Berardino, A. (1970). The superior colliculus and pretectum in visually guided behavior and visual discrimination in the cat. *Brain, Behavior and Evolution, 3,* 285–294.

Stone, J. (1983). *Parallel processing in the visual system.* New York: Plenum Press.

Stryker, M. P., & Zahs, K. R. (1983). On and off sublaminae in the lateral geniculate nucleus of the ferret. *Journal of Neuroscience, 3,* 1943–1951.

Sur, M., Esguerra, M., Garraghty, P. E., Kritzer, M. F., & Sherman, S. M. (1987). Morphology of physiologically identified retinogeniculate X- and Y-axons in the cat. *Journal of Neurophysiology, 58,* 1–32.

Swadlow, H. A., & Gusev, A. G. (2001). The impact of "bursting" thalamic impulses at a neocortical synapse. *Nature Neuroscience, 4*, 402–408.

Swadlow, H. A., Gusev, A. G., & Bezdudnaya, T. (2002). Activation of a cortical column by a thalamocortical impulse. *Journal of Neuroscience, 22*, 7766–7773.

Tehovnik, E. J., Slocum, W. M., & Schiller, P. H. (2003). Saccadic eye movements evoked by microstimulation of striate cortex. *European Journal of Neuroscience, 17*, 870–878.

Theyel, B. B., Llano, D. A., & Sherman, S. M. (2010). The corticothalamocortical circuit drives higher order cortex in the mouse. *Nature Neuroscience, 13*, 84–88.

Trageser, J. C., Burke, K. A., Masri, R., Li, Y., Sellers, L., & Keller, A. (2006). State-dependent gating of sensory inputs by zona incerta. *Journal of Neurophysiology, 96*, 1456–1463.

Trageser, J. C., & Keller, A. (2004). Reducing the uncertainty: Gating of peripheral inputs by zona incerta. *Journal of Neuroscience, 24*, 8911–8915.

Uhlrich, D. J., Cucchiaro, J. B., Humphrey, A. L., & Sherman, S. M. (1991). Morphology and axonal projection patterns of individual neurons in the cat perigeniculate nucleus. *Journal of Neurophysiology, 65*, 1528–1541.

Umeno, M. M., & Goldberg, M. E. (1997). Spatial processing in the monkey frontal eye field. I. Predictive visual responses. *Journal of Neurophysiology, 78*, 1373–1383.

Urbain, N., & Deschênes, M. (2007). Motor cortex gates vibrissal responses in a thalamocortical projection pathway. *Neuron, 56*, 714–725.

Van Essen, D. C., Anderson, C. H., & Felleman, D. J. (1992). Information processing in the primate visual system: An integrated systems perspective. *Science, 255*, 419–423.

Van Horn, S. C., Erişir, A., & Sherman, S. M. (2000). The relative distribution of synapses in the A-laminae of the lateral geniculate nucleus of the cat. *Journal of Comparative Neurology, 416*, 509–520.

Van Horn, S. C., & Sherman, S. M. (2004). Differences in projection patterns between large and small corticothalamic terminals. *Journal of Comparative Neurology, 475*, 406–415.

Van Horn, S. C., & Sherman, S. M. (2007). Fewer driver synapses in higher order than in first order thalamic relays. *Neuroscience, 475*, 406–415.

Varela, C., & Sherman, S. M. (2007). Differences in response to muscarinic agonists between first and higher order thalamic relays. *Journal of Neurophysiology, 98*, 3538–3547.

Varela, C., & Sherman, S. M. (2008). Differences in response to serotonergic activation between first and higher order thalamic nuclei. *Cerebral Cortex, 19*, 1776–1786.

Viaene, A. N., Petrof, I., & Sherman, S. M. (2011a). Properties of the thalamic projection from the posterior medial nucleus to primary and secondary somatosensory cortices in the mouse. *Proceedings of the National Academy of Sciences of the United States of America, 108*, 18156–18161.

Viaene, A. N., Petrof, I., & Sherman, S. M. (2011b). Synaptic properties of thalamic input to layers 2/3 in primary somatosensory and auditory cortices. *Journal of Neurophysiology, 105*, 279–292.

Viaene, A. N., Petrof, I., & Sherman, S. M. (2011c). Synaptic properties of thalamic input to the subgranular layers of primary somatosensory and auditory cortices in the mouse. *Journal of Neuroscience, 31*, 12738–12747.

von Holst, E., & Mittelstaedt, H. (1950). The reafference principle. Interaction between the central nervous system and the periphery. In R. Martin (Ed., Trans.), *Selected Papers of Erich von Holst: The Behavioural Physiology of Animals and Man* (pp. 139–173). Coral Gables: University of Miami Press.

Walker, A. E. (1938). *The primate thalamus.* Chicago: University of Chicago Press.

Walls, G. L. (1953). *University of California Publications in Physiology,* Vol. 9(1), *The lateral geniculate nucleus and visual histophysiology.* Berkeley, CA: University of California Press.

Wang, S., Bickford, M. E., Van Horn, S. C., Erişir, A., Godwin, D. W., & Sherman, S. M. (2001). Synaptic targets of thalamic reticular nucleus terminals in the visual thalamus of the cat. *Journal of Comparative Neurology, 440*, 321–341.

Wang, S., Eisenback, M. A., & Bickford, M. E. (2002). Relative distribution of synapses in the pulvinar nucleus of the cat: Implications regarding the "driver/modulator" theory of thalamic function. *Journal of Comparative Neurology, 454*, 482–494.

Wei, H., Bonjean, M., Petry, H. M., Sejnowski, T. J., & Bickford, M. E. (2011). Thalamic burst firing propensity: A comparison of the dorsal lateral geniculate and pulvinar nuclei in the tree shrew. *Journal of Neuroscience, 31*, 17287–17299.

Wickelgren, B. G., & Sterling, P. (1969). Influence of visual cortex on receptive fields in the superior colliculus of the cat. *Journal of Neurophysiology, 32*, 16–23.

Wilson, J. R., Friedlander, M. J., & Sherman, S. M. (1984). Fine structural morphology of identified X- and Y-cells in the cat's lateral geniculate nucleus. *Proceedings of the Royal Society of London. Series B, Containing Papers of a Biological Character, 221*, 411–436.

Wise, S. P., Boussaoud, D., Johnson, P. B., & Caminiti, R. (1997). Premotor and parietal cortex: Corticocortical connectivity and combinatorial computations. *Annual Review of Neuroscience, 20*, 25–42.

Womelsdorf, T., Fries, P., Mitra, P. P., & Desimone, R. (2006). Gamma-band synchronization in visual cortex predicts speed of change detection. *Nature, 439*, 733–736.

Wong, P. Y., Collins, C. E., Baldwin, M. K. L., & Kaas, J. H. (2009). Cortical connections of the visual pulvinar complex in prosimian galagos (*Otolemur garnetti*). *Journal of Comparative Neurology, 517*, 493–511.

Wong-Riley, M. T. T. (1977). Connections between the pulvinar nucleus and the prestriate cortex in the squirrel monkey as revealed by peroxidase histochemistry and autoradiography. *Brain Research, 134*, 225–236.

Zeki, S. (1993). *A vision of the brain.* Oxford: Blackwell Scientific Publications.

Zhan, X. J., Cox, C. L., Rinzel, J., & Sherman, S. M. (1999). Current clamp and modeling studies of low threshold calcium spikes in cells of the cat's lateral geniculate nucleus. *Journal of Neurophysiology, 81*, 2360–2373.

20 Light Responsiveness and Photic Entrainment of the Mammalian Circadian Clock

JOHANNA H. MEIJER, SAMER HATTAR, AND JOSEPH S. TAKAHASHI

The rotation of the Earth around its axis causes 24-h rhythms in many aspects of the physical environment, while the Earth's revolution around the sun causes seasonal changes. As an adaptation to daily environmental changes, virtually all living systems have developed endogenous circadian clocks with periods of about 24 h to anticipate changes in the solar day. Anticipation of, rather than passive responsiveness to, the daily cycle is deeply embedded in biology and perhaps first evolved in cyanobacteria 3–4 billion years ago (Johnson & Golden, 1999). The photic environment was likely the selective agent both for optimization of photosynthesis as well as escape from the mutagenic effects of ultraviolet irradiation. In animals this evolutionary history persists in circadian rhythms that can be observed at the level of gene expression, cellular metabolism, endocrine function, physiology, and behavior (Takahashi, Turek, & Moore, 2001).

For proper functioning, circadian rhythms have to be synchronized, or entrained, to the day–night cycle. The major synchronizing stimulus in the environment is light. Light, rather than temperature variation or other environmental features, is a reliable signal to indicate whether it is day or night, and changes in day length are a reliable signal to indicate the time of the year. Not surprisingly, circadian clocks are responsive to light. In mammals the circadian clock is located in the suprachiasmatic nuclei (SCN) at the base of the anterior hypothalamus. Light information reaches the SCN via specialized neuronal projections, of which the retinohypothalamic tract (RHT) provides a substantial input.

Circadian rhythms are generated at the cellular level, and in mammals, individual neurons in the SCN are cell autonomous circadian oscillators (Welsh et al., 1995). At the cell level "clock" genes play a role in the generation of circadian rhythms. An autoregulatory transcriptional–translational feedback loop forms the core mechanism of the circadian clock in animals. The positive elements of the oscillator are *Clock* and *BMAL1*

whose gene products (CLOCK and BMAL1) form heterodimers to activate the transcription of the mammalian *Period* and *Cryptochrome* genes. The negative elements of the oscillator are *Period* and *Cryptochrome*, whose gene products (PERIOD and CRYPTOCHROME) accumulate, associate with each other, and translocate into the nucleus to inhibit the CLOCK/BMAL1 activation of their own transcription. As the negative elements turn over, CLOCK and BMAL1 then become active again to begin a new cycle of transcription of the *Period* and *Cryptochrome* genes (Lowrey & Takahashi, 2000; Welsh et al., 1995; Young & Kay, 2001).

This chapter focuses on how light information received by the eyes changes the phase of the circadian oscillator in the SCN to allow mammals to entrain to the light–dark environment. It starts with an overview of the behavioral responsiveness of circadian rhythms to light as it is measured in the behaving animal. This is followed by a discussion of retinal photoreceptors and the light input pathway to the SCN, the neurotransmitters involved, and the light response properties of SCN neurons. The last part of the chapter deals with intracellular responsiveness of SCN neurons to light, including the effects of light on clock gene expression.

RESPONSIVENESS OF BEHAVIORAL CIRCADIAN RHYTHMS TO LIGHT

To understand the physiological aspects of animal behavior, it is important to have robust phenotypic assays, the ability to modify molecular pathways, and precise methods to interrogate circuitry. The circadian field has provided robust behavioral assays to determine precise and quantitative aspects of daily behavior including endogenous circadian period length, circadian photoentrainment, direct light effects on behavior, and short light pulses' effects on the phase of the circadian oscillator. This section discusses how light modifies the circadian system.

Characteristically, the effects of light on the clock change as a function of time of day, and this is essential for photic entrainment. During the night the circadian system responds to light with a phase adjustment, whereas during the day this response is absent. When mammals are kept in a constant environment, their activity pattern is still rhythmic, but the period of the cycle deviates slightly from 24 h. The characteristic light response of the circadian pacemaker is best illustrated when animals are kept in constant darkness and exposed to a short pulse of light (e.g., 5–15 min) at a particular phase of the circadian cycle. A light pulse that is presented shortly after a nocturnal animal's activity onset (which corresponds to the beginning of the animal's "subjective night") will induce a phase delay of the circadian cycle (figure 20.1A). This means that on the next cycle the animal will start its activity later than was expected on basis of its previous activity rhythm. A light pulse presented toward the end of the animal's activity (which corresponds to the end of the animal's "subjective night") will induce a phase advance of the animal's rhythm (figure 20.1B). During the animal's resting phase (which corresponds to the animal's "subjective day"), light does not affect the pacemaker's phase. Importantly, both phase delays and advances are dependent on an intact retinal pathway (figure 20.1C).

The behavioral phase shifts that are induced by light pulses are permanent and an indication that the underlying pacemaker has shifted in phase. The effects of light pulses can be summarized in a phase response curve by plotting the light-induced phase shift as a function of the phase of pulse application (Pittendrigh & Daan, 1976), which shows how the pacemaker's responsiveness changes during the course of the cycle (figure 20.1D).

For diurnal species in which behavioral activity is shifted from night to day, the phase response curve is nearly identical. Light pulses presented at the beginning of the animal's resting phase ("early subjective night") induce phase delays in diurnal animals, and light pulses at the end of the resting phase ("late subjective night") produce phase advances. The phase-delaying and phase-advancing responses at the beginning and end of the subjective night are common features of circadian pacemakers and assure that animals entrain to a light–dark cycle. For example, when nocturnal animals leave their burrow too early in the evening, the evening light will induce a phase delay, causing the animal to leave its burrow later on the next cycle. In contrast, the circadian system is advanced

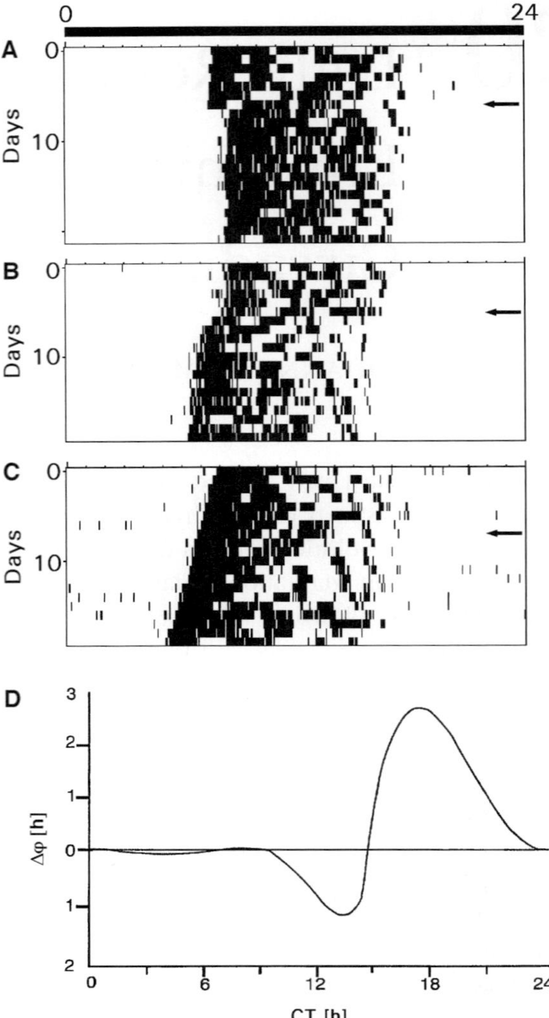

FIGURE 20.1   Phase-shifting effects of light on running wheel activity in the hamster. Consecutive days are plotted beneath one another. The presence of running wheel activity is recorded with a resolution of 1 min. Following entrainment to a light-dark cycle (L:D 14:10), hamsters were released in constant darkness, resulting in a free-running rhythm with a period that is slightly deviating from 24 hours. (A) Presentation of a 15-min light pulse (100 lux) 2 h after activity onset results in a phase delay of the free-running activity rhythm. (B) An identical light pulse that is presented 7 h after activity onset results in a phase advance of the rhythm. Panels a and b illustrate the phase dependence of the effects of light on the pacemaker. (C) Ineffectiveness of light presentation 2 h after activity onset to induce a phase delay in a blinded hamster. Panel c illustrates that the effects of light on the pacemaker are dependent on integrity of retinal input. (Based on Meijer et al., 1999) (D) Phase response curve for the phase-shifting effects of light pulses on circadian running wheel activity rhythm in the hamster. Onset of behavioral activity is defined as circadian time 12 (CT12). The phase response curve shows that presentation of light in the first few hours after activity onset results in phase delays. Delays are maximal around CT14. Toward the end of activity, light results in phase advances with maximal advances around CT19. A dead zone exists in which the animals' circadian clocks are not responding to light with a phase shift. (Based on Takahashi et al., 1984.)

when animals return to their burrow too late in the morning and see the morning light. In conclusion, photic entrainment relies on well-directed phase adjustments of the circadian pacemaker that follow from the phase-dependent responsiveness of the pacemaker to light.

Additional response properties of the circadian pacemaker to light further support the animal's ability to entrain. The magnitude of either a phase delay or a phase advance obtained at a particular phase of the cycle depends on (1) light intensity, (2) duration of the light pulse, and (3) wavelength of the light. For light pulses with durations of 5–15 min, light intensities up to a threshold level of about 0.1 lux or 10^{11} photons cm^{-2} s^{-1} for hamsters and 1 lux for rats do not phase-shift the circadian clock (Meijer, Groos, & Rusak, 1986; Meijer, Rusak, & Ganshirt, 1992; Nelson & Takahashi, 1991). This threshold intensity is very high compared to the thresholds for scotopic vision and is similar to the threshold for photopic vision. Above threshold intensity the magnitude of the phase shifts increases in a nearly linear way with light intensity. At around 100 lux saturation occurs, and further increments in intensity do not result in a further increase in phase shift. The working range of about 2 log units is very small compared to the range of light intensities that occur in the environment in the course of a day. For example on a sunny day, intensity levels are about 10^5 lux. If we transform the environmental variations in light intensity using the intensity dependence of the circadian system, the daily cycle of light in the world would become a rectangular waveform with transitions at dawn and dusk, allowing the system to discriminate between day and night effectively. In view of the function of light in entraining the system to the day–night cycle, this light responsiveness of the pacemaker makes sense.

The duration of light pulses is also an important determinant of light-induced phase shifts (Meijer, Rusak, & Ganshirt, 1992; Nelson & Takahashi, 1991). In hamsters durations ranging from milliseconds to hours can reset circadian rhythms, but the optimal duration is about 5 min (Nelson & Takahashi, 1991). Short pulses on the order of seconds are not effective stimuli and require very high intensities of light, whereas long pulses on the order of hours saturate the system and so are also less efficient on a quantum basis. Indeed, the intensity–response functions for durations between seconds and hours can all be explained by integration of the total quanta during the pulse. Under these conditions reciprocity between intensity and duration holds over the range from 1 to 45 min, suggesting that the circadian photic entrainment system acts as a "photon counter."

Complexity in Light Response

Several complexities exist in an animal's behavioral responses to light. A phase shift is often not completed on the first circadian cycle after a light pulse but can exhibit transients. Especially for phase advances, several transient cycles occur in which the phase shift gradually grows in consecutive days (figure 20.1B). A two-pulse paradigm has been used to investigate whether the transient cycles reflect the pacemaker's position or whether, alternatively, the pacemaker is fully shifted on the first day after the pulse. In the latter case transients would be a reflection of secondary mechanisms that resynchronize to the SCN only slowly (Best et al., 1999). The data indicate that the pacemaker is fully shifted on day 1, at least the light-sensitive part of the pacemaker, while interaction with secondary downstream oscillators either inside or outside the SCN determine the delayed response of the behavioral activity pattern.

Repeated or prolonged light exposure leads to saturation of the phase shift. The pacemaker shows a reduction in light response in terms of its phase-shifting capacity that persists for at least 1 h after a subsaturating light pulse (Nelson & Takahashi, 1999). Two hours after a light pulse the pacemaker's responsiveness seems to recover (Best et al., 1999). The decreased responsiveness to light does not correspond with the pacemaker's ability to track day length and suggests that day length is coded for in a different way.

Light pulses cause period changes in addition to phase shifts (DeCoursey, 1989; Hut, van Oort, & Daan, 1999). The light-induced changes in period depend on the circadian phase of the pulse application. The changes in period occur such that they may contribute to photic entrainment (i.e., period lengthening occurs after light pulses during early night, and period shortening occurs following pulses during the late night).

Light is the most important but not the only phase-resetting stimulus because behavioral activity of the animal can also cause phase shifts (Maywood et al., 1999; Mrosovsky et al., 1989; Reebs & Mrosovsky, 1989). Behavioral activity of the animal causes phase shifts when the activity is triggered by, for instance, cage cleaning or by presenting a dark pulse against an otherwise illuminated background. Forced activity of the animal is less effective in producing a phase shift. Behavioral activity has phase-shifting effects when it occurs during the day but not when it occurs during the night. Nevertheless, it has the capacity at night to inhibit light-induced phase shifts, indicating that physical activity affects the circadian system and its responsiveness to light.

In addition to the light influence on the circadian phase, light can directly affect several parameters, which include activity and hormonal levels. Interestingly, the intensity–response curves for phase shifting and melatonin suppression are different. Melatonin production in the pineal gland is high during the night and low during the day and still cycles when animals are kept in constant darkness (Illnerova, 1991; Illnerova & Sumova, 1997). The nighttime elevation of melatonin can be suppressed by light through a multisynaptic pathway involving the SCN. Light suppression of melatonin appears substantially more sensitive to irradiance as compared to phase shifting (Nelson & Takahashi, 1991). The difference is 1.4 log units when half-saturation values are compared. As light information reaches the pineal gland via the SCN, the increased sensitivity cannot readily be explained and indicates differences in the organization of input pathways.

NEURONAL INPUT PATHWAYS

Despite the well-established role for extraretinal photoreception in many vertebrate species, mammals have lost the ability to perceive light extraretinally, and photoentrainment is mediated exclusively via the retina (Yamazaki, Goto, & Menaker, 1999). Action spectra have indicated a maximal sensitivity of the circadian system in its phase-shifting response to wavelengths of about 480–500 nm (Provencio & Foster, 1995; Takahashi et al., 1984) and a secondary peak in the near-ultraviolet range of the spectrum (Amir & Robinson, 1995). Wavelength sensitivity is determined largely by the retinal pigments that mediate photic entrainment. The maximal sensitivity at about 500 nm predicted a role for opsin-based photopigments, and recent studies allowed us to create an accurate model of how retinal photoreceptors influence circadian photoentrainment. Initially, however, two surprising results were observed in mice that lack both rods and cones: normal phase-shifting responses to 509-nm monochromatic light pulses (Freedman et al., 1999) and normal suppression of pineal melatonin in response to light (Lucas et al., 1999). These results indicated that classical photoreceptors do not fully account for light effects on the circadian pacemaker and suggested the presence of non-rod/non-cone photoreceptors. The race was on to identify the non-rod/non-cone photoreceptors that allow the rodless/coneless animals to phase shift and suppress their pineal melatonin in response to light. A classical study revealed that retinal ganglion cells, usually the output neurons of the retina, that project to the SCN are capable of responding to light in the absence of rod and cone signaling (Berson, Dunn, & Takao, 2002).

These neurons, now called intrinsically photosensitive retinal ganglion cells or ipRGCs, depolarize to light even when synaptic input from rods and cones is blocked or when they are surgically detached from the retina (Berson, Dunn, & Takao, 2002). The spectral sensitivity function of these ganglion cells shows a peak at 484 nm. Constant illumination of the cells caused a tonic depolarization and increase in discharge rate, and the change in membrane potential appears to be a function of light intensity. Latencies in response are long and are on the order of several seconds. Physiological recordings of retrogradely labeled SCN-projecting ganglion cells have shown similar characteristics. These ganglion cells showed sustained responses to light and receptive field sizes of 2° to 5° (Pu, 2000). It has been suggested that converging W (or type III) cells account for the response characteristics of SCN neurons (Cooper, Herbin, & Nevo, 1993; Groos & Mason, 1980; Hattar et al., 2002). The axons of these cells are fine and unmyelinated and make up the retinohypothalamic tract (RHT), which projects directly via the optic chiasm to the SCN (Moore, 1973). The terminal field is in the ventral and often also the lateral part of the SCN in most species of rodents (figure 20.2).

The discovery of ipRGCs raised the next logical question: What imparts photosensitivity to these retinal ganglion cells? The ipRGCs express two probable photopigments: cryptochrome and melanopsin. This expression pattern produced two divergent views about which photopigment may be responsible for the light detection in ipRGCs. A view posited on the basis of a vitamin A deprivation study in mammals (Selby et al., 2000; Thompson et al., 2004) that cryptochrome, which is a flavin-based photopigment in plants and *Drosophila* (Cashmore, 2003), must be responsible. The other view

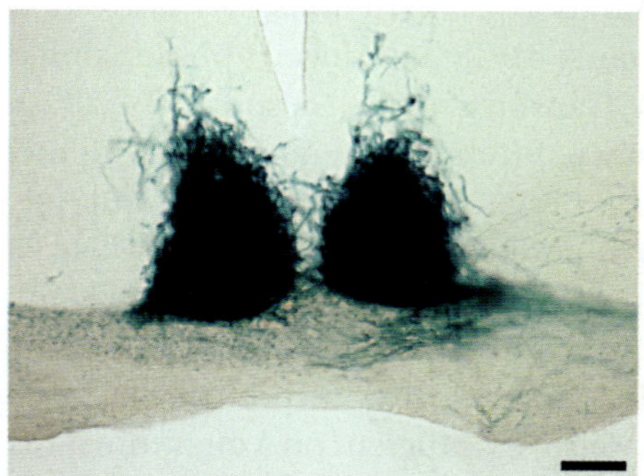

FIGURE 20.2   Retinal innervation of the SCN in the mouse as revealed by X-gal staining from a genetically modified mouse that expresses the *tau-LacZ* gene from the melanopsin locus. The *tau-LacZ* expresses beta-galactosidase, and the tau allows the enzyme to be transported down the axons. ipRGCs expressing melanopsin will stain blue, and their axons can be traced to the SCN (Hattar et al., 2006).

   JOHANNA H. MEIJER, SAMER HATTAR, AND JOSEPH S. TAKAHASHI

posited that melanopsin, which is a vitamin A–based photopigment originally cloned from photosensitive skin melanophores and later found to be expressed in the retina (Provencio et al., 1998, 2000), is responsible for the light sensitivity in ipRGCs. To resolve this issue, two independent groups produced animals that lack the melanopsin protein in addition to lacking outer retinal photoreceptor function either through degeneration (Panda et al., 2003) or through eliminating the phototransduction signaling pathways in rods and cones (Hattar et al., 2003). The data from both these animals were unequivocal: the mice failed to photoentrain (Hattar et al., 2003; Panda et al., 2003) even though cryptochrome levels were unaffected (Hattar et al., 2003). This indicates that cryptochrome does not contribute to circadian photoentrainment. Subsequent studies indicated that mammalian cryptochromes are nonphotoreceptive (Zhu et al., 2005), and their function is confined to the canonical clockwork machinery. The focus now shifts to the contribution of vitamin A–based light signaling to circadian light functions in mammals.

INTERACTIONS BETWEEN CONVENTIONAL RETINAL PHOTORECEPTORS AND IPRGCS

Initially, ipRGCs were thought to be a homogeneous population of cells, but recent evidence shows surprising diversity in morphological, electrophysiological, and behavioral outputs (Ecker et al., 2010; Schmidt, Chen, & Hattar, 2011). In this chapter we concentrate on the originally identified ipRGCs, which are now known as M1 ipRGCs, because exploration of all the ipRGC subtypes is beyond the scope of this chapter, and their contribution to circadian photoentrainment is minimal.

M1 ipRGCs represent a small subset of retinal ganglion cells (about 1% of the total population) and robustly express the photopigment melanopsin (Hattar et al., 2002). The intrinsic photosensitivity of these cells is absolutely dependent on the melanopsin protein (Lucas et al., 2003). In contrast to rods and cones, which have modified cilia that enhance photon capture, ipRGCs do not have modified structures for light absorption, as they are canonical retinal ganglion cells. The lack of modified structures for photon capture is compensated for by the high gain in signaling captured photons (Do et al., 2009). Melanopsin is present in the axons, cell bodies and proximal dendrites of these neurons (Hattar et al., 2002). Based on a reporter mouse line that expresses tau-beta galactosidase under the melanopsin promoter, M1 ipRGCs innervate the SCN bilaterally and also project to the intergeniculate leaflet (IGL), the pretectum, and to a lesser extent the ventral lateral geniculate nucleus (Hattar et al., 2006). It was suggested, therefore, that melanopsin plays a role in transducing light information to non-image-forming visual brain areas (Hattar et al., 2002, 2006). To directly test this hypothesis, several groups eliminated the ipRGCs using three independent yet complementary methods (Göz et al., 2008; Güler et al., 2008; Hatori et al., 2008). Eliminating ipRGCs did not affect visual functions (Güler et al., 2008) but caused major deficits in circadian photoentrainment (Göz et al., 2008; Güler et al., 2008; Hatori et al., 2008). The data showed that rod/cone light input to circadian entrainment requires ipRGCs.

Now that we know that ipRGCs are the only relay to the SCN, the question is what is the contribution of the outer retinal photoreceptors to circadian photoentrainment? By combining a variety of retinal mutant mice, it was shown that cone-based photoreception plays a minor role in circadian photoentrainment (Lall et al., 2010) and that, unexpectedly, rods are the predominant photoreceptors, in addition to ipRGCs, for circadian photoentrainment (Altimus et al., 2010). This was a surprising result because the exquisite sensitivity of rods to light was thought to be inconsistent with the rather low sensitivity of circadian light responses. Additionally, rods are capable of signaling for nonimage functions even at light intensities at which they are bleached and thus incapable of supporting image detection. Finally, it was shown that rods use different retinal circuits to contribute to circadian photoentrainment at both low- and high-light intensities (Altimus et al., 2010).

Although outer retinal photoreceptors contribute to circadian photoentrainment, the absence of the melanopsin protein, which leads to lack of an intrinsic light response in ipRGCs, causes substantial deficits in circadian phase delays (Panda et al., 2002; Ruby et al., 2002). In addition, melanopsin knockout animals exhibit larger deficits when tested on functions that depend on prolonged photon counting. For example, placing animals in constant-light conditions or exposing animals to several hours of light at night leads to lengthening of the circadian period or inhibition of wheel-running activity, respectively. Melanopsin knockout animals particularly fail at lengthening the period in constant-light conditions (Panda et al., 2002; Ruby et al., 2002) or maintaining the inhibition of activity in the dark for a prolonged light pulse (Mrosovsky & Hattar, 2003). This indicates that the intrinsic melanopsin response contributes more robustly than the extrinsic rod/cone input to circadian light functions and provides a possible explanation of why melanopsin was evolutionarily preserved.

As mentioned earlier M1 ipRGCs project to several brain regions in addition to the SCN and IGL that are important for circadian photoentrainment. A surprising result indicated that mice lacking the transcription factor Brn3b (also known as Pou4f2), which show 80% RGC loss, lack innervation of the majority of the M1 ipRGC brain targets but maintain normal innervation of the SCN (Badea et al., 2009). Combining the promoter activities of Brn3b and melanopsin genes to genetically eliminate Brn3b-positive ipRGCs led to a reduction in the number of ipRGCs in the mouse retina to only 200 cells. Remarkably, these 200 ipRGCs predominantly innervated the SCN and to a lesser extent the IGL and were sufficient for circadian photoentrainment (Chen, Badea, & Hattar, 2011).

From the Brn3b studies mentioned above, we conclude that the RHT is predominantly important for photic entrainment, and hence we focus on this pathway in this chapter. Three other photic pathways reach the SCN: via the intergeniculate leaflet of the lateral geniculate nucleus; via the raphe; and via the pretectum (Meijer, 2001). The raphe projects with serotonergic fibers to the SCN, while the intergeniculate nucleus projects with GABA-, NPY-, and enkephalin-containing neurons (Harrington, 1997; Morin, 1994). The dominant effect of the raphe and intergeniculate leaflet on the SCN is inhibitory.

The RHT contains three different neurotransmitters: (1) excitatory amino acids; (2) PACAP (pituitary adenylate cyclase–activating peptide); and (3) substance P. Increased glutamate immunoreactivity has been demonstrated inside the SCN (Castel et al., 1993; de Vries et al., 1993). In situ hybridization and immunocytochemical studies have indicated the presence of ionotropic and metabotropic (mGluR1) receptor subunits. Ionotropic receptors are usually divided into NMDA and non-NMDA receptors, and both are present in the SCN. Intraperitoneal administration of a noncompetitive antagonist of the NMDA receptor (MK-801) blocks the phase-shifting effects of light pulses (Colwell, Foster, & Menaker, 1991) and of electrical stimulation of the RHT (de Vries et al., 1994), whereas application of glutamate or NMDA mimics the effects of light on the pacemaker's phase in vitro (Ding et al., 1994; Shirakawa & Moore, 1994) as well as in vivo (Mintz et al., 1999; Vindlacheruvu et al., 1992). A role for the NMDA receptor in mediating phase shifts is therefore relatively well established. Several non-NMDA receptor antagonists also block phase shifting by light (Ebling, 1996).

PACAP is a member of the vasoactive polypeptide (VIP)/glucagon/secretin/growth-hormone–releasing hormone superfamily, and it is present in the terminals of the RHT. PACAP-containing ganglion cells contain the photopigment melanopsin (Hannibal et al., 2002). PACAP induces phase shifts during the night when applied at physiological dosages (Harrington et al., 1999), probably through potentiation of NMDA currents.

Substance P is contained in retinal ganglion cells projecting to the SCN, and after enucleation, the density of substance P–containing fibers in the SCN is markedly reduced (Mikkelsen & Larsen, 1993). The acute effect of substance P on SCN neurons is excitatory. Substance P potentiates glutamatergic activation (Piggins & Rusak, 1997), and its phase-shifting effect in vitro is similar to that of glutamate or NMDA (Hamada et al., 1999; Kim et al., 2001). Substance P–induced phase shifts can be blocked not only by the substance P receptor antagonists spantide or L-703,606 but also by NMDA and non-NMDA receptor antagonists (Hamada et al., 1999; Kim et al., 2001). Conversely, however, the substance P antagonist L-703,606 does not block glutamate-induced phase delays, suggesting that glutamate is downstream from substance P (Kim et al., 2001).

ELECTROPHYSIOLOGICAL RESPONSE PROPERTIES OF SCN NEURONS

SCN neurons are among the smallest in the central nervous system, making it difficult for electrophysiological recordings using intracellular methods without causing damage to the cells and changes in cellular function. Only a few studies have investigated the membrane properties of retinorecipient SCN neurons by these techniques. The few available studies with intracellular recordings showed that retinorecipient SCN neurons are characterized by a high input resistance and a membrane potential of about –60 mV (Jiang et al., 1997; Kim & Dudek, 1993). A circadian variation in conductance was demonstrated in SCN neurons, with higher conductances observed during subjective day/dawn and lower conductances during subjective dusk, while the membrane potential appeared high during day and low during the night (de Jeu, Hermes, & Pennartz, 1998; Jiang et al., 1997; Schaap et al., 1999). This rhythmic change in membrane potential causes several intracellular oscillations in the SCN, including fluctuations in the levels of intracellular calcium, MAP kinase (MAPK), cAMP, and nitric oxide (NO) (Colwell, 2011). Of particular interest are the rhythmic levels of cAMP, which are now known to be a core component of the mammalian circadian pacemaker (O'Neill et al., 2008). cAMP, through its effector exchange protein activated by cAMP (EPAC), leads to the phosphorylation of CRE-binding protein (CREB) to regulate transcription of

 JOHANNA H. MEIJER, SAMER HATTAR, AND JOSEPH S. TAKAHASHI

clock genes such as *mPer1* and *mPer2* (Hastings, Maywood, & O'Neill, 2008). Recent work has shown that *cryptochromes* can regulate intracellular cAMP signaling, providing a further link between the cAMP oscillator and the clockwork transcriptional translational feedback oscillator (Zhang et al., 2010).

The electrical properties of light-responsive SCN neurons have been studied mainly by extracellular recording methods. These recordings have revealed that a subpopulation of SCN neurons responds to light stimulation with a change in discharge rate. The location of these light-responsive SCN neurons corresponds with the terminal field of the RHT (Meijer, Groos, & Rusak, 1986), indicating that at least part of the neurons may be monosynaptically driven by retinal ganglion cells. In the nocturnal rat, mouse, and hamster, about one-third of the SCN neurons respond to light (Aggelopoulos & Meissl, 2000; Brown et al., 2011; Cui & Dyball, 1996; Groos & Mason, 1980; Meijer, Groos, & Rusak, 1986). The receptive fields of SCN neurons are large, often exceeding 20° in diameter, and are characterized by an absence of antagonistic center-surround organization (Groos & Mason, 1980). The SCN neurons respond with either a suppression or an activation to light spots that are presented throughout large parts of the visual field. Their large receptive fields show that light-responsive SCN neurons integrate light from large areas in the visual field (Groos & Mason, 1980), which is consistent with the location and projection area of M1 melanopsin-containing ganglion cells (Schmidt, Chen, & Hattar, 2011).

Most light-responsive SCN neurons of the rat and hamster show an excitation in response to light, although a smaller fraction are light suppressed (Aggelopoulos & Meissl, 2000; Brown et al., 2011; Cui & Dyball, 1996; Groos & Mason, 1980; Meijer, Groos, & Rusak, 1986) or show transient responses only (Brown et al., 2011). The origin of the suppression is not clear, as glutamate and PACAP are the major neurorotransmitters of the RHT. Electrical stimulation of the SCN-containing brain slice revealed that part of the suppression may be driven by GABAergic interneurons but that other suppressions seem to be driven directly by the RHT (Irwin & Allen, 2009; Jiao & Rusak, 2003).

Studies in diurnal 13-lined ground squirrel (*Spermophilus tridecemlineatus*) and diurnal degu (*Octodon degus*) have demonstrated that in the diurnal species the relative contribution of activated and suppressed neurons is very different and that a much larger proportion of light-responsive SCN neurons are suppressed by light (Jiao, Lee, & Rusak, 1999; Meijer, Rusak, & Harrington, 1989). Response latencies are long and range from 25 ms to several seconds (Brown et al., 2011;

Meijer, Groos, & Rusak, 1986; Sawaki, 1979). Most responses developed fully between 200 ms and 20 s after the onset of a light pulse, so that brief light flashes of less than 1 s are not very effective in stimulating these neurons (Meijer, Groos, & Rusak, 1986; Sawaki, 1979). The time course of the responsiveness is extraordinary slow as compared to responses in visually responsive brain areas and is consistent with the long latencies that have been observed in melanopsin-containing ganglion cells (Berson, Dunn, & Takao, 2002; Dacey et al., 2005). It is also possible that the delayed responsive neurons are driven indirectly via the IGL (Brown et al., 2011). The difference in effectiveness of long versus short stimuli reflects a remarkable property of photically responsive SCN cells.

Responses of SCN neurons are typically sustained despite a short transient "overshoot" at the onset of a light response and a short suppression at light-off (figure 20.3). The sustained response shows no adaptation for light pulses of 30 min or longer (Meijer et al., 1998), which is an exceptional response characteristic of SCN neurons. The sustained light responsiveness of the SCN together with its large receptive-field properties allow the SCN to monitor ambient light levels reliably in the natural environment. The sustained response of SCN neurons is in contrast with the light responses observed in many other visual brain areas, which show transient rather than sustained responses. Other areas that show sustained responses are the intergeniculate leaflet of the geniculate nucleus (Harrington, 1997), the raphe nucleus (Mosko & Jacobs, 1974), and the pretectum (Trejo & Cicerone, 1984); all receive direct input from ipRGCs. The capacity to respond with a relative lack of adaptation is thought to rely strongly on melanopsin-containing retinal ganglion cells. Recordings in the SCN of melanopsin-deficient animals have shown a decrease in, but not elimination of, the sustained discharge level as compared to the response magnitude observed in WT animals (Mure et al., 2007). These results correspond with recordings of ipRGCs in melanopsin-deficient and WT animals (Schmidt & Kofuji, 2010) and with recordings in the thalamocortical visual system (Brown et al., 2010).

Recordings of the SCN of freely moving mice have revealed a special role for UV light in the regulation of sustained discharge rates (Van Oosterhout, 2012). This response is possible because the crystalline lens of the mouse allows transmission of UV radiation. UV light can be detected by all photoreceptors, and specifically by a short-wavelength–sensitive cone (UVS cone), which is present in the murine retina, showing maximal sensitivity (λ_{max}) of 360 nm. The sustained pattern of firing of SCN neurons is unaffected by the loss of melanopsin,

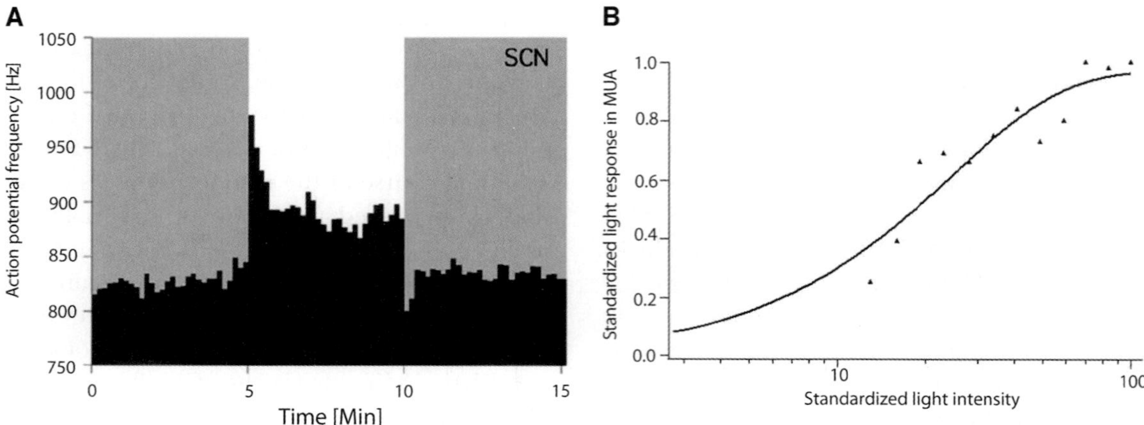

FIGURE 20.3   Light-activated response in the SCN. (A) The example shows a transient overshoot in electrical impulse frequency at the onset of the light pulse, a sustained discharge level during the light presentation, and a transient decrease at the offset of the light. The recording was performed with an implanted stationary electrode in an unanesthetized mouse. The duration of the pulse was 5 min. (B) Intensity–response curve for the sustained level of discharge. White light pulses of 100 s were applied, and neutral density filters were used to dim the light. The maximal light intensity was normalized to 100% and corresponds with 1,100 μW/cm^2.

demonstrating that sustained responses can occur independently of melanopsin phototransduction. A functional role for UV light is underscored by the ability of UV to induce phase shifts (Provencio & Foster, 1995), to suppress pineal melatonin production (Benshoff et al., 1987; Brainard et al., 1994; Podolin, Rollag, & Brainard, 1987), and to induce c-fos expression in the SCN. Future studies should determine why cones then are unable to drive photoentrainment alone (Lall et al., 2010) despite a strong and sustained input from UV light to the SCN (Van Oosterhout, 2012). An intriguing possibility is that long-wave and short-wave cones couple differently to ipRGCs, which are the only relay to the SCN.

The steady-state discharge rate that is obtained during a light pulse increases with light intensity (Brown et al., 2011; Groos & Mason, 1980; Meijer, Groos, & Rusak, 1986; Meijer, Rusak, & Ganshirt, 1992; Nakamura et al., 2004). The intensity response curve for SCN neurons is sigmoid, with a small working range (Meijer, Groos, & Rusak, 1986; Nakamura et al., 2004). These response curves are nearly identical to the intensity–response curve for phase shifting (Meijer, Rusak, & Ganshirt, 1992; Mure et al., 2007; Takahashi et al., 1984).

INTRACELLULAR SIGNALING RESPONSES TO LIGHT

The light responsiveness of SCN neurons changes in the course of the circadian cycle. With implanted electrodes in freely moving animals, it was shown that the light response during the night is larger than the response during the day (Meijer et al., 1998). The rhythm in light responsiveness is superimposed on the rhythm in baseline activity. During the subjective day, when baseline discharge is high, light presentations elicit only small responses. During the subjective night, when baseline discharge is low, the light responses are large and exceed the response levels that are obtained during daytime (figure 20.4). These data are also consistent with in vitro SCN recordings showing enhanced NMDA receptor activity during the night (Pennartz, Hamstra, & Geurtsen, 2001) and enhanced RHT-mediated c-Fos expression and AMPA-stimulated calcium responses during the night (Bennett, Aronin, & Schwartz, 1996; Michel, Itri, & Colwell, 2002). Combined, these studies show how the same light pulse has little effect during the day but large effects during the night.

Recordings in mammalian SCN neurons have now shown that NMDA induces calcium transients that are large during the night and small during the day (Colwell, 2001). Thus, the rhythm in calcium response follows the rhythm in light-induced changes in discharge rate. This is not surprising because calcium influx is triggered by depolarization of the membrane potential. The intracellular calcium response is one of the best-studied intracellular responses to light and stimulates, by itself or in concert with other second messengers, a complex cascade of intracellular events. A few elements of the intracellular signal transduction pathway have been identified that lead to transcriptional activation (Gillette, 1997). Light-induced release of glutamate and activation of both NMDA and non-NMDA

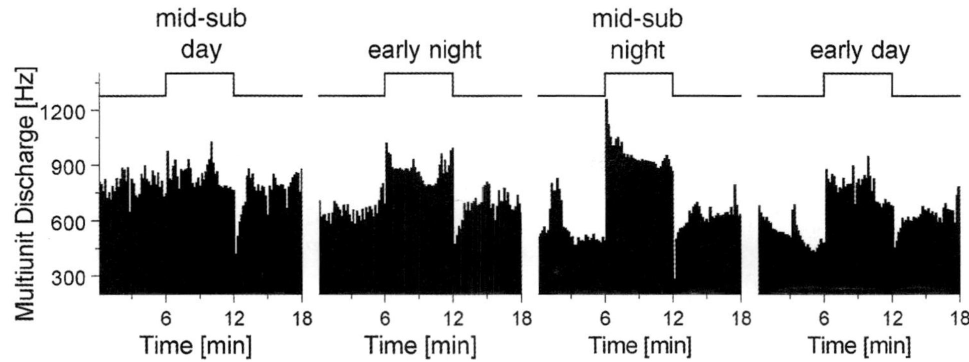

FIGURE 20.4   Light response of SCN neurons as a function of circadian time based on Meijer et al. (1998). The timing of a 6-min light pulse is indicated above the records. Discharge rate of SCN neurons is given per 10 s. Light pulse presentation during mid–subjective day (CT 7) to a dark-adapted animal does not induce an increase in electrical activity. At the offset of the pulse a transient suppression is visible. During early night (CT 12), light induces a clear sustained increase in neuronal discharge rate. Maximal light responses are obtained during mid–subjective night (CT19). Toward the end of the night responses decrease, and only small responses are obtained during early subjective day (CT2).

receptors lead to an increase in intracellular calcium (Tominaga et al., 1994). Intracellular calcium can be increased by calcium influx across the membrane and/or by calcium release from intracellular stores. Cellular organelles such as the endoplasmatic reticulum function in the storage of calcium. Stimulation of calcium channels on intracellular organelles results in phase delays during early night but not in phase advances during late night. Blocking of the ryanodine receptor by dantrolene and ruthenium red blocks light- and glutamate-induced delays during early night but not the glutamate-induced advances during late night (Ding et al., 1998). These data indicate that the ryanodine receptors play a role that is restricted to the early night.

Intracellular increase in calcium levels leads to formation of nitric oxide (NO) (Amir, 1992; Ding et al., 1994) via activation of nitric oxide synthethase (NOS). NO produces phase advances and delays depending on circadian time of NO application. The effects of NO are similar to those of light, and blocking of NO transmission disrupts light transmission to the SCN.

A third intracellular component of importance is cGMP (Prosser, McArthur, & Gillette, 1989). The SCN is reset by cGMP during the night but not during the day. The effect of cGMP is a phase advance, with a maximum phase shift at mid–subjective night. The phase response curve for cGMP shows no delay zone, and thus, it does not resemble the phase response curve for light. However, the sensitive part of the phase response curve is similar to the period in which the pacemaker responds to light with a phase shift (Prosser, McArthur, & Gillette, 1989).

Light has been shown to induce phosphorylation of the transcription factor cAMP response element-binding protein (CREB) in the hamster SCN (Ginty et al., 1993; Kornhauser et al., 1996). Phosphorylation occurs within several minutes after light exposure and occurs only at night and not during the day (Ginty et al., 1993; Kornhauser et al., 1996). Increased phosphorylation of CREB can be blocked by the NMDA receptor antagonist APV, indicating that it results from activation of the NMDA receptor (Ginty et al., 1993). Depolarization of the membrane potential by itself and in the presence of APV is also able to induce CREB, indicating that NMDA receptor activation can be bypassed if membrane depolarization occurs (Ginty et al., 1993).

As discussed earlier in this chapter, low concentrations of PACAP cause phase shifts that resemble the phase response curve for light pulses. Several studies have indicated that cAMP-PKA and also calcium-dependent pathways are involved in PACAP-induced phase shifts during the night (Kopp et al., 1999; Tischkau et al., 2000; von Gall et al., 1998). Moreover, PACAP can stimulate CREB (Gillette, 1997). There is substantial evidence, therefore, that PACAP is involved in photic entrainment by stimulating intracellular components that also respond to light or NMDA.

Intracellular calcium also activates the MAP kinase (MAPK) signaling pathway in the SCN. Once phosphorylated, MAPK is translocated from the cytosol into the nucleus, where it can regulate transcriptional activity (Obrietan, Impey, & Storm, 1998). MAPK shows an endogenous circadian rhythmicity (Obrietan, Impey, & Storm, 1998). Light induces activation of MAPK during the night but not during the day, and MAPK activation depends on the duration of light exposure, with 15-min light pulses being more effective than 5-min pulses (Obrietan, Impey, & Storm, 1998). Glutamate application or membrane depolarization also triggers activation of MAPK. Blocking the MAPK pathway results in a 50% reduction of glutamate-induced CREB phosphorylation in the SCN. Taken together these results show that MAPK is an important signaling component for CREB phosphorylation in the SCN (Obrietan, Impey, & Storm, 1998).

Two circadian clock genes, *mPer1* and *mPer2*, homologues of *Drosophila Per*, are thought to be involved in light-induced resetting of the pacemaker (Albrecht et al., 1997; Field et al., 2000; Hastings et al., 1999; Shearman et al., 1997; Takumi et al., 1998a; Yan et al., 1999; Zylka et al., 1998). The expression of *mPer1* and *mPer2* is regulated not only by light but also by the CREB pathway (Gillette, 1997), by NMDA (Moriya et al., 2000), and by PACAP (Nielsen et al., 2001). Increases in *mPer1* occur within 10 min after light exposure, which is very rapid (Shigeyoshi et al., 1997), and show peaks in expression about 60 min after light exposure (Albrecht et al., 1997; Shearman et al., 1997; Takumi et al., 1998a). Expression of *mPer1* is initially present in the retinorecipient area of the SCN and spreads through the SCN in the course of 60–120 min (Albrecht et al., 1997; Shigeyoshi et al., 1997). In transgenic rats or mice with a reporter gene under the control of the *mPer1* promoter, light results in increased *mPer1* expression (figure 20.5) and in rapid shifts of *mPer1* expression cycle (Kuhlman, Quintero, & McMahon, 2000; Wilsbacher et al., 2002; Yamazaki et al., 2000). In cultured SCN slices with an *mPer1:luc* reporter gene, the cycle in *mPer1* is quickly reset by NMDA (Asai et al., 2001). As with light, *mPer1* induction by NMDA occurs only during the night and not during the day. Evidence for a role of *mPer1* in light-induced delays is provided by experiments that showed that intracerebroventricular injections of antisense oligonucleotide repress light-induced phase shifts in running wheel activity and that direct application in vitro attenuates glutamate-induced delays of neuronal firing rhythms (Akiyama et al., 1999).

As with *mPer1*, *mPer2* is induced by light at a much more delayed time (about 90 min after light exposure), although different studies show somewhat different results in the time required to show maximal expression. Light-induced *mPer2* expression is dependent on circadian time, and maximal responsiveness exists during the end of the subjective day and during the first part of the night. In contrast to *mPer1* and *mPer2*, *mPer3* appears unresponsive to light (Albrecht et al., 1997; Shearman et al., 1997; Takumi et al., 1998b; Zylka et al., 1998). Consistent with the light induction of both *mPer1* and *mPer2*, a complete block of light-induced phase delays is obtained when both *mPer1* and *mPer2* antisense oligonucleotides are injected, indicating that *mPer1* and *mPer2* have additive effects on photic entrainment (Wakamatsu et al., 2001).

Thus far the mammalian cryptochromes have not been shown to be functional photopigments (Hall, 2000; Sancar, 2000). Rather, in mammals, *mCry1* and *mCry2* are essential components of the molecular clockwork (van der Horst et al., 1999; Vitaterna et al., 1999), where they play a role as negative elements in the autoregulatory feedback loop. Indeed *mCry1/mCry2* double knockouts appear to have intact *mPer1* and *mPer2* light responses in the SCN (Okamura et al., 1999).

At a conceptual level a major question to be answered is how the phase response curve, with its bidirectional response to light, can be explained. In *Drosophila* and *Neurospora* there is indication that light-regulated clock components such as *timeless* and *frq* show a unidirectional response to light throughout the circadian cycle. This response is associated with phase delays and advances solely on the basis of the time by which the response occurs. In mammals there is substantial evidence for divergence in the intracellular transduction cascade resulting in specific cues that mediate advances and delays (figure 20.6). Whether divergence in intracellular signals accounts for phase advances and delays remains to be proven.

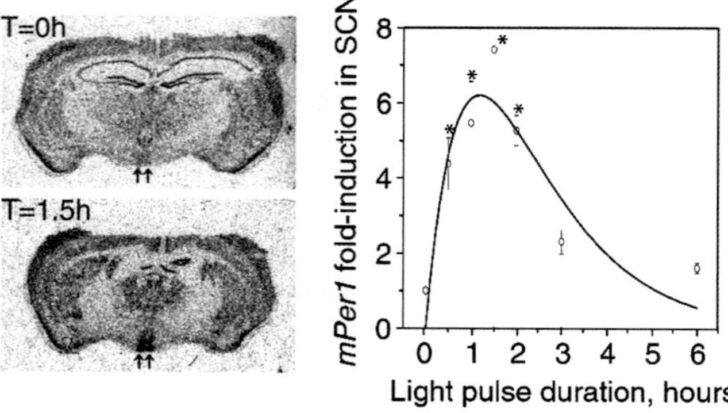

FIGURE 20.5 Light-induced *mPer1* mRNA expression in the SCN. Mice were maintained in constant darkness for at least 7 days prior to the experiment, and the endogenous circadian rhythm of each animal was measured. At CT17, mice were transferred to a light box. The graph to the right shows a time course for *mPer1* induction over 6 h. The *y*-axis indicates the fold induction from baseline expression at CT17 (Wilsbacher et al., 2002).

SUMMARY

It has become evident that the light responsiveness of the SCN differs in many ways from light responsiveness in brain areas involved in vision. The discovery of a set of novel ganglion cells that are intrinsically

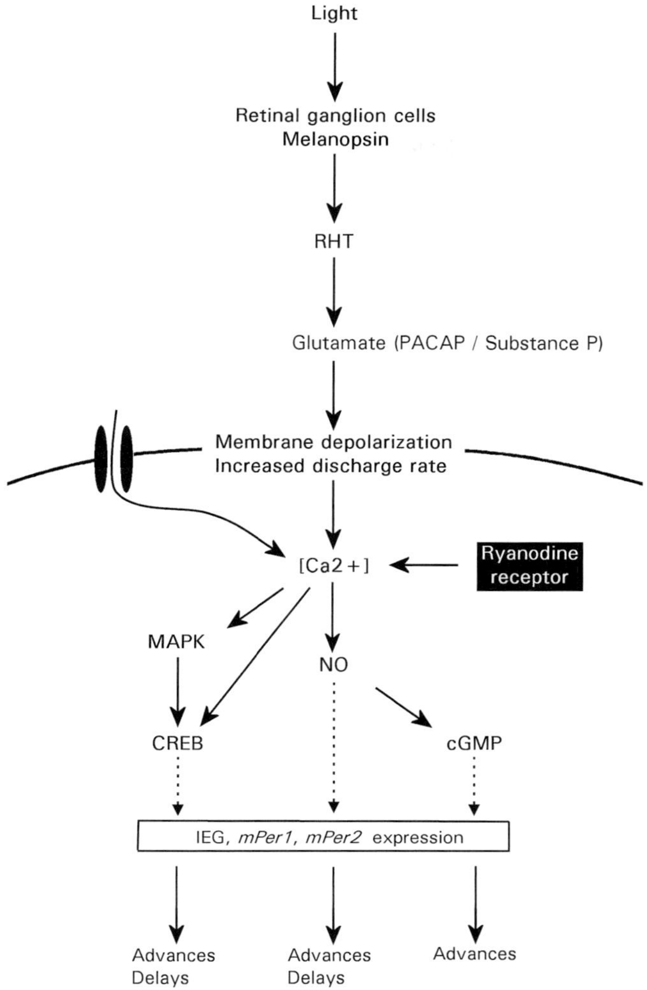

FIGURE 20.6 Schematic overview of light input pathways to the circadian clock. See text for details.

photosensitive has answered some of the most intriguing questions about the ability of SCN neurons to code for luminance and to show sustained responses to light. These melanopsin-containing retinal ganglion cells release glutamate and PACAP to influence the intracellular signaling in the SCN, ultimately affecting clock gene transcription. The daily rhythm of SCN membrane potential provides a time window or gate for modulating neuronal responses, and although the clock is insensitive to light-induced phase shifts during the day, it responds with bidirectional phase shifts during the night in order to promote entrainment.

REFERENCES

Aggelopoulos, N. C., & Meissl, H. (2000). Responses of neurones of the rat suprachiasmatic nucleus to retinal illumination under photopic and scotopic conditions. *Journal of Physiology, 523*(Pt 1), 211–222.

Akiyama, M., Kouzu, Y., Takahashi, S., Wakamatsu, H., Moriya, T., Maetani, M., et al. (1999). Inhibition of light- or glutamate-induced *mPer1* expression represses the phase shifts into the mouse circadian locomotor and suprachiasmatic firing rhythms. *Journal of Neuroscience, 19*, 1115–1121.

Albrecht, U., Sun, Z. S., Eichele, G., & Lee, C. C. (1997). A differential response of two putative mammalian circadian regulators, mper1 and mper2, to light. *Cell, 91*, 1055–1064.

Altimus, C. M., Guler, A. D., Alam, N. M., Arman, A. C., Prusky, G. T., Sampath, A. P., et al. (2010). Rod photoreceptors drive circadian photoentrainment across a wide range of light intensities. *Nature Neuroscience, 13*, 1107–1112. doi:10.1038/nn.2617.

Amir, S. (1992). Blocking NMDA receptors or nitric oxide production disrupts light transmission to the suprachiasmatic nucleus. *Brain Research, 586*, 336–339.

Amir, S., & Robinson, B. (1995). Ultraviolet light entrains rodent suprachiasmatic nucleus pacemaker. *Neuroscience, 69*, 1005–1011.

Asai, M., Yamaguchi, S., Isejima, H., Jonouchi, M., Moriya, T., Shibata, S., et al. (2001). Visualization of *mPer1* transcription in vitro: NMDA induces a rapid phase shift of *mPer1* gene in cultured SCN. *Current Biology, 11*, 1524–1527. doi:10.1016/S0960-9822(01)00445-6.

Badea, T. C., Cahill, H., Ecker, J., Hattar, S., & Nathans, J. (2009). Distinct roles of transcription factors brn3a and brn3b in controlling the development, morphology, and function of retinal ganglion cells. *Neuron, 61*, 852–864.

Bennett, M. R., Aronin, N., & Schwartz, W. J. (1996). In vitro stimulation of c-Fos protein expression in the suprachiasmatic nucleus of hypothalamic slices. *Brain Research. Molecular Brain Research, 42*, 140–144.

Benshoff, H. M., Brainard, G. C., Rollag, M. D., & Lynch, G. R. (1987). Suppression of pineal melatonin in *Peromyscus leucopus* by different monochromatic wavelengths of visible and near-ultraviolet light (UV-A). *Brain Research, 420*, 397–402.

Berson, D. M., Dunn, F. A., & Takao, M. (2002). Phototransduction by retinal ganglion cells that set the circadian clock. *Science, 295*, 1070–1073.

Best, J. D., Maywood, E. S., Smith, K. L., & Hastings, M. H. (1999). Rapid resetting of the mammalian circadian clock. *Journal of Neuroscience, 19*, 828–835.

Brainard, G. C., Barker, F. M., Hoffman, R. J., Stetson, M. H., Hanifin, J. P., Podolin, P. L., et al. (1994). Ultraviolet regulation of neuroendocrine and circadian physiology in rodents. *Vision Research, 34*, 1521–1533. doi:10.1016/0042-6989(94)90154-6.

Brown, T. M., Gias, C., Hatori, M., Keding, S. R., Semo, M., Coffey, P. J., et al. (2010). Melanopsin contributions to irradiance coding in the thalamo-cortical visual system. *PLoS Biology, 8*, e1000558. doi:10.1371/journal.pbio.1000558.

Brown, T. M., Wynne, J., Piggins, H. D., & Lucas, R. J. (2011). Multiple hypothalamic cell populations encoding distinct visual information. *Journal of Physiology, 589*(Pt 5), 1173–1194.

Cashmore, A. R. (2003). Cryptochromes: Enabling plants and animals to determine circadian time. *Cell, 114*, 537–543.

Castel, M., Belenky, M., Cohen, S., Ottersen, O. P., & Storm-Mathisen, J. (1993). Glutamate-like immunoreactivity in retinal terminals of the mouse suprachiasmatic nucleus. *European Journal of Neuroscience, 5*, 368–381.

Chen, S. K., Badea, T. C., & Hattar, S. (2011). Photoentrainment and pupillary light reflex are mediated by distinct populations of ipRGCs. *Nature, 476*, 92–95.

Colwell, C. S. (2001). NMDA-evoked calcium transients and currents in the suprachiasmatic nucleus: Gating by the circadian system. *European Journal of Neuroscience, 13*, 1420–1428.

Colwell, C. S. (2011). Linking neural activity and molecular oscillations in the SCN. *Nature Reviews. Neuroscience, 12*, 553–569.

Colwell, C. S., Foster, R. G., & Menaker, M. (1991). NMDA receptor antagonists block the effects of light on circadian behavior in the mouse. *Brain Research, 554*, 105–110.

Cooper, H. M., Herbin, M., & Nevo, E. (1993). Ocular regression conceals adaptive progression of the visual system in a blind subterranean mammal. *Nature, 361*, 156–159.

Cui, L. N., & Dyball, R. E. (1996). Synaptic input from the retina to the suprachiasmatic nucleus changes with the light-dark cycle in the Syrian hamster. *Journal of Physiology, 497*(Pt 2), 483–493.

Dacey, D. M., Liao, H. W., Peterson, B. B., Robinson, F. R., Smith, V. C., Pokorny, J., et al. (2005). Melanopsin-expressing ganglion cells in primate retina signal colour and irradiance and project to the LGN. *Nature, 433*, 749–754. doi:10.1038/nature03387.

DeCoursey, P. J. (1989). *Photoentrainment of circadian rhythms: An ecologist's viewpoint.* Sapporo: Hokkaido University Press.

de Jeu, M., Hermes, M., & Pennartz, C. (1998). Circadian modulation of membrane properties in slices of rat suprachiasmatic nucleus. *Neuroreport, 9*, 3725–3729. doi:10.1097/00001756-199811160-00028.

de Vries, M. J., Nunes Cardozo, B., van der Want, J., de Wolf, A., & Meijer, J. H. (1993). Glutamate immunoreactivity in terminals of the retinohypothalamic tract of the brown Norwegian rat. *Brain Research, 612*, 231–237.

de Vries, M. J., Treep, J. A., de Pauw, E. S., & Meijer, J. H. (1994). The effects of electrical stimulation of the optic nerves and anterior optic chiasm on the circadian activity rhythm of the Syrian hamster: Involvement of excitatory amino acids. *Brain Research, 642*, 206–212.

Ding, J. M., Buchanan, G. F., Tischkau, S. A., Chen, D., Kuriashkina, L., Faiman, L. E., et al. (1998). A neuronal ryanodine receptor mediates light-induced phase delays of the circadian clock. *Nature, 394*, 381–384. doi:10.1038/28639.

Ding, J. M., Chen, D., Weber, E. T., Faiman, L. E., Rea, M. A., & Gillette, M. U. (1994). Resetting the biological clock: Mediation of nocturnal circadian shifts by glutamate and NO. *Science, 266*, 1713–1717.

Do, M. T., Kang, S. H., Xue, T., Zhong, H., Liao, H. W., Bergles, D. E., et al. (2009). Photon capture and signalling by melanopsin retinal ganglion cells. *Nature, 457*, 281–287. doi:10.1038/nature07682.

Ebling, F. J. (1996). The role of glutamate in the photic regulation of the suprachiasmatic nucleus. *Progress in Neurobiology, 50*, 109–132.

Ecker, J. L., Dumitrescu, O. N., Wong, K. Y., Alam, N. M., Chen, S. K., LeGates, T., et al. (2010). Melanopsin-expressing retinal ganglion-cell photoreceptors: Cellular diversity and role in pattern vision. *Neuron, 67*, 49–60. doi:10.1016/j.neuron.2010.05.023.

Field, M. D., Maywood, E. S., O'Brien, J. A., Weaver, D. R., Reppert, S. M., & Hastings, M. H. (2000). Analysis of clock proteins in mouse SCN demonstrates phylogenetic divergence of the circadian clockwork and resetting mechanisms. *Neuron, 25*, 437–447.

Freedman, M. S., Lucas, R. J., Soni, B., von Schantz, M., Munoz, M., David-Gray, Z., et al. (1999). Regulation of mammalian circadian behavior by non-rod, non-cone, ocular photoreceptors. *Science, 284*, 502–504. doi:10.1126/science.284.5413.502.

Gillette, M. U. (1997). Cellular and biochemical mechanisms underlying circadian rhythms in vertebrates. *Current Opinion in Neurobiology, 7*, 797–804.

Ginty, D. D., Kornhauser, J. M., Thompson, M. A., Bading, H., Mayo, K. E., Takahashi, J. S., et al. (1993). Regulation of CREB phosphorylation in the suprachiasmatic nucleus by light and a circadian clock. *Science, 260*, 238–241. doi:10.1126/science.8097062.

Göz, D., Studholme, K., Lappi, D. A., Rollag, M. D., Provencio, I., & Morin, L. P. (2008). Targeted destruction of photosensitive retinal ganglion cells with a saporin conjugate alters the effects of light on mouse circadian rhythms. *PLoS One, 3*, e3153. doi:10.1371/journal.pone.0003153.

Groos, G., & Mason, R. (1980). The visual properties of rat and cat suprachiasmatic neurons. *Journal of Comparative Physiology, 135*, 349–356.

Güler, A. D., Ecker, J. L., Lall, G. S., Haq, S., Altimus, C. M., Liao, H. W., et al. (2008). Melanopsin cells are the principal conduits for rod-cone input to non-image-forming vision. *Nature, 453*, 102–105. doi:10.1038/nature06829.

Hall, J. C. (2000). Cryptochromes: Sensory reception, transduction, and clock functions subserving circadian systems. *Current Opinion in Neurobiology, 10*, 456–466.

Hamada, T., Yamanouchi, S., Watanabe, A., Shibata, S., & Watanabe, S. (1999). Involvement of glutamate release in substance P-induced phase delays of suprachiasmatic neuron activity rhythm in vitro. *Brain Research, 836*, 190–193.

Hannibal, J., Hindersson, P., Knudsen, S. M., Georg, B., & Fahrenkrug, J. (2002). The photopigment melanopsin is exclusively present in pituitary adenylate cyclase-activating polypeptide-containing retinal ganglion cells of the retinohypothalamic tract. *Journal of Neuroscience, 22*, RC191.

Harrington, M. E. (1997). The ventral lateral geniculate nucleus and the intergeniculate leaflet: Interrelated structures in the visual and circadian systems. *Neuroscience and Biobehavioral Reviews, 21*, 705–727.

Harrington, M. E., Hoque, S., Hall, A., Golombek, D., & Biello, S. (1999). Pituitary adenylate cyclase activating peptide phase shifts circadian rhythms in a manner similar to light. *Journal of Neuroscience, 19*, 6637–6642.

Hastings, M. H., Field, M. D., Maywood, E. S., Weaver, D. R., & Reppert, S. M. (1999). Differential regulation of mPER1 and mTIM proteins in the mouse suprachiasmatic nuclei: New insights into a core clock mechanism. *Journal of Neuroscience, 19*, RC11.

Hastings, M. H., Maywood, E. S., & O'Neill, J. S. (2008). Cellular circadian pacemaking and the role of cytosolic rhythms. *Current Biology, 18*, R805–R815.

Hatori, M., Le, H., Vollmers, C., Keding, S. R., Tanaka, N., Schmedt, C., et al. (2008). Inducible ablation of melanopsin-expressing retinal ganglion cells reveals their central role in non-image forming visual responses. *PLoS One, 3*, e2451. doi:10.1371/journal.pone.0002451.

Hattar, S., Kumar, M., Park, A., Tong, P., Tung, J., Yau, K. W., et al. (2006). Central projections of melanopsin-expressing

   JOHANNA H. MEIJER, SAMER HATTAR, AND JOSEPH S. TAKAHASHI

retinal ganglion cells in the mouse. *Journal of Comparative Neurology, 497*, 326–349. doi:10.1002/cne.20970.

Hattar, S., Liao, H. W., Takao, M., Berson, D. M., & Yau, K. W. (2002). Melanopsin-containing retinal ganglion cells: Architecture, projections, and intrinsic photosensitivity. *Science, 295*, 1065–1070.

Hattar, S., Lucas, R. J., Mrosovsky, N., Thompson, S., Douglas, R. H., Hankins, M. W., et al. (2003). Melanopsin and rod-cone photoreceptive systems account for all major accessory visual functions in mice. *Nature, 424*, 76–81. doi:10.1038/nature01761.

Hut, R. A., van Oort, B. E., & Daan, S. (1999). Natural entrainment without dawn and dusk: The case of the European ground squirrel (*Spermophilus citellus*). *Journal of Biological Rhythms, 14*, 290–299. doi:10.1177/074873099129000704.

Illnerova, H. (1991). *The suprachiasmatic nucleus and rhythmic pineal melatonin production*. New York: Oxford University Press.

Illnerova, H., & Sumova, A. (1997). Photic entrainment of the mammalian rhythm in melatonin production. *Journal of Biological Rhythms, 12*, 547–555. doi:10.1177/07487304970 1200609.

Irwin, R. P., & Allen, C. N. (2009). GABAergic signaling induces divergent neuronal Ca^{2+} responses in the suprachiasmatic nucleus network. *European Journal of Neuroscience, 30*, 1462–1475.

Jiang, Z. G., Yang, Y., Liu, Z. P., & Allen, C. N. (1997). Membrane properties and synaptic inputs of suprachiasmatic nucleus neurons in rat brain slices. *Journal of Physiology, 499*(Pt 1), 141–159.

Jiao, Y. Y., Lee, T. M., & Rusak, B. (1999). Photic responses of suprachiasmatic area neurons in diurnal degus (*Octodon degus*) and nocturnal rats (*Rattus norvegicus*). *Brain Research, 817*, 93–103.

Jiao, Y. Y., & Rusak, B. (2003). Electrophysiology of optic nerve input to suprachiasmatic nucleus neurons in rats and degus. *Brain Research, 960*, 142–151.

Johnson, C. H., & Golden, S. S. (1999). Circadian programs in cyanobacteria: Adaptiveness and mechanism. *Annual Review of Microbiology, 53*, 389–409.

Kim, D. Y., Kang, H. C., Shin, H. C., Lee, K. J., Yoon, Y. W., Han, H. C., et al. (2001). Substance P plays a critical role in photic resetting of the circadian pacemaker in the rat hypothalamus. *Journal of Neuroscience, 21*, 4026–4031.

Kim, Y. I., & Dudek, F. E. (1993). Membrane properties of rat suprachiasmatic nucleus neurons receiving optic nerve input. *Journal of Physiology, 464*, 229–243.

Kopp, M. D., Schomerus, C., Dehghani, F., Korf, H. W., & Meissl, H. (1999). Pituitary adenylate cyclase-activating polypeptide and melatonin in the suprachiasmatic nucleus: Effects on the calcium signal transduction cascade. *Journal of Neuroscience, 19*, 206–219.

Kornhauser, J. M., Ginty, D. D., Greenberg, M. E., Mayo, K. E., & Takahashi, J. S. (1996). Light entrainment and activation of signal transduction pathways in the SCN. *Progress in Brain Research, 111*, 133–146.

Kuhlman, S. J., Quintero, J. E., & McMahon, D. G. (2000). GFP fluorescence reports period 1 circadian gene regulation in the mammalian biological clock. *Neuroreport, 11*, 1479–1482.

Lall, G. S., Revell, V. L., Momiji, H., Al Enezi, J., Altimus, C. M., Guler, A. D., et al. (2010). Distinct contributions of rod, cone, and melanopsin photoreceptors to encoding irradiance. *Neuron, 66*, 417–428. doi:10.1016/j.neuron.2010.04.037.

Lowrey, P. L., & Takahashi, J. S. (2000). Genetics of the mammalian circadian system: Photic entrainment, circadian pacemaker mechanisms, and posttranslational regulation. *Annual Review of Genetics, 34*, 533–562.

Lucas, R. J., Freedman, M. S., Munoz, M., Garcia-Fernandez, J. M., & Foster, R. G. (1999). Regulation of the mammalian pineal by non-rod, non-cone, ocular photoreceptors. *Science, 284*, 505–507.

Lucas, R. J., Hattar, S., Takao, M., Berson, D. M., Foster, R. G., & Yau, K. W. (2003). Diminished pupillary light reflex at high irradiances in melanopsin-knockout mice. *Science, 299*, 245–247.

Maywood, E. S., Mrosovsky, N., Field, M. D., & Hastings, M. H. (1999). Rapid down-regulation of mammalian period genes during behavioral resetting of the circadian clock. *Proceedings of the National Academy of Sciences of the United States of America, 96*, 15211–15216. doi:10.1073/pnas.96.26.15211.

Meijer, J. H. (2001). *Photic entrainment in mammals* (Vol. 12). New York: Kluver Academic/Plenum Press.

Meijer, J. H., Groos, G. A., & Rusak, B. (1986). Luminance coding in a circadian pacemaker: The suprachiasmatic nucleus of the rat and the hamster. *Brain Research, 382*, 109–118.

Meijer, J. H., Rusak, B., & Ganshirt, G. (1992). The relation between light-induced discharge in the suprachiasmatic nucleus and phase shifts of hamster circadian rhythms. *Brain Research, 598*, 257–263.

Meijer, J. H., Rusak, B., & Harrington, M. E. (1989). Photically responsive neurons in the hypothalamus of a diurnal ground squirrel. *Brain Research, 501*, 315–323.

Meijer, J. H., Thio, B., Albus, H., Schaap, J., & Ruijs, A. C. (1999). Functional absence of extraocular photoreception in hamster circadian rhythm entrainment. *Brain Research, 831*, 337–339.

Meijer, J. H., Watanabe, K., Schaap, J., Albus, H., & Detari, L. (1998). Light responsiveness of the suprachiasmatic nucleus: Long-term multiunit and single-unit recordings in freely moving rats. *Journal of Neuroscience, 18*, 9078–9087.

Michel, S., Itri, J., & Colwell, C. S. (2002). Excitatory mechanisms in the suprachiasmatic nucleus: The role of AMPA/KA glutamate receptors. *Journal of Neurophysiology, 88*, 817–828.

Mikkelsen, J. D., & Larsen, P. J. (1993). Substance P in the suprachiasmatic nucleus of the rat: An immunohistochemical and in situ hybridization study. *Histochemistry, 100*, 3–16. doi:10.1007/BF00268873.

Mintz, E. M., Marvel, C. L., Gillespie, C. F., Price, K. M., & Albers, H. E. (1999). Activation of NMDA receptors in the suprachiasmatic nucleus produces light-like phase shifts of the circadian clock in vivo. *Journal of Neuroscience, 19*, 5124–5130.

Moore, R. Y. (1973). Retinohypothalamic projection in mammals: A comparative study. *Brain Research, 49*, 403–409.

Morin, L. P. (1994). The circadian visual system. *Brain Research. Brain Research Reviews, 19*, 102–127.

Moriya, T., Horikawa, K., Akiyama, M., & Shibata, S. (2000). Correlative association between N-methyl-D-aspartate receptor-mediated expression of period genes in the suprachiasmatic nucleus and phase shifts in behavior with photic entrainment of clock in hamsters. *Molecular Pharmacology*, *58*, 1554–1562.

Mosko, S. S., & Jacobs, B. L. (1974). Midbrain raphe neurons: Spontaneous activity and response to light. *Physiology & Behavior*, *13*, 589–593.

Mrosovsky, N., & Hattar, S. (2003). Impaired masking responses to light in melanopsin-knockout mice. *Chronobiology International*, *20*, 989–999.

Mrosovsky, N., Reebs, S. G., Honrado, G. I., & Salmon, P. A. (1989). Behavioural entrainment of circadian rhythms. *Experientia*, *45*, 696–702.

Mure, L. S., Rieux, C., Hattar, S., & Cooper, H. M. (2007). Melanopsin-dependent nonvisual responses: Evidence for photopigment bistability in vivo. *Journal of Biological Rhythms*, *22*, 411–424.

Nakamura, T. J., Fujimura, K., Ebihara, S., & Shinohara, K. (2004). Light response of the neuronal firing activity in the suprachiasmatic nucleus of mice. *Neuroscience Letters*, *371*, 244–248.

Nelson, D. E., & Takahashi, J. S. (1991). Sensitivity and integration in a visual pathway for circadian entrainment in the hamster (*Mesocricetus auratus*). *Journal of Physiology*, *439*, 115–145.

Nelson, D. E., & Takahashi, J. S. (1999). Integration and saturation within the circadian photic entrainment pathway of hamsters. *American Journal of Physiology*, *277*(5 Pt 2), R1351–R1361.

Nielsen, H. S., Hannibal, J., Knudsen, S. M., & Fahrenkrug, J. (2001). Pituitary adenylate cyclase-activating polypeptide induces period1 and period2 gene expression in the rat suprachiasmatic nucleus during late night. *Neuroscience*, *103*, 433–441.

Obrietan, K., Impey, S., & Storm, D. R. (1998). Light and circadian rhythmicity regulate MAP kinase activation in the suprachiasmatic nuclei. *Nature Neuroscience*, *1*, 693–700.

Okamura, H., Miyake, S., Sumi, Y., Yamaguchi, S., Yasui, A., Muijtjens, M., et al. (1999). Photic induction of *mPer1* and *mPer2* in cry-deficient mice lacking a biological clock. *Science*, *286*, 2531–2534. doi:10.1126/science.286.5449.2531.

O'Neill, J. S., Maywood, E. S., Chesham, J. E., Takahashi, J. S., & Hastings, M. H. (2008). cAMP-dependent signaling as a core component of the mammalian circadian pacemaker. *Science*, *320*, 949–953.

Panda, S., Provencio, I., Tu, D. C., Pires, S. S., Rollag, M. D., Castrucci, A. M., et al. (2003). Melanopsin is required for non-image-forming photic responses in blind mice. *Science*, *301*, 525–527. doi:10.1126/science.1086179.

Panda, S., Sato, T. K., Castrucci, A. M., Rollag, M. D., DeGrip, W. J., Hogenesch, J. B., et al. (2002). Melanopsin (Opn4) requirement for normal light-induced circadian phase shifting. *Science*, *298*, 2213–2216. doi:10.1126/science.1076848.

Pennartz, C. M., Hamstra, R., & Geurtsen, A. M. (2001). Enhanced NMDA receptor activity in retinal inputs to the rat suprachiasmatic nucleus during the subjective night. *Journal of Physiology*, *532*(Pt 1), 181–194.

Piggins, H. D., & Rusak, B. (1997). Effects of microinjections of substance P into the suprachiasmatic nucleus region on hamster wheel-running rhythms. *Brain Research Bulletin*, *42*, 451–455.

Pittendrigh, C. S., & Daan, S. (1976). A functional analysis of circadian pacemakers in nocturnal rodents. IV. Entrainment: Pacemaker as a clock. *Journal of Comparative Physiology*, *106*, 291–331. Now this is cited in the text.

Podolin, P. L., Rollag, M. D., & Brainard, G. C. (1987). The suppression of nocturnal pineal melatonin in the Syrian hamster: dose-response curves at 500 and 360 nm. *Endocrinology*, *121*, 266–270.

Prosser, R. A., McArthur, A. J., & Gillette, M. U. (1989). cGMP induces phase shifts of a mammalian circadian pacemaker at night, in antiphase to cAMP effects. *Proceedings of the National Academy of Sciences of the United States of America*, *86*, 6812–6815. doi:10.1073/pnas.86.17.6812.

Provencio, I., & Foster, R. G. (1995). Circadian rhythms in mice can be regulated by photoreceptors with cone-like characteristics. *Brain Research*, *694*, 183–190.

Provencio, I., Jiang, G., De Grip, W. J., Hayes, W. P., & Rollag, M. D. (1998). Melanopsin: An opsin in melanophores, brain, and eye. *Proceedings of the National Academy of Sciences of the United States of America*, *95*, 340–345.

Provencio, I., Rodriguez, I. R., Jiang, G., Hayes, W. P., Moreira, E. F., & Rollag, M. D. (2000). A novel human opsin in the inner retina. *Journal of Neuroscience*, *20*, 600–605.

Pu, M. (2000). Physiological response properties of cat retinal ganglion cells projecting to suprachiasmatic nucleus. *Journal of Biological Rhythms*, *15*, 31–36. doi:10.1177/074873040001500104.

Reebs, S. G., & Mrosovsky, N. (1989). Effects of induced wheel running on the circadian activity rhythms of Syrian hamsters: Entrainment and phase response curve. *Journal of Biological Rhythms*, *4*, 39–48. doi:10.1177/074873048900400103.

Ruby, N. F., Brennan, T. J., Xie, X., Cao, V., Franken, P., Heller, H. C., et al. (2002). Role of melanopsin in circadian responses to light. *Science*, *298*, 2211–2213. doi:10.1126/science.1076701.

Sancar, A. (2000). Cryptochrome: The second photoactive pigment in the eye and its role in circadian photoreception. *Annual Review of Biochemistry*, *69*, 31–67.

Sawaki, Y. (1979). Suprachiasmatic nucleus neurones: Excitation and inhibition mediated by the direct retino-hypothalamic projection in female rats. *Experimental Brain Research*, *37*, 127–138.

Schaap, J., Bos, N. P., de Jeu, M. T., Geurtsen, A. M., Meijer, J. H., & Pennartz, C. M. (1999). Neurons of the rat suprachiasmatic nucleus show a circadian rhythm in membrane properties that is lost during prolonged whole-cell recording. *Brain Research*, *815*, 154–166.

Schmidt, T. M., Chen, S. K., & Hattar, S. (2011). Intrinsically photosensitive retinal ganglion cells: Many subtypes, diverse functions. *Trends in Neurosciences*, *34*, 572–580.

Schmidt, T. M., & Kofuji, P. (2010). Differential cone pathway influence on intrinsically photosensitive retinal ganglion cell subtypes. *Journal of Neuroscience*, *30*, 16262–16271.

Selby, C. P., Thompson, C., Schmitz, T. M., Van Gelder, R. N., & Sancar, A. (2000). Functional redundancy of cryptochromes and classical photoreceptors for nonvisual ocular photoreception in mice. *Proceedings of the National Academy of Sciences of the United States of America*, *97*, 14697–14702. doi:10.1073/pnas.260498597.

Shearman, L. P., Zylka, M. J., Weaver, D. R., Kolakowski, L. F., Jr., & Reppert, S. M. (1997). Two period homologs: Circadian expression and photic regulation in the suprachiasmatic nuclei. *Neuron*, *19*, 1261–1269.

Shigeyoshi, Y., Taguchi, K., Yamamoto, S., Takekida, S., Yan, L., Tei, H., et al. (1997). Light-induced resetting of a mammalian circadian clock is associated with rapid induction of the *mPer1* transcript. *Cell, 91,* 1043–1053. doi:10.1016/S0092-8674(00)80494-8.

Shirakawa, T., & Moore, R. Y. (1994). Responses of rat suprachiasmatic nucleus neurons to substance P and glutamate in vitro. *Brain Research, 642,* 213–220.

Takahashi, J. S., DeCoursey, P. J., Bauman, L., & Menaker, M. (1984). Spectral sensitivity of a novel photoreceptive system mediating entrainment of mammalian circadian rhythms. *Nature, 308*(5955), 186–188.

Takahashi, J. S., Turek, F. W., & Moore, R. Y. (2001). *Handbook of behavioral neurobiology: Circadian clocks* (Vol. 12). New York: Kluwer Academic/Plenum Press.

Takumi, T., Matsubara, C., Shigeyoshi, Y., Taguchi, K., Yagita, K., Maebayashi, Y., et al. (1998a). A new mammalian period gene predominantly expressed in the suprachiasmatic nucleus. *Genes to Cells, 3,* 167–176. doi:10.1046/j.1365-2443.1998.00178.x.

Takumi, T., Taguchi, K., Miyake, S., Sakakida, Y., Takashima, N., Matsubara, C., et al. (1998b). A light-independent oscillatory gene *mPer3* in mouse SCN and OVLT. *European Molecular Biology Organization Journal, 17,* 4753–4759. doi:10.1093/emboj/17.16.4753.

Thompson, C. L., Selby, C. P., Van Gelder, R. N., Blaner, W. S., Lee, J., Quadro, L., et al. (2004). Effect of vitamin A depletion on nonvisual phototransduction pathways in cryptochromeless mice. *Journal of Biological Rhythms, 19,* 504–517. doi:10.1177/0748730404270519.

Tischkau, S. A., Gallman, E. A., Buchanan, G. F., & Gillette, M. U. (2000). Differential cAMP gating of glutamatergic signaling regulates long-term state changes in the suprachiasmatic circadian clock. *Journal of Neuroscience, 20,* 7830–7837.

Tominaga, K., Geusz, M. E., Michel, S., & Inouye, S. T. (1994). Calcium imaging in organotypic cultures of the rat suprachiasmatic nucleus. *Neuroreport, 5,* 1901–1905.

Trejo, L. J., & Cicerone, C. M. (1984). Cells in the pretectal olivary nucleus are in the pathway for the direct light reflex of the pupil in the rat. *Brain Research, 300,* 49–62.

van der Horst, G. T., Muijtjens, M., Kobayashi, K., Takano, R., Kanno, S., Takao, M., et al. (1999). Mammalian Cry1 and Cry2 are essential for maintenance of circadian rhythms. *Nature, 398,*627–630. doi:10.1038/19323.

Van Oosterhout, F., Fisher, S. P., Van Diepen, H. C., Watson, T. S., Houben, T., Vanderleest, H. T., et al. (2012). Ultraviolet light provides a major input to non-image-forming light detection in mice. *Current Biology, 22,* 1397–1402.

Vindlacheruvu, R. R., Ebling, F. J., Maywood, E. S., & Hastings, M. H. (1992). Blockade of glutamatergic neurotransmission in the suprachiasmatic nucleus prevents cellular and behavioural responses of the circadian system to light. *European Journal of Neuroscience, 4,* 673–679.

Vitaterna, M. H., Selby, C. P., Todo, T., Niwa, H., Thompson, C., Fruechte, E. M., et al. (1999). Differential regulation of mammalian period genes and circadian rhythmicity by cryptochromes 1 and 2. *Proceedings of the National Academy of Sciences of the United States of America, 96,* 12114–12119. doi:10.1073/pnas.96.21.12114.

von Gall, C., Duffield, G. E., Hastings, M. H., Kopp, M. D., Dehghani, F., Korf, H. W., et al. (1998). CREB in the mouse SCN: A molecular interface coding the phase-adjusting stimuli light, glutamate, PACAP, and melatonin for clockwork access. *Journal of Neuroscience, 18,* 10389–10397.

Wakamatsu, H., Yoshinobu, Y., Aida, R., Moriya, T., Akiyama, M., & Shibata, S. (2001). Restricted-feeding-induced anticipatory activity rhythm is associated with a phase-shift of the expression of *mPer1* and *mPer2* mRNA in the cerebral cortex and hippocampus but not in the suprachiasmatic nucleus of mice. *European Journal of Neuroscience, 13,* 1190–1196.

Welsh, D. K., Logothetis, D. E., Meister, M., & Reppert, S. M. (1995). Individual neurons dissociated from rat suprachiasmatic nucleus express independently phased circadian firing rhythms. *Neuron, 14,* 697–706.

Wilsbacher, L. D., Yamazaki, S., Herzog, E. D., Song, E. J., Radcliffe, L. A., Abe, M., et al. (2002). Photic and circadian expression of luciferase in *mPeriod1-luc* transgenic mice in vivo. *Proceedings of the National Academy of Sciences of the United States of America, 99,* 489–494. doi:10.1073/pnas.0122 48599.

Yamazaki, S., Goto, M., & Menaker, M. (1999). No evidence for extraocular photoreceptors in the circadian system of the Syrian hamster. *Journal of Biological Rhythms, 14,* 197–201.

Yamazaki, S., Numano, R., Abe, M., Hida, A., Takahashi, R., Ueda, M., et al. (2000). Resetting central and peripheral circadian oscillators in transgenic rats. *Science, 288,* 682–685. doi:10.1126/science.288.5466.682.

Yan, L., Takekida, S., Shigeyoshi, Y., & Okamura, H. (1999). *Per1* and *Per2* gene expression in the rat suprachiasmatic nucleus: Circadian profile and the compartment-specific response to light. *Neuroscience, 94,* 141–150.

Young, M. W., & Kay, S. A. (2001). Time zones: A comparative genetics of circadian clocks. *Nature Reviews. Genetics, 2,* 702–715.

Zhang, E. E., Liu, Y., Dentin, R., Pongsawakul, P. Y., Liu, A. C., Hirota, T., et al. (2010). Cryptochrome mediates circadian regulation of cAMP signaling and hepatic gluconeogenesis. *Nature Medicine, 16,* 1152–1156. doi:10.1038/nm.2214.

Zhu, H., Yuan, Q., Briscoe, A. D., Froy, O., Casselman, A., & Reppert, S. M. (2005). The two CRYs of the butterfly. *Current Biology, 15,* R953–R954.

Zylka, M. J., Shearman, L. P., Weaver, D. R., & Reppert, S. M. (1998). Three period homologs in mammals: Differential light responses in the suprachiasmatic circadian clock and oscillating transcripts outside of brain. *Neuron, 20,* 1103–1110.

21 Inhibitory Circuits in the Visual Thalamus

JUDITH A. HIRSCH, XIN WANG, VISHAL S. VAINGANKAR, AND FRIEDRICH T. SOMMER

The inhibitory circuits of the visual thalamus dominate local processing and influence all information that relay cells transmit from retina to cortex. These circuits divide into two major groups. Each one occupies a different position in the visual pathway and serves a different role in sensory processing. One circuit provides feedforward inhibition and involves local interneurons within the lateral geniculate (LGN). These cells receive input from the retina and connect with relay cells and with each other (Sherman, 2004). The second circuit provides feedback inhibition mediated by the perigeniculate sector (PGN) of the thalamic reticular nucleus, a thin and interconnected sheet of inhibitory neurons that covers the surface of the thalamus (Sillito & Jones, 2008). Reticular neurons receive input from relay cells and project back to them in turn. Notably, the two inhibitory pathways seem separate; reticular neurons do not synapse with local interneurons in the LGN (Wang et al., 2001). This chapter explores both of these circuits (figure 21.1), starting from the perspective of the relay cell.

SPATIAL ARRANGEMENT OF EXCITATION AND INHIBITION IN THE RELAY CELL'S RECEPTIVE FIELD: PUSH–PULL RESPONSES TO STIMULI OF THE OPPOSITE SIGN

The first recordings from the LGN revealed that relay cells have receptive fields built of concentric ON and OFF subregions (Hubel, 1960; Hubel & Wiesel, 1961), much like ganglion cells in retina (Kuffler, 1953). Later on, intracellular recordings showed that each subregion has a push–pull arrangement of excitation and inhibition; for example, bright light flashed in an ON subregion excites, whereas dark stimuli inhibit (Hirsch, 2003; Hirsch et al., 2002; Wang et al., 2007, 2011). For the remainder of the chapter we use the term "push" to describe excitation evoked by the preferred luminance contrast (e.g., bright light presented within an ON subregion). Accordingly, "pull" refers to inhibition evoked by stimuli of the opposite sign (e.g., a dark

stimulus presented within an OFF subregion). This pattern of response is illustrated in figure 21.2B (left), which provides a summary diagram of push–pull circuits in the thalamus.

For thalamic relay cells, it is clear that the push (i.e., excitation evoked by stimuli of the preferred sign) is provided by retinal inputs. Its spatial profile mirrors the receptive fields of presynaptic ganglion cells (Levick, Cleland, & Dubin, 1972; Usrey, Reppas, & Reid, 1999) and is largely unchanged by removing cortex, (Jones & Sillito, 1994), the other major source of visual input to the thalamus (Dankowski & Bickford, 2003; Datskovskaia, Carden, & Bickford, 2001; Erisir et al., 1997; Erisir, Van Horn, & Sherman, 1998; Guillery, 1969a; Hamos et al., 1985, 1987; Montero, 1991; Van Horn, Erisir, & Sherman, 2000). Further, retinal afferents select proximal regions of the dendritic arbor, whereas corticothalamic axons favor more distal arbors (Erisir et al., 1997; Hamos et al., 1985; Van Horn, Erisir, & Sherman, 2000; Wilson, 1989; Wilson, Friedlander, & Sherman, 1984). These anatomical relationships suggest that retinal input drives thalamic firing, whereas cortical input serves a modulatory role (Sherman, 2001a; Sherman & Guillery, 1998). The pull has the same basic shape as the push, but it is generated de novo in the thalamus, as we discuss in later sections.

Push–pull is found at the first three stages in the early visual pathway, including retina, thalamus, and simple cells in the cat's cortex (Hirsch & Martinez, 2006a, 2006b). What advantages might this arrangement of excitation and inhibition provide? Universal roles for push–pull are to extend the dynamic range of neural sensitivity to the stimulus and to restore linearity of response following rectification through the synapse (Werblin, 2010). In addition, as a result of the geometry of neural receptive fields, push–pull also serves other functions in the early visual system. We illustrate these using an early example from the literature. Kuffler (1953) was the first to discover that ganglion cells in the mammalian retina had receptive fields with a center-surround structure. This arrangement of OFF and ON

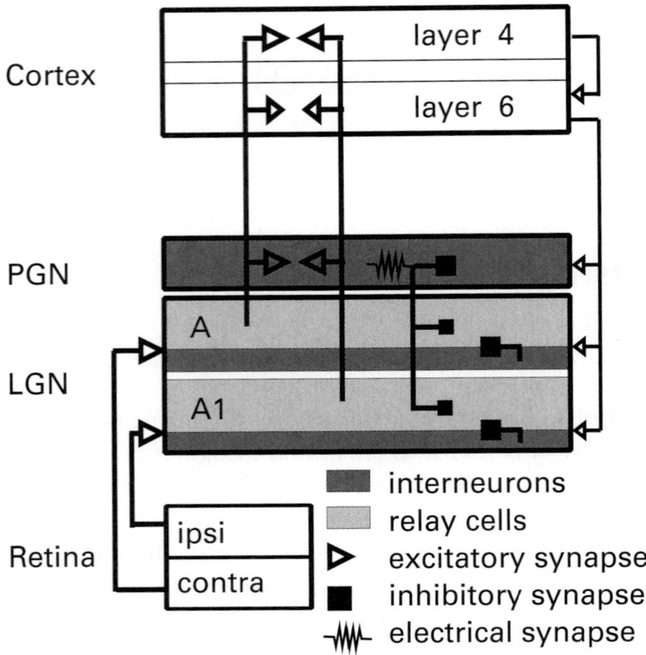

FIGURE 21.1 Block diagram of feedforward and feedback connections in the visual thalamus. The legend in the figure explains all symbols and color codes. Populations of interneurons and relay cells in the LGN are drawn separately for clarity, although they are mixed in situ. Only connections for the primary eye-specific layers of the LGN (A, A1) are shown; however, connections for the C layers are similar.

subregions confers sensitivity to local contrast borders. Kuffler also realized that the center and the surround had an antagonistic relationship with each other, going so far as to use the term push–pull. Thus, ganglion cells respond robustly to stimuli of the preferred luminance contrast (bright or dark) that are confined to the center. A larger disk, one that spills from the center to cover the surround, evokes few spikes, however. This diminished response to large homogeneous stimuli reflects the suppressive effect of pull in one subregion on the push in the other (Wiesel, 1959). Hence, push–pull interactions across subregions enhance the selectivity for local contrast borders. The same principle holds for relay cells as well (Hubel & Wiesel, 1961). Based on this biological pattern of response, the receptive fields of ganglion cells and relay cells have been described mathematically as a difference of Gaussians (Enroth-Cugell & Robson, 1966; Marr & Hildreth, 1980; Shapley & Lennie, 1985). Processing an image with this function, in essence, produces the second (spatial) derivative of the stimulus, which enhances contrast borders.

One can also consider the role of push–pull from the perspective of information theory (Barlow, 1961), following hypotheses about sensory coding advanced shortly after Kuffler's discovery. In a framework called efficient coding, Barlow (1961) proposed that as

patterns of neural activity are recoded from stage to stage, informative signals are selectively extracted to reduce redundancy in the information encoded by the spike train. Observations about the statistics of natural images highlight the importance of these early ideas. Mathematical analyses of visual scenes, whether in nature or in urban settings, show that visual patterns in the environment are highly redundant (Simoncelli & Olshausen, 2001). Simply put, spatial correlations emerge because adjacent points in a picture often have similar luminance values (think of a patch of blue sky or the facade of a building). In more formal terms the visual world has $1/f$ statistics; the power spectrum of a given natural image is skewed toward low spatial frequencies (Field, 1987; Geisler, 2008; Simoncelli & Olshausen, 2001). Because, as above, the receptive field essentially computes the second derivative of the image, higher spatial frequencies (e.g., effectively, edges within the image) are enhanced, and redundancy in the stimulus is reduced.

Push–pull operates in time as well as space. Movements of spatially correlated images across the retina introduce temporal correlations. Thus, luminance values within the receptive field remain similar for long periods of time and then suddenly change (Simoncelli & Olshausen, 2001). For neurons with push–pull, changes in stimulus polarity over time elicit commensurate reversals in the sign of the synaptic response (Alitto, Weyand, & Usrey, 2005; Wang et al., 2011; Werblin, 2010). Accordingly, pronounced transitions from the nonpreferred to the preferred contrast drive relay cells best (Alitto, Weyand, & Usrey, 2005; Wang et al., 2007). Presumably, the pull readies the membrane to fire using the same biophysical principles that Hodgkin and Huxley (1952) described for the squid axon to explain excitation at anode break. Regardless of mechanism, this preference for biphasic stimuli corresponds to selectivity for the first (temporal) derivative of the image and thus reduces redundancy of signals with $1/f$ temporal structure, such as natural stimuli (Dan, Atick, & Reid, 1996; Dong & Atick, 1995).

Analyses of simultaneous recordings from ganglion and relay cells provide a separate line of support for efficient coding in thalamus (Rathbun, Warland, & Usrey, 2010; Sincich, Horton, & Sharpee, 2009; Wang et al., 2011). This work shows that only a fraction of the spikes a ganglion cell fires actually trigger the relay cell, such that each thalamic spike contains more information (more bits per spike) than a retinal one (Rathbun, Warland, & Usrey, 2010; Sincich, Horton, & Sharpee, 2009; Wang et al., 2010). One can think of this transformation as down-sampling across the synapse, a process that reduces redundancy in the coding of slowly

    J. A. HIRSCH, X. WANG, V. S. VAINGANKAR, AND F. T. SOMMER

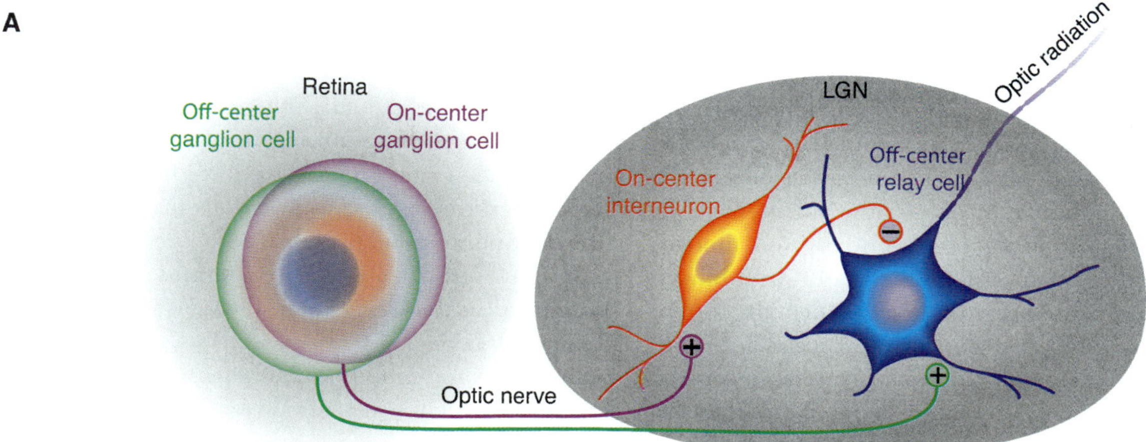

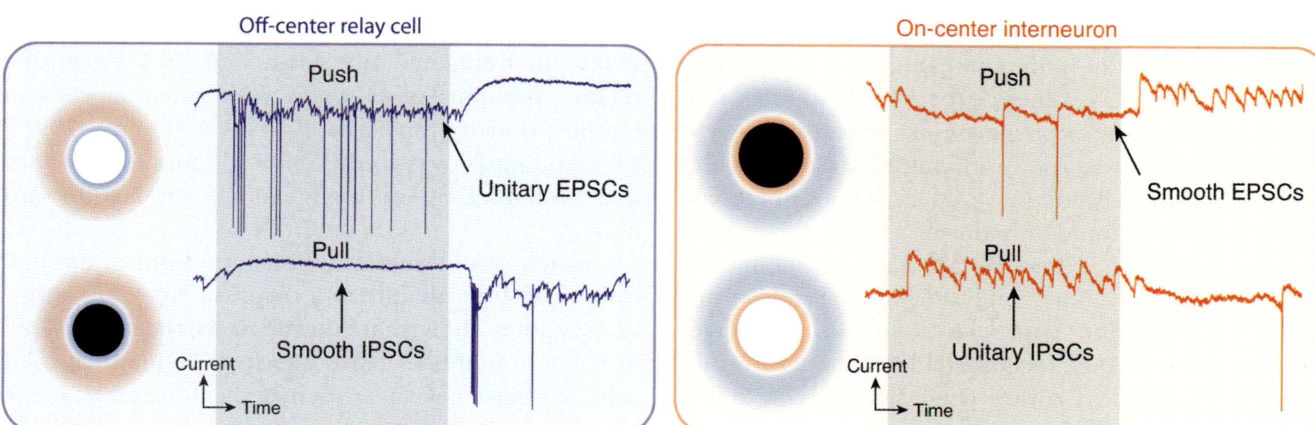

FIGURE 21.2 (A) Wiring diagram of connections between retinal ganglion cells and thalamic neurons. An OFF-center thalamic relay cell (blue) receives excitatory input (push, green) from an OFF-center retinal ganglion cell and inhibitory input (pull, purple) from an ON-center local interneuron (red) driven by an ON-center ganglion cell; retinal ganglion cells are represented as their receptive fields. (B) Relay cells and interneurons have complementary patterns of push–pull responses to stimuli of the opposite contrast. Bright and dark stimuli (indicated by icons positioned within neural receptive fields) are shown alongside synaptic responses evoked from an OFF-center relay cell (left) and ON-center interneuron (right); traces plot membrane current against time. The push is formed by trains of individual excitatory events for the relay cell, whereas the pull is smooth. By contrast, the push recorded from the interneuron is smooth, and the pull is formed by a series of unitary events. See text for an explanation of the functional consequences of these different physiological profiles.

changing signals such as natural stimuli. Although inhibition is likely to contribute to this increase in the economy of the neural code, other synaptic mechanisms, such as temporal summation (Carandini, Horton, & Sincich, 2007; Usrey, Reppas, & Reid, 1998; Wang, Hirsch, & Sommer, 2010), are likely to play a part.

INTERACTIONS BETWEEN PUSH–PULL AND THE INTRINSIC PROPERTIES OF THE NEURAL MEMBRANE

The roles for push–pull described so far apply equally well to retina and thalamus. However, there are differences between the intrinsic membrane properties of

cells in the two structures that permit a specialized role for the pull in the LGN. Specifically, relay cells fire in two modes, depending on the resting potential. They fire in tonic mode when resting levels are near spike threshold and in burst mode when the membrane voltage is low (Jahnsen & Llinas, 1984). The bursts are generated by T-type calcium channels that open at low threshold (below that for the sodium channels that produce spikes) but close shortly thereafter (Huguenard & Prince, 1992; Huguenard & McCormick, 1992). Further, these channels cannot open again unless exposed to strong hyperpolarization.

Relay cells fire in burst mode during sleep because neuromodulators that lower the resting potential are secreted during that time. But the fact that bursts are generated during sleep does not preclude an important role in waking (Sherman, 2001b) and vision (see chapter 19 by Sherman and Guillery). In fact, bursts are predictably driven by natural stimuli (Denning & Reinagel, 2005; Lesica & Stanley, 2004; Wang et al., 2007) and appear to play an active role in vision (Alitto & Usrey, 2005; Sherman, 2001b) and other sensory modalities (Swadlow & Gusev, 2001). One reason that bursts are important is that they are more effective than tonic spikes in activating postsynaptic neurons (Swadlow & Gusev, 2001; Usrey, Reppas, & Reid, 1998) and thus may help propagate activity through cortex. Furthermore, bursts are able to transmit information different from that encoded by single impulses (Denning & Reinagel, 2005; Lesica et al., 2006; Reinagel et al., 1999).

Although most experiments exploring the role of bursts in vision have been done in anesthetized animals, a growing number of studies have employed alert preparations. By and large, these experiments support the view that bursts play a role in sensory processing (Niell & Stryker, 2010; Swadlow, Bezdudnaya, & Gusev, 2005; Swadlow, Gusev, & Bezdudnaya, 2002). This is not to say that bursts occur frequently during vision; they might not. However, even if they are rare, they can be important. For example, bursts might signal special occurrences in the environment, such as the abrupt change in the stimulus over the receptive field, and thus provide novel information to cortex (Lesica & Stanley, 2004; Sherman, 2001c).

What mechanisms evoke bursts during vision? First, the pull is able to hyperpolarize the membrane substantially (Wang et al., 2007). Second, spatiotemporal correlations in the visual environment create instances in which stimuli of the nonpreferred sign (e.g., a dark image patch over an ON center) cover the receptive field for long periods, even hundreds of milliseconds (Denning & Reinagel, 2005; Lesica & Stanley, 2004; Wang et al., 2007). These lingering stimuli evoke prolonged inhibition that is able to deinactivate T-channels. Hence, when stimulus contrast reverses (the dark image patch is replaced with a bright one), retinal input is able to trigger a burst of spikes (Wang et al., 2007). Thus, visually evoked synaptic input engages the intrinsic properties of the membrane to drive relay cells between burst and tonic modes (Wang et al., 2007).

NEURAL CIRCUITS THAT CONTRIBUTE TO PUSH–PULL

Both the push and pull in the relay cell's receptive field have a center-surround structure, much like ganglion cells in the retina (Wang, Sommer, & Hirsch, 2011; Wang et al., 2007, 2011). Although the push can be explained by retinal input, the pull cannot because ganglion cells excite their targets. The simplest explanation for the pull is that it derives from local interneurons with the opposite preference for stimulus contrast (figure 21.2A). Like relay cells, local interneurons also have receptive fields with a center-surround structure (Wang, Sommer, & Hirsch, 2011). Local interneurons even have push–pull responses and process visual signals in a roughly linear fashion (Wang, Sommer, & Hirsch, 2011).

The push and pull should not be thought of as mirror images of each other, however. This conclusion is reached from studies in which the receptive fields of excitatory and inhibitory inputs were mapped separately with individual bright and dark spots. The peaks of the push and of the pull in the resulting maps were displaced (Martinez et al., 2005; Wang et al., 2007, 2011), on average, by the distance predicted from the relative positions of ON and OFF ganglion cells in retina within a given class (Wässle & Boycott, 1991). This finding provides additional support for the idea that push–pull circuits are generated by feedforward input.

Further, it is likely that the spatial extent of the pull will prove larger than that of the push. First, ultrastructural studies show that interneurons receive at least double the number of synapses from retina than relay cells do (Datskovskaia, Carden, & Bickford, 2001; Erisir, Van Horn, & Sherman, 1998; Van Horn, Erisir, & Sherman, 2000), which is not surprising because there are many fewer interneurons than relay cells (Fitzpatrick, Penny, & Schmechel, 1984; LeVay & Ferster, 1979; Lin, Kratz, & Sherman, 1977; Montero, 1991; Montero & Singer, 1985; Weber & Kalil, 1983). If one assumes that higher numbers of presynaptic retinal terminals indicate commensurately greater pools of convergent ganglion cells, then the receptive fields of interneurons should be larger than those of relay cells (note that the receptive fields of ganglion cells within a given class barely overlap (Wässle & Boycott, 1991). Determining the relative sizes of the receptive fields of relay cells and interneurons at a given point in visual space is difficult to test, however, and remains to be shown. A second mechanism that could increase the size of the pull in the relay cell's receptive field involves convergent input from several interneurons. Work in brain slices suggests that this is the case (Crunelli et al., 1988; Ziburkus, Lo, & Guido, 2003).

So far, it seems as if feedforward inhibitory circuits are simple, but they are not. Local interneurons are very different from relay cells at both the anatomical and physiological levels. For example, both the dendrites and axons of local interneurons are presynaptic to other cells (Sherman, 2004). The axons contact all classes of relay cells—X, Y, and W—favoring the soma and proximal dendrites and also synapse with other interneurons (Guillery, 1969b; Hamos et al., 1985; Montero, 1987). Inhibitory axonal terminals are called F1 (flat vesicle, type 1). Connections made by F1 terminals of local interneurons could easily mediate push–pull responses by connecting cells with the opposite preferences for stimulus contrast—an OFF interneuron synapsing with an ON relay cell or ON interneuron, for example. Alternatively, F1 connections are equally able to supply inhibition to targets with the same preference for stimulus contrast. We refer to this arrangement as same-sign (as opposed to push–pull) inhibition; an example is an OFF interneuron connecting to an OFF relay cell or OFF interneuron.

THE SYNAPTIC TRIAD AND OTHER MECHANISMS OF SAME-SIGN INHIBITION

Interneurons make an unusual type of connection with many of their targets; they form dendrodendritic synapses (i.e., the presynaptic and postsynaptic compartments are dendritic). The presynaptic elements are called F2 terminals. Some of these dendrodendritic synapses are made en passant, but many are made via dendritic appendages (Guillery, 1969b; Hamos et al., 1985; Montero, 1986; Pasik, Pasik, & Hámori, 1976). Because dendrites can be depolarized by synaptic input, F2 terminals are able to supply inhibition even in the absence of action potentials (Cox & Sherman, 2000). Thus, in principle each of an interneuron's dendrites is able to operate as an independent unit of processing (Sherman, 2004). That does not mean that the dendritic and axonal outputs of a single interneuron cannot be coordinated in time. Action potentials are able to back-propagate into dendrites, presumably triggering release from F2 and F1 terminals at once (Casale & McCormick, 2011).

Dendrodendritic connections are often made in specialized structures called triads (Famiglietti & Peters, 1972; Fitzpatrick, Penny, & Schmechel, 1984; Guillery, 1969b; Hamos et al., 1985; Montero, 1986; Szentagothai, Hamori, & Tombol, 1966). In these structures, a retinal bouton synapses with the juxtaposed dendrites of a relay cell and of an interneuron. The interneuron synapses onto the relay cell in turn via an F2 terminal (Sherman, 2004). Triads commonly involve X relay cells and are made at "grape-like" appendages, small bulbs that cluster around regions where primary dendrites branch. These complicated, interdigitated synaptic profiles are ensheathed by glial membranes to form structures called glomeruli (Hamos, et al., 1985; Szentagothai, Hamori, & Tombol, 1966).

Triads are suited for same-sign inhibition but unlikely to mediate push–pull because common retinal input yokes inhibition to excitation. Functional roles for triads have been very difficult to establish. The obvious approach is to compare responses of X relay cells with those of other types of relay cells (Y or W) that are not the main targets of F2 terminals. To date, however, the differences in response properties among the three main types of relay cells appear to reflect retinal input and do not give insight into the role of triads. One might at the very least expect that retinogeniculate EPSPs would be shortest in X cells, clipped by feedforward same-sign inhibition. Yet the time courses of retinogeniculate EPSPs seem similar across relay cells in vivo (Carandini, Horton, & Sincich, 2007; Wang, Hirsch, & Sommer, 2010; Wang et al., 2007). Work in vitro, however, shows that activation of a single retinal input jointly evokes an EPSP immediately followed by an IPSP in some cells (Blitz & Regehr, 2005), consistent with triadic or other mechanisms of same-sign inhibition.

Thus, the role of triads remains an open question. A leading hypothesis suggests that triads have a role in gain control (Sherman, 2004). This makes sense because the pharmacology of the connections between retinal afferents and relay cells is different from that between retinal afferents and interneurons. The synapse between retina and interneurons includes a substantial balance of metabotropic receptors, which require sustained exposure to glutamate to influence the membrane voltage (Crandall & Cox, 2012; Govindaiah & Cox, 2004). Thus, metabotropic receptors might provide disproportionate inhibition when stimulus contrast is high and retinal input is strong (Sherman, 2004). Notably the rodent somatosensory thalamus lacks interneurons (Arcelli et al., 1997). Perhaps this species difference can be exploited to form new hypotheses about roles of interneurons in general and triads in particular.

There are many variations on the basic triadic circuit; some involve input from various brainstem regions (Sherman, 2004). In addition, there is evidence that relay cells send sparse collaterals within the LGN (Cox, Reichova, & Sherman, 2003; Lorincz et al., 2009). These collaterals sometimes contribute to triadic arrangements (Bickford, Carden, & Patel, 2008), effectively substituting for retinal afferents. Such profiles occur most frequently in interlaminar zones (Bickford, Carden, & Patel, 2008), specialized regions that lie between the eye-specific layers of the LGN. Last, many other types of dendrodendritic synapses are made outside the glomerulus and can involve different classes of relay cells or even two interneurons (Dankowski & Bickford, 2003; Datskovskaia, Carden, & Bickford, 2001; Pasik, Pasik, & Hámori, 1976).

Recent computational analyses of thalamic spike trains provide evidence for a form of same-sign inhibition that is likely independent of triads because it is recorded from a variety of relay cells. This inhibition is evoked by stimuli of the preferred sign but is delayed relative to excitation (Butts et al., 2007). At present it is not clear whether this form of suppression originates in retina (Alitto & Usrey, 2008), thalamus, or both. Like sequential push–pull, delayed same-sign inhibition provides a mechanism to establish sensitivity for the temporal derivative of the stimulus. Further, sequences of excitation followed by inhibition, whether same-sign or push–pull, are able to increase response precision by narrowing the time window in which spikes can fire (Butts et al., 2007).

SYNAPTIC PHYSIOLOGY OF PUSH–PULL AND THE TRANSMISSION OF INFORMATION DOWNSTREAM

Although both relay cells and interneurons have receptive fields with similar shapes and with push–pull, the two types of cells process their synaptic inputs in very different ways during vision (Wang et al., 2011). The push in relay cells is generated by a series of large EPSPs that have a fast rise, sharp peak, and slow fall. The pull, by contrast, is smooth; under standard conditions single IPSPs are not visible in what appears as a virtually seamless hyperpolarization. The picture for interneurons is markedly different. The push is smooth and graded; individual EPSPs are rarely visible. The pull, however, has a very jagged appearance because it is formed by trains of unitary IPSPs. In other words, for relay cells the push seems high-passed in that the waveforms have sharp peaks. By contrast, the pull seems low-passed, changing smoothly over time. The opposite situation obtains for interneurons (figure 21.2B).

How might the diverse shapes of these waveforms relate to visual processing? In order to address this question, it is useful to remember that relay cells are responsible for transmitting retinal information to the cortex. Theoretical analyses show that retinal spike trains encode information in both high- and low-frequency bands (Koepsell et al., 2009, 2010). The low-frequency band represents variations in spike rate that are evoked by changes in visual pattern over time. In other words the low-frequency band transmits information encoded by changes in firing that follow temporal changes in the stimulus (Nirenberg et al., 2001; Reinagel & Reid, 2000). The high-frequency band encodes information with respect to the timing of ongoing oscillations (Koepsell et al., 2009, 2010) generated by distributed networks of ganglion cells. Thus, the high-frequency band transmits information encoded in spike timing relative to intrinsic rhythms. Remarkably, the spike trains of relay cells are able to preserve information across the frequency spectrum.

Because most cells in the brain encode information in the low-frequency band (Koepsell et al., 2010), unique synaptic mechanisms seem unnecessary for the task. Transmitting information in the high-frequency band, however, calls for extraordinary temporal precision. Recordings of EPSPs and spikes from the same relay cell show that, in fact, retinal input drives spikes with millisecond fidelity (Koepsell et al., 2009). Intuitively, the large sharp EPSPs seem ideal for promoting rapid and reliable firing. This impression is supported by simulations with conductance-based models of relay cells. The results of these simulations show that if the peaks of EPSPs are made smooth (low-passed), then there is a corresponding loss of fine temporal information in the thalamic spike train (Wang et al., 2011). Further, the simulations suggest that smooth inhibition, modeled to reflect the actual pull signal in relay cells, does not interfere with the tight coupling between retinal inputs and the spikes they evoke. This, too, is intuitive. A smooth, or low-passed, waveform would not interfere with the peaked, or high-passed, structure of the retinogeniculate EPSPs. By contrast, if smooth inhibition is replaced with jagged IPSPs in the simulations, the amount of information relayed to cortex is reduced across all time scales.

How might the graded inhibition that relay cells receive take shape? Several different mechanisms are likely to operate in unison. Retinal input to interneurons often targets distal, electrotonically remote dendrites. If these remotely generated EPSPs were to spread passively, they would attenuate greatly or even fully before reaching the soma and therefore fail to trigger an action potential (Bloomfield & Sherman, 1989). In such instances the output of the interneuron would be restricted to F2 synapses. Work in brain slices, however, shows that the dendritic membrane is active and contains L-type calcium channels (Acuna-Goycolea, Brenowitz, & Regehr, 2008). If retinal input is strong enough, these channels generate a depolarizing plateau potential that travels to the soma (Acuna-Goycolea, Brenowitz, & Regehr, 2008; Dilger, Shin, & Guido, 2011) and evokes spikes. Because the time to the peak and the length of the plateau fluctuate, the latency between retinal input and spikes evoked by the plateau currents varies considerably. Hence, the timing of retinal input is decoupled from the interneuron's spike train by the

L-type currents. Other factors that might contribute to temporal blur include attenuation of the EPSP through the dendritic cable and slow excitation mediated by metabotropic glutamate receptors (Govindaiah & Cox, 2006b). Remarkably, early studies of retinogeniculate connectivity support the idea of jittered transmission. Cross-correlations made from spike trains of ganglion cells and putative interneurons (neurons that could not be activated antidromically from cortex) were broader than those for relay cells (Dubin & Cleland, 1977). Because multiple interneurons are likely to innervate a single relay cell (Bickford et al., 2010; Crunelli et al., 1988; Ziburkus, Lo, & Guido, 2003), it seems reasonable that asynchronously arriving inhibitory inputs would blend to form smooth pull.

As yet it is not clear how the jagged pull in interneurons is formed, although there are a few clues. First, the frequency of the individual IPSPs matches retinal firing rates (Wang et al., 2011). This indicates a roughly one-to-one correspondence between spikes in a presynaptic ganglion cell and IPSPs in the interneuron. Second, anatomical studies show that retinal input to interneurons (Montero, 1991; Van Horn, Erisir, & Sherman, 2000) frequently targets F2 terminals and that these terminals are often involved in dendritic synapses with other interneurons (Pasik, Pasik, & Hámori, 1976). Thus, it is possible to construct the following scenario for generating the pull, which we illustrate for an OFF interneuron. Imagine that an F2 terminal of an ON interneuron receives input from an ON ganglion cell and also synapses with the dendrite of the OFF inter-neuron. Each time the retinal afferent spikes, it depolarizes the dendrite of the ON interneuron, which then releases transmitter onto the OFF cell. This is merely a guess, but it is testable. If this scheme were correct, then it would be possible to turn off the pull in the interneuron (and disinhibit postsynaptic relay cells) at the timescale of a single retinal action potential.

CLASSES OF LOCAL INTERNEURONS IN THE LGN

So far, we have discussed interneurons as a homogeneous group of cells. In cat the X, Y, and W pathways originate in retina and continue through the geniculate in the form of three corresponding classes of relay cells. It is not clear that interneurons divide into these three categories, although it has been noted that larger interneurons are found in regions that receive greater input from Y cells and small interneurons are found in areas that are preferentially innervated by X cells (Montero & Zempel, 1985).

This is not to say that there are not different types of local interneurons. In fact there are at least two classes of interneurons within the main layers of the LGN (Bickford, Carden, & Patel, 1999; Famiglietti, 1970; Guillery, 1966; Updyke, 1979). To understand the literature that describes these types of cells, it is useful to review some early work of Guillery (1966). He used the Golgi stain to label neurons in the LGN and divided these into three main morphological classes. Class 1 neurons are large with radial dendritic arbors; these correlate with classical Y relay cells. Class 2 cells have smaller somata and grape-like appendages; these are classical X relay cells. Guillery's last category, class 3, comprises cells that have smaller somata still, stalked dendritic appendages (i.e., F2 terminals), and have axons that ramify locally. This last feature suggested that class 3 cells were inhibitory, an impression that has been confirmed histochemically (Bickford, Carden, & Patel, 1999; Montero, 1986, 1987). Later work revealed a second class of GABAergic neuron, called class 5 (Bickford, Carden, & Patel, 1999; Famiglietti, 1970; Updyke, 1979). There are several differences between class 3 and class 5 cells, including the facts that class 5 cells signal using nitrous oxide, do not synapse within glomeruli (Bickford, Carden, & Patel, 1999), and are few in number. In aggregate these properties suggest a role for class 5 cells in modulating overall levels of activity (Bickford, Carden, & Patel, 1999).

Further, there is a special class of inhibitory cell situated in the interlaminar zones of the LGN. These interlaminar cells are innervated by relay cells but do not receive retinal input (Bickford et al., 2008; Montero, 1989). On the bases of these and other anatomical features, Montero speculated that interlaminar cells might be related to those in reticular nucleus. Later work supported this idea by showing that reticular and interlaminar neurons alike stain with antibodies to parvalbumin, whereas intralaminar interneurons are positive for calbindin (Sanchez-Vives et al., 1996). It has since become clear that several other immunohistochemical markers are selective for reticular neurons but do not label interlaminar cells (Kawasaki et al., 2004). Thus, at present, the special role of interlaminar neurons awaits further experiments.

RECEPTIVE FIELDS AND VISUAL RESPONSE PROPERTIES OF NEURONS IN THE PGN

Neurons in the PGN provide feedback inhibition to relay cells. Because cells in the PGN are spatially segregated from the LGN, it is possible to target reticular cells using extracellular recording. Still, the sensory physiology of the PGN has received sparse attention over the years, perhaps because the nucleus is slim and thus easy to miss, and perhaps also because studies of the thalamic reticular nucleus (of which the PGN is one sector) have focused on sleep (Destexhe & Sejnowski, 2003; Steriade, 2005).

Early studies that compared the LGN with the PGN made the point that the receptive fields of cells in the two structures were markedly different from each other. Recall that relay cells and local interneurons in the LGN have monocular receptive fields with a center-surround structure. By contrast, receptive fields in the PGN are typically described as diffuse, frequently binocular, and with ON and OFF components of various strengths that often overlap (Ahlsén, Lindström, & Lo, 1982; Dubin & Cleland, 1977; Funke & Eysel, 1998; Jones & Sillito, 1994; Levick, Cleland, & Dubin, 1972; Sanderson, 1971; Uhlrich et al., 1991; Wrobel & Bekisz, 1994). As varied as receptive fields in the PGN seem to be, they are all likely to take their shapes from feedforward input from relay cells; removing the cortex has only modest effects (Jones & Sillito, 1994; Xue et al., 1988).

The same group of investigators who first mapped receptive fields in the PGN made a complementary observation about relay cells. They noted that the activity of relay cells was modulated by stimuli of either sign, bright or dark, that fell outside of the classical (center-surround) receptive field (Levick, Cleland, & Dubin, 1972). This component of the receptive field was called the suppressive field (Levick, Cleland, & Dubin, 1972; Bonin, Mante, & Carandini, 2005) to distinguish it from the classical antagonistic surround. Quantitative measurements have shown that this region can span very large distances in visual space (Bonin, Mante, & Carandini, 2005). Because receptive fields in the PGN are often ON–OFF and can be large, at least by some accounts, it was reasonable to assume that the suppressive field was supplied by the PGN. Thus, it seemed as if the PGN might provide a spatially extensive, nonselective form of gain control (Bonin, Mante, & Carandini, 2005; Levick, Cleland, & Dubin, 1972) to relay cells that complemented the local and selective inhibition supplied by interneurons within the LGN proper. A growing body of work, however, challenges the view that the PGN generates the suppressive field. Usrey and colleagues, for example, showed that the spatially extensive suppression recorded from relay cells in monkey actually arises in retina and is merely inherited by the LGN (Alitto & Usrey, 2008). Further, recent quantitative analyses reveal that neurons in the PGN are sensitive to diverse and complex visual features (Vaingankar et al., 2012).

Francis Crick proposed an altogether separate role for the reticular nucleus (Crick, 1984) in sensory processing. He suggested that this structure was ideally positioned—because it receives input from both bottom-up and top-down projections (Bickford et al., 2008; Cucchiaro, Uhlrich, & Sherman, 1991; Ide, 1982)—to help to focus the "searchlight" of attention and thereby improve detection of salient features in the visual field. Recent work in the alert behaving monkey supports his general idea. These experiments showed that the firing rate of a reticular neuron drops when an animal attends to, rather than passively views, a stimulus shown in the receptive field. Simultaneously, the firing rate of a relay cell that shares the same receptive field grows faster (McAlonan, Cavanaugh, & Wurtz, 2008). Thus, it seems as if attention has the net effect of increasing activity in the LGN by means of disinhibition—by reducing the amount of feedback provided by the PGN. Whether attention regulates local interneurons in the LGN, which also receive cortical feedback (Govindaiah & Cox, 2006a; Montero, 1991) and other sources of modulatory input (Sherman & Guillery, 2002), is unknown.

If the PGN plays a role in spatial attention, then one might make the following two assumptions. The first is that the sizes of reticular receptive fields would be localized; searchlights shine on only a small part of the larger terrain. Second, one would expect that connections between the PGN and LGN would be topographic in order to focus the searchlight on the appropriate spot. In support of the first assumption there is some evidence that reticular receptive fields are not very much larger than those in the LGN in monkey (McAlonan, Cavanaugh, & Wurtz, 2008). In the cat, however, there are wide-ranging descriptions of receptive field structure in the PGN, from very large to somewhat localized (Ahlsén, Lindström, & Lo, 1982; Dubin & Cleland, 1977; Funke & Eysel, 1998; Jones & Sillito, 1994; Levick et al., 1972; Sanderson, 1971; Uhlrich et al., 1991; Wrobel & Bekisz, 1994). In addition, tuning for spatial frequency ranges from narrow (Xue et al., 1988) to broad (So & Shapley, 1981; Xue, et al., 1988). A resolution of the spatial scale of receptive fields in the PGN as a whole awaits further work.

There is considerable support for the idea that connections to and from the PGN are organized topographically (Friedlander et al., 1981; Lam, Nelson, & Sherman, 2006; Montero, Guillery, & Woolsey, 1977; Sillito & Jones, 2008; Uhlrich et al., 1991). Anatomical studies show that the axons of single reticular cells usually form dense and spatially restricted terminal fields in the LGN (Uhlrich et al., 1991), although there is a modest degree of scatter (Friedlander et al., 1981; Uhlrich et al., 1991). Cortical projections to the PGN are topographic as well (Murphy & Sillito, 1996; Robson, 1984; Sillito & Jones, 2008). Physiological studies made with pharmacological tools to activate or silence the reticular nucleus (Funke & Eysel, 1998) or associated primary sensory nuclei (Lam, Nelson, & Sherman, 2006; Lam & Sherman, 2011) further support the idea of topographic, though not necessarily reciprocal (Pinault, 2004), connectivity.

It is possible that circuits within the PGN also help to constrain the spread of local activity. Reticular neurons are coupled together by electrical synapses (Cruikshank et al., 2005; Deleuze & Huguenard, 2006; Lam, Nelson, & Sherman, 2006; Landisman et al., 2002; Long, Landisman, & Connors, 2004; Pinault, Smith, & Deschenes, 1997) that seem to tie together local groups of cells (Long, Landisman, & Connors, 2004). The connection strength of these gap junctions also changes as a function of previous activity (Haas, Zavala, & Landisman, 2011; Landisman & Connors, 2005) and so, in principle, might enlarge or reduce the scope of the searchlight as a given situation warrants. Reticular neurons also form GABAergic synapses (Ahlsen & Lindstrom, 1982; Deleuze & Huguenard, 2006; Lam, Nelson, & Sherman, 2006; Sanchez-Vives, Bal, & McCormick, 1997; Shu & McCormick, 2002) either via axonal (F1) terminals (Ide, 1982; Pinault, Smith, & Deschenes, 1997) or dendritic (F2) appendages (Ide, 1982; Montero & Singer, 1984; Pinault, Smith, & Deschenes, 1997). Thus, activity in one region could have the effect of suppressing activity in surrounding regions and limiting the spread of excitation through the nucleus.

SUMMARY

In sum, relay cells in the LGN receive input from two separate types of inhibitory cells. Local interneurons in the LGN provide feedforward inhibition, and reticular neurons in the adjacent PGN supply feedback inhibition. The local interneurons contribute a form of inhibition that is selective for the same sorts of stereotyped visual features that drive relay cells. Coupled with retinal input this inhibition provides the substrate for push–pull responses within the relay cell's receptive field. Push–pull serves the purposes of extending the dynamic range of response, improving sensitivity to luminance contrast in both the spatial and temporal domains, and reducing the amount of redundant information transmitted to the cortex. Also, the pull is strong enough to drive the membrane from tonic to burst modes. In addition, local interneurons provide same-sign inhibition via dendrodendritic synapses that may be involved in gain control. Furthermore, the synaptic physiology of local interneurons is highly specialized and seems adapted to preserve information encoded in the retinal spike train and thus to optimize the amount of information transmitted to the cortex.

Unlike local interneurons, cells in the PGN supply inhibition that appears to be driven by different and diverse patterns of visual input. So little is known about the specific role of the PGN in visual processing per se that even speculation is difficult at present, although recent work highlights an important role in spatial attention.

REFERENCES

Acuna-Goycolea, C., Brenowitz, S. D., & Regehr, W. G. (2008). Active dendritic conductances dynamically regulate GABA release from thalamic interneurons. *Neuron, 57,* 420–431.

Ahlsen, G., & Lindstrom, S. (1982). Mutual inhibition between perigeniculate neurones. *Brain Research, 236,* 482–486.

Ahlsén, G., Lindström, S., & Lo, F. S. (1982). Functional distinction of perigeniculate and thalamic reticular neurons in the cat. *Experimental Brain Research, 46,* 118–126.

Alitto, H. J., & Usrey, W. M. (2005). Dynamic properties of thalamic neurons for vision. *Progress in Brain Research, 149,* 83–90.

Alitto, H. J., & Usrey, W. M. (2008). Origin and dynamics of extraclassical suppression in the lateral geniculate nucleus of the macaque monkey. *Neuron, 57,* 135–146.

Alitto, H. J., Weyand, T. G., & Usrey, W. M. (2005). Distinct properties of stimulus-evoked bursts in the lateral geniculate nucleus. *Journal of Neuroscience, 25,*514–523.

Arcelli, P., Frassoni, C., Regondi, M. C., Biasi, S. D., & Spreafico, R. (1997). GABAergic neurons in mammalian thalamus: A marker of thalamic complexity? *Brain Research Bulletin, 42,* 27–37.

Barlow, H. B. (1961). The coding of sensory messages. In W. H. Thorpe & O. L. Zangwill (Eds.), *Current Problems in Animal Behaviour* (pp. 331–360). Cambridge: Cambridge University Press.

Bickford, M. E., Carden, W. B., & Patel, N. C. (1999). Two types of interneurons in the cat visual thalamus are distinguished by morphology, synaptic connections, and nitric oxide synthase content. *Journal of Comparative Neurology, 413,* 83–100.

Bickford, M. E., Slusarczyk, A., Dilger, E. K., Krahe, T. E., Kucuk, C., & Guido, W. (2010). Synaptic development of the mouse dorsal lateral geniculate nucleus. *Journal of Comparative Neurology, 518,* 622–635.

Bickford, M. E., Wei, H., Eisenback, M. A., Chomsung, R. D., Slusarczyk, A. S., & Dankowsi, A. B. (2008). Synaptic organization of thalamocortical axon collaterals in the perigeniculate nucleus and dorsal lateral geniculate nucleus. *Journal of Comparative Neurology, 508,* 264–285.

Blitz, D. M., & Regehr, W. G. (2005). Timing and specificity of feed-forward inhibition within the LGN. *Neuron, 45,* 917–928.

Bloomfield, S. A., & Sherman, S. M. (1989). Dendritic current flow in relay cells and interneurons of the cat's lateral geniculate nucleus. *Proceedings of the National Academy of Sciences of the United States of America, 86,* 3911–3914.

Bonin, V., Mante, V., & Carandini, M. (2005). The suppressive field of neurons in lateral geniculate nucleus. *Journal of Neuroscience, 25,* 10844–10856.

Butts, D. A., Weng, C., Jin, J., Yeh, C. I., Lesica, N. A., Alonso, J. M., et al. (2007). Temporal precision in the neural code and the timescales of natural vision. *Nature, 449*, 92–95. doi:10.1038/nature06105.

Carandini, M., Horton, J. C., & Sincich, L. C. (2007). Thalamic filtering of retinal spike trains by postsynaptic summation. *Journal of Vision, 7*, 1–11. doi:10.1167/7.14.20.

Casale, A. E., & McCormick, D. A. (2011). Active action potential propagation but not initiation in thalamic interneuron dendrites. *Journal of Neuroscience, 31*, 18289–18302.

Cox, C. L., Reichova, I., & Sherman, S. M. (2003). Functional synaptic contacts by intranuclear axon collaterals of thalamic relay neurons. *Journal of Neuroscience, 23*, 7642–7646.

Cox, C. L., & Sherman, S. M. (2000). Control of dendritic outputs of inhibitory interneurons in the lateral geniculate nucleus. *Neuron, 27*, 597–610.

Crandall, S. R., & Cox, C. L. (2012). Local dendrodendritic inhibition regulates fast synaptic transmission in visual thalamus. *Journal of Neuroscience, 32*, 2513–2522.

Crick, F. (1984). Function of the thalamic reticular complex: The searchlight hypothesis. *Proceedings of the National Academy of Sciences of the United States of America, 81*, 4586–4590. doi:10.1073/pnas.81.14.4586.

Cruikshank, S. J., Landisman, C. E., Mancilla, J. G., & Connors, B. W. (2005). Connexon connexions in the thalamocortical system. *Progress in Brain Research, 149*, 41–57.

Crunelli, V., Haby, M., Jassik-Gerschenfeld, D., Leresche, N., & Pirchio, M. (1988). Cl⁻ and K⁺-dependent inhibitory postsynaptic potentials evoked by interneurones of the rat lateral geniculate nucleus. *Journal of Physiology, 399*, 153–176.

Cucchiaro, J. B., Uhlrich, D. J., & Sherman, S. M. (1991). Electron-microscopic analysis of synaptic input from the perigeniculate nucleus to the A-laminae of the lateral geniculate nucleus in cats. *Journal of Comparative Neurology, 310*, 316–336.

Dan, Y., Atick, J. J., & Reid, R. C. (1996). Efficient coding of natural scenes in the lateral geniculate nucleus: Experimental test of a computational theory. *Journal of Neuroscience, 16*, 3351–3362.

Dankowski, A., & Bickford, M. E. (2003). Inhibitory circuitry involving Y cells and Y retinal terminals in the C laminae of the cat dorsal lateral geniculate nucleus. *Journal of Comparative Neurology, 460*, 368–379.

Datskovskaia, A., Carden, W. B., & Bickford, M. E. (2001). Y retinal terminals contact interneurons in the cat dorsal lateral geniculate nucleus. *Journal of Comparative Neurology, 430*, 85–100.

Deleuze, C., & Huguenard, J. R. (2006). Distinct electrical and chemical connectivity maps in the thalamic reticular nucleus: Potential roles in synchronization and sensation. *Journal of Neuroscience, 26*, 8633–8645.

Denning, K. S., & Reinagel, P. (2005). Visual control of burst priming in the anesthetized lateral geniculate nucleus. *Journal of Neuroscience, 25*, 3531–3538.

Destexhe, A., & Sejnowski, T. J. (2003). Interactions between membrane conductances underlying thalamocortical slow-wave oscillations. *Physiological Reviews, 83*, 1401–1453.

Dilger, E. K., Shin, H.-S., & Guido, W. (2011). Requirements for synaptically evoked plateau potentials in relay cells of the dorsal lateral geniculate nucleus of the mouse. *Journal of Physiology, 589*, 919–937.

Dong, D. W., & Atick, J. J. (1995). Statistics of natural time-varying images. *Network: Computation in Neural Systems, 6*, 345–358.

Dubin, M. W., & Cleland, B. G. (1977). Organization of visual inputs to interneurons of lateral geniculate nucleus of the cat. *Journal of Neurophysiology, 40*, 410–427.

Enroth-Cugell, C., & Robson, J. G. (1966). The contrast sensitivity of retinal ganglion cells of the cat. *Journal of Physiology, 187*, 517–552.

Erisir, A., Van Horn, S. C., Bickford, M. E., & Sherman, S. M. (1997). Immunocytochemistry and distribution of parabrachial terminals in the lateral geniculate nucleus of the cat: A comparison with corticogeniculate terminals. *Journal of Comparative Neurology, 377*, 535–549.

Erisir, A., Van Horn, S. C., & Sherman, S. M. (1998). Distribution of synapses in the lateral geniculate nucleus of the cat: Differences between laminae A and A1 and between relay cells and interneurons. *Journal of Comparative Neurology, 390*, 247–255.

Famiglietti, E. V., Jr. (1970). Dendro-dendritic synapses in the lateral geniculate nucleus of the cat. *Brain Research, 20*, 181–191.

Famiglietti, E. V., Jr., & Peters, A. (1972). The synaptic glomerulus and the intrinsic neuron in the dorsal lateral geniculate nucleus of the cat. *Journal of Comparative Neurology, 144*, 285–334.

Field, D. J. (1987). Relations between the statistics of natural images and the response properties of cortical cells. *Journal of the Optical Society of America. A, Optics and Image Science, 4*, 2379–2394. doi:10.1364/JOSAA.4.002379.

Fitzpatrick, D., Penny, G. R., & Schmechel, D. E. (1984). Glutamic acid decarboxylase-immunoreactive neurons and terminals in the lateral geniculate nucleus of the cat. *Journal of Neuroscience, 4*, 1809–1829.

Friedlander, M. J., Lin, C. S., Stanford, L. R., & Sherman, S. M. (1981). Morphology of functionally identified neurons in lateral geniculate nucleus of the cat. *Journal of Neurophysiology, 46*, 80–129.

Funke, K., & Eysel, U. T. (1998). Inverse correlation of firing patterns of single topographically matched perigeniculate neurons and cat dorsal lateral geniculate relay cells. *Visual Neuroscience, 15*, 711–729.

Geisler, W. S. (2008). Visual perception and the statistical properties of natural scenes. *Annual Review of Psychology, 59*, 167–192.

Govindaiah, G., & Cox, C. L. (2004). Synaptic activation of metabotropic glutamate receptors regulates dendritic outputs of thalamic interneurons. *Neuron, 41*, 611–623.

Govindaiah, G., & Cox, C. L. (2006a). Excitatory actions of synaptically released catecholamines in the rat lateral geniculate nucleus. *Neuroscience, 137*, 671–683.

Govindaiah, G., & Cox, C. L. (2006b). Metabotropic glutamate receptors differentially regulate GABAergic inhibition in thalamus. *Journal of Neuroscience, 26*, 13443–13453.

Guillery, R. W. (1966). A study of Golgi preparations from the dorsal lateral geniculate nucleus of the adult cat. *Journal of Comparative Neurology, 128*, 21–50.

Guillery, R. W. (1969a). The organization of synaptic interconnections in the laminae of the dorsal lateral geniculate nucleus of the cat. *Zeitschrift fur Zellforschung und Mikroskopische Anatomie (Vienna, Austria), 96*, 1–38.

Guillery, R. W. (1969b). A quantitative study of synaptic interconnections in the dorsal lateral geniculate nucleus of the cat. *Zeitschrift fur Zellforschung und Mikroskopische Anatomie (Vienna, Austria), 96*, 39–48.

Haas, J. S., Zavala, B., & Landisman, C. E. (2011). Activity-dependent long-term depression of electrical synapses. *Science, 334*, 389–393.

Hamos, J. E., Van Horn, S. C., Raczkowski, D., & Sherman, S. M. (1987). Synaptic circuits involving an individual retinogeniculate axon in the cat. *Journal of Comparative Neurology, 259*, 165–192.

Hamos, J. E., Van Horn, S. C., Raczkowski, D., Uhlrich, D. J., & Sherman, S. M. (1985). Synaptic connectivity of a local circuit neurone in lateral geniculate nucleus of the cat. *Nature, 317*, 618–621.

Hirsch, J. A. (2003). Synaptic physiology and receptive field structure in the early visual pathway of the cat. *Cerebral Cortex, 13*, 63–69.

Hirsch, J. A., & Martinez, L. M. (2006a). Circuits that build visual cortical receptive fields. *Trends in Neurosciences, 29*, 30–39. doi:10.1016/j.tins.2005.11.001.

Hirsch, J. A., & Martinez, L. M. (2006b). Laminar processing in the visual cortical column. *Current Opinion in Neurobiology, 16*(4), 377–384.

Hirsch, J. A., Martinez, L. M., Alonso, J. M., Desai, K., Pillai, C., & Pierre, C. (2002). Synaptic physiology of the flow of information in the cat's visual cortex in vivo. *Journal of Physiology, 540*(Pt 1), 335–350.

Hodgkin, A. L., & Huxley, A. F. (1952). A quantitative description of membrane current and its application to conduction and excitation in nerve. *Journal of Physiology, 117*, 500–544.

Hubel, D. H. (1960). Single unit activity in lateral geniculate body and optic tract of unrestrained cats. *Journal of Physiology, 150*, 91–104.

Hubel, D. H., & Wiesel, T. N. (1961). Integrative action in the cat's lateral geniculate body. *Journal of Physiology, 155*, 385–398.

Huguenard, J. R., & McCormick, D. A. (1992). Simulation of the currents involved in rhythmic oscillations in thalamic relay neurons. *Journal of Neurophysiology, 68*, 1373–1383.

Huguenard, J., & Prince, D. (1992). A novel T-type current underlies prolonged Ca(2+)-dependent burst firing in GABAergic neurons of rat thalamic reticular nucleus. *Journal of Neuroscience, 12*, 3804–3817.

Ide, L. S. (1982). The fine structure of the perigeniculate nucleus in the cat. *Journal of Comparative Neurology, 210*(4), 317–334.

Jahnsen, H., & Llinas, R. (1984). Voltage-dependent burst-to-tonic switching of thalamic cell activity: An in vitro study. *Archives Italiennes de Biologie, 122*, 73–82.

Jones, H. E., & Sillito, A. M. (1994). The length-response properties of cells in the feline perigeniculate nucleus. *European Journal of Neuroscience, 6*, 1199–1204.

Kawasaki, H., Crowley, J. C., Livesey, F. J., & Katz, L. C. (2004). Molecular organization of the ferret visual thalamus. *Journal of Neuroscience, 24*, 9962–9970.

Koepsell, K., Wang, X., Hirsch, J. A., & Sommer, F. T. (2010). Exploring the function of neural oscillations in early sensory systems. *Frontiers in Neuroscience, 4*, 53–61. doi:10.3389/neuro.01.010.2010.

Koepsell, K., Wang, X., Vaingankar, V., Wei, Y., Wang, Q., Rathbun, D. L., et al. (2009). Retinal oscillations carry visual information to cortex. *Frontiers in Systems Neuroscience, 3*, 1–18. doi:10.3389/neuro.06.004.2009.

Kuffler, S. W. (1953). Discharge patterns and functional organization of the mammalian retina. *Journal of Neurophysiology, 16*, 37–68.

Lam, Y.-W., Nelson, C. S., & Sherman, S. M. (2006). Mapping of the functional interconnections between thalamic reticular neurons using photostimulation. *Journal of Neurophysiology, 96*, 2593–2600.

Lam, Y. W., & Sherman, S. M. (2011). Functional organization of the thalamic input to the thalamic reticular nucleus. *Journal of Neuroscience, 31*, 6791–6799.

Landisman, C. E., & Connors, B. W. (2005). Long-term modulation of electrical synapses in the mammalian thalamus. *Science, 310*, 1809–1813.

Landisman, C. E., Long, M. A., Beierlein, M., Deans, M. R., Paul, D. L., & Connors, B. W. (2002). Electrical synapses in the thalamic reticular nucleus. *Journal of Neuroscience, 22*, 1002–1009.

Lesica, N. A., & Stanley, G. B. (2004). Encoding of natural scene movies by tonic and burst spikes in the lateral geniculate nucleus. *Journal of Neuroscience, 24*, 10731–10740.

Lesica, N. A., Weng, C., Jin, J., Yeh, C. I., Alonso, J. M., & Stanley, G. B. (2006). Dynamic encoding of natural luminance sequences by LGN bursts. *PLoS Biology, 4*, e209. doi:10.1371/journal.pbio.0040209.

LeVay, S., & Ferster, D. (1979). Proportion of interneurons in the cat's lateral geniculate nucleus. *Brain Research, 164*, 304–308.

Levick, W. R., Cleland, B. G., & Dubin, M. W. (1972). Lateral geniculate neurons of cat: Retinal inputs and physiology. *Investigative Ophthalmology, 11*, 302–311.

Lin, C. S., Kratz, K. E., & Sherman, S. M. (1977). Percentage of relay cells in the cat's lateral geniculate nucleus. *Brain Research, 131*, 167–173.

Long, M. A., Landisman, C. E., & Connors, B. W. (2004). Small clusters of electrically coupled neurons generate synchronous rhythms in the thalamic reticular nucleus. *Journal of Neuroscience, 24*, 341–349.

Lorincz, M. L., Kekesi, K. A., Juhasz, G., Crunelli, V., & Hughes, S. W. (2009). Temporal framing of thalamic relay-mode firing by phasic inhibition during the alpha rhythm. *Neuron, 63*, 683–696.

Marr, D., & Hildreth, E. (1980). Theory of edge detection. *Proceedings of the Royal Society of London. Series B, Biological Sciences, 207*, 187–217.

Martinez, L. M., Wang, Q., Reid, R. C., Pillai, C., Alonso, J. M., Sommer, F. T., et al. (2005). Receptive field structure varies with layer in the primary visual cortex. *Nature Neuroscience, 8*, 372–379. doi:10.1038/nn1404.

McAlonan, K., Cavanaugh, J., & Wurtz, R. H. (2008). Guarding the gateway to cortex with attention in visual thalamus. *Nature, 456*, 391–394.

Montero, V. M. (1986). Localization of gamma-aminobutyric acid (GABA) in type 3 cells and demonstration of their source to F2 terminals in the cat lateral geniculate nucleus: A Golgi-electron-microscopic GABA-immunocytochemical study. *Journal of Comparative Neurology, 254*, 228–245.

Montero, V. M. (1987). Ultrastructural identification of synaptic terminals from the axon of type 3 interneurons in the cat lateral geniculate nucleus. *Journal of Comparative Neurology, 264*, 268–283.

Montero, V. M. (1989). The GABA-immunoreactive neurons in the interlaminar regions of the cat lateral geniculate nucleus: Light and electron microscopic observations. *Experimental Brain Research, 75*, 497–512.

Montero, V. M. (1991). A quantitative study of synaptic contacts on interneurons and relay cells of the cat lateral geniculate nucleus. *Experimental Brain Research, 86*, 257–270.

Montero, V. M., Guillery, R. W., & Woolsey, C. N. (1977). Retinotopic organization within the thalamic reticular nucleus demonstrated by a double label autoradiographic technique. *Brain Research, 138*,407–421.

Montero, V. M., & Singer, W. (1984). Ultrastructure and synaptic relations of neural elements containing glutamic acid decarboxylase (GAD) in the perigeniculate nucleus of the cat. A light and electron microscopic immunocytochemical study. *Experimental Brain Research, 56*, 115–125.

Montero, V. M., & Singer, W. (1985). Ultrastructural identification of somata and neural processes immunoreactive to antibodies against glutamic acid decarboxylase (GAD) in the dorsal lateral geniculate nucleus of the cat. *Experimental Brain Research, 59*,151–165.

Montero, V. M., & Zempel, J. (1985). Evidence for two types of GABA-containing interneurons in the A-laminae of the cat lateral geniculate nucleus: A double-label HRP and GABA-immunocytochemical study. *Experimental Brain Research, 60*, 603–609.

Murphy, P. C., & Sillito, A. M. (1996). Functional morphology of the feedback pathway from area 17 of the cat visual cortex to the lateral geniculate nucleus. *Journal of Neuroscience, 16*, 1180–1192.

Niell, C. M., & Stryker, M. P. (2010). Modulation of visual responses by behavioral state in mouse visual cortex. *Neuron, 65*, 472–479.

Nirenberg, S., Carcieri, S. M., Jacobs, A. L., & Latham, P. E. (2001). Retinal ganglion cells act largely as independent encoders. *Nature, 411*, 698–701.

Pasik, P., Pasik, T., & Hámori, J. (1976). Synapses between interneurons in the lateral geniculate nucleus of monkeys. *Experimental Brain Research, 25*, 1–13.

Pinault, D. (2004). The thalamic reticular nucleus: Structure, function and concept. *Brain Research. Brain Research Reviews, 46*, 1–31.

Pinault, D., Smith, Y., & Deschenes, M. (1997). Dendrodendritic and axoaxonic synapses in the thalamic reticular nucleus of the adult rat. *Journal of Neuroscience, 17*, 3215–3233.

Rathbun, D. L., Warland, D. K., & Usrey, W. M. (2010). Spike timing and information transmission at retinogeniculate synapses. *Journal of Neuroscience, 30*, 13558–13566.

Reinagel, P., Godwin, D., Sherman, S. M., & Koch, C. (1999). Encoding of visual information by LGN bursts. *Journal of Neurophysiology, 81*, 2558–2569.

Reinagel, P., & Reid, R. C. (2000). Temporal coding of visual information in the thalamus. *Journal of Neuroscience, 20*, 5392–5400.

Robson, J. A. (1984). Reconstructions of corticogeniculate axons in the cat. *Journal of Comparative Neurology, 225*, 193–200.

Sanchez-Vives, M. V., Bal, T., Kim, U., von Krosigk, M., & McCormick, D. A. (1996). Are the interlaminar zones of the ferret dorsal lateral geniculate nucleus actually part of the perigeniculate nucleus? *Journal of Neuroscience, 16*,5923–5941.

Sanchez-Vives, M. V., Bal, T., & McCormick, D. A. (1997). Inhibitory interactions between perigeniculate GABAergic neurons. *Journal of Neuroscience, 17*, 8894–8908.

Sanderson, K. J. (1971). The projection of the visual field to the lateral geniculate and medial interlaminar nuclei in the cat. *Journal of Comparative Neurology, 143*, 101–108.

Shapley, R., & Lennie, P. (1985). Spatial frequency analysis in the visual system. *Annual Review of Neuroscience, 8*, 547–583.

Sherman, S. M. (2001a). Thalamic relay functions. *Progress in Brain Research, 134*, 51–69.

Sherman, S. M. (2001b). Tonic and burst firing: Dual modes of thalamocortical relay. *Trends in Neurosciences, 24*, 122–126. doi:10.1016/S0166-2236(00)01714-8.

Sherman, S. M. (2001c). A wake-up call from the thalamus. *Nature Neuroscience, 4*(4), 344–346.

Sherman, S. M. (2004). Interneurons and triadic circuitry of the thalamus. *Trends in Neurosciences, 27*, 670–675. doi: 10.1016/j.tins.2004.08.003.

Sherman, S. M., & Guillery, R. W. (1998). On the actions that one nerve cell can have on another: Distinguishing "drivers" from "modulators." *Proceedings of the National Academy of Sciences of the United States of America, 95*, 7121–7126. doi:10.1073/pnas.95.12.7121.

Sherman, S. M., & Guillery, R. W. (2002). The role of the thalamus in the flow of information to the cortex. *Philosophical Transactions of the Royal Society of London. Series B, Biological Sciences, 357*, 1695–1708. doi:10.1098/rstb.2002.1161.

Shu, Y., & McCormick, D. A. (2002). Inhibitory interactions between ferret thalamic reticular neurons. *Journal of Neurophysiology, 87*, 2571–2576.

Sillito, A. M., & Jones, H. E. (2008). The role of the thalamic reticular nucleus in visual processing. *Thalamus & Related Systems, 4*, 1–12. doi:10.1017/S1472928807000295.

Simoncelli, E. P., & Olshausen, B. A. (2001). Natural image statistics and neural representation. *Annual Review of Neuroscience, 24*, 1193–1216.

Sincich, L. C., Horton, J. C., & Sharpee, T. O. (2009). Preserving information in neural transmission. *Journal of Neuroscience, 29*, 6207–6216.

So, Y. T., & Shapley, R. (1981). Spatial tuning of cells in and around lateral geniculate nucleus of the cat: X and Y relay cells and perigeniculate interneurons. *Journal of Neurophysiology, 45*, 107–120.

Steriade, M. (2005). Sleep, epilepsy and thalamic reticular inhibitory neurons. *Trends in Neurosciences, 28*, 317–324. doi:10.1016/j.tins.2005.03.007.

Swadlow, H. A., Bezdudnaya, T., & Gusev, A. G. (2005). Spike timing and synaptic dynamics at the awake thalamocortical synapse. *Progress in Brain Research, 149*, 91–105.

Swadlow, H. A., & Gusev, A. G. (2001). The impact of "bursting" thalamic impulses at a neocortical synapse. *Nature Neuroscience, 4*, 402–408.

Swadlow, H. A., Gusev, A. G., & Bezdudnaya, T. (2002). Activation of a cortical column by a thalamocortical impulse. *Journal of Neuroscience, 22*, 7766–7773.

Szentagothai, J., Hamori, J., & Tombol, T. (1966). Degeneration and electron microscope analysis of the synaptic glomeruli in the lateral geniculate body. *Experimental Brain Research, 2*, 283–301.

Uhlrich, D. J., Cucchiaro, J. B., Humphrey, A. L., & Sherman, S. M. (1991). Morphology and axonal projection patterns of individual neurons in the cat perigeniculate nucleus. *Journal of Neurophysiology, 65*, 1528–1541.

Updyke, B. V. (1979). A golgi study of the class V cell in the visual thalamus of the cat. *Journal of Comparative Neurology, 186*, 603–619.

Usrey, W. M., Reppas, J. B., & Reid, R. C. (1998). Paired-spike interactions and synaptic efficacy of retinal inputs to the thalamus. *Nature, 395*, 384–387.

Usrey, W. M., Reppas, J. B., & Reid, R. C. (1999). Specificity and strength of retinogeniculate connections. *Journal of Neurophysiology, 82*, 3527–3540.

Vaingankar, V., Soto Sanchez, C., Wang, X., Sommer F. T., & Hirsch, J. A. (2012). Neurons in the thalamic reticular nucleus are selective for complex visual features. *Frontiers in Integrative Neuroscience, 6*, 118. doi:10.3389/fnint.2012.00118.

Van Horn, S. C., Erisir, A., & Sherman, S. M. (2000). Relative distribution of synapses in the A-laminae of the lateral geniculate nucleus of the cat. *Journal of Comparative Neurology, 416*, 509–520.

Wang, S., Bickford, M. E., Van Horn, S. C., Erisir, A., Godwin, D. W., & Sherman, S. M. (2001). Synaptic targets of thalamic reticular nucleus terminals in the visual thalamus of the cat. *Journal of Comparative Neurology, 440*, 321–341.

Wang, X., Hirsch, J. A., & Sommer, F. T. (2010). Recoding of sensory information across the retinothalamic synapse. *Journal of Neuroscience, 30*, 13567–13577.

Wang, X., Sommer, F. T., & Hirsch, J. A. (2011). Inhibitory circuits for visual processing in thalamus. *Current Opinion in Neurobiology, 21*,726–733.

Wang, X., Vaingankar, V., Sanchez, C. S., Sommer, F. T., & Hirsch, J. A. (2011). Thalamic interneurons and relay cells use complementary synaptic mechanisms for visual processing. *Nature Neuroscience, 14*, 224–231.

Wang, X., Wei, Y., Vaingankar, V., Wang, Q., Koepsell, K., Sommer, F. T., et al. (2007). Feedforward excitation and inhibition evoke dual modes of firing in the cat's visual thalamus during naturalistic viewing. *Neuron, 55*, 465–478. doi:10.1016/j.neuron.2007.06.039.

Wässle, H., & Boycott, B. B. (1991). Functional architecture of the mammalian retina. *Physiological Reviews, 71*, 447–480.

Weber, A. J., & Kalil, R. E. (1983). The percentage of interneurons in the dorsal lateral geniculate nucleus of the cat and observations on several variables that affect the sensitivity of horseradish peroxidase as a retrograde marker. *Journal of Comparative Neurology, 220*, 336–346.

Werblin, F. S. (2010). Six different roles for crossover inhibition in the retina: Correcting the nonlinearities of synaptic transmission. *Visual Neuroscience, 27*, 1–8. doi:10.1017/S0952523810000076.

Wiesel, T. N. (1959). Recording inhibition and excitation in the cat's retinal ganglion cells with intracellular electrodes. *Nature, 183*, 264–265.

Wilson, J. (1989). Synaptic organization of individual neurons in the macaque lateral geniculate nucleus. *Journal of Neuroscience, 9*, 2931–2953.

Wilson, J. R., Friedlander, M. J., & Sherman, S. M. (1984). Fine structural morphology of identified X- and Y-cells in the cat's lateral geniculate nucleus. *Proceedings of the Royal Society of London. Series B, Biological Sciences, 221*, 411–436.

Wrobel, A., & Bekisz, M. (1994). Visual classification of X and Y perigeniculate neurons of the cat. *Experimental Brain Research, 101*, 307–313.

Xue, J. T., Carney, T., Ramoa, A. S., & Freeman, R. D. (1988). Binocular interaction in the perigeniculate nucleus of the cat. *Experimental Brain Research, 69*, 497–508.

Ziburkus, J., Lo, F. S., & Guido, W. (2003). Nature of inhibitory postsynaptic activity in developing relay cells of the lateral geniculate nucleus. *Journal of Neurophysiology, 90*, 1063–1070.

22 Functional Properties of Cortical Feedback to the Primate Lateral Geniculate Nucleus

FARRAN BRIGGS AND W. MARTIN USREY

The thalamus and cerebral cortex are intimately interconnected by a reciprocal network of feedforward and feedback connections. In the visual system of primates, projections from the lateral geniculate nucleus (LGN) of the thalamus are organized into functionally distinct, parallel streams that relay different types of visual information from the retina to distinct laminar and compartmental targets in the primary visual cortex (V1). The cortex, in turn, gives rise to feedback projections that target the LGN. Although the functional role of cortical feedback in visual processing remains poorly understood, increasing evidence indicates that feedback projections are also organized into parallel streams. This chapter examines the anatomical and physiological organization of corticogeniculate circuits in the primate. This organization provides important clues about the functional role of corticogeniculate feedback in visual processing.

Visual signals leaving the eye do not have direct access to the cerebral cortex but rather are relayed to the cortex primarily by neurons in the LGN. Consequently, the communication of visual information along the retinogeniculocortical pathway is typically characterized as a feedforward process. However, a careful examination of the neural circuits that provide synaptic input to the LGN reveals a myriad of nonretinal sources of input, with the single greatest source of input coming from a select group of V1 neurons called corticogeniculate neurons (Erisir et al., 1997; Erisir, Van Horn, & Sherman, 1997; Fitzpatrick et al., 1994; Gilbert & Kelly, 1975; Guillery, 1969; Hendrickson, Wilson, & Ogren, 1978). The transmission of signals from the LGN to V1 should therefore not be viewed as a process that operates independently of the cortex but rather as a process that includes dynamic interactions with the cortex.

Cortical projections to first-order thalamic nuclei (e.g., the LGN, MGB, and VPM/VPL) are ubiquitous across mammalian species and sensory modalities (vision, auditory, and somatic sensation). Given the evolutionary conservation and anatomical strength of these connections, it is surprising that we know so little about their functional roles in sensory processing. In this chapter we focus on the corticogeniculate pathway in the macaque monkey as a model for studying the structural and functional organization of corticothalamic feedback.[1] The precise and parallel organization of corticogeniculate feedback to the primate LGN provides significant insight into the potential role this pathway serves for vision.

FUNCTIONAL ORGANIZATION OF THE PRIMATE EARLY VISUAL SYSTEM

In primates the visual pathway from the retina to V1 is organized into three parallel-processing streams. These streams are established at the earliest processing stage, the retina, and are maintained as they continue through the LGN and into V1. They are called the magnocellular, parvocellular, and koniocellular streams on the basis of their constituent cells' morphological properties in the LGN (magnocellular, large cell bodies; parvocellular, small cell bodies; koniocellular, dust-like). Other chapters in this volume (chapters 16 by Kaplan and 25 by Callaway) discuss the anatomy and physiology of these parallel-processing streams in detail. Here we provide a brief overview of their functional organization with an emphasis on those features most relevant to our understanding of the organization and physiology of corticogeniculate neurons.

The magnocellular, parvocellular, and koniocellular streams arise from three classes of retinal ganglion cells (RGCs)—the parasol, midget, and small bistratified RGCs, respectively. Compared to midget RGCs, parasol RGCs respond better to low-contrast stimuli and to stimuli moving at high temporal frequencies (reviewed in Briggs & Usrey, 2009a; Lankow & Usrey, 2011). Parasol RGCs also have larger receptive fields that lack a chromatic organization, whereas midget receptive fields are smaller and display a chromatic antagonism that is established by the middle (M)- and long (L)-wavelength-sensitive cones. Less is known about the small bistratified RGCs. However, they are generally believed to have response properties intermediate to those of parasol and midget cells, with the exception that they clearly receive strong input from the short (S)-wavelength-sensitive cones that is antagonistic to input from the M and L cones. Based on these properties, the parasol RGCs provide information important for processing motion, while the midget and small bistratified RGCs provide information important for processing color. The relatively small size of the midget receptive field also indicates that midget RGCs likely play an important role in the processing of fine visual features, a property consistent with their ability to respond well to stimuli containing higher spatial frequencies.

The axons of the parasol, midget, and small bistratified RGCs selectively innervate the magnocellular, parvocellular, and koniocellular layers of the LGN, respectively (reviewed in Briggs & Usrey, 2009a). In Old World primates, including macaque monkeys, the magnocellular layers (layers 1 and 2) are located ventral to the parvocellular layers (layers 3–6), and the koniocellular layers are located between each of these layers and below layer 1 (figure 22.1). The stream specificity of retinogeniculate axons for specific LGN layers, along with a limited convergence of retinal axons onto individual LGN neurons, results in LGN neurons having response properties very similar to those of their retinal inputs (table 22.1). A few noteworthy differences exist, however, in the response properties of LGN neurons and their retinal inputs: the antagonistic receptive-field surrounds of LGN neurons are stronger than those of their retinal inputs (Hubel & Wiesel, 1961; Usrey, Reppas, & Reid, 1999), LGN neurons produce fewer spikes than their retinal inputs (Kaplan, Purpura, & Shapley, 1987; Rathbun, Warland, & Usrey, 2010; Sincich, Horton, & Sharpee, 2009; Usrey, Reppas, & Reid, 1998; Weyand, 2007), and LGN neurons have a greater range of responses to visual stimuli (e.g., burst and tonic spikes; Alitto, Weyand, & Usrey, 2005; Sherman, 1996). These differences likely reflect the integration of inputs from retinal and nonretinal sources, including corticogeniculate feedback (discussed below).

Neurons in the magnocellular, parvocellular, and koniocellular layers of the LGN selectively innervate

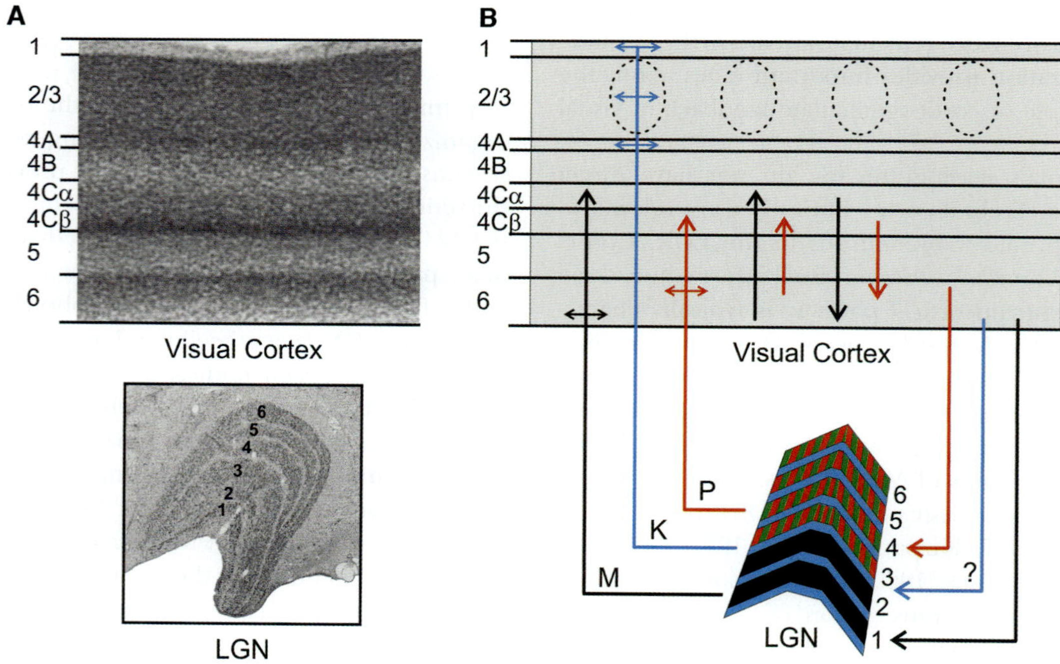

FIGURE 22.1 (A) Nissl-stained sections of primary visual cortex (above) and LGN (below) from the macaque monkey. LGN layers 1 and 2 are the magnocellular layers, layers 3–6 are the parvocellular layers, and in between these layers and below layer 1 are the koniocellular layers. (B) Circuit diagram showing the organization of connections between the LGN and visual cortex. A subset of the connections between cortical layers 4C and 6 is also provided. The magnocellular stream is indicated in black, the parvocellular stream in red, and the koniocellular stream in blue.

     FARRAN BRIGGS AND W. MARTIN USREY

TABLE 22.1

Comparison of response properties for neurons in the magnocellular, parvocellular, and koniocellular layers of the primate LGN

Parallel-processing streams in the feedforward pathway from LGN to V1			
	Magno cells	Parvo cells	Konio cells
Axon velocity	Faster	Slower	?
Firing rate	Higher	Lower	?
Responses to low-contrast stimuli	Very good	Poor	?
Responses to fast-moving stimuli	Very good	Poor	?
Chromatic organization	Broadband	L/M-cone opponent	S-cone input

layer 4Cβ, 4Cβ, and the cytochrome-oxidase-rich blobs in layers 2/3 of V1, respectively. Thus, the parallel streams for visual processing initially established in the retina are maintained in the early stages of cortical processing (figure 22.1). Because the axons of the three classes of LGN neurons differ in their diameters and degrees of myelination, visual signals traveling down magnocellular axons arrive first to V1, followed by parvocellular signals, and then koniocellular signals. Finally, magnocellular and parvocellular axons also give rise to weaker collateral projections that are biased to the lower and upper divisions of layer 6, respectively (Hendrickson, Wilson, & Ogren, 1978). This point will be important in our discussion of the physiological properties of corticogeniculate neurons across the depth of layer 6.

Once visual signals from the magno-, parvo-, and koniocellular streams arrive in V1, there are a multitude of intrinsic circuits that provide further processing. Some of these circuits appear to maintain stream specificity, and others appear to provide a substrate for stream mixing. For instance, photostimulation experiments from living slices of V1 indicate that individual layer 5 neurons primarily receive mixed-stream input from the overlying cortical layers, whereas layer 6 neurons are more diverse with respect to the organization of their inputs, with some neurons receiving mixed-stream input while other neurons are dominated by either magnocellular-stream or parvocellular-stream input (Briggs & Callaway, 2001, 2005). Although our current understanding of the local circuit architecture of V1 is insufficient to resolve the full extent to which magnocellular-, parvocellular-, and koniocellular-stream signals remain segregated or become mixed within and

beyond V1, evidence described in detail below indicates that feedback from V1 to the LGN is organized into parallel-processing streams that align well with the feedforward streams.

STRUCTURAL ORGANIZATION OF V1 FEEDBACK TO LGN

Similar to other mammals, corticogeniculate neurons in V1 of the macaque monkey are all pyramidal neurons with cell bodies located in layer 6. Species-specific differences exist, however, in the relative percentage of layer 6 neurons that project to the LGN as well as in the distribution of corticogeniculate neurons across the depth of layer 6. For instance, although approximately 50% of layer 6 neurons in the carnivore and rodent are corticogeniculate neurons, only 14% of layer 6 neurons in the macaque monkey project to the LGN (Gilbert & Kelly, 1975; Fitzpatrick et al., 1994). Moreover, layer 6 in the macaque monkey can be subdivided into three tiers—an upper, middle, and lower tier—with corticogeniculate neurons restricted to the upper and lower tiers (Fitzpatrick et al., 1994). This three-tier pattern is not present in carnivores and rodents.

Retrograde-tracer experiments demonstrate stream specificity in the projections from cortical layer 6 to the LGN in the macaque. In particular, layer 6 neurons projecting to the magnocellular LGN layers have cell bodies restricted to the lower tier of layer 6, whereas layer 6 neurons projecting to the parvocellular layers have cell bodies in the upper tier (Fitzpatrick et al., 1994). Although less conclusive, anatomical evidence also supports the view that a separate population of neurons in the lower tier of layer 6 projects to the koniocellular LGN layers.

The majority of corticogeniculate neurons in the macaque monkey fall into three morphological classes. Corticogeniculate neurons in the upper tier of layer 6, likely targeting the parvocellular layers of the LGN, match the previously described type Iβ and IβA morphologies with axons and dendrites arborizing primarily in layers 4Cβ and 4Cβ/4A, respectively (Briggs & Callaway, 2001; Wiser & Callaway, 1996). Corticogeniculate neurons located in the lower tier of layer 6, likely projecting to the magnocellular layers, display the type IC morphology with axons and dendrites arborizing across the full extent of layer 4C. The third main morphological class of corticogeniculate neurons constitute a previously unidentified cell class (unpublished data). These cells have apical dendrites that appear tilted as they emerge from the cell soma, and they are located in the deepest zones of layer 6, often encroaching into the white matter. In other species, cells with this morphology frequently project to the layers of the LGN equivalent to the koniocellular layers (Usrey & Fitzpatrick, 1996).

Each of the corticogeniculate cell types described above has axonal projections targeting layer 4C (see chapter 25 by Callaway). This is consistent with the accepted model of reciprocal connectivity between layer 6 and the overlying cortical layers that receive LGN input. Given that the local axonal projections of corticogeniculate neurons are often specific to the magno- or parvocellular-recipient lamina (e.g., type Iβ/IβA cells send local axonal arbors to layer 4Cβ), it is possible that local axonal connections from corticogeniculate neurons also preserve stream specificity.

We currently lack a detailed morphological understanding of corticogeniculate axon arbors within the primate LGN. For example, it is unknown whether individual corticogeniculate neurons target single or multiple LGN layers, perhaps preserving the ocular-dominance property of the projecting neuron. Likewise, it is not known whether individual axons preferentially target LGN relay cells or local interneurons (as a population, corticogeniculate neurons target both). With that said, some general rules are known. For instance, corticogeniculate neurons have axon collaterals that also innervate neurons in the reticular nucleus of the thalamus (RTN) (reviewed in Sherman & Guillery, 2005), a sheet-like structure that wraps around the dorsal surface of the LGN. The RTN contains entirely GABAergic neurons that interconnect locally with each other via chemical and electrical synapses and project extrinsically to the LGN. Because the RTN also receives excitatory input from the collaterals of LGN axons destined for V1, the RTN is able to sample both feedforward and feedback signals and, based on these signals, provide inhibition to the LGN. This inhibition could serve to sculpt the receptive-field properties of LGN neurons as well as gate or influence the transmission of visual signals from the LGN to V1.

In contrast to our limited understanding of the role corticogeniculate neurons serve in visual processing, much is known about corticogeniculate synapses at the cellular and ultrastructural level (reviewed in Briggs & Usrey, 2011). In all mammals examined thus far corticogeniculate neurons preferentially provide synaptic input onto the distal dendrites of LGN neurons, whereas synapses from retinal ganglion cells are located more proximal to the cell body of LGN neurons. Corticogeniculate synapses are type II synapses, smaller in size and containing fewer vesicles compared to the larger type I synapses that arise from retinal axons. Corticogeniculate axons use glutamate for synaptic transmission, and this glutamate activates both ionotropic and metabotropic receptors. The excitatory postsynaptic currents (EPSCs) activated by corticogeniculate terminals are small and facilitating, compared to the EPSCs of retinal inputs, which are large and depressing. Because corticogeniculate axon terminals occupy distal dendritic sites and have type II structure and physiology, corticogeniculate input to the LGN is believed to modulate LGN activity, whereas retinal input is believed to drive LGN activity (Sherman & Guillery, 1998).

In summary, the structural organization of corticogeniculate cells and their projections to the LGN supports the view of parallel streams of feedback that are likely to modulate selectively the responses of magnocellular, parvocellular, and possibly even koniocellular LGN neurons. What remains to be determined is whether the corticogeniculate neurons giving rise to these selective feedback projections exhibit stream-specific visual responses. If so, then visual stimuli that drive a particular class of corticogeniculate neurons will lead to a selective and dynamic modulation of LGN neurons processing visual signals traveling from the LGN to V1.

PHYSIOLOGICAL ORGANIZATION OF V1 FEEDBACK TO THE LGN

Results from recent physiological studies conducted in our laboratory provide strong support for the view that the corticogeniculate pathway is organized into parallel streams with properties that are consistent with the magnocellular, parvocellular, and koniocellular streams (figure 22.2). By electrically stimulating the LGN and antidromically activating corticogeniculate neurons, we identified individual layer 6 neurons with axons targeting the LGN (Briggs & Usrey, 2007a, 2009b). We then used an arsenal of visual stimuli to characterize the response properties of these neurons. Importantly, these studies were conducted in the alert macaque monkey, thereby avoiding the dampening effects of anesthesia on the responses of corticogeniculate neurons.

Similar to other species, across corticogeniculate neurons there was a broad range of conduction times for antidromic spikes traveling from the LGN to V1 (Harvey, 1978; Swadlow & Weyand, 1981, 1987; Tsumoto & Suda, 1980). Within this range, however, there was a strict relationship between axon conduction time and cell classification as simple versus complex. Corticogeniculate neurons with axon latencies less than 7 ms were complex cells, those with latencies between 7 and 15 ms were simple cells, and those with latencies greater than 15 ms were again complex cells. Interestingly, results from past studies in carnivores also suggest corticogeniculate can be separated into three groups on

     FARRAN BRIGGS AND W. MARTIN USREY

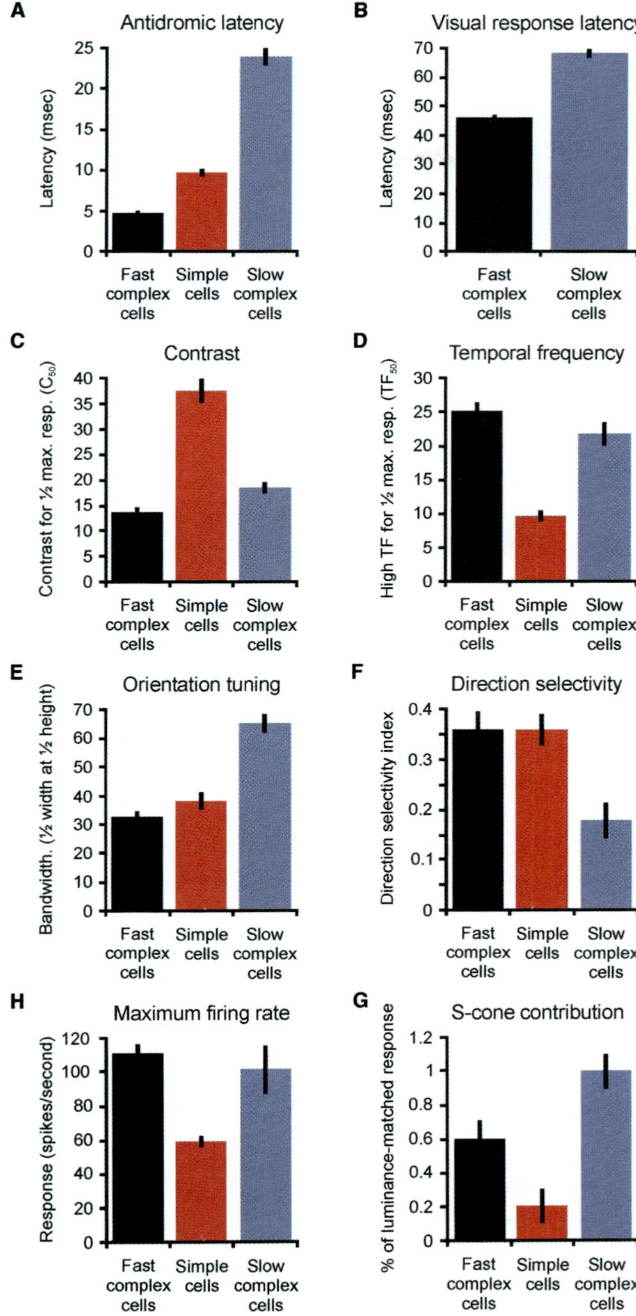

FIGURE 22.2 A comparison of the response properties of corticogeniculate neurons in the macaque monkey. (A–H) Across the three classes of neurons, the fast complex cells, simple cells, and slow complex cells have physiological properties that align well with the magnocellular, parvocellular, and koniocellular streams, respectively (see table 22.1; adapted from Briggs & Usrey, 2009b).

the basis of axon conduction latency (Briggs & Usrey, 2007b; Harvey, 1978; Tsumoto & Suda, 1980). Thus, there appears to be an evolutionary advantage or constraint in having a feedback pathway comprised of neurons with distinct temporal properties.

Of the three groups of corticogeniculate neurons in the macaque monkey, only the fast-conducting complex cells receive suprathreshold "driving" input from the collaterals of LGN axons innervating layer 4 (Briggs & Usrey, 2007a). Although the other two groups of corticogeniculate neurons, those with medium- and slow-conducting axons, likely receive feedforward LGN input as well, this input does not appear sufficient to evoke spikes directly. This finding indicates that fast-conducting corticogeniculate neurons set the lower bound on how quickly signals from the LGN can traverse the thalamocorticothalamic circuit, requiring less than 10 ms. Although it is interesting to speculate on the functional significance of fast, reciprocal communication between the LGN and V1, a specific role for this communication in visual processing remains an open question.

Investigations into additional response properties of corticogeniculate neurons further support the view that there are three classes of feedback neurons (figure 22.2) (Briggs & Usrey, 2009b). These studies demonstrate that, similar to magnocellular neurons in the LGN, the first group of corticogeniculate neurons with fast-conducting axons and complex receptive fields are (1) very sensitive to low-contrast stimuli, (2) able to follow stimuli at high temporal frequencies, (3) less excited by S-cone input from the retina than L- or M-cone input, and (4) suppressed by stimuli that extend into the extraclassical surround (figure 22.2). In addition, fast-conducting, complex corticogeniculate neurons are well tuned for stimulus orientation and have the strongest direction selectivity compared to the other groups of corticogeniculate neurons.

Sharing features with the parvocellular stream, the second group of corticogeniculate neurons with medium-conducting axons and simple receptive fields are (1) less sensitive to low-contrast stimuli, (2) not able to follow stimuli at high temporal frequencies, (3) less excited by S-cone input from the retina than by L- or M-cone input, and (4) weakly suppressed by stimuli that extend into the extraclassical receptive field (figure 22.2). This group of corticogeniculate neurons is also well tuned for stimulus orientation but less selective than the first group (the fast complex cells) for the direction of moving stimuli.

Response properties of the third group of corticogeniculate neurons, those with slow-conducting axons and complex receptive fields, are intermediate to those of the first two groups in terms of contrast sensitivity and temporal frequency tuning (figure 22.2). Significant differences still exist, however, as neurons in this third group have poor orientation tuning and minimal direction selectivity. Consistent with the koniocellular stream, neurons in this third group receive significantly greater input from S-cones in the retina than either the fast- or medium-conducting corticogeniculate neurons.

Taken together with the known anatomy of corticogeniculate projections, the physiological evidence indicates that the corticogeniculate pathway is composed of at least three distinct cell classes that have visual response properties and anatomical specificity that map well onto the magnocellular, parvocellular, and koniocellular streams. Given this organization it is tempting to make predictions about the functional role of corticogeniculate circuits in visual processing.

FUNCTIONAL ROLE OF V1 FEEDBACK TO LGN

Elucidating the functional role(s) of corticogeniculate feedback in visual processing has proven challenging for several reasons. First and foremost, evidence indicates that corticogeniculate feedback is a modulatory, rather than a driving, pathway (Sherman & Guillery, 1998). Consequently, the influence of feedback on LGN responses should be subtle in comparison to the influence of retinal inputs. Second, corticogeniculate synapses experience rate-dependent facilitation, and most studies of feedback function are conducted using anesthetized animals (reviewed in Briggs & Usrey, 2011) where activity levels are generally reduced. As a result the influence of corticogeniculate feedback on LGN responses in these preparations is likely to be greatly diminished and more difficult to quantify than in the unanesthetized state. Third, most studies that have examined the function of corticogeniculate feedback have relied on sequential measurements from individual LGN neurons (but see Sillito et al., 1994; Sillito & Jones, 2002). Although this type of measurement is useful for generating response functions and quantifying features of a neuron's receptive field, it cannot be used to determine whether feedback affects the temporal relationship of activity across neuronal ensembles. Fourth, it is entirely possible that the influence of feedback is stream specific and dynamic, shifting between streams to meet the ongoing physical, biological, and behavioral conditions. Despite these complications and difficulties, previous studies have shed some light on the potential roles of corticogeniculate circuits in vision. Although many of these experiments were conducted in nonprimate species, they nevertheless provide important clues about the role of corticogeniculate feedback in primates.

One of the earliest proposed roles for corticothalamic function in sensory processing was a role in sharpening the tuning of thalamic neurons. This view has received experimental support from work in the cat visual system, the bat auditory system, and the rodent somatosensory system (reviewed in Briggs & Usrey, 2008). In the visual system the sharpening of the spatial properties of LGN receptive fields could come about by a combination of mechanisms involving direct excitation and indirect inhibition supplied by local interneurons and/or RTN circuits. In the carnivore this inhibition likely plays an important role in strengthening the extraclassical/suppressive surround of LGN receptive fields (Murphy & Sillito, 1987). This particular role is less evident in primates, where the extraclassical surround depends to a greater extent on circuits within the retina and is more prominent among magnocellular than parvocellular LGN neurons (Alitto & Usrey, 2008).

Corticothalamic feedback may also modulate the gain of thalamic responses to sensory stimulation. For instance, past studies indicate that corticogeniculate feedback multiplicatively increases parvocellular responses to visual stimuli in a contrast-independent fashion over a wide range of contrasts (Przybyszewski et al., 2000). Similar results are seen for magnocellular neurons but over a more limited range of contrasts. Because this work was performed in anesthetized animals when feedback pathways were likely suppressed, it is entirely possible that this role for feedback is more robust in alert animals and extends to other response properties including those modulated by the spatial and temporal frequencies of a visual stimulus. Somewhat related to an influence on response gain, corticogeniculate feedback has been implicated in adjusting the reliability, timing, and burst/tonic activity profile of LGN responses to visual stimuli (Andolina et al., 2007; Funke et al., 1996; Sherman, 1996). These effects remain to be determined in the alert monkey and have yet to be examined with respect to the parallel streams of corticogeniculate feedback.

Finally, there is growing evidence that the corticogeniculate pathway plays a critical role in influencing LGN activity levels to meet the ongoing behavioral/cognitive demands of the cortex. For example, recent results indicate that the LGN and RTN are modulated by visual spatial attention and that this modulation may be stream specific (McAlonan, Cavanaugh, & Wurtz, 2008; O'Connor et al., 2002; Vanduffel, Tootell, & Orban, 2000). Although these studies have not directly demonstrated that the corticogeniculate pathway is involved in attentional modulation of LGN responses, the corticogeniculate pathway is the major route for cortical information to reach the LGN and therefore is in a strategic position to communicate attentional signals to the LGN.

SUMMARY

Our understanding of the anatomical and physiological organization of the corticogeniculate pathway has increased dramatically in recent years. In primates corticogeniculate feedback is organized into three parallel streams that display characteristics that align with the feedforward magnocellular, parvocellular, and koniocellular streams. This relationship between the feedforward and feedback streams leads to a number of testable hypotheses regarding the functional role of corticogeniculate feedback in visual processing. Future developments in molecular, genetic, electrophysiological, and imaging technologies will undoubtedly make critical contributions toward unlocking the mysteries of this important pathway.

NOTE

1. Readers interested in learning more about the relationship between the visual cortex and the pulvinar nucleus (a higher-order thalamic nucleus) are directed to chapter 19 by Sherman and Guillery.

REFERENCES

Alitto, H. J., & Usrey, W. M. (2008). Origin and dynamics of extraclassical suppression in the lateral geniculate nucleus of the macaque monkey. *Neuron, 57*, 135–146.

Alitto, H. J., Weyand, T. G., & Usrey, W. M. (2005). Distinct properties of visually evoked bursts in the lateral geniculate nucleus. *Journal of Neuroscience, 25*, 514–523.

Andolina, I. M., Jones, H. E., Wang, W., & Sillito, A. M. (2007). Corticothalamic feedback enhances stimulus response precision in the visual system. *Proceedings of the National Academy of Sciences of the United States of America, 104*, 1685–1690. doi:10.1073/pnas.0609318104.

Briggs, F., & Callaway, E. M. (2001). Layer-specific input to distinct cell types in layer 6 of monkey primary visual cortex. *Journal of Neuroscience, 21*, 3600–3608.

Briggs, F., & Callaway, E. M. (2005). Laminar patterns of local excitatory input to layer 5 neurons in macaque primary visual cortex. *Cerebral Cortex, 15*, 479–488.

Briggs, F., & Usrey, W. M. (2007a). A fast, reciprocal pathway between the lateral geniculate nucleus and visual cortex in the macaque monkey. *Journal of Neuroscience, 27*, 5431–5436.

Briggs, F., & Usrey, W. M. (2007b). Temporal properties of feedforward and feedback pathways between the thalamus and visual cortex in the ferret. *Thalamus & Related Systems, 3*, 133–139. doi:10.1017/S1472928807000131.

Briggs, F., & Usrey, W. M. (2008). Emerging views of corticothalamic function. *Current Opinion in Neurobiology, 18*, 403–407.

Briggs, F., & Usrey, W. M. (2009a). Visual system structure. In E. B. Goldstein (Ed.), *The Sage encyclopedia of perception* (pp. 1130–1134). Thousand Oaks, CA: Sage Publications.

Briggs, F., & Usrey, W. M. (2009b). Parallel processing in the corticogeniculate pathway of the macaque monkey. *Neuron, 62*, 135–146.

Briggs, F., & Usrey, W. M. (2011). Corticogeniculate feedback and parallel processing in the primate visual system. *Journal of Physiology, 589*, 33–40.

Erisir, A., Van Horn, S. C., Bickford, M. E., & Sherman, S. M. (1997). Immunocytochemistry and distribution of parabrachial terminals in the lateral geniculate nucleus of the cat: A comparison with corticogeniculate terminals. *Journal of Comparative Neurology, 377*, 535–549.

Erisir, A., Van Horn, S. C., & Sherman, S. M. (1997). Relative numbers of cortical and brainstem inputs to the lateral geniculate nucleus. *Proceedings of the National Academy of Sciences of the United States of America, 94*, 1517–1520.

Fitzpatrick, D., Usrey, W. M., Schofield, B. R., & Einstein, G. (1994). The sublaminar organization of neurons in layer 6 of macaque striate cortex. *Visual Neuroscience, 11*, 307–315.

Funke, K., Nelle, E., Li, B., & Worgotter, F. (1996). Corticofugal feedback improves the timing of retino-geniculate signal transmission. *Neuroreport, 7*, 2130–2134.

Gilbert, C. D., & Kelly, J. P. (1975). The projections of cells in different layers of the cat's visual cortex. *Journal of Comparative Neurology, 163*, 81–106.

Guillery, R. W. (1969). A quantitative study of synaptic interconnections in the dorsal lateral geniculate nucleus of the cat. *Cell and Tissue Research, 96*, 39–48. doi:10.1007/BF00321475.

Harvey, A. R. (1978). Characteristics of corticothalamic neurons in area 17 of the cat. *Neuroscience Letters, 7*, 177–181.

Hendrickson, A. E., Wilson, J. R., & Ogren, M. P. (1978). The neuroanatomical organization of pathways between the dorsal lateral geniculate nucleus and visual cortex in Old World and New World primates. *Journal of Comparative Neurology, 182*, 123–136.

Hubel, D. H., & Wiesel, T. N. (1961). Integrative action in the cat's lateral geniculate body. *Journal of Physiology, 155*, 385–398.

Kaplan, E., Purpura, K., & Shapley, R. M. (1987). Contrast affects the transmission of visual information through the mammalian lateral geniculate nucleus. *Journal of Physiology, 391*, 267–288.

Lankow, B. S., & Usrey, W. M. (2011). Visual processing in the monkey. In R. M. Williams (Ed.), *Monkeys: Biology, behavior and disorders* (pp. 181–197). Hauppauge, NY: Nova Science Publishers.

McAlonan, K., Cavanaugh, J., & Wurtz, R. H. (2008). Guarding the gateway to cortex with attention in visual thalamus. *Nature, 456*, 391–394.

Murphy, P. C., & Sillito, A. M. (1987). Corticofugal feedback influences the generation of length tuning in the visual pathway. *Nature, 329*, 727–729.

O'Connor, D. H., Fukui, M. M., Pinsk, M. A., & Kastner, S. (2002). Attention modulates responses in the human lateral geniculate nucleus. *Nature Neuroscience, 5*, 1203–1209.

Przybyszewski, A. W., Gaska, J. P., Foote, W., & Pollen, D. A. (2000). Striate cortex increases contrast gain of macaque LGN neurons. *Visual Neuroscience, 17*, 485–494.

Rathbun, D. L., Warland, D. K., & Usrey, W. M. (2010). Spike timing and information transmission at retinogeniculate synapses. *Journal of Neuroscience, 30*, 13558–13566.

Sherman, S. M. (1996). Dual response modes in lateral geniculate neurons: Mechanisms and functions. *Visual Neuroscience, 13*, 205–213.

Sherman, S. M., & Guillery, R. W. (1998). On the actions that one nerve cell can have on another: Distinguishing "drivers" from "modulators." *Proceedings of the National Academy of Sciences of the United States of America, 95*, 7121–7126. doi:10.1073/pnas.95.12.7121.

Sherman, S. M., & Guillery, R. W. (2005). *Exploring the thalamus and its role in cortical function* (2nd ed.). Cambridge, MA: MIT Press.

Sillito, A. M., & Jones, H. E. (2002). Corticothalamic interactions in the transfer of visual information. *Philosophical Transactions of the Royal Society of London. Series B, Biological Sciences, 357*, 1739–1752. doi:10.1098/rstb.2002.1170.

Sillito, A. M., Jones, H. E., Gerstein, G. L., & West, D. C. (1994). Feature-linked synchronization of thalamic relay cell firing induced by feedback from the visual cortex. *Nature, 369*, 479–482.

Sincich, L. C., Horton, J. C., & Sharpee, T. O. (2009). Preserving information in neural transmission. *Journal of Neuroscience, 29*, 6207–6216.

Swadlow, H. A., & Weyand, T. G. (1981). Efferent systems of the rabbit visual cortex: Laminar distribution of the cells of origin, axonal conduction velocities, and identification of axonal branches. *Journal of Comparative Neurology, 203*, 799–822.

Swadlow, H. A., & Weyand, T. G. (1987). Corticogeniculate neurons, corticotectal neurons, and suspected interneurons in visual cortex of awake rabbits: Receptive-field properties, axonal properties, and effects of EEG arousal. *Journal of Neurophysiology, 57*, 977–1001.

Tsumoto, T., & Suda, K. (1980). Three groups of corticogeniculate neurons and their distribution in binocular and monocular segments of cat striate cortex. *Journal of Comparative Neurology, 193*, 223–236.

Usrey, W. M., & Fitzpatrick, D. (1996). Specificity in the axonal connections of layer VI neurons in tree shrew striate cortex: Evidence for separate granular and supragranular systems. *Journal of Neuroscience, 16*, 1203–1218.

Usrey, W. M., Reppas, J. B., & Reid, R. C. (1998). Paired-spike interactions and synaptic efficacy of retinal inputs to thalamus. *Nature, 395*, 384–387.

Usrey, W. M., Reppas, J. B., & Reid, R. C. (1999). Specificity and strength of retinogeniculate connections. *Journal of Neurophysiology, 82*, 3527–3540.

Vanduffel, W., Tootell, R. B., & Orban, G. A. (2000). Attention-dependent suppression of metabolic activity in the early stages of the macaque visual system. *Cerebral Cortex, 10*, 109–126.

Weyand, T. G. (2007). Retinogeniculate transmission in wakefulness. *Journal of Neurophysiology, 98*, 769–785.

Wiser, A. K., & Callaway, E. M. (1996). Contributions of individual layer 6 pyramidal neurons to local circuitry in macaque primary visual cortex. *Journal of Neuroscience, 16*, 2724–2739.

23 Superior Colliculus and Visual Attention

RICHARD J. KRAUZLIS

The superior colliculus (SC) is a highly conserved structure in the brainstem that plays a major role in visual processing and sensory–motor integration. In primates the SC is especially well known for its role in the control of orienting movements of the eyes and head, as reviewed in detail elsewhere (Gandhi & Katnani, 2011; Wurtz & Albano, 1980). Here, we focus on the contributions of the SC to visual processing and spatial attention, functions that have been studied in a wide range of species. The functional picture that emerges is very consistent—if a meaningful visual event takes place, the SC plays a crucial role in detecting the event and selecting the visual object.

We begin by briefly summarizing the basic structure and anatomy of the SC, followed by a survey of visual processing in the SC, how it relates to the control of visual attention, and finish by discussing possible circuit mechanisms.

STRUCTURAL FEATURES

The SC is located on the roof of the vertebrate midbrain and consists of a pair of prominent hill-like structures measuring several millimeters across and extending several millimeters deep. In nonmammalian species, such as fish, frogs, and birds, the SC is more often referred to as the "optic tectum" and is one of the largest components of the brain. In mammals the SC forms a smaller fraction of the brain because of the expansion of the neocortex but nonetheless retains a central role in sensory–motor processes, including vision and attention.

In all vertebrates the SC is a laminar structure whose layers are defined by alternating strata of cell bodies and fibers (figure 23.1). The most superficial layers of the SC are strictly visual. In mammals the superficial layers consist of a thin cell-free outermost layer (stratum zonale, SZ), followed by a cell-rich layer (stratum griseum superficiale, SGS), and then by another mostly cell-free layer containing axon fibers (stratum opticum, SO). A similar organization holds true in nonmammals, but the convention is to number the layers rather than give them names; for example, in birds the superficial layers of the optic tectum are identified as layers 1 through 10. In this review we refer to these superficial layers together with the abbreviation sSC.

The intermediate and deep layers of the SC also contain visual signals, but these are mixed with signals from other sensory modalities as well as activity related to the control of orienting movements. Just below the stratum opticum (SO) in mammals is another cell-rich layer, the stratum griseum intermediale (SGI), followed by another fiber-rich layer, the stratum album intermediale (SAI), and then by a somewhat less well-defined layer of cells, the stratum griseum profundum (SGP), which merges with the midbrain reticular formation. In birds the intermediate and deep layers of the optic tectum correspond to layers 11 through 15. The intermediate and deep layers are abbreviated here as dSC.

CONNECTIVITY

The different layers of the SC have distinct anatomical connections and functions. The sSC receives a direct and topographically organized input from retinal ganglion cells, and it is this input that establishes perhaps the most distinctive features of the SC—an orderly map of the visual field. As in many other primary visual areas, the sSC on each side of the brain represents the contralateral visual field, and the representation of the visual field is distorted to reflect the density of retinal ganglion cell inputs. In primates, for example, the central 10° of the visual field occupies about a third of the entire SC map (Pollack & Hickey, 1979) in order to accommodate the high-resolution inputs from the foveal region of the retina (figure 23.2).

In addition to inputs directly from the retina, the sSC receives signals from several other early stages of visual processing, and these inputs are likewise organized to match the retinotopic topography of the SC map. The most prominent sources are visual areas in the forebrain, such as visual cortex in mammals (Lui et al., 1995) or the visual hyperpallium in birds (Karten & Dubbeldam, 1973; Reiner et al., 2004). Less well-known,

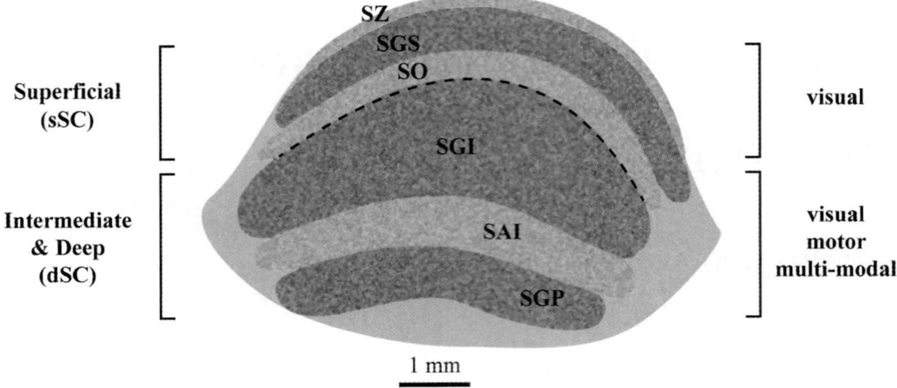

FIGURE 23.1   Laminar structure of the superior colliculus (SC). The upper layers (SZ, SGS, SO) form the superficial layers and are purely visual. The intermediate and deep layers (SGI, SAI, SGP) also have visual responses plus motor and multimodal activity. The sizes of the different layers in this diagram are based on the dimensions of the SC in the macaque monkey.

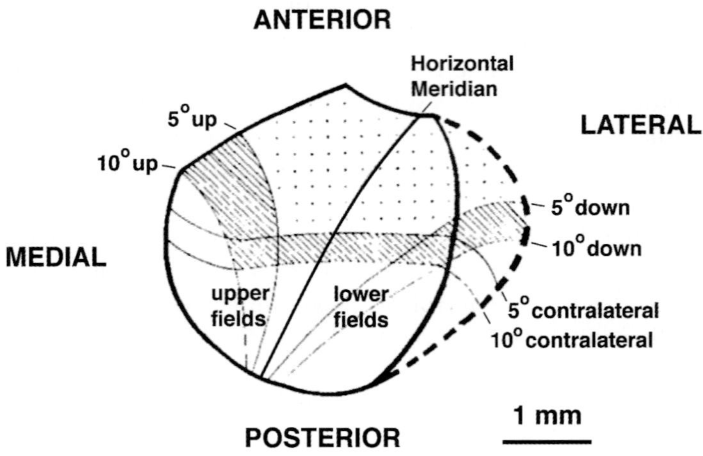

FIGURE 23.2   Representation of the visual field on the surface of the right SC as viewed from above. Stippled area represents contralateral visual field within the central 5°. (Adapted from Cynader & Berman, 1972.)

but also extensive, are projections from components of the pretectal complex (Büttner-Ennever et al., 1996), a set of visual nuclei in the midbrain involved in a range of basic and highly conserved visual functions, including sensing visual motion and controlling the refractive state of the eye lens. A third source of inputs is from the isthmi nuclei, extensively studied in birds (Luksch, 2003) but less so in mammals (where the homologous structure is called the parabigeminal nucleus); these nuclei provide a combination of cholinergic and GABA-ergic inputs to the superficial layers that modulate the transmission of visual signals through the sSC.

The outputs of the sSC are directed to several targets. Some outputs form anatomical loops with structures that also provide inputs to the SC. The sSC projects back to the parabigeminal nucleus and the pretectum (Harting, 1977) and also to thalamic nuclei such as the

lateral geniculate nucleus, which in turn project to visual cortex (Harting et al., 1980). The sSC also has prominent projections to the ventral lateral geniculate nucleus, known in primates as the pregeniculate nucleus, but the functions of this visual nucleus are not yet known.

The sSC also provides an important second pathway for visual information to reach visual cortex, independent of the well-known relay through the dorsal lateral geniculate nucleus. This pathway may be especially important for species such as mice, in which a majority of retinal ganglion cells project to the sSC and not to the lateral geniculate nucleus (Hofbauer & Dräger, 1985). This projection from the sSC, which passes through the inferior pulvinar of the thalamus to extrastriate visual areas (Lyon, Nassi, & Callaway, 2010) may be especially important for processing visual motion signals or for suppressing motion signals caused by saccades (Berman & Wurtz, 2010, 2011). This part of the pulvinar also projects to regions of the striatum, providing a possible route for the sSC to influence how the basal ganglia accomplishes functions such as reinforcement learning and visual selection (Harting, Updyke, & Van Lieshout, 2001).

The dSC has an even more varied set of anatomical connections than the sSC. Many of these are involved in the control of orienting movements of the eyes, head, and body and include connections with the reticular formation, brainstem premotor nuclei, precerebellar nuclei, and the cerebellum (May, 2006; Wurtz & Albano, 1980). The circuits formed between the dSC and these motor-related structures convert signals about the spatial location of important objects into motor commands that make it possible to execute skilled movements toward or away from those objects.

The dSC also receives inputs related to the perceptual and cognitive processing steps that precede the initiation of orienting movements from cortical areas including the frontal eye fields, lateral intraparietal area, supplementary eye fields, and the prefrontal cortex (Kawamura, Sprague, & Niimi, 1974; Kunzle & Akert, 1977). These inputs provide a range of visual, motor, and decision-related signals that play a major role in selecting the targets for orienting movements. The dSC is also one of the major sites in the brain for combining signals from different sensory modalities, including auditory and somatosensory signals. Signals from these other modalities are mapped in the same retinotopic reference frame established by the visual inputs (Groh & Sparks, 1996; Jay & Sparks, 1987; McHaffie, Kao, & Stein, 1989), providing a mechanism for representing the location of salient objects regardless of how the object is detected.

Although historically the emphasis has been on the descending pathways from cortex to the dSC, more recently there has been recognition of the importance of ascending pathways from the dSC to cortex (see chapter 65). The dSC has projections, through the thalamus, back to many of the same areas of the cerebral cortex that provide inputs to the dSC. One route is from the dSC through the medial dorsal thalamic nucleus to prefrontal cortex, including the frontal eye fields. This pathway has been shown to provide corollary discharge signals about saccades—copies of the motor command leaving the SC—that make it possible to compensate for the fact that visual inputs shift every time a saccade is made (Sommer & Wurtz, 2008). Another route is from the dSC through the lateral pulvinar to parietal and visual cortex; this pathway may be important for modulating visual activity in cortex with shifts of attention (Wurtz et al., 2011).

The dSC has several types of connections with the basal ganglia that are implicated in motor and cognitive functions. The best known is the projection from the substantia nigra pars reticulata (SNpr), which provides a GABAergic inhibitory input to the dSC (Deniau & Chevalier, 1992; Huerta, Van Lieshout, & Harting, 1991; Redgrave, Marrow, & Dean, 1992). This pathway provides a mechanism for disinhibiting activity in the dSC—a pause in the tonic inhibition from the SNpr may tip the balance of activity in the dSC sufficiently to initiate an orienting movement (Hikosaka, Takikawa, & Kawagoe, 2000). The tonic inhibition itself may play an important role in regulating the level of activity across the SC map. Also, because activity in the SNpr is partly controlled by cortical inputs through the striatum, this pathway provides another route by which descending cortical signals can influence activity in the SC.

The dSC also has projections back to the basal ganglia. One pathway involves projections to the "nonspecific" intralaminar nuclei in the thalamus, which project to the caudate and putamen in the basal ganglia. The intralaminar nuclei are considered part of the ascending reticular activating system, but the rostral portion of the intralaminar nuclei that receives inputs from the dSC appears to play a more specific role in cognitive functions rather than general arousal (Smith et al., 2004; Van der Werf, Witter, & Groenewegen, 2002). There is also a direct projection from the dSC to the substantia nigra pars compacta, which is well known for containing dopamine neurons involved in reinforcement learning (Comoli et al., 2003; May et al., 2009; McHaffie et al., 2006). The response of dopaminergic neurons to unexpected and salient events is widely believed to provide a "prediction error" signal that guides learning (Niv & Schoenbaum, 2008; Schultz, 2010), and the input from the dSC could be an important source for driving this activity.

BASIC FEATURES OF VISUAL RESPONSES

Visual activity in the SC is related to meaningful events. Some of the first observations about visual responses in the SC were made in frogs, where it was recognized that visual neurons could be broadly divided into "sameness" neurons and "newness" neurons (Gaillard, 1990; Lettvin et al., 1961). "Sameness" neurons respond well to small spots that move across the visual field and continue to discharge if the spot stops within the neuron's receptive field. In contrast, "newness" neurons also respond to moving spots but rapidly decrease their discharge as the stimulus persists and show habituation if the stimulus is presented repeatedly. Most details of the visual stimulus are irrelevant to the response; what matters is the presence of a moving, or suddenly visible, visual object and whether it has just appeared ("newness") or has been present for a while ("sameness").

More detailed examination of the frog SC shows that this dichotomy between "newness" and "sameness" neurons is a simplification, and that there are a range of neuron types, with multiple identifiable classes based on morphology, location in the SC map, receptive field size, level of spontaneous activity, and other properties (Gaillard, 1990). Moreover, across different species, there are notable differences in the detailed properties of visual neurons and responses (Frost & DiFranco, 1976; Humphrey, 1968; Sterling & Wickelgren, 1969). Nonetheless, the function of visual responses in the SC appears to be remarkably conserved, and the fundamental aspects of the "newness" and "sameness" responses from frogs can be found even in primates. SC visual responses do not provide a systematic representation of visual features like that found in visual cortex but instead are involved in the detection of evocative and behaviorally relevant stimuli.

Visual motion is a highly salient cue for detecting danger and identifying prey, and responses to motion are found in the SC of every species (figure 23.3). However, motion-sensitive neurons are often not selective for the direction of motion, and in some species, including rodents and primates, the SC contains only a handful of direction-selective neurons (Humphrey, 1968; Krauzlis, 2004; Updyke, 1974). For species in which direction selectivity is more common, such as cats, rabbits, and frogs, the selectivity for direction is very broadly tuned and not uniformly distributed across possible directions. In cats, for example, most neurons prefer horizontal motion directed away from the center of the visual field (Sterling & Wickelgren, 1969), and even this sensitivity is dependent on input from the cerebral cortex (Wickelgren & Sterling, 1969). Neurons show some degree of selectivity for speed—for example, in rats, cats, and primates, neurons tend to show the largest responses for speeds of about 5–10°/s, continue to respond for speeds up to about 30°/s, but are unresponsive for faster speeds (Cynader & Berman, 1972; Humphrey, 1968; Marrocco & Li, 1977).

The features of motion stimuli that matter most are the ones that determine whether the motion is salient. For example, jerky motion—the type of motion you might generate by waving your hands to attract attention—is an especially effective stimulus. Flashing a stimulus on and off is very effective and often works whether the stimulus is moving or stationary. Some neurons may initially respond to smooth motion but then decrease their activity and only be activated by jerky motion. Some neurons will continue to respond to a moving stimulus that stops and remains stationary in the receptive field, although this tends to less common in the primate. Other neurons show the peculiar property of first being unresponsive to a stimulus in their receptive field but then later responding with a burst of impulses every time it moves. In several species, including birds (Wu et al., 2005), frogs (Kang & Li, 2010; Nakagawa & Hongjian, 2010), and cats (Liu, Wang, & Li, 2011), there are neurons that respond selectively to expanding retinal images and provide signals that can be used to detect whether a visual object is on a collision course.

Aside from motion and stimulus onsets and offsets, most other major visual features tend not to matter. The magnitude of SC visual responses is not appreciably affected by the shape or contrast of the visual stimulus. Neurons show little or no selectivity for orientation, even in species that receive major projections from visual cortex. Changing the polarity of the stimulus—light stimulus against dark background or vice versa—usually does not matter, but when it does, neurons tend to show a preference for dark objects against a lighter background. There is no strong preference for color—individual neurons show some preferences across wavelength, but the tuning is very broad, and the receptive fields do not show color-opponent organization (Marrocco & Li, 1977; Schiller & Malpeli, 1977; White et al., 2009). Whether inputs come from one or both eyes

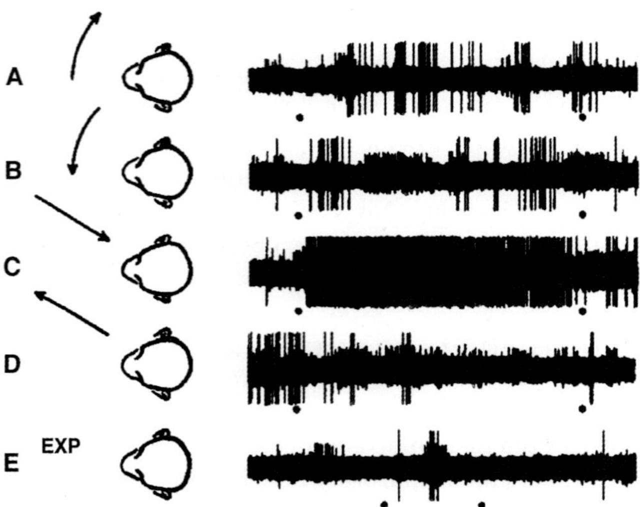

FIGURE 23.3   Response of sSC in monkey to moving stimuli. Neuron responds to motion with an irregular discharge with no preference for direction (A, B), responds vigorously to a visual object approaching the animal (C), stops discharging when object moves away (D), and responds weakly to a projected expanding spot stimulus that simulates an approaching object (E). Similar activity could be found when viewed with either eye alone. (Adapted from Updyke, 1974.)

follows from the degree of binocular overlap in the visual field—in frogs, some neurons are monocular and others binocular, whereas in cat and monkey, most are binocular.

The basic visual properties of neurons in the SC do not change much across its different layers. The most notable change in visual properties is the size of receptive fields, which increase with depth from the surface and with eccentricity in the visual field. For example, in primates, near the center of the visual field map, neurons in the sSC have small receptive fields with diameters less than 2° across, whereas neurons in the dSC at the same eccentricity can have receptive fields with diameters about 10 times larger (Cynader & Berman, 1972; Goldberg & Wurtz, 1972a; Humphrey, 1968).

The largest change in properties from the superficial to the intermediate and deep layers is the addition of other signals. Neurons in the superficial layers are exclusively visual, whereas neurons in the intermediate and deep layers exhibit visual responses combined with sensitivity to other sensory modalities, cognitive signals, and motor-related activity. In the dSC, the responses to other modalities—for example, auditory signals—often interact with visual signals in a superadditive fashion, especially when the visual stimulus by itself evokes only a weak response and when the two inputs are spatially and temporally aligned (Wallace, Wilkinson, & Stein, 1996); these effects may be part of a mechanism for combining cues from different modalities to more accurately infer whether a particular event has taken place (see chapter 74).

MODULATION OF VISUAL RESPONSES WITH SACCADES

Visual images sweep across the visual field with each saccade, generating visual motion that could be a very effective stimulus for driving the activity of visual neurons in the SC and potentially creating ambiguity between the motion of visual objects in the world and motion caused by the animal's own eye movements. However, sSC neurons are able to distinguish between real and self-induced stimulus motion (Robinson & Wurtz, 1976). Many sSC neurons respond to visual motion with speeds like those generated during saccades, but they do not respond when a saccade sweeps a stationary stimulus across their receptive field. The background activity of these neurons is also suppressed during saccades made in complete darkness, showing that saccadic suppression of these neurons is due to an extraretinal signal, perhaps from movement-related activity of other neurons in the SC. Recordings made

in brain slices have identified a possible circuit within the SC for saccadic suppression—premotor neurons in the dSC not only project to downstream motor structures but also provide a collateral to inhibitory interneurons in the intermediate layer that suppress the activity of visual neurons in the sSC (Phongphanphanee et al., 2011).

Another problem is that saccades change the set of neurons representing a particular visual object because the retinotopic location of the object shifts with each saccade. One possible mechanism for maintaining perceptual continuity across saccades is the phenomenon of perisaccadic remapping of receptive fields. This effect was first described for neurons in the lateral intraparietal cortex (Duhamel, Colby, & Goldberg, 1992) and has been observed in several areas of cerebral cortex (Berman & Colby, 2009; Nakamura & Colby, 2002). In perisaccadic remapping, neurons show predictive responses to visual stimuli that will be in their receptive field after the saccade is completed—they respond even before the saccade starts or so quickly after the saccade that the response cannot be due to direct visual stimulation. Similar predictive responses are found for about one-third of the neurons in the intermediate and deep layers of the SC but not the superficial layers (Walker, Fitzgibbon, & Goldberg, 1995). The dSC is also a source of signals that are necessary for predictive remapping of receptive fields in cerebral cortex (Sommer & Wurtz, 2008).

VISUAL TARGET SELECTION

One of the major visual functions of the dSC is the selection of visual targets. The dSC is well known for its role in the motor control of orienting movements, especially in primates, but it also plays a crucial role in the preceding step of determining which visual object, among several competing alternatives, is selected as the goal of the movement (see chapter 64).

Neuronal activity in the dSC is related to saccade target selection, and this activity is related to the selection itself, not just the preparation of the saccade. Neurons show elevated activity for a visual stimulus that will be selected as the endpoint of a saccadic eye movement, compared to distracter visual stimuli that will be ignored. This preference is not evident in the initial visual response but emerges about 150–250 ms after stimulus onset (Glimcher & Sparks, 1992; Krauzlis & Dill, 2002; McPeek & Keller, 2002). Because the timing of the saccade is variable, it is possible to determine whether the neuronal preference for the target is time-locked to the visual onset of the stimuli or to the motor execution of the saccade. For some neurons the timing is correlated with the onset of the saccadic eye movement, suggesting a role in saccade preparation. However, for other neurons, the timing is correlated with the onset of the visual stimuli, suggesting a role in visual target selection (McPeek & Keller, 2002).

A similar distinction between dSC activity related to saccade selection and visual target selection has been found in visual motion discrimination tasks (Horwitz & Newsome, 1999; Horwitz, Batista, & Newsome, 2004). In this task, animals view a patch of visual motion and are rewarded for making a saccade in the direction shown in the patch. One type of dSC neuron shows activity that predicts which saccade will be made. However, another type of dSC neuron shows activity that predicts the direction of motion that will be identified rather than the direction of saccade used to report that choice. These neurons tend to prefer motions directed toward the location of their movement fields; their elevated activity might be used as a mnemonic to translate the visual judgment about the motion patch into the correct saccade motor command.

Other evidence for a role of the dSC in visual target selection comes from experiments using smooth pursuit eye movements, the voluntary eye movement used by primates to track visual objects of interest as they move in the world (Krauzlis, 2005; Lisberger, 2010; see chapter 62). Neurons in the dSC show changes in their activity during smooth pursuit as you would expect from the location of their response fields—they increase their discharge when the retinal location of the pursuit stimulus falls inside their response field and decrease their discharge otherwise; because the retinal stimulus during pursuit tends to be on or near the fovea, neurons with pursuit-related activity are found mainly in the rostral part of the dSC (Krauzlis, 2003; Krauzlis, Basso, & Wurtz, 1997, 2000).

When there are multiple moving visual objects, primates can selectively track one of them using pursuit and ignore the others. Unlike saccades the spatial location of the selection can be dissociated from the direction of the eye movement required to track it. For example, if a target appears on the left and moves rightward, the subject should select the stimulus on the left even though this requires a rightward movement; also, if the starting location is chosen carefully, the subject's response will not include corrective saccades (Rashbass, 1961). Using this approach neurons in the dSC have been found to show preferences for visual targets selected for pursuit as well as for saccades (Krauzlis & Dill, 2002). During pursuit, dSC neurons show enhanced responses for the visual target, even though the eye movement is directed away from the starting position of the target. This distinction shows that neuronal activity in the dSC is related to the selection of the visual target and not just the preparation of the movement.

Other evidence for activity related to target selection comes from recordings in the dSC of the owl (Mysore & Knudsen, 2011; Mysore, Asadollahi, & Knudsen, 2011). Looming visual stimuli are very effective at driving activity of dSC neurons in the owl, and a subset of these neurons show a specialized capacity to identify the most salient stimulus (figure 23.4). Some neurons

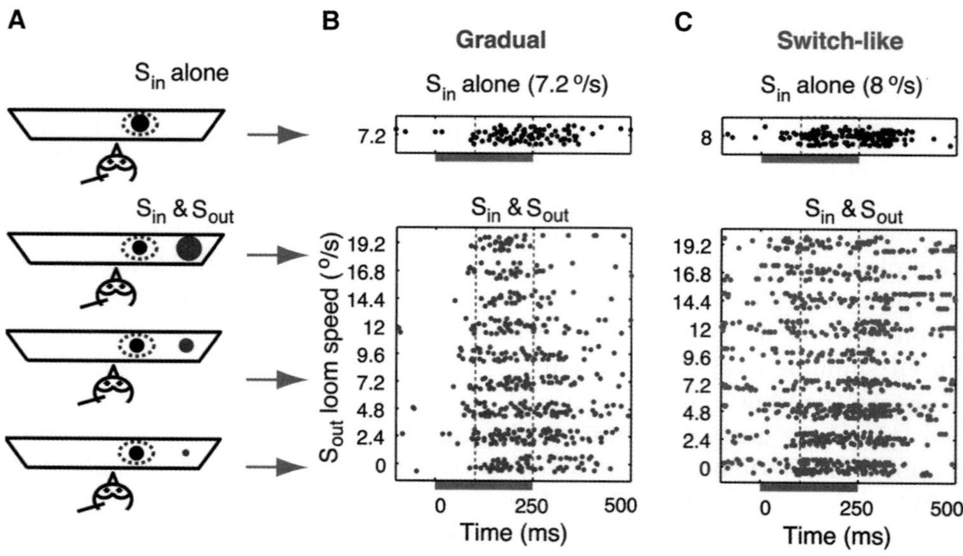

FIGURE 23.4   Gradual and switch-like responses in the SC of the owl. A looming stimulus of fixed salience is presented in the receptive field (S_{in}) while the strength of a competing stimulus outside the receptive field (S_{out}) is varied (A). Some units show a gradual suppression of their response as the strength of S_{out} is increased (B). Other units show an abrupt switch-like suppression (C). (Adapted from Mysore, Asadollahi, & Knudsen, 2011.)

show a graded increase or decrease in their response to a looming stimulus in its receptive field as the strength of competing stimuli outside the receptive field is decreased or increased by varying the speed of motion. However, other neurons show a "switch-like" change in their response—their activity increases abruptly when the stimulus inside the neuron's receptive field is the strongest. This response pattern is an example of flexible categorization like that seen in parietal cortex (Freedman & Assad, 2006) because the switch-like behavior depends on the relative, not absolute, strength of the competing stimuli.

Activity in the dSC not only is correlated with visual target selection but also plays a causal role. For example, saccade target selection is impaired when activity in the dSC is reversibly blocked by injection of pharmacological agents such as lidocaine (McPeek & Keller, 2004). When a pop-out visual stimulus is presented in the part of the visual field affected by the SC inactivation, saccades tend to be directed to one of the distracter stimuli rather than to the target. Conversely, when neuronal activity in the SC is artificially increased by electrical microstimulation, saccade selection is biased in favor of the stimulus located at the activated SC site (Carello & Krauzlis, 2004; Dorris, Olivier, & Munoz, 2007).

Results from experiments using pursuit eye movements demonstrate a causal role in target selection, distinct from saccade selection. When activity in the dSC is focally inactivated using microinjections of muscimol, target selection is strongly biased against the stimulus that is initially located inside the affected part of the visual field and is biased in favor of the stimulus located outside the affected region (Nummela & Krauzlis, 2010). Because the pursuit response required a movement away from, rather than toward, this retinotopic location, this effect shows an influence on the selection of the visual target per se rather than on the selection of the movement direction. Conversely, artificial activation of dSC activity with microstimulation causes a bias in favor of the stimulus placed at the corresponding location in the visual field even though the pursuit response requires a movement in the opposite direction (Carello & Krauzlis, 2004).

Deficits in visual target selection are found during SC inactivation even for manual responses when the subject is required to maintain fixation. In one experiment animals reported their choice with buttons that could be reached with the hand but could not be seen (Nummela & Krauzlis, 2010). Chemical inactivation of the dSC biased choices away from the stimulus placed in the affected part of the visual field, causing effects qualitatively similar to, but weaker than, the effects seen on saccade and pursuit tasks. Larger effects on manual

target selection are found when animals need to directly reach to the visual targets using a touch screen (Song, Rafal, & McPeek, 2011). A possible explanation for these effects is that dSC inactivation causes a motor deficit by affecting neurons with activity related to arm movements (Stuphorn, Bauswein, & Hoffmann, 2000; Werner, Dannenberg, & Hoffmann, 1997); however, the movement preferences of these neurons do not match the retinotopic organization of the SC map, making it unlikely that they are responsible for the spatial deficits found after inactivation. Thus, activity in the dSC plays a causal role in the selection of visual targets, especially when the targets need to be localized in space, regardless of whether the target is acquired with the eyes or hands.

SPATIAL ATTENTION

Visual activity in the sSC provided one of the first known correlates of spatial attention in the primate brain. About half of the visual neurons in the sSC show elevated visual responses when the stimulus in their receptive field will be the target of a saccadic eye movement (Goldberg & Wurtz, 1972b). The effect is not due to general arousal because it shows spatial specificity— when the saccade is directed outside the neuron's receptive field, the neuron does not show enhanced responses.

The overlap between the control of saccades and spatial attention in the superior colliculus has been explored in detail in the dSC. One study used electrical microstimulation of the dSC to artificially trigger saccades and showed that the endpoints of these evoked saccades were shifted in the direction of spatial cues about the likely location of the upcoming target (Kustov & Robinson, 1996). The time course of these changes in saccade endpoint depended on the type of cue. Peripheral flashed cues led to deviations at short delays, consistent with a bottom-up effect driven by the visual stimulus, whereas symbolic color cues led to deviations at longer delays, consistent with a top-down effect driven by cognitive processing of the color stimulus.

The interaction between top-down and bottom-up influences on spatial attention during the preparation of saccades is also evident in the activity of dSC neurons. When a peripheral flash is presented just before the appearance of the saccade target in the same visual field, it both reduces saccade latency and facilitates the response of the neuron to the target (Bell, Fecteau, & Munoz, 2004; Fecteau, Bell, & Munoz, 2004). These effects are time-locked to the onset of the stimulus, consistent with a bottom-up influence of spatial attention. Also, these effects are larger and more prolonged when the flashed stimulus is predictive, consistent with an additional effect of top-down influences. These results support the view of the SC as a priority map for saccade target selection (Fecteau & Munoz, 2006); the term "priority map" is meant to draw a distinction from other terms that emphasize the role of top-down (e.g., attention map) or bottom-up (e.g., salience map) processes.

Demonstrating that dSC activity is related to the allocation of spatial attention, even in the absence of saccades, requires the animal to perform some form of discrimination task so that a behavioral measure of spatial attention can be defined. In one study (Ignashchenkova et al., 2004), animals were trained to discriminate the orientation of a Landolt "C," and the size of the gap was varied to control the difficulty of the task and to measure acuity (figure 23.5). To manipulate spatial attention, the onset of the "C" was preceded by either a spatial cue or a symbolic cue. Spatial and symbolic cues produce significant improvements in discrimination performance in this task and also alter the activity of visually responsive neurons in the dSC. However, the effects of spatial cues depend on the type of neuron. "Visual" neurons have visual responses but

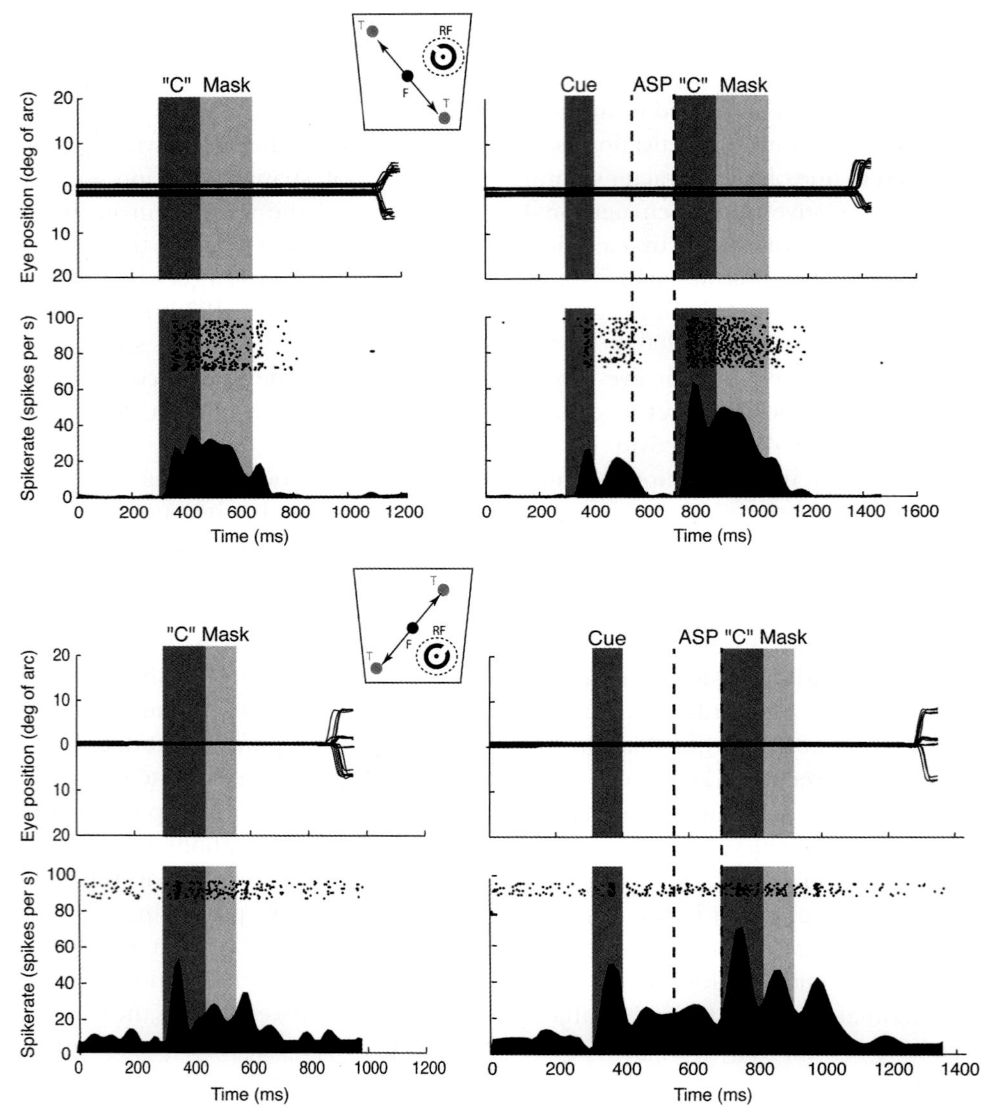

FIGURE 23.5   Activity of dSC neurons is modulated with covert shifts of attention, as shown by this sample visual neuron (top row) and visual-motor neuron (bottom row). The left column shows activity on trials when no cue was presented; the right column shows activity on trials when a spatially precise cue was presented. Both the visual and visual-motor neurons respond to the presentation of the "C," but this activity is higher when the location of the "C" was cued than when it was not cued. Also, the visual-motor neuron shows activity during the time epoch before the presentation of the "C" (attention shift period, ASP), but the visual neuron does not. (From Ignashchenkova et al., 2004.)

no saccade-related activity. These neurons respond to the appearance of the "C" in their receptive field, and this evoked activity is higher when the location was previously cued, regardless of whether the cue was spatial or symbolic. "Visual-motor neurons" have visual responses and also show saccade-related activity. These neurons also respond to the appearance of the "C" in their response fields, and show higher activity when the location was previously cued, but this enhancement occurs for spatial cues and not for symbolic cues. Visual-motor neurons also show elevated activity during the delay period after the cue has been presented and before the appearance of the "C," but only for spatial cues. Because this delay period is presumably when spatial attention is shifted, this distinction suggests that visual-motor neurons in the SC may be especially important for stimulus-driven shifts of attention.

Neuronal recordings in the SC provide correlative evidence of a role in covert attention, but they do not provide a test of causality. One problem is that modulation in the dSC activity could be related to the preparation of saccades, even if the animal refrains from making those saccades during the covert attention task. However, several experiments have tested whether the SC plays a causal role in the control of covert spatial attention.

One of these studies applied subthreshold microstimulation (i.e., too weak to directly evoke saccades) in the SC during a test of change-blindness (Cavanaugh & Wurtz, 2004). Change-blindness refers to the inability to detect changes in a visual scene when those changes are accompanied by a full-field transient, such as a blank screen or blurred visual input during saccades (Rensink, 2002). In this task, animals fixated a central spot and were shown three patches of random-dot motion arranged across the visual display for about a second, then a blank screen, and then the three patches again. When the three patches reappeared, one of them may have changed its direction of motion, and the task of the animal was to detect the change by making a saccade to it. Animals were given a visual cue on half of the trials indicating which motion patch might change; these spatial cues improved the animal's detection performance and reduced their reaction times, providing behavioral evidence for effects on spatial attention. In microstimulation trials, instead of providing the animal with a spatial cue, the SC was microstimulated at a location matching one of the three motion patches. SC microstimulation caused effects similar to those observed with visual cues—the animal's detection performance improved, and reaction times were reduced. Importantly, microstimulation did not lead to an increase in false alarms—they did not

simply become more likely to report a change but became better at detecting the occurrence of a change. Other experiments ruled out the explanation that SC stimulation generated a visual phosphene that only indirectly affects spatial attention (Cavanaugh, Alvarez, & Wurtz, 2006).

Another study used subthreshold microstimulation of the SC during a spatial attention task involving visual motion discrimination (Müller, Philiastides, & Newsome, 2005). In this task, animals judged the direction of motion in a patch containing random-dot motion, which was made more challenging by including flickering distracter dots elsewhere in the display. The fraction of dots that moved coherently in the motion patch was varied across trials, and a psychometric curve was constructed by tabulating the animal's performance for these different stimulus conditions. Subthreshold microstimulation of the dSC improved discrimination performance—it shifted the psychometric curves to the left, so that less visual motion was required to achieve a particular level of performance. This improvement in performance was observed only if the motion patch and the site of dSC stimulation were at spatially coincident locations.

The results from these microstimulation studies provide strong evidence that the dSC plays a causal role in spatial attention. However, they leave open the possibility that the dSC is not crucial for the control of covert attention, but, as part of the neuronal circuits for orienting, it is simply linked to covert processes that mostly take place elsewhere.

To test whether activity in the dSC is necessary for covert attention, behavioral performance has been tested before and after reversible chemical inactivation of the dSC (Lovejoy & Krauzlis, 2010). In these experiments the task of the animals was to discriminate the direction of motion in a random-dot motion patch that appeared at a previously cued location in the visual display. To ensure that spatial attention was necessary to perform the task, the display also included a "foil" stimulus that also contained random-dot motion but appeared at an uncued location and should therefore be ignored.

Performance in this discrimination task is profoundly impaired after dSC inactivation, and this impairment is selective for stimuli placed in the part of the visual field affected by the SC chemical inactivation (figure 23.6). When the cued stimulus is placed in the affected part of the visual field, performance is severely degraded, but the errors are not random—they tend to be based on the irrelevant foil stimulus placed outside the affected region. This pattern of errors shows that the animal still attempts to discriminate the direction of

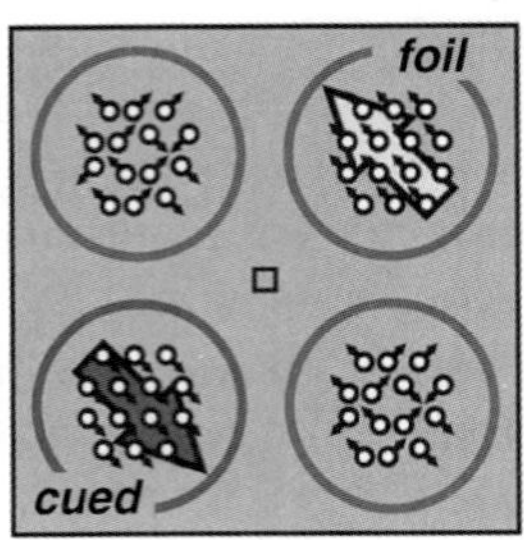
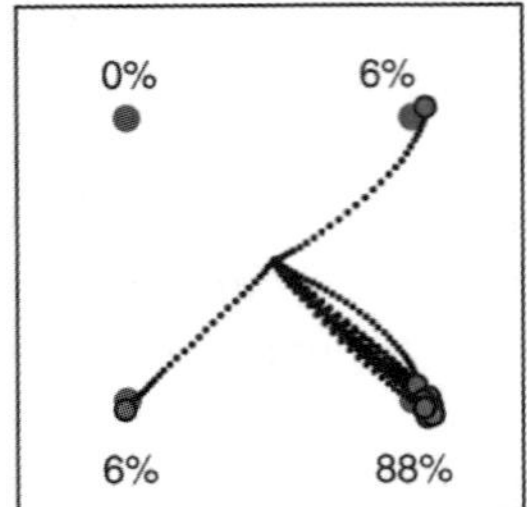
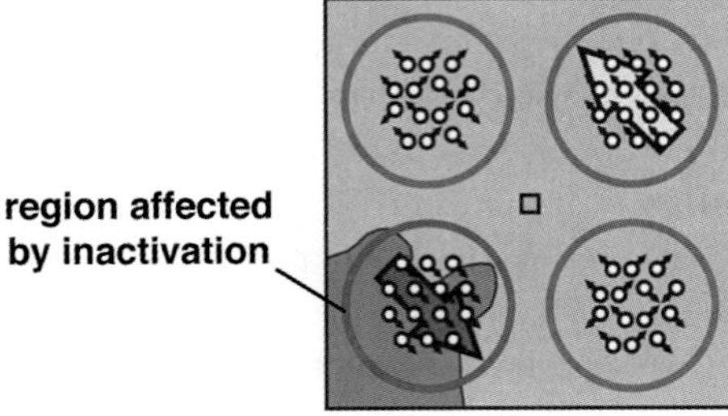
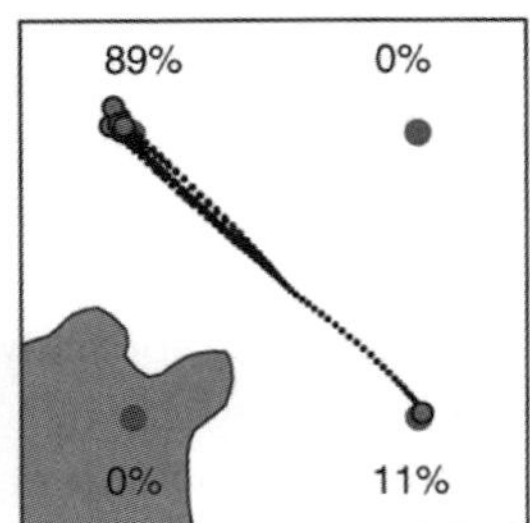

FIGURE 23.6   Impairments in covert selection of signals for perceptual judgments during dSC inactivation. The top-left panel shows the stimulus configuration, with the cued motion stimulus located in the lower left quadrant and the foil stimulus in the upper right quadrant. The top-right panel shows the choices made by the animal in this condition before injection of muscimol into the dSC. Most choices were correctly directed based on the motion in the cued patch. The bottom-left panel shows the stimulus configuration again, with the region affected by the SC inactivation shaded. After SC inactivation, most of the animal's choices are incorrectly directed based on the motion in the irrelevant foil patch. (Adapted from Lovejoy & Krauzlis, 2010.)

motion after SC inactivation, but erroneously bases his choices on the stimulus at the wrong spatial location.

In control experiments, when only a single motion patch is presented in the affected part of the visual field, the animal exhibits only a minor impairment in performance, showing that the processing of visual motion signals per se is mostly intact. Instead, the primary deficit caused by SC inactivation appears to be an inability to filter out distracting or misleading sensory information. The same pattern of results is found regardless of whether the animal responds with its eyes or its hands, indicating that the changes in performance are not due to a motor impairment. Thus, activity in the dSC plays a causal role in the control of covert spatial attention in addition to its role in visual selection for orienting movements.

CIRCUITS FOR VISUAL SELECTION

The circuits used by the SC in target selection and visual attention are not yet known, but several candidate circuit elements have been identified that are important for the selection and gating of visual signals. One of the key circuit properties for visual selection is inhibition, so that weaker inputs can be suppressed and a unique winner can emerge, but the source or sources of inhibition that accomplish this "winner-take-all" selection remain an open issue.

One possibility is that the inhibition is part of the intrinsic circuitry of the dSC. As described in several models of the dSC (Trappenberg et al., 2001; Van Opstal & Van Gisbergen, 1989), if the response fields of dSC neurons had a center-surround organization, this could provide the local excitation combined with long-range inhibition that could implement the winner-take-all selection. There is some experimental evidence for long-range inhibition within the dSC; for example, electrical stimulation in the dSC of the primate shows that activation at one site causes short-latency inhibition of activity at other, remote sites (Munoz & Istvan, 1998). However, pharmacological manipulation of dSC activity, which, unlike electrical stimulation, would not affect fibers of passage, does not produce long-range inhibition (Watanabe et al., 2005). Also, in vitro studies of the dSC in slice preparations have not found evidence for long-range inhibition—excitatory effects are coextensive with the inhibitory effects (Isa & Hall, 2009; Phongphanphanee, Kaneda, & Isa, 2008). Ironically, given the emphasis on winner-take-all selection in the dSC, these same slice studies provide evidence for long-range inhibition in the sSC.

There are also prominent sources of inhibition from outside the dSC that could serve a similar role in visual selection. The best-known of these is the substantia nigra pars reticulata in the basal ganglia, which has an inhibitory GABAergic projection to the dSC implicated in the top-down control of saccades (Hikosaka, Takikawa, & Kawagoe, 2000); alternate models in the primate have shown how this external source of inhibition could support a winner-take-all mechanism (Arai & Keller, 2005). In addition to substantia nigra there are about a dozen other likely sources of inhibitory inputs to the dSC from subcortical structures, many related to control of orienting, but whose specific functions are mostly unknown (Appell & Behan, 1990).

One source of inhibition that has been studied in detail in birds is the nucleus isthmi pars magnocellularis (Imc); the homologous structure in primates may be in the lateral tegmentum. The Imc receives topographically organized inputs from the SC and contains GABAergic neurons that provide a projection back to the SC (Wang, Major, & Karten, 2004). Strikingly, the terminals from Imc neurons spread broadly across the dSC map but avoid the spatial location that provides their input. Consequently, unlike a classic inhibitory surround, this "antitopographic" inhibition does not decrease with distance but acts across the entire visual field (Lai et al., 2011; Mysore, Asadollahi, & Knudsen, 2010). Furthermore, Imc neurons not only provide an inhibitory input to the dSC, they also contact other neurons within the Imc (Wang, Major, & Karten, 2004). Reciprocal inhibition between competing spatial channels in the Imc may add a novel computational feature to the selection process. Lateral inhibition alone implements a fixed criterion for selecting one stimulus over another, but the addition of reciprocal inhibition allows the criterion to shift according to the strength of the competitors, providing a mechanism for flexible categorization (Mysore & Knudsen, 2012; Sharpee, 2012). This type of mechanism could underlie the "switch-like" behavior found for some neurons in the dSC (Mysore, Asadollahi, & Knudsen, 2011).

The nucleus isthmi circuit is also involved in gating visual signals through the SC using another mechanism (figure 23.7). In addition to its projection to the dSC, the Imc sends a GABAergic projection to the cholinergic nucleus isthmi pars parvocellularis (Ipc); the homologous structure in mammals is the parabigeminal nucleus. As with the dSC, the input from the Imc to the Ipc exerts a long-range inhibitory effect, and Ipc neurons also show switch-like properties, increasing their activity when the stimulus in their receptive field becomes the strongest (Asadollahi, Mysore, & Knudsen, 2010, 2011). The Ipc, in turn, forms another loop with

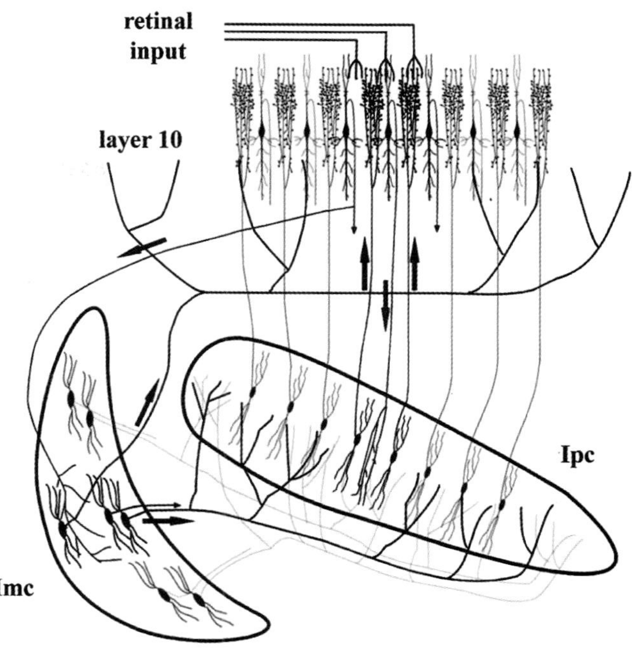

Figure 23.7
Diagram showing the circuit between the SC and the nucleus isthmi in the bird. The Ipc is reciprocally and topographically connected with the SC; the input from the Ipc to the SC forms cholinergic "paintbrush" terminals. The Imc receives inputs from the SC and sends inhibitory projections to both the SC and the Ipc. (Adapted from Marín et al., 2005; data from Wang, Major, & Karten, 2004.)

the SC. The Ipc receives a topographically organized input from the SC, and cholinergic neurons in the Ipc project back to the SC (Wang et al., 2006). The inputs from the Ipc form a very distinctive columnar pattern in the sSC referred to as "paintbrush" axon terminals, which are responsible for the distinctive oscillatory bursts recorded in the sSC (Marín et al., 2005). Each paintbrush is juxtaposed with retinal axon terminals as they make contact with the broad dendritic trees of tectal ganglion cells, a prominent class of neurons in the sSC that respond to visual motion across large parts of the visual field and provide an ascending projection to the thalamic nucleus rotundus (Rt; homologous to pulvinar in mammals).

The cholinergic input from the Ipc modulates the efficacy of the retinal input onto tectal ganglion cells, and, because of the precise topography among the connections between the SC and Ipc, this modulation constitutes a spatial filter determining which retinal inputs get transmitted to the thalamus. When activity in the Ipc is locally blocked, the corresponding visually evoked cholinergic feedback in the dSC is eliminated, and thalamic neurons in Rt receiving input from those tectal ganglion cells no longer respond to visual motion in that subregion of the visual field (Marín et al., 2007). Thus, the cholinergic feedback from the Ipc selectively enhances the salience of stimuli at particular retinotopic locations and helps select which signals get transmitted from SC to thalamus.

Whether a similar functional circuit applies in the primate is not clear, but pathways through thalamus to cortex are believed to be one of the primary routes by which the SC influences visual attention. One pathway originates from the sSC, passes through the inferior pulvinar nucleus, and targets areas in extrastriate cortex specialized for processing visual motion. This pathway has been identified by using electrical stimulation to identify neurons that both project to cortex and receive inputs from the SC (Berman & Wurtz, 2010) and also by using transsynaptic anatomical tracers (Lyon, Nassi, & Callaway, 2010). Neurons in the inferior pulvinar identified as receiving SC input tend to lack direction selectivity, as do sSC neurons, indicating that this pathway is unlikely to provide directional signals to cortex, but they do show suppression of visual activity during saccades (Berman & Wurtz, 2011). Because it provides a direct route to motion areas in cortex, this pathway might also underlie the changes in performance on visual motion tasks caused by SC stimulation (Cavanaugh & Wurtz, 2004; Müller, Philiastides, & Newsome, 2005).

Other possible routes for influencing visual attention originate in the SC and reach areas of prefrontal and parietal cortex through several thalamic nuclei, including the mediodorsal nucleus and lateral pulvinar (Harting et al., 1980; Robinson & Petersen, 1992). The prefrontal and parietal cortices have long been recognized to play a major role in the control of visual attention, and more recent work has described how these areas may contain salience maps that influence the processing of sensory signals elsewhere in cortex (Bisley & Goldberg, 2010; Moore, 2006; see chapter 75). The role of SC-recipient thalamic nuclei in visual attention is less clear, but inactivation of the lateral pulvinar is associated with deficits in visual attention tasks (Desimone et al., 1990). The deficits after pulvinar lesions are especially pronounced when distracters are present, similar to the pattern found with SC inactivation (Lovejoy & Krauzlis, 2010), supporting the conjecture that the role of the SC in controlling visual attention is mediated by this ascending pathway.

CONCLUSIONS

The SC has been the subject of a large amount of research, and for an excellent reason—it plays a highly conserved role in detecting the occurrence of meaningful events and in orchestrating appropriate behavioral responses. In some species, such as frogs, the bulk of processing needed to detect events occurs en route to the SC in a bottom-up manner, whereas in other species, such as primates, these processing steps involve interactions between bottom-up sensory and top-down cognitive factors. As events are detected, a selection mechanism winnows down the field of potential targets so that a particular visual object guides the behavioral response. The circuits involved in selecting the winner are not yet fully understood, but a variety of animal models and experimental techniques, ranging from slice work to behavioral neurophysiology, have provided solid clues about the structures involved and the role of inhibitory feedback. The availability of these diverse approaches, combined with the conserved functions of the SC, suggest that the SC may be one of the first places in the vertebrate brain where the puzzle of how neuronal circuits accomplish higher-order visual functions gets solved.

REFERENCES

Appell, P. P., & Behan, M. (1990). Sources of subcortical GABAergic projections to the superior colliculus in the cat. *Journal of Comparative Neurology, 302,* 143–158.

Arai, K., & Keller, E. L. (2005). A model of the saccade-generating system that accounts for trajectory variations produced by competing visual stimuli. *Biological Cybernetics, 92,* 21–37.

Asadollahi, A., Mysore, S. P., & Knudsen, E. I. (2010). Stimulus-driven competition in a cholinergic midbrain nucleus. *Nature Neuroscience, 13,* 889–895.

Asadollahi, A., Mysore, S. P., & Knudsen, E. I. (2011). Rules of competitive stimulus selection in a cholinergic isthmic nucleus of the owl midbrain. *Journal of Neuroscience, 31,* 6088–6097.

Bell, A. H., Fecteau, J. H., & Munoz, D. P. (2004). Using auditory and visual stimuli to investigate the behavioral and neuronal consequences of reflexive covert orienting. *Journal of Neurophysiology, 91,* 2172–2184.

Berman, R., & Colby, C. (2009). Attention and active vision. *Vision Research, 49,* 1233–1248. doi:10.1016/j.visres.2008.06.017.

Berman, R. A., & Wurtz, R. H. (2010). Functional identification of a pulvinar path from superior colliculus to cortical area MT. *Journal of Neuroscience, 30,* 6342–6354.

Berman, R. A., & Wurtz, R. H. (2011). Signals conveyed in the pulvinar pathway from superior colliculus to cortical area MT. *Journal of Neuroscience, 31,* 373–384.

Bisley, J. W., & Goldberg, M. E. (2010). Attention, intention, and priority in the parietal lobe. *Annual Review of Neuroscience, 33,* 1–21. doi:10.1146/annurev-neuro-060909-152823.

Büttner-Ennever, J. A., Cohen, B., Horn, A. K., & Reisine, H. (1996). Efferent pathways of the nucleus of the optic tract in monkey and their role in eye movements. *Journal of Comparative Neurology, 373,* 90–107.

Carello, C. D., & Krauzlis, R. J. (2004). Manipulating intent: Evidence for a causal role of the superior colliculus in target selection. *Neuron, 43*, 575–583.

Cavanaugh, J., Alvarez, B. D., & Wurtz, R. H. (2006). Enhanced performance with brain stimulation: Attentional shift or visual cue? *Journal of Neuroscience, 26*, 11347–11358.

Cavanaugh, J., & Wurtz, R. H. (2004). Subcortical modulation of attention counters change blindness. *Journal of Neuroscience, 24*, 11236–11243.

Comoli, E., Coizet, V., Boyes, J., Bolam, J. P., Canteras, N. S., Quirk, R. H., et al. (2003). A direct projection from superior colliculus to substantia nigra for detecting salient visual events. *Nature Neuroscience, 6*, 974–980. doi:10.1038/nn1113.

Cynader, M., & Berman, N. (1972). Receptive-field organization of monkey superior colliculus. *Journal of Neurophysiology, 35*, 187–201.

Deniau, J. M., & Chevalier, G. (1992). The lamellar organization of the rat substantia nigra pars reticulata: Distribution of projection neurons. *Neuroscience, 46*, 361–377.

Desimone, R., Wessinger, M., Thomas, L., & Schneider, W. (1990). Attentional control of visual perception: Cortical and subcortical mechanisms. *Cold Spring Harbor Symposia on Quantitative Biology, 55*, 963–971.

Dorris, M., Olivier, E., & Munoz, D. (2007). Competitive integration of visual and preparatory signals in the superior colliculus during saccadic programming. *Journal of Neuroscience, 27*, 5053–5062.

Duhamel, J.-R., Colby, C. L., & Goldberg, M. E. (1992). The updating of the representation of visual space in parietal cortex by intended eye movements. *Science, 255*, 90–92.

Fecteau, J. H., Bell, A. H., & Munoz, D. P. (2004). Neural correlates of the automatic and goal-driven biases in orienting spatial attention. *Journal of Neurophysiology, 92*, 1728–1737.

Fecteau, J. H., & Munoz, D. P. (2006). Salience, relevance, and firing: A priority map for target selection. *Trends in Cognitive Sciences, 10*, 382–390.

Freedman, D. J., & Assad, J. A. (2006). Experience-dependent representation of visual categories in parietal cortex. *Nature, 443*, 85–88.

Frost, B. J., & DiFranco, D. E. (1976). Motion characteristics of single units in the pigeon optic tectum. *Vision Research, 16*, 1229–1234. doi:10.1016/0042-6989(76)90046-8.

Gaillard, F. (1990). Visual units in the central nervous system of the frog. *Comparative Biochemistry and Physiology. A. Comparative Physiology, 96*, 357–371.

Gandhi, N. J., & Katnani, H. A. (2011). Motor functions of the superior colliculus. *Annual Review of Neuroscience, 34*, 205–231.

Glimcher, P. W., & Sparks, D. L. (1992). Movement selection in advance of action in the superior colliculus. *Nature, 355*, 542–545.

Goldberg, M. E., & Wurtz, R. H. (1972a). Activity of superior colliculus in behaving monkey. I. Visual receptive fields of single neurons. *Journal of Neurophysiology, 35*, 542–559.

Goldberg, M. E., & Wurtz, R. H. (1972b). Activity of superior colliculus in behaving monkey. II. Effect of attention on neuronal responses. *Journal of Neurophysiology, 35*, 560–574.

Groh, J. M., & Sparks, D. L. (1996). Saccades to somatosensory targets. II. Motor convergence in primate superior colliculus. *Journal of Neurophysiology, 75*, 428–438.

Harting, J. K. (1977). Descending pathways from the superior collicullus: An autoradiographic analysis in the rhesus monkey (*Macaca mulatta*). *Journal of Comparative Neurology, 173*, 583–612.

Harting, J. K., Huerta, M. F., Frankfurter, A. J., Strominger, N. L., & Royce, G. J. (1980). Ascending pathways from the monkey superior colliculus: An autoradiographic analysis. *Journal of Comparative Neurology, 192*, 853–882.

Harting, J. K., Updyke, B. V., & Van Lieshout, D. P. (2001). The visual-oculomotor striatum of the cat: Functional relationship to the superior colliculus. *Experimental Brain Research, 136*, 138–142.

Hikosaka, O., Takikawa, Y., & Kawagoe, R. (2000). Role of the basal ganglia in the control of purposive saccadic eye movements. *Physiological Reviews, 80*, 953–978.

Hofbauer, A., & Dräger, U. C. (1985). Depth segregation of retinal ganglion cells projecting to mouse superior colliculus. *Journal of Comparative Neurology, 234*, 465–474.

Horwitz, G. D., Batista, A. P., & Newsome, W. T. (2004). Representation of an abstract perceptual decision in macaque superior colliculus. *Journal of Neurophysiology, 91*, 2281–2296.

Horwitz, G. D., & Newsome, W. T. (1999). Separate signals for target selection and movement specification in the superior colliculus. *Science, 284*, 1158–1161.

Huerta, M. F., Van Lieshout, D. P., & Harting, J. K. (1991). Nigrotectal projections in the primate *Galago crassicaudatus*. *Experimental Brain Research, 87*, 389–401.

Humphrey, N. K. (1968). Responses to visual stimuli of units in the superior colliculus of rats and monkeys. *Experimental Neurology, 20*, 312–340.

Ignashchenkova, A., Dicke, P. W., Haarmeier, T., & Thier, P. (2004). Neuron-specific contribution of the superior colliculus to overt and covert shifts of attention. *Nature Neuroscience, 7*, 56–64.

Isa, T., & Hall, W. C. (2009). Exploring the superior colliculus in vitro. *Journal of Neurophysiology, 102*, 2581–2593.

Jay, M. F., & Sparks, D. L. (1987). Sensorimotor integration in the primate superior colliculus. II. Coordinates of auditory signals. *Journal of Neurophysiology, 57*, 35–55.

Kang, H.-J., & Li, X.-H. (2010). Response properties and receptive field organization of collision-sensitive neurons in the optic tectum of bullfrog, *Rana catesbeiana*. *Neuroscience Bulletin, 26*, 304–316.

Karten, H. J., & Dubbeldam, J. L. (1973). The organization and projections of the paleostriatal complex in the pigeon (*Columba livia*). *Journal of Comparative Neurology, 148*, 61–89.

Kawamura, S., Sprague, J. M., & Niimi, K. (1974). Corticofugal projections from the visual cortices to the thalamus, pretectum and superior colliculus in the cat. *Journal of Comparative Neurology, 158*, 339–362.

Krauzlis, R. J. (2003). Neuronal activity in the rostral superior colliculus related to the initiation of pursuit and saccadic eye movements. *Journal of Neuroscience, 23*, 4333–4344.

Krauzlis, R. J. (2004). Activity of rostral superior colliculus neurons during passive and active viewing of motion. *Journal of Neurophysiology, 92*, 949–958.

Krauzlis, R. J. (2005). The control of voluntary eye movements: New perspectives. *The Neuroscientist, 11*, 124–137.

Krauzlis, R. J., Basso, M. A., & Wurtz, R. H. (1997). Shared motor error for multiple eye movements. *Science, 276*, 1693–1695.

Krauzlis, R. J., Basso, M. A., & Wurtz, R. H. (2000). Discharge properties of neurons in the rostral superior colliculus of the monkey during smooth-pursuit eye movements. *Journal of Neurophysiology, 84*, 876–891.

Krauzlis, R., & Dill, N. (2002). Neural correlates of target choice for pursuit and saccades in the primate superior colliculus. *Neuron, 35*, 355–363.

Kunzle, H., & Akert, K. (1977). Efferent connections of cortical, area 8 (frontal eye field) in *Macaca fascicularis*. A reinvestigation using the autoradiographic technique. *Journal of Comparative Neurology, 173*, 147–164.

Kustov, A., & Robinson, D. (1996). Shared neural control of attentional shifts and eye movements. *Nature, 384*, 74–77.

Lai, D., Brandt, S., Luksch, H., & Wessel, R. (2011). Recurrent antitopographic inhibition mediates competitive stimulus selection in an attention network. *Journal of Neurophysiology, 105*, 793–805.

Lettvin, J. Y., Maturana, H. R., Pitts, W. H., & McCulloch, W. S. (1961). Two remarks on the visual system of the frog. In W. Rosenblith (Ed.), *Sensory communication* (pp. 757–776). New York: MIT Press and John Wiley & Sons.

Lisberger, S. G. (2010). Visual guidance of smooth-pursuit eye movements: Sensation, action, and what happens in between. *Neuron, 66*, 477–491.

Liu, Y.-J., Wang, Q., & Li, B. (2011). Neuronal responses to looming objects in the superior colliculus of the cat. *Brain, Behavior and Evolution, 77*, 193–205.

Lovejoy, L. P., & Krauzlis, R. J. (2010). Inactivation of primate superior colliculus impairs covert selection of signals for perceptual judgments. *Nature Neuroscience, 13*, 261–266.

Lui, F., Gregory, K. M., Blanks, R. H., & Giolli, R. A. (1995). Projections from visual areas of the cerebral cortex to pretectal nuclear complex, terminal accessory optic nuclei, and superior colliculus in macaque monkey. *Journal of Comparative Neurology, 363*, 439–460.

Luksch, H. (2003). Cytoarchitecture of the avian optic tectum: Neuronal substrate for cellular computation. *Reviews in the Neurosciences, 14*, 85–106.

Lyon, D. C., Nassi, J. J., & Callaway, E. M. (2010). A disynaptic relay from superior colliculus to dorsal stream visual cortex in macaque monkey. *Neuron, 65*, 270–279.

Marín, G., Mpodozis, J., Mpodozis, J., Sentis, E., Ossandón, T., & Letelier, J. C. (2005). Oscillatory bursts in the optic tectum of birds represent re-entrant signals from the nucleus isthmi pars parvocellularis. *Journal of Neuroscience, 25*, 7081–7089.

Marín, G., Salas, C., Sentis, E., Rojas, X., Letelier, J. C., & Mpodozis, J. (2007). A cholinergic gating mechanism controlled by competitive interactions in the optic tectum of the pigeon. *Journal of Neuroscience, 27*, 8112–8121.

Marrocco, R. T., & Li, R. H. (1977). Monkey superior colliculus: Properties of single cells and their afferent inputs. *Journal of Neurophysiology, 40*, 844–860.

May, P. J. (2006). The mammalian superior colliculus: Laminar structure and connections. *Progress in Brain Research, 151*, 321–378.

May, P. J., McHaffie, J. G., Stanford, T. R., Jiang, H., Costello, M. G., Coizet, V., et al. (2009). Tectonigral projections in the primate: A pathway for pre-attentive sensory input to midbrain dopaminergic neurons. *European Journal of Neuroscience, 29*, 575–587. doi:10.1111/j.1460-9568.2008.06596.x.

McHaffie, J. G., Jiang, H., May, P. J., Coizet, V., Overton, P. G., Stein, B. E., et al. (2006). A direct projection from superior colliculus to substantia nigra pars compacta in the cat. *Neuroscience, 138*, 221–234.

McHaffie, J. G., Kao, C. Q., & Stein, B. E. (1989). Nociceptive neurons in rat superior colliculus: Response properties,

topography, and functional implications. *Journal of Neurophysiology, 62*, 510–525.

McPeek, R. M., & Keller, E. L. (2002). Saccade target selection in the superior colliculus during a visual search task. *Journal of Neurophysiology, 88*, 2019–2034.

McPeek, R. M., & Keller, E. L. (2004). Deficits in saccade target selection after inactivation of superior colliculus. *Nature Neuroscience, 7*, 757–763.

Moore, T. (2006). The neurobiology of visual attention: Finding sources. *Current Opinion in Neurobiology, 16*, 159–165.

Müller, J. R., Philiastides, M. G., & Newsome, W. T. (2005). Microstimulation of the superior colliculus focuses attention without moving the eyes. *Proceedings of the National Academy of Sciences of the United States of America, 102*, 524–529.

Munoz, D. P., & Istvan, P. J. (1998). Lateral inhibitory interactions in the intermediate layers of the monkey superior colliculus. *Journal of Neurophysiology, 79*, 1193–1209.

Mysore, S. P., Asadollahi, A., & Knudsen, E. I. (2010). Global inhibition and stimulus competition in the owl optic tectum. *Journal of Neuroscience, 30*, 1727–1738.

Mysore, S. P., Asadollahi, A., & Knudsen, E. I. (2011). Signaling of the strongest stimulus in the owl optic tectum. *Journal of Neuroscience, 31*, 5186–5196.

Mysore, S. P., & Knudsen, E. I. (2011). Flexible categorization of relative stimulus strength by the optic tectum. *Journal of Neuroscience, 31*, 7745–7752.

Mysore, S. P., & Knudsen, E. I. (2012). Reciprocal inhibition of inhibition: A circuit motif for flexible categorization in stimulus selection. *Neuron, 73*, 193–205.

Nakagawa, H., & Hongjian, K. (2010). Collision-sensitive neurons in the optic tectum of the bullfrog, *Rana catesbeiana*. *Journal of Neurophysiology, 104*, 2487–2499.

Nakamura, K., & Colby, C. L. (2002). Updating of the visual representation in monkey striate and extrastriate cortex during saccades. *Proceedings of the National Academy of Sciences of the United States of America, 99*, 4026–4031. doi:10.1073/pnas.052379899.

Niv, Y., & Schoenbaum, G. (2008). Dialogues on prediction errors. *Trends in Cognitive Sciences, 12*, 265–272. doi:10.1016/j.tics.2008.03.006.

Nummela, S. U., & Krauzlis, R. J. (2010). Inactivation of primate superior colliculus biases target choice for smooth pursuit, saccades, and button press responses. *Journal of Neurophysiology, 104*, 1538–1548.

Phongphanphanee, P., Kaneda, K., & Isa, T. (2008). Spatiotemporal profiles of field potentials in mouse superior colliculus analyzed by multichannel recording. *Journal of Neuroscience, 28*, 9309–9318.

Phongphanphanee, P., Mizuno, F., Lee, P. H., Yanagawa, Y., Isa, T., & Hall, W. C. (2011). A circuit model for saccadic suppression in the superior colliculus. *Journal of Neuroscience, 31*, 1949–1954.

Pollack, J. G., & Hickey, T. L. (1979). The distribution of retino-collicular axon terminals in rhesus monkey. *Journal of Comparative Neurology, 185*, 587–602.

Rashbass, C. (1961). The relationship between saccadic and smooth tracking eye movements. *Journal of Physiology, 159*, 326–338.

Redgrave, P., Marrow, L., & Dean, P. (1992). Topographical organization of the nigrotectal projection in rat: Evidence for segregated channels. *Neuroscience, 50*, 571–595.

Reiner, A., Perkel, D. J., Bruce, L. L., Butler, A. B., Csillag, A., Kuenzel, W., et al. (2004). Revised nomenclature for avian telencephalon and some related brainstem nuclei. *Journal of Comparative Neurology, 473*, 377–414. doi:10.1002/cne.20118.

Rensink, R. A. (2002). Change detection. *Annual Review of Psychology, 53*, 245–277.

Robinson, D. L., & Petersen, S. E. (1992). The pulvinar and visual salience. *Trends in Neurosciences, 15*, 127–132. doi:10.1016/0166-2236(92)90354-B.

Robinson, D. L., & Wurtz, R. H. (1976). Use of an extraretinal signal by monkey superior colliculus neurons to distinguish real from self-induced stimulus movement. *Journal of Neurophysiology, 39*, 852–870.

Schiller, P. H., & Malpeli, J. G. (1977). Properties and tectal projections of monkey retinal ganglion cells. *Journal of Neurophysiology, 40*, 428–445.

Schultz, W. (2010). Dopamine signals for reward value and risk: Basic and recent data. *Behavioral and Brain Functions, 6*, 24.

Sharpee, T. O. (2012). Adaptive switches in midbrain circuits. *Neuron, 73*, 6–7.

Smith, Y., Raju, D. V., Pare, J.-F., & Sidibe, M. (2004). The thalamostriatal system: A highly specific network of the basal ganglia circuitry. *Trends in Neurosciences, 27*, 520–527. doi:10.1016/j.tins.2004.07.004.

Sommer, M. A., & Wurtz, R. H. (2008). Brain circuits for the internal monitoring of movements. *Annual Review of Neuroscience, 31*, 317–338.

Song, J.-H., Rafal, R. D., & McPeek, R. M. (2011). Deficits in reach target selection during inactivation of the midbrain superior colliculus. *Proceedings of the National Academy of Sciences of the United States of America, 108*, E1433–E1440. doi:10.1073/pnas.1109656108.

Sterling, P., & Wickelgren, B. G. (1969). Visual receptive fields in the superior colliculus of the cat. *Journal of Neurophysiology, 32*, 1–15.

Stuphorn, V., Bauswein, E., & Hoffmann, K.-P. (2000). Neurons in the primate superior colliculus coding for arm movements in gaze-related coordinates. *Journal of Neurophysiology, 83*, 1283–1299.

Trappenberg, T. P., Dorris, M. C., Munoz, D. P., & Klein, R. M. (2001). A model of saccade initiation based on the competitive integration of exogenous and endogenous signals in the superior colliculus. *Journal of Cognitive Neuroscience, 13*, 256–271.

Updyke, B. V. (1974). Characteristics of unit responses in superior colliculus of the Cebus monkey. *Journal of Neurophysiology, 37*, 896–909.

Van der Werf, Y. D., Witter, M. P., & Groenewegen, H. J. (2002). The intralaminar and midline nuclei of the thalamus. Anatomical and functional evidence for participation in processes of arousal and awareness. *Brain Research. Brain Research Reviews, 39*, 107–140.

Van Opstal, A. J., & Van Gisbergen, J. A. (1989). A nonlinear model for collicular spatial interactions underlying the metrical properties of electrically elicited saccades. *Biological Cybernetics, 60*, 171–183.

Walker, M. F., Fitzgibbon, E. J., & Goldberg, M. E. (1995). Neurons in the monkey superior colliculus predict the visual result of impending saccadic eye movements. *Journal of Neurophysiology, 73*, 1988–2003.

Wallace, M. T., Wilkinson, L. K., & Stein, B. E. (1996). Representation and integration of multiple sensory inputs in primate superior colliculus. *Journal of Neurophysiology, 76*, 1246–1266.

Wang, Y., Luksch, H., Brecha, N. C., & Karten, H. J. (2006). Columnar projections from the cholinergic nucleus isthmi to the optic tectum in chicks (*Gallus gallus*): A possible substrate for synchronizing tectal channels. *Journal of Comparative Neurology, 494*, 7–35.

Wang, Y., Major, D. E., & Karten, H. J. (2004). Morphology and connections of nucleus isthmi pars magnocellularis in chicks (*Gallus gallus*). *Journal of Comparative Neurology, 469*, 275–297.

Watanabe, M., Kobayashi, Y., Inoue, Y., & Isa, T. (2005). Effects of local nicotinic activation of the superior colliculus on saccades in monkeys. *Journal of Neurophysiology, 93*, 519–534.

Werner, W., Dannenberg, S., & Hoffmann, K.-P. (1997). Arm-movement-related neurons in the primate superior colliculus and underlying reticular formation: Comparison of neuronal activity with EMGs of muscles of the shoulder, arm and trunk during reaching. *Experimental Brain Research, 115*, 191–205.

White, B. J., Boehnke, S. E., Marino, R. A., Itti, L., & Munoz, D. P. (2009). Color-related signals in the primate superior colliculus. *Journal of Neuroscience, 29*, 12159–12166.

Wickelgren, B. G., & Sterling, P. (1969). Influence of visual cortex on receptive fields in the superior colliculus of the cat. *Journal of Neurophysiology, 32*, 16–23.

Wu, L., Niu, Y., Yang, J., & Wang, S. (2005). Tectal neurons signal impending collision of looming objects in the pigeon. *European Journal of Neuroscience, 22*, 2325–2331.

Wurtz, R. H., & Albano, J. E. (1980). Visual-motor function of the primate superior colliculus. *Annual Review of Neuroscience, 3*, 189–226.

Wurtz, R. H., McAlonan, K., Cavanaugh, J., & Berman, R. A. (2011). Thalamic pathways for active vision. *Trends in Cognitive Sciences, 15*, 177–184. doi:10.1016/j.tics.2011.02.004.

24 Attentional Activation in Corticothalamic Loops of the Visual System

ANDRZEJ WRÓBEL

All sensory information in mammals, except for smell, arrives at the cerebral cortex via the thalamus (Jones, 2001, 2007). However, far from being a simple relay, thalamic sensory nuclei receive much more reentrant cortical connections than ascending peripheral ones (Rouiller & Welker, 2000). These thalamocortical loops have been implicated in numerous cognitive functions (Briggs & Usrey, 2008; Saalmann & Kastner, 2009, 2011), including attention (Rees, 2009; Wróbel, 2000), perceptual grouping (Robertson, 2003; Ward et al., 2002; Wilke at al., 2010), and general consciousness (Dehaene & Changeux, 2011; Llinas et al., 1998; Ward, 2011). It is also widely accepted that the momentary state of the neuronal network modifies stimulus information processing (Anderson et al., 2000; Arieli et al., 1996; Briggs & Usrey, 2007; Wróbel & Kublik, 2000) and that the resulting response variability (initially attributed to "noise") might represent one of the foundations of brain function (Bernasconi et al., 2011). The factors governing such variability across different processing states are under intensive investigations (Saalmann & Kastner, 2011; Sherman, 2005) and are not yet fully understood. Much is already known about gross state transitions, such as between sleep and wakefulness (Steriade, 1997), but in order to understand the mechanisms underlying various brain functions it is necessary to investigate subtle, rapid natural state transitions evoked during behavioral/cognitive tests in awake animals (i.e., Bekisz & Wróbel, 1993; Buschman & Miller, 2007; Sobolewski et al., 2010). Proper performance in these tasks is critically dependent on the neuromodulatory actions of brainstem nuclei (Buhl et al., 1998; Roopun et al., 2010; Soma et al., 2012; Steriade et al., 1991; Wróbel & Kublik, 2000; see Harris & Thiele, 2011, for review) and the dynamic neural mechanisms utilized during processing of information within the neuronal network. The purpose of this chapter is to shed light on how different attentive demands reconfigure the functional arrangement of the corticothalamic network of the visual system.

VISUAL THALAMIC RELAYS, THEIR POSTULATED FUNCTIONS, AND STATE MODULATION

In mammals visual information arrives at the thalamus via its two largest afferent pathways. Signals in the retinogeniculate pathway reach the dorsal lateral geniculate nucleus (dLGN), directly from the retina and in turn project further to area 17 of the visual cortex. In the retinotectal pathway retinal signals arrive at the thalamus lateral posterior pulvinar complex (LP-P) of the cat and pulvinar complex of the monkey indirectly (via the superior colliculus), where they project further to higher-order visual cortical areas.

The retinogeniculocortical pathway contains three separate streams conveyed by morphologically and functionally different neurons. The first of them (X in the cat/P in the monkey) provides maximum acuity for detailed vision. The second (Y/M) is chiefly involved in motion detection and processing of high temporal and low spatial frequencies. The function of the third stream (W/K) appears to be involved in heterogeneous functions (Sherman, 2009) and is beyond the scope of this chapter. Most of the retinogeniculate axons of the two latter streams branch to innervate midbrain targets implicated in eye movements, pupillary control, and other functions. Among them the superior colliculus provides a link to the LP-P/pulvinar (Chalupa, 1991; Garey, Dreher, & Robinson, 1991). The function of the dLGN in processing visual information is well documented; however, the role of the LP-P/pulvinar relay is less clear, although many findings suggest an involvement in attention (Baluch & Itti, 2011; Chalupa, 1991; Saalmann & Kastner, 2009; Shipp, 2004) and active vision (Noudoost et al., 2010; Wurtz et al., 2011).

Visual thalamic nuclei project to the primary cortex, albeit in parallel streams, and the LP-P sends additional projections to higher-order visual areas (Berman & Wurtz, 2010; Sherman, 2009; Symonds et al., 1981; Wurtz et al., 2011). Ascending output projections from primary to higher visual areas are less separated than previously thought (Casagrande, 1994), but some cortical areas in the ventral visual stream are predominantly innervated by X/P input and dorsal areas by Y/M input (Nassi & Callaway, 2006; Waleszczyk et al., 2004). Nevertheless, the cortical areas of the ventral stream (like area 21a in the cat and area V4 in the monkey) are functionally specialized for processing of form and pattern information, whereas dorsal stream areas are specialized for analyses of motion and spatial relationships (Casagrande & Xu, 2004; Waleszczyk et al., 2004).

Importantly, the activity of both thalamic relay nuclei can be modulated powerfully by stream-specific corticothalamic feedback from layer 6 (figure 24.1) (Briggs &

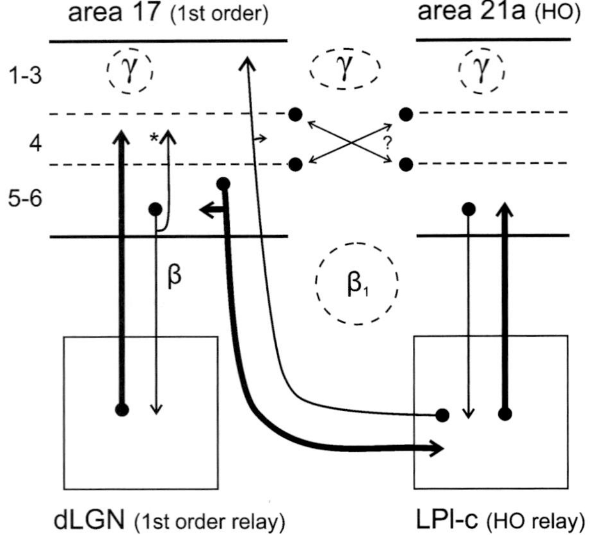

FIGURE 24.1 Schematic diagram showing connections between first- and higher-order relays in the cat visual system. The dLGN (a first-order thalamic relay, left) represents the relay of retinal information to area 17 (first-order cortical area, homologue to V1). The LPl-c (right), a part of the higher-order relay (LP-P), relays information from layer 5 of area 17 to the higher-order cortical area 21a (homologue to V4). The driver inputs (thick lines) show the postulated route of thalamocortico-thalamocortical communication, which involves afferent input to a primary cortex, a projection from its layer 5 to a higher-order thalamic relay and from there to higher-order cortical areas. All these relay centers receive parallel modulatory feedbacks from higher levels (thin lines). The question mark indicates direct corticocortical projections where the function, drive, or modulation is not known. dLGN, dorsal lateral geniculate nucleus; LPl-c, caudal part of the lateral zone of the LP-P, lateral posterior–pulvinar complex; HO, higher order; β, β1, γ, putative circuits with characteristic resonance frequencies as outlined in the text. The asterisk indicates one of the intracortical modulatory feedbacks; see text for details.

Usrey, 2009a; Vanduffel, Tootell, & Orban, 2000; Wróbel et al., 2007). The LP-P complex receives additional recurrent projections from layer 5 of many cortical areas (distinct from modulatory corticothalamic feedback) whose ultrastructure suggests a strong synaptic weight (Guillery, 1995; Huppe-Gourgues et al., 2006). Such corticofugal connections, present in various sensory modalities of mammalian brains, inspired Sherman and Guillery (2002) to put forward a hypothesis dividing thalamic relays into first-order and higher-order types and describing the function of thalamic relays by the source and nature of their strongest input (the driver). In this context first-order thalamic nuclei (such as dLGN) would be predicted to transmit peripheral signals to the primary cortex (e.g., V1), and higher-order nuclei (e.g., LP-P) would process ascending traffic between the cortices (figure 24.1; compare Theyel, Llano, & Sherman, 2010). Note that driving fibers projecting to higher-order centers are paralleled with reciprocally directed modulatory feedback pathways at each level of processing (Harris & Thiele, 2011; Sherman & Guillery, 2002). In support of this hypothesis an increasing number of studies have demonstrated that the complex functions of the thalamus, including the first and higher-order type relays, affect the nature of information relayed in a manner that reflects behavioral state, including attention (Sherman, 2009; Sobolewski et al., 2010, 2011b).

RESONANCE FREQUENCIES IN CORTICOTHALAMIC LOOPS

Alpha and Gamma Bands

The rich network of reentrant corticothalamic loops (figure 24.1) generates oscillations of different frequencies related to its architecture and the properties of the neurons involved (Wróbel, Hedström, & Lindström, 1998; for reviews see Steriade, 2000; Wang, 2010). Indeed, over the years recordings in humans and experimental animals have shown that alpha rhythm (8–12 Hz) is associated with an idle state of sensory networks, and attention reduces the amplitude of this oscillation (Bollimunta et al., 2011; Hanslmayer et al., 2007; Kelly, Gomez-Ramirez, & Foxe, 2009; Sauseng et al., 2005; Snyder & Foxe, 2010; Sobolewski et al., 2011a; Worden, 2000). In contrast, it is well documented that enhanced attention is accompanied by increased amplitude and/or synchrony of gamma rhythms (30–60 Hz) in many visual centers (Doesburg et al., 2008; Fries et al., 2001; Gregoriou et al., 2009; Ray et al., 2008; Taylor et al., 2005. See also topical reviews: Deco & Thiele, 2009; Engel, Fries, & Singer, 2001; Fries, 2009; Harris &

Thiele, 2011). These centers include the primary visual cortex and dLGN, where gamma activity increases in power and becomes synchronized during attention (Briggs & Usrey, 2009a; Bekisz & Wróbel, 1999; Steriade et al., 1991).

Two Beta Frequencies Characterize Distinct Corticothalamocortical Loops

The attentional enhancement related to beta-band (12–30 Hz) activity has remained relatively less studied (reviewed in Engel & Fries, 2010; Saalmann & Kastner, 2011; Wróbel, 2000). Experiments on cats (Bekisz & Wróbel, 1993; Wróbel et al., 1994, 2007; Wróbel, Bekisz, & Waleszczyk, 1994) provided data supporting the hypothesis that beta-band activity plays an important role in both overt and covert attention processes in numerous thalamic (dLGN, LP-P) and cortical (area 17, 18, and various parts of suprasylvian cortex) regions of the visual system (figure 24.2A–E). This hypothesis was based on the finding that beta-band power increased during the anticipatory period of a visual spatial differentiation task only in trials resulting in correct responses (figure 24.2C). Additionally, it was shown that attentional demands increase corticothalamic synchronization of beta oscillations between structures representing consecutive visual processing levels (i.e., dLGN and area 17—Bekisz & Wróbel, 1999, 2003; or area 17, LP-P, and suprasylvian sulcus—Wróbel et al. 2007; shown in figures 24.1 and 24.2F).

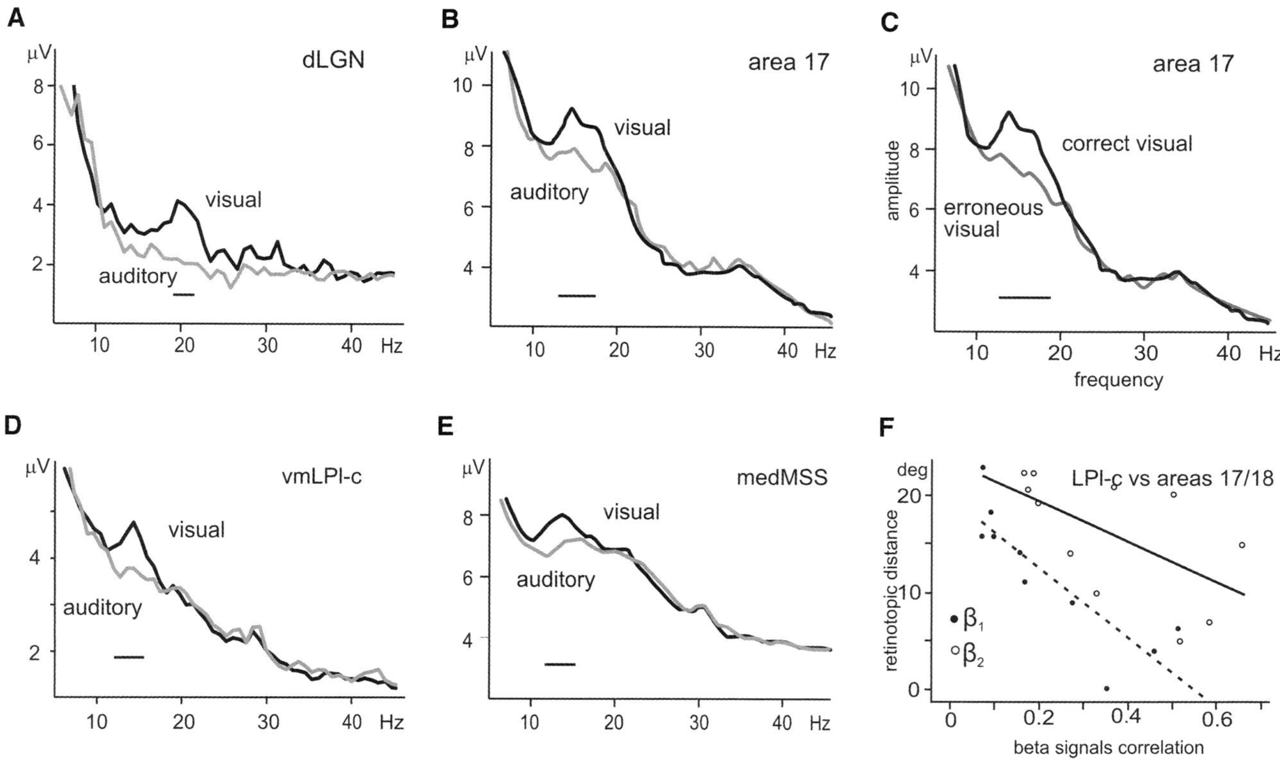

FIGURE 24.2   Beta activity in the visual centers of the cat performing a spatial differentiation task with visual and auditory cues increases during the anticipatory period of the visual trials (adapted from Wróbel et al., 2007). Mean FFT amplitude (example ordinate labeled in C) spectra (in frequency units, abscissa labeled in C) for individual cortical sites in dLGN (A), area 17 (B), vmLPl-c, the ventromedial part of the lateral zone of the LP-P complex (D), medMSS, medial part of middle suprasylvian sulcus (E) of a cat attentively expecting visual (black lines) or auditory (gray lines) target stimuli. The FFTs shown correspond to recordings during the stimulus-free time interval between cessation of either modality cue and onset of the appropriate target stimuli. Averages were taken only from trials that ended with correct behavioral responses. (C) FFT spectra averaged from visual trials ending with a correct response (black line) are compared with those ending with an incorrect conditional response (gray). Short lines above the horizontal axes denote significantly different frequency ranges between the spectra ($p < 0.01$). Note that, in these examples, activity in vmLPl-c, cortical areas 17, and medMSS was enhanced during visual trials in the low beta-frequency range (beta 1), but, at the same time, dlLPl-c, dorsolateral part of the lateral zone of the LP-P complex, and areas 18 and latMSS, lateral part of middle suprasylvian sulcus, recordings exhibited larger amplitudes in higher, beta 2 frequencies (not shown). See text for further details. (F) Beta synchronization depends on the retinotopic distance between cortical and LPl-c recording sites. The signals recorded from correlated pairs were filtered at beta 1 (12–19 Hz) and beta 2 (17–25 Hz) ranges, respectively. Pearson's correlation within the beta 1 pairs, –0.851 ($p = 0.002$); and within beta 2 pairs, –0.603 ($p = 0.05$).

To further define a putative functional role of the higher-order corticothalamocortical pathway in visual processing, Wróbel et al. (2007) analyzed attention-related changes of local field potentials recorded from the LP-P complex and interconnected primary and higher-order cortical visual areas in the cat. Among the latter the suprasylvian region (encompassing the higher-order visual areas) was chosen because of its strong interconnections with the LP-P (Garey, Dreher, & Robinson, 1991; Huppe-Gourgues et al., 2006; Payne & Lomber, 2003) and its involvement in visually guided behavior (Kiefer et al., 1989; Ouellette et al., 2004; Rudolph & Pasternak, 1996). It was observed that solely during visual, but not auditory, anticipatory attentive tasks, the amplitude of beta activity increased in signals recorded from the caudal part of the lateral zone of the LP-P (LPl-c), cortical areas 17 and 18, and the complex located at the middle suprasylvian sulcus (MSS) extending to area 21a (compare figure 24.2, panels B–E). Ventromedial and dorsolateral subregions of LPl-c were distinguished based on the occurrence of attentionally related beta activity of low (12–18 Hz; beta 1) and high (18–25 Hz; beta 2) frequencies, respectively. A selective increase of beta 1 power was observed in the medial bank of the MSS and area 21a, whereas cortical area 18 and the lateral bank of the MSS were activated in the high, beta 2 range. Area 17 exhibited attentional activation in the whole beta range. Phase-correlation analysis revealed that two distinct corticothalamic systems were synchronized by the beta activity of different frequencies. Beta 1 was associated with cortical area 17, ventromedial region of LPl-c, and regions extending to area 21a (and most probably represents the ventral stream), and beta 2 involved area 18 and the dorsolateral LPl-c (regions belonging to the dorsal stream).

These observations suggest that the LPl-c belongs to the wide corticothalamic attentional system, which is functionally segregated by distinct, and each retinotopically specific (figure 24.2F), streams of beta activity (Wróbel et al., 2007). This hypothesis on the functional organization of the cat LP-P is remarkably similar to that proposed by Shipp (2004) for the monkey pulvinar complex based on anatomical connections and retinotopic mapping. Shipp proposed that the primate pulvinar complex, regardless of internal anatomical boundaries, might be divided into two functional parts, which are used by separate corticothalamocortical loops matching reciprocal ventral and dorsal pathways of cortical processing of visual information. Both findings provide functional and anatomical support for the involvement of the pulvinar complex of cats and monkeys in attentional modulation of visual information processing (Olshausen, Anderson, & Van Essen, 1993; Saalmann & Kastner, 2009, 2011; Sherman & Guillery, 2002; Wilke et al., 2010).

Buschman and Miller (2007) recently demonstrated a positive correlation between beta-band power and visual attention in frontal and parietal cortices of the monkey. Importantly, they also found that synchrony between frontal and parietal areas during a paradigm involving top-down (anticipatory) attention was stronger in lower frequencies (beta) and during bottom-up (stimulus-evoked) attention synchrony in higher frequencies (gamma dominated). A spatially selective increase of attentional beta EEG activity during stimulus expectancy period was also found in humans (Basile et al., 2007; Kamiński et al., 2011; Siegel et al., 2008), and a very recent study indicates that the deterioration of attention observed in elderly subjects correlates with changes in beta-band activity (Gola et al., 2011). Similarly, Hanslmayer et al. (2007) demonstrated that brief visual stimuli were perceived by subjects only when preceded by periods of increased phase coupling in both beta and gamma EEG bands. Finally, in a magnetoencephalographic study, Gross et al. (2004) showed that long-range beta-band phase synchronization was a significant indicator of performance in an attentional blink task.

PUTATIVE MECHANISMS OF ATTENTIONAL ENHANCEMENT

The two mechanisms postulated for the neuronal correlates of attention have assumed either biased competition between neuronal ensembles (by increasing the firing rate of neurons encoding the attended stimulus features and decreasing the unattended [Reddy, Kanwisher, & VanRullen, 2009]) or synchronization of their firing by high-frequency oscillations (Fries et al., 2001; Gray et al., 1989; see Harris & Thiele, 2011, for review). Although the two phenomena are independent (Buehlmann & Deco, 2008), they can influence each other. Thus, intrinsic, cellular, and field oscillations during attentive states might self-regulate specific firing generated within the sensory system, as hypothesized by several investigators (for reviews see Schroeder & Lakatos, 2009; Wang, 2010; Ward, 2011; Wróbel, 2000). The long-range, top-down attentional signals may additionally influence the amplitude and phase of synchronous activities in the same frequency range or across different frequencies (Engel, Fries, & Singer, 2001; Kopell et al., 2000; Saalmann, Pigarev, & Vidyasagar, 2007; Siegel, Donner, & Engel, 2012). These descending signals spread to sensory systems from parietal and prefrontal cortical areas (Moore, 2006), and the coherence between them is modulated by attention (Siegel et al., 2008). Note that the flow of activity in these corticocortical circuits may exhibit directionality as shown for beta 2 rhythms between association and primary auditory cortices (Roopun et al., 2010) and between the primary visual cortex and dLGN for the whole beta range (figure 24.1) (Bekisz & Wróbel, 1993; Wróbel, Bekisz, & Waleszczyk, 1994). The reverberatory circuit dynamics in the long- and short-range circuits might also generate location-specific, self-sustained persistent activity (Siegel et al., 2008; Siegel, Donner, & Engel, 2012; Wang, 2010). In general it is widely accepted that enhancement of gamma oscillations often accompanies bottom-up attentional control of stimulus features (Fries et al., 2001; Lima, Singer, & Neuenschwander, 2011; but compare Chalk et al., 2010) and that beta-band activity is involved in top-down attentional signaling (Buschman & Miller, 2007; Gross et al., 2004; see Wang, 2010, for review).

How enhanced synchrony may support selective attention has also been investigated in several modeling studies. A pronounced attention-specific rate enhancement of neuronal activity based on gamma synchronization between sensory and executive cortical areas was reproduced by Ardid, Wang, and Compte (2010). Buia and Tiesinga (2008) postulated that feedforward and recurrent cortical interneurons in area V4 are responsible for setting gamma and beta rhythms according to attentional demands. Grossberg and Versace (2008) investigated a thalamocortical model in which gamma synchronization between various brain regions supported visual attention and learning, whereas a mismatch in synchronization led to inhibition of learning and instead initiated beta oscillations in the deep cortical layers. Similarly, the large multineuronal model of the mammalian thalamocortical sensory system spontaneously exhibited interregional beta oscillation, preferably in basket cells of layer 5, which in turn generated strong local gamma rhythms (Izhikevich & Edelman, 2008). Thus, these modeling results provide additional support for the idea that long-range beta and gamma oscillations might be used in the process of distributing attentional signals in sensory systems, and the local gamma oscillations would be utilized for columnar processing.

Modulation by Feedback Pathways

The idea that descending feedback pathways are involved in attention processing is not new (Adrian, 1953; Hernandez-Peon, 1966). Because of the complicated organization of the interlaminar and corticocortical connections (Douglas & Martin, 2004; Ferster & Lindström, 1983), this was first investigated in cortico-thalamic pathways of the visual system (Lindström & Wróbel, 1990; Wróbel et al., 2007). It is now widely agreed that the descending projections from layer 6 of the visual cortex play a modulatory role in the transmission of retinal information through the dLGN (Sherman, 2005). One putative function of this feedback loop is optimization of the selective segmentation of visual features (Sillito & Jones, 2003). A second possibility is that the loop provides a gain regulatory mechanism (Lindström & Wróbel, 1990; Livingstone & Hubel, 1981; Saalmann & Kastner, 2009; Waleszczyk, Bekisz, & Wróbel, 2005) that could be utilized during attentive visual processing (Bekisz & Wróbel, 1993; Wróbel, 2000).

It was originally proposed that the corticogeniculate projection from layer 6 might form a positive neuronal amplifier and, together with its perigeniculate (PGN) recurrent inhibitory counterpart, be used in attention control (figure 24.3A) (Ahlsén, Lindström, & Lo, 1985). Using intracellular recording techniques, Lindström and Wróbel (1990) showed that corticogeniculate synapses have a built-in frequency potentiation mechanism that reaches optimal (i.e., approximately three times enhancement of EPSPs and below an epileptic threshold at about 50 Hz) values at the beta frequency (figure 24.3B). Based on this observation it was proposed that beta-frequency activity transmitted in bursts via the corticogeniculate modulatory pathway would depolarize geniculate cells for a hundred to several hundred milliseconds (figure 24.3B) and therefore increase the input–output gain of the geniculate relay (figure 24.3C) (Lindström & Wróbel, 1990; Wróbel, Bekisz, & Waleszczyk, 1994). Accordingly, it was recently found that the amplitude of potentials evoked by chiasm stimulation in the primary visual cortex of attending cats increased in synchrony with beta bursts, suggesting a greater number of neurons involved in the population spike due to the lowered threshold in depolarized dLGN cells (figure 24.3E, F) (A. Wróbel, M. Bekisz, W. Bogdan, A. Ghazaryan, unpublished results). Note that, in parallel, spatially specific inhibitory feedback is reduced as the cortically induced increase of geniculate activity was found to be accompanied by a decrease of firing of PGN recurrent inhibitory interneurons (see figure 7 of Wróbel, 2000; Funke & Eysel, 1998; McAlonan, Cavanaugh, & Wurtz, 2008; Waleszczyk, Bekisz, & Wróbel, 2005).

It is plausible to postulate that a similar gain regulatory mechanism may exist in all processing levels of the visual system, thereby enabling each higher center to enhance the activity of its predecessor. Such a mechanism certainly applies for the feedback collaterals of axons of layer 6 cells, which, despite dLGN targets, also

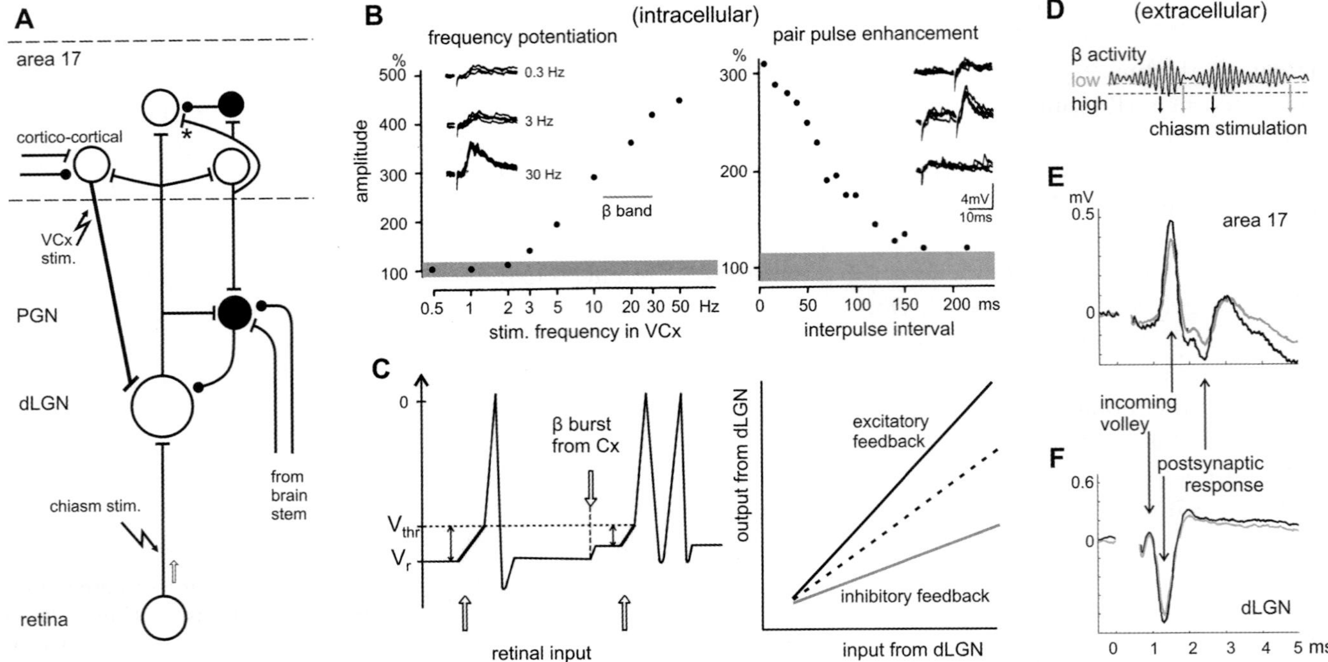

Figure 24.3

The mechanism of gain modulation by recurrent projections from layer 6 cortical cells and PGN (perigeniculate) interneurons in the cat dLGN. (A) Principal dLGN cells monosynaptically excite layer 6 corticogeniculate cells and in turn receive from them modulatory feedback projection (Ahlsén, Lindström, & Lo, 1985). They also send excitatory collaterals to cells in layer 4, the main cortical targets of dLGN fibers, and this synapse (*) also has a pronounced frequency sensitivity (Ferster & Lindström, 1985). Thus, layer 6 cells can influence visual input to the cortex by a combined positive feedback–feedforward action onto the main input neurons in layer 4 of the cortex. To maintain stability in such a system it is probably necessary to keep the resting gain low, which is controlled by the other branch of the recurrent pathway projecting to the PGN, which is also under brainstem control (the ascending modulation of principal dLGN cells is not shown). During visual attention, on the other hand, the gain in the dLGN should be increased in order to significantly influence input to the cortex. This is achieved by frequency enhancement at corticothalamic synapses. On the other hand the intracortical inhibitory loop secures similar balanced activity at layer 4 by a divisive gain modulation (Chance, Abbott, & Reyes, 2002). (B) Frequency enhancement and paired pulse enhancement of corticogeniculate EPSPs in a dLGN principal cell recorded in vivo from an anaesthetized animal (from Lindström & Wróbel, 1990). The EPSPs were evoked by cortical stimulation (as shown in A), and the recorded cell was hyperpolarized to the reversal level for the recurrent IPSP. The left plot shows the mean amplitude of maximally enhanced EPSPs at different frequencies expressed as a percentage of the mean amplitude at 1 Hz. The right plot shows the mean amplitude of the second EPSP in a paired pulse test, expressed as a percentage of the unconditioned test response for increasing interpulse interval. The hatched fields represent the mean test levels' SEM. Calibration bars refer to both sets of representative EPSPs. (C) The left plot shows the scheme of enhancement of principal dLGN cell activity in response to the same retinal input after depolarization evoked by beta burst activity at the corticogeniculate synapse. On the right is shown the input–output function for dLGN principal cells with and without the recurrent circuits operating. (D–F) Recordings from a cat during an anticipatory attention test. (D) Each chiasm stimulation (as shown in A) was triggered at a time when beta-filtered local field potential in the visual cortex reached the arbitrarily set low or high thresholds. (E, F) Evoked potentials were recorded after the lowest beta activity during the auditory task (gray lines) or the highest beta activity during the visual task (black lines). Note the increased postsynaptic wave of dLGN potential and increased amplitudes of the population spike and postsynaptic wave in cortical recordings. VCx, primary visual cortex; PGN, perigeniculate nucleus; dLGN, dorsal lateral geniculate nucleus; stim, electrical stimulation; (intracellular) recordings applies for B and C, and (extracellular) for D–F.

innervate input cortical neurons in layer 4 of the visual cortex (figure 24.1) (Ferster & Lindström, 1985). Interestingly, Briggs and Usrey (2009b) found recently that average gamma activity is higher among output cells of layer 6 as compared to the input neurons in layer 4, as measured in alert monkeys. This may be due to the gain regulatory mechanism utilizing prominent beta activity as observed in layer 6 in this experiment. The balanced excitatory and inhibitory barrage from layer 6 (marked by the asterisk in figure 24.3A) may result in a divisive gain modulation of layer 4 neuronal responses (Chance, Abbott, & Reyes, 2002) with a resulting decrease of their population firing rate (Olsen et al., 2012) and spatially specific increase of the stimulus processing stream (figure 24.3C).

Further research is needed to support generalization of the hypothesis that frequency potentiation in the beta frequency range serves as a gain enhancer in all modulatory feedback pathways (delineated by thin lines in figure 24.1) during processes of visual attention.

GENERAL SUMMARY INCLUDING THE ROLE OF CROSS-FREQUENCY COHERENCE IN COGNITION

In recent years the complex relations between different frequency bands and their functional meanings for corticothalamic systems have been intensively investigated. As outlined in this chapter many recent findings have provided a new framework for understanding many neural mechanisms supporting attentional processes and other cognitive functions (Benchenane, Tiesinga, & Battaglia, 2011; Fan et al., 2007; Grossberg & Versace, 2008; Kopell et al., 2010; Siegel et al., 2008; Steriade, 2006; Wilke, Mueller, & Leopold, 2009).

An increasing amount of data converge in favor of the hypothesis that a top-down covert attentional mechanism is supported by beta oscillation and synchronization (Buschman & Miller, 2007; Gross et al., 2004; Saalmann, Pigarev, & Vidyasagar, 2007). Beta-band oscillations are recorded predominantly in infragranular cortical layers, whereas gamma-band oscillations dominate in supragranular layers (Buffalo, Fries, & Desimone, 2004; Buhl, Tamas, & Fisahn, 1998; Maier et al., 2010; Roopun et al., 2006). Anatomical findings match the physiology: upper layers send rich afferent horizontal connections to higher visual areas (Binzegger, Douglas, & Martin, 2004); the latter mediate feedback projections originating in deep layers to the lower areas as well as to subcortical brain structures (Douglas & Martin, 2004). Thus, beta oscillations originating in deep layers may be especially involved in long-distance signaling along feedback, corticothalamocortical pathways (figure 24.1) (Sherman & Guillery, 2002; see also model by Grossberg & Versace, 2008), whereas gamma oscillations would support local and strong monosynaptic interareal cortical coherence (figure 24.1) (for reviews, see Donner & Siegel, 2011; von Stein & Sarnthein, 2000).

The hypothesis proposed in this chapter is based on the assumption that beta-frequency activity causes subthreshold depolarizations within the regions of visual system selected by attentional requirements. These depolarizations are produced by a frequency potentiation mechanism at the synaptic level of the feedback modulatory pathways and result in lower thresholds

within the required ensemble of neurons (Lindström & Wróbel, 1990; Wróbel, 2000; Wróbel et al., 2007; compare figures 24.1 and 24.3B, C). This would evoke an activation state (attention) allowing for synchronization of higher-frequency activities in local networks (Briggs & Usrey, 2009b; Buschman & Miller, 2010; Lakatos et al., 2008; Steriade, Amzica, & Contreras, 1996) and could be utilized during putative feature-binding processes (Bibbig, Traub, & Whittington, 2002; Eckhorn et al., 1988; Gray et al., 1989). In favor of this hypothesis recordings from the cat dLGN and visual cortex have shown that attention-related bursts of beta activity tend to correlate in time with gamma bursts (Bekisz & Wróbel, 1999).

REFERENCES

Adrian, E. D. (1953). The physiological basis of perception. In E. D. Adrian, F. Bremer, & H. H. Jasper (Eds.), *Brain mechanisms and consciousness* (pp. 237–248). Oxford: Blackwell.

Ahlsén, G., Lindström, S., & Lo, F. S. (1985). Interaction between inhibitory pathways to principal cells in the lateral geniculate nucleus of the cat. *Experimental Brain Research, 58*, 134–143.

Anderson, J., Lampl, I., Reichova, I., Carandini, M., & Ferster, D. (2000). Stimulus dependence of two-state fluctuations of membrane potential in cat visual cortex. *Nature Neuroscience, 3*, 617–621.

Ardid, S., Wang, X. J., & Compte, A. (2010). Reconciling coherent oscillation with modulation of irregular spiking activity in selective attention: Gamma-range synchronization between sensory and executive cortical areas. *Journal of Neuroscience, 30*, 2856–2870.

Arieli, A., Sterkin, A., Grinvald, A., & Aertsen, A. (1996). Dynamics of ongoing activity: Explanation of the large variability in evoked cortical responses. *Science, 273*, 1868–1871.

Baluch, F., & Itti, L. (2011). Mechanism of top-down attention. *Trends in Neurosciences, 34*, 210–224.

Basile, L. F. H., Anghinah, R., Ribeiro, P., Ramos, R. T., Piedade, R., Ballester, G., et al. (2007). Interindividual variability in EEG correlates of attention and limits of functional mapping. *International Journal of Psychophysiology, 65*, 238–251.

Bekisz, M., & Wróbel, A. (1993). 20 Hz rhythm of activity in visual system of perceiving cat. *Acta Neurobiologiae Experimentalis, 53*, 175–182.

Bekisz, M., & Wróbel, A. (1999). Coupling of beta and gamma activity in corticothalamic system of cats attending to visual stimuli. *Neuroreport, 10*, 3589–3594.

Bekisz, M., & Wróbel, A. (2003). Attention-dependent coupling between beta activities recorded in the cat's thalamic and cortical representations of the central visual field. *European Journal of Neuroscience, 17*, 421–426.

Benchenane, K., Tiesinga, P. H., & Battaglia, F. (2011). Oscillations in the prefrontal cortex: A gateway to memory and attention. *Current Opinion in Neurobiology, 21*, 475–485.

Berman, R. A., & Wurtz, R. H. (2010). Functional identification of a pulvinar path from superior colliculus to cortical area MT. *Journal of Neuroscience, 30*, 6342–6354.

Bernasconi, F., De Lucia, M., Tzovara, A., Manuel, A. L., Murray, M. M., & Spierer, L. (2011). Noise in brain activity engenders perception and influences discrimination sensitivity. *Journal of Neuroscience, 31,* 17971–17981.

Bibbig, A., Traub, R. D., & Whittington, M. A. (2002). Long-range synchronization of gamma and beta oscillations and the plasticity of excitatory and inhibitory synapses: A network model. *Journal of Neurophysiology, 88,* 1634–1654.

Binzegger, T., Douglas, R. J., & Martin, K. A. C. (2004). A quantitative map of the circuit of cat primary visual cortex. *Journal of Neuroscience, 24,* 8441–8453.

Bollimunta, A., Mo, J., Schroeder, C. E., & Ding, M. (2011). Neuronal mechanism and attentional modulation of corticothalamic alpha oscillations. *Journal of Neuroscience, 31,* 4935–4943.

Briggs, F., & Usrey, W. M. (2007). Cortical activity influences geniculocortical spike efficacy in the macaque monkey. *Frontiers in Integrative Neuroscience, 1*(3). doi:10.3389/neuro.07.003.2007.

Briggs, F., & Usrey, W. M. (2008). Emerging views of cortico-thalamic function. *Current Opinion in Neurobiology, 18,* 403–407.

Briggs, F., & Usrey, W. M. (2009a). Parallel processing in the corticogeniculate pathway of the macaque monkey. *Neuron, 62,* 135–146.

Briggs, F., & Usrey, W. M. (2009b). Modulation of gamma-band activity across local cortical circuits. *Frontiers in Integrative Neuroscience, 3*(15). Epub.

Buehlmann, A., & Deco, G. (2008). The neuronal basis of attention: Rate versus synchronization modulation. *Journal of Neuroscience, 28,* 7679–7686.

Buffalo, E.A., Fries, P., & Desimone R. (2004). Layer-specific attentional modulation in early visual areas. *Society for Neuroscience Abstracts, 30,* 717.6.

Buhl, E. H., Tamas, G., & Fisahn, A. (1998). Cholinergic activation and tonic excitation induce persistent gamma oscillations in mouse somatosensory cortex in vitro. *Journal of Physiology, 513,* 117–126.

Buia, C. I., & Tiesinga, P. H. (2008). Role of interneuron diversity in the cortical microcircuit for attention. *Journal of Neurophysiology, 99,* 2158–2182.

Buschman, T. J., & Miller, E. K. (2007). Top-down versus bottom-up control of attention in the prefrontal and posterior parietal cortices. *Science, 315,* 1860–1862.

Buschman, T. J., & Miller, E. K. (2010). Shifting the spotlight of attention: Evidence for discrete computations in cognition. *Frontiers in Human Neuroscience, 4,* 194–203. doi:10.3389/fnhum.2010.00194.

Casagrande, V. A. (1994). A third parallel visual pathway to primate area V1. *Trends in Neurosciences, 17,* 305–310.

Casagrande, V. A., & Xu, X. (2004). Parallel visual pathways: A comparative perspective. In L. Chalupa & J. S. Werner (Eds.), *The visual neurosciences* (pp. 494–506). Cambridge, MA: MIT Press.

Chalk, M., Herrero, J. L., Gieselmann, M. A., Delicato, L. S., Gotthardt, S., & Thiele, A. (2010). Attention reduces stimulus-driven gamma frequency oscillations and spike field coherence in V1. *Neuron, 66,* 114–125.

Chalupa, L. M. (1991). Visual function of the pulvinar. In A. G. Leventhal (Ed.), *Vision and visual dysfunction* (Vol. 4, pp. 141–159). Boca Raton, FL: CRC Press.

Chance, F. S., Abbott, L. F., & Reyes, A. D. (2002). Gain modulation from background synaptic input. *Neuron, 35,* 773–782.

Deco, G., & Thiele, A. (2009). Attention—oscillations and neuropharmacology. *European Journal of Neuroscience, 30,* 347–354.

Dehaene, S., & Changeux, J.-P. (2011). Experimental and theoretical approaches to conscious processing. *Neuron, 70,* 201–227.

Doesburg, S. M., Roggeveen, A. B., Kitajo, K., & Ward, L. M. (2008). Large-scale gamma-band phase synchronization and selective attention. *Cerebral Cortex, 18,* 386–396.

Donner, T. H., & Siegel, M. (2011). A framework for local cortical oscillation patterns. *Trends in Cognitive Sciences, 15,* 191–199.

Douglas, R. J., & Martin, K. A. C. (2004). Neuronal circuits of the neocortex. *Annual Review of Neuroscience, 27,* 419–451.

Eckhorn, R., Bauer, R., Jordan, W., Brosch, M., Kruse, W., Munk, M., et al. (1988). Coherent oscillations: A mechanism of feature linking in the visual cortex? *Biological Cybernetics, 60,* 121–130.

Engel, A. K., & Fries, P. (2010). Beta-band oscillations—signalling the status quo? *Current Opinion in Neurobiology, 20,* 156–165. doi:10.1016/j.conb.2010.02.015.

Engel, A., Fries, P., & Singer, W. (2001). Dynamic predictions: Oscillations and synchrony in top-down processing. *Nature Reviews. Neuroscience, 2,* 704–716.

Fan, J., Byrne, J., Worden, M. S., Guise, K. G., McCandliss, B. D., Fossella, J., et al. (2007). The relation of brain oscillations to attentional networks. *Journal of Neuroscience, 27,* 6197–6206.

Ferster, D., & Lindström, S. (1983). An intracellular analysis of geniculo-cortical connectivity in area 17 of the cat. *Journal of Physiology, 342,* 181–215.

Ferster, D., & Lindström, S. (1985). Augmenting responses evoked in area 17 of the cat by intracortical axon collaterals of cortico-geniculate cells. *Journal of Physiology, 367,* 217–232.

Fries, P. (2009). Neuronal gamma-band synchronization as a fundamental process in cortical computation. *Annual Review of Neuroscience, 32,* 209–224.

Fries, P., Reynolds, J., Rorie, A., & Desimone, R. (2001). Modulation of oscillatory neuronal synchronization by selective visual attention. *Science, 291,* 1560–1563.

Funke, K., & Eysel, U. T. (1998). Inverse correlation of firing patterns of single topographically matched perigeniculate neurons and cat dorsal lateral geniculate relay cells. *Visual Neuroscience, 15,* 711–729.

Garey, L. J., Dreher, B., & Robinson, S. R. (1991). The organization of visual thalamus. In B. Dreher & S. R. Robinson (Eds.), *Vision and visual dysfunction* (Vol. 3, pp. 176–234). Houndmills, UK: Macmillan.

Gola, M., Kamiński, J., Brzezicka, A., & Wróbel, A. (2011). Beta band oscillations as a correlate of alertness—changes in aging. *International Journal of Psychophysiology, 85,* 62-67.

Gray, C. M., König, P., Engel, A. K., & Singer, W. (1989). Oscillatory responses in cat visual cortex exhibit inter-columnar synchronization which reflects global stimulus properties. *Nature, 338,* 334–337.

Gregoriou, G. G., Gotts, S. J., Zhou, H., & Desimone, R. (2009). High-frequency, long-range coupling between prefrontal and visual cortex during attention. *Science, 324,* 1207–1210.

Gross, J., Schmitz, F., Schnitzler, I., Kessler, K., Shapiro, K., Hommel, B., et al. (2004). Modulation of long-range neural synchrony reflects temporal limitations of visual attention in humans. *Proceedings of the National Academy of Sciences of the United States of America, 101*, 13050–13055.

Grossberg, S., & Versace, M. (2008). Spikes, synchrony, and attentive learning by laminar thalamocortical circuits. *Brain Research, 1218*, 278–312.

Guillery, R. W. (1995). Anatomical evidence concerning the role of the thalamus in corticocortical communication: A brief review. *Journal of Anatomy, 187*, 583–592.

Hanslmayer, S., Aslan, A., Staudigl, T., Klimesch, W., Herrmann, C. S., & Bauml, K. (2007). Prestimulus oscillations predict visual perception performance between and within subjects. *NeuroImage, 37*, 1465–1473.

Harris, K. D., & Thiele, A. (2011). Cortical state and attention. *Nature Reviews. Neuroscience, 12*, 509–523.

Hernandez-Peon, R. (1966). Physiological mechanisms in attention. In R. W. Russel (Ed.), *Frontiers in physiological psychology* (pp. 121–147). New York: Academic Press.

Huppe-Gourgues, F., Bickford, M. E., Boire, D., Ptito, M., & Casanova, C. (2006). Distribution, morphology and synaptic targets of corticothalamic terminals in the cat lateral posterior-pulvinar complex that originate from the posteromedial lateral suprasylvian cortex. *Journal of Comparative Neurology, 497*, 847–863.

Izhikevich, E. M., & Edelman, G. M. (2008). Large-scale model of mammalian thalamocortical system. *Proceedings of the National Academy of Sciences of the United States of America, 105*, 3593–3598.

Jones, E. G. (2001). The thalamic matrix and thalamocortical synchrony. *Trends in Neurosciences, 24*, 595–601.

Jones, E. G. (2007). *The thalamus* (2nd ed.). New York: Cambridge University Press.

Kamiński, J., Brzezicka, A., Gola, M., & Wróbel, A. (2011). Beta band oscillations engagement in human alertness process. *International Journal of Psychophysiology, 85*(1), 125–128.

Kelly, S. P., Gomez-Ramirez, M., & Foxe, J. J. (2009). The strength of anticipatory spatial biasing predicts target discrimination at attended locations: A high-density EEG study. *European Journal of Neuroscience, 30*, 2224–2234.

Kiefer, W., Kruger, K., Strauss, G., & Berlucchi, G. (1989). Considerable deficits in the detection performance of the cat after lesion of the suprasylvian visual cortex. *Experimental Brain Research, 75*, 208–212.

Kopell, N., Ermentrout, G. B., Whittington, M. A., & Traub, R. D. (2000). Gamma rhythms and beta rhythms have different synchronization properties. *Proceedings of the National Academy of Sciences of the United States of America, 97*, 1867–1872.

Kopell, N., Kramer, M. A., Malerba, P., & Whittington, M. A. (2010). Are different rhythms good for different functions? *Frontiers in Human Neuroscience, 4*, 187. Epub.

Lakatos, P., Karmos, G., Mehta, A. D., Ulbert, I., & Schroeder, C. E. (2008). Entrainment of neuronal oscillation as a mechanism of attentional selection. *Science, 320*, 110–113.

Lima, B., Singer, W., & Neuenschwander, S. (2011). Gamma responses correlate with temporal expectation in monkey primary visual cortex. *Journal of Neuroscience, 31*, 15919–15931.

Lindström, S., & Wróbel, A. (1990). Frequency dependent corticofugal excitation of principal cells in the cat's dorsal lateral geniculate nucleus. *Experimental Brain Research, 79*, 313–318.

Livingstone, M. S., & Hubel, D. H. (1981). Effects of sleep and arousal on the processing of visual information in the cat. *Nature, 291*, 554–561.

Llinas, R., Ribary, U., Contreras, D., & Pedroarena, C. (1998). The neuronal basis for consciousness. *Philosophical Transactions of the Royal Society of London, Series B, 353*, 1841–1849.

Maier, A., Adams, G. K., Aura, C., & Leopold, D. A. (2010). Distinct superficial and deep laminar domains of activity in the visual cortex during rest and stimulation. *Frontiers in Systems Neuroscience, 4*, 31. doi:10.3389/fnsys.2010.00031.

McAlonan, K., Cavanaugh, J., & Wurtz, R. H. (2008). Guarding the gateway to cortex with attention in visual thalamus. *Nature, 456*, 391–395.

Moore, T. (2006). The neurobiology of visual attention: Finding sources. *Current Opinion in Neurobiology, 16*, 159–165.

Nassi, J. J., & Callaway, E. M. (2006). Multiple circuits relying primate parallel visual pathways to the middle temporal area. *Journal of Neuroscience, 26*, 12789–12798.

Noudoost, B., Chang, M. H., Steinmetz, N. A., & Moore, T. (2010). Top-down control of visual attention. *Current Opinion in Neurobiology, 20*, 183–190.

Olsen, S. R., Bortone, D. S., Adesnik, H., & Scanziani, M. (2012). Gain control by layer six in cortical circuits of vision. *Nature, 483*, 47–52.

Olshausen, B. A., Anderson, C. H., & Van Essen, D. C. (1993). A neurobiological model of visual attention and invariant pattern recognition based on dynamic routing of information. *Journal of Neuroscience, 13*, 4700–4719.

Ouellette, B. G., Minville, K., Faubert, J., & Casanova, C. (2004). Simple and complex visual motion response properties in the anterior medial bank of the lateral suprasylvian cortex. *Neuroscience, 123*, 231–245.

Payne, B. R., & Lomber, S. G. (2003). Quantitative analyses of principal and secondary compound parieto-occipital feedback pathways in cat. *Experimental Brain Research, 152*, 420–433.

Ray, S., Niebur, E., Hsiao, S., Sinai, A., & Crone, N. (2008). High-frequency gamma activity (80–150 Hz) is increased in human cortex during selective attention. *Clinical Neurophysiology, 119*, 116–133.

Reddy, L., Kanwisher, N. G., & VanRullen, R. (2009). Attention and biased competition in multi-voxel object representations. *Proceedings of the National Academy of Sciences of the United States of America, 106*, 21447–21452.

Rees, G. (2009). Visual attention: The thalamus at the centre? *Current Biology, 17*(5), R213–R214.

Robertson, L. C. (2003). Binding, spatial attention and perceptual awareness. *Nature Reviews. Neuroscience, 4*, 93–102.

Roopun, A. K., LeBeau, F. E. N., Rammell, J., Cunningham, M. O., Traub, R. D., & Whittington, M. A. (2010). Cholinergic neuromodulation controls directed temporal communication in neocortex in vitro. *Frontiers in Neural Circuits, 4*. doi:10.3389/fncir.2010.00008.

Roopun, A., Middleton, S., Cunningham, M., LeBeau, F., Bibbig, A., Whittington, M., et al. (2006). A beta2-frequency (20–30 Hz) oscillation in nonsynaptic networks of somatosensory cortex. *Proceedings of the National Academy of Sciences of the United States of America, 103*, 15646–15650.

Rouiller, E. M., & Welker, E. (2000). A comparative analysis of the morphology of corticothalamic projections in mammals. *Brain Research Bulletin, 53*, 727–741.

Rudolph, K. K., & Pasternak, T. (1996). Lesions in cat lateral suprasylvian cortex affect perception of complex motion. *Cerebral Cortex, 6*, 814–822.

Saalmann, Y. B., & Kastner, S. (2009). Gain control in the visual thalamus during perception and cognition. *Current Opinion in Neurobiology, 19*, 408–414.

Saalmann, Y. B., & Kastner, S. (2011). Cognitive and perceptual functions of the visual thalamus. *Neuron, 71*, 209–223.

Saalmann, Y., Pigarev, I., & Vidyasagar, T. (2007). Neural mechanisms of visual attention: How top-down feedback highlights relevant locations. *Science, 316*, 1612–1615.

Sauseng, P., Klimesch, W., Stadler, W., Schabus, M., Doppelmayr, M., Hanslmayr, S., et al. (2005). A shift of visual spatial attention is selectively associated with human EEG alpha activity. *European Journal of Neuroscience, 22*, 2917–2926.

Schroeder, C. E., & Lakatos, P. (2009). Low-frequency neuronal oscillations as instruments of sensory selection. *Trends in Neurosciences, 32*, 9–18.

Sherman, S. M. (2005). Thalamic relays and cortical functioning. *Progress in Brain Research, 149*, 107–126.

Sherman, S. M. (2009). Thalamic mechanisms in vision. In L. R. Squire (Ed.), *Encyclopedia of neuroscience* (Vol. 9, pp. 929–944). Oxford: Academic Press.

Sherman, S. M., & Guillery, R. W. (2002). The role of the thalamus in the flow of information to the cortex. *Philosophical Transactions of the Royal Society of London. Series B, Biological Sciences, 357*, 1695–1708.

Shipp, S. (2004). The brain circuitry of attention. *Trends in Cognitive Sciences, 8*, 223–230.

Siegel, M., Donner, T. H., & Engel, A. K. (2012). Spectral fingerprints of large-scale neuronal interactions. *Nature Reviews. Neuroscience, 13*, 121–134.

Siegel, M., Donner, T. H., Oostenveld, R., Fries, P., & Engel, A. K. (2008). Neuronal synchronization along the dorsal visual pathway reflects the focus of spatial attention. *Neuron, 60*, 709–719.

Sillito, A. M., & Jones, H. E. (2003). Feedback systems in visual processing. In L. M. Chalupa & J. S. Werner (Eds.), *The Visual Neurosciences* (Vol. 2, pp. 609–624). Cambridge, MA: MIT Press,

Snyder, A. C., & Foxe, J. J. (2010). Anticipatory attentional suppression of visual features indexed by oscillatory alphaband power increases: A high-density electrical mapping study. *Journal of Neuroscience, 30*, 4024–4032.

Sobolewski, A., Kublik, E., Świejkowski, D., Kamiński, J., & Wróbel, A. (2011b). Function of first and higher-order somato-sensory thalamic relays varies with level of arousal. *Acta Neurobiologiae Experimentalis, 71*(suppl.), 84.

Sobolewski, A., Kublik, E., Świejkowski, D. A., Łęski, S., Kamiński, J. K., & Wróbel, A. (2010). Cross-trial correlation analysis of evoked potentials reveals arousal related attenuation of thalamo-cortical coupling. *Journal of Computational Neuroscience, 29*, 485–493.

Sobolewski, A., Świejkowski, D. A., Wróbel, A., & Kublik, E. (2011a). The 5–12Hz oscillations in the barrel cortex of awake rats—sustained attention during behavioral idling? *Clinical Neurophysiology, 122*, 483–489.

Soma, S., Shimegi, S., Osaki, H., & Sato, H. (2012). Cholinergic modulation of response gain in the primary visual cortex of the macaque. *Journal of Neurophysiology, 107*, 283–291.

Steriade, M. (1997). Synchronized activities of coupled oscillators in the cerebral cortex and thalamus at different levels of vigilance. *Cerebral Cortex, 7*, 583–604.

Steriade, M. (2000). Corticothalamic resonance, states of vigilance and mentation. *Neuroscience, 101*, 243–276.

Steriade, M. (2006). Grouping of brain rhythms in corticothalamic systems. *Neuroscience, 137*, 1087–1106.

Steriade, M., Amzica, F., & Contreras, D. (1996). Synchronization of fast (30–40 Hz) spontaneous cortical rhythms during brain activation. *Journal of Neuroscience, 16*, 392–417.

Steriade, M., Dossi, R. C., Pare, D., & Oakson, G. (1991). Fast oscillations (20–40 Hz) in thalamocortical systems and their potentiation by mesopontine cholinergic nuclei in the cat. *Proceedings of the National Academy of Sciences of the United States of America, 88*, 4396–4400.

Symonds, L. L., Rosenquist, A. C., Edwards, S. B., & Palmer, L. A. (1981). Projections of the pulvinar-lateral posterior complex to visual cortical areas in the cat. *Neuroscience, 6*, 1995–2020.

Taylor, K., Mandon, S., Freiwald, W. A., & Kreiter, A. K. (2005). Coherent oscillatory activity in monkey area V4 predicts successful allocation of attention. *Cerebral Cortex, 15*, 1424–1437.

Theyel, B. B., Llano, D. A., & Sherman, M. (2010). The corticothalamocortical circuit drives higher-order cortex in the mouse. *Nature Neuroscience, 13*, 84–89.

Vanduffel, W., Tootell, R. B. H., & Orban, G. A. (2000). Attention-dependent suppression of metabolic activity in the early stages of the macaque visual system. *Cerebral Cortex, 10*, 109–126.

von Stein, A., & Sarnthein, J. (2000). Different frequencies for different scales of cortical integration: From local gamma to long range alpha/theta synchronization. *International Journal of Psychophysiology, 38*, 301–313.

Waleszczyk, W. J., Bekisz, M., & Wróbel, A. (2005). Cortical modulation of neuronal activity in the cat's lateral geniculate and perigeniculate nuclei. *Experimental Neurology, 196*, 54–72.

Waleszczyk, W. J., Wang, C., Benedek, G., Burke, W., & Dreher, B. (2004). Motion sensitivity in cat's superior colliculus: Contribution of different visual processing channels to response properties of collicular neurons. *Acta Neurobiologiae Experimentalis, 64*, 209–228.

Wang, X.-J. (2010). Neurophysiological and computational principles of cortical rhythms in cognition. *Physiological Reviews, 90*, 1195–1268.

Ward, L. M. (2011). The thalamic dynamic core theory of conscious experience. *Consciousness and Cognition, 20*, 464–486.

Ward, R., Danziger, S., Owen, V., & Rafal, R. (2002). Deficits in spatial coding and feature binding following damage to spatiotopic maps in the human pulvinar. *Nature Neuroscience, 5*, 99–100.

Wilke, M., Mueller, K.-M., & Leopold, D. A. (2009). Neural activity in the visual thalamus reflects perceptual suppression. *Proceedings of the National Academy of Sciences of the United States of America, 106*, 9465–9470.

Wilke, M., Turchi, J., Smith, K., Mishkin, M., & Leopold, D. A. (2010). Pulvinar inactivation disrupts selection of movement plans. *Journal of Neuroscience, 30*, 8650–8659.

Worden, M. S., Foxe, J. J., Wang, N., & Simpson, G. V. (2000). Anticipatory biasing of visuospatial attention indexed by retinotopically specific alpha-band electroencephalography increases over occipital cortex. *Journal of Neuroscience, 20,* RC63.

Wróbel, A. (2000). Beta activity: A carrier for visual attention. *Acta Neurobiologiae Experimentalis, 60,* 247–260.

Wróbel, A., Bekisz, M., Kublik, E., & Waleszczyk, W. (1994). 20 Hz bursting beta activity in the cortico-thalamic system of visually attending cats. *Acta Neurobiologiae Experimentalis, 54,* 95–107.

Wróbel, A., Bekisz, M., & Waleszczyk, W. (1994). 20 Hz bursts of activity in the cortico-thalamic pathway during attentive perception. In C. Pantev, T. H. Elbert, & B. Lutkenhoner (Eds.), *Oscillatory event related brain dynamics. NATO A/Life Sciences, 271* (pp. 311–324). London: Plenum Press.

Wróbel, A., Ghazaryan, A., Bekisz, M., Bogdan, W., & Kamiński, J. (2007). Two streams of attention dependent beta activity in the striate recipient zone of cat's lateral posterior-pulvinar complex. *Journal of Neuroscience, 27,* 2230–2240.

Wróbel, A., Hedström, A., & Lindström, S. (1998). Synaptic excitation of principal cells in the cat's lateral geniculate nucleus during focal epileptic seizures in the visual cortex. *Acta Neurobiologiae Experimentalis, 58,* 271–276.

Wróbel, A., & Kublik, E. (2000). Modification of evoked potentials in the rat's barrel cortex induced by conditioning stimuli. In M. Kossut (Ed.), *The barrel cortex* (pp. 229–239). Johnson City, TN: Graham Publishing.

Wurtz, R. H., McAlonan, K., Cavanaugh, J., & Berman, R. A. (2011). Thalamic pathway for active vision. *Trends in Cognitive Sciences, 15,* 177–184.

IV PROCESSING IN PRIMARY VISUAL CORTEX

25 Cell Types and Local Circuits in Primary Visual Cortex of the Macaque Monkey

EDWARD M. CALLAWAY

A major aim of vision research is to understand how neural circuits mediate the computations that underlie visual perception. Because the visual response properties of cortical neurons arise largely as a consequence of the organization and function of their connections with other neurons, it is believed that understanding how neural circuits give rise to the specific visual response properties of individual cortical neurons contributes significantly to this aim.

Local circuits in the primary visual cortex (V1) of primates provide a unique venue for elucidating cortical computations. This is due largely to the highly specialized organization and function of primate V1. Parallel inputs from multiple functionally specialized pathways impinge on V1 (Nassi & Callaway, 2009; see also chapter 16 by Kaplan), and the local circuits here must simultaneously extract the features from each input pathway that are relevant to the computations required for the creation of output neurons with novel receptive field properties and maintain an ordered substrate for efficiently interacting with distant cortical and subcortical neurons. The result is a highly organized structure of multiple circuits that are intertwined at multiple levels of organization. Nevertheless, these circuits remain embedded within a framework that V1 has in common with other cortical areas.

At present our understanding of relationships between circuits and function in the visual system is strongest at the levels of areas and modules (layers or columns within an area). There is detailed information about the functional properties and interconnections of dozens of cortical areas (e.g., Felleman & Van Essen, 1991), and for some of these areas (particularly V1) we know that different layers or columns receive input from different sources and that the neurons in these locations have corresponding differences in their responses to visual stimuli (see Callaway, 1998, and below for review).

But it is apparent that much of the computation that is carried out by visual cortical circuits will not be understood until we understand circuits and function at a finer level of resolution than cortical modules. This is most apparent from the simple observation that any given cortical module contains many functionally and anatomically distinct neuron types. In particular it appears that the number of anatomically distinct cell types in primate V1 may far exceed that in other cortical areas. And recent studies of functional connectivity have revealed that anatomically distinct neuron types within the same V1 module are uniquely interconnected (see below). It follows that the diversity of visual response properties within a module is likely to reflect the connectional differences between anatomically distinct neuron types. It is therefore possible to evaluate correlations between circuits and function at a finer level of resolution by taking advantage of the unique features of each cell type in V1.

The focus of this chapter is the organization of local circuits in V1 at the level of specific cell types. It is both the recognition and the exploitation, as well as continuing discovery, of the functional, structural, and connectional diversity of cell types that allow circuits to be linked to function at increasingly fine levels. I begin with a brief review of the anatomy and function of parallel pathways from the retina and lateral geniculate nucleus (LGN) to V1. I then turn to the primary focus of the chapter: a detailed description of the cell types that compose V1's local circuits and the specific connectivity of each cell type. Despite extensive diversity of inhibitory cell types (Fishell & Rudy, 2011), this review focuses on excitatory neurons with an emphasis on how V1 circuits mediate interactions between parallel inputs to V1 and the projection neurons that provide output to various extrastriate cortical areas.

THE PATHWAYS FROM LGN TO PRIMARY VISUAL CORTEX

The retinogeniculocortical pathways in the macaque monkey are composed of at least three distinct parallel systems, the M, P, and K pathways, which originate in the retina and remain segregated until they reach V1 (Nassi & Callaway, 2009; see also chapter 16 by Kaplan). Each of these systems seems to be specialized for the detection of a unique set of visual cues. At the other end of the visual hierarchy, there are dozens of extrastriate cortical areas, each of which is also specialized for the detection of particular attributes of visual scenes (cf., Desimone & Ungerleider, 1989; Felleman & Van Essen, 1991). It is believed that the extrastriate visual areas also comprise two more or less parallel systems. One system is specialized for analysis of spatial relationships and involves visual areas in parietal cortex. The other is believed to be responsible for object identification and involves areas in temporal cortex. Because virtually all of the retinal input relayed through the LGN to primate cortical areas converges on area V1 (Benevento & Standage, 1982; Bullier & Kennedy, 1983), V1 provides a unique interface between the M, P, and K pathways and extrastriate cortex. It also provides a unique opportunity for computations requiring early integration of information from more than one stream.

Our understanding of circuits underlying the contributions of the M, P, and K pathways to extrastriate cortical areas is based largely on studies of the flow of information into and out of various functional compartments in V1. These functional compartments include laminar (i.e., layers 4B, 4Cα, and 4Cβ) and columnar (i.e., blob, interblob) divisions. But there are also different cell types within each compartment, which are anatomically and/or connectionally distinct. Over the last several decades there has been considerable progress in relating the functional characteristics of parallel pathways with the functional architecture of V1 and in turn the functional differences between neurons providing output to extrastriate areas (see Callaway, 1998; Nassi & Callaway, 2009, and below for review). But there is still much to be learned about the functional and anatomical heterogeneities within each pathway and how these are related.

The locations in V1 that receive the combined input from the M, P, and K cells of the LGN correlate closely to the pattern revealed by staining for the mitochondrial enzyme cytochrome oxidase (CO) (Livingstone & Hubel, 1982). The densest LGN input is to layer 4C, but there are also inputs to layer 1, layer 2/3 blobs, layer 4A, and layer 6. Each of the M, P, and K pathways has a unique relationship to these afferent termination zones.

The M Pathway

Neurons in the most ventral M layers of the LGN project most strongly to the upper division of layer 4C, layer 4Cα, and secondarily to layer 6 (Hendrickson, Wilson, & Ogren, 1978; Hubel & Wiesel, 1972). Anatomical reconstructions of individual afferent axons show that they are heterogeneous. Most innervate the entire depth of layer 4Cα, but a subset selectively targets the upper portion of 4Cα (Blasdel & Lund, 1983). These termination zones closely mimic the laminar organization of the dendritic arbors of two types of layer 4Cα spiny stellate neurons (Yabuta & Callaway, 1998b; see also below). Cells with dendrites in upper 4Cα, where they would contact both types of M afferents, project axons to layer 4B and selectively to CO blobs in layer 3. Cells with narrowly stratified dendrites in lower 4Cα, where they would contact only one type of M afferent, project axons densely and selectively to CO interblobs in layer 3. These observations suggest that layer 4B and blobs are influenced by both types of M input, whereas interblobs are influenced exclusively by the M input that spans the entire depth of layer 4Cα.

M cells could potentially vary from each other with respect to numerous functional attributes (see below for functional properties of each pathway), and these variations might be systematically related to the anatomically defined termination zones of M afferents (upper versus lower layer 4Cα). We have hypothesized that the functional differences between M afferents that terminate in upper 4Cα and those that terminate throughout 4Cα might be related to their linearity (Yabuta & Callaway, 1998b). Some cells in the M layers of the LGN respond to high spatial frequency counterphase gratings with frequency-doubled responses, irrespective of spatial phase (Kaplan & Shapley, 1982). In this respect they are similar to Y cells, which in the cat's visual system arborize preferentially in upper layer 4 of area 17 (Humphrey et al., 1985). Most M cells in the monkey, however, respond more linearly and in this respect are similar to the cat's X cells, which arborize throughout the depth of layer 4.

Although there may be functional and anatomical similarities between cat X or Y cells and subsets of LGN M cells in the monkey, it is unclear whether these pathways are truly analogous. For example, the parasol cells of the primate retina, which give rise to the M pathway (Leventhal, Rodieck, & Dreher, 1981), are most similar to the cat's alpha retinal ganglion cells (RGCs) and not the beta RGCs, which give rise to the X pathway

(Leventhal, Rodieck, & Dreher, 1985). Furthermore, it is not clear whether there are discrete classes of linear and nonlinear M cells, or whether frequency-doubled cells simply reflect an extreme of nonlinearity (Levitt et al., 2001). On the other hand at least some parasol cells that project to the superior colliculus are nonlinear Y cells (Crook et al., 2008). Thus, differences in linearity are just one of many possible functional differences that may distinguish M afferents in upper versus lower 4Cα.

P and K Pathways

Neurons in the dorsal LGN layers include not only P cells but also K cells. K cells are primarily located in the intercalated zones between the P layers (and M layers) but are also found scattered within the P layers (Hendry & Yoshioka, 1994). This arrangement has made it difficult to ascertain which layers of V1 receive input from P versus K cells or which of the functionally distinct neuron types recorded in the dorsal LGN correspond to P or K cells. For example, early studies describing anterograde labeling following tracer injections (or lesion-induced degeneration) in the dorsal layers of the LGN reported label in superficial layers (layers 1, 2/3, and 4A) as well as layers 4Cβ and 6 (Hendrickson, Wilson, & Ogren, 1978; Hubel & Wiesel, 1972). The superficial labeling was originally believed to arise from the P pathway, but we now know that much of label in the superficial layers must have originated from K cells, not P cells (see below). Similarly, recordings from LGN neurons in and around the P layers were all attributed to the P pathway before it was appreciated that there is a distinct K pathway. And even with this realization, recordings made in the LGN cannot distinguish P cells from the K cells that are scattered within the P layers. More recent studies recording directly from LGN afferents within silenced V1 have directly revealed correlations between afferent terminations and functional properties (Chatterjee & Callaway, 2003; see further discussion below).

Despite uncertainty about the precise anatomy and physiology of the K pathway, the available evidence strongly suggests that the great majority of LGN neurons that project axons to layers 4Cβ or 4A are P cells and that only K cells expressing αCAM kinase II or calbindin (Hendry & Yoshioka, 1994) project to more superficial layers (layer 1 and layer 2/3 blobs). There are four main pieces of evidence that support this configuration. (1) Reconstructions of individual LGN afferents fail to reveal any axons that project densely both to layer 4Cβ and to superficial layers (Blasdel & Lund, 1983; Freund et al., 1989); these inputs therefore appear to arise from anatomically distinct populations. (2) The K pathway originates from RGCs with the smallest-diameter axons, and these axons innervate the intercalated K layers (Conley & Fitzpatrick, 1989). (3) LGN neurons retrogradely labeled from superficial layers above layer 4A of V1 include cells in the intercalated K layers along with scattered cells in the P layers, and these cells stain for αCAM kinase or calbindin (Hendry & Yoshioka, 1994), whereas tracer injections involving layer 4A also include αCAM kinase–negative cells. (4) A correlation between the functional properties of LGN afferents terminating in layer 4A, which have blue-OFF receptive fields (Chatterjee & Callaway, 2003), and anatomical and physiological studies of the retina indicating that at least some blue-OFF cells are midget ganglion cells (the provenance of the P pathway) (Field et al., 2010; Klug et al., 2003) argues that input to layer 4A arises from the P pathway.

Functional Properties of M, P, and K cells

Relative to P or K cells, M cells have excellent contrast sensitivity, prefer lower spatial frequencies and higher temporal frequencies, and lack cone opponent receptive fields (Nassi & Callaway, 2009). The most distinctive property of cells recorded in the dorsal layers of the LGN (P and K layers) is that, unlike M cells, they have cone-opponent receptive fields (e.g., Wiesel & Hubel, 1966). These cells are tuned along the cardinal red-green and blue-yellow color axes (Derrington, Krauskopf, & Lennie, 1984; De Valois et al., 2000). Thus, they either receive input from long-wavelength–sensitive (L) cones opposed to middle-wavelength–sensitive (M) cones ("red-green" color opponency), or they have short-wavelength–sensitive (S) cone input opposed by the L+M cones ("blue-yellow" color opponency).

Recent studies directly correlating the functional properties of LGN neurons with their location of termination within V1 (Chatterjee & Callaway, 2003) provide considerable insight into the relationships between color opponency and the M, P, and K pathways. These studies have demonstrated that LGN afferents terminating in superficial layers (2/3 and 4A) are exclusively blue-yellow color opponent, whereas those terminating in layer 4Cβ are exclusively red-green color opponent, and input to layer 4Cα is not color opponent. These observations confirmed the expectation that the M pathway carries achromatic information to layer 4Cα as well as the expectation that at least some of the input to layer 4Cβ should be red-green opponent. But the lack of red-green opponency in the superficial layers strongly argues that the K pathway, which projects only to more superficial layers, does not include red-green–opponent neurons. The unexpected finding that blue-ON and blue-OFF inputs were segregated, projecting to layers 3 and 4A, respectively, indicated that they come from separate populations.

These functional and anatomical relationships can be more clearly linked to M, P, and K pathways by considering the functional properties of different types of retinal ganglion cells (RGCs) that give rise to each pathway. By definition the M pathway originates from parasol RGCs, the P pathway originates from midget RGCs, and the K pathway originates from a large diversity of other retinal ganglion cell types with small-diameter axons (Dacey, 2000). Parasol cells are achromatic and have properties similar to M cells in the LGN and their afferents that terminate in layer $4C\alpha$ (see above). Most midget RGCs have red-green–opponent receptive fields, as do the majority of cells in P layers of the LGN and the red-green afferents that terminate in layer $4C\beta$ of V1 (see above). Blue-ON RGCs include both large and small bistratified cells (Dacey, 2000). It can therefore be inferred that these form a substantial part of the K pathway and contribute to the blue-ON afferents that terminate in layer 3 blobs. But there is also evidence for blue-ON input to layer 1 (Blasdel & Lund, 1983), suggesting that small and large bistratified cells might create distinct pathways that contribute inputs to layers 3 and 1, respectively. Functional and anatomical evidence indicates that blue cones also connect to a subset of OFF midget RGCs (but not ON midgets) (Field et al., 2010; Klug et al., 2003), indicating that many if not all blue-OFF RGCs are midgets and contribute to the P pathway. Because blue-OFF LGN afferents terminate in layer 4A of V1 (Chatterjee & Callaway, 2003), and retrograde tracer injections involving layer 4A label αCAM kinase–negative P cells in the LGN (Hendry & Yoshioka, 1994), this converging evidence strongly suggests that the blue-OFF input to layer 4A arises from blue-OFF LGN P cells.

Projections to Extrastriate Cortical Areas

The next section of this chapter describes the cell types that receive direct input from the M, P, and K pathways and how they distribute this information to V1's extrastriate cortical projection neurons. In this context it is therefore also helpful to briefly review the organization of V1 neurons that project to various extrastriate cortical areas and also to functionally distinct compartments in area V2.

As a general rule, neurons that project to dorsal visual areas are found in layer 4B, and those that project to ventral visual areas are found in layer 2/3 (Felleman & Van Essen, 1991; chapter 17 by Bell, Pasternak, and Ungerleider, for review). For example, layer 4B neurons provide direct input to areas V3 and MT, and layer 2/3 neurons connect to area V4. But detailed connectivity between areas V1 and V2 has only more recently been studied in the macaque monkey (Nassi & Callaway, 2007; Sincich & Horton, 2002, 2005; Sincich, Jocson, & Horton, 2007). Although it was previously appreciated that neurons throughout the cortical depth, from layers 2 to 4B, contain neurons connecting to area V2, their relationships to CO stripe compartments in V2 and to blobs and interblobs in V1 had not been investigated. Because each of the CO stripe compartments in V2 connects to a unique set of cortical areas, this is particularly relevant to the question of how each type of V1 cortical projection neuron distributes information to dorsal and ventral visual areas. In particular, the V2 thick stripes connect to areas V3 and MT, whereas thin stripes and interstripes connect to V4 (for review, see DeYoe & Van Essen, 1988; Zeki & Shipp, 1988).

Input to the CO compartments in macaque V2 is most closely related to the CO organization in V1, not to V1's laminar organization. Layers 2 to 4B neurons in (and under) interblob regions of V1 connect to V2 thick stripes and interstripes, whereas those in (and under) blob regions connect to V2 thin stripes (Sincich & Horton, 2002, 2005; Sincich, Jocson, & Horton, 2007). Thus, although it had long been thought that layer 4B provided the only input to V2 thick stripes, it is now clear that layer 2/3 and 4A interblob neurons also provide input to the thick stripes (Sincich & Horton, 2002). It had also been thought that V2 thin stripes receive input only from layer 2/3 blob neurons, but newer results indicate that the layer 4A and 4B cells under blobs also connect to thin stripes (Sincich & Horton, 2002, 2005; Sincich, Jocson, & Horton, 2007). Despite strong evidence that V1 blob and interblobs neurons have strong preferences for projecting to V2 thin stripes versus thick stripes and interstripes, respectively, there is also evidence for at least some mixing of these populations so that the location to which a V1 neuron projects can not necessarily be determined unambiguously from its position (Xiao & Felleman, 2004), perhaps more so in the marmoset monkey than in macaques (Federer et al., 2009). It is also noteworthy that the input from layer 4B to V2 arises from a separate population of neurons than the layer 4B neurons that project to MT (Sincich & Horton, 2003) and that the V2 projecting cells are mostly pyramidal neurons, whereas the direct input to MT from V1 arises predominantly from layer 4B spiny stellate neurons (Nassi & Callaway, 2007; Shipp & Zeki, 1989; see further below).

Excitatory Cell Types in V1, Their Local Connectivity, and Relationships to Extrastriate Cortical Areas

Intracellular labeling and anatomical reconstruction of the axonal and dendritic arbors of individual neurons throughout the layers of V1 of macaque monkeys (see Callaway, 1998, for review) have revealed at least 27 distinct excitatory neuron types (see details below). Photostimulation experiments have revealed the sources of local excitatory input to 16 of these cell types (Briggs & Callaway, 2001, 2005; Sawatari & Callaway, 1996, 2000; Yabuta, Sawatari, & Callaway, 2001). Systematic differences in the sources of local input to different cell types have revealed that each cell type receives input from unique sources, even when the locations of their cell bodies and/or dendritic arbors are the same (see also Dantzker & Callaway, 2000; Xu & Callaway, 2009). Thus, connectivity cannot be predicted based solely on light-level anatomy because axons that arborize in a particular location selectively connect to the dendrites of certain cell types within that location. In some cases the sources of functional excitatory input can be highly diverse, even for neurons of the same type (Sawatari & Callaway, 2000). Differences in functional input to each cell type imply that there should be corresponding differences in their visual function in vivo. Future studies will be necessary to reveal any such correlations.

Excitatory Cell Types and Connections from Layer 4C to Superficial Layers

As described above, the most dense LGN input to V1 is to layer 4C. Layer 4C neurons in turn connect directly to extrastriate projection neurons in more superficial layers. It appears that the most salient features of the classical receptive fields of neurons in superficial cortical layers can be attributed largely to the organization of these connections (Callaway, 1998). Therefore, understanding the detailed relationships between these circuits and the visual responses of the neurons that form them is expected to be particularly useful for understanding how V1 circuits function.

Within layer 4Cα three distinct types of spiny stellate neuron have been identified (Yabuta & Callaway, 1998b). The first type has dense axonal arbors only within layer 4Cα and layer 6—it does not arborize in layer 5 or in superficial layers. The second cell type has dense axonal arbors in layers 4Cα and 4B but also makes substantial connections with layers 3B, 5, and 6. The connections in layer 3B are located selectively in the CO blobs. The last type of layer 4Cα spiny stellate

is the most unusual. It has narrowly stratified dendrites at the bottom of layer 4Cα that neither extend into layer 4Cβ nor to upper layer 4Cα. Its M input is therefore limited to those afferents that terminate throughout layer 4Cα and not those that target only upper 4Cα (Blasdel & Lund, 1983; see also above for possible functional significance). These cells have very dense axonal arbors selectively in the CO interblobs of layer 3B. They have moderate axonal arbors within layer 4C. Together these three cell types receive information from LGN M afferents and distribute it to all three of the major compartments that contain extrastriate projection neurons—blobs, interblobs, and layer 4B (see above). But the narrowly stratified cells that provide input to interblobs are likely to receive only one type of M input, whereas the cells connecting to layer 4B and blobs receive both types of M input (see also above). Possible functional differences between these sources of LGN input remain unknown.

There are two distinct types of layer 4Cβ spiny stellate cells, although subtle differences in connections to layers 5 and 6 may also be diagnostic of additional types (Callaway & Wiser, 1996; Yabuta & Callaway, 1998b). Cells near the layer 4Cα/4Cβ border have dendritic arbors that span the border and are therefore likely to receive both M and P input. But most layer 4Cβ spiny stellates confine their dendrites to layer 4Cβ and therefore can receive LGN input only from P cells. Both of these layer 4Cβ spiny stellate cell types make extremely dense connections to layer 3B, and these connections lack specificity for blobs or interblobs.

These anatomical observations of layer 4C neurons suggest distinct differences in the sources of input to the extrastriate projection neurons in layers 3B and 4B. But more detailed photostimulation experiments reveal that the connections are even more precise. Neurons in layer 4B provide the output from V1 to areas in the dorsal visual pathway, including areas V2, V3, and MT (see above). Furthermore, there are differences in the cell types projecting to these areas. About one-third of layer 4B excitatory neurons have a spiny stellate morphology, and the remainder are pyramidal (Callaway & Wiser, 1996). But the less common spiny stellates provide the majority of the input to area MT (Nassi & Callaway, 2007), whereas the pyramids are more likely to connect to V2 and V3. Photostimulation studies (figure 25.1) (Sawatari & Callaway, 1996; Yabuta, Sawatari, & Callaway, 2001) reveal that both cell types receive their strongest local inputs from layer 4Cα, as expected from anatomical observations (see above). But pyramidal cells also receive a substantial excitatory input from layer 4Cβ (see figure 25.1), presumably via their apical dendrites in layer 3. This P-dominated input from layer

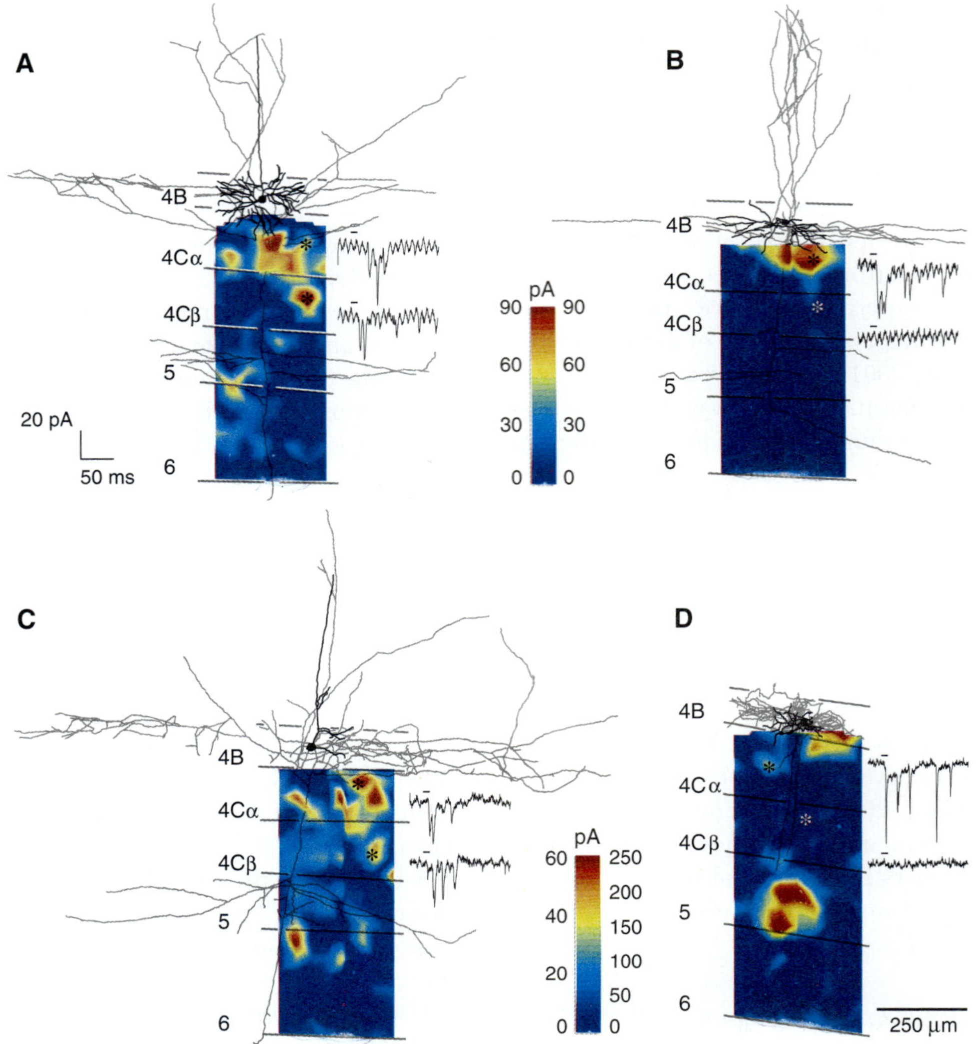

FIGURE 25.1   Laminar excitatory input to layer 4B pyramidal neurons (A, C), a spiny stellate neuron (B), and an inhibitory basket cell (D). The pyramidal cells receive strong excitatory input from both layers 4Cα and 4Cβ, whereas spiny stellate neurons receive input from layer 4Cα but not 4Cβ (see text for further details). The excitatory neurons (spiny stellate and pyramidal) get their strongest input from layer 4C, whereas this inhibitory neuron receives its strongest input from layer 5. These patterns of input to the excitatory neurons versus the inhibitory neuron are reminiscent of input to pyramidal neurons and a subtype of inhibitory basket cell in rat visual cortex (Dantzker & Callaway, 2000). Colored maps indicating patterns of excitatory input to each neuron are linear interpolations of the estimated evoked input (EEI) (see Yabuta et al., 2001) values measured following photostimulation at discrete sites. Colored vertical scale bars indicate the corresponding EEI values for input maps to the left and right. To the right of each input map are example voltage-clamp recordings made while stimulating presynaptic regions (indicated by asterisks) in layer 4Cα (upper traces) or layer 4Cβ (lower traces). Short dashes above each recording show the onset of photostimulation. Horizontal lines crossing the input maps represent the anatomical laminar borders. Anatomical reconstructions of dendritic (black) and axonal arbors (gray) are superimposed over the input maps. Scale bars apply to all panels. (From Yabuta et al., 2001.)

4Cβ is about half the strength of the M-dominated input from layer 4Cα. Layer 4B spiny stellates receive no detectable input from layer 4Cβ. Thus, the direct pathway from layer 4B of V1 to area MT is influenced far less by the P pathway than indirect pathways from layer 4B pyramids via areas V2 or V3 to MT. Further observations argue that there is extremely little if any P input to the layer 4B neurons projecting directly to MT.

In particular, although some layer 4B pyramids project to area MT, these have a distinctly different morphology from those that project to V2 (Nassi & Callaway, 2007), and transsynaptic tracing with rabies virus injected into area MT results in essentially no labeling in the P recipient layer 4Cβ (Nassi & Callaway, 2006). These observations imply that there are likely to be differences in the visual receptive field properties of layer 4B spiny stellate versus pyramidal neurons that reflect the differential contributions from the P pathway and that there are also likely to be correlations between receptive field properties and the area to which each neuron projects. The difficulty of relating receptive fields to corticocortical projections based on antidromic stimulation (Movshon & Newsome, 1996) might be overcome by modern imaging approaches that allow visual responses of projections neurons to be characterized (Jarosiewicz et al., 2012; Osakada et al., 2011; Sato & Svoboda, 2010).

There are also differences in the sources of local excitatory input to different pyramidal cell types in layer 3B. About one-third of the pyramidal neurons in layer 3B of V1 project an axon into the white matter (probably to area V2), but the remaining pyramids have axonal arbors that are completely local to V1 (Callaway & Wiser, 1996; Sawatari & Callaway, 2000). The projecting pyramids have apical dendrites with tufts in layer 1 or layer 2, whereas the local pyramids rarely have tufted apical dendrites (Sawatari & Callaway, 2000). Photostimulation studies revealed that these anatomical differences were closely correlated with sources of local excitatory input (Sawatari & Callaway, 2000). Most notably, functional excitatory connections from layer 4Cβ, which provide the most dense anatomical input to layer 3B, could never be detected onto projecting pyramids. In contrast, most local pyramids received strong excitatory input from layer 4Cβ. There were also systemic differences in the sources of functional excitatory input to layer 3B pyramids in blobs versus interblobs. These differences were consistent with the anatomical differences in local input to blobs and interblobs (see above). It is therefore expected that future in vivo recordings will reveal differences in the visual receptive fields of local versus projecting pyramids and that there may be further correlations with location relative to blobs and/or V2 stripe compartments to which the neurons project. Differences between local and projecting pyramids are likely to reflect the functional influence of direct input from layer 4Cβ. In addition, layer 2/3A also contains both local and projecting pyramids (Callaway & Wiser, 1996) that are also likely to receive different functional inputs and have different receptive fields. And anatomical observations indicate that the sources of input to these cells must differ from those in layer 3B (Callaway & Wiser, 1996; Lachica, Beck, & Casagrande, 1992; Nassi & Callaway, 2006; Yabuta & Callaway, 1998b). Finally, layer 2/3 pyramids can also be separated into two types based on their patterns of horizontal intralaminar axonal arbors (Yabuta & Callaway, 1998a). Although most pyramidal neurons have clustered intralaminar axons projecting for long distances, those located at distances of about 125 μm from the centers of blobs tend to lack long axons. Studies of the relationships between the locations of V1 neurons relative to blob centers and the V2 stripes to which they project (Federer et al., 2009) raise the possibility that these cell types might also project to different types of V2 stripes.

Excitatory Cell Types in Deep Layers and Their Connections

The contributions of V1 to visual processing depend on the patterns of activity that are generated in its output neurons. These include not only the extrastriate projection neurons found in superficial cortical layers (see above) but also subcortically projecting neurons found in layers 5 and 6 (Fitzpatrick et al., 1994; Hendrickson, Wilson, & Ogren, 1978; Lund et al., 1975). Interestingly, however, only a minority of layer 5 and layer 6 neurons in macaque V1 have axonal projections outside of V1 (Briggs & Callaway, 2001, 2005; Callaway & Wiser, 1996; Fitzpatrick et al., 1994; Wiser & Callaway, 1996). The majority make connections that are entirely intrinsic to V1. This suggests that the role of these neurons is to influence the patterns of activity that are generated in the output neurons. Other deep-layer neurons do project outside of V1, and their local axonal projections and input sources are often distinctly different from those that do not.

These observations and analyses of local axonal arbors lead to the hypothesis that there are two distinct classes of deep-layer neurons. The first class has local, recurrent axon collaterals that form extensive arbors in superficial cortical layers. The layer 5 pyramidal neurons of this class have axonal arbors that target layers 2–4B, whereas this class of layer 6 pyramidal neuron targets layer 4C (see figure 25.2). The second class of deep-layer pyramidal neurons has laterally projecting local axon collaterals that generally avoid superficial cortical layers. These same two cell types have also been distinguished on both anatomical and physiological grounds in other species and cortical areas (cf., Chagnac-Amitai, Luhmann, & Prince, 1990).

Because of our extensive knowledge of the functional organization of macaque V1, it has been possible to make inferences about the functional influences of the different types of deep-layer neurons (Callaway, 1998). Specifically, deep-layer neurons with local, recurrent axonal arbors appear to have a modulatory rather than driving influence on the visual response properties of neurons in superficial layers. Local recurrent pyramidal neurons in layers 5 and 6 have widespread recurrent axon collaterals in layers 2–4B and layer 4C, respectively. Despite these widespread projections, the

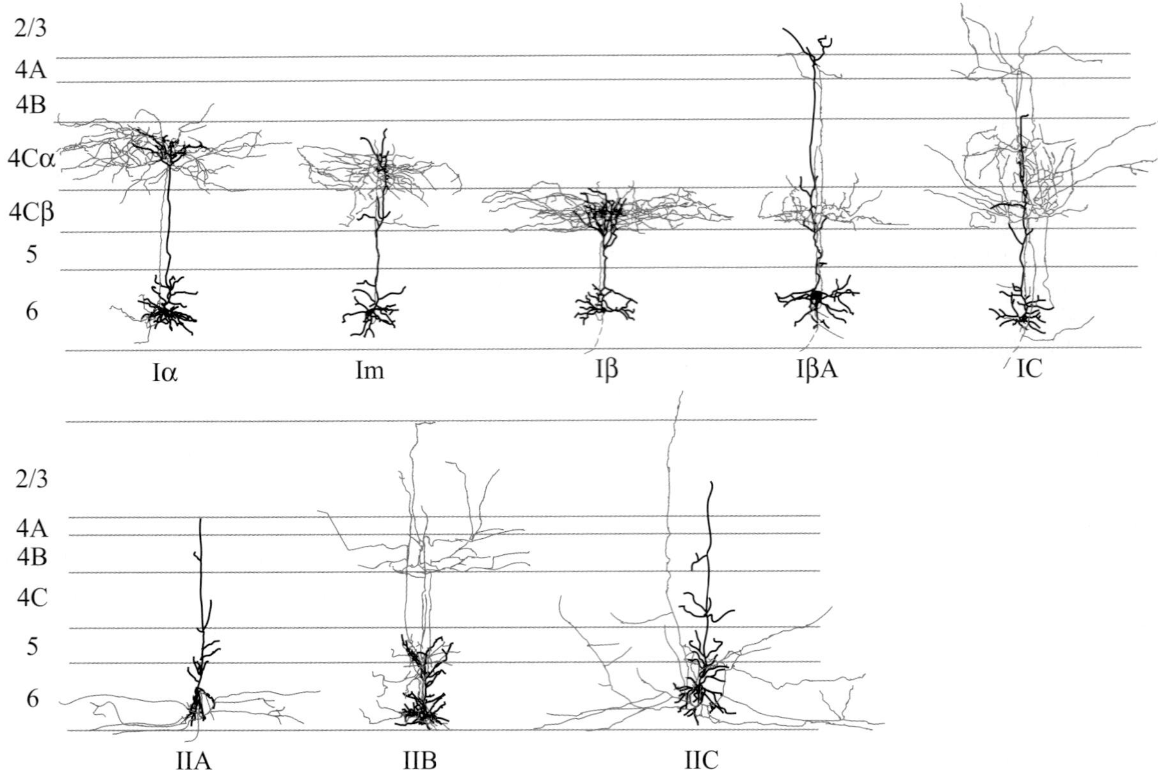

FIGURE 25.2   Eight types of pyramidal neuron identified in layer 6 of macaque V1. Class I neurons (top row) have dense axonal and apical dendritic arbors in layer 4C. Each type of class I neuron has a unique distribution of axons and dendrites within layer 4C and therefore a unique relationship to the magno- and parvocellular streams. Class II neurons (bottom row) have more extensive dendritic arbors in layer 5 and project axons primarily to either layers 2–4B (type IIB) or to deep layers (types IIA and IIC). See text for further details. (From Briggs & Callaway, 2001.)

receptive fields of neurons in layers 2–4C remain relatively small. Furthermore, the projections from deep layers lack specificity for functional compartments (i.e., ocular dominance columns, blobs, interblobs), yet the visual response properties of neurons in the recipient modules appear to be reflective of their inputs from the LGN and layer 4 (see above and Callaway, 1998). For example, layer 6 pyramidal neurons have widespread axonal arbors in layer 4C that are not ocular dominance column–specific (Wiser & Callaway, 1997). This indicates that, under monocular viewing conditions, layer 6 cells connecting to ocular dominance columns corresponding to the closed eye are active, but the recipient layer 4C neurons are not driven above threshold. The layer 6 input cannot by itself activate the layer 4C neurons. The connections are therefore modulatory. A similar argument can be made for the widespread projections from layer 5 to layers 2–4B. The same argument does not hold for the deep-layer neurons that have local, laterally projecting axonal arbors. Consistent with these interpretations, recent optogenetic studies have suggested that layer 6 neurons in the mouse primary visual cortex that project to layer 4 and to LGN have a

net inhibitory influence (Olsen et al., 2012). It remains to be determined if layer 5 neurons that project to superficial cortical layers might make similar functional contributions.

Beyond the local, recurrent versus local, lateral distinction that can be made between deep-layer neurons, they can be further classified into at least three types of layer 5 pyramidal neurons and nine types of layer 6 pyramidal neurons (figure 25.2). The different patterns of axonal and dendritic arborization, as well as differential connectivity of these cell types, suggest that they are likely to be involved uniquely in information processing within V1. Each cell type receives local input from a unique set of sources and provides output to a unique combination of target layers, implying that there are likely to also be corresponding differences in their visual receptive fields.

Neurons in the deep cortical layers, 5 and 6, are even more diverse than those in superficial layers (Briggs & Callaway, 2001, 2005; Callaway & Wiser, 1996; Wiser & Callaway, 1996). Four distinct types of layer 5 pyramidal neuron have been identified in macaque V1 (Briggs & Callaway, 2005; Callaway & Wiser, 1996; Nhan &

Callaway, 2012; Valverde, 1985). One type includes all of the local recurrent neurons found in this layer. They have extensive axonal arbors in superficial layers (2–4B) but not deep layers and do not project out of V1. A second type has a "backbranching" dendritic morphology, and its local axons project laterally within the bottom of layer 5. The backbranching cells project out of V1, perhaps exclusively to the pulvinar nucleus of the thalamus (Callaway & Wiser, 1996; Lund et al., 1975), because their morphology is inconsistent with the morphology of cells that project to the superior colliculus (Nhan & Callaway, 2012). Finally, there are the "tall" pyramids (Lund & Boothe, 1975; Valverde, 1985) and the nontufted Meynert cells, which recent evidence suggests are separate populations (Nhan & Callaway, 2012). The Meynert cells, as originally defined (le Gros Clark, 1942; Meynert, 1867) and more recently confirmed (Nhan & Callaway, 2012), have long (more than 500 mm) lateral basal dendrites and apical dendrites that extend to layer 1. Photostimulation-based investigation of the sources of local excitatory input to layer 5 local pyramids and backbranching pyramids revealed that both cell types receive extensive input across all cortical layers, although not all individual neurons received their input from the same sources (Briggs & Callaway, 2005). Instead, cluster analysis revealed four distinct patterns of input, suggesting that there should be similar diversity in the functional properties of these two types of layer 5 pyramids. Layer 5 tall-tufted and Meynert cells were not sampled.

Nine anatomically distinct types of layer 6 pyramidal neurons have been identified (figure 25.2) (Briggs & Callaway, 2001; Wiser & Callaway, 1996), and photostimulation reveals systematic differences in the sources of local excitatory input to each type (Briggs & Callaway, 2001). Layer 6 pyramids are grouped into two classes, I and II. Class I cells have apical dendritic branches and dense axonal arbors in layer 4C. These are divided into five types based on their specific patterns of local axonal arborization within subdivisions of layer 4. They project specifically to either layer 4Cα (type Iα), 4Cβ (type Iβ), the middle of layer 4C/lower 4Cα (type Im), 4Cβ plus 4A (type IβA), or throughout layer 4C plus 3B/4A (type IC). These primary distinguishing features are correlated with the position of the cell body within the depth of layer 6 and whether the cell projects an axon into the white matter. These features indicate that the class I cells are the source of feedback from V1 to the LGN (see Fitzpatrick et al. 1994; Wiser & Callaway, 1996). The feedback to P layers of the LGN comes from type Iβ, IβA, and perhaps IC cells. Feedback to M layers of the LGN comes from type IC cells. Type Iα and Im cells are found only in the middle of layer 6 and never project axons outside of V1. Thus, each type of class I cell has unique relationships to the M and P pathways, locally in V1, and with respect to input from and feedback to the LGN (see also Briggs & Callaway, 2001). Future studies of the differences in the functional properties of each of these cell types in vivo will provide considerable insight into the functions of these circuits.

Distinct anatomical features and photostimulation studies of local input indicate that class II layer 6 pyramids are likely to have very different functions than class I cells. Only one type of class II cell, type IIA, projects an axon to the white matter. The type IIA cells are found preferentially in the middle of layer 6 and have long, lateral local axons exclusively in deep layers of V1. They are therefore very similar to claustral projecting cells in the cat (Katz, 1987) and are hypothesized to also project to the visual claustrum of the monkey (Wiser & Callaway, 1996). They may therefore have very long receptive fields and play a role in the generation of end-stopped receptive fields (Bolz & Gilbert, 1986; Grieve & Sillito, 1995; Sherk & LeVay, 1983). Type IIB cells provide dense local feedback to superficial layers of V1, whereas type IIC cells are unusual in receiving diffuse input from all layers of V1 and extending sparse local axons throughout the cortical layers (Briggs & Callaway, 2001). Finally, the layer 6 pyramids also include giant cells of Meynert (le Gros Clark, 1942; Winfield, Rivera-Dominguez, & Powell, 1981). They project to area MT, and at least a subset likely projects to the superior colliculus (Fries, Keizer, & Kuypers, 1985; Nhan & Callaway, 2012), and they are likely to have direction-selective receptive fields (Movshon & Newsome, 1996). Area MT–projecting Meynert cells are also likely to receive direct input from both M and P layers of the LGN (Nassi, Lyon, & Callaway, 2006). The distinct inputs and outputs of each type of the other class II cells suggest that they are each likely to also have different visual receptive field properties in vivo.

SUMMARY AND CONCLUSIONS

Understanding how neural circuits give rise to visual perception requires the understanding of numerous details, most of which are necessary but probably not sufficient for solving the problem. It is clear that we must first understand how the neurons that compose the circuits are connected—what is the circuit? But we must also correlate circuits with function and perturb function in vivo to test hypotheses that emerge from observed correlations. As we understand the precision of cortical circuits at increasingly fine levels, it becomes clear that it will also be necessary to develop methods and implement experiments designed to observe correlations between circuitry and function and to perturb function at increasingly fine levels. Studies of the circuitry in V1 have shown that individual cell types are precisely interconnected. Thus, future methods and experiments should be designed to attack function at this same level of resolution. Recent advances in modern genetic and optogenetic methods for tracing circuits, correlating circuitry with function, and manipulating the activity of specific cell types are likely to lead to important advances in the near future (Callaway, 2005; Luo, Callaway, & Svoboda, 2008). But because of the relative difficulty in implementing these approaches in primates (Callaway, 2005; Diester et al., 2011; Han et al., 2009), progress in understanding circuits unique to primate visual processing is likely to lag behind that of circuits that are common between primates and mice. Therefore, in parallel to studies in primates, there is likely to be an increasing importance of studies using the mouse as a model system (see chapter 29 by Niell, Bonin, and Andermann).

In this review I have emphasized our understanding of V1 circuits at the level of individual cell types. In particular, it has become increasingly apparent that V1 of the macaque monkey contains dozens of anatomically distinct excitatory and inhibitory cell types. We now know that each of the distinct cell types whose sources of functional input have been investigated has proven to be connected differently from other cell types. This is the case even for neurons with similar cell body positions or similar laminar patterns of dendritic arborization. Our understanding of the connectivity in macaque V1 at this cell type–specific level of detail is so far confined to excitatory neurons. However, studies of the connectivity of inhibitory neurons in rat and mouse visual cortex (Dantzker & Callaway, 2000; Xu & Callaway, 2009) and the anatomical diversity of inhibitory neurons in macaque V1 (Lund, 1987; Lund, Hawken, & Parker, 1988; Lund & Wu, 1997; Lund & Yoshioka, 1991) strongly suggest that functional connectivity will also be strongly correlated with distinct anatomical features of inhibitory cell types in macaque V1. We are therefore likely to gain a great deal of understanding about the functional roles of inhibitory neurons and their interactions with excitatory neurons from future studies revealing the functional connectivity of each inhibitory cell type.

Regardless of the detail with which we understand the connectivity between distinct neuron types, we are unlikely to be able to understand the mechanisms by which these circuits contribute to visual perception without also correlating cell types with function. Future studies should take advantage of the strong

correlations between anatomical features and connectivity in V1 in order to correlate visual responses with circuits. For example, in vitro studies show that layer 4B spiny stellate cells and projecting pyramids in layer 3B receive strong functional input from layer 4Cα but not from 4Cβ, whereas layer 4B pyramids and layer 3B local pyramids receive considerable layer 4Cβ input. Thus, comparing the visual responses of these four anatomically distinct cell types would provide further insight into the contributions of layers 4Cα and 4Cβ to visual responses in more superficial layers. Furthermore, it is likely that in the near future it will be possible to directly correlate the visual responses of primate V1 neurons with the extrastriate areas and/or V2 stripe compartments to which they project using various combinations of retrograde tracing and calcium imaging that have proven successful in rodents and ferrets (Jarosiewicz et al., 2012; Osakada et al., 2011; Sato & Svoboda, 2010).

It should also be apparent that identifying correlations between circuits and function must be accompanied by perturbations that precisely test any hypotheses that emerge based on correlative studies. The incorporation of modern molecular and genetic methods into the experimental arsenal of primate neurophysiologists is likely to contribute importantly to this process. Although existing methods allow genetic perturbation of activity in primates (Diester et al., 2011; Han et al., 2009; Tan et al., 2006) using replication-incompetent viruses for delivery (Callaway, 2005), it has proven to be far more difficult to restrict gene expression to particular cell types in monkeys than in mice. Progress has been made in using viral vectors to target cells based on their unique projections and to express genes to control activity from these vectors (Lima et al., 2009; Osakada et al., 2011), but these methods have not yet been successfully implemented in primates. Furthermore, although a combination of viral tropism and cell type–specific promoters has allowed restriction of gene expression to excitatory neurons (Han et al., 2009; Nathanson et al., 2009b), the use of cell type–specific promoters in viral vectors to target other cortical cell types has proven difficult (Nathanson et al., 2009a).

REFERENCES

Benevento, L. A., & Standage, G. P. (1982). Demonstration of lack of dorsal lateral geniculate nucleus input to extrastriate areas MT and visual 2 in the macaque monkey. *Brain Research, 252,* 161–166.

Blasdel, G. G., & Lund, J. S. (1983). Termination of afferent axons in macaque striate cortex. *Journal of Neuroscience, 3,* 1389–1413.

Bolz, J., & Gilbert, C. D. (1986). Generation of end-inhibition in the visual cortex via interlaminar connections. *Nature, 320*, 362–365.

Briggs, F., & Callaway, E. M. (2001). Layer-specific input to distinct cell types in layer 6 of monkey primary visual cortex. *Journal of Neuroscience, 21*, 3600–3608.

Briggs, F., & Callaway, E. M. (2005). Laminar patterns of local excitatory input to layer 5 neurons in macaque primary visual cortex. *Cerebral Cortex, 15*, 479–488.

Bullier, J., & Kennedy, H. (1983). Projection of the lateral geniculate nucleus onto cortical area V2 in the macaque monkey. *Experiments in Brain Research, 53*, 168–172.

Callaway, E. M. (1998). Local circuits in primary visual cortex of the macaque monkey. *Annual Review of Neuroscience, 21*, 47–74.

Callaway, E. M. (2005). A molecular and genetic arsenal for systems neuroscience. *Trends in Neurosciences, 28*, 196–201.

Callaway, E. M., & Wiser, A. K. (1996). Contributions of individual layer 2–5 spiny neurons to local circuits in macaque primary visual cortex. *Visual Neuroscience, 13*, 907–922.

Chagnac-Amitai, Y., Luhmann, H. J., & Prince, D. A. (1990). Burst generating and regular spiking layer 5 pyramidal neurons of rat neocortex have different morphological features. *Journal of Comparative Neurology, 296*, 598–613.

Chatterjee, S., & Callaway, E. M. (2003). Parallel colour-opponent pathways to primary visual cortex. *Nature, 426*, 668–671.

Conley, M., & Fitzpatrick, D. (1989). Morphology of retinogeniculate axons in the macaque. *Visual Neuroscience, 2*, 287–296.

Crook, J. D., Peterson, B. B., Packer, O. S., Robinson, F. R., Troy, J. B., & Dacey, D. M. (2008). Y-cell receptive field and collicular projection of parasol ganglion cells in macaque monkey retina. *Journal of Neuroscience, 28*, 11277–11291.

Dacey, D. M. (2000). Parallel pathways for spectral coding in primate retina. *Annual Review of Neuroscience, 23*, 743–775.

Dantzker, J. L., & Callaway, E. M. (2000). Laminar sources of synaptic input to cortical inhibitory interneurons and pyramidal neurons. *Nature Neuroscience, 3*, 701–707.

Derrington, A. M., Krauskopf, J., & Lennie, P. (1984). Chromatic mechanisms in lateral geniculate nucleus of macaque. *Journal of Physiology, 357*, 241–265.

Desimone, R., & Ungerleider, L. (1989). Neural mechanisms of visual processing in monkeys. In F. Boller & J. Grafman (Eds.), *Handbook of neuropsychology* (Vol. 2, pp. 267–299). Amsterdam: Elsevier.

De Valois, R. L., Cottaris, N. P., Elfar, S. D., Mahon, L. E., & Wilson, J. A. (2000). Some transformations of color information from lateral geniculate nucleus to striate cortex. *Proceedings of the National Academy of Sciences of the United States of America, 97*, 4997–5002.

DeYoe, E. A., & Van Essen, D. C. (1988). Concurrent processing streams in monkey visual cortex. *Trends in Neurosciences, 11*, 219–226.

Diester, I., Kaufman, M. T., Mogri, M., Pashaie, R., Goo, W., Yizhar, O., et al. (2011). An optogenetic toolbox designed for primates. *Nature Neuroscience, 14*, 387–397.

Federer, F., Ichida, J. M., Jeffs, J., Schiessl, I., McLoughlin, N., & Angelucci, A. (2009). Four projection streams from primate V1 to the cytochrome oxidase stripes of V2. *Journal of Neuroscience, 29*, 15455–15471.

Felleman, D. J., & Van Essen, D. C. (1991). Distributed hierarchical processing in the primate cerebral cortex. *Cerebral Cortex, 1*, 1–47.

Field, G. D., Gauthier, J. L., Sher, A., Greschner, M., Machado, T. A., Jepson, L. H., et al. (2010). Functional connectivity in the retina at the resolution of photoreceptors. *Nature, 467*, 673–677.

Fishell, G., & Rudy, B. (2011). Mechanisms of inhibition within the telencephalon: "Where the wild things are." *Annual Review of Neuroscience, 34*, 535–567.

Fitzpatrick, D., Usrey, W. M., Schofield, B. R., & Einstein, G. (1994). The sublaminar organization of corticogeniculate neurons in layer 6 of macaque striate cortex. *Visual Neuroscience, 11*, 307–315.

Freund, T. F., Martin, K. A., Soltesz, I., Somogyi, P., & Whitteridge, D. (1989). Arborisation pattern and postsynaptic targets of physiologically identified thalamocortical afferents in striate cortex of the macaque monkey. *Journal of Comparative Neurology, 289*, 315–336.

Fries, W., Keizer, K., & Kuypers, H. G. (1985). Large layer VI cells in macaque striate cortex (Meynert cells) project to both superior colliculus and prestriate visual area V5. *Experimental Brain Research, 58*, 613–616.

Grieve, K. L., & Sillito, A. M. (1995). Differential properties of cells in the feline primary visual cortex providing the corticofugal feedback to the lateral geniculate nucleus and visual claustrum. *Journal of Neuroscience, 15*(Pt 1), 4868–4874.

Han, X., Qian, X., Bernstein, J. G., Zhou, H. H., Franzesi, G. T., Stern, P., et al. (2009). Millisecond-timescale optical control of neural dynamics in the nonhuman primate brain. *Neuron, 62*, 191–198.

Hendrickson, A. E., Wilson, J. R., & Ogren, M. P. (1978). The neuroanatomical organization of pathways between the dorsal lateral geniculate nucleus and visual cortex in Old World and New World primates. *Journal of Comparative Neurology, 182*, 123–136.

Hendry, S. H., & Yoshioka, T. (1994). A neurochemically distinct third channel in the macaque dorsal lateral geniculate nucleus. *Science, 264*, 575–577.

Hubel, D., & Wiesel, T. (1972). Laminar and columnar distribution of geniculo-cortical fibers in the macaque monkey. *Journal of Comparative Neurology, 146*, 421–450.

Humphrey, A. L., Sur, M., Uhlrich, D. J., & Sherman, S. M. (1985). Projection patterns of individual X- and Y-cell axons from the lateral geniculate nucleus to cortical area 17 in the cat. *Journal of Comparative Neurology, 233*, 159–189.

Jarosiewicz, B., Schummers, J., Malik, W. Q., Brown, E. N., & Sur, M. (2012). Functional biases in visual cortex neurons with identified projections to higher cortical targets. *Current Biology, 22*, 269–277.

Kaplan, E., & Shapley, R. M. (1982). X and Y cells in the lateral geniculate nucleus of macaque monkeys. *Journal of Physiology, 330*, 125–143.

Katz, L. C. (1987). Local circuitry of identified projection neurons in cat visual cortex brain slices. *Journal of Neuroscience, 7*, 1223–1249.

Klug, K., Herr, S., Ngo, I. T., Sterling, P., & Schein, S. (2003). Macaque retina contains an S-cone OFF midget pathway. *Journal of Neuroscience, 23*, 9881–9887.

Lachica, E. A., Beck, P. D., & Casagrande, V. A. (1992). Parallel pathways in macaque monkey striate cortex: Anatomically defined columns in layer III. *Proceedings of the National*

Academy of Sciences of the United States of America, 89, 3566–3570.

le Gros Clark, W. E. (1942). The cells of Meynert in the visual cortex of the monkey. *Journal of Anatomy, 76*, 369–376.

Leventhal, A. G., Rodieck, R. W., & Dreher, B. (1981). Retinal ganglion cell classes in the Old World monkey: Morphology and central projections. *Science, 213*, 1139–1142.

Leventhal, A. G., Rodieck, R. W., & Dreher, B. (1985). Central projections of cat retinal ganglion cells. *Journal of Comparative Neurology, 237*, 216–226.

Levitt, J. B., Schumer, R. A., Sherman, S. M., Spear, P. D., & Movshon, J. A. (2001). Visual response properties of neurons in the LGN of normally reared and visually deprived macaque monkeys. *Journal of Neurophysiology, 85*, 2111–2129.

Lima, S. Q., Hromadka, T., Znamenskiy, P., & Zador, A. M. (2009). PINP: a new method of tagging neuronal populations for identification during in vivo electrophysiological recording. *PLoS One, 4*(7), e6099.

Livingstone, M. S., & Hubel, D. H. (1982). Thalamic inputs to cytochrome oxidase-rich regions in monkey visual cortex. *Proceedings of the National Academy of Sciences of the United States of America, 79*, 6098–6101.

Lund, J. S. (1987). Local circuit neurons of macaque monkey striate cortex: I. Neurons of laminae 4C and 5A. [Research Support, U.S. Gov't, P.H.S.]. *Journal of Comparative Neurology, 257*, 60–92.

Lund, J. S., & Boothe, R. G. (1975). Interlaminar connections and pyramidal neuron organisation in the visual cortex, area 17, of the macaque monkey. *Journal of Comparative Neurology, 159*,305–334.

Lund, J. S., Hawken, M. J., & Parker, A. J. (1988). Local circuit neurons of macaque monkey striate cortex: II. Neurons of laminae 5B and 6. *Journal of Comparative Neurology, 276*, 1–29.

Lund, J. S., & Wu, C. Q. (1997). Local circuit neurons of macaque monkey striate cortex: IV. Neurons of laminae 1-3A. *Journal of Comparative Neurology, 384*, 109–126.

Lund, J. S., & Yoshioka, T. (1991). Local circuit neurons of macaque monkey striate cortex: III. Neurons of laminae 4B, 4A, and 3B. *Journal of Comparative Neurology, 311*, 234–258.

Lund, J. S., Lund, R. D., Hendrickson, A. E., Bunt, A. H., & Fuchs, A. F. (1975). The origin of efferent pathways from the primary visual cortex, area 17, of the macaque monkey as shown by retrograde transport of horseradish peroxidase. *Journal of Comparative Neurology, 164*, 287–303.

Luo, L., Callaway, E. M., & Svoboda, K. (2008). Genetic dissection of neural circuits. *Neuron, 57*, 634–660.

Meynert, T. (1867). Der Bau der Grosshirnrinde und seine ortlichen Verschiedenheiten nebsteinem pathologisch-anatomischen Corollarium. *Vjschr Psychiat, 1*, 198–217.

Movshon, J. A., & Newsome, W. T. (1996). Visual response properties of striate cortical neurons projecting to area MT in macaque monkeys. *Journal of Neuroscience, 16*, 7733–7741.

Nassi, J. J., & Callaway, E. M. (2006). Multiple circuits relaying primate parallel visual pathways to the middle temporal area. *Journal of Neuroscience, 26*, 12789–12798.

Nassi, J. J., & Callaway, E. M. (2007). Specialized circuits from primary visual cortex to V2 and area MT. *Neuron, 55*, 799–808.

Nassi, J. J., & Callaway, E. M. (2009). Parallel processing strategies of the primate visual system. *Nature Reviews. Neuroscience, 10*, 360–372.

Nassi, J. J., Lyon, D. C., & Callaway, E. M. (2006). The parvocellular LGN provides a robust disynaptic input to the visual motion area MT. *Neuron, 50*, 319–327.

Nathanson, J. L., Jappelli, R., Scheeff, E. D., Manning, G., Obata, K., Brenner, S., et al. (2009a). Short promoters in viral vectors drive selective expression in mammalian inhibitory neurons, but do not restrict activity to specific inhibitory cell-types. *Front Neural Circuits, 3*, 19.

Nathanson, J. L., Yanagawa, Y., Obata, K., & Callaway, E. M. (2009b). Preferential labeling of inhibitory and excitatory cortical neurons by endogenous tropism of adeno-associated virus and lentivirus vectors. *Neuroscience, 161*, 441–450.

Nhan, H. L., & Callaway, E. M. (2012). Morphology of superior colliculus- and middle temporal area-projecting neurons in primate primary visual cortex. *Journal of Comparative Neurology, 520*, 52–80.

Olsen, S. R., Bortone, D. S., Adesnik, H., & Scanziani, M. (2012). Gain control by layer six in cortical circuits of vision. *Nature, 483*, 47–52.

Osakada, F., Mori, T., Cetin, A. H., Marshel, J. H., Virgen, B., & Callaway, E. M. (2011). New rabies virus variants for monitoring and manipulating activity and gene expression in defined neural circuits. *Neuron, 71*, 617–631.

Sato, T. R., & Svoboda, K. (2010). The functional properties of barrel cortex neurons projecting to the primary motor cortex. *Journal of Neuroscience, 30*, 4256–4260.

Sawatari, A., & Callaway, E. M. (1996). Convergence of magno- and parvocellular pathways in layer 4B of macaque primary visual cortex. *Nature, 380*, 442–446.

Sawatari, A., & Callaway, E. M. (2000). Diversity and cell type specificity of local excitatory connections to neurons in layer 3B of monkey primary visual cortex. *Neuron, 25*, 459–471.

Sherk, H., & LeVay, S. (1983). Contribution of the corticoclaustral loop to receptive field properties in area 17 of the cat. *Journal of Neuroscience, 3*, 2121–2127.

Shipp, S., & Zeki, S. (1989). The organization of connections between areas V5 and V1 in macaque monkey visual cortex. *European Journal of Neuroscience, 1*, 309–332.

Sincich, L. C., & Horton, J. C. (2002). Divided by cytochrome oxidase: A map of the projections from V1 to V2 in macaques. *Science, 295*, 1734–1737.

Sincich, L. C., & Horton, J. C. (2003). Independent projection streams from macaque striate cortex to the second visual area and middle temporal area. *Journal of Neuroscience, 23*, 5684–5692.

Sincich, L. C., & Horton, J. C. (2005). Input to V2 thin stripes arises from V1 cytochrome oxidase patches. *Journal of Neuroscience, 25*, 10087–10093.

Sincich, L. C., Jocson, C. M., & Horton, J. C. (2007). Neurons in V1 patch columns project to V2 thin stripes. *Cerebral Cortex, 17*, 935–941.

Tan, E. M., Yamaguchi, Y., Horwitz, G. D., Gosgnach, S., Lein, E. S., Goulding, M., et al. (2006). Selective and quickly reversible inactivation of mammalian neurons in vivo using the *Drosophila* allatostatin receptor. *Neuron, 51*, 157–170.

Valverde, F. (1985). *The organizing principles of the primary visual cortex in the monkey* (Vol. 3). New York: Plenum Press.

Wiesel, T. N., & Hubel, D. H. (1966). Spatial and chromatic interactions in the lateral geniculate body of the rhesus monkey. *Journal of Neurophysiology, 29*, 1115–1156.

Winfield, D. A., Rivera-Dominguez, M., & Powell, T. P. (1981). The number and distribution of Meynert cells in area 17 of the macaque monkey. *Proceedings of the Royal Society of London, B Biological Science, 213*, 27–40.

Wiser, A. K., & Callaway, E. M. (1996). Contributions of individual layer 6 pyramidal neurons to local circuitry in macaque primary visual cortex. *Journal of Neuroscience, 16*, 2724–2739.

Wiser, A. K., & Callaway, E. M. (1997). Ocular dominance columns and local projections of layer 6 pyramidal neurons in macaque primary visual cortex. *Visual Neuroscience, 14*, 241–251.

Xiao, Y., & Felleman, D. J. (2004). Projections from primary visual cortex to cytochrome oxidase thin stripes and interstripes of macaque visual area 2. *Proceedings of the National Academy of Sciences of the United States of America, 101*, 7147–7151.

Xu, X., & Callaway, E. M. (2009). Laminar specificity of functional input to distinct types of inhibitory cortical neurons. *Journal of Neuroscience, 29*, 70–85.

Yabuta, N. H., & Callaway, E. M. (1998a). Cytochrome-oxidase blobs and intrinsic horizontal connections of layer 2/3 pyramidal neurons in primate V1. *Visual Neuroscience, 15*, 1007–1027.

Yabuta, N. H., & Callaway, E. M. (1998b). Functional streams and local connections of layer 4C neurons in primary visual cortex of the macaque monkey. *Journal of Neuroscience, 18*, 9489–9499.

Yabuta, N. H., Sawatari, A., & Callaway, E. M. (2001). Two functional channels from primary visual cortex to dorsal visual cortical areas. *Science, 292*, 297–300.

Zeki, S., & Shipp, S. (1988). The functional logic of cortical connections. *Nature, 335*, 311–317.

26 The Cortical Assembly of Visual Receptive Fields

SARI ANDONI, ANDREW TAN, AND NICHOLAS J. PRIEBE

A central goal of neuroscience is to understand how the sensory inputs we receive and our expectations about the world around us are integrated to generate appropriate behaviors. As sensory information moves through our nervous system, it undergoes systematic transformations in representation. A signature of this process is the evolution of the receptive field along the sensory pathways of the brain. The concept of the sensory neuron receptive field was originally developed in the visual system, where it has been defined as the intricate and specific visual stimulus space that is able to modulate responses of neurons in retina to thalamus to cortex (Hartline, 1938). Over the last half century the identification and study of the receptive field construct has been extended to all sensory pathways: in the auditory system it corresponds to the combination of sounds that modulate neural responses; in the somatosensory system it corresponds to the location and type of pressure felt across the body that modulate neural responses. Indeed, neuronal receptive fields—across all sensory modalities—represent how information about the world is transformed and processed as it moves from the site of initial neural transduction and into the cerebral cortex. Importantly, it has been our ability to characterize the receptive field structure that has allowed us to probe the neural computations that underlie cortical processing and, ultimately, behavior.

In the visual system, visual information is initially transduced by retinal photoreceptors, each of which has a punctate receptive field corresponding to the specific region of visual space that falls on the corresponding region of the retina. From the photoreceptor the incoming visual information undergoes a number of transformations within the series of neurons of the retina, ultimately progressing to the retinal ganglion cells, the retinal output neurons that have circularly symmetric receptive fields with an antagonistic center-surround organization (Kuffler, 1953). In mammals retinal ganglion cells principally project to the lateral geniculate nucleus (LGN) of the thalamus, which in turn relays visual information to the first cortical station: primary visual cortex (V1).

V1 is the site of dramatic transformations in the neural representation of the visual world. Both retinal ganglion cells and their target LGN relay cells are characterized by circularly symmetric receptive fields and respond to almost any stimulus presented within their receptive fields, whereas V1 neurons are fastidiously sensitive to several complex visual stimulus attributes, including the orientation, direction of motion, size, and binocular disparity of visual stimulus contours (Hubel & Wiesel, 1962). In many cases this selectivity is exquisite: V1 neurons may not respond to visual stimulation at all unless the stimulus features specifically match the neuron's preferences, as defined by its receptive field.

These profound changes in the character of the receptive field along the visual pathway have made V1 a model system for studying neuronal processing by the circuitry of the cerebral cortex. The cerebral cortex is characterized by its relative anatomical uniformity, and the massive expansion of the cerebral cortex is a hallmark of the human brain. The cerebral cortex is thought to be the neural substrate for our cognitive abilities, memories, and perceptual awareness, yet how the circuitry in the cerebral cortex allows us to recognize complex stimuli, to make decisions, and to generate intricate plans is unclear, and the methods to uncover the neural basis for those computations on a circuit level have not been developed. By comparison, the transformations performed by V1 on the inputs it receives from the LGN—the extraction of contour orientation and motion direction, for example—are complex enough to be interesting but simple enough to be tractable. The neuronal circuitry and the computations it makes to allow for the emergence of the intricate receptive fields in V1 neurons have therefore become the focus of intense scrutiny and debate, all the while advancing our understanding of the processing conducted by the cerebral cortex (Ferster & Miller, 2000; Sompolinsky & Shapley, 1997).

The emergence of orientation selectivity is perhaps the most remarkable transformation in V1. In contrast to afferent LGN relay cells, which have circular receptive fields, V1 neurons, including those receiving direct

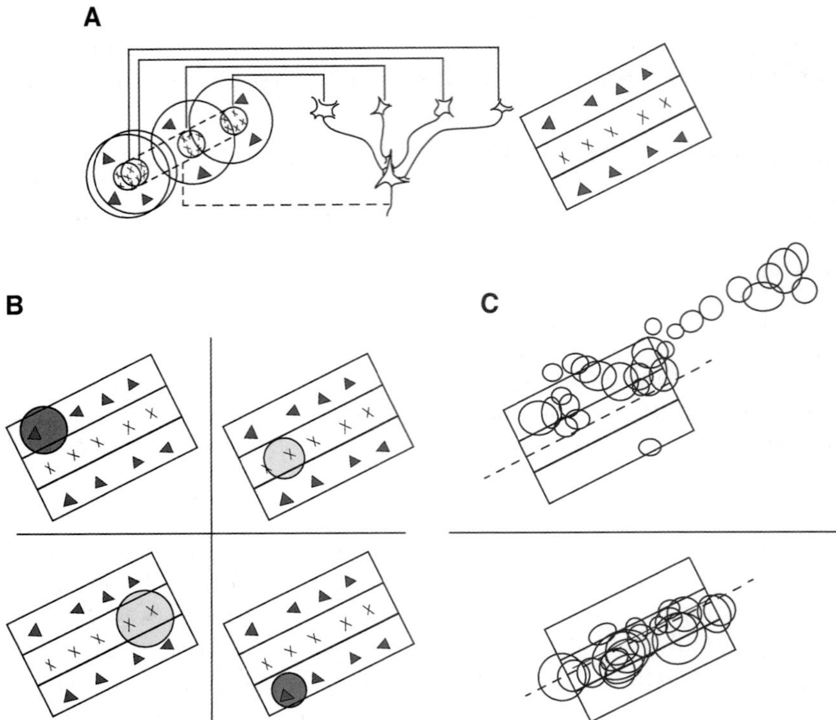

FIGURE 26.1   The feedforward model of orientation selectivity in primary visual cortex. (A) Orientation-selective simple cells in visual cortex have elongated receptive fields with segregated ON regions (marked by X) and OFF regions (marked by filled triangles). A target simple cell receives input from thalamic relay cells with circularly symmetric receptive fields that are spatially offset along the axis of the cortical cells' orientation preference. (Adapted from Hubel & Wiesel, 1962.) (B) Connected pairs of relay cells and cortical cells exhibit the same polarity preference. In each subpanel, the spatial extent of a different relay cell/cortical cell pair are shown. The polarity preference of the thalamic relay cell is indicated by shading. (Adapted from Reid & Alonso, 1995.) (C) The receptive fields of thalamic relay cells that provide input to a single cortical column are dispersed along the preferred axis of the orientation column. In this experiment, the orientation selectivity of the column was first measured, and then the cortical column was inactivated using muscimol. The receptive fields of the remaining thalamocortical projections were then measured. (Adapted from Chapman, Zahs, & Stryker, 1991.)

thalamic inputs, have elongated receptive fields; it is this elongation that endows V1 neurons with orientation selectivity (figure 26.1). V1 neurons can be further distinguished as simple or complex cells (Hubel & Wiesel, 1962; Skottun et al., 1991). Simple cells receive direct input from LGN relay cells and exhibit a receptive field configuration in which regions that prefer light (ON) and dark (OFF) changes in luminance are segregated. Such segregation between ON and OFF receptive field regions allows these neurons to be well described by a single spatiotemporal filter. Complex cells, on the other hand, usually receive input from simple cells and have receptive fields in which ON and OFF regions are not spatially segregated but instead are overlapping. Because complex cells respond to both increases and decreases in luminance at the same location, they require a receptive field description that includes multiple spatiotemporal filters. In this chapter we review the cortical organization that underlies the

emergence of orientation selectivity, with a focus on its emergence in simple cells of V1.

THE EMERGENCE OF ORIENTATION SELECTIVITY IN V1

Few models of neuronal computation have the simplicity and longevity of Hubel and Wiesel's feedforward model for the first emergence of orientation selectivity in V1. In 1962 they proposed that V1 simple cells—the neurons that receive direct thalamic input—are orientation selective by virtue of excitatory input from LGN relay cells whose receptive fields are aligned along the axis of the simple cell's preferred stimulus orientation (figure 26.1). As a result, properly oriented stimuli activate all the presynaptic LGN relay cells simultaneously and lead to a large spiking response from the simple cell; improperly (e.g., orthogonally) oriented stimuli activate only a fraction of the presynaptic inputs, eliciting little to no simple cell spiking response. Note that for moving stimuli, the total excitatory input integrated over the whole stimulus will be nearly identical at all orientations. Because the LGN relay cells are largely orientation insensitive due to their circularly symmetric receptive fields, their individual spiking responses will not vary with stimulus orientation (Hubel & Wiesel, 1962). The relative timing of their responses, however, does vary with stimulus orientation: The spiking responses of localized LGN relay cells will be nearly simultaneous when an oriented stimulus is presented at the target V1 neuron's preferred orientation, but the corresponding LGN relay cell responses will be spread out in time for nonpreferred stimulus orientations. Even for nonpreferred stimuli, however, the total excitatory input from LGN relay cells is nonzero. A threshold is therefore required to render the spike output of the cell perfectly orientation selective, with no response at the null (orthogonal) orientation (Finn, Priebe, & Ferster, 2007; Troyer et al., 1998).

One feature of their spiking records that may have prompted Hubel and Wiesel to propose the feedforward model is the resemblance between the ON and OFF centers and surrounds of retinal ganglion cells and LGN relay cells and the ON and OFF regions of V1 simple cells. Yet this observation was not sufficient to confirm the feedforward model. Rather, complete validation of their proposal would require that the receptive field properties of the LGN relay cells providing direct input to orientation-selective V1 neurons have the correct polarity (ON and OFF preferences) and spatial selectivity to underlie orientation tuning. This prediction was not so readily testable, however, as it would require the experimental identification of pairs of directly connected LGN relay cells and target V1 neurons. Indeed, it took 25 years for this prediction to be tested. In a set of landmark studies two groups (Reid & Alonso, 1995; Tanaka, 1983, 1985) recorded simultaneously from relay cell–cortical cell pairs and compared their receptive field properties. Both research groups found that the polarity and spatial selectivity of the LGN relay cell–V1 neuron pairs have matching polarity and spatial selectivity, supporting Hubel and Wiesel's model for the emergence of orientation selectivity in V1 (figure 26.1B).

Although these seminal studies demonstrated that the polarity and spatial selectivity of LGN relay cell–V1 neuron pairs are in register as required by the feedforward model, another critical feature of Hubel and Wiesel's proposal was that many LGN relay cells with spatially but specifically offset receptive fields converge onto a single V1 neuron to generate its orientation selectivity. To uncover whether this functional anatomy existed, in 1991 Chapman and colleagues took advantage of the columnar organization of orientation selectivity within V1 to analyze the spatial offset of the receptive fields of many LGN neurons (Chapman, Zahs, & Stryker, 1991). The investigators inactivated cortex—and thus cortical neuron spiking responses—using the GABA$_A$ agonist muscimol and measured the selectivity of the still-active LGN relay cell afferents to a single V1 orientation column. They identified LGN relay cells with spatially offset receptive fields such that their aggregate receptive field was oriented along the orientation of the cortical column under study (figure 26.1C). Together with the paired-cell experiments described above, it became evident that the biological underpinnings elegantly speculated by Hubel and Wiesel so many years ago do indeed exist in a manner consistent with the generation of orientation-selective receptive fields in the neurons of V1.

These important results were consistent with Hubel and Wiesel's model for the generation of oriented receptive fields in V1, but they were not sufficient to exclude other cortical influences on the generation of V1 orientation selectivity. In the mid- to late 1990s, to uncover whether the LGN relay cell inputs to V1 are sufficient to support orientation tuning or whether additional cortical network activity would be required, Ferster and co-workers inactivated cortex—via cooling and via shock—and measured its impact on V1 orientation tuning (Chung & Ferster, 1998; Ferster, Chung, & Wheat, 1996). Experimentally, inactivating cortical circuitry posed a problem, as inactivation would also remove the spiking responses of the cortical neurons typically utilized for measuring orientation selectivity. To circumvent this problem Ferster and colleagues performed technically challenging experiments using intracellular recordings in vivo, allowing for the recording of subthreshold (e.g., membrane potential) responses of V1 neurons in order to measure orientation tuning in both the presence and absence of cortical activity. They found that V1 neuron orientation selectivity—defined by the set of stimulus orientations that elicit subthreshold responses—is largely unaffected by cortical inactivation, providing evidence that the aggregate receptive field properties of LGN relay cells that provide input to individual V1 neurons are sufficient to underlie V1 orientation selectivity and also that active cortical circuitry is not required to refine said orientation selectivity.

Today, there is general consensus in the field supporting Hubel and Wiesel's original proposal that the basic organization of V1 simple cell receptive fields, including their preferred orientation selectivity, is derived from the LGN relay cell input received by these V1 neurons. This organization appears to hold across different species such as cat (Reid & Alonso, 1995), ferret (Chapman, Zahs, & Stryker, 1991), and tree shrew (Mooser, Bosking & Fitzpatrick, 2004). (For the macaque monkey, see chapter 25 by Callaway; for rodents, see chapter 29 by Niell, Bonin, and Andermann; for an alternative view of the Hubel and Wiesel model, see chapter 31 by Ringach.) The success of the feedforward model in describing this cortical computation has led to the view that new receptive field profiles emerge at each processing station by combining together excitatory inputs with differing receptive field profiles, thus generating a novel and more complex receptive field. Less certain, however, is whether the feedforward model is sufficient to explain the menagerie of other V1 simple cell response properties beyond orientation selectivity or whether additional circuit elements and mechanisms are required. Indeed, a number of these other V1 simple cell receptive field properties appear to be inconsistent with a purely linear feedforward model and suggest an important role for cortical inhibition in sculpting selectivity. We discuss below two such receptive field profiles that have been the focus of intense study precisely because their appearance appears to be at odds with the simple Hubel and Wiesel framework: contrast-invariant orientation tuning width and cross-orientation suppression.

CONTRAST-INVARIANT ORIENTATION TUNING IN V1

V1 simple cells are able to maintain the shape and width of their tuning curves for orientation despite large changes that may occur in stimulus contrast (Alitto & Usrey, 2004; Sclar & Freeman, 1982; Skottun et al., 1987). Although the amplitude of simple cell spiking responses does change systematically with stimulus contrast, the range of orientations that are able to elicit responses remains the same and does not depend on stimulus contrast.

This contrast-invariant feature of V1 neuron orientation selectivity presents a fundamental challenge to the simple feed-forward models proposed by Hubel and Wiesel, largely as a consequence of what is termed the "iceberg" effect: At any stimulus orientation, increasing stimulus contrast increases the activity of LGN relay cells and thereby increases the synaptic input to the target V1 simple cell across stimulus orientations. As the synaptic (LGN) input increases with contrast at all orientations, more and more of the orientations should evoke cortical responses that rise above threshold, and the width of tuning of the cortical spike output should therefore broaden (figure 26.2A–C). In addition, cells

with a large proportion of their excitation originating in the LGN should depolarize to high-contrast stimuli of the nonpreferred orientation and fire action potentials at the nonpreferred orientation and fire action potentials at all to low-contrast stimuli at the preferred orientation (figure 26.2C), breaking contrast invariance.

Cross-orientation inhibition has long been recognized as a possible solution to this problem. Cortical inhibition tuned to the nonpreferred orientation could suppress any depolarization and spiking evoked by such nonpreferred stimuli. Threshold could then be lowered so that low-contrast stimuli of the preferred orientation evoke spikes, as is observed in V1 simple cells (figure 26.2C). Orientation-independent inhibition (omniorientation inhibition) from inhibitory cells lacking orientation tuning could also create contrast invariance in V1 neuron spike responses (Lauritzen & Miller, 2003; Martinez et al., 2005; Nowak, Sanchez-Vives, & McCormick, 2008). If the amplitude of inhibition increases with contrast, the orientation tuning curve reaches threshold at the same orientation regardless of stimulus contrast (Troyer, Krukowski, & Miller, 2002).

The feedforward model and the cross-orientation inhibition model make very distinct predictions about changes in membrane potential in simple cells evoked by stimuli presented in the null orientation. The feedforward model predicts that because LGN relay cells are not selective for orientation, the mean excitation evoked by null-oriented stimuli should be just as large as that evoked by preferred stimuli (although the peak depolarization at the preferred orientation is much larger). With cross-orientation inhibition (either tuned or untuned), the net change in membrane potential evoked by null-oriented stimuli should be 0 or negative. For a population of 120 V1 simple cells, however, null-oriented stimuli were observed to evoke a significant depolarization, on average 43% as large as that evoked by the preferred orientation stimulus (Finn, Priebe, & Ferster, 2007). Further, the amount of null-evoked depolarization was equal to the proportion of excitatory input V1 simple cells received from the LGN (Finn, Priebe, & Ferster, 2007). The more excitatory input a simple cell received from other cortical neurons, the less the null stimulus evoked depolarization, presumably because cortical neurons are strongly orientation selective and respond weakly at the null orientation. These results appear inconsistent with either untuned inhibition or cross-orientation inhibition serving primary roles in the generation of contrast-invariant orientation tuning.

If inhibition is not responsible for contrast-invariant orientation selectivity, how does contrast-invariant orientation tuning emerge? An alternative proposal to the

     SARI ANDONI, ANDREW TAN, AND NICHOLAS J. PRIEBE

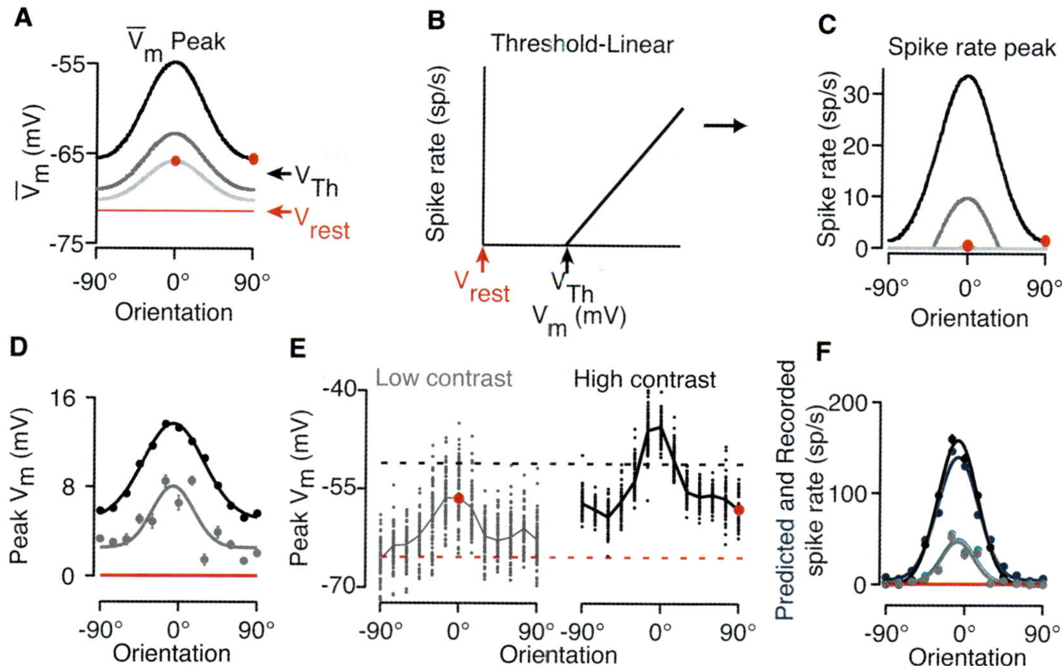

FIGURE 26.2   The emergence of contrast-invariant orientation tuning in primary visual cortex. (A) A model of input to a cortical simple cells assuming it arrives from eight ON-center thalamic relay cells, the behaviors of which are based on recordings from the LGN. Orientation tuning curves at two different contrasts are shown. (B) A threshold-linear transformation is used to convert membrane potential to spike rate. (C) Tuning for spike rate broadens with increasing contrast as more of the tuning curve for membrane potential rises above threshold. (D) As predicted from the simple relay cell model shown in A, membrane potential responses of V1 simple cells are not contrast invariant. (E) The membrane potential responses of cortical neurons are highly variable. Orientation tuning curves are plotted for the peak (F1+DC) response of the cell at low contrast (left panel) and high contrast (right panel). Each point represents the peak response to a single cycle. The dashed line shows a threshold membrane potential that would cause the same set of orientations to elicit spiking responses for high and low contrast. (F) Contrast-invariant orientation selectivity at the spike rate level emerges for the neuron shown in D (black and gray symbols). A simple elaboration of the model that relates mean membrane potential, its variability, and spike rate also predicts contrast invariance (dark and light blue symbols). (Adapted from Finn, Priebe, & Ferster, 2007.)

use of inhibition has arisen that does not depend on inhibition but instead relies on changes in response variability that occur with stimulus contrast. Membrane potential responses vary dramatically on a trial-to-trial basis, and this variability affects the relationship between mean membrane potential and firing rate. Both trial-to-trial variability and moment-to-moment synaptic noise tend to smooth the relationship between average membrane potential and average spike rate so that there is no longer a sharp inflection. Instead, spike rate rises gradually with membrane potential, starting right from the resting membrane potential. This smoothed V_m-to-spiking transformation narrows orientation tuning curves at all contrasts by approximately the same amount (Anderson et al., 2000; Hansel & van Vreeswijk, 2002; Miller & Troyer, 2002).

Even after the smoothing of the relationship between membrane potential and spike rate has been taken into account, however, contrast invariance will still break down in the feedforward model at low spike rates

(figure 26.2C); the predicted response to a high-contrast stimulus of the null orientation, although small, is still larger than the predicted response to a low-contrast stimulus at the preferred orientation. The solution to this problem comes from the observation that trial-to-trial variability of the membrane potential is contrast dependent (Finn, Priebe, & Ferster, 2007). Variability increases with decreasing contrast, and because trial-to-trial variability is partly responsible for carrying the membrane potential above threshold, an increase in variability generates an increase in spikes, even when mean membrane potential is unchanged (figure 26.2E). As a result, even though a low-contrast stimulus of the preferred orientation evokes a smaller depolarization than a high-contrast stimulus of the null orientation, it evokes more spikes. The null stimulus almost never evokes spikes, either because the underlying mean depolarization is too low (low contrast) or because the trial-to-trial variability is too low (high contrast). Contrast invariance therefore appears in the spike output of V1 simple cells without any requirement for lateral cortical inhibition, even when the visually evoked synaptic inputs are themselves not invariant (Finn, Priebe, & Ferster, 2007). Although this explanation for contrast-invariant orientation tuning does not explicitly include a role for inhibition, inhibition may nonetheless be a critical factor in shaping the degree of trial-to-trial variability. The bases for changes in response variability with contrast are as yet not well known, and both excitatory feedforward and inhibitory feedback mechanisms may be involved. The generation of this receptive field property—contrast-invariant orientation tuning—is therefore consistent with Hubel and Wiesel's original feedforward proposal. But the additional cortical circuitry, if any, required to elicit the changes in response variability essential to contrast invariance remains unclear.

CROSS-ORIENTATION SUPPRESSION IN V1

The most compelling evidence that the simple feedforward model for the receptive field organization of simple cells requires additional cortical inhibitory circuitry has come from the strong functional interactions between stimuli of different orientations, called cross-orientation suppression. In psychophysical experiments it has been shown that the detectability of one oriented stimulus ("test") is lowered by superimposing a second stimulus of the orthogonal orientation ("mask") (Campbell & Kulikowski, 1966). At the single-cell level the spike responses of a V1 neuron to a stimulus of the preferred orientation are reduced by superimposing an orthogonal stimulus (Bishop, Coombs, & Henry, 1973). The responses to high-contrast preferred-orientation stimuli can be suppressed by as much as 50%; the responses to low-contrast preferred-orientation stimuli can be suppressed almost entirely. It has long been thought that this suppression arises from inhibition between cells with orthogonal preferred orientations. In support of this interpretation, antagonists of $GABA_A$-mediated inhibition reduce cross-orientation suppression in visual evoked potentials (Morrone, Burr, & Maffei, 1982; Morrone, Burr, & Speed, 1987).

Cross orientation has been considered to be a cortical phenomenon because it appears to be sensitive to the orientation of the mask, but there are aspects of cross-orientation suppression that appear to be at odds with cortical inhibition as its source. First, cross-orientation suppression is largely monocular (Ferster, 1981; Walker, Ohzawa, & Freeman, 1998); a null-oriented mask stimulus presented to one eye has little effect on a preferred test stimulus presented to the other eye,

whereas the majority of cortical neurons, presumably including inhibitory interneurons, are binocular. Second, strong suppression can be evoked by mask stimuli of high temporal frequency, beyond the frequencies to which most cortical neurons can respond (Freeman et al., 2002). Third, suppression is insensitive to contrast adaptation, whereas the responses of most cortical cells—presumably including inhibitory interneurons—are strongly suppressed by adaptation (Freeman et al., 2002). Fourth, the onset of suppression is coincident with the onset of neuronal responses, leaving no time for the activation of the inhibitory circuits (Smith, Bair, & Movshon, 2006). Finally, evidence from intracellular recording for strong inhibition evoked by stimuli of orthogonal orientation is equivocal, and both inhibition and excitation appear to decrease when the test and mask stimuli are presented (Priebe & Ferster, 2006).

All of these properties of cross-orientation suppression are more reminiscent of LGN relay cells than they are of V1 inhibitory interneurons: Relay cells are monocular, respond at high temporal frequency, adapt little to contrast, and respond simultaneously with the excitatory input to the cortex. It has therefore been proposed that cross-orientation suppression arises from nonlinear interactions within the relay-cell pathway itself (Carandini, Heeger, & Senn, 2002; Ferster, 1986). One such nonlinearity is synaptic depression—the mask stimulus could increase the level of depression at the synapses between relay cells and cortical cells, thereby reducing the excitatory drive evoked by the test stimulus. Because thalamocortical depression may not be strong enough to account fully for strong cross-orientation suppression (Boudreau & Ferster, 2005; Li et al., 2006; Reig et al., 2006), it has also been proposed that cross-orientation suppression may arise from two nonlinearities in the responses of LGN relay cells: contrast saturation and firing-rate rectification (Ferster, 1986; Li et al., 2006; Priebe & Ferster, 2006).

To understand how nonlinearities in the feedforward pathway generate cross-orientation suppression, it is useful to consider how LGN relay cells respond to drifting gratings. LGN relay cells modulate their response to drifting gratings, but their responses are not purely sinusoidal. Because LGN relay cells have a low spontaneous firing rate, high-contrast drifting gratings cause response rectification to occur, clipping the neuronal response at 0 spikes/s (figure 26.3A). Further, LGN relay cell responses do not increase linearly with contrast, instead saturating for contrasts above 32% (figure 26.3B, inset). These two nonlinearities have strong effects on the total input that a

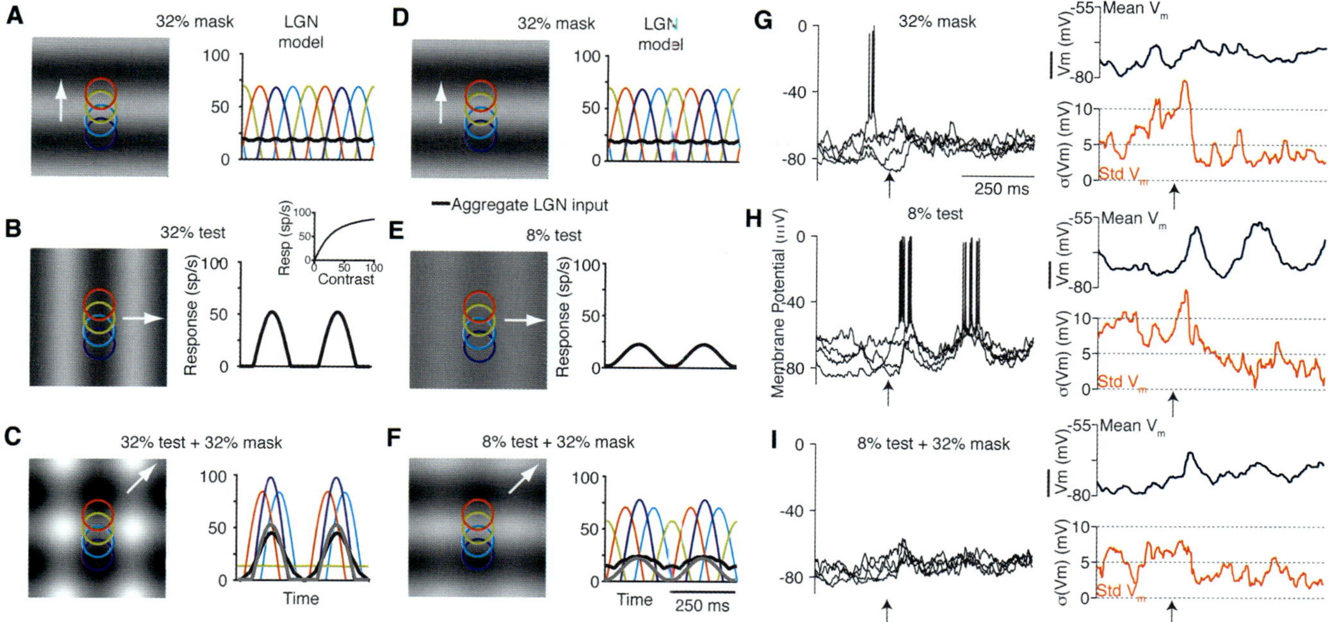

FIGURE 26.3   Cross-orientation suppression in models and responses of simple cells. (A–F) Predicted membrane potential responses are shown for six different grating stimuli. Spiking responses of constituent geniculate relay cells are shown in the colors of their receptive field centers (left). Membrane potential responses of the simple cell are shown in black and are derived from the average of the relay-cell responses. On the right of each panel, spontaneous activity in the relay cells is low, so that the responses rectify, and the amplitude of modulation saturates with increasing contrast (inset in B). The nonlinear model predicts that the mask stimulus induces a 15% reduction in the modulation component of the response to the high-contrast test stimulus (compare black and gray traces in C) and 50% reduction in the response to the low-contrast test stimulus (compare black and gray traces in F). Responses of a V1 simple cell demonstrate a cross-orientation suppression when a 32% mask is presented with an 8% test stimulus. (G–I) The presentation of the mask alone causes both a depolarization and a decrease in the membrane potential variability across trials (G). The decrease in membrane potential variability also occurs for the 8% test stimulus alone (H), but the combination of the test and mask stimulus results in a lower degree of membrane potential variability, as measured by the standard deviation of the membrane potential traces (I). The vertical arrow marks the initiation of the visual stimulus. (Adapted from Priebe & Ferster, 2006.)

target V1 neuron receives when a stimulus composed of preferred (test) and orthogonal (mask) gratings is presented.

The presentation of the superimposed test and mask gratings (*plaid stimulus*) causes systematic changes in the contrast of the stimulus for each relay cell. When the test and mask contrasts are matched in the plaid, some LGN relay cells will encounter locations in the plaid stimulus where the dark bars from the two gratings superimpose, alternating with the locations where the bright bars superimpose. The result is a luminance modulation exactly twice as large as that generated by either grating stimulus alone (figure 26.3C, blue relay cell). For other LGN relay cells their receptive fields lie at a location within the plaid stimulus where the bright bars from one grating superimpose on the dark bars from the other grating. As a result, there is no modulation of luminance in the relay cell's receptive field, and its response falls to zero (figure 26.3C, gold relay cell).

Including the mask stimulus therefore causes systematic shifts in the contrast of the stimulus that falls over each LGN relay cell's receptive field.

For a purely linear system these changes in contrast would not affect the total input to V1, but contrast saturation and response rectification exhibited by LGN relay cells change cortical input significantly. The response of the relay cells that receive no contrast modulation still falls to zero because its stimulus has zero contrast, but the response of those relay cells that receive twice the contrast modulation does not double. Although the mask stimulus doubles the local contrast relative to the test stimulus alone (figure 26.3A, C blue neuron), because the test stimulus was already nearly saturating, the cell's response increases only slightly (right). For low-contrast test gratings (figure 26.3D–F), the effects of the mask grating are to reduce dramatically the amplitude of the modulation input by almost 50%.

This suppression of excitatory relay-cell input to V1 simple cells predicted by the nonlinear LGN model (15% for high-contrast test gratings and 50% for low-contrast test gratings) matches closely what is observed in the membrane potential responses of V1 simple cells: 9% for the high-contrast test gratings and 52% for low-contrast test gratings (Priebe & Ferster, 2006) (figure 26.3G). To account for the even larger effects observed in V1 neuron spiking responses (29% and 89%), the nonlinearity of spike threshold is needed. Threshold amplifies the effects of the mask gratings in the same way it sharpens orientation tuning. Together with the nonlinearity of relay-cell responses, it accounts quantitatively for the cross-orientation effects in V1 simple cells (Priebe & Ferster, 2006).

Note that although the model accounts for the mask-induced reduction in the modulation component of membrane potential, it also predicts a rise in the mean LGN input to V1 neurons and therefore a corresponding rise in mean membrane potential of approximately 50% (figure 26.4F, compare black and gray traces). That a large rise in the mean is not observed experimentally could be explained at least in part by short-term synaptic depression at the thalamocortical synapse (Carandini, Heeger, & Senn, 2002; Freeman et al., 2002) and by the fact that many simple cells receive less than half of their excitatory input from the LGN (Chung & Ferster, 1998; Ferster, Chung, & Wheat, 1996). An additional factor that may also contribute to the lack of a large mean increase to cortical neurons is in the change in cortical response variability that is also seen when the orthogonal stimulus is presented. As shown for the contrast invariance of orientation tuning, trial-to-trial response variability in cortical simple cells is reduced by stimulus contrast, whether the stimulus is at the preferred (test) or orthogonal (mask) orientation. In V1 simple cells we observe a systematic change in the trial-to-trial variability of membrane potential fluctuations when the mask is presented that reduces the gain of the membrane potential to spike rate transformation (figure 26.3G–I). This effect of variability on response gain may thus contribute to a reduction in intracortical activity that would normally reflect the increased mean activity provided by the LGN relay cells. As for contrast-invariant orientation tuning, adjusting response variability with contrast represents a key aspect controlling the gain of cortical responses. As with contrast-invariant orientation tuning, however, it appears that an account for cross-orientation suppression is consistent with the feedforward circuitry proposed by Hubel and Wiesel and that accounting for cross-orientation suppression does not require cortical inhibitory circuitry to sculpt orientation selectivity.

ROLE OF INHIBITION IN V1 ORIENTATION SELECTIVITY

The models described in the previous section are able to generate orientation tuning without cortical inhibition only if some specific assumptions are made. For example, to account for contrast-invariant orientation tuning, the change in variability could be subcortical, and yet cortical inhibition could also play a role in controlling response variability. The change in effective threshold could be caused by changes in membrane potential variance, but it could also result from a change in the balance of excitation and inhibition (Chance, Abbott, & Reyes, 2002; Holt & Koch, 1997). V1 neurons that receive monosynaptic LGN input also receive considerable cortical input, estimated to account for ~40% of the depolarization in V1 simple cells (Chung & Ferster, 1998; Ferster, Chung, & Wheat, 1996). Because a given depolarization could arise from a continuum of possible excitatory-to-inhibitory ratios, the mix of excitation and inhibition that comprises the cortical contribution is not well constrained. Because of this ambiguity, we review theoretical and experimental explorations of how excitatory and inhibitory cortical architectures could contribute to the emergence of cortical receptive fields.

To account for contrast-invariant orientation tuning Troyer, Krukowski, and Miller (1998) began with a purely feed-forward model and extended it to include cortical excitation and inhibition. V1 neurons receive excitatory LGN input that has two components: a modulated component that varies with spatial phase and is orientation tuned and a second temporally unmodulated component that is phase insensitive and entirely untuned for orientation. Troyer and colleagues pointed out that the untuned component of the LGN input could disrupt contrast invariance and suggested that cortical inhibitory neurons that are tuned for orientation but respond to all orientations could counteract this untuned component.

For the inhibitory neurons to suppress the untuned component from the LGN, Troyer and co-workers initially proposed that inhibitory neurons exist that are contrast variant, as opposed to their excitatory counterparts. Alternatively, it is also possible to construct a model in which the excitatory and most inhibitory simple cells are contrast invariant, but an additional set of inhibitory neurons, characterized as non-orientation-tuned complex cells, act to cancel the untuned component of the net thalamic input (Lauritzen & Miller, 2003). Both forms of the Troyer et al. model predict the existence of cortical inhibitory neurons whose spike response properties differ substantially from those of

     SARI ANDONI, ANDREW TAN, AND NICHOLAS J. PRIEBE

cortical excitatory neurons. Evidence has demonstrated distinct receptive field properties for excitatory and inhibitory cortical neurons, but the relative frequency with which excitatory and inhibitory neurons exhibit receptive field differences is still not known (Ahmed et al., 1997; Azouz et al., 1997; Cardin, Palmer, & Contreras, 2007; Hirsch et al., 2003; Martin, Somogyi, & Whitteridge, 1983; Nowak, Sanchez-Vives, & McCormick, 2008).

Troyer, Krukowski, and Miller also suggested a particular architecture for inhibitory and excitatory neurons in terms of spatial preference. They suggested that inhibitory neurons converge synaptically onto excitatory neurons in the push–pull architecture: If a bright bar at a particular location elicits an increase in excitation, a dark bar at the same location elicits an increase in inhibition. Indeed, when spots or bars of light are flashed at different positions in the receptive field, membrane potential records indicate a characteristic push–pull spatial receptive field structure: Where bright spots of light evoke excitation, dark spots evoke inhibition; where dark spots evoke excitation, light spots evoke inhibition (Hirsch et al., 1998). This push–pull architecture in spatial receptive fields is also revealed in the response to sinusoidal moving gratings where excitation and inhibition are temporally out of phase (Anderson, Carandini, & Ferster, 2000; Ferster, 1988; Monier et al., 2003; Priebe & Ferster, 2006). The presence of this push–pull relationship between excitation and inhibition thus places constraints on the circuitry underlying cortical receptive fields.

Further, the Troyer et al. model predicts that sinusoidal moving gratings evoke inhibition that contains an untuned inhibitory component. This is consistent with the imperfect push–pull suggested by Borg-Graham, Monier, and Fregnac (1998) and the clear excitatory and inhibitory responses evoked by gratings oriented orthogonally to those that evoke maximal responses (Borg-Graham, Monier, & Fregnac, 1998; Monier et al., 2003; Priebe, Lisberger, & Movshon, 2006). Finally, as suggested by Troyer et al., excitatory and inhibitory inputs to most cortical neurons appear to have the same preferred orientation and similar tuning widths (Anderson, Carandini, & Ferster, 2000; Ferster, 1988; Marino et al., 2005; Monier et al., 2003).

Although measures of excitation and inhibition do provide support for the Troyer et al. proposal, they do not yet distinguish between alternative models of cortical architecture. For example, it has been suggested that inhibitory input may be more broadly tuned than excitatory input (Ben-Yishai, Bar-Or, & Sompolinsky, 1995; Somers, Nelson, & Sur, 1995). Sharp orientation selectivity in these models is produced by recurrent

horizontal connections between neurons in the same cortical layer. In regard to the horizontal connections, the characteristic two-dimensional pinwheel pattern of orientation selectivity across the cortex has also been used to model the functional organization of V1 (McLaughlin, Shapley, & Shelley, 2003). Whereas preferred orientation preference changes smoothly across the cortical surface, pinwheels are formed by discontinuities in the map. Spike responses, however, are equally well orientation tuned regardless of their proximity to pinwheel centers (Maldonado et al., 1997; Ohki et al., 2006). In contrast the subthreshold inputs to neurons close to pinwheel centers are more broadly tuned and more variable than those to neurons distal from pinwheel centers (Marino et al., 2005; Schummers, Marino, & Sur, 2002, 2004). The cortical functional map therefore influences synaptic input, and either inhibition or intrinsic mechanisms maintain orientation selectivity at the spiking level. The generation of the receptive field in V1 therefore appears to depend critically on the local cortical architecture (Stimberg et al., 2009).

There seems to be considerable diversity in the fashion that orientation selectivity emerges. Monier et al. (2003) demonstrate, for example, that although excitation and inhibition are co-tuned for some neurons, for other neurons it appears that excitation and inhibition have orthogonal orientation preferences (Monier et al., 2003; Volgushev et al., 1993). In stark contrast, the above models posit that the excitatory and inhibitory circuit elements that generate orientation selectivity follow simple and stereotyped architectures. One of the methods that Stimberg et al. (2009) employed to create diversity in the cortical circuitry was to include the spatial distribution of preferred orientations on the cortical surface. Further, diversity in the circuitry may be a result of differences in circuit elements across cortical layers (Martinez et al., 2002). More generally, however, the above-mentioned circuit models do not represent all biochemical mechanisms by which the circuits form and process visual information, and some of the mechanisms not explicitly modeled therefore appear as "noise."

SPONTANEOUS ACTIVITY AND CORTICAL DYNAMICS OF ORIENTATION TUNING

The work detailed thus far has sought to understand a neuron's receptive field via the neural responses it generates to visual stimuli, but other studies have also probed the spontaneous cortical network activity to understand the circuitry underlying the receptive field. Ongoing spontaneous activity during the absence of a visual stimulus may reveal much information on the overall connectivity and functional structure of V1 by demonstrating which aspects of the cortical circuit are synchronously active (Arieli et al., 1995, 1996; Fitzpatrick, 2000; Kenet et al., 2003; Shoham et al., 1999; Tsodyks et al., 1999). Although spontaneous activity could represent a background cortical state to which the cortical network relaxes in the absence of a stimulus, experimental evidence has suggested that ongoing activity may represent dynamic switching among intrinsic cortical states that are also present during stimulus-driven activity. For example, spontaneous cortical activity is often highly correlated to activity elicited by a single orientation when imaged using a voltage-sensitive dye (figure 26.4A–C) (Kenet et al., 2003; Tsodyks et al., 1999). The spatial structure of these spontaneous activity patterns reveals an underlying cortical circuitry in which neurons with similar orientation selectivity are coupled across cortex. Surprisingly, spontaneous maps resembled the evoked maps with a correlation coefficient almost as high as the relationship found between a single stimulus presentation and the averaged orientation map (figure 26.4C). Over the duration of the experiment the relationship between instantaneous spontaneous activity and an averaged orientation map fluctuated between high and low correlations with a Gaussian distribution (figure 26.4D). However, when the spontaneous activity is correlated with an arbitrary orientation map not found during evoked activity, such as the inverse of a stimulus-evoked map, the high correlations go away, resulting in a narrower distribution of correlation coefficients. Because high correlations were found across all stimulus-evoked orientations, Grinvald and colleagues suggested that ongoing activity in V1 reflected dynamic switching among a set of intrinsic states many of which correspond closely to stimulus-evoked orientation maps (Goldberg, Rokni, & Sompolinsky, 2004; Kenet et al., 2003).

To make constraints on the cortical architecture and connectivity that could give rise to the dynamic network behavior observed by Grinvald and colleagues, different models have been proposed that could result in similar dynamics. One proposal that reproduced these dynamics is a cortical network with "balanced amplification," a circuit characterized by strong recurrent excitation and stabilized by feedback inhibition (Murphy & Miller, 2009). The interplay between excitation and inhibition in this architecture could allow the cortical circuit to distinguish and amplify specific activity patterns from unstructured inputs. In this manner the presence of a noisy input results in spontaneous activity that switches among activity modes that are similar to stimulus-evoked orientation maps, as observed by optical imaging.

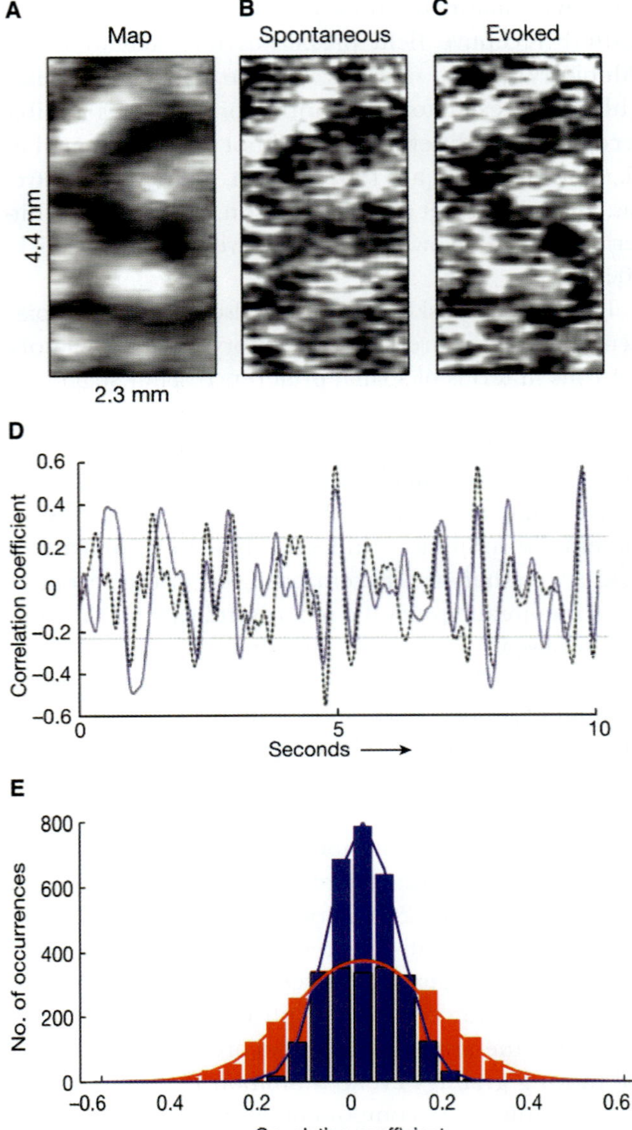

FIGURE 26.4 Similarity between spontaneous and evoked maps in visual cortex. (A) An evoked map extracted by averaging responses to a vertically oriented grating. (B) A single frame of a spontaneous map that is highly correlated with the evoked map. (C) A single frame in response to a vertical grating. (D) Correlation coefficient between evoked and spontaneous maps plotted over time for the top (solid) and bottom (dashed) patches of the cortical area shown in A. (E) Distribution of correlation coefficient between spontaneous maps and an evoked map (red) or an arbitrary (inverted) map (blue). (Adapted from Kenet et al., 2003.)

Other studies have suggested that the similarity between spontaneous and evoked cortical maps could be equivalent to switching among attractor states in a dynamical system (Ben-Yishai, Bar-Or, & Sompolinsky, 1995; Goldberg, Rokni, & Sompolinsky, 2004). In this perspective the cortical dynamics is assumed to settle into attractor states that are manifested as activity patterns across the cortex and are determined by the underlying functional connectivity of the network. Goldberg, Rokni, and Sompolinsky (2004) examined whether spontaneous maps in the visual cortex, as observed by Kenet et al. (2003), were a result of a single background state or multiple attractor states with noise-driven spontaneous switching among them. A comparison of a single-state model to a "ring" attractor, where each location on the ring corresponds to an orientation preference, revealed that a single-state model reproduced the observed Gaussian-distributed correlations more robustly than a ring-attractor model (see figure 26.4C).

Although many open questions remain on the similarity between spontaneous and evoked activity in the visual cortex, the ongoing cortical activity and its relationship and interaction with stimulus-evoked responses could provide insight into the circuit architecture and functional connectivity of cortical networks and their role in sensory perception (see chapter 32 by Swadlow and Alonso).

The high variability present in cortical spontaneous activity could degrade response reliability. Such response unreliability—potentially the root source of creativity— is an essential component for reinforcement learning, which enables organisms to adapt to changing environments (Barto, Sutton, & Anderson, 1983; Mazzoni, Andersen, & Jordan, 1991). Although such a role for spontaneous activity is highly speculative, the requisite synaptic plasticity is present in primary visual cortex (Fregnac et al., 1988; Gilbert, Sigman, & Crist, 2001; Meliza & Dan, 2006). Synaptic plasticity changes cortical circuit organization according to an animal's experience and could contribute to the generation of receptive field diversity (Monier et al., 2003; Rust et al., 2005; Volgushev et al., 1993). Although Hubel and Wiesel's orientation-tuned neurons can be interpreted as the edge detectors necessary for animals to make sense of almost all natural scenes (Daugman, 1985; Hubel & Wiesel, 1962; Marr & Hildreth, 1980; Shapley & Tolhurst, 1973), receptive field diversity beyond the Hubel and Wiesel paradigm may subserve flexible object recognition and is therefore also interpretable in scene analysis terms (Lampl et al., 2004; Riesenhuber & Poggio, 1999; Ullman & Bart, 2004). Balancing these perspectives suggests that although the requirements of specific tasks promote fine-scale diversity, statistical symmetries of natural scenes constrain the large-scale structure of visual cortex (Kaschube et al., 2010; Miller, 2010).

ACKNOWLEDGMENTS

We are grateful to Jessica Hanover for helpful discussions. Supported by grants from the National Institutes of Health (EY-019288) and the Pew Charitable Trusts.

REFERENCES

Ahmed, B., Allison, J. D., Douglas, R. J., & Martin, K. A. (1997). An intracellular study of the contrast-dependence of neuronal activity in cat visual cortex. *Cerebral Cortex, 7,* 559–570.

Alitto, H. J., & Usrey, W. M. (2004). Influence of contrast on orientation and temporal frequency tuning in ferret primary visual cortex. *Journal of Neurophysiology, 91,* 2797–2808.

Anderson, J. S., Carandini, M., & Ferster, D. (2000). Orientation tuning of input conductance, excitation, and inhibition in cat primary visual cortex. *Journal of Neurophysiology, 84,* 909–926.

Anderson, J. S., Lampl, L., Gillespie, D., & Ferster, D. (2000). The contribution of noise to contrast invariance of orientation tuning in cat visual cortex. *Science, 290,* 1968–1971.

Arieli, A., Shoham, D., Hildesheim, R., & Grinvald, A. (1995). Coherent spatiotemporal patterns of ongoing activity revealed by real-time optical imaging coupled with single-unit recording in the cat visual cortex. *Journal of Neurophysiology, 73,* 2072–2093.

Arieli, A., Sterkin, A., Grinvald, A., & Aertsen, A. (1996). Dynamics of ongoing activity: Explanation of the large variability in evoked cortical responses. *Science, 273,* 1868–1871.

Azouz, R., Gray, C., Nowak, L. G., & McCormick, D. A. (1997). Physiological properties of identified inhibitory interneurons in cat striate cortex. *Cerebral Cortex, 7,* 534–545.

Barto, A. G., Sutton, R. S., & Anderson, C. W. (1983). Neuron-like adaptive elements that can solve difficult learning control-problems. *IEEE Transactions on Systems, Man, and Cybernetics, 13,* 834–846.

Ben-Yishai, R., Bar-Or, R. L., & Sompolinsky, H. (1995). Theory of orientation tuning in visual cortex. *Proceedings of the National Academy of Sciences of the United States of America, 92,* 3844–3848. doi:10.1073/pnas.92.9.3844.

Bishop, P. O., Coombs, J. S., & Henry, G. H. (1973). Receptive fields of simple cells in the cat striate cortex. *Journal of Physiology, 231,* 31–60.

Borg-Graham, L. J., Monier, C., & Fregnac, Y. (1998). Visual input evokes transient and strong shunting inhibition in visual cortical neurons. *Nature, 393,* 369–373.

Boudreau, C. E., & Ferster, D. (2005). Short-term depression in thalamocortical synapses of cat primary visual cortex. *Journal of Neuroscience, 25,* 7179–7190.

Campbell, F. W., & Kulikowski, J. J. (1966). Orientational selectivity of the human visual system. *Journal of Physiology, 187,* 437–445.

Carandini, M., Heeger, D. J., & Senn, W. (2002). A synaptic explanation of suppression in visual cortex. *Journal of Neuroscience, 22,* 10053–10065.

Cardin, J. A., Palmer, L. A., & Contreras, D. (2007). Stimulus feature selectivity in excitatory and inhibitory neurons in primary visual cortex. *Journal of Neuroscience, 27,* 10333–10344.

Chance, F. S., Abbott, L. F., & Reyes, A. D. (2002). Gain modulation from background synaptic input. *Neuron, 35,* 773–782.

Chapman, B., Zahs, K. R., & Stryker, M. P. (1991). Relation of cortical cell orientation selectivity to alignment of receptive fields of the geniculocortical afferents that arborize within a single orientation column in ferret visual cortex. *Journal of Neuroscience, 11*, 1347–1358.

Chung, S., & Ferster, D. (1998). Strength and orientation tuning of the thalamic input to simple cells revealed by electrically evoked cortical suppression. *Neuron, 20*, 1177–1189.

Daugman, J. G. (1985). Uncertainty relation for resolution in space, spatial frequency, and orientation optimized by two-dimensional visual cortical filters. *Journal of the Optical Society of America. A, Optics and Image Science, 2*, 1160–1169. doi:10.1364/JOSAA.2.001160.

Ferster, D. (1981). A comparison of binocular depth mechanisms in areas 17 and 18 of the cat visual cortex. *Journal of Physiology, 311*, 623–655.

Ferster, D. (1986). Orientation selectivity of synaptic potentials in neurons of cat primary visual cortex. *Journal of Neuroscience, 6*, 1284–1301.

Ferster, D. (1988). Spatially opponent excitation and inhibition in simple cells of the cat visual cortex. *Journal of Neuroscience, 8*, 1172–1180.

Ferster, D., Chung, S., & Wheat, H. (1996). Orientation selectivity of thalamic input to simple cells of cat visual cortex. *Nature, 380*, 249–252.

Ferster, D., & Miller, K. D. (2000). Neural mechanisms of orientation selectivity in the visual cortex. *Annual Review of Neuroscience, 23*, 441–471.

Finn, I. M., Priebe, N. J., & Ferster, D. (2007). The emergence of contrast-invariant orientation tuning in simple cells of cat visual cortex. *Neuron, 54*, 137–152.

Fitzpatrick, D. (2000). Cortical imaging: Capturing the moment. *Current Biology, 10*, R187–R190. doi:10.1016/S0960-9822(00)00348-1.

Freeman, T. C., Durand, S., Kiper, D. C., & Carandini, M. (2002). Suppression without inhibition in visual cortex. *Neuron, 35*, 759–771.

Fregnac, Y., Bienenstock, E., Shulz, D., & Thorpe, S. (1988). A cellular analog of visual cortical plasticity. *Nature, 333*, 367–370.

Gilbert, C. D., Sigman, M., & Crist, R. E. (2001). The neural basis of perceptual learning. *Neuron, 31*, 681–697.

Goldberg, J. A., Rokni, U., & Sompolinsky, H. (2004). Patterns of ongoing activity and the functional architecture of the primary visual cortex. *Neuron, 42*, 489–500.

Hansel, D., & van Vreeswijk, C. (2002). How noise contributes to contrast invariance of orientation tuning in cat visual cortex. *Journal of Neuroscience, 22*, 5118–5128.

Hartline, H. (1938). The response of single optic nerve fibers of the vertebrate eye to illumination of the retina. *American Journal of Physiology, 121*, 106–154.

Hirsch, J. A., Alonso, J. M., Reid, R. C., & Martinez, L. M. (1998). Synaptic integration in striate cortical simple cells. *Journal of Neuroscience, 18*, 9517–9528.

Hirsch, J. A., Martinez, L. M., Pillai, C., Alonso, J. M., Wang, Q., & Sommer, F. T. (2003). Functionally distinct inhibitory neurons at the first stage of visual cortical processing. *Nature Neuroscience, 6*, 1300–1308.

Holt, G. R., & Koch, C. (1997). Shunting inhibition does not have a divisive effect on firing rates. *Neural Computation, 9*, 1001–1013. doi:10.1162/neco.1997.9.5.1001.

Hubel, D. H., & Wiesel, T. N. (1962). Receptive fields, binocular interaction and functional architecture in the cat's visual cortex. *Journal of Physiology, 160*, 106–154.

Kaschube, M., Schnabel, M., Lowel, S., Coppola, D. M., White, L. E., & Wolf, F. (2010). Universality in the evolution of orientation columns in the visual cortex. *Science, 330*, 1113–1116.

Kenet, T., Bibitchkov, D., Tsodyks, M., Grinvald, A., & Arieli, A. (2003). Spontaneously emerging cortical representations of visual attributes. *Nature, 425*, 954–956.

Kuffler, S. W. (1953). Discharge patterns and functional organization of mammalian retina. *Journal of Neurophysiology, 16*, 37–68.

Lampl, I., Ferster, D., Poggio, T., & Riesenhuber, M. (2004). Intracellular measurements of spatial integration and the MAX operation in complex cells of the cat primary visual cortex. *Journal of Neurophysiology, 92*, 2704–2713.

Lauritzen, T. Z., & Miller, K. D. (2003). Different roles for simple-cell and complex-cell inhibition in V1. *Journal of Neuroscience, 23*, 10201–10213.

Li, B., Thompson, J. K., Duong, T., Peterson, M. R., & Freeman, R. D. (2006). Origins of cross-orientation suppression in the visual cortex. *Journal of Neurophysiology, 96*, 1755–1764.

Maldonado, P. E., Godecke, I., Gray, C. M., & Bonhoeffer, T. (1997). Orientation selectivity in pinwheel centers in cat striate cortex. *Science, 276*, 1551–1555.

Marino, J., Schummers, J., Lyon, D. C., Schwabe, L., Beck, O., Wiesing, P., et al. (2005). Invariant computations in local cortical networks with balanced excitation and inhibition. *Nature Neuroscience, 8*, 194–201.

Marr, D., & Hildreth, E. (1980). Theory of edge detection. *Proceedings of the Royal Society of London. Series B, Biological Sciences, 207*, 187–217. doi:10.1098/rspb.1980.0020.

Martin, K. A., Somogyi, P., & Whitteridge, D. (1983). Physiological and morphological properties of identified basket cells in the cat's visual cortex. *Experiments in Brain Research, 50*, 193–200.

Martinez, L. M., Alonso, J. M., Reid, R. C., & Hirsch, J. A. (2002). Laminar processing of stimulus orientation in cat visual cortex. *Journal of Physiology, 540*(Pt 1), 321–333.

Martinez, L. M., Wang, Q., Reid, R. C., Pillai, C., Alonso, J. M., Sommer, F. T., et al. (2005). Receptive field structure varies with layer in the primary visual cortex. *Nature Neuroscience, 8*, 372–379.

Mazzoni, P., Andersen, R. A., & Jordan, M. I. (1991). A more biologically plausible learning rule for neural networks. *Proceedings of the National Academy of Sciences of the United States of America, 88*, 4433–4437. doi:10.1073/pnas.88.10.4433.

McLaughlin, D., Shapley, R., & Shelley, M. (2003). Large-scale modeling of the primary visual cortex: Influence of cortical architecture upon neuronal response. *Journal of Physiology (Paris), 97*, 237–252. doi:10.1016/j.jphysparis.2003.09.019.

Meliza, C. D., & Dan, Y. (2006). Receptive-field modification in rat visual cortex induced by paired visual stimulation and single-cell spiking. *Neuron, 49*, 183–189.

Miller, K. D. (2010). Neuroscience. π = visual cortex. *Science, 330*, 1059–1060.

Miller, K. D., & Troyer, T. W. (2002). Neural noise can explain expansive, power-law nonlinearities in neural response functions. *Journal of Neurophysiology, 87*, 653–659.

Monier, C., Chavane, F., Baudot, P., Graham, L. J., & Fregnac, Y. (2003). Orientation and direction selectivity of synaptic inputs in visual cortical neurons: A diversity of combinations produces spike tuning. *Neuron, 37*, 663–680.

Mooser, F., Bosking, W. H., & Fitzpatrick, D. (2004). A morphological basis for orientation tuning in primary visual cortex. *Nature Neuroscience, 7*, 872–879.

Morrone, M. C., Burr, D. C., & Maffei, L. (1982). Functional implications of cross-orientation inhibition of cortical visual cells. I. Neurophysiological evidence. [Research Support, Non-U.S. Gov't]. *Proceedings of the Royal Society of London. Series B, Biological Sciences, 216*, 335–354. doi:10.1098/rspb.1982.0078.

Morrone, M. C., Burr, D. C., & Speed, H. D. (1987). Cross-orientation inhibition in cat is GABA mediated. *Experiments in Brain Research, 67*, 635–644.

Murphy, B. K., & Miller, K. D. (2009). Balanced amplification: A new mechanism of selective amplification of neural activity patterns. *Neuron, 61*, 635–648.

Nowak, L. G., Sanchez-Vives, M. V., & McCormick, D. A. (2008). Lack of orientation and direction selectivity in a subgroup of fast-spiking inhibitory interneurons: Cellular and synaptic mechanisms and comparison with other electrophysiological cell types. *Cerebral Cortex, 18*, 1058–1078.

Ohki, K., Chung, S., Kara, P., Hubener, M., Bonhoeffer, T., & Reid, R. C. (2006). Highly ordered arrangement of single neurons in orientation pinwheels. *Nature, 442*, 925–928.

Priebe, N. J., & Ferster, D. (2006). Mechanisms underlying cross-orientation suppression in cat visual cortex. *Nature Neuroscience, 9*, 552–561.

Priebe, N. J., Lisberger, S. G., & Movshon, J. A. (2006). Tuning for spatiotemporal frequency and speed in directionally selective neurons of macaque striate cortex. *Journal of Neuroscience, 26*, 2941–2950.

Reid, R. C., & Alonso, J. M. (1995). Specificity of monosynaptic connections from thalamus to visual cortex. *Nature, 378*, 281–284.

Reig, R., Gallego, R., Nowak, L. G., & Sanchez-Vives, M. V. (2006). Impact of cortical network activity on short-term synaptic depression. *Cerebral Cortex, 16*, 688–695.

Riesenhuber, M., & Poggio, T. (1999). Hierarchical models of object recognition in cortex. *Nature Neuroscience, 2*, 1019–1025.

Rust, N. C., Schwartz, O., Movshon, J. A., & Simoncelli, E. P. (2005). Spatiotemporal elements of macaque V1 receptive fields. *Neuron, 46*, 945–956.

Schummers, J., Marino, J., & Sur, M. (2002). Synaptic integration by V1 neurons depends on location within the orientation map. *Neuron, 36*, 969–978.

Schummers, J., Marino, J., & Sur, M. (2004). Local networks in visual cortex and their influence on neuronal responses and dynamics. *Journal of Physiology (Paris), 98*, 429–441. doi:10.1016/j.jphysparis.2005.09.017.

Sclar, G., & Freeman, R. D. (1982). Orientation selectivity in the cat's striate cortex is invariant with stimulus contrast. *Experiments in Brain Research, 46*, 457–461.

Shapley, R. M., & Tolhurst, D. J. (1973). Edge detectors in human vision. *Journal of Physiology, 229*, 165–183.

Shoham, D., Glaser, D. E., Arieli, A., Kenet, T., Wijnbergen, C., Toledo, Y., et al. (1999). Imaging cortical dynamics at high spatial and temporal resolution with novel blue voltage-sensitive dyes. *Neuron, 24*, 791–802.

Skottun, B. C., Bradley, A., Sclar, G., Ohzawa, I., & Freeman, R. (1987). The effects of contrast on visual orientation and spatial frequency discrimination: A comparison of single cells and behavior. *Journal of Neurophysiology, 57*, 773–786.

Skottun, B. C., De Valois, R. L., Grosof, D. H., Movshon, J. A., Albrecht, D. G., & Bonds, A. B. (1991). Classifying simple and complex cells on the basis of response modulation. *Vision Research, 31*, 1079–1086.

Smith, M. A., Bair, W., & Movshon, J. A. (2006). Dynamics of suppression in macaque primary visual cortex. *Journal of Neuroscience, 26*, 4826–4834.

Somers, D. C., Nelson, S. B., & Sur, M. (1995). An emergent model of orientation selectivity in cat visual cortical simple cells. *Journal of Neuroscience, 15*, 5448–5465.

Sompolinsky, H., & Shapley, R. (1997). New perspectives on the mechanisms for orientation selectivity. *Current Opinion in Neurobiology, 7*, 514–522.

Stimberg, M., Wimmer, K., Martin, R., Schwabe, L., Marino, J., Schummers, J., et al. (2009). The operating regime of local computations in primary visual cortex. *Cerebral Cortex, 19*, 2166–2180. doi:10.1093/cercor/bhn240.

Tanaka, K. (1983). Cross-correlation analysis of geniculostriate neuronal relationships in cats. *Journal of Neurophysiology, 49*, 1303–1318.

Tanaka, K. (1985). Organization of geniculate inputs to visual cortical cells in the cat. *Vision Research, 25*, 357–364.

Troyer, T. W., Krukowski, A. E., & Miller, K. D. (2002). LGN input to simple cells and contrast-invariant orientation tuning: an analysis. *Journal of Neurophysiology, 87*, 2741–2752.

Troyer, T. W., Krukowski, A. E., Priebe, N. J., & Miller, K. D. (1998). Contrast-invariant orientation tuning in cat visual cortex: Thalamocortical input tuning and correlation-based intracortical connectivity. *Journal of Neuroscience, 18*, 5908–5927.

Tsodyks, M., Kenet, T., Grinvald, A., & Arieli, A. (1999). Linking spontaneous activity of single cortical neurons and the underlying functional architecture. *Science, 286*, 1943–1946.

Ullman, S., & Bart, E. (2004). Recognition invariance obtained by extended and invariant features. *Neural Networks, 17*, 833–848. doi:10.1016/j.neunet.2004.01.006.

Volgushev, M., Pei, X., Vidyasagar, T. R., & Creutzfeldt, O. D. (1993). Excitation and inhibition in orientation selectivity of cat visual cortex neurons revealed by whole-cell recordings in vivo. *Visual Neuroscience, 10*, 1151–1155. doi:10.1017/S0952523800010257.

Walker, G. A., Ohzawa, I., & Freeman, R. D. (1998). Binocular cross-orientation suppression in the cat's striate cortex. *Journal of Neurophysiology, 79*, 227–239.

27 The Cortical Organization of Binocular Vision

RALPH D. FREEMAN

BINOCULAR FUNCTION IN THE VISUAL CORTEX

The early visual pathway from ganglion cells in the retina converges on the second-order neurons in the lateral geniculate neurons (LGN) of the thalamus. Although there is some minimal amount of binocular interaction in LGN cells, the main first-stage combination of inputs from left and right eyes occurs in the primary visual cortex. In animals with frontally placed eyes there is considerable visual field overlap of left and right images. This provides a basis for stereoscopic depth processing. Primary visual cortex is therefore the initial stage in the binocular visual processing system.

This review is intended to cover basic physiological characteristics of the binocular system. The focus is almost entirely on physiological mechanisms in the visual cortex. In the field of binocular vision, stereoscopic depth discrimination is of primary importance. Although some topics on stereoscopic depth are covered in this review, other chapters in this volume are more directly relevant (see chapter 28 by Parker and chapter 57 by Schor).

An effort has been made to cover topics that are considered the most important for this subject. However, there was no attempt to be comprehensive or to summarize all reports in the literature. The main intention is to discuss significant work that is directly relevant to the topics of binocular interaction.

OVERVIEW

Inputs from left and right eyes project into the visual cortex and enter specific cortical laminae. Layers 4 and 6 are the main input layers. The input layer cells are largely monocular in the primate but less so in the cat (Hubel & Wiesel, 1962, 1968). The elaboration from input layers into other laminae of the visual cortex delivers binocular characteristics. If we only consider monocular pathways, we obtain a two-dimensional analysis of external visual space. From monocular images we are limited to two-dimensional maps of object space.

These maps are processed in monocular pathways. However, in animals with frontally placed eyes, left and right eye views of visual space are slightly different. This causes left and right images to be relatively displaced from each other. These disparate images cause binocular retinal disparity, which is the necessary and sufficient condition for stereoscopic depth discrimination. From this process, two-dimensional projections of visual space, which result from monocular images, are transformed into a three-dimensional percept. This disparity-based stereopsis was demonstrated in behavioral investigations in the early nineteenth century.

The major original investigations of binocular function in the visual cortex were conducted during systematic single-cell recordings in area 17 in the anesthetized cat. In this work neurons were found that could be activated through either left or right eyes. The assumption was made that if stimulation via either eye was effective in driving a cortical cell, the neuron could then be considered as binocular. From this early work and much that followed, conclusions about binocular function were made based solely on monocular tests. As considered below, monocular testing can yield inaccurate assessments of binocularity. However, it should be mentioned that in the landmark early investigations of binocular interaction in the cat (Hubel & Wiesel, 1962), some binocular tests were conducted, and it was noted for some cells that responses were obtained only for simultaneous activation of both eyes. In spite of this constraint a considerable literature followed in which binocularity was assessed solely on the basis of monocular activation. A field that expanded substantially during this time was concerned with the development and plasticity of vision. A primary assessment in most of these studies had to do with the degree of binocularity in early visual development or following specific regimes of rearing. The problems with this approach are considered below.

The discovery of binocular neurons in the visual cortex was interpreted as a direct physiological connection to stereoscopic depth perception (Hubel & Wiesel

1962). However, no mechanism was proposed to indicate how the system might be organized. Experiments that followed the early work generally focused on cells that responded to stimuli at different depth positions in space. The assumption again was that cortical neurons are directly involved in depth detection. This kind of approach has a long history in which critical features of biological systems are studied in relation to survival requirements. A limitation of this type of approach is that results of a given experiment may be linked exclusively to a specific interpretive conclusion. For example, cortical binocular cells might be organized as if they are responding to depth when it is possible that the observations could be accounted for in terms of receptive field organization without direct cognitive implications. In order to establish conclusively that binocular cells in the visual cortex are in fact part of a depth discrimination system, it is necessary to extend single-cell study further by inclusion of behavioral observations. This required extension of the early work has been made, and it is now possible to conclude that binocular cells in the visual cortex are, in fact, directly related to binocular depth discrimination.

The most basic physiological application of binocular vision concerns the rules by which neurons combine the inputs from both eyes. Approaches to address this area are similar to those that have been used for the monocular system. The overall purpose is to establish how a neuron integrates visual input within receptive fields. The intention of this approach is to unify a consideration of mechanisms of both monocular and binocular processing. This approach is exemplified as follows. In the simplest case symmetrical inputs are combined in a linear manner. This is represented in figure 27.1A. A threshold mechanism that follows binocular convergence can be altered following binocular input. Another fundamental type of combination is multiplicative convergence, which is depicted in figure 27.1B. An example of this kind of combination is seen when a cortical cell is not responsive to input from either eye but responds to simultaneous activation of both eyes (Barlow, Blakemore, & Pettigrew, 1967; Hubel & Wiesel, 1962, 1968). In a symmetrical input postsynaptic or presynaptic inhibition may be included, as illustrated in figure 27.1C and D, respectively. A considerable amount of data in the literature is consistent with processes of linear summation that are combined with threshold mechanisms. There are also data for which multiplicative binocular interaction appears to account for findings in cortical cells (Anzai, Ohzawa, & Freeman, 1999a, 1999b).

Regardless of the rules by which signals from left and right eyes are combined, the primary application, as

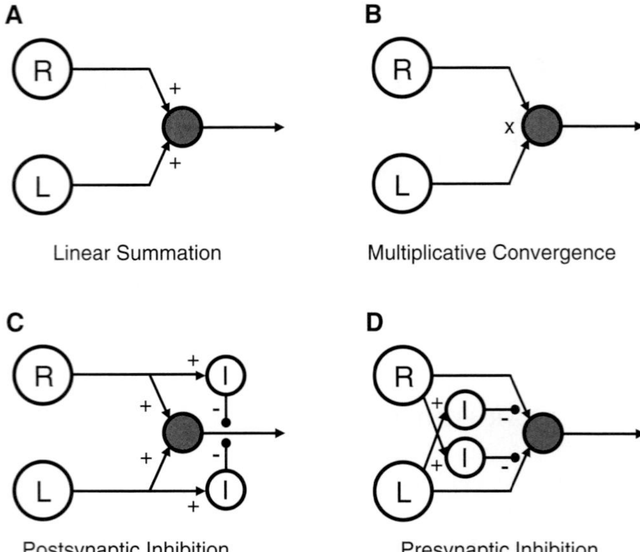

FIGURE 27.1   Models are depicted that represent possible ways by which input from each eye may be combined at the level of visual cortex. Right and left eyes are represented by R. and L., respectively, and inhibitory units are designated as I.

noted above, has been focused on stereoscopic depth discrimination. Chapters 28 by Parker, 57 by Schor, and 58 by Blake in this volume are concerned with the details of this process. An overview of the physiological basis of stereoscopic depth processing is considered in what follows. Although binocular cortical neurons were identified in earlier work, the first direct investigation of the neural basis of stereoscopic depth discrimination was performed in an anesthetized preparation (Barlow, Blakemore, & Pettigrew, 1967). Cells in area 17 of the cat were found to respond selectively to stimuli that produced different retinal disparities. These differences in retinal position of binocular stimuli corresponded to different depths in space from near to far positions. The experimental arrangement is illustrated in figure 27.2. Neurons were studied that responded to a considerable range of both crossed and uncrossed disparities. The range of disparities for vertically oriented stimuli was around three times larger than that for horizontal orientations. This difference is consistent with the horizontal displacement of the two eyes. Subsequent investigations varied in the extent to which differences were found between horizontal and vertical disparities (Ferster, 1981; Joshua & Bishop, 1970; LeVay & Voigt, 1988; Nikara, Bishop, & Pettigrew, 1968; von der Heydt et al., 1978). All of the early experiments and analyses were predicated on the basis that retinal disparity occurred only from shifts in image positions between left and right eyes. As described below, disparity pro-

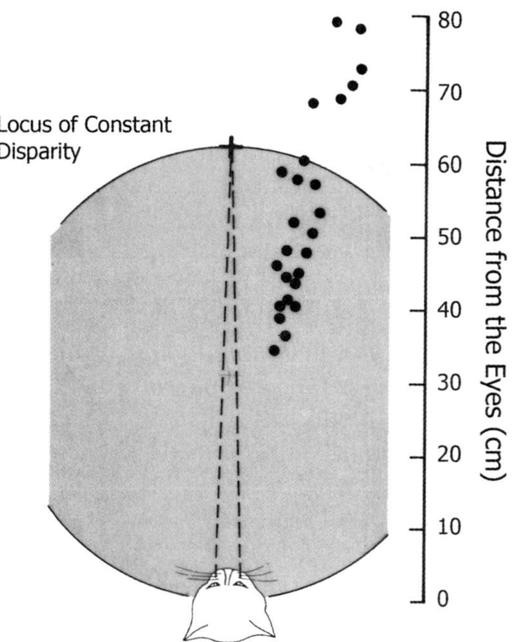

FIGURE 27.2 An animal is depicted viewing a curved tangent screen at a fixed distance. Filled circles denote neurons that are tuned to various depths from near to far positions. The curved arc passing through the fixation point traces constant-disparity positions.

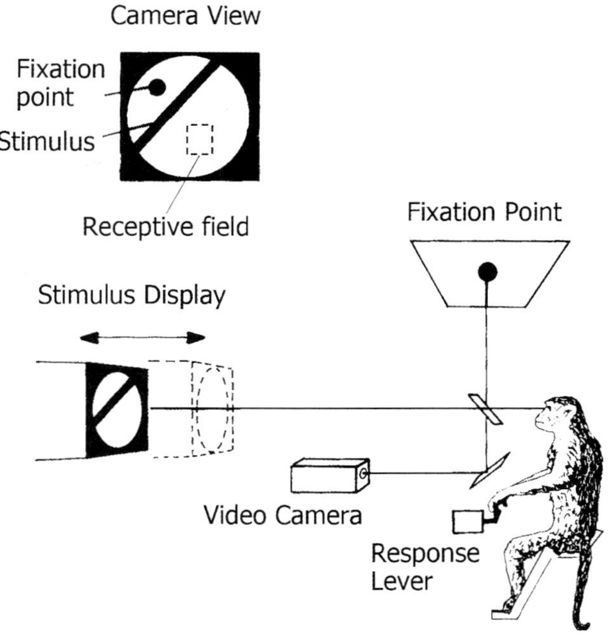

FIGURE 27.3 A testing condition is illustrated for a primate to enable simultaneous neurophysiological and behavioral determinations. A fixation point and a stimulus display are viewed via a beam splitter. Relative depth of the stimulus display may be changed as illustrated. Neurophysiological study of receptive fields of cortical cells is analyzed during different depth positions. The animal is trained to press a button during illumination of the fixation spot. Different visual objects are used while neural activity and behavioral decisions are monitored.

cessing may also occur from dissimilarities in the internal structure of left and right eye receptive fields.

The establishment of disparity-sensitive neurons in the cat's visual cortex was an important advance. However, it was necessary to extend these findings to the case of the primate. And for this purpose investigations were conducted in monkey striate cortex. The initial study of this type produced the remarkable claim that there was an absence of disparity-sensitive cells in V1 but a population of such neurons in V2 (Hubel & Wiesel, 1970). It is not obvious why in this study disparity-sensitive neurons were missed in V1, but subsequent work with behaving monkeys from which neurons were also recorded showed clearly that both V1 and V2 and higher-order areas contained high proportions of cells that responded differentially to changes in relative depth (Cumming & DeAngelis, 2001; Krug & Parker, 2011; Poggio & Fischer, 1977; Poggio, Gonzalez, & Krause, 1988; Poggio & Talbot, 1981; Shiozaki et al., 2012; Tanabe, Haefner, & Cumming, 2011). The study of awake behaving monkeys is important because it allows simultaneous assessment of both behavioral and neurophysiological responses. An experimental setup for this work is depicted in figure 27.3. In this configuration animals were trained to fixate on a location on the screen. The screen's distance from the animal could

be varied, and individual cortical cells were recorded during fixation. Categories of cell types were reported, and in the initial work four types of depth-sensitive neurons were identified: tuned excitatory, tuned inhibitory, near, and far cells as represented in figure 27.4. Note that the figure contains two additional response types—tuned near and tuned far. The two additional categories are the result of a subsequent study (Poggio, Gonzalez, & Krause, 1988), but the idea of distinct classes of depth-sensitive neurons is conceptually problematic, as outlined in other reports (Freeman & Ohzawa, 1990; LeVay & Voigt, 1988).

In the initial investigations no attempt was made to consider a system within which disparity detecting and encoding might be unified. One way to do this is to use an encoding that is scaled to the size or spatial frequency selectivity of cortical cells. For this type of scheme, receptive field internal spatial detail must be considered. Receptive field characteristics such as subunits of left and right eye images may be analyzed to determine if there are differences between the two eyes. If so, they can be utilized to encode relative disparity.

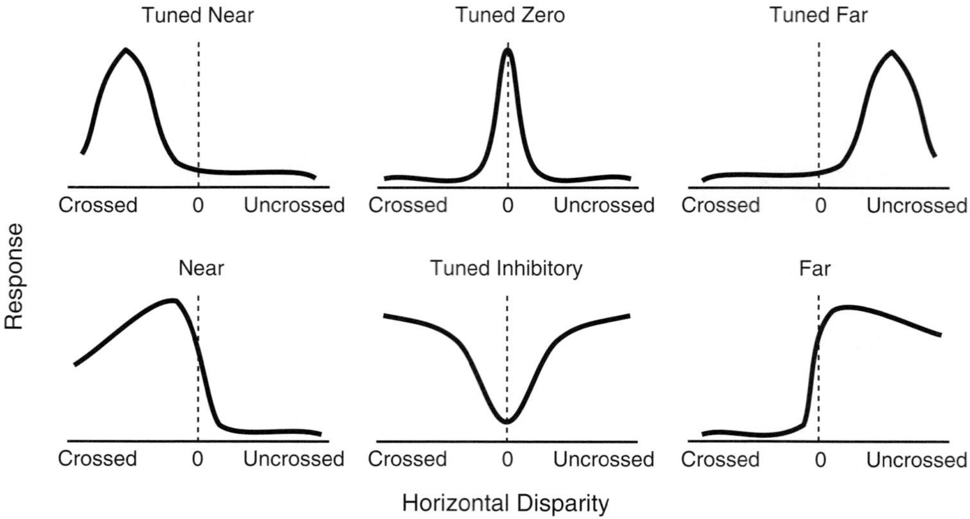

FIGURE 27.4   Six binocular response patterns are represented as typical for cortical neurons. Crossed and uncrossed horizontal disparity positions are indicated on the abscissas. In this scheme tuned zero or tuned inhibitory types respond or are suppressed, respectively, at a disparity of zero. Near and far cells are activated, respectively, at close or distant locations. Tuned-near and tuned-far types represent intermediate location peak responses.

For this approach a quantitative analysis is required of the details of internal receptive field shape or phase. This kind of analysis may provide information on binocular disparity coding. We can then determine if phase-disparity-selective neurons at different spatial frequency scales may be used in disparity processing in addition to information concerning relative position. This type of analysis is considered further on in this chapter.

Binocular Assessment from Monocular Tests

In the early exploratory studies of the visual cortex, it was determined that most cortical neurons responded to stimulation of either eye and were therefore considered to be binocular (Hubel & Wiesel, 1962). The subjective observations were made almost exclusively from monocular tests. Findings included the observation that different cortical cells do not necessarily respond equally to stimulation of each eye. Frequently, stimulation of one eye elicited stronger responses than that of the other. The result of this observation was the creation of a subjective scale of relative ocular dominance (Hubel & Wiesel, 1962). The original scale included seven categories: Groups 1 and 7 represent neurons that are entirely monocular; group 4 includes cells that have equally balanced input from left and right eyes; groups 2, 3, 5, and 6 designate binocular cells with different degrees of ipsilateral or contralateral dominance. Some studies that followed the original ones employed different numbers of categories of ocular dominance. Nearly all of this work was of a subjective nature. In the original study a footnote was included that stated that groups 2 and 6 included some cells for which the nondominant eye was ineffective in activation of a cell but could influence the response to stimulation of the dominant eye (Hubel & Wiesel, 1962).

Soon after the original exploration of the visual cortex, considerable interest was expressed in studies of development and plasticity of the visual system. An extensive literature resulted from these studies in which binocular status of rearing conditions was critical to the conclusions of the investigations. In most of this work subjectively obtained ocular dominance histograms were used to assess effects of visual experience. It is somewhat remarkable that very little concern was given to testing procedures. The problem, as shown below, is that assessment of binocular function depends on whether monoptic or dichoptic stimulation is used. In general, dichoptic activation reveals more binocular function than is assessed by monoptic techniques. This is probably because of a threshold nonlinearity in which a subthreshold input from one eye that appears to be silent can be expressed under dichoptic viewing conditions. Specifically, cells can appear to be monocular when each eye is tested individually but can exhibit clear binocular interaction when eyes are tested dichoptically (Freeman & Ohzawa, 1988, 1990, 1992; Freeman & Robson, 1982; Ohzawa, DeAngelis, & Freeman, 1996; Ohzawa & Freeman, 1986a, 1986b, 1988).

Measurement of Binocular Interaction

Early tests of binocular interaction were conducted in which visual stimuli consisted generally of single bars or spots or random dots with selected disparity varying presentation. To avoid the possibility of missing critical interactive spatial locations, attempts were made to verify disparity-selective neurons in the visual cortex by use of relatively large sinusoidal gratings that were presented to each eye. In these studies relative interocular phase was shifted in order to create different interocular relative disparity conditions (Freeman & Ohzawa, 1990; Freeman & Robson, 1982). This approach follows from early studies of the retina in which gratings with sinusoidal luminance profiles were employed to determine spatial summation characteristics of receptive fields of ganglion cells (Enroth-Cugell & Robson, 1966). The interocular stimulation conditions are illustrated in figure 27.5. Simple cells in the visual cortex are represented by Gabor functions in the figure. The abscissas of these curves designate retinal position, and the ordinates give sensitivity to luminous increments. Negative positions of each curve represent OFF (dark excitatory) subregions, and positive positions denote ON (bright excitatory) subregions. Dashed curves represent Gaussian envelopes of the idealized receptive fields.

For simple cells in the visual cortex dichoptic presentation of phase-varying gratings should cause increases and decreases of firing rates that correspond, respectively, to stimulation of ON and OFF areas of the receptive field. Examples of this type of response are shown in figure 27.6. Initially, monoptic tests are conducted to establish preferred stimulus parameters for each cell. Results for the monocular tests are shown along with those for dichoptic stimulation in the figure. Monoptic, dichoptic, and blank or null conditions are interleaved in each run. Note for this example that dichoptic tests yield modulated bursts of activity, which correspond to ON areas, and silent phases in which ON and OFF regions are counterbalanced. Null conditions establish levels of spontaneous activity.

Complex cells in the visual cortex generally produce a different style of response than do simple cells. Spike rates from monoptically driven complex cells are generally uniform and often vigorous and have discharge activity that is relatively constant across relative interocular phase ranges. Examples of the expected relative interocular phase patterns for complex cells are shown in figure 27.7. There are, however, unexpected responses from complex cells: a large proportion show phase specific interaction. An example of the latter is illustrated in figure 27.8. Standard monocular tuning functions for orientation and spatial frequency show weak and

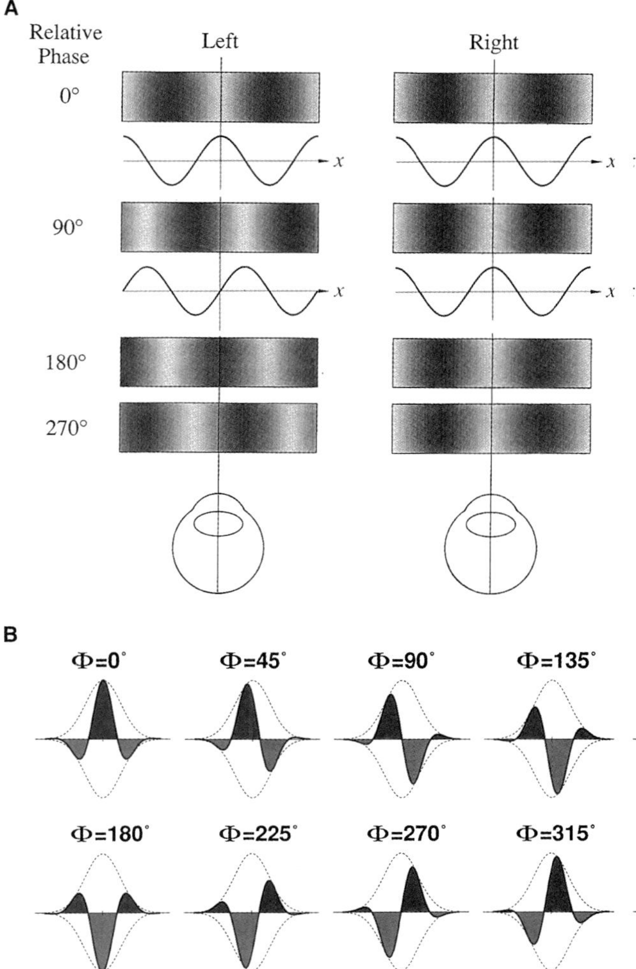

FIGURE 27.5 Conditions are represented here for binocular relative phase shift experiments. (A) Relative phase-shift positions between left and right eyes are shown for a 0° and a 90° phase shift. At the bottom additional phase-shift conditions are shown for 180° and 270°. (B) Various phase angles from 0° to 315° are illustrated for Gabor functions with Gaussian envelopes, which represent receptive field profiles for simple cells.

unmodulated responses. But for dichoptic relative interocular phase runs, post-stimulus-time histograms show changes in response levels. For this example, although monocular responses are weak, relative interocular phase activity is reasonably strong. Moreover, there are clear differences in response values as a function of relative interocular phase so that the relative phase-tuning function shown in figure 27.8 is clearly phase tuned. This is not expected because ON and OFF regions in complex cells are generally considered to be mixed together so that stimulus position in the monocular mode is not critical. Binocular phase tuning of complex cells is possible if subunits are organized as simple cell receptive fields. In this case the subunits

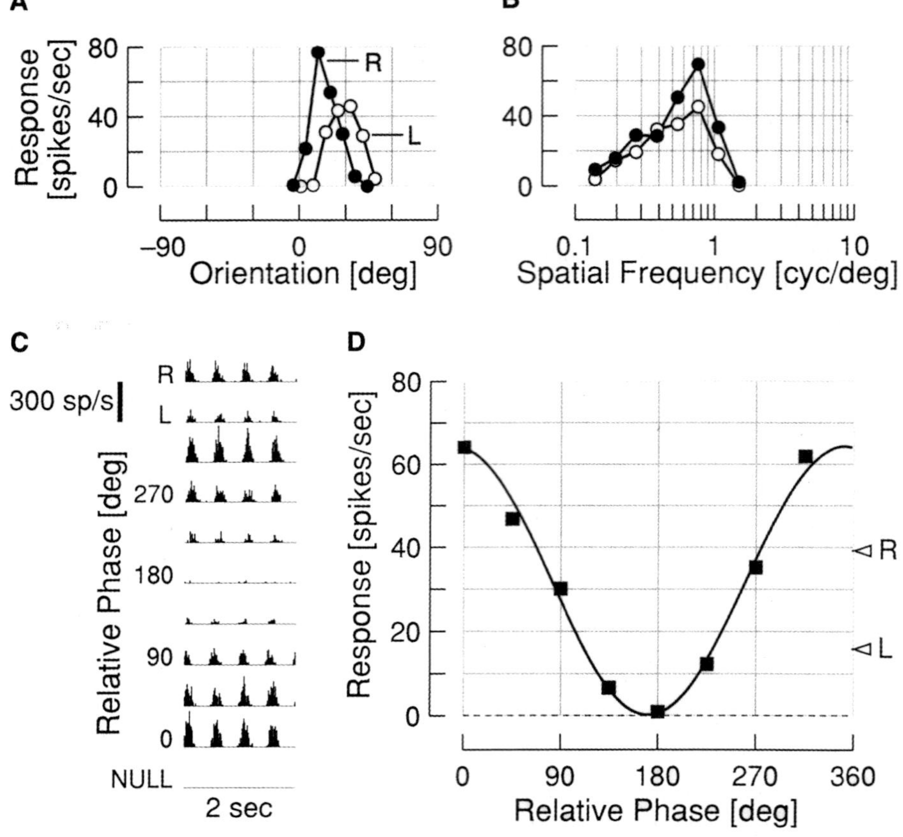

FIGURE 27.6   Response patterns are illustrated for monoptic and dichoptic tests of a simple cell in the visual cortex. (A, B) Tuning functions are shown for left and right eyes for orientation and spatial frequency. (C) Monocular responses and relative interocular phase responses are shown in poststimulus time histogram form. (D) The relative interocular phase profile for this cell is shown together with monocular responses of right and left eyes.

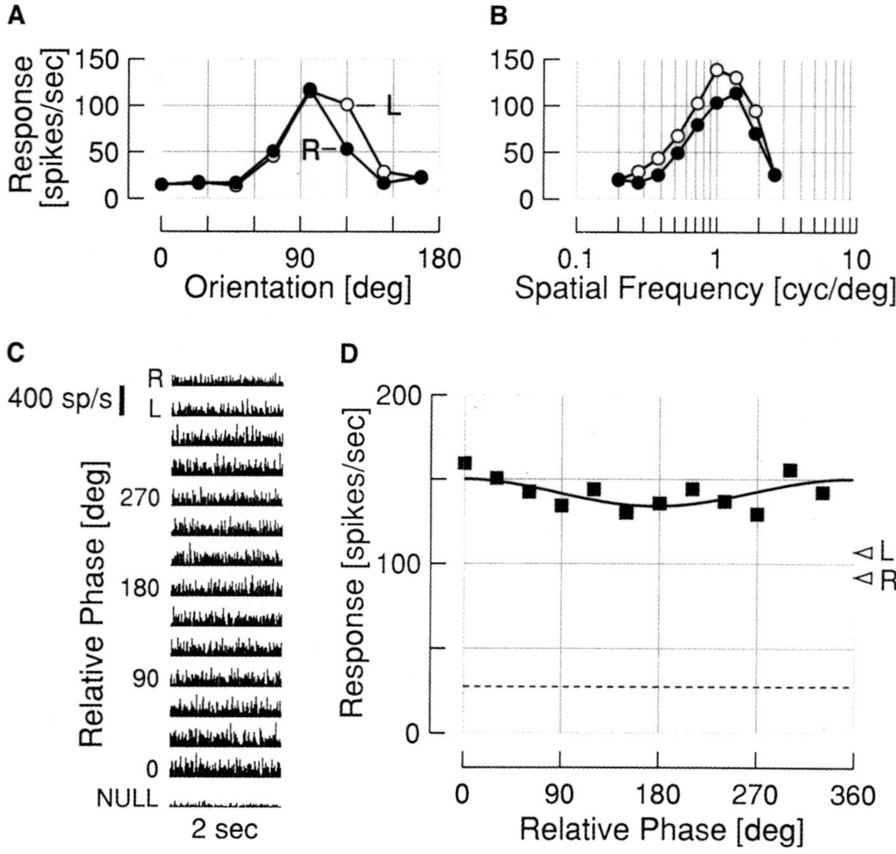

FIGURE 27.7   Response patterns are shown here for monoptic and dichoptic tests of a complex cell in the visual cortex. Conditions and components are the same as in figure 27.6.

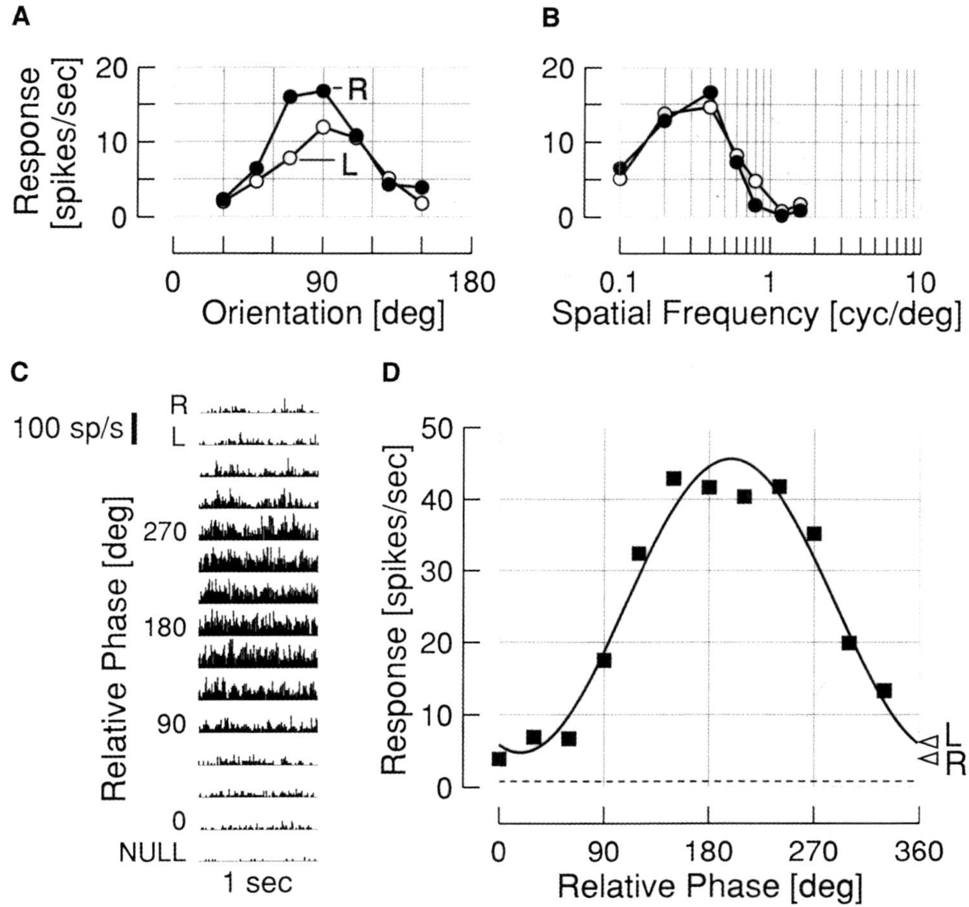

FIGURE 27.8   Another set of response patterns is shown for monoptic and dichoptic conditions for a cortical complex cell with responses that are binocularly phase specific. Conditions and components are the same as in the previous two figures.

would be phase specific, and this would be retained at the output level of the neuron.

A technique may be employed to probe subunit organization as follows. In figure 27.9 a complex cell is represented for three different stimulus protocols. In A, a sinusoidal grating with optimal parameters of orientation, spatial frequency, and temporal frequency is drifted across the receptive field while one eye is occluded. The resulting response of this stimulation is a standard increase in mean discharge. A second condition, shown in figure 27.9B, illustrates a different monoptic stimulation protocol. Instead of being drifted as in the previous case, this grating is counterphased in a single position so that it flickers sinusoidally with dark and bright components. In this case, as shown at the bottom, the cell discharges in a modulated fashion at twice the temporal frequency of the counterphase. In other words there are two bursts of discharge per cycle of flicker—a frequency-doubled response (Movshon, Thompson, & Tolhurst, 1978). The third condition shown in the figure enables an exploration of the cells'

subunits. In this condition the same sine-wave grating used in condition A is repeated with the exception that both eyes are activated instead of one. In addition, the grating drift for left and right eyes is in opposite directions. Because identical sine waves drifted in opposite directions at the same drift rate combine to result in a single flickering sine wave with twice the peak amplitude of each drifting component, the following test is created. If the drifting gratings shown to each eye are combined at an early neural stage, the resulting response should be equivalent to that for a counterphase grating. In this case there will be modulated discharge at twice the frequency of the drifting gratings. This would indicate that the subunits of the complex cell sum the inputs from each eye in a linear manner. If there is instead a nonlinear combination, the expected response of the cell would be a mean discharge increase that is not modulated. These possible response patterns are represented in figure 27.9C.

Results for the test outlined above are shown for an example complex cell in figure 27.10. The top two

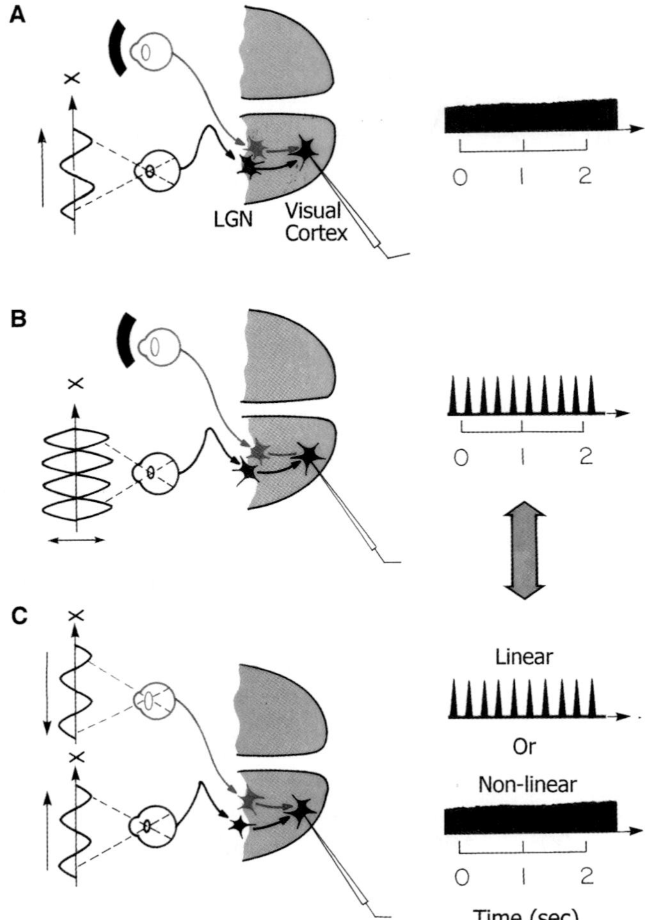

FIGURE 27.9   Three stimulus conditions are shown that are intended to identify the stage at which binocular combination occurs. In the three conditions a complex cell is stimulated as follows. (A) A sinusoidal grating is drifted across the receptive field of one eye while the other eye is occluded. A standard discharge rate increase is obtained for this condition. (B) The drifting grating is replaced by one that is presented in counterphase to the same single eye while the other is occluded. In this case the cell responds with a modulated discharge rate that is double the temporal frequency of the stimulus. The final panel (C) depicts stimulation of both eyes with drifting sinusoidal gratings for which the drift direction is opposite in left and right eyes. The result of this stimulus pattern is either a modulated or average increase discharge that represents linear or nonlinear combinations, respectively.

histograms represent monocular responses for left and right eyes. In this example there is a modest response for the right eye and a weak one for the left eye. Because the stimulation drift direction is opposite for each eye, this response pattern indicates direction selectivity. However, in the dichoptic mode, there is a clear modulated discharge for all interocular phase values that were used. Note that the temporal scale indicates 1 s of response, and each grating is moving at a drift rate of

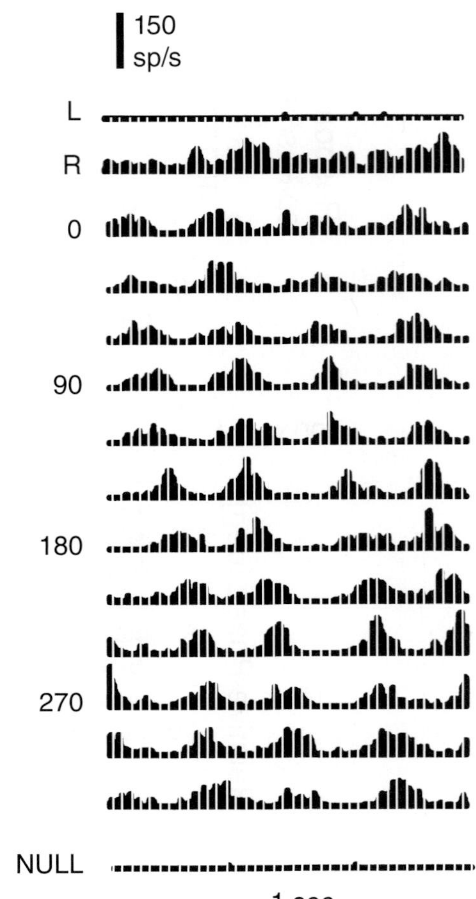

FIGURE 27.10   Poststimulus time histograms are shown as response patterns for a complex cell that is activated by sinusoidal gratings drifting in opposite directions for left and right eyes. For dichoptic presentation there are four bursts of discharge per second for a drift rate of 2 Hz, which indicates a frequency-doubled response.

2 Hz. In other words the cell produces a frequency-doubled discharge. The implication is that there has been a linear combination of signals from left and right eyes at the subunit level. Data for this kind of experiment are available (Freeman & Ohzawa, 1988, 1992; Ohzawa & Freeman, 1986a, 1986b).

Monocular versus Binocular Contrast Sensitivity

It is of interest to consider monocular versus binocular processing of basic visual parameters. Contrast is probably the most straightforward feature of the visual scene. It is likely that changes in contrast levels do not necessarily bring into play different categories of cells. Most of the findings that consider contrast sensitivity of binocular compared to monocular stimulation result from psychophysical investigations. A standard notion is that binocular summation is not simply a statistical event in which two eyes are more effective than one. An active process is thought to occur in the neuronal interaction between left and right eye signals (Blake, 1981; Blake, Sloane, & Fox, 1981; Blake & Wilson, 2011).

Some neurophysiological studies have been focused on this area, and there have been efforts to connect behavioral and physiological findings. The basic neurophysiological assessment of binocular summation may be obtained from measurements of contrast response functions of neurons. Comparisons of tests from monoptic and dichoptic stimulation permit the derivation of neurometric functions that may be applied to single neurons. In one study contrast thresholds and slopes for each neurometric function were compared in order to connect physiological and behavioral assessments (Bradley et al., 1985). Related neurophysiological findings are illustrated in figure 27.11. Results are compared for left, right, and both eyes for a simple cell in A. As shown, responses increase approximately linearly with log contrast above a threshold. Saturation occurs in the binocular test at high contrast levels. There is a steady and regular increase in response strength of binocular data compared to monocular

runs at most contrast levels. Analogous data are shown for a second cell (complex type) in figure 27.11B, where similar patterns may be observed. A considerable amount of data and analysis shows that individual neurons have a clear binocular advantage in the processing of visual data (Anzai et al., 1995; Bradley et al., 1985).

Unequal Contrast in Left and Right Eyes

An interesting property of contrast processing in the visual cortex is analogous to one in the retina. Retinal ganglion cells adapt to prevailing absolute luminance levels, and neurons in visual cortex express a similar mechanism with respect to ambient contrast. The contrast gain control system in visual cortex has been studied and described in previous work (Ohzawa, Sclar, & Freeman, 1982, 1985). Results showed that the contrast gain control mechanism is an effective way to maintain high sensitivity across a range of contrast values. Comparisons of monocular and binocular contrast encoding suggest a mechanism that is primarily monocular (Truchard, Ohzawa, & Freeman, 2000). In most previous work on binocular effects, contrast values

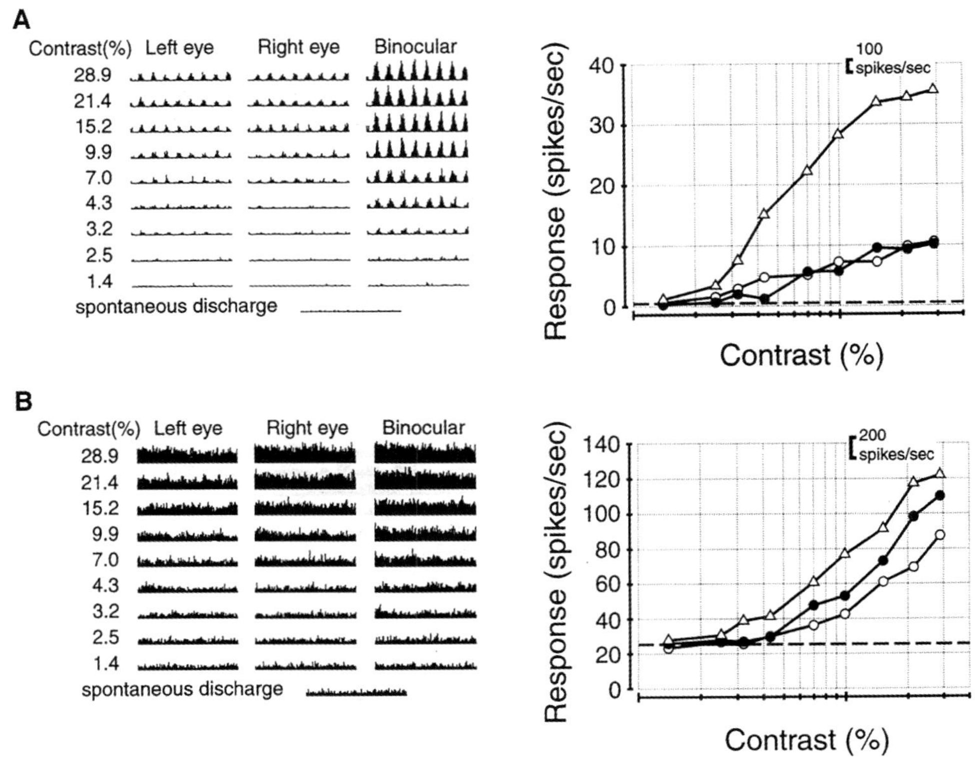

FIGURE 27.11   Results are presented for tests of monocular and binocular contrast sensitivity of cortical cells. Poststimulus time histograms and contrast response functions are shown for monocular and binocular conditions for a simple cell (A) and a complex cell (B). Filled circles, open circles, and open triangles represent mean discharge rates for right eye, left eye, and binocular stimulation, respectively. Spontaneous levels are indicated by dashed lines.

for both eyes have been equal. However, it is of interest to examine the possibility of low contrast for one eye and a high value in the other. There are clinical conditions for which visual function in one eye is relatively reduced. It is also possible in a natural scene for a normal binocular individual to have relatively reduced contrast in one eye. Standard psychophysical tests have shown that a weak signal in one eye can substantially reduce binocular functions such as stereoscopic vision (Schor & Heckmann, 1989). However, there is a striking difference in physiological results involving interocular contrast differences. If one conducts binocular interaction tests in which interocular phase is varied between dichoptic gratings with different contrast values for each eye, there is a compensation process that boosts the weaker eye response so that binocular interaction is not severely affected (Truchard, Ohzawa, & Freeman, 2000).

Results of tests that demonstrate this process are shown in figure 27.12. First, 50% contrast gratings are phase-varied and presented to left and right eyes. Clear phase-specific responses are observed. This example cell was a complex type for which phase-specific responses occur from around half the population. A second run for the same cell was made with reduced contrast (6.3%) to one eye. As seen in the figure, the second run produces a nearly identical result to that of the first. Other runs were made with different contrast values for each eye, and similar findings were obtained. These results are summarized in a quantitative binocular interaction profile in the figure. To do this, one cycle of the sinusoid is fit to the phase-specific response curve, and a depth-of-modulation index is derived. The depth-of-modulation index is equal to the ratio of the amplitude of the fitted sinusoid over the mean of the binocular responses. The larger the index, the stronger will be the degree of binocular interaction. A relatively flat interaction curve yields low index values. At the limit, stimuli presented to one eye have no effect on the other. The depth-of-modulation index therefore provides a quantified indication of the effect one eye has on a given neuron. For this index, excitation through the other eye is used as the reference.

Summary results for the above data are presented in figure 27.12C. Depth of modulation is given on the left vertical axis, and the cell's response to different contrasts is shown for one eye as labeled on the right vertical axis. It is clear that intraocular contrast differences have minimal effects on depth of modulation, which is similar to that for identical contrast values for each eye. Similar findings have been observed for a population of these types of tests. A summary of the findings for the cells tested is given in figure 27.12D. Clearly,

contrast differences between left and right eyes have minimal consequences on depth-of-modulation values. The implication is that there is a sort of contrast gain reduction system at a monocular level (Truchard, Ohzawa, & Freeman, 2000).

The data from these experiments show a clear compensatory mechanism that results in relatively constant binocular interaction. Presentation of a low-contrast grating appears to raise the threshold, which creates a synaptic gain of signals from the weak-stimulus eye. Presumably, gain is high when input is weak and low for high-contrast input. There may be two components in the compensatory process. First, a threshold mechanism may be operational that follows left and right eye input convergence. A strong input level from either or both eyes could raise spike discharge threshold levels. Since monocular input can raise the threshold, and as a result both eyes are affected, this could account for interocular transfer of contrast gain control. The second component may also be a contrast gain control mechanism. It could be monocular and operate at the synaptic interface between one eye and a cortical neuron. The second mechanism could account for constancy of binocular interaction when low- and high-contrast gratings are used, respectively, for left and right eyes.

An interesting implication of the results discussed above concerns the use of ocular dominance designation as a way of assessing binocular vision. The dichoptic presentation of gratings of unequal contrast between left and right eyes alters ocular dominance at the input level. Contrast is a stimulus parameter that affects input strength at the cortical level. Although the gain control mechanism is not easily accessible for experimental control, the overall input to a given cortical cell may be changed by the strength of the stimulus. In effect this changes the ocular dominance of the cortical cell. This is, of course, an artificial change and of a temporary nature. The surprising finding of unequal-contrast stimuli to right and left eyes is that binocular interaction remains relatively constant over a wide range of input strength differences. In a standard neurophysiological assessment the range of ocular dominance of values that are observed could be the consequence of a genetic compromise because it might be relatively difficult to program every cortical cell to have exactly balanced input from the two eyes. Ocular dominance variability, as measured in the laboratory, may be relatively unimportant for binocular visual function. The categorization of ocular dominance could be a laboratory exercise that does not apply in the functional operation of binocular vision in awake behaving animals.

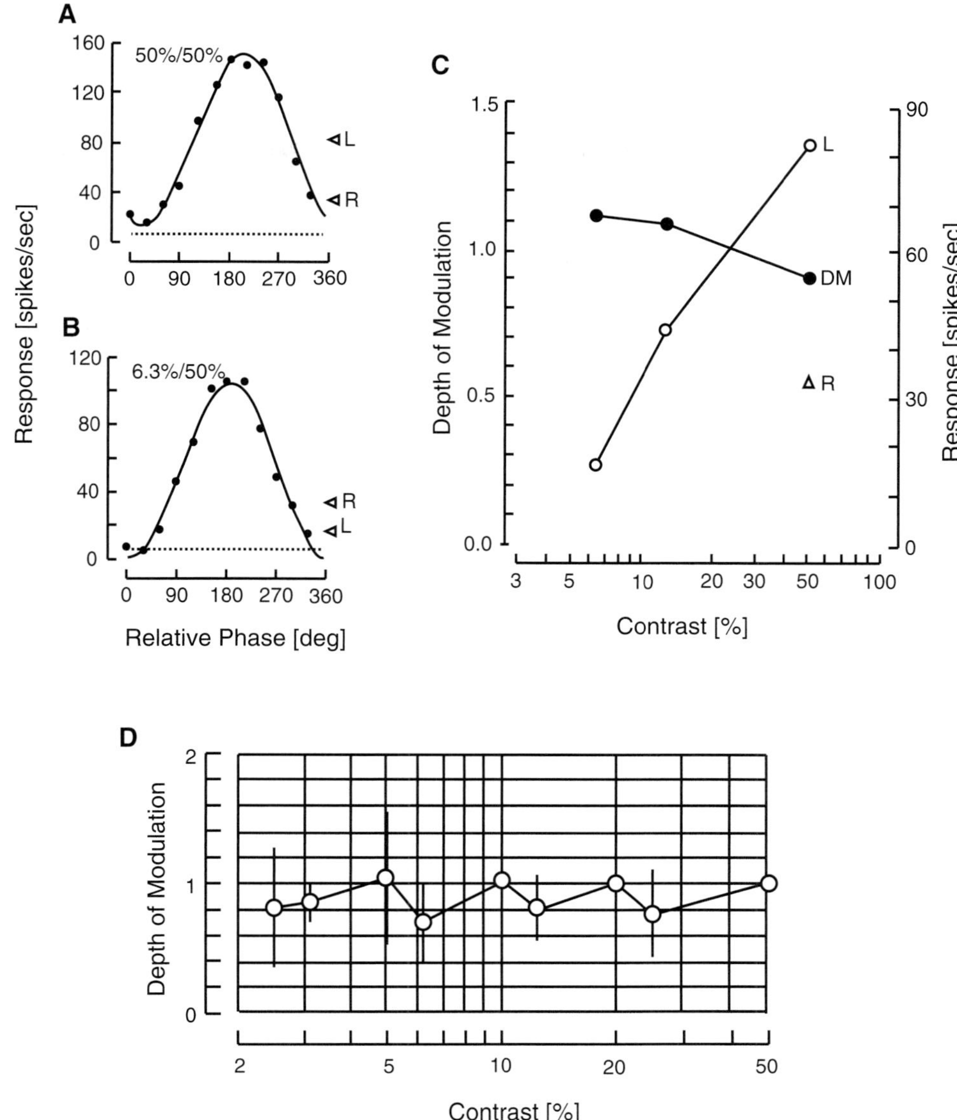

FIGURE 27.12   Results are presented for binocular interaction runs derived from tests in which grating contrast is substantially different in left and right eyes. Binocular interaction tests in A and B were conducted for 50% contrast gratings in left and right eyes (A) and for 50% and 6.3% in left and right eyes (B). Depth of modulation is shown for four contrast difference conditions (C, filled circles). A contrast response function is also shown for one eye with response rate on the right. (D) Results for a cell population are shown indicating that the depth of modulation is relatively constant regardless of contrast differences between left and right eyes.

Plasticity of Binocular Vision

The development and plasticity of the visual system comprise an extensive topic that is covered in other chapters in this volume. However, a brief section on this topic is included here because it emphasizes how important it is to study the physiological organization of binocular vision by observation of response characteristics for dichoptic instead of solely by monoptic experimental procedures. A basic question has to do with developmental onset of functional performance of the binocular apparatus. Quantitative studies in kitten and primate in which dichoptic stimulation was employed show clearly that the physiological apparatus for binocular vision is operational as early as accurate measurement is possible (Freeman & Ohzawa, 1992). The importance of dichoptic stimulation is illustrated in the following example (see figure 27.13). The example is for a cortical neuron recorded from a normally reared 3-week postnatal kitten. Standard tuning curves for orientation and spatial frequency are shown for each eye. For one eye the functions appear

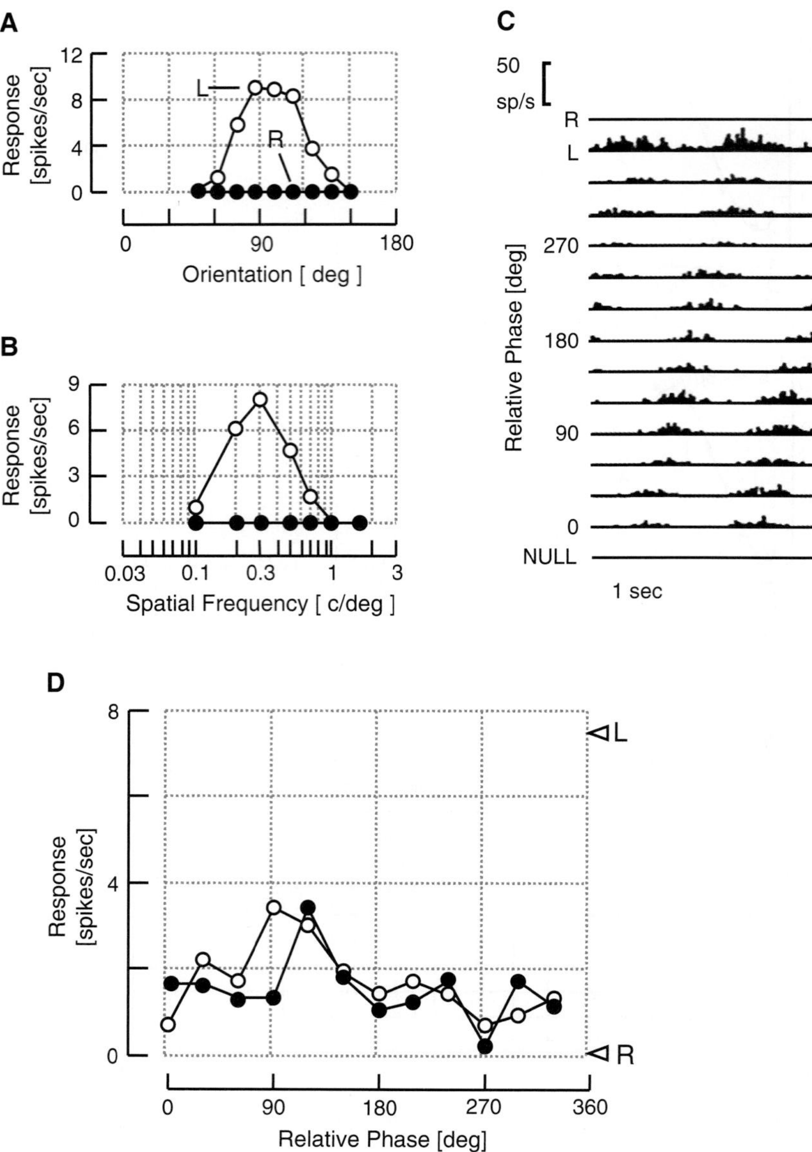

FIGURE 27.13   A response set is shown that is similar to those of previous figures. In this case the findings are from an immature visual cortex for which monocular stimulation shows a lack of right eye response for both orientation and spatial frequency (A, B). However, binocular interaction poststimulus time histograms (C) show clear evidence of suppression by the eye that was silent in monocular tests. (D) Relative binocular phase-response runs are shown. Open and filled symbols represent separate tests, which yield consistent findings.

standard, but for the other, there is a complete lack of response. For gratings presented dichoptically at different relative interocular phases, clear modulated responses are observed. In this case the effect is suppressive. The eye that is silent when tested monoptically exerts a clear suppressive effect in the dichoptic mode. The figure also shows that a repeat of the same stimulation paradigm produces closely similar results to those from the first set of tests. These findings show clearly that an eye that appears silent in the monocular mode is physiologically connected to the neuron under study during dichoptic tests. In other words inaccurate conclusions would be made from monocular tests alone. Many similar examples of these types of data are available (Freeman & Ohzawa, 1988, 1990, 1992; Freeman & Robson, 1982; Ohzawa, DeAngelis, & Freeman, 1990, 1996; Ohzawa & Freeman, 1986a, 1986b). The conclusion from this work is that dichoptic tests are required to establish binocular characteristics of visual neurons.

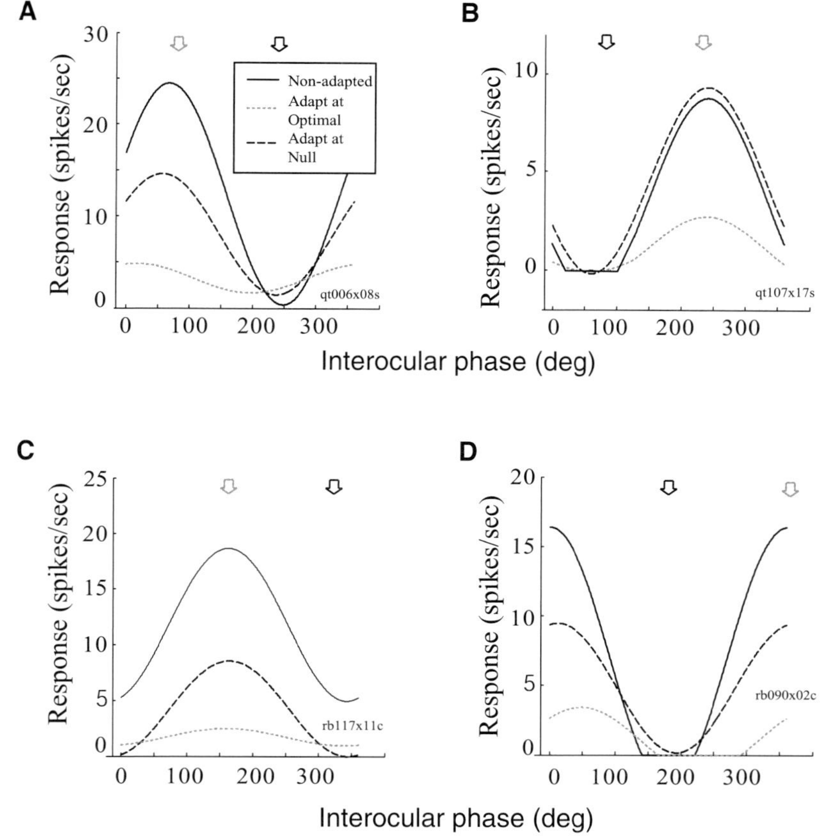

FIGURE 27.14   Results of different types of binocular disparity adaptation are shown for four representative cells (A and B are simple cells, and C and D are complex). The three curves in each panel represent nonadapted (solid curve), optimal phase adaptation (dashed curve), and null-phase adaptation (dotted curve). Black and gray open arrows represent null and optimal phases, respectively.

Adaptation of Neurons in the Visual Cortex That Code Depth Discrimination

Extended or repeated sensory stimulation has been used as an investigative tool of neural connectivity. In general the adaptive process that ensues produces a reduction in sensitivity to similar subsequent stimulation. Different visual stimulus parameters have been investigated with both behavioral and neurophysiological approaches (Blakemore & Campbell, 1969; Blakemore & Hague, 1972; Blakemore & Sutton, 1969; Duong & Freeman, 2007; Sanchez-Vives, Nowak, & McCormick, 2000a, 2000b). In addition to the reduction of responses to subsequent stimuli, there are also changes in specificity to parameters such as orientation, spatial frequency, and direction selectivity (Marlin, Douglas, & Cynader, 1993; Movshon & Lennie, 1979; Muller et al., 1999). A recent study has also explored adaptation of neurons that encode binocular depth discrimination in the primary visual cortex. In this work

cortical cells are adapted to prolonged periods of dichoptically presented sinusoidal gratings that have different relative interocular phase values. Interocular phase-tuning curves are determined before and after adaptation at optimal, near-optimal, and null interocular phase positions. Results show that adaptation at optimal phase values causes marked reductions in responses to subsequent stimuli. For the null phase, adaptation has minimal effects. At intermediate phases, there are changes in tuning curves away from those of the adapting phase values. Examples of typical results for interocular phase-tuning curves before and after adaptation to optimal and null phases are shown in figure 27.14. For the cells shown, there are minimal changes at null phase values but substantial reductions in tuning curve amplitudes for adaptation at optimal phases. These examples represent the general result that adaptation at optimal interocular phase values suppresses subsequent responses, whereas null phase effects are minimal (Duong, Moore, & Freeman, 2011).

Two-Photon Calcium Imaging

Most of the approaches that have been used in the study of the organization of binocular vision have been based on single- or multiple-neuron recordings in the central visual pathway. As in any discipline new approaches may provide insights that were not previously possible. A current example of a relatively new technology is two-photon calcium imaging. In this procedure a calcium indicator is pressure ejected just below the cortical surface in order to provide a population of stained cells. A very large population of cells for which two-photon images are available can then be analyzed. Two-photon calcium imaging may be used to sample monoptic and dichoptic visual stimuli from neighboring cells in a region of the visual cortex. In one report (Kara & Boyd, 2009), local circuits for ocular dominance are shown to have smooth transitions from monocular to binocular zones. For local maps of ocular dominance and binocular disparity, there were measurable gradients at local cortical sites. Results suggest that ocular dominance is not specifically related to binocular disparity tuning. On the other hand there are precise maps of ocular dominance and binocular disparity (Kara & Boyd, 2009).

SUMMARY

Some basic factors have been considered here on the physiological combination of monocular inputs from the two eyes into a binocular system. Left and right eyes view visual space from a slightly different perspective, which varies with interocular distance. The resulting retinal or binocular disparity is the necessary and sufficient condition for stereoscopic depth discrimination. The main focus of the study of binocular vision is related to stereoscopic functions, which are considered in detail in other chapters of this volume (chapters 28 by Parker and 57 by Schor). In this chapter an attempt is made to show the inadequacy of monocular tests to evaluate binocular vision. Ocular dominance measurements can be misleading and clearly inaccurate if they are made with monocular tests. Because of generally subtle influences that each eye has on a given cortical cell, monocular tests will underestimate the extent of binocular connections. The only way this can be avoided is to use dichoptic presentations with quantitative methodology. Examples are given in this chapter of the use of phase-varying sinusoidal gratings presented dichoptically. The advantage of binocular compared to monocular vision is illustrated with reference to contrast processing. One unexpected and interesting result of this study is related to tests of binocular interaction using contrast differences between left and right eye gratings. The surprising result is that relatively constant binocular interaction profiles are found. Gain control mechanisms are considered in relation to this result. A related observation is that the use of left and right eye grating contrast differences provides artificial changes in effective ocular dominance conditions. It is also suggested that compensatory mechanisms may offset ocular dominance variability in the case of conscious perceptual vision. Finally, a relatively new technology, two-photon calcium imaging, may be used to elucidate important interrelationships in the binocular system.

ACKNOWLEDGMENTS

Our work on binocular vision has been supported by grant EY01175 from the National Eye Institute. Major contributions to the subjects covered here were made by Izumi Ohzawa, Greg DeAngelis, and Aki Anzai.

REFERENCES

Anzai, A., Bearse, M. A., Jr., Freeman, R. D., & Cai, D. (1995). Contrast coding by cells in the cat's striate cortex: Monocular vs. binocular detection. *Visual Neuroscience, 12,* 77–93.

Anzai, A., Ohzawa, I., & Freeman, R. D. (1999a). Neural mechanisms for encoding binocular disparity: Receptive field position versus phase. *Journal of Neurophysiology, 82,* 874–890.

Anzai, A., Ohzawa, I., & Freeman, R. D. (1999b). Neural mechanisms for processing binocular information I. Simple cells. *Journal of Neurophysiology, 82,* 891–908.

Barlow, H. B., Blakemore, C., & Pettigrew, J. D. (1967). The neural mechanism of binocular depth discrimination. *Journal of Physiology, 193,* 327–342.

Blake, R. (1981). Binocular rivalry and perceptual interference. *Perception & Psychophysics, 29,* 77–78. doi:10.3758/BF03198843.

Blake, R., Sloane, M., & Fox, R. (1981). Further developments in binocular summation. *Perception & Psychophysics, 30,* 266–276. doi:10.3758/BF03214282.

Blake, R., & Wilson, H. (2011). Binocular vision. *Vision Research, 51,* 754–770.

Blakemore, C., & Campbell, F. W. (1969). Adaptation to spatial stimuli. *Journal of Physiology, 200,* 11P–13P.

Blakemore, C., & Hague, B. (1972). Evidence for disparity detecting neurones in the human visual system. *Journal of Physiology, 225,* 437–455.

Blakemore, C., & Sutton, P. (1969). Size adaptation: A new aftereffect. *Science, 166,* 245–247.

Bradley, A., Skottun, B. C., Ohzawa, I., Sclar, G., & Freeman, R. D. (1985). Neurophysiological evaluation of the differential response model for orientation and spatial-frequency discrimination. *Journal of the Optical Society of America. A, Optics and Image Science, 2,* 1607–1610. doi:10.1364/JOSAA.2.001607.

Cumming, B. G., & DeAngelis, G. C. (2001). The physiology of stereopsis. *Annual Review of Neuroscience, 24,* 203–238.

Duong, T., & Freeman, R. D. (2007). Spatial frequency-specific contrast adaptation originates in the primary visual cortex. *Journal of Neurophysiology, 98,* 187–195.

Duong, T., Moore, B. D., IV, & Freeman, R. D. (2011). Adaptation changes stereoscopic depth selectivity in visual cortex. *Journal of Neuroscience, 31,* 12198–12207.

Enroth-Cugell, C., & Robson, J. G. (1966). The contrast sensitivity of retinal ganglion cells of the cat. *Journal of Physiology, 187,* 517–552.

Ferster, D. (1981). A comparison of binocular depth mechanisms in areas 17 and 18 of the cat visual cortex. *Journal of Physiology, 311,* 623–655.

Freeman, R. D., & Ohzawa, I. (1988). Monocularly deprived cats: Binocular tests of cortical cells reveal functional connections from the deprived eye. *Journal of Neuroscience, 8,* 2491–2506.

Freeman, R. D., & Ohzawa, I. (1990). On the neurophysiological organization of binocular vision. *Vision Research, 30,* 1661–1676.

Freeman, R. D., & Ohzawa, I. (1992). Development of binocular vision in the kitten's striate cortex. *Journal of Neuroscience, 12,* 4721–4736.

Freeman, R. D., & Robson, J. G. (1982). A new approach to the study of binocular interaction in visual cortex: Normal and monocularly deprived cats. *Experiments in Brain Research, 48,* 296–300.

Hubel, D. H., & Wiesel, T. N. (1962). Receptive fields, binocular interaction and functional architecture in the cat's visual cortex. *Journal of Physiology, 160,* 106–154.

Hubel, D. H., & Wiesel, T. N. (1968). Receptive fields and functional architecture of monkey striate cortex. *Journal of Physiology, 195,* 215–243.

Hubel, D. H., & Wiesel, T. N. (1970). Stereoscopic vision in macaque monkey: Cells sensitive to binocular depth in area 18 of the macaque monkey cortex. *Nature, 225,* 41–42.

Joshua, D. E., & Bishop, P. O. (1970). Binocular single vision and depth discrimination. Receptive field disparities for central and peripheral vision and binocular interaction on peripheral single units in cat striate cortex. *Experiments in Brain Research, 10,* 389–416. doi:10.1007/BF02324766.

Kara, P., & Boyd, J. D. (2009). A micro-architecture for binocular disparity and ocular dominance in visual cortex. *Nature, 458,*627–631.

Krug, K., & Parker, A. J. (2011). Neurons in dorsal visual area V5/MT signal relative disparity. *Journal of Neuroscience, 31,* 17892–17904.

LeVay, S., & Voigt, T. (1988). Ocular dominance and disparity coding in cat visual cortex. *Visual Neuroscience, 1,* 395–414.

Marlin, S., Douglas, R., & Cynader, M. (1993). Position-specific adaptation in complex cell receptive fields of the cat striate cortex. *Journal of Neurophysiology, 69,* 2209–2221.

Movshon, J. A., & Lennie, P. (1979). Pattern-selective adaptation in visual cortical neurones. *Nature, 278,* 850–852.

Movshon, J. A., Thompson, I. D., & Tolhurst, D. J. (1978). Receptive field organization of complex cells in the cat's striate cortex. *Journal of Physiology, 283,* 79–99.

Muller, J. R., Metha, A. B., Krauskopf, J., & Lennie, P. (1999). Rapid adaptation in visual cortex to the structure of images. *Science, 285,* 1405–1408.

Nikara, T., Bishop, P. O., & Pettigrew, J. D. (1968). Analysis of retinal correspondence by studying receptive fields of binocular single units in cat striate cortex. *Experiments in Brain Research, 6,* 353–372.

Ohzawa, I., DeAngelis, G. C., & Freeman, R. D. (1990). Stereoscopic depth discrimination in the visual cortex: Neurons ideally suited as disparity detectors. *Science, 249,* 1037–1041.

Ohzawa, I., DeAngelis, G. C., & Freeman, R. D. (1996). Encoding of binocular disparity by simple cells in the cat's visual cortex. *Journal of Neurophysiology, 75,* 1779–1805.

Ohzawa, I., & Freeman, R. D. (1986a). The binocular organization of complex cells in the cat's visual cortex. *Journal of Neurophysiology, 56,* 243–259.

Ohzawa, I., & Freeman, R. D. (1986b). The binocular organization of simple cells in the cat's visual cortex. *Journal of Neurophysiology, 56,* 221–242.

Ohzawa, I., & Freeman, R. D. (1988). Binocularly deprived cats: Binocular tests of cortical cells show regular patterns of interaction. *Journal of Neuroscience, 8,* 2507–2516.

Ohzawa, I., Sclar, G., & Freeman, R. D. (1982). Contrast gain control in the cat visual cortex. *Nature, 298,* 266–268.

Ohzawa, I., Sclar, G., & Freeman, R. D. (1985). Contrast gain control in the cat's visual system. *Journal of Neurophysiology, 54,* 651–667.

Poggio, G. F., & Fischer, B. (1977). Binocular interaction and depth sensitivity in striate and prestriate cortex of behaving rhesus monkey. *Journal of Neurophysiology, 40,* 1392–1405.

Poggio, G. F., Gonzalez, F., & Krause, F. (1988). Stereoscopic mechanisms in monkey visual cortex: Binocular correlation and disparity selectivity. *Journal of Neuroscience, 8,* 4531–4550.

Poggio, G. F., & Talbot, W. H. (1981). Mechanisms of static and dynamic stereopsis in foveal cortex of the rhesus monkey. *Journal of Physiology, 315,* 469–492.

Sanchez-Vives, M. V., Nowak, L. G., & McCormick, D. A. (2000a). Cellular mechanisms of long-lasting adaptation in visual cortical neurons in vitro. *Journal of Neuroscience, 20,* 4286–4299.

Sanchez-Vives, M. V., Nowak, L. G., & McCormick, D. A. (2000b). Membrane mechanisms underlying contrast adaptation in cat area 17 in vivo. *Journal of Neuroscience, 20,* 4267–4285.

Schor, C., & Heckmann, T. (1989). Interocular differences in contrast and spatial frequency: Effects on stereopsis and fusion. *Vision Research, 29,* 837–847.

Shiozaki, H. M., Tanabe, S., Doi, T., & Fujita, I. (2012). Neural activity in cortical area V4 underlies fine disparity discrimination. *Journal of Neuroscience, 32,* 3830–3841.

Tanabe, S., Haefner, R. M., & Cumming, B. G. (2011). Suppressive mechanisms in monkey V1 help to solve the stereo correspondence problem. *Journal of Neuroscience, 31,* 8295–8305.

Truchard, A. M., Ohzawa, I., & Freeman, R. D. (2000). Contrast gain control in the visual cortex: Monocular versus binocular mechanisms. *Journal of Neuroscience, 20,* 3017–3032.

von der Heydt, R., Adorjani, C., Hanny, P., & Baumgartner, G. (1978). Disparity sensitivity and receptive field incongruity of units in the cat striate cortex. *Experiments in Brain Research, 31,* 523–545.

28 Cortical Pathways for Binocular Depth

ANDREW J. PARKER

OVERVIEW

The ability to use the two eyes in coordination to see in depth is a biological development that takes place only to a limited degree in most animal species. Humans share with a few other species the characteristic of having eyes that are directed frontward. The advantage provided by binocular coordination is the ability to use the horizontal separation of the two eyes to acquire slightly different views of the visual scene in front of the animal (Parker, 2004). These different views result in small differences in the visual images of the scene that are received by the left and right eyes. The brain is able to exploit these differences to enable humans and animals to see stereoscopic depth and to segment objects in visually crowded environments, potentially breaking camouflage.

The degree to which eyes point forward varies from species to species. Among mammals, frontal alignment of the eyes is almost complete among cats and the primates (a group that includes humans) but occurs to a lesser degree among other mammals. There is consequently a variation in the size of the visual field that can be monitored with both eyes simultaneously (the "binocular overlap"). It is evident from the organization of the neural structures that support binocular vision that the forms of binocularity that occur in some birds, owls in particular (Nieder & Wagner, 2001), have developed separately from the binocularity that occurs in mammals. Recent reviews of this comparative biology have drawn out some strong similarities in functional characteristics between stereoscopic vision in owls and mammals (van der Willigen et al., 2010). The factors that have brought about these similarities during the course of independent evolutionary histories are as yet poorly understood.

Figure 28.1 shows in an elementary way how the separation of the eyes can lead to the ability to register depth differences between features in the part of the visual field that possesses binocular overlap. The geometry of the lines of sight from left and right eyes and the optical projections of visual features onto the retinas imply that differences in the depth of objects in the visual scene are converted into small differences in the position of a visual feature in the images of the left and right eyes. These small differences are called horizontal disparities. Although the diagram in figure 28.1 is very helpful in conveying the underlying geometric principles of binocular depth and stereopsis, principles that are common to all stereo systems, biological or indeed man-made, the underlying mathematical structure of the optical projections from the visual scene to form binocular images is considerably more complex. The horizontal separation of the eyes also generates small vertical differences: A vertical rod placed close to the left eye projects an image that is slightly more elongated vertically in the left eye than in the right, simply because of the closeness of the rod to the left eye (Bishop, 1989; Mayhew & Longuet-Higgins, 1982). There are further complexities as the eyes move, mainly because each eye rotates about its line of sight as the eyes converge on near or far targets in the upper and lower visual fields. The behavioral and perceptual consequences of these vertical image differences (and related phenomena such as cyclorotation of the eyes during convergence) have been studied extensively (Read, Phillipson, & Glennerster, 2009; Schreiber, Tweed, & Schor, 2006). Surprisingly, these complexities are not of great current concern in reviewing the physiology of visual cortical neurons that respond to stereoscopic depth, particularly the role of different visual cortical areas. But the fact that an appreciation of the full geometry of stereo depth imaging has not impacted on much of the current physiological literature indicates there is a good deal more to understand about the properties of binocular neurons.

Convergence of Signals from Left and Right Eyes

The signals leaving the left and right eyes are carried by separate neural pathways. Convergence of these signals is required for a response to stereoscopic depth and also for a number of other functions, including the control of binocular eye position and the generation of a sense of visual direction with respect to the head. Any of these functions requires that neurons with receptive fields serving a small patch of retina in the left eye should connect to a binocular neuron that is also

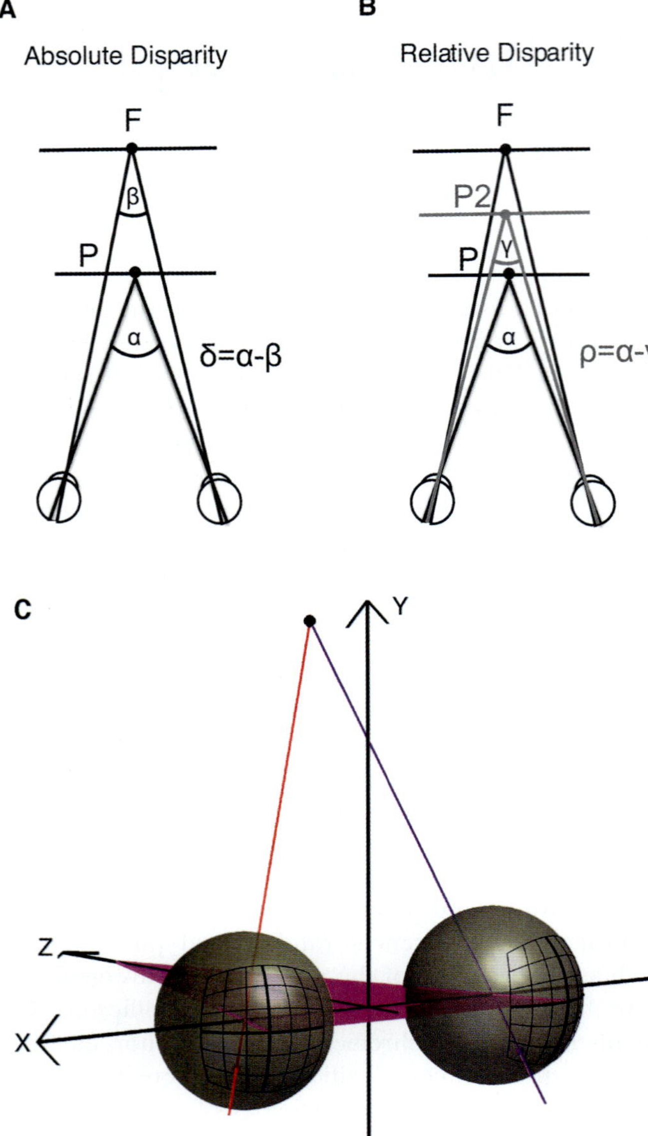

FIGURE 28.1 Stereoscopic disparities at the retinas of left and right eyes. (A, B) Absolute and relative horizontal stereo disparities. The eyes are shown fixating a point straight ahead. (A) The absolute disparity of point P is calculated as the angle between the visual axis of the eye and the point P. (B) The relative disparity between points P1 and P2 does not depend on the alignment of the visual axes of the eyes. Relative disparity can be calculated between the two visual features without reference to the coordinate frame of the eyes themselves. (C) How vertical disparity arises. The eyes are shown fixating a point on the midline, (X,Y,Z) = (0,0,8). An azimuth-longitude/elevation-longitude coordinate system is marked on each retina, with the fovea at its origin. Heavy black lines mark the horizontal and vertical retinal meridians; lighter lines mark lines of longitude drawn every 15°. The pink lines show the optic axes, that is, the line linking the fixation point to each fovea. Colored rays show the retinal projections of an object at (−6,6,10). When the locations where the red and blue rays intersect each retinal coordinate system are compared, it is clear that the object projects lower down in the right retina than the left (from Serrano-Pedraza, Phillipson, & Read, 2010).

connected to neurons serving a similarly located patch of retina in the right eye. The locations of patches in the left and right eyes are typically in correspondence, which means that when the two eyes look naturally at the visual world, the patches on left and right eyes receive optical projections from the same object in the external world (see figure 28.1). Tightening up this definition further, points on the left and right eyes are in geometric correspondence when the projection lines passing through the optical centers of each eye to the points in each eye form the same angle with the visual axis of each eye, where the visual axis is defined as the line leaving the fovea and passing through the optical center of the eye.

Convergence onto Single Neurons

A simple convergence of excitatory connections from neurons serving corresponding points on the left and right eyes onto a more central neuron enables binocular summation. The central binocular neuron will be most strongly stimulated when there is a strong stimulus arriving at the necessary locations on both the left and the right retinas. Strong stimulation of the binocular neuron can be achieved in a number of ways, but under natural viewing a single stimulus at the correct distance from the observer's eyes will arrive simultaneously on the left and right corresponding points. As the stimulus moves forward or backward in depth away from the correct distance, the response of the central binocular neuron will decrease (Barlow, Blakemore, & Pettigrew, 1967; Poggio, Gonzalez, & Krause, 1988) because such movements misalign the stimulus with at least one, and often both, of the corresponding points.

Of course the response of the binocular neuron also decreases when the stimulus in the visual world moves from side to side, rather than in depth. This illustrates a general point about the signals from single sensory neurons, which is that their signals do not unambiguously indicate a particular event or change in the outside world. Almost always the signals indicate that a relevant change has taken place, but to recover exactly what has caused that change requires analysis of the signals from a population of neurons. For our purposes here this means going beyond a simple search for neurons that are sensitive to changes of binocular disparity. The aim must be to discover neurons whose response properties match some of the well-characterized aspects of perceptual behavior.

Functionally speaking, binocular correspondence of this type generates neuronal signals that are sensitive to disparity computed in a coordinate frame organized around the binocular fixation point adopted by the eyes. Consider a binocular neuron that is formed by anatomical projections that brings together signals from a point on the left retina with signals from a point on the right retina. When the two eyes are aligned on a single target, thereby bringing the images of that target to the foveas of the left and right eyes, the binocular neuron is inevitably sensitive to changes in the disparity of a visual stimulus falling within the neuron's receptive field. Such sensitivity has been termed absolute disparity because it depends on exact geometric alignment of a single target with the sensitive areas of the binocular neuron (Cumming & Parker, 1999; Westheimer, 1979). Behaviorally, however, the human perceptual system is not most sensitive to changes in absolute disparity. For our perceptual behavior, the relative disparity between pairs of visible features is a more salient and relevant visual cue. This distinction has been at the center of a large number of studies in the past decade (see Absolute versus Relative Disparity below).

Geniculostriate Pathway

The geniculostriate pathway is formed by axons that leave the lateral geniculate nucleus (LGN) in the thalamus to reach the primary visual cortex (also known as the striate cortex). Within the thalamus and its projections out through this pathway, the primary excitatory signals are specific to a single eye of origin. There are findings to indicate that neuronal firing within the LGN is subject to a binocular influence, but the evidence points to this being a descending modulatory influence from corticofugal fibers (Schmielau & Singer, 1977).

Hence, the arrival of geniculostriate afferent fibers in the primary visual cortex is the first point of binocular combination. As such, although these connections have been well researched over a number of years, this site of combination remains one of the most significant points in the pathway for binocular vision (Hubel & Wiesel, 1962). This combination generates cortical neurons that respond to the binocular visual presentation of features. The binocularity of these neurons has been identified by two distinct types of test, one in which the responses of the neuron to stimulating each eye separately are measured and another that measures the effect of combined stimulation of both eyes.

The result of stimulating each eye separately has been summarized with the concept of ocular dominance. The neuron's response may be dominated by the signals arriving from the left eye alone or from the right eye alone, or the neuron may respond equally strongly to input from either eye. In animals with laterally placed eyes the signals from each eye are sent via neural pathways that cross the midline of the brain from one side to the other. Thus, the left eye connects to the right side of the brain, generating what is termed a contralateral signal, and similarly the right eye connects to the left side of the brain. In animals with forward-pointing eyes both eyes also connect to the same side of the brain on which they lie, with neural pathways that do not cross the midline. These pathways carry ipsilateral signals. Ocular dominance of visual responses was originally assessed on a 7-point scale by Hubel and Wiesel, on which 1 is complete dominance by the ipsilateral eye, 7 is complete dominance by the contralateral eye, and equal strength input from each eye is placed at 4 in the middle of this scale (Hubel & Wiesel, 1962). More recent measurements, based on quantification of the neural responses, have used a continuous scale of ocular dominance index (ODI) from 0 to 1, in which 1 is purely monocular and 0 is purely binocular (LeVay & Voigt, 1988; Prince et al., 2002; Read & Cumming, 2004).

The result of stimulating the eyes in combination has usually been assessed by measuring the ability of the neuron to discriminate different values of stimulus disparity with a disparity discrimination index (DDI) (Prince et al., 2002) or the capacity of signals to summate or cancel when presented in combination to both left and right eyes at the same time, which is quantified by the binocular interaction index (BII) (Ohzawa & Freeman, 1986a, 1986b). These tests for binocular signaling are conceptually different from the ODI. The ODI is assessed purely on the basis of presenting visual stimuli to one eye at a time, whereas the DDI and BII are based on simultaneous stimulation of both eyes (for more details, see chapter 27 by Freeman). What the ODI does not capture is that significant binocular interactions can take place not just when the signals from one eye summate with signals from the other eye but also when the interaction between the eyes is suppressive or inhibitory. Thus, a neuron could easily be monocular with an ODI of 1 but show (1) strong binocularity with the BII and (2) an ability to discriminate different disparities because it receives a suppressive signal from one eye.

These are important distinctions to bear in mind. First, the distinctions are not purely conceptual. There are now several experimental studies in cat and monkey that demonstrate the absence of any consistent relationship between the ODI and either BII or DDI (LeVay & Voigt, 1988; Prince et al., 2002; Read & Cumming, 2004). Also, for those neurons that are tuned to disparity, there appears to be no relationship between ODI and the shape of the disparity-tuning curve (Prince et al., 2002). One recent study (Kara & Boyd, 2009) that examined optical imaging of populations of neurons in the cat primary visual cortex suggests that there are orthogonal representations on the cortical surface for ocular dominance and binocularity (as assessed by BII). Second, the relationship between ocular dominance and identifiable anatomical structures is well established by a variety of different methodologies. No such clear link between structure and function has been established yet for purely binocular signaling. Third, the study of ocular dominance has also influenced the treatment of the human clinical condition of amblyopia and much of the interpretation of the literature about the models of this condition in experimental animals. Clinical amblyopia is the loss of visual acuity in one eye without overt pathological cause, but it is characteristically induced by squint or long-term experience of defocused images in the affected eye. It is typically treated by temporarily removing input to the brain from the stronger eye (e.g., by the wearing of an eye patch) so that the weaker eye can boost its neural connections into the central nervous system. Such treatment, of course, temporarily reduces the level of truly binocular stimulation reaching the cortex. Experimental interventions that model amblyopia in experimental animals typically also seek to study the weakening of one eye's input into the central nervous system rather than to examine directly the disruption of binocular vision. It seems clear that these issues are linked, not least because brief periods of reduced input from one eye may induce a long-term disorganization of binocular function. But the link itself remains unclear.

Disparity and its Neural Signaling

The measurement of selectivity for disparity at the level of neuronal signals has converged onto a consistent methodology in many respects. This has enabled much more rapid progress because it enables reasonably fair comparisons to be made across data sets acquired within different laboratories. In particular, there has been a consistent use of stereoscopic random-dot patterns for testing the selectivity of the neurons, consistent use of similar metrics of neuronal activity, and consistent approaches to data analysis and reporting of results. There are now available summary data showing for several visual cortical areas the proportion of neurons that exhibit binocular interactions, the range of binocular disparities coded by neurons in those areas, and the narrowness of neuronal selectivity for binocular disparity. Data are summarized for V1, V2, V4, and V5/MT in Parker (2007), and similar comparisons are available for V1, V3, V3A, V4, and V5/MT in Anzai, Chowdhury, and DeAngelis (2011).

The most striking aspect of these comparisons emerges when these properties, particularly range and narrowness of tuning, are considered across different cortical areas at similar visual eccentricities. In this regard the responses of neurons in all the extrastriate cortical areas tested so far appear to be similar. Only V1 appears to differ, with a smaller range of disparities represented with correspondingly tighter selectivity for the individual neurons. The reasons for this difference remain unclear. Other differences have been found in the shapes of the tuning for disparity. The tuning of some neurons is odd-symmetric, so these neurons are typically excited by disparities nearer than the fixation point and suppressed by disparities beyond it (or the opposite). Other neurons are even-symmetric in their tuning, so these neurons are maximally excited or inhibited at one disparity with a response that changes systematically as disparity is altered away from this maximum. It has been found that cortical areas V4 and V5/MT tend to have more disparity-selective neurons that are odd-symmetric in profile in comparison with neurons in V1, whose selectivity profiles for disparity have a marked tendency to be even-symmetric (Cumming & DeAngelis, 2001). Neurons in V3 and V3A appear to be intermediate in this regard between those of V4, V5/MT and those of V1 (Anzai, Chowdhury, & DeAngelis, 2011).

Disparity-Specific Pathways and the Functional Localization of Binocular Depth Perception

As the next stage in disparity processing from V1, the secondary visual area V2 contains both anatomical structures and functional specializations of its neurons associated with the further processing of stereo disparity. Anatomically, cytochrome oxidase staining reveals a set of thick and thin stripes within the cortical surface of V2 (Livingstone & Hubel, 1982). These thick stripes have for some time been hypothesized as having a role in binocular processing (Livingstone & Hubel, 1988). In respect of functional measures related to depth perception, some neurons in V2 respond to the relative disparity between adjacent visual surfaces (Thomas, Cumming, & Parker, 2002). When macaque monkeys are performing a stereoscopic depth task, the firing of disparity-specific neurons in V2 but not in V1 is linked to the animal's behavioral decisions about the binocular depth of the visual stimuli (Nienborg & Cumming, 2006). Both of these observations suggest a functional role for V2 in the processing of perceptual signals about stereo depth, consistent with the earlier lesion evidence (Cowey & Wilkinson, 1991). For at least one aspect of functional stereo signal, the link between structure and function has been validated very clearly with recent use of measurements based on intrinsic optical imaging of cortical activations. Activations related to the variation of binocular signals are dominant in the thick stripes (Chen, Lu, & Roe, 2008). However, as yet no link has been established between these anatomical structures and neurons that signal relative disparity or those that are behaviorally relevant for stereo perception.

In understanding the relationship between anatomical structure and physiological or behavioral function in V2, there are some parallels with the discussions that surround the role of ocular dominance columns in V1 in binocular vision. Recently, for experimental models of strabismic amblyopia in macaque monkeys, V2 has been identified as a site where the alteration of binocular function due to the experimental intervention more closely matches the change in perceptual function than the neuronal changes in V1 (Bi et al., 2011). On the other hand, a recent anatomical study finds no evidence for changes in the V2 stripe structures as a result of amblyopia induced by monocular deprivation (Sincich, Jocson, & Horton, 2012). Ideally, research should aim to establish a firm set of links from anatomy through physiology to the behavioral consequences. If such a link exists for cortical area V2, it is not established yet. At present, our best interpretation of existing evidence is that V2 appears to play an important role in establishing perceptually relevant neuronal signals about binocular depth, but its contribution is most probably as a layer within a deep hierarchy of visual areas that begins with V1 and stretches out along both dorsal and ventral visual pathways.

Binocular Depth Perception in Dorsal and Ventral Visual Areas

The identification of perceptually relevant signals for binocular depth perception has gained momentum through the consistent application of specific tests of stereoscopic performance. Three tests have mainly been used. For each of these tests, earlier work has established that the neurons of the primary visual area V1 respond according to the predictions of simple summation or correlation-based models of neuronal computation. As both the experimental results and the computational predictions of these models differ clearly from the way in which the perception of stereoscopic vision performs, the results from V1 have acted as a platform from which to launch a search of extrastriate visual areas for an explanation of perceptual performance in stereoscopic vision. The three tests have been the responsiveness of neurons to binocularly anticorrelated random-dot stereograms that do not give rise to

a consistent percept of stereoscopic depth; the ability of neurons to select out and respond differentially to stimuli presented to the left and right eyes when stereoscopic perception brings these stimuli into a matched relationship to generate a clear sense of stereo depth; and the distinction between responsiveness to absolute or relative disparity (mentioned earlier above and in Parker, 2004). Each of these tests is reviewed in relation to what they have revealed about the function of different elements of the dorsal and ventral visual pathways.

Binocular Anticorrelation and Stereo Depth Perception

Helmholtz first studied the effect of inverting the contrast relationship between the two eyes, arranging for the presentation of a line drawing to one eye and its photographic negative to the other eye in a stereoscopic display. Julesz noted that, specifically with the random-dot stereograms that he invented, inversion of the contrast relationship between the dots between the two eyes' images could completely abolish the perception of stereoscopic depth, unlike the case with simple line drawings (Julesz, 1971). In the case of random-dot stereograms, this contrast inversion is referred to as anticorrelation: The relationship between contrast in the left and right eyes is not uncorrelated (as it would be if two independent samples of random dots were presented to left and right eyes); rather it is consistently negatively correlated. Neurons in V1 respond to anticorrelated random-dot stereograms with a disparity-response curve that is inverted in shape in comparison with the neuron's response to the same patterns presented in their correlated versions (Cumming & Parker, 1997). Thus, each disparity that generates a strong response with a positively correlated random-dot stereogram results in a weaker response when the same disparity is presented in its anticorrelated configuration. Most strikingly, disparities that generate weak or zero response with correlated stereograms often produce strong responses to the same disparity when it is presented in an anticorrelated version.

This inversion of response between correlated and anticorrelated form is an indicator that the neuronal mechanism for disparity detection in V1 includes a device capable of detecting and signaling interocular correlation. The fact that anticorrelated stereograms of sufficiently high dot density do not generate a percept of stereo depth shows that the perceptual system for detecting stereo depth cannot be modeled as a whole with a correlator mechanism. A simple correlator mechanism also fails the other two tests that I discuss in this section. Thus, the response of V1 neurons to anticorrelated random-dot stereograms is already an indicator that these neurons may also fail the other two tests under study.

The search has therefore gone forward for neurons that eliminate the response to binocular anticorrelation. The only brain site meeting this criterion in full is the high-level ventral visual area TEo, for which Janssen and colleagues found that the response to changing disparity under binocular anticorrelation was essentially flat in neurons that showed substantial modulation in their response as the three-dimensional shape of surfaces was altered using correlated stereo disparities (Janssen et al., 2003). Most other results have been less categorical, with some cortical areas showing about the same strength of response to anticorrelation as V1 itself and others showing somewhat weaker responses to anticorrelation, suggestive of a progressive elimination of V1-like responses (Parker, 2007).

The overall pattern in these results has been summarized as a ventral pathway leading from early visual areas, V1 and V2, through V4 and out into inferotemporal cortex and a dorsal pathway leading from early visual areas to V5/MT and onward into MST. The ventral pathway is thought to be primarily concerned with perception of three-dimensional shape of individual objects, whereas the dorsal pathway is concerned with control of eye movements and localization of the observer in the three-dimensional world (Parker, 2007). This view was supported by the observation that anticorrelated random-dot stereograms are capable of inducing rapid alterations of the vergence angle of the eyes in circumstances when the eyes are landing on a new target at the end of a saccade. When this vergence movement is induced by anticorrelated random-dot figures, the pattern of movement is reversed in comparison with correlated random-dot figures (Masson, Busettini, & Miles, 1997). Thus, with anticorrelated random-dot stereo figures, a binocular disparity indicating a surface closer to the observer than the current vergence angle (a near-disparity) actually induces a divergent "correction" of binocular eye position. This is initially counterintuitive, but it follows logically from the fact that a near-disparity in an anticorrelated stereogram is actually increasing the activity of neurons that are excited by far-disparities under normal circumstances (Takemura et al., 2001).

These results also suggest a general distinction between cortical systems for eye-movement control and those for the conscious perception of depth. However, recent work has suggested a close link between the inferotemporal areas involved in the perception of solid shape and regions in the parietal cortex, specifically the anterior intraparietal cortex (AIP), which is concerned with reaching and grasping movements made by the upper limbs in a three-dimensional environment. When AIP neurons are tested with three-dimensional shapes displayed with anticorrelated random-dot figures, the neurons in this parietal region behave similarly to the neurons in the inferotemporal region TEo, which is thought to communicate with AIP (Theys et al., 2012). These results raise a number of issues that need to be investigated experimentally: in particular, it would be of interest to understand exactly how reaching and grasping movements are performed under control of binocular disparity cues and whether reaching and grasping movements show the same pattern of reversal in movement direction when the responses made to binocularly anticorrelated figures are compared against the responses made to binocularly correlated figures.

Resolving Multiple Matches

Julesz argued that the invention of the random-dot stereogram shifted the focus of binocular vision research away from the issue of how depth is generated from the visual cue of binocular disparity toward the question of how the brain sorts out and segregates which features falling on one retina should match up with those falling on the other retina. Julesz recognized that the random-dot stereogram presents this matching problem in an extreme form. For any single dot in one eye's view, there are a large number of identical dots in the other eye's view; given that both eyes are viewing large numbers of dots, there is a combinatorial explosion in the number of alternative matches that might need to be assessed. This led to the fascination with stereo-matching as a showcase problem during the ground-breaking developments of computational vision in the 1970s (Marr & Poggio, 1976, 1979).

The issue of resolving multiple matches is also present with smaller numbers of features. Indeed, the simplest possibility of a pair of closely spaced parallel vertical lines presented to each eye results in four possible ways in which the lines could be matched with each other (Panum, 1858). A version of this was explored neurophysiologically in primary visual cortex, with the outcome that V1 neurons represent all of the possible matches of these configurations (Cumming & Parker, 2000). Considering that when we look at a pair of vertical lines on an otherwise blank page, we typically see just two lines in the depth plane of the page, it is clear that there must be further processing between the stage of V1 and any neural representation that corresponds to reported perception. Exactly how these potential matches that are not seen ("false matches") become eliminated has been the focus of a great deal of psychophysical and computational study (McKee et al., 2007), but so far there are only hints about the neurophysiological sites that might be involved (Bakin, Nakayama, & Gilbert, 2000) and no substantive understanding of the neural mechanisms involved.

When there is only a small number of repeating features or a clear contour surrounding a set of repeating features, the brain appears to use these constraints to help distinguish among the possible pairings of similar features. If extended repeating patterns are presented, there is no single solution to the matching problem: there is more than one solution allowable, and the brain does flip from one solution to another, as manifest by the so-called "wallpaper illusion." In addition to reliance on exterior geometric constraints to solve the problem of multiple matches, identification by visual feature is also a possibility. For binocular vision a great deal of research has taken place on the question of whether discrepant features on the left and right eyes can be matched to provide a sense of stereo depth. Many configurations of different line lengths, orientations, color textures, and so on have been explored experimentally (Braddick, 1979). Here, however, we are more specifically concerned with whether visual feature similarity can be exploited to help resolve the multiple matching problem.

Harris and Parker (1995) used random-dot patterns in a task designed to measure the statistical performance of human vision in detecting small consistent differences in depth against a noisy disparity field generated by randomly adding an extra disparity to each dot forming the stereo figure. The measure of performance is the signal-to-noise ratio for the detection of a step change in depth against this noisy background of varying disparities. Harris and Parker showed that the signal-to-noise ratios improved when half of the dots were white and half were black, in comparison with the circumstances when all were black or all were white. They argued that the feature of contrast sign (black vs. white relative to a gray background) supported the elimination of false matches, effectively reducing the noise level for the discrimination task. Recently, this observation has been pursued in considerably more quantitative detail by (Read, Vaz, & Serrano-Pedraza, 2011), who extended the tasks studied to include detection of binocular correlation as well as detection of stereo depth differences. They pointed out that the conventional and widely accepted model (the "energy" model; Ohzawa, 1998; Ohzawa, DeAngelis, & Freeman, 1990, 1997; Prince et al., 2002) for the performance of disparity-tuned neurons in primary visual cortex eliminates the information about bright and dark features that these experiments demonstrate is used by the

human visual system. The energy model suggests that there would be pooling of information across luminance increments and decrements before the correspondence of multiple matches is sorted out, whereas the experimental results (Harris & Parker, 1995; Read, Vaz, & Serrano-Pedraza, 2011) suggest that information about the sign of luminance contrast is preserved during the process of sorting out multiple matches.

It seems relevant to link this result with other aspects of the binocular matching problem. Recall that for the binocularly anticorrelated stimuli, the construction of the stereogram is such that each white dot is geometrically paired with a dark dot in the other eye's image. Such anticorrelated stimuli will present a strong challenge to a system that relies on the sign of luminance contrast during the matching process (i.e., a system that at some stage separately matches the bright and dark features available in left and right eyes). Future studies will need to test the matching process in more detail to discover which features are most relevant for the establishment of stereoscopic perception.

Absolute versus Relative Disparity

The human visual system is essentially a binocular device with a high-resolution field of interest at the fovea, which is brought to bear successively on a sequence of targets during a cognitive task. The succession of saccadic eye movements, particularly evident during a demanding visual task such as reading text, serve to bring the visual information for analysis onto the highest-resolution part of the visual pathways. Binocular eye movement control is very much organized around the successful *bifoveation* of targets (i.e., bringing a real-world target simultaneously onto the fovea of the left and right eyes). It is from this baseline that the neural structures involved in the analysis of binocular disparity are organized in the human visual system.

When the foveas of the left and right eyes are aligned on a target, this brings the neural structures of the left and right retinas into alignment as well. The only variable under eye movement control is the degree of rotation of the eye around the line of sight (a line joining the center of the fovea of the eye out into the visual world to the point that is being fixated). Rotation about the line of sight has the greatest effect on the alignment of the left and right retinas for points on the retina far away from the fovea. The nineteenth century investigators (Helmholtz, 1863) established that the degree of rotation of the eye about the line of sight is linked to the direction of binocular gaze relative to the head.

More recent analyses under binocular viewing suggest that the oculomotor control of the two eyes is coordinated to ensure that there is minimal variation in the relative rotation of each eye around the line of sight as the eyes take up different fixation directions relative to the head position (Schreiber, Tweed, & Schor, 2006). Thus, if we begin from the simple situation of the two eyes looking directly ahead to a target in the far distance and consider how parts of the left retina and of the right retina need to send anatomical connections to form a binocular structure further in the visual system, the anatomical organization resulting from that mapping of left and right retinas onto a binocular central representation is sufficient to deal with other gaze positions as well.

As mentioned earlier, the visual projections onto the left and right retinas may result in horizontal or vertical disparities or both. Thus, to deal with the possible range of three-dimensional scenes, a point on one retina must be connected to a binocular representation that it is receiving from a neighborhood of points on the other eye's retina. This point-to-neighborhood connectivity has to be reciprocal so that every point on each eye's retina is connected to a set of points on the other eye's retina. These connections are set up during neural development but are refined under visual experience in postnatal development (Blakemore, 1979; Freeman & Ohzawa, 1992). It can be appreciated immediately that if this form of connectivity were applied universally across both retinas, then the number of potential connections would be very large indeed. Of course many of those potential connections are completely unnecessary because they correspond to geometries that are highly unlikely in the natural world. Recent studies of the statistics of natural scenes have begun to provide a quantitative basis for the specifications of the connectivity that would be required (Hibbard, 2007; Liu, Bovik, & Cormack, 2008); as yet, however, the detailed wiring that would be required for an efficient representation of these natural-world constraints has not been followed through.

A connection from a point on one retina and another point on the other retina to form a neuron sensitive to binocular visual stimulation creates a form of binocular disparity sensitivity, which has been called sensitivity to absolute disparity. Neurons with this connectivity are effectively stimulated whenever light simultaneously stimulates the relevant points on the left and right retinas. Neurons of this type were first identified in the visual cortex of the cat (Barlow, Blakemore, & Pettigrew, 1967; Nikara, Bishop, & Pettigrew, 1968; Pettigrew, Nikara, & Bishop, 1968). Such neurons are also found in the primary visual cortex of macaque monkeys, for which there is clear evidence of their sensitivity to absolute disparity (Cumming & Parker, 1999).

In recent years disparity-sensitive neurons with a different type of disparity selectivity have been identified. These respond to the relative disparity between two different visual features within the visual scene (Krug & Parker, 2011; Neri, Bridge, & Heeger, 2004; Shiozaki et al., 2012; Thomas, Cumming, & Parker, 2002; Uka et al., 2005; Umeda, Tanabe, & Fujita, 2007). The initial impetus to search for such neurons was the human psychophysical observation that individuals with typical stereo performance are able to resolve much smaller differences of binocular depth when using relative disparities as opposed to absolute disparities (Mitchison, 1993; Westheimer, 1979), and the recent physiological studies have specifically examined discrimination of fine disparities as an important criterion.

One of the issues that has emerged from these different studies is that the spatial distribution of relative disparities within a neuron's receptive field (or in the neighborhood of the receptive field) is critical not just in setting the responses of the neuron but also whether it displays selectivity for relative disparity at all (Parker, 2007). Specifically, in a careful series of studies (Uka & DeAngelis, 2003, 2004, 2006), it was shown that V5/MT neurons do not signal the relative disparity between a central region of dots and an annular surround region of similar dots, this configuration having been used in several other studies to reveal relative disparity coding in other visual cortical areas. However, when V5/MT neurons are presented with a pair of transparent moving planes, the neurons in this cortical region are sensitive to the relative depth between the two planes (Krug & Parker, 2011; see figure 28.2). These results show that neurons in both the dorsal and ventral visual pathways may exhibit selectivity for relative disparity, depending on the spatial configuration of the stimulus being viewed.

CONCLUSIONS

An emerging picture from recent studies is the integration of information about stereoscopic depth with other sources of information about the three-dimensional structure of the visual world. Sometimes stereoscopic disparity will be used to segregate one region of the image from another, whereas in other cases the stereoscopic depth cue will be integrated with other sources of information about depth and solid shape. The emphasis is shifting away from a view of extrastriate visual processing areas, in which the task of visual neuroscientists should be to discover the specializations of those areas for the extraction and processing of specific visual cues. There has been a greater emphasis on understanding the contribution of neurons during the

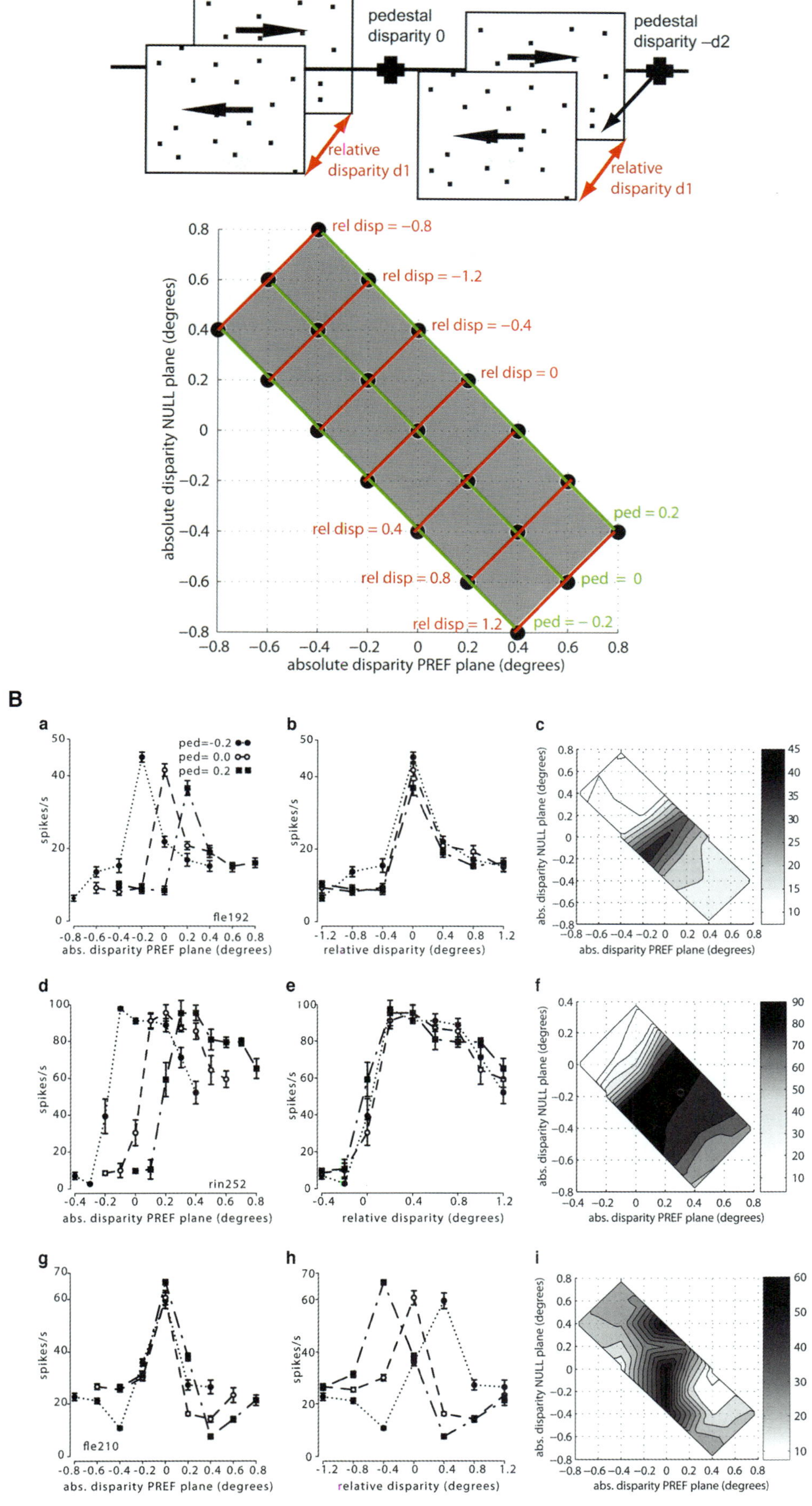

FIGURE 28.2 Binocular sensitivity of neurons in the dorsal visual cortical area V5/MT (from Krug & Parker, 2011) to the relative disparity between overlapping planes of dots. (A) Stimuli consisting of planes of moving random dots, separated in depth by relative disparity and shifted in depth as a pair of planes by changes in pedestal disparity. (B) Each row shows tuning curves from a single neuron in V5/MT. Two neurons (a,b,c and d,e,f) were sensitive to relative disparity, and one neuron (g,h,i) was sensitive to absolute disparity.

Continued

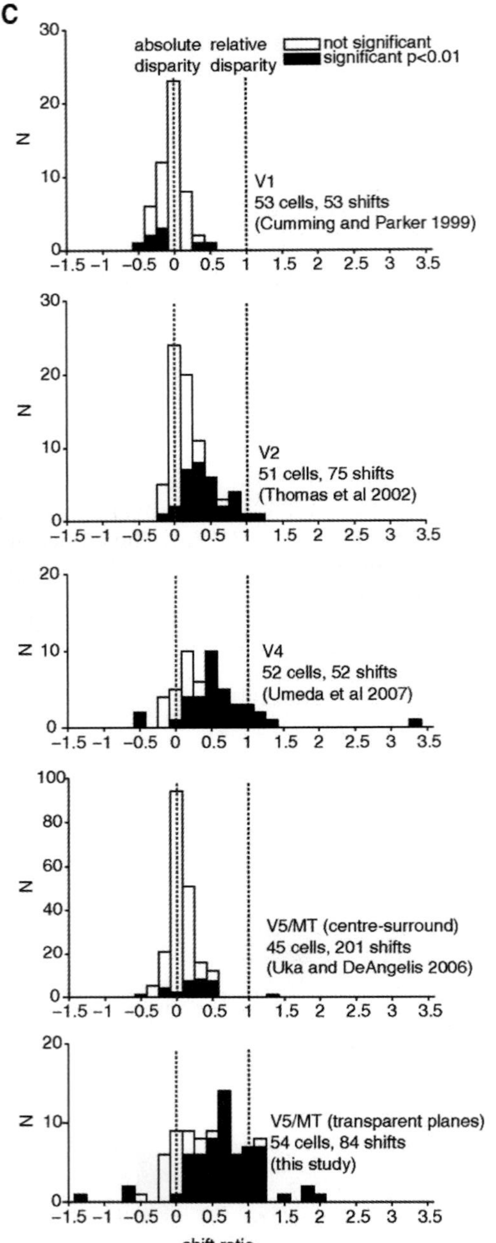

FIGURE 28.2 *Continued* (C) Scaled sensitivity of neurons to absolute and relative disparity. Neurons with a shift ratio of 1 are sensitive to relative disparity, and those with a shift ratio of 0 are primarily sensitive to absolute disparity. Neurons in V5/MT are sensitive to absolute disparity (like V1) when tested with center-surround stimuli, but some of these V5/MT neurons are sensitive to relative disparity when tested with transparent planes.

performance of specific visual tasks, with the design of the task being carefully crafted to probe a particular aspect of visual performance.

In regard to binocular vision, this approach has revealed the importance of different visual areas in different tasks. This allows us to interpret a number of experimental observations that would be difficult to understand on the basis of a single cortical site specialized for the processing of binocular depth. For example, the stereoscopic performance of an individual, a well-studied case of visual agnosia, is highly dependent on the choice of task and visual target (Read et al., 2010). In the much more widely occurring cluster of visual deficits that include amblyopia and associated dysfunctions of binocular performance, it is clear that there are wide variations in visual function that cannot be adequately reconciled with the view that all these losses can be traced to a single locus that governs stereoscopic performance (Kiorpes, 2006; McKee, Levi, & Movshon, 2003).

REFERENCES

Anzai, A., Chowdhury, S. A., & DeAngelis, G. C. (2011). Coding of stereoscopic depth information in visual areas V3 and V3A. *Journal of Neuroscience, 31*, 10270–10282.

Bakin, J. S., Nakayama, K., & Gilbert, C. D. (2000). Visual responses in monkey areas V1 and V2 to three-dimensional surface configurations. *Journal of Neuroscience, 20*, 8188–8198.

Barlow, H. B., Blakemore, C., & Pettigrew, J. D. (1967). The neural mechanism of binocular depth discrimination. *Journal of Physiology, 193*, 327–342.

Bi, H., Zhang, B., Tao, X., Harwerth, R. S., Smith, E. L., & Chino, Y. M. (2011). Neuronal responses in visual area V2 (V2) of macaque monkeys with strabismic amblyopia. *Cerebral Cortex, 21*, 2033–2045.

Bishop, P. O. (1989). Vertical disparity, egocentric distance and stereoscopic depth constancy: A new interpretation. *Proceedings of the Royal Society of London. Series B, Biological Sciences, 237*, 445–469.

Blakemore, C. (1979). The development of stereoscopic mechanisms in the visual cortex of the cat. *Proceedings of the Royal Society of London. Series B, Biological Sciences, 204*, 477–484.

Braddick, O. J. (1979). Binocular single vision and perceptual processing. *Proceedings of the Royal Society of London. Series B, Biological Sciences, 204*, 503–512.

Chen, G., Lu, H. D., & Roe, A. W. (2008). A map for horizontal disparity in monkey V2. *Neuron, 58*, 442–450.

Cowey, A., & Wilkinson, F. (1991). The role of the corpus callosum and extra striate visual areas in stereoacuity in macaque monkeys. *Neuropsychologia, 29*, 465–479.

Cumming, B. G., & DeAngelis, G. C. (2001). The physiology of stereopsis. *Annual Review of Neuroscience, 24*, 203–238.

Cumming, B. G., & Parker, A. J. (1997). Responses of primary visual cortical neurons to binocular disparity without depth perception. *Nature, 389*, 280–283.

Cumming, B. G., & Parker, A. J. (1999). Binocular neurons in V1 of awake monkeys are selective for absolute, not relative, disparity. *Journal of Neuroscience, 19*, 5602–5618.

Cumming, B. G., & Parker, A. J. (2000). Local disparity not perceived depth is signaled by binocular neurons in cortical area V1 of the macaque. *Journal of Neuroscience, 20*, 4758–4767.

Freeman, R. D., & Ohzawa, I. (1992). Development of binocular vision in the kitten striate cortex. *Journal of Neuroscience, 12*, 4721–4736.

Harris, J. M., & Parker, A. J. (1995). Independent neural mechanisms for bright and dark information in binocular stereopsis. *Nature, 374*, 808–811.

Helmholtz, H. (1863). The croonian lecture: On the normal motions of the human eye in relation to binocular vision. *Proceedings of the Royal Society of London, 13*, 186–199.

Hibbard, P. B. (2007). A statistical model of binocular disparity. *Visual Cognition, 15*, 149–165.

Hubel, D. H., & Wiesel, T. N. (1962). Receptive fields, binocular interaction and functional architecture in the cat's visual cortex. *Journal of Physiology, 160*, 106–154.

Janssen, P., Vogels, R., Liu, Y., & Orban, G. A. (2003). At least at the level of inferior temporal cortex, the stereo correspondence problem is solved. *Neuron, 37*, 693–701.

Julesz, B. (1971). *Foundations of cyclopean perception*. Chicago: University of Chicago Press.

Kara, P., & Boyd, J. D. (2009). A micro-architecture for binocular disparity and ocular dominance in visual cortex. *Nature, 458*, 627–631. doi:10.1038/nature07721.

Kiorpes, L. (2006). Visual processing in amblyopia: Animal studies. *Strabismus, 14*, 3–10.

Krug, K., & Parker, A. J. (2011). Neurons in dorsal visual area V5/MT signal relative disparity. *Journal of Neuroscience, 31*, 17892–17904.

LeVay, S., & Voigt, T. (1988). Ocular dominance and disparity coding in cat visual cortex. *Visual Neuroscience, 1*, 395–414.

Liu, Y., Bovik, A. C., & Cormack, L. K. (2008). Disparity statistics in natural scenes. *Journal of Vision, 8*(11), 19. doi:10.1167/8.11.19.

Livingstone, M. S., & Hubel, D. H. (1982). Thalamic inputs to cytochrome oxidase-rich regions in monkey visual cortex. *Proceedings of the National Academy of Sciences of the United States of America–Biological Sciences, 79*, 6098–6101.

Livingstone, M. S., & Hubel, D. H. (1988). Segregation of form, color, movement, and depth—anatomy, physiology, and perception. *Science, 240*, 740–749.

Marr, D., & Poggio, T. (1976). Cooperative computation of stereo disparity. *Science, 194*, 283–287.

Marr, D., & Poggio, T. (1979). A computational theory of human stereo vision. *Proceedings of the Royal Society of London. Series B, Biological Sciences, 204*, 301–328.

Masson, G. S., Busettini, C., & Miles, F. A. (1997). Vergence eye movements in response to binocular disparity without depth perception. *Nature, 389*, 283–286.

Mayhew, J. E. W., & Longuet-Higgins, H. C. (1982). A computational model of binocular depth-perception. *Nature, 297*, 376–378.

McKee, S. P., Levi, D. M., & Movshon, J. A. (2003). The pattern of visual deficits in amblyopia. *Journal of Vision, 3*(5), 380–405. doi:10.1167/3.5.5.

McKee, S. P., Verghese, P., Ma-Wyatt, A., & Petrov, Y. (2007). The wallpaper illusion explained. *Journal of Vision, 7*(14), 10. doi:10.1167/7.14.10.

Mitchison, G. (1993). The neural representation of stereoscopic depth contrast. *Perception, 22*, 1415–1426.

Neri, P., Bridge, H., & Heeger, D. J. (2004). Stereoscopic processing of absolute and relative disparity in human visual cortex. *Journal of Neurophysiology, 92*, 1880–1891.

Nieder, A., & Wagner, H. (2001). Hierarchical processing of horizontal disparity information in the visual forebrain of behaving owls. *Journal of Neuroscience, 21*, 4514–4522.

Nienborg, H., & Cumming, B. G. (2006). Macaque V2 neurons, but not V1 neurons, show choice-related activity. *Journal of Neuroscience, 26*, 9567–9578.

Nikara, T., Bishop, P. O., & Pettigrew, J. D. (1968). Analysis of retinal correspondence by studying receptive fields of binocular single units in cat striate cortex. *Experimental Brain Research, 6*, 353–372.

Ohzawa, I. (1998). Mechanisms of stereoscopic vision: The disparity energy model. *Current Opinion in Neurobiology, 8*, 509–515.

Ohzawa, I., DeAngelis, G. C., & Freeman, R. D. (1990). Stereoscopic depth discrimination in the visual cortex—neurons ideally suited as disparity detectors. *Science, 249*, 1037–1041.

Ohzawa, I., DeAngelis, G. C., & Freeman, R. D. (1997). The neural coding of stereoscopic depth. *Neuroreport, 8*, R3–R12.

Ohzawa, I., & Freeman, R. D. (1986a). The binocular organization of complex cells in the cat's visual cortex. *Journal of Neurophysiology, 56*, 243–259.

Ohzawa, I., & Freeman, R. D. (1986b). The binocular organization of simple cells in the cat's visual cortex. *Journal of Neurophysiology, 56*, 221–242.

Panum, P. L. (1858). *Physiologische Untersuchungen über das Sehen mit zwei Augen*. Kiel: Schwerssche Buchhandlung.

Parker, A. J. (2004). From binocular disparity to the perception of stereoscopic depth. In L. Chalupa & J. S. Werner (Eds.), *The visual neurosciences* (Vol. 1, pp. 779–792). Cambridge, MA: MIT Press.

Parker, A. J. (2007). Binocular depth perception and the cerebral cortex. *Nature Reviews. Neuroscience, 8*, 379–391.

Pettigrew, J. D., Nikara, T., & Bishop, P. O. (1968). Binocular interaction on single units in cat striate cortex: Simultaneous stimulation by single moving slit with receptive fields in correspondence. *Experimental Brain Research, 6*, 391–410.

Poggio, G. F., Gonzalez, F., & Krause, F. (1988). Stereoscopic mechanisms in monkey visual cortex—binocular correlation and disparity selectivity. *Journal of Neuroscience, 8*, 4531–4550.

Prince, S. J. D., Pointon, A. D., Cumming, B. G., & Parker, A. J. (2002). Quantitative analysis of the responses of V1 neurons to horizontal disparity in dynamic random-dot stereograms. *Journal of Neurophysiology, 87*, 191–208.

Read, J. C. A., & Cumming, B. G. (2004). Ocular dominance predicts neither strength nor class of disparity selectivity with random-dot stimuli in primate V1. *Journal of Neurophysiology, 91*, 1271–1281.

Read, J. C. A., Phillipson, G. P., & Glennerster, A. (2009). Latitude and longitude vertical disparities. *Journal of Vision, 9*(13), 11. doi:10.1167/9.13.11.

Read, J. C. A., Phillipson, G. P., Serrano-Pedraza, I., Milner, A. D., & Parker, A. J. (2010). Stereoscopic vision in the absence of the lateral occipital cortex. *PLoS One, 5*, e12608. doi:10.1371/journal.pone.0012608.

Read, J. C. A., Vaz, X. A., & Serrano-Pedraza, I. (2011). Independent mechanisms for bright and dark image features in a stereo correspondence task. *Journal of Vision, 11*(12), 4. doi:10.1167/11.12.4.

Schmielau, F., & Singer, W. (1977). Role of visual cortex for binocular interactions in cat lateral geniculate nucleus. *Brain Research, 120*, 354–361.

Schreiber, K. M., Tweed, D. B., & Schor, C. M. (2006). The extended horopter: Quantifying retinal correspondence across changes of 3D eye position. *Journal of Vision, 6*(1), 64–74.

Serrano-Pedraza, I., Phillipson, G. P., & Read, J. C. A. (2010). A specialization for vertical disparity discontinuities. *Journal of Vision, 10*(3), 2, 1–25. doi:10.1167/10.3.2.

Shiozaki, H. M., Tanabe, S., Doi, T., & Fujita, I. (2012). Neural activity in cortical area V4 underlies fine disparity discrimination. *Journal of Neuroscience, 32*, 3830–3841.

Sincich, L. C., Jocson, C. M., & Horton, J. C. (2012). Neuronal projections from V1 to V2 in amblyopia. *Journal of Neuroscience, 32*, 2648–2656.

Takemura, A., Inoue, Y., Kawano, K., Quaia, C., & Miles, F. A. (2001). Single-unit activity in cortical area MST associated with disparity-vergence eye movements: Evidence for population coding. *Journal of Neurophysiology, 85*, 2245–2266.

Theys, T., Srivastava, S., van Loon, J., Goffin, J., & Janssen, P. (2012). Selectivity for three-dimensional contours and surfaces in the anterior intraparietal area. *Journal of Neurophysiology, 107*, 995–1008.

Thomas, O. M., Cumming, B. G., & Parker, A. J. (2002). A specialization for relative disparity in V2. *Nature Neuroscience, 5*, 472–478.

Uka, T., & DeAngelis, G. C. (2003). Contribution of middle temporal area to coarse depth discrimination: Comparison of neuronal and psychophysical sensitivity. *Journal of Neuroscience, 23*, 3515–3530.

Uka, T., & DeAngelis, G. C. (2004). Contribution of area MT to stereoscopic depth perception: Choice-related response modulations reflect task strategy. *Neuron, 42*, 297–310.

Uka, T., & DeAngelis, G. C. (2006). Linking neural representation to function in stereoscopic depth perception: Roles of the middle temporal area in coarse versus fine disparity discrimination. *Journal of Neuroscience, 26*, 6791–6802.

Uka, T., Tanabe, S., Watanabe, M., & Fujita, I. (2005). Neural correlates of fine depth discrimination in monkey inferior temporal cortex. *Journal of Neuroscience, 25*, 10796–10802.

Umeda, K., Tanabe, S., & Fujita, I. (2007). Representation of stereoscopic depth based on relative disparity in macaque area V4. *Journal of Neurophysiology, 98*, 241–252.

van der Willigen, R. F., Harmening, W. M., Vossen, S., & Wagner, H. (2010). Disparity sensitivity in man and owl: Psychophysical evidence for equivalent perception of shape-from-stereo. *Journal of Vision, 10*(1), 10. doi:10.1167/10.1.10.

Westheimer, G. (1979). Cooperative neural processes involved in stereoscopic acuity. *Experimental Brain Research, 36*, 585–597.

29 Functional Organization of Circuits in Rodent Primary Visual Cortex

CRISTOPHER M. NIELL, VINCENT BONIN, AND MARK L. ANDERMANN

Although the first physiological studies of the mouse visual cortex were performed over three decades ago (Drager, 1975), it was not until recently that the mouse became generally considered as a valuable model for visual neuroscience. Instead, research focused on larger animals such as monkeys and cats, which have larger brains and a visual system that more closely resembles that of humans. Indeed, the wealth of anatomical, physiological, and behavioral data amassed in these species in the time since Hubel and Wiesel made it quite difficult to justify work in the mouse.

Several factors limit the use of mice to study visual function. In addition to being nocturnal animals, mice have extremely low acuity, and many laboratory mouse strains show genetic retinal defects. Furthermore, the central visual system of the mouse lacks many of the anatomical elaborations, such as distinct laminar segregation in the thalamus and large-scale maps of orientation and eye preference in visual cortex, that are canonical features of the visual system of most larger mammals. Finally, awake head-fixed recording methods and behavioral training paradigms for vision-based tasks in mice have lagged far behind those in primates and other species.

However, recent technical advances have made the mouse an appealing model for many aspects of systems neuroscience. Molecular genetic tools allow the identification of cell types, tracing of connectivity, and manipulation of spiking activity in neural circuits (Luo, Callaway, & Svoboda, 2008; O'Connor, Huber, & Svoboda, 2009). Electrophysiological tools such as in vivo whole-cell recording (Chorev et al., 2009) and multisite extracellular recording (Buzsaki, 2004) enable the study of neural activity from synaptic currents to large neuronal ensembles. In vivo functional imaging methods, such as two-photon calcium imaging, allow visualization of activity in identified cells and at known spatial locations (Kerr & Denk, 2008). Although these methods can be applied to other species, they are most accessible in mouse because of both its widespread use as a genetic model system and the experimental advantages of studying smaller brains with lissencephalic cortices.

Furthermore, it is now clear that many aspects of visual processing in larger mammals are similar in mice (Huberman & Niell, 2011). Mice use their visual system for natural behaviors and can be trained to perform behaviors that rely solely on vision. The lower functional acuity of mouse vision results in orders of magnitude fewer neurons in each visual area, which should greatly facilitate the characterization of complete cortical circuits. Beyond these experimental considerations the comparison of visual function in mouse and other rodents (Krubitzer, Campi, & Cooke, 2011) with higher species such as primates is also of value in understanding common evolutionary patterns, similarities and differences in visual strategies, and universal aspects of cortical computation.

In this chapter we assess the functional organization of rodent visual cortex, focusing on the mouse for the reasons described above. We draw on other rodent species as appropriate, particularly where knowledge is lacking in mice. Additionally, although the mouse visual system has been used extensively to study development and plasticity mechanisms, here we focus on functional organization and visual processing. We first give a brief overview of visual pathways leading into visual cortex and then describe the macroscopic organization of mouse primary visual cortex (V1) (For pathways emanating from V1 see chapter 18 by Burkhalter, Sporns, Gao, and Wang.) We then describe the receptive field properties of individual V1 neurons, followed by a discussion of how these properties vary across cell types and of how they are generated. Finally, we conclude by discussing how activity in visual cortex is affected by the animal's behavioral state and how it may be used to guide behavior.

ORGANIZATION OF VISUAL CORTEX

Retinal Inputs to Mouse Cortex

We first review the nature of the visual input coming from the periphery, which imposes significant constraints on the cortical representation. The photoreceptor mosaic in the mouse retina consists of both rods and cones at roughly the same spatial density as in other species (Jeon, Strettoi, & Masland, 1998). The light sensitivity of rods and cones, as measured behaviorally, is also similar to that of humans, although the mouse retina contains only two types of cones, with absorption centered in green (wavelength = 511 nm) and ultraviolet (360 nm) (Calderone & Jacobs, 1995) . However, the small diameter of the mouse eye (3 mm) (Remtulla & Hallett, 1985) results in poor optical resolution (Artal et al., 1998) and fewer total photoreceptors to sample the visual world. Furthermore, the mouse retina lacks a foveal region of increased cone density, which underlies high acuity vision in other species. Together, these factors mean that the retina of the mouse samples the visual field at a resolution that is approximately two orders of magnitude lower than the human fovea. Thus, receptive fields in the retina and at subsequent stages of the visual system are much larger than those in other species, with mean ganglion cells receptive field center diameters in the range of 6° to 10° (Koehler, Akimov, & Renteria, 2011; Stone & Pinto, 1993). As we see below, this does not imply that other visual properties are similarly degraded.

The output of the mouse retina is provided by a diverse array of roughly 20 types of retinal ganglion cells (Volgyi, Chheda, & Bloomfield, 2009). In addition to cells with classical ON and OFF center-surround and X- and Y-like response properties (Stone & Pinto, 1993), there are also a number of other classes with distinct response properties, including at least eight types of direction-selective ganglion cells (Kay et al., 2011; Rivlin-Etzion et al., 2011). Although these diverse cell types may account for over half of all ganglion cells, it is still unknown exactly which of them send signals to mouse V1, although at least some project to the visual thalamus, suggesting a role in cortical processing.

Macroscopic Organization of Visual Cortex

Before addressing the properties of individual neurons, we briefly describe the cytoarchitectural organization of primary visual cortex in the mouse. V1 spans approximately 3.5 mm^2 on the posterior dorsal surface of cortex (figure 29.1A). Because mouse cortex lacks sulci and gyri, the entirety of V1 and higher visual cortical areas are exposed dorsally, making them easily accessible for optical imaging or targeting of electrode penetrations. The mouse visual cortex is approximately 1 mm thick, less than half the thickness of those of primates, and shows the general six-layered cytoarchitecture seen in other species (figure 29.1B). However, the lamination is less clearly defined than in other species. In particular, there is no clear demarcation between cell bodies

in layer 2 and layer 3, and the thalamic recipient layer 4 lacks the sublaminae observed in other species. Although the layers may thus appear more diffuse, a number of molecular markers have been found to both identify and provide genetic access to layer-specific subpopulations (Heintz, 2004; Madisen et al., 2010; Molyneaux et al., 2009).

In addition to cytoarchitectural organization, mouse V1 contains, as in all other species, a topographic representation of visual space (retinotopy) across the surface of the cortex (figure 29.1C, D) (Drager, 1975; Kalatsky & Stryker, 2003; Wagor, Mangini, & Pearlman, 1980). Consistent with the absence of a fovea, the cortical magnification factor (distance on the cortical surface corresponding to a fixed angle in visual space) and receptive field size are relatively constant across the cortex, varying only by a factor of 2 (range ~10–20 μm/°) (Schuett, Bonhoeffer, & Hubener, 2002), in contrast to the dramatic variation with eccentricity found in other species. Cortical magnification also differs significantly between the horizontal and vertical axes of space, with the vertical axis expanded almost twofold.

Each hemisphere of mouse V1 represents the contralateral region of visual space (figure 29.1E). Because the mouse's eyes are oriented laterally, most of this territory receives input from one eye, the contralateral eye, except for a region of binocular overlap approximately 600 μm wide (representing ~30° on either side of the midline), where neurons with different eye preferences are intermingled (Gordon & Stryker, 1996; Mrsic-Flogel et al., 2007), in contrast to the segregation of the neurons' eye preference into ocular dominance columns as in cats and some primates. Thus, the binocular zone largely reflects retinotopy rather than eye-specific segregation. However, the relative strength of input from the two eyes within the binocular zone can be altered by visual experience during a critical period of development, which has proven fruitful as a model for experience-dependent plasticity (see chapters 95 by Nagakura, Mellios, and Sur and 100 by Coleman, Heynen, and Bear).

RECEPTIVE FIELD PROPERTIES

Orientation Selectivity

The hallmark of visual cortex is the transformation from neurons with center-surround receptive fields to neurons with oriented receptive fields (Hubel & Wiesel, 1962). The orientation tuning of V1 responses is robust and holds true independent of other stimulus properties (see chapter 26 by Andoni, Tan, and Priebe). In

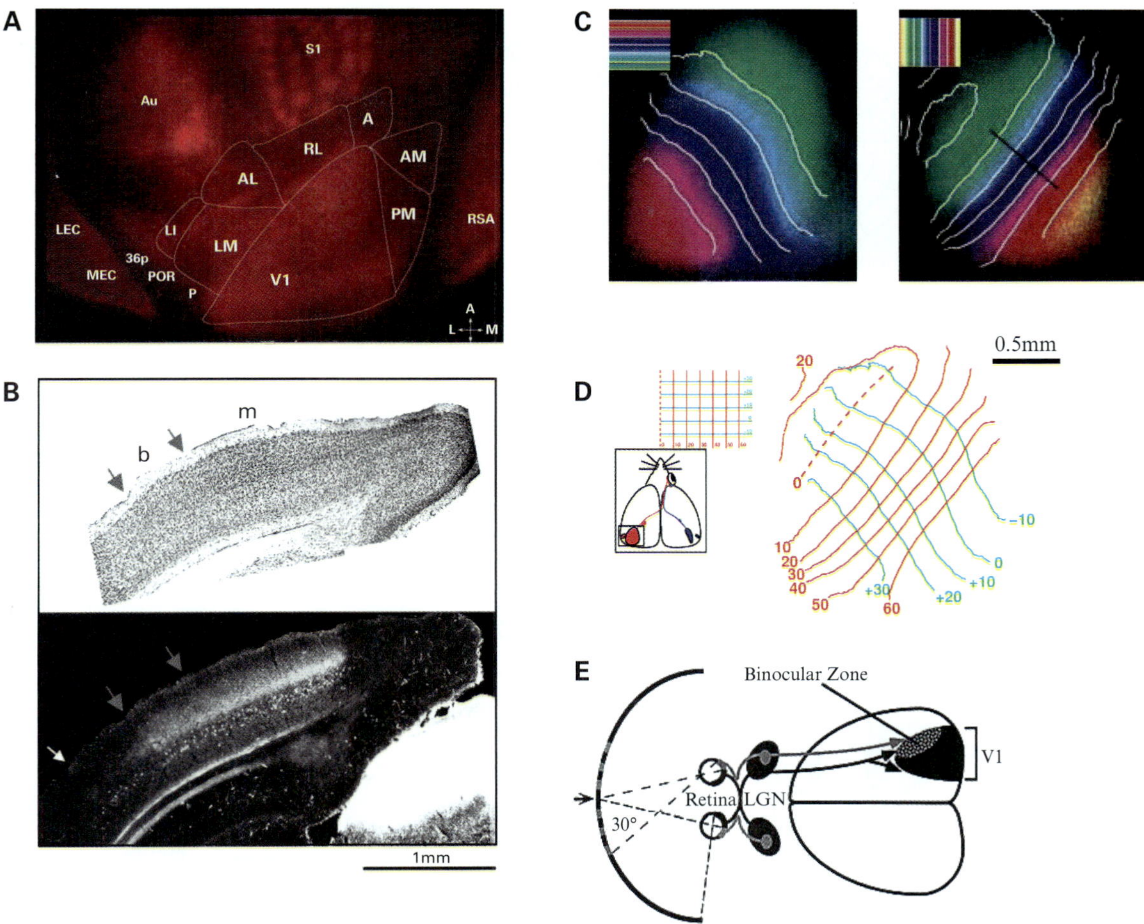

FIGURE 29.1   Organization of mouse V1. (A) Layout of primary visual cortex in relation to extrastriate and other cortical areas (flatmount section of a PV-Cre × tdTomato mouse, courtesy of Demetris Roumis). (B) Cytoarchitecture of primary visual cortex. (Upper) Nissl stain shows distinct cell body density in layers of V1. (Lower) WGA-HRP transneuronal tracing shows visual projections into layer 4 and upper layers, demarcating primary visual cortex. (Reproduced with permission from Antonini, Fagiolini, & Stryker, 1999.) (C, D) Retinotopic maps for elevation (C, left) and azimuth (C, right) as measured with intrinisic signal imaging, and corresponding contour map (D). (Reproduced with permission from Kalatsky & Stryker, 2003.) (E) Organization of retinotopy and binocular zone in the projections to mouse visual cortex. (Reproduced with permission from Gordon & Stryker, 1996.)

higher mammals orientation preference is organized across the cortical surface, with nearby neurons responding to bars of similar orientation. This organization was long thought to play a role in the emergence of the neurons' visual properties (Hubel & Wiesel, 1963). Work in rodents, however, has shown that columnar organization is not necessary for orientation tuning. The visual cortices of the mouse and other rodents do not show orientation maps, yet individual neurons have pronounced orientation selectivity (Drager, 1975; Ohki et al., 2005; Van Hooser et al., 2005). The tuning curves in mouse V1 closely resemble those observed in other species (Kerlin et al., 2010; Metin, Godement, & Imbert, 1988; Niell & Stryker, 2008). In layer 2/3 and layer 4 excitatory neurons, tuning width matches that of

primate and carnivores, with median values of 20°–25° half-width at half-maximum (figure 29.2A,C). This suggests that neither high spatial acuity nor functional organization is necessary for implementing orientation selectivity. As discussed more fully below, however, inhibitory neurons in mouse V1 show broader tuning (figure 29.2B), as do layer 5 excitatory neurons (Niell & Stryker, 2008).

Importantly, as in other species, orientation selectivity in mouse V1 is independent of the contrast of the stimulus (Niell & Stryker, 2008). The lack of broadening with increased drive, known as contrast-invariant tuning, suggests that several mechanisms interact to construct the neurons' receptive fields. We explore some of these mechanisms later in this chapter.

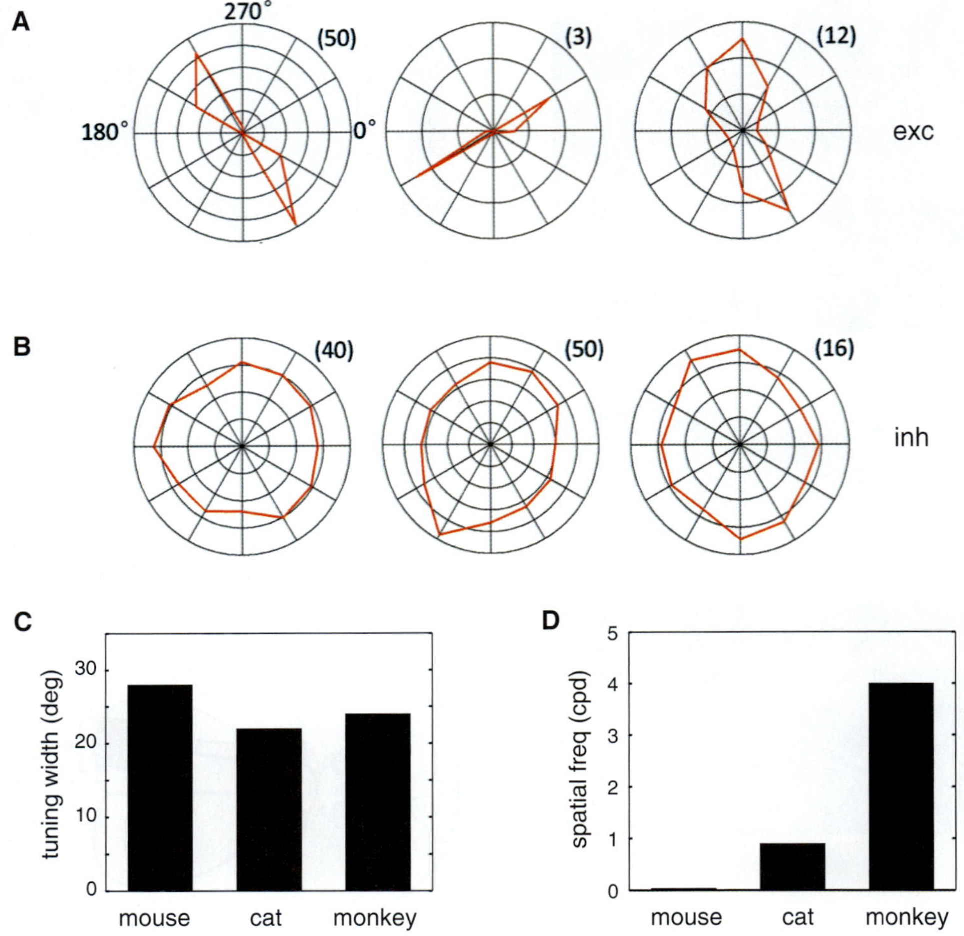

FIGURE 29.2   Tuning properties in mouse V1. (A, B) Sample orientation-tuning curves measured in response to moving gratings for (A) excitatory neurons, demonstrating sharp orientation tuning, and (B) inhibitory neurons, demonstrating nonselective responses. (Reproduced with permission from Liu et al., 2009.) (C, D) Comparison of median orientation-tuning width (C; half-width at half-maximum) and preferred spatial frequency (D) across species. Mouse data from Niell and Stryker (2008); cat and monkey data from Van Hooser (2007).

Spatial and Temporal Frequency

As mentioned above, the most pronounced difference between the receptive field properties of visual neurons in mice and those in other species is spatial scale (figure 29.1D). Studies in anesthetized animals have reported a spatial frequency preference centered on 0.04 cycle/deg (corresponding to a grating with 25° spacing), with a small fraction of neurons responding to full-field luminance changes, and with the highest spatial frequency eliciting responses around 0.5 cycle/deg (2° spacing) (Gao, DeAngelis, & Burkhalter, 2010; Kerlin et al., 2010; Niell & Stryker, 2008). This high cutoff is consistent with the sampling of the RGC mosaic (Jeon, Strettoi, & Masland, 1998) and with behavioral measures of visual acuity (Prusky & Douglas, 2004). Selectivity for low spatial frequencies does not preclude a role

in visually guided behavior, as such large visual angles could correspond either to large features at a distance (as in landmarks for navigation) or small features at close range (as for visual inspection of nearby objects). Spatial frequency bandwidth is similar to that of other species, approximately 2–3 octaves.

By comparison, the selectivity of mouse V1 responses for temporal modulations matches that of other species. Studies in anesthetized mice have found temporal frequency preferences between 0.5 and 4 Hz, with some neurons showing bandpass tuning and some showing lowpass (Gao, DeAngelis, & Burkhalter, 2010; Niell & Stryker, 2008). Interestingly, a recent study in awake mice (Andermann et al., 2011) found that both mean spatial and temporal frequency preferences were approximately twofold higher, suggesting that these values may be underestimates (discussed further below).

Linearity

Hubel and Wiesel initially identified two visual cortical cell types, simple and complex cells, based on how orientation-tuned neurons respond to optimally oriented light and dark bars presented at different positions on their receptive fields (Hubel & Wiesel, 1962). This distinction was later cast as linear or nonlinear spatial summation (Movshon, Thompson, & Tolhurst, 1978a, 1978b; Skottun et al., 1991) and can be characterized by the time course of firing in response to drifting sinusoidal grating stimuli. Simple cells show a periodic response to drifting grating stimuli, indicative of linear responses, whereas complex cells show more tonic responses, indicative of nonlinear spatial summation. As in other species, the degree of response nonlinearity in mouse V1 neurons follows a bimodal distribution (Niell & Stryker, 2008). In addition response linearity strongly depends on the neuron's laminar location and type. Most layer 4 and 2/3 excitatory neurons are linear, whereas layer 5 and inhibitory cells are nonlinear (Bonin et al., 2011; Liu et al., 2009; Niell & Stryker, 2008). However, as discussed further below, mouse V1 neurons that show nonlinear responses tend to have broad orientation tuning and therefore differ from the orientation-tuned complex cells originally described by Hubel and Wiesel, although cells with similar properties have been observed in cat visual cortex (Hirsch et al., 2003). Furthermore, although many species show a progression from simple/linear cells in layer 4 to nonlinear/ complex cells in layer 2/3, nonlinear cells constitute a minority in mouse layer 2/3.

Spatial Structure of Receptive Fields

In addition to tuning properties as measured with oriented stimuli such as bars and gratings, several studies have directly measured the spatial receptive field structure of linear cells using reverse correlation methods (Bonin et al., 2011; Liu et al., 2009; Niell & Stryker, 2008; Smith & Hausser, 2010). These have demonstrated that mouse V1 receptive fields consist of one to three segregated ON and OFF subregions and that this arrangement can predict the preferred orientation and spatial frequency of responses. Consistent with the decreased acuity in mouse, both the overall size and spacing of subregions are one to two orders of magnitude larger than in carnivores or primates, with typical classical receptive fields in the mouse being 10°–20° across. Despite this difference in length scale, the detailed structure of mouse spatial receptive fields is quite similar to those of other species in that they resemble a specific subset of Gabor functions (Jones & Palmer, 1987; Niell & Stryker, 2008; Ringach, 2002). This suggests that V1 receptive field ensembles form a universal cortical representation of visual scenes across species.

Other Tuning Properties

Neurons in mouse V1 show a number of other response properties that have been identified in other species, including direction selectivity and size tuning, contrast gain control, and surround suppression. Binocular interactions have been studied in the context of ocular dominance plasticity (Gordon & Stryker, 1996; Mrsic-Flogel et al., 2007) by measuring the relative strength of inputs and, recently, matching of orientation preference between the two eyes (Wang, Sarnaik, & Cang, 2010). However, no study has yet addressed whether mouse binocular neurons are tuned for disparity, which could be used for stereopsis. Furthermore, although mouse cones do have two types of photopigments, no studies have looked for color-opponent receptive fields in cortex. Because many cones coexpress the two pigments and expression varies across the retina (Calderone & Jacobs, 1995), any such color processing would likely be significantly different than in carnivores or primates.

Local Organization

Recent work has shed light on the local organization of receptive fields in mouse visual cortex. Receptive field position in mouse V1 does not vary smoothly across cortex but rather shows significant deviations between nearby neurons (Drager, 1975). Cellular imaging in the mouse has shown that this scatter is on the order of half the size of the receptive field (Bonin et al., 2011), consistent with electrode recording in cats and monkeys (DeAngelis et al., 1999; Hubel & Wiesel, 1974). Combined with the lack of organization of orientation preference, the scatter causes response properties of nearby visual cortical neurons in rodents to be distributed almost randomly (Bonin et al., 2011; Smith & Hausser, 2010) (figure 29.3). One exception is a tendency for clustering of spatial phase whereby receptive field subregions of the same sign in nearby neurons overlap more often than expected from random sampling (Bonin et al., 2011; Smith & Hausser, 2010). The nearly random distribution of receptive fields in local patches of mouse visual cortex suggests a high degree of specificity in the local connections that underlie the neurons' visual properties (see Patterns of Connectivity).

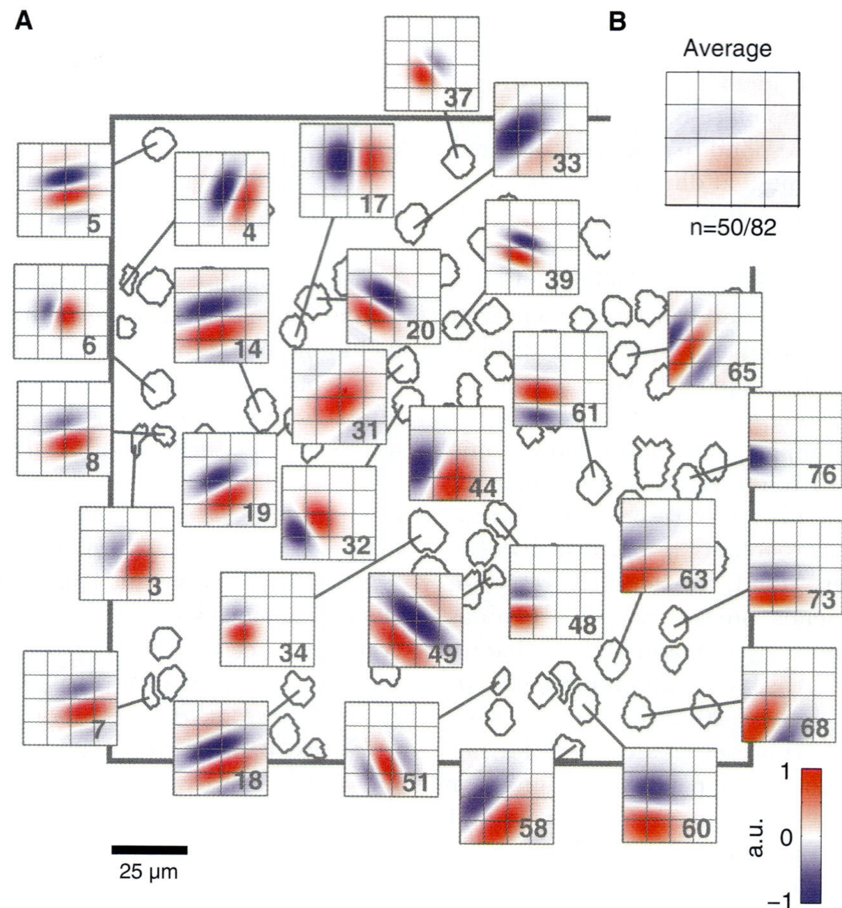

FIGURE 29.3 Local organization of receptive fields. (A) Spatial receptive fields of a population of neurons spanning ~200 × 200 μm in layer 2/3 of cortex, measured by two-photon calcium imaging and response fit to Gabor functions. (B) Average of all receptive fields reveals a slight bias toward shared subunits. (Reproduced with permission from Bonin et al., 2011.)

CELL TYPES, SYNAPTIC MECHANISMS, AND CONNECTIVITY

Thus far we have seen that neurons in mouse visual cortex show many of the basic response properties that have been previously observed in higher mammals. We now turn to the question of how the various neuronal types, together with their synaptic inputs and anatomical connectivity, generate these functional properties.

Composition of the Rodent Visual Cortex

Cortical neurons are highly diverse, yet the function of different cell types is only beginning to be unraveled. The two main cell types in cortex are spiny neurons and smooth neurons. Spiny neurons are excitatory, release the neurotransmitter glutamate, and make up about 85% of all neurons. The remaining neurons are mostly smooth, inhibitory interneurons, which release the neurotransmitter γ-aminobutyric acid (GABA) together with varying sets of cotransmitter molecules among other peptides.

The distribution of excitatory cells in the visual cortex in the mouse has not been characterized in detail but has been described in the rat (Peters & Kara, 1985). As in other species, most excitatory neurons exhibit pyramidal morphology. The main difference in excitatory neuron morphology between visual cortex of rats and that of higher mammals is that few, if any, spiny stellate cells are found in rat visual cortical layer 4. Instead, most layer 4 visual cortical neurons have pyramidal or star pyramidal morphologies.

GABAergic neurons, although less numerous than excitatory neurons, form in all species a highly heterogeneous group (Markram et al., 2004). The mouse is no exception. Studies in rodents have focused on interneuron subclasses for which molecular markers are available. In mouse visual cortex, Ca^{2+}-binding protein parvalbumin (PV), somatostatin (SOM), and vasointestinal peptide (VIP) are expressed in three nonoverlapping populations (Xu, Roby, & Callaway, 2010), although VIP+ neurons may represent a subset of a larger group of neurons expressing the ionotropic serotonin receptor 5HT3R (Rudy et al., 2011). Interestingly PV+, SOM+, and 5HT3+ neurons are not only nonoverlapping but also comprise almost all inhibitory neurons.

As in other species, there exists in rodents an approximate correspondence between molecularly defined and electrophysiologically defined subtypes (Burkhalter, 2008; Gonchar, Wang, & Burkhalter, 2007; Kawaguchi & Kubota, 1997; Xu, Roby, & Callaway, 2006). In rodent V1, PV+ neurons are thought to correspond to fast-spiking basket and chandelier cells, which account for up to 40% of all GABAergic neurons. SOM+ neurons correspond to Martinotti cells, which project their axons to distal dendrites of pyramidal cells in layer 1 and account for about 15% of GABAergic neurons. Last, VIP+ neurons mostly have bipolar morphology and account for about 8% of GABAergic neurons. These proportions, however, vary widely from one layer to another. Numerous Cre driver lines (Madisen et al., 2010; Taniguchi et al., 2011) and other transgenic lines, such as PV+ GAD67-GFP (Tamamaki et al., 2003), PV+ G42 mice (Molyneaux et al., 2007), and SOM+ GIN mice (Oliva et al., 2000), have been generated, allowing labeling of specific interneuron subtypes with fluorescent proteins or reporters of neural activity (Zariwala et al., 2012).

Distinct Functional Properties of Excitatory and Inhibitory Neurons

Until recently only two broad classes of neurons, regular and fast-spiking neurons, could be reliably targeted in vivo (e.g., Niell & Stryker, 2008). Recording from identified neuron types in vivo has become possible through the application of two-photon microscopy to image neurons labeled with fluorescent calcium indicators (Stosiek et al., 2003) or to target them for recording with glass pipettes (Margrie et al., 2003). More recently, extracellular recordings in combination with optogenetic methods have also been used to identify various neural populations (Lima et al., 2009).

Studies in visual cortex of anesthetized mice have revealed distinct response properties in excitatory and inhibitory neurons. Targeted recordings and calcium-imaging studies have shown that excitatory neurons have very diverse receptive field properties. Their responses are sharply tuned for stimulus orientation (figure 29.2A) (Bonin et al., 2011; Hofer et al., 2011; Kerlin et al., 2010; Liu et al., 2009; Niell & Stryker, 2008; Sohya et al., 2007; Tan et al., 2011), and their receptive fields typically have multiple ON and OFF subregions (Bonin et al., 2011; Liu et al., 2009; Liu et al., 2010; Niell & Stryker, 2008; Smith & Hausser, 2010). A majority of mouse visual cortical excitatory neurons have simple-cell receptive fields with nonoverlapping ON and OFF subregions, whereas a minority has complex receptive fields with overlapping ON and OFF subregions.

By comparison, inhibitory neurons have somewhat homogeneous visual properties. The responses of most GABAergic neurons are broadly tuned for stimulus orientation (figure 29.2B) (Atallah et al., 2012; Hofer et al., 2011; Kerlin et al., 2010; Liu et al., 2009; Niell & Stryker, 2008; Sohya et al., 2007; Zariwala et al., 2011; but see Runyan et al., 2010). They have complex-like response properties (Bonin et al., 2011; Niell & Stryker, 2008), and their receptive fields have large overlapping ON and OFF subregions (Liu et al., 2009). The broad tuning of fast-spiking cells seems to exist across all cortical layers (Niell & Stryker, 2008) and also holds for PV+ layer 2/3 regular-spiking neurons (Liu et al., 2009). SOM+ Martinotti cells (Ma et al., 2010) may have more diverse receptive field properties as compared to PV+ neurons (but see Kerlin et al., 2010). These results differ from observations in cat visual cortex, where the responses of fast-spiking inhibitory neurons are generally tuned for orientation and can have simple receptive fields, although they resemble results obtained for a distinct cat visual cortical population, particularly in layer 4 (Cardin, Palmer, & Contreras, 2007; Hirsch et al., 2003; Nowak, Sanchez-Vives, & McCormick, 2008).

These properties have important implications for how the mouse visual cortical network operates. The diverse properties of excitatory neurons are consistent with the presence of a rich representation of the visual world and the capacity for distributing relevant information to the rest of the brain. The properties of inhibitory neurons in mice, as well as in cats and primates, are consistent with a model in which these neurons promiscuously sum the activity of all nearby neurons and axonal inputs. In species that have an orientation map, such that neighboring cells have similar orientation preference, summing the local population would still result in orientation-selective inhibitory cells, whereas in mice summing the local activity results in much broader tuning. This may also explain the broader orientation tuning of layer 5 excitatory neurons, as in vitro studies have shown that layer 5 neurons pool input from a large number of layer 2/3 cells (Fino & Yuste, 2011).

The general similarity in tuning properties of inhibitory subtypes seems at odds with the diversity of morphologies and intrinsic properties. Although there may be more differences in their selectivity that remain to be discovered (see Ma et al., 2010), it is also possible that the diversity is more reflective of differences in temporal firing patterns, sensitivity to neuromodulators, and/or gating of different feedforward, lateral, and feedback input pathways.

Synaptic Mechanisms Underlying Response Selectivity

How does the tuning of visual cortical neurons arise? One way to approach this question is to measure the responses of synaptic inputs to the neurons. In both cat and mouse visual cortex excitatory and inhibitory inputs to simple cells have similar intrinsic properties and tuning preferences (Liu et al., 2011; Tan et al., 2011), so the mechanisms that underlie orientation selectivity are likely to be similar in both species. However, as discussed below, the spatial and temporal alignment of excitation and inhibition do differ between the species.

In cat visual cortex excitatory and inhibitory inputs to simple cells vary in opposite directions (Anderson, Carandini, & Ferster, 2000; Hirsch et al., 1998): Inhibition decreases when excitation increases, and vice versa. As a result, excitation and inhibition have "opposite" receptive fields. This push–pull arrangement is thought to help maintain response linearity in the face of rectification due to the spike threshold. Several studies have tried but failed to find evidence for a similar push–pull phenomenon in mouse visual cortex (Liu et al., 2010, 2011; Tan et al., 2011). Instead, it appears that excitation and inhibition are proportional to one another in mouse visual cortex (figure 29.4A) (Tan et al., 2011): Any visual stimulus evokes balanced levels of both excitatory and inhibitory currents. As a result, excitatory and inhibitory receptive fields are largely overlapping (Liu et al., 2010), and firing rate responses result from the dynamic interplay between excitation and inhibition (figure 29.4B).

Imaging methods have also been used to directly visualize sensory-evoked synaptic inputs. Calcium imaging of subthreshold calcium responses in local dendritic regions of single neurons in vivo indicates that layer 2/3 pyramidal neurons receive inputs from neurons with very different tuning properties (Jia et al., 2010). In the future these and other methods will yield a better understanding of how visual neurons embedded in the network perform their computations.

Patterns of Connectivity

Connections between excitatory neurons in rodent visual cortex are generally sparse with only a fraction of the neurons connected to each other. In rat visual cortex paired recordings in vitro have shown that connectivity among nearby layer 2/3 excitatory neurons (Holmgren et al., 2003; Yoshimura, Dantzker, & Callaway, 2005), as well as between layer 4 and layer 2/3 pyramidal cells (Yoshimura, Dantzker, & Callaway,

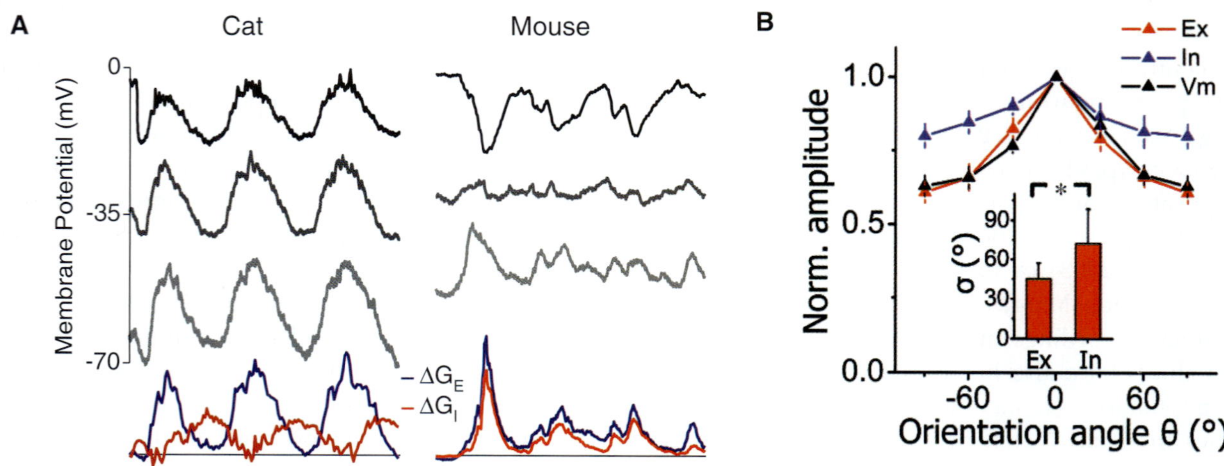

FIGURE 29.4   Synaptic mechanisms underlying cortical selectivity. (A) Excitatory and inhibitory conductances in response to drifting gratings in cat and mouse. By measuring membrane potential at different holding currents (top three traces), it is possible to extract the excitatory and inhibitory conductances (ΔG_E, ΔG_I). In cat these are in an antiphase, push–pull configuration, whereas in the mouse they are largely overlapping. Scale bar: magnitude of conductance (cat, 2 nS; mouse, 5 nS). (Reproduced with permission from Tan et al., 2011.) (B) Excitatory and inhibitory conductances generating orientation selectivity in mice. Broad excitation interacts with even broader inhibition to generate membrane potential tuning, which is then sharpened by the action potential threshold. (Inset) Average subthreshold excitatory and inhibitory tuning bandwidths. (Reproduced with permission from Liu et al., 2011.)

2005), is about 10–20%. Similarly, in mouse visual cortex, probability of connections among excitatory neurons is around 20% (Ko et al., 2011). Pairwise connectivity, however, is a poor predictor of network connectivity. Reciprocal connections within layer 2/3 (Holmgren et al., 2003) and within layer 5 (Song et al., 2005) occur more frequently than would be expected from random connections. Higher-order connectivity patterns (involving more than two neurons) within layer 5 pyramidal neurons are also overrepresented (Song et al., 2005). Thus, visual cortical excitatory connections are highly specific and are poorly modeled by random networks. In fact connections between layer 2/3 pyramidal cells, as well as between layer 4 and layer 2/3 cells, are a strong predictor of the degree of common inputs that the cells receive (Yoshimura, Dantzker, & Callaway, 2005). Together, these results suggest the existence of distinct functional subnetworks among pyramidal cells, which may play distinct roles in visual processing (see also chapter 25 by Callaway for related findings in primate visual cortex).

Recent work combining in vitro recordings with *in vivo* cellular imaging supports this hypothesis. Layer 2/3 neurons with similar functional properties are more likely to be connected (Ko et al., 2011). The probability of local layer 2/3 connections between pyramidal cells increases steeply with the similarity of visual responses, ranging from 50% for neurons with similar visual properties to near zero for neuron pairs with opposite preferences.

In contrast, connections between visual cortical excitatory and inhibitory neurons are denser and more promiscuous. The probability of connections between nearby layer 2/3 excitatory neurons and inhibitory neurons is over 50% (Hofer et al., 2011; Yoshimura & Callaway, 2005). The strength of these connections is also higher by an order of magnitude as compared to excitatory–excitatory connections (Hofer et al., 2011). Ultrastructural reconstruction of functionally identified cells has shown that smooth cells receive a broad range of input that is independent of the neurons' orientation preference (Bock et al., 2011), again consistent with inhibitory neurons pooling the local activity as discussed above.

Together, these studies have begun to shed light on the neural circuit motifs that make up visual cortex. In cases where direct comparison can be made to data in other species, it appears that similar synaptic mechanisms exist in mouse cortex with the most prominent difference in functional output (decreased orientation selectivity in inhibitory neurons) arising as a result of the lack of columnar organization for preferred orientation.

ACTIVITY IN VISUAL CORTEX AND ITS RELATION TO VISUAL BEHAVIORS

Although mice are a nocturnal species, there is substantial evidence that they use vision in their natural environment. They can rely on visual information to perform a wide variety of behaviors including spatial navigation, object recognition, as well as many social behaviors (Latham & Mason, 2004). For example, some species of mice use vision in long-distance homing behavior (Mather & Baker, 1980) and foraging behavior (Zollner & Lima, 1999) as well as "way-marking" by distributing visually conspicuous small objects in new territory as navigational aids (Stopka & Macdonald, 2003). As described below, laboratory mice have also been trained to perform a wide variety of visual behaviors, which allows assessment of visual function. At least some of the visual processing required for many of these natural and trained behaviors takes place in area V1 and likely requires an integration of feedforward visual response tuning with factors relating to behavioral state, including, for example, general arousal, modality- and stimulus-selective attention, and reward expectancy.

Response Properties during Anesthesia, Quiet Wakefulness, and Locomotion

The majority of rodent visual physiology studies have been conducted under various states of anesthesia, but recent studies using head-restrained mice and rats have enabled detailed examination of activity in awake rodent V1 while maintaining precise stimulus control (Andermann et al., 2011; Frenkel et al., 2006; Girman, 1985; Greenberg, Houweling, & Kerr, 2008; Niell & Stryker, 2010). Measurements of spontaneous and evoked activity in rodent V1 are sensitive to anesthesia versus various waking states (Greenberg, Houweling, & Kerr, 2008; Niell & Stryker, 2010), whereas orientation selectivity in V1 neurons is quite similar across states (Niell & Stryker, 2010). However, as in primates (Alitto et al., 2011), spatial and temporal frequency preferences were recently found to be higher in awake mouse V1 neurons (Andermann et al., 2011) than those typically observed in anesthetized mouse V1 (Gao, DeAngelis, & Burkhalter, 2010; Kerlin et al., 2010; Marshel et al., 2011; Niell & Stryker, 2008).

Locomotion involves visual processing of distant objects (Pinto & Enroth-Cugell, 2000) and is a basic component of foraging and other navigation behaviors in mice and other rodents. Acquisition and/or performance of many such behaviors is likely to require an intact primary visual cortex (Shankar & Ellard, 2000).

Surprisingly, most neurons in mouse V1 demonstrate severalfold increases in visual response magnitude during locomotion (Andermann et al., 2011; Niell & Stryker, 2010). Orientation tuning and spatial frequency preferences in V1 remain largely invariant to this change in gain, although temporal frequency preferences may increase slightly (<½ octave). Although the mechanism of this change is unknown, activation of cholinergic signaling in visual cortex has been shown to have similar effects, enhancing responsiveness and reliability (Goard & Dan, 2009; Herrero et al., 2008) and suggesting a potential role for neuromodulators.

V1 Activity and Plasticity during Visually Guided Behaviors

Recently, several studies have monitored responses of single neurons in rodent V1 during foraging and other reward-related behaviors, demonstrating activity in V1 that is not directly driven by visual inputs. A study using chronic recordings in foraging rats found that individual neurons in deeper layers of V1 showed "place preferences" for specific locations in a circular maze, likely corresponding to the site of specific spatial cues (Ji & Wilson, 2007). This study then showed correlated reactivation of these same neurons, together with specific hippocampal neurons, during subsequent periods of slow-wave sleep (Ji & Wilson, 2007; see also Han, Caporale, & Dan, 2008).

Using similar multielectrode recordings in freely moving rats, Shuler and Bear devised a task in which the timing of a reward varied depending on which eye was visually stimulated (Shuler & Bear, 2006). These authors found that neurons that responded preferentially to stimulation of a given eye also demonstrated robust and specific activity indicative of the associated reward timing. In trained rats this induced reward-related spiking lasted for several seconds prior to reward delivery, whereas V1 neurons in naive rats demonstrated only transient stimulus-evoked responses. By contrast, reward- and task-related behavioral signals that are not specifically associated with the presentation of visual stimuli may not influence visual responses in rat V1 neurons (Girman, 1985).

The ability to monitor and manipulate specific cells repeatedly will be of particular value for understanding the neural plasticity underlying learning. To record from the same cells in rodent V1 across days and weeks, recent studies have used chronic two-photon calcium imaging in freely moving rats (Sawinski et al., 2009) and in head-restrained rats (Greenberg, Houweling, & Kerr, 2008) and mice (Andermann, Kerlin, & Reid, 2010; Mank et al., 2008). These approaches are likely complementary: Whereas recordings in freely moving animals have the potential for closer similarity to rodent tasks developed over the last 80 years, head-restrained preparations allow greater stimulus control, field of view, and access for recording and manipulation. Recent studies using head-restrained mice have demonstrated reliable spatial navigation (Harvey et al., 2009) and visual detection/discrimination behaviors providing hundreds of trials per day for several months (Andermann, Kerlin, & Reid, 2010; Histed, Carvalho, & Maunsell, 2012). In combination with genetically encoded calcium indicators (Andermann, Kerlin, & Reid, 2010), high-throughput behavioral studies in head-fixed or freely moving animals (figure 29.5) will provide a means for robust characterization of responses from the same local volume of V1 neurons across time, allowing long-term monitoring of neural activity during learning.

Relating Neural Activity in Awake Mouse V1 to Perceptual Thresholds

Consistent with the narrow orientation tuning observed in awake mouse visual cortex (Frenkel et al., 2006; Niell & Stryker, 2010), head-restrained mice can discriminate orientations as little as 10° apart in a GO-NOGO task (Andermann, Kerlin, & Reid, 2010). Similarly, consistent with C50 contrast thresholds of less than 10% in a subset of neurons in awake mouse V1 (Busse et al., 2011), strongly motivated head-restrained mice can detect grating stimuli with contrasts as low as 2% (Histed, Carvalho, & Maunsell, 2012); but see other studies reporting higher contrast thresholds using different behavioral assays (Busse et al., 2011; Gianfranceschi, Fiorentini, & Maffei, 1999; Prusky et al., 2004; van Alphen, Winkelman, & Frens, 2009). A number of these reports have demonstrated the upper behavioral limit of spatial acuity in mice to be approximately 0.5 cycle/deg, consistent with the upper limit of spatial frequencies that evoke responses in mouse V1 (Andermann et al., 2011; Gao, DeAngelis, & Burkhalter, 2010).

Probing the Involvement of V1 in Aspects of Visual Behavior

Many visual pathways stemming from the retina can be used to guide behavior. Which aspects of rodent visual behavior depend on area V1? Beginning with the studies of Lashley in rats (Lashley, 1939), the most common behavioral deficit reported following localized lesions of rodent V1 is a decrease in the spatial acuity threshold during either detection or pattern discrimination tasks. Recent lesions studies involving discrimination of orthogonal grating stimuli at various spatial frequencies

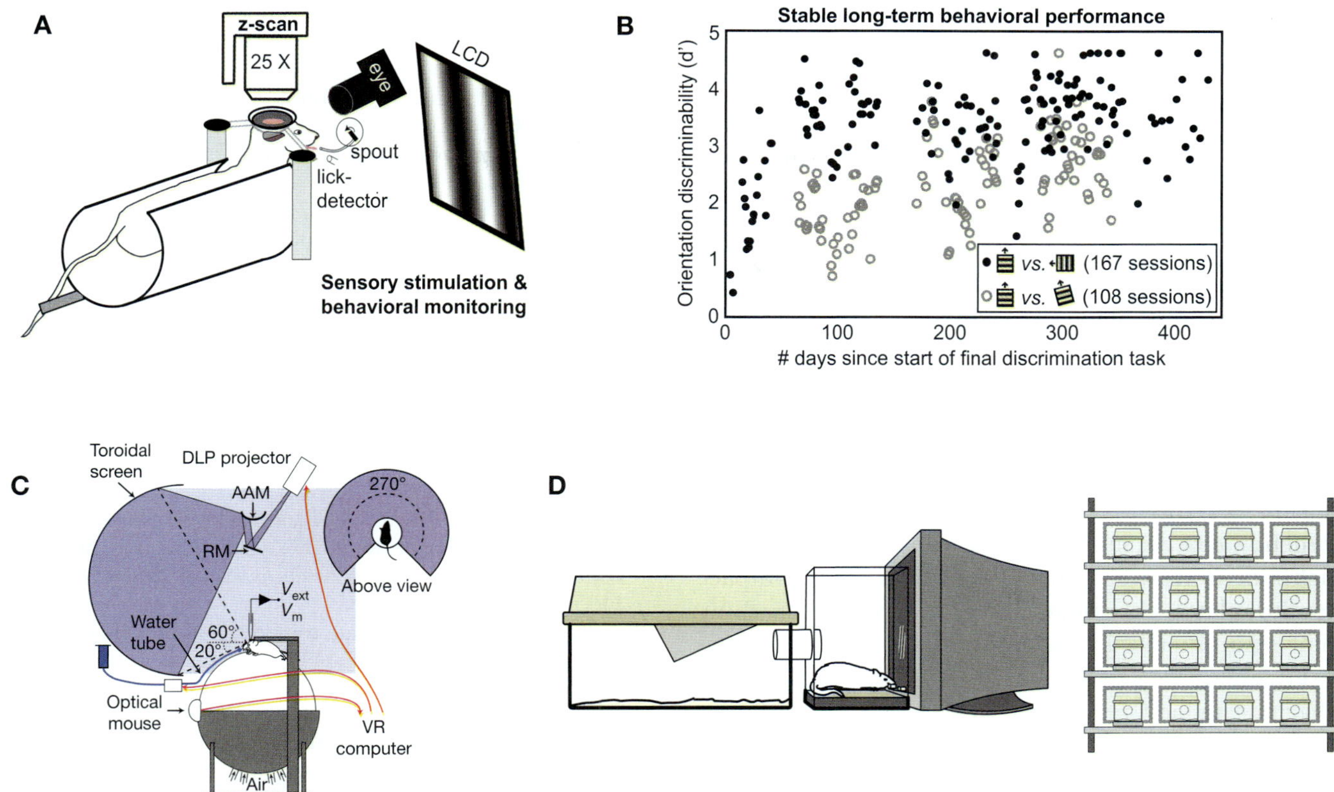

FIGURE 29.5   Behavioral paradigms for vision in rodents. (A) Awake head-fixed mice can be trained in a GO/NOGO orientation task, allowing two-photon calcium imaging of neural responses. (B) Behavioral responses can achieve a high level of accuracy and are stable over months. (A and B reproduced with permission from Andermann, Kerlin, & Reid, 2010.) (C) Awake head-fixed mice running on a spherical treadmill can learn to navigate through a virtual reality using only visual cues generated by projection of a visual image onto a toroidal screen and controlled by the animal's movement on the treadmill. (Reproduced with permission from Harvey et al., 2009.) (D) High-throughput training system for visual discrimination in rats. Animals move from their home cage into a behavior box in which access to water is contingent on participation in visual discrimination tasks. Because of the modular nature of the design, a number of animals can be trained and tested in parallel. (Reproduced with permission from Meier, Flister, & Reinagel, 2011.)

confirmed this effect in rats—decrement in cutoff threshold from 1.0 cycle/deg to 0.7 cycle/deg (Dean, 1981; McDaniel, Coleman, & Lindsay, 1982)—and in mice—from ~0.5 cycle/deg to 0.3 cycle/deg (Prusky & Douglas, 2004).

The absence of visual deficits at lower spatial frequencies following V1-specific lesions may be due to preservation of behaviors by collicular and other subcortical visual pathways (McDaniel, Coleman, & Lindsay, 1982; Prusky & Douglas, 2004). However, because of the inability of these studies of freely moving animals to present stimuli at specific retinotopic locations, it is possible that partial lesions of V1 would result in behavioral compensation by neighboring retinotopic regions of V1 (Lashley, 1939; McDaniel, Coleman, & Lindsay, 1982), whereas total lesions of V1 may result in additional damage to higher visual areas immediately lateral

to V1 (McDaniel, Coleman, & Lindsay, 1982), which have been shown in rodents to receive direct thalamic input in addition to V1 afferent input (Caviness & Frost, 1980; Simmons, Lemmon, & Pearlman, 1982). The irreversible damage and the delay to retesting after lesions may also result in plasticity and subsequent compensation by other areas (cf., Talwar, Musial, & Gerstein, 2001).

Future studies in rodent V1 can address these considerations by use of increasingly specific manipulations of neural activity. In addition to the use of reversible pharmacological manipulations (e.g., muscimol, cf. O'Connor, Huber, & Svoboda, 2010; Talwar, Musial, & Gerstein, 2001), neurons in rodent V1 can be selectively and reversibly silenced for short or long periods of time using optogenetic techniques (Mattis et al., 2011) or pharmacogenetic techniques (Rogan & Roth, 2011).

These techniques will afford spatial resolution via targeted virus injections into local regions of V1 and spatially specific receptor activation, in combination with laminar- and cell-type specificity obtained using transgenic mice (e.g., Atallah et al., 2012; Olsen et al., 2012). Beyond control of neural activity, future studies in mouse V1 can also employ increasingly specific genetic mutant mouse models to constrain models of visual computation (e.g., Heimel et al., 2010) and to understand the molecular underpinnings of visual cortical dysfunction.

CONCLUSION

The number of studies using mouse primary visual cortex as a model system to understand molecular, genetic, and circuit aspects of visual processing and behavior is growing rapidly—so rapidly as to outpace (and outdate) any single chapter focusing on anything more than the basic properties of mouse visual cortex. Consequently, we have focused on aspects of mouse V1 that are already reasonably well understood, and we have outlined the flavor of current and future experiments involving recent genetic, physiological, and behavioral strategies adapted to mice. We suggest that, despite several clear differences in the nature of the primary visual cortex and its peripheral inputs as outlined above, there is now a general consensus that a great deal of visual cortical processing may be quite similar between mice and better-studied mammals including cats and primates. Thus, the growing body of work in mouse visual cortex is quite likely to provide generalizable insight into many aspects of human vision despite an over 1,000-fold reduction in the number of neurons in mouse V1 as compared to humans.

REFERENCES

Alitto, H. J., Moore, B. D. IV, Rathbun, D. L., & Usrey, W. M. (2011). A comparison of visual responses in the lateral geniculate nucleus of alert and anaesthetized macaque monkeys. *Journal of Physiology, 589*(Pt 1), 87–99.

Andermann, M. L., Kerlin, A. M., & Reid, R. C. (2010). Chronic cellular imaging of mouse visual cortex during operant behavior and passive viewing. *Frontiers in Cellular Neuroscience, 4*, 3. doi:10.3389/fncel.2010.00003.

Andermann, M. L., Kerlin, A. M., Roumis, D. K., Glickfeld, L. L., & Reid, R. C. (2011). Functional specialization of mouse higher visual cortical areas. *Neuron, 72*, 1025–1039.

Anderson, J. S., Carandini, M., & Ferster, D. (2000). Orientation tuning of input conductance, excitation, and inhibition in cat primary visual cortex. *Journal of Neurophysiology, 84*, 909–926.

Antonini, A., Fagiolini, M., & Stryker, M. P. (1999). Anatomical correlates of functional plasticity in mouse visual cortex. *Journal of Neuroscience, 19*, 4388–4406.

Artal, P., Herreros de Tejada, P., Munoz Tedo, C., & Green, D. G. (1998). Retinal image quality in the rodent eye. *Visual Neuroscience, 15*, 597–605.

Atallah, B. V., Bruns, W., Carandini, M., & Scanziani, M. (2012). Parvalbumin-expressing interneurons linearly transform cortical responses to visual stimuli. *Neuron, 73*, 159–170.

Bock, D. D., Lee, W. C., Kerlin, A. M., Andermann, M. L., Hood, G., Wetzel, A. W., et al. (2011). Network anatomy and in vivo physiology of visual cortical neurons. *Nature, 471*, 177–182.

Bonin, V., Histed, M. H., Yurgenson, S., & Reid, R. C. (2011). Local diversity and fine-scale organization of receptive fields in mouse visual cortex. *Journal of Neuroscience, 31*, 18506–18521.

Burkhalter, A. (2008). Many specialists for suppressing cortical excitation. *Frontiers in Neuroscience, 2*, 155–167.

Busse, L., Ayaz, A., Dhruv, N. T., Katzner, S., Saleem, A. B., Scholvinck, M. L., et al. (2011). The detection of visual contrast in the behaving mouse. *Journal of Neuroscience, 31*, 11351–11361.

Buzsaki, G. (2004). Large-scale recording of neuronal ensembles. *Nature Neuroscience, 7*, 446–451.

Calderone, J. B., & Jacobs, G. H. (1995). Regional variations in the relative sensitivity to UV light in the mouse retina. *Visual Neuroscience, 12*, 463–468.

Cardin, J. A., Palmer, L. A., & Contreras, D. (2007). Stimulus feature selectivity in excitatory and inhibitory neurons in primary visual cortex. *Journal of Neuroscience, 27*, 10333–10344.

Caviness, V. S., Jr., & Frost, D. O. (1980). Tangential organization of thalamic projections to the neocortex in the mouse. *Journal of Comparative Neurology, 194*, 335–367.

Chorev, E., Epsztein, J., Houweling, A. R., Lee, A. K., & Brecht, M. (2009). Electrophysiological recordings from behaving animals—going beyond spikes. *Current Opinion in Neurobiology, 19*, 513–519.

Dean, P. (1981). Grating detection and visual acuity after lesions of striate cortex in hooded rats. *Experimental Brain Research, 43*, 145–153.

DeAngelis, G. C., Ghose, G. M., Ohzawa, I., & Freeman, R. D. (1999). Functional micro-organization of primary visual cortex: Receptive field analysis of nearby neurons. *Journal of Neuroscience, 19*, 4046–4064.

Drager, U. C. (1975). Receptive fields of single cells and topography in mouse visual cortex. *Journal of Comparative Neurology, 160*, 269–290.

Fino, E., & Yuste, R. (2011). Dense inhibitory connectivity in neocortex. *Neuron, 69*, 1188–1203.

Frenkel, M. Y., Sawtell, N. B., Diogo, A. C., Yoon, B., Neve, R. L., & Bear, M. F. (2006). Instructive effect of visual experience in mouse visual cortex. *Neuron, 51*, 339–349.

Gao, E., DeAngelis, G. C., & Burkhalter, A. (2010). Parallel input channels to mouse primary visual cortex. *Journal of Neuroscience, 30*, 5912–5926.

Gianfranceschi, L., Fiorentini, A., & Maffei, L. (1999). Behavioural visual acuity of wild type and bcl2 transgenic mouse. *Vision Research, 39*, 569–574.

Girman, S. V. (1985). Responses of neurons of primary visual cortex of awake unrestrained rats to visual stimuli. *Neuroscience and Behavioral Physiology, 15*, 379–386.

Goard, M., & Dan, Y. (2009). Basal forebrain activation enhances cortical coding of natural scenes. *Nature Neuroscience, 12,* 1444–1449.

Gonchar, Y., Wang, Q., & Burkhalter, A. (2007). Multiple distinct subtypes of GABAergic neurons in mouse visual cortex identified by triple immunostaining. *Frontiers in Neuroanatomy, 1,* 3. doi:10.3389/neuro.05.003.2007.

Gordon, J. A., & Stryker, M. P. (1996). Experience-dependent plasticity of binocular responses in the primary visual cortex of the mouse. *Journal of Neuroscience, 16,* 3274–3286.

Greenberg, D. S., Houweling, A. R., & Kerr, J. N. (2008). Population imaging of ongoing neuronal activity in the visual cortex of awake rats. *Nature Neuroscience, 11,* 749–751.

Han, F., Caporale, N., & Dan, Y. (2008). Reverberation of recent visual experience in spontaneous cortical waves. *Neuron, 60,* 321–327.

Harvey, C. D., Collman, F., Dombeck, D. A., & Tank, D. W. (2009). Intracellular dynamics of hippocampal place cells during virtual navigation. *Nature, 461,* 941–946.

Heimel, J. A., Saiepour, M. H., Chakravarthy, S., Hermans, J. M., & Levelt, C. N. (2010). Contrast gain control and cortical TrkB signaling shape visual acuity. *Nature Neuroscience, 13,* 642–648.

Heintz, N. (2004). Gene expression nervous system atlas (Gensat). *Nature Neuroscience, 7,* 483.

Herrero, J. L., Roberts, M. J., Delicato, L. S., Gieselmann, M. A., Dayan, P., & Thiele, A. (2008). Acetylcholine contributes through muscarinic receptors to attentional modulation in V1. *Nature, 454,* 1110–1114.

Hirsch, J. A., Alonso, J. M., Reid, R. C., & Martinez, L. M. (1998). Synaptic integration in striate cortical simple cells. *Journal of Neuroscience, 18,* 9517–9528.

Hirsch, J. A., Martinez, L. M., Pillai, C., Alonso, J. M., Wang, Q., & Sommer, F. T. (2003). Functionally distinct inhibitory neurons at the first stage of visual cortical processing. *Nature Neuroscience, 6,* 1300–1308.

Histed, M. H., Carvalho, L. A., & Maunsell, J. H. (2012). Psychophysical measurement of contrast sensitivity in the behaving mouse. *Journal of Neurophysiology, 107,* 758–765.

Hofer, S. B., Ko, H., Pichler, B., Vogelstein, J., Ros, H., Zeng, H., et al. (2011). Differential connectivity and response dynamics of excitatory and inhibitory neurons in visual cortex. *Nature Neuroscience, 14,* 1045–1052.

Holmgren, C., Harkany, T., Svennenfors, B., & Zilberter, Y. (2003). Pyramidal cell communication within local networks in layer 2/3 of rat neocortex. *Journal of Physiology, 551*(Pt 1), 139–153.

Hubel, D. H., & Wiesel, T. N. (1962). Receptive fields, binocular interaction and functional architecture in the cat's visual cortex. *Journal of Physiology, 160,* 106–154.

Hubel, D. H., & Wiesel, T. N. (1963). Shape and arrangement of columns in cat's striate cortex. *Journal of Physiology, 165,* 559–568.

Hubel, D. H., & Wiesel, T. N. (1974). Uniformity of monkey striate cortex: A parallel relationship between field size, scatter, and magnification factor. *Journal of Comparative Neurology, 158,* 295–305.

Huberman, A. D., & Niell, C. M. (2011). What can mice tell us about how vision works? *Trends in Neurosciences, 34,* 464–473. doi:10.1016/j.tins.2011.07.002.

Jeon, C. J., Strettoi, E., & Masland, R. H. (1998). The major cell populations of the mouse retina. *Journal of Neuroscience, 18,* 8936–8946.

Ji, D., & Wilson, M. A. (2007). Coordinated memory replay in the visual cortex and hippocampus during sleep. *Nature Neuroscience, 10,* 100–107.

Jia, H., Rochefort, N. L., Chen, X., & Konnerth, A. (2010). Dendritic organization of sensory input to cortical neurons in vivo. *Nature, 464,* 1307–1312.

Jones, J. P., & Palmer, L. A. (1987). An evaluation of the two-dimensional Gabor filter model of simple receptive fields in cat striate cortex. *Journal of Neurophysiology, 58,* 1233–1258.

Kalatsky, V. A., & Stryker, M. P. (2003). New paradigm for optical imaging: Temporally encoded maps of intrinsic signal. *Neuron, 38,* 529–545.

Kawaguchi, Y., & Kubota, Y. (1997). GABAergic cell subtypes and their synaptic connections in rat frontal cortex. *Cerebral Cortex, 7,* 476–486.

Kay, J. N., De la Huerta, I., Kim, I. J., Zhang, Y., Yamagata, M., Chu, M. W., et al. (2011). Retinal ganglion cells with distinct directional preferences differ in molecular identity, structure, and central projections. *Journal of Neuroscience, 31,* 7753–7762.

Kerlin, A. M., Andermann, M. L., Berezovskii, V. K., & Reid, R. C. (2010). Broadly tuned response properties of diverse inhibitory neuron subtypes in mouse visual cortex. *Neuron, 67,* 858–871.

Kerr, J. N., & Denk, W. (2008). Imaging in vivo: Watching the brain in action. *Nature Reviews. Neuroscience, 9,* 195–205.

Ko, H., Hofer, S. B., Pichler, B., Buchanan, K. A., Sjostrom, P. J., & Mrsic-Flogel, T. D. (2011). Functional specificity of local synaptic connections in neocortical networks. *Nature, 473,* 87–91.

Koehler, C. L., Akimov, N. P., & Renteria, R. C. (2011). Receptive field center size decreases and firing properties mature in ON and OFF retinal ganglion cells after eye opening in the mouse. *Journal of Neurophysiology, 106,* 895–904.

Krubitzer, L., Campi, K. L., & Cooke, D. F. (2011). All rodents are not the same: A modern synthesis of cortical organization. *Brain, Behavior and Evolution, 78,* 51–93.

Lashley, K. S. (1939). The mechanism of vision. XVI. The functioning of small remnants of the visual cortex. *Journal of Comparative Neurology, 70,* 45–67.

Latham, N., & Mason, G. (2004). From house mouse to mouse house: The behavioural biology of free-living *Mus musculus* and its implications in the laboratory. *Applied Animal Behaviour Science, 86,* 261–289.

Lima, S. Q., Hromadka, T., Znamenskiy, P., & Zador, A. M. (2009). PINP: A new method of tagging neuronal populations for identification during in vivo electrophysiological recording. *PLoS One, 4,* e6099. doi:10.1371/journal.pone.0006099.

Liu, B. H., Li, P., Li, Y. T., Sun, Y. J., Yanagawa, Y., Obata, K., et al. (2009). Visual receptive field structure of cortical inhibitory neurons revealed by two-photon imaging guided recording. *Journal of Neuroscience, 29,* 10520–10532.

Liu, B. H., Li, P., Sun, Y. J., Li, Y. T., Zhang, L. I., & Tao, H. W. (2010). Intervening inhibition underlies simple-cell receptive field structure in visual cortex. *Nature Neuroscience, 13,* 89–96.

Liu, B. H., Li, Y. T., Ma, W. P., Pan, C. J., Zhang, L. I., & Tao, H. W. (2011). Broad inhibition sharpens orientation selectivity by expanding input dynamic range in mouse simple cells. *Neuron, 71,* 542–554.

Luo, L., Callaway, E. M., & Svoboda, K. (2008). Genetic dissection of neural circuits. *Neuron, 57*(5), 634–660.

Ma, W. P., Liu, B. H., Li, Y. T., Huang, Z. J., Zhang, L. I., & Tao, H. W. (2010). Visual representations by cortical somatostatin inhibitory neurons—selective but with weak and delayed responses. *Journal of Neuroscience, 30*,14371–14379.

Madisen, L., Zwingman, T. A., Sunkin, S. M., Oh, S. W., Zariwala, H. A., Gu, H., et al. (2010). A robust and high-throughput Cre reporting and characterization system for the whole mouse brain. *Nature Neuroscience, 13*, 133–140.

Mank, M., Santos, A. F., Direnberger, S., Mrsic-Flogel, T. D., Hofer, S. B., Stein, V., et al. (2008). A genetically encoded calcium indicator for chronic in vivo two-photon imaging. *Nature Methods, 5*, 805–811.

Margrie, T. W., Meyer, A. H., Caputi, A., Monyer, H., Hasan, M. T., Schaefer, A. T., et al. (2003). Targeted whole-cell recordings in the mammalian brain in vivo. *Neuron, 39*, 911–918.

Markram, H., Toledo-Rodriguez, M., Wang, Y., Gupta, A., Silberberg, G., & Wu, C. (2004). Interneurons of the neocortical inhibitory system. *Nature Reviews. Neuroscience, 5*, 793–807.

Marshel, J. H., Garrett, M. E., Nauhaus, I., & Callaway, E. M. (2011). Functional specialization of seven mouse visual cortical areas. *Neuron, 72*, 1040–1054.

Mather, J. G., & Baker, R. R. (1980). A demonstration of navigation by small rodents using an orientation cage. *Nature, 284*, 259–262.

Mattis, J., Tye, K. M., Ferenczi, E. A., Ramakrishnan, C., O'Shea, D. J., Prakash, R., et al. (2011). Principles for applying optogenetic tools derived from direct comparative analysis of microbial opsins. *Nature Methods, 9*, 159–172.

McDaniel, W. F., Coleman, J., & Lindsay, J. F., Jr. (1982). A comparison of lateral peristriate and striate neocortical ablations in the rat. *Behavioural Brain Research, 6*, 249–272.

Meier, P., Flister, E., & Reinagel, P. (2011). Collinear features impair visual detection by rats. *Journal of Vision, 11*, 1–16. doi:10.1167/11.3.22.

Metin, C., Godement, P., & Imbert, M. (1988). The primary visual cortex in the mouse: Receptive field properties and functional organization. *Experimental Brain Research, 69*, 594–612.

Molyneaux, B. J., Arlotta, P., Fame, R. M., MacDonald, J. L., MacQuarrie, K. L., & Macklis, J. D. (2009). Novel subtype-specific genes identify distinct subpopulations of callosal projection neurons. *Journal of Neuroscience, 29*, 12343–12354.

Molyneaux, B. J., Arlotta, P., Menezes, J. R., & Macklis, J. D. (2007). Neuronal subtype specification in the cerebral cortex. *Nature Reviews. Neuroscience, 8*, 427–437.

Movshon, J. A., Thompson, I. D., & Tolhurst, D. J. (1978a). Receptive field organization of complex cells in the cat's striate cortex. *Journal of Physiology, 283*, 79–99.

Movshon, J. A., Thompson, I. D., & Tolhurst, D. J. (1978b). Spatial summation in the receptive fields of simple cells in the cat's striate cortex. *Journal of Physiology, 283*, 53–77.

Mrsic-Flogel, T. D., Hofer, S. B., Ohki, K., Reid, R. C., Bonhoeffer, T., & Hubener, M. (2007). Homeostatic regulation of eye-specific responses in visual cortex during ocular dominance plasticity. *Neuron, 54*, 961–972.

Niell, C. M., & Stryker, M. P. (2008). Highly selective receptive fields in mouse visual cortex. *Journal of Neuroscience, 28*, 7520–7536.

Niell, C. M., & Stryker, M. P. (2010). Modulation of visual responses by behavioral state in mouse visual cortex. *Neuron, 65*, 472–479.

Nowak, L. G., Sanchez-Vives, M. V., & McCormick, D. A. (2008). Lack of orientation and direction selectivity in a subgroup of fast-spiking inhibitory interneurons: Cellular and synaptic mechanisms and comparison with other electrophysiological cell types. *Cerebral Cortex, 18*, 1058–1078. doi:10.1093/cercor/bhm137.

O'Connor, D. H., Clack, N. G., Huber, D., Komiyama, T., Myers, E. W., & Svoboda, K. (2010). Vibrissa-based object localization in head-fixed mice. *Journal of Neuroscience, 30*, 1947–1967.

O'Connor, D. H., Huber, D., & Svoboda, K. (2009). Reverse engineering the mouse brain. *Nature, 461*, 923–929.

Ohki, K., Chung, S., Ch'ng, Y. H., Kara, P., & Reid, R. C. (2005). Functional imaging with cellular resolution reveals precise micro-architecture in visual cortex. *Nature, 433*, 597–603.

Oliva, A. A., Jr., Jiang, M., Lam, T., Smith, K. L., & Swann, J. W. (2000). Novel hippocampal interneuronal subtypes identified using transgenic mice that express green fluorescent protein in GABAergic interneurons. *Journal of Neuroscience, 20*, 3354–3368.

Olsen, S. R., Bortone, D. S., Adesnik, H., & Scanziani, M. (2012). Gain control by layer six in cortical circuits of vision. *Nature, 483*, 47–52.

Peters, A., & Kara, D. A. (1985). The neuronal composition of area 17 of rat visual cortex. I. The pyramidal cells. *Journal of Comparative Neurology, 234*, 218–241.

Pinto, L. H., & Enroth-Cugell, C. (2000). Tests of the mouse visual system. *Mammalian Genome, 11*, 531–536. doi:10.1007/s003350010102.

Prusky, G. T., Alam, N. M., Beekman, S., & Douglas, R. M. (2004). Rapid quantification of adult and developing mouse spatial vision using a virtual optomotor system. *Investigative Ophthalmology & Visual Science, 45*, 4611–4616. doi:10.1167/iovs.04-0541.

Prusky, G. T., & Douglas, R. M. (2004). Characterization of mouse cortical spatial vision. *Vision Research, 44*, 3411–3418. doi:10.1016/j.visres.2004.09.001.

Remtulla, S., & Hallett, P. E. (1985). A schematic eye for the mouse, and comparisons with the rat. *Vision Research, 25*, 21–31. doi:10.1016/0042-6989(85)90076-8.

Ringach, D. L. (2002). Spatial structure and symmetry of simple-cell receptive fields in macaque primary visual cortex. *Journal of Neurophysiology, 88*,455–463.

Rivlin-Etzion, M., Zhou, K., Wei, W., Elstrott, J., Nguyen, P. L., Barres, B. A., et al. (2011). Transgenic mice reveal unexpected diversity of ON-OFF direction-selective retinal ganglion cell subtypes and brain structures involved in motion processing. *Journal of Neuroscience, 31*, 8760–8769.

Rogan, S. C., & Roth, B. L. (2011). Remote control of neuronal signaling. *Pharmacological Reviews, 63*, 291–315.

Rudy, B., Fishell, G., Lee, S., & Hjerling-Leffler, J. (2011). Three groups of interneurons account for nearly 100% of neocortical GABAergic neurons. *Developmental Neurobiology, 71*, 45–61.

Runyan, C. A., Schummers, J., Van Wart, A., Kuhlman, S. J., Wilson, N. R., Huang, Z. J., et al. (2010). Response features of parvalbumin-expressing interneurons suggest precise roles for subtypes of inhibition in visual cortex. *Neuron, 67*, 847–857.

  CRISTOPHER M. NIELL, VINCENT BONIN, AND MARK L. ANDERMANN

Sawinski, J., Wallace, D. J., Greenberg, D. S., Grossmann, S., Denk, W., & Kerr, J. N. (2009). Visually evoked activity in cortical cells imaged in freely moving animals. *Proceedings of the National Academy of Sciences of the United States of America, 106*, 19557–19562. doi:10.1073/pnas.0903680106.

Schuett, S., Bonhoeffer, T., & Hubener, M. (2002). Mapping retinotopic structure in mouse visual cortex with optical imaging. *Journal of Neuroscience, 22*, 6549–6559.

Shankar, S., & Ellard, C. (2000). Visually guided locomotion and computation of time-to-collision in the mongolian gerbil (*Meriones unguiculatus*): The effects of frontal and visual cortical lesions. *Behavioural Brain Research, 108*, 21–37.

Shuler, M. G., & Bear, M. F. (2006). Reward timing in the primary visual cortex. *Science, 311*, 1606–1609.

Simmons, P. A., Lemmon, V., & Pearlman, A. L. (1982). Afferent and efferent connections of the striate and extrastriate visual cortex of the normal and reeler mouse. *Journal of Comparative Neurology, 211*, 295–308.

Skottun, B. C., De Valois, R. L., Grosof, D. H., Movshon, J. A., Albrecht, D. G., & Bonds, A. B. (1991). Classifying simple and complex cells on the basis of response modulation. *Vision Research, 31*, 1079–1086. doi:10.1016/0042-6989(91)90033-2.

Smith, S. L., & Hausser, M. (2010). Parallel processing of visual space by neighboring neurons in mouse visual cortex. *Nature Neuroscience, 13*, 1144–1149.

Sohya, K., Kameyama, K., Yanagawa, Y., Obata, K., & Tsumoto, T. (2007). GABAergic neurons are less selective to stimulus orientation than excitatory neurons in layer ii/iii of visual cortex, as revealed by in vivo functional Ca^{2+} imaging in transgenic mice. *Journal of Neuroscience, 27*, 2145–2149.

Song, S., Sjostrom, P. J., Reigl, M., Nelson, S., & Chklovskii, D. B. (2005). Highly nonrandom features of synaptic connectivity in local cortical circuits. *PLoS Biology, 3*, e68. doi:10.1371/journal.pbio.0030068.

Stone, C., & Pinto, L. H. (1993). Response properties of ganglion cells in the isolated mouse retina. *Visual Neuroscience, 10*, 31–39.

Stopka, P., & Macdonald, D. W. (2003). Way-marking behaviour: An aid to spatial navigation in the wood mouse (*Apodemus sylvaticus*). *BMC Ecology, 3*, 3. doi:10.1186/1472-6785-3-3.

Stosiek, C., Garaschuk, O., Holthoff, K., & Konnerth, A. (2003). In vivo two-photon calcium imaging of neuronal networks. *Proceedings of the National Academy of Sciences of the United States of America, 100*, 7319–7324. doi:10.1073/pnas.1232232100.

Talwar, S. K., Musial, P. G., & Gerstein, G. L. (2001). Role of mammalian auditory cortex in the perception of elementary sound properties. *Journal of Neurophysiology, 85*, 2350–2358.

Tamamaki, N., Yanagawa, Y., Tomioka, R., Miyazaki, J., Obata, K., & Kaneko, T. (2003). Green fluorescent protein expression and colocalization with calretinin, parvalbumin, and somatostatin in the gad67-gfp knock-in mouse. *Journal of Comparative Neurology, 467*, 60–79.

Tan, A. Y., Brown, B. D., Scholl, B., Mohanty, D., & Priebe, N. J. (2011). Orientation selectivity of synaptic input to neurons in mouse and cat primary visual cortex. *Journal of Neuroscience, 31*, 12339–12350.

Taniguchi, H., He, M., Wu, P., Kim, S., Paik, R., Sugino, K., et al. (2011). A resource of Cre driver lines for genetic targeting of GABAergic neurons in cerebral cortex. *Neuron, 71*, 995–1013.

van Alphen, B., Winkelman, B. H., & Frens, M. A. (2009). Age- and sex-related differences in contrast sensitivity in C57BL/6 mice. *Investigative Ophthalmology & Visual Science, 50*, 2451–2458.

Van Hooser, S. D. (2007). Similarity and diversity in visual cortex: Is there a unifying theory of cortical computation? *Neuroscientist, 13*, 639–656.

Van Hooser, S. D., Heimel, J. A., Chung, S., Nelson, S. B., & Toth, L. J. (2005). Orientation selectivity without orientation maps in visual cortex of a highly visual mammal. *Journal of Neuroscience, 25*, 19–28.

Volgyi, B., Chheda, S., & Bloomfield, S. A. (2009). Tracer coupling patterns of the ganglion cell subtypes in the mouse retina. *Journal of Comparative Neurology, 512*, 664–687.

Wagor, E., Mangini, N. J., & Pearlman, A. L. (1980). Retinotopic organization of striate and extrastriate visual cortex in the mouse. *Journal of Comparative Neurology, 193*, 187–202.

Wang, B. S., Sarnaik, R., & Cang, J. (2010). Critical period plasticity matches binocular orientation preference in the visual cortex. *Neuron, 65*, 246–256.

Xu, X., Roby, K. D., & Callaway, E. M. (2006). Mouse cortical inhibitory neuron type that coexpresses somatostatin and calretinin. *Journal of Comparative Neurology, 499*, 144–160.

Xu, X., Roby, K. D., & Callaway, E. M. (2010). Immunochemical characterization of inhibitory mouse cortical neurons: Three chemically distinct classes of inhibitory cells. *Journal of Comparative Neurology, 518*, 389–404.

Yoshimura, Y., & Callaway, E. M. (2005). Fine-scale specificity of cortical networks depends on inhibitory cell type and connectivity. *Nature Neuroscience, 8*, 1552–1559.

Yoshimura, Y., Dantzker, J. L., & Callaway, E. M. (2005). Excitatory cortical neurons form fine-scale functional networks. *Nature, 433*, 868–873.

Zariwala, H. A., Borghuis, B. G., Hoogland, T. M., Madisen, L., Tian, L., De Zeeuw, C. I., et al. (2012). A Cre-dependent gcamp3 reporter mouse for neuronal imaging in vivo. *Journal of Neuroscience, 32*, 3131–3141.

Zariwala, H. A., Madisen, L., Ahrens, K. F., Bernard, A., Lein, E. S., Jones, A. R., et al. (2011). Visual tuning properties of genetically identified layer 2/3 neuronal types in the primary visual cortex of Cre-transgenic mice. *Frontiers in Systems Neuroscience, 4*, 162. doi:10.3389/fnsys.2010.00162.

Zollner, P. A., & Lima, S. L. (1999). Illumination and the perception of remote habitat patches by white-footed mice. *Animal Behaviour, 58*, 489–500.

30 Beyond the Classical Receptive Field: Surround Modulation in Primary Visual Cortex

ALESSANDRA ANGELUCCI AND S. SHUSHRUTH

The concept of the classical receptive field (RF) (Barlow, 1953; Hubel & Wiesel, 1959) and hierarchical feedforward models of the visual system (Hubel & Wiesel, 1962; Riesenhuber & Poggio, 2003) have provided a foundation for theories of the neural representation of visual objects. These theories view cells as "filters" or "feature detectors" and visual information as ascending through a hierarchy of cortical areas (Van Essen & Maunsell, 1983), with RFs in higher areas processing information from increasingly larger regions of visual space and coding for increasingly more complex aspects of visual stimuli. These models are feedforward in that the increasingly complex and invariant representation of objects is built by integrating convergent feedforward inputs from lower levels.

In recent years, anatomical, computational and physiological evidence has challenged these theories. Anatomically, in addition to feedforward projections between cortical areas, there is a massive system of feedback connections that is unaccounted for by purely feedforward models. Computationally, feedforward models can perform object recognition in simple environments but not in cluttered environments such as natural scenes. This is because local information processed by neuronal RFs in natural scenes is ambiguous, and the computation of object boundaries requires knowledge about the global properties of a scene for disambiguation. Physiologically, there is evidence that in the visual system global-to-local computations occur even at the lowest level of processing, that is, in the primary visual cortex (V1), where neuronal responses to local features within their RFs are affected by the perceptual organization of the scene as a whole. A fundamental example of such computations is surround modulation—the ability of neurons in V1 (and other areas) to change their response to stimuli within their classical RF depending on visual context, i.e., the stimuli presented in the RF surround (Allman, Miezin, & McGuinness, 1985; Blakemore & Tobin, 1972; Maffei & Fiorentini, 1976; Nelson & Frost, 1978). This phenomenon implies integration of signals across distant visual field locations, well beyond the classical RF of single V1 neurons, and thus cannot be easily explained by feedforward mechanisms and classical RF concepts.

In the present chapter we first review the phenomenology of surround modulation in V1, then examine the circuitry and mechanisms that may generate it, and finally discuss its role in visual processing and perception.

PHENOMENOLOGY OF SURROUND MODULATION

Hubel and Wiesel (1965) first described a class of cells in areas 18 and 19 of the cat visual cortex that were selective for the length of a bar, in that their response was suppressed by stimuli extending beyond a critical length. They named these cells "hypercomplex," believing they were generated by converging feedforward afferents from area 17 complex cells. However, later studies found that even simple and complex cells throughout area 17 showed length tuning and that increasing the width of a stimulus had a similar effect (Gilbert, 1977; Maffei & Fiorentini, 1976; Nelson & Frost, 1978). The terms "end stopping" and "side inhibition" or surround suppression came to replace the term hypercomplex. These early reports suggested that the notion of the classical RF was insufficient to understand the neural substrates of visual perception. However, the concept of a non-classical RF (or surround) did not become established until the mid-1980s (Allman, Miezin, & McGuinness, 1985). A series of studies followed in which surround modulation was characterized quantitatively using circular or annular gratings and varying systematically the stimulation parameters. Overall, these studies indicated that surround modulation is a property of most V1 cells: 56–86% of cells in cat V1 (Sengpiel, Sen, & Blakemore, 1997; Walker, Ohzawa, & Freeman, 2000) and 60–100% in macaque V1 (Cavanaugh, Bair, & Movshon, 2002a; Sceniak, Hawken, & Shapley, 2001; Shushruth et al., 2009). They also showed that the predominant surround effect is *suppression* of the spiking response to RF stimulation, and that this suppression is sensitive to surround stimulus parameters (such as orientation, spatial frequency, speed, and contrast), that is, the suppression is reduced or turns to *facilitation* for specific stimulation parameters. Below we review these studies.

Spatial Properties of Surround Modulation: Defining the Receptive Field and the Surround

The surround is defined as the region outside the RF where presentation of stimuli does not cause the cell to spike but can modulate its response to stimuli inside the RF. Therefore, any study of surround modulation requires defining the limits of the RF. This is not simple because RF size varies depending on how it is measured. A common approach to map the RF has been to use a small stimulus (a light or dark bar or small grating) at the appropriate orientation for the cell to delimit the visual field area that elicits spikes from the cell, a measure of RF size termed *minimum response field* (mRF) or *classical RF* (Barlow, Blakemore, & Pettigrew, 1967; Hubel & Wiesel, 1962). However, intracellular recordings in cat area 17 (Bringuier et al., 1999) have demonstrated that the mRF is surrounded by a larger subthreshold depolarizing field incapable of driving the cell when stimulated alone but capable of increasing the cell's response when stimulated with the mRF. With extracellular recordings, this subthreshold region of the RF can be revealed to some extent by measuring the cell's response to a circular grating of increasing radius centered on the mRF. A typical V1 cell increases its response with grating size up to a peak and is suppressed for further increases in size (figure 30.1, black curve). The RF size thus corresponds to the grating's radius at the cell's peak response. In macaque V1 at parafoveal eccentricities, RF dimensions based on these areal summation measurements using high-contrast gratings are two to three times larger than the mRF of the same cells (Angelucci et al., 2002). Furthermore, when this measure of RF is performed using gratings of low contrast, the stimulus area over which summation occurs increases by about twofold compared to the summation area measured with high-contrast gratings (figure 30.1, gray curve) (Sceniak et al., 1999; Sengpiel, Sen, & Blakemore, 1997). In this chapter we refer to the RF size based on these summation measurements

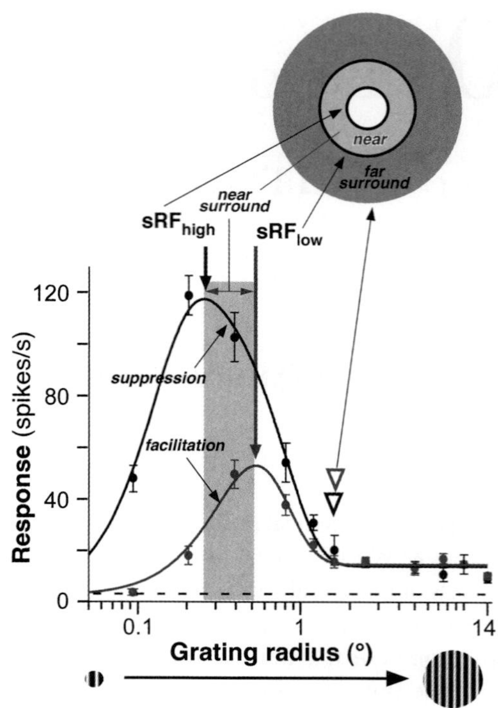

FIGURE 30.1 Size tuning curve of an example cell in macaque V1. Black and gray curves show responses to a grating of high (70%) and low (12%) contrast, respectively. Dashed line shows mean spontaneous firing rate. Thick arrows indicate the radii of the sRF_high (black; 0.26°) and sRF_low (gray; 0.54°). Gray shaded column indicates the near surround. Arrowheads point at surround radius at high (black; 1.41°) and low (gray; 1.41°) contrast. Cartoon illustrates schematics of the different components of the RF and surround of a V1 cell: white area, RF; gray area, surround, consisting of a near and a far region. (Modified from Shushruth et al., 2009.)

at high or low stimulus contrast, as the cell's high- or low-contrast summation RF (sRF_high and sRF_low, respectively) (figure 30.1).

FACILITATORY MODULATIONS: STUDIES WITH DISCRETE STIMULI A consequence of the contrast dependence of the sRF size is that stimuli presented in the region between the sRF_high and the sRF_low (gray column in figure 30.1) can exert facilitatory or suppressive effects, depending on stimulus contrast. Whether this region should be considered part of the surround or part of the RF has been a matter of debate. In particular there is still disagreement as to whether surround effects can be facilitatory under specific stimulation conditions. Several studies have shown that facilitatory modulations occur predominantly when discrete stimuli (bars or Gabor patches) are presented at the end zones of the mRF collinearly aligned in visual space with the orientation of the stimulus inside the mRF (figure 30.2A) (Chisum, Mooser, & Fitzpatrick, 2003; Kapadia et al.,

426   ALESSANDRA ANGELUCCI AND S. SHUSHRUTH

1995; Maffei & Fiorentini, 1976; Nelson & Frost, 1985; Polat et al., 1998). This facilitation dominates when the stimulus in the mRF is presented at low contrast, and can turn to suppression when it is presented at high contrast (Kapadia, Westheimer, & Gilbert, 2000; Mizobe et al., 2001; Polat et al., 1998). This phenomenon, known as *collinear facilitation*, is thought to be the neural correlate of perceptual contour integration (Hess & Field, 1999; Kapadia, Westheimer, & Gilbert, 2000) (discussed below). If one regards the sRF_{low} of a cell as part of the cell's RF, then these facilitatory effects can be viewed as simply occurring within the RF (figure 30.2A), and therefore, one would conclude that there are no facilitatory surround effects (Angelucci & Bullier, 2003; DeAngelis, Freeman, & Ohzawa, 1994; Fitzpatrick, 2000; Walker, Ohzawa, & Freeman, 1999). Collinear facilitation and increased spatial summation at low stimulus contrast would thus simply share similar mechanisms. Consistent with this interpretation, the strength of collinear facilitation decreases with increasing spatial separation of the stimuli inside and outside the mRF (figure 30.2A, bottom). However, because the region between the sRF_{high} and sRF_{low} can be suppressive at high contrast (figure 30.1, black curve), it can also be viewed as part of the surround. This observation together with anatomical studies (discussed below) suggesting a distinctive circuitry for this region have led some (Angelucci & Bressloff, 2006) to refer to it as the *near surround* and to the region beyond the sRF_{low} as the *far surround* (figure 30.1). Moreover, not all facilitatory surround effects can be attributed to stimulation of the RF by a surround stimulus, because facilitatory inputs can arise 12° away from the RF center in both cat (Mizobe et al., 2001) and macaque V1 (Ichida et al., 2007; Shushruth et al., 2012).

A more comprehensive explanation of surround effects, instead, is that the sign of surround modulation depends on the strength of activation of both the RF and its surround. When the RF is strongly activated (e.g., by an optimally oriented high-contrast stimulus fitted to the cell's sRF_{high}), weak or strong surround stimulation evokes predominantly suppression (Cavanaugh, Bair, & Movshon, 2002a; DeAngelis, Freeman, & Ohzawa, 1994; Ichida et al., 2007; Levitt & Lund, 1997; Sceniak, Hawken, & Shapley, 2001; Sengpiel et al., 1998; Shushruth et al., 2009). Instead, surround facilitation is more frequently observed for weak stimulation of both the RF and the surround. For example, it occurs in 34–38% of V1 cells when the RF and surround are stimulated with discrete elements (of optimal orientation for the cell) (Chen et al., 2001; Polat et al., 1998) (figure 30.2A), especially if the element in the RF is at low contrast. It is also frequently observed

(~60% of macaque V1 cells) when the sRF_{low} is stimulated by an optimally oriented grating near the cell's contrast threshold and the surround by an iso-oriented thin annular grating; lowering the contrast of this surround stimulus increases the strength of facilitation, whereas increasing the width of the annular grating turns the facilitation into suppression (Ichida et al., 2007) (figure 30.2B). Similarly, surround facilitation emerges in many cells (~35% in macaque) when both the RF and surround are weakly stimulated with high-contrast gratings of suboptimal orientation for the recorded cell (Shushruth et al., 2012). Importantly, the threshold for suppression versus facilitation is cell specific, so that there is no single contrast level or surround stimulus size that causes facilitation across the entire cell population.

SUPPRESSIVE MODULATIONS: STUDIES WITH LARGE GRATINGS   In most other studies surround modulation was investigated using a central grating fitted to the cell's sRF_{high} surrounded by a large annular grating involving the near and far surround, while varying the center and surround grating parameters (e.g., their relative orientation, spatial frequency and contrast). These studies found that when the sRF_{high} is stimulated by a grating of optimal orientation for the cell, a large iso-oriented surround grating predominantly *suppresses* the cell response (Cavanaugh, Bair, & Movshon, 2002a; DeAngelis, Freeman, & Ohzawa, 1994; Levitt & Lund, 1997, 2002; Sceniak, Hawken, & Shapley, 2001; Sengpiel, Sen, & Blakemore, 1997; Walker, Ohzawa, & Freeman, 2000), even when the center grating contrast is lowered (Cavanaugh, Bair, & Movshon, 2002a; DeAngelis, Freeman, & Ohzawa, 1994; Levitt & Lund, 1997). Suppression is reduced or abolished, occasionally turning to facilitation, as surround stimulus parameters increasingly differ from those of the center grating. Facilitation was rarely observed in these studies because of the strong surround stimulation exerted by large surround annuli.

Furthermore, the surround is not always organized in a concentric and symmetric fashion, as suppression can arise from a single modulatory zone (Walker, Ohzawa, & Freeman, 1999), from only the end or side zones of the sRF, and is often stronger at the end zones (Cavanaugh, Bair, & Movshon, 2002b; DeAngelis, Freeman, & Ohzawa, 1994; Sceniak, Hawken, & Shapley, 2001). Moreover, suppression is stronger in the near than in the far surround.

SPATIAL EXTENT AND STRENGTH OF SURROUND MODULATION   Two different stimulus protocols have been used to measure the extent and strength of

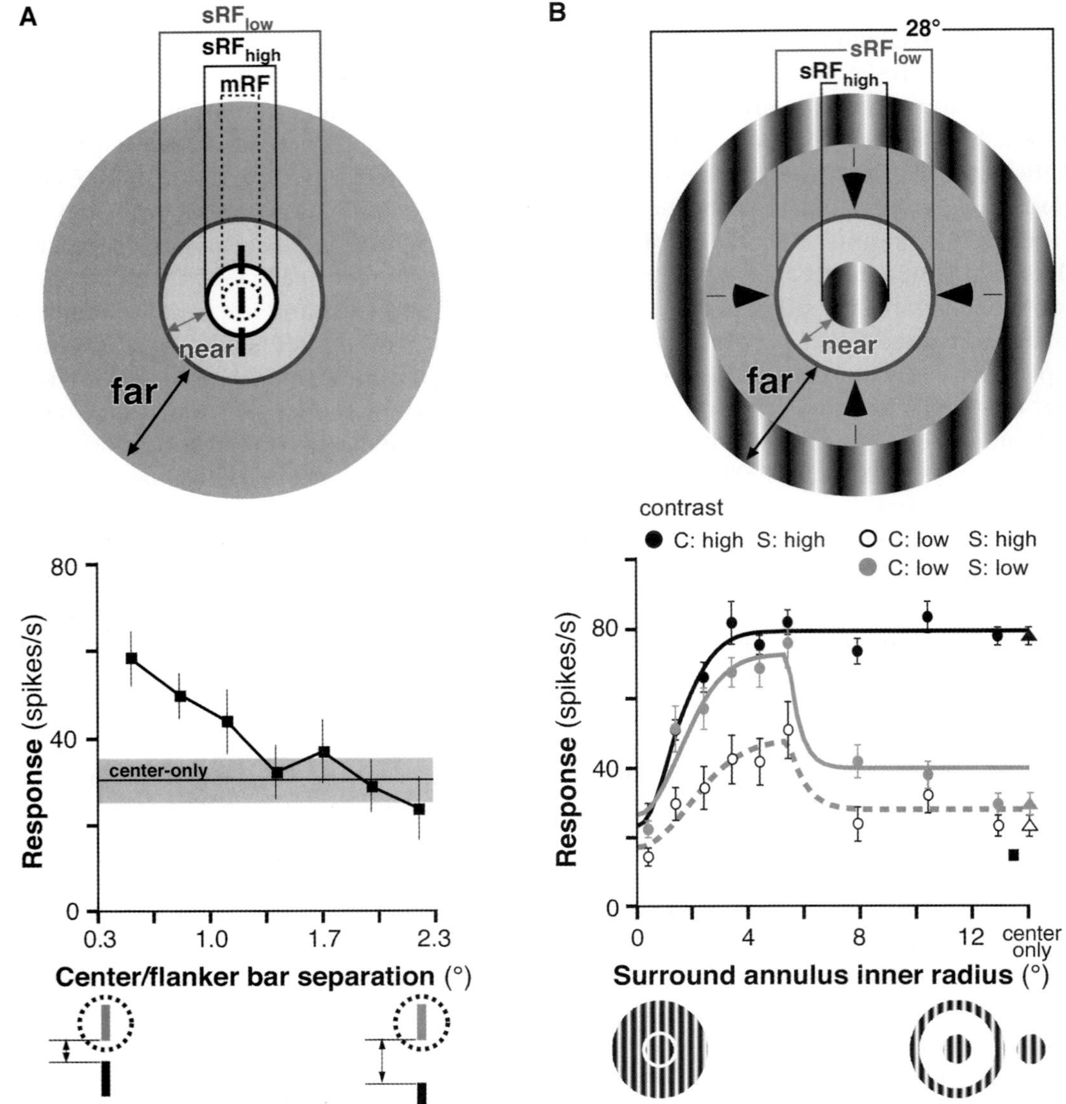

FIGURE 30.2

Center and surround stimuli that can evoke facilitatory modulations. (A) Collinear facilitation. (Top) cartoon indicates different RF and surround components and the bar stimuli used to evoke collinear facilitation. (Bottom) Response of a V1 cell in awake macaque as a function of the spatial separation of the bar stimuli inside and outside the mRF. Horizontal line marks the cell's response to the center bar alone (presented at low contrast). Gray area represents 1 SEM of the center-only response. Iso-oriented and co-aligned discrete stimuli inside and outside the mRF, but within the sRF$_{low}$, can evoke facilitation. (Bottom modified from Kapadia et al., 1995, reproduced with permission from Cell Press.) (B) Facilitation from the far surround. (Top) RF and surround components and the grating stimuli used to evoke far-surround facilitation. The center grating is matched to the size of the sRF$_{high}$, and the inner radius of the annular grating in the surround is decreased (arrows) from 13° to the size of the cell's sRF$_{low}$, thus stimulating only the far surround. (Bottom) Response of a V1 cell in an anesthetized macaque as a function of the inner radius of the surround annular grating. Dashed and solid curves indicate responses to different combinations of center (C) and surround (S) stimulus contrast as follows: 70%/70% (black), 20%/70% (solid gray), 20%/20% (dashed gray). The triangles are responses to center-only stimulation. The square indicates response to a surround stimulus of smallest inner radius presented alone. At high center contrast, surround stimulation always evokes suppression. When the center grating is at low contrast, a small annular stimulus in the far surround evokes facilitation, followed by suppression as more of the surround is stimulated. Thus, large surround stimuli of any contrast evoke suppression. (Modified from Ichida et al., 2007, reproduced with permission from the American Physiological Society.)

surround modulation: the expanding patch (figure 30.1) and expanding annulus (figure 30.2B) methods. The former activates both the near and far surround, but it predominantly reveals the stronger suppressive effects from the near surround. The latter, by masking out the near surround, reveals the weaker modulations from the far surround. In macaque V1 at parafoveal eccentricities, the average surround radius measured with the expanding patch method is ~1.6°, that is, five to six times larger than the sRF$_{high}$ of V1 cells, and can extend up to ~3° (Cavanaugh, Bair, & Movshon, 2002a; Levitt & Lund, 2002; Sceniak, Hawken, & Shapley, 2001; Shushruth et al., 2009). Surround radius measured with the expanding annulus method is larger, averaging 5.5° and extending up to 12.5° (Shushruth et al., 2009).

Near-surround suppression is strong; about 50% of V1 cells in macaque are suppressed by 60% or more (mean 58%, ranging up to 87%) (Sceniak, Hawken, & Shapley, 2001; Shushruth et al., 2009), and ~40% in cat are suppressed by more than 40% (Walker, Ohzawa, & Freeman, 2000). Far-surround suppression (mean 25%, ranging up to 61%) is weaker than near-surround suppression (Shushruth et al., 2009). Surround facilitation has a similar spatial extent as suppression, that is, it can arise up to 12° away from the RF center, and near-surround facilitation is stronger (mean 64%) than far-surround facilitation (mean 32%) (Ichida et al., 2007).

Surround modulation occurs in all V1 layers (Levitt & Lund, 2002; Sceniak, Hawken, & Shapley, 2001; Walker, Ohzawa, & Freeman, 2000). However, subtle laminar differences exist that may reflect laminar differences in connectivity. For example, in macaque V1 larger surrounds are absent in geniculocortical recipient layer 4C (Ichida et al., 2007; Shushruth et al., 2009), a layer lacking long-range intracortical connections. This suggests that the larger far surrounds in other V1 layers are generated by long-range intracortical connections within these layers. Moreover, near- and far-surround suppression in macaque V1 are both stronger in the supragranular layers (4B and above) (Sceniak, Hawken, & Shapley, 2001; Shushruth et al., 2009).

Tuning of Surround Modulation

THE EFFECTS OF CHANGING THE ORIENTATION OF THE SURROUND STIMULUS By stimulating the RF with gratings of optimal parameters for the recorded cell while varying the parameters of the grating in the surround, previous studies concluded that surround modulation is selective for surround stimulus orientation and spatial and temporal frequency and that this selectivity is similar to, but broader than that of the same cells' RF (Cavanaugh, Bair, & Movshon, 2002b; DeAngelis,

Freeman, & Ohzawa, 1994; Li & Li, 1994; Webb et al., 2005). Typically, suppression is maximal when the center and surround gratings have the same orientation, spatial frequency, drift direction (Cavanaugh, Bair, & Movshon, 2002b; DeAngelis, Freeman, & Ohzawa, 1994; Levitt & Lund, 1997; Muller et al., 2003; Sengpiel, Sen, & Blakemore, 1997; Walker, Ohzawa, & Freeman, 1999) and speed (Li & Li, 1994). Suppression is reduced or disappears, sometimes turning to facilitation, as the difference in orientation (or other parameters) between the center and surround stimuli is increased. Instead, suppression is insensitive to the relative spatial phase of the stimuli in the RF and surround (DeAngelis, Freeman, & Ohzawa, 1994; Levitt & Lund, 1997) or to chromatic contrast (Solomon, Peirce, & Lennie, 2004). The orientation tuning of surround modulation also depends on a cell's position in the orientation map, with suppression being more sharply tuned at orientation domains than at pinwheel centers (Hashemi-Nezhad & Lyon, 2011), likely reflecting differences in the orientation specificity of the local and long-range connectivity at these different map locations.

Most previous studies used gratings that stimulated both the near and far surround. Recently, however, Shushruth et al. (2013), using annular gratings confined to the near or far surround, found that near suppression is more sharply tuned than far suppression (figure 30.3). Using similar stimuli and a contrast-matching task, these authors obtained a similar result for near- and far-surround suppression of perceived contrast in human observers. In both V1 cells and human perception, broader tuning of far-surround suppression was due to nonoptimal stimulus orientations exerting stronger suppression in the far than in the near surround. These results suggest different orientation specificities of the circuits underlying near- and far-surround suppression and point to an important relationship between surround suppression in V1 and human perception. The same study also found laminar differences in the orientation tuning of both near- and far-surround suppression, suggesting laminar differences in the orientation specificity of their underlying circuitry (see below). Specifically, near suppression is most sharply tuned in layers 3B, 4B and 4Cα, and far suppression in layer 4B. Below we discuss the idea that the different tuning of near- and far-surround suppression may reflect a statistical bias in the distribution of oriented elements in natural images.

THE EFFECTS OF CHANGING THE ORIENTATION OF THE CENTER STIMULUS Few studies have examined how the orientation tuning of surround modulation is affected by changing the orientation of stimuli inside the RF.

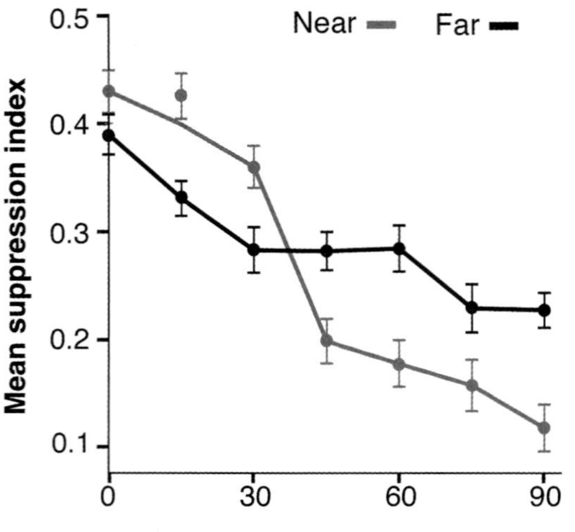

FIGURE 30.3   Orientation tuning of near- and far-surround suppression. Mean suppression index (SI) for a population of macaque V1 cells ($n = 68$), caused by near-(gray) or far-surround (black) stimulation. Far-surround stimulation was achieved using the stimulus in figure 30.2B, with a center grating of 75% contrast at the optimal orientation and a surround grating of 2° inner radius varied in orientation. For near-surround stimulation the surround grating had a 2° outer radius and was separated from the center grating by a 0.25° gap. Thus, complementary surround regions were stimulated in the two conditions. $SI = 1 - (R_{CS}/R_C)$, where R_{CS} is the cell response to the center + surround stimulus and R_C the response to the center-only stimulus. A larger SI indicates stronger suppression. (Modified from Shushruth et al., 2013.)

Sillito et al. (1995) showed in a few V1 cells that maximal suppression occurred when the stimuli in the RF and surround were similarly oriented, irrespective of whether they were presented at the cell's preferred orientation. This result was replicated in a small population of broadly orientation-tuned cells in macaque V1 (Cavanaugh, Bair, & Movshon, 2002b), but the prevalence of this tuning behavior in V1 cells remained unclear. Furthermore, Sillito et al. (1995) reported that any orientation discontinuity in the RF and surround evoked facilitation, even when the RF was stimulated with an orientation orthogonal to that preferred by the cell, which did not activate the cell in the absence of surround stimulation. Others (Cavanaugh, Bair, & Movshon, 2002b; Walker, Ohzawa, & Freeman, 1999) could not replicate this result and attributed it to encroachment of the surround stimuli onto the RF. Recently, Shushruth et al. (2012) reexamined this issue by recording the response of a large population of cells in macaque V1 to high-contrast gratings of changing orientation in the RF and surround; they used the stimulus shown in figure 30.2B, growing the far-surround grating toward the RF without stimulating the near surround. They found that for the majority of V1 cells (including sharply orientation-tuned cells) the orientation specificity of surround modulation is independent of the stimulus orientation preferred by the RF, but changes with the orientation presented to the RF. Strongest suppression occurs when the stimuli in the RF and surround are of the same orientation (figure 30.4A, B), and strongest facilitation occurs when the stimuli are cross-oriented (figure 30.4C), even when the stimulus in the RF is not at the cell's preferred orientation, but provided that this stimulus reliably activates the cell when presented without a surround stimulus. Thus, surround stimulation had no effect when the RF was stimulated by orientations orthogonal to optimal. Unlike previous studies, Shushruth et al. (2012) found that facilitation emerged in many (35%) cells when the RF and surround were both weakly activated, the RF by stimuli of suboptimal orientation for the recorded cell, the surround by small annular gratings. Thus, previous studies failed to observe facilitation likely because they used stimuli that strongly activated the surround and/or the RF. To explain the tuning behavior of surround modulation, Shushruth et al. (2012) proposed a computational model that is discussed below.

Contrast Dependence of Surround Modulation

The contrast of the stimuli in the RF and surround affects the spatial extent, tuning and strength of surround modulation. Surround sizes are larger when measured with low-contrast gratings (Shushruth et al., 2009). The orientation selectivity of surround modulation is reduced when the contrast of the stimulus in the RF is reduced (Cavanaugh, Bair, & Movshon, 2002b; Hashemi-Nezhad & Lyon, 2011), and (as discussed above) surround orientations that typically evoke suppression can facilitate the cell's response when center stimulus contrast is reduced (e.g., figure 30.2B). When the RF and surround are stimulated at low contrast, the strength of surround suppression is reduced (Cavanaugh, Bair, & Movshon, 2002a; Sadakane et al., 2006; Schwabe et al., 2010), and many cells show facilitatory surround effects (Ichida et al., 2007). For any given center stimulus contrast, suppression strength increases as surround stimulus contrast increases. However, there have been variable reports on the effects of changing center stimulus contrast on the strength of surround suppression. In some studies a high-contrast surround stimulus suppressed more strongly the response to a lower-contrast than to a higher-contrast center stimulus (Cavanaugh, Bair, & Movshon, 2002a; Levitt & Lund, 1997). Schwabe et al. (2010), instead, observed weaker surround suppression for lower-contrast than for

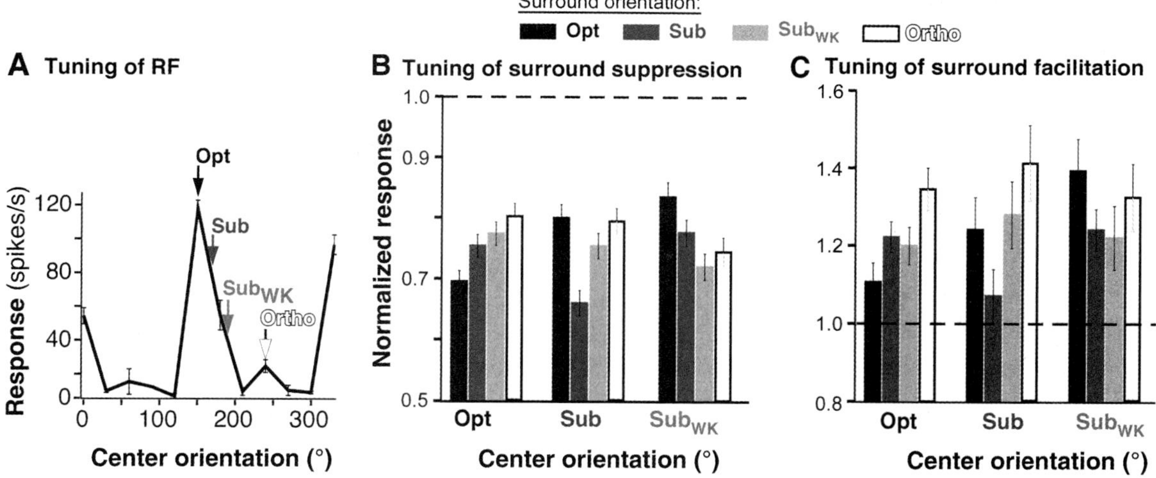

FIGURE 30.4   Orientation tuning of surround suppression and facilitation. (A) Orientation-tuning curve of the RF response for a V1 cell. The arrows indicate four orientations chosen for this cell to characterize the orientation specificity of surround modulation. Opt, optimal orientation; Sub, Sub$_{WK}$, suboptimal orientations evoking weaker responses from the cell, with Sub$_{WK}$ evoking a weaker response than Sub; Ortho, orthogonal. For each cell four orientations were similarly chosen on the basis of the cell's RF tuning curve. (B) Mean normalized population responses to each of three center grating orientations (indicated on the *x* axis) presented together with one of four surround grating orientations (bars of different gray levels). Each center + surround response was measured at the largest surround grating size used and normalized to each respective center-only response. For any center orientation, suppression is maximal for iso-oriented center-surround stimuli. (C) Mean normalized population responses for cells showing surround facilitation. Responses were measured at the surround grating size that evoked maximal facilitation for each cell. Facilitation is weakest in the iso-orientation condition and is strongest at the surround orientation nearest to orthogonal to that presented in the RF. (A–C, modified from Shushruth et al., 2012, reproduced with permission from the Society for Neuroscience.)

higher-contrast center stimuli. Using model simulations, Schwabe et al. (2010) were able to reconcile these discrepancies by demonstrating opposite effects on surround suppression strength depending on both the size and the contrast level of the stimulus presented to the RF. In particular, weaker suppression of lower-contrast than higher-contrast center stimuli is observed when the center stimulus is near the cell's contrast threshold.

Surround stimulation also changes the contrast response function of a V1 cell in a manner that is best described by a change in response gain, that is, a divisive suppression that scales responses equally at all contrasts (Cavanaugh, Bair, & Movshon, 2002a).

Timing and Dynamics of Surround Modulation

Using a variety of stimuli and methodological approaches, several studies have shown that the onset of orientation-specific surround suppression is delayed by 15–60 ms relative to the RF response (Hupé et al., 2001; Knierim & Van Essen, 1992; Lamme, 1995; Lamme, Rodriguez-Rodriguez, & Spekreijse, 1999; Nothdurft, Gallant, & Van Essen, 1999, 2000). Bair, Cavanaugh, and Movshon (2003) measured the onset latency of surround suppression using dynamic center-surround stimuli (160 ms duration), with the center fixed at the preferred orientation for the cell and the surround orientation changed from preferred to orthogonal-to-preferred. They measured the timing of response change when the surround stimulus transitioned from a nonsuppressive to a suppressive orientation, and therefore effectively measured the onset timing of orientation-tuned suppression. The average suppression latency relative to the stimulus transition was 60 ms (range 25–110 ms), with suppression being delayed on average by 9 ms relative to the onset of the RF response. This delay was about 30 ms longer for weakly than for strongly suppressed cells. Moreover, by moving the surround grating progressively farther from the RF, they found that the latency of suppression induced by stimuli in the far surround was similar to that induced by stimuli in the near surround (figure 30.5). These results point to an underlying circuit with fast conduction velocity and high spatial divergence and convergence. As discussed below, interareal feedback connections to V1 show both properties.

Xing et al. (2005) looked at the time course of orientation tuning based on reverse correlation analysis of stimuli two to five times larger than the radius of the neuron's sRF$_{high}$. They found that suppression consists

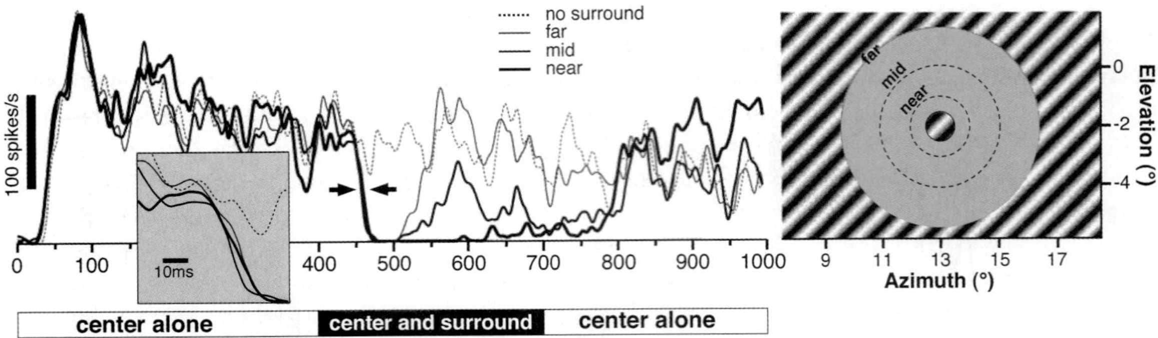

FIGURE 30.5   Time course of surround suppression for stimuli at different distances from the RF. (Left) mean firing rates of an example cell in response to a center-only stimulus, or a center + surround stimulus involving the near, mid, or far surround (lines of different thickness). (Gray inset) Blowup of the curves around the time of suppression onset. For this cell the onset of the RF response was 48 ms, and the onset of near suppression was 65 ms. (Right) The stimulus used consisted of a center grating fitted to the sRF$_{high}$ and a grating in the surround whose inner radius was moved away from the RF, from near to far (dashed circles). The center stimulus appeared first (at $t = 0$) and lasted for 1 s, and the surround stimulus appeared at $t = 400$ ms and lasted for 300 ms. (Modified from Bair, Cavanaugh, & Movshon, 2003; reproduced with permission from the Society for Neuroscience.)

of an early component, which is orientation *un*tuned and peaks at virtually the same time as the stimulus-driven excitation, and a late component, which is orientation tuned and peaks about 17 ms after the peak of excitation. The untuned component was also seen when the stimulus was confined to the sRF$_{high}$ of the neuron, suggesting that it arises from suppression of the feedforward input itself, perhaps as a result of surround suppression of geniculocortical afferents (see below).

ANATOMICAL CIRCUITS FOR SURROUND MODULATION

Feedforward connections to V1 from the lateral geniculate nucleus (LGN), long-range intra-V1 horizontal connections, and feedback connections from extrastriate cortex to V1 have all been implicated in surround modulation.

The Role of Feedforward Connections

V1 receives driving feedforward inputs from the LGN, with afferents from the magnocellular and parvocellular channels terminating primarily in layer 4Cα and 4Cβ, respectively (Lund, 1988). These connections are spatially restricted, that is, they connect corresponding regions of visual field representation in LGN and V1 (Perkel, Bullier, & Kennedy, 1986), and are thought to contribute to the spatial and tuning properties of V1 cells' RFs (Bauer et al., 1999; Hubel & Wiesel, 1962; Reid & Usrey, 2004). However, several lines of evidence indicate that geniculocortical afferents also contribute

to V1 surrounds. First, in addition to their classical center-surround RF, LGN cells have an extraclassical, nonlinear surround that overlaps and extends beyond the classical RF and that can strongly suppress LGN responses (Alitto & Usrey, 2008; Bonin, Mante, & Carandini, 2005; Felisberti & Derrington, 1999; Levick, Cleland, & Dubin, 1972; Sceniak, Chatterjee, & Callaway, 2006; Solomon, White, & Martin, 2002). Median surround radius in macaque LGN at parafoveal eccentricities is 0.8° (up to ~5°, but <2.5° for most cells) (Sceniak, Chatterjee, & Callaway, 2006), thus significantly smaller than surrounds in V1. There is strong evidence that surround suppression in the LGN originates subcortically (Alitto & Usrey, 2008), and therefore, it must influence V1 responses to large stimuli. Second, blockade of intracortical inhibition in cat V1 does not abolish near-surround suppression (Ozeki et al., 2004). Third, two mechanisms contribute to surround suppression in V1, one having broad spatiotemporal tuning (likely originating in the LGN), the other being sharply tuned for orientation, spatial and temporal frequency (likely generated intracortically) (Webb et al., 2005).

However, LGN surround suppression cannot fully account for V1 surrounds. First, the orientation tuning of surround modulation in V1 points to intracortical mechanisms for its generation. Although some have argued that surround suppression in cat LGN is orientation tuned (Naito et al., 2007; Sillito, Cudeiro, & Murphy, 1993), others have disagreed (Bonin, Mante, & Carandini, 2005), and yet others have shown that geniculate surround suppression is substantially less orientation tuned than in V1 (Ozeki et al., 2009). LGN surrounds in primates, instead, are untuned for orientation (Solomon, White, & Martin, 2002; Webb et al., 2002). Therefore, the LGN is unlikely to be the source of strong orientation-selective surround suppression in V1. Second, LGN surrounds are significantly smaller than V1 surrounds. In macaque the narrow visuotopic spread of geniculocortical axons added to the size of the LGN surrounds can account for the size of V1 cells' near surround, but not for their far surround (Angelucci & Sainsbury, 2006). The different RF sizes of LGN and V1 cells further argue that a stimulus of optimal size for a V1 cell is suppressive for most LGN cells. Thus, V1 cells at the peak of their size-tuning curve summate inputs from partly suppressed LGN afferents, indicating that suppression in the LGN does not necessarily lead to surround suppression in V1. In summary, the spatial scale of geniculate afferents to V1 can contribute to the V1 cell's sRF$_{high}$ and to near-surround, but not far-surround, suppression in V1. The absence of far surrounds in V1 layer 4C (Ichida et al., 2007) is also consistent with this notion.

We propose that V1 cells inherit a spatially restricted orientation-untuned component of suppression from the LGN (Webb et al., 2005; Xing et al., 2005) whose spatial scale is determined by the visuotopic spread of geniculocortical afferents and LGN surround sizes. However, intracortical mechanisms must contribute a spatially broader and strongly orientation-tuned component to V1 surround suppression.

The Role of Horizontal Connections

In macaque V1, excitatory neurons in layers 2/3, 4B/upper 4Cα and 5/6 send intralaminar projections linking regions over several millimeters (Rockland & Lund, 1983). Similar projections have been described in V1 layers 2/3 of many other species, for example, tree shrew (Rockland & Lund, 1982) and cat (Gilbert & Wiesel, 1983). Horizontal connections in layers 2/3 were first proposed to generate surround modulation in V1 (Gilbert et al., 1996) because many of their features are well suited to explain many properties of surround modulation. In particular, these connections in layers 2/3 link preferentially neurons of similar orientation preference (Malach et al., 1993) with RFs aligned along an axis in visual space collinear with the orientation preference of the connected neurons (Bosking et al., 1997; Schmidt et al., 1997; Sincich & Blasdel, 2001) (see figure 44.1 in chapter 44 by Field, Golden, and Hayes). This property could generate the orientation tuning of surround modulation and collinear facilitation. Instead, horizontal connections in different V1 layers do not appear to link cortical domains of similar

functional property (eye dominance) (Li et al., 2003; Lund, Angelucci, & Bressloff, 2003), a feature that may explain the weaker orientation tuning of surround modulation in the infragranular V1 layers (see above; Shushruth et al., 2013). Moreover, horizontal connections, at least in layers 2/3, contact both excitatory (80%) and inhibitory neurons (McGuire et al., 1991), a property that is useful to generate both long-range suppression and facilitation. Finally, these connections only elicit subthreshold responses (Hirsch & Gilbert, 1991; Yoshimura et al., 2000) and thus do not drive, but only modulate, the response of their target cells.

However, studies in macaque in which the spatial dimensions of V1 neurons' RFs and surrounds were quantitatively compared with the visuotopic extent of horizontal and feedback connections to the same V1 sites have led to new hypotheses on the role of these connections (Angelucci et al., 2002). By combining neuronal tracer injections with electrophysiological characterization at and around the injected V1 site, these studies demonstrated that the monosynaptic spread of V1 horizontal connections is commensurate with the size of the sRF$_{low}$, or near surround, of V1 cells. Thus, these connections cannot monosynaptically account for the extent of V1 cells' far surrounds.

Polysynaptic chains of horizontal connections are also unlikely to underlie far-surround modulation because of the slow conduction velocity of horizontal axons. A variety of methods have estimated the speed of signal propagation within V1 to be 0.1–0.3 m/s (Bringuier et al., 1999; Grinvald et al., 1994; Slovin et al., 2002), but 0.1 m/s for most horizontal axons (Girard, Hupé, & Bullier, 2001). As discussed above, surround signals in V1 can arise 12.5° from the RF center. In parafoveal V1 this corresponds to a cortical distance of approximately 29 mm (using a magnification factor of 2.3 mm/° at 5° eccentricity) (Van Essen, Newsome, & Maunsell, 1984). Signals traveling at 0.1 m/s along horizontal axons would take 290 ms to reach the RF (97 ms for horizontal axons conducting at 0.3 m/s). Polysynaptic chains of horizontal connections would further add integration times of about 5–20 ms at each synaptic relay. Thus, clearly horizontal connections are too slow to account for the fast onset of far-surround suppression (9–60 ms, see above). These connections, however, could mediate near-surround suppression, as they can relay signals to the RF from a distance of 3 mm (average radius of horizontal connections in macaque V1) (Angelucci et al., 2002) in about 10–30 ms, which is compatible with the onset delay of near-surround suppression.

Thus, the spatiotemporal properties and synaptic physiology of horizontal connections are well suited to underlie near-surround modulation in V1. This includes facilitatory effects from the near surround, such as increased spatial summation at low stimulus contrast (figure 30.1) (Kapadia, Westheimer, & Gilbert, 1999; Sceniak et al., 1999) and collinear facilitation (figure 30.2A), but also near-surround suppression (e.g., size tuning at high contrast; figure 30.1). In principle, both facilitatory and suppressive effects associated with size tuning and its contrast dependence can be accounted for by the interaction of the horizontal networks with local GABAergic neurons (Schwabe et al., 2006; Somers et al., 1998). Recordings from cortical slices have shown that weak electrical stimulation of horizontal circuits elicits purely EPSPs, whereas stronger stimulation elicits EPSPs followed by strong IPSPs (Hirsch & Gilbert, 1991). Therefore, low level of activity in the network, as may be evoked by low-contrast stimuli, results in summation; instead, high-contrast or large-size stimuli would lead to higher level of activity in the horizontal network and subsequent recruitment of inhibition, resulting in suppression. In summary, horizontal connections are likely to contribute to near-surround modulation, its contrast dependence and orientation tuning, but they are unlikely to generate far-surround modulation.

The Role of Feedback Connections

V1 sends feedforward projections to several extrastriate areas, including V2, V3 and V5/MT, which in turn send a dense network of feedback projections to V1. These projections arise from excitatory neurons in layers 2/3A and 5/6 of extrastriate cortex, target both excitatory and inhibitory neurons (Anderson & Martin, 2009; Gonchar & Burkhalter, 2003), and terminate in V1 layer 1 and in the same layers (2/3, 4B and 6) that give rise to feedforward projections from V1. Functionally, feedback connections do not drive V1 cells but enhance their responses to stimulation of their RF (Hupé et al., 1998, 2001; Mignard & Malpeli, 1991; Sandell & Schiller, 1982).

Recent studies on the spatiotemporal properties and functional organization of feedback connections have led to suggest that they underlie far-surround modulation in V1 (Angelucci & Bressloff, 2006; Angelucci & Bullier, 2003). First, feedback connections to V1 have the appropriate spatial extent to underlie far-surround modulation (Angelucci et al., 2002). Feedback connections from V2, V3 and MT convey information to V1 from regions of visual space that are about 5, 10 and 25 times, respectively, the size of the V1 cell's sRF_{high}. Second, inactivation of area MT in macaque (Hupé et al., 1998) and of posterotemporal visual cortex in cat

(presumed homologue of primate inferotemporal cortex) (Bardy et al., 2009) by cooling reduces surround suppression in V1. In contrast, inactivation of area V2 loci by GABA does not affect the modulation of V1 responses generated by static texture patterns in the surround (Hupé et al., 2001). This result, however, does not rule out involvement of feedback connections from other extrastriate areas. Furthermore, the stimulus used by Hupé et al. (2001) was confined to the near surround, thus likely recruiting, in addition to feedback, horizontal connections, which were unperturbed in this study. Third, feedback connections have the appropriate conduction velocities to account for the fast onset of surround modulation. Electrical stimulation studies between macaque areas V1 and V2 have shown that both feedforward and feedback connections conduct at 2–6 m/s, which is about 10 times faster than the conduction velocity of V1 horizontal connections (Girard, Hupé, & Bullier, 2001). Fast and highly divergent feedback connections also explain how the onset latency of surround suppression is almost independent of cortical distance (Bair, Cavanaugh, & Movshon, 2003) (figure 30.5). If chains of horizontal V1 connections mediated far-surround suppression, one would expect the latency of suppression to increase with distance from the RF center, a prediction that was not confirmed by Bair, Cavanaugh, and Movshon (2003). In summary, on the basis of propagation speed and spatial extent, feedback connections are the most likely substrate for far-surround modulation in V1.

Data on the patterning and functional organization of feedback connections relative to the orientation map in V1 are controversial, with reports of both anatomically widespread (Maunsell & Van Essen, 1983; Rockland & Pandya, 1979; Ungerleider & Desimone, 1986) and orientation-unspecific (Stettler et al., 2002) V2-to-V1 feedback connections, and "patchy" and orientation-specific V2-to-V1 feedback connections (Angelucci & Bressloff, 2006; Angelucci et al., 2002; Shmuel et al., 2005) in primates. One hypothesis is that there are multiple feedback systems with differing functional specificities, possibly terminating in different V1 layers (Angelucci & Bressloff, 2006). In particular, the broader orientation tuning of far-surround suppression compared to near-surround suppression (Shushruth et al., 2013) (figure 30.3) suggests that feedback connections are more broadly orientation biased than horizontal connections. Furthermore, sharper tuning of far-surround modulation in V1 layer 4B suggests greater orientation specificity of feedback to this layer.

In summary, it is likely that the RF center and surround of V1 neurons result from integration of signals from feedforward, horizontal and feedback connections operating at different spatiotemporal scales (figure 30.6). The sRF$_{high}$ of V1 neurons is generated by converging feedforward inputs from the classical RF of LGN cells whose response is partially suppressed by stimuli of optimal size for the V1 cell (light green arrows in figure 30.6). Near surround-modulation is generated by all three sets of connections: feedforward inputs from the extraclassical RF of LGN cells (dark green arrows) strongly suppressed by large stimuli, horizontal (red arrows), and feedback (blue arrows) connections. Instead, far-surround modulation is mediated exclusively by feedback.

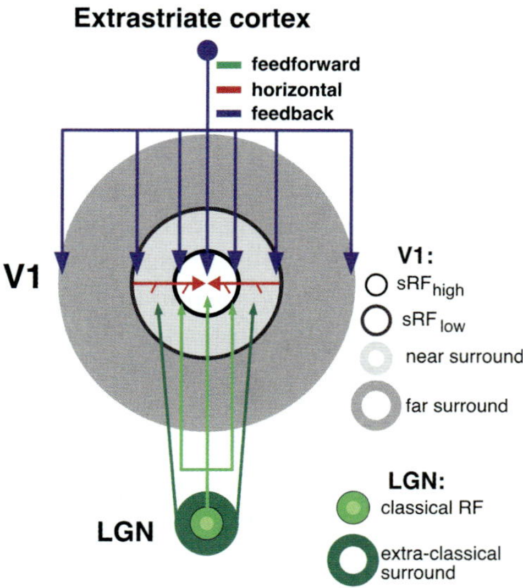

FIGURE 30.6   Circuits for surround modulation. Anatomical circuits (arrows) postulated to generate the RF center (white area) and surround (gray areas) of V1 neurons.

MECHANISMS FOR SURROUND MODULATION

Synaptic Mechanisms

Based on the model in figure 30.6, surround suppression in V1 likely results from multiple mechanisms. Surround suppression of LGN afferents would cause withdrawal of feedforward excitation and therefore fast untuned suppression of V1 cells; increased inhibition via horizontal and feedback connections acting through local inhibitory neurons would then cause slower tuned suppression. In vivo intracellular recordings have provided evidence for both reduced excitation and increased inhibition in the steady-state response of cat V1 cells to stimuli of intermediate length (8°) (Anderson et al., 2001). However, reduction in both excitatory and inhibitory conductances was observed for shorter (4°) and longer stimuli (12° length or 20° diameter) (Anderson et al., 2001; Ozeki et al., 2009). Importantly, this steady-state decrease in excitation and inhibition was temporally preceded by a transient increase in inhibition and could not be fully explained by suppression of inputs from the LGN. Similar stimuli as used in cortex did not evoke strong enough or sufficiently orientation-tuned suppression in LGN to account for suppression in V1, which therefore requires additional intracortical mechanisms. Ozeki et al. (2009) proposed that a steady-state decrease in excitatory and inhibitory conductances can be explained if V1 operates as an inhibition-stabilized network in which strong recurrent excitation is balanced by strong recurrent inhibition (figure 30.7). Within such a network, an increase in external excitatory input (via the surround pathways) to a local inhibitory neuron leads to a transient increase

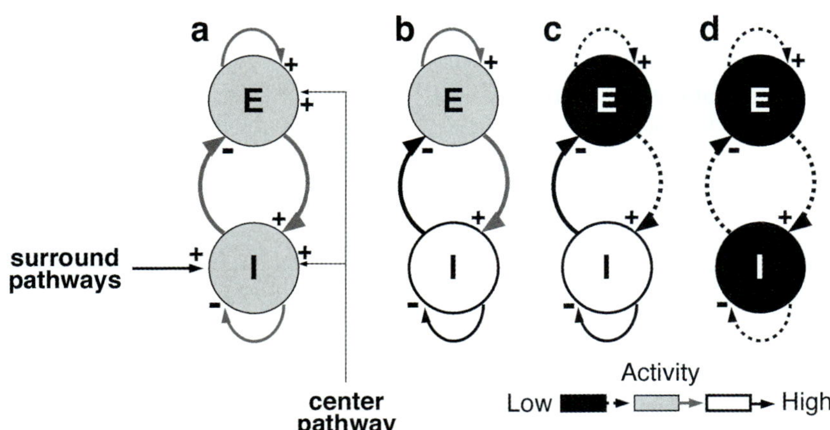

FIGURE 30.7   Inhibition-stabilized network model. Excitatory (E) and inhibitory (I) neurons making recurrent and reciprocal connections receive excitatory (+) feedforward inputs driven by RF stimulation, and lateral excitatory inputs driven by surround stimulation. (a–d) Sequence of events occurring when a surround stimulus is added to a preexisting center stimulus. Gray scale codes activity level. (Modified from Ozeki et al., 2009, with permission.)

in inhibition (figure 30.7b, c), which in turn leads to a withdrawal of recurrent excitation and inhibition at steady state (figure 30.7d).

Intracellular recordings in cat V1 found that natural visual stimulation of the RF and surround compared to stimulation of the RF alone caused an increase in membrane hyperpolarization (stronger IPSPs) and an increase in trial-to-trial reliability of EPSPs and spikes in pyramidal neurons (Haider et al., 2010). Instead, fast-spiking inhibitory neurons increased their firing rates during RF and surround stimulation, possibly causing increased IPSP amplitude in pyramidal cells.

Models

PHENOMENOLOGICAL MODELS   Phenomenological models have focused on understanding the computation underlying surround modulation. They describe surround modulation as resulting from two overlapping Gaussian mechanisms, an excitatory mechanism (representing the RF) and a spatially broader inhibitory one (representing the surround), interacting either subtractively (DoG model) or divisively (RoG model) (Cavanaugh, Bair, & Movshon, 2002a; Sceniak et al., 1999). Both models can satisfactorily fit spatial summation curves measured experimentally. They also accurately describe the contrast dependence of the sRF, either as a contrast-induced change in the size of the center excitatory Gaussian (DoG model) or as a change in the gains of both the center and surround Gaussians (RoG model). In contrast to some mechanistic models (Schwabe et al., 2006; Somers et al., 1998) (see below), the DoG model predicts the strength of surround suppression to be insensitive to stimulus contrast (Sceniak et al., 1999; Schwabe et al., 2010). However, when strength of suppression is calculated directly from neuronal responses rather than from DoG model parameters, it does show a clear contrast dependence, with weaker suppression occurring at low contrast (Schwabe et al., 2010).

Phenomenological models, however, by virtue of their design, do not have the same explanatory and predictive power of mechanistic neuronal network models.

MECHANISTIC MODELS   This second group of models have attempted to understand how cortical circuits generate surround modulation. The models of Stemmler, Usher, and Niebur (1995) and Somers et al. (1998) have focused on understanding how facilitation at low contrast and suppression at high contrast can occur using fixed cortical connections. These models consist of a lattice of orientation hypercolumns, each comprising several orientation columns with two populations of neurons (excitatory, E, and local inhibitory, I)

reciprocally and recurrently connected within each column. Different hypercolumns interact via orientation-specific horizontal connections targeting both local E and I neurons. Both models make the crucial assumption that there is an asymmetry in the response of the E and I neurons, such that for weak visual inputs (e.g., low-contrast stimuli) I neurons are silent, and inputs from the surround increase the E neuron response (causing facilitation). Instead, for strong inputs (e.g., high-contrast stimuli) the I neuron response increases, and surround inputs thus cause suppression. In the model of Stemmler, Usher, and Niebur (1995), the E–I asymmetry is implemented as a lower spontaneous input to the I than to the E neurons, whereas in the model of Somers et al. (1998), it is implemented as I neurons having higher threshold and gain than E neurons, as originally proposed by Lund et al. (1995). This mechanism can also reproduce the contrast dependence of sRF size. This is because at low contrast only the E neurons are active, and as stimulus size is increased, more and more of the horizontal network gets recruited, leading to an increase in the E neuron activity, which saturates at stimulus sizes corresponding to the length of horizontal connections. Instead, at high contrast I neurons are active, and increases in stimulus size recruit horizontal inputs that bring the I neurons to threshold; thus, suppression occurs at smaller stimulus sizes. A similar mechanism was used in the model of Schwabe et al. (2006). The latter, however, extended these previous models, introducing interareal feedback connections to V1 to account for the fast onset and large extent of surround suppression, and using realistic spatial scales and conduction velocities for horizontal and feedback circuits. This model can account for a wider range of physiological data compared to its two predecessor models, including (1) suppression and facilitation from the far surround using annular surround gratings such as those in figure 30.2B (Ichida et al., 2007; Levitt & Lund, 2002; Shushruth et al., 2009), (2) the effects of inactivating feedback connections on RF center and surround responses (Hupé et al., 1998), (3) the timing and dynamics of surround suppression (Bair, Cavanaugh, & Movshon, 2003), and (4) the contrast dependence of suppression strength (Cavanaugh, Bair, & Movshon, 2002a; Levitt & Lund, 1997; Schwabe et al., 2010). With respect to the last point, in fact, this model provided an explanation for apparently contradictory data on the effects on suppression strength of lowering center stimulus contrast (see above).

Recently, there has been theoretical (Bressloff & Cowan, 2002; van Vreeswijk & Sompolinsky, 1996) and experimental (Mariño et al., 2005; Stimberg et al., 2009) support for the idea that the cortex operates in a regime of strong but balanced recurrent excitation and inhibition. Ozeki et al. (2009) have shown that only an inhibition-stabilized network can explain the synaptic physiology of surround modulation (see above, figure 30.7). Shushruth et al. (2012) extended the model of Schwabe et al. (2006) by incorporating orientation tuning and strong local recurrent connections between orientation columns (figure 30.8A). This model could explain the stimulus-dependent orientation specificity of surround modulation (Shushruth et al., 2012) (figure 30.4). In the model this tuning behavior results from the interaction of orientation-specific surround inputs with strong and poorly orientation-specific local recurrent connections. The mechanism is shown in figure 30.8B. When the stimulus in the RF is at the optimal orientation (0°; figure 30.8B, a–c), the E1 cells receive strongest feedforward activation and, thus, provide the strongest recurrent excitation within the center hypercolumn; because of the strong recurrency, the whole hypercolumn becomes strongly activated. Iso-oriented surround stimuli thus suppress the strongest source of recurrent excitation within the hypercolumn, i.e. the E1 cells: recurrent excitation is withdrawn, and the whole hypercolumn becomes suppressed. This withdrawal of excitation is greater when the E1 cells are strongly suppressed by an optimally oriented (0°) surround stimulus (figure 30.8B, c) than when they are weakly suppressed by a nonoptimal one (–22.5°; figure 30.8B, b). When the RF center is stimulated by a suboptimal orientation for the E1cells (–22.5°; figure 30.8B, d–f) instead, these cells receive weaker feedforward excitation and thus contribute weaker recurrent excitation within the hypercolumn. Suppression of the E1 cells by an optimal (0°) stimulus now has little effect on the hypercolumn activity (figure 30.8B, e). Instead, the E2 cells, which are maximally activated by a stimulus of –22.5° orientation, provide the strongest recurrent excitation to the hypercolumn, and suppressing them results in strongest suppression of the whole hypercolumn (figure 30.8B, f). In summary, recurrent excitation is weakest when center and surround stimuli are at the same orientation; because of the strong recurrent regime, the level of recurrent excitation in the hypercolumn has a greater effect on the E1 neuron responses than the direct inhibition from the surround pathways.

ROLE OF SURROUND MODULATION IN VISION

Role in Visual Information Processing

It has been suggested that surround suppression could reflect two important computations performed by sensory neurons—gain control and/or redundancy reduction.

Neurons in V1 maintain their orientation tuning even at high stimulus contrasts, which saturate their response. This property, and other nonlinearities of V1 RF responses, can be explained by a normalization model, according to which the activity of each neuron is divided by the responses of a pool of neighboring neurons (Carandini & Heeger, 1994; Carandini, Heeger, & Movshon, 1997). Recently, this model has been extended by suggesting that for larger stimuli the activated cells in the surround sharing similar functional properties to those of the RF also contribute to the normalization pool (Schwartz & Simoncelli, 2001). Contrast normalization is a desirable computation that allows V1 neurons to handle the range of contrasts in natural scenes, because single neurons lack the dynamic range needed for this task (Heeger, 1992).

A second hypothesis is that surround suppression serves to reduce redundancies in visual inputs. Theoretically, the "efficient coding" hypothesis (Barlow, 1961) states that sensory neurons are tuned to the statistics of natural images (Geisler, 2008; Simoncelli & Olshausen, 2001) and that their role is to reduce redundancies in sensory inputs by maintaining statistical independence in their responses. Because natural scenes contain strong spatial correlations, with neighboring locations being highly similar, it has been suggested that surround suppression serves to reduce information redundancy in natural images by maximizing statistical independence in the response of neurons representing such correlated inputs. Schwartz and Simoncelli (2001) examined the response of oriented filters (as models of V1 RFs) to natural images and found strong dependency between the responses of spatially separated filters. They implemented a form of divisive normalization wherein each filter's response was divided by a weighted sum of the responses of other filters, with the weights derived to maximize the independence of filter responses to an ensemble of natural images. The filter responses to grating stimuli now showed properties similar to those seen in V1 neurons, such as contrast-dependent spatial summation, thus supporting the theory.

The idea that surround modulation subserves a form of efficient coding is also supported by experimental evidence. Stimulation of the RF surround of V1 neurons in awake macaques with natural images increases the selectivity of individual neurons, reduces the correlations between the responses of neuron pairs, and increases the sparseness of neuronal responses (Vinje & Gallant, 2000, 2002). Sparseness is a nonparametric measure of neuronal selectivity. A neuron with increased

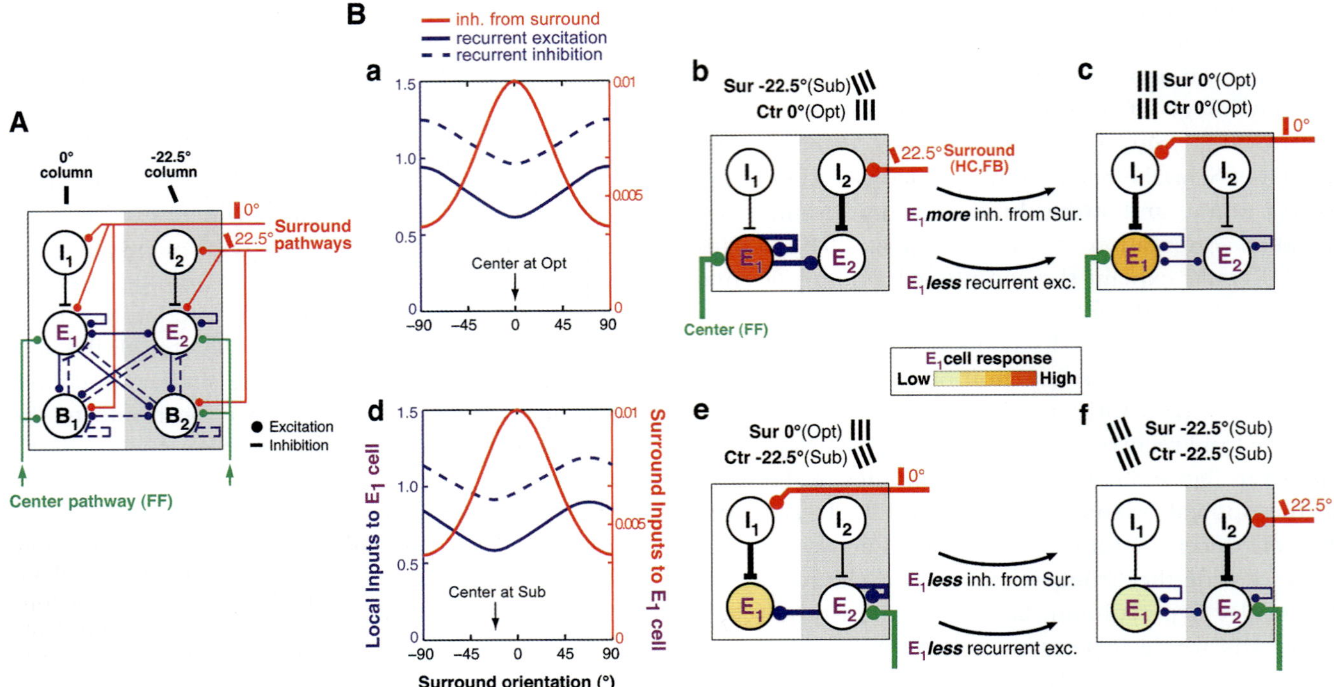

FIGURE 30.8   Network model of V1 with strong recurrent connections. (A) The network architecture consists of one hypercolumn of 32 recurrently connected orientation columns, of which only two (preferring 0° and −22.5° orientation) are shown for simplicity. Each column consists of an excitatory neuron (E) and two kinds of inhibitory neurons, high-threshold inhibitory neurons (I), which receive surround modulation from outside the hypercolumn, and basket neurons (B), which support local recurrent connections within the hypercolumn. The surround modulatory inputs (red) are orientation specific, the local recurrent connections (blue) are not. (B) The mechanism underlying stimulus-dependent orientation tuning of surround suppression. (a, d) Inputs to the E1 cell preferring 0° orientation for varying surround orientations, when the center is stimulated at the neuron's preferred orientation (a) or at a suboptimal, nonpreferred orientation (d). The solid blue curve indicates the local recurrent inputs from E neurons in other orientation columns; the dashed blue curve indicates the local recurrent inputs (negative) from B neurons in the same and other orientation columns; the red curve is the input (negative) from the surround. Note that the surround input (red y-axis) is much smaller than the local recurrent inputs (blue y-axis). (b, c) Diagrams showing the inputs that most affect the E1 cell response for a center stimulus at the optimal orientation (0°) for cell E1, and a surround stimulus at −22.5° (Sub) (b) or at 0° (Opt) (c) orientations. Only the relevant cells and connections are depicted. Line thickness indicates input strength. White and gray shading indicate 0° and −22.5° orientation columns, respectively. Changing the surround orientation from Sub (b) to the iso-orientation condition (c) leads to increased inhibition of the E1 cell via the surround inputs and to less recurrent excitation. (e, f) Same as in b and c, but for a center orientation of −22.5° and surround orientations of 0° (e) or −22.5° (f). This leads to reduced inhibition of the E1 cell via the surround pathways, but also to less recurrent excitation; the latter causes stronger suppression at iso-orientation. (Modified from Shushruth et al., 2012; reproduced with permission from the Society for Neuroscience.)

sparseness responds to a more restricted set of stimuli and therefore is more selective (Olshausen & Field, 2004), and this mechanism has been suggested to increase the efficiency of information transmission about a visual stimulus (Vinje & Gallant, 2002).

Intracellular recordings from cat V1 have shown that for excitatory cells, natural stimulation of the surround increases the sparseness of the spiking response and the trial-to-trial reliability of membrane potential responses (see above) (Haider et al., 2010). An increase in spike reliability allows neurons with sparse responses to overcome trial-to-trial response variability and, thus, to transmit reliable information to downstream neurons.

In natural images there is a statistical relation between edge orientation and distance between edges, such that nearby edges have a higher probability than distant edges of being co-oriented and cocircular and of belonging to the same physical contour (Geisler et al., 2001). The different orientation tuning of near- and far-surround suppression observed in macaque V1 (Shushruth et al., 2013; see above) may reflect this statistical bias; accordingly, suppression should be narrowly tuned for nearby edges and more broadly tuned for distant edges. Sharply orientation-tuned near-surround suppression may serve to detect small orientation differences in nearby edges, whereas broadly tuned far-surround suppression may serve to direct saccades/attention to salient distant locations.

Another class of computational models of vision considers visual processing a form of statistical inference (Yuille & Kersten, 2006). One approach of this class of models is the predictive coding theory, which posits that higher visual areas learn the statistical regularities of natural images and feed back to lower areas predictions based on such regularities. The activity in lower areas reflects deviations from these predictions, that is, from these regularities (Rao & Ballard, 1999); thus, only non-matched predictions are signaled to higher areas. In these models neurons respond higher to unpredictable elements of the visual scene, such as pop-out stimuli and edges. Spratling (2010) has proposed an alternative implementation of predictive coding wherein horizontal connections within V1 take the role ascribed by other models to feedback input.

Role in Visual Perception

Above, we have discussed how surround modulation could serve visual information processing. Although such computations could ultimately serve visual perception, surround modulation in V1 has also been proposed to be the direct neural correlate of some perceptual contextual effects.

The observation that, compared to similar stimuli, dissimilar stimuli in the RF and surround result in higher neuronal responses led to the suggestion that surround modulation in V1 plays a role in visual saliency and pop-out perceptual phenomena (Knierim & Van Essen, 1992; Nothdurft, Gallant, & Van Essen, 1999). More generally, surround modulation enhances neuronal responses in regions of orientation discontinuities, which typically occur at object boundaries (Nothdurft, Gallant, & Van Essen, 2000). This property has been hypothesized to underlie the perceptual ability of delineating figures from background—a process termed "figure–ground segregation." Lamme (1995) and Zipser, Lamme, and Schiller (1996) recorded from V1 while presenting figure–ground stimuli defined by differently oriented line segments (figure 30.9). They reported that neuronal responses were higher when the RF was located on the figural part of the stimulus than when it was on the background, although the local stimulus in the RF was the same in both conditions. The larger response to the figure persisted even when the diameter of the figure was increased to 8°. These results suggested that V1 represents figure–ground relationships in visual scenes. However, using similar stimuli, Rossi, Desimone, and Ungerleider (2001) failed to find any difference in response to the figure versus the

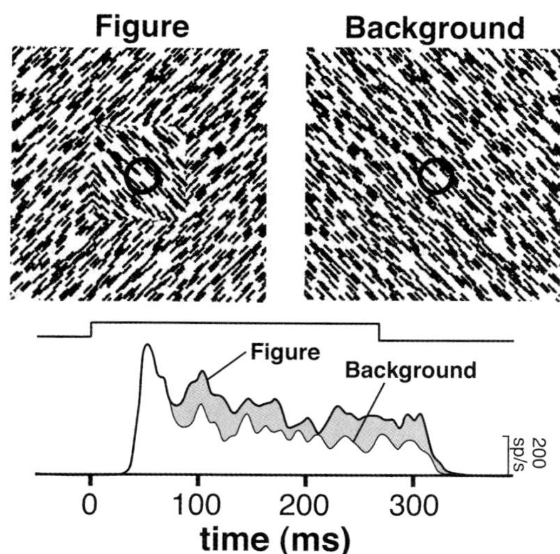

FIGURE 30.9   Figure–ground segregation. (Top) Figure and background stimuli used to stimulate the RF (circles) of V1 cells. (Bottom) V1 cell response to the two stimuli at the top. The response is higher when the RF overlays the figure than when it overlays the background. (Modified from Zipser et al., 1996; reprinted with permission from the Society for Neuroscience.)

background, except when the boundaries of the figure were within 1° of the cells' mRF border and thus likely within their sRF. This suggested that V1 neurons respond to local figure boundaries rather than to the figure per se. Alternatively, given the large size of V1 surrounds, the differential responses of V1 cells to figure and background could simply reflect different strengths of surround suppression evoked by different amounts of surround stimulation in the two conditions. Specifically, compared to the figure condition, in the background condition a larger fraction of the surround is activated by a stimulus of the same orientation as that in the RF, causing stronger suppression. In conclusion, additional studies are needed to determine whether these phenomena are just manifestations of surround modulations or responses to texture boundaries, as opposed to true figure–ground analysis in V1. However, the longer latency of figure–ground signals (30–70 ms after RF response onset) (Lamme, 1995; Zipser, Lamme, & Schiller, 1996), compared to that of surround modulation (9 ms) (Bair, Cavanaugh, & Movshon, 2003) suggests that these may be two separate phenomena and that the latter is at a processing stage prior to figure–ground analysis.

Collinear facilitation in V1 cells (figure 30.2A) is thought to be the neural correlate of perceptual "contour integration," which is the visual system's ability to segregate into a contour collinear line segments from a background of randomly oriented elements (see figure 44.2 in chapter 44 by Field, Golden, and Hayes). Psychophysical studies have demonstrated that individual contour elements can be grouped with other elements based on the Gestalt principles of good continuation (Hess & Field, 1999). Contour integration is the topic of chapter 44 by Field, Golden, and Hayes in this volume and therefore is only briefly discussed here. Kapadia et al. (1995) conducted parallel psychophysical studies in humans and electrophysiological recordings in macaque V1 on the effect of collinearly placed flankers on the perception of target line elements. They reported that the flankers enhanced the detectability of the target in human observers as well as the response of V1 neurons to an iso-oriented line segment inside the RF (figure 30.2A). Li, Piech, and Gilbert (2006) further demonstrated a close correlation among the perceptual saliency of a contour, the animal performance on a contour detection task, and V1 responses. We refer the reader to chapter 70 of this volume by Li and Gilbert for details on these studies.

In summary, surround modulation enhances neuronal responses to perceptually salient aspects of a visual scene such as contours, figure boundaries and texture borders. Accordingly, it has been proposed that V1 serves as a preattentive map of visual saliency in which higher neuronal responses correspond to the perceptual salience of the image location they represent (Li, 2002). This map serves to direct visual attention to the most salient locations in a scene.

CONCLUSIONS AND FUTURE DIRECTIONS

Models of V1 based purely on classical RF concepts and feedforward interactions are insufficient to understand the neuronal basis of visual perception. These models have portrayed a view of V1 neurons as localized and functionally independent windows over the visual world. Studies of surround modulation have, instead, demonstrated that the responses of V1 neurons even to simple visual stimuli reflect integration of signals from distant cortical regions. This process engages a complex network of feedforward, local recurrent, and long-distance horizontal and feedback circuits. The selective properties of surround modulation and the precise connectivity of V1 may serve to provide a meaningful representation of the image structure that goes beyond the computation of contrast gain control and redundancy reduction.

Surround modulation has so far been extensively characterized using highly simplified stimuli and by recording from one neuron at a time. A challenge for future research is to understand how V1 neurons respond to the kind of contextual information that is present in natural scenes, and how the response of individual neurons relates to the large-scale population activity in cortical networks. Multielectrode array recording of V1 neuronal population responses to natural visual stimuli can provide the next step in understanding the role of context in natural vision. Furthermore, causal relationships among neuronal circuits, surround modulation, and visual perception need to be established. Recent advances in the ability to perturb specific neural circuits using viral technology and molecular genetic techniques (Han et al., 2009; Luo, Callaway, & Svoboda, 2008; Osakada et al., 2011), even in nonhuman primates, will open new avenues for addressing this challenge.

REFERENCES

Alitto, H. J., & Usrey, W. M. (2008). Origin and dynamics of extraclassical suppression in the lateral geniculate nucleus of the macaque monkey. *Neuron, 57,* 135–146. doi:10.1016/j.neuron.2007.11.019.

Allman, J., Miezin, F., & McGuinness, E. (1985). Stimulus specific responses from beyond the classical receptive field: Neurophysiological mechanisms for local-global comparisons in visual neurons. *Annual Review of Neuroscience, 8,* 407–430. doi:10.1146/annurev.ne.08.030185.002203.

Anderson, J. C., & Martin, K. A. C. (2009). The synaptic connections between cortical areas V1 and V2 in macaque monkey. *Journal of Neuroscience, 29,* 11283–11293.

Anderson, J. S., Lampl, I., Gillespie, D. C., & Ferster, D. (2001). Membrane potential and conductance changes underlying length tuning of cells in cat primary visual cortex. *Journal of Neuroscience, 21,* 2104–2112.

Angelucci, A., & Bressloff, P. C. (2006). The contribution of feedforward, lateral and feedback connections to the classical receptive field center and extra-classical receptive field surround of primate V1 neurons. *Progress in Brain Research, 154,* 93–121.

Angelucci, A., & Bullier, J. (2003). Reaching beyond the classical receptive field of V1 neurons: Horizontal or feedback axons? *Journal of Physiology, Paris, 97,* 141–154. doi:10.1016/j.jphysparis.2003.09.001.

Angelucci, A., Levitt, J. B., Walton, E., Hupé, J. M., Bullier, J., & Lund, J. S. (2002). Circuits for local and global signal integration in primary visual cortex. *Journal of Neuroscience, 22,* 8633–8646.

Angelucci, A., & Sainsbury, K. (2006). Contribution of feedforward thalamic afferents and corticogeniculate feedback to the spatial summation area of macaque V1 and LGN. *Journal of Comparative Neurology, 498,* 330–351. doi:10.1002/cne.21060.

Bair, W., Cavanaugh, J. R., & Movshon, J. A. (2003). Time course and time–distance relationships for surround suppression in macaque V1 neurons. *Journal of Neuroscience, 23,* 7690–7701.

Bardy, C., Huang, J. Y., Wang, C., Fitzgibbon, T., & Dreher, B. (2009). "Top-down" influences of ispilateral or contralateral postero-temporal visual cortices on the extra-classical receptive fields of neurons in cat's striate cortex. *Neuroscience, 158,* 951–968. doi:10.1016/j.neuroscience.2008.09.057.

Barlow, H. B. (1953). Summation and inhibition in the frog's retina. *Journal of Physiology, 119,* 69–88.

Barlow, H. B. (1961). Possible principles underlying the transformation of sensory messages. In W. A. Rosenblith (Ed.), *Sensory communication* (pp. 217–234). Cambridge, MA: MIT Press.

Barlow, H. B., Blakemore, C., & Pettigrew, J. D. (1967). The neural mechanisms of binocular depth discrimination. *Journal of Physiology, 193,* 327–342.

Bauer, U., Scholz, M., Levitt, J. B., Lund, J. S., & Obermayer, K. (1999). A model for the depth dependence of receptive field size and contrast sensitivity of cells in layer 4C of macaque striate cortex. *Vision Research, 39,* 613–629.

Blakemore, C., & Tobin, E. A. (1972). Lateral inhibition between orientation detectors in the cat's visual cortex. *Experimental Brain Research, 15,* 439–440. doi:10.1007/BF00234129.

Bonin, V., Mante, V., & Carandini, M. (2005). The suppressive field of neurons in lateral geniculate nucleus. *Journal of Neuroscience, 25,* 10844–10856.

Bosking, W. H., Zhang, Y., Schofield, B., & Fitzpatrick, D. (1997). Orientation selectivity and the arrangement of horizontal connections in tree shrew striate cortex. *Journal of Neuroscience, 17,* 2112–2127.

Bressloff, P. C., & Cowan, J. D. (2002). An amplitude equation approach to contextual effects in visual cortex. *Neural Computation, 14,* 493–525. doi:10.1162/089976602317250870.

Bringuier, V., Chavane, F., Glaeser, L., & Frégnac, Y. (1999). Horizontal propagation of visual activity in the synaptic integration field of area 17 neurons. *Science, 283,* 695–699.

Carandini, M., & Heeger, D. J. (1994). Summation and division by neurons in primate visual cortex. *Science, 264,* 1333–1336. doi:10.1126/science.8191289.

Carandini, M., Heeger, D. J., & Movshon, J. A. (1997). Linearity and normalization in simple cells of the macaque primary visual cortex. *Journal of Neuroscience, 17,* 8621–8644.

Cavanaugh, J. R., Bair, W., & Movshon, J. A. (2002a). Nature and interaction of signals from the receptive field center and surround in macaque V1 neurons. *Journal of Neurophysiology, 88,* 2530–2546.

Cavanaugh, J. R., Bair, W., & Movshon, J. A. (2002b). Selectivity and spatial distribution of signals from the receptive field surround in macaque V1 neurons. *Journal of Neurophysiology, 88,* 2547–2556.

Chen, C., Kasamatsu, T., Polat, U., & Norcia, A. M. (2001). Contrast response characteristics of long-range lateral interactions in cat striate cortex. *Neuroreport, 12,* 655–661.

Chisum, H. J., Mooser, F., & Fitzpatrick, D. (2003). Emergent properties of layer 2/3 neurons reflect the collinear arrangement of horizontal connections in tree shrew visual cortex. *Journal of Neuroscience, 23,* 2947–2960.

DeAngelis, G. C., Freeman, R. D., & Ohzawa, I. (1994). Length and width tuning of neurons in the cat's primary visual cortex. *Journal of Neurophysiology, 71,* 347–374.

Felisberti, F., & Derrington, A. M. (1999). Long-range interactions modulate the contrast gain in the lateral geniculate nucleus of cats. *Visual Neuroscience, 16,* 943–956. doi:10.1017/S0952523899165143.

Fitzpatrick, D. (2000). Seeing beyond the receptive field in primary visual cortex. *Current Opinion in Neurobiology, 10,* 438–443. doi:10.1016/S0959-4388(00)00113-6.

Geisler, W. S. (2008). Visual perception and the statistical properties of natural scenes. *Annual Review of Psychology, 59,* 167–192. doi:10.1146/annurev.psych/58/110405.085632.

Geisler, W. S., Perry, J. S., Super, B. J., & Gallogly, D. P. (2001). Edge co-occurrence in natural images predicts contour grouping performance. *Vision Research, 41,* 711–724. doi:10.1016/S0042-6989(00)00277-7.

Gilbert, C. D. (1977). Laminar differences in receptive field properties of cells in cat primary visual cortex. *Journal of Physiology, 268,* 391–421.

Gilbert, C. D., Das, A., Ito, M., Kapadia, M., & Westheimer, G. (1996). Spatial integration and cortical dynamics. *Proceedings of the National Academy of Sciences of the United States of America, 93,* 615–622. doi:10.1073/pnas.93.2.615.

Gilbert, C. D., & Wiesel, T. N. (1983). Clustered intrinsic connections in cat visual cortex. *Journal of Neuroscience, 3,* 1116–1133.

Girard, P., Hupé, J. M., & Bullier, J. (2001). Feedforward and feedback connections between areas V1 and V2 of the monkey have similar rapid conduction velocities. *Journal of Neurophysiology, 85,* 1328–1331.

Gonchar, Y., & Burkhalter, A. (2003). Distinct GABAergic targets of feedforward and feedback connections between lower and higher areas of rat visual cortex. *Journal of Neuroscience, 23,* 10904–10912.

Grinvald, A., Lieke, E. E., Frostig, R. D., & Hildesheim, R. (1994). Cortical point-spread function and long-range lateral interactions revealed by real-time optical imaging of macaque monkey primary visual cortex. *Journal of Neuroscience, 14,* 2545–2568.

Haider, B., Krause, M. R., Duque, A., Yu, Y., Touryan, J., Mazer, J. A., et al. (2010). Synaptic and network mechanisms of sparse and reliable visual cortical activity during nonclassical receptive field stimulation. *Neuron, 65,* 107–121.

Han, X., Qian, X. G., Bernstein, J. G., Zhou, H. H., Franzesi, G. T., Stern, P., et al. (2009). Millisecond-timescale optical control of neural dynamics in the nonhuman primate brain. *Neuron, 62,* 191–198. doi:10.1016/j.neuron.2009.03.011.

Hashemi-Nezhad, M., & Lyon, D. C. (2011). Orientation tuning of the suppressive extraclassical surround depends on intrinsic organization of V1. *Cerebral Cortex, 22,* 308–326.

Heeger, D. J. (1992). Normalization of cell responses in cat striate cortex. *Vision Research, 9,* 181–198.

Hess, R., & Field, D. (1999). Integration of contours: new insights. *Trends in Cognitive Sciences, 3,* 480–486. doi:10.1016/S1364-6613(99)01410-2.

Hirsch, J. A., & Gilbert, C. D. (1991). Synaptic physiology of horizontal connections in the cat's visual cortex. *Journal of Neuroscience, 11,* 1800–1809.

Hubel, D. H., & Wiesel, T. N. (1959). Receptive fields of single neurones in the cat's striate cortex. *Journal of Physiology, 148,* 574–591.

Hubel, D. H., & Wiesel, T. N. (1962). Receptive fields, binocular interaction and functional architecture in the cat's visual cortex. *Journal of Physiology, 160,* 106–154.

Hubel, D. H., & Wiesel, T. N. (1965). Receptive fields and functional architecture in two nonstriate visual areas (18 and 19) of the cat. *Journal of Neurophysiology, 28,* 229–289.

Hupé, J. M., James, A. C., Girard, P., & Bullier, J. (2001). Response modulations by static texture surround in area V1 of the macaque monkey do not depend on feedback connections from V2. *Journal of Neurophysiology, 85,* 146–163.

Hupé, J. M., James, A. C., Payne, B. R., Lomber, S. G., Girard, P., & Bullier, J. (1998). Cortical feedback improves discrimination between figure and background by V1, V2 and V3 neurons. *Nature, 394,* 784–787. doi:10.1038/29537.

Ichida, J. M., Schwabe, L., Bressloff, P. C., & Angelucci, A. (2007). Response facilitation from the "suppressive" receptive field surround of macaque V1 neurons. *Journal of Neurophysiology, 98,* 2168–2181.

Kapadia, M. K., Ito, M., Gilbert, C. D., & Westheimer, G. (1995). Improvement in visual sensitivity by changes in local context: Parallel studies in human observers and in V1 of alert monkeys. *Neuron, 15,* 843–856. doi:10.1016/0896-6273(95)90175-2.

Kapadia, M. K., Westheimer, G., & Gilbert, C. D. (1999). Dynamics of spatial summation in primary visual cortex of alert monkeys. *Proceedings of the National Academy of Sciences of the United States of America, 96,* 12073–12078. doi:10.1073/pnas.96.21.12073.

Kapadia, M. K., Westheimer, G., & Gilbert, C. D. (2000). Spatial distribution of contextual interactions in primary visual cortex and in visual perception. *Journal of Neurophysiology, 84,* 2048–2062.

Knierim, J. J., & Van Essen, D. (1992). Neuronal responses to static texture patterns in area V1 of the alert macaque monkey. *Journal of Neurophysiology, 67,* 961–980.

Lamme, V. A. F. (1995). The neurophysiology of figure-ground segregation in primary visual cortex. *Journal of Neuroscience, 15,* 1605–1615.

Lamme, V. A. F., Rodriguez-Rodriguez, V., & Spekreijse, H. (1999). Separate processing dynamics for texture elements, boundaries and surfaces in primary visual cortex of the macaque monkey. *Cerebral Cortex, 9,* 406–413. doi:10.1093/cercor/9.4.406.

Levick, W. R., Cleland, B. G., & Dubin, M. W. (1972). Lateral geniculate neurons of the cat: Retinal inputs and physiology. *Investigative Ophthalmology, 11,* 302–311.

Levitt, J. B., & Lund, J. S. (1997). Contrast dependence of contextual effects in primate visual cortex. *Nature, 387,* 73–76. doi:10.1038/387073a0.

Levitt, J. B., & Lund, J. S. (2002). The spatial extent over which neurons in macaque striate cortex pool visual signals. *Visual Neuroscience, 19,* 439–452. doi:10.1017/S0952523802194065.

Li, C., & Li, W. (1994). Extensive integration field beyond the classical receptive field of cat's striate cortical neurons: Classification and tuning properties. *Vision Research, 34,* 2337–2355. doi:10.1016/0042-6989(94)90280-1.

Li, H., Fukuda, M., Tanifuji, M., & Rockland, K. S. (2003). Intrinsic collaterals of layer 6 Meynert cells and functional columns in primate V1. *Neuroscience, 120,* 1061–1069. doi:10.1016/S0306-4522(03)00429-9.

Li, W., Piech, V., & Gilbert, C. D. (2006). Contour saliency in primary visual cortex. *Neuron, 50,* 951–962. doi:10.1016/j.neuron.2006.04.035.

Li, Z. (2002). A saliency map in primary visual cortex. *Trends in Cognitive Sciences, 6,* 9–16. doi:10.1016/S1364-6613(00)01817-9.

Lund, J. S. (1988). Anatomical organization of macaque monkey striate visual cortex. *Annual Review of Neuroscience, 11,* 253–288. doi:10.1146/annurev.ne.11.0301188.001345.

Lund, J. S., Angelucci, A., & Bressloff, P. C. (2003). Anatomical substrates for functional columns in macaque monkey primary visual cortex. *Cerebral Cortex, 13,* 15–24. doi:10.1093/cercor/13.1.15.

Lund, J. S., Wu, Q., Hadingham, P. T., & Levitt, J. B. (1995). Cells and circuits contributing to functional properties in area V1 of macaque monkey cerebral cortex: Bases for neuroanatomically realistic models. *Journal of Anatomy, 187,* 563–581.

Luo, L., Callaway, E. M., & Svoboda, K. (2008). Genetic dissection of neural circuits. *Neuron, 57,* 634–660. doi:10.1016/j.neuron.2008.01.002.

Maffei, L., & Fiorentini, L. (1976). The unresponsive regions of visual cortical receptive fields. *Vision Research, 16,* 1131–1139.

Malach, R., Amir, Y., Harel, M., & Grinvald, A. (1993). Relationship between intrinsic connections and functional architecture revealed by optical imaging and in vivo targeted biocytin injections in primate striate cortex. *Proceedings of the National Academy of Sciences of the United States of America, 90,* 10469–10473. doi:10.1073/pnas.90.22.10469.

Mariño, J., Schummers, J., Lyon, D. C., Schwabe, L., Beck, O., Wiesing, P., et al. (2005). Invariant computations in local cortical networks with balanced excitation and inhibition. *Nature Neuroscience, 8,* 194–201. doi:10.1038/nn1391.

Maunsell, J. H. R., & Van Essen, D. C. (1983). The connections of the middle temporal visual area (MT) and their relationship to a cortical hierarchy in the macaque monkey. *Journal of Neuroscience, 3,* 2563–2586.

McGuire, B. A., Gilbert, C. D., Rivlin, P. K., & Wiesel, T. N. (1991). Targets of horizontal connections in macaque primary visual cortex. *Journal of Comparative Neurology, 305,* 370–392. doi:10.1002/cne.903050303.

Mignard, M., & Malpeli, J. G. (1991). Paths of information flow through visual cortex. *Science, 251,* 1249–1251. doi:10.1126/science.1848727.

Mizobe, K., Polat, U., Pettet, M. W., & Kasamatsu, T. (2001). Facilitation and suppression of single striate-cell activity by spatially discrete pattern stimuli presented beyond the receptive field. *Visual Neuroscience, 18,* 377–391. doi:10.1017/S0952523801183045.

Muller, J. R., Metha, A. B., Krauskopf, J., & Lennie, P. (2003). Local signals from beyond the receptive fields of striate cortical neurons. *Journal of Neurophysiology, 90,* 822–831.

Naito, T., Sadakane, O., Okamoto, M., & Sato, H. (2007). Orientation tuning of surround suppression in lateral geniculate nucleus and primary visual cortex of cat. *Neuroscience, 149,* 962–975. doi:10.1016/j.neuroscience.2007.08.001.

Nelson, J. I., & Frost, B. (1978). Orientation selective inhibition from beyond the classical visual receptive field. *Brain Research, 139,* 359–365. doi:10.1016/0006-8993(78)90937-X.

Nelson, J. I., & Frost, B. (1985). Intracortical facilitation among co-oriented, co-axially aligned simple cells in cat striate cortex. *Experimental Brain Research, 61,* 54–61.

Nothdurft, H. C., Gallant, J. L., & Van Essen, D. C. (1999). Response modulation by texture surround in primate area V1: Correlates of "popout" under anesthesia. *Visual Neuroscience, 16,* 15–34.

Nothdurft, H. C., Gallant, J. L., & Van Essen, D. C. (2000). Response profiles to texture border patterns in area V1. *Visual Neuroscience, 17,* 421–436. doi:10.1017/S0952523800173092.

Olshausen, B. A., & Field, D. J. (2004). Sparse coding of sensory inputs. *Current Opinion in Neurobiology, 14*, 481–487. doi:10.1016/j.conb.2004.07.007.

Osakada, F., Mori, T., Cetin, A. H., Marshel, J. H., Virgen, B., & Callaway, E. M. (2011). New rabies virus variants for monitoring and manipulating activity and gene expression in defined neural circuits. *Neuron, 71*, 617–631. doi:10.1016/j.neuron.2011.07.005.

Ozeki, H., Finn, I. M., Schaffer, E. S., Miller, K. D., & Ferster, D. (2009). Inhibitory stabilization of the cortical network underlies visual surround suppression. *Neuron, 62*, 578–592. doi:10.1016/j.neuron.2009.03.028.

Ozeki, H., Sadakane, O., Akasaki, T., Naito, T., Shimegi, S., & Sato, H. (2004). Relationship between excitation and inhibition underlying size tuning and contextual response modulation in the cat primary visual cortex. *Journal of Neuroscience, 24*, 1428–1438.

Perkel, D. J., Bullier, J., & Kennedy, H. (1986). Topography of the afferent connectivity of area 17 in the macaque monkey: A double-labelling study. *Journal of Comparative Neurology, 253*, 374–402. doi:10.1002/cne.902530307.

Polat, U., Mizobe, K., Pettet, M. W., Kasamatsu, T., & Norcia, A. M. (1998). Collinear stimuli regulate visual responses depending on cell's contrast threshold. *Nature, 391*, 580–584. doi:10.1038/35372.

Rao, R. P., & Ballard, D. H. (1999). Predictive coding in the visual cortex: A functional interpretation of some extra-classical receptive-field effects. *Nature Neuroscience, 2*, 79–87. doi:10.1038/4580.

Reid, R. C., & Usrey, W. M. (2004). Functional connectivity in the pathway from retina to striate cortex. In L. M. Chalupa & J. S. Werner (Eds.), *The visual neurosciences* (Vol. 1, pp. 673–679). Cambridge, MA: MIT Press.

Riesenhuber, M., & Poggio, T. (2003). How the visual cortex recognizes objects: The tale of the standard model. In L. M. Chalupa & J. S. Werner (Eds.), *The visual neurosciences* (Vol. 2, pp. 1640–1653). Cambridge, MA: MIT Press.

Rockland, K. S., & Lund, J. S. (1982). Widespread periodic intrinsic connections in the tree shrew visual cortex. *Science, 215*, 1532–1534. doi:10.1126/science.7063863.

Rockland, K. S., & Lund, J. S. (1983). Intrinsic laminar lattice connections in primate visual cortex. *Journal of Comparative Neurology, 216*, 303–318. doi:10.1002/cne.902160307.

Rockland, K. S., & Pandya, D. N. (1979). Laminar origins and terminations of cortical connections of the occcipital lobe in the Rhesus monkey. *Brain Research, 179*, 3–20. doi:10.1016/0006-8993(79)90485-2.

Rossi, A. F., Desimone, R., & Ungerleider, L. G. (2001). Contextual modulation in primary visual cortex of macaques. *Journal of Neuroscience, 21*, 1698–1709.

Sadakane, O., Ozeki, H., Naito, T., Akasaki, T., Kasamatsu, T., & Sato, H. (2006). Contrast-dependent, contextual response modulation in primary visual cortex and lateral geniculate nucleus of the cat. *European Journal of Neuroscience, 23*, 1633–1642.

Sandell, J. H., & Schiller, P. H. (1982). Effect of cooling area 18 on striate cortex cells in the squirrel monkey. *Journal of Neurophysiology, 48*, 38–48.

Sceniak, M. P., Chatterjee, S., & Callaway, E. M. (2006). Visual spatial summation in macaque geniculocortical afferents. *Journal of Neurophysiology, 96*, 3474–3484.

Sceniak, M. P., Hawken, M. J., & Shapley, R. M. (2001). Visual spatial characterization of macaque V1 neurons. *Journal of Neurophysiology, 85*, 1873–1887.

Sceniak, M. P., Ringach, D. L., Hawken, M. J., & Shapley, R. (1999). Contrast's effect on spatial summation by macaque V1 neurons. *Nature Neuroscience, 2*, 733–739. doi:10.1038/11197.

Schmidt, K. E., Goebel, R., Löwell, S., & Singer, W. (1997). The perceptual grouping criterion of colinearity is reflected by anisotropies of connections in the primary visual cortex. *European Journal of Neuroscience, 9*, 1083–1089.

Schwabe, L., Ichida, J. M., Shushruth, S., Mangapathy, P., & Angelucci, A. (2010). Contrast-dependence of surround suppression in macaque V1: Experimental testing of a recurrent network model. *NeuroImage, 52*, 777–792. doi:10.1016/j.neuroimage.2010.01.032.

Schwabe, L., Obermayer, K., Angelucci, A., & Bressloff, P. C. (2006). The role of feedback in shaping the extra-classical receptive field of cortical neurons: A recurrent network model. *Journal of Neuroscience, 26*, 9117–9129.

Schwartz, O., & Simoncelli, E. P. (2001). Natural signal statistics and sensory gain control. *Nature Neuroscience, 4*, 819–825. doi:10.1038/90526.

Sengpiel, F., Baddley, R. J., Freeman, T. C. B., Harrad, R., & Blakemore, C. (1998). Different mechanisms underlie three inhibitory phenomena in cat area 17. *Vision Research, 38*, 2067–2080. doi:10.1016/S0042-6989(97)00413-6.

Sengpiel, F., Sen, A., & Blakemore, C. (1997). Characteristics of surround inhibition in cat area 17. *Experimental Brain Research, 116*, 216–228. doi:10.1007/PL00005751.

Shmuel, A., Korman, M., Sterkin, A., Harel, M., Ullman, S., Malach, R., et al. (2005). Retinotopic axis specificity and selective clustering of feedback projections from V2 to V1 in the owl monkey. *Journal of Neuroscience, 25*, 2117–2131.

Shushruth, S., Ichida, J. M., Levitt, J. B., & Angelucci, A. (2009). Comparison of spatial summation properties of neurons in macaque V1 and V2. *Journal of Neurophysiology, 102*, 2069–2083.

Shushruth, S., Mangapathy, P., Ichida, J. M., Bressloff, P. C., Schwabe, L., & Angelucci, A. (2012). Strong recurrent networks compute the orientation-tuning of surround modulation in primate primary visual cortex. *Journal of Neuroscience, 4*, 308–321.

Shushruth, S., Nurminen, L., Bijanzadeh, M., Ichida, J. M., Vanni, S., & Angelucci, A. (2013). Different orientation-tuning of near and far surround suppression in macaque primary visual cortex mirrors their tuning in human visual perception. *Journal of Neuroscience, 33*, 106–119.

Sillito, A. M., Cudeiro, J., & Murphy, P. C. (1993). Orientation sensitive elements in the corticofugal influence on centre-surround interactions in the dorsal lateral geniculate nucleus. *Experimental Brain Research, 93*, 6–16.

Sillito, A. M., Grieve, K. L., Jones, H. E., Cudeiro, J., & Davis, J. (1995). Visual cortical mechanisms detecting focal orientation discontinuities. *Nature, 378*, 492–496.

Simoncelli, E. P., & Olshausen, B. A. (2001). Natural image statistics and neural representation. *Annual Review of Neuroscience, 24*, 1193–1216. doi:10.1146/annurev.neuro.24.1.1193.

Sincich, L. C., & Blasdel, G. G. (2001). Oriented axon projections in primary visual cortex of the monkey. *Journal of Neuroscience, 21*, 4416–4426.

Slovin, H., Arieli, A., Hildesheim, R., & Grinvald, A. (2002). Long-term voltage-sensitive dye imaging reveals cortical dynamics in behaving monkeys. *Journal of Neurophysiology, 88*, 3421–3438.

Solomon, S. G., Peirce, J. W., & Lennie, P. (2004). The impact of suppressive surrounds on chromatic properties of cortical neurons. *Journal of Neuroscience, 24*, 148–160.

Solomon, S. G., White, A. J. R., & Martin, P. R. (2002). Extraclassical receptive field properties of parvocellular, magnocellular, and koniocellular cells in the primate lateral geniculate nucleus. *Journal of Neuroscience, 22*, 338–349.

Somers, D. C., Todorov, E. V., Siapas, A. G., Toth, L. J., Kim, D. S., & Sur, M. (1998). A local circuit approach to understanding integration of long-range inputs in primary visual cortex. *Cerebral Cortex, 8*, 204–217. doi:10.1093/cercor/8.3.204.

Spratling, M. W. (2010). Predictive coding as a model of response properties in cortical area V1. *Journal of Neuroscience, 30*, 3531–3543.

Stemmler, M., Usher, M., & Niebur, E. (1995). Lateral interactions in primary visual cortex: a model bridging physiology and psychophysics. *Science, 269*, 1877–1880. doi:10.1126/science.7569930.

Stettler, D. D., Das, A., Bennett, J., & Gilbert, C. D. (2002). Lateral connectivity and contextual interactions in macaque primary visual cortex. *Neuron, 36*, 739–750. doi:10.1016/S0896-6273(02)01029-2.

Stimberg, M., Wimmer, K., Martin, R., Schwabe, L., Marino, J., Schummers, J., et al. (2009). The operating regime of local computations in primary visual cortex. *Cerebral Cortex, 19*, 2166–2180. doi:10.1093/cercor/bhn240.

Ungerleider, L. G., & Desimone, R. (1986). Cortical connections of visual area MT in the macaque. *Journal of Comparative Neurology, 248*, 190–222. doi:10.1002/cne.902480204.

Van Essen, D. C., & Maunsell, J. H. R. (1983). Hierarchical organization and functional streams in the visual cortex. *Trends in Neurosciences, 6*, 370–375. doi:10.1016/0166-2236(83)90167-4.

Van Essen, D. C., Newsome, W. T., & Maunsell, J. H. (1984). The visual field representation in striate cortex of the macaque monkey: Asymmetries, anisotropies, and individual variability. *Vision Research, 24*, 429–448. doi:10.1016/0042-6989(84)90041-5.

van Vreeswijk, C., & Sompolinsky, H. (1996). Chaos in neuronal networks with balanced excitatory and inhibitory activity. *Science, 274*, 1724–1726. doi:10.1126/science.274.5293.1724.

Vinje, W. E., & Gallant, J. L. (2000). Sparse coding and decorrelation in primary visual cortex during natural vision. *Science, 287*, 1273–1276. doi:10.1126/science.287.5456.1273.

Vinje, W. E., & Gallant, J. L. (2002). Natural stimulation of the nonclassical receptive field increases information transmission efficiency in V1. *Journal of Neuroscience, 22*, 2904–2915.

Walker, G. A., Ohzawa, I., & Freeman, R. D. (1999). Asymmetric suppression outside the classical receptive field of the visual cortex. *Journal of Neuroscience, 19*, 10536–10553.

Walker, G. A., Ohzawa, I., & Freeman, R. D. (2000). Suppression outside the classical cortical receptive field. *Visual Neuroscience, 17*, 369–379. doi:10.1017/S0952523800173055.

Webb, B. S., Dhruv, N. T., Solomon, S. G., Taliby, C., & Lennie, P. (2005). Early and late mechanisms of surround suppression in striate cortex of macaque. *Journal of Neuroscience, 25*, 11666–11675.

Webb, B. S., Tinsley, C. J., Barraclough, N. E., Easton, A., Parker, A., & Derrington, A. M. (2002). Feedback from V1 and inhibition from beyond the classical receptive field modulates the responses of neurons in the primate lateral geniculate nucleus. *Visual Neuroscience, 19*, 583–592.

Xing, D., Shapley, R. M., Hawken, M. J., & Ringach, D. L. (2005). Effect of stimulus size on the dynamics of orientation selectivity in macaque V1. *Journal of Neurophysiology, 94*, 799–812.

Yoshimura, Y., Sato, H., Imamura, K., & Watanabe, Y. (2000). Properties of horizontal and vertical inputs to pyramidal cells in the superficial layers of the cat visual cortex. *Journal of Neuroscience, 20*, 1931–1940.

Yuille, A., & Kersten, D. (2006). Vision as Bayesian inference: Analysis by synthesis? *Trends in Cognitive Sciences, 10*, 301–308. doi:10.1016/j.tics.2006.05.002.

Zipser, K., Lamme, V. A., & Schiller, P. H. (1996). Contextual modulation in primary visual cortex. *Journal of Neuroscience, 16*, 7376–7389.

31 Peripheral Guidance of Cortical Organization

DARIO L. RINGACH

A salient feature of primary visual cortex in higher mammals is its organization into maps of visual space, ocular dominance, and orientation preference among others. The two-dimensional structure of these maps has been revealed on a large scale by optical imaging methods, and down to single-cell resolution by two-photon microscopy (Basole, White, & Fitzpatrick, 2003; Blasdel, 1992; Blasdel & Campbell, 2001; Blasdel & Salama, 1986; Grinvald et al., 1986; Ohki et al., 2005, 2006; Stosiek et al., 2003). Much effort has been devoted to studying the relationship among cortical maps (Blasdel, Obermayer, & Kiorpes, 1995; Carreira-Perpinan & Goodhill, 2002; Hubener et al., 1997, 2000; Kara & Boyd, 2009; Kim et al., 1999; Matsuda et al., 2000; Muller et al., 2000; Swindale, 1991, 2004), how a neuron's visual response properties depend on the local structure of the orientation map it is embedded in (Maldonado et al., 1997; Nauhaus et al., 2008; Schummers, Marino, & Sur, 2002, 2004), and how long-range horizontal connections relate to the underlying orientation map (Bosking et al., 1997; Buzas, Eysel, & Kisvarday, 1998; Chisum, Mooser, & Fitzpatrick, 2003; Das & Gilbert, 1999; Fitzpatrick, 1996; Stettler et al., 2002; Yousef et al., 1999). Despite decades of research we cannot help but feel justifiable discomfort at the recognition that we still lack an explanation for what function cortical maps play, if any, in cortical visual processing.

Expert opinion on the importance of maps in cortical processing has spanned the entire range of possibilities. Some have argued that cortical maps are a "key building block in the infrastructure of information processing by the nervous system" (Knudsen, du Lac, & Esterly, 1987), whereas others have suggested that they may be nothing more than a "by-product of synaptic development," not substantially different than the "spots, stripes, whorls and other markings on the skin and fur of various mammals" (Purves, Riddle, & Lamantia, 1992). This latter view has been reinforced by the remarkable similarity between mathematical models of cortical development and those used to account for skin markings and other biological patterns (Bard, 1981; Kondo, 2002;

Kondo, Iwashita, & Yamaguchi, 2009; Morelli et al., 2012; Swindale, 1996; Turing, 1952; Young, 1984).

A third possibility is that cortical maps evolved to support functions other than cortical computation per se. As an engineer would readily point out, randomly repositioning neurons in the cortex while keeping their connectivity intact ought to have no influence on function other than making the wiring of the brain more complex. This would be akin to shuffling the locations of transistors on an integrated circuit while keeping the circuit diagram unchanged. Indeed, minimizing wiring length may be one important reason for organizing neurons in two-dimensional maps (Chklovskii & Koulakov, 2004; Chklovskii, Schikorski, & Stevens, 2002; Koulakov & Chklovskii, 2001). However, if wiring optimization is the sole reason for the existence of maps, we would be hard pressed to accept the notion that they play a critical function in the computations carried out by the cortex.

HOW CAN WE OBTAIN EVIDENCE THAT CORTICAL MAPS PLAY A ROLE IN VISUAL PROCESSING?

One strategy has been to take advantage of the fact that not all species express the same cortical maps in primary visual cortex. Ocular dominance columns and orientation maps are present in some species but not others (Horton & Adams, 2005; Niell & Stryker, 2008; Ohki et al., 2005; Van Hooser et al., 2005). In some cases maps are highly variable even among individuals of the same species (Adams & Horton, 2003). Thus, in order to reveal a function of cortical maps, one could search for differences in visual performance that correlate with their presence or absence. Unfortunately, such a search has yet to produce a single compelling example of a correlation between behavioral performance and map expression (Adams & Horton, 2006, 2009; Horton & Adams, 2005) (see also chapter 29 by Niell, Bonin, and Andermann).

An indirect strategy has been to show that a particular cortical area performs a nontrivial computation on incoming information and that the result is represented spatially on the cortical surface. This is, in fact, how some define the concept of a computational map (Knudsen, du Lac, & Esterly, 1987). Under this definition, maps representing visual space or the surface of the body are not computational maps because they "simply reproduce the peripheral representation of the sensory information" (Knudsen, du Lac, & Esterly, 1987). On the other hand, the orientation map in primary visual cortex is deemed to be a computational map as it appears to result from cells computing a nontrivial stimulus property that is not explicitly represented by the input signals and because the outcome is represented spatially as a map on the cortical surface. The reasoning behind this approach is that it would be unlikely for nature to have evolved a complex structure like a computational map that has no function at all. Thus, the mere existence of computational maps ought to indicate they play some function, even if that function is unknown to us at present.

But is the orientation map a bona fide computational map according to this definition? In what way is orientation tuning a nontrivial computation? And what evidence is there to support the existence of an active process that organizes the resulting calculations in a two-dimensional map?

IS ORIENTATION TUNING A NONTRIVIAL COMPUTATION?

Orientation tuning arises for the first time in primary visual cortex (see chapter 26 by Andoni, Tan, and Priebe). Receptive fields in the retina and the lateral geniculate nucleus (LGN) have concentric, center-surround receptive fields that are not tuned for orientation. How difficult would it be to wire an orientation-tuned, simple-cell receptive field with side-by-side ON and OFF subregions (Hubel & Wiesel, 1962, 1968)? The first point I would like to make is that the answer to this question depends on set of receptive fields represented by the afferent fibers.

In the classic model by Hubel and Wiesel (Ferster, Chung, & Wheat, 1996; Ferster & Miller, 2000; Hirsch et al., 1998; Hubel & Wiesel, 1962; Reid & Alonso, 1995) there is an implicit assumption that a large number of overlapping ON- and OFF-center inputs cover the same region of visual space (figure 31.1, top). An advantage of having such a rich input is that it allows for the possibility of wiring receptive fields with an arbitrary preferred set of orientations (figure 31.1, top right). The same richness poses important challenges for the wiring of simple cells. How can simple cells select their thalamic inputs to match the sign of ON and OFF subregions (Alonso, Usrey, & Reid, 2001; Reid & Alonso, 1995)? How can cells in a single cortical column

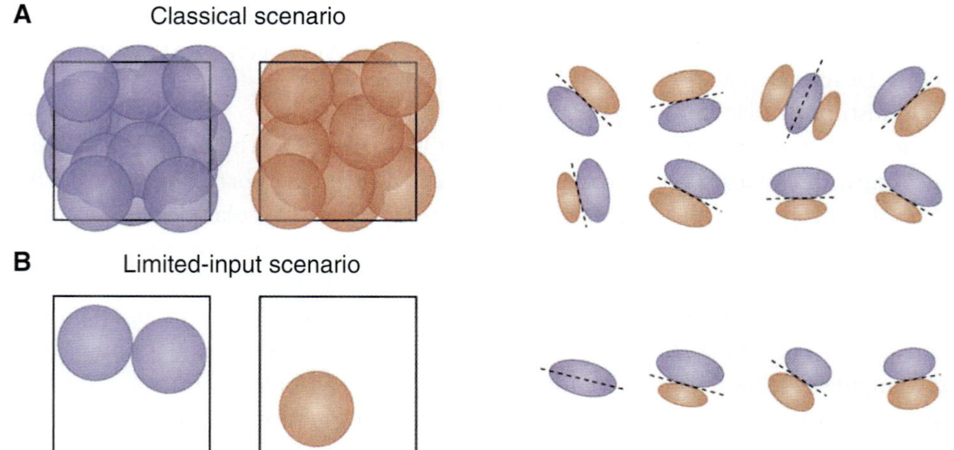

FIGURE 31.1   Classical and limited-input scenarios (Ringach, 2011). (A) In the classic scenario a very large number of overlapping ON- and OFF-center inputs, covering the same area of the visual field (outlined squares), provide cortical cells with the ability to wire themselves in nearly arbitrary ways. (B) In the limited-input scenario only a handful of receptive fields can be combined to generate a cortical receptive field. As shown on the right, only a limited range of orientations (clustering near the horizontal) can be produced with the given resources on the left. In this case, given the input, we should be able to predict the preferred orientation of the cortical column.

coordinate their wiring so that they all prefer the same orientation? How can the brain develop a quasi-periodic orientation map to ensure that all orientations are represented close to one another on the cortical surface?

Solving all these problems with a rich set of inputs is indeed nontrivial. Spontaneous retinal waves and their propagation through the thalamus, along with synaptic learning rules, have been offered as a possible solution (Cang et al., 2005; Chiu & Weliky, 2001, 2002; Eglen, 1999; Erwin & Miller, 1998; Firth, Wang, & Feller, 2005; McLaughlin et al., 2003; Miller, 1994; Miller & Erwin, 2001; Miller, Erwin, & Kayser, 1999; Stellwagen & Shatz, 2002; Warland, Huberman, & Chalupa, 2006; Weliky, 1999). Another approach has been to invoke molecular guidance, where a blueprint of the cortical organization is imprinted on the cortex by markers that direct geniculate afferents to their appropriate targets (Crowley & Katz, 2000, 2002; Hubener & Bonhoeffer, 1999; Huberman, 2007; Lambot et al., 2005; Price et al., 2006; Shatz, 1997; Skaliora, Adams, & Blakemore, 2000). Although these mechanisms have been shown to play a role in development, they have failed to provide a satisfactory explanation for the origin of receptive fields and the orientation map (Crowley & Katz, 2000, 2002; Hubener & Bonhoeffer, 1999; Katz & Crowley, 2002; Ohshiro & Weliky, 2006; Ringach, 2007).

There is an alternative scenario that, surprisingly, does not generate such difficult challenges. Assume that a cortical column receives a limited set of ON-center and OFF-center inputs (figure 31.1A) (Ringach, 2011). In such a limited-input scenario, cells in the recipient cortical column have few options for the wiring of their receptive fields. Moreover, the set of orientation-tuned receptive fields that is possible to construct relying on such inputs (assuming binary synaptic weights for simplicity) is restricted—in this example the set of feasible orientations attainable is biased to the horizontal (figure 31.1B, right). In other words in the limited-input scenario the set of preferred orientations cortical cells can achieve is constrained, or biased, by the spatial arrangement of a small number of ON- and OFF-center inputs (Jin et al., 2011; Ringach, 2011).

COST AND BENEFITS OF THE LIMITED-INPUT SCENARIO

The limited-input scenario offers some important benefits, as many difficulties the brain faces in wiring receptive fields and orientation columns in the classic scenario are alleviated or completely solved. First, simple isotropic pooling of geniculate receptive fields, which are circularly symmetric and untuned for orientation, can yield oriented receptive fields that resemble the structure of simple cells. This explains how orientation tuning can emerge without the need of connectivity rules that invoke synaptic specificity (Ringach, 2004a, 2007). Second, the tricky problem of wiring orientation columns is mitigated by the fact that all neurons within a column are biased in the same way by their shared input. Thus, in the limited-input scenario, orientation columns emerge naturally, even if neurons within a column make independent decisions about the wiring of their inputs. As a consequence there is no need for cells within a column to coordinate their wiring so they all develop the same preferred orientation. There is a cost to these benefits in that not all orientations can be represented equally well by cortical receptive fields at one location. Thus, the limited-input scenario predicts the existence of what one may term "orientation scotomas" (Paik & Ringach, 2011).

THE CASE FOR THE LIMITED-INPUT SCENARIO

Which of the two extreme scenarios most closely represents the actual cortical input? I will argue next that key findings on the statistics of RGC mosaics support the view that the cortex receives input that more closely resembles the limited-input scenario.

The 2σ Rule

The first property of RGC mosaics, called the 2σ rule, is a restatement about the coverage ratio of retinal ganglion cell (RGC) mosaics in terms of the size of their receptive field centers (Borghuis et al., 2008; DeVries & Baylor, 1997; Gauthier et al., 2009). When RGC receptive fields are mapped simultaneously by means of dense electrode arrays using reverse-correlation methods (Chichilnisky, 2001; Ringach & Shapley, 2004; Ringach, 2004b), one finds that the average distance between nearest neighbors of the same center sign is ~2σ, where σ represents the standard deviation of a Gaussian fit to the receptive field spatial profile (figure 31.2).

There are some important attributes of the 2σ rule that are worth mentioning. First, this property is universal in the sense that it is consistently observed in retinal mosaics from different ganglion cell types and across different species (Anishchenko et al., 2010; Borghuis et al., 2008; DeVries & Baylor, 1997; Gauthier et al., 2009). Second, the 2σ rule is invariant with eccentricity (figure 31.2, right). Third, the 2σ rule is optimal in the sense that it allows the retina to convey the maximum amount of information per receptive field when stimulated with

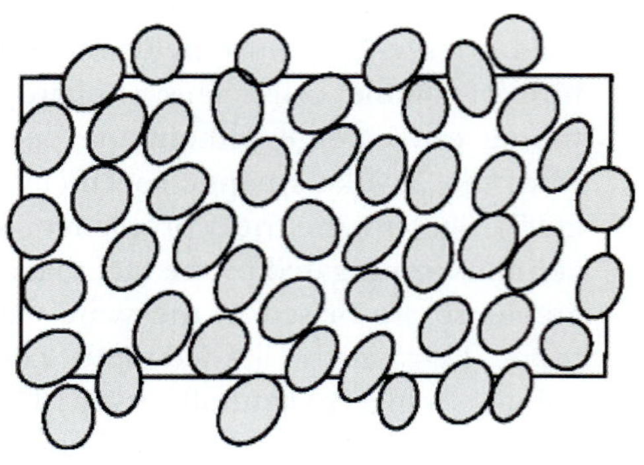
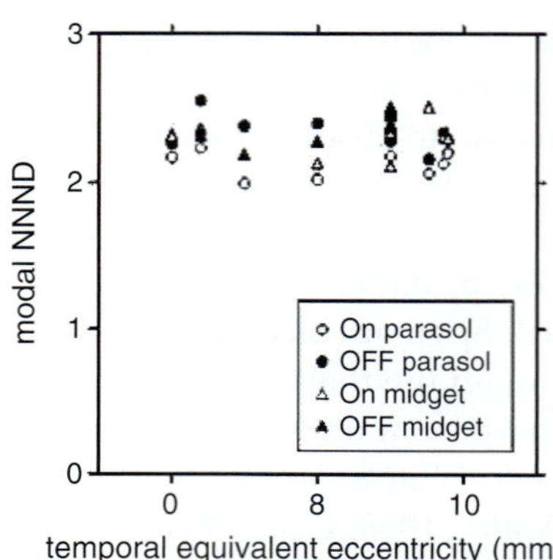

FIGURE 31.2   The 2σ rule. (Left) σ-levels sets of Gaussian fits to the RFs of a RGC mosaic. The fact that the contours of nearby fields nearly "kiss" each other implies that the average distance between nearest neighbors is about ~2σ. (Right) Mode of the nearest-neighbor distances as a function of eccentricity. The 2σ rule is constant with eccentricity. (Figure adapted from Gauthier et al., 2009.)

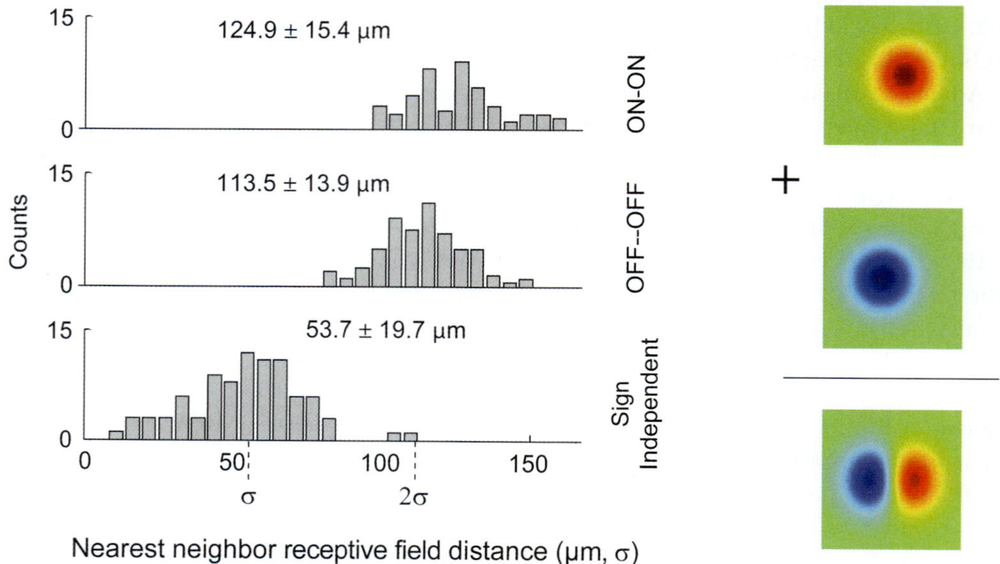

FIGURE 31.3   The nearest-neighbor rule. (Left) Distribution of nearest-neighbor distances for the ON-center mosaic (top), OFF-center mosaics (middle), and nearest-neighbor distance distribution independent of the center sign (bottom). Distances are shown in millimeters of retina and in terms of mean receptive field size, σ. (Right) ON/OFF-center receptive fields 1σ apart can converge onto a cortical cell to generate an oriented receptive field with side-by-side subregions.

natural scenes having $1/f$ spectra (Borghuis et al., 2008). Fourth, 2σ is the maximal distance two Gaussian receptive fields can be separated while their linear combination would result in a smooth, elongated subregion without a saddle point in between.

The Dipole Rule

A second key property of RGC mosaics is the so-called *dipole rule*. It states that the nearest neighbor of an ON-center receptive field is, with very high probability, an OFF-center receptive field, and vice versa (Wässle, Boycott, & Illing, 1981) (figure 31.3). In other words ON/OFF inputs originating from one ganglion cell class naturally arrange themselves in pairs that we call dipoles. When expressed in terms of center size, the average distance between ON/OFF receptive fields forming a dipole is 1σ (figure 31.3). If the input to a cortical cell were to be dominated by a single dipole, then its receptive field would resemble that of a simple cell, with side-by-side ON and OFF subregions (figure 31.3, right). Could this organization of the periphery bias cortical cells to prefer a particular orientation?

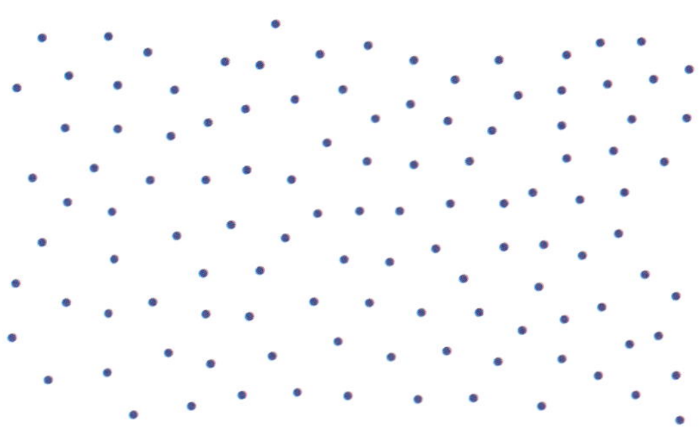

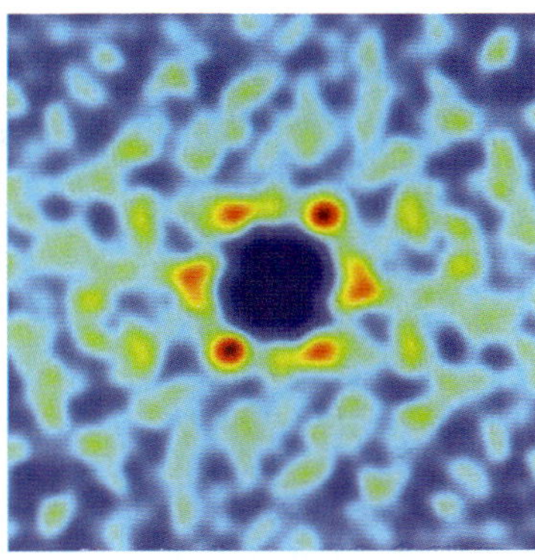

FIGURE 31.4   Retinal mosaics have hexagonal structure. (Left) Sample reconstructions of an OFF-center receptive field mosaic. (Data from Gauthier et al., 2009.) (Right) Autocorrelation of the center locations in these reconstructions (with the center peak suppressed for better visualization of the secondary peaks. Red hues represent positive values; blue hues represent negative ones.

THE NUMBER OF RETINAL GANGLION CELLS CONTRIBUTING TO A SIMPLE-CELL RECEPTIVE FIELD

To assess this possibility we could start by asking if there are any compelling reasons to believe that a simple cell in the cortex might be dominated by input from a handful of RGCs? The statistics of monosynaptic connections between geniculate and layer 4 cells in cat, along with the relative sizes of geniculate and cortical receptive fields, suggests this might be so (Alonso, Usrey, & Reid, 2001; Reid & Alonso, 1995).

We know the subregions of layer 4 simple cells have widths comparable to (in fact slightly smaller than) those of geniculate receptive field centers (cf. figure 31.4, left, in Alonso, Usrey, & Reid, 2001). Moreover, along the length of each subregion, thalamic inputs rarely make a connection if the distance from their receptive field centers to that of the cortical receptive field subregion is larger than the width of the subregion itself (see figure 31.4, right, in Alonso, Usrey, & Reid, 2001). Finally, we know that the average nearest-neighbor distance between the centers of two receptive fields of the same sign is 2σ. Given these data we must conclude that a given subregion of a simple-cell receptive field is likely to receive input from just one or two retinal ganglion cells. The same reasoning applies to a flanking subregion of the opposite sign.

Altogether the relative size of simple-cell receptive fields in terms of geniculate centers and the coverage factor of RGC mosaics expressed in terms of the 2σ rule imply that only a handful of retinal ganglion cell receptive fields converge onto a layer 4 simple cell. This conclusion is consistent with the limited-input scenario, but see Alonso, Usrey, and Reid (2001) for a different estimate based on anatomy that yields estimates 10 times larger, and the discussion of such a calculation by Ringach (2004a).

WHY ARE ORIENTATION MAPS QUASI-PERIODIC?

The limited input scenario may explain in part how orientation tuned receptive fields and columns may emerge, but it does not offer, by itself, an explanation for why the resulting orientation map would be quasi-periodic.

Here we need to invoke one additional property of the retinal mosaics. On a local scale the centers of their receptive fields lie on the vertices of a noisy, hexagonal lattice. The first experimental observation in support of this fact came from the measurements of lattice angles in reconstructions of cell-body mosaics by Wässle and colleagues, who reported a mode at 60° (Wässle et al., 1981). We have also established the hexagonal structure of RGC arrays on more solid footing by calculating the two-dimensional autocorrelations of the actual receptive field arrays (figure 31.4, right) (Paik & Ringach, 2011).

When two periodic retinal mosaics with slightly different densities and orientations are superimposed, they generate a moiré interference pattern (figure 31.5). Within the interference pattern we observe the dipole rule at work—the nearest neighbor of a cell is one of the opposite sign. Moreover, the orientation of the dipoles changes smoothly across space, is periodic, and has a hexagonal structure. If inputs from ON/OFF dipoles bias the preferred orientation of their target cortical columns, then the global moiré pattern generates a periodic input to the cortex that, we hypothesize, seeds the structure of the orientation map during the earliest stages of development. Activity-dependent mechanisms are expected to later maintain and refine these initial structures during the critical period until they reach maturity.

Orientation Maps Have Hexagonal Structure

A prediction of moiré interference can be readily put to the test. The theory predicts that, under the assumption of an isotropic magnification factor, cortical sites with a preference for a given orientation should also lie on a hexagonal lattice on the cortex (figure 31.5, circles). Indeed, we recently confirmed this is the case in four different species: ferret, tree-shrew, cat, and monkey (Paik & Ringach, 2011) (figure 31.6). This result is remarkable, as it points to a universal feature shared among orientation maps of various species,

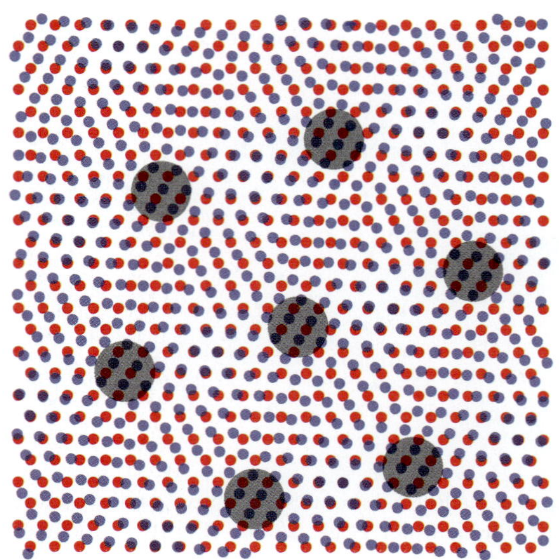

FIGURE 31.5   Example of a moiré interference pattern produced by the superposition of two hexagonal ON/OFF lattices with slightly different densities and orientations. The circles show that dipoles with similar orientation are predicted to arrange themselves in a hexagonal lattice too.

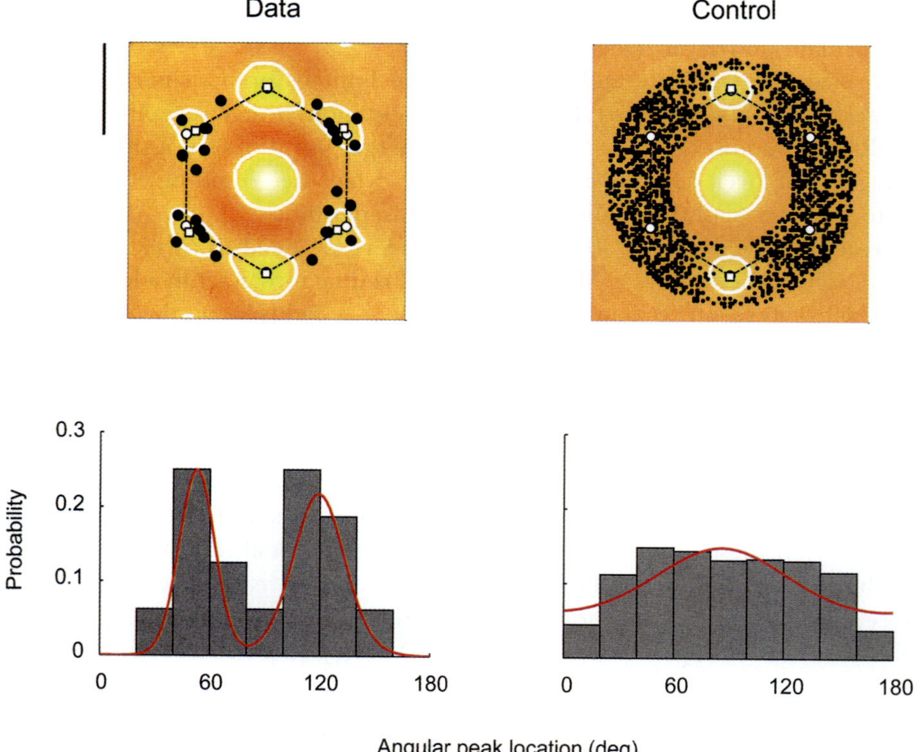

FIGURE 31.6   Orientation maps have hexagonal structure. (Left) Autocorrelation of orientation maps show hexagonal structure, with peaks located at multiples of 60°. (Right) Control maps do not show hexagonal structure. (Adapted from Paik & Ringach, 2011.)

suggesting they share a common generating mechanism and one that is consistent with the spatial interference of ON/OFF mosaics.

Orientation and Retinotopic Maps Are Linked

Another prediction of the model is that there must be a link between the retinotopic and orientation maps. In other words the two maps are not independent of each other. Specifically, the layout of orientation preferences around orientation singularities (Bonhoeffer & Grinvald, 1991) of opposite signs must differ. We have recently confirmed the predicted relationship in tree shrew primary visual cortex (Paik & Ringach, 2012), providing additional support for the model.

DISCUSSION

The goal of this review was to put forward the hypothesis that orientation tuning, columns, and maps arise as a natural consequence of a structured input from the periphery. Of course, the idea still remains to be conclusively proven, but it is encouraging that the resulting theory is supported by the parsimonious explanation it offers to a diverse set of phenomena.

Despite its simplicity, the model explains some key features of the statistics of monosynaptic connections between the thalamus and cortex (Ringach, 2004a). The sign rule (Alonso, Usrey, & Reid, 2001), which refers to the tendency for ON/OFF-center inputs to connect to simple-cell subregions of the same sign, is explained by the limited overlap of inputs present in the mosaics (see figure 31.4, left, in Ringach, 2004a). The limited overlap of ON/OFF inputs is also consistent with the clustering of ON- and OFF-center afferents in layer 4 (Jin et al., 2008) and with the finding that the average input from thalamic afferents is biased with a preferred orientation that matches that of the target cortical column (Jin et al., 2011; Ringach, 2011). The model also explains a tendency for simple-cell receptive fields to have odd-symmetric profiles (Movshon, Thompson, & Tolhurst, 1978; Ringach, 2002), which is a consequence of their inputs being initially dominated by a single dipole that generates an odd-symmetric receptive field (cf. figures 1A and 5 of Paik & Ringach, 2011), and the sharing of subregions of the same sign by nearby neurons (Smith & Hausser, 2010) (see also chapter 29 by Niell, Bonin, and Andermann). Moreover, the model suggests how cortical orientation columns can arise in a straightforward way—a set of neurons within a cortical column share the same inputs and are biased toward the same preferred orientation. The global structure of the interference pattern accounts for the hexagonal symmetry of orientation maps (Muir et al., 2011; Paik & Ringach, 2011) and a tendency for cocircularity in their organization (Braitenberg & Braitenberg, 1979; Hunt et al., 2009; Lee & Kardar, 2006; Lee, Yahyanejad, & Kardar, 2003; Sigman et al., 2001). Altogether, the ability of the model to account for these diverse findings lends support to the notion that spatially structured and limited input from the contralateral retina may seed receptive fields and the orientation map during the earliest stages of development.

One may be surprised that the organization of map structure visualized via optical imaging of signals in layer 2/3, even across species with different laminar architecture (Paik & Ringach, 2011), would show remnants of a spatially structured retinal input to the cortex. However, the model makes a general statement about the class of linear receptive fields that may be implemented at any one point in the visual field given a limited set of retinal inputs. So long as we restrict ourselves to linear combinations of the retinal signals, the particular anatomical organization of the early visual pathways is immaterial. The answer to the question is thus determined by the structure of the RGC mosaics and the assumption that the input to the first neurons exhibiting orientation selectivity can be well approximated as a linear combination of signals from the retina.

Finally, many important questions emerge in relation to the proposed scheme that require further research. Are the retinal mosaics sufficiently regular to allow for the proposed spatial interference? What mosaic class is responsible for establishing the orientation map? How precise should the retinotopic map be? How are binocular receptive fields with matching orientation preferences established? More broadly, one may conjecture that spatially structured input from the periphery may also help seed the organization of primary sensory cortices in other modalities. It would then be of interest to explore if the model is applicable in other systems. If these ideas are eventually confirmed, they could profoundly influence the way we view cortical maps and their development and function.

REFERENCES

Adams, D. L., & Horton, J. C. (2003). Capricious expression of cortical columns in the primate brain. *Nature Neuroscience, 6,* 113–114.

Adams, D. L., & Horton, J. C. (2006). Monocular cells without ocular dominance columns. *Journal of Neurophysiology, 96,* 2253–2264.

Adams, D. L., & Horton, J. C. (2009). Ocular dominance columns: Enigmas and challenges. *Neuroscientist, 15,* 62–77.

Alonso, J. M., Usrey, W. M., & Reid, R. C. (2001). Rules of connectivity between geniculate cells and simple cells in cat primary visual cortex. *Journal of Neuroscience, 21,* 4002–4015.

Anishchenko, A., Greschner, M., Elstrott, J., Sher, A., Litke, A. M., Feller, M. B., et al. (2010). Receptive field mosaics of retinal ganglion cells are established without visual experience. *Journal of Neurophysiology, 103,* 1856–1864. doi:10.1152/jn.00896.2009.

Bard, J. B. L. (1981). A model for generating aspects of zebra and other mammalian coat patterns. *Journal of Theoretical Biology, 93,* 363–385.

Basole, A., White, L. E., & Fitzpatrick, D. (2003). Mapping multiple features in the population response of visual cortex. *Nature, 423,* 986–990.

Blasdel, G. G. (1992). Differential imaging of ocular dominance and orientation selectivity in monkey striate cortex. *Journal of Neuroscience, 12,* 3115–3138.

Blasdel, G., & Campbell, D. (2001). Functional retinotopy of monkey visual cortex. *Journal of Neuroscience, 21,* 8286–8301.

Blasdel, G., Obermayer, K., & Kiorpes, L. (1995). Organization of ocular dominance and orientation columns in the striate cortex of neonatal macaque monkeys. *Visual Neuroscience, 12,* 589–603.

Blasdel, G. G., & Salama, G. (1986). Voltage-sensitive dyes reveal a modular organization in monkey striate cortex. *Nature, 321,* 579–585.

Bonhoeffer, T., & Grinvald, A. (1991). Iso-orientation domains in cat visual cortex are arranged in pinwheel-like patterns. *Nature, 353,* 429–431.

Borghuis, B. G., Ratliff, C. P., Smith, R. G., Sterling, P., & Balasubramanian, V. (2008). Design of a neuronal array. *Journal of Neuroscience, 28,* 3178–3189.

Bosking, W. H., Zhang, Y., Schofield, B., & Fitzpatrick, D. (1997). Orientation selectivity and the arrangement of horizontal connections in tree shrew striate cortex. *Journal of Neuroscience, 17,* 2112–2127.

Braitenberg, V., & Braitenberg, C. (1979). Geometry of orientation columns in the visual-cortex. *Biological Cybernetics, 33,* 179–186.

Buzas, P., Eysel, U. T., & Kisvarday, Z. F. (1998). Functional topography of single cortical cells: an intracellular approach combined with optical imaging. *Brain Research Protocols, 3,* 199–208.

Cang, J. H., Renteria, R. C., Kaneko, M., Liu, X. R., Copenhagen, D. R., & Stryker, M. P. (2005). Development of precise maps in visual cortex requires patterned spontaneous activity in the retina. *Neuron, 48,* 797–809.

Carreira-Perpinan, M. A., & Goodhill, G. J. (2002). Are visual cortex maps optimized for coverage? *Neural Computation, 14,* 1545–1560.

Chichilnisky, E. J. (2001). A simple white noise analysis of neuronal light responses. *Network, 12,* 199–213.

Chisum, H. J., Mooser, F., & Fitzpatrick, D. (2003). Emergent properties of layer 2/3 neurons reflect the collinear arrangement of horizontal connections in tree shrew visual cortex. *Journal of Neuroscience, 23,* 2947–2960.

Chiu, C., & Weliky, M. (2001). Spontaneous activity in developing ferret visual cortex in vivo. *Journal of Neuroscience, 21,* 8906–8914.

Chiu, C., & Weliky, M. (2002). Relationship of correlated spontaneous activity to functional ocular dominance columns in the developing visual cortex. *Neuron, 35,* 1123–1134.

Chklovskii, D. B., & Koulakov, A. A. (2004). Maps in the brain: What can we learn from them? *Annual Review of Neuroscience, 27,* 369–392.

Chklovskii, D. B., Schikorski, T., & Stevens, C. F. (2002). Wiring optimization in cortical circuits. *Neuron, 34,* 341–347.

Crowley, J. C., & Katz, L. C. (2000). Early development of ocular dominance columns. *Science, 290,* 1321–1324.

Crowley, J. C., & Katz, L. C. (2002). Ocular dominance development revisited. *Current Opinion in Neurobiology, 12,* 104–109.

Das, A., & Gilbert, C. D. (1999). Topography of contextual modulations mediated by short-range interactions in primary visual cortex. *Nature, 399,* 655–661.

DeVries, S. H., & Baylor, D. A. (1997). Mosaic arrangement of ganglion cell receptive fields in rabbit retina. *Journal of Neurophysiology, 78,* 2048–2060.

Eglen, S. J. (1999). The role of retinal waves and synaptic normalization in retinogeniculate development. *Philosophical Transactions of the Royal Society of London. Series B, Biological Sciences, 354,* 497–506.

Erwin, E., & Miller, K. D. (1998). Correlation-based development of ocularly matched orientation and ocular dominance maps: Determination of required input activities. *Journal of Neuroscience, 18,* 9870–9895.

Ferster, D., Chung, S., & Wheat, H. (1996). Orientation selectivity of thalamic input to simple cells of cat visual cortex. *Nature, 380,* 249–252.

Ferster, D., & Miller, K. D. (2000). Neural mechanisms of orientation selectivity in the visual cortex. *Annual Review of Neuroscience, 23,* 441–471.

Firth, S. I., Wang, C. T., & Feller, M. B. (2005). Retinal waves: Mechanisms and function in visual system development. *Cell Calcium, 37,* 425–432.

Fitzpatrick, D. (1996). The functional organization of local circuits in visual cortex: Insights from the study of tree shrew striate cortex. *Cerebral Cortex, 6,* 329–341.

Gauthier, J. L., Field, G. D., Sher, A., Shlens, J., Greschner, M., Litke, A. M., et al. (2009). Uniform signal redundancy of parasol and midget ganglion cells in primate retina. *Journal of Neuroscience, 29,* 4675–4680. doi:10.1523/JNEUROSCI.5294-08.2009.

Grinvald, A., Lieke, E., Frostig, R. D., Gilbert, C. D., & Wiesel, T. N. (1986). Functional architecture of cortex revealed by optical imaging of intrinsic signals. *Nature, 324,* 361–364.

Hirsch, J. A., Alonso, J. M., Reid, R. C., & Martinez, L. M. (1998). Synaptic integration in striate cortical simple cells. *Journal of Neuroscience, 18,* 9517–9528.

Horton, J. C., & Adams, D. L. (2005). The cortical column: A structure without a function. *Philosophical Transactions of the Royal Society B: Biological Sciences, 360,* 837–862.

Hubel, D. H., & Wiesel, T. N. (1962). Receptive fields, binocular interaction and functional architecture in the cat's visual cortex. *Journal of Physiology, 160,* 106–154.

Hubel, D. H., & Wiesel, T. N. (1968). Receptive fields and functional architecture of monkey striate cortex. *Journal of Physiology (London), 195,* 215–243.

Hubener, M., & Bonhoeffer, T. (1999). Eyes wide shut. *Nature Neuroscience, 2,* 1043–1045.

Hubener, M., Grinvald, A., Shoham, D., Bonhoeffer, T., & Swindale, N. V. (2000). Coverage optimization as a principle for the arrangement of functional maps in the visual cortex. *European Journal of Neuroscience, 12,* 194.

Hubener, M., Shoham, D., Grinvald, A., & Bonhoeffer, T. (1997). Spatial relationships among three columnar systems in cat area 17. *Journal of Neuroscience, 17*, 9270–9284.

Huberman, A. D. (2007). Mechanisms of eye-specific visual circuit development. *Current Opinion in Neurobiology, 17*, 73–80.

Hunt, J. J., Giacomantonio, C. E., Tang, H., Mortimer, D., Jaffer, S., Vorobyov, V., et al. (2009). Natural scene statistics and the structure of orientation maps in the visual cortex. *NeuroImage, 47*, 157–172. doi:10.1016/j.neuroimage.2009.03.052.

Jin, J., Wang, Y., Swadlow, H. A., & Alonso, J. M. (2011). Population receptive fields of ON and OFF thalamic inputs to an orientation column in visual cortex. *Nature Neuroscience, 14*, 232–238. doi:10.1038/nn.2729.

Jin, J. Z., Weng, C., Yeh, C. I., Gordon, J. A., Ruthazer, E. S., Stryker, M. P., et al. (2008). ON and OFF domains of geniculate afferents in cat primary visual cortex. *Nature Neuroscience, 11*, 88–94. doi:10.1038/nn2029.

Kara, P., & Boyd, J. D. (2009). A micro-architecture for binocular disparity and ocular dominance in visual cortex. *Nature, 458*,627–631.

Katz, L. C., & Crowley, J. C. (2002). Development of cortical circuits: Lessons from ocular dominance columns. *Nature Reviews. Neuroscience, 3*, 34–42.

Kim, D. S., Matsuda, Y., Ohki, K., Ajima, A., & Tanaka, S. (1999). Geometrical and topological relationships between multiple functional maps in cat primary visual cortex. *Neuroreport, 10*, 2515–2522. doi:10.1097/00001756-199908200-00015.

Knudsen, E. I., du Lac, S., & Esterly, S. D. (1987). Computational maps in the brain. *Annual Review of Neuroscience, 10*, 41–65.

Kondo, S. (2002). The reaction-diffusion system: A mechanism for autonomous pattern formation in the animal skin. *Genes to Cells, 7*, 535–541.

Kondo, S., Iwashita, M., & Yamaguchi, M. (2009). How animals get their skin patterns: Fish pigment pattern as a live Turing wave. *International Journal of Developmental Biology, 53*, 851–856.

Koulakov, A. A., & Chklovskii, D. B. (2001). Orientation preference patterns in mammalian visual cortex: A wire length minimization approach. *Neuron, 29*, 519–527.

Lambot, M. A., Depasse, F., Noel, J. C., & Vanderhaeghen, P. (2005). Mapping labels in the human developing visual system and the evolution of binocular vision. *Journal of Neuroscience, 25*, 7232–7237.

Lee, H. Y., & Kardar, M. (2006). Patterns and symmetries in the visual cortex and in natural images. *Journal of Statistical Physics, 125*, 1247–1270.

Lee, H. Y., Yahyanejad, M., & Kardar, M. (2003). Symmetry considerations and development of pinwheels in visual maps. *Proceedings of the National Academy of Sciences of the United States of America, 100*, 16036–16040. doi:10.1073/pnas.2531343100.

Maldonado, P. E., Godecke, I., Gray, C. M., & Bonhoeffer, T. (1997). Orientation selectivity in pinwheel centers in cat striate cortex. *Science, 276*, 1551–1555.

Matsuda, Y., Ohki, K., Saito, T., Ajima, A., & Kim, D. S. (2000). Coincidence of ipsilateral ocular dominance peaks with orientation pinwheel centers in cat visual cortex. *Neuroreport, 11*, 3337–3343.

McLaughlin, T., Torborg, C. L., Feller, M. B., & O'Leary, D. D. (2003). Retinotopic map refinement requires spontaneous retinal waves during a brief critical period of development. *Neuron, 40*, 1147–1160.

Miller, K. D. (1994). Models of activity-dependent neural development. Self-organizing. *Progress in Brain Research, 102*, 303–318.

Miller, K. D., & Erwin, E. (2001). Effects of monocular deprivation and reverse suture on orientation maps can be explained by activity-instructed development of geniculocortical connections. *Visual Neuroscience, 18*, 821–834.

Miller, K. D., Erwin, E., & Kayser, A. (1999). Is the development of orientation selectivity instructed by activity? *Journal of Neurobiology, 41*, 44–57.

Morelli, L. G., Uriu, K., Ares, S., & Oates, A. C. (2012). Computational approaches to developmental patterning. *Science, 336*, 187–191.

Movshon, J. A., Thompson, I. D., & Tolhurst, D. J. (1978). Spatial summation in the receptive-fields of simple cells in cats striate cortex. *Journal of Physiology (London), 283*, 53–77.

Muir, D. R., Da Costa, N. M., Girardin, C. C., Naaman, S., Omer, D. B., Ruesch, E., et al. (2011). Embedding of cortical representations by the superficial patch system. *Cerebral Cortex, 21*, 2244–2260. doi:10.1093/cercor/bhq290.

Muller, T., Stetter, M., Hubener, M., Sengpiel, E., Bonhoeffer, T., Godecke, I., et al. (2000). An analysis of orientation and ocular dominance patterns in the visual cortex of cats and ferrets. *Neural Computation, 12*, 2573–2595. doi:10.1162/089976600300014854.

Nauhaus, I., Benucci, A., Carandini, M., & Ringach, D. L. (2008). Neuronal selectivity and local map structure in visual cortex. *Neuron, 57*, 673–679.

Niell, C. M., & Stryker, M. P. (2008). Highly selective receptive fields in mouse visual cortex. *Journal of Neuroscience, 28*, 7520–7536.

Ohki, K., Chung, S., Ch'ng, Y. H., Kara, P., & Reid, R. C. (2005). Functional imaging with cellular resolution reveals precise micro-architecture in visual cortex. *Nature, 433*, 597–603.

Ohki, K., Chung, S. Y., Kara, P., Hubener, M., Bonhoeffer, T., & Reid, R. C. (2006). Highly ordered arrangement of single neurons in orientation pinwheels. *Nature, 442*, 925–928.

Ohshiro, T., & Weliky, M. (2006). Simple fall-off pattern of correlated neural activity in the developing lateral geniculate nucleus. *Nature Neuroscience, 9*, 1541–1548.

Paik, S. B., & Ringach, D. L. (2011). Retinal origin of orientation maps in visual cortex. *Nature Neuroscience, 14*, 919–925.

Paik, S. B., & Ringach, D. L. (2012). Link between orientation and retinotopic maps in primary visual cortex. *Proceedings of the National Academy of Sciences of the United States of America, 109*, 7091–7096. doi:10.1073/pnas.1118926109.

Price, D. J., Kennedy, H., Dehay, C., Zhou, L. B., Mercier, M., Jossin, Y., et al. (2006). The development of cortical connections. *European Journal of Neuroscience, 23*, 910–920. doi:10.1111/j.1460-9568.2006.04620.x.

Purves, D., Riddle, D. R., & Lamantia, A. S. (1992). Iterated patterns of brain circuitry (or how the cortex gets its spots). *Trends in Neurosciences, 15*, 362–368. doi:10.1016/0166-2236(92)90180-G.

Reid, R. C., & Alonso, J. M. (1995). Specificity of monosynaptic connections from thalamus to visual-cortex. *Nature, 378*, 281–284.

Ringach, D. L. (2002). Spatial structure and symmetry of simple-cell receptive fields in macaque primary visual cortex. *Journal of Neurophysiology, 88,* 455–463.

Ringach, D. L. (2004a). Haphazard wiring of simple receptive fields and orientation columns in visual cortex. *Journal of Neurophysiology, 92,* 468–476.

Ringach, D. L. (2004b). Mapping receptive fields in primary visual cortex. *Journal of Physiology (London), 558,* 717–728.

Ringach, D. L. (2007). On the origin of the functional architecture of the cortex. *PLoS One, 2,* e251. doi:10.1371/journal.pone.0000251.

Ringach, D. L. (2011). You get what you get and you don't get upset. *Nature Neuroscience, 14,* 123–124.

Ringach, D., & Shapley, R. (2004). Reverse correlation in neurophysiology. *Cognitive Science, 28,* 147–166.

Schummers, J., Marino, J., & Sur, M. (2002). Synaptic integration by V1 neurons depends on location within the orientation map. *Neuron, 36,* 969–978.

Schummers, J., Marino, J., & Sur, M. (2004). Local networks in visual cortex and their influence on neuronal responses and dynamics. *Journal of Physiology (Paris), 98,* 429–441.

Shatz, C. J. (1997). Emergence or order in visual system development. *American Journal of Medical Genetics, 74,* 556.

Sigman, M., Cecchi, G. A., Gilbert, C. D., & Magnasco, M. O. (2001). On a common circle: Natural scenes and Gestalt rules. *Proceedings of the National Academy of Sciences of the United States of America, 98,* 1935–1940. doi:10.1073/pnas.031571498.

Skaliora, I., Adams, R., & Blakemore, C. (2000). Morphology and growth patterns of developing thalamocortical axons. *Journal of Neuroscience, 20,* 3650–3662.

Smith, S. L., & Hausser, M. (2010). Parallel processing of visual space by neighboring neurons in mouse visual cortex. *Nature Neuroscience, 13,* 1144–1149.

Stellwagen, D., & Shatz, C. J. (2002). An instructive role for retinal waves in the development of retinogeniculate connectivity. *Neuron, 33,* 357–367.

Stettler, D. D., Das, A., Bennett, J., & Gilbert, C. D. (2002). Lateral connectivity and contextual interactions in macaque primary visual cortex. *Neuron, 36,* 739–750.

Stosiek, C., Garaschuk, O., Holthoff, K., & Konnerth, A. (2003). In vivo two-photon calcium imaging of neuronal networks. *Proceedings of the National Academy of Sciences of the United States of America, 100,* 7319–7324. doi:10.1073/pnas.1232232100.

Swindale, N. V. (1991). Coverage and the design of striate cortex. *Biological Cybernetics, 65,* 415–424.

Swindale, N. V. (1996). The development of topography in the visual cortex: A review of models. *Network (Bristol, England), 7,* 161–247.

Swindale, N. V. (2004). How different feature spaces may be represented in cortical maps. *Network (Bristol, England), 15,* 217–242.

Turing, A. M. (1952). The chemical basis of morphogenesis. *Philosophical Transactions of the Royal Society of London. Series B, Biological Sciences, 237,* 37–72.

Van Hooser, S. D., Heimel, J. A. F., Chung, S., Nelson, S. B., & Toth, L. J. (2005). Orientation selectivity without orientation maps in visual cortex of a highly visual mammal. *Journal of Neuroscience, 25,* 19–28.

Warland, D. K., Huberman, A. D., & Chalupa, L. M. (2006). Dynamics of spontaneous activity in the fetal macaque retina during development of retinogeniculate pathways. *Journal of Neuroscience, 26,* 5190–5197.

Wässle, H., Boycott, B. B., & Illing, R. B. (1981). Morphology and mosaic of ON-beta and OFF-beta cells in the cat retina and some functional considerations. *Proceedings of the Royal Society of London. Series B, Biological Sciences, 212,* 177–195.

Weliky, M. (1999). Recording and manipulating the in vivo correlational structure of neuronal activity during visual cortical development. *Journal of Neurobiology, 41,* 25–32.

Young, D. A. (1984). A local activator–inhibitor model of vertebrate skin patterns. *Mathematical Biosciences, 72,* 51–58.

Yousef, T., Bonhoeffer, T., Kim, D. S., Eysel, U. T., Toth, E., & Kisvarday, Z. F. (1999). Orientation topography of layer 4 lateral networks revealed by optical imaging in cat visual cortex (area 18). *European Journal of Neuroscience, 11,* 4291–4308.

32 Brain State and Geniculocortical Communication

HARVEY A. SWADLOW AND JOSE MANUEL ALONSO

It has been known for nearly half a century that dorsal thalamic nuclei serve to filter sensory information passing to sensory neocortex and that this filter is related to brain state. Early studies showed that transmission through the dorsal lateral geniculate nucleus (LGN) was strongly modulated by the activity of the mesencephalic reticular formation (Doty et al., 1973; Singer, 1977) during the sleep–wake cycle (Livingstone & Hubel, 1981; Maffei, Moruzzi, & Rizzolatti, 1965; Mukhametov & Rizzolatti, 1970) and fluctuations in alertness (Bartlett et al., 1973; Coenen & Vendrik, 1972; Swadlow & Weyand, 1985). The mechanisms underlying these changes are still under investigation, as is the manner in which visual perceptions and receptive field properties change with state. It is important to distinguish "global" brain states such as arousal, drowsiness, and slow-wave sleep, which reflect changes occurring throughout much of the brain, from processes such as selective visual attention, which are local and affect a restricted retinotopic portion of the visual brain (Desimone & Duncan, 1995; Kastner & Ungerleider, 2000; Reynolds & Chelazzi, 2004). This chapter focuses on global changes in state. More local, retinotopic changes that result from selective visual attention are addressed in other chapters of this volume.

Studies of visual perception have been largely performed on alert and attentive subjects. However, under real-world conditions, mammals shift between alert and nonalert/drowsy states tens or even hundreds of times per day. Importantly, nonalert subjects must be able to perceive their environment and respond appropriately to potentially interesting or dangerous stimuli. Drowsiness is not simply a brief, transient state that precedes sleep. Many people spend much of their working day in a sleep-deprived, drowsy state sometimes performing demanding visual tasks (Carrier & Monk, 2000; Mitler et al., 1988; Torsvall & Akerstedt, 1987). The hazards of "drowsy driving," caused by either extended sleep deprivation or monotonous long trips, are beginning to reach public awareness (http://drowsydriving.org/about/facts-and-stats/). Drowsiness is thought to be associated with a shrinking peripheral visual field

(Roge, Kielbasa, & Muzet, 2002) and increased reaction times (Philip et al., 1999). However, determining the extent to which these changes in performance are due to perceptual mechanisms, per se, presents a challenge, as does understanding the neural mechanisms of such perceptual changes. Visual perception and its neural mechanisms are difficult to study in nonalert subjects because the eyes tend to drift, and the visual stimulus is, therefore, difficult to control. However, rabbits have been used successfully for these studies because their eyes remain remarkably stable (and open) during transitions between EEG-defined alert and nonalert states (figure 32.1). This chapter reviews how brain state affects the geniculocortical processing of visual information. It reviews specific effects of brain state on neuronal firing patterns, response properties, and synaptic transmission. In addition, it discusses a potential role of thalamic bursting in sensory processing during nonalert states.

Several facts about the thalamus are worth noting in these introductory paragraphs. The first is that LGN neurons, like neurons of many thalamic sensory nuclei, receive only a small percentage of their synaptic inputs (~10%) from the sensory periphery, and these inputs form powerful, driving synapses (Sherman & Guillery, 1998; Usrey, Reppas, & Reid, 1998; Wilson, Friedlander, & Sherman, 1984). Most of the remaining inputs come largely from layer 6 of the visual cortex, from GABAergic neurons (both intrinsic to the LGN, and from the thalamic reticular nucleus), and from cholinergic neurons from the brainstem reticular formation (Erisir, Van Horn, & Sherman, 1997). Many of these synapses are mediated by metabotropic receptor mechanisms that are well suited to modulate the powerful retinogeniculate responses (Sherman, 2001a; chapter 19 by Sherman and Guillery). Another important discovery is that thalamic neurons have two modes of firing, "burst mode" and "relay tonic mode," which are controlled by a low-threshold, voltage- and time-dependent Ca^{2+} conductance (Jahnsen & Llinas, 1984a, 1984b). This discovery provided an important key to understanding some of the mechanisms underlying the thalamic filtering of

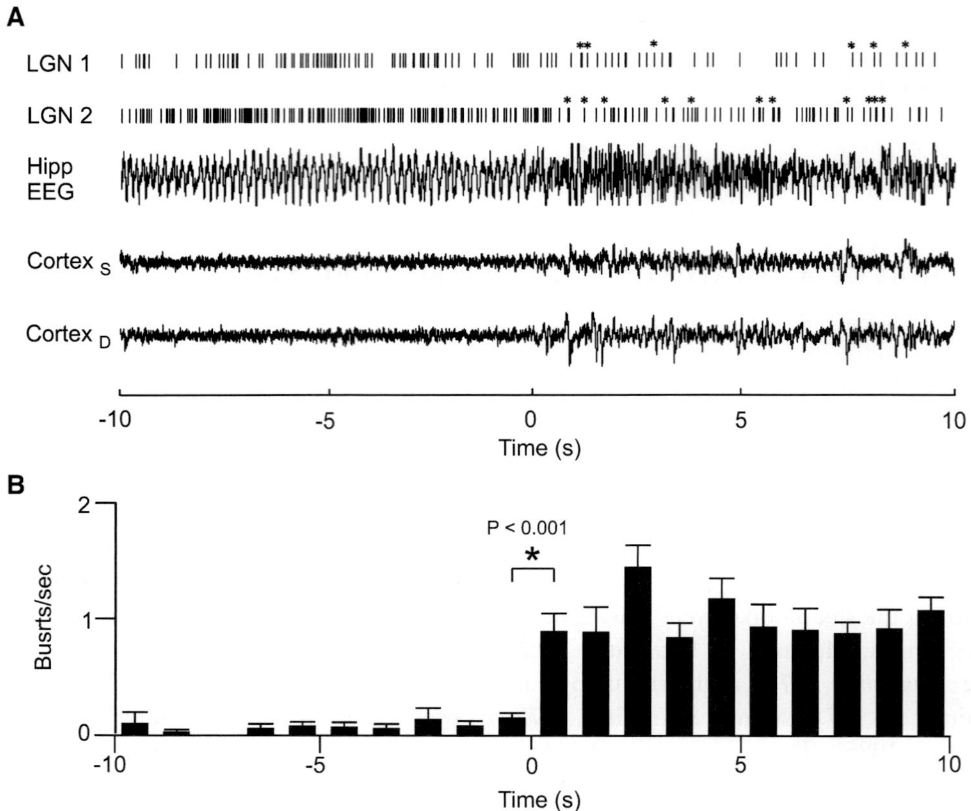

FIGURE 32.1   Profound changes in LGN bursting within seconds of EEG shifts from alert to nonalert state. (A) Spontaneous activity of two LGN cells for 10 s before and 10 s after the transition from alert (left) to nonalert states (right). EEG was recorded from the hippocampus and from the superficial (S) and deep (D) layers of cortex. Asterisks mark bursts. Bursts were defined as two or more spikes with interspike intervals of <4 ms and an interval of >100 ms preceding the initial spike of the burst (Lu, Guido, & Sherman, 1992). The state-transition point is shown at the zero value. (B) Average burst rates (bursts/s) for 10 cells, 10 s before and after the state-transition point. The difference in burst rate between 1 s before and after the transition is highly significant (asterisk). (Adapted from Bezdudnaya et al., 2006.)

sensory information and the nonlinear nature of this filter (McCormick, 1992; McCormick & Feeser, 1990; Sherman & Guillery, 1996; Sherman & Koch, 1986; Steriade & Llinas, 1988). Burst and tonic modes of thalamic firing depend on the background membrane potential of thalamic cells: Arousal is associated with depolarization and sleep/drowsiness with hyperpolarization (Llinas & Steriade, 2006). Because of this relationship, state-dependent changes in thalamocortical processing have been frequently attributed to the changes in thalamic firing mode.

BRAIN STATE EFFECTS ON LGN FIRING PATTERNS

In one of the earliest single-unit studies of LGN activity, where recordings were obtained from awake cats that were free to drift into sleep states, Hubel (1960) noted that, "In the awake animal . . . impulses occurred at more or less random intervals . . . ," but as ". . . the animal became drowsy and finally slept, there developed an increasing tendency to firing in characteristic brief, high-frequency clusters of impulses." He pointed out that such "bursts of repetitive firing . . . were seldom if ever seen in alert animals . . . ," and so began a prolonged controversy concerning the role, if any, of thalamic "bursts" in awake, perceiving subjects (Sherman, 2001b; Steriade, 2001). Importantly, Hubel noted that such state-related bursting was not seen in axons of the optic tract, thereby localizing this property to the LGN.

The basic observations by Hubel were extensively replicated during the 1960s and 1970s (Maffei, Moruzzi, & Rizzolatti, 1965; Mukhametov & Rizzolatti, 1970; Sakakura, 1968), and it was then generally agreed that thalamic burst firing was largely restricted to slow-wave sleep and general anesthesia (i.e., to brain states not generally associated with visual perception). More recently, considerable evidence has accrued that thalamic bursting also occurs in awake subjects, albeit at a lower frequency than is seen during sleep or anesthesia, and that bursting can be reliably elicited by some types of visual stimulation. Before exploring this it will be useful to briefly review the intrinsic cellular membrane properties that are responsible for the generation of two distinct modes of thalamic activity. Thalamic burst firing is dependent on T-type Ca^{2+} channels found in the somatic and dendritic membranes of thalamic neurons (Jahnsen & Llinas, 1984a). These channels mediate a voltage-gated Ca^{2+} conductance that is inactivated by prolonged membrane depolarization. Importantly, this inactivation is removed (referred to as deinactivation) when the membrane is hyperpolarized for a suitable length of time. Deinactivation of these Ca^{2+} channels is both voltage and time dependent, generally requiring ~100 ms to occur. Thus, when thalamic neurons have been hyperpolarized for >100 ms, these Ca^{2+} channels are deinactivated, and a rapid depolarization will elicit an all-or-none calcium spike with a burst of action potentials riding on its crest. If the same depolarization occurs while the membrane is already depolarized, the cell will fire in a tonic mode, where spikes are more linearly related to the depolarizing signal (Lu, Guido, & Sherman, 1992).

Studies in awake subjects have made it clear that thalamic burst mode is not limited to deep sleep and anesthetic states. It is known that the membrane potential of thalamic neurons is under state-related neuromodulatory control and that thalamic neurons are more depolarized in an awake, alert state than during slow-wave sleep or drowsiness (Llinas & Steriade, 2006). Given this, and the above properties of thalamic neurons, it is not surprising that thalamic bursting is very prevalent during slow-wave sleep and anesthesia. However, the above facts also suggest that (1) some thalamic bursting should be seen in awake subjects, and (2) the prevalence of bursting can, to some extent, be controlled by the nature of stimulus presentation. Indeed, any conditions that generate membrane hyperpolarization for >100 ms and subsequent depolarization may trigger a burst in thalamic neurons. Weyand, Boudreaux, and Guido (2001) studied LGN neurons in awake cats trained to fixate a target and found that, although burst frequency was low during alertness, most cells did show some level of bursting that was primarily associated with the onset of visual stimulation or with saccadic eye movements. Importantly, they showed that bursting could be preferentially generated by some types of natural sensory stimulation (see also Alitto, Weyand, & Usrey, 2005; Lesica & Stanley, 2005; Wang et al., 2007; also see below).

Although bursting of LGN neurons may be relatively rare in alert, attentive subjects (Ramcharan et al., 2000; Ruiz et al., 2006; Weyand, Boudreaux, & Guido, 2001), it is quite common in awake animals that are nonalert. Awake rabbits shift between alert and nonalert states very quickly, and this shift is evident in the hippocampal and neocortical EEG. The shift from nonalert to alert states can either be spontaneous or elicited by novel, low-intensity sensory stimulation. By contrast, the shift from alert to nonalert states is always spontaneous. Figure 32.1A (from Bezdudnaya et al., 2006) shows 10 s before and after one such spontaneous shift from an alert to a nonalert state, as indicated by hippocampal and cortical EEG. In the alert state the hippocampal EEG is dominated by 5- to 7-Hz theta activity and the neocortical EEG by desynchronous activity. When the animal shifts into a nonalert state, the hippocampal theta activity switches to high-voltage, irregular activity, and the neocortical desynchronous activity switches to high-voltage slow-wave activity. Note that the state shift from alert to nonalert is quite abrupt and obvious in the EEG records and that the state shift causes a dramatic reduction in the spontaneous activity of the LGN neurons and an increase in bursting (indicated by asterisks). Notably, the changes in LGN bursting and spontaneous activity occur within a single second of the shift from alert to nonalert states (figure 32.1B), a finding that also holds in ventrobasal somatosensory thalamus (Swadlow & Gusev, 2001; Stoelzel, Bereshpolova, and Swadlow, 2009).

BRAIN STATE EFFECTS ON LGN VISUAL RESPONSE PROPERTIES

For a variety of reasons it has proved difficult to make quantitative comparisons of visual response properties across differing global brain states. A major problem in the past has been the control of the visual stimulus. Hubel's (1960) early work in sleeping cats could only use diffuse stimulation through closed eyelids, but other preparations allowed a better control of the visual stimulus as subjects drifted between slow-wave sleep and arousal. For example, using a "midpontine pretrigeminal" preparation, Maffei, Moruzzi, and Rizzolatti (1965) showed that slow-wave sleep nearly abolished the linear responses of cat LGN neurons to low-frequency stimulation seen during arousal. Coenen and Vendrik (1972) also noted a reduced gain (transfer ratio) at the cat retinogeniculate synapse during sleep and drowsiness (also see Sherman & Koch, 1986). Livingstone and Hubel (1981) studied cats immobilized by a paralytic agent and reported that, as compared with slow-wave sleep, arousal enhanced both the excitatory center and

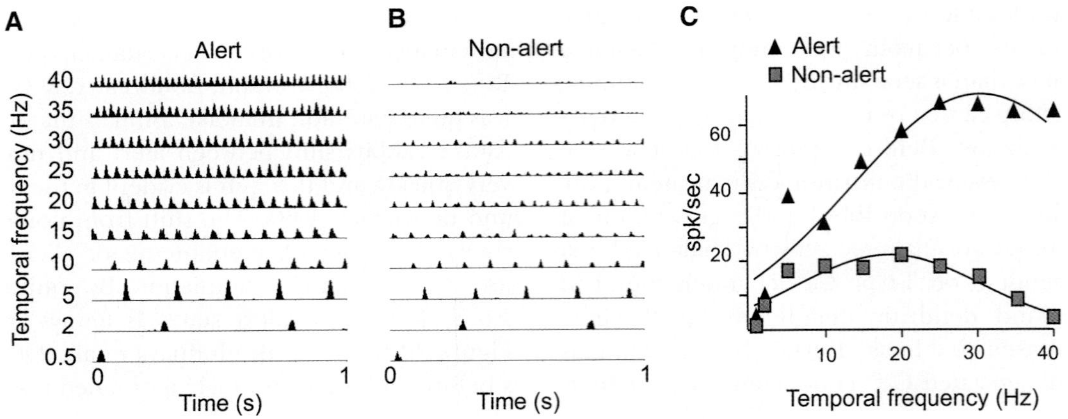

FIGURE 32.2   Temporal tuning in an LGN neuron during alert and nonalert states. (A, B) Peristimulus time histograms (PSTHs) for 10 temporal frequencies in alert (A) and in the nonalert (B) states for a transient cell. PSTHs are shown for a period of 1 s. (C) Temporal frequency tuning. Triangles, alert; squares, nonalert. (Adapted from Bezdudnaya et al., 2006.)

the suppressive surround mechanism but "produced no obvious changes in receptive field size, as mapped by small spots."

Lu, Guido, and Sherman (1992) demonstrated that LGN neurons generate bursts in response to visual stimulation when the membrane potential is hyperpolarized. At depolarized membrane potentials, when the voltage-gated Ca^{2+} conductance is inactivated, the spiking output is more linear and more faithfully related to the drifting sinusoidal grating than when the membrane is hyperpolarized and the Ca^{2+} conductance is deinactivated. Lu, Guido, and Sherman also developed important criteria for identifying, in extracellular recordings, bursts of action potentials that resulted from deinactivation of Ca^{2+} channels and a subsequent calcium spike. They found that two conditions had to be met to reliably identify such bursts. The first was an interspike interval of <4 ms, and the second was a silent period of >100 ms (50 ms under some circumstances) that preceded the first spike of the burst. These criteria have been used in numerous extracellular studies investigating burst firing in thalamic neurons.

An increase in bursting activity in the thalamus is associated with pronounced changes in temporal frequency tuning and response gain. In awake rabbits one class of LGN concentric neurons respond in a sustained manner to maintained standing contrast over the receptive field center but only when subjects display an alert EEG. When the EEG indicates a nonalert state, the sustained response disappears (Swadlow & Weyand, 1985). Importantly, this pronounced change in temporal tuning is not present in optic tract axons. Further quantitative studies (Bezdudnaya et al., 2006) reported pronounced changes in LGN temporal frequency tuning and burst frequency that began within 1 s of the shift from the alert to the nonalert state. Figure 32.2

shows the temporal tuning of a concentric "transient" LGN cell that responded at considerably higher temporal frequencies during alert than nonalert states. The cell responded to flickering stimulus of up to 40 Hz only in the alert state (figure 32.2A and C). During the nonalert state (figure 32.2B and C) responses to the high frequencies were severely attenuated, and the peak response shifted toward the lower temporal frequencies. In agreement with Livingstone and Hubel (1981), no significant state-related changes in the size of the receptive field centers were seen. Similar results in anesthetized cats were reported by Mukherjee and Kaplan (1995). These authors found that the temporal tuning of the LGN cell, but not the retinal input, was strongly related to the LGN bursting activity induced by anesthesia. As the LGN cells became more "bursty," temporal tuning became more "bandpass," with sharp reductions in firing at both high and low frequencies. Based on these results, Mukherjee and Kaplan (1995) concluded that the filtering of sensory information by the LGN is akin to a "tunable temporal filter," a conclusion that is consistent with the results from Bezdudnaya et al. (2006) in awake rabbits.

The contrast response functions of LGN neurons are also strongly influenced by state. Cano et al. (2006) found that the gain of the contrast response function increased considerably when rabbits shifted from the nonalert to the alert state. Notably, despite this large change in response gain, the contrast sensitivity (as indicated by the contrast that generated half of the maximal response) remained relatively constant. In addition, and in agreement with results of Lu, Guido, and Sherman (1992), Cano et al. (2006) found that the visual responses to sinusoidal drifting gratings were considerably more linear in alert than nonalert states. Figure 32.3 shows responses of an LGN "transient" cell

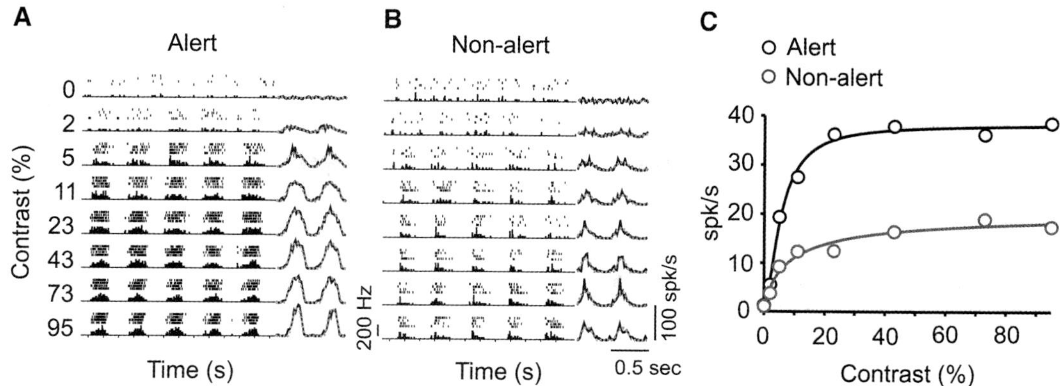

FIGURE 32.3   Contrast–response function in an LGN neuron during alert and nonalert states. (A, B) Response to drifting gratings presented at eight increasing contrasts (from top to bottom). The responses are illustrated as rasters and peristimulus time histograms (PSTHs) for the first five stimulus cycles and as PSTHs superimposed on rectified sinusoidal fits for the last two cycles. (C) Contrast response function. Black circles, alert; gray circles, nonalert. (Adapted from Cano et al., 2006.)

to a sinusoidal drifting grating of various contrasts presented at the optimal velocity and spatial frequency. Note that the visual responses are stronger and more linear (appear more like a rectified sinusoid) in the alert than the nonalert state (figure 32.3A, B). However, the contrast that generated half of the maximum response is roughly the same as that in the alert state (figure 32.3C). It should be noted that this multiplicative effect caused by alertness in response gain is very different from the multiplicative effect caused by other processes such as selective visual attention. Depending on the size of the stimulus and attention field (Lee & Maunsell, 2009; Reynolds & Heeger, 2009), visual attention can either multiplicatively scale orientation/directional tuning (McAdams & Maunsell, 1999; Treue & Martinez-Trujillo, 1999) or increase neuronal sensitivity by shifting the contrast-response function toward lower contrast (Martinez-Trujillo & Treue, 2002; Reynolds, Pasternak, & Desimone, 2000; Williford & Maunsell, 2006). Conversely, alertness acts consistently by increasing the response gain with relatively minor shifts in contrast sensitivity.

The changes in temporal tuning and response gain described above are associated with changes in the frequency of thalamic bursts. However, such association does not imply that changes in tuning are caused by the low-threshold calcium current that generates thalamic bursts. For example, noradrenergic activation by itself can increase the signal-to-noise ratio and reduce the receptive field size of neurons in the somatosensory thalamus, and cholinergic activation can cause the opposite effect (Hirata, Aguilar, & Castro-Alamancos, 2006). Also, the simple depolarization caused by stimulation of the mesencephalic reticular formation or direct application of ACh to somatosensory thalamic

neurons eliminates the response suppression to high-frequency stimulation seen in anesthetized subjects (Castro-Alamancos, 2002). Temporal frequency tuning can also be affected by synaptic filtering (Krukowski & Miller, 2001) and changes in neuronal and network-induced time constants (Alitto & Usrey, 2004; Carandini, Heeger, & Movshon, 1997). Similarly, changes in contrast response gain can result from variations in membrane depolarization (Murphy & Miller, 2003; Sanchez-Vives, Nowak, & McCormick, 2000) and/or conductance (Chance & Abbott, 2000; Mitchell & Silver, 2003), variables that are influenced by the shift between thalamic tonic mode and burst mode. Neuromodulators released during arousal act directly to depolarize thalamic relay neurons by control of K^+ conductances. Moreover, layer-6 corticogeniculate input also has a depolarizing influence, and this may be the reason that cooling of macaque visual cortex results in a reduction in the response gain of parvocellular LGN neurons (Przybyszewski et al., 2000). It should be noted, however, that cortical layer 6 can also have a powerful suppressive effect in LGN (Bolz & Gilbert, 1986; Olsen et al., 2012; Sillito, Cudeiro, & Jones, 2006).

The effect of brain state on visual response properties has often been studied using electrical stimulation of the reticular formation in anesthetized subjects. However, the appropriateness of this model for alert awake subjects is questionable. Electrical stimulation of the reticular formation activates many ascending systems (diffuse thalamic, noradrenergic, cholinergic, serotoninergic, dopaminergic) that are unlikely to be synchronously activated in awake subjects. Moreover, the effect of reticular stimulation is complicated by the use of anesthetic agents that bind to different receptor sites (e.g., pentobarbital at the $GABA_A$ site).

For example, electrical stimulation of the reticular formation facilitates transmission through the LGN (Doty et al., 1973; Singer, 1977), a finding that has been attributed to reduction of both intrinsic and extrinsic inhibitory inputs from the thalamic reticular nucleus (Singer, 1973). However, when studying LGN neurons in cats that drifted from slow-wave sleep to awake states, Livingstone and Hubel (1981) reported that "both the excitatory and inhibitory components of [LGN] responses to spatially restricted visual stimuli were enhanced" instead of reduced. In agreement with Livingstone and Hubel (1981), Swadlow and Weyand (1985) demonstrated that EEG arousal results in an enhancement of both excitatory responses and feedback inhibitory responses in the LGN of awake rabbits. Similarly, at the level of the visual cortex, Singer, Tretter, and Cynader (1976) reported a loss in response selectivity of V1 neurons following stimulation of the mesencephalic reticular formation, but Livingstone and Hubel (1981) found no reduced directional selectivity, orientation selectivity, or end-stopping in awake cats as they fell asleep, and similar results were found by Swadlow and Weyand (1987) in awake rabbits. These discrepancies indicate that stimulation of the reticular formation in anesthetized subjects is not a faithful mimic of the alert state.

A SPECIAL ROLE FOR THALAMIC BURSTING IN SENSORY PROCESSING?

An important point made by Lu, Guido, and Sherman (1992) is that LGN bursts can be reliably triggered by visual stimulation, a finding that was further developed by a number of subsequent studies from Sherman's lab. Thus, using signal detection theory, Guido et al. (1995) showed that visual responses during burst mode are less linear but more "detectable" than spikes during tonic mode. Information theoretic studies (Reinagel et al., 1999) also concluded that both tonic spikes and bursts encode stimulus information efficiently, but bursts (considered as unitary events) carry considerably more information. These findings led Sherman and Guillery (1996) to propose that LGN bursts serve as a "wake-up call" to the cortex that is fed back, through the corticogeniculate pathway, to a restricted region of the LGN. As a consequence of this feedback signal, the restricted LGN region depolarizes and switches to tonic mode, which makes visual responses more linear and the visual stimulus analysis more precise. This provocative proposal followed a related suggestion by Crick (1984) as well as theoretical notions that bursts are well suited to powerfully activate postsynaptic targets (Lisman, 1997).

The notion that bursts may encode different visual information than tonic spikes is interesting and provocative. One strategy to further examine this issue has been to compare the stimulus conditions that generate bursts versus single spikes, usually in a stable anesthetic state. Thus, Rivadulla et al. (2003) reported a shrinkage of the receptive field center for bursts when compared with single spikes. Grubb and Thompson (2005) demonstrated that bursts are driven more effectively by lower temporal frequencies than single spikes. Alitto, Wayand, and Usrey (2005) showed that burst spikes occur more reliably and at shorter latencies than single spikes and had a greater dependence on stimuli that quickly transition from suppressive to excitatory portions of the visual field. Finally, two other studies demonstrated that LGN bursts are frequently triggered by stimuli in the natural environment, especially those in which the spatial and temporal correlations of natural scenes included sequences of prolonged inhibitory stimuli followed by the sudden appearance of an excitatory stimulus (Lesica & Stanley, 2005; Wang et al., 2007).

Together, the above findings are very compatible with the notion that thalamic bursts serve some special function in perception, perhaps providing a kind of "attentional searchlight" (Crick, 1984) or wake-up call (Sherman & Guillery, 1996) to the cortex for the detection of novel, potentially interesting, or dangerous stimuli. To serve such a function, however, it is essential that thalamic bursts effectively activate cortical circuits. Although bursts may, in principle, be well suited to activate postsynaptic targets (Lisman, 1997), the cortex may not be effectively driven when modulatory cholinergic inputs are reduced during burst thalamic mode. To address this issue the efficacy of synaptic thalamocortical transmission was measured in layer 4 putative inhibitory interneurons in the somatosensory cortex of awake rabbits (Swadlow & Gusev, 2001). The results from this study demonstrated that thalamic bursts powerfully activate these cortical layer 4 neurons. Moreover, they showed that the initial impulse of each burst has about two times greater efficacy in generating a postsynaptic spike than does a single tonic spike and that the efficacy is raised even further by subsequent impulses in the burst. Notably, it was shown that the potency of the initial spike in the burst was due to the interval that necessarily precedes each burst. Thalamocortical synapses are depressing, and the long interval that precedes each burst allows for recovery from this depression and the generation of a potent postsynaptic response. This result, which was originally predicted by Ramcharan et al. (2000), has now been replicated with different methods in both the visual and somatosensory thalamocortical systems (Stoelzel, Bereshpolova, & Swadlow, 2009; Stoelzel et al., 2008; Swadlow, Gusev, & Bezdudnaya, 2002).

BRAIN STATE EFFECTS ON THALAMOCORTICAL SYNAPTIC TRANSMISSION

Nicotinic ACh receptors are found on thalamic afferents to sensory cortex (Lavine, Reuben, & Clarke, 1997) and may be involved in state-related modulation of thalamic input to sensory neocortex. In vitro work by Gil, Connors, and Amitai (1997) showed that the activation of nicotinic receptors on thalamic afferents enhances the amplitude of the synaptic drive onto first-order cortical neurons, probably because of an enhanced presynaptic release of transmitter. These results led Gil and colleagues to speculate that thalamic synapses would be selectively enhanced, as compared to corticocortical inputs, during arousal. This selective enhancement would shift the balance of synaptic inputs onto first-order cortical neurons by increasing the weight of thalamic inputs (also see Hasselmo & Bower, 1992; Kimura, 2000). Consistent with the above findings, Disney, Aoki, and Hawken (2007) found evidence for presynaptic nicotinic acetylcholine receptors on thalamocortical terminals in layer 4c of macaque V1. The receptors were found on the thalamic terminals that were synapsing onto excitatory cortical neurons but not inhibitory neurons. Disney and co-workers also studied, in anesthetized subjects, the responses of layer 4c neurons to visual stimulation before and after iontophoretic application of nicotine. Importantly, they showed that nicotine enhances the gain of these neurons to visual stimulation by increasing the responsiveness and lowering the contrast threshold.

The above study by Disney, Aoki, and Hawken (2007) was done in anesthetized monkeys but suggests that thalamocortical synaptic transmission could be increased during the alert state following a cholinergic release in the cortical layers that receive thalamic input. However, it is not at all clear that state shifts in awake animals can be emulated by the direct application of cholinergic agonists in anesthetized animals. Indeed, two recent studies in awake rabbits (Stoelzel, Bereshpolova, & Swadlow, 2009; Swadlow & Gusev, 2001) found no difference in thalamocortical synaptic transmission between alert and nonalert states. In both studies burst frequency increased greatly in the nonalert state, and bursts were more effective in generating postsynaptic responses than single spikes. However, in both studies, the greater efficacy of bursts was attributable to the long interval preceding the initial spike of each burst, which allowed recovery from synaptic depression. The

synaptic impact generated by single spikes with comparable preceding interspike intervals was not state dependent.

BRAIN STATE EFFECTS ON VISUAL CORTEX

The neocortex receives diffuse, state-related noradrenergic and cholinergic neuromodulatory inputs that are implicated in arousal, vigilance, spatial attention, spatial working memory, stimulus discrimination/selection, and neural plasticity (Aston-Jones & Bloom, 1981; Bentley, Husain, & Dolan, 2004; Kirkwood et al., 1999; Krnjevic, 2004; Robbins, 1998; Sarter & Bruno, 1997). However, much of the cortical work on the neuromodulatory control of brain state has focused on the cholinergic inputs, which originate in basal forebrain and act on different nicotinic and muscarinic receptor types in sensory cortices (Kimura & Baughman, 1997; Kuczewski et al., 2005). Because the cholinergic input to the cortex is reduced in nonalert states, cortical neurons could amplify the state-related changes in response gain and temporal tuning that are relayed by LGN inputs.

ACh has diverse direct and indirect actions on cortical neurons, some of which are cell-type specific and voltage dependent. Thus, McCormick and Prince (1985, 1986) reported that pressure pulses of ACh onto layer 5 pyramidal neurons generate a short-latency inhibition mediated by a fast muscarinic activation of local GABAergic interneurons. The short-latency inhibition is followed by a prolonged increase in excitability and depolarization, which is mediated by a direct, muscarinic, voltage-dependent increase in input resistance. The increase in input resistance is caused by a reduction of two types of potassium currents: the M-type potassium current and the Ca^{2+}-activated potassium conductance that underlies the afterhyperpolarization. Notably, the slow, excitatory effects of ACh were prolonged, lasting for 10–60 s. Xiang, Huguenard, and Prince (1998) have reexamined the action of ACh on cortical GABAergic interneurons and found distinctly different effects on two major classes of these cells in layer 5 of rat visual cortex. Whereas low-threshold spiking interneurons were directly excited (via nicotinic receptors), fast-spike interneurons responded with a direct, muscarinic-mediated hyperpolarization to ACh. Because low-threshold spiking interneurons have axons that are mainly vertically directed (e.g., Martinotti cells), and axons of fast-spike interneurons are mainly horizontally directed (basket cells), Xiang, Huguenard, and Prince (1998) suggested that behavioral arousal might serve to "switch" the neocortical network toward enhanced intracolumnar (and reduced intercolumnar) inhibition.

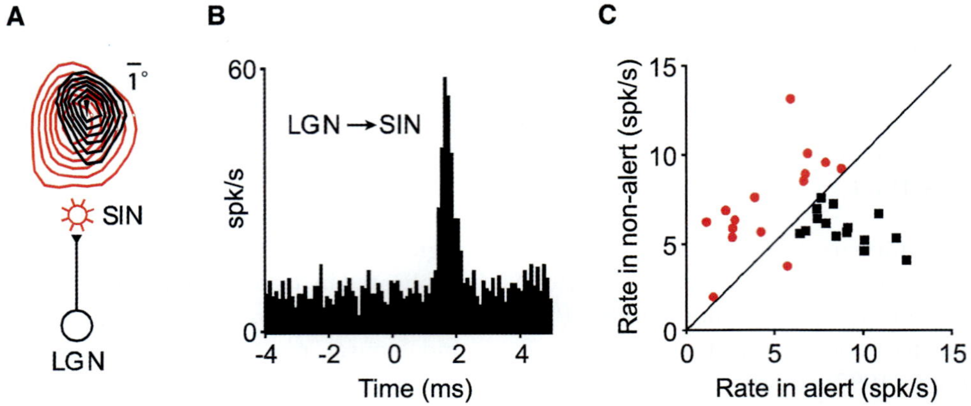

FIGURE 32.4   State-dependent opposing shifts in spontaneous rates between synaptically connected LGN neurons and layer 4 suspected inhibitory neurons (SINs). (A) Precisely aligned receptive fields of an LGN neuron (black) and a SIN in visual cortical layer 4 (red). (B) Cross-correlogram of the spike trains of these cells. Time "0" represents the time of the LGN spike. (C) Scatterplot for the firing rate of the LGN cell (black) and the SIN (red) during the 5 s before and after the transition from alert to nonalert states. Fifteen 5-s transitions are shown. (Adapted from Bereshpolova et al., 2011.)

The cell-specific actions of ACh in cortex emphasize the need for cell-class identification in studies of brain state and visual response properties. The need for such identification is illustrated by recent measurements from two classes of visual cortical layer 4 neurons and their LGN afferents during different brain states (Bereshpolova et al., 2011). Whereas the LGN neurons showed marked reduction in spontaneous firing rate and increase in bursting following the transition from alert to nonalert states, the layer 4 simple cells (putative excitatory neurons) showed no such changes. Even more surprising, neighboring fast-spike neurons (suspected inhibitory neurons, SINs) of layer 4 showed a paradoxical increase in their spontaneous firing rates during the nonalert state even though the rates of their LGN inputs decreased. These surprising findings were replicated in pairs of thalamic and cortical neurons that were monosynaptically connected. Figure 32.4 shows one cell pair, where a LGN cell was studied simultaneously with a layer 4 SIN. The receptive fields of these cells were in precise retinotopic alignment (figure 32.4A), and cross-correlation analysis of their spike train indicated a strong synaptic connectivity (figure 32.4B). Spontaneous firing rates during the 5-s periods just before and after 15 state transitions (alert to nonalert) were studied. Even though this SIN received a strong excitatory synaptic input from this LGN neuron, it shifted its spontaneous rate in the opposite direction than the LGN input (figure 32.4C).

Importantly, although the behavior of layer 4 simple cells does not reflect the state-related changes in spontaneous activity of their LGN inputs, they do show changes in their contrast response gain that mirror the changes described in the LGN (Zhuang et al., 2011). However, although the response gains of both LGN neurons and cortical simple cells increase during the alert state, there are many response properties that remain remarkably constant. Thus, layer 4 simple cells show no state-related changes in linearity of spatial summation (F1/F0 ratio), orientation tuning width, or spatial frequency tuning width. Similarly, Livingstone and Hubel (1981) reported that, in cats drifting between sleep and arousal, most visual cortical neurons showed a stronger response during arousal (i.e., a higher gain) but little change in either orientation tuning, directional specificity, or end-stopping. Also, Swadlow and Weyand (1987) found pronounced changes in the temporal tuning of V1 simple cells in alert versus nonalert awake rabbits but no changes in orientation or directional tuning. By contrast, however, Worgotter et al. (1998), studying anesthetized cats, reported large increases in receptive field size during periods of highly synchronous EEG activity.

WHAT IS "DROWSY VISION"?

The above review has touched, all too briefly, on many studies that compare geniculocortical processing of visual information during different brain states. Most of the review has focused on the awake brain, but we also included studies performed under anesthesia and sleep. Studies under anesthesia and sleep are of fundamental interest but do not directly relate to the question of how (or if) visual perceptions are altered by brain state. To address this question it is necessary to study changes that occur in awake subjects, during

   HARVEY A. SWADLOW AND JOSE MANUEL ALONSO

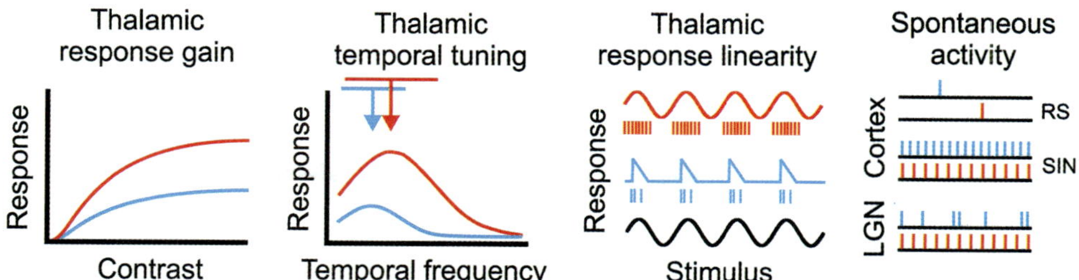

FIGURE 32.5   Summary diagram. Main changes in thalamic and cortical activity caused by shifts from alert (red) to nonalert (blue) states in awake rabbits. (A) The thalamic contrast response function decreases its gain, but there are no major changes in contrast sensitivity (defined by the contrast that generates the half-maximum response). (B) The thalamic temporal frequency tuning becomes narrower, and its peak shifts toward lower frequencies. (C) The thalamic responses become less linear, and the frequency of bursts increases. (D) Spontaneous activity is reduced and is less regular in LGN (bottom), increased in cortical layer 4 SINs (suspected inhibitory neurons), and does not change in layer 4 simple cells (putative excitatory neurons).

distinct and clearly defined brain states in which perception is possible, for example, alert and nonalert awake states. There are other awake brain states that we have not reviewed here. Niell and Stryker (2010), for example, described differences in visual response properties of cortical neurons in awake mice that were either running or balancing on top of a ball. Perception during hyperaroused states is also in need of further study (Stetson, Fiesta, & Eagleman, 2007).

We have shown that many aspects of thalamocortical visual processing change considerably within seconds of shifting between alert and nonalert states (figures 32.1 and 32.5). What is, perhaps, most surprising about the results during the nonalert state is that they are very similar to those found during slow-wave sleep or deep anesthesia, states in which perception is not possible. The EEG-defined nonalert state that abruptly follows an alert state in rabbits is clearly an awake state. First, rabbits remain very responsive to innocuous novel stimuli (visual, tactile, or auditory) that can make them shift readily to an alert state. Second, the eyes remain very stable and open. Third, changes in temporal tuning, response gain, and bursting occur within a single second following the transition from alert to nonalert states. Finally, sleep spindles never occur within 5 s of the transition from alert to nonalert states (Bereshpolova et al., 2011). This is important because sleep spindles are widely viewed as indicative of the earliest stage of sleep in rodents (Gervasoni et al., 2004).

Given the above considerations, it is reasonable to suspect that some of the changes in thalamocortical visual processing observed in awake but nonalert subjects may be related to reports of impaired visual performance in drowsy or sleep-deprived human subjects (Philip et al., 1999; Roge et al., 2004). Unfortunately, there are very few well-controlled psychophysical studies in nonalert human subjects and few comparisons across states. It is possible, for example, that driving while tired is dangerous because visual processes shift fundamentally, and our sensory abilities and perceptions change. Alternatively, the increased risks associated with drowsiness/nonalertness may be primarily or solely due to an increased probability of falling asleep or to associated nonvisual (e.g., cognitive or motor) impairment. At present, we do not know.

REFERENCES

Alitto, H. J., & Usrey, W. M. (2004). Influence of contrast on orientation and temporal frequency tuning in ferret primary visual cortex. *Journal of Neurophysiology, 91,* 2797–2808.

Alitto, H. J., Weyand, T. G., & Usrey, W. M. (2005). Distinct properties of stimulus-evoked bursts in the lateral geniculate nucleus. *Journal of Neuroscience, 25,* 514–523.

Aston-Jones, G., & Bloom, F. E. (1981). Activity of norepinephrine-containing locus coeruleus neurons in behaving rats anticipates fluctuations in the sleep-waking cycle. *Journal of Neuroscience, 1,* 876–886.

Bartlett, J. R., Doty, R. W., Pecci-Saavedra, J., & Wilson, P. D. (1973). Mesencephalic control of lateral geniculate nucleus in primates. 3. Modifications with state of alertness. *Experimental Brain Research, 18,* 214–224.

Bentley, P., Husain, M., & Dolan, R. J. (2004). Effects of cholinergic enhancement on visual stimulation, spatial attention, and spatial working memory. *Neuron, 41,* 969–982.

Bereshpolova, Y., Stoelzel, C. R., Zhuang, J., Amitai, Y., Alonso, J. M., & Swadlow, H. A. (2011). Getting drowsy? Alert/nonalert transitions and visual thalamocortical network dynamics. *Journal of Neuroscience, 31,* 17480–17487.

Bezdudnaya, T., Cano, M., Bereshpolova, Y., Stoelzel, C. R., Alonso, J. M., & Swadlow, H. A. (2006). Thalamic burst mode and inattention in the awake LGNd. *Neuron, 49,* 421–432.

Bolz, J., & Gilbert, C. D. (1986). Generation of end-inhibition in the visual cortex via interlaminar connections. *Nature, 320,* 362–365.

Cano, M., Bezdudnaya, T., Swadlow, H. A., & Alonso, J. M. (2006). Brain state and contrast sensitivity in the awake visual thalamus. *Nature Neuroscience, 9*, 1240–1242.

Carandini, M., Heeger, D. J., & Movshon, J. A. (1997). Linearity and normalization in simple cells of the macaque primary visual cortex. *Journal of Neuroscience, 17*, 8621–8644.

Carrier, J., & Monk, T. H. (2000). Circadian rhythms of performance: New trends. *Chronobiology International, 17*, 719–732.

Castro-Alamancos, M. A. (2002). Different temporal processing of sensory inputs in the rat thalamus during quiescent and information processing states in vivo. *Journal of Physiology, 539*, 567–578.

Chance, F. S., & Abbott, L. F. (2000). Divisive inhibition in recurrent networks. *Network, 11*, 119–129.

Coenen, A. M., & Vendrik, A. J. (1972). Determination of the transfer ratio of cat's geniculate neurons through quasi-intracellular recordings and the relation with the level of alertness. *Experimental Brain Research, 14*, 227–242.

Crick, F. (1984). Function of the thalamic reticular complex: The searchlight hypothesis. *Proceedings of the National Academy of Sciences of the United States of America, 81*, 4586–4590. doi:10.1073/pnas.81.14.4586.

Desimone, R., & Duncan, J. (1995). Neural mechanisms of selective visual attention. *Annual Review of Neuroscience, 18*, 193–222.

Disney, A. A., Aoki, C., & Hawken, M. J. (2007). Gain modulation by nicotine in macaque V1. *Neuron, 56*, 701–713.

Doty, R. W., Wilson, P. D., Bartlett, J. R., & Pecci-Saavedra, J. (1973). Mesencephalic control of lateral geniculate nucleus in primates. I. Electrophysiology. *Experimental Brain Research, 18*, 189–203.

Erisir, A., Van Horn, S. C., & Sherman, S. M. (1997). Relative numbers of cortical and brainstem inputs to the lateral geniculate nucleus. *Proceedings of the National Academy of Sciences of the United States of America, 94*, 1517–1520.

Gervasoni, D., Lin, S. C., Ribeiro, S., Soares, E. S., Pantoja, J., & Nicolelis, M. A. (2004). Global forebrain dynamics predict rat behavioral states and their transitions. *Journal of Neuroscience, 24*, 11137–11147.

Gil, Z., Connors, B. W., & Amitai, Y. (1997). Differential regulation of neocortical synapses by neuromodulators and activity. *Neuron, 19*, 679–686.

Grubb, M. S., & Thompson, I. D. (2005). Visual response properties of burst and tonic firing in the mouse dorsal lateral geniculate nucleus. *Journal of Neurophysiology, 93*, 3224–3247.

Guido, W., Lu, S. M., Vaughan, J. W., Godwin, D. W., & Sherman, S. M. (1995). Receiver operating characteristic (ROC) analysis of neurons in the cat's lateral geniculate nucleus during tonic and burst response mode. *Visual Neuroscience, 12*, 723–741.

Hasselmo, M. E., & Bower, J. M. (1992). Cholinergic suppression specific to intrinsic not afferent fiber synapses in rat piriform (olfactory) cortex. *Journal of Neurophysiology, 67*, 1222–1229.

Hirata, A., Aguilar, J., & Castro-Alamancos, M. A. (2006). Noradrenergic activation amplifies bottom-up and top-down signal-to-noise ratios in sensory thalamus. *Journal of Neuroscience, 26*, 4426–4436.

Hubel, D. H. (1960). Single unit activity in lateral geniculate body and optic tract of unrestrained cats. *Journal of Physiology, 150*, 91–104.

Jahnsen, H., & Llinas, R. (1984a). Ionic basis for the electro-responsiveness and oscillatory properties of guinea-pig thalamic neurones in vitro. *Journal of Physiology, 349*, 227–247.

Jahnsen, H., & Llinas, R. (1984b). Electrophysiological properties of guinea-pig thalamic neurones: an in vitro study. *Journal of Physiology, 349*, 205–226.

Kastner, S., & Ungerleider, L. G. (2000). Mechanisms of visual attention in the human cortex. *Annual Review of Neuroscience, 23*, 315–341.

Kimura, F. (2000). Cholinergic modulation of cortical function: A hypothetical role in shifting the dynamics in cortical network. *Neuroscience Research, 38*, 19–26.

Kimura, F., & Baughman, R. W. (1997). Distinct muscarinic receptor subtypes suppress excitatory and inhibitory synaptic responses in cortical neurons. *Journal of Neurophysiology, 77*, 709–716.

Kirkwood, A., Rozas, C., Kirkwood, J., Perez, F., & Bear, M. F. (1999). Modulation of long-term synaptic depression in visual cortex by acetylcholine and norepinephrine. *Journal of Neuroscience, 19*, 1599–1609.

Krnjevic, K. (2004). Synaptic mechanisms modulated by acetylcholine in cerebral cortex. *Progress in Brain Research, 145*, 81–93.

Krukowski, A. E., & Miller, K. D. (2001). Thalamocortical NMDA conductances and intracortical inhibition can explain cortical temporal tuning. *Nature Neuroscience, 4*, 424–430.

Kuczewski, N., Aztiria, E., Gautam, D., Wess, J., & Domenici, L. (2005). Acetylcholine modulates cortical synaptic transmission via different muscarinic receptors, as studied with receptor knockout mice. *Journal of Physiology, 566*, 907–919.

Lavine, N., Reuben, M., & Clarke, P. B. (1997). A population of nicotinic receptors is associated with thalamocortical afferents in the adult rat: Laminal and areal analysis. *Journal of Comparative Neurology, 380*, 175–190.

Lee, J., & Maunsell, J. H. (2009). A normalization model of attentional modulation of single unit responses. *PLoS One, 4*, e4651. doi:10.1371/journal.pone.0004651.

Lesica, N., & Stanley, G. (2005). Signal detection of salient visual features by the early visual pathway. *Annual International Conference of the IEEE Engineering in Medicine and Biology Society, 1*, 425–428.

Lisman, J. E. (1997). Bursts as a unit of neural information: Making unreliable synapses reliable. *Trends in Neurosciences, 20*, 38–43. doi:10.1016/S0166-2236(96)10070-9.

Livingstone, M. S., & Hubel, D. H. (1981). Effects of sleep and arousal on the processing of visual information in the cat. *Nature, 291*, 554–561.

Llinas, R. R., & Steriade, M. (2006). Bursting of thalamic neurons and states of vigilance. *Journal of Neurophysiology, 95*, 3297–3308.

Lu, S. M., Guido, W., & Sherman, S. M. (1992). Effects of membrane voltage on receptive field properties of lateral geniculate neurons in the cat: Contributions of the low-threshold Ca^{2+} conductance. *Journal of Neurophysiology, 68*, 2185–2198.

Maffei, L., Moruzzi, G., & Rizzolatti, G. (1965). Influence of sleep and wakefulness on the response of lateral geniculate units to sinewave photic stimulation. *Archives Italiennes de Biologie, 103*, 596–608.

Martinez-Trujillo, J., & Treue, S. (2002). Attentional modulation strength in cortical area MT depends on stimulus contrast. *Neuron, 35*, 365–370.

McAdams, C. J., & Maunsell, J. H. (1999). Effects of attention on orientation-tuning functions of single neurons in macaque cortical area V4. *Journal of Neuroscience, 19*, 431–441.

McCormick, D. A. (1992). Cellular mechanisms underlying cholinergic and noradrenergic modulation of neuronal firing mode in the cat and guinea pig dorsal lateral geniculate nucleus. *Journal of Neuroscience, 12*, 278–289.

McCormick, D. A., & Feeser, H. R. (1990). Functional implications of burst firing and single spike activity in lateral geniculate relay neurons. *Neuroscience, 39*, 103–113.

McCormick, D. A., & Prince, D. A. (1985). Two types of muscarinic response to acetylcholine in mammalian cortical neurons. *Proceedings of the National Academy of Sciences of the United States of America, 82*, 6344–6348. doi:10.1073/pnas.82.18.6344.

McCormick, D. A., & Prince, D. A. (1986). Mechanisms of action of acetylcholine in the guinea-pig cerebral cortex in vitro. *Journal of Physiology, 375*, 169–194.

Mitchell, S. J., & Silver, R. A. (2003). Shunting inhibition modulates neuronal gain during synaptic excitation. *Neuron, 38*, 433–445.

Mitler, M. M., Carskadon, M. A., Czeisler, C. A., Dement, W. C., Dinges, D. F., & Graeber, R. C. (1988). Catastrophes, sleep, and public policy: Consensus report. *Sleep, 11*, 100–109.

Mukhametov, L. M., & Rizzolatti, G. (1970). The responses of lateral geniculate neurons to flashes of light during the sleep-waking cycle. *Archives Italiennes de Biologie, 108*, 348–368.

Mukherjee, P., & Kaplan, E. (1995). Dynamics of neurons in the cat lateral geniculate nucleus: In vivo electrophysiology and computational modeling. *Journal of Neurophysiology, 74*, 1222–1243.

Murphy, B. K., & Miller, K. D. (2003). Multiplicative gain changes are induced by excitation or inhibition alone. *Journal of Neuroscience, 23*, 10040–10051.

Niell, C. M., & Stryker, M. P. (2010). Modulation of visual responses by behavioral state in mouse visual cortex. *Neuron, 65*, 472–479.

Olsen, S. R., Bortone, D. S., Adesnik, H., & Scanziani, M. (2012). Gain control by layer six in cortical circuits of vision. *Nature, 483*, 47–52.

Philip, P., Taillard, J., Quera-Salva, M. A., Bioulac, B., & Akerstedt, T. (1999). Simple reaction time, duration of driving and sleep deprivation in young versus old automobile drivers. *Journal of Sleep Research, 8*, 9–14. doi:10.1046/j.1365-2869.1999.00127.x.

Przybyszewski, A. W., Gaska, J. P., Foote, W., & Pollen, D. A. (2000). Striate cortex increases contrast gain of macaque LGN neurons. *Visual Neuroscience, 17*, 485–494.

Ramcharan, E. J., Cox, C. L., Zhan, X. J., Sherman, S. M., & Gnadt, J. W. (2000). Cellular mechanisms underlying activity patterns in the monkey thalamus during visual behavior. *Journal of Neurophysiology, 84*, 1982–1987.

Reinagel, P., Godwin, D., Sherman, S. M., & Koch, C. (1999). Encoding of visual information by LGN bursts. *Journal of Neurophysiology, 81*, 2558–2569.

Reynolds, J. H., & Chelazzi, L. (2004). Attentional modulation of visual processing. *Annual Review of Neuroscience, 27*, 611–647.

Reynolds, J. H., & Heeger, D. J. (2009). The normalization model of attention. *Neuron, 61*, 168–185.

Reynolds, J. H., Pasternak, T., & Desimone, R. (2000). Attention increases sensitivity of V4 neurons. *Neuron, 26*, 703–714.

Rivadulla, C., Martinez, L., Grieve, K. L., & Cudeiro, J. (2003). Receptive field structure of burst and tonic firing in feline lateral geniculate nucleus. *Journal of Physiology, 553*, 601–610.

Robbins, T. W. (1998). Arousal and attention: Psychopharmacological and neuropsychological studies in experimental animals. In R. Parasuramen (Ed.), *The attentive brain* (pp. 189–220). Cambridge, MA: MIT Press.

Roge, J., Kielbasa, L., & Muzet, A. (2002). Deformation of the useful visual field with state of vigilance, task priority, and central task complexity. *Perceptual and Motor Skills, 95*, 118–130. doi:10.2466/PMS.95.4.118-130.

Roge, J., Pebayle, T., Lambilliotte, E., Spitzenstetter, F., Giselbrecht, D., & Muzet, A. (2004). Influence of age, speed and duration of monotonous driving task in traffic on the driver's useful visual field. *Vision Research, 44*, 2737–2744.

Ruiz, O., Royal, D., Sary, G., Chen, X., Schall, J. D., & Casagrande, V. A. (2006). Low-threshold Ca^{2+}-associated bursts are rare events in the LGN of the awake behaving monkey. *Journal of Neurophysiology, 95*, 3401–3413.

Sakakura, H. (1968). Spontaneous and evoked unitary activities of cat lateral geniculate neurons in sleep and wakefulness. *Japanese Journal of Physiology, 18*, 23–42. doi:10.2170/jjphysiol.18.23.

Sanchez-Vives, M. V., Nowak, L. G., & McCormick, D. A. (2000). Membrane mechanisms underlying contrast adaptation in cat area 17 in vivo. *Journal of Neuroscience, 20*, 4267–4285.

Sarter, M., & Bruno, J. P. (1997). Cognitive functions of cortical acetylcholine: Toward a unifying hypothesis. *Brain Research. Brain Research Reviews, 23*, 28–46. doi:10.1016/S0165-0173(96)00009-4.

Sherman, S. M. (2001a). Tonic and burst firing: Dual modes of thalamocortical relay. *Trends in Neurosciences, 24*, 122–126. doi:10.1016/S0166-2236(00)01714-8.

Sherman, S. M. (2001b). A wake-up call from the thalamus. *Nature Neuroscience, 4*, 344–346.

Sherman, S. M., & Guillery, R. W. (1996). Functional organization of thalamocortical relays. *Journal of Neurophysiology, 76*, 1367–1395.

Sherman, S. M., & Guillery, R. W. (1998). On the actions that one nerve cell can have on another: Distinguishing "drivers" from "modulators." *Proceedings of the National Academy of Sciences of the United States of America, 95*, 7121–7126. doi:10.1073/pnas.95.12.7121.

Sherman, S. M., & Koch, C. (1986). The control of retinogeniculate transmission in the mammalian lateral geniculate nucleus. *Experimental Brain Research, 63*, 1–20. doi:10.1007/BF00235642.

Sillito, A. M., Cudeiro, J., & Jones, H. E. (2006). Always returning: Feedback and sensory processing in visual cortex and thalamus. *Trends in Neurosciences, 29*, 307–316. doi:10.1016/j.tins.2006.05.001.

Singer, W. (1973). The effect of mesencephalic reticular stimulation on intracellular potentials of cat lateral geniculate neurons. *Brain Research, 61*, 35–54. doi:10.1016/0006-8993(73)90514-3.

Singer, W. (1977). Control of thalamic transmission by corticofugal and ascending reticular pathways in the visual system. *Physiological Reviews, 57*, 386–420.

Singer, W., Tretter, F., & Cynader, M. (1976). The effect of reticular stimulation on spontaneous and evoked activity in the cat visual cortex. *Brain Research, 102,* 71–90. doi:10.1016/0006-8993(76)90576-X.

Steriade, M. (2001). To burst, or rather, not to burst. *Nature Neuroscience, 4,* 671.

Steriade, M., & Llinas, R. R. (1988). The functional states of the thalamus and the associated neuronal interplay. *Physiological Reviews, 68,* 649–742.

Stetson, C., Fiesta, M. P., & Eagleman, D. M. (2007). Does time really slow down during a frightening event? *PLoS One, 2,* e1295. doi:10.1371/journal.pone.0001295.

Stoelzel, C. R., Bereshpolova, Y., Gusev, A. G., & Swadlow, H. A. (2008). The impact of an LGNd impulse on the awake visual cortex: Synaptic dynamics and the sustained/transient distinction. *Journal of Neuroscience, 28,* 5018–5028.

Stoelzel, C. R., Bereshpolova, Y., & Swadlow, H. A. (2009). Stability of thalamocortical synaptic transmission across awake brain states. *Journal of Neuroscience, 29,* 6851–6859.

Swadlow, H. A., & Gusev, A. G. (2001). The impact of "bursting" thalamic impulses at a neocortical synapse. *Nature Neuroscience, 4,* 402–408.

Swadlow, H. A., Gusev, A. G., & Bezdudnaya, T. (2002). Activation of a cortical column by a thalamocortical impulse. *Journal of Neuroscience, 22,* 7766–7773.

Swadlow, H. A., & Weyand, T. G. (1985). Receptive-field and axonal properties of neurons in the dorsal lateral geniculate nucleus of awake unparalyzed rabbits. *Journal of Neurophysiology, 54,* 168–183.

Swadlow, H. A., & Weyand, T. G. (1987). Corticogeniculate neurons, corticotectal neurons, and suspected interneurons in visual cortex of awake rabbits: Receptive-field properties, axonal properties, and effects of EEG arousal. *Journal of Neurophysiology, 57,* 977–1001.

Torsvall, L., & Akerstedt, T. (1987). Sleepiness on the job: Continuously measured EEG changes in train drivers. *Electroencephalography and Clinical Neurophysiology, 66,* 502–511.

Treue, S., & Martinez Trujillo, J. C. (1999). Feature-based attention influences motion processing gain in macaque visual cortex. *Nature, 399,* 575–579.

Usrey, W. M., Reppas, J. B., & Reid, R. C. (1998). Paired-spike interactions and synaptic efficacy of retinal inputs to the thalamus. *Nature, 395,* 384–387.

Wang, X., Wei, Y., Vaingankar, V., Wang, Q., Koepsell, K., Sommer, F. T., et al. (2007). Feedforward excitation and inhibition evoke dual modes of firing in the cat's visual thalamus during naturalistic viewing. *Neuron, 55,* 465–478.

Weyand, T. G., Boudreaux, M., & Guido, W. (2001). Burst and tonic response modes in thalamic neurons during sleep and wakefulness. *Journal of Neurophysiology, 85,* 1107–1118.

Williford, T., & Maunsell, J. H. (2006). Effects of spatial attention on contrast response functions in macaque area V4. *Journal of Neurophysiology, 96,* 40–54.

Wilson, J. R., Friedlander, M. J., & Sherman, S. M. (1984). Fine structural morphology of identified X- and Y-cells in the cat's lateral geniculate nucleus. *Proceedings of the Royal Society of London. Series B, Biological Sciences, 221,* 411–436.

Worgotter, F., Suder, K., Zhao, Y., Kerscher, N., Eysel, U. T., & Funke, K. (1998). State-dependent receptive-field restructuring in the visual cortex. *Nature, 396,* 165–168.

Xiang, Z., Huguenard, J. R., & Prince, D. A. (1998). Cholinergic switching within neocortical inhibitory networks. *Science, 281,* 985–988.

Zhuang J., Bereshpolova Y., Stoelzel C., Huff J., Alonso J.-M., & Swadlow H. A. (2011). Getting drowsy? Alert/non-alert transitions and response properties in layer 4 of V1. *Society for Neuroscience Abstracts.* Program 484.15.

V BRIGHTNESS AND COLOR

33 Color Vision and the Retinal Mosaic

HEIDI J. HOFER AND DAVID R. WILLIAMS

The numerosity and spatial arrangement of the three cone classes in the human retina determine its ability to encode spatial and spectral variations in the retinal image and constrain the neural circuitry for spatial and color vision. In this chapter we describe the organization of the human cone mosaic and discuss its impact on visual processing.

THE TRICHROMATIC CONE MOSAIC

Relative Numerosity and Arrangement of L, M, and S Cones

The human cone mosaic is a spatially interleaved array of long-wavelength-sensitive (L), middle-wavelength-sensitive (M), and short-wavelength-sensitive (S) cones. S cones have been more easily studied than L and M cones due to their unique spectral (Williams, MacLeod, & Hayhoe, 1981), histochemical (Curcio et al., 1991; de Monasterio et al., 1985), and morphological (Ahnelt, Kolb, & Pflug, 1987) properties. They serve distinct neural circuitry (reviewed in Lee, Martin, & Grünert, 2010) and form a sparse and spatially independent submosaic, with peak density less than 10% of the total cone population (Ahnelt, Kolb, & Pflug, 1987; Bumsted & Hendrickson, 1999; Curcio et al., 1991; Hofer et al., 2005; Roorda & Williams, 1999). These features are highly conserved across both different individuals (Curcio et al., 1991; Hofer et al., 2005; Roorda et al., 2001) and different species due to the evolutionarily ancient nature of this subsystem (Ahnelt & Kolb, 2000; Neitz & Neitz, 2011). In most nonhuman primates S cones are distributed in a quasi-regular fashion (Ahnelt et al., 1987; Bumsted & Hendrickson, 1999; de Monasterio et al., 1985; Lee, Martin, & Grünert, 2010; Roorda et al., 2001; Shapiro, Schein, & de Monasterio, 1985), whereas in humans this regular arrangement becomes more disordered near the fovea (Bumsted & Hendrickson, 1999; Curcio et al., 1991; Hofer et al., 2005; Roorda et al., 2001), where there is typically an S cone–free region of about one-third of a degree (Curcio et al., 1991; Williams, MacLeod, & Hayhoe, 1981).

Characterizing the L and M cone submosaics has been more challenging because they exhibit no known morphological or histochemical differences and their pigments are 96% identical (Nathans, Thomas, & Hogness, 1986). Early studies using indirect methods suggested large variability in the ratio of L to M cones with L cones approximately twice as numerous as M cones on average (Carroll, Neitz, & Neitz, 2002; Cicerone & Nerger, 1989; Deeb et al., 2000; DeVries, 1946; Hagstrom, Neitz, & Neitz, 1998; Kremers et al., 2000; Otake & Cicerone, 2000; Pokorny, Smith, & Wesner, 1991; Rushton & Baker, 1964; Yamaguchi, Motulsky, & Deeb, 1997). Microspectrophotometry (Dartnall, Bowmaker, & Mollon, 1983) allows direct measurement of the numbers and arrangements of L and M cones—but only for small numbers of cones in postmortem tissue. With this method Mollon and Bowmaker (1992) observed a random arrangement of L and M cones in patches of talapoin monkey retina, and Bowmaker, Parry, and Mollon (2003) reported a random arrangement of L and M cones in patches of two human retinas near the fovea. Packer, Williams, and Bensinger (1996), using a transmittance imaging technique with excised tissue, reported a weak tendency toward L and M cone clumping in patches of peripheral macaque retina. This tendency has been confirmed in more recent experiments using multiunit electrophysiology to reconstruct the cone topography of the peripheral macaque retina (Field et al., 2010).

Adaptive optics retinal imaging allows direct measurements of L, M, and S cone locations in large patches of living retina (Roorda & Williams, 1999). Measurements with this technique have confirmed the large variability in L- to M-cone ratio for humans with normal color vision (Hofer et al., 2005; Roorda & Williams, 1999). Figure 33.1 shows false color images of the arrangement of cones in patches of retina at ~1° eccentricity from 18 color-normal human subjects with L to M ranging from 0.37:1 to 16.5:1. L and M cones near the fovea are generally randomly interleaved, although a small degree of clumping of like-type cones occurs in some individuals (Hofer et al., 2005; Roorda et al., 2001).

Within an individual retina, the high correlation of large-field ERG-derived estimates with those obtained in small patches with adaptive optics suggests that L- to

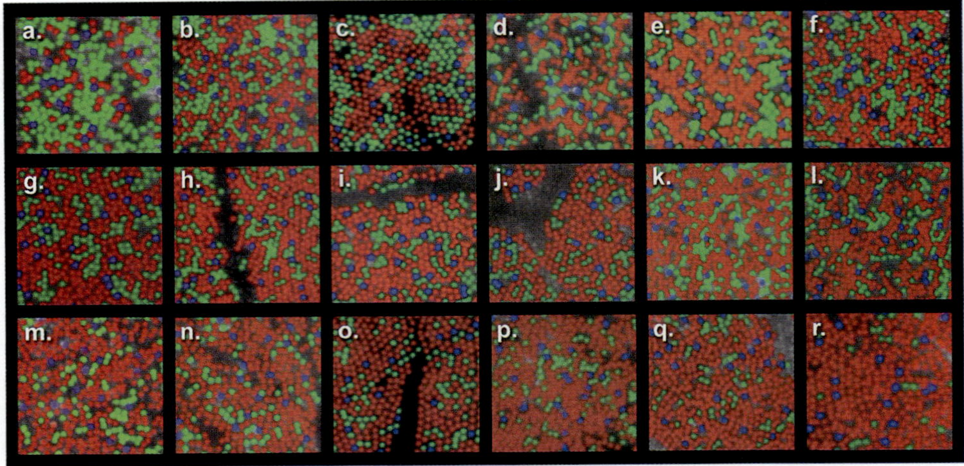

FIGURE 33.1 False-color images showing the distribution and relative numerosity of L (red), M (green), and S (blue) cones in patches of retina at ~1° eccentricity for 18 color-normal human subjects. L-to-M ratio varies from 0.37:1 to 16.5:1, whereas S-cone numerosity is relatively constant and ranges from 4% to 7% of total cone number. (c and o, Roorda & Williams, 1999; a, b, f–j, and r, Hofer et al., 2005; d, e, k–n, p, and q courtesy of Heidi Hofer and Osamu Masuda.)

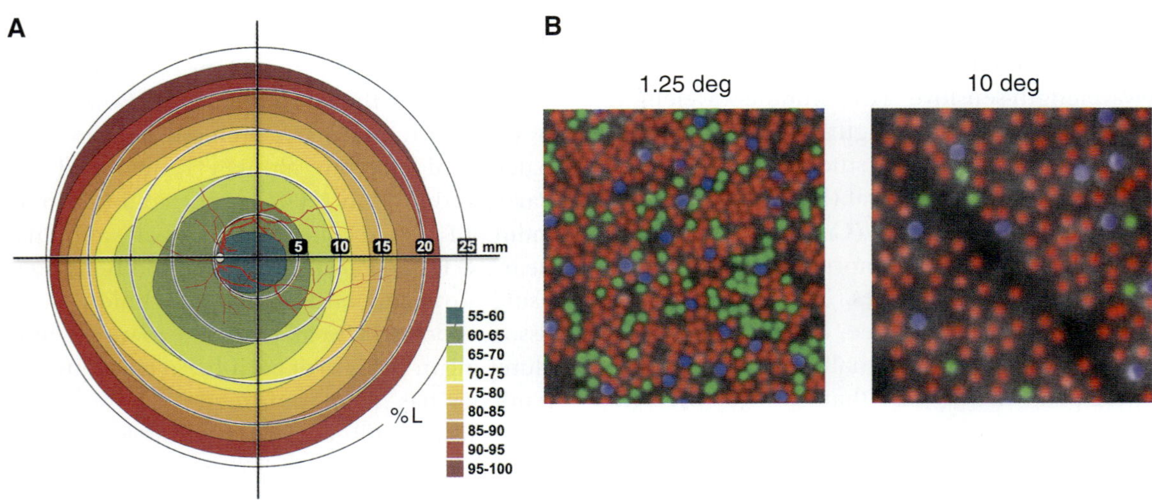

FIGURE 33.2 (A) L- to M-cone ratio across the retina of a representative human male donor retina (from Neitz et al., 2006). L- to M-cone ratio starts to increase steadily beyond ~5 mm (~15°) from the central fovea. (B) Some subjects display significant differences in L- to M-cone ratio at closer eccentricities; for example, this subject (shown in panel m in figure 33.1) has a significantly higher L- to M-cone ratio at 10° than at 1.25°. (Courtesy of Heidi Hofer and Osamu Masuda.)

M-cone ratio is relatively constant throughout the mid-peripheral retina (Brainard et al., 2000; Hofer et al., 2005), although L cones become progressively more numerous farther in the periphery (Bowmaker, Parry, & Mollon, 2003; Hagstrom, Neitz, & Neitz, 1998, 2000; Kuchenbecker et al., 2008; Neitz et al., 2006), as shown in figure 33.2. Cone topography measurements in peripheral macaque retina and adaptive optics imaging in near peripheral human retina suggest that weak clumping of L and M cones may be more prevalent in peripheral retina as well. Within the central retina adaptive optics imaging has provided direct cone numerosity information in patches of nasal versus temporal retina for two human subjects, one of whom displayed a significant difference in L- to M-cone ratio (1.24:1 and 1.77:1) (Hofer et al., 2005) while the other did not (3.67:1 and 3.90:1) (Roorda et al., 2001). This indicates that at least some humans exhibit a nasal-temporal asymmetry in the local L- to M-cone ratio, similar to that observed in macaque (Deeb et al., 2000), but whether this is the exception or the rule is yet to be established.

In summary, the notable features of the organization of the L- and M-cone submosaic are (1) large variability in L- to M-cone ratio across individuals, (2) a random or slightly clumped arrangement, and (3) an increasing L- to M-cone ratio in the periphery. In the following section we briefly consider developmental mechanisms that might account for these properties before we move on to discuss their impact on spatial and color vision.

Retinal Development and the L- and M-Cone Submosaic

The L- and M-cone pigment genes sit head to tail on the X chromosome, and their expression is controlled by a single locus control region (LCR) so that only one pigment gene can be expressed at any time. Whether an individual cone expresses the L- or M-cone pigment gene is believed to be the result of a stochastic process by which the LCR interacts and forms a complex with one of the pigment gene promoters (Nathans, 1999). Both genetic differences in nonopsin sequences of the LM pigment gene array and the distance of the opsin genes from the LCR have been shown to influence the L- to M-cone ratio, presumably by altering the LCR's probability of binding with the promoter regions (Gunther, Neitz, & Neitz, 2008; McMahon, Neitz, & Neitz, 2004). However, these genetic differences cannot explain the extent of variation in L- to M-cone ratio observed in humans with normal color vision. Knoblauch, Neitz, and Neitz (2006) have suggested that each developing cone randomly alternates between expressing the L- or M-cone pigment, with each expression increasing the probability of future expression, perhaps through epigenetic changes (Neitz & Neitz, 2011), until ultimately each cone is "locked in" to a single choice. This type of model could explain the large variation in L- to M-cone ratio across individuals, and also variation in L- to M-cone ratio across an individual retina, under the assumption that mechanisms responsible for altering the probability of gene expression are passed on to daughter cells when the immature photoreceptors divide. For example, neighboring cones descending from the same progenitor may have similar "epigenetic memories," resulting in local nonuniformities in the probability of expressing the L and M pigment genes (Neitz & Neitz, 2011) consistent with the slight L- and M-cone clumping observed in some retinas. Similarly, combining this type of mechanism with the differing time course of receptor development from the central to peripheral retina (Bumsted et al., 1997; Xiao & Hendrickson, 2000) provides a plausible explanation for the increase in L- to M-cone ratio in the peripheral retina (Knoblauch, Neitz, & Neitz, 2006).

In females, who have two X chromosomes, L- and M-cone organization is further influenced by the process of X-inactivation. X-inactivation is expected to lead to patches of retina expressing genes from the same X-chromosome (Smallwood et al., 2003), which would be accompanied by L- and M-cone clumping in human females who carry color vision deficiencies. L- and M-cone clumping would also be expected in trichromatic New World female monkeys whose X chromosomes contain only a single L or M pigment gene. Despite these expectations, L- and M-cone clumping has not been observed either in a human protan carrier (Hofer et al., 2005; figure 33.1a), female marmoset (Bowmaker, Parry, & Mollon, 2003), or female squirrel monkey (Mollon, Bowmaker, & Jacobs, 1984), although weak clumping would not have been detectable in these last two studies because of the small numbers of cones measured. One possibility is that cone migration during foveal formation (Diaz-Araya & Provis, 1992; Hendrickson & Yuodelis, 1984; Yuodelis & Hendrickson, 1986) disturbs L and M clumping due to X-inactivation. In this case one would expect significant clumping in the peripheral retinas of human female carriers and trichromatic New World female monkeys. No data have been acquired to date that would allow this hypothesis to be tested.

Foveal migration might be expected to decrease any degree of order in the organization of the cone mosaic in the central retina. This may account for the fact that L- and M-cone clumping has been observed more frequently in the periphery than near the fovea, although in humans data have only been acquired out to $10°$ eccentricity. Even greater clumping might be expected in the human retina at more peripheral locations that are less affected by foveal migration. Foveal cone migration may also explain the irregularity in human S-cone arrangement right outside the fovea. However, because the S-cone distribution becomes regular at eccentricities that are well within those impacted by migration, this would imply that either S cones outside the foveal pit remain at fixed locations while L and M cones migrate around them or that there is a mechanism that preserves the regularity of the S-cone distribution as they migrate toward the fovea.

IMPACT OF THE TRICHROMATIC MOSAIC ON SPATIAL AND COLOR VISION

Sampling and Information-Encoding Ability

The numbers and arrangement of the three cone classes have a fundamental impact on the information-coding abilities of the retinal mosaic because

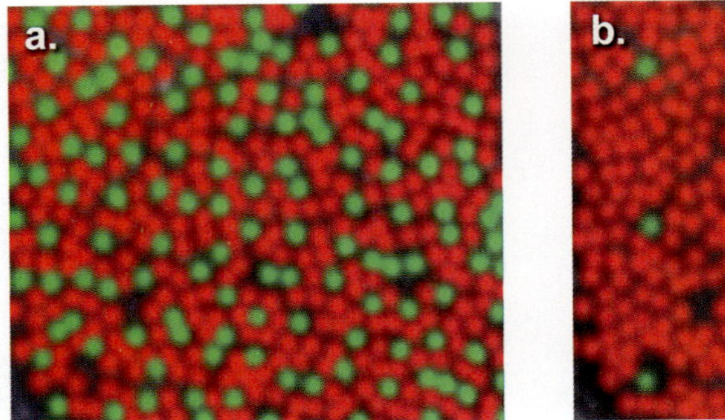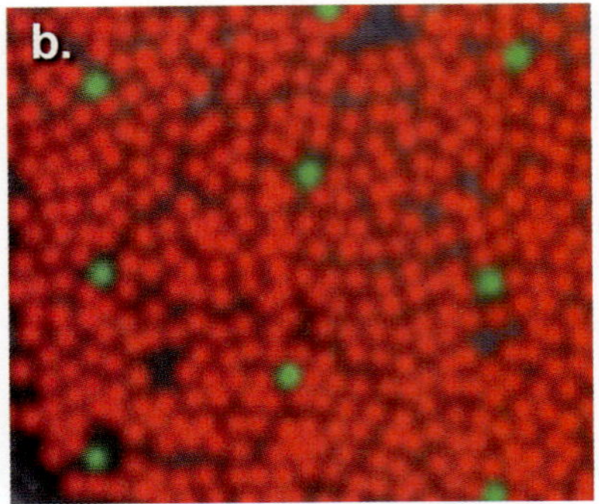

FIGURE 33.3   The hypothetical cone mosaic on the left (a) will be less effective at encoding spatial variation but more effective at encoding spectral variation than the mosaic on the right (b). Red and green denote L and M cones, respectively.

information cannot be encoded about image variations that occur between, or entirely within, receptors. For the two hypothetical cone mosaics shown in figure 33.3, the first would be expected to be capable of encoding much richer patterns of spectral variation than the second, although the second mosaic would have the advantage for images with purely spatial variation. Changing the spatial characteristics of the cone mosaic, for example by increasing its density or altering its packing geometry, will also impact the information it can encode. We refer the reader to the first edition of this encyclopedia (Williams & Hofer, 2004) for a discussion of these effects and here focus on the unique issues confronted when faced with a spatially interleaved array of different chromatic receptors.

With respect to information coding capability, the human trichromatic cone mosaic would appear to exhibit a number of puzzling features. For example, why is there such large variability in L- to M-cone ratio in different individuals with normal color vision? Is it because there is not yet an effective mechanism to constrain this ratio given the similarity of the two cone types and their relatively recent evolutionary divergence? Or is the retina's ability to encode ecologically relevant information little impacted by variations in L- to M-cone ratio? Similarly, why are L and M cones arranged in such a disordered fashion, unlike the orderly and highly conserved organization of the S-cones submosaic? Is this due to the similarity of the L and M cones precluding a plausible mechanism to promote order? Or are there advantages to the disordered arrangement or at least minimal negative consequences? Similar questions have been asked about the optimality or suitability of the number of cone pigments and their spectral sensitivities (reviewed by Osorio & Vorobyev, 2008). Here we consider the problem of how to best organize a retina composed of the existing three classes of human cones.

Different approaches can be used to address the problem of how to optimally arrange a cone mosaic.

For example, one approach is to consider the arrangement that maximizes the number of potential different images that can be encoded, and another is to consider the arrangement that minimizes the amount of sampling-related information loss. Another approach might be to consider the arrangement resulting in best discrimination, or highest signal-to-noise ratio, along some ecologically relevant stimulus dimension. Regardless of the particular method it is important to consider only *relevant* visual information because having a retina theoretically capable of encoding more visual scenes is of little use if those extra scenes are never encountered in daily life (or alternatively, if they do not represent ecologically meaningful differences).

The sparseness of the S-cone mosaic is perhaps the least-puzzling aspect of the arrangement of the human cone mosaic. The general consensus is that only a small number of S cones are needed to adequately sample the retinal image because chromatic aberration blurs the short wavelengths for which they are most sensitive (Yellott, Wandell, & Cornsweet, 1984; Mollon, 1989; Williams, Sekiguchi, & Brainard, 1993). More recently McLellan et al. (2002) suggested that the eye's optical aberrations normalize optical blur across the different cone types despite chromatic aberration; however, they based these conclusions on only a small number of eyes. Autrusseau, Thibos, and Shevell (2011) have since demonstrated that ocular aberrations do not equalize blur across cone types, consistent with the earlier literature, in the vast majority of individuals. However, as pointed out by Garrigan et al. (2010), because the spectral range most sharply focused on the retina is not fixed but is under accommodative control, it is not obvious that longitudinal chromatic aberration alone can adequately explain observed S-cone density. For example, if our retinas were to contain a dense S-cone submosaic (and a sparse L- and M-cone submosaic), we could presumably choose to accommodate so that short wavelengths were in focus. Whether this would be less

472   HEIDI J. HOFER AND DAVID R. WILLIAMS

advantageous than the current state of affairs depends on additional factors such as the signal-to-noise ratio in the different receptor classes and the spatial and spectral characteristics of typical retinal images. By considering these factors Garrigan et al. (2010) were able to show that the observed sparseness of the S-cone submosaic is optimal, or nearly so, given the reduction in S-cone signal to noise due to short-wavelength filtering by the ocular media in addition to longitudinal chromatic aberration.

Unlike with S cones, there is enormous variation in L- to M-cone ratio across different individuals with normal color vision. This is surprising because intuition holds that there ought to be a unique ratio of L to M cones that maximizes the information the retina can encode. For example, it seems reasonable that equal numbers of L and M cones would provide the best spectral discrimination. Modeling studies have confirmed this intuition to an extent. If one assumes that L and M cones are equally efficient, equally noisy, and that equal amounts of visual information are present in their two spectral bands, then equal numbers of L and M cones are indeed ideal (Garrigan et al., 2010; Manning & Brainard, 2009). However, if there are asymmetries in any of these factors, the ideal balance can shift (Manning & Brainard, 2009). Garrigan et al. (2010) further demonstrate that chromatic aberration and the eye's optical blur should confer a tolerance to variations in the L- to M-cone ratio, with a large range of cone ratios expected to result in nearly maximally informative retinal mosaics. It seems then that the large variation observed in L- to M-cone ratio is of little consequence for the retina's ability to *encode* visually relevant information. Assuming our visual machinery for *extracting* this information is sufficiently flexible to accommodate these variations in L- to M-cone ratio (a topic we discuss further below), this suggests that the relative numerosity of L, M, and S cones observed in the human retina is optimal, or nearly so, given existing environmental and optical constraints.

The information-coding ability of the retina depends on the arrangement, as well as relative numbers, of the L, M, and S cones. One might expect that the best way to arrange multiple receptor types would be to alternate them or interleave them to the extent biologically possible. This principle underlies retinal organization in many fish (Bowmaker & Kunz, 1987; Fernald, 1981; Scholes, 1975), chicken (Kram, Mantey, & Corbo, 2010), and the S-cone submosaic in humans and most nonhuman primates. However, the human L- and M-cone submosaic is not regularly interleaved but disordered, or slightly clumped, as a general rule. Two studies have quantitatively investigated the optimal manner of interleaving L and M cones. Osorio, Ruderman, and Cronin (1998) suggest that an alternating, or regular, arrangement will lead to the least luminance noise in reconstructing the retinal image. Their work incorporated various assumptions on how the visual system performs the image reconstruction process and, as such, is silent as to whether a regularly arranged retina will outperform a randomly arranged, or slightly clumped, retina if served by optimal reconstruction mechanisms. Manning and Brainard (2009) recently began to address this question using a Bayesian analysis incorporating estimates of the spatial and spectral properties of retinal images—but with a highly simplified retina of only one spatial dimension and two cone pigments with nonoverlapping spectra. Under these conditions they also found the regularly interleaved arrangement to be optimal, in the sense of minimizing sampling-related information loss, regardless of cone ratio.

The observed irregularity in the arrangement of L and M cones in the human retina thus appears contrary to optimal organizational principles. It may be that there is simply no physical mechanism that could create a more ideal, interleaved arrangement given the similarity of the L and M cones and the stochastic nature of their pigment gene expression choice. Additionally (or perhaps consequently), the existing retinal circuitry may not be poised to take advantage of a more optimally arranged cone mosaic. On the other hand it is possible that the limited modeling work done to date has overlooked some critical feature that would render the observed organization more advantageous. For example, it has been suggested that a disordered arrangement might prevent the formation of regularly patterned aliases (we discuss aliasing more below), perhaps making this form of reconstruction noise easier to filter from the visual signal (Yellott, 1982). Neither Osorio, Ruderman, and Cronin (1998) nor Manning and Brainard (2009) addressed this issue, as they did not consider higher-order scene statistics (for example, edges) or the nature of the error in the reconstructed visual signal.

Visual Consequences of Retinal Sampling

To the extent that the interleaved nature of the cone mosaic results in sampling-related information loss, there must be attendant visual consequences. Because only one of the three cone types is present at any retinal location, spatial and spectral information are necessarily confounded at small spatial scales. This is demonstrated in figure 33.4, which illustrates how a difference in the excitation of a neighboring L and M cone could

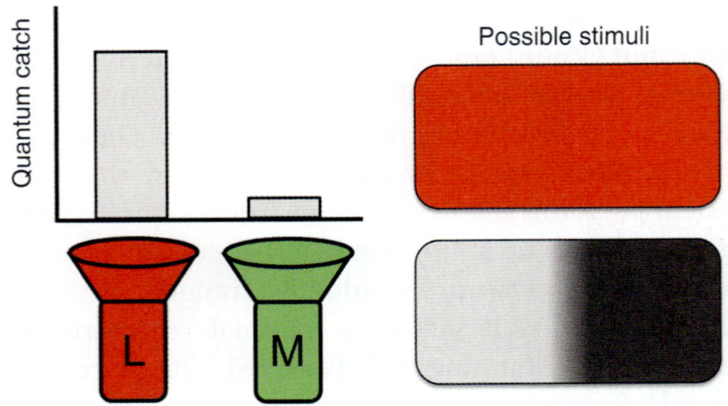

FIGURE 33.4   A difference in cone activity between neighboring L and M cones could indicate either a difference in light intensity between the two receptors or uniform illumination of the two receptors with long-wavelength light. Spatial and spectral information are necessarily confounded on small spatial scales because of the interleaved nature of the cone mosaic.

be consistent with a spatial variation in image intensity, or uniform illumination of both cones with long-wavelength light. This is an example of the general phenomenon referred to as aliasing, defined as two or more physically distinct images that result in identical patterns of cone activity (i.e., that result in identical sampled images). The visual system is then faced with the conundrum of which interpretation to choose.

Figure 33.4 illustrates an example of spatiochromatic aliasing, where stimuli that differ in both their chromatic and spatial content become indistinguishable once sampled by the retinal mosaic. Although spatiochromatic aliasing is a common source of artifacts in digital imaging, its effects are oddly absent from our everyday visual experience. Spatiochromatic aliasing can be induced under carefully controlled experimental conditions (Sekiguchi, Williams, & Brainard, 1993; Williams & Collier, 1983; Williams et al., 1991), but even then the effects are subtle and fleeting. Our immunity to the perceptual consequences of aliasing is typically explained by our eye's optics because they tend to blur the high spatial frequencies that are most often involved. However, this is true only in considering purely spatial aliases (i.e., one spatial pattern being confusable with a higher-frequency spatial pattern), and as pointed out by Williams (1990; Williams & Hofer, 2004), spatiochromatic aliases involve lower spatial frequencies that are not prevented from reaching the retina by the eye's optical blur. That we rarely, if ever, suffer sampling-related perceptual artifacts suggests that our visual system employs a highly sophisticated strategy when reconstructing images from the information encoded in the cone mosaic.

To accurately reconstruct the spatial and spectral information in a visual scene, the visual system requires a measure of the L-, M-, and S-cone spectral signals at each point. As discussed above, the interleaved nature of the mosaic makes this impossible. One solution is to make the cones very small and pack them very densely so that all spectral and spatial variation occurs on a scale much larger than a cone, but this solution is biologically unfeasible and energetically costly (due to the need for a much greater number of much smaller cones). If we accept the mosaic as it is, then there are numerous ways the missing samples could be estimated. For example the L, M, and S signals can be interpolated with some method independently. Williams et al. (1991) demonstrate that this strategy can qualitatively predict aspects of the appearance of spatiochromatic aliases but cannot account for their perceptual subtlety. A more sophisticated approach would be to incorporate some knowledge of the spatial and spectral characteristics of retinal images of visual scenes. For example, with respect to figure 33.4, it may be that the chromatic stimulus is more likely to be formed on the retina than the achromatic stimulus, in which case this information could be used to guide the choice of image interpretation.

Brainard, Williams, and Hofer (2008) recently formalized this approach as a Bayesian estimation problem in which the most likely visual stimulus given a particular pattern of photoreceptor activity can be computed by taking into account prior knowledge of the spatial and spectral statistics of natural scenes, the eye's optical blur, and the identities and locations of the cones in the retinal mosaic. This approach essentially specifies the method of combining photoreceptor responses that will maximize the average veridicality, or statistical likelihood, of the reconstructed visual percept given the types of images that are likely to be formed on the retina when viewing real-world visual scenes. Figure 33.5 shows low- (6 cycles/deg) and high-spatial-frequency (24 cycles/deg) achromatic gratings reconstructed with this method when sampled with two individuals' cone mosaics (Brainard, Williams, & Hofer, 2008). The achromatic reconstructions with both cone mosaics are qualitatively similar to visual experience: There are no traces of visual artifacts in the low-spatial-frequency reconstructions and only subtle chromatic artifacts in the high-spatial-frequency reconstructions. This subtlety is consistent with subject reports when viewing high-spatial-frequency achromatic gratings but quite unlike the strong colored artifacts that are predicted with methods that do not incorporate a priori knowledge of the cone mosaic and retinal image statistics (Williams et al., 1991).

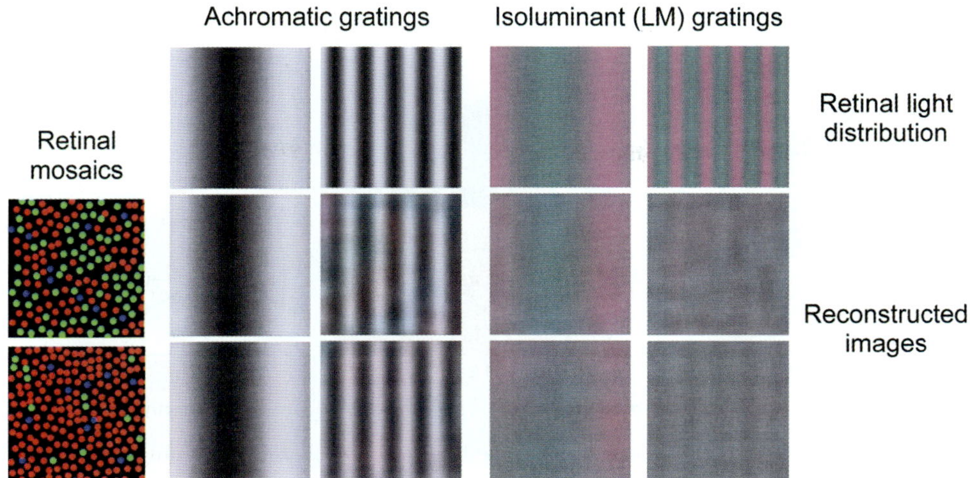

FIGURE 33.5   Low (6 cycles/deg) and high (24 cycles/deg) achromatic and isoluminant (LM) grating stimuli and their predicted reconstructions for two individuals with different cone mosaics using a Bayesian image reconstruction algorithm that incorporates knowledge of the cone mosaic, eye's optics, and spatial and spectral statistics of natural scenes (Brainard, Williams, & Hofer, 2008). Red, green, and blue denote L, M, and S cones, respectively.

Figure 33.5 also shows reconstructions for low- (6 cycles/deg) and high-spatial-frequency (24 cycles/deg) LM isoluminant gratings. Interestingly, the predicted appearance is more distorted for the observer with the most imbalanced cone ratio (the individual whose mosaic is shown in figure 33.1r), consistent with reports that individuals with extreme L- to M-cone ratios have compromised chromatic contrast sensitivity (Gunther & Dobkins, 2002). In addition, the reconstructed high-spatial-frequency gratings are of substantially lower contrast than the achromatic reconstructions. This is consistent with the more rapid falloff in chromatic contrast sensitivity than achromatic contrast sensitivity that has previously been described (Mullen, 1985). Importantly though, here this behavior arises as a fundamental consequence of the statistics of natural scenes combined with the information loss imposed by interleaved trichromatic sampling and not as an unfortunate consequence of subsequent retinal or neural processing.

Another example of the consequences of sampling-related information loss is that subjects routinely misjudge the color appearance of tiny flashes of monochromatic light (Cicerone & Nerger, 1989; Hartridge, 1946; Hofer, Singer, & Williams, 2005; Holmgren, 1886; Krauskopf, 1978; Koenig & Hofer, 2012). Hofer, Singer, and Williams (2005) investigated this phenomenon using adaptive optics to create stimuli the size of individual cones in five subjects with known cone mosaics. They found a large diversity in color reports, with reported color depending significantly on the ratio of L to M cones in each subject's mosaic (see figure 33.6). They were able to show that the diversity in response was great enough to preclude uniform roles for cones within the same class in color perception (discussed further below). Subjects also tended to categorize a large fraction of stimuli as white, with the most white responses made by those subjects with the most extreme cone ratios. Hofer, Singer, and Williams (2005) suggested that this may reflect a failure of cones situated in clumps of cones of like type to contribute to color, as they are evidently unable to contribute to a spectrally opponent signal. This intuition was largely supported by the strong concordance between subject's color responses and the predictions of the Bayesian image reconstruction model developed by Brainard, Williams, and Hofer (2008) (figure 33.6). The success of this model in accounting for subjects' behavior when viewing small spots of light produced with adaptive optics, as well as the low salience of aliasing artifacts when viewing high-spatial-frequency achromatic gratings, suggests that our visual system employs an image reconstruction strategy specifically tailored to our eye's own cone mosaic and optics in addition to the statistics of our visual environment.

Retinal Microcircuitry and the Trichromatic Mosaic

Up to this point we have considered the consequences of the organization of the trichromatic cone mosaic without regard to any constraints imposed by the neural architecture it serves. We have examined the optimality of the organization of the cone mosaic and its impact on vision from an information-coding perspective with

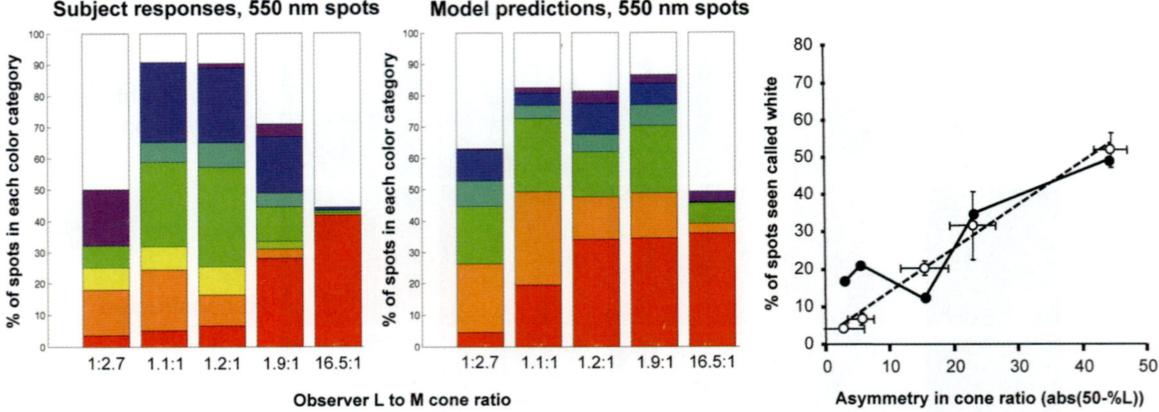

FIGURE 33.6   Color reports made by subjects when viewing threshold (50% probability of seeing) point stimuli produced with adaptive optics (left panel; Hofer, Singer, & Williams, 2005) depend on cone ratio and are broadly consistent with predictions made for those individuals with a Bayesian image reconstruction algorithm that incorporates knowledge of the cone mosaic, eye's optics, and spatial and spectral statistics of natural scenes (middle panel; Brainard, Williams, & Hofer, 2008). The Bayesian model also correctly predicts that subjects with the most extreme cone ratios will label the highest fraction of spots as "white" (right panel). Data (open circles) and predictions (solid circles) are for the fraction of threshold flashes called "white" averaged over three wavelengths (500, 550, and 600 nm). (Data from Hofer, Singer, & Williams, 2005; Brainard, Williams, & Hofer, 2008.)

the assumption that subsequent processing stages are ideal. However, in reality, visual processing must be carried out by a real physiological substrate, and it is not a priori certain that the transformations required for optimal image extraction are compatible with biological constraints. If not, the constraints imposed by the neural architecture would present additional visual consequences and might create a situation in which a theoretically optimal mosaic no longer represents the best organizational choice. Here we discuss aspects of retinal microcircuitry relevant to L and M cone organization; for general descriptions of the neural retina we refer the reader to reviews by Field & Chichilnisky (2007), Wässle (2004), and Dacey (2000).

The neural retina transforms the visual signal into roughly three streams, a luminance signal, and the L–M and LM–S spectrally opponent signals. This is an efficient recombination of cone inputs in that it removes redundancy due to the high correlation in L-, M- (and, to a lesser extent) S-cone responses when viewing a typical scene (Buchsbaum & Gottschalk, 1983). Although LM–S cone opponency is served by dedicated retinal circuitry, the midget ganglion cell is the only known candidate for encoding the retinal L–M opponent signal (Lennie, 2000; Lennie, Haake, & Williams, 1991; Paulus & Kröger-Paulus, 1983). In the central few degrees of the retina, midget ganglion cell centers are driven by individual cones (Boycott & Wassle, 1991; Dacey, 1999), although several neighboring cones drive the center of a peripheral midget ganglion cell due to increasing neural convergence (Dacey, 1999). Whether or not the midget cell is selectively or nonselectively

wired with respect to its L- and M-cone inputs (S-cone input appears functionally absent; Sun et al., 2006) has been contentious (Buzás et al., 2006; Diller et al., 2004; Field et al., 2010; Martin et al., 2001, 2011; Reid & Shapley, 1992; Solomon et al., 2005). In the fovea, due to the "private-line" arrangement and the midget cell's spatial opponency, nonselective wiring is sufficient for L–M opponent encoding. However, in peripheral retina, neural convergence should result in weakened chromatic opponency. Anatomical work suggests nonselective connections (Calkins & Sterling, 1996; Jusuf, Martin, & Grünert, 2006) but does not rule out functional bias (i.e., differential weighting of L- and M-cone inputs) that could result in a degree of cone-type specificity.

A limitation of many prior studies is that conclusions depended in part on the assumption of random L- and M-cone arrangement. However, L and M cones are often slightly clumped, which would confer stronger opponency with random wiring than would be expected if L and M cones were arranged randomly (see figure 33.7). Recent work by Field et al. (2010) has overcome this limitation by using a multiunit array technique to measure both cone weights to midget receptive fields as well as the identities and locations of the cones that feed them. They observed a degree of opponency in peripheral macaque midget cells that was slightly larger than could be explained by random wiring, even when they took into account the clumped nature of the L- and M-cone submosaic in the retina from which they recorded. Consequently, there is a growing consensus that although midget cell receptive fields mostly

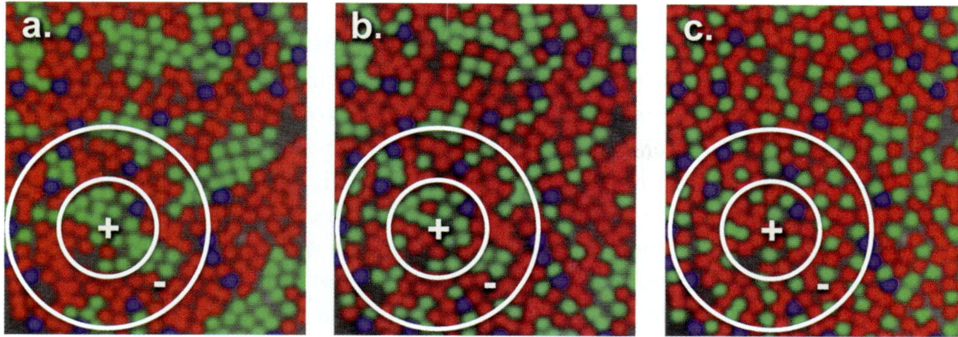

FIGURE 33.7 Because of neural convergence, significant L–M opponency in peripheral midget ganglion cells is more likely to be conferred when L and M cones are clumped (a) or randomly (b) arranged than if they were regularly interleaved (c). The concentric white circles illustrate the cones contributing to the center and surround of a theoretical peripheral midget ganglion cell for three mosaics constructed by taking one individual's actual cone locations (b) and manually permuting the L and M cone assignments to produce a clumped (a) or regularly interleaved (c) L- and M-cone submosaic. Red, green, and blue denote L, M, and S cones, respectively.

conform to the random-wiring model, a small amount of functional specificity likely exists (Buzás et al., 2006; Field et al., 2010; Martin et al., 2011). Adaptive optics may eventually allow similarly detailed analysis of the receptive fields of foveal midget cells (Sincich et al., 2009).

The midget cell also plays a critical role in fine spatial vision. Importantly, with respect to ideal cone arrangement, these two roles are somewhat incompatible. Although deviations from a regularly interleaved arrangement may be suboptimal in terms of information-encoding ability (Manning & Brainard, 2009; Osorio, Ruderman, & Cronin, 1998), a randomly, or slightly clumped, cone mosaic may present advantages given the existing structure of midget cell receptive fields (which incidentally appears ideally suited to extracting information from a random cone mosaic; Wachtler et al., 2007). In the fovea the random arrangement may strike a balance between the need to encode both spatial and spectral variations. For example, a midget ganglion cell drawing from cones of a single type will clearly be best at encoding spatial information but will be unable to encode spectral information. On the other hand, a midget cell whose central cone is of opposite type to those in the surround will encode a robust spectral signal but will be less effective at encoding spatial variations. A random arrangement may be an effective compromise between these two extremes. It has also been suggested that the patchiness resulting from a random (or slightly clumped) cone arrangement may help to ameliorate the spectral blurring that occurs as a result of electrical coupling between neighboring cones (DeVries et al., 2002; Hornstein, Verweij, & Schnapf, 2004; Hsu et al., 2000).

In the periphery where neural convergence is high, clumping of like-type cones may be advantageous because it would increase the number of midget cells whose receptive field centers would be driven by a majority of cones of the same class, as shown in figure 33.7. This could explain why many investigators have found remarkably good color discrimination in the periphery provided large enough stimuli are used (Abramov, Gordon, & Chan, 1991; Gordon & Abramov, 1977; Noorlander et al., 1983; van Esch et al., 1984), although peripheral color vision in general is significantly disadvantaged (Mullen & Kingdom, 1996; Noorlander et al., 1983). Testing these hypotheses—that a disordered L- and M-cone arrangement may be optimal given the constraints imposed by cone coupling and the structure of foveal and peripheral receptive fields—could be accomplished by mechanistically modeling these features and computing the information encoded in the ganglion cell array given typical retinal images under different cone arrangement strategies.

Extracting Color from Cone Signals

Extracting spectral information from the retinal image requires comparing the relative absorption rates of the L, M, and S cones. The interleaved nature of the cone mosaic makes this comparison difficult for stimuli with fine spatial or spectral variation, but for large stimuli it is straightforward to tabulate the relationship between the triad of cone absorptance values and various spectral compositions, which to us appear as different colors (at least for simple stimuli where the effects of context on perceived color can be ignored). This has resulted in basic equations that link cone mechanisms to different hues (Kaiser & Boynton, 1996; Stockman & Brainard, 2010; Walraven, 1962; Wuerger, Atkinson, & Cropper, 2005), with the unique hues—red, green, yellow, and blue—traditionally enjoying a privileged status given Hering's observations of their opponent role in color appearance (Hering, 1878/1964).

The relationships between relative cone excitations and perceived hues were largely specified before much of the knowledge of the human retinal mosaic and its variation among individuals was available. It became common to take these relationships a step further and associate the hue labels directly with the outputs of individual retinal opponent neurons or individual cones of each type. For example excitation and inhibition of midget ganglion cells have been associated with unique red and green, and L, M, and S cones have been associated with the unique red, green, and blue, respectively (with unique yellow thought to reflect equal L- and M-cone excitation in the absence of S-cone excitation). This linkage creates the expectation that an individual's color experience ought to depend on the relative numerosity of cones in the retinal mosaic (for example, people with many L cones seeing stimuli as more red) (Cicerone, 1987; Krauskopf, 2000). Early measurements of unique yellow in subjects believed to have diverse ratios of L to M cones suggested this may not be the case (Jordan & Mollon, 1997; Miyahara et al., 1998; Pokorny, Smith, & Wesner, 1991), with later direct tests confirming that unique yellow is essentially constant across different color-normals despite even larger variation in L- to M-cone ratio than previously suspected (Brainard et al., 2000; Hofer et al., 2005; Carroll et al., 2000). Although cone ratio may have a modest impact on some aspects of color vision (anomaloscope matching range: Jordan & Mollon, 1993; chromatic constrast sensitivity: Gunther & Dobkins, 2002), it evidently does not determine our perception of hue. Hue perception instead appears to be largely determined by the environment, with neural mechanisms adapting so as to signal only departures from the mean environmental chromaticity (Delahunt et al., 2004; Mollon, 1982; Neitz et al., 2002; Pokorny et al., 1991; Webster & Leonard, 2008).

The fact that the hues that maximally excite (or inhibit) a midget cell are not red and green (but rather orangish and bluish-green) and similarly for the small bistratified cell involved in LM–S opponency, also challenges the association of L, M, and S cones directly with the red, green, and blue unique hues. This has encouraged a search for methods of combining or modifying the postreceptoral retinal mechanism to create new mechanisms with spectral signatures reflecting the unique hues (DeValois & DeValois, 1993; Guth, 1991;

Neitz & Neitz, 2011) as well as a search for their potential neural basis (Stoughton & Conway, 2008). Despite these known limitations, the tendency to assume that individual cones within a class play a fixed role in color vision has been pervasive, and surprisingly, the association of L, M, and S with red, green, and blue still sometimes occurs.

The data of Hofer, Singer, and Williams (2005) indicate that not only is it incorrect to associate the output of individual L, M, and S cones with unique hues but that individual cones within the same class do not always contribute in the same way to color even in the absence of contextual factors or excitation in other classes of cones (i.e., as described by Knoblauch & Shevell, 2001). For example, it must be possible for excitation of a single cone to sometimes fail to contribute to a chromatic percept. Work by Brainard, Williams, and Hofer (2008) suggests that this behavior is a natural consequence of an optimal image reconstruction strategy where cone contributions are determined so as to provide the most accurate visual reconstructions given the cone mosaic and the nature of real world scenes. Although this model does not include physiological constraints or details of specific visual mechanisms, it still provides some neural insight, as its linearity results in a predicted association of a localized color percept with each individual retinal cone.

A key feature of these predictions is that ideal image reconstruction requires that the contributions of individual cones depend on the local structure of the cone mosaic. Figure 33.8 shows examples of the predicted impact of the local cone mosaic on perceptual responses to excitation of individual cones. For example, stimulating an L cone surrounded by other L cones is not expected to contribute a chromatic percept (the same is true for an M cone surrounded by M cones). Intuitively this arises because there is only information in one spectral band, so no chromatic comparison can be made, and the most likely interpretation is a very small spot of average chromaticity (although, in general, spots as small as single cones are very unlikely given the eye's optical blur, in this case it is the only reasonable image interpretation). On the other hand, stimulating a single L cone surrounded by M cones (or some mix of L and M cones) is expected to result in a reddish percept because in this case the lack of signal in the neighboring M cones is consistent with a (larger) red spot. The contributions of individual M cones are predicted to be similarly sensitive to the local retinal mosaic, with one of the more salient features that they are expected to contribute to blue sensations when there is no S cone in the immediate neighborhood and to greenish sensations when there is. This is consistent

    HEIDI J. HOFER AND DAVID R. WILLIAMS

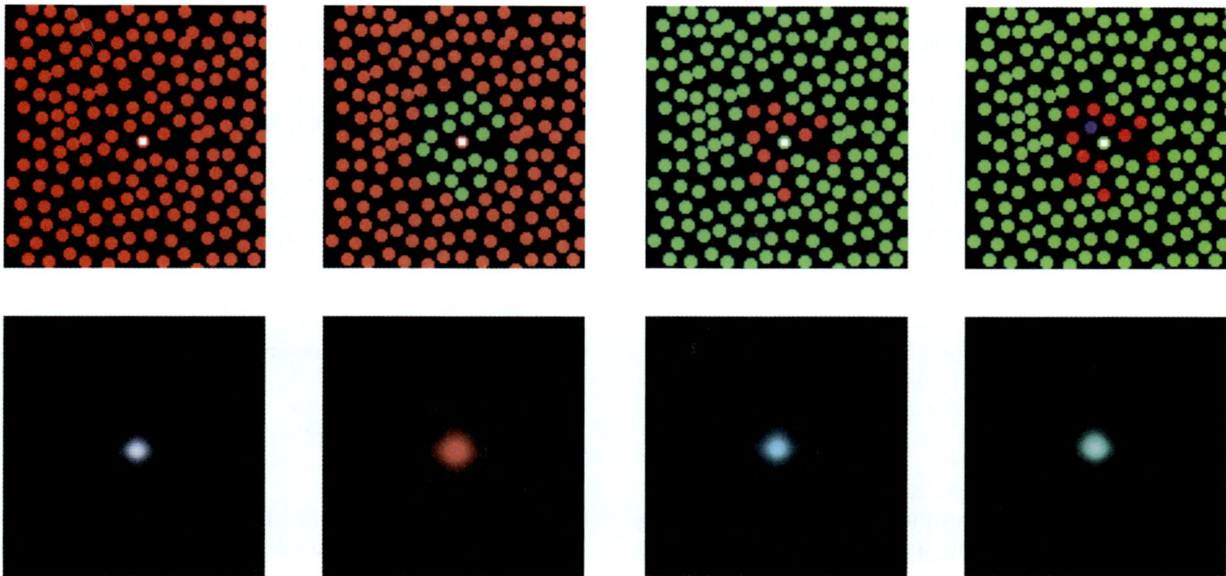

FIGURE 33.8   Percepts predicted with a Bayesian image reconstruction algorithm that incorporates knowledge of the cone mosaic, eye's optics, and spatial and spectral statistics of natural scenes (bottom row) when exciting individual L or M cones situated in different local cone mosaics (top row). Red, green, and blue denote L, M, and S cones, respectively (Brainard, Williams, & Hofer, 2008).

with prior suggestions of M-cone contributions to blueness (DeValois & DeValois, 1993; Drum, 1989; Hofer, Singer, & Williams, 2005b; Schirillo & Reeves, 2001) and makes intuitive sense because the highest ratio of M- to L-cone excitation occurs with bluish lights. In the future these predictions may be directly testable by delivery of tiny adaptive optics stimuli to the retinas of characterized observers and with concomitant retinal imaging (Arathorn et al., 2007; Putnam et al., 2005).

It has long been appreciated that, despite biological constraints, many features of our sense organs and neural system are exquisitely adapted to our environment to allow nearly optimal signal transduction and processing. Our cone mosaic and visual processing strategies appear to be no different. Where not optimal, our limitations appear complementary so as to minimize the overall impact. For example, it appears that the eye's optical blur decreases sensitivity to variability in L- to M-cone ratio. Similarly, it seems possible that the irregularity in the arrangement of L and M cones in the retinal mosaic is uniquely suited to the structure of retinal microcircuitry and vice versa. However tolerance to this diversity comes at a cost because it requires flexible and adaptive visual mechanisms to effectively extract the information encoded in a set of diverse cone mosaics. The similarity of color experience across individuals with very different cone mosaics and our insusceptibility to mosaic-related visual artifacts suggest that we rely on nearly optimal image reconstruction strategies that take into account the structure of our own cone mosaics in addition to the statistics of our visual environment.

REFERENCES

Abramov, I., Gordon, J., & Chan, H. (1991). Color appearance in the peripheral retina: Effects of stimulus size. *Journal of the Optical Society of America. A, Optics and Image Science, 8,* 404–414.

Ahnelt, P. K., & Kolb, H. (2000). The mammalian photoreceptor mosaic-adaptive design. *Progress in Retinal and Eye Research, 19,* 711–777.

Ahnelt, P. K., Kolb, H., & Pflug, R. (1987). Identification of a subtype of cone photoreceptor, likely to be blue sensitive, in the human retina. *Journal of Comparative Neurology, 255,* 18–34.

Arathorn, D. W., Yang, Q., Vogel, C. R., Zhang, Y., Tiruveedhula, P., & Roorda, A. (2007). Retinally stabilized cone-targeted stimulus delivery. *Optics Express, 15,* 13731–13744.

Autrusseau, F., Thibos, L., & Shevell, S. K. (2011). Chromatic and wavefront aberrations: L-, M- and S-cone stimulation with typical and extreme retinal image quality. *Vision Research, 51,* 2282–2294.

Bowmaker, J. K., & Kunz, Y. W. (1987). Ultraviolet receptors, tetrachromatic colour vision and retinal mosaics in the brown trout (*Salmo trutta*): Age-dependent changes. *Vision Research, 27,* 2101–2108.

Bowmaker, J. K., Parry, J. W. L., & Mollon, J. D. (2003). The arrangement of L and M cones in human and a primate retina. In J. D. Mollon, J. Pokorny, & K. Knoblauch (Eds.), *Normal and defective colour vision* (pp. 39–50). New York: Oxford University Press.

Boycott, B. B., & Wassle, H. (1991). Morphological classification of bipolar cells of the primate retina. *European Journal of Neuroscience, 3,* 1069–1088.

Brainard, D. H., Roorda, A., Yamauchi, Y., Calderone, J. B., Metha, A., Neitz, M., et al. (2000). Functional consequences of the relative numbers of L and M cones. *Journal of the Optical Society of America. A, Optics, Image Science, and Vision, 17,* 607–614. doi:10.1364/JOSAA.17.000607.

Brainard, D. H., Williams, D. R., & Hofer, H. (2008). Trichromatic reconstruction from the interleaved cone mosaic: Bayesian model and the color appearance of small spots. *Journal of Vision, 8,* 15, 1–23. doi:10.1167/8.5.15.

Buchsbaum, G., & Gottschalk, A. (1983). Trichromacy, opponent colours coding and optimum colour information transmission in the retina. *Proceedings of the Royal Society of London. Series B, Biological Sciences, 220,* 89–113. doi:10.1098/rspb.1983.0090.

Bumsted, K., Jasoni, C., Szel, A., & Hendrickson, A. (1997). Spatial and temporal expression of cone opsins during monkey retinal development. *Journal of Comparative Neurology, 378,* 117–134.

Bumsted, K., & Hendrickson, A. (1999). Distribution and development of short-wavelength cones differ between *Macaca* monkey and human fovea. *Journal of Comparative Neurology, 403,* 502–516.

Buzás, P., Blessing, E. M., Szmajda, B. A., & Martin, P. R. (2006). Specificity of M and L cone inputs to receptive fields in the parvocellular pathway: Random wiring with functional bias. *Journal of Neuroscience, 26,* 11148–11161.

Calkins, D. J., & Sterling, P. (1996). Absence of spectrally specific lateral inputs to midget ganglion cells in primate retina. *Nature, 381,* 613–615.

Carroll, J., McMahon, C., Neitz, M., & Neitz, J. (2000). Flicker-photometric electroretinogram estimates of L:M cone photoreceptor ratio in men with photopigment spectra derived from genetics. *Journal of the Optical Society of America. A, Optics, Image Science, and Vision, 17,* 499–509.

Carroll, J., Neitz, J., & Neitz, M. (2002). Estimates of L:M cone ratio from ERG flicker photometry and genetics. *Journal of Vision, 2,* 531–542. doi:10.1167/2.8.1.

Cicerone, C. M. (1987). Constraints placed on color vision models by the relative numbers of different cone classes in human fovea centralis. *Die Farbe, 34,* 59–66.

Cicerone, C. M., & Nerger, J. L. (1989). The relative numbers of long-wavelength-sensitive to middle-wavelength-sensitive cones in the human fovea centralis. *Vision Research, 29,* 115–128.

Curcio, C. A., Allen, K. A., Sloan, K. R., Lerea, C. L., Hurley, J. B., Klock, I. B., et al. (1991). Distribution and morphology of human cone photoreceptors stained with anti-blue opsin. *Journal of Comparative Neurology, 312,* 610–624.

Dacey, D. M. (1999). Primate retina: Cell types, circuits and color opponency. *Progress in Retinal and Eye Research, 18,* 737–763.

Dacey, D. M. (2000). Parallel pathways for spectral coding in primate retina. *Annual Review of Neuroscience, 23,* 743–775.

Dartnall, H. J., Bowmaker, J. K., & Mollon, J. D. (1983). Human visual pigments: Microspectrophotometric results from the eyes of seven persons. *Proceedings of the Royal Society of London. Series B, Biological Sciences, 220,* 115–130. doi:10.1098/rspb.1983.0091.

Deeb, S. S., Diller, L. C., Williams, D. R., & Dacey, D. M. (2000). Interindividual and topographical variation of L:M cone ratios in monkey retinas. *Journal of the Optical Society of America. A, Optics, Image Science, and Vision, 17,* 538–544.

Delahunt, P., Webster, M. A., Ma, L., & Werner, J. S. (2004). Color appearance changes after cataract surgery reveal a long-term chromatic adaptation mechanism. *Visual Neuroscience, 21,* 301–307.

de Monasterio, F. M., McCrane, E. P., Newlander, J. K., & Schein, S. J. (1985). Density profile of blue-sensitive cones along the horizontal meridian of macaque retina. *Investigative Ophthalmology & Visual Science, 26,* 289–302.

De Valois, R. L., & De Valois, K. K. (1993). A multi-stage color model. *Vision Research, 33,* 1053–1065.

DeVries, H. (1946). Luminosity curve of trichromats. *Nature, 157,* 736–737.

DeVries, S. H., Qi, X., Smith, R., Makous, W., & Sterling, P. (2002). Electrical coupling between mammalian cones. *Current Biology, 12,* 1900–1907. doi:10.1016/S0960-9822(02)01261-7.

Diaz-Araya, C., & Provis, J. M. (1992). Evidence of photoreceptor migration during early foveal development: A quantitative analysis of human fetal retinae. *Visual Neuroscience, 8,* 505–514.

Diller, L., Packer, O. S., Verweij, J., McMahon, M. J., Williams, D. R., & Dacey, D. M. (2004). L and M cone contributions to the midget and parasol ganglion cell receptive fields of macaque monkey retina. *Journal of Neuroscience, 24,* 1079–1088.

Drum, B. (1989). Hue signals from short- and middle-wavelength-sensitive cones. *Journal of the Optical Society of America. A, Optics and Image Science, 6,* 153–157.

Fernald, R. D. (1981). Chromatic organization of a cichlid fish retina. *Vision Research, 21,* 1749–1753.

Field, G. D., & Chichilnisky, E. J. (2007). Information processing in the primate retina: Circuitry and coding. *Annual Review of Neuroscience, 30,* 1–30. doi:10.1146/annurev.neuro.30.051606.094252.

Field, G. D., Gauthier, J. L., Sher, A., Greschner, M., Machado, T. A., Jepson, L. H., et al. (2010). Functional connectivity in the retina at the resolution of photoreceptors. *Nature, 467,* 673–677. doi:10.1038/nature09424.

Garrigan, P., Ratliff, C. P., Klein, J. M., Sterling, P., Brainard, D. H., & Balasubramanian, V. (2010). Design of a trichromatic cone array. *PLoS Computational Biology, 6,*e1000677. doi:10.1371/journal.pcbi.1000677.

Gordon, J., & Abramov, I. (1977). Color vision in the peripheral retina II: Hue and saturation. *Journal of the Optical Society of America, 67,* 202–207.

Gunther, K. L., & Dobkins, K. R. (2002). Individual differences in chromatic (red/green) contrast sensitivity are constrained by the relative number of L- versus M-cones in the eye. *Vision Research, 42,* 1367–1378.

Gunther, K. L., Neitz, J., & Neitz, M. (2008). Nucleotide polymorphisms upstream of the X-chromosome opsin gene array tune L:M cone ratio. *Visual Neuroscience, 25,* 265–271.

Guth, S. L. (1991). Model for color vision and light adaptation. *Journal of the Optical Society of America. A, Optics and Image Science, 8,* 976–993.

Hagstrom, S. A., Neitz, J., & Neitz, M. (1998). Variations in cone populations for red-green color vision examined by analysis of mRNA. *Neuroreport, 9,* 1963–1967. doi:10.1097/00001756-199806220-00009.

Hagstrom, S. A., Neitz, M., & Neitz, J. (2000). Cone pigment gene expression in individual photoreceptors and the chromatic topography of the retina. *Journal of the Optical Society of America. A, Optics, Image Science, and Vision, 17,* 527–537.

Hartridge, H. (1946). Colour receptors of the human fovea. *Nature, 158,* 97–98.

Hendrickson, A. E., & Yuodelis, C. (1984). The morphological development of the human fovea. *Ophthalmology, 91,* 603–612.

Hering, E. (1964). *Outlines of a theory of the light sense* (L. M. Hurvich & D. Jameson, D., Trans.). Cambridge, MA: Harvard University Press. (Original work published 1878.)

Hofer, H., Carroll, J., Neitz, J., Neitz, M., & Williams, D. R. (2005). Organization of the human trichromatic cone mosaic. *Journal of Neuroscience, 25,* 9669–9679.

Hofer, H., Singer, B., & Williams, D. R. (2005). Different sensations from cones with the same photopigment. *Journal of Vision, 5,* 444–454. doi:10.1167/5.5.5.

Holmgren, F. (1886). Hr. A. Konig verlas vor Eintritt in die Tagesordnung folgende ihm von Hrn. Frithiof Holmgren (in Upsala) unter Beziwhung auf der Sitzungsbericht des internationalen medicinischen Congresses (Kopenhagen, August 1884) eingesandte Mittheilung. *Verhandlungen der Physiologischen Gesellschaft zu Berlin 11 Jahrgang,* 4–6.

Hornstein, E. P., Verweij, J., & Schnapf, J. L. (2004). Electrical coupling between red and green cones in primate retina. *Nature Neuroscience, 7,* 745–750.

Hsu, A., Smith, R. G., Buchsbaum, G., & Sterling, P. (2000). Cost of cone coupling to trichromacy in primate fovea. *Journal of the Optical Society of America. A, Optics, Image Science, and Vision, 17,* 635–640.

Jordan, G., & Mollon, J. D. (1993). A study of women heterozygous for colour deficiencies. *Vision Research, 33,* 1495–1508.

Jordan, G., & Mollon, J. D. (1997). Unique hues in heterozygotes for protan and deutan deficiencies. In C. R. Cavonius (Ed.) *13th Symposium of the International Research Group on Colour Vision Deficiencies* (pp. 67–76). Dordrecht: Kluwer Academic Publishers.

Jusuf, P. R., Martin, P. R., & Grünert, U. (2006). Random wiring in the midget pathway of primate retina. *Journal of Neuroscience, 26,* 3908–3917.

Kaiser, P. K., & Boynton, R. M. (1996). *Human color vision.* Washington, DC: Optical Society of America.

Knoblauch, K., Neitz, M., & Neitz, J. (2006). An urn model of the development of L/M cone ratios in human and macaque retinas. *Visual Neuroscience, 23,* 387–394.

Knoblauch, K., & Shevell, S. K. (2001). Relating cone signals to color appearance: Failure of monotonicity in yellow/blue. *Visual Neuroscience, 18,* 901–906.

Koenig, D. E., & Hofer, H. J. (2012). Do color appearance judgments interfere with detection of small threshold stimuli? *Journal of the Optical Society of America. A, Optics, Image Science, and Vision, 29,* A258–A267. doi:10.1364/JOSAA.29.00A258.

Kram, Y. A., Mantey, S., & Corbo, J. C. (2010). Avian cone photoreceptors tile the retina as five independent, self-organizing mosaics. *PLoS One, 5,* e8992. doi:10.1371/journal.pone.0008992.

Krauskopf, J. (1978). On identifying detectors. In J. C. Armington, J. Krauskopf, & B. R. Wooten (Eds.), *Visual psychophysics and physiology* (pp. 283–298). New York: Academic Press.

Krauskopf, J. (2000). Relative number of long- and middle-wavelength-sensitive cones in the human fovea. *Journal of the Optical Society of America. A, Optics, Image Science, and Vision, 17,* 510–516.

Kremers, J., Scholl, H. P. N., Knau, H., Berendschot, T. T. J. M., Usui, T., & Sharpe, L. T. (2000). L/M cone ratios in human trichromats assessed by psychophysics, electroretinography, and retinal densitometry. *Journal of the Optical Society of America. A, Optics, Image Science, and Vision, 17,* 517.

Kuchenbecker, J., Sahay, M., Tait, D. M., Neitz, M., & Neitz, J. (2008). Topography of the long- to middle-wavelength sensitive cone ratio in the human retina assessed with a wide-field color multifocal electroretinogram. *Visual Neuroscience, 25,* 301–306.

Lee, B. B., Martin, P. R., & Grünert, U. (2010). Retinal connectivity and primate vision. *Progress in Retinal and Eye Research, 29,* 622–639.

Lennie, P. (2000). Color vision: Putting it together. *Current Biology, 10,* R589–R591. doi:10.1016/S0960-9822(00)00632-1.

Lennie, P., Haake, W., & Williams, D. R. (1991). The design of chromatically opponent receptive fields. In M. Landy & A. Movshon (Eds.), *Computational models of visual processing* (pp. 71–82). Cambridge, MA: MIT Press.

Manning, J. R., & Brainard, D. H. (2009). Optimal design of photoreceptor mosaics: Why we do not see color at night. *Visual Neuroscience, 26,* 5–19. doi:10.1017/S095252380808084X.

Martin, P. R., Blessing, E. M., Buzás, P., Szmajda, B. A., & Forte, J. D. (2011). Transmission of colour and acuity signals by parvocellular cells in marmoset monkeys. *Journal of Physiology, 589*(Pt 11), 2795–2812.

Martin, P. R., Lee, B. B., White, A. J., Solomon, S. G., & Rüttiger, L. (2001). Chromatic sensitivity of ganglion cells in the peripheral primate retina. *Nature, 410,* 933–936.

McLellan, J. S., Marcos, S., Prieto, P. M., & Burns, S. A. (2002). Imperfect optics may be the eye's defence against chromatic blur. *Nature, 417,* 174–176.

McMahon, C., Neitz, J., & Neitz, M. (2004). Evaluating the human X-chromosome pigment gene promoter sequences as predictors of L:M cone ratio variation. *Journal of Vision, 4,* 203–208. doi:10.1167/4.3.7.

Miyahara, E., Pokorny, J., Smith, V. C., Baron, R., & Baron, E. (1998). Color vision in two observers with highly biased LWS/MWS cone ratios. *Vision Research, 38,* 601–612.

Mollon, J. D. (1982). Color vision. *Annual Review of Psychology, 33,* 41–85. doi:10.1146/annurev.ps.33.020182.000353.

Mollon, J. D. (1989). "Tho' she kneel'd in that place where they grew..." The uses and origins of primate colour vision. *Journal of Experimental Biology, 146,* 21–38.

Mollon, J. D., & Bowmaker, J. K. (1992). The spatial arrangement of cones in the primate fovea. *Nature, 360,* 677–679.

Mollon, J. D., Bowmaker, J. K., & Jacobs, G. H. (1984). Variations of colour vision in a New World primate can be explained by polymorphism of retinal photopigments. *Proceedings of the Royal Society of London. Series B, Biological Sciences, 222,* 373–399. doi:10.1098/rspb.1984.0071.

Mullen, K. T. (1985). The contrast sensitivity of human colour vision to red-green and blue-yellow chromatic gratings. *Journal of Physiology, 359,* 381–400.

Mullen, K. T., & Kingdom, F. A. (1996). Losses in peripheral colour sensitivity predicted from "hit and miss" post-receptoral cone connections. *Vision Research, 36,* 1995–2000.

Nathans, J. (1999). The evolution and physiology of human review color vision: Insights from molecular genetic studies of visual pigments. *Neuron, 24*, 299–312.

Nathans, J., Thomas, D., & Hogness, D. (1986). Molecular genetics of human color vision: The genes encoding blue, green, and red pigments. *Science, 232*, 193–202.

Neitz, J., & Neitz, M. (2011). The genetics of normal and defective color vision. *Vision Research, 51*, 633–651.

Neitz, J., Carroll, J., Yamauchi, Y., Neitz, M., & Williams, D. R. (2002). Color perception is mediated by a plastic neural mechanism that is adjustable in adults. *Neuron, 35*, 783–792.

Neitz, M., Balding, S. D., McMahon, C., Sjoberg, S. A., & Neitz, J. (2006). Topography of long- and middle-wavelength sensitive cone opsin gene expression in human and Old World monkey retina. *Visual Neuroscience, 23*, 379–385.

Noorlander, C., Koenderink, J. J., Den Ouden, R. J., & Edens, B. W. (1983). Sensitivity to spatiotemporal colour contrast in the peripheral visual field. *Vision Research, 23*, 1–11. doi:10.1016/0042-6989(83)90035-4.

Osorio, D., Ruderman, D. L., & Cronin, T. W. (1998). Estimation of errors in luminance signals encoded by primate retina resulting from sampling of natural images with red and green cones. *Journal of the Optical Society of America. A, Optics, Image Science, and Vision, 15*, 16–22. doi:10.1364/JOSAA.15.000016.

Osorio, D., & Vorobyev, M. (2008). A review of the evolution of animal colour vision and visual communication signals. *Vision Research, 48*, 2042–2051.

Otake, S., & Cicerone, C. M. (2000). L and M cone relative numerosity and red–green opponency from fovea to mid-periphery in the human retina. *Journal of the Optical Society of America. A, Optics, Image Science, and Vision, 17*, 615–627.

Packer, O. S., Williams, D. R., & Bensinger, D. G. (1996). Photopigment transmittance imaging of the primate photoreceptor mosaic. *Journal of Neuroscience, 16*, 2251–2260.

Paulus, W., & Kröger-Paulus, A. (1983). A new concept of retinal colour coding. *Vision Research, 23*, 529–540.

Pokorny, J., Smith, V. C., & Wesner, M. F. (1991). Variability in cone populations and implications. In A. Valberg & B. B. Lee (Eds.), *From pigments to perception* (pp. 23–34). New York: Plenum Press.

Putnam, N. M., Hofer, H. J., Doble, N., Chen, L., Carroll, J., & Williams, D. R. (2005). The locus of fixation and the foveal cone mosaic. *Journal of Vision, 5*, 632–639. doi:10.1167/5.7.3.

Reid, R. C., & Shapley, R. M. (1992). Spatial structure of cone inputs to receptive fields in primate lateral geniculate nucleus. *Nature, 356*, 716–718.

Roorda, A., Metha, A. B., Lennie, P., & Williams, D. R. (2001). Packing arrangement of the three cone classes in primate retina. *Vision Research, 41*, 1291–1306.

Roorda, A., & Williams, D. R. (1999). The arrangement of the three cone classes in the living human eye. *Nature, 397*, 520–522.

Rushton, W. A. H., & Baker, H. D. (1964). Red/green sensitivity in normal vision. *Vision Research, 4*, 75–85.

Schirillo, J. A., & Reeves, A. (2001). Color-naming of M-cone incremental flashes. *Color Research and Application, 26*, 132–140.

Scholes, J. H. (1975). Colour receptors, and their synaptic connexions, in the retina of a cyprinid fish. *Philosophical Transactions of the Royal Society of London. Series B, Biological Sciences, 270*, 61–118. doi:10.1098/rstb.1975.0004.

Sekiguchi, N., Williams, D. R., & Brainard, D. H. (1993). Efficiency in detection of isoluminant and isochromatic interference fringes. *Journal of the Optical Society of America. A, Optics and Image Science, 10*, 2118–2133. doi:10.1364/JOSAA.10.002118.

Shapiro, M. B., Schein, S. J., & de Monasterio, F. M. (1985). Regularity and structure of the spatial pattern of blue cones of the macaque retina. *Journal of the American Statistical Association, 80*, 803–812. doi:10.2307/2288535.

Smallwood, P. M., Olveczky, B. P., Williams, G. L., Jacobs, G. H., Reese, B. E., Meitster, M., & Nathans, J. (2003). Genetically engineered mice with an additional class of cone photoreceptors: implications for the evolution of color vision. *Proceedings of the National Academy of Sciences of the United States of America, 100*, 11706–11711.

Sincich, L. C., Zhang, Y., Tiruveedhula, P., Horton, J. C., & Roorda, A. (2009). Resolving single cone inputs to visual receptive fields. *Nature Neuroscience, 12*, 967–969.

Solomon, S. G., Lee, B. B., White, A. J. R., Rüttiger, L., & Martin, P. R. (2005). Chromatic organization of ganglion cell receptive fields in the peripheral retina. *Journal of Neuroscience, 25*, 4527–4539.

Stockman, A., & Brainard, D. H. (2010). Color vision mechanisms. In M. Bass (Ed.), *The OSA handbook of optics* (3rd ed., 11.1–11.104.). New York: McGraw-Hill.

Stoughton, C. M., & Conway, B. R. (2008). Neural basis for unique hues. *Current Biology, 18*, R698–R699. doi:10.1016/j.cub.2008.06.018.

Sun, H., Smithson, H. E., Zaidi, Q., & Lee, B. B. (2006). Do magnocellular and parvocellular ganglion cells avoid short-wavelength cone input? *Visual Neuroscience, 23*, 441–446.

van Esch, J. A., Koldenhof, E. E., van Doom, A. J., & Koenderink, J. J. (1984). Spectral sensitivity and wavelength discrimination of the human peripheral visual field. *Journal of the Optical Society of America. A, Optics and Image Science, 1*, 443–450.

Wachtler, T., Doi, E., Lee, T.-W., & Sejnowski, T. J. (2007). Cone selectivity derived from the responses of the retinal cone mosaic to natural scenes. *Journal of Vision, 7*, 6, 1–14. doi:10.1167/7.8.6.

Walraven, P. L. (1962). *On the mechanisms of colour vision*. Soesterberg: University of Utrecht.

Wässle, H. (2004). Parallel processing in the mammalian retina. *Nature Reviews. Neuroscience, 5*, 747–757.

Webster, M. A., & Leonard, D. L. (2008). Adaptation and perceptual norms in color vision. *Journal of the Optical Society of America. A, Optics, Image Science, and Vision, 25*, 2817–2825.

Williams, D. R. (1990). The invisible cone mosaic. In *Advances in Photoreception: Proceedings of a Symposium on Frontiers of Visual Science* (pp. 135–148). Washington, DC: National Academy Press.

Williams, D. R., & Collier, R. (1983). Consequences of spatial sampling by a human photoreceptor mosaic. *Science, 221*, 385–387.

Williams, D. R., & Hofer, H. (2004). Formation and acquisition of the retinal image. In L. M. Chalupa & J. S. Werner (Eds.), *The visual neurosciences* (pp. 795–810). Cambridge, MA: MIT Press.

Williams, D. R., MacLeod, D. I., & Hayhoe, M. M. (1981). Foveal tritanopia. *Vision Research, 21*(9), 1341–1356.

Williams, D., Sekiguchi, N., & Brainard, D. (1993). Color, contrast sensitivity, and the cone mosaic. *Proceedings of the*

National Academy of Sciences of the United States of America, 90, 9770–9777. doi:10.1073/pnas.90.21.9770.

Williams, D. R., Sekiguchi, N., Haake, W., Brainard, D., & Packer, O. (1991). The cost of trichromacy for spatial vision. In A. Valberg & B. B. Lee (Eds.), *From pigments to perception* (pp. 11–22). New York: Plenum Press.

Wuerger, S. M., Atkinson, P., & Cropper, S. (2005). The cone inputs to the unique-hue mechanisms. *Vision Research, 45,* 3210–3223.

Xiao, M., & Hendrickson, A. (2000). Spatial and temporal expression of short, long/medium, or both opsins in human fetal cones. *Journal of Comparative Neurology, 425,* 545–559.

Yamaguchi, T., Motulsky, A. G., & Deeb, S. S. (1997). Visual pigment gene structure and expression in human retinae. *Human Molecular Genetics, 6,* 981–990.

Yellott, J. I. (1982). Spectral analysis of spatial sampling by photoreceptors: Topological disorder prevents aliasing. *Vision Research, 22,* 1205–1210.

Yellott, J., Wandell, B., & Cornsweet, T. (1984). The beginnings of visual perception: The retinal image and its initial encoding. In I. Darian-Smith (Ed.), *Handbook of physiology, Section 1: The nervous system III, Part 2* (pp. 257–316). Bethesda, MD: American Physiological Society.

Yuodelis, C., & Hendrickson, A. (1986). A qualitative and quantitative analysis of the human fovea during development. *Vision Research, 26,* 847–855.

34 The Interaction of Rod and Cone Signals: Pathways and Psychophysics

STEVEN L. BUCK

The activity of rod photoreceptors has ubiquitous influence on cone-photoreceptor-mediated visual performance and perception and on the multiple parallel neural pathways that mediate the full range of visual function. As detailed by Buck (2004) and updated in the following pages, rod activity interacts with cone activity in determining the spatial, temporal, and spectral sensitivities of human vision as well as motion and color perception and chromatic discrimination. The goal of this chapter is to review some of the important advances over the past decade in our understanding of the effects of rod–cone interactions and of the neural pathways shared by rod and cone signals in human and primate visual systems. My sincere apologies are extended to those whose work is not cited here because of the limited space for this chapter.

NEURAL PATHWAYS OF ROD–CONE INTERACTION

In what postreceptoral pathways do rod and cone signals combine? What neural mechanisms support those combinations, and under what conditions do they operate? These are all long-standing questions that both psychophysicists and physiologists have attempted to answer. This section highlights some of the key advances made over the past decade.

Rod–Cone Coupling

Signals from rod photoreceptors have at least two routes to influence both ON and OFF retinal pathways that also carry cone signals. In addition to the "classical" pathway for rod signals via rod bipolar cells and AII amacrine cells, which synapse onto ON and OFF cone bipolar cells (see figure 34.1), it is now understood that rod photoreceptors make electrical gap-junction synapses with outgrowths (telodendrites) of cone photoreceptors (shown schematically in figure 34.1). Gap junctions are low-resistance intercellular channels that span the plasma membranes of adjacent cells. Hornstein et al., (2005) showed that L and M cones are

coupled with multiple rods but found no evidence of coupling between S cones and rods.

Recent work has revealed a wide range of fundamental roles rod–cone coupling may play in vision and retinal neural function. Coupling extends the range of scotopic vision by alleviating saturation at the synapses of rods and rod bipolar cells (Hornstein et al., 2005), keeps signals from more numerous rods from saturating the retinal network and impairing postreceptoral cone signaling during mesopic vision (Seeliger et al., 2011), increases the sensitivity and receptive field size of cone pathways, slows the temporal response of cones, and broadens cone spectral sensitivity (Ribelayga, Cao, & Mangel, 2008), and regulates light adaptation (Cameron & Lucas, 2009).

Another consequence of rod–cone coupling is that H1 horizontal cells will receive input from rods and L and M cones and will exert this mixed influence on bipolar-cell surrounds, potentially affecting both magnocellular and parvocellular pathways (Verweij et al., 1999). Trümpler et al. (2008) note that coupling could allow horizontal cell axon terminals to integrate signals over the entire visual intensity range and over large areas of the retina to modulate retinal processing and visual function.

A key finding is that the extent and strength of rod–cone coupling are regulated by the retinal 24-h circadian clock, so that coupling is much enhanced at night and weak during the day, at least in goldfish and mice (Ribelayga, Cao, & Mangel, 2008). This allows rod input to reach cone pathways at night when light levels are below threshold for cone phototransduction. During the day rod input to cones is weak for these species, helping to reduce rod interference with the spatial resolution acuity of mesopic vision (Schneeweis & Schnapf, 1995).

Ribelayga, Cao, and Mangel (2008) suggest that the increased nighttime rod–cone coupling may (1) improve detection of large dim stimuli, (2) support daily synchronization of both neural and metabolic activity, and (3) permit exchange of intracellular signaling molecules, nutrients, and small metabolites between

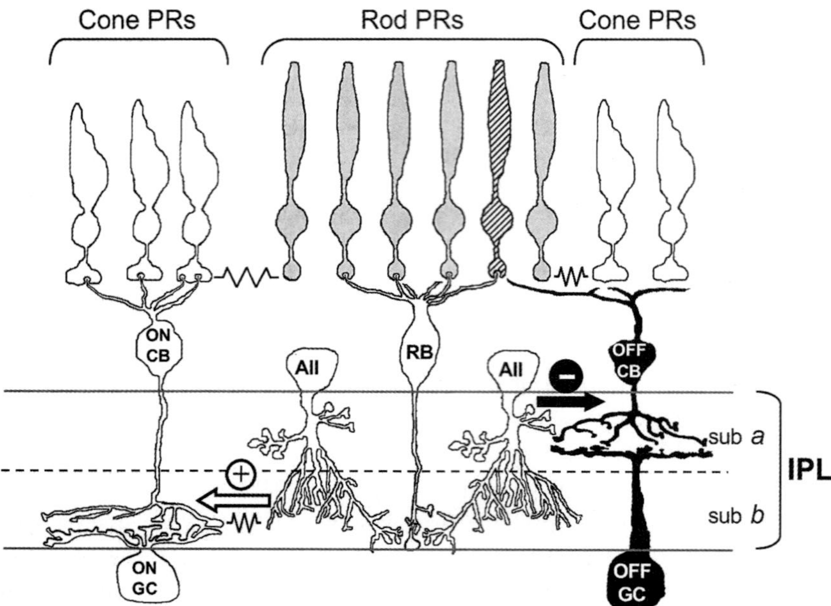

FIGURE 34.1   Pathways of rod and cone signals in early retina include the classical pathway via AII amacrine cells, rod–cone coupling, and direct contact of rods with OFF-cone bipolar cells. (Reprinted from Protti et al., 2005, by permission of the American Physiological Society.)

coupled rods and cones, possibly helping to maintain the health of or survival of cones. Similarly, Camacho et al. (2010) suggest that a coupled network may be essential for controlling the rhythmic shedding and renewal cycle of both rod and cone photoreceptors.

More work is needed to better understand the importance of rod-cone coupling for visual processing, possible differences among mammalian species, and the role of coupling in maintaining the health and function of photoreceptors.

Other Retinal Pathways of Rod–Cone Interaction

Over the past 15 years, investigators have identified more retinal pathways sharing rod and cone signals and more functions for previously known pathways such as those involving AII amacrine cells.

A third pathway by which rod and cone signals can merge was shown in mouse (Soucy et al., 1998). A subset of rod photoreceptors is contacted directly by dendrites of one or more types of OFF cone bipolar cells (see figure 34.1). The significance of this pathway apparently varies among species. Protti et al. (2005) showed that it functioned to provide strong signals in rabbit but, in their hands, was weak in mouse and absent in rat. To date, no anatomical or physiological evidence for this pathway has not been found in primates.

Because the AII has bidirectional gap-junction coupling with ON cone bipolar cells, signals from cones and the rods coupled with them can enter the AII via these ON cone bipolars (Stone, Buck, & Dacey, 1997; Trexler, Li, & Massey, 2005). Recently, it has been shown that this coupling allows the AII to play an important role in a rapid inhibitory circuit that responds to looming dark features at daytime light levels (Münch et al., 2009). It is unclear what role rod signals might play in this high-light-level function.

Yet another role played by the coupling between ON cone bipolar and AII amacrine cells appears to be to allow cone and rod signals to extend the operating range of OFF visual pathways in daylight (Manookin et al., 2008). Mixed rod and cone input to the AII via the coupling with ON cone bipolars ultimately modulates inhibition of OFF ganglion cells on which the AII synapses. Thus, light decrements cause excitation of the OFF ganglion cell by release of inhibition.

Because the expression of gap junctions—whether between rods and cones or later neurons—is labile and rapidly controlled by factors such as light adaptation and circadian rhythms, electrical coupling provides tremendous capacity for plasticity and modulatory control in retinal circuits. We undoubtedly will learn more in the years ahead about the ways that coupling allows rod and cone signals to jointly influence retinal processing.

Rod and Cone Inputs to Melanopsin Ganglion Cells

Another major advance of the past dozen years is the discovery of intrinsically photosensitive, melanopsin-containing ganglion cells (melanopsin cells) that also receive input from rods and cones. These very large, sparse ganglion cells have been shown to project to LGN as well as to subcortical areas (Dacey et al., 2003, 2005) and play a role in controlling pupil size (Gamlin et al., 2007) and circadian sensitivity (reviewed by Berson, 2003). The role of melanopsin cells in conscious visual perception is as yet unclear, but they display highly sensitive rod input at scotopic light levels and spectrally opponent cone input at photopic levels (Dacey et al., 2005) as well as intrinsic melanopsin sensitivity. Interestingly, the cone opponency is S OFF, L+M ON, which is rare among retinal ganglion cells. The convergence of rod, cone, and melanopsin photoresponses positions the melanopsin cell to encode overall luminance over the entire range of light sensitivity (Dacey et al., 2005). Further work is needed to understand the role of these cells and their rod and cone inputs in conscious visual perception and the control of retinal sensitivity.

Rod Input to S-Cone/Koniocellular Pathways

The best known retinal output for S-cone signals is the small bistratified ganglion (SBG) cell, which shows largely coextensive S-ON and L+M-OFF responses (Dacey & Lee, 1994) and projects to koniocellular layers of the LGN (Szmajda, Grünert, & Martin, 2008). Psychophysical studies of rod influence on hue originally suggested that rod signals could add with S-cone signals to influence color perception (Buck, 2001; Buck et al., 1998), but it was not until recently that this prediction was confirmed. Earlier studies were contradictory about rod influence to S-ON cells (Virsu, Lee, & Creutzfeldt, 1987; Lee et al., 1997). However, Crook et al. (2009) and Field et al. (2009) demonstrated definitively that rod signals add with the same sign as S-cone signals and provide a high-sensitivity input to SBG cells and their subsequent koniocellular pathway. The light level of maximal rod influence on the opponent spectral tuning of SBG cells is roughly at the light levels showing psychophysical rod influence (Field et al., 2009).

Rod Input to Magnocellular and Parvocellular Pathways

It has long been known that rod signals strongly influence parasol ganglion cells and magnocellular LGN pathways. However, the role of rod signals in midget ganglion (MG) cells and parvocellular pathways has been less clear, with some indications that rod influence is weak, at best (Lee et al., 1997; Purpura, Kaplan, & Shapley, 1988). Nevertheless, other physiological (reviewed below) and psychophysical (next section) studies suggest a stronger role for rod signals in parvocellular pathways.

Rudvin and Valberg (2006) find that visually evoked potentials (VEPs) provide a sensitive measure of human parvocellular activity, even down to very low levels of activation. They concluded that parvocellular-mediated responses are the dominant source of high-contrast isochromatic flicker VEPs at all light levels, including scotopic and mesopic levels.

Several studies have found strong rod input to parvocellular as well as magnocellular pathways in New World dichromatic primates (Yeh et al., 1995; Kremers, Weiss, & Zrenner, 1997). Weiss, Kremers, and Maurer (1998) demonstrated that rod and cone signals combine by vector summation in both pathways in marmosets. Rod input strengthened with eccentricity and weakened with increased light level but was often measurable up to the human equivalent of 700 trolands (Td). More work is needed to determine if there is a difference between New World and Old World primates in the strength of rod signals in parvocellular pathways.

If rod signals were excluded from some postreceptoral pathways, one might expect the characteristics of cortical responses to change with transition from cone to rod vision. However, Duffy and Hubel (2007) found that dark adaptation and low light levels did not alter the basic receptive-field properties of V1 cortical cells in awake behaving macaques. They found that most cells in V1 with receptive fields centered beyond 2° eccentricity get rod input but that most closer to the fovea do not. Consistent with this eccentricity effect, Hadjikhani and Tootell (2000) found rod activation selectively absent from fMRI images of foveal representations but robust in the peripheral representations of multiple areas. Thus, eccentricity of testing is another parameter that could influence the finding of strong or weak rod influence.

Questions about the role of rod signals in the responses of MG cells and parvocellular pathways, especially among primates, and about the cortical pathways influenced by mixed rod and cone signals remain among the most important unanswered research challenges related to rod–cone interactions.

ROD INFLUENCE ON VISUAL PERCEPTION AND PERFORMANCE

Additivity of Rod and Cone Signals

Over the past decade the use of stimuli that allow variation of the relative phase of separate modulations of selected cone types and rods has clarified our understanding of how well rod and cone signals combine in specific visual pathways.

As noted above, it has long been appreciated that rods provide strong input to parasol ganglion cells and magnocellular pathways. With use of sinusoidally modulated stimuli, both physiological (Cao, Lee, & Sun, 2010) and psychophysical (Sun, Pokorny, & Smith, 2001b) studies have now shown that rod and cone signals sum linearly in the magnocellular pathway. The latter study concluded that rod signals added with near perfect summation with M- and L-cone signals at both 1 and 10 photopic trolands when stimuli were modulated at 10 Hz and presumably carried in magnocellular pathways. However, rod signals showed only probability summation with M- and L-cone signals when stimuli were modulated at 2 Hz, when cone signals were presumably carried by parvocellular pathways. In contrast, Kilavik and Kremers (2006) found that rod and L-cone signals combined with complete vector summation over the entire range of temporal frequencies from 1 to 15 Hz in deuteranopes. More work will be needed to reconcile these results.

The evidence about combination of rod and S-cone signals is also complicated. Zele, Kremers, and Feigl (2012) demonstrated that rod and S-cone signals combine linearly in some individuals but only by probability summation in one other individual. In addition, they also found a complex nonlinear mutual reinforcement of rod and S-cone signals with a different procedure. More work is needed to sort out how these effects may generalize to other conditions.

Finally, Cao et al. (2008) have shown that rods contribute to all three pathways—parvocellular, koniocellular, and magnocellular—and do so linearly with rod stimulus contrast.

Rod Influence on Luminance and Brightness

There has been long-standing interest in characterizing how rod and cone signals jointly contribute to brightness and luminance of a mesopic stimulus that excites both receptor classes (see Buck, 2004, and Stockman & Sharpe, 2006, for reviews). Standard spectral luminosity functions have been determined for both photopic vision—CIE 1924 $V(\lambda)$ and CIE 1964 $V_{10}(\lambda)$—and for scotopic vision, CIE 1951 $V'(\lambda)$. However, there is no single mesopic luminosity function. The relative contributions of rods and cones depend on stimulus parameters such as size, timing, and retinal location and the adaptation state of the observer in addition to the overall light level and other factors (reviewed by Stockman & Sharpe, 2006).

Despite these complexities, attempts have been made to determine generalizable rules relating the rod and cone contributions to mesopic luminosity and brightness. A challenge for the specification of *mesopic luminance* is that common tasks for isolating photopic luminance such as minimally distinct border or flicker photometry are unsuitable because they disadvantage scotopic sensitivity, which has poor spatial and temporal sensitivity. A variety of alternative means of specifying mesopic luminance include measures of reaction time (Rea et al., 2004), search time (Walkey, Harlow, & Barbur, 2006), contrast and acuity thresholds (Varady & Bodrogi, 2006), and motion photometry (Raphael & MacLeod, 2011). Of these, reaction time and similar performance measures have particular relevance for real-world tasks such as driving.

The use of minimum-motion photometry by Raphael and MacLeod (2011) is particularly promising because the technique has a history of use in the measurement of photopic luminance but does not disadvantage scotopic sensitivity. They found a sigmoidal shift from rod to cone dominance as light levels rose from purely scotopic to purely photopic conditions. Importantly, the light level range of mesopic luminosity depended dramatically on eccentricity: The midpoint of the mesopic transition increased by a factor of 10 between 2° and 18° eccentricity. This suggests that the transition to purely photopic vision is determined by the relative strengths of rod and cone signals and not by a fixed ceiling (saturation) for rods or by a level of cone stimulation that blocks rod influence. This underscores the complexity of the processes that mediate the rod–cone transition.

Attempts have also continued to develop a system to specify *mesopic brightness*. The CIE developed a supplementary system of photometry to measure and predict brightness at any light level that is based on the specification of a metric they called equivalent luminance (CIE, 2001), which is the luminance of a standard reference light (540-THz frequency or ~555 nm wavelength) that matches the brightness of a test light or object. Such a system allows precise *specification* of brightness for a specific light or object, but for *prediction* of brightness of an arbitrary light or object, one must have a model of how mesopic brightness differs from luminance, with contributions from rods and chromatic mechanisms. Efforts to refine such photometric models are ongoing but outside the scope of this review.

Another fundamental difference between luminance and brightness is that brightness can be influenced by the stimuli that surround and precede it (as in simultaneous and successive brightness contrast). Sun, Pokorny, and Smith (2001a) showed that modulation of rod

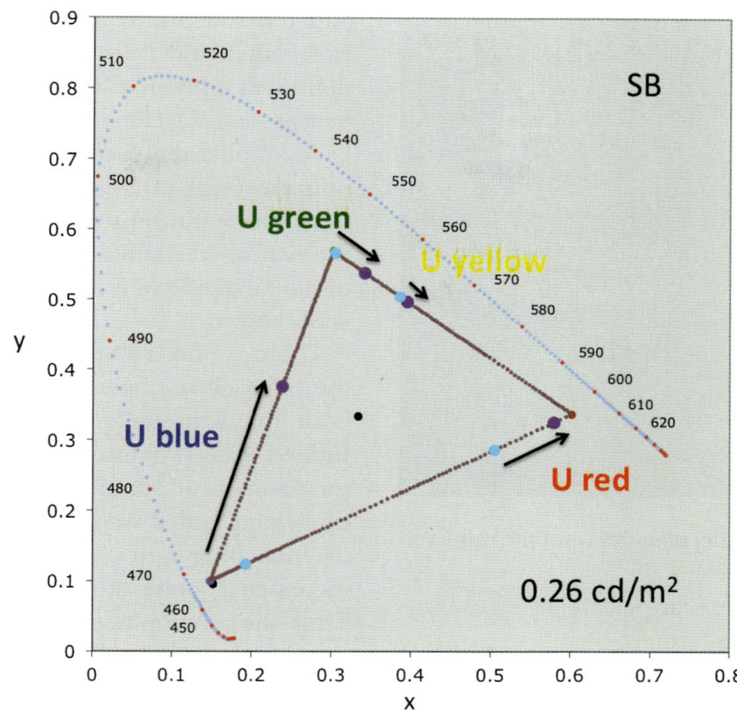

FIGURE 34.2   Rod hue shifts for unique-hue loci observed on CRT monitor in CIE chromaticity space. (Based on Buck et al., 2012.)

signals in an annular surround could induce brightness contrast in a test field when stimuli were presented at 1, 10, or 100 photopic trolands.

Rod Influence on Hue

Rod stimulation can affect all three perceptual dimensions of color—hue, brightness, and saturation—as well as chromatic discrimination. This review focuses on recent progress in understanding rod influences on hue and chromatic discrimination.

One way that rod stimulation can influence hue perception is to shift or bias the hue that would otherwise be determined by cones alone. There are three distinct rod hue biases that reflect rod involvement in different portions of the pathways that serve hue perception (reviewed in Buck, 2004). All can be seen in figure 34.2.

1. Rods can enhance green relative to red. This *rod green bias* is most easily seen at longer wavelengths, where it can be revealed as a shift of unique yellow (a red-green balance) to longer wavelengths, as the cone contribution to redness has to be increased to balance the rod contribution to greenness.

2. Rods can also have the opposite effect and enhance red relative to green. This *rod red bias* is most prominent at shorter wavelengths, where it can be revealed as a

shift of unique blue (another red-green balance) to longer wavelengths.

3. Rods can enhance blue relative to yellow. This rod blue bias is revealed as a shift of unique green (a blue-yellow balance) to longer wavelengths and as an analogous shift of the extraspectral unique red toward greater long-wavelength energy.

MECHANISMS OF ROD HUE BIAS   We have suggested that the rod signals combine with same sign as M- and L-cone signals in midget-ganglion-cell pathways, but with a stronger weighting on M-cone signals than on L-cone signals, to produce the *rod green bias* (Buck, 2001; Buck et al., 1998) (see figure 34.3). The neural basis for this differential weighting is not known, but an attractive possibility is that rod/M-cone signals are differentially strengthened as part of normalization of signals between M cones and the typically more numerous L cones. One explicit test (Cao, Pokorny, & Smith, 2005) did not find an association between L-/M-cone ratio, as inferred from flicker photometry, and the magnitude of rod green bias. However, increases of rod stimulation in mesopic fields produced hue changes that were matched by stimuli that increased M-cone excitation more than L-cone excitation, consistent with greater rod effect associated with M-cone pathways. Also, modulation of rods in an annular surround produced

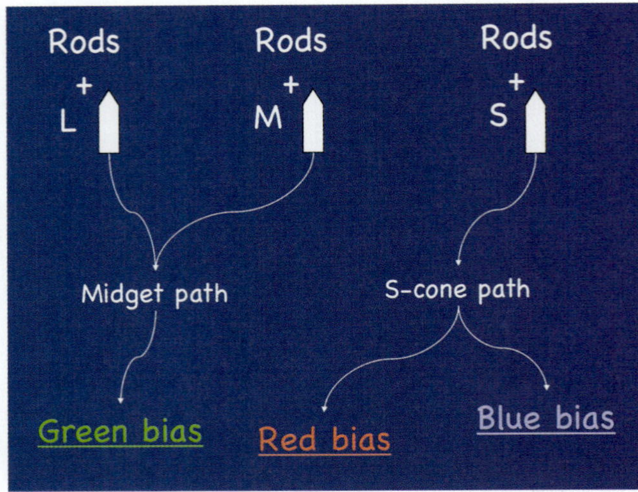

FIGURE 34.3   Conceptual model of pathways of the three rod hue biases. (Based on Buck, 2001.)

chromatic contrast effects in the center that mimicked those of annular M-cone modulation. The origin of the rod green bias and the differential rod effect on M- and L-cone pathways remains among the most intriguing puzzles of the neural substrate of rod hue biases.

We have also suggested that rod signals combine with the same sign as S-cone signals, presumably in small bistratified ganglion-cell pathways, to produce both the *rod red bias* and the *rod blue bias* (see figure 34.3). Fundamental color models link increases of S-cone stimulation to increases of both blue and red, consistent with the violet percept associated with very-short-wavelength lights. Thus, rod signals added to S-cone pathways mimic the effect of S-cone signals and enhance both redness and blueness.

Because the rod red bias and the rod green bias have separate early neural substrates, they can end up being pitted against each other, as occurs in their influence on unique blue. This can be demonstrated because the red and green rod hue biases have different time courses. Normally the sluggish rod red bias is stronger for longer-duration stimuli (≥200 ms), but if stimuli are very brief (30 ms), the brisk rod green bias is revealed (Buck & Knight, 2003; Buck et al., 2008). The other situation in which the rod green bias can dominate at unique blue appears to be for small foveally centered stimuli, possibly because of the sparseness of foveal S cones needed to mediate the rod red bias (Buck et al., 2006, 2012; Thomas & Buck, 2006).

The pitting of red and green rod hue biases against each other also provides a potential explanation of early findings of a single rod blue bias (reviewed in Buck, 2004). The use of both longer- and shorter-wavelength mixture lights would tend to produce both green

and red rod hue biases, which would tend to cancel each other and leave the rod blue bias as the single dominant influence on hue. The absence of the rod red bias at long wavelengths, when S cones are not stimulated, implies that S-cone activity gates rod influence: A nonzero level of S-cone activity is necessary for rod signals to be transmitted by the S-cone pathways. Neither the neural mechanism of this gating nor the existence of similar gating of rod influence in L–M cone pathways is yet known.

There is only a single direction of rod influence on the blue-yellow dimension, a *rod blue bias*, which shifts both unique green and unique red (Buck et al., 2012). Indeed, Cao, Pokorny, and Smith (2005) found that increases of rod stimulation in mesopic fields produced hue changes that required increased S-cone excitation in a matching light when they could not be matched by increased M-cone stimulation alone (see figure 34.4). Thus, the psychophysics and retinal physiology (see previous section) dovetail in implicating SBG-cell pathways in mediating the rod blue bias.

GENERALITY OF ROD HUE BIASES   One of the areas of greatest progress over the past few years has been in the expansion of our understanding of the range of conditions in which rod hue biases may be observed. Early studies mostly used single monochromatic stimulus disks that were turned on and off in isolation in dark fields. Recent studies have begun to explore stimuli that are more spatially complex, temporally dynamic, and chromatically desaturated in order to help understand how rods influence hue in real-world visual environments.

In the spatial domain several studies have used large background or surround stimuli to modulate rod hue biases. Knight and Buck (2001, 2002) used large concentric scotopic backgrounds to demonstrate that it is the duration of rod stimulation, and not the duration of cone stimulation, that drives the emergence of the rod red bias at short wavelengths. Cao, Pokorny, and Smith (2005) showed that temporal modulation of rods in a concentric surround induced chromatically contrasting hues in an enclosed disk that paralleled the rod hue biases seen when rod stimulation is modulated in a mesopic field.

Buck and DeWenter (2010) showed that the magnitude of rod blue and green biases was little affected by colored surrounds that induced simultaneous chromatic contrast into the test stimulus. They also showed that rod hue biases persisted for a rectangular patch in the center of a multi-colored 5×5-patch Mondrian display. Mondrian displays have been shown to stabilize hue perception, presumably by means of processes

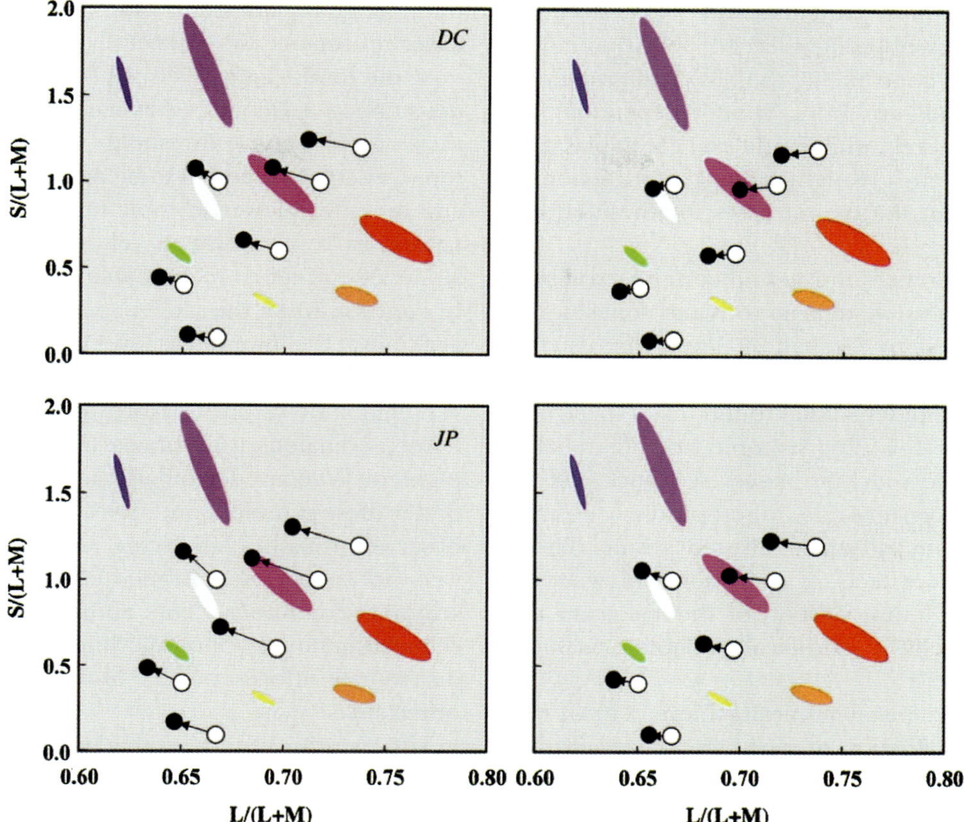

FIGURE 34.4   Rod hue shifts (points and arrows) seen with four-channel photostimulator at 2 Td (left) and 10 Td (right) in relation to basic color categories. (Reprinted from Cao, Pokorny, & Smith, 2005; copyright 2005; with permission from Elsevier.)

involved in maintaining color constancy and desensitizing chromatic sensitivity (Brown & MacLeod, 1997). To whatever extent the Mondrian display used for this study produced such effects, it did not eliminate rod hue biases. Further study is needed to better assess possible interactions of rod hue effects with other spatial and temporal influences on hue.

Other studies have begun to assess the prevalence of rod hue biases for more desaturated stimuli. Buck et al. (2012) found all three rod hue biases with a CRT display that provided less saturated stimuli (because of the mixture of broader-band phosphors) and reduced spatial contrast (because of the background veiling luminance) compared to a Maxwellian-view apparatus presenting spectral stimuli. Rod hue biases were found at light levels as high as 2.6 cd/m^2 (but not at 26 cd/m^2). However, critical light levels and magnitudes of rod hue biases varied considerably among observers. The possibility that this variation is related to the relative desaturation of the CRT stimuli receives support from another study of rod hue biases using desaturated stimuli (Buck & Cunningham, 2009). Observers showed all three rod hue biases for desaturated stimuli, in some cases with as little as ~25% excitation purity, but with

overall differences in persistence across unique hues and observers. In general, rod hue biases persisted to lower purity levels for blue and green stimuli than with yellow stimuli, which also showed the greatest variation among observers.

Although qualitatively similar rod hue biases have been identified by means of unique-hue-shift and hue-scaling tasks, adjustment, forced-choice, and staircase methods and use of stimuli that are equated for either photopic or scotopic luminance, it is less clear how each of these variations may affect the details of the resulting rod hue biases.

In one study Volbrecht and colleagues (2010) found that the method of measuring unique green influenced both the locus of unique green and the magnitude of its shift under rod influence. A hue-scaling task showed longer-wavelength loci for unique green and greater rod influence than a staircase task, although the measured rod blue bias was relatively small.

The dependence of rod hue biases on stimulus light level has both theoretical and practical importance—and vexing complexities. In general rod hue biases clearly occur under conditions of relatively strong rod signals and relatively weak cone signals. However, it is less clear whether this represents simple signal-strength influence or more subtle light-level-dependent retinal neural changes. It appears that the blue and green rod hue biases generally are more strongly dependent on low mesopic light levels, whereas the rod green bias may persist to higher levels (Knight & Buck, 2001; Thomas & Buck, 2004), but the neural bases for these effects remain unknown.

Because of differences in rod and cone spectral sensitivities, it is not possible to keep rods and cones simultaneously at constant excitation levels across the spectrum, so choices must be made in specifying light levels. Several older studies found that a spectrum-wide light level of 1–1.5 log scotopic trolands reliably produced all three rod hue biases (see Buck, 2004). The use of spectrum-wide photopically constant light levels may not yield such consistent results. There may also be some effects of the choice of whether stimuli *within* a relevant portion of the spectrum are equated photopically or scotopically (Thomas & Buck, 2004).

Arguably the premier technical achievement of this period is the introduction of a new four-channel photostimulator that allows simultaneous control of all four photoreceptor types (Pokorny, Smithson, & Quinlan, 2004). The photostimulator allows independent control of two stimulus fields (e.g., disk and concentric surround) presented in Maxwellian view at mesopic light levels that stimulate all three cones and rods. Within each field any number of photoreceptor types may be modulated over time with the remainder held constant. The photostimulator has allowed direct demonstration and validation of the previously described green and blue rod hue biases without the need for comparison of settings made in bleached and dark-adapted states (Cao, Pokorny, & Smith, 2005).

SURFACE-COLOR PERCEPTION An intriguing recent study investigated how rods influence the naming of hues among varied ensembles of real paper samples (Pokorny et al., 2006). Most studies of rod hue biases have presented single lighted stimulus patches "floating" in a dark field with no stimuli of other hues for simultaneous comparison. Instead, these investigators had observers judge the "related" hues of individual pieces of paper, presented as part of a 24-piece ensemble from the eight basic nondark hue categories. As light level was reduced below purely photopic levels, rods and L cones mediated detection of most samples. Long-wavelength samples with higher L-cone stimulation tended to be labeled in the same way they were labeled photopically: as either orange or red. Short- and middle-wavelength stimuli of higher reflectance were generally labeled as blue or green, and those of lower reflectance were generally labeled black.

At the lowest light levels, where only rods mediated detection, a variety of color names were still used for those samples above threshold: Those with higher scotopic reflectance tended to be named as green or blue, and those with lower scotopic reflectance tended to be named as red or orange. Thus, stimulation of rods alone led to reports of a reliable range of related hues. In contrast, when the same papers were presented one at a time at the lowest light level, all papers were labeled as blue or green if they were above threshold.

A companion study (Pokorny et al., 2008) found that deuteranomalous trichromats (having shifted M-cone pigments) behaved similarly to normal trichromats under these stimulus conditions. However, dichromatic observers (missing either the M- or L-cone pigment) were not so consistent in their use of hue names under scotopic conditions. The authors suggest that the reduced gamut of photopic hues experienced by the dichromats affords poorer association with scotopic reflectance.

These studies remind us that the brain constructs hue on the basis of many factors in addition to the actual photoreceptor signals generated by a stimulus. In this case the simultaneous comparison of multiple samples enables the brain to infer a range of probable related hues, based on comparisons either of rods and L cones or of rods alone, and informed by prior natural experience. This study underscores the importance of expanding the investigation of rod influence on hue to conditions that are as naturalistic as possible, where rods may play an even greater role in hue perception than that suggested by studies of unrelated hues.

CHROMATIC PERCEPTIVE FIELDS Another measure of rod influence on color vision is how rod signals influence the size of chromatic perceptive fields (PF). PFs are the psychophysically measured equivalents of neural receptive fields. *Chromatic* PFs can be defined as the stimulus diameter at which the strength of a specified color percept, either a specific basic hue or saturation, stops increasing with increasing stimulus size.

Jan Nerger, Vicki Volbrecht, and colleagues have worked extensively to reveal the influence of rod signals on chromatic PF sizes over a wide range of light levels and retinal locations having different degrees of rod and cone convergence on ganglion cell pathways. The general finding is that rod stimulation increases the size of chromatic PFs for all four basic hues, particularly at lower retinal illuminance levels of 2–200 photopic Td (Pitts et al., 2005; Troup et al., 2005; Volbrecht et al., 2009). Overall, rods increase PF size most when rods are stimulated most strongly relative to cones and in retinal areas where rods are most numerous relative to cones. However, this leaves unclear what specific retinal and/or postretinal pathways are actually involved. These investigators argue that rod influences on PF size do not result from simple rod desaturation but could result from cortical interactions of inputs from parvocellular and magnocellular pathways.

Rod Influence on Chromatic Discrimination

Understanding how rod signals influence chromatic discrimination is of considerable theoretical and practical importance, given the increasing use of displays that use color to convey information. Recent studies have underscored previous findings that, for normal trichromats, rod stimulation typically impairs discriminations mediated by either L- plus M-cone (Nagy & Doyal, 1993; Stabell & Stabell, 1977) or S-cone (Knight et al., 1998) chromatic pathways. However, the recent studies have shown that rods do so in different ways and that there are asymmetries in the rod effects for each type of discrimination.

Consistent with prior results on L–M discriminations, Volbrecht, Nerger, and Trujillo (2011) found that rods progressively reduced hue discrimination in the longer-wavelength half of the spectrum as rods gained sensitivity during dark adaptation. The authors suggested that rods impair chromatic discrimination via magnocellular pathways, interacting in visual cortex with cone signals, possibly L-cone signals in particular, conveyed by parvocellular pathways. Unfortunately, the wavelength range used by Volbrecht, Nerger, and Trujillo stopped short of providing a test of the puzzling and unreplicated finding of Stabell and Stabell (1977) that rods *improved* wavelength discrimination for very-long-wavelength stimuli.

Cao, Zele, and Pokorny (2008) showed asymmetrical effects of rod stimulation on L–M discrimination using a four-channel photostimulator that allowed manipulation of rod and cone excitations without the need for comparison of dark-adapted and bleached adaptation conditions. They found that rod excitation impaired discrimination of L–M cone increments (increases of L relative to M) but had no effect on discrimination of L–M cone decrements (increases of M relative to L). Further comparisons of these rod effects with perceptually matching cone stimuli suggested that the L–M increment impairment resulted from rods acting in chromatic pathways, a different conclusion from that of Volbrecht, Nerger, and Trujillo (2011). The absence of rod influence on L–M decrements by Cao, Zele, and Pokorny also stands in contrast to earlier work that

found rod impairment of both increments and decrements of L–M, although the increment effects were larger than the decrement effects (Nagy & Doyal, 1993). Further work is needed to determine what aspect of the different stimuli and methods among these studies may have led to the different results.

Although Volbrecht, Nerger, and Trujillo (2011) did not find an effect of rods on short-wavelength hue discrimination, other studies have. Knight et al. (1998) found that dark adaptation selectively impaired chromatic discrimination on the FM100-hue test along an S-cone axis and that this effect increased at lower light levels. Knight, Buck, and Pereverzeva (2001) found much stronger rod impairment of S-cone decrements than S-cone increments for small extrafoveal stimuli but no consistent rod effect with large extrafoveal stimuli. This size dependence is opposite in direction to that demonstrated by Nagy & Doyal (1993) for L–M discrimination, underscoring the different ways that rods affect the two types of chromatic discrimination but leaving us unsure as to why this is.

Asymmetrical rod effects on S-cone increments and decrements were also found in two other studies. Shepherd & Wyatt (2008) used a simultaneous color contrast task and interpreted their result as suggesting that rod signals have preferential access to S-OFF pathways. Cao, Zele, and Pokorny (2008) also found that rods impaired discrimination of S-cone decrements but had no effect on S-cone increments (see figure 34.5). The anatomical basis for these effects is not known.

Finally, a study by Walkey and colleagues (2001) is important for *not* showing any rod influence on chromatic discriminations measured in multiple directions in color space on a luminance contrast masking field, as used in the Cambridge Color Test to eliminate detection of scotopic and photopic luminance contrast signals and isolate chromatic pathways. No consistent differences of chromatic discrimination ellipses measured in 18 directions in color space were found between cone plateau and dark-adapted conditions for stimuli centered 3.5° from fixation. Possibly the relatively small stimulus eccentricity may have produced only weak rod signals whereas a greater eccentricity would have revealed rod effects on chromatic discrimination. However, a more intriguing possibility is that the luminance-contrast masking may have masked rod influence, perhaps by saturating the pathway(s) carrying rod signals.

We still have an incomplete understanding of the range of conditions under which rods affect chromatic discrimination and the nature of the neural pathways mediating those effects. An incentive to improve that understanding comes from the fact that the

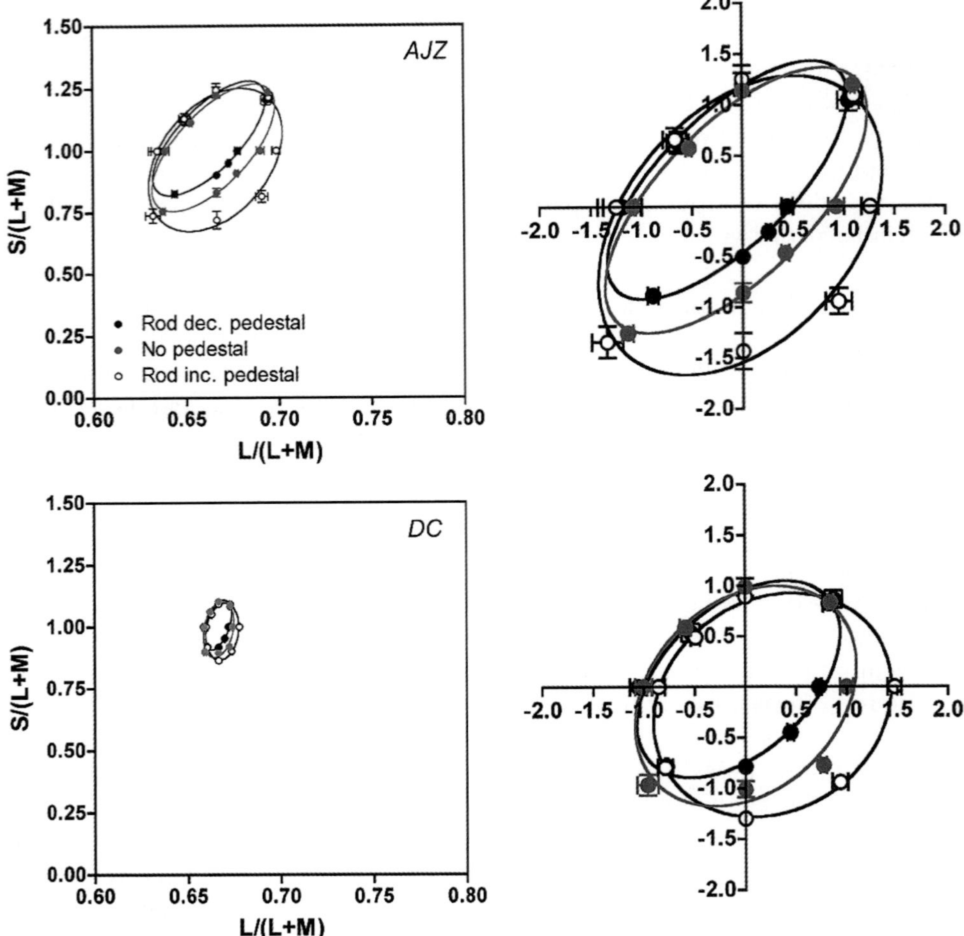

FIGURE 34.5   Asymmetrical influence of rods on chromatic discrimination: rods affect S-cone decrements but not increments and L-cone increments but not decrements. (Reprinted from Cao, Zele, & Pokorny, 2008, copyright Cambridge University Press, reproduced with permission.)

abovementioned studies showing rod *impairments* of chromatic discrimination in normal trichromats stand in contrast to demonstrations of rod *enhancement* of chromatic discrimination for dichromats (e.g., Pokorny & Smith, 1977) and monochromats (Reitner, Sharpe, & Zrenner, 1991), for whom rods seem able to provide a missing dimension of chromatic discrimination. The reasons for this, and the search for any comparable rod enhancements for normal trichromats, stand among the most intriguing unresolved issues concerning rod–cone interactions related to color vision.

THE CHALLENGES AHEAD

Among the most important unanswered questions that will challenge researchers in the immediate future are the following:

• How does rod–cone coupling serve to maintain the health and survival of photoreceptors?

• What is the role of rod signals on the responses of MG cells and parvocellular pathways?

• What cortical pathways process signals that started out in rods?

• Why does rod activation exert a net perceptual green-versus-red hue bias?

• Why are there asymmetries of rod influence on chromatic discriminations that involve S-cone increments and decrements?

• Are there circumstances in which rod activation can improve human chromatic discrimination?

REFERENCES

Berson, D. M. (2003). Strange vision: Ganglion cells as circadian photoreceptors. *Trends in Neurosciences, 26*, 314–320.

Brown, R. O., & MacLeod, D. I. A. (1997). Color appearance depends on the variance of surround colors. *Current Biology, 7*, 844–849.

Buck, S. L. (2001). What is the hue of rod vision? *Color Research and Application, 26*(Suppl.), S57–S59.

Buck, S. L. (2004). Rod-cone interactions. In L. M. Chalupa & J. S. Werner (Eds.), *The visual neurosciences* (Vol. 2, pp. 863–878). Boston: MIT Press.

Buck, S. L., & Cunningham, C. (2009). Rod influence on desaturated color mixtures. *Journal of Vision, 9*, 55. doi:10.1167/9.14.55.

Buck, S. L., & DeWenter, R. (2010). Rod influence on complex backgrounds. *Journal of Vision, 10*, 50. doi:10.1167/10.15.50.

Buck, S. L., Juve, R., Wisner, D., & Concepcion, A. (2012). Rod hue biases produced on CRT displays. *Journal of the Optical Society of America. A, Optics, Image Science, and Vision, 29*, A36–A43. doi:10.1364/JOSAA.29.000A36.

Buck, S. L., & Knight, R. F. (2003). Stimulus duration affects rod influence on hue perception. In J. D. Mollon, J. Pokorny, & K. Knobluach (Eds.), *Normal and defective colour vision* (pp. 177–184). Oxford: Oxford University Press.

Buck, S. L., Knight, R. F., Fowler, G. A., & Hunt, B. (1998). Rod influence on hue-scaling functions. *Vision Research, 38*, 3259–3263. doi:10.1016/S0042-6989(97)00436-7.

Buck, S. L., Thomas, L., Connor, C., Green, K., & Quintana, T. (2008). Time-course of rod influences on hue perception. *Visual Neuroscience, 25*, 517–520.

Buck, S. L., Thomas, L., Hillyer, N., & Samuelson, E. (2006). Do rods influence the hue of foveal stimuli? *Visual Neuroscience, 23*, 519–523.

Camacho, E. T., Colon Velez, M. A., Hernandez, D. J., Bernier, U. R., Van Laarhoven, J., & Wirkus, S. (2010). A mathematical model for photoreceptor interactions. *Journal of Theoretical Biology, 267*, 638–646.

Cameron, M. A., & Lucas, R. J. (2009). Influence of the rod photoresponse on light adaptation and circadian rhythmicity in the cone ERG. *Molecular Vision, 15*, 2209–2216.

Cao, D., Lee, B. B., & Sun, H. (2010). Combination of rod and cone inputs in parasol ganglion cells of the magnocellular pathway. *Journal of Vision, 10*, 1–15. doi:10.1167/10.11.4.

Cao, D., Pokorny, J., & Smith, V. C. (2005). Matching rod percepts with cone stimuli. *Vision Research, 45*, 2119–2128. doi:10.1016/j.visres.2005.01.034.

Cao, D., Pokorny, J., Smith, V. C., & Zele, A. J. (2008). Rod contributions to color perception: Linear with rod contrast. *Vision Research, 48*, 2586–2592. doi:10.1016/j.visres.2008.05.001.

Cao, D., Zele, A. J., & Pokorny, J. (2008). Chromatic discrimination in the presence of incremental and decremental rod pedestals. *Visual Neuroscience, 25*, 399–404.

CIE. (2001). *Testing of Supplementary Systems of Photometry*, CIE Publication No. 141. Vienna: Central Bureau of the Commission Internationale de l'Eclairage.

Crook, J. D., Davenport, C. M., Peterson, B. B., Packer, O. S., Detwiler, P. B., & Dacey, D. M. (2009). Parallel ON and OFF cone bipolar inputs establish spatially coextensive receptive field structure of blue-yellow ganglion cells in primate retina. *Journal of Neuroscience, 29*, 8372–8387.

Dacey, D. M., & Lee, B. B. (1994). The blue-ON opponent pathway in primate retina originates from a distinct bistratified ganglion cell type. *Nature, 367*, 731–735.

Dacey, D. M., Liao, H. W., Peterson, B. B., Robinson, F. R., Smith, V. C., Pokorny, J., et al. (2005). Melanopsin-expressing ganglion cells in primate retina signal colour and irradiance and project to the LGN. *Nature, 433*, 749–754.

Dacey, D. M., Peterson, B. B., Robinson, F. R., & Gamlin, P. D. (2003). Fireworks in the primate retina: In vitro photodynamics reveals diverse LGN-projecting ganglion cell types. *Neuron, 37*, 15–27.

Duffy, K. R., & Hubel, D. H. (2007). Receptive field properties of neurons in the primary visual cortex under photopic and scotopic lighting conditions. *Vision Research, 47*, 2569–2574. doi:10.1016/j.visres.2007.06.009.

Field, G. D., Greschner, M., Gauthier, J. L., Rangel, C., Shlens, J., Sher, A., et al. (2009). High-sensitivity rod photoreceptor input to the blue–yellow color opponent pathway in macaque retina. *Nature Neuroscience, 12*, 1159–1164. doi:10.1038/nn.2353.

Gamlin, P. D., McDougal, D. H., Pokorny, J., Smith, V. C., Yau, K. W., & Dacey, D. M. (2007). Human and macaque pupil responses driven by melanopsin-containing retinal ganglion cells. *Vision Research, 47*, 946–954. doi:10.1016/j.visres.2006.12.015.

Hadjikhani, N., & Tootell, R. B. (2000). Projection of rods and cones within human visual cortex. *Human Brain Mapping, 9*, 55–63.

Hornstein, E. P., Verweij, J., Li, P. H., & Schnapf, J. L. (2005). Gap-junctional coupling and absolute sensitivity of photoreceptors in macaque retina. *Journal of Neuroscience, 25*, 11201–11209.

Kilavik, B. E., & Kremers, J. (2006). Interactions between rod and L-cone signals in deuteranopes: Gains and phases. *Visual Neuroscience, 23*, 201–207.

Knight, R. F., & Buck, S. L. (2001). Rod influences on hue perception: Effect of background light level. *Color Research and Application, 26*(Suppl.), S60–S64.

Knight, R. F., & Buck, S. L. (2002). Time-dependent changes of rod influence on hue perception. *Vision Research, 42*, 1651–1662. doi:10.1016/S0042-6989(02)00087-1.

Knight, R. F., Buck, S. L., Fowler, G. A., & Nguyen, A. (1998). Rods affect S-cone discrimination on the Farnsworth-Munsell 100-Hue Test. *Vision Research, 38*, 3477–3481. doi:10.1016/S0042-6989(97)00414-8.

Knight, R. F., Buck, S. L., & Pereverzeva, M. (2001). Stimulus size affects rod influence on tritan chromatic discrimination. *Color Research and Application, 26*(Suppl.), S65–S68.

Kremers, J., Weiss, S., & Zrenner, E. (1997). Temporal properties of marmoset lateral geniculate cells. *Vision Research, 37*, 2649–2660. doi:10.1016/S0042-6989(97)00090-4.

Lee, B. B., Smith, V. C., Pokorny, J., & Kremers, J. (1997). Rod inputs to macaque ganglion cells. *Vision Research, 37*, 2813–2828. doi:10.1016/S00042-6989(97)00108-9.

Manookin, M. B., Beaudoin, D. L., Ernst, Z. R., Flagel, L. J., & Demb, J. B. (2008). Disinhibition combines with excitation to extend the operating range of the OFF visual pathway in daylight. *Journal of Neuroscience, 28*, 4136–4150.

Münch, T. A., da Silveira, R. A., Siegert, S., Viney, T. J., Awatramani, G. B., & Roska, B. (2009). Approach sensitivity in the retina processed by a multifunctional neural circuit. *Nature Neuroscience, 12*, 1308–1316. doi:10.1038/nn.2389.

Nagy, A. L., & Doyal, J. A. (1993). Red-green color discrimination as a function of stimulus field size in peripheral vision. *Journal of the Optical Society of America. A, Optics and Image Science, 10*, 1147–1156.

Pitts, M. A., Troup, L. J., Volbrecht, V. J., & Nerger, J. L. (2005). Chromatic perceptive field sizes change with retinal illuminance. *Journal of Vision, 5*, 435–443. doi:10.1167/5.5.4.

Pokorny, J., Lutze, M., Cao, D., & Zele, A. J. (2006). The color of night: Surface color perception under dim illuminations. *Visual Neuroscience, 23*, 525–530.

Pokorny, J., Lutze, M., Cao, D., & Zele, A. J. (2008). The color of night: Surface color categorization by color defective observers under dim illuminations. *Visual Neuroscience, 25*, 475–480.

Pokorny, J., & Smith, V. C. (1977). Evaluation of single pigment shift model of anomalous trichromacy. *Journal of the Optical Society of America, 67*, 1196–1209.

Pokorny, J., Smithson, H., & Quinlan, J. (2004). Photostimulator allowing independent control of rods and the three cone types. *Visual Neuroscience, 21*, 263–267.

Protti, D. A., Flores-Herr, N., Li, W., Massey, S. C., & Wassle, H. (2005). Light signaling in scotopic conditions in the rabbit, mouse and rat retina: A physiological and anatomical study. *Journal of Neurophysiology, 93*, 3479–3488.

Purpura, K., Kaplan, E., & Shapley, R. M. (1988). Background light and the contrast gain of primate P and M retinal ganglion cells. *Proceedings of the National Academy of Sciences of the United States of America, 85*, 4534–4537. doi:10.1073/pnas.85.12.4534.

Raphael, S., & MacLeod, D. I. A. (2011). Mesopic luminance assessed with minimum motion photometry. *Journal of Vision, 11*, 1–21. doi:10.1167/11.9.14.

Rea, M. S., Bullough, J. D., Freyssinier-Nove, J. P., & Bierman, A. (2004). A proposed unified system of photometry. *Lighting Research & Technology, 36*, 85–111.

Reitner, A., Sharpe, L. T., & Zrenner, E. (1991). Is colour vision possible with only rods and blue-sensitive cones? *Nature, 352*, 798–800.

Ribelayga, C., Cao, Y., & Mangel, S. (2008). The circadian clock in the retina controls rod-cone coupling. *Neuron, 59*, 790–801.

Rudvin, I., & Valberg, A. (2006). Flicker VEPs reflecting multiple rod and cone pathways. *Vision Research, 46*, 699–717.

Schneeweis, D. M., & Schnapf, J. L. (1995). Photovoltage of rods and cones in the macaque retina. *Science, 268*, 1053–1056.

Seeliger, M. W., Brombas, A., Weiler, R., Humphries, P., Knop, G., Tanimoto, N., et al. (2011). Modulation of rod photoreceptor output by HCN1 channels is essential for regular mesopic cone vision. *Nature Communications, 2*, 532. doi:10.1038/ncomms1540.

Shepherd, A. J., & Wyatt, G. (2008). Changes in induced hues at low luminance and following dark adaptation suggest rod-cone interactions may differ for luminance increments and decrements. *Visual Neuroscience, 25*, 387–394.

Soucy, E., Wang, Y., Nirenberg, S., Nathans, J., & Meister, M. (1998). A novel signaling pathway from rod photoreceptors to ganglion cells in mammalian retina. *Neuron, 21*, 481–493.

Stabell, U., & Stabell, B. (1977). Wavelength discrimination of peripheral cones and its change with rod intrusion. *Vision Research, 17*, 423–426.

Stockman, A., & Sharpe, L. T. (2006). Into the twilight zone: The complexities of mesopic vision and luminous efficiency. *Ophthalmic & Physiological Optics, 26*, 225–239.

Stone, S., Buck, S. L., & Dacey, D. (1997). Pharmacological dissection of rod and cone bipolar input to the AII amacrine in macaque retina. *Investigative Ophthalmology & Visual Science, 38*(Suppl.), S689.

Sun, H., Pokorny, J., & Smith, V. C. (2001a). Brightness induction from rods. *Journal of Vision, 1*, 32–41. doi:10.1167/1.1.4.

Sun, H., Pokorny, J., & Smith, V. C. (2001b). Rod–cone interaction assessed in inferred postreceptoral pathways. *Journal of Vision, 1*, 42–54. doi:10.1167/1.1.5.

Szmajda, B. A., Grünert, U., & Martin, P. R. (2008). Retinal ganglion cell inputs to the koniocellular pathway. *Journal of Comparative Neurology, 510*, 251–268.

Thomas, L., & Buck, S. L. (2004). Generality of rod hue biases with smaller, brighter, and photopically specified stimuli. *Visual Neuroscience, 21*, 257–262.

Thomas, L., & Buck, S. L. (2006). Foveal vs. extra-foveal contributions to rod hue biases. *Visual Neuroscience, 23*, 539–542.

Trexler, E. B., Li, W., & Massey, S. C. (2005). Simultaneous contribution of two rod pathways to AII amacrine and cone bipolar cell light responses. *Journal of Neurophysiology, 93*, 1476–1485.

Troup, L. J., Pitts, M. A., Volbrecht, V. J., & Nerger, J. L. (2005). Effect of stimulus intensity on the sizes of chromatic perceptive fields. *Journal of the Optical Society of America. A, Optics, Image Science, and Vision, 22*, 2137–2142. doi:10.1364/JOSAA.22.002137.

Trümpler, J., Dedek, K., Schubert, T., de Sevilla Müller, L. P., Seeliger, M., Humphries, P., et al. (2008). Rod and cone contributions to horizontal cell light responses in the mouse retina. *Journal of Neuroscience, 28*, 6818–6825. doi:10.1523/JNEUROSCI.1564-08.2008.

Varady, G., & Bodrogi, P. (2006). Mesopic spectral sensitivity functions based on visibility and recognition contrast thresholds. *Ophthalmic & Physiological Optics, 26*, 246–253.

Verweij, J., Dacey, D., Peterson, B., & Buck, S. L. (1999). Sensitivity and dynamics of rod signals in macaque H1 horizontal cells. *Vision Research, 39*, 3662–3672. doi:10.1016/S0042-6989(99)00093-0.

Virsu, V., Lee, B. B., & Creutzfeldt, O. D. (1987). Mesopic spectral responses and the Purkinje shift of macaque lateral geniculate cells. *Vision Research, 27*, 191–200. doi:10.1016/0042-6989(87)90181-7.

Volbrecht, V. J., Clark, C. L., Nerger, J. L., & Randall, C. E. (2009). Chromatic perceptive field sizes measured at 10° along the horizontal and vertical meridians. *Journal of the Optical Society of America. A, Optics, Image Science, and Vision, 26*, 1167–1177.

Volbrecht, V. J., Nerger, L., Baker, L. S., Trujillo, A. R., & Youngpeter, K. (2010). Unique hue loci differ with methodology. *Ophthalmic & Physiological Optics, 30*, 545–552.

Volbrecht, V. J., Nerger, J. L., & Trujillo, A. R. (2011). Middle- and long-wavelength discrimination declines with rod photopigment regeneration. *Journal of the Optical Society of America. A, Optics, Image Science, and Vision, 28*, 2600–2606.

Walkey, H. C., Barbur, J. L., Harlow, J. A., & Makous, W. (2001). Measurements of chromatic sensitivity in the mesopic range. *Color Research and Application, 26*, S36–S42.

Walkey, H. C., Harlow, J. A., & Barbur, J. L. (2006). Characterising mesopic spectral sensitivity from reaction times. *Vision Research, 46*, 4232–4243. doi:10.1016/j.visres.2006.08.002.

Weiss, S., Kremers, J., & Maurer, J. (1998). Interaction between rod and cone signals in responses of lateral geniculate neurons in dichromatic marmosets (*Callithrix jacchus*). *Visual Neuroscience, 15*, 931–943.

Yeh, T., Lee, B. B., Kremers, J., Cowing, J. A., Hunt, D. M., Martin, P. R., et al. (1995). Visual responses in the lateral geniculate nucleus of dichromatic and trichromatic marmosets (*Callithrix jacchus*). *Journal of Neuroscience, 15*, 7892–7904.

Zele, A. J., Kremers, J., & Feigl, B. (2012). Mesopic rod and S-cone interactions revealed by modulation thresholds. *Journal of the Optical Society of America. A, Optics, Image Science, and Vision, 29*, A19–A26.

35 Brightness and Lightness

FREDERICK A. A. KINGDOM

This chapter concerns the visual sensations of achromatic color. The first part defines basic terms and outlines the relationships among lightness, brightness, contrast, and illumination. The second part deals with the brain areas where brightness and lightness are thought to be encoded. The third part reviews current models of brightness and lightness perception. A more detailed exposition of all these topics can be found in a recent review by Kingdom (2011).

BRIGHTNESS, LIGHTNESS, CONTRAST, AND ILLUMINATION

Image Decomposition

Any image can be decomposed into *layers* or *intrinsic images*. To understand brightness and lightness, and in particular their relationship, it is helpful to consider two of these layers: reflectance and illumination. The reflectance layer represents material variations such as paint and pigment and is itself decomposable into sublayers of spectral reflectance (*color*) and intensive reflectance (*albedo*). Intensive reflectance, from now on simply called *reflectance*, is defined as the proportion of light incident on a surface that is reflected from it. The second, illumination layer comprises the incident light. It is neither temporally nor spatially uniform. Temporally, the most dramatic changes in illumination occur in the ambient level as a result of the diurnal cycle. Spatially, the visual world is replete with various types of nonuniform illumination: shadows, shading, spotlights, interreflections, highlights, and light sources, to name the main varieties. The reflectance and illumination layers can be appreciated from the photograph in figure 35.1.

Brightness

Brightness is perceived luminance. Although brightness could in principle be derived from the sensory measurement of a single image point, it depends on context and mainly on contrast. The precise relationship between brightness and contrast is discussed later in this section and in the section on models.

Lightness

Lightness is perceived reflectance. If x and y are Cartesian coordinates in the two-dimensional image plane, the luminance $L(x,y)$, reflectance $R(x,y)$, and illumination $I(x,y)$ layers are related by the equation:

$$L(x,y) = I(x,y) \; R(x,y) \qquad (35.1)$$

Equation 35.1 reveals that it is impossible to determine from the luminance of any one image point whether it is part of a light surface that is shaded or a dark surface that is brightly illuminated. Pixel luminance is inherently ambiguous because an infinite combination of the two unknowns, $R(x,y)$ and $I(x,y)$ can produce the same $L(x,y)$. It is therefore axiomatic that only by examining the *relations* among image points is it possible to determine the perceptual correlate of $R(x,y)$, that is, lightness. The capacity to recognize that surfaces under different illuminations have the same reflectance is termed *lightness constancy*. Lightness constancy in natural scenes is generally thought to be quite good: The white walls in figure 35.1 appear white even though their luminances vary considerably over space and time. The visual system achieves lightness constancy in part by *comparing* the luminances of different parts of the image. Although there is no consensus as to the details of how this is done, the basic strategy works because luminance ratios, or contrasts, between surfaces remain invariant with changes in illumination (Wallach, 1976; Jacobsen & Gilchrist, 1988). This fact follows from equation 35.1: If two surfaces have reflectances Ra and Rb and are under a common illumination I, their ratio of luminances, La/Lb remains constant irrespective of I. The idea that lightness involves the computation of contrast has its origins in Hering's (1874/1964) ideas on the role of reciprocal neural interactions in the perception of brightness.

Lightness Anchoring

Although the computation of contrast is in principle sufficient to generate a scale of *relative* lightness values (e.g., this surface is lighter than that one but only slightly darker than that other one), *absolute* lightness values (this surface is light gray, that one is white),

FIGURE 35.1 Perceptual layer decomposition enables us to distinguish among reflectance (the white walls) shading (underside of roof, inside of left wall) and shadows (on front-facing walls).

require that the contrasts be "anchored" to some value. Traditionally, the two contenders for the lightness anchor have been the average luminance, designated as midgray, and the highest luminance, designated as white. Gilchrist (2006) argues that most evidence supports the latter contender. However, as he also points out, surfaces can appear white even in rooms where the light source, for example, a fluorescent light, is both visible and the highest luminance. Moreover, the shadowed area in the photograph in figure 35.1 looks white, even though it is not the highest luminance. These observations point to an alternative: White is determined not by the highest *luminance* but by the highest *lightness*, as suggested by Rudd and Zemach (2005).

Brightness versus Lightness

Brightness and lightness are distinct percepts when there are visible illumination boundaries, as in figure 35.1. The surfaces of the walls in the photograph appear uniformly white—a lightness judgment—yet are brighter in some places than others—a brightness judgment—due to the presence of shading and shadows. The ability of observers to distinguish between brightness and lightness has been confirmed in numerous psychophysical experiments using laboratory stimuli that contain depictions of nonuniform illumination (Arend & Spehar, 1993a, 1993b).

In displays that are devoid of visible illumination boundaries, however, such as in the simultaneous contrast displays on the left of figure 35.3, the percepts of brightness and lightness become synonymous. Thus, models of the photometric qualities of such stimuli can be considered as models of either brightness or lightness.

Whereas lightness can be estimated only by discounting the effects of nonuniform illumination, brightness can in principle be computed without recourse to layer decomposition. Yet there is strong evidence that nonuniform illumination influences brightness. For example, shadows appear brighter than their equal-in-luminance reflectance counterparts (Logvinenko, 2005), light sources appear brighter than their equal-in-luminance reflectance counterparts (Agostini & Galmonte, 2002; Correani, Scott-Samuel, & Leonard, 2006; Zavagno & Caputo, 2001), and surfaces appearing to lie in shadow or behind dark transparencies appear brighter than if presented on equivalent reflectance backgrounds (Adelson, 1993; Anderson, 1997; Kingdom, Blakeslee, & McCourt, 1997; Logvinenko, 1999). These findings have been influential in persuading a number of researchers that brightness is encoded not only by mechanisms sensitive to contrast but also by mechanisms sensitive to nonuniform illumination. However, the extent of the involvement of perceived nonuniform illumination in brightness perception is controversial (see Models of Brightness and Lightness below).

　　FREDERICK A. A. KINGDOM

Brightness and Contrast

Given that brightness/lightness is intimately related to contrast, what, precisely, is the relationship? For complex scenes the answer is not straightforward, but for relatively simple stimuli such as a patch on a uniform background, the problem is more tractable, and Whittle's seminal studies using such stimuli have gone a long way to providing the answer (summarized in Whittle, 1994). Whittle showed that for a variety of tasks using patch-on-background stimuli, a linear relationship exists between brightness and contrast if contrast is measured by the metric log W. In log W, $W = \Delta L / (L_{min} + k)$, where ΔL is the difference in luminance between patch and background, L_{min} is the smaller of the patch and background luminances, and k is a constant that prevents W approaching infinity when L_{min} approaches zero (k can be regarded as a measure of internal neural noise when luminance is zero, but if L_{min} is not close to zero the constant can be safely omitted). As figure 35.2 shows, W is calculated differently for an incremental and a decremental patch. W differs from conventional metrics such as Weber ($\Delta L / L_b$) and Michelson ($L_{max} - L_{min}) / (L_{max} + L_{min}$) contrast ($L_{max}$ is the maximum luminance, equal to $L_{min} + \Delta L$ for the increment and L_b for the decrement in the figure).

It has been argued that log W succeeds where other metrics fail because it encapsulates two visual processes: local light adaptation and a compressive, specifically logarithmic, contrast nonlinearity (Kingdom & Whittle, 1996; McIlhagga & Peterson, 2006). In all metrics of contrast the light-adaptation level is embodied in the equation's denominator, and with the denominator set to L_{min}, W embodies the idea that neurons sensitive to patch contrast light-adapt to the lower of the two luminances that fall within their receptive field. The logarithmic transformation of W arguably embodies the "true" shape of the function relating brightness to contrast and is notably different from the power exponent typically used to model contrast behavior when using Michelson contrast (see Kingdom & Whittle, 1996).

Increments versus Decrements

The distinction between increments and decrements inherent in W finds expression in an enduring home truth: Increments and decrements are processed by distinct mechanisms. The evidence for separate pathways for increments and decrements is both physiological, specifically the ON and OFF pathways beginning in the retinas of the mammalian visual system (Schiller, 1982; Fiorentini et al., 1990), and psychophysical (reviewed by Huang, Kingdom, & Hess, 2006). Phenomenologically, it is striking how difficult it is to find a luminance setting of an increment that matches the brightness of a decrement and vice versa: They simply never look the same (Vladusich, 2012). Increments invariably appear brighter than decrements whatever their luminance or contrast (e.g., top left figure 35.3).

The Neural Loci of Lightness and Brightness

Cornsweet's (1970) early influential book on vision championed the idea that reciprocal interactions among *retinal* neurons were responsible for simultaneous brightness contrast and, by implication, lightness. More recently, however, opinion has shifted in favor of a cortical locus for lightness perception. Nevertheless,

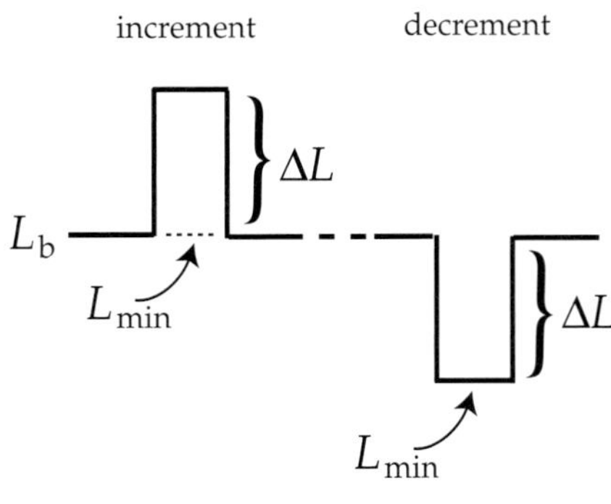

$$W = \Delta L / (L_{min} + k)$$

FIGURE 35.2   Calculation of W for both an increment and a decrement.

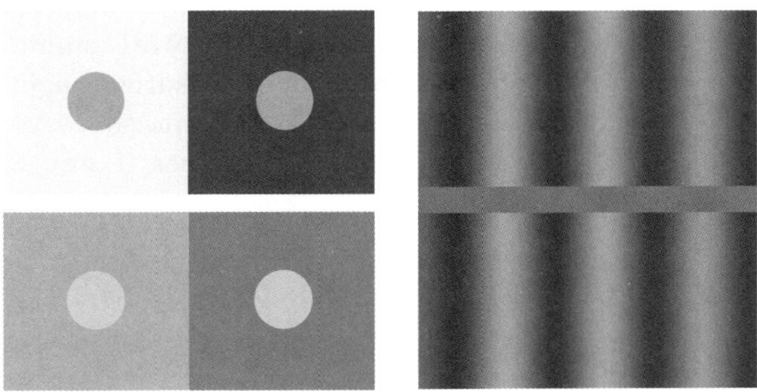

FIGURE 35.3   (Left) Two forms of classic simultaneous contrast. The two patches in each figure have the same luminance. (Top left) The left patch is an increment, the right patch a decrement. (Bottom left) Both patches are increments. (Right) Grating induction stimulus; the narrow stripe running horizontally through the middle is uniform in luminance.

the retina plays a critical role. Light adaptation, the process whereby the visual system adjusts to the local average light level, is universally believed to result from gain changes among *retinal* neurons. Thus, it is the retina that normalizes luminance differences to the local light level, converting those differences into the contrasts (or ratios) that are carried in the signals from retinal ganglion cells and that form an essential component of lightness constancy (Shapley & Enroth-Cugell, 1984; Walraven et al., 1990).

The evidence for the involvement of the cortex in brightness and lightness perception has emerged from a number of quarters. Psychophysical studies have shown that perceived depth relations, which are encoded in the cortex, influence lightness and brightness perception (reviewed by Gilchrist, 2006, pp. 120–122 and 159–172; the notable exception to the rule is Schirillo, Reeves, & Arend's 1990 study, which found an effect of depth on lightness but not brightness). Studies have also shown that changes to the perceived three-dimensional structure of a stimulus can cause edges to switch in appearance from being material to nonuniform illumination (Knill & Kersten, 1991; Buckley, Frisby, & Freeman, 1994; Logvinenko & Menshikova, 1994), consistent with a cortical involvement in lightness and brightness perception.

Physiological evidence for a cortical involvement includes the presence of cells or brain areas that respond to changes in surface lightness but that are unresponsive either to the borders of the surfaces or to changes in illumination. For example, Roe, Lu, and Hung (2005) reported cells in monkey V2 that responded to the purely illusory lightness variations in the Cornsweet illusion, and MacEvoy and Paradiso (2001) described cells in cat V1 whose responses were modulated by interactions from outside the classical receptive field in such a way as to make them immune to changes in illumination. Human fMRI studies report strong responses to changes in surface lightness but not surface borders in retinotopic areas V1, V2, and V3 (Boyaci et al., 2007; Haynes, Lotto, & Rees, 2004).

MODELS OF BRIGHTNESS AND LIGHTNESS

Types of Error

The models discussed in this section are motivated primarily by the errors in lightness and brightness perception. Although in everyday vision we rarely encounter lightness/brightness errors in the form of blatantly contradictory percepts, a plethora of laboratory stimuli have been carefully fashioned to flush them out. The

FIGURE 35.4   White's effect. The two sets of gray bars are the same luminance. (Based on the original stimulus in White, 1979.)

best known "error" is the classic simultaneous contrast display shown top left of figure 35.3. The two equal-luminance patches set against different-luminance backgrounds appear different in lightness/brightness. This stimulus, together with its innumerable variants (such as the double-increment figure shown bottom left), continues to play a big part in the scientific investigation of lightness and brightness. A form of simultaneous contrast that has also been influential is the grating induction stimulus shown on the right of figure 35.3 (McCourt, 1982).

Another class of lightness/brightness error is assimilation. Assimilation is the opposite of contrast in that lightness/brightness shifts *toward* rather than away from that of the background. The conventional form of White's effect (White, 1979), shown here in figure 35.4, is an example of assimilation. The two sets of gray bars have the same luminance but differ markedly in lightness/brightness. In this case, however, the lightnesses/brightnesses of the test bars are shifted toward those of the flanking phases of the grating, which abut more with the test bars than the coaxial phases.

I now describe five classes of models of brightness/lightness that have been influential in recent years,

   FREDERICK A. A. KINGDOM

termed *edge integration, Gestalt anchoring, multiscale filtering, intrinsic image,* and *empirical.*

Edge Integration Models

Land and McCann (1971) first championed the idea that lightness was computed by integrating local edge contrasts across the image. Their Retinex algorithm was designed to recover the lightnesses of Mondrian-like patterns—these are arrays of rectangles with quasi-random spatial dimensions and reflectances—that were subject to gradual illumination gradients such as shading. Lightness computation was achieved by a four-stage process: (1) the detection of edges; (2) removal of any weak edges, such as those due to the shading; (3) reintegration of the remaining edges to recreate the original rectangles and compute their relative lightnesses; and (4) anchoring to convert the relative lightnesses into absolute ones. The Retinex, as well as the models it spawned (e.g., Hurlbert & Poggio, 1988; Land, 1986) would fail, however, with the photograph in figure 35.1 because sharp-bordered shadows would be incorrectly identified as reflectance changes. The Retinex also fails to predict the illusory lightness differences in simultaneous contrast and White's effect because it is designed to generate accurate estimates of lightness.

The most recent expression of the lightness integration approach is the class of models associated with Rudd and colleagues (Rudd & Popa, 2007; Rudd & Zemach, 2007). These models combine edge integration with the physiological process known as *contrast gain control.* The contrast gain control acts between neighboring edge detectors, with the gain of each edge detector being a positive function of the response magnitudes of neighboring edge detectors and a negative function of their distances from each other. These models have overcome the limitations of previous lightness integration models in that they are capable of accounting for lightness errors such as simultaneous contrast and assimilation.

Gestalt-Anchoring Models

Gestalt-anchoring models are models of lightness perception that posit multiple anchors within nested image frameworks. The frameworks are derived from classic Gestalt grouping principles such as "belongingness" as well as principles such as common depth (Bressan, 2006a, 2006b, 2007; Gilchrist, 2006; Gilchrist et al., 1999).

The principle of the best known of these models, that of Gilchrist and colleagues, can be grasped by considering how it applies to the simultaneous contrast display in the top left of figure 35.3. In the model the stimulus is divided into two local and one global perceptual framework, with the lightness of each test patch computed as a weighted average of two lightness values—one derived from the local, the other from the global, framework. The global framework comprises the stimulus as a whole, whereas the local frameworks consist of the two surrounds and their respective test patches. Within each framework the anchor is the highest luminance, which is assigned white, and all other regions within the framework are assigned grays according to their luminance ratio with respect to the white anchor. Within the global framework, the white background of the entire stimulus is the highest luminance and therefore assigned white, and the two test patches are assigned identical lower lightnesses because their luminance ratios with respect to the background are the same. However, for the two local frameworks, the patch lightness assignments are different. For the patch on the dark surround, the patch is the highest luminance and therefore assigned white. For the patch on the light surround, the surround is the highest luminance and therefore assigned white, and the patch is assigned a value relative to it—a midgray. The net patch lightnesses are the averages of the global framework values (both patches equal and midgray) and the local framework values (patch on dark surround white, patch on light surround midgray). The result is a higher lightness value for the patch on the dark compared to the light surround. Thus, although the global framework has an important influence on the lightnesses of the patches, it is the influence of the two local frameworks that causes the difference in patch lightness and hence the illusion.

One notable failure of the model is that it predicts that increments of equal luminance on different backgrounds should appear equal in lightness. However, as the bottom left of figure 35.3 shows, and as confirmed by psychophysical studies (Blakeslee, Reetz, & McCourt, 2009; Bressan, 2006a, 2006b; Bressan & Actis-Grosso, 2001; Rudd & Zemach, 2005), they do not appear equal. Bressan's (2006a, 2006b, 2007) "double-anchoring" model solves this problem by positing two anchors per framework: highest-luminance-is-white and surround-luminance-is-white.

Kingdom (2011) has argued that in their present form both edge-integration and Gestalt-anchoring models are limited to dealing with a Mondrian world—one with regions of uniform lightness demarcated by sharp edges—and would have difficulty handling the complexity of natural scenes. Rudd and colleagues are aware of this limitation and anticipate incorporating multiscale filtering into their edge-integration model (Rudd & Zemach, 2005).

Multiscale-Filtering Models

Multiscale-filtering models seek to explain brightness errors in terms of the known physiological response properties of early cortical neurons that are selective for orientation and spatial frequency, such as simple cells. The best-known of these models is the ODOG (oriented difference-of-Gaussian) model of Blakeslee, McCourt, and colleagues (Blakeslee & McCourt, 2004; Blakeslee, Pasieka, & McCourt, 2005; Blakeslee, Reetz, & McCourt, 2009).

In the ODOG model two processes conspire to produce brightness errors. First, very low spatial frequencies are attenuated, and this accounts for a variety of contrast errors such as simultaneous contrast and grating induction, as noted also by others (Perna & Morrone, 2007; Shapiro & Lu, 2011). The second process is contrast normalization, which in the ODOG model equates the filter responses across the six orientation channels, each of which is itself a weighted linear sum of seven spatial-frequency channels. The contrast normalization stage is the key to understanding assimilation phenomena such as White's effect. The most responsive filters to White's stimulus are the relatively high-spatial-frequency vertically-oriented filters tuned to the inducer grating. Contrast normalization has the effect of reducing the responses of these filters relative to those sensitive to horizontal orientations. The low-frequency horizontally oriented filters pool the luminances of the flanking bars with those of the test patches, and it is their relative enhancement that delivers the illusion.

Robinson, Hammon, and de Sa (2007) have argued that the predictive power of the ODOG model is improved if the contrast normalization stage is restricted to nearby receptive fields and nearby spatial frequencies. In a related model by Dakin and Bex (2003), the contrast normalization equates filter responses across spatial frequency, not orientation, and this appears to provide a good explanation for the Craik-Cornsweet-O'Brien illusion (see figure 35.6). "Contrast" theories of brightness and lightness perception, of which multiscale-filtering models are the most modern expression, have traditionally been criticized for failing to take into account the effects of remote context, on the grounds that they only deal with contrast in the immediate vicinity of the edge. However, this is a red herring: In multiscale spatial-filtering models remote context makes its impact via the coarse-scale (low-spatial-frequency) filters.

One of the undoubted strengths of multiscale-filtering models is that they are potentially applicable to modeling brightness variations in everyday scenes because any image can be passed through an array of model filters, and the combined filter responses require no rules of interpretation. It has been argued that a present limitation of multiscale-filtering models is that they do not categorize luminance variations as either reflectance or illumination and hence fail to take into account the specific effects of nonuniform illumination on brightness mentioned earlier.

Intrinsic-Image Models

The influence of nonuniform illumination on brightness underpins an approach—termed here the intrinsic-image model—that elevates layer decomposition to the forefront of explaining lightness and brightness perception (Anderson, 2001; Arend, 1994; Bergström, 1994). This approach has its origins in the ideas of Helmholtz (1866/1962) (see Kingdom, 1997). Intrinsic-image models do not in general make quantitative predictions of brightness: their proponents focus on the creation of captivating demonstrations of the effects of perceived nonuniform illumination and transparency on brightness. Two examples are shown in figure 35.5. The allure of these figures is the sheer magnitude of their illusory brightness differences, which appear to far surpass those of standard simultaneous contrast. Each appears to demonstrate that depictions of nonuniform illumination or transparency cause brightness to shift toward the "true" lightnesses of the test regions "beneath" the transparencies/shadows.

Impressive as these demonstrations are, a number of researchers have urged caution in their interpretation (Blakeslee & McCourt, 2012; Kingdom, 2011; Todorovic, 2006). For example, in the snake/antisnake figure, although the difference in brightness between the two rows of equiluminant diamonds is bigger in the snake compared to the antisnake figure, there are significant differences in the contrast composition of the two figures, and this might be the cause. In the snake figure, the upper row of diamonds is surrounded by a larger area of black, and the lower row of diamonds by a larger area of white, than the corresponding diamonds in the antisnake figure. Recent studies have suggested that it is these contrast differences, rather than the process of perceptual layer decomposition, that underlie the difference in the magnitude of the illusion size between the two figures (Shapiro & Lu, 2011; Blakeslee & McCourt, 2012).

Notwithstanding the above caveats, intrinsic-image models have provided a lush new terrain for exploring both the visual cues that facilitate layer decomposition (reviewed by Kingdom, 2008) and the impact of nonuniform illumination and transparency on our perception of brightness and lightness.

 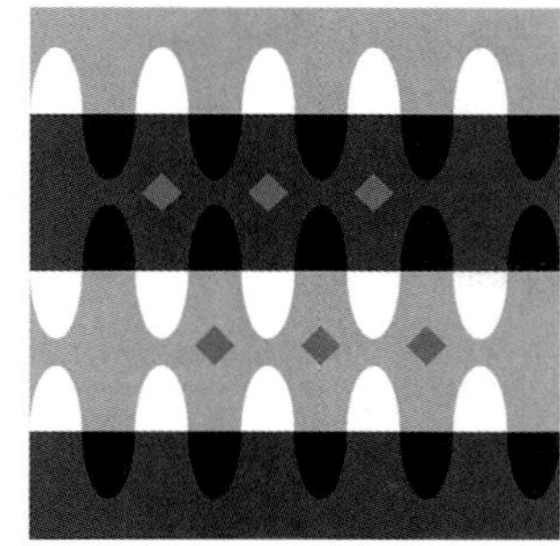 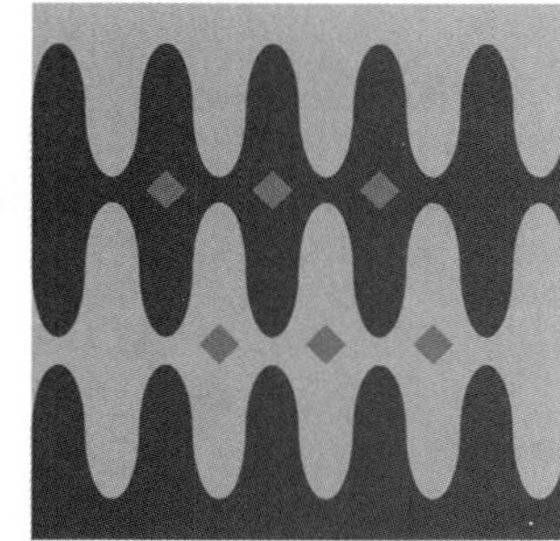

FIGURE 35.5   (Left) Logvinenko's (1999) figure of a wall of blocks with shading. All rows of diamonds have the same luminance, but alternating rows differ dramatically in both brightness and lightness. (Figure supplied by the author.) (Middle and Right) Adelson's "snake" (middle) and "antisnake" (right) figures. The small diamonds in both figures all have the same luminance. The brightness difference between the upper and lower rows of diamonds is much bigger in the snake compared to the antisnake figure. (From Adelson, 2000, reproduced with permission.)

Empirical Models

In their empirical approach to lightness perception, Purves & Lotto (2003) suggest that when organisms are confronted with the need to identify reflectances in the context of spatially nonuniform illumination, they estimate the most likely reflectance values based on the pattern of luminances observed, together with their knowledge of image statistics learned through goal-directed behavior. Lightness illusions occur because in any given situation the most likely value of reflectance will often differ from its true value. For example, in the case of simultaneous contrast, Purves and Lotto argue that patches on dark backgrounds are more likely to be lying in shadow compared to patches on bright backgrounds. Hence, two equal-in-luminance patches, one on a dark the other on a bright background, will likely have different reflectances, and this is how they are perceived (Purves & Lotto, 2003). Other illusions such as the Craik-Cornsweet-O'Brien illusion (figure 35.6) are similarly explained: The illusory percept matches physical illumination patterns that arise because of the manner in which nondiffuse illumination is reflected from three-dimensional objects.

Although the empirical approach appears to echo many of the themes of intrinsic-image models, it does not posit the need for an explicit stage of perceptual layer decomposition. Rather, image statistics learned through goal-directed behavior are used to make lightness estimates. To illustrate the operation of the empirical approach as it relates to lightness errors, Corney & Lotto (2007) trained an artificial back-propagation neural network to identify the reflectances of target surfaces in synthetic images. The synthetic images consisted of three-dimensional arrangements of multiple-sized reflectance patches subject to simulated nonuniform illumination. The only sense data available to

the network were the luminances of the patches, so the network had to learn to identify surface reflectance in spite of the inherent ambiguity of patch luminance. Having learned to identify the target reflectances in the synthetic images to a criterion level of accuracy, the network was then required to estimate the reflectances of target patches in classic lightness-illusion displays such as simultaneous contrast, White's effect, and Mach bands. The network made very similar lightness errors to those reported by human observers.

It has been argued that the empirical approach is not so different from mechanistic approaches such as spatial-filtering and intrinsic-image models as might at first be thought (Kingdom, 2011). For example, the responses of many cortical neurons, such as simple cells, are designed to be largely invariant to the ambient level of illumination, so the output of such filters will, on average, more closely correlate with the pattern of image reflectances than with the pattern of image luminances. In other words cortical neurons serve to reduce the range of lightness values from which the visual system must choose. By the same token, a mechanism that discounts spatially varying illumination via a process of layer decomposition also reduces the potential range of lightness choices. In short, the mechanisms deployed by vision for coding lightness (multiscale filtering, contrast gain control, layer decomposition, etc.) have been honed during evolution and/or development to maximize the probability of judging lightness correctly, but being imperfect, they nevertheless make errors. The difference between the empirical and mechanistic approaches to lightness perception may in the end come down to a difference in the type of image statistics exploited by vision for judging lightness. For example, perceptual layer decomposition is believed to exploit relatively high-order statistical relations such as X-junctions (these are the points in the middle figure in figure 35.5 where two borders cross) and three-dimensional shape (Kingdom, 2008), whereas the neural networks employed in Corney and Lotto's study presumably capture relatively low-order statistical relations. Sensitivity to high-order image statistics is a defining property of human vision, so one should not be surprised to find that these statistics are exploited by vision for lightness perception.

"Filling-In" Models

One of the enduring themes in vision science is that the lightness/brightness of uniform regions results from the propagation or spreading of neural activity from the edges of the region inward—termed "filling in" (Grossberg & Todorovic, 1988; Paradiso & Nakayama, 1991; Rossi & Paradiso, 1996). The term "filling in" is not, however, synonymous with neural spreading—it is also a metaphor for the representation of uniform areas by low spatial frequencies (Dakin & Bex, 2003). However, it is the idea that filling in involves neural spreading that concerns us here.

Although there is psychophysical (Paradiso & Hahn, 1996; Paradiso & Nakayama, 1991; Rossi & Paradiso, 1996), neurophysiological (Huang & Paradiso, 2008), and brain-imaging (Pereverzeva & Murray, 2008) evidence in favor of the neural-spreading concept, it has recently been challenged (Blakeslee & McCourt, 2008; Cornelissen et al., 2006; Dakin & Bex, 2003). Blakeslee & McCourt (2008) measured the time course of brightness induction in the grating-induction stimulus (figure 35.3, right). They employed a highly sensitive method that exploited the visual system's keen motion sensitivity in order to leverage tiny temporal differences into conspicuous changes in motion direction. They found that the temporal response of the induced brightness variations in the uniform stripe in the grating induction stimulus lagged by less than 1 ms and, furthermore, was constant across wide variations in the stripe's height. The fact that a method capable of measuring the time course of brightness induction with millisecond precision showed that it is virtually instantaneous constitutes a serious difficulty for the neural spreading idea because neural spreading must presumably have a time course.

Dakin and Bex (2003) took a new look at what is often heralded as the flagship of neural spreading: the Craik-Cornsweet-O'Brien (CCOB) illusion. In the CCOB stimulus a step function in luminance is connected to uniform, equal-in-luminance regions on either side by sharp ramps. Yet the stimulus appears to be a step edge (figure 35.6, top). The neural spreading idea is that neural signals propagate outward from the edge to fill in the regions on either side and deliver the

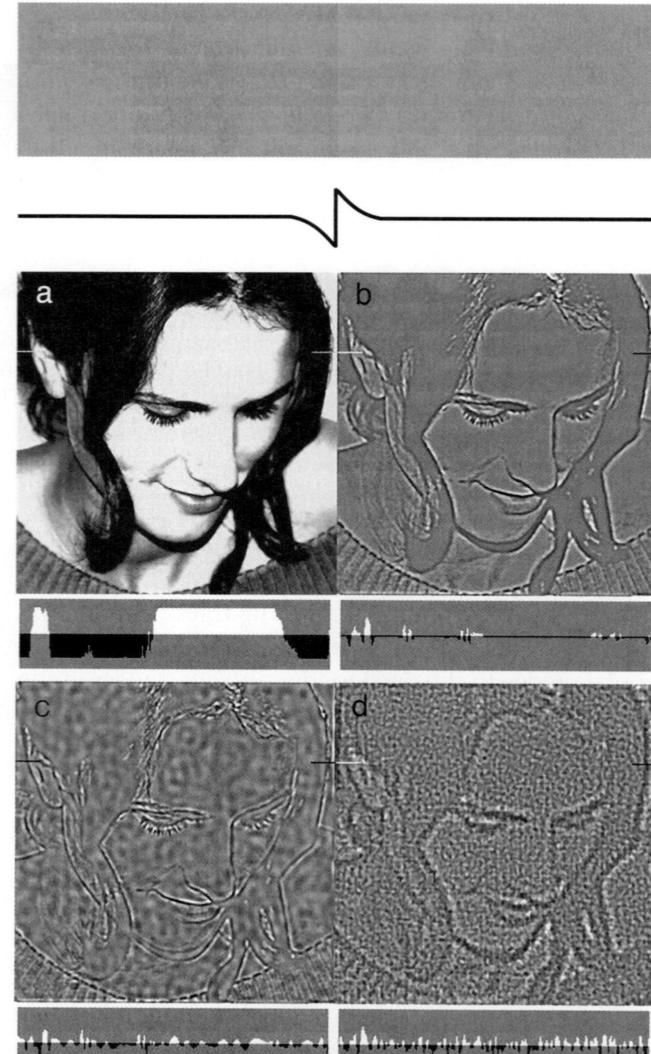

FIGURE 35.6　(Top) Craik-Cornsweet-O'Brien illusion. The regions either side of the edge have the same luminance but differ in brightness. (Bottom) Illusion applied to an image of a face. (a) Input image; (b) attenuation of low spatial frequencies by high-pass filtering produces the illusion; (c) if the low frequencies are completely removed rather than attenuated, the illusion disappears; (d) adding luminance noise to b. does not destroy the illusion. Below each figure is the cross-section luminance profile at the point marked by the short horizontal lines. (From Dakin & Bex, 2003, reproduced with permission.)

illusion. Dakin & Bex (2003), however, show that the CCOB is critically dependent on the presence of residual low-spatial-frequency information in sine-phase at the edge: If the low frequencies are phase-scrambled, the illusion disappears (figure 35.6, bottom). They propose that contrast normalization, which in their model tends to equalize the energy across equal log bands of spatial frequency, boosts the residual low-frequency information, producing the illusion. They also show, as did Burr (1987), that filling the CCOB stimulus with luminance noise, which would be expected to prevent any propagation of neural signals from the edges, fails to destroy the illusion.

SUMMARY

To compute lightness, or perceived reflectance, the visual system must discount the effects of ambient as well as spatial nonuniform illumination such as shadows and shading. Although the detection of contrast followed by anchoring is believed to be critical to the computation of lightness, there is no consensus as to how these processes are instantiated. Models of lightness perception such as edge-integration and Gestalt-anchoring models successfully account for lightness errors in only a restricted set of laboratory stimuli and do not lend themselves easily to more complex stimuli such as everyday scenes. Multiscale-filtering models of brightness perception, on the other hand, have potentially wider applicability but have been criticized for failing to take into account the effects of spatially nonuniform illumination on brightness perception. Intrinsic-image models of brightness perception, using captivating demonstrations of the effects of nonuniform illumination on brightness, have highlighted the importance of perceptual layer decomposition in brightness and lightness perception. However, intrinsic-image models are not yet sufficiently developed to make quantitative predictions and have been criticized for exaggerating the importance of perceived nonuniform illumination and transparency on brightness perception. The traditional idea that the brightness of uniform areas results from the spreading of edge signals to the interior has been recently challenged by studies showing that brightness induction is near instantaneous and involves low-spatial-frequency information. Finally, Whittle's studies with simple patch–background displays have revealed a metric, log W, that provides a good account of the relationship between brightness and contrast.

ACKNOWLEDGMENTS

This chapter was supported by a Canadian Institute of Health (CIHR) research grant No.11554 given to the author. Special thanks to Mike Webster and Arthur Shapiro for their support and comments on an earlier draft, and thanks to all those who provided figures.

REFERENCES

Adelson, E. H. (1993). Perceptual organization and the judgement of brightness. *Science, 262,* 2042–2044.

Adelson, E. H. (2000). Lightness perception and lightness illusions. In M. Gazzaniga (Ed.), *The new cognitive neuroscience* (2nd ed., pp. 339–351). Cambridge, MA: MIT Press.

Agostini, T., & Galmonte, A. (2002). A new effect of luminance gradient on achromatic simultaneous contrast. *Psychonomic Bulletin & Review, 9,* 264–269.

Anderson, B. L. (1997). A theory of illusory lightness and transparency in monocular and binocular images: The role of contour junctions. *Perception, 26,* 419–453.

Anderson, B. L. (2001). Contrasting theories of White's illusion. *Perception, 30,* 1499–1501.

Arend, L. E. (1994). Surface colors, illumination, and surface geometry: Intrinsic-image models of human color perception. In A. L. Gilchrist (Ed.), *Lightness, brightness, and transparency* (pp. 159–213). Hillsdale, NJ: Lawrence Erlbaum Associates.

Arend, L. R., & Spehar, B. (1993a). Lightness, brightness and brightness contrast: 1. Illuminance variation. *Perception & Psychophysics, 54,* 446–456.

Arend, L. R., & Spehar, B. (1993b). Lightness, brightness and brightness contrast: 2. Reflectance variation. *Perception & Psychophysics, 54,* 457–468.

Bergström, S. S. (1994). Color constancy: Arguments for a vector model for the perception and illumination, color and depth. In A. L. Gilchrist (Ed.), *Lightness, brightness, and transparency* (pp. 215–255). Hillsdale, NJ: Lawrence Erlbaum Associates.

Blakeslee, B., & McCourt, M. E. (2004). A unified theory of brightness contrast and assimilation incorporating oriented multiscale spatial filtering and contrast normalization. *Vision Research, 44,* 2483–2503. doi:10.1016/j.visres.2004.05.015.

Blakeslee, B., & McCourt, M. E. (2008). Nearly instantaneous brightness induction. *Journal of Vision, 8,* 1–8. doi:10.1167/8.2.15.

Blakeslee, B., & McCourt, M. E. (2012). When is spatial filtering enough? Investigation of brightness and lightness perception in stimuli containing a visible illumination component. *Vision Research, 60,* 40–50. doi:10.1016/j.visres.2012.03.006.

Blakeslee, B., Pasieka, W., & McCourt, M. E. (2005). Oriented multiscale spatial filtering and contrast normalization: A parsimonious model of brightness induction in a continuum of stimuli including White, Howe and simultaneous brightness contrast. *Vision Research, 45,* 607–615. doi:10.1016/j.visres.2004.09.027.

Blakeslee, B., Reetz, D., & McCourt, M. E. (2009). Spatial filtering versus anchoring accounts of brightness/lightness perception in staircase and simultaneous brightness/lightness contrast stimuli, *Journal of Vision, 9,* 1–17. doi:10.1167/9.3.22.

Boyaci, H., Fang, F., Murray, S. O., & Kersten, D. (2007). Responses to lightness variations in early visual cortex. *Current Biology, 17,* 989–993.

Bressan, P. (2006a). The place of white in a world of grays: A double-anchoring theory of lightness perception. *Psychological Review, 113,* 526–553.

Bressan, P. (2006b). Inhomogenous surrounds, conflicting frameworks, and the double-anchoring theory of lightness.

Psychonomic Bulletin & Review, 13, 22–32. doi:10.3758/BF03193808.

Bressan, P. (2007). Dungeons, gratings, and black rooms: A defense of double-anchoring theory and a reply to Howe et al. (2007). *Psychological Review, 114,* 1111–1115. doi:10.1037/0033-295X.114.4.1111.

Bressan, P., & Actis-Grosso, R. (2001). Simultaneous lightness contrast with double increments. *Perception, 30,* 889–897.

Buckley, D., Frisby, J. P., & Freeman, J. (1994). Lightness perception can be affected by surface curvature from stereopsis. *Perception, 23,* 869–881.

Burr, D. C. (1987). Implications of the Craik-O'Brien illusion for brightness perception. *Vision Research, 27,* 1903–1913. doi:10.1016/0042-6989(87)90056-3.

Cornelissen, F. W., Wade, A. R., Vladusich, T., Dougherty, R. F., & Wandell, B. A. (2006). No functional magnetic resonance imaging evidence for brightness and color filling-in in early human visual cortex. *Journal of Neuroscience, 26,* 3633–3641.

Corney, D., & Lotto, B. (2007). What are lightness illusions and why do we see them? *PLoS Computational Biology, 3,* 1790–1800. doi:10.1371/journal.pcbi.0030180.

Cornsweet, T. N. (1970). *Visual Perception.* London: Academic Press.

Correani, A., Scott-Samuel, N. E., & Leonard, U. (2006). Luminosity—a perceptual "feature" of light-emitting objects? *Vision Research, 46,* 3915–3925. doi:10.1016/j.visres.2006.05.001.

Dakin, S. C., & Bex, P. J. (2003). Natural image statistics mediate brightness "filling-in." *Proceedings. Biological Sciences, 270,* 2341–2348. doi:10.1371/journal.pcbi.0030180.

Fiorentini, A., Baumgartner, G., Magnusson, S., Schiller, P. H., & Thomas, J. P. (1990). The perception of brightness and darkness. In L. Spillman & J. S. Werner (Eds.), *Visual perception: The neurophysiological foundations* (pp. 129–161). San Diego: Academic Press.

Gilchrist, A. (2006). *Seeing black and white* (pp. 180–187). Oxford: Oxford University Press.

Gilchrist, A., Kossyfidis, C., Bonato, F., Agostini, T., Cataliotti, J., Li, X., et al. (1999). An anchoring theory of lightness perception. *Psychological Review, 106,* 795–834.

Grossberg, S., & Todorovic, D. (1988). Neural dynamics of 1-D and 2-D brightness perception: A unified model of classical and recent phenomena. *Perception & Psychophysics, 43,* 241–277.

Haynes, J., Lotto, R. B., & Rees, G. (2004). Responses of human visual cortex to uniform surfaces. *Proceedings of the National Academy of Sciences of the United States of America, 101,* 4286–4291. doi:10.1098/rspb.2003.2528.

Helmholtz, H. von. (1962). *Treatise on physiological optics* (Vol. II, pp. 264–301). (J. P. L. Southall, Trans.). New York: Dover Publications. (Original work published 1866.)

Hering, E. (1964). *Outlines of a theory of the light sense* (L. M. H. D. Jameson, Trans.). Cambridge, MA: Harvard University Press. (Original work published 1874.)

Huang, P.-C., Kingdom, F. A. A., & Hess, R. F. (2006). Only two phase mechanisms, ±cosine, in human vision. *Vision Research, 46,* 2069–2081. doi:10.1016/j.visres.2005.12.020.

Huang, X., & Paradiso, M. A. (2008). V1 response timing and surface filling-in. *Journal of Neurophysiology, 100,* 539–547.

Hurlbert A., & Poggio, T. A. (1988). Synthesizing a color algorithm from examples. *Science, 239,* 484–485.

Jacobsen, A., & Gilchrist, A. (1988). The ratio principle holds over a million-to-one range of illumination. *Perception & Psychophysics, 43,* 1–6. doi:10.3758/BF03208966.

Kingdom, F. A. A. (1997). Simultaneous contrast: The legacies of Hering and Helmholtz. *Perception, 26,* 673–677.

Kingdom, F. A. A. (2008). Perceiving light versus material. *Vision Research, 48,* 2090–2105. doi:10.1016/j.visres.2008.03.020.

Kingdom, F. A. A. (2011). Lightness, brightness and transparency: A quarter century of new ideas, captivating demonstrations and unrelenting controversy [Invited review]. *Vision Research, 51,* 652–673. doi:10.1016/j.visres.2010.09.012.

Kingdom, F. A. A., Blakeslee, B., & McCourt, M. E. (1997). Brightness with and without perceived transparency: When does it make a difference? *Perception, 26,* 493–506.

Kingdom, F. A. A., & Whittle, P. (1996). Contrast discrimination at high contrasts reveals the influence of local light adaptation on contrast processing. *Vision Research, 36,* 817–829.

Knill, D. C., & Kersten, D. (1991). Apparent surface curvature affects lightness perception. *Nature, 351,* 228–230.

Land, E. H. (1986). An alternative technique for the computation of the designator in the retinex theory of color vision. *Proceedings of the National Academy of Sciences of the United States of America, 83,* 3078–3080. doi:10.1073/pnas.83.10.3078.

Land, E. H., & McCann, J. J. (1971). Lightness and retinex theory. *Journal of the Optical Society of America, 61,* 1–11. doi:10.1364/JOSA.61.000001.

Logvinenko, A. D. (1999). Lightness induction revisited. *Perception, 28,* 803–816.

Logvinenko, A. D. (2005). On achromatic colour appearance. In J. L. Nieves and J. Hernanderez-Andres (Eds.), *AIC Colour 05. The 10th Congress of the International Colour Association. May 8–13, 2005, Granada, Spain* (Part 1, pp. 639–642).

Logvinenko, A., & Menshikova, G. (1994). Trade-off between achromatic colour and perceived illumination as revealed by the use of pseudoscopic inversion of apparent depth. *Perception, 23,* 1007–1023.

MacEvoy, S. P., & Paradiso, M. A. (2001). Lightness constancy in primary visual cortex. *Proceedings of the National Academy of Sciences of the United States of America, 98,* 8827–8831. doi:10.1073/pnas.161280398.

McCourt, M. E. (1982). A spatial frequency dependent grating induction effect. *Vision Research, 22,* 119–134.

McIlhagga, W., & Peterson, R. (2006). Sinusoid = light bar + dark bar? *Vision Research, 46,* 1934–1945. doi:10.1016/j.visres.2005.12.004.

Paradiso, M. A., & Hahn, S. (1996). Filling-in percepts produced by luminance modulation. *Vision Research, 36,* 2657–2663. doi:10.1016/0042-6989(96)00033-8.

Paradiso, M. A., & Nakayama, K. (1991). Brightness perception and filling-in. *Vision Research, 31,* 1221–1236. doi:10.1016/0042-6989(91)90047-9.

Pereverzeva, M., & Murray, S. O. (2008). Neural activity in human V1 correlates with dynamic lightness induction. *Journal of Vision, 8,* 1–10. doi:10.1167/8.15.8.

Perna, A., & Morrone, M. C. (2007). The lowest spatial frequency channel determines brightness perception. *Vision Research, 47,* 1282–1291. doi:10.1016/j.visres.2007.01.011.

Purves, D., & Lotto, R. B. (2003). *Why we see what we do: An empirical theory of vision.* Sunderland, MA: Sinauer Associates.

Roe, A. W., Lu, H. D., & Hung, C. P. (2005). Cortical processing of a brightness illusion. *Proceedings of the National Academy of Sciences of the United States of America, 102,* 3869–3874. doi:10.1073/pnas.0500097102.

Robinson, A. E., Hammon, P. S., & de Sa, V. R. (2007). Explaining brightness illusions using spatial filtering and local response normalization. *Vision Research, 47,* 1631-1644. doi:10.1016/j.visres.2007.02.017

Rossi, A. F., & Paradiso, M. A. (1996). Temporal limits of brightness induction and mechanisms of brightness perception. *Vision Research, 36,* 1391–1398. doi:10.1016/0042-6989(95)00206-5.

Rudd, M. E., & Popa, D. (2007a). Stevens's brightness law, contrast gain control, and edge integration in achromatic color perception: A unified model. *Journal of the Optical Society of America. A, Optics, Image Science, and Vision, 24,* 2766–2782. doi:10.1364/JOSAA.24.002766.

Rudd, M. E., & Popa, D. (2007b). Stevens's brightness law, contrast gain control, and edge integration in achromatic color perception: A unified model. Errata. *Journal of the Optical Society of America. A, Optics, Image Science, and Vision, 24,* 3335. doi:10.1364/JOSAA.24.003335.

Rudd, M. E., & Zemach, I. K. (2005). The highest luminance rule in achromatic color perception: Some counterexamples and an alternative theory. *Journal of Vision, 5,* 983–1003. doi:10.1167/5.11.5.

Rudd, M. E., & Zemach, I. K. (2007). Contrast polarity and edge integration in achromatic color perception. *Journal of the Optical Society A. Optics, Image Science, and Vision, 24,* 2134–2156. doi:10.1364/JOSAA.24.002134.

Schiller, P. H. (1982). Central connections of the retinal ON and OFF pathways. *Nature, 297,* 580–583.

Schirillo, J., Reeves, A., & Arend, L. (1990). Perceived lightness, but not brightness, of achromatic surfaces depends on perceived depth information. *Perception & Psychophysics, 48,* 82–90.

Shapiro, A., & Lu, Z.-L. (2011). Relative brightness in natural images can be accounted for by removing blurry content. *Psychological Science, 22*(11), 1452–1459.

Shapley, R., & Enroth-Cugell, C. (1984). Visual adaptation and retinal gain controls. In N. N. Osbourne & G. J. Chader (Eds.), *Retinal research* (pp. 263–346). Oxford: Pergamon Press.

Todorovic, D. (2006). Lightness, illumination, and gradients. *Spatial Vision, 19,* 219–261.

Vladusich, T. (2012). Simultaneous contrast and gamut relativity in achromatic color perception. *Vision Research, 69,* 49–63. doi:10.1016/j.visres.2012.07.022

Wallach, H. (1976). *On Perception.* New York: Quadrangle/The New York Times Book Co.

Walraven, J., Enroth-Cugell, C., Hood, D. C., MacLeod, D. I. A., & Schnapf, J. L. (1990). The control of visual sensitivity: Receptoral and postreceptoral processs. In L. Spillman & J. S. Werner (Eds.), *Visual perception: The neurophysiological foundations* (pp. 53–101). San Diego, CA: Academic Press.

White, M. (1979). A new effect on perceived lightness. *Perception, 8,* 413–416.

Whittle, P. (1994). The psychophysics of contrast brightness. In A. L. Gilchrist (Ed.), *Lightness, Brightness, and Transparency* (pp. 35–110). Hillsdale, NJ: Lawrence Erlbaum Associates.

Zavagno, D., & Caputo, G. (2001). The glare effect and the perception of luminosity. *Perception, 30,* 209–222.

36 Color Appearance, Language, and Neural Coding

DELWIN T. LINDSEY AND ANGELA M. BROWN

Color appearance refers to our perceptual experiences that are linked to the spectral composition of the light entering the eye. In simple terms color appearance refers to what the nonspecialist would call the *color* of something, as embodied in its perceived qualities such as hue, saturation, and brightness. Color appearance is also related to color *qualia*, a term used by philosophers when referring to the quality of experiencing the "coloredness" of something (for example, the redness of a red tomato).

Although color is related to the spectral composition of the light contained in the visual stimulus, color itself does not exist outside the human mind. Rather, it is an emergent property of the mapping of the spectral composition of the incoming light onto a series of internal *color representations* within the visual nervous system. These representations start with the responses of the photoreceptors to the light falling onto the retina and continue with the increasingly complex patterns of connectivity among neurons at higher and higher levels of visual information processing. On one hand our understanding of some of these representations is properly in the domain of neurophysiology. On the other hand other representations are known only from behavioral experiments. The challenge in the study of color appearance is to relate these two types of color representations to our phenomenal experience of color and to other perceptual and cognitive capabilities.

The subjective nature of color appearance is clear from the lack of any strictly one-to-one mapping between the spectral composition of the light reaching the eye and the color appearance of the stimulus the light comes from. The color appearance of a stimulus that is fixed in spectral composition can vary greatly, depending (among other things) on the colors of the stimuli located elsewhere in the visual field, a one-to-many mapping. Some spectrally dissimilar lights look identical when they are metamerically matched, and they can look similar or identical when each is viewed in a different color context, examples of many-to-one mappings. Finally, a particular many-to-one mapping occurs when the huge range of discriminable colors is reduced to a few named color categories. Each of these mappings occurs in the mind of the observer and cannot be deduced from the physical nature of light.

COLOR MATCHING AND COLOR APPEARANCE

The best-understood and most rigorous treatment of color vision is based on color matching, a procedure in which the observer adjusts two spectrally dissimilar mixtures of lights until they are perceptually indistinguishable from one another. The indistinguishable lights are called *metameric* lights, and the measurement of color based on the matching procedure is called *colorimetry* (see Brainard & Stockman, 2010, for a contemporary review). The basic result of a color-matching experiment is that a fixed standard light, contained in one patch of color, can be matched by adjusting the intensity of no more than three primary lights of fixed spectral composition but variable intensity contained in a second, adjacent patch of color (in some cases, one of the primaries must be moved over and added to the standard light). When the two patches of color are matched in this way, they cannot be distinguished by any means: they are physiologically identical because the color-matching procedure equates the two lights for the quantum catch rates in the three cone types. In this sense normal human color vision is said to be trichromatic, and the representation of normal color matching is three-dimensional.

Every visible light can be represented as a triplet a, where the elements, a_i ($i = 1 \ldots 3$), of the triplet correspond to the intensities of each of the three primaries required to establish the match. This immediately suggests that the lights could be represented in a three-dimensional Euclidean space, and indeed, several important color spaces are based on a or linear transformations of a. The values of a_i that match a set of monochromatic lights across the visible spectrum are called the *color-matching functions* and are a linear transformation of the spectral sensitivities of the cones (König & Dieterici, 1893).

Despite its quantitative rigor, color matching cannot specify how lights actually appear. All that color matching can provide is information about the conditions under which lights of different spectral composition are indiscriminable from one another. Nonetheless, the trichromacy of color matching establishes a fundamental constraint on the information available to the visual system at all subsequent levels of processing.

COLOR APPEARANCE OF ISOLATED COLORS

The color appearance of isolated lights has long been conceptualized within the context of a geometric framework (see Kuehni, 2003, for a historical review). One simple three-dimensional representation maps colors into a cylindrical coordinate frame in which the perceptual dimensions are hue, saturation, and brightness (figure 36.1). *Hue* is represented as an angular dimension that captures the change in color appearance of monochromatic lights as wavelength is varied. Additional nonspectral hues, the purples, are formed by mixing together lights from the spectral extremes in various proportions (e.g., a deep red light plus a violet light). All visible lights are contained within the convex region bounded by the spectrum locus and purple line. *Brightness* is the dimension of color appearance most closely associated with the intensity of the light. If two lights differ only in intensity, the one with the higher intensity will be brighter. At a given brightness level the origin of the color space is the *white point,* the locus of the color with zero hue. *Saturation* is the perceptual dimension corresponding to the similarity of a color to white and is therefore represented as its distance from white. Lights mapping farther away from the white point appear more "colored" and thus more saturated and more different from white. Lights that are equally saturated fall on a circle centered at the white point. Brightness is represented on an axis perpendicular to

the plane swept out by all equally bright mixtures of monochromatic lights.

There is no simple relationship between the colorimetric representations of color determined by color matching and the perceptual dimensions of hue, saturation, and brightness just described. Brightness and saturation cannot be represented as simple linear transformations of colorimetric color space (see Shevell, 2003). Moreover, the hue of a monochromatic light typically varies when its luminance changes (the Bezold–Brücke effect) (Purdy, 1937) as well as when it is desaturated by another light of fixed spectral composition (the Abney effect) (Burns et al., 1984).

Mizokami and colleagues (2006) have recently suggested that the Abney effect may reflect an evolutionary adaptation that preserves hue when colors in the real world are desaturated. In this view desaturation occurs when the spectrum of a light is broadened rather than when it is mixed to varying degrees with other lights of fixed spectral composition (as is typically done in the lab). Indeed, Mizokami et al. have shown that the hue of lights with Gaussian-shaped energy spectra remains invariant as spectral bandwidth is varied, provided that the peak wavelength of the spectrum is held constant. Crucially, the constant-hue loci in colorimetric color space produced by varying bandwidth closely follow those obtained in classical measurements of the Abney effect.

Despite these caveats it is tempting to relate the space of lights defined by color matching to the space of lights based on color appearance. Both spaces are three-dimensional, and both spaces represent all physically realizable colors within the convex hull formed by the purples and the monochromatic lights ordered according to wavelength. Thus, the structure of color space, as determined by color matching, can be identified readily with that based on the psychological attributes of color appearance (see Koenderink & van Doorn, 2003).

COLOR OPPONENCY

The classic framework for understanding the appearance of colors as outlined above is Hering's theory of color vision. It is based on two fundamental principles (Hering, 1878/1964). The first principle is that there are only four perceptually unitary colors—red, green, blue, and yellow—which correspond to four *elemental color sensations.* The visual system combines these elemental sensations in varying degrees to create the appearances of the other colors. For example, the appearance of purple can be represented as a combination of the elemental sensations of redness and

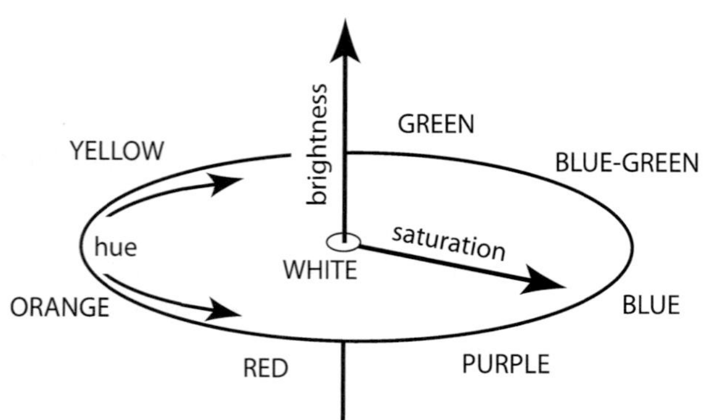

FIGURE 36.1 A diagram of a hue, saturation, brightness color space.

blueness. The second principle is that these elemental sensations are grouped into two color-opponent pairs: red/green and blue/yellow. This second principle stems from the observation that the opponent colors are both mutually exclusive and mutually antagonistic. Opponent colors are said to be mutually exclusive because no color normally contains the sensations corresponding to both poles of an opponent color pair. For example, a light that possesses any degree of redness cannot also have perceived greenness. Opponent colors are said to be mutually antagonistic because they cancel one another when they are added together. For example, a light that, when added to another, increases the apparent redness of the mixture, also reduces its greenness. Hering proposed that color opponency arises from independent, bipolar neural responses. Hering also recognized elemental sensations related to the unitary colors black and white, although he supposed that the black/white dimension is mediated by a more complex neural substrate than the other two. With the addition of this third opponent pair of sensations, each light can be expressed as a triplet of opponent neural responses.

Modern psychophysical evidence supports this view of the appearance of colors. Boynton, Schafer, and Neun (1964) assessed color appearance directly by simply showing observers colored lights and asking them to indicate what fraction of the lights' color appearance can be attributed to red, green, blue, and yellow and to overall saturation. This method has proven to be quite reliable despite the purely introspective methodology. Boynton, Neun, and Schafer (1964) confirmed the fact that no ordinary light contained both redness and greenness, or blueness and yellowness, at the same time. This method has been used successfully to study the appearance of colors over a variety of manipulations (e.g., Abramov & Gordon, 1977; Alpern, Krantz, & Kitahara, 1983; Gordon & Abramov, 1965).

An appealing aspect of the opponent process account of color appearance is that it is at heart trichromatic. Thus, it suggests a link between the coordinates of a light as specified by color matching, *a*, and another set of coordinates, the red/green, blue/yellow, and black/white bipolar neural responses that specify its appearance (figure 36.2).

In the mid-twentieth century investigators sought to place color opponency on a firm theoretical and empirical footing by using hue cancellation as a means of determining the coordinates of monochromatic lights (Hurvich & Jameson, 1957; Jameson & Hurvich, 1951). In this paradigm (see also Bruckner, 1927), the two hue dimensions of a light—its redness or greenness and its

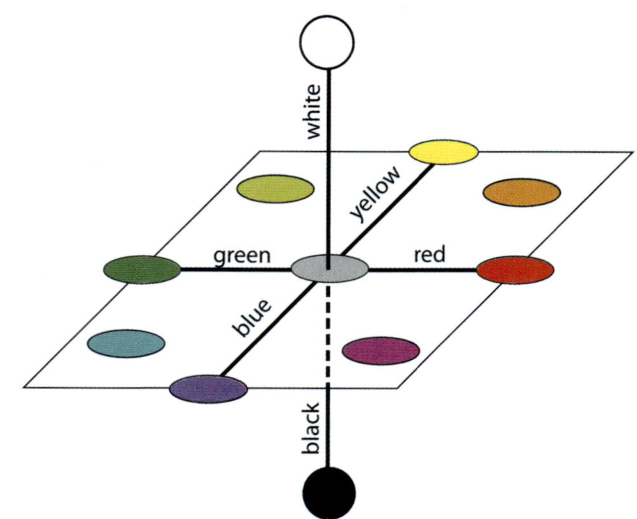

FIGURE 36.2   A diagram of the Hering color space.

blueness or yellowness—can be measured by determining the amounts of each of a fixed set of standard lights—red, green, blue, and yellow—that are needed to cancel each of the fundamental color-opponent hue sensations in the test light. For example, the redness of any test light can be specified by the amount of a standard green light that, when added to the test light, produces a mixture light that is neither reddish nor greenish. At this point the response of the red/green opponent process due to the test has been *nulled*, and the mixture is said to be in red-green equilibrium. The remaining residual colors (blue, yellow, or white) are called red-green *equilibrium lights* (Larimer, Krantz, & Cicerone, 1974). By a similar procedure, blue-yellow equilibrium can be determined (Larimer, Krantz, & Cicerone, 1974). In this way one can empirically establish the magnitudes (the *valences*) of the signed responses in the red/green and blue/yellow opponent color systems (figure 36.3a). Of particular interest are the equilibrium lights, called *unique hues* (Jameson & Hurvich, 1951), that elicit the unitary red, green, blue, or yellow color sensations without need for additional hue cancellation. Three isolated monochromatic lights that satisfy these criteria appear at about 478 nm (unique blue), 517 nm (unique green), and 576 nm (unique yellow), with considerable variation from observer to observer (see Kuehni, 2003, for a review). Unique red is not a monochromatic light, but it can be created by mixing a long-wavelength light (which will look slightly yellowish) with a short-wavelength light (which cancels the yellowness of the long-wavelength light). The resulting pinkish-looking light is neither bluish nor yellowish. The logic underlying this paradigm does not work very well for measuring the black/

white dimension because there is no light whose color can be described as "black" that can be added to a whitish light to reduce its whiteness. Rather, blackness requires the introduction of an inducing field that creates a black stimulus by simultaneous contrast. Furthermore, it is not straightforward to identify a light that is in black-white equilibrium, that is, whose blackness has been fully canceled by adding whiteness. We discuss these issues in more detail below.

Assuming that the visual system is linear up to the point where color appearance is specified (Hurvich & Jameson, 1957; Krantz, 1975), and assuming that the valence spectrum of the black-white dimension follows the spectral luminous efficiency function $V(\lambda)$, Hurvich and Jameson expressed the color-opponent systems as a linear transformation of the color-matching functions. Because these are a linear transformation of the spectral sensitivities of the cones,

$$\begin{pmatrix} (r-g)_\lambda \\ (b-y)_\lambda \\ (bl-w)_\lambda \end{pmatrix} = A \begin{pmatrix} L_\lambda \\ M_\lambda \\ S_\lambda \end{pmatrix} \qquad (36.1)$$

where the coefficients on the left-hand side of the equation represent responses in the three opponent systems, L, M, and S are cone responses, and A is a 3 × 3 matrix specifying the mapping from photoreceptor to opponent responses. The color cancellation data in figure 36.3a are obtained from equation 36.1, and figure 36.3b is a diagram of the constant-brightness plane, according to their model (Hurvich & Jameson, 1956).

In modern times, we refer to this overall view of color vision as a *zone theory*, as it is based on two zones of color representation: The first zone consists of the cone photoreceptors, and the second zone is a subsequent site where the cone signals are recombined to create the bipolar neural responses posited by Hering. Thus, the color-cancellation method and the theory associated with it were crucial for establishing the second, color-opponent zone of color representation.

One important consequence of the linearity of equation 36.1 is scalar invariance (Krantz, 1975; Suppes et al., 1989). If light **a** is in equilibrium for one of the opponent processes (red/green or blue/yellow), then a light t times as intense ($t*$**a**) must also be in equilibrium for that process. Some tests of scalar invariance have confirmed that prediction. For example, the wavelengths of the spectral blue, green, and yellow unique hues, unlike other lights, are approximately invariant with light level (e.g., Larimer, Krantz, & Cicerone, 1974, 1975). A second consequence of linearity is additivity: If lights **a** and **b** are each in equilibrium for one of the opponent processes, say in red/green equilibrium, then the mixture **a b** is also in equilibrium for the red/green process. Critical tests involving mixtures of lights generally have failed to confirm the additivity prediction, particularly for mixtures involving lights in blue/yellow equilibrium (e.g., Larimer, Krantz, & Cicerone, 1975).

One might suppose that the color opponency theory embodied in Hurvich and Jameson's model could be made to work by introducing a suitable nonlinearity. For example, the photoreceptor inputs into the

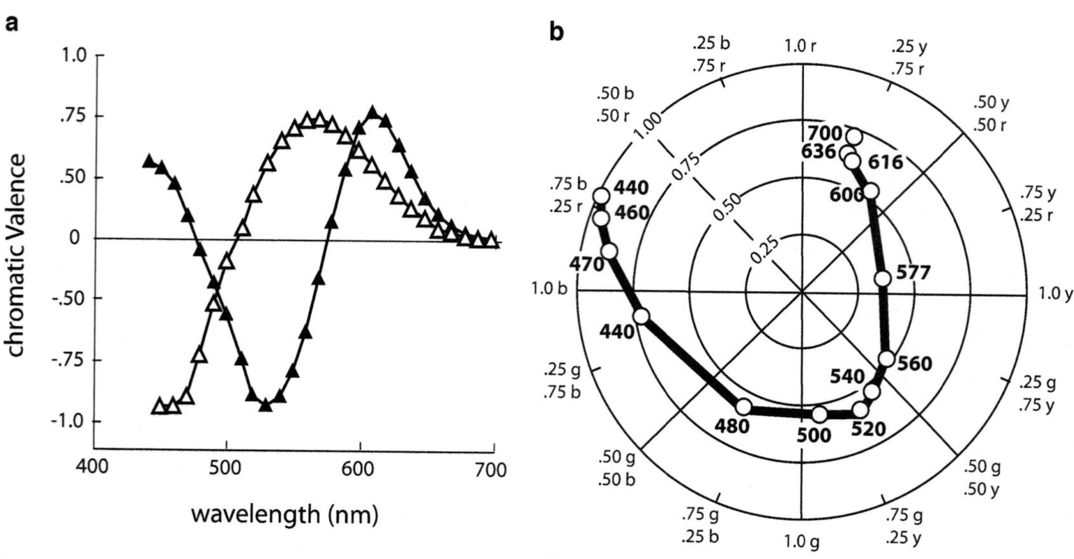

FIGURE 36.3 (a) Chromatic valences, which are a linear transformation of the color-matching functions of the standard observer. (b) The R/G, B/Y constant-brightness plane defined by the color cancellation data. The bold line is the spectrum locus, computed from the chromatic valences on the left.

514 DELWIN T. LINDSEY AND ANGELA M. BROWN

color-opponent channels might vary nonlinearly with quantal catch (Larimer, Krantz, & Cicerone, 1975; reviewed in Knoblauch & Shevell, 2001). In that class of models the signals contributed to the color-opponent system by each cone type remain fixed in polarity. For example, the S cones can contribute to redness, but if they do, they cannot contribute to greenness. In order for this constraint to prevail, receptor signals must act monotonically on the target color-opponent channels, a weaker relation than the linearity prescribed in the original Hurvich and Jameson model.

However, even the monotonicity assumption does not hold. Knoblauch and Shevell (2001) showed that even the monotonicity constraint does not strictly hold for human color vision. In their experiments subjects created equilibrium lights by adjusting the contribution of one cone type at a time. For red/green equilibria, subjects controlled the L-cone contribution for different levels of M-cone response while the S-cone responses were held constant. The L- and M-cone values at red/green equilibrium were linearly related, as predicted by equation 36.1. In contrast, the amount of S-cone response needed to maintain blue/yellow equilibrium varied nonmonotonically with the level of L-cone responses for a fixed value of M-cone response. Knoblauch and Shevell point out that nonmonotonicity in the blue/yellow equilibrium may arise from receptor-dependent gain changes or from separate color-opponent transforms for each polarity of the visual system's responses along the blue/yellow dimension of color appearance.

SURFACE COLORS AND COLORS IN CONTEXT

In our everyday life, colors are generally not seen in isolation. Rather, they are more often associated with pigmented surfaces under varying conditions of illumination and color context, which can greatly affect color appearance. Generally, the perceived properties of an isolated light, when viewed under only one stimulus condition, cannot predict the color appearance of a natural stimulus with the same spectral composition viewed under other, more diverse conditions. The warning that color matching cannot predict color appearance is even more important here because color matching, unlike the appearance of a natural stimulus, is based only on the spatially local computations of the retinal excitation arising from the matching lights themselves, regardless of the state of adaptation or the context in which they are viewed. In this section, we address this domain of more complex stimulus configurations.

The Munsell Color Order System

The *Munsell Book of Color* is a set of reflective color samples that instantiate a color order system that was devised by Alfred Munsell in 1905 (reviewed in Nickerson, 1940), and was later revised and expanded to admit a larger gamut of colors (see: Newhall, Nickerson & Judd, 1943). Originally intended to standardize the specification of colors for artists, the Munsell system is in wide use today in science and industry.

The colors in the present Munsell system consist of a large number of reflective samples that are intended to be viewed against a standard mid-gray background, illuminated with a specific illuminant (Standard Illuminant C). The three-dimensional structure of the Munsell color diagram is in some ways similar to the cylindrical space described above, although the Munsell diagram is more irregular in shape because of limitations in printing a wide gamut of reflective samples (figure 6.4).

The dimensions of the Munsell color diagram are hue, chroma, and value. *Hue* is specified by the angular separation of colors falling on concentric hue circles of discrete radii emanating from the *value* axis. The angular separations of colors are constant and reflect the approximately equal perceptual spacing of colors (40 per circle) on each circle. *Value* (also called *lightness*) is the perceived intensity of a stimulus, relative to white. Lightness differs from brightness in that brightness is a property of an isolated light and can vary from dim to brilliant, whereas lightness is a property of a light seen in context and varies from dark through neutral to light. Both differ from *luminance* in that they are phenomenal characteristics of lights, whereas luminance is a photometric quantity based on the spectral sensitivity of the eye (Lennie, Pokorny, & Smith, 1993; Wyszecki & Stiles, 1982). At the center of each circle of hues is an equally light neutral color, which looks gray rather than white because these are reflective samples seen in a standard context rather than isolated lights. The radius of each color circle is the dimension of *chroma*; thus, colors falling on a line emanating from the value axis differ in chroma, but have the same hue. Chroma differs from saturation in that chroma is the difference between a color and an equally light neutral color, whereas saturation is the diference of a light from white.

The Munsell system is called a color order system because there is no metric for specifying the perceptual differences among arbitrary pairs of colors. Moreover, since the Munsell colors are chosen to be equally perceptually spaced along any single dimension—hue, chroma, or value—the perceptual difference between any two colors that differ along more than one

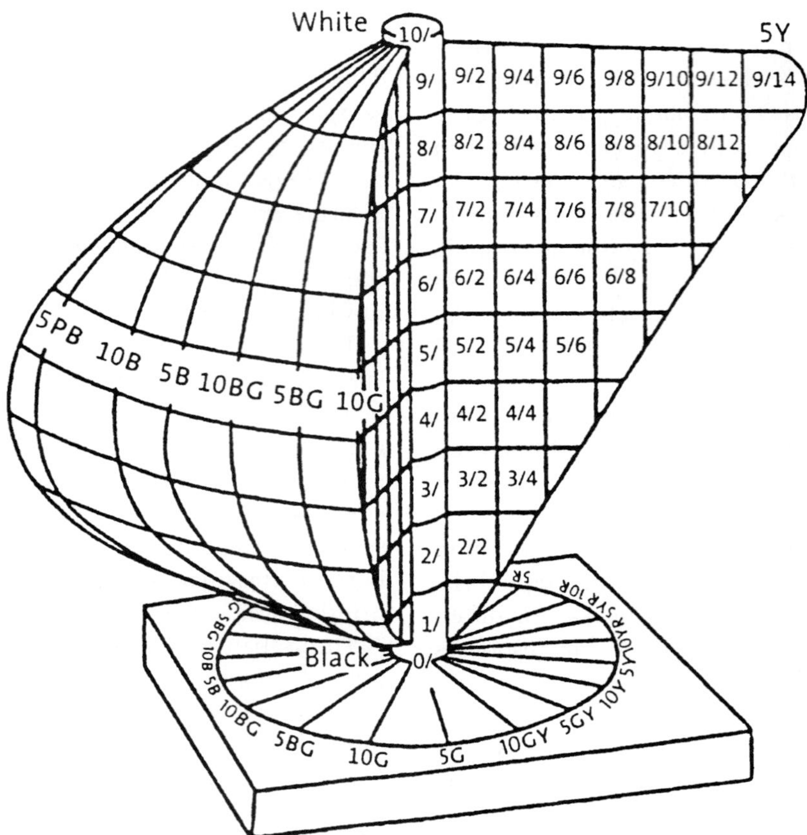

FIGURE 36.4   A diagram of the Munsell color order system. (From D'Andrade & Romney, 2003.) The vertical dimension is value, with the lightest samples at the top and the darkest at the bottom. The angular dimension is hue. The cutaway at hue = 5Y shows the radial dimension, chroma, with the highest chroma samples farthest from the center. See text for further details.

dimension in this space is not defined. Finally, modern empirical color scaling studies have shown that the "equal-spacing" assumption only holds approximately (Indow, 1988).

We saw in our discussion of the Hering theory of color perception that even that theory, which is grounded in the appearance of isolated lights, nevertheless requires the context of an inducing field to establish black-white as the third dimension of color-opponent processing. Next, we discuss this and related color appearance phenomena.

Chromatic Induction

COLOR CONTRAST   The color of a stimulus presented in one location of visual space can be affected by a surrounding light of a different color (*chromatic induction*). When the shift in color appearance is away from that of a nearby region, induction is called *simultaneous color contrast* (figure 36.5a–d) (Krauskopf, Zaidi, & Mandler, 1986; Ware & Cowan, 1982; Zaidi et al., 1992). For example, a white test surrounded by a red inducing field will appear greenish in hue. Color contrast effects are often large and can change the hue, the chroma, and the lightness of the test color and must be the result of spatial interactions among visual responses to the colored regions. Ware and Cowan (1982) proposed a two-stage model of simultaneous color contrast: a multiplicative gain change at the level of the photoreceptors followed by an additive change at a subsequent, color-opponent stage. In contrast, Krauskopf, Zaidi, and Mandler (1986) showed that a simple two-stage model is not sufficient to account for simultaneous color contrast and implicated even higher-level chromatic mechanisms in simultaneous color induction.

THE DARK COLORS   A broadband light that is viewed in an otherwise dark field can appear white, but it is never seen as gray or black regardless of its luminance. The range of colors that we perceive as achromatic, ranging from black, through gray, to white, are a consequence of interactions occurring between the test light and its context. This is important to understanding the black/white dimension of the Hering model. Particularly, the

   DELWIN T. LINDSEY AND ANGELA M. BROWN

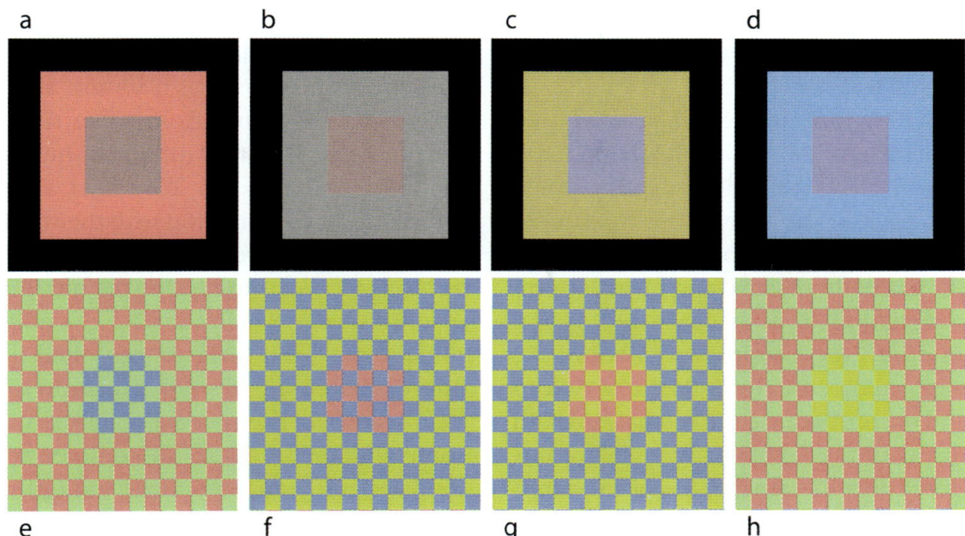

FIGURE 36.5   Chromatic induction (after Stockman & Brainard, 2009, figure 1). (a–d) Simultaneous color contrast. The spectral compositions of the light from the central squares of a and b are identical, as are those of c and d. In each case the surrounding square pushes the color appearance to the opposite direction from its own color. The pink surround makes the central square look grayish (a), whereas the gray surround makes the central square look more pinkish (b). A similar effect in c and d makes the central blue square look more bluish in c and more purplish in d. (e, f) Color assimilation. The contrasting colored checks are drawn toward the color appearance of the inducing checks. The central blue checks in e and f are spectrally identical, as are the central red checks in f and g and the central yellow checks in g and h. However, the blue checks an e look more greenish, and those in f look more reddish, because of the greenish and reddish inducing checks, respectively. Similarly, the red checks in f look more bluish, and those in g look more yellowish, because of the inducing checks, and the yellow checks in f and g look more reddish in g and more greenish in f because of the colors of the inducing checks.

color black is induced in a neutral target by a brighter surrounding context, which can be any color. The action spectrum of blackness induction corresponds closely to the spectral luminous efficiency function (Chichilnisky & Wandell, 1999; Volbrecht, Werner, & Cicerone, 1990; Werner et al., 1984) (but see Shinomori, Schefrin, & Werner, 1997, for evidence of more complex spatial interactions).

In a similar vein the presence of a moderately bright white surrounding field induces darkness in an adjacent region containing a chromatic color. The hue associated with dark colors is usually similar to the hue of the isolated light; for example, maroon and navy blue stimuli are perceived as darker versions of their isolated red and blue counterparts. However, brown, which is produced by surrounding a saturated yellow or orange light with a more luminous white surround, is a qualitatively different percept than either yellow or orange presented in isolation in the dark. Thus brown, like black and gray, is experienced only in the context of brighter lights in nearby parts of the visual field.

COLOR ASSIMILATION   When the color of a test field shifts toward the hue of the inducer, the phenomenon is called *assimilation* (figure 36.5e, f) (Helson, 1926).

Assimilation tends to occur in high-spatial-frequency stimuli. In stimuli consisting of alternating test and inducing strips, simultaneous color contrast is seen at spatial frequencies below 3–6 cycles/deg, and assimilation occurs above this value (Fach & Sharpe, 1986). Optical factors, such as chromatic aberration and diffraction, which blur the retinal image more at some wavelengths than at others, may be important to assimilation. However, some of the effects in assimilation are clearly due to spatially antagonistic neural processing in the cerebral cortex (Cao & Shevell, 2005; Monnier & Shevell, 2003; Shevell, 2012).

CHROMATIC ADAPTATION AND COLOR APPEARANCE

A light that looks yellow when viewed in isolation will appear more greenish when it is superimposed on a uniform red background. Inasmuch as the light reaching the eye from the test location consists of a mixture of the middle-wavelength (yellow) test light and the long-wavelength (red) background lights, one would predict the test to appear more reddish rather than greenish if color appearance were determined solely by the chromaticity of the light from the test location.

Instead, chromatic adaptation to the red background shifts the appearance of the yellow test light toward green.

The classic explanation for this phenomenon is based on the von Kries coefficient law (Brill, 2005; Ives, 1912; Shevell, 1978), which states that the adapting background desensitizes each of the three cone types independently, without changing their spectral sensitivities. Von Kries adaptation is multiplicative and inversely proportional to the excitation of each cone type by the adapting background. In the example given above the change in color appearance of the yellow test due to the red background occurs because the red background reduces the sensitivity of the L cones to the yellow test light more than it reduces the sensitivities of the other two cone types.

The coefficient law does a good job of capturing many color-appearance-related features of adaptation to uniform colored backgrounds of moderate photopic intensity. However, it is only an approximation, and some of its predictions are not supported by empirical results, especially in the case of asymmetric color matching. For example, consider two lights, one viewed on a red background and the other viewed on an achromatic background. Suppose that the subject has adjusted the two test lights in spectral composition and intensity to match in color appearance under these asymmetric viewing conditions. The coefficient law predicts that if the test lights' intensities are changed proportionately, they should still match. A classic study by Jameson and Hurvich (1959) showed that this is not the case. Vimal, Pokorny, and Smith (1987) reported a similar result for briefly presented lights and lights viewed for a prolonged period.

Models of chromatic adaptation for the simple case of a test stimulus superimposed on a large, uniform background, generally feature two adaptation stages that mediate, respectively, multiplicative and additive effects of the background on the test stimuli (Jameson & Hurvich, 1972; Pugh & Mollon, 1979; Shevell, 1978; Werner & Walraven, 1982). The multiplicative stage typically involves a von Kries–like receptoral desensitization, and the second, subtractive stage shifts the baseline of the hypothesized color-opponent neural responses toward the chromaticity of the test plus background mixture. More complex two-stage models are possible (see Shevell, 2003) in which each stage consists of additive and/or multiplicative processes.

The foregoing discussion of chromatic adaptation and color appearance is based on large, steadily presented uniform backgrounds, which are rarely encountered in nature. This is a great simplification compared to the changing, chromatically variegated backgrounds such as those encountered in the natural environment. In the spatial domain the obvious simplification is to treat a complex scene as if it were a uniform *equivalent background* computed from some simple global scene statistic—for example, the mean luminance and chromaticity of the scene. To a first approximation this approach, in combination with the von Kries coefficient law, often works reasonably well. For example, color appearance in computer displays depicting variegated colored scenes, which simulate the effects of lights illuminating colored surfaces, can often be accounted for approximately in this way (Bäuml, 1995; Brainard & Wandell, 1992; Brenner & Cornelissen, 1998). However, some adaptation effects in complex scenes cannot be modeled with an equivalent background (Jenness & Shevell, 1995). For example, in color induction, either assimilation or induction can occur depending on the spatial configuration, as described above. Moreover, chromatic adaptation alone cannot explain color constancy (see below).

In the temporal domain the effects of chromatic adaptation on color appearance appear to occur on several different timescales. There are one or more adaptation processes with fast time constants (less than 1 s) and at least one with a slower time constant (between 1 s and 1 min; see Rinner & Gegenfurtner, 2000). Furthermore, adaptation can continue to change color appearance on even longer timescales, on the order of a few minutes (Shevell, 2001), or even of days to weeks or longer (Belmore & Shevell, 2008; Nietz et al., 2002). The slower among these processes are thought to be involved in setting the chromaticity of the equilibrium white (Pokorny & Smith, 1987) and thus calibration of the opponent channel responses. Such adjustments could partly compensate for variations in the average color of the environment over the course of a day, over the seasons of the year (Juricevic & Webster, 2009; Webster, Mizokami, & Webster, 2007), or over the lifetime of the observer, to compensate for the yellowing of the ocular lens as we age (Schefrin & Werner, 1990).

COLOR CONSTANCY

The color appearance of a surface is remarkably little affected by the spectral content of the illuminant, a phenomenon called *color constancy*. Unless the stimulus is self-luminous (such as a signal light or a video monitor), the spectrum of light entering the eye is the product of the spectral reflectance of the stimulus surface times the spectral composition of the illuminant. Thus, if color appearance depended only on the spectral characteristics of light, we would experience large fluctuations in the color appearance of surfaces as environmental illumination varies. Contrary to that prediction, the surface color that we experience is approximately constant, even as the surface is viewed indoors under tungsten illumination or outdoors during the daily phases of sunlight or in the laboratory under a range of broadband illuminants. It is a difficult (and classic) problem to understand how this could occur (for reviews, see chapter 38 of this volume and Brainard & Maloney, 2011; Foster, 2011).

From a computational perspective, the stable property of the material is its reflectance function, $R(\lambda)$. But how can the visual system estimate this function, based on only three pieces of information at each location in the image: the responses of the L, M, and S cones, ρ_i:

$$\rho_L = \int_\lambda R(\lambda)E(\lambda)S_L(\lambda)d\lambda, \quad \rho_M = \int_\lambda R(\lambda)E(\lambda)S_M(\lambda)d\lambda,$$

$$\rho_S = \int_\lambda R(\lambda)E(\lambda)S_S(\lambda)d\lambda$$

where $E(\lambda)$ is the illuminant spectrum and S_i are the spectral sensitivities of the cone photoreceptors? Mathematically, estimation of $R(\lambda)$ at every location in the retinal image is an ill-posed problem. However, the problem becomes more computationally tractable if the compositions of all illuminant and material reflectance spectra are approximated as linear sums of a small number of intrinsic basis functions e_i and r_i, respectively (Cohen, 1964; Judd, MacAdam, & Wyszecki, 1964; Maloney & Wandell, 1986). Suppose, for example, there are three basis functions for each. Then $R(\lambda) = \sum_{i=1\ldots3} a_i r_i$ and $E(\lambda) = \sum_{i=1\ldots3} b_i e_i$, and the problem of color constancy boils down to estimating the values of three parameters, a_i, for each surface in the scene. These are reasonable approximations (Judd, MacAdam, & Wyszecki, 1964; Maloney & Wandell, 1986; Solomon & Lennie, 2005) to environmental illuminants and reflectances, and it can be shown that recovery of the parameters specifying $R(\lambda)$ from ρ_i is possible (Buchsbaum, 1980; Sallstrom, 1973) (see Maloney, 1999, for a review). A number of related approaches have also been proposed; all rely on a low-dimension reparameterization of the representation of $R(\lambda)$, and recovery of an estimate of the illuminant. In general, all these approaches require a global assessment of the photoreceptor signals, ρ_i, from multiple surfaces in a scene. This requirement is consistent with human color constancy: Surfaces are only color-constant if they are seen in the color context of other surfaces that are lit by the same illuminant; the appearance of an isolated, uniform patch of color is not color-constant, even if it is created by illuminating a reflective surface (Helson, 1943).

Other investigators have framed color constancy as a form of visual adaptation to the illuminant (D'Zmura & Lennie, 1986; Foster & Nascimento, 1994; Webster & Mollon, 1995; West & Brill, 1982), including visual adaptation to complex naturalistic scenes (reviewed in Webster, 2011), while Golz and MacLeod (2002) combine the computational and adaptation approaches. Color constancy is not perfect, and empirical studies of the patterns of errors in human color constancy can be used to evaluate candidate models of color constancy (Brainard & Maloney, 2011).

THE PHYSIOLOGY OF COLOR APPEARANCE

How is the spectral composition of a stimulus encoded by the visual system, as related to color appearance? In this section we review the neural representations of color that might underlie color appearance in the major ascending visual pathways beyond the retina.

The Dorsal Lateral Geniculate Nucleus of the Thalamus

The dorsal lateral geniculate nucleus (dLGN) serves primarily as a relay nucleus between retina and cortex. Early recordings in macaque dLGN provided encouraging support for a Hering-type encoding of spectral information early in the visual processing chain (DeValois, Abramov, & Jacobs, 1966; Wiesel & Hubel, 1966) (see Mollon, 2003, for a historical discussion). In the experiments of DeValois et al. and Wiesel and Hubel, most dLGN cells displayed spatially antagonistic *center-surround* receptive field organization. Some of these cells were spectrally broadband, and some were spectrally antagonistic in ways that roughly corresponded to the Hering red/green and blue/yellow components of color perception. However, subsequent work has failed to support the idea that the dLGN is the place in the brain where a Hering-style color representation encodes color appearance.

More recent work indicates different responses in three distinct cell types in dLGN: respectively, parvocellular (p-), koniocellular (k-), and magnocellular (m-) cells. Signals in each cell type arise from distinct retinal ganglion cell types and are transmitted to distinct regions of primary visual cortex. The p-cells encode spatially and chromatically antagonistic interactions between L- and M-cones (Derrington, Krauskopf, & Lennie, 1984). These are by far the most numerous cells in dLGN, and they have been closely linked to spatial as well as color vision (see Lennie & Movshon, 2005). However, very few p-cells receive S-cone signals, and their chromatic signatures do not conform to a

Hering-type red/green encoding, which requires S-cone as well as L- and M-cone signals.

The k-cells encode signal differences between S-cones and the sum of the L- and M-cone responses in a spatially coextensive rather than center-surround configuration. Although the resulting spectral properties of k-cells match the requirements for a blue-yellow perceptual color channel, their linear responses do not match the nonlinear behavior associated with blue-yellow color vision. Some k-cells with an S-OFF spectral signature also receive signals from melanopsin, a photopigment that is present in the plasma membranes of intrinsically photoresponsive retinal ganglion cells (Dacey et al., 2005). The contribution of these signals to visual perception remains to be determined.

The m-cells possess spatially antagonistic center-surround receptive fields composed of additive weightings of L and M cones. The m-cells thus have a broadband spectral signature that closely matches the photopic luminance efficiency function, as determined by flicker photometry (Lee, Martin, & Valberg, 1988) and minimally distinct border measurements (Kaiser et al., 1990). The m-cells might seem to be a likely substrate for Hering's "black/white" color-opponent processing. However, they have a number of characteristics (for example, their very low sampling density as compared to p-cells) that make m-cells ill-suited for this role (see Lennie & Movshon, 2005; Lennie, Pokorny, & Smith, 1993). Instead, it is believed that luminance information relevant to form and information relevant to the L–M dimension of color vision are both conveyed to visual cortex via p-cells. Because of their receptive field structure, p-cells co-encode chromatic information (for spatially uniform stimuli) as well as luminance information (at high spatial frequencies) (Ingling & Martinez-Uriegas, 1983; Wiesel & Hubel, 1966). These two forms of information are thought to be demultiplexed in visual cortex.

Representation of Color Appearance in Visual Cortex

Single-cell recordings in macaque cortex indicate that a number of different cortical areas respond selectively to chromatic stimuli. Functional MRI studies that look at population responses among visually responsive cells confirm this basic finding in humans. These areas include V1 (the primary visual cortex) as well as V2, V3, V4, and parts of inferotemporal cortex (see, for example, Brouwer & Heeger, 2009; Engel, Zhang, & Wandell, 1997; Kleinschmidt et al., 1996; Wade et al., 2008). However, an understanding of precisely how the different color-responsive cortical areas are related functionally has remained elusive.

PRIMARY VISUAL CORTEX CORTEX (V1)   By far the most comprehensive study of color neurophysiology in cortex has involved macaque V1. Early work suggested that the color-responsive cells in V1 are concentrated within cytochrome oxidase-staining *blobs* in layer 2/3 (Livingstone & Hubel, 1984; Ts'o & Gilbert, 1988) and may even be anatomically organized with respect to hue (Xiao, Wang, & Felleman, 2003). However, many other studies report a more diffuse anatomical distribution of color-selective cells in V1 (Lennie, Krauskopf, & Sclar, 1990; Leventhal et al., 1995). In humans, fMRI studies involving multivoxel pattern analysis (MVPA) show color-selective patterns of responses in V1 to different colors, possibly related to the unique hues (Parkes et al., 2009). However, evidence for a homomorphism between the neural representation of color in V1 cortex and a Hering-type psychological representation has not been found either at the fMRI or single-unit level of analysis. (An in-depth discussion of the response characteristics of V1 cells is presented in chapter 40 of this volume; see also reviews by Gegenfurtner, 2003, and Shapley & Hawken, 2011.)

Single-unit studies in macaque indicate that although the majority of cells in V1 are at least somewhat color selective, only about 10% of cells are strongly so (Johnson, Hawken, & Shapley, 2004; Lennie & Movshon, 2005). The remaining cells respond to varying degrees to modulation in both luminance and color. Functional MRI responses to color in V1, however, are very robust (Engel, Zhang, & Wandell, 1997; Kleinschmidt et al., 1996; Wade et al., 2008). This apparent discrepancy between single-unit and fMRI studies of V1 probably reflects differences between individual and population responses of cells in V1 that vary in degree of color selectivity (Schluppeck & Engel, 2002). In contrast to the distinctive spectral signatures of cells in dLGN, the overall spectral tuning characteristics of V1 cells are much more diverse, with no general tendency to prefer certain directions in color space (DeValois et al., 2000; Lennie, Krauskopf, & Sklar, 1990; Wachtler, Seijnowski, & Albright, 2003). Moreover, many cells in V1, unlike those in dLGN, respond nonlinearly to visual stimulation (e.g., Lennie, Krauskopf, & Sklar, 1990).

Cells in macaque V1 generally show distinctive spatial as well as spectral tuning characteristics, suggesting they may be involved primarily in the coanalysis of both form and color. Some V1 cells, like their retinal and dLGN counterparts, are "single-opponent" (e.g., L-ON-center, M-OFF-surround) and respond well to large, uniform areas of color. However, other cells have "double-opponent" receptive fields (e.g., adjacent areas of L-ON, M-OFF and M-ON, L-OFF sensitivity), which must be computed within V1. Many double-opponent cells respond well to color differences across a border (Johnson, Hawken, & Shapley, 2001), suggesting that they could be the basis for some of the simultaneous color contrast and color constancy effects observed behaviorally.

There is some evidence that cells in V1 process chromatic information in close association with the spatial context within which it appears, which suggests a relationship with the behavioral data on color seen in context (discussed above). For example, Wachtler, Sejnowski, and Albright (2003) have reported that chromatic tuning of cells in macaque V1 was affected by the spectral characteristics of stimuli located well outside the classic receptive fields of V1 cells. Moreover, these stimuli produced color induction effects in humans that were well predicted by the average responses of the macaque V1 cells. These results, among others, suggest that a color representation in V1 exists at the level of populations of neurons rather than in the spectral signatures of a few anatomically segregated cell types. Little else is currently known, however, about how population responses of cells like those found in macaque V1 might contribute to the representation of color appearance in humans.

extrastriate cortex Color information is also represented in the extrastriate visual pathways: V2–V4 and inferotemporal cortex (see also chapter 41 of this volume). Cells in V2 share many of the characteristics of V1 cells (reviewed by Conway, 2009), although there is some evidence for an elaboration in hue processing (Xiao, Wang, & Felleman, 2003). Recent work has suggested that there may be a lateralized fMRI response in human left-hemisphere V2/V3 that depends on the color category membership of the stimuli (Siok et al., 2009). Those authors linked this result to the still-controversial claim that activation of the nearby language centers can alter the physiological representation of color differences in the left but not the right cerebral hemisphere (Gilbert et al., 2006; but see Brown, Lindsey, & Guckes, 2011; Witzel & Gegenfurtner, 2011; and Color Naming below).

Area V4 in macaque was once thought to be devoted exclusively to the processing of color (Zeki, 1983a, 1983b), but the current view is that this area is more likely to be involved primarily in mediating form perception and color-form interactions (reviewed by Shapley & Hawken, 2011). Interestingly, an MVPA fMRI study by Brouwer and Heeger (2009) suggests that the neural representation of color in human V4 approximates the ordering of colors in perceptual color space. They did not, however, test for any special relationship between the neural representation of color in V4 and the unique hues defined in behavioral experiments.

Further downstream, an area currently known as posterior inferotemporal cortex (PIT) is just anterior to V4 in macaque, although humans and macaques may differ importantly in the topographical organization of these areas (Wade et al., 2008). Recent data suggest that the representation of color in the macaque PIT/V4 area may consist of patches of cells (called *globs*) interspersed among non-color-selective patches of cortex (Conway, Moeller, & Tsao, 2007). Single-unit recordings revealed that the interglob regions responded best to luminance. In contrast, glob cells were narrowly tuned to specific ranges of color, especially to directions in color space corresponding to the red, green, and blue unique hues typically measured in humans (Stoughton & Conway, 2008). This suggests that color processing in V4/PIT, as in V1, is probably more related to the population responses of groups of neurons than to the tuning characteristics of just a few specific cell types. The apparent predominance of cells tuned to the unique hues is the closest empirical link found to date to an explicit physiological basis for Hering's fundamental sensations. However, this interpretation of Conway's data remains controversial (Conway & Stoughton, 2009; Mollon, 2009) because the stimuli that elicited the strongest, most selective responses also corresponded to stimuli that modulated the L, M, and S cone types maximally and also to the stimuli that had the greatest amplitude modulation under the conditions in which the experiment was run. Moreover, the uniqueness of unique hues might not arise from a strict isomorphism between the responses of individual cells and the red, green, blue or yellow directions of color space. Rather, uniqueness might be related more closely to other, specific patterns of population responses among chromatically diverse neurons (MacLeod, 2010).

Other single-unit studies of IT may also be specifically relevant to color cognition. For example, Komatsu and colleagues (1992) have found some cells in macaque anterior IT that respond to hue. Furthermore, later studies by this group revealed that response magnitudes of anterior IT cells depended on whether the monkey's behavioral task was to discriminate or categorize colors (Koida & Komatsu, 2007). This line of work may ultimately prove important to understanding the physiological basis of color categorical behavior.

In the meantime it is remarkable that, to date, so many studies of the ascending visual pathways and visual cortex have described so many interesting aspects of color processing but have failed to establish a clear understanding of the neural basis of color appearance in humans.

The preceding discussion of color appearance has been framed within the context of trichromacy at the photo-receptor level and trichromatic color opponency at higher levels of visual processing. The naming of colors reveals a very different aspect of color appearance. Humans can perceptually distinguish approximately 10^6 physically distinct colors, a number far greater than what a person can possibly remember or have distinct words for. On the other hand, if colors are grouped into lexically defined categories, where each group contains colors that are "similar" to one another on one or more perceptual color dimensions, then color cognition becomes more tractable. The need to categorize and name the stimuli in the environment is not restricted to color, of course. However, color naming is a classic topic of research on the relation between cognition and perception because the spectral composition of visible lights is continuously variable, yet most world languages partition the color continuum into distinct color categories, each with its own color term.

Two Traditions: Universalism and Relativism

How exactly is this continuum of colors to be partitioned into categories? A significant finding is that people differ greatly in the sizes of their color lexicons. In some cases only two color names are used (e.g., speakers of Dani in West Papua), in other cases as many as 12 color terms (e.g., speakers of Russian), although 5–7 color names are more typical. For the last 150 years (Gladstone, 1858) the major accounts of this wide range of color-naming behavior have centered on two issues. The first is whether color naming is due to factors, possibly related to human visual neurophysiology, that are common to all people, the *universalist* view, or whether the governing factors are particular to each person's culture, the *relativist* view (see Dedrick, 1998, for a review). The second issue concerns the effects of language on color cognition. Deeply embedded in the relativist view is the idea that the language a person speaks determines broadly how a person thinks, his/her worldview. This idea is called *linguistic relativity* or the *Sapir-Whorf hypothesis*, after the two linguists who first formally advanced it (see Brown, 1976). As applied to color, linguistic relativity asserts that two individuals with different color lexicons should have fundamentally different high-level representations for color, which are a result of their distinctive color categorical systems.

The relativist/universalist dichotomy has guided much of the work in the field of color naming (e.g., Saunders & van Brakel, 1997, and 28 associated essays).

To anticipate, the evidence as a whole does not strongly favor one view over the other. In recent years some investigators (e.g., Kay & Regier, 2006; Regier & Kay, 2009) have attempted to create a hybrid position, with elements of each side of the dichotomy applying, depending on the conditions and task. Furthermore, other theoretical frameworks—some involving the evolution of color terms to facilitate communication (e.g., Jameson & Komarova, 2009; Steels & Belpaeme, 2005) and others involving regularities in the color environment (e.g., Yendrikhovskij, 2001)—are motivating some of the more current work on the subject. We discuss those below. However, because the relativist/universalist dichotomy has so greatly influenced the work in this field over the last 150 years, we use it to organize our discussion here.

Berlin and Kay and Universalism

Most modern work on these issues has been framed in the context of three proposals advanced by Berlin and Kay (1969). They collected data from native-speaking informants on 20 languages (figure 36.6) and augmented that data set with a scholarly analysis of the color vocabularies of 78 additional languages. Berlin and Kay's first proposal was that the vocabularies of most languages have *basic color terms* (BCTs), single-word units that are used to name colors. A set of BCTs is a core color vocabulary that is both necessary and sufficient to name the colors discussed in everyday conversation among speakers of a language. BCTs are used in a consistent way across speakers of a language, and their usage is general: They can be used to describe the color of any kind of object. Eleven BCTs are recognized in English[1]: "black," "white." "red," "green," "yellow," "blue," "brown," "purple," "pink," "orange," and "gray." English-speaking subjects often use other names as well—"turquoise," "lavender," and "peach," for instance—but subjects are generally able to substitute one of the eleven BCTs for the nonbasic word if asked to do so.

The second proposal of Berlin and Kay is that focal colors—the "best examples" of the named color categories—tend to cluster in color space when compared across languages. This result suggested to Berlin and Kay that the locations of named color categories in color space, as specified by focal colors, are subject to innate universal constraints. They argued that if this were not so, then the focal colors would have to be located more or less randomly throughout color space.

Berlin and Kay's third proposal was that languages differ in the number of BCTs in their vocabularies because they are at different stages along a tightly

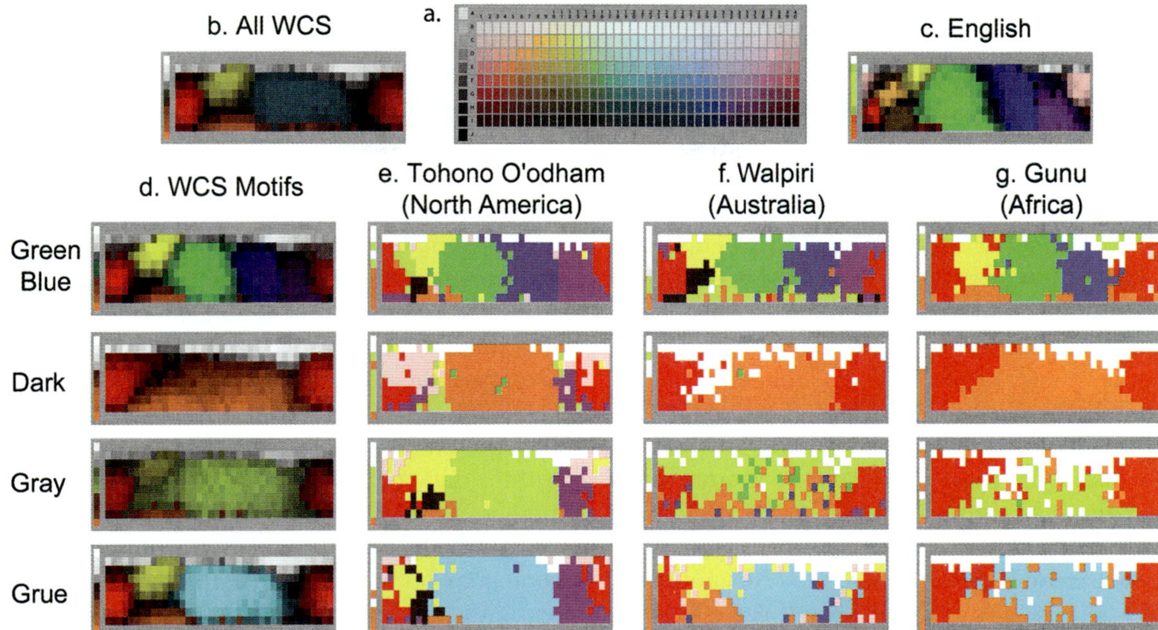

FIGURE 36.6   Color naming in the World Color Survey (WCS). (a) The stimulus set included 330 Munsell samples. Each tiny rectangle in the color-naming diagrams (b–g) corresponds to a sample in a. False hue codes refer to glossed color terms: orange codes the color term BLACK; chartreuse codes GRAY; cyan codes GRUE (GREEN-or-BLUE); red codes RED; similarly, green, blue, yellow, white, pink, purple, and brown code their corresponding colors. (b, d) Consensus diagrams for groups of informants. False hues are the glossed color terms used by the majority of informants, false lightness codes the percentage of informants using the majority term. Consensus is lowest in the darkest areas. (b) All informants in the WCS. (c) 25 English-speaking informants. (d) Informants using each of the motifs (motif names on left). (e–g) Individual informants speaking each of three languages (top of each column), using each of the four motifs. Individuals from different continents use the same motifs (notice the rows of similar data sets), but individuals speaking the same language use different motifs (notice the diversity within each column).

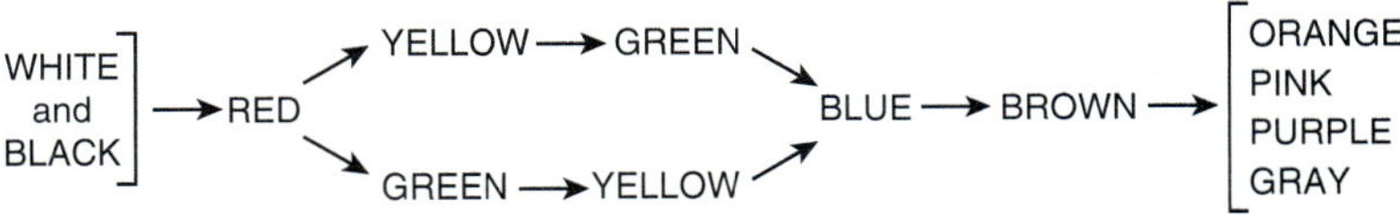

constrained evolutionary sequence defined by the BCTs, which emerge in the following order:

The simplest color lexicons have only two color terms, BLACK and WHITE, where BLACK is generally used to name the dark and cool colors and WHITE is used to name the light and warm colors. As languages evolve, they add color terms one at a time, starting with RED, then either YELLOW then GREEN, or GREEN then YELLOW, and so forth, as their societies become more complex and distinctions among the colors become more crucial in the everyday lives of the people who speak them.

Subsequent Evidence in Support of the Universalist View

Since the publication of Berlin and Kay's monograph, there has been repeated evidence for the fundamental correctness of their proposals about the BCTs. For example, Boynton and Olson (1987, 1990) showed that English-speaking subjects use the English BCTs faster and with greater consensus and consistency than non-basic terms. Uchikawa and Boynton (1987) extended this successfully to Japanese, and a lot more work has been done extending the list of languages (e.g., Davies et al., 1995; Jameson & Alvarado, 2003).

In the late 1970s in response to concerns raised by critics regarding the scope of their original study, Berlin and Kay, in collaboration with Maffi, Merrifield, and Cook, conducted a large prospective color-naming study called the World Color Survey (Kay et al., 2009). The WCS studied 2,616 informants of 110 unwritten languages spoken in preindustrial societies. All informants were tested *en scène* using 330 colored and achromatic color samples from the *Munsell Book of Color*. Each of about 25 informants speaking each language provided a one-word color term for each sample. Informants were then asked to select a focal color for each of their named color categories. Inasmuch as the WCS data were collected to test the predictions of the Berlin and Kay view of language development prospectively, the results of the WCS are particularly compelling.

The WCS dataset has been analyzed by several investigators. Kay and Regier (2003) examined the centroid of the region in color space defined by each color word deployed by each informant as well as each informant's focal colors (Regier, Kay, & Cook, 2005). They found that both the centroids and the focal colors from the WCS tended to cluster near the centroids of the 11 English BCTs regardless of the total number of words that the informant deployed, just as Berlin and Kay had found in 1969. In view of the fact that the 110 languages were not generally related to one another by historical linguistic ties, and speakers of these 110 languages have not had contact with one another over historical time, the clustering of these centroids and focal colors argues in favor of a universalist view. Kay and Regier also examined the concordance across the languages in the WCS by rotating the data around the color circle by varying amounts and comparing the color term boundaries with those of the reference language at each step. They found that the color boundaries agreed best when they were in their correct alignment, suggesting that the division of the hues into color categories is universal and not idiosyncratic from language to language.

Lindsey and Brown (2006) performed a cluster analysis of the full WCS data set rather than reducing each color term to a single number (a centroid or a focal color). They found 11 universal patterns of color term usage, which generally resembled the color-naming patterns that occur in English, except that YELLOW-or-ORANGE was one color term, and there was a very common term GRUE (meaning GREEN-or-BLUE) that is not present in English. These 11 clusters provided the common glossary for Lindsey and Brown's second analysis (2009). They assembled each informant's dataset into a glossed color-naming system and compared all of these to each other. This second cluster analysis showed that WCS color naming fell into approximately four color-naming systems, or *motifs*, a term that emphasizes the recurrence of patterns of color naming, with minor variations, across informants. The four motifs differ from one another mainly in the blue, green, and purple regions of the color chart (figure 36.6d), where the colors are variously BLACK (the Dark motif), GRAY (the Gray motif), or GRUE (the Grue motif), or else are divided into BLUE, GREEN, and sometimes PURPLE (the Green/Blue/Purple or GBP motif), as in English. In good general agreement with Berlin and Kay, the results of Lindsey and Brown suggest a limited universal color glossary and a limited number of color-naming motifs within which the BCTs of a language occur.

Further evidence in favor of the universalist view comes from an examination of the motifs as they occur within and across the WCS languages. The motifs show striking similarities across traditional cultures that are very unlikely ever to have had contact with one another (compare the motifs across the rows in figure 36.6e–g). Furthermore, there is great diversity among the speakers of single languages in the WCS (compare the motifs within the columns in figure 36.6e–g) (Lindsey & Brown, 2009; Webster & Kay, 2007). The motifs of individuals living on different continents are often more similar to one another than those among individuals speaking the same language! These results present a serious difficulty for the relativist view. If color naming is entirely determined by a person's culture, why is there as much diversity within languages as between languages in color motif usage? And, if color terms within a language facilitate communication about color among the people who speak it, then one would expect all members of a culture to express the same motif.

WHAT MIGHT THE UNIVERSAL CONSTRAINTS BE? An intuitively appealing conjecture is that the standard color-opponent perceptual representation—consisting of Hering's six elemental sensations—plays an essential role in constraining color categorical structure (Kay & McDaniel, 1978). In this view the elemental sensations provide internal reference points around which color categories are built. When a person names a color he/she evaluates the similarity of that color to each of the internal reference points and chooses the BCT corresponding to the "closest" focal color. Colors close to a reference point are easily classified and named; those that are midway between two or more reference points will be more difficult to name. In Kay and McDaniel's view category membership of colors varies on a continuum, much like color itself. However, it has been difficult to establish a causal link between the Hering sensations and color categories because the neural substrate for Hering's six elemental color sensations has not yet been found, as explained above. Moreover, theories of color category formation based on the Hering primaries do not explain how other nonelemental sensations give rise to such universal color categories as BROWN, PINK, PURPLE, and GRAY.

  DELWIN T. LINDSEY AND ANGELA M. BROWN

Several other intriguing alternatives to the idea that the Hering elemental color sensations serve as perceptual anchors for color category formation have been developed. Levinson (2000) has argued that languages build their color names around reference points, but the reference points may not be related to the Hering primaries. In at least some languages colors that are not close to these reference points are not named at all.

Xiao and colleagues (2011) have recently suggested that the correct link between color naming and the universal color physiology of all people bypasses the Hering primaries altogether. They suggested that the warm/cool distinction, which appears in essentially all WCS datasets and therefore appears as a deep gash of dark in the WCS-wide consensus diagram (figure 36.6b), is related to the responses of the sign of the L–M cone contrast in the ascending visual pathway, especially in specific locations within the "hue maps" that Xiao, Wang, and Felleman (2003) found in V1.

Another approach is to treat color naming as an optimization problem in color communication. The *interpoint distance model* (Jameson, 2005; Jameson & D'Andrade, 1997) supposes that, for a given number of color categories, boundaries between these categories are drawn to maximize the similarity between colors within categories while minimizing the similarity between colors in different categories. Regier, Kay, and Khetarpal (2007) have applied this approach to the WCS and have successfully predicted many of the color-naming patterns in that dataset. Yendrikhovskij (2001) has proposed a model of color category formation based on the clustering of perceptually similar colors that occur in an observer's environment. Yendrikhovskij based his estimates of similarity on the metric structure of the CIELUV uniform color space. However, Steels and Belpaeme (2005) have shown that randomly chosen colors from either CIELUV or CIELAB color spaces are statistically similar to those obtained by carefully sampled "real-world" colors, suggesting that Yendrikhovskij's results may have more to do with the metric properties of the color space he used in his analysis and less to do with the statistical structure of color in the environment. Nonetheless, the approaches of Yendrichovskij and of Regier and co-workers are attractive because they avoid the need for explicitly assuming the primacy of the Hering unitary hues. However, it is important to recognize that these approaches do rely critically on a different set of assumptions, namely, the metric structure of the color space used to specify perceptual distances between colors.

Lindsey and Brown (2002) proposed that the geographical distribution of Dark and Grue motifs might be related to the phototoxic effects of sunlight on the lens and retina. Solar ultraviolet light (UV-B) is known to accelerate the natural yellowing of the ocular lens (lens brunescence) and to damage the S cones. Both of these effects reduce the sensitivity of the eye to short-wavelength light (Schefrin et al., 1992). Stimuli that are predominantly short wavelength are predicted to become either less visible or less blue-looking or both, perhaps reducing the salience of these stimuli and diminishing the utility of lexical distinctions involving blue. However, we also know that the visual system can "recalibrate" itself to diminished short-wavelength light stimulation over the range of loss seen during senescence at temperate latitudes (Hardy et al., 2005; Schefrin & Werner, 1990). However, for extreme changes in optical density, recalibration is incomplete (Delahunt et al., 2004; Schefrin & Werner, 1990). It remains to be seen, quantitatively, how these issues play out for more drastic changes in density that might occur in observers who live near the equator.

The Relativist View Revisited

Despite the empirical support for universal constraints on color naming, there are several aspects of color naming that clearly depend on culture. The most obvious aspect is that color terms have to be learned as each generation of children learns to speak. This is not necessarily evidence against the existence of universal constraints on color naming, but it does provide an opportunity for color terms, like other aspects of language, to change over time (Jameson & D'Andrade, 1997; Steels & Belpaeme, 2005). The second aspect is simply that people around the world use different numbers of BCTs, which on its face contradicts the supposition that BCTs are strictly governed by factors that are universal to all humans. Subsequent analyses of the WCS data suggest much greater diversity in color term evolution (Kay & Maffi, 1999; Kay & McDaniel, 1978; Lindsey & Brown, 2009) than Berlin and Kay's original, tightly constrained evolutionary sequence. This diversity suggests that cultural factors may affect color term evolution in complex, possibly unpredictable ways.

DEVELOPMENT OF COLOR NAMING   If the Hering sensations had special status in guiding color category formation within a culture, then one might expect that the colors corresponding to the elemental sensations would be among the first to be named during language development. Even though infants generally have color vision by age 3 months (reviewed by Brown, 1990; Teller,

1997), and even though there is evidence for some categorical color perception in infants as young as 4 months old (Bornstein, Kessen, & Weiskopf, 1976; Franklin & Davies, 2004), it is not clear what role the Hering sensations play as a child learns to name colors between 36 and 40 months of age. Particularly, developmental studies provide little support for view that BCTs corresponding to the fundamental Hering sensations, RED, GREEN, BLUE, YELLOW, BLACK, and WHITE are acquired first, before other BCTs such as PINK, ORANGE, and PURPLE (reviewed by Bornstein, 1985; Pitchford & Mullen, 2002). Moreover, accurate color naming is not acquired in children until ages 4–7 years, often only after extensive training (Rice, 1980) (cited in Pitchford & Mullen, 2001). Finally, studies of color term acquisition in non-English-speaking children whose language has only five BCTs have yielded contradictory results, which could be interpreted as support for either the universalist or the relativist view (Franklin et al., 2005; Roberson et al., 2004).

COLOR MEMORY   Some of the classic literature on the relationship between color names and color cognition challenges a strictly universalist account of color naming in favor of linguistic relativity. For example, Brown and Lenneberg (1954) studied color memory in English speakers using a color recognition task. The focal colors corresponding to the eight chromatic BCTs (black, white, and gray were not tested) were better remembered than the colors of samples of intermediate hues. Moreover, focal colors were also more "codable" than the intermediate colors; that is, they were more easily named because they were "best examples" of the subjects' color categories. The authors concluded that the named color categories improve memory by facilitating the efficiency with which an internal lexical representation of the colored samples could be generated and stored in memory.

A compelling way of testing Brown and Lenneberg's claim would be to look for cross-cultural differences in color memory. If color memory and the highly codable focal colors were tightly related to a person's language, then people with few color terms in their language should have few codable focal colors, and their color memory should be best for only those colors. Rosch Heider (Heider, 1972; Heider & Olivier, 1972) tested that prediction in a series of experiments on color memory with the Dani people of Papua New Guinea. Color memory was best for the 11 English focal colors (compared to similar, nonfocal colors) even though the Dani informants named only two color categories. This remarkable result suggested that the English color categories are innately present in human cognition, and

their primacy in color memory is independent of the observer's color lexicon. However, Heider's conclusions have been questioned in recent years on statistical grounds as well as on the basis of new cross-cultural studies on the Berinmo people of Irian Jaya, who do not use separate words for BLUE and GREEN. These studies report that color memory can be strongly influenced by the observer's color lexicon (Roberson, Davies, & Davidoff, 2000). At present, the available evidence best supports the view that language influences color memory.

CATEGORICAL COLOR PERCEPTION   Classic work by Kay and Kempton (1984) on categorical color perception (CCP) also provides support for linguistic relativity. In that experiment U.S. observers and observers who spoke the Mayan language, Tarahumara, judged the similarity of pairs of colors that either fell within color categories named by their respective languages or else straddled a color boundary between two such named categories. The colors from within named categories looked more similar than pairs of colors that straddled color categorical boundaries, where the colors were chosen to be separated by an equal number of steps in the Munsell color order system. Kay and Kempton further showed, in a color-grouping experiment, that the locations of the boundaries between color categories differed according to the observer's language. They interpreted their result as evidence that the color categorical structure of a language can distort the perceptual metric structure of color space. Winawer et al. (2007) reported a similar result for Russian, which may have distinct terms for LIGHT BLUE and DARK BLUE and therefore may have 12 BCTs.

More recent attempts to generalize CCP to other tasks have produced inconsistent results. Some investigators have reported that speeded discrimination of colors was categorical in a reaction-time (RT) experiment or that visual search for colors was categorical (e.g., Daoutis et al., 2006), with the categories closely related to the color terms in the subject's language. Several studies have suggested an even tighter relation between CCP and language because they reported that CCP effects in visual search were confined to the right visual field, suggesting that visual search RT might benefit from the colocation of language and right visual field processing in the left cerebral hemisphere (Gilbert et al., 2006; see also Siok et al., 2009, discussed above). This proposal remains controversial, however, because several laboratories have failed to replicate the laterality effect (Brown, Lindsey, & Guckes, 2011; Witzel & Gegenfurtner, 2011). Others have questioned the evidence for CCP itself, citing several methodological problems (Brown, Lindsey, & Guckes, 2011). Most studies up to now, including Kay & Kempton (1984) and Gilbert et al. (2006), have relied heavily on the metric properties of Munsell color space. This is problematical, as the colors are only approximately equally spaced (as outlined above) and equally spaced hues are not equally discriminable from one another (Bellamy & Newhall, 1942). Furthermore, the limited metric properties of the Munsell color order system are based on aggregate data from 41 observers (Newhall, 1940) and will not apply typically to any individual person.

UNIVERSALISM VERSUS RELATIVISM

As of today neither universalism nor relativism has provided a compelling, unified view of the relationship among color naming and color perception and cognition. The universalists have not yet provided an explicit statement about what it is that is universal (physiology or something else), an explanation of why trichromatic color appearance space is decimated into so many universal categories or why physiologically similar people from these diverse cultures around the world use such a wide range of color term vocabularies. The cultural relativists have not yet provided an explanation of why people around the world name color in such a similar way or how culture determines exactly which color terms are used or why people within a given language or culture differ so greatly from each other in their color term usage. We expect that constructive theoretical work that is not driven by either the universalist or the relativist view or the debate between them will be most likely to make important progress in this interesting field.

ACKNOWLEDGMENTS

This chapter was supported by NSF BCS-1152841 to D.T.L. and NIH/NEI R21 EY018321–0251 to A.M.B. The authors wish to thank Michael A. Webster and Kimberly A. Jameson for their helpful comments.

NOTE

1. In this discussion, the color itself is in lowercase letters, the English color term is in quotation marks, the international gloss is in capital letters, and the name of the motif has the first letter capitalized.

REFERENCES

Abramov, I., & Gordon, J. (1977). Color-vision in peripheral retina. II. Hue and saturation. *Journal of the Optical Society of America, 67*, 202–207.

Alpern, M., Krantz, D. H., & Kitahara, K. (1983). Perception of colour in unilateral tritanopia. *Journal of Physiology (London), 335*, 683–697.

Bäuml, K. H. (1995). Illuminant changes under different surface collections: Examining some principles of color appearance. *Journal of the Optical Society of America. A, 12*, 261–271.

Bellamy, B. R., & Newhall, S. M. (1942). Attributive limens in selected regions of the Munsell color solid. *Journal of the Optical Society of America, 32*, 465–473.

Belmore, S. C., & Shevell, S. K. (2008). Very-long-term chromatic adaptation: Test of gain theory and a new method. *Visual Neuroscience, 25*, 411–414.

Berlin, B., & Kay, P. (1969). *Basic color terms: Their universality and evolution.* Berkeley, Los Angeles: University of California Press.

Bornstein, M. H. (1985). On the development of color naming in young children: Data and theory. *Brain and Language, 26*, 72–93.

Bornstein, M. H., Kessen, W., & Weiskopf, S. (1976). Color-vision and hue categorization in young human infants. *Journal of Experimental Psychology. Human Perception and Performance, 2*, 115–129.

Boynton, R. M., Schafer, W., & Neun, M. E. (1964). Hue-wavelength relation measured by color-naming method for three retinal locations. *Science, 146*, 666–668.

Boynton, R. M., & Olson, C. X. (1987). Locating basic colors in the OSA space. *Color Research and Application, 12*, 94–105.

Boynton, R. M., & Olson, C. X. (1990). Salience of chromatic basic color terms confirmed by three measures. *Vision Research, 30*, 1311–1317.

Brainard, D. H., & Maloney, L. T. (2011). Surface color perception and equivalent illumination models. *Journal of Vision, 11*, 1. doi:10.1167/11.5.1.

Brainard, D. H., & Stockman, A. (2010). Colorimetry. In M. Bass (Ed.), *The Optical Society of America handbook of optics: Volume III. Vision and vision optics* (pp. 10.1–10.56). New York: McGraw-Hill.

Brainard, D. H., & Wandell, B. A. (1992). Asymmetric color matching: How color appearance depends on the illuminant. *Journal of the Optical Society of America. A, 9*, 1433–1448.

Brenner, E., & Cornelissen, F. W. (1998). When is a background equivalent? Sparse chromatic context revisited. *Vision Research, 38*, 1789–1793.

Brill, M. H. (2005). The relation between the color of the illuminant and the color of the illuminated object by Herbert E. Ives [Commentary]. *Color Research and Application, 20*(1), 70–76.

Brouwer, G. J., & Heeger, D. J. (2009). Decoding and reconstructing color from responses in human visual cortex. *Journal of Neuroscience, 29*, 13992–14003.

Brown, A. M. (1990). Development of visual sensitivity to light and color vision in human infants—a critical review. *Vision Research, 30*, 1159–1188.

Brown, A. M., Lindsey, D. T., & Guckes, K. M. (2011). Color names, color categories, and color-cued visual search: Sometimes, color perception is not categorical. *Journal of Vision, 11*, 2. doi:10.1167/11.12.2.

Brown, R. W. (1976). In memorial tribute to Eric Lenneberg. *Cognition, 4*, 125–153.

Brown, R. W., & Lenneberg, E. H. (1954). A study in language and cognition. *Journal of Abnormal and Social Psychology, 49*, 454–462.

Bruckner, A. (1927). Zur frage der eichung von farbsystemen. *Zeitschrift fur Sinnesphysiologie, 58,* 322–362.

Buchsbaum, G. (1980). A spatial processor model for object colour perception. *Journal of the Franklin Institute, 310,* 1–26.

Burns, S. A., Elsner, A. E., Pokorny, J., & Smith, V. C. (1984). The Abney effect: Chromaticity coordinates of unique and other constant hues. *Vision Research, 24,* 479–489.

Cao, D., & Shevell, S. K. (2005). Chromatic assimilation: Spread light or neural mechanism? *Vision Research, 45,* 1031–1045.

Chichilnisky, E. J., & Wandell, B. A. (1999). Trichromatic opponent color classification. *Vision Research, 39,* 3444–3458.

Cohen, J. (1964). Dependency of the spectral reflectance curves of the Munsell color chips. *Psychonomic Science, 1,* 369–370.

Conway, B. R. (2009). Color vision, cones, and color-coding in the cortex. *Neuroscientist, 15,* 274–290.

Conway, B. R., Moeller, S., & Tsao, D. Y. (2007). Specialized color modules in macaque extrastriate cortex. *Neuron, 56,* 560–573.

Conway, B. R., & Stoughton, C. M. (2009). Response: Towards a neural representation for unique hues. *Current Biology, 19,* R442–R443.

Dacey, D. M., Liao, H. W., Peterson, B. B., Robinson, F. R., Smith, V. C., Pokorny, J., et al. (2005). Melanopsin-expressing ganglion cells in primate retina signal colour and irradiance and project to the LGN. *Nature, 433,* 749–754. doi:10.1038/nature03387.

D'Andrade, R. G., & Romney, A. K. (2003). A quantitative model for transforming reflectance spectra into the Munsell color space using cone sensitivity functions and opponent process weights. *Proceedings of the National Academy of Sciences of the United States of America, 100,* 6281–6286.

Daoutis, C. A., Franklin, A., Riddett, A., Clifford, A., & Davies, I. R. L. (2006). Categorical effects in children's colour search: A cross-linguistic comparison. *British Journal of Developmental Psychology, 24,* 373–400.

Davies, I. R. L., Corbett, G., Mtenje, A., & Sowden, P. (1995). The basic color terms of Chichewa. *Lingua, 95*(4), 259–278.

Dedrick, D. (1998). *Naming the rainbow: Colour language, colour science, and culture.* Dordrecht: Kluwer.

Delahunt, P. B., Webster, M. A., Ma, L., & Werner, J. S. (2004). Long-term renormalization of chromatic mechanisms following cataract surgery. *Visual Neuroscience, 21,* 301–307.

Derrington, A. M., Krauskopf, J., & Lennie, P. (1984). Chromatic mechanisms in lateral geniculate nucleus of macaque. *Journal of Physiology, 357,* 241–265.

DeValois, R. L., Abramov, I., & Jacobs, G. H. (1966). Analysis of response patterns of LGN cells. *Journal of the Optical Society of America, 56,* 966–977.

DeValois, R. L., Cottaris, N. P., Elfar, S. D., Mahon, L. E., & Wilson, A. J. (2000). Some transformations of color information from lateral geniculate nucleus to striate cortex. *Proceedings of the National Academy of Sciences of the United States of America, 97,* 4997–5002.

D'Zmura, M., & Lennie, P. (1986). Mechanisms of color constancy. *Journal of the Optical Society of America. A, 3,* 1662–1672.

Engel, S., Zhang, X. M., & Wandell, B. A. (1997). Colour tuning in human visual cortex measured with functional magnetic resonance imaging. *Nature, 388,* 68–71.

Fach, C., & Sharpe, L. T. (1986). Assimilative hue shifts in color gratings depend on bar width. *Perception & Psychophysics, 40,* 412–418.

Foster, D. H. (2011). Color constancy. *Vision Research, 51,* 674–700.

Foster, D. H., & Nascimento, S. M. C. (1994). Relational colour constancy from invariant cone-excitation ratios. *Proceedings. Biological Sciences, 257,* 115–121.

Franklin, A., Clifford, A., Williamson, E., & Davies, I. R. L. (2005). Color term knowledge does not affect categorical perception of color in toddlers. *Journal of Experimental Child Psychology, 90,* 114–141.

Franklin, A., & Davies, I. R. L. (2004). New evidence for infant colour categories. *British Journal of Developmental Psychology, 22,* 349–377.

Gegenfurtner, K. T. (2003). Cortical mechanisms of colour vision. *Nature Reviews. Neuroscience, 4,* 563–572.

Gilbert, A. L., Regier, T., Kay, P., & Ivry, R. B. (2006). Whorf hypothesis is supported in the right visual field but not the left. *Proceedings of the National Academy of Sciences of the United States of America, 103,* 489–494.

Gladstone, W. E. (1858). *Studies on Homer and the Homeric age III* (Sec. IV, pp. 457–499). London: Oxford University Press.

Golz, J., & MacLeod, D. I. A. (2002). Influence of scene statistics on colour constancy. *Nature, 415,* 637–640.

Gordon, J., & Abramov, I. (1965). Scaling procedures for specifying color appearance. *Color Research and Application, 13,* 146–152.

Hardy, J. L., Frederick, C. M., Kay, P., & Werner, J. S. (2005). Color naming, lens aging, and grue: What the optics of the aging eye can teach us about color language. *Psychological Science, 16,* 321–327.

Heider, E. R. (1972). Universals in color naming and memory. *Journal of Experimental Psychology, 93,* 10–20.

Heider, E. R., & Olivier, D. C. (1972). The structure of the color space in naming and memory for two languages. *Cognitive Psychology, 3,* 337–354.

Helson, H. (1926). The psychology of "Gestalt." *American Journal of Psychology, 37,* 25–62.

Helson, H. (1943). Some factors and implications of color constancy. *Journal of the Optical Society of America, 33,* 555–567.

Hering, E. (1964). *Grundzuge der Lehre vom Lichtsinn (Outlines of a theory of the light sense)* (L. M. Hurvich & D. Jameson, Trans.). Cambridge, MA: Harvard University Press. (Original work published 1878.)

Hurvich, L. M., & Jameson, D. (1956). Some quantitative aspects of an opponent-colors theory. IV. A psychological color specification system. *Journal of the Optical Society of America, 46,* 416–421.

Hurvich, L. M., & Jameson, D. (1957). An opponent-process theory of color vision. *Psychological Review, 64,* 384–404.

Indow, T. (1988). Multidimensional studies of Munsell color solid. *Psychological Review, 95,* 456–470.

Ingling, C. R., & Martinez-Uriegas, E. (1983). The relationship between spectral sensitivity and spatial sensitivity for the primate r-g X-channel. *Vision Research, 23,* 1495–1500.

Ives, H. E. (1912). The relation between the color of the illuminant and the color of the illuminated object. *Transactions of the Illumination Engineering Society, 7,* 62–72.

Jameson, D., & Hurvich, L. M. (1951). Use of spectral hue-invariant loci for the specification of white stimuli. *Journal of Experimental Psychology, 41*, 455–463.

Jameson, D., & Hurvich, L. M. (1959). Perceived color and its dependence on focal, surrounding, and preceding stimulus variables. *Journal of the Optical Society of America, 49*, 890–897.

Jameson, D., & Hurvich, L. M. (1972). Color adaptation: Sensory control, contrast, afterimages. In D. Jameson & L. M. Hurvich (Eds.), *Handbook of sensory physiology* (Vol. II/4, pp. 568–581). Berlin: Springer-Verlag.

Jameson, K. A. (2005). Why GRUE? An interpoint distance model analysis of composite color categories. *Cross-Cultural Research, 39*, 159–194.

Jameson, K. A., & Alvarado, N. (2003). Differences in color naming and color salience in Vietnamese and English. *Color Research and Application, 28*, 113–138.

Jameson, K. A., & D'Andrade, R. G. (1997). It's not really red, green, yellow, blue: An inquiry into cognitive color space. In C. L. Hardin & L. Maffi (Eds.), *Color categories in thought and language* (pp. 295–319). Cambridge: Cambridge University Press.

Jameson, K. A., & Komarova, N. L. (2009). Evolutionary models of color categorization. I. Population categorization systems based on normal and dichromat observers. *Journal of the Optical Society of America. A, 26*, 1414–1423.

Jenness, J. W., & Shevell, S. K. (1995). Color appearance with sparse chromatic context. *Vision Research, 35*, 797–805.

Johnson, E. N., Hawken, M. J., & Shapley, R. (2001). The spatial transformation of color in the primary visual cortex of the macaque monkey. *Nature Neuroscience, 4*(4), 409–416.

Johnson, E. N., Hawken, M. J., & Shapley, R. (2004). Cone inputs in macaque primary visual cortex. *Journal of Neurophysiology, 91*, 2501–2514.

Judd, D. B., MacAdam, D. L., & Wyszecki, G. (1964). Spectral distribution of typical daylight as a function of correlated color temperature. *Journal of the Optical Society of America, 54*, 1031–1040.

Juricevic, I., & Webster, M. A. (2009). Variations in normal color vision: V. Simulations of adaptation to natural color environments. *Visual Neuroscience, 26*, 133–145.

Kaiser, P. K., Lee, B. B., Martin, P. R., & Valberg, A. (1990). The physiological basis of the minimally distinct border demonstrated in the ganglion cells of the macaque retina. *Journal of Physiology, 422*, 153–183.

Kay, P., Berlin, B., Maffi, L., Merrifield, W. R., & Cook, R. S. (2009). *The world color survey*. Stanford, CA: CSLI.

Kay, P., & Kempton, W. (1984). What is the Sapir-Whorf hypothesis? *American Anthropologist, 86*, 65–79.

Kay, P., & Maffi, L. (1999). Color appearance and the emergence and evolution of basic color lexicons. *American Anthropologist, 101*, 743–760.

Kay, P., & McDaniel, K. (1978). The linguistic significance of the meanings of basic color terms. *Language, 54*, 610–646.

Kay, P., & Regier, T. (2003). Resolving the question of color naming universals. *Proceedings of the National Academy of Sciences of the United States of America, 100*, 9085–9089.

Kay, P., & Regier, T. (2006). Language, thought and color: Recent developments. *Trends in Cognitive Sciences, 10*, 51–54.

Kleinschmidt, A., Lee, B. B., Requardt, M., & Frahm, J. (1996). Functional mapping of color processing by magnetic resonance imaging of responses to selective P- and M-pathway stimulation. *Experimental Brain Research, 110*, 279–288.

Knoblauch, K., & Shevell, S. K. (2001). Relating cone signals to color appearance: Failure of monotonicity in yellow/blue. *Visual Neuroscience, 18*, 901–906.

Koenderink, J. J., & van Doorn, A. J. (2003). Perspectives on color space. In R. Mausfeld & D. Heyer (Eds.), *Colour perception; Mind and the physical world* (pp. 1–56). Oxford: Oxford University Press.

Koida, K., & Komatsu, H. (2007). Effects of task demands on the responses of color-selective neurons in the inferior temporal cortex. *Nature Neuroscience, 10*, 108–116. doi:10.1038/nn1823.

Komatsu, H., Ideura, Y., Kanji, S., & Yamane, S. (1992). Color selectivity of neurons in the inferior temporal cortex of the awake macaque monkey. *Journal of Neuroscience, 12*, 408–424.

König, A., & Dieterici, C. (1893). Die Grundempfindungen in normalen und anomalen Farben Systemen und ihre Intensitats-Verteilung im Spectrum. *Zeitschrift für Psychologie und Physiologie der Sinnesorgane, 4*, 241–247.

Krantz, D. H. (1975). Color measurement and color theory: II. Opponent-colors theory. *Journal of Mathematical Psychology, 12*, 304–327.

Krauskopf, J., Zaidi, Q., & Mandler, M. (1986). Mechanisms of simultaneous color induction. *Journal of the Optical Society of America. A, 3*, 1752–1757.

Kuehni, R. G. (2003). *Color space and its divisions: color order from antiquity to the present*. New York: Wiley Interscience.

Larimer, J., Krantz, D. H., & Cicerone, C. M. (1974). Opponent-process additivity—I: Red/green equilibria. *Vision Research, 14*, 1127–1140.

Larimer, J., Krantz, D. H., & Cicerone, C. M. (1975). Opponent process additivity—II: Yellow/blue equilibria and nonlinear models. *Vision Research, 15*, 723–731.

Lee, B. B., Martin, P. R., & Valberg, A. (1988). The physiological-basis of heterochromatic flicker photometry demonstrated in the ganglion-cells of the macaque retina. *Journal of Physiology, 404*, 323–347.

Lennie, P., Krauskopf, J., & Sclar, G. (1990). Chromatic mechanisms in striate cortex of macaque. *Journal of Neuroscience, 10*, 649–669.

Lennie, P., & Movshon, J. A. (2005). Coding of color and form in the geniculostriate visual pathway (invited review). *Journal of the Optical Society of America. A, 22*, 2013–2033.

Lennie, P., Pokorny, J., & Smith, V. C. (1993). Luminance. *Journal of the Optical Society of America. A, 10*, 1283–1293.

Leventhal, A. G., Thompson, K. G., Liu, D., Zhou, H., & Ault, S. J. (1995). Concomitant sensitivity to orientation, direction, and color of cells in layers 2, 3, and 4 of monkey striate cortex. *Journal of Neuroscience, 15*, 1808–1818.

Levinson, S. C. (2000). Yélî Dnye and the theory of basic color terms. *Journal of Linguistic Anthropology, 10*, 3–55.

Lindsey, D. T., & Brown, A. M. (2002). Color naming and the phototoxic effects of sunlight on the eye. *Psychological Science, 13*, 506–512.

Lindsey, D. T., & Brown, A. M. (2006). Universality of color names. *Proceedings of the National Academy of Sciences of the United States of America, 103*, 16608–16613.

Lindsey, D. T., & Brown, A. M. (2009). World color survey color naming reveals universal motifs and their within-language diversity. *Proceedings of the National Academy of Sciences of the United States of America, 106*, 19785–19790.

Livingstone, M. S., & Hubel, D. H. (1984). Anatomy and physiology of a color system in the primate visual cortex. *Journal of Neuroscience, 4*, 309–356.

MacLeod, D. I. A. (2010). Into the neural maze. In J. Cohen & M. Matthen (Eds.), *Color ontology and color science* (pp. 151–178). Cambridge, MA: MIT Press.

Maloney, L. T. (1999). Physics-based approaches to modeling surface color perception. In K. T. Gegenfurtner & L. T. Sharpe (Eds.), *Color vision: From genes to perception* (pp. 387–421). Cambridge: Cambridge University Press.

Maloney, L. T., & Wandell, B. A. (1986). Color constancy: A method for recovering surface spectral reflectance. *Journal of the Optical Society of America. A, 3*, 29–33.

Mizokami, Y., Werner, J. S., Crognale, M. A., & Webster, M. A. (2006). Nonlinearities in color coding: Compensating color appearance for the eye's spectral sensitivity. *Journal of Vision, 6*, 996–1007. doi:10.1167/6.9.12.

Mollon, J. D. (2003). The origins of modern color science. In S. K. Shevell (Ed.), *The science of color* (2nd ed., pp. 1–39). Amsterdam: Elsevier.

Mollon, J. D. (2009). A neural basis for unique hues? *Current Biology, 19*, R441–R442.

Monnier, P., & Shevell, S. K. (2003). Large shifts in color appearance from patterned chromatic backgrounds. *Nature Neuroscience, 6*, 801–802.

Newhall, S. M. (1940). Preliminary report of the O.S.A. subcommittee on the spacing of the Munsell colors. *Journal of the Optical Society of America, 30*, 617–645.

Newhall, S. M., Nickerson, D., & Judd, D. B. (1943). Final report of the O.S.A. Subcommittee on the Spacing of the Munsell Colors. *Journal of the Optical Society of America, 33*, 385–411.

Nickerson, D. (1940). History of the Munsell color system and its scientific application. *Journal of the Optical Society of America, 30*, 575–586.

Nietz, J., Carroll, J., Yamauchi, Y., Neitz, M., & Williams, D. R. (2002). Color perception is mediated by a plastic neural mechanism that is adjustable in adults. *Neuron, 35*, 783–792.

Parkes, L. M., Marsman, J.-B., Oxley, D. C., Goulermas, J. Y., & Wuerger, S. M. (2009). Multivoxel fMRI analysis of color tuning in human primary visual cortex. *Journal of Vision, 9*, 1–13. doi:10.1167/9.1.1.

Pitchford, N. J., & Mullen, K. T. (2001). Conceptualization of perceptual attributes: A special case for color? *Journal of Experimental Child Psychology, 80*, 289–314.

Pitchford, N. J., & Mullen, K. T. (2002). Is the acquisition of basic-colour terms in young children constrained? *Perception, 31*, 1349–1370.

Pokorny, J., & Smith, V. C. (1987). L/M cone ratios and the null point of the perceptual red/green opponent system. *Die Farbe, 34*, 53–57.

Pugh, E. N., & Mollon, J. D. (1979). A theory of the Pi_1 and Pi_3 color mechanisms of Stiles. *Vision Research, 19*, 293–312.

Purdy, D. M. (1937). The Bezold-Brücke phenomenon and contours for constant hue. *American Journal of Psychology, 49*, 313–315.

Regier, T., & Kay, P. (2009). Language, thought, and color: Whorf was half right. *Trends in Cognitive Sciences, 13*, 439–446.

Regier, T., Kay, P., & Cook, R. S. (2005). Focal colors are universal after all. *Proceedings of the National Academy of Sciences of the United States of America, 102*, 8386–8391.

Regier, T., Kay, P., & Khetarpal, N. (2007). Color naming reflects optimal partitions of color space. *Proceedings of the National Academy of Sciences of the United States of America, 104*, 1436–1441.

Rice, M. (1980). *Cognition to language: Categories, word meanings, and training.* Baltimore: University Park Press.

Rinner, O., & Gegenfurtner, K. R. (2000). Time course of chromatic adaptation for color appearance and discrimination. *Vision Research, 40*, 1813–1826.

Roberson, D., Davidoff, J., Davies, I. R. L., & Shapiro, L. R. (2004). The development of color categories in two languages: A longitudinal study. *Journal of Experimental Psychology. General, 133*, 554–571.

Roberson, D., Davies, I. R. L., & Davidoff, J. (2000). Colour categories are not universal: Replications and new evidence from a stone-age culture. *Journal of Experimental Psychology. General, 129*, 369–398.

Sallstrom, P. (1973). *Color and physics: Some remarks concerning the physical aspects of human color vision.* Stockholm: University of Stockholm, Institute of Physics Report 73–09.

Saunders, B. A. C., & van Brakel, J. (1997). Are there non-trivial constraints on colour categorization? *Behavioral and Brain Sciences, 20*, 167–179.

Schefrin, B. E., & Werner, J. S. (1990). Loci of spectral unique hues throughtout the life-span. *Journal of the Optical Society of America. A, 7*, 305–311.

Schefrin, B. E., Werner, J. S., Plach, M., Utlaut, N., & Switkes, E. (1992). Sites of age-related sensitivity loss in a short-wave cone pathway. *Journal of the Optical Society of America. A, 9*, 355–363.

Schluppeck, D., & Engel, S. A. (2002). Color opponent neurons in V1: A review and model reconciling results from imaging and single-unit recording. *Journal of Vision, 2*, 480–492. doi:10.1167/2.6.5.

Shapley, R., & Hawken, M. J. (2011). Color in the cortex: Single- and double-opponent cells. *Vision Research, 51*, 701–717.

Shevell, S. K. (1978). The dual role of chromatic backgrounds in color-perception. *Vision Research, 18*, 1649–1661.

Shevell, S. K. (2001). The time course of chromatic adaptation. *Color Research and Application, 26*, S170–S173.

Shevell, S. K. (2003). Color appearance. In S. K. Shevell (Ed.), *The science of color* (pp. 149–190). Oxford: Elsevier.

Shevell, S. K. (2012). The Verriest Lecture: Color lessons from space, time and motion. *Journal of the Optical Society of America. A, 29*, A337–A345.

Shinomori, K., Schefrin, B. E., & Werner, J. S. (1997). Spectral mechanisms of spatially induced blackness: Data and quantitative model. *Journal of the Optical Society of America. A, 14*, 372–387.

Siok, W. T., Kay, P., Wang, W. S. Y., Chan, A. H. D., Chen, L., Luke, K. K., & Tan, L. H. (2009). Language regions of brain are operative in color perception. *Proceedings of the National Academy of Sciences of the United States of America, 106*, 8140–8145. doi:10.1073/pnas.0903627106.

Solomon, S. G., & Lennie, P. (2005). Chromatic gain controls in visual cortical neurons. *Journal of Neuroscience, 25*, 4779–4792.

Steels, L., & Belpaeme, T. (2005). Coordinating perceptually grounded categories through language: A case study for colour. *Behavioral and Brain Sciences, 28*, 469–489.

Stockman, A., & Brainard, D. H. (2009). Color vision mechanisms. In M. Bass (Ed.), *The Optical Society of America handbook of optics III. Vision and vision optics* (pp. 11.1–11.104). New York: McGraw-Hill.

Stoughton, C. M., & Conway, B. R. (2008). Neural basis for unique hues. *Current Biology, 18*(16), R698–R699.

Suppes, P., Luce, R. D., Krantz, D. H., & Tversky, A. (1989). *Foundations of measurement.* Vol. 2: *Geometrical, threshold, and probabilistic representations.* New York: Academic Press.

Teller, D. Y. (1997). First glances: The vision of infants—The Friedenwald Lecture. *Investigative Ophthalmology and Visual Science, 38*, 2183–2203.

Ts'o, D. Y., & Gilbert, C. D. (1988). The organization of chromatic and spatial interactions in the primate striate cortex. *Journal of Neuroscience, 8*, 1712–1727.

Uchikawa, K., & Boynton, R. M. (1987). Catergorical color perception of Japanese observers: Comparison with that of Americans. *Vision Research, 27*, 1825–1833.

Vimal, R. L. P., Pokorny, J., & Smith, V. C. (1987). Appearance of steadily viewed lights. *Vision Research, 27*, 1309–1318.

Volbrecht, V. J., Werner, J. S., & Cicerone, C. M. (1990). Additivity of spatially induced blackness. *Journal of the Optical Society of America. A, 7*, 106–112.

Wachtler, T., Seijnowski, T. J., & Albright, T. D. (2003). Representation of color stimuli in awake macaque primary visual cortex. *Neuron, 37*, 681–691.

Wade, A., Augath, M., Logothetis, N., & Wandell, B. A. (2008). fMRI measurements of color in macaque and human. *Journal of Vision, 8*, 1–19. doi:10.1167/8.10.6.

Ware, C., & Cowan, W. B. (1982). Changes in perceived color due to chromatic interactions. *Vision Research, 22*, 1353–1362.

Webster, M. A. (2011). Adaptation and visual coding. *Journal of Vision, 11*(5), 3. doi:10.1167/11.5.3

Webster, M. A., & Kay, P. (2007). Individual and population differences in focal colors. In R. E. MacLaury, G. V. Paramei, & D. Dedrick (Eds.), *Anthropology of color: Interdisciplinary multilevel modeling* (pp. 29–53). Amsterdam: John Benjamins.

Webster, M. A., Mizokami, Y., & Webster, S. M. (2007). Seasonal variations in the color statistics of natural images. *Network: Computation in Neural Systems, 18*, 213–233.

Webster, M. A., & Mollon, J. D. (1995). Color constancy influenced by contrast adaptation. *Nature, 373*, 694–698.

Werner, J. S., Cicerone, C. M., Kliegl, R., & DellaRosa, D. (1984). Spectral efficiency of blackness induction. *Journal of the Optical Society of America. A, 1*, 981–986.

Werner, J. S., & Walraven, J. (1982). Effect of chromatic adaptation on the achromatic locus: The role of contrast, luminance, and background color. *Vision Research, 22*, 929–943.

West, G., & Brill, M. H. (1982). Necessary and sufficient conditions for von Kries chromatic adaptation to give color constancy. *Journal of Mathematical Biology, 15*, 249–258.

Wiesel, T. N., & Hubel, D. H. (1966). Spatial and chromatic properties in the lateral geniculate body of the rhesus monkey. *Journal of Neurophysiology, 29*, 1115–1156.

Winawer, J., Witthoft, N., Frank, M. C., Wu, L., Wade, A. R., & Boroditsky, L. (2007). Russian blues reveal effects of language on color discrimination. *Proceedings of the National Academy of Sciences of the United States of America, 104*, 7780–7785.

Witzel, C., & Gegenfurtner, K. R. (2011). Is there a lateralized category effect for color? *Journal of Vision, 11*, 16. doi:10.1167/11.12.16.

Wyszecki, G., & Stiles, W. S. (1982). *Color science: Concepts and methods, quantitative data and formulae* (2nd ed.). New York: Wiley.

Xiao, Y. P., Kavanau, C., Bertin, L., & Kaplan, E. (2011). The biological basis of a universal constraint on color naming: Cone contrasts and the two-way categorization of colors. *PLoS One, 6*, e24994.

Xiao, Y. P., Wang, Y., & Felleman, D. J. (2003). A spatially organized representation of colour in macaque cortical area V2. *Nature, 421*, 535–539.

Yendrikhovskij, S. (2001). Computing color categories from statistics of natural images. *Journal of Imaging Science and Technology, 45*, 409–417.

Zaidi, Q., Yoshimi, B., Flanigan, N., & Canova, A. (1992). Lateral interactions within color mechanisms in simultaneous induced contrast. *Vision Research, 32*, 1695–1707.

Zeki, S. (1983a). Colour coding in the cerebral cortex: The reaction of cells in monkey visual-cortex to wavelengths and colours. *Neuroscience, 9*, 741–765.

Zeki, S. (1983b). Colour coding in the cerebral cortex: The responses of wavelength-selective and colour-coded cells in monkey visual cortex to changes in wavelength composition. *Neuroscience, 9*, 767–781.

37 Adaptation in Color and Form Perception

MICHAEL A. WEBSTER

It is a laboured truth that all things and experiences are comparative.

—Slavomir Rawicz, *The Long Walk*

In *The Long Walk*, Rawicz recounts escaping a Soviet prison camp during the Second World War by traveling on foot from Siberia to India (Rawicz, 1997). Throughout he endured hardships few of us can imagine (and which some have claimed he only imagined). Yet, as the quote suggests, faced with such experiences we might come to find them routine. This chapter reviews how comparative judgments driven by our experiences are built into the very fabric of seeing. The mechanisms of perception are highly dynamic and constantly adjusting—or adapting—to make comparisons relative to the current stimulus context. These adjustments profoundly affect visual experience, and how they alter sensitivity and perception has provided a penetrating window into the neural mechanisms of vision. In the review of adaptation in the previous edition, a focus was on how these adjustments have been used to dissect the visual channels encoding information about color and form. The present chapter concentrates on how our understanding of adaptation itself has changed since then. Over this time, explorations of adaptation have grown steadily and have revealed a far broader impact of sensitivity regulation at both higher and lower levels of the visual system, over a wider range of timescales, and for a much more diverse range of perceptual attributes. The similar effects of adaptation across a wide range of stimulus dimensions have pointed to common coding principles and to the central role that adaptation plays in this coding. A detailed account of these developments can be found in Webster (2011). There are a number of other recent reviews on adaptation, some with greater emphasis on the physiological and computational mechanisms (Clifford et al., 2007; Clifford & Rhodes, 2005; Demb, 2008; Kohn, 2007; Rieke & Rudd, 2009; Wark, Lundstrom, & Fairhall, 2007; Webster & MacLeod, 2011).

ADAPTATION AND VISUAL PLASTICITY

Stare at the cross in the image on the left side of figure 37.1 and then shift your gaze to the square on the right. You should briefly experience a visual impression of a face (which blinking may help to revive). The afterimage is a consequence of local light adaptation in the retina (Zaidi et al., 2012). Receptors exposed to dark parts of the image increase their sensitivity and thus respond more to the blank field, while cells stimulated by the bright regions become less sensitive. As a result the phantom face is a negative afterimage of the adapting image. Adaptation aftereffects also occur for more complex stimulus attributes. For example, after looking at clockwise bars, a vertical bar appears counterclockwise to the viewer, and viewing the downward flow of a waterfall makes the rocks to the side appear to ooze upward. Such "pattern-selective" aftereffects occur even when the time-averaged luminance levels across the retina remain constant (e.g., by moving or counterphasing the stimulus) and thus implicate additional adaptable mechanisms sensitive to particular properties of the stimulus.

Visual adaptation is typically measured and defined in terms of these brief exposures and the aftereffects they induce (Thompson & Burr, 2009). However, the visual system exhibits many forms of plasticity and many experience-dependent and context-dependent adjustments. As a result it is difficult to give a functional definition of adaptation that uniquely distinguishes it. As the above examples illustrate, aftereffects begin with adjustments as early as the receptors but can extend to the sensory-motor corrections coordinating perception and action (Shadmehr, Smith, & Krakauer, 2010). Numerous distinct mechanisms contribute to this cascade of calibrations, and multiple mechanisms are involved even in adjusting to a single stimulus property such as the average light level in retinal neurons (Rieke & Rudd, 2009) or contrast in cortical neurons (Dhruv et al., 2011). Moreover, perception is also modulated by

FIGURE 37.1   An aftereffect of light adaptation. Fixate the cross in the left image for several seconds and then view the square on the right. The face that appears is a negative afterimage of the adapting image and results from local sensitivity changes in the retina. (From Webster & MacLeod, 2011.)

processes such as priming, perceptual learning, and attention. These can be differentiated from adaptation by a number of criteria, but the boundaries between them remain blurred. For example, adaptation has been described as a form of learning (Barlow, 1990a) and may involve adjustments that can themselves be learned (Yehezkel et al., 2010). Similarly, adaptation is clearly distinct from attention, in part because we can adapt to patterns we cannot even see (Blake & He, 2005). Yet many perceptual aftereffects can be modulated by attention, especially at higher levels of visual coding, and adaptation and attention could reflect complementary modulations of neural responses (Barlow, 1997; Pestilli, Viera, & Carrasco, 2007; Rezec, Krekelberg, & Dobkins, 2004).

It is also difficult to isolate adaptation based on the timescale of the adjustment. Sensitivity is calibrated not only by the recent past but also nearly instantaneously to changes in the spatial context. Like adaptation, this "normalization" appears ubiquitous in sensory coding (Carandini & Heeger, 2011). It is not clear to what extent these spatial and temporal adjustments are functionally distinct (Schwartz, Hsu, & Dayan, 2007), but they are often described by similar terms. At the other extreme, adaptation effects are also being extended to increasingly longer durations, but it remains unknown when these reflect a qualitative switch in function or mechanism [e.g., from intracellular (Sanchez-Vives, Nowak, & McCormick, 2000) to synaptic changes (Massey & Bashir, 2007) to long-term developmental changes].

A final problem is how adaptation effects probed at different levels (e.g., single cells or behavior) are related (Krekelberg, Boynton, & van Wezel, 2006). Neural responses show a number of adaptive changes (e.g., in their tuning functions or dynamic range) that are not evident behaviorally (Kohn, 2007). An important development in neural imaging has been the advent of fMRI adaptation (Grill-Spector, Henson, & Martin, 2006; Weigelt, Muckli, & Kohler, 2008). The BOLD response declines when the stimulus is repeated, and thus, the selectivity of the response can be measured by determining how the stimulus must change to release the suppression. Studies of fMRI adaptation have demonstrated response changes paralleling many classic perceptual aftereffects. Yet some accounts of repetition effects in fMRI have emphasized the connections to priming (Schacter, Wig, & Stevens, 2007) or behavioral habituation (Turk-Browne, Scholl, & Chun, 2008) rather than adaptation.

EARLY STAGES OF ADAPTATION

The pattern selectivity and interocular transfer of many visual aftereffects implied a cortical locus and suggested precortical mechanisms might adjust only to simple features such as the mean luminance and chromaticity. However, recent studies of the retina continue to reveal increasingly complex computations in early vision (Gollisch & Meister, 2010), and, paralleling this, the range of ways in which the retina adapts has increased dramatically. For example, it is now apparent that the retina of many species adapts not only to the average light level but also to contrast (Baccus & Meister, 2002; Brown & Masland, 2001; Chander & Chichilnisky, 2001; Rieke, 2001; Smirnakis et al., 1997). The contrast adjustments include not only a rapid contrast gain control (Shapley & Enroth-Cugell, 1984) but also sluggish

534   MICHAEL A. WEBSTER

sensitivity changes with a time course similar to that of contrast aftereffects measured psychophysically (Baccus & Meister, 2002), and these dynamics can even adjust to the rate of variations in the stimulus (Wark, Fairhall, & Rieke, 2009). Adaptation within the retina can also adjust to spatiotemporal patterns to which the cells are not directly selective. For example, ganglion cells can adapt to the orientation of patterns (Hosoya, Baccus, & Meister, 2005) or to movement (Olveczky, Baccus, & Meister, 2007). The extent to which these early adjustments contribute to behavioral contrast aftereffects is not certain. In primate retina contrast adaptation is confined largely to the magnocellular pathway and thus may primarily affect sensitivity at high temporal frequencies and lower test contrasts (Camp, Tailby, & Solomon, 2009; Solomon et al., 2004).

ADAPTATION AND HIGH-LEVEL AFTEREFFECTS

Adaptation has also been extended to higher levels of visual coding and more abstract perceptual attributes. For example, adaptation affects not only perceived tilt or curvature but also higher-order shape properties such as aspect ratio (Suzuki & Cavanagh, 1998) or three-dimensional viewpoint (Fang & He, 2005). Tilt aftereffects can also be selective for figure–ground relationships (von der Heydt, Macuda, & Qiu, 2005) and for spatiotopic rather than retinotopic coordinates (Melcher, 2008). Similarly, some color aftereffects may depend on surface reflectance (independent of the illuminant) (Goddard, Solomon, & Clifford, 2010) as well as material properties such as whether they appear glossy or matte (Motoyoshi et al., 2007) and can be contingent on extraretinal cues such as gaze direction (Bompas & O'Regan, 2006; Richters & Eskew, 2009). Adaptation can also influence the perceived layout and affordances of scenes (Greene & Oliva, 2010).

A number of unique motion aftereffects have been distinguished based on the types of stimuli that induce them (e.g., static vs. dynamic or translating vs. expanding) or over which they transfer (e.g., to new retinal locations) (Mather et al., 2008). Motion aftereffects can also be generated by attentional tracking (Culham et al., 2000) or by imagining motion or viewing still photographs that depict movement (Winawer, Huk, & Boroditsky, 2008, 2010). Aftereffects can also occur for attributes that are inferred from motion, such as the perception of gender derived from biological motion (Jordan, Fallah, & Stoner, 2006; Troje et al., 2006).

Many studies have explored adaptation and face perception (Webster & MacLeod, 2011). After viewing a distorted face, an undistorted face appears distorted in the opposite direction (Webster & MacLin, 1999). These aftereffects occur for many of the natural dimensions defining faces including identity, gender and ethnicity, expression, age, and attractiveness (Hsu & Young, 2004; Leopold et al., 2001; O'Neil & Webster, 2011; Rhodes et al., 2003; Schweinberger et al., 2010; Webster et al., 2004). For example, adapting to a female face causes an androgynous face to appear more male, and these aftereffects have therefore been used as a potential tool for probing how faces are encoded and represented in the visual system.

But to what extent does adaptation to a high-level attribute reflect response changes at high levels in the visual stream? This is difficult to assess because sensitivity regulation occurs at many levels, so that higher levels may inherit response changes from earlier stages. For example, contrast adaptation in cells in MT is selective for the stimulated subregion within their receptive field, suggesting that some sensitivity changes are passed on from V1 (Kohn & Movshon, 2003), whereas others arise within MT (Priebe, Churchland, & Lisberger, 2002). Similarly, face aftereffects can be induced by adaptation to local shapes (Xu et al., 2008) or orientation gradients (Dickinson et al., 2010) that are not themselves face-like. Thus, the issue is whether there is evidence for additional stages of adaptation that are linked to the explicit coding of the attribute. A variety of procedures have been developed to try to isolate these stages by maintaining the perceptual similarity of the images as faces while reducing their low-level featural similarity. Face aftereffects show strong transfer across changes in the size, position, or orientation of the adapt and test stimuli, and this transfer has suggested that the adaptation at least partly reflects response changes at high levels of visual processing (Webster & MacLeod, 2011).

The finding that adaptation can affect many complex and abstract visual judgments suggests that most aspects of perception are adaptable. Thus, adaptation appears to be a central and inherent mechanism in visual processing and might warrant the status of a general law (Helson, 1964). Moreover, the basic pattern of high-level aftereffects shows a number of striking similarities to the aftereffects for simpler stimulus attributes. For example, the buildup and decay of face aftereffects follow the same time course as contrast adaptation in gratings (Leopold et al., 2005) and, as discussed below, may calibrate the representation of faces in the same way that adaptation sets the reference level for stimuli such as color. This suggests that how the visual system adapts and represents different perceptual dimensions may often draw on common coding strategies (Clifford, 2002; Webster & MacLeod, 2011).

The plethora of newly discovered aftereffects further suggest that adaptation is constantly engaged by the patterns of stimuli we encounter in everyday viewing. This raises the question of how the processes of adaptation are matched to the properties of natural scenes. The range of light levels and contrasts at different points within a scene are often far greater than could be coded by the dynamic range of neurons, and this strongly constrains how retinal mechanisms adjust as scenes are sampled with eye movements (Rieke & Rudd, 2009). Variations in mean luminance and contrast are also uncorrelated within natural scenes, which predicts that light and contrast adaptation should operate independently, and this independence is seen in both single cells (Mante et al., 2005) and perception (Webster & Wilson, 2000). The statistics of natural color signals have also been central to understanding adaptation and mechanisms of color constancy (Smithson, 2005).

A further important issue is what the "natural" adaptation states of the visual system might be, for these are the states that would be most relevant for understanding the normal operating characteristics of our visual system. One way to explore this has been to examine how we adapt to properties that are characteristic of natural scenes (Wainwright, 1999; Webster & Miyahara, 1997; Webster & Mollon, 1997). For example, natural images have more energy at low spatial and temporal frequencies, with amplitude spectra that roughly fall as $1/f$ (Field, 1987). Adaptation to this structure selectively reduces sensitivity to low spatial frequencies, resulting in marked changes in the shape of the contrast sensitivity function (CSF) (Bex, Solomon, & Dakin, 2009; Webster & Miyahara, 1997) or in the tuning functions of individual cortical cells (Sharpee et al., 2006). Similar structure is found for spatial variations in the chromatic contrast (Parraga, Troscianko, & Tolhurst, 2002), and adaptation to this chromatic structure can cause the normally low-pass chromatic CSF to become nearly band pass (Webster et al., 2006). Conventional measures of contrast sensitivity, based on adaptation and presentation on a uniform gray field, thus may fail to capture important features of visual sensitivity under natural viewing conditions. Natural scenes also have characteristic color gamuts, which vary across specific environments or over time with the seasons. These variations are large enough so that specific environments may hold observers in different states of color adaptation (Webster, Mizokami, & Webster, 2007; Webster & Mollon, 1997).

Adaptation also plays an important role in adjusting to changes within the observer. These changes are often gradual and thus again may involve many forms of plasticity. However, the visual system changes throughout the lifespan and thus must be continuously recalibrated to maintain the match between visual coding and the visual environment. These calibrations may also be critically important for matching visual coding across different retinal locations or stimulus attributes (e.g., so that size coding is consistent across orientation or vice versa).

Most studies of how observers adapt to changes in their own visual system have explored optical changes or the filtering effects of preretinal pigments. As we age the lens progressively yellows so that less short-wavelength light reaches the retina. Without adapting, the world should thus appear increasingly yellower, yet color appearance instead remains stable across the lifespan (Hardy et al., 2005; Schefrin & Werner, 1990; Werner & Schefrin, 1993). This correction could begin with gain changes in the cones to match their sensitivity to the average spectrum we are exposed to. The same processes may also help correct for differences in spectral sensitivity across the visual field (Webster & Leonard, 2008).

An important source of variation in spatial sensitivity is from optical aberrations of the eye. A number of studies have explored how perception adapts to retinal image blur. Viewing a blurred or sharpened image produces rapid changes in the level of physical blur that appears focused (Webster, Georgeson, & Webster, 2002) (figure 37.2). With longer exposures, adaptation to defocus can also lead to improvements in visual acuity (Mon-Williams et al., 1998; Pesudovs & Brennan, 1993). The adaptation is also induced by the actual patterns of blur resulting from the eye's optics, including low-order aberrations of defocus and astigmatism as well as high-order aberrations (Sawides et al., 2010, 2011a). Moreover, observers may be naturally adapted to the level of blur normally present in their retinal image, so that this blur appears subjectively in focus (Artal et al., 2004; Sawides et al., 2011b).

A common consequence to these adjustments is that they tend to compensate visual appearance for the spectral or spatial sensitivity of the observer, and thus, the adaptation helps to maintain perceptual constancy (Walraven & Werner, 1991). This has important implications for visual experience (Webster, Werner, & Field, 2005) because it predicts that the world will "look" less different than predicted by the interobserver

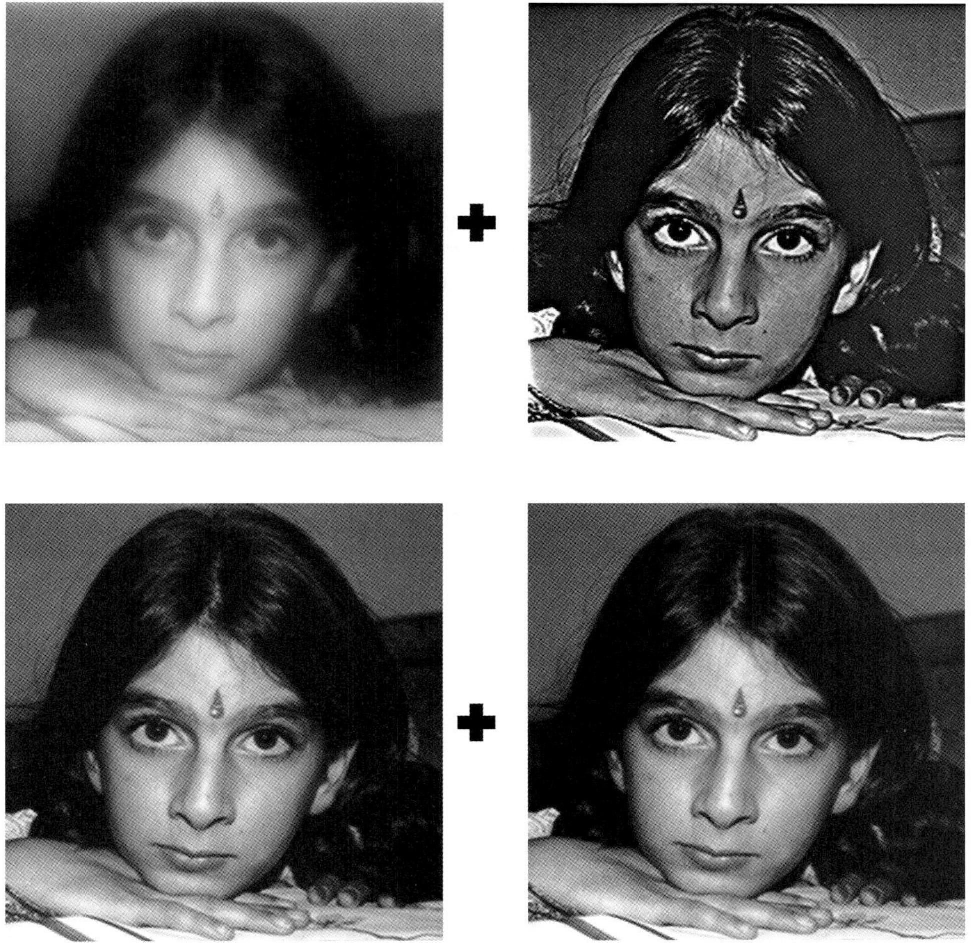

FIGURE 37.2 Blur aftereffects. Fixate the cross between the blurred and sharpened images at the top for several seconds and then switch to the lower cross between the two focused images. The right image will briefly appear blurrier because of adaptation to the preceding blur. (From Webster, 2011.)

differences in threshold sensitivity (Webster, Juricevic, & McDermott, 2010). Moreover, this may tend to mask sensitivity losses with progressive disease, so that observers may be less aware of a developing visual impairment. Finally, these compensations highlight an important asymmetry of visual adaptation—that it is the observer that is adjusted to match the world (Clifford & Rhodes, 2005). Thus, environment variations may be more important than variations in visual sensitivity for controlling some aspects of perception. For example, whether two observers experience the same stimulus as white may depend more on whether they are exposed to the same color environment than whether their eyes filter the environment in the same way.

Surprisingly little is known about how adaptation itself varies across observers, although large individual differences have been documented for some visual aftereffects (Vera-Diaz, Woods, & Peli, 2010). The integrity of adaptation across the lifespan also remains poorly understood. The development of light adaptation, contrast sensitivity, and contrast gain control has been well characterized in infant vision (Brown & Lindsey, 2009), and pattern-selective adaptation appears within the first weeks (Suter et al., 1994). Yet it is not clear whether there are significant developmental changes in the mechanisms of cortical adaptation. Similarly, correcting age-related losses requires that adaptation remains functional throughout life. Senescent changes in adaptability have been found for dark adaptation (Jackson, Owsley, & McGwin, 1999) and chromatic adaptation (Werner et al., 2010) and for high-level shape aftereffects (Rivest et al., 2004). However, the strength of some cortical aftereffects shows little decline with age (Elliott et al., 2007). Thus, some aspects of adaptation remain stable despite dramatic age-related neural changes, perhaps because they are important for compensating for these changes.

Adaptation remains widely used as a tool for characterizing visual representations. The logic is that aftereffects will only transfer between stimuli to the extent that they are encoded by a common mechanism or channel. The degree of selectivity or spread of the adaptation can thus be used to infer how broadly or narrowly tuned the adapted mechanisms are for the stimulus dimension as well as the number of mechanisms that span the dimension. A classic example is the multiple-channel model of spatial frequency encoding, which was inspired in part by the finding that adapting to a given spatial frequency reduced contrast sensitivity only for nearby frequencies (Graham, 1989).

Recently a number of studies have focused on using adaptation to distinguish between multiple-channel models and norm-based or opponent models. Figure 37.3 shows an illustration of these different models and the pattern of aftereffects they predict. In figure 37.3A, there are many narrowly tuned channels encoding different levels of the stimulus, which might be represented by something like the peak response among the channels. Adapting to one level produces a local sensitivity loss, and this biases the response distribution at other levels away from the adapting stimulus. This produces "repulsion" aftereffects, in which stimuli appear less like the adapting stimulus (and thus shift in opposite directions above or below the adapting level, while the adapting level itself appears unaltered). This pattern is characteristic of the aftereffects for a number of attributes, including size or spatial frequency (Blakemore & Sutton, 1969), hue angle (Webster & Mollon, 1994), and object viewpoint and gaze direction (Calder et al., 2008; Fang & He, 2005). In the second model (figure 37.3B) the dimension is instead represented by only two overlapping channels with broad sensitivities, and the stimulus level might therefore be coded by the ratio of responses across the two channels. Aftereffects thus transfer over a wider range and are instead

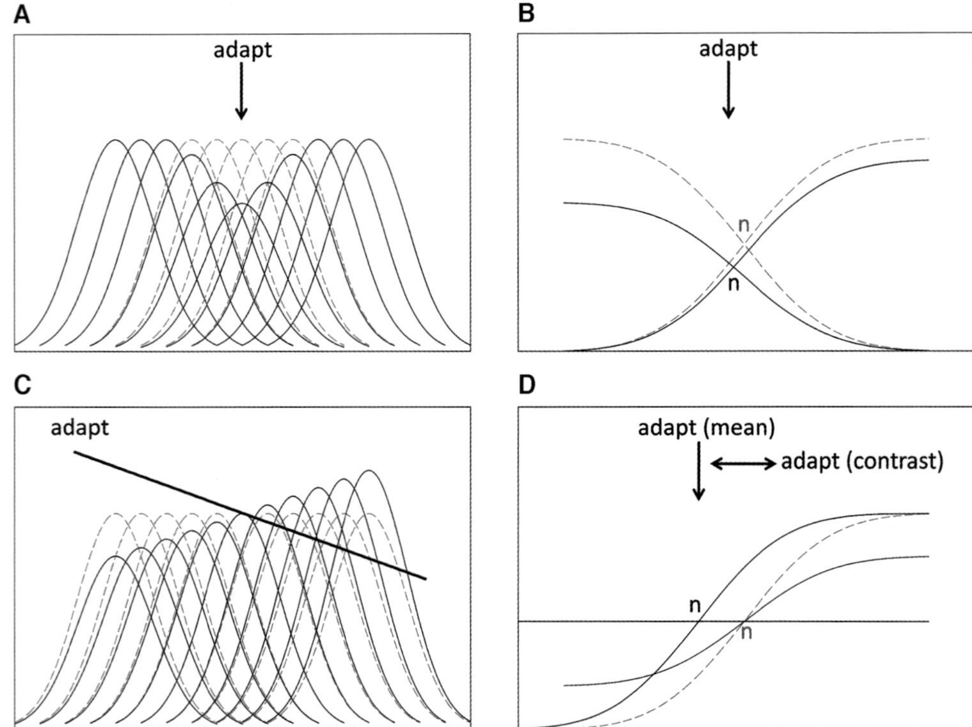

FIGURE 37.3   Adaptation and visual channels. Curves show the sensitivity of a set of channels before (dashed) or after (solid) adaptation. (A) Multiple, narrowly tuned channels. Adaptation reduces sensitivity in the channels sensitive to the adapting level, biasing the modal response to flanking stimuli away from the response to the adaptor. (B) Broadly tuned channels. Adaptation reduces sensitivity in the more strongly stimulated channel, shifting the balance point or norm (n) toward the adapting level. (C) When the stimulus is broadband, narrowly tuned channels can also exhibit normalization-like aftereffects corresponding to balanced activity across the channels. (D) An opponent mechanism in which the norm corresponds to the null between excitation and inhibition. Adaptation at a preopponent site can change the balance of inputs, resulting in a mean shift in the norm toward the adapt level. Adaptation within the mechanism (e.g., to contrast) instead alters sensitivity without shifting the null point.

characterized by perceptual shifts of the same sign across all levels, including the adapting level, a pattern described as "normalization." This model has been widely used to account for face aftereffects (Webster & MacLeod, 2011), although the actual form of the response changes—and the underlying channels they imply—remains uncertain for faces (Robbins, McKone, & Edwards, 2007; Storrs & Arnold, 2012; Zhao et al., 2011). In the final case (figure 37.3D), the dimension is encoded by a single opponent mechanism that responds in opposite ways (e.g., excitation or inhibition) to levels above or below its neutral point. Adaptation within this mechanism should reduce its sensitivity and thus weaken both positive and negative responses so that, after adaptation, all stimuli appear weaker or shifted toward the neutral point. This pattern is evident in the changes in perceived saturation after adaptation to color contrast (Webster & Mollon, 1994), yet beyond color there are surprisingly few examples of adaptation within an explicit opponent mechanism (Webster & MacLeod, 2011).

Renormalization is predicted not only when the stimulus dimension is encoded by broadly tuned channels (figure 37.3B) but also when the stimulus spectrum itself is broad (figure 37.3C). For example, color is sampled by only three broadly tuned mechanisms, and thus, adaptation even to a narrowband wavelength will normalize the responses across the three cones. Conversely, multiple spatial frequency channels might generate a repulsion aftereffect when adapting to a single grating, but natural stimuli such as edges have broad amplitude spectra, and thus, adaptation to these should instead normalize the responses across a broad range of channels (Elliott, Georgeson, & Webster, 2011). Thus, the models and aftereffects differ only by whether the stimulus and channels are narrow or broad. Moreover, normalization also occurs within multiple-channel models. If each channel adjusts independently, then this adaptation will level the playing field across the channel array for the prevailing stimulus. Consequently norms do not have to be prebuilt in the system because they arise through adaptation. In fact, norms are synonymous with adaptation states.

An important assumption in these models is that perceptual norms appear neutral or unique because they reflect a unique null or balance in the underlying neural responses. Consistent with this, stimuli that "look" neutral may be special in not producing an aftereffect, presumably because adaptation to them does not alter the neural balance. For example, distortion aftereffects do not occur for undistorted faces (Webster & MacLin, 1999), and the stimulus that appears white or in focus to an individual does not induce a color or blur aftereffect (Sawides et al., 2011b; Webster & Leonard, 2008). Although this suggests a close relationship between perceptual norms and the normalized responses of visual mechanisms, how the visual system normalizes is clearly more complex than the models of figure 37.3 suggest. For example, tilt aftereffects are several times stronger than predicted by the changes in the perceived tilt of the adapting stimulus (Müller et al., 2009). Further, adaptation may change not only the sensitivities of the channels but also their tuning functions (Clifford et al., 2007; Clifford, Wenderoth, & Spehar, 2000) and could also involve interactions between channels (Barlow, 1990b). Consequently, the interpretation of selective aftereffects remains uncertain (Benton & Burgess, 2008; Hegde, 2009; Mur et al., 2010).

THE TIMESCALES OF ADAPTATION

As noted, adaptation is normally defined by changes in sensitivity over brief timescales, from milliseconds to minutes. Multiple rates of adaptation operate within this interval (Kohn, 2007; Wark, Lundstrom, & Fairhall, 2007). Yet there is growing evidence that these brief aftereffects may ride atop sensitivity adjustments also acting at far longer times, from hours to weeks or perhaps even years. Long-term effects have been documented most clearly in color vision. For example, in the McCollough effect (color afterimages contingent on orientation), aftereffects may persist permanently until reset by a new pattern (McCollough-Howard & Webster, 2009; Vul, Krizay, & MacLeod, 2008). Persistent hue changes also occur when observers wear tinted lenses (Belmore & Shevell, 2008, 2010; Eisner & Enoch, 1982; Neitz et al., 2002) or when the "tinted lens" is removed in cataract surgery (Delahunt et al., 2004). Recent studies have also revealed long-term adaptation to contrast (Kwon et al., 2009; Zhang et al., 2009).

Multiple timescales of adaptation may afford a number of advantages. The dynamics of even a single adaptation process should itself be adaptable to cope with speed–accuracy trade-offs. Rapid adjustments are required to preserve coding when the signal changes but will induce the wrong states if they are driven by noise (Wark, Fairhall, & Rieke, 2009). Further, changes in the signal itself can arise from multiple sources with different time constants, and the timescales of adaptation should be matched to these events. Optimal adaptation should be rapid but also decay rapidly when the source of change is short-term and should be slow yet persistent when the source is a long-term one (Kording, Tenenbaum, & Shadmehr, 2007; Shadmehr, Smith, & Krakauer, 2010; Smith, Ghazizadeh, & Shadmehr,

2006). Models based on this premise, with adaptation in at least two processes tracking different timescales, can predict many properties of habituation and motor adaptation including spacing effects, spontaneous recovery, and relearning (Kording, Tenenbaum, & Shadmehr, 2007; Shadmehr, Smith, & Krakauer, 2010; Smith, Ghazizadeh, & Shadmehr, 2006; Staddon & Higa, 1996). Finally, the source of stimulus changes may require qualitatively different forms of response change and thus distinct mechanisms at different timescales. For example, adaptation of contrast coding may require rapid changes in contrast gain to match the current stimulus level but changes in response gain to adjust to persistent changes in the maximum contrast (Kwon et al., 2009).

THE FUNCTIONS OF ADAPTATION

Despite the expanding interest in adaptation over the last decade, the reasons the visual system adapts remain uncertain. This problem might in part be because adaptation actually serves multiple purposes, and these may change depending on the nature of the stimulus and the constraints on visual coding. Moreover, as noted at the outset, the term adaptation is applied to a diverse range of adjustments, and just as these may differ in mechanism, they may also differ in function. Finally, the function may also depend critically on the timescale, for the calibrations required to establish visual coding when we are born, or to maintain it as we age, may have to cope with very different problems from the daily fine-tuning typically studied in the lab.

Functional accounts of adaptation at early visual stages have focused on coding efficiency (Clifford et al., 2007; Wainwright, 1999; Wark, Lundstrom, & Fairhall, 2007), where it is clearly critical to scale sensitivity to cope with the enormous variation in light levels within and across scenes. A second related function is to enhance efficiency by predictive coding, so that the mean or expected stimulus is represented only implicitly (Srinivasan, Laughlin, & Dubs, 1982), a scheme that is also metabolically efficient (Lennie, 2003). This is effectively a norm-based code and allows the full response capacity to signal deviations from the norm, or errors in the prediction, thus enhancing the salience of novel information (Barlow, 1990a; McDermott et al., 2010; Ranganath & Rainer, 2003). A third set of benefits involves error correction (Andrews, 1967; Kording, Tenenbaum, & Shadmehr, 2007) and constancy. As discussed above, adaptation is important for maintaining stable percepts by compensating for sensitivity variations in the observer and also by discounting "irrelevant" variations in the stimulus. For example, color

constancy depends in part on adapting out or discounting the mean chromaticity of the illuminant (Foster, 2011).

Importantly, there are also costs to adapting. One is that adaptation can be deleterious if it adjusts to or amplifies the noise rather than the signal (Rieke & Rudd, 2009; Wark, Fairhall, & Rieke, 2009). A second potential cost is if changes in the neural responses are attributed to a change in the stimulus rather than to the state of adaptation. This "coding catastrophe" is in fact the basis of most perceptual aftereffects (Schwartz, Hsu, & Dayan, 2007). Finally, although many of the proposed functions of adaptation are complementary, they are not always compatible. For example, adjustments that enhance efficiency or salience may not help to maintain perceptual constancy (McDermott et al., 2010; Webster & Mollon, 1995). As a result it is perhaps not surprising that beneficial effects of adaptation in one context seem hidden or even maladaptive in another. Yet despite this, the repeating pattern of aftereffects across diverse dimensions and visual levels remains striking. Deciphering the role they play may depend on understanding why these calibrations are manifest throughout the visual system and often in such similar ways, even though the constraints on sensory processing may be very different.

ACKNOWLEDGMENTS

Supported by EY-10834.

REFERENCES

Andrews, D. P. (1967). Perception of contour orientation in the central fovea part 1: Short lines. *Vision Research, 7,* 975–997.

Artal, P., Chen, L., Fernandez, E. J., Singer, B., Manzanera, S., & Williams, D. R. (2004). Neural compensation for the eye's optical aberrations. *Journal of Vision, 4,* 281–287. doi:10.1167/4.4.4.

Baccus, S. A., & Meister, M. (2002). Fast and slow contrast adaptation in retinal circuitry. *Neuron, 36,* 909–919.

Barlow, H. B. (1990a). Conditions for versatile learning, Helmholtz's unconscious inference, and the task of perception. *Vision Research, 30,* 1561–1571.

Barlow, H. B. (1990b). A theory about the functional role and synaptic mechanism of visual aftereffects. In C. Blakemore (Ed.), *Visual coding and efficiency* (pp. 363–375). Cambridge: Cambridge University Press.

Barlow, H. (1997). Adaptation by hyperpolarization. *Science, 276,* 913–914.

Belmore, S. C., & Shevell, S. K. (2008). Very-long-term chromatic adaptation: Test of gain theory and a new method. *Visual Neuroscience, 25,* 411–414.

Belmore, S. C., & Shevell, S. K. (2010). Very-long-term and short-term chromatic adaptation: Are their influences cumulative? *Vision Research, 51,* 362–366.

Benton, C. P., & Burgess, E. C. (2008). The direction of measured face aftereffects. *Journal of Vision, 8,* 1–6. doi:10.1167/8.15.1.

Bex, P. J., Solomon, S. G., & Dakin, S. C. (2009). Contrast sensitivity in natural scenes depends on edge as well as spatial frequency structure. *Journal of Vision, 9,* 1–19. doi:10.1167/9.10.1.

Blake, R., & He, S. (2005). Adaptation as a tool for probing the neural correlates of visual awareness: progress and precautions. In C. W. G. Clifford & G. Rhodes (Eds.), *Fitting the mind to the world: Adaptation and aftereffects in high-level vision* (pp. 281–308). Oxford: Oxford University Press.

Blakemore, C., & Sutton, P. (1969). Size adaptation: A new aftereffect. *Science, 166,* 245–247.

Bompas, A., & O'Regan, J. K. (2006). Evidence for a role of action in colour perception. *Perception, 35,* 65–78. doi:10.1068/p5356.

Brown, A. M., & Lindsey, D. T. (2009). Contrast insensitivity: The critical immaturity in infant visual performance. *Optometry and Vision Science, 86,* 572–576.

Brown, S. P., & Masland, R. H. (2001). Spatial scale and cellular substrate of contrast adaptation by retinal ganglion cells. *Nature Neuroscience, 4,* 44–51.

Calder, A. J., Jenkins, R., Cassel, A., & Clifford, C. W. G. (2008). Visual representation of eye gaze is coded by a nonopponent multichannel system. *Journal of Experimental Psychology. General, 137,* 244–261.

Camp, A. J., Tailby, C., & Solomon, S. G. (2009). Adaptable mechanisms that regulate the contrast response of neurons in the primate lateral geniculate nucleus. *Journal of Neuroscience, 29,* 5009–5021.

Carandini, M., & Heeger, D. J. (2011). Normalization as a canonical neural computation. *Nature Reviews. Neuroscience, 13,* 51–62.

Chander, D., & Chichilnisky, E. J. (2001). Adaptation to temporal contrast in primate and salamander retina. *Journal of Neuroscience, 21,* 9904–9916.

Clifford, C. W. G. (2002). Perceptual adaptation: Motion parallels orientation. *Trends in Cognitive Sciences, 6,*136–143. doi:10.1016/S1364-6613(00)01856-8.

Clifford, C. W. G., & Rhodes, G. (2005). *Fitting the mind to the world: Adaptation and aftereffects in high-level vision.* Advances in Visual Cognition Series (Vol. 2). Oxford: Oxford University Press.

Clifford, C. W., Webster, M. A., Stanley, G. B., Stocker, A. A., Kohn, A., Sharpee, T. O., et al. (2007). Visual adaptation: Neural, psychological and computational aspects. *Vision Research, 47,* 3125–3131. doi:10.1016/j.visres.2007.08.023.

Clifford, C. W. G., Wenderoth, P., & Spehar, B. (2000). A functional angle on some after-effects in cortical vision. *Proceedings. Biological Sciences, 267,* 1705–1710. doi:10.1098/rspb.2000.1198.

Culham, J. C., Verstraten, F. A., Ashida, H., & Cavanagh, P. (2000). Independent aftereffects of attention and motion. *Neuron, 28,* 607–615.

Delahunt, P. B., Webster, M. A., Ma, L., & Werner, J. S. (2004). Long-term renormalization of chromatic mechanisms following cataract surgery. *Visual Neuroscience, 21,* 301–307.

Demb, J. B. (2008). Functional circuitry of visual adaptation in the retina. *Journal of Physiology, 586*(Pt 18), 4377–4384.

Dhruv, N. T., Tailby, C., Sokol, S. H., & Lennie, P. (2011). Multiple adaptable mechanisms early in the primate visual pathway. *Journal of Neuroscience, 31,* 15016–15025.

Dickinson, J. E., Almeida, R. A., Bell, J., & Badcock, D. R. (2010). Global shape aftereffects have a local substrate: A tilt aftereffect field. *Journal of Vision, 10,* 1–12. doi:10.1167/10.13.5.

Eisner, A., & Enoch, J. M. (1982). Some effects of 1 week's monocular exposure to long-wavelength stimuli. *Perception & Psychophysics, 31,* 169–174.

Elliott, S. L., Georgeson, M. A., & Webster, M. A. (2011). Response normalization and blur adaptation: Data and multi-scale model. *Journal of Vision, 11,* 1–18. doi:10.1167/11.2.7.

Elliott, S. L., Hardy, J. L., Webster, M. A., & Werner, J. S. (2007). Aging and blur adaptation. *Journal of Vision, 7,* 1–9. doi:10.1167/7.6.8.

Fang, F., & He, S. (2005). Viewer-centered object representation in the human visual system revealed by viewpoint aftereffects. *Neuron, 45,* 793–800.

Field, D. J. (1987). Relations between the statistics of natural images and the response properties of cortical cells. *Journal of the Optical Society of America. A, Optics and Image Science, 4,* 2379–2394.

Foster, D. H. (2011). Color constancy. *Vision Research, 51,* 674–700.

Goddard, E., Solomon, S., & Clifford, C. (2010). Adaptable mechanisms sensitive to surface color in human vision. *Journal of Vision, 10,* 1–13. doi:10.1167/10.9.17.

Gollisch, T., & Meister, M. (2010). Eye smarter than scientists believed: Neural computations in circuits of the retina. *Neuron, 65,* 150–164.

Graham, N. V. (1989). *Visual pattern analyzers.* Oxford: Oxford University Press.

Greene, M. R., & Oliva, A. (2010). High-level aftereffects to global scene properties. *Journal of Experimental Psychology. Human Perception and Performance, 36,* 1430–1442.

Grill-Spector, K., Henson, R., & Martin, A. (2006). Repetition and the brain: Neural models of stimulus-specific effects. *Trends in Cognitive Sciences, 10,* 14–23. doi:10.1016/j.tics.2005.11.006.

Hardy, J. L., Frederick, C. M., Kay, P., & Werner, J. S. (2005). Color naming, lens aging, and grue: What the optics of the aging eye can teach us about color language. *Psychological Science, 16,* 321–327.

Hegde, J. (2009). How reliable is the pattern adaptation technique? A modeling study. *Journal of Neurophysiology, 102,* 2245–2252.

Helson, H. (1964). *Adaptation-level theory.* New York: Harper & Row.

Hosoya, T., Baccus, S. A., & Meister, M. (2005). Dynamic predictive coding by the retina. *Nature, 436,* 71–77.

Hsu, S. M., & Young, A. W. (2004). Adaptation effects in facial expression recognition. *Visual Cognition, 11,* 871–899.

Jackson, G. R., Owsley, C., & McGwin, G., Jr. (1999). Aging and dark adaptation. *Vision Research, 39,* 3975–3982.

Jordan, H., Fallah, M., & Stoner, G. R. (2006). Adaptation of gender derived from biological motion. *Nature Neuroscience, 9,* 738–739.

Kohn, A. (2007). Visual adaptation: Physiology, mechanisms, and functional benefits. *Journal of Neurophysiology, 97,* 3155–3164.

Kohn, A., & Movshon, J. A. (2003). Neuronal adaptation to visual motion in area MT of the macaque. *Neuron, 39,* 681–691.

Kording, K. P., Tenenbaum, J. B., & Shadmehr, R. (2007). The dynamics of memory as a consequence of optimal adaptation to a changing body. *Nature Neuroscience, 10,* 779–786.

Krekelberg, B., Boynton, G. M., & van Wezel, R. J. A. (2006). Adaptation: From single cells to BOLD signals. *Trends in Neurosciences, 29,* 250–256. doi:10.1016/j.tins.2006.02.008.

Kwon, M., Legge, G. E., Fang, F., Cheong, A. M., & He, S. (2009). Adaptive changes in visual cortex following prolonged contrast reduction. *Journal of Vision, 9,* 21–16. doi:10.1167/9.2.20.

Lennie, P. (2003). The cost of cortical computation. *Current Biology, 13,* 493–497.

Leopold, D. A., O'Toole, A. J., Vetter, T., & Blanz, V. (2001). Prototype-referenced shape encoding revealed by high-level aftereffects. *Nature Neuroscience, 4,* 89–94.

Leopold, D. A., Rhodes, G., Muller, K. M., & Jeffery, L. (2005). The dynamics of visual adaptation to faces. *Proceedings. Biological Sciences, 272,* 897–904.

Mante, V., Frazor, R. A., Bonin, V., Geisler, W. S., & Carandini, M. (2005). Independence of luminance and contrast in natural scenes and in the early visual system. *Nature Neuroscience, 8,* 1690–1697.

Massey, P. V., & Bashir, Z. I. (2007). Long-term depression: Multiple forms and implications for brain function. *Trends in Neurosciences, 30,* 176–184. doi:10.1016/j.tins.2007.02.005.

Mather, G., Pavan, A., Campana, G., & Casco, C. (2008). The motion aftereffect reloaded. *Trends in Cognitive Sciences, 12,*481–487. doi:10.1016/j.tics.2008.09.002.

McCollough-Howard, C., & Webster, M. A. (2009). McCollough effect. *Scholarpedia: Encyclopedia of Computational Neuroscience, 6,* 8175. doi:10.4249/scholarpedia.8175.

McDermott, K. C., Malkoc, G., Mulligan, J. B., & Webster, M. A. (2010). Adaptation and visual salience. *Journal of Vision, 10,* 1–32. doi:10.1167/10.13.17.

Melcher, D. (2008). Dynamic, object-based remapping of visual features in trans-saccadic perception. *Journal of Vision, 8,* 1–17. doi:10.1167/8.14.2.

Mon-Williams, M., Tresilian, J. R., Strang, N. C., Kochhar, P., & Wann, J. P. (1998). Improving vision: Neural compensation for optical defocus. *Proceedings. Biological Sciences, 265,* 71–77.

Motoyoshi, I., Nishida, S., Sharan, L., & Adelson, E. H. (2007). Image statistics and the perception of surface qualities. *Nature, 447,* 206–209.

Müller, K. M., Schillinger, F., Do, D. H., & Leopold, D. A. (2009). Dissociable perceptual effects of visual adaptation. *PLoS One, 4,* e6183. doi:10.1371/journal.pone.0006183.

Mur, M., Ruff, D. A., Bodurka, J., Bandettini, P. A., & Kriegeskorte, N. (2010). Face-identity change activation outside the face system: "Release from adaptation" may not always indicate neuronal selectivity. *Cerebral Cortex, 20,* 2027–2042.

Neitz, J., Carroll, J., Yamauchi, Y., Neitz, M., & Williams, D. R. (2002). Color perception is mediated by a plastic neural mechanism that is adjustable in adults. *Neuron, 35,* 783–792.

Olveczky, B. P., Baccus, S. A., & Meister, M. (2007). Retinal adaptation to object motion. *Neuron, 56,* 689–700.

O'Neil, S., & Webster, M. A. (2011). Adaptation and the perception of facial age. *Visual Cognition, 19,* 534–550. doi:10.1080/13506285.2011.561262.

Parraga, C. A., Troscianko, T., & Tolhurst, D. J. (2002). Spatiochromatic properties of natural images and human vision. *Current Biology, 12,* 483–487.

Pestilli, F., Viera, G., & Carrasco, M. (2007). How do attention and adaptation affect contrast sensitivity? *Journal of Vision, 7,* 1–12. doi:10.1167/7.7.9.

Pesudovs, K., & Brennan, N. A. (1993). Decreased uncorrected vision after a period of distance fixation with spectacle wear. *Optometry and Vision Science, 70,* 528–531.

Priebe, N. J., Churchland, M. M., & Lisberger, S. G. (2002). Constraints on the source of short-term motion adaptation in macaque area MT. I. The role of input and intrinsic mechanisms. *Journal of Neurophysiology, 88,* 354–369.

Ranganath, C., & Rainer, G. (2003). Neural mechanisms for detecting and remembering novel events. *Nature Reviews. Neuroscience, 4,* 193–202.

Rawicz, S. (1997). *The long walk.* New York: Lyons Press.

Rezec, A., Krekelberg, B., & Dobkins, K. R. (2004). Attention enhances adaptability: Evidence from motion adaptation experiments. *Vision Research, 44,* 3035–3044.

Rhodes, G., Jeffery, L., Watson, T. L., Clifford, C. W. G., & Nakayama, K. (2003). Fitting the mind to the world: Face adaptation and attractiveness aftereffects. *Psychological Science, 14,* 558–566.

Richters, D. P., & Eskew, R. T., Jr. (2009). Quantifying the effect of natural and arbitrary sensorimotor contingencies on chromatic judgments. *Journal of Vision, 9,* 1–11. doi:10.1167/9.4.27.

Rieke, F. (2001). Temporal contrast adaptation in salamander bipolar cells. *Journal of Neuroscience, 21,* 9445–9454.

Rieke, F., & Rudd, M. E. (2009). The challenges natural images pose for visual adaptation. *Neuron, 64,* 605–616.

Rivest, J., Kim, J. S., Intriligator, J., & Sharpe, J. A. (2004). Effect of aging on visual shape distortion. *Gerontology, 50,* 142–151. doi:10.1159/000076776.

Robbins, R., McKone, E., & Edwards, M. (2007). Aftereffects for face attributes with different natural variability: Adapter position effects and neural models. *Journal of Experimental Psychology. Human Perception and Performance, 33,* 570–592.

Sanchez-Vives, M. V., Nowak, L. G., & McCormick, D. A. (2000). Membrane mechanisms underlying contrast adaptation in cat area 17 in vivo. *Journal of Neuroscience, 20,* 4267–4285.

Sawides, L., de Gracia, P., Dorronsoro, C., Webster, M., & Marcos, S. (2011a). Adapting to blur produced by ocular high-order aberrations. *Journal of Vision, 11,* 1–11. doi:10.1167/11.7.21.

Sawides, L., de Gracia, P., Dorronsoro, C., Webster, M. A., & Marcos, S. (2011b). Vision is adapted to the natural level of blur present in the retinal image. *PLoS One, 6,* e27031. doi:10.1371/journal.pone.0027031.

Sawides, L., Marcos, S., Ravikumar, S., Thibos, L., Bradley, A., & Webster, M. (2010). Adaptation to astigmatic blur. *Journal of Vision, 10,* 1–15. doi:10.1167/10.12.22.

Schacter, D. L., Wig, G. S., & Stevens, W. D. (2007). Reductions in cortical activity during priming. *Current Opinion in Neurobiology, 17,* 171–176.

Schefrin, B. E., & Werner, J. S. (1990). Loci of spectral unique hues throughout the life span. *Journal of the Optical Society of America. A, Optics and Image Science, 7,* 305–311.

Schwartz, O., Hsu, A., & Dayan, P. (2007). Space and time in visual context. *Nature Reviews. Neuroscience, 8,* 522–535.

Schweinberger, S. R., Zaske, R., Walther, C., Golle, J., Kovacs, G., & Wiese, H. (2010). Young without plastic surgery: Perceptual adaptation to the age of female and male faces. *Vision Research, 50,* 2570–2576.

Shadmehr, R., Smith, M. A., & Krakauer, J. W. (2010). Error correction, sensory prediction, and adaptation in motor control. *Annual Review of Neuroscience, 33*, 89–108.

Shapley, R. M., & Enroth-Cugell, C. (1984). Visual adaptation and retinal gain controls. *Progress in Retinal Research, 3*, 263–346.

Sharpee, T. O., Sugihara, H., Kurgansky, A. V., Rebrik, S. P., Stryker, M. P., & Miller, K. D. (2006). Adaptive filtering enhances information transmission in visual cortex. *Nature, 439*, 936–942.

Smirnakis, S. M., Berry, M. J., Warland, D. K., Bialek, W., & Meister, M. (1997). Adaptation of retinal processing to image contrast and spatial scale. *Nature, 386*, 69–73.

Smith, M. A., Ghazizadeh, A., & Shadmehr, R. (2006). Interacting adaptive processes with different timescales underlie short-term motor learning. *PLoS Biology, 4*, e179. doi:10.1371/journal.pbio.0040179.

Smithson, H. E. (2005). Sensory, computational and cognitive components of human colour constancy. *Philosophical Transactions of the Royal Society of London. Series B, Biological Sciences, 360*, 1329–1346. doi:10.1098/rstb.2005.1633.

Solomon, S. G., Peirce, J. W., Dhruv, N. T., & Lennie, P. (2004). Profound contrast adaptation early in the visual pathway. *Neuron, 42*, 155–162.

Srinivasan, M. V., Laughlin, S. B., & Dubs, A. (1982). Predictive coding: A fresh view of inhibition in the retina. *Proceedings of the Royal Society of London. Series B, Biological Sciences, 216*, 427–459. doi:10.1098/rspb.1982.0085.

Staddon, J. E., & Higa, J. J. (1996). Multiple time scales in simple habituation. *Psychological Review, 103*, 720–733.

Storrs, K. H., & Arnold, D. K. (2012). Not all face aftereffects are equal. *Vision Research, 64*, 7. doi:10.1016/j.visres.2012.04.020.

Suter, P. S., Suter, S., Roessler, J. S., Parker, K. L., Armstrong, C. A., & Powers, J. C. (1994). Spatial-frequency-tuned channels in early infancy: VEP evidence. *Vision Research, 34*, 737–745.

Suzuki, S., & Cavanagh, P. (1998). A shape-contrast effect for briefly presented stimuli. *Journal of Experimental Psychology. Human Perception and Performance, 24*, 1315–1341.

Thompson, P., & Burr, D. (2009). Visual aftereffects. *Current Biology, 19*, R11–R14.

Troje, N. F., Sadr, J., Geyer, H., & Nakayama, K. (2006). Adaptation aftereffects in the perception of gender from biological motion. *Journal of Vision, 6*, 850–857. doi:10.1167/6.8.7.

Turk-Browne, N. B., Scholl, B. J., & Chun, M. M. (2008). Babies and brains: Habituation in infant cognition and functional neuroimaging. *Frontiers in Human Neuroscience, 2*(16), 1–11. doi:10.3389/neuro.09.016.2008.

Vera-Diaz, F. A., Woods, R. L., & Peli, E. (2010). Shape and individual variability of the blur adaptation curve. *Vision Research, 50*, 1452–1461.

von der Heydt, R., Macuda, T., & Qiu, F. T. (2005). Border-ownership-dependent tilt aftereffect. *Journal of the Optical Society of America. A, Optics, Image Science, and Vision, 22*, 2222–2229.

Vul, E., Krizay, E., & MacLeod, D. I. (2008). The McCollough effect reflects permanent and transient adaptation in early visual cortex. *Journal of Vision, 8*, 1–12. doi:10.1167/8.12.4.

Wainwright, M. J. (1999). Visual adaptation as optimal information transmission. *Vision Research, 39*, 3960–3974.

Walraven, J., & Werner, J. S. (1991). The invariance of unique white; a possible implication for normalizing cone action spectra. *Vision Research, 31*, 2185–2193.

Wark, B., Fairhall, A., & Rieke, F. (2009). Timescales of inference in visual adaptation. *Neuron, 61*, 750–761.

Wark, B., Lundstrom, B. N., & Fairhall, A. (2007). Sensory adaptation. *Current Opinion in Neurobiology, 17*, 423–429.

Webster, M. A. (2011). Adaptation and visual coding. *Journal of Vision 11*, 1–23. doi:10.1167/11.5.3.

Webster, M. A., Georgeson, M. A., & Webster, S. M. (2002). Neural adjustments to image blur. *Nature Neuroscience, 5*, 839–840.

Webster, M. A., Juricevic, I., & McDermott, K. C. (2010). Simulations of adaptation and color appearance in observers with varying spectral sensitivity. *Ophthalmic & Physiological Optics, 30*, 602–610.

Webster, M. A., Kaping, D., Mizokami, Y., & Duhamel, P. (2004). Adaptation to natural facial categories. *Nature, 428*, 557–561.

Webster, M. A., & Leonard, D. (2008). Adaptation and perceptual norms in color vision. *Journal of the Optical Society of America. A, Optics, Image Science, and Vision, 25*, 2817–2825.

Webster, M. A., & MacLeod, D. I. A. (2011). Visual adaptation and face perception. *Philosophical Transactions of the Royal Society of London. Series B, Biological Sciences, 366*, 1702–1725.

Webster, M. A., & MacLin, O. H. (1999). Figural aftereffects in the perception of faces. *Psychonomic Bulletin & Review, 6*, 647–653.

Webster, M. A., & Miyahara, E. (1997). Contrast adaptation and the spatial structure of natural images. *Journal of the Optical Society of America. A, Optics, Image Science, and Vision, 14*, 2355–2366.

Webster, M. A., Mizokami, Y., Svec, L. A., & Elliott, S. L. (2006). Neural adjustments to chromatic blur. *Spatial Vision, 19*, 111–132.

Webster, M. A., Mizokami, Y., & Webster, S. M. (2007). Seasonal variations in the color statistics of natural images. *Network, 18*, 213–233.

Webster, M. A., & Mollon, J. D. (1994). The influence of contrast adaptation on color appearance. *Vision Research, 34*, 1993–2020.

Webster, M. A., & Mollon, J. D. (1995). Colour constancy influenced by contrast adaptation. *Nature, 373*, 694–698.

Webster, M. A., & Mollon, J. D. (1997). Adaptation and the color statistics of natural images. *Vision Research, 37*, 3283–3298.

Webster, M. A., Werner, J. S., & Field, D. J. (2005). Adaptation and the phenomenology of perception. In C. Clifford & G. Rhodes (Eds.), *Fitting the mind to the world: Adaptation and aftereffects in high-level vision*. Advances in Visual Cognition Series (Vol. 2, pp. 241–277). Oxford: Oxford University Press.

Webster, M. A., & Wilson, J. A. (2000). Interactions between chromatic adaptation and contrast adaptation in color appearance. *Vision Research, 40*, 3801–3816.

Weigelt, S., Muckli, L., & Kohler, A. (2008). Functional magnetic resonance adaptation in visual neuroscience. *Reviews in the Neurosciences, 19*, 363–380.

Werner, A., Bayer, A., Schwarz, G., Zrenner, E., & Paulus, W. (2010). Effects of ageing on postreceptoral short-wavelength gain control: transient tritanopia increases with age. *Vision Research, 50*, 1641–1648.

Werner, J. S., & Schefrin, B. E. (1993). Loci of achromatic points throughout the life span. *Journal of the Optical Society of America. A, Optics and Image Science, 10*, 1509–1516.

Winawer, J., Huk, A. C., & Boroditsky, L. (2008). A motion aftereffect from still photographs depicting motion. *Psychological Science, 19*, 276–283.

Winawer, J., Huk, A. C., & Boroditsky, L. (2010). A motion aftereffect from visual imagery of motion. *Cognition, 114*, 276–284.

Xu, H., Dayan, P., Lipkin, R. M., & Qian, N. (2008). Adaptation across the cortical hierarchy: Low-level curve adaptation affects high-level facial-expression judgments. *Journal of Neuroscience, 28*, 3374–3383.

Yehezkel, O., Sagi, D., Sterkin, A., Belkin, M., & Polat, U. (2010). Learning to adapt: Dynamics of readaptation to geometrical distortions. *Vision Research, 50*, 1550–1558.

Zaidi, Q., Ennis, R., Cao, D., & Lee, B. (2012). Neural locus of color afterimages. *Current Biology, 22*, 220–224.

Zhang, P., Bao, M., Kwon, M., He, S., & Engel, S. A. (2009). Effects of orientation-specific visual deprivation induced with altered reality. *Current Biology, 19*, 1956–1960.

Zhao, C., Series, P., Hancock, P. J., & Bednar, J. A. (2011). Similar neural adaptation mechanisms underlying face gender and tilt aftereffects. *Vision Research, 51*, 2021–2030.

38 Color Constancy

DAVID H. BRAINARD AND ANA RADONJIĆ

Vision is useful because it informs us about the physical environment. In the case of color, two distinct functions are generally emphasized (e.g., Jacobs, 1981; Mollon, 1989). First, color helps to segment objects from each other and the background. A canonical task for this function is locating fruit in foliage (Regan et al., 2001; Sumner & Mollon, 2000). Second, color provides information about object properties (e.g., fresh fish versus old fish; figure 38.1). Using color for this second function is enabled to the extent that perceived object color correlates well with object reflectance properties. Achieving such correlation is a nontrivial requirement, however, because the spectrum of the light reflected from an object confounds variation in object surface reflectance with variation in the illumination (figure 38.2; Brainard, Wandell, & Chichilnisky, 1993; Hurlbert, 1998; Maloney, 1999). In particular, the spectrum of the reflected light is the wavelength-by-wavelength product of the spectrum of the illumination and the object's surface reflectance function (figure 38.2B). The dependence of the reflected light on illumination leads to ambiguity about object reflectance because changes in the spectrum of the reflected light can arise from changes in object reflectance, changes in illumination, or both. Despite this ambiguity, color appearance is often quite stable across illumination changes (e.g., Helmholtz, 1896/2000; Katz, 1935; Brainard, 2004). The approximate invariance of object color appearance is called color constancy. When color constancy fails, as in the case of Uluru (figure 38.2A), it is a phenomenon to remark on.

Ambiguity occurs pervasively in perception (e.g., Gregory, 1978; Helmholtz, 1896/2000; Wolfe et al., 2006) and at higher levels of processing (for example, language) (Foder, Bever, & Garrett, 1974; Miller, 1973). Characterizing how biological information processing resolves informational ambiguity is a priority for vision science (e.g., Brainard, 2009; Knill & Richards, 1996; Purves & Lotto, 2003; Rust & Stocker, 2010). Object color perception provides a model system for moving such characterization forward.

This chapter provides an update on the treatment of color constancy that appeared in the first edition of this volume (Brainard, 2004). Our emphasis here is to provide a self-contained introduction, to highlight some more recent results, and to outline what we see as important challenges for the field. We focus on key concepts and avoid much in the way of technical development. Other recent treatments complement the one provided here and provide a level of technical detail beyond that introduced in this chapter (Brainard, 2009; Brainard & Maloney, 2011; Foster, 2011; Gilchrist, 2006; Kingdom, 2008; Shevell & Kingdom, 2008; Smithson, 2005; Stockman & Brainard, 2010).

MEASURING CONSTANCY

The study of constancy requires methods for measuring it. The classic approach to such measurement is to assess the extent to which the color appearance of objects varies as they are viewed under different illuminations. Helson (1938; Helson & Jeffers, 1940), for example, trained observers to use a color-naming system (Munsell notation) and then had them name the colors of surfaces viewed under different illuminants. He found very good constancy as long as the illumination was not too close to monochromatic.

More recent work has focused on continuous measures. For example, in a cross-illumination *asymmetric matching* experiment, the observer adjusts the color of a matching stimulus seen under one illuminant so that it appears to have the same color as a reference stimulus presented under another illuminant (e.g., Arend & Reeves, 1986; Brainard, Brunt, & Speigle, 1997; Burnham, Evans, & Newhall, 1957). The conceptual idea is illustrated in figure 38.3. The top of the figure illustrates two contexts, shown as two collections of flat colored rectangles (so-called *mondrians*), each uniformly illuminated. One context is defined as the standard context, and the other is defined as the test context. In the figure the two contexts differ because the collection of surfaces in each is seen under a different illuminant. In parallel with our nomenclature for the contexts, we refer to these as the standard and test illuminants. A reference surface is presented in the standard context, and the observer's task is to adjust a matching surface, presented in the test context, so that its color appearance matches that of the reference.

FIGURE 38.1   Color tells us about object properties: The sushi on the left plate looks more appetizing than the sushi on the right plate. (Image on left courtesy of Michael Eisenstein. Image on right manipulated by hand in Photoshop to simulate aging of the fish. After a demonstration by Bei Xiao.)

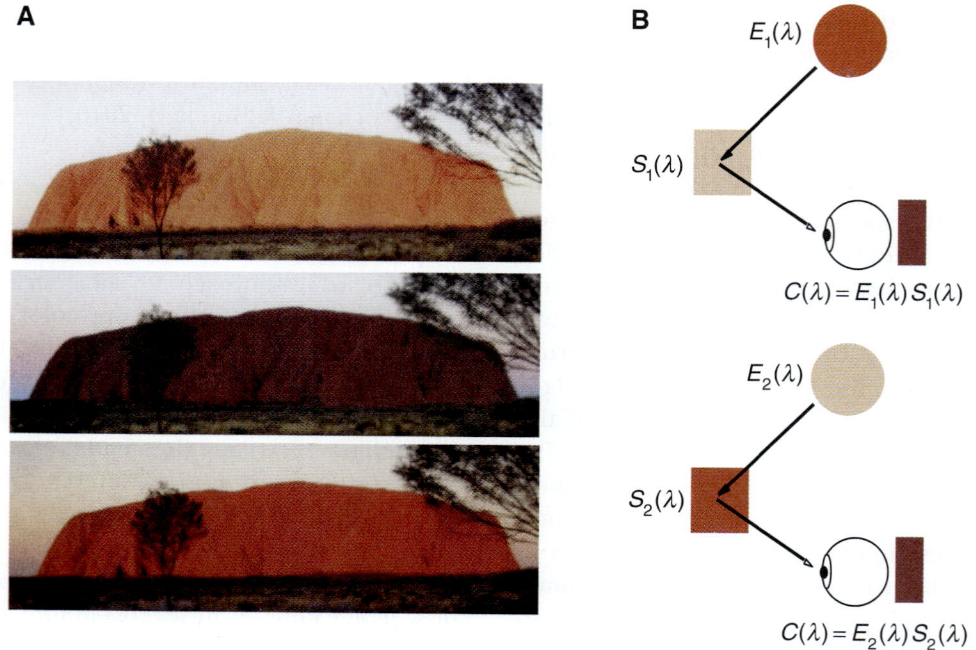

FIGURE 38.2   (A) Uluru, a sandstone formation in the Australian outback famous for its large changes in color appearance. These changes occur because of variation in the spectrum of the light impinging on the rock, which in turn affects the reflected spectrum. What is unusual about Uluru is not the illumination changes—these are pervasive in natural viewing—but that they are perceived as large changes in object color. Photographs copyright © Anya Hurlbert and reproduced with permission. (B) The spectrum of the light reflected from an object, $C(\lambda)$, is the wavelength-by-wavelength product of the illuminant spectrum $E(\lambda)$ and the object's surface reflectance $S(\lambda)$. The same spectrum reaching the eye can arise from many different combinations of illuminant and surface. The figure illustrates two such combinations.

Various techniques may be used to produce and manipulate the stimuli for this type of experiment (for examples, see Arend & Reeves, 1986; Brainard, Brunt, & Speigle, 1997; Delahunt & Brainard, 2004; Xiao et al., 2012).

To connect asymmetric matches to color constancy, consider possible experimental outcomes. One possibility is that the observer will set the matching surface to have the same reflectance as the reference surface despite the change in illuminant. This surface will reflect different light to the observer under the test illuminant than under the standard illuminant. An observer who sets such a match would be considered perfectly color constant (100% constant, figure 38.3) because for such an observer we can infer that the reference surface retains its appearance across the change in illuminant.

A second possibility is that the observer will set the matching surface so that it reflects the same spectrum to the eye under the test illuminant as does the

546   DAVID H. BRAINARD AND ANA RADONJIĆ

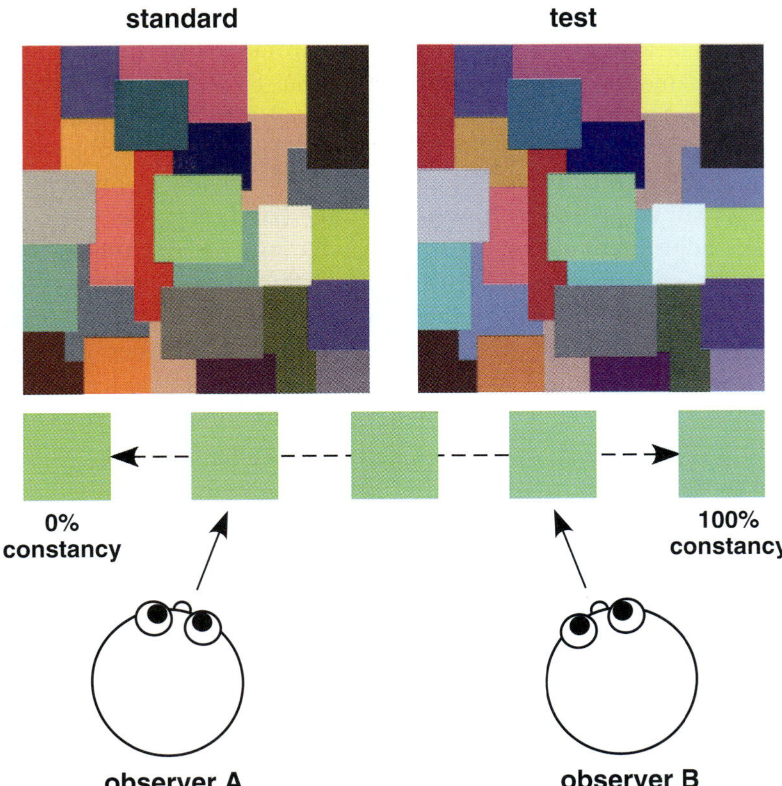

FIGURE 38.3   In a cross-illumination asymmetric matching task, an observer is presented with two contexts. In the figure these are shown as two collections of colored rectangles (so-called mondrians). One context is the standard, and the other is the test, with the illumination differing between the two. The experimenter presents a reference surface in the standard context, and the subject adjusts a matching surface presented in the test context until its color matches that of the reference. In the figure the reference and matching surfaces are at the center of their corresponding mondrians. Below the mondrians are shown five example matches that might be set by an observer. One possibility is that the observer will set a match that has the same surface reflectance as the reference surface. This is illustrated on the right and labeled 100% constancy. Another possibility is that the observer will set a match that corresponds to a different surface reflectance, but that reflects the same spectrum as the reference. This is shown schematically on the left and labeled 0% constancy. Observer matches typically fall between these two theoretical endpoints, and a constancy index may be assigned according to where between the endpoints the match lies. For example, if observer A sets the match indicated by the leftmost arrow, then A's constancy index would be approximately 25%. Similarly, observer B's constancy index would be approximately 75%.

reference surface under the standard illuminant. This possibility is depicted at the left-hand side of the set of sample squares in the figure. Note that the matches set by an observer of this type are the same as those that would be set on the basis of physical measurements of the reflected spectrum. Thus, we say that an observer who sets this type of match has no constancy (0% constant).

In actual experiments observers' matches tend to lie between the two possibilities described above: The matches are neither 0% constant nor 100% constant. Based on this it is common to use asymmetric matching data to assign a constancy index value, typically ranging between 0% and 100%, that characterizes how close the observer's matches are to the idealized endpoint matches. Such indices summarize performance and allow consideration of what factors influence the degree of constancy exhibited by observers. A number of sources discuss the quantitative analysis used to compute constancy indices from asymmetric matches (Arend et al., 1991; Brainard, Brunt, & Speigle, 1997; Brainard & Wandell, 1991; Troost & de Weert, 1991).

A second method that has been used to assess constancy is achromatic adjustment. This method is conceptually similar to asymmetric matching except that instead of adjusting a test surface to match a physical reference, the observer adjusts the chromaticity of a test so that it appears achromatic and then repeats this adjustment with the test embedded in a variety of contexts. The resulting achromatic chromaticities obtained in different contexts are then considered to be matched in appearance, and the data can be analyzed to produce constancy indices in much the same way as achromatic matching data (Brainard, 1998). Speigle and Brainard (1999) established that asymmetric matching and achromatic adjustment lead to similar conclusions about constancy when the two tasks are compared with well-matched stimuli. Achromatic adjustment allows one to study performance across contexts presented in isolation without introducing a heavy memory load on the observer.

BASIC OBSERVATIONS

Given an understanding of how constancy may be measured, we can turn to making some basic empirical generalizations from over 100 years of experimental literature. Much of the experimental literature has considered a simplified laboratory model in which a spatially diffuse light source homogeneously illuminates a set of flat matte surfaces (*flat-matte-diffuse conditions*; for reviews see Brainard, 2004; Brainard & Maloney, 2011; Maloney, 1999; for a selection of experiments see Arend & Reeves, 1986; Brainard, 1998; Brainard, Brunt, & Speigle, 1997; Delahunt & Brainard, 2004; Granzier et al., 2005; Helson & Jeffers, 1940; Kraft & Brainard, 1999; McCann, McKee, & Taylor, 1976; Olkkonen, Hansen, & Gegenfurtner, 2009). Such simplification is reasonable. Although one might view a characterization of constancy for the richer conditions of natural viewing as the ultimate end goal, it is also true that natural scenes are very complicated. To make experimental progress some degree of simplification is necessary, and actual experiments represent a compromise between an attempt to capture key aspects of natural viewing and an attempt to provide enough experimental control that the data are interpretable.

For flat-matte-diffuse conditions constancy across illuminant changes can be very good. Brainard (1998), for example, used real illuminated surfaces and the method of achromatic adjustment and found constancy indices of about 85%. High degrees of constancy can also be observed with stimuli that consist of graphics simulations of illuminated objects (Delahunt & Brainard, 2004, experiment 1, average constancy index 73%). Foster (2011) provides the constancy indices found in a large number of studies many of which are similarly high. Such high degrees of constancy are consistent with the intuition that the color appearance of objects does not change much from one situation to another, and data of this type lead to the empirical generalization that human color constancy is very good when all that is changed is the scene illumination.

In most studies of constancy the experimental manipulation is as described above—the illuminant is changed while the surfaces surrounding the test and reference are held fixed. This is the manipulation that is illustrated by the comparison between figure 38.4A and figure 38.4B. Although the motivation for studying this type of manipulation is obvious, note that for a collection of scenes where most of the surfaces never change, constancy is not a challenging computational problem. Indeed, changing only the illuminant is not the only way to vary the context in which an object is seen. The comparison between figure 38.4A and figure 38.4C illustrates a case in which the illuminant changes and the reflectance of the surface that forms the immediate background of the test also changes. In this example the change in background reflectance was chosen to exactly cancel the effect of the illuminant change for that background surface so that the light reflected from the background remained the same. The effect of this manipulation is to silence local contrast as a cue to constancy. For stimulus manipulations of this type constancy is reduced but not eliminated (Delahunt & Brainard, 2004; Kraft & Brainard, 1999; Kraft, Maloney, & Brainard, 2002; McCann, 1992; McCann, Hall, & Land, 1977; Werner, 2006). For example, Delahunt and

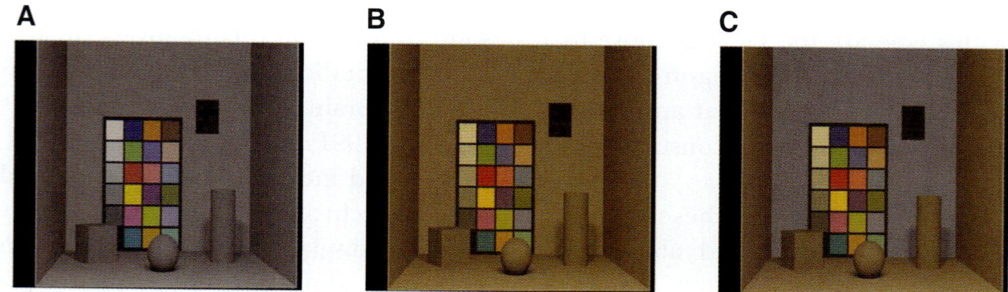

FIGURE 38.4   Experimental manipulations used in studying constancy. Between A and B the illuminant has changed. Between B and C the illuminant is held constant, but the reflectance of the background surface has been changed so that the light reflected from the background is the same as in A. Images depict stimuli used by Delahunt and Brainard (2004). (Reprinted with permission from Brainard & Maloney, 2011.)

Brainard (2004, experiment 2) found an average constancy index of 22% across context changes of this type. Kraft and Brainard (1999) showed similarly that manipulations that silence the change in the most luminous surface in the contextual image or the change in the spatial average of the reflected light have a similar effect: Constancy is reduced but not completely eliminated.

The results outlined above highlight the point that blanket statements about the degree of human constancy are not useful because the measured degree is highly dependent on the stimulus manipulation chosen for study. Any successful model of constancy must account not only for the fact that constancy is sometimes very good but also for the systematic failures of constancy that can be observed. This point appears underappreciated in the literature and is critical to bear in mind when one is comparing the degree of constancy reported across studies.

MODELING CONSTANCY

One approach to understanding how constancy varies with different stimulus manipulations is to connect human performance to a model of how an ideal visual system would estimate and correct for illuminant changes so as to stabilize object color appearance. Early efforts along these lines focused on the information carried by the spatial mean of the reflected light or on the information carried from the most luminous surface in the scene (Judd, 1940; Land, 1964; von Kries, 1902, 1905). Models of this sort capture a number of regularities in the data for experiments when only the illuminant is changed (McCann, McKee, & Taylor, 1976), but their predictions deviate from the data when the cues they rely on are silenced experimentally (Kraft & Brainard, 1999).

A more general approach is to employ a Bayesian algorithm to estimate the illuminant (Brainard & Freeman, 1997; D'Zmura, Iverson, & Singer, 1995; Gehler et al., 2008) and to use this to predict performance. The Bayesian approach has the advantage that it is sensitive to all the information about the illuminant carried by the initial visual representation of color. This information is provided by the isomerization rates of photopigment in three classes of cone photoreceptors (e.g., Brainard & Stockman, 2010). Indeed, there is useful information not only in the spatial average of the cone isomerization rates but also in their covariance and higher-order moments. Note that the observation that the covariance of the isomerization rates carries information about the illuminant precedes Bayesian treatments of color constancy (MacLeod & Golz, 2003;

Maloney & Wandell, 1986; Webster & Mollon, 1995; Zaidi, 1998).

Brainard and colleagues (Brainard et al., 2006) used the Bayesian approach to model the data of Delahunt and Brainard (2004) and showed that it provided a good account of performance. The essence of their model was to assume that observers, in effect, use a specific Bayesian algorithm to estimate the illuminant and then discount the effect of this estimated illuminant (see also Brainard, Brunt, & Speigle, 1997; Brainard & Wandell, 1991; Speigle & Brainard, 1996). Brainard (2009) and Brainard and Maloney (2011) provide a conceptual review of this approach and how it may be generalized to more complex viewing situations.

A key feature of Brainard et al.'s (2006) Bayesian model is that it correctly predicts variation in the degree of measured constancy across different stimulus manipulations and connects this variation to the information about the illuminant available in natural images. Although the efficacy of this model has not been tested extensively even within the restricted flat-matte-diffuse stimulus ensemble, it provides both a framework for thinking about constancy at Marr's (1982) computational level of analysis and theory that links the computations to measurements of human performance. In addition, a successful computational model is important for understanding the neural mechanisms that subserve constancy in that such a model characterizes the phenomena a neural theory must account for. That said, an important challenge for theories of color constancy remains elucidation of the mechanisms that implement it in the human visual pathways. In this regard the Bayesian analysis does not provide any direct insight about the mechanisms that accomplish constancy.

The earlier version of this chapter (Brainard, 2004; see also Stockman & Brainard, 2010) provides an overview of mechanistic ideas as they relate to color constancy, and we briefly review these ideas here. At a general level, mechanistic accounts of constancy focus on the notion of adaptation. This is the idea that the transformation between the initial representation of the stimulus and responses later in the visual pathways depends on context. As noted above, the initial representation of a color stimulus is given by the isomerization rates of photopigment in three classes of cone photoreceptors (e.g., Brainard & Stockman, 2010). These are referred to as the L (long-wavelength-sensitive), M (middle-wavelength-sensitive), and S (short-wavelength-sensitive) cones. Subsequent retinal and cortical processing then transforms the LMS cone representation into one that consists of sums and differences of signals from the different cone classes. At various sites along the pathways that implement this transformation, the signals are subject to multiplicative gain control as well as a subtractive form of adaptation (e.g., Blakeslee & McCourt, 2004; Chen, Foley, & Brainard, 2000; Engel & Furmanski, 2001; Jameson & Hurvich, 1964; Poirson & Wandell, 1993; Shevell, 1978; Stiles, 1967; Valeton, 1983; von Kries, 1902; Wade & Wandell, 2002; Walraven, 1976; Webster & Mollon, 1994; for reviews see Smithson, 2005; Stockman & Brainard, 2010; Walraven & Werner, 1982; Webster, 1996). Key is that the exact values of the multiplicative and subtractive transformations depend on the viewing context, so the representation corresponding to any triplet of LMS cone isomerization rates varies with context. Adaptation of this sort supports constancy to the extent that its overall effect is to stabilize the postreceptoral representation of the light reflected from objects across changes of illumination (as well as other contextual changes).

A challenge for connecting our mechanistic understanding of color adaptation to color constancy is that the stimuli that have been used to study constancy are generally more complex than those that have been used to study adaptation. For example, we know that constancy as measured under flat-matte-diffuse conditions is well described by changes in multiplicative gain (e.g., Bauml, 1995; Brainard, Brunt, & Speigle, 1997; Brainard & Wandell, 1992; McCann, McKee, & Taylor, 1976). What we do not know is how to independently derive the values of the multiplicative gain from a specification of a spatially rich stimulus. Two lines of research that aim to close this gap are worth note. First, Webster and Mollon (1994, 1995) analyzed how illuminant variation changes the mean and covariance of cone isomerization rates and showed that measured adaptation to such changes could support color constancy with respect to illumination changes. Second, Blakeslee and McCourt (2004; Blakeslee, Reetz, & McCourt, 2009) developed a model of lightness based on an abstracted model of mechanistic processing from retina to early visual cortex and showed that such a model can account for a variety of constancy-related lightness effects. Pushing the connections between mechanistic models, computational models, and the empirical data represents an important research frontier.

FUTURE DIRECTIONS

It is now possible to imagine a fairly complete account of color constancy for flat-matte-diffuse stimuli, although there is still much to be done. Even within the flat-matte-diffuse stimulus ensemble, we can go further toward incorporating a growing knowledge about the statistical structure of natural scenes as it pertains to color (surface reflectance functions: Jaaskelainen, Parkkinen & Toyooka, 1990; Krinov, 1947; Nickerson, 1957; Parkkinen, Hallikainen, & Jaaskelainen, 1989; Vrhel, Gershon, & Iwan, 1994; illuminant spectral power distributions: DiCarlo & Wandell, 2000; Judd, MacAdam, & Wyszecki, 1964; calibrated color images: Burge & Geisler, 2011; Ciurea & Funt, 2003; Gehler et al., 2008; Geisler & Perry, 2011; Olmos & Kingdom, 2004; Tkacik et al., 2011; hyperspectral image datasets: Chakrabarti & Zickler, 2011; Foster, Nascimento, & Amano, 2004; Longère & Brainard, 2001; Parraga et al., 1998; Ruderman, Cronin, & Chiao, 1998). Nonetheless, one can envision how a focused research program could bring together the computational and mechanistic ideas outlined above and test these with experiments that are within current technical reach.

Thus, it is worth asking, suppose that enterprise of understanding color constancy for flat-matte-diffuse conditions is brought to a successful conclusion, what would remain to be done within the field of color constancy? Below we outline what we see as some important directions.

Real-World Scenes

Real scenes do not consist of flat matte surfaces under diffuse illumination. First, real scenes exist in three-dimensional space so that objects differ in their depth as well as their position on the retinal projection. Second, real objects have three-dimensional shape and are made of nonmatte materials. Third, real objects often have texture of some sort, so that surface reflectance can vary across an object. Finally, real illumination has geometric structure.

Figure 38.5 illustrates richness that is introduced when we depart from flat-matte-diffuse. Figure 38.5A shows the effect of varying the three-dimensional pose of a flat matte surface when the light is directional rather than diffuse. As illustrated by the figure, the amount of light reflected to an observer depends strongly on the pose. Although the figure illustrates the effect for a change in intensity, similar effects occur in color (Bloj, Kersten, & Hurlbert, 1999; Boyaci, Doerschner, & Maloney, 2004). If the visual system did not compensate for such effects, objects would change their appearance as a function of their pose in the three-dimensional environment.

Constancy with respect to surface pose has received attention in the literature, although primarily in the domain of lightness (Bloj & Hurlbert, 2002; Bloj, Kersten, & Hurlbert, 1999; Boyaci, Doerschner, &

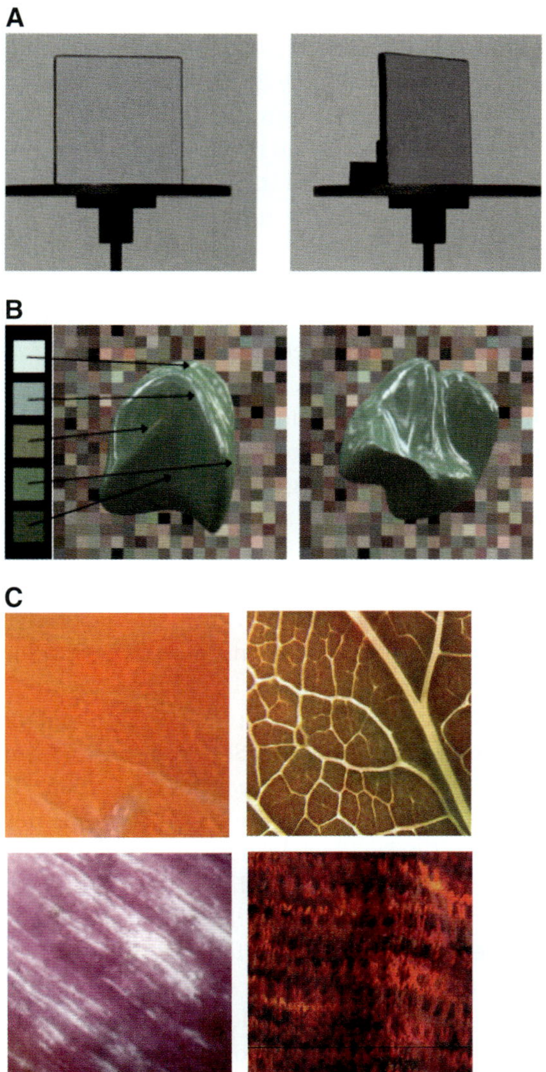

FIGURE 38.5 (A) Photograph of the same surface in two different poses with respect to a directional light source. The intensity of the light reflected from the surface varies with pose. (Reprinted with permission from Ripamonti et al., 2004.) (B) Two objects ("blob" and "pepper") made from the same material and rendered under the same spatially complex illumination field. The pattern of light reflected to the eye across the object varies with object shape. This variation is shown explicitly by the colored squares (left), which are drawn from individual pixels on the "blob." (Figure courtesy of Maria Olkkonen.) (C) Illustrative image patches from polychromatic objects. Clockwise from upper left: salmon (original photograph copyright © James Bowe); leaf (original photograph copyright © Peter Shanks); sweater (original photograph copyright © "orchidgalore"); eggplant (original photograph copyright © Liz West). Image patches obtained from photographs posted at flickr.com; all were posted under the creative commons attribution 2.0 generic license (http://creativecommons.org/licenses/by/2.0/deed.en).

Maloney, 2004; Boyaci, Maloney, & Hersh, 2003; Doerschner, Boyaci, & Maloney, 2004; Epstein, 1961; Gilchrist, 1980; Hochberg & Beck, 1954; Radonjić, Todorović, & Gilchrist, 2010; Ripamonti et al., 2004). The bulk of this work indicates that the human visual system indeed compensates for changes in reflected light that arise from changes of surface pose within three-dimensional scenes, although the exact stimulus conditions that support such compensation are less well understood. Outlines of theories that could be deployed to understand these effects are also available. One line of thought seeks to connect the result of three-dimensional manipulations of surface pose to results for flat-matte-diffuse conditions by invoking grouping principles that segment the three-dimensional scene into separate frameworks and applying principles derived from studies of simplified stimuli to understand what happens within each framework (Gilchrist, 2006; Gilchrist et al., 1999; see also Adelson, 2000; Anderson & Winawer, 2005). A second approach is to extend the computational approach described above to include geometric aspects of the illumination (Bloj et al., 2004; Boyaci, Doerschner, & Maloney, 2004; Boyaci, Maloney, & Hersh, 2003; Doerschner, Boyaci, & Maloney, 2004; reviewed in Brainard & Maloney, 2011).

Figure 38.5B shows how the light reflected from a three-dimensional object can vary across the surface of the object. For a fixed surface reflectance the pattern of reflected light can depend strongly on object shape. Variation in the pose of a three-dimensional object, the geometry of the illumination, and the material out of which the object is made affects the pattern of reflected light. A challenge is thus to measure and understand how we judge the color of objects in the face of these effects. The challenge arises because once variations in object shape, pose, illumination geometry, and object material are introduced into the experimental paradigm, the number of stimulus dimensions available for exploration grows to a point where systematic study via fully crossed designs is not feasible. Therefore, some type of simplifying theory becomes critical for guiding experiments. At present, such a theory is not available.

One theoretical suggestion, explored primarily in the lightness literature, is that simple summary statistics extracted from the luminance histogram of light reflected from an object might provide direct predictors of its lightness (and also of its perceived glossiness). Although initial results were promising (Motoyoshi et al., 2007; Nishida & Shinya, 1998; Sharan et al., 2008), this approach has not generalized well (Anderson & Kim, 2009; Kim & Anderson, 2010; Kim, Marlow, & Anderson, 2011; Olkkonen & Brainard, 2010, 2011). It remains possible, however, that theories that combine such statistics with other information, such as object shape, will be useful. Another theoretical approach has been to test for simplifying empirical regularities (e.g., separability of effects of shape and illumination geometry/spectra), but these efforts have not to date led to successful accounts of the empirical data (Olkkonen & Brainard, 2010, 2011; Xiao et al., 2012). Work aimed at generalizing our understanding toward the full richness of three-dimensional objects and scenes thus remains exploratory.

Many objects (e.g., leaves, skin, rock, fabric) exhibit reflectance variation with varying degrees of spatial regularity. This is a form of texture, although the variation often does not have the quasi-periodic spatial structure of stimuli typically studied in the texture literature. Introspection suggests that reflectance variation need not prevent us from perceiving objects as having a characteristic color appearance (figure 38.5C). We currently know little about color constancy for such polychromatic objects, or indeed about how their overall color is extracted even under a single illuminant, although interest in this question is growing (Beeckmans, 2004; Hurlbert, Ling, & Vurro, 2008; Hurlbert et al., 2009; Hurlbert, Vurro, & Ling, 2008; Ling, Pietta, & Hurlbert, 2009; Olkkonen, Hansen, & Gegenfurtner, 2008; Vurro, Ling, & Hurlbert, 2009; Yoonessi & Zaidi, 2010). As with the geometric effects considered above, progress toward understanding color appearance and color constancy for polychromatic objects will require identification and confirmation of simplifying regularities.

Real-World Tasks

The discussion of constancy above focused on color appearance as the key dependent measure, and indeed, most of what we know about constancy is based on measurements of color appearance. These measurements, however, do not directly characterize performance in real-life tasks that require the use of color to choose among objects. As Zaidi (1998; see also Abrams, Hillis, & Brainard, 2007) has emphasized, an object could look different across a change in illumination, but it still might be possible to use color effectively for object selection or identification. This would, for example, be the case if a person perceived that the illumination had changed and used this fact to reason about object identity. Only a few papers report studies of how well subjects can use lightness or color to identify objects across changes in illumination (Robilotto & Zaidi, 2004, 2006; Zaidi & Bostic, 2008). Related work has considered the extent to which subjects can discriminate between image changes that result from changes in illumination and those that result from changes in surface reflectance (Craven & Foster, 1992; D'Zmura & Mangalick, 1994; Foster & Nascimento, 1994; Gerhard & Maloney, 2010). Additional study of the relation between perceived color appearance and performance on color-based tasks is a priority for understanding how color is used in the real world.

A second challenge for understanding the role of color in real-world tasks is that the results of color appearance experiments can depend on the instructions given to subjects (Arend & Reeves, 1986; Arend et al., 1991; Bauml, 1999; Reeves, Amano, & Foster, 2008; Troost & de Weert, 1991). In an influential asymmetric matching study Arend and Reeves (1986) found that when subjects are asked to "match the hue, saturation, and brightness of the [target color], while disregarding as much as possible other areas in the screen," the data show very little constancy across changes in illuminant. On the other hand, when they are asked to "to make the [match] look as if were cut from the same piece of paper [as the target]," constancy is considerably better.

There is no agreement about the fundamental nature of such instructional effects. Some authors (Arend & Spehar, 1993; Logvinenko & Maloney, 2006; Rock, 1975) posit that subjects have available multiple perceptual modes of color appearance and that instructions modulate which mode is used when making a match. Other authors are of the view that there is a single perceptual representation and that instructional effects modulate the degree to which subjects use explicit reasoning (Blakeslee, Reetz, & McCourt, 2008; MacLeod, 2011; see also Gibson, 1966; Koffka, 1935). Distinguishing between these and other related accounts has been difficult (see Wagner, 2012, for a recent review of the broad issues in the context of size judgments). This is in part because the varying views do not make discernibly different predictions for the outcome of most of the relevant experiments (but see Logvinenko & Maloney, 2006, for an interesting exception) and in part because delineating the precise experimental conditions (method, class of stimuli, and instructional wording) that lead to instructional effects has proved elusive (Cornelissen & Brenner, 1995; Delahunt & Brainard, 2004; Logvinenko & Tokunaga, 2011; Ripamonti et al., 2004). Independent of the theoretical position one adopts, understanding the relation between laboratory experiments and how color is used in real life clearly requires better control and characterization of instructional (as well as related individual difference) effects. Our sense is that these issues will become more acute as we continue to extend our experiments and thinking to richer and more naturalistic stimuli.

  DAVID H. BRAINARD AND ANA RADONJIĆ

REFERENCES

Abrams, A. B., Hillis, J. M., & Brainard, D. H. (2007). The relation between color discrimination and color constancy: When is optimal adaptation task dependent? *Neural Computation, 19,* 2610–2637.

Adelson, E. H. (2000). Lightness perception and lightness illusions. In M. Gazzaniga (Ed.), *The new cognitive neurosciences* (2nd ed., pp. 339–351). Cambridge, MA: MIT Press.

Anderson, B. L., & Kim, J. (2009). Image statistics do not explain the perception of gloss and lightness. *Journal of Vision, 9,* 10–26. doi:10.1167/9.11.10.

Anderson, B. L., & Winawer, J. (2005). Image segmentation and lightness perception. *Nature, 434,* 79–83.

Arend, L. E., & Reeves, A. (1986). Simultaneous color constancy. *Journal of the Optical Society of America. A, Optics and Image Science, 3,* 1743–1751.

Arend, L. E., Reeves, A., Schirillo, J., & Goldstein, R. (1991). Simultaneous color constancy: Papers with diverse Munsell values. *Journal of the Optical Society of America. A, Optics and Image Science, 8,* 661–672.

Arend, L. E., & Spehar, B. (1993). Lightness, brightness, and brightness contrast. 1. Illuminance variation. *Perception & Psychophysics, 54,* 446–456.

Bauml, K. H. (1995). Illuminant changes under different surface collections: Examining some principles of color appearance. *Journal of the Optical Society of America. A, Optics, Image Science, and Vision, 12,* 261–271.

Bauml, K. H. (1999). Simultaneous color constancy: How surface color perception varies with the illuminant. *Vision Research, 39,* 1531–1550.

Beeckmans, J. (2004). Chromatically rich phenomenal percepts. *Philosophical Psychology, 17*(1), 27–44.

Blakeslee, B., & McCourt, M. E. (2004). A unified theory of brightness contrast and assimilation incorporating oriented multiscale filtering and contrast normalization. *Vision Research, 44,* 2483–2503. doi:10.1016/j.visres.2004.05.015.

Blakeslee, B., Reetz, D., & McCourt, M. E. (2008). Coming to terms with lightness and brightness: Effects of stimulus configuration and instructions on brightness and lightness judgments. *Journal of Vision, 8,* 1–14. doi:10.1167/8.11.3.

Blakeslee, B., Reetz, D., & McCourt, M. E. (2009). Spatial filtering versus anchoring accounts of brightness/lightness perception in staircase and simultaneous brightness/lightness contrast stimuli. *Journal of Vision, 9,* 1–17. doi:10.1167/9.3.22.

Bloj, M. G., & Hurlbert, A. C. (2002). An empirical study of the traditional Mach card effect. *Perception, 31,* 233–246.

Bloj, M., Kersten, D., & Hurlbert, A. C. (1999). Perception of three-dimensional shape influences colour perception through mutual illumination. *Nature, 402,* 877–879.

Bloj, M., Ripamonti, C., Mitha, K., Greenwald, S., Hauck, R., & Brainard, D. H. (2004). An equivalent illuminant model for the effect of surface slant on perceived lightness. *Journal of Vision, 4,* 735–746. doi:10.1167/4.9.6.

Boyaci, H., Doerschner, K., & Maloney, L. T. (2004). Perceived surface color in binocularly viewed scenes with two light sources differing in chromaticity. *Journal of Vision, 4,* 664–679. doi:10.1167/4.9.1.

Boyaci, H., Maloney, L. T., & Hersh, S. (2003). The effect of perceived surface orientation on perceived surface albedo in binocularly viewed scenes. *Journal of Vision, 3,* 541–553. doi:10.1167/3.8.2.

Brainard, D. H. (1998). Color constancy in the nearly natural image. 2. Achromatic loci. *Journal of the Optical Society of America. A, Optics, Image Science, and Vision, 15,* 307–325.

Brainard, D. H. (2004). Color constancy. In L. M. Chalupa & J. S. Werner (Eds.), *The visual neurosciences* (pp. 948–961). Cambridge, MA: MIT Press.

Brainard, D. H. (2009). Bayesian approaches to color vision. In M. S. Gazzaniga (Ed.), *The cognitive neurosciences* (4th ed., pp. 395–408). Cambridge, MA: MIT Press.

Brainard, D. H., Brunt, W. A., & Speigle, J. M. (1997). Color constancy in the nearly natural image. 1. Asymmetric matches. *Journal of the Optical Society of America. A, Optics, Image Science, and Vision, 14,* 2091–2110.

Brainard, D. H., & Freeman, W. T. (1997). Bayesian color constancy. *Journal of the Optical Society of America. A, Optics, Image Science, and Vision, 14,* 1393–1411.

Brainard, D. H., Longere, P., Delahunt, P. B., Freeman, W. T., Kraft, J. M., & Xiao, B. (2006). Bayesian model of human color constancy. *Journal of Vision, 6,* 1267–1281. doi:10.1167/6.11.10.

Brainard, D. H., & Maloney, L. T. (2011). Surface color perception and equivalent illumination models. *Journal of Vision, 11,* 1–18. doi:10.1167/11.5.1.

Brainard, D. H., & Stockman, A. (2010). Colorimetry. In M. Bass, C. DeCusatis, J. Enoch, V. Lakshminarayanan, G. Li, C. Macdonald, V. Mahajan, & E. van Stryland (Eds.), *The Optical Society of America handbook of optics, 3rd ed., Volume III: Vision and vision optics* (pp. 10.11–10.56). New York: McGraw-Hill.

Brainard, D. H., & Wandell, B. A. (1991). A bilinear model of the illuminant's effect on color appearance. In M. S. Landy & J. A. Movshon (Eds.), *Computational models of visual processing* (pp. 171–186). Cambridge, MA: MIT Press.

Brainard, D. H., & Wandell, B. A. (1992). Asymmetric color-matching: How color appearance depends on the illuminant. *Journal of the Optical Society of America. A, Optics and Image Science, 9*(9), 1433–1448.

Brainard, D. H., Wandell, B. A., & Chichilnisky, E. J. (1993). Color constancy: From physics to appearance. *Current Directions in Psychological Science, 2,* 165–170.

Burge, J., & Geisler, W. S. (2011). Optimal defocus estimation in individual natural images. *Proceedings of the National Academy of Sciences of the United States of America, 108,* 16849–16854.

Burnham, R. W., Evans, R. M., & Newhall, S. M. (1957). Prediction of color appearance with different adaptation illuminations. *Journal of the Optical Society of America, 47,* 35–42.

Chakrabarti, A., & Zickler, T. (2011). Statistics of real-world hyperspectral images. Paper presented at Proceedings of the IEEE Computer Society Conference on Computer Vision and Pattern Recognition (pp. 193–200).

Chen, C. C., Foley, J. M., & Brainard, D. H. (2000). Detection of chromoluminance patterns on chromoluminance pedestals II: model. *Vision Research, 40,* 789–803.

Ciurea, F., & Funt, B. (2003). A large image database for color constancy research. Paper presented at Proceedings of the IS&T 11th Color Imaging Conference, Scottsdale, AZ (pp. 160–164).

Cornelissen, F. W., & Brenner, E. (1995). Simultaneous colour constancy revisited—an analysis of viewing strategies. *Vision Research, 35*(17), 2431–2448. doi:10.1016/0042-6989(94)00318-1.

Craven, B. J., & Foster, D. H. (1992). An operational approach to colour constancy. *Vision Research, 32,* 1359–1366. doi:10.1016/0042-6989(92)90228B.

Delahunt, P. B., & Brainard, D. H. (2004). Does human color constancy incorporate the statistical regularity of natural daylight? *Journal of Vision, 4,* 57–81. doi:10.1167/4.2.1.

DiCarlo, J. M., & Wandell, B. A. (2000). Illuminant estimation: beyond the bases. Paper presented at IS&T/SID Eighth Color Imaging Conference, Scottsdale, AZ (pp. 91–96).

Doerschner, K., Boyaci, H., & Maloney, L. T. (2004). Human observers compensate for secondary illumination originating in nearby chromatic surfaces. *Journal of Vision, 4,* 92–105. doi:10.1167/4.2.3.

D'Zmura, M., Iverson, G., & Singer, B. (1995). Probabilistic color constancy. In R. D. Luce, M. D'Zmura, D. Hoffman, G. Iverson, & A. K. Romney (Eds.), *Geometric representations of perceptual phenomena: Papers in honor of Tarow Indow's 70th birthday* (pp. 187–202). Mahwah, NJ: Lawrence Erlbaum Associates.

D'Zmura, M., & Mangalick, A. (1994). Detection of contrary chromatic change. *Journal of the Optical Society of America. A, Optics, Image Science, and Vision, 11*(2), 543–546.

Engel, S. A., & Furmanski, C. S. (2001). Selective adaptation to color contrast in human primary visual cortex. *Journal of Neuroscience, 21*(11), 3949–3954.

Epstein, W. (1961). Phenomenal orientation and perceived achromatic color. *Journal of Psychology, 52,* 51–53.

Foder, J. A., Bever, T. G., & Garrett, M. F. (1974). *The psychology of language.* New York: McGraw-Hill.

Foster, D. H. (2011). Color constancy. *Vision Research, 51,* 674–700.

Foster, D. H., & Nascimento, S. M. C. (1994). Relational colour constancy from invariant cone-excitation ratios. *Proceedings of the Royal Society of London. Series B, Biological Sciences, 257,* 115–121.

Foster, D. H., Nascimento, S. M. C., & Amano, K. (2004). Information limits on neural identification of colored surfaces in natural scenes. *Visual Neuroscience, 21,* 331–336. doi:10.1017/S0952523804213335.

Gehler, P., Rother, C., Blake, A., Minka, T., & Sharp, T. (2008). Bayesian color constancy revisited. Paper presented at IEEE Computer Society Conference on Computer Vision and Pattern Recognition, Anchorage, Alaska.

Geisler, W. S., & Perry, J. S. (2011). Statistics for optimal point prediction in natural images. *Journal of Vision, 11*(12), 14, 1–17. doi:10.1167/11.12.14.

Gerhard, H. E., & Maloney, L. T. (2010). Detection of light transformations and concomitant changes in surface albedo. *Journal of Vision, 10,* 1–14. doi:10.1167/10.9.1.

Gibson, J. J. (1966). *The senses considered as perceptual systems.* Boston: Houghton Mifflin.

Gilchrist, A. L. (1980). When does perceived lightness depend on perceived spatial arrangement? *Perception & Psychophysics, 28,* 527–538.

Gilchrist, A. (2006). *Seeing black and white.* Oxford: Oxford University Press.

Gilchrist, A., Kossyfidis, C., Bonato, F., Agostini, T., Cataliotti, J., Li, X., et al. (1999). An anchoring theory of lightness perception. *Psychological Review, 106,* 795–834.

Granzier, J. J. M., Brenner, E., Cornelissen, F. W., & Smeets, J. B. J. (2005). Luminance-color correlation is not used to estimate the color of the illumination. *Journal of Vision, 5,* 20–27. doi:10.1167/5.1.2.

Gregory, R. L. (1978). *Eye and brain.* New York: McGraw-Hill.

Helmholtz, H. von. (2000). *Handbuch der Physiologischen Optik* (J. P. C. Southall, Trans., 1924). Reprinted London: Thoemmes Press. (Original work published 1896.)

Helson, H. (1938). Fundamental problems in color vision. I. The principle governing changes in hue, saturation and lightness of non-selective samples in chromatic illumination. *Journal of Experimental Psychology, 23,* 439–476.

Helson, H., & Jeffers, V. B. (1940). Fundamental problems in color vision. II. Hue, lightness, and saturation of selective samples in chromatic illumination. *Journal of Experimental Psychology, 26,* 1–27. doi:10.1037/h0060515.

Hochberg, J. E., & Beck, J. (1954). Apparent spatial arrangement and perceived brightness. *Journal of Experimental Psychology, 47,* 263–266.

Hurlbert, A. C. (1998). Computational models of color constancy. In V. Walsh & J. Kulikowski (Eds.), *Perceptual constancy: Why things look as they do* (pp. 283–322). Cambridge: Cambridge University Press.

Hurlbert, A., Ling, Y. Z., & Vurro, M. (2008). Polychromatic colour constancy [Abstract]. *Perception, 37*(2), 311.

Hurlbert, A., Pietta, I., Vurro, M., & Ling, Y. (2009). Surface discrimination of natural objects: When is a blue kiwi off-colour? *Journal of Vision, 9,* 330. doi:10.1167/9.8.330.

Hurlbert, A., Vurro, M., & Ling, Y. (2008). Colour constancy of polychromatic surfaces. *Journal of Vision, 8,* 1101. doi:10.1167/8.6.1101.

Jaaskelainen, T., Parkkinen, J., & Toyooka, S. (1990). A vector-subspace model for color representation. *Journal of the Optical Society of America. A, Optics and Image Science, 7,* 725–730.

Jacobs, G. H. (1981). *Comparative color vision.* New York: Academic Press.

Jameson, D., & Hurvich, L. M. (1964). Theory of brightness and color contrast in human vision. *Vision Research, 4,* 135–154.

Judd, D. B. (1940). Hue saturation and lightness of surface colors with chromatic illumination. *Journal of the Optical Society of America, 30,* 2–32. doi:10.1364/JOSA.30.000002.

Judd, D. B., MacAdam, D. L., & Wyszecki, G. W. (1964). Spectral distribution of typical daylight as a function of correlated color temperature. *Journal of the Optical Society of America, 54,* 1031–1040.

Katz, D. (1935). *The world of colour.* London: Kegan, Paul, Trench, Trübner & Co.

Kim, J., & Anderson, B. L. (2010). Image statistics and the percepton of surface gloss and lightness. *Journal of Vision, 10,* 3. doi:10.1167/10.9.3.

Kim, J., Marlow, P., & Anderson, B. L. (2011). The perception of gloss depends on highlight congruence with surface shading. *Journal of Vision, 11,* 4. doi:10.1167/11.9.4.

Kingdom, F. A. A. (2008). Perceiving light versus material. *Vision Research, 48,* 2090–2105.

Knill, D. C., & Richards, W. (1996). *Perception as Bayesian inference.* Cambridge: Cambridge University Press.

Koffka, K. (1935). *Principles of Gestalt psychology.* New York: Harcourt, Brace.

Kraft, J. M., & Brainard, D. H. (1999). Mechanisms of color constancy under nearly natural viewing. *Proceedings of the National Academy of Sciences of the United States of America, 96,* 307–312.

Kraft, J. M., Maloney, S. I., & Brainard, D. H. (2002). Surface-illuminant ambiguity and color constancy: Effects of scene complexity and depth cues. *Perception, 31*, 247–263.

Krinov, E. L. (1947). *Surface Reflectance Properties of Natural Formations*. National Research Council of Canada: Technical Translation, TT-439.

Land, E. H. (1964). The retinex. In A. V. S. de Reuck & J. Knight (Eds.), *CIBA Foundation Symposium on colour vision* (pp. 217–227). Boston: Little, Brown and Company.

Ling, Y., Pietta, I., & Hurlbert, A. (2009). The interaction of colour and texture in an object classification task. *Journal of Vision, 9*, 788. doi:10.1167/9.8.788.

Logvinenko, A. D., & Maloney, L. T. (2006). The proximity structure of achromatic surface colors and the impossibility of asymmetric lightness matching. *Perception & Psychophysics, 68*, 76–83.

Logvinenko, A. D., & Tokunaga, R. (2011). Lightness constancy and illumination discounting. *Attention, Perception & Psychophysics, 73*, 1886–1902.

Longère, P., & Brainard, D. H. (2001). Simulation of digital camera images from hyperspectral input. In C. J. van den Branden Lambrecht (Ed.), *Vision models and applications to image and video processing* (pp. 123–150). Boston: Kluwer Academic.

MacLeod, D. I. A. (2011). Into the neural maze. In J. Cohen & M. Matthen (Eds.), *Color ontology and color science* (pp. 151–178). Cambridge, MA: MIT Press.

MacLeod, D. I. A., & Golz, J. (2003). A computational analysis of colour constancy. In R. Mausfeld & D. Heyer (Eds.), *Colour perception: Mind and the physical world* (pp. 205–242). Oxford: Oxford University Press.

Maloney, L. T. (1999). Physics-based approaches to modeling surface color perception. In K. R. Gegenfurtner & L. T. Sharpe (Eds.), *Color vision: From genes to perception* (pp. 387–416). Cambridge: Cambridge University Press.

Maloney, L. T., & Wandell, B. A. (1986). Color constancy: A method for recovering surface spectral reflectances. *Journal of the Optical Society of America. A, Optics and Image Science, 3*, 29–33.

Marr, D. (1982). *Vision*. San Francisco: W. H. Freeman.

McCann, J. J. (1992). Rules for colour constancy. *Ophthalmic & Physiological Optics, 12*, 175–179.

McCann, J. J., Hall, J. A., & Land, E. H. (1977). Color Mondrian experiments: The study of average spectral distributions. *Journal of the Optical Society of America, 67*, 1380.

McCann, J. J., McKee, S. P., & Taylor, T. H. (1976). Quantitative studies in retinex theory: A comparison between theoretical predictions and observer responses to the "Color Mondrian" experiments. *Vision Research, 16*, 445–458.

Miller, G. A. (1973). *Communication, language, and meaning*. New York: Basic Books.

Mollon, J. D. (1989). "Tho' she kneel'd in that place where they grew ..." The uses and origins of primate color vision. *Journal of Experimental Biology, 146*, 21–38.

Motoyoshi, I., Nishida, S., Sharan, L., & Adelson, E. H. (2007). Image statistics and the perception of surface qualities. *Nature, 447*, 206–209.

Nickerson, D. (1957). *Spectrophotometric data for a collection of munsell samples*. Washington, D.C.: U.S. Department of Agriculture.

Nishida, S., & Shinya, M. (1998). Use of image-based information in judgments of surface-reflectance properties. *Journal*

of the Optical Society of America. A, Optics, Image Science, and Vision, 15, 2951–2965. doi:10.1364/JOSAA.15.002951.

Olkkonen, M., & Brainard, D. H. (2010). Perceived glossiness and lightness under real-world illumination. *Journal of Vision, 10*, 5. doi:10.1167/10.9.5.

Olkkonen, M., & Brainard, D. H. (2011). Joint effects of illumination geometry and object shape in the perception of surface reflectance. *Perception, 2*, 1014–1034.

Olkkonen, M., Hansen, T., & Gegenfurtner, K. R. (2008). Color appearance of familiar objects: Effects of object shape, texture, and illumination changes. *Journal of Vision, 8*, 13–28. doi:10.1167/8.5.13.

Olkkonen, M., Hansen, T., & Gegenfurtner, K. R. (2009). Categorical color constancy for simulated surfaces. *Journal of Vision, 9*, 6–23. doi:10.1167/9.12.6.

Olmos, A., & Kingdom, F. A. A. (2004). A biologically inspired algorithm for the recovery of shading and reflectance images. *Perception, 33*, 1463–1473.

Parkkinen, J. P. S., Hallikainen, J., & Jaaskelainen, T. (1989). Characteristic spectra of Munsell colors. *Journal of the Optical Society of America, 6*, 318–322.

Parraga, C. A., Brelstaff, G., Troscianko, T., & Moorehead, I. R. (1998). Color and luminance information in natural scenes. *Journal of the Optical Society of America. A, Optics, Image Science, and Vision, 15*, 563–569.

Poirson, A. B., & Wandell, B. A. (1993). Appearance of colored patterns—pattern color separability. *Journal of the Optical Society of America. A, Optics and Image Science, 10*(12), 2458–2470.

Purves, D., & Lotto, R. B. (2003). *Why we see what we do: An empirical theory of vision*. Sunderland, MA: Sinauer Associates.

Radonjić, A., Todorović, D., & Gilchrist, A. (2010). Adjacency and surroundedness in the depth effect on lightness. *Journal of Vision, 10*, 12–27. doi:10.1167/10.9.12.

Reeves, A., Amano, K., & Foster, D. H. (2008). Color constancy: Phenomenal or projective? *Perception & Psychophysics, 70*, 219–228.

Regan, B. C., Julliot, C., Simmen, B., Vienot, F., Charles-Dominique, P., & Mollon, J. D. (2001). Fruits, foliage and the evolution of primate color vision. *Philosophical Transactions of the Royal Society of London. Series B, Biological Sciences, 356*, 229–283.

Ripamonti, C., Bloj, M., Hauck, R., Mitha, K., Greenwald, S., Maloney, S. I., et al. (2004). Measurements of the effect of surface slant on perceived lightness. *Journal of Vision, 4*, 7–23. doi:10.1167/4.9.4.

Robilotto, R., & Zaidi, Q. (2004). Perceived transparency of neutral density filters across dissimilar backgrounds. *Journal of Vision, 4*, 5–17. doi:10.1167/4.3.5.

Robilotto, R., & Zaidi, Q. (2006). Lightness identification of patterned three-dimensional, real objects. *Journal of Vision, 6*, 3–21. doi:10.1167/6.1.3.

Rock, I. (1975). *An introduction to perception*. New York: Macmillan.

Ruderman, D. L., Cronin, T. W., & Chiao, C. C. (1998). Statistics of cone responses to natural images: Implications for visual coding. *Journal of the Optical Society of America. A, Optics, Image Science, and Vision, 15*, 2036–2045.

Rust, N. C., & Stocker, A. A. (2010). Ambiguity and invariance: Two fundamental challenges for visual processing. *Current Opinion in Neurobiology, 20*, 382–388.

Sharan, L., Li, Y., Motoyoshi, I., Nishida, S., & Adelson, E. H. (2008). Image statistics for surface reflectance perception. *Journal of the Optical Society of America. A, Optics, Image Science, and Vision, 25*, 846–865.

Shevell, S. K. (1978). The dual role of chromatic backgrounds in color perception. *Vision Research, 18*, 1649–1661. doi:10.1016/0048-6989(78)90257-2.

Shevell, S. K., & Kingdom, F. A. A. (2008). Color in complex scenes. *Annual Review of Psychology, 59*, 143–166.

Smithson, H. E. (2005). Sensory, computational, and cognitive components of human color constancy. *Philosophical Transactions of the Royal Society of London. Series B, Biological Sciences, 360*, 1329–1346.

Speigle, J. M., & Brainard, D. H. (1996). Luminosity thresholds: Effects of test chromaticity and ambient illumination. *Journal of the Optical Society of America. A, Optics, Image Science, and Vision, 13*, 436–451.

Speigle, J. M., & Brainard, D. H. (1999). Predicting color from gray: The relationship between achromatic adjustment and asymmetric matching. *Journal of the Optical Society of America. A, Optics, Image Science, and Vision, 16*, 2370–2376.

Stiles, W. S. (1967). Mechanism concepts in colour theory. *Journal of the Colour Group, 11*, 106–123.

Stockman, A., & Brainard, D. H. (2010). Color vision mechanisms. In M. Bass, C. DeCusatis, J. Enoch, V. Lakshminarayanan, G. Li, C. Macdonald, V. Mahajan, & E. van Stryland (Eds.), *The Optical Society of America handbook of optics, 3rd ed, Volume III: Vision and vision optics* (pp. 11.11–11.104). New York: McGraw-Hill.

Sumner, P., & Mollon, J. D. (2000). Catarrhine photopigments are optimized for detecting targets against a foliage background. *Journal of Experimental Biology, 203*, 1963–1986.

Tkacik, G., Garrigan, P., Ratliff, C., Milcinski, G., Klein, J. M., Sterling, P., et al. (2011). Natural images from the birthplace of the human eye. *PLoS One, 6*, e20409). doi:10.1371/journal.pone.0020409.

Troost, J. M., & de Weert, C. M. M. (1991). Naming versus matching in color constancy. *Perception & Psychophysics, 50*, 591–602.

Valeton, J. M. (1983). Photoreceptor light adaptation models: An evaluation. *Vision Research, 23*, 1549–1554. doi:10.1016/0042-6989(83)90168-2.

von Kries, J. (1902). Chromatic adaptation. In *Sources of color vision* (pp. 109–119), reprinted 1970, Cambridge, MA: MIT Press.

von Kries, J. (1905). Influence of adaptation on the effects produced by luminous stimuli. In D. L. MacAdam (Ed.), *Sources of color science* (1970 reprint, pp. 120–127. Cambridge, MA: MIT Press.

Vrhel, M. J., Gershon, R., & Iwan, L. S. (1994). Measurement and analysis of object reflectance spectra. *Color Research and Application, 19*, 4–9.

Vurro, M., Ling, Y., & Hurlbert, A. (2009). Memory colours of polychromatic objects. *Journal of Vision, 9*, 333. doi:10.1167/9.8.333.

Wade, A. R., & Wandell, B. A. (2002). Chromatic light adaptation measured using functional magnetic resonance imaging. *Journal of Neuroscience, 22*, 8148–8157.

Wagner, M. (2012). Sensory and cognitive explanations for a century of size constancy research. In S. R. Allred & G. Hatfield (Eds.), *Visual experience: Sensation, cognition, and constancy* (pp. 63–86). Oxford: Oxford University Press.

Walraven, J. (1976). Discounting the background: The missing link in the explanation of chromatic induction. *Vision Research, 16*, 289–295.

Walraven, J., & Werner, J. S. (1982). Chromatic adaptation and π mechanisms. *Color Research and Application, 7*, 50–53.

Webster, M. A. (1996). Human colour perception and its adaptation. *Network (Bristol, England), 7*, 587–634.

Webster, M. A., & Mollon, J. D. (1994). The influence of contrast adaptation on color appearance. *Vision Research, 34*, 1993–2020.

Webster, M. A., & Mollon, J. D. (1995). Colour constancy influenced by contrast adaptation. *Nature, 373*, 694–698.

Werner, A. (2006). The influence of depth segregation on colour constancy. *Perception, 35*, 1171–1184.

Wolfe, J. M., Kluender, K. R., Levi, D. M., Bartoshuk, L. M., Herz, R. S., Klatzky, R. L., et al. (2006). *Sensation and perception*. Sunderland, MA: Sinauer Associates.

Xiao, B., Hurst, B., MacIntyre, L., & Brainard, D. H. (2012). The color constancy of three-dimensional objects. *Journal of Vision, 12*, 6. doi:10.1167/12.4.6.

Yoonessi, A., & Zaidi, Q. (2010). The role of color in recognizing material changes. *Ophthalmic & Physiological Optics, 30*, 626–631.

Zaidi, Q. (1998). Identification of illuminant and object colors: Heuristic-based algorithms. *Journal of the Optical Society of America. A, Optics, Image Science, and Vision, 15*, 1767–1776.

Zaidi, Q., & Bostic, M. (2008). Color strategies for object identification. *Vision Research, 48*, 2673–2681.

39 Recent Developments in Comparative Color Vision

GERALD H. JACOBS

For a wide range of natural conditions, the overall photon flux, the plane of polarization, and the distribution of spectral power of light reaching the eye vary both spatially and temporally. Such variations are the raw materials animals use in the process of converting light into sight. Depending on the nature of the recipient visual system, different features of light can be exploited to yield vision. Arrangements that allow nervous systems to analyze specific differences in the distribution of spectral power have evolved in many different taxa. The usual result is that such animals have color vision, but there are striking variations in the nature and utility of color vision across species, and it is those variations and their consequences that form the targets for comparative studies of color vision. In an earlier edition of this volume I reviewed basic aspects of the nature, distribution, evolution, and utility of color vision in different animals (Jacobs, 2004). In the decade since a number of new facts and important issues relevant to these topics have emerged. Those recent advances are the focus of this chapter.

EVALUATIONS OF ANIMAL COLOR VISION

Defining Animal Color Vision

For most people, the experience of color seems self-evident—lights and objects are either colored or they are not—and we seem naturally to appreciate those environmental conditions where color is prevalent in our worlds and those where it is not and, additionally, also often learn to attach meaning to the presence of particular colors. The scientific study of color, however, requires definitions, and that is particularly a concern for studies of color vision in other species. A standard definition of color vision is that it represents the ability to discriminate between lights of different spectral compositions irrespective of their relative intensities (Wyszecki & Stiles, 1982). A large number of color vision experiments with human subjects have been carried out in such a discrimination context, exhaustively probing the boundaries of what lights or combinations of lights appear the same or different. The results from such investigations form the basis for much of what has been learned about the formal features of human color vision (Stockman & Brainard, 2010). Over the years it has also proven possible to use behavioral paradigms to conduct analogous discrimination experiments on quite a number of other species. But, as others have noted (Shevell, 2003; Wandell, 1995), discrimination experiments tell us in great detail what lights look the same and what look different, but they do not tell us what the lights actually look like. The latter falls in the province of studies of color appearance, where the experimental techniques employed normally require the use of the medium of language in one form or another. Given that, it is not surprising there have been only a relatively small number of studies that sought to infer what color appearance may be like for animal subjects (Jacobs, 2004).

Recently, there has been renewed discussion about the nature of the definitions of color vision as they apply to different species. No one questions that satisfying the formal criterion listed above is a necessary condition for demonstrating the presence of color vision, but there is concern whether that, by itself, is sufficient. Skorupski and Chittka (2011) have argued forcefully that it is not. In doing so they listed various examples where plants and bacteria, and even some machines, are shown capable of responding to wavelength differences in lights independent of their relative intensities. They suggest that such instances do not qualify as cases of color vision simply because they are not instances of vision. For them, as for others, vision necessarily "provides information about the properties and identity of objects" (Brainard & Maloney, 2011). According to that definition a demonstration of color vision would require that an animal must be shown able to use spectral cues in a task that involves some form of image segmentation. This is a high standard, so far technically achieved in only a relatively small number of animal studies.

Kelber and Osorio (2010) offer a considerably broader approach to the issue of definition, suggesting that color vision can be thought to exist as a series of four different levels or grades. The four share in common the basic criterion that the animal has to be able to disentangle information about wavelength and intensity differences. Beyond that, they ascend from (seemingly) more simple to more complex, as follows: (1) Photokinesis and phototaxis, where animal movements are directed by the spectral properties of ambient light; for example, the small aquatic crustacean *Daphnia* moves toward algae-rich yellowish water. Image-forming eyes are not required for such behaviors. (2) Innate (unlearned) preferences triggered by the spectral properties of objects. These behaviors require the presence of image-forming eyes, and they serve to allow reliable recognition of critical objects, for example, food or mates. Over the years many such examples have been described, and, appropriately, they are usually referred to as *wavelength-specific behaviors* (Goldsmith, 1990; Menzel, 1979). (3) Learned associations to the spectral properties of objects. This is the classic approach to establishing the presence of color vision in animals—the criterion here is that there has to be a demonstration of the modifiability of behavior in response to alterations in the spectral properties of targets. Earlier reviews list numerous examples of the application of this technique to a wide variety of species (Jacobs, 1981, 1993, 2010; Kelber, Vorobyev, & Osorio, 2003). (4) Color categorization and color appearance. Humans can reliably group colors into categories (e.g., reds, yellows, greens) and can discern subtleties in the appearances of color; for instance, we see separable dimensions of hue and saturation.

Although the domains of color categorization and color appearance have not often been studied in nonhuman subjects, there have been a few attempts to establish whether other animals categorize colors in a manner analogous to the way people do, and there is evidence to indicate that some birds, for example both pigeons and chickens, form color categories, and perhaps some fish do as well (Kelber & Osorio, 2010). Given the commonalities in the organization of visual systems across primates, it is somewhat surprising to find that there is debate as to whether nonhuman primates form color categories. Early experiments, as well as a range of less-systematic observations, strongly implied that Old World nonhuman primates experience color categories similar to those apparent to humans (reviewed in Jacobs, 1981), but a recent study conducted with baboons (*Papio papio*) failed to find evidence for the presence of a categorical boundary between the colors "blue" and "green"—a boundary that was detected by human observers that were tested in the same way—despite the fact that wavelength discrimination in this region of the spectrum is as good

for the baboon as it is for humans (Fagot et al., 2006). Which of these conclusions is correct may well reflect variations in the experimental paradigms that have so far been utilized, but, in any case, given that many nonhuman primates can use chromatic cues to guide food choices (for example, selecting "reddish" ripe fruits among "greenish" less-ripe fruits), it seems hard to imagine that capacity does not depend to some extent on the consistent utilization of color categories. Another aspect of color appearance is that humans rather automatically perceive a difference between scenes that contain color and those that do not, irrespective of other details of those scenes, and a recent experiment suggests that monkeys share that capacity. In that investigation marmosets (*Callithrix jacchus*) learned to discriminate images containing color from various grayscale images, and that skill then readily transferred to new images, thus strongly suggesting that color per se is as distinctive to nonhuman primates as it is to humans (Derrington et al., 2002).

In the end whether the presence of color vision in animals requires the demonstration of a cognitive component in spectrally based discriminations or can instead be thought to exist as a series of graded levels running from simple to more complex behaviors is unlikely to be settled by argument alone. Kelber and Osorio (2010) make the apt point that, in any case, the ultimate goal of studies of color vision in other species is to better understand how they exploit spectral information to form their own unique view of the world. Reaching that goal will require, in addition to laboratory-derived descriptions of the basic features of their color vision and relevant visual system biology, a much deeper understanding of visually guided natural behaviors, in effect a species-specific exploration of the visual ecology of color.

Using Proxies to Infer Animal Color Vision

For a relatively small number of well-studied species—including some mammals, birds, fishes, and insects—there is a clear understanding both of the nature of their color vision and various aspects of the biology that underlies that capacity. From those cases we have learned, among other things, that the dimensionality of their color vision can be compellingly linked to the number of types of cone pigment they possess. My chapter in the earlier edition of this book (Jacobs, 2004) illustrated that fact by showing that the presence of two classes of cone photopigment allows for dichromatic color vision in the case of the domestic dog, that three (very different) classes of cone pigment underlie color vision in both the trichromatic macaque monkey and the trichromatic honeybee, and that the goldfish expresses four classes of cone pigment and has tetrachromatic color vision . From examples such as these, as well as from detailed examinations of human color vision variations, it long ago became clear that the presence of various biological markers can serve as proxies for inferring the presence and nature of color vision.

Over the past few years there has been a significant expansion of information about the photopigment complements for many different animals. This comes from direct measurements of photopigments (Bowmaker, 2008; Hart, 2001), from antibody labeling of cone photoreceptors (Ahnelt & Kolb, 2000; Peichl, 2005; Szel et al., 2000), and from comparative examinations of the structure of the genes that specify the photopigment proteins (opsins) (Davies, Collin, & Hunt, 2012; Hunt et al., 2009; Yokoyama, 2008). In the face of this flood of new information it has become common practice to use these various indices as direct indicators of color vision; thus, for instance, detection of three types of cone opsin gene usually triggers the assertion that the animal has "trichromatic color vision."

Although these various biological proxies are indeed compellingly linked to color vision in many animals that have been well studied, there are a number of reasons for being cautious in assuming they can be universally applied across the animal kingdom. To consider just one of these, note that documentation of the structure and number of cone opsin genes offers no insight into the number and spatial distribution of the cones that contain the various pigment types or, indeed, of anything about the organization of the visual system needed to support the appropriate processing of the cone signals. All these features can critically influence the presence and nature of color vision. To cite one specific example, opsin genes specifying two classes of cone photopigments were detected in Old World fruitbats (Pteropodidae) and taken as evidence for presence of dichromatic color vision (Zhao et al., 2009). However, an examination of one such species (*Pteropus rufus*—the flying fox) reveals that the retina of this animal contains only a very sparse population of cones (~0.5% of all photoreceptors), and of those, fewer than 10%, a scant 200–300 cones/mm^2, express the short-wavelength pigment (Muller, Goodman, & Peichl, 2007). Such an array provides, at best, an arrangement allowing only very inefficient photon capture, and it seems quite possible that might prove insufficient to support the signals required to yield color vision. In this particular animal that difficulty is exacerbated still further because this species is principally nocturnal and thus mostly active when there is little light available. So whether this animal (and others like it) derives much color vision

from the presence of two types of cone pigment is very much in question. The essential point to remember in evaluating evidence derived from these various proxies is that, in the end, the establishment of color vision requires a clear documentation of appropriate behavioral responses, responses that necessarily reflect the operation of the entire visual system evaluated in the context of the environmental conditions that are appropriate for the animal in question.

EVOLUTION OF COLOR VISION

Probably no topic in comparative color vision has received greater attention over the past few years than this one. That situation largely reflects significant advances in understanding the comparative distribution and the structure/function relationships of opsin genes. From such sources of information it has proven possible to build up phylogenies that provide plausible pictures of the evolution of opsin genes. Although the cautions noted above are clearly relevant here, this work also then allows insights into the evolution of color vision. We consider three aspects of this still-developing story.

Natural History of Vertebrate Opsin Genes

Opsin genes are members of a large family of cell-surface receptor genes (linked also, for example, to receptors for pheromones, hormones, and neurotransmitters), all of which probably derived from a single ancestor in this family. Sequence comparisons of opsin genes from contemporary jawed and jawless vertebrates suggest that four classes of cone opsin genes (designated as SWS1, SWS2, Rh2, LWS) emerged at least 540 million years ago; the rod opsin genes (Rh1) appeared later (Collin et al., 2009). All the photopigments of contemporary vertebrates are derived from opsin genes drawn from these five families. A phylogenetic tree illustrating the distribution of these gene families for a small sample of vertebrate species is shown in figure 39.1. Detailed results for additional vertebrates obtained from similar comparisons have been presented elsewhere (Collin et al., 2009; Hunt et al., 2009; Shichida & Matsuyama, 2009; Yokoyama, 2000), and corresponding accounts of the evolution of various invertebrate opsins have also been published (Briscoe & Chittka, 2001; Cronin et al., 2010; Kashiyama et al., 2009). The ranges of the spectral absorption peaks (λ_{max}) for the photopigments derived from each of the gene families are shown to the right of the brackets in figure 39.1. Note that whereas the spectral peaks of vertebrate rod photopigments vary only modestly across species, the cone photopigment

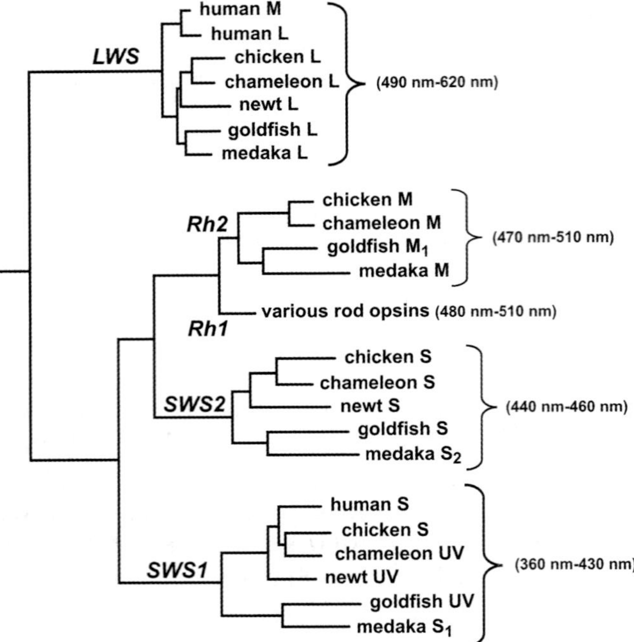

FIGURE 39.1 Opsin-gene family tree. Shown are a few representative vertebrate species that derive photopigments from the five opsin-gene families. The ranges of λ_{max} values for the pigments derived from each of these families are listed to the right of the brackets. (The tree construction was derived from Hisatomi & Tokunaga, 2002.)

peaks span nearly the entire visible spectrum (from ~360 nm to 620 nm). The structural basis for these variations in spectral tuning has been studied intensively and found to result principally from the presence of amino acid substitutions occurring at a restricted number of sites in the opsin molecule, while also reflecting the control that chromophore composition exerts on the spectral absorption properties of photopigments (Hunt et al., 2004, 2009; Neitz & Neitz, 2011; Yokoyama, 2008).

Another important lesson learned is that opsin genes have been both gained and lost during vertebrate evolution. The loss is particularly obvious (figure 39.1); for example, although all four families of cone opsin genes are represented in some contemporary birds and fishes, one of these (Rh2) is not found in modern amphibians (Bowmaker, 2008). The evolution of opsin genes among mammals mainly is a story of loss, as is outlined in figure 39.2. As for the amphibians, no contemporary mammals have Rh2 opsin genes so that gene family must have disappeared prior to the initial mammalian diversification. SWS2 genes are not found in either marsupials or eutherian mammals, and although monotremes have SWS2 genes, they do not express SWS1-derived genes. The loss of cone-opsin gene representation among the mammals is usually attributed to the long period of

nocturnality that characterized the early history of mammals, a time during which the value of having keen daylight vision would likely have been diminished (Heesy & Hall, 2010). To the extent that opsin genes predict the dimensionality of color vision, these evolutionary sketches suggest that some birds and fishes could realize tetrachromatic color vision, that amphibians might achieve trichromatic color vision, and that the basal pattern for mammals should be no better than dichromatic color vision. In general these predictions align with what has been learned from actual color-vision measurements (Jacobs, 2012).

Not surprisingly the evolution of primate color vision has been a particular target for study. That topic has been reviewed on several occasions and so merits only a brief summary here (for coverage of this topic see Hunt, Jacobs, & Bowmaker, 2005; Jacobs, 2008; Jacobs & Nathans, 2009; Surridge, Osorio, & Mundy, 2003). Similar to other eutherian mammals, most primates derive a single short-wavelength-sensitive (S) cone pigment based on an SWS1-derived opsin, but there are three separate patterns of representation for LWS genes. (1) Some primates are like other eutherian mammals in having only a single cone pigment from this gene family. (2) Others have evolved two or more spectrally discrete cone pigments from the LWS gene family. In the case of Old World primates (humans, apes, Old World monkeys), an X-chromosome gene duplication occurring some 35–40 mya provided two structurally distinct opsin genes from the LWS family, and their presence allowed for the emergence of two separate types (M and L) of cone pigment and, along with them, trichromatic color vision. By contrast, almost all New World monkeys have X-chromosome opsin-gene polymorphisms, typically yielding three spectrally unique versions of the LWS pigments, thus providing heterozygous females with both M and L pigments and trichromatic color vision while all the male monkeys, along with homozygous females, have only a single M/L pigment type and dichromatic color vision. (3) The third major group of primates, the more primitive strepsirrhines, features species that either show polymorphisms similar to that of the New World monkeys or have a total of two cone pigments similar to that of most nonprimate mammals. Finally, there are also a few primate species whose retinas contain only a single type of cone pigment. How that came about is considered next.

Opsin Pseudogenes

Pseudogenes are DNA sequences that are similar to those of known genes but lack their protein-coding

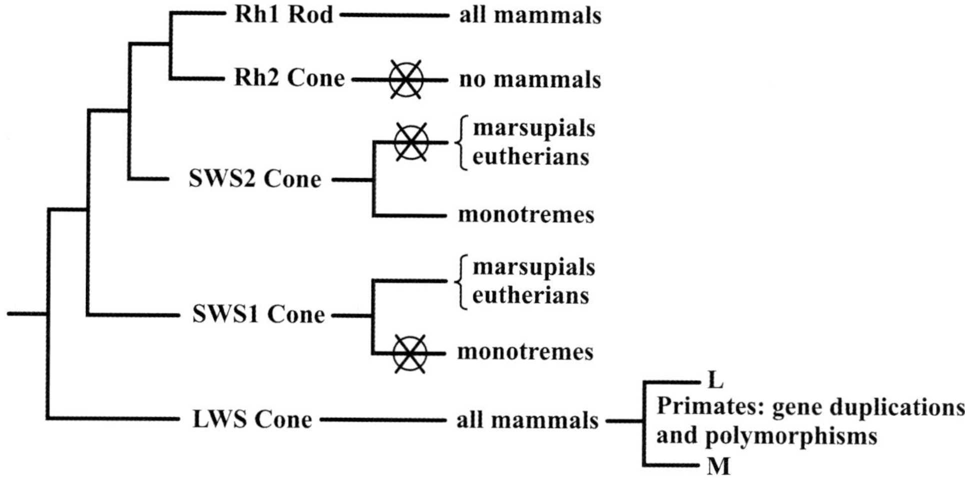

FIGURE 39.2   Retention and loss of representation from the five opsin gene families in mammals. The Xs indicate stages at which there was a loss of representation from that gene family. The gene-duplication event that occurred in some of the primates is described in the text.

capability. The transition of a gene to pseudogene status is usually thought to occur when the function(s) that gene serves becomes dispensable, although there are a few instances where inactivation is believed to have been adaptive (Podlaha & Zhang, 2010). Pseudogenes turn out to be of particular interest because they can help to illuminate the evolutionary history of the organism. About 15 years ago it was discovered that the SWS1 cone-opsin genes in two species of primate, owl monkeys (*Aotus trivirgatus*) and bushbabies (*Galago garnetti*), are pseudogenes, as in each of these cases the opsin gene sequences contain mutational changes that are sufficient to obviate their expression (Jacobs, Neitz, & Neitz, 1996). Because both of these species have only a single LWS-derived cone pigment, each expresses only a single type of cone, and therefore, these primates must lack ordinary color vision.

Since they were first found in these two types of primate, SWS1 opsin pseudogenes have been detected in more than 70 additional mammalian species, and there are likely many more yet to be identified (Jacobs, 2013). Those known to belong to this group include a number of other strepsirrhine primates as well as assorted carnivores, cetaceans, rodents, and bats (Jacobs, 2010; Peichl, 2005). In each instance some ancestral species in these lineages must have possessed intact SWS1 opsin genes and functional S cones that subsequently were lost through gene mutation. The relative timing of the occurrence of loss seems to have varied from case to case; for example, the SWS1 opsin genes of all the cetaceans (whales, dolphins, porpoises) contain the identical mutational change, implying that the pseudogene emerged early in the history of this

order, prior to any diversification of the lineage (Levenson & Dizon, 2003), whereas among the procyonids (raccoons, coatis, kinkajous, and the like) the loss is seen in some but not all contemporary species, indicating that the change must have been relatively more recent (Jacobs & Deegan, 1992; Peichl & Pohl, 2000).

S cones constitute only a small percentage of all cones in most mammalian retinas, and it is believed that the principal function subserved by signals derived from these receptors is the support of a dimension of color vision (Calkins, 2001). Because most mammals have only two cone opsin genes, the transition of functional SWS1 opsin genes to pseudogene status would thus have eliminated color vision. It is not obvious why S cones were lost in such a diverse group of mammals. One obvious trait that links all of the terrestrial species in which such pseudogenes have been detected is that they are nocturnal. By definition such animals are predominantly active when ambient light levels are insufficient to support much cone-based vision, and, consequently, color vision might be thought to offer only minimal advantages in such animals. The marine mammals (cetaceans and pinnipeds) that have lost functional SWS1 opsin genes are often classified as being arrhythmic, rather than nocturnal, but they too inhabit photic environments that often feature low light levels. Although nocturnality may have been a necessary condition for gene loss, it must not be sufficient because there are many mammals known to be nocturnal, some stringently so, who still maintain functional opsin genes and S cones. And because ancestral mammals are assumed to have been nocturnal, then, if nocturnality were key, why were not all SWS1 cones lost early in mammalian history? In sum, what we know at this point is that a significant number of mammals suffered an evolutionary loss of their SWS1 cone opsin genes and, consequently, a capacity for color vision. No single explanation for these losses has yet emerged (Jacobs, 2013).

Tuning Color Vision with Differential Gene Expression

Although all vertebrate cone opsin genes are members of four gene families, there are numerous variations on that theme. As we have seen there has been frequent loss of representation of cone opsin genes during the course of evolution, particularly among the mammals. One instance of a gain in gene representation was also noted, the case where the LWS gene in Old World primates duplicated, in the process generating two opsin genes that specify spectrally discrete M- and L-cone pigments. In recent years it has become clear that there have also been numerous instances of opsin-gene duplication in teleost fish, genes from each of the four families having been duplicated at least once and, in some cases, several times. Probably the best examples come from detailed studies of the cichlid fish that occupy the Rift Valley Lakes of East Africa (Carleton, 2009 Parry et al., 2005). These fish are renowned for their great diversity, their often colorful appearances, and the rapidity with which hundreds of species have evolved. Although there are typically only single versions of the SWS1 and LWS genes in these fish, gene duplication has resulted in the presence of two SWS2 cone-opsin genes and three Rh2 cone-opsin genes. Absorption spectra for the six photopigments that can potentially be specified by such a gene array are shown at the top of figure 39.3 (Hofmann & Carleton, 2009).

Although there are some exceptions, individual species of cichlid generally express only three of the photopigments from this array, but the set that is expressed varies across species, as is illustrated for three such species at the bottom of figure 39.3. The principal mechanism for determining the cone-pigment complement in each of these three types of fish is differential gene expression (Spady et al., 2006). To complicate the picture further not all of these pigments are expressed to same level; for example, there might be very low expression of the SWS1 opsin gene, and thus restricted representation of the pigment it specifies, but, at the same time, a severalfold higher expression of an Rh2 gene and of the pigment it specifies. Such variation can greatly impact both the spectral sensitivity of the fish and the quality of its color vision. Additionally, there may be large changes in the expression patterns of these genes over the period of development, such that, for example, a cone starts out expressing SWS1, next expresses an SWS2 gene, and then, finally, expresses a second version of an SWS2 gene (in the course, shifting the peak of the pigment found in that cone from the ultraviolet to violet and then to blue) (Carleton, 2009). All of these changes can occur during the first few months of the life of the fish. Finally, within each of these gene types spectral tuning can be also be altered by substitutions in the opsin-gene sequences. The sum of all these pigment possibilities is that there is the potential for enormous variations in visual sensitivity and in color vision in these cichlids, both within and between species.

The functional consequences of the pigment variations seen in the cichlids have been the topic of much recent discussion. One striking aspect of these fish is that the coloration patterns of male and female cichlids are much the same except during breeding season, when male cichlids become more intensely colored (so-called nuptial colors). At those times females are said to prefer males displaying more saturated colors and, during that same time period, color also is a factor in triggering aggressive interactions between males defending territories (Pauers, McKinnon, & Ehlinger, 2004; Pauers et al., 2008). A recent study started with measurements of the spectral absorption properties of the cone pigments, of the light environment, and of the reflectivity of fish color patterns and then modeled the conspicuousness of cichlid color patterns asking whether the nuptial colors displayed by males are in fact more conspicuous to other fish than are the colors of females (Dalton et al., 2010). If they are, it would suggest that sexual selection may be driving the evolution of male color patterns. Of the nine species examined males were indeed predicted to be more conspicuous to other fish than were the females, supporting the sexual selection idea. However, in two species the reverse was found, so other factors must also be at play in the evolution of male color patterns.

Recent studies have also examined the role that variations in cone-opsin-gene expression in cichlids plays in matching visual capacity to the photic environment. The photic environments in these lakes vary considerably depending on depth and quality of the water, both features serving to bias significantly the spectral distributions of ambient light. Reassuringly, the gene-expression pattern often seems to follow what might be expected. For example, in some locations the water greatly attenuates short-wavelength light, and fish that inhabit those waters show correspondingly very low expression of the short-wavelength photopigments (Hofmann et al., 2009). However, a subsequent study

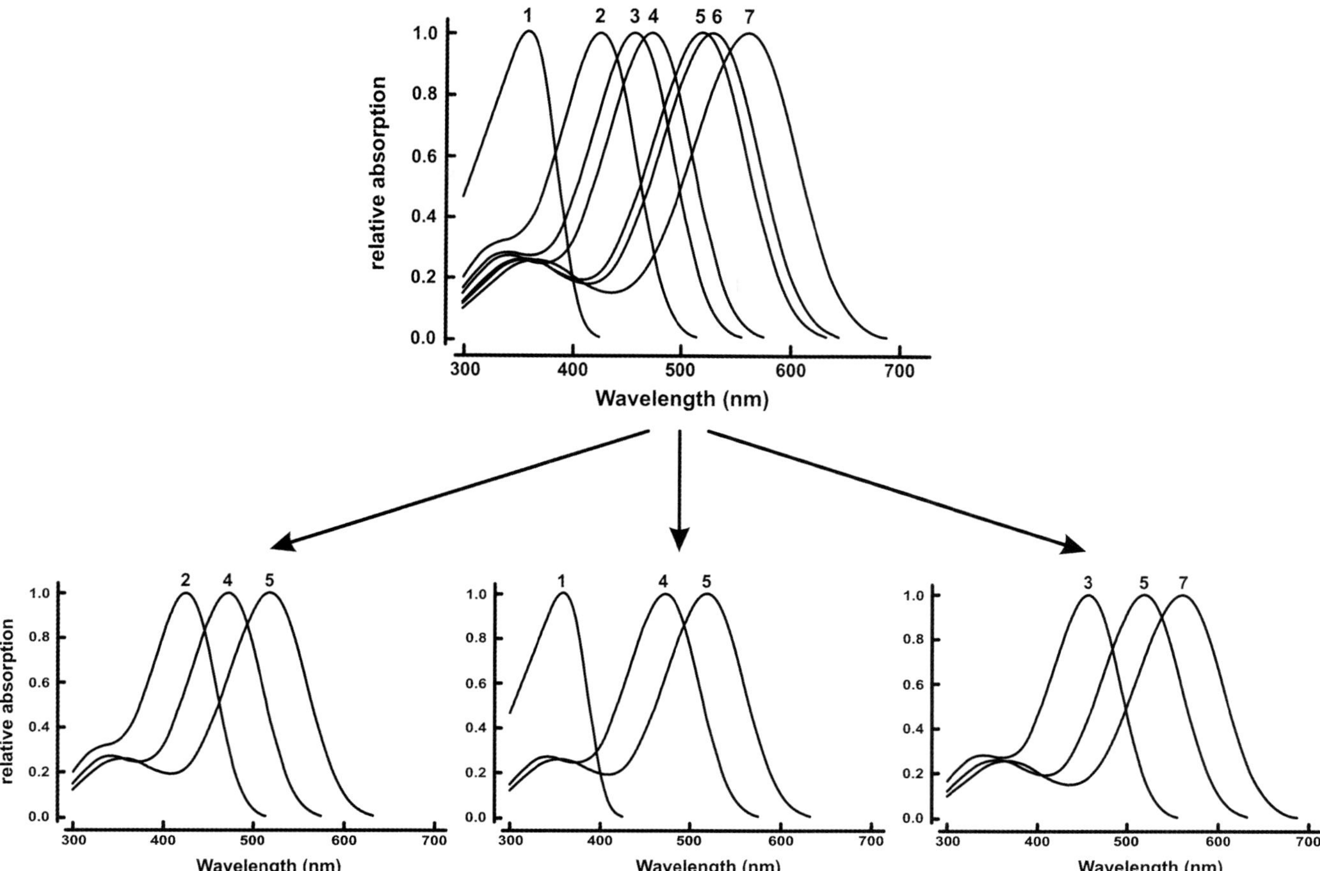

FIGURE 39.3 The top panel shows the absorption curves for the full palette of photopigments available in cichlid fish from the Rift Lakes of East Africa. The numbers at the top indicate the gene family from which the pigment is derived, keyed as follows: 1 (SWS1), 2 (SWS2B), 3 (SWS2A), 4 (Rh2B), 5 (Rh2Aβ), 6 (Rh2Aα), 7 (LWS). The photopigment absorption curves shown at the bottom are for the sets of cone photopigments that are differentially expressed in each of three species of fish. (These curves were derived from Hofmann & Carleton, 2009.)

revealed that there is still much to be understood about photopigment variation and seeing in these fish, for it turns out that there are also considerable intraspecific variations in gene expression in some cichlids, and in those cases the pattern of gene expression does not appear to vary in a manner consistent with providing sensitivity advantages that would be predicted by the local photic environment (Smith et al., 2011). If all that were not enough, it also seems that some of these fish express not three but four photopigments—in some cases perhaps even five (Sabbah et al., 2010). This raises the prospect that the color vision of these fish may be immensely variegated. To appreciate the functional consequences of variations in opsin genes and photopigments in cichlids and to understand how they link to the photic environment will require eventually direct studies of color vision. That will be a most challenging task.

COLOR VISION UNDER DIM ILLUMINATION

Classical duplicity theory links the perception of color to the activation of cone photoreceptors, but in duplex visual systems rods and cones operate in common across a considerable range of illuminances—some four log units in the case of human vision—and numerous studies with human subjects make it clear that across this interval ("mesopic" vision) rod signals can influence the perception of color in a wide variety of ways that are dependent on the conditions of stimulation (Buck, 2004). Although strict color discrimination must fail under conditions where only rods are active, it is clear that, if they are allowed to make use of differences in relative scotopic lightness, both normal and color-defective human subjects are capable of reliably categorizing surface colors under illumination conditions where only rods are activated (Pokorny et al., 2006,

2008). The ability to do this apparently reflects the experience observers accumulate through repeated viewing of familiar objects in natural environments under both bright and dim illuminations. Although there seems to be no evidence one way or another, it certainly seems reasonable to suppose that at least some nonhuman species may learn to behave in an analogous fashion.

For human subjects there has long been evidence to show that at mesopic light levels rod and cone signals can be contrasted to yield an additional dimension of color vision, allowing, for example, dichromatic individuals to make trichromatic color matches under some lighting conditions (Smith & Pokorny, 1977). Such possibilities are much harder to investigate in nonhuman subjects, but examples have emerged that strongly imply that other species have this same capacity. In one such case California sea lions (*Zalophus californaus*) were tested in a behavioral experiment and found capable of making color discriminations (Griebel & Schmid, 1992), yet a later genetic examination showed convincingly that this species has only a single type of cone photopigment (Levenson et al., 2006), implying that in the behavioral experiment the animals must have been able to exploit signals from their single cone type in combination with rod signals to support a dimension of color vision. Animals with visual adaptations similar to that of the sea lion, and that would include at the least all those mammals identified above as having SWS1 opsin pseudogenes, could presumably be without color vision at both high and low light levels, yet still able to derive some dichromatic color capabilities at intermediate light levels.

In recent years some surprising new examples of the presence of color vision at low light levels have emerged from comparative studies. The hawkmoth (*Deilephilia elpenor*) is a nocturnal species that feeds on nectar harvested from visits made to flowers during the dark of night. The compound eyes of this insect contain three different types of photopigment (Hoglund, Hamdorf, & Rosner, 1973), and behavioral tests show that this species is able to make color discriminations at very low light levels, at an ambient illuminance that is equivalent to dim starlight—this is a light level at which human color vision is absent entirely (Kelber, Balkenius, & Warrant, 2002). Several features of the eyes of hawkmoths appear to support this unusual capacity. First, this insect has a superposition eye in which light focused through many lens facets combines to yield a single image, which, in conjunction with a reflective tapetum, increases significantly the photon efficiency of the eye. Second, there are also mechanisms in this visual system that allow for enhanced spatial and temporal

summation. All of these features combine to allow the hawkmoth true color vision at low light levels (Land & Osorio, 2003; Kelber & Roth, 2006). In another case, a vertebrate species, the gecko (*Tarentola chazaliae*), has also been shown to enjoy color vision at low light levels (Roth & Kelber, 2004). Although it is a nocturnal animal, this lizard has an all-cone retina featuring three classes of cone pigment (Loew et al., 1996) and, similar to the hawkmoth, has evolved a number of structural adaptations that enhance the abilities of its cones to capture scarce photons. These adaptations included unusually large photoreceptors and an eye that has a short focal length and a large pupil, features that combine to provide a device that is highly efficient at gathering light. Behavioral tests show that this gecko is capable of making color discriminations at light levels equivalent to those available in the dim moonlight (Roth & Kelber, 2004). Although color vision in these two species may in the end prove to be quite unusual, these cases serve as stark reminders that our understanding of human vision provides a less-than-perfect model for thinking about the possibilities for color vision in other species.

ENGINEERING COLOR VISION

In recent years the advent of new molecular genetics technologies has transformed research on the visual system. In addition to providing animal models for studying various disorders of vision, these tools have also been used to ask basic questions about visual system organization (Chalupa & Williams, 2008). Because the process of evolution continually alters the organization of the visual system, these same techniques also provide a means for evaluating hypotheses about the evolution of vision.

As described above, opsin genes were both lost and gained during the course of evolution. The addition of a novel opsin gene can produce a new photopigment, and one central question for understanding the evolution of color vision is whether the mere appearance of a new retinal photopigment yields a sufficient change in visual capacity to allow the new gene to evolve, or whether some additional change(s) in the nervous system may also be required. A first attempt to examine this issue involved the production of a transgenic mouse that had incorporated a human L-cone opsin gene (Shaaban et al., 1998). The pigment specified by the transgene (having λ_{max} = 556 nm) was abundantly expressed in the cones of this transgenic mouse, and subsequent electrophysiological and behavioral measurements showed that these animals experienced a significant increase in sensitivity to long-wavelength

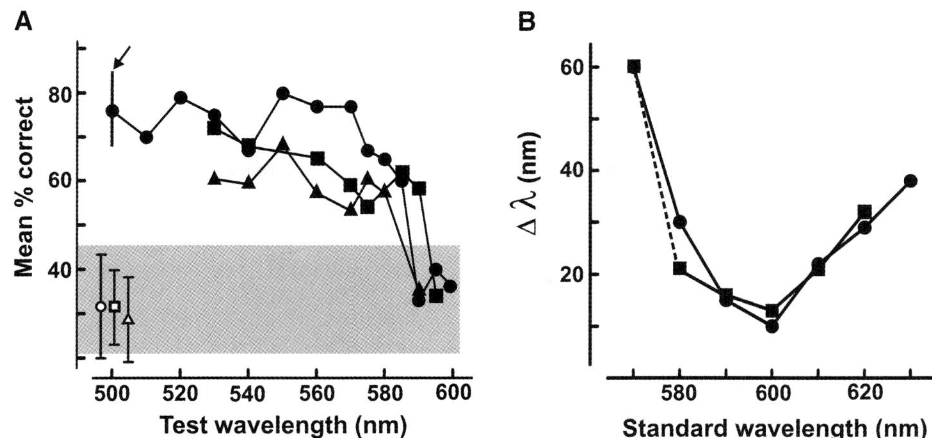

FIGURE 39.4   Results obtained from two tests of color vision conducted on knock-in mice (see text). (A) Wavelength discrimination. In a three-choice discrimination experiment mice were required to discriminate various test wavelengths from an equally bright 600-nm light. The shaded area indicates chance performance levels. The asymptotic performance levels achieved by three control animals are shown to the lower left, and the results for three heterozygous mice are given by the three dark symbols. Note that the latter successfully discriminated a wide range of wavelengths from the 600-nm light. (B) Results from a second wavelength-discrimination test in which two heterozygous mice were tested for the ability to discriminate differences in wavelength. The $\Delta\lambda$ values plot how much wavelength had to change (in nm) for the mouse to reliably discriminate it from the standard wavelength. The results from both tests demonstrate that these knock-in mice acquired color vision that was novel to their species. (Data from Jacobs et al., 2007.)

lights. This result implies that simple addition of a new pigment can, by itself, yield immediate changes in behavioral capacity and thus at least provide the potential for an adaptive advantage (Jacobs et al., 1999). The retina of this transgenic mouse featured three types of cone photopigment: native UV (λ_{max} = 360 nm), native M (λ_{max} = 510 nm), and the human L. The human L pigment was coexpressed in cones together with the native mouse pigments, a fact that very likely obviated the possibility of these transgenic mice acquiring any new color vision (Jacobs et al., 1999).

A better approximation of the arrangement through which color vision probably evolved in primates was achieved by developing knock-in mice in which the normal (X-chromosome-linked) mouse M-opsin gene was replaced with constructs that contained the human L-cone opsin gene (Onishi et al., 2005; Smallwood et al., 2003). Subsequent breeding of these animals yielded mice of three different cone-pigment phenotypes. All of them had the normal UV cone pigment; in addition, hemizygous males and homozygous females expressed either the mouse M or the human L pigment, whereas heterozygous female mice had both the M and the L pigments. As a result of early random X-chromosome inactivation, the two pigments were individually expressed in the cones of the heterozygous females, just as they are in the polymorphic New World monkeys.

Results from behavioral tests (figure 39.4) show that these heterozygous mice are capable of acquiring a color-vision capacity that is entirely novel to the species (Jacobs et al., 2007). A similar outcome was later obtained in an experiment in which a gene-therapy technique was used to add a third cone pigment to the eyes of dichromatic squirrel monkeys that subsequently were shown capable of making trichromatic color discriminations (Mancuso et al., 2009). It is notable that in this latter experiment the gene transfer was accomplished in adult animals, thus demonstrating that no correlative changes in the visual system were required beyond the mere presence of a new cone pigment. Experiments of these kinds demonstrate that, under the right set of circumstances, the mere addition of a novel photopigment gene is sufficient to trigger a capacity for new color vision. That is likely what happened during the evolution of primate color vision, and it seems probable that the same process also occurred at many other stages during the evolution of animal color vision. These experiments also provide nice demonstrations of the potential that molecular genetics approaches hold for deriving a deeper understanding of the evolution of color vision and, in the process, furthering our understanding of why this remarkable capacity appears in such a myriad of forms across the animal kingdom.

REFERENCES

Ahnelt, P. K., & Kolb, H. (2000). The mammalian photoreceptor mosaic-adaptive design. *Progress in Retinal and Eye Research, 19,* 711–770. doi:10.1016/S1350-9462(00)00012-4.

Bowmaker, J. K. (2008). Evolution of vertebrate visual pigments. *Vision Research, 48,* 2022–2041. doi:10.1016/j.visres.2008.03.025.

Brainard, D. H., & Maloney, L. T. (2011). Surface color perception and equivalent illumination models. *Journal of Vision, (11)5,* 1–10. doi:10.11167/11.5.1.

Briscoe, A. D., & Chittka, L. (2001). The evolution of color vision in insects. *Annual Review of Entomology, 46,* 471–510.

Buck, S. L. (2004). Rod-cone interactions in human vision. In L. M. Chalupa & J. S. Werner (Eds.), *The visual neurosciences* (Vol. 1, pp. 863–878). Cambridge, MA: MIT Press.

Calkins, D. J. (2001). Seeing with S cones. *Progress in Retinal and Eye Research, 20,* 255–287.

Carleton, K. (2009). Cichlid fish visual systems: Mechanisms of spectral tuning. *Integrative Zoology, 4,* 75–86.

Chalupa, L. M., & Williams, R. W. (Eds.). (2008). *Eye, retina and visual system of the mouse.* Cambridge, MA: MIT Press.

Collin, S. P., Davies, W. L., Hart, N. S., & Hunt, D. M. (2009). The evolution of early vertebrate photoreceptors. *Philosophical Transactions of the Royal Society of London. Series B, Biological Sciences, 364,* 2925–2940.

Cronin, T. W., Porter, M. L., Bok, M. J., Wolf, J. B., & Robinson, P. R. (2010). The molecular genetics and evolution of colour and polarization vision in stomatopod crustaceans. *Ophthalmic & Physiological Optics, 30,* 460–469.

Dalton, B. E., Cronin, T. W., Marshall, N. J., & Carleton, K. L. (2010). The fish eye view: Are cichlids conspicuous? *Journal of Experimental Biology, 213,* 2243–2255.

Davies, W. I. E. L., Collin, S. P., & Hunt, D. M. (2012). Molecular ecology and adaptation of visual pigments in craniates. *Molecular Ecology, 21,* 3121–3158.

Derrington, A. M., Parker, A. R., Barraclough, N. E., Easton, A., Goodson, G. R., Parker, K. S., et al. (2002). The uses of colour vision: Behavioural and physiological distinctiveness of colour stimuli. *Philosophical Transactions of the Royal Society of London. Series B, Biological Sciences, 357,* 975–985.

Fagot, J., Goldstein, J., Davidoff, J., & Pickering, A. (2006). Cross-species differences in color categorization. *Psychonomic Bulletin & Review, 13,* 275–280.

Goldsmith, T. H. (1990). Optimization, constraint, and history in the evolution of eyes. *Quarterly Review of Biology, 65,* 281–322.

Griebel, U., & Schmid, A. (1992). Color vision in the California sea lion (*Zalophus californianus*). *Vision Research, 32,* 477–482. doi:10.1016/0042-6989(92)90239-F.

Hart, N. S. (2001). Variations in cone photoreceptor abundance and the visual ecology of birds. *Journal of Comparative Physiology. A, Neuroethology, Sensory, Neural, and Behavioral Physiology, 187,* 685–698.

Heesy, C. P., & Hall, M. I. (2010). The nocturnal bottleneck and the evolution of mammalian vision. *Brain, Behavior and Evolution, 75,* 195–203.

Hisatomi, O., & Tokunaga, F. (2002). Molecular evolution of proteins involved in vertebrate phototransduction. *Comparative Biochemistry and Physiology. B, Comparative Biochemistry, 133,* 509–522.

Hofmann, C. M., & Carleton, K. L. (2009). Gene duplication and differential gene expression play an important role in the diversification of visual pigments in fish. *Integrative and Comparative Biology, 49,* 630–643.

Hofmann, C. M., O'Quin, K. E., Marshall, N. J., Cronin, T. W., Seehausen, O., & Carleton, K. L. (2009). The eyes have it: Regulatory and structural changes both underlie cichlid visual pigment diversity. *PLoS Biology, 7,* e1000266. doi:10.1371/journal.pbio.1000266.

Hoglund, G., Hamdorf, K., & Rosner, G. (1973). Trichromatic visual system in an insect and its sensitivity control by blue light. *Journal of Comparative Physiology, 86,* 265–279.

Hunt, D. M., Carvalho, L. S., Cowing, J. A., & Davies, W. L. (2009). Evolution and spectral tuning of visual pigments in birds and mammals. *Philosophical Transactions of the Royal Society of London. Series B, Biological Sciences, 364,* 2941–2956. doi:10.1098/rstb.2009.0044.

Hunt, D. M., Cowing, J. A., Wilkie, S. E., Parry, J. W. L., Poopalasundaram, S., & Bowmaker, J. K. (2004). Divergent mechanisms for the tuning of shortwave sensitive visual pigments in vertebrates. *Photochemical & Photobiological Sciences, 3,* 713–720.

Hunt, D. M., Jacobs, G. H., & Bowmaker, J. K. (2005). The genetics and evolution of primate visual pigments. In J. Kremers (Ed.), *The primate visual system: A comparative approach* (pp. 73–126). Chichester: John Wiley & Sons.

Jacobs, G. H. (1981). *Comparative color vision.* New York: Academic Press.

Jacobs, G. H. (1993). The distribution and nature of colour vision among the mammals. *Biological Reviews of the Cambridge Philosophical Society, 68,* 413–471.

Jacobs, G. H. (2004). Comparative color vision. In L. M. Chalupa & J. S. Werner (Eds.), *The visual neurosciences* (Vol. 2, pp. 962–973). Cambridge, MA: MIT Press.

Jacobs, G. H. (2008). Primate color vision: A comparative perspective. *Visual Neuroscience, 25,* 619–633.

Jacobs, G. H. (2010). Recent progress in understanding mammalian color vision. *Ophthalmic & Physiological Optics, 30,* 422–434.

Jacobs, G. H. (2012). The evolution of vertebrate color vision. *Advances in Experimental Medicine and Biology, 739,* 156–172.

Jacobs, G. H. (2013). Losses of functional opsin genes, short-wavelength cone photopigments and color vision—a significant trend in the evolution of mammalian vision. *Visual Neuroscience, 30,* 39–53. doi:10.1017/S0952523812000429.

Jacobs, G. H., & Deegan, J. F. II. (1992). Cone photopigments in nocturnal and diurnal procyonids. *Journal of Comparative Physiology. A, Neuroethology, Sensory, Neural, and Behavioral Physiology, 171,* 351–358.

Jacobs, G. H., Fenwick, J. C., Calderone, J. B., & Deeb, S. S. (1999). Human cone pigment expressed in transgenic mice yields altered vision. *Journal of Neuroscience, 19,* 3258–3265.

Jacobs, G. H., & Nathans, J. (2009). The evolution of primate color vision. *Scientific American, 300,* 40–47.

Jacobs, G. H., Neitz, M., & Neitz, J. (1996). Mutations in S-cone pigment genes and the absence of colour vision in two species of nocturnal primate. *Proceedings. Biological Sciences, 263,* 705–710.

Jacobs, G. H., Williams, G. A., Cahill, H., & Nathans, J. (2007). Emergence of novel color vision in mice engineered to express a human cone photopigment. *Science, 315,* 1723–1725.

Kashiyama, K., Seki, T., Numata, H., & Goto, S. G. (2009). Molecular characterization of visual pigments in branchiopoda and the evolution of opsins in arthropoda. *Molecular Biology and Evolution, 26*, 299–311.

Kelber, A., Balkenius, A., & Warrant, E. J. (2002). Scotopic colour vision in nocturnal hawkmoths. *Nature, 419*, 922–925.

Kelber, A., & Osorio, D. (2010). From spectral infromation to animal colour vision: Experiments and concepts. *Proceedings. Biological Sciences, 277*, 1617–1625.

Kelber, A., & Roth, L. S. V. (2006). Nocturnal colour vision—not as rare as we might think. *Journal of Experimental Biology, 209*, 781–788.

Kelber, A., Vorobyev, M., & Osorio, D. (2003). Animal colour vision—behavioural tests and physiological concepts. *Biological Reviews of the Cambridge Philosophical Society, 78*, 81–118.

Land, M. F., & Osorio, D. (2003). Colour vision: Colouring the dark. *Current Biology, 13*, R83–R85.

Levenson, D. H., & Dizon, A. (2003). Genetic evidence for the ancestral loss of SWS cone pigments in mysticetee and odontocete cetaceans. *Proceedings. Biological Sciences, 270*, 673–679.

Levenson, D. H., Ponganis, P. J., Crognale, M. A., Deegan, J. F., II, Dizon, A., & Jacobs, G. H. (2006). Visual pigments of marine carnivores: Pinnipeds, polar bear, and sea otter. *Journal of Comparative Physiology. A, Neuroethology, Sensory, Neural, and Behavioral Physiology, 192*, 823–843.

Loew, E. R., Govardovskii, V. I., Rohlich, P., & Szel, A. (1996). Microspectrophotometric and immunocytochemical identification of ultraviolet photoreceptors in geckos. *Visual Neuroscience, 13*, 247–256.

Mancuso, K., Hauswirth, W. W., Li, Q., Connor, T. B., Kuckenbecker, J. A., Mauck, M. C., et al. (2009). Gene therapy for red-green color blindness in adult primates. *Nature, 461*, 784–787.

Menzel, R. (1979). Spectral sensitivity and color vision in invertebrates. In H. Autrum (Ed.), *Handbook of sensory physiology, Vol VII/6A* (pp. 503–580). Berlin: Springer-Verlag.

Muller, B., Goodman, S. M., & Peichl, L. (2007). Cone photoreceptor diversity in the retinas of fruit bats (Megachiroptera). *Brain, Behavior and Evolution, 70*, 90–104.

Neitz, J., & Neitz, M. (2011). The genetics of normal and defective color vision. *Vision Research, 51*, 633–651.

Onishi, A., Hasegawa, J., Imai, H., Chisaka, O., Ueda, Y., Honda, Y., et al. (2005). Generation of knock-in mice carrying third cones with spectral sensitivity different from S and L cones. *Zoological Science, 22*, 1145–1156. doi:10.2108/zsj.22.1145.

Parry, J. W., Carleton, K. L., Spady, T., Carboo, A., Hunt, D. M., & Bowmaker, J. K. (2005). Mix and match color vision: Tuning spectral sensitivity by differential gene expression in Lake Malawi cichlids. *Current Biology, 15*, 1734–1739.

Pauers, M. J., Kapfer, J. M., Fendos, C. E., & Berg, C. S. (2008). Aggressive biases towards similarly coloured males in Lake Malawi cichlid fishes. *Biology Letters, 4*, 156–159.

Pauers, M. J., McKinnon, J. S., & Ehlinger, T. J. (2004). Directional sexual selection on chroma and within-pattern colour contrast in *Labeotropheus fuelleborni. Proceedings of the Royal Society of London. Series B, Biological Sciences, 271*, S444–S447.

Peichl, L. (2005). Diversity of mammalian photoreceptor properties: Adaptations to habitat and lifestyle? *Anatomical Record. Part A, Discoveries in Molecular, Cellular, and Evolutionary Biology, 287A*, 1001–1012.

Peichl, L., & Pohl, B. (2000). Cone types and cone/rod ratios in the crab-eating raccoon and coati (Procyonidae). *Investigative Ophthalmology & Visual Science, 41*, S494.

Podlaha, O., & Zhang, J. (2010). *Pseudogenes and their evolution. Encyclopedia of life sciences.* Chichester: John Wiley & Sons.

Pokorny, J., Lutze, M., Cao, D., & Zele, A. J. (2006). The color of night: Surface color perception under dim illuminations. *Visual Neuroscience, 23*, 525–530.

Pokorny, J., Lutze, M., Cao, D. C., & Zele, A. J. (2008). The color of night: Surface color categorization by color defective observers under dim illuminations. *Visual Neuroscience, 25*, 475–480.

Roth, L. S. V., & Kelber, A. (2004). Nocturnal colour vision in geckos. *Proceedings. Biological Sciences, 271*(Suppl.), S485–S487.

Sabbah, S., Laria, R. L., Gray, S. M., & Hawryshyn, C. W. (2010). Functional diversity in the color vision of cichlid fishes. *BMC Biology, 8.* doi:133 10.1186/1741-7007-8-133

Shaaban, S. A., Crognale, M. A., Calderone, J. B., Huang, J., Jacobs, G. H., & Deeb, S. S. (1998). Transgenic mice expressing a functional human photopigment. *Investigative Ophthalmology & Visual Science, 39*, 1036–1043.

Shevell, S. K. (2003). Color appearance. In S. K. Shevell (Ed.), *The science of color* (2nd ed., pp. 149–190). Oxford: Elsevier.

Shichida, Y., & Matsuyama, T. (2009). Evolution of opsins and phototransduction. *Philosophical Transactions of the Royal Society B-Biological Sciences, 364*, 2881–2895. doi:10.1098/rstb.2009.0051.

Skorupski, P., & Chittka, L. (2011). Is colour cognitive? *Optics & Laser Technology, 43*, 251–260.

Smallwood, P. M., Olveczky, B. P., Williams, G. A., Jacobs, G. H., Reese, B. E., Meister, M., et al. (2003). Genetically engineered mice with an additional class of cone photoreceptors: Implications for the evolution of color vision. *Proceedings of the National Academy of Sciences of the United States of America, 100*, 11706–11711. doi:10.1073/pnas.1934712100.

Smith, A. R., D'Annunzio, L., Smith, A. E., Sharma, A., Hofmann, C. M., Marshall, N. J., et al. (2011). Intraspecific cone opsin expression variation in the cichlids of Lake Malawi. *Molecular Ecology, 20*, 299–310.

Smith, V. C., & Pokorny, J. (1977). Large-field trichromacy in protanopes and deuteranopes. *Journal of the Optical Society of America, 67*, 213–220.

Spady, T. C., Parry, J. W. L., Robinson, F. R., Hunt, D. M., Bowmaker, J. K., & Carleton, K. L. (2006). Evolution of the cichlid visual palette through ontogenetic subfunctionalization of the opsin gene arrays. *Molecular Biology and Evolution, 23*, 1538–1547. doi:10.1093/molbev/msl014.

Stockman, A., & Brainard, D. H. (2010). Color vision mechanisms. In M. Bass (Ed.), *OSA handbook of optics* (pp. 11.11–11.104). New York: McGraw-Hill.

Surridge, A. K., Osorio, D., & Mundy, N. I. (2003), Evolution and selection of trichromatic vision in primates. *Trends in Ecology & Evolution, 18*, 198–206.

Szel, A., Lukats, A., Fekete, T., Szepessy, Z., & Rohlich, P. (2000). Photoreceptor distribution in the retinas of subprimate mammals. *Journal of the Optical Society of America. A, Optics, Image Science, and Vision, 17*, 568–579.

Wandell, B. A. (1995). *Foundations of vision.* Sunderland, MA: Sinauer Associates.

Wyszecki, G., & Stiles, W. S. (1982). *Color science* (2nd ed.). New York: John Wiley & Sons.

Yokoyama, S. (2000). Molecular evolution of vertebrate visual pigments. *Progress in Retinal and Eye Research, 19,* 385–419.

Yokoyama, S. (2008). Evolution of dim-light and color vision pigments. *Annual Review of Genomics and Human Genetics, 9,* 259–282.

Zhao, H., Rossiter, S. J., Teeling, E. C., Li, C., Cotton, J. A., & Zhang, S. (2009). The evolution of color vision in nocturnal mammals. *Proceedings of the National Academy of Sciences of the United States of America, 106,* 8980–8985. doi:10.1073/pnas.0813201106.

40 Color in the Primary Visual Cortex

ROBERT SHAPLEY, MICHAEL HAWKEN, AND ELIZABETH JOHNSON

In this chapter we review how the primary visual cortex (V1) processes color signals. There is increasing evidence that color and form are linked inextricably in visual cortical processing. V1 has a large role to play in the linkage of form and color. The reevaluation of the important role of V1 in color vision was caused in part by investigations of human V1 responses to color, measured with psychophysics and with functional magnetic resonance imaging (fMRI) and in part by the results of studies of single-unit neurophysiology in nonhuman primates. The neurophysiological results have highlighted the importance of double-opponent cells in V1 responses to color surfaces.

One major philosophical view is that color is an objective material property (Hyman, 2006), not a subjective experience. However, those who study visual perception know that surrounding colors have a great influence on color perception (Brainard, 2004; Katz, 1935), a fact that implies that color is not simply objective. The brain needs to construct a color signal to recover, as well as it can, the reflective properties of a surface, independent of illumination. To accomplish this task the neural mechanisms of color perception must make comparative computations that take into account the spatial layout of the scene as well as the spectral reflectances of the target surfaces (Brainard, 2004; Shevell & Kingdom, 2008). It is not known how the visual system integrates form and color, but it is now widely believed that the primary visual cortex, V1, plays an important role (Friedman, Zhou, & von der Heydt, 2003; Hurlbert and Wolf, 2004; Johnson, Hawken, & Shapley, 2001; Wachtler, Sejnowski, & Albright, 2003).

THE "SEGREGATED-COLOR" VIEW

The most widely publicized view about color in the cortex used to be that color, motion, and form were analyzed in parallel by separate visual cortical modules. This *segregated-color* view was in part a legacy of the ideas of Hering transmitted by Hurvich and Jameson (1957). The segregated-color or modular view was also influenced by the work of Krauskopf and colleagues in their

work on cardinal directions in color space (Derrington, Krauskopf, & Lennie, 1984; Krauskopf, Williams, & Heeley, 1982), although the interpretation of the earlier work has been modified subsequently (Tailby et al., 2008). The modular view was also influenced by the shape of the contrast sensitivity function for red-green equiluminant patterns, which is low-pass, as opposed to the spatial bandpass contrast sensitivity function for luminance patterns (Mullen, 1985). We first review briefly the neurophysiological results on the visual cortex from the late 1980s that were interpreted in terms of what we call the segregated-color or modular view (after Livingstone and Hubel, 1987; Zeki and Shipp, 1988) and then turn to more recent research on color signals in the cortex that either still reflects the modular view or supports a different, more integrated view of the neural mechanisms of color and form perception (cf. reviews by Gegenfurtner, 2003; Gegenfurtner & Kiper, 2003; see chapter 41 by Kiper and Gegenfurtner).

De Valois (1965) first provided evidence that color-opponent neurons existed in the primate visual system and could be important for color vision. Wiesel and Hubel (1966) then found that there were color-opponent cells in the parvocellular layers of the monkey lateral geniculate nucleus (LGN). LGN magnocellular layer neurons were largely color-blind. Later work on the LGN supported the idea that the parvocellular stream from retina to LGN carried signals about color while the magnocellular stream conveyed signals about low-luminance contrast (Benardete & Kaplan, 1999; Chatterjee & Callaway, 2003; Crook et al., 1987; Kaplan & Shapley, 1982, 1986; Lee, Martin, & Valberg, 1989; Lee et al., 2012; Reid & Shapley 1992, 2002). Subsequent work revealed that there was a third parallel retina → cortex stream, the koniocellular pathway (Casagrande, 1994; Chatterjee & Callaway, 2003; Hendry & Reid, 2000) that conveyed color-opponent signals where the opponency was between the S cones and the longer-wavelength-sensitive M and L cones. Koniocellular cells innervated the cytochrome oxidase (CO) patches in layer 3 of V1 cortex directly (figure 40.1B;

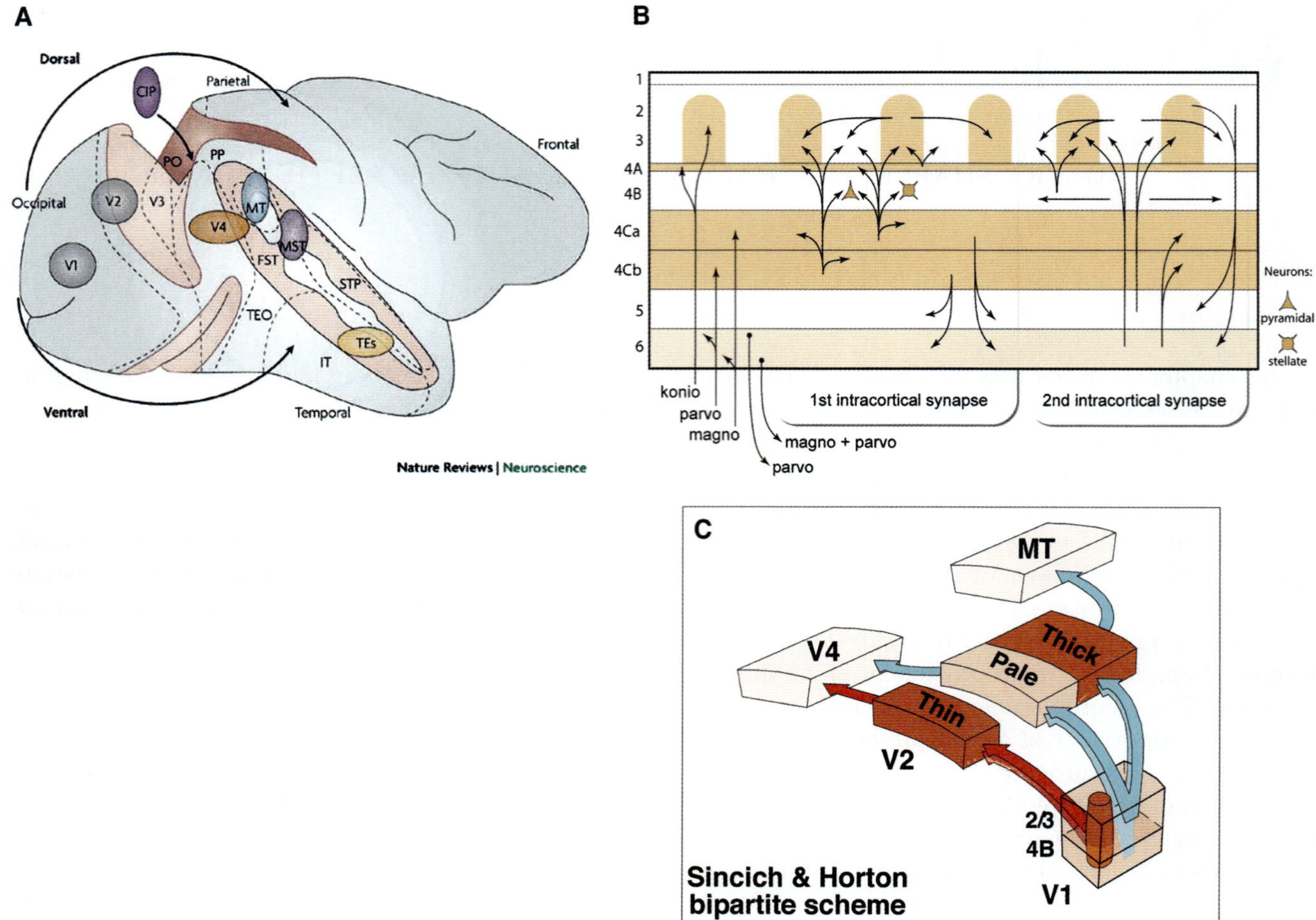

FIGURE 40.1   (A) Lateral view of the Old World monkey brain. The striate (V1) and extrastriate regions are shown on a surface where the sulci have been "opened" to allow viewing of the borders between regions that are otherwise embedded in a sulcus. The regions in the upper parietal region of extrastriate cortex form the dorsal stream; those in the lower temporal region form the ventral stream. (B) A schematic view of the processing in V1. The darker brown regions indicate the cytochrome oxidase (CO)-rich regions of V1 that correspond to layers receiving the stronger input from the lateral geniculate nucleus (LGN); the light brown shows an intermediate level of CO staining. The input streams from the different divisions of the LGN magnocellular (magno or M), parvocellular (parvo or P), and koniocellular (konio or K) are shown at the lower left of the figure. The main axonal projections within the cortex (intracortical projections) are shown as the first and second intracortical projections in the center and right of the figure. The cytochrome-rich regions of V1 in layer 3 and lower 2 are referred to as CO patches (or "blobs"). The regions between CO patches are the interpatches (or interblobs). (C) The axonal projections to the first extrastriate visual area V2 from neurons in the CO-rich patches (dark brown barrels in V1) project preferentially to the thin CO-rich stripes. The axonal projections from the interpatch regions in V1 (turquoise) preferentially project to the CO-poor (pale) and thick CO-rich stripes in V2. The further projections from the V2 regions are shown to visual area V4 and MT (V5). These two extrastriate areas are those that were identified by Zeki as the "color" area (V4) and the "motion" area (MT or V5).

Casagrande et al., 2007; Hendry & Yoshioka, 1994) and also provided weak additional direct input to V1 outside the CO patches (Casagrande et al., 2007).

The modular view of cortical color processing was emphasized in the work of Semir Zeki on the functions of extrastriate visual cortex (Zeki 1973, 1978a,b). Zeki reported that different extrastriate regions in the macaque monkey cortex (figure 40.1A) were specialized for different visual features. For instance, V5 (also called MT) cortex was reported to be comprised of neurons almost all of which were directionally selective. Thus, V5 was viewed as the motion area, the region of the visual cortex designed to respond to motion in the visual scene. On the other hand, cortical area V4 was reported to contain mainly neurons that were color responsive.

Also supporting modularity, Zeki's later work analyzed the differences between V4 and V1 (Zeki, 1983a,b), comparing color responses in the monkey cortex with human color perception. In those experiments wavelength distribution and perceived color were dissociated by surrounding context. Zeki (1983a,b) reported that a fraction of V1 cells responded to the wavelength distribution of a target in their receptive fields, but only V4 cells responded to the perceived color of the target in experiments in which context was manipulated. Based on these results and his earlier work on the functional specificity of anatomical areas, Zeki proposed that V4 was a color center in the monkey brain.

The famous studies of V1 and V2 cortex in monkeys by Livingstone and Hubel (1984, 1988; Hubel & Livingstone, 1987) supported Zeki's modular concept and linked it to parallel processing of color and form in the retina and LGN. Livingstone and Hubel (1984, 1987, 1988) proposed a tripartite model in which signals from parvocellular LGN were further subdivided in V1 cortex into two separate streams, one for color processing, localized in the CO patches (Horton & Hubel, 1981; Wong-Riley, 1979) or "blobs," and one for form processing in the interpatch regions of layers 2/3—layers that provide a substantial part of the output of V1 to extrastriate visual areas (figure 40.1C). The Livingstone-Hubel hypothesis projected the parallel parvo and magno streams out of V1 into the three CO stripe compartments in V2 cortex (see Federer et al., 2009; Sincich & Horton, 2005; Sincich, Jocson, & Horton, 2010, for anatomical studies on compartmentalized V1-to-V2 projections).

One specific component of the Livingstone-Hubel tripartite model that we focus on in this chapter is the proposed *double-opponent* cell. Such a hypothesized neuron is double-opponent because it adds up opposite-signed inputs from different cones (cone opponency) and also opposite-signed inputs from different locations in the cell's receptive field (spatial opponency). The defining characteristic of a double-opponent cell is that it is strongly responsive to color patterns but weakly or nonresponsive to full-field color stimuli or shallow gradients of color.

Livingstone and Hubel (1984, 1988) reported that most of the color-responsive cells in the CO patches were double-opponent cells strongly responsive to localized color bars but insensitive to full-field color stimuli. The segregated-color channel from CO patches to V2 thin stripes was therefore conceived as carrying color-contrast signals from the double-opponent cells in CO-rich patches. Besides their sensitivity to color contrast, another feature of the double-opponent cells in the CO patches described by Livingstone and Hubel (1984; Hubel & Livingstone, 1987) was that the cells' receptive fields were roughly circularly symmetric and that the cells responded at all orientations of a (colored) bar. The lack of orientation tuning of the double-opponent color cells in the CO patches was an important part of the role they were hypothesized to play in the tripartite model (cf. Sincich & Horton, 2005).

NEW DIRECTIONS

More recently, there has been a shift away from the segregated-color view toward the view that neural signals about color are more integrated with form-related neural signals in the cerebral cortex (Gegenfurtner, 2003; Lennie, 1999; Shapley & Hawken, 2002, 2011; see chapter 41 by Kiper and Gegenfurtner). Neurophysiological investigations of this problem were strongly influenced by psychophysical and perceptual results that showed there were strong reciprocal interactions between color and form, and form and color.

Color and Form

Color has been found to influence the perception of form. One example is given by psychophysical results on orientation-dependent and spatial-frequency-dependent masking of equiluminant red–green sine grating patterns and black–white luminance gratings. The masking results suggest that the neural mechanisms of color detection are selective for spatial frequency and orientation, almost as much as the spatial channels for luminance detection (Beaudot & Mullen, 2005; Losada & Mullen, 1994; Pandey-Vimal, 1997; Switkes, Bradley, & De Valois, 1988). Analogous experiments on spatial-frequency adaptation revealed spatial-frequency-tuned adaptable detection mechanisms for both luminance and color patterns (Bradley, Switkes, & De Valois, 1988). Thus, even though the contrast sensitivity function for equiluminant colored patterns was found to be spatially low-pass (Mullen, 1985), the adaptation and masking results implied that the low-pass contrast sensitivity function was an envelope of many narrow-band, tuned, spatial–chromatic mechanisms. Additional compelling evidence for chromatic signals carrying spatial information comes from results on orientation discrimination, which was found to be almost as fine with red–green equiluminant patterns as with luminance patterns (Webster, De Valois, & Switkes, 1990). Moreover, the tilt illusion was found to be a strong effect for equiluminant as well as luminance patterns (Clifford et al., 2003). Even the Ebbinghaus, Müller-Lyer, Poggendorff, Ponzo, and Zöllner illusions were present with red–green equiluminant stimuli as long as the thickness of the colored lines in the illusory figures was large enough (Hamburger, Hansen, & Gegenfurtner, 2007).

Blue–yellow signals about visual patterns may be processed differently from red–green and black–white signals (Brainard & Williams, 1993). The S-cone array provides a large component of blue–yellow perception, and the S-cone array is much sparser than the L- and M-cone arrays in the retina. Also, blue–yellow signals have a parallel route into V1 through the koniocellular pathway, and distinct intracortical processing of blue–yellow signals is a possibility discussed below. For these reasons there needs to be more investigation in the future about the role of blue–yellow signals in form perception.

Form and Color

There is extensive evidence for the reciprocal influence of form on color perception. One example is filling-in of color in stabilized images across long distances from an unstabilized boundary (Yarbus, 1967). Color filling-in in the periphery of the visual field can be seen even with voluntary fixation (Krauskopf, 1963). These perceptual results suggest that the color appearance of a region may be more dependent on color contrast at the boundary of the region than it is on the spectral reflectance of the region's interior. The linkage between color and form helps to clarify how the spatial and chromatic receptive field organizations of cortical neurons work together to produce the perception of color.

A familiar and very important property of color perception is its sensitivity to spatial context. The strong influence of surrounding regions on the color of a target region has been known for a very long time (as in the case of Monge in the eighteenth century; see Mollon, 2006). Color and form interact through the contrast that surrounding forms exert on the target. Color contrast effects are assumed to be involved in the important subject of color constancy (Brainard, 2004). It has been suggested that the neural substrate for color constancy is the population of orientation-selective double-opponent neurons in V1 cortex (see the review by Gegenfurtner, 2003).

SINGLE- AND DOUBLE-OPPONENT CELLS IN V1

More recent research on the neurophysiology of the visual cortex has led to an understanding of cortical color processing very different from the modular view (reviewed in Gegenfurtner, 2003; Shapley & Hawken, 2002, 2011). It is important to state up front one major set of results that has guided the new way of thinking:

namely that V1 plays an important role in color perception through the combined activity of two kinds of color-sensitive cortical neurons, single- and double-opponent cells.

The receptive field organization that we refer to as double-opponent (Shapley & Hawken, 2002) was put forward as a result of findings that many neurons in V1 were spatially selective to chromatic patterns, a spatial selectivity that included orientation selectivity (Johnson, Hawken, & Shapley, 2001, 2004, 2008; Lennie, Krauskopf, & Sclar, 1990; Solomon & Lennie 2005; Thorell, De Valois, & Albrecht, 1984; also reviewed in Shapley & Hawken, 2011; Solomon & Lennie, 2007). Most orientation-selective double-opponent neurons respond well to luminance and to chromatic edges (Johnson, Hawken, & Shapley, 2008; Thorell, De Valois, & Albrecht, 1984). Many of the neurons that strongly prefer chromatic targets rather than achromatic targets are single-opponent in their chromatic receptive field organization, often with balanced antagonistic regions (Johnson, Hawken, & Shapley, 2001, 2008).

Single-opponent and double-opponent cells have different functions. Single-opponent cells respond to large areas of color and to the interiors of large patches. Double-opponent cells respond to color patterns, textures, and color boundaries. For perceiving the color of objects and for understanding color pictures, the double-opponent cells are probably very important. For responding to color illumination and color atmospheres, single-opponent cells have an advantage. Color contrast is an aspect of perception that likely depends on double-opponent cells. Double-opponent cells in V1 are likely to be the neural substrate of the spatially tuned chromatic channels that were revealed in psychophysical masking and adaptation studies with equiluminant patterns (Beaudot & Mullen, 2005; Bradley, Switkes, & De Valois, 1988; Losada & Mullen, 1994; Pandey-Vimal, 1997; Switkes, Bradley, & De Valois, 1988). Color assimilation and color-spreading (De Valois & De Valois, 1988; Jameson & Hurvich, 1975) are perceptions that may derive from the activity of single-opponent cells. Now we consider major issues that have led to new directions in research on the response to color in the primary visual cortex, V1, and the role of V1 in color vision.

THE RESPONSIVENESS OF V1 CORTEX TO
COLOR AND TO BLACK AND WHITE

The first issue we consider is the relative responsiveness of the primary visual cortex, V1, to color versus black–white stimuli. De Valois (1965) and Wiesel and Hubel (1966), among others, had established that a large

   ROBERT SHAPLEY, MICHAEL HAWKEN, AND ELIZABETH JOHNSON

signal about color was coming into V1 via the parvocellular pathway. Derrington, Krauskopf, and Lennie (1984) studied parvocellular LGN cells in detail and found that almost all parvocellular cells were cone-opponent and color-responsive. However, it was known that parvocellular cells could respond to black and white, especially to patterned black–white stimuli such as bars, edges, and grating patterns, as shown by De Valois and Pease (1971).

Some early studies of V1 indicated that most V1 cells responded better to black–white stimuli than to color stimuli. For example, Hubel and Wiesel (1968) reported that over 80% of V1 cells responded to spatial patterns, independent of color. Lennie, Krauskopf, and Sclar (1990) reported that most V1 cells they studied preferred achromatic stimuli, in agreement with Hubel and Wiesel (1968). However, other early studies revealed that a substantial fraction of V1 cells were sensitive to the color of the stimulus (Dow & Gouras, 1973; Vautin & Dow, 1985). Thorell, De Valois, and Albrecht (1984) found that almost 80% of macaque V1 cells were color-selective.

More recently, many different investigators reported that there were many color-responsive neurons in V1 cortex. Victor et al. (1994), recording local field potentials (LFPs) in macaque V1 with multielectrode probes, found LFP responses to equiluminant red–green modulation both in upper and lower cortical layers at a majority of recording sites. Leventhal et al. (1995) found that a large fraction of cells in the upper layers of macaque V1 were color sensitive. Single-unit studies undertaken in the twenty-first century benefited from the insights derived from fMRI studies of human V1 that indicated that human V1 responded very strongly to red–green color contrast (Engel, Zhang, & Wandell, 1997; reviewed below). In our own studies we estimated that about 40% of all macaque V1 cells were color selective, but this percentage rose to 60% in layer 2/3 (Johnson, Hawken, & Shapley, 2001, 2004, 2008). Friedman, Zhou, and von der Heydt (2003) found a very similar percentage, 64%, of color-selective cells in layer 2/3 of macaque V1.

Studies of human V1 with fMRI had a big influence on how scientists think about color processing in V1. Many fMRI results comparing color and luminance responses in human V1 indicated that color responses were comparable to achromatic responses. For instance, Engel, Zhang, and Wandell (1997) found that the strongest fMRI modulation in human V1 was caused by red–green modulation that evoked opposite-signed L- and M-cone responses (figure 40.2). The data were plotted as criterion response contours in cone contrast space. V1 response contours were approximately elliptical with the minor axis of the ellipse oriented in the L–M direction (figure 40.2), meaning that response/cone contrast was greatest in the opponent L–M direction and least in the nonopponent L+M direction. These results supported the concept that human V1 was strongly responsive for color, a result that has been replicated in many different laboratories (Beauchamp et al., 1999; Brewer et al., 2005; Hadjikhani et al., 1998; Kleinschmidt et al., 1996; McKeefry & Zeki, 1997; Mullen et al., 2007; Wade et al., 2008). One might question what the fMRI signal is measuring because the relation between fMRI and neuronal activity is still a controversial topic. But there is a good argument for taking these fMRI results seriously because it has been established that the magnitudes of the fMRI signals are highly correlated with human behavioral performance in pattern detection (Engel, Zhang, & Wandell, 1997) and adaptation to color contrast (Engel & Furmanski, 2001). Another issue is whether cone contrast is the correct metric for comparing black–white versus color sensitivity. Recently, McDermott and Webster (2012) showed that if one uses multiples of discrimination threshold as a metric, then black–white and color responsiveness are roughly balanced in human observers, and they suggest their results may be explained by contrast adaptation to the gamut of color contrast and black–white contrast in the natural environment.

COLOR AND ACHROMATIC CELLS IN V1 CORTEX

How specialized are V1 neurons for color? The modular viewpoint would predict that the only cells that contribute to color perception are cells highly specialized for color detection and highly selective for different colors (reviewed in Gegenfurtner & Kiper, 2003). An exemplar of this approach is the study of color responses in V1 and V4 neurons by Zeki (1983a,b). Zeki selected from the population of V1 neurons only those cells that gave a bigger response to color than to black–white stimuli. Then he found that the population of cells he sampled was not really color selective but, rather, wavelength selective, as described earlier. This is the line of reasoning that led Zeki to conclude that V1 was not really responding to color.

More recent studies also focused on V1 cells that respond much more to color than to achromatic stimuli (Conway 2001; Conway, Hubel, & Livingstone, 2002; Conway & Livingstone, 2006). Here, as in Zeki's (1983a,b) studies, neurons were selected for investigation if and only if they responded best to color. The criterion used to screen cells for color selectivity in Conway (2001) and Conway and Livingstone (2006) was

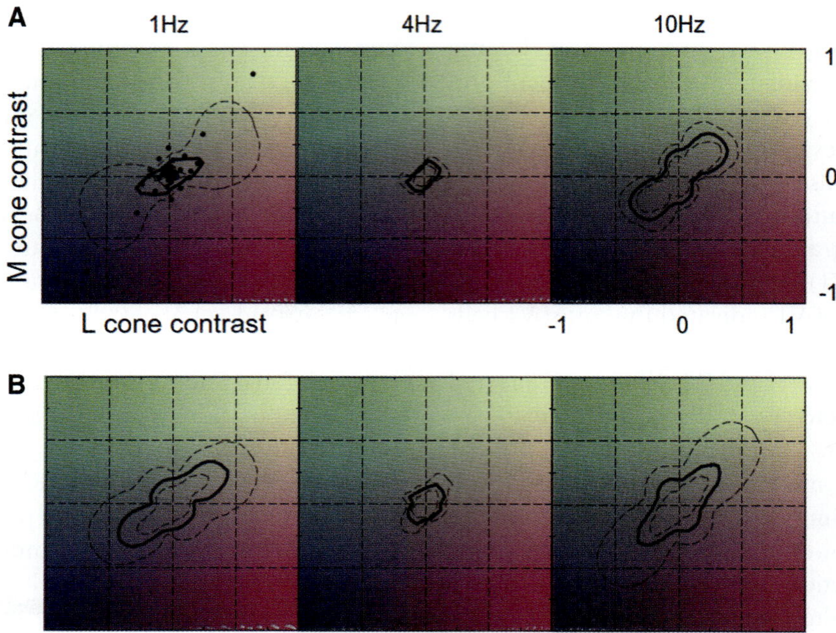

FIGURE 40.2 Human fMRI responses in cone contrast space (from Engel et al., 1997). The amount of contrast required to reach a criterion response is plotted for 18 different stimulus color directions—shown by the small dots in the panel labeled 1 Hz in A. Each of the panels shows the responses modulated at three different temporal frequencies: 1, 4, and 10 Hz for two subjects. The data for BW are shown in the row marked A, for SE in the row marked B. The shape of the threshold contour is reciprocal to the relative strength of the response in different color directions. When the contour is close to the origin, it means that the amount of contrast required to evoke a criterion response was low. If the points on the negative diagonal are closer to the origin than the points on the positive diagonal, it means that the opponent L–M mechanism is more sensitive than the luminance L+M mechanism. The L–M opponent mechanism is more sensitive than the luminance mechanism for 1 and 4 Hz. Similar results were obtained in psychophysical experiments.

opposite signs of response to the same type of cone-isolating stimuli used in the main experiments: so-called sparse noise (Reid, Victor, & Shapley, 1997), that is, spots flashed briefly at random locations in the visual field. The selection of neurons for study by Conway (2001), Conway, Hubel, and Livingstone (2002), and Conway and Livingstone (2006) were motivated by the modular point of view. Their idea was that the only cells that contribute to color perception were those that were highly specific in responding to color and not to achromatic visual stimuli.

Other investigators found many color-responsive neurons in V1 that were also selective for spatial patterns. Thorell, De Valois, and Albrecht (1984) reported the existence of many V1 neurons that were responsive to equiluminant color stimuli and also tuned for spatial frequency. Lennie, Krauskopf, and Sclar (1990) replicated and extended Thorell, De Valois, and Albrecht's results, studying V1 neuron responses to sinusoidal grating patterns that were formed by achromatic or color contrast. Their stimuli varied over a range of directions in DKL color space (Derrington, Krauskopf, & Lennie, 1984), a space in which black–white (achromatic) is much higher modulation in cone contrast

than is red–green (equiluminant) modulation. Lennie, Krauskopf, and Sclar studied the properties of all neurons they could record in V1 cortex. They found many neurons that were spatial-frequency-tuned and that also responded to both chromatic and achromatic stimuli. Most of the neurons they studied that were spatial-frequency-tuned were classified as cone-opponent. They found a small population that responded much more strongly to equiluminant color than to black–white grating patterns, and these were most responsive at low spatial frequencies. Lennie, Krauskopf, and Sclar (1990) hypothesized, as did Zeki (1983a,b), that the only V1 neurons that were important for color perception were the small percentage of neurons that strongly preferred equiluminant stimuli. The idea that color perception depends on the minority of strongly color-preferring neurons was inconsistent with the integrated-color viewpoint articulated by Lennie (1999), although it resurfaced later in the review by Lennie and Movshon (2005).

The integrated-color viewpoint was represented unambiguously by Leventhal et al. (1995). They tested all neurons they encountered in the upper layers (2–4) of macaque V1 with a battery of visual stimuli to test sensitivity for color, direction of motion, orientation, and spatial selectivity. As part of their investigation they compared orientation selectivity and direction selectivity between neurons that were more or less color selective. In retrospect this seems a logical requirement for understanding how specialized the "color" neurons were. The well-known result of Leventhal et al. (1995) was that the neurons they studied were selective on many dimensions. Cells sensitive to color were orientation-selective approximately as much as the cells that were unselective for color (see figures 9 and 11 in Leventhal et al., 1995).

Friedman, Zhou, and von der Heydt (2003) used flashed, uniformly colored, geometric figures as stimuli to study color coding in awake macaque monkeys. They concluded that "contrary to the idea of feature maps, colour, orientation, and edge polarity are multiplexed in cortical signals." They studied the response of cells in V1 upper layers and V2 to squares or bars of color (including neutral gray, white, and black), flashed in different positions with respect to cell receptive fields, on neutral gray backgrounds. Using several different indices of color selectivity, Friedman, Zhou, and von der Heydt (2003) found a substantial proportion (64%) of color-selective neurons in the upper layers of V1, and most of these were edge sensitive. Friedman, Zhou, and von der Heydt (2003) also found a smaller proportion of what they termed color-surface-responsive cells, so-called because these cells were not especially sensitive to edges. They found that the edge-responsive color cells were mostly orientation selective, and that the surface-responding cells were not selective for orientation. They stated, "the vast majority of colour-coded cells are orientation tuned." Friedman, Zhou, and von der Heydt's findings seem quite congruent with Leventhal's about the large numbers of color-responsive cells in layers 2/3 of striate cortex. Also like Leventhal et al. (1995), Friedman, Zhou, and von der Heydt (2003) specifically analyzed the correlation between orientation selectivity and sensitivity to color and found little correlation in upper layers of V1 or in V2 cortex, as illustrated in figure 40.3 (from Friedman, Zhou, & von der Heydt, 2003). In fact, as the authors noted, the weak correlation they found (figure 40.3, top) was slightly positive, meaning that high selectivity for orientation was slightly more likely when there was high sensitivity for color, a result opposite from that expected on the segregated-color or modular hypothesis.

In our own work (Johnson, Hawken, & Shapley, 2001), we studied all visually responsive single neurons that we recorded in macaque monkey V1 and compared color-sensitive and color-insensitive neurons' responses to spatial patterns of color and luminance.

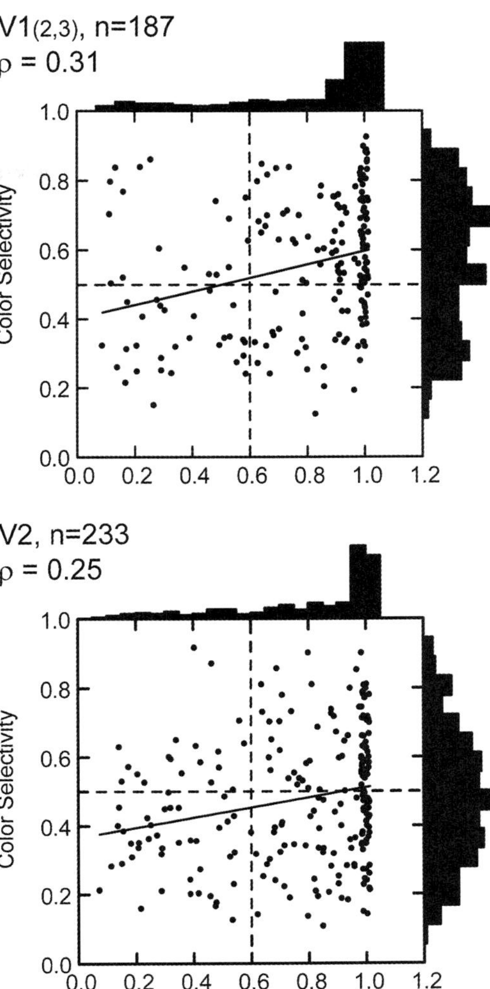

FIGURE 40.3   Color and orientation selectivity in V1 and V2 (from Friedman, Zhou, & von der Heydt, 2003). Selectivity for stimulus orientation (*x*-axis) and for color (*y*-axis) is shown for neurons recorded in V1 and V2 of the awake monkey. The orientation modulation was measured with bars of the optimal color oscillating across the receptive field at 1 Hz. The orientation modulation index was calculated as $[R_{max} - R_{min}]/[R_{max} + R_{min}]$. An orientation modulation index of 1 indicates that there was no response at the orientation 90° to the optimal orientation, and an index close to zero means that the neuron was untuned for orientation. The color selectivity index was calculated as the relative response to 15 flashed bars of different colors. A color selectivity index near one indicates a response to only one of the colored bars; an index close to zero indicates an equal response across all colors. There are many orientation-selective neurons that are also color-selective in both V1 and V2, shown by points in the upper right quadrants of both panels.

We attempted to equate colored and black–white stimuli for average cone contrast so that their effectiveness in driving V1 neurons could be compared quantitatively (in this way following a procedure similar to Thorell, De Valois, & Albrecht, 1984, and one that was also adopted in a more recent study from Solomon & Lennie, 2005). In order to compare relative color sensitivity across the population of neurons, we assigned to each neuron a single number, its sensitivity index, defined as a ratio:

$$I = \max\{\text{equilum response}\}/\max\{\text{lum response}\}$$

The index I was distributed broadly, ranging from 0 to 64. High values indicated preference for colored stimuli compared to achromatic. We divided the population into three groups: luminance-preferring ($I < 0.5$); color-luminance ($0.5 < I < 2$); and color-preferring ($I > 2$) cells. A majority ($60\% = 100/167$) of V1 cells sampled were luminance-preferring, and color-luminance cells were about 29% of the total. The color-preferring cells were only 11% of the cells recorded, in agreement with previous studies. The percentage of color-luminance cells was higher in layer 2/3, where more than 50% of the cells we recorded were color-luminance cells. Solomon and Lennie (2005) later replicated the parcellation of V1 cells into these three groups.

Color-luminance cells were spatially tuned for equiluminant and also for black–white grating patterns (Johnson, Hawken, & Shapley, 2001). In fact, the spatial frequency preference and bandwidth for a color-luminance cell was approximately the same for black–white or red–green equiluminant patterns (cf. Thorell, De Valois, & Albrecht, 1984). Most color-preferring cells were not spatially tuned for equiluminant grating patterns; they preferred the lowest spatial frequency (as found in Lennie, Krauskopf, & Sclar, 1990, and replicated in Solomon & Lennie, 2005). The population-averaged spatial frequency tuning curves for luminance-preferring, color-preferring, and color-luminance cells are drawn in figure 40.4 (from Schluppeck & Engel, 2002). There are (at least) three quantitative observations one can glean from figure 40.4: (1) Color-preferring cells did not respond to red–green grating patterns at higher spatial frequencies (>3 cycles/deg). (2) Spatial frequency tuning of luminance-preferring cells was similar to that of color-luminance cells in preference and bandwidth. (3) Color-luminance cells responded poorly to color (and luminance) patterns at low spatial frequencies (<0.5 cycles/deg). Later, Johnson, Hawken, and Shapley (2004) proposed that most color-luminance cells were in fact double-opponent cells.

Proponents of the modular color view might argue that color-luminance cells could not be important for color perception because they send mixed signals about color and brightness contrast (Conway, Hubel, & Livingstone, 2002; Lennie & Movshon, 2005). Our own

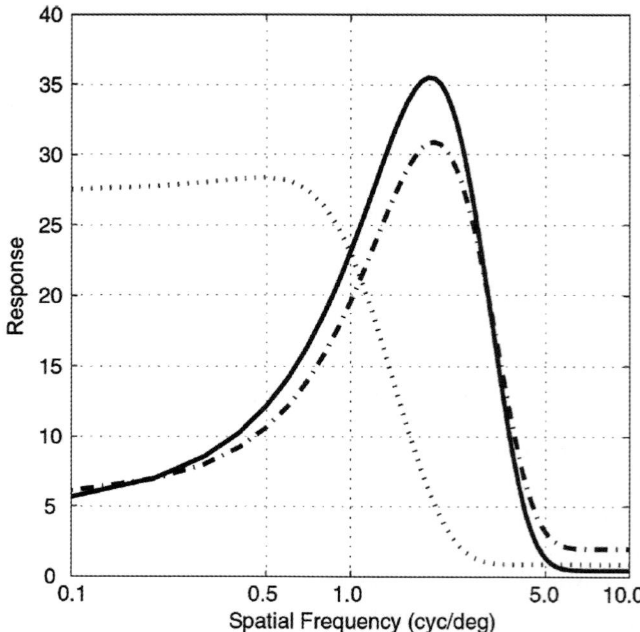

FIGURE 40.4   The average spatial frequency tuning for three populations of V1 neurons. The tuning functions were estimated by Schluppeck and Engel (2002) from 230 neurons recorded by Johnson, Hawken, and Shapley. The dotted line represents the responses of the color-preferring neurons. It shows the characteristic low-pass spatial frequency tuning reported in most studies (Johnson, Hawken, & Shapley, 2001, 2004, 2008; Lennie, Krauskopf, & Sclar, 1990; Solomon, Peirce, & Lennie, 2004; Solomon et al., 2005; Thorell, De Valois, & Albrecht, 1984). The dashed line shows the responses of color-luminance neurons—cells classified as having robust responses to equiluminant color and to black–white luminance when the stimuli are matched for cone contrast. Most of the chromatically opponent color-luminance simple cells are double-opponent in that they have spatially separated chromatically opponent responses to L and M cones (Johnson, Hawken, & Shapley, 2008; see figure 40.5). The solid line is the average spatial frequency tuning of the luminance-preferring neurons. The maximum responses of luminance cells to luminance patterns are more than twice the amplitude of the best response to equiluminance. The tuning of the color-luminance and luminance cells are bandpass and similar in both preferred spatial frequency (2.56 ± 1.26 cycles/deg and 2.09 ± 1.00 cycles/deg, respectively) and in bandwidth (2.05 ± 0.70 octaves full-width, half-height and 2.96 ± 0.69 octaves, respectively).

view is that color and brightness are not as segregated in perception as they are in the modular hypothesis. We base our view on the interaction between brightness contrast and color. For example, measurements of perceived saturation of chromatic induction reach a peak at minimal brightness contrast (Gordon & Shapley, 2006; Kirschmann, 1891). The results on brightness–color interactions do not prove the hypothesis that color-luminance cells are involved in color perception, but they do not rule out the hypothesis. There also are more general considerations that indicate color and brightness contrast might not be analyzed independently in separate modules. Most edges are defined both by color and luminance contrast with a wide variation in the proportions of color and luminance (Hansen & Gegenfurtner, 2009). Also it is quite reasonable to suppose that single cells are not the basis of color perception but rather that there is a population code for color that could extract and decipher what the color is without being "confused" by a multiplexed signal from a single neuron (Lehky & Sejnowski, 1999; Wachtler, Sejnowski, & Albright, 2003). The issue of population coding is discussed below.

Another line of work entirely, the study of human visual evoked potentials (VEPs), also implied that color responses in human V1 cortex were produced by color-sensitive neural mechanisms that were spatially tuned. The biggest VEP evoked by a contrast-reversed equiluminant grating pattern was evoked by a 3–4 cycles/deg grating rather than full-field or low spatial frequency (Rabin et al., 1994; Tobimatsu, Tomoda, & Kato, 1995). The VEP tuning results also suggested that the low-pass color-preferring cells did not contribute a larger VEP signal than the color-luminance cells because if they had, the VEP would have had its maximal response at the lowest spatial frequency. Another VEP experiment on the spatial symmetry of chromatic and achromatic neurons yielded the important result that color-responsive neurons should have the same diversity of spatial symmetry in their receptive fields as non-color-responsive cells, and, in particular, that there ought to be odd-symmetric color-responsive cells (Girard & Morrone, 1995). This prediction from VEP experiments on human observers was completely confirmed in studies of the receptive field properties of single- and double-opponent cells in macaque V1, as discussed next.

SPATIAL RECEPTIVE FIELD PROPERTIES OF COLOR-RESPONSIVE NEURONS IN PRIMARY VISUAL CORTEX

The simplest color receptive field model is the single-opponent cell model (De Valois, 1965; Wiesel & Hubel, 1966), which has been used to explain the properties of LGN cells and retinal ganglion cells that respond to color. As in the LGN, the first tests of the single- and double-opponent receptive field models in V1 cortex came from experiments with grating patterns (Johnson, Hawken, & Shapley, 2001; Thorell, De Valois, & Albrecht, 1984). There were single-opponent cells in V1, and they were the color-preferring cells described earlier (figure 40.4). These cells had low-pass spatial frequency responses to equiluminant red–green grating patterns—like the single-opponent LGN cells studied by De Valois and Pease (1971)—and also low-pass responses to cone-isolating gratings as in the P (midget) ganglion cells and parvocellular LGN cells studied by Lee et al. (2012). However, unlike many single-opponent LGN cells, the V1 single-opponent cells responded poorly to achromatic patterns of higher spatial frequency (Johnson, Hawken, & Shapley, 2001). V1 single-opponent cells, like LGN single opponent cells, had nearly equal but opposite inputs from L and M cones, and some received S-cone input (Johnson, Hawken, & Shapley, 2004). When the receptive fields of V1 single-opponent cells were mapped with cone-isolating stimuli, the result was that their receptive fields appeared to be type II (Wiesel & Hubel, 1966) and roughly circularly symmetric, consistent with their weak to nonexistent orientation selectivity (Johnson, Hawken, & Shapley, 2008).

For double-opponent cells each cone mechanism in the neuron's visual receptive field can be approximated as a difference-of-Gaussians function. That is, each cone input could be positive or negative at different visual field positions. The double-opponent cells in V1 cortex were almost all color-luminance cells. Spatial frequency analysis with cone-isolating stimuli, and with equiluminant stimuli, revealed that most color-luminance cells were tuned for spatial frequency in the 1–3 cycles/deg range (Johnson, Hawken, & Shapley, 2001), and this can be accounted for with double-opponent receptive fields. Furthermore, when the receptive fields of V1 double-opponent cells were mapped with cone-isolating stimuli, we obtained results like those in figure 40.5B, C, where the receptive field subregions were highly elliptical and not circularly symmetric (redrawn from Johnson, Hawken, & Shapley, 2008). The L- and M-cone inputs had elongated, spatially separated increment-excitatory and decrement-excitatory subregions, indicating that there was spatial opponency for each cone. Furthermore, the cone inputs were of opposite sign at each location—therefore cone-opponent everywhere. A cell with such a receptive field would be expected to respond strongly to color patterns and edges but would respond poorly to extended areas of color or to color patterns of low spatial frequency. This was precisely how the color-luminance double-opponent cells responded. Most double-opponent cells were orientation-selective for both achromatic and chromatic stimuli (Johnson, Hawken, & Shapley, 2008). There were a few color-preferring double-opponent cells that responded only weakly to achromatic stimuli (Johnson, Hawken, & Shapley, 2004, 2008), although most had a significant response to achromatic patterns.

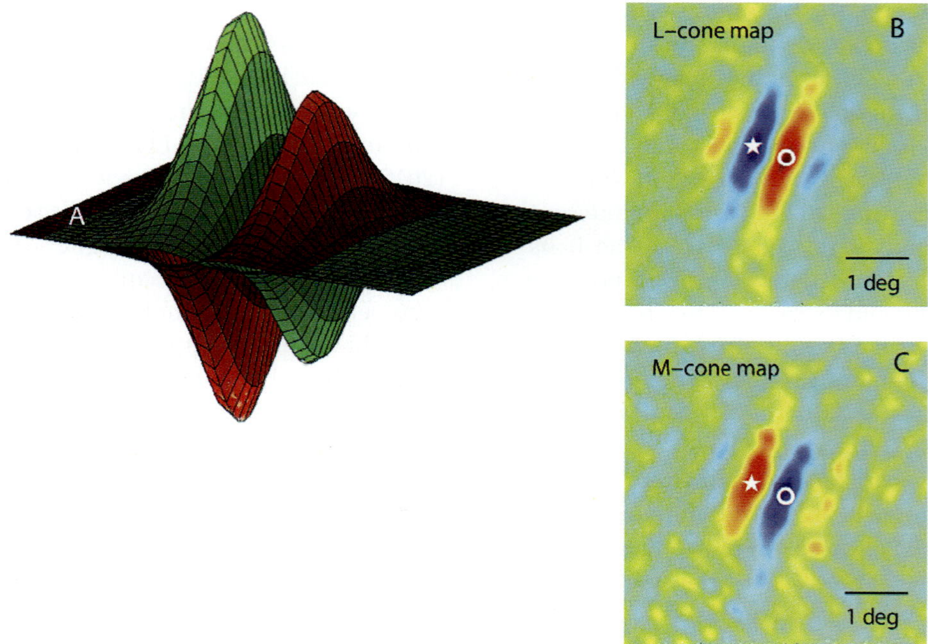

FIGURE 40.5   Double-opponent cells in V1 (from Johnson, Hawken, & Shapley, 2008). The spatial organization of an orientation-selective, spatial-frequency-bandpass, double-opponent neuron's receptive field (after Johnson, Hawken, & Shapley, 2008). (A) A schematic receptive field with side-by-side spatially antagonistic regions with opponent cone weights. The weighting above the horizontal plane is ON, where an increment of light will evoke an increase in response; the weighting below the line is OFF, where a decrement will result in a response. (B) Two-dimensional spatial map obtained from a neuron in V1 by means of the subspace reverse correlation technique (Ringach et al., 1997) with L-cone isolating grating stimuli. (C) Map obtained with M-cone isolating stimuli. At the starred location in B, the L-cone map is decrement excitatory. At the same location in C, the M-cone map is increment excitatory, and vice versa for the locations marked by the open circles. The schematic in A is a three-dimensional representation of the overlay of the two cone maps to give an overall profile. A is not to scale with respect to B and C.

Conway and Livingstone (2006) used a different approach and reported roughly circularly symmetric, roughly even-symmetric, double-opponent cells in macaque cortex. Their experiments involve mapping receptive fields by reverse correlation with cone-isolating stimuli that were flashed colored squares on a gray background. They selected neurons for study that responded well to the flashed colored squares and did not report the results on any of the other cells in their large sample. They stated that some of their double-opponent cells were "weakly orientation-selective."

Johnson, Hawken, and Shapley (2008) reported the orientation selectivity distributions of double-opponent, single-opponent, and nonopponent cells in V1; the double-opponent cells as a population were orientation-selective but not as selective as the nonopponent cells. Most cells that Conway and Livingstone (2006) called double-opponent had very weak surround effects (see their figure 7A). One possible explanation for the weak surrounds is that most of the cells Conway and Livingstone studied were what Johnson, Hawken, and Shapley (2001, 2004, 2008) would have classified as color-preferring, single-opponent cells.

The receptive field models we have been discussing only apply to neurons that act in a quasi-linear manner. Nonlinearities could cause edge responses and grating responses to fail to correspond to predictions from receptive field maps. Although color scientists often have favored linear models of cortical function (reviewed in Gegenfurtner, 2003), cortical neurophysiologists have been cautious about assuming linearity without proof (Friedman, Zhou, & von der Heydt, 2003). It would be desirable in future research to compare responses to multiple stimulus sets to check for the linearity of the cortical network in experiments on color in the cortex because very large nonlinear effects in V1 have been observed in other experiments (Yeh et al., 2009). A majority of double-opponent cells were classified as complex cells (Johnson, Hawken, & Shapley, 2004), but a significant fraction were classified as simple cells, meaning their responses were modulated by the position of the visual pattern in the visual field. Such spatial-phase sensitivity would seem to be a requirement for the double-opponent cells to contribute to perception of the color of a target. However, the fact that a cell is put in the complex cell category does not mean it has no spatial phase sensitivity. Moreover, many cells classified as complex can have robust responses to the contrast sign and position of a visual stimulus (Yeh et al., 2009). Future work on population coding of color (and brightness) by V1 cell populations will be needed to assess the nature and the magnitude of signals about color in the double-opponent population.

SEGREGATION WITHIN THE V1/V2 NETWORK OF NEURONS THAT RESPOND TO COLOR FROM THOSE THAT RESPOND TO FORM

One component of the modular view of cortical color processing was the idea that color-responsive and/or color-selective cells would be segregated within V1 and V2 cortex, within CO patches in V1 and within thin CO stripes in V2 (Livingstone & Hubel, 1984, 1988). This is an area of research that has produced controversial results. Studies that have been done with optical imaging of cortical activity have found clusters of red–green color-selective and color-responsive cells in early visual cortex. Optical imaging results on the clustering of blue–yellow cells in V1 and V2 were negative. Studies employing electrophysiological techniques have yielded mixed results on cortical clustering of red–green neurons. As yet, there has been no definitive resolution of the apparent contradictions.

Results in favor of clustering were obtained in a number of optical imaging studies. Landisman and Ts'o (2002a, 2002b) used optical imaging and microelectrode recording combined in a study of macaque V1. They found a significant degree of overlap between CO patches and regions of heightened color sensitivity, but there were many regions of color response outside CO patches. Moreover, Landisman and Ts'o reported that color domains were usually larger than the CO patches, sometimes covering two neighboring patches and the interpatch region between. Their electrophysiological recordings were targeted to the color domains identified in optical imaging experiments and confirmed that there tended to be more color-selective neurons near CO patches.

Lu and Roe (2008) used similar optical imaging techniques and confirmed the presence of patches of color-selective neurons in V1, but their results implied the color patches were more localized to CO patches than Landisman and Ts'o had found. Lu and Roe (2008) argued that differences between their study and the earlier results of Landisman and Ts'o might have been caused by signal/noise of the optical signal and by how optical signals were averaged. Lu and Roe (2008) also concluded that color selectivity and orientation selectivity were not localized together in macaque V1. However, this conclusion needs to be reconsidered because of the possibility that orientation preference might be changing more rapidly near CO patches than far from them, an idea proposed by Edwards et al. (1995). The low spatial resolution of the optical imaging technique could confound regions of rapid orientation-preference change with regions of low orientation selectivity. As reviewed below, single-unit studies, which have high spatial resolution, usually have not confirmed that neurons in CO patches have poor orientation selectivity (Economides et al., 2011; Lennie, Krauskopf, & Sclar, 1990; Leventhal et al., 1995).

Two related papers by Xiao and colleagues reinforce the idea of segregation and localization of color in V1 and V2 (Xiao, Wang, & Felleman, 2003; Xiao et al., 2007). In their experiments Xiao and colleagues studied the dependence on the hue of the color stimulus of the peak of the evoked response measured by optical imaging. They found systematic hue maps, first in V2 and then in V1 cortex, using differential optical imaging of responses to large squares or full fields of color versus the responses to achromatic grating patterns.

There are issues to consider about optical imaging studies. First, the investigators usually used low-spatial-frequency grating patterns to evoke color responses. They calculated optical images as the difference signal between the response to a low-spatial-frequency color pattern and the response to a high-contrast black–white grating pattern. The choice of spatial frequency probably meant the color stimuli activated the single-opponent color cells selectively, much more than the double-opponent cells. Subtracting the responses to color and achromatic stimuli also biases the measurement of activity to cells that are not equally responsive to both types of stimuli. Second, the image contrast in the optical image could have been a thresholded, nonlinear function of neuronal activity. It is possible that the differences in neuronal response between color and achromatic patterns could have been quite small, and the imaging may have exaggerated small differences.

Single-unit recording studies in V1 have provided evidence against segregation of color cells in different regions of V1 cortex. For instance, Lennie, Krauskopf, and Sclar (1990) asserted that there was no correlation between color preference and location of a cell in a CO patch. Leventhal et al. (1995) stated that there was no relationship between receptive field properties they studied and CO staining in upper layers of V1.

Leventhal et al. (1995) systematically explored CO compartments; they show records of electrode tracks through layer 2/3 of V1, and there was no evidence of clustering of color-sensitive cells or of non-orientation-selective cells in CO patches. Recently Economides et al. (2011) reported that, in array recordings in V1 cortex, there was only a very small difference in orientation selectivity between neurons located in CO patches and those in interpatches, confirming and extending the original findings of Leventhal et al. (1995).

BLUE–YELLOW COLOR SIGNALS IN THE PRIMARY VISUAL CORTEX

Most research on color in the cortex has involved studies of red–green color vision that is very salient in Old World primates. However, the color pathway that is found most often in mammalian species is the neuronal channel that carries blue–yellow signals from eye to cortex (see review by Jacobs, 2008). The koniocellular pathway in primates was proposed as the vehicle for blue–yellow signals to reach cortex (Hendry & Reid, 2000), and direct proof that S–(L+M) signals are carried by the koniocellular cells was provided for the marmoset LGN by Martin et al. (1997) and for the macaque LGN by Roy et al. (2009). Chatterjee and Callaway (2003) measured the blue–yellow input to layer 4A/3B in their study of the laminar pattern of afferent LGN input to V1 in macaques: anatomically the afferent input from the LGN to layer 4A is associated with the honeycomb pattern of dense afferent terminals in layer 4A that is also evident in CO-stained cortex. In human primary visual cortex there is no prominent layer 4A in CO (Horton & Hedley-Whyte, 1984) or vGluT2-stained sections (Garcia-Marin et al., 2013) in contrast to the pattern observed in monkey. Currently it is not known how the blue–yellow input to human cortex is organized, but it is unlikely to be similar to the pattern observed in Old World monkeys. Buzas et al. (2008) studied the laminar distribution of S-cone-driven responses in marmoset V1 and found no evidence for clustering of S-cone-driven cells in layer 3 CO patches; rather, the spatial distribution was uniform throughout layer 3.

There is no consensus on the relative contribution of S-cone-driven color signals in V1. Some single-unit studies in macaque V1 report relatively weak S-cone input, commensurate with the relative frequency of recording S-cone single-opponent cells in the LGN (Johnson, Hawken, & Shapley, 2004; Solomon & Lennie, 2005), but De Valois et al. (2000) reported an enhanced S-cone input in V1. fMRI studies of human V1 also disagree on this point, with Liu and Wandell (2005) reporting relatively weak S-cone-driven responses while Mullen et al. (2007) report stronger S-cone activity that was approximately as strong as the L–M signal in V1.

Johnson, Van Hooser, and Fitzpatrick (2010) reported studies of S-cone-driven signals in the V1 cortex of tree shrews with the technique of intrinsic signal optical imaging and the higher-resolution method of two-photon imaging. Tree shrews, like many nonprimate mammals, have dichromatic vision mainly supported by signals from cells that compute S–L signals from the two cone photoreceptors available. Johnson, Van Hooser, and Fitzpatrick (2010) found that tree shrew V1 cells that received S-cone input could be color-opponent or not and could be orientation-selective or not (figures 40.6 and 40.7). The degree of color selectivity was not correlated with orientation selectivity, consistent with the results of Friedman, Zhou, and von der Heydt (2003). The results of Johnson, Van Hooser, and Fitzpatrick (2010) are consistent with the idea that S-cone opponent signals are combined with achromatic signals in cortical double-opponent cells. Support for this conclusion can also be found in the cone weight plots from macaque V1 in Johnson, Hawken, and Shapley (2004).

COLOR CONTRAST IN THE RESPONSES OF NEURONS IN PRIMARY VISUAL CORTEX

Contrast at the edge between target and surroundings has a big influence on color appearance (Brainard, 2004; Gordon & Shapley, 2006; Katz, 1935; Krauskopf, 1963; Shevell & Kingdom, 2008). Zeki (1983a,b) compared V1 and V4 neurons and their sensitivity to color contrast. He concluded that V1 neurons were not responding to color but rather to wavelength. However, because we now have a better understanding of how color-sensitive neurons in V1 can be divided into single-opponent and double-opponent groups, it is worth reconsidering these conclusions. Zeki selected the V1 neurons on the basis that their response to color was stronger than their response to luminance. Probably these were what we term color-preferring single-opponent neurons. Such a neuron will most likely have a single-opponent receptive field with a low-pass spatial frequency response like that shown in figure 40.4 (dotted line). Therefore, it would likely not be affected by color contrast. However, if Zeki had also examined the responses of orientation-selective double-opponent cells, we conjecture he would have found the responses of some of them linked to the color of the illuminated targets as were the V4 cells he investigated. The edge-sensitive color cells studied by Friedman, Zhou, and von der Heydt (2003) should also respond to color contrast at the edge of a target, based on the conjecture that

    ROBERT SHAPLEY, MICHAEL HAWKEN, AND ELIZABETH JOHNSON

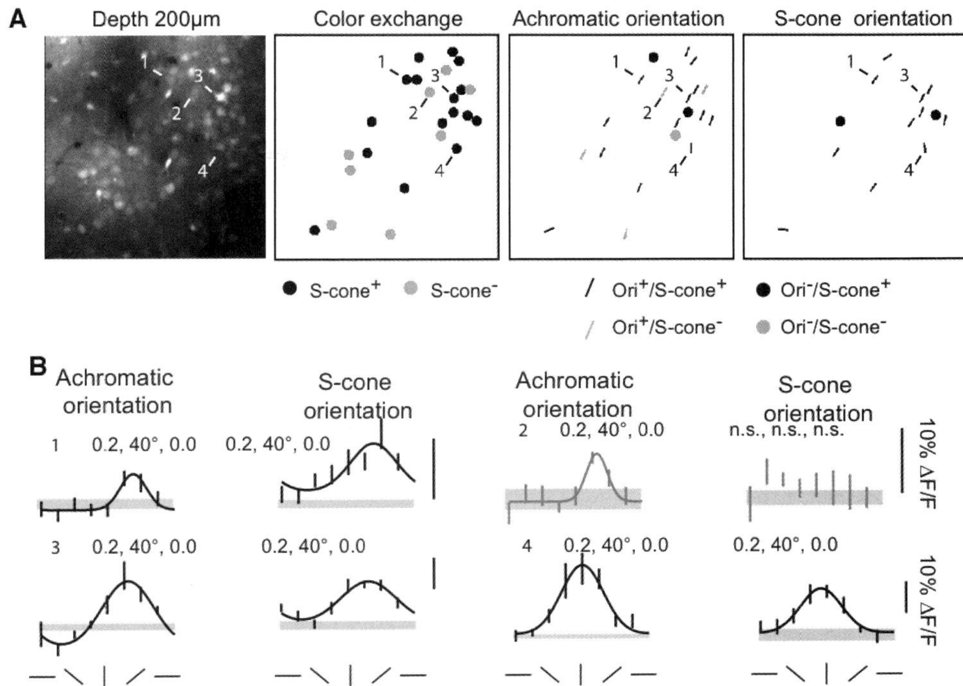

FIGURE 40.6   Tree shrew V1 cells receive S-cone opponent inputs, and these cells are often orientation-selective (after Johnson, Van Hooser, & Fitzpatrick, 2010). Tree shrews are a highly visual dichromatic mammal with cone-dominated retina, strong color vision capabilities, and a well-characterized V1 functional architecture. (A left) Two-photon image of an example field of cells labeled with calcium dye Oregon green BAPTA-1. (A, middle left) Schematic of responses to an S-cone isolating stimulus. Dark circles are neurons that exhibited significant S-cone responses, and light circles are neurons that did not respond to S-cone isolating stimuli. (A middle right) Schematic of responses to achromatic orientation. Orientation of the bar matches the orientation preference of each cell. (A right) S-cone isolating orientation responses. (B) Achromatic and S-cone isolating orientation tuning curves for the four cells indicated in A. Circular variance (CV), tuning width (half-width at half-height), and orthogonal-to-preferred (O/P) ratio are shown for each curve.

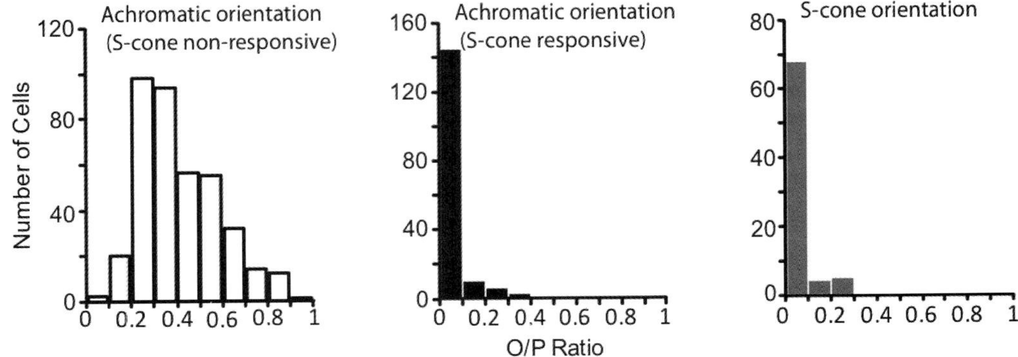

FIGURE 40.7   Histograms of orthogonal-to-preferred (O/P) ratio of orientation responses for tree shrew V1 cells (after Johnson, Van Hooser, & Fitzpatrick, 2010). This ratio indicates how the orientation tuning curve flanks compare to the peak response. The majority of S-cone-responsive neurons were orientation-selective, whether measured with S-cone isolating or achromatic stimuli ($N = 164$, S-cone responsive, achromatic orientation; $N = 383$, S-cone nonresponsive, achromatic orientation; $N = 77$, S-cone responsive, S-cone isolating orientation).

most edge-sensitive color cells were from the same population that we termed double-opponent cells. In other words it is plausible that the neural basis of color contrast perception (including color constancy) begins in V1, not in extrastriate cortex. This plausible prediction is testable.

Wachtler, Sejnowski, and Albright (2003) found that V1 neurons' responses could be affected by color contrast. They studied V1 in the awake monkey with single-cell recording. Their stimuli were colored squares centered on a cell's receptive field, with the squares chosen to be at least two times wider and longer than the receptive field. The stimuli were stationary and flashed for 500 ms. Wachtler, Sejnowski, and Albright (2003) measured the effect of changing the surroundings of the stimulus square from midgray to different colors systematically arranged around the color circle and found small but significant shifts in color preference of the cells that followed the shift in surrounding color. The shifts were in the same direction as perceptual hue shifts caused by color induction from a surround. This was a very significant finding of color contrast in V1, but we do not yet know whether or not the chromatic shifts might have been much bigger if stimuli had been optimized. For instance, if instead of centering a big square on the receptive field, Wachtler and colleagues had measured the response of a double-opponent cell near the boundary of the target square, perhaps the response would have been much more affected by color contrast. However, Solomon, Peirce, and Lennie (2004) reported a negative result for a similar experiment in which they measured tuning direction in DKL space as a function of the color of a surrounding region and found no effect.

One innovation introduced by Wachtler, Sejnowski, and Albright (2003) may be important for future research. They calculated the neuronal basis for perceived hue as the response of a population of V1 neurons in analogy with the population vector approach of Georgopoulos, Schwartz, and Kettner (1986). This was a different way of thinking about what the cortex was doing in its representation of a hue and moved us away from a strict single-neuron viewpoint. In studying color representation in the cortex many neuroscientists have adopted the single-neuron viewpoint only, and that may have prevented us from seeing the forest for the trees. Visual perception could tap many populations of neurons in the visual cortex, and it could make use of many possible computations to extract percepts. The vector average computation is perhaps not the only way population activity can be used to reckon hue or saturation or brightness. For instance, Lehky and Sejnowski (1999) suggested that nonlinear combination of population signals might be utilized to compute the hue of a colored region. Further exploration of the possibilities of population coding for color perception may lead to a new understanding of color in the cortex and of visual perception as an outcome of cortical population activity.

COLOR AND NATURAL IMAGE STATISTICS

We conclude not with a summary but with a review of some theoretical work that illustrates the complexity and controversy of studying color vision. This is work in the spirit of the times, about the statistical nature of the world and how the visual system is matched to the statistics of natural scenes. Two major studies have been done, one by Tailor, Finkel, and Buchsbaum (2000) and one by Caywood, Willmore, and Tolhurst (2004). The earlier Tailor study concluded that the optimal independent components that match color and pattern in natural scenes would have separate and independent color and spatial channels, rather like the ideas of the modularists. The later work by Caywood, Willmore, and Tolhurst (2004) offered technical criticism of Tailor and colleagues and concluded that optimal filters in V1 would resemble the population of single- and (orientation-selective) double-opponent cells that we have reported (Johnson, Hawken, and Shapley, 2004, 2008).

CONCLUSION

Vision scientists have studied color perception, often under conditions where form is minimal. It is certainly possible to perceive colors under such reduced conditions, and such colors have even been given a name, "aperture colors" (Katz, 1935). The idea that color is analyzed separately by the brain is deeply ingrained. It is natural for us as human beings to think of color as separate and apart.

The main point of this chapter is that color is not separate and apart. Rather, color and form and motion are linked inextricably in V1 signal processing, as properties of objects in visual perception (cf. Lennie, 1999; Wallach, 1935, translated as Wuerger, Shapley, & Rubin, 1996). The famous psychologist Gaetano Kanizsa was an eloquent advocate of this viewpoint; he wrote ". . . space and color are not distinct elements but, rather, are interdependent aspects of a unitary process of perceptual organization" (Kanizsa, 1979).

ACKNOWLEDGMENTS

The preparation of this chapter was supported in part by grants from the U.S. National Eye Institute, EY01472,

    ROBERT SHAPLEY, MICHAEL HAWKEN, AND ELIZABETH JOHNSON

EY08300, and EY17945, and from the National Science Foundation, 0745253.

REFERENCES

Beauchamp, M. S., Haxby, J. V., Jennings, J. E., & DeYoe, E. A. (1999). An fMRI version of the Farnsworth–Munsell 100-Hue test reveals multiple color-selective areas in human ventral occipitotemporal cortex. *Cerebral Cortex, 9,* 257–263.

Beaudot, W. H., & Mullen, K. T. (2005). Orientation selectivity in luminance and color vision assessed using 2-d bandpass filtered spatial noise. *Vision Research, 45,* 687–696.

Benardete, E. A., & Kaplan, E. (1999). Dynamics of primate P retinal ganglion cells: Responses to chromatic and achromatic stimuli. *Journal of Physiology, 519,* 775–790.

Bradley, A., Switkes, E., & De Valois, K. (1988). Orientation and spatial frequency selectivity of adaptation to color and luminance gratings. *Vision Research, 28,* 841–856.

Brainard, D. (2004). Color constancy. In L. Chalupa & J. Werner (Eds.), *Visual neuroscience* (pp. 948–961). Cambridge, MA: MIT Press.

Brainard, D. H., & Williams, D. R. (1993). Spatial reconstruction of signals from short-wavelength cones. *Vision Research, 33,* 105–116.

Brewer, A. A., Liu, J., Wade, A. R., & Wandell, B. A. (2005). Visual field maps and stimulus selectivity in human ventral occipital cortex. *Nature Neuroscience, 8,* 1102–1109.

Buzas, P., Szmajda, B. A., Hashemi-Nezhad, M., Dreher, B., & Martin, P. R. (2008). Color signals in the primary visual cortex of marmosets. *Journal of Vision, 8*(10), 7, 1–16. doi:10.1167/8.10.7.

Casagrande, V. A. (1994). A third parallel visual pathway to primate area V1. *Trends in Neurosciences, 17,* 305–310. doi:10.1016/0166-2236(94)90065-5.

Casagrande, V. A., Yazar, F., Jones, K. D., & Ding, Y. (2007). The morphology of the koniocellular axon pathway in the macaque monkey. *Cerebral Cortex, 17,* 2334–2345.

Caywood, M., Willmore, B., & Tolhurst, D. (2004). Independent components of color natural scenes resemble V1 neurons in their spatial and color tuning. *Journal of Neurophysiology, 91,* 2859–2873.

Chatterjee, S., & Callaway, E. M. (2003). Parallel colour-opponent pathways to primary visual cortex. *Nature, 426,* 667–668.

Clifford, C. W., Spehar, B., Solomon, S. G., Martin, P. R., & Zaidi, Q. (2003). Interactions between color and luminance in the perception of orientation. *Journal of Vision, 3,* 106–115. doi:10.1167/3.2.1.

Conway, B. R. (2001). Spatial structure of cone inputs to color cells in alert macaque primary visual cortex (V1). *Journal of Neuroscience, 21,* 2768–2783.

Conway, B. R., Hubel, D. H., & Livingstone, M. S. (2002). Color contrast in macaque V1. *Cerebral Cortex, 12,* 915–925.

Conway, B. R., & Livingstone, M. S. (2006). Spatial and temporal properties of cone signals in alert macaque primary visual cortex. *Journal of Neuroscience, 26,* 10826–10846.

Crook, J. M., Lee, B. B., Tigwell, D. A., & Valberg, A. (1987). Thresholds to chromatic spots of cells in the macaque geniculate nucleus as compared to detection sensitivity in man. *Journal of Physiology, 392,* 193–211.

Derrington, A. M., Krauskopf, J., & Lennie, P. (1984). Chromatic mechanisms in lateral geniculate nucleus of macaque. *Journal of Physiology, 357,* 241–265.

De Valois, R. L. (1965). Analysis and coding of color vision in the primate visual system. *Cold Spring Harbor Symposia on Quantitative Biology, 38,* 567–580. doi:10.1101/SQB.1965.030.01.055.

De Valois, R. L., Cottaris, N. P., Elfar, S. D., Mahon, L. E., & Wilson, J. A. (2000). Some transformations of color information from lateral geniculate nucleus to striate cortex. *Proceedings of the National Academy of Sciences of the United States of America, 97*(9), 4997–5002. doi:10.1073/pnas.97.9.4997.

De Valois, R. L. & De Valois, K. K. (1988). *Spatial vision* (pp. 228–230). New York: Oxford University Press.

De Valois, R. L. & Pease, P. L. (1971). Contours and contrast: Responses of monkey lateral geniculate nucleus cells to luminance and colour figures. *Science, 171,* 694–696.

Dow, B. M., & Gouras, P. (1973). Color and spatial specificity of single units in rhesus monkey foveal striate cortex. *Journal of Neurophysiology, 36,* 79–100.

Economides, J. R., Sincich, L. C., Adams, D. L., & Horton, J. C. (2011). Orientation tuning of cytochrome oxidase patches in macaque primary visual cortex. *Nature Neuroscience, 14,* 1574–1580.

Edwards, D. P., Purpura, K. P., & Kaplan, E. (1995). Contrast sensitivity and spatial frequency response of primate cortical neurons in and around the cytochrome oxidase blobs. *Vision Research, 35,* 1501–1523.

Engel, S. A., & Furmanski, C. S. (2001). Selective adaptation to color contrast in human primary visual cortex. *Journal of Neuroscience, 21,* 3949–3954.

Engel, S., Zhang, X., & Wandell, B. (1997). Colour tuning in human visual cortex measured with functional magnetic resonance imaging. *Nature, 388,* 68–71.

Federer, F., Ichida, J. M., Jeffs, J., Schiessl, I., McLoughlin, N., & Angelucci, A. (2009). Four projection streams from primate V1 to the cytochrome oxidase stripes of V2. *Journal of Neuroscience, 29,* 15455–15471.

Friedman, H. S., Zhou, H., & von der Heydt, R. (2003). The coding of uniform colour figures in monkey visual cortex. *Journal of Physiology, 548,* 593–613.

Garcia-Marin, V., Ahmed, T., Afzal, Y., & Hawken, M. J. (2013) The distribution of the vesicular glutamate transporter (VGluT2) in the primary visual cortex of macaque and human. *Journal of Comparative Neurology. 521,* 130–151.

Gegenfurtner, K. R. (2003). Cortical mechanisms of colour vision. *Nature Reviews Neuroscience, 4,* 563–572.

Gegenfurtner, K. R., & Kiper, D. C. (2003). Color vision. *Annual Review of Neuroscience, 26,* 181–206.

Georgopoulos, A. P., Schwartz, A. B., & Kettner, R. E. (1986). Neuronal population coding of movement direction. *Science, 233,* 1416–1419.

Girard, P., & Morrone, M. C. (1995). Spatial structure of chromatically opponent receptive fields in the human visual system. *Visual Neuroscience, 12,* 103–116.

Gordon, J., & Shapley, R. (2006). Brightness contrast inhibits color induction: Evidence for a new kind of color theory. *Spatial Vision, 19,* 133–146.

Hadjikhani, N., Liu, A. K., Dale, A. M., Cavanagh, P., & Tootell, R. B. (1998). Retinotopy and color sensitivity in visual cortical area V8. *Nature Neuroscience, 1,* 235–241.

Hamburger, K., Hansen, T., & Gegenfurtner, K. R. (2007). Geometric-optical illusions at isoluminance. *Vision Research, 47*, 3276–3285.

Hansen, T., & Gegenfurtner, K. R. (2009). Independence of color and luminance edges in natural scenes. *Visual Neuroscience, 26*, 35–49.

Hendry, S. H., & Reid, R. C. (2000). The koniocellular pathway in primate vision. *Annual Review of Neuroscience, 23*, 127–153.

Hendry, S. H., & Yoshioka, T. (1994). A neurochemically distinct third channel in the macaque dorsal lateral geniculate nucleus. *Science, 264*, 575–577.

Horton, J. C., & Hedley-Whyte, E. T. (1984). Mapping of cytochrome oxidase patches and ocular dominance columns in human visual cortex. *Philosophical Transactions of the Royal Society of London. Series B, Biological Sciences, 304*, 255–272. doi:10.1098/rstb.1984.0022.

Horton, J. C., & Hubel, D. H. (1981). Regular patchy distribution of cytochrome oxidase staining in primary visual cortex of macaque monkey. *Nature, 292*, 762–764.

Hubel, D. H, & Livingstone, M. S. (1987). Segregation of form, color, and stereopsis in primate area 18. *Journal of Neuroscience, 7*, 3378–3415.

Hubel, D. H., & Wiesel, T. N. (1968). Receptive fields and functional architecture of monkey striate cortex. *Journal of Physiology, 195*, 215–243.

Hurlbert, A., & Wolf, K. (2004). Color contrast: A contributory mechanism to color constancy. *Progress in Brain Research, 144*, 147–160.

Hurvich, L. M., & Jameson, D. (1957). An opponent-process theory of color vision. *Psychological Review, 64*, 384–404.

Hyman, J. (2006). *The objective eye.* Chicago: University of Chicago Press.

Jacobs, G. H. (2008). Primate color vision: A comparative perspective. *Visual Neuroscience, 25*, 619–633.

Jameson, D., & Hurvich, L. M. (1975). From contrast to assimilation; in art and in the eye. *Leonardo, 8*, 125–131. doi:10.2307/1572954.

Johnson, E. N., Hawken, M. J., & Shapley, R. (2001). The spatial transformation of color in the primary visual cortex of the macaque monkey. *Nature Neuroscience, 4*, 409–416.

Johnson, E. N., Hawken, M. J., & Shapley, R. (2004). Cone inputs in macaque primary visual cortex. *Journal of Neurophysiology, 91*, 2501–2514.

Johnson, E. N., Hawken, M. J., & Shapley, R. (2008). The orientation selectivity of color-responsive neurons in macaque V1. *Journal of Neuroscience, 28*, 8096–8106.

Johnson, E. N., Van Hooser, S. D., & Fitzpatrick, D. (2010). The representation of S-cone signals in primary visual cortex. *Journal of Neuroscience, 30*, 10337–10350.

Kanizsa, G. (1979). *Organization in perception.* New York: Praeger.

Kaplan, E., & Shapley, R. (1982). X and Y cells in the lateral geniculate nucleus of the macaque monkey. *Journal of Physiology, 330*, 125–143.

Kaplan, E., & Shapley, R. (1986). Two types of ganglion cell in the monkey retina with different contrast sensitivity. *Proceedings of the National Academy of Sciences of the United States of America, 83*(8), 2755–2757. doi:10.1073/pnas.83.8.2755.

Katz, D. (1935). *The world of colour* (R. B. MacLeod & C. W. Fox, Trans.). London: Kegan, Paul, Trench, Truebner and Co.

Kirschmann, A. (1891). Ueber die quantitativen Verhaeltnisse des simultanen Helligkeits- und Farben-contrastes. *Philosophische Studien, 6*, 417–491.

Kleinschmidt, A., Lee, B. B., Requardt, M., & Frahm, J. (1996). Functional mapping of color processing by magnetic resonance imaging of responses to selective P- and M-pathway stimulation. *Experimental Brain Research, 110*, 279–288. doi:10.1007/BF00228558.

Krauskopf, J. (1963). Effect of retinal image stabilization on the appearance of heterochromatic targets. *Journal of the Optical Society of America, 53*, 741–744.

Krauskopf, J., Williams, D. R., & Heeley, D. W. (1982). Cardinal directions of color space. *Vision Research, 22*, 1123–1131.

Landisman, C. E., & Ts'o, D. Y. (2002a). Color processing in macaque striate cortex: Relationships to ocular dominance, cytochrome oxidase, and orientation. *Journal of Neurophysiology, 87*, 3126–3137.

Landisman, C. E., & Ts'o, D. Y. (2002b). Color processing in macaque striate cortex: Electrophysiological properties. *Journal of Neurophysiology, 87*, 3138–3151.

Lee, B. B., Martin, P. R., & Valberg, A. (1989). Sensitivity of macaque retinal ganglion cells to chromatic and luminance flicker. *Journal of Physiology, 414*, 223–243.

Lee, B. B., Shapley, R. M., Hawken, M. J., & Sun, H. (2012). Spatial distributions of cone inputs to cells of the parvocellular pathway investigated with cone-isolating gratings. *Journal of the Optical Society of America. A, Optics, Image Science, and Vision, 29*, A223–A232. doi:10.1364/JOSAA.29.00A223.

Lehky, S. R., & Sejnowski, T. J. (1999). Seeing white: Qualia in the context of decoding population codes. *Neural Computation, 11*, 1261–1280. doi:10.1162/089976699300016232.

Lennie, P. (1999). Color coding in the cortex. In K. Gegenfurtner & L. Sharpe (Eds.), *Color: From genes to perception* (pp. 235–248). Cambridge: Cambridge University Press.

Lennie, P., Krauskopf, J., & Sclar, G. (1990). Chromatic mechanisms in striate cortex of macaque. *Journal of Neuroscience, 10*, 649–669.

Lennie, P., & Movshon, J. A. (2005). Coding of color and form in the geniculostriate visual pathway. *Journal of the Optical Society of America. A, Optics, Image Science, and Vision, 22*, 2013–2033. doi:10.1364/JOSAA.22.002013.

Leventhal, A. G., Thompson, K. G., Liu, D., Zhou, Y., & Ault, S. J. (1995). Concomitant sensitivity to orientation, direction, and color of cells in layers 2, 3, and 4 of monkey striate cortex. *Journal of Neuroscience, 15*, 1808–1818.

Liu, J., & Wandell, B. A. (2005). Specializations for chromatic and temporal signals in human visual cortex. *Journal of Neuroscience, 25*, 3459–3468.

Livingstone, M. S., & Hubel, D. H. (1984). Anatomy and physiology of a color system in the primate visual cortex. *Journal of Neuroscience, 4*, 309–356.

Livingstone, M. S., & Hubel, D. H. (1987). Psychophysical evidence for separate channels for the perception of form, color, movement, and depth. *Journal of Neuroscience, 7*, 3416–3468.

Livingstone, M., & Hubel, D. (1988). Segregation of form, color, movement, and depth: Anatomy, physiology, and perception. *Science, 240*, 740–749.

Losada, M. A., & Mullen, K. T. (1994). The spatial tuning of chromatic mechanisms identified by simultaneous masking. *Vision Research, 34*, 331–341.

Lu, H. D., & Roe, A. W. (2008). Functional organization of color domains in V1 and V2 of macaque monkey revealed by optical imaging. *Cerebral Cortex, 18*, 516–533.

Martin, P. R., White, A. J., Goodchild, A. K., Wilder, H. D., & Sefton, A. E. (1997). Evidence that blue-ON cells are part of the third geniculocortical pathway in primates. *European Journal of Neuroscience, 9*, 1536–1541.

McDermott, K. C., & Webster, M. A. (2012). Uniform color spaces and natural image statistics. *Journal of the Optical Society of America. A, Optics, Image Science, and Vision, 29*, A182–A187. doi:10.1364/JOSAA.29.00A182.

McKeefry, D. J., & Zeki, S. (1997). The position and topography of the human colour centre as revealed by functional magnetic resonance imaging. *Brain, 120*, 2229–2242. doi:10.1093/brain/120.12.2229.

Mollon, J. D. (2006). Monge: The Verriest Lecture, Lyon, July 2005. *Visual Neuroscience, 23*, 297–309. doi:10.1017/S0952523806233479.

Mullen, K. T. (1985). The contrast sensitivity of human colour vision to red-green and blue-yellow chromatic gratings. *Journal of Physiology, 359*, 381–400.

Mullen, K. T., Dumoulin, S. O., McMahon, K. L., de Zubicaray, G. I., & Hess, R. F. (2007). Selectivity of human retinotopic visual cortex to S-cone-opponent, L/M-cone-opponent and achromatic stimulation. *European Journal of Neuroscience, 25*, 491–502.

Pandey Vimal, R. L. (1997). Orientation tuning of the spatial-frequency-tuned mechanisms of the red-green channel. *Journal of the Optical Society of America. A, Optics, Image Science, and Vision, 14*, 2622–2632. doi:10.1364/JOSAA.14.002622.

Parker, A. J. (2007). Binocular depth perception and the cerebral cortex. *Nature Reviews Neuroscience, 8*, 379–391.

Rabin, J., Switkes, E., Crognale, M., Schneck, M. E., & Adams, A. J. (1994). Visual evoked potentials in three-dimensional color space: Correlates of spatio-chromatic processing. *Vision Research, 34*, 2657–2671.

Reid, R. C., & Shapley, R. (1992). Spatial structure of cone inputs to receptive fields in primate lateral geniculate nucleus. *Nature, 356*, 716–718.

Reid, R. C., & Shapley, R. M. (2002). Space and time maps of cone photoreceptor signals in macaque lateral geniculate nucleus. *Journal of Neuroscience, 22*, 6158–6175.

Reid, R. C., Victor, J., & Shapley, R. (1997). The use of m-sequences in the analysis of visual neurons. *Visual Neuroscience, 14*, 1015–1027.

Ringach, D., Carandini, M., Sapiro, G., & Shapley, R. (1997). A subspace reverse correlation method for the study of visual neurons. *Vision Research, 37*, 2455–2464.

Roy, S., Jayakumar, J., Martin, P. R., Dreher, B., Saalmann, Y. B., Hu, D., et al. (2009). Segregation of short-wavelength-sensitive (S) cone signals in the macaque dorsal lateral geniculate nucleus. *European Journal of Neuroscience, 30*, 1517–1526.

Schluppeck, D., & Engel, S. A. (2002). Color opponent neurons in V1: A review and model reconciling results from imaging and single-unit recording. *Journal of Vision, 2*, 480–492. doi:10.1167/2.6.5.

Shapley, R., & Hawken, M. (2002). Neural mechanisms for color perception in the primary visual cortex. *Current Opinion in Neurobiology, 12*, 426–432.

Shapley, R., & Hawken, M. (2011). Color in the cortex, VR50 issue. *Vision Research, 51*, 701–717.

Shevell, S. K., & Kingdom, F. A. (2008). Color in complex scenes. *Annual Review of Psychology, 59*, 143–166.

Sincich, L. C., & Horton, J. C. (2005). The circuitry of V1 and V2: Integration of color, form, and motion. *Annual Review of Neuroscience, 28*, 303–326.

Sincich, L. C., Jocson, C. M., & Horton, J. C. (2010). V1 inter-patch projections to V2 thick stripes and pale stripes. *Journal of Neuroscience, 30*, 6963–6974.

Solomon, S. G., Lee, B. B., White, A. J., Ruttiger, L., & Martin, P. R. (2005). Chromatic organization of ganglion cell receptive fields in the peripheral retina, *Journal of Neuroscience, 25*, 4527–4539.

Solomon, S. G., & Lennie, P. (2005). Chromatic gain controls in visual cortical neurons. *Journal of Neuroscience, 25*, 4779–4792.

Solomon, S. G., & Lennie, P. (2007). The machinery of colour vision. *Nature Reviews Neuroscience, 8*, 276–286.

Solomon, S. G., Peirce, J. W., & Lennie, P. (2004). The impact of suppressive surrounds on chromatic properties of cortical neurons. *Journal of Neuroscience, 24*, 148–160.

Switkes, E., Bradley, A., & De Valois, K. K. (1988). Contrast dependence and mechanisms of masking interactions among chromatic and luminance gratings. *Journal of the Optical Society of America. A, Optics and Image Science, 5*, 1149–1162. doi:10.1364/JOSAA.5.001149.

Tailby, C., Solomon, S. G., Dhruv, N. T., & Lennie, P. (2008). Habituation reveals fundamental chromatic mechanisms in striate cortex of macaque. *Journal of Neuroscience, 28*, 1131–1139.

Tailor, D., Finkel, L., & Buchsbaum, G. (2000). Color opponent receptive fields derived from independent component analysis of natural images. *Vision Research, 40*, 2671–2676.

Thorell, L. G., De Valois, R. L., & Albrecht, D. G. (1984). Spatial mapping of monkey V1 cells with pure color and luminance stimuli. *Vision Research, 24*, 751–769.

Tobimatsu, S., Tomoda, H., & Kato, M. (1995). Parvocellular and magnocellular contributions to visual evoked potentials in humans: Stimulation with chromatic and achromatic gratings and apparent motion. *Journal of the Neurological Sciences, 134*, 73–82. doi:10.1016/0022-510X(95)00222-X.

Vautin, R. G., & Dow, B. M. (1985). Color cell groups in foveal striate cortex of the behaving macaque. *Journal of Neurophysiology, 54*, 273–292.

Victor, J. D., Purpura, K., Katz, E., & Mao, B. (1994). Population encoding of spatial frequency, orientation, and color in macaque V1. *Journal of Neurophysiology, 72*, 2151–2166.

Wachtler, T., Sejnowski, T. J., & Albright, T. D. (2003). Representation of color stimuli in awake macaque primary visual cortex. *Neuron, 37*, 681–691.

Wade, A., Augath, M., Logothetis, N., & Wandell, B. (2008). FMRI measurements of color in macaque and human. *Journal of Vision, 8*, 1–19. doi:10.1167/8.10.6.

Wallach, H. (1935). Ueber visuell wahrgenommene Bewegungsrichtung. *Psychologische Forschung, 20*, 325–380.

Webster, M. A., De Valois, K. K., & Switkes, E. (1990). Orientation and spatial-frequency discrimination for luminance and chromatic gratings. *Journal of the Optical Society of America. A, Optics and Image Science, 7*, 1034–1049. doi:10.1364/JOSAA.7.001034.

Wiesel, T. N., & Hubel, D. H. (1966). Spatial and chromatic interactions in the lateral geniculate body of the rhesus monkey. *Journal of Neurophysiology, 29*, 1115–1156.

Wong-Riley, M. (1979). Changes in the visual system of monocularly sutured or enucleated cats demonstrable with cytochrome oxidase histochemistry. *Brain Research, 171,* 11–28. doi:10.1016/0006-8993(79)90728-5.

Wuerger, S., Shapley, R., & Rubin, N. (1996). "On the visually perceived direction of motion" by Hans Wallach, 1935: Translation. *Perception, 25,* 1317–1368. doi:10.1068/p251317.

Xiao, Y., Casti, A., Xiao, J., & Kaplan, E. (2007). Hue maps in primate striate cortex. *NeuroImage, 35,* 771–786.

Xiao, Y., Wang, Y., & Felleman, D. J. (2003). A spatially organized representation of colour in macaque cortical area V2. *Nature, 421,* 535–539.

Yarbus, A. L. (1967). *Eye movements and vision.* New York: Plenum Press.

Yeh, C. I., Xing, D., Williams, P. E., & Shapley, R. M. (2009). Stimulus ensemble and cortical layer determine V1 spatial receptive fields. *Proceedings of the National Academy of Sciences of the United States of America, 106,* 14652–14657. doi:10.1073/pnas.0907406106.

Zeki, S. M. (1973). Colour coding in rhesus monkey prestriate cortex. *Brain Research, 53,* 422–427.

Zeki, S. M. (1978a). Uniformity and diversity of structure and function in rhesus monkey prestriate visual cortex. *Journal of Physiology, 277,* 273–290.

Zeki, S. M. (1978b). Functional specialisation in the visual cortex of the rhesus monkey. *Nature, 274,* 423–428.

Zeki, S. (1983a). Colour coding in the cerebral-cortex—the reaction of cells in monkey visual-cortex to wavelengths and colours. *Neuroscience, 9,* 741–765.

Zeki, S. (1983b). Colour coding in the cerebral cortex: The responses of wavelength-selective and colour-coded cells in monkey visual cortex to changes in wavelength composition. *Neuroscience, 9,* 767–781.

Zeki, S., & Shipp, S. (1988). The functional logic of cortical connections. *Nature, 335,* 311–317.

41 The Processing of Color in Primate Extrastriate Cortex

DANIEL C. KIPER AND KARL R. GEGENFURTNER

The processing of chromatic signals in the retina, lateral geniculate nucleus (LGN), and primary visual cortex (V1) has been the focus of numerous studies (see Shapley & Hawken, 2011, for a recent review). Surprisingly, much less is known about the fate of color signals in extrastriate cortex. Here we first review well-established knowledge of chromatic processing in extrastriate cortex and then focus on recent developments. In the first part of the chapter we briefly review the distribution of color-selective neurons within extrastriate areas, the chromatic properties of individual neurons in several of these areas, and describe how color selectivity relates to the processing of other visual attributes such as orientation, size, and motion (see Gegenfurtner & Kiper, 2003). In the second part we describe several topics that have been subject to significant recent developments and promise to remain areas of interest in the future. These are the possible clustering of color-selective cells within individual cortical areas, the question of the existence of a color center in the brain, and the nature of chromatic processing in the primate brain as revealed by functional magnetic resonance imaging (fMRI), including the role of color signals in synesthesia.

CHROMATIC PROPERTIES OF INDIVIDUAL NEURONS IN EXTRASTRIATE CORTEX

Although several studies showed that individual neurons in the dorsal visual pathway, in particular in area MT of the macaque monkey (Croner & Albright, 1999; Dobkins & Albright, 1994; Gegenfurtner et al., 1994; Thiele, Dobkins, & Albright, 1999), can significantly respond to chromatic variations, these responses are typically smaller than those obtained with luminance stimuli (Conway et al., 2010) and do not account for the animal's behavioral performance. Our discussion is thus restricted to the areas of the ventral pathway that are known to play a critical role in color processing.

Proportion of Color-Selective Cells

In extrastriate areas of the ventral pathway the number of neurons whose responses are affected by the chromatic properties of the stimulus (i.e., neurons that respond to chromatic contrast in addition to or instead of luminance contrast) remains surprisingly constant despite the variability in the criteria used to classify neurons. This proportion reaches 50% in area V2 (Gegenfurtner, Kiper, & Fenstemaker, 1996) and 54% in V3 (Gegenfurtner, Kiper, & Levitt, 1997). Estimates in later areas of the ventral pathways are more variable. In area V4, often considered a "color" area of the primate brain (but see below), original estimates ranged from less than 20% (Schein, Marrocco, & de Monasterio, 1982) to 100% (Zeki, 1983a). A more recent estimate of 66% has been reported by Kotake et al. (2009). In IT cortex it has been estimated between 48% (Gross, Rocha-Miranda, & Bender, 1972) and 70% (Komatsu et al., 1992).

Chromatic Properties of Individual Neurons

Compared to the relay cells in the retina or LGN, the chromatic properties of cortical neurons seem to differ in two important respects. First, a significant proportion of neurons in each area possess a high degree of color selectivity. These neurons show a narrow tuning in color space (figure 41.1). Indeed, although retinal and LGN neurons are well described by models that postulate linear combination of cone signals (Derrington, Krauskopf, & Lennie, 1984), the selectivity of many cortical neurons is too narrow to be explained by this model. Narrowly tuned neurons have been reported, in different proportions, within areas V1 (Cottaris & De Valois, 1998), V2 (Kiper, Fenstemaker, & Gegenfurtner, 1997), V3 (Gegenfurtner, Kiper, & Levitt, 1997), and V4 (Zeki, 1983b).

A second major difference between precortical and cortical processing concerns individual neurons'

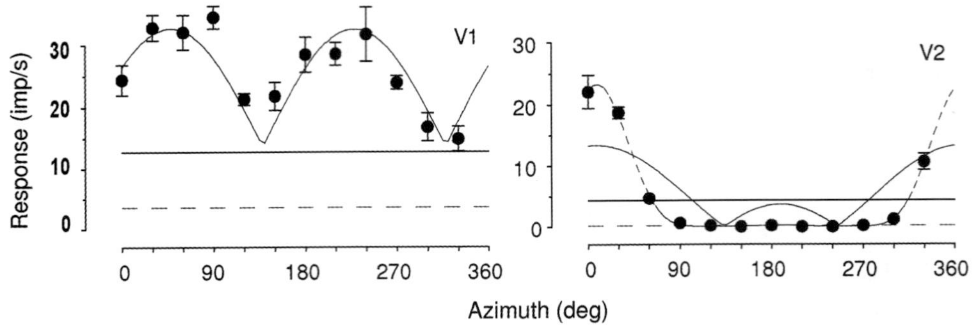

FIGURE 41.1   Color tuning of two individual neurons expressed as a function of color angle (azimuth) in DKL color space. The left panel shows a typical V1 neuron that combines its input signals linearly (solid curve). The horizontal lines show the cell's response to an achromatic stimulus (solid line) and the cell's spontaneous firing rate (dashed line). The right panel shows a narrowly tuned V2 neuron. (From Kiper, Fenstemaker, & Gegenfurtner, 1997.)

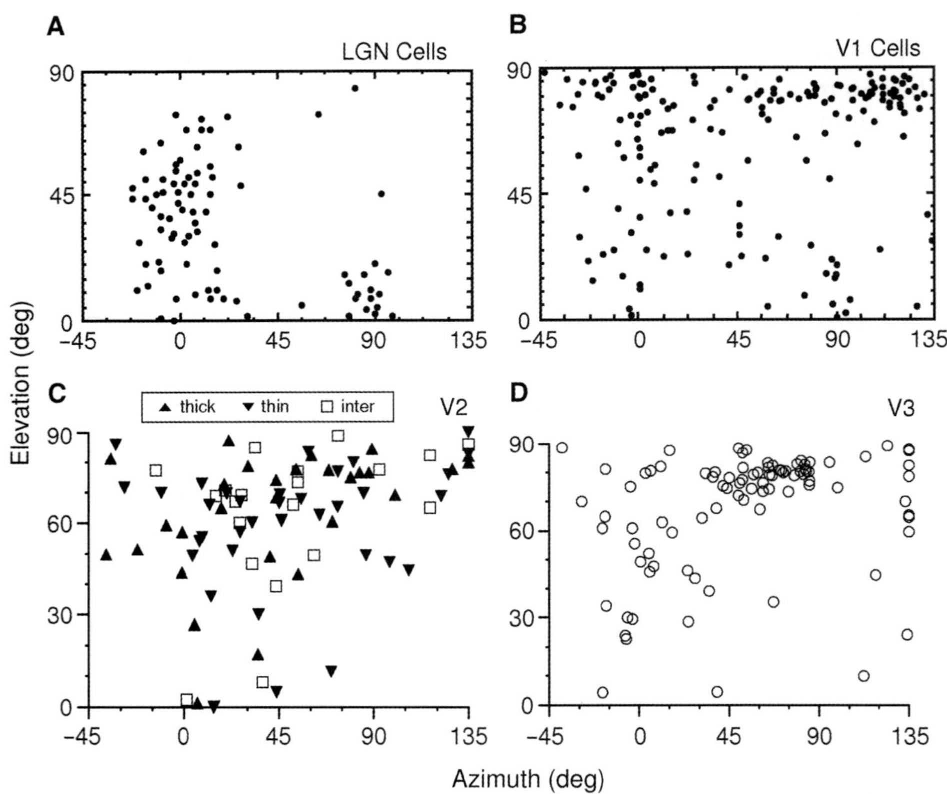

FIGURE 41.2   Distribution of preferred colors in the LGN (A), V1 (B), different CO compartments of V2 (C), and V3 (D). Azimuth (analogous to hue) and elevation (luminance) refer to angle within the Derrington-Krauskopf-Lennie (DKL) color space. (From Gegenfurtner & Kiper, 2004.)

preferred colors. The vast majority of color-selective neurons in the retina and LGN oppose L and M cones or oppose the S- to the sum of L+M-cone signals. In the cortex the distribution of preferred colors is not clustered in these regions of color space but is much more uniformly distributed (figure 41.2). The existence of the red–green and blue–yellow channels in subcortical processing is likely due to the bottleneck created by the retinocortical pathway: It was shown that coding the chromatic content of natural objects within a restricted-capacity channel is best done using this coding scheme (Buchsbaum & Gottschalk, 1983). In the cortex, where the number of available neurons and connections is vastly larger, coding of all colors into distinct red–green and yellow–blue channels is not necessary.

Color versus Other Visual Attributes

A number of studies (Livingstone & Hubel, 1984; Roe & Ts'o, 1999; Shipp & Zeki, 2002) had proposed that cortical cells selective for color (i.e., neurons that respond to chromatic in addition to luminance modulations or to chromatic modulations only) are not selective for other visual attributes such as stimulus motion or orientation. In that view color is processed by segregated populations of neurons that can be found in the cytochrome oxidase (CO)-rich patches in layer 2/3 of V1 or the thin CO bands of V2. This view, however, has been challenged. Numerous studies in the last decade have shown that cells within V1 (Friedman, Zhou, & von der Heydt, 2003; Johnson, Hawken, & Shapley, 2001; Leventhal et al., 1995), V2 (Friedman, Zhou, & von der Heydt, 2003; Gegenfurtner, Kiper, & Fenstemaker, 1996), V3 (Gegenfurtner, Kiper, Levitt, 1997; Seymour et al., 2009), as well as V4 (Yoshioka & Dow, 1996) can be concomitantly tuned for several dimensions of the visual stimulus. It thus appears that color is not processed independently but by the same neuronal populations that also code orientation or size. Note, however, that a recent study reported the existence of a significant subpopulation of V4 cells that respond to chromatic but not luminance variations (Bushnell et al., 2011), leading the authors to suggest that color and luminance might be treated by different channels within area V4.

RECENT AND PROBABLE FUTURE DEVELOPMENTS

Clustering of Color-Selective Cells

The debate concerning the organization of color-selective cells into clusters within a given area has extended to several visual areas. In V1 the CO-rich patches have been thought to represent the location of clusters of color-selective cells that are relatively unselective for orientation (see Conway et al., 2010), but this has been contested by a number of anatomical and physiological studies (see Economides et al., 2011; Gegenfurtner & Kiper, 2003).

Several studies reported clustering of color-selective neurons in extrastriate visual areas. In V2 the thin bands defined by CO staining have been reported to represent clusters of color-selective cells (Hubel & Livingstone, 1987) and to be the source of the color signals sent to area V4 and areas of the inferotemporal cortex (Conway et al., 2010; Zeki & Shipp, 1989). Moreover, optical imaging studies of area V2 concluded that color is represented in an orderly fashion within the thin

stripes in the form of well-defined color maps that resemble those based on human color perception (Lim et al., 2009; Xiao, Wang, & Felleman, 2003). As in V1, however, the clustering of color selectivity within the thin CO stripes has been challenged (Gegenfurtner, Kiper, & Fenstemaker, 1996).

In areas more posterior within the temporal pathway, color-selective cells have been reported to cluster into subregions. This clustering is in fact often offered as an explanation for the large variance in estimates of the proportion of color-selective cells in temporal cortex. Microelectrode recordings in dorsal V4 are thought to encounter many color clusters, which seem to be much less prevalent in ventral parts of V4. In a recent study Tanigawa, Lu, and Roe (2010) reported that area V4 might be divided into a number of subregions that would treat orientation and color independently. As for earlier visual areas, this issue is currently not resolved.

In the most anterior parts of the ventral pathways, distinction between cortical areas is much less clear, and the relationship between areas reported in the macaque brain with those of the human is still controversial (see Shapley & Hawken, 2011). It has nonetheless been suggested (Conway & Tsao, 2006) that color might be treated within a specialized pathway that extends across several of the ventral visual areas including V2, V4, and the dorsal portion of posterior inferotemporal cortex (PITd). Within PIT, color-selective cells would be clustered into islands, themselves containing orderly, columnar color maps (Conway & Tsao, 2009), reminiscent of the organization previously reported in V2 (Lim et al., 2009; Xiao, Wang, & Felleman, 2003).

Is There a Color Center in the Primate Brain?

Several of the issues discussed above are intimately related to the question of whether the primate brain contains one color center, a cortical area whose main function would be to support most or all aspects of color perception. The notion of a color center is a natural consequence for the proponents of a strictly modular view of cortical organization (Zeki, 1993). It has received support from clinical reports of patients' selective impairments in color vision (cerebral achromatopsia) following lesion(s) to the ventral temporal cortex. Although somewhat obscured by controversies concerning the equivalence of human and macaque cortical areas (for example hV4—human V4—versus macaque V4), most researchers agree that there are areas in ventral cortex that are highly activated by color stimuli, but none of these areas has been established as predominant over the others. A meta-study of clinical cases of cerebral achromatopsia (Bouvier & Engel,

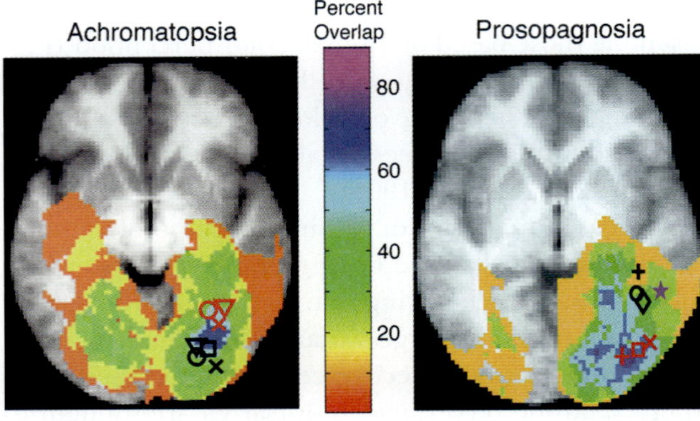

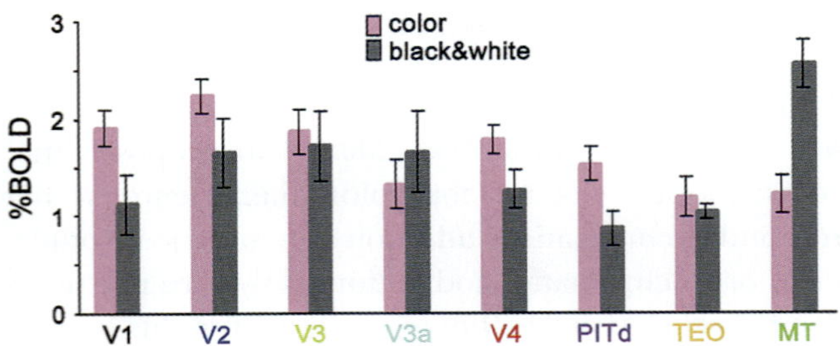

FIGURE 41.4   Comparison of the BOLD responses of cortical regions to 99% contrast luminance images (gray bars) and to saturated equiluminant images (pink bars). Error bars are standard error of the means. (From Conway & Tsao, 2006.)

FIGURE 41.3   Lesion overlaps with neuroimaging results. Achromatopsia and prosopagnosia lesion overlaps are shown with representative neuroimaging peak activations superimposed. On achromatopsia lesion overlap (left), black symbols indicate posterior color-sensitive findings, and red symbols indicate anterior color-sensitive findings in studies reporting multiple responses. On prosopagnosia overlap (right), black symbols indicate responses near the face-sensitive area FFA, red symbols indicate responses near the face-sensitive area OFA, and the purple symbol indicates a response near the face-sensitive area STS. (From Bouvier & Engel, 2006.)

2006) showed that lesions that produce achromatopsia overlap with those that result in other perceptual deficits such a prosopagnosia (figure 41.3).

Moreover, experimental lesions in macaque V4 (the most popular candidate for the role of color center) do not result in a complete loss of color vision (Schiller, 1993). These results cast doubts on the notion of a unique color center and support the idea that color, like many other visual attributes, is treated within a network of neuronal populations distributed within the ventral occipitotemporal pathway (Shapley & Hawken, 2011).

fMRI Studies of Color Processing in the Primate Brain

Over the last two decades many studies using fMRI have focused on color processing, both in the macaque monkey and the human brain. These studies have highlighted some of the difficulties that exist in equating cortical areas originally described in the monkey to those revealed by human experiments. In particular Lueck et al. (1989) and Zeki and Bartels (1999) have reported the existence of two human cortical areas that respond preferentially to chromatic modulation in the ventral occipitotemporal cortex. They called these areas hV4, for human V4, and V4α. Note that the similarity between monkey and human V4 has been questioned (Goddard et al., 2011). Others reported on another

color-specific area anterior to hV4, which they called V8 (Hadjikhani et al., 1998) or VO-1 and VO-2 (Brewer et al., 2005). As a whole these studies convincingly showed two important points: First, chromatic modulations can induce large BOLD signals in primate visual cortex, including areas V1 and V2. This result contrasts with the relative scarcity of color-selective cells identified with electrophysiological techniques (Engel, Zhang, & Wandell, 1997). Second, they showed that color is not processed within a single area but, as mentioned above, by a network of areas along the ventral occipital-temporal pathway (figure 41.4).

This multitude of areas responding to color stimuli has also been observed in the macaque brain, where BOLD responses to color equal or exceed those to luminance stimuli in areas V1, V2, V3, V4, PITd, and TEO (Conway & Tsao, 2006).

In the macaque brain fMRI studies revealed a number of cell clusters that contain color-selective neurons (Conway, Moeller, & Tsao, 2007), particularly in area V4, PITd, and the posterior part of area TEO. Studies combining fMRI with single-cell recordings concluded that each of these regions contains cells selective for different colors and that adjacent regions tend to have similar color preferences (Conway & Tsao, 2009). These results, if confirmed, argue for a fine-scale, orderly representation of color in occipitotemporal cortex. Although the functional significance of these color islands remains largely unknown, it has been proposed that the responses of cells in IT cortex support perceptual color judgments (Koida & Komatsu, 2007; Murphey, Yoshor, & Beauchamp, 2008) and could also be the neuronal correlate of the perceptual unique hues (Stoughton & Conway, 2008; but see Mollon, 2009, for a response).

In recent years a number of studies have used fMRI to study the role of color signals in synesthesia, whose most common form binds graphemes with specific colors. It has been reported that this form of synesthesia could be due to abnormal connectivity between color-selective regions of the temporal lobe and areas involved in grapheme processing (Hubbard & Ramachandran, 2005; Rouw & Scholte, 2007; but see Jäncke et al., 2009). A number of fMRI studies reported activation of area V4 in human synesthetes exposed to graphemes (Nunn et al., 2002; Weiss et al., 2001). A more recent and careful study (Hupé, Bordier, & Dojat, 2012), however, concluded that putative color-selective areas of the human brain are not activated during grapheme–color association in synesthesia.

What Does the Future Hold?

As the preceding sections have emphasized, the processing of color signals in extrastriate cortex remains poorly understood, and specialization is not only not resolved but often highly controversial. In general studies performed in the last few years have converged to show that the hypothesis of a single, specialized color center in the primate brain has been refuted. Instead, color is now known to be processed within a network of areas mostly contained within the ventral visual processing streams. Within these areas color-processing cells seem to cluster into subregions whose organization and functional role are still relatively unknown. Future studies will undoubtedly focus on the individual properties of these extrastriate color-processing cells and reveal their contributions to perceptual phenomena.

Indeed, important perceptual phenomena such as color constancy (see Foster, 2011), unique hues, or color categorization remain largely unexplained. Further studies reconciling imaging or lesion results in the human brain with those obtained by methods revealing single activity in the macaque brain are thus necessary to fill these gaps. Moreover, the relationship between color signals and those associated with other visual attributes such as object shape or motion needs further scrutiny, particularly in extrastriate areas. Today, it is safe to say that a full understanding at the neuronal level of perceptual phenomena associated with color is still eluding the vision science community.

REFERENCES

Bouvier, S. E., & Engel, S. A. (2006). Behavioral deficits and cortical damage loci in cerebral achromatopsia. *Cerebral Cortex, 16,* 183–191.

Brewer, A. A., Liu, J., Wade, A. R., & Wandell, B. A. (2005). Visual field maps and stimulus selectivity in human ventral occipital cortex. *Nature Neuroscience, 8,* 1102–1109.

Buchsbaum, G., & Gottschalk, A. (1983). Trichromacy, opponent colours coding and optimum information transmission in the retina. *Proceedings of the Royal Society of London. Series B, Biological Sciences, 220,* 89–113.

Bushnell, B. N., Harding, P. J., Kosai, Y., Bair, W., & Pasupathy, A. (2011). Equiluminance cells in visual cortical area V4. *Journal of Neuroscience, 31,* 12398–12412.

Conway, B. R., Chatterjee, S., Field, G. D., Horwitz, G. D., Johnson, E. N., Koida, K., et al. (2010). Advances in color science: From retina to behavior. *Journal of Neuroscience, 30,* 14955–14963.

Conway, B. R., Moeller, S., & Tsao, D. Y. (2007). Specialized color modules in macaque extrastriate cortex. *Neuron, 56,* 560–573.

Conway, B. R., & Tsao, D. Y. (2006). Color architecture in alert macaque cortex revealed by fMRI. *Cerebral Cortex, 16,* 1604–1613. doi:10.1093/cercor/bhj099.

Conway, B. R., & Tsao, D. Y. (2009). Color-tuned neurons are spatially clustered according to color preference within alert macaque posterior inferior temporal cortex. *Proceedings of the National Academy of Sciences of the United States of America, 106,* 18034–18039. doi:10.1073/pnas.0810943106.

Cottaris, N. P., & De Valois, R. L. (1998). Temporal dynamics of chromatic tuning in macaque primary visual cortex. *Nature, 395,* 896–900.

Croner, L. J., & Albright, T. D. (1999). Segmentation by color influences responses of motion-sensitive neurons in the cortical middle temporal visual area. *Journal of Neuroscience, 19,* 3935–3951.

Derrington, A. M., Krauskopf, J., & Lennie, P. (1984). Chromatic properties of neurons in macaque LGN. *Journal of Physiology, 357,* 241–265.

Dobkins, K. R., & Albright, T. D. (1994). What happens if it changes color when it moves? The nature of chromatic input to macaque visual area MT. *Journal of Neuroscience, 14,* 4854–4870.

Economides, J. R., Sincich, L. C., Adams, D. L., & Horton, J. C. (2011). Orientation tuning of cytochrome oxidase patches in macaque primary visual cortex. *Nature Neuroscience, 14,* 1574–1580.

Engel, S., Zhang, X., & Wandell, B. (1997). Colour tuning in human visual cortex measured with functional magnetic resonance imaging. *Nature, 388,* 68–71.

Foster, D. H. (2011). Color constancy. *Vision Research, 51,* 674–700.

Friedman, H. S., Zhou, H., & von der Heydt, R. (2003). The coding of uniform colour figures in monkey visual cortex. *Journal of Physiology, 548*(Pt 2), 593–613.

Gegenfurtner, K. R., & Kiper, D. C. (2003). Color vision. *Annual Review of Neuroscience, 26,* 181–206.

Gegenfurtner, K. R., & Kiper, D. C. (2004). The processing of color in extrastriate cortex. In L. M. Chalupa & J. S. Werner (Eds.), *The visual neurosciences* (Vol. 2, pp. 1017–1028). Cambridge, MA: MIT Press. doi:10.1167/11.4.3.

Gegenfurtner, K. R., Kiper, D. C., Beusmans, J. M., Carandini, M., Zaidi, Q., & Movshon, J. A. (1994). Chromatic properties of neurons in macaque MT. *Visual Neuroscience, 11,* 455–466.

Gegenfurtner, K. R., Kiper, D. C., & Fenstemaker, S. B. (1996). Processing of color, form, and motion in macaque area V2. *Visual Neuroscience, 13,* 161–172.

Gegenfurtner, K. R., Kiper, D. C., & Levitt, J. B. (1997). Functional properties of neurons in macaque area V3. *Journal of Neurophysiology, 77,* 1906–1923.

Goddard, E., Mannion, D. J., McDonald, J. S., Solomon, S. G., & Clifford, C. W. (2011). Color responsiveness argues against a dorsal component of human V4. *Journal of Vision, 11*, 3. doi:10.1167/11.4.3.

Gross, C. G., Rocha-Miranda, C. E., & Bender, D. B. (1972). Visual properties of neurons in inferotemporal cortex of the macaque. *Journal of Neurophysiology, 35*, 96–111.

Hadjikhani, N., Liu, A. K., Dale, A. M., Cavanagh, P., & Tootell, R. B. (1998). Retinotopy and color sensitivity in human visual cortical area V8. *Nature Neuroscience, 1*, 235–241.

Hubbard, E. M., & Ramachandran, V. S. (2005). Neurocognitive mechanisms of synesthesia. *Neuron, 48*, 509–520.

Hubel, D. H., & Livingstone, M. S. (1987). Segregation of form, color, and stereopsis in primate area 18. *Journal of Neuroscience, 7*, 3378–3415.

Hupé, J. M., Bordier, C., & Dojat, M. (2012). The neural bases of grapheme–color synesthesia are not localized in real color-sensitive areas. *Cerebral Cortex, 22*, 1622–1633.

Jäncke, L., Beeli, G., Eulig, C., & Hänggi, J. (2009). The neuroanatomy of grapheme-color synesthesia. *European Journal of Neuroscience, 29*, 1287–1293.

Johnson, E. N., Hawken, M. J., & Shapley, R. (2001). The spatial transformation of color in the primary visual cortex of the macaque monkey. *Nature Neuroscience, 4*, 409–416.

Kiper, D. C., Fenstemaker, S. B., & Gegenfurtner, K. R. (1997). Chromatic properties of neurons in macaque area V2. *Visual Neuroscience, 14*, 1061–1072.

Koida, K., & Komatsu, H. (2007). Effects of task demands on the responses of color-selective neurons in the inferior temporal cortex. *Nature Neuroscience, 10*, 108–116.

Komatsu, H., Ideura, Y., Kaji, S., & Yamane, S. (1992). Color selectivity of neurons in the inferior temporal cortex of the awake macaque monkey. *Journal of Neuroscience, 12*, 408–424.

Kotake, Y., Morimoto, H., Okazaki, Y., Fujita, I., & Tamura, H. (2009). Organization of color-selective neurons in macaque visual area V4. *Journal of Neurophysiology, 102*, 15–27.

Leventhal, A. G., Thompson, K. G., Liu, D., Zhou, Y., & Ault, S. J. (1995). Concomitant sensitivity to orientation, direction, and color of cells in layers 2, 3, and 4 of monkey striate cortex. *Journal of Neuroscience, 15*(Pt 1), 1808–1818.

Lim, H., Wang, Y., Xiao, Y., Hu, M., & Felleman, D. J. (2009). Organization of hue selectivity in macaque V2 thin stripes. *Journal of Neurophysiology, 102*, 2603–2615.

Livingstone, M. S., & Hubel, D. H. (1984). Anatomy and physiology of a color system in the primate visual cortex. *Journal of Neuroscience, 4*, 309–356.

Lueck, C. J., Zeki, S., Friston, K. J., Deiber, M. P., Cope, P., Cunningham, V. J., et al. (1989). The colour centre in the cerebral cortex of man. *Nature, 340*, 386–389.

Mollon, J. D. (2009). A neural basis for unique hues? *Current Biology, 19*, R441. doi:10.1016/j.cub.2009.05.008.

Murphey, D. K., Yoshor, D., & Beauchamp, M. S. (2008). Perception matches selectivity in the human anterior color center. *Current Biology, 18*, 216–220.

Nunn, J. A., Gregory, L. J., Brammer, M., Williams, S. C., Parslow, D. M., Morgan, M. J., et al. (2002). Functional magnetic resonance imaging of synesthesia: Activation of V4/V8 by spoken words. *Nature Neuroscience, 5*, 371–375.

Roe, A. W., & Ts'o, D. Y. (1999). Specificity of color connectivity between primate V1 and V2. *Journal of Neurophysiology, 82*, 2719–2730.

Rouw, R., & Scholte, H. S. (2007). Increased structural connectivity in grapheme-color synesthesia. *Nature Neuroscience, 10*, 792–797.

Schein, S. J., Marrocco, R. T., & de Monasterio, F. M. (1982). Is there a high concentration of color-selective cells in area V4 of monkey visual cortex? *Journal of Neurophysiology, 47*, 193–213.

Schiller, P. H. (1993). The effects of V4 and middle temporal (MT) area lesions on visual performance in the rhesus monkey. *Visual Neuroscience, 10*, 717–746.

Seymour, K., Clifford, C. W., Logothetis, N. K., & Bartels, A. (2009). The coding of color, motion, and their conjunction in the human visual cortex. *Current Biology, 19*, 177–183.

Shapley, R. M., & Hawken, M. J. (2011). Color in the cortex: Single- and double-opponent cells. *Vision Research, 51*, 701–717.

Shipp, S., & Zeki, S. (2002). The functional organization of area V2, I: Specialization across stripes and layers. *Visual Neuroscience, 19*, 187–210.

Stoughton, C. M., & Conway, B. R. (2008). Neural basis for unique hues. *Current Biology, 18*, R698. doi:10.1016/j.cub.2008.06.018.

Tanigawa, H., Lu, H. D., & Roe, A. W. (2010). Functional organization for color and orientation in macaque V4. *Nature Neuroscience, 12*, 1542–1548.

Thiele, A., Dobkins, K. R., & Albright, T. D. (1999). The contribution of color to motion processing in macaque middle temporal area. *Journal of Neuroscience, 19*, 6571–6587.

Weiss, P. H., Shah, N. J., Toni, I., Zilles, K., & Fink, G. R. (2001). Associating colours with people: A case of chromatic-lexical synaesthesia. *Cortex, 37*, 750–753.

Xiao, Y., Wang, Y., & Felleman, D. J. (2003). A spatially organized representation of colour in macaque cortical area V2. *Nature, 421*, 535–539.

Yoshioka, T., & Dow, B. M. (1996). Color, orientation and cytochrome oxidase reactivity in areas V1, V2 and V4 of macaque monkey visual cortex. *Behavioural Brain Research, 76*, 71–88.

Zeki, S. (1983a). The distribution of wavelength and orientation selective cells in different areas of the monkey visual cortex. *Proceedings of the Royal Society of London, 217*, 449–470.

Zeki, S. (1983b). Colour coding in the cerebral cortex: The reaction of cells in monkey visual cortex to wavelengths and colours. *Neuroscience, 9*, 741–765.

Zeki, S. (1993). The visual association cortex. *Current Opinion in Neurobiology, 2*, 155–159.

Zeki, S., & Bartels, A. (1999). The clinical and functional measurement of cortical (in)activity in the visual brain, with special reference to the two subdivisions (V4 and V4 alpha) of the human colour centre. *Philosophical Transactions of the Royal Society of London. Series B, Biological Sciences, 354*, 1371–1382. doi:10.1098/rstb.1999.0485.

Zeki, S., & Shipp, S. (1989). Modular connections between areas V2 and V4 of macaque monkey visual cortex. *European Journal of Neuroscience, 1*, 494–506.

VI PATTERN, SURFACE, AND SHAPE

42 Spatial Scale in Visual Processing

ROBERT F. HESS

In the early 1970s our views began to change about the nature of visual processing. The default idea, that visual processing produced an image description based on a local feature representation, was being challenged. The new suggestion, which had its origins in visual psychophysics (Campbell & Robson, 1968), was that visual information is analyzed in terms of the amplitude of different Fourier components. According to this approach, information about spatial scale was available to the higher stages of perception and used in a very specific way. Some found this latter view particularly objectionable because, unlike its applicability to physical optics, neural processes possessed a number of characteristics (nonlinearities and changes in spatial grain with eccentricity) that would potentially render such an analysis inappropriate (Petrov, Pigarev, & Zenkin, 1980; Westheimer, 1973). However, the issue was developing into one of a more general nature. A slow transition occurred in thinking from the original Fourier proposal to a more general one involving the importance of a local, scale-based spatial analysis. This was essentially a debate about whether perception had access to information separately at each of a number of spatial scales or whether, at an early stage, visual information was rigidly combined across scale for a feature-based analysis. The emerging neurophysiology showed that neurons in the striate cortex have receptive fields that could be considered as spatial and temporal filters (Campbell et al., 1968; Campbell, Cooper, & Enroth-Cugell, 1969; Glezer, Cooperman, & Tscherbach, 1973; Maffei & Fiorentini, 1973), suggesting that, at least at the level of the striate cortex, information could be separately accessed at a range of different spatial scales. Thus, the balance tipped toward the independent scale proposal. However, the more we learn about the properties of cells in extrastriate cortex (Kourtzi & Kanwisher, 2001), the more we realize that scale combination is an essential part of vision. Subsequent work reviewed here suggests a role for both types of analysis, an initial scale-independent process followed by, in some cases, scale combination. The relative roles of each depend on the particular task and stimuli.

HISTORY OF THE MULTISCALE VIEW

The emergence of spatial scale as an important aspect of visual processing came at about the same time from neurophysiological studies of animals and psychophysical studies of humans. In humans, following from the early aerial reconnaissance work of Selwyn (1948) in England and the applied work of Schade (1956) in the United States, Campbell and Green (1965) began to develop and apply the measurement of contrast sensitivity as a function of spatial frequency to better specify human vision. This approach emphasized the importance of visual processing for object sizes larger than the resolution limit.

In particular, a contrast sensitivity function describes the relationship between the size of a stimulus and the contrast necessary to just detect it (i.e., its *contrast threshold*). The stimulus of choice is a sinusoidal grating (black and white bars with a sinusoidal luminance profile, modulated about a mean light level) specified in units of spatial frequency (in cycles per degree subtended at the eye). Such a stimulus allows contrast to be altered without affecting the mean adaptational state of the eyes, and in addition, the retinal image will also be sinusoidal in form because the optics are linear. The form of this relationship is shown in figure 42.1. Human sensitivity is best at intermediate object sizes (or spatial frequencies) and is reduced at both higher and lower frequencies. Although the optics contribute to the reduction in contrast sensitivity at high spatial frequencies, the majority of the falloff at high spatial frequencies and all of the falloff at low spatial frequencies is due to the sensitivity of neural processes. The contrast sensitivity curve depicted in figure 42.1 is for foveal viewing and photopic light levels. If stimuli are imaged on more peripheral parts of the field or under scotopic light levels, there is preferential loss of sensitivity at high spatial frequencies as a consequence of the reduced neural sensitivity under these conditions.

The next major advance came when Campbell and Robson (1968) and later Pantle and Sekular (1968) and Blakemore and Campbell (1969) provided evidence

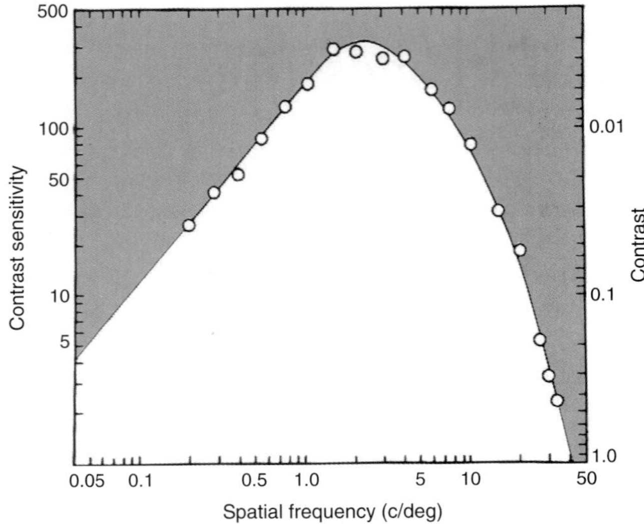

FIGURE 42.1 Contrast sensitivity function for foveal vision under photopic conditions. Contrast sensitivity (the reciprocal of the contrast needed for threshold detection) is plotted against the spatial frequency of a sinusoidal grating stimulus. This overall sensitivity function is itself composed of a set of more spatially restricted mechanisms termed "spatial channels."

that the contrast sensitivity function was itself composed of a number of more narrowly tuned, independent spatial mechanisms. In a series of psychophysical studies that followed, the degree to which these "spatial channels," as they were called, were independent (Graham, Robson, & Nachmias, 1978) and the ways in which they interacted (Graham & Nachmias, 1971) were outlined.

During this same period single-cell measurements from different parts of the visual pathway showed that neuronal receptive fields were of various sizes (Hubel & Wiesel, 1959, 1962) and that there was a systematic scaling of receptive field size with eccentricity in the retina and, to some extent, in the cortex (Hubel & Weisel, 1968). Up until this time, neurons were not really considered as filters (in the sense that they simply only transmitted information of a particular spatial scale, just as a green filter attenuates blue and red light and only allows light of intermediate wavelengths to pass); the prevailing idea at this time was that neurons encoded certain stimulus features and that they needed to do this at a range of different sizes. The idea that neurons could be considered as linear filters emerged from the retinal work of Enroth-Cugell and Robson (1966), which was very controversial at the time and only began to have its main impact a decade later. The primary visual cortex contains cells that are grouped together along a number of key processing dimensions: orientation, ocular dominance, and spatial frequency.

Cells with similar spatial frequency preference are grouped together into domains whose map is locally continuous across V1 (Issa, Trepel, & Stryker, 2000).

The impact of spatial scale on our thinking in vision research is reflected in the different computational approaches (Marr, 1982; Marr & Hildreth, 1980; Marr & Poggio, 1979; Morrone & Burr, 1990; Watt & Morgan, 1985; Wilson & Gelb, 1984; Wilson & Richards, 1989) that have developed subsequent to the work described above. All the models assume an initial spatial decomposition. They differ in the extent of this decomposition and in the level at which the information from these different spatial filters is combined and analyzed. The two extreme versions of this were captured in Watt and Morgan's MIRAGE model and Wilson's line element model. In the MIRAGE model spatial filters were combined at an early stage, and only symbolic descriptors subsequent to this combination were used (Watt & Morgan, 1985). In the Wilson line element model later stages of processing had independent access to the output of individual spatial filters, and their outputs were flexibly combined to solve different tasks (Wilson & Gelb, 1984; Wilson & Richards, 1989).

In this chapter I give examples of the advantages of accessing information at different spatial scales; these include foveal specialization and visual stability under different light levels. After detailing the evidence for independent access to scale information, scale selection rules are discussed. Finally, I give examples of situations where information at different scales is not kept separate but combined in specific ways (scale combination rules).

SCALE AND THE FUNCTIONAL SPECIALIZATION OF CENTRAL VISION

The relationship between spatial scale and eccentricity has important consequences for visual processing, yet it is poorly understood. Our traditional view of this relationship has been dominated by the measurement of visual acuity and its relationship with eccentricity. Acuity is best at the fovea and progressively deteriorates at more peripheral loci. This is similar to how receptive field size of retinal ganglion cells changes from fovea to periphery (Crook et al., 1988; De Monastereo & Gouras, 1975; Hubel & Wiesel, 1960). Small receptive fields are confined to the center of the visual field, and progressively larger ones are found in the periphery. In the cortex this situation changes. True, there is gradual enlarging of the size of the average receptive field in regions of the cortex representing more peripheral parts of the visual field. However, at any eccentricity there is a range of different sizes of receptive field.

Hubel and Wiesel's (1968) data suggest that this range is about 4:1 in the fovea. A better description is that, in more peripheral parts of the visual field, the range of different-sized receptive fields, and by implication the range of spatial scales, reduces because the number of cells with small receptive fields declines with eccentricity (De Valois, Albrecht, & Thorell, 1982).

Our understanding of how spatial scale varies across eccentricity has been greatly extended by the use of contrast sensitivity measurements for targets of different spatial scales. Numerous investigators have contributed to this understanding. Particularly noteworthy was the study of Robson and Graham (1981), who used stimuli of fixed spatiotemporal bandwidth that were well localized in space. They showed that for a wide range of spatial scales, the decline in sensitivity with eccentricity (plotted in absolute units, i.e., in degrees) was linear and depended on spatial scale; the finer the spatial scale the more rapidly sensitivity fell off with absolute eccentricity. This result is seen in figure 42.2B, where contrast sensitivity is plotted against eccentricity in degrees for a range of different spatial

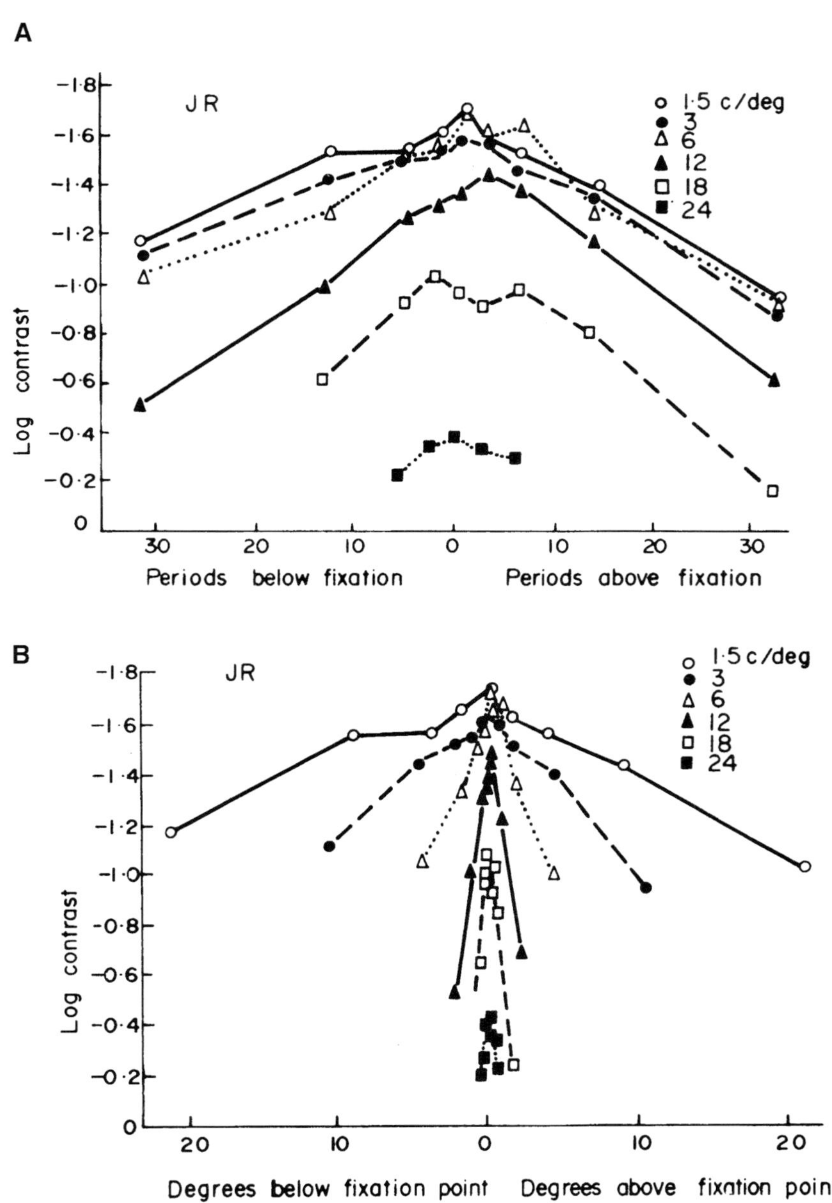

FIGURE 42.2   The regional variation of contrast sensitivity. (A and B) Results of Robson and Graham (1981) showing how contrast sensitivity for a range of different spatial frequencies declines with eccentricity along the vertical meridian. (A) Eccentricity is in relative units (periods of the particular spatial frequency). (B) Results plotted against eccentricity in degrees. Note that the relative slope in A is the same for all spatial frequencies in the range they tested. (Reproduced with permission from Robson & Graham, 1981.)

frequencies. However, if a relative eccentricity metric (figure 42.2A) was used (i.e., measured in periods of the particular spatial scale), then all spatial frequencies within the range 1–20 cycles/deg exhibited a similar falloff; for the vertical meridian this was 60 periods/decade (contrast sensitivity fell by a decade at an eccentricity equal to 60 periods of a particular spatial frequency).

In a subsequent study Pointer and Hess (1989) extended Robson and Graham's approach to spatial frequencies below 1 cycle/deg. Their results showed that there were three different rules for how spatial scale varied with eccentricity depending on the spatial range involved. At high to mid spatial frequencies (above 1.6 cycles/deg), they replicated the previous results of Robson and Graham (1981); the sensitivity gradient was 60 periods/decade. At mid to low spatial frequencies (0.2–0.8 cycles/deg) the decline of sensitivity with relative eccentricity increases (30 periods/decade), and at very low spatial frequencies (0.1 cycle/deg and below), the decline in sensitivity is much more gradual (90 periods/decade). From a purely psychophysical viewpoint the relationship between spatial scale and eccentricity is fairly straightforward. At most spatial scales sensitivity is best in the fovea, the extent of this superiority depending on spatial scale: The finer the spatial scale the greater the superiority of the fovea. At very coarse scales, sensitivity is the same in the center of the field as it is in the periphery. There is no evidence for either a peripheral superiority or a foveal inferiority at coarse scales. Thus, at more eccentric loci, the range of different spatial filters available to perception is reduced. Most neurophysiologists would not be comfortable with the above view because this is not reflected in how the receptive field properties of neurons vary with eccentricity in primary cortex, where, as I have already said, the size of the average receptive field increases with eccentricity. This is best seen in the population response using functional imaging. Marrett et al. (1997), using an adaptation of the peripheral phase-encoding method, showed that in human V1, spatial scale varies inversely with eccentricity—the fovea containing higher spatial scales, the periphery lower spatial scales. However, what one needs to appreciate is that the psychophysical data reflect the potential contribution of all visual areas, not just V1. Area V2 in primate (so too for area 18 in cat) is known to contain neurons responding to lower spatial scales than V1 (De Valois, Albrecht, & Thorell, 1982; Foster et al., 1985; Issa, Trepel, & Stryker, 2000; Movshon, Thompson, & Tolhurst, 1978). Extrastriate visual areas generally have larger receptive fields than V1 and sizable foveal representations. What this means is that by virtue of the additional information provided by these different extrastriate areas, the fovea has access to additional levels of spatial processing at coarser scales as well as the extrafine scales provided by V1. When we think of foveal specialization we often think in terms of the magnified areal representation of the fovea in the primary visual cortex. Another way in which the fovea is specialized is that it has access, via the contribution of different visual areas, to a larger range of spatial filters than its peripheral counterpart. This provides a great deal of potential flexibility and computational power. It also

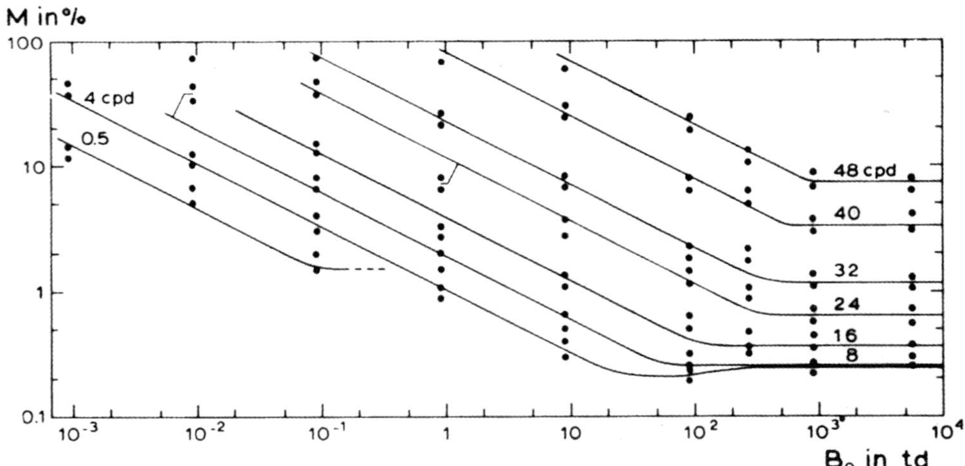

FIGURE 42.3   The scale-independent way in which contrast sensitivity varies with mean light level. The results are from Van Nes and Bouman (1967). Contrast sensitivity is plotted against the mean retinal illuminance in Trolands for a range of different spatial frequencies (listed in top inset). Each spatial frequency response exhibits two different rules depending on the mean illuminance: a Weber rule at high light levels and a DeVries–Rose rule at low light levels. The transition from the former to the latter also depends on the mean light level. (Reproduced with permission from Van Nes & Bouman, 1967.)

represents an important economy in terms of processing bandwidth because not all scales need to be represented at all eccentricities.

SCALE AND LIGHT ADAPTATION

One of the most impressive aspects of mammalian vision is the range of light levels over which it operates without greatly sacrificing sensitivity. The usual explanation for this is duplex retinal function of the photoreceptors together with gain control in postreceptoral retinal neurons. The fact that information is processed independently across scale, a mainly postretinal phenomenon, also helps extend our visual range and makes a significant contribution to perceptual stability across different light levels. The relationship between spatial scale and mean light level is best seen in the contrast threshold results of Van Nes and Bouman (1967). They showed that at any one spatial scale there was a characteristic relationship between sensitivity and light level. At low light levels sensitivity varies as the square root of the light level (so-called Rose–DeVries law) (DeVries, 1943; Rose, 1942, 1948), whereas at higher light levels, sensitivity is independent of the average light level (so-called Weber's law). The light level at which one behavior gives way to the other varies with spatial scale. This is shown in figure 42.3, in which, for a range of different spatial frequencies, contrast sensitivity is plotted against the mean retinal illuminance (in Trolands). These results suggest that neurons with smaller receptive fields reach this "transition luminance" at higher light levels than those with larger receptive fields. The psychophysics shows that this Rose–DeVries/Weber behavior occurs at all spatial scales.

At the level of the retina the degree of spatial selectivity also varies with the light level. Enroth-Cugell and Robson (1966) showed that the position of peak responsivity of a neuron shifts to lower spatial frequencies at lower light levels. Curiously, the low-frequency limb of the response function can be in its Weber region when the high-frequency limb is in its Rose–DeVries region. In the cortex this does not happen. Here the spatial tuning does not alter (either position of peak responsivity or bandwidth); only sensitivity varies with light level. The way in which sensitivity varies is just how Van Nes and Bouman (1967) described it in humans. This is true for both simple and complex cells in area 17 of cat (Hess, 1990; Kaufman & Palmer, 1990). Thus, by the time one gets to the cortex, reducing the light level seems to affect only sensitivity, not the spatial and temporal tuning of individual neurons. Processing information independently at different spatial scales provides an opportunity for a greater stability in our perceptions

at reduced light levels. Our perception of spatial and temporal structure, information derived from the output of many filters, is not greatly affected by reducing the light level. For a change in the average light level by a factor of between 3 and 6 orders of magnitude, our perceptions of spatial and temporal frequency are influenced by as little as 10% (Hess, 1990). That sort of perceptual stability is remarkable and makes a significant contribution to our ability to operate effectively over a wide range of light levels.

EVIDENCE FOR INDEPENDENT SCALE PROCESSING

Scale and Space

CONTRAST PERCEPTION   Many studies have shown that the perceived contrast of a stimulus depends on the contrast of surrounding structures, a phenomenon known as *contrast induction* (Cannon & Fullenkamp, 1991; Chubb, Sperling, & Solomon, 1989; Ejima & Takahashi, 1985; Ellemberg et al., 1998; Klein, Stromeyer, & Ganz, 1974; Mackay, 1973). Two previous studies have shown that the magnitude of contrast induction depends on the spatial frequency difference between test and inducing stimuli (Chubb, Sperling, & Solomon, 1989; Ellemberg et al., 1998). Chubb, Sperling, and Solomon (1989) showed that if the spatial frequency composition of test and inducing isotropic textures were 1 octave apart, the apparent contrast dropped from 40% to 15%. Ellemberg and colleagues (1998) showed a similar effect for single one-dimensional Gabor elements. Solomon, Sperling, and Chubb (1993) showed that this contrast induction is also tuned for orientation. Thus, it appears that the perceived contrast of an isolated region is in part determined by the contrast in adjacent regions and that this lateral influence is mediated within spatial scale. This is a curious result because one would intuitively expect contrast normalization to be mediated by a spatially and temporally broadband mechanism.

CONTRAST-DEFINED   STRUCTURE   Objects can be defined in a variety of ways, such as by modulations in luminance, contrast, chromaticity, motion, and disparity. Human vision is best at detecting luminance modulation (so-called first-order or *carrier modulation*) but can also detect objects defined by contrast modulation (second-order modulation). Indeed, there is evidence that cortical cells can process both types of information in early cortex (Baker, 1999; Zhou & Baker, 1993, 1994, 1996), although there are different ways of modeling the detection of contrast-defined objects. When the carrier spatial frequency is close to that of the modulation spatial frequency (e.g., a factor of about 2), then these stimuli can be detected by striate neurons with surround suppression responding to the luminance information in the stimulus (Tanaka & Ohzawa, 2009). When the carrier is of much higher spatial frequency than that of the envelope (specifically outside the spatial frequency range that luminance stimuli evoke responses, i.e., the luminance spatial frequency passband), then neurons in extrastriate cortex respond to the contrast modulation (second-order detectors) (Li et al., 2011; Rosenberg & Issa, 2009). The standard model involves two stages: a linear first stage composed of bandpass spatial filters and a second stage of linear filtering preceded by a rectifying nonlinearity. It has been of interest to know the relationship between the first and second stage of linear filters; for example, do second-stage filters of a particular spatial scale receive the summed input of all first-stage filters? If they do, they would act as generic "texture-grabbers" by detecting contrast modulation irrespective of the spatial composition of the texture. This would provide an obvious economy because fewer second-order neurons would be needed to cover all the possibilities of the first-order input. The results, however, suggest the opposite, namely that spatial information is not collapsed prior to the second-stage detectors (Dakin & Mareschal, 2000; Graham, Sutter, & Venkatesan, 1993; Langley, Fleet, & Hibbard, 1996). The current neurophysiology also supports this. Mareschal and Baker (1998) have shown that second-order detectors are tuned for the spatial frequency and, to a lesser extent, orientation of their input carrier frequencies. This means that more second-order detectors are required to cover all possibilities for the spatial tuning of their first-order input, but this may not be quite the problem it was once thought (Baker, 1999). This is an example of spatial scale being important at two different serial stages of visual analysis.

CONTOUR   Recent attempts at understanding contour processing have used spatially narrowband elements. Such stimuli allow tests of whether the integration along a contour occurs only between cells with the same receptive field properties or also between cells with different receptive field properties. In one approach (Field, Hayes, & Hess, 1993), each presentation comprises an array of spatially and spatial-frequency-localized elements. In one of the two presentations a subset of these elements are arranged, by virtue of their orientation alignment, to define a contour. When subjects are asked to detect which presentation contains the contour, using a standard 2AFC psychophysical procedure, performance is found to be at ceiling for straight contours but declines as contour curvature increases (Field, Hayes, & Hess, 1993).

Another reason for using spatially narrowband elements and limiting the processing to just one scale concerns the properties of natural images (Field, 1993). Consider the orientation structure of a fractal edge at a number of different spatial scales (Field, Hayes, & Hess, 1993). Unlike smooth edges, fractal edges do not exhibit the same consistency of local orientation across scale, making a number of previously proposed edge-detection strategies (Canny, 1983; Lowe, 1988; Marr & Hildreth, 1980) problematic. A fractal edge is continuous at each scale, but the precise position and orientation of the edge change across scale. At any particular position the edge may show a particular orientation at only one particular scale. This may be one good reason for having cortical neurons with bandpass properties (Field, 1987, 1993; Hayes, 1989).

The above argument based on the properties of natural images suggests that it may be important for the visual system to solve the continuity problem separately at each scale, but is there any direct psychophysical support for this? The stimuli displayed in frame B of figure 42.4 provide such a test. Here we are using the same paradigm as described above except that the display is composed of equal numbers of Gabors and phase-scrambled edges. Phase-scrambled edges are micropatterns composed of $f + 3f + 5f$ compound Gabors in which the relative phases are randomized. When the components are phase-aligned they produce local edge stimuli. The results in figure 42.4C and D show that when contour elements alternate between Gabors and phase-scrambled edges (as is also the case for the background elements), good performance (triangles) was maintained across different spatial scales for the Gabor, scales that were common to the phase-scrambled elements. This suggests that the visual system solves the continuity problem independently at each of a number of spatial scales. Interestingly, this was not the case when Gabors were alternated with phase-aligned edges (circular symbols), suggesting the importance of phase alignment in broadband stimuli for scale combination (Dakin & Hess, 1999). Textures composed of orientation flows have also been shown to exhibit spatial frequency selectivity. Kingdom and Keeble (2000) have shown that texture segmentation for such a task depends on the spatial frequency composition of the local elements, suggesting a scale-selective analysis.

Scale and Motion

Historically, there have been two different models of human motion detection, one using information within

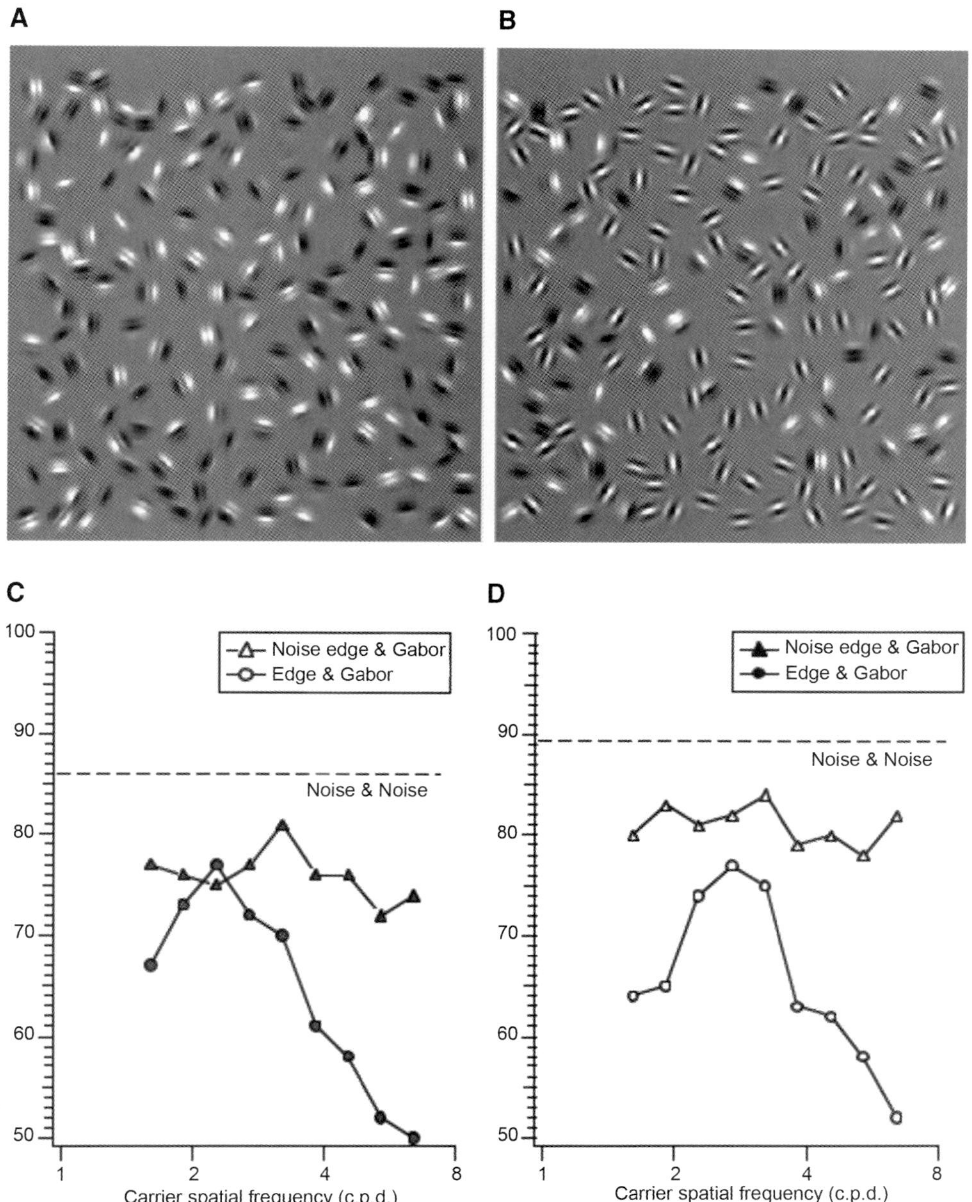

FIGURE 42.4 The scale-independent nature of contour integration. (A and B) Examples of the stimuli in which only a subset of the elements have their local orientation aligned along a notional curved contour. (A) All the elements are composed of phase-randomized $f + 3f + 5f$ compound Gabor micropatterns. (B) Half of the elements are phase-randomized $f + 3f + 5f$ compound Gabor micropatterns, and half are Gabor micropatterns having a spatial frequency corresponding to the abscissa in C and D. (C and D) Psychophysical detection results for a number of different stimulus conditions. Of particular note is the fact that performance is good across a range of different spatial frequencies of Gabor when the linking is between the phase-randomized $f + 3f + 5f$ compound Gabor micropatterns and Gabors of a single frequency, indicating that the task can be solved independently within each of a number of different spatial scales (triangles). Note that this is not the case when the compound Gabors exhibit phase coherence (i.e., edges: circles). (These results were reproduced with permission from Dakin & Hess, 1999.)

different spatial scales and the other involving information after it has been combined across spatial scales. The most popular example of the former is the motion energy model (i.e., the oriented energy in the spatio-temporal spectrum) utilizing neurons with receptive fields narrowly tuned for spatial frequency and orientation (Adelson & Bergen, 1985; van Santen & Sperling, 1985; Watson & Ahumada, 1985). A good test of this involves the motion of a spatially filtered broadband stimulus (spatial noise). For two-flash apparent motion, according to the first proposal above, the direction of displacement would be detected independently at a number of different spatial scales. D_{min} (the smallest displacement detected) would be signaled by the detectors working at the finest spatial scale and would be ultimately limited by the signal/noise ratio within these detectors. D_{max} (the largest displacement detected) would be signaled by the detectors working at the largest scale and would be ultimately limited by their half-cycle limit. An altogether different view (Morgan, 1992; Morgan & Mather, 1994; also see Johnston, McOwen, & Buxton, 1992) is that the visual system works on the displacements of image features that only exist when information across different spatial scales has been collapsed together and combined nonlinearly by a single spatial filter prior to motion detection.

Initially the independent scale model of motion was supported by the finding that, at least over a part of the range (not at low spatial scales), D_{max} scaled with the center frequency of bandpass-filtered noise images (Bischof & di Lollo, 1990, 1991; Chang & Julesz, 1983, 1985; Cleary, 1990; Cleary & Braddick, 1990; Morgan, 1992; Morgan & Mather, 1994). However, this result was equally well described within the feature model of motion because the average separation between image features in bandpass-filtered images also scales with center frequency (Morgan, 1992; Morgan & Mather, 1994). The failure at very low spatial scales was due to the use of "white" rather than a "fractal" noise for the broadband stimuli. Fractal noise images (a $1/f$ amplitude spectrum), unlike white-noise images (a flat amplitude spectrum), provide comparable energy for octave-wide detectors tuned to different spatial scales (Field, 1987). Stimuli composed of white noise do not

adequately stimulate visual detectors tuned to low spatial scales compared with those tuned to higher spatial scales; this resulted in the belief that high spatial frequencies interacted with low spatial frequencies for motion (Chang & Julesz, 1983, 1985; Cleary & Braddick, 1990).

Two subsequent results, both involving differential spatial filtering between the frames of a two-flash motion sequence, suggested that the detection of image motion occurs within, rather than across, spatial scale. The first was the finding that motion can be reliably detected between two image frames so long as there is motion energy within a common spatial frequency band, suggesting that the visual system has equal access to information across spatial scale (Brady, Bex, & Fredericksen, 1997; Ledgeway, 1996). The second finding is that motion can be detected between image sequences of fractal noise that are differentially low- or highpass filtered (Bex et al., 1995). One frame can be unfiltered, the other either highpass or lowpass filtered. This destroys the correlation between edge structures but maintains information at common scales (Hess et al., 1998). These results are displayed in figure 42.5 (A, symbols show results for lowpass filtering of one frame; B, symbols show results for highpass filtering of one frame). The multichannel prediction was generated by assuming that D_{max} was determined by information carried by the lowest scale supported by the stimulus within the constraints of the half-cycle limit. The spectral characteristics of the spatial noise are again important here. Initial attempts (Morgan & Mather, 1994) with white noise failed because low-frequency

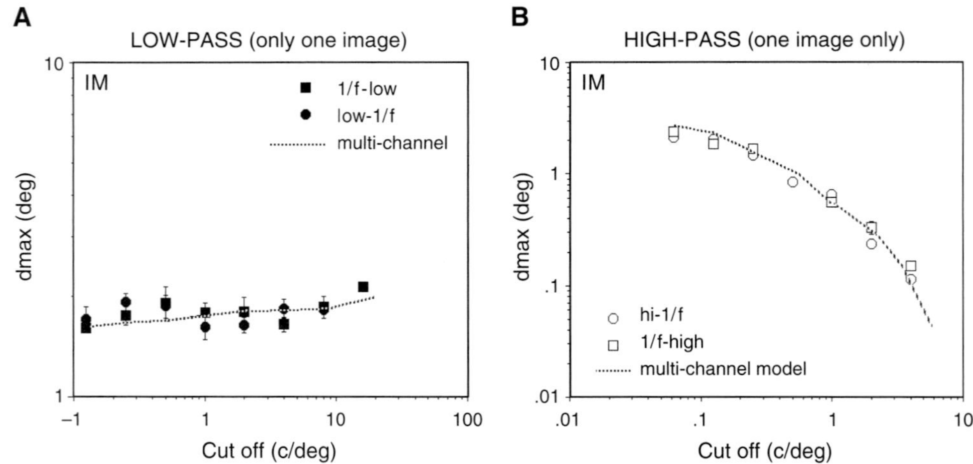

FIGURE 42.5   The effect of spatial scale on the discrimination of motion direction. Direction discrimination data (symbols) for fractal noise images seen in two-flash apparent motion when only one of the two frames is subject to spatial filtering. Results are compared with a multichannel model in which motion direction is processed independently by channels, each at a number of different spatial scales, and the channel with the highest signal/noise determines performance. (Reproduced with permission from Hess et al., 1998.)

mechanisms were severely disadvantaged by the use of white-noise stimuli. The use of fractal noise stimuli, which affords similar stimulation of spatial detectors tuned to fine and coarse scales, demonstrates that the detection of motion direction can occur independently at each of a number of spatial scales. This of course does not bear on exactly what is computed within each of these scales. It could be motion energy (Adelson & Bergen, 1985) or some other scale-localized feature (Eagle, 1996). This is still an open question, one that may be resolved by investigating temporal summation aspects of motion detection.

Scale and Stereo

The relationship between spatial scale and stereo processing, like that of motion processing, has been controversial. The initial receptive field positional disparity model advanced by Barlow, Blakemore, and Pettigrew (1967) and Pettigrew, Nikara, and Bishop (1968) did not have any specific role for receptive fields of different size. A much later model in which disparity was encoded not by positional displacements of receptive fields but by phase disparities within receptive fields driven by the right and left eyes (Ohzawa, DeAngelis, & Freeman, 1990, 1996; Ohzawa & Freeman, 1986) did have a specific link to the spatial properties of individual cells. It relied on high-spatial-frequency-tuned cells processing only fine disparities and low-spatial-frequency-tuned-cells processing only coarse disparities (the so-called size-disparity correlation).

Support for such a size-disparity correlation in human stereo processing has not been clear cut. For example, Schor and Woods (1983) provided the first psychophysical evidence for a size-disparity correlation by measuring the relationship between stereo sensitivity (D_{min} and D_{max}) and luminance spatial frequency. For D_{min} (the lower disparity limit), below a spatial frequency of 2.4 cycles/deg, stereo thresholds depend directly on the peak luminance spatial frequency of the stimulus, representing a constant phase limit of around 1/36th of a spatial cycle (figure 42.6A). For D_{max} (the upper disparity limit), they found a square-root relationship over approximately the same spatial frequency range with an asymptote at around 2.4 cycles/deg (figure 42.6C). This suggested that at least for D_{min}, disparity may have been computed within each of a number of independent spatial scales (Schor, Woods, & Ogawa, 1984).

Later work (Smallman & MacLeod, 1994, 1997) suggested that such a correlation may occur, at least for low-contrast targets, across the whole spatial frequency range including that above 2.4 cycles/deg. The interpretation of these results in terms of the role of spatial channels in stereo processing has been controversial. The above interpretation, that there are spatial frequency mechanisms processing stereoscopic information only below 2.4 cycles/deg (Schor, Woods, & Ogawa, 1984), has been challenged by the results of Yang and Blake (1991) and Kontsevich and Tyler (1994). Yang and Blake's (1991) masking results and Kontsevich and Tyler's (1994) modeling results argue that only those spatial channels above 2.4 cycles/deg process stereo information. More recently Glennerster and Parker (1997) questioned the conclusions of Yang and Blake's results on the basis that they had not taken into account the overall visibility of the stimuli and argued instead for multiple spatial mechanisms processing stereo information below 2 cycles/deg, a result supported by subsequent work (Prince, Eagle, & Rogers, 1998).

In the majority of the above studies (e.g., figure 42.6A, C), the stimulus of choice was a one-dimensional spatially and spatial-frequency-localized stimulus, a difference of Gaussians (DOGs). One property of such a stimulus is that peak spatial frequency and overall size of the stimulus covary, leaving one to wonder whether the relationship described in figure 42.6A is due to stimulus spatial frequency or size. A more general test of the size-disparity correlation would involve the use of broadband fractal stimuli for the reasons outlined above for the comparable case in motion. Such stimuli contain a range of different spatial scales and are not only representative of natural images but also optimal in stimulating different spatial-frequency-tuned neurons (Field, 1987). If we assume that the visual system has independent access to an array of spatial-frequency-tuned detectors, each responding up to its individual phase-disparity limit, there are some clear predictions for how D_{min} and D_{max} should vary with low- and high-pass filtering of such a broadband stimulus. D_{min} should be determined by disparity neurons with the highest spatial frequency tuning (assuming each spatial detector has comparable internal noise), and D_{max} by the lowest. For D_{min}, low-pass filtering should reduce stereo performance in a manner corresponding to some fixed fraction of a spatial cycle of the highest spatial-frequency channel supported by the stimulus. This fraction will depend on factors such as stimulus contrast because it represents not only a spatial limit but also a signal/noise limit. For the same reason high-pass filtering should have no effect on performance. D_{max}, on the other hand, which is thought to be a predominately spatial limit, should be limited by the half-cycle limit of the lowest spatial-frequency channel supported by the stimulus. Therefore, it should display a falloff with filter cutoff frequency corresponding to a fixed phase relationship in the case of high-pass filtering but no effect

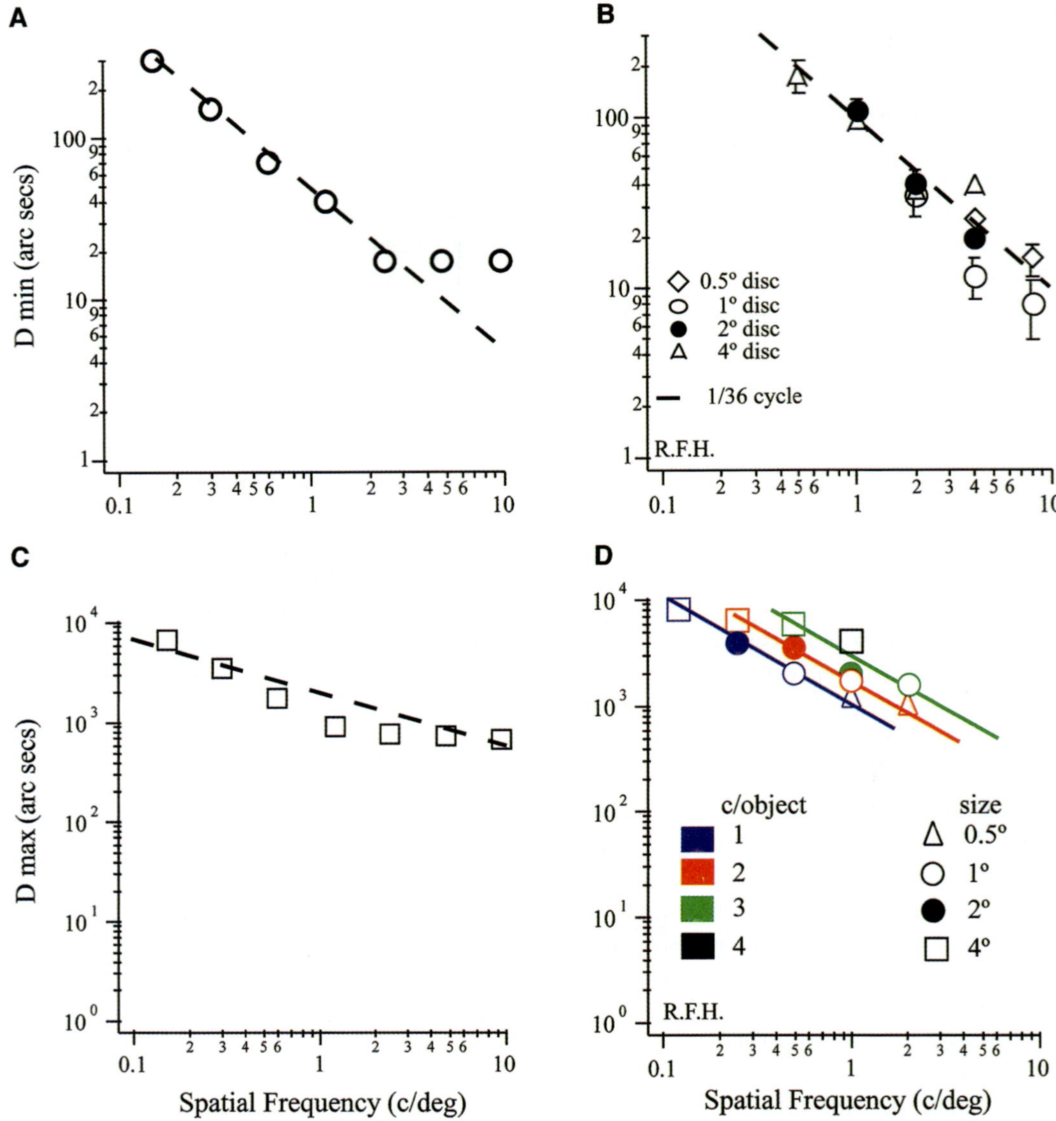

FIGURE 42.6   The scale-dependent nature of local stereo processing. (A and C) Results replotted from Schor and Woods (1983) for $D_{\min}$ and $D_{\max}$, respectively, for a one-dimensional bandpass DOG stimulus in which the overall size and peak spatial frequency covary. (B and D) Results for $D_{\min}$ and $D_{\max}$, respectively, for two-dimensional bandpass fractal noise stimuli in which overall size and peak spatial frequency have been decoupled (Hess, Liu, & Wang, 2002). $D_{\min}$ does scale with the peak spatial frequency of the image irrespective of stimulus size; $D_{\max}$ does not. $D_{\max}$ scales with object frequency (common colors) not retinal frequency (D).

for low-pass filtering. These predictions follow from possibly the simplest view of the relationship between spatial-frequency-tuned disparity mechanisms that the channel with the highest signal/noise ratio will determine performance. There are, of course, many other possibilities. For example, disparity mechanisms may receive input from the combined output of many spatial channels, and other factors may limit $D_{\max}$ and $D_{\min}$ (e.g., the type of local primitive derived from such a multiscale analysis).

The effect of various types of spatial filtering on stereopsis is seen in figure 42.6B and D. Figure 42.6B shows how $D_{\min}$ for a broadband fractal image varies with low-pass filtering (high-pass filtering has no effect, data not presented). Irrespective of the stimulus size (here spatial frequency and stimulus size have been studied separately), $D_{\min}$ gets progressively worse with progressive low-pass filtering (slope = 1/36th of the period of the highest frequency), bearing out the conclusions of Schor and Woods (1983). Figure 42.6D

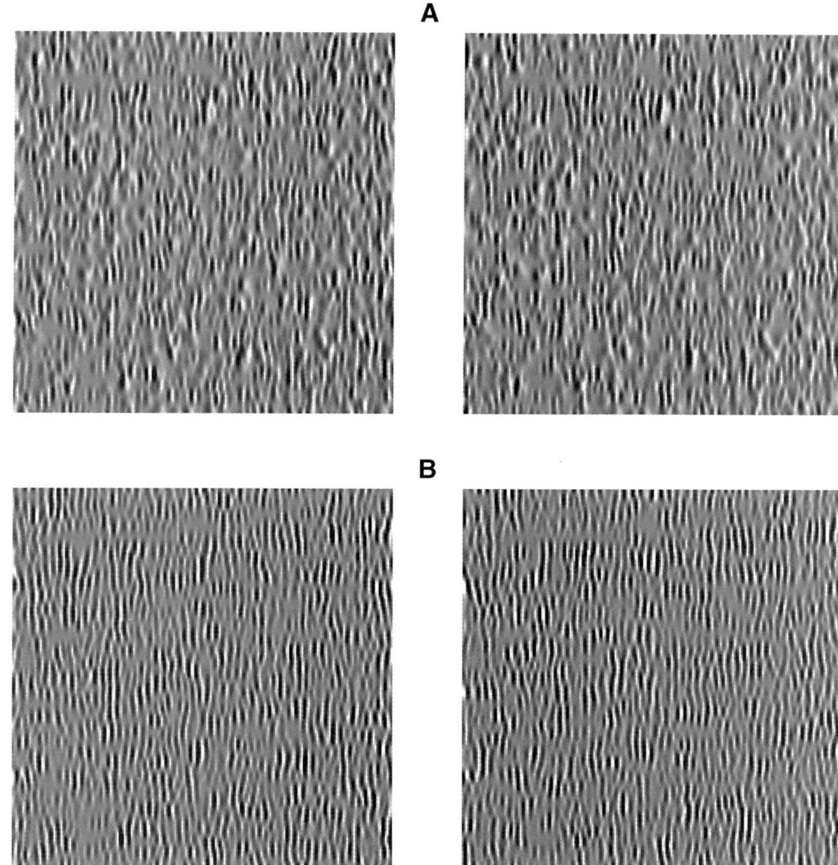

FIGURE 42.7   A scale-dependent example of transparency for global stereopsis. (A and B) Example of dual-surface disparity gratings. Fusion of the two stereo pairs in A and B results in the percept of an obliquely-oriented corrugated structure. This structure is actually composed of two interwoven sinusoidal disparity surfaces added out of phase. (A) Each surface is composed of Gabor elements of a single scale, although there is an octave difference in the scale that represents each surface. In this case, the transparent nature of the two surfaces is perceived. (B) Gabors of different scale are not segregated with respect to the surfaces. In this case, transparency is not perceived. (Reproduced with permission from Kingdom, Ziegler, & Hess, 2001.)

shows similar results for D_{max} but this time in terms of high-pass filtering (low-pass filtering has no effect, data not presented). However, the relationship is now much shallower (square root) than the independent scale prediction (slope of unity), and there is an effect of stimulus size. In fact, D_{max} seems to show the scale-dependent prediction (slope of unity) when we compared performance for stimuli not with the same spatial frequency (i.e., cycles per degree) but the same object frequency (cycles per object), suggesting it follows the information content of the stimulus rather than a spatial scale limit imposed by the visual system. Although D_{min} can be thought of as reflecting the activity of the highest spatial frequency tuned disparity neurons supported by a particular stimulus, D_{max} can not. It must involve information combined across scale. The Ohzawa and Freeman (1986) model would work for D_{min} but not for D_{max}. A combination of the model of Ohzawa and Freeman (1986) and the implied model from the work

of Barlow, Blakemore, and Pettigrew (1967) might be able to explain both limits.

There is more to stereopsis than the measures D_{max} and D_{min} for surfaces in the frontoparallel plane. For example, local stereopsis can support the perception of complex three-dimensional surfaces (Tyler, 1974). Does spatial scale maintain its relevance at this higher level? The answer is *yes*, at least in some cases. Take the example of two identical stereodefined sinusoidal surfaces added 180° out of phase (figure 42.7). Such a stimulus is difficult to disambiguate when each surface is made up of an array of micropatterns comprising the mixture of two spatial frequencies (figure 42.7A) an octave apart. Yet these surfaces can be disambiguated (figure 42.7B) when each surface contains its own exclusive spatial scale (Kingdom, Ziegler, & Hess, 2001). This segregation by spatial scale has obvious ecological relevance because similar objects (e.g., foliage) at different depths will be represented at different spatial

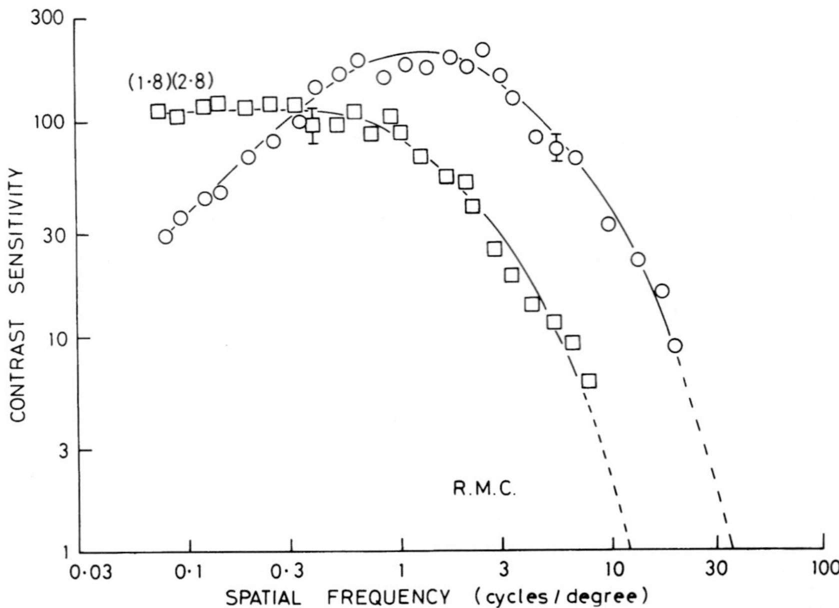

FIGURE 42.8   Spatial scale differences between achromatic and chromatic vision. Contrast sensitivity functions for achromatic and chromatic stimuli show that whereas the former excels at high spatial frequencies, the latter excels at low spatial frequencies. (Reproduced with permission from Mullen, 1985.)

scales, and this difference in scale by itself can aid segregation.

Scale and Color

Any understanding of the relation between chromatic and achromatic processing necessitates an understanding of spatial scale. A comprehensive comparison of the spatial processing in chromatic and achromatic vision was done using the contrast sensitivity function (Granger & Heurtley, 1973; Mullen, 1985; Sekiguchi, Williams, & Brainard, 1993). This showed that the overall spatial sensitivity of mechanisms that process red–green and blue–yellow are very different from that of the achromatic system. Chromatic sensitivity is low pass compared with the bandpass response of the achromatic system. Furthermore, although chromatic sensitivity is worse at higher spatial frequencies compared with its achromatic counterpart, its low spatial frequency sensitivity is better. These results are shown in figure 42.8. Chromatic processes and achromatic processes work best at complementary spatial scales. Color vision does not provide us with the sort of detail that achromatic vision is capable of, but it does provide us with the ability to segment large regions based on subtle color differences.

The spatial properties of the individual mechanisms underlying our chromatic sensitivity are similar to those previously described for achromatic vision, namely narrowly tuned bandpass mechanisms extending over the entire range (Bradley, Switkes, & De Valois, 1988; Losada & Mullen, 1994, 1995; Mullen & Losada, 1999). The scale-dependent rules for how sensitivity falls off across the visual field that have been described above for luminance gratings are also relevant to chromatic sensitivity with the proviso that red–green chromatic sensitivity falls off much more rapidly than its achromatic counterpart (Mullen, 1991). Blue–yellow and red–green cone opponencies are distributed differently across the visual field, suggesting different underlying neural constraints (Mullen & Kingdom, 2002). However, this difference, like the above chromatic/achromatic difference, does not itself depend on spatial scale (K. T. Mullen, personal communication).

SCALE SELECTION

The ability to process information independently at different scales has some obvious advantages. These involve situations where scales are differentially affected, for example, in reduced luminance levels, peripheral viewing, motion, stereo, and color. Under these situations, common information derived from multiple scales will be less corrupted, and, as a consequence, perceptions are more stable. This benefit comes at a cost. The cost is, first, that some of the more interesting image features occur only after scale combination, and second, for information that is carried independently within multiple scales, one has to decide which scales to select and which scales to ignore: the problem of

606   ROBERT F. HESS

scale selection. "Intelligent" selection of spatial scale is always better than a "dumb" combination, but the question is "What constitutes intelligent selection?" A number of different rules have been proposed, each with application to a specific task. These range from selection of the scale with the best signal/noise ratio or the minimum variability to the tracking of filter output features across scale.

Initially, at least for texture tasks, rules were delineated for the selection of features for segregation (Julesz, 1981). Some of these feature rules for textures can be recast as filter selection, especially where the contrast polarity of features is important. A number of different suggestions have been made for the selection of the appropriate scale of analysis. For example, within the image-processing field, features such as zero crossings from the output of filters of different spatial scale were identified as important markers of image features. One scale selection rule was proposed whereby these zero crossings were tracked across scale, and the highest scale at which they persist was selected (Witkin & Tennenbaum, 1983). This initial approach, which was successful for one-dimensional features, ran into difficulty in its application to two-dimensional features (Yuille & Poggio, 1986). Another suggestion is to use the scale with the best signal/noise ratio. A number of studies on motion (Bex et al., 1995; Brady, Bex, & Fredericksen, 1997; Eagle, 1996; Hess et al., 1998) and stereo (Hess, Liu, & Wang, 2002) have assumed this. Malik and Perona (1990) proposed a "leader-takes-all" rule for a texture discrimination task. Elder and Zucker (1998) proposed the use of the minimum reliable scale for edge detection. Dakin (1997) proposed a statistical rule, the minimization of local orientation variance for a global orientation task.

EVIDENCE FOR SCALE COMBINATION

Useful information in an image occurs at different scales within one spatial region as well as at one scale across different regions. Scale combination may help provide important object-based information in the presence of surface noise (Marr, 1982). Below is a list of specific situations where a rigid (i.e., indiscriminate) scale combination has been shown to occur. In some cases it is not the stimuli but the task that determines whether the visual system combines information across scale or analyzes it separately at each of a number of scales. An example of the task-dependent nature of scale combination is in "local" versus "global" discrimination tasks. For both spatial (Dakin & Bex, 2001) and motion detection (Bex & Dakin, 2000) at a local level, elements are grouped in a scale-dependent way but at a global level; these local groupings are integrated across scale. On the other hand, proposals have been made for a rigid scale combination rule operating from coarse to fine scales. This was originally proposed to solve the correspondence problem for stereo (Marr & Poggio, 1979) but later adapted into a model of spatio-temporal analysis, where progressively finer scales are analyzed at progressively later times (Watt, 1987; Parker, Lishman, & Hughes, 1997).

SPACE

Harmon and Julesz (1973) were the first to show that images that are block-quantized are difficult to recognize. Compare figure 42.9 (top row). The low-spatial-frequency components that are shown in rightmost panel on the top row are also present in middle panel of the top row but cannot be used to aid recognition. Although at first this was thought to be due to "masking" of the low spatial frequencies by the high spatial frequencies introduced by the quantization, it was later realized that it had more to do with phase continuity across scale (Canny, 1983; Hayes, 1989; Morrone & Burr, 1983). Compare figure 42.9 panels on the bottom row: In the middle panel of the bottom row, additional high spatial frequencies similar to those introduced by blocking are added, but recognition is not so impaired as it is in the leftmost panel of the bottom row. Image identification is restored when the phase relationships are disrupted between the (blocking) high frequencies and the (original) low-spatial-frequency components of the image. This is seen in figure 42.9 in the rightmost panel of the bottom row, where the phase of the high-frequency components introduced by blocking have been shifted by 180°—now recognition is easier. If the visual system carried out an independent scale analysis without rigid combination, a block-quantized image should be able to be recognized via its low-spatial-frequency components. For example, faces that are blurred are not unrecognizable (figure 42.9, rightmost panel of the top row). That this is not the case suggests that although processing may initially occur within scale, it is only by combining information across scale that interesting image features can be revealed.

Texture Perception

Glass patterns are visual textures composed of a field of element pairs (dipoles) whose global orientation is determined by a simple geometric transformation (Glass, 1969). These patterns are ideal for looking at scale combination because it has been shown that this task can be performed equally well at a number of

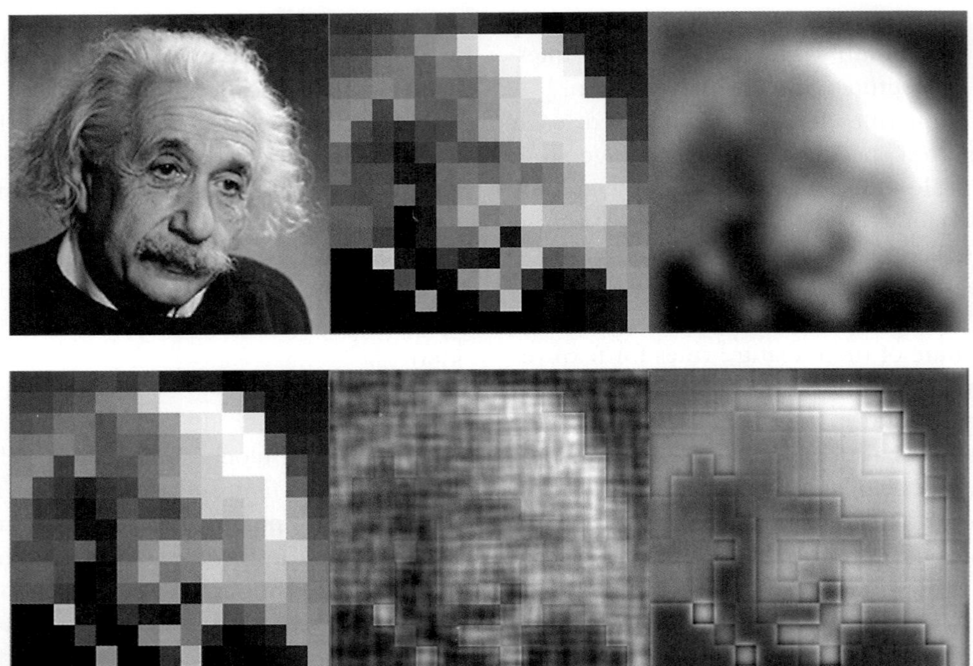

FIGURE 42.9   Object recognition does not occur independently within different scales. In the top frame, recognition is impaired when the image is block-quantized (top, middle) and to a much worse extent than expected on the basis of the available low-spatial-frequency information (top, right). This effect was originally thought to be due to critical band masking (Harmon & Julesz, 1973), but adding high frequencies (bottom, middle) does not impair performance (Morrone & Burr, 1983) to the same extent as block-quantization (bottom, left). Flipping the luminance polarity (bottom, right) of the original block-quantized edges does aid recognition (Hayes, 1989). (Reproduced with permission of S. C. Dakin.)

independent spatial scales (Dakin, 1997). To detect the global structure in these multielement arrays, observers must perform local grouping at the level of the dipoles in order to detect the local oriented structure and global grouping at the level of the array as a whole in order to combine these local orientation estimates to derive global structure. Dakin and Bex (2001) have shown that the local grouping operation occurs independently at each of a number of spatial scales (i.e., has a bandpass tuning), whereas the global grouping combines information across spatial scale (i.e., has a low-pass tuning). The spatial tuning results for these two processes are seen in figure 42.10 for the geometric transformations of rotation, translation, and expansion.

Blur Perception

Our understanding of how we discriminate blur is still unresolved. There have been a number of different proposals. In some cases (Elder & Zucker, 1998; Field & Brady, 1997; Mather, 1997; Watt & Morgan, 1983), the neural mechanism involved would need to collapse information across spatial scale, whereas in others

(Georgeson, 1994) multiscale or single-scale (Elder & Zucker, 1998) processing would suffice.

The straightforward suggestion that blur is signaled by the degree of activity at the highest spatial scale is not correct because we are better at discriminating a change in blur from a reference edge that is already blurred compared with a reference edge that is perfectly sharp (Watt & Morgan, 1983). This implies that either we can not independently access the highest spatial frequency mechanisms that we possess to signal blur or that we choose not to use it for other reasons (i.e., if the finest scale was too noisy). Watt and Morgan (1983) proposed that there was a rigid combination of filter outputs (MIRAGE model) prior to blur analysis. They suggested that blur was signaled by the distance between adjacent peaks and troughs in the output of a second-derivative filter. Field and Brady (1997) and Mather (1997) proposed that images look blurred when there is a change in the relative amplitude of detectable structure at different frequencies. To determine this would require a comparison of the activity across scales rather than solely within. Georgeson (1994) proposed (see also Kayargadde & Martens, 1996) another way of computing blur for aperiodic and periodic sinusoids

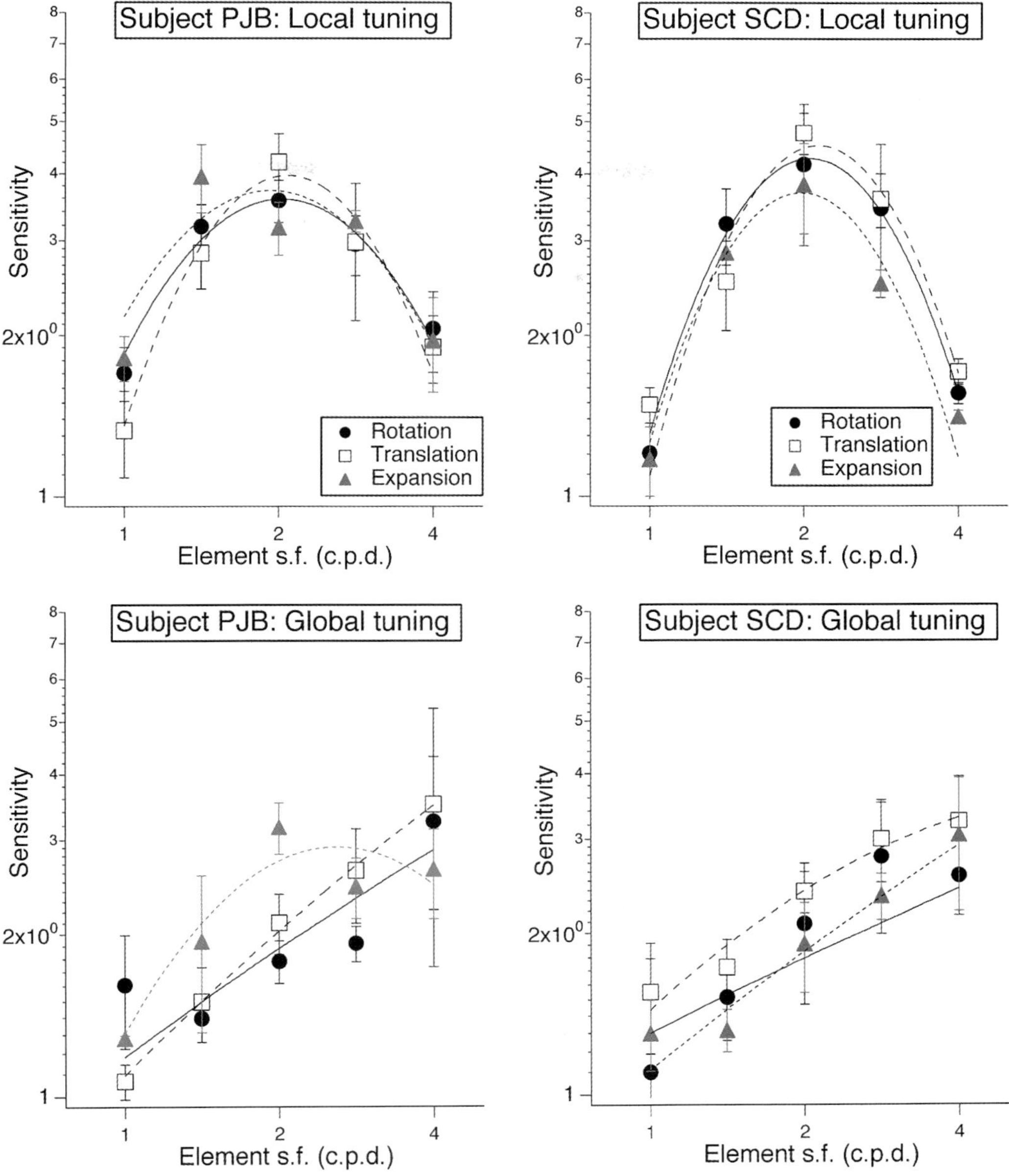

FIGURE 42.10 Spatial frequency tuning of local and global grouping for the recognition of structure in Glass patterns. In the top two frames, local sensitivity (the reciprocal of the signal/noise ratio at threshold) is plotted as a function of the spatial frequency of the element paired with a 2 cycles/deg dipole element. Here, sensitivity directly reflects the sensitivity of the underlying mechanism. In the bottom two frames, global sensitivity is plotted as a function of the spatial frequency of a masking stimulus. Here, sensitivity is related to the sensitivity of the underlying mechanism. Local grouping occurs within scale, whereas global grouping occurs across scale. (Reproduced with permission from Dakin & Bex, 2001.)

based on a multiscale template model (Georgeson, 2001) in which blur is encoded by finding which of a set of multiscale Gaussian derivative templates best fits the second-derivative profile of the edge; this model is simple, multiscale, fits well and (for one-dimensional edges) has no free parameters, and could be readily implemented by simple cells. An important contribution was also made by Elder and Zucker (1998), who used adaptive scale filtering—at each location the

system uses the "smallest reliable scale," defined by signal-to-noise ratio.

Motion Velocity

Even though the early stages of visual processing involve the activity of filters that are space-time oriented but separable in terms of their spatial and temporal frequency dependence, there is evidence that our discrimination of image motion is in terms of velocity rather than temporal frequency. To achieve this it is necessary to have filters that are nonseparable (i.e., oriented) in spatial frequency and temporal frequency, necessitating the combination of information across spatial and temporal scales prior to the processing of image velocity. Psychophysical studies have provided evidence for the importance of velocity in motion processing. Thompson (1981) showed that a common metric for motion adaptation was in terms of velocity rather than spatial and temporal frequency. McKee, Silverman, and Nakayama (1986) showed the importance of velocity for the discrimination of image motion. More recently, it has been shown that the visual system integrates information across different orientation bands in a nonselective manner, leading to suboptimal performance in some situations (Schrater, Knill, & Simoncelli, 2000). Perrone and Thiele (2001) and more recently Priebe, Lisberger, and Movshon (2006) and Nishimoto and Gallant (2006) have provided neurophysiological support for the receptive field properties of neurons at the level of area MT in the extrastriate cortex being oriented in spatial frequency and temporal frequency. This suggests that information previously contained within individual spatial and temporal scales in area V1 have been combined at the level of MT to encode velocity. A computational model describing the extraction of image velocity by putative MT cells based on the responses of oriented and directionally selective V1 cells has been proposed by Simoncelli and Heeger (1998).

COMBINATION RULES

At some stage within the visual process, for particular tasks, it is advantageous to combine information across spatial scale to derive interesting image features. The type of combination will critically depend on the nature of the image feature as many different combination rules have been proposed; these range from "blind" (i.e., indiscriminate) pooling to optimal pooling strategies. The main determinants are the particular feature and task involved.

For example, if the feature is the global orientation in an image then it has been shown that information is first collapsed across scale or spatial frequency. Similarly, if the selected feature is the local spatial frequency structure, combination across orientation occurs (Dakin & Bex, 2001; Olzak & Thomas, 1991, 1992; Thomas & Olzak, 1996). There is also evidence that the visual system contains higher-level mechanisms that sum across spatial scale to process the orientation of image features but do not sum across scale to process either

the spatial frequency or contrast of image features (Olzak & Wickens, 1997). If the feature is local velocity, Schrater, Knill, and Simoncelli (2000) have shown that the visual system uses a fixed combination rule over orientation rather than one adapted to the spatial properties of a particular stimulus. Their result is consistent with velocity tuned detectors that measure the energy in fixed (i.e., blind pooling), orientationally broadband, planar regions of spatiotemporal frequency (see also Nishimoto & Gallant, 2006; Simoncelli & Heeger, 1998). If the feature is the global structure in either Glass patterns (Dakin & Bex, 2001) or random-dot kinematograms (Bex & Dakin, 2000), information is combined across spatial scale with greater weight given to lower spatial frequencies. If the feature is the global perception of plaids composed of two oblique sinusoidal gratings, information is collapsed across orientation prior to edge extraction (Georgeson & Meese, 1997).

A special case of interscale combination occurs for edge features where the combination process is phase sensitive. A number of computational schemes for edge detection exploit this by assessing the correlated activity between filter outputs across scale (Canny, 1983; Georgeson, 1992, 1994; Lowe, 1988; Marr, 1982; Marr & Hildreth, 1980; Morrone & Burr, 1988; Torre & Poggio, 1986). Georgeson and Meese (1997) report that the addition of a $3f$ component, if added in square-wave phase to one component of their plaid stimuli, breaks down the previous combination across orientation. The perception of the plaid is now in terms of its components. This suggests that a phase-sensitive combination of scale to extract edge features may be a special case and may take precedence over interorientation linking. Dakin and Hess (1999), using a contour integration task between elements with broad and narrow band spatial properties, also argue that edge extraction may represent a special case and occur concurrent with or prior to contour integration.

SPATIAL SCALE INVARIANCE

Some visual sensitivities display a viewing distance invariance and are termed scale invariant. This can occur for a number of different reasons. For example, D_{max} for random-dot kinematograms was initially found to exhibit a form of scale invariance (Lappin & Bell, 1976) This was probably because as the viewing distance is reduced and the stimulus extends to greater eccentricities, lower-spatial-frequency components in the now-magnified noise stimulus could then determine the largest jump size for motion (Baker & Braddick, 1985). Another form of scale invariance can occur for tasks that do not exhibit any dependence on element spatial frequency, such as a global form task (Achtman, Hess, & Wang, 2003). A different and much more interesting form of scale invariance occurs for some types of so-called second-order stimuli. The processing of these stimuli is thought to involve two serial stages of visual processing. The first stage involves bandpass filters. The second stage, which also involves bandpass filters, combines the input from the first-stage filters. These stimuli comprise modulations in contrast, motion, orientation, disparity, and others. The general finding is that the sensitivity for detecting these modulations is not dependent on viewing distance. What this means is that modulation sensitivity does not depend on retinal spatial frequency (i.e., cycles per degree) but rather on object spatial frequency (cycles per object). This form of scale invariance is seen for contrast modulation (Jamar & Koenderink, 1985), motion modulation (Meso & Hess, 2010), and orientation flow-field modulation (Kingdom & Keeble, 1999). However, it is not universally seen for second-order stimuli. For example, it is not seen for all types of orientation modulation (Meso & Hess, 2011) or for disparity modulation (N. Wirtz & R. F. Hess, unpublished data). In the cases where scale invariance is exhibited (i.e., contrast modulation, motion modulation, and orientation flow fields), there is also an optimal ratio between carrier (tuning of first-stage filters) and modulator (tuning of second-stage filters) spatial frequencies.

CONCLUSION

It has been argued by some that the visual system processes information only within scale, and by others that information is processed only across scale. However, one can find examples of both within- as well as across-scale analysis in vision, and this is quite task dependent. For some tasks, scale combination occurs at a later stage in visual processing than in others. One example is the efficient coding of edges. At the level of the striate cortex we know that we have neurons with the appropriate spatial tuning properties to encode edge structure successfully via its harmonic components. However, if edges are an important feature of everyday images, and if we want to achieve as sparse a code as possible (Olshausen & Field, 1996, 1997; see also chapter 87 by Olshausen and Lewicki), it might be advantageous to also have some hardwired edge-detecting neurons (combination across scale) that would more sparsely encode such common natural features. Having such detectors would not preclude the use of multiple scale processing for broadband images in which the phase relations were random (e.g., noise). This is an example of where within- and combined-scale analysis could coexist at the same stage of processing and even be capable of competing to accomplish similar tasks.

Initially the within-scale processing idea was taken to extreme limits by assuming that the visual system did a global Fourier analysis of the retinal image as a whole. This view has now given way to the more moderate proposal that visual information can be processed separately at each of a number of scales within a localized region of the field, at least in early parts of the cortical pathway. Understanding this gives predictive power about what limits various types of visual processes. It is, of course, not the full story. In some cases, for example, image velocity encoding, spatial combination is a fundamental part of the extraction of the relevant visual information. Spatial scale is now a fundamental part of our thinking in vision, particularly relevant to the early stages of visual processing. It is as hazardous to neglect it as it is not to be aware of its limitations.

ACKNOWLEDGMENTS

I am grateful to all my collaborators throughout the years for their insight, patience, and understanding and to the CIHR and NSERC for their support. I am grateful to Rebecca Achtman for helping with the illustrations.

REFERENCES

Achtman, R. L., Hess, R. F., & Wang, Y. Z. (2003). Sensitivity for global shape detection. *Journal of Vision (Charlottesville, Va.)*, *3*, 616–624. doi:10.1167/3.10.4.

Adelson, E. H., & Bergen, J. R. (1985). Spatio-temporal energy models for the perception of motion. *Journal of the Optical Society of America. A, Optics and Image Science*, *2*, 284–299.

Baker, C. L., Jr. (1999). Central neural mechanisms for detecting second-order motion. *Current Opinion in Neurobiology*, *9*, 461–466.

Baker, C. L., Jr., & Braddick, O. J. (1985). Eccentricity-dependent scaling of the limits for short-range apparent motion perception. *Vision Research*, *25*, 803–812.

Barlow, H. B., Blakemore, C., & Pettigrew, J. D. (1967). The neural mechanism of binocular depth discrimination. *Journal of Physiology*, *193*, 327–342.

Bex, P. J., Brady, N., Fredericksen, R. E., & Hess, R. F. (1995). Energetic motion detection. *Nature*, *378*, 670–672. doi:10.1038/378670b0.

Bex, P. J., & Dakin, S. C. (2000). Narrowband local and broadband global spatial frequency selectivity for motion perception. *Investigative Ophthalmology & Visual Science*, *41*(Suppl.), s545.

Bischof, E. H., & di Lollo, V. (1990). Perception of directional sampled motion in relation to displacement and spatial frequency: Evidence for a unitary motion system. *Vision Research*, *9*, 1341–1362. doi:10.1016/0042-6989(90)90008-9.

Bischof, W. F., & di Lollo, V. (1991). On the half-cycle displacement limit of sampled directional motion. *Vision Research*, *31*, 649–660. doi:10.1016/0042-6989(91)90006-Q.

Blakemore, C., & Campbell, F. W. (1969). On the existence of neurones in the human visual system selectively sensitive to the orientation and size of retinal images. *Journal of Physiology, 203*, 237–260.

Bradley, A., Switkes, E., & De Valois, K. (1988). Orientation and spatial frequency selectivity of adaptation to color and luminance gratings. *Vision Research, 28*, 841–856. doi:10.1016/0042-6989(88)90031-4.

Brady, N., Bex, P. J., & Fredericksen, R. E. (1997). Independent coding across spatial scales in moving fractal images. *Vision Research, 37*, 1873–1884. doi:10.1016/S0042-6989(97)00007-2.

Campbell, F. W., Cleland, B., Cooper, G. F., & Enroth-Cugell, C. (1968). The angular selectivity of visual cortical cells to moving gratings. *Journal of Physiology, 198*, 237–250.

Campbell, F. W., Cooper, G. F., & Enroth-Cugell, C. (1969). The spatial selectivity of the visual cells of the cat. *Journal of Physiology, 203*, 223–235.

Campbell, F. W., & Green, D. G. (1965). Optical and retinal factors affecting visual resolution. *Journal of Physiology, 181*, 576–593.

Campbell, F., & Robson, J. (1968). Application of Fourier analysis to the visibility of gratings. *Journal of Physiology, 197*, 551–566.

Cannon, W. M., & Fullenkamp, S. C. (1991). Spatial interactions in apparent contrast: Inhibitory effects among grating patterns of different spatial frequencies, spatial positions and orientations. *Vision Research, 31*, 1985–1998. doi:10.1016/0042-6989(91)90193-9.

Canny, J. F. (1983). *Finding Edges and Lines in Images* (AI-TR-720). Boston: MIT AI Laboratory Technical Report.

Chang, J. J., & Julesz, B. (1983). Displacement limits, directional anisotropy and direction versus form discrimination in random dot cinematograms. *Vision Research, 23*, 639–646.

Chang, J. J., & Julesz, B. (1985). Cooperative and non-cooperative processes of apparent motion of random dot cinematograms. *Spatial Vision, 1*, 39–41. doi:10.1163/156856885X00062.

Chubb, C., Sperling, G., & Solomon, J. A. (1989). Texture interactions determine perceived contrast. *Proceedings of the National Academy of Sciences of the United States of America, 86*, 9631–9635. doi:10.1073/pnas.86.23.9631.

Cleary, R. (1990). Contrast dependence of short range apparent motion. *Vision Research, 30*, 463–478.

Cleary, R., & Braddick, O. J. (1990). Direction discrimination for bandpass filtered random dot kinematograms. *Vision Research, 30*, 303–316.

Crook, J. M., Lange-Malecki, B., Lee, B. B., & Valbert, A. (1988). Visual resolution of macaque retinal ganglion cells. *Journal of Physiology, 396*, 205–224.

Dakin, S. C. (1997). The detection of structure in Glass patterns: Psychophysics and computational models. *Vision Research, 37*, 2227–2259. doi:10.1016/S0042-6989(97)00038-2.

Dakin, S. C., & Bex, P. J. (2001). Local and global visual grouping: Tuning for spatial frequency and contrast. *Journal of Vision (Charlottesville, Va.), 1*, 99–112. doi:10.1167/1.2.4.

Dakin, S. C., & Hess, R. F. (1999). Contour integration and scale combination processes in visual edge detection. *Spatial Vision, 1*, 309–327. doi:10.1163/156856899X00184.

Dakin, S. C., & Mareschal, I. (2000). Sensitivity to contrast modulation depends on carrier spatial frequency and orientation. *Vision Research, 4*, 311–329.

De Monastereo, F. M., & Gouras, P. (1975). Functional properties of ganglion cells in the rhesus monkey retina. *Journal of Physiology, 25*, 167–195.

De Valois, R. L., Albrecht, D. G., & Thorell, L. G. (1982). Spatial frequency selectivity of cells in macaque visual cortex. *Vision Research, 22*, 545–559.

DeVries, H. (1943). The quantum nature of light and its bearing upon the threshold of vision, the differential sensitivity and visual acuity of the eye. *Physica, 10*, 553–564.

Eagle, R. A. (1996). What determines the maximum displacement limit for spatially broadband kinematograms. *Journal of the Optical Society of America. A, Optics, Image Science, and Vision, 13*, 408–418.

Ejima, Y., & Takahashi, S. (1985). Apparent contrast of a sinusoidal grating in the simultaneous presence of peripheral gratings. *Vision Research, 25*, 1223–1232.

Elder, J. H., & Zucker, S. W. (1998). Local scale control for edge detection and blur estimation. *IEEE Transactions on Pattern Analysis and Machine Intelligence, 20*, 699–716.

Ellemberg, D., Wilkinson, F., Wilson, H. R., & Arsenault, A. S. (1998). Apparent contrast and spatial frequency of local texture elements. *Journal of the Optical Society of America. A, Optics, Image Science, and Vision, 15*, 1733–1739. doi:10.1364/JOSAA.15.001733.

Enroth-Cugell, C., & Robson, J. G. (1966). The contrast sensitivity of retinal ganglion cells of the cat. *Journal of Physiology, 187*, 517–552.

Field, D. J. (1987). Relations between the statistics of natural images and the response properties of cortical cells. *Journal of the Optical Society of America. A, Optics and Image Science, 4*, 2379–2394. doi:10.1364/JOSAA.4.002379.

Field, D. J. (1993). Scale-invariance and self-similar "wavelet" transforms: An analysis of natural scenes and mammalian visual systems. In M. Marge, J. C. R. Hunt, & J. C. Vassilicos (Eds.), *Wavelets, fractals and Fourier transforms* (pp. 151–193). Oxford: Clarendon Press.

Field, D. J., & Brady, N. (1997). Visual sensitivity, blur and the sources of variability in the amplitude spectra of natural scenes. *Vision Research, 37*, 3367–3384. doi:10.1016/S0042-6989(97)00181-8.

Field, D. J., Hayes, A., & Hess, R. F. (1993). Contour integration by the human visual system: evidence for a local "association field." *Vision Research, 33*, 173–193. doi:10.1016/0042-6989(93)90156-Q.

Foster, K. H., Gaska, J. P., Nagler, M., & Pollen, D. A. (1985). Spatial and temporal frequency selectivity of neurones in visual cortical areas V1 and V2 of the macaque monkey. *Journal of Physiology, 365*, 331–363.

Georgeson, M. (1992). Human vision combines oriented filters to compute edges. *Proceedings. Biological Sciences, 249*, 235–245.

Georgeson, M. A. (1994). From filters to features: Location, orientation, contrast and blur. In M. J. Morgan (Ed.), *Higher-Order Processing in the Visual System: CIBA Foundation Symposium 184* (pp. 147–165). Chichester, UK: John Wiley & Sons.

Georgeson, M. A. (2001). *Seeing edge blur: Receptive fields as multi-scale neural templates*. Vision Sciences Conference, Sarasota, Florida.

Georgeson, M. A., & Meese, T. S. (1997). Perception of stationary plaids: The role of spatial filters in edge analysis. *Vision Research, 37*(23), 3255. doi:10.1016/S0042-6989(97)00124-7.

Glass, L. (1969). Moiré effects from random dots. *Nature, 243*, 578–580.

Glennerster, A., & Parker, A. J. (1997). Computing stereo channels from masking. *Vision Research, 37*, 2143–2152.

Glezer, V. D., Cooperman, A. M., & Tscherbach, T. A. (1973). Investigation of complex and hyper-complex receptive fields of visual cortex of the cat as spatial frequency filters. *Vision Research, 13*, 1875–1904.

Graham, N., & Nachmias, J. (1971). Detection of grating patterns containing two spatial frequencies: A comparison of single channel and multi-channel models. *Vision Research, 11*, 251–259.

Graham, N., Robson, J. G., & Nachmias, J. (1978). Grating summation in fovea and periphery. *Vision Research, 18*, 816–825.

Graham, N., Sutter, A., & Venkatesan, C. (1993). Spatial frequency and orientation selectivity of simple and complex channels in regional segregation. *Vision Research, 33*, 1893–1911. doi:10.1016/0042-6989(93)90017-Q.

Granger, E. M., & Heurtley, J. C. (1973). Visual chromaticity-modulation transfer function. *Journal of the Optical Society of America, 63*, 1173–1174.

Harmon, L. D., & Julesz, B. (1973). Masking in visual recognition: Effects of two dimensional filtered noise. *Science, 180*, 1194–1197.

Hayes, A. (1989). *Representation by images restricted in resolution and intensity range* (PhD thesis). Department of Psychology, University of Western Australia, Perth, Australia.

Hess, R. F. (1990). Vision at low light levels: Role of spatial, temporal and contrast filters. *Ophthalmic & Physiological Optics, 10*, 351–359.

Hess, R. F., Liu, H. C., & Wang, Y. Z. (2002). Luminance spatial scale and local stereo-sensitivity. *Vision Research, 42*(3), 331. doi:10.1016/S0042-6989(01)00285-1.

Hubel, D. H., & Wiesel, T. N. (1959). Receptive fields of single neurons in the cat's striate cortex. *Journal of Physiology, 148*, 574–591.

Hubel, D. H., & Wiesel, T. N. (1960). Receptive fields of optic nerve fibres in the spider monkey. *Journal of Physiology, 154*, 572–580.

Hubel, D. H., & Wiesel, T. N. (1962). Receptive fields, binocular interaction and functional architecture in the cat's visual cortex. *Journal of Physiology, 160*, 106–154.

Hubel, D. H., & Wiesel, T. N. (1968). Receptive fields and functional architecture of monkey striate cortex. *Journal of Physiology, 195*, 215–243.

Issa, N. P., Trepel, C., & Stryker, M. P. (2000). Spatial frequency maps in cat visual cortex. *Journal of Neuroscience, 20*(22), 8504–8514.

Jamar, J. H., & Koenderink, J. J. (1985). Contrast detection and detection of contrast modulation for noise gratings. *Vision Research, 25*, 511–521.

Johnston, A., McOwen, P., & Buxton, H. (1992). A computational model of the analysis of some first-order and second-order motion patterns by simple and complex cells. *Proceedings of the Royal Society of London, B250*, 297–306.

Julesz, B. (1981). Textons, the elements of texture perception, and their interactions. *Nature, 29*, 91–97.

Kaufman, D. A., & Palmer, L. A. (1990). The luminance dependence of the spatiotemporal response of cat striate cortical cells. *Investigative Ophthalmology & Visual Science, 31*(Suppl), 398.

Kayargadde, V., & Martens, J. B. (1996). Perceptual characterization of images degraded by blur and noise: experiments. *Journal of the Optical Society of America A-Optics & Image Science, 13*, 1166–1177.

Kingdom, F. A., & Keeble, D. R. (1999). On the mechanism for scale invariance in orientation-defined textures. *Vision Research, 39*, 1477–1489. doi:10.1016/S0042-6989(98)00217-X.

Kingdom, F. A. A., & Keeble, D. R. T. (2000). Luminance spatial frequency differences facilitate the segmentation of superimposed textures. *Vision Research, 40*, 1077–1087.

Kingdom, F. A. A., Ziegler, L. R., & Hess, R. F. (2001). Luminance spatial scale facilitates depth segmentation. *Journal of the Optical Society of America. A, Optics, Image Science, and Vision, 18*, 993–1002. doi:10.1364/JOSAA.18.000993.

Klein, S. A., Stromeyer, C. F. III, & Ganz, L. (1974). The simultaneous spatial frequency shift: A dissociation between the detection and perception of gratings. *Vision Research, 14*, 1421–1432.

Kontsevich, L. L., & Tyler, C. W. (1994). Analysis of stereo-thresholds for stimuli below 2.5 c/deg. *Vision Research, 34*, 2317–2329. doi:10.1016/0042-6989(94)90110-4.

Kourtzi, Z., & Kanwisher, N. (2001). Representation of perceived object shape by the human lateral occipital complex. *Science, 293*, 1506–1509.

Langley, K., Fleet, D. J., & Hibbard, P. B. (1996). Linear filtering precedes non-linear processing in early vision. *Current Biology, 6*, 891–896.

Lappin, J. S., & Bell, H. H. (1976). The detection of coherence in moving random dot patterns. *Vision Research, 16*(2), 161–168.

Ledgeway, T. (1996). How similar must the Fourier spectra of the frames of a random dot kinematogram be to support motion perception? *Vision Research, 36*, 2489–2495. doi:10.1016/0042-6989(95)00315-0.

Li, G., Wang, Z., Yao, Z., Yuan, N., Talebi, V., Tan, J., et al. (2011). Form-cue invariant second-order contrast envelope responses in macaque V2. *Society for Neuroscience Abstract*, Program # 271.08/ Poster II27.

Losada, M. A., & Mullen, K. T. (1994). The spatial tuning of chromatic mechanisms identified by simultaneous masking. *Vision Research, 34*, 331–341.

Losada, M. A., & Mullen, K. T. (1995). Color and luminance spatial tuning estimated by noise masking in the absence of off-frequency looking. *Journal of the Optical Society of America. A, Optics, Image Science, and Vision, 12*, 250–260.

Lowe, D. G. (1988). *Proceedings from the second international conference on computer vision: Organization of smooth image curves at multiple spatial scales*. New York: IEEE Computer Society Press.

Mackay, D. M. (1973). Lateral interaction between neural channels sensitive to texture density. *Nature, 245*, 159–161.

Maffei, L., & Fiorentini, A. (1973). The visual cortex as a spatial frequency analyser. *Vision Research, 13*, 1255–1267.

Malik, J., & Perona, P. (1990). Preattentive texture discrimination with early visual mechanisms. *Journal of the Optical Society of America. A, Optics and Image Science, 7*, 923–932.

Marr, D. (1982). *Vision: A computational investigation into the human representation and processing of visual information*. New York: W. H. Freeman & Co.

Marr, D., & Hildreth, E. (1980). Theory of edge detection. *Proceedings of the Royal Society of London. Series B, Biological Sciences, 207*, 187–217.

Marr, D., & Poggio, T. (1979). A computational theory of human stereo vision. *Proceedings of the Royal Society of London. Series B, Biological Sciences, 204*, 301–328.

Marrett, S., Dale, A. M., Mendela, J. D., Sereno, M. I., Liu, A. K., & Tootell, R. B. H. (1997). Preferred spatial frequency varies with eccentricity in human visual cortex. *NeuroImage, 3*, s157.

Mather, G. (1997). The use of image blur as a depth cue. *Perception, 26*, 1147–1158.

McKee, S. P., Silverman, G. H., & Nakayama, K. (1986). Precise velocity discrimination despite random variations in temporal frequency and contrast. *Vision Research, 26*, 609–619.

Meso, A. I., & Hess, R. F. (2010). Visual motion gradient sensitivity shows scale invariant spatial frequency and speed tuning properties. *Vision Research, 50*, 1475–1485.

Meso, A. I., & Hess, R. F. (2011). Orientation gradient detection exhibits variable coupling between 1st and 2nd stage filtering mechanisms. *Journal of the Optical Society of America. A, Optics, Image Science, and Vision, 28*, 1721–1731.

Morgan, M. J. (1992). Spatial filtering precedes motion detection. *Nature, 335*, 344–346.

Morgan, M. J., & Mather, G. (1994). Motion discrimination in 2-frame sequences with differing spatial frequency content. *Vision Research, 34*, 197–208.

Morrone, M. C., & Burr, D. C. (1983). Added noise restores recognition of coarse quantized images. *Nature, 305*, 226–228.

Morrone, M. C., & Burr, D. C. (1990). Feature detection in human vision: A phase-dependent energy model. *Proceedings of the Royal Society of London. Series B, Biological Sciences, 235*, 221–245.

Movshon, J. A., Thompson, I. D., & Tolhurst, D. J. (1978). Spatial and temporal contrast sensitivity in areas 17 and 18 of the cat's visual cortex. *Journal of Physiology, 283*, 101–130.

Mullen, K. T. (1985). The contrast sensitivity of human colour vision to red-green and blue-yellow chromatic gratings. *Journal of Physiology, 359*, 381–400.

Mullen, K. T. (1991). Colour vision as a post-receptoral specialization of the central visual field. *Vision Research, 31*, 119–130.

Mullen, K. T., & Kingdom, F. A. A. (2002). Differential distribution of red-green and blue-yellow cone opponency across the visual field. *Visual Neuroscience, 19*, 108–118. doi:10.1017/S0952523802191103.

Mullen, K. T., & Losada, M. A. (1999). The spatial tuning of color and luminance peripheral vision measured with notch filtered noise. *Vision Research, 39*, 721–731.

Nishimoto, S., & Gallant, J. L. (2006). A three-dimensional spatiotemporal receptive field model explains responses of area MT neurons to naturalistic movies. *Journal of Neuroscience, 31*(41), 14552–14564.

Ohzawa, I., DeAngelis, G. C., & Freeman, R. D. (1990). Stereoscopic depth discrimination in the visual cortex: Neurones ideally suited as disparity detectors. *Science, 249*, 1037–1041. doi:10.1126/science.2396096.

Ohzawa, I., DeAngelis, G. C., & Freeman, R. D. (1996). Encoding of binocular disparity by simple cells in cat's visual cortex. *Journal of Neurophysiology, 75*, 1779–1805.

Ohzawa, I., & Freeman, R. D. (1986). The binocular organization of simple cells in cat striate cortex. *Journal of Neurophysiology, 56*, 221–242.

Olshausen, B. A., & Field, D. J. (1996). Emergence of simple-cell receptive field properties by learning a sparse code for natural images. *Nature, 381*, 607–609.

Olshausen, B. A., & Field, D. J. (1997). Sparse coding with an overcomplete basis set: A strategy employed by V1? *Vision Research, 37*, 3311–3325. doi:10.1016/S0042-6989(97)00169-7.

Olzak, L. A., & Thomas, J. P. (1991). When orthogonal orientations are not processed independently. *Vision Research, 31*, 51–57.

Olzak, L. A., & Thomas, J. P. (1992). Configural effects constrain Fourier models of pattern discrimination. *Vision Research, 32*, 1885–1898. doi:10.1016/0042-6989(92)90049-O.

Olzak, L. A., & Wickens, T. D. (1997). Discrimination of complex patterns: Orientation information is integrated across spatial scale; spatial frequency and contrast information are not. *Perception, 26*, 1101–1120.

Pantle, A., & Sekular, R. (1968). Size detecting mechanisms in human vision. *Science, 62*, 1146–1148.

Parker, D. M., Lishman, J. R., & Hughes, J. (1997). Evidence for the view that spatiotemporal integration in vision is temporally anisotropic. *Perception, 26*, 1169–1180. doi:10.1068/p261169.

Perrone, J. A., & Thiele, A. (2001). Speed skills: Measuring the visual speed analyzing properties of primate MT neurons. *Nature Neuroscience, 4*, 526–532.

Petrov, A. P., Pigarev, I. N., & Zenkin, G. M. (1980). Some evidence against Fourier analysis as a function of the receptive fields in cat's striate cortex. *Vision Research, 20*, 1023–1025. doi:10.1016/0042-6989(80)90087-5.

Pettigrew, J. D., Nikara, T., & Bishop, P. O. (1968). Binocular interaction on single units in cat striate cortex: Simultaneous stimulation by single moving slit with receptive fields in correspondence. *Experimental Brain Research, 6*, 394–410. doi:10.1007/BF00233186.

Pointer, J. S., & Hess, R. F. (1989). The contrast sensitivity gradient across the human visual field: Emphasis on the low spatial frequency range. *Vision Research, 29*, 1133–1151. doi:10.1016/0042-6989(89)90061-8.

Priebe, N. J., Lisberger, S. G., & Movshon, J. A. (2006). Tuning for spatiotemporal frequency and speed in directionally selective neurons of macaque striate cortex. *Journal of Neuroscience, 26*(11), 2941–2950.

Prince, S. J. D., Eagle, R. A., & Rogers, B. J. (1998). Contrast masking reveals spatial-frequency channels in stereopsis. *Perception, 27*, 1345–1355. doi:10.1068/p271345.

Robson, J. G., & Graham, N. (1981). Probability summation and regional variation in contrast sensitivity across the visual field. *Vision Research, 21*, 409–418.

Rose, A. (1942). Quantum and noise limitations of the visual process. *Journal of the Optical Society of America, 43*, 715–725.

Rose, A. (1948). The sensitivity performance of the human eye on an absolute scale. *Journal of the Optical Society of America, 38*, 196–208.

Rosenberg, A., & Issa, N. P. (2009). The Y cell visual pathway implements a demodulating nonlinearity. *Neuron, 71*, 348–361.

Schade, O. H. (1956). Optical and photo-electric analog of the eye. *Journal of the Optical Society of America, 46*, 721–739.

Schor, C. M., & Woods, I. (1983). Disparity range for local stereopsis as a function of luminance spatial frequency. *Vision Research, 23*, 1649–1654. doi:10.1016/0042-6989(83)90179-7.

Schor, C., Woods, I., & Ogawa, J. (1984). Binocular sensory fusion is limited by spatial resolution. *Vision Research, 24,* 661–665. doi:10.1016/0042-6989(84)90207-4.

Schrater, P. R., Knill, D. C., & Simoncelli, E. P. (2000). Mechanisms of visual motion detection. *Nature Neuroscience, 3,* 64–68. doi:10.1038/71134.

Sekiguchi, N., Williams, D. R., & Brainard, D. H. (1993). Aberration-free measurements of the visibility of isoluminant gratings. *Journal of the Optical Society of America. A, Optics and Image Science, 10,* 2105–2117. doi:10.1364/JOSAA.10.002105.

Selwyn, E. W. H. (1948). The photographic and visual resolving power of lenses. *Photographic Journal, 88,* 6–12, 46–57.

Simoncelli, E. P., & Heeger, D. J. (1998). A model of neuronal responses in visual area MT. *Vision Research, 38*(5), 743–761.

Smallman, H. S., & MacLeod, D. I. A. (1994). Size-disparity correlation in stereopsis at contrast threshold. *Journal of the Optical Society of America. A, Optics, Image Science, and Vision, 1*(11), 2169–2183. doi:10.1364/JOSAA.11.002169.

Smallman, H. S., & MacLeod, D. I. A. (1997). Spatial scale interactions in stereosensitivity and the neural representation of binocular disparity. *Perception, 26,* 977–994.

Solomon, J. A., Sperling, G., & Chubb, C. (1993). The lateral inhibition of perceived contrast is indifferent to on-center/off-center segregation, but specific to orientation. *Vision Research, 33,* 2671–2683. doi:10.1016/0042-6989(93)90227-N.

Tanaka, H., & Ohzawa, I. (2009). Neuronal responses to texture-defined form in macaque visual area V2. *Journal of Neurophysiology, 101,* 1444–1462.

Thomas, J. P., & Olzak, L. A. (1996). Uncertainty experiments support the roles of second order mechanisms in spatial frequency and orientation discriminations. *Journal of the Optical Society of America. A, Optics, Image Science, and Vision, 13,* 689–696.

Thompson, P. (1981). Velocity after-effects: The effects of adapting to moving stimuli on the perception of subsequently seen moving stimuli. *Vision Research, 21,* 337–345.

Torre, V., & Poggio, T. A. (1986). On edge detection. *IEEE Transactions on Pattern Analysis and Machine Intelligence, 8,* 147–163.

Tyler, C. W. (1974). Depth perception in disparity gratings. *Nature, 251,* 140–142.

Van Nes, F. L., & Bouman, M. A. (1967). Spatial modulation transfer in the human eye. *Journal of the Optical Society of America, 57,* 401–406.

van Santen, J. P. H., & Sperling, G. (1985). Elaborated Reichardt detectors. *Journal of the Optical Society of America. A, Optics and Image Science, 2,* 300–321.

Watson, A. B., & Ahumada, A. J. (1985). Model of human visual motion sensing. *Journal of the Optical Society of America. A, Optics and Image Science, 2,* 322–341.

Watt, R. J. (1987). Scanning from coarse to fine spatial-scales in the human visual system after the onset of the stimulus. *Journal of the Optical Society of America. A, Optics and Image Science, 4,* 2006–2021. doi:10.1364/JOSAA.4.002006.

Watt, R. J., & Morgan, M. J. (1983). The recognition and representation of edge blur: Evidence for spatial primitives in human vision. *Vision Research, 23,* 1465–1477. doi:10.1016/0042-6989(83)90158-X.

Watt, R. J., & Morgan, M. J. (1985). A theory of the primitive spatial code in human vision. *Vision Research, 25,* 1661–1674. doi:10.1016/0042-6989(85)90138-5.

Westheimer, G. (1973). Fourier analysis of vision. *Investigative Ophthalmology, 12,* 86–87.

Wilson, H., & Gelb, D. (1984). Modified line-element theory for spatial-frequency and width discrimination. *Journal of the Optical Society of America. A, Optics and Image Science, 1,* 124–131.

Wilson, H. R., & Richards, W. A. (1989). Mechanisms of contour curvature discrimination. *Journal of the Optical Society of America. A, Optics and Image Science, 6,* 106–115.

Witkin, A., & Tennenbaum, J. (1983). On the role of structure in vision. In J. Beck, B. Hope, & A. Rosenfeld (Eds.), *Human and machine vision* (pp. 481–543). London: Academic Press.

Yang, Y., & Blake, R. (1991). Spatial frequency tuning of human stereopsis. *Vision Research, 31,* 1177–1189. doi:10.1016/0042-6989(91)90043-5.

Yuille, A. L., & Poggio, T. (1986). Scaling theorems for zero-crossings. *IEEE Transactions on Pattern Analysis and Machine Intelligence, 8,* 15–25.

Zhou, Y. X., & Baker, C. L., Jr. (1993). A processing stream in mammalian visual cortex neurons for non-Fourier responses. *Science, 261,* 98–101.

Zhou, Y. X., & Baker, C. L., Jr. (1994). Envelope-responsive neurons in areas 17 and 18 of cat. *Journal of Neurophysiology, 72,* 2134–2150.

Zhou, Y. X., & Baker, C. L., Jr. (1996). Spatial properties of envelope-responsive cells in area 17 and 18 neurons of the cat. *Journal of Neurophysiology, 75,* 1038–1050.

43 Configural Pooling in the Ventral Pathway

HUGH R. WILSON AND FRANCES WILKINSON

Primary visual cortex, or V1, receives information from the retina via the lateral geniculate nucleus (LGN). Following processing in V1 there exist two major processing pathways for vision: ventral and dorsal. These have variously been characterized as the *what* and *where* pathways (Mishkin, Ungerleider, & Macko, 1983) or the *vision for object recognition* and *vision for action* (Milner & Goodale, 1995) pathways, respectively. In this chapter we focus on the transformations inherent in the ventral pathway, which we refer to as either the *form* or *object vision* pathway. We also briefly consider changes in the ventral pathway under clinical conditions and during healthy aging. As a prelude to this we first briefly review work on spatial channels in V1, and the reader is referred to our chapter in the first edition of *The Visual Neurosciences* for a more detailed description (Wilson & Wilkinson, 2004) that remains largely accurate.

Beginning with the pioneering work of Hubel and Wiesel (1968), it is now firmly established that primate V1 analyzes objects in terms of local contour and edge orientations on a range of different spatial or spatial frequency scales. Psychophysical evidence for both orientation and spatial frequency tuning in humans has been derived from both pattern adaptation studies (Blakemore & Campbell, 1969; Blakemore, Carpenter, & Georgeson, 1970) and from pattern masking (Phillips & Wilson, 1984; Wilson, McFarlane, & Phillips, 1983). A convenient summary of the two-dimensional spatial tuning profiles of psychophysically measured orientation-selective units in humans is provided in Wilson (1991). The psychophysically measured tuning agrees well with both spatial frequency and orientation tuning of macaque V1 neurons (DeValois, Albrecht, & Thorell, 1982; Wilson, 1991).

Receptive fields or spatial filter shapes were originally thought to be constructed in an exclusively feedforward manner from small arrays of LGN inputs as suggested by Hubel and Wiesel (1968), and they were also assumed to operate independently and in parallel. However, more recent evidence suggests a more dynamic and interactive organization. First, the discovery of contrast gain control pools, epitomized by the influential Heeger (1992) model, clearly demonstrated that a divisive normalization occurs across all orientations. This gain control normalization serves to alter the contrast response of units and to narrow their orientation bandwidths. Such inhibitory gain controls were first reported in cats (Bonds, 1989). Second, the discovery of collinear facilitation both in humans (Field, Hayes, & Hess, 1993; Polat & Sagi, 1993) and macaques (Malach et al., 1993) demonstrated another mechanism contributing to orientation tuning. These long-range connections between neurons with similar orientations processing information from adjacent, roughly collinear spatial locations both enhance contour continuity and alter the effective orientation tuning of cells. Recent neural models of V1 have incorporated these effects (McLaughlin et al., 2000; Vidyasagar, Pei, & Volgushev, 1996). In addition to inhibitory gain controls and collinear facilitation, it is now known that pattern adaptation involves much more than simple neuronal fatigue. In macaques at least three adaptation sites have been documented: a broadly tuned precortical stage, an orientation-selective cortical stage, and adaptation of gain control mechanisms themselves (Dhruv et al., 2011). A more detailed summary of these V1 mechanisms may be found elsewhere (Wilson & Wilkinson, 2004).

VENTRAL PATHWAY: PHYSIOLOGY

A major role of V1 contrast gain control pools is to produce a normalized response independent of contrast once pattern contrast is modestly above threshold. In addition, collinear facilitation will serve to enhance the salience of straight or smoothly curving contours. This information (along with information about color and texture) is then passed into the ventral pathway with the ultimate purpose of object detection and recognition. To gain some insight into how this is accomplished, let us first examine the ventral pathway and its receptive field characteristics.

In macaques the ventral pathway hierarchy is generally considered to comprise V1, V2, V4, posterior occipitotemporal cortex (TEO), and anterior temporal cortex (TE) (Van Essen, Anderson, & Felleman, 1992). As depicted in figure 43.1, there are two sets of bidirectional connections among these areas. The first set interconnect adjacent areas: V1–V2, V2–V4, V4–TEO, and TEO–TE. The second set of connections skip one area and are therefore known as skipping connections: V1–V4, V2–TEO, and V4–TE. In all these cases there are reciprocal feedback connections. Relatively little is known about the role of most of these feedback connections, although it has been suggested that feedback is responsible for context effects (Lamme & Roelfsema, 2000) and for aspects of selective attention (Tsotsos, 1993). Indeed, there is some evidence that

rapid feedforward processing of low spatial frequencies is followed by feedback that facilitates recognition of high spatial frequencies in FFA, the fusiform face area (Bar et al., 2006). However, detailed descriptions of these feedback roles remain elusive.

In humans the situation becomes less clear beyond the V1–V2–V4 sequence. Certainly, these provide input into the occipital face area (OFA), fusiform face area (FFA) (Gauthier et al., 2000; Haxby, Hoffman, & Gobbini, 2000; Kanwisher, McDermott, & Chun, 1997), and lateral occipital complex (LOC) containing many subregions for object recognition (Haxby et al., 2001). A similar picture has emerged in macaques (Tsao et al., 2003). Thus, areas TEO and TE in the ventral pathway diagram must be viewed as standing for an incomplete representation of a more complex interaction among subareas to be elucidated by further research.

One key characteristic of ventral pathway receptive fields is their systematic increase in diameter from lower to higher areas. Figure 43.2 represents our compilation of data from four different studies (Boussaoud, Desimone, & Ungerleider, 1991; Elston & Rosa, 1998; Kobatake & Tanaka, 1994; Op de Beeck & Vogels, 2000), which, when combined, provide measurements for all five cortical areas. All data are from central visual field. Here we have plotted the logarithm of mean receptive field diameter as a function of cortical area. The solid black line represents a progressive increase in diameter by a factor of 2.7 from area to area. This striking regularity has a natural interpretation. If receptive fields in

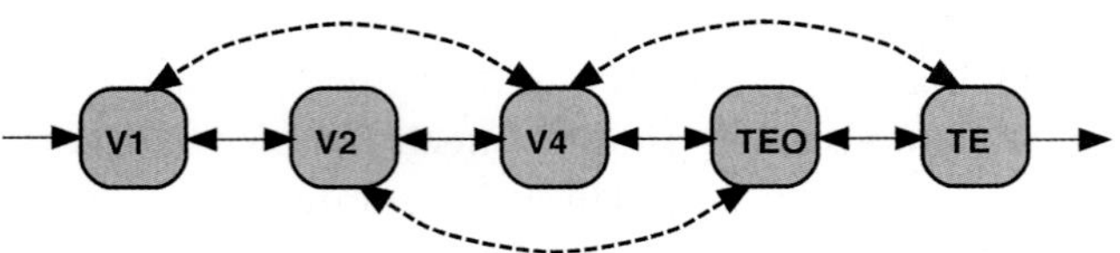

FIGURE 43.1 Cortical areas in the ventral, form vision pathway. Solid arrows depict area-to-area interconnections, and dashed arrows show skipping connections that leapfrog an intervening area. All arrows are double-headed to indicate that all forward connections between areas are accompanied by reciprocal feedback, the function of which is largely unknown.

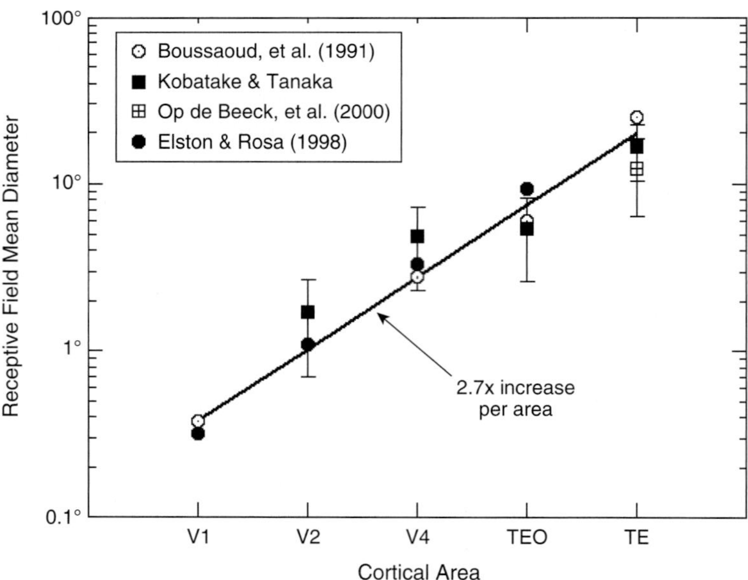

FIGURE 43.2 Increasing receptive field width along the ventral pathway in macaques: data from four studies (see text). As shown by the solid line, width increases by approximately 2.7 times from area to area. This implies neighborhood processing and increasing globality progressing through the pathway.

successive ventral areas pooled responses from a central unit and its adjacent, nonoverlapping neighbors, receptive fields would increase by approximately 3.0 times in diameter. The slightly smaller factor of 2.7 presumably reflects modest overlap of pooled neighbors. Thus, it appears that the transition from local contour orientation detection in V1 to global object representation in TE entails intermediate levels of increasing globality for its realization.

Given the evidence that increasing ventral pathway receptive field sizes reflect pooling over about three neighboring spatial locations, it becomes possible to provide structure to neurophysiological response properties along the pathway. Analysis of natural scene contours has demonstrated that the two dominant local properties are collinearity and curvature or cocircularity (Olshausen & Field, 1996; Sigman et al., 2001). On the assumption that the adult ventral pathway reflects these properties of natural scenes as a combined result of evolution and postnatal visual experience, then it would be natural to expect ventral pathway areas to contain neurons selective for curvature, extended lines, and angles. In agreement with this many V2 neurons do indeed respond optimally to curved arcs, angles, and line intersections (Anzai, Peng, & Van Essen, 2007; Hegdé & Van Essen, 2000). In order to do this it is necessary to pool different orientations at adjacent spatial locations, and this is consistent with the observed increase in receptive field size. Importantly, this pooling must be configural.

Area V4 contains even larger receptive fields, and for this area there is currently a wealth of physiological data. Although V4 was first claimed to be a color area, it is now understood to be a form vision area that incorporates both object shape and color visual characteristics. Early studies indicated that V4 contains neurons sensitive to closed circular shapes as well as radial and hyperbolic patterns (Gallant, Braun, & Van Essen, 1993; Gallant et al., 1996). Human fMRI data have also shown that V4 responds strongly to both quasi-circular shapes (RF patterns, see below and figure 43.3) and to radial patterns (Wilkinson et al., 2000). A subsequent fMRI study confirmed that human V4 is indeed most strongly activated by closed curved shapes rather than simply by randomly oriented collections of curved arcs (Dumoulin & Hess, 2007). The combination of local curves from V2 into closed curved shapes in V4 again requires significantly larger receptive field sizes in agreement with the data in figure 43.2.

Extensive recent research into the responses of macaque V4 receptive fields has elucidated the role of V4 in encoding closed shapes (Pasupathy & Connor, 2001, 2002). These studies demonstrated that most V4

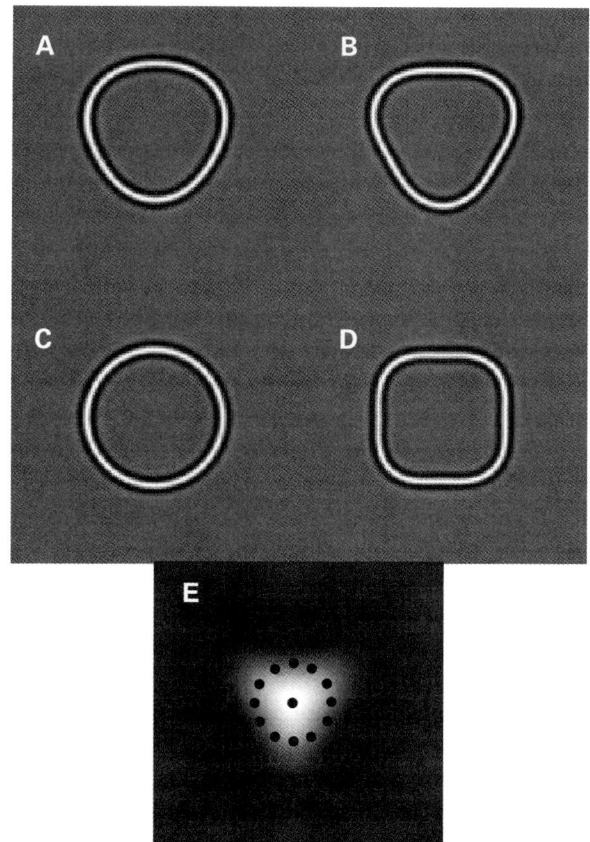

FIGURE 43.3 Examples of radial frequency patterns and model responses. (A) RF3 pattern with amplitude $A = 0.05$. (B) RF3, $A = 0.10$. (C) Circle, $A = 0$. (D) RF4, $A = 0.07$. (E) Response of model V4 neurons to pattern in B. Lightest grays indicate largest responses. The largest occurs at the pattern center (center dot), and this results in selection of 12 surrounding units (circle of dots) for sine- and cosine-weighted summation (see text).

neurons are selective to particular curvatures at a particular location *relative to the center of a curved shape*. In addition, the preponderance of cells were sensitive to convex curvatures relative to the shape's center. A clever simulation based on their data showed that there was sufficient information available in the V4 neural responses to provide a population code for the overall shape of closed curved contours (Pasupathy & Connor, 2002). Further results suggest that V4 encodes curved shapes using a sparse population code for curvature extrema (Carlson et al., 2011). As shown below, these results are consistent with our neural model of V4 shape encoding.

Beyond V4, details of shape encoding become less precise aside from the general observation that faces and objects are represented by the level of TE in macaques or FFA and LOC in humans. Human fMRI has demonstrated that FFA responds strongly to quasi-circular shapes (Wilkinson et al., 2000). A later study corroborated that responses of neurons in macaque face areas respond to faces and also circular contours but no other shapes (Tsao et al., 2006). There is even evidence from multivoxel pattern analysis that FFA contains cortical columns sensitive to the shape of human heads (Nichols, Betts, & Wilson, 2010). It is reasonable to assume that this representation of curved head shapes derives from V4 inputs, as V4 responds very effectively to the same images.

There is one study that points to TEO responses being the result of V4 object-centered curvature pooling (Brincat & Connor, 2004). This neurophysiological study demonstrated that TEO neurons pool multiple V4 curvature responses in either a linear or multiplicative manner. In addition, TEO neurons were found to exhibit size constancy over a fairly broad range.

PSYCHOPHYSICS

Concentric Glass patterns (Glass, 1969) provided some of the first evidence for configural orientation pooling in the visual system. In a psychophysical study that restricted the signal dot pairs to only part of the pattern, evidence was found for global linear summation of orientations tangent to concentric Glass contours (Wilson, Wilkinson, & Asaad, 1997). This result was interpreted with a simple neural model of orientation pooling that accurately predicted Glass pattern thresholds (see

figure 43.4 for an updated version of this model). An additional study demonstrated that radial and other Glass pattern shapes could be explained with analogous models (Wilson & Wilkinson, 1998).

In order to study global configural curvature pooling further, we devised a novel class of visual stimuli: radial frequency (RF) patterns (Wilkinson, Wilson, & Habak, 1998). These are defined in polar coordinates as patterns in which the radius R varies sinusoidally with the polar angle θ:

$$R(\theta) = R_0 \left[1 + A \cdot \cos(\omega\theta + \varphi) \right] \qquad (43.1)$$

In this expression ω is the radial frequency or number of cycles around the circumference, and the phase φ determines the rotation of the pattern. If RF amplitude $A = 0$, this is just the equation for a circle of constant radius R_0. Examples of RF3–RF4 are illustrated in figure 43.3 for several amplitudes. For equation 43.1 to describe a closed curve, it is necessary that ω assume only integer values and that $A < 1$. For at least the range RF3–RF12, threshold amplitudes A for discriminating an RF pattern from a circle fall well into the hyperacuity range, averaging about 7.0 arc sec for $R_0 = 0.5°$, with thresholds obeying Weber's law and thus exhibiting size constancy for larger R_0.

Several studies have demonstrated that, for low RFs, optimum detection performance must involve global pooling across multiple cycles of the pattern, as thresholds exceed those predicted by probability summation alone for patterns between 2 and 5–8 radial cycles (Day

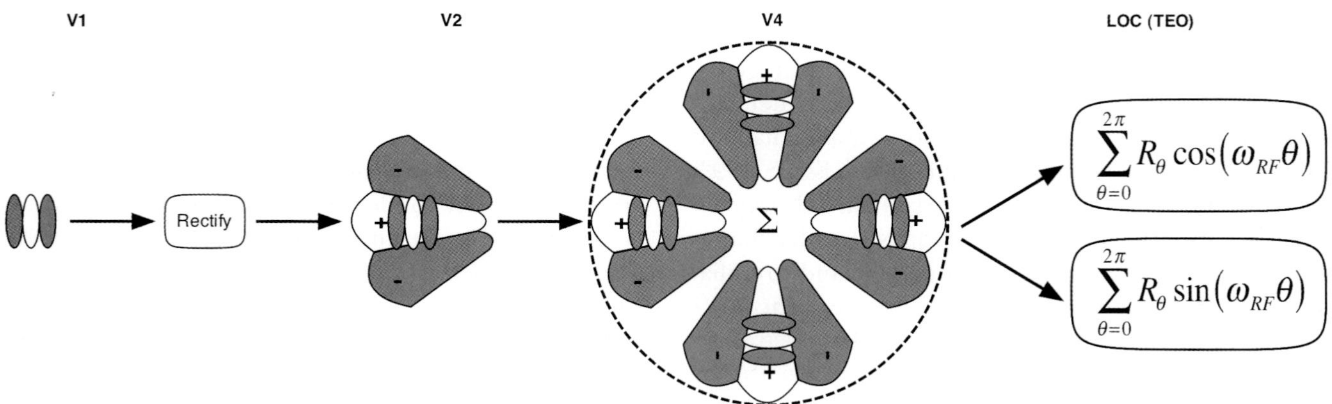

FIGURE 43.4   Schematic of a model for configural processing along the ventral pathway based on Poirier and Wilson (2006). Orientation processing (12 orientations) occurs in V1. Following rectification, V2 extracts curvature using center-surround filters that are orthogonal to the V1 orientations being pooled. The pie-wedge shape of these filters enhances size constancy (see text for details). The subsequent V4 stage involves concentric pooling of V2 curvature responses. This results in detection of the curved pattern center and selection of 12 surrounding points for further curvature processing (see figure 43.3E). The final stage, assumed to reflect LOC in humans or TEO in macaques, comprises pairs of units that receive cosine- and sine-weighted sums of inputs from the 12 units selected by V4. These units produce bandpass channels for each radial frequency (ω_{RF} varies for different unit pairs), and these in turn encode RF amplitude and phase. More complex shapes, such as heads and fruit, will be represented by a sparse population code of LOC responses.

& Loffler, 2009; Hess, Wang, & Dakin, 1999; Jeffrey, Wang, & Birch, 2002; Loffler, Wilson, & Wilkinson, 2003; Wilkinson, Wilson, & Habak, 1998). In an interesting alternative conceptualization of the task, Jeffrey, Wang, and Birch (2002) have reported that the spatial limit for this task, which is a hyperacuity, may be set by a minimum length along the circular (unmodulated) contour (CCL) such that a radial cycle of the pattern covers not more than about 0.5–1° (this measurement, which is linear degrees, not polar angle, is somewhat counterintuitive because it is not based on a straight line). This does not negate the concept of global pooling at low RFs, which the Jeffrey, Wang, and Birch study in fact replicates, but most likely is related to a sampling limitation on orientation/curvature processing at the local level. The one direct test of the CCL model using high-amplitude RF patterns in an adaptation paradigm showed no evidence of cross-frequency adaptation being enhanced by matching circular contour length (Bell, Dickinson, & Badcock, 2008), which supports the argument that the CCL limit is related to local sampling issues that obtain only at modulation detection threshold.

In follow-up work using RF stimuli, sensitivity to curvature increments has been examined, along with interactions between frequencies using a combination of adaptation and masking paradigms. Adaptation both reduces sensitivity to the adapting stimulus configuration and also alters the appearance of similar but not identical patterns. Effects of this sort are also seen for orientation in adaptation-induced contrast threshold elevation and in the tilt aftereffect (Gibson & Radner, 1937). The first study to indicate that complex curvature-based forms could produce shape-specific threshold elevations was by Alter and Schwartz (1988) using Fourier descriptor stimuli. These are much like RF patterns except that they are not band-limited in spatial frequency. These authors reported frequency-tuned adaptation such that the modulation amplitude required to differentiate the adapted frequency from a circle was elevated although the threshold of adjacent frequencies was unaffected. Anderson et al. (2007) examined the perceptual aspect of RF adaptation and found that adaptation to a high-amplitude RF pattern (e.g., RF5) altered the appearance of a circle, such that it resembled an RF5 180° out of phase with the adapting stimulus. The appearance of nearby RFs was little affected by adaptation, implicating separate shape-level mechanisms rather than low-level adaptation effects. Similar effects have been reported with other global patterns (Regan & Hamstra, 1992; Suzuki, 2001, 2003). More recent adaptation studies provide direct evidence that the perception of RF patterns of high deformation amplitude is also supported by mechanisms involving global pooling (Bell et al., 2010; Bell & Kingdom, 2009), that RF amplitude must be reflected in this coding, and that all parts of the RF contour (concavities, convexities, and zero-crossings or inflection points) contribute roughly equally to the encoding of these shapes (Bell et al., 2010).

Masking studies using spatially (Habak et al., 2004) and temporally (Habak, Wilkinson, & Wilson, 2006) separated masks have demonstrated shape-specific interactions that extend over a spatial zone of at least 1° and over a broad range of temporal offsets peaking at 80–110 ms. A variety of findings, including the observation that shapes defined by first-order and by second-order contours can show masking interactions (Bell & Badcock, 2008; Habak et al., 2004), provide strong evidence that these spatiotemporal interactions occur at the level after form abstraction rather than at early stages of the pathway where local features are extracted.

There is strong psychophysical evidence that an intermediate curvature coding stage precedes the final global pooling (Bell & Badcock, 2008; Bell et al., 2011); contour components as long as or longer than one cycle can be integrated into a global form showing the pooling gains at threshold and adaptation effects described above when components alternate between light and dark. Shorter segments of alternating local structure show progressive decrements in performance as the number of sequential same-polarity elements decreases (Bell et al., 2011).

Other recent work has attempted to characterize interactions among the configural mechanisms coding these shapes by examining compound shapes created by adding RF patterns of different frequencies, phases, and amplitudes to produce complex shapes (Bell et al., 2009). When thresholds for detection of one component of a two-component RF pattern are assessed in the presence of a high-amplitude second component, masking is found to occur between neighboring RFs. This is strongly suggestive of narrow RF-band channels with inhibitory interactions among channels. This was further supported by the observation that reducing the sensitivity of one component of such compounds by selective adaptation improved thresholds for the other component RF. Again, adaptation to second-order adaptors had the same threshold-lowering effect on first-order complex contours as did first-order RF adaptors, and adaptors did not have to be the same radius as the test stimuli. This provides further evidence that inhibitory interactions between higher-level shape-specific mechanisms mediate the coding of complex curved closed contours.

VENTRAL PATHWAY NEURAL MODEL

We have developed a series of quantitative neural models for concentric orientation pooling in the ventral pathway that have been applied to concentric Glass patterns (Wilson, Wilkinson, & Asaad, 1997), radial Glass patterns (Wilson & Wilkinson, 1998), RF patterns (Poirier & Wilson, 2006), and symmetry perception (Poirier & Wilson, 2010). Here we briefly summarize a version of the Poirier and Wilson (2006) model that is extended to incorporate a final stage in which different bandpass RF channels are explicitly represented. As schematized in figure 43.4, the model comprises four stages that can be speculatively linked to areas in the ventral pathway. Stage 1, associated with V1, involves orientation-selective filtering at 12 orientations using model receptive fields derived from psychophysical masking studies (Wilson, 1991). Only one spatial frequency is represented in the model, as RFs have a narrow spatial frequency spectrum, but additional frequencies can easily be incorporated.

The second stage involves rectification and combination of orientations at adjacent spatial locations to extract a contour curvature signal, an operation assumed to occur in V2 (Anzai, Peng, & Van Essen, 2007; Hegdé & Van Essen, 2000). Curvature computation involves pooling of rectified V1 orientations using an orthogonal center-surround filter. This computation has been shown to produce no response to extended straight lines but a response proportional to curvature over a considerable range (Wilson, 1999). The sole embellishment here is that the second-stage filter assumes a pie-wedge shape in which the space constant of the filter increases linearly from the center of the pattern:

$$RF(x, y) = \left[3\exp\left(-y^2 / \sigma_y^2\right) - \exp\left(-y^2 / \left(3\sigma_y\right)^2\right) \right]$$
$$\exp\left(-x^2 / \sigma_x^2\right)$$
$$\sigma_y = \sigma_0 \left(a + bx\right) \tag{43.2}$$

This equation describes the V2 filter orientation shown in figure 43.4 and can be rotated for other orientations. This filter shape reflects the observation that size constancy implies that a greater contour length is used for curvature computations as distance from the center of curvature increases. Note that the model V2 receptive field is approximately 2.7 times the linear dimensions of the V1 oriented filter input stage in agreement with figure 43.2.

The third stage incorporates concentric summation of V2 wedge filters at 12 orientations (only four are depicted in figure 43.4 for clarity) around the center of the receptive field. Based on data cited earlier this is assumed to reflect complex curvature processing in V4. As shown in figure 43.3E, the maximum response of a spatial array of V4 filters occurs at the center of an RF pattern, and this maximum can be determined by a spatially regional winner-take-all mechanism such as has been proposed for perception of Marroquin patterns (Wilson, Krupa, & Wilkinson, 2000). Units such as these can provide the critical information for encoding convex curvature extrema relative to the center of a closed contour, as has been reported in macaque V4 (Pasupathy & Connor, 2001, 2002).

After the center of a closed contour has been located by these V4 filters, the final stage of the computation involves selection of a fixed set of 12 filter responses at equal 30° polar angles and at a constant distance from the contour center. These are shown by the ring of 12 dots around the center in figure 43.3E. Note that these are selected automatically by the network after determination of the pattern center, which suggests a role for interarea feedback. If these 12 responses R_θ are separately cosine- and sine-weighted by a radial frequency ω_{RF}, the units doing the summation will be tuned to only the pattern defined by ω_{RF}. Given evidence for V4 curvature pooling in macaque TEO (Brincat & Connor, 2004), which is presumptively homologous to human LOC, this weighted summation is hypothesized to reflect processing in those areas. These presumptive LOC units for RF2–RF6 provide separate channels for each ω_{RF}, with the cosine- and sine-weighted pair encoding both RF amplitude and phase (A and φ in equation 43.1). More complex curved shapes composed of several RFs would be described by a sparse population code of several active units with appropriate inhibitory interactions. Furthermore, simulations show that the wedge-shaped V2 receptive fields in the model confer size constancy over a 2:1 range of radii. Inclusion of several spatial scales can extend this (Poirier & Wilson, 2006). Thus, the model provides a good quantitative approximation of configural processing for curved shapes, including head shapes and many fruit, throughout much of the ventral pathway.

LIFESPAN CHANGES IN GLOBAL POOLING

The development of global processing in early life has been examined using spatial as well as temporal pooling. Using radial frequency patterns in a preferential looking paradigm, Birch, Swanson, and Wang (2000) reported a rapid improvement in sensitivity to radial deformations over the early months of life. The same group has recently extended this study across the lifespan (Wang et al., 2009). By studying children of several age groups they have demonstrated that the rapid early improvement is followed by a more gradual improvement up to 21 years of age, whereas resolution acuity reaches its asymptotic adult value by 11 years. Glass patterns, the perception of which also depends on global pooling of local orientation signals, showed marked improvement between 6 and 9 years of age and did not show statistically significant differences from adults by 9 years (Lewis et al., 2004).

Although threshold sensitivity for these global processing tasks seems only to be achieved in adolescence or early adulthood, once achieved, those skills that have been tested endure longer in healthy aging than lower-level functions such as spatial resolution. Wang (2001) reported that seniors show very little sensitivity loss for RF4 patterns compared to younger adults, whereas spatial resolution shows much more marked degradation by the same age. This finding of preserved RF sensitivity was confirmed using RF5 by Habak, Wilkinson, and Wilson (2009), who also showed that there was no change in processing speed for this task nor any abnormality in lateral interactions among shapes. However, slightly reduced sensitivity was seen for RF patterns defined by second-order rather than first-order contours. In their more recent longitudinal study Wang et al. (2009) report that the decline in both RF sensitivity and spatial resolution begins at about 55 years, but the slope of the loss is steeper for resolution acuity than for RF hyperacuity. In an unpublished study in our laboratory (C. Habak, F. Wilkinson, & H. R. Wilson, unpublished data), we have found that, provided changes in contrast sensitivity are compensated for by using stimuli that were a fixed multiple of contrast threshold, Glass pattern coherence thresholds are also normal in older participants.

CLINICAL ABNORMALITIES IN GLOBAL POOLING

Examination of global contour-processing deficits has been extensively done in amblyopia and autism. Amblyopia (severely impaired visual acuity) can arise through visual deprivation (e.g., cataract), strabismus, or markedly different refractive errors in the two eyes (anisometropia). Intermediate visual pooling measures have been used in both deprivation amblyopia (Jeffrey, Wang, & Birch, 2004; Lewis et al., 2002) and in strabismic amblyopia (Dallala, Wang, & Hess, 2010; Hess et al., 1999). Individuals with either unilateral or bilateral cataract removal were found to have deficits in concentric Glass pattern detection in their amblyopic eye(s) (Lewis et al., 2004), which was more marked following binocular than unilateral cataract. Testing with a range of RF frequencies and radii also revealed marked deficits in thresholds for detecting deformations of circular contours (Jeffrey, Wang, & Birch, 2004); however, this deficit did not differ in degree between unilateral and bilateral deprivation.

Strabismic amblyopes have also been tested with RF patterns (Dallala, Wang, & Hess, 2010; Hess et al., 1999) and have been reported to show impairments in RF detection and RF increment thresholds that cannot be explained through acuity deficits. The latter study used modified RF patterns that allowed the authors to separate the contributions of pooling of position versus orientation cues over the pattern; although both were found to contribute to the impairment, orientation was found to have the larger role. It remains to be determined how undersampling and positional variance contribute to this impairment.

In autism and in individuals with mild autistic traits, unusually good performance is seen in the ability to detect embedded figures, leading to the suggestion that this superability at local analysis might reflect weak global processing. Recent studies testing this hypothesis have demonstrated that children with autism spectrum disorder have elevated thresholds compared to typically developing age-matched controls for detection of RF3 patterns, which have been shown to incorporate global processing, but are comparable to controls on RF24 patterns, which depend on local processing (Grinter et al., 2010). However, when RF patterns well above detection threshold are employed in a visual search task, university students scoring at the high end of the range of autistic-like traits (Almeida et al., 2010a, 2010b) show better than control visual search (lower slopes as a function of set size), which these authors relate to their superior performance on embedded figures. These seemingly inconsistent findings have not yet been reconciled.

CONCLUSIONS

We have summarized data on increasing receptive field size and physiological response properties moving up the ventral, form vision pathway. All neurophysiological evidence points to configural pooling between spatially adjacent orientations extracted in V1 to compute angles and local contour curvatures in V2 (Anzai, Peng, & Van Essen, 2007). Following this, global pooling in V4 leads to the computation of pattern centers and curvatures relative to those centers (Pasupathy & Connor, 2001, 2002). Finally, concentric V4 curvature responses are combined in TEO to represent curved shapes (Brincat & Connor, 2004).

These neurophysiological findings have been paralleled by psychophysical studies of form vision for closed contours in humans. One of the major stimuli used in these studies has been RF patterns (Wilkinson, Wilson, & Habak, 1998). As sums of RFs describe many complex shapes, including heads and many fruit, they provide a biologically relevant pattern set for studying form vision. Psychophysical data show hyperacuity for detection of RFs, size constancy (Wilkinson, Wilson, & Habak, 1998), and evidence for visual channels tuned to individual RFs (Anderson et al., 2007; Habak et al., 2004). In addition, RF patterns are proving increasingly useful in a clinical context. The neural model illustrated in figure 43.4 provides an explanation of these psychophysical data and is consistent with primate physiology.

The configural pooling data discussed in this chapter include contours describing head shapes, many types of fruit, and many other curved, concentric forms. However, this is hardly a complete inventory of configural contour processing but only the beginning. Many shape contours, particularly of human artifacts, have angular contours. Examples include most house shapes, knives, tepees, tables, and chairs, among others. We suggest that a parallel approach to configural processing of curved shapes starting with angle extraction in V2 (Loffler, 2008) will lead to a similar level of understanding of the neural representation of angular and other shapes.

Other aspects of configural processing in the ventral pathway that have scarcely been alluded to in the literature and are ignored in this chapter include the role of skipping connections that bypass an intermediate area and the role of feedback in general. Hypotheses have been proposed (Lamme & Roelfsema, 2000) for feedback and perhaps for skipping connections (Bar et al., 2006), but there is a lack of concrete data to support or refute them. Thus, there is now a substantial, organized corpus on many aspects of configural pooling in form vision, but years of challenging and exciting research lie ahead.

ACKNOWLEDGMENT

This work was supported in part by CIHR grant 172103 to both authors, NSERC grant 7551 to F.W., and a Canadian Institute for Advanced Research grant to H.R.W.

REFERENCES

Almeida, R. A., Dickinson, J. E., Maybery, M. T., Badcock, J. C., & Badcock, D. R. (2010a). A new step towards understanding embedded figures test performance in the autism spectrum: The radial frequency search task. *Neuropsychologia, 48*, 374–381.

Almeida, R. A., Dickinson, J. E., Maybery, M. T., Badcock, J. C., & Badcock, D. R. (2010b). Visual search performance in the autism spectrum II: The radial frequency search task with additional segmentation cues. *Neuropsychologia, 48*, 4117–4124.

Alter, I., & Schwartz, E. L. (1988). Psychophysical studies of shape with Fourier descriptor stimuli. *Perception, 17*, 191–202.

Anderson, N. D., Habak, C., Wilkinson, F., & Wilson, H. R. (2007). Evaluating shape aftereffects with radial frequency patterns. *Vision Research, 47*, 298–308.

Anzai, A., Peng, X., & Van Essen, D. C. (2007). Neurons in monkey visual area V2 encode combinations of orientations. *Nature Neuroscience, 10*, 1313–1321.

Bar, M., Kassam, K. S., Ghuman, A. S., Boshyan, J., Schmid, A. M., Dale, A. M., et al. (2006). Top-down facilitation of visual recognition. *Proceedings of the National Academy of Sciences of the United States of America, 103*, 449–452. doi: 10.1073/pnas.0507062103.

Bell, J., & Badcock, D. R. (2008). Luminance and contrast cues are integrated in global shape detection with contours. *Vision Research, 48*, 2336–2344.

Bell, J., Dickinson, J. E., & Badcock, D. R. (2008). Radial frequency adaptation suggests polar-based coding of local shape cues. *Vision Research, 48*, 2293–2301.

Bell, J., Gheorghiu, E., Hess, R. F., & Kingdom, F. A. (2011). Global shape processing involves a hierarchy of integration stages. *Vision Research, 51*, 1760–1766.

Bell, J., Hancock, S., Kingdom, F. A. A., & Peirce, J. W. (2010). Global shape processing: Which parts form the whole? *Journal of Vision, 10*(6), 16. doi: 10.1167/10.6.16.

Bell, J., & Kingdom, F. A. A. (2009). Global contour shapes are coded differently from their local components. *Vision Research, 49*, 1702–1710.

Bell, J., Wilkinson, F., Wilson, H. R., Loffler, G., & Badcock, D. R. (2009). Radial frequency adaptation reveals interacting contour shape channels. *Vision Research, 49*, 2306–2317.

Birch, E. E., Swanson, W. H., & Wang, Y. Z. (2000). Infant hyperacuity for radial deformation. *Investigative Ophthalmology & Visual Science, 41*, 3410–3414.

Blakemore, C., & Campbell, F. W. (1969). On the existence of neurones in the human visual system selectively sensitive to the orientation and size of retinal images. *Journal of Physiology, 203*, 237–260.

Blakemore, C., Carpenter, R. H. S., & Georgeson, M. A. (1970). Lateral inhibition between orientation detectors in the human visual system. *Nature, 228*, 37–39.

Bonds, A. B. (1989). Role of inhibition in the specification of orientation selectivity of cells in the cat striate cortex. *Visual Neuroscience, 2*, 41–55.

Boussaoud, D., Desimone, R., & Ungerleider, L. G. (1991). Visual topography of area TEO in the macaque. *Journal of Comparative Neurology, 306*, 554–575.

Brincat, S. L., & Connor, C. E. (2004). Underlying principles of visual shape selectivity in posterior inferotemporal cortex. *Nature Neuroscience, 7*, 880–886.

Carlson, E. T., Rasquinha, R. J., Zhang, K., & Connor, C. E. (2011). A sparse object coding scheme in area V4. *Current Biology, 21*, 288–293.

Dallala, R., Wang, Y.-Z., & Hess, R. F. (2010). The global shape detection deficit in strabismic amblyopia: Contribution of local orientation and position. *Vision Research, 50*, 1612–1617.

Day, M., & Loffler, G. (2009). The role of orientation and position in shape perception. *Journal of Vision, 9*(10), 14, 1–17. doi: 10.1167/9.10.14.

DeValois, R. L., Albrecht, D. G., & Thorell, L. G. (1982). Spatial frequency selectivity of cells in macaque visual cortex. *Vision Research, 22*, 545–559.

Dhruv, N. T., Tailby, C., Sokol, S. H., & Lennie, P. (2011). Multiple adaptable mechanisms early in the primate visual pathway. *Journal of Neuroscience, 31*, 15016–15025.

Dumoulin, S. O., & Hess, R. F. (2007). Cortical specialization for concentric shape processing. *Vision Research, 47*, 1608–1613.

Elston, G. N., & Rosa, M. G. P. (1998). Morphological variation of layer III pyramidal neurones in the occipitotemporal pathway of the macaque monkey visual cortex. *Cerebral Cortex, 8*, 278–294.

Field, D. J., Hayes, A., & Hess, R. F. (1993). Contour integration by the human visual system: Evidence for a local "association field." *Vision Research, 33*, 173–193.

Gallant, J. L., Braun, J., & Van Essen, D. C. (1993). Selectivity for polar, hyperbolic, and Cartesian gratings in macaque visual cortex. *Science, 259*, 100–103.

Gallant, J. L., Connor, C. E., Rakshit, S., Lewis, J. W., & Van Essen, D. C. (1996). Neural responses to polar, hyperbolic, and Cartesian gratings in area V4 of the macaque monkey. *Journal of Neurophysiology, 76*, 2718–2739.

Gauthier, I., Tarr, M. J., Moylan, J., Skudlarski, P. G. J. C., & Anderson, A. W. (2000). The fusiform "face area" is part of a network that processes faces at the individual level. *Journal of Cognitive Neuroscience, 12*, 495–504.

Gibson, J. J., & Radner, M. (1937). Adaptation, aftereffect and contrast in the perception of tilted lines. I. Quantitative studies. *Journal of Experimental Psychology, 20*, 453–467.

Glass, L. (1969). Moiré effect from random dots. *Nature, 223*, 578–580.

Grinter, E. J., Maybery, M. T., Pellicano, E., Badcock, J. C., & Badcock, D. R. (2010). Perception of shapes targeting local and global processes in autism spectrum disorders. *Journal of Child Psychology and Psychiatry, and Allied Disciplines, 51*, 717–724.

Habak, C., Wilkinson, F., & Wilson, H. R. (2006). Dynamics of shape interaction in human vision. *Vision Research, 46*, 4305–4320.

Habak, C., Wilkinson, F., & Wilson, H. R. (2009). Preservation of shape discrimination in aging. *Journal of Vision, 9*(12), 1–8. doi: 10.1167/9.12.18.

Habak, C., Wilkinson, F., Zakher, B., & Wilson, H. R. (2004). Curvature population coding for complex shapes in human vision. *Vision Research, 44*, 2815–2823.

Haxby, J. V., Gobbini, M. I., Furey, M. L., Ishai, A. S. J. L., & Pietrini, P. (2001). Distributed and overlapping representations of faces and objects in ventral temporal cortex. *Science, 293*, 2425–2430.

Haxby, J. V., Hoffman, E. A., & Gobbini, M. I. (2000). The distributed human neural system for face perception. *Trends in Cognitive Sciences, 4*, 223–233.

Heeger, D. J. (1992). Normalization of cell responses in cat striate cortex. *Visual Neuroscience, 9*, 181–197.

Hegdé, J., & Van Essen, D. C. (2000). Selectivity for complex shapes in primate visual area V2. *Journal of Neuroscience, 20*, RC61–66.

Hess, R. F., Wang, Y. A., & Dakin, S. C. (1999). Are judgements of circularity local or global? *Vision Research, 39*, 4354–4360.

Hess, R. F., Wang, Y. Z., Demanins, R., Wilkinson, F., & Wilson, H. R. (1999). A deficit in strabismic amblyopia for global shape detection. *Vision Research, 39*, 901–914.

Hubel, D. H., & Wiesel, T. N. (1968). Receptive fields and functional architecture of monkey striate cortex. *Journal of Physiology, 195*, 215–243.

Jeffrey, B. G., Wang, Y.-Z., & Birch, E. E. (2002). Circular contour frequency in shape discrimination. *Vision Research, 42*, 2773–2779.

Jeffrey, B. G., Wang, Y. Z., & Birch, E. E. (2004). Altered global shape discrimination in deprivation amblyopia. *Vision Research, 44*, 167–177.

Kanwisher, N., McDermott, J., & Chun, M. M. (1997). The fusiform face area: A module in human extrastriate cortex specialized for face recognition. *Journal of Neuroscience, 17*, 4302–4311.

Kobatake, E., & Tanaka, K. (1994). Neuronal selectivities to complex object features in the ventral visual pathway of the macaque cerebral cortex. *Journal of Neurophysiology, 71*, 856–867.

Lamme, V. A. F., & Roelfsema, P. R. (2000). The distinct modes of vision offered by feedforward and recurrent processing. *Trends in Neurosciences, 23*, 571–577.

Lewis, T. L., Ellemberg, D., Maurer, D., Dirks, M., Wilkinson, F., & Wilson, H. R. (2004). A window on the normal development of sensitivity to global form in Glass patterns. *Perception, 33*, 409–418.

Lewis, T. L., Ellemberg, D., Maurer, D., Wilkinson, F., Wilson, H. R., Dirks, M., et al. (2002). Sensitivity to global form in glass patterns after early visual deprivation in humans. *Vision Research, 42*, 939–948.

Loffler, G. (2008). Perception of contours and shapes: Low and intermediate stage mechanisms. *Vision Research, 48*, 2106–2127.

Loffler, G., Wilson, H. R., & Wilkinson, F. (2003). Local and global contributions to shape discrimination. *Vision Research, 43*, 519–530.

Malach, R., Amir, Y., Harel, M., & Grinvald, A. (1993). Relationship between intrinsic connections and functional architecture revealed by optical imaging and in vivo targeted biocytin injections in primate striate cortex. *Proceedings of the National Academy of Sciences of the United States of America, 90*, 10469–10473.

McLaughlin, D. C., Shapley, R., Shelley, J., & Wielaard, D. J. (2000). A neuronal network model for macaque primary visual cortex (V1): Orientation selectivity and dynamics in the input layer 4Ca. *Proceedings of the National Academy of Sciences of the United States of America, 97*, 8087–8092.

Milner, A. D., & Goodale, M. A. (1995). *The visual brain in action.* Oxford: Oxford University Press.

Mishkin, M., Ungerleider, L. G., & Macko, K. A. (1983). Object vision and spatial vision: Two cortical pathways. *Trends in Neurosciences, 6*, 414–417.

Nichols, D. F., Betts, L. R., & Wilson, H. R. (2010). Decoding of faces and face components in face-sensitive human visual cortex. *Frontiers in Psychology, 1*, 1–13.

Olshausen, B. A., & Field, D. J. (1996). Emergence of simple-cell receptive field properties by learning a sparse code for natural images. *Nature, 381*, 607–609.

Op de Beeck, H., & Vogels, R. (2000). Spatial sensitivity of macaque inferior temporal neurons. *Journal of Comparative Neurology, 426*, 505–518.

Pasupathy, A., & Connor, C. E. (2001). Shape representation in area V4: Position-specific tuning for boundary conformation. *Journal of Neurophysiology, 86*, 2505–2519.

Pasupathy, A., & Connor, C. E. (2002). Population coding of shape in area V4. *Nature Neuroscience, 5*, 1332–1338.

Phillips, G. C., & Wilson, H. R. (1984). Orientation bandwidths of spatial mechanisms measured by masking. *Journal of the Optical Society of America. A, Optics and Image Science, 1*, 226–232.

Poirier, F. J., & Wilson, H. R. (2006). A biologically plausible model of human radial frequency perception. *Vision Research, 46*, 2443–2455.

Poirier, F., & Wilson, H. R. (2010). A biologically plausible model of human shape symmetry. *Journal of Vision, 10*(1), 9, 1–16. doi: 10.1167/10.1.9.

Polat, U., & Sagi, D. (1993). Lateral interactions between spatial channels: Suppression and facilitation revealed by lateral masking experiments. *Vision Research, 33*, 993–999.

Regan, D., & Hamstra, S. J. (1992). Shape discrimination and the judgement of perfect symmetry: Dissociation of shape from size. *Vision Research, 32*, 1845–1864.

Sigman, M., Cecchi, G. A., Gilbert, C. D., & Magnasco, M. O. (2001). On a common circle: Natural scenes and gestalt rules. *Proceedings of the National Academy of Sciences of the United States of America, 98*, 1935–1940.

Suzuki, S. (2001). Attention-dependent brief adaptation to contour orientation: A high-level aftereffect for convexity? *Vision Research, 41*, 3883–3902.

Suzuki, S. (2003). Attentional selection of overlapped shapes: A study using brief shape aftereffects. *Vision Research, 43*, 549–561.

Tsao, D. Y., Freiwald, W. A., Knutsen, T. A., Mandeville, J. B., & Tootell, R. B. (2003). Faces and objects in macaque cerebral cortex. *Nature Neuroscience, 6*, 989–995.

Tsao, D. Y., Freiwald, W. A., Tootell, R. B., & Livingstone, M. S. (2006). A cortical region consisting entirely of face-selective cells. *Science, 311*, 670–674.

Tsotsos, J. K. (1993). An inhibitory beam for attentional selection. In L. Harris & M. Jenkin (Eds.), *Spatial vision in humans and robots* (pp. 313–331). New York: Cambridge University Press.

Van Essen, D. C., Anderson, C. H., & Felleman, D. J. (1992). Information processing in the primate visual system: An integrated systems perspective. *Science, 255*, 419–423.

Vidyasagar, T. R., Pei, X., & Volgushev, M. (1996). Multiple mechanisms underlying the orientation selectivity of visual cortical neurones. *Trends in Neurosciences, 19*, 272–277.

Wang, Y.-Z. (2001). Effects of aging on shape discrimination. *Optometry and Vision Science, 78*, 447–454. doi: 10.1097/00006324-200106000-00019.

Wang, Y.-Z., Morale, S. E., Cousins, R., & Birch, E. E. (2009). Course of development of global hyperacuity over lifespan. *Optometry and Vision Science, 86*, 695–700.

Wilkinson, F., James, T. W., Wilson, H. R., Gati, J. S., Menon, R. S., & Goodale, M. A. (2000). Radial and concentric gratings selectively activate human extrastriate form areas: An fMRI study. *Current Biology, 10*, 1455–1458.

Wilkinson, F., Wilson, H. R., & Habak, C. (1998). Detection and recognition of radial frequency patterns. *Vision Research, 38*, 3555–3568.

Wilson, H. R. (1991). Psychophysical models of spatial vision and hyperacuity. In D. Regan (Ed.), *Spatial form vision* (pp. 64–86). New York: Macmillan.

Wilson, H. R. (1999). Non-Fourier cortical processes in texture, form, and motion perception. In P. S. Ulinski & E. G. Jones (Eds.), *Cerebral cortex, 13: Models of cortical circuitry* (pp. 445–477). New York: Plenum Press.

Wilson, H. R., Krupa, B., & Wilkinson, F. (2000). Dynamics of perceptual oscillations in form vision. *Nature Neuroscience, 3*, 170–176.

Wilson, H. R., McFarlane, D. K., & Phillips, G. C. (1983). Spatial frequency tuning of orientation selective units estimated by oblique masking. *Vision Research, 23*, 873–882.

Wilson, H. R., & Wilkinson, F. (1998). Detection of global structure in Glass patterns: Implications for form vision. *Vision Research, 38*, 2933–2947.

Wilson, H. R., & Wilkinson, F. (2004). Spatial channels in vision & spatial pooling. In L. M. Chalupa & J. S. Werner (Eds.), *The visual neurosciences* (pp. 1060–1068). Cambridge, MA: MIT Press.

Wilson, H. R., Wilkinson, F., & Asaad, W. (1997). Concentric orientation summation in human form vision. *Vision Research, 37*, 2325–2330.

44 Contour Integration and the Association Field

DAVID J. FIELD, JAMES R. GOLDEN, AND ANTHONY HAYES

Edges and extended contours are key components of any object or natural scene. Mechanisms that are sensitive to edges and lines were among the first to be described with the invention of techniques that permitted the recording of the responses of cells in visual cortex. However, the steps between the coding of edges and the recognition of objects have remained elusive. In this chapter we focus on one of those steps: the integration of edge or line elements into a contour. The idea that continuity is important to visual perception was a central idea of the Gestalt psychologists who, in the first half of the twentieth century, described a set of perceptual grouping principles that included the "law of good continuation." Over the past 20 years there has been renewed interest in the representation of contours and continuity. Researchers in visual anatomy, neurophysiology, computer science, and visual psychophysics have combined their approaches to develop models of how contours are perceived and integrated by the visual system. In the late 1950s Hubel and Wiesel (see Hubel, 1988, for review) provided the first clear evidence that neurons in the primary visual cortex (V1) responded to local regions of space and were selective to properties such as orientation, position, spatial frequency, and direction of motion. However, there remained the question of how this information, encoded by different neurons, is used in the perception of whole objects and scenes.

In this chapter we review recent work that suggests that the response properties of neurons in the visual pathway depend on a complex relationship among the input, the activity of neighboring neurons, and feedback from higher levels of processing. In particular, neurons in primary visual cortex make use of long-range lateral connections that allow integration of information from far beyond the classical receptive field, and the evidence suggests that these connections are involved in associating neurons that respond along the length of a contour.

The classical receptive field of a visual neuron is defined as the area of the visual field that has the capacity to modify the resting potential of the neuron. However, although this basic feedforward linear model of the simple cell receptive field has been invoked to explain a wide variety of perceptual phenomena—and is at the heart of a wide range of modeling studies—it is essentially wrong. Some of the earliest studies that measured receptive-field properties of cortical neurons recognized that stimuli presented outside the classical receptive field can modify the activity of the neuron even if those regions by themselves cannot effect a response (e.g., Maffei & Fiorentini, 1976). Neurons in primary visual cortex show a variety of interesting nonlinearities, with many occurring within the classical receptive field. However, the nonlinearities that are of interest to us here are the responses to regions outside the classical receptive field. Stimulation of these areas typically does not produce a response, but it can modulate the activity of the neuron. This modulation in activity was commonly described as inhibitory, and a variety of theories have been proposed (e.g., Allman, Miezin, & McGuinness, 1985). One popular account has argued that modulation can serve to normalize the neuron's response and make greater use of the neuron's relatively limited dynamic range (Carandini & Heeger, 2012).

We concentrate here on one component of these nonlinear effects. This approach proposes that the nonclassical surrounds of receptive fields are intimately involved in a process called "contour integration." We do not mean to imply that contour integration is their only role; however, the evidence suggests that it is one role. Indeed, some of the effects that have given rise to the notion of nonclassical surrounds may be generated by the active grouping or "association" of cells in neighboring regions of the visual field. In accord with the term "receptive field," we used the term "association field" to describe the region of associated activity (Field, Hayes, & Hess, 1993), a term that has become relatively popular (e.g., Li, Piëch, & Gilbert, 2008; McManus, Li, & Gilbert, 2011). Others have used the term "integration field" (e.g., Chavane et al., 2000) or "contextual field" (e.g., Phillips & Singer, 1997) or "extension field" (Papari & Petkov, 2011).

We address four questions and explore some of the research that is providing new insights. The questions are as follows: (1) What is contour integration, and why is it important? (2) What do anatomy and physiology suggest about the underlying mechanism? (3) What does the behavior of individuals (i.e., psychophysics) suggest about the underlying mechanism? (4) What insights are provided by computational models of the process?

We note that when putting this review together we discovered over 1,000 papers published in the last 20 years that bear directly on these issues, including a number of excellent reviews and discussions published on the topic or on associated topics. We recommend Fitzpatrick (2000), Gilbert (1998), and Callaway (1998) for discussions of the anatomy and physiology; Polat (1999), Hess and Field (1999), and Graham (2011) for reviews of the psychophysics; and the papers by Zhaoping (2011), Yen and Finkel (1998), and Papari and Petkov (2011) for their comprehensive discussions of the computational issues.

WHAT IS CONTOUR INTEGRATION?

Because reflection and illumination vary across different surfaces, occlusions between surfaces commonly produce a luminance discontinuity (i.e., an edge). However, edges in scenes do not occur only at occlusions. They may also arise from different textures within surfaces as well as from luminance and shading discontinuities. In the 1980s a number of modeling studies were published that proposed computational strategies that would help to identify which of the edges in a scene made up the principal boundaries of an object. Under the assumption that boundary edges were likely to extend over large regions of the visual field, the computations were designed to extract only those edges that were continuous over an extended area. The algorithms that were developed were based on the assumption that the problem could be at least partially solved by integrating over neighboring regions that had similar orientations. Although some of these integration models included, or were derived from, known physiology (e.g., Grossberg & Mingolla, 1985; Parent & Zucker, 1989), the evidence that an integration algorithm of this kind was actually performed by the visual system was not widely accepted.

Two lines of research helped to support the plausibility of a scheme as described above. The first line comes from a series of anatomical and physiological studies that used both cat and primate and suggest that there exist long-range connections between neurons in primary visual cortex that link neurons with similar orientations. The second line consists of psychophysical studies that have provided evidence for the sorts of associations implied by the physiological and anatomical results (e.g., Dakin & Baruch, 2009; Field et al., 1993; Polat & Sagi, 1993). The results of these studies converge on an account that suggests that neurons in primary visual cortex integrate information from outside the classical receptive field in a way that helps with the integration of contours. Below we review some of the studies from both lines of research.

PHYSIOLOGY AND ANATOMY OF LATERAL CONNECTIONS

It is now well understood that stimuli outside of the classical receptive field of a neuron in visual cortex can modulate that neuron's activity. The sources of modulation potentially originate from feedforward connections, feedback connections from neurons further along the visual pathway, and lateral projections from neighboring neurons. Although we concentrate here on lateral connections, the modulation activity is almost certainly dependent on a more complex circuit involving all three. What has been remarkable, however, has been the close ties found between both the anatomy and physiology of the lateral connections and the visual behavior of humans and macaques when completing appropriate psychophysical tasks. However, we also note some exceptions to this general rule (Dakin & Baruch, 2009; May & Hess, 2007).

Early studies exploring the horizontal connections in visual cortex discovered that pyramidal neurons have connections that extend laterally for 2 mm to 5 mm parallel to the surface and have terminations that are patchy and selective (Gilbert & Wiesel, 1979; Rockland & Lund, 1982). Studies on the extent and specificity of lateral projections have been completed on tree shrew (e.g., Rockland & Lund, 1982; Bosking et al., 1997), primate (e.g., Malach et al., 1993; Sincich & Blasdel, 2001), ferret (e.g., Ruthazer & Stryker, 1996), and cat (e.g., Gilbert & Wiesel, 1989) with largely good agreement between species but also some important differences.

Figure 44.1A illustrates an impressive combination of optical imaging and anatomical techniques that reveal the specificity of V1 projections. These results from Bosking et al. (1997) show an overlay of the orientation columns revealed from optical imaging with the lateral projections of pyramidal neurons near the injection site synapsing onto the surrounding regions. The lateral projections are revealed through extracellular injections of biocytin, which label a small number of neurons near the injection site along with their projections. The orientation tuning of a particular neuron is estimated by its location within an orientation column.

As figure 44.1A shows, the orientation column of the injection (shown by the dark areas) is the same orientation as the majority of the long-range projections (i.e., they synapse onto neurons that are also in the dark regions). The short-range projections do not show such specificity. Bosking et al. (1997) also found that in tree shrew the extent of the long-range projections was significantly greater along the axis corresponding to the orientation of the central neuron.

Other work has demonstrated that the lateral projections of these pyramidal neurons are quite specific, projecting to regions of the cortex with iso-orientation columns, as well as similar ocular dominance columns and cytochrome oxidase blobs (e.g., Malach et al., 1993; Yoshioka et al., 1996). Pyramidal cells that are tuned to an orientation aligned with the axes of the projections are shown to project primarily to iso-orientation columns. That is, neurons project primarily to neurons of similar orientation preference.

In some species, as exemplified in figure 44.1A, the projections are considerably longer along the primary axis than orthogonal to the primary axis. For tree shrew (Bosking et al., 1997), owl monkey, and squirrel monkey (Sincich & Blasdel, 2001), the axis of projections is a factor or 2 to 4 longer along the primary axis than the orthogonal axis. Much of the anatomical work has been conducted in layers III and V of primary visual cortex. We should note here that layer IV neurons do not show

the high degree of connectivity to neurons with similar orientations that has been found in layers III and V (Chisum, Mooser, & Fitzpatrick, 2003). Karube and Kisvárday (2011) have found that a large fraction of neurons in layer IV of cat primary visual cortex synapse with neurons in orientation columns with a preferred orientation shifted 60°–90° from the parent cell, providing connections between orientation columns, which they speculate could aid in coding image features such as discontinuities and junctions.

In addition to anatomical studies, neurophysiological studies have explored the effects of co-oriented stimuli presented outside of the classical receptive field (Ito & Gilbert, 1999; Kapadia et al., 1995; Kapadia, Westheimer, & Gilbert, 2000; Li, Piëch, & Gilbert, 2006; McManus, Li, & Gilbert, 2011; Polat et al., 1998). These studies demonstrate that when a neuron is presented with an oriented stimulus within its receptive field, a second collinear stimulus can increase the response rate of the neuron, whereas the same oriented stimulus presented orthogonal to the main axis (displaced laterally) will produce inhibition or at least less facilitation. Kapadia, Westheimer, and Gilbert (2000) attempt to map out these inhibitory and facilitatory effects in awake behaving macaques with the results showing good agreement with the anatomy and with human behavior.

Figure 44.1B shows our theoretical depiction of these lateral projections, which we have called an "association field" (Field, Hayes, & Hess, 1993). This depiction incorporates results from our psychophysical measurements with the particular example from figure 44.1A. First we summarize what we see as some of the principal

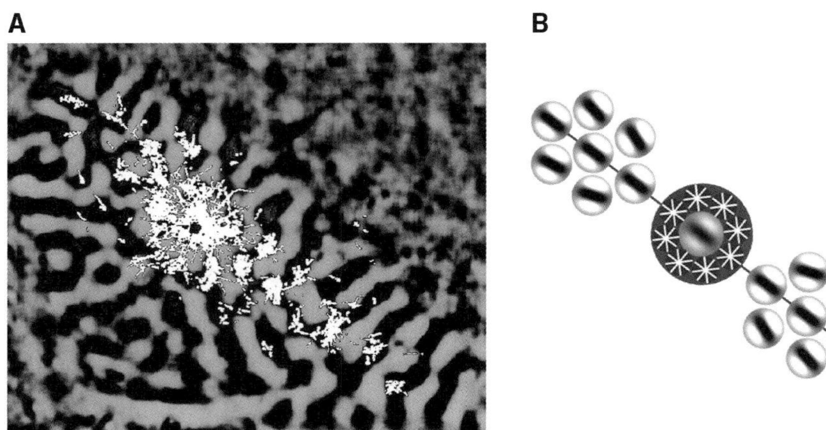

FIGURE 44.1   (A) Results (modified from Bosking et al., 1997) demonstrating the orientation-specific projections of a set of V1 neurons in the tree shrew. Optical imaging is used to reveal the orientation columns, and injections of biocytin are used to map the projections of a set of neurons taking up the biocytin (shown in white). As can be seen, the location of the orientation column of the injection is the same in most cases as the orientation column of the projection. (B) An experimentally and theoretically derived "association field" (Field, Hayes, & Hess, 1993) summarizing our beliefs regarding the underlying projections. Short-range connections are theorized to be largely inhibitory and independent of orientation, whereas long-range connections are orientation specific and largely excitatory.

anatomical and neurophysiological findings that are important to our discussion of contour integration.

1. Long-range projections of pyramidal cells are "patchy," projecting primarily to neurons in iso-orientation columns (i.e., with similar orientation tuning) (Gilbert & Wiesel, 1989; Malach et al., 1993).

2. Long-range projections commonly extend to distances two to four times the size of classical receptive fields and extend primarily in a direction collinear with orientation tuning of the cell (Bosking et al., 1997; McManus, Li, & Gilbert, 2011; Sincich & Blasdel, 2001).

3. Long-range projections to collinear neurons appear to be largely facilitatory (Kapadia et al., 1995; Nelson & Frost, 1985; Polat et al., 1998); however, the neurophysiology results also suggest that facilitation is largely dependent on contrast, with excitation predominant at low contrasts (or with high-contrast cluttered backgrounds) and inhibition predominant at high contrasts (Kapadia, Westheimer, & Gilbert, 1999).

4. Long-range connections appear to be reciprocal (Kisvarday & Eysel, 1992).

5. Short-range projections appear to be largely independent of orientation and have been argued to be predominantly inhibitory (Das & Gilbert, 1999).

These conclusions are not unequivocal. For example, as noted above, not all species show elongation of projections along the primary axis. Furthermore, there is some debate as to whether these long-range connections are the source of facilitatory effects. Kapadia's results (Kapadia, Westheimer, & Gilbert, 2000) imply that regions orthogonal to the main axis will produce inhibitory modulation. However, Walker, Ohzawa, and Freeman (1999), using patches of gratings at positions adjacent to the classical receptive field, found little evidence for facilitation. Although they did find inhibition, that inhibition was rarely symmetric and was typically distributed unevenly. Kapadia et al. (1995) and Polat et al. (1998) have also shown that when a second line segment is presented collinearly outside of the classical receptive field and is aligned with the preferred orientation, a small majority of neurons produce a stronger response when the line segments are separated by a small gap. Their response is stronger to a discontinuous line than to a continuous line. These results might be expected if we assume that short-range inhibition could, in some neurons, cancel out facilitatory collinear effects.

The diagram in figure 44.1B is a simplified representation of what we believe to be the underlying mechanism of contour integration. We have made some assumptions that lack clear support in the anatomy and neurophysiology. For example, in our model we imply that projections to regions offset from the primary axis will project to orientations that are offset in a regular manner. Figure 44.1A provides a weak suggestion that the off-axis projections are not as centered in the orientation column as those along the main axis, but this suggestion lacks quantitative physiological or anatomical data (the hypothesis has not been adequately tested). The motivation for the arrangement shown in the figure comes not from the anatomy and physiology but from behavioral data discussed below.

McManus, Li, and Gilbert (2011) may provide the most thorough physiological evidence yet demonstrating the role of an association field in contour integration. They show that V1 neurons in macaque are selective to contours that extend over the cells' nonclassical receptive fields. They believe this selectivity is achieved through the top-down gating of the lateral connections. In their study, neurons in V1 were recorded with an awake, behaving, macaque. The macaque was cued with a contour composed of three line elements. It then performed a match-to-sample task by observing two fields of random line segments and saccading to the one containing a contour identical to the cue. To identify a cell's preferred contour, a number of contours composed of three segments were presented in the receptive field, and the cell's response was recorded. An optimization algorithm was used to find the preferred contour, which then became a starting point for finding the optimal contour with five segments. This procedure was then carried out for extended contours made up of seven segments—several times the width of the classical receptive field.

McManus and colleagues argue that the association field provides a good account of their V1 responses to these multibar stimuli. For smooth contours the association fields match closely with theory in that they are collinear or cocircular. However, McManus and co-workers not only observed selectivity in response to line segments but found selectivity to sinusoidal and closed circular contours as well. They observed a number of individual neurons that could shift between linear and circular selectivity depending on the cue given in the matching task. By aggregating the single-unit recordings, they determined that the contour selectivity of populations is also context dependent. The contour selectivity of V1 neurons is therefore a dynamic property that McManus and colleagues speculate is wholly dependent on top-down, task-specific control signals from higher processing levels that may functionally turn on or off sets of lateral connections between V1 neurons. They note that this is distinct from the conventional view of top-down control of V1 through minor amplitude modulation of neuronal responses (e.g., contrast gain control).

Li, Piëch, and Gilbert (2008) addressed the role of top-down influences on contour integration in a different manner. They examined the time course of contour integration in macaque and found results that strongly suggested a form of perceptual learning. There was gradual improvement up to an asymptote, and the improved accuracy was restricted to the trained location in the visual field. To directly investigate the role of attention, before training, the animal was given a distracter task and simultaneously shown contour stimuli in the trained areas of the visual field. Before training, recordings of V1 neurons revealed that responses were no different from the responses to noise patterns. However, after training, responses of V1 neurons were directly related to the length of the contours during the contour task and were weakly related in the distracter task, demonstrating the necessity of task-specific feedback. Contour-related responses wholly disappeared with no top-down input when the macaque was anesthetized. This result is largely consistent with evidence of Angelucci and Bressloff (2006) that suggests that a full understanding of the dynamics of contour integration (facilitation) and surround suppression requires a model with feedback from higher levels in addition to feedforward and lateral connections.

PSYCHOPHYSICS

Psychophysical research using different methodologies has demonstrated effects that correspond to the anatomical and physiological data discussed above.

An example of a stimulus used in psychophysical research is demonstrated in figure 44.2. Human observers are presented with arrays of high-contrast-oriented elements in which a subset has been aligned according to one of several alignment rules. Human observers attempt to identify the presence of this subset of elements as a function of the alignment. Figure 44.2 provides examples of two different rules. As is readily observed from the figure, observers are more sensitive to collinear alignment than to orthogonal alignment even though the stimuli are equated for their information content (Field, Hayes, & Hess, 1993).

A raft of studies that use the contour integration task have been published over the last 20 years to investigate the conditions under which integration occurs. For example, studies have demonstrated that integration is possible with elements that have multiple depth planes (Hess & Field, 1995; Hess, Hayes, & Kingdom, 1997), with elements that have different phase or polarity (Field, Hayes, & Hess, 2000), with elements that have different bandwidths, and rather weakly with elements at multiple scales (Dakin & Hess, 1998, 1999). Hayes (2000) has demonstrated that when a local-motion signal induces an apparent displacement of each whole element, integration is stronger when the alignment corresponds to the elements' perceptual location as opposed to their physical location. Mullen, Beaudot, and McIlhagga (2000) demonstrated that although integration across multiple hues is possible, integration is more effective between similar hues. Lee and Blake (2001) have shown a similar effect for movement. Hess and Dakin (1997) have suggested that there is a decline in contour integration accuracy in the periphery and

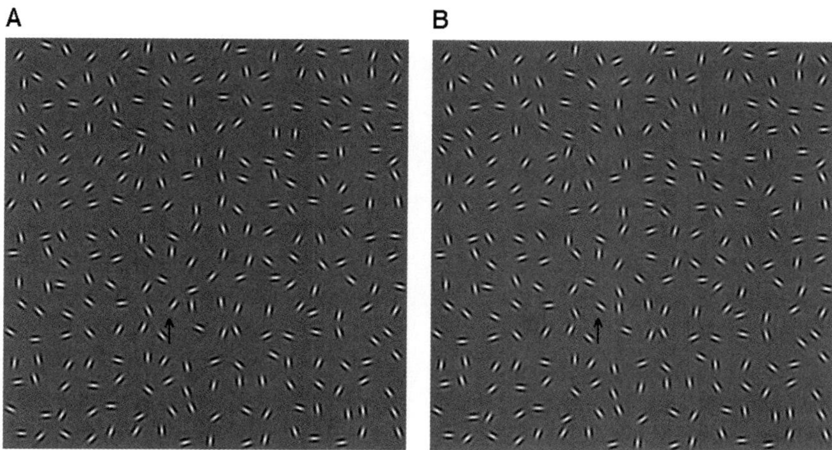

FIGURE 44.2   An example of two stimuli used to measure human sensitivity to contours. The two images contain a path at the same location in each image (as marked by arrows) but created according to two different rules (after Field, Hayes, & Hess, 1993). The reader should be able to see that the contour or "path" contained in the image on the left (A) is more visible than that on the right (B). This figure is one example that demonstrates the human visual system's increased sensitivity to collinear arrangements.

argue that this is related to an overreliance on large receptive fields. Others have found that contour integration declines in the periphery at a similar rate to other visual functions, such as acuity (Nugent et al., 2003).

Other psychophysical techniques have demonstrated intriguing results. Chavane and colleagues (2000) have demonstrated that the speed of an oriented element appears higher when it moves in a direction collinear to its axis than when it moves in a direction orthogonal to its axis. They argue that long-range connections may be responsible. Kapadia, Westheimer, and Gilbert (2000) have demonstrated that elements placed along the ends of a central element can induce a perceived change in the orientation of the central element toward the orientation of the flanking element. However, when the flanking elements are placed along the opposite axis (adjacent to the central element), the central element can be shifted away from the orientation of the flanking elements. They also demonstrated that the spatial distribution of this effect showed good agreement to the neurophysiology of cortical facilitation produced by the flanking lines. Mareschal, Sceniak, and Shapley (2001) have also demonstrated that, with a collinear arrangement, flanking grating patches can significantly increase the orientation discrimination thresholds of the central patch. Furthermore, the threshold increase is significantly higher in the collinear arrangement than when the orientation of the elements is perpendicular to the positions of the three patches.

Fulvio, Singh, and Maloney (2008) tested the ability of subjects to interpolate contours between two line segments. In their experiment, the two contours were spatially separated by several degrees of visual angle by an occlusion, with a short fragment visible between them. The subjects were required to align this fragment to match a continuous contour between the two segments. This is used to test the "relatability hypothesis" of Kellman and Shipley (1991), which claims that for contour elements to be linked, they must have an intersection and that the angle that one must turn is less than 90°. This view is slightly less strict than the collinearity/cocircularity linkages of the association field. Kellman and Shipley predict a sharp dropoff in the ability of subjects to link contours with a turning angle greater than 90°. Fulvio and colleagues tested subjects for a continuous range of turning angles below and above 90° and found a smooth dropoff instead in their abilities to interpolate contours.

Kovacs and Julesz (1993) have demonstrated that when the visibility of a path of elements is measured as a function of the density of the surrounding elements, the path is significantly more visible if the contour forms a closed figure. Pettet, McKee, and Grzywacz (1998) argue that this effect may be related to the directional smoothness of the contours (i.e., a circular figure has all the elements changing orientation in a consistent direction). In either case both results demonstrate that sensitivity depends on elements further away than the immediate neighbors. The simple model, based on excitatory effects between neighbors, will not produce this effect. Of course, there is no reason to assume that these psychophysical effects necessarily occur in V1, and these psychophysical results may be an indication of the direction of attention toward features that undergo predictable change. Nonetheless, the results suggest that thresholds for perceiving contours depend on complex relationships.

Dakin and Baruch (2009) investigated the interplay between these psychophysical areas: how contextual modulation affects contour integration directly. They sought to test this phenomenon under the flanking conditions described by Polat and Sagi (1993). They created Gabor-element contours in noise backgrounds where the orientation of distracter elements to contour elements was systematically varied. A strong effect was observed for "snakes" (see figure 44.2A), such that parallel distracter elements significantly inhibited integration of the actual contours compared with randomly aligned flankers, whereas distracter elements aligned perpendicularly with the path facilitated integration. Interestingly, all alignments of flankers facilitated the integration of "ladders" (see figure 44.2B) equally over randomly aligned flankers, but not nearly as much as that found with snakes. Dakin and Baruch note that this result is not predicted by the original association field model, nor has the physiological support for this been investigated (although see Meirovithz et al., 2010). Dakin and Baruch also propose a detailed computational model, which is discussed below.

The prevailing idea behind why the visual system integrates contours in the way it does is that it is optimized to the types of contours that occur most frequently in natural scenes. Snake-like elements usually indicate contours, ladder-like elements sometimes do so, and arrays of elements aligned somewhere between snakes and ladders usually do not. Extensions of the Field, Hayes, and Hess (1993) design have been applied to task-dependent learning scenarios to show this optimization might not be immutable. Schwarzkopf and Kourtzi (2008) trained naive observers on the contour integration task with ladders and demonstrated significant improvements in their performance. They found that this improved performance persisted in a follow-up test 3–5 months after training. Additionally, they were able to train subjects to a high degree of accuracy (75%) with stimuli having Gabor elements offset at 30°–45°. This high level of performance required a longer training period than for ladder stimuli, but it is surprising considering that image elements with this offset usually signal discontinuities, not contours; additionally, it contradicts the conventional belief that contour integration abilities have evolved to match the contour statistics present in natural images. This study shows that the cortical connections that allow for contour integration can be thoroughly modified by short-term training and lend support to the idea of dynamic association fields of McManus et al. (2011).

COMPUTATIONAL MODELING

Many computational models are simple reflections of the data found experimentally. They can be considered as "existence proofs" that demonstrate that it is at least possible to perform the desired task with the proposed architecture. They cannot demonstrate that the visual system necessarily uses the architecture of the model, but they can demonstrate that such a model would work if that architecture did underlie the task. However, at times, these models are most useful when they fail, and that may well be the case in some of the following studies that we discuss.

Although the physiology and anatomy that underlie contour integration are becoming clearer, the range of potentially plausible models is narrowing. In this section we briefly review natural-scene work, but for the most part we concentrate on recent computational strategies that focus on behavioral phenomena of contour integration whose underlying mechanisms have not been elucidated. Most of these computational explanations attempt to describe observed psychophysical phenomena such as crowding, the effect of flankers, or short-term remapping of cortical connections.

To integrate contours, a variety of algorithms have been proposed that use the technique of integrating similar orientations along collinear directions. Part of the argument for using a collinearity algorithm appears to be that the task demands it. However, these early studies also went to some lengths to explain how such an algorithm might fit with the known physiology and anatomy (e.g., Grossberg & Mingolla, 1985; Parent & Zucker, 1989; Sha'ashua & Ullman, 1988). In recent years, as our understanding of the underlying physiology has increased, so has the sophistication of computational models (e.g., Zhaoping, 2011; McManus, Ullman, & Gilbert, 2008; Yen & Finkel, 1998). These models have demonstrated that the architecture revealed by the physiology and anatomy can be used to

provide an efficient means of extracting contours in natural scenes and can be used to account for a significant amount of the psychophysical data.

As noted above, early computational algorithms for contour detection operated on the principle that nearby image patches with the same orientation were likely elements of the same contour. A model that embodies this assumption has intuitive appeal as the way of performing contour integration. Some early models attempted to incorporate the known physiology (e.g., Grossberg & Mingolla, 1985; Parent & Zucker, 1989; Sha'ashua & Ullman, 1988); many more recent models hew closely to the physiology as it is now understood, although some forgo this approach to simplify the exercise. We proposed the original association field model (Field, Hayes, & Hess, 1993) in part because it clearly was an efficient method to encode the practically infinite space of possible contours that could occur in natural scenes. An alternative model may propose neurons that each encode a particular contour, but this scheme would require a vast number of "contour feature detectors."

Geisler et al. (2001) and Sigman et al. (2001) took the ecological approach further and asked whether the contour integration model is an efficient means of coding natural scene contours. They measured the co-occurrence statistics of edge elements in natural scenes and found that the relative orientations of neighboring contour segments match well with those predicted physiologically and with psychophysically defined association fields. The results of Geisler et al. are particularly interesting because of the requirements needed to measure these co-occurrence statistics. As these authors argue, these statistics are multidimensional in nature. Given an edge at a particular location with a particular orientation, the region around that location is a three-dimensional probability map of x-position by y-position by orientation. Only by mapping out this full probability map does one see the full set of statistical dependencies. And it is in these conditional probabilities that one finds the orientation dependencies that map onto the "association field" properties. The probability map is much higher in dimension if we include the additional dependencies across scale, chromaticity, motion, and disparity.

One approach proposes that continuity is represented by a temporal code, presumably tied to the synchronous activity of neighboring neurons. This approach to "binding" has received considerable recent attention and has some experimental support (e.g., Gray et al., 1989). The difficulty with this model is that it requires a mechanism to detect the synchrony. Hess, Dakin, and Field (1998) suggest a rather different and more basic version of a temporal code. They suggest that contrast information is represented by the initial response generated by the feedforward activity with the later response determined by the lateral connections and the context of the surrounding regions. The contrast signal could then be extracted from the collinearity signal by simply tracking the sustained response. This hypothesis was derived from the neurophysiological work of Zipser, Lamme, and Schiller (1996), which found results for textures consistent with this theory. However, Kapadia, Westheimer, and Gilbert (1999) provide data that are supportive in some ways but also make the story more complex. As noted in the previous section, Kapadia, Westheimer, and Gilbert found that collinear facilitation for neurons in V1 occurs only at low contrasts or in complex backgrounds. They also noted that this facilitation occurs after the initial transient response of the neuron during the "sustained" component of the response. This aspect of the response fits the model proposed by Hess, Dakin, and Field (1998). At high contrasts, however, the neurons do not show this sustained response but only the sharp transient response. What sort of model predicts this high-contrast behavior? It may involve some degree of contrast normalization (Heeger, 1992), but at present we are not aware of any model that predicts both the timing of responses and the lack of facilitation at high contrasts.

One area of recent research where lateral connections play an important role is in the phenomenon of "crowding." Crowding describes a perceptual experience in which the flanking stimuli (often letters) reduce the ability to recognize a central stimulus (e.g., a letter). Although crowding research has a large literature (see Levi, 2008, for a review), much of the recent work has focused on the importance of lateral interactions in visual cortex (e.g., Dakin et al., 2011; Greenwood, Bex, & Dakin, 2009; Saarela et al., 2009).

May and Hess (2007) observed a crowding effect that interfered with detection of "ladder" stimuli in the periphery. They describe a computational model that replicates the contour integration performance of their human subjects and that is also susceptible to crowding in the periphery. They note that they are not explicitly modeling the physiological mechanisms underlying contour integration but merely the computations that are carried out in physiology. The model receives as input only the coordinates and orientations of the Gabor elements in the input images; May and Hess claim that this is an accurate representation of the association field in physiology, as contour integration ability does not increase with contrast of Gabor elements (Hess, Dakin, & Field, 1998). Association fields were created using methods similar to those of Pelli, Palomares, and Majaj (2004), growing larger in the periphery. The association of each element with every other element is calculated based on whether the element is within an association field, how similar their orientations are, and whether the pair is more snake-like or ladder-like. A rule-based approach was used to sort these association strengths and form the longest possible snake and ladder contours. The model was applied to the same 2AFC experiment the subjects performed, and results were very similar, with ceiling accuracy for snakes at all eccentricities but a sharp dropoff as a function of eccentricity for ladders. The model was also applied to letter stimuli, and the linking of contours across different letters, as would be expected when crowding occurs, was also observed.

Pelli (2008) investigated the effects of the relative scale and distance of the flankers on crowding as a function of eccentricity. His work suggests that the critical spacing at all eccentricities corresponds to approximately 6 mm on the surface of the cortex. This roughly corresponds to the spatial extent of the lateral connections, which he calls the "integration field." Dakin and Baruch (2009) found that human subjects' contour integration abilities were facilitated or impaired by surrounding Gabor elements, and they attempted to replicate the results of their experiment with a computational model. Their approach used a technique they call "opponent-orientation filtering." Images were filtered with oriented Gabors, and then the difference of the outputs of orthogonal filters was calculated. At this stage the response of the model for contours of any length was strongly positive and was inhibited by the surrounding orthogonal elements. Dakin and Baruch found that this inhibitory model provided a better model for the contextual effects found when searching for contours in noisy backgrounds.

McManus, Ullman, and Gilbert (2008) used an association field model to investigate the phenomenon of perceptual fill-in, which occurs when a subset of photoreceptors in both retinas are destroyed by a disease such as macular degeneration. After some time, patients are able to see features within the patches of their visual field that had been blind spots. This recovery is due to a reorganization of the horizontal connections of the association field in V1 (see Gilbert, 1998, for review), where axons from V1 cells with normal input grow into the lesion-projection zone. The association field is used as the basis for a computational model of how perceptual fill-in occurs. McManus and colleagues developed a model that recurrently feeds input to V1 complex cells within the lesion projection zone input from cocircular horizontal connections. The model, shown in

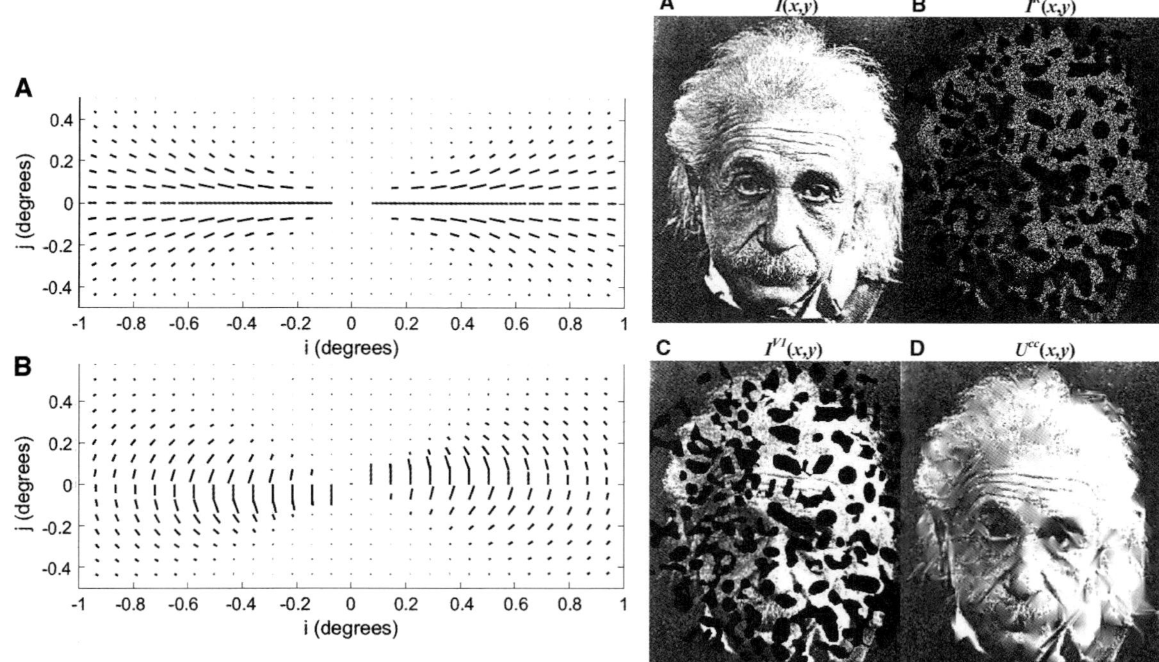

FIGURE 44.3   McManus, Ullman, and Gilbert (2008) modeled the association field using a recurrent neural network. When the model was applied to images with patches missing to simulate retinal damage, the association field was somewhat able to "fill in" the missing information. This approach has the potential to successfully fill in a missing patch when the information in the patch corresponds to contours found in the association field. The figure shows (A) the original image, (B) the modeled response of a defective retina, (C) the modeled signal transmitted to V1, and (D) the estimate of the original image with defective areas recovered.

figure 44.3, outputs images that simulate the "holes" in the visual field and shows the approach has some limited success at filling in these holes.

Kovacs et al. (1996) demonstrated a binocular version of the above finding by presenting to each eye two completely different natural scenes, such that each eye received patches from both images: the left eye received the complement of the right (e.g., right eye gets 1,2,1,2,2,1 with the left eye receiving 2,1,2,1,1,2). Kovacs and co-workers found that observers commonly see complete images (1,1,1,1,1,1 or 2,2,2,2,2). The contours and other visual information were successfully integrated between the two eyes into a single perceptual whole. This result implies that the association field is less sensitive to ocular origin than it is sensitive to continuity.

SUMMARY AND CONCLUSIONS

In the last two decades a large number of studies in visual anatomy, neurophysiology, psychophysics, and computational modeling have provided significant new insights into how the information from different V1 neurons is integrated. Our original proposal in 1993 was that lateral connections between V1 neurons played a key role in the process of contour integration. Evidence for this proposal has been supported by a wide range of studies. However, the simple "association field" model is incomplete and cannot account for many aspects of contour perception: as described above, higher-level processing is certainly involved. We know that feedback from higher levels can alter the activity of V1 neurons. The work of McManus, Li, and Gilbert (2011) suggests that feedback may have a greater effect on the nonclassical receptive field than it does on the classical receptive field. When the animal is involved in a search for a particular class of stimulus (e.g., curved lines or straight lines), lateral connections help tune the neuron to the particular stimulus. This work suggests that even at the level of V1, the visual system is much more dynamic and flexible than most of us have thought.

We also do not wish to argue that the only role of lateral connections in V1 is to integrate contours. The communication between neighboring neurons certainly plays a number of roles in pattern perception that range from spatial selectivity to contrast normalization to texture segregation. Contour integration represents just one component of early visual processing. Furthermore, the visual system performs various forms of integration (e.g., motion, stereo, color). The general concept of an association field should perhaps include all aspects of the stimulus that are "associated."

We should also note that it is likely that few "computational problems" are solved in V1. Lateral connections certainly play a significant role. However, we expect to see that a full account will include more complex dynamical interaction between neurons and other regions of the brain.

In this chapter we can address a fraction of the studies that bear on contour integration. This body of research is the result of remarkable feedback among a broad range of disciplines, where the advances in one discipline have very rapidly influenced the research in a second, and which have led to new advances in a third. There remain a number of fundamental questions that have yet to be answered. Exactly how does lateral input mediate the firing of V1 cells in response to contours? What is the physiological explanation for why some flanking stimuli facilitate and others inhibit? The phenomenon of crowding has had a number of different explanations over the past few decades, and now the association field seems to provide an explanation for some pieces of that puzzle. The exciting new finding pointing toward dynamic, task-dependent gating of the association field (McManus, Li, & Gilbert, 2011) brings attention into the picture, which is itself a high-level mechanism of great complexity. Characterizing the physiology behind this interaction will be a daunting task.

ACKNOWLEDGMENTS

This work was partially supported by research grant SUG 58100047 from Nanyang Technological University to A. Hayes, and a Department of Psychology grant to David J. Field and James R. Golden.

REFERENCES

Allman, J., Miezin, F., & McGuinness, E. (1985). Stimulus specific responses from beyond the classical receptive field: Neurophysiological mechanisms for local-global comparisons in visual neurons. *Annual Review of Neuroscience, 8,* 407–430.

Angelucci, A., & Bressloff, P. C. (2006). The contribution of feedforward, lateral and feedback connections to the classical receptive field center and extra-classical receptive field surround of primate V1 neurons. *Progress in Brain Research, 154,* 93–121.

Bosking, W. H., Zhang, Y., Schofield, B., & Fitzpatrick, D. (1997). Orientation selectivity and the arrangement of horizontal connections in tree shrew striate cortex. *Journal of Neuroscience, 17,* 2112–2127.

Callaway, E. M. (1998). Local circuits in primary visual cortex of the macaque monkey. *Annual Review of Neuroscience, 21,* 47–74.

Carandini, M., & Heeger, D. J. (2012). Normalization as a canonical neural computation. *Nature Reviews. Neuroscience, 13,* 51–62.

Chavane, F., Monier, C., Bringuier, V., Baudot, P., Borg-Graham, L., Lorenceau, J., et al. (2000). The visual cortical association field: A Gestalt concept or a psychological entity? *Journal of Physiology, Paris, 94,* 333–342. doi:10.1016/S0928-4257(00)01096-2.

Chisum, H. J., Mooser, F., & Fitzpatrick, D. (2003). Emergent properties of layer 2/3 neurons reflect the collinear arrangement of horizontal connections in tree shrew visual cortex. *Journal of Neuroscience, 23,* 2947–2960.

Dakin, S. C., & Baruch, N. J. (2009). Context influences contour integration. *Journal of Vision, 9*(2), 13, 1–13. doi:10.1167/9.2.13.

Dakin, S., Greenwood, J., Carlson, T., & Bex, P. (2011). Crowding is tuned for perceived (not physical) location. *Journal of Vision (Charlottesville, Va.), 11*(2), 1–13. doi:10.1167/11.9.2.

Dakin, S. C., & Hess, R. F. (1998). Spatial-frequency tuning of visual contour integration. *Journal of the Optical Society of America, Series A, 15,* 1486–1499. doi:10.1364/JOSAA.15.001486.

Dakin, S. C., & Hess, R. F. (1999). Contour integration and scale combination processes in visual edge detection. *Spatial Vision, 12,* 309–327.

Das, A., & Gilbert, C. D. (1999). Topography of contextual modulations mediated by short-range interactions in primary visual cortex. *Nature, 399,* 655–661.

Field, D. J., Hayes, A., & Hess, R. F. (1993). Contour integration by the human visual system: Evidence for a local "association field." *Vision Research, 33,* 173–193.

Field, D. J., Hayes, A., & Hess, R. F. (2000). The roles of polarity and symmetry in contour integration. *Spatial Vision, 13,* 51–66.

Fitzpatrick, D. (2000). Seeing beyond the receptive field in primary visual cortex. *Current Opinion in Neurobiology, 10,* 438–443.

Fulvio, J., Singh, M., & Maloney, L. (2008). Precision and consistency of contour interpolation. *Vision Research, 48,* 831–849.

Geisler, W. S., Perry, J. S., Super, B. J., & Gallogly, D. P. (2001). Edge co-occurrence in natural images predicts contour grouping performance. *Vision Research, 41,* 711–724.

Gilbert, C. D. (1998). Adult cortical dynamics. *Physiological Reviews, 78,* 467–485.

Gilbert, C. D., & Wiesel, T. N. (1979). Morphology and intracortical projections of functionally characterised neurones in the cat visual cortex. *Nature, 280,* 120–125.

Gilbert, C. D., & Wiesel, T. N. (1989). Columnar specificity of intrinsic horizontal and corticocortical connections in cat visual cortex. *Journal of Neuroscience, 9,* 2432–2442.

Graham, N. (2011). Beyond multiple pattern analyzers modeled as linear filters (as classical V1 simple cells): Useful additions of the last 25 years. *Vision Research, 51,* 1397–1430. doi:10.1016/j.visres.2011.02.007.

Gray, C. M., Konig, P., Engel, A. K., & Singer, W. (1989). Oscillatory responses in cat visual cortex exhibit inter-columnar synchronization which reflects global stimulus properties. *Nature, 338,* 334–337.

Greenwood, J., Bex, P., & Dakin, S. (2009). Positional averaging explains crowding with letter-like stimuli. *Proceedings of the National Academy of Sciences of the United States of America, 106,* 13130–13135. doi:10.1073/pnas.0901352106.

Grossberg, S., & Mingolla, E. (1985). Neural dynamics of perceptual grouping: Textures, boundaries, and emergent segmentations. *Perception & Psychophysics, 38,* 141–171.

Hayes, A. (2000). Apparent position governs contour-element binding by the visual system. *Proceedings of the Royal Society, Series B, 267,* 1341–1345.

Heeger, D. J. (1992). Normalization of cell responses in cat striate cortex. *Visual Neuroscience, 9,* 181–198.

Hess, R. F., & Dakin, S. C. (1997). Absence of contour linking in peripheral vision. *Nature, 390,* 602–604.

Hess, R. F., Dakin, S. C., & Field, D. J. (1998). The role of "contrast enhancement" in the detection and appearance of visual contours. *Vision Research, 38,* 783–787.

Hess, R. F., & Field, D. J. (1995). Contour integration across depth. *Vision Research, 35,* 1699–1711. doi:10.1016/0042-6989(94)00261-J.

Hess, R., & Field, D. (1999). Integration of contours: New insights. *Trends in Cognitive Sciences, 12,* 480–486.

Hess, R. F., Hayes, A., & Kingdom, F. A. A. (1997). Integrating contours within and through depth. *Vision Research, 3,* 691–696.

Hubel, D. H. (1988). *Eye, brain, and vision.* New York: Scientific American Library.

Ito, M., & Gilbert, C. D. (1999). Attention modulates contextual influences in the primary visual cortex of alert monkeys. *Neuron, 22,* 593–604.

Kapadia, M. K., Ito, M., Gilbert, C. D., & Westheimer, G. (1995). Improvement in visual sensitivity by changes in local context: Parallel studies in human observers and in V1 of alert monkeys. *Neuron, 15,* 843–856.

Kapadia, M. K., Westheimer, G., & Gilbert, C. D. (1999). Dynamics of spatial summation in primary visual cortex of alert monkeys. *Proceedings of the National Academy of Sciences of the United States of America, 96,* 12073–12078. doi:10.1073/pnas.96.21.12073.

Kapadia, M. K., Westheimer, G., & Gilbert, C. D. (2000). Spatial distribution of contextual interactions in primary visual cortex and in visual perception. *Journal of Neurophysiology, 84,* 2048–2062.

Karube, F., & Kisvárday, Z. (2011). Axon topography of layer IV spiny cells to orientation map in the cat primary visual cortex (area 18). *Cerebral Cortex, 21,* 1443.

Kellman, P. J., & Shipley, T. F. (1991). A theory of visual interpolation in object perception. *Cognitive Psychology, 23,* 141–221.

Kisvarday, Z. F., & Eysel, U. T. (1992). Cellular organization of reciprocal patchy networks in layer III of cat visual cortex (area 17). *Neuroscience, 46,* 275–286.

Kovacs, I., & Julesz, B. (1993). A closed curve is much more than an incomplete one: Effect of closure in figure-ground segmentation. *Proceedings of the National Academy of Sciences of the United States of America, 90,* 7495–7497. doi:10.1073/pnas.90.16.7495.

Kovacs, I., Papathomas, T. V., Yang, M., & Feher, A. (1996). When the brain changes its mind: Interocular grouping during binocular rivalry. *Proceedings of the National Academy of Sciences of the United States of America, 93,* 15508–15511. doi:10.1073/pnas.93.26.15508.

Lee, S. H., & Blake, R. (2001). Neural synergy in visual grouping: When good continuation meets common fate. *Vision Research, 41,* 2057–2064. doi:10.1016/S0042-6989(01)00086-4.

Levi, D. (2008). Crowding—an essential bottleneck for object recognition: A mini-review. *Vision Research, 48,* 635–654.

Li, W., Piëch, V., & Gilbert, C. D. (2006). Contour saliency in primary visual cortex. *Neuron, 50,* 951–962.

Li, W., Piëch, V., & Gilbert, C. D. (2008). Learning to link visual contours. *Neuron, 57,* 442–451.

Maffei, L., & Fiorentini, A. (1976). The unresponsive regions of visual cortical receptive fields. *Vision Research, 16,* 1131–1139. doi:10.1016/0042-6989(76)90253-4.

Malach, R., Amir, Y., Harel, M., & Grinvald, A. (1993). Relationship between intrinsic connections and functional architecture revealed by optical imaging and in vivo targeted biocyting injections in primate striate cortex. *Proceedings of the National Academy of Sciences of the United States of America, 90,* 10469–10473. doi:10.1073/pnas.90.22.10469.

Mareschal, I., Sceniak, M. P., & Shapley, R. M. (2001). Contextual influences on orientation discrimination: Binding local and global cues. *Vision Research, 41,* 1915–1930. doi:10.1016/S0042-6989(01)00082-7.

May, K., & Hess, R. (2007). Ladder contours are undetectable in the periphery: A crowding effect? *Journal of Vision, 7*(13), 9, 1–15. doi:10.1167/7.13.9.

McManus, J., Li, W., & Gilbert, C. (2011). Adaptive shape processing in primary visual cortex. *Proceedings of the National Academy of Sciences of the United States of America, 108,* 9739–9746. doi:10.1073/pnas.1105855108.

McManus, J., Ullman, S., & Gilbert, C. (2008). A computational model of perceptual fill-in following retinal degeneration. *Journal of Neurophysiology, 99,* 2086–2100.

Meirovithz, E., Ayzenshtat, I., Bonneh, Y., Itzhack, R., Werner-Reiss, U., & Slovin, H. (2010). Population response to contextual influences in the primary visual cortex. *Cerebral Cortex, 20,* 1293–1304.

Mullen, K. T., Beaudot, W. H., & McIlhagga, W. H. (2000). Contour integration in color vision: A common process for the blue-yellow, red-green and luminance mechanisms? *Vision Research, 40,* 639–655.

Nelson, J. I., & Frost, B. J. (1985). Intracortical facilitation among co-oriented, co-axially aligned simple cells in cat striate cortex. *Experimental Brain Research, 61,* 54–61.

Nugent, A. K., Keswani, R. N., Woods, R. L., & Peli, E. (2003). Contour integration in peripheral vision reduces gradually with eccentricity. *Vision Research, 43,* 2427–2437. doi:10.1016/S0042-6989(03)00434-6.

Papari, G., & Petkov, N. (2011). Edge and line oriented contour detection: State of the art. *Image and Vision Computing, 29,* 79–103. doi:10.1016/j.imavis.2010.08.009.

Parent, P., & Zucker, S. (1989). Trace inference, curvature consistency and curve detection. *IEEE Transactions on Pattern Analysis and Machine Intelligence, 11,* 823–839.

Pelli, D. (2008). Crowding: A cortical constraint on object recognition. *Current Opinion in Neurobiology, 18,* 445–451.

Pelli, D. G., Palomares, M., & Majaj, N. J. (2004). Crowding is unlike ordinary masking: Distinguishing feature integration from detection. *Journal of Vision, 4*(12), 12. doi:10.1167/4.12.12.

Pettet, M. W., McKee, S. P., & Grzywacz, N. M. (1998). Constraints on long range interactions mediating contour detection. *Vision Research, 38,* 865–879.

Phillips, W. A., & Singer, W. (1997). In search of common foundations for cortical computation. *Behavioral and Brain Sciences, 20,* 657–722.

Polat, U. (1999). Functional architecture of long-range perceptual interactions. *Spatial Vision, 12*, 143–162.

Polat, U., Mizobe, K., Pettet, M. W., Kasamatsu, T., & Norcia, A. M. (1998). Collinear stimuli regulate visual responses depending on cell's contrast threshold. *Nature, 391*, 580–584.

Polat, U., & Sagi, D. (1993). Lateral interactions between spatial channels: Suppression and facilitation revealed by lateral masking experiments. *Vision Research, 33*, 993–999.

Rockland, K. S., & Lund, J. S. (1982). Widespread periodic intrinsic connections in the tree shrew visual cortex. *Science, 215*, 1532–1534.

Ruthazer, E. S., & Stryker, M. P. (1996). The role of activity in the development of long-range horizontal connections in area 17 of the ferret. *Journal of Neuroscience, 16*, 7253–7269.

Saarela, T. P., Sayem, B., Westheimer, G., & Herzog, M. H. (2009). Global stimulus configuration modulates crowding. *Journal of Vision, 9*(2), 5. doi:10.1167/9.2.5.

Schwarzkopf, D., & Kourtzi, Z. (2008). Experience shapes the utility of natural statistics for perceptual contour integration. *Current Biology, 18*, 1162–1167.

Sha'ashua, A., & Ullman, S. (1988). Structural saliency. In *Proceedings of the International Conference on Computer Vision, Tampa, Florida* (pp. 482–488).

Sigman, M., Guillermo, G. A., Gilbert, C. D., & Magneasco, M. O. (2001). On a common circle: Natural scenes and Gestalt rules. *Proceedings of the National Academy of Sciences of the United States of America, 98*, 1935–1940. doi:10.1073/pnas.031571498.

Sincich, L. C., & Blasdel, G. G. (2001). Oriented axon projections in primary visual cortex of the monkey. *Journal of Neuroscience, 21*, 4416–4426.

Walker, G. A., Ohzawa, I., & Freeman, R. D. (1999). Asymmetric suppression outside the classical receptive field of the visual cortex. *Journal of Neuroscience, 19*, 10536–10553.

Yen, S. C., & Finkel, L. H. (1998). Extraction of perceptually salient contours by striate cortical networks. *Vision Research, 38*, 719–741.

Yoshioka, T., Blasdel, G. G., Levitt, J. B., & Lund, J. S. (1996). Relation between patterns of intrinsic lateral connectivity, ocular dominance, and cytochrome oxidase-reactive regions in macaque monkey striate cortex. *Cerebral Cortex, 6*, 297–310.

Zhaoping, L. (2011). Neural circuit models for computations in early visual cortex. *Current Opinion in Neurobiology, 21*, 808–815.

Zipser, K., Lamme, V. A. F., & Schiller, P. H. (1996). Contextual modulation in primary visual cortex. *Journal of Neurophysiology, 16*, 7376–7389.

45 Texture Analysis and Perception

MICHAEL S. LANDY

What is visual texture? *The Oxford English Dictionary* provides several definitions of the word. The first definition comes from the etymology: The word comes from Latin *textūra,* meaning "a weaving." Later definitions elaborate and extend this theme, including "any natural structure having an appearance or consistence as if woven; a tissue; a web, e.g., of a spider," and, finally, "the constitution, structure, or substance of anything with regard to its constituents or formative elements." In other words, texture is "stuff" (Adelson & Bergen, 1991) that is composed of smaller elements. This latter definition thus encompasses all the sorts of images that we typically describe as visual texture, whether regular and clearly human-made (wallpaper, fabric, floor tiles), irregular deriving from natural variations in surface reflectance (wood grain), resulting from the interplay of illumination and a variegated three-dimensional surface (plaster) or from the separate elements in a scene (a field of wheat), or combinations of these factors. In each case the image consists of repeating smaller elements (individual design elements in a tile, single small bumps in a plaster wall, single growth lines in wood) that repeat, possibly with variation, across the textured image. One can even extend this definition to encompass the idea of auditory "texture" (McDermott & Simoncelli, 2011).

In this chapter I concentrate on the broad array of research on the visual coding and perception of texture that has appeared in the last decade or so since my previous review of this topic (Landy & Graham, 2004). For earlier reviews, see Bergen (1991) and Landy (1996). As we will see, there has been a tremendous amount of work in this area in the intervening years, requiring major revisions in our understanding of the processing of visual texture and its neural underpinnings. I begin by reviewing psychophysical work on the detection of texture-defined boundaries ("texture segregation") and its importance for the development of biologically grounded models of texture coding. I relate models of texture segregation to models of texture appearance based on image statistics. Then, I discuss recent work on the use of surface texture for visual estimation of object shape and surface properties in three-dimensional scenes. The chapter concludes with a survey of studies of the neural processing of texture. In recent years neurophysiological work on texture coding has ranged from single-cell physiology and optical imaging to visual evoked potentials (VEP) and functional magnetic resonance imaging (fMRI).

MODELS OF TEXTURE SEGREGATION

The Back-Pocket Model

Consider the image of the boardwalk in figure 45.1. Simple cells in primary visual cortex have receptive fields that can be modeled as applying an orientation-selective linear spatial filter to the retinal image. This computation emphasizes oriented lines in an image, such as those between neighboring boards in the walkway. But linear filtering cannot specifically respond to the texture-defined borders in this image, such as the boundaries between differently oriented boardwalk sections. This is because, on a coarse scale, the average luminance is identical on either side of these boundaries. What differs across such a boundary is the average local orientation (i.e., a local average statistic of the image). What kind of computation can reveal such boundaries, which are clearly salient to human observers?

A number of researchers, beginning with Bergen and Adelson (1988), have suggested that texture-defined boundaries may be computed by the sort of mechanism we have called the "back-pocket" model of texture segregation (Chubb & Landy, 1991). In its simplest form the back-pocket model (figure 45.2) consists of a sequence of three feedforward stages: filter, rectify, filter (FRF; sometimes also called LNL for linear, nonlinear, linear). To make this concrete, figure 45.3 illustrates this sequence of operations on a section of the boardwalk image. The first linear spatial filter is tuned for orientation and spatial frequency so that it responds preferentially to one of the two constituent textures. In figure 45.3B a vertically tuned filter has been applied. However, a linear filter will typically respond both with strong positive values (when the receptive field is located so that excitatory regions align with brighter parts of the image) and negative values (when aligned

FIGURE 45.1   The boardwalk at Coney Island, Brooklyn, NY. Note the salient boundaries between boardwalk sections each consisting of parallel boards, with the sections' boards in different orientations.

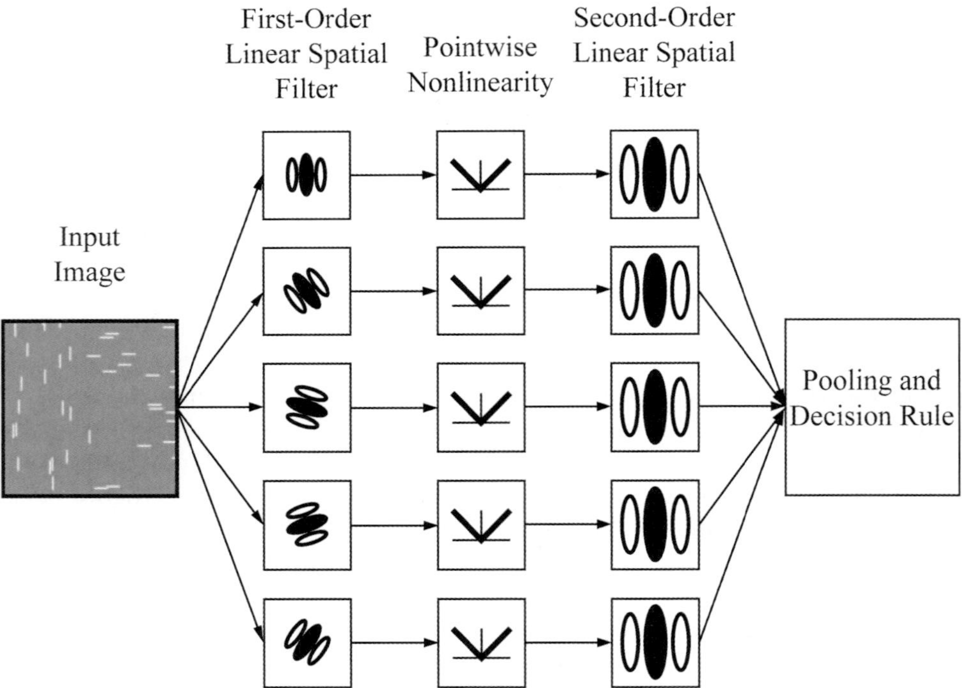

FIGURE 45.2   The back-pocket model of texture segregation, consisting of a linear filter, a pointwise nonlinearity, and a second-order linear filter at a coarser scale.

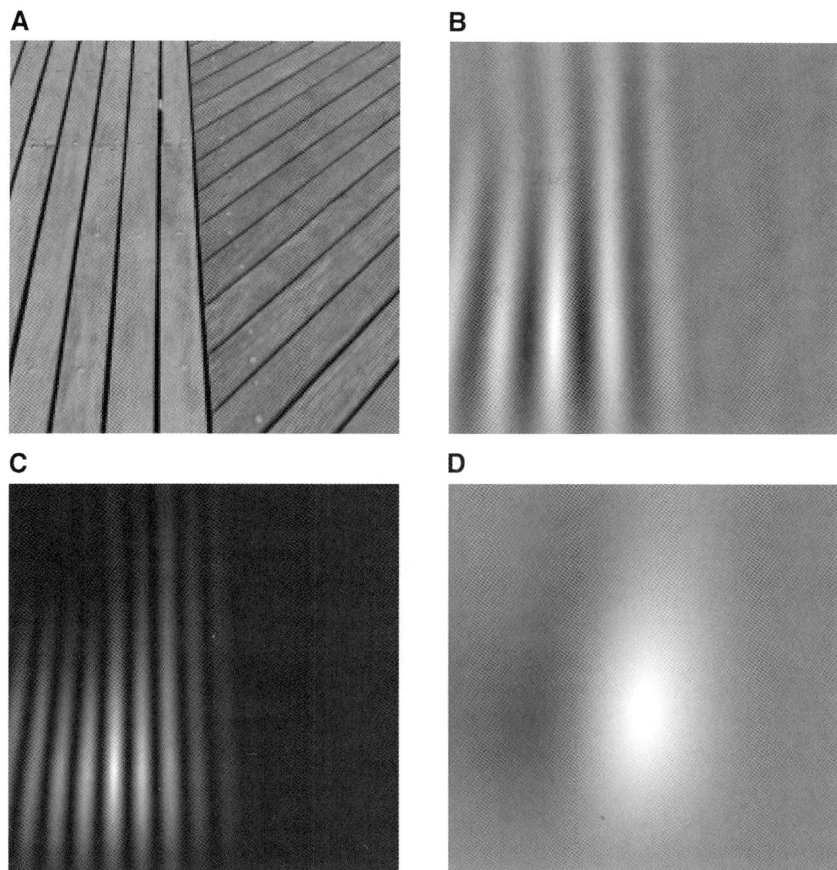

FIGURE 45.3 The back-pocket model of figure 45.2 applied to the image of figure 45.1. (A) Portion of figure 45.1 with an orientation-defined edge. (B) Convolved with a vertically oriented filter. (C) Full-wave rectified by squaring. (D) Convolved with a second, coarser-scale, vertically oriented filter. In B and D, a filter response of zero is represented as midgray; in C zero is represented as black.

with darker image regions). Thus, a spatial average of these linear filter responses will be identical on either side of the texture edge. In figure 45.3B the averages on the left and right sides are identical (gray, indicating a filter output of zero). The nonlinearity (e.g., a threshold or pointwise squaring operation) converts these regions of high variance in response to regions producing higher average response (figure 45.3C, where now black represents a value of zero). After this stage a "second-order" linear filter, typically of a substantially larger scale, will respond strongly to the texture-defined edge (figure 45.3D). The kinds of image structure delineated by FRF models are prevalent in natural images and are not correlated with first-order, luminance-defined structure (Schofield, 2000).

Some recent work serves to support the FRF model and establish its basic properties. In much of this work researchers have developed psychophysical techniques that seek to measure the variety and filtering properties of the second-order filters using experimental methods analogous to those classically used to measure the properties of first-order filters (i.e., of spatial frequency channels [Graham, 1989]). In these studies second-order stimuli involve modulations of texture that observers must detect or discriminate. Contrast sensitivity for the detection of orientation modulation (OM; see figure 45.4 for example stimuli) is relatively flat over a large range of spatial frequency (Landy & Oruç, 2002). Discrimination of second-order modulation depth with OM textures results in subthreshold summation; that is, thresholds decrease in the presence of pedestal contrast compared to no pedestal (the classic "dipper function" of first-order contrast discrimination). This improvement in threshold indicates either a further nonlinearity (i.e., FRFR) or a reduction in spatial uncertainty due to the presence of the pedestal (Landy & Oruç, 2002). Subthreshold summation (i.e., dipper functions) has been observed using OM stimuli as well as spatial-frequency-modulated (FM) and contrast-modulated (CM) textures with no facilitation across texture types (Kingdom, Prins, & Hayes, 2003). These results are reasonably consistent with FRF

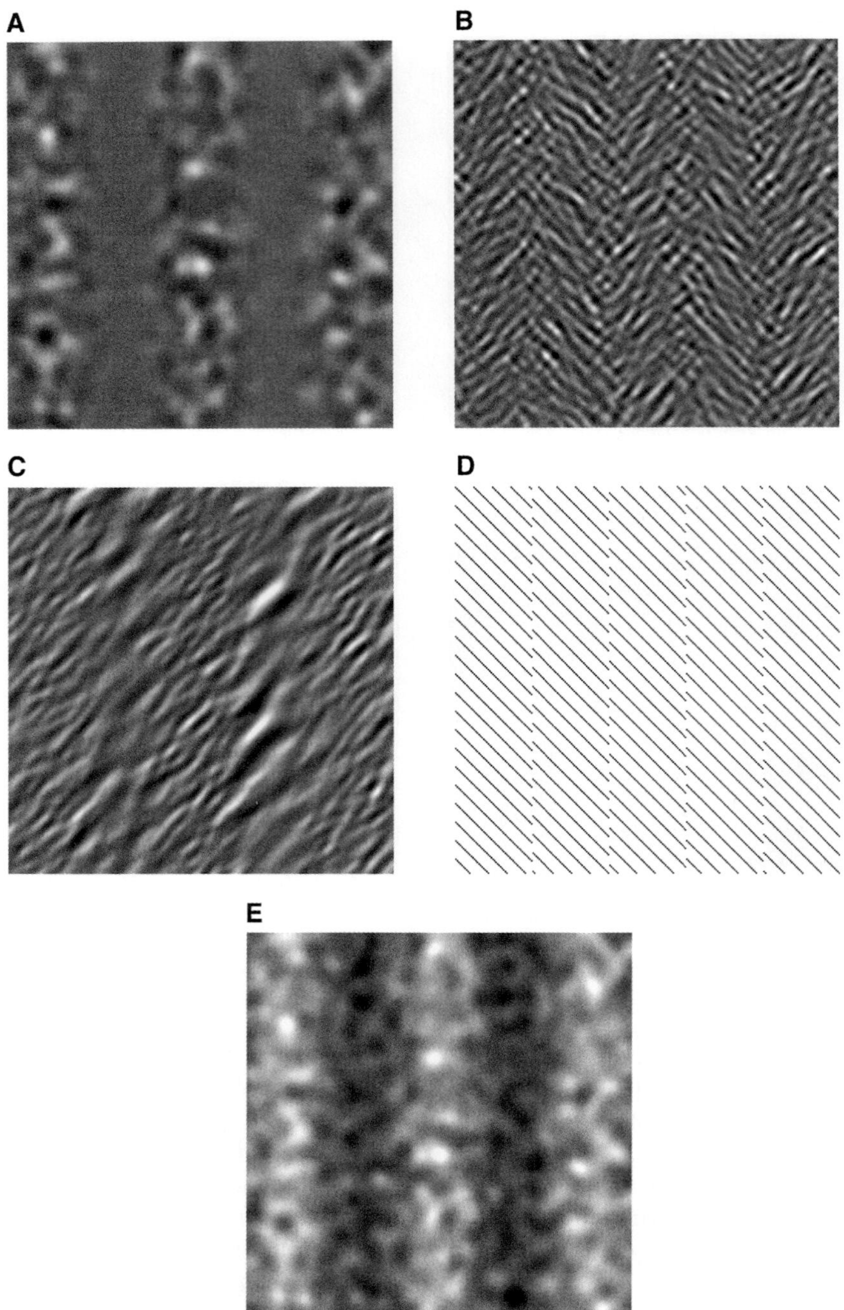

FIGURE 45.4   Typical second-order stimulus modulations used in texture experiments in which a noise carrier is modulated. (A) Contrast modulation—CM. (B) Orientation modulation—OM. (C) Frequency modulation—FM. (D) Illusory contour—IC. (E) First-order luminance modulation—LM. In A–C, a modulator, here a vertical sine wave, controls the contrast of one (A) or two (B, C) carrier patterns, here consisting of filtered noise.

predictions, further supporting the FRF model. Kingdom and colleagues also argued that normalization (e.g., Heeger, 1992), sometimes referred to as surround suppression, was needed to explain their results (see following section for further discussion of normalization processes in texture perception). As in the first-order case, the second-order contrast sensitivity function is an envelope over multiple, narrower-bandwidth, second-order spatial frequency channels. This has been demonstrated, for example, using a second-order version of the summation technique originally used by Graham and Nachmias (1971) to infer the existence of first-order channels. Using this technique with OM stimuli, Landy and Oruç (2002) found that second-order channels had a similar bandwidth (~1 octave) to first-order channels. Data from a critical-band masking experiment also suggest that second-order channels of modest bandwidth (~1.5 octaves) are used to identify texture-defined letters (Oruç, Landy, & Pelli, 2006). Prins and Kingdom (2003) tested the FRF model using a "2×2 task" (two tasks × two intervals) with OM, FM, or CM stimuli. Subjects were required to both detect the interval in which there was second-order modulation and identify the type of modulation in that stimulus (OM vs. CM, OM vs. FM, or CM vs. FM). For two of the three pairings (except CM vs. FM) the thresholds for detection and identification were identical. This is consistent with a model in which FRF channels are "labeled" with the first-order spatial filter they use, so that knowledge of the specific channel that detected the stimulus also identifies which type of modulation it contained. Ellemberg, Allen, and Hess (2006) also used the 2×2 task in the manner of Watson and Robson (1981). The two discriminanda differed in second-order spatial frequency or orientation, and they increased the difference in spatial frequency or orientation until detection and discrimination modulation thresholds were identical, indicating that the two stimuli were detected by distinct channels. Ellemberg and colleagues found that the number of distinct second-order orientation and spatial-frequency channels was similar to the number of first-order channels (at least for lower spatial frequencies). Consistent with the FRF model, adaptation to first-order (grating) patterns is most effective at impeding texture segregation when the adapter spatial frequency matches the tuning of the first-order filter for the FRF channel that is the most informative for the task (Prins & Kingdom, 2006).

In the basic FRF model the second-order linear filter uses the same input (i.e., from the same first-order filter) for both its excitatory and inhibitory regions. Both Prins, Nottingham, and Mussap (2003) and Graham and Wolfson (2004) provide evidence against the existence of more complex orientation-opponent and double-opponent arrangements in which two orthogonal first-order channels, after rectification, contribute to the responses of a second-order channel.

Extensions of the Model

Recent psychophysical work in texture discrimination has sought to flesh out and extend the basic FRF model. For example, an FRF model that computes local texture "energy" (with a square-law nonlinearity) should be insensitive to the local phase of the carrier patterns. But textures with a preponderance of even-phase (0° or 180°, i.e., positive or negative cosine phase) patterns segregate when placed on a random-phase background (Hansen & Hess, 2006), suggesting that the sign of cosine phase is encoded (although a local luminance nonlinearity might provide an alternative explanation of these results). A number of studies have followed up the observation that orientation-defined texture edges are easiest to detect when one of the bordering textures shares a local (carrier) orientation with the second-order texture edge (Wolfson & Landy, 1995). This effect may be due to texture elements aligned with and abutting the texture edge, whereas texture elements interior to a texture-defined figure aid in segregation when orthogonal to the boundary (Giora & Casco, 2007). Segregation improves further when interior texture elements align, grouping to form long chains (Harrison & Keeble, 2008). Alberti and colleagues (2010) confirmed this finding but also found that collinear texture elements are detrimental to motion-defined segregation.

In addition to linear spatial filtering and point nonlinearities, normalization, in which the response of one neuron is normalized by a local, pooled response of nearby neurons (e.g., Heeger, 1992), is likely a widespread computation used throughout cortical circuits (Carandini & Heeger, 2011). This leads to an extension of the FRF model to $F_1R_1N_1F_2R_2N_2$, in which "N" stands for a normalization stage. Normalization has been suggested as an important component of texture processing (e.g., Graham, 1991; Graham & Sutter, 2000). Several recent papers suggest that first-order normalization (N_1) is needed to explain psychophysical performance using second-order stimuli (Kingdom, Prins, & Hayes, 2003; Li & Zaidi, 2009; Motoyoshi & Nishida, 2004). The Motoyoshi and Nishida (2004) result also suggests that there are FRF (or FRNF) channels in which the first-order filter is not orientation-tuned; several other studies are consistent with this observation (Motoyoshi & Kingdom, 2007; Prins, 2008). Schade and Meinecke (2009, 2011) demonstrate a kind of second-order metacontrast masking. They found that detection of a second-order pattern was impeded if the stimulus was followed by presentation of a masker containing a task-irrelevant texture-defined border. This might be modeled as normalization with a normalization pool that is extended in time. Finally, evidence is beginning to emerge for second-order normalization as well, both in terms of surround suppression's effects on sensitivity (Wang, Heeger, & Landy, 2012) and on perceived contrast (Ellemberg, Allen, & Hess, 2004).

Texture segregation experiments have used a wide variety of natural and computer-generated stimuli in which texture-defined edges are signaled by modulations of contrast, orientation, spatial frequency, line terminations, and others. Thus, modelers are motivated to find simple explanations for these complex phenomena that do not require positing a new texture "channel" for each newly defined type of stimulus (Prins & Kingdom, 2003). This efficiency might result from channels that are "carrier invariant," responding to several forms of texture modulation (e.g., CM, OM, and FM) or, in the case of CM stimuli, responding to contrast modulation of a range of carrier patterns. Two studies found cross-adaptation between different kinds of contours including both first-order (luminance-modulated or LM) and second-order (CM, OM, and illusory contour or IC), suggesting a common mechanism is used to encode all of these types (Filangieri & Li, 2009; Hawley & Keeble, 2006). However, other studies (e.g., Larsson, Landy, & Heeger, 2006) suggest that LM mechanisms are separate from those coding second-order patterns.

The FRF model is consistent with the observation that, for OM patterns, detectability of an edge is a function of the "orientation gradient" (Landy & Bergen, 1991; Nothdurft, 1985), that is, the change in orientation across the edge divided by the distance over which it occurs. However, in an interesting series of papers (Ben-Shahar, 2006; Ben-Shahar & Zucker, 2004; Ben-Yosef & Ben-Shahar, 2008), Ben-Shahar showed that perception of boundaries in OM patterns is more complex than this. He examined orientation flow patterns in which orientation changes continuously across the image and found that contours are perceived in locations in which the orientation gradient is constant across the image (figure 45.5). He suggested a theory for the encoding of second-order spatial patterns in which two local measures of orientation change are computed: normal and tangential curvature of the orientation flow. This appears consistent with the finding that structure in textures made of random pairs of oriented texture elements is most salient when the pairs consist of cocircular elements, that is, support a local orientation flow (Motoyoshi & Kingdom, 2010). It may also relate to the many behavioral results suggesting that texture-defined edges are most salient when a texture on one side of the edge has orientation content oriented parallel to the edge (Casco et al., 2005; Giora & Casco, 2007; Wolfson & Landy, 1995).

Substantial work has been done on developing biologically inspired models of texture segregation. Two modeling efforts take distinct approaches to this task. In the model of Thielscher and Neumann (2003, 2005, 2007), a feedforward chain of filtering and rectification, representing computations in cortical areas V1, V2, and V4, is modulated by feedback signals from V4 back to V1. Thielscher and Neumann show that the feedback signals are necessary to account for human performance. On the other hand, the models of Zhaoping Li (Jingling & Zhaoping, 2008; Zhaoping, 2003; Zhaoping, Guyader, & Lewis, 2009; Zhaoping & May, 2007) are primarily feedforward, with some interaction between model neurons within a single cortical area.

Texture and Visual Attention

It is well known that visual attention can improve performance in low-level visual tasks. A series of studies by Carrasco and colleagues (Barbot, Landy, & Carrasco, 2011; Carrasco & Yeshurun, 2009; Yeshurun & Carrasco, 2000, 2008; Yeshurun, Montagna, & Carrasco, 2008) indicates that both exogenous attention, induced involuntarily by sudden stimulus change in the periphery, and endogenous attention, the top-down allocation of spatial attention to a region of interest, have strong effects on second-order processing. For example, consider the central performance drop, in which performance at a texture-segregation task is best at a near-peripheral location and worse at the fovea (Gurnsey, Pearson, & Day, 1996; Kehrer, 1989). It has been suggested that this task requires a particular scale of processing for optimal performance. At each eccentricity there is a scale or spatial frequency to which the visual system is most sensitive. This preferred spatial frequency is highest at the fovea and drops with increasing eccentricity, so that it is "just right" for this texture-segregation task in the near-periphery. With exogenous attention, performance improves in this task at peripheral locations and worsens at the fovea consistent with the view that exogenous attention improves spatial resolution wherever it is directed (Yeshurun & Carrasco,

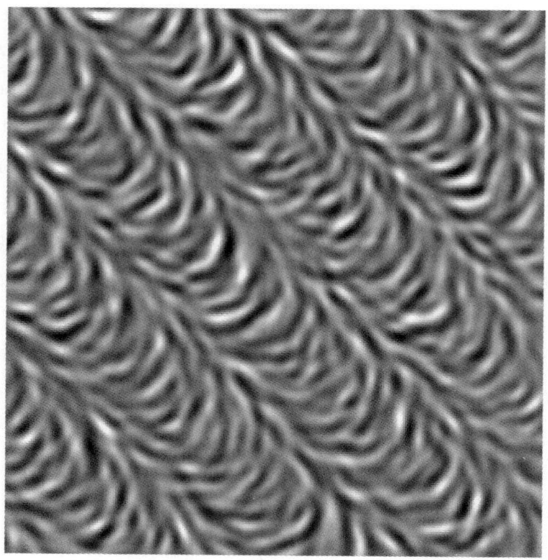

FIGURE 45.5   Example of a texture stimulus that has constant orientation gradient throughout the pattern and yet has salient contours inconsistent with typical FRF models. (Modeled after Ben-Yosef & Ben-Shahar, 2008.)

2000). In locations where the optimal scale is already too high for the task (near the fovea), the increase in resolution makes performance worse, but where it is too low (farther in the periphery), performance improves with attention. However, endogenous attention is more flexible and improves performance at all eccentricities in this task (Yeshurun, Montagna, & Carrasco, 2008). In addition to improving resolution, attention also improves second-order contrast sensitivity (Barbot, Landy, & Carrasco, 2011).

Development

The developmental sequence for perception of texture-defined patterns lasts throughout childhood. Global orientation leads to significant VEPs in 2- to 5-month-olds, and global orientation patterns (e.g., circles vs. pinwheels) lead to responses by 6–13 months (Norcia et al., 2005). On the other hand, behavioral performance in simple spatial discrimination tasks reaches adult levels by age 12 for luminance-defined patterns but continues to improve significantly after age 12 for second-order, contrast-defined patterns (Bertone et al., 2008, 2010).

TEXTURE CODING AND STATISTICS

Early work on texture segregation was concerned with what "order" of statistics was encoded by the visual system (Caelli & Julesz, 1978; Caelli, Julesz, & Gilbert, 1978; Julesz et al., 1973; Julesz, Gilbert, & Victor, 1978). In this work the term "second-order statistics" meant dipole statistics, in which one computes the distribution of gray values seen at the ends of a dipole (a line segment of a particular length and orientation) positioned at a random location on the texture, with one distribution per choice of dipole length and orientation. Third-order statistics are analogous, except that one randomly positions triangles (of a particular shape and size) on the texture. Counterexamples were found to the hypotheses that second-order or third-order statistics determined texture appearance (Caelli & Julesz, 1978; Julesz, Gilbert, & Victor, 1978).

Understanding texture perception by looking at image statistics remains an important approach. Even the encoding of first-order statistics (graylevel distributions) is complex. Chubb and colleagues (Chubb, Econopouly, & Landy, 1994; Chubb, Landy, & Econopouly, 2004) showed that graylevel distributions are encoded by human observers using only three statistics: mean (average luminance), variance (i.e., contrast), and a third statistic they term "blackshot" because it is primarily sensitive to distinctions among the darkest

graylevels in the scene. Both first- and fourth-order statistics are encoded in a continuous, graded fashion (Victor & Conte, 2004). Differences in image statistics that are most salient are also those that are most informative in natural images (Tkačik et al., 2010); that is, they are principal components of the first-order statistics or are maximum-entropy components for fourth-order statistics. Segregation of texture-defined borders between natural image textures requires third- or higher-order statistics because performance differs for phase-scrambled textures that preserve second- but not higher-order statistics (Arsenault, Yoonessi, & Baker, 2011). The encoding of phase was also examined by Emrith and colleagues (2010). They varied the degree of phase scrambling and used the maximum-likelihood difference-scaling method (Knoblauch & Maloney, 2008) to determine perceptual sensitivity to the degree of phase scrambling. Observers were most sensitive to intermediate degrees of phase scrambling, and the authors suggested that a model with a third stage of nonlinearity and linear filtering (i.e., FRFRF) would be required to account for this behavior.

Many natural textures can be characterized by the distribution of local orientation across the texture (cf. figure 45.1), just as experimenters have often used OM textures in their experiments. When observers are asked to judge the mean orientation of textures consisting of collections of oriented local texture elements (oriented Gabor patches), sampling efficiency (the fraction of the texture elements observers used for the judgment) depends only on the number of texture elements in the stimulus (Dakin, 2001). It is independent of the density of texture elements or overall size of the texture image. Girshick, Landy, and Simoncelli (2011) also asked observers to discriminate the mean orientation of patterns containing oriented texture elements. In their study, orientation uncertainty (the variance) was varied to measure biases in orientation estimation and infer the underlying orientation prior (in the Bayesian-estimation sense) used by observers. They found that human observers use a prior that peaks at the cardinal orientations (horizontal and vertical) similar to the distribution of orientations in natural images, and a neural population code that emphasized the cardinal orientations (with more neurons tuned to these orientations possessing narrower orientation bandwidth) could account for their results. If the stimuli to be discriminated include both first-order (luminance-defined) and second-order (contrast-defined) texture elements, observers are not able to combine the two cues, and they use first-order luminance information even when it is less informative (Allen et al., 2003). However, combinations of spatial frequency and orientation cues do lead to improved psychophysical performance from cue integration—much as Landy and Kojima (2001) had found previously using a localization task—and stronger VEP signal during the task (Bach et al., 2000).

How are all these studies of image statistics related to the FRF and similar models? The FRF computation is an image transformation. A pixel in the output image is a linear combination of the pixels in the intermediate image that results from the first linear filter and rectification. The pixels of this intermediate image are a nonlinear combination of input image pixels. For example, if the nonlinearity is a square law, then output pixels are a quadratic polynomial function of input pixels. For more severe nonlinearities this function will have higher polynomial order. Thus, the task of determining the number and types of FRF channels in the visual system is tantamount to determining the number and types of image statistics we encode. This connection is clearest in the work on texture-synthesis models that take an input image, compute a collection of statistics on that image, and then try to synthesize new images that match those statistics. If the statistics correspond to those human observers use to determine texture appearance, then the synthesized images should look as though they are made of the same "stuff" as the original image. Two such texture-synthesis models have been developed that begin with a multiscale oriented image pyramid, in analogy to the multiple spatial frequency and orientation channels in human vision. The first of these (Heeger & Bergen, 1995) uses only first-order statistics (the histogram of values in each spatial channel and the pixel histogram). A more recent such model (Portilla & Simoncelli, 2000) also includes various correlational statistics between each pyramid value and its spatial, scale, and orientation neighbors. Figure 45.6 shows example syntheses by this model, which works quite well on a wide variety of textures.

The Portilla and Simoncelli (2000) model (PS model) has been examined as a potential model for human visual coding in several psychophysical studies. Balas (2006) asked observers to discriminate the odd texture in a group of three synthesized textures, asking whether various subsets of the statistics in the PS model are necessary for describing human performance. Humans can detect differences in textures when any subclass of statistics is left out of the model, but the salience of the difference depends on the class of statistics that is omitted—the marginal statistics, similar to those in the Heeger/Bergen (HB) model, are the most salient—and on the particular natural texture being synthesized. Surprisingly, the power spectrum is a better predictor of human texture classification than either the HB or the

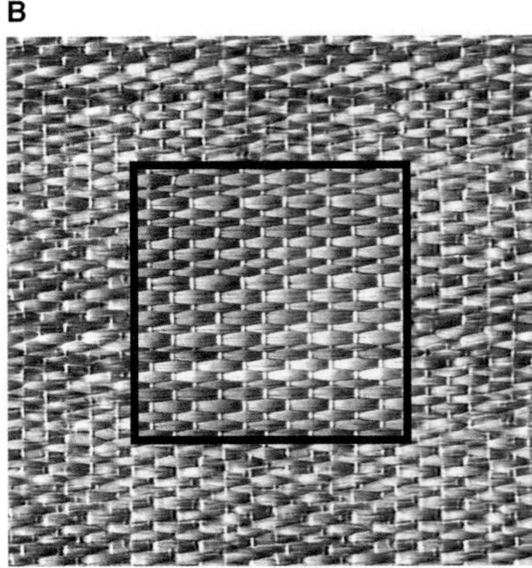

FIGURE 45.6 Texture representation based on image statistics. In each panel the inset square is an original texture, and the rest of the image is new texture using the technique of Portilla and Simoncelli (2000).

PS model (Balas, 2008). This shows that the ability of a model to synthesize textures that humans perceive as matching an original texture does not imply that the model can predict human texture classification. In a particularly interesting paper Balas, Nakano, and Rosenholtz (2009) suggested that the sorts of statistics in the PS model are all that is encoded in peripheral vision. They did this by showing that performance in a peripheral visual crowding task was similar to performance in which the peripheral search patterns were replaced by "mongrel" patterns (i.e., resynthesized images using the PS model, figure 45.7) but observed foveally. Following this logic a step further, Freeman and Simoncelli (2011) created image "metamers" by replacing peripheral portions of an image with

646   MICHAEL S. LANDY

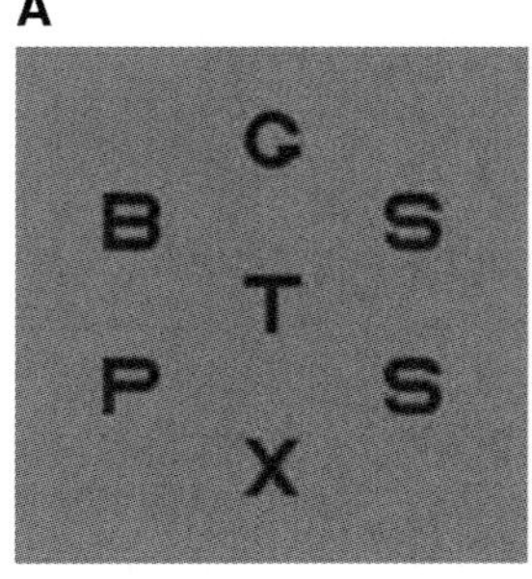 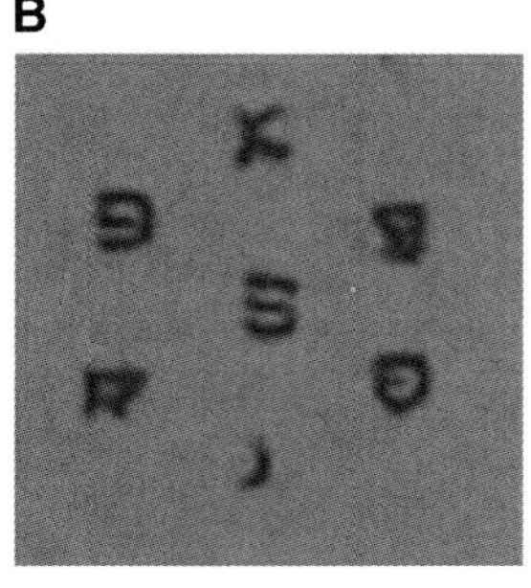

FIGURE 45.7 (A) Sample stimulus for visual crowding study. (B) "Mongrel" image based on the Portilla and Simoncelli (2000) model as described by Balas, Nakano, and Rosenholtz (2009).

PS-model resyntheses. To do this, they had to choose an optimal image patch size for measuring local image statistics in each region of the image to drive the texture-synthesis process so that the full synthetic image appeared identical to the original image. By using a cortical magnification model in which the pooling region for statistics grows with eccentricity, they found that the cortical scaling that could be supported with this manipulation matched that of cortical area V2, suggesting that texture statistics might be encoded in that brain area.

SHAPE FROM TEXTURE

Surface texture is one of the many pictorial cues to three-dimensional shape, and work in this area has often centered on the assumptions observers make about surface textures to compute shape from texture and what computation is used (e.g., Knill, 1998). Li and Zaidi (2000, 2001a, 2001b, 2003, 2004; Zaidi & Li, 2002) have developed a simple idea concerning the aspects of texture that support three-dimensional perception. Originally working with developable (i.e., singly curved) surfaces and textures synthesized as sums of sine-wave gratings, they found that veridical perception of qualitative aspects of three-dimensional shape (e.g., concave vs. convex) required that the texture include contours along the orientation of maximum curvature and that those spectral components need to be relatively isolated (without additional components at nearby orientations). In later work this idea was confirmed with natural textures and with doubly curved three-dimensional shapes. Todd and colleagues have produced a series of counterexamples and experimental data they see as contradicting this theory (Thaler, Todd, & Dijkstra, 2007; Todd & Oomes, 2002; Todd & Thaler, 2010; Todd et al., 2004, 2007), resulting in an extended argument between the two groups played out in the journals that

does not seem to have been resolved to either group's satisfaction. Todd and colleagues have been concerned with the precise shapes observers perceive and, in particular, with the circumstances under which observers misperceive shape. They have suggested several heuristic computations that accord with these patterns of nonveridical shape perception, although these heuristics mainly apply to textures that are collections of individual, nonoverlapping texture elements. This class does not include the sorts of noise textures examined by Li and Zaidi. On the other hand, Li and Zaidi's work has been concerned mainly with qualitative shape perception (e.g., the sign of perceived slant or curvature) and not on perceived depth per se.

SURFACE PROPERTIES FROM TEXTURE

The definition of texture suggests that texture should be regarded as a surface property. Indeed, one can speak of three-dimensional texture as a property of a surface akin to shape but at a far smaller scale. Perception of one such surface property, roughness, depends on viewing conditions. It seems that observers use the amount of cast shadow in an image of a rough surface as a cue to surface roughness, even though cast shadow can vary with such things as the direction of illumination for a constant amount of surface roughness (Ho, Landy, & Maloney, 2006; Ho, Maloney, & Landy, 2007; Landy et al., 2011). Surface roughness (or bumpiness) and gloss are not coded independently in the sense that varying one affects perception of the other surface property (Ho, Landy, & Maloney, 2008). Padilla and colleagues (2008) looked at perceived surface roughness of three-dimensional surfaces with a height function that was fractal (i.e., had a $1/f^\beta$ spectrum) and found that roughness increased with decreasing value of spectral slope β.

NEURAL CODING OF TEXTURE

Physiological Studies in Animals

There has been a modest effort to determine how second-order patterns are encoded in cortex and in which brain areas the computation takes place. Baker and colleagues have suggested in a series of papers (Song & Baker, 2006, 2007; Zhan & Baker, 2006, 2008) that in the cat second-order patterns (both CM and IC) are encoded in area 18. In single-cell experiments, they found that a subpopulation of area 18 cells is tuned for the orientation of the modulator and that many cells behave in a carrier-invariant fashion. Using optical imaging, they found that there are similar first- and second-order orientation maps in area 18. These findings only occur for second-order patterns in which the carrier pattern has a spatial frequency above the linear (first-order) passband of the cell or cortical area. They suggest these results are consistent with the behavior of an FRF model. However, these results remain controversial. In rhesus macaque area V2 (perhaps a region homologous to cat area 18), cue-invariant second-order responses are nearly absent (El-Shamayleh & Movshon, 2011), and in a small set of cat area 18 cells these authors were unable to replicate Baker's results. They suggested that tuning for second-order patterns might emerge gradually across several cortical regions rather than all at once in area V2. In area 18 of the cat, cells have similar disparity tuning to the contrast envelope of second-order patterns as they have to luminance disparity (Tanaka & Ohzawa, 2006). Many cells in cat area 17 are tuned for the second-order orientation and spatial frequency of CM gratings, and this second-order tuning arises simply by the cell having a nonclassical surround that is asymmetrically placed or organized relative to the classical receptive field (Tanaka & Ohzawa, 2009). Finally, in awake macaque, population responses from area V4 appear to be capable of representing three-dimensional textures (Arcizet, Jouffrais, & Girard, 2008). In this study macaques viewed images from a database of three-dimensional surface textures (the CURET database) while responses of V4 cells were recorded. Many cells were tuned to subsets of these images, and a clustering algorithm on the population responses was able to cluster individual textures independent of the illumination direction as well as a similar algorithm applied to a pyramid Gabor representation of these images.

Visual Evoked Potential Studies

Several groups have studied the VEP response in humans corresponding to the segregation of textures (the tsVEP). The tsVEP has two components: The early (100-ms) component is task-independent, whereas the later (230-ms) component is abolished if the subject performs an attention-distracting task (Heinrich, Andrés, & Bach, 2007). The tsVEP is stronger for more salient textures, such as when the texture elements are parallel to the texture edge (Casco et al., 2005). The tsVEP really is a response to texture per se and is not just due to the first-order content of the image because a change in orientation content leads to a VEP signal that cuts off at 17 Hz, whereas the tsVEP cuts off at 12 Hz (Lachapelle et al., 2008). A series of studies by Appelbaum and colleagues (2006, 2008) used a variety of second-order patterns in which they tagged responses

to the figure and ground textures by cycling different versions of each at different frequencies ("frequency tagging"). Responses to the figure begin in area V1 and continue on to the lateral occipital complex (LOC) independent of the texture cue, whereas responses to the background appear in more medial regions. Nonlinear interaction responses (e.g., at frequency $f_1 + f_2$, where f_1 and f_2 are the figure and ground frequencies, respectively) signal the texture-defined boundary and disappear when a luminance-defined gap is introduced between figure and ground.

fMRI Studies

There have been a number of studies that attempted to link particular brain areas to texture processing using fMRI. In the simplest of these, subjects are shown textured versus "equated" nontextured objects, and a search is done for areas that respond more strongly to texture or adapt to recently viewed texture (Cant, Arnott, & Goodale, 2009; Cavina-Pratesi et al., 2010a, 2010b; Stylianou-Korsnes et al., 2010) or for areas that respond more strongly to shape defined by texture (Georgieva et al., 2008). A variety of brain areas are implicated for texture processing by these studies, including the posterior collateral sulcus and right inferior parietal lobule, consistent with some neurophysiological findings. For shape from texture, several brain regions respond more strongly to stimuli that appear three-dimensional due to texture cues, including LOC, lateral occipital sulcus, and the inferior temporal gyrus.

In studies such as these, especially those using subtraction contrasts, it is hard to determine what function is carried out by the revealed brain areas, that is, what aspects of the textured stimuli lead to stronger responses. Single-unit studies suggest that only a small subset of neurons in early visual cortical regions encode second-order patterns, so fMRI methods are required that can demonstrate the function of a potentially small subset of the neurons in a voxel. fMRI adaptation is an experimental technique suited to providing evidence for tuning of a subset of neurons in a voxel, and this technique has demonstrated second-order orientation and spatial frequency tuning in several visual cortical areas (Hallum, Landy, & Heeger, 2011; Larsson et al., 2006; Montaser-Kouhsari et al., 2007). In all of these studies the adaptation index (the reduction of response due to adaptation to the same orientation or frequency of modulation compared to adaptation to a different modulation) increases steadily from V1 through downstream visual areas. Consistent orientation-selective adaptation to LM, CM, and OM patterns was found in V1, V2, V3, V3A/B, LO1, hV4, and VO1 (Larsson et al.,

2006). For first-order (LM) patterns adaptation was no stronger in downstream areas than in V1, suggesting that the adaptation effect derived from V1 processing and was inherited by downstream areas. In contrast, stronger adaptation was found in VO1 for OM patterns and additionally in V3A/B and LO1 for CM patterns, suggesting further downstream processing, consistent with the "gradualist encoding" suggestion of El-Shamayleh and Movshon (2011). A similar gradual increase in adaptation was found for illusory contours with the addition of significant adaptation in areas V7 and LO2 (Montaser-Kouhsari et al., 2007). Significant tuning for second-order spatial frequency was found in V1, V2, V3, and V4 (Hallum, Landy, & Heeger, 2011). However, a model of these responses suggests that first-order normalization was involved in second-order spatial frequency tuning and that the bulk of the adaptation may have occurred in V1. Finally, fMRI responses increase in several cortical areas with increasing salience of the texture-defined border (achieved by increasing the orientation gradient), especially in areas hV4, V3, and LOC (Thielscher et al., 2008).

REFERENCES

Adelson, E. H., & Bergen, J. R. (1991). The plenoptic function and the elements of early vision. In M. S. Landy & J. A. Movshon (Eds.), *Computational models of visual processing* (pp. 3–20). Cambridge, MA: MIT Press.

Alberti, C. F., Pavan, A., Campana, G., & Casco, C. (2010). Segmentation by single and combined features involves different contextual influences. *Vision Research, 50*, 1065–1073.

Allen, H. A., Hess, R. F., Mansouri, B., & Dakin, S. C. (2003). Integration of first- and second-order orientation. *Journal of the Optical Society of America. A, Optics, Image Science, and Vision, 20*, 974–986.

Appelbaum, L. G., Wade, A. R., Pettet, M. W., Vildavski, V. Y., & Norcia, A. M. (2008). Figure-ground interaction in the human visual cortex. *Journal of Vision, 8*(9), 8, 1–19. doi:10.1167/8.9.8.

Appelbaum, L. G., Wade, A. R., Vildavski, V. Y., Pettet, M. W., & Norcia, A. M. (2006). Cue-invariant networks for figure and background processing in human visual cortex. *Journal of Neuroscience, 26*, 11695–11708.

Arcizet, F., Jouffrais, C., & Girard, P. (2008). Natural textures classification in area V4 of the macaque monkey. *Experimental Brain Research, 189*, 109–120.

Arsenault, E., Yoonessi, A., & Baker, C. (2011). Higher order texture statistics impair contrast boundary segmentation. *Journal of Vision, 11*(10), 14, 1–15. doi:10.1167/11.10.14.

Bach, M., Schmitt, C., Quenzer, T., Meigen, T., & Fahle, M. (2000). Summation of texture segregation across orientation and spatial frequency: electrophysiological and psychophysical findings. *Vision Research, 40*, 3559–3566.

Balas, B. J. (2006). Texture synthesis and perception: Using computational models to study texture representations in the human visual system. *Vision Research, 46*, 299–309.

Balas, B. (2008). Attentive texture similarity as a categorization task: Comparing texture synthesis models. *Pattern Recognition, 41*, 972–982.

Balas, B., Nakano, L., & Rosenholtz, R. (2009). A summary-statistic representation in peripheral vision explains visual crowding. *Journal of Vision, 9*(12), 13, 1–18. doi:10.1167/9.12.13.

Barbot, A., Landy, M. S., & Carrasco, M. (2011). Exogenous attention enhances 2nd-order contrast sensitivity. *Vision Research, 51*, 1086–1098.

Ben-Shahar, O. (2006). Visual saliency and texture segregation without feature gradient. *Proceedings of the National Academy of Sciences of the United States of America, 103*, 15704–15709.

Ben-Shahar, O., & Zucker, S. W. (2004). Sensitivity to curvatures in orientation-based texture segmentation. *Vision Research, 44*, 257–277.

Ben-Yosef, G., & Ben-Shahar, O. (2008). Curvature-based perceptual singularities and texture saliency with early vision mechanisms. *Journal of the Optical Society of America. A, Optics, Image Science, and Vision, 25*, 1974–1993.

Bergen, J. R. (1991). Theories of visual texture perception. In D. Regan (Ed.), *Vision and visual dysfunction* (Vol. 10B, pp. 114–134). New York: Macmillan.

Bergen, J. R., & Adelson, E. H. (1988). Early vision and texture perception. *Nature, 333*, 363–364.

Bertone, A., Hanck, J., Cornish, K. M., & Faubert, J. (2008). Development of static and dynamic perception for luminance-defined and texture-defined information. *Neuroreport, 19*, 225–228.

Bertone, A., Hanck, J., Guy, J., & Cornish, K. (2010). The development of luminance- and texture-defined form perception during the school-aged years. *Neuropsychologia, 48*, 3080–3085.

Caelli, T., & Julesz, B. (1978). On perceptual analyzers underlying visual texture discrimination: Part I. *Biological Cybernetics, 28*, 167–175.

Caelli, T., Julesz, B., & Gilbert, E. N. (1978). On perceptual analyzers underlying visual texture discrimination: Part II. *Biological Cybernetics, 29*, 201–214.

Cant, J. S., Arnott, S. R., & Goodale, M. A. (2009). fMR-adaptation reveals separate processing regions for the perception of form and texture in the human ventral stream. *Experimental Brain Research, 192*, 391–405.

Carandini, M., & Heeger, D. J. (2011). Normalization as a canonical neural computation. *Nature Reviews Neuroscience, 13*, 51–62.

Carrasco, M., & Yeshurun, Y. (2009). Covert attention effects on spatial resolution. In N. Srinivasan (Ed.), *Progress in brain research: Attention* (Vol. 176, pp. 65–86). Amsterdam: Elsevier.

Casco, C., Grieco, A., Campana, G., Corvino, M. P., & Caputo, G. (2005). Attention modulates psychophysical and electrophysiological response to visual texture segmentation in humans. *Vision Research, 45*, 2384–2396.

Cavina-Pratesi, C., Kentridge, R. W., Heywood, C. A., & Milner, A. D. (2010a). Separate processing of texture and form in the ventral stream: Evidence from fMRI and visual agnosia. *Cerebral Cortex, 20*, 433–446.

Cavina-Pratesi, C., Kentridge, R. W., Heywood, C. A., & Milner, A. D. (2010b). Separate channels for processing form, texture, and color: Evidence from fMRI adaptation and visual object agnosia. *Cerebral Cortex, 20*, 2319–2332.

Chubb, C., Econopouly, J., & Landy, M. S. (1994). Histogram contrast analysis and the visual segregation of IID textures. *Journal of the Optical Society of America. A, Optics, Image Science, and Vision, 11*, 2350–2374.

Chubb, C., & Landy, M. S. (1991). Orthogonal distribution analysis: A new approach to the study of texture perception. In M. S. Landy & J. A. Movshon (Eds.), *Computational models of visual processing* (pp. 291–301). Cambridge, MA: MIT Press.

Chubb, C., Landy, M. S., & Econopouly, J. (2004). A visual mechanism tuned to black. *Vision Research, 44*, 3223–3232.

Dakin, S. C. (2001). Information limit on the spatial integration of local orientation signals. *Journal of the Optical Society of America. A, Optics, Image Science, and Vision, 18*, 1016–1026.

Ellemberg, D., Allen, H. A., & Hess, R. F. (2004). Investigating local network interactions underlying first- and second-order processing. *Vision Research, 44*, 1787–1797.

Ellemberg, D., Allen, H. A., & Hess, R. F. (2006). Second-order spatial frequency and orientation channels in human vision. *Vision Research, 46*, 2798–2803.

El-Shamayleh, Y., & Movshon, J. A. (2011). Neuronal responses to texture-defined form in macaque visual area V2. *Journal of Neuroscience, 31*, 8543–8555.

Emrith, K., Chantler, M. J., Green, P. R., Maloney, L. T., & Clarke, A. D. F. (2010). Measuring perceived differences in surface texture due to changes in higher order statistics. *Journal of the Optical Society of America. A, Optics, Image Science, and Vision, 27*, 1232–1244.

Filangieri, C., & Li, A. (2009). Three-dimensional shape from second-order orientation flows. *Vision Research, 49*, 1465–1471.

Freeman, J., & Simoncelli, E. P. (2011). Metamers of the ventral stream. *Nature Neuroscience, 14*, 1195–1201.

Georgieva, S. S., Todd, J. T., Peeters, R., & Orban, G. A. (2008). The extraction of 3D shape from texture and shading in the human brain. *Cerebral Cortex, 18*, 2416–2438.

Giora, E., & Casco, C. (2007). Region- and edge-based configurational effects in texture segmentation. *Vision Research, 47*, 879–886.

Girshick, A. R., Landy, M. S., & Simoncelli, E. P. (2011). Cardinal rules: Visual orientation perception reflects knowledge of environmental statistics. *Nature Neuroscience, 14*, 926–932.

Graham, N. V. S. (1989). *Visual pattern analyzers.* New York: Oxford University Press.

Graham, N. (1991). Complex channels, early local nonlinearities, and normalization in perceived texture segregation. In M. S. Landy & J. A. Movshon (Eds.), *Computational models of visual processing* (pp. 273–290). Cambridge, MA: MIT Press.

Graham, N., & Nachmias, J. (1971). Detection of grating patterns containing two spatial frequencies: A comparison of single-channel and multiple-channels models. *Vision Research, 11*, 251–259.

Graham, N., & Sutter, A. (2000). Normalization: Contrast-gain control in simple (Fourier) and complex (non-Fourier) pathways of pattern vision. *Vision Research, 40*, 2737–2761.

Graham, N., & Wolfson, S. S. (2004). Is there opponent-orientation coding in the second-order channels of pattern vision. *Vision Research, 44*, 3145–3175.

Gurnsey, R., Pearson, P., & Day, D. (1996). Texture segmentation along the horizontal meridian: Nonmonotonic changes in performance with eccentricity. *Journal of Experimental Psychology. Human Perception and Performance, 22*, 738–757.

Hallum, L. E., Landy, M. S., & Heeger, D. J. (2011). Human primary visual cortex (V1) is selective for second-order spatial frequency. *Journal of Neurophysiology, 105*, 2121–2131.

Hansen, B. C., & Hess, R. F. (2006). The role of spatial phase in texture segmentation and contour integration. *Journal of Vision, 6*, 594–615. doi:10.1167/6.5.5.

Harrison, S. J., & Keeble, D. R. T. (2008). Within-texture collinearity improves human texture segmentation. *Vision Research, 48*, 1955–1964.

Hawley, S. J., & Keeble, D. R. T. (2006). Tilt aftereffect for texture edges is larger than in matched subjective edges, but both are strong adaptors of luminance edges. *Journal of Vision, 6*, 37–52. doi:10.1167/6.1.4.

Heeger, D. J. (1992). Normalization of cell responses in cat striate cortex. *Visual Neuroscience, 9*, 181–197.

Heeger, D. J., & Bergen, J. R. (1995). Pyramid-based texture analysis/synthesis. In *Proceedings of the 22nd Annual Conference on Computer Graphics & Interactive Techniques, 30*, 229–238.

Heinrich, S. P., Andrés, M., & Bach, M. (2007). Attention and visual texture segregation. *Journal of Vision, 7*, 1–10. doi:10.1167/7.6.6.

Ho, Y.-X., Landy, M. S., & Maloney, L. T. (2006). How direction of illumination affects visually perceived surface roughness. *Journal of Vision, 6*, 634–648. doi:10.1167/6.5.8.

Ho, Y.-X., Landy, M. S., & Maloney, L. T. (2008). Conjoint measurement of gloss and surface texture. *Psychological Science, 19*, 196–204.

Ho, Y.-X., Maloney, L. T., & Landy, M. S. (2007). The effect of viewpoint on perceived visual roughness *Journal of Vision, 7*(1), 1, 1–16. doi:10.1167/7.1.1.

Jingling, L., & Zhaoping, L. (2008). Change detection is easier at texture border bars when they are parallel to the border: Evidence for V1 mechanisms of bottom-up salience. *Perception, 37*, 197–206.

Julesz, B., Gilbert, E. N., Shepp, L. A., & Frisch, H. L. (1973). Inability of humans to discriminate between visual textures that agree in second-order statistics—revisited. *Perception, 2*, 391–405.

Julesz, B., Gilbert, E. N., & Victor, J. D. (1978). Visual discrimination of textures with identical third-order statistics. *Biological Cybernetics, 31*, 137–140.

Kehrer, L. (1989). Central performance drop on perceptual segregation tasks. *Spatial Vision, 4*, 45–62.

Kingdom, F. A. A., Prins, N., & Hayes, A. (2003). Mechanism independence for texture-modulation detection is consistent with a filter-rectify-filter mechanism. *Visual Neuroscience, 20*, 65–76.

Knill, D. C. (1998). Ideal observer perturbation analysis reveals human strategies for inferring surface orientation from texture. *Vision Research, 38*, 2635–2656.

Knoblauch, K., & Maloney, L. T. (2008). MLDS: Maximum likelihood difference scaling in R. *Journal of Statistical Software, 25*, 1–26.

LaChapelle, J., McKerral, M., Jauffret, C., & Bach, M. (2008). Temporal resolution of orientation-defined texture segregation: A VEP study. *Documenta Ophthalmologica, 117*, 155–162.

Landy, M. S. (1996). Texture perception. In G. Adelmen (Ed.), *Encyclopedia of neuroscience.* Amsterdam: Elsevier.

Landy, M. S., & Bergen, J. R. (1991). Texture segregation and orientation gradient. *Vision Research, 31,* 679–691.

Landy, M. S., & Graham, N. V. G. (2004). Visual perception of texture. In L. M. Chalupa & J. S. Werner (Eds.), *The visual neurosciences* (pp. 1106–1118). Cambridge, MA: MIT Press.

Landy, M. S., Ho, Y.-X., Serwe, S., Trommershäuser, J., & Maloney, L. T. (2011). Cues and pseudocues in texture and shape perception. In J. Trommershäuser, K. Körding, & M. S. Landy (Eds.), *Sensory cue integration* (pp. 263–278). New York: Oxford University Press.

Landy, M. S., & Kojima, H. (2001). Ideal cue combination for localizing texture-defined edges. *Journal of the Optical Society of America. A, Optics, Image Science, and Vision, 18,* 2307–2320.

Landy, M. S., & Oruç, İ. (2002). Properties of second-order spatial frequency channels. *Vision Research, 42,* 2311–2329.

Larsson, J., Landy, M. S., & Heeger, D. J. (2006). Orientation-selective adaptation to first- and second-order patterns in human visual cortex. *Journal of Neurophysiology, 95,* 862–881.

Li, A., & Zaidi, Q. (2000). Perception of three-dimensional shape from texture is based on patterns of oriented energy. *Vision Research, 40,* 217–242.

Li, A., & Zaidi, Q. (2001a). Information limitations in perception of shape from texture. *Vision Research, 41,* 1519–1533.

Li, A., & Zaidi, Q. (2001b). Veridicality of three-dimensional shape perception predicted from amplitude spectra of natural textures. *Journal of the Optical Society of America. A, Optics, Image Science, and Vision, 18,* 2430–2447.

Li, A., & Zaidi, Q. (2003). Observer strategies in perception of 3-D shape from isotropic textures: Developable surfaces. *Vision Research, 43,* 2741–2758.

Li, A., & Zaidi, Q. (2004). Three-dimensional shape from non-homogeneous textures: Carved and stretched surfaces. *Journal of Vision, 4,* 860–878. doi:10.1167/4.10.3.

Li, A., & Zaidi, Q. (2009). Release from cross-orientation suppression facilitates 3D shape perception. *PLoS One, 4,* e8333. doi:10.1371/journal.pone.0008333.

McDermott, J. H., & Simoncelli, E. P. (2011). Sound texture perception via statistics of the auditory periphery: Evidence from sound synthesis. *Neuron, 71,* 926–940.

Montaser-Kouhsari, L., Landy, M. S., Heeger, D. J., & Larsson, J. (2007). Orientation-selective adaptation to illusory contours in human visual cortex. *Journal of Neuroscience, 27,* 2186–2195.

Motoyoshi, I., & Kingdom, F. A. A. (2007). Differential roles of contrast polarity reveal two streams of second-order visual processing. *Vision Research, 47,* 2047–2054.

Motoyoshi, I., & Kingdom, F. A. A. (2010). The role of co-circularity of local elements in texture perception. *Journal of Vision, 10*(1), 3, 1–8. doi:10.1167/10.1.3.

Motoyoshi, I., & Nishida, S. (2004). Cross-orientation summation in texture segregation. *Vision Research, 44,* 2567–2576.

Norcia, A. M., Pei, F., Bonneh, Y., Hou, C., Sampath, V., & Pettet, M. W. (2005). Development of sensitivity to texture and contour information in the human infant. *Journal of Cognitive Neuroscience, 17,* 569–579.

Nothdurft, H. C. (1985). Sensitivity for structure gradient in texture discrimination tasks. *Vision Research, 25,* 551–560.

Oruç, İ., Landy, M. S., & Pelli, D. G. (2006). Noise masking reveals channels for second-order letters. *Vision Research, 46,* 1493–1506.

Padilla, S., Drbohlav, O., Green, P. R., Spence, A., & Chantler, M. J. (2008). Perceived roughness of $1/f^\beta$ noise surfaces. *Vision Research, 48,* 1791–1797.

Portilla, J., & Simoncelli, E. P. (2000). A parametric texture model based on joint statistics of complex wavelet coefficients. *International Journal of Computer Vision, 40,* 49–71.

Prins, N. (2008). Texture modulation detection by probability summation among orientation-selective and isotropic mechanisms. *Vision Research, 48,* 2751–2766.

Prins, N., & Kingdom, F. A. A. (2003). Detection and discrimination of texture modulations defined by orientation, spatial frequency, and contrast. *Journal of the Optical Society of America. A, Optics, Image Science, and Vision, 20,* 401–410.

Prins, N., & Kingdom, F. A. A. (2006). Direct evidence for the existence of energy-based texture mechanisms. *Perception, 35,* 1035–1046.

Prins, N., Nottingham, N. K., & Mussap, A. J. (2003). The role of local grouping and global orientation contrast in perception of orientation-modulated textures. *Vision Research, 43,* 2315–2331.

Schade, U., & Meinecke, C. (2009). Spatial distance between target and irrelevant patch modulates detection in a texture segmentation task. *Spatial Vision, 22,* 511–527.

Schade, U., & Meinecke, C. (2011). Texture segmentation: Do the processing units on the saliency map increase with eccentricity? *Vision Research, 51,* 1–12.

Schofield, A. J. (2000). What does second-order vision see in an image? *Perception, 29,* 1071–1086.

Song, Y., & Baker, C. L. (2006). Neural mechanisms mediating responses to abutting gratings: Luminance edges vs. illusory contours. *Visual Neuroscience, 23,* 181–199.

Song, Y., & Baker, C. L. (2007). Neuronal response to texture- and contrast-defined boundaries in early visual cortex. *Visual Neuroscience, 24,* 65–77.

Stylianou-Korsnes, M., Reiner, M., Magnussen, S. J., & Feldman, M. W. (2010). Visual recognition of shapes and textures: An fMR study. *Brain Structure & Function, 214,* 355–359.

Tanaka, H., & Ohzawa, I. (2006). Neural basis for stereopsis from second-order contrast cues. *Journal of Neuroscience, 26,* 4370–4382.

Tanaka, H., & Ohzawa, I. (2009). Surround suppression of V1 neurons mediates orientation-based representation of high-order visual features. *Journal of Neurophysiology, 101,* 1444–1462.

Thaler, L., Todd, J. T., & Dijkstra, T. M. H. (2007). The effects of phase on the perception of 3D shape from texture: Psychophysics and modeling. *Vision Research, 47,* 411–427.

Thielscher, A., Kölle, M., Neumann, H., Spitzer, M., & Grön, G. (2008). Texture segmentation in human perception: A combined modeling and fMRI study. *Neuroscience, 151,* 730–736.

Thielscher, A., & Neumann, H. (2003). Neural mechanisms of cortico-cortical interaction in texture boundary detection: A modeling approach. *Neuroscience, 122,* 921–939.

Thielscher, A., & Neumann, H. (2005). Neural mechanisms of human texture processing: Texture boundary detection and visual search. *Spatial Vision, 18,* 227–257.

Thielscher, A., & Neumann, H. (2007). A computational model to link psychophysics and cortical cell activation patterns in human texture processing. *Journal of Computational Neuroscience, 22,* 255–282.

Tkačik, G., Prentice, J. S., Victor, J. D., & Balasubramanian, V. (2010). Local statistics in natural scenes predict the saliency of synthetic textures. *Proceedings of the National Academy of Sciences of the United States of America, 107,* 18149–18154.

Todd, J. T., & Oomes, A. H. J. (2002). Generic and non-generic conditions for the perception of surface shape from texture. *Vision Research, 42,* 837–850.

Todd, J. T., Oomes, A. H. J., Koenderink, J. J., & Kappers, A. M. L. (2004). The perception of doubly curved surfaces from anisotropic textures. *Psychological Science, 15,* 40–46.

Todd, J. T., & Thaler, L. (2010). The perception of 3D shape from texture based on directional width gradients. *Journal of Vision, 10,* 17, 1–13. doi:10.1167/10.5.17.

Todd, J. T., Thaler, L., Dijkstra, T. M. H., Koenderink, J. J., & Kappers, A. M. L. (2007). The effects of viewing angle, camera angle, and sign of surface curvature on the perception of three-dimensional shape from texture. *Journal of Vision, 7*(12), 9, 1–16. doi:10.1167/7.12.9.

Victor, J. D., & Conte, M. M. (2004). Visual working memory for image statistics. *Vision Research, 44,* 541–556.

Wang, H. X., Heeger, D. J., & Landy, M. S. (2012). Responses to second-order texture modulations undergo surround suppression. *Vision Research, 62,* 192–200.

Watson, A. B., & Robson, J. G. (1981). Discrimination at threshold: Labelled detectors in human vision. *Vision Research, 21,* 1115–1122.

Wolfson, S. S., & Landy, M. S. (1995). Discrimination of orientation-defined texture edges. *Vision Research, 35,* 2863–2877.

Yeshurun, Y., & Carrasco, M. (2000). The locus of attentional effects in texture segmentation. *Nature Neuroscience, 3,* 622–627.

Yeshurun, Y., & Carrasco, M. (2008). The effects of transient attention on spatial resolution and the size of the attentional cue. *Perception & Psychophysics, 70,* 104–113.

Yeshurun, Y., Montagna, B., & Carrasco, M. (2008). On the flexibility of sustained attention and its effects on a texture segmentation task. *Vision Research, 48,* 80–95.

Zaidi, Q., & Li, A. (2002). Limitations on shape information provided by texture cues. *Vision Research, 42,* 815–835.

Zhan, C. A., & Baker, C. L. (2006). Boundary cue invariance in cortical orientation maps. *Cerebral Cortex, 16,* 896–906.

Zhan, C. A., & Baker, C. L. (2008). Critical spatial frequencies for illusory contour processing in early visual cortex. *Cerebral Cortex, 18,* 1029–1041.

Zhaoping, L. (2003). V1 mechanisms and some figure-ground and border effects. *Journal of Physiology, Paris, 97,* 503–515.

Zhaoping, L., Guyader, N., & Lewis, A. (2009). Relative contributions of 2D and 3D cues in a texture segmentation task, implications for the roles of striate and extrastriate cortex in attentional selection. *Journal of Vision, 9*(11), 20, 1–22. doi:10.1167/9.11.20.

Zhaoping, L., & May, K. A. (2007). Psychophysical tests of the hypothesis of a bottom-up saliency map in primary visual cortex. *PLoS Computational Biology, 3,* e62. doi:10.1371/journal.pcbi.0030062.

46 The Perceptual Organization of Depth, Lightness, Color, and Opacity

BARTON L. ANDERSON

The world is filled with objects and materials that interact with light, generating the image structure the visual system uses to recover scene structure. Our experience reflects certain aspects of the organization that exists in the material world, which allow us to perform the behaviors needed to survive and reproduce. The world is filled with "stuff" distributed in space and time, and vision provides some of the most important information about the behaviorally relevant properties of the physical world. The issue confronting vision scientists is to understand what that information is, how it is extracted from retinal images, and how it is used to guide behavior.

There are two complementary problems that arise in recovering scene properties from images: segmentation (or decomposition), and synthesis (or grouping and/or interpolation), which both have associated sub-problems. The light reflected from materials is a conflated mixture of surface optics (color, lightness, translucency, specularity, etc.), the illumination field (the distribution of primary light sources and light reflects from different materials), and three-dimensional (3D) shape. These different sources of image variation must be sorted appropriately to recover scene structure. Another segmentation problem involves the problem of segmenting images into distinct objects and materials. This is not merely a problem of identifying "edges" in an image, since any luminance or chromatic discontinuity can arise in a variety of different ways. Image discontinuities can be generated by changes in pigmentation, depth discontinuities, 3D folds or corners of objects, changes in illumination, or specular reflections, which all contribute different kinds of information about scene structure. Thus, something much more than the mere detection of edges is needed to understand how objects are segmented from their surroundings. The complementary grouping and interpolation problem involves understanding how the visual system fills in missing information. The partial occlusion and camouflage of surfaces in natural scenes generates fragmented image data, which must be somehow unified to generate coherent representations of surfaces and materials.

In this chapter, I describe some of the background and recent progress into how the visual system sorts information into a stratified representation of surfaces in depth, with a focus on the problem of image segmentation. Although much research in vision is pursued with a "divide and conquer" strategy, I will argue that many of these ostensibly distinct problems are intimately coupled, and argue that a full understanding of one domain requires understanding how sources of information for one kind of property are distinguished from another.

THE PERCEPTUAL ORGANIZATION OF DEPTH

One of the general areas of vision research is depth perception, which refers to the transformation of 2D images into an experience of 3D scenes. One vivid and extensively studied source of depth information is stereopsis, which is based on the information provided by the slight differences in viewpoint generated by our frontally placed eyes. These different viewpoints generate binocular parallax, which is the apparent difference in position of an object when viewed from different positions. In order for the visual system to extract parallax, it must determine what counts as the "same object" in the two eyes' views, which is known as the *correspondence problem*. One of the main ongoing areas of research in stereopsis focuses on understanding how the visual system determines binocular correspondence and computes the positional shifts of common world features (binocular disparity) (Anderson & Nakayama, 1994; Kumano, Tanabe, & Fujita, 2008; Schreiber, Tweed, & Schor, 2006; Tanabe, Umeda, & Fujita, 2004; van Ee & Anderson, 2001). The prevailing assumption is that the "problem" of stereoscopic depth perception is essentially solved by the computation of disparity, which is predicated on the assumption that there is a simple one-to-one relationship between disparity and perceived depth. But it is not this simple.

The complexity arises because disparity and perceived depth do not always have a simple one-to-one relationship. Part of the problem arises as a consequence of the "stuff" that the brain uses to establish correspondence. There is currently an abundance of data demonstrating that disparity is computed by measuring the positional shifts of some local measure of image contrast. In its most general form, "contrast" refers to a normalized luminance difference, such as that generated by a local edge. Unfortunately, there is currently no single measure of image contrast that adequately captures perceived contrast, and it remains unclear precisely how disparity is computed from binocular image contrast (or what aspects of contrast are used as primitives for matching). Nonetheless, the mere fact that *some* form of luminance *difference* is used to compute disparity is sufficient to demonstrate that the relationship between disparity and depth can be a one-to-many mapping, which implies that there is more to the perceptual organization of stereoscopic depth than the computation of binocular disparity.

Consider a simple luminance discontinuity that is present in both eyes' views, and assume that the visual system has correctly identified the discontinuity as arising from the same source in the two images (i.e., it is binocularly matched). This local discontinuity, or "edge," would give rise to a single disparity, and hence be assigned a single depth. But what if this edge was generated by an occluding contour? In such contexts, one side of the edge (the occluder) is nearer than the other (the occluded background), a fact that is not captured by the single-value of disparity generated by the edge. For the visual system to recover scene geometry in the vicinity of occluding edges, it must map a single value of disparity onto two values of depth (Anderson, 2003b; Anderson, Singh, & Fleming, 2002; Fleming & Anderson, 2003).

This simple geometric fact implies that disparity does not always map onto depth (perceived or physical) in a simple one-to-one manner. This, in turn, implies that the "problem" of stereoscopic depth perception cannot be reduced to establishing a map of local disparity estimates, despite the fact that a large body of work in this field continues to embrace this view. This view has been driven in part by Julesz's seminal invention of the computer-generated random-dot stereogram (RDS). In RDSs, depth can be manipulated by interocularly shifting the positions of individual dots, which results in a corresponding shift in perceived depth of those dots. But RDSs are special cases, whose rich texture obscures some of the complexities of the relationship between disparity and depth that arise in natural scenes. Understanding how the disparity of a local contrast relationship is translated into perceived depth requires a broader understanding of the different ways that contrast can be generated by the world, and how these different sources of image contrast are mapped onto physical and perceived depth.

The relationship between contrast and perceived depth depends on the kinds of surfaces and material properties that generate the pattern of local contrasts in the two eyes. One source of image contrast is local surface texture or variations in reflectance or pigmentation along an opaque surface. In such conditions, the relationship between disparity and depth can be largely considered a simple one-to-one mapping, as is exemplified by the depth experienced in RDSs. However, the same is not true for something as simple as a luminance discontinuity ("edge"); how depth is assigned at an edge requires understanding what generated the edge in the world (Anderson, 2003b). If the edge is an occluding contour, then the occluding edge must be assigned a different depth than the occluded side of the edge. If the edge is a boundary of a transparent overlay (a form of partial occlusion), then depth must be split in *two* ways: at the location of the edge's boundary, in the same manner as an occluding contour; and underneath the transparent surface, into a set of overlapping surfaces. The former decomposition cuts the image like a cookie cutter, splitting the image into pieces like a jigsaw puzzle; the latter decomposition splits the image into layers, like the layers of an onion. All of this may transpire in an image region containing only a single depth estimate or disparity value, which means that the full range of depth experienced in these regions must arise from other sources of information.

Although the geometry of occlusion introduces a source of complexity in interpreting local disparity signals, it also imposes a significant, inviolable constraint on how depth is organized in the neighborhood of a local contrast signal. I have previously dubbed this contain the "contrast depth asymmetry principle" (or CDAP; see Anderson, 2003b). This constraint is most easily understood by considering local luminance discontinuities. It expresses something quite simple and seemingly innocuous: in a world of opaque surfaces, the depth assignment of the individual luminances that define a luminance discontinuity (i.e., a local "edge") are constrained such that they must appear *at least as distant* as the depth of the edge, or one side of the edge can appear more distant (as an occluded region). This seemingly benign constraint can explain a host of phenomena that receives no coherent explanation within conventional theories of binocular vision or depth perception. For example, the CDAP explains a large variety of "depth spreading" asymmetries in stereopsis. For

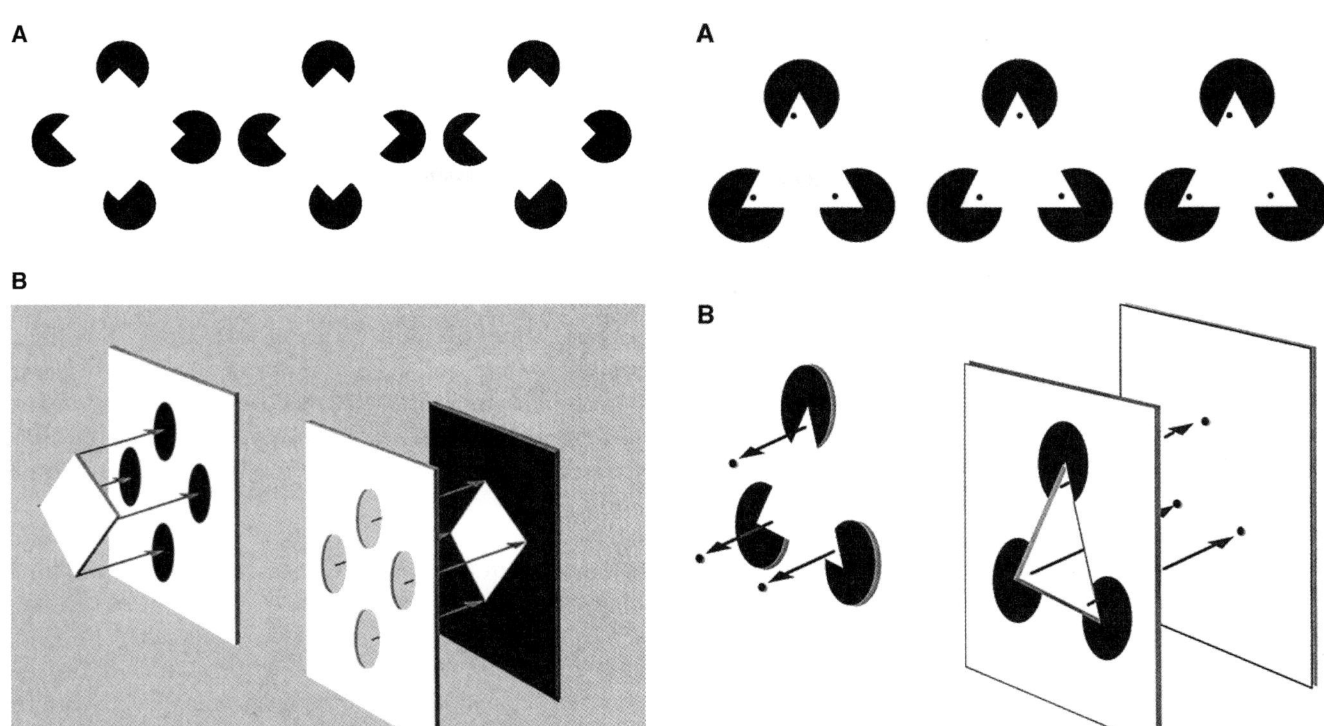

FIGURE 46.1   A stereo Kanizsa figure (A), and the depth and surface organization the two different depth organizations evoke (B). The image on the left in B corresponds to the depth experienced when the two left images are cross fused, or the two right images are divergently fused. The image on the right of B corresponds to the other pair of fused images. Note that the black regions appear as disconnected discs when the diamond appears in front, but appear as a unified coherent surface when the inducing elements of the diamond are placed behind. (From Anderson, 2003b.)

FIGURE 46.2   A "depth spreading" developed by Takeichi et al. The schematic in B depicts the two experiences that arise when the dots are given a near disparity (left) or far disparity (right). Note the asymmetric spread in depth for the near and far depth configurations. (From Anderson, 2003b.)

example, the inducing elements of a stereoscopic Kanizsa figures can be transformed from appearing as disconnected disks that are partially occluded by an illusory figure, into four illusory "portholes," in which the interior of the disks appear as a unified background surface or void (figure 46.1). Takeichi, Watanabe, and Shimojo (1992) reported a related effect, wherein a few small dots that are placed inside and in front of a Kanizsa figure appear as isolated dots floating in front of the illusory figure, but transform the that figure to a vivid illusory hole when the depth of the dots is inverted, dragging all of the surround back in depth with the dots (figure 46.2). Such effects can be seen in even simpler displays: if a few dots are stereoscopically placed in front of the edges of a computer monitor, they appear to float in empty space; but if their depth is inverted, the background on which they are placed recede in depth and appear behind the boundaries of the monitor.

All of these phenomena involve asymmetries in how depth is assigned to local image contrast, or local "edges." Although this idea may seem completely trivial and obvious when expressed so simply, it is far from universally understood or appreciated. Indeed, there are a large number of papers that purport to manipulate the depth of targets in ways that are simply impossible because they violate the constraints imposed by the CDAP. For example, some early studies of simultaneous contrast attempted to compare the effects of stereoscopically placing targets in front or behind their inducing fields (surrounds). The former is possible; the latter is not. Similar attempts to place targets behind their adjacent inducing fields were made with other forms of lightness illusions, such as White's effect, which is also precluded by the CDAP (Spehar, Gilchrist, & Arend, 1995). Similar errors have occurred in visual search experiments (Davis & Driver, 1998), the role of familiarity in figure-ground displays (Peterson & Gibson, 1994), and host of other experiments that attempted to use binocular disparity a tool for inverting depth. Indeed, figure schematics have even been produced that are both physically impossible, and non-representative of what is actually perceived in the displays they are intended to depict.

In general, the CDAP captures a constraint on how local image contrast can be mapped onto possible world events. Our discussion has focused on the role of the CDAP in explaining a variety of asymmetries in how surface properties are organized in depth in displays in which disparity relationships are simply inverted. In such contexts, the CDAP imposes the constraint that both sides of an edge must appear at least as distant in depth as the edge. But this simple statement fails to capture the full range of ways in which local edges (or local contrast more generally) can be generated, and the complexity of the percepts that can be evoked by simply inverting depth information. A deeper understanding of this constraint requires considering a broader class of world events, and how they relate to the local image data that they generate.

SCISSION FROM DEPTH

The preceding section articulated a constraint on how the brain maps local image structure (i.e., local image contrast) into a representation of surfaces in the world. The fact that our perceptual experience mirrors this constraint indicates that the visual system "understands" this constraint imposed by occlusion. Occlusion can, however, be a matter of *degree*, surfaces can vary in their opacity or "hiding power," which occurs in conditions of transparency. This fact introduces a new set of ambiguities in mapping local image structure (contrast) onto a representation of world properties.

Consider again the possible interpretations of a local image contrast (like an edge) with an associated depth signal (such as binocular disparity). For opaque surfaces, the edge could have been generated by a variation in surface reflectance, an illumination boundary, a fold in the 3D geometry, or an occluding and occluded surface. The problem is further complicated if the effects of transparent surfaces or media are considered. In this case, either side of the edge could be a transparent surface. It is also possible that *both* sides of the edge are partially obscured by a transparency overlay. In such cases, multiple depths would have to be assigned to one or both sides of the local image contrast to recover scene geometry (Anderson, 2003b).

How do these considerations impact on the interpretive constraints imposed by the CDAP? Transparency requires that depth be partitioned into multiple depths along the same line of sight, a process historically described as *scission* (Koffka, 1935). There are three possible ways that this can happen: Either side of the edge may be split into a transparent surface overlying another surface, or both sides of the edge are interpreted as containing multiple layers. In the former case, one side of the edge is still treated as an occluding surface, but one that does not completely obscure the more distant, underlying layer. The latter case is more interesting, and provides a refinement of the CDAP as it has been expressed to this point. This refinement arises from considering how a local contrast (such as an edge) would be affected by an overlying transparent surface. Transparent overlays (surfaces or media) affect underlying surfaces in a number of ways: they reduce the overall amount of light coming from the underlying surface, and they typically add an additional source of reflected light from the pigments within the media or filter (Adelson & Anandan, 1990; Anderson, 1997; Beck, 1985; Metelli, 1970, 1974a, 1974b). The former effect is essentially identical to a change in illumination; the latter, additive effect reduces the contrast of the underlying surface. Note, however, that transparent surfaces cannot reverse the *sign* (or polarity) of contrast generated by underlying surface structure; they can only reduce or preserve the contrast that arises in a given direction.

These facts provide additional insights into the mapping between local image contrast and scene geometry. The constraint imposed by the CDAP on the interpretation of local image contrast applies to its *polarity*, or sign, not to its magnitude (or "strength"). In its unrefined form, it states that the component luminances along both sides of an edge must appear at least as distant as the depth of the edge. In its more refined form, when transparency is also considered, it states that *some* of the luminance on either side of the edge must appear at least as distant as the depth of the edge in a manner that preserves the *polarity* of the edge. The polarity constraints on transparency have been described in detail by a variety of authors, but the particular way in which polarity constraints are coupled to perceived depth is described more precisely in the CDAP.

The existence of transparency also induces a general source of ambiguity in the interpretation of all images. Consider a photograph of an arbitrary scene. The structure in the image is defined by the pattern of local contrasts that collectively form the image. How does the visual system determine whether a given image, or image region, is in plain view or partially obscured by a transparent surface or medium? Consider figure 46.3. The dark and light gray bars in the surround appear as portions of a surface in plain view; only the central ellipse appears transparent. Yet it is theoretically possible that the image of the surround was generated by a higher contrast set of stripes that is partially obscured by a contrast—reducing transparent layer. Yet this

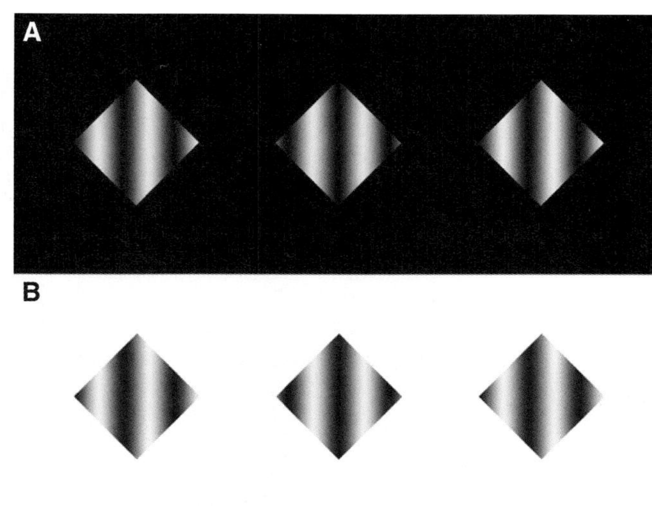

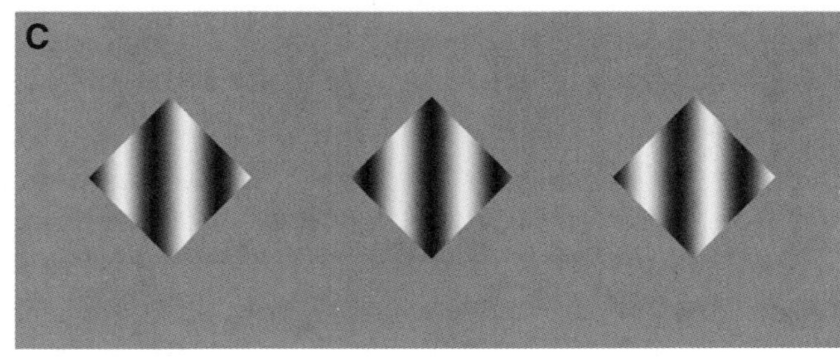

FIGURE 46.4 Stereoscopic grating stimuli. The stereo images
were created by embedding a sine wave grating in diamond
shaped apertures. Disparity was introduced by shifting the
aperture boundaries relative to the grating. When the grating
has a near disparity, it appears as (A) black stripes in front of
a white diamond, (B) white stripes in front of a black diamond,
and (C) unstable and incoherent. When the grating has a far
disparity, no scission is experienced, and the grating appears
as a coherent surface behind a diamond-shaped aperture.
(From Anderson, 1999.)

percept is never experienced, which implies that some
additional constraint is needed to explain when transparent scene interpretations are invoked.

I proposed that the visual system imposes a visual
form of "Occam's razor" when computing transparency
(Anderson, 1999; 2003b; Anderson & Winawer, 2005,
2008). In particular, I argued that transparency is only
inferred when there is visual evidence of a *perturbation*
in local luminance and/or contrast along continuous
contours or textures. This constraint was dubbed the
"transmittance anchoring principle," or TAP. The TAP
states that the highest contrast region of a scene serves
as an "anchor" point against which contrast variations
are assessed; reductions in contrast along continuous
contours or textures, which preserve contrast polarity,
can induce percepts of transparency. The intuition
motivating this principle is that changes in illumination
preserve contrast, whereas transparent surfaces generi-
cally reduce contrast (the only exception being for
transparent surfaces that reflect no discernible light).

These two principles—the TAP and CDAP—can
provide a unified explanation of how the visual system
partitions images into layered representations of surfaces in depth in a broad range of stimuli. Two examples of their predictive power are presented in figures
46.4 and 46.5 (from Anderson, 1999). Figure 46.4
depicts a sine wave grating placed within a diamond
shaped aperture boundary. Binocular disparity is introduced by phase-shifting the grating relative to the aperture boundary in the two eyes. When the two images
on the top right are cross-fused (or two left fused

divergently), the grating appears in a single depth
plane as a single surface. However, when disparity is
inverted (cross fusing the two top left images, or divergently fusing the two top right images), the percept is
dramatically transformed: the image now appears as
uniform white diamond behind a series of fuzzy black
stripes which vary in opacity. A similar effect can be
observed in the image on the white background in the
middle of figure 46.4, but the attribution of lightness
to the two layers is inverted: a uniform black diamond
now appears behind a series of fuzzy white stripes which
vary in opacity. A similar transformation can be experienced when the two left columns in figure 46.5 are cross
fused: dark clouds in front of light disks (top), and light
clouds in front of dark disks on the bottom. Like the
grating images, the luminance modulations within the
near transparent surface appear to vary in opacity.
Importantly, no clear percept of transparency or layers

is experienced when the surround color falls between the luminance values of the texture inside the aperture; the intermediate luminance of the surround causes the contrast polarity along the aperture-texture boundary to reverse, which violates the conditions for coherent scission (bottom of figure 46.4).

There are a number of aspects of the perceptual organization of these displays that need to be explained. The two depth configurations use exactly the same images as input; they simply exchange which image the left and right eye sees. Thus, the disparity relationships in the two depth configurations are simply inverted, yet the perceptual organization of the texture is dramatically different in the two cases. Any cogent theory of these effects must explain: (1) what causes the dramatic transformation in the appearance of the texture when depth is inverted, and (2) the specific pattern of lightness, transparency, and depth that is experienced.

Consider the case in which the texture appears behind the aperture boundaries. The CDAP requires that the light and dark "sides" of a local texture must appear at least as distant as its disparity defined depth to account for its polarity. In this depth configuration, the texture is the most distant contrast in the image, so it appears in the same far depth behind the aperture boundary for both the grating and cloud textures, as

conventional theories of stereopsis would predict. The contrast variations along the aperture boundaries occur in in the near surface, and hence do not induce scission within the more distant texture (recall that the magnitude and polarity constraints apply to the more distant image contrast). However, when the texture is given a near disparity, the aperture boundary is now the far contrast in the scene. The CDAP requires that there must be a luminance relationship at this far depth that causes the edge to have its particular sign (polarity). When the surround is dark (top rows of figures 46.4 and 46.5), this means that there must be something at, or more distant than, the depth of the aperture boundary that is lighter than the surround, which accounts for the polarity of the contour. The converse holds for light surrounds: there must be something darker within the aperture, at this more distant depth, that is darker than the aperture boundary.

The constraint imposed by the CDAP is, however, purely local, and only requires that depth be assigned in the immediate vicinity of the contrast generated by the relatively distant contour. But the transformations experienced in figures 46.4 and 46.5 are *global*: the entire *texture* appears to split into two layers, and the distant layer appears to take on the lightness defined by either the most extreme light values in the texture

658   BARTON L. ANDERSON

(top of figures 46.4 and 46.5), or the extreme dark values in the texture (second rows of figures 46.4 and 46.5). Thus, something else is needed to explain the percepts of transparency that arise in these images, and the way that lightness is attributed to the multiple layers. One key component of that explanation is the TAP.

To understand the role of the TAP in these images, note that the textures on the dark surround appear as light diamonds (figure 46.4) or disks (figure 46.5), and that the lightest components of these textures appear in plain view (i.e., unobscured by a transparent surface). The converse holds for the light surrounds: now the darkest regions in the texture appear as portions of a distant surface in plain view. This is the constraint imposed by the TAP: The lightest components of the textures create the highest contrast aperture boundary when the surround is dark, and the darkest components create the highest contrast regions along the aperture boundary when the surround is light. These are the regions that are seen in plain view. Note further that the contrast along the aperture boundary varies continuously in strength, which provides a contrast cue signaling the possible presence of a transparent surface or medium. These contrast variations are, by hypothesis, scaled by the highest contrast contour segment, and used to infer the relative opacity of the transparent layer: The transparent surfaces appear most opaque in the low contrast regions along the aperture boundary, and takes on intermediate values in opacity when the contrast of the aperture boundary falls between the highest and lowest contrast values. Note further that the CDAP and TAP predict an incoherent or inhomogeneous percept of transparency when the texture-surround boundary reverses polarity (bottom of figure 46.4), requiring the color of the underlying surface to undergo polarity alternations, which is what is experienced.

Taken together, the CDAP and TAP can provide a principled account of a number of properties of these images. It should be noted, however, that this explanation is not complete. The constraint imposed by the CDAP is purely local, but the images in figures 46.4 and 46.5 generate global percepts of transparency. Something more is needed to explain what causes these textures to split into coherent, global percepts of transparent surfaces. One possible explanation for the global organization of these images is provided in Anderson (1999, 2003b).

DEPTH FROM SCISSION

The preceding sections focused on how the visual system uses local depth signals to induce layered representations of surfaces in depth. Binocular disparity was used as a source of local information about depth ordering, which can have a profound effect on how global depth and surface structure is experienced. Although the CDAP and the TAP were developed to explain how local depth determines the layout and organization of surfaces in images containing local depth information (like binocular disparity), their utility extends beyond images in which local depth is locally specified. More generally, these principles articulate general *contingencies* between perceived depth and surface properties, whether the depth is given or inferred.

Since the seminal work of Metelli, it has been known that there are basic geometric and photometric constraints that must be satisfied for transparency to be experienced (Metelli, 1970, 1974a, 1974b, 1985; Metelli, Da Pos, & Cavedon, 1985). Metelli's own work focused on forms of transparency that can be understood with a simple physical devise called an episcotister—a rapidly rotating disk that had open sectors through which an underlying, two-tone background is visible (similar to figure 46.3). For such simple stimuli, it is possible to generate an algebraic solution to the transmittance and reflectance of the transparent layer (the fan blade). This solution was only possible because of the simple physical conditions under consideration, where the transparent surface had a uniform lightness and transmittance, and the underlying surfaces had a uniform reflectance. Nonetheless, some basic constraints on the conditions for perceived transparency can be derived from this special case. *Geometric* constraints involve the continuity of the underlying and overlying (transparent) layers, and *photometric* constraints involve the changes in the luminance and contrast between regions in plain view and regions of transparency.

The geometric and photometric constraints embodied in the CDAP and TAP suggest that it should be possible to generate layered image representations in non-stereoscopic stimuli. Consider figure 46.6. The textures within the boundaries of the chess pieces are identical in the top and bottom, but appear dramatically different: The top image appears as a series of white chess pieces visible through black smoke, and the bottom appears as a series of black chess pieces visible through light fog. These images were created by changing the overall luminance and contrast range of a single "seed" texture (see figure 46.7). The target regions were created by increasing the contrast of the noise texture so that it spanned the full luminance range of the monitor. The luminance range of the surrounds was restricted to span from black to mid-gray on the top, and mid-gray to white on the bottom (note that the

FIGURE 46.6   Transformations in perceived lightness induced by placing identically textured images on textured surrounds with different luminance ranges. The textures within the chess pieces are identical in the top and bottom images, but are perceptually decomposed in opposite ways: as black clouds overlying white chess pieces (top), and white clouds overlying black chess pieces (bottom). (From Anderson & Winawer, 2005.)

lightest and darkest components of the surround occur in the same locations in the top and bottom images; only their absolute intensities differ). The continuity of the texture within the chess pieces and their surrounds is assured by using the same seed texture for both the targets and surrounds. Moreover, by choosing an appropriate luminance range for the light and dark surrounds, the contrast polarity around the edges of the chess pieces all had a consistent, but opposite sign in the top and bottom images: dark–light in the top image, and light–dark in the bottom (referenced from surround to chess pieces). Only the magnitude of contrast along the surround—target borders varies in both of the images. Thus, the geometric and photometric constraints for transparency are satisfied in these images; the issue is explaining the particular pattern of depth, lightness, and opacity that they evoke.

First, note that the lightest regions in the top image's chess pieces appear as the highest contrast regions and appear in plain view (e.g., the top of the left Bishop in the top), whereas the darkest regions within the bottom chess pieces appear as the highest contrast regions and appear in plain view (the top of the rightmost Rook or King on the bottom image). This is consistent with the TAP, and explains why the chess pieces appear to have the specific lightness that they do (white on top, black on the bottom). Second, note that the lowest-contrast regions along the boundaries of the chess pieces appear the most opaque (i.e., occluded), which is consistent with the scaling of opacity by the strength of the edge in plain view (but see below for some complexities with this computation). Third, the chess pieces of both surrounds appear to lie on an approximately mid-gray background, which are the regions where the contrast

FIGURE 46.7   The component images used to create the stimuli depicted in figure 46.6. See text for details.

of chess pieces appears strongest against the two surrounds. Thus, the CDAP and TAP can explain which regions are perceived in plain view, the depth and lightness associated with the two layers, and the relative opacity of the transparent surface or medium.

One complexity that was avoided in the previous analysis is the definition of local contrast used to determine how to apply the CDAP and TAP to these images. In the stereo images depicted in figures 46.4 and 46.5, the particular form of normalization used to define contrast could be largely avoided because the surrounds used in these images were a fixed, uniform luminance. Thus, the ordering of the contrast variations along the target—surround boundary was determined solely by the luminance differences between the targets and the surround; any normalizing factor would generate the same order of "edge strengths" (i.e., edge contrast) as those defined by the luminance difference. The same is not true for the images depicted in figure 46.6 because both the luminance within the targets (chess pieces) and the luminance in the surround varies. The concept of "local contrast" is currently ill defined because there is no definition of contrast that can capture perceived contrast in arbitrary images. This ambiguity has generated some controversy as to how to interpret the contrast constraints of the CDAP in these images. Although there is still no definition of contrast that can account for the perceived edge contrast in these images, the CDAP and TAP are still nonetheless predictive of the

percepts in these images if *perceived* contrast is used to determine the regions in plain view.

COUPLED COMPUTATIONS OF DEPTH AND SURFACE QUALITIES

The phenomena and organizational principles articulated in the preceding reveal that there are systematic constraints on how local image data is organized into percepts of surfaces in depth. One of the core insights is that depth per se is not something that is seen. Rather, some *thing* (or *"stuff"*) is seen at as *having* a particular depth. In order to understand the relationship between how local image structure provides information about the layout of surfaces and materials—in short, the world—it is necessary to understand how different world properties constrain the local image data, and the extent to which visual processes embody these constraints when inferring scene structure from the images.

The CDAP expresses a constraint on the organization of depth imposed by the geometry of occlusion, which can have large transformative effects on perceived lightness and color. This suggests that scission may play a general role in lightness and color perception. This idea has not, however, been widely embraced. Indeed, a number of authors have argued that the transformations in perceived lightness that emerge in figures 46.4–46.6 are of a different kind (i.e., driven by different mechanisms) than other forms of contextual effects in

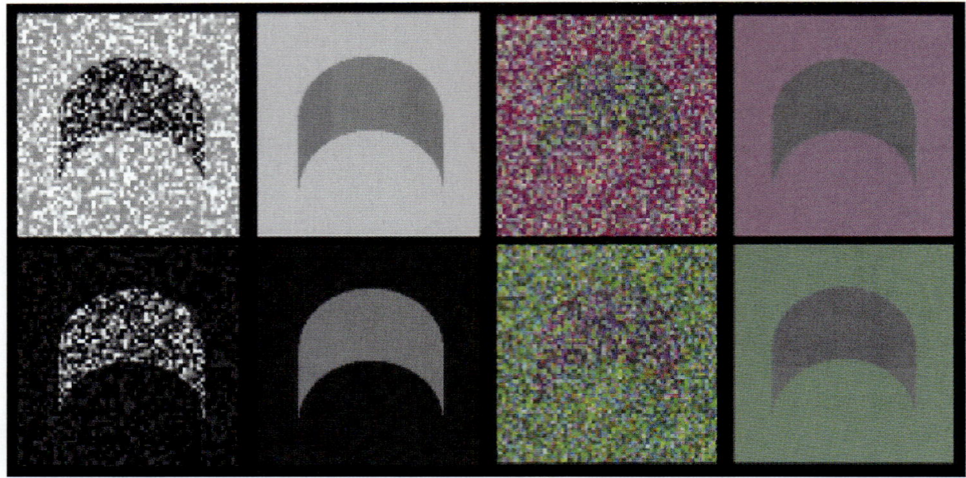

FIGURE 46.8   Similar forms of scission can be observed using white-noise textures that do not contain any explicit junctions or occlusion cues, in both achromatic images (left) and chromatic images (right). The homogenous targets on homogeneous surrounds are presented to demonstrate the size of the effects expects from simultaneous contrast. (Adapted from Wollschläger & Anderson, 2009.)

perceived lightness and color. The issue at stake concerns whether (and when) the process of layered image decompositions contributes to our experience of surface reflectance. At one extreme, some authors (Albert, 2007; Kingdom, 2008) have suggested that the lightness transformations experienced in figure 46.6 are simply a type of figure-ground reversal. The logic of this view is that the regions that appear as the occluded regions that appear in plain view in the two images occur on complimentary places in the two images, i.e., on the two "sides" of the luminance gradient. The link between the transformations in perceived surface reflectance and occlusion was one of the primary inspirations behind the development of these displays, and forms the basis of the CDAP and TAP. Nonetheless, there are a number of problems with the assertion that these effects can be *reduced* to a description as a figure-ground reversal. First, such accounts provide no explanation of the fact that the textures in these images contain a continuous distribution of luminance values that are decomposed into a percept of overlying layers. There are no "edges" in these images where the putative figure-ground reversal occurs; the regions that appear in plain view are perceptual outcomes that require explanation. Any proposed explanation must articulate where the occluding (transparent) layer begins, and why the luminance variations are experienced as variations in the opacity of a transparent surface, rather than variations in surface reflectance, 3D shape, or any other source in luminance variation.

A second problem with figure-ground "explanations" is their failure to appreciate the full range of stimuli that elicit percepts of scission and its transformative effects on perceived lightness and color (Anderson, 2003a; Anderson & Winawer, 2008; Wollschläger & Anderson, 2009). Some examples are presented the images in figure 46.8. The central targets in the top and bottom images of all three columns are identical, yet appear dramatically different (particularly when played as movies); see (Wollschläger & Anderson, 2009). For example, the random textures of the central patch in the third column are composed of the exact same distribution of chromaticities, but the top image appears green, and the bottom image appears magenta. When played as dynamic movies, these regions generate a vivid percept of multiple layers: a random noise pattern identical to the surround, and a uniform, saturated surface, which appears as approximately the same color as the most saturated element (complementary to the surround) within the target. Similar effects can be seen for the lightness version of these effects in the first column: the top image appears blackish, and the bottom appears whitish. Neither pair of images contains image structures to which local figure-ground computations can be applied.

Whereas some authors have suggested that the effects of scission on lightness and color are special cases, other authors have argued that scission plays a crucial and general role in a variety of well-known induction phenomena. Recent work by Ekroll, Faul, and colleagues have argued that the phenomena of simultaneous contrast, "gamut expansion," and "crispening" are all versions of the same effect, in which scission plays a crucial role (Ekroll & Faul, 2009; Ekroll,

Faul, & Niederee, 2004; Ekroll, Faul, Niederee, & Richter, 2002; Ekroll, Faul, & Wendt, 2011; Faul & Ekroll, 2002; Faul, Ekroll, & Wendt, 2008). Simultaneous contrast is well known, and refers to the changes in perceived lightness or color of targets placed on surrounds with a different lightness or color. Gamut expansion refers to the fact that the perceived range of chromaticities appears greater on a uniform background that is close to the chromaticity of the targets, than when the same targets are placed on a variegated surround. "Crispening" refers to the fact that the perceived *difference* between two targets is greatest when the target just changes from being an increment to a decrement relative to the chromaticity or lightness of its surround. In an elegant series of papers, these authors have shown that all of these phenomena can be modeled as the consequence of two sources of induction: a crispening component (which they attribute to scission), and a multiplicative scaling component (von Kries scaling), which treats the effect of context as a multiplicative transformation. In contradistinction to those suggesting that the scission experienced in textured displays may represent a special form of induction, the results by Ekroll, Faul, and colleagues suggest that layered image decompositions play a general and potentially ubiquitous role in computations of lightness and color; just how ubiquitous is an active an open question of ongoing research (see, e.g., (Anderson, Khang, & Kim, 2011)).

CONCLUSION

The problem of depth perception has often been divorced from other problems of surface and material perception, such as perceived lightness and color. The phenomena and organizational principles described in the preceding chapter suggest that this is a false division, and that any understanding of depth perception requires an understanding of what is assigned a given depth. I have shown that simple inversions of depth can cause dramatic transformations in the perceptual organization of depth and material properties, such as their lightness, color, and opacity. I have also argued that these asymmetries can be understood with a few general principles derived from constraints on image interpretation that arise from two sources: the geometry of occlusion, and the photometric constraints imposed by transparent overlays. Future work is needed to understand how other material properties, such as translucency and gloss, shape the structure in the image and constrain how depth and material qualities are computed.

REFERENCES

Adelson, E. H., & Anandan, P. (1990). Ordinal characteristics of transparency *Proceedings of the AAAI-90 Workshop on Qualitative Vision* (pp. 77–81). Boston: AAAI Press.

Albert, M. K. (2007). Occlusion, transparency, and lightness. *Vision Research, 47*, 3061–3069. doi:10.1016/j.visres.2007.06.004.

Anderson, B. L. (1997). A theory of illusory lightness and transparency in monocular and binocular images: the role of contour junctions. *Perception, 26*, 419–453.

Anderson, B. L. (1999). Stereoscopic surface perception. *Neuron, 24*, 919–928.

Anderson, B. L. (2003a). Perceptual organization and White's illusion. *Perception, 32*, 269–284. doi:10.1068/p3216.

Anderson, B. L. (2003b). The role of occlusion in the perception of depth, lightness, and opacity. *Psychological Review, 110*, 785–801. doi:10.1037/0033-295x.110.4.785.

Anderson, B. L., Khang, B. G., & Kim, J. (2011). Using color to understand perceived lightness. *Journal of Vision, 11*, 19. doi:10.1167/11.13.19.

Anderson, B. L., & Nakayama, K. (1994). Toward a general theory of stereopsis: Binocular matching, occluding contours, and fusion. *Psychological Review, 101*, 414–445.

Anderson, B. L., Singh, M., & Fleming, R. W. (2002). The interpolation of object and surface structure. *Cognitive Psychology, 44*, 148–190. doi:10.1006/cogp.2001.0765.

Anderson, B. L., & Winawer, J. (2005). Image segmentation and lightness perception. *Nature, 434*, 79–83. doi:10.1038/nature03271.

Anderson, B. L., & Winawer, J. (2008). Layered image representations and the computation of surface lightness. *Journal of Vision, 8*, 18 11–22. doi: 10.1167/8.7.18

Beck, J. (1985). Perception of transparency in man and machine. In A. Rosenfeld (Ed.), *Human and machine vision II* (pp. 1–12). New York: Academic Press.

Davis, G., & Driver, J. (1998). Kanizsa subjective figures can act as occluding surfaces at parallel stages of visual search. *Journal of Experimental Psychology. Human Perception and Performance, 24*, 169–184.

Ekroll, V., & Faul, F. (2009). A simple model describes large individual differences in simultaneous colour contrast. *Vision Research, 49*, 2261–2272. doi:10.1016/j.visres.2009.06.015.

Ekroll, V., Faul, F., & Niederee, R. (2004). The peculiar nature of simultaneous colour contrast in uniform surrounds. *Vision Research, 44*, 1765–1786. doi:10.1016/j.visres.2004.02.009.

Ekroll, V., Faul, F., Niederee, R., & Richter, E. (2002). The natural center of chromaticity space is not always achromatic: A new look at color induction. *Proceedings of the National Academy of Sciences of the United States of America, 99*, 13352–13356. doi:10.1073/pnas.192216699.

Ekroll, V., Faul, F., & Wendt, G. (2011). The strengths of simultaneous colour contrast and the gamut expansion effect correlate across observers: evidence for a common mechanism. *Vision Research, 51*, 311–322. doi:10.1016/j.visres.2010.11.009.

Faul, F., & Ekroll, V. (2002). Psychophysical model of chromatic perceptual transparency based on substractive color mixture. *Journal of the Optical Society of America. A, Optics, Image Science, and Vision, 19*, 1084–1095.

Faul, F., Ekroll, V., & Wendt, G. (2008). Color appearance: The limited role of chromatic surround variance in the "gamut expansion effect." *Journal of Vision, 8*, 30 31–20. doi: 10.1167/8.3.30

Fleming, R. W., & Anderson, B. L. (2003). The perceptual organization of depth. In L. Chalupa & J. S. Werner (Eds.), *The visual neurosciences* (pp. 1284–1299). Cambridge, MA: MIT Press.

Kingdom, F. A. A. (2008). Perceiving light versus material. *Vision Research, 48*, 2090–2105. doi:10.1016/j.visres.2008. 03.020.

Koffka, K. (1935). *Principles of Gestalt psychology.* New York: Harcourt, Brace, & World.

Kumano, H., Tanabe, S., & Fujita, I. (2008). Spatial frequency integration for binocular correspondence in macaque area V4. *Journal of Neurophysiology, 99*, 402–408. doi:10.1152/ jn.00096.2007.

Metelli, F. (1970). An algebraic development of the theory of perceptual transparency. *Ergonomic, 13*, 59–66.

Metelli, F. (1974a). Achromatic color conditions in the perception of transparency. In R. B. MacLeod & H. L. Pick (Eds.), *Perception: Essays in honor of J.J. Gibson* (pp. 95–116). Ithaca, NY: Cornell University Press.

Metelli, F. (1974b). The perception of transparency. *Scientific American, 230*, 90–98.

Metelli, F. (1985). Stimulation and perception of transparency. *Psychological Research, 47*, 185–202.

Metelli, F., Da Pos, O., & Cavedon, A. (1985). Balanced and unbalanced, complete and partial transparency. *Perception & Psychophysics, 38*, 354–366.

Peterson, M. A., & Gibson, B. S. (1994). Object recognition contributions to figure-ground organization: operations on outlines and subjective contours. *Perception & Psychophysics, 56*, 551–564.

Schreiber, K. M., Tweed, D. B., & Schor, C. M. (2006). The extended horopter: quantifying retinal correspondence across changes of 3D eye position. *Journal of Vision, 6*, 64–74. doi:10.1167/6.1.6.

Spehar, B., Gilchrist, A., & Arend, L. (1995). The critical role of relative luminance relations in White's effect and grating induction. *Vision Research, 35*, 2603–2614.

Takeichi, H., Watanabe, T., & Shimojo, S. (1992). Illusory occluding contours and surface formation by depth propagation. *Perception, 21*, 177–184.

Tanabe, S., Umeda, K., & Fujita, I. (2004). Rejection of false matches for binocular correspondence in macaque visual cortical area V4. *Journal of Neuroscience, 24*, 8170–8180. doi:10.1523/JNEUROSCI.5292-03.2004.

van Ee, R., & Anderson, B. L. (2001). Motion direction, speed, and orientation in binocular matching. *Nature, 410*, 690–694.

Wollschläger, D., & Anderson, B. L. (2009). The role of layered scene representations in color appearance. *Current Biology, 19*, 430–435.

47 Image-Parsing Mechanisms of the Visual Cortex

RÜDIGER VON DER HEYDT

Vision is, of course, processing of the information that is available in images, processing that allows us to search for and recognize things, to compare, to plan actions, and to perform many other tasks. But vision is not just processing; it also involves the representation of visual information. The system must transform the image information and code it in a way that is suitable for performing the various tasks. "The study of vision must therefore include not only the study of how to extract from images the various aspects of the world that are useful to us, but also an inquiry into the nature of the internal representations by which we capture this information and thus make it available as a basis for decisions about our thoughts and actions" (Marr, 1982).

In this chapter I discuss visual processes and representations that are often labeled as *intermediate-level vision*. Area V1 transforms the image information into a "feature map," a representation of local features. Like the retina, V1 encodes position by the location of receptive fields, but each neuron here represents not a pixel but a small patch of the image. Several new dimensions are added to the three color dimensions, such as orientation, spatial frequency, direction of motion, and binocular disparity. Although the local feature representation of V1 is known in detail, we still do not have a good understanding of what the extrastriate areas do and represent. At first glance the visual properties of the neurons in area V2 are rather similar to those of their input neurons of V1 except that V2 neurons have larger receptive fields (Burkhalter & Van Essen, 1986; Zeki, 1978). Many neurons of area V4 again have properties that can be found already in V1 and V2, such as selectivity for color, orientation, and spatial frequency (Desimone et al., 1985; Schein & Desimone, 1990; Zeki, 1978). Some neurons in V4 show selectivity for increasingly complex features (Roe et al., 2012).

Here I review evidence that the extrastriate areas perform a function that may be described as "image parsing." This stage mediates between the local feature representation of V1 and higher processing centers by making figure–ground relationships explicit, assigning contours to the figure regions, and providing a structure for selective attention. From the enormous amount of information that streams in through the optic nerves at each moment, the visual system selects a small fraction and, in general, precisely what is relevant for a given task. This amazing performance indicates powerful mechanisms for organizing the incoming information. The Gestalt psychologists first pointed out that the visual system tends to organize elemental visual units (such as points and lines) into larger perceptual units, or figures, according to certain rules called "Gestalt laws" (Kanizsa, 1979; Spillmann & Ehrenstein, 2003; Wagemans et al., 2012). Much of this organization occurs independently of what the subject knows or thinks about the visual stimulus. Gaetano Kanizsa has created wonderful illustrations to demonstrate this "autonomy of perception." One is a graphic supposed to show a knife behind a glass. Everybody knows that glass is transparent and knifes are not, but perception has the knife transparent and passing in front of the glass (Kanizsa, 1979, figure 2.19).

Grouping together features that belong to an object is a general task of perception. Specific visual problems arise from the fact that vision is based on two-dimensional projections of a three-dimensional world. Due to spatial interposition, parts of the scene are occluded, and features of near and far objects are cluttered in the image. The true three-dimensional shape of the objects and their relations in space can only be inferred from the images. In principle any image has an infinite number of possible interpretations in three dimensions; vision is an "ill-posed problem" (Poggio & Koch, 1985). In spite of this fundamental ambiguity vision is the most reliable of our senses. Apparently, through evolution and experience, biological vision systems have learned to make efficient use of the regularities present in images and to infer the missing information (Attneave, 1954; Barlow, 1961; Helmholtz, 1866; Marr, 1982; Poggio & Koch, 1985; Ullman, 1996).

ILLUSORY CONTOURS: CREATIVE MECHANISMS

A prominent example of this creative process is the phenomenon of illusory contours (figure 47.1). In A, the system "visibly" fills in the missing contours of an overlying triangle. Note that the illusory contours are not just interpolations between given contrast borders, as it might seem in A, but form also in the absence of contrast borders that could be interpolated (C). In fact, when the corners of the overlying triangle *are* defined by lines that could be interpolated, illusory contours *do not* form (B). What all illusory contour figures have in common is the presence of occlusion cues (Coren, 1972) such as terminations of lines and edges. Thus, the system seems to infer an occluding object. However, this is not an inference in abstract terms. The mere expectation of a contour does not lead to perception of illusory contours (figure 47.1D). Apparently, in forming the contours the system combines evidence from occlusion cues with rules such as the Gestalt principle of good continuation.

Interestingly, illusory contours are represented in the visual cortex at a relatively early stage. In monkey area V2 many cells respond to illusory contour stimuli as if the contours were contrast borders (von der Heydt, Peterhans, & Baumgartner, 1984). Figure 47.2 shows an example of a cell that was tested with a moving illusory bar. The raster plot in B shows that the cell responds when the illusory bar traverses a small region that was determined beforehand as the cell's minimum response field (ellipse, see legend). Figure 47.2C shows a control

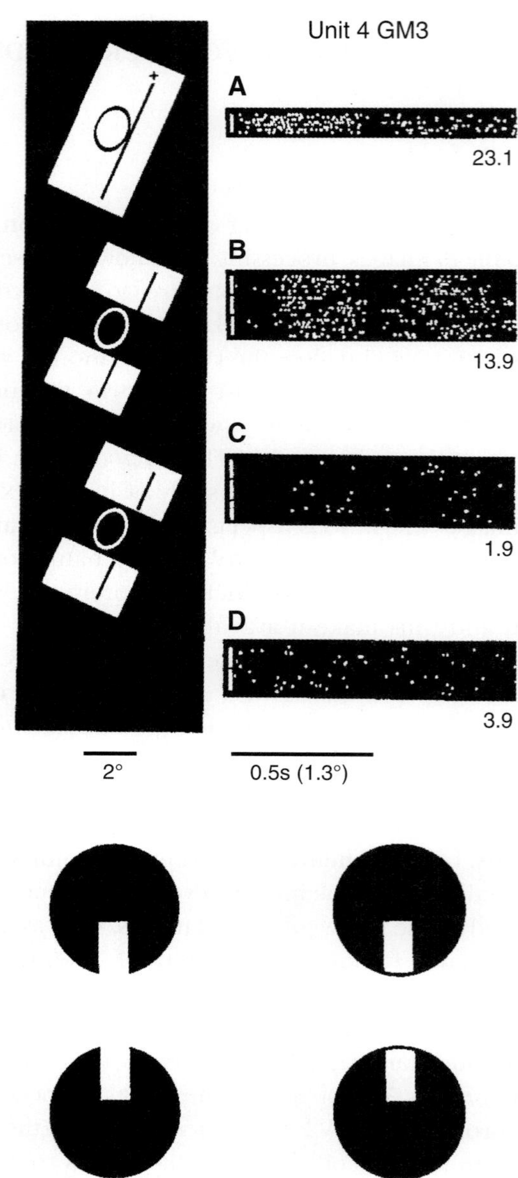

FIGURE 47.2   Illusory contour responses in a neuron of area V2. Each line of dots in the raster plots represents a sequence of action potentials fired in response to the stimulus shown on the left. Responses (A) to a moving dark bar, (B) to a figure in which a moving illusory bar is perceived, (C) to a modified figure in which the illusion is abolished by adding line segments. Note the reduction of responses. Figures at the bottom illustrate the perceptual effect of adding lines. (D) Spontaneous activity. Ellipses indicate the minimum response field of the neuron (that is, the minimum region outside which a bar does not evoke a response); cross indicates fixation point. (From Peterhans & von der Heydt, 1989.)

FIGURE 47.1   Perception of illusory contours. (A and C after Kanizsa, 1979; D, drawing by René Magritte.)

in which the two bars were moved exactly as in B, but the open ends were closed off with thin lines. Closing lines weaken the perceptual illusion (see figure at the bottom), and they also reduce the responses of the neuron. Cells in V2 respond not only to figures with illusory bars but also to other figures that produce illusory contours, such as a pattern of two abutting line gratings (figure 47.3, right). It can be seen that the cell of figure 47.3 responds at the same orientations for the illusory contour as for the bar stimulus; thus, it signals the orientation of an illusory contour. Using the criteria of consistent orientation tuning and response reduction by the closing lines, researchers found that 30–40% of the cells of V2 signal illusory contours of one or the other type, and the results obtained with the two types of contour were highly correlated (Peterhans & von der Heydt, 1989; von der Heydt & Peterhans, 1989).

As shown in figure 47.2, illusory contour responses can be evoked by stimuli that are devoid of contrast over the excitatory center of the receptive field. The inducing contrast features can be restricted to regions from which an optimized bar stimulus would not evoke any response. The cells seem to integrate occlusion features over a region larger than the conventional receptive field (Peterhans & von der Heydt, 1989). Nevertheless, the extent of spatial integration is limited; for neurons with near-foveal receptive fields, the responses declined if the gap of the stimulus (figure 47.2 B) was made wider than about 3° visual angle.

The fact that so many cells in V2 respond this way indicates that illusory contour stimuli probe a basic function of the visual cortex. V2 is an early stage of processing where responses are fast and highly reproducible. Illusory contour responses arise as early as 70 ms after stimulus onset (Lee & Nguyen, 2001; von der Heydt & Peterhans, 1989). This indicates that illusory contours are probably not the result of object recognition processes at higher levels but are generated within the visual cortex. Computational models have shown how such contours might be generated (e.g., Finkel & Sajda, 1992; Grossberg & Mingolla, 1985; Heitger et al., 1998).

Representation of illusory contours has also been demonstrated in V1 of cat (Redies, Crook, & Creutzfeldt, 1986; Sheth et al., 1996) and monkey (Grosof, Shapley, & Hawken, 1993; Lee & Nguyen, 2001; Ramsden, Hung, & Roe, 2001). However, it is not clear if cells in V1 also generalize over the different types of illusory contour figures and if they signal the orientation of the contour. Sheth et al. (1996) and Ramsden, Hung, and Roe (2001) used a combination of optical imaging and single-unit recording to identify the illusory contour representation with the abutting-grating type of stimulus. Sheth et al. found cells with consistent orientation tuning for illusory and real contours in cat V1. Ramsden, Hung, and Roe found that, in monkey V1, illusory contours *reduced* the activity in columns of the corresponding orientation and *increased* activity in columns of the

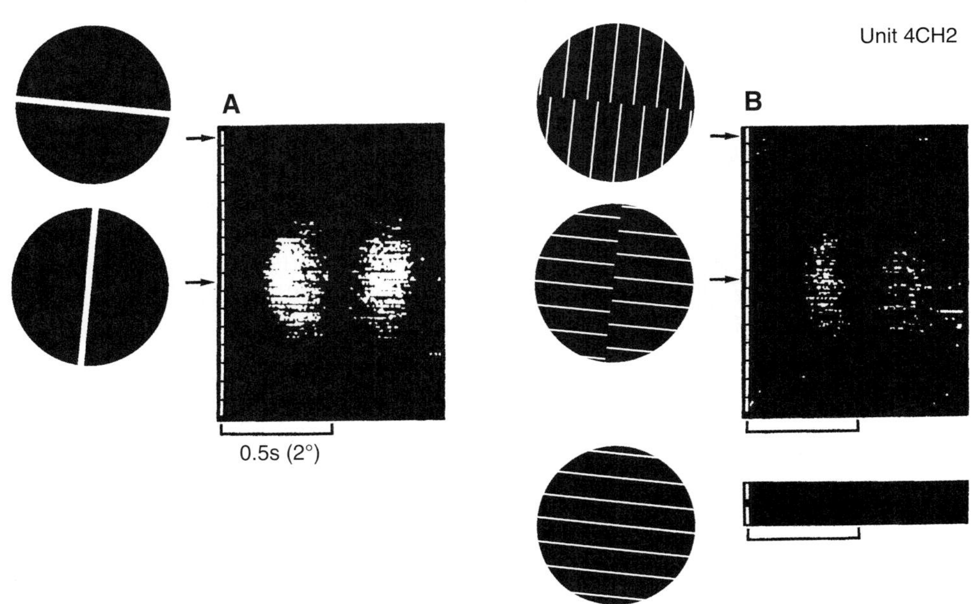

FIGURE 47.3   Illusory contour responses in another neuron of V2. Bars (A) and the border between two gratings (B) were moved across the receptive field at 16 different orientations spanning 180°. The neuron responds at the same orientations for bar and illusory contour. (Bottom right) Control: grating without border of discontinuity. (Modified from von der Heydt & Peterhans, 1989.)

orthogonal orientation, in contrast to V2, where the same columns were activated by illusory contours and contrast lines. They concluded that V1 deemphasizes illusory contours. Studies that compared V1 and V2 invariably found marked differences in the proportion of cells that signaled illusory contours, the orientation tuning, and the degree of cue invariance (Bakin, Nakayama, & Gilbert, 2000; Leventhal et al., 1995; Ramsden, Hung, & Roe, 2001; Sheth et al., 1996; von der Heydt & Peterhans, 1989).

Correlation of Physiology and Perception

Varying the configurations and spatial parameters of the abutting grating displays (figure 47.3B), Soriano, Spillmann, and Bach (1996) found a tight correspondence between human perception and the neuronal responses in monkey V2 (von der Heydt & Peterhans, 1989). However, in discriminating the shape of illusory figures, the human visual system shows larger spatial integration (Ringach & Shapley, 1996). Because neurons that signal illusory contours are only a subset of the cells that signal contrast edges, orientation-dependent adaptation aftereffects should transfer from contrast-defined to illusory contours, but not in the reverse direction, and the discrimination of orientation should be less accurate for illusory contours than for contrast-defined contours. Both predictions were borne out in psychophysical experiments (Paradiso, Shimojo, & Nakayama, 1989; Westheimer & Li, 1996). Illusory contours are usually associated with perception of overlay (Coren, 1972), and some neurons in V2 are selective for the implied direction of occlusion of illusory contours (Baumann, van der Zwan, & Peterhans, 1997). Thus, the illusory contour mechanisms may be related to the coding of border ownership discussed below.

Illusory Contours Are Universal

Perception of illusory contours has been demonstrated in a variety of nonhuman species, including bee, cat, and barn owl (Bravo, Blake, & Morrison, 1988; De Weerd et al., 1990; Srinivasan, Lehrer, & Wehner, 1987; for a review see Nieder, 2002; Nieder & Wagner, 1999). Most elegant is the combination of behavioral experiments with single-cell recordings (Nieder & Wagner, 1999).

BORDER OWNERSHIP: CONTEXT INTEGRATION

Illusory contours and related visual phenomena are only the tip of an iceberg of cortical processes involved in perceptual organization. Kanizsa's figure (figure 47.1A) suggests that illusory contours are the product of mechanisms in figure–ground segregation. The system takes the peculiar arrangement of the black elements as evidence for an occluding triangle and hence creates a representation of its contours (Gregory, 1972). In fact, it also creates the representation of a white opaque surface, as one can see from the subtle difference in brightness relative to the background. The illusory contours appear as the edges of this surface. Perception generally tends to interpret sharp contrast borders as occluding contours and tries to assign them to a surface on one or the other side. This compulsion is demonstrated by Rubin's vase figure (figure 47.4A), where the borders are perceived either as the contours of a vase or as the contours of two faces. In the case of a simple figure such as the white square of figure 47.4B, the contrast borders are "of course" perceived as the contours of the square. They seem to belong to the enclosed light-textured region. The surrounding gray, which does not "own" these borders, is perceived as extending behind the square, forming the "background." The phenomenon of border ownership remained long unnoticed until it was discovered by the Gestalt psychologists (Koffka, 1935; Rubin, 1915).

One could argue that even the display of figure 47.4B is ambiguous. With some effort, the square can also be perceived as a window, and the border then appears as the edge of the frame. Unambiguous displays can be produced by means of random-dot stereograms, as shown in figure 47.4C. When binocularly fused by crossing the eyes (see legend), the top pair shows a tipped square floating in front of a background plane, and the bottom pair shows a square window through which a background plane can be seen. In the first case the stereoscopic borders are perceived as the edges of the square; in the second case they are perceived as edges of the window frame. Perception of border ownership cannot be reversed in this stereogram. In general, stereoscopic depth and the assignment of border ownership profoundly influence the way an image is interpreted in terms of objects and how surfaces are perceived (Gregory & Harris, 1974; Idesawa, 1991; Nakayama, Shimojo, & Silverman, 1989; Shimojo, Silverman, & Nakayama, 1989).

In perceptual experiments we observe the tip of the iceberg. By recording signals in visual cortex we should be able to explore the depth of the iceberg as well. Contrast borders are represented in the visual cortex by signals of the orientation-selective cells discovered by Hubel and Wiesel. Do these signals also represent the relationship between border and surface? This idea can be tested with a simple experiment (Zhou, Friedman,

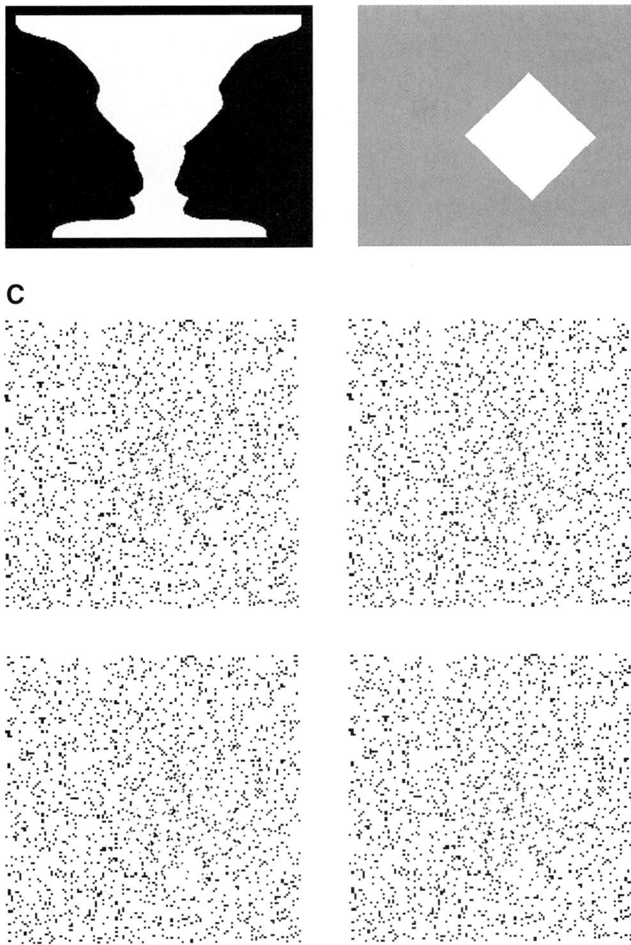

FIGURE 47.4   Perception of border ownership. (A) Physiologist's version of Rubin's vase. The black–white borders are perceived either as contours of a vase or as contours of faces. (B) White square. The contrast borders are generally perceived as contours of the square. (C) Stereograms. Left and right textured square fields can be fused, for example, by squinting. (Try crossing the lines of sight of the two eyes until three fields are perceived instead of two; the center field then shows the result of binocular fusion.) On fusion with crossed eyes, the top pair shows a square figure, whereas the bottom pair shows a square window. In the former, the three-dimensional edges belong to the figure; in the latter, to the surround.

& von der Heydt, 2000). Light–dark borders are placed in the receptive field of a neuron at optimal orientation (figure 47.5), and the same border is presented either as the right side of a light square (e.g., A1) or as the left side of a dark square (B1). Column 2 shows a similar test with displays of reversed contrast, and columns 3–4 and 5–6 show the same kind of test with squares of larger sizes. The bar graph at the bottom represents the responses of a cell of V2. If we compare the responses to the corresponding displays in A and B, we see that in every case the neuron responds more strongly when the edge in the receptive field belongs to a square to the left than a square to the right, despite locally identical stimulation.

Note that the corresponding displays in rows A and B are identical over the entire region occupied by the two squares (as one can see by superimposing them). Thus, if a neuron responds differently, it must have information from outside this region. Thus, by varying the square size we can reveal the extent of image context integration. In this example square sizes of 4°, 10°, and 15° were tested, and in each case the responses differed depending on the location of the figure. By contrast, the size of the classical receptive field of this cell was only 0.4°, which is typical for V2 neurons of the foveal representation. Thus, although the cell can "see" only a small piece of contrast border through the aperture of its receptive field, its responses reveal processing of an area at least 15° in diameter. Using fragmented figures, Zhang and von der Heydt (2010) explored the context integration mechanism. A contour fragment in the classical receptive field was necessary to evoke responses, but most of the fragments outside this field modulated the responses, producing enhancement on one side and suppression on the other.

What might be the mechanism of side-of-figure selectivity? For a single square figure on a uniform background relatively simple algorithms would be able to discriminate figure and ground. The convexity of the figure area could be used or simply the orientation of the L-junctions (corners) on either side of the receptive field or the fact that the figure is a region of one color enclosed by a region of a different color (*surroundedness*).

Any of these strategies would work for the isolated square. However, for other displays in which border ownership is also perceptually clear, mechanisms based on one simple strategy would fail to produce the right answer. Zhou, Friedman, and von der Heydt (2000) used two other configurations besides squares to see how well the neural responses correlated with perception, a C-shaped figure as shown in columns 3 and 4 of figure 47.6 and a pair of overlapping squares as shown in columns 5 and 6 of the same figure. For the C-shape, convexity is not valid, and the L-junctions next to the receptive field are reflected to the background side, but surroundedness would still be a valid cue. For the overlapping squares, surroundedness is violated, while convexity and orientation of L-junctions are valid.

The bar graph at the bottom represents the responses of an example neuron of V2. The test with C-shaped figures shows that the neuron "correctly" preferred the display in which the C-figure was located on the left

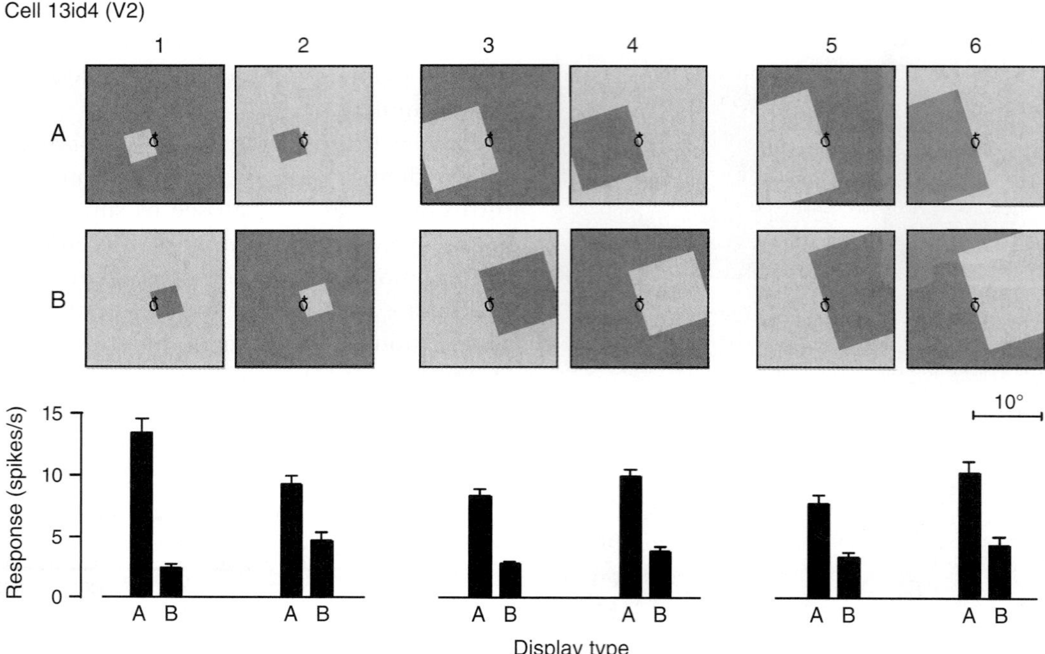

FIGURE 47.5 Selectivity for side of figure in a neuron of area V2. Edges of squares were tested so that the square was either on the left side (A) or on the right side (B) of the receptive field (ellipses show minimum-response field; cross marks fixation point). Note that corresponding displays in A and B are identical over the combined area of the two squares. Tests with square sizes of 4°, 10°, and 15° are shown. Bar graph represents mean firing rates with standard errors. In every case the neuron responds more strongly when the figure is on the left side despite locally identical stimulation. (From Zhou, Friedman, & von der Heydt, 2000.)

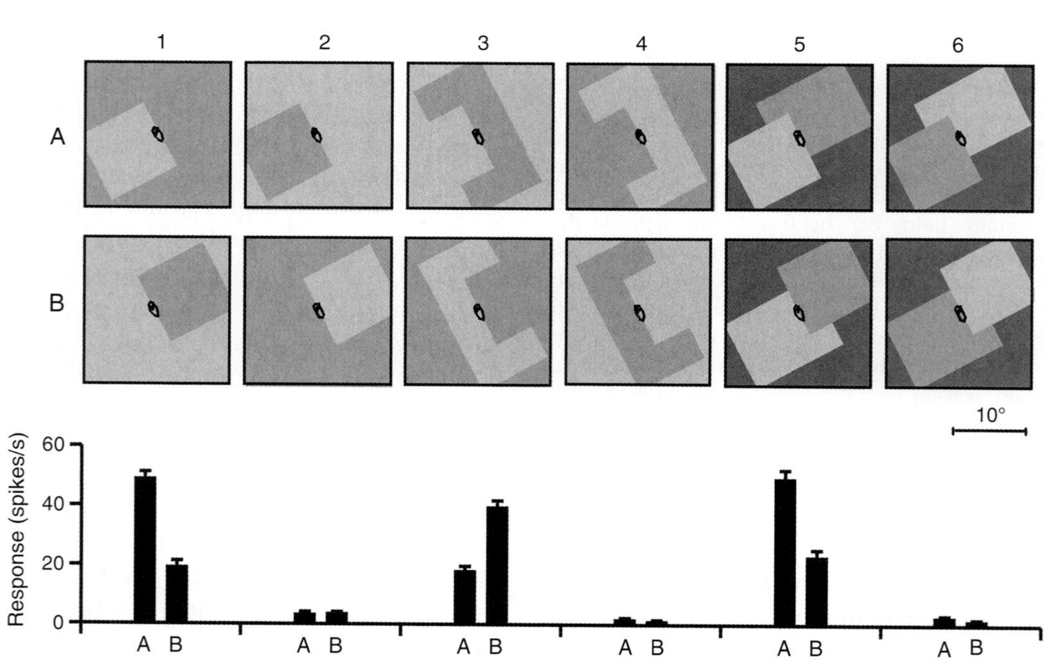

FIGURE 47.6 Generalization of side-of-figure selectivity. Responses of an example V2 neuron to squares, C-shaped figures, and overlapping figures. This neuron was color selective with a preference for violet (depicted here as light gray) and selective for local contrast edge polarity (compare responses to odd- and even-numbered displays). The tests with single squares (1, 2) show preference for figure on the lower left-hand side of the receptive field (A1). With C-shaped figures (3, 4), the neuron responds better to B3 than to A3, again pointing to the lower left as the figure side, in agreement with perception. With overlapping figures (5, 6), the neuron responds better to A5 than B5, assigning the edge to the figure that is perceived as overlaying. (From Zhou, Friedman, & von der Heydt, 2000.)

(display B3), although the L-junctions next to the receptive field would, rather, suggest a figure on the opposite side. Also in the test with overlapping figures the neuron preferred display A5 in which the border in the receptive field belongs to the lower left figure. In this case the T-junctions might account for the emergence of the occluding square as a figure, but convexity might also contribute because the overlapped region has a concavity, whereas the overlapping region does not. Thus, the responses of this cell are entirely consistent with the perception of border ownership. Not all cells tested showed this pattern, but the example is not unusual. About half of the cells with a side-of-figure effect for single squares exhibited the corresponding side preference for overlapping figures, whereas the others showed no significant response difference. When tested with the concave side of C-figures, only about one third of the cells with a side-of-figure effect for single squares showed preference for the C on the same side; the others were indifferent.

Cells with a response preference for one or the other side of the figure were found for any location and orientation of receptive field, and side-of-figure preference was invariant throughout the recording period. The responses of these cells seem to carry information not only about the location and orientation of the contours but also about the side to which they belong. About half of the orientation-selective cells of areas V2 and V4 were found to be side-of-figure selective by the test of figure 47.5. In about 30% of the V2 cells the ratio of the responses to preferred and nonpreferred sides was greater than 2, and ratios as high as 10 were not unusual. Side-of-figure selectivity was also found in V1, but it was in a smaller proportion of the cells. Figure 47.7 shows the time course of the population responses to preferred and nonpreferred sides of a square in the three areas. The difference appears shortly after response onset, reaching half-maximal strength around 70 ms after stimulus onset.

Inferring Depth Order

Figure–ground organization and border ownership assignment obviously relate to the task of interpreting images in terms of a three-dimensional world, specifically to the problem of resolving situations of occlusion. In this sense "figure" and "ground" correspond to foreground and background. The results illustrated in figures 47.5 and 47.6, especially the test with overlapping figures, suggest that side-of-figure selectivity relates to the task of inferring depth order from the two-dimensional configuration of contours in the image. A crucial test of this hypothesis is to examine the responses

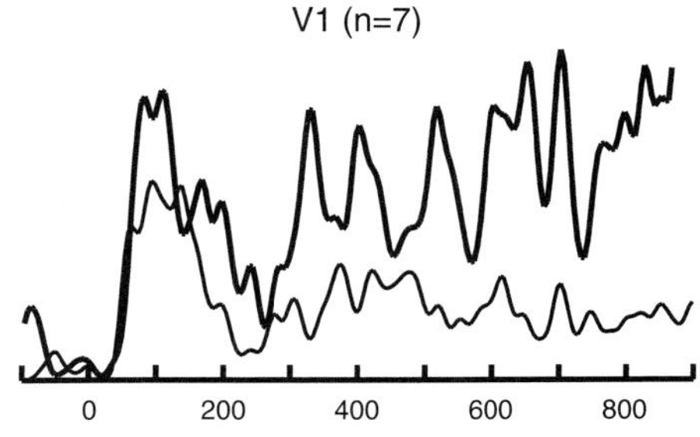
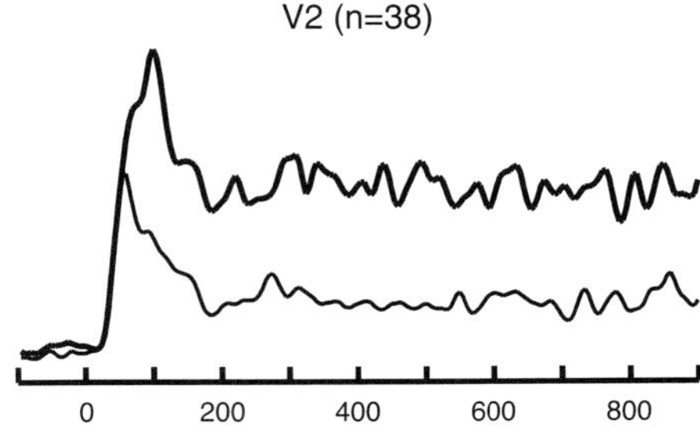
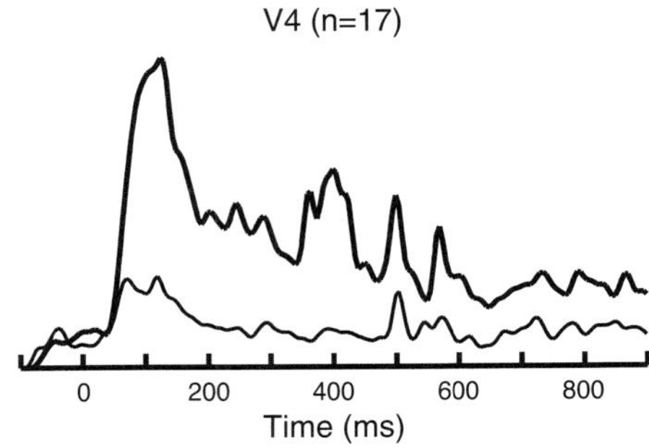

Figure 47.7   The time course of border ownership signals. The figure shows the averaged normalized responses of neurons in three cortical areas. Squares of 4° or 6° size were presented as in figure 47.5. Zero on the time scale refers to the onset of the display. Thick and thin lines represent responses to preferred and nonpreferred sides, averaged over both contrast polarities. The delay between onset of response and differentiation of side of figure was less than 25 ms. (From Zhou, Friedman, & von der Heydt, 2000.)

of these neurons with displays in which depth is defined stereoscopically. A contrast-defined square is generally perceived as a figure, and the contrast borders as its contours, but a corresponding region in a random-dot stereogram is perceived either as a figure or as a window, depending on the disparities (figure 47.4). In the stereogram the nearer surface always owns the border. Thus, the random-dot stereogram is the "gold standard" of border ownership perception.

Binocular disparity is represented extensively in the monkey visual cortex (Cumming & DeAngelis, 2001; Poggio, 1995), and cells that signal edges in random-dot stereograms exist in area V2 (von der Heydt, Zhou, & Friedman, 2000). These cells are orientation selective

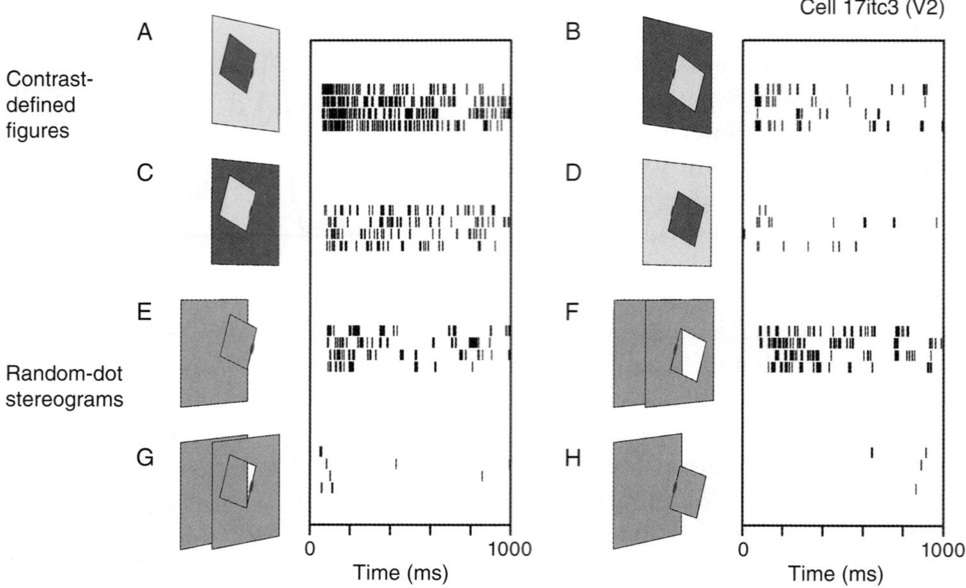

FIGURE 47.8   Inferring depth order. Comparison of side-of-figure selectivity and stereoscopic edge selectivity in a V2 neuron. (A–D) Contrast-defined figures without depth. (E, F) Stereoscopic figures without contrast. With the stereoscopic displays the neuron responded when the surface to the left of the receptive field was in front of a background surface (E, F) but not to edges of the reversed depth order (G, H) or flat surfaces at any depth. With contrast displays the neuron responded more strongly when the figure region was to the left of the receptive field (A, C) compared to the right (B, D). This shows that contrast figures are interpreted as objects in front of a background. (From Qiu & von der Heydt, 2005.)

and respond to disparity-defined edges as well as to contrast borders. Most of them are selective for the depth order of the stereoscopic edge, responding, for example, to a vertical edge if the front surface is on the right side but not if the front surface is on the left side. If the side-of-figure–selective cells in V2 represent border ownership, then these neurons should also be selective for stereoscopic depth order, and the side-of-figure preference should agree with the preferred depth order. That is, the preferred figure side should be the "near" side of the preferred step edge. A test of this hypothesis in a V2 neuron is illustrated in figure 47.8. With the random-dot stereograms, the cell is activated by the right edge of the figure and the left edge of the window (E, F) but not by the opposite edges (G, H). Thus, the cell responds when the surface to the left of its receptive field owns the border. Therefore, the responses to the contrast-defined square (A and C stronger than B and D) show that the cell "correctly" assigns the border to the square, suggesting that the cortex interprets the square as an object occluding a background. The vast majority of neurons that were selective for side-of-figure and stereo edge polarity showed this combination of preferences (the "object" interpretation of the square), and the opposite combination (the "window" interpretation of the square) was rare (Qiu & von der Heydt, 2005).

The influence of stereoscopic edge polarity as just described is consistent with the idea that border ownership–selective neurons signal the depth ordering at the local contrast border. Orientation-selective neurons signal the borders between image regions, and the differential activity in pairs of neurons with opposite border ownership preference indicates which region is in front and which is in back (figure 47.9). This is the familiar opponent coding scheme used by the nervous system in many instances, such as for coding light and dark or direction of motion. Opponent coding also allows multiplexing of dimensions, as many border ownership–selective cells are also selective for the direction of color or luminance contrast at the border (Zhou, Friedman, & von der Heydt, 2000). This may explain why color and depth order interact in perception (von der Heydt & Pierson, 2006). One specific prediction from this kind of coding is the possibility of selective adaptation. Indeed, by use of the adaptation paradigm, it was possible to demonstrate the existence of border ownership–selective neurons in the human visual cortex, both psychophysically (von der Heydt, Macuda, & Qiu, 2005) and by functional magnetic resonance imaging (Fang, Boyaci, & Kersten, 2009). By coding the depth ordering of surfaces, the system can represent the occlusion structure of three-dimensional scenes of any complexity.

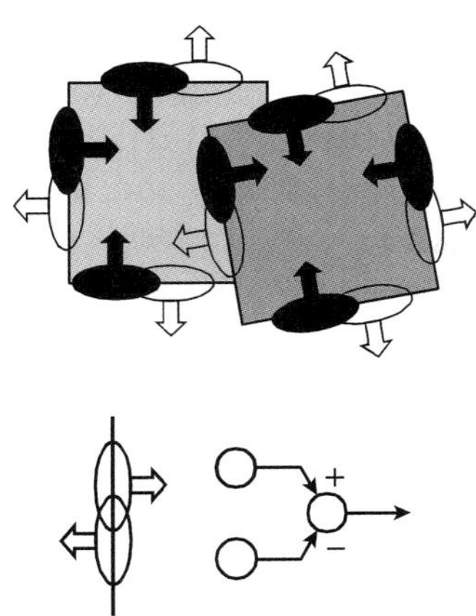

FIGURE 47.9   Schematic illustration of the cortical representation of a pair of overlapping squares. Ellipses indicate receptive fields. Each piece of contour is represented by two pools of orientation-selective neurons with opposite border ownership preference as indicated by arrows. Filled symbols indicate the neurons whose activity would be enhanced for this stimulus. Border ownership is thought to be encoded in the relative activation of the two pools.

Models for Border Ownership Assignment

Before it was discovered in neurons, border ownership coding was proposed in computational models of figure–ground organization (Finkel & Sajda, 1992; Sajda & Finkel, 1995). Other figure–ground organization models make depth stratification explicit rather than border ownership (e.g., Domijan & Setic, 2008; Grossberg, 1994). The existence of neurons with fixed border ownership preference was not anticipated. Recently, a number of models have been proposed to explain the neurophysiological finding of border ownership selectivity (Baek & Sajda, 2005; Craft et al., 2007; Jehee, Lamme, & Roelfsema, 2007; Kikuchi & Akashi, 2001; Kogo et al., 2010; Sakai & Nishimura, 2006; Zhaoping, 2005).

There are two important constraints on models of border ownership selectivity, the wide range of context integration (figure 47.6) and the short latency (figure 47.7). V1 and V2 are large, in fact, the two largest areas in the cerebral cortex of macaques and humans (Adams, Sincich, & Horton, 2007; Felleman & Van Essen, 1991). Because they are retinotopically organized, the relevant context information in the border ownership tests is typically represented far from the recorded neuron. There are two ways to spread information across the cortex, lateral propagation within the area and feedback loops through another area. Because lateral propagation relies on intracortical "horizontal" fibers, which conduct slowly, delays are inevitable, and the expected delays increase with the distance to be bridged. In

contrast, feedback loops use white-matter fibers, which conduct an order of magnitude faster than horizontal fibers (Girard, Hupe, & Bullier, 2001). Therefore, in the visual cortex a loop through another area with large receptive fields may propagate the contextual information faster than lateral propagation (Angelucci et al., 2002; Bullier, 2001). This important insight is often overlooked when the latency argument is used in discussing the flow of information in the cortex.

Among the models that consider neurophysiological implementations, some are based on lateral signal propagation (e.g., Baek & Sajda, 2005; Kogo et al., 2010; Zhaoping, 2005) and others on feedback loops through higher-level areas (e.g., Craft et al., 2007; Jehee, Lamme, & Roelfsema, 2007). Measurements of the latency of border ownership signals and calculation of the cortical distances indicate that the lateral propagation scheme cannot explain the neurophysiological results, whereas feedback (e.g., from other prestriate areas) appears to be a plausible way to achieve the fast context integration (Sugihara, Qiu, & von der Heydt, 2011; Zhang & von der Heydt, 2010).

It is important to distinguish the feedback scheme from what is generally referred to as top-down processing, that is, processing that involves long-term memory or volitional deployment of attention. It is often argued, rightfully, that it would take considerable time for such processes to develop and influence signals in the visual cortex. Clearly, object memory, as represented in IT cortex, cannot be involved in generating the border ownership signals, at least in the initial phase, because the signals emerge (Sugihara, Qiu, & von der Heydt, 2011) before IT neurons even begin to respond (Brincat & Connor, 2006; Bullier, 2001). A role of volitional attention in generating these signals has also been ruled out by specific experiments (Qiu, Sugihara, & von der Heydt, 2007). The feedback in the models (Craft et al., 2007; Jehee, Lamme, & Roelfsema, 2007) is thought to be stimulus driven, involving neither attention nor memory.

EMERGING OBJECT REPRESENTATIONS

Mechanisms that infer depth ordering and add figure–ground pointers to the cortical contour map seem like a plausible explanation for the observation of border ownership selectivity that is so common in V2. However, the phenomenon of figure–ground organization, as described by the Gestalt psychologists (Koffka, 1935; Rubin, 1915), implies more than just depth ordering. Figure regions have a distinct quality: Shapes of figures are easily recognized, whereas the shapes of ground regions are hard to see; figures attract attention, but the ground does not. Figure–ground perception seems to reveal something fundamental, a process that defines what is an object even before attention comes into play and before something is recognized. Several recent findings indicate that border ownership selectivity reflects this fundamental process.

One observation is that border ownership signals reflect the perceptual object interpretation also in situations where it is not dictated by depth ordering. Qiu and von der Heydt (2007) used a configuration that looks like two crossed bars, one light and one dark, in transparent overlay (figure 47.10, inset b) (for perceptual and theoretical aspects of transparent overlay see Adelson, 1993). The stimulus (b) actually consists of two light and two dark squares on a medium gray background. Qiu and von der Heydt found that when only one of the squares is presented the border ownership signals at the edges point to the interior of the square (a), but when all four squares are presented, the signals on the edges bounding the region of apparent overlay are reversed, pointing to the interior of the perceived bars (b). When the corners of the squares are rounded

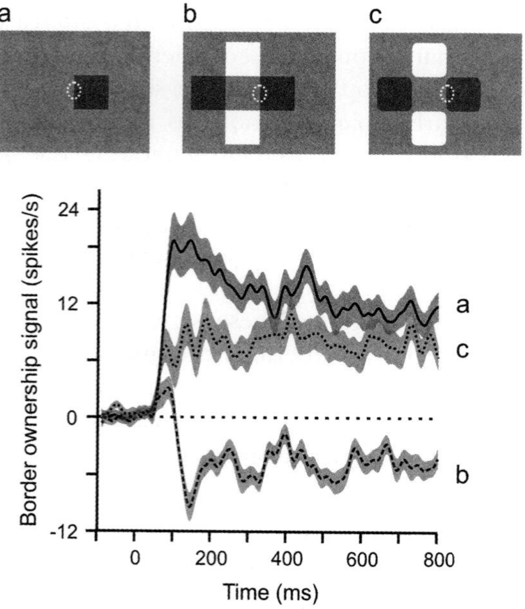

FIGURE 47.10 Representing object structure. An isolated dark square "owns" its borders (a), but when two white and a dark square are added, a cross of two bars in transparent overlay is perceived, and one edge of the original square is now owned by the vertical bar (b). When the corners of the squares are rounded, four individual objects are perceived, and border ownership returns to the dark square (c). The neural signals show the corresponding reversals. Curves show the time course of the mean differential firing rates of a population of V2 neurons; shading indicates ±1SEM. (Modified from Qiu & von der Heydt, 2007.)

off, these signals point again to the interior of the square (c). Thus, the context leads to a reorganization in the perceptual object interpretation, and the same reorganization is observed in the border ownership signals.

Another indication that border ownership signals relate to object coding is their persistence: Generally, responses in the visual cortex rise and decay quickly, within a few tens of milliseconds, and border ownership signals emerge within less than 30 ms after response onset (figure 47.7). But when an ambiguous edge is substituted for the edge of a figure, the border ownership signals persist for a second or more (O'Herron & von der Heydt, 2009). This long persistence is reminiscent of the slow rates of perceptual alternation observed in ambiguous displays. The persistence is not a result of the ambiguous situation (the absence of figure–ground cues): O'Herron and von der Heydt (2011) created displays in which one object is moved relative to another, thereby reversing the figure–ground cues at the border between the objects while preserving their identities. They found that the border ownership signals persisted despite the reversal of figure–ground cues. Thus, persistence seems to reflect continuity of object representations.

A third group of observations relates to the role of attention. In situations where several objects are displayed and the subject (monkey) needs to select one of them for a shape discrimination task, the selective attention modulates the firing rate in many of the border ownership–selective neurons (Qiu, Sugihara, & von der Heydt, 2007). However, the border ownership signals emerge at the attended as well as the ignored objects. This shows that border ownership modulation is not a product of attention but reflects autonomous, preattentive mechanisms. When the task involves one of two overlapping objects, the responses to the occluding edge are suppressed when the object in back is attended compared to the object in front, in agreement with the perceptual phenomenon of "extrinsic edge suppression" (Nakayama, Shimojo, & Silverman, 1989). A spatial attention filter (spotlight model) would produce the opposite, namely a relative enhancement, because when the background object was attended, the occluding edge was closer to the focus of attention than when the foreground object was attended. Thus, the observation of extrinsic edge suppression in neurons (Qiu, Sugihara, & von der Heydt, 2007) reveals mechanisms of object-based attention. Responses to the border between two overlapping figures reveal an asymmetry of the attention influence: In each neuron, attention on one side of the receptive field produces enhancement relative to attention on the other side. The side of enhancement tends to be the same as the preferred side of border ownership.

This asymmetry and the extrinsic edge suppression can be understood easily within the feedback model by Craft et al. (2007). This model introduced the principle of grouping cells that sum the responses from edge cells in multiple locations and, by feedback, set the gain of the same edge cells. Each of them communicates unilaterally with grouping cells on one side of its receptive field, which makes it side-of-figure selective. Mihalas et al. (2011) showed that a purely spatial attention field propagating backward in this network will be reshaped, repositioned, and sharpened to match the object's shape and scale. As yet, the existence of grouping cells is hypothetical.

CONCLUDING COMMENTS

In the three sections of this chapter I discussed different aspects of contour representation in the visual cortex. Illusory contour mechanisms might serve to make contours explicit and "fill in the gaps." Border ownership mechanisms seem to make the three-dimensional structure of a scene explicit by representing occluding contours and showing how they belong to surfaces. Finally, border ownership signals might be related to the emergence of object representations.

Representing occluding contours is perhaps the most important first step on the way from image to object representation because it specifies the assignment of borders to regions and the depth ordering of surfaces. The convergence of stereoscopic and shape-based border assignment mechanisms may explain the striking effects of binocular disparity on the perception of surfaces (Gregory & Harris, 1974; Idesawa, 1991; Nakayama, Shimojo, & Silverman, 1989; Shimojo, Silverman, & Nakayama, 1989). A number of elegant studies have demonstrated the influence of stereoscopic depth on cortical neuron responses with paradigms that reveal the relation between neuronal and perceptual effects: figure–ground modulation of surface responses in V1 (Zipser, Lamme, & Schiller, 1996), illusory contour responses in V2 (Bakin, Nakayama, & Gilbert, 2000; Heider, Spillmann, & Peterhans, 2002), and the integration of motion signals in area MT (Duncan, Albright, & Stoner, 2000). In general, the neural representation was consistent with the figure (foreground object) dominating, but a representation of occluded object parts (amodal completion) has also been described (Sugita, 1999).

As mentioned, illusory contour cells can also be border ownership selective. Perceptual demonstrations often conflate the two phenomena. Some models try to account for both by the same mechanism (e.g., Kogo et al., 2010). However, little is known about how the neural mechanisms underlying the two phenomena are related. One striking difference is the range of context integration. Illusory contour responses decline rapidly when the gap between the inducing elements is increased beyond a few degrees (Peterhans & von der Heydt, 1989), whereas side-of-figure selectivity shows context influences over distances of 10° and more (figure 47.5; Zhang & von der Heydt, 2010; Zhou, Friedman, & von der Heydt, 2000). This discrepancy suggests that there are two stages of processing, a semilocal mechanism for contour completion followed by global shape processing for border ownership. Psychophysical measurements of the span of illusory contours (Ringach & Shapley, 1996) might involve both stages.

A neural correlate of figure–ground segregation was first discovered by Lamme (1995) in the form of enhancement of responses over the region of a figure compared to the ground. This figure enhancement, found in V1, emerges somewhat later than the border ownership signals in V2 (Zhou, Friedman, & von der Heydt, 2000). How these two findings are related has not been clarified. Lamme's paradigm compares texture-evoked responses (in most cortical neurons uniform regions do not evoke surface responses; Friedman, Zhou, & von der Heydt, 2003). It has been studied in V1 and V4 (Poort et al., 2012), mainly with multiunit recording. Border ownership coding can be found in V1 as well as V2, and the small proportion of V1 cells that are border ownership selective show the same short signal latencies and broad context integration as their counterparts in V2 (Sugihara, Qiu, & von der Heydt, 2011; Zhang & von der Heydt, 2010).

The recent findings summarized in the third section indicate that border ownership coding implies more than representation of occluding contours and depth order. Border ownership signals reflect perceptual object interpretation even in situations where depth order is irrelevant; they vary under the influence of attention in a way that is consistent with object-based attention; and they show persistence—a surprising observation in the visual cortex. These new findings pose many new questions. What is the mechanism of persistence? If border ownership signals are related to object representation, as suggested, do they transfer across cortex when an object moves? Do they persist across eye movements? Perhaps further research along these lines will bring us closer to understanding how the visual system defines what is an object, how it links elemental features to objects, how it maintains these links over time, and how object-based attention works.

ACKNOWLEDGMENTS

Preparation of this chapter was supported by NIH grants EY02966 and EY016281.

REFERENCES

Adams, D. L., Sincich, L. C., & Horton, J. C. (2007). Complete pattern of ocular dominance columns in human primary visual cortex. *Journal of Neuroscience, 27,* 10391–10403.

Adelson, E. H. (1993). Perceptual organization and the judgment of brightness. *Science, 262,* 2042–2044.

Angelucci, A., Levitt, J. B., Walton, E. J., Hupe, J. M., Bullier, J., & Lund, J. S. (2002). Circuits for local and global signal integration in primary visual cortex. *Journal of Neuroscience, 22,* 8633–8646.

Attneave, F. (1954). Some informational aspects of visual perception. *Psychological Review, 61,* 183–193.

Baek, K., & Sajda, P. (2005). Inferring figure-ground using a recurrent integrate-and-fire neural circuit. *IEEE Transactions on Neural Systems and Rehabilitation Engineering, 13,* 125–130.

Bakin, J. S., Nakayama, K., & Gilbert, C. D. (2000). Visual responses in monkey areas V1 and V2 to three-dimensional surface configurations. *Journal of Neuroscience, 20,* 8188–8198.

Barlow, H. B. (1961). Possible principles underlying the transformations of sensory messages. In W. A. Rosenblith (Ed.), *Sensory communication* (pp. 217–257). Cambridge, MA: MIT Press.

Baumann, R., van der Zwan, R., & Peterhans, E. (1997). Figure-ground segregation at contours: A neural mechanism in the visual cortex of the alert monkey. *European Journal of Neuroscience, 9,* 1290–1303.

Bravo, M., Blake, R., & Morrison, S. (1988). Cats see subjective contours. *Vision Research, 28,* 861–865.

Brincat, S. L., & Connor, C. E. (2006). Dynamic shape synthesis in posterior inferotemporal cortex. *Neuron, 49,* 17–24.

Bullier, J. (2001). Integrated model of visual processing. *Brain Research. Brain Research Reviews, 36,* 96–107.

Burkhalter, A., & Van Essen, D. C. (1986). Processing of color, form and disparity information in visual areas VP and V2 of ventral extrastriate cortex in the macaque monkey. *Journal of Neuroscience, 6,* 2327–2351.

Coren, S. (1972). Subjective contours and apparent depth. *Psychological Review, 79,* 359–367.

Craft, E., Schuetze, H., Niebur, E., & von der Heydt, R. (2007). A neural model of figure-ground organization. *Journal of Neurophysiology, 97,* 4310–4326.

Cumming, B. G., & DeAngelis, G. C. (2001). The physiology of stereopsis. *Annual Review of Neuroscience, 24,* 203–238.

Desimone, R., Schein, S. J., Moran, J., & Ungerleider, L. G. (1985). Contour, color and shape analysis beyond the striate cortex. *Vision Research, 25,* 441–452.

De Weerd, P., Vandenbussche, E., Debruyn, B., & Orban, G. A. (1990). Illusory contour orientation discrimination in the cat. *Behavioural Brain Research, 39,* 1–17. doi:10.1016/0166-4328(90)90117-W.

Domijan, D., & Setic, M. (2008). A feedback model of figure-ground assignment. *Journal of Vision, 8,* 1–27. doi:10.1167/8.7.10.

Duncan, R. O., Albright, T. D., & Stoner, G. R. (2000). Occlusion and the interpretation of visual motion: Perceptual and neuronal effects of context. *Journal of Neuroscience, 20,* 5885–5897.

Fang, F., Boyaci, H., & Kersten, D. (2009). Border ownership selectivity in human early visual cortex and its modulation by attention. *Journal of Neuroscience, 29,* 460–465.

Felleman, D. J., & Van Essen, D. C. (1991). Distributed hierarchical processing in the primate cerebral cortex. *Cerebral Cortex, 1,* 1–47. doi:10.1093/cercor/1.1.1-a.

Finkel, L. H., & Sajda, P. (1992). Object discrimination based on depth-from-occlusion. *Neural Computation, 4,* 901–921. doi:10.1162/neco.1992.4.6.901.

Friedman, H. S., Zhou, H., & von der Heydt, R. (2003). The coding of uniform color figures in monkey visual cortex. *Journal of Physiology, 548,* 593–613.

Girard, P., Hupe, J. M., & Bullier, J. (2001). Feedforward and feedback connections between areas V1 and V2 of the monkey have similar rapid conduction velocities. *Journal of Neurophysiology, 85,* 1328–1331.

Gregory, R. L. (1972). Cognitive contours. *Nature, 238,* 51–52.

Gregory, R. L., & Harris, J. P. (1974). Illusory contours and stereo depth. *Percept. Psychophysics, 15,* 411–416.

Grosof, D. H., Shapley, R. M., & Hawken, M. J. (1993). Macaque-V1 neurons can signal illusory contours. *Nature, 365,* 550–552.

Grossberg, S. (1994). 3-D vision and figure-ground separation by visual cortex. *Perception & Psychophysics, 55,* 48–120.

Grossberg, S., & Mingolla, E. (1985). Neural dynamics of form perception: Boundary completion, illusory figures, and neon color spreading. *Psychological Review, 92,* 173–211.

Heider, B., Spillmann, L., & Peterhans, E. (2002). Stereoscopic illusory contours—cortical neuron responses and human perception. *Journal of Cognitive Neuroscience, 14,* 1018–1029.

Heitger, F., von der Heydt, R., Peterhans, E., Rosenthaler, L., & Kübler, O. (1998). Simulation of neural contour mechanisms: Representing anomalous contours. *Image and Vision Computing, 16,* 409–423.

Helmholtz, H. v. (1866). *Handbuch der physiologischen Optik.* Hamburg: Voss.

Idesawa, M. (1991). Perception of 3-D illusory surface with binocular viewing. *Japanese Journal of Applied Physics, 30,* 751–754. doi:10.1143/JJAP.30.L751.

Jehee, J. F., Lamme, V. A., & Roelfsema, P. R. (2007). Boundary assignment in a recurrent network architecture. *Vision Research, 47,* 1153–1165.

Kanizsa, G. (1979). *Organization in vision. Essays on Gestalt perception.* New York: Praeger.

Kikuchi, M., & Akashi, Y. (2001). A model of border-ownership coding in early vision. In G. Dorffner, H. Bischof, & K. Hornik (Eds), *Artificial neural networks—ICANN 2001, lecture notes in computer science* (pp. 1069–1074). Berlin: Springer.

Koffka, K. (1935). *Principles of Gestalt psychology.* New York: Harcourt, Brace & World.

Kogo, N., Strecha, C., Van Gool, L., & Wagemans, J. (2010). Surface construction by a 2-D differentiation-integration process: A neurocomputational model for perceived border ownership, depth, and lightness in Kanizsa figures. *Psychological Review, 117,* 406–439.

Lamme, V. A. F. (1995). The neurophysiology of figure-ground segregation in primary visual cortex. *Journal of Neuroscience, 15,* 1605–1615.

Lee, T. S., & Nguyen, M. (2001). Dynamics of subjective contour formation in the early visual cortex. *Proceedings of the National Academy of Sciences of the United States of America, 9*, 1907–1911. doi:10.1073/pnas.031579998.

Leventhal, A. G., Thompson, K. G., Liu, D., Zhou, Y., & Ault, S. J. (1995). Concomitant sensitivity to orientation, direction, and color of cells in layers 2, 3, and 4 of monkey striate cortex. *Journal of Neuroscience, 15*, 1808–1818.

Marr, D. (1982). *Vision. A computational investigation into the human representation and processing of visual information.* San Francisco: Freeman.

Mihalas, S., Dong, Y., von der Heydt, R., & Niebur, E. (2011). Mechanisms of perceptual organization provide auto-zoom and auto-localization for attention to objects. *Proceedings of the National Academy of Sciences of the United States of America, 108*, 7583–7588. doi:10.1073/pnas.1014655108.

Nakayama, K., Shimojo, S., & Silverman, G. H. (1989). Stereoscopic depth: Its relation to image segmentation, grouping, and the recognition of occluded objects. *Perception, 18*, 55–68.

Nieder, A. (2002). Seeing more than meets the eye: Processing of illusory contours in animals. *Journal of Comparative Physiology. A, Neuroethology, Sensory, Neural, and Behavioral Physiology, 188*, 249–260.

Nieder, A., & Wagner, H. (1999). Perception and neuronal coding of subjective contours in the owl. *Nature Neuroscience, 2*, 660–663.

O'Herron, P. J., & von der Heydt, R. (2009). Short-term memory for figure-ground organization in the visual cortex. *Neuron, 61*, 801–809.

O'Herron, P.J., & von der Heydt, R. (2011). Representation of object continuity in the visual cortex. *Journal of Vision, 11*(2), 12, 1–12. doi:10.1167/11.2.12.

Paradiso, M. A., Shimojo, S., & Nakayama, K. (1989). Subjective contours, tilt aftereffects, and visual cortical organization. *Vision Research, 29*, 1205–1213.

Peterhans, E., & von der Heydt, R. (1989). Mechanisms of contour perception in monkey visual cortex. II. Contours bridging gaps. *Journal of Neuroscience, 9*, 1749–1763.

Poggio, G. F. (1995). Mechanisms of stereopsis in monkey visual cortex. *Cerebral Cortex, 5*, 193–204.

Poggio, T., & Koch, C. (1985). Ill-posed problems in early vision: From computational theory to analogue networks. *Proceedings of the Royal Society of London. Series B, Biological Sciences, 226*, 303–323. doi:10.1098/rspa.1985.0124.

Poort, J., Raudies, F., Wannig, A., Lamme, V., Neumann, H., & Roelfsema, P. (2012). The role of attention in figure-ground segregation in areas V1 and V4 of the visual cortex. *Neuron, 75*, 143–156. doi:10.1016/j.neuron.2012.04.032.

Qiu, F. T., Sugihara, T., & von der Heydt, R. (2007). Figure-ground mechanisms provide structure for selective attention. *Nature Neuroscience, 10*, 1492–1499.

Qiu, F. T., & von der Heydt, R. (2005). Figure and ground in the visual cortex: V2 combines stereoscopic cues with Gestalt rules. *Neuron, 47*, 155–166.

Qiu, F. T., & von der Heydt, R. (2007). Neural representation of transparent overlay. *Nature Neuroscience, 10*, 283–284.

Ramsden, B. M., Hung, C. P., & Roe, A. W. (2001). Real and illusory contour processing in area V1 of the primate: A cortical balancing act. *Cerebral Cortex, 11*, 648–665.

Redies, C., Crook, J. M., & Creutzfeldt, O. D. (1986). Neuronal responses to borders with and without luminance gradients in cat visual cortex and dorsal lateral geniculate nucleus. *Experiments in Brain Research, 61*, 469–481.

Ringach, D. L., & Shapley, R. (1996). Spatial and temporal properties of illusory contours and amodal boundary completion. *Vision Research, 36*, 3037–3050.

Roe, A. W., Chelazzi, L., Connor, C. E., Conway, B. R., Fujita, I., Gallant, J. L., et al. (2012). Toward a unified theory of visual area V4. *Neuron, 74*, 12–29.

Rubin, E. (1915). *Synsoplevede Figurer.* Copenhagen: Gyldendals.

Sajda, P., & Finkel, L. H. (1995). Intermediate-level visual representations and the construction of surface perception. *Journal of Cognitive Neuroscience, 7*, 267–291.

Sakai, K., & Nishimura, H. (2006). Surrounding suppression and facilitation in the determination of border ownership. *Journal of Cognitive Neuroscience, 18*, 562–579.

Schein, S. J., & Desimone, R. (1990). Spectral properties of V4 neurons in the macaque. *Journal of Neuroscience, 10*, 3369–3389.

Sheth, B. R., Sharma, J., Rao, S. C., & Sur, M. (1996). Orientation maps of subjective contours in visual cortex. *Science, 274*, 2110–2115.

Shimojo, S., Silverman, G. H., & Nakayama, K. (1989). Occlusion and the solution to the aperture problem for motion. *Vision Research, 29*, 619–626.

Soriano, M., Spillmann, L., & Bach, M. (1996). The abutting grating illusion. *Vision Research, 36*, 109–116.

Spillmann, L., & Ehrenstein, W. H. (2003). Gestalt factors in the visual neurosciences. In L. M. Chalupa & J. S. Werner (Eds.), *The visual neurosciences* (pp. 1573–1589). Cambridge, MA: MIT Press.

Srinivasan, M., Lehrer, M., & Wehner, R. (1987). Bees perceive illusory contours induced by movement. *Vision Research, 27*, 1285–1290.

Sugihara, T., Qiu, F. T., & von der Heydt, R. (2011). The speed of context integration in the visual cortex. *Journal of Neurophysiology, 106*, 374–385.

Sugita, Y. (1999). Grouping of image fragments in primary visual cortex. *Nature, 401*, 269–272.

Ullman, S. (1996). *High-level vision.* Cambridge, MA: MIT Press.

von der Heydt, R., Macuda, T. J., & Qiu, F. T. (2005). Border-ownership dependent tilt aftereffect. *Journal of the Optical Society of America, 22*, 2222–2229. doi:10.1364/JOSAA.22.002222.

von der Heydt, R., & Peterhans, E. (1989). Mechanisms of contour perception in monkey visual cortex. I. Lines of pattern discontinuity. *Journal of Neuroscience, 9*, 1731–1748.

von der Heydt, R., Peterhans, E., & Baumgartner, G. (1984). Illusory contours and cortical neuron responses. *Science, 224*, 1260–1262.

von der Heydt, R., & Pierson, R. (2006). Dissociation of color and figure-ground effects in the watercolor illusion. *Spatial Vision, 19*, 323–340. doi:10.1163/156856806776923416.

von der Heydt, R., Zhou, H., & Friedman, H. S. (2000). Representation of stereoscopic edges in monkey visual cortex. *Vision Research, 40*, 1955–1967.

Wagemans, J., Elder, J. H., Kubovy, M., Palmer, S. E., Peterson, M. A., Singh, M., et al. (2012). A century of Gestalt psychology in visual perception: I. Perceptual grouping and figure-ground organization. *Psychological Bulletin, 138*, 1172–1217.

Westheimer, G., & Li, W. (1996). Classifying illusory contours by means of orientation discrimination. *Journal of Neurophysiology, 75,* 523–528.

Zeki, S. M. (1978). Uniformity and diversity of structure and function in rhesus monkey prestriate visual cortex. *Journal of Physiology, 277,* 273–290.

Zhang, N. R., & von der Heydt, R. (2010). Analysis of the context integration mechanisms underlying figure-ground organization in the visual cortex. *Journal of Neuroscience, 30,* 6482–6496.

Zhaoping, L. (2005). Border ownership from intracortical interactions in visual area V2. *Neuron, 47,* 147–153.

Zhou, H., Friedman, H. S., & von der Heydt, R. (2000). Coding of border ownership in monkey visual cortex. *Journal of Neuroscience, 20,* 6594–6611.

Zipser, K., Lamme, V. A. F., & Schiller, P. H. (1996). Contextual modulation in primary visual cortex. *Journal of Neuroscience, 16,* 7376–7389.

VII OBJECTS AND SCENES

48 Visual Crowding

DENNIS M. LEVI

In peripheral vision, objects that can be readily recognized when viewed in isolation become unrecognizable in clutter. This is the interesting phenomenon known as *visual crowding*. You can experience this by looking directly at the plus in figure 48.1a. The isolated letter N on the left is easy to read, while the same letter N in the word "LUNCH" is quite difficult despite it being the same distance from the cross, that is, at the same retinal eccentricity. Interestingly, the viewing distance does not matter much because changing the viewing distance scales both the size and spacing of the letters and their eccentricity.

It has long been known that visual acuity is reduced in peripheral vision. The threshold size for an isolated letter in peripheral vision is ≈1/50 of the eccentricity. However, for the letter to be recognized unimpaired, the distance between neighboring letters may have to be as large as half of the eccentricity. This difference between the threshold size of an object required for recognition (visual acuity) and the threshold spacing is shown in figure 48.1b. At an eccentricity of 10° features need to be separated by more than 20 times their threshold size. Crowding is not restricted to letters—it occurs with a wide range of objects including simple features such as lines and gratings and complex objects, faces, and shapes. We have known about crowding for decades. The term was first used by the Scandinavian ophthalmologist Ehlers in 1936 (H. Strasburger, personal communication), and it has been studied in waves. In the 1960s Flom, Heath, and Takahashi (1963) discovered that crowding occurs when target and flankers were presented to different eyes, suggesting a cortical locus. In the 1970s Bouma (1970, 1973) discovered that the extent of crowding is a more or less constant fraction of the target eccentricity and that crowding occurs in a variety of tasks (Andriessen & Bouma, 1976). In the 1990s He, Cavanagh, and Intrilligator (1996) demonstrated that orientation-specific adaptation is not affected by crowding, implying that the influence of crowding on spatial resolution may take place beyond the primary visual cortex, and in the last 5 years or so there has been an explosion of interest in crowding, as evidenced by the publication of more than 200 articles on the topic since the beginning of 2007. All of this effort has led to substantial progress in our understanding of the neural basis of visual crowding and to the development of new models. Despite this surge in activity, however, there remain many unknowns.

Crowding represents an essential bottleneck, setting limits on object perception, eye movements, visual search, reading, and perhaps other functions in peripheral, amblyopic, and developing vision (Whitney & Levi, 2011). It is generally defined as the deleterious influence of nearby contours on visual discrimination, but the effects of crowding go well beyond impaired discrimination. Crowding impairs the ability to recognize and respond appropriately to objects in clutter. Thus, studying crowding may lead to a better understanding of the processes involved in object recognition. Crowding also has important clinical implications for patients with macular degeneration, amblyopia, and dyslexia.

There have been several recent reviews on the topic of peripheral vision in general (Strasburger, Rentschler, & Juttner, 2011) and crowding in particular (Levi, 2008, 2011; Pelli & Tillman, 2008; Whitney & Levi, 2011), so the focus of this chapter is on providing a general account of crowding as we currently understand it, emphasizing what is known and where the field seems to be headed.

WHAT IS CROWDING?

The hallmark of crowding is reduced performance for object recognition and identification under clutter. This is not simply a result of reduced peripheral acuity because the object can be easily recognized when presented in isolation (figure 48.1), and it is not just a reduction in visibility because detection thresholds are relatively unimpaired (Levi, Hariharan, & Klein, 2002a, 2002b; Pelli, Palomares, & Majaj, 2004; Levi & Carney, 2009). Rather, crowding alters object appearance. Figure 48.1a provides the reader with an opportunity to experience crowding directly, and it is obvious that crowding does not result in reduced apparent contrast—rather, crowded objects are perceived as having high contrast but are indistinct or jumbled together.

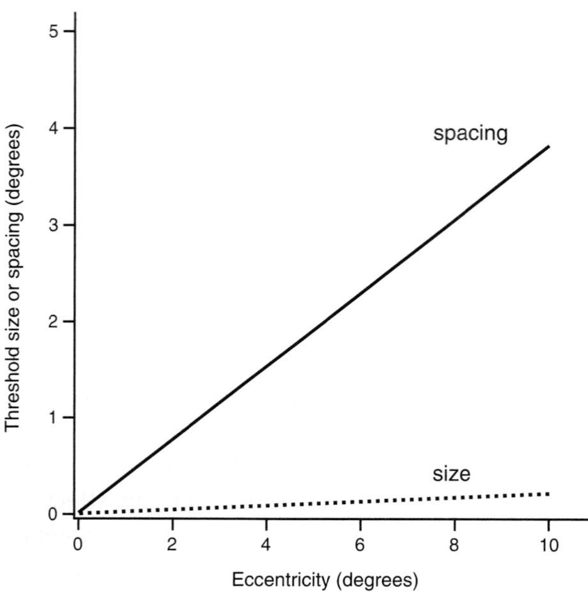

FIGURE 48.1   Visual crowding. (a) Crowding renders a target that is easily recognized in isolation unrecognizable. When one looks directly at the plus, the isolated letter N on the left is easily read, but the same letter N in the word "LUNCH" is quite difficult. (b) Threshold size (visual acuity—black dotted line) and threshold spacing (gray solid line) versus eccentricity.

How Is Crowding Measured?

Many studies have quantified crowding by using a fixed-size target that, in isolation, is correctly identified about 90% of the time. In this paradigm systematically varying the flanker distance and plotting percentage correct versus flanker distance allows the experimenter to quantify both the strength and spatial extent of crowding (Bouma, 1970; Chung, 2007; Flom, Weymouth, & Kahneman, 1963; and many others). The classical work of Bouma (1970) suggests that the spatial extent or critical spacing for crowding is proportional to eccentricity. Specifically, "for complete visual isolation of a letter presented at an eccentricity of $\phi°$, it follows that no other letters should be present (roughly) within $0.5\phi°$ distance." Bouma's observation has sometimes been afforded the status of a law; however, as discussed below, ϕ depends on a number of stimulus, task, and attentional factors (Whitney & Levi, 2011).

Another approach, used to assess the influence of the flanks on target perception, is to measure a "threshold" (e.g., orientation or contrast) for identifying the target (Chung, Levi, & Legge, 2001; Levi, Hariharan, & Klein,

2002a; Pelli, Palomares, & Majaj, 2004; Strasburger, Harvey, & Rentschler, 1991). This method enables the experimenter to independently vary target and flank size, distance, and other parameters. It is time consuming but informative for disentangling the effects of various stimulus and flanker parameters and for disentangling variations in the extent (critical spacing) and the strength (e.g., threshold elevation produced by flanks at a given distance) of crowding.

A third approach is to make two parametric measurements of target size threshold (e.g., Levi, Song, & Pelli, 2007; Petrov & Meleschkevich, 2011a). One measure is with a flanked target, where the flank spacing is a multiple of the target size (e.g., spacing equals 1.1 times size). By itself, this measurement would confound size and spacing; however, also making a comparable measurement with an isolated target—one with infinite spacing—allows one to disentangle the two if the flanked threshold size is much larger than, and thus not limited by, unflanked threshold size (as it is in peripheral vision and in the central field of strabismic amblyopes). This method has some practical advantages. It saves time by reducing the two-dimensional space of size versus spacing to just one dimension, spacing equals 1.1 times size. Moreover, it allows one to use the same procedure for isolated and flanked targets. It enables measurement of small critical spacings without overlapping the targets. For letter targets it allows testing with normal text spacing (i.e., the spacing used in normal print is about 1.1 times the letter size) over the whole range, reinforcing its relevance to reading. The cost of covariation is that it confounds the variables. A number of studies have suggested that peripheral and amblyopic letter identification is limited by the center-to-center spacing, *not* the letter size (Hariharan, Levi, & Klein, 2005; Levi, Hariharan, & Klein, 2002a, 2002b; Pelli, Palomares, & Majaj, 2004; Pelli & Tillman, 2008; Strasburger, Rentschler, & Juttner, 1991; Tripathy & Cavanagh, 2002). However, it should be noted that the maximum threshold elevation (for a particular task) increases with eccentricity and that the "critical spacing" is not strictly proportional to eccentricity (Gurnsey, Roddy, & Chanab, 2011).

What Are the Conditions under Which Crowding Takes Place?

CROWDING IS NEARLY UBIQUITOUS IN PERIPHERAL VISION   In peripheral vision, crowding can be seen in a wide variety of tasks, including letter recognition (Bouma, 1970; Flom, Weymouth, & Kahneman, 1963; Toet & Levi, 1992; see figure 48.1), Vernier acuity (Levi, Klein, & Aitsebaomo, 1985; Westheimer & Hauske,

1975), orientation discrimination (Andriessen & Bouma, 1976; Westheimer, Shimamura, & McKee, 1976), stereoacuity (Butler & Westheimer, 1978), object recognition (Wallace & Tjan, 2011), face recognition (Louie, Bressler, & Whitney, 2007; Martelli, Majaj, & Pelli, 2005), and Mooney (two-tone) face recognition. Crowding occurs for chromatic stimuli with equiluminant backgrounds, with similar extents to crowding in the luminance domain (Tripathy & Cavanagh, 2002). Thus, crowding is not simply a property of "luminance channels." Additionally, crowding also occurs for moving stimuli (Bex & Dakin, 2005; Bex, Dakin, & Simmers, 2003). However, there are notable exceptions. For example, there is little or no effect of crowding on simple detection of a target (Andriessen & Bouma, 1976; Levi, Hariharan, & Klein, 2002a; Livne & Sagi, 2007; Pelli, Palomares, & Majaj, 2004).

In foveal vision, crowding typically only occurs over very small distances (4–6 arc minutes, e.g., Flom, Weymouth, & Kahneman,1963 ; Liu & Arditi, 2000; Toet & Levi, 1992) or is reported not to occur at all (Strasburger, Harvey, & Rentschler, 1991), and there is still some question as to whether genuine crowding actually occurs in the fovea (Levi, 2008). In contrast, crowding in peripheral vision occurs over very large distances (up to 0.5 times the eccentricity of the target—Bouma, 1970; Kooi et al., 1994; Toet & Levi, 1992) where the retinal point spread functions of the target and flanks are clearly separate. Extensive crowding also occurs in the central visual field of strabismic amblyopes.

CROWDING IS "TUNED" TO TARGET/FLANKER SIMILARITY Crowding is selective along a number of dimensions. It is strongest and most extensive when target and flankers are similar. These dimensions include shape and size (Kooi et al., 1994; Nazir, 1992), orientation (Andriessen & Bouma, 1976; Hariharan, Levi, & Klein, 2005; Levi & Carney, 2009; Levi, Hariharan, & Klein, 2002b; see also Livne & Sagi, 2011), spatial frequency (Chung, Levi, & Legge, 2001), depth (Kooi et al., 1994), color (Kooi et al., 1994, in most but not all observers), and motion (i.e., motion-defined targets do not interact with luminance-defined flanks in a Vernier task; Banton & Levi, 1993). For crowded letters the error rate also increases with target flanker similarity (Bernard & Chung, 2011).

Tuning along many of these dimensions would be expected based on low-level considerations and also on the basis of grouping, but the story is made a bit more complex by the details. For example, for discriminating the orientation of a near-threshold T (Kooi et al., 1994) or Landolt C (Hess, Dakin, & Kapoor, 2000; Hess et al.,

2000), crowding is both stronger and more extensive when target and flankers are the same contrast polarity (e.g., both black or both white) than when they are opposite polarity (e.g., target black, flankers white). This strong polarity tuning has been an important piece of evidence used to support both low-level and high-level explanations for crowding. However, when the stimuli are larger than the resolution limit, the polarity tuning evaporates, and crowding is similar in strength and magnitude when target and flankers are the same polarity or the opposite polarity, at least in foveal vision (Ehrt et al., 2003; Hariharan, Levi, & Klein, 2005; Hess, Williams, & Chaudhry, 2001).

The polarity selectivity also depends on the temporal properties of the stimulus. Polarity tuning is strong at low temporal frequencies but is absent beyond 6–8 Hz (Chakravarthi & Cavanagh, 2007). Interestingly, Chakravarthi and Cavanagh found no polarity effect at any temporal frequency for lateral masking of target detection.

An exception to the similarity rule is target–flanker contrast. Crowding is not tuned to the target contrast; rather, the strength of crowding depends monotonically on the flank contrast (Chung, Levi, & Legge, 2001; Levi & Carney, 2009) rather than being strongest when target and flank contrasts are the same. Importantly, at any target-to-flank spacing, the threshold and saturation contrasts of the flanks to affect the signal are the same (Pelli, Palomares, & Majaj, 2004). Similarly, the threshold contrast of the flanks is independent of the target contrast (Levi & Carney, 2009, 2011). This threshold-versus-contrast (TvC) function is quite different from the usual TvC function for masking. Based on the spacing independence of the threshold and saturation contrasts, Pelli, Palomares, and Majaj conclude that "the fixed threshold and saturation contrasts of the mask are determined not at the variously distant sensor that detects the signal, but, instead, at a sensor local to the mask. In other words, the effects of signal and mask are mediated by separate feature detections."

Another case in which tuning fails relates to the properties or order that define the target and flanks. Although there is no interaction between motion-defined targets and luminance-defined flankers for a Vernier task (Banton & Levi, 1993), there is substantial crossover crowding between static targets and flanks that are of different order types. Specifically, Chung, Li, and Levi (2007) reported substantial interaction between first- and second-order targets and flanks, suggesting that the processing of these stimuli is not independent at the stage of processing at which crowding occurs.

As noted above, crowded objects do not simply disappear. Tyler and Likova (2007) noted their strong subjective impression of a crowded letter as a "gray, or inchoate, smudge between the two outer letters, including the inner parts of those letters" (see their figure 2). The reader can draw his or her own impressions from figure 48.1a; however, it is clear that some target information survives crowding, and this can provide some clues about the nature of crowding. For example, observers can easily detect the appearance of a feature under conditions that render identification or discrimination of a change in the feature impossible (Levi & Carney, 2009, 2011; Levi, Hariharan, & Klein, 2002a, 2002b; Pelli, Palomares, & Majaj, 2004).

Although observers are unable to correctly report the orientation of an individual patch under conditions of crowding, they can reliably report the average orientation, suggesting that the local orientation signals are combined rather than lost (Parkes et al., 2001). Their results led to the widely held notion that crowded signals undergo a form of compulsory pooling or averaging of signals rather than being lost or suppressed. This inappropriate pooling or ensemble processing of the target and flanker signals has now been demonstrated under a variety of different conditions (e.g., Dakin et al., 2010; Greenwood, Bex, & Dakin, 2009; Levi & Carney, 2009) and forms the basis of the "faulty integration" theory. To the extent that peripheral vision is based on statistical inference (Balas, Nakano, & Rosenholtz, 2009; Dakin et al., 2010; van den Berg, Roerdink, & Cornelissen, 2010), crowding may simply be a regularization process that results in a consistent appearance (a uniform texture) among neighboring objects.

Observers frequently mistakenly report a flanker rather than the target in crowded displays. Whether this reflects positional uncertainty (i.e., the observer confuses the position of the flanker with that of the target) or simply the fact that the observer has to report something, and if unable to see the crowded target, the observer simply reports what he or she could see (a flanker), is not clear. However, when an observer is required to report all the letters (i.e., give a full report of target and flanker letters), the proportion of correct "target" responses is much higher when the correct order (position) is not required (Eriksen & Rohrbaugh, 1970; Popple & Levi, 2005; Strasburger, 2005). That is, given a crowded display BTH, an observer may respond "BHT." Recent work suggests that mislocation errors increase with increasing target–flanker similarity (Bernard & Chung, 2011). They argue that "common features shared by a target and a flanker will serve as anchors for other features to be combined. A pair of letters that do not share any common features will not have any anchoring features, and thus, even in the presence of feature mislocation, perceptual errors will not occur." Clearly some information about the target (or the features comprising the target) is preserved, but the location information may be lost or uncertain.

Importantly, crowded targets "look" like flankers. In an elegant series of experiments, Greenwood, Bex, and Dakin (2010) showed that crowded objects assume characteristics of the flankers. Specifically, they showed that crowding can induce an orientation-specific change in the target representation. They conclude, in agreement with Levi and Carney (2009), that crowding is a regularization process that simplifies the appearance of the peripheral array by promoting consistent appearance among adjacent objects.

The results of Greenwood, Bex, and Dakin are consistent with a number of earlier studies suggesting that information about crowded objects is not lost. For example, Parkes et al. (2001) concluded that crowding reflects compulsory averaging of signals (but see Baldassi, Megna, & Burr, 2006; Livne & Sagi, 2007) and that crowding is the term we use to define texture perception "when we do not wish it to occur." Under conditions of crowding, orientation perception is characterized by strong perceptual assimilation (i.e., the flanker orientation captures the target) near the target and perceptual repulsion (i.e., "anticrowding") farther from the target (Felisberti, Solomon, & Morgan, 2005; Mareschal, Morgan, & Solomon, 2010; Song & Levi, 2010). Assimilation would be useful in regularizing perception of the peripheral array, whereas repulsion would be most useful in highlighting salient differences among visual signals (making different stimuli "pop out"). Assimilation and repulsion reflect opponent influences on orientation perception, and a recent study (Mareschal, Morgan, & Solomon, 2010) suggests that the switch from assimilation (crowding) to repulsion depends on cortical distance.

Is Crowding Compulsory?

There are certain conditions under which crowding is reduced or eliminated completely. For example, as noted above, when targets and flankers are dissimilar, they "ungroup," and the target pops out. Thus, crowding is reduced when targets and flankers are dissimilar in shape and size (Kooi et al., 1994; Nazir, 1992), orientation (Andriessen & Bouma, 1976; Hariharan, Levi, & Klein, 2005; Levi, Hariharan, & Klein, 2002b), polarity (Chakravarthi & Cavanagh, 2007; Hess, Dakin, & Kapoor, 2000; Hess et al., 2000; Kooi et al., 1994), spatial frequency (Chung, Levi, & Legge, 2001), depth (Kooi et al., 1994), color (Kooi et al., 1994), and motion (Banton & Levi, 1993).

Similarly, with multielement flankers, when the flankers group together separately from the target, crowding may be reduced (Banks, Larson, & Prinzmetal, 1979; Banks & White, 1984; Estes, Allmeyer, & Reder, 1976; Levi & Carney, 2009; Livne & Sagi, 2007, 2010; Malania, Herzog, & Westheimer, 2007; Põder, 2006; Saarela et al., 2009; Saarela, Westheimer, & Herzog, 2010). Thus, when target and flanker look like a regular texture, crowding is strong, whereas when the target appears distinct from the flankers, crowding is weak or absent (Saarela, Westheimer, & Herzog, 2010). Indeed, observations of this type have led to the suggestion that "crowding is grouping" (Rosen, Chakravarthi, & Pelli, 2011).

There are also configural effects in crowding of high-level representations of objects. For example, recognition of an upright target face is more strongly impaired when surrounded by a crowd of nearby upright faces than by a crowd of inverted faces. This high-level configuration effect only occurred in peripheral vision and was not found for nonface objects (Louie, Bressler, & Whitney, 2007). The inversion effect is not caused by similarity or grouping and suggests that crowding may operate redundantly at multiple stages (Whitney & Levi, 2011).

Detection of a change, when cued, may escape crowding. Freeman and Pelli (2007) showed that closely spaced flankers impair uncued change detection performance while they have no effect on cued change detection. The cue to the target location eliminated the effects of crowding. Although this finding is consistent with a high-level explanation, Freeman and Pelli argue for a bottom-up account. Specifically, they suggest that crowded letters look less familiar, so we must use longer internal descriptions to remember them. Thus, fewer fit into working memory. However, in the cued condition the observer need remember only the cued letter, so the memory limit does not apply. It is difficult to see how this explanation could apply to Yeshurun and Rashal's (2010) finding that cuing can reduce the critical spacing in an identification task.

Crowding may be "released" when the flankers are masked (Chakravarthi & Cavanagh, 2009). However, this release occurred only with noise and metacontrast masks but not with object substitution masks. Chakravarthi and Cavanagh argue that noise and metacontrast masks act early in the visual processing cascade, degrading the features, whereas object substitution masks do not interfere with feature encoding but act much later by replacing the representation of the stimulus.

Crowding is also "released" when the flankers are suppressed from visual awareness. Wallis and Bex (2011) used "adaptation-induced blindness" to render flankers perceptually invisible and used a dual report paradigm in order to obtain a trial-by-trial assessment of awareness and crowding. They found that target identification was dependent on the number of flanking letters that were perceived on a given trial, independent of the number that were physically present, and concluded that crowding is "released" when the flankers are suppressed from visual awareness.

THE SIZE AND SHAPE OF CROWDING

As noted above, both the magnitude and extent of crowding increase with eccentricity (figure 48.1). Importantly, it is the *perceived* rather than the physical distance of the flankers from the target that determines the magnitude of crowding (Dakin et al., 2011; Maus, Fischer, & Whitney, 2011). However, the size and shape of the crowding zone depend on the location in the visual field, the arrangement of the stimulus array (Bouma, 1970; Feng, Jiang, & He, 2007; Pelli & Tillman, 2008; Petrov & Meleshkevich, 2011a; Petrov, Popple, & McKee, 2007; Toet & Levi, 1992), and the locus of the observer's window of attention (Petrov & Meleshkevich, 2011b). Figure 48.2 illustrates some of these asymmetries and anisotropies in the size and shape of crowding: (1) the strong radial arrangement of the crowding zones, which are elongated along lines radiating from

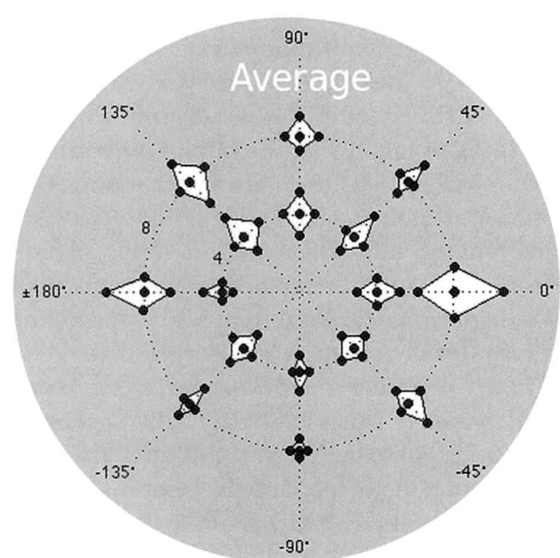

Figure 48.2 The size and shape of crowding in peripheral vision. "Average" results from Petrov and Meleshkevich (2011a) illustrate some of the asymmetries and anisotropies in the size and shape of crowding.

the fovea to the target; the extent of crowding, which is about two to three times larger in the radial direction than in the tangential direction (Toet & Levi, 1992). (2) The extent of crowding is also field dependent: It is larger in the upper visual field than in the lower field (He, Cavanagh, & Intrilligator, 1996), consistent with the lower "resolution of attention" in the upper field (the minimum spacing at which observers can select individual items) (Intriligator & Cavanagh, 2001). (3) Crowding is stronger when the target and distractors are horizontally rather than vertically arranged (Feng, Jiang, & He, 2007). (4) There is an inward–outward asymmetry (Bouma, 1970) in which crowding is stronger with a single flanker at an eccentric locus greater than the target compared to a single flanker at an eccentric locus nearer to the fovea (at the same angular separation from the target) but only along the horizontal meridian (Petrov & Meleshkevich, 2011a). Not evident in this figure are the very substantial individual differences also noted by Toet and Levi (1992).

Interestingly, outer (more peripheral) letters have *higher* recognition scores than inner (less peripheral) letters, whether in the left or right visual field (Bouma, 1973). This result is counterintuitive because one might expect that an inner letter would be more visible than an outer one (which is at a larger eccentricity) and thus be a more potent mask. Bouma proposed a form of visual interference "of a masking type, characteristic for eccentric vision and predominantly acting towards the fovea." This inner/outer asymmetry has been replicated a number of times for recognition of letters (Banks, Larson, & Prinzmetal, 1979; Bex, Dakin, & Simmers, 2003; Chastain, 1982; Krumhansl & Thomas, 1977) and for the identification of Gabor patch orientation (Petrov & Meleshkevich, 2011a, 2011b; Petrov, Popple, & McKee, 2007). Indeed, Petrov, Popple, and McKee (2007) suggest that the inward–outward anisotropy can be used as a litmus test for crowding.

Motter and Simoni (2007) suggested a very simple explanation for this asymmetry in terms of cortical geometry: "Although the angular separations for near and far flankers are the same in visual space, the far flanker is actually closer to the target than the near flanker after mapping to cortical space" (see also Pelli, 2008). However, this view has been challenged by Petrov and his colleagues on the grounds that the inward–outward asymmetry is too large to be accounted for by cortical magnification (Petrov, Popple, & McKee, 2007) and that it can be altered by attention (Petrov & Meleshkevich, 2011b—discussed further below). They argue additionally that along the vertical meridian the radial–tangential anisotropy is in the wrong direction to be explained by cortical geometry.

As noted earlier, Bouma's observation that the extent of crowding is about 0.5 $\phi°$, suggesting that the notion of a "critical spacing" that is proportional to eccentricity (ϕ) has been very influential (Pelli, 2008; Pelli & Tillman, 2008) because it has important implications. Because of the log conformal mapping of the visual world onto the retinotopic visual cortex (Schwartz, 1980), it suggests that, at any eccentricity, the critical distance represents a fixed distance on the cortex. Thus, objects can be recognized only when they are sufficiently separated on retinotopically organized cortex (Levi, Klein, & Aitsebaomo, 1985; Pelli, 2008). Pelli argues the critical spacing (Bouma's constant, $b = 0.5 \phi°$) corresponds to a cortical distance of ~6 mm in area V1. This is also the range of dichoptic interactions in the region of the cortex corresponding to the blind spot (Tripathy & Levi, 1994). Indeed, several lines of evidence suggest that crowding is determined by cortical rather than physical separation (Dakin et al., 2011; Mareschal, Morgan, & Solomon, 2010; Maus, Fischer, & Whitney, 2011), and several studies have confirmed that the critical spacing for crowding depends on target eccentricity, not target size (Levi, Hariharan, & Klein, 2002a, 2002b; Pelli, Palomares, & Majaj, 2004; Tripathy & Cavanagh, 2002; but see Gurnsey, Roddy, & Chanab, 2011).

The notion that crowding depends simply on cortical distance is simple and appealing. But the evidence shows that Bouma's constant, b, is substantially larger when the polarity, color, or shape of targets and flankers are the same than when they differ (Kooi et al., 1994); b is also larger when the complexity of targets and flankers are the same compared to when they differ (Zhang et al., 2009). Recent work also suggests that b is larger when the target and flankers are letter-like symbols than when they are letters (Grainger, Tydgat, & Issele, 2010) and that b is larger when the target location is cued versus not cued (Yeshurun & Rashal, 2010). Bouma's constant also depends on the orientation of the target and flankers and their location in the visual field (figure 48.2): b is about two to three times larger in the radial direction than in the tangential direction (Petrov & Meleshkevich, 2011b; Toet & Levi, 1992). It is also noteworthy that it is possible to shrink b through perceptual learning (Chung, 2007; Sun, Chung, & Tjan, 2010) and that attention can either engage crowding or release it in dynamic displays (Cavanagh & Holcombe, 2007).

MODELS OF CROWDING

There is no shortage of ideas about crowding, but until recently, few were computational or made specific quantitative predictions. The large number of different notions (see Levi, 2008, for a review) maybe distilled down to three basic classes: masking models, pooling models, and substitution models. Within each class many different architectures and algorithms have been proposed. Below we discuss the extant models.

Masking Models

This class of models treats crowding as a disruptive process in which information about the feature or object is suppressed. However, it is now clear that crowding does not simply suppress the target information. The substantial threshold-elevating effect of flankers on orientation discrimination with much smaller (or even opposite-signed) effects on detection (Levi & Carney, 2009, 2011; Levi, Hariharan, & Klein, 2002a, 2002b) is not consistent with a masking or disruptive model. We note that although low-level masking models cannot easily explain why detection escapes crowding, a masking-like process (e.g., contrast normalization) at a stage beyond that of feature detection could (Chung, Levi, & Legge, 2001; see Levi, 2008, and Pelli & Tillman, 2008, for reviews). However, the finding that crowding alters the appearance of the target is difficult to reconcile with crowding being a disruptive process.

Pooling Models

There are many variations of this model, based on the suggestion by Parkes et al. (2001) that signals from low-level features are pooled or integrated over an inappropriately large area by compulsory averaging. Thus, when observers are asked to report the orientation (Parkes et al., 2001) or position (Greenwood, Bex, & Dakin, 2009; Dakin et al., 2010) of a crowded target, they will often report the average orientation of the target and flanker. Other studies show that observers shown a red target letter with green flankers will sometimes report "illusory conjunctions," reporting the correct color but wrong letter, or vise-versa. These illusory conjunctions are also considered to be an outcome of crowding due to pooling (see Pelli, Palomares, & Majaj, 2004). One variant of pooling is the "faulty integration" model in which the visual system samples a limited discrete number of independent features, objects, or positions before integration (Levi & Carney, 2009). Another variant is that each feature (target and flanks) is analyzed independently by local feature detectors that are perturbed by independent noise, but only the detector shouting the loudest gets passed along to the next level. This "maximum" combination rule has been shown to account nicely for the high confidence errors that observers make in judging target tilt in a crowded display (Baldassi, Megna, & Burr, 2006).

May and Hess (2007) have argued that the integration field for crowding may actually be the "association field" proposed to mediate contour integration (Field, Hayes, & Hess, 1993). Specifically, the association field integrates information across neighboring first-stage filters tuned to similar orientations and is thought to be involved in texture segmentation.

WHAT IS POOLED? There is psychophysical evidence for pooling of orientation (Mareschal, Morgan, & Solomon, 2010; Parkes et al., 2001; Solomon, Felisberti, & Morgan, 2004; Song & Levi, 2010) and position (Greenwood, Bex, & Dakin, 2009; Dakin et al., 2010). Indeed, Balas, Nakano, and Rosenholtz (2009) propose a model in which visual stimuli in the periphery are represented by the joint statistics of responses of neurons sensitive to position, phase, orientation, and scale, which are pooled over regions corresponding to Bouma's proportionality constant. Pooling of image statistics results in information loss and can lead to the jumbled, "textural" representation of peripheral stimuli. To the extent that peripheral vision is based on statistical inference, crowding may simply be a regularization process that results in a consistent appearance (a uniform texture) among neighboring objects. Whether this pooling occurs "preattentively" or by attentional mechanisms and where it occurs in the brain are subjects of much current research and debate. The key to successfully modeling crowding will ultimately depend on a better understanding of the pooling process and how it interacts with image segregation, attention, and learning.

WHERE DOES POOLING TAKE PLACE? Current thinking places the pooling beyond the stage of feature detection. Whether this pooling occurs "preattentively" or by attentional mechanisms is the subject of much current research and debate. There are two different attentional models, attentional allocation and attentional resolution. The first attributes crowding to a failure to deploy or allocate appropriate attentional resources to the location of the target. One prediction of this model is that the deployment of attention to the target location should release or at least reduce crowding. A number of studies have shown either no effect or a small effect of cuing on crowding (Nazir, 1992; Scolari et al., 2007; Wilkinson, Wilson, & Ellemberg, 1997). However, several recent studies (Freeman & Pelli, 2007; Petrov & Meleshkevich, 2011b; Yeshurun & Rashal, 2010) showed substantial effects of precuing, and Põder (2007) showed a substantial effect of simultaneous location cuing, which he attributed to exogenous attention. Nonetheless, crowding still occurs with cuing and when the observer knows precisely when and where the target will be, so it seems unlikely that crowding can be fully explained on the basis of a lack of attentional deployment. Yeshurun and Rashal (2010) suggest that one way to reconcile the pooling and attentional deployment models is to have the stage of integration take place at an intermediate stage of processing, where attention can influence the size of receptive fields.

Another attentional model argues that peripheral crowding results from limitations set by attentional resolution (He, Cavanagh, & Intriligator, 1996). This is effectively a pooling model, with pooling taking place in attentional brain areas. He, Cavanagh, and Intriligator (1996) showed that orientation-specific adaptation occurs under conditions where crowding reduces the observers' ability to report the adapting grating's orientation to chance performance. Although this is not direct evidence for an attentional locus, it did seem to provide strong evidence that crowding occurs beyond V1 (where adaptation effects are thought to occur). However, this interpretation has been called into question by Blake and colleagues (2006). Their results showed that the threshold-elevation aftereffect was reduced substantially during crowding and that the strong aftereffect reported by He, Cavanagh, and Intriligator could be explained by the response saturation that occurred at their very high adapting contrast level. Blake's group suggest that their findings indicate that the neural events that underlie crowding are inaugurated at least in part at an early stage of visual processing.

There are other lines of evidence that have been used to argue for an attentional explanation for crowding. Some are circumstantial rather than direct. These include the report that crowding, like other attentional effects, is stronger in the upper than in the lower visual field (He, Cavanagh, & Intriligator, 1996), the fact that the polarity effect in crowding has a temporal resolution similar to that for attention (Chakravarthi & Cavanagh, 2007), and the demonstration that crowding is specific to the attentional selection region and does not occur outside it (Cavanagh & Holcombe, 2007). More direct evidence comes from studies showing that manipulating attention cuing can alter the size and/or strength of crowding (Freeman & Pelli, 2007; Petrov & Meleshkevich, 2011b; Yeshurun & Rashal, 2010).

The pooling/faulty integration hypothesis is appealing because it predicts a number of known aspects of crowding. These include changes in target appearance induced by flankers (Greenwood, Bex, & Dakin, 2010). For access to statistical properties such as average orientation or position, see Parkes et al. (2001), Greenwood, Bex, and Dakin (2009), and Dakin et al. (2010). It also predicts that crowding would impair discrimination and recognition but not detection if the pooling takes place at a site beyond that of the initial feature extraction (Chung, Levi, & Legge, 2001; Levi, Hariharan, & Klein, 2002a, 2002b; Pelli, Palomares, & Majaj, 2004). Precisely what is pooled and where the pooling takes place remain open questions; however, to the extent that only similar features are pooled (i.e., the pooling unit is tuned), it is consistent with the reduction of crowding when targets and flankers are dissimilar (e.g., Kooi et al., 1994).

Substitution Models

Observers frequently mistakenly report a flanker instead of a target (Bernard & Chung, 2011; Krumhansl & Thomas, 1977; Popple & Levi, 2005; Strasburger, 2005; Wolford, 1975), leading to the notion that crowding is a result of substitution errors. In this substitution model the observer's responses are randomly drawn from the location of the target or from one of the flankers so that correct identification of the target (percentage correct) reflects the average of trials drawn from the target and flanker locations. Peripheral vision is characterized by high degrees of spatial uncertainty (Pelli, 1985), and this uncertainty results in a loss of position information. For example, in crowded displays there is a proportion of confusions between neighboring positions (higher than predicted by chance performance), and these errors have been attributed to a noisy stimulus representation, with both object and positional uncertainty under crowded conditions (Popple & Levi, 2005). In addition, in peripheral vision there is a strong correlation between observers' errors and "parts of the flanks that resemble target features" (Nandy & Tjan, 2007). Interestingly, no such correlation was observed in the normal fovea. Both the peripheral and amblyopic visual systems are known to have a high degree of positional uncertainty and a high degree of crowding.

Although this model readily explains the high proportion of flanker errors, it is not so clear that it would predict the reduced crowding with dissimilar flankers or the fact that only about one-third of the errors are substitution errors (Bernard & Chung, 2011).

COMPUTATIONAL INSTANTIATIONS Wilkinson, Wilson, and Ellemberg (1997) proposed the first quantitative computational model for crowding. In this model isolated visual contours are processed by simple cells, which suppress weak complex cell responses. However, in the presence of nearby similarly oriented flanking contours, information is present in a small area, complex cells respond vigorously because of spatial pooling, and they then suppress simple cell activity within their receptive field area. This texture model nicely predicts several aspects of their data; however, the pooling parameter was based on the simulations that best fit the data, rather than on physiology or some other principled approach.

There are several more recent approaches to modeling crowding. As noted above, Balas, Nakano, and Rosenholtz (2009) take the approach that the visual system represents peripheral targets by the joint statistics of responses of neurons that are sensitive to different features of the stimulus array, such as their positions, orientations, and sizes (Portilla & Simoncelli, 2000). These joint statistics are computed over a local "pooling" area, resulting in a "texture-like" representation. The putative local pooling regions overlap, tile the visual field, and grow with eccentricity (Balas, Nakano, & Rosenholtz, 2009; Freeman & Simoncelli, 2011). In the Balas model the size of the pooling regions is fixed to coincide with the critical spacing Bouma's "law." Human observers, when required to identify peripheral targets, show similar performance to identifying "mongrels" (i.e., model texture summaries of the same stimuli) in the fovea.

A recent model with a very similar flavor (Freeman & Simoncelli, 2011) has pooling regions that are determined based on estimates of receptive field size in V2, and their experiments showed that stimuli that resulted in similar model responses were "metameric," that is, indistinguishable to human observers.

Van den Berg, Roerdink, and Cornelissen (2010) proposed a quantitative model for spatial integration of orientation signals based on the principles of population coding. Their model nicely predicts several properties of crowding, including critical spacing, "compulsory averaging," and the inner/outer asymmetry. However, in its current form it fails to predict the effect of target flank similarity (Kooi et al., 1994; Nazir, 1992) and the configuration effects (Levi & Carney, 2009; Livne & Sagi, 2007, 2010).

Dayan and Solomon (2010) take a very different approach, in which spatial selection of a target among flankers emerges through a process of Baysian inference, in a computational form. "Interference" (i.e., crowding) in this model results from spatial uncertainty inherent in large receptive fields, and receptive field size is assumed to increase with eccentricity according to the cortical magnification factor. The model was developed to explain the Eriksen flanker task. It remains to be seen how well it accounts for many of the key features of crowding.

None of the models naturally accounts for the radial/ tangential anisotropy in a principled way. Van den Berg, Roerdink, and Cornelissen (2010) simply use different parameters to define the radial and tangential integration fields. Freeman and Simoncelli (2011) assumed an aspect ratio of radial width to circumferential width of approximately 2. In contrast, Nandy and Tjan (2012) begin with a physiologically plausible model of cortical area V1 and its geometry and lateral connections, quite similar to the model of Neri and Levi (2006), combined with the important role of natural image statistics (Balas, Nakano, & Rosenholtz, 2009). However, the novel insight and advance in this paper are the ideas that image statistics are acquired primarily at attended spatial locations via a gating mechanism and that spatial attention and any subsequent eye movement that it elicits overlap in time. Nandy and Tjan further argue that learning image statistics during development leads to the formation of lateral connections that distort the true image statistics in the peripheral field, leading in turn to the radial–tangential anisotropy of crowding. The model not only accounts for many of the known phenomena of crowding but also makes specific and testable predictions that will keep the field busy for years to come.

THE LOCUS OF CROWDING

Precisely where crowding takes place is a long-standing and still open question. We know that it resides in the cortex because crowding occurs when the target is presented to one eye and the flanks to the other, so crowding must occur at or beyond the site of binocular combination (Flom, Heath, & Takahashi, 1963). However, there is surprisingly little physiology or fMRI that bears directly on the problem (discussed further below). One approach to this question by psychophysicists has been *psychoanatomy*. For example, Flom, Weymouth, and Kahneman (1963) suggested "Larger central and overall diameters of receptive fields with greater retinal eccentricity in man would be consistent with this regional variation in resolution and would lead to the expectation of interaction over greater distances in the retinal periphery." However, they did not specify which receptive fields might be responsible for crowding.

Crowding occurs when the stimuli are closer than a fixed cortical distance (Levi, Klein, & Aitsebaomo, 1985). For Vernier acuity, crowding occurs when the critical spacing is "less than a cortical distance of about 1 mm—approximately the size of a human cortical ocular dominance column" (Hitchcock & Hickey, 1980). Note that the critical spacing for Vernier acuity is ~0.1 E, much smaller than for many other tasks, where the critical spacing is 0.4–0.5 E. Indeed, for letter-like stimuli, Tripathy and Levi (1994) estimated the cortical extents of crowding to be on the order of ~6 mm and suggested that this may reflect the role of long-range horizontal connections. Pelli (2008) came to a similar conclusion that "objects in the world, unless they are very dissimilar, can be recognized only if they are sufficiently separated in visual cortex: specifically, in V1, at least 6 mm apart in the radial direction (increasing eccentricity) or 1 mm apart in the circumferential direction (equal eccentricity). Objects closer together than this critical spacing are perceived as an unidentifiable jumble."

Crowding almost certainly resides beyond the retina or LGN because crowding is orientation specific (Levi & Carney, 2009; Levi, Hariharan, & Klein, 2002b) and occurs when the target is presented to one eye and the flanks to the other. Thus, crowding must take place in the visual cortex at or beyond the site of binocular combination (Flom, Heath, & Takahashi, 1963). Thus, it is reasonable to ask whether the large extent of peripheral crowding reflects pooling of target and flanks by large peripheral receptive fields in visual cortex. Figure 48.3 (filled symbols) shows several estimates of average or population receptive field (RF) size, expressed as a fraction of eccentricity, in cortical areas V1, V2, and V4, obtained from both physiology and fMRI. For comparison, the open gray symbols show a range of estimates of the spacing limit from psychophysical studies of crowding, and the thin black line shows the uncrowded acuity limit. The figure reveals a

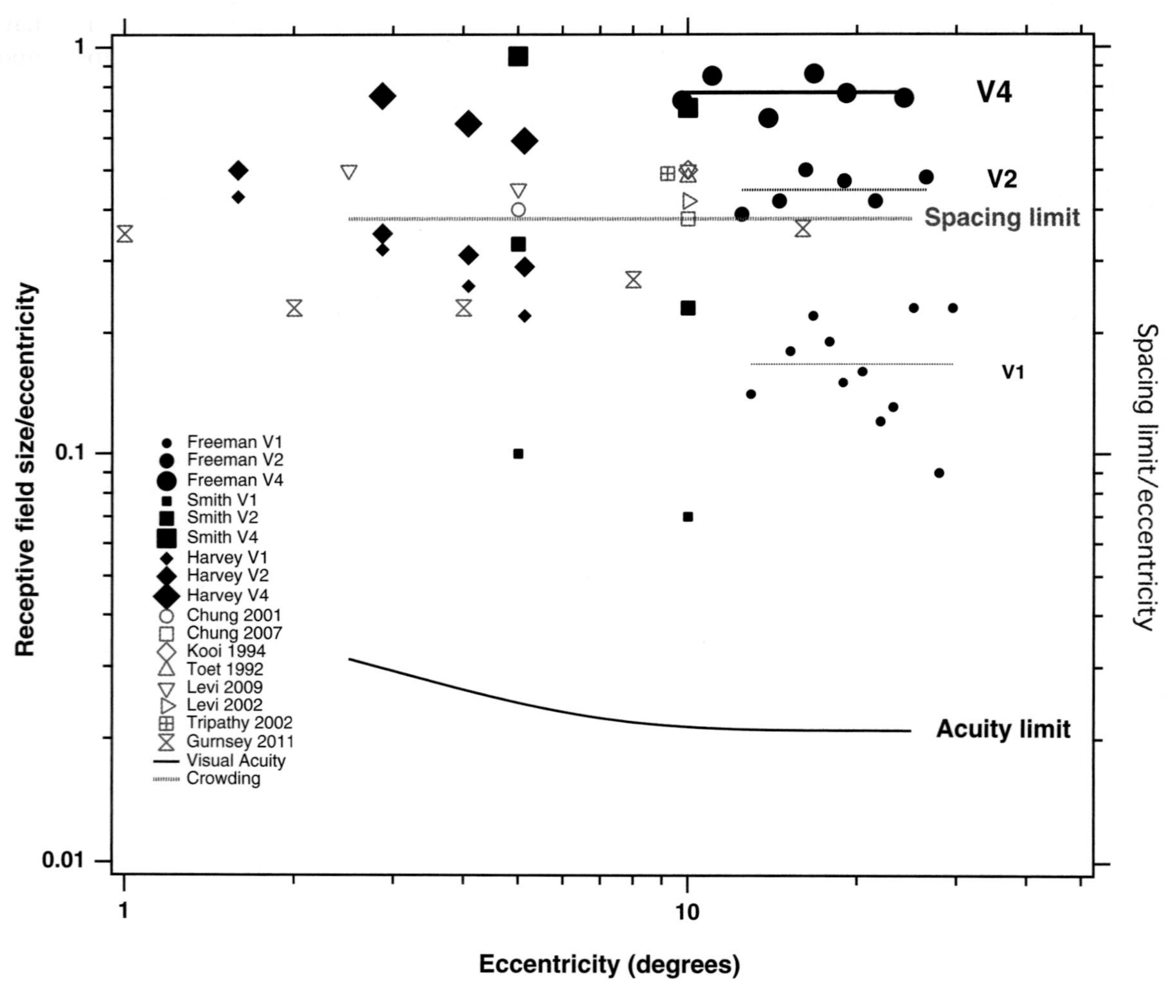

FIGURE 48.3   Receptive field size and the spacing limit for crowding versus eccentricity. Solid black symbols show several estimates of average or population receptive field (RF) size, expressed as a fraction of eccentricity, in cortical areas V1 (small black symbols), V2 (medium black symbols), and V4 (large black symbols), obtained from both physiology and fMRI. (Circles from Freeman & Simoncelli, 2011; squares from Smith et al., 2001; and diamonds from Harvey & Dumoulin, 2011.) For comparison, the open gray symbols show a range of estimates of the spacing limit from psychophysical studies of crowding, and the thin black line shows the uncrowded acuity limit.

number of interesting points. First, the acuity or size limit is very much smaller than the crowding or spacing limit, by about a factor of 16! Second, although the acuity limit is considerably smaller than the average or population V1 RF size, the spacing limit is close to the average or population V2 RF size. Third, the average or population V4 RF is larger than the crowding region. There are several important caveats: There is likely a large spread of receptive field sizes at each eccentricity, and they are not reflected in this figure.

A few neurophysiological studies have compared crowding with RFs in different areas, with mixed conclusions. For example, Motter (2002) suggests that the RFs of neurons in V4 are consistent with the large extent of crowding. However, figure 48.3 seems to indicate that V4 RFs are too large. Attempts to link crowding to specific visual areas in humans have been mixed. fMRI studies suggest that the effects of crowding are not evident in the BOLD response in V1 but are evident as early as V2 and beyond (Arman, Chung, & Tjan, 2006; Bi et al., 2009; Fang & He, 2008; Freeman, Donner, & Heeger, 2011). However, most of these studies have introduced flankers into the stimulus to induce crowding, making the results difficult to interpret. A very recent study (Anderson et al., 2012) used an elegant approach based on the change in appearance as a signature of crowding and reports that crowding induced changes in BOLD signal in areas V1, V2, V3, and V4, but weaker in V1 than in areas downstream.

Freeman and Simoncelli (2011) have taken a rather different approach to the question. Specifically, they developed a population model for "mid–ventral stream" processing, in which nonlinear combinations of V1 responses are averaged in receptive fields that grow with eccentricity. To test their model they generated visual "metamers" that were indistinguishable to human observers but physically different. They used these metameric stimuli to estimate scaling constants and found them to be consistent with average V2 RF sizes (as shown in figure 48.3). They argue that their V2 model can account for the spacing and eccentricity dependence of crowding. Freeman and Simoncelli "fix" the aspect ratio of their RFs to be 2:1, so there is not a natural account for the radial–tangential anisotropy in crowding.

A most intriguing aspect of figure 48.3 is that despite the fact that there are clearly RFs considerably smaller than the spacing limit, we do not appear to have access to them under conditions of clutter or crowding.

It is now clear that flankers have distinct effects that depend on the visual task (Levi & Carney, 2011) and the nature of the stimuli, suggesting that crowding may occur at multiple stages of visual processing. As Whitney

and Levi (2011) note, "If crowding occurs at multiple levels of visual analysis, or if different 'channels' (chromatic, spatial frequency, object, etc.) each possess their [*sic*] own unique crowding bottleneck, then one would expect that the gradient of crowding as a function of eccentricity might be channel or stimulus specific." There is some evidence to support this view, but more work is needed.

DEVELOPMENT, AGING, AND PLASTICITY

Recent work suggests that the development of crowding is quite protracted, such that it operates over much larger distances in children as old as 11 years of age than in adults—well after the development of visual acuity (Atkinson, 1991; Jeon et al., 2010). This protracted development of crowding has important implications for developmental abnormalities such as amblyopia and dyslexia, in which crowding may be differentially impaired. Much less is known about the effects of aging (at the other end of the spectrum) on crowding, although the report by Habak, Wilkinson, and Wilson (2009) suggests that, at least in central vision, the effects of flanking masks on a shape discrimination task are similar in old and young observers.

Importantly, several studies have shown that it is possible to reduce crowding through perceptual learning (Chung, 2007; Maniglia et al., 2011; Sun, Chung, & Tjan, 2010). This reduction in the extent of crowding places a strong constraint on any model for crowding. For example, it is not clear how a simple "fixed RF size" model can accommodate this kind of shrinkage without additional assumptions.

REFERENCES

Anderson, E. J., Dakin, S. C., Schwarzkopf, D. S., Reese, G., & Greenwood, J. (2012). The neural correlates of crowding-induced changes in appearance. *Current Biology, 22,* 1199–1206.

Andriessen, J. J., & Bouma, H. (1976). Eccentric vision: Adverse interactions between line segments. *Vision Research, 16,* 71–78.

Arman, A. C., Chung, S. T. L., & Tjan, B. S. (2006). Neural correlates of letter crowding in the periphery [Abstract]. *Journal of Vision, 6*(6), 804.

Atkinson, J. (1991). Review of human visual development: Crowding and dyslexia. In J. Stein (Ed.), *Vision and visual dyslexia* (pp. 44–57). New York: Macmillan.

Balas, B., Nakano, L., & Rosenholtz, R. (2009). A summary-statistic representation in peripheral vision explains visual crowding. *Journal of Vision, 9*(13), 11–18. doi:10.1167/9.12.13.

Baldassi, S., Megna, N., & Burr, D. C. (2006). Visual clutter causes high-magnitude errors. *PLoS Biology, 4*(3), e56. doi:10.1371/journal.pbio.0040056.

Banks, W. P., Larson, D. W., & Prinzmetal, W. (1979). Asymmetry of visual interference. *Perception & Psychophysics, 25,* 447–456.

Banks, W. P., & White, H. (1984). Lateral interference and perceptual grouping in visual detection. *Perception & Psychophysics, 36,* 285–295.

Banton, T., & Levi, D. M. (1993). Spatial localization of motion-defined and luminance-defined contours. *Vision Research, 33,* 2225–2237.

Bernard, J. B., & Chung, S. T. (2011). The dependence of crowding on flanker complexity and target-flanker similarity. *Journal of Vision, 11*(8), 1–16. doi:10.1167/11.8.1.

Bex, P. J., & Dakin, S. C. (2005). Spatial interference among moving targets. *Vision Research, 45,* 1385–1398.

Bex, P. J., Dakin, S. C., & Simmers, A. J. (2003). The shape and size of crowding for moving targets. *Vision Research, 43,* 2895–2904.

Bi, T., Cai, P., Zhou, T., & Fang, F. (2009). The effect of crowding on orientation-selective adaptation in human early visual cortex. *Journal of Vision, 9*(11), 13, 1–10. doi:10.1167/9.11.13.

Blake, R., Tadin, D., Sobel, K. V., Raissian, T. A., & Chong, S. C. (2006). Strength of early visual adaptation depends on visual awareness. *Proceedings of the National Academy of Sciences of the United States of America, 103,* 4783–4788.

Bouma, H. (1970). Interaction effects in parafoveal letter recognition. *Nature, 226,* 177–178.

Bouma, H. (1973). Visual interference in the parafoveal recognition of initial and final letters of words. *Vision Research, 13,* 767–782.

Butler, T. W., & Westheimer, G. (1978). Interference with stereoscopic acuity: Spatial, temporal, and disparity tuning. *Vision Research, 18,* 1387–1392.

Cavanagh, P., & Holcombe, A. O. (2007). Non-retinotopic crowding. *Vision Sciences Society Abstract.* http://journalofvision.org/7/9/338/.

Chakravarthi, R., & Cavanagh, P. (2007). Temporal properties of the polarity advantage effect in crowding. *Journal of Vision, 7,* 1–13. doi:10.1167/7.2.11.

Chakravarthi, R., & Cavanagh, P. (2009). Recovery of a crowded object by masking the flankers: Determining the locus of feature integration. *Journal of Vision, 9*(10), 4, 1–9. doi:10.1167/9.10.4.

Chastain, G. (1982). Confusability and interference between members of parafoveal letter pairs. *Perception & Psychophysics, 32,* 576–580.

Chung, S. T. L. (2007). Learning to identify crowded letters: Does it improve reading speed? *Vision Research, 47*(25), 3150–3159. doi:10.1016/j.visres.2007.08.017.

Chung, S. T., Levi, D. M., & Legge, G. E. (2001). Spatial-frequency and contrast properties of crowding. *Vision Research, 41,* 1833–1850.

Chung, S. T. L., Li, R. W., & Levi, D. M. (2007). Crowding between first- and second-order letter stimuli in normal foveal and peripheral vision. *Journal of Vision, 7,* 1–13. doi:10.1167/7.2.10.

Dakin, S. C., Cass, J., Greenwood, J. A., & Bex, P. J. (2010). Probabilistic, positional averaging predicts object-level crowding effects with letter-like stimuli. *Journal of Vision, 10,* 1–16. doi:10.1167/10.10.14.

Dakin, S. C., Greenwood, J. A., Carlson, T. A., & Bex P. J. (2011). Crowding is tuned for perceived (not physical) location. *Journal of Vision, 11*(9), 2, 1–13. doi:10.1167/11.9.2.

Dayan, P., & Solomon, J. A. (2010). Selective Bayes: Attentional load and crowding. *Vision Research, 50,* 2248–2260.

Ehrt, O., Hess, R. F., Williams, C. B., & Sher, K. (2003). Foveal contrast thresholds exhibit spatial- frequency- and polarity-specific contour interactions. *Journal of the Optical Society of America, 20*(1), 11–17.

Eriksen, C. W., & Rohrbaugh, J. W. (1970). Some factors determining efficiency of selective attention. *American Journal of Psychology, 83,* 330–343.

Estes, W. K., Allmeyer, D. H., & Reder, S. M. (1976). Serial position functions for letter identification at brief and extended exposure durations. *Perception & Psychophysics, 19,* 1–15.

Fang, F., & He, S. (2008). Crowding alters the spatial distribution of attention modulation in human primary visual cortex. *Journal of Vision, 8*(9), 6, 1–9. doi:10.1167/8.9.6.

Felisberti, F. M., Solomon, J. A., & Morgan, M. J. (2005). The role of target salience in crowding. *Perception, 34,* 823–833.

Feng, C., Jiang, Y., & He, S. (2007). Horizontal and vertical asymmetry in visual spatial crowding effects. *Journal of Vision, 7*(2), 13, 1–10. doi:10.1167/7.2.13.

Field, D. J., Hayes, A., & Hess, R. F. (1993). Contour integration by the human visual system: Evidence for a local "association field." *Vision Research, 33*(2), 173–193.

Flom, M. C., Heath, G. G., & Takahashi, E. (1963). Contour interaction and visual resolution: Contralateral effect. *Science, 142,* 979–980.

Flom, M. C., Weymouth, F. W., & Kahneman, D. (1963). Visual resolution and contour interaction. *Journal of the Optical Society of America, 53,* 1026–1032.

Freeman, J., Donner, T. H., & Heeger, D. J. (2011). Inter-area correlations in the ventral visual pathway reflect feature integration. *Journal of Vision, 11*(4), 15, 1–23. doi:10.1167/11.4.15.

Freeman, J., & Pelli, D. G. (2007). An escape from crowding. *Journal of Vision, 7*(2), 22, 1–14. doi:10.1167/7.2.22.

Freeman, J., & Simoncelli, E. P. (2011). Metamers of the ventral stream. *Nature Neuroscience, 14,* 1195–1201.

Grainger, J., Tydgat, I., & Issele, J. (2010). Crowding affects letters and symbols differently. *Journal of Experimental Psychology. Human Perception and Performance, 36,* 673–688.

Greenwood, J. A., Bex, P. J., & Dakin, S. C. (2009). Positional averaging explains crowding with letter-like stimuli. *Proceedings of the National Academy of Sciences of the United States of America, 106*(31), 13130–13135. doi:10.1073/pnas.0901352106.

Greenwood, J. A., Bex, P. J., & Dakin, S. C. (2010). Crowding changes appearance. *Current Biology, 20,* 496–501.

Gurnsey, R., Roddy, G., & Chanab, W. (2011). Crowding is size and eccentricity dependent. *Journal of Vision, 11*(7), 15, 1–17. doi:10.1167/11.7.15.

Habak, C., Wilkinson, F., & Wilson, H. R. (2009). Preservation of shape discrimination in aging. *Journal of Vision, 9*(12), 18, 1–8. doi:10.1167/9.12.18.

Hariharan, S., Levi, D. M., & Klein, S. A. (2005). "Crowding" in normal and amblyopic vision assessed with Gaussian and Gabor C's. *Vision Research, 45,* 617–633.

Harvey, B. M., & Dumoulin, S. O. (2011). The relationship between cortical magnification factor and population receptive field size in human visual cortex: Constancies in cortical architecture. *Journal of Neuroscience, 31,* 13604–13612.

He, S., Cavanagh, P., & Intrilligator, J. (1996). Attentional resolution and the locus of visual awareness. *Nature, 383,* 334–337.

Hess, R. F., Dakin, S. C., & Kapoor, N. (2000). The foveal "crowding" effect: Physics or physiology? *Vision Research, 40,* 365–370.

Hess, R. F., Dakin, S. C., Kapoor, N., & Tewfik, M. (2000). Contour interaction in fovea and periphery. *Journal of the Optical Society of America. A, Optics, Image Science, and Vision, 17,* 1516–1524.

Hess, R. F., Williams, C. B., & Chaudhry, A. (2001). Contour interaction for easily resolvable stimulus. *Journal of the Optical Society of America. A, Optics, Image Science, and Vision, 18,* 2414–2418.

Hitchcock, P. F., & Hickey, T. L. (1980). Ocular dominance columns: Evidence for their presence in humans. *Brain Research, 182,* 176–179.

Intriligator, J., & Cavanagh, P. (2001). The spatial resolution of visual attention. *Cognitive Psychology, 43,* 171–216.

Jeon, S. T., Hamid, J., Maurer, D., & Lewis, T. L. (2010). Developmental changes during childhood in single-letter acuity and its crowding by surrounding contours. *Journal of Experimental Child Psychology, 107,* 423–437.

Kooi, F. L., Toet, A., Tripathy, S. P., & Levi, D. M. (1994). The effect of similarity and duration on spatial interaction in peripheral vision. *Spatial Vision, 8,* 255–279.

Krumhansl, C. L., & Thomas, E. A. (1977). Effect of level of confusability on reporting letters from briefly presented visual displays. *Perception & Psychophysics, 21,* 269–279.

Levi, D. M. (2008). Crowding—an essential bottleneck for object recognition: A mini-review. *Vision Research, 48,* 635–654.

Levi, D. M. (2011). Visual crowding. *Current Biology, 21,* R678–R679.

Levi, D. M., & Carney, T. (2009). Crowding in peripheral vision: Why bigger is better. *Current Biology, 19,* 1988–1993.

Levi, D.M., & Carney T. (2011). The effects of flankers on three tasks in central, peripheral and amblyopic vision. *Journal of Vision, 11*(1), 10, 1–23. doi:10.1167/11.1.10.

Levi, D. M., Hariharan, S., & Klein, S. A. (2002a). Suppressive and facilitatory spatial interactions in peripheral vision: Peripheral crowding is neither size invariant nor simple contrast masking. *Journal of Vision, 2,* 167–177. doi:10.1167/2.2.3.

Levi, D. M., Hariharan, S., & Klein, S. A. (2002b). Suppressive and facilitatory spatial interactions in amblyopic vision. *Vision Research, 42,* 1379–1394.

Levi, D. M., Klein, S. A., & Aitsebaomo, A. P. (1985). Vernier acuity, crowding and cortical magnification. *Vision Research, 25,* 963–977.

Levi, D. M., Song, S., & Pelli, D. G. (2007). Amblyopic reading is crowded. *Journal of Vision, 7*(2), 21, 1–17. doi:10.1167/7.2.21.

Liu, L., & Arditi, A. (2000). Apparent string shortening concomitant with letter crowding. *Vision Research, 40,* 1059–1067.

Livne, T., & Sagi, D. (2007). Configuration influence on crowding. *Journal of Vision, 7*(4), 1–12. doi:10.1167/7.2.4.

Livne, T., & Sagi, D. (2010). How do flankers' relations affect crowding? *Journal of Vision, 10*(3), 1, 1–14. doi:10.1167/10.3.1.

Livne, T., & Sagi, D. (2011). Multiple levels of orientation anisotropy in crowding with Gabor flankers. *Journal of Vision, 11*(13), 18. doi:10.1167/11.13.18.

Louie, E. G., Bressler, D. W., & Whitney, D. (2007). Holistic crowding: Selective interference between configural representations of faces in crowded scenes. *Journal of Vision, 7*(2), 24, 1–11. doi:10.1167/7.2.24.

Malania, M., Herzog, M. H., & Westheimer, G. (2007). Grouping of contextual elements that affect vernier thresholds. *Journal of Vision, 7*(2), 1, 1–7. doi:10.1167/7.2.1.

Maniglia, M., Pavan, A., Cuturi, L. F., Campana, G., Sato, G., & Casco, C. (2011). Reducing crowding by weakening inhibitory lateral interactions in the periphery with perceptual learning. *PLoS One, 6*(10), e25568.

Mareschal, I., Morgan, M. J., & Solomon, J.A. (2010). Cortical distance determines whether flankers cause crowding or the tilt illusion. *Journal of Vision, 10*(18), 13, 1–14. doi:10.1167/10.8.13.

Martelli, M., Majaj, N. J., & Pelli, D. G. (2005). Are faces processed like words? A diagnostic test for recognition by parts. *Journal of Vision, 5*(1), 6, 58–70. doi:10.1167/5.1.6.

Maus, G. W., Fischer, J., & Whitney, D. (2011). Perceived positions determine crowding. *PLoS One, 6*(5), e19796.

May, K. A., & Hess, R. F. (2007). Ladder contours are undetectable in the periphery: A crowding effect? *Journal of Vision, 7*(13), 9, 1–15. doi:10.1167/7.13.9.

Motter, B. C. (2002). Crowding and object integration within the receptive field of V4 neurons. *Journal of Vision, 2*(7), 274. doi:10.1167/2.7.274.

Motter, B. C., & Simoni, D. A. (2007). The roles of cortical image separation and size in active visual search performance. *Journal of Vision, 7*(2), 6, 1–15. doi:10.1167/7.2.6.

Nandy, A. S., & Tjan, B. S. (2007). The nature of letter crowding as revealed by first- and second-order classification images. *Journal of Vision, 7,* 1–26. doi:10.1167/7.2.5.

Nandy, A. S., & Tjan, B. S. (2012). Saccade-confounded image statistics explain visual crowding. *Nature Neuroscience, 15,* 463–469. doi:10.1038/nn.3021.

Nazir, T. A. (1992). Effects of lateral masking and spatial precueing on gap-resolution in central and peripheral vision. *Vision Research, 32,* 771–777.

Neri, P., & Levi, D. M. (2006). Spatial resolution for feature binding is impaired in peripheral and amblyopic vision. *Journal of Neurophysiology, 96,* 42–53.

Parkes, L., Lund, J., Angelucci, A., Solomon, J. A., & Morgan, M. (2001). Compulsory averaging of crowded orientation signals in human vision. *Nature Neuroscience, 4,* 739–744.

Pelli, D. G. (1985). Uncertainty explains many aspects of visual contrast detection and discrimination. *Journal of the Optical Society of America. A, Optics and Image Science, 2*(9), 1508–1532. doi:10.1364/JOSAA.2.001508.

Pelli, D. G. (2008). Crowding: A cortical constraint on object recognition. *Current Opinion in Neurobiology, 18,* 445–451.

Pelli, D. G., Palomares, M., & Majaj, N. J. (2004). Crowding is unlike ordinary masking: Distinguishing feature integration from detection. *Journal of Vision, 4*(12), 1136–1169. doi:10.1167/4.12.12.

Pelli, D. G., & Tillman, K. A. (2008). The uncrowded window for object recognition. *Nature Neuroscience, 11,* 1129–1135.

Petrov, Y., & Meleshkevich, O. (2011a). Asymmetries and idiosyncratic hot spots in crowding. *Vision Research, 51,* 1117–1123.

Petrov, Y., & Meleshkevich, O. (2011b). Locus of spatial attention determines inward-outward anisotropy in crowding. *Journal of Vision, 11*(4), 1, 1–11. doi:10.1167/11.4.1

Petrov, Y., Popple, A.V., & McKee, S.P. (2007). Crowding and surround suppression: Not to be confused. *Journal of Vision, 7*(2), 12, 1–9. doi:10.1167/7.2.12.

Põder, E. (2006). Crowding, feature integration, and two kinds of "attention." *Journal of Vision, 6*(2), 7, 163–169. doi:10.1167/6.2.7.

Põder, E. (2007). Effect of colour pop-out on the recognition of letters in crowding conditions. *Psychological Research, 71*, 615–715.

Popple, A. V., & Levi, D. M. (2005). The perception of spatial order at a glance. *Vision Research, 45*, 1085–1090.

Portilla, J., & Simoncelli, E. (2000). A parametric texture model based on joint statistics of complex wavelet coefficients. *International Journal of Computer Vision, 40*, 49–71.

Rosen, S., Chakravarthi, R., & Pelli, D. G. (2011). Crowding reveals a third stage of object recognition. *Journal of Vision, 11*(11), 1142. doi:10.1167/11.11.1142.

Saarela, T. P., Sayim, B., Westheimer, G., & Herzog, M. H. (2009). Global stimulus configuration modulates crowding. *Journal of Vision, 9*(2), 5, 1–11. doi:10.1167/9.2.5.

Saarela, T. P., Westheimer, G., & Herzog, M. H. (2010). The effect of spacing regularity on visual crowding. *Journal of Vision, 10*(10), 17, 1–7. doi:10.1167/10.10.17.

Schwartz, E. L. (1980). Computational anatomy and functional architecture of striate cortex: A spatial mapping approach to perceptual coding. *Vision Research, 20*, 645–669.

Scolari, M., Kohnen, A., Barton, B., & Awh, E. (2007). Spatial attention, preview, and popout: Which factors influence critical spacing in crowded displays? *Journal of Vision, 7*(2), 7, 1–23. doi:10.1167/7.2.7.

Smith, A. T., Singh, K. D., Williams, A. I., & Greenlee, M. W. (2001). Estimating receptive field size from fMRI data in human striate and extrastriate visual cortex. *Cerebral Cortex, 11*, 1182–1190.

Solomon, J. A., Felisberti, F. M., & Morgan, M. J. (2004). Crowding and the tilt illusion: Toward a unified account. *Journal of Vision, 4*(6), 9, 500–508. doi:10.1167/4.6.9.

Song, S., & Levi, D. M. (2010). Spatiotemporal mechanisms for simple image feature perception in normal and amblyopic vision. *Journal of Vision, 10*(13), 21, 1–22. doi:10.1167/10.13.21.

Strasburger, H. (2005). Unfocused spatial attention underlies the crowding effect in indirect form vision. *Journal of Vision, 5*(11), 8, 1024–1037. doi:10.1167/5.11.8.

Strasburger, H., Harvey, L. O., Jr., & Rentschler, I. (1991). Contrast thresholds for identification of numeric characters in direct and eccentric view. *Perception & Psychophysics, 49*, 495–508.

Strasburger, H., Rentschler, I., & Juttner, M. (2011). Peripheral vision and pattern recognition: A review. *Journal of Vision, 11*(5), 13. doi:10.1167/11.5.13.

Sun, G. J., Chung, S. T. L., & Tjan, B. (2010). Ideal observer analysis of crowding and the reduction of crowding through learning. *Journal of Vision, 10*(5), 16, 1–14. doi:10.1167/10.5.16.

Toet, A., & Levi, D. M. (1992). The two-dimensional shape of spatial interaction zones in the parafovea. *Vision Research, 32*, 1349–1357.

Tripathy, S. P., & Cavanagh, P. (2002). The extent of crowding in peripheral vision does not scale with target size. *Vision Research, 42*, 2357–2369.

Tripathy, S. P., & Levi, D. M. (1994). Long-range dichoptic interactions in the human visual cortex in the region corresponding to the blind spot. *Vision Research, 34*, 1127–1138.

Tyler, C. W., & Likova, L. T. (2007). Crowding: A neuroanalytic approach. *Journal of Vision, 7*, 1–9. doi:10.1167/7.2.16.

van den Berg, R., Roerdink, J. B., & Cornelissen, F. W. (2010). A neurophysiologically plausible population code model for feature integration explains visual crowding. *PLoS Computational Biology, 6*, e1000646.

Wallace, J. M., & Tjan, B. S. (2011). Object crowding. *Journal of Vision, 11*(6), 19, 1–17. doi:10.1167/11.6.19.

Wallis, T. S., & Bex, P. J. (2011). Visual crowding is correlated with awareness. *Current Biology, 21*(3), 254–258.

Westheimer, G., & Hauske, G. (1975). Temporal and spatial interference with vernier acuity. *Vision Research, 15*, 1137–1141.

Westheimer, G., Shimamura, K., & McKee, S. P. (1976). Interference with line-orientation sensitivity. *Journal of the Optical Society of America, 66*, 332–338.

Whitney, D., & Levi, D. M. (2011). Visual crowding: A fundamental limit on conscious perception and object recognition. *Trends in Cognitive Sciences, 15*, 160–168.

Wilkinson, F., Wilson, H. R., & Ellemberg, D. (1997). Lateral interactions in peripherally viewed texture arrays. *Journal of the Optical Society of America, 14*, 2057–2068.

Wolford, G. (1975). Perturbation model for letter identification. *Psychological Review, 82*, 184–199.

Yeshurun, Y., & Rashal, E. (2010). Precueing attention to the target location diminishes crowding and reduces the critical distance. *Journal of Vision, 10*, 1–12. doi:10.1167/10.10.16.

Zhang, J. Y., Zhang, T., Xue, F., Liu, L., & Yu, C. (2009). Legibility of Chinese characters in peripheral vision and the top-down influences on crowding. *Vision Research, 49*, 44–53.

49 From Textures to Crowds: Multiple Levels of Summary Statistical Perception

DAVID WHITNEY, JASON HABERMAN, AND TIMOTHY D. SWEENY

Billions of bits of information arrive at the retina every moment, but our awareness is limited to a small fraction of this information. There are many bottlenecks in visual processing, ranging from physiological and iconic bottlenecks (Nakayama, 1990), attentional bottlenecks in space (Rensink, O'Regan, & Clark, 1997; Simons & Levin, 1997; Whitney & Levi, 2011) and in time (Battelli, Pascual-Leone, & Cavanagh, 2007; Franconeri, Alvarez, & Enns, 2007; Marois, Yi, & Chun, 2004; Raymond, Shapiro, & Arnell, 1992), capacity limits in attention and memory (e.g., Franconeri, in press; Luck & Vogel, 1997; Scholl & Pylyshyn, 1999), and many others. One way in which the brain surmounts some of these bottlenecks, which occur at multiple levels of visual processing, is by representing the statistical regularity of the world in the form of condensed ensemble representations. The leaves of a tree, the blades of grass, and the tiles of the floor are similar and redundant, giving rise to the percepts of "treeness," "lawnness," and "floorness," respectively. The individual components of each texture are lost in favor of a concise, summary statistical representation: an ensemble percept.

In this chapter we review summary statistical perception. Most of the work on this topic has focused on perception of average and variance in displays (e.g., the average length of blades of grass, the average size of a tree's leaves, or the average expression in a crowd of faces). Several other chapters in this volume discuss in more detail texture perception (chapter 45 by Landy), crowding (chapter 48 by Levi), and scene and gist perception (chapter 51 by Oliva). Although these are very clearly related, our focus is on the intersection between these topics, on the nature of ensemble representations, what can form an ensemble and at what levels of visual processing, and how ensembles might be represented in the brain.

The concept of summary representation has recently generated significant interest and debate within the vision science community (Alvarez, 2011; Alvarez & Oliva, 2008, 2009; Ariely, 2001, 2008; Chong & Treisman, 2003, 2005a, 2005b; de Fockert & Marchant, 2008; Haberman & Whitney, 2007, 2009; Koenderink, van Doorn, & Pont, 2004; Myczek & Simons, 2008; Simons & Myczek, 2008). Also sometimes called ensemble coding or ensemble perception, summary representation refers to the idea that the visual system naturally and directly represents an emergent quality (i.e., the gist) of a set of similar items (such as blades of grass). Such a system is intuitively appealing in terms of computational efficiency, and it has far-reaching implications for understanding awareness. For example, Chong and Treisman (2003) and, more recently, we (Haberman & Whitney, 2009) and other authors have suggested that summary representation can provide coarse information from sources across our entire field of view, driving the compelling impression that we have a complete and accurate grasp of our visual world (Haberman & Whitney, 2009). Thus, the "grand illusion" (Noe, Pessoa, & Thompson, 2000) may not be an illusion at all but rather a noisy summary representation of all that we survey. Put another way, although many of the individual details of a scene are inaccessible, ensemble coding may provide a viable algorithm to keep the gist ever present.

EARLY CONCEPTUALIZATIONS OF SUMMARY STATISTICAL PERCEPTION

The concept of ensemble representation is not a new one. Aristotle described perception as a mean of sensory inputs, which could be used to identify stimulus changes as the sense organ gathered more information. Empirical examination of this phenomenon began centuries later with investigations of Gestalt grouping (Wertheimer, 1923), although this early conceptualization was not referred to as ensemble or summary statistical perception, per se. The Gestaltists viewed emergent object perception as a synergy of lower-level inputs; the final percept was more than the sum of its parts. Researchers argued that the grouped object was the favored percept and that the individual features were (at worst) lost or (at best) difficult to perceive (Koffka, 1935). Although Gestaltists outlined several basic heuristics by which the visual system groups features (similarity, proximity, common fate, etc.), the underlying mechanism(s) driving this grouping, as well as the algorithm that supports it, remained elusive. It may be that Gestalt grouping amounts to a summary statistical representation, and the mechanism of ensemble coding may provide an explanation for several Gestalt phenomena.

Although Gestalt phenomenology helped to define some elemental principles of object perception, researchers in this area were not explicitly thinking in terms of ensemble perception or summary statistical representation. Some of the earliest explicit work on ensemble coding was done from a social psychology perspective. In an extensive line of research Norman Anderson outlined a simple yet flexible model called "integration theory" (Anderson, 1971). His work demonstrated that a weighted mean more precisely captured how information is integrated than a summation model. For example, subjects rated another individual more favorably when that person was described by two extremely positive terms compared to when that person was described by two extremely positive terms in addition to two moderately positive terms (Anderson, 1965). Integration theory was extended to numerous other social contexts, including group attractiveness (Anderson, Lindner, & Lopes, 1973), shopping preferences (Levin, 1974), and even the perceived "badness" of criminals (Leon, Oden, & Anderson, 1973). Thus, it appears that humans readily integrate semantic as well as social information, although the mechanism behind this process remains largely unknown. The implication is clear, however: Social perceptions and attitudes may hinge on the same sort of underlying summary computations that allow us to perceive the gist of sets of visual features like the average direction of snow blowing in a blizzard.

The modern era of summary statistical research can be divided into two stages. Psychophysical work in the 1980s and 1990s demonstrated that humans integrate low-level motion into something akin to an ensemble percept (Watamaniuk & Duchon, 1992; Watamaniuk, Sekuler, & Williams, 1989; Williams & Sekuler, 1984). These researchers proposed straightforward mechanisms for perceiving the average; local information may be pooled across a population of low-level motion detectors operating in parallel (Watamaniuk & McKee, 1998). Although these early accounts did not explicitly refer to ensembles or summary statistics, they laid the groundwork for a flood of modern work across an

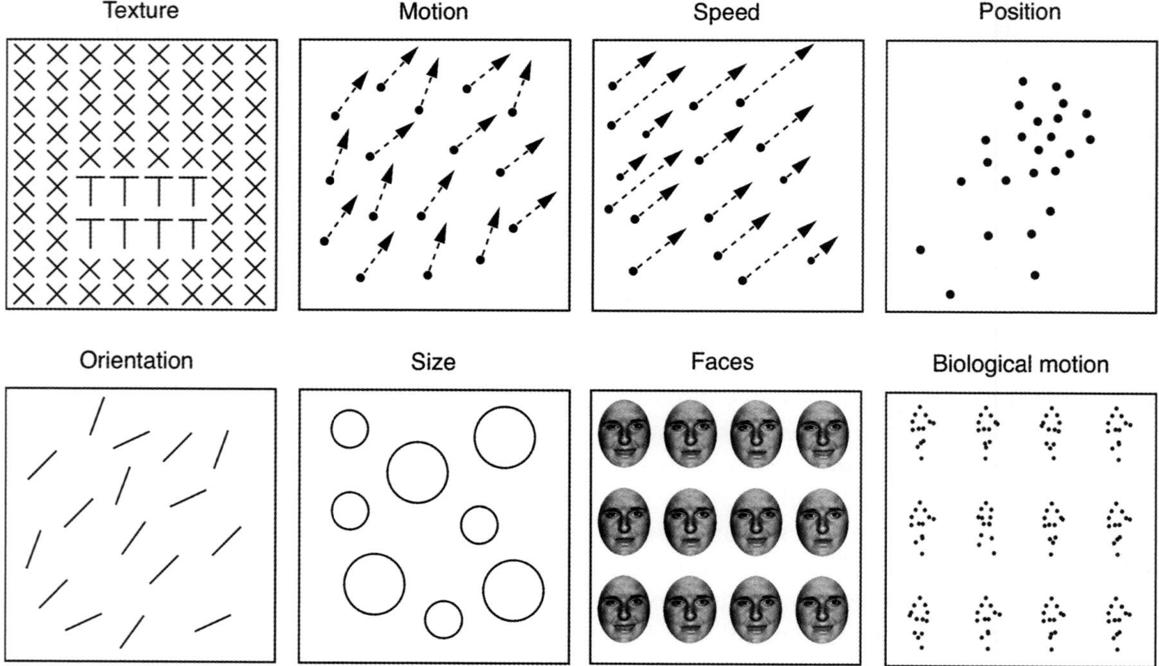

FIGURE 49.1 Summary statistical perception occurs across a wide range of stimuli. The flexibility of ensemble representation suggests that it occurs across multiple levels along the visual hierarchy.

 DAVID WHITNEY, JASON HABERMAN, AND TIMOTHY D. SWEENY

impressive range of low- to high-level visual features (figure 49.1).

The current era in the study of ensemble statistics began with the discovery that humans also derive a summary representation for the *size* of a set of arbitrary objects (Ariely, 2001; Chong & Treisman, 2003, 2005a, 2005b) and that this summary representation is favored over a representation of the individual items composing the set. The striking aspects of this research are twofold. First, it suggested that we perceive ensembles implicitly, perhaps through parallel mechanisms. Second, it showed that summary statistical perception could occur for objects. This raises several interesting questions, including these: Are there low-level feature detectors designed to operate on object size in a manner akin to motion or orientation? If not, how does average size perception, if it is indeed parallel, bypass traditional limitations of serial attention? Does ensemble coding extend beyond low-level stimuli to complex objects important for social interactions (e.g., faces and moving bodies)?

Although open questions remain (some of which are addressed below), it is clear that ensemble coding is connected to several areas of vision science, and this, in part, explains the growing interest in summary statistical perception. In addition to providing a window on gist, ensemble perception has implications for the way we understand visual search, texture, depth, scene perception, object recognition, spatial vision, attention, and awareness. The remainder of this chapter surveys the history of this subfield, highlights in greater detail some of the more influential work, and speculates as to where future work should be directed.

MULTIPLE LEVELS OF SUMMARY STATISTICAL PERCEPTION

We begin our survey by discussing which kinds of visual features are known to form ensemble percepts and at what levels of visual processing these summary representations might be formed. The purpose of this approach is to highlight the incredible versatility of ensemble coding for perceiving various feature dimensions and to reveal the possible stages of visual processing in which ensemble codes are likely to be generated. Throughout our review we highlight evidence that ensemble percepts are formed implicitly and automatically but are nonetheless susceptible to manipulations of attention. As with the study of almost any visual process, exploring the role of attention will help to reveal the underlying mechanisms of summary statistical perception and refine our understanding of awareness.

Perceiving Average Motion and Speed

As mentioned above, low-level motion was one of the first visual features shown to produce an ensemble or gist percept. Humans precisely perceive the average direction of a group of dots moving along unique local vectors (Watamaniuk, Sekuler, & Williams, 1989; Williams & Sekuler, 1984). Similar results are found for a group of dots that vary in speed (Watamaniuk & Duchon, 1992). Rather than perceiving each moving dot individually, the dominant percept is the average direction or speed of motion. Although interesting on their own, these demonstrations also provide descriptions of pooling mechanisms that might underlie summary statistical perception with other features. For example, perceiving the average direction of motion from a set of moving dots (or blowing snow) is consistent with established physiological mechanisms of motion perception (Britten & Heuer, 1999; Britten et al., 1992; Newsome & Pare, 1988); information may be pooled across low-level motion detectors operating in parallel, potentially obviating the involvement of serial attention (Watamaniuk & McKee, 1998; but see also Bulakowski, Bressler, & Whitney, 2007). These early accounts were not referred to as ensemble perception per se, but they clearly provide some of the first evidence of summary statistical perception.

Perceiving Average Position

Several psychophysical experiments have shown that humans are sensitive to average or centroid position (Hess & Holliday, 1992; Morgan & Glennerster, 1991; Whitaker et al., 1996). Recent work shows that this sensitivity is based on a real statistical decision, much like a *t*-test. Fouriezos, Rubenfeld, and Capstick (2008) found that in an attempt to judge which of two crowds of vertically oriented bars had the greater average height (which could rely in part on a judgment of the average position of the top of the bars), performance was improved when groups with more bars were visible, but it was impaired when these groups had higher variability.

Perceiving average position can occur when awareness is compromised, such as in crowding. Greenwood, Bex, and Dakin (2009) asked observers to indicate whether a horizontal bar intersected above or below the midpoint of a peripherally located vertical bar. Similar flankers with above- or below-the-midpoint intersections accompanied the target. Position information in the flankers influenced where observers perceived the intersection in the target; observers based their decisions on the pooled position information across the flankers and target—*position averaging*. Complementary work by Alvarez and Oliva (2008) further suggests that selective attention may play a minimal role in ensemble position perception. Using a multiple object-tracking task (Intriligator & Cavanagh, 2001; Pylyshyn & Storm, 1988), Alvarez and Oliva (2008) found that even when observers were unable to localize individual unattended objects, they could localize the centroid of those objects, and their performance was perfectly predicted by averaging the noisy representations of the individual objects. Although Chong and Treisman (2005b; discussed below) demonstrated that distributed attention could improve an estimate of the mean, this work (Alvarez & Oliva, 2008) suggested that ensemble position might be derived even beyond the focus of attention.

Perceiving Average Orientation

Humans readily perceive average orientation. Dakin (2001) was one of the first to demonstrate this when he showed that humans pool the orientations of multiple Gabor patches to estimate the mean orientation of a texture composed of heterogeneous orientations. This pooling occurs over large areas of space, and the number of samples used to estimate the average is approximately the square root of the size of the set. This suggests that, unlike tracking multiple objects (Franco-neri, in press), perceiving ensembles may not be limited by a fixed number of features. In fact, ensemble orientation perception becomes more precise when more items are in a set; Robitaille and Harris (2011) showed higher precision and reduced response times when larger sets were available to make mean orientation and size judgments. The efficiency of this pooling, however, can be compromised when attention is overloaded; Dakin and colleagues (2009) showed that an attentionally demanding task reduced the effective number of local orientations observers used to estimate the mean.

There is both psychophysical and physiological evidence suggesting that representing average orientation is a parallel process. Some of the strongest evidence for this comes from Parkes and colleagues (2001), who showed that the orientation of a Gabor patch crowded out of awareness (i.e., observers were unable to discriminate its orientation) nonetheless influenced the perceived average orientation of an entire set of surrounding patches. Even though observers could not consciously individuate or scrutinize the target Gabor patch, orientation detectors could process the set in parallel and subsequently pool the information into a summary percept. A similar conclusion was reached by Alvarez and Oliva (2009). These results with crowding suggest that an orientation averaging system is not directly dependent on mechanisms of selective attention. This is consistent with the notion that average orientation representation reflects an automatic, low-level physiological mechanism (Bosking, Crowley, & Fitzpatrick, 2002; Victor et al., 1994; Vogels, 1990). Several accounts have advocated a back-pocket model of visual texture (see chapter 45 by Landy), in which a second-stage mechanism pools the outputs of local filters, as a plausible mechanism for ensemble orientation perception.

Although it is clear that crowding is not necessary for the extraction of ensemble information, one intriguing possibility is that it enhances the precision of the summary representation. Similar to distributed attention that improved average size representation (Chong & Treisman, 2005a), crowding (Levi, 2008; Pelli, Palomares, & Majaj, 2004; Whitney & Levi, 2011) by definition disrupts any serial attentive process (Intriligator & Cavanagh, 2001), which may force observers into an attentional strategy more conducive to summary representation. Thus, crowding might facilitate the condensation of (even consciously inaccessible) information into efficient "chunks."

Perhaps even more intriguing is the notion that people may actually *perceive* the mean orientation. For example, Morgan, Chubb, and Solomon (2008) as well as Ross and Burr (2008) showed that orientation pooling allows people to gauge the variance in a texture of orientations, and as long as the overall variability across the set is below a certain threshold, each local element takes on the appearance of the mean orientation of the group (Parkes et al., 2001). In other words, we actually perceive the mean even when it is physically absent from the set.

Perceiving Average Size

The current era in the study of ensemble statistics began when Ariely (2001) provided evidence that observers could implicitly derive the average of a set of differently sized dots. In fact, this summary representation was the favored representation. Observers viewed sets of dots for 2 s and then indicated whether a subsequently viewed test dot was a member of the set. The striking aspect of these data was not just that observers performed poorly at the member identification task. As the size of the test dot approached the average size of the array of dots, observers were much more likely to respond that the test dot was a member of the set. Even though observers were instructed to attend to the individual members, they instead represented the summary of the set constituents. When explicitly asked, observers were nearly as precise in discriminating the mean size

 DAVID WHITNEY, JASON HABERMAN, AND TIMOTHY D. SWEENY

of several dots as they were in discriminating the size of a single dot. As with orientation (Dakin, 2001), mean discrimination performance seemed invariant to the number of dots in the set (up to 16), possibly suggesting that serial attention mechanisms were not required.

The impact of this seminal work is probably responsible for the fact that several, if not the majority of, investigations of ensemble perception have regarded the perception of size. Accordingly, this subsection of our review is more extensive than the others. This is not to say that summarizing information about size is more important than summarizing other features (e.g., orientation or faces). It merely reflects the abundance of relevant work.

Seeing average size is not just a physical calculation. Im and Chong (2009) harnessed the Ebbinghaus illusion to create sets of circles that differed in their perceived size and physical size. The ensemble percept followed the perceived size, which is probably encoded in V1 at the earliest (Arnold, Birt, & Wallis, 2008; Murray, Boyaci, & Kersten, 2006). Choo and Franconeri (2010) provided further evidence that ensemble size is computed in early stages of visual processing. When they used masking to truncate the representation of the size of a subset of circles to lower-level stages of processing, these circles continued to influence the perception of the mean size.

As with average orientation perception, perception of average size follows statistical rules. de Gardelle and Summerfield (2011) found that observers discounted extreme values (i.e., outliers) when computing the mean shape of a set of circles and squares, suggesting that variance may be encoded in addition to the average (similar outlier exclusion has been found with faces [Haberman & Whitney, 2010]). Solomon, Morgan, and Chubb (2011) provide complementary findings in which observers were able to perceive the variance of a set of circles (or oriented Gabors [Solomon, 2010]) more efficiently than they could perceive the mean size (or orientation). The notion that we are all statisticians appears to be more than an anecdote.

Most investigations of summary statistical perception involve estimating the mean of a static array of features, but the world is dynamic, and recent research shows that the visual system accounts for this by pooling information across time. Albrecht and Scholl (2010) showed that a pooling mechanism takes multiple samples across time to precisely represent the average size of a continuously changing circle.

Average size perception is also robust and appears to occur automatically and without intention. Chong and Treisman (2003) showed that perceiving which of two sets of 12 circles had a larger mean size was immune to changes in presentation (simultaneously versus successively) and duration (even with 50-ms presentations, although see Whiting & Oriet, 2011, for evidence that 200 ms is a more appropriate lower limit). Moreover, observers' discrimination of the average size of the set was nearly as precise as their discrimination of the size of a single circle. The precision of mean representation (at least for size) is best when attention is spread over a large spatial extent (Chong & Treisman, 2005a). Nevertheless, Demeyere and colleagues found that a patient with simultanagnosia (Balint syndrome) could perceive ensemble size (and color) in an array of stimuli despite severely limited spatial attention (Demeyere et al., 2008).

Average size is even computed across multiple sets, in parallel, preceding or perhaps bypassing limitations imposed by the attentional bottleneck. Chong and Treisman (2005b) found that when observers discriminated the average size of a subset of an array of circles that was segregated from the rest of the array by color, average size perception did not depend on whether the color cue preceded or followed the array of circles and was no worse even when only a single color was presented. Ensemble size perception is even possible when attention is divided across stimulus modalities. Albrecht, Scholl, and Chun (2011) asked observers to listen to a sequence of tones while simultaneously viewing a sequence of differently sized disks. Depending on a cue, they made a subsequent judgment about the mean of one set or the other. Ensemble tone or size judgments were unaffected by whether or not the cue preceded or followed the sequences. In other words, dividing attention across the two modalities had no cost on the efficiency of perceiving the means. There is, however, a cost in ensemble precision when attention is divided between two feature dimensions (e.g., size and speed) (Emmanouil & Treisman, 2008).

Attentional manipulations may not only affect the precision or efficiency of the ensemble code but, under certain conditions, can also bias the representation in predictable ways. For example, priming observers to a dot of a particular size (either the largest or the smallest dot) biased estimates of mean size in the direction of the prime (de Fockert & Marchant, 2008). One interpretation is that observers allocated more resources to the primed dot, resulting in a biased estimate of mean size or a representation that reflected only a spatially proximal subset of the dots. A complementary result found by Brady and Alvarez (2011) is that the mean size estimate reflects ready access to one of several hierarchical means within the set. Observers were biased by the mean size of a set even though they were asked to attend to an individual member. Interestingly, this bias emerged only when attention was directed to a particular feature dimension (i.e., whether the member came from a red or a blue set, which were simultaneously presented). These studies suggest that observers represent multiple means simultaneously, but that those representations are predictably biased by attention.

Although the role of attention in average size representation is an ongoing debate (Ariely, 2008; Chong et al., 2008; Myczek & Simons, 2008; Simons & Myczek, 2008), these studies provide support for the existence of an automatic mechanism responsible for average size computation.

Perceiving Ensembles of Faces

For many years, the focus of research on summary statistical perception has been on low-level stimuli (motion, orientation, position, size, etc.). However, given our effortless interaction with highly complex scenes and our subjective impression of a rich and complete visual world, it is reasonable to think that the ensemble coding heuristic might operate on a processing level beyond that of orientation, size, or texture. Haberman and Whitney (2007, 2009) and Haberman, Harp, and Whitney (2009) explored the possibility that observers could extract an average representation from high-level stimuli, including faces. The authors created a series of morphs, varying the expression of faces ranging from extremely happy to extremely sad. Observers viewed sets of these emotionally varying faces and were asked whether a subsequent test face was happier or sadder than the mean expression of the previous set. Remarkably, observers could discriminate the average expression of the whole set as well as they could discriminate the expression of a single face. This phenomenon proved to be robust and flexible, operating implicitly and explicitly (Haberman & Whitney, 2009), across a variety of expressions as well as gender morphs (Haberman & Whitney, 2007), at short exposure durations (as low as 50 ms, although with reduced precision (Haberman & Whitney, 2009), and on sets containing as many as 20 faces (Haberman, Harp, & Whitney, 2009; see figure 49.2). Control experiments demonstrated that the mean discrimination of expression declined when subjects viewed sets of inverted or scrambled faces, suggesting that the visual system extracts summary statistical information about the configural or holistic properties of faces, not just about low-level visual cues such as spatial frequency (Oliva & Torralba, 2001; Torralba & Oliva, 2003) or orientation.

Perceived facial expression also rapidly integrates over time (Haberman, Harp, & Whitney, 2009). Observers viewed sequences of different faces presented at various temporal frequencies and made judgments about the mean expression of those sequences. Observers were able to accurately derive a mean expression in a sequence of 20 faces presented at 20 Hz. The ensemble required 800 ms of temporal integration. Although this integration time is higher than that for low-level motion (Burr, 1981; Nakayama, 1985; Snowden & Braddick, 1989), it compares favorably with the time it takes the visual system to perceive biological motion (Neri, Morrone, & Burr, 1998) and suggests the work of a parallel mechanism.

Although the results from many of these investigations of summary perception and countless accounts of crowding suggest that access to information about individuals (e.g., the size of a particular circle or the expression of a particular face) is lost when the summary statistic is calculated (e.g., Rosenholtz, 2011), there is also good reason to suspect that holistic object-level representations remain intact but are somehow blocked from conscious access in favor of the ensemble. Fischer and Whitney (2011) demonstrated a result that built on that from Parkes and colleagues (2001) in which the expression of a face crowded out of awareness influenced the average expression perceived in a crowd. Crucially, this pooling did not occur when the crowded face was inverted; the ensemble only incorporated holistic information from the face, indicating that a high-level object representation was intact and influencing the average but was not accessible to awareness.

High-level ensemble coding is further supported by other work showing that observers can rapidly perceive the mean identity of sets of faces (de Fockert & Wolfenstein, 2009; Yamanashi-Leib et al., 2012) as well as research showing rapid within-hemifield emotional averaging predicted by properties of neural averaging (Sweeny et al., 2009).

Perceiving Ensembles of Biological Motion

If ensemble perception occurs for high-level forms such as faces, then it is reasonable to believe that it might also occur for perception of biological motion, a high-level visual feature defined by the integration of form and motion. Sweeny, Haroz, and Whitney (2012a) used a design reminiscent of that used by Dakin (2001) to examine this possibility. Observers estimated the headings of briefly presented crowds of point-light walkers that differed in the number and headings of their members (i.e., people in differently sized crowds had identical or increasingly variable directions of walking). They found that observers rapidly pooled information from multiple walkers to precisely estimate the overall direction of the crowd. The striking aspect of these

     DAVID WHITNEY, JASON HABERMAN, AND TIMOTHY D. SWEENY

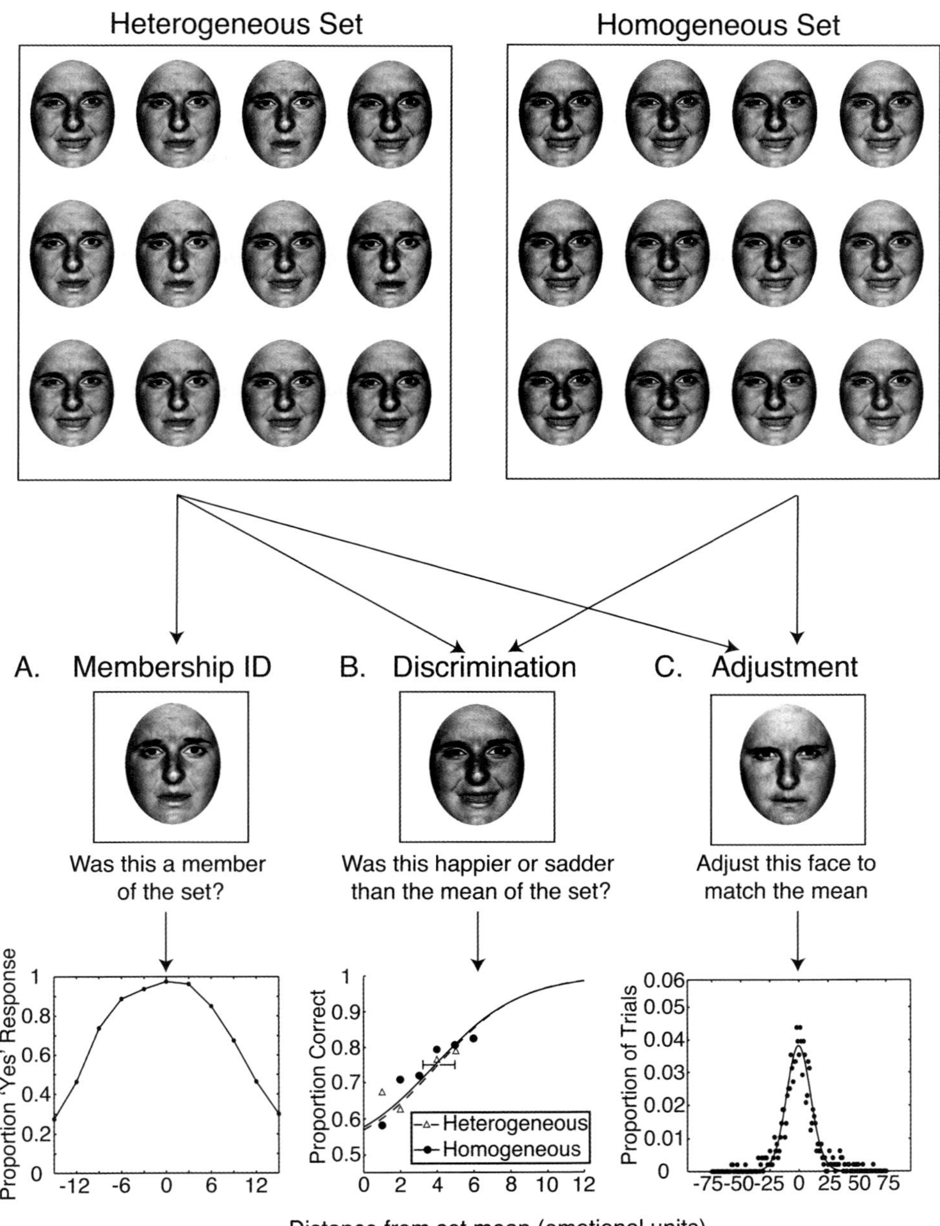

FIGURE 49.2   Several ensemble perception paradigms. Observers view sets of stimuli (e.g., faces). (A) In one experiment observers had to identify whether a test face was a member of the previously displayed set (membership ID). Observers were most likely to indicate a test face was a set member when it approached the mean expression of the set (0 indicates the mean expression). Thus, observers were unable to represent the individual set constituents but instead favored the ensemble. (B) Observers were explicitly asked about the average expression in a set (discrimination task). Surprisingly, they could discriminate the mean expression as well as they could discriminate any single face. (C) Observers used the mouse to adjust the test face to match the mean expression of the set. This provided the full error distribution of the mean representation (0 indicates the mean expression). Responses tended to cluster around the mean expression of the set. (Adapted from Haberman & Whitney, 2011b.)

results is that pooling across noisy individual walker directions allowed observers to perceive the direction of a crowd better than the direction of an individual. This precise ensemble percept required upright orientations and human configurations, suggesting that the code was formed in high-level visual areas where form and motion are integrated. Moreover, it occurred within a surprisingly brief amount of time (200 ms), showing that ensemble percepts of even the most complex visual stimuli can be formed in parallel.

This brief survey is necessarily incomplete, but it provides a glimpse at the history of ensemble perception. The robust summary of statistical representations found across domains suggests that ensembles are calculated at multiple levels in both the dorsal and ventral streams. Some ensembles, such as average brightness, color, and orientation, may be generated at the earliest cortical (and possibly even subcortical) stages. Others, such as motion and position, may be generated along the dorsal stream. High-level shape and face ensembles are likely generated along the ventral, object-processing stream. Finally, biological motion ensembles are likely generated after the convergence of the dorsal and ventral pathways. Because ensemble percepts can emerge at independent levels of analysis, for example on holistic representations of faces independent of the ensemble brightness, orientation, or facial features in a scene, no single visual or cortical area is likely to be responsible for ensemble perception. Consequently, although there are several physiologically inspired models that might generate ensemble representations at single levels of visual processing (Balas, Nakano, & Rosenholtz, 2009; Freeman & Simoncelli, 2011; Rosenholtz et al., 2012), they are not prepared to capture the repetitive and independent nature of ensemble representations at multiple stages along the visual hierarchy. This holds especially true for high-level objects such as crowds of walking humans or faces.

Despite the distinct object properties processed at each level, the uniting commonality is that any set may be represented by a single ensemble percept. This percept is created and maintained for conscious access, while the individual constituents are lost (via limitations of visual working memory, crowding, etc.). Because the visual system creates a representation of many of the items within a set, (conscious) loss of the individual is inconsequential. Many unanswered questions remain, such as how many concurrent ensemble percepts can be maintained, whether there is interference between different levels of ensemble analysis (e.g., average facial expression, brightness, and orientation), and whether the ensembles bypass the limited capacity of attention and visual short-term memory or instead simply act as "chunks" of information, increasing processing efficiency while still drawing on the finite resources of attention and memory.

It is easy to imagine how summary statistics explain texture appearance—the graniteness, stucconess, and so on of surfaces. Although textures have been extensively studied (Beck, 1983; Landy & Graham, 2004;

Malik & Rosenholtz, 1997; Nothdurft, 1991), and summary statistical representation of low-level features holds for typical textures, the finding that groups of faces or walking people are perceived as an ensemble—as a texture—suggests that textures can occur at multiple, distinct levels of the visual processing hierarchy.

Is Ensemble Perception Just a Prototype?

The demonstration of summary statistical representation for faces may raise the concern that the results are simply due to a prototype effect (Solso & McCarthy, 1981). Indeed, there has been significant research providing evidence that observers implicitly develop statistical sensitivities to arbitrary patterns over time (Fiser & Aslin, 2001; Posner & Keele, 1968). However, unlike the prototype effect, ensemble coding requires no learning; summary statistical representation is a perceptual process, and observers are sensitive to it after only a single trial. Prototypes suggest that observers falsely recognize an average face due to predominant exposure to specific facial features over an extended period (Solso & McCarthy, 1981). The average face (or size, orientation, etc.) in ensemble coding, though, changes on a trial-by-trial basis and is immediately recognizable. Ensemble perception is therefore a much more flexible pooling of important information into computationally palatable chunks. Observers never actually see the average face of a set, and yet they favor the ensemble percept over the individuals.

ENSEMBLES AS AN EXPLICIT CODE

Given the explosion of work convincingly showing that the visual system is sensitive to summary statistics, and the ease with which they are represented, there has been surprisingly little work exploring the supporting mechanisms. Although there have been a few studies that have addressed this directly, as described below, there are many fundamental questions left unanswered. For example, how are ensembles computed in the brain? Are there multiple levels of representation corresponding to the level at which each exemplar is analyzed (e.g., is average orientation computed in early visual cortex, and average emotion computed in face selective regions)?

Recent work exploring how ensembles are represented demonstrated an aftereffect specific to the average size. Aftereffects occur as a result of neural adaptation and reveal that a given feature is directly represented in the brain (e.g., Suzuki, 2005). In other words the existence of an aftereffect demonstrates dedicated neural coding for a given feature dimension. In their study Corbett and colleagues (2012) had observers adapt to sets of dots varying in size and then judge which of two test circles was larger. The test dots were perceived as smaller after adaptation to a set with a large mean size, and vice versa, suggesting that average size is an explicitly represented feature dimension. However, one concern with the study is that the authors did not distinguish between local adaptation to the individual elements (which must occur) and adaptation to the average. To clearly establish that these aftereffects indeed reflected adaptation to the mean size, the results must demonstrate adaptation independent of what is predicted by local adaptation effects. Nevertheless, this finding makes a clear prediction for future research; it should be possible to identify and characterize an ensemble representation in the brain.

So while evidence is accumulating to support the notion that summary statistics are directly represented in the brain, the algorithm for computing this summary value is not entirely clear. A linear pooling mechanism is probably the most plausible and popular candidate (see figure 49.3). This type of mechanism is fairly straightforward (representations of individual features are integrated, and the average is determined), and modeling illustrates that it can provide a reasonable approximation for perception of average orientation (Parkes et al., 2001) and average biological motion (Sweeny, Haroz, & Whitney, 2012b). These pooling

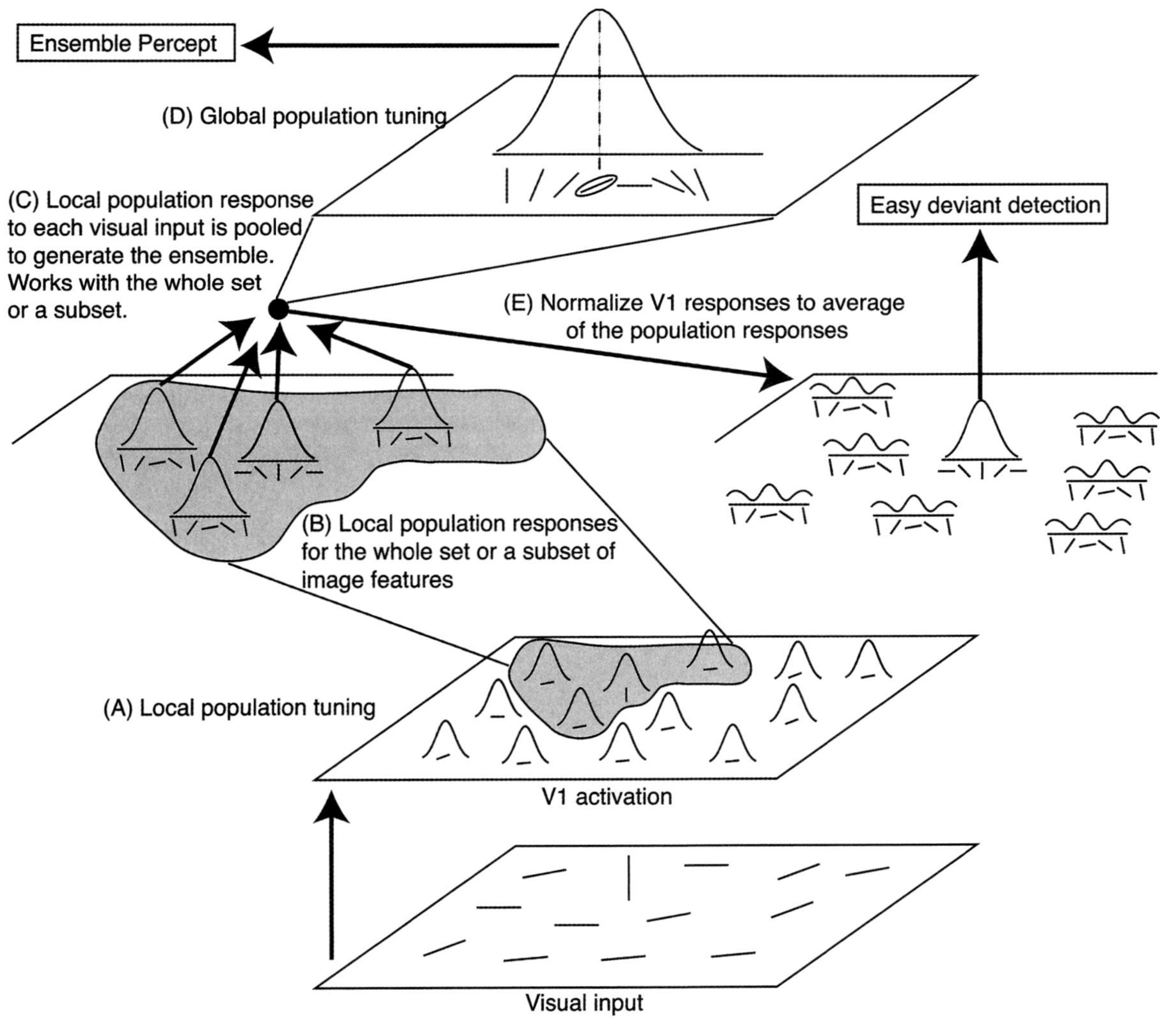

FIGURE 49.3   One possible physiological mechanism driving pop-out. (A, B) Orientation-selective cells (possibly in V1) fire in response to visual input. (C, D) The activity from some or all of the orientation-selective cells is combined to create the ensemble. (E) Via feedback or horizontal connections, the activity from orientation-selective cells is normalized to the population response (i.e., ensemble). Any cell activity remaining will correspond to the deviant. One of the strengths of this model is that it can operate in parallel, negating the computationally inefficient method of comparing each item with every other one. (Adapted from Haberman & Whitney, 2011b.)

models include Gaussian-shaped early-stage channel noise in the encoding of each individual feature and a subsequent stage in which the outputs of these channels are combined, averaged, and perturbed by late-stage Gaussian-shaped noise. For example, the outputs of several orientation-tuned channels could be integrated into a population code. The centroid of this population of noisy individual feature representations would determine the average orientation. Population coding such as this is common (Suzuki, 2005), and it is reasonable to assume that it could apply to perception of ensembles.

Pooling, especially across large subsets of features, affords the prospect of averaging out noisy estimates of local features (Dakin, 2001; Ross & Burr, 2008; Morgan, Chubb, & Solomon, 2008). Moreover, linear pooling predicts that when encoding is particularly noisy, the average should be as precise as (or even more precise than) perception of an individual. Such precision has been borne out in several investigations (Alvarez & Oliva, 2008; Ariely, 2001; Bulakowski, Bressler, & Whitney, 2007; Haberman & Whitney, 2009; Sweeny, Haroz, & Whitney, 2012a). For example, a recent study by Yamanashi-Leib et al. (2012) found that prosopagnosics, who have difficulty recognizing individual faces (i.e., have noisy face recognition), nevertheless recognize crowds surprisingly well.

Choo and Franconeri (2010) provide a compelling explanation of how pooling might be implemented in the brain. They note that the mandatory pooling that occurs when objects are nearby in space (i.e., in what are referred to as integration fields by Pelli, Palomares, & Majaj, 2004, or areas of minimal attentional resolution by Intriligator & Cavanagh, 2001) is consistent with the length of horizontal connections in V1 and the size of receptive fields in V4. They suggest that averaging is a result of integration through these connections in lower-level areas and pooling within receptive fields in high-level areas. This hypothesis is consistent with the fact that ensemble perception is better when attention is diffusely spread (Chong & Treisman, 2005a). Furthermore, Sweeny and colleagues (2009) provide direct empirical support for this speculation. In their investigation observers viewed a briefly and simultaneously presented pair of faces with different emotional expressions and were postcued to rate the emotional expression of just one face in the pair. Critically, the locations of the faces were varied such that both faces either fell within large receptive fields of high-level neurons (both within a hemifield) or in separate receptive fields (each in a separate hemifield). Perceptual averaging occurred (the expression of a given face appeared more like the average of the pair), but only when the faces fell within

what would be expected to be the same receptive fields of high-level face-tuned neurons.

Although these studies hint that ensembles may be a fundamental representation of the visual system, much work remains to be done to determine whether multiple levels of representation exist and to characterize the algorithm more precisely. In our view these represent some of the most challenging and exciting avenues for future research in the field.

IMPLICATIONS OF ENSEMBLE CODING

Overall, it is clear that although ensemble percepts may not be completely independent of attention, they do occur implicitly. What are the broader implications of implicit statistical summarization of the environment? How does this knowledge inform traditional notions of awareness? We explore these questions in this section.

Bypassing the Bottleneck

The discovery that ensembles could be represented implicitly led several researchers to speculate that summary perception might drive the sense of visual completeness in spite of limited awareness (Cavanagh, 2001; Chong & Treisman, 2003; Haberman & Whitney, 2009). Even though the visual system can explicitly represent just a few items simultaneously (e.g., Luck & Vogel, 1997; Franconeri, in press), the world beyond the focus of attention does not fade to black. In fact, the objects and scenes we are not attending to seem remarkably rich. It is reasonable to speculate that this "grand illusion" may be due, in part, to summary statistical perception. The statistics of natural scenes (e.g., Simoncelli & Olshausen, 2001) are, indeed, quite stable (Oliva, 2005; Torralba & Oliva, 2003), and the visual system efficiently exploits this natural redundancy by generating summary percepts. Several recent findings support this hypothesis by showing (1) how ensemble representations provide strikingly precise percepts in spite of noisy encoding of individual details and (2) that moment-to-moment awareness of visual scenes more closely follows ensemble representations than abrupt changes in individual features.

Sweeny, Haroz, and Whitney (2012a) demonstrated that pooling across multiple noisy features produces an ensemble percept that surpasses the precision with which we can perceive an individual. Observers viewed a crowd composed of individual people with different directions of walking. Because the crowd spanned a large spatial extent and was only visible for 200 ms, the encoding of each walker was noisy. Nevertheless, observers perceived the crowd's average direction of walking more precisely than they perceived a single foveally presented person's direction. This shows that even though perception of a given feature in the periphery may be poor, perception of the group (or the whole scene) truly is precise.

Similar high-resolution pooling occurs outside the focus of attention. Alvarez and Oliva (2008) found that observers were just as good at reporting the average position of a set of dots they had been tracking as with dots they had not been tracking (i.e., beyond the focus of attention). Modeling showed that, although the position representation of the individual dots beyond the focus of attention was noisy, as expected, the average of these noisy representations nonetheless accurately predicted performance on the average position task. This suggests that ensemble information was preserved in spite of limited awareness, and it supports the assertion that ensembles provide an efficient means to maintain perceptual stability (i.e., deriving a precise ensemble code even though only noisy information was available).

Moment-to-moment awareness of a scene closely follows ensemble representations even when abrupt changes in individual features go unnoticed. Alvarez and Oliva (2009) showed that while engaged in an attentionally demanding tracking task, observers were explicitly aware of changes in the background that altered the average orientation of the top and bottom halves of the screen. However, observers were oblivious to changes of the same magnitude that preserved the overall ensemble (i.e., the average orientation). Thus, information regarding global scene statistics remains available even in the face of exhausted attentional resources.

Complementary evidence that ensembles provide low-cost perceptual stability comes from a dual-task paradigm employed by Haberman and Whitney (2011a). Observers viewed two sets of 16 successively presented faces on each trial. Within a given trial, 4 of the 16 faces changed from one emotional extreme to another (e.g., four happy faces turned sad). This shift created a change in the overall mean emotion of the set. Observers were instructed to identify (1) which of the two sets was on average happier (ensemble task) and (2) the location of just one of the four changes (change localization task). In trials in which change localization failed (i.e., when the observer could not report where on the screen the change occurred), observers were nonetheless significantly above chance in identifying which set was on average happier. Although change localization reflects the limitations of explicit awareness, ensemble codes seem to bypass these limitations. Taken together, these studies point to

a robust and efficient heuristic at work, one that can maintain the stability of our visual world in the face of limited information.

Visual Search

The possible connection between ensemble coding and visual search is appealing. Despite the rich literature on the properties of visual search (Treisman, 1982; Verghese, 2001; Wolfe, Cave, & Franzel, 1989), a physiologically plausible mechanism (e.g., an algorithm or neural implementation; Marr, 1982) that generates popout is still debated (Eckstein, 1998; Itti & Koch, 2000; Wolfe, 2003). Summary statistical representations—ensemble coding—may serve as a computationally efficient means of calculating deviance. Several models have made similar suggestions (e.g., Callaghan, 1984; Duncan & Humphreys, 1989). Often these models suggest that similarity influences popout (Duncan & Humphreys, 1989). However, what counts as "similar" or "dissimilar" is unclear. Summary statistical representations, per se, could provide the underlying metric of similarity, one that affords deviance detection (Rosenholtz et al., 2012). Recent accounts of visual search also acknowledge the possibility that much of the periphery may be represented as an ensemble and processed preattentively and that this nonselective ensemble pathway generates a gist impression that guides a selective pathway, leading to more efficient real-world search (Wolfe et al., 2011).

How might a very simple, physiologically plausible, population-coding algorithm extract ensemble information and generate popout? Figure 49.3A shows an example of an array of oriented lines that might stimulate many local populations of orientation-selective cells (e.g., in V1). If a subset of *locally* tuned receptive fields is sampled (figure 49.3B), and its output is pooled (figure 49.3C), a *global* population tuning curve is represented [only a subset of the items needs to be sampled]; cf. Dakin & Watt, 1997; Morgan, Chubb, & Solomon, 2008; Myczek & Simons, 2008). This global population curve is the average of local tuning curves and ultimately produces an ensemble percept (figure 49.3D). Note that the impact of any deviant orientation is mitigated in the global population curve, as most of the inputs are of similar orientations. The global population response then normalizes the local tuning (via feedback or horizontal connections; figure 49.3E). Most of the local population responses are reduced to near 0, and what is left is activity corresponding to the deviant orientation. Although low-level normalization or contextually dependent procedures have been implemented in other models (e.g., Itti, Koch, & Niebur,

1998; Li, 1999), this model implicates ensemble coding and the generation of ensemble percepts as the basis for popout. A particular strength of this model is that the normalization operation may be carried out in parallel, without repetitive comparisons across local population responses.

CONCLUSION

Despite many bottlenecks in visual processing and the limited nature of awareness, humans rapidly extract an enormous amount of information from scenes (Oliva & Torralba, 2001; Potter, 1976; Thorpe, Fize, & Marlot, 1996; Torralba & Oliva, 2003). It is becoming clear that much of this information may take the form of condensed summary statistics—computationally efficient ensemble representations of similar features and objects in scenes. Ensembles are encoded from the lowest levels of feature processing to the highest levels of object and face perception. Ensemble perception occurs quickly, automatically, and outside the focus of attention, although it is also modulated by attention. More broadly, ensemble perception may underlie much of our impression of perceiving a complete and rich visual world.

REFERENCES

Albrecht, A. R., & Scholl, B. (2010). Perceptually averaging in a continuous visual world: Extracting statistical summary representations over time. *Psychological Science, 21,* 560–567.

Albrecht, A. R., Scholl, B., & Chun, M. M. (2011). Perceptual averaging by eye and ear: Computing summary statistics from multimodal stimuli. *Journal of Vision, 11*(11), 1210. doi:10.1167/11.11.1210.

Alvarez, G. A. (2011). Representing multiple objects as an ensemble enhances visual cognition. *Trends in Cognitive Neurosciences, 15,* 122–131. doi:10.1016/j.tics.2011.01.003.

Alvarez, G. A., & Oliva, A. (2008). The representation of simple ensemble visual features outside the focus of attention. *Psychological Science, 19,* 392–398.

Alvarez, G. A., & Oliva, A. (2009). Spatial ensemble statistics are efficient codes that can be represented with reduced attention. *Proceedings of the National Academy of Sciences of the United States of America, 106,* 7345–7350.

Anderson, N. H. (1965). Averaging versus adding as a stimulus-combination rule in impression-formation. *Journal of Experimental Psychology, 70*(4), 394–400.

Anderson, N. H. (1971). Integration theory and attitude change. *Psychological Review, 78,* 171–206.

Anderson, N. H., Lindner, R., & Lopes, L. L. (1973). Integration theory applied to judgments of group attractiveness. *Journal of Personality and Social Psychology, 26,* 400–408.

Ariely, D. (2001). Seeing sets: Representation by statistical properties. *Psychological Science, 12,* 157–162.

Ariely, D. (2008). Better than average? When can we say that subsampling of items is better than statistical summary representations? *Perception & Psychophysics, 70,* 1325–1326.

Arnold, D. H., Birt, A., & Wallis, T. S. A. (2008). Perceived size and spatial coding. *Journal of Neuroscience, 28,* 5954–5958.

Balas, B., Nakano, L., & Rosenholtz, R. (2009). A summary-statistic representation in peripheral vision explains visual crowding. *Journal of Vision, 9*(12), 1–18. doi:10.1167/9.12.13.

Battelli, L., Pascual-Leone, A., & Cavanagh, P. (2007). The "when" pathway of the right parietal lobe. *Trends in Cognitive Sciences, 11,* 204–210.

Beck, J. (1983). Textural segmentation, 2nd-order statistics, and textural elements. *Biological Cybernetics, 48,* 125–130.

Bosking, W. H., Crowley, J. C., & Fitzpatrick, D. (2002). Spatial coding of position and orientation in primary visual cortex. *Nature Neuroscience, 5,* 874–882.

Brady, T. F., & Alvarez, G. A. (2011). Hierarchical encoding in visual working memory: Ensemble statistics bias memory for individual items. *Psychological Science, 22,* 384–392.

Britten, K. H., & Heuer, H. W. (1999). Spatial summation in the receptive fields of MT neurons. *Journal of Neuroscience, 19,* 5074–5084.

Britten, K. H., Shadlen, M. N., Newsome, W. T., & Movshon, J. A. (1992). The analysis of visual-motion—a comparison of neuronal and psychophysical performance. *Journal of Neuroscience, 12,* 4745–4765.

Bulakowski, P. F., Bressler, D. W., & Whitney, D. (2007). Shared attentional resources for global and local motion processing. *Journal of Vision, 7*(10), 810–817. doi:10.1167/7.10.10.

Burr, D. C. (1981). Temporal summation of moving images by the human visual-system. *Proceedings of the Royal Society of London. Series B, Biological Sciences, 211,* 321–339.

Callaghan, T. C. (1984). Dimensional interaction of hue and brightness in preattentive field segregation. *Perception & Psychophysics, 36*(1), 25–34.

Cavanagh, P. (2001). Seeing the forest but not the trees. *Nature Neuroscience, 4,* 673–674.

Chong, S. C., Joo, S. J., Emmanouil, T. A., & Treisman, A. (2008). Statistical processing: Not so implausible after all. *Perception & Psychophysics, 70,* 1327–1334.

Chong, S. C., & Treisman, A. (2003). Representation of statistical properties. *Vision Research, 43,* 393–404.

Chong, S. C., & Treisman, A. (2005a). Attentional spread in the statistical processing of visual displays. *Perception & Psychophysics, 67,* 1–13.

Chong, S. C., & Treisman, A. (2005b). Statistical processing: Computing the average size in perceptual groups. *Vision Research, 45,* 891–900.

Choo, H., & Franconeri, S. L. (2010). Objects with reduced visibility still contribute to size averaging. *Attention, Perception & Psychophysics, 72,* 86–99.

Corbett, J. E., Wurnitsch, N., Schwartz, A., & Whitney, D. (2012). An aftereffect of adaptation to mean size. *Visual Cognition, 20,* 211–231.

Dakin, S. C. (2001). Information limit on the spatial integration of local orientation signals. *Journal of the Optical Society of America. A, Optics, Image Science, and Vision, 18,* 1016–1026.

Dakin, S. C., Bex, P. J., Cass, J. R., & Watt, R. J. (2009). Dissociable effects of attention and crowding on orientation averaging. *Journal of Vision, 9*(11), 1–16. doi:10.1167/9.11.28.

Dakin, S. C., & Watt, R. J. (1997). The computation of orientation statistics from visual texture. *Vision Research, 37,* 3181–3192.

de Fockert, J. W., & Marchant, A. P. (2008). Attention modulates set representation by statistical properties. *Perception & Psychophysics, 70,* 789–794.

de Fockert, J., & Wolfenstein, C. (2009). Rapid extraction of mean identity from sets of faces. *Quarterly Journal of Experimental Psychology, 62*, 1716–1722.

de Gardelle, V., & Summerfield, C. (2011). Robust averaging during perceptual judgment. *Proceedings of the National Academy of Sciences of the United States of America, 108*, 13341–13346.

Demeyere, N., Rzeskiewicz, A., Humphreys, K. A., & Humphreys, G. W. (2008). Automatic statistical processing of visual properties in simultanagnosia. *Neuropsychologia, 46*, 2861–2864.

Duncan, J., & Humphreys, G. W. (1989). Visual-search and stimulus similarity. *Psychological Review, 96*, 433–458.

Eckstein, M. P. (1998). The lower visual search efficiency for conjunctions is due to noise and not serial attentional processing. *Psychological Science, 9*, 111–118.

Emmanouil, T. A., & Treisman, A. (2008). Dividing attention across feature dimensions in statistical processing of perceptual groups. *Perception & Psychophysics, 70*, 946–954.

Fischer, J., & Whitney, D. (2011). Object-level visual information gets through the bottleneck of crowding. *Journal of Neurophysiology, 106*, 1389–1398.

Fiser, J., & Aslin, R. N. (2001). Unsupervised statistical learning of higher-order spatial structures from visual scenes. *Psychological Science, 12*, 499–504.

Fouriezos, G., Rubenfeld, S., & Capstick, G. (2008). Visual statistical decisions. *Perception & Psychophysics, 70*, 456–464.

Franconeri, S. L. (in press). The nature and status of visual resources. In D. Resiberg (Ed.), *Oxford handbook of cognitive psychology*. Oxford: Oxford University Press.

Franconeri, S. L., Alvarez, G. A., & Enns, J. T. (2007). How many locations can be selected at once? *Journal of Experimental Psychology. Human Perception and Performance, 33*, 1003–1012.

Freeman, J., & Simoncelli, E. P. (2011). Metamers of the ventral stream. *Nature Neuroscience, 14*(9), 1195–1201.

Greenwood, J. A., Bex, P. J., & Dakin, S. C. (2009). Positional averaging explains crowding with letter-like stimuli. *Proceedings of the National Academy of Sciences of the United States of America, 106*, 13130–13135.

Haberman, J., Harp, T., & Whitney, D. (2009). Averaging facial expression over time. *Journal of Vision, 9*(11), 1–13. doi:10.1167/9.11.1.

Haberman, J., & Whitney, D. (2007). Rapid extraction of mean emotion and gender from sets of faces. *Current Biology, 17*, R751–R753.

Haberman, J., & Whitney, D. (2009). Seeing the mean: Ensemble coding for sets of faces. *Journal of Experimental Psychology. Human Perception and Performance, 35*(3), 718–734.

Haberman, J., & Whitney, D. (2010). The visual system discounts emotional deviants when extracting average expression. *Attention, Perception & Psychophysics, 72*, 1825–1838.

Haberman, J., & Whitney, D. (2011a). Efficient summary statistical representation when change localization fails. *Psychonomic Bulletin & Review, 18*, 955–959.

Haberman, J., & Whitney, D. (2011b). Ensemble perception: Summarizing the scene and broadening the limits of visual processing. In J. Wolfe & L. Robertson (Eds.), *A festschrift in honor of Anne Treisman*. Oxford: Oxford University Press.

Hess, R. F., & Holliday, I. E. (1992). The coding of spatial position by the human visual-system—effects of spatial scale and contrast. *Vision Research, 32*, 1085–1097.

Im, H. Y., & Chong, S. C. (2009). Computation of mean size is based on perceived size. *Attention, Perception & Psychophysics, 71*, 375–384.

Intriligator, J., & Cavanagh, P. (2001). The spatial resolution of visual attention. *Cognitive Psychology, 43*, 171–216.

Itti, L., & Koch, C. (2000). A saliency-based search mechanism for overt and covert shifts of visual attention. *Vision Research, 40*, 1489–1506.

Itti, L., Koch, C., & Niebur, E. (1998). A model of saliency-based visual attention for rapid scene analysis. *IEEE Transactions on Pattern Analysis and Machine Intelligence, 20*, 1254–1259.

Koenderink, J. J., van Doorn, A. J., & Pont, S. C. (2004). Light direction from shad(ow)ed random Gaussian surfaces. *Perception, 33*, 1405–1420.

Koffka, K. (1935). *The principles of Gestalt psychology*. London: Routledge and Kegan Paul.

Landy, M., & Graham, N. (2004). Visual perception of texture. In L. M. Chalupa & J. S. Werner (Eds.), *The visual neurosciences* (Vol. 2, pp. 1106–1118). Cambridge, MA: MIT Press.

Leon, M., Oden, G. C., & Anderson, N. H. (1973). Functional measurement of social values. *Journal of Personality and Social Psychology, 27*, 301–310.

Levi, D. M. (2008). Crowding—an essential bottleneck for object recognition: A mini-review. *Vision Research, 48*, 635–654.

Levin, I. P. (1974). Averaging processes in ratings and choices based on numerical information. *Memory & Cognition, 2*, 786–790.

Li, Z. (1999). Contextual influences in V1 as a basis for pop out and asymmetry in visual search. *Proceedings of the National Academy of Sciences of the United States of America, 96*, 10530–10535.

Luck, S. J., & Vogel, E. K. (1997). The capacity of visual working memory for features and conjunctions. *Nature, 390*(6657), 279–281.

Malik, J., & Rosenholtz, R. (1997). Computing local surface orientation and shape from texture for curved surfaces. *International Journal of Computer Vision, 23*, 149–168.

Marois, R., Yi, D. J., & Chun, M. M. (2004). The neural fate of consciously perceived and missed events in the attentional blink. *Neuron, 41*, 465–472.

Marr, D. (1982). *Vision: A computational investigation into the human representation and processing of visual information*. San Francisco: W. H. Freeman.

Morgan, M., Chubb, C., & Solomon, J. A. (2008). A "dipper" function for texture discrimination based on orientation variance. *Journal of Vision, 8*(11), 9. doi:10.1167/8.11.9.

Morgan, M. J., & Glennerster, A. (1991). Efficiency of locating centres of dot-clusters by human observers. *Vision Research, 31*, 2075–2083.

Murray, S. O., Boyaci, H., & Kersten, D. (2006). The representation of perceived angular size in human primary visual cortex. *Nature Neuroscience, 9*, 429.

Myczek, K., & Simons, D. J. (2008). Better than average: Alternatives to statistical summary representations for rapid judgments of average size. *Perception & Psychophysics, 70*, 772–788.

Nakayama, K. (1985). Biological image motion processing—a review. *Vision Research, 25*, 625–660.

Nakayama, K. (1990). The iconic bottleneck and the tenuous link between early visual processing and perception.

In C. Blakemore (Ed.), *Vision: Coding and efficiency* (pp. 411–422). Cambridge: Cambridge University Press.

Neri, P., Morrone, M. C., & Burr, D. C. (1998). Seeing biological motion. *Nature, 395*, 894–896.

Newsome, W. T., & Pare, E. B. (1988). A selective impairment of motion perception following lesions of the middle temporal visual area (MT). *Journal of Neuroscience, 8*, 2201–2211.

Noe, A., Pessoa, L., & Thompson, E. (2000). Beyond the grand illusion: What change blindness really teaches us about vision. *Visual Cognition, 7*, 93–106.

Nothdurft, H. C. (1991). Texture segmentation and pop-out from orientation contrast. *Vision Research, 31*, 1073–1078.

Oliva, A. (2005). Gist of the scene. In L. Itti, G. Rees, & J. K. Tsotsos (Eds.), *Neurobiology of attention* (pp. 251–256). San Diego, CA: Elsevier.

Oliva, A., & Torralba, A. (2001). Modeling the shape of the scene: A holistic representation of the spatial envelope. *International Journal of Computer Vision, 42*(3), 145–175.

Parkes, L., Lund, J., Angelucci, A., Solomon, J. A., & Morgan, M. (2001). Compulsory averaging of crowded orientation signals in human vision. *Nature Neuroscience, 4*(7), 739–744.

Pelli, D. G., Palomares, M., & Majaj, N. J. (2004). Crowding is unlike ordinary masking: Distinguishing feature integration from detection. *Journal of Vision, 4*(12), 1136–1169. doi:10.1167/4.12.12.

Posner, M. I., & Keele, S. W. (1968). On genesis of abstract ideas. *Journal of Experimental Psychology, 77*, 353–363.

Potter, M. C. (1976). Short-term conceptual memory for pictures. *Journal of Experimental Psychology. Human Learning and Memory, 2*, 509–522.

Pylyshyn, Z. W., & Storm, R. W. (1988). Tracking multiple independent targets: Evidence for a parallel tracking mechanism. *Spatial Vision, 3*(3), 179–197.

Raymond, J. E., Shapiro, K. L., & Arnell, K. M. (1992). Temporary suppression of visual processing in an RSVP task: An attentional blink? *Journal of Experimental Psychology. Human Perception and Performance, 18*, 849–860.

Rensink, R. A., O'Regan, J. K., & Clark, J. J. (1997). To see or not to see: The need for attention to perceive changes in scenes. *Psychological Science, 8*, 368–373.

Robitaille, N., & Harris, I. M. (2011). When more is less: Extraction of summary statistics benefits from larger sets. *Journal of Vision, 11*(12), 1–8. doi:10.1167/11.12.18.

Rosenholtz, R. (2011). What your visual system sees where you are not looking. In B. E. Rogowitz & T. N. Pappas (Eds.), *Proceedings of the SPIE 7865, Human Vision and Electronic Imaging, XVI*, 786510, San Francisco, CA.

Rosenholtz, R., Huang, J., Raj, A., Balas, B. J., & Llie, L. (2012). A summary statistic representation in peripheral vision explains visual search. *Journal of Vision, 12*(4), 1–17. doi:10.1167/12.4.14.

Ross, J., & Burr, D. (2008). The knowing visual self. *Trends in Cognitive Sciences, 12*, 363–364. doi:10.1016/j.tics.2008.06.007.

Scholl, B. J., & Pylyshyn, Z. W. (1999). Tracking multiple items through occlusion: Clues to visual objecthood. *Cognitive Psychology, 38*, 259–290.

Simoncelli, E. P., & Olshausen, B. A. (2001). Natural image statistics and neural representation. *Annual Review of Neuroscience, 24*, 1193–1216.

Simons, D. J., & Levin, D. T. (1997). Change blindness. *Trends in Cognitive Sciences, 1*, 261–267. doi:10.1016/S1364-6613(97)01080-2.

Simons, D. J., & Myczek, K. (2008). Average size perception and the allure of a new mechanism. *Perception & Psychophysics, 70*(7), 1335–1336.

Snowden, R. J., & Braddick, O. J. (1989). The combination of motion signals over time. *Vision Research, 29*, 1621–1630.

Solomon, J. A. (2010). Visual discrimination of orientation statistics in crowded and uncrowded arrays. *Journal of Vision, 10*(14), 1–16. doi:10.1167/10.14.19.

Solomon, J. A., Morgan, M., & Chubb, C. (2011). Efficiencies for statistics of size discrimination. *Journal of Vision, 12*(12), 1–11. doi:10.1167/11.12.13.

Solso, R. L., & McCarthy, J. E. (1981). Prototype formation of faces—a case of pseudo-memory. *British Journal of Psychology, 72*, 499–503.

Suzuki, S. (2005). High-level pattern coding revealed by brief shape aftereffects. In C. Clifford & G. Rhodes (Eds.), *Advances in visual cognition: Vol 2. Fitting the mind to the world: Adaptation and aftereffects in high-level vision* (pp. 135–172). New York: Oxford University Press.

Sweeny, T. D., Grabowecky, M., Paller, K., & Suzuki, S. (2009). Within-hemifield perceptual averaging of facial expressions predicted by neural averaging. *Journal of Vision, 9*(3), 1–11. doi:10.1167/9.3.2.

Sweeny, T. D., Haroz, S., & Whitney, D. (2012a). Perceiving group behavior: Sensitive ensemble coding mechanisms for biological motion of human crowds. *Journal of Experimental Psychology: Human Perception and Performance, 32*, 329–337. doi:10.1037/a0028712.

Sweeny, T. D., Haroz, S., & Whitney, D. (2012b). Reference repulsion in the categorical perception of biological motion. *Vision Research, 64*, 26–34. doi:10.1016/j.visres.2012.05.008.

Thorpe, S., Fize, D., & Marlot, C. (1996). Speed of processing in the human visual system. *Nature, 381*, 520–522.

Torralba, A., & Oliva, A. (2003). Statistics of natural image categories. *Network (Bristol, England), 14*(3), 391–412.

Treisman, A. (1982). Perceptual grouping and attention in visual-search for features and for objects. *Journal of Experimental Psychology. Human Perception and Performance, 8*(2), 194–214.

Verghese, P. (2001). Visual search and attention: A signal detection theory approach. *Neuron, 31*, 523–535.

Victor, J. D., Purpura, K., Katz, E., & Mao, B. Q. (1994). Population encoding of spatial-frequency, orientation, and color in macaque V1. *Journal of Neurophysiology, 72*(5), 2151–2166.

Vogels, R. (1990). Population coding of stimulus orientation by striate cortical-cells. *Biological Cybernetics, 64*, 25–31.

Watamaniuk, S. N. J., & Duchon, A. (1992). The human visual-system averages speed information. *Vision Research, 32*, 931–941.

Watamaniuk, S. N. J., & McKee, S. P. (1998). Simultaneous encoding of direction at a local and global scale. *Perception & Psychophysics, 60*, 191–200.

Watamaniuk, S. N. J., Sekuler, R., & Williams, D. W. (1989). Direction perception in complex dynamic displays—the integration of direction information. *Vision Research, 29*, 47–59.

Wertheimer, M. (1923). Untersuchungen zur Lehre von der Gestalt. *Psychologische Forschung, 4*, 301–350.

Whitaker, D., McGraw, P. V., Pacey, I., & Barrett, B. T. (1996). Centroid analysis predicts visual localization of first- and second-order stimuli. *Vision Research, 36*, 2957–2970.

Whiting, B. F., & Oriet, C. (2011). Rapid averaging? Not so fast! *Psychonomic Bulletin & Review, 18*, 484–489.

Whitney, D., & Levi, D. M. (2011). Visual crowding: A fundamental limit on conscious perception and object recognition. *Trends in Cognitive Sciences, 15*, 160–168. doi:10.1016/j.tics.2011.02.005.

Williams, D. W., & Sekuler, R. (1984). Coherent global motion percepts from stochastic local motions. *Vision Research, 24*, 55–62.

Wolfe, J. M. (2003). Moving towards solutions to some enduring controversies in visual search. *Trends in Cognitive Sciences, 7*, 70–76. doi:10.1016/S1364-6613(02)00024-4.

Wolfe, J. M., Cave, K. R., & Franzel, S. L. (1989). Guided search—an alternative to the feature integration model for visual-search. *Journal of Experimental Psychology. Human Perception and Performance, 15*, 419–433.

Wolfe, J. M., Võ, M. L. H., Evans, K. K., & Greene, M. R. (2011). Visual search in scenes involves selective and nonselective pathways. *Trends in Cognitive Sciences, 15*, 77–84.

Yamanashi-Leib, A., Puri, A. M., Fischer, J., Bentin, S., Whitney, D., & Robertson, L. (2012). Crowd perception in prosopagnosia. *Neuropsychologia, 50*, 1698–1707. doi:10.1016/j.neuropsychologia.2012.03.026.

50   Face Perception

GILLIAN RHODES AND ANDREW J. CALDER

We use a wealth of cues from human faces to guide our interactions with others. With little apparent effort, we read subtle cues to identity, emotional state, gender, ethnicity, age, attractiveness, and focus of attention. This fluency is remarkable considering the similarity of faces as visual patterns and the difficult perceptual discriminations required. In this chapter we review the neural and computational mechanisms that allow us to recognize thousands of faces and their more fleeting changes of emotional expression. We first outline a face-coding network in the brain. We then consider the computational mechanisms used to code identity and expression and how these may be implemented in that network.

A FACE-CODING NETWORK

Face perception relies on a network of face-selective regions in extrastriate visual cortex (Haxby & Gobbini, 2011; Haxby, Hoffman, & Gobbini, 2000) (figure 50.1). The core network consists of an occipital face area (OFA) in inferior occipital cortex, a fusiform face area (FFA) in the lateral fusiform gyrus, and an area of the posterior superior temporal sulcus (STS). Although bilateral, these areas are often larger and more reliably found in the right hemisphere, consistent with older evidence for a right hemisphere advantage in face perception (De Renzi et al., 1994). This core system derives a visual representation of the face. Although the precise functional role of these areas and their interconnections are still being worked out, the OFA may code an initial representation that is projected in parallel to the FFA and STS (Fairhall & Ishai, 2007; Haxby, Hoffman, & Gobbini, 2000; Pitcher, Walsh, & Duchaine, 2011). The FFA is thought to code invariant properties such as identity, sex, and race (Gauthier et al., 2000; George et al., 1999; Grill-Spector, Knouf, & Kanwisher, 2004; Haxby, Hoffman, & Gobbini, 2000; Kanwisher & Barton, 2011; Kanwisher & Yovel, 2006; Mazard, Schiltz, & Rossion, 2006; Rotshtein et al., 2005; Winston et al., 2004), whereas the STS processes changeable aspects related to emotional expression, gaze direction, and facial speech (Andrews & Ewbank, 2004; Calder & Young, 2005; Haxby, Hoffman, & Gobbini, 2002;

Hoffman & Haxby, 2000). In addition, extrastriate regions outside the core system that respond significantly (but not maximally) to faces may help code facial appearance (for a review, see Haxby & Gobbini, 2011). The core areas project to an extended system that extracts various types of social information (e.g., person knowledge, emotional states) from the visual face representations.

There has been considerable debate about whether the FFA is specialized for processing faces (Kanwisher, McDermott, & Chun, 1997; Kanwisher & Yovel, 2006) or objects of expertise more generally (Gauthier & Nelson, 2001). However, the core prediction of the expertise hypothesis, that FFA responses to nonface objects should increase with expertise, has not been consistently supported (for a review, see McKone, Kanwisher, & Duchaine, 2007). Rather, the FFA appears to be relatively selective for faces (Kanwisher & Barton, 2011).

CODING IDENTITY

Here we consider two computational mechanisms, holistic coding and adaptive norm-based coding, that may help us discriminate and recognize thousands of faces despite their similarity as visual patterns.

Holistic Coding

Unlike many other visual objects faces can seldom be recognized from diagnostic parts or the first-order arrangement of parts that is shared by all faces (eyes above nose above mouth). Instead, successful recognition entails sensitivity to more subtle, second-order variations in the spatial relations between internal features or between those features and the rest of the face (Diamond & Carey, 1986). More generally, face recognition seems to entail integration of information across the face (for reviews, see Farah et al., 1998; Maurer, Le Grand, & Mondloch, 2002; McKone, Kanwisher, & Duchaine, 2007; McKone & Robbins, 2011; Peterson & Rhodes, 2003). Various terms have been used to characterize the kind of coding that produces such representations: "holistic," "configural," "(second-order)

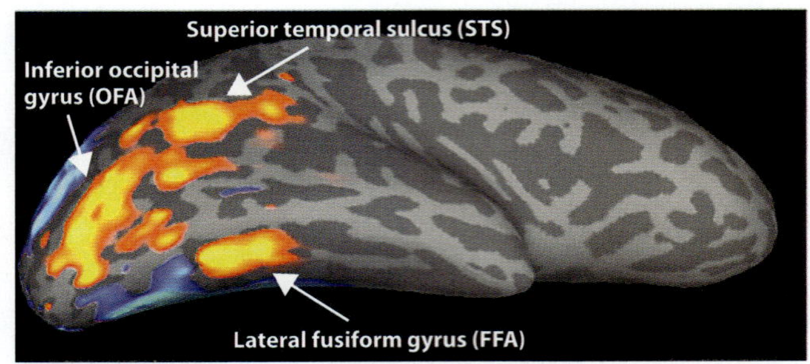

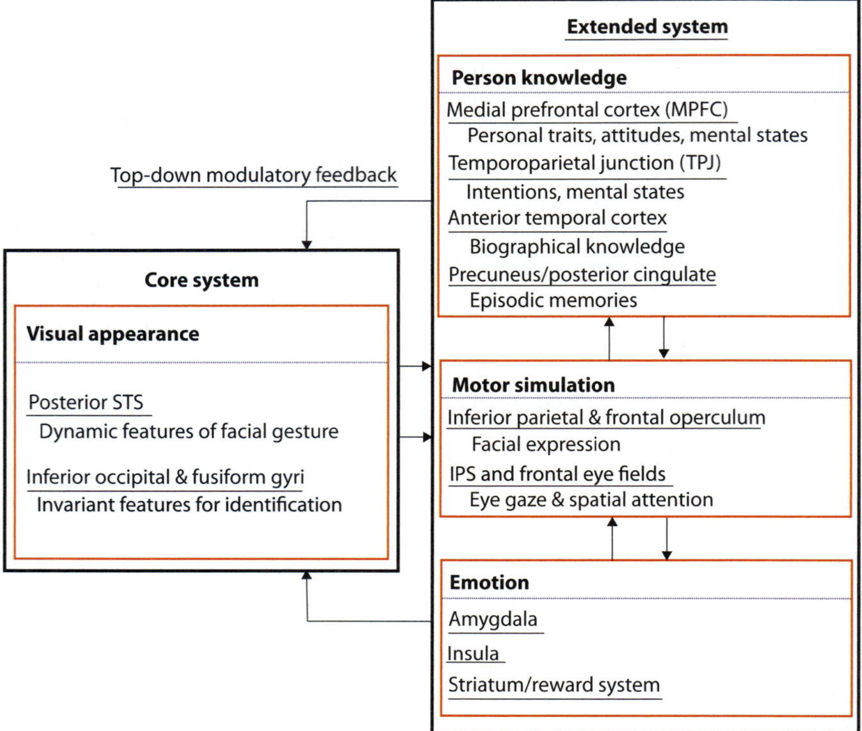

FIGURE 50.1   (Top) Three visual extrastriate regions in occipitotemporal cortex that responded more strongly to faces than houses in a single participant in Haxby et al. (1999), shown on an inflated cortical surface. (Reproduced from Haxby & Gobbini, 2011, with permission.) (Bottom) Haxby's model of a distributed neural system for face perception. (Reproduced from Haxby & Gobbini, 2011, with permission.)

relational," "coarse," and "global." Although these differ in precise meaning and/or operational definition (for reviews, see Farah et al., 1998; Peterson & Rhodes, 2003), they all contrast with more local, feature-based coding (variously referred to as "local," "piecemeal," "part-based," "componential," "fine-grained," or "analytic"). Here we use *holistic* as an umbrella term for coding that integrates information across the face to represent the appearance of features and their spatial relations, making it difficult to selectively attend to isolated parts (cf. McKone, Kanwisher, & Duchaine, 2007; McKone & Robbins, 2011; Rossion, 2008; Tanaka & Gordon, 2011).

Evidence for holistic coding of faces comes from three effects, as explained in figure 50.2: the composite effect, the wholes advantage (or part–whole effect), and the disproportionate inversion effect for faces. Holistic processing appears to be a hallmark of face processing rather than expert processing more generally (McKone & Robbins, 2011), although spatial context effects are certainly not unique to faces (Schwartz, Hsu, & Dayan, 2007).

Several lines of evidence indicate that holistic coding helps us identify faces. First, it is greatly reduced or absent for inverted faces, which are poorly recognized (e.g., Tanaka & Farah, 1993). Second, it is reduced for

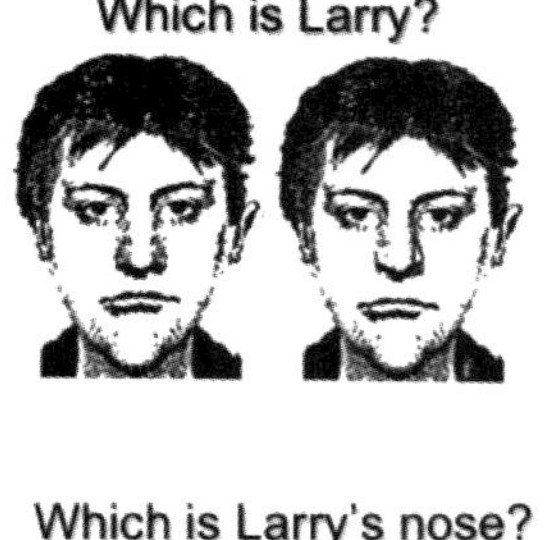
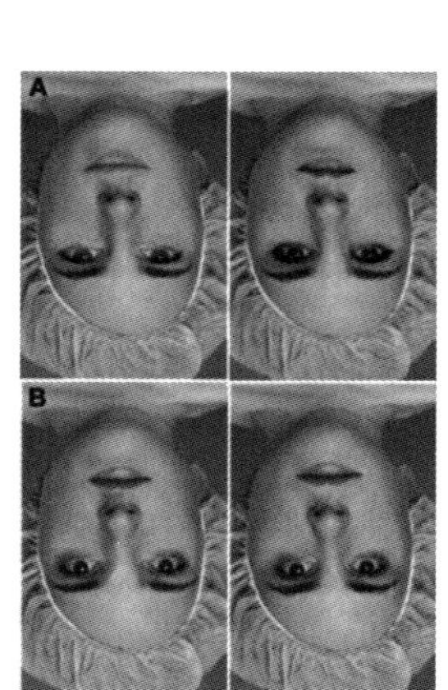

FIGURE 50.2 (A) Composite effect. Pairing the top half of one face with the bottom half of another makes it harder to recognize the top face (top left), relative to a misaligned condition (top right) (Young, Hellawell, & Hay, 1987). The faces used to make the composites are shown below. (B) Wholes advantage (part–whole effect). After seeing Larry, it is easier to recognize Larry's nose in the context of the whole face than in isolation (Tanaka & Farah, 1993). (C) Inversion effect. Inversion disrupts perception of faces, particularly their spatial relations (row B faces differ in eye spacing). Sensitivity to feature appearance (row A faces have different eyes) is often less affected, but not always (for a review see McKone & Yovel, 2009). (Adapted from Le Grand et al., 2001 by permission from Macmillan Publishers Ltd., *Nature*, copyright 2001.)

other-race faces, which are recognized more poorly than own-race faces (Hancock & Rhodes, 2008; Hayward, Rhodes, & Schwaninger, 2008; Michel, Caldara, & Rossion, 2006; Michel, Corneille, & Rossion, 2007; Michel et al., 2006; Mondloch et al., 2010; Rhodes, Hayward, & Winkler, 2006; Rhodes et al., 1989, 2009; Tanaka, Kiefer, & Bukach, 2004). Third, it is reduced in individuals with acquired prosopagnosia, who have difficulty recognizing faces following focal brain damage (e.g., Barton et al., 2002; Barton, Zhao, & Keenan, 2003), and in some individuals with congenital prosopagnosia (Le Grand et al., 2006; Palermo et al., 2011b). Finally, stable individual differences in face-specific recognition ability (Wilmer et al., 2010; Zhu et al., 2010) appear to be linked to individual differences in holistic coding of faces (Wang et al., 2012). Moreover, individual differences in holistic coding (for some, but not all, measures) predict performance on The Cambridge Face Memory Test (Richler, Cheung, & Gauthier, 2011). The evidence for a link with more perceptually based tasks is less clear, with some studies finding a link (Richler, Cheung, & Gauthier, 2011) and others failing to do so (Konar, Bennet, & Sekuler, 2010; Mondloch & Desjarlais, 2010). It remains possible that a clearer link would be found if face-specific performance was isolated in these tasks, but this has not yet been done.

There are many open questions and unresolved controversies in this area. Indeed there is no clear consensus about what holistic coding is or how it should be measured. Different measures do not always yield consistent results (e.g., Mondloch et al., 2010), even variants of the same basic measure (Richler, Cheung, & Gauthier, 2011; but see DeGutis et al., 2013). Moreover, the notion of qualitative differences between coding of upright and inverted faces, which is central to claims that face coding is special, has not gone unchallenged (Gold, Mundy, & Tjan, 2012; Hayward, Rhodes, & Schwaninger, 2008; Rhodes, Hayward, & Winkler, 2006; Sekuler et al., 2004). Certainly, large inversion effects do not directly index configural coding of spatial relations, as often assumed, because large inversion effects can occur for the perception of features as well as spatial relations (Hayward, Rhodes, & Schwaninger, 2008; Rhodes, Brake, & Atkinson, 1993; Rhodes, Hayward, & Winkler, 2006). Recently, ideal observer analysis of contrast thresholds suggests that neither upright nor inverted faces are perceived better than would be expected from perception of their local parts (Gold, Mundy, & Tjan, 2012). It remains to be seen how this result can be reconciled with other evidence for holistic processing of upright faces and whether it will generalize to tasks using clearly visible faces, such as face recognition.

Norm-Based Coding

Many theorists have suggested that an elegant and economical way for the brain to represent faces is to code how each face deviates from a perceptual norm or prototype, which constitutes the central tendency (average) of a distribution of faces represented in a multidimensional "face space" (Diamond & Carey, 1986; Goldstein & Chance, 1980; Hebb, 1949; Hochberg, 1978; Leopold et al., 2001; Rhodes, 1988, 1996; Rhodes, Brennan, & Carey, 1987; Rhodes & Jeffery, 2006; Tsao & Freiwald, 2006; Valentine, 1991). This form of coding may allow the visual system to see past the shared structure of faces to the subtle variations that define individuals. Because the norm is adaptively updated by experience, it can also ensure that our face-coding mechanisms are well calibrated to our diet of faces (for reviews, see Armann et al., 2011; Rhodes & Leopold, 2011; Webster & MacLeod, 2011). For simplicity we will talk about a single norm, although different norms are used to code faces of different genders and races (for reviews, see Armann et al., 2011; Rhodes & Leopold, 2011).

Compelling evidence for norm-based coding comes from face identity aftereffects, in which viewing a face for a few seconds selectively biases us to see the "opposite" identity in a subsequently presented face (Anderson & Wilson, 2005; Armann et al., 2011; Leopold et al., 2001, 2005; Rhodes & Jeffery, 2006; Rhodes et al., 2007; Rhodes & Leopold, 2011). For example, after antiDan, who lies diametrically opposite Dan in face space, has been viewed, the average face is likely to be identified as Dan (figure 50.3). These identity aftereffects cannot be explained by adaptation to low-level image features because they survive changes in the size and/or retinal position between adapt and test faces (Rhodes & Leopold, 2011) and are larger for upright than inverted faces (Rhodes, Evangelista, & Jeffery, 2009). Rather, viewing a face seems to temporarily shift some higher-level representation of the average (norm) toward that face so that low-identity-strength versions of Dan become more distinctive and easier to identify as Dan. The selectivity of the bias for the identity that lies opposite the adapting face in face space strongly suggests that the average face functions as a perceptual norm for coding identity.

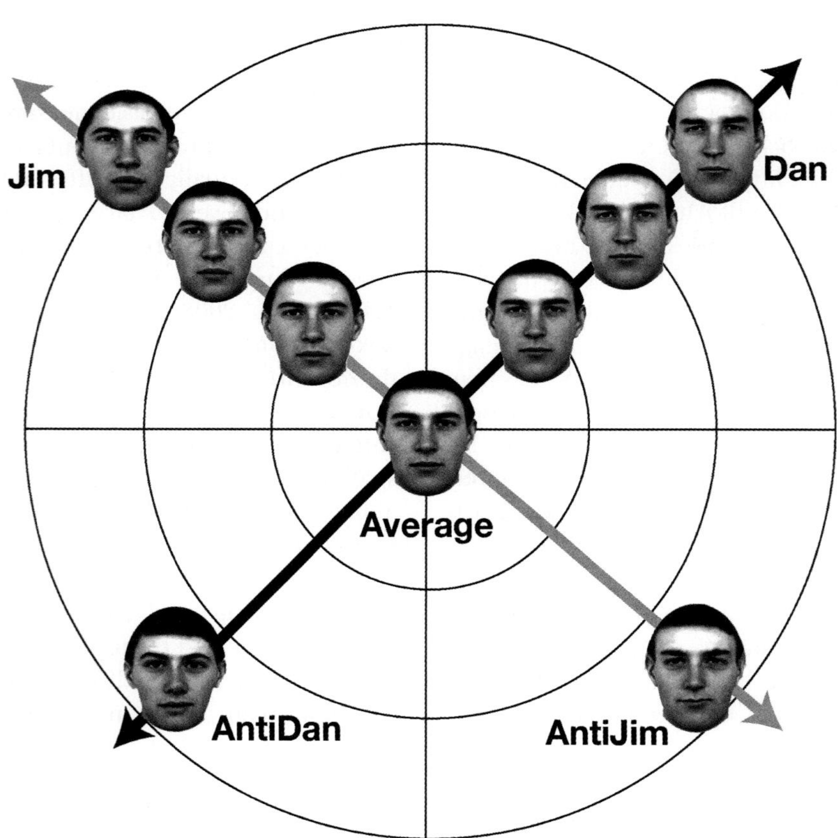

FIGURE 50.3   A simple two-dimensional face space with two faces, Dan and Jim, and an average face, created by morphing 20 male Caucasian faces, at the center. For each face we can construct a corresponding antiface (antiDan and antiJim, also shown) with opposite properties by morphing the original face toward the average and beyond. Reduced identity strength versions (anticaricatures) of Dan and Jim, created by morphing those identities toward the average, are also shown. Identity aftereffects occur when exposure to a face biases subsequent perception toward a face with opposite properties. For example, after viewing antiDan for a few seconds, we are biased (briefly) to see Dan. This identity aftereffect is measured as the reduction in identity strength needed to successfully identify a face (e.g., Dan) after viewing its opposite (antiDan).

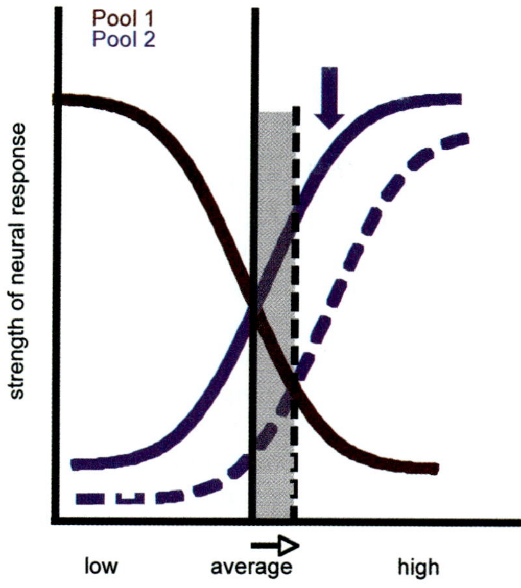

FIGURE 50.4   A simple norm-based coding model in which each face dimension is opponently coded by a pair of neural populations, one responding to below-average and the other to above-average values. Average values are coded implicitly, by equal (and low) activation of the two populations. Following adaptation to a face with a particular dimension value (large arrow), the relative contribution of the population that responds most strongly to that stimulus is diminished. As a result the balance point (perceived average) shifts toward (small arrow, shaded area), and biases subsequent perception away from, the adapting value.

Norm-based coding of faces could be implemented by a simple opponent coding model (Rhodes & Jeffery, 2006; Rhodes et al., 2005; Robbins, McKone, & Edwards, 2007; Tsao & Freiwald, 2006) similar to those proposed for simpler properties such as color and aspect ratio (Regan & Hamstra, 1992; Suzuki, 2005; Webster & MacLeod, 2011). In the model, pairs of neural populations are tuned to above-average (e.g., large eyes) and below-average (e.g., small eyes) values, respectively, on each dimension (e.g., eye size) of face space (Rhodes & Jeffery, 2006; Rhodes et al., 2005; Robbins, McKone, & Edwards, 2007; Susilo, McKone, & Edwards, 2010a; Tsao & Freiwald, 2006) (figure 50.4). The norm is signaled by equal (and low) activation in both members of each pair. The model predicts that more extreme adaptors will produce stronger identity aftereffects than less extreme adaptors, as found in several studies (Jeffery et al., 2010, 2011; Robbins, McKone, & Edwards, 2007; Susilo, McKone, & Edwards, 2010b).

Unlike holistic coding, norm-based coding is widely used in perception, where it may help us discriminate subtle differences in brightness, contrast, color, motion direction, shape, and other simple stimulus attributes (Bartlett, 2007; MacLeod & von der Twer, 2003; Mather, 1980; Regan & Hamstra, 1992; Sutherland, 1961; Suzuki, 2005; Webster, 2003; Webster & Leonard, 2008). It may also help us discriminate and identify faces, with better performance sometimes (but not always) reported around the norm or adapted state (Armann et al., 2011). Moreover, face adaptation is reduced in children with autism spectrum disorders (ASDs), who have difficulty discriminating and recognizing faces (Ewing, Pellicano, & Rhodes, 2013; Pellicano et al., 2007), and in congenital prosopagnosics (Palermo et al., 2011a). Finally, adult face recognition ability correlates positively with the size of identity-related (eye height) face aftereffects (Dennett et al., 2012). These results suggest that adaptive, norm-based coding contributes to our face expertise.

Implementation in the Face Network

Several studies implicate the FFA in holistic face coding (Kanwisher & Yovel, 2006). It is the neural source of behavioral face inversion effects (Yovel & Kanwisher, 2005), it is sensitive to variations in spatial relations between features (Rhodes et al., 2009), and it shows composite effects reflecting integration of information across the top and bottom halves of the face (Schiltz & Rossion, 2006). The right FFA also responds more strongly during matching of whole faces that differ on a single part than matching of the parts in isolation (Rossion et al., 2000). Interestingly, there is little selectivity for spatial relations over features (or vice versa) in face-selective cortex, consistent with integration of features and relations in holistic face representations (Maurer et al., 2007; Yovel & Kanwisher, 2004). Importantly, however, there is strong sensitivity to identity-related information (Fox et al., 2009; Rotshtein et al., 2005).

The right OFA may also play a role in holistic coding because it is sensitive to spatial relations (Rhodes et al., 2009; but see Pitcher et al., 2007) and composite effects (Schiltz & Rossion, 2006). More generally, both lesion studies (Bouvier & Engel, 2006; see also Rossion et al., 2003; Steeves et al., 2006) and ERP studies indicating rapid access to individuating information (Corentin & Rossion, 2006; Eimer, 2011) suggest a role for the right OFA in representing the facial appearance. However, unlike the FFA, it does not discriminate between face changes that are perceived as different identities and those that are not (Fox et al., 2009; Rotshtein et al., 2005). It likely represents an earlier (and more retinotopic) stage with less explicit coding of identity than the FFA.

Little is known about how adaptive, norm-based coding is implemented in the face network. Certainly, face-selective cortex adapts, and fMR adaptation is widely used to assess its sensitivity to various stimulus parameters (Grill-Spector et al., 1999; Grill-Spector & Malach, 2001); but see Ewbank et al. (2011, 2013) for evidence that these effects may reflect top-down influences from other regions rather than neuronal fatigue in the area showing adaptation. In addition, adaptation to distorted faces affects electrophysiological correlates of perceived face normality, but the source of these effects remains unclear (Burkhardt et al., 2010). Responses in human FFA increase with distance of faces from the average (Loffler et al., 2005), as do responses of face-selective cells in monkeys (Freiwald, Tsao, & Livingstone, 2009; Leopold, Bondar, & Giese, 2006). Moreover, face-selective neurons tuned to one or another extreme on a variety of face dimensions have been found in monkeys (Freiwald, Tsao, & Livingstone, 2009). Taken together, these findings are consistent with norm-based coding. However, activation in the human face network has yet to be linked directly to the face identity aftereffects that indicate norm-based coding.

CODING EXPRESSION

Next we consider the role of holistic coding in perception of facial expressions, evidence that a norm-based multidimensional framework provides a suitable method of coding facial expressions, and how both might be implemented in the face network.

Holistic Coding

The vast majority of research addressing holistic and configural coding of faces has focused on their role in facial identity recognition, or basic face perception assessed with "faceness" decisions (i.e., is it a face?) and face matching (Maurer, Le Grand, & Mondloch, 2002). However, work also points to the role of holistic coding in processing facial characteristics besides identity, including facial expressions. For example, in an expression composite task participants found it harder to identify the expression shown in one-half of composite facial expressions (comprising the top and bottom halves of two different emotional expressions posed by the same person) when the faces were aligned than when misaligned (Calder et al., 2000b; White, 2000). Similarly, in a matching variant of the task, participants found it harder to decide that the top halves of two composite expressions were the same when their bottom halves showed different expressions (Calder & Jansen,

2005). The effects in both experiments were significantly reduced by stimulus inversion, which impairs the recognition of whole veridical facial expressions (McKelvie, 1995; Prkachin, 2003).

An obvious question is whether holistic coding of facial expressions is processed separately from holistic coding of facial identity, as proposed by influential models (Bruce & Young, 1986; Haxby, Hoffman, & Gobbini, 2000). To address this question Calder and colleagues (2000b) used a variant of the composite effect in which only composite faces were used. The top and bottom halves of the composite faces in this experiment showed (1) different expressions posed by the same identity (different expression/same identity), (2) the same expression posed by different identities (same expression/different identity), or (3) different expressions posed by different identities (different expression/different identity) (figure 50.5A). Participants found it harder to report the expression in the top face half when the two halves contained different expressions (i.e., different expression/same identity and different expression/different identity conditions) than when they contained the same expressions posed by different identities (i.e., same expression/different identity condition; figure 50.5B, top graph). Importantly, when the two halves contained different expressions, there was no added cost if the two halves also showed different identities. In other words, incongruent identity information in the two face halves did not interfere with holistic processing of facial expressions. A corresponding effect was found when participants were asked to report the identity of one face half. These results appear to support the dissociable processing of facial identity and facial expression, as suggested by authors positing separate visual routes for these two facial characteristics (Bruce & Young, 1986; Haxby, Hoffman, & Gobbini, 2000). However, as we go on to discuss, they can be accounted for by a single multidimensional framework coding both facial identity and facial expression.

It is interesting to consider how holistic processing of facial expressions contributes to facial expression recognition in general. We are not aware of work addressing this in healthy or typical participants. However, individuals with congenital or acquired face recognition impairments can show apparently normal, or close to normal, facial expression recognition despite showing significantly reduced composite effects for facial expressions (Baudouin & Humphreys, 2006; Palermo et al., 2011b). These results potentially question the contribution of holistic processing to facial expression recognition. Alternatively, and as suggested by both sets of authors, the preserved expression recognition may

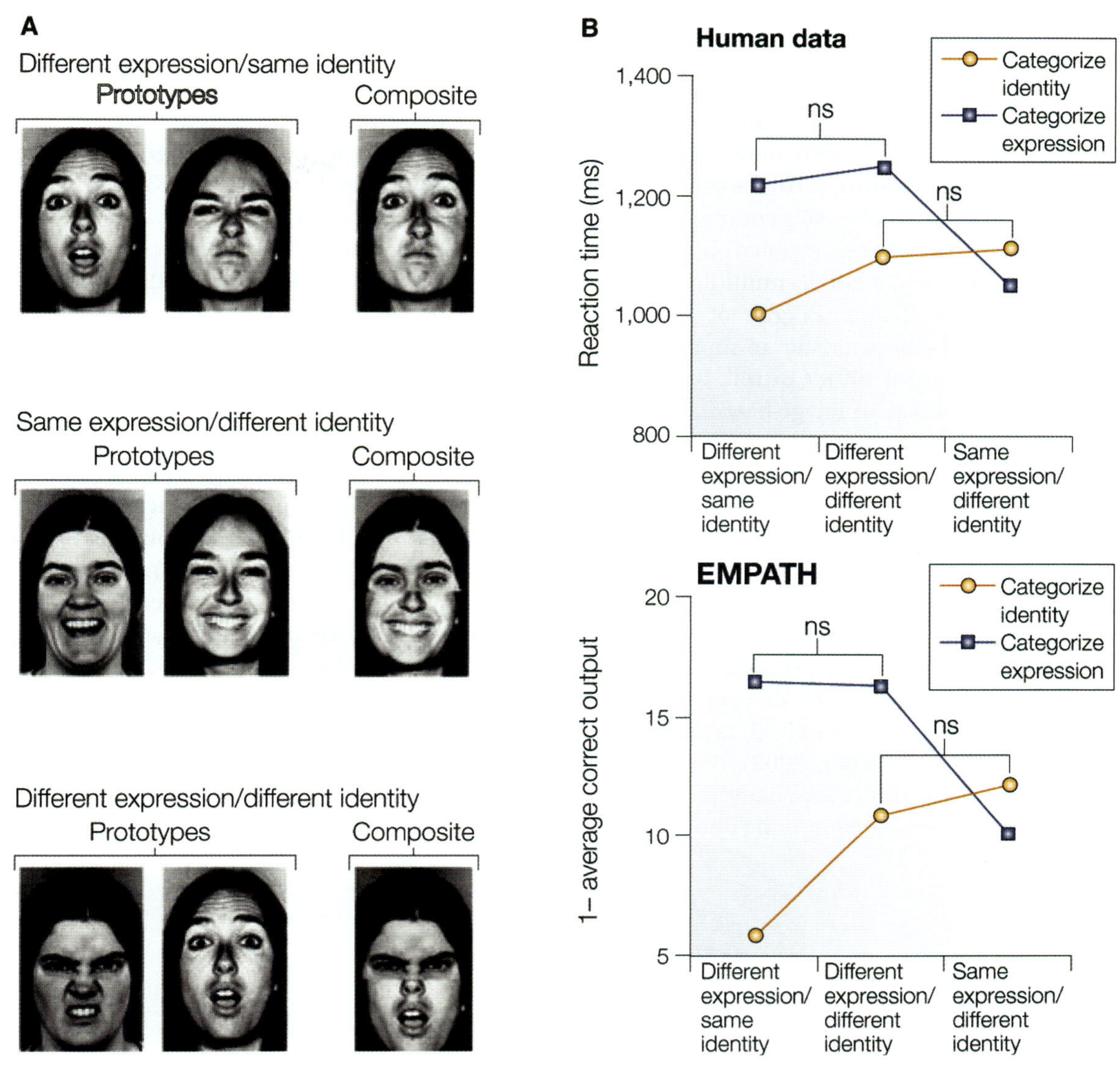

FIGURE 50.5 (A) Composite facial expressions that combine the top and bottom halves of two different facial expressions posed by the same identity (top), the same expression posed by different identities (middle), and different expressions posed by different identities (bottom). (B) The top graph shows participants' mean correct reaction times to categorize the identity or expression shown in the top half of the composite faces. The bottom graph shows a simulation of this effect in a variant of the EMPATH model (Cottrell, Branson, & Calder, 2002). (Reprinted from Calder & Young, 2005, with permission from *Nature* publishing group.)

reflect the use of compensatory strategies. The latter interpretation has important implications for previous work in which preserved recognition of facial expression in individuals with acquired or congenital face recognition impairments was interpreted as support for a dissociation between the neural mechanisms coding facial identity and facial expression (Bruyer et al., 1983; Duchaine, Paerker, & Nakayama, 2003; Humphreys, Avidan, & Behrmann, 2007; Tranel, Damasio, & Damasio, 1988). Instead, these results may reflect the more ready application of compensatory strategies to recognizing facial expressions than facial identities.

This may be due to expressions being associated with salient individual features and that, as adults, we encounter fewer facial expressions than facial identities.

Norm-Based Coding

Although the face-space metaphor was originally developed to represent invariant facial properties—initially identity and then race and sex (O'Toole et al., 1995; Valentine, 1995), it has been extended to include facial expressions (Calder et al., 2000a, 2001). A single face space coding facial identity and facial expression goes against the proposal that these facial characteristics are processed by entirely separate visual routes (Bruce & Young, 1986; Haxby, Hoffman, & Gobbini, 2000). However, Calder and Young (2005) have argued that the neuropsychological literature provides considerably less support for separate visual routes coding these two facial characteristics than is often assumed. In addition, work has shown that such a single multidimensional system can display a sufficient degree of separable coding of the two facial characteristics to support these sorts of effects (Calder et al., 2001; Cottrell, Branson, & Calder, 2002). For example, an image-based analysis of a set of facial expressions (Ekman & Friesen, 1976) showed both separate *and* common dimensions coding the physical image properties associated with the identity and expression of these faces (Calder, 2011; Calder et al., 2001).

Moreover, computational modeling has shown that the dissociable processing of the holistic information for facial identity and facial expression (Calder & Young, 2005; Calder et al., 2000b) can be simulated within a single face space that codes both facial properties (Cottrell, Branson, & Calder, 2002) (figure 50.5, bottom graph). Note that this research addressed identity and expression processing from static photographic images. Hence, the proposal is that the *visual form* of facial identity and facial expression are coded in a single multidimensional space. Dynamic information is thought to be coded separately and is more relevant for expression than identity, although again both facial characterstics may be coded by a single system representing dynamic information. For a more detailed discussion, see Calder (2011).

If a single framework codes facial identity and expression, then we might expect facial expressions to be coded by the sort of norm-based, opponent-coded system proposed for facial identity. Support for this proposal has recently been found in work showing that adapting to an "antiexpression" (e.g., antianger) caused the average or prototype expression to be seen as the corresponding (i.e. "opposite") facial expression (e.g., anger) (Cook, Matei, & Johnston, 2011; Skinner & Benton, 2010, 2012) (figure 50.6). Moreover, the size of the effect was related to how extreme the antiface was, consistent with the predictions of an opponent-coded system (Skinner & Benton, 2010, 2012). Also consistent with the single-face-space concept, a number of studies have reported significantly less facial expression adaptation when the identities of the adapt and test faces are different versus the same (Campbell & Burke, 2009; Ellamil, Susskind, & Anderson, 2008; Fox

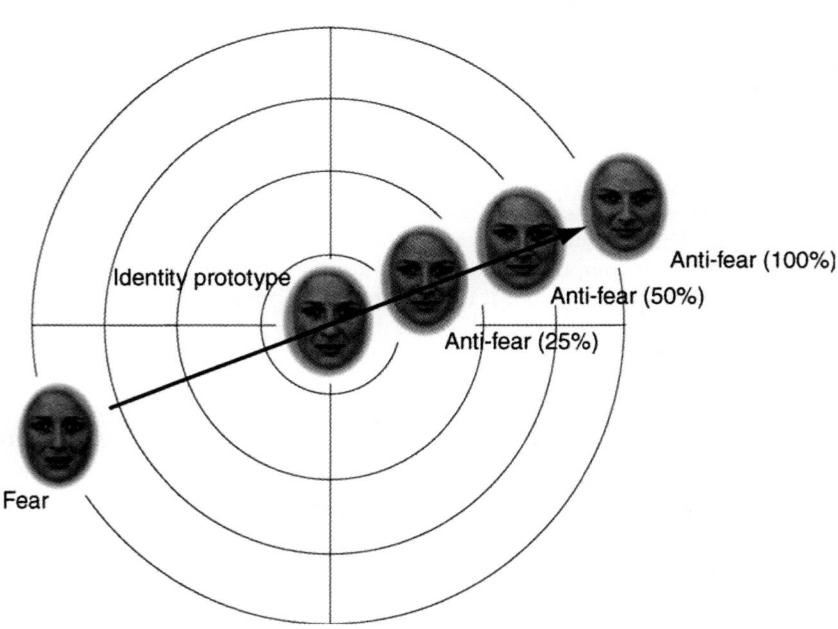

FIGURE 50.6   A two-dimensional representation showing how the three levels of antifear expressions were prepared from a fear expression. The identity prototype (or average) is prepared by averaging seven expressions (happy, sad, angry, fear, disgust, surprise, and neutral) posed by multiple identities. Similarly, the fear face is prepared from the average of multiple fear expressions posed by the same identities. In the example shown, identity is kept constant by applying the deformations associated with the prototype fear expression and average face to a single facial identity. Antifear expressions were prepared by morphing the fear face toward the prototype and beyond. (Reprinted from Skinner & Benton, 2012, with permission.)

& Barton, 2007; Skinner & Benton, 2012). Similarly, to-be-ignored changes in facial identity interfere with participants' ability to categorize facial expressions (Schweinberger, Burton, & Kelly, 1999; Schweinberger & Soukup, 1998). Interestingly, these interference effects are typically asymmetrical, such that to-be-ignored changes in facial identity affect categorization and adaptation of facial expression, but not vice versa (Fox & Barton, 2007; Fox, Oruc, & Barton, 2008; Schweinberger, Burton, & Kelly, 1999; Schweinberger & Soukup, 1998) (but see Ganel & Goshen-Gottstein, 2004, for symmetric interference). Important issues for future research are to provide an explanation of the asymmetry and to explore whether computational models can simulate these effects.

Implementation in the Face Network

Although some have suggested that holistic processing of both facial identity and facial expression occur at the same level of the face perception system (Calder & Jansen, 2005; Palermo et al., 2011b), it remains to be shown whether the FFA and OFA are involved in holistic coding of both facial properties. Indeed, we are not aware of any neuroimaging research that has examined the neural correlates of holistic coding of facial expressions.

Few studies have addressed adaptive coding of facial expressions in the face network, and, as far as we are aware, none has specifically addressed norm-based coding. Winston et al. (2004) used a functional magnetic resonance (fMR)-adaptation procedure (also known as repetition suppression) to examine the effects of repeating the same facial identity or facial expression. The results showed that the fusiform gyrus and posterior STS were sensitive to the repetition of facial identity, whereas a more anterior section of STS was sensitive to repetition of facial expression. Using a similar procedure a second study found that both the posterior STS and the fusiform gyrus showed adaptation of both facial properties, but that study replicated the role of anterior STS in facial expression adaptation alone (Fox et al., 2009). A third, MEG study found adaptation to facial expressions in the posterior STS around 300 and 400 ms post test face onset (Furl et al., 2007). The authors suggest that this region signals the dissimilarity between the adapting and test expressions and that its late timing argues against the traditional fatigue account of adaptation effects; see also Ewbank et al. (2013) for evidence against a simple fatigue explanation of adaptation of facial identity. Together these results suggest that the FFA is involved in both facial identity and facial expression perception, whereas the

anterior STS may have a more specific role in facial expression perception. The contribution of the posterior STS is less clear but may relate more to facial expression perception, given that it is not typically seen in work examining neural adaptation of facial identity alone (e.g. Rhodes et al., 2009).

CONCLUSIONS AND FUTURE DIRECTIONS

Face expertise lies at the heart of human social interaction. We have reviewed evidence that holistic coding and adaptive norm-based coding mechanisms contribute to two aspects of our face expertise: recognition of identity and emotional expressions. That the same computational mechanisms are implicated in both suggests that a common representation of face form may underlie the perception of identity and expression, in contrast with traditional models with distinct processing streams (Bruce & Young, 1986; Haxby, Hoffman, & Gobbini, 2000). However, a detailed understanding of how these computational mechanisms are implemented in the neural face network remains to be worked out. More extensive reviews of face perception and its underlying mechanisms can be found in several recent volumes (Bruce & Young, 2012; Calder et al., 2011).

ACKNOWLEDGMENTS

This work was supported by the Australian Research Council Centre of Excellence for Cognition and its Disorders (project number CE110001021), by an Australian Research Council Professorial Fellowship to G.R. (project number DP0877379), and by the UK Medical Research Council (project number MC-A060–5PQ50).

REFERENCES

Anderson, N. D., & Wilson, H. R. (2005). The nature of synthetic face adaptation. *Vision Research, 45*, 1815–1828. doi:10.1016/j.visres.2005.01.012.

Andrews, T. J., & Ewbank, M. P. (2004). Distinct representations for facial identity and changeable aspects of faces in the human temporal lobe. *NeuroImage, 23*, 905–913.

Armann, R., Jeffery, L., Calder, A. J., Bülthoff, I., & Rhodes, G. (2011). Race-specific norms for coding face identity and a functional role for norms. *Journal of Vision, 11*(13), 9, 1–14. doi:10.1167/11.13.9.

Bartlett, M. S. (2007). Information maximization in face processing. *Neurocomputing, 70*, 2204–2217. doi:10.1016/j.neucom.2006.02.025.

Barton, J. J. S., Press, D. Z., Keenan, J. P., & O'Connor, M. (2002). Lesions of the fusiform face area impair perception of facial configuration in prosopagnosia. *Neurology, 58*, 71–78.

Barton, J. J. S., Zhao, J., & Keenan, J. P. (2003). Perception of global facial geometry in the inversion effect and prosopagnosia. *Neuropsychologia, 41,* 1703–1711.

Baudouin, J.-Y., & Humphreys, G. W. (2006). Compensatory strategy in processing facial emotions: Evidence from prosopagnosia. *Neuropsychologia, 44,* 1364–1369.

Bouvier, S. E., & Engel, S. A. (2006). Behavioral deficits and cortical damage loci in cerebral achromatopsia. *Cerebral Cortex, 16,* 183–191.

Bruce, V., & Young, A. W. (1986). Understanding face recognition. *British Journal of Psychology, 77,* 305–327.

Bruce, V., & Young, A. W. (2012). *Face perception.* New York: Psychology Press.

Bruyer, R., Laterre, C., Seron, X., Feyereisen, P., Strypstein, E., Pierrard, E., et al. (1983). A case of prosopagnosia with some preserved covert remembrance of familiar faces. *Brain and Cognition, 2,* 257–284. doi:10.1016/0278-2626(83)90014-3.

Burkhardt, A., Blaha, L. M., Schneider Jurs, B., Rhodes, G., Jeffery, L., Wyatte, D., et al. (2010). Adaptation modulates the electrophysiological substrates of perceived facial distortion: Support for opponent coding. *Neuropsychologia, 48,* 3743–3756. doi:10.1016/j.neuropsychologia.2010.08.016.

Calder, A. J. (2011). Does facial identity and facial expression recognition involve separate visual routes? In A. J. Calder, G. Rhodes, M. H. Johnson, & J. V. Haxby (Eds.), *The Oxford handbook of face perception* (pp. 427–448). Oxford: Oxford University Press.

Calder, A. J., Burton, A. M., Miller, P., Young, A. W., & Akamatsu, S. (2001). A principal component analysis of facial expressions. *Vision Research, 41,* 1179. doi:10.1016/S0042-6989(01)00002-5.

Calder, A. J., & Jansen, J. (2005). Configural coding of facial expressions: The impact of inversion and photographic negative. *Visual Cognition, 12,* 495–518. doi:10.1080/13506280444000418.

Calder, A. J., Rhodes, G., Johnson, M. H., & Haxby, J. V. (Eds.). (2011). *The Oxford handbook of face perception.* Oxford: Oxford University Press.

Calder, A. J., Rowland, D., Young, A. W., Nimmo-Smith, I., Keane, J., & Perrett, D. I. (2000a). Caricaturing facial expressions. *Cognition, 76,* 105–146.

Calder, A. J., & Young, A. W. (2005). Understanding the recognition of facial identity and facial expression. *Nature Reviews. Neuroscience, 6,* 641–651.

Calder, A. J., Young, A. W., Keane, J., & Dean, M. (2000b). Configural information in facial expression perception. *Journal of Experimental Psychology. Human Perception and Performance, 26,* 527–551.

Campbell, J., & Burke, D. (2009). Evidence that identity-dependent and identity-independent neural populations are recruited in the perception of five basic emotional facial expressions. *Vision Research, 49,* 1532–1540. doi:10.1016/j.visres.2009.03.009.

Cook, R., Matei, M., & Johnston, A. (2011). Exploring expression space: Adaptation to orthogonal and anti-expressions. *Journal of Vision, 11*(4), 1–9. doi:10.1167/11.4.2.

Corentin, J., & Rossion, B. (2006). The speed of individual face categorization. *Psychological Science, 16,* 485–492.

Cottrell, G. W., Branson, K. M., & Calder, A. J. (2002). Do expression and identity need separate representations? Paper presented at the 24th Annual Meeting of the Cognitive Science Society, Fairfax, Virginia.

DeGutis, J., Wilmer, J., Mercado, R. J., & Cohan, S. (2013). Using regression to measure holistic face processing reveals a strong link with face recognition ability. *Cognition, 126,* 87–100. doi:10.1016/j.cognition.2012.09.004.

Dennett, H. W., McKone, E., Edwards, M., & Susilo, T. (2012). Face aftereffects predict individual differences in face recognition ability. *Psychological Science, 23,* 1279–1287.

De Renzi, E., Perani, D., Carlesimo, G. A., Silveri, M., & Fazio, F. (1994). Prosopagnosia can be associated with damage confined to the right hemisphere—MRI and PET study and a review of the literature. *Neuropsychologia, 32,* 893–902.

Diamond, R., & Carey, S. (1986). Why faces are and are not special: An effect of expertise. *Journal of Experimental Psychology. General, 115,* 107–117.

Duchaine, B. C., Paerker, H., & Nakayama, K. (2003). Normal recognition of emotion in a prosopagnosic patient. *Perception, 32,* 827–839.

Eimer, M. (2011). The face-sensitive N170 component of the event-related brain potential. In A. J. Calder, G. Rhodes, M. H. Johnson, & J. V. Haxby (Eds.), *The Oxford handbook of face perception* (pp. 329–344). Oxford: Oxford University Press.

Ekman, P., & Friesen, W. V. (1976). *Pictures of facial affect.* Palo Alto, CA: Consulting Psychologists Press.

Ellamil, M., Susskind, J. M., & Anderson, A. K. (2008). Examinations of identity invariance in facial expression adaptation. *Cognitive, Affective & Behavioral Neuroscience, 8,* 273–281.

Ewbank, M. P., Henson, R. N., Rowe, J. B., Stoyanova, R. S., & Calder, A. J. (2013). Different neural mechanisms within occipitotemporal cortex underlie repetition suppression across same and different size faces. *Cerebral Cortex, 23,* 1073–1084. doi:10.1093/cercor/bhs070.

Ewbank, M. P., Lawson, R. P., Henson, R. N., Rowe, J. B., Passamonti, L., & Calder, A. J. (2011). Changes in "top-down" connectivity underlie repetition suppression in the ventral visual pathway. *Journal of Neuroscience, 31,* 5635–5642.

Ewing, L., Pellicano, E., & Rhodes, G. (2013). Atypical updating of face representations with experience in children with autism. *Developmental Science, 16,* 116–123. doi:10.1111/desc.12007.

Fairhall, S. L., & Ishai, A. (2007). Effective connectivity within the distributed cortical network for face perception. *Cerebral Cortex, 17,* 2400–2406.

Farah, M. J., Wilson, K. D., Drain, M., & Tanaka, J. N. (1998). What is "special" about face perception? *Psychological Review, 105,* 482–498.

Fox, C. J., & Barton, J. J. S. (2007). What is adapted in face adaptation? The neural representations of expression in the human visual system. *Brain Research, 1127,* 80–89.

Fox, C. J., Moon, S. Y., Iaria, G., & Barton, J. J. (2009). The correlates of subjective perception of identity and expression in the face network: An fMRI adaptation study. *NeuroImage, 44,* 569–580.

Fox, C. J., Oruc, I., & Barton, J. J. S. (2008). It doesn't matter how you feel. The facial identity aftereffect is invariant to changes in facial expression. *Journal of Vision, 8*(11), 1–13. doi:10.1167/8.3.11.

Freiwald, W. A., Tsao, D. Y., & Livingstone, M. S. (2009). A face feature space in the macaque temporal lobe. *Nature Neuroscience, 12,* 1187–1196.

Furl, N., van Rijsbergen, N. J., Treves, A., Friston, K. J., & Dolan, R. J. (2007). Experience-dependent coding of facial expression in superior temporal sulcus. *Proceedings of the*

National Academy of Sciences of the United States of America, 104, 13485–13489. doi:10.1073/pnas.0702548104.

Ganel, T., & Goshen-Gottstein, Y. (2004). Effects of familiarity on the perceptual integrity of the identity and expression of faces: the parallel-route hypothesis revisited. *Journal of Experimental Psychology: Human Perception & Performance, 30,* 583–597.

Gauthier, I., & Nelson, C. (2001). The development of face expertise. *Current Opinion in Neurobiology, 11,* 219–224.

Gauthier, I., Tarr, M. J., Moylan, J., Skudlarski, P., Gore, J. C., & Anderson, A. W. (2000). The fusiform "face area" is part of a network that processes faces at the individual level. *Journal of Cognitive Neuroscience, 12,* 495–504.

George, N., Dolan, R. J., Fink, G. R., Baylis, G. C., Russell, C., & Driver, J. (1999). Contrast polarity and face recognition in the human fusiform gyrus. *Nature Neuroscience, 2,* 574–580.

Gold, J. M., Mundy, P. J., & Tjan, B. S. (2012). The perception of a face is no more than the sum of its parts. *Psychological Science, 23,* 427–434.

Goldstein, A. G., & Chance, J. E. (1980). Memory for faces and schema theory. *Journal of Psychology, 105,* 47–59.

Grill-Spector, K., Knouf, N., & Kanwisher, N. (2004). The fusiform face area subserves face perception, not generic within-category identification. *Nature Neuroscience, 7,* 555–562.

Grill-Spector, K., Kushnir, T., Edelman, S., Avidan, G., Itzchak, Y., & Malach, R. (1999). Differential processing of objects under various viewing conditions in the human lateral occipital complex. *Neuron, 24,* 187–203.

Grill-Spector, K., & Malach, R. (2001). fMR-adaptation: A tool for studying the functional properties of human cortical neurons. *Acta Psychologica, 107,* 293–321.

Hancock, K., & Rhodes, G. (2008). Contact, configural coding and the other-race effect in face recognition. *British Journal of Psychology, 99,* 45–56.

Haxby, J. V., & Gobbini, M. I. (2011). Distributed neural systems for face perception. In A. J. Calder, G. Rhodes, M. H. Johnson, & J. V. Haxby (Eds.), *The Oxford handbook of face perception* (pp. 93–110). Oxford: Oxford University Press.

Haxby, J. V., Hoffman, E. A., & Gobbini, M. I. (2000). The distributed human neural system for face perception. *Trends in Cognitive Sciences, 4,* 223–233.

Haxby, J. V., Hoffman, E. A., & Gobbini, M. I. (2002). Human neural systems for face recognition and social communication. *Biological Psychiatry, 51,* 59–67.

Haxby, J. V., Ungerleider, L. G., Clark, V. P., Schouten, J. L., Hoffman, E. A., & Martin, A. (1999). The effect of face inversion on activity in human neural systems for face and object perception. *Neuron, 22,* 189–199.

Hayward, W. G., Rhodes, G., & Schwaninger, A. (2008). An own-race advantage for components as well as configurations in face recognition. *Cognition, 106,* 1017–1027.

Hebb, D. O. (1949). *The organisation of behaviour; A neuropsychological theory.* New York: John Wiley & Sons.

Hochberg, J. (1978). *Perception* (2nd ed.). Englewood Cliffs, NJ: Prentice Hall.

Hoffman, E. A., & Haxby, J. V. (2000). Distinct representations of eye gaze and identity in the distributed human neural system for face perception. *Nature Neuroscience, 3,* 80–84.

Humphreys, K., Avidan, G., & Behrmann, M. (2007). A detailed investigation of facial expression processing in congenital prosopagnosia as compared to acquired prosopagnosia. *Experimental Brain Research, 176,* 356–373.

Jeffery, L., McKone, E., Haynes, R., Firth, E., Pellicano, E., & Rhodes, G. (2010). Four-to-six-year-old children use norm-based coding in face-space. *Journal of Vision, 10*(5), 1–19. doi:10.1167/10.5.18.

Jeffery, L., Rhodes, G., McKone, E., Pellicano, E., Crookes, K., & Taylor, E. (2011). Distinguishing norm-based from exemplar-based coding of identity in children: Evidence from face identity aftereffects. *Journal of Experimental Psychology. Human Perception and Performance, 37,* 1824–1840.

Kanwisher, N., & Barton, J. J. S. (2011). The functional architecture of the face system: Integrating evidence from fMRI and patient studies. In A. J. Calder, G. Rhodes, M. H. Johnson, & J. V. Haxby (Eds.), *The Oxford handbook of face perception* (pp. 111–130). Oxford: Oxford University Press.

Kanwisher, N., McDermott, J., & Chun, M. M. (1997). The fusiform face area: A module in human extrastriate cortex specialized for face perception. *Journal of Neuroscience, 17,* 4302–4311.

Kanwisher, N., & Yovel, G. (2006). The fusiform face area: A cortical region specialized for the perception of faces. *Philosophical Transactions of the Royal Society of London, Series B, 361,* 2109–2128. doi:10.1098/rstb.2006.1934.

Konar, Y., Bennet, P. J., & Sekuler, A. B. (2010). Holistic processing is not correlated with face-identification accuracy. *Psychological Science, 21,* 38–43.

Le Grand, R., Cooper, P. A., Mondloch, C. J., Lewis, T. L., Sagiv, N., de Gelder, B., et al. (2006). What aspects of face processing are impaired in developmental prosopagnosia? *Brain and Cognition, 61,* 139–158. doi:10.1016/j.bandc.2005.11.005.

Le Grand, R., Mondloch, C. J., Maurer, D., & Brent, H. P. (2001). Neuroperception: Early visual experience and face processing. *Nature, 410,* 890.

Leopold, D. A., Bondar, I. V., & Giese, M. A. (2006). Norm-based face encoding by single neurons in the monkey inferotemporal cortex. *Nature, 442,* 572–575.

Leopold, D. A., O'Toole, A. J., Vetter, T., & Blanz, V. (2001). Prototype-referenced shape encoding revealed by high-level aftereffects. *Nature Neuroscience, 4,* 89–94.

Leopold, D. A., Rhodes, G., Muller, K. M., & Jeffery, L. (2005). The dynamics of visual adaptation to faces. *Proceedings. Biological Sciences, 272,* 897–904.

Loffler, G., Yourganov, G., Wilkinson, F., & Wilson, H. R. (2005). fMRI evidence for the neural representation of faces. *Nature Neuroscience, 8,* 1386–1391.

MacLeod, D. I. A., & von der Twer, T. (2003). The pleisto-chrome: Optimal opponent codes for natural colours. In R. Mausfield & D. Heyer (Eds.), *Colour perception: Mind and the physical world* (pp. 155–184). Oxford: Oxford University Press.

Mather, G. (1980). The movement aftereffect and a distribution-shift model for coding the direction of visual movement. *Perception, 9,* 379–392.

Maurer, D., Le Grand, R., & Mondloch, C. J. (2002). The many faces of configural processing. *Trends in Cognitive Sciences, 6,* 255–260.

Maurer, D., O'Craven, K. M., Le Grand, R., Mondloch, C. J., Springer, M. V., Lewis, T. L., et al. (2007). Neural correlates of processing facial identity based on features versus their spacing. *Neuropsychologia, 45,* 1438–1451. doi:10.1016/j. neuropsychologia.2006.11.016.

Mazard, A., Schiltz, C., & Rossion, B. (2006). Recovery from adaptation to facial identity is larger for upright than inverted faces in the human occipito-temporal cortex. *Neuropsychologia, 44,* 912–922.

McKelvie, S. J. (1995). Emotional expression in upside-down faces: Evidence for configurational and componential processing. *British Journal of Social Psychology, 34,* 325–334.

McKone, E., Kanwisher, N., & Duchaine, B. (2007). Can generic expertise explain special processing for faces? *Trends in Cognitive Sciences, 11,* 8–15. doi:10.1016/j.tics.2006.11.002.

McKone, E., & Robbins, R. (2011). Are faces special? In A. J. Calder, G. Rhodes, M. H. Johnson, & J. V. Haxby (Eds.), *The Oxford handbook of face perception* (pp. 149–176). Oxford: Oxford University Press.

McKone, E., & Yovel, G. (2009). Why does picture-plane inversion sometimes dissociate perception of features and spacing in faces, and sometimes not? Toward a new theory of holistic processing. *Psychonomic Bulletin & Review, 16,* 778–797.

Michel, C., Caldara, R., & Rossion, B. (2006). Same-race faces are perceived more holistically than other-race faces. *Visual Cognition, 14,* 55–73.

Michel, C., Corneille, O., & Rossion, B. (2007). Race-categorization modulates holistic face encoding. *Cognitive Science, 31,* 911–924.

Michel, C., Rossion, B., Han, J., Chung, C. H., & Caldara, R. (2006). Holistic processing is finely tuned for faces of one's own race. *Psychological Science, 17,* 608–615.

Mondloch, C. J., & Desjarlais, M. (2010). The function and specificity of sensitivity to cues to facial identity: An individual differences approach. *Perception, 39,* 819–829.

Mondloch, C. J., Elms, N., Maurer, D., Rhodes, G., Hayward, W. G., Tankana, J. W., et al. (2010). Processes underlying the cross-race effect: An investigation of holistic, featural and relational processing of own- versus other-race faces. *Perception, 39,* 1065–1085. doi:10.1068/p6608.

O'Toole, A. J., Abdi, H., Deffenbacher, K. A., & Valentin, D. (1995). A perceptual learning theory of the information in faces. In T. Valentine (Ed.), *Cognitive and computational aspects of face recognition: Explorations in face space* (pp. 159–182). London: Routledge.

Palermo, R., Rivolta, D., Wilson, C. E., & Jeffery, L. (2011a). Adaptive face space coding in congenital prosopagnosia: Typical figural aftereffects but abnormal identity aftereffects. *Neuropsychologia, 49,* 3801–3812.

Palermo, R., Willis, M. L., Rivolta, D., McKone, E., Wilson, C. E., & Calder, A. J. (2011b). Impaired holistic coding of facial expression and facial identity in congenital prosopagnosia. *Neuropsychologia, 49,* 1226–1235.

Pellicano, E., Jeffery, L., Burr, D., & Rhodes, G. (2007). Abnormal adaptive face-coding mechanisms in children with autism spectrum disorder. *Current Biology, 17,* 1508–1512.

Peterson, M. P., & Rhodes, G. (Eds.). (2003). *Perception of faces, objects and scenes: Analytic and holistic processing.* Cambridge, MA: Oxford University Press.

Pitcher, D., Walsh, V., & Duchaine, B. (2011). The role of the occipital face area in the cortical face perception network. *Experimental Brain Research, 209,* 481–493.

Pitcher, D., Walsh, V., Yovel, G., & Duchaine, B. (2007). TMS evidence for the involvement of the right occipital face area in early face processing. *Current Biology, 17,* 1568–1573.

Prkachin, G. C. (2003). The effects of orientation on detection and identification of facial expressions of emotion. *British Journal of Psychology, 94,* 45–62.

Regan, D., & Hamstra, S. J. (1992). Shape discrimination and the judgement of perfect symmetry: Dissociation of shape from size. *Vision Research, 32,* 1845–1864. doi:10.1016/0042-6989(92)90046-L.

Rhodes, G. (1988). Looking at faces: First-order and second-order features as determinants of facial appearance. *Perception, 17,* 48–63.

Rhodes, G. (1996). *Superportraits: Caricatures and Recognition.* Hove: Psychology Press.

Rhodes, G., Brake, S., & Atkinson, A. P. (1993). What's lost in inverted faces? *Cognition, 47,* 25–57.

Rhodes, G., Brennan, S., & Carey, S. (1987). Identification and ratings of caricatures: Implications for mental representations of faces. *Cognitive Psychology, 19,* 473–497.

Rhodes, G., Evangelista, E., & Jeffery, L. (2009). Orientation-sensitivity of face identity aftereffects. *Vision Research, 49,* 2379–2385. doi: 10.1016/j.visres.2009.07.010.

Rhodes, G., Hayward, W. G., & Winkler, C. (2006). Expert face coding: Configural and component coding of own-race and other-race faces. *Psychonomic Bulletin & Review, 13,* 499–505.

Rhodes, G., & Jeffery, L. (2006). Adaptive norm-based coding of facial identity. *Vision Research, 46,* 2977–2987. doi:10.1016/j.visres.2006.03.002.

Rhodes, G., Jeffery, L., Clifford, C. W. G., & Leopold, D. A. (2007). The timecourse of higher-level aftereffects. *Vision Research, 47,* 2291–2296. doi:10.1016/j.visres.2007.05.012.

Rhodes, G., & Leopold, D. A. (2011). Adaptive norm-based coding of face identity. In A. J. Calder, G. Rhodes, M. H. Johnson, & J. V. Haxby (Eds.), *The Oxford handbook of f ace perception* (pp. 263–286). Oxford: Oxford University Press.

Rhodes, G., Michie, P. T., Hughes, M. E., & Byatt, G. (2009). FFA and OFA show sensitivity to spatial relations in faces. *European Journal of Neuroscience, 30,* 721–733.

Rhodes, G., Robbins, R., Jaquet, E., McKone, E., Jeffery, L., & Clifford, C. W. G. (2005). Adaptation and face perception: How aftereffects implicate norm-based coding of faces. In C. W. G. Clifford & G. Rhodes (Eds.), *Fitting the mind to the world: Adaptation and aftereffects in high-level vision* (pp. 213–240). Oxford: Oxford University Press.

Rhodes, G., Tan, S., Brake, S., & Taylor, K. (1989). Expertise and configural coding in face recognition. *British Journal of Psychology, 80,* 313–331.

Richler, J. J., Cheung, O. S., & Gauthier, I. (2011). Holistic processing predicts face recognition. *Psychological Science, 22,* 464–471.

Robbins, R., McKone, E., & Edwards, M. (2007). Aftereffects for face attributes with different natural variability: Adapter position effects and neural models. *Journal of Experimental Psychology. Human Perception and Performance, 33,* 570–592.

Rossion, B. (2008). Picture-plane inversion leads to qualitative changes of face perception. *Acta Psychologica, 128,* 274–289.

Rossion, B., Caldara, R., Seghier, M., Schuller, A.-M., Lazeyras, F., & Mayer, E. (2003). A network of occipito-temporal face-sensitive areas besides the right middle fusiform gyrus is necessary for normal face processing. *Brain, 126,* 2381–2395. doi:10.1093/brain/awg241.

Rossion, B., Dricot, L., Devolder, A., Bodart, J.-M., Crommelinck, M., de Gelder, B., et al. (2000). Hemispheric asymmetries for whole-based and part-based face processing in the human fusiform gyrus. *Journal of Cognitive Neuroscience, 12,* 793–802. doi:10.1162/089892900562606.

Rotshtein, P., Henson, R. N., Treves, A., Driver, J., & Dolan, R. J. (2005). Morphing Marilyn into Maggie dissociates physical and identity face representations in the brain. *Nature Neuroscience, 8,* 107–113.

Schiltz, C., & Rossion, B. (2006). Faces are represented holistically in the human occipito-temporal cortex. *NeuroImage, 32,* 1385–1394.

Schwartz, O., Hsu, A., & Dayan, P. (2007). Space and time in visual context. *Nature Reviews Neuroscience, 8,* 522–535.

Schweinberger, S. R., Burton, A. M., & Kelly, S. W. (1999). Asymmetric dependencies in perceiving identity and emotion: Experiments with morphed faces. *Perception & Psychophysics, 6,* 1102–1115.

Schweinberger, S. R., & Soukup, G. R. (1998). Asymmetric relationships among perceptions of facial identity, emotion, and facial speech. *Journal of Experimental Psychology. Human Perception and Performance, 24,* 1748–1765.

Sekuler, A. B., Gaspar, C. M., Gold, J. M., & Bennett, P. J. (2004). Inversion leads to quantitative, not qualitative, changes in face processing. *Current Biology, 14,* 391–396.

Skinner, A. L., & Benton, C. P. (2010). Anti-expression aftereffects reveal prototype-referenced coding of facial expressions. *Psychological Science, 21,* 1248–1253.

Skinner, A. L., & Benton, C. P. (2012). The expressions of strangers: Our identity-independent representation of facial expression. *Journal of Vision, 12*(2), 1–13. doi:10.1167/12.2.12.

Steeves, J. K., Culham, J. C., Duchaine, B. C., Pratesi, C. C., Valyear, K. F., Schindler, I., et al. (2006). The fusiform face area is not sufficient for face recognition: Evidence from a patient with dense prosopagnosia and no occipital face area. *Neuropsychologia, 44,* 594–609. doi:10.1016/j.neuropsychologia.2005.06.013.

Susilo, T., McKone, E., & Edwards, M. (2010a). Solving the upside-down puzzle: Why do upright and inverted face aftereffects look alike? *Journal of Vision, 10*(13), 1–16. doi:10.1167/10.13.1.

Susilo, T., McKone, E., & Edwards, M. (2010b). What shape are the neural response functions underlying opponent coding in face space? A psychophysical investigation. *Vision Research, 50,* 300–314.

Sutherland, N. S. (1961). Figural aftereffects and apparent size. *Quarterly Journal of Experimental Psychology, 13,* 222–228.

Suzuki, S. (2005). High-level pattern coding revealed by brief shape aftereffects. In C. W. G. Clifford & G. Rhodes (Eds.), *Fitting the mind to the world: Adaptation and aftereffects in high level vision* (pp. 135–172). Oxford: Oxford University Press.

Tanaka, J. W., & Farah, M. J. (1993). Parts and wholes in face recognition. *Quarterly Journal of Experimental Psychology, 46A,* 225–245.

Tanaka, J. W., & Gordon, I. (2011). Features, configuration, and holistic face processing. In A. J. Calder, G. Rhodes, M. H. Johnson, & J. V. Haxby (Eds.), *The Oxford handbook of face perception* (pp. 177–194). Oxford: Oxford University Press.

Tanaka, J. W., Kiefer, M., & Bukach, C. (2004). A holistic account of the own-race effect in face recognition: Evidence from a cross-cultural study. *Cognition, 93,* B1–B9. doi:10.1016/j.cognition.2003.09.011.

Tranel, D., Damasio, A. R., & Damasio, H. (1988). Intact recognition of facial expression, gender, and age in patients with impaired recognition of face identity. *Neurology, 38,* 690–696.

Tsao, D. Y., & Freiwald, W. A. (2006). What's so special about the average face? *Trends in Cognitive Sciences, 10,* 391–393. doi:10.1016/j.tics.2006.07.009.

Valentine, T. (1991). A unified account of the effects of distinctiveness, inversion, and race in face recognition. *Quarterly Journal of Experimental Psychology, 43A,* 161–240.

Valentine, T. (Ed.). (1995). *Cognitive and computational aspects of face recognition: Explorations in face space.* London: Routledge.

Wang, R., Li, J., Fang, H., Tian, M., & Liu, J. (2012). Individual differences in holistic processing predict face recognition ability. *Psychological Science, 23,* 169–177.

Webster, M. A. (2003). Light adaptation, contrast adaptation, and human vision. In R. Mausfield & D. Heyer (Eds.), *Colour perception: Mind and the physical world.* Oxford: Oxford University Press.

Webster, M. A., & Leonard, D. (2008). Adaptation and perceptual norms in color vision. *Journal of the Optical Society of America. A, Optics, Image Science, and Vision, 25,* 2817–2825. doi:10.1364/JOSAA.25.002817.

Webster, M. A., & MacLeod, D. I. A. (2011). Visual adaptation and face perception. *Philosophical Transactions of the Royal Society: Biological Sciences, 366,* 1702–1725. doi:10.1098/rstb.2010.0360.

White, M. (2000). Parts and wholes in expression recognition. *Cognition and Emotion, 40,* 39–60.

Wilmer, J. B., Germine, L., Chabris, C. F., Chatterjee, G., Williams, M., Loken, E., et al. (2010). Human face recognition ability is specific and highly heritable. *Proceedings of the National Academy of Sciences of the United States of America, 107,* 5238–5241. doi:10.1073/pnas.0913053107.

Winston, J. S., Henson, R. N. A., Fine-Goulden, M. R., & Dolan, R. J. (2004). fMRI-adaptation reveals dissociable neural representations of identity and expression in face perception. *Journal of Neurophysiology, 92,* 1830–1839.

Young, A. W., Hellawell, D., & Hay, D. C. (1987). Configurational information in face perception. *Perception, 16,* 747–759.

Yovel, G., & Kanwisher, N. (2004). Face perception: Domain-specific, not process-specific. *Neuron, 44,* 889–898.

Yovel, G., & Kanwisher, N. (2005). The neural basis of the behavioural face-inversion effect. *Current Biology, 15,* 2256–2262.

Zhu, Q., Song, Y., Hu, S., Li, X., Tian, M., Zhen, Z., et al. (2010). Heritability of the specific cognitive ability of face perception. *Current Biology, 20,* 137–142. doi:10.1126/science.1077091.

51 Scene Perception

AUDE OLIVA

Visual scene perception is the gateway to many of our most valued behaviors, including navigation, recognition, and reasoning with the world around us. What is a "visual scene"? What are its properties? Is scene perception different from object perception? Operationally, a visual scene can be defined as a view in which objects and surfaces are arranged in a meaningful way, for example a kitchen, a street, or a forest. Scenes contain elements arranged in a *spatial layout* and can be viewed at a variety of spatial scales (e.g., the up-close view of an office desk or the view of the entire office). As a rough distinction, one generally takes action *on* an object, whereas one usually acts *within* a scene.

One paradoxical feature of visual scene analysis is that the complex arrangement of objects and surfaces in the world creates the impression that there is too much to see at once. How can so much visual information be processed and understood in a timely manner? Remarkably, we are able to interpret the meaning of multifaceted and complex scene images—a wedding, a birthday party, or a stadium crowd—in a fraction of a second (Potter, 1975)! This is about the same time it takes a person to identify that a single object is a face, a dog, or a car (Grill-Spector & Kanwisher, 2005; Intraub, 1981; Thorpe, Fize, & Marlot, 1996). An unmistakable demonstration of the brain's prowess in visual scene understanding can be experienced at the movies: With a few rapid scene cuts from a movie to form a trailer, it seems as if we have perceived and understood much more of the story in a few seconds than could be described later in the same amount of time. Perceiving scenes in a glance is like looking at an abstract painting of a landscape and recognizing that a "forest" is depicted before seeing the "trees" that create it (Navon, 1977).

This chapter reviews research in the behavioral, computational, and cognitive neuroscience domains that describe how the human visual system analyses real-world scenes. Although we typically experience scenes in a three-dimensional physical world, most studies are conducted using two-dimensional pictures. There are likely important differences between perceiving the world and perceiving visual scenes via pictures, and this chapter describes principles that are likely to apply to both mediums (for a review, see Cutting, 2003).

BEHAVIORAL STUDIES OF SCENE PERCEPTION

Historical Perspective

Pioneering work done by Mary Potter (1975), David Navon (1977), and Irving Biederman (1981) has shown that the overall scene meaning is invariably captured within a glance regardless of the complexity of the image. These landmark studies have laid down much of the empirical and theoretical foundation for modern inquiries into how the human brain perceives complex, real-world scenes.

Figure 51.1 illustrates what scenes are made of: surfaces of different materials laid out within a three-dimensional physical space. Some surfaces define the boundaries of the space (e.g., walls, tall objects), and other surfaces are created by objects of specific identities and functions (e.g., a dining room table). Importantly, scene perception is not merely the sum of its parts; it is paramount to consider the scene as a whole. In figure 51.1, for example, scenes can be similar in layout (B, D) or contain similar objects (see C, D) yet belong to different semantic categories. Scene analysis involves perceiving the type of surfaces, objects, their placement, and their quantity.

In their original study Potter and Levy (1969) allowed observers a single brief glance at a series of real-world images and subsequently tested their memory of these images. They observed that understanding happens fast, very fast: When presented alone for 100 ms, each image was easily remembered and described. Together with other experimental evidence, these results showed that most of the information from an image could be captured in tenths of a second. So what type of representation could possibly be built so fast?

Theoretical Perspectives

Biederman (1981) proposed three levels of representations observers could build quickly, either sequentially or in parallel, to reach a rich description of the scene by the end of a glance. According to this framework, scene understanding may follow from (1) a prominent object or surface, (2) a multiple-component

Different objects, different spatial layout   Different objects, same spatial layout

Same objects, different spatial layout   Same objects, same spatial layout

FIGURE 51.1 Each image pair contains scenes of different semantic categories. Yet notice how the scenes can differ in spatial layout (A, C) or object content (A, B) or have similar layouts (B, D) and similar objects (C, D).

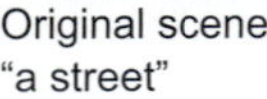

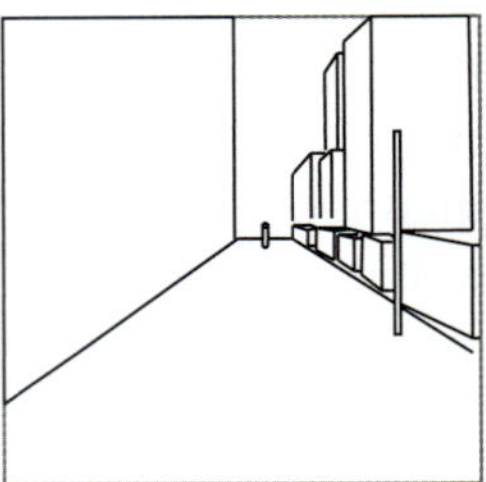

Original scene "a street"   Suggestive contours (Biederman, 1981)   Geometric forms (Biederman, 1981)   Blobs in relations (Schyns & Oliva, 1994)   Sketch of textures (Oliva & Torralba, 2001)

FIGURE 51.2 Illustrations of global scene emergent features. These representations have formed the basis of object-free computational models of scene perception.

representation incorporating a layout of distinct objects or surfaces, and (3) a more global representation of scene-emergent features, which does not necessarily depend on object and surface identification. The *prominent object* representation capitalizes on a region of the scene (Friedman, 1979)—often, a large and diagnostic object (a bed in a bedroom, a sofa in a living room, a large region of grass in a field)—to quickly activate contextual information from stored knowledge. The *multiple-component* representation is based on objects and regions segmented from the background and organized in a coherent layout. Finally, the *global scene-emergent features* representation is built from grasping components from all over the available percept that do not necessarily correspond to segmented objects or meaningful regions. Figure 51.2 illustrates some of the forms these emergent global scene features haven taken in various works.

Inspired by this early proposal studies in behavioral, computational, and cognitive neuroscience of the past decade have converged to describe two complementary paths of scene perception: an *object-centered approach* in which components are segmented and function as the scene descriptors (i.e., this is a street because there are buildings and cars); and a *scene or space-centered approach* in which spatial layout and global properties of the whole image or place act as the scene descriptors (i.e., this is a street because it is an outdoor, urban environment flanked with tall frontal vertical surfaces with squared patterned textures). How do these different levels of scene information unfold over the course of a glance?

Although a complete picture of the time course of the components contributing to scene perception has yet to emerge, most experimental work distinguishes between early stages (before 100 ms) and late stages (from 200 to 300 ms, before the observer moves her eye) of scene analysis. When an image is briefly presented, there is a temporal progression to how an observer perceives a scene's content: As image exposure increases, observers are better able to fully perceive the details of an image such as high spatial frequencies (Schyns & Oliva, 1994), texture (Walker-Renninger & Malik, 2004), or object identities (Fei-Fei et al., 2007; Rayner et al., 2009). This is known as global-to-local (Navon, 1977) or coarse-to-fine (Schyns & Oliva, 1994) scene analysis. For instance, in Fei-Fei et al. (2007), observers were presented with briefly masked pictures depicting various events and scenery (e.g., a soccer game, a busy hair salon, a choir, a dog playing fetch) and later asked to describe in detail what they saw in the picture. The authors found that observers perceived global scene information, such as whether the picture was outdoor or indoor, well above chance with less than 100 ms of exposure, whereas details about objects were reported with a couple of hundred milliseconds more exposure time. Similarly, Greene and Oliva (2009a) found that global scene properties (i.e., the volume, openness, naturalness of a scene) explain early scene categorization better than representation of a scene solely by its objects (i.e., trees, grass, rock). In fact, an exposure of only 20 to 30 ms is sufficient to know whether the scene depicts a natural or an urban place (Greene & Oliva, 2009b; Joubert et al., 2007) or whether the scene has a small or large volume (e.g., cave vs. lake). Yet it takes twice that much time to determine the basic level category of the scene (Greene & Oliva, 2009b), for example mountain versus beach. It follows that, during early visual processing, there is a time point at which a scene may be classified as a large or navigable landscape but not yet as a mountain or lake.

One important question in visual analysis is the role of temporal history in perception: How does what we have recently perceived influence what is presently perceived? This refers to the notion of *adaptation*: When observers are overexposed to certain visual features, the adaptation of those features affects the conscious perception of a subsequently presented stimulus. Which representations of scene information, if any, adapt? Our daily life experiences suggest that spatial layout characteristics might be susceptible to aftereffects. For example, after spending all day working in a small office or cubicle, the first sight of the expansive world right outside of the office building might appear much larger than it did on entry.

Using an adaptation paradigm Greene and Oliva (2010) found that prolonged exposure to scenes with shared global properties can influence how observers perceive that property in later scenes, and, interestingly, these aftereffects even influence semantic categorization of the scene. In their experiment observers adapted to a stream of open and panoramic views of natural images (openness) or to a stream of pictures depicting scenes enclosed with frontal and lateral surfaces (closeness). When ambiguous probe images (e.g., a field with trees) were then tested, they were more likely to be judged as a *closed* space if the observer had adapted to openness or as an *open* space if adapted to closeness. Similar aftereffects are found after adapting to natural versus urban spaces (see also Kaping, Tzvetanov, & Treue, 2007). Furthermore, adapting to an open or closed spatial layout even shifted the categorical recognition of a briefly perceived scene. This design took advantage of the fact that fields are usually open scenes, whereas forests are typically closed scenes, but importantly there is a continuum between field and forest scenes, with some scenes existing ambiguously between the two categories that can be perceived both as a field or a forest (see figure 51.3). Indeed, Greene and Oliva (2010) found that adapting to a stream of closed natural

FIGURE 51.3   Continuum between fields and forests. Scenes in the middle of continuum have an ambiguous category and can be perceived as either a field or a forest, after adapting to closed versus open scenes, respectively. (From Greene & Oliva, 2010.)

images caused an ambiguous field/forest to be judged as more likely to be a field. In summary, the observation of aftereffects for global scene properties implies that, during natural scene processing, the visual system extracts statistics and becomes attuned to recently processed global properties such as scene layout and volume.

Discovering what scene properties become available over the course of a glance and learning what representational role global properties have in scene recognition provide critical insights into the possible computations underlying scene perception.

COMPUTATIONAL FRAMEWORK OF SCENE PERCEPTION

Just as external shape and internal features are separable dimensions of face encoding, Oliva and Torralba (2001, 2002) proposed a framework in which a scene, whether a physical space or its projection onto a two-dimensional image, can be represented by two separable and complementary descriptors: its *spatial boundary* (i.e., the external shape, size, and scope of the space the scene represents) and its *content* (the internal elements, encompassing textures, colors, materials, and objects). As illustrated in figure 51.4, the shape of an outdoor scene may be expansive and open to the horizon, as in field and parking lot, or closed and bounded by frontal and lateral surfaces, as in forests and streets. Importantly, the spatial boundary is independent of the scene's content, which may contain natural or manufactured elements. This framework is general enough so that it does not make assumptions regarding the type of features used to represent the scene. For instance, one can identify a scene as a landscape by using colors, textures, materials, or objects.

The size or scope of a scene can be captured by geometrical relations between its boundaries or by low-level image features that are correlated with scene size (Torralba & Oliva, 2002, 2003) and semantic categories (Oliva & Torralba, 2001; Xiao et al., 2010). Therefore, this framework is orthogonal to object- and scene-centered views to scene perception: Both boundaries and content descriptors can be built up, in theory, from segmented objects, from relations between components, or from global "scene-emergent" features (Oliva & Torralba, 2001; Ross & Oliva, 2010). Given that visual scene analysis generates a rich percept with multiple representations of description, how does the brain accomplish these diverse functions of scene understanding?

NEUROIMAGING STUDIES OF SCENE PERCEPTION

A growing body of research spanning behavioral, computational, and neuroimaging methods has shown that scene representations are not unitary per se. Rather, natural scene processing appears to involve a combination of many visual feature and scene property representations that, in parallel, create the scene image representation. Vision scientists have made significant progress in identifying several candidate brain regions responsible for processing different scene properties. These brain regions are distributed across low-, mid-, and high-level visual areas, with different areas supporting different types of scene information.

For a given semantic scene category (e.g., forest), images are correlated with a plethora of low- and midlevel visual features that help to distinguish that category from other semantic categories. Recent neuroimaging studies have shown that the activity from

FIGURE 51.4 A schematic illustration of how pictures of real-world scenes can be represented uniquely by their spatial boundaries and content. Keeping the enclosed spatial layout, if we strip off the natural content of a forest and fill the space with urban contents, then the scene becomes a street. Keeping the open spatial layout, if we strip off the natural content of a field and fill the space with urban contents, then the scene becomes a parking lot. (Adapted from Park et al., 2011.)

early visual areas (V1 to V4) can be used to predict which particular image was being viewed by an observer and, even more impressively, to reconstruct some of the visual feature content of the scenes themselves (Kay et al., 2008; Naselaris et al., 2009). Furthermore, the pattern of responses in these regions also contains the information to allow scene classification into a handful of semantic categories (Naselaris et al., 2009; Walther et al., 2009).

Which constituent representations are necessary and sufficient to support our remarkable ability to rapidly understand complex real-world scenes? Although this remains an open question, human functional neuroimaging investigations of the last decade have made significant progress in identifying brain regions important for higher-level aspects of scene and space perception (figure 51.5). Akin to the empirical and theoretical proposals described in the previous sections, recent neuroimaging studies (Kravitz, Peng, & Baker, 2011; MacEvoy & Epstein, 2011; Park et al., 2011) have found that visual scene analysis recruits distinct and complementary high-level representations, indicating distinct neural pathways for the representation of scene/space-centered versus content/object-centered information.

Scene- and Space-Centered Cortical Regions

The two most studied scene-selective regions so far have been the parahippocampal place area (PPA), a region of the collateral sulcus near the parahippocampal–lingual boundary (Epstein & Kanwisher, 1998) and the retrosplenial complex (RSC) (Bar & Aminoff, 2003). Both of these regions respond preferentially to pictures depicting scenes, spaces (like those shown in figures 51.1 and 51.3), and landmarks more than to pictures of faces or single movable objects (for a review, see Epstein & MacEvoy, 2011). A third scene-selective functional region, the occipital place area (OPA) (Dilks, Julian, Paunov, & Kanwisher, 2013) is found around the transverse occipital sulcus (see chapter 52 by Kanwisher and Dilks). Interestingly, neither the PPA nor the RSC responses are modulated by the quantity of objects in the scene (i.e., both regions are equally active when viewing an empty room or a room with clutter) (Epstein & Kanwisher, 1998); however, they both show selectivity to the spatial layout of the scene in various tasks (Aguirre, Zarahn, & D'Esposito, 1998; Epstein & Kanwisher, 1998; Janzen & Van Turennout, 2004; Park et al., 2011). So what distinguishes PPA from RSC?

Several studies have shown that the PPA and RSC have different selectivities to perceiving a scene from an observer's viewpoint or perceiving the place the scene is embedded in. For instance, Park and Chun (2009) showed observers different views of scenes that were part of the same panoramic scene, simulating the perception of an observer moving her head, taking snapshots of views in a spatiotemporally continuous way. They found that the PPA treated each view of the panoramic scene as a different "image," suggesting a view-specific representation in PPA (see also Epstein, Graham, & Downing, 2003; Epstein, Parker, & Feiler 2007). By contrast, Park and Chun found that RSC treated different views of a panorama as the same stimulus, suggesting that this region may hold a larger representation of the place beyond the current view (see also Park et al., 2007; Park, Chun, & Johnson, 2010; Epstein, Parker, & Feiler, 2007, for supporting evidence; see Baumann & Mattingley, 2010, for a related topic in heading direction). Interestingly, however, another recent study (Dilks et al., 2011) found that scene representations in PPA were in fact tolerant to more severe transformations (i.e., reflections about the vertical axis—a transformation of 180°). Thus, the further question of whether PPA representations are only tolerant to mirror reversals is an interesting one.

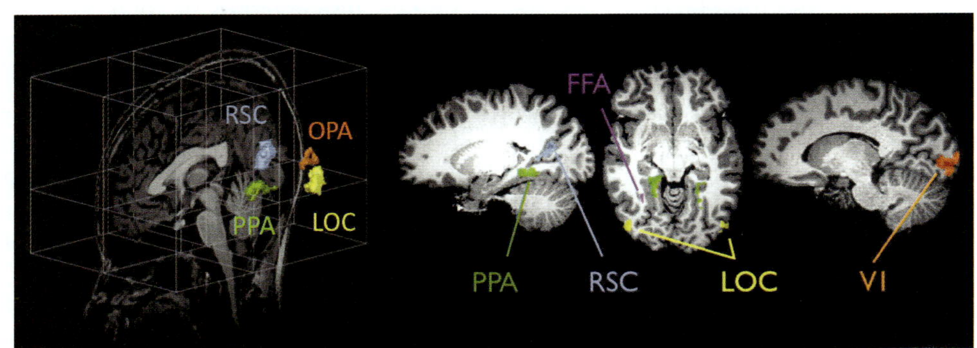

FIGURE 51.5   Several functionally defined regions involved in scene perception are shown for two individuals. PPA, parahippocampal place area; RSC, retrosplenial complex; OPA, occipital place area; LOC, lateral occipital complex. FFA (fusiform face area) and V1 (primary visual cortex) are shown here for comparison. (See chapter 52 by Kanwisher and Dilks.)

Beyond spatial layout information, perceiving objects is an important part of scene processing. Much of the time identifying the objects in a scene will dictate the scene's function (figure 51.1). For that reason the lateral occipital complex (LOC) is another candidate region of the scene perception network. The LOC is an area of the ventral visual pathway that specializes in representing object shapes and object categories (Grill-Spector et al., 1998; see chapter 52 by Kanwisher and Dilks for a complete review). Among the many neuro-imaging studies examining the nature of object representations in the LOC, the section below describes findings with implications for an object-centered pathway to scene understanding.

Because scenes typically contain several objects rather than single isolated objects, it stands to reason that object-processing brain regions should be encoding the content of a scene. Recent studies have shown that the presence of multiple objects in the visual field can be decoded from averaging the activity for individual objects in LOC (MacEvoy & Epstein, 2009, 2011). Furthermore, the pattern of neural responses in the LOC is sufficient to distinguish among a few scene categories (e.g., beach and city) (Walther et al., 2009) as well as to decode whether certain objects were present within the scenes (Peelen, Fei-Fei, & Kastner, 2009).

The LOC is not the only brain region involved in object processing. Large objects—landmarks and buildings, for example—have been shown to activate the PPA (Epstein & Kanwisher, 1998). Importantly, certain physical and experience-based properties of real-world objects evoke selective brain responses based on properties such as their real-world size (e.g., paperclip vs. car) (Konkle & Oliva, 2012), their contextual strength within a scene (e.g., a fire hydrant vs. a book) (Bar, 2004), and whether the objects define a local space in a larger scene (e.g., a sofa) (Mullally & Maguire, 2011). An example of selective activity for object properties was shown by Konkle and Oliva (2012), in which the authors identified a region of interest in the parahippocampal cortex that showed peaks of activity for large objects that our bodies typically interact with (e.g., by physically interacting with that object as with a bed or a sofa or walking toward it as to a piano or a dishwasher). Similarly, a left-lateralized region in the occipitotemporal sulcus showed peaks of selectivity for objects of a small physical size in the world that can be typically handled (e.g., a strawberry or a hat). Thus, as multiple regions of the brain encode different spatial aspects of a scene, multiple regions may represent different types of content and objects encountered in a scene.

Complementary Representations of Spatial Boundary and Object Content

How can these multiple regions, and perhaps others, work together to form a complete representation of a scene? Although this remains an open question for future neuroimaging investigation, some recent lines of research have started to shed light on how the brain analyzes an input natural image using a visual feature representation and ultimately creates a higher-level representation of the space and content of the scene. Inspired by the fact that spatial boundaries and scene content are orthogonal properties of a real-world scene (see figure 51.4), Park et al. (2011) designed an experimental paradigm to empirically test the underlying nature of the representations in the PPA and LOC. As illustrated in figure 51.4 the shape of a scene may be expansive and open to the horizon, as in a field or parking lot, or closed and bounded by frontal and lateral surfaces, as in forests or streets. Furthermore, a scene may be comprised of natural or urban (manufactured) objects, independent of its spatial boundary.

Park et al. (2011) used pattern analysis of neural responses in the PPA and LOC to classify whether a scene belonged to a particular class (i.e., open, closed, natural, or urban space). By analyzing the types of classification errors that occurred when each region's response was used, the authors could probe questions relating to whether the regions represent a scene in an *overlapping* fashion (e.g., both produce similar errors when classifying different scenes) or in a *complementary* fashion (e.g., each region shows a specialization in representing either the boundaries or content of a scene). They found a dissociation between the types of classification errors made in two brain regions. The PPA confused scenes with similar spatial boundaries, regardless of the type of content, whereas the LOC made the opposite errors (confusing scenes with the same content, independent of their spatial layout). In the PPA, scenes representing closed urban environments, such as streets and buildings, were most confused with closed natural scenes, such as forests and canyons (see also Kravitz, Peng, & Baker, 2011). On the other hand, the LOC confused scenes that contained similar objects and surfaces, such as mistaking fields, deserts, or ocean with forest, mountain, or canyons; LOC accuracy was insensitive to the spatial layout.

Using multivoxel pattern analysis as well, MacEvoy and Epstein (2011) found a similar result. In their study neural responses evoked in the LOC by pictures of scenes (e.g., a kitchen, a street) were well predicted by the response patterns elicited by their sole prominent objects (e.g., refrigerator, traffic light), although this was not the case for PPA activity. Altogether, the current state of the art in neuroimaging studies suggests that scene analysis in the brain recruits distinct regions that perform different and complementary computations to create a singular and unique representation of the scene in view.

CONCLUSION

Altogether, studies across disciplines and methods provide evidence for the existence of a distributed scene representation, encoding different levels of information, as a basis for scene perception. The existence of neural pathways for the representation of scene/space-centered versus content/object-centered information corroborates the experimental and computational frameworks of scene perception that have emerged in the last decade. Whereas many discoveries remain to be found regarding how the brain computes this immediate "understanding" of the world, and which levels of features are used to perform particular operations, the common theoretical framework in scene perception between disciplines should provide fast-track progress in the years to come.

ACKNOWLEDGMENTS

The writing of this chapter was supported by a National Eye Institute grant EY020484 to A.O. The author would like to thank Barbara Hidalgo-Sotelo, Daniel Dilks, and Soojin Park for insightful comments.

REFERENCES

Aguirre, G. K., Zarahn, E., & D'Esposito, M. (1998). An area within human ventral cortex sensitive to "building" stimuli: Evidence and implications. *Neuron, 21*, 373–383.

Bar, M. (2004). Visual objects in context. *Nature Reviews. Neuroscience, 5*, 617–629.

Bar, M., & Aminoff, E. (2003). Cortical analysis of visual context. *Neuron, 38*, 347–358.

Baumann, O., & Mattingley, J. B. (2010). Medial parietal cortex encodes perceived heading direction in humans. *Journal of Neuroscience, 30*, 12897–12901.

Biederman, I. (1981). On the semantics of a glance at a scene. In M. Kubovy & J. R. Pomerantz (Eds.), *Perceptual organization* (pp. 213–263). Hillsdale, NJ: Lawrence Erlbaum Associates.

Cutting, J. E. (2003). Reconceiving perceptual space. In H. Hecht, M. Atherton, & R. Schwartz (Eds.), *Perceiving pictures: An interdisciplinary approach to pictorial space* (pp. 215–238). Cambridge, MA: MIT Press.

Dilks, D. D., Julian, J. B., Kubilius, J., Spelke, E. S., & Kanwisher, N. (2011). Mirror-image sensitivity and invariance in object and scene processing pathways. *Journal of Neuroscience, 31*(31), 11305–11312.

Dilks, D. D., Julian, J. B., Paunov, A., & Kanwisher, N. (2013). The occipital place area (OPA) is causally and selectively involved in scene perception. *Journal of Neuroscience, 33*, 1331–1336.

Epstein, R., Graham, K. S., & Downing, P. E. (2003). Viewpoint-specific scene representations in human parahippocampal cortex. *Neuron, 37*, 865–876.

Epstein, R., & Kanwisher, N. (1998). A cortical representation of the local visual environment. *Nature, 392*, 598–601.

Epstein, R. A., & MacEvoy, S. P. (2011). Making a scene in the brain. In L. Harris & M. Jenkin (Eds.), *Vision in 3D environments* (pp. 255–279). Cambridge: Cambridge University Press.

Epstein, R. A., Parker, W. E., & Feiler, A. M. (2007). Where am I now? Distinct roles for parahippocampal and retrosplenial cortices in place recognition. *Journal of Neuroscience, 27*, 6141–6149.

Fei-Fei, L., Iyer, A., Koch, C., & Perona, P. (2007). What do we perceive in a glance of a real-world scene? *Journal of Vision, 7*(1), 1–29.

Friedman, A. (1979). Framing pictures: The role of knowledge in automatized encoding and memory for gist. *Journal of Experimental Psychology. General, 108*, 316–355.

Greene, M. R., & Oliva, A. (2009a). Recognition of natural scenes from global properties: Seeing the forest without representing the trees. *Cognitive Psychology, 58*, 137–176.

Greene, M. R., & Oliva, A. (2009b). The briefest of glances: The time course of natural scene understanding. *Psychological Science, 20*, 464–472.

Greene, M. R., & Oliva, A. (2010). High-level aftereffects to global scene property. *Journal of Experimental Psychology. Human Perception and Performance, 36*, 1430–1442.

Grill-Spector, K., & Kanwisher, N. (2005). Visual recognition: As soon as you know it is there, you know what it is. *Psychological Science, 16*, 152–160.

Grill-Spector, K., Kushnir, T., Edelman, S., Itzchak, Y., & Malach, R. (1998). Cue-invariant activation in object-related areas of the human occipital lobe. *Neuron, 21*, 191–202.

Intraub, H. (1981). Rapid conceptual identification of sequentially presented pictures. *Journal of Experimental Psychology. Human Perception and Performance, 7*, 604–610.

Janzen, G., & Van Turennout, M. (2004). Selective neural representation of objects relevant for navigation. *Nature Neuroscience, 7*, 673–677.

Joubert, O., Rousselet, G., Fize, D., & Fabre-Thorpe, M. (2007). Processing scene context: Fast categorization and object interference. *Vision Research, 47*, 3286–3297.

Kaping, D., Tzvetanov, T., & Treue, S. (2007). Adaptation to statistical properties of visual scenes biases rapid categorization. *Visual Cognition, 15*, 12–19.

Kay, K. N., Naselaris, T., Prenger, R. J., & Gallant, J. L. (2008). Identifying natural images from human brain activity. *Nature, 452*, 352–355.

Konkle, T., & Oliva, A. (2012). A real-world size organization of object responses in occipito-temporal cortex. *Neuron, 74*, 1114–1124.

Kravitz, D. J., Peng, C. S., & Baker, C. I. (2011). Real-world scene representations in high-level visual cortex: It's the spaces more than the places. *Journal of Neuroscience, 31*, 7322–7333.

MacEvoy, S. P., & Epstein, R. A. (2009). Decoding the representation of multiple simultaneous objects in human occipitotemporal cortex. *Current Biology, 19*, 943–947.

MacEvoy, S. P., & Epstein, R. A. (2011). Constructing scenes from objects in human occipitotemporal cortex. *Nature Neurosciences, 14,* 1323–1329.

Mullally, S. L., & Maguire, E. A. (2011). A new role for the parahippocampal cortex in representing space. *Journal of Neuroscience, 31,* 7441–7449.

Naselaris, T., Prenger, R. J., Kay, K. N., Oliver, M., & Gallant, J. L. (2009). Bayesian reconstruction of natural images from human brain activity. *Neuron, 53,* 902–915.

Navon, D. (1977). Forest before the trees: The precedence of global features in visual perception. *Cognitive Psychology, 9,* 353–383.

Oliva, A., & Torralba, A. (2001). Modeling the shape of the scene: A holistic representation of the spatial envelope. *International Journal of Computer Vision, 42,* 145–175.

Oliva, A., & Torralba, A. (2002). Scene-centered description from spatial envelope properties. In H. Bulthoff, S. W. Lee, T. Poggio, & C. Wallraven (Eds.), *Computer Science Series Procedure, Second International Workshop on Biologically Motivated Computer Vision* (pp. 263–272). Tübingen: Springer-Verlag.

Park, S., Brady, T. F., Greene, M. R., & Oliva, A. (2011). Disentangling scene content from its spatial boundary: Complementary roles for the PPA and LOC in representing real-world scenes. *Journal of Neuroscience, 31*(4), 1333–1340.

Park, S., & Chun, M. M. (2009). Different roles of the parahippocampal place area (PPA) and retrosplenial cortex (RSC) in panoramic scene perception. *NeuroImage, 47,* 1747–1756.

Park, S., Chun, M. M., & Johnson, M. K. (2010). Refreshing and integrating visual scenes in scene-selective cortex. *Journal of Cognitive Neuroscience, 22,* 2813–2822.

Park, S., Intraub, H., Widders, D., Yi, D. J., & Chun, M. M. (2007). Beyond the edges of a view: Boundary extension in human scene-selective visual cortex. *Neuron, 54,* 335–342.

Peelen, M. V., Fei-Fei, L., & Kastner, S. (2009). Neural mechanisms of rapid natural scene categorization in human visual cortex. *Nature, 460,* 94–97.

Potter, M. C. (1975). Meaning in visual scenes. *Science, 187,* 965–966.

Potter, M. C., & Levy, E. I. (1969). Recognition memory for a rapid sequence of pictures. *Journal of Experimental Psychology, 81,* 10–15.

Rayner, K., Smith, T. J., Malcolm, G. L., & Henderson, J. M. (2009). Eye movements and visual encoding during scene perception. *Psychological Science, 20,* 6–10.

Ross, M. G., & Oliva, A. (2010). Estimating perception of scene layout properties from global image features. *Journal of Vision, 10*(1), 2, 1–25. doi:10.1167/10.1.2.

Schyns, P. G., & Oliva, A. (1994). From blobs to boundary edges: Evidence for time- and spatial-scale-dependent scene recognition. *Psychological Science, 5,* 195–200.

Thorpe, S., Fize, D., & Marlot, C. (1996). Speed of processing in the human visual system. *Nature, 381,* 520–522.

Torralba, A., & Oliva, A. (2002). Depth estimation from image structure. *IEEE Pattern Analysis and Machine Intelligence, 24,* 1226–1238.

Torralba, A., & Oliva, A. (2003). Statistics of natural images categories. *Network (Bristol, England), 14,* 391–412.

Walker Renninger, L., & Malik, J. (2004). When is scene identification just texture recognition? *Vision Research, 44,* 2301–2311.

Walther, D. B., Caddigan, E., Fei-Fei, L., & Beck, D. M. (2009). Natural scene categories revealed in distributed patterns of activity in the human brain. *Journal of Neuroscience, 29,* 10573–10581.

Xiao, J., Hayes, J., Ehinger, K., Oliva, A., & Torralba, A. (2010). SUN Database: Large-scale scene recognition from abbey to zoo. In *Proceedings of the 23rd IEEE Conference on Computer Vision and Pattern Recognition* (pp. 3485–3492). San Francisco, CA: IEEE Computer Society.

52 The Functional Organization of the Ventral Visual Pathway in Humans

NANCY KANWISHER AND DANIEL D. DILKS

We can recognize an object within a fraction of a second, even if we have never seen that particular object before, and even if we have no advance information about what kind of object it might be (Potter, 1976; Thorpe, Fize, & Marlot, 1996). The cognitive and neural mechanisms underlying this remarkable ability are not well understood, and current computer vision algorithms still lag far behind human performance. One promising strategy for attempting to understand human visual recognition is to characterize the neural system that accomplishes it, the ventral visual pathway (VVP), which extends from the occipital lobe into inferior and lateral regions of the temporal lobe. Here we describe key results from the last 15 years of neuroimaging research on humans that have begun to elucidate the general organization and functional properties of the cortical regions involved in visually perceiving people, places, and things.

The central finding from this now-substantial body of work is that the VVP is not homogeneous but is instead a highly differentiated structure containing a set of regions each with its own distinct functional profile. These regions include the fusiform face area (FFA), which responds selectively to faces (Kanwisher, McDermott, & Chun, 1997a; McCarthy et al., 1997), the parahippocampal place area (PPA), which responds selectively to places (Epstein & Kanwisher, 1998), the extrastriate body area (EBA), which responds selectively to bodies (Downing et al., 2001), the lateral occipital complex (LOC), which responds to object shape (Kanwisher et al., 1997b; Malach et al., 1995) largely independent of object category, and the visual word form area (VWFA), which responds selectively to both visually presented words and consonant strings (Baker, Hutchison, & Kanwisher, 2007; Cohen et al., 2000). Each of these regions is present in approximately the same location in virtually every healthy subject. These regions and their cohorts (e.g., the occipital face area or OFA) constitute the fundamental machinery of high-level visual recognition in humans. An understanding of the function of each of these regions is likely to provide important clues about how visual recognition works.

In this chapter we first describe the best-established functionally specific regions in the VVP and then the ongoing controversies about the degree of such specialization. We then review the available evidence on the functional organization of the entire VVP, and finally, we consider the computational advantages that may be afforded by having specialized regions in the first place.

THE BEST-ESTABLISHED FUNCTIONALLY SPECIFIC REGIONS IN THE VENTRAL VISUAL PATHWAY

Figure 52.1 shows the main functionally distinct regions of the VVP. We begin by focusing on the FFA, PPA, EBA, and LOC because these are the best-established regions in this pathway in several respects. First, each of these regions has been found consistently in a large number of studies and labs; although the theoretical significance of these regions can be debated, their existence cannot. Second, the category selectivity by which each region is defined is not merely statistically significant but also large in effect size: Each of these regions responds about twice as strongly to any stimulus from its preferred category as to any nonpreferred stimulus category. Effect size is often ignored in the brain imaging literature, but it should not be, as it determines the strength of the inference you can draw: If you know how to double the response of a region, you generally have a better handle on its function than if you merely know how to change its response by a small amount. Third, the fact that these regions can be found easily and now algorithmically (Julian et al., 2012) in any healthy subject has made possible a "region of interest" (ROI) research strategy whereby the region is first functionally identified in each subject individually in a short "localizer" scan and then characterized in rich detail in subsequent experiments, including not just its response profile but the information it represents.

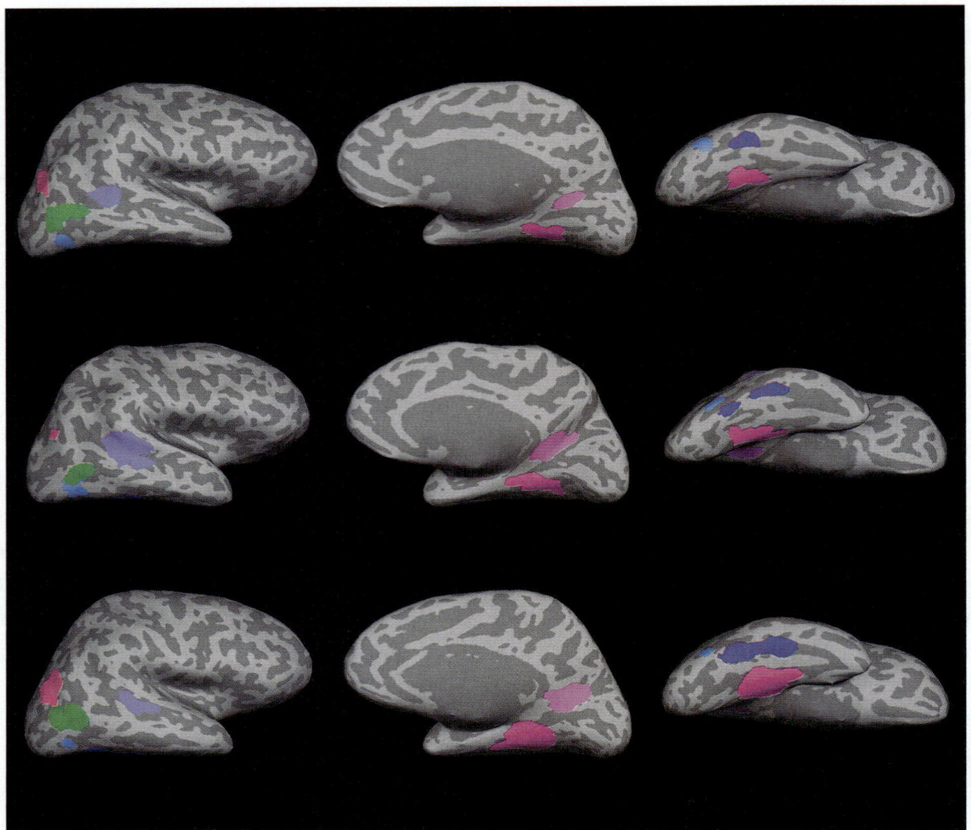

FIGURE 52.1   Inflated brain from three representative subjects showing regions specifically involved in the perception of faces (blues), places (pinks), and bodies (green). Dark blue, FFA; purplish blue, pSTS; light blue, OFA; magenta, PPA; light purple/pink, RSC; reddish pink, OPA; green, EBA. Each of these regions can be found in a short functional scan in essentially every healthy subject. LOC is not shown here; it is a very large region generally responsive to any object shape, and hence both of its subregions (LO and pFs) overlap partially with some of the regions shown here. The VWFA (left hemisphere) and FBA (partially overlapping with FFA) are also not pictured.

The approach to neuroimaging in which cognitive functions are assigned to brain regions has been widely maligned as "mere phrenology," as if the label itself is an argument against the whole enterprise. But name-calling is not argumentation any more in science than in the schoolyard. There is nothing fundamentally wrongheaded with the effort to characterize the cognitive functions of particular brain regions—it is an empirical question how successful this research program will turn out to be. The problem is rather that most neuroimaging studies neither robustly establish the existence of a functionally specific region nor precisely identify its function. When a brain region is identified based on just a few conditions in a single study, the evidence will necessarily be weak, and the functional characterization of the region will be sparse and unsatisfying. But when the same brain region has been identified in many labs, and when each of those labs has measured the functional response of that region in numerous conditions, each testing a different hypothesis about the representations it contains, then we can begin to approach the kind of rich cognitive (and perhaps some day computational) characterization of that brain region that will be of real theoretical significance to cognitive science. Thus, brain imaging can discover not just the *locations* of cognitive functions (who cares?) but, more fundamentally, it can discover and characterize those cognitive functions themselves. That is, brain imaging can exploit the functional specificity of the brain to discover the cognitive architecture of the mind. Although still very much a work in progress, this enterprise is most advanced for the regions we begin with here: the FFA, PPA, EBA, and LOC.

The Fusiform Face Area

The FFA is the region of the midfusiform gyrus (on the bottom surface of the cerebral cortex, just above the cerebellum) that responds significantly more strongly when subjects view faces than when they view objects

(Kanwisher, McDermott, & Chun, 1997a; McCarthy et al., 1997; Puce et al., 1996). The precise shape of the FFA varies across subjects, and consists of two regions in some subjects (Weiner & Grill-Spector, 2012), but is present in nearly all subjects and highly reproducible within each (Peelen & Downing, 2005). Earlier neuroimaging work (Haxby et al., 1994; Puce et al., 1995) had shown activations in this general region for faces (compared to scrambled faces, textures, and letter strings), but those contrasts left open the possibility that this region was engaged in processing generic object shape. The subsequent evidence for the *specificity* of this region for faces per se came from the discovery that this region responds similarly and strongly to a wide variety of face images (Kanwisher & Yovel, 2006), including photos of familiar and unfamiliar faces, schematic faces, cartoon faces, and cat faces (Kanwisher & Barton, 2010) as well as faces presented in different sizes, locations, and viewpoints (Axelrod & Yovel, 2012; Grill-Spector et al., 1999; Schwarzlose et al., 2008) and much less strongly to nonface stimuli. Extensive evidence now rejects alternative hypotheses proposed earlier that the FFA is more generally engaged in fine-grained discrimination of exemplars of any category, or of any category for which the subject has gained substantial expertise (McKone & Robbins, 2010; Tarr & Gauthier, 2000; Yovel & Kanwisher, 2004). Consistent with the evidence from fMRI, face-selective responses have been observed in approximately the same location in subdural electrode recordings from the brains of subjects undergoing presurgical mapping for epilepsy treatment (Allison et al., 1999; McCarthy et al., 1999; Puce, Allison, & McCarthy, 1999), and lesions in approximately this location can produce selective deficits in face perception (Kanwisher & Barton, 2010). Thus, as discussed further below, the FFA appears to be quite selectively engaged during the perception of faces.

Answering the question of what exactly the FFA does with faces has been more difficult. Importantly, the magnitude of the FFA response is correlated trial-by-trial with success both in detecting the presence of faces and in identifying individual faces (Grill-Spector & Malach, 2004) but not with detection or identification of nonface objects, and the extent of face selectivity in the fusiform is correlated with behavioral ability at face (but not object) identification across subjects (Furl et al., 2011). In terms of the kind of information represented in the FFA, current evidence indicates that the FFA is sensitive to multiple aspects of face stimuli including face parts (eyes, noses, and mouths), the T-shaped configuration of those features, and external features of faces (Liu, Harris, & Kanwisher, 2010). Further, FFA responses show some invariance across changes in stimulus position and image size (Schwarzlose et al., 2008; but see Yue et al., 2011). Using multivoxel pattern analysis (MVPA), a recent study (Axelrod & Yovel, 2012) found similar neural representations for mirror-symmetric views of faces but not for other changes in viewpoint, and other studies using fMRI adaptation have found a moderate invariance for viewpoint changes (up to 30°) for familiar (celebrity) faces (e.g., Ewbank & Andrews, 2008). Finally, the FFA exhibits neural correlates of well-established behavioral signatures of face perception (McKone & Robbins, 2010), including sensitivity to differences in face identity for upright but not inverted faces (Yovel & Kanwisher, 2005) and sensitivity to holistic information in upright but not inverted faces (Schiltz & Rossion, 2006). Thus, the FFA appears to represent perceptual information about face shape in a fashion partially invariant to image changes and to reflect the well-known behavioral signatures of face-specific processing.

The Parahippocampal Place Area

The PPA is defined functionally as the region adjacent to the collateral sulcus in parahippocampal cortex that responds significantly more strongly to images of scenes than objects (Epstein & Kanwisher, 1998). The PPA responds to a wide variety of scenes, including indoor and outdoor scenes, familiar and unfamiliar scenes, and even abstract "scenes" made of Legos (Aguirre et al., 1996; Epstein, 2005, 2008). The PPA is primarily responsive to the spatial layout of one's surroundings: Its response is not reduced when all of the objects are removed from an indoor scene, leaving just the floor and walls (Epstein & Kanwisher, 1998). The PPA has also been shown to respond selectively to high-spatial-frequency geometric shapes in humans and monkeys (Rajimehr et al., 2011), which suggests that the PPA may use such information for detecting scene details during place perception and navigation. By contrast, Bar and colleagues have proposed that the parahippocampal response to scenes does not reflect spatial layout but rather the activation of a "context frame" representation that includes information about which objects typically appear in that context and where they are likely to be located relative to each other (Bar, 2004). However, the evidence for the context hypothesis is weak: The finding of greater response to strong- versus weak-context objects only replicates at slow presentation rates, is reliable in only a minority of subjects, and can be alternatively explained in terms of scene imagery (Epstein & Ward, 2010). On the other hand, consistent with the spatial layout hypothesis, studies using MVPA found that the PPA contains significantly more information about spatial layout—that is, whether a scene is "open" versus "closed" (Oliva & Torralba, 2001)—than information about whether a scene is manmade (e.g., a city) or natural (e.g., a forest) (Kravitz, Peng, & Baker, 2011; Park et al., 2011). Evidence that the PPA is not only activated when information about spatial layout is processed but that it is further *necessary* for this function comes from patients with damage in or near the PPA, who have deficits in simple identification of scenes or landmarks (Aguirre & D'Esposito, 1999; Mendez & Cherrier, 2003) and difficulty more generally in knowing where they are (Epstein et al., 2001; Habib & Sirigu, 1987). This high response to spatial layout information is tantalizingly reminiscent of the "geometric module" (Cheng & Gallistel, 1984; Hermer & Spelke, 1996), inferred from behavioral data, in which rats and human infants (and adults whose language system is tied up by a concurrent verbal task) rely exclusively on the layout of space, not on objects or landmarks, to reorient themselves in an environment once they are disoriented. However, we recently found that representations in the PPA are largely invariant to mirror-image reversals, a result that challenges its role in navigation and reorientation (Dilks et al., 2011). The precise role of the PPA in place perception and navigation is a topic of ongoing investigation (see chapter 51 by Oliva).

The Extrastriate Body Area

The EBA is a region on the lateral surface of the brain adjacent to (and sometimes partly overlapping with) visual motion area MT, which responds significantly more strongly to images of bodies than to images of objects or faces (Downing et al., 2001). This region responds equally to visually different images of bodies, from a photograph of a hand, to a photograph of a body (human or animal), to a line drawing, or even a schematic stick figure of a person. The EBA is more involved in perceiving other people's bodies, in a viewpoint-dependent manner (Taylor, Wiggett, & Downing, 2010), than one's own (Chan, Peelen, & Downing, 2004; Saxe, Jamal, & Powell, 2006) and is more engaged in the perception of the form/identity of bodies than in the actions they are carrying out (Downing & Peelen, 2011; Moro et al., 2008). Evidence that this region not only is activated during, but is also necessary for, the perception of bodies comes from studies in which disruption of the EBA by a brain lesion (Moro et al., 2008) or by transcranial magnetic stimulation (TMS) (Pitcher et al., 2009; Urgesi, Berlucchi, & Aglioti, 2004; van Koningsbruggen, Peelen, & Downing, 2013) impairs the perception of body form but not the perception of

faces or of object shape (Pitcher et al., 2009). Recent evidence indicates that the EBA may consist of several subregions and that these subregions of the EBA and neighboring cortex may respond differentially to distinct body parts (Bracci et al., 2010; Op de Beeck et al., 2010; Orlov, Makin, & Zohary, 2010; Weiner & Grill-Spector, 2011). A recent review (Downing & Peelen, 2011) argues that the EBA and another more ventral body-specific region, the fusiform body area or FBA (Schwarzlose, Baker, & Kanwisher, 2005), are not primarily engaged in a higher-level interpretation of the individual identity, emotional content, motion, or action goals but rather extract a "cognitively unelaborated" visual representation of the shape and posture of the people in the current percept.

The Lateral Occipital Complex

In addition to the category-specific regions just described, a large region of lateral and inferior occipital cortex just anterior to retinotopic cortex (the lateral occipital complex or LOC), and partially overlapping with some of the regions described above, responds more strongly to stimuli depicting shapes than stimuli with similar low-level features that do not depict shapes (Kanwisher et al., 1997b; Malach et al., 1995). Common areas within LOC are activated by shapes defined by motion, texture, and luminance contours (Grill-Spector et al., 1998), showing that representations of shape in LOC are quite abstract. Importantly, the LOC responds similarly to familiar and unfamiliar shapes (Kanwisher et al., 1997b; Malach et al., 1995), suggesting that this region is not involved in matching to stored object representations or to semantic or verbal coding of the stimuli. Further support for the idea that LOC represents object shape, not semantic information about objects, comes from the fact that fMRI adaptation is not found in LOC across objects that are similar in meaning but differ in shape (Kim et al., 2009). Several studies have implicated the LOC in visual object recognition by showing that activity in this region is correlated with success on a variety of object recognition tasks (Bar et al., 2001; Grill-Spector et al., 2000), and indeed the location of LOC matches very nicely the location of the lesion in the famous ventral pathway agnosic patient DF (James et al., 2003). Thus, an investigation of the response properties of the LOC may provide important clues about the nature of the representations underlying object recognition.

MVPA and fMRI adaptation studies show that LOC contains fairly abstract representations of object shape. First, fMRI adaptation studies have shown that representations in the vicinity of LOC are partially invariant to changes in size and position but largely specific to viewpoint and direction of illumination (Grill-Spector et al., 1999). Evidence on the nature of the shape representations in LOC comes from the fact that fMRI adaptation occurs in LOC between stimulus pairs that have different contours but the same perceived shape (because of changes in occlusion) but not between stimulus pairs with identical contours but different perceived shape (because of a figure–ground reversal) (Kourtzi & Kanwisher, 2001). Further, MVPA analyses show that a posterior subregion of LOC (often called LO) contains representations that are more tied to the stimulus, whereas a more anterior subregion (called pFs) contains representations that are correlated with observer-specific perceptions of shape similarity (Haushofer, Livingstone, & Kanwisher, 2008). Finally, another recent study found that the more posterior region LO showed sensitivity to mirror-image reversals of objects, although the more anterior region pFs did not, suggesting a hierarchy of object processing in which left–right information is represented at earlier (more posterior) stages in the hierarchy and invariance is then computed at later (more anterior) stages (Dilks et al., 2011).

In sum, the VVP contains a large multipart region, the LOC, that responds strongly to object structure but that exhibits little selectivity for specific object categories, along with a small number of category-specific regions (for faces, places, and bodies). Efforts to date have not found regions of the VVP robustly selective for other categories (Downing et al., 2006; Lashkari et al., 2011), with the exception of the VWFA (Baker et al., 2007; Cohen et al., 2000).

Specificity: Do Category-Selective Regions Contribute Only to the Perception of Their Preferred Stimuli?

So far, we have argued that the FFA, PPA, and EBA are primarily if not exclusively engaged in processing their "preferred" stimuli (the stimulus class they respond most strongly to). However, each of these regions responds significantly (albeit weakly) to objects that are not in the preferred category ("nonpreferred" objects). Further, in what we consider the most important challenge to the claimed specificity of these regions, Haxby and colleagues (Haxby et al., 2001) reported that the spatial *pattern* of response across the FFA contains information about nonfaces and that the pattern of response within the PPA contains information about nonscenes, and hence, that "Regions such as the 'PPA' and 'FFA' are not dedicated to representing only spatial arrangements or human faces, but, rather, are part of a more extended representation for all objects."

We too find that the FFA and PPA contain information about nonpreferred stimuli (Reddy & Kanwisher, 2007), and current physiological evidence (Tsao et al., 2006) also indicates that face-selective regions carry a small but significant amount of information about nonpreferred stimuli. However, the information that category-selective regions contain is much weaker for nonpreferred than preferred stimuli. As Haxby and colleagues (O'Toole et al., 2005) noted, "preferred regions for faces and houses are not well suited to object classifications that do not involve faces and houses, respectively." These findings raise two questions, which we address next.

First, is the information about nonpreferred objects in category-selective regions still present when subjects view cluttered displays more typical of real-world vision? Most analyses of the spatial pattern of the fMRI response within the VVP have been based on responses elicited by single cut-out objects on a blank background, presented at the fovea. Of course real-world visual stimuli are not this simple: A typical visual scene contains multiple objects and complex background textures (i.e., "clutter"). A recent study (Reddy & Kanwisher, 2007) tested whether information about nonpreferred objects is present in a minimalist case of clutter, with two objects present simultaneously in the visual field (both on blank backgrounds). When single cut-out stimuli were shown one at a time, the pattern of response in the FFA contained considerable information about faces and significant though weaker information about nonfaces. Similarly, the pattern of response in the PPA contained robust information about houses (which activate the PPA strongly, although not as strongly as a full scene) and significant but weak information about nonscenes. Crucially, however, when two objects were present at once, information about preferred stimuli was virtually undiminished from the single-object case, but information about nonpreferred stimuli dropped to insignificance (Reddy & Kanwisher, 2007). This study and later related studies suggest that category-selective regions may have little or no information about nonpreferred stimuli under natural (i.e., cluttered) viewing conditions.

Still, given that fMRI is bound to underestimate the information present in the full neural population code, it is possible that future physiological studies will reveal some information about nonpreferred stimuli in the FFA, the PPA, and similarly selective regions, even for the cluttered stimuli typical of real-world viewing. Thus, the second and most important question is whether such information is *used in the perception of those stimuli*, or whether it is epiphenomenal (Williams, Berberovic, & Mattingley, 2007). Some relevant evidence is available for the case of the FFA from the study of individuals with focal brain damage. Some of these individuals exhibit deficits only in face perception (i.e., prosopagnosia), with little or no deficit in object recognition, after damage to regions in or near the FFA, suggesting that even if the FFA contains information about nonfaces, this information is not necessary for object perception. Although no published case of acquired prosopagnosia has completely ruled out the existence of any other deficits beyond face perception (Garrido, Duchaine, & Nakayama, 2008) using the most sensitive tests of object perception such as reaction time measures (Gauthier, Behrmann, & Tarr, 1999), some cases come close (Sergent & Signoret, 1992; Wada & Yamamoto, 2001).

However, because the locus and extent of lesions in humans are not under our control, an important complementary method for testing the functional specificity and causal role of cortical regions in perception is TMS. In TMS a brief magnetic pulse is delivered to the scalp through a coil held next to the scalp, disrupting neural processing in the cortical region immediately beneath the coil. We can now precisely position the TMS coil to directly target specific cortical regions defined functionally within individual subjects. Although the FFA and PPA are too medial to be reached by TMS, other more lateral category-selective regions (e.g., the face-selective OFA and the scene-selective OPA, previously known as TOS—Dilks et al., 2013—discussed in more detail below) can be. Using this method, Pitcher and colleagues (Pitcher et al., 2009) showed that TMS to the EBA disrupted perception of bodies (Urgesi, Berlucchi, & Aglioti, 2004) *but not faces or objects*, TMS to the OFA disrupted perception of faces *but not objects or bodies* (but see Silvanto et al., 2010), and TMS to LO disrupted perception of objects *but not bodies or faces*. Similarly, Dilks and colleagues (Dilks et al., 2013) recently showed that TMS to OPA impaired discrimination of scenes *but not faces*, while TMS to OFA impaired discrimination of faces *not scenes*. In a second experiment, Dilks et al. (2013) showed that TMS over OPA impaired categorization accuracy of scenes *but not objects*, while TMS over LOC impaired categorization accuracy of objects *not scenes* (see also Ganaden, Mullin, & Steeves, in press; Mullin & Steeves, 2011). These striking triple and double dissociations, respectively, suggest that category-selective regions play a causal role in the perception of their preferred stimulus class, but not their nonpreferred stimulus class. Thus, even if the pattern of response across these regions contains some information about nonpreferred stimulus categories, the available evidence suggests that such information plays no detectable causal role in perception.

In sum, current evidence suggests that category-selective regions sometimes contain weak but significant information about nonpreferred stimuli, which is likely to be underestimated by fMRI. Nonetheless, results from neuropsychology and TMS are consistent with the hypothesis that any information about nonpreferred stimuli in category-selective regions is epiphenomenal (i.e., not causally involved in perception of those stimuli). It will be important in the future to test this hypothesis further with new data from patients, TMS, and other disruption methods, such as electrical microstimulation in macaque monkeys and humans (Afraz, Kiani, & Esteky, 2006; Puce, Allison, & McCarthy, 1999).

THE FUNCTION AND STRUCTURE OF THE WHOLE VENTRAL VISUAL PATHWAY

Of course no complex cognitive process is accomplished in a single brain region, and arguments for the specificity of the regions described above in no way preclude an important role for other brain regions. "Earlier" cortical regions such as primary visual cortex are obviously crucial in the perception of faces, places, and bodies, and "higher" areas (e.g., in parietal and frontal regions) are also probably necessary for information in the FFA, PPA, and EBA to be used by other cognitive systems and to reach awareness (Kanwisher, 2001). Further, none of these regions is the only one with its defining selectivity. Other category-selective regions have not been studied in the same detail as the FFA, PPA, and EBA, so their functions are less clear. However, the existence of multiple selective regions, and the growing evidence for a functional division of labor between them, raises the exciting possibility that we may ultimately understand how face recognition, for example, emerges from the joint activity of a number of functionally distinct regions. Next, we briefly review the literature on other functionally distinct regions engaged in face and scene perception.

For faces, selective responses are found not only in the FFA but in many subjects also in a nearby but more posterior OFA (Gauthier et al., 2000) as well as other more anterior regions, such as the posterior superior temporal sulcus (pSTS) (Puce et al., 1998), and anterior temporal pole (Rajimehr, Young, & Tootell, 2009). Based on the more posterior location and generally lower selectivity of the OFA, it is often assumed to constitute an early stage of face perception (or face detection), which is then followed by continued processing in more anterior regions (e.g., the FFA and pSTS) (Haxby, Hoffman, & Gobbini, 2000; but see Rossion, Hanseeuw, & Dricot, 2012). Consistent with this picture, (1) the right OFA has been more implicated in the representation of the parts of a face, including the eyes, nose, and mouth, than the configuration of those parts (Liu, Harris, & Kanwisher, 2010; Pitcher et al., 2007), whereas the FFA is sensitive to both face parts and their overall configuration in the face (Liu, Harris, & Kanwisher, 2010), and (2) the OFA is more sensitive to mirror-image reversals than is the FFA or pSTS (Axelrod & Yovel, 2012). By contrast, the face-selective region in the pSTS has been implicated in the representation of more dynamic high-level face and social information, including eye, mouth, and head movements (Carlin et al., 2011; Fox et al., 2009; Haxby, Hoffman, & Gobbini, 2000; Pitcher et al., 2011; Puce et al., 1998) and facial expression (Phillips & David, 1997; Winston et al., 2004). In one of the most striking functional dissociations within the face system, the face-selective region in the pSTS responds about three times as strongly to movies of faces (but not movies of bodies or objects) as to static snapshots taken from those face movies, whereas the FFA responds the same to movies and snapshots (Pitcher et al., 2011; see also Puce et al., 1998).

For scenes, selective responses are found not only in the PPA but also in retrosplenial complex (RSC) and a region near the transverse occipital sulcus, formerly referred to as "TOS" (Grill-Spector, 2003), now called the occipital place area (OPA) (Dilks et al., 2013). Like PPA, RSC is primarily responsive to the spatial layout of one's surroundings (Dilks et al., 2011; Epstein, 2008; Kravitz, Peng, & Baker, 2011; Park & Chun, 2009), with a recent study reporting only spatial layout information, not object information, in RSC (Harel, Kravitz, & Baker, 2013). However, in contrast to the ongoing debate about the PPA's precise role in place perception and navigation, most studies find clear evidence that RSC plays a role in navigation. For example, one fMRI adaptation study (Baumann & Mattingley, 2010) has shown that this region encodes heading direction. Further, RSC is sensitive to left–right information in scenes (i.e., mirror-image reversals of a scene), which is presumably important for navigation (Dilks et al., 2011). Finally, patients with RSC damage have been reported to recognize salient landmarks but not to use these landmarks to orient themselves or to navigate through a larger environment (Takahashi et al., 1997). OPA is the least-studied scene-selective region, but a recent study from our lab found that it is causally and selectively involved in scene perception: TMS over OPA impaired discrimination of scenes but not faces or objects (Dilks et al., 2013).

The existence of multiple face-selective regions and multiple scene-selective regions offers the exciting prospect of taking apart the process of face and scene perception by understanding the functional division of labor among the various regions within each system. An important part of this story concerns the connectivity of the different regions within each system and between those regions and the rest of the brain. Evidence on this important question is sparse, however, because neither of the two methods currently available in humans can answer these questions definitively. Resting-state fMRI is intriguing but can reveal strong correlations between regions known not to be directly connected (Tian et al., 2007). Diffusion tractography is subject to ambiguities both in tracing connectivity from specific functionally identified gray matter regions into the underlying white matter and in tracing specific connections through white matter (rather than simply following known major fiber bundles that run through the VVP, e.g., the ILF and IFOF). Indeed, preliminary evidence from these methods does not fully agree. Although both diffusion tractography and resting-state fMRI agree that the nearby OFA and FFA are connected, connections between the FFA and the face-selective region in the pSTS have been found with resting-state fMRI (Turk-Browne, Norman-Haignere, & McCarthy, 2010) but not with diffusion methods (Gschwind et al., 2011). It remains to be determined whether methods for discovering specific anatomical connections in humans will ultimately be able to discover the precise connectivity of specific subregions of the VVP.

Perhaps the biggest open question concerning the functional organization of the VVP is whether the functionally distinctive regions identified here are best thought of as discrete processors or whether it makes more sense to consider the broader region that contains them as a single processor in which each of these regions simply constitutes a local peak in the functional response. On the latter view, the question would still remain as to why that landscape would contain the particular replicable configuration it does across the VVP, and what if any are the dimensions represented by axes of this broader "map" (Op de Beeck, Haushofer, & Kanwisher, 2008; Kanwisher & Schwarzlose, 2008). Some of the locations of particular regions in the VVP may be explained in terms of a center–periphery map (Hasson et al., 2002) or a representation of real-world object size (Konkle, 2011). Further widely noted and intriguing aspects of the structure of the VVP are that many of the object- and category-selective regions come in pairs, with one on the ventral surface and one on the lateral surface (Hasson et al., 2003; Schwarzlose et al., 2008), and that body-selective and face-selective regions tend to be close and sometimes overlapping with each other (Kanwisher & Schwarzlose, 2008; Weiner & Grill-Spector, 2011). One clue to the question of whether the ventral pathway is best thought of as a single representational space or a set of at least partially distinct processors may come from anatomy: To the extent that the different regions discussed above have distinctive connectivity and cytoarchitecture, that would support the interpretation of these regions as distinct entities. Indeed, recent evidence indicates that face selectivity within the fusiform gyrus can be predicted from connectivity with the rest of the brain (Saygin et al., 2011) and that at least one face-selective region in the fusiform gyrus may have a distinctive cytoarchitecture (Caspers et al., 2013).

WHY HAVE SELECTIVE REGIONS IN THE FIRST PLACE?

Perhaps the deepest question raised by the work on the VVP is this: Why do some visual categories get their own private piece of real estate in the VVP, whereas others apparently do not (Downing et al., 2006; Lashkari et al., 2011)? To think clearly about this question, we need to consider what computational advantages are afforded by functional specialization in the first place. To be detected by fMRI, functional specializations must have two properties: (1) selectivity of the response of neurons to the relevant information (e.g., face selectivity) and (2) spatial clustering of selective neurons. These phenomena are related but distinct (Ohki et al., 2005) and will be discussed in turn in this final, highly speculative section.

Selectivity/Sparseness

The advantages of selectivity, or "sparseness," in neural coding have been widely noted (Barlow, 1995; Foldiak & Young, 1995; Olshausen & Field, 2004). If a given object is coded by the activity of a small subset of the available neurons, then interference is minimized in two important senses. First, it is possible to represent multiple objects simultaneously with minimal ambiguity because the neural codes for different objects are unlikely to overlap. Thus, we can perceive a face and a place simultaneously without the two representations colliding (Reddy & Kanwisher, 2007). Perhaps this is one reason we have neural populations selectively responsive to faces, places, and bodies: to provide "private lines" of communication about particularly important classes of stimuli that are protected from crosstalk of other irrelevant information.

Second, the use of sparse codes can also reduce interference from learning, enabling us to learn new exemplars of one class of objects without altering stored information about another class of objects. With one neural population to represent faces and a nonoverlapping neural population to represent the spatial layout of places, we can learn new faces without disrupting our memories of places and vice versa. From this perspective we may expect to find relatively sparse codes for classes of information characterized by continual lifelong learning (such as faces and places).

Third, building specialized brain regions and precise connectivity linking them to other brain regions could bootstrap development by essentially hardwiring constraints on inductive inference. For example, if information in faces provides the key input required for learning about other people's minds, then perhaps the most efficient way to construct the machinery for thinking about other minds is to hardwire a face area and connect it to another available region, which will then have the constrained input it needs to construct the circuits necessary for social cognition. Evidence against this particular hypothesis comes from the recent finding that congenitally blind individuals show the same location and pattern of activation as sighted subjects when thinking about other people's thoughts, even though input from the FFA is likely very different or nonexistent in these people (Bedny, Pascual-Leone, & Saxe, 2009). Nevertheless, the general idea that specialized brain regions and their connections may serve as constraints on development is worth considering in other cases.

A fourth possible advantage of relatively sparse codes is metabolic rather than computational: Less energy is required if fewer neurons are firing. From this perspective the greatest lifelong energy savings would come about if sparse codes were available for classes of stimuli that occur most frequently (Foldiak & Young, 1995). Thus, even from a purely metabolic perspective, it makes sense to use relatively sparse codes for faces, places, and bodies because they are among the most frequently encountered visual stimuli.

In sum, sparse codes, in which information is represented by a relatively small percentage of the available neurons, each with relatively high selectivity, have certain advantages. At the same time, sparse codes have well-known disadvantages, such as greater susceptibility to damage (because of the smaller number of neurons involved in any given representation) and a smaller number of possible patterns that can be held (one at a time) by a fixed number of neurons. The speculation here is that these disadvantages are outweighed by the particular advantages in the coding of biologically important stimuli such as faces, places, and bodies: (1) reduction of interference or crosstalk when multiple stimuli must be represented simultaneously, (2) the ability to learn new information about one stimulus class without disrupting stored information about another class, (3) bootstrapping the development of other regions, and (4) the potential energy efficiency of coding the most frequently encountered stimuli through the activity of the smallest number of neurons.

Spatial Clustering

The second property implied by functionally selective regions detected by fMRI, after selectivity of neurons, is spatial clustering of those neurons. Spatial clustering of functional properties is a familiar phenomenon in the brain, found not only in retinotopic, somatotopic, tonotopic, and other-topic maps that follow the organization of the receptor surface, but also in the organization of functional information that is computed de novo, such as orientation columns in primary visual cortex and chromotopic maps in posterior inferotemporal cortex (Conway & Tsao, 2009). Spatial organization is such a pervasive and familiar property of the cortex that we can easily forget to ask ourselves why it occurs. This mystery has been articulated most clearly (Chklovskii & Koulakov, 2004) as follows: "Imagine taking a cortical area containing a map and scrambling neurons in that area while preserving all the connections between neurons. Because the circuit remains unchanged, the functional properties of the neurons remain intact. Then the scrambled region without a map is functionally identical to the original one with the map." Given that the identical circuit can be constructed in a spatially clustered or spatially scrambled version, why does spatial clustering occur?

This question is sharpened by the facts that the strong spatial clustering seen in some systems, such as orientation-selective cells in cat visual cortex, is not found in other very similar systems, such as orientation-selective cells in rodent visual cortex (Ohki & Reid, 2007), and further by the fact that in the rodent olfactory processing pathway, the precise spatial clustering (and odorant specificity) constructed in the olfactory bulb is thrown away in the next stage of processing, the piriform cortex (Stettler & Axel, 2009).

Chklovskii and Koulakov (2004) argue that the need to minimize wiring length (for developmental, metabolic, and conduction delay reasons) must be a fundamental constraint in the nervous system that produces spatial clustering of neurons that are densely connected to each other. To the extent that this wiring-length minimization principle is an important determinant of cortical organization, it suggests that we may find functional specialization in focal cortical regions for functions that are implemented in circuits for which the neurons have to be densely connected to each other. A testable prediction of this idea is that neurons within face-selective patches of monkey cortex must be richly interconnected, either directly or via webs of inhibitory interneurons found in those same regions. A further prediction of the axon-length-minimization principle is that to the extent that readout of a neural code (by the next stage of processing) requires convergence of multiple inputs on a particular neuron, it may be easier to read out a population code represented in a focal region of cortex where those inputs can all conveniently converge on a common output neuron. In a different vein the functional significance of spatial clustering in the cortex may derive from the requirement to selectively modulate a given functional circuit by way of nonsynaptic diffusible messenger molecules that can spread a few millimeters through the cortex.

Functionally Specific Cortical Regions for Computationally Different Problems?

Of course, the classic argument for functional specialization is that efficiency can be gained by division of labor (Rueffler, Hermisson, & Wagner, 2012) when the computational requirements differ across tasks (Marr, 1982). Indeed, especially for the case of faces and places, both theoretical considerations and extensive empirical evidence suggest that different kinds of representations are extracted from these stimulus classes, and different uses are made of the resulting information (Cheng & Gallistel, 1984; Hermer & Spelke, 1996; McKone & Robbins, 2010). On the other hand, the observed clustered selectivity for faces, places, and bodies does not in itself imply qualitative differences in the computations and representations entailed in the perception of one of these categories versus another. We might have functional selectivity of the relevant neuronal responses for each category without any fundamental differences in the kinds of computations conducted for each, just as we see in retinotopic cortex, where completely nonoverlapping pools of neurons code for visual information in one visual field location versus another, but fundamentally similar computations are conducted by each. A crucial question for the enterprise of using functional specificity of the brain to infer fundamental components of the mind will therefore be: Which cortical selectivities reflect fundamentally different underlying cognitive processes, and which simply reflect convenient compartmentalization of similar processes?

CONCLUSIONS

Over the past 15 years, fMRI studies have taught us a great deal about the functional organization of the VVP in humans. We have learned that the machinery that conducts visual object recognition in humans is not a homogeneous mass of tissue but instead a richly structured system composed of functionally distinct regions, each found in approximately the same location in every healthy subject. Exciting directions for future research will exploit a suite of powerful new methods, including the ability to relate the "representational dissimilarity matrices" extracted in each region (or the whole ventral pathway) via fMRI with representational spaces derived from behavioral and monkey single-unit data (Kriegeskorte et al., 2008) and the ability to link functionally defined regions and functional profiles with cytoarchitecture (Caspers et al., in press) and connectivity (Saygin et al., 2011).

Yet many fundamental questions have proven to be difficult or impossible to answer with current methods available in humans. What information is represented in each region at the spatial and temporal resolution of actual neural responses (Freiwald & Tsao, 2010)? What neural circuits extract this information? What is the causal role of each region in perception? What is the connectivity among these various regions, and between each of them and the rest of the brain (Moeller, Freiwald, & Tsao, 2008)? Further, how do these regions get wired up in the brain in development, and what are the relative roles of experience (Baker et al., 2007; Srihasam et al., 2012) and genes (Duchaine, Germine, & Nakayama, 2007; Sugita, 2008; Turati, Bulf, & Simion, 2008; Wilmer et al., 2010; Zhu et al., 2010) in this process? One of the most exciting developments of the last decade has been the discovery of face-selective (Tsao et al., 2006) and place-selective (Nasr et al., 2011) regions in the temporal lobes of macaques. These discoveries offer the possibility that the questions that have proven most intractable in humans will be answered by work on macaques, where it is possible to provide much richer characterizations of neural representations and their underlying circuits, the causal roles of these circuits in behavior, the structural correlates of functionally defined regions, the interplay of genes and experience that wire those regions up during development, and the interactions among those regions during task performance.

ACKNOWLEDGMENTS

The writing of this chapter was supported by EY13455, by a grant from the Ellison Medical Foundation to N.K., and by a Simons Foundation Autism Research Initiative (SFARI) postdoctoral fellowship to D.D.D.

REFERENCES

Afraz, S. R., Kiani, R., & Esteky, H. (2006). Microstimulation of inferotemporal cortex influences face categorization. *Nature, 442,* 692–695.

Aguirre, G. K., & D'Esposito, M. (1999). Topographical disorientation: A synthesis and taxonomy. *Brain, 122,* 1613–1628. doi:10.1093/brain/122.9.1613.

Aguirre, G. K., Detre, J. A., Alsop, D. C., & D'Esposito, M. (1996). The parahippocampus subserves topographical learning in man. *Cerebral Cortex, 6,* 823–829. doi:10.1093/cercor/6.6.823.

Allison, T., Puce, A., Spencer, D. D., & McCarthy, G. (1999). Electrophysiological studies of human face perception. I: Potentials generated in occipitotemporal cortex by face and non-face stimuli. *Cerebral Cortex, 9,* 415. doi:10.1093/cercor/9.5.415.

Axelrod, V., & Yovel, G. (2012). Hierarchical processing of face viewpoint in human visual cortex. *Journal of Neuroscience, 32,* 2442–2452.

Baker, C. I., Hutchison, T. L., & Kanwisher, N. (2007). Does the fusiform face area contain subregions highly selective for nonfaces? *Nature Neuroscience, 10,* 3–4.

Baker, C. I., Liu, J., Wald, L. L., Kwong, K. K., Benner, T., & Kanwisher, N. (2007). Visual word processing and experiential origins of functional selectivity in human extrastriate cortex. *Proceedings of the National Academy of Sciences of the United States of America, 104,* 9087–9092. doi:10.1073/pnas.0703300104.

Bar, M. (2004). Visual objects in context. *Nature Reviews. Neuroscience, 5,* 617–629.

Bar, M., Tootell, R. B., Schacter, D. L., Greve, D. N., Fischl, B., Mendola, J. D., et al. (2001). Cortical mechanisms specific to explicit visual object recognition. *Neuron, 29,* 529–535. doi:10.1016/S0896-6273(01)00224-0.

Barlow, H. B. (1995). The neuron doctrine in perception. *Cognitive Neurosciences, 1,* 415–436.

Baumann, O., & Mattingley, J. B. (2010). Medial parietal cortex encodes perceived heading direction in humans. *Journal of Neuroscience, 30,* 12897–12901.

Bedny, M., Pascual-Leone, A., & Saxe, R. (2009). Growing up blind does not change the neural bases of theory of mind. *Proceedings of the National Academy of Sciences of the United States of America, 106,* 11312–11317. doi:10.1073/pnas.0900010106.

Bracci, S., Ietswaart, M., Peelen, M. V., & Cavina-Pratesi, C. (2010). Dissociable neural responses to hands and non-hand body parts in human left extrastriate visual cortex. *Journal of Neurophysiology, 103,* 3389–3397.

Carlin, J. D., Rowe, J. B., Kriegeskorte, N., Thompson, R., & Calder, A. J. (2011). Direction-sensitive codes for observed head turns in human superior temporal sulcus. *Cerebral Cortex, 22,* 735–744. doi:10.1093/cercor/bhr061.

Caspers, J., Zilles, K., Eickhoff, S. B., Schleicher, A., Mohlberg, H., & Amunts, K. (2013). Cytoarchitectonical analysis and probabilistic mapping of two extrastriate areas of the human posterior fusiform gyrus. *Brain Structure & Function. 218,* 511–526. doi:10.1007/s00429-012-0411-8.

Chan, A. W., Peelen, M. V., & Downing, P. E. (2004). The effect of viewpoint on body representation in the extrastriate body area. *Neuroreport, 15,* 2407–2410.

Cheng, K., & Gallistel, C. R. (1984). Testing the geometric power of an animal's spatial representation. In H. L. Roitblatt, T. G. Bever, & H. S. Terrace (Eds.), *Animal cognition* (pp. 409–423). London: Lawrence Erlbaum Associates.

Chklovskii, D. B., & Koulakov, A. A. (2004). Maps in the brain: What can we learn from them? *Annual Review of Neuroscience, 27,* 369–392.

Cohen, L., Dehaene, S., Naccache, L., Lehericy, S., Dehaene-Lambertz, G., Henaff, M. A., et al. (2000). The visual word form area: Spatial and temporal characterization of an initial stage of reading in normal subjects and posterior split-brain patients. *Brain, 123*(Pt.2), 291–307. doi:10.1093/brain/123.2.291.

Conway, B. R., & Tsao, D. Y. (2009). Color-tuned neurons are spatially clustered according to color preference within alert macaque posterior inferior temporal cortex. *Proceedings of the National Academy of Sciences of the United States of America, 106*, 1–6. doi:10.1073/pnas.0810943106.

Dilks, D. D., Julian, J. B., Kubilius, J., Spelke, E. S., & Kanwisher, N. (2011). Mirror-image sensitivity and invariance in object and scene processing pathways. *Journal of Neuroscience, 31*, 11305–11312.

Dilks, D. D., Julian, J. B., Paunov, A., & Kanwisher, N. (2013). The occipital place area (OPA) is causally and selectively involved in scene perception. *Journal of Neuroscience, 33*, 1331–1336. doi:10.1523/JNEUROSCI.4081-12.2013.

Downing, P. E., Chan, A. W., Peelan, M. V., Dodds, C. M., & Kanwisher, N. (2006). Domain specificity in visual cortex. *Cerebral Cortex, 16*, 1453–1461.

Downing, P. E., Jiang, Y., Shuman, M., & Kanwisher, N. (2001). A cortical area selective for visual processing of the human body. *Science, 293*, 2470–2473.

Downing, P. E., & Peelen, M. V. (2011). The role of occipito-temporal body-selective regions in person perception. *Cognitive Neuroscience, 2*, 186–203.

Duchaine, B., Germine, L., & Nakayama, K. (2007). Family resemblance: Ten family members with prosopagnosia and within-class object agnosia. *Cognitive Neuropsychology, 24*, 419–430.

Epstein, R. A. (2008). Parahippocampal and retrosplenial contributions to human spatial navigation. *Trends in Cognitive Sciences, 12*, 388–396. doi:10.1016/j.tics.2008.07.004.

Epstein, R. (2005). The cortical basis of visual scene processing. *Visual Cognition, 12*, 954–978.

Epstein, R., De Yoe, E., Press, D., & Kanwisher, N. (2001). Neuropsychological evidence for a topographical learning mechanism in parahippocampal cortex. *Cognitive Neuropsychology, 18*, 481–508.

Epstein, R., & Kanwisher, N. (1998). A cortical representation of the local visual environment. *Nature, 392*, 598–601.

Epstein, R. A., & Ward, E. J. (2010). How reliable are visual context effects in the parahippocampal place area? *Cerebral Cortex, 20*, 294–303.

Ewbank, M. P., & Andrews, T. J. (2008). Differential sensitivity for viewpoint between familiar and unfamiliar faces in human visual cortex. *NeuroImage, 40*, 1857–1870. doi:10.1016/j.neuroimage.2008.01.049.

Foldiak, P., & Young, M. P. (1995). *Handbook of brain theory and neural networks.* Cambridge, MA: MIT Press.

Fox, C. J., Moon, S.-Y., Iaria, G., & Barton, J. J. S. (2009). The correlates of subjective perception of identity and expression in the face network: An fMRI adaptation study. *NeuroImage, 44*, 569–580.

Freiwald, W. A., & Tsao, D. Y. (2010). Functional compartmentalization and viewpoint generalization within the macaque face-processing system. *Science, 330*, 845–851.

Furl, N., Garrido, L., Dolan, R. J., Driver, J., & Duchaine, B. (2011). Fusiform gyrus face selectivity relates to individual differences in facial recognition ability. *Journal of Cognitive Neuroscience, 23*, 1723–1740.

Ganaden, R. E., Mullin, C. R., & Steeves, J. K. E. (2013). Transcranial magnetic stimulation to the transverse occipital sulcus affects scene but not object processing. *Journal of Cognitive Neuroscience* [epub ahead of print]. doi:10.1162/jocn_a_00372.

Garrido, L., Duchaine, B., & Nakayama, K. (2008). Face detection in normal and prosopagnosic individuals. *Journal of Neuropsychology, 2*(Pt 1), 119–140.

Gauthier, I., Behrmann, M., & Tarr, M. J. (1999). Can face recognition really be dissociated from object recognition? *Journal of Cognitive Neuroscience, 11*, 349–370.

Gauthier, I., Tarr, M. J., Moylan, J., Skudlarski, P., Gore, J. C., & Anderson, A. W. (2000). The fusiform "face area" is part of a network that processes faces at the individual level. *Journal of Cognitive Neuroscience, 12*, 495–504.

Grill-Spector, K. (2003). The neural basis of object perception. *Current opinion in neurobiology, 13*, 159–166.

Grill-Spector, K., Kushnir, T., Edelman, S., Avidan, G., Itzchak, Y., & Malach, R. (1999). Differential processing of objects under various viewing conditions in the human lateral occipital complex. *Neuron, 24*, 187–203.

Grill-Spector, K., Kushnir, T., Edelman, S., Itzchak, Y., & Malach, R. (1998). Cue-invariant activation in object-related areas of the human occipital lobe. *Neuron, 21*, 191–202.

Grill-Spector, K., Kushnir, T., Hendler, T., & Malach, R. (2000). The dynamics of object-selective activation correlate with recognition performance in humans. *Nature Neuroscience, 3*, 837–843.

Grill-Spector, K., & Malach, R. (2004). The human visual cortex. *Annual Review of Neuroscience, 27*, 649–677.

Gschwind, M., Pourtois, G., Schwartz, S., Van De Ville, D., & Vuilleumier, P. (2011). White-matter connectivity between face-responsive regions in the human brain. *Cerebral Cortex, 22*, 1564–1576. doi:10.1093/cercor/bhv226.

Habib, M., & Sirigu, A. (1987). Pure topographical disorientation: A definition and anatomical basis. *Cortex, 23*, 73–85.

Harel, A., Kravitz, D. J., & Baker, C. I. (2013). Deconstructing visual scenes in cortex: Gradients of object and spatial layout information. *Cerebral Cortex, 2*, 947–957. doi:10.1093/cercor/bhs091.

Hasson, U., Harel, M., Levy, I., & Malach, R. (2003). Large-scale mirror-symmetry organization of human occipito-temporal object areas. *Neuron, 37*, 1027–1041.

Hasson, U., Levy, I., Behrmann, M., Hendler, T., & Malach, R. (2002). Eccentricity bias as an organizing principle for human high-order object areas. *Neuron, 34*, 479–490.

Haushofer, J., Livingstone, M. S., & Kanwisher, N. (2008). Multivariate patterns in object-selective cortex dissociate perceptual and physical shape similarity. *PLoS Biology, 6*, e187. doi:10.1371/journal.pbio.0060187.

Haxby, J. V., Gobbini, M. I., Furey, M. L., Ishai, A., Schouten, J. L., & Pietrini, P. (2001). Distributed and overlapping representations of faces and objects in ventral temporal cortex. *Science, 293*, 2425–2430.

Haxby, J. V., Hoffman, E. A., & Gobbini, M. I. (2000). The distributed human neural system for face perception. *Trends in Cognitive Sciences, 4*, 223–233. doi:10.1016/S1364-6613(00)01482-0.

Haxby, J. V., Horwitz, B., Ungerleider, L. G., Maisog, J. M., Pietrini, P., & Grady, C. L. (1994). The functional organization of human extrastriate cortex: A PET-RCBF study of selective attention to faces and locations. *Journal of Neuroscience, 14*, 6336–6353.

Hermer, L., & Spelke, E. (1996). Modularity and development: The case of spatial reorientation. *Cognition, 61*, 195–232.

James, T. W., Culham, J. C., Humphrey, G. K., Milner, A. D., & Goodale, M. A. (2003). Ventral occipital lesions impair object recognition but not object-directed grasping: An fMRI study. *Brain, 126*, 2463–2475. doi:10.1093/brain/awg248.

Julian, J. B., Fedorenko, E., Webster, J., & Kanwisher, N. (2012). An algorithmic method for functionally defining regions of interest in the ventral visual pathway. *NeuroImage, 60*, 2357–2364. doi:10.1016/j.neuroimage.2012.02.055.

Kanwisher, N. & Barton, J. J. S. (2011). The functional architecture of the face system: Integrating evidence from fMRI and patient studies. In J. Haxby, M. Johnson, G. Rhodes, & A. Calder (Eds.), *The handbook of face perception* (pp. 111–130). Oxford: Oxford University Press.

Kanwisher, N. (2001). Neural events and perceptual awareness. *Cognition, 79*, 89–113.

Kanwisher, N. G., McDermott, J., & Chun, M. M. (1997a). The fusiform face area: A module in human extrastriate cortex specialized for face perception. *Journal of Neuroscience, 17*, 4302–4311.

Kanwisher, N., & Schwarzlose, R. F. (2008). *Principles governing the large-scale organization of object selectivity in ventral visual cortex.* Ph.D. Thesis, Massachusetts Institute of Technology.

Kanwisher, N., Woods, R. P., Iacoboni, M., & Mazziotta, J. C. (1997b). A locus in human extrastriate cortex for visual shape analysis. *Journal of Cognitive Neuroscience, 9*, 133–142.

Kanwisher, N., & Yovel, G. (2006). The fusiform face area: A cortical region specialized for the perception of faces. *Philosophical Transactions of the Royal Society of London. Series B, Biological Sciences, 361*, 2109–2128. doi:10.1098/rstb.2006.1934.

Kim, J. G., Biederman, I., Lescroart, M. D., & Hayworth, K. J. (2009). Adaptation to objects in the lateral occipital complex (LOC): Shape or semantics? *Vision Research, 49*, 2297–2305. doi:10.1016/j.visres.2009.06.020.

Konkle, T. (2011). *The role of real-world size in object representation.* Ph.D. Thesis, Massachusetts Institute of Technology.

Kourtzi, Z., & Kanwisher, N. (2001). Representation of perceived object shape by the human lateral occipital complex. *Science, 293*, 1506–1509.

Kravitz, D. J., Peng, C. S., & Baker, C. I. (2011). Real-world scene representations in high-level visual cortex: It's the spaces more than the places. *Journal of Neuroscience, 31*, 7322–7333.

Kriegeskorte, N., Mur, M., Ruff, D. A., Kiani, R., Bodurka, J., Esteky, H., et al. (2008). Matching categorical object representations in inferior temporal cortex of man and monkey. *Neuron, 60*, 1126–1141.

Lashkari, D., Sridharan, R., Vul, E., Hsieh, P. J., Kanwisher, N., & Golland, P. (2011). Search for patterns of functional specificity in the brain: A nonparametric hierarchical Bayesian model for group fMRI data. *NeuroImage, 59*, 1348–1368. doi:10.1016/j.neuroimage.2011.08.031.

Liu, J., Harris, A., & Kanwisher, N. (2010). Perception of face parts and face configurations: An fMRI study. *Journal of Cognitive Neuroscience, 22*, 203–211.

Malach, R., Reppas, J. B., Benson, R. R., Kwong, K. K., Jiang, H., Kennedy, W. A., et al. (1995). Object-related activity revealed by functional magnetic resonance imaging in human occipital cortex. *Proceedings of the National Academy of Sciences of the United States of America, 92*, 8135–8139. doi:10.1073/pnas.92.18.8135.

Marr, D. (1982). *Vision: A computational investigation into the human representation and processing of visual information.* San Francisco: W. H. Freeman.

McCarthy, G., Puce, A., Belger, A., & Allison, T. (1999). Electrophysiological studies of human face perception. II: Response properties of face-specific potentials generated in occipitotemporal cortex. *Cerebral Cortex, 9*, 431.

McCarthy, G., Puce, A., Gore, J. C., & Allison, T. (1997). Face-specific processing in the human fusiform gyrus. *Journal of Cognitive Neuroscience, 9*, 605–610.

McKone, E. M., & Robbins, R. R. (2011). Are faces special? In J. Haxby, M. Johnson, G. Rhodes, & A Calder (Eds.), *The handbook of face perception* (pp. 149–176). Oxford: Oxford University Press.

Mendez, M. F., & Cherrier, M. M. (2003). Agnosia for scenes in topographagnosia. *Neuropsychologia, 41*, 1387–1395.

Moeller, S., Freiwald, W. A., & Tsao, D. Y. (2008). Patches with links: A unified system for processing faces in the macaque temporal lobe. *Science, 320*, 1355–1359.

Moro, V., Urgesi, C., Pernigo, S., Lanteri, P., Pazzaglia, M., & Aglioti, S. M. (2008). The neural basis of body form and body action agnosia. *Neuron, 60*, 235–246.

Mullin, C. R., & Steeves, J. K. E. (2011). TMS to lateral occipital cortex disrupts object processing but facilitates scene processing. *Journal of Cognitive Neuroscience, 23*, 4174–4184.

Nasr, S., Liu, N., Devaney, K. J., Yue, X., Rajimehr, R., Ungerleider, L. G., et al. (2011). Scene-selective cortical regions in human and nonhuman primates. *Journal of Neuroscience, 31*, 13771–13785. doi:10.1523/JNEUROSCI.2792-11.2011.

Ohki, K., Chung, S., Ch'ng, Y. H., Kara, P., & Reid, R. C. (2005). Functional imaging with cellular resolution reveals precise micro-architecture in visual cortex. *Nature, 433*, 597–603.

Ohki, K., & Reid, R. C. (2007). Specificity and randomness in the visual cortex. *Current Opinion in Neurobiology, 17*, 401–407.

Oliva, A., & Torralba, A. (2001). Modeling the shape of the scene: A holistic representation of the spatial envelope. *International Journal of Computer Vision, 42*, 145–175.

Olshausen, B. A., & Field, D. J. (2004). Sparse coding of sensory inputs. *Current Opinion in Neurobiology, 14*, 481–487.

Op de Beeck, H. P., Brants, M., Baeck, A., & Wagemans, J. (2010). Distributed subordinate specificity for bodies, faces, and buildings in human ventral visual cortex. *NeuroImage, 49*, 3414–3425. doi:10.1016/j.neuroimage.2009.11.022.

Op de Beeck, H. P., Haushofer, J., & Kanwisher, N. G. (2008). Interpreting fMRI data: Maps, modules and dimensions. *Nature Reviews. Neuroscience, 9*, 123–135.

Orlov, T., Makin, T. R., & Zohary, E. (2010). Topographic representation of the human body in the occipitotemporal cortex. *Neuron, 68*, 586–600.

O'Toole, A. J., Jiang, F., Abdi, H., & Haxby, J. V. (2005). Partially distributed representations of objects and faces in ventral temporal cortex. *Journal of Cognitive Neuroscience, 17*, 580–590.

Park, S., Brady, T. F., Greene, M. R., & Oliva, A. (2011). Disentangling scene content from spatial boundary: Complementary roles for the parahippocampal place area and lateral occipital complex in representing real-world scenes. *Journal of Neuroscience, 31*, 1333–1340.

Park, S., & Chun, M. M. (2009). Different roles of the parahippocampal place area (PPA) and retrosplenial cortex (RSC) in panoramic scene perception. *NeuroImage, 47*, 1747–1756. doi:10.1016/j.neuroimage.2009.04.058.

Peelen, M. V., & Downing, P. E. (2005). Within-subject reproducibility of category-specific visual activation with functional MRI. *Human Brain Mapping, 25*, 402–408.

Phillips, M. L., & David, A. S. (1997). Viewing strategies for simple and chimeric faces: An investigation of perceptual bias in normals and schizophrenic patients using scan paths. *Brain and Cognition, 35*, 225–238.

Pitcher, D., Charles, L., Devlin, J. T., Walsh, V., & Duchaine, B. (2009). Triple dissociation of faces, bodies, and objects in extrastriate cortex. *Current Biology, 19*, 319–324.

Pitcher, D., Dilks, D. D., Saxe, R. R., Triantafyllou, C., & Kanwisher, N. (2011). Differential selectivity for dynamic versus static information in face-selective cortical regions. *NeuroImage, 56*, 2356–2363. doi:10.1016/j.neuroimage.2011.03.067.

Pitcher, D., Walsh, V., Yovel, G., & Duchaine, B. (2007). TMS evidence for the involvement of the right occipital face area in early face processing. *Current Biology, 17*, 1568–1573. doi:10.1016/j.cub.2007.07.063.

Potter, M. C. (1976). Short-term conceptual memory for pictures. *Journal of Experimental Psychology. Human Learning and Memory, 5*, 509–522.

Puce, A., Allison, T., Asgari, M., Gore, J. C., & McCarthy, G. (1996). Differential sensitivity of human visual cortex to faces, letterstrings, and textures: A functional magnetic resonance imaging study. *Journal of Neuroscience, 16*, 5205–5215.

Puce, A., Allison, T., Bentin, S., Gore, J. C., & McCarthy, G. (1998). Temporal cortex activation in humans viewing eye and mouth movements. *Journal of Neuroscience, 18*, 2188–2199.

Puce, A., Allison, T., Gore, J. C., & McCarthy, G. (1995). Face-sensitive regions in human extrastriate cortex studied by functional MRI. *Journal of Neurophysiology, 74*, 1192–1199.

Puce, A., Allison, T., & McCarthy, G. (1999). Electrophysiological studies of human face perception. III: Effects of top-down processing on face-specific potentials. *Cerebral Cortex, 9*, 445.

Rajimehr, R., Devaney, K. J., Bilenko, N. Y., Young, J. C., & Tootell, R. B. (2011). The "parahippocampal place area" responds preferentially to high spatial frequencies in humans and monkeys. *PLoS Biology, 9*, e1000608.

Rajimehr, R., Young, J. C., & Tootell, R. B. (2009). An anterior temporal face patch in human cortex, predicted by macaque maps. *Proceedings of the National Academy of Sciences of the United States of America, 106*, 1995–2000. doi:10.1073/pnas.0807304106.

Reddy, L., & Kanwisher, N. (2007). Category selectivity in the ventral visual pathway confers robustness to clutter and diverted attention. *Current Biology, 17*, 2067–2072. doi:10.1016/j.cub.2007.10.043.

Rossion, B., Hanseeuw, B., & Dricot, L. (2012). Defining face perception areas in the human brain: A large-scale factorial fMRI face localizer analysis. *Brain and Cognition, 79*, 138–157.

Rueffler, C., Hermisson, J., & Wagner, G. P. (2012). Evolution of functional specialization and division of labor. *Proceedings of the National Academy of Sciences of the United States of America, 109*, E326–E335. doi:10.1073/pnas.1110521109.

Saxe, R., Jamal, N., & Powell, L. (2006). My body or yours? The effect of visual perspective on cortical body representations. *Cerebral Cortex, 16*, 178.

Saygin, Z. M., Osher, D. E., Koldewyn, K., Reynolds, G., Gabrieli, J. D. E., & Saxe, R. R. (2011). Anatomical connectivity patterns predict face selectivity in the fusiform gyrus. *Nature Neuroscience, 15*, 321–327.

Schiltz, C., & Rossion, B. (2006). Faces are represented holistically in the human occipito-temporal cortex. *NeuroImage, 32*, 1385–1394.

Schwarzlose, R. F., Baker, C. I., & Kanwisher, N. (2005). Separate face and body selectivity on the fusiform gyrus. *Journal of Neuroscience, 25*, 11055–11059.

Schwarzlose, R. F., Swisher, J. D., Dang, S., & Kanwisher, N. (2008). The distribution of category and location information across object-selective regions in human visual cortex. *Proceedings of the National Academy of Sciences of the United States of America, 105*, 4447–4452. doi:10.1073/pnas.0800431105.

Sergent, J., & Signoret, J. L. (1992). Implicit access to knowledge derived from unrecognized faces in prosopagnosia. *Cerebral Cortex, 2*, 389–400.

Silvanto, J., Schwarzkopf, D. S., Gilaie-Dotan, S., & Rees, G. (2010). Differing causal roles for lateral occipital cortex and occipital face area in invariant shape recognition. *European Journal of Neuroscience, 32*, 165–171.

Srihasam, K., Mandeville, J. B., Morocz, I. A., Sullivan, K. J., & Livingstone, M. S. (2012). Behavioral and anatomical consequences of early versus late symbol training in macaques. *Neuron, 73*, 608–619.

Stettler, D. D., & Axel, R. (2009). Representations of odor in the piriform cortex. *Neuron, 63*, 854–864.

Sugita, Y. (2008). Face perception in monkeys reared with no exposure to faces. *Proceedings of the National Academy of Sciences of the United States of America, 105*, 394–398.

Takahashi, N., Kawamura, M., Shiota, J., Kasahata, N., & Hirayama, K. (1997). Pure topographic disorientation due to right retrosplenial lesion. *Neurology, 49*, 464–469.

Tarr, M. J., & Gauthier, I. (2000). FFA: A flexible fusiform area for subordinate-level visual processing automatized by expertise. *Nature Neuroscience, 3*, 764–769.

Taylor, J. C., Wiggett, A. J., & Downing, P. E. (2010). fMRI—adaptation studies of viewpoint tuning in the extrastriate and fusiform body areas. *Journal of Neurophysiology, 103*, 1467–1477.

Thorpe, S., Fize, D., & Marlot, C. (1996). Speed of processing in the human visual system. *Nature, 381*, 520–522.

Tian, L., Jiang, T., Liu, Y., Yu, C., Wang, K., Zhou, Y., et al. (2007). The relationship within and between the extrinsic and intrinsic systems indicated by resting state correlational patterns of sensory cortices. *NeuroImage, 36*, 684–690. doi:10.1016/j.neuroimage.2007.03.044.

Tsao, D. Y., Freiwald, W. A., Tootell, R. B., & Livingstone, M. S. (2006). A cortical region consisting entirely of face-selective cells. *Science, 311*, 670–674.

Turati, C., Bulf, H., & Simion, F. (2008). Newborns' face recognition over changes in viewpoint. *Cognition, 106*, 1300–1321. doi:10.1016/j.cognition.2007.06.005.

Turk-Browne, N. B., Norman-Haignere, S. V., & McCarthy, G. (2010). Face-specific resting functional connectivity between the fusiform gyrus and posterior superior temporal sulcus. *Frontiers in Human Neuroscience, 4*, 176. doi:10.3389/fnhum.2010.00176.

Urgesi, C., Berlucchi, G., & Aglioti, S. M. (2004). Magnetic stimulation of extrastriate body area impairs visual processing of nonfacial body parts. *Current Biology, 14*, 2130–2134. doi:10.1016/j.cub.2004.11.031.

van Koningsbruggen, M. G., Peelen, M. V., & Downing, P. E. (2013). A causal role for the extrastriate body area in detecting people in real-world scenes. *Journal of Neuroscience, 33*, 7003–7010. doi:10.1523/JNEUROSCI.2853-12.2013.

Wada, Y., & Yamamoto, T. (2001). Selective impairment of facial recognition due to a haematoma restricted to the right fusiform and lateral occipital region. *Journal of Neurology, Neurosurgery, and Psychiatry, 71*, 254–257.

Weiner, K. S., & Grill-Spector, K. (2011). Neural representations of faces and limbs neighbor in human high-level visual cortex: Evidence for a new organization principle. *Psychological Research* [epub ahead of print December 3, 2011].

Weiner, K. S., & Grill-Spector, K. (2012). The improbable simplicity of the fusiform face area. *Trends in Cognitive Sciences, 16*, 251–254. doi:10.1016/j.tics.2012.03.003.

Williams, M. A., Berberovic, N., & Mattingley, J. B. (2007). Abnormal FMRI adaptation to unfamiliar faces in a case of developmental prosopamnesia. *Current Biology, 17*, 1259–1264. doi:10.1016/j.cub.2007.06.042.

Wilmer, J. B., Germine, L., Chabris, C. F., Chatterjee, G., Williams, M., Loken, E., et al. (2010). Human face recognition ability is specific and highly heritable. *Proceedings of the National Academy of Sciences of the United States of America, 107*, 5238–5241. doi:10.1073/pnas.0913053107.

Winston, J. S., Henson, R. N., Fine-Goulden, M. R., & Dolan, R. J. (2004). fMRI-adaptation reveals dissociable neural representations of identity and expression in face perception. *Journal of Neurophysiology, 92*, 1830–1839.

Yovel, G., & Kanwisher, N. (2004). Face perception domain specific, not process specific. *Neuron, 44*, 889–898.

Yovel, G., & Kanwisher, N. (2005). The neural basis of the behavioral face-inversion effect. *Current Biology, 15*, 2256–2262. doi:10.1016/j.cub.2005.10.072.

Yue, X., Cassidy, B. S., Devaney, K. J., Holt, D. J., & Tootell, R. B. H. (2011). Lower-level stimulus features strongly influence responses in the fusiform face area. *Cerebral Cortex, 21*, 35–47. doi:10.1093/cercor/bhq050.

Zhu, Q., Song, Y., Hu, S., Li, X., Tian, M., Zhen, Z., et al. (2010). Heritability of the specific cognitive ability of face perception. *Current Biology, 20*, 137–142. doi:10.1016/j.cub.2009.11.067.

VIII TIME, MOTION, AND DEPTH

53 Visual Time Perception

ALAN JOHNSTON

The question "How is time represented in the brain?" is readily asked but not so readily answered. To make progress with this issue we will need to constrain our inquiry. In particular we need to address some more concrete questions that offer more chance of success. We can ask what it is we need to know about the temporal world. We can ask what properties of the world we need to encode, and by what mechanisms we can gain this knowledge while continuing to consider how algorithms to accomplish these tasks can be instantiated in neural systems.

Time perception should tell us about events, including the temporal order of events, the relative timing of events, in a metric rather than ordinal sense, and the duration of events. At a higher level we need to be able to encode temporal pattern, temporal overlap, and temporal continuity and discontinuity and use temporal information to infer causality. In the hope that the mechanisms of time perception share some common ground, independent of the medium by which the information is provided, we concentrate here on the timing of visual events unless drawn to compare vision and other senses.

There has been concentrated activity in recent years to identify brain areas that are involved in timing tasks using functional imaging. The variety of timing tasks, including perceptual and motor tasks, the variety of temporal scales investigated, and the different requirements placed on central control and decision processes depending on the task, require too much detail to be effectively summarized here. However these studies have been extensively and carefully reviewed elsewhere (Coull, Cheng, & Meck, 2010) and have also been subject to meta-analysis (Wiener, Turkeltaub, & Coslett, 2010). In general terms, functional imaging has shown that a network of areas are involved in timing tasks. This contrasts with the functional specialization of visual areas such as V5/MT or FFA, which are active when we are engaged in motion or face processing. There does not appear to be one particular area that is specifically engaged in time judgments, although the supplementary motor area (SMA) and basal ganglia (BG) are often active, and no single brain lesion results in a loss of ability to make temporal judgments (Coull, Cheng, & Meck, 2010).

It is tempting to think that the time at which events appear to occur simply reflects the time of neural activation of the representations that unambiguously signal the presence of some external feature. However, this leads to a number of difficulties. There is the homunculus problem, familiar from pictorial theories of spatial vision. Who is inspecting the brain and distinguishing and timing neural events? How do we explain the temporal perception of the inspector? In any case, neural latency data are just too noisy to be a useful timing signal. Given the latency differences between areas and between neurons within areas (Bullier, 2001) and the dependencies on stimulus strength (Lee, Williford, & Maunsell, 2007), which brain events can be relied on to accurately reflect the timing of events in the physical world?

Communication delay between different brain areas makes brain time relativistic. If simultaneous events occur in connected but distant areas A and B, from the perspective of A, B will occur later, and from the perspective of B, A will occur later. This leads to a significant computational problem for neural systems. Many neural computations must access information from multiple representations in the brain that refer to the same physical environmental event. In order to generate synchronous computation for time-varying data, there needs to be some form of compensation for delays, or the computation may suffer from temporal mismatch, temporal blurring, or misbinding. Note that the need for temporal alignment of neural signals is a neural processing issue and may or may not give rise to perceived temporal misalignment, although it may lead to perceptual errors. Physical clocks mark the objectively measurable passage of time. The timing of brain events can be assessed in relation to these clocks; however, we need to consider the nature of these brain events, not their timing as such, to relate neural processing and the perceived time course of external events (Johnston & Nishida, 2001). Fortunately, we know quite a lot about visual temporal mechanisms from a different context.

Any modern discussion of visual mechanisms for spatial localization, relative position of points, or mechanisms for the perception of the length of a line would start with a shared understanding of the importance of spatial filters and their properties in encoding spatial information. Temporal filters have largely been explored in the context of motion perception, but they may perform double duty in the context of temporal encoding of visual events. This perspective embeds time perception within a specific modality. It means that there may be vision-specific and general mechanisms involved in a timing task and, by analogy, other modality-specific and general mechanisms. We explicitly encode the color of a surface or its motion without preference or effort. Stimulus qualities and magnitudes are available locally and are generally constant through an inspection interval in a psychophysical task. However, we cannot encode spatial relations automatically, as even pairwise relations would lead to a combinatorial explosion. Perceptual judgments about duration can be made only at the end of the interval to be timed, and relations such as temporal order or synchrony require specific perceptual routines (Johnston & Nishida, 2001; Ullman, 1984) to be set up to make explicit information that may be held implicitly in the sensory system. It is useful to distinguish between the automatic processes that exist in sensory systems that provide information on which judgments may be made and the more generic, but not necessarily anatomically intersecting, routines that are needed to extract task-relevant information.

TEMPORAL MECHANISMS

Visual information is initially encoded through channels that differ in their sensitivity to image change over time. Unlike spatial channels, temporal channels are few in number. It is generally agreed there are no more than three (Cass & Alais, 2006; Hess & Plant, 1985; Mandler & Makous, 1984; Watson & Robson, 1981). Hess and Snowden (1992), using a masking paradigm, found three channels: one low-pass and one band-pass peaking around 10 Hz, and a third temporal channel peaking around 16 Hz, which was only evident for low-spatial-frequency stimuli. Although, the two-pulse method for deriving the temporal impulse response suggested chromatic information was processed by a single low-pass temporal channel (Burr & Morrone, 1993, 1996), a number of other paradigms have revealed at least two channels for isoluminant chromatic patterns (Cass et al., 2009; Eskew, Stromeyer, & Kronauer, 1994; Metha & Mullen, 1996). Temporal filters modulate the sensitivity of the visual system to temporal

patterns; however, their functional role may go beyond this characterization of their action. Johnston and Clifford (1995) showed that a Gaussian in log time and its first and second derivatives provided an excellent fit to the temporal filters measured by Hess and Snowden (1992), reducing the number of parameters required to fit their data from six to two (figure 53.1). Using a Gaussian in log time for the blur kernel of the filters ensures the derivative filters are causal (Koenderink, 1988); that is, they do not act on image brightness values in negative (future) time. The fact that temporal filters are causal means they inevitably introduce some phase delay as the blur kernel averages over some time in the recent past. Neural filters have been interpreted to function as signal modulators, matched templates, or parts of a basis set, which would allow the content of a temporal segment to be encoded in an efficient manner as the proportions in which the basis filters need to be added to reconstruct the signal. The more tightly constrained derivative property allows us to go beyond these ideas to the idea that neurons could reconstruct the signal outside the scope of their temporal receptive fields, allowing prediction (and postdiction).

TEMPORAL FREQUENCY

Contrast sensitivity to temporal frequency peaks around 8–10 Hz and has an upper limit exceeding 60 Hz in optimal conditions (Kelly, 1961, 1979). The response to high temporal frequency improves at high luminance levels, shifting the shape of the temporal contrast sensitivity function (CSF) from low-pass at low luminance levels to be more bandpass at high luminance levels. The CSF in the temporal domain can be considered to be the envelope of two or three temporal channels. Although the form of the temporal CSF and component channels has been studied extensively, the mechanisms underlying the perception of temporal frequency and temporal patterning are not well understood. Some indication about how these filters may interact to code temporal frequency comes from studies of the velocity aftereffect. After adaptation to motion, perceived velocity decreases for motion in the same direction that has the same or a lower velocity than the adapting velocity (Carlson, 1962; Thompson, 1981). If the test velocity is higher than the adapt velocity and is moving in the same direction, it can appear to be increased (Smith & Edgar, 1994). This shift in perceived speed or drift rate away from the adapt velocity or temporal frequency also occurs for adapting gratings that oscillate in direction during the adaptation phase (Ayhan et al., 2009; Johnston, 2010; Johnston, Arnold, & Nishida, 2006). This temporal frequency shift effect can be explained

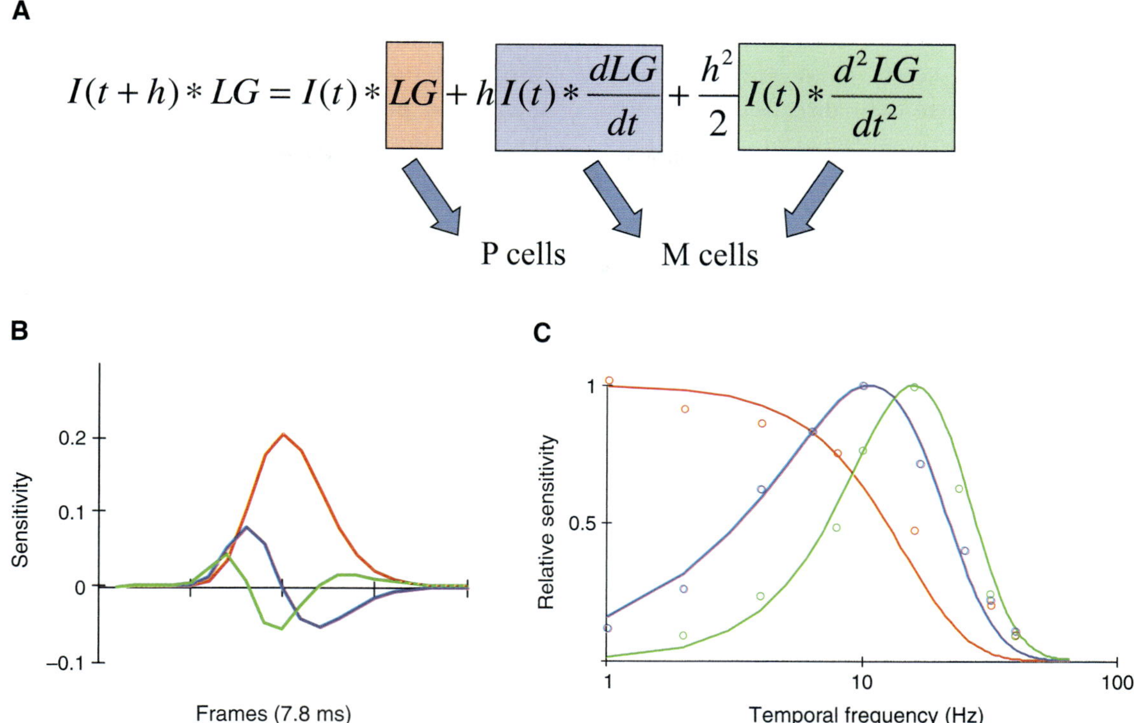

FIGURE 53.1   (A) Taylor's series can be used to predict the value of a function (the image brightness, *I*) in the neighborhood around a point in time *t*. Temporal derivatives of the image brightness are calculated at a point *t* by convolving (∗) the image brightness with the derivatives of Gaussians in log time. The excursion parameter, *h*, indicates the point (*t* + *h*) at which the value of the function is to be predicted. The parameter *h* can take on a range of values. (B) The impulse responses for the filters, calculated as the inverse Fourier transform of the corresponding frequency sensitivity curves in C. The first-derivative operator is biphasic, and the second-derivative operator is triphasic. (C) The temporal tuning functions for human vision. The data are from Hess and Snowden (1992), and the fitted functions, the Gaussian in log time (low pass) and its first (bandpass, peaking around 8 Hz) and second (bandpass, peaking around 16 Hz) derivatives are from Johnston and Clifford (1995). All data for each of the three functions are fitted by adjusting just one parameter (see Johnston & Clifford, 1995, for details). The bandpass filters correspond to the tuning curves of magnocellular neurons, and the low-pass temporal filter is representative of parvocellular neuron's temporal tuning curves.

by changes in the relative sensitivity of a small set of broadband filters (Smith & Edgar, 1994). Smith and Edgar (1994) proposed that temporal frequency could be coded as the relative activity of two or three broadly tuned mechanisms. Note that if temporal frequency channels can be modeled as differentiating filters, their relative activation is proportional to temporal frequency as the differentiation in the time domain corresponds to multiplication by $i\omega$ (where i is $\sqrt{-1}$ and ω is temporal frequency) in the frequency domain (Johnston, 2010). The multiplication by i reflects the fact that a sine function will be shifted in phase by 90° (cosine) by differentiation. Although an algorithm based on the relative magnitude of the zero-, first-, and second-order derivative filters could recover the temporal frequency of a sinusoidal modulation of brightness, how a more complex temporal pattern might be encoded by just three broadband overlapping filters remains an open question.

PERCEPTUAL FLUCTUATION

The critical flicker-fusion threshold is the temporal frequency above which flicker can no longer be detected. It is impossible to distinguish between a sampled and a continuous signal if the frequency components that differ are above the CFF frequency. Temporal sampling of continuous motion leads to the classical wagon-wheel effect seen in the cinema in which stagecoach wheels appear to rotate in a direction opposite to their true direction. This is simply a reflection of the fact that in the sampled version of the rotation the shortest path between the spokes of the wheel is physically in the reversed direction. Cinema projectors update the image at a rate of 24 frames per second. If we saw reversed rotation of wagon wheels in the unmediated viewing of a rotating wagon wheel, this would be evidence of temporal sampling in the human visual system. Real wagon wheels do not typically appear to reverse at a particular rate of rotation, so it is unlikely that the retinal image is temporally sampled in any direct sense (Kline, Holcombe, & Eagleman, 2004). However, rotating patterns can appear to reverse their direction of rotation in some conditions (Purves, Paydarfar, & Andrews, 1996; Schouten, 1967; VanRullen, Reddy, & Koch, 2005). These reversals are not sustained but rather alternate with periods of forward motion. VanRullen, Reddy, and Koch (2005) showed the experience of reversal did not occur if attention was directed to a primary RSVP task, indicating that focused attention was necessary and that sampling may have been accomplished through attentional snapshots. More recently Busch and VanRullen (2010) have shown that the effects of attention on target detection fluctuate with the phase of a 7-Hz periodic EEG component. These results indicate a temporal variation in visual performance linked to a physiological response, which Busch and VanRullen attribute to attention-driven periodic sampling, even in conditions of sustained attention.

Arnold and Johnston (2003) reported an example of a motion display that introduced time-varying perceptual fluctuations where none exists in the physical stimulus. Isoluminant chromatic stimuli appear to move more slowly than stimuli defined by luminance contrast (Cavanagh, Tyler, & Favreau, 1984). Arnold and Johnston drifted a red–green isoluminant bullseye pattern over a dark background. The red–green border appeared to jitter when isoluminant even though the bullseye was moving rigidly. This as not related to the use of CRT displays or artificial lighting, as the jitter could be seen for a printed stimulus rotated mechanically in daylight and disappeared when the target was tracked on CRT monitors. The rate of jitter was around 22 Hz when measured by matching to a flickering LED. The match frequency varied more between subjects than when velocity varied by a factor of two and did not change as a function of border separation (Arnold & Johnston, 2005), again suggesting the perceptual fluctuation was internally generated. We obtained an estimate of around 10 Hz when the jitter was measured by matching to a physically jittering display (Amano et al., 2008). This was roughly half the frequency measured in the previous studies, suggesting that subjects matched each apparent change in position to the bright phase of LED flicker. MEG recordings showed an enhanced peak around 10 Hz for an isoluminant condition that generated illusory jitter compared to real jitter, luminance difference, and isoluminant color on an isoluminant background condition. This enhanced rhythmic activity was of a similar frequency to that reported to modulate attention, 7 Hz (Busch & VanRullen, 2010), and accompanies the

continuous wagon-wheel illusion, 13Hz (VanRullen, Reddy, & Koch, 2006). Arnold and Johnston (2003) attributed the jitter to a motion-induced spatial conflict. They proposed that the velocity signals at the boundary were used to shift the spatial pattern forward. The benefit of this was considered to be that a predicted signal could be used to calibrate the motion analysis system. More recently Roach, McGraw, and Johnston (2011) have provided convincing psychophysical evidence that spatial pattern is predicted forward at the leading edge of motion. A match between the predicted signal and the incoming signal would indicate a perfectly calibrated motion computation. However the difference between the speed calculated at the isoluminant chromatic edge and the luminance edge leads to an erroneous prediction, which is discarded in favor of the new data, leading to perceived jitter. This process might also be considered an example of the use of predictive coding (Rao & Ballard, 1999) in the early visual system. The persistence of this systems-level illusion results from the fact that the velocity difference between isoluminant and luminance boundaries cannot easily be calibrated away. However, the reason this mechanism delivers a constant temporal rate is not obvious—the predict-and-compare strategy outlined above could be continuous.

TEMPORAL ALIGNMENT

A real-time image-processing pipeline can operate asynchronously if a downstream stage in the pipeline simply requires access to the current state of its inputs. However, if the process requires comparisons, such as those involved in error checking, or optionally bringing together signals, then the timing of those processes becomes critical. Consider the operations required to set up the motion computation check outlined above. A measure of speed specifies a rate. To check the rate of spatial displacement one needs to specify a particular location at which to test and a specific time at which to compare the prediction for that location and the representation of the spatial pattern that arises at that time. The image motion refers to events in the world, and so we can predict the spatial pattern in the image say 100 ms ahead in the world event. However, the neural mechanisms checking the prediction against the dynamic changes in the spatial representation at the checking point need to take account of the time course of the operations within the neural implementation of the routine; that is, the neural processes need to be temporally regulated alongside the representations of event time; otherwise it would not be possible to generate an accurate estimate of the error between prediction and predicted. It would seem that the requirement to set up a perceptual routine necessitates accurate relative neural timing, which may necessitate transient coordinated brain activity that would appear in the electrical or magnetic activity of the brain as a rhythmic signal at a particular frequency determined by the perceptual routine.

To generate some temporal alignment across components of a neural circuit implementing a perceptual routine, it may be necessary for the brain to adjust processes into temporal alignment for a neural assembly to compute some function across those inputs accurately. If it is possible to adjust the time course of neural events to bring events into apparent temporal alignment, then it may be possible to generate an apparent misalignment of physically synchronous events.

Fujisaki et al. (2004) showed that adaptation to a 235-ms lag between a flash and a tone pip caused a subsequent shift in the point of subjective equality of the timing of the tone and pip of around 30 ms. In the stream bounce illusion the likelihood of seeing two converging dots bounce rather than stream through one another is maximized when a tone pip is presented at the time the dots overlap. In a subsequent experiment Fujisaki and Nishida (2010) showed that the relative time lag between convergence and tone pip that maximized the bounce percept was also shifted by adapting to a time lag by about the same amount of time (approximately 24 ms). The fact that the stream bounce illusion does not require an explicit encoding of the temporal relationship between the auditory and visual stimulus is good evidence that adaptation to time lags affects sensory processing.

Hogendoorn, Verstraten, and Johnston (2010) demonstrated a shift in apparent temporal alignment within a single modality. They introduced a display in which subjects made judgments about the relative time on running clocks after motion adaptation. On a given trial eight clocks located at regular intervals within an annulus centered on fixation were set running. At some point six clocks were removed and subjects had to say which of the two remaining clocks was showing a later time. On each trial prior to the presentation of the clocks, four of the stations were adapted to a bullseye grating pattern whose motion drifted at either 5 Hz or 20 Hz and oscillated between expansion and contraction. The motion of the adaptor was orthogonal to the direction of the clock. The clocks rotated at 1 Hz. After 20-Hz adaptation, the clock in the adapted region appeared to be around 10 ms ahead of the clock in the unadapted region when the clocks were physically aligned. This corresponds to an angular difference of around 3.6°. Note that this was not because the

20-Hz-adapted clock appeared to move faster, as, paradoxically, it appeared to move more slowly than the unadapted clock. There was also no difference in choice reaction time between an adapted and unadapted region, indicating that the adaptation paradigm did not alter neural processing time.

The apparent temporal order of events can even appear to be reversed (Morrone, Ross, & Burr, 2005) although not for all subjects (Kitazawa et al., 2007; Terao et al., 2008). Subjects were asked to judge which of a pair of near-isoluminant horizontal bars, located above and below fixation and saccadic targets, was presented first, around the time of a horizontal saccade. They found that the interval defined by the presentation of the bars appeared compressed and that when bars with short-interval separations were presented just before a saccade (−70 to −30 ms before the saccade), the temporal order appeared reversed. Interestingly, temporal order reversal also occurs in the tactile domain (Kitazawa et al., 2007; Yamamoto & Kitazawa, 2001). In this paradigm subjects can accurately determine the temporal order of stimuli presented to the two hands but report temporal order reversals when the hands are crossed. This finding shows that temporal order of tactile stimulation depends on spatial processing rather than on the order of stimulation on the receptor surface, which is not changed by crossing hands. Kitazawa et al. attribute the reversal to a requirement that the spatial position of the hands needs to be represented before the temporal order of stimulation. For crossed hands, in some cases participants report which spatial location was stimulated first, not which hand was stimulated first. For a temporal order difference of around 100 ms, the stimulation of the right crossed hand in left space is mapped to the left hand for some period of time before the correct attribution of hand and spatial location. Interestingly, Kitazawa et al. (2007) also report a temporal order reversal for tactile stimulation around the time of a saccade. Terao et al. (2008) asked whether the apparent temporal interval compression in the Morrone, Ross, and Burr paradigm might be related to saccadic suppression, thought to primarily influence the magnocellular pathway (Burr, Morrone, & Ross, 1994). They adapted the locations to be filled with the horizontal bars to high-frequency dynamic temporal noise. This reduced the sensitivity to the near-isoluminant chromatic bars. In this case they reported temporal compression without saccades. Terao et al. (2008) concluded that reduction of sensitivity of magnocellular mechanisms through the use of isoluminant targets, high temporal frequency adaptation, or low contrast could lead to temporal compression.

The rate at which two features can be compared can vary dramatically (Fujisaki & Nishida, 2010; Holcombe, 2009). When two oriented patterns are superimposed and alternated quickly, it is possible for the visual system to identify texture boundaries defined by orientation (Motoyoshi & Nishida, 2001) or identify which orientation is linked with one color rather than another (Holcombe & Cavanagh, 2001) at reversal rates of over 20 Hz. This suggests the possibility that a local representation of orientation contrast or orientation and color can be arrived at fairly automatically, perhaps by neurons that are simultaneously sensitive to orientation and chromatic contrast in the case of the color orientation task (Holcombe & Cavanagh, 2001). However, when subjects are asked to report which color was presented in combination with a particular grating orientation for spatially separated alternating blocks of color and orientation, then subjects find the task impossible for patterns alternating above 3 Hz (Holcombe & Cavanagh, 2001). Fujisaki and Nishida (2010) distinguished between *binding* tasks, in which subjects had to identify a feature in one domain that was coincident with a feature in another domain, and *synchrony* tasks, in which subjects had to report whether features across domains were synchronous or not. They studied a range of comparisons within and across sensory modalities. They found over a range of cross-attribute and cross-modality comparisons that the limits on the binding tasks were universally around 2–3 Hz. It is clear that individuating a feature from one sequence, shifting attention, and identifying the value of a corresponding feature in another sequence requires the instantiation of a specific task-relevant perceptual routine—an attention-demanding and time-limited process.

COLOR-MOTION ASYNCHRONY

The limitations imposed on the perceptual system by requiring it to orchestrate a bespoke neural routine to perform a particularly difficult combinatorial task can lead to profound apparent asynchronies for physically synchronous events. Moutoussis and Zeki (1997) discovered that observers judged changes in motion direction and color to be synchronous when the color change preceded the motion direction change by around 100 ms. They used a binding task, but a similar apparent temporal shift can be demonstrated in a synchrony task (Nishida & Johnston, 2002). Although Moutoussis and Zeki (1997) attributed this effect to differences in the time at which the motion and color changes were represented in the brain, this

interpretation could not account for subsequent manipulations. Nishida and Johnston (2002) showed that motion color asynchrony did not occur for motion direction reversals and color changes when the time of the fifth transition in the sequence was compared to the timing of a single transition of either the same or an alternate submodality, showing that the temporal patterning of the stimulus was not the key feature. In addition they reversed the temporal properties of the position and color signals. When participants were asked to compare the timing of a reversal in the direction of change of a color temporal gradient, red to green and back again, with a transient jump in position they reported synchrony when the color gradient direction change led by around 100 ms, indicating that the temporal patterning was the key factor not the submodality. Nishida and Johnston proposed that, when under time pressure, observers tend to link time markers with similar properties (Nishida & Johnston, 2002, 2010). In the case of color–motion asynchrony first-order color changes are linked to first-order changes in position, motion segments, rather than second-order features, motion direction changes, which are difficult to individuate and match to first-order features at alternation rates above about 2–3 Hz. More recently it has been shown that when an attentional cue is presented with the change in color or motion direction for an array of eight stimuli alternating in color and direction within an annulus around fixation, or when an attentional window is guided around the array, motion asynchrony can be reduced or abolished (Cavanagh, Holcombe, & Chou, 2008; Holcombe & Cavanagh, 2008). This provides further evidence that color–motion asynchrony is not a reflection of processing latency.

DURATION PERCEPTION

Judgment of the duration of a visual event requires the computation of extent rather than simultaneity versus asynchrony or temporal order. The computational problem of recovering duration is sufficiently different from judging temporal order that it is likely that different mechanisms are required for each. Judging the temporal order of events in a restricted spatial neighborhood could be done fairly automatically. We can, for example, determine the direction of apparent motion, which requires little cognitive effort. Judging duration is different, however, from judging other properties of an event in that is it is not specified until the end (Morgan, Giora, & Solomon, 2008). Typically duration judgments require attention and are cognitively taxing.

The standard model proposes that duration is encoded by an amodal cognitive clock (Creelman,

1962; Treisman, 1963). A clock requires a signal of constant rate, such as a pacemaker, and a means of integrating the output of the pacemaker over the duration. Typically the output of the pacemaker is gated into an accumulator, and the value in the accumulator reflects the elapsed time between gate opening and closing. This mechanism could apply generally to all events and temporal scales and is independent of the content of the interval that is being timed.

However, there is considerable evidence that the content of the interval influences its apparent duration. Kanai and colleagues (2006) showed that the duration of intervals containing low-temporal-frequency expanded gratings and flickering Gaussian blobs were underestimated compared to high-frequency versions of these stimuli. The compression effect saturated at around 4–8 Hz. This may be stimulus dependent. In another study using drifting Gabors (Kaneko & Murakami, 2009), duration compression was found to be speed dependent. The effects of content on duration have been linked to the number of change events in the interval (Brown, 1995); however, it is not clear what constitutes an enumerable event for a moving grating. Bruno, Ayhan, and Johnston (2012) reported that the duration of an interval made up of moving and static segments appeared compressed relative to a constant-motion condition and no different from a static condition. This result is at odds with the idea that apparent duration is a reflection of the number of events.

Johnston, Arnold, and Nishida (2006) introduced an adaptation-based approach to time perception. They adapted a region of the near visual periphery on one side of fixation to drifting sine gratings or Gaussian blobs whose luminance varied sinusoidally over time. After a 15-s adaptation period, a subsecond interval of 10-Hz drift (motion) or luminance modulation (flicker) was presented sequentially to the adapted and nonadapted sides of the visual field. Observers were asked to report which interval lasted longer. The duration of the interval on the nonadapted side was varied from trial to trial to generate a psychometric function. The point of subjective equality provided a measure of relative duration. Johnston and colleagues found that apparent duration of a 10-Hz motion stimulus on the adapted side was reduced after adaptation to a 20-Hz drift but only slightly reduced or unaffected after a 5-Hz adaptation. A similar result was found for the flicker stimulus. Adapting to motion or flicker may have nonspecific effects, such as a change in attention or arousal, but any nonspecific effect will not contribute to the comparison between adapted and nonadapted regions of the visual field.

The change in perceived duration of identical intervals after adaptation of a localized region of the visual field cannot be explained with reference to a generic central clock or state changes, such as changes in arousal or attention on the clock rate. If these changes in apparent duration were a result of attending more to the adapted region, then we should expect the same amount of apparent time compression in the 5-Hz condition as in the 20-Hz condition, but this was not the case. The 5-Hz and 20-Hz adaptors shift the apparent frequency of a 10-Hz test pattern by about the same amount (around 3-Hz shift away from the adaptor). In addition, we have demonstrated that time compression occurs even in the case of invisible (60-Hz) flicker (Johnston et al., 2008). In this paradigm, because the adapter is invisible, it is not possible for observers to tell which side of the visual field is being adapted and attend more to the adapted side.

Adaptation can also change the apparent temporal frequency of the test patterns (Johnston, Arnold, & Nishida, 2006; Smith & Edgar, 1994; Thompson, 1981). Adapting to a high-temporal-frequency (20-Hz) drifting grating reduced the apparent frequency of a 10-Hz test grating in the adapted region of visual space, and adapting to low temporal frequency (5 Hz) had the reverse effect, increasing the apparent temporal frequency of the 10-Hz test. This raises the issue of whether the changes in apparent duration are mediated by changes in apparent temporal frequency. However, there are a number of reasons to discount this idea of mediation via temporal frequency. Changes in apparent temporal frequency are bidirectional, whereas adaptation introduces only a reduction in apparent duration, except in dyslexia (Johnston et al., 2008). There is only a very small, <1 Hz, decrease in apparent temporal frequency after adaptation to the same stimulus; nevertheless, there is still a compression of apparent duration (Johnston, Arnold, & Nishida, 2006). Time compression occurs for stimuli matched in apparent temporal frequency (Bruno, Ayhan, & Johnston, 2010; Johnston, Arnold, & Nishida, 2006) or speed (Curran & Benton, 2012). Adaptation to 20 Hz only reduces the apparent temporal frequency of 10 Hz to 7 Hz, and Kanai et al. (2006) showed that the effects of temporal frequency on duration saturated at around 4–8 Hz. Dyslexics show a different pattern from controls when asked to judge duration but not when asked to judge temporal frequency after the same exposure to the adapter (Johnston et al., 2008). Duration compression has also been demonstrated for interleaved 5-Hz and 20-Hz adapters, with duty cycles chosen to null any changes in apparent temporal frequency (Ayhan et al., 2009, 2010; Bruno, Ayhan, & Johnston, 2010).

Adaptation may alter the perception of the onset and offset of the interval. For example, the onset might appear to be delayed after adapting to high temporal frequencies with little effect on offset. However, apparent compression was found to be proportional to the length of the interval (Johnston, Arnold, & Nishida, 2006) rather than being a subtraction from the length of the interval, which is what would be expected if adaptation introduced some processing delay at onset. Typically apparent onset and offset times measured relative to an auditory tone are little affected by adaptation (Johnston, Arnold, & Nishida, 2006).

THE MAGNOCELLULAR THEORY OF DURATION PERCEPTION

There are a number of reasons to believe that adaptation-based time compression may derive from changes occurring early in the visual pathway. First, adaptation-based compression of apparent duration is not orientation specific. Rotating the test grating 90° with respect to the adapting grating has virtually no effect on the strength of the apparent compression (Johnston, Arnold, & Nishida, 2006). Neurons in the visual pathway up to the input layers of visual cortex have a center–surround receptive-field geometry. Thus, the lack of orientation specificity is consistent with a precortical site for adaptation. Adaptation-based time compression also appears to be well localized to the adapted area, in some cases vanishing within 1° of visual angle (Ayhan et al., 2009). Duration compression does not show interocular transfer (Bruno, Ayhan, & Johnston, 2010; but see Burr, Tozzi, & Morrone, 2007). Compression also occurs for invisible (60-Hz) flicker (Johnston et al., 2008). Neurons in the retina and lateral geniculate nucleus (LGN) respond to frequencies about 20 Hz higher than cortical neurons (Hawken, Shapley, & Grosof, 1996). The absent or poor response of cortical cells to 60-Hz stimulation indicates a subcortical locus for the adaptation effect. Cells in the magnocellular pathway are less responsive to isoluminant chromatic contrast. If adaptation-based time compression results from the adaptation of M cells, then we should not see duration compression for isoluminant stimuli. Indeed, duration compression disappears at isoluminance for red–green chromatic gratings, but it can be restored by introducing green and orange backgrounds, which previously were found by Stromeyer and colleagues (1997) to generate maximal phase lags between the L and M cone signals, thereby generating a luminance response to isoluminant gratings (Ayhan et al., 2010).

Solomon and co-workers (2004) demonstrated that M cells in the LGN show slow adaptation at high temporal frequencies (45 Hz) but not at low temporal frequencies (1 Hz). Concurrently recorded S potentials, action potentials that reflect ganglion cell input, show this slow adaptation to be retinal in origin. Magnocellular cells also show a fast contrast-gain adaptation (Mante, Bonin, & Carandini, 2008; Shapley & Victor, 1978) in which the contrast response function is shifted to the right (toward higher contrasts). This results in a higher threshold contrast but greater variation in response to contrast at high contrasts (i.e., reduced saturation). Contrast gain reduces the responsivity of the cell and sharpens and advances the temporal impulse response (Benardete & Kaplan, 1999). Parvocellular (P) neurons in the LGN and retina do not show any substantive fast or slow contrast adaptation. Both contrast adaptation in the magnocelluar pathway up to the LGN and adaptation-based time compression occur selectively at high temporal frequency and can occur for temporal frequencies beyond those likely to stimulate the visual cortex. Both are independent of the relative orientation of adapter and test, and both are induced by luminance modulation and drifting gratings.

Slow adaptation induces a drop in the maintained discharge rate of neurons after adaptation rather than a decline in firing rate during the adaptation period. There is also a change in the contrast gain as indicated by a right shift in the neuron's contrast response function. The size of the postadaptation drop in the discharge rate and the magnitude of the contrast-gain shift tend to be correlated (Solomon et al., 2004). Slow adaptation delivers time compression. If this is mediated by a change in the temporal impulse response associated with contrast-gain control, then one would also expect to get effects of fast adaptation (contrast-gain control) on time perception. Bruno and Johnston (2010) presented subjects with a sequence that contained five intervals in total. The first and last intervals contained a high-contrast (90%), and the middle interval contained a low-contrast (10%) drifting sine grating. In other blocks the first and last intervals contained a low-contrast (10%) grating, and the middle interval was occupied by a high-contrast (90%) grating. Two intermediate-contrast (50%) intervals, in positions 2 and 4, were the test intervals. Subjects reported which of the two 50% contrast intervals lasted longer. The duration of one of the two intermediate contrast intervals was varied over trials to generate a psychometric function. The point of subjective equality provided a measure of relative perceived duration. The gratings drifted at 2, 4, 10, and 20 Hz in separate blocks. On each trial the stimulus location varied within an annulus centered on fixation to avoid slow adaptation. The adaptation intervals were 1.5 s, and the standard test interval was 500 ms. An apparent temporal compression of the interval following the high-contrast adapter relative to the low-contrast context was found, but only for high (10- or 20-Hz) temporal frequencies. Contrast gain-induced compression also disappeared at isoluminance for chromatic stimuli, confirming its dependence on magnocellular adaptation. In addition, we know that the temporal impulse response broadens at low light levels. The prediction that duration should appear expanded at low luminance was confirmed by Bruno, Ayhan, and Johnston (2011).

Although there are now a range of findings that support the proposals that changes in M cells' temporal tuning critically determine adaptation-based compression and that M-cell responsivity may be a factor in other effects such as improved temporal discrimination after adaptation (Terao et al., 2008), there are some results that challenge these views. Curran and Benton (2012) reported adaptation-based duration compression when the test stimulus was in the same direction as the adaptation but not when the test was in the opposite direction. They also showed that compression occurred for plaid adapters but not for transparent random dot adapters moving with the same velocity as the plaid components, indicating compression can occur after motion integration. These findings suggest the involvement of higher cortical areas in the motion pathway. However, although direction selectivity is a common property of cortical cells, around 20% of LGN neurons show a direction bias (Xu et al., 2002), and LGN cells with S-cone inputs show strong direction selectivity to achromatic gratings (Tailby et al., 2010; Tailby, Solomon, & Lennie, 2008). In cats the orientation dependence of LGN cells can be removed by inactivation of the cortex (Ye et al., 2009), and it remains a possibility that direction-contingent effects on duration perception may result from cortical feedback to the LGN. The drift velocities used by Curran and Benton (2012) were considerably lower than those used by Johnston, Arnold, and Nishida (2006), opening up the possibility that the absence of duration compression for test stimuli moving in the opposite direction to the adapter in Curran and Benton's paradigm may be specific to low-velocity adapters and test patterns.

The other challenges are the proposal that retinotopic duration compression can be attributed solely to the effects of temporal frequency adaptation and the observation that duration compression occurs within a spatiotopic frame of reference (Burr, Tozzi, & Morrone, 2007). The evidence that adaptation-based duration can occur in the absence of temporal frequency adaptation is outlined above. To confirm that retinotopic adaptation can occur in the absence of spatiotopic adaptation, Bruno, Ayhan, and Johnston (2010) adapted to interleaved (5-Hz and 20-Hz) adapters with a duty cycle chosen to eliminate any effects on the apparent temporal frequency of a 10-Hz test grating. In one condition one adapter was placed above a pursuit target and yoked to the eye movement while another adapter was placed below the pursuit target and moved in the opposite direction to the target but covered the same amount of screen as the yoked target. A test pattern placed in the retinotopic location of the adapter showed apparent duration compression relative to a comparison placed in the location that had the same degree of spatiotopic adaptation as the standard, demonstrating retinotopic adaptation in the absence of temporal frequency adaptation. The exploration of spatiotopic adaptation-based duration compression has provided mixed results (Bruno, Ayhan, & Johnston, 2010; Burr, Cicchini, & Arrighi, 2011; Burr, Tozzi, & Morrone, 2007; Johnston, Bruno, & Ayhan, 2011).

A CONTENT-DEPENDENT CLOCK

A content-independent timer as exemplified by the standard stopwatch model cannot explain the effects of spatially localized adaptation. The manipulations outlined above are designed to alter the properties of temporal filters in the visual system. However, changes to linear filters can only alter the phase and contrast of drifting sine gratings, not their durations. Changes in perceived temporal frequency can be explained by changes in filter sensitivity, but a different approach is necessary to explain changes in apparent duration.

We have characterized the action of temporal filters as blurring and differentiating the temporal image. Temporal differentiation can be accomplished by convolving a temporal signal with the temporal differentiating filters. The zero-order blur kernel has the low-pass temporal property characteristic of parvocellular neurons. The derivative filters have the bandpass property associated with magnocellular neurons (figure 53.1B, C). The relationship between the temporal impulse response and the frequency response of the filters is shown in figure 53.1B, C.

One can exploit the derivative properties of the temporal filters to construct a Taylor-series approximation (figure 53.1A) of the image brightness. The Taylor series allows the reconstruction of a function from a sum of terms based on the derivatives of the function at a point. The advantage of this procedure for biological vision is that it extends the representation of visual information to before and after the immediate present. The visual system can also use it as a simple way

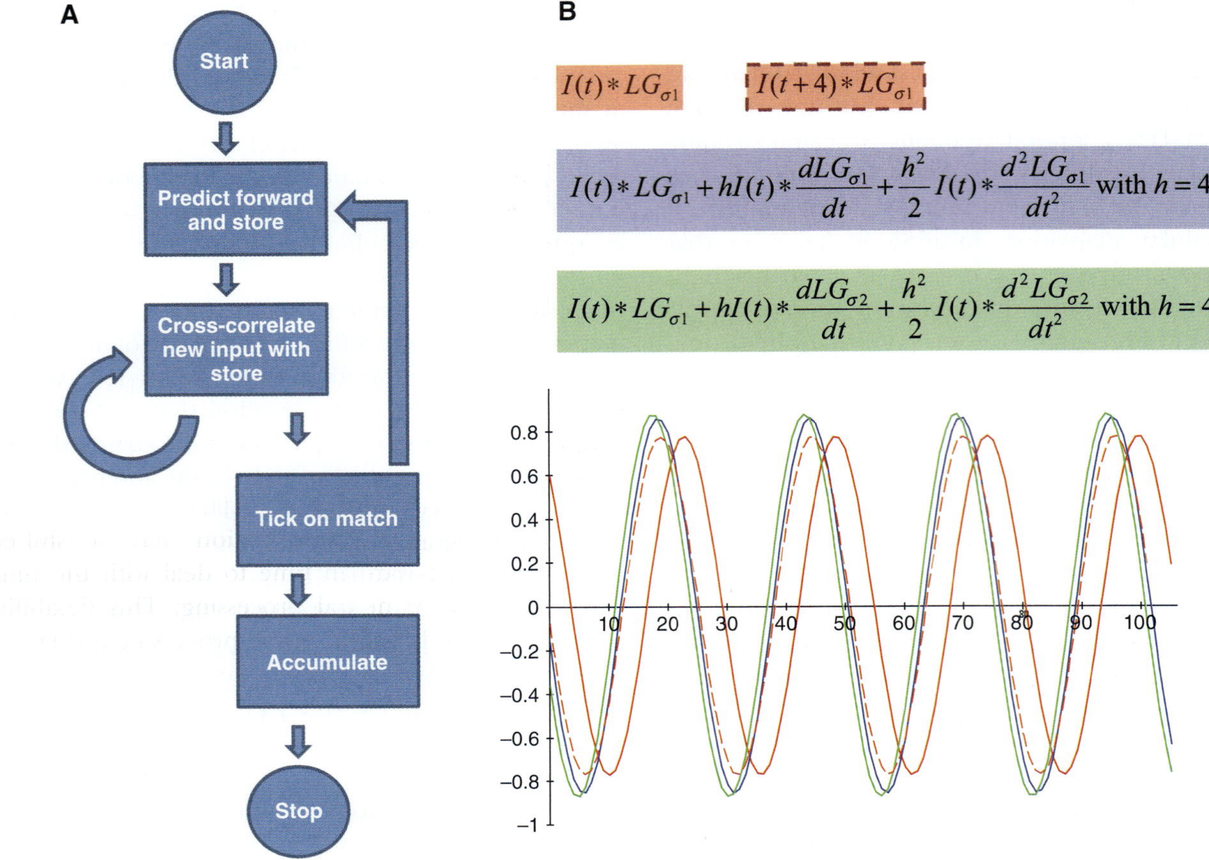

FIGURE 53.2   (A) A content-dependent clock. The free-running pacemaker from the standard clock model is replaced with a "predict and compare" circuit. The Taylor series is used to predict an image-based temporal sequence at some future time at a point or region in space. This forward prediction is stored. The new input is cross-correlated with the stored data in a temporal buffer. When the cross-correlation peaks, the clock ticks, the tick is accumulated, and the prediction is reset. The interval over which the prediction is made could vary, and if it does vary, the number of ticks may be scaled by the length of the interval over which the prediction is made (e.g., 10×30-ms ticks). (B) The bold solid line is the profile of a filtered (by the log Gaussian) sine function. The bold dashed line is the solid line time shifted forward by 4 temporal units. The faint solid line is the prediction constructed from the Taylor expansion. The predicted signal is slightly increased in amplitude at this frequency; however, the temporal shift is quite accurate. The faint dotted line is the prediction, with a sharpened temporal impulse response in the M-cell filters, as a response to temporal adaptation. The advance in the filter leads to a phase advance in the derivative functions, which would have the same effect as predicting forward with a larger excursion parameter.

of predicting forward (or backward) in time. The parameter h in the Taylor series determines the direction and amount of predicted displacement. We can substitute a range for h instead of an instantaneous value to predict a displaced temporal sequence (figure 53.2B). This construction would allow a forward or reverse prediction of the image brightness through time. Spatial differentiating filters allow the same trick to be played in the spatial domain. Note that the prediction is just the weighted sum of parvocellular and magnocellular outputs.

We can now build a content-sensitive clock in the following way (figure 53.2A). First, predict the current image brightness sequence forward for, let us say, 30 ms using the magnocellular signal and store it in a temporal buffer. Next, cross-correlate the current parvocellular-based sequence with the stored sequence. The new input has to be filtered by the zero-order kernel, otherwise, the phase of the signal will be quite different from the stored signal. After 30 ms of comparing prediction and input, the cross-correlation should peak and, at that point, we determine that 30 ms has passed and reset the prediction. The number of 30-ms resettings, which we could think of as clock ticks, are then accumulated as in the standard stopwatch model until the stimulus has completed, at which point the stimulus duration is read out. Note that the parameter that determines the time shift is h, which only multiplies the magnocellular outputs. The magnocellular pathway controls the temporal prediction.

Only magnocellular neurons show significant adaptation to contrast. However, changes in magnocellular sensitivity after adaptation cannot explain time compression. A relatively lower magnocellular signal, as a result of adaptation, would have the same effect as a smaller value of the excursion parameter h, leading to more resetting, more ticks, and a longer duration. A higher magnocellular signal should lead to compression. Thus, an analysis based on sensitivity appears to provide a prediction that is opposite to our observations.

A change in sensitivity is just one aspect of contrast-gain control; the other is that the phase of signals carried by magnocellular neurons is advanced though adaptation (Benardete & Kaplan, 1999). This phase shift is seen in magnocellular but not parvocellular neurons (Benardete & Kaplan, 1999). The phase shift will shift the predicted sine wave forward in time, which is equivalent to having a larger h parameter. This should lead to resetting after a longer delay, leading to fewer ticks, and therefore time compression. The greater the magnocellular phase advance relative to the parvocellular phase, the greater will be the time compression. Although this forward model, or "predict and compare" strategy, could be applied generally to time perception, in the current context the key theoretical elements that link duration perception to the magnocellular pathway are the identification of a role for the magnocellular system in generating a forward prediction through a Taylor expansion, the recognition of the role of magnocellular cells as temporal differentiators, which is also central to their role in motion computation (Johnston, McOwan, & Benton, 1999; Johnston, McOwan, & Benton, 2003; Johnston, McOwan, & Buxton, 1992), the magnocellular system's greater sensitivity to adaptation in comparison to the parvocellular system, the dual change in sensitivity and sharpening of the impulse response, and, specifically, the phase advance caused by the shortening of the temporal impulse response as a result of temporal frequency adaptation and contrast gain control.

An alternative content-based approach to timing has recently been proposed by Ahrens and Sahani (2011). In their model the size of the difference in some image measure is expected to increase with elapsed time between the samples. Given sufficient processes and an assumption of smoothness of natural images the distribution of these differences should increase with time. They show that duration discrimination is better for intervals with dynamic content than for intervals with static content, which supports the proposal that image content can be used to judge duration. Network dynamics has also been considered as providing a timing signal (Buonomano & Laje, 2010; Buonomano & Merzenich, 1995). In this case the pattern over network nodes is taken as a signature of the elapsed time. The advantage of stochastic and network models is that timing information is intrinsic and distributed; however, the challenge for this approach is to identify how the information might be made explicit depending on task requirements.

SUMMARY AND CONCLUSIONS

The characterization of visual temporal filters as differential operators points to a role for early visual processes in prediction as well as shaping contrast sensitivity. Prediction allows a more labile flexible view of temporal representation and leads us to consider mechanisms by which information may be shifted forward and backward in time to deal with the time delays inherent in neural processing. This flexibility may require synchronous neural processing giving rise to temporal fluctuations in perceptual representation. Prediction also has a role to play in duration encoding through a "predict and compare" content-dependent clock. Temporal prediction just requires a measure of the current value of some representation and its rate of change, making the ideas developed here generalizable to many domains and levels of visual processing. The theory provides a role for bandpass temporal filters like those in the magnocellular system in predictive vision as well as contrast detection and motion processing.

REFERENCES

Ahrens, M. B., & Sahani, M. (2011). Observers exploit stochastic models of sensory change to help judge the passage of time. *Current Biology, 21*, 200–206. doi:10.1016/j.cub.2010.12.043.

Amano, K., Arnold, D. H., Takeda, T., & Johnston, A. (2008). Alpha band amplification during illusory jitter perception. *Journal of Vision, 8*(10), 1–8. doi:10.1167/8.10.3.

Arnold, D. H., & Johnston, A. (2003). Motion-induced spatial conflict. *Nature, 425*, 181–184.

Arnold, D. H., & Johnston, A. (2005). Motion induced spatial conflict following binocular integration. *Vision Research, 45*, 2934–2942.

Ayhan, I., Bruno, A., Nishida, S., & Johnston, A. (2009). The spatial tuning of adaptation-based time compression. *Journal of Vision, 9*(11), 1–12. doi:10.1167/9.11.2.

Ayhan, I., Bruno, A., Nishida, S., & Johnston, A. (2010). The effect of luminance signal on adaptation-based duration compression. *Journal of Vision, 10*(7), 1412. doi:10.1167/10.7.1412.

Benardete, E. A., & Kaplan, E. (1999). The dynamics of primate M retinal ganglion cells. *Visual Neuroscience, 16*, 355–368.

Brown, S. W. (1995). Time, change, and motion: The effects of stimulus movement on temporal perception. *Perception & Psychophysics, 57*, 105–116.

Bruno, A., Ayhan, I., & Johnston, A. (2010). Retinotopic adaptation-based visual duration compression. *Journal of Vision, 10*(10), 30. doi:10.1167/10.10.30.

Bruno, A., Ayhan, I., & Johnston, A. (2011). Duration expansion at low luminance levels. *Journal of Vision, 11*(14), 13. doi:10.1167/11.14.13.

Bruno, A., Ayhan, I., & Johnston, A. (2012). Effects of temporal features and order on the apparent duration of a visual stimulus. *Frontiers in Psychology, 3*. doi:10.3389/fpsyg.2012.00090.

Bruno, A., & Johnston, A. (2010). Contrast gain shapes visual time. *Frontiers in Psychology, 1*, 12. doi:10.3389/fpsyg.2010.00170.

Bullier, J. (2001). Integrated model of visual processing. *Brain Research. Brain Research Reviews, 36*, 96–107.

Buonomano, D. V., & Laje, R. (2010). Population clocks: Motor timing with neural dynamics. *Trends in Cognitive Sciences, 14*, 520–527. doi:10.1016/j.tics.2010.09.002.

Buonomano, D. V., & Merzenich, M. M. (1995). Temporal information transformed into a spatial code by a neural network with realistic properties. *Science, 267*, 1028–1030.

Burr, D. C., Cicchini, G. M., & Arrighi, R. (2011). Spatiotopic selectivity of adaptation-based compression of event duration. *Journal of Vision, 11*(2), 21. doi:10.1167/11.2.21.

Burr, D. C., & Morrone, M. C. (1993). Impulse-response functions for chromatic and achromatic stimuli. *Journal of the Optical Society of America. A, Optics and Image Science, 10*, 1706–1713. doi:10.1364/JOSAA.10.001706.

Burr, D. C., & Morrone, M. C. (1996). Temporal impulse response functions for luminance and colour during saccades. *Vision Research, 36*, 2069–2078.

Burr, D. C., Morrone, M. C., & Ross, J. (1994). Selective suppression of the magnocellular visual pathway during saccadic eye movements. *Nature, 371*, 511–513.

Burr, D. C., Tozzi, A., & Morrone, M. C. (2007). Neural mechanisms for timing visual events are spatially selective in real-world coordinates. *Nature Neuroscience, 10*, 423–425.

Busch, N. A., & VanRullen, R. (2010). Spontaneous EEG oscillations reveal periodic sampling of visual attention. *Proceedings of the National Academy of Sciences of the United States of America, 107*, 16048–16053. doi:10.1073/pnas.1004801107.

Carlson, V. R. (1962). Adaptation in the perception of visual velocity. *Journal of Experimental Psychology, 64*, 192–197.

Cass, J., & Alais, D. (2006). Evidence for two interacting temporal channels in human visual processing. *Vision Research, 46*, 2859–2868. doi:10.1016/j.visres.2006.02.015.

Cass, J., Clifford, C. W., Alais, D., & Spehar, B. (2009). Temporal structure of chromatic channels revealed through masking. *Journal of Vision, 9*(5), 11–15. doi:10.1167/9.5.17.

Cavanagh, P., Holcombe, A. O., & Chou, W. (2008). Mobile computation: Spatiotemporal integration of the properties of objects in motion. *Journal of Vision, 8*(12), 1–23. doi:10.1167/8.12.1.

Cavanagh, P., Tyler, C. W., & Favreau, O. E. (1984). Perceived velocity of moving chromatic gratings. *Journal of the Optical Society of America. A, Optics and Image Science, 1*, 893–899.

Coull, J. T., Cheng, R. K., & Meck, W. H. (2010). Neuroanatomical and neurochemical substrates of timing. *Neuropsychopharmacology, 36*, 3–25.

Creelman, C. (1962). Human discrimination of auditory duration. *Journal of the Acoustical Society of America, 34*, 582–593.

Curran, W., & Benton, C. P. (2012). The many directions of time. *Cognition, 122*, 252–257. doi:10.1016/j.cognition.2011.10.016.

Eskew, R. T., Jr., Stromeyer, C. F., III, & Kronauer, R. E. (1994). Temporal properties of the red-green chromatic mechanism. *Vision Research, 34*, 3127–3137.

Fujisaki, W., & Nishida, S. (2010). A common perceptual temporal limit of binding synchronous inputs across different sensory attributes and modalities. *Proceedings of the Royal Society B: Biological Sciences, 277*, 2281–2290. doi:10.1098/rspb.2010.0243.

Fujisaki, W., Shimojo, S., Kashino, M., & Nishida, S. (2004). Recalibration of audiovisual simultaneity. *Nature Neuroscience, 7*, 773–778.

Hawken, M. J., Shapley, R. M., & Grosof, D. H. (1996). Temporal-frequency selectivity in monkey visual cortex. *Visual Neuroscience, 13*, 477–492.

Hess, R. F., & Plant, G. T. (1985). Temporal frequency discrimination in human vision: Evidence for an additional mechanism in the low spatial and high temporal frequency region. *Vision Research, 25*, 1495–1500.

Hess, R. F., & Snowden, R. J. (1992). Temporal properties of human visual filters: Number, shapes and spatial covariation. *Vision Research, 32*, 47–60.

Hogendoorn, H., Verstraten, F., & Johnston, A. (2010). Spatially localised time shifts of the perceptual stream. *Frontiers in Psychology, 1*, 12. doi:10.3389/fpsyg.2010.00181.

Holcombe, A. O. (2009). Seeing slow and seeing fast: two limits on perception. *Trends in Cognitive Sciences, 13*, 216–221. doi:10.1016/j.tics.2009.02.005.

Holcombe, A. O., & Cavanagh, P. (2001). Early binding of feature pairs for visual perception. *Nature Neuroscience, 4*(2), 127–128. doi:10.1038/83945.

Holcombe, A. O., & Cavanagh, P. (2008). Independent, synchronous access to color and motion features. *Cognition, 107*(2), 552–580. doi:10.1016/j.cognition.2007.11.006.

Johnston, A. (2010). Modulation of time perception by visual adaptation. In A. C. Nobre & J. T. Coull (Eds.), *Attention and time* (pp. 187–200). Oxford: Oxford University Press.

Johnston, A., Arnold, D., & Nishida, S. (2006). Spatially localized distortions of event time. *Current Biology, 16*, 472–479.

Johnston, A., Bruno, A., & Ayhan, I. (2011). Retinotopic selectivity of adaptation-based compression of event duration: Reply to Burr, Cicchini, Arrighi, and Morrone. *Journal of Vision, 11*(2), 21a. doi:10.1167/11.2.21a.

Johnston, A., Bruno, A., Watanabe, J., Quansah, B., Patel, N., Dakin, S., et al. (2008). Visually-based temporal distortion in dyslexia. *Vision Research, 48*, 1852–1858. doi:10.1016/j.visres.2008.04.029.

Johnston, A., & Clifford, C. W. (1995). A unified account of three apparent motion illusions. *Vision Research, 35*, 1109–1123.

Johnston, A., McOwan, P. W., & Benton, C. P. (1999). Robust velocity computation from a biologically motivated model of motion perception. *Proceedings. Biological Sciences, 266*, 509–518.

Johnston, A., McOwan, P. W., & Benton, C. P. (2003). Biological computation of image motion from flows over boundaries. *Journal of Physiology, Paris, 97*, 325–334.

Johnston, A., McOwan, P. W., & Buxton, H. (1992). A computational model of the analysis of some first-order and second-order motion patterns by simple and complex cells. *Proceedings. Biological Sciences, 250,* 297–306.

Johnston, A., & Nishida, S. (2001). Time perception: Brain time or event time? *Current Biology, 11*(11), R427–R430.

Kanai, R., Paffen, C. L. E., Hogendoorn, H., & Verstraten, F. A. J. (2006). Time dilation in dynamic visual display. *Journal of Vision, 6*(12), 1421–1430. doi:10.1167/6.12.8.

Kaneko, S., & Murakami, I. (2009). Perceived duration of visual motion increases with speed. *Journal of Vision, 9*(7), 14. doi:10.1167/9.7.14.

Kelly, D. H. (1961). Visual responses to time-dependent stimuli. I. Amplitude sensitivity measurements. *Journal of the Optical Society of America, 51,* 422–429. doi:10.1364/JOSA.51.000422.

Kelly, D. H. (1979). Motion and vision. II. Stabilized spatiotemporal threshold surface. *Journal of the Optical Society of America, 69,* 1340–1349.

Kitazawa, S., Moizumi, S., Okuzumi, A., Saito, F., Shibuya, S., Takahashi, T., et al. (2007). Reversal of subjective temporal order due to sensory and motor interactions. In P. Haggard, M. Kawato, & Y. Rosetti (Eds.), *Attention and performance XXII* (pp. 73–97). Oxford: Oxford University Press.

Kline, K., Holcombe, A. O., & Eagleman, D. M. (2004). Illusory motion reversal is caused by rivalry, not by perceptual snapshots of the visual field. *Vision Research, 44,* 2653–2658. doi:10.1016/j.visres.2004.05.030.

Koenderink, J. J. (1988). Scale-time. *Biological Cybernetics, 58,* 159–162.

Lee, J., Williford, T., & Maunsell, J. H. (2007). Spatial attention and the latency of neuronal responses in macaque area V4. *Journal of Neuroscience, 27,* 9632–9637. doi:10.1523/JNEUROSCI.2734-07.2007.

Mandler, M. B., & Makous, W. (1984). A three channel model of temporal frequency perception. *Vision Research, 24,* 1881–1887.

Mante, V., Bonin, V., & Carandini, M. (2008). Functional mechanisms shaping lateral geniculate responses to artificial and natural stimuli. *Neuron, 58,* 625–638. doi:10.1016/j.neuron.2008.03.011.

Metha, A. B., & Mullen, K. T. (1996). Temporal mechanisms underlying flicker detection and identification for red-green and achromatic stimuli. *Journal of the Optical Society of America. A, Optics, Image Science, and Vision, 13,* 1969–1980.

Morgan, M. J., Giora, E., & Solomon, J. A. (2008). A single "stopwatch" for duration estimation, a single "ruler" for size. *Journal of Vision, 8*(2), 11–18. doi:10.1167/8.2.14/8/2/14.

Morrone, M. C., Ross, J., & Burr, D. C. (2005). Saccadic eye movements cause compression of time as well as space. *Nature Neuroscience, 8,* 950–954.

Motoyoshi, I., & Nishida, S. (2001). Temporal resolution of orientation-based texture segregation. *Vision Research, 41,* 2089–2105.

Moutoussis, K., & Zeki, S. (1997). Functional segregation and temporal hierarchy of the visual perceptive systems. *Proceedings. Biological Sciences, 264,* 1407–1414.

Nishida, S., & Johnston, A. (2002). Marker correspondence, not processing latency, determines temporal binding of visual attributes. *Current Biology, 12,* 359–368.

Nishida, S., & Johnston, A. (2010). Time marker theory of cross-channel temporal binding. In R. Nijhawan & B. Khurana (Eds.), *Problems of space and time in perception and action* (pp. 278–300). Cambridge: Cambridge University Press.

Purves, D., Paydarfar, J. A., & Andrews, T. J. (1996). The wagon wheel illusion in movies and reality. *Proceedings of the National Academy of Sciences of the United States of America, 93,* 3693–3697.

Rao, R. P., & Ballard, D. H. (1999). Predictive coding in the visual cortex: A functional interpretation of some extra-classical receptive-field effects. *Nature Neuroscience, 2,* 79–87. doi:10.1038/4580.

Roach, N. W., McGraw, P. V., & Johnston, A. (2011). Visual motion induces a forward prediction of spatial pattern. *Current Biology, 21,* 740–745. doi:10.1016/j.cub.2011.03.031.

Schouten, J. F. (1967). Subjective stroboscopy and a model of visual movement detectors. In W. Wathen-Dunn (Ed.), *Models for the perception of speech and visual form* (pp. 44–45). Cambridge, MA: MIT Press.

Shapley, R. M., & Victor, J. D. (1978). The effect of contrast on the transfer properties of cat retinal ganglion cells. *Journal of Physiology, 285,* 275–298.

Smith, A. T., & Edgar, G. K. (1994). Antagonistic comparison of temporal frequency filter outputs as a basis for speed perception. *Vision Research, 34,* 253–265.

Solomon, S. G., Peirce, J. W., Dhruv, N. T., & Lennie, P. (2004). Profound contrast adaptation early in the visual pathway. *Neuron, 42,* 155–162.

Stromeyer, C. F. III, Chaparro, A., Tolias, A. S., & Kronauer, R. E. (1997). Colour adaptation modifies the long-wave versus middle-wave cone weights and temporal phases in human luminance (but not red-green) mechanism. *Journal of Physiology, 499*(Pt 1), 227–254.

Tailby, C., Dobbie, W. J., Solomon, S. G., Szmajda, B. A., Hashemi-Nezhad, M., Forte, J. D., et al. (2010). Receptive field asymmetries produce color-dependent direction selectivity in primate lateral geniculate nucleus. *Journal of Vision, 10*(8), 1. doi:10.1167/10.8.1.

Tailby, C., Solomon, S. G., & Lennie, P. (2008). Functional asymmetries in visual pathways carrying S-cone signals in macaque. *Journal of Neuroscience, 28,* 4078–4087. doi:10.1523/JNEUROSCI.5338-07.2008.

Terao, M., Watanabe, J., Yagi, A., & Nishida, S. (2008). Reduction of stimulus visibility compresses apparent time intervals. *Nature Neuroscience, 11,* 541–542. doi:10.1038/nn.2111.

Thompson, P. (1981). Velocity after-effects: The effects of adaptation to moving stimuli on the perception of subsequently seen moving stimuli. *Vision Research, 21,* 337–345. doi:0042–6989(81)90161–9.

Treisman, M. (1963). Temporal discrimination and the indifference interval. Implications for a model of the "internal clock." *Psychological Monographs, 77,* 1–31.

Ullman, S. (1984). Visual routines. *Cognition, 18,* 97–159.

VanRullen, R., Reddy, L., & Koch, C. (2005). Attention-driven discrete sampling of motion perception. *Proceedings of the National Academy of Sciences, 102,* 5291–5296.

VanRullen, R., Reddy, L., & Koch, C. (2006). The continuous wagon wheel illusion is associated with changes in electroencephalogram power at approximately 13 Hz. *Journal of Neuroscience, 26,* 502–507.

Watson, A. B., & Robson, J. G. (1981). Discrimination at threshold: Labelled detectors in human vision. *Vision Research, 21,* 1115–1122.

Wiener, M., Turkeltaub, P., & Coslett, H. B. (2010). The image of time: A voxel-wise meta-analysis. *NeuroImage, 49*, 1728–1740. doi:10.1016/j.neuroimage.2009.09.064.

Xu, X., Ichida, J., Shostak, Y., Bonds, A. B., & Casagrande, V. A. (2002). Are primate lateral geniculate nucleus (LGN) cells really sensitive to orientation or direction? *Visual Neuroscience, 19*, 97–108.

Yamamoto, S., & Kitazawa, S. (2001). Reversal of subjective temporal order due to arm crossing. *Nature Neuroscience, 4*, 759–765. doi:10.1038/89559.

Ye, X., Li, G., Yang, Y., & Zhou, Y. (2009). The effect of orientation adaptation on responses of lateral geniculate nucleus neurons with high orientation bias in cats. *Neuroscience, 164*, 760–769. doi:10.1016/j.neuroscience.2009.08.016.

54 Motion Perception: Human Psychophysics

DAVID BURR

Perceiving motion is a fundamental skill of any visual system: to analyze the form and velocity of moving objects; to avoid collision with moving masses; to navigate through our environment; to analyze the three-dimensional structure of the world we move through; and much more. Much progress has been made over the past few decades to learn how humans and other animals analyze visual signals of objects in motion. This chapter concentrates primarily on advances in human psychophysics. For an excellent review of imaging studies of human and nonhuman primates—and the homologies between them—the interested reader is referred to the excellent chapter (chapter 55) by Orban and Jastorff.

Over the last few decades many important conceptual and empirical advances have enormously expanded our understanding of the principles behind motion perception. Advances in the psychophysics of motion have been accompanied by important breakthroughs in physiology. This chapter concentrates on the main advances in motion psychophysics that have contributed to our understanding of human visual motion perception.

MOTION DETECTORS CONSIDERED AS SPATIOTEMPORAL FILTERS

One of the more important conceptual leaps for motion research was to apply the powerful technique of Fourier analysis to show how suitably tuned spatiotemporal filters can model motion perception (Adelson & Bergen, 1985; Burr, Ross, & Morrone, 1986; van Santen & Sperling, 1985; Watson & Ahumada, 1985). The details of the various models are probably less important than the general message they all convey: that many aspects of motion, thought to be mysterious, are well explained in the frequency domain.

Figure 54.1 illustrates the three stages of one representative model (Adelson & Bergen, 1985). It starts with spatiotemporal filters that integrate the motion input over space (A) and time (B). The outputs of these filters are combined by "quadrature pairing," a technique that produces direction selectivity (C). The outputs of the two classes of filters (sine and cosine) are then squared and summed together (C), to yield a smooth response, selective for direction and also weakly selective for speed. Figure 54.1D shows the resulting frequency response, clearly selective to a specific range of velocities.

This model describes perception of real motion and also accounts for many other phenomena, including *apparent* or *sampled* motion, previously thought to reflect separate processes (e.g., Kolers, 1972): The integration in space and in time causes the discrete motion sequence to become continuous. It also explains some motion illusions, including the "fluted square wave" illusion (Adelson & Bergen, 1985) and the reverse-phi illusion of Anstis (1970). In both cases the explanation of the illusions is that the stimuli contain motion energy in the direction in which they are perceived, even though this is not obvious without analyzing the spatiotemporal frequency spectrum. Interestingly, the reverse-phi illusion has recently been extended to demonstrate different transmission times of ON- and OFF-luminance channels (Del Viva, Gori, & Burr, 2006), again taking advantage of the fact that this illusion has spatiotemporal energy corresponding to the perceived direction of motion. Frequency-based models also provide the basis for explaining many illusions discovered more recently, such as Pinna and Brelstaff's (2000) powerful illusion.

Considering motion as spatiotemporal energy was an important conceptual breakthrough, but it is important to note that these more recent models build on the pioneering work of Werner Reichardt (1957, 1961), which compares the output from one part of space with the delayed output of another. Two such units operate together, mutually inhibiting each other to eliminate a response to flashes. The original Reichardt detector had no filters, just sampling two points of the retina and a simple delay line, although later adaptations included spatiotemporal filtering (Egelhaaf et al., 1988).

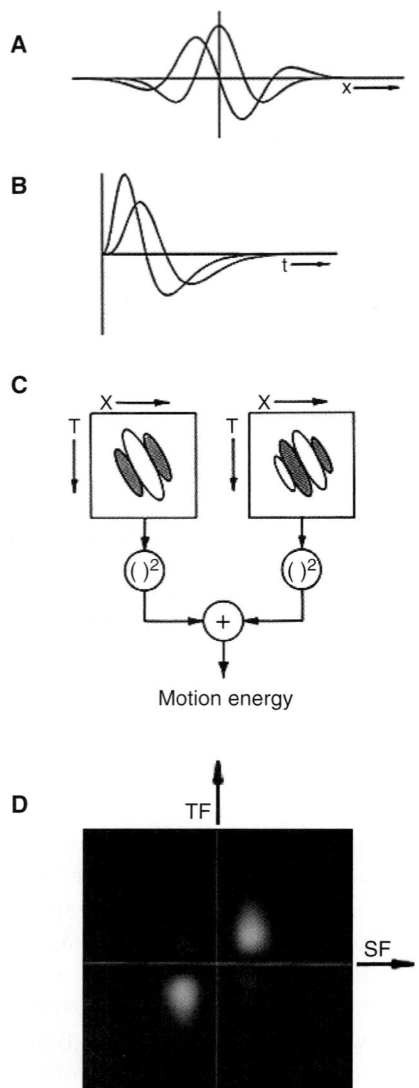

FIGURE 54.1   Constructing a spatiotemporally tuned motion detector. (A, B) The models of Adelson and Bergen (1985), Watson and Ahumada (1985), and van Santen and Sperling (1985) all start with separable operators (or impulse response functions) tuned in space (A) and in time (B), each both in sine and in cosine phase. Each spatial operator is multiplied with each temporal operator to yield four separate spatiotemporal impulse response functions of different phases. (C) Appropriate subtractive combination of these separable spatiotemporal impulse response functions yields two "quadrature pairs" of linear filters (Watson & Ahumada, 1985), oriented in space-time (hence selective to motion direction). In Adelson and Bergen's model these are combined after squaring to yield a phase-independent measure that is known as "motion energy." The full detector has another quadrature pair tuned to the opposite direction, which combines subtractively to enhance direction selectivity (and inhibition responsiveness to nondirected flashes). (D) The spatiotemporal energy spectrum of the motion detector in C. Responding only to one quadrant of spatiotemporal frequency gives the direction selectivity and broad selectivity to speed. (Reproduced with permission from Adelson & Bergen, 1985.)

Burr and colleagues (Burr & Ross, 1986; Burr, Ross, & Morrone, 1986) measured the characteristics of the spatiotemporal filters by the psychophysical technique of masking and used these results to account for how the form of moving objects is perceived. To make the results more intuitive they inverse-Fourier transformed the filter from frequency space to space-time, introducing the concept of the spatiotemporal receptive field, oriented in space-time (see figure 54.2). This representation makes obvious many of the phenomena that seemed mysterious, such as "motion smear" (Burr, 1980), "spatiotemporal interpolation" (Burr, 1979), and seemingly unrelated phenomena such as metacontrast (Burr, 1984). Interestingly, many similar issues have been reemerging recently (e.g., Boi et al., 2009), and it seems that these illusions too can be explained, both qualitatively and quantitatively, by receptive fields oriented in space-time (Pooresmaeili et al., 2012).

SECOND-ORDER, HIGHER-ORDER, AND FEATURE-TRACKING MOTION

"Second-order" motion was first demonstrated by David Badcock and colleagues (Badcock & Derrington, 1985; Derrington & Badcock, 1985; Derrington & Henning, 1987) with complex gratings comprising two drifting harmonics that caused "beats" as they came into and out of phase. The apparent direction of motion of these stimuli could vary, either in the physical direction of motion, as predicted by energy models, or in the direction of the beats (which contain no energy in Fourier space that would excite the energy models), and could not be explained by trivial nonlinearities such as distortion products (Badcock & Derrington, 1989). This class of motion stimulus, which contains no energy in the Fourier plane describing the direction of perceived motion, has variously been called "non-Fourier motion," "second-order motion" (a more correct term than had prevailed), higher-order motion, and sometimes "feature motion."

Zanker (1990, 1993) devised another motion stimulus, which he coined "theta motion," motion of motion-defined forms, for example, leftward drifting dots confined to a rectangular region that was itself drifting rightward. But second-order motion is most often associated with Chubb and Sperling (1988), who devised a clever series of "drift-balanced" stimuli that have no directed motion energy that energy detectors would pick up but are perceived clearly to move in one direction or another. They developed a simple model that will detect second-order motion, mainly because of a nonlinear rectifying stage after the linear filters, which renders the output visible to an energy-extraction stage.

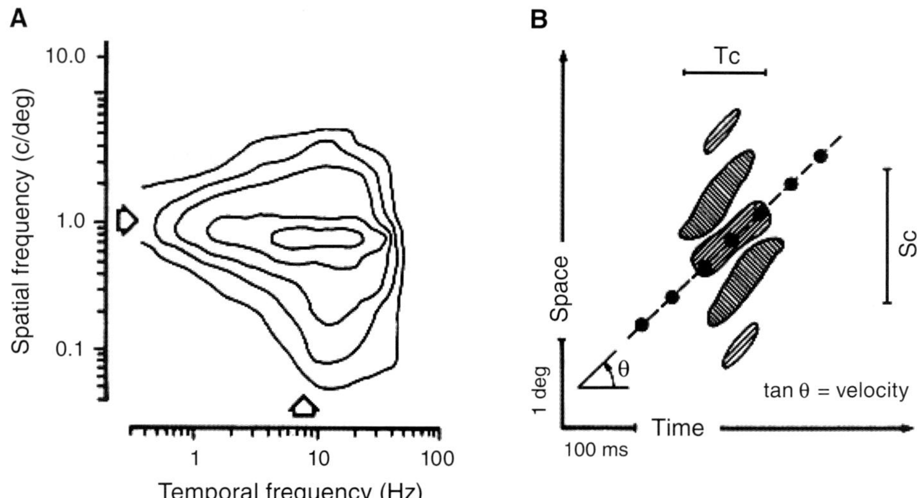

FIGURE 54.2   (A) Spatiotemporal tuning of a hypothetical unit of the human motion system measured by the technique of "masking" (Burr, Ross, & Morrone, 1986). The function is tuned to 1 cycle/deg, 8 Hz, and falls off steadily away from the peak (contour lines represent 0.5 log-unit attenuation). (B) Spatiotemporal receptive field derived from the filter (assuming linear phase). Forward cross-hatching represents excitatory regions; back cross-hatching represents inhibitory regions. The orientation in space-time means it has a preferred velocity, both direction and speed. Spatiotemporal operators of this sort (inferred from all the filter-based motion models of the mid-1980s) go a long way toward explaining many phenomena such as integrating the path of sampled motion (indicated by the series of dots) so it is perceived as smooth and "spatiotemporal interpolation" (see Burr & Ross, 1986). They also help to explain why we do not see the world to be as smeared as may be expected from a "camera" analogy. The field extends for over 100 ms in time (indicated by symbol T_C) and may be expected to smear targets by this amount. However, the analysis is not in this direction but orthogonal to the long axis of the receptive field, where the spread in space-time is considerably less.

There is still some debate on whether second-order motion requires a functionally distinct system for its analysis or whether both could be subserved by the same system. For example, Taub, Victor, and Conte (1997) claim that the most parsimonious explanation is that both types of motion are detected by a common mechanism, with a simple rectifying nonlinearity at the front end to convert the "non-Fourier" into "Fourier" motion energy (see also Cavanagh & Mather, 1989).

But evidence also exists for separate systems. Animation sequences that require integration of first-order and second-order frames do not give rise to unambiguous motion (Ledgeway & Smith, 1994; Mather & West, 1993). There are also qualitative differences between the two types of motion: Contrast thresholds for identifying motion direction are higher (relative to detection) for second-order than for first-order motion (Smith, Snowden, & Milne, 1994), as are temporal-frequency thresholds (Derrington, Badcock, & Henning, 1993; Smith & Ledgeway, 1998). Perhaps the strongest evidence is neuropsychological, as several patients have been described with selective impairment of either first- or second-order motion (Greenlee & Smith, 1997; Vaina & Cowey, 1996; Vaina & Soloviev, 2004).

In addition, another class of motion has been proposed, variously termed "third-order" (Lu & Sperling, 1995a, 1995b, 2001) or "attentional" motion (Cavanagh, 1992; Verstraten, Cavanagh, & Labianca, 2000). Third-order motion is thought to depend on psychological attributes of the stimuli, such as attention or "salience" (the probability that the image will be perceived as "figure" rather than "ground"; Lu & Sperling, 2001), so a perceptually salient figure is seen to move over a background. Examples can be constructed to which both the first- and second-order systems are blind, such as a moving stimulus that continually changes in orientation, contrast, or chromaticity. Interestingly, changing the salience of equiluminant drifting gratings causes activation of the inferior parietal lobule (IPL), implicating a different area in the analysis of third-order motion, one that is involved in attention (Claeys et al., 2003).

Attention has also been implicated in describing higher-order motion. Cavanagh introduced a new class of motion stimuli that he termed *attentional motion stimuli* (in many respects similar to Lu and Sperling's third-order motion stimuli). A typical example could be a luminance-modulated grating drifting in one direction with a superimposed chromatic-modulated grating drifting in the opposite direction: Attending to one or the other determines the direction of drift. Whether motion of this type is functionally distinct from third-order motion, or indeed whether either type really defines a unique class of motion, is, of course, subject to debate.

One question that many perplexed readers may wish to ask at this stage is, "What purpose does this second- and higher-order motion serve? When may we normally encounter a second-order motion stimulus modulated, say, in contrast but not luminance?" One approach has been to suggest that the higher-order motions represent a form of *feature tracking*, a system specialized to monitor the motion of salient features. This is reminiscent of Lu and Sperling's third-order motion but may in fact be a more general goal of motion mechanisms.

The early motion models of the Marr group were designed to track edges in two-dimensional motion (Hildreth, 1984; Marr & Ullman, 1981), and much experimental evidence is consistent with edge tracking or, more generally, feature tracking (Cavanagh & Mather, 1989; Derrington & Ukkonen, 1999; Morgan, 1992; Morgan & Mather, 1994; Seiffert & Cavanagh, 1998). Del Viva and Morrone (1998, 2006) developed a feature-tracking algorithm based on the "local energy" feature-detection algorithm (Morrone & Burr, 1988), which first detects salient features in scenes and then searches for peaks in space-time corresponding to the motion of these features. In some respects the model resembles Chubb and Sperling's (1988), in that the early nonlinearity converts the contrast features into energy detectable by basic Reichardt-type models. They show that their algorithm can predict qualitatively and quantitatively human perceptual performance on many interesting examples of motion stimuli that defy many other motion models. One key factor is "phase congruence" between harmonics of compound gratings in determining whether the harmonics will move as a block or be seen in transparency. Phase, also important for Fleet and Langley's (1994) model, has little effect on Fourier power but is fundamental in the formation of visually salient features.

SEGMENTATION AND INTEGRATION OF MOTION SIGNALS

A particularly challenging problem for motion perception is to understand when to integrate motion signals and when to segregate them (see Braddick, 1993, for an excellent discussion of this issue). Much evidence shows that motion mechanisms can integrate over a wide area. One clear example is what has been termed "motion capture" (Mackay, 1961; Ramachandran & Inada, 1984, 1985): A field of dynamic random dots moving in no clear direction can be "captured" by a moving frame or low-frequency grating or even a subjective contour to appear to move coherently. But motion mechanisms can also segregate, so that shapes defined solely by motion can stand out on stationary or reverse-moving backgrounds (Dick, Ullman, & Sagi, 1987; Julesz, 1971).

Nowhere is the conflicting requirement for segregation and integration more apparent than in the "aperture problem." Figure 54.3A illustrates the point. As a circle moves horizontally to the right, the local changes in the image can occur over a wide range of directions. Local measurements of motion (by neurons with small receptive fields) will all indicate motion perpendicular to the orientation of the edge passing through its field. To determine the true global motion of the object, local motions must be combined. The real problem here is for the system to know when to combine motions to yield a global percept of a moving object and when to segregate these motions to resolve a moving pattern from its background.

To tackle this problem Adelson and Movshon (1982) introduced the "plaid" stimulus—two sinusoidal gratings of different orientations moving, therefore, in different directions—and asked under what conditions the two gratings slide one over the other transparently, and when do they cohere into a single plaid pattern. They showed that, in vector space, the motion of each grating is consistent with a family of motions that lie along a line. Each motion has such a constraint line, and these two lines cross one another at "the intersection of constraints," which determines the single direction and speed of motion that can satisfy both components of the plaid (figure 54.3B). This notion predicted rather well the perceived direction of the plaids, even in the case where the intersection of constraints prediction is at odds with the vector sums of the components.

The intersection-of-constraints model has been questioned, with some evidence that the vector sum or vector average of the components gives a more accurate estimation of the direction of movement of the resulting plaid (Ferrera & Wilson, 1990; Wilson & Kim, 1994; Yo & Wilson, 1992). Other researchers have also suggested a role for local features such as blobs in the plaid patterns (e.g., Alais, Wenderoth, & Burke, 1994; Bowns & Alais, 2006). Although the details are still much under dispute, this research clearly exemplifies one of the real problems of motion perception: when to integrate components into a single motion and when to keep them segregated.

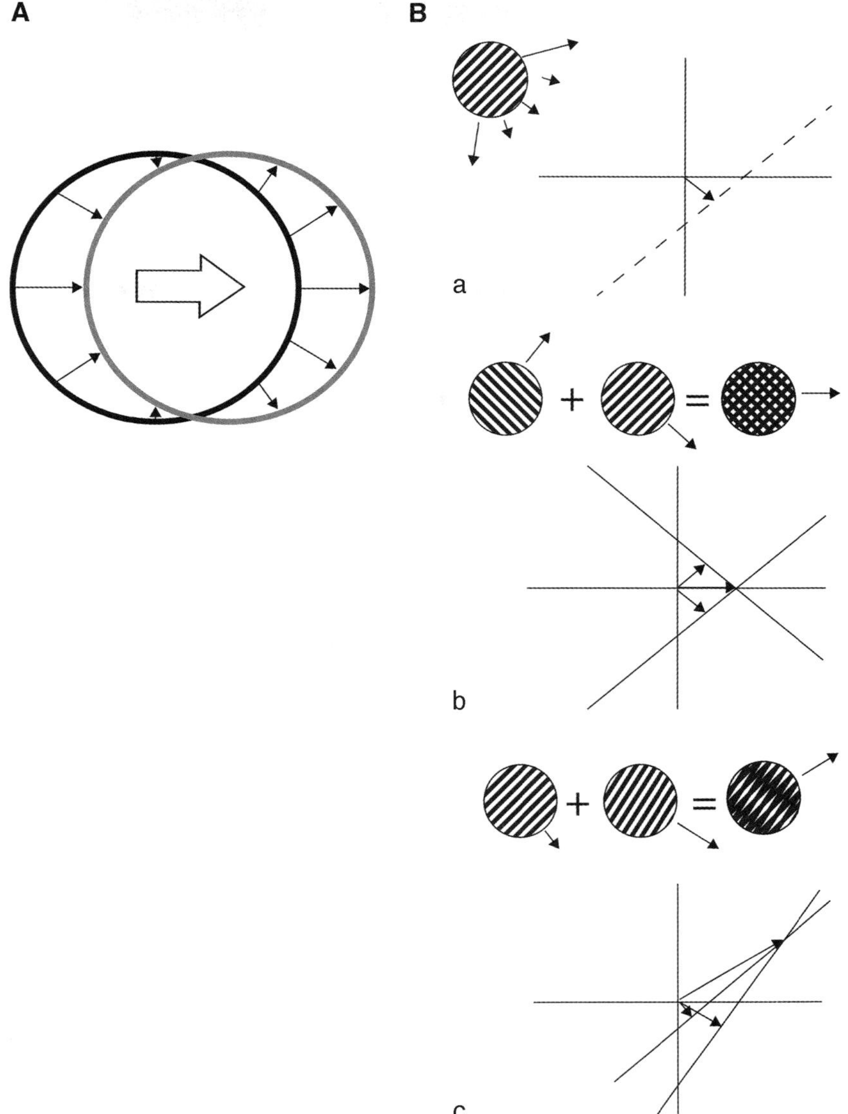

FIGURE 54.3   (A) Illustration of the aperture problem. When a circle moves horizontally, the local movement of the contours may be in a wide range of directions. If only the vector of motion perpendicular to the local edge orientation is seen, the range of motions of the circle will extend from vertically downward through rightward movement to vertically upward. This gamut of motion directions must be integrated to give the global movement. (B) The intersection of constraints model of plaid motion (from Adelson & Movshon, 1982). (a) A 45° grating with a motion vector perpendicular to its orientation is ambiguous in that the size of the vector of motion parallel to its orientation is not knowable. The dotted constraint line provides the locus of all the motion vectors. (b) When added to a second grating orthogonal to the first and moving upward to the right, a single point marks the intersection of the two constraint lines, which predicts correctly the perceived horizontal movement of the plaid. (c) A so-called type 2 plaid in which the intersection of constraints prediction of motion lies outside the component vectors. This prediction is, therefore, very different from a vector sum or vector average model of plaid motion.

The simplest mechanism for integration is the simple linear filtering incorporated into most modern models of motion energy detection, which blur together all signals falling within their receptive fields (figures 54.1 and 54.2). Psychophysical studies suggest that the size of the receptive fields of motion detectors increases with velocity and spatial frequency preference, can be quite large, up to 8° for low-frequency, fast-moving gratings (Anderson & Burr, 1987, 1991), and can extend over around 100 ms in time (Burr, 1981). But the situation is more complex than predicted by the spatial and temporal extent of the front-end filters. Motion signals from these front-end Reichardt-like detectors are combined at a later, intermediate stage of motion processing to mediate global perception of coherent motion of random-dot and other complex patterns (Bex & Dakin, 2002; Yang & Blake, 1994).

Coherence thresholds for random-dot patterns seem to tap into a higher level of processing. For example, detection thresholds for discriminating motion coherence improve with exposure duration up to 3 s, compared with 100–300 ms for contrast detection thresholds. The 100–300 ms agrees well with the temporal properties of neurons in primary visual cortex (Duysons et al., 1985; Tolhurst & Movshon, 1975), whereas integration times beyond 1 s are quite beyond what would be expected in primary cortex, implying the action of higher mechanisms such as prefrontal cortex and the functional link with area MT (Zaksas & Pasternak, 2006). Random-dot patterns also reveal spatial summation fields much larger than that revealed by contrast sensitivity measurements, up to 70° (Burr, Morrone, & Vaina, 1998), particularly for flow motion.

It is important to note that although the motion systems can summate over large regions, up to 70° for complex optic-flow motion (Burr, Morrone, & Vaina, 1998), the summation is not obligatory but under clear attentional control (Burr et al., 2009). When regions of moving stimuli are cued, the noncued regions can be ignored, even when the cued regions are not contiguous in space. This shows that the summation does not reflect a large, hard-wired receptive field of a high-level mechanism but flexible summation under attentional control. Indeed, there is evidence that summation between patches of motion stimuli is more effective than within a single contiguous patch of comparable area (Verghese & Stone, 1995, 1996).

Motion is a powerful cue for scene segregation: A field of dots moving coherently stands out clearly from a background comprised of stationary dots, forming a clear shape defined by motion information alone (Julesz, 1971). The resolution of motion as a cue to segregation is less than that of luminance but is nevertheless quite fine, in the order of 2' arc (Loomis & Nakayama, 1973; Nakayama et al., 1985; Regan & Hong, 1990). Motion-defined resolution varies with both filter frequency and image speed, with performance for unfiltered patterns moving at 1–4°/s yielding a stripe resolution of about 3' (Burr, McKee, & Morrone, 2006), similar to the smallest receptive size of motion units. However, "vernier acuity" under similar conditions was about 2' arc, only slightly better than grating acuity (whereas standard luminance-based vernier acuity is typically 3–10 times better than resolution). Imaging studies show that many areas, including motion areas V3a and V5, as well as V4, show shape selectivity for motion-defined contours (Braddick et al., 2000; Mysore et al., 2008).

Not only is motion integration under voluntary control, but it is strongly subject to contextual influences. Lorenceau and Alais (2001) devised a stimulus from a diamond figure orbiting behind an occluding surface. The local signals within the separate apertures have completely ambiguous direction, but the global impression is of an orbiting diamond outline. However, when the stimuli within the local windows are swapped (leaving the local stimulation patterns unaltered), the pattern is perceived as a sliding motion rather than as a rotating diamond. This demonstration provides a clear example of the resourcefulness of the system in integrating motion signals appropriately and also demonstrates the tight links between form and motion perception (often considered to be independent "modules"), where form provides a clear veto for motion integration in the absence of closure.

Tadin and colleagues (2003) have described a clever technique for investigating the neural mechanisms underlying the segregation of motion signals. They use a variant of the summation technique, varying the size of visual stimuli and measuring direction discrimination thresholds (by varying exposure duration). Their counterintuitive result is that, for high-contrast stimuli, increasing the size of the stimulus (over about 3°) decreases sensitivity for direction discrimination. Large stimuli are also less effective for inducing the motion-after effect. They suggest that these results reflect the action of center–surround neural mechanisms like those that have been described for area MT (Born et al., 2000; Born & Tootell, 1992; Raiguel et al., 1995) and MSTl (Eifuku & Wurtz, 1998): Large stimuli activate the inhibitory surround, weakening the response of these units.

MOTION TRANSPARENCY

An important practical example of selective segregation and integration is *motion transparency*, where a foreground field slides over a stationary or differently moving background. Here the visual system has to represent multiple motions in the same part of the visual field. However, not all stimuli with opposing local-motion signals are seen as transparent. Qian, Andersen, and Adelson (1994) devised a stimulus with two patterns of pseudorandomly positioned dots moving in opposite directions over the same region. When the patterns were constrained so the opposing motion signals were *locally* coincident (paired), there was no perceptual impression of transparency. To produce transparency the displays have to have locally unbalanced motion signals with some microregions containing motion in one direction, others in the other direction. Provided there are some regions with net motion in a given direction, the system can segregate these from those moving in the other direction and then integrate these disparate regions to yield one or more coherent surfaces moving in particular directions.

In a companion paper Qian and Andersen (1994) show how cells in V1 and MT respond to these patterns. In general, V1 cells do not distinguish between the stimulus conditions in which the dots of opposite motion direction were constrained to fall within a local region (paired) and those in which the patterns contained locally unbalanced signals (unpaired). MT cells, on the other hand, reliably distinguished between the two conditions, responding well only to the nonpaired stimuli. They suggest that this is consistent with a two-stage model. The first stage, like a simple Reichardt detector responding only to motion energy, corresponds well to the behavior of V1 cells. The second stage introduces local inhibition between opposing directions of motion within a local region, presumably for noise suppression and preventing flicker producing a sense of motion. fMRI studies reveal similar differences in humans: V1 responds more strongly to counterphase flicker (the sum of two opposed drifting gratings) than to a single drifting grating, but for MT the reverse is true (Heeger et al., 1999).

What remains to be explained, of course, is how the signals of directed motion—some leftward, others rightward—combine appropriately with each other to yield the impression of a surface in motion. This clearly recalls the idea of "common fate" of Wertheimer (1912). What it points to, however, is a very clear example of how the visual system needs to segregate stimuli on the basis of their direction of motion and then to integrate these same signals. No linear system can achieve both at the same time. Some intermediate nonlinearity—which we can describe as a *feature extraction*—is necessary.

There is also evidence (Del Viva & Morrone, 2006; Meso & Zanker, 2009) that transparency is determined by *phase congruency*, which to a large extent governs visually salient features (Morrone & Burr, 1988). When two extended patterns with clear features drift in opposite directions (for example, two square waves), those Fourier components in the composite, bidirectional stimulus that are not coherent in phase are seen to drift in transparency. To model the effect it is necessary to introduce an oriented spatiotemporal filter that operates after feature extraction (discussed above) and is selective to phase congruency. With this scheme, pooling of motion signals occurs between components

that produce features and segregation of the different transparent surfaces by analyzing along fixed directions the trajectories of the features.

APPEARANCE OF OBJECTS IN MOTION: MOTION BLUR AND SPEEDLINES

Many of the *modular* models of vision assume that form and motion are processed separately by different brain areas (e.g., Marr, 1982; Mishkin, Ungerleider, & Macko, 1983; Zeki, 1993). Although this may or may not be to some extent true (see, for example, Burr, 1999; Lennie, 1998), motion and form are clearly interconnected. The most obvious example is "biological motion" (see Blake & Shiffrar, 2007), where it is the motion itself that defines the form. But even for simple objects in motion, the mechanisms that analyze their form must be capable of taking the motion into account.

One very basic aspect of form analysis of moving objects is that they do not seem to be as smeared as would be expected on a simple "camera analogy" (Burr, 1980). Early visual mechanisms integrate information for around 100 ms, whether stationary (Barlow, 1958) or in motion (Burr, 1981). This integration may be expected to smear the images over time, like opening the shutter of a camera for this period. However, as mentioned earlier, motion mechanisms are tuned to the motion and hence are oriented in space-time (figure 54.2). This means that they do not simply integrate over time, but they integrate in the direction of the receptive field in space-time. The spatial structure of the image in motion is analyzed not normal to the space axis (as static objects would be) but normal to the axis of slant of the spatiotemporal receptive field (see also Burr & Ross, 1986, and figure 54.3). That is, they rotate space-time, effectively annulling the smearing effects of the motion. The relevant smear is not given by the duration over which these detectors spread but by the width normal to their axis. Detectors not tuned to the motion cannot do this and will cause smear much the same as a still camera will. Since these initial experiments a great deal of work has been done on motion smear, largely by Beddel and his group, showing that many factors contribute to smear, such as the presence of multiple rather than single targets (Chen, Bedell, & Ogmen, 1995) and pursuit eye movements (Bedell, Chung, & Patel, 2004; Tong, Stevenson, & Bedell, 2008).

There is another side to the motion-smear coin, one that has come to be known as *motion streaks* or *speedlines*. Geisler (1999) pointed out that the motion streaks left behind by moving stimuli provide potentially important information about the direction of motion, particularly in conditions where direction is made ambiguous by

MOTION PERCEPTION   769

the aperture problem. Moving objects of finite size will stimulate two classes of cell: those tuned to the direction of motion and also cells without motion tuning, oriented orthogonally (in space-space) to the direction of motion. Wilson proposed a simple model in which the broad direction tuning of motion units could combine with the fine orientation tuning of other units to enhance direction selectivity.

Ross, Badcock, and Hayes (2000) reported a motion illusion in which random sequences of Glass patterns (pairs of dots all aligned in a coherent fashion) appear to move in coherent directions following the direction of the dot pairs. There is no actual motion energy in this direction, and it is easy to show that the motion energy is completely random. A likely explanation for this illusion is that it results from the dot pairs stimulating the hypothetical motion-streak or speed-line mechanisms. The randomly positioned dot pairs should generate a strong but noncoherent sense of motion, equally strong in all directions, exciting many broadly tuned motion detectors. However, only very limited classes of static, orientation-selective neurons will be stimulated, those parallel to the dot-pair alignment. This mechanism will signal local motion parallel to the direction of dot alignment (in both directions), which will lead to global coherent motion that follows the coherent Glass pattern. Interestingly, the apparent direction of motion of these Glass patterns is not fixed but alternates, as would be expected by the random changes in average motion energy.

Much evidence, both psychophysical and neurophysiological, has accumulated in favor of motion streaks. Noise or Glass patterns oriented near the direction of motion strongly degrade motion discrimination thresholds (Burr & Ross, 2002). Furthermore, motion induced by Glass patterns adds vectorially with real motion, suggesting that common mechanisms are being stimulated (Krekelberg et al., 2003). The streaks left by fast motion interact with stationary oriented patterns in interesting ways, causing motion aftereffects and tilt illusions (Apthorp & Alais, 2009), raising contrast thresholds in an orientation-specific manner (Apthorp, Cass, & Alais, 2010) and even causing orientation-selective suppression in rivalry (Apthorp, Wenderoth, & Alais, 2009).

There is also good electrophysiological evidence that motion streaks activate neurons in early visual cortex. Geisler and colleagues (2001) reported that cells in V1 of cat and monkey respond to dot motion orthogonal to their preferred direction (producing motion streaks parallel to their preferred orientation) and that the relative strength of the response to this direction increases with stimulus speed. Just as dynamic Glass patterns (that contain no coherent motion energy) are

seen by humans to move coherently (Ross, Badcock, & Hayes, 2000), they also stimulate cells in MT and MST of monkey (Krekelberg et al., 2003). The direction preference of these STS cells was tuned for both real and "implied" motion and to combinations of them, suggesting that these cells did not distinguish between them. Taken together these results suggest that the implied motion streaks of dynamic Glass patterns generate motion signals in early visual cortex, to which the cells in STS respond, in the same way that they do to real motion signals.

The studies on motion streaks have clearly illustrated the *resourcefulness* of the visual motion system and its capacity to use all available information—even a cue that might normally be thought of as a hindrance to motion rather than a feature—to help it uncover the direction of moving objects and solve the aperture problem.

INFLUENCE OF MOTION ON POSITION AND SPACE

The previous section described interactions between motion and form processing in human vision. This section explores how motion also has a profound influence on perceived position of objects.

Perhaps the clearest and best-known example of motion influencing the perceived position of a target is the *flash-lag illusion*: a stimulus moving continuously seems to lead a briefly flashed light. It is one of the more robust visual illusions, easily demonstrated in the classroom by mounting a translucent card in front of a photographic flash and moving it around in normal lighting, periodically setting off the flash: the flash lags behind the moving card.

The illusion dates back at least to the 1930s, when Metzger (1931) reported that rotating stimuli seemed to move ahead of brief flashes of the stimulus moving behind adjacent slits. Donald Mackay rediscovered the effect by observing that, under strobe lighting, the glowing head of a moving cigarette moved ahead of the base (Mackay, 1958). But Nijhawan's (1994) recent rediscovery and new interpretation of the illusion has spurred a surge of interest. Nijhawan's original explanation was that the illusion compensated for the various delays in processing visual stimuli by extrapolating the motion trajectory forward in time, so a moving target seems to lead a stationary target. Certainly an interesting idea, but it has not stood up to rigorous testing. For example, if a moving stimulus is abruptly stopped or reversed, the extrapolated trajectory should go beyond the reversal point, but this was not observed experimentally (Brenner & Smeets, 2000; Whitney, Cavanagh, &

Murakami, 2000; Whitney & Murakami, 1998; Whitney, Murakami, & Cavanagh, 2000).

Whitney and co-workers put forward a simpler explanation for the flash-lag effect, the "differential latency hypothesis," suggesting that the visual system responds with shorter latency to moving than to flashed stimuli. Although this explanation has the appeal of simplicity, it again fails to account for many of the complexities of the flash-lag phenomenology. For example, increasing the number of flashes (in a repetitive sequence) or the duration of the flash leads to a reduction in the magnitude of flash lag (difficult to reconcile with a simple latency). Furthermore, the flash-lag effect is far more general than was originally thought. Indeed, it does not require that objects actually move in space but can change in other dimensions, such as color or luminance (Sheth, Nijhawan, & Shimojo, 2000) and even works for streams of changing letters (Bachmann & Poder, 2001). Indeed, it is not even restricted to vision. Analogous phenomena of even greater magnitude occur in audition, both for moving sound sources and "chirps," sounds that increase or decrease in pitch over time (Alais & Burr, 2003), with leads of up to 200 ms (compared with the far more modest 20 ms in vision). Flash-lag phenomena also occur cross-modally, probing auditory motion with a visual flash and vice versa. For these effects differential latencies seem particularly implausible. Indeed Arrighi, Alais, and Burr (2005) tested the latency hypothesis directly and showed not only that the latencies are insufficient to explain the measured flash-lag results but actually go in the wrong direction.

Murakami (2001) devised a particularly clever adaptation of the flash-lag effect. Rather than using continuous motion, he presented bars in random positions over time, and subjects judged whether they appeared to the left or right of a marker; again, this produced a robust flash-lag effect, with the additional advantage of being an objective technique, not possible to predict by cognitive reasoning. The results with this were difficult to reconcile with interpolation or spatial averaging but did seem reasonably consistent with differential latencies.

Despite the enormous research effort expended on the flash-lag effect, no single clear explanation has emerged. In the end the flash-lag effect has probably opened more problems than it has solved, in particular the general question of how time and temporal order are encoded in the brain. This has proven to be a profitable line of research, outside of the scope of this chapter, but discussed in chapter 53 by Johnston. And whatever the explanation of the flash-lag effect may be, it appears to have one crucial consequence in everyday life: Baldo, Ranvaud, and Morya (2002) provided convincing evidence that soccer assistant referees' errors in flagging offsides are consistent with the flash-lag effect influencing their decisions.

When a grating drifts behind a stationary window, the window appears to be displaced in the direction of the motion. The effect, described by De Valois and De Valois (1991), is extremely compelling, with shifts up to 15 min for low-frequency gratings drifting at 4–8 Hz. Ramachandran and Anstis (1990) also reported that random dots moving within a stationary window displace the position of the window and that the effect is strongest when the patterns are equiluminant. These demonstrations go to show that motion affects space perception: position and motion are not completely independent for the brain. It is still not exactly clear how this occurs, but presumably it is related to the signal that spatiotemporal receptive fields give about the location in space of objects stimulating them.

Snowden (1998) and Nishida and Johnston (1999) have shown independently that motion can distort position indirectly, via the motion aftereffect. After viewing a drifting grating (or rotating windmill) for some seconds, a grating patch displayed to the adapted region seems to be displaced in the direction of the motion aftereffect. Interestingly, the spatial distortions caused by motion extend beyond the range of the moving stimulus. Whitney and Cavanagh (2000, 2002) showed that moving stimuli affect the perceived position of stimuli briefly flashed to positions quite remote from the motion; they also influence fast-reaching movements (Whitney, 2002; Whitney, Westwood, & Goodale, 2003; Yamagishi, Anderson, & Ashida, 2001) and saccades (Zimmermann, Morrone, & Burr, 2012). Very brief motion displays are sufficient to create large spatial distortions, maximum at motion onset, suggesting very rapidly adapting mechanisms (Roach & McGraw, 2009). Interestingly, the spatial distortions produced by motion and by adapting to motion are clearly distinguishable from the classical motion aftereffects. Whitney and Cavanagh (2003) have demonstrated clear shifts in spatial position, with no corresponding aftereffect. McKeefry, Laviers, and McGraw (2006) have more convincing evidence: Although the motion aftereffect is chromatically selective, motion-induced spatial distortions were completely insensitive to chromatic composition. The dissociation between chromatic selectivity of aftereffects suggested that chromatic inputs are segregated during initial analysis but are later integrated, before the site where motion affects spatial position. Along these lines, Turi and Burr (2012) have shown that whereas the motion aftereffect is *retinotopic* (moving with the eyes), the positional motion aftereffect is *spatiotopic*, specific to position in external space, not to the retinal region stimulated.

The studies reviewed in this section show that form, motion, and position cannot be thought of in isolation. Form can influence motion—most clearly shown in the motion-streak studies—and motion can influence form, in reducing blur in moving objects and in strongly affecting the perceived position of objects in motion and objects flashed near moving stimuli.

CONCLUDING REMARKS

One clear conclusion to emerge from the wealth of studies is that the visual motion system is extremely resourceful in the face of very challenging problems. Another is that motion and position and form interact strongly with each other and are difficult to study in isolation. We need to segregate moving objects from their background, but this segregation involves the integration of local movement signals, many of which have, seemingly, little in common; sometimes we need to keep these local signals separate so that we can distinguish separate objects, as, for example, in transparency. It seems certain that our motion system uses any and all the information it can to make the best bet of what is out there in the real world. For example, the "speed lines" familiar from comics and cartoons can be used to disambiguate and refine our estimation of motion direction. That such a novel mechanism of motion perception can be discovered so recently, after it seemed that much of what could be learned of motion perception had already been learned, suggests that there may well be many other surprises around the corner.

ACKNOWLEDGMENTS

This work has been supported by ERC grant "STANIB" and by the Italian Ministry of Universities and Research.

REFERENCES

Adelson, E. H., & Bergen, J. R. (1985). Spatio-temporal energy models for the perception of motion. *Journal of the Optical Society of America. A, Optics and Image Science, 2*, 284–299.

Adelson, E. H., & Movshon, J. A. (1982). Phenomenal coherence of moving visual patterns. *Nature, 300*, 523–525.

Alais, D., & Burr, D. (2003). The "flash-lag" effect occurs in audition and cross-modally. *Current Biology, 13*, 59–63.

Alais, D., Wenderoth, P., & Burke, D. (1994). The contribution of one-dimensional motion mechanisms to the perceived direction of drifting plaids and their after effects. *Vision Research, 34*, 1823–1834.

Anderson, S. J., & Burr, D. C. (1987). Receptive field sizes of human motion detectors. *Vision Research, 27*, 621–635.

Anderson, S. J., & Burr, D. C. (1991). Receptive field length and width of human motion detector units: Spatial summation. *Journal of the Optical Society of America. A, Optics and Image Science, 8*, 1330–1339.

Anstis, S. M. (1970). Phi movement as a subtractive process. *Vision Research, 10*, 1411–1430.

Apthorp, D., & Alais, D. (2009). Tilt aftereffects and tilt illusions induced by fast translational motion: Evidence for motion streaks. *Journal of Vision, 9*(1), 27, 21–11. doi:10.1167/9.1.27.

Apthorp, D., Cass, J., & Alais, D. (2010). Orientation tuning of contrast masking caused by motion streaks. *Journal of Vision, 10*, 11. doi:10.1167/10.10.11.

Apthorp, D., Wenderoth, P., & Alais, D. (2009). Motion streaks in fast motion rivalry cause orientation-selective suppression. *Journal of Vision, 9*, 11–14.

Arrighi, R., Alais, D., & Burr, D. (2005). Neural latencies do not explain the auditory and audio-visual flash-lag effect. *Vision Research, 45*, 2917–2925.

Bachmann, T., & Poder, E. (2001). Change in feature space is not necessary for the flash-lag effect. *Vision Research, 41*, 1103–1106.

Badcock, D. R., & Derrington, A. M. (1985). Detecting the displacement of periodic patterns. *Vision Research, 25*, 1253–1258.

Badcock, D. R., & Derrington, A. M. (1989). Detecting the displacements of spatial beats: No role for distortion products. *Vision Research, 29*, 731–739.

Baldo, M. V., Ranvaud, R. D., & Morya, E. (2002). Flag errors in soccer games: The flash-lag effect brought to real life. *Perception, 31*, 1205–1210.

Barlow, H. B. (1958). Temporal and spatial summation in human vision at different background intensities. *Journal of Physiology, 141*, 337–350.

Bedell, H. E., Chung, S. T., & Patel, S. S. (2004). Attenuation of perceived motion smear during vergence and pursuit tracking. *Vision Research, 44*, 895–902.

Bex, P. J., & Dakin, S. C. (2002). Comparison of the spatial-frequency selectivity of local and global motion detectors. *Journal of the Optical Society of America. A, Optics, Image Science, and Vision, 19*, 670–677.

Blake, R., & Shiffrar, M. (2007). Perception of human motion. *Annual Review of Psychology, 58*, 47–73.

Boi, M., Ogmen, H., Krummenacher, J., Otto, T. U., & Herzog, M. H. (2009). A (fascinating) litmus test for human retino- vs. non-retinotopic processing. *Journal of Vision, 9*, 1–11.

Born, R. T., Groh, J. M., Zhao, R., & Lukasewycz, S. J. (2000). Segregation of object and background motion in visual area MT: Effects of microstimulation on eye movements. *Neuron, 26*, 725–734.

Born, R. T., & Tootell, R. B. (1992). Segregation of global and local motion processing in primate middle temporal visual area. *Nature, 357*, 497–499.

Bowns, L., & Alais, D. (2006). Large shifts in perceived motion direction reveal multiple global motion solutions. *Vision Research, 46*, 1170–1177.

Braddick, O. (1993). Segmentation versus integration in visual motion processing. *Trends in Neurosciences, 16*, 263–268.

Braddick, O. J., O'Brien, J. M., Wattam-Bell, J., Atkinson, J., & Turner, R. (2000). Form and motion coherence activate independent, but not dorsal/ventral segregated, networks in the human brain. *Current Biology, 10*, 731–734. doi:10.1016/S0960-9822(00)00540-6.

Brenner, E., & Smeets, J. B. (2000). Motion extrapolation is not responsible for the flash-lag effect. *Vision Research, 40*, 1645–1648.

Burr, D. C. (1979). Acuity for apparent vernier offset. *Vision Research, 19*, 835–837.

Burr, D. C. (1980). Motion smear. *Nature, 284*, 164–165.

Burr, D. C. (1981). Temporal summation of moving images by the human visual system. *Proceedings of the Royal Society of London. Series B, Biological Sciences, 211*, 321–339.

Burr, D. C. (1984). Summation of target and mask metacontrast stimuli. *Perception, 13*, 183–192.

Burr, D. (1999). Vision: Modular analysis—or not? *Current Biology, 9*, R90–R92.

Burr, D. C., Baldassi, S., Morrone, M. C., & Verghese, P. (2009). Pooling and segmenting motion signals. *Vision Research, 49*, 1065–1072.

Burr, D. C., McKee, S., & Morrone, M. C. (2006). Resolution for spatial segregation and spatial localization by motion signals. *Vision Research, 46*, 932–939.

Burr, D. C., Morrone, M. C., & Vaina, L. (1998). Large receptive fields for optic flow direction in humans. *Vision Research, 38*, 1731–1743.

Burr, D. C., & Ross, J. (1986). Visual processing of motion. *Trends in Neurosciences, 9*, 304–306.

Burr, D. C., & Ross, J. (2002). Direct evidence that "speed-lines" influence motion mechanisms. *Journal of Neuroscience, 22*, 8661–8664.

Burr, D. C., Ross, J., & Morrone, M. C. (1986). Seeing objects in motion. *Proceedings of the Royal Society of London, B227*, 249–265.

Cavanagh, P. (1992). Attention-based motion perception. *Science, 257*, 1563–1565.

Cavanagh, P., & Mather, G. (1989). Motion: The long and short of it. *Spatial Vision, 4*, 103–129.

Chen, S., Bedell, H. E., & Ogmen, H. (1995). A target in real motion appears blurred in the absence of other proximal moving targets. *Vision Research, 35*, 2315–2328.

Chubb, C., & Sperling, G. (1988). Drift-balanced random stimuli: A general basis for studying non-Fourier motion perception. *Journal of the Optical Society of America. A, Optics and Image Science, 5*, 1986–2007.

Claeys, K. G., Lindsey, D. T., De Schutter, E., & Orban, G. A. (2003). A higher order motion region in human inferior parietal lobule: Evidence from fMRI. *Neuron, 40*, 631–642. doi:10.1016/S0896-6273(03)00590-7.

Del Viva, M. M., Gori, M., & Burr, D. C. (2006). Powerful motion illusion caused by temporal asymmetries in ON and OFF visual pathways. *Journal of Neurophysiology, 95*, 3928–3932.

Del Viva, M. M., & Morrone, M. C. (1998). Motion analysis by feature tracking. *Vision Research, 38*, 3633–3653.

Del Viva, M. M., & Morrone, M. C. (2006). A feature-tracking model simulates the motion direction bias induced by phase congruency. *Journal of Vision, 6*,179–195. doi:10.1167/6.3.1.

Derrington, A. M., & Badcock, D. R. (1985). Separate detectors for simple and complex grating patterns? *Vision Research, 25*, 1869–1878.

Derrington, A. M., Badcock, D. R., & Henning, G. B. (1993). Discriminating the direction of second-order motion at short stimulus durations. *Vision Research, 33*, 1785–1794.

Derrington, A. M., & Henning, G. B. (1987). Errors in direction-of-motion discrimination with complex stimuli. *Vision Research, 27*, 61–75.

Derrington, A. M., & Ukkonen, O. I. (1999). Second-order motion discrimination by feature-tracking. *Vision Research, 39*, 1465–1475.

De Valois, R. L., & De Valois, K. K. (1991). Vernier acuity with stationary moving Gabors. *Vision Research, 31*, 1619–1626.

Dick, M., Ullman, S., & Sagi, D. (1987). Parallel and serial processes in motion detection. *Science, 237*, 400–402.

Duysons, J., Orban, G. A., Cremieux, J., & Maes, H. (1985). Visual cortical correlates of visual persistence. *Vision Research, 25*, 171–178.

Egelhaaf, M., Hausen, K., Reichardt, W., & Wehrhahn, C. (1988). Visual course control in flies relies on neuronal computation of object and background motion. *Trends in Neurosciences, 11*, 351–358.

Eifuku, S., & Wurtz, R. H. (1998). Response to motion in extrastriate area MSTL: Center-surround interactions. *Journal of Neurophysiology, 80*, 282–296.

Ferrera, V. P., & Wilson, H. R. (1990). Perceived direction of moving two-dimensional patterns. *Vision Research, 30*, 273–287.

Fleet, D. J., & Langley, K. (1994). Computational analysis of non-Fourier motion. *Vision Research, 34*, 3057–3079.

Geisler, W. S. (1999). Motion streaks provide a spatial code for motion direction. *Nature, 400*, 65–69.

Geisler, W. S., Albrecht, D. G., Crane, A. M., & Stern, L. (2001). Motion direction signals in the primary visual cortex of cat and monkey. *Visual Neuroscience, 18*, 501–516.

Greenlee, M. W., & Smith, A. T. (1997). Detection and discrimination of first- and second-order motion in patients with unilateral brain damage. *Journal of Neuroscience, 17*, 804–818.

Heeger, D. J., Boynton, G. M., Demb, J. B., Seidemann, E., & Newsome, W. T. (1999). Motion opponency in visual cortex. *Journal of Neuroscience, 19*, 7162–7174.

Hildreth, E. C. (1984). The computation of the velocity field. *Proceedings of the Royal Society of London. Series B, Biological Sciences, 221*, 189–220.

Julesz, B. (1971). *Foundations of cyclopean perception.* Chicago: University of Chicago Press.

Kolers, P. A. (1972). *Aspects of motion perception.* New York: Pergamon Press.

Krekelberg, B., Dannenberg, S., Hoffmann, K. P., Bremmer, F., & Ross, J. (2003). Neural correlates of implied motion. *Nature, 424*, 674–677.

Ledgeway, T., & Smith, A. T. (1994). Evidence for separate motion-detecting mechanisms for first- and second-order motion in human vision. *Vision Research, 34*, 2727–2740.

Lennie, P. (1998). Single units and visual cortical organization. *Perception, 27*, 889–935.

Loomis, J. M., & Nakayama, K. (1973). A velocity analogue of brightness contrast. *Perception, 2*, 425–427.

Lorenceau, J., & Alais, D. (2001). Form constraints in motion binding. *Nature Neuroscience, 4*, 745–751.

Lu, Z. L., & Sperling, G. (1995a). Attention-generated apparent motion. *Nature, 377*, 237–239.

Lu, Z. L., & Sperling, G. (1995b). The functional architecture of human visual motion perception. *Vision Research, 35*, 2697–2722.

Lu, Z. L., & Sperling, G. (2001). Three-systems theory of human visual motion perception: Review and update. *Journal of the Optical Society of America. A, Optics, Image Science, and Vision, 18*, 2331–2370.

Mackay, D. M. (1958). Perceptual stability of a stroboscopically lit visual field containing self-luminous objects. *Nature, 181*, 507–508.

Mackay, D. M. (1961). Visual effects of non-redundant stimulation. *Nature, 192*, 739–740.

Marr, D. (1982). *Vision*. San Fransisco: Freeman.

Marr, D., & Ullman, S. (1981). Directional selectivity and its use in early visual processing. *Proceedings of the Royal Society of London. Series B, Biological Sciences, 211*, 151–180.

Mather, G., & West, S. (1993). Evidence for second-order motion detectors. *Vision Research, 33*, 1109–1112.

McKeefry, D. J., Laviers, E. G., & McGraw, P. V. (2006). The segregation and integration of colour in motion processing revealed by motion after-effects. *Proceedings of the Royal Society of London. Series B, Biological Sciences, 273*, 91–99.

Meso, A. I., & Zanker, J. M. (2009). Perceiving motion transparency in the absence of component direction differences. *Vision Research, 49*, 2187–2200.

Metzger, W. (1931). Versuch einer gemeinsamen Theorie der Phänomene fröhlichs und Hazelhoffs und Kritik ihrer Verfahren zur Messung der Empfindungszeit. *Psychologische Forschung, 16*, 176–200.

Mishkin, M., Ungerleider, L. G., & Macko, K. A. (1983). Object vision and spatial vision: Two cortical pathways. *Trends in Neurosciences, 6*, 414–417.

Morgan, M. J. (1992). Spatial filtering precedes motion detection. *Nature, 355*, 344–346.

Morgan, M. J., & Mather, G. (1994). Motion discrimination in two-frame sequences with differing spatial frequency content. *Vision Research, 34*, 197–208.

Morrone, M. C., & Burr, D. C. (1988). Feature detection in human vision: A phase dependent energy model. *Proceedings of the Royal Society of London, B235*, 221–245.

Murakami, I. (2001). A flash-lag effect in random motion. *Vision Research, 41*, 3101–3119.

Mysore, S. G., Vogels, R., Raiguel, S. E., & Orban, G. A. (2008). Shape selectivity for camouflage-breaking dynamic stimuli in dorsal V4 neurons. *Cerebral Cortex, 18*, 1429–1443. doi:10.1093/cercor/bhm176.

Nakayama, K., Silverman, G. H., MacLeod, D. I., & Mulligan, J. (1985). Sensitivity to shearing and compressive motion in random dots. *Perception, 14*, 225–238.

Nijhawan, R. (1994). Motion extrapolation in catching. *Nature, 370*, 256–257.

Nishida, S., & Johnston, A. (1999). Influence of motion signals on the perceived position of spatial pattern. *Nature, 397*, 610–612.

Pinna, B., & Brelstaff, G. J. (2000). A new visual illusion of relative motion. *Vision Research, 40*, 2091–2096.

Pooresmaeili, A., Cicchini, G. M., Morrone, M. C., & Burr, D. C. (2012). "Non-retinotopic processing" in ternus motion displays modelled by spatio-temporal filters. *Journal of Vision, 12*, 12745–12758. doi:10.1167/12.1.10.

Qian, N., & Andersen, R. A. (1994). Transparent motion perception as detection of unbalanced motion signals. II. Physiology. *Journal of Neuroscience, 14*, 7367–7380.

Qian, N., Andersen, R. A., & Adelson, E. H. (1994). Transparent motion perception as detection of unbalanced motion signals. I. Psychophysics. *Journal of Neuroscience, 14*, 7357–7366.

Raiguel, S., Van Hulle, M. M., Xiao, D. K., Marcar, V. L., & Orban, G. A. (1995). Shape and spatial distribution of

receptive fields and antagonistic motion surrounds in the middle temporal area (V5) of the macaque. *European Journal of Neuroscience, 7*, 2064–2082.

Ramachandran, V. S., & Anstis, S. M. (1990). Illusory displacement of equiluminous kinetic edges. *Perception, 19*, 611–616.

Ramachandran, V. S., & Inada, V. (1984). Motion capture in random-dot patterns. *Optics News, 10*, 77.

Ramachandran, V. S., & Inada, V. (1985). Spatial phase and frequency in motion capture of random-dot patterns. *Spatial Vision, 1*, 57–67.

Regan, D., & Hong, X. H. (1990). Visual acuity for optotypes made visible by relative motion. *Optometry and Vision Science, 67*, 49–55.

Reichardt, W. (1957). Autokorrelationsauswertung als Funktionsprinzip des Zentralnervensystems. *Zeitschrift für Naturforschung, 12b*, 447–457.

Reichardt, W. (1961). Autocorrelation, a principle for evaluation of sensory information by the central nervous system. In W. Rosenblith (Ed.), *Sensory communications* (pp. 303–317). New York: John Wiley.

Roach, N. W., & McGraw, P. V. (2009). Dynamics of spatial distortions reveal multiple time scales of motion adaptation. *Journal of Neurophysiology, 102*, 3619–3626.

Ross, J., Badcock, D. R., & Hayes, A. (2000). Coherent global motion in the absence of coherent velocity signals. *Current Biology, 10*, 679–682.

Seiffert, A. E., & Cavanagh, P. (1998). Position displacement, not velocity, is the cue to motion detection of second-order stimuli. *Vision Research, 38*, 3569–3582.

Sheth, B. R., Nijhawan, R., & Shimojo, S. (2000). Changing objects lead briefly flashed ones. *Nature Neuroscience, 3*, 489–495.

Smith, A. T., & Ledgeway, T. (1998). Sensitivity to second-order motion as a function of temporal frequency and eccentricity. *Vision Research, 38*, 403–410.

Smith, A. T., Snowden, R. J., & Milne, A. B. (1994). Is global motion really based on spatial integration of local motion signals? *Vision Research, 34*, 2425–2430.

Snowden, R. J. (1998). Shifts in perceived position following adaptation to visual motion. *Current Biology, 8*, 1343–1345.

Tadin, D., Lappin, J. S., Gilroy, L. A., & Blake, R. (2003). Perceptual consequences of centre-surround antagonism in visual motion processing. *Nature, 424*, 312–315.

Taub, E., Victor, J. D., & Conte, M. M. (1997). Nonlinear preprocessing in short-range motion. *Vision Research, 37*, 1459–1477.

Tolhurst, D. J., & Movshon, J. A. (1975). Spatial and temporal contrast sensitivity of striate cortical neurones. *Nature, 257*, 674–675.

Tong, J., Stevenson, S. B., & Bedell, H. E. (2008). Signals of eye-muscle proprioception modulate perceived motion smear. *Journal of Vision, 8*, 1–6. doi:10.1167/8.14.7.

Turi, M., & Burr, D. C. (2012). Spatiotopic perceptual maps in humans: Evidence from motion adaptation. *Proceedings of the Royal Society of London B Biological Science, 279*, 3091–3097. doi:10.1098/rspb.2012.0637.

Vaina, L. M., & Cowey, A. (1996). Impairment of the perception of second-order motion but not first-order motion in a patient with unilateral focal brain damage. *Proceedings of the Royal Society of London. Series B, Biological Sciences, 263*, 1225–1232.

Vaina, L. M., & Soloviev, S. (2004). First-order and second-order motion: Neurological evidence for neuroanatomically distinct systems. *Progress in Brain Research, 144,* 197–212.

van Santen, J. P., & Sperling, G. (1985). Elaborated Reichardt detectors. *Journal of the Optical Society of America. A, Optics and Image Science, 2,* 300–321.

Verghese, P., & Stone, L. S. (1995). Combining speed information across space. *Vision Research, 35,* 2811–2823.

Verghese, P., & Stone, L. S. (1996). Perceived visual speed constrained by image segmentation. *Nature, 381,* 161–163.

Verstraten, F. A., Cavanagh, P., & Labianca, A. T. (2000). Limits of attentive tracking reveal temporal properties of attention. *Vision Research, 40,* 3651–3664.

Watson, A. B., & Ahumada, A. J. (1985). Model of human visual-motion sensing. *Journal of the Optical Society of America. A, Optics and Image Science, 2,* 322–341.

Wertheimer, M. (1912). Experiementelle Studien uber das Sehen von Bewegung. *Zeitschrift fur Psychologie mit Zeitschrift fur Angewandte Psychologie, 61,* 151–265.

Whitney, D. (2002). The influence of visual motion on perceived position. *Trends in Cognitive Sciences, 6,* 211–216.

Whitney, D., & Cavanagh, P. (2000). Motion distorts visual space: Shifting the perceived position of remote stationary objects. *Nature Neuroscience, 3,* 954–959.

Whitney, D., & Cavanagh, P. (2002). Surrounding motion affects the perceived locations of moving stimuli. *Visual Cognition, 9,* 139–152.

Whitney, D., & Cavanagh, P. (2003). Motion adaptation shifts apparent position without the motion aftereffect. *Perception & Psychophysics, 65,* 1011–1018.

Whitney, D., Cavanagh, P., & Murakami, I. (2000). Temporal facilitation for moving stimuli is independent of changes in direction. *Vision Research, 40,* 3829–3839.

Whitney, D., & Murakami, I. (1998). Latency difference, not spatial extrapolation. *Nature Neuroscience, 1,* 656–657.

Whitney, D., Murakami, I., & Cavanagh, P. (2000). Illusory spatial offset of a flash relative to a moving stimulus is caused by differential latencies for moving and flashed stimuli. *Vision Research, 40,* 137–149.

Whitney, D., Westwood, D. A., & Goodale, M. A. (2003). The influence of visual motion on fast reaching movements to a stationary object. *Nature, 423,* 869–873.

Wilson, H. R., & Kim, J. (1994). Perceived motion in the vector sum direction. *Vision Research, 34,* 1835–1842.

Yamagishi, N., Anderson, S. J., & Ashida, H. (2001). Evidence for dissociation between the perceptual and visuomotor systems in humans. *Proceedings of the Royal Society of London. Series B, Biological Sciences, 268,* 973–977.

Yang, Y., & Blake, R. (1994). Broad tuning for spatial frequency of neural mechanisms underlying visual perception of coherent motion. *Nature, 371,* 793–796.

Yo, C., & Wilson, H. R. (1992). Perceived direction of moving two-dimensional patterns depends on duration, contrast and eccentricity. *Vision Research, 32,* 135–147.

Zaksas, D., & Pasternak, T. (2006). Directional signals in the prefrontal cortex and in area MT during a working memory for visual motion task. *Journal of Neuroscience, 26,* 11726–11742.

Zanker, J. M. (1990). Theta motion: A new psychophysical paradigm indicating two levels of visual motion perception. *Naturwissenschaften, 77,* 243–246.

Zanker, J. M. (1993). Theta motion: A paradoxical stimulus to explore higher order motion extraction. *Vision Research, 33,* 553–569.

Zeki, S. (1993). *A vision of the brain.* Oxford: Blackwell Scientific.

Zimmermann, E., Morrone, M. C., & Burr, D. C. (2012). Visual motion distorts perceptual and motor space. *Journal of Vision, 12,* 1–8. doi:10.1167/12.2.10.

55 Functional Mapping of Motion Regions in Human and Nonhuman Primates

GUY A. ORBAN AND JAN JASTORFF

CONCEPTUAL FRAMEWORK

Lower- and Higher-Order Motion-Selective Neurons

Knowledge concerning the functional properties of motion-selective neurons (Orban, 2008) provides an extremely useful framework for the functional imaging studies of the motion-sensitive regions in the human brain. We use the term motion-sensitive for cortical regions defined using functional imaging to stress the fact that imaging, unlike single-cell studies, provides no direct evidence of neuronal selectivity. Subtractive single-voxel techniques cannot provide such data, and more advanced techniques either overestimate selectivity, as repetition suppression does (Mur et al., 2010; Sawamura, Orban, & Vogels, 2006), or have to invoke additional assumptions that are not necessarily verified, as the grouping of selective neurons for multivoxel pattern analysis.

Lower-order motion-selective neurons are the standard direction-selective neurons described in monkey V1 (Hubel & Wiesel, 1968) or area MT/V5 (Albright, 1984), which can also be speed selective (Hawken, Parker, & Lund, 1988; Lagae, Raiguel, & Orban, 1993; Maunsell & Van Essen, 1983; Mikami, Newsome, & Wurtz, 1986; Orban, Kennedy, & Bullier, 1986). Many higher-order motion-selective neurons are selective for patterns of motion vectors. The typical examples are the MSTd neurons selective for flow components, specified by smooth variations of motion direction (Duffy & Wurtz, 1991; Graziano, Andersen, & Snowden, 1994; Lagae et al., 1994; Tanaka et al., 1986). A more recent example is provided by the neurons selective for smooth variations of speeds, the so-called speed-gradient-selective neurons, which have been documented in MT/V5, MSTd, and FST (Mysore et al., 2010a,b; for review see Orban, 2011; Sugihara et al., 2002; Xiao et al., 1997). Finally, sudden changes in direction or speed are indicative of boundaries, the so-called non-luminance-defined or higher-order boundaries. It was long ago established that IT neurons can be shape selective

for such boundaries (Sary, Vogels, & Orban, 1993), a property often overlooked (El-Shamayleh & Movshon, 2011). Selectivity for orientation of kinetic boundaries is weak or absent in V1 and V2 (Marcar et al., 2000) but present in V4 (An et al., 2010; Mysore et al., 2006). A similar hierarchical processing seems to take place in the case of texture-defined or illusory boundaries (El-Shamayleh & Movshon, 2011; Pan et al., 2010). One type of relatively simple higher-order motion is the so-called global motion, based on integrating motion pulses in static Gabor patches and related to apparent motion, for which the neuronal mechanism is still unknown (Hedges et al., 2011).

Finally, a number of neurons in the monkey STS are selective for actions performed by conspecifics (Perrett et al., 1985; Singer & Sheinberg, 2010; Vangeneugden et al., 2011; Vangeneugden, Pollick, & Vogels, 2009). Although in principle such selectivity can arise from the spatial pattern of motion vectors (Giese & Poggio, 2003; Lange & Lappe, 2006), and thus action-selective neurons can be considered higher-order motion-selective neurons, recent studies have emphasized temporal changes in the body shape as the source of this selectivity (Singer & Sheinberg, 2010; Vangeneugden, Pollick, & Vogels, 2009). It is noteworthy that in the action stimuli the various body parts move relatively one to another. Thus, rigidity constraints cannot apply, but constraints are provided by the biological kinematics, the body shape, and its permissible deformations.

Parallel Imaging in Humans and Monkeys

Whereas single-cell studies and functional imaging studies in humans are extremely complementary, their direct comparison is difficult, a problem that is frequently overlooked. Indeed, action potentials and MR signals are of very different origin and cannot simply be equated, even if often correlated, and monkey and human brains, although similar, are different. Old World monkeys diverged from humans more than 23 million years ago, cortical surfaces in the two species differ by a factor of almost 10 (Van Essen et al., 2011), and the behavioral repertoires show marked differences. Thus, a monkey brain cannot simply be considered a miniature human brain, and relating human fMRI studies to monkey single-cell studies amounts to solving a single equation with two unknowns. The problem becomes tractable when the intermediate step, fMRI in alert monkeys, is added to the two other experimental strategies (Orban, 2002, 2011). Hence, the present review emphasizes comparative imaging studies in human and nonhuman primates.

Definition and Properties of Cortical Areas

fMRI studies typically link a function, here the processing of motion signals, to anatomical regions in the brain. Because the precise parcellation of the human brain into cortical areas is unknown, the anatomy is usually described in terms of coordinates such as the Talairach or MNI coordinates or makes use of vague anatomical terms such as Brodmann areas. In the monkey cortical areas are defined by a combination of morphological and functional criteria. These include the topographic organization and the functional properties of neurons. Initially, attempts were made to define human cortical areas by function, such as the motion or color areas (Zeki et al., 1991). With time, however, it has become clear that what was initially described as the motion area (MT or V5) was in fact a complex of areas, hMT/V5+ (DeYoe et al., 1996), analogous to the situation in the monkey (figure 55.1), triggering

attempts to parcel this complex (Huk, Dougherty, & Heeger, 2002). As it became clear that functional subtractions or localizers cannot define single cortical areas, determining the topographic organization of cortical areas has regained interest. This interest is bolstered by the fact that fMRI, unlike single-neuron recordings, is very adept at detecting weak overall topographic organizations, which involve only a fraction of neurons in a given area—those with small receptive fields (RFs). Because the mapping is performed in individual subjects, there is no need to smooth the data to overcome the variability between subjects. This yields far sharper distinctions in functional properties in neighboring cortical areas (functional grid analysis, see Kolster, Peeters, & Orban, 2010).

LOWER-ORDER MOTION-SENSITIVE REGIONS

The Occipitotemporal Motion-Sensitive Region of hMT/V5+

Single-cell and anatomical studies have indicated that area MT/V5 is surrounded by several satellite areas to which it projects (for review see Orban, 1997). Their number is a subject of debate: three (MSTv, MSTd, and FST) according to Tanaka et al. (1993) or four according to Lewis and Van Essen (2000). Initially these areas were defined in the fMRI by their motion sensitivity as indicated by a significant difference in responses to moving or static random dots (Vanduffel et al., 2001). In keeping with these authors, three motion-sensitive

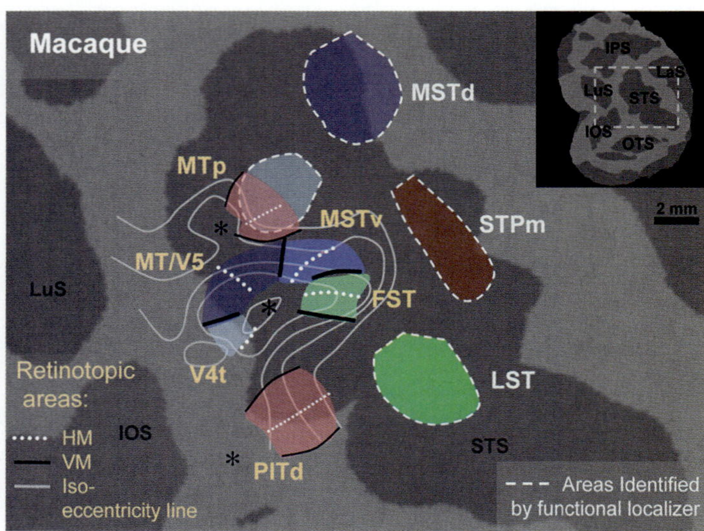

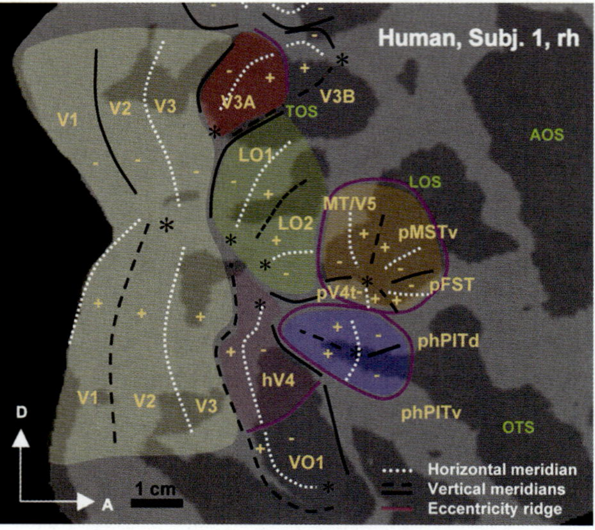

FIGURE 55.1   The MT/V5 cluster in macaque (left) and human (right). The four areas included in the cluster are shown on a part of the flattened right hemisphere in relation to other cortical regions: motion-sensitive regions MTp, MSTd, STPm, and LST and retinotopic field PITd in the monkey, and retinotopically defined regions V1–3, V3A, LO1-2, phPITd and phPITv, hV4, and VO1 in humans. Projections of meridians and spatial scales are indicated. (From Kolster, Peeters, & Orban, 2010.)

areas (MT/V5, MSTv, and FST) were described in the chapter of the previous handbook edition (Orban & Vanduffel, 2004). Nelissen, Vanduffel, and Orban (2006) described three additional motion-sensitive regions in the caudal STS: MSTd, STPm, and LST (figure 55.1). This procedure identified regions by their local maxima, not by their boundaries. Therefore, the retinotopic mapping of the MT/V5 cluster (Kolster et al., 2009) achieved considerable progress. This cluster included the three regions MT/V5, MSTv, and FST, joined at their central representations, plus a fourth area labeled V4t. Comparison with the regions defined by Nelissen, Vanduffel, and Orban (2006) (figure 55.1) suggests that what Nelissen, Vanduffel, and Orban (2006) considered to be MSTv, which was located relatively caudal in the STS, may correspond to a seventh motion-sensitive region, MTp, as defined originally by Ungerleider and Desimone (1986).

As indicated in the 2004 chapter (Orban & Vanduffel, 2004), the human motion-sensitive region in the ascending ITS, or hMT/V5+, is likely to include multiple motion-sensitive regions. Huk, Dougherty, and Heeger (2002) described only two regions based on the ipsilateral field representation in the anterior region labeled MST. Amano, Wandell, and Dumoulin (2009), using polar angle mapping, also described two areas, TO1 and TO2. Using complete retinotopic mapping techniques (polar and eccentricity mapping) similar to those used in the monkey, Kolster, Peeters, and Orban (2010) showed that the posterior two-thirds of hMT/V5+ corresponded to the human MT/V5 cluster including, just as in monkeys, four retinotopically defined areas, MT/V5 and the homologues of MSTv, FST and V4t (figure 55.1). This way, area MT/V5 has become the first extrastriate area for which the homology is reasonably certain, insofar as most criteria (Orban &

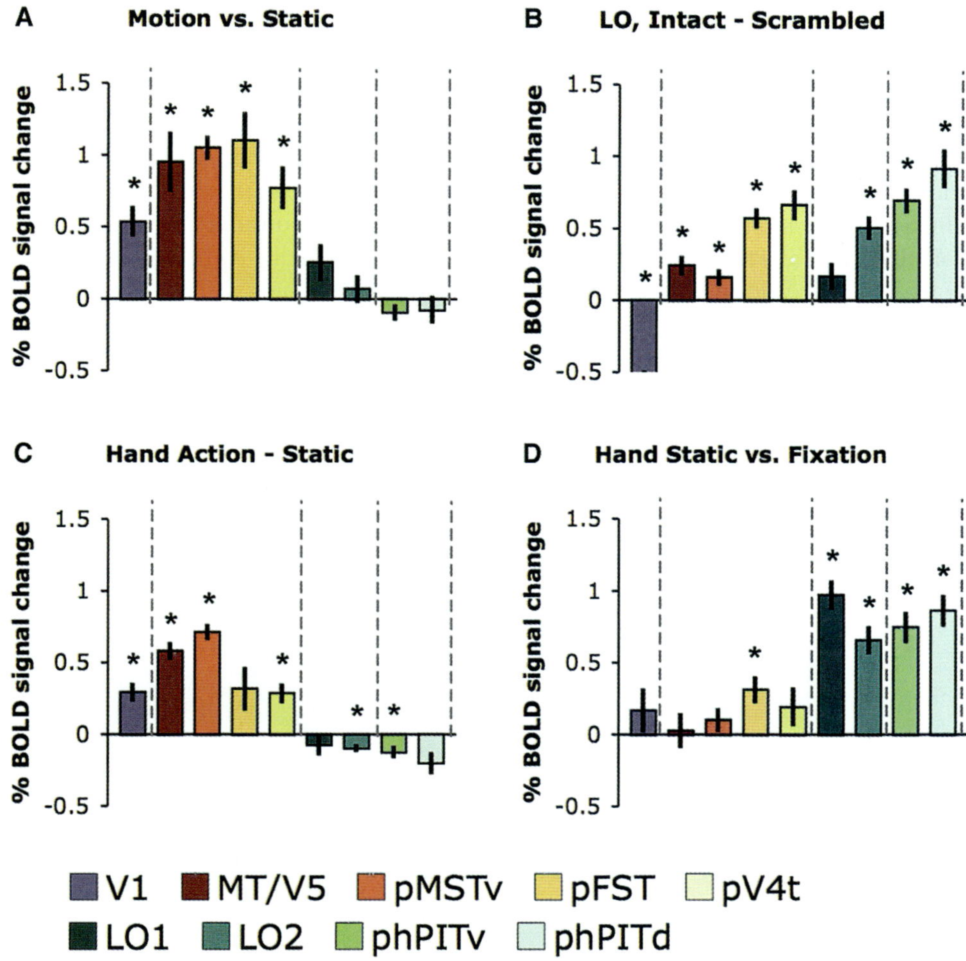

FIGURE 55.2   Percentage BOLD signal change in the members of the MT/V5 cluster and neighboring areas for different subtractions indicated; asterisks indicate significant changes. (From Kolster, Peeters, & Orban, 2010.)

Vanduffel, 2004) are fulfilled. Whether additional retinotopic areas corresponding to the remaining motion-sensitive regions in hMT/V5+ can be described must await further experimentation. Also, the relationship with the region anterior to hMT/V5+ sensitive to motion in depth (Likova & Tyler, 2007) will have to be clarified.

The use of retinotopic mapping to define cortical areas has the additional benefit that responses in a given retinotopic area can be averaged across subjects without the need for smoothing. Hence, functional differences between neighboring areas are much sharper than expected from earlier studies, such as those between LO2 and MT/V5 or between pFST and phPITd (figure 55.2A, C, D) or between pMSTv and pFST (figure 55.2B). The activation of pFST and pV4t by the shape localizer confirmed an earlier observation of an overlap between the ventral part of hMT/V5+ and the lateral occipital complex (LOC; Kourtzi et al., 2002). It is noteworthy that human MT/V5 is strongly activated by hand action observation (Kolster, Peeters, & Orban, 2010), as it is in the monkey (Nelissen, Vanduffel, & Orban, 2006).

The Dorsal Motion Area: V6

Galletti et al. (1999) described a retinotopically organized area V6, located in the depth of the parieto-occipital sulcus (POS) of the monkey and housing a high proportion of direction-selective neurons (figure 55.3). V6 only partially corresponds to the earlier defined region PO (Colby et al., 1988). In contrast to the ventral motion area MT/V5, area V6 does not overrepresent the central visual field. A similar region, hV6, which in flatmaps abuts dorsal V3 (figure 55.3), has been described in the POS of humans (Pitzalis et al.,

2006; but see Van Essen et al., 2011). This area is also motion sensitive (Pitzalis et al., 2010), but concentric expanding rings do not stimulate hV6 as well as the hMT/V5+ complex, whereas, conversely, coherently moving, compared to scrambled, flow fields activate hV6 and far less so hMT/V5+. Because flow-induced activation extended beyond hV6, it is premature to conclude that flow-field responses are an adequate substitute for retinotopic mapping of hV6.

A Species Difference: V3A and the Parietal Regions

There is at present disagreement about the exact definition of V3A in humans (Georgieva et al., 2009; Larsson & Heeger, 2006; Smith et al., 1998; Tootell et al., 1997) and the extent to which it is motion sensitive. In our definition (Georgieva et al., 2009), V3A is motion sensitive, as in the original definition of Tootell et al. (1997), and unlike monkey V3A (Vanduffel et al., 2001). Orban et al. (2003, 2006) have attributed the motion sensitivity of human parietal regions, such as ventral intraparietal sulcus (VIPS), parieto-occipital intraparietal sulcus (POIPS), dorsal intraparietal sulcus medial (DIPSM), and dorsal intraparietal sulcus anterior (DIPSA) regions (Sunaert et al., 1999), to input from human V3A. Comparison of a range of functional properties provides evidence that human DIPSM corresponds to anterior LIP, which is motion sensitive in the monkey, and that DIPSA is the equivalent of posterior AIP (Durand et al., 2009). Originally, two neighboring motion-sensitive regions, DIPSL and DIPSM, were described (Sunaert et al., 1999), which over time have come to be considered as a single region (DIPSM). It could be, however, that the more lateral of the two regions now labeled DIPSM in fact corresponds to anterior LIP, and the more medial region to VIP (see below). In humans and

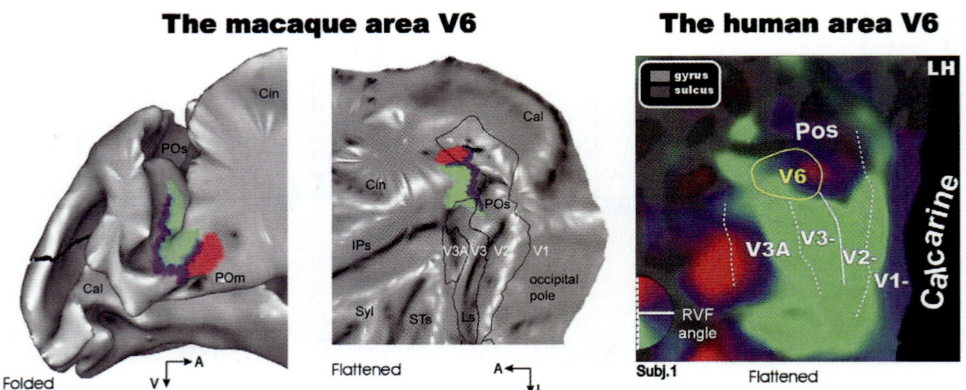

FIGURE 55.3   Cortical area V6 shown as colored voxels on part of folded and flattened macaque left hemisphere (left) and as yellow outline on a region of flattened human cortex (right). Polar angle color code is indicated in right panel (Fattori, Pitzalis, & Galletti, 2009). Note that the similarity of topological relationship between V6 and dorsal V3 is more visible in the flatmaps.

monkeys a retroinsular motion region, posterior insular cortex (PIC), has been described (Claeys et al., 2003; Sunaert et al., 1999; Vanduffel et al., 2001). This region may correspond to the visual posterior Sylvian (VPS) area, a region housing optic-flow-selective neurons (Chen, DeAngelis, & Angelaki, 2011), as PIC and VPS have a similar relationship to parietoinsular vestibular cortex (PIVC).

HIGHER-ORDER RIGID-MOTION-SENSITIVE REGIONS

Kinetic Boundary-Sensitive Regions

Van Oostende et al. (1997) defined the kinetic occipital region as a region located in between V3A and hMT/V5+ more responsive to kinetic boundary gratings than to a uniform field or transparent motion (see white outlines in figure 55.7). In the monkey a similar subtraction activates dorsal V4 (Fize et al., 2001), in which neurons tuned to kinetic boundary orientation have been reported (Mysore et al., 2006). The region in dorsal V4 activated by the kinetic boundary contrast was also activated by other non-luminance-defined contours. Because the activation was restricted to dorsal V4, which may reflect technical limitations such as susceptibility artifacts at 1.5 T, only a preliminary abstract was published. Two studies in humans disputed the KO (kinetic occipital) label applied to the kinetic-boundary responsive region. Zeki, Perry, and Bartels (2003) disputed the KO label as they showed that color-defined boundaries also activated this region. Tyler et al. (2006) proposed the term "occipital depth structure region" as a replacement because KO was also responsive to disparity-defined boundaries, which evoke the percept of depth structure, as do kinetic gratings.

Using the original KO localizer, Larsson and Heeger (2006) showed that several retinotopic regions, particularly LO1 and LO2, responded to this functional localizer. In a subsequent study (Larsson, Heeger, & Landy, 2010), they found that most of these retinotopic areas (LO1, LO2, V3A/B, and V7) showed adaptation to the orientation of the kinetic boundaries. Thus, KO provides another example suggesting that a single functional localizer is unlikely to define a unique cortical area, a reservation already put forward by Van Oostende et al. (1997). It is possible that some of the contradictions noted by Zeki, Perry, and Bartels (2003) and Tyler et al. (2006) in the properties of KO reflect the differential involvement of these different areas. Obviously, more work is needed to determine to what extent non-luminance-defined contours and depth structure

activate the different retinotopic areas posterior to hMT/V5+.

Optic-Flow-Sensitive Regions

Although at least four cortical areas are known to house optic-flow-selective neurons in the monkey—MSTd (Tanaka et al., 1986), VIP (Schaafsma & Duysens, 1996), VPS (Chen, DeAngelis, & Angelaki, 2011), and PEc (Squatrito et al., 2001)—no activations of these areas have been reported in monkey fMRI, perhaps due to overly small stimulus sizes (but see Guipponi et al., 2011). Nelissen, Vanduffel, and Orban (2006) even reported stronger activation of their motion-sensitive region MSTd by translation than by rotation.

In humans, Wall and Smith (2008) devised a subtraction isolating self-motion-sensitive regions which compared the viewing a large expanding flow field with the viewing of a composite of nine small expanding flow fields. They reported activation for this contrast (1) in MST as defined according to Huk, Dougherty, and Heeger (2002) in a location different from that reported by Morrone et al. (2000), (2) in a site identified as VIP, and (3) in a novel visual area located in the cingulate sulcus (CSv). The visual nature of this latter site and its involvement in motion analysis has recently been confirmed by Fischer et al. (2011). A subsequent study (Cardin & Smith, 2010) using the same stimuli added PIVC, a site labeled pV6 (see above), the precuneus, and p2v to the list of self-motion sensitive regions (figure 55.4).

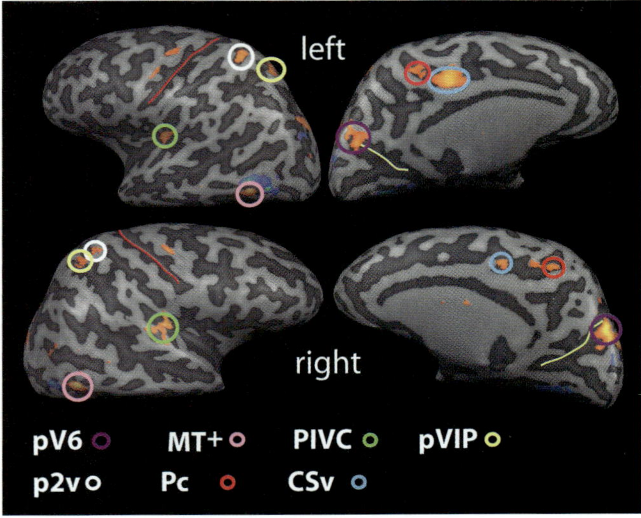

FIGURE 55.4   Location of sites activated by the comparison of a large expanding flow field to a set of nine small fields, shown on the inflated hemispheres of a single subject (Cardin & Smith, 2010).

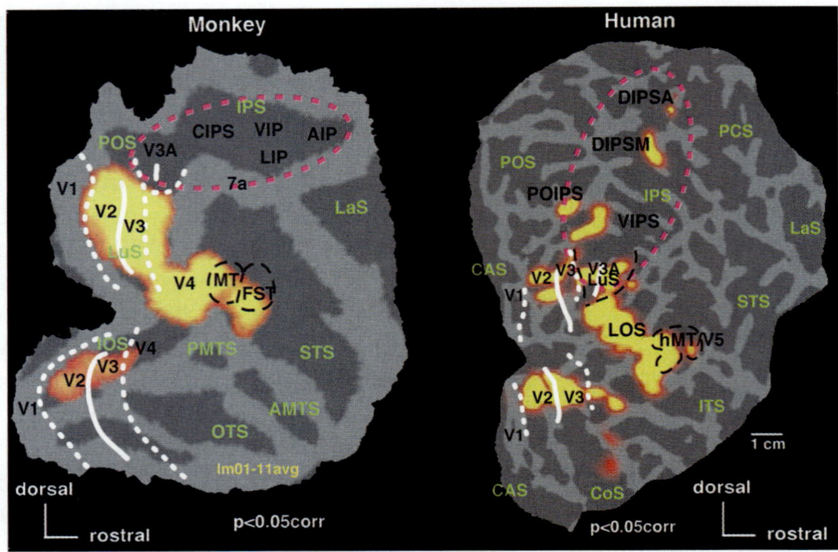

FIGURE 55.5   Comparison of voxels significantly activated by random lines rotating in depth versus random lines translating in the plane projected onto parts of the flattened right hemisphere of a macaque (left) or human (right). Dashed stippled red lines indicate parietal regions (Vanduffel et al., 2002).

Great care is needed in using identical labels for the human and monkey areas in the absence of monkey fMRI data, which are indispensable for validating the MR paradigms used in human imaging, or applying a battery of functional imaging tests identifying cortical areas in both species. In this respect Bartels, Zeki, and Logothetis (2008) provided some additional features indicating that the global-motion-sensitive site labeled mPPC might indeed be homologous to monkey VIP. Its coordinates (5, –65, 59) are relatively similar to those (18, –60, 62) listed initially for DIPSM by Sunaert et al. (1999), and DIPSM was shown to be involved in heading judgments by Peuskens et al. (2001). On the other hand, the coordinates of DIPSL (25, –54, 62) are very similar to those (25, –55, 49) listed by Cardin & Smith (2010) for VIP. Yet comparison of the location of these three sites on a flatmap (see green dots in figure 55.7), together with two other sites claimed to be the homologue of monkey VIP (Bremmer et al., 2001; Sereno & Huang, 2006) shows a wide dispersion for these candidate homologous areas. This underscores the need for monkey fMRI to identify homologous regions in humans and monkeys.

The role of hMT/V5+, VIP, precuneus, and anterior cingulate sulcus in self-motion perception is supported by their activation in a contrast between periods during which subjects reported self-motion versus periods during which they reported object motion when viewing a three-dimensional expanding flow field (see Klein-schmidt et al., 2002, for a related study; Kovacs, Raabe, & Greenlee, 2008). Although self-motion-sensitive regions included many vestibular regions such as PIVC or p2v, only the anterior part of hMST and CSv have been found to respond to galvanic stimulation (Smith, Wall, & Thilo, 2011).

Three-Dimensional Shape from Motion-Sensitive Regions

The comparison between kinetic stimulus conditions yielding a three-dimensional shape perception and those eliciting a flat two-dimensional shape perception yielded a different activation pattern in humans and monkeys (Vanduffel et al., 2002, figure 55.5). The human results that we obtained (Orban et al., 1999, 2006; Peuskens et al., 2004; Vanduffel et al., 2002) are in general agreement with those of Beer et al. (2009) or Yamamoto et al. (2008). Several regions in human parietal cortex, in or dorsal to the IPS, were activated in humans but not in monkeys (Vanduffel et al., 2002). This functional difference was related to the use of tools, which is much more extensive in humans than in monkeys. Indeed, we later discovered a region concerned with observing tool use in left anterior supra-marginal gyrus of humans, apparently without a functional counterpart in monkeys (Peeters et al., 2009).

On the other hand, three-dimensional shape-from-motion activations in lateral occipitotemporal cortex of the two species were similar, involving MT/V5 and FST in monkeys and the hMT/V5+ region in humans (figure 55.5). Because the latter complex has now been shown to include the homologues of MT/V5 and probably also FST, it is likely that these human areas house speed-gradient-selective neurons similar to those described in the monkey areas (Orban, 2011). Finally, ventral occipital regions are also involved in the processing of three-dimensional shape from motion, especially when stimuli portraying three-dimensional surfaces are presented to humans (Kriegeskorte et al., 2003; Orban et al., 2006) and monkeys (Mysore et al., 2010a,b). Further work is needed to clarify these ventral regions and their overlap with regions extracting three-dimensional shape from disparity (Georgieva et al., 2009; Joly, Vanduffel, & Orban, 2009) and from static texture or shading cues (Georgieva et al., 2008; Nelissen et al., 2009).

Attention-Driven Motion-Sensitive Regions

One other type of higher-order motion is based on the displacement of spatial regions marked by attention rather than by energy in the image (Cavanagh, 1992; Lu & Sperling, 1995). Claeys et al. (2003) used four types of gratings: isoluminant gratings in which either (1) green or (2) red bars were more salient, compared to (3) isoluminant and isosalient gratings with equal saturation of red and green and (4) isosalient gratings with luminance differences. Motion of the first two types of gratings would be captured by the higher-order system, that of the fourth type, by lower-order regions such as hMT/V5+, and motion of the third type grating should be invisible as it drives neither system. One region in the human brain in the right inferior parietal lobule (IPL) was found to be specifically activated by motion of the isoluminant gratings with salient color. Figure 55.6A shows the location of this higher-order IPL (HM-IPL) region, its activity in the eight individual conditions, and its differential activity for higher- and lower-order motion, compared to that of hMT/V5+. Its location on the flattened hemisphere is indicated in figure 55.7 (ellipse 6). It did not react to second-order motion, which has been shown to activate a temporal region (Noguchi et al., 2005). The same HM-IPL region was also found to be activated bilaterally by long-range apparent motion (figure 55.6B), perhaps because the onset of the dots also makes them salient. Note that the left HM-IPL region is located just behind the tool-use region (Peeters et al., 2009). Activation of these IPL regions would be necessary for the apparent motion stimuli to appear to be moving. Feedback signals could be sent from HM-IPL, perhaps via hMT/V5+ (Muckli et al., 2002; Sterzer & Kleinschmidt, 2005; Sterzer et al., 2002), to V1, where the path of the stimulus is represented (Muckli et al., 2005). Whether this attention-based mechanism associated with linear apparent motion might also operate in apparent rotation is

unclear. Stimuli appearing to rotate in depth evoke a very different activation pattern (Weigelt et al., 2007).

It has recently been shown that macaque MT/V5 neurons respond to local motion but not to global motion produced by multiple local motion pulses, creating long-range apparent motion (Hedges et al., 2011). If a similar area exists in the monkey, right HM-IPL is certainly a candidate area for the extraction of global motion, perhaps more so than the anterior temporal region defined by Zhuo et al. (2003), involved in apparent shape change rather than apparent motion.

ACTION OBSERVATION AND BIOLOGICAL MOTION-SENSITIVE REGIONS

Observation of Actions

Observation of the actions of others, compared to control conditions, activates a network of cortical areas that span three levels (figure 55.7): occipitotemporal, parietal, and premotor cortex (Buccino et al., 2001, 2004; Gazzola & Keysers, 2009; Jastorff et al., 2010; Oosterhof, Tipper, & Downing, 2012). Because action includes both shape and motion aspects (Giese & Poggio, 2003), static controls with the body (part) shape do not suffice. The most efficient controls for local motion are the moving small-object control as in Jastorff et al. (2010), appropriate for actions involving the distal parts of limbs, or flow extracted from the video superimposed on random noise with temporal shuffling (Jastorff et al., 2011).

The occipito-temporal activation by action observation typically starts from the hMT/V5+ complex and spreads forward along two pathways: rostrally over the posterior ITS, MTG, STS, and sometimes STG, and ventrorostrally into the OTS and fusiform gyrus, overlapping with the LOC. For the observation of manipulation the main parietal activation overlaps with DIPSA and parts of phAIP and DIPSM, and a small posterior activation overlaps VIPS (figure 55.7). Also, in the monkey, the observation of grasping involves three levels: the STS, AIP and neighboring PFG, and F5 in ventral premotor cortex (figure 55.7). These areas are anatomically connected so that by combining tracing and fMRI data the complete circuit can be reconstructed (Nelissen et al., 2011).

Generally, subjects remain passive in these studies. When subjects prepare to perform an action, even one not matched with that shown in the video, additional action-processing regions are recruited (Jastorff, Abdollahi, & Orban, 2012). Some of these are nonspecific for the effector, such as those in parieto-occipital cortex, but others, such as those in fusiform and posterior

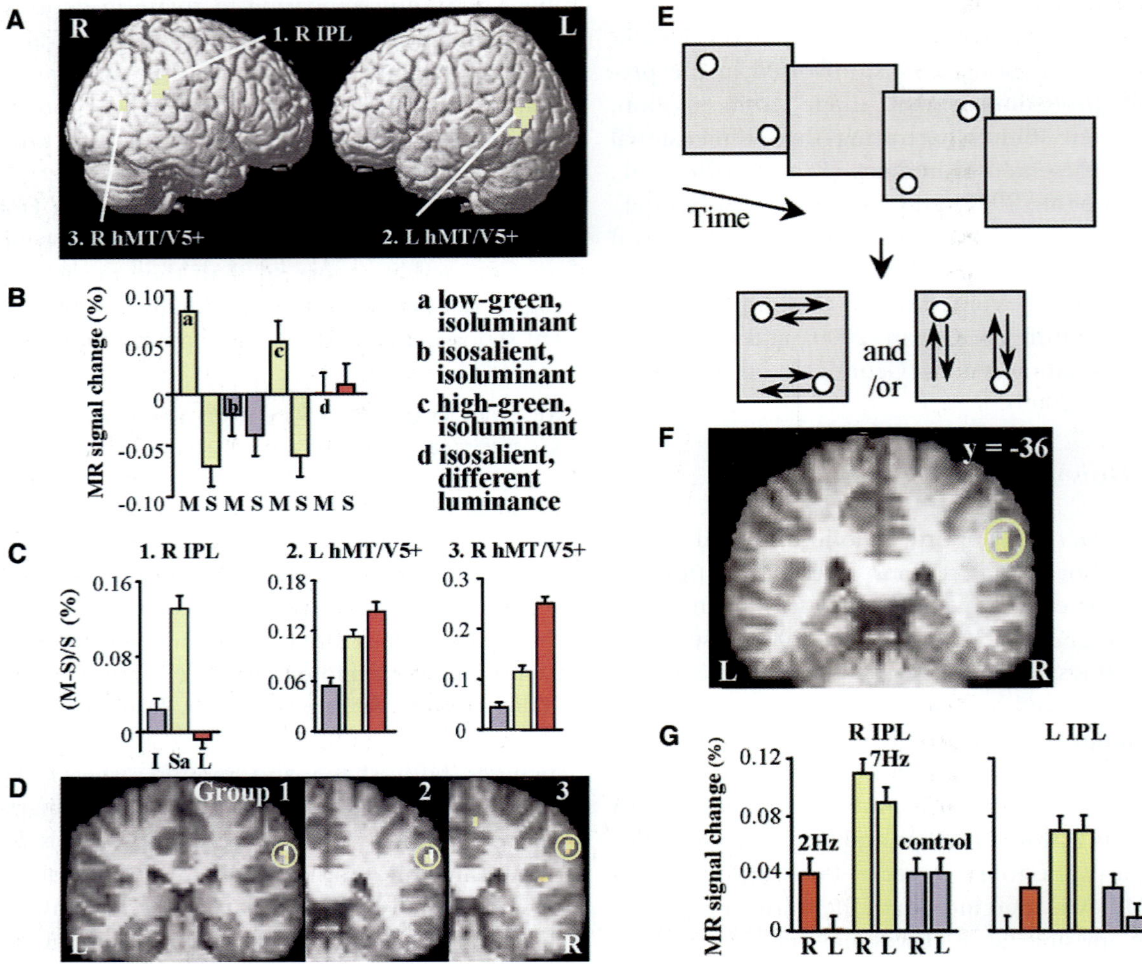

FIGURE 55.6   The higher-order motion IPL region in the right hemisphere. (A) Significant voxels in the difference between moving and stationary "different saliency" stimuli (experiment 1) on rendered brain. (B) Percentage signal change in right HM-IPL with respect to fixation for the eight stimulus conditions. (C) Percentage signal change in moving versus static conditions in the three saliency–luminance combinations in right HM-IPL and both hMT/V5+. (D) Location of HM-IPL on coronal section in three subgroups of scans. (E) Apparent motion stimuli and percept in experiment 2. (F) Location of HM-IPL on coronal section in experiment 2. (G) Percentage signal change with respect to fixation in the six different stimulus conditions testing apparent motion (Claeys et al., 2003).

middle temporal gyrus (pMTG), are effector specific. It is noteworthy that this latter selectivity was observed only in single-subject analysis of unsmoothed data, not in smoothed group data. This indicates how cautious one must be in deciding whether regions overlap when using group data.

There is growing evidence that the premotor level of the action observation network is somatotopically organized (Buccino et al., 2001; Jastorff et al., 2010), with foot actions projecting dorsally and mouth and hand actions ventrally. On the other hand, the parietal level is organized according to the type of action observed (Jastorff et al., 2010). Indeed, the latter authors found that observing actions bringing the object toward the actor activated the rostral part of phAIP, whereas actions

moving the object away from the actor activated the caudal part of phAIP. The results of Filimon et al. (2007) and Abdollahi, Jastorff, and Orban (2012) extend this view to include a wider range of actions than object manipulation. Finally, the occipitotemporal level seems to be sensitive to the context of the action (Pelphrey, Morris, & McCarthy, 2004; Pelphrey et al., 2003; Pelphrey, Viola, & McCarthy, 2004). In particular, it is activated when the actions observed are not adapted to the context (irrational actions; Jastorff et al., 2011).

Biological Motion

Influenced by monkey electrophysiology (Oram & Perrett, 1994), initial studies of biological motion (BM)

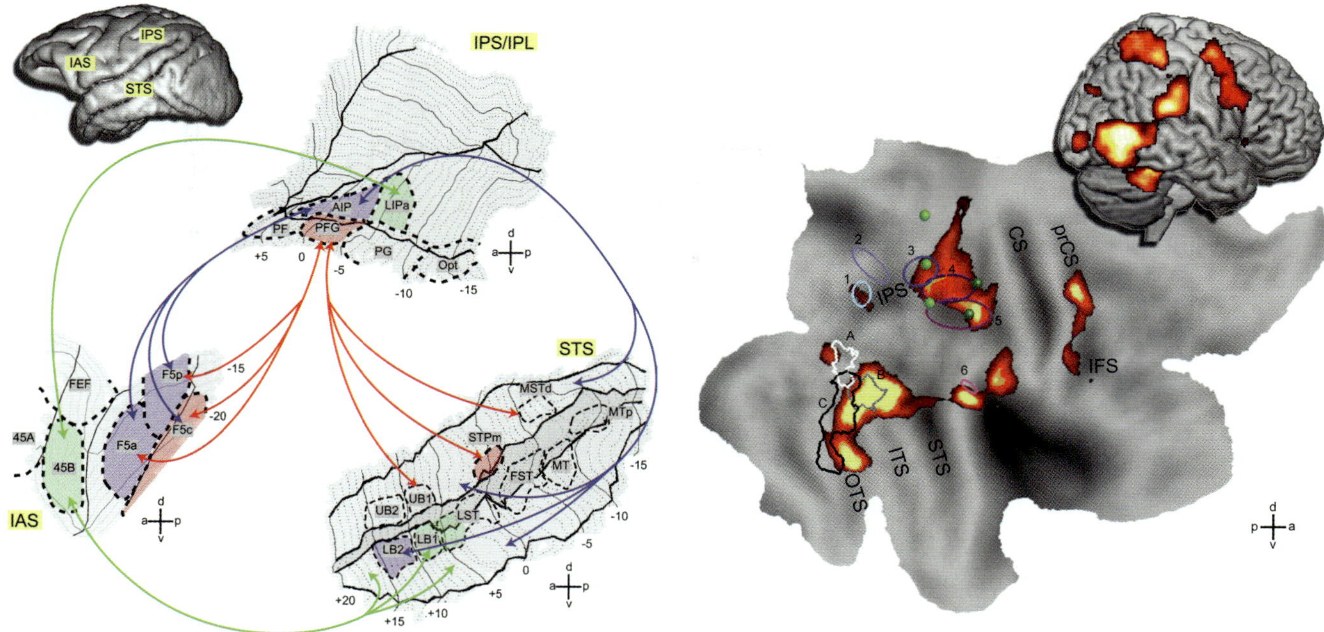

Figure 55.7   Action observation in macaque (left) and human (right). On the left, hatched areas indicate regions activated by observation of grasping in the inferior arcuate sulcus, intraparietal sulcus/ inferior parietal lobule, and superior temporal sulcus (STS). Colored arrows indicate the anatomical connections defining three distinct circuits, two of which link premotor cortex with STS. On the right, voxels significantly more activated by observing manipulation than control stimuli on flattened right hemisphere. Colored ellipses numbered 1–6 indicate VIPS (1), POIPS (2), DIPSM (3), DIPSA (4), phAIP (5), and HM-IPL (6) in clockwise direction from the back; green dots are candidate homologues of VIP; gray, white, and black outlines indicate hMT/V5+, KO, and LOC, respectively (Jastorff et al., 2010; Nelissen et al., 2011).

processing focused on activations in the posterior STS region of humans (Beauchamp et al., 2003; Grossman et al., 2000; Peelen, Wiggett, & Downing, 2006; Safford et al., 2010; Thompson et al., 2005). This was in line with the activation of this region by a range of body movements such as facial and eye movements (Allison, Puce, & McCarthy, 2000; Pelphrey et al., 2005). This restrictive view overlooked the finding that the comparison of biological with scrambled motion also typically activates the inferior temporal gyrus (Grossman & Blake, 2002; Jastorff, Kourtzi, & Giese, 2009; Jastorff & Orban, 2009; Peuskens et al., 2005), as shown in figure 55.8. Thus, the occipitotemporal activations in action observation and in biological motion are rather similar.

Jastorff and Orban (2009) used factorial designs to disentangle the contribution of various components of biological motion processing (Giese & Poggio, 2003; Lange & Lappe, 2006; Troje, 2002). They showed that the *kinematics* factor activated mainly the pMTG branch of the occipitotemporal activation, whereas the *configuration* factor drove the ITG branch. On the other hand the factor *opposed motion* supposedly extracted in KO (figure 55.7) contributed little to the biological motion responses. Finally they observed that the two main factors, kinematics and configuration, were integrated

in the two body areas: the extrastriate body area (EBA) and the fusiform body area (FBA) (figure 55.8). In combining their findings with results from monkey fMRI (Nelissen, Vanduffel, & Orban, 2006) and electrophysiology (Singer & Sheinberg, 2010; Vangeneugden, Pollick, & Vogels, 2009), Jastorff and Orban (2009) proposed that the posterior MTG branch of the BM activation corresponds to the upper bank of monkey STS and that the ITG branch of the activation would correspond to the lower bank of monkey STS.

To test this hypothesis, Jastorff et al. (2012) performed an identical study in the monkey and found that most of their predictions were verified: The *kinematics* factor activated predominantly the upper bank, whereas the *configuration* factor activated predominantly the lower bank of monkey STS. As in humans, both factors interacted in the STS in cortical areas also responsive to the presentation of static bodies. However, the kinematics effect also engaged the fundus of the STS, and the configuration effect was localized in the lateral part of the lower bank (figure 55.8). Three additional observations of Jastorff et al. (2012) are worth mentioning. First, the BM activations are more symmetric in monkeys than in humans (figure 55.8); second, they observed a small species effect in the monkey, whereby the actions

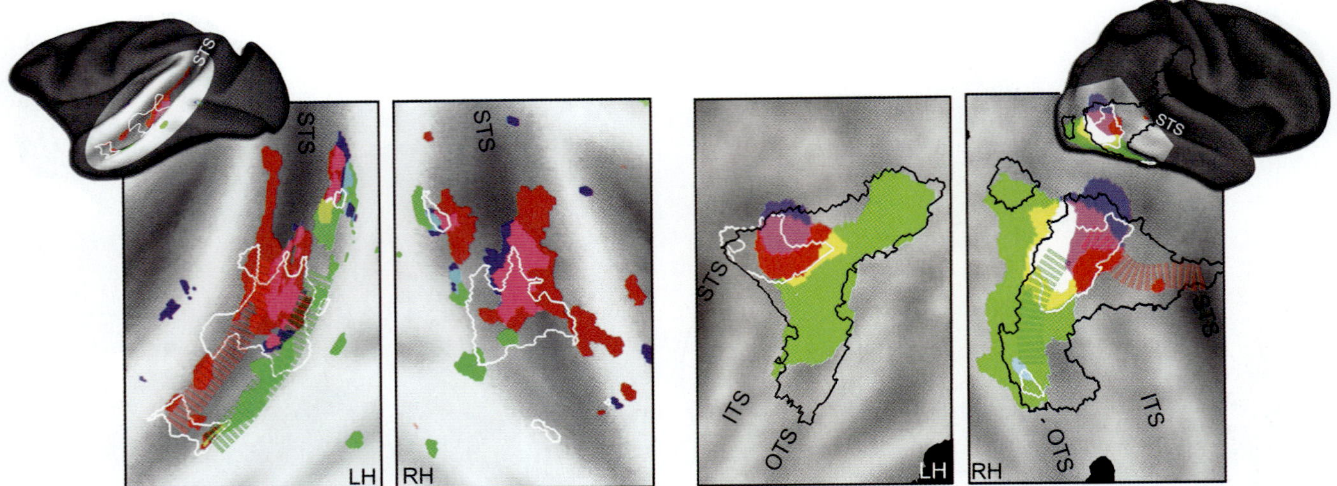

FIGURE 55.8   Voxels significantly activated by the factors kinematics (red) and configuration (green) of biological motion and their interaction (blue), indicated on flattened right and left STS of the monkey (left) and on part of flattened left and right hemispheres in human (right). White outlines EBA, FBA (humans), and anterior, middle, and posterior body patches (monkey); black outlines on the right indicate activation limits for intact versus scrambled biological motion. Striped lines indicate the correspondence between upper-bank STS of monkeys and posterior ITS/MTG/STS in humans and between lower bank STS and posterior ITG/OTS (Jastorff & Orban, 2009; Jastorff et al., 2012).

of monkeys were more efficient than those of humans; and third, the STS regions activated by BM largely overlapped with those activated by the full videos of the same actions. These results not only establish a direct link with single-cell results, although further studies such as these are certainly warranted given the small number of BM-selective neurons observed by Oram and Perrett (1994), but also indicate that the processing of BM is relatively similar in humans and nonhuman primates (Vangeneugden et al., 2010). One interpretation of these human and macaque fMRI results is that the motion and shape cues of BM are integrated in the middle body patch and its likely human counterpart, EBA. This is the level where actions of others are extracted from the retinal array. The configuration- and kinematics-specific activations in front of these body regions have then to be interpreted as representing the postural and kinematic properties of the actions observed, whereas the anterior body regions (anterior body patch and FBA) may process the identities of the actors.

Biological motion was initially considered yet another aspect of higher-order motion somewhat akin to three-dimensional structure from motion (Peuskens et al., 2005), but the recent results reviewed here make it increasingly clear that BM is little more than a very reduced version of action observation (Oram & Perrett, 1994; Saygin et al., 2004). Indeed, the occipitotemporal levels of processing are relatively similar. It is therefore intriguing that the two other levels of the action

observation network have not yet been observed for BM. Saygin et al. (2004), Jastorff, Kourtzi, and Giese (2009), and Jastorff and Orban (2009) have reported inferior frontal activation sites, but those regions do not precisely match the premotor activation found with action observation (figure 55.7). Further work is needed to clarify this discrepancy.

CONCLUSION

The studies reviewed in the present chapter clearly indicate the need for monkey fMRI to link single-cell studies in monkeys with human fMRI and to try to ascertain homologies between cortical areas in the two species. They also clearly indicate that a single functional characteristic such as responsiveness to motion does not uniquely define a single cortical area.

More specifically, these studies indicate that the human cortex includes multiple motion-processing pathways. Each of these pathways runs deeper than initially anticipated and includes several processing levels, as is illustrated by the optic-flow-sensitive areas, which extend far beyond MSTd into dorsal parietal cortex, where they are most likely used to control locomotion. These multiple pathways provide a rich variety of exit points from the visual system, underscoring the importance of considering the visual system from the back end (Orban, 2007) rather than from the retinal starting point (Hubel & Wiesel, 1962) when attempting to understand visual processing.

REFERENCES

Abdollahi, R.O., Jastorff, J., & Orban, G.A. (2012). Common and segregated processing of observed actions in human SPL. *Cerebral Cortex ,* epub. doi:10.1093/cercor/bhs264.

Albright, T. D. (1984). Direction and orientation selectivity of neurons in visual area MT of the macaque. *Journal of Neurophysiology, 52,* 1106–1130.

Allison, T., Puce, A., & McCarthy, G. (2000). Social perception from visual cues: role of the STS region. *Trends in Cognitive Sciences, 4,* 267–278. doi:10.1016/S1364-6613(00)01501-1.

Amano, K., Wandell, B. A., & Dumoulin, S. O. (2009). Visual field maps, population receptive field sizes, and visual field coverage in the human MT+ complex. *Journal of Neurophysiology, 102,* 2704–2718.

An, X., Yin, J., Pan, Y., Zhang, X., Gong, H., Yang, Y., et al. (2010). The reversal visual cortical mechanism of functional organizations underlying kinetic contour processing in V1, V2 and V4 of rhesus macaques. *Society for Neuroscience Abstracts, 372,* 15.

Bartels, A., Zeki, S., & Logothetis, N. K. (2008). Natural vision reveals regional specialization to local motion and to contrast-invariant, global flow in the human brain. *Cerebral Cortex, 18,* 705–717.

Beauchamp, M. S., Lee, K. E., Haxby, J. V., & Martin, A. (2003). fMRI responses to video and point-light displays of moving humans and manipulable objects. *Journal of Cognitive Neuroscience, 15,* 991–1001.

Beer, A. L., Watanabe, T., Ni, R., Sasaki, Y., & Andersen, G. J. (2009). 3D surface perception from motion involves a temporal-parietal network. *European Journal of Neuroscience, 30,* 703–713.

Bremmer, F., Schlack, A., Shah, N. J., Zafiris, O., Kubischik, M., Hoffmann, K. P., et al. (2001). Polymodal motion processing in posterior parietal and premotor cortex: A human fMRI study strongly implies equivalencies between humans and monkeys. *Neuron, 29,* 287–296. doi:10.1016/S0896-6273(01)00198-2.

Buccino, G., Binkofski, F., Fink, G. R., Fadiga, L., Fogassi, L., Gallese, V., et al. (2001). Action observation activates premotor and parietal areas in a somatotopic manner: An fMRI study. *European Journal of Neuroscience, 13,* 400–404. doi:10.1111/j.1460-9568.2001.01385.x.

Buccino, G., Lui, F., Canessa, N., Patteri, I., Lagravinese, G., Benuzzi, F., et al. (2004). Neural circuits involved in the recognition of actions performed by nonconspecifics: An fMRI study. *Journal of Cognitive Neuroscience, 16,* 114–126. doi:10.1162/089892904322755601.

Cardin, V., & Smith, A. T. (2010). Sensitivity of human visual and vestibular cortical regions to egomotion-compatible visual stimulation. *Cerebral Cortex, 20,* 1964–1973.

Cavanagh, P. (1992). Attention-based motion perception. *Science, 257,* 1563–1565.

Chen, A., DeAngelis, G. C., & Angelaki, D. E. (2011). Convergence of vestibular and visual self-motion signals in an area of the posterior Sylvian fissure. *Journal of Neuroscience, 31,* 11617–11627.

Claeys, K. G., Lindsey, D. T., De Schutter, E., & Orban, G. A. (2003). A higher order motion region in human inferior parietal lobule: Evidence from fMRI. *Neuron, 40,* 631–642.

Colby, C. L., Gattass, R., Olson, C. R., & Gross, C. G. (1988). Topographical organization of cortical afferents to extrastriate visual area PO in the macaque: A dual tracer study. *Journal of Comparative Neurology, 269,* 392–413.

DeYoe, E. A., Carman, G. J., Bandettini, P., Glickman, S., Wieser, J., Cox, R., et al. (1996). Mapping striate and extrastriate visual areas in human cerebral cortex. *Proceedings of the National Academy of Sciences of the United States of America, 93,* 2382–2386. doi:10.1073/pnas.93.6.2382.

Duffy, C. J., & Wurtz, R. H. (1991). Sensitivity of MST neurons to optic flow stimuli. I. A continuum of response selectivity to large-field stimuli. *Journal of Neurophysiology, 65,* 1329–1345.

Durand, J. B., Peeters, R., Norman, J. F., Todd, J. T., & Orban, G. A. (2009). Parietal regions processing visual 3D shape extracted from disparity. *NeuroImage, 46,* 1114–1126.

El-Shamayleh, Y., & Movshon, J. A. (2011). Neuronal responses to texture-defined form in macaque visual area V2. *Journal of Neuroscience, 31,* 8543–8555.

Fattori, P., Pitzalis, S., & Galletti, C. (2009). The cortical visual area V6 in macaque and human brains. *Journal of Physiology, Paris, 103,* 88–97.

Filimon, F., Nelson, J. D., Hagler, D. J., & Sereno, M. I. (2007). Human cortical representations for reaching: mirror neurons for execution, observation, and imagery. *NeuroImage, 37,* 1315–1328.

Fischer, E., Bulthoff, H. H., Logothetis, N. K., & Bartels, A. (2012. Visual motion responses in the posterior cingulate sulcus: A comparison to V5/MT and MST. *Cerebral Cortex, 22,* 865–876.

Fize, D., Vanduffel, W., Nelissen, K., Van Hecke, P., Mandeville, J. B., Tootell, R. B., & Orban, G. A. (2001). Distributed processing of kinetic boundaries in monkeys investigated using fMRI. *Society for Neuroscience Abstracts 27,* 11.19.

Galletti, C., Fattori, P., Gamberini, M., & Kutz, D. F. (1999). The cortical visual area V6: Brain location and visual topography. *European Journal of Neuroscience, 11,* 3922–3936.

Gazzola, V., & Keysers, C. (2009). The observation and execution of actions share motor and somatosensory voxels in all tested subjects: Single-subject analyses of unsmoothed fMRI data. *Cerebral Cortex, 19,* 1239–1255.

Georgieva, S., Peeters, R., Kolster, H., Todd, J. T., & Orban, G. A. (2009). The processing of three-dimensional shape from disparity in the human brain. *Journal of Neuroscience, 29,* 727–742.

Georgieva, S. S., Todd, J. T., Peeters, R., & Orban, G. A. (2008). The extraction of 3D shape from texture and shading in the human brain. *Cerebral Cortex, 18,* 2416–2138.

Giese, M. A., & Poggio, T. (2003). Neural mechanisms for the recognition of biological movements. *Nature Reviews. Neuroscience, 4,* 179–192.

Graziano, M. S., Andersen, R. A., & Snowden, R. J. (1994). Tuning of MST neurons to spiral motions. *Journal of Neuroscience, 14,* 54–67.

Grossman, E. D., & Blake, R. (2002). Brain areas active during visual perception of biological motion. *Neuron, 35,* 1167–1175. doi:10.1016/S0896-6273(02)00897-8.

Grossman, E., Donnelly, M., Price, R., Pickens, D., Morgan, V., Neighbor, G., et al. (2000). Brain areas involved in perception of biological motion. *Journal of Cognitive Neuroscience, 12,* 711–720. doi:10.1162/089892900562417.

Guipponi, O., Wardak, C., Pinede, S., Comte, J. C., Sappey-Marinier, D., & Hamed, S. B. (2011). Identification of ventral intraparietal area (VIP) with multiple sensory stimulations: A functional magnetic resonance imaging (fMRI) study in awake monkeys. *Society for Neuroscience Abstracts, 575*, 04.

Hawken, M. J., Parker, A. J., & Lund, J. S. (1988). Laminar organization and contrast sensitivity of direction-selective cells in the striate cortex of the Old World monkey. *Journal of Neuroscience, 8*, 3541–3548.

Hedges, J. H., Gartshteyn, Y., Kohn, A., Rust, N. C., Shadlen, M. N., Newsome, W. T., et al. (2011). Dissociation of neuronal and psychophysical responses to local and global motion. *Current Biology, 21*, 2023–2028. doi:10.1016/j.cub.2011.10.049.

Hubel, D. H., & Wiesel, T. N. (1962). Receptive fields, binocular interaction and functional architecture in the cat's visual cortex. *Journal of Physiology, 160*, 106–154.

Hubel, D. H., & Wiesel, T. N. (1968). Receptive fields and functional architecture of monkey striate cortex. *Journal of Physiology, 195*, 215–243.

Huk, A. C., Dougherty, R. F., & Heeger, D. J. (2002). Retinotopy and functional subdivision of human areas MT and MST. *Journal of Neuroscience, 22*, 7195–7205.

Jastorff, J., Abdollahi, R. O., & Orban, G. A. (2012). Acting alters visual processing: Flexible recruitment of visual areas by one's own actions. *Cerebral Cortex, 22*, 2930–2942.

Jastorff, J., Begliomini, C., Fabbri-Destro, M., Rizzolatti, G., & Orban, G. A. (2010). Coding observed motor acts: Different organizational principles in the parietal and premotor cortex of humans. *Journal of Neurophysiology, 104*, 128–140.

Jastorff, J., Clavagnier, S., Gergely, G., & Orban, G. A. (2011). Neural mechanisms of understanding rational actions: Middle temporal gyrus activation by contextual violation. *Cerebral Cortex, 21*, 318–329.

Jastorff, J., Kourtzi, Z., & Giese, M. A. (2009). Visual learning shapes the processing of complex movement stimuli in the human brain. *Journal of Neuroscience, 29*, 14026–14038.

Jastorff, J., & Orban, G. A. (2009). Human functional magnetic resonance imaging reveals separation and integration of shape and motion cues in biological motion processing. *Journal of Neuroscience, 29*, 7315–7329.

Jastorff, J., Popivanov, I. D., Vogels, R., Vanduffel, W., & Orban, G. A. (2012). Integration of shape and motion cues in biological motion processing in the monkey STS. *NeuroImage, 60*, 911–921.

Joly, O., Vanduffel, W., & Orban, G. A. (2009). The monkey ventral premotor cortex processes 3D shape from disparity. *NeuroImage, 47*, 262–272.

Kleinschmidt, A., Thilo, K. V., Buchel, C., Gresty, M. A., Bronstein, A. M., & Frackowiak, R. S. (2002). Neural correlates of visual-motion perception as object- or self-motion. *NeuroImage, 16*, 873–882.

Kolster, H., Mandeville, J. B., Arsenault, J. T., Ekstrom, L. B., Wald, L. L., & Vanduffel, W. (2009). Visual field map clusters in macaque extrastriate visual cortex. *Journal of Neuroscience, 29*, 7031–7039.

Kolster, H., Peeters, R., & Orban, G. A. (2010). The retinotopic organization of the human middle temporal area MT/V5 and its cortical neighbors. *Journal of Neuroscience, 30*, 9801–9820.

Kourtzi, Z., Bulthoff, H. H., Erb, M., & Grodd, W. (2002). Object-selective responses in the human motion area MT/MST. *Nature Neuroscience, 5*, 17–18.

Kovacs, G., Raabe, M., & Greenlee, M. W. (2008). Neural correlates of visually induced self-motion illusion in depth. *Cerebral Cortex, 18*, 1779–1787. doi:10.1093/cercor/bhm203.

Kriegeskorte, N., Sorger, B., Naumer, M., Schwarzbach, J., van den Boogert, E., Hussy, W., et al. (2003). Human cortical object recognition from a visual motion flowfield. *Journal of Neuroscience, 23*, 1451–1463.

Lagae, L., Maes, H., Raiguel, S., Xiao, D. K., & Orban, G. A. (1994). Responses of macaque STS neurons to optic flow components: A comparison of areas MT and MST. *Journal of Neurophysiology, 71*, 1597–1626.

Lagae, L., Raiguel, S., & Orban, G. A. (1993). Speed and direction selectivity of macaque middle temporal neurons. *Journal of Neurophysiology, 69*, 19–39.

Lange, J., & Lappe, M. (2006). A model of biological motion perception from configural form cues. *Journal of Neuroscience, 26*, 2894–2906.

Larsson, J., & Heeger, D. J. (2006). Two retinotopic visual areas in human lateral occipital cortex. *Journal of Neuroscience, 26*, 13128–13142.

Larsson, J., Heeger, D. J., & Landy, M. S. (2010). Orientation selectivity of motion-boundary responses in human visual cortex. *Journal of Neurophysiology, 104*, 2940–2950.

Lewis, J. W., & Van Essen, D. C. (2000). Mapping of architectonic subdivisions in the macaque monkey, with emphasis on parieto-occipital cortex. *Journal of Comparative Neurology, 428*, 79–111.

Likova, L. T., & Tyler, C. W. (2007). Stereomotion processing in the human occipital cortex. *NeuroImage, 38*, 293–305.

Lu, Z. L., & Sperling, G. (1995). Attention-generated apparent motion. *Nature, 377*, 237–239.

Marcar, V. L., Raiguel, S. E., Xiao, D., & Orban, G. A. (2000). Processing of kinetically defined boundaries in areas V1 and V2 of the macaque monkey. *Journal of Neurophysiology, 84*, 2786–2798.

Maunsell, J. H., & Van Essen, D. C. (1983). Functional properties of neurons in middle temporal visual area of the macaque monkey. II. Binocular interactions and sensitivity to binocular disparity. *Journal of Neurophysiology, 49*, 1148–1167.

Mikami, A., Newsome, W. T., & Wurtz, R. H. (1986). Motion selectivity in macaque visual cortex. I. Mechanisms of direction and speed selectivity in extrastriate area MT. *Journal of Neurophysiology, 55*, 1308–1327.

Morrone, M. C., Tosetti, M., Montanaro, D., Fiorentini, A., Cioni, G., & Burr, D. C. (2000). A cortical area that responds specifically to optic flow, revealed by fMRI. *Nature Neuroscience, 3*, 1322–1328.

Muckli, L., Kohler, A., Kriegeskorte, N., & Singer, W. (2005). Primary visual cortex activity along the apparent-motion trace reflects illusory perception. *PLoS Biology, 3*, e265. doi:10.1371/journal.pbio.0030265.

Muckli, L., Kriegeskorte, N., Lanfermann, H., Zanella, F. E., Singer, W., & Goebel, R. (2002). Apparent motion: Event-related functional magnetic resonance imaging of perceptual switches and states. *Journal of Neuroscience, 22*, RC219.

Mur, M., Ruff, D. A., Bodurka, J., Bandettini, P. A., & Kriegeskorte, N. (2010). Face-identity change activation outside the face system: "Release from adaptation" may not always indicate neuronal selectivity. *Cerebral Cortex, 20*, 2027–2042. doi:10.1093/cercor/bhp272.

Mysore, S. G., Vogels, R., Kolster, H., Vanduffel, W., & Orban, G. (2010). Cortical network of 3D structure from motion (3DSFM) in the macaque: A functional imaging study. *Society for Neuroscience Abstract Online* program No 776.9.

Mysore, S. G., Vogels, R., Raiguel, S. E., & Orban, G. A. (2006). Processing of kinetic boundaries in macaque V4. *Journal of Neurophysiology, 95*, 1864–1880.

Mysore, S. G., Vogels, R., Raiguel, S. E., Todd, J. T., & Orban, G. A. (2010). The selectivity of neurons in the macaque fundus of the superior temporal area for three-dimensional structure from motion. *Journal of Neuroscience, 30*, 15491–15508.

Nelissen, K., Borra, E., Gerbella, M., Rozzi, S., Luppino, G., Vanduffel, W., et al. (2011). Action observation circuits in the macaque monkey cortex. *Journal of Neuroscience, 31*, 3743–3756. doi:10.1523/JNEUROSCI.4803-10.2011.

Nelissen, K., Joly, O., Durand, J. B., Todd, J. T., Vanduffel, W., & Orban, G. A. (2009). The extraction of depth structure from shading and texture in the macaque brain. *PLoS One, 4*, e8306. doi:10.1371/journal.pone.0008306.

Nelissen, K., Vanduffel, W., & Orban, G. A. (2006). Charting the lower superior temporal region, a new motion-sensitive region in monkey superior temporal sulcus. *Journal of Neuroscience, 26*, 5929–5947.

Noguchi, Y., Kaneoke, Y., Kakigi, R., Tanabe, H. C., & Sadato, N. (2005). Role of the superior temporal region in human visual motion perception. *Cerebral Cortex, 15*, 1592–1601.

Oosterhof, N. N., Tipper, S. P., & Downing, P. E. (2012). Viewpoint (in)dependence of action representations: An MVPA study. *Journal of Cognitive Neuroscience, 24*, 975–989.

Oram, M. W., & Perrett, D. I. (1994). Responses of anterior superior temporal polysensory (STPa) neurons to "biological motion" stimuli. *Journal of Cognitive Neuroscience, 6*, 99–116.

Orban, G. A. (1997). Visual processing in macaque area MT/V5 and its satellites (MSTd and MSTv). In K. S. Rockland, J. H. Kaas, & A. Peters (Eds.), *Extrastriate cortex in primates* (Vol. 12, pp. 359–434). New York: Plenum Press.

Orban, G. A. (2002). Functional MRI in the awake monkey: The missing link. *Journal of Cognitive Neuroscience, 14*, 965–969.

Orban, G. A. (2007). *La vision, mission du cerveau* (Vol. 10). Paris: Collège de France/Fayard.

Orban, G. A. (2008). Higher order visual processing in macaque extrastriate cortex. *Physiological Reviews, 88*(1), 59–89.

Orban, G. A. (2011). The extraction of 3D shape in the visual system of human and nonhuman primates. *Annual Review of Neuroscience, 34*, 361–388.

Orban, G. A., Claeys, K., Nelissen, K., Smans, R., Sunaert, S., Todd, J. T., et al. (2006). Mapping the parietal cortex of human and non-human primates. *Neuropsychologia, 44*, 2647–2667. doi:10.1016/j.neuropsychologia.2005.11.001.

Orban, G. A., Fize, D., Peuskens, H., Denys, K., Nelissen, K., Sunaert, S., et al. (2003). Similarities and differences in motion processing between the human and macaque brain: Evidence from fMRI. *Neuropsychologia, 41*, 1757–1768. doi:10.1016/S0028-3932(03)00177-5.

Orban, G. A., Kennedy, H., & Bullier, J. (1986). Velocity sensitivity and direction selectivity of neurons in areas V1 and V2 of the monkey: Influence of eccentricity. *Journal of Neurophysiology, 56*, 462–480.

Orban, G. A., Sunaert, S., Todd, J. T., Van Hecke, P., & Marchal, G. (1999). Human cortical regions involved in extracting depth from motion. *Neuron, 24*, 929–940.

Orban, G. A., & Vanduffel, W. (2004). Functional mapping of motion regions. In L. M. Chalupa & J. S. Werner (Eds.), *The visual neurosciences* (Vol. 2, pp. 1229–1246). Cambridge, MA: MIT Press.

Pan, Y., An, X., Yin, J., Zhang, X., Gong, H., Yang, Y., et al. (2010). The functional organizations underlying illusory contour processing in extrastriate cortex V2 and V4d in macaques. *Society for Neuroscience Abstracts, 580*, 12.

Peelen, M. V., Wiggett, A. J., & Downing, P. E. (2006). Patterns of fMRI activity dissociate overlapping functional brain areas that respond to biological motion. *Neuron, 49*, 815–822.

Peeters, R., Simone, L., Nelissen, K., Fabbri-Destro, M., Vanduffel, W., Rizzolatti, G., et al. (2009). The representation of tool use in humans and monkeys: Common and uniquely human features. *Journal of Neuroscience, 29*, 11523–11539. doi:10.1523/JNEUROSCI.2040-09.2009.

Pelphrey, K. A., Morris, J. P., & McCarthy, G. (2004). Grasping the intentions of others: The perceived intentionality of an action influences activity in the superior temporal sulcus during social perception. *Journal of Cognitive Neuroscience, 16*, 1706–1716.

Pelphrey, K. A., Morris, J. P., Michelich, C. R., Allison, T., & McCarthy, G. (2005). Functional anatomy of biological motion perception in posterior temporal cortex: An fMRI study of eye, mouth and hand movements. *Cerebral Cortex, 15*, 1866–1876.

Pelphrey, K. A., Singerman, J. D., Allison, T., & McCarthy, G. (2003). Brain activation evoked by perception of gaze shifts: The influence of context. *Neuropsychologia, 41*,156–170.

Pelphrey, K. A., Viola, R. J., & McCarthy, G. (2004). When strangers pass: Processing of mutual and averted social gaze in the superior temporal sulcus. *Psychological Science, 15*, 598–603.

Perrett, D. I., Smith, P. A., Mistlin, A. J., Chitty, A. J., Head, A. S., Potter, D. D., et al. (1985). Visual analysis of body movements by neurones in the temporal cortex of the macaque monkey: A preliminary report. *Behavioural Brain Research, 16*, 153–170. doi:10.1016/0166-4328(85)90089-0.

Peuskens, H., Claeys, K. G., Todd, J. T., Norman, J. F., Van Hecke, P., & Orban, G. A. (2004). Attention to 3-D shape, 3-D motion, and texture in 3-D structure from motion displays. *Journal of Cognitive Neuroscience, 16*, 665–682.

Peuskens, H., Sunaert, S., Dupont, P., Van Hecke, P., & Orban, G. A. (2001). Human brain regions involved in heading estimation. *Journal of Neuroscience, 21*, 2451–2461.

Peuskens, H., Vanrie, J., Verfaillie, K., & Orban, G. A. (2005). Specificity of regions processing biological motion. *European Journal of Neuroscience, 21*, 2864–2875.

Pitzalis, S., Galletti, C., Huang, R. S., Patria, F., Committeri, G., Galati, G., et al. (2006). Wide-field retinotopy defines human cortical visual area V6. *Journal of Neuroscience, 26*, 7962–7973. doi:10.1523/JNEUROSCI.0178-06.2006.

Pitzalis, S., Sereno, M. I., Committeri, G., Fattori, P., Galati, G., Patria, F., et al. (2010). Human V6: The medial motion area. *Cerebral Cortex, 20*, 411–424. doi:10.1093/cercor/bhp112.

Safford, A. S., Hussey, E. A., Parasuraman, R., & Thompson, J. C. (2010). Object-based attentional modulation of biological motion processing: Spatiotemporal dynamics using functional magnetic resonance imaging and electro-encephalography. *Journal of Neuroscience, 30*, 9064–9073.

Sary, G., Vogels, R., & Orban, G. A. (1993). Cue-invariant shape selectivity of macaque inferior temporal neurons. *Science, 260*, 995–997.

Sawamura, H., Orban, G. A., & Vogels, R. (2006). Selectivity of neuronal adaptation does not match response selectivity: A single-cell study of the fMRI adaptation paradigm. *Neuron, 49*, 307–318.

Saygin, A. P., Wilson, S. M., Hagler, D. J., Jr., Bates, E., & Sereno, M. I. (2004). Point-light biological motion perception activates human premotor cortex. *Journal of Neuroscience, 24*, 6181–6188.

Schaafsma, S. J., & Duysens, J. (1996). Neurons in the ventral intraparietal area of awake macaque monkey closely resemble neurons in the dorsal part of the medial superior temporal area in their responses to optic flow patterns. *Journal of Neurophysiology, 76*, 4056–4068.

Sereno, M. I., & Huang, R. S. (2006). A human parietal face area contains aligned head-centered visual and tactile maps. *Nature Neuroscience, 9*, 1337–1343.

Singer, J. M., & Sheinberg, D. L. (2010). Temporal cortex neurons encode articulated actions as slow sequences of integrated poses. *Journal of Neuroscience, 30*, 3133–3145.

Smith, A. T., Greenlee, M. W., Singh, K. D., Kraemer, F. M., & Hennig, J. (1998). The processing of first- and second-order motion in human visual cortex assessed by functional magnetic resonance imaging (fMRI). *Journal of Neuroscience, 18*, 3816–3830.

Smith, A. T., Wall, M. B., & Thilo, K. V. (2011). Vestibular inputs to human motion-sensitive visual cortex. *Cerebral Cortex, 22*, 1068. doi:10.1093/cercor/bhr179.

Squatrito, S., Raffi, M., Maioli, M. G., & Battaglia-Mayer, A. (2001). Visual motion responses of neurons in the caudal area PE of macaque monkeys. *Journal of Neuroscience, 21*, RC130.

Sterzer, P., & Kleinschmidt, A. (2005). A neural signature of colour and luminance correspondence in bistable apparent motion. *European Journal of Neuroscience, 21*, 3097–3106.

Sterzer, P., Russ, M. O., Preibisch, C., & Kleinschmidt, A. (2002). Neural correlates of spontaneous direction reversals in ambiguous apparent motion. *NeuroImage, 15*, 908–916.

Sugihara, H., Murakami, I., Shenoy, K. V., Andersen, R. A., & Komatsu, H. (2002). Response of MSTd neurons to simulated 3D orientation of rotating planes. *Journal of Neurophysiology, 87*, 273–285.

Sunaert, S., Van Hecke, P., Marchal, G., & Orban, G. A. (1999). Motion-responsive regions of the human brain. *Experimental Brain Research, 127*, 355–370.

Tanaka, K., Hikosaka, K., Saito, H., Yukie, M., Fukada, Y., & Iwai, E. (1986). Analysis of local and wide-field movements in the superior temporal visual areas of the macaque monkey. *Journal of Neuroscience, 6*, 134–144.

Tanaka, K., Sugita, Y., Moriya, M., & Saito, H. (1993). Analysis of object motion in the ventral part of the medial superior temporal area of the macaque visual cortex. *Journal of Neurophysiology, 69*, 128–142.

Thompson, J. C., Clarke, M., Stewart, T., & Puce, A. (2005). Configural processing of biological motion in human superior temporal sulcus. *Journal of Neuroscience, 25*, 9059–9066.

Tootell, R. B., Mendola, J. D., Hadjikhani, N. K., Ledden, P. J., Liu, A. K., Reppas, J. B., et al. (1997). Functional analysis of V3A and related areas in human visual cortex. *Journal of Neuroscience, 17*, 7060–7078.

Troje, N. F. (2002). Decomposing biological motion: A framework for analysis and synthesis of human gait patterns. *Journal of Vision, 2*, 371–387. doi:10.1167/2.5.2.

Tyler, C. W., Likova, L. T., Kontsevich, L. L., & Wade, A. R. (2006). The specificity of cortical region KO to depth structure. *NeuroImage, 30*, 228–238.

Ungerleider, L. G., & Desimone, R. (1986). Projections to the superior temporal sulcus from the central and peripheral field representations of V1 and V2. *Journal of Comparative Neurology, 248*, 147–163.

Van Essen, D. C., Glasser, M. F., Dierker, D. L., Harwell, J., & Coalson, T. (2011). Parcellations and Hemispheric Asymmetries of Human Cerebral Cortex Analyzed on Surface-Based Atlases. *Cereb Cortex*. doi:10.1093/cercor/bhr291.

Van Oostende, S., Sunaert, S., Van Hecke, P., Marchal, G., & Orban, G. A. (1997). The kinetic occipital (KO) region in man: an fMRI study. *Cereb Cortex, 7*(7), 690–701.

Vanduffel, W., Fize, D., Mandeville, J. B., Nelissen, K., Van Hecke, P., Rosen, B. R., et al. (2001). Visual motion processing investigated using contrast agent-enhanced fMRI in awake behaving monkeys. *Neuron, 32*, 565–577. doi:10.1016/S0896-6273(01)00502-5.

Vanduffel, W., Fize, D., Peuskens, H., Denys, K., Sunaert, S., Todd, J. T., et al. (2002). Extracting 3D from motion: Differences in human and monkey intraparietal cortex. *Science, 298*, 413–415. doi:10.1126/science.1073574.

Van Essen, D. C., Glasser, M. F., Dierker, D. L., Harwell, J., & Coalson, T. (2012). Parcellations and hemispheric asymmetries of human cerebral cortex analyzed on surface-based atlases. *Cerebral Cortex, 22*, 2241–2261.

Vangeneugden, J., De Maziere, P. A., Van Hulle, M. M., Jaeggli, T., Van Gool, L., & Vogels, R. (2011). Distinct mechanisms for coding of visual actions in macaque temporal cortex. *Journal of Neuroscience, 31*, 385–401.

Vangeneugden, J., Pollick, F., & Vogels, R. (2009). Functional differentiation of macaque visual temporal cortical neurons using a parametric action space. *Cerebral Cortex, 19*, 593–611.

Vangeneugden, J., Vancleef, K., Jaeggli, T., VanGool, L., & Vogels, R. (2010). Discrimination of locomotion direction in impoverished displays of walkers by macaque monkeys. *Journal of Vision, 10*, 1–19. doi:10.1167/10.4.22.

Van Oostende, S., Sunaert, S., Van Hecke, P., Marchal, G., & Orban, G. A. (1997). The kinetic occipital (KO) region in man: An fMRI study. *Cerebral Cortex, 7*, 690–701.

Wall, M. B., & Smith, A. T. (2008). The representation of egomotion in the human brain. *Current Biology, 18*, 191–194.

Weigelt, S., Kourtzi, Z., Kohler, A., Singer, W., & Muckli, L. (2007). The cortical representation of objects rotating in depth. *Journal of Neuroscience, 27*, 3864–3874.

Xiao, D. K., Marcar, V. L., Raiguel, S. E., & Orban, G. A. (1997). Selectivity of macaque MT/V5 neurons for surface orientation in depth specified by motion. *European Journal of Neuroscience, 9*, 956–964.

Yamamoto, T., Takahashi, S., Hanakawa, T., Urayama, S., Aso, T., Fukuyama, H., & Ejima, Y. (2008). Neural correlates of the stereokinetic effect revealed by functional magnetic resonance imaging. *Journal of Vision, 8*, 1–17. doi:10.1167/8.10.14.

Zeki, S., Perry, R. J., & Bartels, A. (2003). The processing of kinetic contours in the brain. *Cerebral Cortex, 13*, 189–202.

Zeki, S., Watson, J. D., Lueck, C. J., Friston, K. J., Kennard, C., & Frackowiak, R. S. (1991). A direct demonstration of functional specialization in human visual cortex. *Journal of Neuroscience, 11*, 641–649.

Zhuo, Y., Zhou, T. G., Rao, H. Y., Wang, J. J., Meng, M., Chen, M., et al. (2003). Contributions of the visual ventral pathway to long-range apparent motion. *Science, 299*, 417–420. doi:10.1126/science.1077091.

56 The Cortical Analysis of Optic Flow: Mechanism, Function, and Dysfunction

CHARLES J. DUFFY

Geniculostriate and colliculopulvinar projections carrying signals about self-movement converge onto dorsal extrastriate cortex. The net effect is a highly interconnected, distributed system for visuospatial processing (Felleman & Van Essen, 1991; Lewis & Van Essen, 2000).

Neurons in the dorsal medial superior temporal area (MSTd) combine visual direction, size, and rotation responses with response preferences for very large stimuli (circular diameters >40°), with their strongest responses to the movement of random dot patterns and not to small moving objects. These findings are the foundation of the notion that MSTd pattern motion responses contribute to self-movement processing (Duffy & Wurtz, 1991b; Graziano, Andersen, & Snowden, 1994; Lappe, 1996; Tanaka, Fukuda, & Saito, 1989).

Many models of MST heading selectivity have been developed. A structural model using larger, partially overlapping, direction-selective, excitatory and inhibitory receptive field gradients accounts for the hierarchy of increasingly selective units (Duffy & Wurtz, 1991a; Yu et al., 2010). A directional template-matching model, using middle temporal (MT)-like direction and speed-tuned visual field subunits, suggests there is a mosaic of sensors across the visual field (Perrone & Stone, 1994). A subset of these direction-speed sensors projected onto each MST-like neuron, conveying Gaussian distributions of FOE selectivity and replicating optic flow position invariance (Perrone & Stone, 1998).

OPTIC FLOW ANALYSIS, PERCEPTION, AND BEHAVIOR

Motion perception is linked to activity in the superior temporal sulcus (STS) cortex (Pasternak & Merigan, 1994). MT and MST neuronal responses to their preferred and antipreferred motion directions are related to the monkey's motion coherence discrimination thresholds for those stimuli. Furthermore, both behavioral and neuronal responses show similar effects of varying motion parameters (Britten et al., 1992; Celebrini & Newsome, 1994; Thiele & Hoffmann, 1996)

and of microstimulation in MSTd (Britten, 1998; Britten & van Wezel, 1998). These findings suggest a correspondence between neuronal activity in STS cortex and perceptual decisions about heading direction from visual motion.

MSTd neuronal activity has been linked to the illusory perception of optic flow. When a radial pattern of optic flow is superimposed on a uniform pattern of planar translational movement, the focus of expansion (FOE) appears to be displaced in the direction of the planar movement (Duffy & Wurtz, 1993). Some MSTd neurons show similar effects of overlapping radial and planar stimuli with responses that shift to match those evoked by radial stimuli with shifted FOEs (Duffy & Wurtz, 1997b). This behavior is mimicked by neurons in a computational model of MSTd optic flow analysis with that correspondence including the spectrum of neurons with and without the illusory shift of FOE tuning. This supports the notion that MSTd neuronal population encoding is linked to optic flow perception (Lappe, 1998).

Optic flow from natural observer movement presents a sequence of motion patterns that reflect heading changes. MSTd neurons respond to continuously changing optic flow patterns by transitioning their responses from those evoked by the preceding pattern to those evoked by the subsequent pattern (Duffy & Wurtz, 1997a). These transitions are not simple linear interpolations between responses. Rather, all MSTd neurons show nonlinear changes in activity during transitions between some stimuli. These effects may be attributable to the temporal context created by a sequence of optic flow stimuli rather than unique responses to particular motion patterns seen during the transition between other patterns (Paolini et al., 2000).

Stimulus sequence effects are more evident in MSTd neuronal responses to whole-body translational movement on clockwise (CW) and counterclockwise (CC) circular paths around a room while the animal views a wall-mounted light array (figure 56.1). Heading selectivity is seen in 35% of the neurons as a preference for

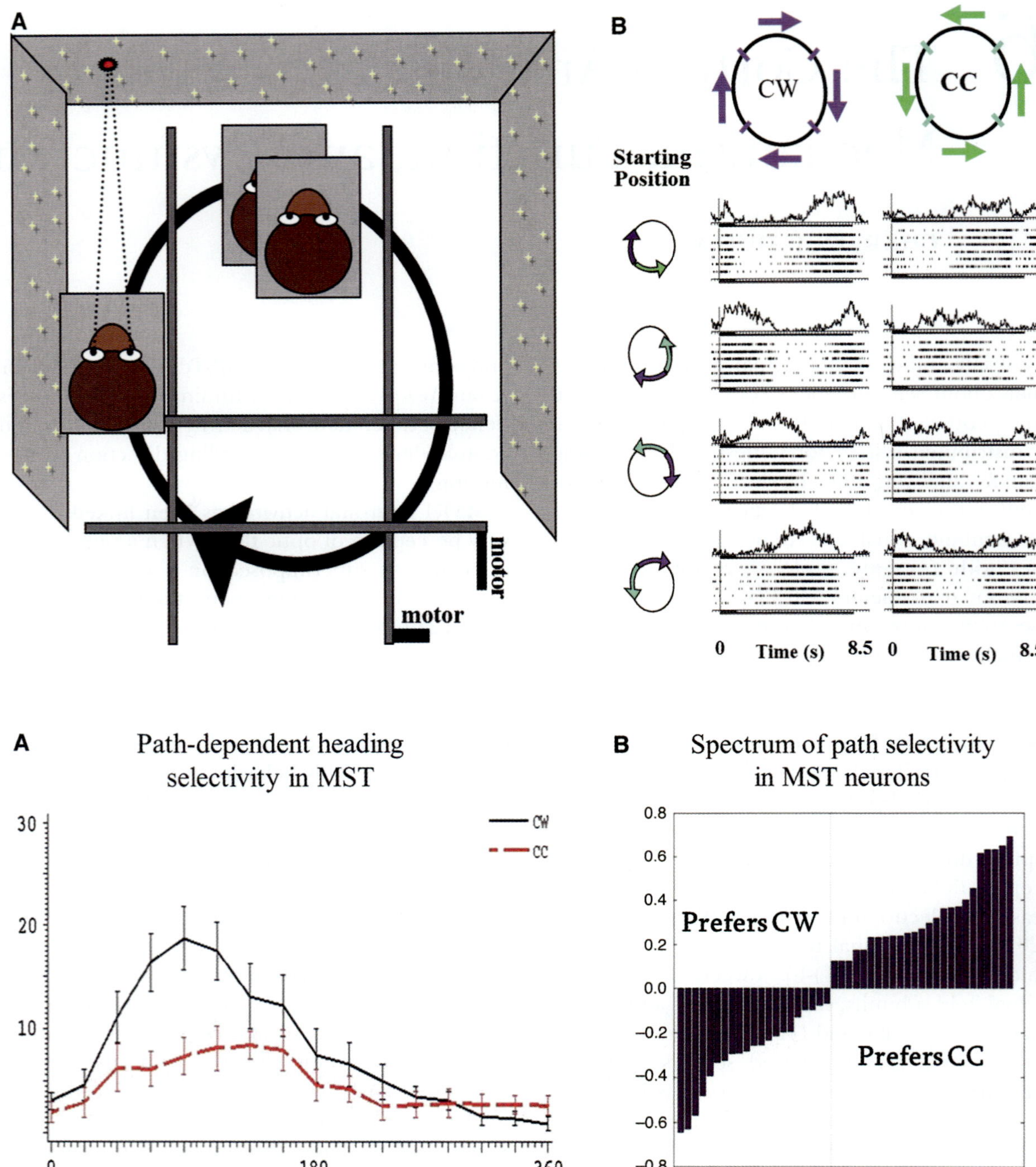

FIGURE 56.1   Path-selective responses of MSTd neurons. (A) The monkey viewed either a light array or a video projection screen during translational sled movement along a circular path. (B) Spike-density histograms and raster displays of a neuronal response showing a left-forward heading preference that is stronger with CW motion (left). (C) A neuron's responses to 16 heading intervals revealing enhanced heading selectivity on the CW path. (D) Contrast ratios of CW and CC response amplitudes at the preferred heading for each neuron; 73% showed significant directionality (Z-statistic) on at least one path (filled bars).

the same heading during CW and CC movement. Another 45% of the neurons show significant heading selectivity on only the CW or the CC path, and 20% of the neurons reversed their heading preference on the CC and CW paths. The latter group respond at the same place in the room regardless of the path to that place. A combination of heading sequence effects and location-specific activity contributes to these path-dependent heading responses and path-independent place responses. Thus, MSTd might contribute to

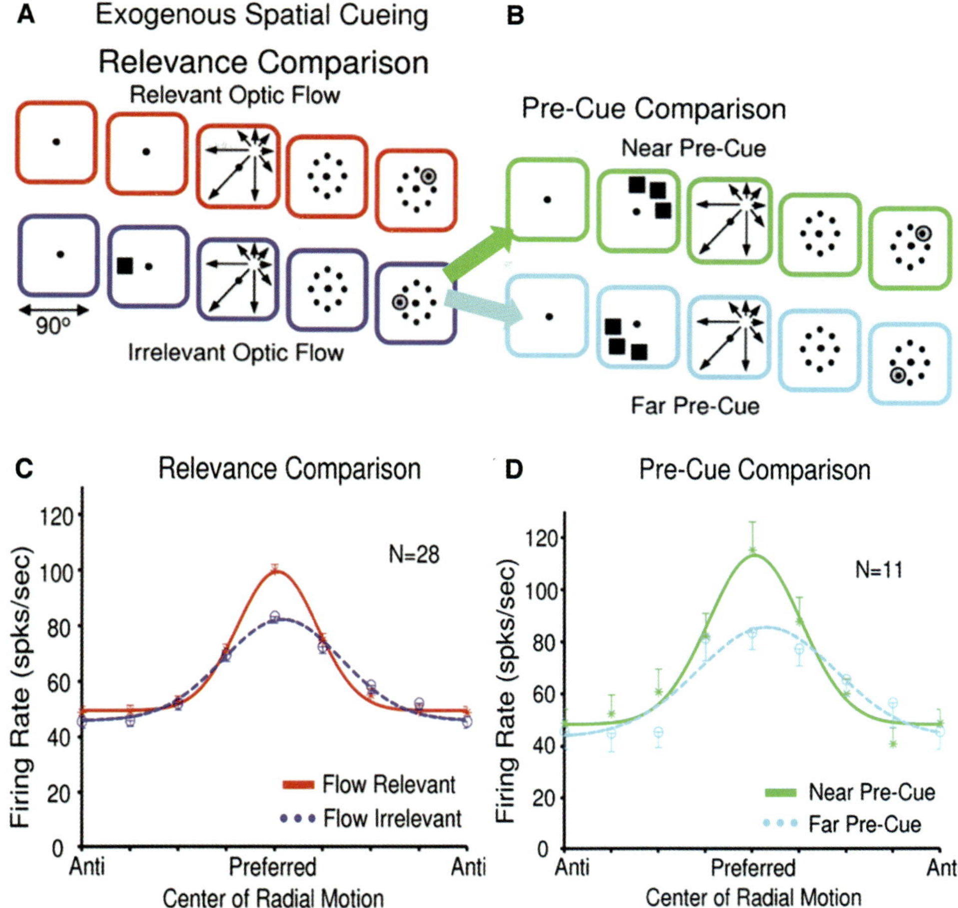

FIGURE 56.2   The effects of exogenous spatial cuing on neuronal responses. (A) In relevant optic flow trials (red), centered fixation is followed by optic flow visual motion with one of eight eccentric radial centers of motion and then an eight-spot array requiring a saccade to the remembered location of the radial center. In irrelevant optic flow trials (blue), a flashed square preceded the optic flow, and a saccade was required to the remembered square location. (B) Precue effects were seen in the irrelevant optic flow condition by comparing trials in which the precue was nearby (green) versus far from (light blue) the radial center. (C) Average tuning profiles of neurons with significant differences between responses to relevant (solid red) and irrelevant (dashed blue) optic flow and with good fits in the relevant and irrelevant conditions. The responses of each neuron were normalized to the amplitude of its largest radial center response. These curves show a larger magnitude (21%) and narrower tuning curve width (32%) in response to relevant optic flow. (D) Average responses of neurons with significant precue proximity effects and with good fits in the near (solid green) and far (dashed light blue) conditions. There are nonsignificantly larger responses and narrower tuning to optic flow following precues that were nearer to the subsequent radial center.

spatial orientation by integrating self-movement cues over time (Froehler & Duffy, 2002). This could be achieved through interactions between MSTd and hippocampal place neurons for mapping extrapersonal space (McNaughton et al., 1994; O'Keefe & Conway, 1978).

BEHAVIORAL INFLUENCES ON OPTIC FLOW ANALYSIS

MSTd neuronal responses to optic flow are affected by working memory and attention. Working memory effects have been seen in a memory-guided saccade task in which optic flow is presented as a distractor between a spatial cue and the saccade. Both cue position and cue proximity effects occur: In some neurons, the absolute position of the cue enhances the optic flow responses (figure 56.2, left). In these neurons the distance from the cue to the FOE determines the cue's influence on the optic flow responses (figure 56.2, right). Attentional effects are seen in these studies when optic flow is alternately presented as the cue in an FOE-guided saccade task and as a distractor in an object-shape-guided saccade task. The optic flow responses differ between the two tasks, mainly with greater response amplitude and narrower FOE tuning during trials in which optic flow guides the subsequent saccade (Dubin & Duffy, 2007). Thus, MSTd's optic flow

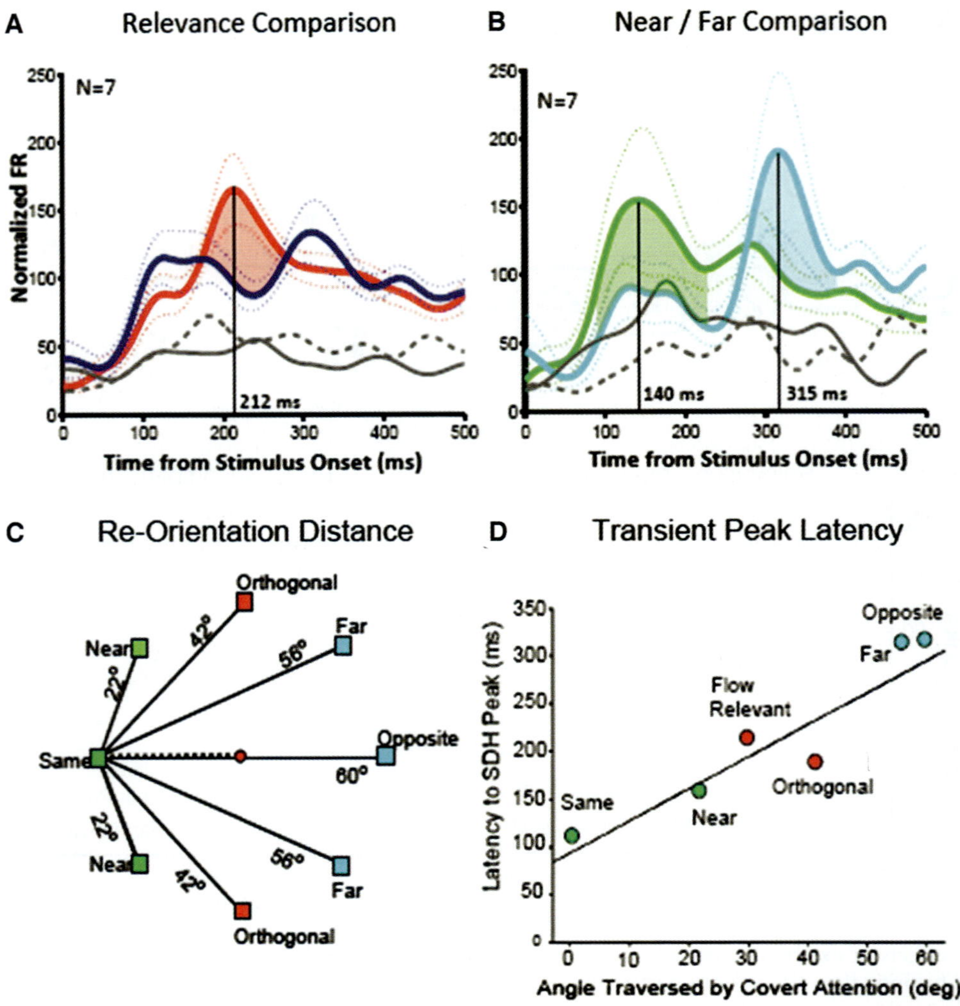

FIGURE 56.3   Effects of exogenous cuing on the latency of a transient response component. (A) Effects of 30° traversal of covert spatial attention (red) from centered fixation to the location of the optic flow's radial center of motion. Average normalized responses (±s.e.) as spike density functions during flow-relevant (red) and flow-irrelevant (blue) trials diverge from 120–300 ms, reaching maximal flow-relevant enhancement in a transient response component peaking at 216 ms. (B) Effects with sustained covert spatial attention (green) when the flashed precue (square) is at the same location as that of the radial center of motion. Effects of 60° traversal of covert spatial attention (blue) when the flashed precue (square) is opposite the location of the radial center of motion. Spike density functions of responses in the near-precue trials (green) peak at 135 ms after stimulus onset, whereas responses in the far-precue trials (light blue) peak at 312 ms. (C, D) There is a linear relationship between transient peak latency and distance traversed by covert spatial attention in the exogenous cuing paradigm. (C) This relationship is seen in the visual angle between the eight precues and a radial center of motion on the horizontal meridian in the left visual field. The dashed line indicates the 30° of visual angle traversed by covert spatial attention in flow-relevant trials. (D) Relationship between latency to the peak of the transient response and the distance (visual angle) between the flashed precue and the location of the radial center of motion in the subsequent optic-flow stimulus. Spike density functions were derived for each of the six attentional shift distances created by these stimuli. The best-fit line through these data has slope of 3.64 and intercept of 94.6, with $r^2 = .89$ and $p < .005$.

responses show effects of both working memory and attentional task demands.

An additional feature of MSTd's responses to precued optic flow is the appearance of transient increments in neuronal firing rate during those responses. In neurons with selective responses to precued optic flow, when the precue is at a moderate distance from the FOE in the subsequent optic flow, the response transient occurs in the middle of the evoked activity (figure 56.3A). When the precue is at the same location as the subsequent FOE, the transient occurs earlier. When the precue is far from the location of the subsequent FOE, the transient occurs later (Fig. 56.3B). Overall, there is a linear relationship between the latency of the optic flow response transient and the angular distance between the precue and the FOE in the subsequent optic flow (Figure 56.3C, D). The varying latency of these response transients appears to reflect the time it takes the monkeys to covertly reorient spatial attention from the site of the precue to that of the FOE (Dubin & Duffy, 2009).

Spatial attentional influences on optic flow processing were further explored in studies of monkey MSTd neuronal responses during simulated driving (Page & Duffy, 2008). Monkeys were trained to use a joystick to steer their simulated heading from one of eight eccentric FOEs in outward radial optic flow and to steer the FOE to the screen-centered fixation point (figure 56.4A–C). We found that the amplitude of the neuronal responses to the eight headings were decreased during active steering, compared to responses during the passive viewing of the same stimuli (figure 56.4D). We considered that decreased active steering responses could mean that the monkey was not using the full-field pattern of optic flow but was instead using only local motion in a part of the pattern.

We defeated the local motion strategy for steering optic flow by randomly interleaving inward and outward radial patterns at each of the eight eccentric headings. With inward optic flow, the monkey initially made 100% errors in 180° the wrong direction, confirming that it was using a local motion cue. Retraining the monkey on interleaved inward and outward patterns encouraged its use of the global pattern of radial optic flow. Repeating the neuronal recording experiments then showed increased responses during active steering compared to passive viewing (figure 56.4E). Thus, MSTd neuronal responses to optic flow depend not only on whether the monkeys used the stimuli to guide behavior but also on the perceptual strategy employed to use the stimuli.

Featural attention provides potent response selectivity in MSTd, selecting optic flow or discrete moving object cues in naturalistically combined stimuli. We trained monkeys in joystick steering based on either optic flow (figure 56.5, top row) or on the simulated relative motion of a discrete object (figure 56.5, second row). In both cases the heading simulated by the motion stimulus was randomly deflected to the left or right as the monkeys used the joystick to steer the heading back to the screen-centered fixation point. The optic flow and object motion stimuli were presented alone at the beginning of each multiresponse trial, so the monkeys knew which cue it was to use in the subsequent set of combined optic flow and object motion stimuli. Combined stimuli were randomly either congruent (optic flow and object depicting the same heading, so when one was centered, they both were) or noncongruent (optic flow and object motion depicting different headings, so centering on the designated cue was rewarded, and steering by the other cue was not).

When the optic flow and object motion stimuli were presented alone, individual MSTd neurons mainly responded preferentially to either the optic flow or the object motion (figure 56.5A, optic flow preferring). When the optic flow and object motion were combined, many MSTd neurons responded as if the cue that was guiding the monkeys' steering was the only cue present. This was particularly striking in the final 1-s period of congruent steering (figure 56.5B, filled segments of responses) when the stimuli on the screen were physically identical in the steering by optic flow (figure 56.5B, left) and steering by object motion (figure 56.5B, right) conditions, but many neurons showed responses that were much like those evoked by the designated cue when it was presented alone. The incongruent combined conditions (figure 56.5C) less often showed the same effects, but the stimuli were not identical in the two steering conditions (Kishore et al., 2011).

These findings lead us to conclude that spatial working memory, spatial attention, and featural attention can all have profound effects on MSTd neuronal responses to optic flow during naturalistic stimulus and task conditions.

THE COGNITIVE NEUROLOGY OF OPTIC FLOW ANALYSIS

The neurological foundations of visuospatial impairments in Alzheimer disease (AD) are evident in the distribution of pathological signs of AD. In mild to moderate AD, neurofibrillary tangles, and to a lesser extent senile plaques, are concentrated in extrastriate visual association areas (Braak & Braak, 1991; Brun & Englund, 1981). In patients with the visuospatial variant of AD, there is a consonant focal hypometabolism in that same distribution (Bradley et al., 2002; Mendez, 2001; Mendez, Ghajarania, & Perryman, 2002). These areas include those activated by large-field optic flow stimuli (Dukelow et al., 2001; Peuskens et al., 2001; Vanduffel et al., 2001).

We pursued links between navigational impairments in AD patients and optic flow perception by comparing dot motion coherence thresholds for horizontal motion and radial optic flow in young adult, older adult, and early AD subjects. Left/right, two-alternative forced choice (2AFC) paradigms presented leftward and rightward directions of horizontal motion and left and right FOEs in radial optic flow. Subjects viewed these stimuli during centered fixation while we algorithmically varied

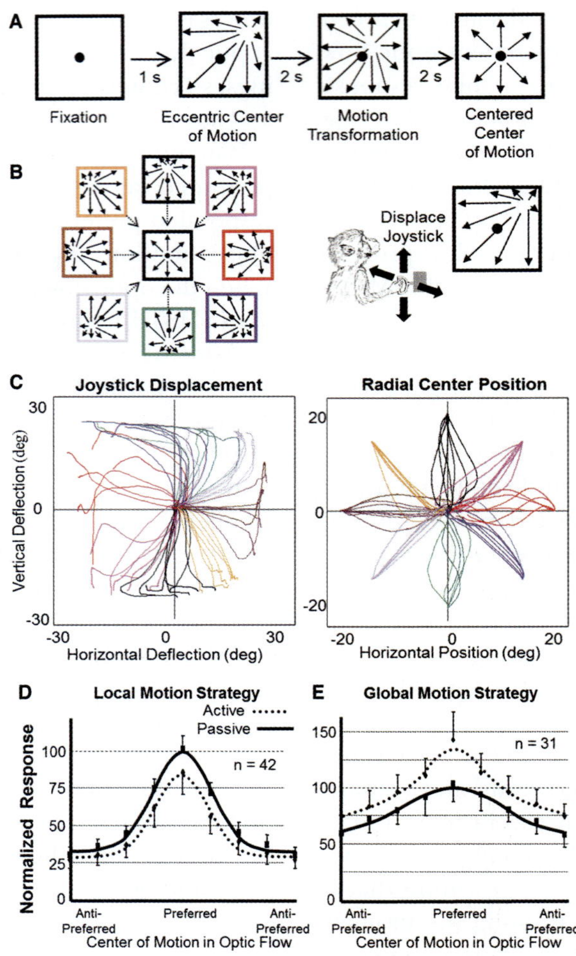

FIGURE 56.4   Behavioral paradigm for comparing optic flow responses during active steering and passive viewing. (A) Both active steering and passive viewing trials began with centered fixation followed by an eccentric radial center of motion optic-flow stimulus. The radial center of motion gradually moved in toward the center of the screen, either by active steering or in passive viewing. (B, left) In all trials the optic flow display initially contained one of eight eccentric radial centers of motion (outer frames) that moved toward the center of the screen (center frame) over the 4-s stimulus period. (B, right) The monkey maintained centered fixation in all trials. In active steering trials the monkey used horizontal and vertical joystick displacement to control the movement of the radial center of motion. (C, left) Joystick deflection during the first 1 s of active steering as averaged traces from trials recorded in six studies. Joystick deflection added a vector of corresponding direction and magnitude to the simulated heading in the optic flow display. (C, right) Screen position of the radial center of motion over the complete 4-s stimulus period as controlled by the monkey in active steering trials. Radial center position recorded during active steering trials was played back during passive viewing trials to create matching visual stimuli for the two recording conditions. Trace colors indicate initial center of motion position as indicated in B. (D) Averaged firing rate across all neurons in the passive-viewing (solid line) and active-steering (dashed line) trials during the 500- to 1,000-ms interval after onset of the radial optic flow pattern. (D) Gaussian fits to center of motion response curve for passive and active trials from the first study averaged across neurons with significant differences between active and passive trials ($n = 31$). Active/passive trials: amplitude = 37.0/49.7 spks/s, baseline = 8.5/9.4 spks/s, tuning width = 0.95/1.04, mean directional difference = 1.44°. (E) Gaussian fits to center of motion tuning curves from the second study averaged across neurons with significant differences between active and passive trials shows larger responses in the active steering condition ($n = 28$). Active/passive trials: amplitude = 11.8/10.1 spks/s, baseline = 17.4/11.4 spks/s, tuning width =1.12/2.04, mean directional difference = 2.7°.

the amount of random dot motion in each stimulus to determine each subject's horizontal and radial dot motion coherence thresholds, high thresholds meaning the subject required a high percentage of the dots moving in the pattern to accurately discriminate the left and right stimuli (figure 56.6A). Young and older adult subjects had similarly low thresholds for horizontal and radial motion, whereas early AD patients showed selectively elevated thresholds for radial optic flow (figure 56.6B) (Tetewsky & Duffy, 1999).

Changing the nature of the 2AFC stimulus set had a substantial impact on the profiles of impairment in

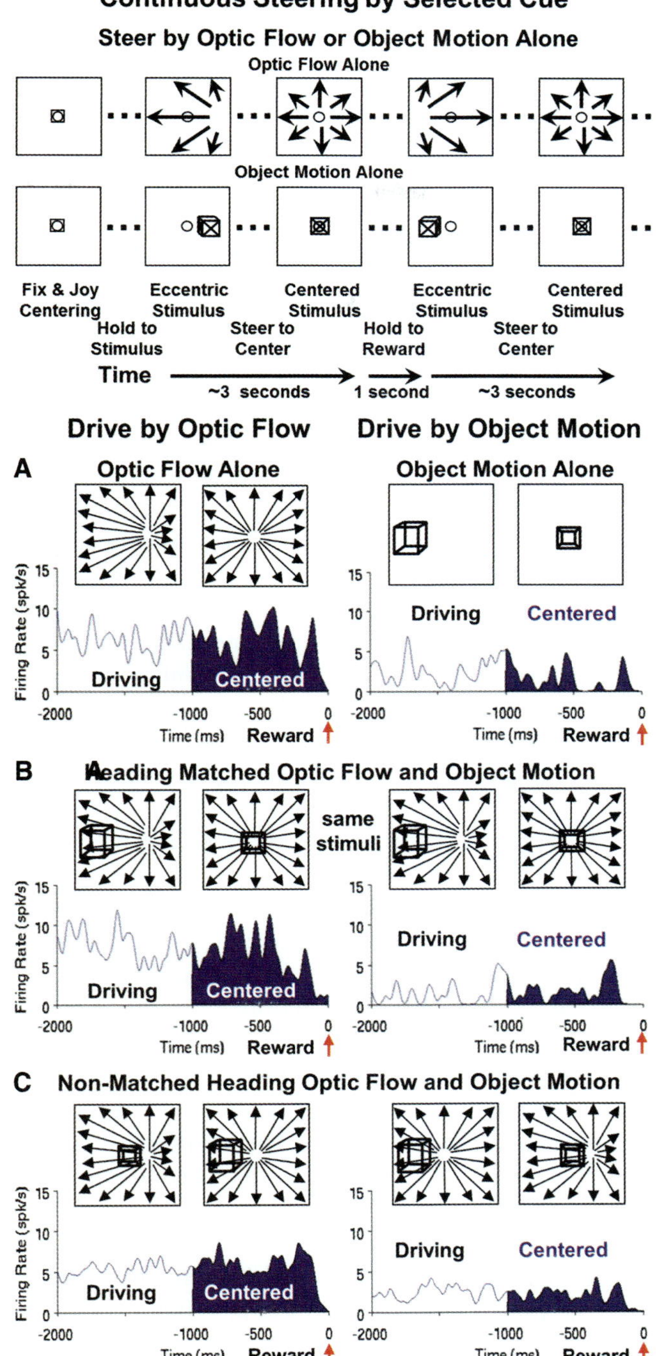

FIGURE 56.5   Diagram of stimuli and tasks showing two consecutive trials in each of the task conditions. The monkey initiated a block of trials by centering the joystick cursor (small square in stimulus frames) and establishing fixation on the centered fixation target (small open circle in stimulus frames). The monkeys then maintained centered gaze and moved the joystick to recenter their simulated heading in a continuous driving task using optic flow and/or object motion. Each block of trials (total ~1.5 min) began with either optic flow (top row) or object motion alone (second row), cuing the monkeys to use that cue throughout the trials (3–4 s) of that block. Here the simulated heading is randomly deviated to the left by imposing a drift velocity in that direction. The monkeys drive to center the heading to earn liquid reward. When centering leftward-deviated optic flow, the monkeys make a rightward joystick deflection. When centering leftward-deviated object motion, the monkeys make a left joystick deflection. The optic flow and moving object are then superimposed such that the movements are along matched headings shown here by the heading displacement to the right, resulting in the earth-fixed object moving to the left. In other blocks of trials, the optic flow and object motion are nonmatched headings, and the irrelevant stimulus is a distracter. Average spike density histograms presenting the final 2 s of the trials in each condition, the final 1 s of which (filled) presented continually centered targets. These responses illustrate the influence of the monkeys' driving strategy on neuronal activation. (A) The responses of a neuron showing much stronger activation to driving optic flow alone (left) compared to driving object motion alone (right). (B) These differences are also evident when the optic flow and object motion are combined in the matched heading stimuli despite the presentation of identical stimuli during driving by optic flow (left) and driving by object motion (right). (C) The same effects of the selected driving cue are evident in responses to the nonmatched heading stimuli even though they contain somewhat different stimulus configurations in the two driving conditions.

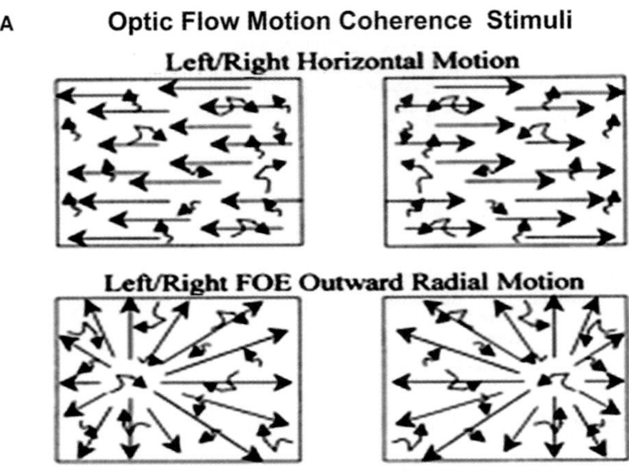

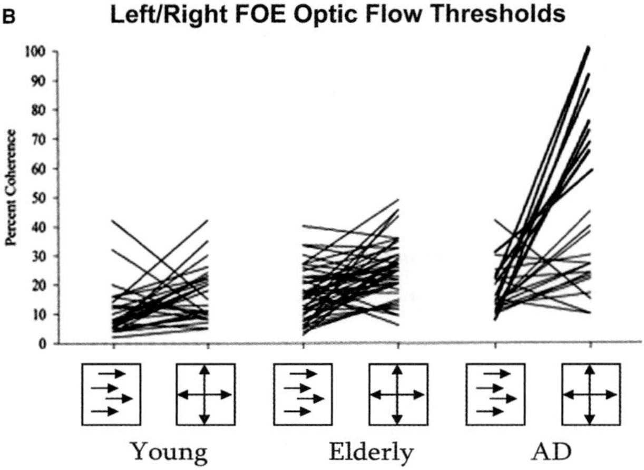

FIGURE 56.6   Optic flow motion coherence threshold determination. (A) Stimuli used to assess motion coherence thresholds. (Top) Horizontal motion stimuli contained either leftward or rightward moving dots superimposed on randomly moving dots. (Bottom) Radial optic flow stimuli consisted of an outward radial pattern with a FOE 15° to the left or right of center with superimposed randomly moving dots. (B) Plots relating horizontal motion and radial optic flow discrimination thresholds. Each subject's thresholds are connected by a solid line. Young normal and elderly normal subjects showed relatively small differences between their horizontal motion and radial optic flow thresholds (mean ± s.e.: young = 1.5 ± 3.5, elderly = 2.2 ± 2.2). Alzheimer subjects showed substantially larger differences between their horizontal motion and radial optic flow thresholds (mean ± s.e. = 13.6 ± 4.23). This difference is attributable to the fact that six of the subjects (55%, 6/11) showed much larger radial optic flow thresholds with average differences of 25.3 ± 4.0 (bold lines), whereas the remaining five subjects showed small differences between those thresholds (–0.4 ± 3.9).

these subjects. We interleaved inward and outward radial optic flow patterns to ambiguate the local motion cues in the left/right 2AFC stimuli, as in our monkey studies of local versus global optic flow processing (Page & Duffy, 2008). We found little effect of whether only

outward optic flow (local motion strategy) or interleaved inward and outward optic flow (global processing required) was presented in young subjects. In contrast, almost all early AD subjects had very high thresholds with interleaved inward and outward optic flow, even those who had low thresholds with outward optic flow alone. Although most older adults responded as the young subjects did, some responded similarly to the early AD patients (figure 56.7). Thus, we considered that optic flow processing impairments might be an early sign of visuospatial AD (O'Brien et al., 2001).

We examined interactions between optic flow and object motion cues about self-movement in aging and AD. In these studies we first determined optic flow dot motion coherence thresholds (figure 56.8A) and object motion path duration thresholds for each subject (figure 56.8B), a longer path being more readily interpreted as implying a particular heading of relative self-movement. We then presented stimuli at each subject's threshold plus one confidence interval in a heading estimation task in which each subject viewed a stimulus and then pointed at his or her perceived simulated heading. The optic flow and object motion stimuli were presented alone and in naturalistic combinations of congruent and noncongruent cues to assess the impact of their interactions on heading estimation (figure 56.8C), as in our studies of interactions between optic flow and object motion in monkeys (Logan & Duffy, 2006). Congruently combined optic flow and object motion suggests the same heading; noncongruent cues imply the heading simulated by the optic flow as the noncongruent object should be interpreted as moving independently (Mapstone & Duffy, 2010).

Young, middle-aged, and older normal control subjects show comparably accurate performance with each stimulus presented alone and in congruent combination, whereas the early AD patients showed particular impairment with heading discrimination by object motion (figure 56.8D, black-and-white bars). The young, middle-aged, and older normal subjects all showed a lateral deflection of their perceived heading when the object was on a heading more lateral to that in the optic flow; when the object was heading more medially, they showed a medial deflection of their perceived heading (figure 56.8D, colored bars). However, the noncongruent cues caused the early AD patients to consistently misestimate their heading as being straight ahead, toward the fixation point (Mapstone & Duffy, 2010; Mapstone, Logan, & Duffy, 2006). Thus, AD is associated with a vulnerability to heading misestimation when patients view the naturalistic combination of self-movement in the presence of independently moving objects, a common circumstance in driving.

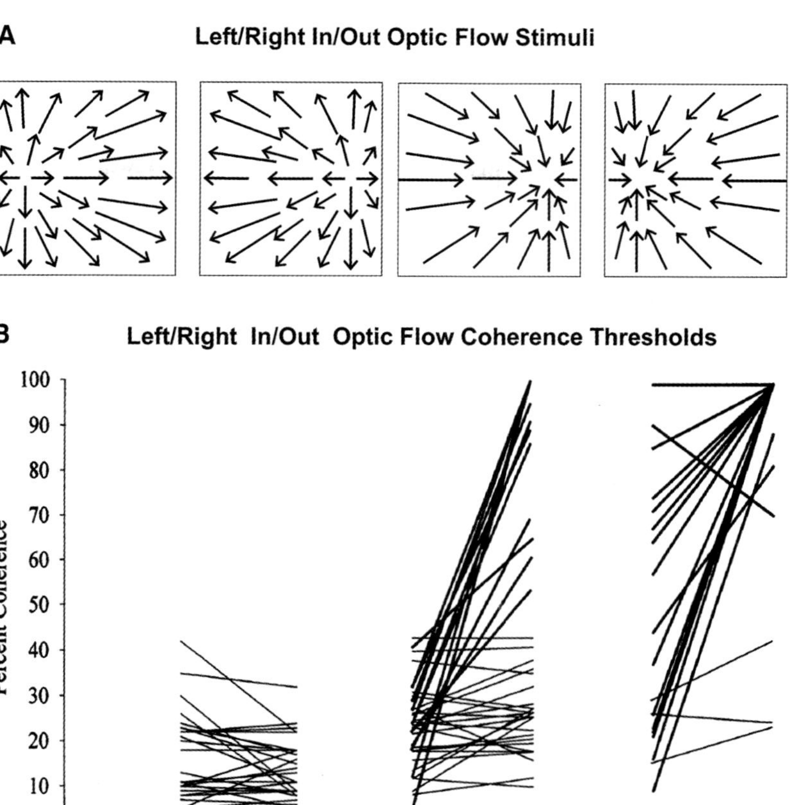

FIGURE 56.7   Global pattern radial motion perception is impaired in AD and in some older adults. (A) Left/right FOE in-out radial motion stimuli were created to remove the utility of local motion cues by interleaving stimuli having radial motion outward from (upper) or inward to (lower) left- or right-sided FOEs. (B) Left/right FOE radial motion discrimination thresholds for subjects (ordinate) in each group (abscissa) are shown for left/right FOE outward radial motion stimuli (left side for each subject group) and for left/right FOE in–out radial motion stimuli (right side for each subject group). Almost all AD ($n = 20$) subjects (85%, 17/20), and some EN ($n = 38$) subjects (32%, 12/38), showed thresholds >50% coherence when local motion cues were removed by interleaving in and out radial motion.

To better understand the impact of aging and AD on real-world navigation, we developed a test battery to characterize our subjects' navigational abilities. This test battery begins with a wheelchair excursion along a fixed path through our hospital lobby and then a series of 80 questions about the route and the environment (figure 56.9A). We found that older adults showed mild deficits in navigation, but early AD patients showed much larger deficits across a variety of subtest domains (figure 56.9B) (Monacelli et al., 2003). We also developed a virtual reality version of this navigational test battery that yields much the same results across subject groups (Cushman, Stein, & Duffy, 2008).

We related navigational performance to optic flow analysis by testing subjects on navigation, optic flow perception, and optic flow neurophysiological responsiveness. We measured neurophysiological responsiveness by developing an optic flow evoked potential triggered by the transition from stationary dots to a radial pattern of optic flow (figure 56.10A). These stimuli evoked robust responses in older adults, with a visual pattern response negative wave that peaks ~200 ms after stimulus onset (N200) that can be modulated by attentional control (Tata et al., 2010). In early AD patients the optic flow N200 was of significantly smaller amplitude, although it peaked at a latency comparable to that in older adults and shared the same distribution across posterior leads as seen in older adults (figure 56.10B) (Kavcic et al., 2006) and showed attentional modulation (Tales et al., 2002).

We combined our psychophysical and neurophysiological measures with the neuropsychological test scores and basic visual functional measures collected on our subjects in a multiple linear regression model of total

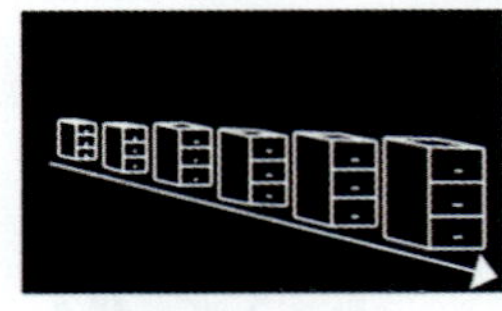

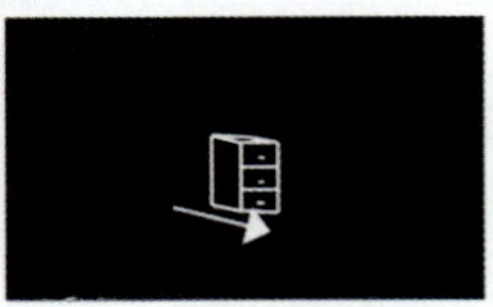

A Sample Optic Flow Coherence Levels

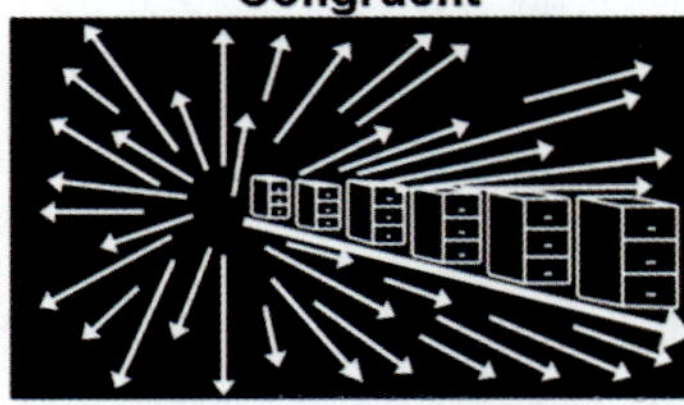

B Sample Object Path Lengths

C Sample Superimposed Heading Cues

D

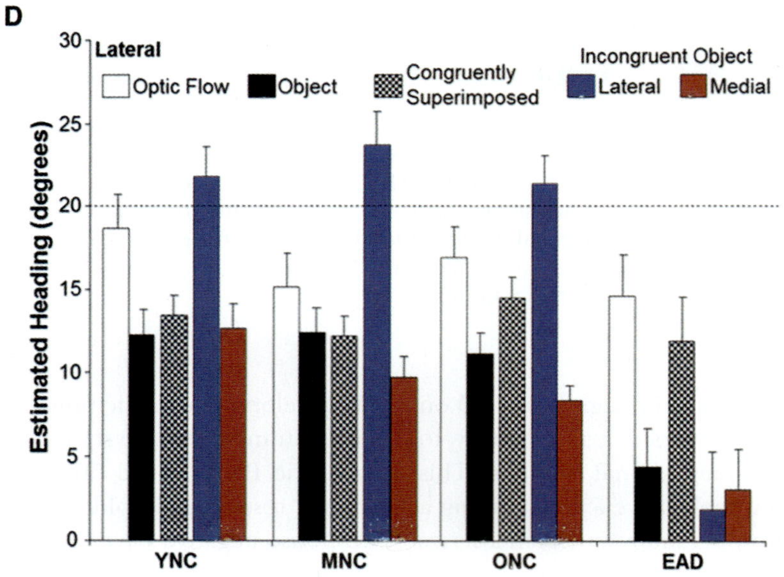

FIGURE 56.8   Visual motion stimuli used in the cue conflict heading discrimination tasks. (A) Outward radial optic flow having a 30° left- or right-sided FOE was formed by a randomly distributed pattern of white dots presented on a black background. A selected percentage of the dots were drawn from the pattern to undergo randomly directed motion to alter the strength of the simulated heading direction cue. The PEST algorithm controlled the percentage of randomly moving dots to determine each subject's motion coherence thresholds. (B) Looming object motion simulated a filing cabinet viewed during the observer's straight-line path of approaching self-movement. The percentage of the full path length presented was selected to alter simulated heading direction signal strength and determine each subject's path length thresholds. (C) Optic-flow and object-motion stimuli were superimposed to create two conditions with respect to a moving observer: two cues simulating the same heading direction in congruent combination (left) as with an earth-fixed object; and two cues simulating different heading directions in incongruent combination (right), as with an independently moving animate object. (D) Heading estimation across five cue configurations and four subject groups. Heading estimation accuracy was determined by varying the simulated heading direction along the horizontal meridian in optic flow and object motion stimuli. Four optic flow or object motion headings were presented alone (±5°, ±10°, ±15°, ±20°, ±25°) or in a limited set of combination stimuli in which optic flow (±10° or ±20°) was combined with object motion either with matched headings (congruently superimposed) or with object motion heading displaced from the optic flow headings by ±5°, ±10°, or ±15° (laterally or medially incongruently superimposed relative to the heading in optic flow). Black and white bars depict results from optic flow and object motion alone or in congruent superimposition, included for comparison to results from incongruently superimposed cues (blue and red bars). Heading discrimination was most accurate using the optic-flow stimuli alone compared to the object alone in all groups. Congruently combined optic flow and (simulated earth-fixed) objects allowed for heading discrimination, which tended to be intermediate between the two single stimuli in all groups. With incongruent stimuli (simulating an independently moving object with respect to a moving observer), all normal control groups showed relatively uniform bias toward the object when it was more lateral to the simulated observer heading in optic flow. In contrast, the AD group did not show a bias toward the lateral object but indicated self-movement heading was generally straight ahead in the direction of central visual fixation. On the other hand, a simulated object with a more medial heading than the observers' simulated heading also led to bias toward the object, but with a pronounced aging and disease effect AD > ONC > MNC > YNC.

---

real-world navigational test scores. This analysis yielded a significant model with an $R^2 = 0.95$ in which the significant regressors were, in descending order of beta weights, N200 amplitude, the difference between radial and planar motion thresholds, and contrast sensitivity (Kavcic et al., 2006). We conclude that factors related to deficits in posterior cortical optic flow analysis play a substantial role in the navigational impairments of early AD.

CONCLUSIONS

Optic flow analysis relies on a distributed temporoparietal system integrating sensory and motor signals about self-movement. The radial pattern of visual motion is the most critical stimulus contributing to optic flow selective cortical neuronal responses.

Area MST supports the successive refinement of these signals across neurons to create optic flow selective responses that are integrated with vestibular signals about self-movement. These multicue signals about heading direction are integrated with working memory, spatial attentional, and featural attentional signals related to ongoing behavioral tasks. Furthermore, the perceptual strategy used in those tasks has a substantial impact on optic flow processing. All of these factors may contribute to the role of optic flow processing in heading estimation and navigation. The importance of optic flow in such tasks is highlighted by the relationship between optic flow processing deficits and navigational impairments in aging and early AD.

The cortical analysis of optic flow provides a model of multisensory integration and its inseparability from reciprocal interactions with behavior. Optic flow processing supports adaptation to the demands of complex environments and to the opportunities afforded by a rich behavioral repertoire while also creating a potential vulnerability to the breakdown of behavioral capacities in neurodegenerative disease.

ACKNOWLEDGMENTS

This chapter is dedicated to my wife and our sons for the light in my life, and to my close colleagues, who have made years of collaborative productivity both fun and exciting. I am grateful for the support of the University of Rochester and of the American people in the form of grants from the National Eye Institute (EY10287), National Institute on Aging (AG17596), and the Office of Naval Research (N000141110525).

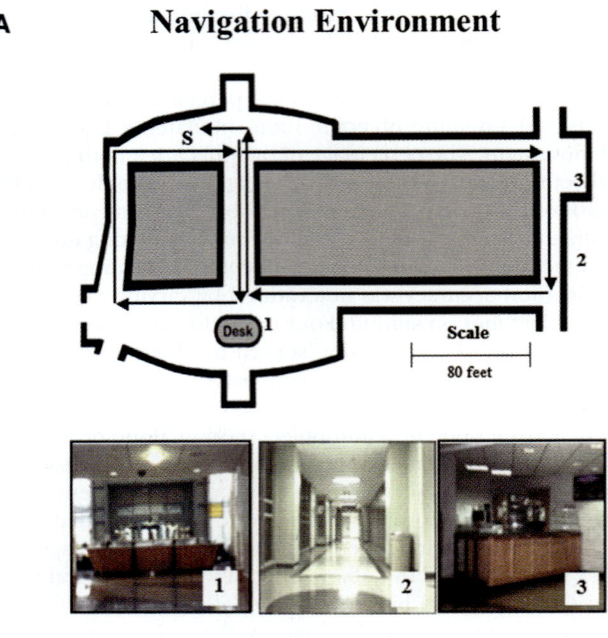

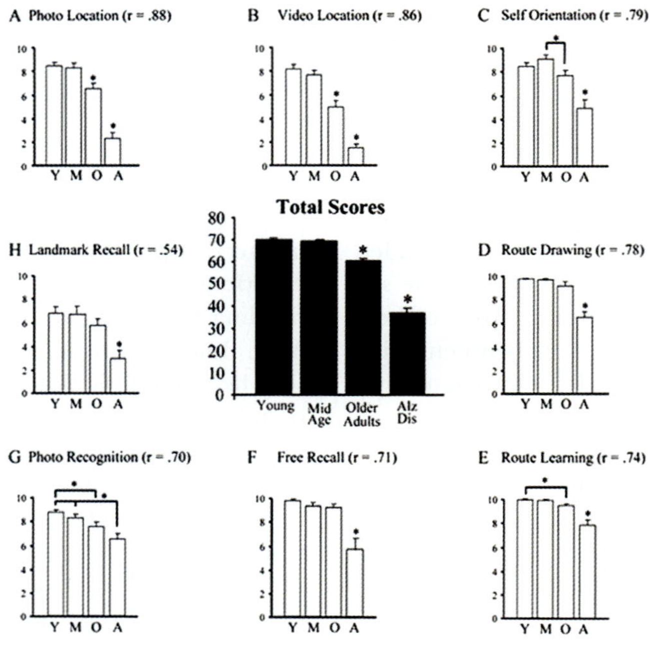

FIGURE 56.9   We created a spatial navigation test given in a single session of 90 min. (A) Bird's eye view of navigational test environment and route (arrows) and representative scenes from the numbered locations in that environment. Tasks were presented in order as: (1) Experimenter-directed tour of the environment, (2) Subject-directed repetition of the tour path, (3) Free-recall of discrete objects seen along path, (4) Drawing of route on map representation, (5) Naming of objects subject used to navigate, (6) Recognition of object photos as being from route, (7) Matching photos to labeled sites on map of route, and (8) Video-clip representation on a map of the route. (B) Group performance on the spatial orientation test. (Center) Total scores for each subject group. The OA group showed significantly poorer performance than the YN or MA groups. The AD group showed significantly poorer performance than all other groups. (A–H) Subtest scores for each group are arranged CW in order of decreasing magnitude of the correlation (r) between that subtest score and the total score. All subtests showed significant correlations with the total score, reflecting their coherence in testing spatial orientation. Asterisks indicate significant group differences (HSD, $p < .05$). Y, young normal; M, middle age; O, older adult; A, Alzheimer's disease.

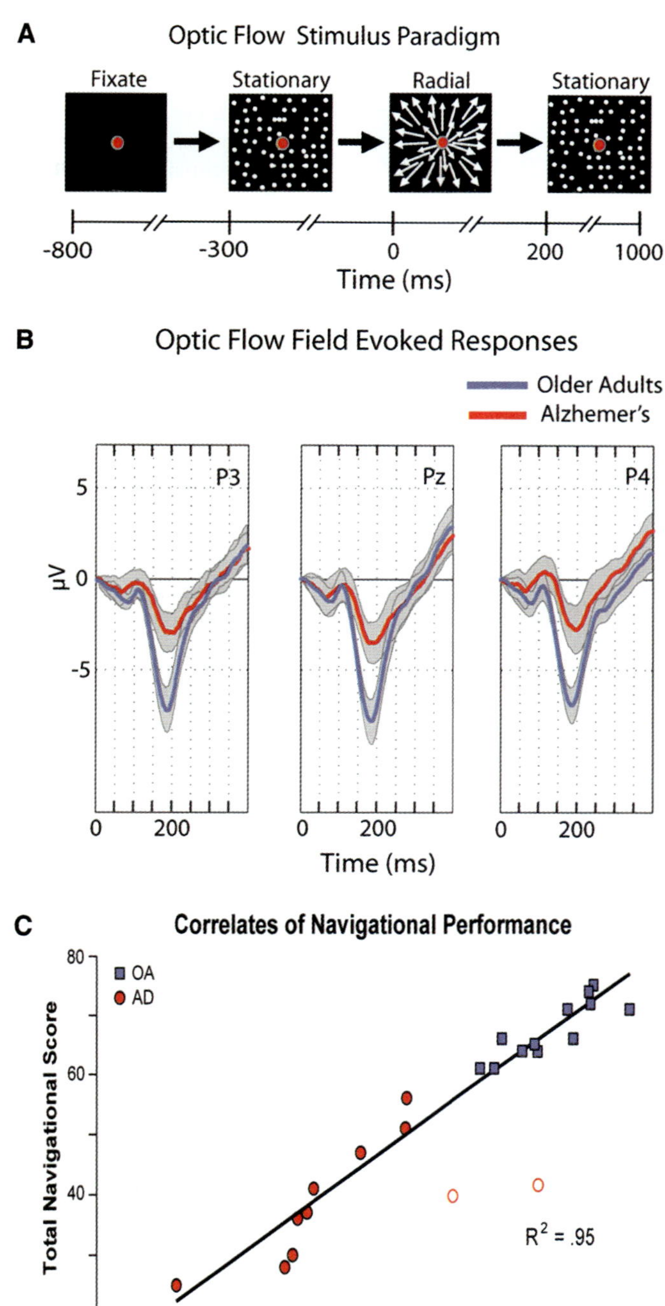

FIGURE 56.10   Stationary dots to radial optic flow-evoked responses in older adults (OA) and Alzheimer disease subjects. (A) Stationary dots preceded radial optic flow stimulation. (B) Averaged waveforms (±1 S.E.M.) showed much larger N200 responses for OA subjects (blue lines) than for Alzheimer disease subjects (red lines), especially at parietal sites. (C) Perceptual and neurophysiological correlates of navigational impairment. Scatterplot of total navigational test score (ordinate) by composite multiple linear regression predictor (abscissa) for all OA (blue squares) and Alzheimer disease (red circles) subjects. Stepwise regression selected the difference between horizontal and radial thresholds (β = .71), the stationary-to-radial N200 response amplitude (β = .38), and visual contrast sensitivity (β = .19) as significant predictors to yield a composite R^2 = .95. The regression analysis rejected the inclusion of the other neuropsychological, perceptual, and neurophysiological variables. The analysis identified two Alzheimer disease subjects as being very impaired outliers (open red circles).

REFERENCES

Braak, H., & Braak, E. (1991). Neuropathological staging of Alzheimer-related changes. *Acta Neuropathologica, 82,* 239–259.

Bradley, K. M., O'Sullivan, V. T., Soper, N. D. W., Nagy, Z., King, E. M. R., Smith, A. D., et al. (2002). Cerebral perfusion SPET correlated with Braak pathological stage in Alzheimer's disease. *Brain, 125,* 1772–1781.

Britten, K. H. (1998). Clustering of response selectivity in the medial superior temporal area of extrastriate cortex in the macaque monkey. *Visual Neuroscience, 15,* 553–558.

Britten, K. H., Shadlen, M. N., Newsome, W. T., & Movshon, J. A. (1992). The analysis of visual motion: A comparison of neuronal and psychophysical performance. *Journal of Neuroscience, 12,* 4745–4765.

Britten, K. H., & van Wezel, R. J. (1998). Electrical microstimulation of cortical area MST biases heading perception in monkeys. *Nature Neuroscience, 1,* 59–63.

Brun, A., & Englund, E. (1981). Regional pattern of degeneration in Alzheimer's disease: Neuronal loss and histopathological grading. *Histopathology, 5,* 549–564.

Celebrini, S., & Newsome, W. T. (1994). Neuronal and psychophysical sensitivity to motion signals in extrastriate area MST of the macaque monkey. *Journal of Neuroscience, 14,* 4109–4124.

Cushman, L. A., Stein, K., & Duffy, C. J. (2008). Detecting navigational deficits in cognitive aging and Alzheimer's disease using virtual reality. *Neurology, 71,* 888–895. doi:71/12/888[pii]10.1212/01.wnl.0000326262.67613.fe.

Dubin, M. J., & Duffy, C. J. (2007). Behavioral influences on cortical neuronal responses to optic flow. *Cerebral Cortex, 17,* 1722–1732.

Dubin, M. J., & Duffy, C. J. (2009). Neuronal encoding of the distance traversed by covert shifts of spatial attention. *Neuroreport, 20,* 49–55.

Duffy, C. J., & Wurtz, R. H. (1991a). Sensitivity of MST neurons to optic flow stimuli. I. A continuum of response selectivity to large-field stimuli. *Journal of Neurophysiology, 65,* 1329–1345.

Duffy, C. J., & Wurtz, R. H. (1991b). Sensitivity of MST neurons to optic flow stimuli. II. Mechanisms of response selectivity revealed by small-field stimuli. *Journal of Neurophysiology, 65,* 1346–1359.

Duffy, C. J., & Wurtz, R. H. (1993). An illusory transformation of optic flow fields. *Vision Research, 33,* 1481–1490.

Duffy, C. J., & Wurtz, R. H. (1997a). Multiple temporal components of optic flow responses in MST neurons. *Experimental Brain Research, 114,* 472–482.

Duffy, C. J., & Wurtz, R. H. (1997b). Planar directional contributions to optic flow responses in MST neurons. *Journal of Neurophysiology, 77,* 782–796.

Dukelow, S. P., DeSouza, J. F. X., Culham, J. C., van den Berg, A. V., Menon, R. S., & Vilis, T. (2001). Distinguishing subregions of the human MT+ complex using visual fields and pursuit eye movements. *Journal of Neurophysiology, 86,* 1991–2000.

Felleman, D. J., & Van Essen, D. C. (1991). Distributed hierarchical processing in the primate cerebral cortex. *Cerebral Cortex, 1,* 1–47.

Froehler, M. T., & Duffy, C. J. (2002). Cortical neurons encoding path and place: Where you go is where you are. *Science, 295,* 2462–2465.

Graziano, M. S. A., Andersen, R. A., & Snowden, R. J. (1994). Tuning of MST neurons to spiral motion. *Journal of Neuroscience, 14,* 54–67.

Kavcic, V., Fernandez, R., Logan, D. J., & Duffy, C. J. (2006). Neurophysiological and perceptual correlates of navigational impairment in Alzheimer's disease. *Brain, 129,* 736–746.

Kishore, S., Hornick, N., Sato, N., Page, W. K., & Duffy, C. J. (2011). Driving strategy alters neuronal responses to self-movement: Cortical mechanisms of distracted driving. *Cerebral Cortex, 22,* 201–208. doi:10.1093/cercor/bhr115.

Lappe, M. (1996). Functional consequences of an integration of motion and stereopsis in area MT of monkey extrastriate visual cortex. *Neural Computation, 8,* 1449–1461.

Lappe, M. (1998). A model of the combination of optic flow and extraretinal eye movement signals in primate extrastriate visual cortex: Neural model of self-motion from optic flow and extraretinal cues. *Neural Networks, 11,* 397–414.

Lewis, J. W., & Van Essen, D. C. (2000). Mapping of architectonic subdivisions in the macaque monkey, with emphasis on parieto-occipital cortex. *Journal of Comparative Neurology, 428,* 79–111.

Logan, D. J., & Duffy, C. J. (2006). Cortical area MSTd combines visual cues to represent 3-D self-movement. *Cerebral Cortex, 16,* 1494–1507.

Mapstone, M., & Duffy, C. J. (2010). Approaching objects cause confusion in patients with Alzheimer's disease regarding their direction of self-movement. *Brain, 133,* 2690–2701. doi:awq140 [pii]10.1093/brain/awq140.

Mapstone, M., Logan, D., & Duffy, C. J. (2006). Cue integration for the perception and control of self-movement in ageing and Alzheimer's disease. *Brain, 129*(Pt 11), 2931–2944.

McNaughton, B. L., Mizumori, S. J. Y., Barnes, C. A., Leonard, B. J., Marquis, M., & Green, E. J. (1994). Cortical representation of motion during unrestrained spatial navigation in the rat. *Cerebral Cortex, 4,* 27–39.

Mendez, M. F. (2001). Visuospatial deficits with preserved reading ability in a patient with posterior cortical atrophy. *Cortex, 37,* 535–543.

Mendez, M. F., Ghajarania, M., & Perryman, K. M. (2002). Posterior cortical atrophy: Clinical characteristics and differences compared to Alzheimer's disease. *Dementia and Geriatric Cognitive Disorders, 14,* 33–40.

Monacelli, A. M., Cushman, L. A., Kavcic, V., & Duffy, C. J. (2003). Spatial disorientation in Alzheimer's disease: The remembrance of things passed. *Neurology, 61,* 1491–1497.

O'Brien, H. L., Tetewsky, S. J., Avery, L. M., Cushman, L. A., Makous, W., & Duffy, C. J. (2001). Visual mechanisms of spatial disorientation in Alzheimer's disease. *Cerebral Cortex, 11,* 1083–1092.

O'Keefe, J., & Conway, D. M. (1978). Hippocampal place units in the freely moving rat: Why they fire where they fire. *Experimental Brain Research, 31,* 573–590.

Page, W. K., & Duffy, C. J. (2008). Cortical neuronal responses to optic flow are shaped by visual strategies for steering. *Cerebral Cortex, 18,* 727–739. doi:bhm109[pii]10.1093/cercor/bhm109.

Paolini, M., Distler, C., Bremmer, F., Lappe, M., & Hoffmann, K. P. (2000). Responses to continuously changing optic flow in area MST. *Journal of Neurophysiology, 84,* 730–743.

Pasternak, T., & Merigan, W. H. (1994). Motion perception following lesions of the superior temporal sulcus in the monkey. *Cerebral Cortex, 4,* 247–259.

Perrone, J. A., & Stone, L. S. (1994). A model of self-motion estimation within primate extrastriate visual cortex. *Vision Research, 34,* 2917–2938.

Perrone, J. A., & Stone, L. S. (1998). Emulating the visual receptive-field properties of MST neurons with a template model of heading estimation. *Journal of Neuroscience, 18,* 5958–5975.

Peuskens, H., Sunaert, S., Dupont, P., Van Hecke, P., & Orban, G. A. (2001). Human brain regions involved in heading estimation. *Journal of Neuroscience, 21,* 2451–2461.

Tales, A., Muir, J. L., Bayer, A., & Snowden, R. J. (2002). Spatial shifts in visual attention in normal ageing and dementia of the Alzheimer type. *Neuropsychologia, 40,* 2000–2012.

Tanaka, K., Fukuda, Y., & Saito, H. (1989). Underlying mechanisms of the response specificity of expansion/contraction and rotation cells in the dorsal part of the medial superior temporal area of the macaque monkey. *Journal of Neurophysiology, 62,* 642–656.

Tata, M. S., Alam, N., Mason, A. L., Christie, G., & Butcher, A. (2010). Selective attention modulates electrical responses to reversals of optic-flow direction. *Vision Research, 50,* 750–760. doi:10.1016/j.visres.2010.01.012.

Tetewsky, S. J., & Duffy, C. J. (1999). Visual loss and getting lost in Alzheimer's disease. *Neurology, 52,* 958–965.

Thiele, A., & Hoffmann, K. P. (1996). Neuronal activity in MST and STPp, but not MT, changes systematically with stimulus-independent decisions. *Neuroreport, 7,* 971–976.

Vanduffel, W., Fize, D., Mandeville, J. B., Nelissen, K., Van Hecke, P., Rosen, B. R., et al. (2001). Visual motion processing investigated using contrast agent-enhanced fMRI in awake behaving monkeys. *Neuron, 32,* 565–577.

Yu, C. P., Page, W. K., Gaborski, R., & Duffy, C. J. (2010). Receptive field dynamics underlying MST neuronal optic flow selectivity. *Journal of Neurophysiology, 103,* 2794–2807. doi:10.1152/jn.01085.2009.

57 Stereopsis

CLIFTON M. SCHOR

Stereopsis is one of our primary senses of spatial layout. Slightly different viewpoints of the two eyes produce binocular retinal image disparities that are used to perceive *relative depth* between objects as well as *surface slant* (orientation) and percepts of *object volume*. Stereopsis also provides information about forms hidden by texture camouflage (such as branches in tree foliage) and direction of object motion in depth relative to the head. Stereopsis is an extremely acute mechanism with one of the lowest visual thresholds—less than the width of a single retinal photoreceptor—which includes it as one of the hyperacuities (Westheimer, 1979b).

VISUAL DIRECTIONS

Stereopsis is stimulated by horizontal disparity that results from the slightly different perspective views of the two eyes. When viewed with one eye at a time, the retinal images of the three-dimensional scene have slightly different locations relative to the fovea of each eye, and they are seen in slightly different directions relative to the line of sight (oculocentric directions). However, each eye perceives objects within the scene in a common direction relative to the head (egocentric direction). Perceived egocentric direction of an object is equal to the average of its oculocentric directions in the two eyes combined with the average of right and left eye positions (conjugate eye position). Movements of the eyes in opposite directions (convergence) have no influence on perceived egocentric direction. Thus, when the two eyes fixate on near objects that lie to the left or right of the midline (straight ahead) in asymmetrical convergence, only the conjugate component of the two eyes' positions (produced by movements of the two eyes in the same direction) contributes to perceived direction. Ewald Hering summarized these facets of egocentric direction as five rules of visual direction (1868), and Howard (1982) has restated them. The laws are mainly concerned with targets imaged on corresponding retinal regions (i.e., stimulation of corresponding retinal points in the two eyes results in percepts in identical visual directions or single vision).

How are oculocentric visual directions of the two eyes combined in a common visual space, sometimes referred to as the cyclopean eye? When the target lies near the plane of fixation, its egocentric direction is based on the similar retinal image locations of the two eyes. When the target is slightly nearer or farther than the plane of fixation, its monocular images are formed on non-corresponding retinal points at different horizontal eccentricities from the foveas; however, its averaged direction is seen as single (allelotropia). The range of disparities that produce singleness is referred to as Panum's fusional area (Schor & Tyler, 1981). The consequence of averaging monocular visual directions of disparate targets is that binocular visual directions are mislocalized by half their retinal image disparity. Binocular visual directions can only be judged accurately for targets with identical monocular visual directions. When targets are positioned far in front or behind the plane of fixation, their retinal image disparity becomes large, and the disparate targets appear diplopic (i.e., they are perceived in two separate directions). The directions of monocular components of the diplopic pair are perceived as though each of the images had an invisible paired image formed on a corresponding retinal point in the other eye. There are ambiguous circumstances in which a target in the peripheral region of a binocular field is only seen by one eye because of partial occlusion of the other eye by the nose. The monocular target could lie at a range of viewing distances; however, its direction is judged as though it were located in the plane of fixation such that if it were seen binocularly, its images would be formed on corresponding retinal points.

Several violations of Hering's rules for visual direction have been observed in both abnormal and normal binocular vision. A common binocular abnormality is strabismus (an eye turn or deviation of one eye from the intended binocular fixation point). In violation of Hering's rules, people who have a constant turn of one eye (unilateral strabismus) can have constant double vision (diplopia), and they use the position of their preferred fixation eye to judge visual direction regardless of whether they fixate a target with their preferred or deviating eye. People who have an alternating strabismus (either eye is used for fixation, but only one at a time) use the position of the fixating eye to judge direction of objects (Mann, Hein, & Diamond, 1979). Both classes of strabismus use the position of only one eye to judge direction, whereas people with normal binocular eye alignment use the average position of the two eyes to judge direction by either eye alone.

An exception of Hering's rules occurs in normal binocular vision when the retinal image of a target that is seen monocularly is imaged near a disparate target seen by both eyes (binocular view). Monocular views occur naturally in the peripheral visual field when one eye's view is occluded by the nose. The direction of the monocular target is judged as though it were positioned at the same depth as any nearby disparate binocular target rather than at the plane of fixation. The visual system assumes there is an occluded counterpart of the monocular target in the contralateral eye that has the same disparity as the nearby binocular target, even though the image is seen only by one eye (Erkelens & van Ee, 1997).

Another exception of Hering's rules is demonstrated by the biased visual direction of a fused disparate target when its monocular image components have unequal contrast (Banks, van Ee, & Backus, 1997). Greater weight is given to the retinal locus of the image that has the higher contrast. The average location in the cyclopean eye of the two disparate retinal sites is biased toward the monocular direction of the higher-contrast image. Interestingly, saccades to binocularly fused unequal-contrast targets are biased toward the mislocalized perceived target location and not toward the physical target location (Weiler, Maxwell, & Schor, 2007). These are minor violations that mainly occur for targets lying nearer or farther than the plane of fixation or distance of convergence. Because visual directions of targets placed before and behind the plane of fixation are mislocalized, even when Hering's rules of visual direction are obeyed, the violations have only minor consequences.

BINOCULAR DISPARITY AND BINOCULAR CORRESPONDENCE

Retinal Disparity and the Horopter

We perceive space with two eyes as though they were merged into a single cyclopean eye located midway between them. This merger of the two ocular images is made possible by a sensory linkage between the two eyes that is based on the anatomical combination of homologous regions of the two retinas in the primary visual cortex. Unique pairs of retinal regions in the two eyes (corresponding points) must receive images of the same object so that these objects can be perceived as single, that is, in a common visual direction. This is made possible by the binocular overlap of the two eyes'

visual fields. Hering defined binocular correspondence operationally as retinal locations in the two eyes that, when stimulated by real points in space, result in a percept in identical visual directions. For a fixed angle of convergence, optical projections into space, from corresponding retinal points located along the equator and midline of each eye, converge on real points in space. In other cases such as oblique eccentric (tertiary) locations on the retina, corresponding points have visual directions that do not intersect in real space. The horopter is the locus of object points in space whose images can be formed on corresponding retinal points. The horopter serves as a reference throughout the visual field for the same disparity as at the fixation point (zero disparity). To appreciate the shape of the horopter, consider a theoretical case in which corresponding points are defined as homologous locations on the two retinas (i.e., corresponding points, which are equidistant from and in the same direction relative to their respective foveas). Consider binocular combinations of points imaged on the horizontal retinal meridians or equators that pass through the foveas. The egocentric visual directions of corresponding points located on the retinal equator intersect in space at real points that define the longitudinal horopter. This theoretical horopter is a circle whose points will be imaged at equal eccentricities from the two foveas on corresponding points except for the small arc of the circle that lies between the two eyes (Ogle, 1962) (figure 57.1). Although the theoretical horopter is always a circle, its radius of curvature increases with viewing distance. This means that its curvature decreases as viewing distance increases. In the limit, the theoretical horopter is a straight line at infinity that is parallel to the face and interocular axis (frontoparallel). Thus, a surface representing zero disparity has many different shapes depending on the viewing distance. The consequence of this spatial variation in horopter curvature is that the spatial pattern of horizontal disparities is insufficient information to specify depth magnitude or even depth ordering or surface shape and curvature. Information about target distance and direction relative to the head is needed to interpret surface shape and orientation from horizontal retinal image disparity (Garding et al., 1995).

Head-Centric Disparity and Perisaccadic Stereopsis

For almost 200 years binocular disparity has remained synonymous with retinal disparity (Wheatstone, 1838), which is computed by subtracting the distance of each half-image from its respective fovea (Hering, 1861). However, binocular disparity could also be coded in

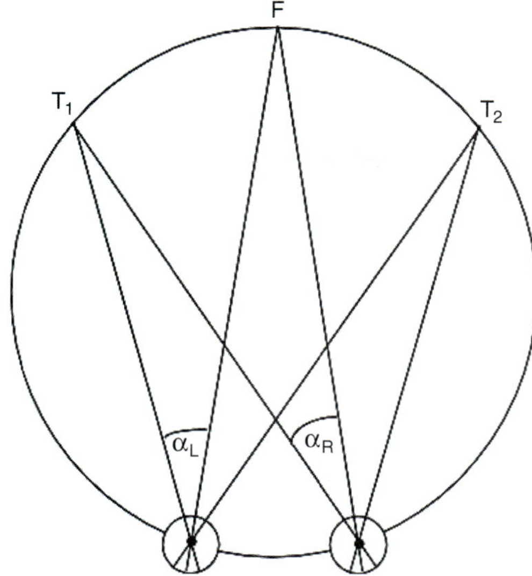

FIGURE 57.1   The theoretical horopter. The theoretical horopter or Vieth-Muller circle passes through the entrance pupils of the two eyes and the point of fixation (F). All points on this circle (e.g., T1 and T2) subtend equal angles at the entrance pupils of the two eyes ($\alpha_L = \alpha_R$). The Vieth-Muller circle represents the geometric locus of targets that subtend zero retinal image disparity.

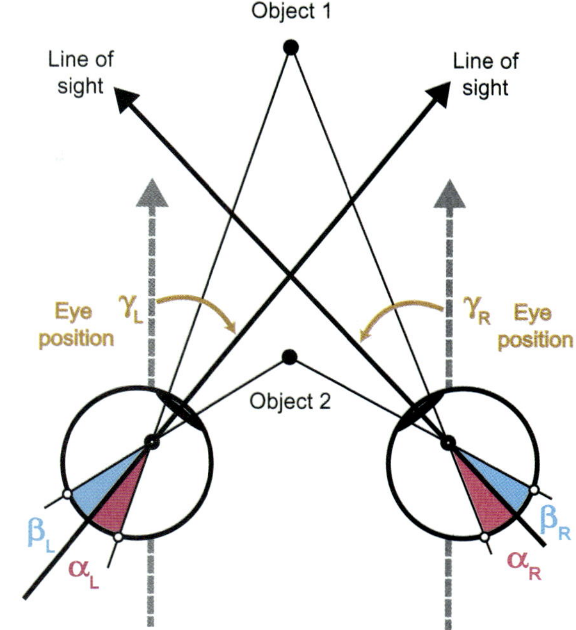

	Retinal Disparity	Head-centric Disparity
Object 1	$\alpha_L - \alpha_R$	$(\alpha_L + \gamma_L) - (\alpha_R + \gamma_R)$
Object 2	$\beta_L - \beta_R$	$(\beta_L + \gamma_L) - (\beta_R + \gamma_R)$
Relative Disparity	$(\alpha_L - \alpha_R) - (\beta_L - \beta_R)$	$(\alpha_L - \alpha_R) - (\beta_L - \beta_R)$

FIGURE 57.2   Retinal versus head-centric disparity coding. Top-down (plan) view of two eyes fixated in between a pair of objects (filled circles) that are separated in depth. The absolute disparity of each object is defined as the difference in the visual directions of its two half-images (open circles). Retinal disparity is coded relative to the line of sight (shaded angles), whereas head-centric disparity is coded relative to the head (taking eye positions into account). Stereo depth perception relies on relative disparity, the difference between the absolute disparities produced by the two objects. In most viewing situations the head-centric visual directions for both objects' half-images include the same eye-position signals, and these cancel out when the two absolute-disparity values are subtracted to form a relative disparity. However, a binocular disparity can be produced in a novel stimulus that presents the half-images of an object asynchronously. If the eye-position signals sampled (at different times) in the head-centric disparity coding are unequal for the two objects, they may not cancel when the disparities are subtracted. (From Zhang, Cantor, & Schor, 2010.)

head-centric instead of retinal coordinates by combining eye position and retinal image position in each eye and representing disparity as differences between visual directions of half images relative to the head (Zhang, Cantor, & Schor, 2010).

The computation of absolute binocular disparity requires a spatial frame of reference for specifying visual directions (figure 57.2). Retinal disparity is computed relative to the line of sight by subtracting the retinocentric visual directions, which are the angles subtended on the retinas between the visual image and the fovea (Wheatstone, 1838). Head-centric disparity is computed by subtracting the head-centric visual directions, which are computed relative to the perceived "straight ahead," by combining (adding) each eye's position signal with the retinocentric visual direction (Zhang, Cantor, & Schor, 2010). Objects at different distances in the scene produce different absolute disparity values, and their difference—relative disparity— affords a metric measure of the depth in the scene (when scaled by viewing distance). Unlike absolute disparity, this relative disparity does not depend on eye position or on the spatial frame of reference for computing disparity. Under almost all viewing conditions, relative head-centric disparity and relative retinal disparity remain identical for a pair of objects.

Although these two disparity-coding schemes suggest very different neural mechanisms, both offer identical predictions for stereopsis in almost every viewing condition, making it difficult to empirically distinguish between them. The use of head-centric disparity in stereopsis has been illustrated with an illusion called

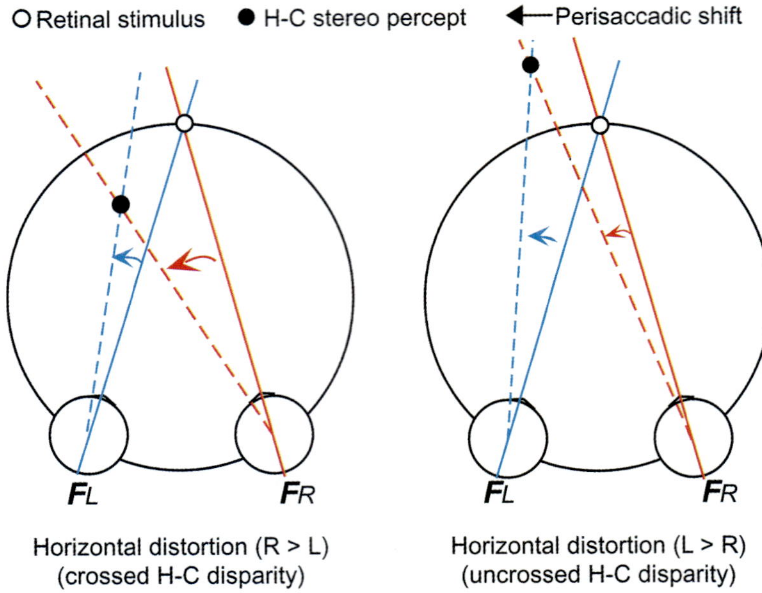

FIGURE 57.3   Producing a head-centric disparity. Head-centric disparity (crossed and uncrossed) is generated by the unequal perisaccadic distortion (blue and orange arrows) of asynchronous, flashed half images. Open circles mark the physical location of both half images. They are extinguished prior to the horizontal saccade and thus produce zero retinal disparity. For a leftward saccade, crossed (uncrossed) head-centric disparity results from greater perceived displacement of the right (left) eye's half-image. (From Zhang, Cantor, & Schor, 2010.)

perisaccadic stereopsis that is produced with binocular disparities that have different magnitudes when represented in retinal versus head-centric coordinates (Zhang, Cantor, & Schor, 2010). In this illusion retinal disparity is always zero, but the head-centric disparity is nonzero, and interestingly, the perisaccadic stimulus produces a stereo-depth effect consistent with the nonzero disparity. The illusion is produced by perisaccadic spatial distortions of targets flashed near the onset of a saccade. When a foveal target is flashed immediately before a horizontal saccade, it appears to move in the direction of the saccade because of its visual persistence. When foveal targets are flashed to the two eyes with a small time delay (50 ms) near the onset of a horizontal saccade, the foveal flashes to the left and right eye appear to move by different amounts (figure 57.3). This produces a binocular head-centric horizontal disparity with a zero retinal disparity, yet it produces a sensation of depth consistent with a nonzero head-centric disparity. Furthermore, this head-centric disparity can cancel and reverse the perceived depth stimulated with nonzero retinal disparity, demonstrating that a coding scheme other than retinal disparity has a role in human stereopsis.

On the basis of reports of physiological coding of binocular disparity, retinal and head-centric coordinates are used by two different disparity systems. The first system is composed of binocular cells in the early visual cortex that process small binocular disparities (<1°) that are represented in retinal coordinates (DeAngelis, Cumming, & Newsome, 1999). On the other hand, larger binocular disparities are represented in a second system by binocular cells in sensorimotor cortical regions, such as MST, LIP, FEF, and the frontal cortex (Gamlin & Yoon, 2000), areas that are important for the representation of eye position.

Although the majority of the literature on stereopsis assumes a retinal-disparity coding scheme, there is indirect evidence that a head-centric coding scheme exists. Additional support comes from patients with strabismus, who appear to remap egocentric space for their deviating (turned) eye using extraretinal (eye position) signals from that eye (Ramachandran, Cobb, & Levi, 1994). These patients do not perceive diplopia of fixated targets even though they have eye turns exceeding several degrees, and neither eye is suppressed. Furthermore, they can have a coarse sense of stereoscopic depth while their eyes are misaligned (Dengler & Kommerell, 1993). Another example is the demonstration of stereopsis in lateral-eyed animals such as the horse (Timney & Keil, 1999). Finally, the possibility of a head-centric disparity coding scheme in humans has been explored in a computational model (Erkelens & van Ee, 1998).

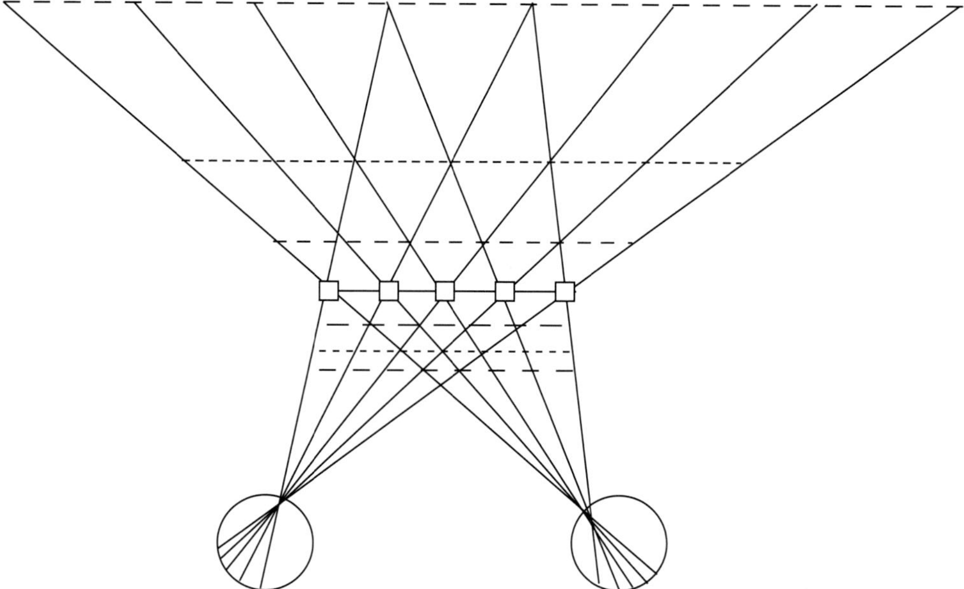

FIGURE 57.4   The wallpaper illusion. A repetitive pattern of vertical bars or an equally spaced horizontal row of dots can be matched in the two eyes to form one of many different depth planes.

BINOCULAR MATCHING: THE CORRESPONDENCE PROBLEM

Alignment of the two retinal images depends on eye position and the optical center of the eye (nodal point). Two separate components of eye position include the direction of gaze relative to the head (version) and the distance of gaze where the two visual axes intersect (vergence). If both version and vergence eye position are known, the three-dimensional scene can be reconstructed geometrically by projecting images from the retina through the nodal points of the two eyes, and the target lies at the points of intersection. However, the success of this exercise relies on selecting image points that correspond to the same target in space. In complex natural scenes there are many textured patterns that have similar forms in different locations, and potential matches are possible between a given feature in one eye and several features in the other eye. This problem is exacerbated by camouflage surface texture such as tree foliage where there is little spatial uniqueness of any particular texture element in the pattern. For example false matches can occur in viewing a vertically oriented repetitive pattern (figure 57.4). The vertical bars in wallpaper patterns can appear to float out of the depth plane of the wall (the wallpaper illusion). Matching could occur at various levels of visual processing including at a high level, after form perception, locally at a low level, prior to form perception, and globally in context with other depth solutions in the visual scene.

Local Retinal Disparity Analysis, Prior to Form Perception

In natural scenes each texture element subtends a specific binocular pair of retinal images; however, this image could correspond to many other textured patterns that have similar forms in different locations, and potential matches are possible between a given feature in one eye and several features in the other eye (the inverse mapping and correspondence problem) (Pizlo, 2001). This problem is exacerbated by camouflage surface texture such as tree foliage, where there is little spatial uniqueness of any particular texture element in the pattern. For example, false matches can occur when one is viewing a vertically oriented repetitive pattern. As illustrated in figure 57.4, the vertical bars in wallpaper patterns can appear to float out of the depth plane of the wall (the wallpaper illusion).

Binocular Disparity Derived from High-Level Form Perception

In the classic view binocular disparity is computed from the difference in visual directions of line elements in individual perceived forms. In this high-level analysis, form perception would precede disparity processing and depth perception (Helmholtz, 1909/1962). However, Julesz (1971) demonstrated low-level disparity matching, prior to form perception, with the random-dot stereogram. In the high-level classic analysis it is clear which monocular components of the perceived binocular images are to be matched because of their uniqueness. The local feature analysis is much more difficult because many similar texture elements must be matched, and the correct pairs of images to match are not obvious.

Cross-Correlation Model

In this local analysis luminance properties of the scene, such as texture and other token elements, could be analyzed to code a disparity map from which depth was estimated. The resulting disparity map would contribute to the perception of form. Formation of the disparity map has been described mathematically as a cross-correlation analysis between the two ocular images (Stevenson, Cormack, & Schor, 1991). The cross-correlation occurs over limited regions of the visual field so that a unique disparity solution is not blurred by variations of depth in the visual scene. Ultimately, the size of these correlation patches has to be large enough to minimize false matches and small enough to avoid blurring of spatial variations in depth (Banks, Gepshtein, & Landy, 2004; Nienborg et al., 2004).

The cross-correlation can be simplified by limiting matches between features with similar contrast, spatial frequencies, and orientations. Disparity is processed with spatial filters that operate at the early stages of visual processing. Center–surround receptive fields and simple and complex cells in the visual cortex have optimal sensitivity to limited ranges of luminance periodicity that can be described in terms of spatial frequency. The tuning or sensitivity profiles of these cells have been modeled with various mathematical functions (difference of Gaussian, Gabor patches, Cauche functions, etc.), all of which have bandpass characteristics. They are sensitive to a limited range of spatial frequencies referred to as a *channel*, and there is some overlap in the sensitivity range of adjacent channels. These channels are also sensitive or tuned to limited ranges of different orientations. Thus, they encode both the size and orientation of contours in space. These filters serve to decompose a complex image into discrete ranges of its spatial frequency components. In the Pollard, Mayhew, and Frisby (PMF) model (Frisby & Mayhew, 1980), three channels tuned to different spatial scales filter different spatial scale image components. The binocular cells within channels can be tuned to either phase or positional disparity (see chapter 28 by Parker), and they can be sensitive to disparity anywhere within their receptive field using quadrature pairing of simple cells that are rectified and squared before summing to form complex cells (see disparity-energy model described in chapter 28 by Parker).

Smoothness and Uniqueness Constraints

Binocular matching is further constrained by global or contextual factors that limit the number of possible matches. Computational models of stereopsis employ a number of algorithms that constrain stereo matches to produce the smallest disparity offset from the plane of fixation (nearest-neighbor match) or to minimize the disparity differences between nearby features (Papathomas & Julesz, 1989; Zhang, Edwards, & Schor, 2001). In some natural scenes, such as a slanted textured plane, the nearest-neighbor match and minimum disparity difference match are in conflict when disparities in regions away from the point of fixation exceed one cycle of the texture spacing. In figure 57.5 the center stimulus in the middle panel is surrounded by upper and lower flanking stimuli with the same disparity amplitude but in opposite depth sign as the center strip. The flank and center are shown in isolation in the top and bottom panels, respectively. These patterns have multiple depth matches as described above in the wallpaper illusion. The minimum depth difference (relative disparity) between the flank and center is smaller than the difference between the absolute disparities of the overall stimulus. When the top panel is fused by crossing the eyes (cross-fused), the flank appears nearer than the fixation lines. When the bottom panel is cross-fused, the center appears farther than the fixation lines. When the center panel is cross-fused, both the center and flanks appear to have the same depth signs either nearer or farther than the fixation lines, and their depth ordering is reversed compared to the top and bottom panels. Figure 57.5 demonstrates a priority of minimum relative depth over minimum absolute disparity matching solutions.

Matching algorithms assume that each point has a single or unique binocular match. Panum's limiting case appears to be a violation of this constraint, in which fusion of a single vertical bar presented to one eye with two adjacent vertical bars presented to the other eye is seen as two bars at different stereo depths. However, this phenomenon can be interpreted as a special case of occlusion or DaVinci stereopsis in which depth differences between the two lines could be estimated from the spatial geometry of an implied occluder (Harris & Wilcox, 2009).

The smoothness and uniqueness constraints and other contextual cues to depth have been described by low-level computational models (Blake & Wilson, 1991). Most of these models employ both mutual facilitation

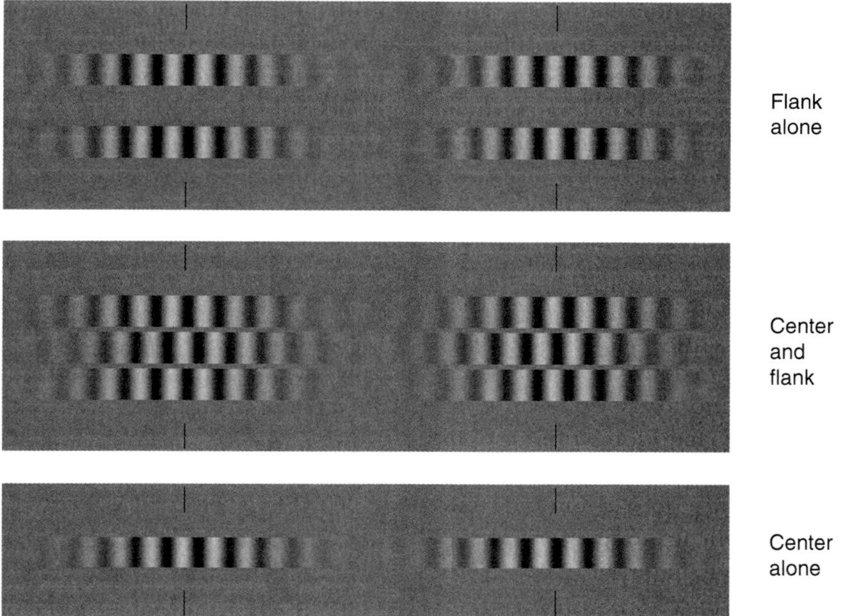

FIGURE 57.5   The center row in the middle panel has the same absolute disparity as the figure shown in the bottom panel, and the flanking rows in the middle panel have the same disparity as the rows shown in the upper panel. When the center panel is fused by crossing your eyes to merge the vertical lines on the left and right, the depth ordering of the center panel is reversed from its counterparts in the upper and lower panel. This stereogram demonstrates that the minimum relative disparity has a higher priority than the minimum absolute disparity in the binocular matching process.

and inhibition; however, several stereo phenomena suggest a lesser role for inhibition. For example, we are able to see depth in transparent planes, such as views of a streambed through a textured water surface or views of a distant scene through a spotted window or intervening plant foliage. In addition, depth of transparent surfaces can be averaged, seen as filled in or as two separate surfaces as their separation increases (Stevenson, Cormack, & Schor, 1989). Inhibition between dissimilar disparity detectors would make these percepts impossible. These contextual global phenomena have also been interpreted with statistical models (van Ee, Adams, & Mamassian, 2003; Geiger, Landendorf, & Yuille, 1995; Nakayama & Shimojo, 1992).

Edge Biases and Second-Order Stereopsis

Binocular matching is most difficult with repetitive textures that are presented in large fields located either in front or behind the plane of fixation. A coarse-to-fine strategy reduces this matching ambiguity by first seeking matches of unambiguous edges (Mitchison & McKee, 1987). Edges or overall target shape can be described as a contrast variation of surface texture. Contrast variations are coded by a nonlinear or second-order process that is equivalent to rectifying the neural representation of the retinal image (Wilson, Ferrera, & Yo, 1991).

Contrast-defined edges (contrast envelope) can be used to make coarse matches in stereo tasks involving large disparities that are near the upper disparity limits for stereopsis (Hess & Wilcox, 1994). Contrast envelope size (Schor, Edwards, & Sato, 2001) and, to a greater extent, temporal synchrony of the two eyes' stimuli (Cogan et al., 1995) are the primary means for selecting matched binocular inputs for transient stereopsis. Second-order contrast-defined stimuli work well in concert with first-order stimuli to assist in binocular matching using a coarse-to-fine strategy. Second-order information from the contrast-defined edges provides coarse information to guide matches of small disparities in the luminance-defined texture by the fine system (Wilcox & Hess, 1997). This is not a function reserved exclusively for second-order stimuli because coarse first-order luminance information also guides disparity matches (Wilson, Ferrera, & Yo, 1991). This coarse-to-fine strategy utilizing contrast and luminance information is effective because spatial information represented by contrast and luminance is usually highly correlated in images formed in natural environments.

Local and High-Level Interactions

There is an interaction between the local and high-level analysis, which simplifies the matching process. Unique shapes are often visible within textured fields. For example there are clumps of leaves in foliage patterns or target edges that are clearly visible before one can sense depth. Vergence eye movements can align these unique monocular patterns on or near corresponding retinal points and reduce the overall range of disparity subtended by the stimulus (Marr & Poggio, 1976). Once this has been done the local system can begin to match tokens based on certain attributes such as luminance, spatial frequency, and orientation.

STEREO ACUITY

Relative and Absolute Stereopsis

Stereopsis is the sense of relative depth between two or more features. It is stimulated by differences between absolute disparities subtended by these features (Westheimer, 1979a). Absolute disparity of an object quantifies its retinal disparity relative to the horopter. It equals the difference in the angular locations of its retinal images in the two eyes, referenced to corresponding retinal points. Stereopsis is stimulated by differences among several absolute disparities, that is, relative disparity. Relative disparity describes the disparity difference between features. Relative disparity thresholds for stereopsis are lowest when targets are imaged on or near the horopter. Stereosensitivity to relative disparity varies dramatically with distance of these targets from the horopter or plane of fixation (figure 57.6). The average disparity of targets from the fixation plane is described as their *depth pedestal*. Stereo depth discrimination thresholds, measured as a function of the depth pedestal, describe a depth-discrimination threshold function that is summarized by a Weber fraction (stereo threshold/depth pedestal). The Weber fraction for stereopsis indicates that the noise or variability of the absolute disparities subtended by the

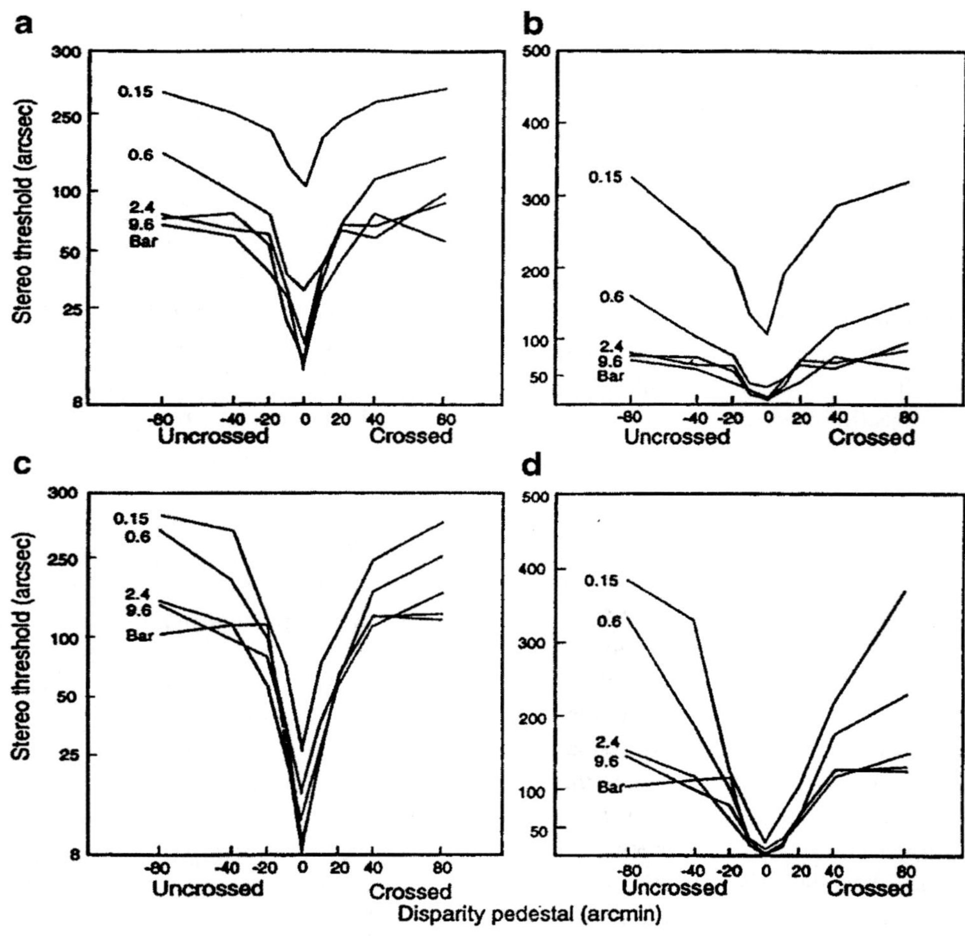

FIGURE 57.6   Threshold for depth discrimination between a test stimulus and comparison stimulus as a function of the disparity offset (pedestal) of the comparison stimulus. Each curve shows the results using bandpass-filtered bar stimuli whose center spatial frequencies range from 0.15 to 9.6 cycles/deg. Stereo depth is most sensitive for targets located on the horopter, and threshold increases exponentially with disparity pedestal.

816   CLIFTON M. SCHOR

targets is less than 5% over a range of disparity pedestals up to 0.5° (Badcock & Schor, 1985). Judgments of depth beyond this range of pedestals are clouded by the appearance of diplopia.

Spatial Pooling and Uncertainty

Stereosensitivity can be either enhanced or reduced by nearby targets. The threshold for detecting depth corrugations of a surface, such as the folds in a hanging curtain, decreases with depth-modulation frequency (reciprocal of spacing between the folds), up to 0.3 cycles/deg) where threshold is lowest (Tyler, 1975). At depth-modulation frequencies lower than 0.3 cycles/deg, the threshold for stereopsis is elevated and appears to be limited by a disparity gradient (i.e., a minimal rate of change of depth/degree of target separation). At depth-modulation frequencies higher than 0.3 cycles/deg, stereo threshold is elevated as a result of depth averaging of adjacent targets. Similar effects are seen with separations between point stimuli for depth (Westheimer, 1986; Westheimer & McKee, 1979). Other factors that could limit spatial stereoresolution include sampling properties of the stimulus, low-pass spatial filtering by mechanisms early in visual processing, the smallest useful correlation window in the visual system, and an assumption that disparity is constant across a correlation patch (Banks, Gepshtein, & Landy, 2004).

When a few isolated targets are viewed foveally, changes in the binocular disparity of one target produce suprathreshold depth changes or depth biases in other targets when their separation is less than 4' arc. This depth attraction illustrates a pooling of disparity signals. When targets are separated by more than 4–6', the depth bias is in the opposite direction, and features appear to repel one another in depth. Attraction and repulsion also occur with cyclopean targets (i.e., targets that can only be revealed by stereo depth such as clumps of tree foliage) (Stevenson, Cormack, & Schor, 1991), showing that they are not based simply on positional effects at the monocular level. The enhancement of depth differences by repulsion might be considered a depth-contrast phenomenon that is analogous to Mach bands in the luminance domain (enhanced perceived contrast of edges), which are thought to result from lateral inhibitory interactions. Depth distortions that are analogous to Mach bands have been demonstrated with horizontal disparity variations between vertically displaced contours (Lunn & Morgan, 1996). Mach bands have been modeled as luminance-based center–surround inhibition; however, disparity-based center–surround inhibition has not been observed in disparity-tuned receptive fields. Rather, it appears that spatial disparity interactions are determined by a weighted mean of the disparity composition of the stimulus (Nienborg et al., 2004).

Influence of Early Spatial Filters on Stereo Acuity

Stereo acuity depends on spatial scale (range of spatial frequencies). When tested with a limited range of spatial frequencies (narrowband luminance stimulus; 1.75 octaves) such as produced by a difference of two Gaussians, stereo threshold becomes elevated when tested with spatial frequencies below 2.5 cycles/deg. (Schor, Wood, & Ogawa, 1984; Smallman & MacLeod, 1994; Kontsevich & Tyler, 1994). The dependence of stereopsis on luminance spatial frequency suggests that disparity is processed early in the visual system by linear channels that are tuned to discrete bands of spatial frequency.

STEREO DEPTH SCALING AND INTERPRETATION

Mapping Disparity from Retinal Angles to Linear Depth in Head-Centric Coordinates

Stereoscopic depth is stimulated by two-dimensional retinal disparities that can be described in angular units. Horizontal retinal image disparity is insufficient to reconstruct the depth and orientation of a three-dimensional space percept. The same horizontal retinal image disparity can correspond to a centimeter at an arm's length and a meter at a remote 10-m viewing distance, and its slant or orientation depends on its azimuth or horizontal eccentricity. Additional information is needed to map the disparity of two retinal images into absolute depth in head-centric space coordinates. Interpretation of disparity for stereo depth perception depends on estimates of both distance and direction (azimuth) of targets relative to the head.

Stereo Depth Scaling

Three independent variables involved in scaling stereo-depth into head-centric coordinates are retinal image disparity, viewing distance, and the separation in space of the two view-points (i.e., the baseline or interpupillary distance). In stereopsis the relationship between the linear depth interval between two objects and their retinal image disparity pattern is approximated by the following expression:

$$\Delta d = \eta \times d^2/2a \qquad (57.1)$$

where η is retinal image disparity in radians, d is viewing distance, $2a$ is the interpupillary distance, and Δd is the

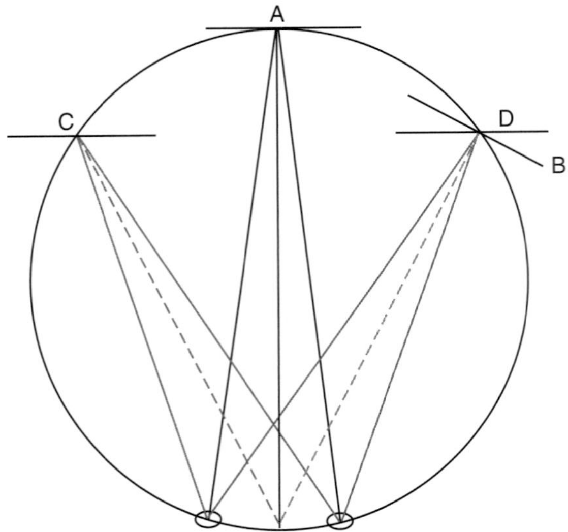

FIGURE 57.7   A top-down view of several slanted surfaces in different locations. Surfaces A and B are gaze normal. They subtend similar horizontal retinal image disparities, yet they have very different slant angles with respect to the head. Surfaces A and C both have the same head-centric frontoparallel orientation, but they subtend very different horizontal disparity patterns.

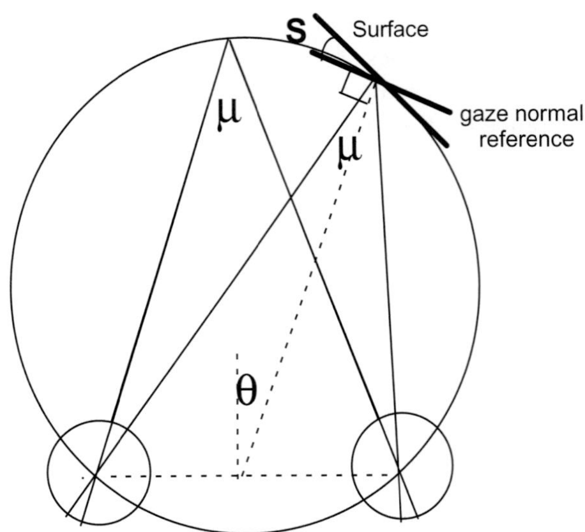

FIGURE 57.8   Slant $(S) = -\tan^{-1} (1/\mu \ln \mathrm{HSR} - \tan \phi)$. Apart from extraretinal eye position signals, vertical disparity cues can also be used to estimate target azimuth when computing slant. Vertical disparities occur naturally with targets in tertiary directions from the point of fixation (Garding et al., 1995; Gillam & Lawergren, 1983; Liu, Stevenson, & Schor, 1994) (see figure 57.9). Vertical disparity can be described as the vertical size ratio of the two retinal images (left/right vertical image size = VSR) (Rogers & Bradshaw, 1995). The relationship among slant angle and combined cues of horizontal and vertical retinal disparity and convergence angle is given by Backus et al. (1999).

linear depth interval between two targets; $2a$, d, and Δd are all expressed in the same units (e.g., meters). The formula implies that in order to perceive depth in units of absolute distance (e.g., meters), the visual system utilizes information about the interpupillary distance and the viewing distance. The equation illustrates that, for a fixed retinal image disparity, the corresponding linear depth interval increases with the square of viewing distance and that information about viewing distance is used to scale the horizontal disparity into a linear depth interval.

Stereo Slant Interpretation

Disparity cues for surface shape and orientation also depend on target location relative to the head (distance and azimuth). For example, a surface that is parallel to the face has a similar horizontal retinal image disparity pattern when viewed straight ahead as a gaze-normal surface that is tilted toward the right eye when viewed in rightward gaze (compare surfaces A and B in figure 57.7). Information about target distance and direction can be obtained from extraretinal and retinal cues. Extraretinal cues include information about version and horizontal vergence eye position. Surface orientation or slant can be estimated relative to the head and body (head-centric coordinates) with a combination of horizontal disparity and eye position cues. Retinal

image disparity for horizontal slant about a vertical axis can be described by the ratio of horizontal size of the two retinal images (left/right horizontal image size = HSR) (Ogle, 1962; Rogers & Bradshaw, 1995). For a single surface the relationship between slant angle and combined cues of horizontal retinal image disparity (HSR), horizontal gaze angle (azimuth, ϕ), and convergence angle (μ) is given by (Backus et al., 1999) (figure 57.8)

$$\text{Slant } (S) = -\tan^{-1} (1/\mu \ln \mathrm{HSR} - \tan \phi) \quad (57.2)$$

$$\text{Slant } (S) = -\tan^{-1} (1/\mu \ln \mathrm{HSR}/\mathrm{VSR}) \quad (57.3)$$

This is a gaze-normal description. Adding the azimuth (ϕ) yields slant relative to a frontoparallel plane. The use of vertical disparity for slant perception is demonstrated by tilt of an apparent frontoparallel plane that is induced by a vertical magnifier before one eye (i.e., the induced effect) (Ogle, 1962). The effect is opposite to the tilt produced by placing a horizontal magnifier before the same eye (the geometric effect). If both a horizontal and a vertical magnifier are placed before one eye, in the form of an overall magnifier, no tilt of the plane occurs. In eccentric gaze, binocular viewing geometry causes the image of the near target to be

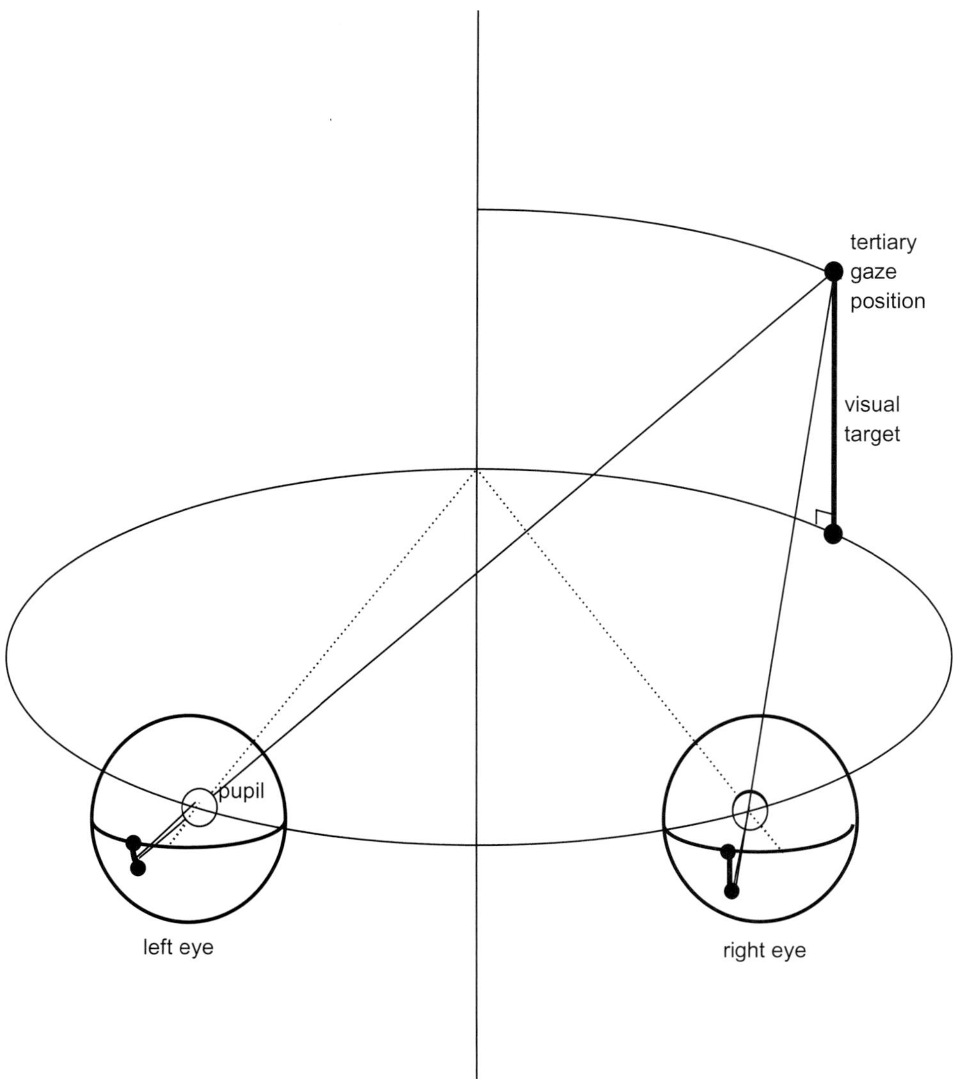

FIGURE 57.9   Vertical disparities occur naturally when targets are placed in tertiary positions while the eyes are aimed in another direction of gaze. Here the target is closer to the left than the right eye, and its image is larger in the left eye. Vertical disparity increases with target azimuth, elevation, and proximity to the observer.

magnified more in the nearer eye (figure 57.9). This unequal magnification could provide a sufficient cue for azimuth to make accurate frontoparallel settings in eccentric gaze. Together, vertical and horizontal magnification provide sufficient information to distinguish a tilted plane that is viewed straight ahead from a frontoparallel plane viewed in eccentric gaze.

TEMPORAL PROPERTIES OF TRANSIENT AND SUSTAINED STEREOPSIS

Second-order contrast cues that define target edges stimulate both sustained and transient stereopsis. Sustained stereopsis is stimulated by small disparities (within the singleness range) that are presented for long durations. Transient stereopsis has a brief (transient) impression of depth, and it is stimulated by briefly presented images (Kumar & Glaser, 1994) whose disparity can range from small values within the sensory fusion range (0.5°) to highly diplopic values (>8°). Both sustained and transient stereopsis can be stimulated by first-order (luminance) and second-order (contrast) forms of spatial information. Transient stereopsis relies more heavily on second-order information than does sustained stereopsis. Vivid stereoscopic depth can be perceived with brief stimuli, even when the detailed structure of the two retinal images differs markedly. The transient system is more broadly tuned than the sustained system to differences between the texture (carrier) properties of the stimulus presented to the two eyes (i.e., spatial frequency, orientation, contrast, and contrast polarity) (Edwards, Pope, & Schor, 1999; Schor, Edwards, & Pope, 1998). For example, when presented for several seconds, a vertical grating patch presented to one eye and a horizontal grating patch presented to the other eye will appear to alternate in appearance (binocular rivalry), but when briefly flashed, they can appear simultaneously in depth arising from the disparity subtended by their contrast defined second-order borders.

Simultaneous versus Sequential Stereopsis

One of the primary functions of the visual system is to provide a spatial sense of our surrounding environment. Tasks such as reaching or grasping involve motor interactions between ourselves and objects about us, and they require knowledge about the location and orientation of objects relative to ourselves in head-centric coordinates. Stereopsis is one of our primary senses of spatial layout. Stereopsis is most sensitive to relative disparity between two or more adjacent targets that can be imaged simultaneously within the high-resolution region of the central retina (Westheimer, 1979a, 1979b). Stereo depth differences are more difficult to estimate when targets are imaged in the periphery such as when comparing depth of widely separated targets, and only one target can be imaged on the fovea at a time. Under these conditions stereoscopic depth discrimination is less sensitive than when both targets can be imaged on the fovea (Schor & Badcock, 1985; McKee et al., 1990). Perceiving the depth differences among several targets in different directions requires either that some of the objects be viewed peripherally at the same time (*simultaneous stereopsis*) or that the targets be viewed sequentially with high-velocity gaze shifts (saccades), at different times, with successive saccadic-foveal fixations (*sequential stereopsis*).

With large target separations simultaneous stereopsis is impaired by elevated thresholds for disparity discrimination in the retinal periphery. Similar effects are found for stereo depth discrimination between targets that are presented sequentially in different directions (Enright, 1991; McKee et al., 1990). One target disparity is imaged briefly (150 ms) on the fovea, and when it is extinguished, another target disparity is imaged briefly on an eccentric retinal location while gaze remains fixed in one direction. The task is to compare differences in depth between sequential views of the two targets that are presented one at a time. Simultaneous and sequential stereopsis have identical thresholds when tested with large target separations (>2°) during steady fixation (Enright, 1991; McKee et al., 1990).

Thresholds increase proportionally with target eccentricity (Berry, 1948; McKee et al., 1990). Saccadic eye movements can improve perceptual performance in both sequential stereo slant and depth estimation tasks when objects are widely separated in space (Zhang, Berends, & Schor, 2003). Stereosensitivity to depth differences between widely separated targets (>5°) can be improved by as much as 50% with sequential gaze shifts that image the two targets sequentially onto the fovea (Enright, 1991; Wright, 1951). Retinal position uncertainty is believed to increase with retinal eccentricity (Levi, Klein, & Aitsebaomo, 1984), and this source of noise is thought to account for the increase of stereo threshold with target separation (McKee et al., 1990).

Sensory–Motor Interactions

Sequential stereo-depth comparisons between targets in different locations are represented in head-centric coordinates, and stereo depth perception depends on target location relative to the head as well as horizontal retinal image disparity. In particular, vergence information is essential for estimating target distance from the head when vertical disparity information is sparse or absent (Brenner & Van Damme, 1998; Wright, 1951; Backus & Matza-Brown, 2003). In the extreme case, vergence aligns the eyes precisely during fixation so that absolute retinal image disparity is reduced to zero. Yet depth differences between the sequentially fixated targets that subtend zero disparity can still be discriminated because efferent signals that align the eyes can be used to estimate target distance from the head (Brenner & Van Damme, 1998). Stereo threshold is limited by the uncertainty with which eye position is sensed (Backus et al., 1999), the duration or dwell time of each fixation, and also by time delays between sequential views (Kumar & Glaser, 1994) that are related to the dynamics of saccadic eye movements (Bahill, Clark, & Stark, 1975). Even small to moderate saccades take 25–50 ms (Bahill, Clark, & Stark, 1975), and this introduces an interstimulus interval (ISI) during which information about the first slant angle may be lost or corrupted. The ISI could be extended beyond the duration of the saccade by saccadic suppression (reduced visual sensitivity during saccades), which occurs while changes in visual direction are updated to correspond with abrupt changes in eye position (Matin, 1986).

Temporal Masking

Temporal masking (Alpern, 1953) also elevates thresholds for sequential stereopsis (Butler & Westheimer,

1978; Tyler & Foley, 1974). Saccadic gaze shifts produce time delays between successively fixated targets. The transient onset of the second stimulus could mask the appearance of the first stimulus (backward masking) or vice versa, especially when the two stimuli are large textured surfaces (Kahneman, 1968). Backward masking has been shown to influence sensitivity to apparent motion with random dot patterns (Braddick, 1973). An example of backward masking in stereopsis was reported by Butler and Westheimer (1978). They observed that stereo threshold was elevated by adjacent contours presented 100 ms after the onset of the stereo test stimulus. The spatiotemporal masking is greatest when the mask and stereo stimulus have the same disparity. Masking is reduced by half when the disparity of the mask differs from the stimulus by only 15" of arc.

Motion in Depth

As we move toward a stationary object in space, as in the approach of a car to a stop sign, there are several cues for estimating this motion, including size looming and changing disparity. The temporal limits for sensing motion in stereo depth can be quantified using perceived depth amplitude of sinusoidal temporal variations in disparity. The stereo system has a low-pass temporal response function, where motion in depth becomes attenuated at temporal frequencies above 2 Hz (Nienborg et al., 2005). In contrast, lateral motion of luminance-defined contours can be sensed at much higher temporal frequencies (>20 Hz). The limited temporal frequency response of the stereo system is a consequence of the cross-correlation process for deriving disparity from two monocular images (see disparity energy model in chapter 28 by Parker). This process involves temporal bandpass filtering of monocular components before computing binocular correlation, and this filtering limits the temporal resolution of the disparity-coding neurons as early as in the primary visual cortex (Nienborg et al., 2005).

CONCLUSION

Space perception is based on several sources of information that are sensed by the visual system. The perception of direction, distance, surface shape and orientation, and object volume can be obtained from monocular sources of information including motion parallax, relative size, texture gradients, perspective cues, and stimulus overlap (partial occlusion) and from the binocular source of retinal image disparity (stereopsis). The subject of stereopsis has been used to investigate spatial resolution limits of the visual system, to learn how three-dimensional spatial information is represented in the brain, and to understand how perceptual systems reduce information content to a manageable level, and also reduce information ambiguity by employing ecological inferences (contextual cues) as well as computational strategies.

Disparity by itself is an ambiguous depth cue because it is an angle represented in retinal or oculocentric coordinates. To yield a depth percept, it must be mapped into head-centric coordinates and scaled using information about direction and distance to interpret relative depth magnitude and depth ordering. Currently there is a concentrated effort in visual media to incorporate stereopsis in rendering realistic views of visual space. Views of natural or "real-world" scenes contain a rich set of consistent cues for depth perception including binocular disparity, shading, and perspective that can be presented with high fidelity in most displays; however, some cues can not. Currently, flat-screen visual displays that simulate depth contain inconsistent depth cues, such as for blur and motion parallax. The fixed screen distance presents a constant defocus cue for focusing the retinal image, while binocular disparities render a wide range of depths in the simulated scene. Motion of the observer does not produce consistent motion parallax cues, and off-axis views have incorrect viewpoint distortions. An important question is how much information is required to render a realistic depth scene? The problem is much like the one faced by the printing industry that determined how small a bandwidth could produce acceptable prints. Current studies that include variable focus cues to distance find that stereo vision is faster and more accurate when focus cues are consistent with, rather than in conflict with, spatially and temporally varying disparity cues (Banks et al., 2008). The question remains about how many other cues need to be varied to render realistic simulated depth scenes that are acceptable to most people. Undoubtedly, there will be different answers for different applications that depend on the number of observers, range of viewing distances, eccentric viewing angles, and tasks involved such as in gaming and in medical analysis. One thing for certain is that stereo depth will be a major component in many future applications of visual displays.

REFERENCES

Alpern, M. (1953). Metacontrast. *Journal of the Optical Society of America, 43,* 648–657.

Backus, B. T., Banks, M. S., van Ee, R., & Crowell, J. A. (1999). Horizontal and vertical disparity, eye position, and stereoscopic slant perception. *Vision Research, 39,* 1143–1170.

Backus, B. T., & Matza-Brown, D. (2003). The contribution of vergence change to the measurement of relative disparity. *Journal of Vision, 3*, 737–750

Badcock, D. R., & Schor, C. M. (1985). Depth-increment detection function for individual spatial channels. *Journal of the Optical Society of America. A, Optics and Image Science, 2*, 1211–1215.

Bahill, A. T., Clark, M. R., & Stark, L. (1975). The main sequence, a tool for studying human eye movements. *Mathematical Biosciences, 24*, 191–204.

Banks, M. S., Akeley, K., Hoffman, D. M., & Girshick, A. R. (2008). Consequences of incorrect focus cues in stereo displays. *Journal of the Society for Information Display, 24*, 7.

Banks, M. S., Gepshtein, S., & Landy, M. S. (2004). Why is spatial stereoresolution so low? *Journal of Neuroscience, 24*, 2077–2089.

Banks, M. S., van Ee, R., & Backus, B. T. (1997). The computation of binocular visual directions: Re-examination of Mansfield and Legge. *Vision Research, 37*, 1605–1610.

Berry, R. N. (1948). Quantitative relations among vernier, real depth and stereoscopic depth acuities. *Journal of Experimental Psychology, 38*, 708–721.

Blake, R., & Wilson, H. R. (1991). Neural models of stereoscopic vision. *Trends in Neurosciences, 14*, 445–452.

Braddick, O. (1973). The masking of apparent motion in random-dot patterns. *Vision Research, 13*, 355–369.

Brenner, E., & Van Damme, W. J. M. (1998). Judging distance from ocular convergence. *Vision Research, 38*, 493–498.

Butler, T. W., & Westheimer, G. (1978). Interference with stereoscopic acuity: Spatial, temporal and disparity tuning. *Vision Research, 18*, 1387–1392.

Cogan, A. I., Konstevich, L. L., Lomakin, A. J., Halpern, D. L., & Blake, R. (1995). Binocular disparity processing with opposite-contrast stimuli. *Perception, 24*, 33–47.

DeAngelis, G. C., Cumming, B. G., & Newsome, W. T. (1999). A new role for cortical area MT: The perception of stereoscopic depth. In M. S. Gazzaniga (Ed.), *The new cognitive neurosciences* (pp. 305–314). Cambridge, MA: MIT Press.

Dengler, B., & Kommerell, G. (1993). Stereoscopic cooperation between the fovea of one eye and the periphery of the other eye at large disparities. Implications for anomalous retinal correspondence in strabismus. *Graefes Archive for Clinical and Experimental Ophthalmology, 231*, 199–206.

Edwards, M., Pope, D. R., & Schor, C. M. (1999). Orientation tuning of the transient-stereopsis system. *Vision Research, 39*, 2717–2727.

Enright, J. T. (1991). Exploring the third dimension with eye movements: Better than stereopsis. *Vision Research, 31*, 1549–1562.

Erkelens, C. J., & van Ee, R. (1997). Capture of the visual direction of monocular objects by adjacent binocular objects. *Vision Research, 37*, 1193–1196.

Erkelens, C. J., & van Ee, R. (1998). A computational model of depth perception based on head-centric disparity. *Vision Research, 38*, 2999–3018.

Frisby, J. P., & Mayhew, J. E. W. (1980). Spatial frequency tuned channels: Implications for structure and function from psychophysical and computational studies of stereopsis. *Philosophical Transactions of the Royal Society of London, 290*, 95–116.

Gamlin, P. D., & Yoon, K. (2000). An area for vergence eye movement in primate frontal cortex. *Nature, 407*, 1003–1007.

Garding, J., Porrill, J., Mayhew, J. E., & Frisby, J. P. (1995). Stereopsis, vertical disparity and relief transformations. *Vision Research, 35*, 703–722.

Geiger, D., Landendorf, B., & Yuille, A. (1995). Occlusions and binocular stereo. *International Journal of Computer Vision, 14*, 211–226.

Gillam, B. J., & Lawergren, B. (1983). The induced effect, vertical disparity and stereoscopic theory. *Perception & Psychophysics, 34*, 121–130.

Harris, J., & Wilcox, L. M. (2009). The role of monocular visible regions in depth and surface perception. *Vision Research, 49*, 2666–2685.

Helmholtz, H. von. (1962). *Handbuch der Physiologischen Optik* (3rd ed.), J. P. C Southall, Trans. Menasha, WI: Optical Society of America (original work published 1909).

Hering, E. (1861). *Beiträge zur Physiologie* (Vol. 5). Leipzig: Engelmann.

Hering, E. (1868). *Die Lehre vom binokularen Sehen*. Leipzig: Verlag Von Wilhelm Engelmann.

Hess, R. F., & Wilcox, L. M. (1994). Linear and non-linear filtering in stereopsis. *Vision Research, 34*, 2431–2438.

Howard, I. P. (1982). *Human visual orientation*. Chichester: Wiley.

Julesz, B. (1971). *Foundations of cyclopean perception*. Chicago: University of Chicago Press.

Kahneman, D. (1968). Method, findings, and theory in the study of visual masking. *Psychological Bulletin, 70*, 404–425.

Kontsevich, L. L., & Tyler, C. W. (1994). Analysis of stereo thresholds for stimuli below 2.5c/deg. *Vision Research, 34*, 2317–2329.

Kumar, T., & Glaser, D. A. (1994). Some temporal aspects of stereoacuity. *Vision Research, 34*, 913–925.

Levi, D. M., Klein, A., & Aitsebaomo, P. (1984). Detection and discrimination of the direction of motion in central and peripheral vision of normal and amblyopic observers. *Vision Research, 24*, 789–800.

Liu, L., Stevenson, S. B., & Schor, C. M. (1994). A polar coordinate system for describing binocular disparity. *Vision Research, 34*(9), 1205–1222.

Lunn, P. D., & Morgan, M. J. (1996). The analogy between stereo depth and brightness. *Perception, 24*, 901–904.

Mann, V. A., Hein, A., & Diamond, R. (1979). Localization of targets by strabismic subjects: Contrasting patterns in constant and alternating suppressors. *Perception & Psychophysics, 25*, 29–34.

Marr, D., & Poggio, T. (1976). Cooperative computation of stereo disparity. *Science, 194*, 283–287.

Matin, L. (1986). Visual localization and eye movements. In K. R. Boff, L. Kaufman, & J. P. Thomas (Eds.), *Handbook of perception and human performance; Vol. 1: Sensory processes and perception* (pp. 1–40). New York: Wiley-Interscience.

McKee, S. P., Welch, L., Taylor, D. G., & Bowne, S. F. (1990). Finding the common bond: Stereoacuity and the other hyperacuities. *Vision Research, 30*, 879–891.

Mitchison, G. J., & McKee, S. P. (1987). The resolution of ambiguous stereoscopic matches by interpolation. *Vision Research, 27*, 285–294.

Nakayama, K., Shimojo, S., (1992) Experiencing and perceiving visual surfaces. *Science, 257*, 1357–1363.

Nienborg, H., Bridge, H., Parker, A. J., & Cumming, B. G. (2004). Receptive field size in V1 neurons limits acuity for perceiving disparity modulation. *Journal of Neuroscience, 24*(9), 2065–2076.

Nienborg, H., Bridge, H., Parker, A. J., & Cumming, B. G. (2005). Neuronal computation of disparity in V1 limits temporal resolution for detecting disparity modulation. *Journal of Neuroscience, 25*(44), 10207–10219.

Ogle, K. N. (1956). Stereoscopic acuity and the role of convergence. *Journal of the Optical Society of America, 46*, 269–273.

Ogle, K. N. (1962). Spatial localization through binocular vision. In H. Davson (Ed.), *The eye* (Vol. 4, pp. 271–324). New York: Academic Press.

Papathomas, T. V., & Julesz, B. (1989). Stereoscopic illusion based on the proximity principle. *Perception, 18*, 589–594.

Pizlo, Z. (2001). Perception viewed as an inverse problem. *Vision Research, 41*, 3145–3161.

Ramachandran, V. S., Cobb, S., & Levi, L. (1994). The neural locus of binocular rivalry and monocular diplopia in intermittent exotropes. *Neuroreport, 5*, 1141–1144.

Rogers, B. J., & Bradshaw, M. F. (1995). Disparity scaling and the perception of frontoparallel surfaces. *Perception, 24*, 155–179.

Schor, C. M., & Badcock, D. (1985). A comparison of stereo and vernier acuity within spatial channels as a function of distance from fixation. *Vision Research, 25*, 1113–1119.

Schor, C. M., Edwards, M., & Pope, D. (1998). Spatial-frequency tuning of the transient-stereopsis system. *Vision Research, 38*, 3057–3068.

Schor, C. M., Edwards, M., & Sato, M. (2001). Envelope size tuning for stereo-depth perception of small and large disparities. *Vision Research, 41*, 2555–2567.

Schor, C. M., & Tyler, C. W. (1981). Spatio-temporal properties of Panum's fusional area. *Vision Research, 21*, 683–692.

Schor, C. M., Wood, I. C., & Ogawa, J. (1984). Spatial tuning of static and dynamic local stereopsis. *Vision Research, 24*, 573–578.

Smallman, H. S., & MacLeod, D. I. (1994). Size-disparity correlation in stereopsis at contrast threshold. *Journal of the Optical Society of America. A, Optics, Image Science, and Vision, 11*, 2169–2183.

Stevenson, S., Cormack, L., & Schor, C. M. (1989). Hyperacuity, superresolution, and gap resolution in human stereopsis. *Vision Research, 29*, 1597–1605.

Stevenson, S. B., Cormack, L. K., & Schor, C. M. (1991). Depth attraction and repulsion in random dot stereograms. *Vision Research, 31*, 805–813.

Timney, B., & Keil, K. (1999). Local and global stereopsis in the horse. *Vision Research, 39*, 1861–1867.

Tyler, C. W. (1975). Spatial organization of binocular disparity sensitivity. *Vision Research, 15*, 583–590.

Tyler, C. W., & Foley, J. M. (1974). Stereomovement suppression for transient disparity changes. *Perception, 3*, 287–296.

van Ee, R., Adams, W. J., & Mamassian, P. (2003). Bayesian modeling of cue interaction: Bistability in stereoscopic slant perception. *Journal of the Optical Society of America. A, Optics, Image Science, and Vision, 20*, 1398–1406.

Weiler, J. A., Maxwell, J. S., & Schor, C. M., (2007). Illusory contrast-induced shifts in binocular visual direction bias saccadic eye movements toward the perceived target position. *Journal of Vision, 7*, 1–18. doi: 10.1167/7.5.3.

Westheimer, G. (1979a). Cooperative neural processes involved in stereoscopic acuity. *Experimental Brain Research, 36*, 585–597.

Westheimer, G. (1979b). The spatial sense of the eye. *Investigative Ophthalmology & Visual Science, 18*, 893–912.

Westheimer, B. (1986). Spatial interaction in the domain of disparity signals in human stereoscopic vision. *Journal of Physiology, 370*, 619–629.

Westheimer, G., & McKee, S. P. (1979). What prior uniocular processing is necessary for stereopsis? *Investigative Ophthalmology & Visual Science, 18*, 614–621.

Wheatstone, C. (1838). Contributions to the physiology of vision—Part the first. On some remarkable, and hitherto unobserved, phenomena of binocular vision. *Philosophical Transactions of the Royal Society of London, 128*, 371.

Wilcox, L. M., & Hess, R. F. (1997). Scale selection for second-order (non-linear) stereopsis. *Vision Research, 37*, 2981–2992.

Wilson, H. R., Ferrera, V. P., & Yo, C. (1991). A psychophysically motivated model for two-dimensional motion perception. *Visual Neuroscience, 9*, 79–97.

Wright, W. D. (1951). The role of convergence in stereoscopic vision. *Proceedings of the Physical Society of London B, 64*, 289–297.

Zhang, Z., Cantor, C., & Schor, C. M. (2010). Perisaccadic stereo depth with zero retinal disparity. *Current Biology, 20*, 1176–1181.

Zhang, Z., Edwards, M., & Schor, C. M. (2001). Spatial interactions minimize relative disparity between adjacent surfaces. *Vision Research, 41*, 2995–3007.

Zhang, Z. L., Berends, E. M., & Schor, C. M. (2003). Thresholds for stereo-slant discrimination between spatially separated targets are influenced mainly by visual and memory factors but not oculomotor instability. *Journal of Vision, 3*, 710–724.

58 Binocular Rivalry Updated

RANDOLPH BLAKE

Binocular rivalry, the endogenously triggered, unpredictable fluctuations in perceptual awareness triggered by dissimilar stimulation of the two eyes, continues to beguile and baffle us. Indeed, the number of published papers on rivalry has exploded since the appearance of the earlier chapter on binocular rivalry in the first edition of this volume (Blake, 2004): According to Google Scholar, 427 articles with the phrase "binocular rivalry" in the title have been published in the past 8 years. Moreover, we see new themes emerging in this literature while, at the same time, some old controversies seem on their way to being resolved. This chapter provides an overview of recent developments—empirical and theoretical—in the field of binocular rivalry and, as well, points out puzzles and inconsistencies in this emerging literature that remain to be resolved. The chapter's emphasis is on work published during the last 8 years following the 2004 appearance of the first edition of *Visual Neuroscience*. Several recent reviews of rivalry will get the reader new to this field up to speed (Alais, 2012; Blake & Wilson, 2011; Blake & O'Shea, 2009).

RIVALRY DYNAMICS

Conventional binocular rivalry typically involves presentation of dissimilar monocular images to the two eyes for an extended viewing period. Aside from an initial, very brief period of time when both stimuli may be dominant simultaneously, most of the time is spent perceiving either one stimulus or the other, along with brief periods of mixed dominance occurring mostly during transitions between states of exclusive dominance. The following sections survey what has been learned recently about factors governing the dynamics of rivalry, starting with the initial state of exclusive dominance.

What Governs Initial Dominance in Rivalry?

Initial dominance is very sensitive to differences in stimulus strength between two rival stimuli, with the stronger of the two nearly always dominating initially (Song & Yao, 2009). Even when both rival stimuli are equal in strength, initial dominance of one can be promoted by focusing attention on that stimulus prior to the onset of both rival stimuli (Chong & Blake, 2006; Mitchell, Stoner, & Reynolds, 2004), an effect that works probably because attention boosts the effective contrast of that stimulus (Carrasco, Ling, & Read, 2004). Onset dominance also can be induced when one of the two rival targets is an implicit cue that boosts efficiency on a concurrent visual search task (Chopin & Mamassian, 2010), and this effect occurs even though observers remain unaware of the cue's significance in that task. Surprisingly, the propensity of a given stimulus to dominate upon rivalry onset varies from location to location within the visual field, with the regional patterns of bias being stable over time and idiosyncratic among observers (Carter & Cavanagh, 2007). These regional onset biases are highly sensitive to small imbalances in the luminance contrast of the rival targets, and when the target contrast is adjusted to be isoluminant between the eyes, a strong dependence on local eye dominance becomes evident in some observers (Stanley, Carter, & Forte, 2011). In all situations showing reliable onset bias, there is never a strong tendency for the favored stimulus to predominate during extended viewing of rivalry; the bias is present only at onset, implicating different processes in the initial registration of rivalry and the subsequent unfolding of rivalry over time (e.g., Mamassian & Goutcher, 2005; Kalisvaart, Rampersad, & Goossens, 2011). A comprehensive review of this literature is available (Stanley, Carter, & Forte, 2011).

Another factor governing initial dominance is the immediate history of stimulation prior to onset of rivalry. We have known for decades that the initially dominant stimulus in rivalry can be biased by prior exposure, such as adaptation to one of the two rival stimuli, a maneuver that encourages initial dominance of the unadapted stimulus (Blake & Overton, 1979; Wade & de Weert, 1986; Walker & Powell, 1979). Pelekanos and colleagues (2012) found that this biasing effect also works when the adapting stimulus (e.g., the image of the face of one particular person) is physically different from either rival stimulus just so long as that adapting stimulus is categorically related to one of the two rival stimuli (i.e., an image of the face of another person). This is not particularly surprising, of course, when we consider that face images contain many features in common including those registered within early stages of visual processing. Moreover, adaptation's biasing effect on initial dominance spreads beyond the retinal region of adaptation when the adapting and rival stimuli are meaningful, complex figures such as faces, whereas adaptation is strictly retinotopic when those stimuli are simple gratings (van Boxtel, Alais, & van Ee, 2008). Pelekanos et al. also found that adaptation has no influence on initial dominance when the adapting stimulus is a word (e.g., "face") related to one of the two rival stimuli, implying that semantics per se cannot influence initial perceptual selection in rivalry. This finding implies that initial dominance includes influences from higher levels of processing where objects are represented within a spatiotopic framework. In a similar vein initial dominance of a given rival target can also be biased by procedures that establish a visual context consistent with one rival stimulus but not the other immediately prior to onset of rivalry (Denison, Piazza, & Silver, 2011). But whatever particular maneuver is employed, a biasing effect on initial dominance seems to work only if observers are actually aware of the stimulus presented just prior to the onset of rivalry and

does not happen when that stimulus is rendered unrecognizable owing to visual crowding (Hancock, Whitney, & Andrews, 2008).

One sure-fire way to promote initial dominance of a given rival stimulus is first to present one rival stimulus followed very closely in time by presentation of the other eye's stimulus (see figure 58.1A). Called flash suppression (FS), this presentation sequence reliably promotes dominance of the second of the two stimuli (Wolfe, 1984). This, too, could be construed as an influence of adaptation to the initially presented stimulus, although the lead time of that stimulus is very brief with FS and would instigate minimal adaptation at best. FS has been exploited successfully in situations where investigators needed to control which one of two dissimilar monocular stimuli was initially dominant, and this includes psychophysical studies (e.g., Brascamp & Blake, 2012), fMRI studies (Lee, Blake, & Heeger, 2007), and neurophysiological studies (Keliris, Logothetis, & Tolias, 2010). One might object that FS is fundamentally different from conventional binocular rivalry in that suppression of the initially presented monocular stimulus arises from interocular masking (e.g., Turvey, 1973), not from suppression underlying rivalry. But, in fact, there is now psychophysical evidence suggesting that rivalry suppression and dichoptic masking may emerge from common neural

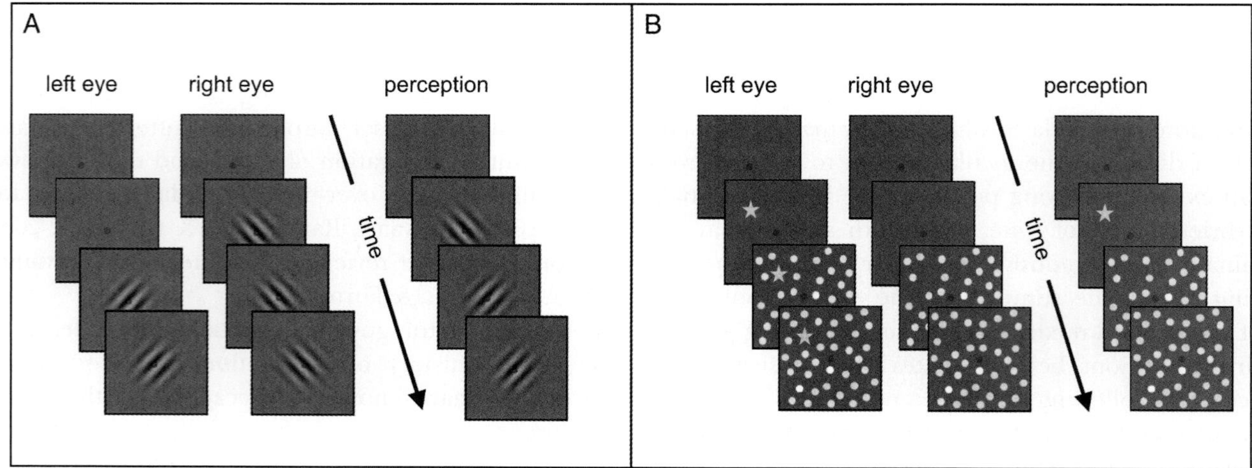

FIGURE 58.1   Schematics of two techniques for inducing predictable states of interocular suppression. (A) Flash suppression. The slight asynchrony in onset time for right-eye and left-eye rival patterns nearly always causes the second pattern to dominate initially. If both patterns continue to be viewed, rivalry lapses into the usual, unpredictable alternations in dominance; flash suppression works, in other words, to promote initial dominance of one particular pattern. (B) General flash suppression. An initially dominant monocular figure (star in this example) is quickly suppressed by binocular presentation of a dense array of "camouflage" elements (white circles in this example) appearing randomly everywhere within the stimulus except at and near the location where the monocular figure is located. Although that figure has no competing elements within the other eye's view, the figure nonetheless is suppressed from vision for several seconds following onset of the camouflage elements. The camouflage elements are even more effective if they are moving but still avoiding the area occupied by the monocular figure.

mechanisms (Baker & Graf, 2009a; van Boxtel, van Ee, & Erkelens, 2007).

In a variant of Wolfe's FS technique one can present a parafoveally viewed monocular figure followed very closely thereafter by a binocular figure with a blank region corresponding to the retinal area where the other eye's monocular target is located (figure 58.1B). Despite the absence of explicit interocular conflict within corresponding areas of the two eyes, the monocular target nonetheless vanishes from perception for several seconds, leaving behind an empty region where the target actually exists (Wilke, Logothetis, & Leopold, 2003). Like its progenitor FS this variant, dubbed generalized flash suppression (GFS), is critically dependent on the timing of the stimulus sequence and, in the case of GFS, on the spatial separation between the outer border of the monocular stimulus and the nearest edge defining the blank region. This technique is useful for testing conditions where suppression must be achieved without presentation of a competing stimulus on the corresponding area of the other eye; one such situation would be brain imaging based on retinotopically defined voxels.

Another effective way to encourage dominance of one rival stimulus is to turn off both rival targets once a given stimulus has achieved dominance, wait a couple of seconds, and then present the pair of rival stimuli again—with appropriate timing, it is possible to force the previously dominant stimulus again to be dominant on reappearance, as if the visual system remembered what was dominant at the end of the previous viewing episode (Brascamp et al., 2009a; Leopold et al., 2002). With optimal timing of intermittent presentations, this forced dominance at onset of rival stimulation can be maintained for minutes at a time. It is noteworthy that the technique works to stabilize dominance of a given stimulus only when the two rival stimuli are reinstated in the same ocular configuration that existed prior to their removal; repeatedly swapping the rival targets between eyes stabilizes dominance in a given eye and thus promotes alternations in dominance of the two rival stimuli (Chen & He, 2004). Pearson and Brascamp (2008) provide a comprehensive review of this stabilization technique and the factors that influence its effectiveness.

Rivalry Alternations with Extended Viewing

Everyone knows, of course, what happens when rival patterns are viewed for an extended period of time: Periods of exclusive perceptual dominance of one stimulus or the other tend to alternate unpredictably over time, meaning that rivalry alternations are not strictly periodic (e.g., Brascamp et al., 2005). Alternation rates slow with aging (Hudak et al., 2011; Ukai, Ando, & Kuze, 2003), but within any given age group there are large individual differences in the rate of alternations in perceptual dominance (Carter & Pettigrew, 2003; Hancock et al., 2012; Miller et al., 2010). This individual variability in alternation rate is correlated with frequency of saccadic eye movements (Hancock et al., 2012) as well as with alternation rates measured for other types of visual bistability (Carter & Pettigrew, 2003). Twin studies point to a strong genetic component to alternation rate (Miller et al., 2010; Shannon et al., 2011), which dovetails with the proposal that rivalry dynamics may be governed by specific neurotransmitter systems (Carter et al., 2005a).

The 2004 chapter summarized stimulus characteristics (e.g., contrast) of rival stimuli that influence dominance and suppression durations and, hence, govern the dynamics of rivalry. Nothing has been discovered since then to overthrow what was believed 8 years ago, but new developments have refined our understanding of the determinants of rivalry alternations. One of those is the growing realization that ecological constraints are embodied in the processes governing competition during rivalry (Ooi & He, 2005, 2006). One simple but compelling example is the substantial boost in predominance enjoyed by a textured surface portraying a ground plane over a rival textured surface portraying a ceiling plane, even though the two rival targets are essentially identical in terms of low-level image properties (Ozkan & Braunstein, 2009). This ground-plane bias befits our being terrestrial creatures oriented to the ground and to the objects resting on it (Gibson, 1950). An even more direct example of the influence of natural scene statistics on rivalry dynamics is provided by Baker and Graf's (2009b) study showing that predominance of a given rival stimulus is maximized when its amplitude spectrum and its phase spectrum match those of natural images (see figure 58.2A, B). These findings provide a plausible reason why pictures of recognizable objects such as faces and houses are more robust rival targets than their phase-scrambled counterparts or than one-dimensional gratings (Alais & Melcher, 2007).

A second advance, also pertaining to stimulus determinants of rivalry dynamics, was Lankheet's (2006) use of reverse correlation to track fluctuations in stimulus strength between two variable-coherence random-dot motion arrays engaged in clear binocular rivalry. The coherence strengths of the two motion arrays were modulated randomly over time and independently in the two eyes, and by looking backward in time relative to switches in dominance, Lankheet was able to correlate those dominance switches to signal strength

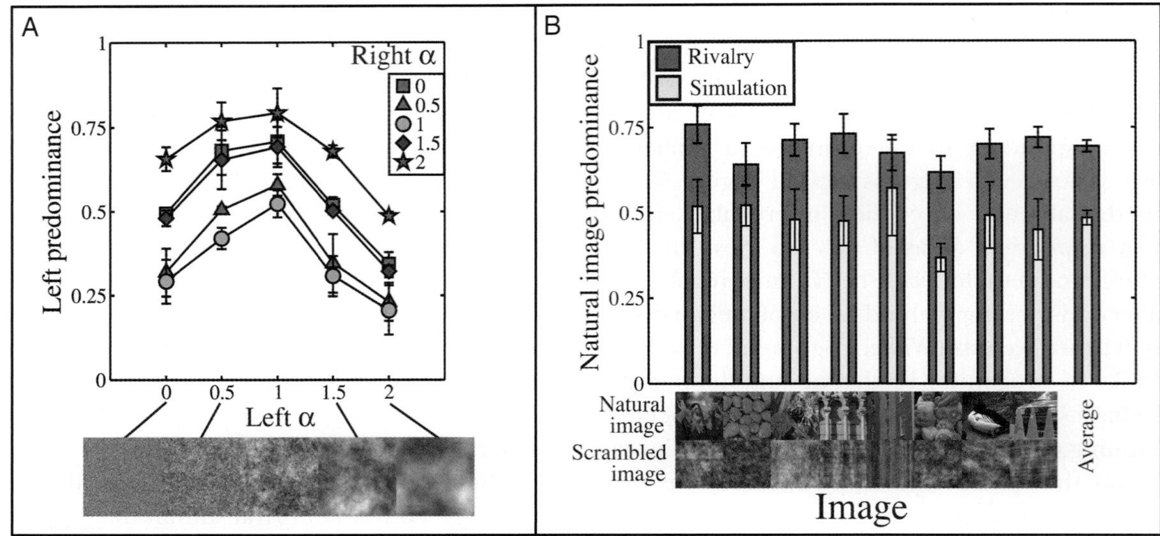

FIGURE 58.2   Images whose amplitude and phase spectra match the spectra characteristic of natural scenes predominate in rivalry. (A) Variations in predominance (total percentage time dominant during entire viewing period) for rival figures comprising static noise whose amplitude spectra were varied parametrically between the two eyes; different noise stimuli are indicated by the slope parameter α (the value $\alpha = 1$ most nearly approximates the spectral slope of natural images). (B) Predominance of natural images in rivalry with their phase-scrambled counterparts. Dark gray bars show predominance values for the natural images, and light gray bars show results from a control condition ruling out bias based on recognition per se as the basis for the strong tendency for natural scenes to predominate. Example images are shown below the histograms. (Reproduced with permission from Baker & Graf, 2009b. In the experiments, color versions of these images were used.)

differences between the two stimuli. As exemplified by the results shown in figure 58.3, immediately before a switch in dominance, motion coherence tended to increase in the suppressed eye's stimulus and to decrease in coherence in the dominant eye's stimulus. The relative stimulus strengths, in other words, were modulating the competitive interactions between the two motion arrays. Lankheet was able to simulate the irregular durations of dominance produced by these stimuli using a model that incorporated reciprocal inhibition and selective neural adaptation between left- and right-eye inputs directly driven by the actual coherence fluctuations used in the psychophysical experiment. Lankheet's findings, incidentally, square with another recent result showing that switches in rivalry state tend to occur just following retinal image transients, including those accompanying saccades (van Dam & van Ee, 2006).

At about the same time, Kim, Grabowecky, and Suzuki (2006) performed an elegant experiment and computational analysis designed to search for stochastic resonance in rivalry dynamics. Their experiment used rival patterns undergoing weak, periodic contrast modulations in anti-phase within the two patterns, with the frequency of these weak modulations varied over a wide range. Kim and colleagues reasoned that if internal, noise-based dynamics is responsible for perceptual

switches, the external modulations should potentiate that process when the modulation frequency matched the internal dynamics, with this signature of stochastic resonance showing up as peaks in the dominance duration histograms at specific multiples of the period of modulation and the coefficient of variation being minimized when the modulation frequency matches the average frequency of spontaneous perceptual alternation. This was, in fact, what they observed, and they went on to perform computational simulations to infer the magnitude and locus of the putative internal noise and to evaluate how this noise would have to be incorporated into several extant neural models of rivalry.

Another notable advance was the introduction by Pastukhov and Braun (2011) of a promising new statistical measure, dubbed cumulative history, to analyze sequential dependencies in successive state durations of rivalry. This measure, built on the mathematical notion of a leaky integrator, assumes that the impact of prior dominance of a given stimulus decays exponentially, with those effects weighted by time combining additively. Measures of cumulative history are constructed for the two alternative perceptual states, and linear correlation coefficients based on those sequences are computed to identify time constants that produce the highest correlation. This analysis is able to uncover sequential dependencies that more conventional measures fail to detect; moreover, the cumulative history can reliably predict the duration of the forthcoming dominance duration. Pastukhov and Braun make a cogent case for the time constant being a reflection of the buildup of adaptation during dominance. Of relevance to this conjecture are results from a study by Alais and colleagues focusing on changes in visual sensitivity measured at many time points throughout states of dominance and suppression (Alais et al., 2010a). By using reverse correlation to analyze test probe sensitivity backward in time from transition points, they were able to document changes in visual sensitivity unfolding during the course of dominance phases and suppression phases, one

example of which is shown in figure 58.3B. The complementary changes in sensitivity revealed by their reverse correlation analysis are consistent with the idea that gradual changes in visual adaptation play a role in triggering reversals in rivalry state, as embodied in many neural models of rivalry (see review by Roumani & Moutoussis, 2012) and as implicated by Pastukhov and Braun's cumulative history index.

We have known for decades that spatial interactions play a significant role in the dynamics of binocular rivalry (e.g., Fukuda & Blake, 1992), but new, revealing instances of these kinds of contextual influences have continued to appear. Thus, for example, the predominance of a circular rival patch containing moving

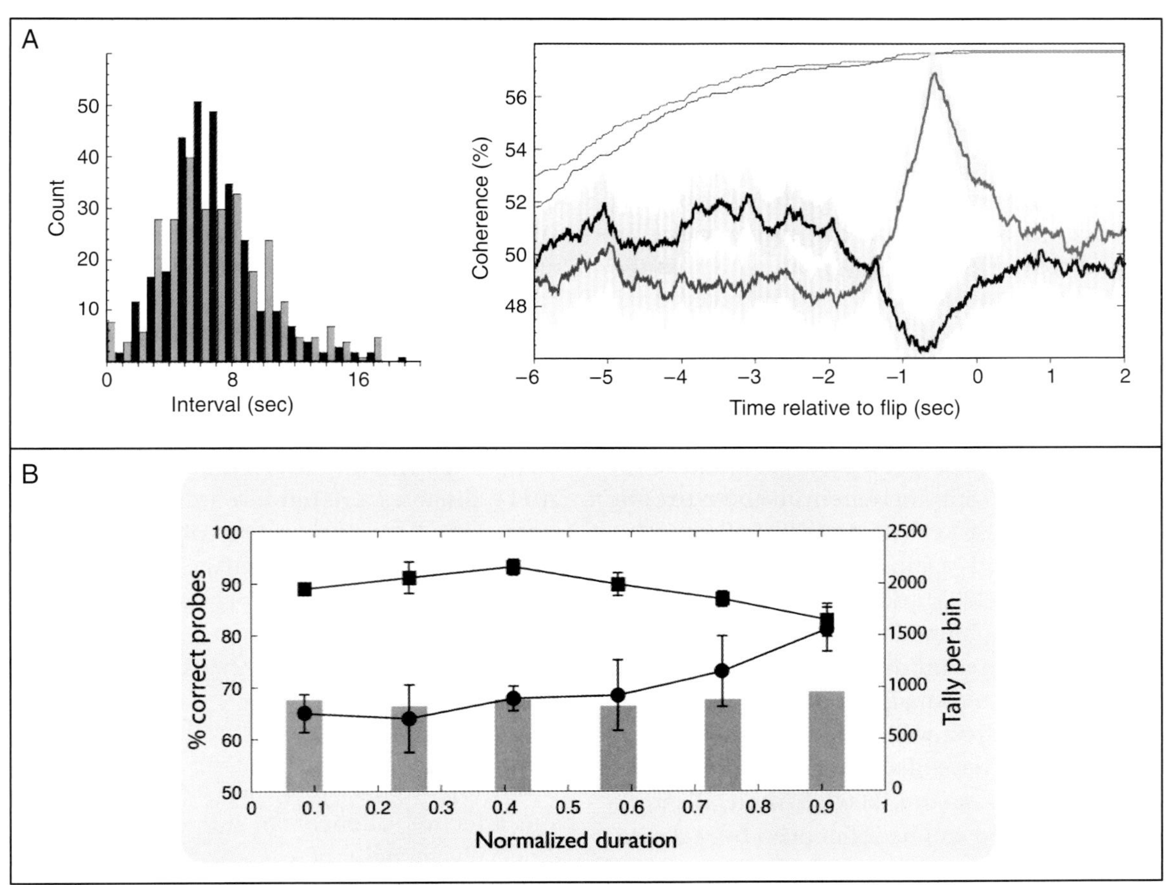

FIGURE 58.3   Microstructure of dominance and suppression phases as revealed by reverse correlation (panel A) and by probe sensitivity measurements. (A) Panel on left shows frequency distributions for dominance durations for rival motion stimuli viewed by the left eye and the right eye. Panel on right shows moment-to-moment random variations in motion coherence (percentage of dots moving in given direction) for the two eyes' motion animations, time-locked to reported switches in rivalry state (denoted by 0 along the horizontal axis). (Figure modified from the original version published in Lankheet, 2006, with permission of the author and the publisher, Association for Research in Vision and Ophthalmology.) (B) Variations in percentage correct on a two-alternative, spatial forced choice task on which probes were presented to a pattern while it was dominant (filled squares) and while it was suppressed (filled circles); performance is plotted relative to switches in rivalry state (with a state change corresponding to the value of 1.0 along the horizontal axis). Histograms indicate the number of observations (Tally per bin) contributing to each set of data points. (Figure modified from the original published in Alais et al., 2010a, with permission of the first author and the publisher, Elsevier Press.)

contours is robustly dependent on the presence and direction of motion in an adjacent surround annulus that itself is not engaged in rivalry (Paffen et al., 2004); similar center/surround interactions have been reported for orientation and color (Paffen et al., 2006). One particularly intriguing spatial effect involving rivalry was discovered by Pearson and Clifford (2005), who created a hybrid rival stimulus that contained three juxtaposed regions each containing a different form of rivalry: conventional binocular rivalry, flicker-and-swap rivalry (Logothetis, Leopold, & Sheinberg, 1996), and monocular rivalry (Breese, 1899). Rival contours of a given orientation within each region tended to become synchronized in their dominance and suppression, promoting unitary perception over the entire composite. This pattern coherence implicates robust spatial interactions among the neural mechanisms involved in these distinct forms of rivalry. Alais et al., (2006) have also provided strong evidence in support of such interactions and, moreover, have formalized the idea within the context of an association field of the sort implicated in other visual phenomena (e.g., Field, Hayes, & Hess, 1993). Whatever the neural bases of these grouping effects in rivalry, it appears that the corpus callosum plays a crucial role in promoting grouping across the vertical midline, for no evidence for coordinated dominance is seen in a split-brain observer unless all components of the rival target are imaged in the same visual hemifield (O'Shea & Corballis, 2005).

It is possible to exert control over these otherwise unpredictable changes in perceptual dominance by introducing a brief, strong increment in the currently suppressed stimulus (Blake, Westendorf, & Fox, 1990) or in the background against which that stimulus appears (Kanai et al., 2005); these kinds of visual transients trigger an almost immediate switch to dominance of that previously suppressed stimulus. Moreover, if that transient is confined to a small region of a larger suppressed stimulus, its effect will spread throughout that stimulus, creating a wave-like spread of dominance (Arnold, James, & Roseboom, 2009; Wilson, Blake, & Lee, 2001) whose speed can be influenced by real stimulus motion (Knapen, van Ee, & Blake, 2007). With appropriate timing of pulses delivered in antiphase to the two eyes' stimuli, one can set up predictable waves of dominance between the two stimuli (Kang, Heeger, & Blake, 2009). These traveling waves of dominance and suppression can be modeled by neural circuitry in which inhibition propagates along lateral connections among neurons with neighboring receptive fields (Wilson, Blake, & Lee, 2001; see refinements to this model by Kang et al., 2010 and by Bressloff & Webber, 2012). A variant of this trigger procedure will produce traveling waves of suppression in an initially dominant stimulus, with suppression spreading well beyond the region of explicit binocular rivalry (Maruya & Blake, 2009). Even without visual transients, spontaneous switches from suppression to dominance in a stimulus tend to originate at a locale in that stimulus where its strength is stronger than elsewhere (Paffen, Naber, & Verstraten, 2008; Stuit, Verstraten, & Paffen, 2010), and even without local hot spots in a rival stimulus, different individuals exhibit reliable endogenously governed hot spots where transitions are most likely to originate (van Ee, 2011).

As mentioned above, periods of mixed dominance are often experienced when viewing rival stimuli, particularly relatively large ones. Alais and Melcher (2007) found that the incidence of mixed dominance is substantially greater when the rival stimuli are simple, one-dimensional gratings than when they are recognizable objects such as houses or faces. The objects more successfully maintain their global coherence, these authors speculate, through top-down–generated feedback signals that influence local feature integration at early stages of processing. Another factor that influences the incidence of mixed dominance is viewing time: With prolonged viewing the incidence of mixture states actually increases substantially (including for houses and faces), and these mixtures do not abate until observers have an opportunity to view matched binocular stimuli that effectively reset the mechanisms of rivalry to their initial state (Klink et al., 2010). This finding, along with others (Lunghi, Burr, & Morrone, 2011), discloses a heretofore unrealized form of short-term, experience-dependent plasticity in the neural mechanisms underlying rivalry.

Top-Down Influences on Rivalry Dynamics

The 2004 chapter on binocular rivalry opined that it was time to examine in greater detail the extent to which top-down factors such as knowledge, meaning, and affective salience can influence rivalry dynamics, especially in light of evidence implicating such top-down influences in other phenomena of visual perception (e.g., Carrasco, Ling, & Read, 2004; Schyns & Oliva, 1999). As anticipated, this issue has indeed become a growing focus of work on rivalry these past 8 years; here is a sample drawn from the roster of intriguing, occasionally baffling findings from that work:

• Moving a computer mouse laterally back and forth favors dominance in rivalry of a visual object whose movements mirror that action, such as rotation of a globe back and forth over time, and this influence of voluntary action impacts both dominance durations and suppression durations of that object (Maruya, Yang, & Blake, 2007).

• Imagining a specific visual pattern impacts the subsequent likelihood of dominance of that stimulus when it is engaged in binocular rivalry, a result implying that processes supporting formation and retention of mental images can influence neural representations of stimuli at levels of processing where those representations compete for dominance (Pearson, Clifford, & Tong, 2008).

• Meditation of a particular kind ("one-point" meditation that involves focused attention) can slow rivalry alternations profoundly, at least in monks well practiced in the technique (Carter et al., 2005b). This slowing of rivalry alternations could be related to modulations of activity in prefrontal cortex known to accompany meditation (Lutz et al., 2004) and to be associated with transition states of binocular rivalry (Knapen et al., 2011; Lumer, Friston, & Rees, 1998).

• Emotionally arousing pictures—both positive and negative—dominate longer in rivalry than do pictures deemed to be nonarousing even though the low-level image properties of the two classes of images are comparable (Sheth & Pham, 2008). The same kind of result has been reported when observers track rivalry alternations between pictures of neutral faces and pictures of faces expressing emotions—the emotional faces, both positive and negative ones, predominate during extending viewing and, for that matter, are highly likely to be seen first at onset of rivalry (Alpers & Pauli, 2006).

• Structurally neutral faces associated through prior conditioning with negative social behaviors ("threw a chair at his classmate") dominate significantly longer in rivalry than do faces associated with positive or neutral social behaviors (Anderson et al., 2011). This top-down influence could arise from enhanced attention drawn to the affective tone acquired by a face through learning.

• When one eye views spatially distributed moving dots that together form a human figure engaged in an activity (point-light animation), those dots remain visible as a group when engaged in rivalry with a different array of dots viewed by the other eye (Watson, Pearson, & Clifford, 2004). The dynamic global human configuration implied by the dots, in other words, promotes their conjoint dominance, a form of grouping that almost certainly transpires at higher stages of processing (Blake & Shiffrar, 2007).

• Manipulations of attention affect rivalry dynamics in ways dependent on the extent to which attention is available to rivalry. When sustained attention is intentionally focused on one of two rival stimuli, the durations of dominance of that stimulus increase,

but the durations of the unattended stimulus do not (Chong, Tadin, & Blake, 2005). When attention is exogenously cued to a given location, alternations in dominance occur more frequently there (Paffen & Van der Stigchel, 2010). On the other hand, directing attention somewhere other than the location of rival stimulation slows the alternation rate between those rivaling stimuli (Alais et al., 2010b; Paffen, Alais, & Verstraten, 2006). In the extreme, when an exceedingly difficult task demands intensely focused attention, rivalry alternations cease altogether (Brascamp & Blake, 2012). These are just three examples from the growing literature on attention's top-down influence on rivalry, a literature that has been comprehensively covered in other recent reviews (Paffen & Alais, 2011; Dieter & Tadin, 2011; Miller, Ngo, & van Swinderen, 2012).

• Nonwords remain dominant in rivalry longer than do meaningful words, and impossible figures show longer dominance durations than do possible objects when both of those figures are relatively simple in terms of total contour length (Wolf & Hochstein, 2011). This outcome seems counterintuitive from the standpoint of resolution of perceptual conflict based on likelihood (Denison, Piazza, & Silver, 2011; Hohwy, Roepstorff, & Friston, 2008) because likelihood should be biased in favor of objects that are more familiar or more plausible. On the other hand, this outcome makes sense from the standpoint of information theory because unfamiliar or implausible objects, when encountered, constitute high-entropy items that plausibly attract greater interest/attention.

• Symbolic magnitude affects predominance in rivalry in the same way as does visual intensity or contrast: Predominance increases with symbolic magnitude, and this is true when magnitude is signified by numerical characters or by objects of different implied size (Paffen, Plukaard, & Kanai, 2011).

In addition to the kinds of top-down influences on rivalry dynamics illustrated above, another category of influences also underscores rivalry's susceptibility to forces beyond those embodied in the rival stimuli themselves: influences emerging from cross-modal sensory interactions. In the last few years several laboratories have shown that the states of dominance during binocular rivalry can be influenced by sensory information from other modalities. So, for example, sound congruent with one of two rival stimuli can bias dominance in favor of that stimulus (Chen, Yeh, & Spence, 2011; Conrad et al., 2010; Kang & Blake, 2005; van Ee et al., 2009), an odor congruent with one of two pictured objects engaged in rivalry can bias dominance in favor of the congruent visual object (W. Zhou et al., 2010), and tactile stimulation from a textured surface can promote dominance of a visual rival stimulus whose texture matches that of the tactile stimulus (Lunghi, Binda, & Morrone, 2010). There is abundant psychophysical and neurophysiological evidence for cross-modal neural interactions (e.g., see review by Spence, 2011), but just how those interactions participate in resolving visual conflict during rivalry remains a mystery. One possibility is that a nonvisual stimulus focuses visual attention on one of two rival stimuli, thereby modulating rivalry dynamics (van Ee et al., 2009).

Are we surprised that rivalry is influenced by top-down influences and by contextual cues from other modalities? In recent years evidence for top-down influences on vision, including influences associated with affective content, has steadily accumulated (e.g., Carrasco, Ling, & Read, 2004; Lupyan, Thompson-Schill, & Swingley, 2010; Phelps, Ling, & Carrasco, 2006; Schyns & Oliva, 1999), and, moreover, a number of recent studies point to the involvement of top-down influences under conditions of subliminal stimulation (e.g., Radel & Clément-Guillotin, 2012). There is no reason to believe that rivalry should be immune from those influences for, after all, a dominant stimulus in rivalry appears to engage all of the same processes ordinarily involved in vision under nonrivalry conditions. At the same time, it is worth noting that the magnitude of the changes in rivalry dynamics produced by context, meaning, and these other top-down influences are quite modest compared to the effects of low-level stimulus manipulations such as contrast or spatial frequency content of rival stimulation. Although not cognitively impenetrable, binocular rivalry appears to be especially sensitive to stimulus features registered within early stages of visual processing.

Before moving to the next topic I wanted to end this overview of rivalry dynamics with a word of caution. In nearly all studies of rivalry, fluctuations in perceptual state and the durations of those states are measured by having observers press different buttons to indicate exclusive dominance of one stimulus or the other as well as mixtures of the two. This is tantamount to requiring observers to perform a three-alternative categorization task that perforce requires establishing criteria for selection of responses. Compounding the problem, rivalry transitions can unfold unpredictably over time, with perception characterized by confusing piecemeal mixtures (Knapen et al., 2011) that sometimes culminate in exclusive dominance of the previously dominant stimulus and not reversals in dominance (Brascamp et al., 2006). Moreover, for complex rival targets that differ along multiple dimensions (e.g., color and form), rivalry dominance can assume a state in which the form

from one eye's stimulus is combined with the color from the other eye's stimulus to create a percept corresponding to neither stimulus (Holmes, Hancock, & Andrews, 2006; Hong & Shevell, 2009). Rivalry, in other words, is not binary. Consequently, observers must adopt some criterion for deciding when to declare a particular stimulus dominant. Faced with this uncertainty, observers' decisions could easily be influenced by extraneous factors outside of the stimuli themselves, including their expectations about the purpose of the study. This is an inherent liability in the tracking procedure, but, fortunately, there are ways to corroborate subjective tracking records. Some investigators believe that indirect measures of rivalry dynamics based on oculomotor reflexes may obviate the problems associated with categorization using subjective report and, in fact, may capture nuances to rivalry not available from subjective report (Naber, Frässle, & Einhäuser, 2011). It is true that perceptual states of rivalry can be inferred from pupil dilation (Einhäuser et al., 2008), ocular following response (Zhu et al., 2008), microsaccades (van Dam & van Ee, 2006), and optokinetic nystagmus (Fox, Todd, & Bettinger, 1975; Hayashi & Tanifuji, 2012[1]; Wei & Sun, 1998). Moreover, some of these oculomotor reflexes apparently predict when a switch is about to happen because the reflex occurs just prior to the observer's perceptual report, and, in the case of pupil dilation, the relative magnitude of the reflex seems to predict the duration of the upcoming perceptual state. However, there are situations in which an objective indirect measure, eye movements, provides a very different outcome relative to concurrent perceptual reports (Spering, Pomplun, & Carrasco, 2011). Given their technical requirements, indirect measures are unlikely to supplant introspective tracking reports in the study of rivalry. In general, investigators of rivalry should remain mindful of the possibility that the complexity of rivalry dynamics may not be fully captured when observers are given just a couple of buttons to indicate what they are seeing during rivalry.

TECHNIQUES FOR INDUCING PROLONGED RIVALRY SUPPRESSION

For some purposes the relatively brief durations of periods of exclusive dominance and the unpredictability of those durations are nuisances because they complicate attempts to make changes in a currently suppressed stimulus, to introduce new stimulation during suppression, or to maintain a given stimulus in the suppressed state for durations longer than a few seconds. For those reasons investigators have turned to several new techniques for generating binocular rivalry that promote relatively complete and long-lasting phases of interocular suppression. The following are summaries of those techniques.

Continuous Flash Suppression

Students of binocular rivalry were awed when Tsuchiya and Koch (2005) introduced what has come to be known as continuous flash suppression (CFS). This potent technique involves presenting to one eye a montage comprising different sized rectangles, each of which resembles a Mondrian pattern, with different montages appearing one after another at a rapid rate (you can experience CFS for yourself by viewing the animation at this website through red/green anaglyphic filters: http://www.klab.caltech.edu/~naotsu/CFS_color_demo.html). This steady stream of densely contoured figures will very effectively suppress an entire monocular stimulus presented to the other eye, with durations of suppression lasting up to an order of magnitude longer than ordinary suppression durations associated with conventional rival displays. Moreover, test probe measurements reveal that the depth of suppression generated by CFS can be much greater than that produced by a conventional binocular rivalry display (Tsuchiya et al., 2006).

Why does CFS work so well? One reason is that the typical CFS display comprises a rich array of spatial frequencies and orientations (see figure 58.4A) that, when randomized in spatial phase once every 100 ms or so, creates a robust stimulus that presumably eludes visual adaptation thought to be a central cause of alternations in dominance when viewing conventional rival stimuli. Still, it is important to realize that suppression produced by CFS is not uniformly effective: Because of anisotropies in the Fourier spectrum of CFS (see figure 58.4B), it exerts a relatively weaker effect on higher spatial frequencies contained within the monocular stimulus viewed by the other eye (Yang & Blake, 2012).

An even more potent variant of CFS was devised by Maruya, Watanabe, and Watanabe (2008), wherein the Mondrian rectangles forming each briefly presented montage themselves contained moving contours, creating what the authors called dynamic CFS. Technically speaking, this maneuver introduces additional motion energy into the conventional 10-Hz CFS stimulus, and, as a consequence, a dynamic CFS stimulus can suppress for an indefinitely long duration even a robust moving stimulus presented to the other eye. Moreover, this dynamic CFS can be confined to an annulus with a blank region in its center and still will exert complete, effective interocular suppression of a large stimulus that includes portions within the corresponding retinal area of the blank region (Watanabe et al., 2011). The spread of interocular suppression, in other words, is spatially extensive with dynamic CFS. A demonstration of this form of interocular suppression can be downloaded from this website: http://visiome.neuroinf.jp/modules/xoonips/detail.php?item_id=6798.

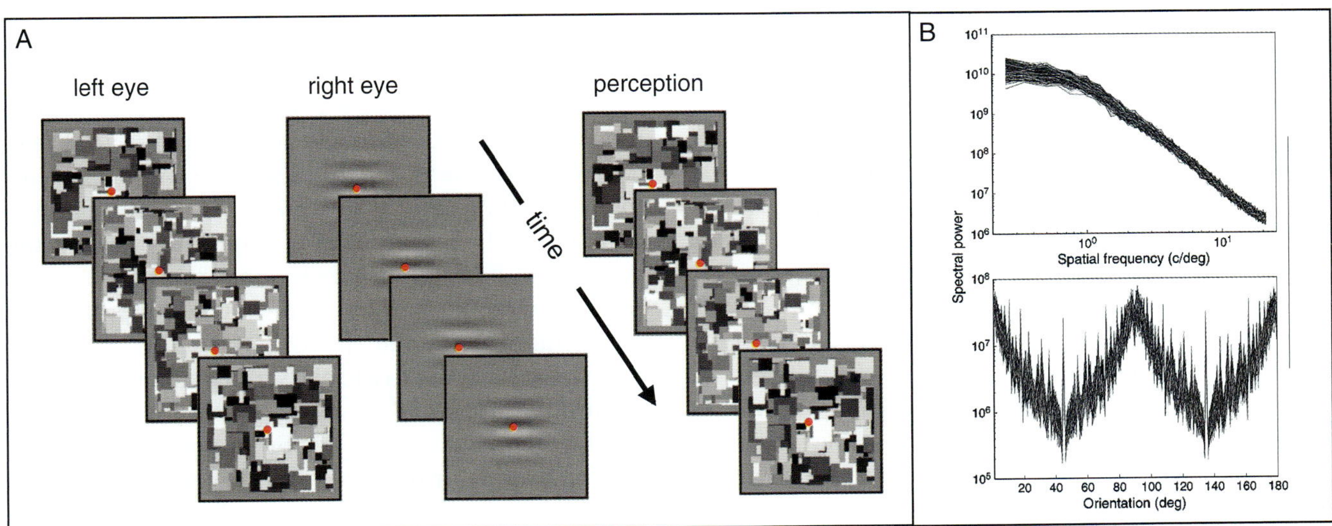

FigURE 58.4   Continuous flash suppression produces potent interocular suppression. (A) Sequence of Mondrian-like arrays presented rapidly one after the other to one eye erases all traces of a rival stimulus presented to the other eye, with suppression of that stimulus lasting for long periods of time. (B) Amplitude spectra of six different versions of CFS displays, showing $1/f$ falloff in power and variations in power associated with different orientations within the two-dimensional spectrum derived using Fourier analysis. See Yang and Blake (2012) for details. (Panel B modified from original version published in Yang & Blake, 2012.)

Because of its effectiveness, CFS was quickly adopted for studying visual processing outside of awareness, and it has been exploited using several different psychophysical strategies. For example, CFS has been combined with visual cuing and priming procedures to ask whether provocative or arousing pictures can capture attention or otherwise affect cognitive judgments when those pictures are never consciously experienced owing to CFS. When it comes to erotic pictures of nudes (Jiang et al., 2006) and pictures (e.g., a bomb exploding) portraying threat (Yamada & Kawabe, 2011), the answer appears to be "yes," those pictures can serve as effective cues. But when the suppressed priming stimuli constitute words, there is no evidence for semantic priming (Kang, Blake, & Woodman, 2011). Also using a priming procedure, Almeida and colleagues (2008) obtained what they construed as evidence that cortical areas in the dorsal stream continue to process information about tools even when pictures portraying exemplars from that category were suppressed from awareness by CFS. That conclusion has been questioned, however, by a recent study showing that this result was attributable to the elongated shape of the tool pictures used by Almeida and colleagues, not the tool category specifically (Sakuraba et al., 2012). Using a location-cuing procedure Chou and Yeh (2012) found a small but statistically significant decrease in reaction times to visible targets appearing at locations cued by the presence of an object that was itself suppressed from awareness by CFS. It remains to be seen how this finding can be reconciled with earlier work failing to find location cuing when the cue was suppressed using conventional binocular rivalry and not CFS (Schall et al., 1993).

Another strategy is to ask whether a stimulus suppressed from awareness by CFS remains sufficiently effective to generate visual adaptation aftereffects. This strategy has been utilized in conjunction with several aftereffects, including perceptual fading of contrast (Yang, Hong, & Blake, 2010), static and dynamic motion aftereffects (Maruya, Watanabe, & Watanabe, 2008), the face identification aftereffect (Moradi, Koch, & Shimojo, 2005), the face shape aftereffect (Stein & Sterzer, 2011), and the emotional facial expression aftereffect (Adams et al., 2010; Yang, Hong, & Blake, 2010). In general, interocular suppression from CFS tends to attenuate but not completely abolish the buildup of the low-level aftereffects and tends to severely reduce or even eliminate the high-level, face aftereffects, with some disagreement concerning the bases of the residual aftereffect in the case of emotion (Adams et al., 2011).

A third strategy assesses the extent to which different visual features are impacted by CFS using test probe procedures. With that procedure, forced-choice thresholds for detection or identification of a stimulus presented during suppression are measured, with the elevation in threshold providing an index of the depth of suppression for a particular stimulus–CFS combination. Test probe studies reveal a significant degree of selectivity in the depth of suppression depending on stimulus properties such as color, motion, spatial frequency, and temporal frequency (Hong & Blake, 2009; Yang & Blake, 2012; Zadbood, Lee, & Blake, 2011). In this respect, interocular suppression produced by CFS resembles suppression associated with conventional rivalry, where the elevations of test probe thresholds are graded depending on the similarity of the probe to the pattern producing suppression (Alais & Parker, 2006; O'Shea & Crassini, 1981; Stuit et al., 2011). As an aside, the question of the selectivity of suppression has generated controversy over the years (Blake & Logothetis, 2002), but the emerging view is that suppression produces a general impairment for all classes of new stimulation introduced to the suppressed eye together with an additional, selective component tied to the stimulus features of the rivalry stimuli themselves (Stuit et al., 2011; Yang & Blake, 2012); this latter component may also be involved in dichoptic masking (Baker & Graf, 2009a).

Finally, one strategy that has really caught on is the so-called breaking of suppression technique devised by Jiang, Costello, and He (2007). On each trial a full-strength CFS display is presented to one eye, and a test figure is presented to the other eye at one of several possible spatial locations. Starting at a very low level, the contrast of the test figure is gradually increased until the observer makes a forced-choice response about the location of the test figure, thus signifying emergence of the test figure from suppression. The time elapsing between trial onset and this response provides the measure of interest, and these times can be measured for different classes of test figures. Using the breaking suppression technique, Jiang and colleagues (2007) found that upright faces broke suppression sooner than did inverted faces and that words printed in script from an observer's native language broke suppression faster than did words printed in unfamiliar script. They also included a monocular control condition in which the test figure was physically superimposed on the CFS display to assess possible masking effects. The authors concluded that information from the suppressed image activated high-level processes specifying the details of that image in a way that potentiated its increasing stimulus strength and speeded its transition from suppression to dominance.

Following the lead of Jiang, Costello, and He (2007), others have devised variants of this procedure to study visual processing outside of awareness in relation to emotional facial expressions (Tsuchiya et al., 2009; Yang, Blake, & Zald, 2007), incongruity among objects in natural scenes (Mudrik et al., 2011), object inversion effects (G. Zhou et al., 2010), affective connotation of words (Yang & Yeh, 2011), and eye contact as conveyed by gaze direction (Stein et al., 2011). Without going into details, suffice it to say that results from these studies have been interpreted to indicate that aspects of a visual stimulus including its meaning, affective connotation, and contextual relevance undergo unconscious processing that empowers that stimulus to emerge from suppression. Stein, Hebart, and Sterzer (2011) have questioned, however, whether that conclusion is warranted based on their doubts about the adequacy of control measures typically employed to rule out alternative explanations of results produced by CFS. At a minimum the critique offered by Stein and colleagues underscores the risks of drawing conclusions based on average suppression durations for different stimulus conditions without also taking into account the distributional properties of those durations. In general, this technique would benefit from the deployment of more sophisticated analytic modeling borrowed from other fields, such as the diffusion model of reaction times that has been so influential in several areas of cognitive science (Ratcliff & McKoon, 2008).

Binocular Switch Suppression

Another hybrid technique for producing persistent suppression of one of two rival targets was devised by Arnold, Law, and Wallis (2008). With their technique, called binocular switch suppression (BSS), dissimilar monocular stimuli—one sharply focused and the other blurred—are repetitively swapped between the eyes at a relatively slow rate in the neighborhood of 1 Hz. These authors present evidence that BSS differs from CFS in several ways, one being that BSS produces even longer suppression durations than does CFS and deeper levels of suppression as indexed by thresholds measured using a probe technique. They also acknowledge that BSS, unlike CFS, may not work well when using the popular anaglyphic method to achieve dichoptic stimulation. And by its very nature BSS successively exposes both eyes to both rival stimuli, a consequence that could complicate attempts to distinguish properties of eye-based rivalry (Blake, 1989) from stimulus-based

rivalry (Logothetis, Leopold, & Sheinberg, 1996). Still, it is intriguing that perception can track the better-focused stimulus for long durations, transcending many eye swaps, a finding that Arnold, Grove, and Wallis (2007) attribute to a functional adaptation serving to promote perceptual stability in cluttered environments.

MODELS OF BINOCULAR RIVALRY

Accompanying these empirical developments is the emergence of new theoretical ideas concerning binocular rivalry. For years the modal view was one in which neurons encoding the two stimuli engage in reciprocal inhibition such that neurons representing one stimulus inhibit those representing the other, thereby promoting perceptual dominance of one over the other. According to this so-called reciprocal inhibition model, neurons responsive to the dominant stimulus adapt, weakening their activity and the inhibition they impose on the neurons responsive to the suppressed stimulus. Eventually the imbalance in activity between the two populations of neurons reverses, triggering an alternation in dominance. This model successfully explains many aspects of rivalry including the dependence of rivalry dynamics on the relative strengths of the two rival stimuli (Kang, 2009; Lankheet, 2006; Levelt, 1965; Mueller & Blake, 1989), the effects of monocular adaptation on rivalry dynamics (Kang & Blake, 2010; van Boxtel, Alais, & van Ee, 2008), the systematic variation in probe thresholds during the individual phases of dominance and suppression (Alais et al., 2010a), the slowing of rivalry alternation rate (Hollins, 1980), and the reduced incidence of exclusive perceptual dominance (Klink et al., 2010) with extended viewing of rivalry. It is worth noting that reciprocal inhibition models make no assumption about the site within the visual pathways where neural competition transpires; reciprocal inhibition and adaptation work equally well for eye-based rivalry (e.g., Matsuoka, 1984) and object-based rivalry (e.g., Dayan, 1998). The reciprocal inhibition model has also been extended to explain the propagation of rivalry dominance, that is, so-called traveling waves that are experienced during transition states between spatially extended rival targets (Kang, Heeger, & Blake, 2009; Wilson, Blake, & Lee, 2001) and to account for the tendency for dominance to be maintained over brief periods of removal of both rival stimuli (Wilson, 2007). Moreover, single-stage versions of reciprocal inhibition recently have been supplanted by versions in which suppression and adaptation are distributed over multiple stages within the visual hierarchy (e.g., Freeman, 2005; Wilson, 2003), the motive being to account for psychophysical, neurophysiological, and brain-imaging data pointing to the distributed nature of rivalry (Roumani & Moutoussis, 2012; Tong, Meng, & Blake, 2006). Finally, top-down influences including attention and semantic content can be incorporated into reciprocal inhibition, with those influences adding to the excitatory drive associated with a given stimulus (Kang & Blake, 2011).

At the same time, there has been a growing realization that intrinsic neural noise must play a role in the dynamics of binocular rivalry, too (see Brascamp et al., 2006; Kang & Blake, 2011; Roumani & Moutoussis, 2012). Earlier accounts of rivalry acknowledged this role, but in recent years noise has been explicitly incorporated into models of rivalry (Laing, Frewen, & Kevrekidis, 2010; Moreno-Bote, Rinzel, & Rubin, 2007; van Ee, 2009). A useful framework for portraying the relative contributions of adaptation and neural noise is provided by the concept of the double-well energy landscape (Kim, Grabowecky, & Suzuki, 2006; Seely & Chow, 2011), as depicted schematically in figure 58.5. Adaptation governs the variable depth of the well associated with the currently dominant stimulus, with well depth decreasing during a dominance phase, thereby increasing the likelihood of a state transition (figure 58.5A). As a deterministic process, adaptation alone should produce sequential dependencies of successive state durations, but in fact those successive state durations are variable. This variability implicates noise in the alternation process, and within the energy landscape scheme noise is portrayed by random fluctuations in the location of the dominant state token within a given well, with those random fluctuations eventually pushing the token into the previously unoccupied well (figure 58.5B). Evidence for the involvement of both adaptation and noise is now compelling, leading to the following conceptualization of rivalry alternations. Just after a perceptual state change, the newly dominant stimulus is likely to remain dominant because the depth of the associated energy well is deep relative to the random, noise-produced fluctuations in the activity level of that stimulus; a state transition is unlikely. But adaptation progressively reduces the depth of the well and, therefore, increases the probability that noise-induced fluctuations can trigger a state change by propelling the state token into the other well. This framework, together

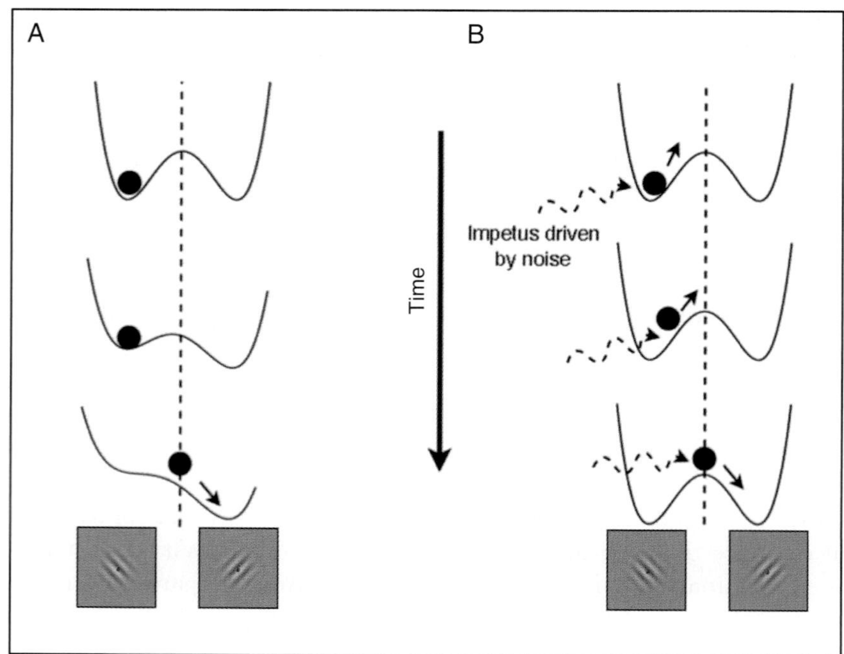

FIGURE 58.5   Energy landscapes depicting two possible states of dominance (left-tilted contours or right-tilted contours) signified by the dips in the landscape profiles, only one of which contains the token denoting dominance at any given time. (A) State changes triggered solely by adaptation of the well currently occupied by the state token, signified by a steady decrease in the depth of that well. On the assumption that the time constant of adaptation is constant, this model predicts strictly periodic switches in state. (B) State changes triggered solely by random fluctuations in the moment-to-moment position of the state token, which by chance will sooner or later move the token into the other energy well. This model predicts an exponential distribution of dominance durations. In fact, neither outcome comports with the measured distributions of dominance during rivalry, implying that rivalry alternations result from both adaptation and noise. (Figure modified from original published in Kang & Blake, 2011.)

with the possible role of noise in the adaptation process itself (Kalarickal & Marshall, 2000; van Ee, 2009), can thus account for the weak correlation between successive dominance durations that posed a challenge for models based on adaptation alone. According to the adaptation-plus-noise model, short- and intermediate-duration dominance states are governed primarily by adaptation, with a significant contribution from noise reflected in the less frequent long-duration dominance states. Several particularly elegant instantiations of these ideas have been advanced by several groups (Shpiro et al., 2009; Theodoni et al., 2011) who devised neural network models in which spiking neurons operating within a regime bordering on the brink of noise-driven switches and adaptation-driven oscillations can successfully mimic experimental data on rivalry dynamics. It is interesting to note that this theoretical framework has previously been used to model working memory, decision making, and perceptual adaptation aftereffects (see review by Theodoni et al., 2011).

To be sure, there are other ways to think of rivalry alternations besides reciprocal inhibition and attractor networks. For example, the Australian group headed by Pettigrew and Miller have championed the view that rivalry alternations result from alternating hemispheric activation, with each hemisphere having its own independent attentional resources that compete for dominance when an observer is confronted with conflicting sensory input. Called the interhemispheric switch hypothesis (IHS), this model receives support from studies employing unilateral caloric stimulation of the vestibular organ, a procedure thought to activate frontoparietal areas in the contralateral hemisphere that control allocation of attentional resources (see Ngo et al., 2007, for an overview of the IHS and the evidence in support of the model). Specific testable hypotheses have been advanced, but what is also needed at this stage is a more formalized account of the IHS that accommodates aspects of binocular rivalry that should serve as acid tests for any comprehensive theory of rivalry. Those include suppression's graded effect on different forms of visual adaptation (Blake et al., 2006; Yang, Hong, & Blake, 2010), the selective effects of suppression on test probe thresholds (e.g., Stuit et al., 2011), the marked individual differences in initial dominance bias within different regions of the visual field (Carter & Cavanagh, 2007), and the influence of predictive context (e.g., Denison, Piazza, & Silver, 2011) and semantic meaning (Wolf & Hochstein, 2011) on rivalry predominance. Whether those aspects of rivalry are explicable within the framework of differential hemispheric activation remains to be learned. Regardless how those developments play out, advocates of the

IHS deserve credit for highlighting intrinsic factors governing individual differences in rivalry dynamics, factors including neurotransmitter systems (Carter et al., 2005a) and genetics (Miller et al., 2010; Miller, Ngo, & van Swinderen, 2012).

Another theoretical twist on rivalry that has gained traction recently characterizes alternations in perceptual dominance as the brain's adaptive reaction to competing perceptual interpretations of ambiguous sensory input (e.g., Sterzer, Jalkanen, & Rees, 2009). Echoing an idea voiced decades ago by Walker (1978), advocates of this viewpoint see rivalry as a high-level, inference-like process that selects from among alternative interpretations consistent with ambiguous sensory information. This account of rivalry has been formalized in several models, including ones grounded within a Bayesian framework (Dayan, 1998; Hohwy, Roepstorff, & Friston, 2008; Sundareswara & Schrater, 2008). In these Bayesian-inspired accounts the dissimilar monocular images instigating rivalry preclude the existence of a single hypothesis about the state of the visual environment that could plausibly give rise to those images, the result being temporary dominance of one of the two stimuli. Consider rivalry between images of a face and a house within the context of this predictive coding scheme. The state of dominance of the face fully accounts for the low-level neural signals evoked by that stimulus (i.e., there is no residual error variance left unexplained for that particular object/event), but the low-level signals associated with the temporarily suppressed house stimulus constitute residual, unexplained error that undermines stable perceptual inference and eventually culminates in a change in brain state. In a nutshell, perception alternates so long as there exist large unexplained but potentially explainable error signals; inference is unstable because neither hypothesis enjoys both high likelihood and high prior probability, and the conjunctive hypothesis that both coexist has an even lower prior probability (presumably because the simultaneous presence of two different monocular images implies the impossible situation of two different objects occupying the same location in visual space at the same time). This view situates the origins of rivalry competition within conflicting top-down cortical explanations for the left and right eyes' inputs and not within competition between those inputs themselves (Dayan, 1998). Hohwy, Roepstorff, and Friston (2008) review psychophysical evidence on rivalry that fits within this predictive coding scheme. This view currently offers tantalizing hope for an integrated account of rivalry that incorporates some of the key concepts characterizing reciprocal inhibition models (e.g., Noest et al., 2007; Wilson, 2007), but within a richer dynamical framework that inherits the glow associated with the idea of hierarchical Bayesian inference and brain function (Friston, 2012; Knill & Pouget, 2004).

For any of these theoretical accounts to remain viable, an important next step in their evolution will be incorporation of stereopsis into the picture, for there are reasons to believe that neural mechanisms involved in rivalry could be related to neural mechanisms underlying stable single vision and stereopsis (Blake & Boothroyd, 1985; Grossberg et al., 2008; Harrad et al., 1994; Hayashi et al., 2004; Li & Atick, 1994; Nichols & Wilson, 2009; Shimojo & Nakayama, 1994). Also not to be ignored is the novel idea that rivalry is involved in resolving binocular conflict when viewing cluttered scenes in which objects at different distances partially occlude one another (Arnold, 2011; Arnold, Grove, & Wallis, 2007).

NEURAL BASES OF BINOCULAR RIVALRY

What are the actual neural concomitants of binocular rivalry? Highlights of recent studies pertinent to that question are summarized in this section, organized by technique.

Neurophysiological Studies in Awake, Behaving Monkeys

• Recording from visual area MT, Maier, Logothetis, and Leopold (2007) found that percept-related activity of a given neuron depended on the specific stimulus being viewed. Nearly all MT neurons showed percept-related modulations to one rivalry configuration but not to others. This implies that it is oversimplified to categorize neurons as either "perceptual" (meaning their activities predict perception) or "sensory" (meaning their activity remains unchanged regardless of perceptual state). Responses, in other words, are highly context dependent, at least in visual area MT. To what extent this reinterpretation of neural responses and perceptual state applies to other visual areas remains to be learned.

• Keliris, Logothetis, and Tolias (2010) recorded spike activity and local field potentials from a large sample of V1 neurons in monkeys while they experienced binocular rivalry induced by FS. A minority of cells showed neural responses (both spikes and LFP) that varied depending on whether the evoking rival stimulus was dominant or suppressed, and, as found in earlier work by the Logothetis group, the percept-related modulations were substantially smaller than those produced by physical presentation and removal of the stimuli. Interestingly, among those percept-related

cells about half were monocular and half were binocular, a more balanced distribution than reported in earlier findings by this group. This and related work (Wilke, Logothetis, & Leopold, 2006) open the door for more detailed analyses of neural responses including different frequency bands comprising the LFP (Gail, Brinksmeyer, & Eckhorn, 2004).

Functional Magnetic Resonance Imaging (fMRI) in Humans

• Human brain-imaging studies have confirmed and expanded the network of visual areas implicated in rivalry by discovering percept-related modulations in neural responses inferred from blood oxygen level–dependent (BOLD) measures. Those areas include the lateral geniculate nucleus (Haynes, Deichmann, & Rees, 2005; Wunderlich et al., 2005), primary visual cortex (Lee, Blake, & Heeger, 2007; Meng, Remus, & Tong, 2005; Wunderlich, Schneider, & Kastner, 2005; but see Watanabe et al., 2011, for an alternative account of response modulations in V1 during rivalry), motion-sensitive visual areas such as V5/human MT+ complex (Moutoussis et al., 2005), the lateral occipital complex in the ventral pathway (Fang & He, 2005), the superior temporal sulcus (Jiang & He, 2006), and the fusiform face area (Fang & He, 2005; Pasley, Mayes, & Schultz, 2004; Sterzer, Haynes, & Rees, 2008; Sterzer & Rees, 2008; Williams et al., 2004). Moreover, varying patterns of responses in different visual areas during prolonged viewing of rivalry contain information sufficient to predict fluctuations in rivalry state using multivariate pattern classification of the fMRI signals (Haynes & Rees, 2005). At the same time fMRI has revealed other visual areas where BOLD responses remain strong, and even visual stimuli that selectively activate those areas are suppressed from awareness during rivalry. Those areas include the amygdala (Jiang & He, 2006; Pasley, Mayes, & Schultz, 2004; Williams et al., 2004) and the intraparietal sulcus in the dorsal pathway (Fang & He, 2005).

• Another, related aspect of rivalry concerns the involvement of frontoparietal areas in controlling state changes during binocular rivalry, an idea first advanced by Lumer, Friston, and Rees (1998) based on fMRI results. Subsequent fMRI papers have confirmed the existence of event-related changes in BOLD signals coinciding with switches during rivalry (e.g., Wilcke, O'Shea, & Watts, 2009), but one of those papers provided additional evidence that these activations are not trigger signals but instead are related to decisional uncertainty around the time of transitions that often include confusing mixtures (Knapen et al., 2011).

Studies showing modulations in rivalry alternation rate following application of repetitive transcranial magnetic stimulation to parietal portions of this putative network (Carmel et al., 2010; Zaretskaya et al., 2010) do not distinguish between these alternative accounts of the role of this network in rivalry.

EEG/MEG

• Recent studies using magnetoencephalography (MEG) and electroencephalography (EEG) reiterate what earlier studies have shown, namely that modulations in amplitude of signals associated with stimuli competing in rivalry can be identified throughout a network of brain areas (Cosmelli & Thompson, 2007; Srinivasan & Petrovic, 2006; Sterzer, Jalkanen, & Rees, 2009), although several of those studies implicate early visual areas as a prime source of those percept-related signal fluctuations (Roeber & Schröger, 2004; Kamphuisen, Bauer, & van Ee, 2008). Studies using EEG have also shown that manipulations of visual attention during rivalry impact signals generated by the rival stimuli evoking those signals, although those studies disagree over whether residual signals persist when attention is completely withdrawn from rivalry (Roeber et al., 2011; Zhang et al., 2011). One recent study using EEG found that the amplitude of the N400, a component of the EEG signal related to semantic context, shrank and ultimately disappeared as the target words evoking that component were reduced in visibility and ultimately abolished from awareness by interocular suppression (Kang, Blake, & Woodman, 2011). Finally, one notable methodological advance was the use of multivariate pattern analysis of EEG signals generated by pictures that were either visible or rendered invisible by CFS, the aim being to learn whether information sufficient for object categorization survived CFS; the answer to that question turned out to be "no" (Kaunitz et al., 2011).

FINAL THOUGHTS

This update on binocular rivalry has merely touched on highlights of a literature that burgeoned since this chapter's precursor appeared in 2004. Because of space limitations several hot topics could not be covered, including rivalry's possible differential reliance on processing within ventral-stream versus dorsal-stream pathways (Denison & Silver, 2012; van Boxtel et al., 2008a), the relation of binocular rivalry to stereopsis (e.g., Andrews & Holmes, 2011; Buckthought, Kim, & Wilson,

2007; Buckthought & Mendola, 2012; Nichols & Wilson, 2009; Su, He, & Ooi, 2009), and the degree to which rivalry resembles other forms of visual bistability such as monocular rivalry (Bhardwaj et al., 2008; Buckthought, Jessula, & Mendola, 2011; Maier, Logothetis, & Leopold, 2005), rivalry within an object-based reference frame (van Boxtel & Koch, 2012), and stimulus rivalry that occurs when dissimilar patterns are swapped rapidly and repetitively between the eyes (e.g., Knapen et al., 2007; Pearson, Tadin, & Blake, 2007; Silver & Logothetis, 2007; van Boxtel et al., 2008b). Nor was I able to touch on possible implications of rivalry for disciplines outside of vision science, such as clinical psychiatry (Nagamine et al., 2009), ophthalmology (Handa et al., 2006), gerontology (Norman et al., 2007), and human factors (Patterson et al., 2007). Suffice it to say, rivalry has captured the imagination of a broad audience.

In the coming years expect to see binocular rivalry maintain its prominent foothold among phenomena able to render ordinarily visible stimuli invisible (Kim & Blake, 2005); it has certainly become a favorite tool for the study of visual processing outside of awareness. But rivalry's continued popularity needs to be accompanied by an expansion of methods utilized for assessing perception during rivalry, methods beyond having people simply track alternations in dominance or indicate when a stimulus first emerges from suppression— more can be extracted from the phenomenon of rivalry than just the percentage of time and the average duration that a given stimulus is dominant or the length of time needed for a stimulus to break suppression. Several promising new approaches to the study of rivalry were introduced recently, including stochastic resonance (Kim, Grabowecky, & Suzuki, 2006), cumulative history analysis (Pastukhov & Braun, 2011), reverse correlation (Lankheet, 2006), and CFS (Tsuchiya & Koch, 2005). In addition, new versions of several inferential strategies were created by combining rivalry with other phenomena such as metacontrast masking (Breitmeyer et al., 2008), phantom motion (Meng, Ferneyhough, & Tong, 2007), afterimage nulling (Brascamp et al., 2009b), and repetition priming (Barbot & Kouider, 2012). In addition, we now have a more enlightened understanding of rivalry's dependence on image properties related to natural scene statistics (Baker & Graf, 2009b; Yang & Blake, 2012), thus setting the stage for a more nuanced approach when studying the impact of different categories of stimuli (e.g., objects versus tools) on rivalry and the differential impact of rivalry suppression on various categories of stimuli. So the future looks bright, and the third edition of this chapter will likely have plenty to say.

ACKNOWLEDGMENTS

During preparation of this chapter R.B. was supported by grants from the NIH (EY13358) and from the World Class University program through the Korea Science and Engineering Foundation funded by the Ministry of Education, Science and Technology (R31–10089). I am grateful to Jan Brascamp, Jochen Braun, Satoru Suzuki, and Sheng He for comments on parts of an earlier draft of this chapter. David Bloom provided invaluable help with referencing and formatting.

NOTE

1. This noteworthy study introduced a new binocular rivalry display consisting of different object images phase-shifting in opposite directions without changing position; the display can create crisp rivalry and robust optokinetic nystagmus whose slow phase correlates with the motion portrayed in the currently dominant stimulus.

REFERENCES

Adams, W. J., Gray, K. L. H., Garner, M. J., & Graf, E. W. (2010). High-level face adaptation without awareness. *Psychological Science, 21*, 205–210. doi:10.1177/0956797609359508.

Adams, W. J., Gray, K. L. H., Garner, M., & Graf, E. W. (2011). On the "special" status of emotional faces: Comment on Yang, Hong, and Blake (2010). *Journal of Vision, 11*, 1–4. doi:10.1167/11.3.10.

Alais, D. (2012). Binocular rivalry: Competition and inhibition in visual perception. *Cognitive Science, 3*, 87–103. doi:10.1002/wcs.151.

Alais, D., Cass, J., O'Shea, R. P., & Blake, R. (2010a). Visual sensitivity underlying changes in visual consciousness. *Current Biology, 20*, 1362–1367. doi:10.1016/j.cub.2010.06.015.

Alais, D., Lorenceau, J., Arrighi, R., & Cass, J. R. (2006). Contour interactions between pairs of Gabors engaged in binocular rivalry reveal a map of the association field. *Vision Research, 46*, 1473–1487. doi:10.1016/j.visres.2005.09.029.

Alais, D., & Melcher, D. (2007). Strength and coherence of binocular rivalry depends on shared stimulus complexity. *Vision Research, 47*, 269–279. doi:10.1016/j.visres.2006.09.003.

Alais, D., & Parker, A. (2006). Independent binocular rivalry processes for motion and form. *Neuron, 52*, 911–920. doi:10.1016/j.neuron.2006.10.027.

Alais, D., van Boxtel, J. J., Parker, A., & van Ee, R. (2010b). Attending to auditory signals slows visual alternations in binocular rivalry. *Vision Research, 50*, 929–935. doi:10.1016/j.visres.2010.03.010.

Almeida, J., Mahon, B. Z., Nakayama, K., & Caramazza, A. (2008). Unconscious processing dissociates along categorical lines. *Proceedings of the National Academy of Sciences of the United States of America, 105*, 15214–15218. doi:10.1073/pnas.0805867105.

Alpers, G. W., & Pauli, P. (2006). Emotional pictures predominate in binocular rivalry. *Cognition and Emotion, 20*, 596–607. doi:10.1080/02699930500282249.

Anderson, E., Siegel, E. H., Bliss-Moreau, E., & Barrett, L. F. (2011). The visual impact of gossip. *Science, 332*, 1446–1448. doi:10.1126/science.1201574.

Andrews, T. J., & Holmes, D. (2011). Stereoscopic depth perception and binocular rivalry. *Frontiers in Human Neuroscience, 5*, 1–6. doi:10.3389/fnhum.2011.00099.

Arnold, D. H. (2011). Why is binocular rivalry uncommon? Discrepant monocular images in the real world. *Frontiers in Human Neuroscience, 5*, 1–7. doi:10.3389/fnhum.2011.00116.

Arnold, D. H., Grove, P. M., & Wallis, T. S. A. (2007). Staying focused: A functional account of perceptual suppression during binocular rivalry. *Journal of Vision, 7*, 1–8. doi:10.1167/7.7.7.

Arnold, D. H., James, B., & Roseboom, W. (2009). Binocular rivalry: Spreading dominance through complex images. *Journal of Vision, 9*, 1–9. doi:10.1167/9.13.4.

Arnold, D. H., Law, P., & Wallis, T. S. A. (2008). Binocular switch suppression: A new method for persistently rendering the visible "invisible." *Vision Research, 48*, 994–1001. doi:10.1016/j.visres.2008.01.020.

Baker, D. H., & Graf, E. W. (2009a). On the relation between dichoptic masking and binocular rivalry. *Vision Research, 49*, 451–459. doi:10.1016/j.visres.2008.12.002.

Baker, D. H., & Graf, E. W. (2009b). Natural images dominate in binocular rivalry. *Proceedings of the National Academy of Sciences of the United States of America, 106*, 5436–5441. doi:10.1073/pnas.0812860106.

Barbot, A., & Kouider, S. (2012). Longer is not better: Nonconscious overstimulation reverses priming influences under interocular suppression. *Attention, Perception & Psychophysics, 74*, 174–184. doi:10.3758/s13414-011-0226-3.

Bhardwaj, R., O'Shea, R. P., Alais, D., & Parker, A. (2008). Probing visual consciousness: Rivalry between eyes and images. *Journal of Vision, 8*, 1–13. doi:10.1167/8.11.2.

Blake, R. (1989). A neural theory of binocular rivalry. *Psychological Review, 96*, 145–167. doi:10.1037/0033-295X.96.1.145.

Blake, R. (2004). Binocular Rivalry. In L. M. Chalupa & J. S. Werner (Eds.), *The Visual Neurosciences* (pp. 1313–1323). Cambridge, MA: MIT Press.

Blake, R., & Boothroyd, K. (1985). The precedence of binocular fusion over binocular rivalry. *Perception & Psychophysics, 37*, 114–124. doi:10.3758/BF03202845.

Blake, R., & Logothetis, N. (2002). Visual competition. *Nature Reviews. Neuroscience, 3*, 13–21. doi:10.1038/nrn701.

Blake, R., & O'Shea, R. P. (2009). Binocular rivalry. In L. Squire (Ed.), *Encyclopedia of Neuroscience* (Vol. 2, pp. 179–187). Oxford: Academic Press.

Blake, R., & Overton, R. (1979). The site of binocular rivalry suppression. *Perception, 8*, 143–152. doi:10.1068/p080143.

Blake, R., & Shiffrar, M. (2007). Perception of human motion. *Annual Review of Psychology, 58*, 47–73. doi:10.1146/annurev.psych.57.102904.190152.

Blake, R., Tadin, D., Sobel, K. V., Raissian, T. A., & Chong, S. C. (2006). Strength of early visual adaptation depends on visual awareness. *Proceedings of the National Academy of Sciences of the United States of America 103*, 4783–4788. doi:10.1073/pnas.0509634103.

Blake, R., Westendorf, D., & Fox, R. (1990). Temporal perturbations of binocular rivalry. *Perception & Psychophysics, 48*, 593–602. doi:10.3758/BF03211605.

Blake, R., & Wilson, H. R. (2011). Binocular vision. *Vision Research, 51*, 754–770. doi:10.1016/j.visres.2010.10.009.

Brascamp, J., & Blake, R. (2012). Inattention abolishes binocular rivalry: Perceptual evidence. *Psychological Science, 23,* 1159–1167. doi:10.1177/0956797612440100.

Brascamp, J. W., Pearson, J., Blake, R., & van den Berg, A. V. (2009a). Intermittent ambiguous stimuli: Implicit memory causes periodic perceptual alternations. *Journal of Vision, 9,* 1–23. doi:10.1167/9.3.3.

Brascamp, J. W., van Boxtel, J. J. A., Knapen, T. H. J., & Blake, R. (2009b). A dissociation of attention and awareness in phase-sensitive but not phase-insensitive visual channels. *Journal of Cognitive Neuroscience, 22,* 2326–2344. doi:10.1162/jocn.2009.21397.

Brascamp, J. W., van Ee, R., Noest, A. J., Jacobs, R. H. A. H., & van den Berg, A. V. (2006). The time course of binocular rivalry reveals a fundamental role of noise. *Journal of Vision, 6,* 1244–1256. doi:10.1167/6.11.8.

Brascamp, J. W., van Ee, R., Pestman, W. R., & van den Berg, A. V. (2005). Distributions of alternation rates in various forms of bistable perception. *Journal of Vision, 5,* 287–298. doi:10.1167/5.4.1.

Breese, B. B. (1899). On inhibition. *Psychological Monographs, 3,* 1–65. doi:10.1037/h0092990.

Breitmeyer, B. G., Koç, A., Öğmen, H., & Ziegler, R. (2008). Functional hierarchies of nonconscious visual processing. *Vision Research, 48,* 1509–1513. doi:10.1016/j.visres.2008.03.015.

Bressloff, P. C., & Webber, M. A. (2012). Neural field model of binocular rivalry waves. *Journal of Computational Neuroscience, 32,* 233–252. doi:10.1007/s10827-011-0351-y.

Buckthought, A., Kim, J., & Wilson, H. R. (2007). Hysteresis effects in stereopsis and binocular rivalry. *Vision Research, 48,* 819–830. doi:10.1016/j.visres.2007.12.013.

Buckthought, A., Jessula, S., & Mendola, J. D. (2011). Bistable percepts in the brain: fMRI contrasts monocular pattern rivalry and binocular rivalry. *PLoS One, 6,* e20367. doi:10.1371/journal.pone.0020367.

Buckthought, A., & Mendola, J. D. (2012). How simultaneous is the perception of binocular depth and rivalry in plaid stimuli? *Perception, 3,* 305–315. doi:10.1068/i0491.

Carmel, D., Walsh, V., Lavie, N., & Rees, G. (2010). Right parietal TMS shortens dominance durations in binocular rivalry. *Current Biology, 20,* R799–R800. doi:10.1016/j.cub.2010.07.036.

Carrasco, M., Ling, S., & Read, S. (2004). Attention alters appearance. *Nature Neuroscience, 7,* 308–313. doi:10.1038/nn1194.

Carter, O., & Cavanagh, P. (2007). Onset rivalry: Brief presentation isolates an early independent phase of perceptual competition. *PLoS One, 2,* e343. doi:10.1371/journal.pone.0000343.

Carter, O. L., & Pettigrew, J. D. (2003). A common oscillator for perceptual rivalries? *Perception, 32,* 295–305. doi:10.1068/p3472.

Carter, O. L., Pettigrew, J. D., Hasler, F., Wallis, G. M., Liu, G. B., Hell, D., et al. (2005a). Modulating the rate and rhythmicity of perceptual rivalry alternations with the mixed 5–HT2A and 5–HT1A agonist psilocybin. *Neuropsychopharmacology, 30,* 1154–1162. doi:10.1038/sj.npp.1300621.

Carter, O. L., Presti, D. E., Callistemon, C., Ungerer, Y., Liu, G. B., & Pettigrew, J. D. (2005b). Meditation alters perceptual rivalry in Tibetan Buddhist monks. *Current Biology, 15,* R412–R413. doi:10.1016/j.cub.2005.05.043.

Chen, X., & He, S. (2004). Local factors determine the stabilization of monocular ambiguous and binocular rivalry stimuli. *Current Biology, 14,* 1013–1017. doi:10.1016/j.cub.2004.05.042.

Chen, Y.-C., Yeh, S.-L., & Spence, C. (2011). Crossmodal constraints on human perceptual awareness: Auditory semantic modulation of binocular rivalry. *Frontiers in Psychology, 2,* 1–13. doi:10.3389/fpsyg.2011.00212.

Chong, S. C., & Blake, R. (2006). Exogenous attention and endogenous attention influence initial dominance in binocular rivalry. *Vision Research, 46,* 1794–1803. doi:10.1016/j.visres.2005.10.031.

Chong, S. C., Tadin, D., & Blake, R. (2005). Endogenous attention prolongs dominance durations in binocular rivalry. *Journal of Vision, 5,* 1004–1012. doi:10.1167/5.11.6.

Chopin, A., & Mamassian, P. (2010). Task usefulness affects perception of rivalrous images. *Psychological Science, 21,* 1886–1893. doi:10.1177/0956797610389190.

Chou, W. L., & Yeh, S. L. (2012). Object-based attention occurs regardless of object awareness. *Psychonomic Bulletin & Review, 19,* 225–231. doi:10.3758/s13423-011-0207-5.

Conrad, V., Bartels, A., Kleiner, M., & Noppeney, U. (2010). Audiovisual interactions in binocular rivalry. *Journal of Vision, 10,* 1–15. doi:10.1167/10.10.27.

Cosmelli, D., & Thompson, E. (2007). Mountains and valleys: Binocular rivalry and the flow of experience. *Consciousness and Cognition, 16,* 623–641. doi:10.1016/j.concog.2007.06.013.

Dayan, P. (1998). A hierarchical model of binocular rivalry. *Neural Computation, 10,* 1119–1135. doi:10.1162/089976698300017377.

Denison, R. N., Piazza, E. A., & Silver, M. A. (2011). Predictive context influences perceptual selection during binocular rivalry. *Frontiers in Human Neuroscience, 5,* 1–11. doi:10.3389/fnhum.2011.00166.

Denison, R. N., & Silver, M. A. (2012). Distinct contributions of the magnocellular and parvocellular visual streams to perceptual selection. *Journal of Cognitive Neuroscience, 24,* 246–259. doi:10.1162/jocn_a_00121.

Dieter, K. C., & Tadin, D. (2011). Understanding attentional modulation of binocular rivalry: A framework based on biased competition. *Frontiers in Human Neuroscience, 5,* 1–12. doi:10.3389/fnhum.2011.00155.

Einhäuser, W., Stout, J., Koch, C., & Carter, O. (2008). Pupil dilation reflects perceptual selection and predicts subsequent stability in perceptual rivalry. *Proceedings of the National Academy of Sciences of the United States of America, 105,* 1704–1709. doi:10.1073/pnas.0707727105.

Fang, F., & He, S. (2005). Cortical responses to invisible objects in the human dorsal and ventral pathways. *Nature Neuroscience, 8,* 1380–1385. doi:10.1038/nn1537.

Field, D. J., Hayes, A., & Hess, R. F. (1993). Contour integration by the human visual system: Evidence for a local "association field." *Vision Research, 33,* 173–193. doi:10.1016/0042-6989(93)90156-Q.

Fox, R., Todd, S., & Bettinger, L. A. (1975). Optokinetic nystagmus as an objective indicator of binocular rivalry. *Vision Research, 15,* 849–853. doi:10.1016/0042-6989(75)90265-5.

Freeman, A. W. (2005). Multistage model for binocular rivalry. *Journal of Neurophysiology, 94,* 4412–4420. doi:10.1152/jn.00557.2005.

Friston, K. (2012). The history of the future of the Bayesian brain. *NeuroImage*, *62*, 1230–1233. doi:10.1016/j.neuroimage.2011.10.004.

Fukuda, H., & Blake, R. (1992). Spatial interactions in binocular rivalry. *Journal of Experimental Psychology. Human Perception and Performance*, *18*, 362–370. doi:10.1037/0096-1523.18.2.362.

Gail, A., Brinksmeyer, H. J., & Eckhorn, R. (2004). Perception-related modulations of local field potential power and coherence in primary visual cortex of awake monkey during binocular rivalry. *Cerebral Cortex*, *14*, 300–313. doi:10.1093/cercor/bhg129.

Gibson, J. J. (1950). *The perception of the visual world.* Boston: Houghton Mifflin.

Grossberg, S., Yazdanbakhsh, A., Cao, Y., & Swaminathan, G. (2008). How does binocular rivalry emerge from cortical mechanisms of 3-D vision? *Vision Research*, *48*, 2232–2250. doi:10.1016/j.visres.2008.06.024.

Hancock, S., Gareze, L., Findlay, J. M., & Andrews, T. J. (2012). Temporal patterns of saccadic eye movements predict individual variation in alternation rate during binocular rivalry. *Perception*, *3*, 88–96. doi:10.1068/i0486.

Hancock, S., Whitney, D., & Andrews, T. J. (2008). The initial interactions underlying binocular rivalry require visual awareness. *Journal of Vision*, *8*, 1–9. doi:10.1167/8.1.3.

Handa, T., Uozato, H., Higa, R., Nitta, M., Kawamorita, T., Ishikawa, H., et al. (2006). Quantitative measurement of ocular dominance using binocular rivalry induced by retinometers. *Journal of Cataract and Refractive Surgery*, *32*, 831–836. doi:10.1016/j.jcrs.2006.01.082.

Harrad, R. A., McKee, S. P., Blake, R., & Yang, Y. (1994). Binocular rivalry disrupts stereopsis. *Perception*, *23*, 15–28. doi:10.1068/p230015.

Hayashi, R., Maeda, T., Shimojo, S., & Tachi, S. (2004). An integrative model of binocular vision: A stereo model utilizing interocularly unpaired points produces both depth and binocular rivalry. *Vision Research*, *44*, 2367–2380. doi:10.1016/j.visres.2004.04.017.

Hayashi, R., & Tanifuji, M. (2012). Which image is in awareness during binocular rivalry? Reading perceptual status from eye movements. *Journal of Vision*, *12*, 1–11. doi:10.1167/12.3.5.

Haynes, J. D., Deichmann, R., & Rees, G. (2005). Eye-specific effects of binocular rivalry in the human lateral geniculate nucleus. *Nature*, *438*, 496–499. doi:10.1038/nature04169.

Haynes, J. D., & Rees, G. (2005). Predicting the orientation of invisible stimuli from activity in human primary visual cortex. *Nature Neuroscience*, *8*, 686–691. doi:10.1038/nn1445.

Hohwy, J., Roepstorff, A., & Friston, K. (2008). Predictive coding explains binocular rivalry: An epistemological review. *Cognition*, *108*, 687–701. doi:10.1016/j.cognition.2008.05.010.

Hollins, M. (1980). The effect of contrast on the completeness of binocular rivalry suppression. *Perception & Psychophysics*, *27*, 550–556. doi:10.3758/BF03198684.

Holmes, D. J., Hancock, S., & Andrews, T. J. (2006). Independent binocular integration of form and colour. *Vision Research*, *46*, 665–677. doi:10.1016/j.visres.2005.05.023.

Hong, S.-W., & Blake, R. (2009). Interocular suppression differentially affects achromatic and chromatic mechanisms. *Attention, Perception & Psychophysics*, *71*, 405–411. doi:10.3758/APP.71.2.403.

Hong, S. W., & Shevell, S. K. (2009). Color-binding errors during rivalrous suppression of form. *Psychological Science*, *20*, 1084–1091. doi:10.1111/j.1467-9280.2009.02408.x.

Hudak, M., Gervan, P., Friedrich, B., Pastukhov, A., Braun, J., & Kovacs, I. (2011). Increased readiness for adaptation and faster alternation rates under binocular rivalry in children. *Frontiers in Human Neuroscience*, *5*, 1–7. doi:10.3389/fnhum.2011.00128.

Jiang, Y., Costello, P., Fang, F., Huang, M., & He, S. (2006). A gender- and sexual orientation-dependent spatial attentional effect of invisible images. *Proceedings of the National Academy of Sciences of the United States of America*, *103*, 17048–17052. doi:10.1073/pnas.0605678103.

Jiang, Y., Costello, P., & He, S. (2007). Processing of invisible stimuli: Advantage of upright faces and recognizable words in overcoming interocular suppression. *Psychological Science*, *18*, 349–355. doi:10.1111/j.1467-9280.2007.01902.x.

Jiang, Y., & He, S. (2006). Cortical responses to invisible faces: Dissociating subsystems for facial-information processing. *Current Biology*, *16*, 2023–2029. doi:10.1016/j.cub.2006.08.084.

Kalarickal, G. J., & Marshall, J. (2000). Neural model of temporal and stochastic properties of binocular rivalry. *Neurocomputing*, *32–33*, 843–853. doi:10.1016/S0925-2312(00)00252-6.

Kalisvaart, J. P., Rampersad, S. M., & Goossens, J. (2011). Binocular onset rivalry at the time of saccades and stimulus jumps. *PLoS One*, *6*, e20017. doi:10.1371/journal.pone.0020017.

Kamphuisen, A., Bauer, M., & van Ee, R. (2008). No evidence for widespread synchronized networks in binocular rivalry: MEG frequency tagging entrains primarily early visual cortex. *Journal of Vision*, *8*, 1–8. doi:10.1167/8.5.4.

Kanai, R., Moradi, F., Shimojo, S., & Verstraten, F. A. J. (2005). Perceptual alternation induced by visual transients. *Perception*, *34*, 803–822. doi:10.1068/p5245.

Kang, M. S. (2009). Size matters: A study of binocular rivalry dynamics. *Journal of Vision*, *9*, 1–11. doi:10.1167/9.1.17.

Kang, M. S., & Blake, R. (2005). Perceptual synergy between seeing and hearing revealed during binocular rivalry. *Psichologija*, *32*, 7–15. Retrieved from http://www.leidykla.eu/fileadmin/Psichologija/32/7-15.pdf.

Kang, M. S., & Blake, R. (2010). What causes alternations in dominance during binocular rivalry? *Attention, Perception & Psychophysics*, *72*, 179–186. doi:10.3758/APP.72.1.179.

Kang, M. S., & Blake, R. (2011). An integrated framework of spatiotemporal dynamics of binocular rivalry. *Frontiers in Human Neuroscience*, *5*, 1–9. doi:10.3389/fnhum.2011.00088.

Kang, M. S., Blake, R., & Woodman, G. F. (2011). Semantic analysis does not occur in the absence of awareness induced by interocular suppression. *Journal of Neuroscience*, *31*, 13535–13545. doi:10.1523/JNEUROSCI.1691-11.2011.

Kang, M. S., Heeger, D., & Blake, R. (2009). Periodic perturbations producing phase-locked fluctuations in visual perception. *Journal of Vision*, *9*, 1–12. doi:10.1167/9.2.8.

Kang, M. S., Lee, S. H., Kim, J., Heeger, D., & Blake, R. (2010). Modulation of spatiotemporal dynamics of binocular rivalry by collinear facilitation and pattern-dependent adaptation. *Journal of Vision*, *10*, 1–15. doi:10.1167/10.11.3.

Kaunitz, L. N., Kamienkowski, J. E., Olivetti, E., Murphy, B., Avesani, P., & Melcher, D. P. (2011). Intercepting the first pass: Rapid categorization is suppressed for unseen stimuli. *Frontiers in Psychology, 2*, 1–10. doi:10.3389/fpsyg.2011.00198.

Keliris, G. A., Logothetis, N. K., & Tolias, A. S. (2010). The role of the primary visual cortex in perceptual suppression of salient visual stimuli. *Journal of Neuroscience, 30*, 12353–12365. doi:10.1523/JNEUROSCI.0677-10.2010.

Kim, C. Y., & Blake, R. (2005). Psychophysical magic: Rendering the visible "invisible." *Trends in Cognitive Sciences, 9*, 381–388. doi:10.1016/j.tics.2005.06.012.

Kim, Y. J., Grabowecky, M., & Suzuki, S. (2006). Stochastic resonance in binocular rivalry. *Vision Research, 46*, 392–406. doi:10.1016/j.visres.2005.08.009.

Klink, P. C., Brascamp, J. W., Blake, R., & van Wezel, R. J. A. (2010). Experience-driven plasticity in binocular vision. *Current Biology, 20*, 1464–1469. doi:10.1016/j.cub.2010.06.057.

Knapen, T., Brascamp, J., Pearson, J., van Ee, R., & Blake, R. (2011). The role of frontal and parietal brain areas in bistable perception. *Journal of Neuroscience, 31*, 10293–10301. doi:10.1523/JNEUROSCI.1727-11.2011.

Knapen, T., Kanai, R., Brascamp, J., van Boxtel, J., & van Ee, R. (2007). Distance in feature space determines exclusivity in visual rivalry. *Vision Research, 47*, 3269–3275. doi:10.1016/j.visres.2007.09.005.

Knapen, T., van Ee, R., & Blake, R. (2007). Stimulus motion propels traveling waves in binocular rivalry. *PLoS One, 2*, e739. doi:10.1371/journal.pone.0000739.

Knill, D. C., & Pouget, A. (2004). The Bayesian brain: The role of uncertainty in neural coding and computation. *Trends in Neurosciences, 27*, 712–719. doi:10.1016/j.tins.2004.10.007.

Laing, C. R., Frewen, T., & Kevrekidis, I. G. (2010). Reduced models for binocular rivalry. *Journal of Computational Neuroscience, 28*, 459–476. doi:10.1007/s10827-010-0227-6.

Lankheet, M. J. M. (2006). Unraveling adaptation and mutual inhibition in perceptual rivalry. *Journal of Vision, 6*, 304–310. doi:10.1167/6.4.1.

Lee, S. H., Blake, R., & Heeger, D. (2007). Hierarchy of cortical responses underlying binocular rivalry. *Nature Neuroscience, 10*, 1048–1054. doi:10.1038/nn1939.

Leopold, D. A., Wilke, M., Maier, A., & Logothetis, N. (2002). Stable perception of visually ambiguous patterns. *Nature Neuroscience, 5*, 605–609. doi:10.1038/nn851.

Levelt, W. J. M. (1965). *On binocular rivalry*. Soesterberg, The Netherlands: Institute for Perception RVO-TNO.

Li, Z., & Atick, J. J. (1994). Efficient stereo coding in the multiscale representation. *Network, 5*, 157–174. doi:10.1088/0954-898X/5/2/003.

Logothetis, N. K., Leopold, D. A., & Sheinberg, D. L. (1996). What is rivalling during binocular rivalry? *Nature, 380*, 621–624. doi:10.1038/380621a0.

Lumer, E. D., Friston, K. J., & Rees, G. (1998). Neural correlates of perceptual rivalry in the human brain. *Science, 280*, 1930–1934. doi:10.1126/science.280.5371.1930.

Lunghi, C., Binda, P., & Morrone, M. C. (2010). Touch disambiguates rivalrous perception at early stages of visual analysis. *Current Biology, 20*, R143–R144. doi:10.1016/j.cub.2009.12.015.

Lunghi, C., Burr, D. C., & Morrone, C. (2011). Brief periods of monocular deprivation disrupt ocular balance in human adult visual cortex. *Current Biology, 21*, R538–R539. doi:10.1016/j.cub.2011.06.004.

Lupyan, G., Thompson-Schill, S., & Swingley, D. (2010). Conceptual penetration of visual processing. *Psychological Science, 21*, 682–691. doi:10.1177/0956797610366099.

Lutz, A., Greischar, L. L., Rawlings, N. B., Ricard, M., & Davidson, R. J. (2004). Long-term meditators self-induce high-amplitude gamma synchrony during mental practice. *Proceedings of the National Academy of Sciences of the United States of America, 101*, 16369–16373. doi:10.1073/pnas.0407401101.

Maier, A., Logothetis, N., & Leopold, D. A. (2005). Global competition dictates local supression in pattern rivalry. *Journal of Vision, 5*, 668–677. doi:10.1167/5.9.2.

Maier, A., Logothetis, N. K., & Leopold, D. A. (2007). Context-dependent perceptual modulation of single neurons in primate visual cortex. *Proceedings of the National Academy of Sciences of the United States of America, 104*, 5620–5625. doi:10.1073/pnas.0608489104.

Mamassian, P., & Goutcher, R. (2005). Temporal dynamics in bistable perception. *Journal of Vision, 5*, 361–375. doi:10.1167/5.4.7.

Maruya, K., & Blake, R. (2009). Spatial spread of interocular suppression is guided by stimulus configuration. *Perception, 38*, 215–231. doi:10.1068/p6157.

Maruya, K., Watanabe, H., & Watanabe, M. (2008). Adaptation to invisible motion results in low-level but not high-level aftereffects. *Journal of Vision, 8*, 1–11. doi:10.1167/8.11.7.

Maruya, K., Yang, E., & Blake, R. (2007). Voluntary action influences visual competition. *Psychological Science, 18*, 1090–1098. doi:10.1111/j.1467-9280.2007.02030.x.

Matsuoka, K. (1984). The dynamic model of binocular rivalry. *Biological Cybernetics, 49*, 201–208. doi:10.1007/BF00334466.

Meng, M., Ferneyhough, E., & Tong, F. (2007). Dynamics of perceptual filling-in of visual phantoms revealed by binocular rivalry. *Journal of Vision, 7*, 1–15. doi:10.1167/7.13.8.

Meng, M., Remus, D. A., & Tong, F. (2005). Filling-in of visual phantoms in the human brain. *Nature Neuroscience, 8*, 1248–1254. doi:10.1038/nn1518.

Miller, S. M., Hansell, N. K., Ngo, T. T., Liu, G. B., Pettigrew, J. D., Martin, N. G., et al. (2010). Genetic contribution to individual variation in binocular rivalry rate. *Proceedings of the National Academy of Sciences of the United States of America, 107*, 2664–2668. doi:10.1073/pnas.0912149107.

Miller, S. M., Ngo, T. T., & van Swinderen, B. (2012). Attentional switching in humans and flies: Rivalry in large and miniature brains. *Frontiers in Human Neuroscience, 5*, 1–17. doi:10.3389/fnhum.2011.00188.

Mitchell, J. F., Stoner, G. R., & Reynolds, J. H. (2004). Object-based attention determines dominance in binocular rivalry. *Nature, 429*, 410–413. doi:10.1038/nature02584.

Moradi, F., Koch, C., & Shimojo, S. (2005). Face adaptation depends on seeing the face. *Neuron, 45*, 169–175. doi:10.1016/j.neuron.2004.12.018.

Moreno-Bote, R., Rinzel, J., & Rubin, N. (2007). Noise-induced alternations in an attractor network model of perceptual bistability. *Journal of Neurophysiology, 98*, 1125–1139. doi:10.1152/jn.00116.2007.

Moutoussis, K., Keliris, G., Kourtzi, Z., & Logothetis, N. (2005). A binocular rivalry study of motion perception in the human brain. *Vision Research, 45*, 2231–2243. doi:10.1016/j.visres.2005.02.007.

Mudrik, L., Breska, A., Lamy, D., & Deouell, L. Y. (2011). Integration without awareness: Expanding the limits of unconscious processing. *Psychological Science, 22,* 764–770. doi:10.1177/095679761140836.

Mueller, T. J., & Blake, R. (1989). A fresh look at the temporal dynamics of binocular rivalry. *Biological Cybernetics, 61,* 223–232. doi:10.1007/BF00198769.

Naber, M., Frässle, S., & Einhäuser, W. (2011). Perceptual rivalry: Reflexes reveal the gradual nature of visual awareness. *PLoS One, 6,* e20910. doi:10.1371/journal.pone.0020910.

Nagamine, M., Yoshino, A., Miyazaki, M., Takahashi, Y., & Nomura, S. (2009). Difference in binocular rivalry rate between patients with bipolar I and bipolar II disorders. *Bipolar Disorders, 11,* 539–546. doi:10.1111/j.1399-5618.2009.00719.x.

Ngo, T. T., Liu, G. B., Tilley, A. J., Pettigrew, J. D., & Miller, S. M. (2007). Caloric-vestibular stimulation reveals discrete neural mechanisms for coherence rivalry and eye rivalry: A meta-rivalry model. *Vision Research, 47,* 2685–2699. doi:10.1016/j.visres.2007.03.024.

Nichols, D. F., & Wilson, H. R. (2009). Stimulus specificity in spatially-extended interocular suppression. *Vision Research, 49,* 2110–2120. doi:10.1016/j.visres.2009.06.001.

Noest, A. J., van Ee, R., Nijs, M. M., & van Wezel, R. J. A. (2007). Percept-choice sequences driven by interrupted ambiguous stimuli: A low-level neural model. *Journal of Vision, 7*(8), 10, 11–14. doi:10.1167/7.8.10.

Norman, J. F., Norman, H. F., Pattison, K., Taylor, M. J., & Goforth, K. E. (2007). Aging and the depth of binocular rivalry suppression. *Psychology and Aging, 22,* 625–631. doi:10.1037/0882-7974.22.3.625.

Ooi, T. L., & He, Z. J. (2005). Surface representation and attention modulation mechanisms in binocular rivalry. In D. Alais & R. Blake (Eds.), *Binocular rivalry* (pp. 117–135). Cambridge, MA: MIT Press.

Ooi, T. L., & He, Z. J. (2006). Binocular rivalry and surface-boundary processing. *Perception, 35,* 581–603. doi:10.1068/p5489.

O'Shea, R. P., & Corballis, P. M. (2005). Visual grouping on binocular rivalry in a split-brain observer. *Vision Research, 45,* 247–261. doi:10.1016/j.visres.2004.08.009.

O'Shea, R. P., & Crassini, B. (1981). The sensitivity of binocular rivalry suppression to changes in orientation assessed by reaction-time and forced-choice techniques. *Perception, 10,* 283–293. doi:10.1068/p100283.

Ozkan, K., & Braunstein, M. L. (2009). Predominance of ground over ceiling surfaces in binocular rivalry. *Attention, Perception & Psychophysics, 71,* 1305–1312. doi:10.3758/APP.71.6.1305.

Paffen, C. L. E., & Alais, D. (2011). Attentional modulation of binocular rivalry. *Frontiers in Human Neuroscience, 5,* 1–10. doi:10.3389/fnhum.2011.00105.

Paffen, C. L. E., Alais, D., & Verstraten, F. A. J. (2006b). Attention speeds binocular rivalry. *Psychological Science, 17,* 752–756. doi:10.1111/j.1467-9280.2006.01777.x.

Paffen, C. L. E., Naber, M., & Verstraten, F. A. J. (2008). The spatial origin of a perceptual transition in binocular rivalry. *PLoS One, 3,* e2311. doi:10.1371/journal.pone.0002311.

Paffen, C. L. E., Plukaard, S., & Kanai, R. (2011). Symbolic magnitude modulates perceptual strength in binocular rivalry. *Cognition, 119,* 468–475. doi:10.1016/j.cognition.2011.01.010.

Paffen, C. L. E., Tadin, D., te Pas, S. F., Blake, R., & Verstraten, F. A. J. (2006a). Adaptive center-surround interactions in human vision revealed during binocular rivalry. *Vision Research, 46,* 599–604. doi:10.1016/j.visres.2005.05.013.

Paffen, C. L. E., te Pas, S. F., Kanai, R., van der Smagt, M. J., & Verstraten, F. A. J. (2004). Center-surround interactions in visual motion processing during binocular rivalry. *Vision Research, 44,* 1635–1639. doi:10.1016/j.visres.2004.02.007.

Paffen, C. L. E., & Van der Stigchel, S. (2010). Shifting spatial attention makes you flip: Exogenous visual attention triggers perceptual alternations during binocular rivalry. *Attention, Perception & Psychophysics, 72,* 1237–1243. doi:10.3758/APP.72.5.1237.

Pasley, B. N., Mayes, L. C., & Schultz, R. T. (2004). Subcortical discrimination of unperceived objects during binocular rivalry. *Neuron, 42,* 163–172. doi:10.1016/S0896-6273(04)00155-2.

Pastukhov, A., & Braun, J. (2011). Cumulative history quantifies the role of neural adaptation in multistable perception. *Journal of Vision, 11,* 1–10. doi:10.1167/11.10.12.

Patterson, R., Winterbottom, M., Pierce, B., & Fox, R. (2007). Binocular rivalry and head-worn displays. *Human Factors, 49,* 1083–1096. doi:10.1518/001872007X249947.

Pearson, J., & Brascamp, J. (2008). Sensory memory for ambiguous vision. *Trends in Cognitive Sciences, 12,* 334–341. doi:10.1016/j.tics.2008.05.006.

Pearson, J., & Clifford, C. W. G. (2005). When your brain decides what you see: Grouping across monocular, binocular, and stimulus rivalry. *Psychological Science, 16,* 516–519. doi:10.1111/j.0956-7976.2005.01566.x.

Pearson, J., Clifford, C. W. G., & Tong, F. (2008). The functional impact of mental imagery on conscious perception. *Current Biology, 18,* 982–986. doi:10.1016/j.cub.2008.05.048.

Pearson, J., Tadin, D., & Blake, R. (2007). The effects of transcranial magnetic stimulation (TMS) on visual rivalry. *Journal of Vision, 7,* 1–11. doi:10.1167/7.7.2.

Pelekanos, V., Roumani, D., & Moutoussis, K. (2012). The effects of categorical and linguistic adaptation on binocular rivalry initial dominance. *Frontiers in Human Neuroscience, 5,* 1–8. doi:10.3389/fnhum.2011.00187.

Phelps, E. A., Ling, S., & Carrasco, M. (2006). Emotion facilitates perception and potentiates the perceptual benefits of attention. *Psychological Science, 17,* 292–299. doi:10.1111/j.1467-9280.2006.01701.x.

Radel, R., & Clément-Guillotin, C. (2012). Evidence of motivational influences in early visual perception: Hunger modulates conscious access. *Psychological Science, 23,* 232–234. doi:10.1177/0956797611427920.

Ratcliff, R., & McKoon, G. (2008). The diffusion decision model: Theory and data for two-choice decision tasks. *Neural Computation, 20,* 873–922. doi:10.1162/neco.2008.12-06-420.

Roeber, U., & Schröger, E. (2004). Binocular rivalry is partly resolved at early processing stages with steady and with flickering presentation: A human event-related brain potential study. *Neuroscience Letters, 371,* 51–55. doi:10.1016/j.neulet.2004.08.038.

Roeber, U., Veser, S., Schröger, E., & O'Shea, R. P. (2011). On the role of attention in binocular rivalry: Electrophysiological evidence. *PLoS One, 6,* e22612. doi:10.1371/journal.pone.0022612.

Roumani, D., & Moutoussis, K. (2012). Binocular rivalry alternations and their relation to visual adaptation. *Frontiers in Human Neuroscience, 6*, 1–5. doi:10.3389/fnhum.2012. 00035.

Sakuraba, S., Sakai, S., Yamanaka, M., Yokosawa, K., & Hirayama, K. (2012). Does the human dorsal stream really process a category for tools? *Journal of Neuroscience, 32*, 3949–3953. doi:10.1523/JNEUROSCI.3973-11.2012.

Schall, J. D., Nawrot, M., Blake, R., & Yu, K. (1993). Visually guided attention is neutralized when informative cues are visible but unperceived. *Vision Research, 33*, 2057–2064. doi:10.1016/0042-6989(93)90004-G.

Schyns, P., & Oliva, A. (1999). Dr. Angry and Mr. Smile: When categorization flexibly modifies the perception of faces in rapid visual presentations. *Cognition, 69*, 243–265. doi:10.1016/S0010-0277(98)00069-9.

Seely, J., & Chow, C. C. (2011). The role of mutual inhibition in binocular rivalry. *Journal of Neurophysiology, 106*, 2136–2150. doi:10.1152/jn.00228.2011.

Shannon, R. W., Patrick, C. J., Jiang, Y., Bernat, E., & He, S. (2011). Genes contribute to the switching dynamics of bistable perception. *Journal of Vision, 11*, 1–7. doi:10.1167/ 11.3.8.

Sheth, B. R., & Pham, T. (2008). How emotional arousal and valence influence access to awareness. *Vision Research, 48*, 2415–2424. doi:10.1016/j.visres.2008.07.013.

Shimojo, S., & Nakayama, K. (1994). Interocularly unpaired zones escape local binocular matching. *Vision Research, 34*, 1875–1882. doi:10.1016/0042-6989(94)90311-5.

Shpiro, A., Moreno-Bote, R., Rubin, N., & Rinzel, J. (2009). Balance between noise and adaptation in competition models of perceptual bistability. *Journal of Computational Neuroscience, 27*, 37–54. doi:10.1007/s10827-008-0125-3.

Silver, M. A., & Logothetis, N. K. (2007). Temporal frequency and contrast tagging bias the type of competition in interocular switch rivalry. *Vision Research, 47*, 532–543. doi:10.1016/ j.visres.2006.10.011.

Song, C., & Yao, H. (2009). Duality in binocular rivalry: Distinct sensitivity of percept sequence and percept duration to imbalance between monocular stimuli. *PLoS One, 4*, e6912. doi:10.1371/journal.pone.0006912.

Spence, C. (2011). Crossmodal correspondences: A tutorial review. *Attention, Perception & Psychophysics, 73*, 971–995. doi:10.3758/s13414-010-0073-7.

Spering, M., Pomplun, M., & Carrasco, M. (2011). Tracking without perceiving: A dissociation between eye movements and motion perception. *Psychological Science, 22*, 216–225. doi:10.1177/0956797610394659.

Srinivasan, R. J., & Petrovic, S. (2006). MEG phase follows conscious perception during binocular rivalry induced by visual stream segregation. *Cerebral Cortex, 16*, 597–608. doi:10.1093/cercor/bhj016.

Stanley, J., Carter, O., & Forte, J. (2011). Color and luminance influence, but can not explain, binocular rivalry onset bias. *PLoS One, 6*, e18978. doi:10.1371/journal.pone. 0018978.

Stein, T., Hebart, M. N., & Sterzer, P. (2011). Breaking continuous flash suppression: A new measure of unconscious processing during interocular suppression? *Frontiers in Human Neuroscience, 5*, 1–17. doi:10.3389/fnhum.2011. 00167.

Stein, T., Senju, A., Peelen, M. V., & Sterzer, P. (2011). Eye contact facilitates awareness of faces during interocular suppression. *Cognition, 119*, 307–311. doi:10.1016/j.cognition. 2011.01.008.

Stein, T., & Sterzer, P. (2011). High-level face shape adaptation depends on visual awareness: Evidence from continuous flash suppression. *Journal of Vision, 11*, 1–14. doi:10.1167/ 11.8.5.

Sterzer, P., Haynes, J. D., & Rees, G. (2008). Fine-scale activity patterns in high-level visual areas encode the category of invisible objects. *Journal of Vision, 8*, 1–12. doi:10.1167/ 8.15.10.

Sterzer, P., Jalkanen, L., & Rees, G. (2009). Electromagnetic responses to invisible face stimuli during binocular suppression. *NeuroImage, 46*, 803–808. doi:10.1016/j.neuroimage. 2009.02.046.

Sterzer, P., & Rees, G. (2008). A neural basis for percept stabilization in binocular rivalry. *Journal of Cognitive Neuroscience, 20*, 389–399. doi:10.1162/jocn.2008.20039.

Stuit, S. M., Paffen, C. L. E., van der Smagt, M. J., & Verstraten, F. A. J. (2011). Suppressed images selectively affect the dominant percept during binocular rivalry. *Journal of Vision, 11*, 1–11. doi:10.1167/11.10.7.

Stuit, S. M., Verstraten, F. A. J., & Paffen, C. L. E. (2010). Saliency in a suppressed image affects the spatial origin of perceptual alternations during binocular rivalry. *Vision Research, 50*, 1913–1921. doi:10.1016/j.visres.2010.06.014.

Su, Y., He, Z. J., & Ooi, T. L. (2009). Coexistence of binocular integration and suppression determined by surface border information. *Proceedings of the National Academy of Sciences of the United States of America, 106*, 15990–15995. doi:10.1073/ pnas.0903697106.

Sundareswara, R., & Schrater, P. R. (2008). Perceptual multistability predicted by search model for Bayesian decisions. *Journal of Vision, 8*, 1–19. doi:10.1167/8.5.12.

Theodoni, P., Panagiotaropoulos, T. I., Kapoor, V., Logothetis, N. K., & Deco, G. (2011). Cortical microcircuit dynamics mediating binocular rivalry: The role of adaptation in inhibition. *Frontiers in Human Neuroscience, 5*, 1–19. doi:10.3389/ fnhum.2011.00145.

Tong, F., Meng, M., & Blake, R. (2006). Neural bases of binocular rivalry. *Trends in Cognitive Sciences, 10*, 502–511. doi:10.1016/j.tics.2006.09.003.

Tsuchiya, N., & Koch, C. (2005). Continuous flash suppression reduces negative afterimages. *Nature Neuroscience, 8*, 1096–1101. doi:10.1038/nn1500.

Tsuchiya, N., Koch, C., Gilroy, L. A., & Blake, R. (2006). Depth of interocular suppression associated with continuous flash suppression, flash suppression, and binocular rivalry. *Journal of Vision, 6*, 1068–1078. doi:10.1167/6.10.6.

Tsuchiya, N., Moradi, F., Felsen, C., Yamazaki, M., & Adolphs, R. (2009). Intact rapid detection of fearful faces in the absence of the amygdala. *Nature Neuroscience, 12*, 1224–1225. doi:10.1038/nn.2380.

Turvey, M. T. (1973). On peripheral and central processes in vision: Inferences from an information-processing analysis of masking with patterned stimuli. *Psychological Review, 80*, 1–52. doi:10.1037/h0033872.

Ukai, K., Ando, H., & Kuze, J. (2003). Binocular rivalry alternation rate declines with age. *Perceptual and Motor Skills, 97*, 393–397. doi:10.2466/PMS.97.5.393-397.

van Boxtel, J. J. A., Alais, D., & van Ee, R. (2008). Retinotopic and non-retinotopic stimulus encoding in binocular rivalry and the involvement of feedback. *Journal of Vision, 8*, 1–10. doi:10.1167/8.5.17.

van Boxtel, J. J. A., Alais, D., Erkelens, C. J., & van Ee, R. (2008). The role of temporally coarse form processing during binocular rivalry. *PLoS One, 3,* e1429. doi:10.1371/journal.pone.0001429.

van Boxtel, J. J. A., Knapen, T., Erkelens, C. J., & van Ee, R. (2008c). Removal of monocular interactions equates rivalry behavior for monocular, binocular, and stimulus rivalries. *Journal of Vision, 8,* 1–17. doi:10.1167/8.15.13.

van Boxtel, J. J. A., & Koch, C. (2012). Visual rivalry without spatial conflict. *Psychological Science, 23,* 410–418. doi:10.1177/0956797611424165.

van Boxtel, J. J. A., van Ee, R., & Erkelens, C. J. (2007). Dichoptic masking and binocular rivalry share common perceptual dynamics. *Journal of Vision, 7,* 1–11. doi:10.1167/7.14.3.

van Dam, L. C. J., & van Ee, R. (2006). Retinal image shifts, but not eye movements per se, cause alternations in awareness during binocular rivalry. *Journal of Vision, 6,* 1172–1179. doi:10.1167/6.11.3.

van Ee, R. (2009). Stochastic variations in sensory awareness are driven by noisy neuronal adaptation: Evidence from serial correlations in perceptual bistability. *Journal of the Optical Society of America. A, Optics, Image Science, and Vision, 26,* 2612–2622. doi:10.1364/JOSAA.26.002612.

van Ee, R. (2011). Percept-switch nucleation in binocular rivalry reveals local adaptation characteristics of early visual processing. *Journal of Vision, 11,* 1–12. doi:10.1167/11.2.13.

van Ee, R., van Boxtel, J. J. A., Parker, A. L., & Alais, D. (2009). Multisensory congruency as a mechanism for attentional control over perceptual selection. *Journal of Neuroscience, 29,* 11641–11649. doi:10.1523/JNEUROSCI.0873-09.2009.

Wade, N. J., & de Weert, C. M. M. (1986). Aftereffects in binocular rivalry. *Perception, 15,* 419–434. doi:10.1068/p150419.

Walker, P. (1978). Binocular rivalry: Central or peripheral selective processes? *Psychological Bulletin, 85,* 376–389. doi:10.1037/0033-2909.85.2.376.

Walker, P., & Powell, D. J. (1979). The sensitivity of binocular rivalry to changes in the nondominant stimulus. *Vision Research, 19,* 247–249. doi:10.1016/0042-6989(79)90169-X.

Watanabe, M., Cheng, K., Murayama, Y., Ueno, K., Asamizuya, T., Tanaka, K., et al. (2011). Attention but not awareness modulates the BOLD signal in the human V1 during binocular suppression. *Science, 334,* 829–831. doi:10.1126/science.1203161.

Watson, T. L., Pearson, J., & Clifford, C. W. G. (2004). Perceptual grouping of biological motion promotes binocular rivalry. *Current Biology, 14,* 1670–1674. doi:10.1016/j.cub.2004.08.064.

Wei, M., & Sun, F. (1998). The alternation of optokinetic responses driven by moving stimuli in humans. *Brain Research, 813,* 406–410. doi:10.1016/S0006-8993(98)01046-4.

Wilcke, J. C., O'Shea, R. P., & Watts, R. (2009). Frontoparietal activity and its structural connectivity in binocular rivalry. *Brain Research, 1305,* 96–107. doi:10.1016/j.brainres.2009.09.080.

Wilke, M., Logothetis, N. K., & Leopold, D. A. (2003). Generalized flash suppression of salient visual targets. *Neuron, 39,* 1043–1052. doi:10.1016/j.neuron.2003.08.003.

Wilke, M., Logothetis, N. K., & Leopold, D. A. (2006). Local field potential reflects perceptual suppression in monkey visual cortex. *Proceedings of the National Academy of Sciences of the United States of America, 103,* 17507–17512. doi:10.1073/pnas.0604673103.

Williams, M. A., Morris, A. P., McGlone, F., Abbott, D. F., & Mattingley, J. B. (2004). Amygdala responses to fearful and happy facial expressions under conditions of binocular suppression. *Journal of Neuroscience, 24,* 2898–2904. doi:10.1523/JNEUROSCI.4977-03.2004.

Wilson, H. R. (2003). Computational evidence for a rivalry hierarchy in vision. *Proceedings of the National Academy of Sciences of the United States of America, 100,* 14499–14503. doi:10.1073/pnas.2333622100.

Wilson, H. R. (2007). Minimal physiological conditions for binocular rivalry and rivalry memory. *Vision Research, 47,* 2741–2750. doi:10.1016/j.visres.2007.07.007.

Wilson, H. R., Blake, R., & Lee, S. H. (2001). Dynamics of travelling waves in visual perception. *Nature, 412,* 907–910. doi:10.1038/35091066.

Wolf, M., & Hochstein, S. (2011). High-level binocular rivalry effects. *Frontiers in Human Neuroscience, 5,* 1–9. doi:10.3389/fnhum.2011.00129.

Wolfe, J. M. (1984). Reversing ocular dominance and suppression in a single flash. *Vision Research, 24,* 471–478. doi:10.1016/0042-6989(84)90044-0.

Wunderlich, K., Schneider, K. A., & Kastner, S. (2005). Neural correlates of binocular rivalry in the human lateral geniculate nucleus. *Nature Neuroscience, 8,* 1595–1602. doi:10.1038/nn1554.

Yamada, Y., & Kawabe, T. (2011). Emotion colors time perception unconsciously. *Consciousness and Cognition, 20,* 1835–1841. doi:10.1016/j.concog.2011.06.016.

Yang, E., & Blake, R. (2012). Deconstructing continuous flash suppression. *Journal of Vision, 12,* 1–14. doi:10.1167/12.3.8.

Yang, E., Blake, R., & Zald, D. H. (2007). Fearful expressions gain preferential access to awareness during continuous flash suppression. *Emotion, 7,* 882–886. doi:10.1037/1528-3542.7.4.882.

Yang, E., Hong, S. W., & Blake, R. (2010). Adaptation aftereffects to facial expressions suppressed from visual awareness. *Journal of Vision, 10,* 1–13. doi:10.1167/10.12.24.

Yang, Y.-H., & Yeh, S. L. (2011). Accessing the meaning of invisible words. *Consciousness and Cognition, 20,* 223–233. doi:10.1016/j.concog.2010.07.005.

Zadbood, A., Lee, S. H., & Blake, R. (2011). Stimulus fractionation by interocular suppression. *Frontiers in Human Neuroscience, 5,* 1–9. doi:10.3389/fnhum.2011.00135.

Zaretskaya, N., Thielscher, A., Logothetis, N. K., & Bartels, A. (2010). Disrupting parietal function prolongs dominance durations in binocular rivalry. *Current Biology, 20,* 2106–2111. doi:10.1016/j.cub.2010.10.046.

Zhang, P., Jamison, K., Engel, S., He, B., & He, S. (2011). Binocular rivalry requires visual attention. *Neuron, 71,* 362–369. doi:10.1016/j.neuron.2011.05.035.

Zhou, G., Zhang, L., Liu, J., Yang, J., & Qu, Z. (2010). Specificity of face processing without awareness. *Consciousness and Cognition, 19,* 408–412. doi:10.1016/j.concog.2009.12.009.

Zhou, W., Jiang, Y., He, S., & Chen, D. (2010). Olfaction modulates visual perception in binocular rivalry. *Current Biology, 20,* 1356–1358. doi:10.1016/j.cub.2010.05.059.

Zhu, M., Hertle, R. W., Kim, C. H., Shi, X., & Yang, D. (2008). Effect of binocular rivalry suppression on initial ocular following responses. *Journal of Vision, 8,* 1–11. doi:10.1167/8.4.19.

IX EYE MOVEMENTS

59 Natural Eye Movements and Vision

MICHAEL B. MCCAMY, STEPHEN L. MACKNIK, AND
SUSANA MARTINEZ-CONDE

As we navigate our lives, we process a vast amount of visual information. Our visual system sifts the wheat from the chaff with surprising ease, giving higher priority to information relevant to the task at hand. Hold out your thumb forward at arm's length, and your thumbnail will subtend about 1–2° of visual angle in the horizontal plane, which is the approximate area of visual space corresponding to the highest resolution part of your retina, the fovea. Outside of this tiny region, your visual acuity is so low that you are legally blind. Aiming your foveal vision toward informative locations in a scene is thus fundamental to visual perception. Knowledge of the dynamics of eye movements, their generation mechanisms, and their perceptual and physiological impact will not only help in deciphering the neural code of vision but also inform the diagnosis and treatment of ophthalmic and neurological disease.

In this chapter, we will discuss natural eye movements in primate vision (with an emphasis on human eye movements during everyday visual tasks) and their effects on perception and neural activity. First, we introduce briefly the structure of the retina and the types of natural eye movements that primates make.

STRUCTURE OF THE RETINA

Visual processing begins inside the back of the eyeball in a thin five-layer stack of neurons called the retina. Two main classes of photoreceptors, cones and rods, transduce incoming photons into electrochemical signals. These signals are further processed by the horizontal, bipolar, and amacrine cell layers before arriving at the ganglion cell layer, the final stage of retinal processing. Ganglion cell axons extend into the brain as a bundle called the optic nerve, which is the sole output of the retina.

The fovea, consisting solely of cones, has a much higher density of photoreceptors dedicated to processing incoming information than anywhere else in the retina. There are three to four ganglion cells per foveal cone, and most foveal ganglion cells process information from only one cone, whereas there is only one ganglion cell per cone at 15–20° eccentricity, and each ganglion cell at these eccentricities integrates information from many cones (Wässle et al., 1990). Visual acuity drops off as the size of ganglion cell receptive fields grows in the peripheral retina. Eye movements in foveate animals such as humans and primates align visual information of potential interest with the fovea. To keep our foveas on the region of interest, we continuously move our eyes, using five main classes of eye movements: saccades, smooth pursuit, reflex eye movements, vergence eye movements, and fixational eye movements.

SACCADES

Saccades are fast ballistic eye movements that carry the fovea across the visual scene at high speeds (i.e., ~400°/s for a 15° saccade). Saccadic peak velocity has a strong linear relationship with saccadic magnitude; this relationship is known as the "main sequence" (see figure 59.1C). Saccades are usually conjugate in both eyes, that is, they have similar magnitudes and directions in the two eyes. Humans normally make up to four saccades per second depending on factors such as the nature of the task and stimulus (Otero-Millan et al., 2008). Saccades smear the retinal image, but the visual system suppresses perception around the time of a saccade so we do not perceive the blur. That is, the brain does not acquire new visual information during saccades, but only during the periods of fixation between saccades, where the eyes are relatively still. This *saccadic suppression* contributes to our perception of a clear and stable world in spite of retinal image motion due to eye movements. For details on these mechanisms, see chapter 66.

SMOOTH PURSUIT

Smooth pursuit is a slow, continuous, and voluntary movement of the eyes that allows primates to track a moving object in a scene (Barnes, 2008; Spering & Montagnini, 2011). Smooth pursuit movements keep the moving object inside or near the fovea and reduce the motion of the object's image on the retina, thus

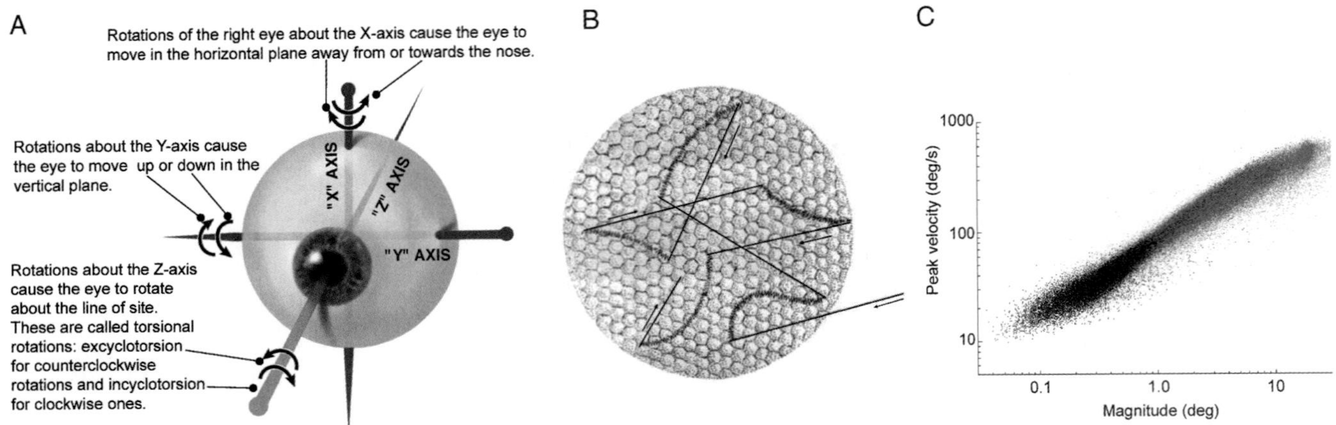

FIGURE 59.1   Eye movements. (A) Rotational axes of the right eye. Modified from Wilson-Pauwels et al. (2010). (B) Fixational eye movements, including microsaccades (straight lines), drifts (curvy lines) and tremor (zigzags superimposed on drifts), transport the visual image over the retinal photoreceptor mosaic (from Pritchard, 1961). (C) Saccades and microsaccades during free-viewing (gray) follow the same main sequence as those during attempted fixation (black). (Modified from Otero-Millan et al., 2008.)

preventing blur and perceptual impairment. During smooth pursuit, retinal and cortical mechanisms read the retinal image velocity of the object of interest and control the magnitude and timing of the smooth pursuit eye movement so that the velocity of the eye matches that of the object. If the object moves too rapidly and falls outside of the fovea, the visual system usually generates a catch-up saccade to correct the error. The onset of smooth pursuit has a latency of ~80–120 ms, depending on target and task properties such as luminance, size, speed, position, motion expectation, and the number of potential targets to be tracked. Humans can track objects with speeds up to ~100°/s, but it becomes increasingly difficult to keep up beyond 30°/s. The target's retinal image velocity controls the first ~100 ms of smooth pursuit. Afterwards, the visual system compares the retinal motion signal to an efference copy signal sent from the motor system, in order to stabilize the image of the target on the fovea.

REFLEX EYE MOVEMENTS

Two types of reflex eye movements work in concert to ensure image stabilization during self-generated head and body movements and whole or partial field movements: the vestibulo-ocular reflex (VOR) and the opto-kinetic reflex (OKR) (Fetter, 2007; Masseck & Hoffmann, 2009). If you turn your head while reading a book, you can keep reading because the VOR compensates for your head motion by rotating your eyes in the opposite direction, thus maintaining your gaze stable on the text. OKR compensates for continuous movements of the visual field (i.e., looking at a tree passing by the window of your car). During OKR movements, our eyes pursue the moving scene ("slow phase") until they no longer can and then quickly saccade in the opposite direction of the motion ("fast phase") before engaging in pursuit once again. Whereas VOR is driven by the inner ear's semicircular canals and the otolith organs that detect acceleration and changes with respect to gravity (Fetter, 2007), OKR relies solely on visual input, and it kicks in whenever the whole visual field, or a part of it, is in continuous motion (usually due to our own movement through the environment) (Tian, Zee, & Walker, 2007).

VERGENCE MOVEMENTS

When we shift our gaze between objects at different distances (or follow an object which changes its distance from us), we make slow vergence movements to place (or keep) the object on the foveas of both eyes. Vergence movements move the left and right eye in opposition to each other, causing the intersection point of their visual axes to move either closer (convergence) or farther away (divergence). Because vergence movements are disconjugate, many researchers believe that separate neural pathways control vergence and saccadic movements, but there is still much debate (Cullen & Van Horn, 2011; Mays, 1984; Robinson, 1968).

FIXATIONAL EYE MOVEMENTS

Our eyes are never still. We produce so called "fixational eye movements" during the fixation periods

    MICHAEL B. MCCAMY, STEPHEN L. MACKNIK, AND SUSANA MARTINEZ-CONDE

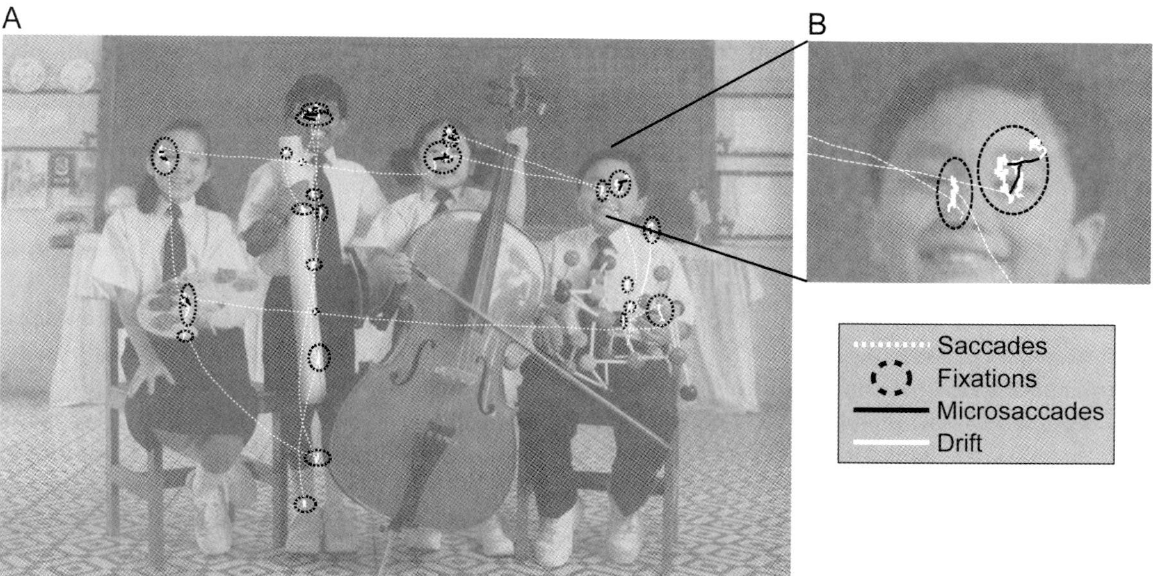

FIGURE 59.2   Eye movements during free-viewing. (A) Each ellipse represents a fixation. The dashed white lines outside of the ellipses are saccades between fixations. (B) Two fixations, one with two microsaccades (solid black lines) and the other with none. Solid white lines indicate drift. (Modified from Otero-Millan et al., 2008.)

between saccades, smooth pursuit, and reflex eye movements. Indeed, when we remove all retinal image motions due to eye movements, visual fading ensues; see the functions and effects of fixational eye movements section for further details. Most material on fixational eye movements in this chapter comes from Martinez-Conde, Macknik, and Hubel (2004), Martinez-Conde et al. (2009), and Rolfs (2009).

We spend ~80% of our free-viewing time fixating our gaze, and because vision is suppressed during saccades, most of our visual perception occurs during fixation. Primate fixational eye movements include microsaccades, drift, and tremor (see figures 59.1B and 59.2). Microsaccades are the fastest and largest of the three and have equivalent physical characteristics to saccades (see figure 59.1C) on a smaller scale (microsaccades are typically <1°). Converging behavioral and physiological evidence indicates that microsaccades and saccades share the same oculomotor circuitry.

Microsaccades occur at a rate of one to two per second during prolonged fixation. During free-viewing of natural scenes, microsaccades occur at a rate of ~0.5/s, and ~14% of fixations contain a microsaccade. Microsaccades generate retinal motions large enough that we should be able to perceive them, but we do not, due to microsaccadic suppression.

Drift is a slow (<2°/s) curved motion that occurs between saccades and/or microsaccades. Drift movements are typically not conjugate and resemble random walks.

Tremor occurs simultaneously with drift and is the smallest fixational eye movement (~1 photoreceptor width, <0.5 arcmin), with a frequency of ~60 Hz. Due to its minute nature, it is difficult to measure tremor noninvasively; thus, almost nothing is known about its role in the visual system.

NATURAL EYE MOVEMENTS

All the eye movements discussed so far occur naturally in everyday vision. Because saccades and fixational eye movements are the most frequent eye movements, we will now focus on their proposed functions, their neural and perceptual effects, the cognitive and perceptual factors that influence them, and their dynamics during ecologically valid (i.e., real-world) tasks.

MEASURING EYE MOVEMENTS

The eye is a sphere that rotates horizontally (about the X-axis in figure 59.1A), vertically (about the Y-axis in figure 59.1A), and about the line of sight (about the Z-axis in figure 59.1A, also called torsional rotation). Eye tracking devices report eye positions in two main ways: (1) point-of-gaze measurements, which describe where the observer is looking, and (2) the degree of visual angle that the horizontal, vertical, and in some cases torsional positions deviate from an arbitrary reference position. Some of the most important considerations when choosing an eye tracking device are (1) the amount of *physical contact* required between the device and the subject (including the amount of visual interference caused by the device), (2) the *physical restraint* required of the subject (head/body restrained vs. unrestrained), (3) the *accuracy* of the device (the expected difference between measured and actual eye position), (4) the *precision* of the device (the reproducibility of measurements), (5) the *resolution* of the device (the smallest eye movement it can measure), and (6) the *range* of the device (the span of eye positions it can measure). Some methods require a lot of physical contact/restraint but can reliably measure very small eye movements, as well as the subject's point of gaze. Others require minimal physical contact/restraint but cannot measure the subject's point of gaze or eye movements smaller than 1°. Also, because some eye movements are mostly conjugate (saccades, microsaccades, and smooth pursuit) and others are not (drift), it is important to assess the experimental need for binocular versus monocular recordings. The three main eye tracking techniques in use today are contact lens methods (coil, mirror), electro-oculography (EOG), and optical/feature recognition methods; see table 59.1. Most of the information describing the various eye tracking methods in the remainder of this section comes from Borah (2006), Eggert (2007), and Young and Sheena (1975).

Contact Lens Methods

The scleral search coil (SSC) method, currently the only contact lens method in use, is considered the gold standard for eye tracking as far as precision and accuracy are concerned. Its main disadvantage is its invasiveness, as it involves a topical anesthetic and a tight fitting contact lens on the subject's eye (SSC are usually surgically implanted in animal models). Human subjects are typically comfortable with the contact lens for only 20–30 min, thus constraining experimental time. The contact lens is a silicon annulus containing coils of thin copper wire (one coil for horizontal and vertical movements and an optional second coil for torsional movements). The subject sits inside an area that contains constant-amplitude, sinusoidally varying (AC) magnetic fields that induce electric current in the coils; the amount of current depends on the angle of the coil relative to the direction of its magnetic field. A thin wire that leaves the corner of the eye sends the induced current signal; this wire may cause additional discomfort to the subject. The subject's head and chin are usually restrained, but no device obscures the field of view. Another coil and magnet may track the head

position, in which case the subject's head can be unrestrained but must stay inside a limited area where the magnetic field strength is close to uniform. The SSC method is not affected by blinks, and it measures horizontal, vertical, and torsional eye movements with accuracies of ~0.2°, precisions of ~0.25 arcmin, resolutions of ~0.5–1 arcmin in each dimension (Robinson, 1963), and a range of ~30° in the horizontal and vertical directions (Remmel, 1984); thus, it can measure microsaccades and drift reliably. The SSC method can also measure point of gaze. Tremor, with a magnitude of <0.5 arcmin (Heckenmueller, 1965), cannot be distinguished from noise, however. Some negative (albeit temporary) consequences of wearing the coil are potential drying of the eye from inadequate blinking, deformations of the cornea, and reduced visual acuity.

An earlier contact lens method, called the "optical lever" method, involved imbedding one or two reflecting mirrors in the contact lens, shining a light source on them, and recording the reflected light, which led to a signal about the orientation of the eye (Ratliff & Riggs, 1950). The optical lever method was more precise than the SSC and allowed for measurement of ocular tremor, but it has gone out of use due to its complexity and invasiveness, in addition to its reduced range (~5 degrees horizontal and vertical).

Electro-oculography

The difference in electric potential between the cornea and the retina creates an electrostatic dipole that rotates with the eyeball. EOG methods use the position of the dipole (detected via skin electrodes placed around the eye) to determine the orientation of the eye in the head. EOG does not require head or chin restraints, is not affected by eye closure, does not obstruct the visual field, and is inexpensive. The accuracy of EOG is ~3–7° with a resolution of ~0.5°. EOG has a range of ~70°, but linearity worsens progressively with movements larger than 25–30°, especially in the vertical axis. It reliably measures larger saccades, smooth pursuit, and nystagmus (a biphasic ocular oscillation alternating a slow smooth pursuit in one direction and a fast saccadic movement in the other direction), but not fixational eye movements, point of gaze, or torsional eye movements. EOG's main problem is its high susceptibility to many types of noise, such as muscle potentials and electrocardiogram activity (Haslwanter & Clarke, 2010). It is typically used only in sleep and infant studies, due to its noninvasiveness and ability to measure eye movements with the subject's eyes closed (Haslwanter & Clarke, 2010).

    MICHAEL B. MCCAMY, STEPHEN L. MACKNIK, AND SUSANA MARTINEZ-CONDE

The human eye has distinct landmarks and optical reflections that one can use to determine its orientation. The main landmarks and reflections used in contemporary eye tracking are the pupil, iris, limbus (boundary between the colored iris and white sclera), the first Purkinje image (reflection from the outer surface of the cornea, also called the corneal reflex), and the fourth Purkinje image (reflection from the inner surface of the lens). We describe the two main methods for optical/feature recognition eye tracking below (though there are many other kinds).

DUAL PURKINJE IMAGE TRACKERS The dual Purkinje image (DPI) method uses the fourth Purkinje image and the corneal reflex to track the eye. The device has no contact with the human subject, but it requires a chin rest and headrest. For higher precision, a bite bar is typically used. The DPI method uses infrared lighting so it does not interfere with normal vision (Cornsweet & Crane, 1973), but the optics need to be fairly close to the eye and thus can obstruct the field of view. The DPI method accounts for small head movements of ~0.5 cm in all three axes, has an accuracy of ~1 arcmin, a typical resolution of <1 arcmin, and a range of 10–20° (Cornsweet & Crane, 1973). It can measure saccades, smooth pursuit, nystagmus, microsaccades, drift, and point of gaze, and it allows for stabilization of the image on the retina, but it cannot measure torsional movements. Although DPI is the most accurate and precise method from this group, it is rarely used due to its elaborate setup (Haslwanter & Clarke, 2010).

VIDEO-OCULOGRAPHY Video-oculography (VOG) methods involve using a camera to digitally film the eye. Landmarks, sometimes combined with optical reflections from the eye, are used to determine the eye's orientation. The most common VOG method uses the pupil with or without the corneal reflex. The camera for recording the eye is either remote (i.e., not attached to the subject) or attached to the subject with a headband. When tracking the pupil only, one must stabilize the subject's head or the subject must use a headband; otherwise, the system will mistake head motions for eye movements. If the corneal reflex is also tracked, then the camera can be remote with the head stabilized or not (depending on the desired resolution/accuracy/ precision). Headbands can produce discomfort and slipping errors, often limiting the length of the experiments to around 30 min. To obtain eye position, image processing algorithms are applied to the images to determine the center of the pupil (and the corneal reflex if tracked). VOG systems use infrared illumination generally, so they do not interfere with normal vision. The cameras, headrests, and chin rests may obstruct the field of view with some VOG devices, however. There are a wide range of VOG devices with large differences in quality. Typical accuracies are ~0.25°–1°, with resolutions between ~0.05–0.2°, and ranges of ~30–50°. VOG systems can measure point of gaze, saccades, smooth pursuit, nystagmus, and eyelid blinks. Some systems can reliably measure microsaccades, such as the Eyelink 1000 with the head and chin support, which has an accuracy of 0.25–0.5° and a resolution of 0.05°. Some systems, such as the Chronos CE-MDD, can record torsional eye movements by tracking natural markers in the iris.

Future of Eye Tracking

The ability to measure eye movements less invasively with more resolution, accuracy, and precision has increased dramatically due to advances in sensor, computer, and image processing technology. Most researchers currently use VOG systems because they are easy to set up, require little if any physical restraint/contact, and can measure most eye movements. The increased reliability and decreased invasiveness of contemporary eye tracking methods have propelled eye movements research (see Martinez-Conde et al., 2004; Rayner, 1998, 2009). Today, tremor is the only eye movement type that noninvasive methods cannot measure reliably. Devices capable of measuring tremor must have a minimum resolution of 0.45 arcsec (Ryle et al., 2009); one example is a piezoelectric probe in contact with the eye, thus requiring local anesthesia and taping of the eyelid. Research on tremor has thus been limited. New noncontact methods to measure tremor are in the developmental phase and not ready for widespread use. As technology continues to improve, we expect to see higher accuracy, precision, and resolution in noninvasive eye tracking with little or no physical restraint.

SCENE PERCEPTION

We will discuss scene perception in the context of the viewing task performed, such as free viewing (i.e., visual exploration) of a scene with no particular goal, visually searching for objects in a scene, or memorizing a scene.

In scene perception, observers obtain information mostly during fixation. To process a visual scene, viewers make fixations of at least 150 ms (Rayner et al., 2009). Saccade rates (two to four saccades per second on average) and fixation durations (~234–300 ms) during

	Scleral search coil	Electro-oculography	Dual Purkinje image	Video-oculography
Physical contact	Contact lens on one eye with anesthetic. Wire exits eye.	Electrodes around eyes.	No contact.	Headband or no contact required.
View obstruction	No obstruction.	No obstruction.	Device obstructs view.	Little or no obstruction of view with headband. No obstruction without headband.
Physical restraint	Usually head and chin rest.	None.	Head and chin rest (bite bar for higher precision).	Usually head and/or chin rest but can be unrestrained sacrificing accuracy and precision.
Accuracy	~0.2°	~3–7°	~1 arcmin	~0.25–1°
Precision	~0.25 arcmin		~1 arcmin	
Resolution	~0.5–1 arcmin	~0.5°	~1 arcmin	~0.05–0.2°
Range	~30°	~70°	~10–20°	~25–50°
Bandwidth	~200 Hz	~100 Hz	~400 Hz	Up to 400 Hz (device dependent)
Eye movements measured (all methods record horizontal and vertical rotations)	All eye movements except for tremor. Torsion and point-of-gaze measurements possible.	Large saccades, smooth pursuit, and nystagmus. Torsion and point-of-gaze measurements not possible.	All eye movements except for tremor. Torsion measurements not possible. Point of gaze is possible.	Device dependent. Torsion and point-of-gaze measurements possible.
Ability to measure eye movements with eye closed	Yes, but not for long periods.	Yes, even during sleep.	No.	No.

scene perception vary with task, viewer, and scene characteristics. Eye movements during scene perception are subject to exogenous, or bottom-up (i.e., stimulus-based) influences and endogenous, or top-down (i.e., from within the brain) influences.

Gist

Observers obtain the "gist" of a scene, including information about objects, scene schema, and even some semantic information, during the first 40–200 ms of the first fixation (Biederman, 1981; Fei-Fei et al., 2007), acquiring sensory or feature-level information such as shading and shape before semantic-level information. Castelhano and Henderson (2007) propose that observers generate an abstract (i.e., size invariant) visual representation in the initial glimpse and retain this representation in memory to guide subsequent eye movements for more detailed/local analysis. This proposal is compatible with studies suggesting that visual information processing proceeds in a global-to-local manner (Navon, 1977; Tatler & Vincent, 2008). After the observer obtains the gist of a scene, both bottom-up and top-down factors guide eye movements.

Bottom-Up Influences on Eye Movements

Visual stimulus properties have a large influence on saccades and fixations. For instance, stimulus size controls saccadic amplitude (Wartburg et al., 2007), and some objects, such as faces, receive long and frequent fixations (Otero-Millan et al., 2008). Certain statistical properties of the scene, such as curved lines and edges, occlusions, isolated spots, high spatial contrast, and regions of uncorrelated intensities also correlate with fixation locations (Tatler et al., 2011).

The study of "saliency maps" is a popular approach to understanding gaze control during scene perception ("salient" here means that it stands out conspicuously). Saliency map models are based on the hypothesis of a

   MICHAEL B. MCCAMY, STEPHEN L. MACKNIK, AND SUSANA MARTINEZ-CONDE

neural representation (which may be distributed throughout the brain) of a scene that encodes the saliency of a given location in that scene (Itti & Koch, 2001). The visual properties of an image, combined with knowledge about features detected by the early visual system (color opponency in the retina, orientation specificity in V1, etc.), are the basis for several feature maps that specify where local features are different from their surround (i.e., more salient). For instance, a lone red apple on a green tree will be very salient in the red–green color opponency feature map. Similarly, a lone vertical bar surrounded by many horizontal bars will be salient in the orientation feature map. Multiple feature maps are combined into a single saliency map that represents the salience of any region in the scene. In Itti and Koch's model, a winner-take-all approach determines where an observer will look, that is, the observer will fixate the most salient region. To move from one location to the next, the current fixation region is inhibited (i.e., the salience is set to zero for some time; see the paragraph about inhibition of return [IOR] below), and the winner-take-all approach determines again the next fixation location. Several studies have proposed a distributed neural representation of the saliency map across several brain locations, such as the lateral intraparietal sulcus of the posterior parietal cortex, the frontal eye fields, the inferior and lateral subdivisions of the pulvinar and the superior colliculus, and area V1, but there is no conclusive proof about its existence or location (Itti & Koch, 2001).

Saliency map models predict fixation locations better than chance, suggesting that salience at least partly drives gaze control (Foulsham & Underwood, 2008; Itti, 2005) and visual search (Itti & Koch, 2000; Nothdurft, 2006). Evidence against the importance of saliency in gaze control is that scan paths predicted by saliency map models differ from actual scan paths (Foulsham & Underwood, 2008), that eye movement biases (i.e., the fact that fixations are biased toward the center of the screen) predict fixation locations better than image salience does (Tatler & Vincent, 2009), and that salience has no effect when searching for a target defined by category or exemplar (Foulsham & Underwood, 2007).

Thus, the evidence for how much saliency influences gaze control is mixed. Stimulus-driven saliency map models do not incorporate the observer's past experience, knowledge, or expectations in predicting eye movements (Itti & Koch, 2000), but other saliency models include some form of top-down modulation (Tsotsos et al., 1995; Wolfe, 1994). Overall, the published literature suggests that both top-down and bottom-up factors influence gaze control, with top-down influences playing the more dominant role (Tatler et al., 2011).

Top-Down Influences on Eye Movements

Top-down influences on eye movements include (Henderson, 2003) the following: (1) *episodic scene knowledge*, that is, the knowledge about a specific scene that one can learn over the short term in the current perceptual encounter or over the longer term across multiple encounters—for instance, a viewer will tend to look at an empty scene region if that region contained a task-relevant object previously; (2) *scene-schema knowledge*, which refers to the generic semantic and spatial knowledge about a particular type of scene, including information about objects likely to be found in a specific category of scene (i.e., bathrooms contain toilets) and spatial regularities associated with a scene category (i.e., car passengers do not ride on the hood); scene-schema knowledge can limit initial fixations to scene regions likely to contain an object relevant to the current task; (3) *task-related knowledge*, which can involve a strategy relevant to a given task, such as periodically checking the inbox of your e-mail to see if new mail has arrived; task-related knowledge exerts a profound influence on gaze control (Ballard & Hayhoe, 2009; Tatler et al., 2011); during task performance, essentially all the fixations fall on task-relevant objects, whereas pretask fixations fall on task-relevant and -irrelevant objects about equally (Tatler et al., 2011). Reward also influences gaze control during scene perception (Hayhoe & Ballard, 2005; Tatler et al., 2011).

Saccadic latencies to recently attended locations are longer than latencies to locations not yet attended (a phenomenon known as inhibition of return [IOR]) (Klein & MacInnes, 1999). IOR may facilitate visual search by inhibiting orienting to previously examined locations (Klein & MacInnes, 1999, but see Tatler & Vincent, 2008). Many saliency map models use IOR as a method to not return immediately to the most salient region in an image for some time after it has been fixated (Itti & Koch, 2001).

Challenges in Scene Perception Studies

Is looking at a picture on a monitor representative of how we perceive scenes in the real world? (Henderson, 2003; Tatler et al., 2011). Two areas of concern are the potential biases introduced by the framing of the scene and the effects of sudden scene onsets. Physical differences between framed static scenes and real environments include the reduced dynamic range and limited field of view of a photograph in comparison to a real scene, the absence of motion cues and many depth cues in photographs, and the fixed viewpoint of the observer in a still image, which is defined by the photographer's viewpoint and typically reflects compositional biases. Also, observers tend to fixate the center of a monitor irrespective of visual content (Tatler et al., 2011).

Further, scene perception studies typically involve the sudden onset of an image, which may influence both eye movement dynamics and the response of neurons involved in target selection (Tatler et al., 2011). Long-duration trials can ameliorate the effects of sudden stimulus onsets on eye movements and neural responses, however, because the onset's impact is limited in time to a small portion of the trial, which can be discarded later in the data analysis (Otero-Millan, Macknik, & Martinez-Conde, 2012).

Other issues include the interference with normal vision caused by eye tracking devices, the use of unnatural stimuli (i.e., "artificial" scenes" as opposed to "real-world" scenes), and task relevance to real life. Also, many scene perception studies are conducted in a dark room. These conditions all contribute to what may be considered unnatural viewing scenarios.

ECOLOGICAL TASKS

Eye movements are intrinsically involved in ecologically valid real-life actions such as making a sandwich, baseball playing, or driving. Unlike standard scene perception research, the studies below require coordinated movements of the hands, body, and eyes.

Driving

Driving is an extremely complex task that involves integration of visual information from many sources, including other drivers, road signs, and pedestrians. Drivers fixate on the tangent point on the inside of each curve ~1–2 s before each bend and refixate it throughout the bend. The tangent point is the moving point on the inside of the bend where the driver's line of sight is tangent to the road edge. It is the point that protrudes furthest into the road and is thus highly visible (Land, 2006; Land & Lee, 1994). Loon et al. (2010) investigated how drivers avoid collisions by recording their eye movements as they made timing judgments about approaching vehicles at an intersection. Drivers made saccadic eye movements back and forth between the approaching car and the road ahead. They spent most of the time fixating the approaching car (37%), followed by the side line of the road on the side of the approaching vehicle (21%). The strategy was independent of the trajectory of the observer or the angle of

the approaching vehicle. Novice drivers have a reduced functional field of view compared to experienced drivers (Crundall, 2005; Di Stasi et al., 2011; Land, 2006; Mourant & Rockwell, 1970; Underwood, 2007).

Domestic Tasks

Land (2009) has drawn two main conclusions from studies that tracked subjects' eyes as they made a cup of tea (Land et al., 1999) or a peanut butter and jelly sandwich (Hayhoe, 2000; Hayhoe et al., 2003): (1) Subjects made saccades almost exclusively to objects involved in the task, even though there were many other objects around; thus, bottom-up salience has little relevance to these kinds of tasks; and (2) the eyes handled one object at a time, corresponding roughly to the duration of the manipulation of the object (which may have involved a number of fixations on different parts of the object).

Land et al. (1999) described four categories of fixations. *Locating fixations* established the locations of objects without accompanying motor activity. These may be used later to quickly locate necessary items. *Directing fixations* accompanied a hand movement to contact a fixated object or to move the object to a fixated new set-down position. This usually required a single fixation, and the eyes moved away from the object or set-down point right before the hand arrived. The main function of directing fixations may be to provide fovea-centered goal-position information to guide the arm. *Guiding fixations* occurred during manipulations of more than one object—for example, a kettle and its lid—where the objects had to be guided relative to each other to interact in an appropriate way. These actions usually involved a number of fixations that alternated between the objects. *Checking fixations* determined when certain conditions were met. In these cases, the eyes might dwell on some region of an object, either in one long fixation or in a series of repeated fixations. Once the condition was met, a new action started. For example, when the kettle was full, the tap was turned off. Subjects rarely fixated their hands, or objects acquired by their hands, during sequences of this kind, indicating that vision is a scarce and valuable resource which disengages from action as soon as another sense can take over (Land, 2006).

Baseball and Cricket

In baseball and cricket, a ball travels from its delivery point to the batter in ~400–500 ms (Land, 2009). In this time, the batter, who must estimate the trajectory of the ball and then hit it, has time for only one, maybe two, saccades (Land, 2006). Angular velocities of the ball reach 500°/s whereas humans cannot usually use smooth pursuit to track targets moving faster than 70°/s (Bahill & LaRitz, 1984; Schalén, 1980). Furthermore, the accuracy required to hit the ball is a few centimeters in space and a few milliseconds in time (Regan, 1992). Hitting a baseball at 90 mph is thus a very complex task that pushes human ability to its limit. Bahill and LaRitz (1984) tracked the eyes of batters facing a simulated fastball and found that they used smooth pursuit to track the ball to about 9 ft in front of them, after which the angular speed of the ball was too fast to track. A professional batter, who had faster smooth pursuit than any reported in the literature, tracked the ball to about 5.5 ft in front of him. The researchers proposed that it is this extraordinary ability, combined with the capacity to suppress VOR and the occasional use of an anticipatory saccade, that makes a good batter. Expert and novice batters also have different fixation strategies. Experts fixate around the pitcher's shoulder–trunk region to pay attention to the entire body of the pitcher, relying on their peripheral vision to efficiently evaluate the pitcher's motion and anticipate the ball's trajectory, whereas novices scatter their fixations throughout the pitcher's head and body and cannot pick up relevant information for batting (Takaakikato & Fukuda, 2002).

Cricket is similar to baseball, except that the ball bounces in front of the batter before it is struck. Cricket batters keep their gaze stationary for a period after delivery, as the ball drops into their field of view. Then, they make an anticipatory saccade to bring the fovea below the ball, close to the point where the ball will subsequently bounce. Good batters make the first anticipatory saccade with shorter latency than poor batters (Land & McLeod, 2000).

Magic

Magic has recently emerged as an interesting tool in neuroscience research. Magic shows are a manifestation of magicians' deep intuition for and understanding of human perception, attention, and awareness. By studying the techniques of magicians, neuroscientists can learn powerful methods to manipulate perceptual and cognitive processes in the laboratory. Most of the information in this section comes from Macknik et al. (2008).

Magicians use many tools to misdirect the audience's attention away from the method (i.e., the secret behind the effect) and toward the magical effect. In overt misdirection, the magician redirects the spectator's *gaze* away from the method, whereas in covert misdirection, the magician redirects the spectator's *attention* without redirecting his or her gaze. Magicians may manipulate the attention and/or gaze of an audience by controlling an object's salience via bottom-up and/or top-down mechanisms. One way a magician might control bottom-up attention is by suddenly producing a flying dove. The spectators' gaze and attention will follow the dove's flight, which will give the magician a few unattended moments to conduct a secret maneuver. A magician can also increase the top-down salience of an object by actively directing attention to it. For instance, the magician may ask a volunteer to perform a task with one specific object onstage, so that the audience will miss changes occurring to a second object.

Magicians also use social cues to redirect gaze: A magician's axiom is that the audience will look where the magician is looking. Not all magic illusions benefit from social misdirection, however. Different magic tricks may be enhanced, lessened, or unchanged by social misdirection, in ways that remain to be explored (Cui et al., 2011).

Magicians can redirect gaze in additional ways. Otero-Millan et al. (2011a) found that a curved motion trajectory of the magician's hand (i.e., in a classical coin vanish) generated stronger misdirection than rectilinear motion, and that the two types of motion trajectories led to differential engagement of the smooth pursuit and the saccadic systems.

THE FUNCTIONS AND EFFECTS OF FIXATIONAL EYE MOVEMENTS

Saccades bring the fovea from one region of interest to another. The beginning of a saccade thus terminates one fixation, and the end of the saccade starts a new fixation. Recall that during each fixation, our eyes are never still but produce three types of fixational eye movements: microsaccades, drift, and tremor. Here we will discuss the proposed functions of fixational eye movements, their perceptual and physiological effects, and the factors that influence their dynamics. Most of the following comes from Martinez-Conde et al. (2004, 2009) and Rolfs (2009).

In the early 1950s, researchers showed that when all eye movements were eliminated, visual perception faded to a homogeneous field. This suggested that fixational eye movements are necessary to prevent and restore loss of vision during fixation, and it led to numerous studies during the 1950s and 1960s. In the late 1970s, the incipient fixational eye movements field arrived at an impasse due to difficulties in data collection, discrepancies between results from different laboratories, and disagreements over interpretation of the data. Kowler and

Steinman (1980) concluded that microsaccades were a laboratory artifact: that is, that microsaccades did not occur in natural viewing conditions but resulted from artificial conditions in which subjects were forced to hold their gaze for very long periods of time while their head was restrained (for instance, with a bite bar). Largely due to this conclusion, research on microsaccades essentially died in the 1980s. In the late 1990s and early 2000s, we and other researchers found that microsaccades modulate the activity of visual neurons and proposed that they have a perceptual impact as well. These findings, combined with technological advances in eye movement recording techniques, computing power, data analysis, and computational modeling, revived and propelled research on fixational eye movements in the 2000s. Contemporary consensus is that fixational eye movements, including microsaccades, are important to natural vision.

Conflicting Results on Microsaccades

Microsaccade research has been polarized, with some researchers concluding that microsaccades serve no function (especially in earlier studies) and others asserting that microsaccades are critical to perception. The discrepancy is likely due in part to methodological differences between early and contemporary studies—for instance, the current standard use of objective algorithms (which are adaptive to noise and base microsaccade characterization on parameters derived from the distribution of involuntary saccades during attempted fixation) versus the earlier reliance on arbitrary magnitude or velocity thresholds, in which microsaccades were typically picked by hand by experts, posing the potential difficulty of replication by other groups. It is worth noting that studies conducted in the 1970s often defined microsaccades arbitrarily as saccades with magnitudes of <12 arcmin. This very stringent parameter is well below the average magnitude of microsaccades found in contemporary human and primate studies.

By definition, microsaccades are small-magnitude saccades that occur when one attempts to fixate. Humans can make voluntary saccades the same size as microsaccades, however, and so it is usually impossible to distinguish a microsaccade from a voluntary saccade, unless the subject is explicitly asked to fixate. Thus, microsaccade studies are often conducted under conditions of instructed and prolonged fixation, which may be considered unrepresentative of natural viewing tasks. One additional complication is that trained subjects can make smaller and less frequent microsaccades than inexperienced subjects. Thus, the use of highly trained fixators as subjects (usually the authors themselves) in the earlier studies could account, at least partly, for the much smaller microsaccade magnitudes in those studies than in current research.

Neural Consequences of Fixational Eye Movements

Neural responses to microsaccades have been studied in the primate lateral geniculate nucleus, area V1, and extrastriate cortex (see figure 59.3A). There is general consensus that microsaccades generate neural responses in early visual neurons by displacing their receptive fields over otherwise stationary stimuli. Some studies have also found microsaccade-driven extraretinal modulations in a minority of neurons in area V1, but the evidence is conflicting as to the sign of such modulations (inhibitory, excitatory, or both) and their timing with respect to the primary retinal responses generated by microsaccades. In every study, the apparent extraretinal responses had smaller magnitudes than the responses explained by retinal activation. Modeling studies suggest that microsaccades might significantly enhance sensitivity to edges, resharpen the image, and improve spatial resolution in retinal neurons.

Very few studies have addressed the neural activity evoked by drift. A subset of neurons in area V1 may respond to drift, but no research to date has precisely quantified the strength of these responses or how they may vary for different types of drift. Tremor's neural consequences have not been studied in the primate.

Functions and Perceptual Consequences of Fixational Eye Movements

MICROSACCADES Microsaccades can counteract (i.e., reverse) perceptual fading during fixation (see figure 59.3B). Martinez-Conde et al. (2006) found that microsaccade rates increased before a peripheral target became perceptually visible and decreased before the target faded. These findings indicated that microsaccades enhance visibility during fixation and supported earlier qualitative reports (Ditchburn & Drysdale, 1977; Gerrits & Vendrik, 1970; Riggs et al., 1953; Sharpe, 1972). More recently, McCamy et al. (2012) have reported that microsaccades are the greatest eye movement contributor to the visual restoration of faded targets, both in the periphery and in the fovea.

Microsaccade production has been linked to target visibility in several fading paradigms, such as the filling-in of artificial scotomas and motion-induced blindness, and to perceptual transitions in other visual phenomena, such as binocular rivalry and illusory motion (Otero-Millan et al., 2012).

Microsaccades are also involved in fixation correction (Otero-Millan et al., 2011b) and other oculomotor and perceptual functions, such as performance in high acuity tasks (Ko, Poletti, & Rucci, 2010) and visual scanning of small regions.

DRIFT Drift also plays a corrective role in fixation, and it can improve discrimination of high (but not low) spatial frequency stimuli. Not many studies have focused on drift, however, due to difficulties in measuring it and its persistent nature. The perceptual effects of tremor are unknown.

Stimulus and Cognitive Factors That Modulate Microsaccades

Task and stimulus modulate microsaccade rates and magnitudes. For instance, subjects make ~0.2 microsaccades per second while free-viewing a blank scene, ~0.6 microsaccades per second while free-viewing natural images, and ~1.3 microsaccades per second near identified targets in a visual search task.

Microsaccade rate abruptly drops to a minimum shortly after the presentation of a sudden visual or auditory stimulus (~100–200 ms after cue onset), followed by a period of enhancement (~300–400 ms after cue onset) and then a return to baseline (see figure 59.3C). This phenomenon, called microsaccadic inhibition, may be related to the role of microsaccades in counteracting fading; that is, a sudden visual stimulus induces transient sensory activity, thus reducing the immediate need for microsaccades to cause such transients (Hafed, 2011).

The primate superior colliculus is crucial for the normal control of selective attention, even without eye movements (Lovejoy & Krauzlis, 2010), and it is also responsible for saccade and microsaccade generation. This common neural substrate suggests that attention shifts can modulate microsaccade production (Laubrock et al., 2010; Pastukhov & Braun, 2010). Most studies have found that microsaccade directions are biased toward and/or away from the spatial location suggested by an attentional cue ~200 ms following the cue, depending on the engagement of endogenous versus exogenous attention in the different experimental tasks, but their conclusions are subject to some debate.

FINAL REMARKS

The statistics of a scene, the task at hand, and past experience can influence the dynamics of saccades and fixations. A common theme among eye movement strategies is that specific task demands dominate the control of saccades and fixations. Learning is critical to improving these strategies (i.e., experts have better eye movement strategies than novices). Saliency can also play a role in gaze control when no specific task is at hand, or when someone such as a magician successfully manipulates someone's attention. Finally, microsaccade and saccade production is higher in the presence than in the absence of visual information. Thus, the overall strategy of the visual system is to optimally sample the visual scene for the specific task at hand, under the constraints of the nervous system and the visual world.

Optimal Sampling Strategy

Vision is an extremely complex computational problem (Tsotsos, 1988, 1990), and yet the primate visual systems solves it with surprising ease and efficiency. Finely processing all of a visual scene in parallel is not computationally tractable, which may explain why the primate visual system evolved to finely process only information from a small region of the eye, the fovea, while processing peripheral information at a more abstract (i.e., filtered) level (Tsotsos, 1988). The anatomical machinery dedicated to processing foveal information is disproportionate in the cortex, taking up 8% of the anatomical matter in V1, whereas the foveal area is only 0.01% of the total retina (Azzopardi & Cowey, 1993; Talbot & Marshall, 1941). If all the information from the visual field was processed in the way it is in the fovea, the skull could not contain the anatomical matter required. Thus, instead of processing the whole visual field as it does in the fovea, the primate visual system makes saccades and fixations to finely process the regions of the visual field which successively fall on the fovea. As previously discussed, a multitude of top-down and bottom-up attentional mechanisms smartly control saccadic and fixation dynamics (i.e., the sampling strategy); for instance, the scan paths of humans while free-viewing natural images are geometrically similar to a class of random walks known as Levy flights. Levy flights are an efficient random search process which minimizes the time required to scan a scene. Levy flights are also scale invariant, which is an advantageous feature given that scale changes are common in vision. The sampling solution of making optimally controlled saccadic movements is not due to the specific mechanics/constraints of the human oculomotor system (Brockmann & Geisel, 2000). Indeed, patients unable to make eye movements make head saccades that have comparable characteristics to eye saccades in normal observers, performing complicated visuomotor tasks, such as making a cup of

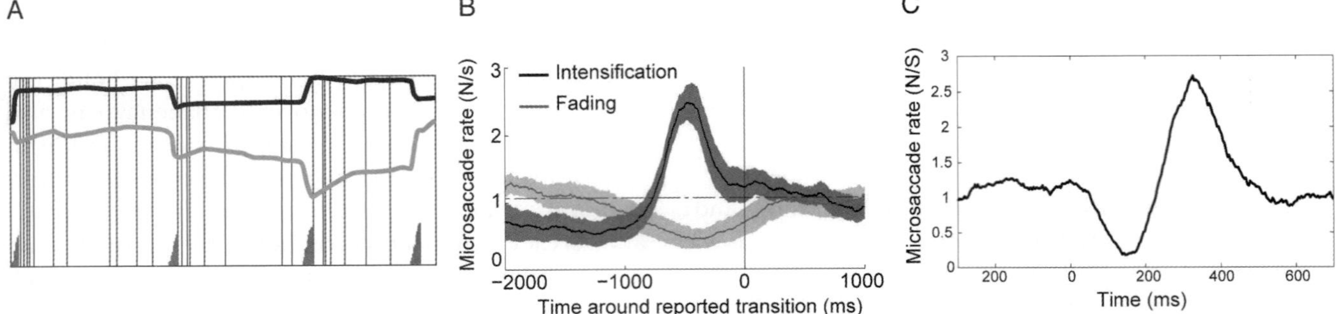

FIGURE 59.3   Microsaccade dynamics and effects. (A) Microsaccades trigger bursts of spikes in area V1. The black and gray traces represent horizontal and vertical eye positions. The triangles at the bottom indicate microsaccade occurrences (the height of the triangles represents microsaccade amplitudes). The vertical lines represent the spikes of a V1 neuron. (Modified from Martinez-Conde, Macknik, and Hubel, 2000 and Martinez-Conde et al., 2004.) (B) Microsaccades counteract visual fading. Microsaccade rates increase before perceptual transitions towards "target intensification" (black) and decrease before transitions to "target fading" (gray). The horizontal line indicates the average rate of microsaccades during the recording session. Shadows indicate the SEM between subjects. (Modified from McCamy et al., 2012.) (C) Microsaccadic inhibition after changes in peripheral stimuli. Zero indicates the sudden change onset. (Modified from Engbert and Kliegl, 2003.)

tea, without problems. Thus, saccadic movements (of the head or the eye), which are smartly controlled via top-down and bottom-up attentional mechanisms, may be a more generic optimal sampling strategy for vision than smooth scanning.

During each fixation, the visual system must also solve the problem of sampling the scene locally. Again, the constraints of the visual system (adaptation, optimal locus of fixation, receptive field structure, etc.) and statistics of visual environment force a particular strategy: using fixational eye movements. Fixational eye movements are well modeled by a self-avoiding random walk. A self-avoiding random walk serves as an optimal strategy "to refresh the perceptual input to the retinal receptor systems" while keeping "excursions of the eye's movements during fixation within the foveal region of the visual field" (Engbert et al., 2011). The visual system has thus solved two sampling problems with certain anatomical and environmental constraints at a global and local level: Smartly controlled saccades and fixations sample the global scene, and smartly controlled fixational eye movements sample local scene regions.

Pathological Eye Movements

Abnormalities in eye movements occur in many different diseases, including autism, schizophrenia, Friedreich's ataxia, and progressive supranuclear palsy (PSP) to name a few; see Leigh and Zee (2006) for an in-depth review. Eye tracking techniques and characterization algorithms may identify deficits in eye movements to help diagnose ophthalmic and neurological disease.

For instance, autistic subjects fixate on social stimuli (such as the eyes on a face) less often than nonautistic observers (Boraston & Blakemore, 2007). Patients with PSP have more saccadic intrusions (horizontal saccades that "intrude on" or interrupt accurate fixation) than do healthy observers, and they produce normal microsaccades very rarely (Otero-Millan et al., 2011b). Studying eye movements has great potential both in the differential and early diagnosis of neurological disease and in the evaluation and development of treatments.

Conclusions

A combination of top-down and bottom-up mechanisms adaptively controls eye movements to sample the visual world optimally. Noninvasive diagnostics of neurological disease based on eye movements will continue to grow as a field and increase in importance. Determining the brain's algorithms for selecting eye movement targets, and the underlying neural mechanisms that carry out those algorithms, is perhaps the most critical unknown information to elucidate the extremely powerful sampling strategy that humans and primates use in perceiving the visual world.

ACKNOWLEDGMENTS

We thank Jorge Otero-Millan and Leandro Luigi Di Stasi for their comments and the following funding agencies for supporting this work: the National Science Foundation (awards 0852636 and 1153786 to SMC and award 0726113 to SLM) and Barrow Neurological Foundation (awards to SMC and SLM).

   MICHAEL B. MCCAMY, STEPHEN L. MACKNIK, AND SUSANA MARTINEZ-CONDE

REFERENCES

Azzopardi, P., & Cowey, A. (1993). Preferential representation of the fovea in the primary visual cortex. *Nature, 361*, 719–721. doi:10.1038/361719a0.

Bahill, A. T., & LaRitz, T. (1984). Why can't batters keep their eyes on the ball? *American Scientist, 72*, 249–253.

Ballard, D. H., & Hayhoe, M. M. (2009). Modelling the role of task in the control of gaze. *Visual Cognition, 17*, 1185–1204. doi:10.1080/13506280902978477.

Barnes, G. R. (2008). Cognitive processes involved in smooth pursuit eye movements. *Brain and Cognition, 68*, 309–326. doi:10.1016/j.bandc.2008.08.020.

Biederman, I. (1981). On the semantics of a glance at a scene. In M. Kubovy & J. R. Pomerantz (Eds.), *Perceptual organization* (pp. 213–263). Hillsdale, NJ: Erlbaum.

Borah, J. (2006). Eye movement, measurement techniques for. *Encyclopedia of medical devices and instrumentation.* doi:10.1002/0471732877.emd112/abstract.

Boraston, Z., & Blakemore, S. (2007). The application of eye-tracking technology in the study of autism. *Journal of Physiology, 581*, 893–898. doi:10.1113/jphysiol.2007.133587.

Brockmann, D., & Geisel, T. (2000). The ecology of gaze shifts. *Neurocomputing, 2*, 643–650.

Castelhano, M. S., & Henderson, J. M. (2007). Initial scene representations facilitate eye movement guidance in visual search. *Journal of Experimental Psychology. Human Perception and Performance, 33*, 753–763. doi:10.1037/0096-1523.33.4.753.

Cornsweet, T. N., & Crane, H. D. (1973). Accurate two-dimensional eye tracker using first and fourth Purkinje images. *Journal of the Optical Society of America, 63*, 921–928. doi:10.1364/JOSA.63.000921.

Crundall, D. (2005). The integration of top-down and bottom-up factors in visual search during driving. In G. Underwood (Ed.), *Cognitive processes in eye guidance* (pp. 283–302). Oxford, England: Oxford University Press. doi:10.1093/acprof:oso/9780198566816.003.0012.

Cui, J., Otero-Millan, J., Macknik, S. L., King, M., & Martinez-Conde, S. (2011). Social misdirection fails to enhance a magic illusion. *Frontiers in Human Neuroscience, 5*, 103. doi:10.3389/fnhum.2011.00103.

Cullen, K. E., & Van Horn, M. R. (2011). The neural control of fast vs. slow vergence eye movements. *European Journal of Neuroscience, 33*, 2147–2154. doi:10.1111/j.1460-9568.2011.07692.x.

Di Stasi, L. L., Contreras, D., Cándido, A., Cañas, J. J., & Catena, A. (2011). Behavioral and eye-movement measures to track improvements in driving skills of vulnerable road users: First-time motorcycle riders. *Transportation Research Part F: Traffic Psychology and Behaviour, 14*, 26–35. doi:10.1016/j.trf.2010.09.003.

Ditchburn, R. W., & Drysdale, A. E. (1977). The effect of retinal-image movements on vision. I. Step-movements and pulse-movements. *Proceedings of the Royal Society of London. Series B. Biological Sciences, 197*, 131–144. doi:10.1098/rspb.1977.0062.

Eggert, T. (2007). Eye movement recordings: Methods. *Developments in Ophthalmology, 40*, 15–34.

Engbert, R., & Kliegl, R. (2003). Microsaccades uncover the orientation of covert attention. *Vision Research, 43*, 1035–1045. doi:10.1016/S0042-6989(03)00084-1.

Engbert, R., Mergenthaler, K., Sinn, P., & Pikovsky, A. (2011). An integrated model of fixational eye movements and microsaccades. *Proceedings of the National Academy of Sciences of the United States of America, 108*, E765–E770. doi:10.1073/pnas.1102730108.

Fei-Fei, L., Iyer, A., Koch, C., & Perona, P. (2007). What do we perceive in a glance of a real-world scene? *Journal of Vision, 7*(1), 10. doi:10.1167/7.1.10.

Fetter, M. (2007). Vestibulo-ocular reflex. In A. Straube & U. Büttner (Eds.), *Developments in ophthalmology* (Vol. 40, pp. 35–51). Basel, Switzerland: Karger.

Foulsham, T., & Underwood, G. (2007). How does the purpose of inspection influence the potency of visual salience in scene perception? *Perception, 36*, 1123–1138. doi:10.1068/p5659.

Foulsham, T., & Underwood, G. (2008). What can saliency models predict about eye movements? Spatial and sequential aspects of fixations during encoding and recognition. *Journal of Vision, 8*(2), 6. doi:10.1167/8.2.6.

Gerrits, H. J. M., & Vendrik, A. J. H. (1970). Simultaneous contrast, filling-in process and information processing in man's visual system. *Experimental Brain Research, 11*, 411–430. doi:10.1007/BF00237914.

Hafed, Z. M. (2011). Mechanisms for generating and compensating for the smallest possible saccades. *European Journal of Neuroscience, 33*, 2101–2113. doi:10.1111/j.1460-9568.2011.07694.x.

Haslwanter, T., & Clarke, A. H. (2010). Eye movement measurement: Electro-oculography and video-oculography. In S. D. Z. Eggers & D. S. Zee (Eds.), *Vertigo and imbalance: Clinical neurophysiology of the vestibular system* (Vol. 9, pp. 61–79). Amsterdam: Elsevier. Retrieved from http://www.sciencedirect.com/science/article/pii/S1567423110090052.

Hayhoe, M. (2000). Vision using routines: A functional account of vision. *Visual Cognition, 7*, 43–64. doi:10.1080/135062800394676.

Hayhoe, M., & Ballard, D. (2005). Eye movements in natural behavior. *Trends in Cognitive Sciences, 9*, 188–194. doi:10.1016/j.tics.2005.02.009.

Hayhoe, M., Shrivastava, A., Mruczek, R., & Pelz, J. B. (2003). Visual memory and motor planning in a natural task. *Journal of Vision, 3*(1), 6. doi:10.1167/3.1.6.

Heckenmueller, E. G. (1965). Stabilization of the retinal image: A review of method, effects, and theory. *Psychological Bulletin, 63*, 157–169. doi:10.1037/h0021743.

Henderson, J. M. (2003). Human gaze control during real-world scene perception. *Trends in Cognitive Sciences, 7*, 498–504. doi:10.1016/j.tics.2003.09.006.

Itti, L. (2005). Quantifying the contribution of low-level saliency to human eye movements in dynamic scenes. *Visual Cognition, 12*, 1093–1123. doi:10.1080/13506280444000661.

Itti, L., & Koch, C. (2000). A saliency-based search mechanism for overt and covert shifts of visual attention. *Vision Research, 40*, 1489–1506. doi:10.1016/S0042-6989(99)00163-7.

Itti, L., & Koch, C. (2001). Computational modelling of visual attention. *Nature Reviews. Neuroscience, 2*, 194–203. doi:10.1038/35058500.

Klein, R. M., & MacInnes, W. J. (1999). Inhibition of return is a foraging facilitator in visual search. *Psychological Science, 10*, 346–352. doi:10.1111/1467-9280.00166.

Ko, H., Poletti, M., & Rucci, M. (2010). Microsaccades precisely relocate gaze in a high visual acuity task. *Nature Neuroscience, 13,* 1549–1553. doi:10.1038/nn.2663.

Kowler, E., & Steinman, R. M. (1980). Small saccades serve no useful purpose: Reply to a letter by R. W. Ditchburn. *Vision Research, 20,* 273–276.

Land, M. F. (2006). Eye movements and the control of actions in everyday life. *Progress in Retinal and Eye Research, 25,* 296–324. doi:10.1016/j.preteyeres.2006.01.002.

Land, M. F. (2009). Vision, eye movements, and natural behavior. *Visual Neuroscience, 26,* 51–62. doi:10.1017/S0952523808080899.

Land, M. F., & Lee, D. N. (1994). Where we look when we steer. *Nature, 369,* 742–744. doi:10.1038/369742a0.

Land, M. F., & McLeod, P. (2000). From eye movements to actions: How batsmen hit the ball. *Nature Neuroscience, 3,* 1340–1345. doi:10.1038/81887.

Land, M. F., Mennie, N., & Rusted, J. (1999). The roles of vision and eye movements in the control of activities of daily living. *Perception, 28,* 1311–1328. doi:10.1068/p2935.

Laubrock, J., Kliegl, R., Rolfs, M., & Engbert, R. (2010). When do microsaccades follow spatial attention? *Attention, Perception & Psychophysics, 72,* 683–694. doi:10.3758/APP.72.3.683.

Leigh, R. J., & Zee, D. S. (2006). *The neurology of eye movements.* New York: Oxford University Press.

Loon, E. M. V., Khashawi, F., & Underwood, G. (2010). Visual strategies used for time-to-arrival judgments in driving. *Perception, 39,* 1216–1229.

Lovejoy, L. P., & Krauzlis, R. J. (2010). Inactivation of primate superior colliculus impairs covert selection of signals for perceptual judgments. *Nature Neuroscience, 13,* 261–266. doi:10.1038/nn.2470.

Macknik, S. L., King, M., Randi, J., Robbins, A., Teller, T., Thompson, J., et al. (2008). Attention and awareness in stage magic: Turning tricks into research. *Nature Reviews. Neuroscience, 9,* 871–879. doi:10.1038/nrn2473.

Martinez-Conde, S., Macknik, S. L., & Hubel, D. H. (2000). Microsaccadic eye movements and firing of single cells in the striate cortex of macaque monkeys. *Nature Neuroscience, 3,* 251–258.

Martinez-Conde, S., Macknik, S. L., & Hubel, D. H. (2004). The role of fixational eye movements in visual perception. *Nature Reviews. Neuroscience, 5,* 229–240.

Martinez-Conde, S., Macknik, S. L., Troncoso, X. G., & Dyar, T. A. (2006). Microsaccades counteract visual fading during fixation. *Neuron, 49,* 297–305. doi:10.1016/j.neuron.2005.11.033.

Martinez-Conde, S., Macknik, S. L., Troncoso, X. G., & Hubel, D. H. (2009). Microsaccades: A neurophysiological analysis. *Trends in Neurosciences, 32,* 463–475. doi:10.1016/j.tins.2009.05.006.

Masseck, O. A., & Hoffmann, K. (2009). Comparative neurobiology of the optokinetic reflex. *Annals of the New York Academy of Sciences, 1164,* 430–439. doi:10.1111/j.1749-6632.2009.03854.x.

Mays, L. E. (1984). Neural control of vergence eye movements: Convergence and divergence neurons in midbrain. *Journal of Neurophysiology, 51,* 1091–1108.

McCamy, M. B., Otero-Millan, J., Macknik, S. L., Yang, Y., Troncoso, X. G., Baer, S. M., et al. (2012). Microsaccadic efficacy and contribution to foveal and peripheral vision. *Journal of Neuroscience, 32,* 9194–9204.

Mourant, R. R., & Rockwell, T. H. (1970). Mapping eye-movement patterns to the visual scene in driving: An exploratory study. *Human Factors: The Journal of the Human Factors and Ergonomics Society, 12,* 81–87.

Navon, D. (1977). Forest before trees: The precedence of global features in visual perception. *Cognitive Psychology, 9,* 353–383. doi:10.1016/0010-0285(77)90012-3.

Nothdurft, H.-C. (2006). Salience and target selection in visual search. *Visual Cognition, 14,* 514–542. doi:10.1080/13506280500194162.

Otero-Millan, J., Macknik, S. L., & Martinez-Conde, S. (2012). Microsaccades and blinks trigger illusory rotation in the "rotating snakes" illusion. *Journal of Neuroscience, 32,* 6043–6051. doi:10.1523/JNEUROSCI.5823-11.2012.

Otero-Millan, J., Macknik, S. L., Robbins, A., McCamy, M. B., & Martinez-Conde, S. (2011a). Stronger misdirection in curved than in straight motion. *Frontiers in Human Neuroscience, 5,* 133. doi:10.3389/fnhum.2011.00133.

Otero-Millan, J., Serra, A., Leigh, R. J., Troncoso, X. G., Macknik, S. L., & Martinez-Conde, S. (2011b). Distinctive features of saccadic intrusions and microsaccades in progressive supranuclear palsy. *Journal of Neuroscience, 31,* 4379–4387. doi:10.1523/JNEUROSCI.2600-10.2011.

Otero-Millan, J., Troncoso, X. G., Macknik, S. L., Serrano-Pedraza, I., & Martinez-Conde, S. (2008). Saccades and microsaccades during visual fixation, exploration and search: Foundations for a common saccadic generator. *Journal of Vision, 8,* 14–21.

Pastukhov, A., & Braun, J. (2010). Rare but precious: Microsaccades are highly informative about attentional allocation. *Vision Research, 50,* 1173–1184. doi:10.1016/j.visres.2010.04.007.

Pritchard, R. M. (1961). Stabilized images on the retina. *Scientific American, 204,* 72–78.

Ratliff, F., & Riggs, L. A. (1950). Involuntary motions of the eye during monocular fixation. *Journal of Experimental Psychology, 40,* 687–701.

Rayner, K. (1998). Eye movements in reading and information processing: 20 years of research. *Psychological Bulletin, 124,* 372–422. doi:10.1037/0033-2909.124.3.372.

Rayner, K. (2009). Eye movements and attention in reading, scene perception, and visual search. *Quarterly Journal of Experimental Psychology, 62,* 1457–1506. doi:10.1080/17470210902816461.

Rayner, K., Smith, T. J., Malcolm, G. L., & Henderson, J. M. (2009). Eye movements and visual encoding during scene perception. *Psychological Science, 20,* 6–10. doi:10.1111/j.1467-9280.2008.02243.x.

Regan, D. (1992). Visual judgements and misjudgements in cricket, and the art of flight. *Perception, 21,* 91–115. doi:10.1068/p210091.

Remmel, R. S. (1984). An inexpensive eye movement monitor using the scleral search coil technique. *IEEE Transactions on Bio-Medical Engineering, BME-31,* 388–390. doi:10.1109/TBME.1984.325352.

Riggs, L. A., Ratliff, F., Cornsweet, J. C., & Cornsweet, T. N. (1953). The disappearance of steadily fixated visual test objects. *Journal of the Optical Society of America, 43,* 495–500. doi:10.1364/JOSA.43.000495.

Robinson, D. A. (1963). A method of measuring eye movement using a scleral search coil in a magnetic field. *IEEE Transactions on Bio-medical Electronics, 10,* 137–145. doi:10.1109/TBMEL.1963.4322822.

Robinson, D. A. (1968). The oculomotor control system: A review. *Proceedings of the IEEE, 56*, 1032–1049. doi:10.1109/PROC.1968.6455.

Rolfs, M. (2009). Microsaccades: Small steps on a long way. *Vision Research, 49*, 2415–2441. doi:10.1016/j.visres.2009.08.010.

Ryle, J. P., Al-Kalbani, M., Collins, N., Gopinathan, U., Boyle, G., Coakley, D., & Sheridan, J. T. (2009). Compact portable ocular microtremor sensor: Design, development and calibration. *Journal of Biomedical Optics, 14*, 014021. doi:10.1117/1.3083435.

Schalén, L. (1980). Quantification of tracking eye movements in normal subjects. *Acta Oto-Laryngologica, 90*(1), 404–413.

Sharpe, C. R. (1972). The visibility and fading of thin lines visualized by their controlled movement across the retina. *The Journal of Physiology, 222*, 113–134.

Spering, M., & Montagnini, A. (2011). Do we track what we see? Common versus independent processing for motion perception and smooth pursuit eye movements: A review. *Vision Research, 51*, 836–852. doi:10.1016/j.visres.2010.10.017.

Takaakikato, K., & Fukuda, T. (2002). Visual search strategies of baseball batters: Eye movements during the preparatory phase of batting. *Perceptual and Motor Skills, 94*, 380–386. doi:10.2466/pms.2002.94.2.380.

Talbot, S., & Marshall, W. (1941). Physiological studies on neural mechanisms of visual localization and discrimination. *American Journal of Ophthalmology, 24*, 1255–1264.

Tatler, B. W., Hayhoe, M. M., Land, M. F., & Ballard, D. H. (2011). Eye guidance in natural vision: Reinterpreting salience. *Journal of Vision, 11*(5), 5. doi:10.1167/11.5.5.

Tatler, B. W., & Vincent, B. T. (2008). Systematic tendencies in scene viewing. *Journal of Eye Movement Research, 2*, 5.

Tatler, B. W., & Vincent, B. T. (2009). The prominence of behavioural biases in eye guidance. *Visual Cognition, 17*, 1029–1054. doi:10.1080/13506280902764539.

Tian, J., Zee, D. S., & Walker, M. F. (2007). Rotational and translational optokinetic nystagmus have different kinematics. *Vision Research, 47*, 1003–1010. doi:10.1016/j.visres.2006.12.011.

Tsotsos, J. K. (1988). A "complexity level" analysis of immediate vision. *International Journal of Computer Vision, 1*, 303–320. doi:10.1007/BF00133569.

Tsotsos, J. K. (1990). Analyzing vision at the complexity level. *Behavioral and Brain Sciences, 13*, 423–469.

Tsotsos, J. K., Culhane, S. M., Kei Wai, W. Y., Lai, Y., Davis, N., & Nuflo, F. (1995). Modeling visual attention via selective tuning. *Artificial Intelligence, 78*, 507–545. doi:10.1016/0004-3702(95)00025-9.

Underwood, G. (2007). Visual attention and the transition from novice to advanced driver. *Ergonomics, 50*, 1235–1249. doi:10.1080/00140130701318707.

Wartburg, R. V., Wurtz, P., Pflugshaupt, T., Nyffeler, T., Lüthi, M., & Müri, R. M. (2007). Size matters: Saccades during scene perception. *Perception, 36*, 355–365. doi:10.1068/p5552.

Wässle, H., Grünert, U., Röhrenbeck, J., & Boycott, B. B. (1990). Retinal ganglion cell density and cortical magnification factor in the primate. *Vision Research, 30*, 1897–1911. doi:10.1016/0042-6989(90)90166-I.

Wilson-Pauwels, L., Stewart, P., Aesson, E. J., & Spacey, S. (2010). *Cranial nerves: Function and dysfunction* (3rd ed.). Shelton, CT: PMPH-USA.

Wolfe, J. M. (1994). Guided Search 2.0: A revised model of visual search. *Psychonomic Bulletin & Review, 1*, 202–238. doi:10.3758/BF03200774.

Young, L. R., & Sheena, D. (1975). Survey of eye movement recording methods. *Behavior Research Methods and Instrumentation, 7*, 397–429. doi:10.3758/BF03201553.

60 Neural Mechanisms of Fixations and Saccades: The Eye Plant and Low-Level Control

LANCE M. OPTICAN, PIERRE M. DAYE, AND CHRISTIAN QUAIA

Clear vision requires that images fall within a small region of the retina (roughly within ±1° of the center of the fovea) and that they be essentially static (moving at less than ±3°/s) (Barnes & Smith, 1981; Westheimer & McKee, 1975). Thus, when we move, or a target of interest moves, either reflexive or voluntary eye movements are required for clear vision. Reflexive movements are used to stabilize gaze, automatically keeping the eyes on target as the body moves around. The neural controllers that underlie reflexive movements ensure this by properly processing vestibular and/or visual information (see chapter 59 by McCamy, Macknik, and Martinez-Conde). Voluntary eye movements are instead used to shift our gaze to point the fovea at a different visual target or to track a moving target. The fastest refixation movements are called saccades. Combined head and eye movements are required for large gaze changes. However, when reading, scanning a picture, or walking around in the natural world, more than 85% of eye movements are less than 15° in amplitude, and head movements are negligible (Bahill et al., 1975). Thus, this chapter focuses on how saccades are generated when the head is not moving (eye-only saccades).

Naturally, the same visual constraints that led to the development of reflexive eye movements also hold for their voluntary counterparts, and consequently saccades must be accurate and end abruptly. Since chapters 63 (Schall) and 64 (Jagadisan and Gandhi) deal with how targets for saccades are selected, we will assume that this important step has been carried out, and treat the retinotopic location of the desired target as the input to the saccadic system. The spatial location of the target relative to the observer, and the observer's current gaze direction, determine where on each retina the image of the target falls, providing an internal measure of the location to be foveated next. The retina is a 2-D map, but each eye has three degrees of freedom,

thus the choice of the twist around the line of sight is undetermined. As will be explained in chapter 61 (Klier, Blohm, and Crawford), the brain solves this problem by always choosing a specific twist (Donder's law), according to a simple rule (Listing's law).

Obviously, the common ability to execute accurate eye movements, which we take for granted in our daily lives, is not innate and must be preserved in spite of mechanical and neural changes associated with growth, aging, and disease. It is commonly accepted that the cerebellum plays a major role in learning to perform appropriate eye movements (Kojima, Soetedjo, & Fuchs, 2011; Optican & Quaia, 2002; Pelisson et al., 2010; Prsa & Thier, 2011). This is, however, beyond our scope, and this all important aspect of saccade generation will not be addressed here.

What we do address here are the neural processes (collectively known as the saccade generator) that take a command to shift gaze and generate the appropriate saccadic eye movements. The output of the saccadic system is not the orientation of the eyes, but rather the innervations carried by the ocular motor neurons of the six muscles controlling each eye. The relationship between these innervations and the orientation of the eyes depends upon the mechanical properties of the orbital tissues that form the oculomotor plant. Thus, any study of the saccade generator must begin with an analysis of the oculomotor plant.

OCULOMOTOR PLANT

The eyeball can be seen, at least in a first-order approximation, as a spherical joint capable of rotation but not translation (Carpenter, 1977). Each eye sits in a bony orbit, suspended by tissues and muscles and cushioned by fat. The term *oculomotor plant* is then used to describe collectively all the tissues within an orbit that

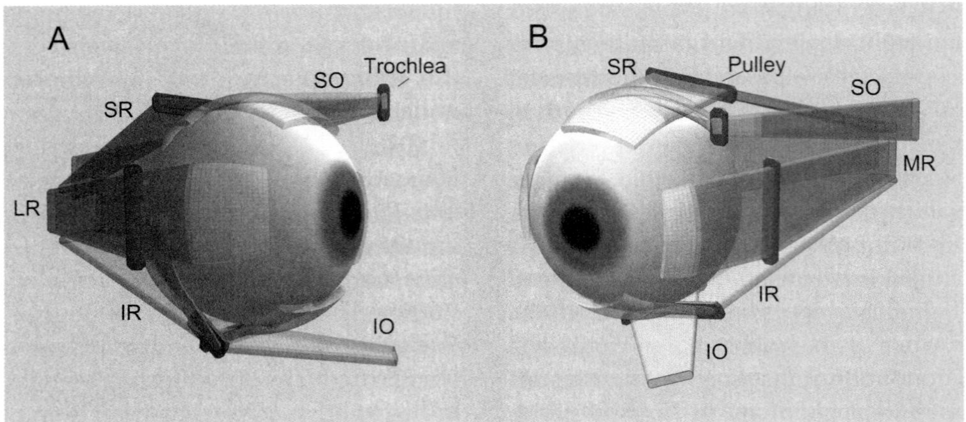

FIGURE 60.1 Two views of a human-scale right eye looking straight ahead. (A) Viewed from the right side, the superior, lateral, and inferior recti muscles (SR, LR, IR) are visible, as well as the inferior oblique muscle (IO) and the tendon of the superior oblique (SO). (B) Viewed from the left side, the medial rectus muscle (MR) and the body of the SO are now visible. Muscles (medium gray) extend from their origin through a pulley (dark gray) to their insertion on the globe. The light gray part of the muscle indicates the tendon. Five of the muscles originate in the back of the orbit, but the IO originates anteriorly on the nasal side of the orbit. Although pulleys are formed from extensive networks of tissues (except the cartilaginous trochlea), the small, dark bars indicate the effective site of the pulleys (i.e., where a point-like pulley would need to be located to produce the observed muscle inflections). The SO (because of the forward position of the trochlea) and the IO pull from in front of the equator of the eye; all other muscles pull from behind the equator.

affect the relationship between the activity of the ocular motor neurons and the orientation of the eye. Obviously, the most prominent elements are the extraocular muscles (EOMs), six in each orbit (see figure 60.1). The EOMs can be regarded as having two layers (global and orbital); the global layers of the EOMs insert, with short tendons, on the sclera, whereas the orbital layers insert on collagenous rings that surround each muscle (Demer, Oh, & Poukens, 2000). Since the eye is a spherical joint, the force delivered by each muscle to its tendon is converted into a torque on the globe. A torque is characterized by a magnitude and an axis of action. In this case the former is proportional to the muscle force, and the latter is the axis around which the muscle, when contracted, rotates the eye. Traditionally it was thought that the axes of action of the muscles simply follow from the location of the muscle's origin and insertion, and the globe's center. However, recent evidence suggests otherwise. Miller (1989) used magnetic resonance imaging of human orbits to show that the posterior half of the muscle does not move as the eye turns, and that there is an inflection point in the eye muscle path (e.g., the anterior, but not the posterior part of the LR moves when the eye looks up). Further research by Demer and colleagues (1995) found that a set of orbital tissues, whose position is possibly under the active control of the orbital layer of EOMs (Demer, Oh, & Poukens, 2000), act as pulleys.

Pulley location determines the location of the inflection point and thus determines the axis of action of the muscle.

The EOMs are not the only mechanically relevant tissues in the orbit. Other orbital tissues apply a passive torque on the eyeball, primarily distributed across its posterior half, which tends to bring the eye to a resting orientation (roughly corresponding to the straight ahead orientation in normal subjects). The orientation of each eye is thus determined by the balance between the forces generated by the orbital tissues and those generated by the six EOMs (inertial forces are usually negligible).

As noted above, the neural controller must solve two problems: (1) It must accurately point the fovea to the desired target (i.e., achieve a certain eye orientation), and (2) it must do so quickly while avoiding postmovement drifts. If EOMs and orbital tissues were mainly elastic structures, both problems would be easily solved. Unfortunately, viscosity plays a dominant role in the eye plant, making control much less straightforward. First, muscles are poor actuators, so that a large fraction of the generated force is actually lost to the muscles' own viscosity, which makes the force delivered to the tendon a heavily smoothed version of the innervational signal (Quaia & Optican, 2011). Second, the passive forces exerted by both muscles and other orbital tissues are dominated by viscous behavior (Anderson et al., 2009;

Quaia et al., 2009a; Quaia, Ying, & Optican, 2009b, 2010).

An additional problem is brought about by the rotational nature of eye movements: although torques are vectors and thus commutative entities (i.e., the order in which torques are applied to different axes does not matter), rotations do not obey commutativity. Thus, the order in which two rotations (around different axes) are executed affects the final orientation (unless they are very tiny). Mathematically, this means that the change in orientation is not equal to the integral of angular rotation. While this issue would be irrelevant for an elastic plant, it becomes a significant nuisance when viscosity enters the picture (Quaia & Optican, 1998; Raphan, 1998; Schnabolk & Raphan, 1994; Tweed, Misslisch, & Fetter, 1994). However, Quaia and Optican (1998) demonstrated mathematically that if the pulleys described by Miller and Demer are in the right place (about halfway between the equator and the posterior pole of the globe), the amount by which the axis of action tips when the eye turns is just enough to compensate for the noncommutativity of 3-D rotations, making the 3-D angular velocity vector approximately equal to the derivative of the 3-D orientation vector. Fortuitously, properly placed pulleys make the plant rotational dynamics appear commutative and thus simpler for the brain to control (Raphan, 1998). Demer and colleagues (2000) pointed out that as the eye turns, the position of the pulleys should also change (*active pulley hypothesis*), so as to keep the geometrical relationship between the pulley and muscle insertion constant. It turns out that this simply means that the pulleys should move back as much as the insertion point does when the globe rotates; this would also prevent the insertion of the contracting muscle from running into its pulley (Quaia & Optican, 2003). Thus, as the muscle contracts, it needs to move both the eye and the pulley.

As a first-order approximation, the presence of the pulleys allows us to ignore the complications brought about by the noncommutativity of 3-D rotations and to treat eye movements as if they were translations (a thorough discussion of 3-D rotations is deferred to chapter 61 by Klier, Blohm, and Crawford). This makes the common practice of considering eye movements around a single, fixed axis (e.g., horizontal or vertical saccades) much less limiting than it would otherwise be. Since a full model of rotational kinematics and dynamics (Quaia & Optican, 2003) is outside the scope of this chapter, we will instead briefly outline a simplified model of the eye plant that is sufficient for comparing neural innervations and rotations about a single axis (e.g., horizontal or vertical rotations).

A Lumped, Single-Axis, Plant Model

D. A. Robinson (1964) measured the mechanical properties of the plant for horizontal movements and found that its dynamics were dominated by viscosity. The dominant time constant of the plant was ~285 ms. Thus, if the innervation to the EOMs were changed in a step-like fashion, the eyes would drift exponentially toward the target (having a 5% error after three time constants, or 855 ms). As we normally make a new saccade every 250–300 ms, our eyes would never be in the eccentricity/slip window for clear vision if they depended upon a step-change in innervation. Robinson suggested that this problem could be avoided by designing the innervation sent to the muscles to compensate for the viscosity of the plant.

The needed compensation depends upon the equations describing the plant, which can be simplified to a second-order linear transfer function (Goldstein, 1983; Optican & Miles, 1985). For this simplified plant, the compensating innervation consists of three parts: a short *pulse*, a long *slide*, and a constant *step*. The pulse generates the force required to overcome viscous drag, the slide compensates for some of the viscosity and the relaxation of tissues that occurs after the rapid part of the movement is over, and the step holds the globe in its final orientation against the orbital restoring forces.

The ratio of the output to the input of a linear model can be expressed as a transfer function (the ratio of two complex polynomials in the Laplace, or complex frequency, domain). A second-order plant model consists of a first-order polynomial in the numerator with a real root called a *zero* (representing the sensitivity of eye movements to the rate of change of the firing rate) and a quadratic polynomial in the denominator with two real roots called *poles* (representing the sensitivity to time integrals of the firing rate), and is thus called the *2p1z* plant model (Goldstein, 1983):

$$\frac{E}{R} = \frac{(sT_z + 1)}{(sT_1 + 1)(sT_2 + 1)}, \qquad (60.1)$$

where E is eye orientation, R is motor neuron firing rate, s is the complex Laplace transform variable, and T_1, T_2, and T_z are time constants.

To find the time constants of the 2p1z plant, we have to consider both the passive orbital tissues and the EOMs. The passive tissues also can be represented as a second-order system. The time constants for the poles in the passive tissues are 20 ms and 1 s, and that for the zero is 0.615 s. A lumped time constant for both muscles of about 0.2 s is commonly used. Combining the passive and active tissues, we can approximate the total plant model with two poles (time constants $T_1 = 0.136$ s, $T_2 = 0.726$ s) and one zero ($T_z = 0.615$ s). The zero approximately cancels the second pole, making the dominant time constant about 0.15 s. Thus, in models where the details of rapid eye movements and the waveform of the innervation are not important, the plant model can be reduced to a single pole:

$$\frac{E}{R} = \frac{1}{(sT_e + 1)} \tag{60.2}$$

with $T_e = 0.15$ s. When time delays between innervation and action need to be taken into account, a delay of about 8 ms is added to the plant (Robinson, 1970). It is important to remember that these linear models are all approximations of a much more complex, nonlinear system (Anderson et al., 2010; Quaia, Ying, & Optican, 2009b, 2011; Robinson, 1981). Care must be taken not to rely on them in clinical cases where the orbital plant may be abnormal.

In one dimension the compensatory innervation needed to drive the 2p1z plant so that it follows the desired step-change in orientation has three components ($R = A \cdot step + B \cdot pulse + C \cdot slide$). To generate an eye movement, we assume that the brain programs the pulse component, which is essentially the desired eye velocity. Then, integrating the pulse gives the step, and low-pass filtering the pulse (with time constant T_s) gives the slide. With perfect plant compensation, the transfer function ratio of the eye movement to the pulse of innervation would simply be $1/s$ (a pure integrator). For normal saccades, the gain of the step should be one, which fixes $A = 1$. This would make the gain of the pulse $B = T_1 T_2 / T_z$, which is about 0.1605. To compensate for the zero in the plant, $T_s = T_z$, or 0.615 s. The gain of the slide would be $C = T_1 + T_2 - T_z - T_1 T_2 / T_z$, which is 0.0865. In a properly compensated plant, the time course of the eye movement will thus simply follow the time course of the step signal. If the size of the pulse, slide, and step are not matched correctly, the eye can drift exponentially after the rapid part of the movement, or even miss the target; this would keep the retinal image outside the eccentricity/slip window for clear vision. The task of computing ocular motor innervations is thus reduced to generating an appropriate pulse of innervation (i.e., a velocity signal), which is then filtered to obtain the appropriate slide and step.

MOVEMENT GENERATION

Figure 60.2 shows a diagram representing the connectivity of neurons in the brainstem involved during horizontal saccadic movements.

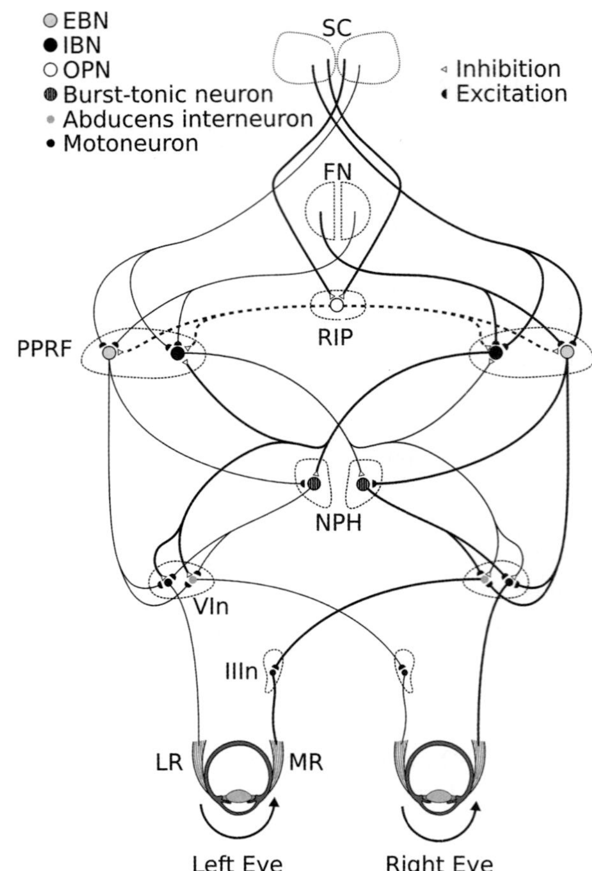

FIGURE 60.2   Schematic diagram (seen from below) of functional connectivity, from the superior colliculus (SC) to the eye muscles, for horizontal saccadic eye movements. Rightward saccades are generated by the activation of the right lateral rectus (LR) and the reciprocal inhibition of the right medial rectus (MR), and vice versa on the left. Left side SC and fastigial nucleus neurons (FN) become active for rightward saccades (thick lines). Omnipause neurons (OPNs) in the nucleus raphe interpositus (RIP) are normally active and pause during saccades (dashed lines). Excitatory burst neurons (EBNs) drive the ipsilateral abducens nucleus (VIn), which shortens the agonist muscle (right lateral rectus, LR). The right abducens interneuron transfers that signal to the left oculomotor nucleus (IIIn), which shortens the left medial rectus (MR). Inhibitory burst neurons (IBNs) reciprocally inhibit the contralateral side and thus relax the antagonist muscles (left LR and right MR). EBNs and IBNs also drive burst-tonic neurons in the nucleus prepositus hypoglossi (NPH). Thin lines represent inactivated axons during a rightward saccade. PPRF, paramedian pontine reticular formation.

Motor Neurons

The purpose of the innervation sent to the muscles is to generate force, which is determined by (1) the firing rate of the motoneurons, (2) which motor units are recruited, and (3) the length of the muscle. To turn the eye, the activity of motoneurons innervating the agonist muscle is increased, and more motor units are recruited. These events lead to an increase in force. However, as the eye turns, the agonist muscle shortens, and its efficiency in converting innervation into force decreases. It is the balance between these two nonlinear phenomena that determines the instantaneous muscle force delivered at the tendon and thus the orientation of the eye.

Horizontal rotations of the eye are mostly controlled by the medial and lateral recti muscles (MR and LR). Thus, the inferior division of the oculomotor nucleus (IIIn) and the abducens nucleus (VIn), which contain the motoneurons that innervate these muscles, can be considered as controlling horizontal eye movements. Vertical rotations require both the vertical (inferior rectus, IR, and superior rectus, SR) and the oblique (inferior oblique, IO, superior oblique, SO) EOMs. The trochlear nucleus (IVn) innervates the SO, and subdivisions of the IIIn innervate the other muscles. Innervation to the IO and IR project to the ipsilateral muscles, but innervation to the SR and SO cross the midline and project to the contralateral muscles. To make an upward movement requires inhibition of the IVn and the IIIn (IR subdivision) and excitation to the IIIn (SR and IO subdivisions). A downward movement requires excitation of the IVn and IIIn (IR subdivision) and inhibition of the IIIn (SR and IO subdivisions) (Leigh & Zee, 2006).

Motor neurons have a preferred direction and a threshold. The firing rate of a motor neuron in its preferred direction can be related to eye rotations by a second-order, ordinary differential equation (Sylvestre & Cullen, 1999):

$$R(t) = \left[m\ddot{E}(t-\delta) + r\dot{E}(t-\delta) + k(E(t-\delta) - E_T) - c\dot{R}(t) \right]^+ ,$$

$$(60.3)$$

where R is the firing rate, $\dot{R}$ is its rate of change, E is the eye's eccentricity in the preferred direction of this neuron, $\dot{E}$ is the eye's angular velocity, and $\ddot{E}$ is the eye's angular acceleration. The constants are defined as follows: E_T is the threshold for this motor unit, δ is the time delay between innervation and action, m is the globe's inertia, r is the orbital viscosity, k is the stiffness of the orbital restoring force, and c determines the phase lead of E relative to R. The bracket symbol ($[\ ]^+$) indicates that negative values are truncated to zero. In general, the firing rate is also sensitive to other factors, including the tension developed in the muscles (Davis-Lopez de Carrizosa et al., 2011), but we will ignore those details in this presentation. As a first-order approximation, equation 60.3 can be reduced to:

$$R(t) = \left[r\dot{E}(t-\delta) + k(E(t-\delta) - E_T) \right]^+ \quad (60.4)$$

When one is modeling both the motor neurons and the plant, the order of the approximations used must be matched (either equation 60.3 is used with equation 60.1, or equation 60.4 is used with equation 60.2).

Final Common Integrator

As noted above, once an appropriate pulse of innervation (i.e., a desired velocity signal) is generated, it is then filtered to obtain the appropriate slide and step. This pulse–slide–step waveform follows from an analysis of the eye plant and was confirmed by recording the firing rate of motor neurons; thus this waveform is not limited to saccadic eye movements. Any eye movement controller (e.g., vestibular, pursuit, or saccade) need only generate an eye velocity command and feed it into the circuits that form the slide and step to drive the eye (Goldman et al., 2002). Thus, everything downstream from the pulse generator can be considered a final common path shared by all oculomotor controllers. The circuitry that generates the step and the slide from the pulse has not been worked out completely yet. However, lesion and single-unit recording studies (Leigh & Zee, 2006) indicate that the vestibular nuclei, the midbrain interstitial nucleus of Cajal (INC, for vertical and torsional components), the brainstem nucleus prepositus hypoglossi (NPH, for horizontal components), and corresponding parts of the cerebellum are important elements. NPH lesions alone cause a 10-fold reduction in the time constant of the neural integrator; this is however still about 10-fold greater than the time constant of the eye plant (Kaneko, 1997). Thus, considerable redundancy must be a prominent feature of this very important circuit. Kaneko (1997) also noted that there were some changes near the end of the saccade, which may have been due to disruption of the mechanism for creating the slide from the pulse, but more research on this topic is needed.

Burst and Pause Neurons

Two populations of neurons play a key role in the generation of the pulse of innervation for saccadic eye movements (Luschei & Fuchs, 1972; Sparks & Travis, 1971). One population discharges vigorously with saccades and is silent during fixation (*burst neurons*). The other fires tonically during fixation but stops discharging during saccades (*pause neurons*). Their activity is tightly coupled with the duration of the saccade (Delgado-Garcia et al., 1988; Keller, 1974; Luschei & Fuchs, 1972; Moschovakis, Scudder, & Highstein, 1991a; Moschovakis et al., 1991b; Villis et al., 1989). Additionally, the amplitude of the burst neurons' activity is correlated with the speed and direction of the saccade, whereas the majority of the pause neurons simply cease their activity, irrespective of the direction and speed of the saccade (Keller, 1974), and are thus called omnipause neurons (OPNs).

Burst neurons can be further subdivided into two functional populations: inhibitory burst neurons (IBNs) and excitatory burst neurons (EBNs) (Hikosaka et al., 1978). IBNs project contralaterally to both EBNs and IBNs (Strassman, Highstein, & McCrea, 1986), thus creating the potential for ocular oscillations due to reciprocal inhibition (Ramat et al., 2005). EBNs and IBNs receive excitatory projections from the cerebellum (Asanuma, Thach, & Jones, 1983; Noda et al., 1990) and the superior colliculus (SC; Nakao et al., 1990; Raybourn & Keller, 1977). Additionally, both EBN and IBN populations are inhibited by the OPNs (Keller, 1977; Langer & Kaneko, 1983; Nakao, Curthoys, & Markham, 1980; Nakao et al., 1988; Ohgaki et al., 1989).

During fixation, EBNs and IBNs are inactive because of OPN inhibition (Ohgaki, Curthoys, & Markham, 1987; Ohgaki et al., 1989), and saccades cannot occur. At the time of a saccade OPNs are deactivated. The EBNs controlling the agonist muscle and the IBNs controlling the antagonist muscle then discharge vigorously; the difference between the two bursts generates the pulse of innervation for the movement. At the end of the saccade the IBNs controlling the agonist muscle discharge briefly (Van Gisbergen, Robinson, & Gielen 1981), possibly slowing down the eye movement (Lefevre, Quaia, & Optican, 1998; Quaia, Lefevre, & Optican, 1999; Scudder, Kaneko, & Fuchs, 2002). When the OPNs are reactivated, they inhibit IBNs and EBNs and fixation ensues.

HORIZONTAL SACCADIC EYE MOVEMENTS The paramedian pontine reticular formation (PPRF) has long been associated with horizontal saccades and quick phases of nystagmus (Cohen & Feldman, 1968). It contains EBNs and IBNs serving the control of horizontal saccades. IBNs project to the contralateral VIn, and EBNs project to the ipsilateral VIn (Grantyn, Baker, & Grantyn, 1980; Strassman, Highstein, & McCrea, 1986).

VERTICAL SACCADIC EYE MOVEMENTS The rostral interstitial nucleus of the medial longitudinal fasciculus (riMLF) in the midbrain contains EBNs related to vertical and torsional saccades (Büttner, Büttner-Ennever, & Henn, 1977; King & Fuchs, 1979) whereas IBNs are located in the INC (Horn, 2006). Although horizontal movements can be easily associated with ipsilateral and contralateral innervation commands, the vertical channel is more complicated. Stimulation of the riMLF

neurons on the right side of the brain induces a clockwise torsional eye movement (right eye extorts, left eye intorts) whereas stimulation of the left side induces counterclockwise rotations (Vilis et al., 1989). Thus, neurons on both sides must be activated simultaneously to induce purely vertical eye movements (i.e., the torsional components must cancel out).

OMNIPAUSE NEURONS A small population of large neurons is found in a small midline area of the brainstem called the nucleus raphe interpositus (RIP) (Büttner-Ennever et al., 1988). These cells fire with a steady background rate (about 100–200 spikes/s) but pause whenever a saccade in any direction is made (Keller, 1974). Stimulation in the RIP stops saccades (Keller, 1974). Thus, to make a saccade, the OPNs must be shut off. Since both the SC (through the contralateral IBNs) and the cerebellum inhibit OPNs (for a review, see Shinoda et al., 2011), a collicular and/or a cerebellar discharge might theoretically turn them off. A hypothesized *latch circuit* would then keep the OPN activity off during the saccade (Van Gisbergen, Robinson, & Gielen, 1981).

Exactly how the OPNs are turned off at the beginning of the saccade, kept off during the movement, and then turned on at the end remains a mystery. Intracellular recordings in cat show that the pause in OPNs begins with a short, sharp hyperpolarization that causes a decrease in firing, followed by a sustained hyperpolarization proportional to eye velocity (Yoshida et al., 1999). If the SC is stimulated, inhibition of OPNs occurs with disynaptic latency (Yoshida et al., 2001). A plausible explanation for these findings is that the sharp hyperpolarization at the beginning of a saccade is caused by an initial burst of activity in the IBNs caused by direct inputs from the caudal SC, whereas the sustained hyperpolarization is caused by EBN activation of the IBNs (Shinoda et al., 2011).

How saccades start and stop is also not yet completely understood. For example, if IBNs are held off by OPNs, how can they become active to stop the OPNs? One theory stems from the observation that there are two populations of IBNs, long- and short-lead (Scudder, Fuchs, & Langer, 1988). The long-lead IBNs begin firing before the OPNs go off, but the short-lead IBNs fire only after the OPNs go off. Thus, it is possible that the long-lead IBNs are part of the pathway that stops the OPNs. However, single-pulse stimulation of the SC inhibits the OPNs, and also drive short-lead but not long-lead burst neurons (Raybourn & Keller, 1977), so more research on this topic is needed.

Latencies between pause beginning and end are also not as tightly coupled to saccade onset and end as

  LANCE M. OPTICAN, PIERRE M. DAYE, AND CHRISTIAN QUAIA

expected if the OPNs are the only factor controlling saccades (Busettini & Mays, 2003). Studies in patients have shown that OPNs can be silent even when saccades stop (Rucker et al., 2011). Also, when lesions are made in the region of the OPNs, saccades still occur with normal latency and amplitude, albeit with lower peak velocities (Miura & Optican, 2006; Soetedjo, Kaneko, & Fuchs, 2002). Much remains to be discovered before we understand how saccades start and stop.

GAZE COMMAND CENTERS

Saccades are generated by specifying a goal, that is, a location in the visual scene to be foveated next. Chapters 63 (Schall) and 64 (Jagadisan and Gandhi) deal with how that target is chosen. Here we focus on how this information is converted into a motor command. Lesions in both the frontal eye fields (FEFs) and the SC obliterate saccades, but lesions in either area alone affect, but do not eliminate, saccades (Schiller, True, & Conway, 1979). Also, lesions in the cerebellum lead to enduring deficits in saccadic accuracy (Dichgans & Jung, 1975; Optican & Robinson, 1980; Ritchie, 1976; Zee et al., 1976). It thus appears that the FEF and the SC carry information about the goal, and the cerebellum plays a critical role in generating a saccade of the appropriate magnitude and direction to foveate that goal. Since either the FEF or the SC alone is sufficient to provide goal information, here we will focus on the network that includes the SC and the cerebellum. Other eye movement areas that are important for choosing targets, such as the basal ganglia, are outside the scope of this chapter.

Superior Colliculus

Since the early 1970s single-unit recordings (Schiller & Stryker, 1972; Wurtz & Goldberg, 1971, 1972) and electrical stimulation experiments (Robinson, 1972; Schiller & Stryker, 1972) have indicated that the intermediate layers of the SC must play an important role in producing saccades. The SC has been reviewed many times (Gandhi & Sparks, 2004; Krauzlis, 2005; McIlwain, 1991; Wurtz & Albano, 1980). Here we give a brief overview of the elements important for saccades. Cells in the SC are characterized by fairly large movement fields (i.e., the range of goal locations/saccade vectors associated with activation of a neuron) (Sparks, Holland, & Guthrie, 1976), which are topographically organized (i.e., cells close together have similar movement fields). Neurons that discharge in correspondence with parafoveal targets (small saccades) are located rostrally whereas eccentric targets (large saccades) are associated with more caudal sites. Accordingly, electrical stimulation at rostral sites results in small saccades, whereas at more caudal sites larger saccades are evoked. These results indicate that the target (or saccade) vector is spatially, and not temporally, encoded in the SC; movements toward targets in the left visual hemifield are encoded in the right SC and vice versa.

Saccade-related neurons in the SC have been divided into three classes, according to their pattern of activity and location: burst, buildup, and fixation neurons (Munoz & Wurtz, 1992, 1993, 1995; Wurtz & Optican, 1994). The burst neurons are characterized by a brisk discharge synchronized with saccade onset, have a closed movement field (i.e., they discharge only for saccades around an optimal vector), and are probably the same cells described by Sparks and colleagues as saccade-related burst neurons (Sparks, 1978; Sparks & Mays, 1980).

The second class of cells is represented by the so-called buildup neurons (located among and just below the burst neurons), which are characterized by a small buildup of activity preceding saccades (hence their name) and have an open movement field (i.e., they discharge, albeit with different intensities, for all saccades in one direction larger than a certain amplitude) (see figure 60.3). Some, but not all, buildup cells are characterized by a burst occurring at saccade onset, similar to that of the burst cells. In the majority of buildup cells this burst component has a closed movement field, similar to that of the burst neurons (see Munoz & Wurtz, 1995, their figures 7B and 8).

The third class of cells, the fixation neurons, located in the rostral pole of the SC, usually behave in a manner opposite to burst neurons, that is, they discharge during active fixation and pause during saccades in any direction. However, sometimes they do not pause, or can even produce a burst, for small, contraversive saccades. For large saccades, these cells pause immediately before the onset of a saccade and resume firing at the time of saccade termination, much like OPNs (Munoz & Wurtz, 1993). Electrical stimulation delivered to this rostral area interrupts ongoing saccades, while stimulation in the rest of the SC induces contralateral saccades. More recently, it has however been suggested that this class of neurons is involved in controlling the execution of small movements, such as microsaccades, and should be seen as part of the buildup class of neurons (Everling et al., 1998; Gandhi & Katnani, 2011; Hafed, Goffart, & Krauzlis, 2009; Otero-Millan et al., 2011).

So far, we have not addressed the issue of whether the SC site activated encodes the location of the target (in retinal coordinates) or the saccadic vector (in motor

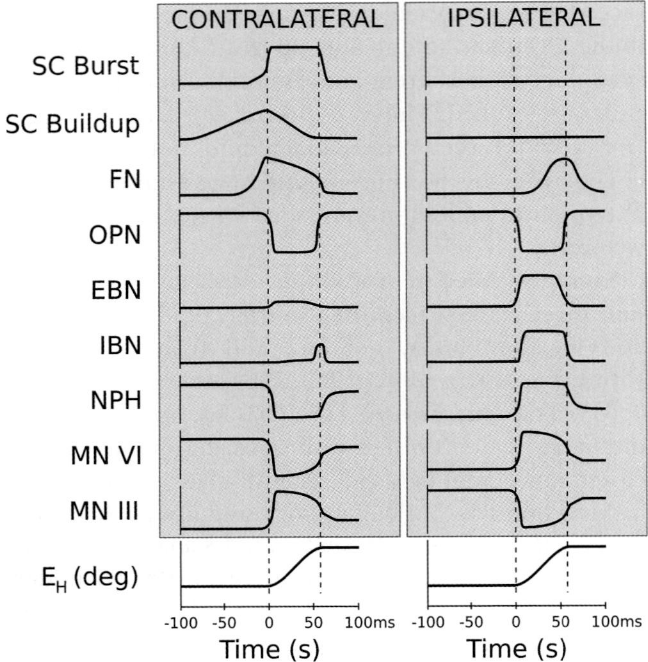

FIGURE 60.3 Schematic diagram of firing rates of neurons in figure 60.2. Each row represents the discharge of each area of figure 60.2 during a rightward saccade. Left (right) column represents the discharge of a typical contralateral (ipsilateral) neuron of the corresponding structure. The last row represents the corresponding left and right eye displacements generated by the discharges. Vertical dashed lines represent onset and offset of the saccadic eye movement. Time origin is set to saccade onset. SC, superior colliculus; FN, fastigial nucleus; OPN, omnipause neuron; EBN, excitatory burst neuron; IBN, inhibitory burst neuron; NPH, nucleus prepositus hypoglossi; MN VI, abducens nucleus motor neuron; MN III, oculomotor nucleus motor neuron; E_H, horizontal eye position.

coordinates) required to foveate it. Traditionally the latter had been assumed. However, recent research points toward the former interpretation; when the two vectors are dissociated, the SC appears to encode the location of the target, not the size of the movement. For example, when a saccade is made to a moving target, the brain anticipates where the target will be after the delay incurred in planning the saccade. Thus, the amplitude of the saccade reflects both the position error of the target when it appears on the retina and the target's velocity. However, the locus of activity on the SC map corresponds to the position error of the target, without regard to its velocity component (Keller, Gandhi, & Weir, 1996; Optican & Quaia, 2002); this is especially obvious when the two are in different hemifields (Carello & Krauzlis, 2004). Furthermore, experiments by Port and Wurtz (2003) looked at strongly curved saccades made in response to two flashed targets. The eyes first turned toward one target, then changed direction to acquire the second target. The final

direction of the saccade was thus different from either of the initial directions to the two targets, yet they did not find activity at the site on the SC corresponding to this redirected movement. Overall, these findings indicate that the SC locus activated is more closely related to the location of a target of interest than to the saccadic vector required to foveate it.

Cerebellum

A great deal of evidence points toward lobuli VIc and VII of the cerebellar vermis as being involved in the control of horizontal saccadic eye movements. Only very small currents are needed to evoke saccades from this region (Fujikado & Noda, 1987; Noda & Fujikado, 1987) whereas much higher currents are needed to evoke saccades from nearby lobuli (Keller, Slakey, & Crandall, 1983; Ron & Robinson, 1973). Ablations of this area result in dysmetric movements (Ritchie, 1976; Takagi, Zee, & Tamargo, 1998). Finally, neurons in this area show saccade-related activity (Helmchen & Büttner, 1995; Ohtsuka & Noda, 1995; Sato & Noda, 1992), whereas activity in neurons belonging to other vermal lobuli is not modulated during saccades (Sato & Noda, 1992). The Purkinje cells in these lobuli project to an ellipsoidal region in the caudal fastigial nucleus (Yamada & Noda, 1987), the so-called fastigial oculomotor region (FOR). These projections are strictly ipsilateral and topographically organized (Carpenter & Batton, 1982; Courville & Diakiw, 1976; Noda et al., 1990).

Consequently, any model that is concerned with the control of saccades by the cerebellum has to give strong importance to the saccade-related discharge of the FOR neurons. They produce an early burst of activity for movements in one direction (preferred direction) and a late burst, near the end of the movement, for saccades in the opposite direction (Fuchs, Robinson, & Straube, 1993; Helmchen, Straube, & Büttner, 1994; Ohtsuka & Noda, 1990, 1991). The preferred direction always has a contralateral horizontal component.

Neurons in the FOR burst early for contraversive and late for ipsiversive saccades. It has been proposed that the early burst in the FOR for contraversive movements helps accelerate the saccade, by providing input to the EBNs ipsilateral to the movement. In contrast, the late ipsilateral burst might help to decelerate the saccade (through the contralateral IBNs) (Dean, 1995; Fuchs, Robinson, & Straube, 1993). Others have pushed this idea further, proposing that the cerebellum generates a drive signal that not only speeds up the saccade but actually steers it toward the target. At the appropriate time, the cerebellum then generates a signal to choke off the combined cerebellar and collicular drives, ensuring that the saccade lands on target (Lefevre, Quaia, & Optican, 1998; Optican, 2005; Optican & Quaia, 2002; Quaia, Lefevre, & Optican, 1999). In this scheme the cerebellum doesn't simply act as a side pathway adaptively tuning the amplitude of movements, as hypothesized by classic models of cerebellar function, but rather keeps track of movement progress toward the target and provides a differently tailored contribution to each individual saccade. Support for this view comes from experiments in which the cerebellum has been subjected to long-term damage: While initially saccades become dysmetric and more variable, over time the average saccade amplitude recovers, but the variability remains (Barash et al., 1999).

A FUNCTIONAL VIEW OF SACCADE GENERATION

Having reviewed the components of the saccade generation circuitry, we are now in a position to understand how a command to look at something is converted into the appropriate eye movement. The prelude to a saccade starts when buildup neurons in the SC, at the site corresponding to the target's retinal location, are activated at a low level. This site is in the hemifield contralateral to the target location (and ensuing movement direction). When the decision to start the movement, which is partly under voluntary control and partly subordinated to an internal trigger mechanism, is issued, the collicular burst neurons become active. At the same time, the neurons in the rostral SC become silent. A large inhibitory drive shuts off the OPNs in the brainstem. The collicular burst neurons can now drive the premotor burst neurons (EBNs and IBNs) ipsilateral to the movement, which carry the pulse signal for the saccade. In addition, the collicular signal is relayed, via the pontine nuclei, to the cerebellum, leading to an activation of the contralateral fastigial nucleus, which also contributes to the activation of ipsilateral EBNs and IBNs. Their pulse of activity is passed on to the appropriate motoneurons; it is also integrated and low-pass filtered by the slide–step circuitry, which also feeds the motoneurons. The activation of a set of motoneurons, and the inhibition of their antagonist counterparts, leads to the contraction of some muscles and the relaxation of others, rotating the eye. As the target is approached, stable fixation must be quickly reestablished. Thus the ipsilateral fastigial nucleus sends a signal to the contralateral IBNs, which generate a signal that chokes off the pulse signal. At this point rostral collicular neurons and OPNs reactivate, stabilizing the circuit and preventing further pulse signals from reaching the motoneurons. The slide takes care of the relaxing passive forces, allowing for a stable fixation.

INTO THE FUTURE

One important question remains: How does the system know when the target is reached and the movement should stop? Obviously, saccadic eye movements are too fast to use visual feedback, so either movements are preprogrammed (i.e., ballistic), or internal feedback must be used. Experiments have revealed that artificial perturbations can be corrected in-flight, indicating that movements are not ballistic. The neural implementation of internal feedback is however still a matter of debate. Some authors (Lefevre, Quaia, & Optican, 1998; Quaia, Lefevre, & Optican, 1999; Xu-Wilson et al., 2011) argue that the cerebellum plays a central role in feedback control whereas others (Dean, 1995; Scudder, Kaneko, & Fuchs, 2002; van Opstal & Goossens, 2008) favor Robinson's (1975) original hypothesis, that it is all taken care of in the brainstem, with the cerebellum acting as a long-term gain controller. That, after 50 years of oculomotor research, such an important function is still the subject of heated debates is a good indication that there is still much to learn about eye-movement generation.

REFERENCES

Anderson, S. R., Lepora, N. F., Porrill, J., & Dean, P. (2010). Nonlinear dynamic modeling of isometric force production in primate eye muscle. *IEEE Transactions on Bio-Medical Engineering, 57,* 1554–1567.

Anderson, S. R., Porrill, J., Sklavos, S., Gandhi, N. J., Sparks, D. L., & Dean, P. (2009). Dynamics of primate oculomotor plant revealed by effects of abducens microstimulation. *Journal of Neurophysiology, 101,* 2907–2923.

Asanuma, C., Thach, W. T., & Jones, E. G. (1983). Brainstem and spinal projections of the deep cerebellar nuclei in the monkey, with observations on the brainstem projections of the dorsal column nuclei. *Brain Research, 286,* 299–322.

Bahill, A. T., Adler, D., & Stark, L. (1975). Most naturally occurring human saccades have magnitudes of 15 degrees or less. *Investigative Ophthalmology, 14,* 468–469.

Barash, S., Melikyan, A., Sivakov, A., Zhang, M., Glickstein, M., & Thier, P. (1999). Saccadic dysmetria and adaptation after lesions of the cerebellar cortex. *Journal of Neuroscience, 19,* 10931–10939.

Barnes, G. R., & Smith, R. (1981). The effects of visual discrimination of image movement across the stationary retina. *Aviation, Space, and Environmental Medicine, 52,* 466–472.

Busettini, C., & Mays, L. E. (2003). Pontine omnipause activity during conjugate and disconjugate eye movements in macaques. *Journal of Neurophysiology, 90,* 3838–3853.

Büttner, U., Büttner-Ennever, J. A., & Henn, V. (1977). Vertical eye movement related unit activity in the rostral mesencephalic reticular formation of the alert monkey. *Brain Research, 130,* 239–252.

Büttner-Ennever, J. A., Cohen, B., Pause, M., & Fries, W. (1988). Raphe nucleus of the pons containing omnipause neurons of the oculomotor system in the monkey, and its homologue in man. *Journal of Comparative Neurology, 267,* 307–321.

Carello, C. D., & Krauzlis, R. J. (2004). Manipulating intent: Evidence for a causal role of the superior colliculus in target selection. *Neuron, 43,* 575–583.

Carpenter, M. B., & Batton, R. R., III. (1982). Connections of the fastigial nucleus in the cat and monkey. In S. L. Palay & V. Chan-Palay (Eds.), *The cerebellum—New vistas* (pp. 250–295). Berlin: Springer-Verlag.

Carpenter, R. H. S. (1977). *Movements of the eyes.* London: Pion.

Cohen, B., & Feldman, M. (1968). Relationship of electrical activity in pontine reticular formation and lateral geniculate body to rapid eye movements. *Journal of Neurophysiology, 31,* 806–817.

Courville, J., & Diakiw, N. (1976). Cerebellar corticonuclear projection in the cat: The vermis of the anterior and posterior lobes. *Brain Research, 110,* 1–20. doi:10.1016/0006-8993(76)90205-5.

Davis-Lopez de Carrizosa, M. A., Morado-Diaz, C. J., Miller, J. M., de la Cruz, R. R., & Pastor, A. M. (2011). Dual encoding of muscle tension and eye position by abducens motoneurons. *Journal of Neuroscience, 31,* 2271–2279.

Dean, P. (1995). Modelling the role of the cerebellar fastigial nuclei in producing accurate saccades: The importance of burst timing. *Neuroscience, 68,* 1059–1077.

Delgado-Garcia, J. M., Vidal, P. P., Gomez, C., & Berthoz, A. (1988). Vertical eye movements related signals in antidromically identified medullary reticular formation neurons in the alert cat. *Experimental Brain Research. Experimentelle Hirnforschung. Experimentation Cerebrale, 70,* 585–589.

Demer, J. L., Miller, J. M., Poukens, V., Vinters, H. V., & Glasgow, B. J. (1995). Evidence for fibromuscular pulleys of the recti extraocular muscles. *Investigative Ophthalmology & Visual Science, 36,* 1125–1136.

Demer, J. L., Oh, S. Y., & Poukens, V. (2000). Evidence for active control of rectus extraocular muscle pulleys. *Investigative Ophthalmology & Visual Science, 41,* 1280–1290.

Dichgans, J., & Jung, R. (1975). Oculomotor abnormalities due to cerebellar lesions. In G. Lennerstrand & P. Bach-y-Rita (Eds.), *Basic mechanisms of ocular motility and their clinical implications* (pp. 281–298). Oxford: Pergamon.

Everling, S., Pare, M., Dorris, M. C., & Munoz, D. P. (1998). Comparison of the discharge characteristics of brain stem omnipause neurons and superior colliculus fixation neurons in monkey: Implications for control of fixation and saccade behavior. *Journal of Neurophysiology, 79,* 511–528.

Fuchs, A. F., Robinson, F. R., & Straube, A. (1993). Role of the caudal fastigial nucleus in saccade generation: I. Neuronal discharge pattern. *Journal of Neurophysiology, 70,* 1723–1740.

Fujikado, T., & Noda, H. (1987). Saccadic eye movements evoked by microstimulation of lobule VII of the cerebellar vermis of macaque monkeys. *Journal of Physiology, 394,* 573–594.

Gandhi, N. J., & Katnani, H. A. (2011). Motor functions of the superior colliculus. *Annual Review of Neuroscience, 34,* 205–231.

Gandhi, N. J., & Sparks, D. L. (2004). Changing views of the role of superior colliculus in the control of gaze. In L. M. Chalupa & J. S. Werner (Eds.), *The visual neurosciences* (pp. 1449–1465). Cambridge, MA: MIT Press.

Goldman, M. S., Kaneko, C. R., Major, G., Aksay, E., Tank, D. W., & Seung, H. S. (2002). Linear regression of eye velocity on eye position and head velocity suggests a common oculomotor neural integrator. *Journal of Neurophysiology, 88,* 659–665.

Goldstein, H. P. (1983). *The neural encoding of saccades in the rhesus monkey.* Doctoral Dissertation, Johns Hopkins University, Baltimore.

Grantyn, R., Baker, R., & Grantyn, A. (1980). Morphological and physiological identification of excitatory pontine reticular neurons projecting to the cat abducens nucleus and spinal cord. *Brain Research, 198,* 221–228.

Hafed, Z. M., Goffart, L., & Krauzlis, R. J. (2009). A neural mechanism for microsaccade generation in the primate superior colliculus. *Science, 323,* 940–943.

Helmchen, C., & Büttner, U. (1995). Saccade-related Purkinje cell activity in the oculomotor vermis during spontaneous eye movements in light and darkness. *Experimental Brain Research. Experimentelle Hirnforschung. Experimentation Cerebrale, 103,* 198–208.

Helmchen, C., Straube, A., & Büttner, U. (1994). Saccade-related activity in the fastigial oculomotor region of the macaque monkey during spontaneous eye movements in light and darkness. *Experimental Brain Research. Experimentelle Hirnforschung. Experimentation Cerebrale, 98,* 474–482.

Hikosaka, O., Igusa, Y., Nakao, S., & Shimazu, H. (1978). Direct inhibitory synaptic linkage of pontomedullary reticular burst neurons with abducens motoneurons in the cat. *Experimental Brain Research. Experimentelle Hirnforschung. Experimentation Cerebrale, 33,* 337–352.

Horn, A. K. (2006). The reticular formation. *Progress in Brain Research, 151,* 127–155.

Kaneko, C. R. (1997). Eye movement deficits after ibotenic acid lesions of the nucleus prepositus hypoglossi in monkeys: I. Saccades and fixation. *Journal of Neurophysiology, 78,* 1753–1768.

Keller, E. L. (1974). Participation of medial pontine reticular formation in eye movement generation in monkey. *Journal of Neurophysiology, 37,* 316–332.

Keller, E. L. (1977). Control of saccadic eye movements by midline brain stem neurons. In R. Baker & A. Berthoz (Eds.), *Control of gaze by brain stem neurons* (pp. 327–336). Amsterdam: Elsevier.

Keller, E. L., Gandhi, N. J., & Weir, P. T. (1996). Discharge of superior collicular neurons during saccades made to moving targets. *Journal of Neurophysiology, 76,* 3573–3577.

Keller, E. L., Slakey, D. P., & Crandall, W. F. (1983). Microstimulation of the primate cerebellar vermis during saccadic eye movements. *Brain Research, 288,* 131–143.

King, W. M., & Fuchs, A. F. (1979). Reticular control of vertical saccadic eye movements by mesencephalic burst neurons. *Journal of Neurophysiology, 42,* 861–876.

Kojima, Y., Soetedjo, R., & Fuchs, A. F. (2011). Effect of inactivation and disinhibition of the oculomotor vermis on saccade adaptation. *Brain Research, 1401,* 30–39.

Krauzlis, R. J. (2005). The control of voluntary eye movements: New perspectives. *Neuroscientist, 11,* 124–137.

Langer, T. P., & Kaneko, C. R. (1983). Efferent projections of the cat oculomotor reticular omnipause neuron region: An autoradiographic study. *Journal of Comparative Neurology, 217*, 288–306.

Lefevre, P., Quaia, C., & Optican, L. M. (1998). Distributed model of control of saccades by superior colliculus and cerebellum. *Neural Networks, 11*, 1175–1190.

Leigh, R. J., & Zee, D. S. (2006). *The neurology of eye movements* (4th ed.). Oxford: Oxford University Press.

Luschei, E. S., & Fuchs, A. F. (1972). Activity of brain stem neurons during eye movements of alert monkeys. *Journal of Neurophysiology, 35*, 445–461.

McIlwain, J. T. (1991). Distributed spatial coding in the superior colliculus: A review. *Visual Neuroscience, 6*, 3–13.

Miller, J. M. (1989). Functional anatomy of normal human rectus muscles. *Vision Research, 29*, 223–240.

Miura, K., & Optican, L. M. (2006). Membrane channel properties of premotor excitatory burst neurons may underlie saccade slowing after lesions of omnipause neurons. *Journal of Computational Neuroscience, 20*, 25–41.

Moschovakis, A. K., Scudder, C. A., & Highstein, S. M. (1991a). Structure of the primate oculomotor burst generator: I. Medium-lead burst neurons with upward on-directions. *Journal of Neurophysiology, 65*, 203–217.

Moschovakis, A. K., Scudder, C. A., Highstein, S. M., & Warren, J. D. (1991b). Structure of the primate oculomotor burst generator: II. Medium-lead burst neurons with downward on-directions. *Journal of Neurophysiology, 65*, 218–229.

Munoz, D. P., & Wurtz, R. H. (1992). Role of the rostral superior colliculus in active visual fixation and execution of express saccades. *Journal of Neurophysiology, 67*, 1000–1002.

Munoz, D. P., & Wurtz, R. H. (1993). Fixation cells in monkey superior colliculus: I. Characteristics of cell discharge. *Journal of Neurophysiology, 70*, 559–575.

Munoz, D. P., & Wurtz, R. H. (1995). Saccade-related activity in monkey superior colliculus: I. Characteristics of burst and buildup cells. *Journal of Neurophysiology, 73*, 2313–2333.

Nakao, S., Curthoys, I. S., & Markham, C. H. (1980). Direct inhibitory projection of pause neurons to nystagmus-related pontomedullary reticular burst neurons in the cat. *Experimental Brain Research, 40*, 283–293.

Nakao, S., Shiraishi, Y., Li, W. B., & Oikawa, T. (1990). Mono- and disynaptic excitatory inputs from the superior colliculus to vertical saccade-related neurons in the cat Forel's field H. *Experimental Brain Research, 82*, 222–226.

Nakao, S., Shiraishi, Y., Oda, H., & Inagaki, M. (1988). Direct inhibitory projection of pontine omnipause neurons to burst neurons in the Forel's field H controlling vertical eye movement-related motoneurons in the cat. *Experimental Brain Research, 70*, 632–636.

Noda, H., & Fujikado, T. (1987). Involvement of Purkinje cells in evoking saccadic eye movements by microstimulation of the posterior cerebellar vermis of monkeys. *Journal of Neurophysiology, 57*, 1247–1261.

Noda, H., Sugita, S., & Ikeda, Y. (1990). Afferent and efferent connections of the oculomotor region of the fastigial nucleus in the macaque monkey. *Journal of Comparative Neurology, 302*, 330–348.

Ohgaki, T., Curthoys, I. S., & Markham, C. H. (1987). Anatomy of physiologically identified eye-movement-related pause neurons in the cat: Pontomedullary region. *Journal of Comparative Neurology, 266*, 56–72.

Ohgaki, T., Markham, C. H., Schneider, J. S., & Curthoys, I. S. (1989). Anatomical evidence of the projection of pontine omnipause neurons to midbrain regions controlling vertical eye movements. *Journal of Comparative Neurology, 289*, 610–625.

Ohtsuka, K., & Noda, H. (1990). Direction-selective saccadic-burst neurons in the fastigial oculomotor region of the macaque. *Experimental Brain Research, 81*, 659–662.

Ohtsuka, K., & Noda, H. (1991). Saccadic burst neurons in the oculomotor region of the fastigial nucleus of macaque monkeys. *Journal of Neurophysiology, 65*, 1422–1434.

Ohtsuka, K., & Noda, H. (1995). Discharge properties of Purkinje cells in the oculomotor vermis during visually guided saccades in the macaque monkey. *Journal of Neurophysiology, 74*, 1828–1840.

Optican, L. M. (2005). Sensorimotor transformation for visually guided saccades. *Annals of the New York Academy of Sciences, 1039*, 132–148.

Optican, L. M., & Miles, F. A. (1985). Visually induced adaptive changes in primate saccadic oculomotor control signals. *Journal of Neurophysiology, 54*, 940–958.

Optican, L. M., & Quaia, C. (2002). Distributed model of collicular and cerebellar function during saccades. *Annals of the New York Academy of Sciences, 956*, 164–177.

Optican, L. M., & Robinson, D. A. (1980). Cerebellar-dependent adaptive control of primate saccadic system. *Journal of Neurophysiology, 44*, 1058–1076.

Otero-Millan, J., Macknik, S. L., Serra, A., Leigh, R. J., & Martinez-Conde, S. (2011). Triggering mechanisms in microsaccade and saccade generation: A novel proposal. *Annals of the New York Academy of Sciences, 1233*, 107–116.

Pelisson, D., Alahyane, N., Panouilleres, M., & Tilikete, C. (2010). Sensorimotor adaptation of saccadic eye movements. *Neuroscience and Biobehavioral Reviews, 34*, 1103–1120.

Port, N. L., & Wurtz, R. H. (2003). Sequential activity of simultaneously recorded neurons in the superior colliculus during curved saccades. *Journal of Neurophysiology, 90*, 1887–1903.

Prsa, M., & Thier, P. (2011). The role of the cerebellum in saccadic adaptation as a window into neural mechanisms of motor learning. *European Journal of Neuroscience, 33*, 2114–2128.

Quaia, C., Lefevre, P., & Optican, L. M. (1999). Model of the control of saccades by superior colliculus and cerebellum. *Journal of Neurophysiology, 82*, 999–1018.

Quaia, C., & Optican, L. M. (1998). Commutative saccadic generator is sufficient to control a 3-D ocular plant with pulleys. *Journal of Neurophysiology, 79*, 3197–3215.

Quaia, C., & Optican, L. M. (2003). Dynamic eye plant models and the control of eye movements. *Strabismus, 11*, 17–31. doi:10.1076/stra.11.1.17.14088.

Quaia, C., & Optican, L. M. (2011). Three-dimensional rotations of the eye. In L. A. Levin, S. F. E. Nilsson, J. V. Hoeve, S. M. Wu, P. L. Kaufman, & A. Alm (Eds.), *Adler's physiology of the eye* (11 ed., pp. 208–219). Edinburgh, Scotland: Saunders Elsevier.

Quaia, C., Ying, H. S., Nichols, A. M., & Optican, L. M. (2009a). The viscoelastic properties of passive eye muscle in primates: I. Static forces and step responses. *PLoS ONE, 4*, e4850. doi:10.1371/journal.pone.0004850.

Quaia, C., Ying, H. S., & Optican, L. M. (2009b). The viscoelastic properties of passive eye muscle in primates: II. Testing the quasi-linear theory. *PLoS ONE, 4*, e6480. doi:10.1371/journal.pone.0006480.

Quaia, C., Ying, H. S., & Optican, L. M. (2010). The viscoelastic properties of passive eye muscle in primates: III. Force elicited by natural elongations. *PLoS ONE, 5,* e9595. doi:10.1371/journal.pone.0009595.

Quaia, C., Ying, H. S., & Optican, L. M. (2011). The nonlinearity of passive extraocular muscles. *Annals of the New York Academy of Sciences, 1233,* 17–25. doi:10.1111/j.1749-6632.2011.06111.x.

Ramat, S., Leigh, R. J., Zee, D. S., & Optican, L. M. (2005). Ocular oscillations generated by coupling of brainstem excitatory and inhibitory saccadic burst neurons. *Experimental Brain Research, 160,* 89–106.

Raphan, T. (1998). Modeling control of eye orientation in three dimensions: I. Role of muscle pulleys in determining saccadic trajectory. *Journal of Neurophysiology, 79,* 2653–2667.

Raybourn, M. S., & Keller, E. L. (1977). Colliculoreticular organization in primate oculomotor system. *Journal of Neurophysiology, 40,* 861–878.

Ritchie, L. (1976). Effects of cerebellar lesions on saccadic eye movements. *Journal of Neurophysiology, 39,* 1246–1256.

Robinson, D. A. (1964). The mechanics of human saccadic eye movement. *Journal of Physiology, 174,* 245–264.

Robinson, D. A. (1970). Oculomotor unit behavior in the monkey. *Journal of Neurophysiology, 33,* 393–403.

Robinson, D. A. (1972). Eye movements evoked by collicular stimulation in the alert monkey. *Vision Research, 12,* 1795–1808. doi:10.1016/0042-6989(72)90070-3.

Robinson, D. A. (1975). Oculomotor control signals. In G. Lennerstrand & P. Bach-y-Rita (Eds.), *Basic mechanisms of ocular motility and their clinical implications* (pp. 337–374). Oxford, England: Pergamon.

Robinson, D. A. (1981). Models of the mechanics of eye movements. In B. L. Zuber (Ed.), *Models of oculomotor behavior and control* (pp. 21–41). Boca Raton, FL: CRC Press.

Ron, S., & Robinson, D. A. (1973). Eye movements evoked by cerebellar stimulation in the alert monkey. *Journal of Neurophysiology, 36,* 1004–1022.

Rucker, J. C., Ying, S. H., Moore, W., Optican, L. M., Büttner-Ennever, J., Keller, E. L., et al. (2011). Do brainstem omnipause neurons terminate saccades? *Annals of the New York Academy of Sciences, 1233,* 48–57.

Sato, H., & Noda, H. (1992). Posterior vermal Purkinje cells in macaques responding during saccades, smooth pursuit, chair rotation and/or optokinetic stimulation. *Neuroscience Research, 12,* 583–595.

Schiller, P. H., & Stryker, M. (1972). Single-unit recording and stimulation in superior colliculus of the alert rhesus monkey. *Journal of Neurophysiology, 35,* 915–924.

Schiller, P. H., True, S. D., & Conway, J. L. (1979). Effects of frontal eye field and superior colliculus ablations on eye movements. *Science, 206,* 590–592.

Schnabolk, C., & Raphan, T. (1994). Modeling three-dimensional velocity-to-position transformation in oculomotor control. *Journal of Neurophysiology, 71,* 623–638.

Scudder, C. A., Fuchs, A. F., & Langer, T. P. (1988). Characteristics and functional identification of saccadic inhibitory burst neurons in the alert monkey. *Journal of Neurophysiology, 59,* 1430–1454.

Scudder, C. A., Kaneko, C. S., & Fuchs, A. F. (2002). The brainstem burst generator for saccadic eye movements: A modern synthesis. *Experimental Brain Research, 142,* 439–462.

Shinoda, Y., Sugiuchi, Y., Takahashi, M., & Izawa, Y. (2011). Neural substrate for suppression of omnipause neurons at the onset of saccades. *Annals of the New York Academy of Sciences, 1233,* 100–106.

Soetedjo, R., Kaneko, C. R. S., & Fuchs, A. F. (2002). Evidence that the superior colliculus participates in the feedback control of saccadic eye movements. *Journal of Neurophysiology, 87,* 679–695.

Sparks, D. L. (1978). Functional properties of neurons in the monkey superior colliculus: coupling of neuronal activity and saccade onset. *Brain Research, 156,* 1–16.

Sparks, D. L., Holland, R., & Guthrie, B. L. (1976). Size and distribution of movement fields in the monkey superior colliculus. *Brain Research, 113,* 21–34.

Sparks, D. L., & Mays, L. E. (1980). Movement fields of saccade-related burst neurons in the monkey superior colliculus. *Brain Research, 190,* 39–50.

Sparks, D. L., & Travis, R. P., Jr. (1971). Firing patterns of reticular formation neurons during horizontal eye movements. *Brain Research, 33,* 477–481.

Strassman, A., Highstein, S. M., & McCrea, R. A. (1986). Anatomy and physiology of saccadic burst neurons in the alert squirrel monkey: II. Inhibitory burst neurons. *Journal of Comparative Neurology, 249,* 358–380.

Sylvestre, P. A., & Cullen, K. E. (1999). Quantitative analysis of abducens neuron discharge dynamics during saccadic and slow eye movements. *Journal of Neurophysiology, 82,* 2612–2632.

Takagi, M., Zee, D. S., & Tamargo, R. J. (1998). Effects of lesions of the oculomotor vermis on eye movements in primate: Saccades. *Journal of Neurophysiology, 80,* 1911–1931.

Tweed, D., Misslisch, H., & Fetter, M. (1994). Testing models of the oculomotor velocity-to-position transformation. *Journal of Neurophysiology, 72,* 1425–1429.

Van Gisbergen, J. A., Robinson, D. A., & Gielen, S. (1981). A quantitative analysis of generation of saccadic eye movements by burst neurons. *Journal of Neurophysiology, 45,* 417–442.

van Opstal, A. J., & Goossens, H. H. (2008). Linear ensemble-coding in midbrain superior colliculus specifies the saccade kinematics. *Biological Cybernetics, 98,* 561–577.

Vilis, T., Hepp, K., Schwarz, U., & Henn, V. (1989). On the generation of vertical and torsional rapid eye movements in the monkey. *Experimental Brain Research, 77,* 1–11. doi:10.1007/BF00250561.

Westheimer, G., & McKee, S. P. (1975). Visual acuity in the presence of retinal-image motion. *Journal of the Optical Society of America, 65,* 847–850.

Wurtz, R. H., & Albano, J. E. (1980). Visual–motor function of the primate superior colliculus. *Annual Review of Neuroscience, 3,* 189–226.

Wurtz, R. H., & Goldberg, M. E. (1971). Superior colliculus cell responses related to eye movements in awake monkeys. *Science, 171,* 82–84.

Wurtz, R. H., & Goldberg, M. E. (1972). The primate superior colliculus and the shift of visual attention. *Investigative Ophthalmology, 11,* 441–450.

Wurtz, R. H., & Optican, L. M. (1994). Superior colliculus cell types and models of saccade generation. *Current Opinion in Neurobiology, 4,* 857–861.

Xu-Wilson, M., Tian, J., Shadmehr, R., & Zee, D. S. (2011). TMS perturbs saccade trajectories and unmasks an internal feedback controller for saccades. *Journal of Neuroscience, 31,* 11537–11546.

Yamada, J., & Noda, H. (1987). Afferent and efferent connections of the oculomotor cerebellar vermis in the macaque monkey. *Journal of Comparative Neurology, 265,* 224–241.

Yoshida, K., Iwamoto, Y., Chimoto, S., & Shimazu, H. (1999). Saccade-related inhibitory input to pontine omnipause neurons: An intracellular study in alert cats. *Journal of Neurophysiology, 82*, 1198–1208.

Yoshida, K., Iwamoto, Y., Chimoto, S., & Shimazu, H. (2001). Disynaptic inhibition of omnipause neurons following electrical stimulation of the superior colliculus in alert cats. *Journal of Neurophysiology, 85*, 2639–2642.

Zee, D. S., Yee, R. D., Cogan, D. G., Robinson, D. A., & Engel, W. K. (1976). Ocular motor abnormalities in hereditary cerebellar ataxia. *Brain, 99*, 207–234.

61 Neural Mechanisms of Eye Movements: Three-Dimensional Control and Perceptual Consequences

ELIANA M. KLIER, GUNNAR BLOHM, AND J. DOUGLAS CRAWFORD

Most body parts (the head, torso, arms, and legs) move in three dimensions (3-D). However, the third dimension is often ignored when it comes to movements of the eye. For example, we know that our eyes move horizontally and vertically, but we are unaware that they also rotate in a third, torsional dimension, which closely corresponds to movement around the line of sight (i.e., gaze) when looking straight ahead. Although these torsional components are of smaller amplitude than their horizontal and vertical counterparts, the brain possess torsion-specific nuclei and muscles, and their control leads to well-defined, torsionally constrained behaviors. Moreover, a complete description of 3-D rotations involves motion control properties and perceptual consequences that disappear when one applies only abstract descriptions based on the mathematical properties of two-dimensional translations. Thus it is worth spending some time examining the neural control of 3-D eye and gaze (combined eye and head) movements.

BEHAVIOR

Measuring 3-D Eye Movements, Listing's Law and the Half-Angle Rule

Understanding 3-D kinematics first involves describing how one measures these movements. Rather than describing eye positions in Cartesian or polar coordinates (e.g., 10° to the left), 3-D eye movement researchers have found it useful to describe eye orientation as an axis of rotation that takes the eye from some reference position (e.g., straight ahead) to its current location (see figure 61.1A). The pointing direction of the axis describes the direction of rotation according to the right hand rule (i.e., align the thumb with the axis and the fingers curl in the direction of eye rotation), and the length of the axis describes the amplitude of rotation. Together, these define an orientation vector.

Interestingly, when the head is upright and motionless, all these orientation vectors lie in a plane—here, that plane corresponds to the paper on which the figure is drawn. If one looks at these same vectors from a different perspective—a side view—one can appreciate its relative thinness (see figure 61.1B). This plane is known as Listing's plane, and eye movements that adhere to it obey Listing's law. Note that the eye is mechanically capable of rotating about the line of sight at any position, and this could produce systematic or variable distortions of Listing's plane without affecting gaze direction. However, the brain chooses not to do this. Thus Listing's law is one particular solution to the "degrees of freedom" problem, as it is called in motor control studies.

To apply these conventions to real eye movements, scientists place a 3-D eye coil (a contact lens with embedded conductive wire) on the eye and ask subjects to move their eyes randomly while seated inside a set of three magnetic fields. Using the rotation vectors described above and, for simplicity, only indicating the tip of each vector, each possible eye position can be seen to lie in this plane (figure 61.1C—behind view; figure 61.1D—side view), which is typically ~1° in thickness (Straumann et al., 1995; Tweed & Vilis, 1990), and the line of site perpendicular to it is called primary position. Note that primary position is typically different from straight ahead eye position as its location depends on the orientation of Listing's plane in the head.

In its simplest form, Listing's law states that each and every horizontal/vertical gaze direction (e.g., 10° right and 5° up) is associated with one unique torsional component, and that component is zero (Helmholtz, 1867; Westheimer, 1957). This formulation holds under two conditions: (1) Eye positions must be described using the conventions described above, and (2) torsion is defined as rotation about a head-fixed axis parallel to

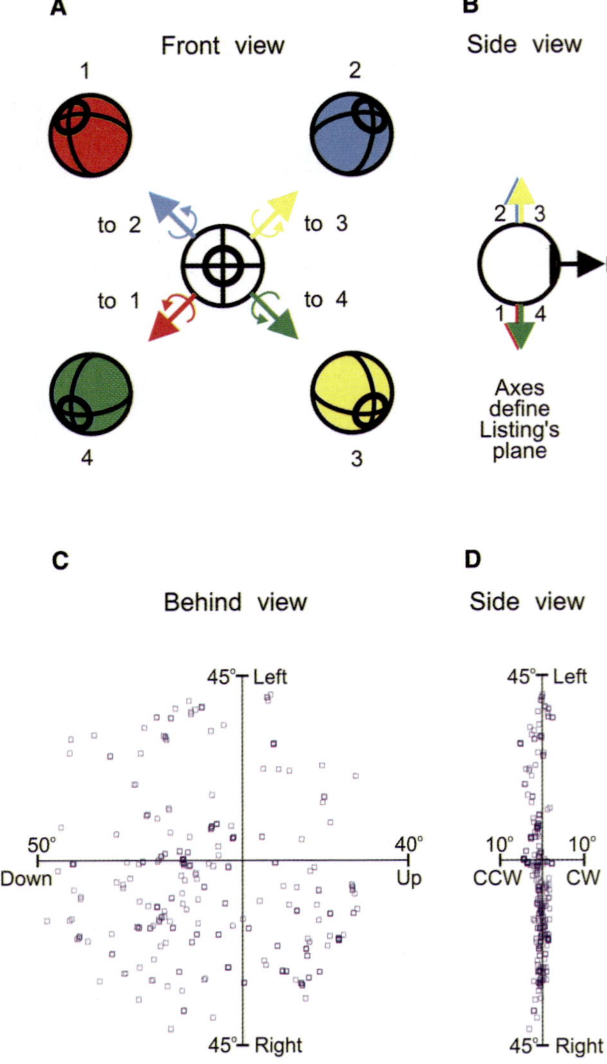

FIGURE 61.1   Measuring 3-D eye movements and Listing's law. (A) Eye orientations are described by axes of rotation (arrows) that take the eye from primary position (center eye) to any other position (1–4). The orientation of each axis describes the direction of rotation according to the right hand rule (align right thumb with the arrow and right fingers curl in the direction of rotation), and the length of the axis describes the amplitude of rotation. (B) A side view of these vectors reveals that they all lie in one plane which is perpendicular to primary position. (A and B adapted with permission from Crawford and Vilis, 1995.) (C) A behind view of real human data, in which only the tips of the rotation vectors are shown (squares). (D) A side view of the data points in (C) illustrate how these data points are confined to Listing's plane. (C and D adapted with permission from Klier and Crawford, 1998.)

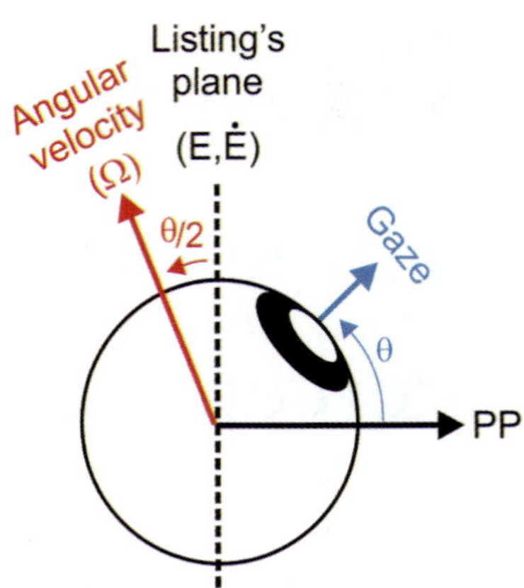

FIGURE 61.2   The half-angle rule. In order for eye position to remain in Listing's plane, angular eye velocity (Ω) must tilt out of Listing's plane by half the amount of gaze eccentricity from primary position (Θ). Note that the derivative of eye position (Ė = dE/dt) remains in Listing's plane (dashed vertical line), but Ė does not correctly describe the velocity of rotating objects (see text).

primary position. This turns out to be a remarkable and tightly controlled motor phenomenon that holds for fixations, saccades, and pursuit eye movements (Ferman, Collewijn, & Van den Berg, 1987). Incidentally, other body parts like the head and arm obey a similar, more general version of Listing's law known as Donders' law (head—Radau, Tweed, & Vilis, 1994; arm—Hore, Watts, & Vilis, 1992). Donders' law states that when a body part points in particular direction, it will always assume the same torsional component (although not necessarily zero) (Donders, 1848).

While Listing's law describes how eye orientation vectors lie in Listing's plane, the rotation vectors that take the eye from one orientation to another tilt out of Listing's plane. Seemingly counterintuitive at first, this becomes clear when examining the underlying mathematics of rotations. Because rotations are noncommutative (rotation A then B ≠ rotation B then A), one cannot simply subtract orientation vectors to obtain the relative rotation. Consequently, the rotation axis (measurable instantaneously as the angular velocity vector) must tilt out of Listing's plane by half the (orthogonal) angle between current gaze and primary position (see figure 61.2). For example, a horizontal saccade made at a gaze elevation of 10° requires that the velocity vector tilt out of Listing's plane by 5°. This is known as the half-angle rule (Tweed & Vilis, 1987). Without this rule, a sequence of two or more rotations about axes in Listing's plane would lead to eye orientations with torsional components out of Listing's plane. Thus the half-angle rule provides the proper compensation of torsional axis tilts to keep eye orientations in Listing's plane throughout a saccade or pursuit eye movement.

Listing's Law Modifications during Vergence and Static Head Tilt

Listing's law, or some variant, is obeyed by all movements that redirect gaze from one object to another. Vergence movements do this for objects in depth and require that both eyes move in opposite directions (converging for nearer objects or diverging for farther objects). A variation of Listing's law, called L2, states that while Listing's plane is frontoparallel when the two eyes are parallel, the planes of the two eyes tilt outwards (like saloon doors) when the two eyes converge (Mok et al., 1992; Van Rijn & Van den Berg, 1993). Thus as both eyes look nasally, Listing's plane of the right eye rotates clockwise and Listing's plane of the left eye rotates counterclockwise (from an above view).

Listing's law is obeyed by eye movements when the head is fixed. However, if the head is fixed and tilted, Listing's plane is still realized, but it exhibits a torsional offset (Crawford & Vilis, 1991; Haslwanter et al., 1992). If the head is rotated clockwise (i.e., right ear down), Listing's plane shifts in a counterclockwise direction (i.e., an offset to the left in figure 61.1D), and vice versa. This shift is due to ocular counterroll which causes a ~10% counterrotation of the eyes whenever the head is rolled (Collewijn et al., 1985).

Violations of Listing's Law during VOR and OKR

Other eye movements are accompanied by head movements and/or involve movements whose goal is to keep the image of the object stable on the retina. To attain these goals, these movements must violate Listing's law and allow the eyes to assume nonzero torsional components. The vestibular–ocular reflex (VOR) is one such example (Crawford & Vilis, 1991; Fetter et al., 1992). This simple reflex causes the eyes to rotate in an equal and opposite direction to head rotation. Thus, torsional head movements (e.g., right ear down to the right shoulder) obviously cause violations of Listing's law. However, even horizontal and vertical VOR movements do not follow the half-angle rule, so they also cause position-dependent violations of Listing's law. Similarly, the optokinetic reflex (OKR) stabilizes the images of moving objects on the retina. Thus if an image is moving in a circular pattern in the frontoparallel plane, the eyes will also rotate torsionally (for as long as possible before resetting with a nystagmus-like quick phase). As a rule of thumb, eye movements engaged in visual tasks that do not specify the required torsion (e.g., saccades, pursuit, vergence) obey Listing's law whereas eye movements where torsion is required for the visual task (e.g., visual stabilization during head rotation; VOR, OKR)

do not obey Listing's law (Crawford, Tweed, & Vilis, 2003).

Listing's Law during Head-Free Movements

When the head is free to move, as is the case in everyday life, torsional constraints are complicated because more parameters need to be controlled. The eyes move in the head, the head moves in space, and both of these contribute to eye motion in space (i.e., gaze). Listing's law is obeyed by the eye-in-head at the end of each movement. In contrast, the head-in-space obeys Donders' law in a specific way known as the Fick strategy (see figure 61.3) (Glenn & Vilis, 1992; Radau et al., 1994). Here, instead of orientation vectors maintaining zero torsion on a flat plane, they form a twisted surface, with torsion at the corners (i.e., clockwise [CW] when up/left and down/right, and counterclockwise [CCW] when up/right and down/left). This is how it looks when torsion is defined using the conventions in figure 61.1 (if Fick

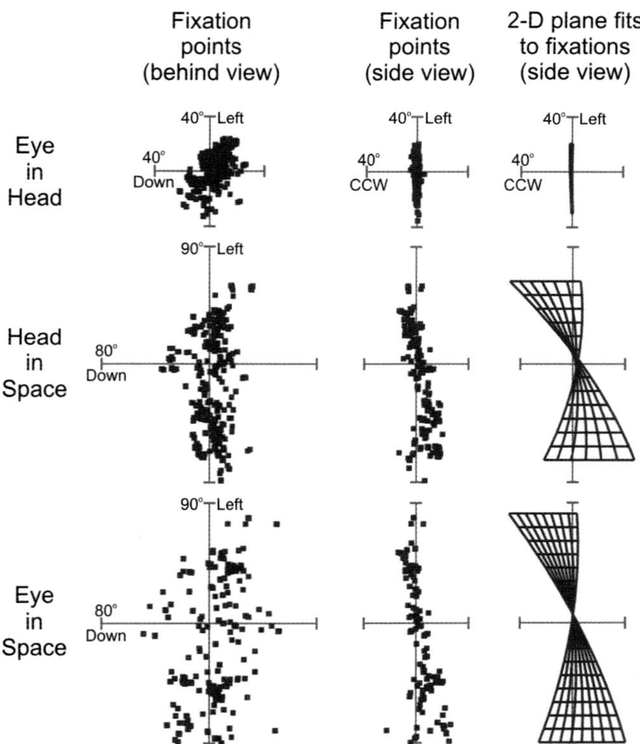

FIGURE 61.3 Three-dimensional behavior of the eyes and head. Tips of rotation vectors of the eye-in-head (top row), head-in-space (middle row), and eye-in-space (bottom row) are illustrated from a behind view (left column) and side view (center column). A 3-D plane is fit to the side view data in the right column that better illustrates the 3-D shape of these data. Notice that the eye-in-head data are planar (Listing's plane) whereas the head- and eye-in-space data look like a twisted plane known as a Fick surface. CCW, counterclockwise. (Adapted with permission from Klier et al., 2003.)

coordinates are used, i.e., where the vertical axis is fixed in the body and the horizontal axis is fixed in the head, then the data would again resemble a zero-torsion plane). The eye-in-space also obeys Donders' law using this Fick strategy. It is thought that the Fick strategy is implemented by the head-in-space and that the eye-in-space follows suit because of the way eye and head rotations naturally combine (Crawford et al., 1999; Glenn & Vilis, 1992).

At the end of a head-free gaze shift, the eye-in-head lands in Listing's plane, but this is not the case during the movement. These movements can be divided into two distinct parts (see figure 61.4A). The first phase (1) begins with the initiation of a gaze shift and ends when the eye-in-space reaches its intended target. For most of this phase, gaze is carried by the eye-in-head and there is little head movement. Here one might expect Listing's law to be obeyed, but it is not for the time being. Instead, the oculomotor system adds a systematic torsional component out of Listing's plane. To understand why, we look to the second phase (2). It begins when the eye-in-space lands on target and the head, now in motion, rotates toward the target until the eyes become centered in the orbits. Due to head motion in this latter phase, and because the eye-in-space must remain on target, the eye-in-head counter rotates according to the VOR. Consequently, Listing's law cannot be obeyed in this second phase. Thus, the oculomotor system must first add torsion to the saccade, in an anticipatory fashion, so that saccade (1) and VOR (2) torsion cancel and movements end up back in Listing's plane (Crawford et al., 1999). The following section describes the physiological implementation of these 3-D rules.

NEURAL AND MECHANICAL CONTROL

Cortical, 2-D Signals

Voluntary gaze shifts (eye only or eye + head) are programmed in the cortex and superior colliculus (SC) and subsequently transmitted to the brain stem for processing and to the oculomotor plant (i.e., the eye and its surrounding muscles and tissues) for execution. The vast majority of oculomotor cortical areas encode only the horizontal and vertical components of desired gaze direction. This has been demonstrated in two ways. First, cortical eye fields, like the frontal and supplementary eye fields, were artificially stimulated with an electrical pulse with the head free to move. This pulse generates a command that is transmitted downstream until an eye-head gaze shift is made. If each cortical site encodes one, unique, nonzero torsional value, then repeated stimulations should elicit gaze shifts with the same, unique, nonzero torsional eye-in-head component. This would lead to violations of Listing's and Donders' laws since final eye and head positions would lie out of their respective surfaces. On the other hand, if each site encodes only a 2-D command and the torsional component is added on downstream, then repeated stimulations should elicit gaze shifts that end on their respective 3-D surfaces, much like normally elicited eye and head movements. The latter was the case (supplementary eye field: Martinez-Trujillo, Wang, & Crawford, 2003; frontal eye field: Monteon et al., 2010). Stimulation of parietal cortex did produce eye movements with inappropriate torsion, but this was because the normal, phase (2) head movements were not elicited (Constantin et al., 2009).

Second, a number of experiments have shown that the SC, the gateway for cortical gaze signals to the brain stem, also carries 2-D commands. Stimulation of the SC also produced normally coordinated eye and head movements that landed in Listing's and Fick planes (Klier, Wang, & Crawford, 2003). Again, this implies that 3-D commands are added on downstream. Other stimulation studies have also backed up the finding that the SC has a 2-D motor map (Hepp et al., 1993; Van Opstal et al., 1991). In addition, bilateral inactivation of the SC does not result in violations of Listing's law during fixations or saccades, and this inactivation does not affect the torsional components of VOR fast phases (Hepp et al., 1993).

The fact that the cortex and SC encode gaze in 2-D makes sense since higher cortical functions are more concerned with redirecting the line of site to a particular location and less concerned about the correct torsional geometry once that 2-D gaze location is reached. However, torsional eye-in-head orientations must be added on somewhere because of the unique 3-D behavior observed during head-free movements and the VOR. Also, a 3-D signal is necessary to move the six extraocular muscles which rotate the eye horizontally, vertically, and torsionally. Thus the SC's 2-D, spatial code must be translated into a 3-D, temporal code. This transformation could be accomplished either neurally, by adding a 3-D signal onto the 2-D command, or by mechanical factors like the anatomical arrangement of the ocular plant. Both these factors are at play.

Brain Stem, 3-D Signal

Output of the SC reaches the burst generators, which produce a 3-D, temporal velocity signal that drives the eye from one location to the next. Interestingly, there are separate burst generators for the horizontal components of eye movements and the vertical/torsional

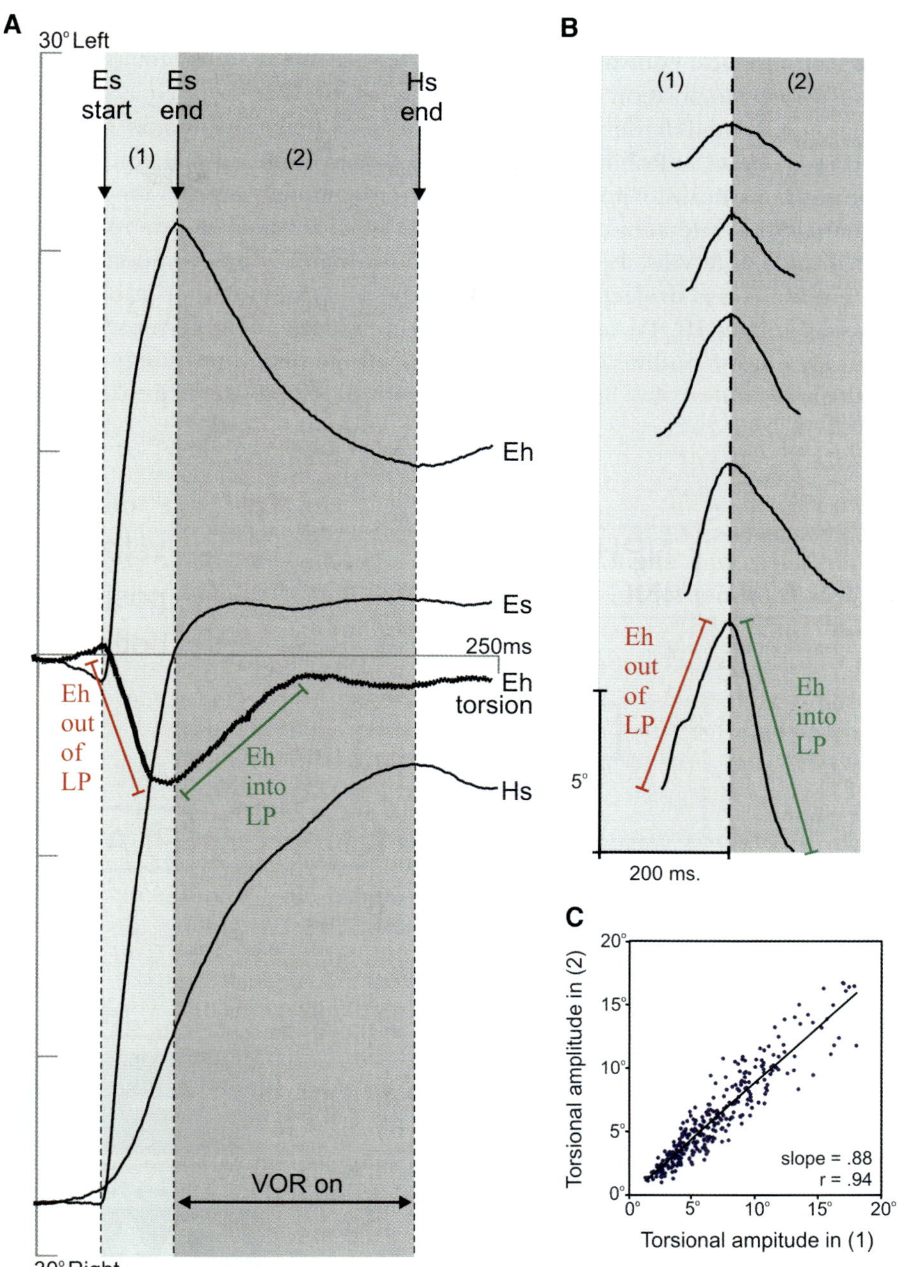

FIGURE 61.4   Torsion during a gaze shift. (A) A plot of eye-in-head (Eh), eye-in-space (Es), and head-in-space (Hs) position as a function of time. In phase 1 (light gray), gaze (Es) is carried to the target by the eye (Eh) while the head (Hs) is mostly stationary. In phase 2 (dark gray), the head (Hs) has overcome inertia and begins to move while gaze (Es) remains steadily fixed on target via the vestibular–ocular reflex (VOR), which rotates the eye (Eh) in an equal and opposite direction to the head (Hs). Eh torsion relative to 0° torsion (abscissa) is superimposed on the same plot. Eh torsion goes out of Listing's plane (LP) in phase 1 in a predictive manner such that it is returned to ~0° torsion in phase 2. Thus the net torsion on the eye (Eh) is zero. (B) Five examples of Eh torsion during gaze shifts of increasing amplitude. The torsional amplitudes in phases 1 and 2 appear to be the same. (C) Quantification of Eh torsional amplitude during phase 1 versus phase 2 for multiple gaze shifts (circles). The amount of torsion in both phases is comparable. (Adapted with permission from Klier et al., 2003.)

components. The former are found in the paramedian pontine reticular formation (PPRF) (Luschei & Fuchs, 1972) while the latter are located in the rostral interstitial nucleus of the medial longitudinal fasciculus (riMLF) (Büttner, Büttner-Ennever, & Henn, 1977; King & Fuchs, 1979). These velocity signals are sufficient to drive the eyes, but once at their new location, a second, position command is necessary to hold them there (otherwise the eyes would drift back toward a more central, resting position). This position signal is generated by groups of cells known as neural integrators that perform the mathematical equivalent of integration (i.e., converting velocity to position). Again, the horizontal neural integrator, located in the nucleus prepositus hypoglossi (NPH) (Canon & Robinson, 1987), is found separately from the vertical/torsional integrators, which are located in the interstitial nucleus of Cajal (INC) (Crawford, Cadera, & Vilis, 1991). Finally, all these four brain stem structures project to the oculomotor neurons (cranial nerves III, IV, and VI) that drive the eye muscles. The classic outline of the brain stem saccade generator is shown in figure 61.5A (Robinson, 1981), and the anatomical locations of the burst neurons and neural integrators are shown in figure 61.5B.

Note that the division of labor between horizontal and vertical/torsional components found in the brain stem mimics the division of labor found in the eye muscles themselves, where the lateral and medial recti control horizontal components of eye movements while the superior and inferior oblique muscles and the superior and inferior recti muscles control vertical/torsional components. Interestingly, a similar division is also found in the semicircular canals that detect head

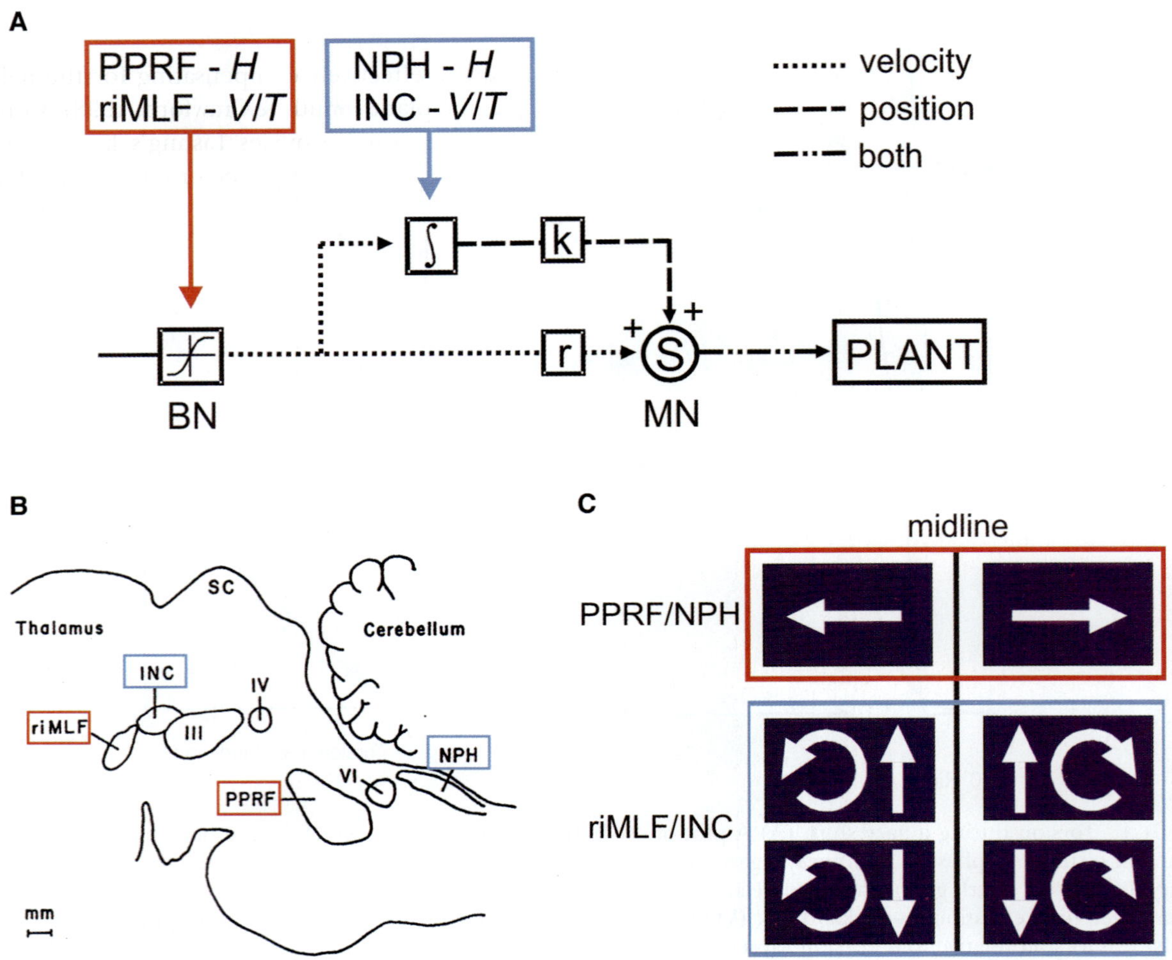

FIGURE 61.5 The three-dimensional, brain stem saccade generator. (A) The Robinson (1981) model of the brain stem saccade generator. Burst neurons (BN) output a velocity command (dotted line) that overcomes the eye's viscosity (r) and is sent to the motoneurons (MN). This velocity command is also sent to the neural integrators ($\int$) that output a position command (dashed line), which counters the eye's elasticity (k), and is also sent to the MNs. Thus the MNs send both position and velocity commands to the oculomotor plant. Horizontal (*H*) BNs are located in the paramedian pontine reticular formation (PPRF), and vertical/torsional (*V/T*) BNs are found in the rostral interstitial nucleus of the medial longitudinal fasciculus (riMLF). Horizontal $\int$s are located in the nucleus prepositus hypoglossi (NPH), and vertical/torsional $\int$s are found in the interstitial nucleus of Cajal (INC; see text). (B) A midsagittal section through the primate brain stem reveals the anatomical locations of the components of the brain stem saccade generator. Adapted with permission from Henn, Hepp, and Büttner-Ennever (1982). (C) Three-dimensional control of the torsional, vertical, and horizontal components of gaze shifts across the midline. (Adapted with permission from Crawford and Vilis, 1992.)

acceleration (the horizontal canals detect motion about the yaw axis, while the left/right-anterior and left/right-posterior canals detect motion about the pitch/roll axes).

Furthermore, directional segregation has been examined within the neural integrators (NPH and INC). The right NPH encodes rightward eye positions, and the left NPH encodes leftward eye positions. The right INC encodes clockwise torsional eye positions, and the left INC encodes counterclockwise torsional eye positions, but upward and downward neuron pools are present bilaterally (Crawford et al., 1991; Crawford & Vilis, 1992). With this unique setup (see figure 61.5C), obeying Listing's law (i.e., zero torsion) can be accomplished by balancing torsional INC neuron pools on either side of the brain. And movements that do not obey Listing's law, like those associated with the VOR and head-free gaze shifts, can be accomplished by activating one side of the INC more than the other (Crawford et al., 1991; Crawford & Vilis, 1992). Also, because these nuclei encode both eye and head movements, disorders such as the ocular tilt reaction and torticollis may be caused by an imbalance in INC activity across the midline (Klier et al., 2002).

Mechanical Factors

Evidence also exists for a role of the oculomotor plant in implementing Listing's law (Demer et al., 1995; Demer, Oh, & Poukens, 2000; Quaia & Optican, 1998). Anatomical and imaging studies have shown that eye muscles are segregated into two distinct layers: global and orbital. The former inserts onto the eyeball while the latter passes through adjacent orbital tissue and acts as a pulley, effectively changing the eye's pulling direction. The placement of the pulleys, and their own differential innervations, influences the eye's axis of rotation to implement Listing's law and the half-angle rule. These observations, made in static preparations, have been successfully modeled and explain eye movements that obey Listing's law. However, they have yet to be realized in dynamic, in vivo preparations, and these mechanical properties would need to be undone neurally in order to produce an ideal VOR (Smith & Crawford, 1998).

Additional evidence supporting a role for the eye plant in 3-D control comes from neural recordings of torsionally related oculomotor neurons. They only appear to encode the derivative of eye position (i.e., changes in eye orientation divided by time) rather than the angular eye velocity axes illustrated in figure 61.2 (Ghasia & Angelaki, 2005). Furthermore, stimulation of the abducens nerve produces eye movements that obey

Listing's law (Klier, Meng, & Angelaki, 2006). Since stimulation so late in the oculomotor pathway bypasses the brain, this finding supports the idea that the plant alone can generate the half-angle rule. However, similar stimulation results are found when the monkey is statically tilted (Klier, Meng, & Angelaki, 2011) and dynamically rotated sinusoidally in the roll plane (Klier, Meng, & Angelaki, 2012), indicating that the plant is obligated to implement the half-angle rule no matter what the situation.

So how can such a plant provide all the behaviors described above? It can still give any 3-D oculomotor behavior if it receives the right inputs. For example, theoretical simulations have shown that the plant can still give an ideal VOR if angular velocity signals from the vestibular system are converted into eye position derivatives, effectively compensating for the half-angle rule in the plant (Smith & Crawford, 1998). Conversely such a plant only provides Listing's law if it receives orientation and derivative vector commands that align with Listing's plane (Crawford & Guitton 1997; Quaia, Lefevre, & Optican, 1999). Finally, violations of Listing's law, like ocular counterroll or the transient torsion observed during head-free gaze shifts, require torsional orientation and derivative commands orthogonal to Listing's plane. Thus, the plant is ideally coupled to the neural mechanisms described above.

Two-Dimensional to Three-Dimensional Transformation

Where does the 2-D signal transform into a 3-D command to drive the eye muscles? Note that this question does not ask about the mechanism responsible for tilting saccade axes out of Listing's plane in a position-dependent manner, as that seems to be largely done by orbital mechanics (Demer et al., 1995, 2000). Instead, the question asks how is zero torsion in Listing's or Donders' coordinates selected? And how is the position range modified for behaviors that follow variations of Donders' law? Finally, how does the brain generate torsional eye-in-head commands that predict and nullify VOR-related, torsional components of head-free gaze shifts?

The 2-D to 3-D transformations must happen downstream of the SC and upstream of the burst neurons. It is possible that there is a direct mapping from the SC onto the burst neurons in such a way that CW and CCW signals on either side of the brain cancel, leading to zero torsion (figure 61.5C—activate the riMLF/INC bilaterally). However, this cannot explain how torsional strategies are modified or how these ranges are maintained after necessary violations of Listing's law (like those associated with head-free gaze shifts). To accomplish this, the system must have a modifiable set point or, more appropriately, a set surface, and a comparator. For example, when the head is tilted torsionally, the eye and corresponding Listing surface tilts in the opposite direction and thus the set surface is changed via vestibular inputs (Bockisch & Haslwanter, 2001; Haslwanter et al., 1992). Similarly, when the INC is inactivated unilaterally, Listing's plane is shifted toward the unaffected side and saccades are still generated in the direction of the shifted plane (even though final eye positions cannot be held in that plane) (Crawford et al., 2003). Here this set surface appears to be maintained by the saccade generator.

Returning to anatomy, neural activity in the nucleus reticularis tegmenti pontis (NRTP) has been correlated with the small corrective movements that are sometimes necessary to bring saccades back into Listing's plane (Van Opstal, Hepp, Suzuki, & Henn, 1996). The NRTP is a nucleus that inputs to the cerebellum, and patients with cerebellar damage show an increase in the thickness of Listing's plane and torsional offsets of the plane (Briar & Dieterich, 2009; Straumann, Zee, & Solomon, 2000). Also, the cerebellar flocculus and paraflocculus have been implicated in contributing an inhibitory torsional eye velocity component to the vestibular nuclei (Ghasia, Meng, & Angelaki, 2008). Finally, the central mesencephalic reticular formation has been implicated in the control of torsional head movements (Pathmanathan et al., 2006).

VISUAL CONSEQUENCES OF 3-D EYE (AND HEAD) MOVEMENTS

The patterns of 3-D eye movement described above have an enormous impact on monocular and binocular vision for action and perception. In addition, they also build and update accurate internal representations of the external world. It has been argued that Listing's law and its variants optimize motor and perceptual factors (Tweed, 1997a). However, while Listing's law may simplify eye movement control and aid binocular vision, it does not trivialize the brain's interpretation of visual signals. The brain must still account for the actual 3-D orientation of the eyes for both perception and action.

Monocular Consequences

When the eyes and head move to explore the visual environment, the resulting retinal images change. This might seem trivial, but the way retinal projections change across eye movements is not intuitive. These changes will be illustrated based on the "cyclopean eye,"

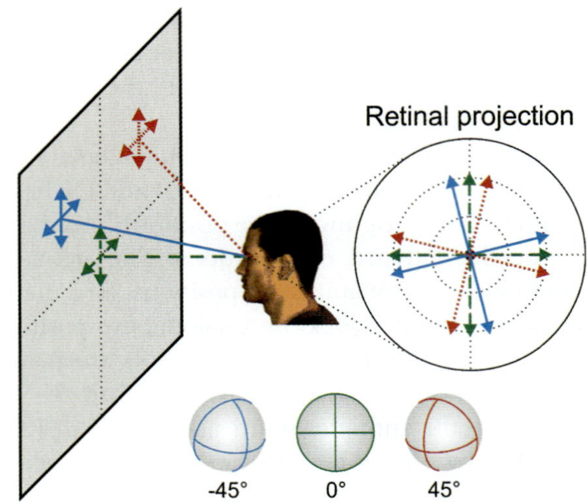

FIGURE 61.6   Retinal-spatial misalignment for oblique eye orientations. Single-axis eye rotations as described by Listing's law result in retinal orientations that are not aligned with space. For example, 45° left–up rotations (blue) result in a misalignment of the retinal axes relative to space so that spatial horizontal and vertical axes are projected onto the retina in a tilted way, that is, tilted counterclockwise relative to when the eyes are straight (green). Same argument holds for right–up fixations with clockwise rotated retinal projections (red). (Adapted with permission from Blohm and Lefevre, 2010.)

that is, a single representation of visual eccentricity based on the combination (e.g., averaging) of both retinal images (Ding & Sperling, 2006; Ono, Mapp, & Howard, 2002). For example, with the head straight, oblique eye movements result in a misalignment of the cyclopean retinal and spatial axes, partly due to eye tilt about the line of sight (Crawford & Guitton, 1997; Henriques & Crawford, 2000). Figure 61.6 shows how the retinal projections change across oblique eye movements. This happens due to the geometry of 3-D rotations, even in the absence of any net torsional eye movement component, because the eyes rotate around a single axis of rotation, which can cause a twist of the retinal axes relative to space. The misalignment angle can reach 15° when the eyes are in eccentric oblique positions (e.g., 45° up and right). And although this seems like a small effect, moving a hammer in the wrong direction by a few degrees can have undesirable consequences. Therefore, the brain must take this retinal-spatial misalignment into account when generating motor commands from visual inputs (Blohm & Crawford, 2007).

The geometry of retinal projections becomes more complex once the head is involved. Now the net rotation of the eye-in-space is determined by both 3-D eye-in-space and 3-D head-in-space orientation (Blohm &

Crawford, 2007; Tweed, 1997a). One simple consequence of head movements on retinal projection geometry can be illustrated for ocular counterroll. Here, the retinal projection pattern is rotated relative to space by the sum of head roll and (negative) ocular torsion. This is important for the visual system because correct interpretation of the visual inputs requires the brain to incorporate knowledge about ocular torsion (or head movements using an internal model of ocular counterroll) into the visual signals (Blohm & Crawford, 2007; Crawford, Henriques, & Medendorp, 2011). Head movements around other axes also influence ocular torsion because these movements alter Listing's plane. Therefore both eye and head orientations are important factors determining the retinal-spatial misalignment.

Moreover, even without torsion, properties of 3-D rotational geometry can produce large mismatches between target displacements in visual and motor coordinates, for example, every time the movement and eye orientation have orthogonal components (Blohm & Crawford 2007; Crawford & Guitton, 1997). In the head-free range, the resulting mismatch between retinal direction and directions in body coordinates can result in huge gaze shift or arm pointing errors. For example, a target that appears 90° left on the retina simply requires a leftward movement from straight ahead, but if gaze and the hand are pointing straight up, the same retinal stimulus requires motion that is equally left and down (Crawford & Guitton, 1997; Klier, Wang, & Crawford, 2001). Moreover, depth and direction become conflated when sensory information relative to the eye must be converted into motor commands relative to the body (Blohm & Crawford, 2007; Crawford et al., 2011). In practice, all these effects interact with the torsional effects described above, and the brain must account for these eye and head orientations (see below).

Consequences for Binocular Vision

Another important consequence of 3-D eye (and head) movements concerns binocular vision, in particular, stereopsis (i.e., depth perception from binocular vision). Stereopsis requires neurons to respond to small differences between both retinal images. And while no binocular orientation strategy results in perfect binocular retinal correspondence (Van Rijn & Van den Berg, 1993), L2 (described above) minimizes overall eye rotation and simultaneously aligns images within the visual plane by adjusting ocular torsion (Tweed, 1997b). Therefore, L2 may have evolved to maximize retinal correspondence and consequently binocularity across 3-D eye movements, thus narrowing the range of

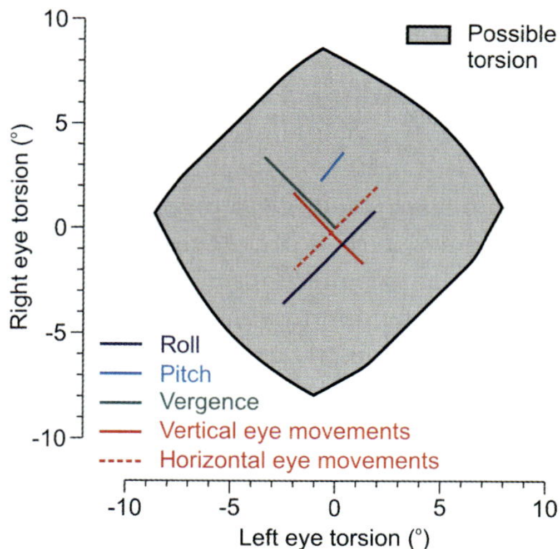

FIGURE 61.7   Possible binocular torsion states. The range of left (x-axis) and right (y-axis) eye torsion is shown across different static head orientations and vergence angles. This range increases dramatically once the head is moved dynamically. The effect of horizontal/vertical eye orientations, ocular vergence, and head roll/pitch on the torsional states is illustrated. For example, changing vergence angles has an opposite effect on left and right eye torsion as shown by the negative slope of the green line. Each eye's torsion here is the 3-D rotational component along the depth axis that brings the eyes from primary position to a given orientation. (Adapted with permission from Blohm et al., 2008.)

disparities that must be handled by striate cortex neurons.

For different ocular vergence angles, both eyes move differently as determined both by the vergence angle and by the head movement (Tweed, 1997b), and both have an interaction effect on Listing's plane (Bockisch & Haslwanter, 2001; Tweed, 1997b). The range of different right and left eye torsion combinations is shown in figure 61.7 and depicts how different eye and head movements influence the combination of both eyes' torsional states. The difference between both eyes' 3-D orientations in space yields two slightly different retinal images. This difference—called retinal disparity—is crucial to depth vision because it informs the brain about the distance of an object relative to the fixation distance, as determined by the vergence angle. Therefore, the way the eyes move has important consequences for depth vision (Blohm et al., 2008; Schreiber, Tweed, & Schor, 2006).

Figure 61.8 illustrates some of the effects eye movements can have on the retinal projection pattern of nonfoveated objects. When the eyes look straight ahead and the head is upright, objects in the visual field have a certain horizontal and vertical disparity (figure 61.8B).

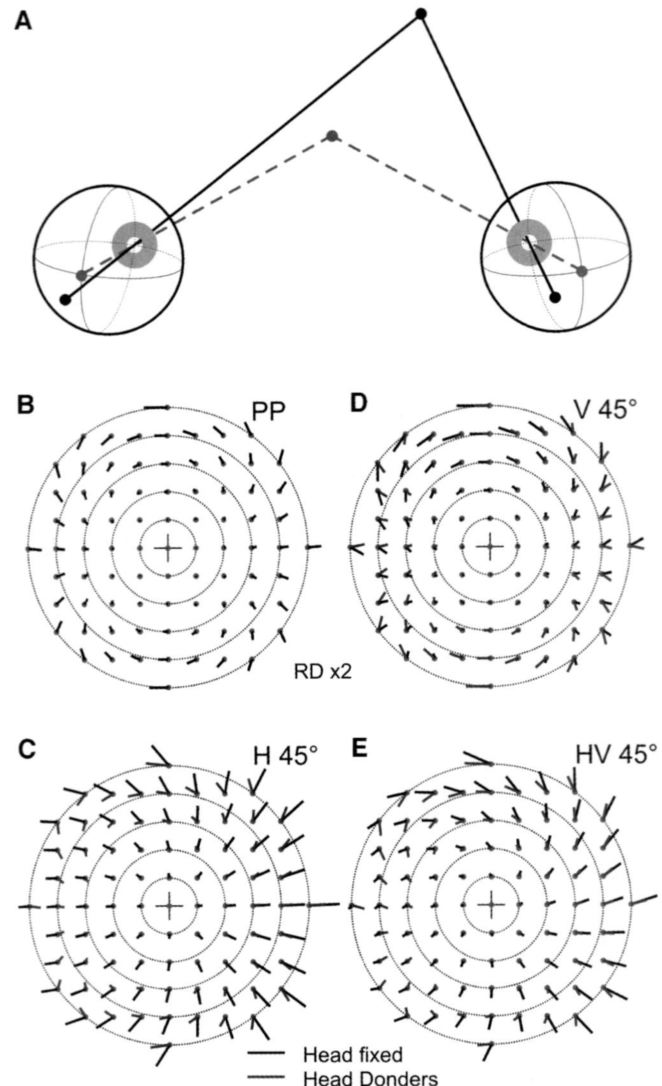

FIGURE 61.8   Consequence of L2 on retinal disparity. (A) Schematic showing the retinal projection geometry. Fixation (gray, dashed lines) and object projection lines (black) are shown. (B) Retinal projection pattern for primary position (PP) at 50-cm distance. Gray dots correspond to different cyclopean-eye-fixed targets (in 10° horizontal and vertical intervals arranged on a hemisphere at 50-cm distance), and the bars attached to them correspond to the disparity of the right and left eye's retinal images. The black bars show the direction and amplitude (length) of the retinal disparity (RD) associated with the cyclopean retinal target positions to which the bars are attached (magnified by a factor 2 for visibility). Target and fixation distance from the cyclopean eye was always 50 cm. Dotted circles are 10° intervals of retinal eccentricity. The central cross indicates the fixation position and fovea. Note that even at primary position, the interocular distance and natural tilt of Listing's law generates a nonzero retinal disparity pattern for targets on an isodistant sphere. (C) Retinal projection pattern when gaze is directed 45° horizontally (rightward), either with the head straight (black bars) or when the head accompanies the eyes in a natural fashion (gray bars), as described by Donder's law. (D) Same for the eyes oriented 45° vertically (up). (E) Same for 45° oblique eye orientations (up–right). Note that the retinal disparity pattern changes dramatically across eye and head orientations. (Adapted with permission from Blohm et al., 2008.)

When the eyes move, this pattern changes dramatically (figure 61.8C–E, black bars). Furthermore, this change depends on the contribution of the head (figure 61.8C–E, gray bars—for simplicity, only one example head orientation is shown). However, all these objects are located at the same distance from the cyclopean eye. Therefore, the brain must interpret these different binocular retinal inputs in an eye/head orientation-dependent fashion.

Which 3-D eye (and head) orientation signals does the brain need to uniquely compute object depth from retinal images? Indeed, it has previously been shown that the same binocular retinal stimulation can result from objects being located at different distances (Blohm

888   ELIANA M. KLIER, GUNNAR BLOHM, AND J. DOUGLAS CRAWFORD

et al., 2008). So how does the brain determine the object's true distance? Theoretically, the 3-D orientations of both eyes are needed to reconstruct 3-D location from binocular retinal inputs, in the case of a single, small, isolated object in darkness (Blohm et al., 2008). Vision of extended objects can provide additional (visual) information about the 3-D orientation of both eyes that are mathematically sufficient to estimate object depth (Horn, 1990). However, Blohm et al. (2008) have shown that 3-D eye or head signals are also used. Since there is no evidence that the brain has direct knowledge about both eyes' torsional angles (Banks, Hooge, & Backus, 2001), the brain must estimate these from other extraretinal signals (Schreiber et al., 2001; van Ee & van Dam, 2003), such as 2-D eye orientations, vergence and 3-D head orientation, as well as an internal model of L2 (Blohm et al., 2008).

Accounting for 3-D Eye Orientation in Vision, Memory, and Movement

To use spatial relationships of objects for perception and action, the brain creates internal representations of relevant locations. This can be seen as a type of internal working memory. Spatial updating is the process of adjusting internal representations of target location to reflect intervening eye, head, or body motion (reviewed in Klier & Angelaki, 2008). Much research has been devoted to uncovering the code (or reference frame) used by the brain during working memory (reviewed in Buneo & Andersen, 2006; Crawford et al., 2011). Early visual striate and extrastriate areas code visual information in a retinal frame of reference. This picture becomes slightly less clear for parietal areas, which are dominated by retinal codes (e.g., Batista et al., 1999; Khan et al., 2005) but include other coding schemes, for example, relative to the head (Battaglia-Mayer et al., 2001, 2003). In prefrontal and frontal areas there seem to be a wide variety of overlapping coding schemes for visual working memory (Martinez-Trujillo et al., 2004). This is important because any nonretinal code has to be generated by integrating retinal information with eye/head (or other) orientation signals. Thus knowledge about what these orientations are is critical. The advantage of these codes is that they may be independent of current gaze, which makes them robust to intervening movements.

Then what happens to the retinal codes in visual, parietal, and other areas when intervening movements of the eyes, head, and so forth occur? These codes must actively account for changes in gaze and remap (or update) the memory accordingly so that it remains in spatial register with the world. This time, absolute orientations of the eyes, head, and so forth are not required, but only changes in orientation and 3-D updating have been shown to account for changes in torsion during active (Medendorp et al., 2002) or passive (Klier, Angelaki, & Hess, 2005) head roll paradigms. Here however, the brain must also deal with the noncommutativity of rotations, which is indeed the case for eye (Smith & Crawford, 2001) and whole body (Klier, Angelaki, & Hess, 2007) movements. Finally, visual motion signals for saccades and perception are also updated across head roll (Ruiz-Ruiz & Martinez-Trujillo, 2008). Therefore, for action planning, updating has been shown to incorporate 3-D eye movement control signals, including ocular torsion.

Further, during movement execution, the brain must account for eye and head orientation to compensate for the geometric effects described above. The brain implements reference frame transformations to compute the correct depth and direction of gaze and reach movements from any initial eye and head orientation (Blohm & Crawford, 2007; Klier & Crawford, 1998). The neural mechanisms for these transformations can be theorized via neural network models that tend to develop appropriate "gain fields" for this purpose (e.g., Blohm, Keith, & Crawford, 2009), but few experiments have tested these models. For the gaze control system, some of these signals are present at the level of the SC (DeSouza et al., 2011), but otherwise the signals and the movements evoked by SC stimulation suggest that it simply uses a retinal code and leaves much of the transformation for later stages (Klier, Wang, & Crawford, 2001).

Less is known about neural mechanisms that account for 3-D eye orientation for perception. For example, there are varied findings concerning the brain's ability to account for ocular torsion. As described above, sensory-to-motor transformations tend to account for 3-D eye orientation when programming saccades (Klier & Crawford, 1998; Medendorp et al., 2002), reaches (Blohm & Crawford, 2007; Medendorp et al., 2002), and smooth pursuit eye movements (Blohm & Lefevre, 2010). However, perceptual experiments have provided mixed results (Wade & Curthoys, 1997). Some have shown that ocular torsion is not accounted for when judging subjective visual verticality (Baier, Bense, & Bieterich, 2008; Brandt, Dieterich, & Danek, 1994), although intersubject differences are large (Clemens et al., 2011; De Vrijer, Medendorp, & Van Gisbergen, 2009). Other studies report that the perceptual system has access to ocular torsion under certain circumstances when judging line orientation under different eye and head orientations (Haustein, 1992; Poljac, Lankheet, & Van Den Berg, 2005). It remains unclear why different perceptual and motor systems may or may not have access to 3-D eye orientation signals, and with very few exceptions (e.g., DeSouza et al., 2011), almost no neurophysiological investigations have even attempted to account for this.

SUMMARY

The central nervous system and peripheral anatomy have evolved with 3-D rotations in mind. Both the brain and body have adapted to restrict eye and head positions to Listing's and Donders' surfaces, respectively, and to implement the eye-in-head choreography necessary to exit and reenter Listing's plane with each and every gaze shift. And while the properties of 3-D eye rotation may seem subtle, they allow one to distinguish between fundamentally different models of neural and mechanical control. In addition, 3-D eye movements are not just a control problem, as they have important consequences on monocular and binocular vision as well as the way we remember and update memorized locations. Three-dimensional eye movement strategies such as Listing's law aid vision by restricting torsional possibilities, but still, these 3-D eye and head movements must be accurately accounted for to perform actions and perceive the external world. Much remains to be understood about the neural mechanisms and perceptual consequences of Listing's law and its variants.

REFERENCES

Baier, B., Bense, S., & Bieterich, M. (2008). Are signs of ocular tilt reaction in patients with cerebellar lesions mediated by the dentate nucleus? *Brain, 131,* 1445–1454.

Banks, M. S., Hooge, I. T., & Backus, B. T. (2001). Perceiving slant about a horizontal axis from stereopsis. *Journal of Vision, 1,* 55–79. doi:10.1167/1.2.1.

Batista, M. S., Buneo, C. A., Snyder, L. H., & Andersen, R. A. (1999). Reach plans in eye-centered coordinates. *Science, 285,* 257–260.

Battaglia-Mayer, A., Caminiti, R., Lacquaniti, F., & Zago, M. (2003). Multiple levels of representation of reaching in the parieto–frontal network. *Cerebral Cortex, 13,* 1009–1022.

Battaglia-Mayer, A., Ferraina, S., Genovesio, A., Marconi, B., Squatrito, S., Molinari, M., et al. (2001). Eye–hand coordination during reaching: II. An analysis of the relationships between visuomanual signals in parietal cortex and parieto–frontal association projections. *Cerebral Cortex, 11,* 528–544.

Blohm, G., & Crawford, J. D. (2007). Computations for geometrically accurate visually guided reaching in 3-D space. *Journal of Vision, 7*(5), 4, 1–22. doi:10.1167/7.5.4.

Blohm, G., Keith, G. P., & Crawford, J. D. (2009). Decoding the cortical transformations for visually guided reaching in 3D space. *Cerebral Cortex, 19,* 1372–1393.

Blohm, G., Khan, A. Z., Ren, L., Schreiber, K. M., & Crawford, J. D. (2008). Depth estimation from retinal disparity requires eye and head orientation signals. *Journal of Vision, 8*(16), 3, 1–23. doi:10.1167/8.16.3.

Blohm, G., & Lefevre, P. (2010). Visuomotor velocity transformations for smooth pursuit eye movements. *Journal of Neurophysiology, 104,* 2103–2115.

Bockisch, C. J., & Haslwanter, T. (2001). Three-dimensional eye position during static roll and pitch in humans. *Vision Research, 41,* 2127–2137. doi:10.1016/S0042-6989(01)00094-3.

Brandt, T., Dieterich, M., & Danek, A. (1994). Vestibular cortex lesions affect the perception of verticality. *Annals of Neurology, 35,* 403–412.

Briar, B., & Dieterich, M. (2009). Ocular tilt reaction: A clinical sign of cerebellar infarctions? *Neurology, 72,* 572–573.

Buneo, C. A., & Andersen, R. A. (2006). The posterior parietal cortex: Sensorimotor interface for the planning and online control of visually guided movements. *Neuropsychologia, 44,* 2594–2606.

Büttner, U., Büttner-Ennever, J. A., & Henn, V. (1977). Vertical eye unit related activity in the rostral mesencephalic reticular formation of the alert monkey. *Brain Research, 130,* 239–252.

Canon, S. C., & Robinson, D. A. (1987). Loss of the neural integrator of the oculomotor system from brain stem lesions in monkey. *Journal of Neurophysiology, 57,* 1383–1409.

Clemens, I. A., De Vrijer, M., Selen, L. P., Van Gisbergen, J. A., & Medendorp, W. P. (2011). Multisensory processing in spatial orientation: An inverse probabilistic approach. *Journal of Neuroscience, 31,* 5365–5377.

Collewijn, H., Van der Steen, J., Ferman, L., & Jansen, T. C. (1985). Human ocular counterroll: Assessment of static and dynamic properties from electromagnetic sclera coil recordings. *Experimental Brain Research, 59,* 185–196.

Constantin, A. G., Wang, H., Monteon, J. A., Martinez-Trujillo, J. C., & Crawford, J. D. (2009). 3-dimensional eye–head coordination in gaze shifts evoked during stimulation of the lateral intraparietal cortex. *Neuroscience, 164,* 1284–1302.

Crawford, J. D., Cadera, W., & Vilis, T. (1991). Generation of torsional and vertical eye position signals by the interstitial nucleus of Cajal. *Science, 252,* 1551–1553.

Crawford, J. D., Ceylan, M. Z., Klier, E. M., & Guitton, D. (1999). Three-dimensional eye–head coordination during gaze saccades in the primate. *Journal of Neurophysiology, 81,* 1760–1782.

Crawford, J. D., & Guitton, D. (1997). Visual–motor transformations required for accurate and kinematically correct saccades. *Journal of Neurophysiology, 78,* 1447–1467.

Crawford, J. D., Henriques, D. Y., & Medendorp, W. P. (2011). Three-dimensional transformations for goal-directed action. *Annual Review of Neuroscience, 34,* 309–331.

Crawford, J. D., Tweed, D. B., & Vilis, T. (2003). Static ocular counterroll is implemented through the 3-D neural integrator. *Journal of Neurophysiology, 90,* 2777–2784.

Crawford, J. D., & Vilis, T. (1991). Axes of eye rotation and Listing's law during rotations of the head. *Journal of Neurophysiology, 65,* 407–423.

Crawford, J. D., & Vilis, T. (1992). Symmetry of oculomotor burst neuron coordinates about Listing's plane. *Journal of Neurophysiology, 68,* 432–448.

Crawford, J. D., & Vilis, T. (1995). How do motor systems deal with the problems of controlling three-dimensional rotations. *Journal of Motor Behavior, 27,* 89–99.

Demer, J. L., Miller, J. M., Poukens, V., Vinters, H. V., & Glasgow, B. J. (1995). Evidence for fibromuscular pulleys of the recti extraocular muscles. *Investigative Ophthalmology & Visual Science, 36,* 1125–1136.

Demer, J. L., Oh, S. Y., & Poukens, V. (2000). Evidence for active control of rectus extraocular muscle pulleys. *Investigative Ophthalmology & Visual Science, 41,* 1280–1290.

DeSouza, J. F., Keith, G. P., Yan, X., Blohm, G., Wang, H., & Crawford, J. D. (2011). Intrinsic reference frames of superior colliculus visuomotor receptive fields during head-unrestrained gaze shifts. *Journal of Neuroscience, 31,* 18313–18326.

De Vrijer, M., Medendorp, W. P., & Van Gisbergen, J. A. (2009). Accuracy–precision trade-off in visual orientation constancy. *Journal of Vision, 9*(2), 9, 1–15. doi:10.1167/9.2.9.

Ding, J., & Sperling, G. (2006). A gain-control theory of binocular combination. *Proceedings of the National Academy of Sciences of the United States of America, 103,* 1141–1146. doi:10.1073/pnas.0509629103.

Donders, F. C. (1848). Beitrag zur lehre von den bewegungen des menschichen auges [translation: The movements of the human eye]. *Holländ Beitr Anat Physiol Wiss, 1,* 104–145.

Ferman, L., Collewijn, H., & Van den Berg, A. V. (1987). A direct test of Listing's law: II. Human ocular torsion measured under dynamic conditions. *Vision Research, 27,* 939–951.

Fetter, M., Tweed, D., Misslisch, H., Fischer, D., & Koenig, E. (1992). Multidimensional descriptions of the optokinetic and vestibuloocular reflexes. *Annals of the New York Academy of Sciences, 656,* 841–842.

Ghasia, F. F., & Angelaki, D. E. (2005). Do motoneurons encode the noncommutativity of ocular rotations? *Neuron, 47,* 281–293.

Ghasia, F. F., Meng, H., & Angelaki, D. E. (2008). Neural correlates of forward and inverse models for eye movements: Evidence from three-dimensional kinematics. *Journal of Neuroscience, 28,* 5082–5087.

Glenn, B., & Vilis, T. (1992). Violations of Listing's law after large eye and head gaze shifts. *Journal of Neurophysiology, 68,* 309–318.

Haslwanter, T., Straumann, D., Hess, B. J., & Henn, V. (1992). Static roll and pitch in the monkey: Shift and rotation of Listing's plane. *Vision Research, 32,* 1341–1348. doi:10.1016/0042-6989(92)90226-9.

Haustein, W. (1992). Head-centric visual localization with lateral body tilt. *Vision Research, 32,* 669–673. doi:10.1016/0042-6989(92)90183-J.

Helmholtz, H. (1867). *Handbuch der Physiologischen Optik* [Treatise of optical physiology] Treatise on Physiological Optics 3(1). Hamburg, Germany: Voss. [English translation, Vol. 3 (Trans. J. P. C. Southall). Rochester, NY: Optical Society of America (1925) pp. 44–51.]

Henn, V., Hepp, K., & Büttner-Ennever, J. A. (1982). The primate oculomotor system: II. Premotor system: A synthesis of anatomical, physiological, and clinical data. *Human Neurobiology, 1,* 87–95.

Henriques, D. Y., & Crawford, J. D. (2000). Direction-dependent distortions of retinocentric space in the visuomotor transformation for pointing. *Experimental Brain Research, 132,* 179–194.

Hepp, K., Van Opstal, A. J., Straumann, D., Hess, B. J., & Henn, V. (1993). Monkey superior colliculus represents rapid eye movements in a two-dimensional motor map. *Journal of Neurophysiology, 69,* 965–979.

Hore, J., Watts, S., & Vilis, T. (1992). Constraints on arm position when pointing in three dimensions: Donders' law and the Fick gimbal strategy. *Journal of Neurophysiology, 68,* 374–383.

Horn, B. K. P. (1990). Relative orientation. *International Journal of Computer Vision, 4,* 59–78.

Khan, A. Z., Pisella, L., Vighetto, A., Cotton, F., Luaute, J., Boisson, D., et al. (2005). Optic ataxia errors depend on remapped, not viewed, target location. *Nature Neuroscience, 8,* 418–420. doi:10.1038/nn1425.

King, W. M., & Fuchs, A. F. (1979). Reticular control of vertical saccadic eye movements by mesencephalic burst neurons. *Journal of Neurophysiology, 42,* 861–876.

Klier, E. M., & Angelaki, D. E. (2008). Spatial updating and the maintenance of visual constancy. *Neuroscience, 156,* 801–818.

Klier, E. M., Angelaki, D. E., & Hess, B. J. (2005). Roles of gravitational cues and efference copy signals in the rotational updating of memory saccades. *Journal of Neurophysiology, 94,* 468–478.

Klier, E. M., Angelaki, D. E., & Hess, B. J. (2007). Human visuospatial updating after noncommutative rotations. *Journal of Neurophysiology, 98,* 537–544.

Klier, E. M., & Crawford, J. D. (1998). Human oculomotor system accounts for 3-D eye orientation in the visual–motor transformation for saccades. *Journal of Neurophysiology, 80,* 2274–2294.

Klier, E. M., Meng, H., & Angelaki, D. E. (2006). Three-dimensional kinematics at the level of the oculomotor plant. *Journal of Neuroscience, 26,* 2732–2737.

Klier, E. M., Meng, H., & Angelaki, D. E. (2011). Revealing the kinematics of the oculomotor plant with tertiary eye positions and ocular counterroll. *Journal of Neurophysiology, 105,* 640–649.

Klier, E. M., Meng, H., & Angelaki, D. E. (2012). Reaching the limit of the oculomotor plant: 3D kinematics after abducens nerve stimulation during the torsional VOR. *Journal of Neuroscience, 32,* 13237–13243.

Klier, E. M., Wang, H., Constantin, A. G., & Crawford, J. D. (2002). Midbrain control of three-dimensional head orientation. *Science, 295,* 1314–1316.

Klier, E. M., Wang, H., & Crawford, J. D. (2001). The superior colliculus encodes gaze commands in retinal coordinates. *Nature Neuroscience, 4,* 627–632.

Klier, E. M., Wang, H., & Crawford, J. D. (2003). Three-dimensional eye–head coordination is implemented downstream from the superior colliculus. *Journal of Neurophysiology, 89,* 2839–2853.

Luschei, E. S., & Fuchs, A. F. (1972). Activity of brain stem neurons during eye movements of alert monkeys. *Journal of Neurophysiology, 35,* 445–461.

Martinez-Trujillo, J. C., Medendorp, W. P., Wang, H., & Crawford, J. D. (2004). Frames of reference for eye–head gaze commands in primate supplementary eye fields. *Neuron, 44,* 1057–1066.

Martinez-Trujillo, J. C., Wang, H., & Crawford, J. D. (2003). Electrical stimulation of the supplementary eye fields in the head-free macaque evokes kinematically normal gaze shifts. *Journal of Neurophysiology, 89*, 2961–2974.

Medendorp, W. P., Smith, M. A., Tweed, D. B., & Crawford, J. D. (2002). Rotational remapping in human spatial memory during eye and head motion. *Journal of Neuroscience, 22*(RC196), 1–4.

Mok, D., Ro, A., Cadera, W., Crawford, J. D., & Vilis, T. (1992). Rotation of Listing's plane during vergence. *Vision Research, 32*, 2055–2064. doi:10.1016/0042-6989(92)90067-S.

Monteon, J. A., Constantin, A. G., Wang, H., Martinez-Trujillo, J. C., & Crawford, J. D. (2010). Electrical stimulation of the frontal eye fields in the head-free macaque evokes kinematically normal 3D gaze shifts. *Journal of Neurophysiology, 104*, 3462–3475.

Ono, H., Mapp, A. P., & Howard, I. P. (2002). The cyclopean eye in vision: The new and old data continue to hit you right between the eyes. *Vision Research, 42*, 1307–1324. doi:10.1016/S0042-6989(01)00281-4.

Pathmanathan, J. S., Presnell, R., Cromer, J. A., Cullen, K. E., & Waitzman, D. M. (2006). Spatial characteristics of neurons in the central mesencephalic reticular formation (cMRF) of head-unrestrained monkeys. *Experimental Brain Research, 168*, 455–470.

Poljac, E., Lankheet, M. J., & Van Den Berg, A. V. (2005). Perceptual compensation for eye torsion. *Vision Research, 45*, 485–496.

Quaia, C., Lefevre, P., & Optican, L. M. (1999). Model of the control of saccades by superior colliculus and cerebellum. *Journal of Neurophysiology, 82*, 999–1018.

Quaia, C., & Optican, L. M. (1998). Commutative saccadic generator is sufficient to control a 3-D ocular plant with pulleys. *Journal of Neurophysiology, 79*, 3197–3215.

Radau, P., Tweed, D., & Vilis, T. (1994). Three-dimensional eye, head, and chest orientations after large gaze shifts and underlying neural strategies. *Journal of Neurophysiology, 72*, 2840–2852.

Robinson, D. A. (1981). The use of control systems analysis in the neurophysiology of eye movements. *Annual Review of Neuroscience, 4*, 463–503.

Ruiz-Ruiz, M., & Martinez-Trujillo, J. C. (2008). Human updating of visual motion direction during head rotations. *Journal of Neurophysiology, 99*, 2558–2576.

Schreiber, K., Crawford, J. D., Fetter, M., & Tweed, D. (2001). The motor side of depth vision. *Nature, 410*, 819–822.

Schreiber, K. M., Tweed, D. B., & Schor, C. M. (2006). The extended horopter: Quantifying retinal correspondence across changes of 3D eye position. *Journal of Vision, 6*, 64–74. doi:10.1167/6.1.6.

Smith, M. A., & Crawford, J. D. (1998). Neural control of rotational kinematics within realistic vestibuloocular coordinate systems. *Journal of Neurophysiology, 80*, 2295–2315.

Smith, M. A., & Crawford, J. D. (2001). Self-organizing task modules and explicit coordinate systems in a neural network model for 3-D saccades. *Journal of Computational Neuroscience, 10*, 127–150.

Straumann, D., Zee, D. S., & Solomon, D. (2000). Three-dimensional kinematics of ocular drift in humans with cerebellar atrophy. *Journal of Neurophysiology, 83*, 1125–1140.

Straumann, D., Zee, D. S., Solomon, D., Lasker, A. G., & Roberts, D. C. (1995). Transient torsion during and after saccades. *Vision Research, 35*, 33321–33334. doi:10.1016/0042-6989(95)00091-R.

Tweed, D. (1997a). Visual–motor optimization in binocular control. *Vision Research, 37*, 1939–1951. doi:10.1016/S0042-6989(97)00002-3.

Tweed, D. (1997b). Three-dimensional model of the human eye–head saccadic system. *Journal of Neurophysiology, 77*, 654–666.

Tweed, D., & Vilis, T. (1987). Implications of rotational kinematics for the oculomotor system in three dimensions. *Journal of Neurophysiology, 58*, 832–849.

Tweed, D., & Vilis, T. (1990). Geometric relations of eye position and velocity vectors during saccades. *Vision Research, 30*, 111–127.

Van Ee, R., & Van Dam, L. C. (2003). The influence of cyclovergence on unconstrained stereoscopic matching. *Vision Research, 43*, 307–319.

Van Opstal, A. J., Hepp, K., Hess, B. J., Straumann, D., & Henn, V. (1991). Two- rather than three-dimensional representation of saccades in monkey superior colliculus. *Science, 252*, 1313–1315.

Van Opstal, A. J., Hepp, K., Suzuki, Y., & Henn, V. (1996). Role of the monkey nucleus reticularis tegmenti pontis in the stabilization of Listing's plane. *Journal of Neuroscience, 16*, 7284–7296.

Van Rijn, L. J., & Van den Berg, A. V. (1993). Binocular eye orientation during fixations: Listing's law extended to include eye vergence. *Vision Research, 33*, 691–708.

Wade, S. W., & Curthoys, I. S. (1997). The effect of ocular torsional position on perception of the roll-tilt of visual stimuli. *Vision Research, 37*, 1071–1078. doi:10.1016/S0042-6989(96)00252-0.

Westheimer, G. (1957). Kinematics of the eye. *Journal of the Optical Society of America, 47*, 967–974.

62 Neural Mechanisms for Smooth Pursuit Eye Movements

MICHAEL J. MUSTARI AND SEIJI ONO

PURPOSE OF SMOOTH PURSUIT EYE MOVEMENTS

The primate visual system is specialized for central vision. To examine an object of interest in detail, its image must be located on or near the fovea and kept relatively stable. This is achieved by different oculomotor subsystems including rapid or step-like movements called saccades and smooth eye movements such as vestibular ocular, optokinetic, ocular following, and smooth pursuit. Optokinetic and vestibular ocular eye movements are reflexive and play a role in maintaining the image of the visual world stable on the retina during movement of the observer (Leigh & Zee, 2006).

Smooth pursuit is a volitional behavior that requires a moving visual target for optimal performance (Leigh & Zee, 2006). Visual target position, motion, and depth are decoded in dorsal stream visual areas, especially middle temporal visual cortex (MT) and medial superior temporal (MST) areas (Born & Bradley, 2005). These visual signals are distributed to the smooth pursuit part of the frontal eye fields (FEFsem), ventral intraparietal (VIP) and lateral intraparietal (LIP) cortical areas (Lynch & Tian, 2006). Cortical visual and eye movement related signals are sent to distal brainstem and cerebellar circuits to generate smooth pursuit.

The pursuit system is a controlled feedback system where eye movements alter the visual sensory driving input (Leigh & Zee, 2006). A consequence of this closed-loop control system is that the retinal error by itself would not be sufficient to drive the eye movements because retinal image motion approaches zero as eye speed reaches target speed. Figure 62.1 captures the fundamental organization of negative and positive feedback loops for smooth pursuit control. Several control models have been proposed that utilize different signals to maintain the drive for the pursuit system once steady state tracking is achieved (Leigh & Zee, 2006). Models featuring position and acceleration of the target, perceived target velocity, efference copy of the motor command sent to the eye muscles, or

a velocity storage mechanism implemented as a local cortical feedback loop have all been proposed. The actual neural sites involved at different stages of processing encompassed in various control system models are still being explored (Nuding et al., 2008). The control of different aspects of smooth pursuit depends on behavioral context and the complexity of the overall tracking environment. Future modeling studies should be able to incorporate metric and cognitive aspects of smooth pursuit.

The smooth pursuit systems of humans and nonhuman primates share many similarities (Leigh & Zee, 2006), making the behaving monkey an ideal model system for studying neural mechanisms supporting smooth pursuit. Smooth pursuit behavior evinces many of the properties found in other sensorimotor behaviors including target selection, prediction, initiation, maintenance, and plasticity. In this review, we will consider some of the evidence indicating where different aspects of pursuit might be represented. Our understanding of neural mechanisms associated with different aspects of smooth pursuit has advanced significantly in the last decade, including how visual motion information is processed at cortical, brainstem, and cerebellar levels and converted into commands for smooth pursuit eye movements (Krauzlis, 2004; Lisberger, 2010; Ono & Mustari, 2009).

PATHWAYS FOR VISUAL FOLLOWING RESPONSES AND SMOOTH PURSUIT

Smooth pursuit evolved with foveal vision and relies, in part, on regions of the brain involved in optokinetic eye movements. The optokinetic response (OKR) is the phylogenetically oldest visual following response in mammals (Leigh & Zee, 2006). It plays a role in compensating for motion of the entire visual scene as often occurs with head movements. Optokinetic eye movements reflexively follow the direction of visual motion. Continuous visual motion leads to optokinetic nystagmus (OKN), comprised of slow compensatory eye

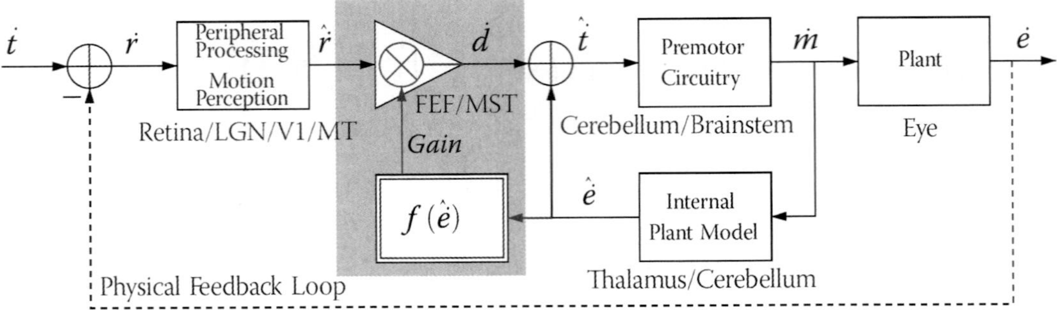

FIGURE 62.1   Extension of the basic model by Robinson et al. (1986) to incorporate dynamic gain control. The symbol $\dot{r}$ denotes retinal error velocity, as defined by the difference between target velocity $\dot{t}$ and eye velocity $\dot{e}$, and $\hat{\dot{r}}$ is the internal estimate of retinal error velocity. Similarly, $\hat{\dot{e}}$ is the internal estimate of eye velocity as provided by the internal plant model, which receives an efference copy of the eye motor command $\dot{m}$. The shaded box represents the dynamic gain control: The feedforward retinal slip signal that constitutes the primary input to the pursuit system is multiplied with a function of estimated eye velocity, that is, the current internal estimate of eye velocity regulates the feedforward gain of the overall system to generate the driving signal $\dot{d}$ for the premotor circuitry (Nuding et al., 2008). LGN, lateral geniculate nucleus; MT, middle temporal cortex; FEF, frontal eye field; MST, medial superior temporal cortex.

movements in the direction of visual motion and quick phases (saccades) that bring eye position back into a usable oculomotor range. OKN plays an important role in maintaining maximal periods of clear vision during continuing rotation or low-frequency movement of the head when the vestibulo-ocular reflex (VOR) is not able to provide full compensation. The OKR can also enhance smooth pursuit when target and background motion occur in the same direction. The OKR of primates is particularly sensitive to visual motion in the central 10° of the retina, although the remaining retina also makes a significant contribution. Two components of the OKR can be distinguished and are attributed to different neural pathways (see figure 62.2). These pathways also play a role in smooth pursuit. The indirect component (figure 62.2, dashed lines) depends on retinal inputs to nuclei in the accessory optic system (AOS) including the nucleus of the optic tract (NOT) (Büttner & Büttner-Ennever, 2006). The NOT also receives visual inputs from areas MT and MST. Horizontal OKN depends on visual motion signals reaching the NOT and its projection to the vestibular nuclei and nucleus prepositus hypoglossi, which contribute to the gradual buildup in slow phase (compensatory) eye movements during extended periods of unidirectional visual motion.

Activation of the direct component pathways by sudden movement of the visual world elicits the short-latency ocular following response (OFR), which moves the eyes in the direction of visual motion (Miles, Kawano, & Optican, 1986). The short-latency OFR is associated with a preceding saccade and has a latency of 50 ms in monkeys. Therefore the OFR has a substantially shorter latency than the typical 100–150 ms latency for smooth pursuit. The short-latency response, the spatial extent of the visual stimulus, and the involuntary character of the OFR eye movements are major features that distinguish it from smooth pursuit.

There is strong evidence for involvement of common neural circuits for the OFR and smooth pursuit. For example, both pursuit and optokinetic eye movements are impaired following lesions of MT and MST areas (Dürsteler & Wurtz, 1988), dorsolateral pontine nucleus (DLPN), and flocculus (Leigh & Zee, 2006). Neurons in MST have been indicated to play a role in the direct component of OKR because their response latencies change in parallel with the simultaneously observed OFR eye movement latencies elicited by different visual stimuli (Kawano et al., 1994). Furthermore, MST neurons reflect the postsaccadic enhancement of the OKR/OFR in their neuronal response. For extended unidirectional visual motion stimulation, OKN can reach velocities above 100°/s in monkeys (Leigh & Zee, 2006), which exceeds foveal smooth pursuit range. These higher velocities are the sum of the direct and indirect components of OKN. Also, when a subject is requested to actively track the optokinetic stimulus, the gain of eye movement increases due to volitional smooth pursuit contributions.

INTERACTION OF SMOOTH PURSUIT, OKR, AND VOR

Smooth pursuit, optokinetic, and vestibular-ocular signals interact during normal behavior. For example,

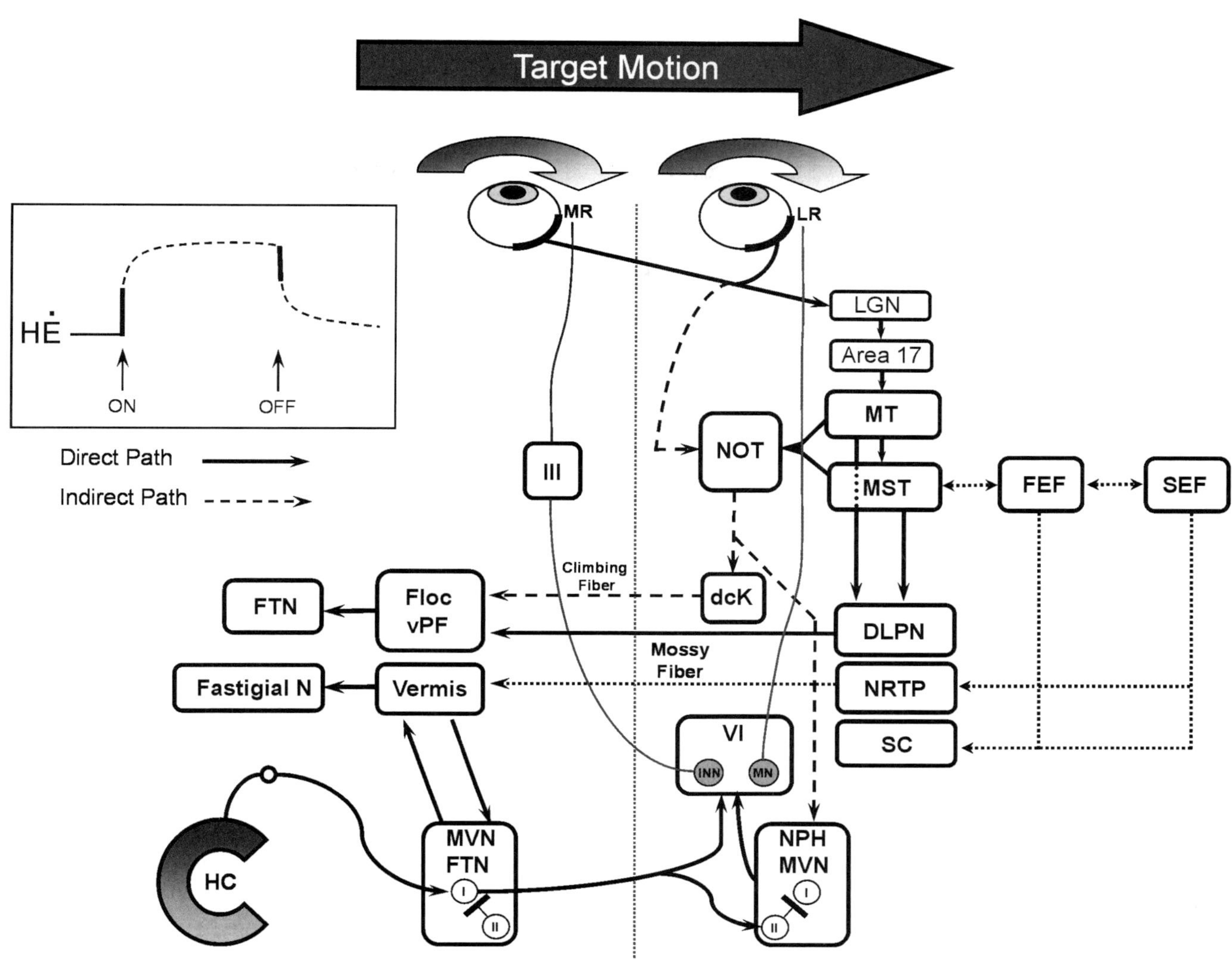

FIGURE 62.2   Organization of smooth pursuit eye movement pathways. Direct and indirect pathways associated with the optokinetic response are indicated in solid and dashed lines. Major feedforward and some feedback pathways for cortical–cortical and cortical–brainstem signals traveling in parallel pathways for smooth pursuit. Different cortical areas contact pontine centers and the superior colliculus, which target different regions of the cerebellum. Different regions of the cerebellum influence eye movements by way of connections with distal premotor and oculomotor neurons. dCK, dorsal cap of Kooy inferior olive; DLPN, dorsolateral pontine nucleus; FEF, frontal eye field; Floc, Flocculus; FTN, floccular target neuron; INN, interneuron; HC, horizontal semicircular canal; HE, horizontal eye velocity; LGN, lateral geniculate nucleus; MN, motor neuron; MST, medial superior temporal cortex; MT, middle temporal cortex; MVN, medial vestibular nucleus; NOT, pretectal nucleus of the optic tract; NPH, nucleus prepositus hypoglossi; NRTP, nucleus reticularis tegmenti pontis; SC, superior colliculus; SEF, supplemental eye field; vPFloc, ventral paraflocculus; III, oculomotor nucleus; VI, abducens nucleus.

the VOR needs to be canceled when a target of interest moves with the head. Cancellation of the VOR is thought to involve smooth pursuit pathways as discussed below (Cullen, 2012; Leigh & Zee, 2006). For example, when the head moves left, the VOR generates a rightward compensatory eye movement (e.g., figure 62.3, VORd). A leftward pursuit response would be required to cancel the right VOR. Consistent with this idea is that cancellation of the VOR is impaired following lesions that impair smooth pursuit (Leigh & Zee, 2006).

Common experience in the laboratory is that monkeys or humans find tracking a small target over a stationary, patterned background difficult (Collewijn & Tamminga, 1984). Even following extensive training, pursuit gains are less in this condition than when tracking a single target over a featureless background. The difference is thought to be due to an optokinetic drive (i.e., large-field visual motion on the retina) that is opposite in direction to current pursuit direction. In contrast, when the smooth pursuit system is supported by the optokinetic system, gain of tracking is improved even for tasks where prediction is difficult such as when the target motion is driven with band limited white noise (Nuding et al., 2008). Tracking a target whose image has foveal and parafoveal extent is a common condition in natural settings and would enhance tracking.

In the laboratory, target and background stimuli are usually coplanar on a tangent screen, but this may not be typical during natural viewing. For example, during binocular viewing where disparity information is available for near targets, pursuit may be less distracted by large-field patterns that oppose target motion. In natural scenes there is often separation of the target from the background based on relative depth and contrast of the target and background. New studies are beginning to define the interaction of eye movements and natural scenes, and this will be important for understanding how pursuit interacts with optokinesis and other eye movement systems during natural behavior.

THE CORTICAL SMOOTH PURSUIT SYSTEM

Smooth pursuit is supported by interconnected regions of cerebral cortex including MT, dorsal and lateral aspects of MST (MSTd, MSTl), FEFsem, supplementary eye fields (SEF), and potentially other areas in the frontal and parietal lobes (see figure 62.2). These areas comprise the cortical pursuit system (Krauzlis, 2004), which is responsible for beginning the process of converting visual motion information into commands for eye movements and also more cognitive aspects of smooth pursuit (Mahaffy & Krauzlis, 2011). Some of these areas such as MT carry pure visual motion signals to specific brainstem regions, and other areas including MST and FEF contain neurons with eye and vestibular sensitivity.

Directional visual error signals are available in MT, MST, and NOT to initiate and adjust smooth pursuit (Lisberger, 2010). Here we consider the organization of cortical-ponto-cerebellar and cortical–brainstem pathways for smooth pursuit beginning with posterior cortical areas (MT, MST) and then the frontal pursuit areas. Some regions of the cortical pursuit systems (MT, MSTd, MSTl, FEF, and SEF) have roles beyond controlling the metrics of smooth pursuit, and only some of those roles are considered here.

Smooth Pursuit Role of Areas MT, MSTd, and MSTl

Visual motion sensitive areas in the dorsal stream including MT and MST are important for smooth pursuit. Early studies showed that lesions in these areas produce retinotopic and directional deficits in smooth pursuit, respectively (Newsome et al., 1985).

Both MT and MST have reciprocal connections with the FEF and SEF regions and other areas including LIP area 7A and VIP (Lynch & Tian, 2006). Neurons in MT have been shown to play a crucial role in visual motion perception, which is essential for volitional smooth pursuit. Individual MT neurons code the direction, speed, and depth of a moving visual target (Born & Bradley, 2005). Such signals are distributed to other regions of the cortical pursuit system including MST and FEF (Lynch & Tian, 2006). If retinotopically organized MT is lesioned, a visual motion scotoma occurs where target motion will be inadequate to drive pursuit (Newsome, Wurtz, & Komatsu, 1988). Thus MT is an obligatory part of the cortical smooth pursuit system and provides directional visual motion information to other cortical areas, the AOS, NOT (Gamlin, 2006; Ono & Mustari, 2010), and the DLPN for smooth pursuit.

The sources of visual motion memory or efference copy of smooth pursuit required to maintain tracking during unity gain smooth pursuit have not been fully identified. However, we can find evidence for extraretinal, visual motion memory or efference copy activity in neurons in MST (Newsome, Wurtz, & Komatsu, 1988; Ono & Mustari, 2006) but not MT (Newsome, Wurtz, & Komatsu, 1988). The neurons illustrated in figure 62.3 had direction-selective responses during smooth pursuit that were best related to eye speed. Furthermore, these neurons continued their response when the target was briefly (100–400 ms) extinguished during smooth pursuit over a dark background (see figure 62.3B). A continuing pursuit related neuronal discharge during a target blink (the target is extinguished for 100–400 ms) indicates extraretinal signals, which could play a role in maintaining tracking. Extraretinal signals might be associated with different mechanisms including efference copy (Tanaka, 2005), visual motion memory (Fukushima, Fukushima, & Warabi, 2011), or eye muscle proprioception (Wang et al., 2007). The latencies of these different extraretinal signals will differ from leading (predictive) to lagging (efference copy or eye position) with respect to the onset of smooth pursuit eye movements. An effective way to examine whether eye movement signals in MST are related to volitional smooth pursuit per se or to slow eye movement in general is to examine neuronal response during smooth eye movements evoked in different behavioral paradigms. In figure 62.3A we show examples of MSTd neurons tested during smooth pursuit and three different VOR behaviors. In these studies the eye moves through the same oculomotor range during smooth pursuit and the VOR. MSTd smooth neuronal discharge was not modulated during rotational VOR testing in

   MICHAEL J. MUSTARI AND SEIJI ONO

darkness (VORd) or in the light (VOR×1). However, MSTd neurons were modulated during cancellation of the VOR (VOR × 0). MSTd neuronal sensitivity during volitional smooth pursuit and cancellation of the VOR fall close to the equal sensitivity line (see figure 62.3C) and had matching directions. We still do not know if the extraretinal signal carried in MST neurons represents an efference copy of the pursuit command or the origin of this signal. Tanaka (2005) has reported neurons in the intralaminar thalamic nuclei that carry pursuit related signals with latency distributions similar to those of MSTd. This region of the thalamus has widespread projections to different regions of the cortical pursuit system. Further studies are needed to

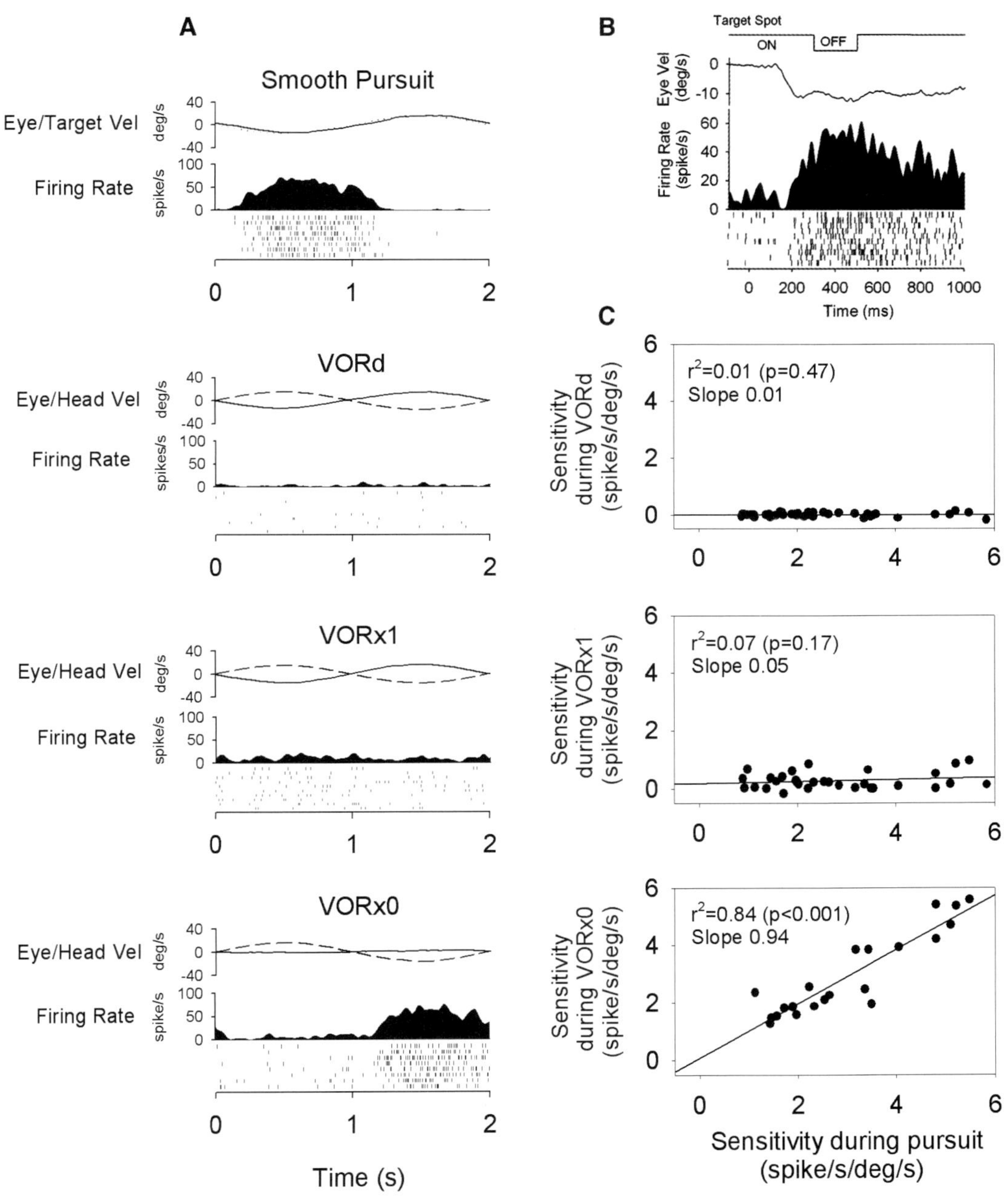

FIGURE 62.3   (A) Smooth pursuit and vestibulo-ocular reflex (VOR) related responses of dorsal medial superior temporal cortex (MSTd) neurons. MSTd neurons with extraretinal signals were modulated during smooth pursuit and cancellation of the VOR (×0). These neurons were not modulated during VOR in complete darkness (VORd) or VOR in light (VORl). Example neuron during four conditions showing response during leftward smooth pursuit and during VORx0 when the head moves leftward. Solid and dashed lines indicate eye and head velocity (vel), respectively. (B) Response of an MSTd neuron during step-ramp smooth pursuit. This and other MSTd neurons continue their smooth pursuit related response even when the target is briefly extinguished. (C) Sensitivity plots for head movement compared to eye movement in four conditions.

determine the type of information carried in these thalamo-cortical connections and their actual role in smooth pursuit.

Some of the parietal cortical areas carry signals related to visual, eye movement, vestibular, and somatosensory information (Angelaki & Cullen, 2008). Coordination of these signals is important for spatial orientation and navigation (Britten, 2008) but not necessarily for smooth pursuit per se. We need more studies to define how smooth pursuit eye movements are taken into account in areas that appear to play a role in gaze pursuit and spatial perception and determine whether any of these areas contribute to more cognitive aspects of smooth pursuit (Barnes, 2008; Mahaffy & Krauzlis, 2011).

When a target moves in a highly predictable fashion such as sinusoidal motion, the smooth pursuit system is able to maintain the eye on target with little or no phase lag within a wide range of frequencies and speeds (e.g., 0.1–1.0 Hz, <60°/s). This sort of phase locking can be seen in the behavior illustrated in figure 62.3A (top). Predictive tracking cannot be derived by visual mechanisms alone because the latency of even the earliest visual response to eye motion (OFR) is about 50 ms (Leigh & Zee, 2006). The location of predictive signals in the pursuit system has not been fully defined, but there is evidence that at least some neurons in FEFsem cortex show phase shifting in their response as the monkey's eye obtains registration with target motion. There may also be different forms of predictive or anticipatory signals with different locations in the smooth pursuit system.

Neurons in MSTl have properties that are similar to those in MSTd. Some MSTl neurons have center–surround organized visual receptive fields. It is possible that such neurons could play a role in extracting target motion information in more complex tracking environments including cases where second order visual motion provides inputs for smooth pursuit. There is a major difference in the neuronal response latency of neurons in MSTd and MSTl, especially related to eye velocity sensitivity. MSTl neurons have early latencies that lead or are coincident with smooth pursuit onset. In contrast, neurons in MSTd often have eye velocity responses that only begin after smooth pursuit has started. Such late-onset responses could help maintain smooth pursuit. MSTl neurons often have rather small visual receptive fields compared to those in MSTd. We know that both MSTd and MSTl regions project to the brainstem including DLPN and NOT, but we do not have specific information about the information these neurons convey to distal centers. A challenge in our field is to determine how these multimodal neurons participate in different behaviors and how they interact with neurons in other cortical areas.

Because some neurons in MST carry multimodal signals including visual, eye motion, and translational vestibular information (Page & Duffy, 2003), it is important to conduct studies that allow differentiation of response components of individual neurons during complex behaviors. Current evidence indicates that MST could play a role not only in smooth pursuit and OFR but also in other sensorimotor behaviors related to computing optic flow, navigation, and spatial orientation during locomotion (Angelaki & Cullen, 2008; Britten, 2008; Cullen, 2012). A major area requiring further study is to determine how different regions of the cortical pursuit system contribute to tracking in different behavioral contexts.

Frontal Eye Fields

The FEF region is located in association with the arcuate sulcus of the frontal lobe. The smooth pursuit region of the FEF is located in the fundus of the arcuate sulcus and is often referred to as the frontal pursuit area, or FEFsem (Lynch & Tian, 2006). This general region contains subregions containing neurons related to fixation, saccades, or smooth pursuit. These FEF subregions also have mostly separate projection fields to the caudate nucleus, superior colliculus (SC), and pontine nuclei (Lynch & Tian, 2006).

Early lesion studies using ablation showed that unilateral damage to the FEF region produced the strongest deficits in ipsilesional smooth pursuit including predictive tracking of periodic target motion. More recent studies that used unilateral injection of muscimol to reversibly inactivate the FEFsem region confirmed and extended earlier findings by showing that pursuit initiation and maintenance were both impaired with the strongest effect for ipsilesional tracking. However, unilateral FEFsem lesions actually reduce the gain of pursuit for all directions of tracking. This omnidirectional effect is consistent with evidence that the FEFsem plays a role in controlling the gain of smooth pursuit in addition to its role in initiation and maintenance (Lisberger, 2010; Mahaffy & Krauzlis, 2011; Nuding et al., 2008).

Electrical stimulation of the FEFsem provides confirmatory results and offers further perspectives. Low-current electrical stimulation of the FEFsem elicits pursuit-like eye movements that often have an ipsilateral component (Gottlieb, Bruce, & MacAvoy, 1993). When electrical stimulation was paired with ongoing pursuit, the evoked eye movements were much larger than those observed during stimulation while the monkey was fixating (Tanaka & Lisberger, 2001). Even at sites where electrical stimulation evoked ipsilateral eye movements, gain was enhanced for all directions of smooth pursuit. The pursuit system requires gain control to maintain tracking targets moving at different speeds especially when eye motion matches target motion (see figure 62.1). The FEFsem appears to support this role. Further studies are needed to determine the relative neural contributions of the FEFsem and other areas to gain control.

Early single-unit recording studies found that neurons in the FEFsem discharge in relation to smooth pursuit and visual motion. The majority of pursuit related FEF neurons begin their response before the onset of pursuit and continue to respond during active tracking. Therefore, FEFsem neurons could contribute to the initiation of pursuit, which is characterized by high retinal slip and eye acceleration (Fukushima, Fukushima, & Warabi, 2011; Mahaffy & Krauzlis, 2011; Ono & Mustari, 2009; Tanaka & Lisberger, 2001). Some neurons in FEFsem show sensitivity to visual, eye movement, vestibular, and somatosensory information (Fukushima, Fukushima, & Warabi, 2011). Multiple linear regression models incorporating visual and eye motion parameters (position, velocity, and acceleration) demonstrate that FEFsem neurons code eye motion variables related to eye acceleration and eye velocity (Mahaffy & Krauzlis, 2011; Ono & Mustari, 2009). Models were slightly improved by inclusions of retinal error terms for some neurons consistent with their visual motion sensitivity, but fits were quite good when only eye motion parameters were used.

FEFsem neurons identified as projecting to the nucleus reticularis tegmenti pontis (NRTP) were often most sensitive to smooth pursuit eye acceleration or eye velocity, consistent with playing a role in pursuit initiation (see figures 62.4 and 62.5). Times to peak firing of individual FEFsem neurons that were antidromically activated following electrical stimulation of the pursuit region of the NRTP had relatively tight onset times and time to peak firing compared to FEF neurons of the same animals that were not activated (see figure 62.4).

Although most studies of smooth pursuit have used single targets, some studies have included target choice paradigms. Recent studies have examined the potential contributions of FEFsem to target selection during smooth pursuit (Mahaffy & Krauzlis, 2011). Recent studies by Fukushima and colleagues (2011) found that some FEFsem neurons not only carry smooth pursuit related signals but were also sensitive to vestibular and neck somatosensory information. These findings support the suggestion that the FEFsem area may also play a role in gaze pursuit. It is likely that FEFsem

neurons carry signals to the brainstem that require further processing in the floccular complex (Lisberger, 2010) and vermis for production of a pursuit.

Supplementary Eye Fields

Some cortical areas such as the SEF region appear to play a role in more cognitive aspects of smooth pursuit behavior rather than generating the metric aspects of smooth pursuit. Recent studies by Fukushima and colleagues (Yang at al., 2010) used a cue-based task to explore the relative roles of FEF and SEF in different aspects of smooth pursuit. In their task neuronal response related to visual motion memory, decision making, movement preparation, and pursuit could be separated. By testing neurons in SEF and FEF of the same monkeys, these investigators have shown that the SEF contains a higher proportion of neurons that code the go and no-go conditions than neurons in FEFsem. Furthermore, when muscimol was delivered into SEF, both directional and no-go errors occurred. These errors did not increase following inactivation of the FEF. Although some FEF neurons showed modulation in the delay periods, most FEF neurons seemed best related to smooth pursuit per se.

Yang and colleagues (2010) examined the role of SEF in encoding rules for smooth pursuit. Their earlier studies found no specific role for SEF in controlling the metrics of smooth pursuit. For example, one of their paradigms used a cue-based go and no-go decision to pursue a target based on its relative trajectory toward a central ($8 \times 8°$) region on the tangent screen. The monkey had to observe the trajectory of a target that would either cross (go) or not cross (no-go) the central square. Because the targets began their motion well away from the edge of the square, monkeys had sufficient time to make a choice to pursue a given target. Easy targets passed well away from the edges (no-go) or through the central square (go). In contrast, difficult trials had target motion that just missed or just traversed the edge of the square. The main finding was that SEF neurons were modulated in relation to the rule that was encoded, not the decision to pursue or pursuit metrics.

Other Cortical Areas

Several different areas including VIP, LIP, 7, FST, and dorsolateral prefrontal cortex (dLPF) that contain neurons modulated during smooth pursuit require further study. Some of these areas of cortex are likely to play a modulatory role in smooth pursuit or other sensorimotor behaviors rather than driving smooth pursuit per se. For example, recent studies show that

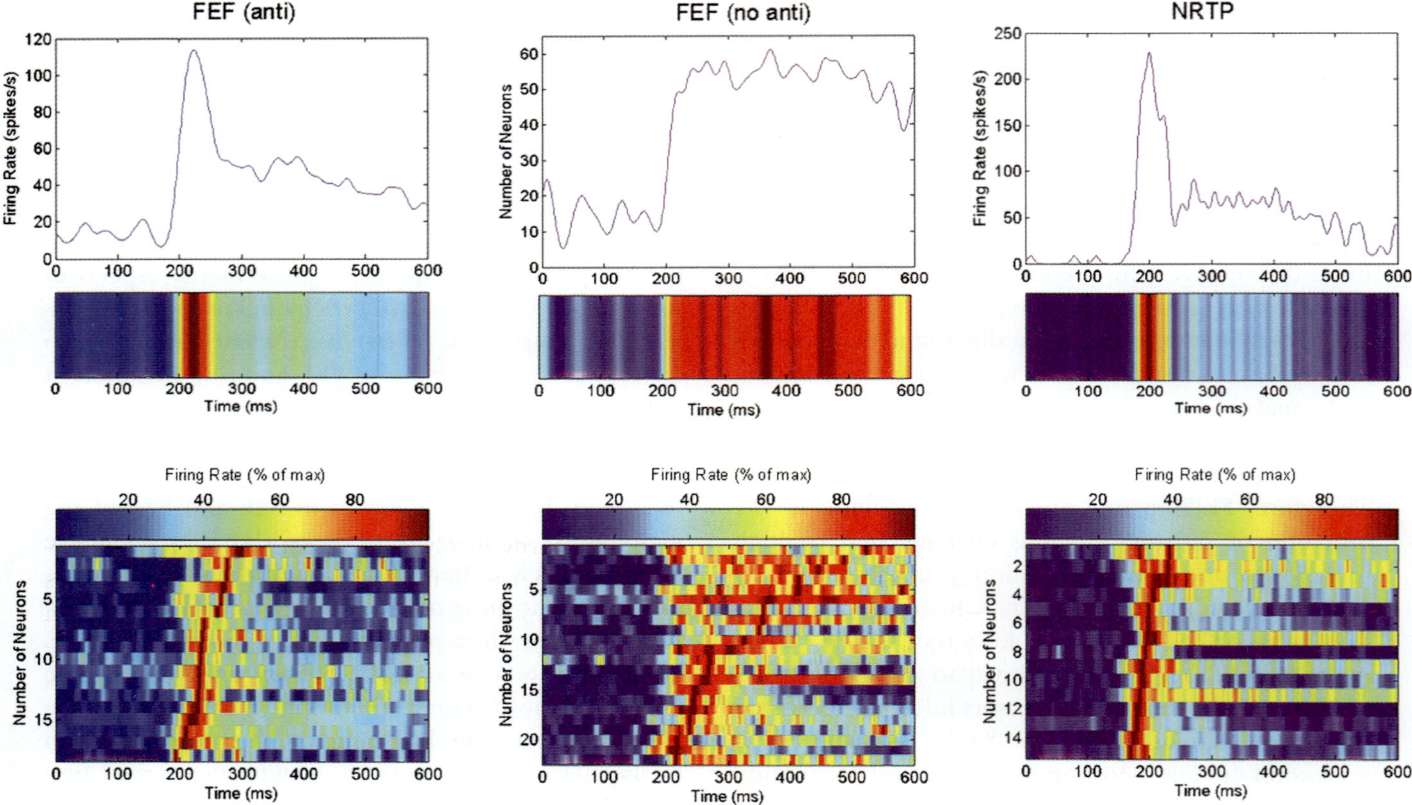

FIGURE 62.4   Comparison of frontal eye field (FEF) and nucleus reticularis tegmenti pontis (NRTP) neuronal response during step-ramp smooth pursuit. FEF neurons that were antidromically activated (anti, left panel) following electrical stimulation of rNRTP have similar times to peak firing during step-ramp smooth pursuit as NRTP neurons (right panel) during similar testing. FEF neurons not activated (no anti, middle panel) following rNRTP stimulation show more spread in the distribution of time to peak firing. Top panel shows the firing rate envelope of neurons aligned on target start at time 0. Bottom row shows individual time to peak firing for each neuron (rows) in a population rank ordered by increasing lead times before pursuit onset.

dLPF and related basal ganglia regions carry go and no-go information for smooth pursuit (Burke & Barnes, 2011). These areas were discovered using functional imaging methods employing magnetoencephalography. Other aspects of smooth pursuit that require cognitive processing for target selection and reward could involve other regions of frontal cortex and basal ganglia (caudate nucleus and substantia nigra). The contribution of these areas to smooth pursuit has to be investigated, but they could play a role in target selection and reward aspects of smooth pursuit (Krauzlis, 2004). Future imaging and neurophysiology studies could help to identify cortical regions and their targets that are active during well-defined smooth pursuit behaviors.

IDENTIFYING SIGNALS IN THE CORTICAL–PONTO–CEREBELLAR SYSTEM

For the cortical pursuit system to produce smooth pursuit, signals must be delivered to appropriate brainstem centers including the NOT, DLPN, rostral NRTP (rNRTP), and SC. Current evidence indicates that these brainstem centers play different roles in smooth pursuit and gaze (Krauzlis, 2004; Ono, Das, & Mustari, 2004). These brainstem centers project to different regions of the cerebellum (flocculus, paraflocculus, and vermis), which play complementary roles in smooth pursuit (Leigh & Zee, 2006; Thier & Möck, 2006).

The actual cortical signals that are delivered to a specific brainstem center and then to the cerebellum remain to be fully defined. Most studies have defined signals in a given cortical region and then determined how those signals might be modified in downstream centers for pursuit (Lisberger, 2010). For example, signals in MT and FEFsem show significant amounts of variability on a trial-by-trial basis compared to that observed in smooth pursuit neurons of the floccular complex. Lisberger (2010) and colleagues have shown that the correlation of MT or FEFsem firing to

     MICHAEL J. MUSTARI AND SEIJI ONO

moment-by-moment smooth pursuit is not as strong as that observed in floccular complex Purkinje cells. This approach will continue to provide important information to constrain hypotheses regarding how information must be processed in the different channels of the pursuit system. However, we still lack information about the specific signals carried in a given pathway. One approach to this problem is to use antidromic activation to identify projection neurons in a given region and then characterize their smooth pursuit related signals (Ono & Mustari, 2009).

If there are separate functional channels leaving the cortex destined for different regions of the cerebellum, we should be able to characterize differences in the information coded in each channel during smooth pursuit. In figure 62.5, we show examples of multiple linear regression models that considered eye or visual motion (position, velocity, and acceleration) for neurons in three potential cortical–brainstem pathways. Eye movement variables make the largest contributions to fits for MSTd and FEF neuronal response. Some FEFsem neurons that project to the rNRTP carry eye acceleration and eye velocity information to their target neurons in the rNRTP. The rNRTP also shows strong eye acceleration and velocity signals during smooth pursuit. The neurons illustrated in figure 62.5 provide an estimate of the overall sensitivity of neurons in different cortical–brainstem channels (Mahaffy & Krauzlis, 2011; Ono & Mustari, 2009). Some neurons in MSTd and DLPN carry eye velocity and visual velocity information. In contrast, other neurons in MT, MSTl, DLPN, and NOT carry retinal image motion information. There is evidence for interactions between variables (e.g., eye velocity and visual velocity) in these areas that reflect possible gain fields (Brostek et al., 2011). We need further studies to determine how different neuronal response components interact during smooth pursuit and other sensorimotor behaviors.

Pontine Nuclei

There are a number of brainstem regions that are known to contribute differentially to smooth pursuit. All of these areas send signals to different regions of the cerebellum where motor commands for smooth pursuit are refined and delivered to distal sites. Recent studies from Lisberger (2010) show that signals in the floccular complex are highly correlated with smooth pursuit eye velocity. Signals in some MT, MST, and FEF cortical neurons are less well correlated with smooth pursuit on a moment-by-moment basis. One possibility is that some of the variability and noise in the cortical afferents to the cerebellum is removed by signal integration in the

pontine nuclei. The DLPN and the NRTP are major components of the cortico-ponto-cerebellar pathway supporting smooth pursuit and gaze control (Ono, Das, & Mustari, 2004). The DLPN receives visual inputs from the MT and MST and other parts of the cortical pursuit system (Thier & Möck, 2006) and sends mossy fiber projections to the contralateral ventral paraflocculus and dorsal paraflocculus and vermal lobule VI and VII. A major open question is whether the pontine nuclei (DLPN, NRTP) serve mainly as relay nuclei or signal integration centers.

The rostral NRTP receives inputs from the pursuit region of the FEFsem and SEF (Lynch & Tian, 2006). NRTP provides mossy fiber projections to the vermal lobules VI and VII. Previous lesion and electrical stimulation studies established the DLPN and rNRTP as being primarily involved in smooth pursuit eye movements. Further evidence for postulating different roles of DLPN and rNRTP in smooth pursuit comes from single-unit studies (Ono, Das, & Mustari, 2004). These single-unit-recording studies show that different DLPN and rNRTP neurons could be classified as preferentially sensitive to smooth pursuit eye velocity, position, or acceleration. We still need to define the specific information carried in different channels to the cerebellum.

Because pontine neurons often evince multiple sensitivities, potential contributions of position, velocity, and acceleration due to eye or retinal image motion during smooth pursuit tasks should be considered. We found that DLPN and rNRTP neurons carry different combinations of position, velocity, and acceleration signals (see figure 62.5). The overall balance of sensitivities in different cortical–ponto–cerebellar pathways could be associated with different roles in initiation (rNRTP) and maintenance (DLPN) of smooth pursuit. Complementary studies by Kawano and colleagues (1992) show that DLPN neurons that respond to large-field visual motion during the OFR are strongly related to retinal image velocity. Overall, single-unit studies in rNRTP and DLPN indicate that these areas carry partially formed smooth pursuit or OFR commands to the cerebellum. For smooth pursuit, some neurons in rNRTP and DLPN carry eye movement signals. For the OFR, signals in MSTd and DLPN appear to be converted into motor commands in the floccular complex.

Nucleus of the Optic Tract

Neurons in the pretectal NOT are best modeled using retinal-error parameters during either ocular following or smooth pursuit. Neurons in MT and NOT lack extraretinal (eye movement) signals but are highly sensitive to visual motion directions and speed (see figure 62.5).

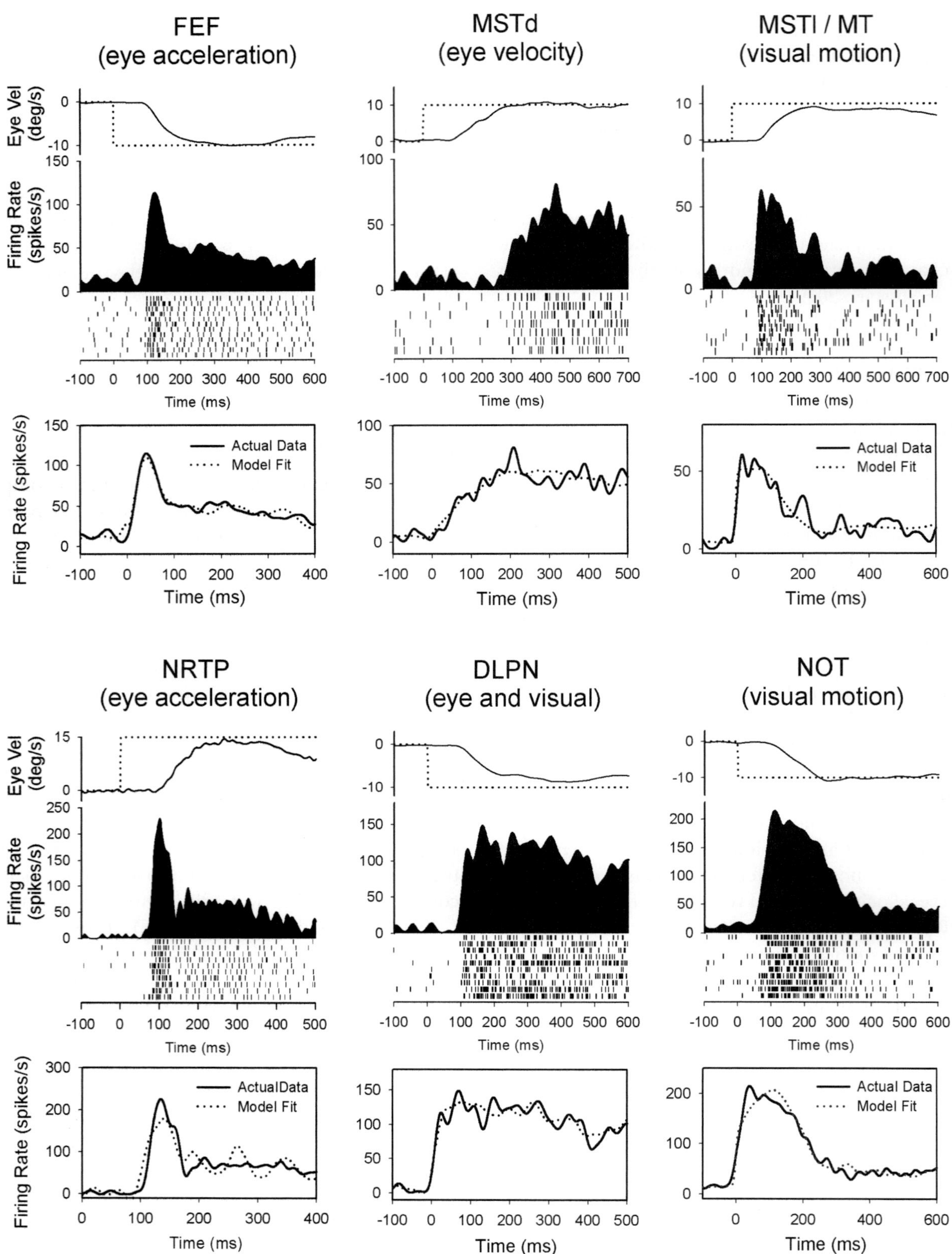

FIGURE 62.5 Examples of step-ramp smooth pursuit responses for cortical and brainstem neurons in different brain regions. Data averaged from at least 10 step-ramp smooth pursuit trials were used to identify coefficients in a model $FR(t) = A + BE(t) + CE'(t) + DE''(t) + ER(t) + FR'(t) + GR''(t)$, where $FR(t)$ is the estimated value of the unit spike density function (solid line, actual data) at time t, $E(t)$ the eye motion at time t, $R(t)$ the retinal error (position, velocity, and acceleration) at time t, and A–G are constants that specify the coefficients in each model. Traces from top, target speed (dotted line), eye speed (solid line), rasters for multiple trials, spike-density function, and model fit (dotted line). There are several possible separate smooth pursuit channels, some carrying only eye or visual information and other channels providing mixed eye and visual signals to the cerebellum. FEF, frontal eye field; MSTd, dorsal medial superior temporal cortex; MSTl, lateral medial superior temporal cortex; MT, middle temporal cortex; NRTP, nucleus reticularis tegmenti pontis; DLPN, dorsolateral pontine nucleus; NOT, nucleus of the optic tract.

Direction sensitive neurons in MT could provide error information to change pursuit output in the presence of consistent errors in tracking. MT could serve this function through connections with the AOS and NOT. Learning in the smooth pursuit can be studied using a step-ramp smooth pursuit paradigm where the target begins moving at one speed and then, 100 ms after pursuit onset, the target speed steps up or down to a new speed. Over the course of 100 or more trials, smooth pursuit gain increases or decreases according to the imposed visual error. The primate NOT contains neurons with horizontal (ipsiversive) direction selectivity, and these neurons are activated during smooth pursuit when residual visual motion exists. For example, during leftward smooth pursuit in a step-up paradigm, an additional leftward visual error would be produced by the second step in target speed and activate neurons in the left NOT. These signals are delivered to the dorsal cap of the inferior olive, which projects to the contralateral cerebellum (see below). Electrical stimulation of the NOT could substitute for the actual visual target motion associated with the second step in speed to produce gain increase or decrease (Ono & Mustari, 2010). Figure 62.6 shows a typical experiment with initial step-ramp target motion of $10°/s$. The second speed of target motion was replaced with electrical stimulation of the left NOT, which produces a leftward error signal (see figure 62.6A, black bar). Over the course of 100 trials, early eye acceleration increases during tracking to the left but decreases for tracking to the right (see figure 62.6A and 62.6B). We need further studies to determine whether visual error information for other directions of pursuit arises in the AOS or other brain regions.

Superior Colliculus

The SC is well-known for its role in saccadic behavior, but it also plays a role in smooth pursuit. There is clear retinotopic organization of visual and saccadic eye movement maps in the superficial and intermediate layers of the SC (Gamlin, 2006). Earlier studies showed that when electrical stimulation was used to activate the rostral SC that contraversive pursuit was enhanced (Basso, Krauzlis, & Wurtz 2000). Inactivation of the SC with muscimol had opposite effects (Basso, Krauzlis, & Wurtz, 2000). Furthermore, electrical stimulation of the rostral SC biases the monkey's choice for targets in the contralateral visual field. The neural bases for these effects have only been partially explained. Single-unit studies have shown that buildup neurons in the intermediate layers of the SC play a role in gating smooth pursuit. The SC receives inputs from the frontal pursuit area where

smooth pursuit related neurons with buildup activity before pursuit onset are found (Mahaffy & Krauzlis, 2011). Whether these neurons provide pursuit related buildup activity with position information from the FEF region remains to be determined. There is strong evidence that the SC also plays a general role in target selection for smooth pursuit and other sensorimotor behaviors. Target selection is critical for pursuit in natural conditions where multiple targets may be present. The neural mechanisms supporting this role could involve retinal error position information in SC. The SC also projects to the medial accessory olivary nucleus, which provides climbing fiber inputs to the contralateral cerebellum that could play a role in plasticity.

Floccular Complex

The floccular complex of the cerebellum is comprised of the flocculus and paraflocculus. Lesions of both flocculus and ventral paraflocculus lead to deficits in initiation and maintenance of smooth pursuit (Leigh & Zee, 2006). The floccular complex plays a role in smooth pursuit and other visual following behaviors like ocular following and visual–vestibular interactions (Ito, 2006). Purkinje cells provide the sole output of the cerebellar cortex, and their response properties are shaped by two primary afferent sources derived from mossy and climbing fibers. Mossy fibers from the NRTP and DLPN carry visual, eye movement, and vestibular information to the floccular complex (Thier & Möck, 2006). Those signals are derived from different regions of the cortical pursuit system. Mossy fiber inputs produce simple-spike activity of Purkinje cells, which provide the moment-by-moment drive for smooth pursuit (Lisberger, 2010) and ocular following. Subsequent studies showed that the firing rate of Purkinje cells was highly correlated with smooth pursuit eye velocity. The floccular complex receives visual motion information representing all directions of eye and visual motion from the DLPN and possibly NRTP (Ono, Das, & Mustari, 2004). It is still not clear how these different signal modalities are combined in the cerebellum to produce commands for eye movements. In studies of the OFR, it was hypothesized that visual signals in MST and DLPN are converted into commands for eye movements in the floccular Purkinje cell. The floccular complex is thought to carry a "forward-model" for the OFR (Ito, 2006; Wolpert & Kawato, 1998) and smooth pursuit (Lisberger, 2010).

Complex spike activity produced by climbing fibers provides error signals that guide motor learning in smooth pursuit and other sensorimotor behavior (e.g., Ito, 2006). Kahlon and Lisberger (2000) described that at least some ventral parafloccular Purkinje cells showed

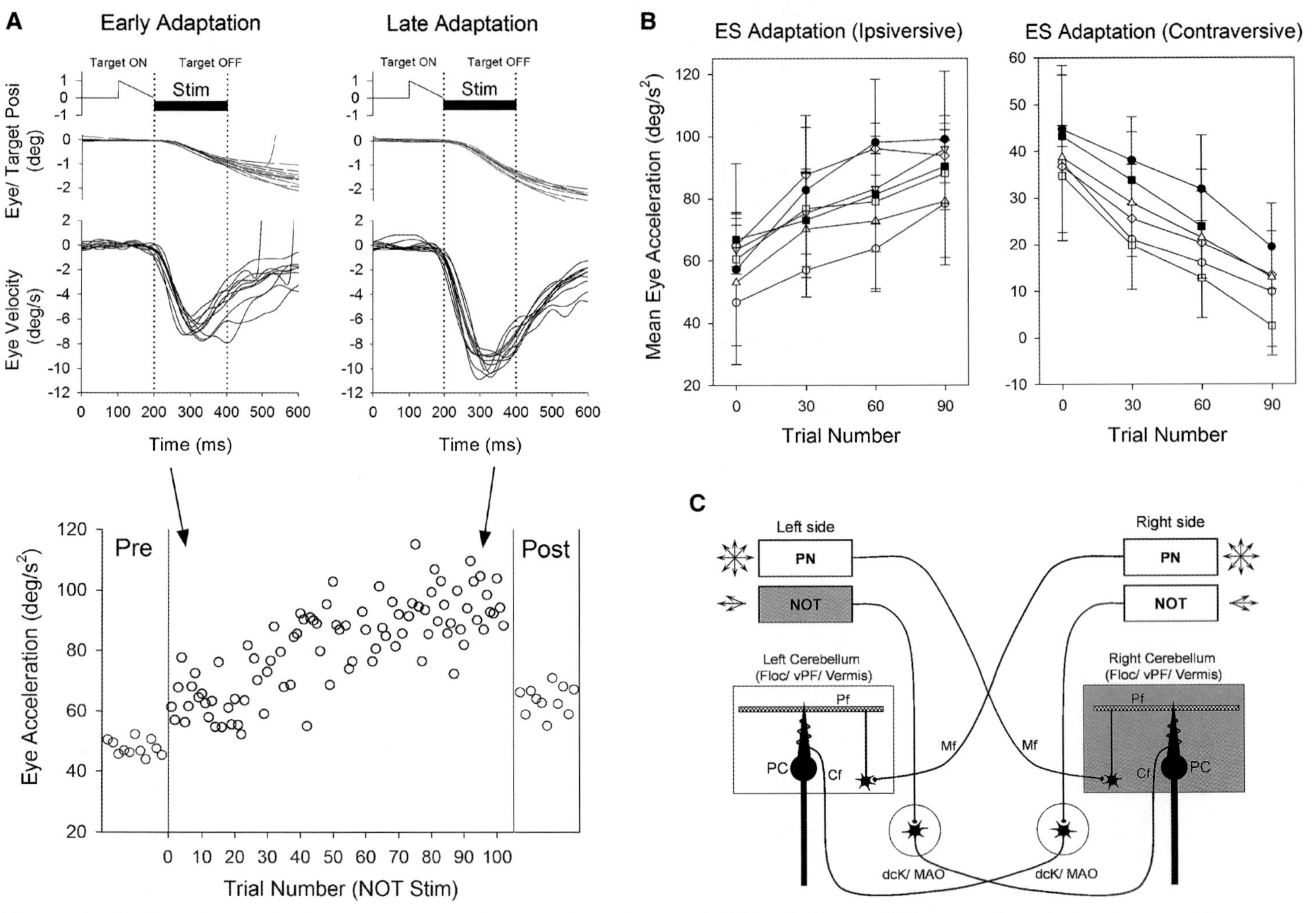

FIGURE 62.6 Experimental evidence for the nucleus of the optic tract (NOT) playing a role in smooth pursuit adaption. The second step of target speed was replaced by a train of electrical stimulation (ES) pulses (stim, solid bar) in the left NOT. Plasticity was estimated by measuring changes in early eye acceleration over the course of 100 trials. (A) Trials early and late in adaptation. (B) Early eye acceleration increases during leftward trials but decreases during right smooth pursuit. (C) A possible neural circuit to support visual adaptation of smooth pursuit involving interaction between mossy fiber (MF) derived simple spikes and climbing fiber (CF) derived complex spikes. Each NOT carries only ipsilateral-direction-selective visual neurons. PN, pontine nucleus; PC, Purkinje cell; dcK, dorsal cap of Kooy; MAO, medial accessory olive; vPF, ventral paraflocculus.

changes in simple-spike and complex-spike firing during smooth pursuit adaptation in the double-step paradigm described earlier. Complex spikes in the floccular complex are known to have contraversive direction selectivity during horizontal tracking (Stone & Lisberger, 1990). For example, during leftward smooth pursuit in the step-up condition, the direction of retinal error is leftward, which is signaled by the left NOT, which projects heavily to the left inferior olive (Büttner & Büttner-Ennever, 1996), thus providing leftward complex spikes to the right floccular complex Purkinje cells (see figure 62.6C). Kahlon and Lisberger (2000) also suggested that eye motion signals in mossy fiber pathways that drive simple spikes could play a role in motor learning in step-up paradigms. Therefore, it seems likely that interaction of simple and complex spikes might be necessary for step-up and step-down adaptation for both pursuit directions.

Vermis

Recent lesion studies have demonstrated that the oculomotor region of the cerebellar vermis plays an essential role in early phase of eye acceleration during step-ramp tracking (Takagi, Zee, & Tamargo, 2000). This is consistent with the properties of neurons in FEFsem projections to the rNRTP described above. In addition lesions of the oculomotor vermis produce deficits in smooth pursuit adaptation in a double-step paradigm (Takagi, Zee, & Tamargo, 2000). Bilateral vermal lesions resulted in deficits in the initial phase of smooth pursuit adaptation in both step-up and step-down paradigms. The symmetry of the reported deficits in step-up and step-down paradigms may have depended on the bilaterality of the vermal lesions. This is because unilateral lesions of the caudal fastigial nucleus, which provides the cerebellar output from the oculomotor vermis, produced asymmetrical smooth pursuit deficits. In recent studies Fukushima and colleagues (2011) found that neurons in oculomotor vermis play a role in go and no-go modulated smooth pursuit behavior. Neurons with go or no-go signals in SEF and FEF may influence the vermis through connections involving their NRTP, but further studies are needed to define the information carried in different FEFsem pathways to the brainstem. In summary, different regions of the cerebellum play complementary roles in smooth pursuit behavior. Most evidence supports a role for the MSTd–DLPN–floccular complex pathway playing a role in smooth pursuit initiation and maintenance (Lisberger, 2010). In contrast, the FEF–rNRTP–vermis pathway plays a role in initial smooth pursuit acceleration (Ono & Mustari, 2009), gain control, learning (Takagi, Zee,

& Tamargo, 2000), and perhaps go/no-go behavior (Fukushima, Fukushima, & Warabi, 2011). Other regions of the cerebellum such as the hemispheres deserve further study to determine the relative roles in smooth pursuit behavior.

FUTURE STUDIES

Our understanding of neural mechanisms for smooth pursuit has advanced considerably during the last 10 years. A number of important areas of investigation remain. For example, we know relatively little about the specific signals carried in different projection neurons from cortex to brainstem and cerebellum during different smooth pursuit and associated behaviors. Determining these signals will help constrain existing models and refine our understanding of neural mechanisms for sensorimotor behavior. Future studies combining traditional neurophysiological methods with new optogenetic techniques could advance our understanding of smooth pursuit circuits. Similarly, studies that examine the interaction of smooth pursuit with other eye movement systems during a variety of sensorimotor behaviors could significantly advance our understanding of how different neural circuits interact during natural behaviors.

ACKNOWLEDGMENTS

This work was supported by grants from the National Institutes of Health (NEI EY06069; EY013308; EY019266; ORIP P51 OD01425; Research to Prevent Blindness).

REFERENCES

Angelaki, D. E., & Cullen, K. E. (2008). Vestibular system: The many facets of a multimodal sense. *Annual Review of Neuroscience, 31,* 125–150.

Barnes, G. R. (2008). Cognitive processes involved in smooth pursuit eye movements. *Brain and Cognition, 68,* 309–326.

Basso, M. A., Krauzlis, R. J., & Wurtz, R. H. (2000). Activation and inactivation of rostral superior colliculus neurons during smooth-pursuit eye movements in monkeys. *Journal of Neurophysiology, 84,* 892–908.

Born, R. T., & Bradley, D. C. (2005). Structure and function of visual area MT. *Annual Review of Neuroscience, 28,* 157–189.

Britten, K. H. (2008). Mechanisms of self-percption of motion. *Annual Review of Neuroscience, 31,* 389–410.

Brostek, L., Eggert, T., Ono, S., Mustari, M. J., Büttner, U., & Glasauer, S. (2011). An information-theoretic approach for evaluating probabilistic tuning functions of single neurons. *Frontiers in Computational Neuroscience, 5,* 1–11.

Burke, M. R., & Barnes, G. R. (2011). The neural correlates of inhibiting pursuit to smoothly moving targets. *Journal of Cognitive Neuroscience, 23,* 3294–3303.

Buttner, U., & Buttner-Ennever, J. A. (2006). Present concepts of oculomotor organization. *Progress in Brain Research, 151,* 1–42.

Collewijn, H., & Tamminga, E. P. (1984). Human smooth and saccadic eye movements during voluntary pursuit of different target motions on different backgrounds. *Journal of Physiology, 351,* 217–250.

Cullen, K. E. (2012). The vestibular system: Multimodal integration and encoding of self-motion for motor control. *Trends in Neurosciences, 35,* 185–196.

Dürsteler, M. R., & Wurtz, R. H. (1988). Pursuit and optokinetic deficits following chemical lesions of cortical areas MT and MST. *Journal of Neurophysiology, 60,* 940–965.

Fukushima, K., Fukushima, J., & Warabi, T. (2011). Vestibular-related frontal cortical areas and their roles in smooth-pursuit eye movements: Representation of neck velocity, neck–vestibular interactions, and memory-based smooth-pursuit. *Frontiers in Neurology, 78,* 1–26.

Gamlin, P. D. R. (2006). The pretectum: connections and oculomotor-related roles. *Progress in Brain Research, 151,* 379–405.

Gottlieb, J. P., Bruce, C. J., & MacAvoy, M. G. (1993). Smooth eye movements elicited by microstimulation in the primate frontal eye field. *Journal of Neurophysiology, 69,* 786–799.

Ito, M. (2006). Cerebellar circuitry as a neuronal machine. *Progress in Neurobiology, 78,* 272–303.

Kahlon, M., & Lisberger, S. G. (2000). Changes in the responses of Purkinje cells in the flocular complex of monkeys after motor learning in smooth pursuit eye movements. *Journal of Neurophysiology, 84,* 2945–2960.

Kawano, K., Shidara, M., Watanabe, Y., & Yamane, S. (1994). Neural activity in cortical area MST of alert monkey during ocular following responses. *Journal of Neurophysiology, 71,* 2305–2324.

Kawano, K., Shidara, M., & Yamane, S. (1992). Neural activity in dorsolateral pontine nucleus of alert monkey during ocular following responses. *Journal of Neurophysiology, 67,* 680–703.

Krauzlis, R. J. (2004). Recasting the smooth pursuit eye movement system. *Journal of Neurophysiology, 91,* 591–603.

Leigh, R. J., & Zee, D. S. (2006). The neurology of eye movements (4th ed.). Oxford, England: Oxford Unversity Press.

Lisberger, S. G. (2010). Visual guidance of smooth-pursuit eye movements: Sensation, action, and what happens in between. *Neuron, 66,* 477–491.

Lynch, J. C., & Tian, J. R. (2006). Cortico–cortical networks and cortico–subcortical loops for the higher control of eye movements. *Progress in Brain Research, 151,* 461–501.

Mahaffy, S., & Krauzlis, R. J. (2011). Neural activity in the frontal pursuit area does not underlie pursuit target selection. *Vision Research, 51,* 853–866.

Miles, F. A., Kawano, K., & Optican, L. M. (1986). Short-latency ocular following responses of monkey. I. Dependence on temporospatial properties of visual input. *Journal of Neurophysiology, 56,* 1321–1354.

Newsome, W. T., Wurtz, R. H., Dursteler, M. R., & Mikami, A. (1985). Deficits in visual motion processing following ibo-tenic acid lesions of the middle temporal visual area of the macaque monkey. *Journal of Neuroscience, 5,* 825–840.

Newsome, W. T., Wurtz, R. H., & Komatsu, H. (1988). Relation of cortical areas MT and MST to pursuit eye movements: II. Differentiation of retinal from extraretinal inputs. *Journal of Neurophysiology, 60,* 604–620.

Nuding, U., Ono, S., Mustari, M. J., Buttner, U., & Glasauer, S. (2008). A theory of the dual pathways for smooth pursuit based on dynamic gain control. *Journal of Neurophysiology, 99,* 2798–2808.

Ono, S., Das, V. E., & Mustari, M. J. (2004). Gaze-related response properties of DLPN and NRTP neurons in the rhesus macaque. *Journal of Neurophysiology, 91,* 2484–2500.

Ono, S., & Mustari, M. J. (2006). Extraretinal signals in MSTd neurons related to volitional smooth pursuit. *Journal of Neurophysiology, 96,* 2819–2825.

Ono, S., & Mustari, M. J. (2009). Smooth pursuit-related information processing in frontal eye field neurons that project to the NRTP. *Cerebral Cortex, 19,* 1186–1197.

Ono, S., & Mustari, M. J. (2010). Visual error signals from the pretectal nucleus of the optic tract guide motor learning for smooth pursuit. *Journal of Neurophysiology, 103,* 2889–2899.

Page, W. K., & Duffy, C. J. (2003). Heading representation in MST: Sensory interactions and population encoding. *Journal of Neurophysiology, 89,* 1994–2013.

Robinson, D. A., Gordon, J. L., & Gordon, S. E. (1986). A model of the smooth pursuit eye movement system. *Biological Cybernetics, 55,* 43–57.

Stone, L. S., & Lisberger, S. G. (1990). Visual responses of Purkinje cells in the cerebellar flocculus during smooth-pursuit eye movements in monkeys. II. Complex spikes. *Journal of Neurophysiology, 63,* 1262–1275.

Takagi, M., Zee, D. S., & Tamargo, R. J. (2000). Effects of lesions of the oculomotor cerebellar vermis on eye movements in primate: Smooth pursuit. *Journal of Neurophysiology, 83,* 2047–2062.

Tanaka, M. (2005). Involvement of the central thalamus in the control of smooth pursuit eye movements. *Journal of Neuroscience, 25,* 5866–5876.

Tanaka, M., & Lisberger, S. G. (2001). Regulation of the gain of visually guided smooth-pursuit eye movements by frontal cortex. *Nature, 409,* 191–194.

Thier, P., & Mock, M. (2006). The oculomotor role of the pontine nuclei and the nucleus reticularis tegmenti pontis. *Progress in Brain Research, 151,* 293–320..

Wang, X., Zhang, M., Cohen, I. S., & Goldberg, M. E. (2007). The proprioceptive representation of eye position in monkey primary somatosensory cortex. *Nature Neuroscience, 10,* 640–646.

Wolpert, D. M., & Kawato, M. (1998). Multiple paired forward and inverse models for motor control. *Neural Networks, 11,* 1317–1329.

Yang, S. N., Hwang, H., Ford, J., & Heinen, S. (2010). Supplementary eye field activity reflects a decision rule governing smooth pursuit but not the decision. *Journal of Neurophysiology, 103,* 2458–2469.

63 Selection of Targets for Saccadic Eye Movements: An Update

JEFFREY D. SCHALL

Vision is an active process. Because primate vision is acute only at the fovea, gaze must shift to identify objects in the scene. However, vision is dramatically impaired during rapid gaze shifts. Therefore, the visual and ocular motor systems must be coordinated judiciously. Because gaze can be directed to only one item at a time, some process must distinguish among possible locations to select the target for a saccade. The outcome of the selection process is defined by the goals of visually guided behavior. Analyses of the pattern of eye movements have revealed regularities such as concentrating on conspicuous and informative features of an image during scrutiny of simple geometric stimuli (Liversedge & Findlay, 2000), natural images (Henderson, 2011) or text (Rayner & Liversedge, 2011), leading to the view that we can direct gaze in a statistically optimal manner (Najemnik & Geisler, 2009). Saccade target selection is coordinated with movements of other parts of the body during natural behaviors (Hayhoe & Ballard, 2011; Land & Tatler, 2009). Research focused on the neural mechanisms of visual search and saccade target selection began 20 years ago (Schall & Hanes, 1993) and is now undertaken by many research groups. This topic has been reviewed thoroughly both in the previous edition of this book (Schall, 2004) and in subsequent publications (Bichot & Desimone, 2006; Bisley & Goldberg, 2010; Constantinidis, 2006; Gottlieb & Balan, 2010; Paré & Dorris, 2011; Schall & Cohen, 2011; Schiller & Tehovnik, 2005; Wardak, Olivier, & Duhamel, 2011), so this chapter will orient the reader to the general issues, highlight more recent findings, and frame the major remaining questions. The citations will be selective and recent; the interested student can find classic references in the previous version of this chapter.

VISUAL SEARCH: SALIENCE, ATTENTION, AND STAGES OF PROCESSING

To investigate how the brain selects the target for an eye movement, multiple stimuli that can be distinguished in some way must be presented. Referred to as visual search, this experimental design has been used extensively to investigate visual selection and attention (reviewed by Geisler & Cormack, 2011; Wolfe & Horowitz, 2004). Search is efficient (with fewer errors and faster response times) if stimuli differ along basic visual feature dimensions, such as color, form, or direction of motion. In contrast, if the distractors resemble the target or no single feature clearly distinguishes the stimuli, then search becomes less efficient (more errors, longer response times).

These observations have been explained most commonly by postulating the existence of a map of salience[1] derived from converging bottom-up and top-down influences (Bundesen, Habekost, & Kyllingsbaek, 2011; Tsotsos, 2011). Salience refers to how distinct one element of the image is from surrounding elements. This distinctness can occur because the element has visual features that are very different from the surroundings (a ripe, red berry in green leaves). The distinctness can also occur because the element is more important than others (the face of a friend among strangers). The distinctness derived from visual features and importance confers upon that part of the image greater likelihood of receiving enhanced visual processing and a gaze shift. In the models of visual search referred to above, one major input to the salience map is the maps of the features (color, shape, motion, depth) of elements of the image. Another major input is top-down modulation based on goals and expectations. The representation of likely targets that is implicit in and dependent on the feature maps becomes explicit in the salience map. Peaks of activation in the salience map that develop as a result of competitive interactions represent locations that have been selected for further processing and thus covert orienting of attention.

Saccade target selection coincides with the allocation of visual attention that has been the focus of considerable research (e.g., chapters 23, 24, 71, 75, 76–78). Attentional allocation and saccade production interact in various ways. Some investigators have explained the connection between saccade production and attention allocation by proposing that the allocation of attention amounts to a subthreshold command to shift gaze. This view is known as the oculomotor readiness hypothesis (Klein, 1980) or the premotor theory of attention (Rizzolatti, 1983). Although an influential guiding hypothesis, numerous observations are inconsistent with a strict interpretation of this hypothesis (reviewed by Awh, Armstrong, & Moore, 2006), and I will review more below. Indeed, I will emphasize the very basic fact that the focus of visual attention can be directed away from the focus of gaze showing that the link between shifting gaze and directing attention is not obligatory. Various lines of evidence demonstrate that the neural process of selecting a target for orienting is functionally distinct from the neural process of preparing a saccade. This distinction has been confirmed by experimental manipulations that independently influence the distinct and successive stages (Sternberg, 2011).

OVERVIEW OF NEURAL SUBSTRATES

As readers of this volume know, saccades are ultimately produced by a network of neurons in the brainstem (chapters 60, 61; see also Cullen & Van Horn 2011). This network that shapes the pattern of activation of the motor neurons innervating extraocular muscles requires two inputs—where to shift gaze and when to initiate the movement. Key brain structures controlling the initiation of saccades by the brainstem include the superior colliculus (SC) (chapter 64; see also White & Munoz, 2011b) and the frontal eye field (FEF; Johnston & Everling, 2011; Schall & Boucher, 2007) operating in a recurrent network through the basal ganglia (Vokoun, Mahamed, & Basso, 2011) and thalamus (Tanaka & Kunimatsu, 2011). Saccade endpoints are specified by the spatial pattern of activation in these structures, which can be appreciated most clearly in the SC (see figure 63.1). Presaccadic neurons that lead to initiation of saccades have movement fields wherein they are most active before saccades of a particular direction and amplitude and progressively less active before saccades deviating from the optimal. Thus, before each saccade, neurons over a rather broad extent of SC are activated in a graded manner. The space of saccade direction and amplitude is mapped in a topographic fashion corresponding to the map of visual space in the superficial layers of the SC. To produce a signal leading to a saccade of a particular direction and amplitude, the activation of many cells in a region of the motor map are combined as vectors. Through this vector combination saccadic endpoints can be more precise than the size of individual neuron movement fields.

Evidence for the vector combination hypothesis is obtained either by electrically stimulating two points in the SC or FEF (Robinson, 1972) or by presenting a pair of visual targets close together (Findlay, 1982). This results in a larger zone of activation occurring across the map (see figure 63.1B). A saccade resulting from the vector average of the zone of activation directs gaze to a location between the targets. The problem of saccade target selection is highlighted, though, when a circular array of stimuli is presented, resulting in a spatial distribution of activation in the SC with multiple peaks that are balanced around the map (see figure 63.1C). The vector combination of this distribution of activity amounts to a resultant of no net length—not a very useful outcome. To produce a useful saccade, the activation in the map of the SC and associated structures must be limited to the neurons contributing to generating just that saccade. Thus, if more than one location in the SC map becomes activated, then additional processing must resolve which of the peaks of activation should become dominant among the rest for an accurate saccade to just one among alternative stimuli. This additional processing is selection of the target.

A network of structures in the visual pathway contributes to selecting targets for saccades. Neurons in primary visual cortex and extrastriate areas represent a variety of more or less elaborated features, surfaces and objects (chapters 25–28, 40, 41, 47, 55, 56), and these representations are influenced by the presence and nature of surrounding stimuli (chapter 30) and experience (chapters 69 and 70). Visual processing is not concluded in the parietal and temporal lobes, for extensive convergence of signals from numerous areas occurs in FEF (Markov et al., 2011; Schall et al., 1995) and SC (May, 2006). In fact, the latency of visual responses in FEF are comparable to those in middle temporal visual cortex (MT) and even precede the latencies of some neurons in V1 (Schmolesky et al., 1998). The density of neurons in the supragranular layers that project to area V4 resembles a feedforward connection (Barone et al., 2000) with terminals on dendritic spines, mainly in supragranular layers of V4 (Anderson, Kennedy, & Martin, 2011). Thus, FEF is positioned anatomically and temporally to influence neural processes in extrastriate visual areas. The influence conveyed by FEF to visual cortex is a central feature of network models of visual attention (e.g., Hamker & Zirnsak, 2006). However, recent evidence indicates that areas V4 and MT receive a different quality of influences from the frontal lobe (Ninomiya et al., 2012); thus, "top-down" modulation is not a unitary process.

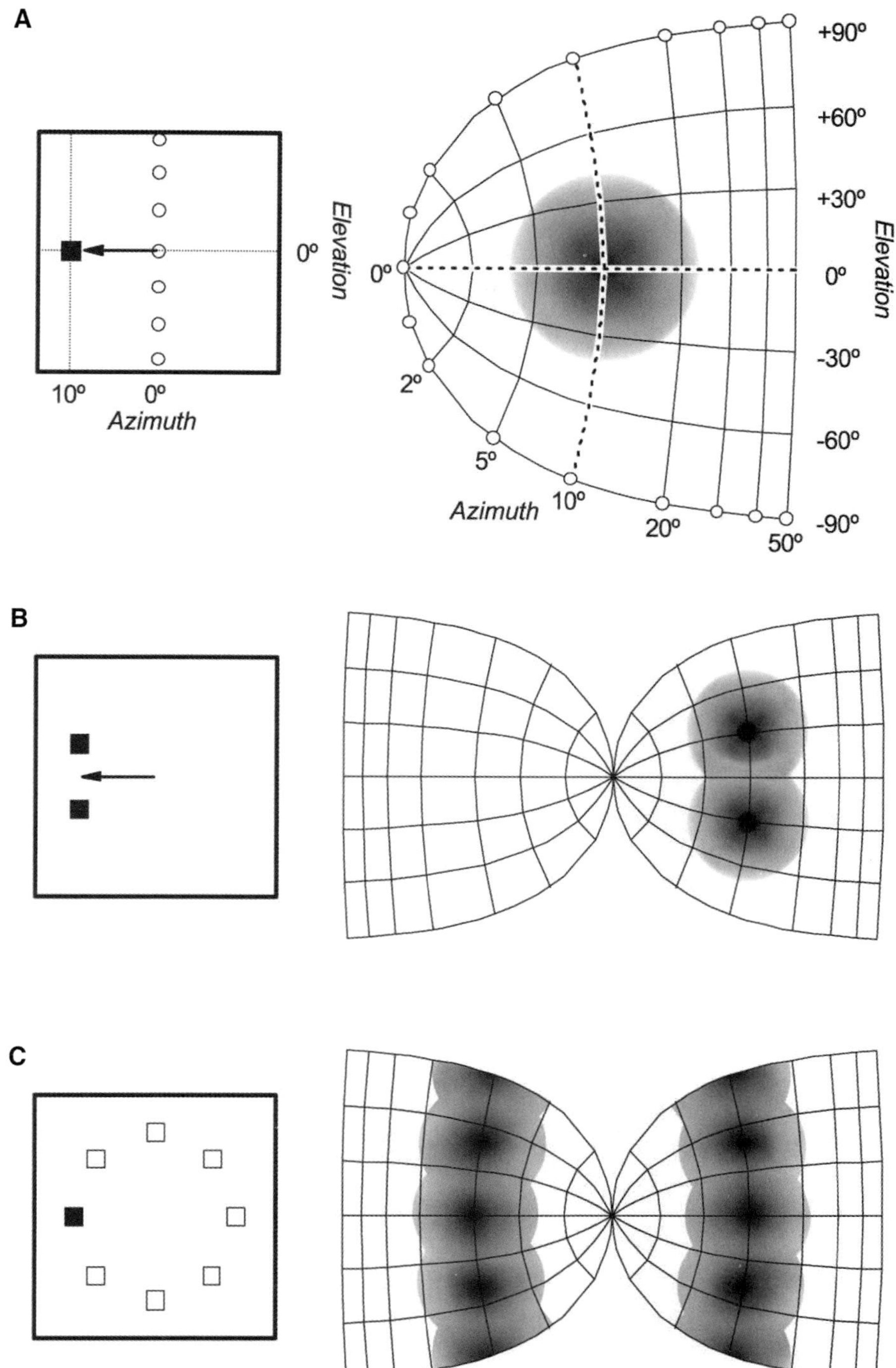

FIGURE 63.1 Superior colliculus activation and saccade target selection. (A) Left panel illustrates the visual field with a saccade (arrow) produced from the center to a target at 10° azimuth and 0° elevation. The vertical meridian is indicated by the row of circles. Right panel illustrates the map of the superior colliculus derived from a systematic survey of the direction and amplitude of saccades evoked by electrical stimulation of various parts of the superior colliculus. A visually guided saccade to a point 10° eccentric on the horizontal meridian is preceded by graded activation over an extended part of the superior colliculus map. The peak of activation (black) is centered at the appropriate point in the map, and surrounding neurons are progressively less active further from the center of activation (gray). The direction and amplitude of the saccade that is produced corresponds to that specified by the location of the center of gravity of the activation in the superior colliculus map. (B) Presentation of two nearby stimuli result in a broader zone of activation in the superior colliculus with two peaks. The center of gravity of the activation is a location in between the two stimuli. This results in a saccade directing gaze to neither stimulus. The map of the superior colliculus representing each hemifield is shown. (C) Presentation of eight stimuli produces activation balanced around the superior colliculus map. The center of gravity of such balanced activation with multiple peaks amounts to a saccade with no amplitude. Thus, to produce a particular saccade, the activation in the superior colliculus map must evolve so that a single peak is present, corresponding to the desired saccade direction and amplitude.

Extensive research has demonstrated how neurons in cortical areas that represent stimulus features are modulated by target and surrounding nontarget features under various task demands (Bichot, Rossi, & Desimone, 2005; Buracas & Albright, 2009; David et al., 2008; Mirabella et al., 2007; Mruczek & Sheinberg, 2007; Ogawa & Komatsu, 2006; Saruwatari, Inoue, & Mikami, 2008). Another major input is top-down modulation based on goals and expectations enabled by neural circuits in the frontal lobe (Everling et al., 2006; Rossi et al., 2007; Zhou & Thompson, 2009).

We can define the functional salience map as the population of neurons that are not intrinsically feature selective but receive input from feature-selective neurons so that they signal the location of objects that are the target or are target-like in a manner that can be used to guide an action like an eye movement. According to this definition, compelling evidence obtained in multiple laboratories supports the conclusion that the neural representation of the salience map is distributed among multiple cortical areas and subcortical structures including FEF, parietal areas LIP and 7a, as well as the SC, basal ganglia, and associated thalamic nuclei. The heterogeneity of neural function within and diversity of connectivity between these areas makes clear that this salience representation is instantiated by an interconnected circuit built from some but not all of the neurons in these structures. Evidence that the selection process observed in these sensorimotor structures can be identified with a salience representation includes the following observations.

When a search array appears (either by flashing on during fixation or after a previous scanning saccade), activation increases at all locations in the map[2] corresponding to the potential saccade targets. This happens because these neurons are not naturally selective for visual features. Following the initial volley, activation becomes relatively lower at locations that would produce saccades to nontarget objects and is sustained or grows at locations corresponding to more conspicuous or important potential targets (see figure 63.2). This process has been observed independently by multiple laboratories in FEF (Cohen et al., 2009b; Lee & Keller, 2008; McPeek, 2006; Ogawa & Komatsu, 2006; Schall & Hanes, 1993), posterior parietal cortex (Balan et al., 2008; Buschman & Miller, 2007; Constantinidis & Steinmetz, 2005; Ipata et al., 2006; Ogawa & Komatsu, 2009; Thomas & Paré, 2007), SC (Kim & Basso, 2008; McPeek & Keller, 2002; Shen & Paré, 2007; White & Munoz, 2011a), substantia nigra pars reticulata (Basso & Wurtz, 2002), and ocular motor thalamic nuclei

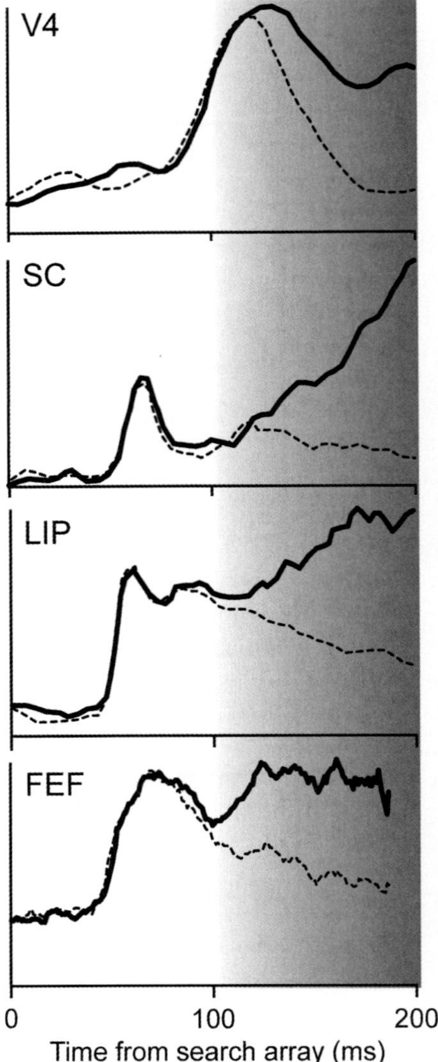

FIGURE 63.2 Illustration of visual and saccade target selection of representative single neurons in area V4 (adapted from Ogawa & Komatsu, 2004), superior colliculus (SC) (adapted from McPeek & Keller, 2002), lateral intraparietal area (LIP) (adapted from Thomas & Paré, 2007), and frontal eye field (FEF) (adapted from Thompson et al., 1996). The average discharge rate on trials when the target appeared in the response field (thick line) is plotted with the average discharge rate on trials when distractors appeared in the response field and the target was elsewhere (thin dashed line). Saccade target selection occurs in a distributed network of cortical and subcortical neurons.

(Schall & Thompson, 1994; Wyder, Massoglia, & Stanford, 2004). In these studies monkeys are responding to one among multiple alternatives for the purpose of earning reinforcement, usually with a single saccade. The target selection process has also been observed during natural scanning eye movements (Bichot, Rossi,

& Desimone, 2005; David et al., 2008; Phillips & Seg-raves, 2010; Zhou & Desimone, 2011). Microstimulation and inactivation have demonstrated causal roles in target selection of FEF (Monosov & Thompson, 2009; Wardak et al., 2006), SC (Lovejoy & Krauzlis, 2010; McPeek, 2008), and LIP (Balan & Gottlieb, 2009; Mirpour, Ong, & Bisley, 2010; Wardak, Olivier, Duhamel, 2004).

Manipulations that influence attention allocation in humans influence in parallel monkey performance and concomitant modulation of neural activity. For example, when search is less as compared to more efficient because target and distractor stimuli are more difficult to discriminate, then the selection process occupies more time and accounts for a greater proportion of the variability of reaction time (RT; Balan et al., 2008; Cohen et al., 2009b; Hayden & Gallant, 2005; Sato et al., 2001; Sato & Schall, 2003; Woodman et al., 2008). The well-known effects of target–distractor similarity on search performance that are expressed in response times and choices by macaque monkeys are paralleled in the magnitude and timing of the visual selection process measured in FEF neurons (Cohen et al., 2009b). When the target is more similar to distractors through either feature similarity or recent stimulus history, the level of neural activity in FEF representing the alternative stimuli is less distinct, leading to a higher likelihood of treating a distractor as if it were the target (Heitz et al., 2010; Thompson, Bichot, & Sato, 2005). This parallel suggests that the statement "less efficient allocation of attention" describes a state of the network in which the activity representing a target and distractors is less capable of being distinguished by either a neurophysiologist or a read-out circuit. Another influence believed to be mediated through the salience map is inhibition of return, the decreased likelihood of directing gaze to a location previously fixated. Neural correlates of this have been described in FEF (Bichot & Schall, 2002), LIP (Mirpour et al., 2009), and SC (Fecteau & Munoz, 2005).

The representation of salience is regarded to guide covert as well as overt orienting independent of effector. The neural selection of the target as a visual location to which to orient attention does not inevitably and immediately lead to reorienting of the eyes. It occurs if no overt response at all is made (Arcizet, Mirpour, & Bisley, 2011; Thompson, Bichot, & Schall, 1997) or if the saccade is directed away from a color singleton (Murthy et al., 2009; Sato & Schall, 2003). The selection process occurs as well if target location or property is signaled by a manual response (Ipata et al., 2009; Monosov & Thompson, 2009; Oristaglio et al., 2006; Thompson, Biscoe, & Sato, 2005).

POPULATION SIGNALS FOR TARGET SELECTION

Having identified key nodes in the network representing visual salience, further investigation of the mechanism has been accomplished. All of the results described above were based entirely on modulation of discharge rates of individual neurons. It is clear, though, that saccade target selection is accomplished by pools of neurons (Bichot et al., 2001; Kim & Basso, 2008, 2010) and probably entails more than just modulation of spike rate because cooperation and competition between pairs of neurons is modulated during target selection (Cohen et al., 2010). Indeed, correlation in discharge rates of FEF neurons over longer time scales has been reported even before stimulus presentation (Ogawa & Komatsu, 2010). Other researchers have measured local field potentials (LFP) in V4, LIP, and FEF during visual search and attention tasks and described increased coherence in the gamma band between spikes and LFP within and across areas such as V4, LIP, and FEF (Bichot, Rossi, & Desimone, 2005; Buschman & Miller, 2007; Gregoriou et al., 2009). Although argued to enhance the representation of attended objects, the functional utility of such signals is not undisputed (Ray & Maunsell, 2010).

An alternative analysis of LFPs is simply to measure the time course of differences in polarization when the target is in or out of the receptive field. This approach corresponds to the measurement of an event-related potential (ERP) on the scalp known as the N2pc that is a signature of the locus and time of attention allocation (Woodman & Luck, 1999). The N2pc has been found in macaque monkeys (Woodman et al., 2007). Source localization procedures indicate that the N2pc arises from parietal and occipitotemporal sources in humans (Boehler et al., 2011) and macaques (Young et al., 2010). In both efficient and inefficient search conditions the target is selected significantly earlier in neural spike rate modulation than in LFP polarization (Cohen et al., 2009a; Monosov, Trageser, & Thompson, 2008), and the delay varies with search efficiency. It appears that local processing within FEF mediated by spike rates results in delayed changes of synaptic potentials manifested in the LFP.

INTERACTIONS BETWEEN THE FRONTAL LOBE AND VISUAL CORTEX DURING TARGET SELECTION

I have described a target selection process that occurs more or less concurrently in multiple cortical areas and subcortical structures. Recent studies in macaque monkeys have investigated interactions between FEF and LIP (Buschman & Miller, 2007), V4 (Gregoriou et al., 2009, 2012; Zhou & Desimone, 2011), and inferior temporal (IT) cortex (Monosov, Sheinberg, & Thompson, 2010; Monosov & Thompson, 2009) as well as an ERP component recorded over visual cortex that indexes attention (Cohen et al., 2009a). While firm conclusions are premature because results were obtained with different tasks, neural signals, measurement procedures, and areas, some results seem consistent across laboratories. First, when search is inefficient, neural signals of attention allocation in FEF precede those in extrastriate visual areas. For example, a recent study demonstrated that spatial selection of a location in FEF precedes object recognition by IT neurons at that location (Monosov, Sheinberg, & Thompson, 2010) and the selection in FEF is necessary for detection and identification of the target (Monosov & Thompson, 2009). Similarly, the target selection observed in spike rate and LFP in FEF preceded the N2pc (Cohen et al., 2009a), and the delay between selection in FEF and visual cortex increased with the number of distractor stimuli, demonstrating that the delay is not due simply to conduction lags. These results expose a puzzling question—if different times of target selection are measured in different nodes of the network and scales of signal, then when would we say that attention has been allocated? Given the variation in selection time across neurons even within an area, can we say that the target is selected when the earliest, the latest, or some intermediate population of neurons resolve target location? Such a basic question highlights our profound uncertainty about how signals arise in and are conveyed between the areas representing features, objects, and salience.

This influence of FEF on visual cortex can influence the quality of attentive visual processing (Monosov & Thompson, 2009; Moore & Fallah, 2004). Weak electrical stimulation of FEF influences extrastriate visual cortex activity in a manner similar to what is observed when attention is allocated (Armstrong, Fitzgerald, & Moore, 2006; Ekstrom et al., 2009; Taylor, Nobre, & Rushworth, 2007; Walker, Techawachirakul, & Haggard, 2009).

FROM SALIENCE TO SACCADE

Explaining how sensory representations lead to accurate movements is a classic problem. One approach to this problem is based on the premise that noisy evidence guiding a response is accumulated over time until a threshold is achieved at which time the response is initiated (Ratcliff & McKoon, 2008; Usher & McClelland, 2001). A recent model inspired by this approach provides an explanation for how signals from neurons that represent target salience can be transformed into a saccade command (Purcell et al., 2010, 2012) (see figure 63.3). The model uses the activity of visually responsive neurons in the FEF representing object salience as evidence for stimulus salience that is accumulated in a network of deterministic accumulators producing saccades to each possible target location to generate accurate and timely saccades during visual search. Response times are specified by the time at which the integrated signal reaches a threshold. The model included leak in the integration process and lateral inhibition between the ensemble of accumulators as well as a form of inhibition that gates the flow of perceptual evidence to the accumulators. Alternative model architectures were excluded because they did not fit the actual distributions of response times nor produce activation profiles corresponding to the form of actual movement neuron activity. At present, this is the only model of visual search that accounts for the range and form of response time distributions (Wolfe, Horowitz, & Palmer, 2010). This union of cognitive modeling and neurophysiology indicates how the visual–motor transformation can occur and provides a concrete mapping between neuron function and specific cognitive processes.

The picture that emerges is that the process of visual selection occupies a certain amount of time that can be shorter and less variable if the target is conspicuous, or it can be longer and more variable if the target is less conspicuous. If subjects wish to prevent a saccade to a nontarget stimulus, then the preparation of the saccade can be delayed until the visual selection process has proceeded to a high degree of resolution. Neural activity mediating saccade preparation begins to grow as the selection process is completed and the rate of growth of activity leading to the movement varies apparently randomly such that sometimes gaze shifts sooner and sometimes gaze shifts later. Systematic adjustments of saccade latency, though, appear to arise through changes in the time that the accumulation of activity begins (Pouget et al., 2011).

STIMULUS–RESPONSE MAPPING

The gated feedforward cascade model assumes that saccade production is guided entirely by the visual salience representation. Thus, errant saccades would be explained by failure to represent evidence correctly. While this has been observed in some testing conditions (Heitz et al., 2010; Thompson, Bichot, & Sato, 2005), several other lines of research demonstrate that the salience representation can be correct even if responses

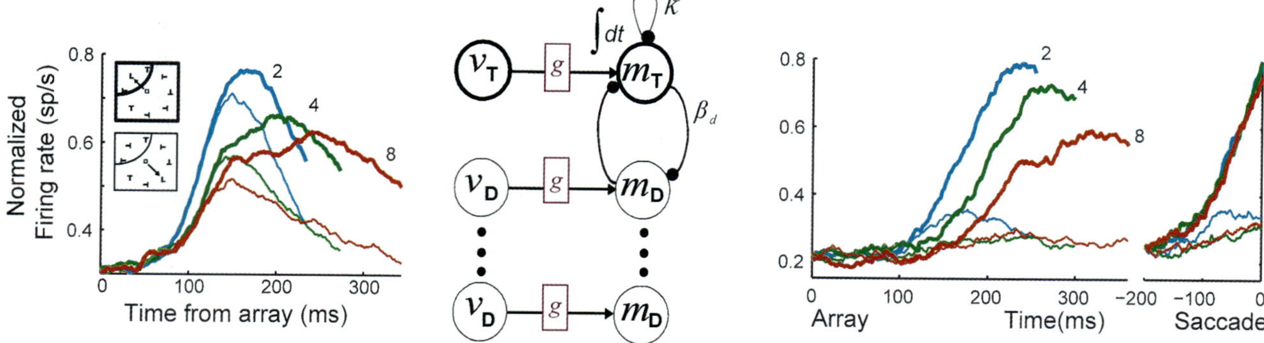

FIGURE 63.3   Object salience can be converted into saccade command through a gated accumulator network. Left: Data showing the evolving selective responses of visual salience neurons to the target "L" (thick) versus distractor "T" (thin) when presented in arrays of two (blue), four (green), or eight (red) objects. Middle: Gated accumulator model architecture in which the visual salience representation at the location of the actual target (vT) and distractors (vD) cascades into a network of units that produce saccades to the actual target location (mT) or distractor locations (mD) when their activation reaches a threshold. The activation accumulates through integration ($\int$dt) of the salience input limited by leak (k) and lateral inhibition (β). Right: Data showing accumulating activity of presaccadic movement neurons before saccades into (thick) and out of (thin) the movement field aligned on array presentation (left) and saccade initiation (right). The model replicated the dynamics of this accumulation process. (Adapted from Purcell et al., 2012.)

are incorrect. For example, in monkeys performing a saccade double-step task with visual search, visual neurons in the FEF locate the new location of the oddball in the search array correctly even when monkeys incorrectly shift gaze to the old location (Murthy et al., 2009). Similarly, when manual response errors occur, the selection process in FEF locates the singleton in the search array correctly (Trageser et al., 2008). However, if the brain located the new location of the oddball correctly, why was an error made? A plausible answer appeals to the hypothesis that the response production stage, although guided by the perceptual stage can operate independently of the perceptual stage. Further evidence for this is the fact that these errors can be corrected very rapidly, even before the brain can register that the gaze shift was an error (Murthy et al., 2007; see also Phillips & Segraves, 2010).

Saccade target selection has also been investigated under conditions that dissociate visual target location from saccade endpoint explicitly. Monkeys were trained to make a prosaccade to a color singleton or an antisaccade to the distractor located opposite the singleton; the shape of the singleton cued the direction of the saccade (Sato & Schall, 2003). As observed in previous studies, the response time for antisaccades was greater than that for prosaccades. A goal of this experiment was to account for this difference in terms of the neural processes that locate the singleton, encode its shape, map the stimulus onto the response, select the endpoint of the saccade, and finally initiate the saccade. Two types of visually responsive neurons were found in FEF. The first, called type I, exhibited the typical pattern

of initially indiscriminant activity followed by selection of the singleton in the response field through elevated discharge rate regardless of whether the singleton's features cue a prosaccade or an antisaccade. Some of these type I neurons maintained the representation of singleton location in antisaccade trials until the saccade was produced. However, the majority of the type I neurons exhibited a dramatic modulation of discharge rate before the antisaccade was initiated (see figure 63.4A). After producing higher discharge rates for the singleton as compared to a distractor in the receptive field, the firing rates changed such that higher discharge rates were observed for the endpoint of the antisaccade relative to the singleton location. This modulation could be described as the focus of attention shifting from one location to the other before the saccade. The second type of neuron, called type II, resembled qualitatively the form of modulation of type I neurons in prosaccade trials, but in antisaccade trials, these neurons did not select the location of the singleton and only selected the endpoint of the saccade (see figure 63.4B).

This experiment revealed a sequence of processes that can be distinguished in the modulation of different populations of neurons in FEF. The time course of these processes can be measured and compared across stimulus–response mapping rules (see figure 63.4C). To summarize, type I neurons selected the singleton earlier than did type II neurons, and the time of this selection did not vary with stimulus–response mapping or account for the difference in RT. However, the singleton selection time of type II neurons in prosaccade trials was less synchronized with array presentation and more related

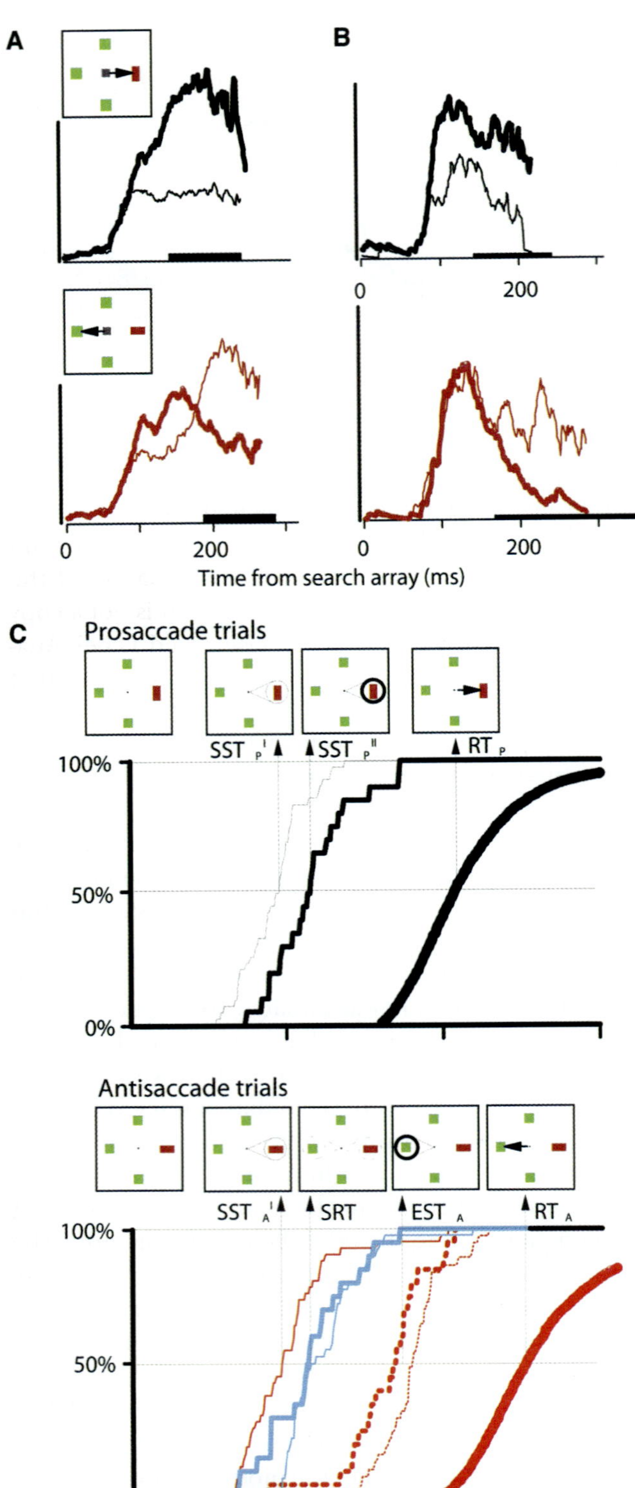

FIGURE 63.4   Elaboration of target selection process when mapping between location of visual target and endpoint of saccade is varied. (A) Activity of FEF neuron with activity indexing allocation of attention (type I). Average spike density function when the singleton fell in the neuron's receptive field (thick line) and when the singleton was located opposite the receptive field (thin line) in prosaccade (top) and antisaccade (bottom) trials. Bracket on abscissa marks range of reaction time in prosaccade (RT_P) and in antisaccade (RT_A) trials. Scale bar represents 100 spikes/s. (B) Activity of FEF neuron with activity indexing selection of the saccade endpoint (type II). (C) Cumulative distributions of modulation times in prosaccade (top) and antisaccade (bottom) trials for type I (thin) and type II (thicker) neurons with corresponding RT (thickest). The inset arrays indicate hypothesized functional correlates. After presentation of the array, the singleton location is selected after a delayed labeled the singleton selection time (SST_P^I and SST_A^I) of type I neurons (indicated by the spotlight on the singleton); this occurs at the same time in prosaccade and antisaccade trials and does not relate to whether or when gaze shifts. In prosaccade but not antisaccade trials type II neurons select the singleton at a later time (indicated by SST_P^{II}), which accounts for some of the variability of RT. A comparison of activation in prosaccade and antisaccade trials reveals the time at which the shape of the singleton is encoded to specify the correct saccade direction, labeled the stimulus-response time (SRT). This follows singleton selection and coincides for type I (thin blue) and type II (thicker blue) neurons in antisaccade trials. At the moment marked by SRT in antisaccade trials the representation of the singleton decreases, and the representation of the location opposite the singleton, the endpoint of the antisaccade, increases (indicated by the weaker spotlight on the singleton and growing spotlight on the saccade endpoint). At this same time in prosaccade trials the representation of the saccade endpoint is enhanced by the selection that occurs in the type II neurons (indicated by the highlighted spotlight on the singleton). In antisaccade trials, further modulation selects the endpoint of the saccade after a delay labeled endpoint selection time (EST_A) (indicated by the highlighted spotlight on the antisaccade endpoint). This is accomplished concomitantly by type I (thin, red, dashed) and type II (thicker red, dashed) neurons. The time taken to select the endpoint of the saccade predicts some of the delay and variability of RT. (Adapted from Sato and Schall, 2003.)

to the time of saccade initiation. In antisaccade trials the time of endpoint selection by type I neurons was significantly later than that of type II neurons. This result is as if the endpoint of a saccade must be identified before attention can shift to that location. The endpoint selection time of type I neurons in antisaccade trials was too late to explain the increase in RT relative to prosaccade trials. In contrast, the endpoint selection time of type II neurons in antisaccade trials accounted for some but not all of the delay and variability of RT. The results of this experiment demonstrate that the process of saccade target selection requires a number of representations and transformations beyond simply representing stimulus salience and producing a saccade.

TESTING THE PREMOTOR THEORY OF ATTENTION

If shifting visual spatial attention corresponds to preparing a saccade, then it should be impossible to dissociate saccade preparation from the focus of attention even if the endpoint of a saccade is directed opposite the attended stimulus. This was tested by probing the evolution of saccade preparation using electrical stimulation of the FEF (Juan, Shorter-Jacobi, & Schall, 2004). The focus of attention was dissociated momentarily from the endpoint of a saccade by training monkeys to perform visual search for an attention-capturing color singleton and then shift gaze either toward (prosaccade) or opposite (antisaccade) this color singleton according to its orientation (Sato & Schall, 2003). Saccade preparation was probed by measuring the direction of saccades evoked by intracortical microstimulation of the FEF at different times following the search array. Eye movements evoked on prosaccade trials deviated progressively toward the singleton that was the endpoint of the saccade. Eye movements evoked on antisaccade trials deviated not toward the singleton but only toward the saccade endpoint opposite the singleton. The interpretation of these results is framed by the findings described above, showing that on antisaccade trials most visually responsive neurons in FEF initially select the singleton while attention is allocated to distinguish its shape. Evidence consistent with these observations has been obtained in human participants using transcranial magnetic stimulation (Juan et al., 2008) and in a study probing explicitly the locus of attention (Smith & Schenk, 2007). Thus, the brain can covertly orient attention without preparing a saccade to the locus of attention. In other words, target selection and saccade preparation are distinct processes because they can be modified separately (Sternberg, 2011). This separate modifiability occurs because different populations of neurons carry out different functions as reviewed above.

Testing the premotor theory requires specifying the anatomical level at which the mechanism maps onto the brain. If shifting attention is accomplished by the same neurons that are preparing a saccade, and if saccade commands are issued by layer 5 pyramidal neurons in FEF, and if FEF influences attention by projections to areas V4 and TEO, then numerous layer 5 neurons must be double labeled by tracer injections in SC and V4/TEO. A recent study found, though, that whereas only pyramidal neurons in layer 5 projected to the SC, the large majority of neurons in FEF projecting to extrastriate visual cortex are located in layers 2 and 3, and no neurons were found projecting to both SC and visual cortex (Pouget et al., 2009). Thus, we can reject the premise that shifting attention is accomplished by the population of neurons that prepare saccades. This conclusion is based on a strict mapping between populations of specific types of neurons and the cognitive processes of attention allocation and saccade preparation. However, a theory formulated too generally to map onto specific neural types loses the relevance of mechanism and force of falsifiability. This result entails that FEF delivers different signals to the visual and ocular motor systems. What, then, is the nature of the influence of FEF on visual processing? If it is not a copy of the saccade command, what else could it be? Anatomical reconstruction of recording sites shows that neurons located in the supragranular layers of FEF are active during the process of attentional target selection (Thompson et al., 1996). Therefore, the kind of signal that extrastriate cortex receives from FEF is the target selection process described above that corresponds to the allocation of attention.

GENERAL SUMMARY

Vision occurs naturally in a continuous cycle of fixations interrupted by gaze shifts. The guidance of these eye movements requires information about what is where in the image. The identity of objects is derived mainly from their visible features. Single neurons in the visual pathway represent the presence of specific features by the level of activation. Each point in the visual field is represented by populations of neurons activated by all types of features. Topographic representations are found throughout the visual and oculomotor systems; neighboring neurons tend to represent similar visual field locations or saccades.

When confronted by an image with many possible targets, the visual system compares the features of elements across the visual field. The retinotopic maps of the visual field facilitate local interactions to implement such comparisons; in particular, a network of lateral inhibition can extract the locations of the most conspicuous stimuli in the visual field. The process of these comparisons can be influenced by knowledge so that inconspicuous but important elements in the image can be the focus of gaze. This selection process results in a state of activation in which neurons with potential targets in their receptive field are more active, and neurons with nontargets in their receptive field are less active.

The outcome of this selection process can be represented at a level of abstraction distinct from the representation of the features themselves. This is why the hypothetical construct of a salience map is useful. The state of neural selection of a salient target relative to surrounding nontarget elements amounts to the covert allocation of attention that usually precedes overt shifts of gaze. The time taken for the brain to achieve an explicit representation of the location of a target varies predictably according to how distinct the target appears in relation to nontarget elements.

Coordinated with this visual processing is activation in a network including the FEF and SC that is responsible for producing the saccade. A saccade is produced when the activation at one location within the motor map reaches a critical threshold. One job of visual processing, influenced by memory and goals, is to ensure that only one site—the best site—within the map of movements becomes activated. This is done when the neurons signaling the location of the desired target develop enhanced activation while the neurons responding to other locations are attenuated. When confronted with ambiguous images having multiple potential targets, partial activation can occur in parts of the motor map representing saccades to nontarget elements that resemble the target. Saccade target selection converts an initially ambiguous pattern of neural activation into a pattern that reliably signals one target location in a winner-take-all fashion. However, the representation of likely targets for orienting does not automatically and unalterably produce a saccade. Sometimes potential targets are perceived without an overt gaze shift, or gaze can shift to locations not occupied by salient stimuli. The explanation of this flexible coupling between target selection and saccade production requires separate stages or modules that select a target for orienting and that produce gaze shifts. The flexible relationship between target selection and saccade production also affords the ability to emphasize speed or accuracy. Accuracy in fixating correctly can be emphasized at the expense of speed by allowing the visual selection process to resolve alternatives before producing a saccade. On the other hand, accuracy can be sacrificed for speed, allowing the visuomotor system to produce a saccade that may be inaccurate because it is premature relative to the target selection process.

Our understanding of how the visual system guides saccadic eye movements has improved considerably since the first version of this chapter appeared. While we continue to pursue remaining questions, we should retain our sense of marvel at the nimble and flexible manner of movements of these shiny globes of gristle that are called the windows to the soul.

ACKNOWLEDGMENTS

Research in the author's laboratory has been supported by the National Eye Institute, the National Institute of Mental Health, the National Science Foundation, the McKnight Endowment Fund for Neuroscience, the Air Force Office of Scientific Research, and by Robin and Richard Patton through the E. Bronson Ingram Chair in Neuroscience.

NOTES

1. Some authors prefer the term "salience" to refer only to conspicuity in the image, and the additional influence of goals is enabled through a "priority" map. In other formulations the salience map combines bottom-up and top-down influences; we will use the latter formulation.
2. The quality of the visual field and saccade vector representation varies across these areas from very precise in SC (chapter 64) to clear but less precise in FEF (Bruce et al., 1985; Suzuki & Azuma 1983) to quite imprecise in parietal areas (Ben Hamed et al., 2001).

REFERENCES

Anderson, J. C., Kennedy, H., & Martin, K. A. (2011). Pathways of attention: Synaptic relationships of frontal eye field to V4, lateral intraparietal cortex, and area 46 in macaque monkey. *Journal of Neuroscience, 31,* 10872–10881.

Arcizet, F., Mirpour, K., & Bisley, J. W. (2011). A pure salience response in posterior parietal cortex. *Cerebral Cortex, 21,* 2498–2506.

Armstrong, K. M., Fitzgerald, J. K., & Moore, T. (2006). Changes in visual receptive fields with microstimulation of frontal cortex. *Neuron, 50,* 791–798.

Awh, E., Armstrong, K. M., & Moore, T. (2006). Visual and oculomotor selection: Links, causes and implications for spatial attention. *Trends in Cognitive Sciences, 10,* 124–130.

Balan, P. F., & Gottlieb, J. (2009). Functional significance of nonspatial information in monkey lateral intraparietal area. *Journal of Neuroscience, 29,* 8166–8176.

Balan, P. F., Oristaglio, J., Schneider, D. M., & Gottlieb, J. (2008). Neuronal correlates of the set-size effect in monkey lateral intraparietal area. *PLoS Biology, 6,* e158.

Barone, P., Batardiere, A., Knoblauch, K., & Kennedy, H. (2000). Laminar distribution of neurons in extrastriate areas projecting to visual areas V1 and V4 correlates with the hierarchical rank and indicates the operation of a distance rule. *Journal of Neuroscience, 20,* 3263–3281.

Basso, M. A., & Wurtz, R. H. (2002). Neuronal activity in substantia nigra pars reticulata during target selection. *Journal of Neuroscience, 22,* 1883–1894.

Ben Hamed, S., Duhamel, J. R., Bremmer, F., & Graf, W. (2001). Representation of the visual field in the lateral intraparietal area of macaque monkeys: A quantitative receptive field analysis. *Experimental Brain Research, 140,* 127–144.

Bichot, N. P., & Desimone, R. (2006). Finding a face in the crowd: Parallel and serial neural mechanisms of visual selection. *Progress in Brain Research, 155*, 147–156.

Bichot, N. P., Rossi, A. F., & Desimone, R. (2005). Parallel and serial neural mechanisms for visual search in macaque area V4. *Science, 308*, 529–534.

Bichot, N. P., & Schall, J. D. (2002). Priming in macaque frontal cortex during popout visual search: Feature-based facilitation and location-based inhibition of return. *Journal of Neuroscience, 22*, 4675–4685.

Bichot, N. P., Thompson, K. G., Rao, S. C., & Schall, J. D. (2001). Reliability of macaque frontal eye field neurons signaling saccade targets during visual search. *Journal of Neuroscience, 21*, 713–725.

Bisley, J. W., & Goldberg, M. E. (2010). Attention, intention, and priority in the parietal lobe. *Annual Review of Neuroscience, 33*, 1–21.

Boehler, C. N., Tsotsos, J. K., Schoenfeld, M. A., Heinze, H. J., & Hopf, J. M. (2011). Neural mechanisms of surround attenuation and distractor competition in visual search. *Journal of Neuroscience, 31*, 5213–5224.

Bruce, C. J., Goldberg, M. E., Bushnell, M. C., & Stanton, G. B. (1985). Primate frontal eye fields. II. Physiological and anatomical correlates of electrically evoked eye movements. *Journal of Neurophysiology, 54*, 714–734.

Bundesen, C., Habekost, T., & Kyllingsbaek, S. (2011). A neural theory of visual attention and short-term memory (NTVA). *Neuropsychologia, 49*, 1446–1457.

Buracas, G. T., & Albright, T. D. (2009). Modulation of neuronal responses during covert search for visual feature conjunctions. *Proceedings of the National Academy of Sciences of the United States of America, 106*, 16853–16858.

Buschman, T. J., & Miller, E. K. (2007). Top-down versus bottom-up control of attention in the prefrontal and posterior parietal cortices. *Science, 315*, 1860–1862.

Cohen, J. Y., Crowder, E. A., Heitz, R. P., Subraveti, C. R., Thompson, K. G., Woodman, G. F., et al. (2010). Cooperation and competition among frontal eye field neurons during visual target selection. *Journal of Neuroscience, 30*, 3227–3238.

Cohen, J. Y., Heitz, R. P., Schall, J. D., & Woodman, G. F. (2009a). On the origin of event-related potentials indexing covert attentional selection during visual search. *Journal of Neurophysiology, 102*, 2375–2386.

Cohen, J. Y., Heitz, R. P., Woodman, G. F., & Schall, J. D. (2009b). Neural basis of the set-size effect in the frontal eye field: Timing of attention during visual search. *Journal of Neurophysiology, 101*, 1699–1704.

Constantinidis, C. (2006). Posterior parietal mechanisms of visual attention. *Reviews in the Neurosciences, 17*, 415–427.

Constantinidis, C., & Steinmetz, M. A. (2005). Posterior parietal cortex automatically encodes the location of salient stimuli. *Journal of Neuroscience, 25*, 233–238.

Cullen, K. E., & Van Horn, M. R. (2011). Brainstem pathways and premotor control. In S. Liversedge, I. Gilchrist, & S. Everling (Eds.), *Oxford handbook of eye movements* (pp. 151–172). Oxford, England: Oxford University Press.

David, S. V., Hayden, B. Y., Mazer, J. A., & Gallant, J. L. (2008). Attention to stimulus features shifts spectral tuning of V4 neurons during natural vision. *Neuron, 59*, 509–521.

Ekstrom, L. B., Roelfsema, P. R., Arsenault, J. T., Kolster, H., & Vanduffel, W. (2009). Modulation of the contrast response function by electrical microstimulation of the macaque frontal eye field. *Journal of Neuroscience, 29*, 10683–10694.

Everling, S., Tinsley, C. J., Gaffan, D., & Duncan, J. (2006). Selective representation of task-relevant objects and locations in the monkey prefrontal cortex. *European Journal of Neuroscience, 23*, 2197–2214.

Fecteau, J. H., & Munoz, D. P. (2005). Correlates of capture of attention and inhibition of return across stages of visual processing. *Journal of Cognitive Neuroscience, 17*, 1714–1727.

Findlay, J. M. (1982). Global visual processing for saccadic eye movements. *Vision Research, 22*, 1033–1045.

Geisler, W. S., & Cormack, L. K. (2011). Models of overt attention. In S. P. Liversedge, I. P. Gilchrist, & S. Everling (Eds.), *Oxford handbook of eye movements* (pp. 439–454). Oxford, England: Oxford University Press.

Gottlieb, J., & Balan, P. (2010). Attention as a decision in information space. *Trends in Cognitive Sciences, 14*, 240–248.

Gregoriou, G. G., Gotts, S. J., & Desimone, R. (2012). Cell-type-specific synchronization of neural activity in FEF with V4 during attention. *Neuron, 73*, 581–594.

Gregoriou, G. G., Gotts, S. J., Zhou, H., & Desimone, R. (2009). High-frequency, long-range coupling between prefrontal and visual cortex during attention. *Science, 324*, 1207–1210.

Hamker, F. H., & Zirnsak, M. (2006). V4 receptive field dynamics as predicted by a systems-level model of visual attention using feedback from the frontal eye field. *Neural Networks, 19*, 1371–1382.

Hayden, B. Y., & Gallant, J. L. (2005). Time course of attention reveals different mechanisms for spatial and feature-based attention in area V4. *Neuron, 47*, 637–643.

Hayhoe, M. M., & Ballard, D. H. (2011). Mechanisms of gaze control in natural vision. In S. P. Liversedge, I. P. Gilchrist, & S. Everling (Eds.), *Oxford handbook of eye movements* (pp. 607–620). Oxford, England: Oxford University Press.

Heitz, R. P., Cohen, J. Y., Woodman, G. F., & Schall, J. D. (2010). Neural correlates of correct and errant attentional selection revealed through N2pc and frontal eye field activity. *Journal of Neurophysiology, 104*, 2433–2441.

Henderson, J. M. (2011). Eye movements and scene perception. In S. Liversedge, I. Gilchrist, & S. Everling (Eds.), *Oxford handbook of eye movements* (pp. 593–606). Oxford, England: Oxford University Press.

Ipata, A. E., Gee, A. L., Bisley, J. W., & Goldberg, M. E. (2009). Neurons in the lateral intraparietal area create a priority map by the combination of disparate signals. *Experimental Brain Research, 192*, 479–488.

Ipata, A. E., Gee, A. L., Goldberg, M. E., & Bisley, J. W. (2006). Activity in the lateral intraparietal area predicts the goal and latency of saccades in a free-viewing visual search task. *Journal of Neuroscience, 26*, 3656–3661.

Johnston, K., & Everling, S. (2011). Frontal cortex and flexible control of saccades. In S. Liversedge, I. Gilchrist, & S. Everling (Eds.), *Oxford handbook of eye movements* (pp. 279–302). Oxford, England: Oxford University Press.

Juan, C. H., Muggleton, N. G., Tzeng, O. J., Hung, D. L., Cowey, A., & Walsh, V. (2008). Segregation of visual selection and saccades in human frontal eye fields. *Cerebral Cortex, 18*, 2410–2415.

Juan, C. H., Shorter-Jacobi, S. M., & Schall, J. D. (2004). Dissociation of spatial attention and saccade preparation. *Proceedings of the National Academy of Sciences of the United States of America, 101*, 15541–15544.

Kim, B., & Basso, M. A. (2008). Saccade target selection in the superior colliculus: A signal detection theory approach. *Journal of Neuroscience, 28,* 2991–3007.

Kim, B., & Basso, M. A. (2010). A probabilistic strategy for understanding action selection. *Journal of Neuroscience, 30,* 2340–2355.

Klein, R. (1980). Does oculomotor readiness mediate cognitive control of visual attention? In R. Nickerson (Ed.), *Attention and Performance* (pp. 259–276). New York: Academic Press.

Land, M., & Tatler, B. (2009). *Looking and acting: Vision and eye movements in natural behaviour.* Oxford, England: Oxford University Press.

Lee, K. M., & Keller, E. L. (2008). Neural activity in the frontal eye fields modulated by the number of alternatives in target choice. *Journal of Neuroscience, 28,* 2242–2251.

Liversedge, S. P., & Findlay, J. M. (2000). Saccadic eye movements and cognition. *Trends in Cognitive Science, 4,* 6–14.

Lovejoy, L. P., & Krauzlis, R. J. (2010). Inactivation of primate superior colliculus impairs covert selection of signals for perceptual judgments. *Nature Neuroscience, 13,* 261–266.

Markov, N. T., Misery, P., Falchier, A., Lamy, C., Vezoli, J., Quilodran, R., et al. (2011). Weight consistency specifies regularities of macaque cortical networks. *Cerebral Cortex, 21,* 1254–1272.

May, P. J. (2006). The mammalian superior colliculus: Laminar structure and connections. *Progress in Brain Research, 151,* 321–378.

McPeek, R. M. (2006). Incomplete suppression of distractor-related activity in the frontal eye field results in curved saccades. *Journal of Neurophysiology, 96,* 2699–2711.

McPeek, R. M. (2008). Reversal of a distractor effect on saccade target selection after superior colliculus inactivation. *Journal of Neurophysiology, 99,* 2694–2702.

McPeek, R. M., & Keller, E. L. (2002). Saccade target selection in the superior colliculus during a visual search task. *Journal of Neurophysiology, 88,* 2019–2034.

Mirabella, G., Bertini, G., Samengo, I., Kilavik, B. E., Frilli, D., Libera, C. D., et al. (2007). Neurons in area V4 of the macaque translate attended visual features into behaviorally relevant categories. *Neuron, 54,* 303–318.

Mirpour, K., Arcizet, F., Ong, W. S., & Bisley, J. W. (2009). Been there, seen that: A neural mechanism for performing efficient visual search. *Journal of Neurophysiology, 102,* 3481–3491.

Mirpour, K., Ong, W. S., & Bisley, J. W. (2010). Microstimulation of posterior parietal cortex biases the selection of eye movement goals during search. *Journal of Neurophysiology, 104,* 3021–3028.

Monosov, I. E., Sheinberg, D. L., & Thompson, K. G. (2010). Paired neuron recordings in the prefrontal and inferotemporal cortices reveal that spatial selection precedes object identification during visual search. *Proceedings of the National Academy of Sciences of the United States of America, 107,* 13105–13110.

Monosov, I. E., & Thompson, K. G. (2009). Frontal eye field activity enhances object identification during covert visual search. *Journal of Neurophysiology, 102,* 3656–3672.

Monosov, I. E., Trageser, J. C., & Thompson, K. G. (2008). Measurements of simultaneously recorded spiking activity and local field potentials suggest that spatial selection emerges in the frontal eye field. *Neuron, 57,* 614–625.

Moore, T., & Fallah, M. (2004). Microstimulation of the frontal eye field and its effects on covert spatial attention. *Journal of Neurophysiology, 91,* 152–162.

Mruczek, R. E., & Sheinberg, D. L. (2007). Activity of inferior temporal cortical neurons predicts recognition choice behavior and recognition time during visual search. *Journal of Neuroscience, 27,* 2825–2836.

Murthy, A., Ray, S., Shorter, S. M., Priddy, E. G., Schall, J. D., & Thompson, K. G. (2007). Frontal eye field contributions to rapid corrective saccades. *Journal of Neurophysiology, 97,* 1457–1469.

Murthy, A., Ray, S., Shorter, S. M., Schall, J. D., & Thompson, K. G. (2009). Neural control of visual search by frontal eye field: Effects of unexpected target displacement on visual selection and saccade preparation. *Journal of Neurophysiology, 101,* 2485–2506.

Najemnik, J., & Geisler, W. S. (2009). Simple summation rule for optimal fixation selection in visual search. *Vision Research, 49,* 1286–1294.

Ninomiya, T., Sawamura, H., Inoue, K., & Takada, M. (2012). Segregated pathways carrying frontally derived top-down signals to visual areas MT and V4 in macaques. *Journal of Neuroscience, 32,* 6851–6858.

Ogawa, T., & Komatsu, H. (2004). Target selection in area V4 during a multidimensional visual search task. *Journal of Neuroscience, 24,* 6371–6382.

Ogawa, T., & Komatsu, H. (2006). Neuronal dynamics of bottom-up and top-down processes in area V4 of macaque monkeys performing a visual search. *Experimental Brain Research, 173,* 1–13.

Ogawa, T., & Komatsu, H. (2009). Condition-dependent and condition-independent target selection in the macaque posterior parietal cortex. *Journal of Neurophysiology, 101,* 721–736.

Ogawa, T., & Komatsu, H. (2010). Differential temporal storage capacity in the baseline activity of neurons in macaque frontal eye field and area V4. *Journal of Neurophysiology, 103,* 2433–2445.

Oristaglio, J., Schneider, D. M., Balan, P. F., & Gottlieb, J. (2006). Integration of visuospatial and effector information during symbolically cued limb movements in monkey lateral intraparietal area. *Journal of Neuroscience, 26,* 8310–8319.

Paré, M., & Dorris, M. C. (2011). The role of posterior parietal cortex in the regulation of saccadic eye movements. In S. P. Liversedge, I. P. Gilchrist, & S. Everling (Eds.), *Oxford handbook of eye movements* (pp. 257–278). Oxford, England: Oxford University Press.

Phillips, A. N., & Segraves, M. A. (2010). Predictive activity in macaque frontal eye field neurons during natural scene searching. *Journal of Neurophysiology, 103,* 1238–1252.

Pouget, P., Logan, G. D., Palmeri, T. J., Boucher, L., Paré, M., & Schall, J. D. (2011). Neural basis of adaptive response time adjustment during saccade countermanding. *Journal of Neuroscience, 31,* 12604–12612.

Pouget, P., Stepniewska, I., Crowder, E. A., Leslie, M. W., Emeric, E. E., Nelson, M. J., et al. (2009). Visual and motor connectivity and the distribution of calcium-binding proteins in macaque frontal eye field: Implications for saccade target selection. *Frontiers in Neuroanatomy, 3,* 2.

Purcell, B. A., Schall, J. D., Logan, G. D., & Palmeri, T. J. (2012). From salience to saccades: Multiple-alternative gated stochastic accumulator model of visual search. *Journal of Neuroscience, 32*, 3433–3446.

Purcell, B. A., Heitz, R. P., Cohen, J. Y., Schall, J. D., Logan, G. D., & Palmeri, T. J. (2010). Neurally constrained modeling of perceptual decision making. *Psychological Review, 117*, 1113–1143.

Ratcliff, R., & McKoon, G. (2008). The diffusion decision model: Theory and data for two-choice decision tasks. *Neural Computation, 20*, 873–922.

Ray, S., & Maunsell, J. H. (2010). Differences in gamma frequencies across visual cortex restrict their possible use in computation. *Neuron, 67*, 885–896.

Rayner, K., & Liversedge, S. P. (2011). Linguistic and cognitive influences on eye movements during reading. In S. Liversedge, I. Gilchrist, & S. Everling (Eds.), *Oxford handbook of eye movements* (pp. 751–766). Oxford, England: Oxford University Press.

Rizzolatti, G. (1983). Mechanisms of selective attention in mammals. In J. Ewert, R. Capranica, & D. Ingle (Eds.), *Advances in Vertebrate Neuroethology* (pp. 261–297). New York: Elsevier.

Robinson, D. A. (1972). Eye movements evoked by collicular stimulation in the alert monkey. *Vision Research, 12*, 1795–1808.

Rossi, A. F., Bichot, N. P., Desimone, R., & Ungerleider, L. G. (2007). Top down attentional deficits in macaques with lesions of lateral prefrontal cortex. *Journal of Neuroscience, 27*, 11306–11314.

Saruwatari, M., Inoue, M., & Mikami, A. (2008). Modulation of V4 shifts from dependent to independent on feature during target selection. *Neuroscience Research, 60*, 327–339.

Sato, T., Murthy, A., Thompson, K. G., & Schall, J. D. (2001). Search efficiency but not response interference affects visual selection in frontal eye field. *Neuron, 30*, 583–591.

Sato, T. R., & Schall, J. D. (2003). Effects of stimulus–response compatibility on neural selection in frontal eye field. *Neuron, 38*, 637–648.

Schall, J. D. (2004). Selection of targets for saccadic eye movements. In L. M. Chalupa & J. S. Werner (Eds.), *The visual neurosciences* (pp. 1369–1390). Cambridge, MA: MIT Press.

Schall, J. D., & Boucher, L. (2007). Executive control of gaze by the frontal lobes. *Cognitive, Affective & Behavioral Neuroscience, 7*, 396–412.

Schall, J. D., & Cohen, J. Y. (2011). The neural basis of saccade target selection. In S. P. Liversedge, I. P. Gilchrist, & S. Everling (Eds.), *Oxford handbook of eye movements* (pp. 357–382). Oxford, England: Oxford University Press.

Schall, J. D., & Hanes, D. P. (1993). Neural basis of saccade target selection in frontal eye field during visual search. *Nature, 366*, 467–469.

Schall, J. D., Morel, A., King, D. J., & Bullier, J. (1995). Topography of visual cortical afferents to frontal eye field in macaque: Functional convergence and segregation of processing streams. *Journal of Neuroscience, 15*, 4464–4487.

Schall, J. D., & Thompson, K. G. (1994). Macaque oculomotor thalamus: Saccade target selection. *Society for Neuroscience Abstracts, 20*, 145.

Schiller, P. H., & Tehovnik, E. J. (2005). Neural mechanisms underlying target selection with saccadic eye movements. *Progress in Brain Research 149*, 157–171.

Schmolesky, M. T., Wang, Y., Hanes, D. P., Thompson, K. G., Leutgeb, S., Schall, J. D., et al. (1998). Signal timing across the macaque visual system. *Journal of Neurophysiology, 79*, 3272–3278.

Shen, K., & Paré, M. (2007). Neuronal activity in superior colliculus signals both stimulus identity and saccade goals during visual conjunction search. *Journal of Vision, 15*, 1–13.

Smith, D. T., & Schenk, T. (2007). Enhanced probe discrimination at the location of a colour singleton. *Experimental Brain Research, 181*, 367–375.

Sternberg, S. (2011). Modular processes in mind and brain. *Cognitive Neuropsychology, 28*, 156–208.

Suzuki, H., & Azuma, M. (1983). Topographic studies on visual neurons in the dorsolateral prefrontal cortex of the monkey. *Experimental Brain Research, 53*, 47–58.

Tanaka, M., & Kunimatsu, J. (2011). Thalamic roles in eye movements. In S. Liversedge, I. Gilchrist, & S. Everling (Eds.), *Oxford handbook of eye movements* (pp. 235–256). Oxford, England: Oxford University Press.

Taylor, P. C., Nobre, A. C., & Rushworth, M. F. (2007). FEF TMS affects visual cortical activity. *Cerebral Cortex, 17*, 391–399.

Thomas, N. W. D., & Paré, M. (2007). Temporal processing of saccade targets in parietal cortex area LIP during visual search. *Journal of Neurophysiology, 97*, 942–947.

Thompson, K. G., Bichot, N. P., & Sato, T. R. (2005). Frontal eye field activity before visual search errors reveals the integration of bottom-up and top-down salience. *Journal of Neurophysiology, 93*, 337–351.

Thompson, K. G., Bichot, N. P., & Schall, J. D. (1997). Dissociation of target selection from saccade planning in macaque frontal eye field. *Journal of Neurophysiology, 77*, 1046–1050.

Thompson, K. G., Biscoe, K. L., & Sato, T. R. (2005). Neuronal basis of covert spatial attention in the frontal eye field. *Journal of Neuroscience, 25*, 9479–9487.

Thompson, K. G., Hanes, D. P., Bichot, N. P., & Schall, J. D. (1996). Perceptual and motor processing stages identified in the activity of macaque frontal eye field neurons during visual search. *Journal of Neurophysiology, 76*, 4040–4055.

Trageser, J. C., Monosov, I. E., Zhou, Y., & Thompson, K. G. (2008). A perceptual representation in the frontal eye field during covert visual search that is more reliable than the behavioral report. *European Journal of Neuroscience, 28*, 2542–2549.

Tsotsos, J. K. (2011). *A computational perspective on visual attention.* Cambridge, MA: MIT Press.

Usher, M., & McClelland, J. L. (2001). The time course of perceptual choice: The leaky, competing accumulator model. *Psychological Review, 108*, 550–592.

Vokoun, C. R., Mahamed, S., & Basso, M. A. (2011). Saccadic eye movements and the basal ganglia. In S. Liversedge, I. Gilchrist, & S. Everling (Eds.), *Oxford handbook of eye movements* (pp. 215–234). Oxford, England: Oxford University Press.

Walker, R., Techawachirakul, P., & Haggard, P. (2009). Frontal eye field stimulation modulates the balance of salience between target and distractors. *Brain Research, 1270*, 54–63.

Wardak, C., Ibos, G., Duhamel, J. R., & Olivier, E. (2006). Contribution of the monkey frontal eye field to covert visual attention. *Journal of Neuroscience, 26*, 4228–4235.

Wardak, C., Olivier, E., & Duhamel, J. R. (2004). A deficit in covert attention after parietal cortex inactivation in the monkey. *Neuron, 42*, 501–508.

Wardak, C., Olivier, E., & Duhamel, J. R. (2011). The relationship between spatial attention and saccades in the frontoparietal network of the monkey. *European Journal of Neuroscience, 33*, 1973–1981.

White, B. J., & Munoz, D. P. (2011a). Separate visual signals for saccade initiation during target selection in the primate superior colliculus. *Journal of Neuroscience, 31*, 1570–1578.

White, B. J., & Munoz, D. P. (2011b). The superior colliculus. In S. P. Liversedge, I. Gilchrist, & S. Everling (Eds.), *Oxford handbook of eye movements* (pp. 195–214). Oxford, England: Oxford University Press.

Wolfe, J. M., & Horowitz, T. S. (2004). What attributes guide the deployment of visual attention and how do they do it? *Nature Reviews Neuroscience. 5*, 495-501.

Wolfe, J. M., Horowitz, T. S., & Palmer, E. M. (2010). Reaction time distributions constrain models of visual search. *Vision Research, 50*, 1304–1311.

Woodman, G. F., Kang, M. S., Rossi, A. F., & Schall, J. D. (2007). Nonhuman primate event-related potentials indexing covert shifts of attention. *Proceedings of the National Academy of Sciences of the United States of America, 104*, 15111–15116.

Woodman, G. F., Kang, M. S., Thompson, K., & Schall, J. D. (2008). The effect of visual search efficiency on response preparation: Neurophysiological evidence for discrete flow. *Psychological Science, 19*, 128–136.

Woodman, G. F., & Luck, S. J. (1999). Electrophysiological measurement of rapid shifts of attention during visual search. *Nature, 400*, 867–869.

Wyder, M. T., Massoglia, D. P., & Stanford, T. R. (2004). Contextual modulation of central thalamic delay-period activity: Representation of visual and saccadic goals. *Journal of Neurophysiology, 91*, 2628–2648.

Young, M. S., Heitz, R. P., Schall, J. D., & Woodman, G. F. (2010). Modeling the neural generators of monkey event-related potentials indexing covert shift of attention. *Program No. 304.1 2010 Neuroscience Meeting Planner*. San Diego, CA: Society for Neuroscience, Online.

Zhou, H. H., & Desimone, R. (2011). Feature-based attention in the frontal eye field and area V4 during visual search. *Neuron, 70*, 1205–1217.

Zhou, H. H., & Thompson, K. G. (2009). Cognitively directed spatial selection in the frontal eye field in anticipation of visual stimuli to be discriminated. *Vision Research, 49*, 1205–1215.

64 Neural Mechanisms of Target Selection in the Superior Colliculus

UDAY K. JAGADISAN AND NEERAJ J. GANDHI

The visual environment is filled with objects that contain potentially useful information for the survival of an organism. In order to best extract this information, the animal needs to orient itself to the part of the visual world most relevant for its immediate behavior while ignoring the unwanted parts. This is especially true for foveating animals such as humans and other primates, where only a small part of the retina (the fovea) is able to resolve the visual world with greatest detail. The brain is therefore faced with a sampling problem—how to decide, among the visual clutter, where to look next? In other words, how does the selection of one or more *targets* for future foveation emerge in the visual–oculo-motor system?

A conceptual approach to this problem is illustrated by the following example. Imagine that you are searching for your keys that were earlier misplaced in your room. You know what your keys look like, and the search ends when you find an object that matches the mental image. You begin the search by casting your glance at a random location, bringing a new set of objects into your visual field. The new visual stimuli are not homogeneous in either their intrinsic properties (brightness, color) or their relevance (likely location for keys). During the process of evaluating the stimulus at the current location and deciding where to look next, your attention can be captured by salient stimuli such as a flashing or ringing phone, a bird flying past your window, or bright and shiny objects such as your computer screen or a bunch of loose change. You also implicitly give weight to locations where you are more likely to have left the keys, such as your desk or your jacket as opposed to under the bed. Moreover, you are less likely to look again in a place you just spent some time focusing on—an exercise in futility. All these attributes of the visual objects in your room and their respective locations must be tied together on a continual basis to enable a thorough evaluation of where to look next. Because gaze is localized to a single location at any given time, the competition between multiple locations must be resolved before a decision selecting the next target is made. Finally, this evolving decision must be relayed to premotor structures to actuate the gaze shift. Although the description of this example goes through a series of stages, it is also possible that they are implemented in a parallel fashion in the ongoing activity in the brain.

Numerous studies in the past couple of decades have attempted to uncover the neural correlates of such visual target selection. Although early studies focused on the role of the frontoparietal cortical network, including the frontal eye fields (FEF) (Schall, 2001; chapter 63, this volume) and lateral intraparietal area (LIP) (Bisley & Goldberg, 2010), more recent work has elucidated the role of the superior colliculus (SC) in selecting targets for eye movements (Krauzlis, Liston, & Carello, 2004; chapter 23, this volume). We have parsed this wealth of information into this chapter as follows. The first section presents a brief overview of the SC, its anatomical and physiological properties, and its role in the control of gaze. In the next section, we review target selection in the SC specifically in the context of visual search. In the third section, we discuss the role of SC in selecting targets in other domains, including smooth pursuit, nonvisual inputs, and nonoculomotor effectors, pointing to a more general function for the midbrain structure. In the fourth section, we present an interpretation of any gaze state as target selection and look at the role of SC in fixation and defixation. This is followed by a discussion of motor preparation and its relation to target-selection-related activity in the SC. Finally, we conclude by discussing a framework that interprets activity in the SC network as a priority map for action.

ANATOMICAL AND PHYSIOLOGICAL PROPERTIES OF THE SC

The SC (and its homologue in nonmammals—the optic tectum, or OT) is an evolutionarily ancient structure whose main function seems to be to direct or orient the attention of an animal, primarily by controlling its gaze.

Located at the roof (tectum) of the brainstem, the SC can be anatomically divided into seven distinct layers (for in depth reviews, see Huerta & Harting, 1984; Isa & Hall, 2009; May, 2006; Sparks & Hartwich-Young, 1989). These layers can be further classified based on their physiological and functional properties into two levels. The superficial layers (SCs) receive inputs directly from the retina as well as primary and extrastriate visual cortices (V1, V2, V4). The pretectum and parabigeminal nucleus (the cholinergic isthmic nucleus in nonmammals) also project to SCs. Neurons in the SCs, in turn, project down to the intermediate and deep layers in the SC, the lateral geniculate nucleus (LGN) in the thalamus, and reciprocally to the pretectal and parabigeminal nuclei. The intermediate and deep layers (SCid; henceforth just called intermediate layers) receive inputs from SCs, the (dorsal) frontoparietal cortical network involved in the processing of visuospatial information, including area LIP and the FEF, the dorsolateral prefrontal cortex (dlPFC), and the inferotemporal cortex. Furthermore, the basal ganglia also project to the SCid, primarily in the form of GABAergic projections from the substantia nigra pars reticulata (SNpr). There is also evidence that the SCid is the recipient of cholinergic inputs from the parabrachial nucleus in the pons. It also receives information related to nonvisual modalities. In turn, the intermediate layer neurons project to brainstem nuclei involved in the control of gaze—the mesencephalic reticular formation (vertical component of gaze) and the paramedian pontine reticular formation (PPRF; horizontal component of gaze) (Moschovakis, Scudder, & Highstein, 1996). Finally, the intermediate layer neurons are also part of an interlaminar network that ascends back onto the superficial layers as well as providing feedback projections to the FEF via the mediodorsal thalamus (Sommer & Wurtz, 2004) and to the parietal and temporal cortices via the pulvinar (Berman & Wurtz, 2010; Clower et al., 2001).

Another important property of the SC network, across layers, is that each colliculus represents a topographic map of the animal's contralateral hemifield in a retinotopic reference frame (see review by Gandhi & Katnani, 2011b). Accordingly, neurons in the superficial layers exhibit responses that are time locked to the onset of visual stimuli (visual neurons) in their receptive field. Neurons in the intermediate layers fire in response to gaze shifts (motor neurons) into their movement field, or both to a gaze shift and visual stimulus onset in their response field (visuomotor neurons). Each class of neurons exhibits a finer spectrum of activity depending on whether their responses are phasic (transient), tonic (sustained), or a combination of the two (for a brief overview of the various response types,

see figure 1 in McPeek & Keller, 2002). The response field locations, to a large extent, overlap across the two layers as one proceeds dorsoventrally through the SC. The topography on the SC tissue is as follows. The rostral region of the SC maps onto the central visual field or small stimulus eccentricities and gaze shift amplitudes. The caudal region maps onto the peripheral visual field—eccentric locations and large movements. Similarly, the lateral halves of the colliculi represent the downward hemifield while the medial halves represent the upward hemifield. The mapping from visual space to SC tissue space is nonlinear, and in fact logarithmic, along the rostrocaudal axis such that significantly more neural tissue is dedicated to the central field compared to the extremities.

This rich diversity in the physiological properties of the SC, along with its anatomical location, supports the idea that it is a critical node in the process of integrating sensory information to produce overt orienting behavior (e.g., saccades). In the following sections, we will look at the neural signatures that identify stimulus selection.

TARGET SELECTION IN VISUAL SEARCH

Since as early as the first studies on the SC, it has been known that stimulus-induced activity on the SC map is modulated by whether the stimulus is the target of an upcoming saccade or not (Goldberg & Wurtz, 1972; Mohler & Wurtz, 1976). The "visual enhancement" is seen even in the superficial layers (SCs) that don't explicitly show movement related activity. Although the effect was originally attributed to peripheral attention, it provided the first indication of a potential role for the SC in target selection. Nevertheless, these results are limited in their interpretability in terms of target selection because the task involved just a single visual stimulus. In order to truly study target selection, it is important that multiple stimuli be present on the screen at the same time.

How is the next target chosen from a field of multiple alternatives? Studies focusing on the SC have used an "oddball visual search" paradigm to probe this question (Kim & Basso, 2008; McPeek & Keller, 2002; Shen & Paré, 2007). The task is fairly simple—a unique visual stimulus embedded in a search array of any number of identical distractors serves as the saccadic target. Various groups have used a similar paradigm to study the relationship between stimulus selection and saccadic target or response selection in the FEF and LIP as well (Schall & Hanes, 1993; Thomas & Paré, 2007; Thompson et al., 1996). In the SC, this task reveals a spectrum of physiological responses across the population, as reported by

McPeek and Keller (2002). Neurons in SCs exhibit no differential activity depending on whether a target stimulus is in their receptive field or a distractor. In contrast, neurons in SCid discriminate the target from distractors to varying degrees depending on their functional properties. Visuomovement neurons with a biphasic visual response (such as in figure 64.1B, left) display enhanced activity in the second peak when a stimulus is the target, and this differential encoding of the target relative to distractors persists during the prelude activity that follows the transient burst. It has therefore been hypothesized that the first peak of the biphasic response reflects direct sensory input whereas the second peak reflects recurrent processing of the initial activity and/or input from top-down sources. For this subset of SCid neurons, the time at which neural activity "selects" the target is independent of movement selection time, highlighting their role in the conceptual stage of target selection. In other visuomovement neurons and purely motor neurons, in contrast, differential encoding of target and distractors occurs after the visual burst and at a time that is correlated with saccade latency, pointing to their role in movement preparation (McPeek & Keller, 2002).

Further insight into the precise nature of the target selection signal comes from studies recording simultaneously from multiple sites on the SC map. Kim and Basso (2008) recorded the activity of neurons corresponding to the four stimulus locations in an oddball visual search task. Because the visual stimuli were constrained by electrode placement in the SC and not vice versa, the asymmetrical nature of the search array produced sessions with varying levels of performance accuracy. The key interpretation is that instead of activity at the target site alone, the level of discriminability between target and distractor-related activity across the entire SC predicted the monkey's performance on individual trials. Furthermore, the discriminability evolved with time from onset of the search array, echoing the accumulation of a decision-like signal reported in numerous studies and psychophysical models (Hanes & Schall, 1996; Kim & Basso, 2008; Ratcliff, Cherian, & Segraves, 2003; Reddi, Asrress, & Carpenter, 2003). Consistent with the idea of target selection based on a relative or competitive signal, lidocaine- or muscimol-induced inactivation of SC selectively affects performance when the target is part of a search array compared to when presented as an isolated stimulus (McPeek & Keller, 2004). The saccades made under inactivation on single-target trials, although slower, were always to the correct (and only) target location. In contrast, when the target was the singleton in an oddball task, inactivation of the SC region corresponding to the target

significantly increased the proportion of erroneous gaze shifts to one of the irrelevant distractors. The effect was target and location specific in that performance was unaffected when a distractor was in the inactivation field. These results highlight the competitive nature of target selection when multiple stimuli are vying for future foveation.

Is target selection purely a posteriori, that is, do selection mechanisms kick in only after all the necessary information has been presented to the visual system? Basso and Wurtz (1998) showed that activity in the SC network is modulated by contextual information about target probability, by systematically varying the number of possible locations at which a single target could appear on a given trial. Neurons with low-level premotor activity had higher activation when the number of potential locations was low, and this was correlated with a reduction in saccade reaction times. As seen before, neurons with these characteristics are also modulated by information about target/distractor identity in visual search (McPeek & Keller, 2002). However, activity in the phasic visual–motor burst neurons was unmodulated by target probability. Thus, a "preparatory set" or "motor set" in the SC population activity indicates that at least part of the selection signal can arise from prior information and can enable a faster implementation of the selection decision (Basso & Wurtz, 1998).

Thus far, we have looked at cases where the location of the target also serves as the eventual saccade goal. This introduces a potential confound between purely target-selection-related activity and activity in preparation of the upcoming movement, especially in the SCid, which is thought to encode both processes. Observing a dissociation of target selection time from saccadic reaction time is by itself insufficient in resolving the confound. A better way to tease apart the two processes is to employ a task design where the saccade endpoint is spatially dissociated from the stimulus cueing the movement—the antisaccade task has been widely used to do precisely this (e.g., Juan, Shorter-Jacobi, & Schall, 2004; Munoz & Everling, 2004; Sato & Schall, 2003). In this task, the subject is required to saccade to an object or location diametrically opposite to a cue stimulus that is presented alone or embedded in a search array. Everling and colleagues looked at activity in the SC following a central instruction cue that indicated to the animal whether the trial required a pro- (toward the peripheral stimulus) or an antisaccade (Everling et al., 1999). Stimulus-related activity, in response to the same visual stimulus, was lower on anti- trials than on pro- trials. Moreover, the activity of neurons in the rostral part of SC—those that respond to the fixated instructional cue—was also modulated; they were higher on anti- trials. A putative explanation for these observations is that the SC receives stronger inhibition from top-down areas, for example, directly from the prefrontal cortex or via the basal ganglia, to prevent unwanted triggering of stimulus-induced prosaccades on antisaccade trials (Everling & Munoz, 2000; Munoz & Everling, 2004). We will revisit the idea of motor preparation and the antisaccade task in a following section.

TARGET SELECTION IN OTHER DOMAINS

Do the aforementioned properties of SC extend to scenarios beyond a static visual scene, such as when the target is in motion (e.g., smooth pursuit, looming stimulus) or where the selection is effected by means other than a gaze shift (e.g., reach, locomotion)? In this section we review evidence that points to a more general function of the SC in target selection.

During natural behavior, our gaze shifts are not limited to saccades or coordinated eye–head movements. The presence of a moving stimulus can also lead to smooth pursuit behavior. Hence, closely linked to the notion of selecting a static target for orientation is the notion of selecting a moving target for tracking. It is possible that the selection of targets for smooth pursuit is coordinated with the network that participates in target selection for saccades. One set of studies (Carello & Krauzlis, 2004; Krauzlis & Dill, 2002) used a behavioral task in which two potential targets appeared at diametrically opposite locations in the visual field. These stimuli could either remain stationary or begin to move horizontally toward the midline of the display, and the animal was required to orient to the stimulus cued by the fixation point. Neural activity in the SCid exhibited selectivity for the target identified by the cue, but presumably only for the duration the target remains in the retinotopically defined receptive field of the neuron. Furthermore, subthreshold microstimulation of SCid during the same task biased the selection of the target contralateral to the stimulus site. In both experiments, target selection was independent of whether the required movement was a saccade or pursuit. Nummela and Krauzlis (2011) used a modified paradigm, in which the two targets move along directions that allow for averaging behavior during initial pursuit. Under normal conditions, the initial pursuit traces an approximately equally weighted average of the two target velocities. Inactivation of the SC strongly biased this weighting against the target corresponding to the inactivated site. The results mirror those observed for the selection of stationary targets, providing evidence for a causal role for SC in selecting "targets" on a more general level.

In a series of studies in the barn owl, Knudsen and colleagues demonstrated the role of the OT in the selection of stimuli based on relative salience (Mysore, Asadollahi, & Knudsen, 2011; Mysore & Knudsen, 2011a, 2011b). The OT is the nonmammalian homologue of the SC and contains topographic representations of auditory as well as visual space (Knudsen, 1982). In one experiment, barn owls were presented with visual stimuli looming in at various speeds (naturalistic stimuli for birds) within the receptive field of an OT neuron. The loom stimuli were presented either alone or in the presence of a competitive distractor at another location. One class of OT neurons was shown to represent relative salience—there is a gradual reduction in their responses to the loom stimulus as the strength (loom speed) of the distractor is varied. Other neurons exhibit switch-like responses when the stimulus in their receptive field becomes the stronger of the two stimuli. Similar experiments with combined auditory and visual stimuli showed that the switch neurons represent the strongest stimulus regardless of the modality of the distractor stimulus (strength of auditory stimulus was modulated by the intensity of broadband noise). Further, the switch-like property is also seen in the population activity, and this code changes as a function of the strength of the strongest stimulus, indicating that the OT may be involved in computing a categorical yet flexible representation of salience (Mysore & Knudsen, 2011a).

Effects of SC inactivation on target selection for reaching movements have been reported to be similar to the effects on target selection for saccades. Muscimol-induced inactivation of a location on the SC map produces deficits in selecting the corresponding stimulus as a target for reach, even in the absence of perceptual or motoric deficits (Song, Rafal, & McPeek, 2011). Analogous to the case for saccades (McPeek & Keller, 2004), the deficits were present only when the selection was under competitive conditions, that is, in the presence of distractors. The strikingly similar role played by the SC in saccade and reach target selection makes a strong case for a general, effector-independent representation of target "priority" in the SC.

Further support for this claim comes from experiments in freely moving rats performing an odor discrimination task (Felsen & Mainen, 2008). The rats were trained to turn and walk toward one of two reward ports on either side of an odor port depending on the identity of an odor or odor mixture that was presented. SC activity was selective for direction (leftward or rightward) during locomotion. In many neurons, the selectivity emerged in advance of movement onset, indicating an active role in selection of the movement. Consistent with the observed pattern of activity, unilateral inactivation of SC also produced a profound movement bias toward the ipsiversive location. Some of the deficits caused by inactivation of the SC seem to share a signature similar to spatial neglect, albeit one of a localized nature and strictly under competition from distracting stimuli. SC has a well-known role in directing spatial attention (Cavanaugh & Wurtz, 2004; Krauzlis, Liston, & Carello, 2004; see also chapter 23 by Krauzlis), and attentional deficits are wont to cause selection deficits because of their tight conceptual linkage.

MECHANISMS OF TARGET SELECTION

Since gaze can be directed to only one location at any time, the distribution of activity in the SC corresponding to different stimuli must eventually result in one "selected" locus of activity. Various mechanisms have been proposed to explain how the competition between multiple stimuli is resolved and to account for the evolution of the target selection signal. In the simple winner-take-all mechanism, in which the locus with the highest activity at a critical time is selected as the target. The results of many studies seem to fall under this category. On the other hand, if the SC is thought to contribute a population code to selection of the target and the corresponding movement, an appropriately weighted vector averaging or summation mechanism could be in place (Goossens & Van Opstal, 2006; Katnani, van Opstal, & Gandhi, 2012; Lee, Rohrer, & Sparks, 1988; Van Opstal & Van Gisbergen, 1990). This mechanism is also seen under reflexive conditions such as express saccades (Chou, Sommer, & Schiller, 1999). Another possibility is that SC activity at a given locus represents the likelihood of selecting that location (Kim & Basso, 2010). The distribution of activity across the population can then be read out by a decoder using Bayesian computations. Using the population activity recorded at four sites corresponding to the stimulus locations in a visual search task, Kim and Basso (2010) explicitly tested the predictions of each of these models on a trial-by-trial basis. The Bayesian maximum a posteriori probability (MAP) -based estimate best predicted the saccade selected on a given trial based on the neural activity at the four locations, outperforming population vector averaging and winner-take-all mechanisms.

How does the distribution of activity on the SC map emerge in the first place? The canonical model relies on a lateral recurrent network with local excitatory projections and long-range inhibitory projections, similar to the organization in primary sensory cortices. Although the basic physiological properties of SC neurons are consistent with this model, recent findings highlight the lack of anatomical evidence for long-range connectivity in the collicular network (Isa & Hall, 2009). The putative function of recurrent lateral inhibition could be served instead by extracollicular projections, such as from the nuclei isthmii—in birds—and its mammalian homologues. The cholinergic nucleus isthmi pars parvocellularis (Ipc, cf. parabrachial and parabigeminal nuclei in mammals) has reciprocal topographic connections with the OT (SC) and may contribute to feedback amplification of target-related activity. Analogously to the OT, the Ipc is also known to exhibit switch-like behavior to represent the location of the most salient stimulus (Asadollahi, Mysore, & Knudsen, 2011). The GABAergic nucleus isthmi pars magnocellularis (cf. lateral tegmental nucleus in mammals) receives input from the OT and has antitopographic projections to the Ipc and OT. This circuit is thought to provide inhibition that enhances discrimination between target and distractor-related activity (Mysore & Knudsen, 2011b). Inputs from the basal ganglia, notably global GABAergic inhibition by the SNpr, may also be a part of the network that facilitates target selection (Hikosaka, Takikawa, & Kawagoe, 2000; Isa & Hall, 2009). Recently, an inhibition-of-inhibition circuit motif has been proposed as a general mechanism for flexible stimulus categorization based on relative stimulus salience (Mysore & Knudsen, 2012). Inhibitory interneurons in the SC could be an important part of a network that implements this computational strategy. Further experiments and models are needed to identify and delineate the precise role of connections within and outside the SC in target selection.

FIXATION AS CONTINUOUS
TARGET SELECTION

As mentioned in the introduction, the visual system is faced with the problem of selecting a target for most of an animal's waking life. This problem can be divided into two steps (not necessarily sequential): (1) whether to deselect the currently foveated target and select another target and (2) if the previous decision is in the affirmative, which target to select from among the menu of options available to the system. We have already discussed the neural correlates of the latter in the previous sections. The two can be thought as belonging to the same class of target selection problems if the following conceptual leap is made: On a moment-by-moment basis the brain implements (or is implementing) a decision to either continue to "select" the current stimulus or to select a new target. With this insight, we can now interpret activity in the SC related to fixation within the target selection framework.

What are the neural correlates of unfixating and shifting gaze in the SC? Early studies reported the existence of neurons in the rostral pole of SCid that seemed to mediate gaze withholding or the maintenance of fixation (Munoz & Wurtz, 1993). These so-called fixation neurons fire at a tonic rate when the eyes are fixated on a spot and lower their activity during saccades, when neurons in the caudal SCid show elevated firing. Microstimulation of the rostral SCid stops saccades in midflight (Munoz, Waitzman, & Wurtz, 1996) and can also suppress buildup activity in the caudal region (Munoz & Istvan, 1998). These observations gave rise to the fixation zone–saccade zone model, where long-range reciprocal inhibition between the rostral and caudal "zones" control the alternating pattern of fixations and saccades seen during typical oculomotor behavior.

However, numerous lines of evidence have since emerged that dispute this hypothesis. First, the nature of perturbations in saccades caused by rostral SCid microstimulation fall along a continuum, along with those caused by stimulation of the caudal SCid, and are qualitatively different from interruptions caused by stimulation of the omnipause neurons (OPNs) in the PPRF (Gandhi & Keller, 1999). The OPNs exhibit activity that is much more closely linked to the onset and offset of fixation and therefore make a more qualified candidate for withholding gaze. Second, experiments in the cat employing multistep gaze shifts have shown that the locus of activity across SCid is correlated with the distance to the eventual goal of the sequence of gaze shifts—or goal-based long-term motor error—with rostral SCid neurons returning to their tonic firing only after the final target is fixated (and not during intermediate fixations) (Bergeron, Matsuo, & Guitton, 2003). Third, recent work has demonstrated a causal role for rostral SC neurons in the generation of microsaccades—tiny movements (<1') of the eyes during fixation (Hafed, Goffart, & Krauzlis, 2009). These findings are in line with the idea that the SC map represents a natural continuum of movements, with large movements represented toward the caudal end and small movements represented near the rostral end. Moreover, the rostrocaudal distribution of activity may serve as an evolving population code for selecting the final target (or "goal") of a gaze shift, ignoring intermediate states that are purely motoric in nature (Krauzlis Liston, & Carello, 2004). This important distinction is unappreciated in purely single-step behavior.

Thus, the transition from withholding gaze to shifting gaze and vice versa is better explained by a model that considers the shifts of balance between the rostral portion of SCid encoding small movements (or target errors) and the caudal portion encoding large movements. In other words, reduction of activity in the rostral neurons along with a synchronous rise in the caudal neurons signifies the deselection of the currently fixated target and the associated microscale movements while selecting the location represented by the locus of activity in the caudal population as the target of the next fixation.

Does the time of deselection of the foveal target necessarily coincide with an impending movement to a new target, or can it be temporally dissociated from the actual saccade? This can be tested in a modified delayed saccade paradigm, in which the subject is presented with a cue during early fixation to indicate an upcoming presentation of a peripheral target. We induced blinks during the initial fixation period in monkeys performing the delayed saccade task, *prior* to the presentation of the eventual saccade target (Gandhi & Jagadisan, 2011). The animals were trained to withhold saccades during the delay period, even in the presence of a peripheral target, and were allowed to look at the target once the fixation spot was extinguished (GO cue). As shown in figure 64.1A, the activity of rostral SC neurons showed a sustained reduction following the blink, even when the eyes remained stable and no gaze shift was produced during this period. Moreover, the reduction of activity in rostral neurons allowed for increased activation in neurons that had the peripheral target appear in their receptive field (figure 64.1B). This result highlights a putative mechanism of foveal target deselection in advance of peripheral target selection and subsequent gaze shift.

It remains to be seen whether the blink plays a special role in forcing the deselection of the fixated stimulus or if any associative cue can produce the same effect. It is well-known that the activity of a subpopulation of rostral SC neurons is driven by the presence of a visual stimulus at the site of fixation; these neurons have lower activity when fixating in the dark (Munoz & Wurtz, 1993). Since an eye blink transiently occludes the fixated stimulus, it is possible that the activity of these neurons is reduced due to the lack of visual input, and subsequent, possibly top-down, mechanisms maintain the activity stable at that level.

TARGET SELECTION AND MOTOR PREPARATION

As discussed above, neurons in the intermediate, but not superficial, layers distinguish the oddball target in a visual search paradigm as early as the second burst of the biphasic visual response. When studied with clever behavioral tasks designed to probe other processes like decision making, confidence evaluation, spatial

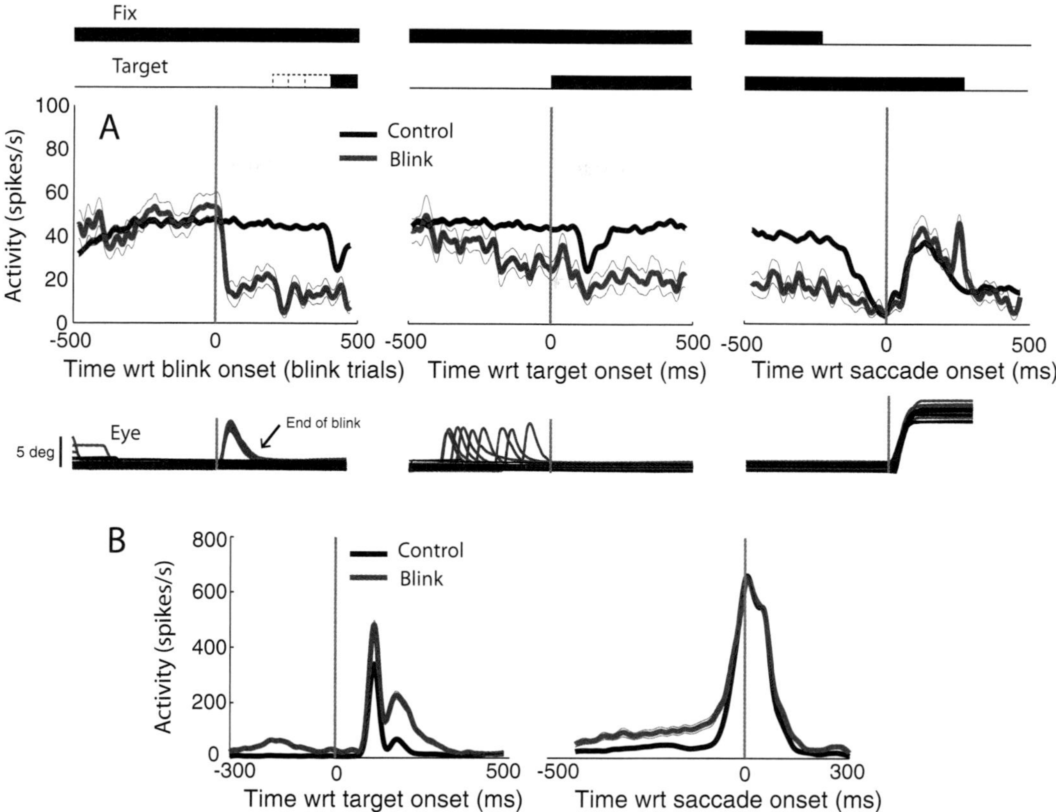

FIGURE 64.1 Blink-induced early deselection of fixated stimulus. (A) The timeline of the task is shown in the header rows. First row: Activity of a representative neuron in rostral superior colliculus (SC) on control (black) and blink (gray) trials. The animal performed in the delayed saccade task, and the blink was evoked by an air puff delivered during fixation, before the peripheral stimulus was illuminated. The activity is aligned on onset of the blink prior to peripheral target onset (left), target onset (middle), and onset of saccade (right). For illustration purposes, the control trace in the left panel (no blink to align on) is a shifted version of the trace from the middle panel. Second row: Eye position traces aligned on the same three events for blink and control trials. The blink is accompanied by a characteristic loopy movement that brings the eyes back to the starting point. All eye movements have reached conclusion by the time of target presentation. (B) Activity of a representative visuomovement neuron in caudal SC aligned on target onset (left) and saccade onset (right). Task is the same as in A. Note the increased activity during the visual response and presaccadic buildup on blink trials.

attention, and reward anticipation, the same class of SCid neurons also exhibits differential modulation, asserting a contribution to multiple cognitive mechanisms. Of course, the typical SCid neuron discharges a high-frequency burst prior to a saccade in its movement field, and it also projects to the brainstem burst generator elements that execute saccades. Thus, studies have proposed a preparatory premotor component to the low-level response (Dorris, Paré, & Munoz, 1997; Glimcher & Sparks, 1992; Hanes & Schall, 1996; Mazzoni et al., 1996; Steinmetz & Moore, 2010), and results assigning a cognitive role to the neural discharge have been subject to the criticism that the low-level activity may instead reflect a preparatory command for a movement (usually a saccade) that is planned but not necessarily executed (Gandhi & Sparks, 2004; Ignashchenkova et al., 2004; Krauzlis, Liston, & Carello, 2004). Of course, it is also likely that both cognitive and premotor

signals are reflected in the low-level discharge, a hypothesis that is the basis of the "premotor theory of attention" (Rizzolatti et al., 1987). While links between visual attention and saccade preparation have been implied by psychophysical (Hoffman & Subramaniam, 1995; Rizzolatti et al., 1987) and neurophysiological (Awh, Armstrong, & Moore, 2006; Cavanaugh & Wurtz, 2004; Corbetta et al., 1998; McPeek & Keller, 2002; Moore & Fallah, 2001; Muller, Philiastides, & Newsome, 2005) studies, evidence also exists for distinct attention and preparation processes, particularly at the level of FEFs (Gregoriou, Gotts, & Desimone, 2012; Juan, Shorter-Jacobi, & Schall, 2004; Schall, 2004; Thompson, Biscoe, & Sato, 2005).

The motor preparation hypothesis states that the low-level discharge in SCid neurons accumulates gradually toward a cell-specific threshold, at which point (1) it converts into a high-frequency burst, (2) the brainstem

OPNs become quiescent, and (3) a saccade is triggered (Dorris, Paré, & Munoz, 1997; Hanes & Schall, 1996; Paré & Hanes, 2003). It has been hypothesized that the low-frequency discharge represents a motor preparation signal that encodes both timing and metrics of the desired saccade. Indeed, the firing rate level in the preparatory period is negatively correlated with the saccade reaction time (the higher the activity, the earlier the movement occurs), and the locus of activity in the SC dictates the saccade vector. Basic computational models that simulate the accumulation or drift rate as either noisy (Lo & Wang, 2006; Ratcliff, Cherian, & Segraves, 2003) or ballistic (Carpenter & Williams, 1995; Reddi, Asrress, & Carpenter, 2003) can sufficiently describe the trial-to-trial variability in the discharge patterns of SCid neurons and the distribution of reaction times. This functional assessment of motor preparation, however, is gauged by correlating neural activity with movement features that are observed hundreds of milliseconds later after the animal is granted permission to generate a response. A stronger foundation for motor preparation, and its time course, could be established if a behavioral output can be revealed as the low-frequency activity is evolving.

Prior to triggering the movement, the saccade generation circuitry must overcome the potent inhibition imposed to preserve fixation. A potential method to reveal the presence and evolution of motor preparation exploits the antagonistic relationship between motor preparation and saccade inhibition. Located in the pons, the OPNs inhibit the saccade burst generator circuit that innervates the extraocular motoneurons. The OPNs discharge at a tonic rate during fixation and cease activity during all saccades. Consider the scenario in which the eyes are stable so that the OPNs are active at a tonic rate, and an experimental manipulation is available to transiently inhibit them at different times after object(s) are presented in the visual periphery. This early and transient withdrawal of inhibition could, in principle, "trick" the saccadic system into prematurely executing an eye movement if the underlying low-frequency discharge in SCid and the brainstem burst generator reflects a premotor signal; note that this reasoning remains agnostic about cognitive signals present simultaneously. If successful, the timing of the earliest saccade will indicate when motor preparation commences, and the kinematics will reveal how it develops. Moreover, concurrently recording activity in an oculomotor structure like the SC in monkeys should provide a neural correlate of timing, kinematics, and direction of prematurely triggered saccades.

To test this hypothesis, Gandhi and Bonadonna (2005a) utilized the observation that OPNs also become quiescent during blinks (Schultz, Williams, & Busettini, 2010). An actual connection between blinks and OPNs may not exist because the pause appears to be associated with the small, loopy blink-associated eye movement (Schultz, Williams, & Busettini, 2010). What is important is that blinks offer a means to remove OPN inhibition. Thus, they delivered an air puff to one eye to invoke the trigeminal blink reflex in nonhuman primates performing various saccade tasks (Gandhi & Bonadonna, 2005a). Figure 64.2A plots temporal traces of eye and eyelid positions of four reflexive saccades with blinks (solid traces) and of an averaged nonblink trial (dashed-black trace). A blink evoked shortly after target onset but before the typical saccade reaction time triggers a saccade of shorter latency. Figure 64.2B illustrates the robust correlation between saccade latency and blink time across many blink trials. It particularly highlights the time course of motor preparation, with the shortest reaction times reflecting express saccade latencies.

The blink perturbation can also be used to probe the time course of motor preparation with paradigms geared to understand target selection. In the standard visual search task, the singleton stimulus also serves as the saccade target. Knowledge of the task structure could lead to parallel processing of target selection and motor preparation, but properties of dissociation or simultaneity in their time course would be better appreciated in a task designed to divorce the singleton location and the required movement vector. In a preliminary study (Gandhi & Katnani, 2011a), we implemented a visual search task that closely resembles that used by Schall and colleagues (Juan, Shorter-Jacobi, & Schall, 2004; Sato & Schall, 2003). Each trial began with several hundred milliseconds of fixation on a central visual target, which was extinguished when four stimuli were presented spaced apart by 90°. One target was bestowed a unique feature (color) that made it "pop out" from the other, identical "distractors" (McPeek & Keller, 2002; Shen & Paré, 2007; Thompson et al., 1996). The color of the singleton indicated whether the correct response was a prosaccade to it or an antisaccade to the distractor separated by 180°. Blinks were evoked at a random time on a small percentage of the trials. Figure 64.2C plots the relationship between blink time and saccade latency for the subset of trials requiring the generation of an antisaccade. The different symbols identify saccades that were directed incorrectly to the singleton (medium gray circles), incorrectly to an orthogonal distractor (light gray circles) or correctly to the opposite location (black circles). When the saccade direction of each movement is plotted as a function of its latency (see figure 64.2D), it becomes clear that

    UDAY K. JAGADISAN AND NEERAJ J. GANDHI

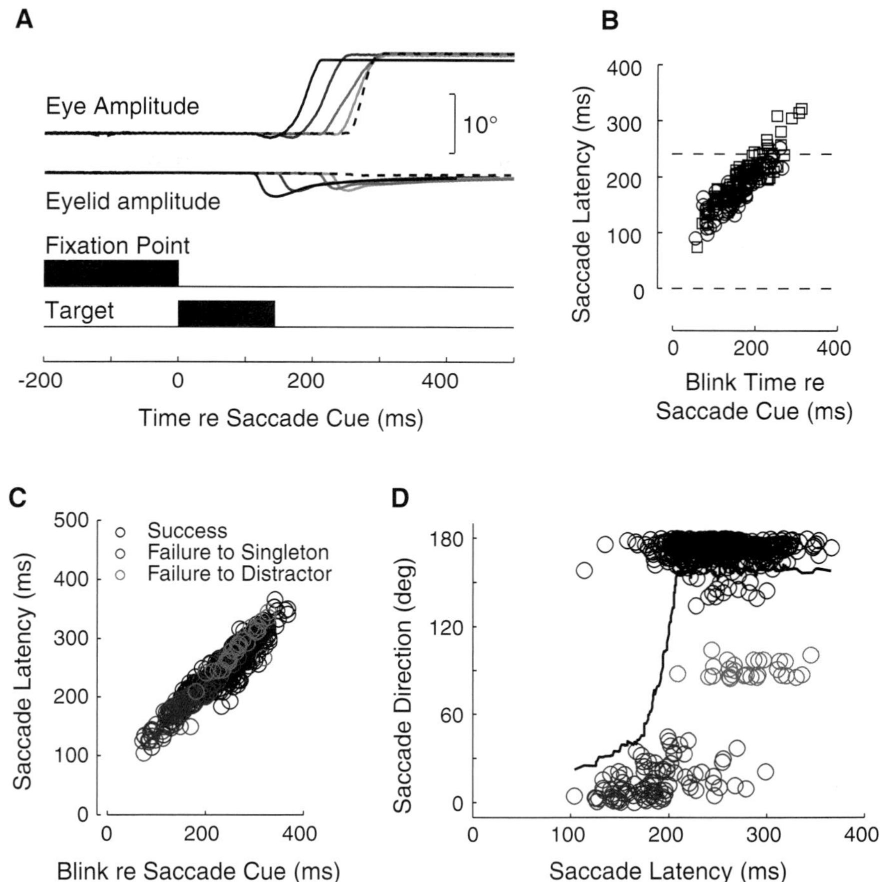

FIGURE 64.2 Behavioral evaluation of the relationship between target selection and motor preparation. (A) Temporal waveforms of eye and eyelid amplitude as a monkey performed visually guided saccades (bottom part of the panel). The dashed traces represent a control movement (no blink). The four solid traces denote individual trials in which a blink was evoked by an air puff to one eye. The blink (downward deflection in eyelid trace) and the associated combined blink–saccade waveform can be paired by matching the shades of gray. It can be appreciated that blinks triggered shortly after the presentation of the peripheral target (and saccade initiation cue) but before a typical saccade reaction time resulted in a prematurely triggered movement. (B) Saccade latency is plotted as a function of blink time relative to saccade cue (time zero in panel A) for many trials, each denoted by one symbol (circles and squares identify leftward and rightward saccades, respectively). The figure highlights the systematic reduction in saccade latency with blink time. (C) Animals performed a visual search task (one singleton, three distractors), in which the color of the singleton indicated whether the correct response was a pro- or antisaccade. Saccade latency is plotted as a function of blink time for only the subset of trials for which an antisaccade was the correct response. The darkest circles identify trials with a correct response to the opposite distractor. The lightest circles represent trials with an incorrect response to an orthogonal distractor. The circles in medium grayscale mark the trials with an incorrect response to the singleton. (D) The direction of the saccade (0°, singleton; 180°, opposite distractor) is plotted as a function of saccade latency for the same data shown in panel C. The solid trace represents a moving average of the data. These data indicate that blink-triggered movements of the shortest latency were directed to the singleton even though the correct response was to the opposite distractor (180°).

saccades with the shortest latencies, comparable to the time when spatial attention is allocated to the singleton, were incorrectly driven to the odd-ball stimulus. The transition to the correct location separated by 180° is observed as the reaction time increases. The blink-perturbation method therefore demonstrates that the motor preparation signal evolves as early as the neural modulation associated with target selection.

The FEFs have been studied exhaustively for correlates of target selection and motor preparation in a paradigm that requires antisaccade generation in the context of a visual search task. The current perspective is spatial attention and motor preparation are dissociated within this cortical node, particularly within visual neurons. It remains to be determined whether the two processes can be simultaneously reflected in visuomotor neurons that exhibit both visual/cognitive selectivity as well as premotor activity. A comparable study has not been conducted on SC neurons, yet it seems prudent to search for such signals in this subcortical structure given its preference for motor-related discharge within the oculomotor neuraxis (Wurtz et al., 2001).

A PRIORITY MAP IN THE SUPERIOR COLLICULUS

We hope to have convinced the reader that the SC contains signals that can be used to select a target from nontargets and, indeed, that the structure plays a causal role in target selection. The studies discussed in this chapter suggest that the function of SC extends beyond its established role in controlling gaze and gaze-related target selection, toward a more general role in selecting stimuli important for transmodal action. Is there a common principle to be gleaned from the activity pattern on the SC map across experimental paradigms and task requirements?

The usage of the term "target" implicitly refers to the fact that a stimulus is the target of a future motor act. This definition does not make any assumptions about the nature of the stimulus itself, beyond its utility for action. However, individual stimuli in the environment are not homogeneous in their intrinsic properties—some are more luminant than others, some present better contrast against their surrounding stimuli, and they come in myriad colors. A simplified, topographic representation of the visual world in terms of the relative strengths of various stimuli has been proposed as a salience (or saliency) map (Koch & Ullman, 1985). This representation is considered bottom-up, independent of the internal state of the animal. On the other hand, the internal state, including top-down factors such as expectations and goals, can be used to assign significance to different objects and locations in the visual field, creating a relevance map. An appropriately weighted, combined representation of bottom-up salience and top-down relevance produces a priority map, high values on the map indicating high priority for action to the corresponding location (such as an eye movement or reach). Such maps have been proposed in various brain regions, including the parietal lobe, FEF, prefrontal cortex, basal ganglia, and indeed, the SC (for a review, see Fecteau & Munoz, 2006). Since the SC is not known to play a direct role in movements other than gaze shifts (stimulation does not induce other types of movements, but see Werner, Dannenberg, & Hoffmann, 1997), it has been hypothesized that a generalized, modality-independent priority map

influences action execution in other domains by relaying the priority information back to the cortex through feedback loops, where it can be parsed into the appropriate command for effector-specific movements (Song, Rafal, & McPeek, 2011).

If the SCid does in fact house a priority map, it is natural to assume that it does so at all times, rather than switching to priority map mode under specific circumstances such as visual search or when competing stimuli are present. Therefore, it is useful to interpret SC activity in simple tasks such as the single-step task within the framework of a priority map. As seen earlier, during fixation, the population activity in SCid is concentrated toward the rostral end. This can be seen as a priority map with a single locus in the parafoveal region. Following initial processing of a stimulus that appears in a peripheral location, the population activity, and thus the priority map, shifts to that location (figure 64.3). During the gap period in a gap task, the priority map smoothly shifts from the foveal to a peripheral location, and this is reflected in the inversely correlated pattern of low-frequency buildup activity in rostral and caudal neurons. Thus, the map evolves as a putative "priority function" of each location in visual space changes in time. The distribution of activity across the SC population during multistep gaze shifts is consistent with an interpretation of this nature. The evolution of the priority signal can also be likened to an emerging oculomotor "decision" signal observed in perceptual discrimination tasks (Gold & Shadlen, 2000).

The premotor theory of attention can also be tied to the notion of a priority map. One consequence of viewing ongoing activity in SCid as a priority map is that the motor machinery in a subpopulation of SCid neurons (motor neurons) and downstream structures has concurrent access to information about the location with highest priority. Does this imply a direct relationship between forming a priority map and motor preparation, or are they serial stages in guiding behavior? In other words, does the priority map map directly onto a space of possible actions? The evidence presented in the previous section suggests that there may indeed be some overlap between the two processes.

How are the individual salience and relevance components communicated to the priority map in the SC? One possibility is that the information about stimulus salience arrives directly or via other cortical areas from the visual cortex, which is thought to contain a representation of bottom-up salience and has direct projections to the SC. Recent findings also implicate the isthmic nuclei (or parabigeminal nuclei) in the

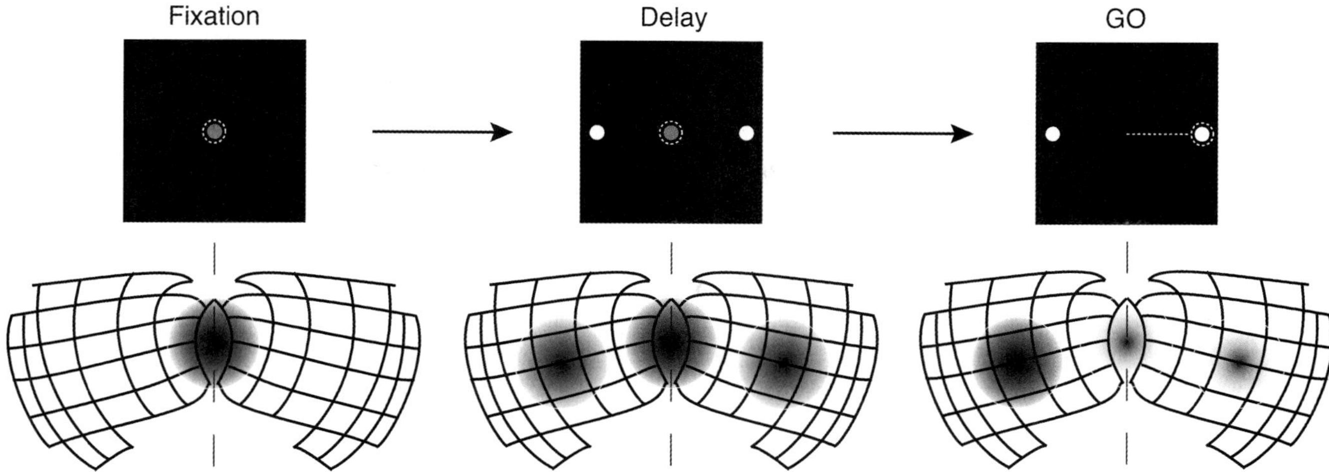

FIGURE 64.3   Evolution of priority in superior colliculus (SC) over time. Top row: Delayed visual search task. Only two peripheral stimuli are used for simplicity. One of the two stimuli is selected as the target based on some feature (not shown). Gaze location and saccade are shown by dotted lines. Bottom row: SC population activity during the task. During fixation, the mound of activity is centered around the rostral poles of the two colliculi. At this time, the highest priority location is the central target. The onset of the search stimuli introduces two new loci of activity on the SC map. There is a transient shift in priority to these locations around the time of the visual response (time course not shown), but the priority of the foveal location is maintained higher than the peripheral targets during the delay period. After permission to saccade, the activity shifts in favor of the selected target (represented in the contralateral SC) and dies down in the other parts of the map.

classification of stimuli based on their relative salience. Information about stimulus relevance is expected to arrive from top-down sources including the respective association cortices and prefrontal cortex. Finally, part of the priority information could be directly relayed from other cortical areas that are postulated to contain modality-specific as well as independent representations of priority, including LIP, FEF, and dlPFC.

In a recent midbrain-centric account of consciousness (Merker, 2007), the SC is proposed to play a critical role in a "selection triangle" that involves target selection, action selection, and internal motivational state. The placement of the colliculus in the command structure of the brain is a key factor behind the proposal. Of course, it is likely that the SC is part of an extended network of regions involved in the selection of targets and actions, albeit with some degree of autonomy considering its diverse functional properties. The idea of a continuously evolving priority map in the SC, together with associated maps in the cortex, basal ganglia, and other regions in the midbrain, fits in neatly within this picture.

ACKNOWLEDGMENTS

This work was supported by NIH grant EY015485 and DC025205.

REFERENCES

Asadollahi, A., Mysore, S. P., & Knudsen, E. I. (2011). Rules of competitive stimulus selection in a cholinergic isthmic nucleus of the owl midbrain. *Journal of Neuroscience, 31,* 6088–6097.

Awh, E., Armstrong, K. M., & Moore, T. (2006). Visual and oculomotor selection: Links, causes and implications for spatial attention. *Trends in Cognitive Sciences, 10,* 124–130. doi:10.1016/j.tics.2006.01.001.

Basso, M. A., & Wurtz, R. H. (1998). Modulation of neuronal activity in superior colliculus by changes in target probability. *Journal of Neuroscience, 18,* 7519–7534.

Bergeron, A., Matsuo, S., & Guitton, D. (2003). Superior colliculus encodes distance to target, not saccade amplitude, in multi-step gaze shifts. *Nature Neuroscience, 6,* 404–413.

Berman, R. A., & Wurtz, R. H. (2010). Functional identification of a pulvinar path from superior colliculus to cortical area MT. *Journal of Neuroscience, 30,* 6342–6354.

Bisley, J. W., & Goldberg, M. E. (2010). Attention, intention, and priority in the parietal lobe. *Annual Review of Neuroscience, 33,* 1–21.

Carello, C. D., & Krauzlis, R. J. (2004). Manipulating intent: Evidence for a causal role of the superior colliculus in target selection. *Neuron, 43,* 575–583.

Carpenter, R. H., & Williams, M. L. (1995). Neural computation of log likelihood in control of saccadic eye movements. *Nature, 377,* 59–62.

Cavanaugh, J., & Wurtz, R. H. (2004). Subcortical modulation of attention counters change blindness. *Journal of Neuroscience, 24,* 11236–11243.

Chou, I. H., Sommer, M. A., & Schiller, P. H. (1999). Express averaging saccades in monkeys. *Vision Research, 39*, 4200–4216. doi:10.1016/S0042-6989(99)00133-9.

Clower, D. M., West, R. A., Lynch, J. C., & Strick, P. L. (2001). The inferior parietal lobule is the target of output from the superior colliculus, hippocampus, and cerebellum. *Journal of Neuroscience, 21*, 6283–6291.

Corbetta, M., Akbudak, E., Conturo, T. E., Snyder, A. Z., Ollinger, J. M., Drury, H. A., et al. (1998). A common network of functional areas for attention and eye movements. *Neuron, 21*, 761–773. doi:10.1016/S0896-6273(00)80593-0.

Dorris, M. C., Paré, M., & Munoz, D. P. (1997). Neuronal activity in monkey superior colliculus related to the initiation of saccadic eye movements. *Journal of Neuroscience, 17*, 8566–8579.

Everling, S., Dorris, M. C., Klein, R. M., & Munoz, D. P. (1999). Role of primate superior colliculus in preparation and execution of anti-saccades and pro-saccades. *Journal of Neuroscience, 19*, 2740–2754.

Everling, S., & Munoz, D. P. (2000). Neuronal correlates for preparatory set associated with pro-saccades and anti-saccades in the primate frontal eye field. *Journal of Neuroscience, 20*, 387–400.

Fecteau, J. H., & Munoz, D. P. (2006). Salience, relevance, and firing: A priority map for target selection. *Trends in Cognitive Sciences, 10*, 382–390.

Felsen, G., & Mainen, Z. F. (2008). Neural substrates of sensory-guided locomotor decisions in the rat superior colliculus. *Neuron, 60*, 137–148.

Gandhi, N. J., & Bonadonna, D. K. (2005a). Temporal interactions of air-puff-evoked blinks and saccadic eye movements: Insights into motor preparation. *Journal of Neurophysiology, 93*, 1718–1729.

Gandhi, N. J., & Jagadisan, U. K. (2011). A neural test bed for simulating executive control deficits in saccade generation. *Society for Neurosciences Abstract, Program 18.09.*

Gandhi, N. J., & Katnani, H. A. (2011a). Interactions of eye and eyelid movements. In S. P. Liversedge, I. D. Gilchrist, & S. Everling (Eds.), *Oxford handbook of eye movements* (pp. 323–338). Oxford, England: Oxford University Press.

Gandhi, N. J., & Katnani, H. A. (2011b). Motor functions of the superior colliculus. *Annual Review of Neuroscience, 34*, 205–231.

Gandhi, N. J., & Keller, E. L. (1999). Comparison of saccades perturbed by stimulation of the rostral superior colliculus, the caudal superior colliculus, and the omnipause neuron region. *Journal of Neurophysiology, 82*, 3236–3253.

Gandhi, N. J., & Sparks, D. L. (2004). Changing views of the role of the superior colliculus in the control of gaze. In L. M. Chalupa & J. S. Werner (Eds.), *The visual neurosciences* (Vol. 2, pp. 1449–1465). Cambridge, MA: MIT Press.

Glimcher, P. W., & Sparks, D. L. (1992). Movement selection in advance of action in the superior colliculus. *Nature, 355*, 542–545.

Gold, J. I., & Shadlen, M. N. (2000). Representation of a perceptual decision in developing oculomotor commands. *Nature, 404*, 390–394.

Goldberg, M. E., & Wurtz, R. H. (1972). Activity of superior colliculus in behaving monkey: I. Visual receptive fields of single neurons. *Journal of Neurophysiology, 35*, 542–559.

Goossens, H. H., & Van Opstal, A. J. (2006). Dynamic ensemble coding of saccades in the monkey superior colliculus. *Journal of Neurophysiology, 95*, 2326–2341.

Gregoriou, G. G., Gotts, S. J., & Desimone, R. (2012). Cell-type-specific synchronization of neural activity in FEF with V4 during attention. *Neuron, 73*, 581–594.

Hafed, Z. M., Goffart, L., & Krauzlis, R. J. (2009). A neural mechanism for microsaccade generation in the primate superior colliculus. *Science, 323*, 940–943.

Hanes, D. P., & Schall, J. D. (1996). Neural control of voluntary movement initiation. *Science, 274*, 427–430.

Hikosaka, O., Takikawa, Y., & Kawagoe, R. (2000). Role of the basal ganglia in the control of purposive saccadic eye movements. *Physiological Reviews, 80*, 953–978.

Hoffman, J. E., & Subramaniam, B. (1995). The role of visual attention in saccadic eye movements. *Perception & Psychophysics, 57*, 787–795.

Huerta, M. F., & Harting, J. K. (1984). The mammalian superior colliculus: Studies of its morphology and connections. In H. Vanegas (Ed.), *Comparative neurology of the optic tectum* (pp. 687–773). New York: Plenum.

Ignashchenkova, A., Dicke, P. W., Haarmeier, T., & Thier, P. (2004). Neuron-specific contribution of the superior colliculus to overt and covert shifts of attention. *Nature Neuroscience, 7*, 56–64.

Isa, T., & Hall, W. C. (2009). Exploring the superior colliculus in vitro. *Journal of Neurophysiology, 102*, 2581–2593.

Juan, C. H., Shorter-Jacobi, S. M., & Schall, J. D. (2004). Dissociation of spatial attention and saccade preparation. *Proceedings of the National Academy of Sciences of the United States of America, 101*, 15541–15544. doi:10.1073/pnas.0403507101.

Katnani, H. A., van Opstal, A. J., & Gandhi, N. J. (2012). A test of spatial temporal decoding mechanisms in the superior colliculus. *Journal of Neurophysiology, 107*, 2442–2452.

Kim, B., & Basso, M. A. (2008). Saccade target selection in the superior colliculus: A signal detection theory approach. *Journal of Neuroscience, 28*, 2991–3007.

Kim, B., & Basso, M. A. (2010). A probabilistic strategy for understanding action selection. *Journal of Neuroscience, 30*, 2340–2355.

Knudsen, E. I. (1982). Auditory and visual maps of space in the optic tectum of the owl. *Journal of Neuroscience, 2*, 1177–1194.

Koch, C., & Ullman, S. (1985). Shifts in selective visual attention: Towards the underlying neural circuitry. *Human Neurobiology, 4*, 219–227.

Krauzlis, R. J., & Dill, N. (2002). Neural correlates of target choice for pursuit and saccades in the primate superior colliculus. *Neuron, 35*, 355–363.

Krauzlis, R. J., Liston, D., & Carello, C. D. (2004). Target selection and the superior colliculus: Goals, choices and hypotheses. *Vision Research, 44*, 1445–1451. doi:10.1016/j.visres.2004.01.005.

Lee, C., Rohrer, W. H., & Sparks, D. L. (1988). Population coding of saccadic eye movements by neurons in the superior colliculus. *Nature, 332*, 357–360.

Lo, C. C., & Wang, X. J. (2006). Cortico–basal ganglia circuit mechanism for a decision threshold in reaction time tasks. *Nature Neuroscience, 9*, 956–963.

May, P. J. (2006). The mammalian superior colliculus: Laminar structure and connections. *Progress in Brain Research, 151*, 321–378.

Mazzoni, P., Bracewell, R. M., Barash, S., & Andersen, R. A. (1996). Motor intention activity in the macaque's lateral intraparietal area: I. Dissociation of motor plan from sensory memory. *Journal of Neurophysiology, 76*, 1439–1456.

McPeek, R. M., & Keller, E. L. (2002). Saccade target selection in the superior colliculus during a visual search task. *Journal of Neurophysiology, 88*, 2019–2034.

McPeek, R. M., & Keller, E. L. (2004). Deficits in saccade target selection after inactivation of superior colliculus. *Nature Neuroscience, 7*, 757–763.

Merker, B. (2007). Consciousness without a cerebral cortex: A challenge for neuroscience and medicine. [Review]. *Behavioral and Brain Sciences, 30*, 63–81, discussion 81–134.

Mohler, C. W., & Wurtz, R. H. (1976). Organization of monkey superior colliculus: Intermediate layer cells discharging before eye movements. *Journal of Neurophysiology, 39*, 722–744.

Moore, T., & Fallah, M. (2001). Control of eye movements and spatial attention. *Proceedings of the National Academy of Sciences of the United States of America, 98*, 1273–1276. doi:10.1073/pnas.021549498.

Moschovakis, A. K., Scudder, C. A., & Highstein, S. M. (1996). The microscopic anatomy and physiology of the mammalian saccadic system. *Progress in Neurobiology, 50*, 133–254.

Muller, J. R., Philiastides, M. G., & Newsome, W. T. (2005). Microstimulation of the superior colliculus focuses attention without moving the eyes. *Proceedings of the National Academy of Sciences of the United States of America, 102*, 524–529.

Munoz, D. P., & Everling, S. (2004). Look away: The antisaccade task and the voluntary control of eye movement. *Nature Reviews. Neuroscience, 5*, 218–228.

Munoz, D. P., & Istvan, P. J. (1998). Lateral inhibitory interactions in the intermediate layers of the monkey superior colliculus. *Journal of Neurophysiology, 79*, 1193–1209.

Munoz, D. P., Waitzman, D. M., & Wurtz, R. H. (1996). Activity of neurons in monkey superior colliculus during interrupted saccades. *Journal of Neurophysiology, 75*, 2562–2580.

Munoz, D. P., & Wurtz, R. H. (1993). Fixation cells in monkey superior colliculus: I. Characteristics of cell discharge. *Journal of Neurophysiology, 70*, 559–575.

Mysore, S. P., Asadollahi, A., & Knudsen, E. I. (2011). Signaling of the strongest stimulus in the owl optic tectum. *Journal of Neuroscience, 31*, 5186–5196.

Mysore, S. P., & Knudsen, E. I. (2011a). Flexible categorization of relative stimulus strength by the optic tectum. *Journal of Neuroscience, 31*, 7745–7752.

Mysore, S. P., & Knudsen, E. I. (2011b). The role of a midbrain network in competitive stimulus selection. *Current Opinion in Neurobiology, 21*, 653–660.

Mysore, S. P., & Knudsen, E. I. (2012). Reciprocal inhibition of inhibition: A circuit motif for flexible categorization in stimulus selection. *Neuron, 73*, 193–205.

Nummela, S. U., & Krauzlis, R. J. (2011). Superior colliculus inactivation alters the weighted integration of visual stimuli. *Journal of Neuroscience, 31*, 8059–8066.

Paré, M., & Hanes, D. P. (2003). Controlled movement processing: Superior colliculus activity associated with countermanded saccades. *Journal of Neuroscience, 23*, 6480–6489.

Ratcliff, R., Cherian, A., & Segraves, M. (2003). A comparison of macaque behavior and superior colliculus neuronal activity to predictions from models of two-choice decisions. *Journal of Neurophysiology, 90*, 1392–1407.

Reddi, B. A., Asrress, K. N., & Carpenter, R. H. (2003). Accuracy, information, and response time in a saccadic decision task. *Journal of Neurophysiology, 90*, 3538–3546.

Rizzolatti, G., Riggio, L., Dascola, I., & Umilta, C. (1987). Reorienting attention across the horizontal and vertical meridians: Evidence in favor of a premotor theory of attention. *Neuropsychologia, 25*(1A), 31–40.

Sato, T. R., & Schall, J. D. (2003). Effects of stimulus–response compatibility on neural selection in frontal eye field. *Neuron, 38*, 637–648.

Schall, J. D. (2001). Neural basis of deciding, choosing and acting. *Nature Reviews. Neuroscience, 2*, 33–42.

Schall, J. D. (2004). On the role of frontal eye field in guiding attention and saccades. *Vision Research, 44*, 1453–1467. doi:10.1016/j.visres.2003.10.025.

Schall, J. D., & Hanes, D. P. (1993). Neural basis of saccade target selection in frontal eye field during visual search. *Nature, 366*, 467–469.

Schultz, K. P., Williams, C. R., & Busettini, C. (2010). Macaque pontine omnipause neurons play no direct role in the generation of eye blinks. *Journal of Neurophysiology, 103*, 2255–2274.

Shen, K., & Paré, M. (2007). Neuronal activity in superior colliculus signals both stimulus identity and saccade goals during visual conjunction search. *Journal of Vision, 7*(5), 15. doi:10.1167/7.5.15.

Sommer, M. A., & Wurtz, R. H. (2004). What the brain stem tells the frontal cortex: I. Oculomotor signals sent from superior colliculus to frontal eye field via mediodorsal thalamus. *Journal of Neurophysiology, 91*, 1381–1402.

Song, J. H., Rafal, R. D., & McPeek, R. M. (2011). Deficits in reach target selection during inactivation of the midbrain superior colliculus. *Proceedings of the National Academy of Sciences of the United States of America, 108*, E1433–E1440. doi:10.1073/pnas.1109656108.

Sparks, D. L., & Hartwich-Young, R. (1989). The deep layers of the superior colliculus. *Reviews of Oculomotor Research, 3*, 213–255.

Steinmetz, N. A., & Moore, T. (2010). Changes in the response rate and response variability of area V4 neurons during the preparation of saccadic eye movements. *Journal of Neurophysiology, 103*, 1171–1178.

Thomas, N. W., & Paré, M. (2007). Temporal processing of saccade targets in parietal cortex area LIP during visual search. *Journal of Neurophysiology, 97*, 942–947.

Thompson, K. G., Biscoe, K. L., & Sato, T. R. (2005). Neuronal basis of covert spatial attention in the frontal eye field. *Journal of Neuroscience, 25*, 9479–9487.

Thompson, K. G., Hanes, D. P., Bichot, N. P., & Schall, J. D. (1996). Perceptual and motor processing stages identified in the activity of macaque frontal eye field neurons during visual search. *Journal of Neurophysiology, 76*, 4040–4055.

Van Opstal, A. J., & Van Gisbergen, J. A. (1990). Role of monkey superior colliculus in saccade averaging. *Experimental Brain Research, 79*, 143–149.

Werner, W., Dannenberg, S., & Hoffmann, K. P. (1997). Arm-movement-related neurons in the primate superior colliculus and underlying reticular formation: Comparison of neuronal activity with EMGs of muscles of the shoulder, arm and trunk during reaching. *Experimental Brain Research, 115*, 191–205.

Wurtz, R. H., Sommer, M. A., Paré, M., & Ferraina, S. (2001). Signal transformations from cerebral cortex to superior colliculus for the generation of saccades. *Vision Research, 41*, 3399–3412. doi:10.1016/S0042-6989(01)00066-9.

65 The Dialogue between Cerebral Cortex and Superior Colliculus: Multiple Ascending Pathways for Corollary Discharge

MARC A. SOMMER AND ROBERT H. WURTZ

Vision requires the integration of two main sources of information: what the eyes are sensing and what they are doing. At any point in time that an image is detected, the eyes may be still, darting across the scene to look at something new, or tracking a mobile object. Hence the brain must evaluate not only inflow from the retina but also movements of the eyes. The retinal and motoric influences on the visual system are intertwined from early in the visual pathway, but the influence of self-movement information becomes more prominent in higher-level visual areas of frontal and parietal–occipital cortex. The output from cortex courses primarily from those areas, with the primary target being the superior colliculus (SC) on the roof of the midbrain. And as we will describe, the SC reciprocates with pathways back to frontal and parietal–occipital cortex (see figure 65.1A).

In the previous edition of this book (Sommer & Wurtz, 2003), we summarized salient aspects of the functioning of descending pathways to SC, along with emerging information about pathways that ascend from SC back to cortex. In the present chapter we build on the latter topic because during the past decade there has been considerable success in understanding two of the specific ascending pathways. One originates from saccade-related neurons in the intermediate layers of the SC, is relayed by neurons in the medial dorsal nucleus (MD) of the thalamus, and targets neurons in the frontal eye field (FEF; figure 65.1B). The second pathway originates from visual neurons in the SC superficial layers, is relayed by neurons in the inferior nucleus of the pulvinar (PI), and targets area MT (see figure 65.1C). While sufficient information is available to consider only these two specific pathways, an evaluation of them may provide insight into pathways that ascend more broadly from the SC to diverse regions of frontal and parietal–occipital cortex.

The ascending pathways from SC may contribute to multiple aspects of behavior and perception. Most of the work to date has focused on one visual function: our maintenance of a stable visual percept in spite of our incessant eye movements. Saccades occur several times per second, displacing the retinal image with each saccade, and yet our visual percept is almost perfectly protected from disruptions. The underlying mechanism for producing this stability is hypothesized to be corollary discharge (CD). Motor plans for an impending saccade would be relayed as CD to regions of the brain where visual processing will be disrupted, providing those regions with information that the movement is about to occur (Wurtz, 2008). A major question asked about ascending pathways has been whether they convey CD of saccades. We will begin by considering how such a signal might be identified.

IDENTIFYING A COROLLARY DISCHARGE

The basic logic of a CD signal is illustrated in figure 65.2A. At the same time that activity is sent downward in the brain to produce a movement, a copy of the activity is sent upward in the brain to inform other regions that the movement is about to occur. The copy of the movement command is used by the other brain regions to compensate for the disruption that will be produced by the movement. This copy of the movement signal has been referred to as a "sense of will" by von Helmholtz (1925) in the nineteenth century, and as a "corollary discharge" by Sperry (1950) or an "efference copy" by von Holst and Mittelstaedt (1950) in the twentieth century. We will use Sperry's terminology. For the compensation required for stable vision, a CD for saccades would be expected to originate in an area where neurons discharge before saccades. One such region is

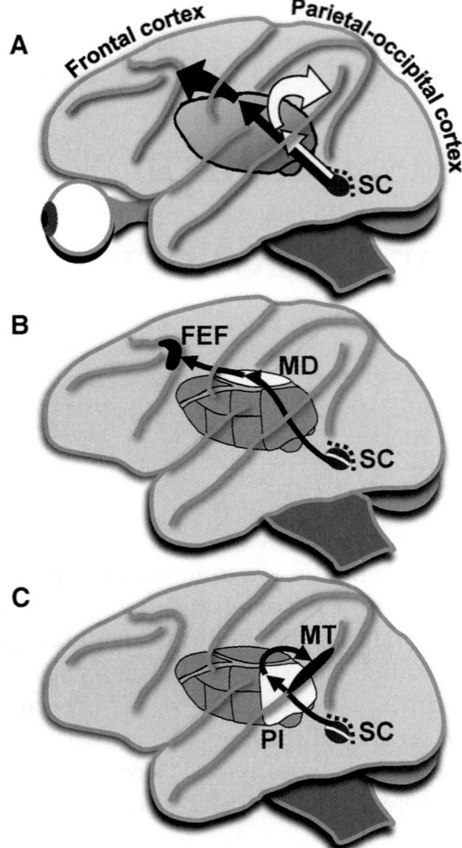

FIGURE 65.1   Ascending pathways from the superior colliculus (SC) to cortex. Lateral view of rhesus monkey brain, depicting the two pathways that are the focus of this chapter. (A) SC neurons send disynaptic pathways, relayed by the thalamus (depicted in bold outline in the middle of the brain), to broad areas of frontal and parietal–occipital cortex. We are beginning to understand the signal content and the visual-related functions of two specific pathways: (B) the SC–MD–FEF pathway (MD, medial dorsal nucleus of the thalamus; FEF, frontal eye field), which originates in the intermediate layers of the SC (depicted below the white line on the SC), is relayed by neurons in the MD, and targets the FEF; and (C) the SC–PI–MT pathway (PI, inferior nucleus of the pulvinar; MT, middle temporal visual cortex), which originates in the superficial layers of the SC (above the white line), is relayed by the PI nucleus of the thalamus, and targets area MT.

the intermediate layers of the SC (see figure 65.2B), where activity begins at a low level several hundred milliseconds before a saccade and then culminates in a burst of activity starting about 50 ms before a saccade. These saccade-related neurons would not only initiate a saccade via their descending connections but also provide information to cerebral cortex via ascending pathways relayed by thalamus. A critical characteristic of a CD signal is its timing: Just like the command signal it emulates, a CD signal must occur before the saccade.

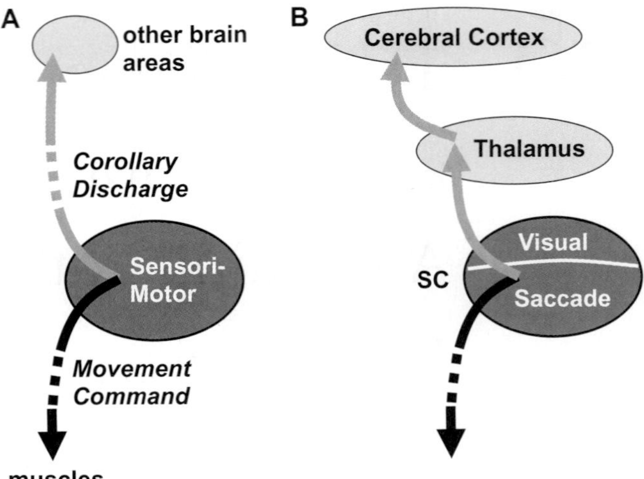

FIGURE 65.2   Role of the superior colliculus (SC) in producing corollary discharge (CD). (A) Principles of CD. A sensorimotor structure generates a movement command that ultimately contracts muscles to produce behavior. At the same time, the structure relays a correlate of the movement command, the CD signal, to other brain areas. The CD provides information about the imminent behavior but does not contribute to generating that behavior. Dotted segments of arrows indicate that, in general, the sensorimotor structure may be connected to the other brain areas and muscles through polysynaptic circuits. (B) The CD concept as it applies to the SC. The intermediate, saccade-related layers of the SC are the origin of the CD signal. Neurons in the intermediate SC produce saccadic commands and, simultaneously, convey CD information about those commands to thalamus and on to cerebral cortex. The CD information may be used within the SC, as well, to affect visual neurons of the superficial SC (not shown).

This timing contrasts with proprioceptive feedback from contracting eye muscles or from the visual input resulting from the saccade (reafference), both of which are available only after the saccade. In sum, a CD signal would provide information about a saccade before the eye even moves.

A major challenge in studying potential CD signals in the primate brain is that they are likely imbedded in an extraordinarily intertwined series of pathways. From this complicated circuitry, how can specific pathways that carry CD be dissected? A related challenge is to know a CD signal when one sees it. How can a CD signal be distinguished from signals that are used to move the eyes? Both types of signals would carry the same information; it is just that they are used by different networks for different purposes. These questions were answered for the SC–MD–FEF pathway (see figure 65.1B) using neurophysiological approaches in behaving monkeys (Sommer & Wurtz, 2002; 2004a, 2004b). That there was a circuit of connected neurons from SC to FEF was

   MARC A. SOMMER AND ROBERT H. WURTZ

known from anatomical studies (Lynch, Hoover, & Strick, 1994). But how could one isolate neurons within this candidate CD pathway? The techniques of anti-dromic and orthodromic electrical stimulation proved successful. By applying these techniques to known points in the circuit (SC, MD, and FEF), it was possible to identify source neurons in the SC, recipient neurons in the FEF, and, most critically, relay neurons in MD. The identified MD relay neurons discharged before saccades, and only before saccades of a given amplitude and direction, just like neurons in the intermediate SC. This led to the second question: How could one distinguish whether the MD neurons convey CD information about the saccade, as opposed to helping to produce the saccade? This was answered with pharmacological manipulation. Reversible inactivation of MD relay neurons had no effect on saccade generation; neither the amplitude nor the direction of saccades was impaired. MD inactivation did, however, disrupt a monkey's ability to perform a double-step saccade task that required the use of CD information (Hallett & Lightstone, 1976).

Thus the activity along the SC–MD–FEF circuit is consistent with a CD, and the activity is necessary for a task that requires CD but not for the generation of saccades. More detailed descriptions of the experiments that established the circuit and its CD information content, including discussions on limitations of interpreting the data, are available in the previous edition (Sommer & Wurtz, 2003), in recent reviews (Crapse & Sommer, 2008a; Sommer & Wurtz, 2008; Wurtz, 2008), and in the original research papers (Sommer & Wurtz, 2002, 2004a, 2004b, 2006).

AN ASCENDING PATHWAY TO FRONTAL CORTEX

Having established that the pathway from SC to FEF conveys CD of saccades, the next step was to assess whether the pathway contributes to visual perception. As we have indicated, a major function of CD might be its contribution to our perception of a stable visual world: The CD anticipates the saccade, and that anticipation alters the processing of visual events that result from the saccade (Crapse & Sommer, 2008a; Sommer & Wurtz, 2008; Wurtz, 2008). Anticipatory visual activity is in fact exhibited by some neurons in parietal cortex (Duhamel, Colby, & Goldberg, 1992), the FEF (Sommer & Wurtz, 2006; Umeno & Goldberg, 1997, 2001), and other cortical and subcortical areas (Dunn, Hall, & Colby, 2010; Nakamura & Colby, 2002; Walker, Fitzgibbon, & Goldberg, 1995). The anticipatory effect was shown in experiments that measured neuronal responses to visual stimuli in receptive field (RF) and other parts of visual space (see figure 65.3A). Long before a saccade, the neurons responded only to stimuli in their RF. Just before a saccade, however, the neurons responded as well, or only, to stimuli presented in a portion of the visual field (which we call the future field; FF) that becomes the neuron's RF after the saccade. This "shifted" response thus anticipates the visual result of the impending saccade. Duhamel, Colby, and Goldberg (1992) argued that it is this anticipatory shift that performs a remapping of the visual field with each saccade, and that this is a possible mechanism for our perception of visual stability.

In order for a neuron to anticipate a future location after the saccade, it must receive information about where the saccade will go before the saccade occurs. That is exactly the information a CD conveys. In the FEF, the CD could arrive from the SC–MD–FEF pathway, and this hypothesis was tested by again inactivating the pathway at the level of the MD relay neurons (Sommer & Wurtz, 2006). The inactivation had no effect on RF visual responses (see figure 65.3B, top row), but it impaired the FF response by more than 70% in the example neuron (bottom row) and by ~50% on average. Hence the CD provided by the SC–MD–FEF pathway does in fact influence visual processing in cerebral cortical neurons. Specifically, it contributes to presaccadic shifting of RFs, the main candidate mechanism for creating visual stability across saccades.

Very recent work has answered four additional questions regarding the role of the SC–MD–FEF pathway in creating visual stability. First, it was known that FEF neurons can shift their RFs in association with a broad range of saccadic directions, including saccades made into ipsilateral space. The SC, however, represents only saccades into contralateral space. So how do FEF neurons know about saccades made everywhere? The hypothesis was that the FEF on one side of the brain receives pathways from both SCs (see figure 65.4A). This was confirmed in an experiment in which FEF neurons on one side of the brain were orthodromically activated by SC stimulation on the same or the opposite side (Crapse & Sommer, 2009). As predicted by the connectivity, FEF neurons that received input from the same-side SC had contralaterally directed visual and movement fields while neighboring FEF neurons that received input from the opposite-side SC had ipsilaterally directed visual and movement fields. Second, it is not necessary for the entire visual scene to be stabilized across saccades; it would be sufficient for perception if only the attended or particularly noticeable stimuli—the *salient* stimuli—remain stable. Are shifting RFs, the presumed mechanism for visual stability, affected by salience? Joiner, Cavanaugh, and Wurtz

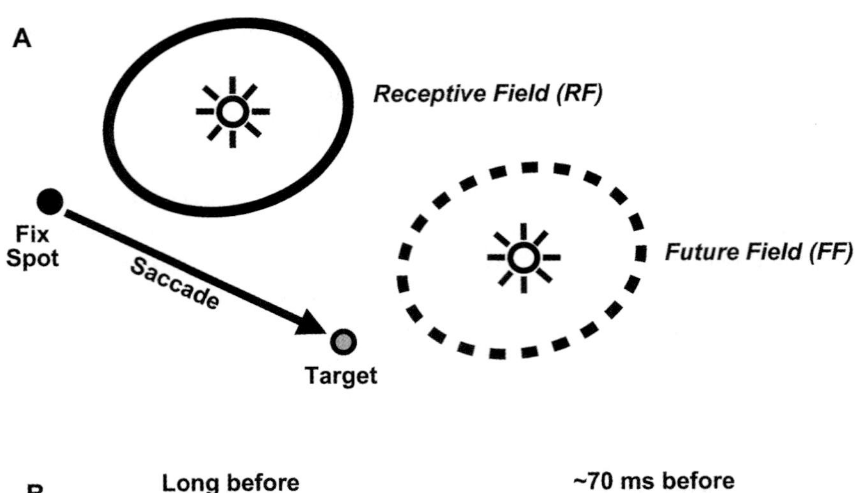

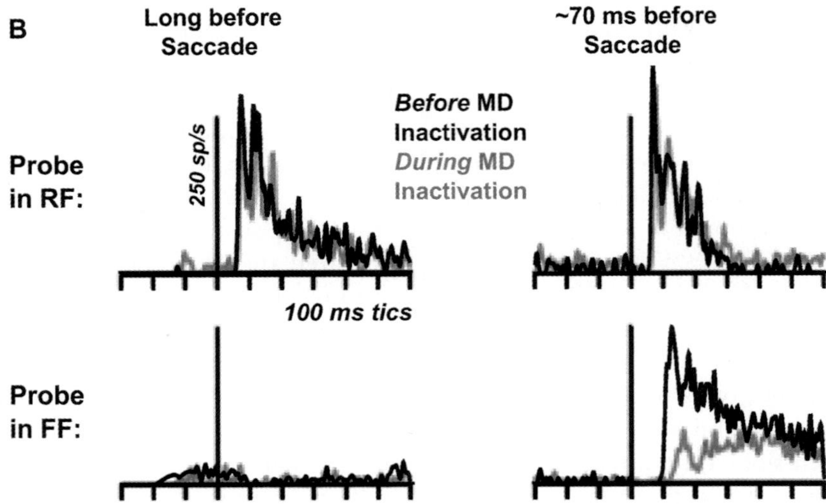

FIGURE 65.3   Shifting receptive fields (RFs) in the frontal eye field (FEF). (A) Task for assessing visual responses of neurons around the time of a saccade. A monkey fixates (Fix Spot) and then makes a saccade to a Target that appears. During fixation, that is, before the saccade, a probe is flashed (circles with radiating lines) at one of two locations: in the center of the neuron's RF or at the center of its future field (FF). The FF is where the RF will reside after the saccade. (B) Example data from an FEF neuron, before and during inactivation of the medial dorsal nucleus of the thalamus (MD). Spike density functions are aligned to probe onset (vertical line). *Before inactivation*: Long before a saccade (left column), probes in the RF (upper panel), but not the FF (lower panel), evoke a visual response. Just before a saccade (right column), probes in the RF still evoke a visual response, but now probes in the FF do as well; this is the shift. *During inactivation*: Visual responses in the RF are unaffected, but the shifted visual response in the FF is impaired. Sp/s, spikes per second. (Modified from Sommer & Wurtz, 2006.)

(2011) found that for many neurons in the FEF, the intensity of the RF shift decreased as the saliency of the stimulus in the FF decreased (see figure 65.4B). Hence RF shifts are not just an obligatory mechanism; they are affected by visual context in a manner that is appropriate for a role in creating the percept of visual stability across saccades. Third, all of this work implies that FEF neurons should distinguish whether the visual world is stable or not as saccades are made. Do they? This was tested by examining the reafferent responses of FEF visual neurons, that is, the responses produced when saccades move the RF onto an existing visual stimulus (Crapse & Sommer, 2012). The postsaccadic stimulus was always at the center of the postsaccadic

RF, so in principle it should have always elicited the same reafferent response. The reafferent responses depended, however, on how the stimulus attained that location (see figure 65.4C). FEF neurons responded differently for stable stimuli than for stimuli that moved to the postsaccadic location during the saccade. The FEF thus acts as a comparator of the visual scene across saccades. Fourth, in all of this work, a link to perception was missing. Do FEF neurons predict whether a monkey will perceive the visual scene as stable or not as it makes saccades? This was tested by training monkeys to make scanning saccades between two fixation spots and report if a peripheral visual stimulus moved during a saccade (Crapse & Sommer,

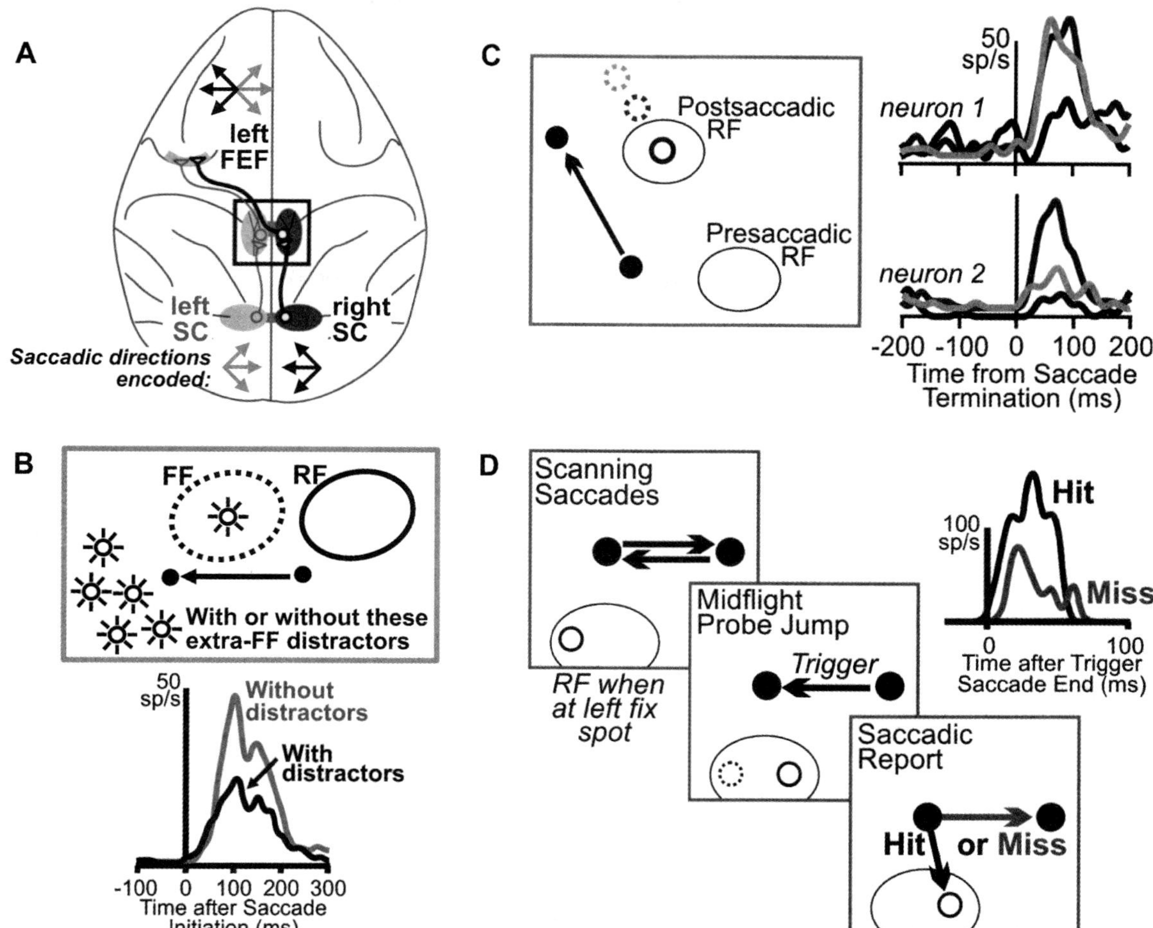

FIGURE 65.4   Recent advances in understanding corollary discharge (CD)–related visual processing in the frontal eye field (FEF). (A) A single FEF receives CD not only from the same-side superior colliculus (SC), representing saccades into contralateral space, but also from the opposite-side SC, representing saccades into ipsilateral space. This may explain why FEF neurons are able to shift their receptive fields (RFs) in association with any direction of saccade. (Modified from Crapse & Sommer, 2009.) (B) The shift depends on the salience of the visual stimulus. Shifting RFs were measured during two conditions: using a future field (FF) probe by itself (high-salience condition) or the same probe accompanied by identical distractors (low-salience condition). The distracting stimuli were positioned so that they would not evoke visual responses in the neuron. Data from an example neuron shows the result: Shifting RF activity was lower when salience of the probe was lower (i.e., when the distractors were presented). (Modified from Joiner, Cavanaugh, & Wurtz, 2011.) (C) FEF neurons assess visual stability of the scene. Monkeys made a saccade in the presence of a peripheral visual stimulus. After the saccade, the stimulus was always at the center of the postsaccadic RF (circle). Before the saccade, the stimulus could be at a different location (dashed circles), in which case it would step to the constant postsaccadic location during the saccade. As shown for two example neurons at right, reafferent visual responses to the same postsaccadic stimulus varied as a function of whether the stimulus remained stable across the saccade (black traces) or moved during the saccade. The grayscale code corresponds to the step amounts shown in task diagram. FEF neurons were tuned for many transaccadic visual changes, including higher activity for steps versus stable stimuli (top) or higher activity for midrange step sizes (bottom). (Modified from Crapse & Sommer, 2012.) (D) Reafferent responses of FEF neurons predict a monkey's percept of visual stability. Monkeys made scanning saccades in the presence of a peripheral stimulus in the RF that eventually moved during one of the saccades (the "Trigger" saccade). The monkey was required to report when the stimulus moved by making a saccade to it. As shown by the example data, reafferent visual responses predicted whether the monkey reported the stimulus as having moved ("Hit") or reported that it remained stable ("Miss"). sp/s, spikes per second. Unpublished data from Crapse & Sommer (2010).

2010). Reafferent responses to the peripheral stimulus were correlated with the behavioral report, and thus presumably with the percept (see figure 65.4D). In sum, the FEF seems to serve as a comparator of the visual scene across saccades, and this is likely an end result of its CD input, shifting RFs, and contextual modulation. FEF neurons are well positioned for informing the rest of the brain about whether the visual scene is stable or not as saccades are made; whether they do so, however, has yet to be established.

Just as there is a pathway from SC to FEF that is part of a much larger pathway to frontal cortex, there are specific pathways from SC to visual areas of parietal–occipital cortex (see figure 65.1A). While the frontal pathways are relayed by the MD nucleus, the parietal–occipital pathways are relayed by the pulvinar complex. In this section we review recent evidence for a functional pathway from SC to one region of parietal–occipital cortex and discuss its potential contribution to cortical visual processing.

The first attempt to demonstrate a functional circuit from SC to parietal–occipital cortex concentrated on a known anatomical pathway from the superficial visual layers of SC through the PI to the cortical motion processing area MT (see figure 65.1C). While the disynaptic structure of this pathway (SC-to-thalamus-to-cortex) parallels that of the circuit from SC to FEF, there are two key differences. First, the origin of the circuit to parietal–occipital cortex originates in the superficial layer of SC, where neurons respond to visual stimuli but show no burst of activity before saccades. This is in striking contrast to the saccade-related neurons in the intermediate layers of the SC, the origin of the pathway to FEF. Second, the target of the pathway to parietal–occipital cortex, area MT, is very different from FEF. MT is a region specialized for detecting visual motion (Dubner & Zeki, 1971; Maunsell & Van Essen, 1983) and likely for mediating the perception of motion (Britten et al., 1992). It is an area of visual processing with limited, if any, relation to saccade generation. This is in contrast to the FEF, where neuronal activity spans the spectrum from purely visual related to purely saccade related. Hence the frontal and posterior pathways share an anatomical resemblance, but they are strikingly different in function at their origin and termination and thus may play different roles in vision.

In studying the SC–PI–MT pathway, a major challenge was to find the area of the thalamic relay neurons because its location was uncertain from existing anatomical data. That such a pathway may exist had been proposed long ago (Diamond & Hall, 1969), and several anatomical studies had suggested that subregions of PI probably served as its relay station (Cusick et al., 1993; Harting et al., 1980). Other anatomical studies, however, suggested that linkage between the SC-to-PI projection and the PI-to-MT projection was uncertain; that is, a continuous pathway might not exist (Stepniewska, Qi, & Kaas, 2000). As was the case with the circuit through MD (Sommer & Wurtz, 2004a), functional continuity from SC to PI to MT was confirmed with antidromic

and orthodromic stimulation. Using these techniques, PI neurons were shown to receive input from SC, project to MT, or both. The critical PI neurons that both received input from SC and projected to MT were the relay neurons of the pathway. Other neurons with either SC input or MT output may have been relay neurons as well although the experiment failed to identify the second connection. All of the neurons were shown to be concentrated in and around a histologically verified subregion of PI, the medial division (PIm). For our current discussion of pathways to cortex, the key point is that this work established a functional circuit from SC to MT through PI.

This identification of the circuit from SC to PI to MT laid the groundwork for answering two major questions about this pathway: What information does it convey, and how does this information differ from that sent from SC to frontal cortex? The first important observation was that PI neurons likely to be in the circuit (relay and singly connected neurons) did not exhibit bursts of activity before saccades (Berman & Wurtz, 2011). The neurons therefore could not be carrying a CD for saccades. Instead, they invariably had visual responses to small spot stimuli and were essentially like the superficial SC neurons from which the circuit originated. This is reminiscent of the similarity in saccade-related activity between MD relay neurons and neurons in the intermediate layers of the SC (Sommer & Wurtz, 2004a).

Interestingly, many PI neurons showed orthodromic, rather than antidromic, activation from stimulation in MT. Consistent with receiving input from MT, these PI neurons typically exhibited directional selectivity to moving stimuli. Such motion tuning is rare in SC visual neurons. Hence some PI neurons seem to be driven by MT, rather than superficial SC, and may not participate in the pathway up to cortex. They may instead participate in the relay of information between cortical regions.

While neuronal activity in the PI relay area showed visual responses rather than saccade-related bursts, the visual activity of many PI neurons *was* modulated by saccades. This modulation was a suppression of neuronal activity with each saccade. Such an effect in PI is reminiscent of the saccade-related suppression demonstrated previously for some visual neurons in the SC (Goldberg & Wurtz, 1972; Robinson & Wurtz, 1976) as illustrated in figure 65.5. In the SC, the suppression had been shown to result not from proprioception but from a central, CD signal (Richmond & Wurtz, 1980).

Modulation of SC visual superficial layer neurons had been hypothesized to result from input from saccade-related neurons in the intermediate layers (Mohler & Wurtz, 1976). The idea was that presaccadic bursts of

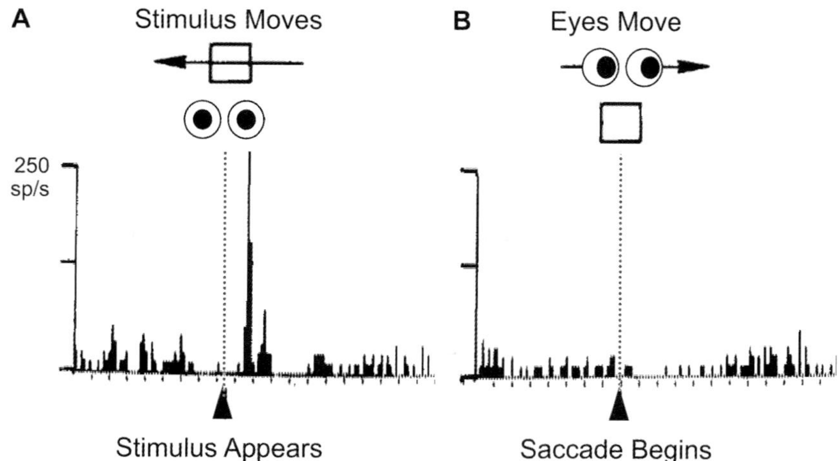

FIGURE 65.5   Suppression of visual responses in superficial superior colliculus (SC) during saccades. (A) Histogram of an example neuronal response as the monkey was fixating. Movement of a visual stimulus across the receptive field produced a sharp increase of visual activity. (B) Activity of the same neuron as the monkey made a saccade in the presence of a stationary stimulus. Because of the eye movement, the image of the stimulus on the retina moved at about the same speed and in the same direction as it did during the moving-stimulus condition of panel A. However, in this condition, there was no visual-related increase in activity; on the contrary, there was a suppression of activity. sp/s, spikes per second. (Modified from Robinson & Wurtz, 1976.)

activity in the intermediate layer acted, via collateral projections, on superficial layer visual neurons, specifically to inhibit them (see figure 65.6A). This was a difficult hypothesis to test because reversible inactivation of the intermediate layers would almost certainly affect the superficial layer as well. This impediment was recently overcome by Phongphanphanee et al. (2011) using a rodent SC slice preparation that allowed them to stimulate and record from a number of neurons at the same time (see figure 65.6B). In a slice through the mouse SC, they identified a putative saccade-related neuron by antidromically activating it from the predorsal bundle (which projects to gaze-orienting centers in the brainstem). They then identified the collateral activation of adjacent neurons caused by this antidromic stimulation and showed that collaterally activated neurons were SC interneurons of the intermediate layers. Finally, by activating those interneurons, they showed that superficial neurons were inhibited. Thus, the combination of establishing the relation of the SC neurons to behavior in the monkey and the interrelation of neurons in a local SC circuit in the rodent strongly supports the hypothesis of an intrinsic CD circuit within the SC, in which its saccade-related neuronal activity inhibits, via interneurons, its visual activity (see figure 65.6C).

Note that these results support the view that this posterior pathway from SC to MT is also related to a CD signal. It is not the CD itself that is conveyed along the pathway, however; rather, it is the *effect* of the CD. The

suppression of the visual response in the SC presumably is conveyed to the pulvinar where the suppression has been observed (Wurtz & Berman, 2012) and then to MT cortex where the suppression is well documented (Bremmer et al., 2009; Ibbotson et al., 2008; Thiele et al., 2002). Hence, the posterior circuit from SC to cortex may convey the suppression of visual activity that results from intracollicular CD, in contrast to the frontal circuit that conveys presaccadic activity that is itself a CD signal.

The next question is whether this suppression along the SC to PI to MT circuit has anything to do with the suppression of visual input that is observed in human psychophysical experiments (Ross et al., 2001). This saccadic suppression contributes to reducing the blur of the visual scene during saccades and thereby reduces the disruption of vision created by saccades. One requirement for any neuronal correlate of this suppression is that the neuronal suppression must start before saccade onset, a hallmark of human perceptual suppression (Latour, 1962). That such neuronal suppression in the SC-to-MT circuit does precede saccade onset was provided by the finding that visual responses to a stimulus flashed just before a saccade were reduced in amplitude relative to responses to stimuli flashed long before a saccade (Wurtz & Berman, 2012). This presaccadic suppression was seen all along the circuit: in SC, PI, and MT. Taken together, all of these experiments imply that the cortical neuronal suppression in MT likely results from relayed, suppressed input from the

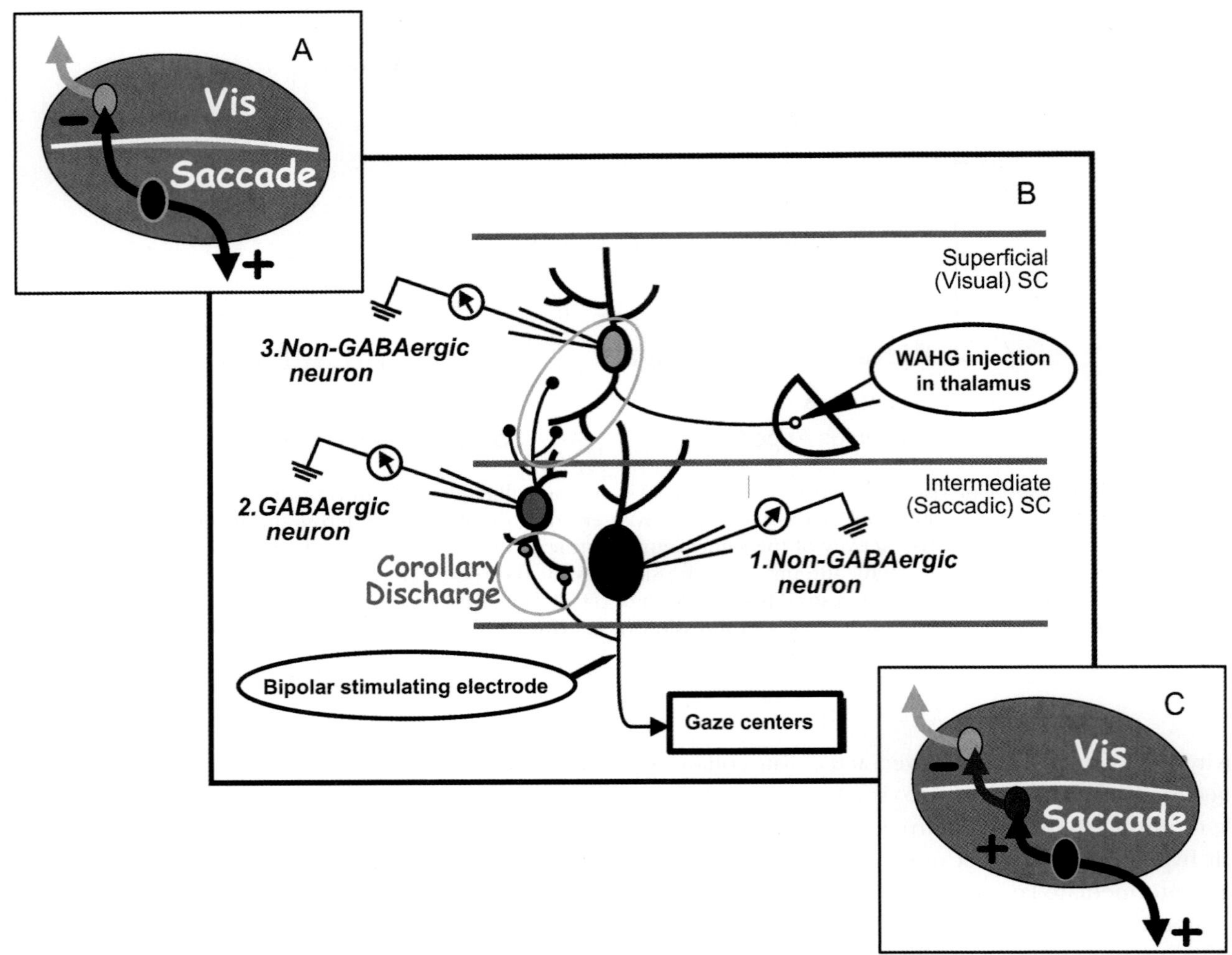

FIGURE 65.6 Microcircuitry of corollary discharge (CD) conveyed within the superior colliculus (SC). (A) Schematic of the intracollicular CD circuit proposed by Robinson and Wurtz (1976). Saccade-related activity in the intermediate layers would inhibit visual activity in the superficial layers. (B) Modified diagram from Phongphanphanee et al. (2011) that summarizes their approach and findings. They used a slice preparation to study functional connectivity of neurons within the rodent SC. Output neurons of the intermediate SC activate GABAergic interneurons via collaterals. The interneurons, in turn, inhibit superficial layer neurons. In intact animals, the collateral would likely carry CD, the superficial layer neurons would likely have visual responses, and the projection targets of the superficial layer neurons would include the pulvinar. The outcome would be saccade-related inhibition of visual activity in the superficial SC. (C) Summary of the putative microcircuit for CD within the SC that incorporates findings from both rodent and monkey work. Vis, visual neurons of the superficial SC; WAHG, wheat germ agglutinin apo–horseradish peroxidase gold.

SC. The next critical step is to determine whether the suppression in MT is modified following inactivation of SC.

CONCLUSION

We have considered the two best understood pathways from the brainstem through thalamus to cerebral cortex that are not part of the classical sensory pathways (see figure 65.7). One runs from the intermediate layers of the SC through the lateral edge of MD to the FEF of frontal cortex. It conveys activity—a CD signal—that encodes the amplitude and direction of the impending saccade. The second pathway runs from the superficial layers of the SC through the PIm region of pulvinar to area MT in parietal cortex. It conveys visual activity that, for many neurons in the pathway, is suppressed during saccades. The saccade-synchronized suppression in parietal cortex is likely carried over from suppression of visual activity within the superficial SC. Intriguingly, the suppression in superficial SC seems to arise from CD signals relayed up from saccade-related neurons of the intermediate SC. Thus the intermediate SC seems to play a dual role, providing CD both to the superficial

    MARC A. SOMMER AND ROBERT H. WURTZ

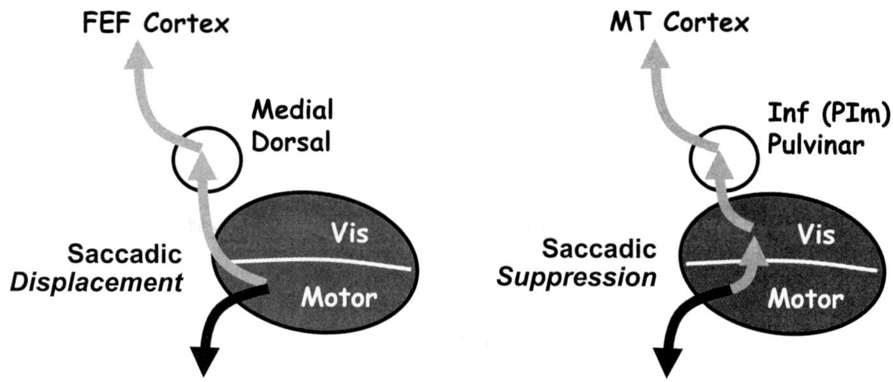

FIGURE 65.7   Summary of the putative visual roles for the two corollary discharge (CD) pathways. Left: The SC–MD–FEF pathway (SC, superior colliculus; MD, medial dorsal nucleus of the thalamus; FEF, frontal eye field) conveys CD of saccades that is produced by the intermediate, motor layers of the SC. In the FEF, the CD signal is used to shift receptive fields just before a saccade, a putative mechanism for the compensation of saccadic displacement of the retinal image across saccades. Right: The SC–PI–MT pathway (PI, inferior nucleus of the pulvinar; MT, middle temporal visual cortex) conveys the effect of CD of saccades. This pathway originates in the superficial, visual SC. The visual responses of neurons there are inhibited by CD from the intermediate, motor SC. These decreased visual responses would be relayed up to MT, where similar, saccade-related decreases in visual responses are found, providing a putative mechanism for the suppression of perceived blur during saccades. Inf, inferior; PIm, medial division of the PI.

SC and, via MD thalamus, to the frontal cortex. Both of the pathways emanating from SC up to cerebral cortex could contribute to our visual stability during saccadic eye movements. The pathway to frontal cortex may compensate for displacements of the retinal image across saccades while the pathway to parietal cortex may suppress the visual smear during saccades. Because the signals at the origin of both pathways, the intermediate SC, are presaccadic, both the compensation and the saccadic suppression are already under way when a saccade begins. This predictive element of the circuitry offers a clear advantage over alternate mechanisms that would rely on proprioceptive signals or reafferent visual input, both of which lag saccade initiation.

We do not know what other types of information are conveyed to cerebral cortex by these two pathways; what has been documented thus far may be just a fraction of the whole. A salient contribution of all of this work, regardless, is to establish a way, at long last, to study transthalamic pathways that are conveying information from place to place within the primate brain. What the pathways carry is hidden until they are probed with microelectrodes during experiments that involve carefully controlled behavioral tasks. The study of CD signals has thus become feasible in primates, as it has been for a long time in the simpler nervous systems of invertebrates (e.g., Delcomyn, 1977; Poulet & Hedwig, 2006; reviewed by Crapse & Sommer, 2008b). By documenting the neuronal information that is related to movement while simultaneously measuring that movement,

and by establishing where the information comes from, where it goes, and what it does to recipient neurons, we can begin to study CD circuits of the primate brain in detail. We have achieved a glimpse at what the pathways from SC might contribute to frontal and parietal–occipital cortex, and using the same experimental strategies, we could begin to study the many other intrinsic circuits of the brain that are relayed by the thalamus.

REFERENCES

Berman, R. A., & Wurtz, R. H. (2011). Signals conveyed in the pulvinar pathway from superior colliculus to cortical area MT. *Journal of Neuroscience, 31*, 373–384.

Bremmer, F., Kubischik, M., Hoffmann, K. P., & Krekelberg, B. (2009). Neural dynamics of saccadic suppression. *Journal of Neuroscience, 29*, 12374–12383.

Britten, K. H., Shadlen, M. N., Newsome, W. T., & Movshon, J. A. (1992). The analysis of visual motion: A comparison of neuronal and psychophysical performance. *Journal of Neuroscience, 12*, 4745–4765.

Crapse, T. B., & Sommer, M. A. (2008a). Corollary discharge circuits in the primate brain. *Current Opinion in Neurobiology, 18*, 552–557.

Crapse, T. B., & Sommer, M. A. (2008b). Corollary discharge across the animal kingdom. *Nature Reviews. Neuroscience, 9*, 587–600.

Crapse, T. B., & Sommer, M. A. (2009). Frontal eye field neurons with spatial representations predicted by their subcortical input. *Journal of Neuroscience, 29*, 5308–5318.

Crapse, T. B., & Sommer, M. A. (2010). Frontal eye field activity predicts performance in a visual stability judgment task. Program No. 280.20. *2010 Neuroscience Meeting Planner*, San Diego, CA: Society for Neuroscience (online).

Crapse, T. B., & Sommer, M. A. (2012). Frontal eye field neurons assess visual stability across saccades. *Journal of Neuroscience, 32*, 2835–2845.

Cusick, C. G., Scripter, J. L., Darensbourg, J. G., & Weber, J. T. (1993). Chemoarchitectonic subdivisions of the visual pulvinar in monkeys and their connectional relations with the middle temporal and rostral dorsolateral visual areas, MT and DLr. *Journal of Comparative Neurology, 336*, 1–30.

Delcomyn, F. (1977). Corollary discharge to cockroach giant interneurons. *Nature, 269*, 160–162.

Diamond, I. T., & Hall, W. C. (1969). Evolution of neocortex. *Science, 164*, 251–262.

Dubner, R., & Zeki, S. M. (1971). Response properties and receptive fields of cells in an anatomically defined region of the superior temporal sulcus in the monkey. *Brain Research, 35*, 528–532.

Duhamel, J. R., Colby, C. L., & Goldberg, M. E. (1992). The updating of the representation of visual space in parietal cortex by intended eye movements. *Science, 255*, 90–92.

Dunn, C. A., Hall, N. J., & Colby, C. L. (2010). Spatial updating in monkey superior colliculus in the absence of the forebrain commissures: Dissociation between superficial and intermediate layers. *Journal of Neurophysiology, 104*, 1267–1285.

Goldberg, M. E., & Wurtz, R. H. (1972). Activity of superior colliculus in behaving monkey: I. Visual receptive fields of single neurons. *Journal of Neurophysiology, 35*, 542–559.

Hallett, P. E., & Lightstone, A. D. (1976). Saccadic eye movements toward stimuli triggered by prior saccades. *Vision Research, 16*, 88–106.

Harting, J. K., Huerta, M. F., Frankfurter, A. J., Strominger, N. L., & Royce, G. J. (1980). Ascending pathways from the monkey superior colliculus: An autoradiographic analysis. *Journal of Comparative Neurology, 192*, 853–882.

Ibbotson, M. R., Crowder, N. A., Cloherty, S. L., Price, N. S., & Mustari, M. J. (2008). Saccadic modulation of neural responses: Possible roles in saccadic suppression, enhancement, and time compression. *Journal of Neuroscience, 28*, 10952–10960.

Joiner, W. M., Cavanaugh, J., & Wurtz, R. H. (2011). Modulation of shifting receptive field activity in frontal eye field by visual salience. *Journal of Neurophysiology, 106*, 1179–1190.

Latour, P. L. (1962). Visual threshold during eye movements. *Vision Research, 2*, 261–262.

Lynch, J. C., Hoover, J. E., & Strick, P. L. (1994). Input to the primate frontal eye field from the substantia nigra, superior colliculus, and dentate nucleus demonstrated by transneuronal transport. *Experimental Brain Research, 100*, 181–186.

Maunsell, J. H., & Van Essen, D. C. (1983). Functional properties of neurons in middle temporal visual area of the macaque monkey: I. Selectivity for stimulus direction, speed, and orientation. *Journal of Neurophysiology, 49*, 1127–1147.

Mohler, C. W., & Wurtz, R. H. (1976). Organization of monkey superior colliculus: Intermediate layer cells discharging before eye movements. *Journal of Neurophysiology, 39*, 722–744.

Nakamura, K., & Colby, C. L. (2002). Updating of the visual representation in monkey striate and extrastriate cortex during saccades. *Proceedings of the National Academy of Sciences of the United States of America, 99*, 4026–4031. doi:10.1073/pnas.052379899.

Phongphanphanee, P., Mizuno, F., Lee, P. H., Yanagawa, Y., Isa, T., & Hall, W. C. (2011). A circuit model for saccadic suppression in the superior colliculus. *Journal of Neuroscience, 31*, 1949–1954.

Poulet, J. F. A., & Hedwig, B. (2006). The cellular basis of a corollary discharge. *Science, 311*, 518–522.

Richmond, B. J., & Wurtz, R. H. (1980). Vision during saccadic eye movements: II. A corollary discharge to monkey superior colliculus. *Journal of Neurophysiology, 43*, 1156–1167.

Robinson, D. L., & Wurtz, R. H. (1976). Use of an extraretinal signal by monkey superior colliculus neurons to distinguish real from self-induced stimulus movement. *Journal of Neurophysiology, 39*, 852–870.

Ross, J., Morrone, M. C., Goldberg, M. E., & Burr, D. C. (2001). Changes in visual perception at the time of saccades. *Trends in Neurosciences, 24*, 113–121.

Sommer, M. A., & Wurtz, R. H. (2002). A pathway in primate brain for internal monitoring of movements. *Science, 296*, 1480–1482.

Sommer, M. A., & Wurtz, R. H. (2003). The dialogue between cerebral cortex and superior colliculus: Implications for saccadic target selection and corollary discharge. In L. M. Chalupa & J. S. Werner (Eds.), *The visual neurosciences* (pp. 1466–1484). Cambridge, MA: MIT Press.

Sommer, M. A., & Wurtz, R. H. (2004a). What the brain stem tells the frontal cortex: I. Oculomotor signals sent from superior colliculus to frontal eye field via mediodorsal thalamus. *Journal of Neurophysiology, 91*, 1381–1402.

Sommer, M. A., & Wurtz, R. H. (2004b). What the brain stem tells the frontal cortex: II. Role of the SC–MD–FEF pathway in corollary discharge. *Journal of Neurophysiology, 91*, 1403–1423.

Sommer, M. A., & Wurtz, R. H. (2006). Influence of the thalamus on spatial visual processing in frontal cortex. *Nature, 444*, 374–377.

Sommer, M. A., & Wurtz, R. H. (2008). Brain circuits for the internal monitoring of movements. *Annual Review of Neuroscience, 31*, 317–338.

Sperry, R. W. (1950). Neural basis of the spontaneous optokinetic response produced by visual inversion. *Journal of Comparative and Physiological Psychology, 43*, 482–489.

Stepniewska, I., Qi, H. X., & Kaas, J. H. (2000). Projections of the superior colliculus to subdivisions of the inferior pulvinar in New World and Old World monkeys. *Visual Neuroscience, 17*, 529–549.

Thiele, A., Henning, P., Kubischik, M., & Hoffmann, K. P. (2002). Neural mechanisms of saccadic suppression. *Science, 295*, 2460–2462.

Umeno, M. M., & Goldberg, M. E. (1997). Spatial processing in the monkey frontal eye field: I. Predictive visual responses. *Journal of Neurophysiology, 78*, 1373–1383.

Umeno, M. M., & Goldberg, M. E. (2001). Spatial processing in the monkey frontal eye field: II. Memory responses. *Journal of Neurophysiology, 86*, 2344–2352.

von Helmholtz, H. (1925). *Helmholtz's treatise on physiological optics* (3rd ed., 1910, J. P. C. Southall, Trans.). New York: Optical Society of America.

von Holst, E., & Mittelstaedt, H. (1950). Das Reafferenzprinzip. Wechselwirkungen zwischen Zentralnervensystem und Peripherie. *Naturwissenschaften, 37*, 464–476.

Walker, M. F., Fitzgibbon, E. J., & Goldberg, M. E. (1995). Neurons in the monkey superior colliculus predict the visual result of impending saccadic eye movements. *Journal of Neurophysiology, 73*, 1988–2003.

Wurtz, R. H. (2008). Neuronal mechanisms of visual stability. *Vision Research, 48*, 2070–2089.

Wurtz, R. H., & Berman, R. A. (2012). One message the pulvinar sends to cortex. *Journal of Vision, 12*(9), 1373. doi:10.1167/12.9.1373.

66 Interaction between Eye Movements and Vision: Perception during Saccades

M. CONCETTA MORRONE

Vision is always clear and stable, despite continual saccadic eye movements that actively reposition our gaze, two to three times a second. Saccades may be made deliberately, but normally they are automatic and pass unnoticed. Not only does the actual movement of the eyes escape notice, so too does the motion of images as they sweep across the retina and the fact that gaze itself has been repositioned. The world seems to stay put. Comparable image motion produced externally, rather than by movements of the observer's own eyes, has an alarming effect on the observer's sense of stability. The problem of visual stability is an old one that has fascinated many scientists, including Descartes, Helmholtz, Mach, and Sherrington, and indeed goes back to the eleventh-century Persian scholar Alhazen: "For if the eye moves in front of visible objects while they are being contemplated, the form of every one of the objects facing the eye … will move on the eyes as the latter moves. But sight has become accustomed to the motion of the objects' forms on its surface when the objects are stationary, and therefore does not judge the objects to be in motion" (Alhazen, 1083). However, only recently have the tools become available to monitor eye movements accurately and to measure their effects qualitatively.

The problem of visual stability during eye movements can be broadly divided into two separate issues: Why do we not perceive the *motion* of the retinal image produced as the eye sweeps over the visual field—and how do we cope dynamically "online" with the continual changes in the retinal image produced by each saccade to construct a stable representation of the world from the successive "snapshots" of each fixation? Although the problem of visual stability is far from solved, tantalizing progress has been made over the last few years. This chapter highlights the progress made in understanding visual perception during large saccades: For small saccade, drift and pursuit eye movements, please refer to chapters 60–65.

SACCADIC SUPPRESSION

Part of the general problem of visual stability is why the fast motion of the retinal image generated by the movement of the eyes completely escapes notice: Comparable wide-field motion generated externally is highly visible and somewhat disturbing (Allison et al., 2010; Burr et al., 1982). It has long been suspected that vision is somehow suppressed during saccades (Holt, 1903), but the nature of the suppression has remained elusive. Now it is clear that the suppression is neither a "central anaesthesia" of the visual system (Holt, 1903), nor a "grey-out" of the world due to fast motion (Campbell & Wurtz, 1978; Dodge, 1900; Woodworth, 1906), as this motion is actually visible, extremely so at low spatial frequencies (Burr & Ross, 1982). What happens is that some stimuli are actively suppressed by saccades while others are not: Stimuli of low spatial frequencies are very difficult to detect if flashed just prior to a saccade while stimuli of high spatial frequencies remain equally visible (Burr et al., 1982; Volkmann et al., 1978). *Equiluminant* stimuli (varying in color but not luminance) are not suppressed during saccades and can even be enhanced (Burr, Morrone, & Ross, 1994), implying that the parvocellular pathway, essential for chromatic discrimination, is left unimpaired while the magnocellular pathway is specifically suppressed (Castet & Masson, 2000; Shioiri & Cavanagh, 1989).

Saccadic suppression follows a specific and very tight time course, illustrated in figure 66.1A (replotted from Diamond, Ross, & Morrone, 2000), very different from saccadic enhancement for equiluminant stimuli (figure 66.1C, replotted from Knoll et al., 2011). Sensitivity for seeing low-spatial-frequency, luminance-modulated stimuli declines 25 ms before saccadic onset, reaching a minimum at the onset of the saccade, then rapidly recovering to normal levels 50 ms afterwards. The suppression effect is multiplicative and is homogeneous at all eccentricities (Knoll et al., 2011), in contrast to what was previously postulated (Mitrani, Mateeff, & Yaki-moff, 1970). Does the suppression result from a central nonvisual *"corollary discharge"* signal (discussed in chapter 66), or could it result simply from visual "masking" effects? This would seem unlikely, as great care was taken to ensure a uniform surround. However, the question is important. In order to be certain that the saccade itself was essential for the suppression, we simulated saccadic eye movements by viewing the stimulus setup through a mirror that could be rotated at saccadic speeds. When the background was uniform, with minimal visual references, the simulated saccades had little or no effect on sensitivity (open symbols of figure 66.1A).

However, that is not to say that under more natural conditions masking does not occur. When the test stimulus is embedded within a textured screen, simulated saccades do decrease contrast sensitivity (see figure 66.1B). Indeed, the maximum suppression is nearly as great as that caused by real saccades and lasts for much longer. This suggests that after the saccade, sensitivity is greater than that expected with comparable motion without the saccade, possibly implying a postsaccadic facilitation. The timing for this postsaccadic facilitation is very similar to that observed for equiluminant stimuli, but for these stimuli the simulated and real saccade produce the same effects (closed and open symbols of figure 66.1C), indicating that the facilitation is related

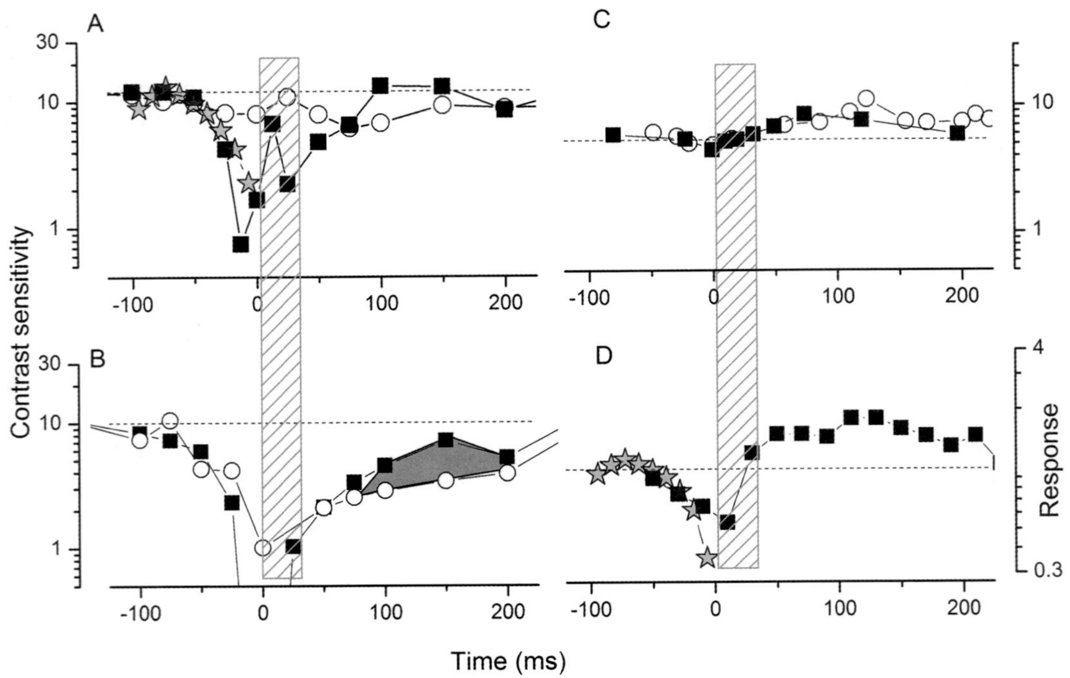

FIGURE 66.1   The effect of saccades on human contrast sensitivity, human blood-oxygen-level dependent (BOLD) response, and firing rate in monkey middle temporal visual cortex (MT; reproduced with permission from Diamond, Ross, & Morrone, 2000; Ibbotson & Cloherty, 2009; Knoll et al., 2011; Vallines & Greenlee, 2006). (A) Filled squares show contrast sensitivity for discriminating (in two-alternative forced choice) the brightness of a brief, low-frequency, luminance-modulated grating patch, as a function of time relative to saccadic onset. The background was of mean luminance, with very few visual referents present. Sensitivity is severely reduced (by more than a log-unit) at saccadic onset. The open circles show measurements made in identical conditions, but instead of making a saccade, a mirror moved at the same speed and amplitude as the saccade; this had very little effect on sensitivity. The stars plot the exponential of the BOLD amplitude of a portion of V1 representing a small patch of grating presented briefly before a saccade. The exponential transformation is applied under the hypothesis that BOLD amplitude is proportional to log contrast sensitivity. (B) As for (A) except that the background was a high-contrast random check pattern. With a structured background, the simulated saccade did reduce visibility, presumably by masking, with the effect lasting longer than it did for a real saccade. The gray shaded area indicates the region where sensitivity was greater during the saccade than in fixation. (C) As for (A) except that the stimulus was a small patch of equiluminant red–green sinusoidal grating presented 5° rightwards and 2° upward of presaccadic fixation. Contrast sensitivity is the inverse of threshold cone contrast. The open symbols report data for the simulated saccade. Note the twofold facilitation in both sets of data. (D) Firing rate of a typical MT neuron in awake monkey in response to stimulation with a brief stimulus (squares) and human V1 BOLD activity (stars), as a function of time relative to saccadic onset. The pattern of the response is similar to the psychophysical results of (A). The enhancement after the saccade may allow for the more rapid recovery from masking during real rather than simulated saccades (difference between filled and open symbols in part B).

948   M. CONCETTA MORRONE

to the spurious retinal motion and not to an active mechanism associated with the saccade.

That real saccades cause a different pattern of luminance-contrast sensitivities from simulated saccades shows that suppression results at least in part from an active, extraretinal signal. Interestingly the amount of suppression varies with age, being much stronger in adolescent children than in adults (Bruno et al., 2006), even though motion perception and masking are largely adult-like by that age (Maurer, Lewis, & Mondloch, 2005; Parrish et al., 2005). This indicates that the mechanisms mediating suppression are still developing into late adolescence. As the saccadic motor system is also not completely mature during adolescence (Fischer, Biscaldi, & Gezeck, 1997), this is further evidence that the extraretinal signal responsible for mediating the saccadic suppression may be linked to the motor system.

Psychophysical studies indicate that saccadic suppression occurs early in the visual system (Burr, Morrone, & Ross, 1994), at or before the site of contrast masking (Watson & Krekelberg, 2011), and before low-level motion processing (Burr, Morgan, & Morrone, 1999). Thilo et al. (2003) addressed this question more directly with a clever electrophysiological technique. Replicating an old study by Riggs, Merton, and Morton (1974), they showed that visual phosphenes produced by electrical stimulation of the eye are suppressed during saccades. However, phosphenes of cortical origin—V1 or V2—generated with the technique of transcranial magnetic stimulation (TMS) were not suppressed. This strongly suggests that saccadic suppression occurs early, before the site of generation of cortical phosphenes, probably within the lateral geniculate nucleus (LGN) or perhaps within V1 itself. A functional magnetic resonance imaging (fMRI; Sylvester, Haynes, & Rees, 2005) study that measured blood-oxygen-level-dependent (BOLD) activity of LGN while subjects made saccades over a field of constant illumination (to avoid the generation of spurious retinal motion) showed a clear suppression in both LGN and V1, reinforcing early suggestions of saccadic suppression in the dark in V1 (Bodis-Wollner, Bucher, & Seelos, 1999; Paus et al., 1995). Interestingly, the amplitude of the BOLD responses in V1 decreased as the stimuli were presented closer to the saccadic onset, following a dynamic similar to that observed psychophysically (see the stars in figure 66.1A, taken from Vallines & Greenlee, 2006), again suggesting an early site of action.

There is also fMRI evidence for postthalamic modulation by saccades. The BOLD response to luminance stimuli is relatively suppressed compared with that to chromatic stimuli during saccades, but the attenuation varies across areas (Kleiser, Seitz, & Krekelberg, 2004),

strong in middle temporal visual cortex (MT)—as expected—but also strong in V4, a cortical area receiving more parvocellular than magnocellular input. That some form of suppression can take place after form analysis is also suggested by the psychophysical result that a suppressed line can influence perception of a form presented postsaccadically (Watson & Krekelberg, 2009). There is also evidence of suppression in higher neural levels, in areas normally associated with attention (Bristow et al., 2005; Kleiser, Seitz, & Krekelberg, 2004). This is interesting, as it could be the suppression of the high-order "attention related" areas that prevents the sense of motion from entering into awareness, causing startle (Allison et al., 2010; Burr et al., 1982).

The electrophysiology of saccadic suppression is more complex. Electrophysiological studies show that the majority of cells in V1 respond vigorously to the movement created by saccades; however, some cells do not respond to saccade-generated motion, but only to real motion in the external world. These cells are the minority, about 10% in V1, 15% in V2, and 40% in V3A (Galletti & Fattori, 2003; Wurtz, 2008). Reppas, Usrey, and Reid (2002) have shown that voluntary saccades induce profound changes in the response of LGN cells, particularly magnocellular cells: Activity is depressed around the time of the saccade, and there is also a larger and long-lasting enhancement after the saccade. There is also clear evidence for strong suppression in the colliculus and pulvinar, which may be important for the suppression of fast motion (see also chapter 65, Wurtz, 2008).

Perhaps the data that can be most readily compared with the psychophysical sensitivities are those in Ibbotson, Crowder, Cloherty, Price and Mustari (2008) and Bremmer et al. (2009), who measured responsiveness of MT/medial superior temporal (MT/MST) cells to a brief stimulus, similar to that used in the psychophysics experiments. Data from a cell (taken from Ibbotson & Cloherty, 2009) are replotted in figure 66.1D. This cell showed a very strong and robust suppression before the start of the saccade, followed by a clear enhancement lasting some 200 ms after the termination of the saccade. While it is difficult to make a quantitative comparison between psychophysical threshold measurements (figure 66.1A) and firing rates of one representative MT cell (figure 66.1D), it is interesting that modulation of MT/MST response follows a similar time course to sensitivity for a brief low-spatial-frequency stimulus. It also follows V1 BOLD activity (stars) (Vallines & Greenlee, 2006), presumably reflecting responses from magnocellular/MT–MST pathways (Bremmer et al., 2009) before the saccade. The very strong postsaccadic enhancement of MT cells, which is present also in total darkness, could explain the relatively higher sensitivity after real saccades compared with after simulated saccades (the difference between open and closed symbols of figure 66.1B). Another interesting result reported for MT neurons is that, in addition to being suppressed, many neurons seem to reverse their preferred direction selectivity (Thiele et al., 2002). This odd behavior could be important in "canceling" the spurious motion information generated by the eye movement, helping to keep the world still. Other areas like ventral intraparietal and lateral intraparietal (LIP) cortex do not show a suppression with similar time course to that observed in MT (Bremmer et al., 2009), pointing again to a specific suppression of the M-pathways and of motion perception (Allison et al., 2010; Burr et al., 1982; Burr, Morrone, & Ross, 1994; Shioiri & Cavanagh, 1989).

To conclude, it is not surprising that saccadic suppression should occur at different levels. Many basic sensory phenomena, such as gain control, do not occur at a single site but at virtually every possible location: photoreceptors, retinal ganglion cells, LGN cells, and cortex (Shapley & Enroth-Cugell, 1984). Indeed, the parallels between saccadic suppression and contrast gain control are strong, suggesting that they may share similar mechanisms. During saccades, the temporal impulse response to luminance, but not to equiluminant stimuli, becomes faster and more transient (Burr & Morrone, 1996). Both LGN (Reppas, Usrey, & Reid, 2002) and MT/MST (Ibbotson et al., 2008) cells show a similar response pattern, with faster and more transient impulse response functions during saccades. These results suggest that saccadic suppression may act by attenuating the contrast gain of the neuronal response, causing a faster impulse response (Shapley & Victor, 1981). Changing contrast gain makes neurons less responsive to low-contrast stimuli, decreasing the effectiveness of the spurious signals caused by the saccade, hence facilitating the recovery to normal sensitivity. That saccadic suppression operates via gain-control mechanisms is consistent with the selective suppression of the magnocellular pathway, as M-cells have much stronger gain control than P-cells (Sclar, Maunsell, & Lennie, 1990). This would certainly be an elegant and economical solution to the problem of saccadic suppression, taking advantage of mechanisms already in place.

The idea that gain control explains both the suppression and the rapid recovery during saccades has been implemented in a model that simulates quantitatively the time course of contrast sensitivity in normal and simulated saccades (Diamond, Ross, & Morrone, 2000). Interestingly, changing response gain is one of the few mechanisms that can explain simultaneously many saccadic suppression properties. It can account for the similarity in sensitivities of real and simulated saccades in the presence of a noise background, but not with a homogeneous background; and it can explain the dependence of suppression from input noise (Watson & Krekelberg, 2011). It can also explain the postsaccadic enhancement, the change of the impulse response function (Burr & Morrone, 1996), and also the change in the strength of masking between brief pre- and postsaccadic stimuli (Burr, Morrone, & Ross, 1994). By operating on gain-control mechanisms, saccadic suppression would serve two important roles: the suppression of image motion, which would otherwise be disturbing, and the rapid return to normal sensitivity after the saccade.

DYNAMIC UPDATING OF INTERNAL SPATIAL MAPS

Besides the (relatively) simple problem of suppressing the motion caused by the fast-moving image on the retina, the brain must also take into account the saccadic movement when determining the instantaneous position of objects in space. Like Alhazen, Helmholtz (1866) also recognized that "the effort of will involved in trying to alter the adjustment of the eyes" could be used to help stabilize perception. Models based on similar ideas of compensation of eye movements were proposed by Sperry (1950) in the 1950s with the concept of corollary discharge and by Von Holst and Mittelstädt (1954) of efference copy: The effort of will of making the eye movements (corollary discharge or efference copy) is subtracted from the retinal signal, to cancel the motion produced by the eye movement and stabilize perception. Now we know that retinal motion signals cannot be easily compensated, given the sophisticated analysis performed by motion detectors. However, there is evidence for the existence of a corollary discharge signal that must be instrumental in maintaining visual stability (for detailed discussion of corollary discharge see chapter 65).

Considerable psychophysical evidence exists for a corollary discharge in humans, going back to the 1960s when Leonard Matin and others reported large transient changes in spatial localization at the time of saccades. When asked to report the position of a target flashed during a saccade, subjects mislocalized it, primarily in the direction of the saccade (Honda, 1991; Mateeff, 1978; Matin & Pearce, 1965). The localization error is typically in the order of half the saccadic size. Later, Mateeff and Honda measured the time course of this effect and showed that the error starts about 50 ms before the saccadic onset and continues well after fixation is regained. The error before the saccadic onset has been taken as an indication of the existence of a slow and sluggish corollary discharge signal that compensates partly for the eye movement: The internal representation of the position and the actual position of the gaze do not match, resulting in errors in the localization of a briefly presented visual target.

We have examined saccadic mislocalization in photopic conditions using equiluminant stimuli (that remain visible during saccades). This approach revealed a bizarre result: At the time of saccades, visual space is not so much shifted in the direction of the saccade but *compressed* toward the saccadic target (Morrone, Ross, & Burr, 1997; Ross, Morrone, & Burr, 1997)—see figure 66.2A and figure 66.2C. Objects flashed at saccadic onset to a range of positions, from close to fixation to positions well beyond the saccadic target, are all perceived at or near the saccadic target. The effect is primarily parallel to the saccade direction (Ross, Morrone, & Burr, 1997) although a small compression is also observed in the orthogonal direction (Kaiser & Lappe, 2004). These results are intriguing because they indicate that the process described mathematically as a simple translation of the internal coordinate system is not plausible: Perhaps the system cannot perform the transformation of space without additional perceptual costs.

The perisaccadic compression of space is so strong that four bars, spread over 20°, are perceived as fused into a single bar (Ross, Morrone, & Burr, 1997). Discrimination of shape (Matsumiya & Uchikawa, 2001) or colors (Lappe et al., 2006) of the bars is still possible, but counting them and perceiving them in separate positions is not. Sometimes the shape or orientation of the flashed object can also change, appearing smaller and more vertical (for horizontal saccades), although these effects have been harder to quantify. The fact that the feature itself is not lost or compressed suggests that the mislocalization occurs at a relatively high level of analysis, after feature extraction.

It has been suggested that saccadic compression occurs only when visual references are present and is absent in the dark (Lappe, Awater, & Krekelberg, 2000). However, subsequent studies (Awater & Lappe, 2006) have shown that this is not necessarily true. In the dark, or in transient dark conditions achieved by a brief blackout immediately following the saccadic onset, compression does occur but can be obscured because in these condition there is also a mislocalization of the saccadic target (Morrone, Ma-Wyatt, & Ross, 2005). When this is taken into account, compression occurs in both light and dark, with and without visual references.

Several studies have shown that visual references per se (like scattered points on the monitor) do not affect compression. However, presenting the same brief stimulus twice perisaccadically, even to different retinal locations, greatly reduces mislocalization (Morrone, Ross, & Burr, 1997; Park, Schlag-Rey, & Schlag, 2003; Pola, 2007; Zhang et al., 2004), and an even stronger reduction of the mislocalization occurs for a flash preceded by a prolonged, continuous or flickering stimulus (Sogo & Osaka, 2001; Watanabe et al., 2005). These findings suggest that the visual system has a mechanism for maintaining position constancy of objects across saccades. On the other hand, relative independence was found between clearly distinct stimuli, such as different shapes presented for different temporal intervals (Hamker, Zirnsak, & Lappe, 2008) or displaced orthogonally to the saccade (Morrone, Ross, & Burr, 1997; Sogo & Osaka, 2002).

Recently, we measured the interaction between stimuli pairs flashed asynchronously in the proximity of a saccade (Cicchini et al., 2013). We found that the localization of a perisaccadic stimulus can be strongly affected by a similar pre- or postsaccadic stimulus, with the two perceived at similar positions even when actually separated by several degrees of visual angle. Figures 66.2B and D show the perceptual localization of probe bars presented at the same physical locations as in figure 66.2C, but followed (after 80 ms) by a similar "reference" bar flashed at screen center (0°): The pattern of probe mislocalization changed dramatically. Under these conditions probe positions closer to the reference bar were attracted by the reference; those closer to the saccadic target and those equidistant (like position 5) had a mixed attraction between the two visual signals. For probe bars presented around 0° (near the screen location of the reference bar) this resulted in virtually veridical localization across the time course. We tested multiple combinations of probe and reference locations and presentation times, and we found that the two bars interacted with each other and were seen together at the same position when they were displayed within a very broad acceptance range, spanning some 20° of horizontal space (the amplitude of the saccade) and some 200–300 ms of time. However, not all stimuli were found to interact: Bars of distinctly different orientation, or displaced orthogonally to saccade direction, have little or no effect on the probe, indicating that the phenomenon takes place late in the visual analysis after basic features have been processed. These two instances of spatial compression, usually toward saccadic target and here toward a postsaccadic reference, may have a common origin: The stronger visual reference (the saccadic target or the additional bar) attracts

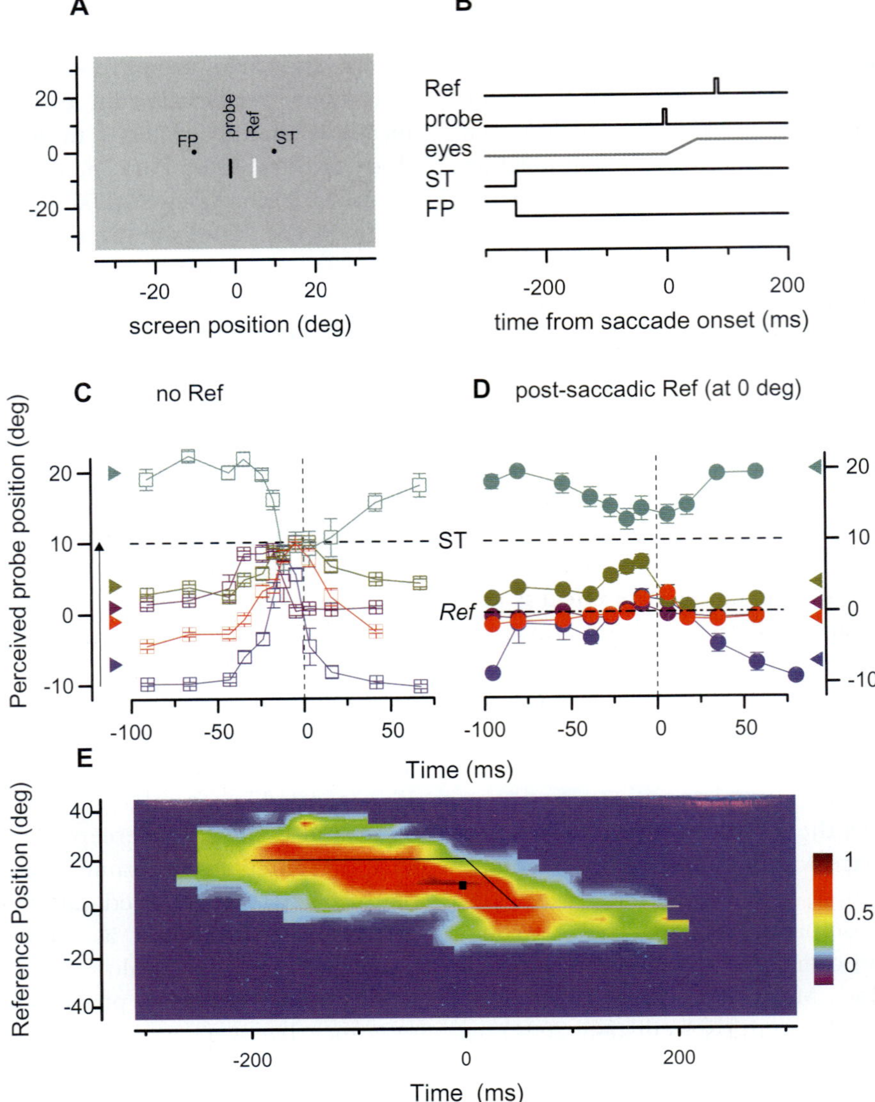

FIGURE 66.2   Effect of saccades on apparent bar position (reproduced from Cicchini et al., 2013). (A) Perceived position of narrow light or black bars, briefly flashed on a red background at various times relative to the onset of a saccade from −10 to +10°. Stimulus display, with the fixation point (FP), the saccade target (ST), and the two flashed bars: reference (Ref) and probe. (B) time course of presentations. (C–D) Perisaccadic compression for a probe bar either flashed alone (C) or followed (80 ms) by another briefly flashed bar at screen center (D). The dashed-dot horizontal line shows the location of the reference stimulus. The black horizontal dashed line marks the location of the saccade target. The different curves refer to different screen positions of the probe (indicated by the color-matched triangles attached to the y-axes). The effect of the saccades (maximal at saccadic onset) is to shift the apparent position of the bar towards the saccadic target, where the eyes land in C (absence of reference bar) and for many bar positions towards the reference position in D. (E) Spatiotemporal map of interactions between a perisaccadic probe bar presented between −20 and 0 ms at screen position 0° (black symbol). The abscissa shows the time of the reference bars and the ordinate the horizontal position of the reference bar in retinal coordinates. The black square shows the location and timing of the probe. The gray and black lines show, respectively, the position of the fovea and of the saccade target. The interaction field extends over 20° of visual angle and over 300 ms across the saccade, but more importantly is elongated in space-time along the trajectory of spurious retinal motion.

the other brief stimuli presented perisaccadically and determines its localization.

Interestingly the spatiotemporal interaction field is quite extensive and, in retinal coordinates, slanted along the trajectory of spurious retinal motion. Figure 66.2E shows a sketch of this perisaccadic interaction that we suggest reflects the action of transiently altered neuronal receptive fields. Visual detectors with spatiotemporally oriented receptive fields are very common in the primate brain. They are fundamental for the computation of motion trajectories, particularly for perceiving the form of the moving object, which would otherwise be subject to heavy motion smear (Burr, 1980). All models of motion perception, from Reichardt's (1957) classic proposal, involve nonlinear combination of systematically delayed signals, which generates a spatiotemporal orientation of receptive fields. A similar strategy may be used to stabilize visual images during saccades, as transiently oriented receptive fields could serve to effectively eliminate the spurious motion signals caused by the movement of the eyes. This profound alteration of neural receptive fields may be crucial to achieve perceptual stability. The perisaccadic extension in space and in time is so large that pre- and postsaccadic information can both activate the same detector, allowing for the integration of images from the two successive fixations. Importantly, only congruent information, concerning similar features, will take place.

But where in the brain do these transiently elongated receptive fields reside? Is there any evidence that a transient craniotopicity is actually implemented physiologically? Electrophysiological studies have reported several transient perisaccadic phenomena. In the LIP, receptive fields change spatial selectivity (Duhamel, Colby, & Goldberg, 1992) just before a monkey makes a saccadic eye movement, anticipating the change in gaze. This is illustrated in figure 66.3A, showing the response of an LIP cell to stimuli flashed within its usual receptive field position (hollow symbols) and within what will became the receptive field after the saccade has been made ("future receptive field," filled symbols). Note that the response in the current receptive field starts to decline, and that in the future receptive field to increase, long before the eye has actually moved to the new fixation. This is termed *predictive remapping*.

This phenomenon occurs not only in LIP but in many other visual areas, including superior colliculus (Walker, Fitzgibbon, & Goldberg, 1995) and area V3 (Nakamura & Colby, 2002), with area V4 showing somewhat different behavior (Tolias et al., 2001). It has even been suggested that a small fraction of neurons in primary visual cortex (V1) might have dynamic updating of receptive fields (Nakamura & Colby, 2002), and predictive remapping can be observed in human cortex using fMRI (Merriam, Genovese, & Colby, 2003, 2007). The origin of the phenomenon has been studied in the frontal eye fields (FEF), and firm evidence demonstrates that it is mediated by a corollary discharge signal, probably originating in the superior colliculus and mediodorsal nucleus of the thalamus (Sommer & Wurtz, 2002, 2006). Deactivation of this nucleus abolishes the predictive updating of the receptive field. The corollary discharge signal arrives nearly 100 ms before

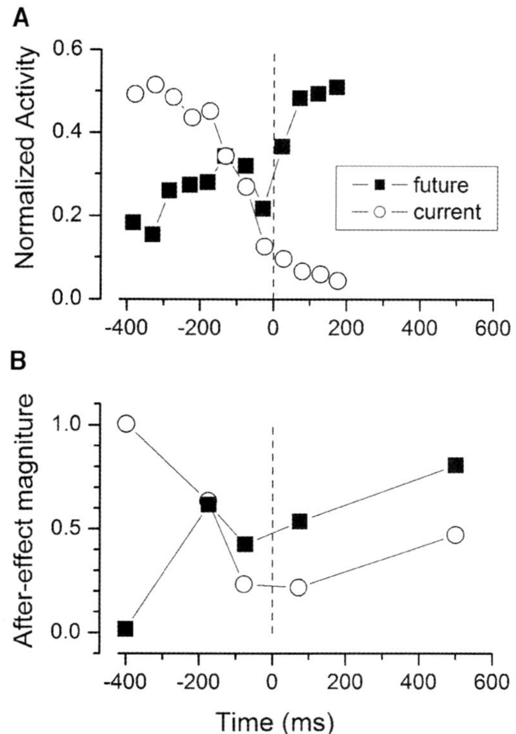

FIGURE 66.3   Predictive remapping in a lateral intraparietal (LIP) cell and human observers. (Reproduced with permission from Kusunoki & Goldberg, 2003; Melcher, 2007; Morrone & Burr, 2009.) (A) The response of a "remapping" cell of area LIP of the macaque around the time of the saccade, to brief stimuli displayed in the "current" (presaccadic) receptive field (open circles) and to stimuli flashed in what will become its receptive field after making the saccade. The response to stimuli in the current receptive field begins to decrease before the eyes actually move. Around the same time, the response in the "future" position begins to increase long before the eyes have actually displaced the receptive field. (B) An experiment showing analogous behavior in human psychophysics. Subjects adapted to a tilted grating, then measured the aftereffect to a grating presented in the same (retinal) position ("current": open circles) or to the position that will correspond to the retinal position of the adaptor after a saccade had been made ("future": filled squares). Long before the saccade, there is no adaptation in the future field and full adaptation in the current field (normalized to unity). Like the cell firing rate, adaptation effects in the current field begin to decrease, and those in the future field to increase, before the eyes have actually moved. Well after the saccade is terminated, the effects do not drop completely to zero, because this position corresponds to the spatiotopic position of the adaptor, and orientation adaptation has a spatiotopic component (Melcher, 2005).

the updating starts in the FEF, indicating the complexity of the reorganization.

In order to build a receptive field oriented along the saccadic direction, it is necessary to introduce delays in the visual processing that vary with position. Interestingly, there is evidence from many studies that early responses at the "remapped" position are indeed delayed (Nakamura & Colby, 2002) and also possibly locked to the time of the saccade execution (Sommer & Wurtz, 2006; Wang, Zhang, & Goldberg, 2008).

Despite these recent efforts, there are several aspects of the remapping phenomenon that remain unclear. For example, between the time when the neuron starts to respond to stimuli in the remapped position and when, after the saccade, it regains its retinotopic specificity, are receptive fields anchored in a transiently craniotopic map as the psychophysical results of figure 66.2 suggest? Do the receptive fields undergo changes in size during the remapping?

The concept of a receptive field oriented in space-time, illustrated in figure 66.2E, provides a simple explanation of how predictive remapping can lead to perceptual stability. The initial presaccadic lobe of the elongated receptive field is consistent with the action of the remapping cells, being shifted in the direction of the saccade (*future receptive field*). However, this "predictive remapping" merely sets the stage for the receptive field to return to its rest position, and it is this return that is important for perceptual stability. The initial shift can be considered a "virtual saccade," before the actual eye movement, at a time when the regular retinotopic cells (which always coexist with predictive remapping cells: Duhamel, Colby, & Goldberg, 1992; Nakamura & Colby, 2002) still respond to a spatially stable visual region. The strategy is to anticipate the problem by shifting the receptive field of detectors in the direction of the saccade, thereby arming the receptive field to return to its resting position, as if loading a spring. The return in position that accompanies the eye movement causes the receptive field to become oriented in space-time, parallel to the spurious retinal motion induced by the saccade, and therefore effectively annulling it. We have no information yet on the dynamics of the return from the future to regular receptive field of visual neurons; no electrophysiological data are yet available. However, the simple concept of a transiently oriented space-time receptive field may not only elucidate the functional role of predicting remapping but also explain many perisaccadic perceptual phenomena, at first sight incongruent. The first is transsaccadic integration and transsaccadic masking.

Clever psychophysical studies also support the existence of transsaccadic integration in humans (Burr & Morrone, 2005; Melcher, 2005, 2007; Turi & Burr, 2012) by examining the spatial selectivity of visual aftereffects or input integration (Melcher & Morrone, 2003). Most aftereffects are spatially selective in retinotopic and/or in spatiotopic coordinates, and the effects can extend for as long as 3 s in time. The degree to which adaptation is spatiotopic varies with the complexity of the signals: Contrast sensitivities (thought to be mediated by primary visual cortex) are primarily retinotopic while more complex signals (like faces or motion coherence sensitivities) are primarily spatiotopic. Adaptation techniques with briefly presented probes (Melcher, 2007) revealed the dynamic of the predictive remapping consequence in perception. Long before the saccade, adaptation is maximal when test and adaptor are presented at the same position, at fixation, with very little adaptation at the position of saccadic target. However, when the test is presented perisaccadically, before and during the eye movement, the maximum adaptation occurs for tests near saccadic target, the position that will correspond to the adapted retina after the eyes have moved (see figure 66.3B). The similarity of the time courses of the adaptation and the response of the LIP neuron (see figure 66.3A) strongly suggests that the brain utilizes the period in which the receptive field regains postsaccadic retinotopic specificity to bridge the perception between the two fixations. That the integration receptive field extends over a large area (see figure 66.2E) is also supported by the recent finding by Zirnsak et al. (2011), who repeated the experiment of Melcher for stimuli positioned in the future receptive field. Interestingly, TMS on human FEF (whose neurons show prominent predictive remapping) reduces transsaccadic memory (Prime, Vesia, & Crawford, 2011) when delivered close to saccadic onset, probably interfering with the organization of the transsaccadic receptive fields.

Recent studies (De Pisapia, Kaunitz, & Melcher, 2010; Hunt & Cavanagh, 2011) show that making saccadic eye movements can actually enhance (rather than degrade) the visibility of a brief perisaccadic stimulus. They presented a brief visual target, followed at various intervals by a "mask," which impedes recognition of the test by "backward masking." The most interesting condition was when test and mask were separated by a brief (12-ms) interval, both presented to stationary eyes, at the same retinal position. When presented 20–30 ms before saccadic onset, visibility of the test improved considerably, particularly for trials where it was perceived as displaced. The results imply that the perisaccadic mislocalization of the test shifts it away from the mask, effectively *demasking* it. In another condition, they used a long test–mask separation with the test and mask straddling the saccade, therefore stimulating distinctly different retinal positions: Yet the masking was strong, suggesting that the representation had been transferred to a spatiotopically corresponding position. Both effects can be simulated by applying filters of the kind illustrated in figure 66.2E.

A well-known perisaccadic phenomenon is that if the saccadic target is displaced after the saccade has been initiated, the displacement (of up to 30% saccade size) is not noted (Bridgeman, Hendry, & Stark, 1975). However, if there is a brief gap in the reappearance of the target in the displaced position, the displacement is immediately apparent (Deubel, Schneider, & Bridgeman, 1996). This observation led to the idea that the system assumes object stability in the absence of contrary information, probably by comparing pre- and postsaccadic positions with some form of short-term memory buffer. These results suggest that the visual system takes advantage of static visual references to help maintain stability across saccades, but the details of how these are selected and stored in some form of memory buffer of limited capacity has been elusive. It is possible that the effect may be mediated by predictive remapping, given that the displacement suppression thresholds are altered by TMS delivered in the perisaccadic epoch over the posterior parietal cortex (Chang & Ro, 2007).

It has recently been argued that the insensitivity to saccadic target displacement (Bridgeman, Hendry, & Stark, 1975) may be explained by optimal sensorimotor integration between the retinal signal and extraretinal corollary discharge signals (Niemeier, Crawford, & Tweed, 2003). At the time of saccades, spatial information about eye position, which is necessary to localize objects in external space, is unreliable. Therefore spatial information during this period is given less weight than information before and after the saccade. The transient distortions of the kind shown in figure 66.2C may also be consistent with statistically optimal, or "Bayesian," integration of information. A recent study has shown how this may be the case by examining audiovisual integration during saccades. Auditory stimuli are usually far more difficult to localize in space than visual stimuli: When vision and sound are in conflict, vision dominates (the "ventriloquist effect") as predicted by optimal integration. However, when visual stimuli are artificially degraded by blurring, audition can dominate (Alais & Burr, 2004), again consistent with optimal integration. As saccades have little effect on auditory space perception (Harris & Lieberman, 1996), they are a useful tool to study saccadic mislocalization. Indeed audiovisual stimuli (bars and beeps presented together in the same spatial position) are mislocalized much less than visual stimuli presented alone, suggesting that visual information is given a low weight during saccades, and this can lead to

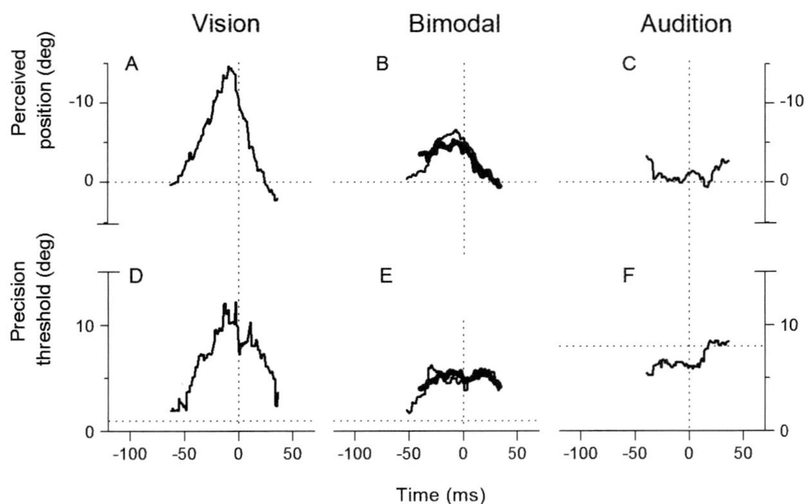

FIGURE 66.4   Illustration of how saccadic mislocalization may result from optimal "Bayesian" fusion. (Reproduced with permission from Binda et al., 2007.) In a two-alternative forced-choice procedure, subjects were asked to report whether a perisaccadic test bar displayed midway between fixation and saccadic target seemed to be located right or left of a presaccadic probe bar (for full details see Binda et al., 2007). Psychometric functions were fitted to these data, to give an estimate of perceived position and also of precision of localization. The upper curves show how perceived position varied with time (relative to saccadic onset). Visual stimuli presented on their own (A) showed the characteristic mislocalization, like that of figure 66.1A. Auditory stimuli, however, were not at all affected by the saccade (C). However, when the sound was played contemporaneously with the bar display, the mislocalization of the bar was reduced (B). The lower curves show the localization thresholds. Again, sound was unaffected by saccades, but the precision of visual localization reduced drastically near saccadic onset. During the bimodal audiovisual presentation, precision improved and was better than either the visual or auditory unimodal localization precision. Indeed this performance, both for perceived position and for precision thresholds, was very close to the Bayesian prediction, indicated by the thick gray line. The dotted horizontal lines indicate performance during fixation.

mislocalization of transient stimuli (Binda et al., 2007). Not only does the idea explain qualitatively the mislocalization, it explains quantitatively the mislocalization of bimodal audiovisual stimuli over the whole time course relative to saccade onset (see figure 66.4).

Binda et al. (2007) go on to develop a Bayesian model of saccadic mislocalization, simply assuming, like Niemeier, Crawford, and Tweed (2003), an increase in noisiness of the eye-position signal at the time of saccades. However this kind of model accounts for only the shift in the direction of the saccade, not the accompanying compression. This would require a further assumption, such as a "prior" or "default rule" for objects to be seen at the fovea. Cicchini and colleagues (2013) advanced the idea of the transient receptive field of the kind illustrated in figure 66.2E, which transiently merges stimuli over an oriented spatiotemporal profile, effectively *implementing* a prior for spatial constancy across saccades.

It is interesting that saccadic compression is positively correlated with peak saccadic velocity (Ostendorf et al., 2007): Individuals with high saccadic velocity show large compression while subjects with slow saccadic velocity show mainly a shift in the saccadic direction (but the effect is not related to the spurious visual motion). This suggests a strong link between perception at the time of saccades and the motor system, probably mediated by the corollary discharge signal (see chapter 65). The receptive field of figure 66.2E is aligned with the saccadic trajectory and may well vary in extension and orientation depending on peak saccadic velocities, explaining this surprising correlation.

Although saccades cause dramatic perceptual localization illusions, when subjects are required to indicate their response by a motor action—secondary saccades or blind hammering—their responses are near veridical (Hallett & Lightstone, 1976a, 1976b; Hansen & Skavenski, 1977, 1985). Other studies (e.g., Bridgeman et al., 1979) also reported that subjects can point accurately to targets that were displaced perisaccadically, even though the subject did not perceive the change in target position. However, a few experiments have failed to replicate the original dissociation between motor accuracy and perceptual error during saccades, reporting localization errors for both tasks (Bockisch & Miller, 1999; Dassonville, Schlag, & Schlag-Rey, 1992, 1995; Honda, 1991; Miller, 1996; Schlag & Schlag-Rey, 1995). Burr, Morrone, and Ross (2001) and Morrone, MaWyatt, and Ross (2005) reported a clear dissociation between verbal reports and blind pointing for saccadic compression. The plot of figure 66.5 shows that briefly flashed stimuli were perceived clearly in false positions, causing the characteristic compression (filled symbols); but when asked to point blindly at the stimuli, with the screen temporally obscured by liquid crystal shutter, observers did so veridically (open symbols).

Interestingly, analogous effects have been reported in audition. Although saccadic eye movements do not affect the localization of tones, saccadic head movements do (Leung, Alais, & Carlile, 2008). Sounds are

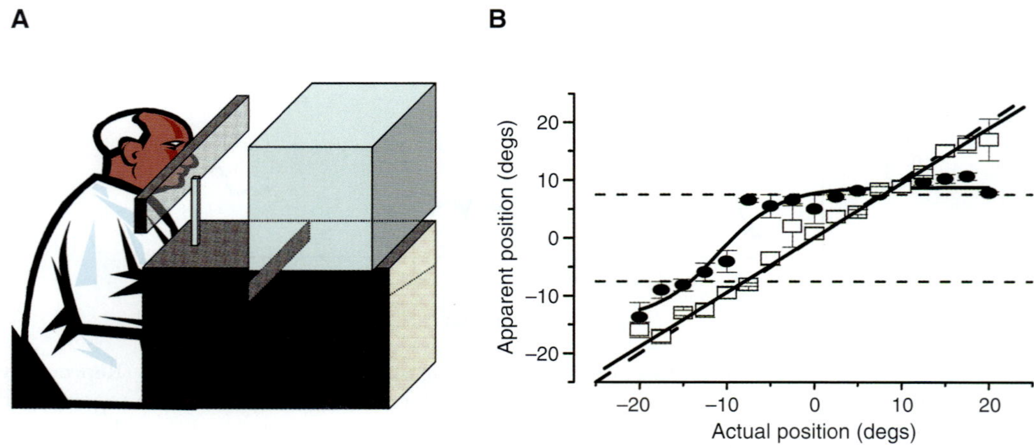

FIGURE 66.5   No spatial compression occurs for rapid motor responses. (Reproduced with permission from Burr, Morrone, & Ross, 2001.) (A) Subjects viewed a CRT monitor through a liquid crystal shutter. On command they made a 15° saccade from −7.5 to +7.5° (dashed lines in B), and a bar was briefly displayed just prior to saccadic onset. Shortly after the saccade was completed, the shutter closed and subjects responded by jabbing at the touch screen with a brisk ballistic movement, the hand hidden from view. (B) The open squares show the results for the jabbing response, for stimuli presented just prior to saccadic onset (−30 < t < 0 ms). The responses are near veridical. The filled circles show results for verbal reports under identical conditions. As shown in figure 66.2, there is a very strong compression, with all stimuli within 10° (degs, degrees) of the saccadic target seen at saccadic target.

compressed toward the end point of the head turn. However, if subjects are ask to point to the apparent sound source (by head turn), the compression disappears, as it does for vision (Burr, Morrone, & Ross, 2001).

However, for visual judgments, introducing clearly visible postsaccadic references under normal lighting conditions causes both verbal report and pointing to show compression. This suggests that vision has access to two maps, one subject to distortion where the receptive fields transiently change form and the other not: The motor map shows no compression except when visual references remain in view for a substantial time after saccade, indicating that these maps are updated postsaccadically, while for perceptual judgments the updating occurs before and during the actual saccade. Both maps contribute to determining the weight given to each map. Perhaps the popular distinction between conscious perception and action (Goodale & Milner, 1992; Trevarthen, 1968) is at best an oversimplification.

Space and time are generally studied separately and thought of as separate and independent dimensions. However, we have observed that not only space but also time undergoes severe transient distortions at the time of saccades: Objects become compressed toward the saccadic target (Ross, Morrone, & Burr, 1997), and perceived temporal durations are severely shrunk (Morrone, Ross, & Burr, 2005). When asked to compare the perceived duration of a temporal interval presented around the time of a saccade with one presented 2 s afterwards, subjects judged it much shorter, about half the duration (see figure 66.6B). Interestingly, the precision of the judgment is higher perisaccadically than at fixation, obeying the general rule of constant Weber fraction (Morrone, Ross, & Burr, 2005). Again the time course of the temporal distortion is quite tight and, after taking into account the duration of the stimuli and the effect of contraction, similar to that of the spatial compression. As the transient changes both in space and in time follow very similar dynamics (compare figure 66.2C with figure 66.6B—continuous curve), they may well be manifestations of a common neural cause, possibly a distortion in the space-time metric induced by the transient orientation of the neuronal receptive field as shown in figure 66.2E.

Alteration of the sense of time can also be more dramatic: Saccades can even cause an inversion of the perceived temporal sequence (Morrone, Ross, & Burr, 2005), anticipating or delaying temporal process.

Binda et al. (2009) showed that the perceived time at saccadic onset (measured by matching an auditory tone) is delayed by about 100 ms, while 50 ms before saccade the latency is reduced by only 20 ms, a small

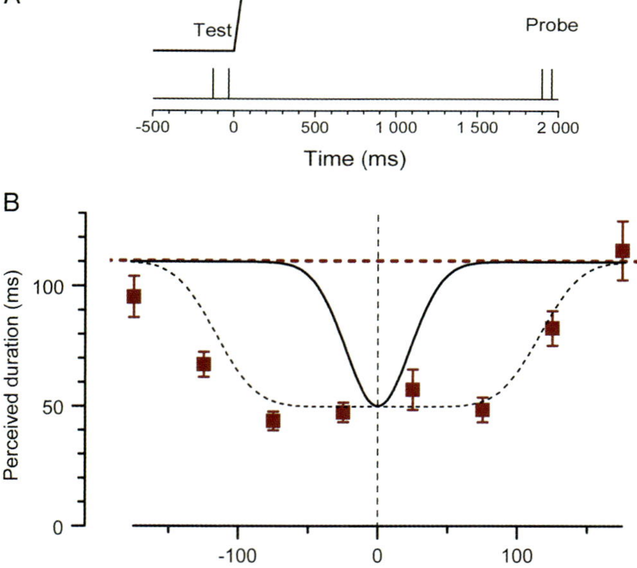

FIGURE 66.6 Time is also compressed during saccades. (Reproduced with permission from Burr et al., 2010.) (A) The subject was asked to compare the duration of the interval of two test flashes (separated by 100 ms) with a postsaccadic probe of variable duration that appeared 2 s later. (B) The apparent duration was then calculated from psychometric functions. Around the time of saccadic onset, apparent duration was about half the physical duration. The dashed line shows the duration match during fixation. Note that the predicted time course is much more tightly tuned than the data (continuous curve), because the data were collected with a broad temporal stimulus (100 ms long) that necessarily blurs the effects over time. A good approximation is obtained by deconvolving the dashed curves with a temporal interval given by the physical stimulus separation and a broad temporal impulse response such as that illustrated in figure 66.2E.

effect, but one sufficient to produce an inversion (Morrone, Ross, & Burr, 2005). This result is consistent with the fact that during remapping the latencies of neurons in areas MT and MST are shorter in response to real than simulated saccades (Price et al., 2005). On the other hand, the postsaccadic delay of the visual stimulus is consistent with the delay of the response of the future receptive field observed by Nakamura and Colby (2002) and Wang, Zhang, and Goldberg (2008). In addition, the delay also indicates that stimuli presented at saccadic onset are coincident with stimuli presented soon after saccades, facilitating the interpretation of their position in the postsaccadic coordinate system and decoding after that the saccade is complete, in a form of *postdiction* (Eagleman & Sejnowski, 2000). Again this is consistent with the long temporal extension of the perisaccadic receptive field (figure 66.2C), which extends well after the saccade is completed.

CONCLUSION

Seeing is usually believing: For about two thirds of our waking lives, we perceive objects where vision tells us they are—which, more often than not, coincides with their actual position. In the remaining time, the visual system sends us erroneous spatial information, presumably because it is engaged in correcting the troublesome consequences of eye movements on retinal afferences. When this happens, we disbelieve visual information; if available, spatial cues from other senses become dominant; if we have to act, we use the robust postsaccadic representation without attempting to update it transsaccadically. If vision is the only signal available, we deform our concept of space and of time to make sense of it and do not miss visual information for more than one third of our waking time.

ACKNOWLEDGMENTS

Supported by European Union PF7 IDEAS of European Research Council: Space, Time and Number in the Brain (STANIB) and by PRIN 2010 from Italian Ministry of University and Research (MIUR). I am grateful for comments from my collaborator and beloved husband, David Burr.

REFERENCES

Alais, D., & Burr, D. (2004). The ventriloquist effect results from near-optimal bimodal integration. *Current Biology, 14,* 257–262.

Alhazen, I. (1989/1083). Book of optics. In A. I. Sabra (Ed.), *The optics of Ibn al-Haytham* (Sabra, A. I., Trans.). London: Warburg Institute.

Allison, R. S., Schumacher, J., Sadr, S., & Herpers, R. (2010). Apparent motion during saccadic suppression periods. *Experimental Brain Research, 202,* 155–169.

Awater, H., & Lappe, M. (2006). Mislocalization of perceived saccade target position induced by perisaccadic visual stimulation. *Journal of Neuroscience, 26,* 12–20.

Binda, P., Bruno, A., Burr, D. C., & Morrone, M. C. (2007). Fusion of visual and auditory stimuli during saccades: A Bayesian explanation for perisaccadic distortions. *Journal of Neuroscience, 27,* 8525–8532.

Binda, P., Cicchini, G. M., Burr, D. C., & Morrone, M. C. (2009). Spatiotemporal distortions of visual perception at the time of saccades. *Journal of Neuroscience, 29,* 13147–13157.

Bockisch, C., & Miller, J. (1999). Different motor systems use similar damped extraretinal eye position information. *Vision Research, 39,* 1025–1038. doi:10.1016/S0042-6989(98)00205-3.

Bodis-Wollner, I., Bucher, S. F., & Seelos, K. C. (1999). Cortical activation patterns during voluntary blinks and voluntary saccades. *Neurology, 53,* 1800–1805.

Bremmer, F., Kubischik, M., Hoffmann, K. P., & Krekelberg, B. (2009). Neural dynamics of saccadic suppression. *Journal of Neuroscience, 29,* 12374–12383.

Bridgeman, B., Hendry, D., & Stark, L. (1975). Failure to detect displacement of visual world during saccadic eye movements. *Vision Research, 15,* 719–722.

Bridgeman, B., Lewis, S., Heit, G., & Nagle, M. (1979). Relation between cognitive and motor-oriented systems of visual position perception. *Journal of Experimental Psychology. Human Perception and Performance, 5,* 692–700.

Bristow, D., Haynes, J. D., Sylvester, R., Frith, C. D., & Rees, G. (2005). Blinking suppresses the neural response to unchanging retinal stimulation. *Current Biology, 15,* 1296–1300.

Bruno, A., Brambati, S. M., Perani, D., & Morrone, M. C. (2006). Development of saccadic suppression in children. *Journal of Neurophysiology, 96,* 1011–1017.

Burr, D. (1980). Motion smear. *Nature, 284,* 164–165.

Burr, D. C., Holt, J., Johnstone, J. R., & Ross, J. (1982). Selective depression of motion sensitivity during saccades. *Journal of Physiology, 333,* 1–15.

Burr, D. C., Morgan, M. J., & Morrone, M. C. (1999). Saccadic suppression precedes visual motion analysis. *Current Biology, 9,* 1207–1209.

Burr, D. C., & Morrone, M. C. (1996). Temporal impulse response functions for luminance and colour during saccades. *Vision Research, 36,* 2069–2078. doi:10.1016/0042-6989(95)00282-0.

Burr, D., & Morrone, M. C. (2005). Eye movements: Building a stable world from glance to glance. *Current Biology, 15,* R839–R840. doi:10.1016/j.cub.2005.10.003.

Burr, D. C., Morrone, M. C., & Ross, J. (1994). Selective suppression of the magnocellular visual pathway during saccadic eye movements. *Nature, 371,* 511–513.

Burr, D. C., Morrone, M. C., & Ross, J. (2001). Separate visual representations for perception and action revealed by saccadic eye movements. *Current Biology, 11,* 798–802.

Burr, D. C., & Ross, J. (1982). Contrast sensitivity at high velocities. *Vision Research, 23,* 3567–3569. doi:10.1016/0042-6989(82)90196-1.

Burr, D. C., Ross, J., Binda, P., & Morrone, M. C. (2010). Saccades compress space, time and number. *Trends in Cognitive Sciences, 14,* 528–533. doi:10.1016/j.tics.2010.09.005.

Campbell, F. W., & Wurtz, R. H. (1978). Saccadic ommision: Why we do not see a greyout during a saccadic eye movement. *Vision Research, 18,* 1297–1303. doi:10.1016/0042-6989(78)90219-5.

Castet, E., & Masson, G. S. (2000). Motion perception during saccadic eye movements. *Nature Neuroscience, 3,* 177–183.

Chang, E., & Ro, T. (2007). Maintenance of visual stability in the human posterior parietal cortex. *Journal of Cognitive Neuroscience, 19,* 266–274.

Cicchini, M., Binda, P., Burr, D., & Morrone, M. (2013). Transient spatiotopic integration across saccadic eye movements mediates visual stability. *Journal of Neurophysiology, 109,* 1117–1125.

Dassonville, P., Schlag, J., & Schlag-Rey, M. (1992). Oculomotor localization relies on a damped representation of saccadic eye movement displacement in human and nonhuman primates. *Visual Neuroscience, 9,* 261–269.

Dassonville, P., Schlag, J., & Schlag-Rey, M. (1995). The use of ego-centric and exocentric location cues in saccadic programming. *Vision Research, 35,* 2191–2199.

De Pisapia, N., Kaunitz, L., & Melcher, D. (2010). Backward masking and unmasking across saccadic eye movements. *Current Biology, 20*, 613–617.

Deubel, H., Schneider, W. X., & Bridgeman, B. (1996). Postsaccadic target blanking prevents saccadic suppression of image displacement. *Vision Research, 36*, 985–996.

Diamond, M. R., Ross, J., & Morrone, M. C. (2000). Extraretinal control of saccadic suppression. *Journal of Neuroscience, 20*, 3442–3448.

Dodge, R. (1900). Visual perception during eye movements. *Psychological Review, 7*, 454–465.

Duhamel, J. R., Colby, C. L., & Goldberg, M. E. (1992). The updating of the representation of visual space in parietal cortex by intended eye movements. *Science, 255*, 90–92.

Eagleman, D. M., & Sejnowski, T. J. (2000). Motion integration and postdiction in visual awareness. *Science, 287*, 2036–2038.

Fischer, B., Biscaldi, M., & Gezeck, S. (1997). On the development of voluntary and reflexive components in human saccade generation. *Brain Research, 754*, 285–297.

Galletti, C., & Fattori, P. (2003). Neuronal mechanisms for detection of motion in the field of view. *Neuropsychologia, 41*, 1717–1727.

Goodale, M. A., & Milner, A. D. (1992). Separate pathways for perception and action. *Trends in Neurosciences, 15*, 20–25. doi:10.1016/0166-2236(92)90344-8.

Hallett, P. E., & Lightstone, A. D. (1976a). Saccadic eye movements towards stimuli triggered by prior saccades. *Vision Research, 16*, 99–106.

Hallett, P. E., & Lightstone, D. (1976b). Saccadic eye movements to flashed targets. *Vision Research, 16*, 107–114.

Hamker, F. H., Zirnsak, M., & Lappe, M. (2008). About the influence of post-saccadic mechanisms for visual stability on peri-saccadic compression of object location. *Journal of Vision, 8*(14), 1, 1–13. doi:10.1167/8.14.1.

Hansen, R. M., & Skavenski, A. A. (1977). Accuracy of eye position information for motor control. *Vision Research, 17*, 919–926.

Hansen, R. M., & Skavenski, A. A. (1985). Accuracy of spatial locations near the time of saccadic eye movments. *Vision Research, 25*, 1077–1082.

Harris, L. R., & Lieberman, L. (1996). Auditory stimulus detection is not suppressed during saccadic eye movements. *Perception, 25*, 999–1004.

Helmholtz, H. v. (1866/1963). Handbuch der Physiologischen Optik. In J. P. C. Southall (Ed.), *A treatise on physiological optics*. New York: Dover.

Holt, E. B. (1903). Eye movements and central anaesthesia. *Psychological Review, 4*, 3–45.

Honda, H. (1991). The time courses of visual mislocalization and of extra-retinal eye position signals at the time of vertical saccades. *Vision Research, 31*, 1915–1921.

Hunt, A. R., & Cavanagh, P. (2011). Remapped visual masking. *Journal of Vision, 11*(1), 13. doi:10.1167/11.1.13.

Ibbotson, M. R., & Cloherty, S. L. (2009). Visual perception: Saccadic omission—Suppression or temporal masking? [Comment]. *Current Biology, 19*, R493–R496.

Ibbotson, M. R., Crowder, N. A., Cloherty, S. L., Price, N. S., & Mustari, M. J. (2008). Saccadic modulation of neural responses: Possible roles in saccadic suppression, enhancement, and time compression. *Journal of Neuroscience, 28*, 10952–10960.

Kaiser, M., & Lappe, M. (2004). Perisaccadic mislocalization orthogonal to saccade direction. *Neuron, 41*, 293–300.

Kleiser, R., Seitz, R. J., & Krekelberg, B. (2004). Neural correlates of saccadic suppression in humans. *Current Biology, 14*, 386–390.

Knoll, J., Binda, P., Morrone, M. C., & Bremmer, F. (2011). Spatiotemporal profile of peri-saccadic contrast sensitivity. *Journal of Vision, 11*(14). doi:10.1167/11.14.15.

Kusunoki, M., & Goldberg, M. E. (2003). The time course of perisaccadic receptive field shifts in the lateral intraparietal area of the monkey. *Journal of Neurophysiology, 89*, 1519–1527.

Lappe, M., Awater, H., & Krekelberg, B. (2000). Postsaccadic visual references generate presaccadic compression of space. *Nature, 403*, 892–895.

Lappe, M., Kuhlmann, S., Oerke, B., & Kaiser, M. (2006). The fate of object features during perisaccadic mislocalization. *Journal of Vision, 6*(11), 1282–1293. doi:10.1167/6.11.11.

Leung, J., Alais, D., & Carlile, S. (2008). Compression of auditory space during rapid head turns. *Proceedings of the National Academy of Sciences of the United States of America, 105*, 6492–6497. doi:10.1073/pnas.0710837105.

Mateeff, S. (1978). Saccadic eye movements and localization of visual stimuli. *Perception & Psychophysics, 24*, 215–224.

Matin, L., & Pearce, D. G. (1965). Visual perception of direction for stimuli flashed during voluntary saccadic eye movments. *Science, 148*, 1485–1487.

Matsumiya, K., & Uchikawa, K. (2001). Apparent size of an object remains uncompressed during presaccadic compression of visual space. *Vision Research, 41*, 3039–3050. doi:10.1016/S0042-6989(01)00174-2.

Maurer, D., Lewis, T. L., & Mondloch, C. J. (2005). Missing sights: Consequences for visual cognitive development. *Trends in Cognitive Sciences, 9*, 144–151. doi:10.1016/j.tics.2005.01.006.

Melcher, D. (2005). Spatiotopic transfer of visual-form adaptation across saccadic eye movements. *Current Biology, 15*, 1745–1748. doi:10.1016/j.cub.2005.08.044.

Melcher, D. (2007). Predictive remapping of visual features precedes saccadic eye movements. *Nature Neuroscience, 10*, 903–907.

Melcher, D., & Morrone, M. C. (2003). Spatiotopic temporal integration of visual motion across saccadic eye movements. *Nature Neuroscience, 6*, 877–881.

Merriam, E. P., Genovese, C. R., & Colby, C. L. (2003). Spatial updating in human parietal cortex. *Neuron, 39*, 361–373.

Merriam, E. P., Genovese, C. R., & Colby, C. L. (2007). Remapping in human visual cortex. *Journal of Neurophysiology, 97*, 1738–1755.

Miller, J. (1996). Egocentric localization of a perisaccadic flash by manual pointing. *Vision Research, 36*, 837–851.

Mitrani, L., Mateeff, S., & Yakimoff, N. (1970). Temporal and spatial characteristics of visual suppression during voluntary saccadic eye movement. *Vision Research, 10*, 417–422.

Morrone, C., & Burr, D. (2009). Visual stability during saccadic eye movements. In M. Gazzaniga (Ed.), *The cognitive neurosciences* (4th ed., pp. 511–523). Cambridge, MA: MIT Press.

Morrone, M. C., Ma-Wyatt, A., & Ross, J. (2005). Seeing and ballistic pointing at perisaccadic targets. *Journal of Vision, 5*(9), 741–754. doi:10.1167/5.9.7.

Morrone, M. C., Ross, J., & Burr, D. (2005). Saccadic eye movements cause compression of time as well as space. *Nature Neuroscience, 8*, 950–954.

Morrone, M. C., Ross, J., & Burr, D. C. (1997). Apparent position of visual targets during real and simulated saccadic eye movements. *Journal of Neuroscience, 17*, 7941–7953.

Nakamura, K., & Colby, C. L. (2002). Updating of the visual representation in monkey striate and extrastriate cortex during saccades. *Proceedings of the National Academy of Sciences of the United States of America, 99*, 4026–4031. doi:10.1073/pnas.052379899.

Niemeier, M., Crawford, J. D., & Tweed, D. B. (2003). Optimal transsaccadic integration explains distorted spatial perception. *Nature, 422*, 76–80.

Ostendorf, F., Fischer, C., Finke, C., & Ploner, C. J. (2007). Perisaccadic compression correlates with saccadic peak velocity: Differential association of eye movement dynamics with perceptual mislocalization patterns. *Journal of Neuroscience, 27*, 7559–7563.

Park, J., Schlag-Rey, M., & Schlag, J. (2003). Spatial localization precedes temporal determination in visual perception. *Vision Research, 43*, 1667–1674. doi:10.1016/S0042-6989(03)00217-7.

Parrish, E. E., Giaschi, D. E., Boden, C., & Dougherty, R. (2005). The maturation of form and motion perception in school age children. *Vision Research, 45*, 827–837.

Paus, T., Marrett, S., Worsley, K. J., & Evans, A. C. (1995). Extraretinal modulation of cerebral blood flow in the human visual cortex: Implications for saccadic suppression. *Journal of Neurophysiology, 74*, 2179–2183.

Pola, J. (2007). A model of the mechanism for the perceived location of a single flash and two successive flashes presented around the time of a saccade. *Vision Research, 47*, 2798–2813. doi:10.1016/j.visres.2007.07.005.

Price, N. S., Ibbotson, M. R., Ono, S., & Mustari, M. J. (2005). Rapid processing of retinal slip during saccades in macaque area MT. *Journal of Neurophysiology, 94*, 235–246.

Prime, S. L., Vesia, M., & Crawford, J. D. (2011). Cortical mechanisms for trans-saccadic memory and integration of multiple object features. *Philosophical Transactions of the Royal Society of London. Series B, Biological Sciences, 366*, 540–553.

Reichardt, W. (1957). Autokorrelationsauswertung als Funktionsprinzip des Zentralnervensystems. *Zeitschrift für Naturforschung, 12b*, 447–457.

Reppas, J. B., Usrey, W. M., & Reid, R. C. (2002). Saccadic eye movements modulate visual responses in the lateral geniculate nucleus. *Neuron, 35*, 961–974.

Riggs, L. A., Merton, P. A., & Morton, H. B. (1974). Suppression of visual phosphenes during saccadic eye movements. *Vision Research, 14*, 997–1011.

Ross, J., Morrone, M. C., & Burr, D. C. (1997). Compression of visual space before saccades. *Nature, 384*, 598–601.

Schlag, J., & Schlag-Rey, M. (1995). Illusory localization of stimuli flashed in the dark before saccades. *Vision Research, 35*, 2347–2357. doi:10.1016/0042-6989(95)00021-Q.

Sclar, G., Maunsell, J. H., & Lennie, P. (1990). Coding of image contrast in central visual pathways of the macaque monkey. *Vision Research, 30*, 1–10. doi:10.1016/0042-6989(90)90123-3.

Shapley, R., & Enroth-Cugell, C. (1984). Visual adaptation and retinal gain controls. In G. J. Chader & N. N. Osborne (Eds.), *Progress in retinal research* (Vol. 3, pp. 263–346). Oxford, England: Pergamon Press.

Shapley, R. M., & Victor, J. D. (1981). How the contrast gain control modifies the frequency responses of cat retinal ganglion cells. *Journal of Physiology, 318*, 161–179.

Shioiri, S., & Cavanagh, P. (1989). Saccadic suppression of low-level motion. *Vision Research, 29*, 915–928.

Sogo, H., & Osaka, N. (2001). Perception of relation of stimuli locations successively flashed before saccade. *Vision Research, 41*, 935–942.

Sogo, H., & Osaka, N. (2002). Effects of inter-stimulus interval on perceived locations of successively flashed perisaccadic stimuli. *Vision Research, 42*, 899–908.

Sommer, M. A., & Wurtz, R. H. (2002). A pathway in primate brain for internal monitoring of movements. *Science, 296*, 1480–1482.

Sommer, M. A., & Wurtz, R. H. (2006). Influence of the thalamus on spatial visual processing in frontal cortex. *Nature, 444*, 374–377.

Sperry, R. W. (1950). Neural basis of the spontaneous optokinetic response produced by visual inversion. *Journal of Comparative and Physiological Psychology, 43*, 482–489.

Sylvester, R., Haynes, J. D., & Rees, G. (2005). Saccades differentially modulate human LGN and V1 responses in the presence and absence of visual stimulation. *Current Biology, 15*, 37–41.

Thiele, A., Henning, P., Kubischik, M., & Hoffmann, K. P. (2002). Neural mechanisms of saccadic suppression. *Science, 295*, 2460–2462.

Thilo, K. V., Santoro, L., Walsh, V., & Blakemore, C. (2003). The site of saccadic suppression. *Nature Neuroscience, 7*, 13–14.

Tolias, A. S., Moore, T., Smirnakis, S. M., Tehovnik, E. J., Siapas, A. G., & Schiller, P. H. (2001). Eye movements modulate visual receptive fields of V4 neurons. *Neuron, 29*, 757–767.

Trevarthen, C. B. (1968). Two mechanisms of vision in primates. *Psychologische Forschung, 31*, 299–348.

Turi, M., & Burr, D. C. (2012). Spatiotopic perceptual maps in humans: Evidence from motion adaptation. *Proceedings of the Royal Society B (London), 279*, 3091–3097.

Vallines, I., & Greenlee, M. W. (2006). Saccadic suppression of retinotopically localized blood oxygen level–dependent responses in human primary visual area V1. *Journal of Neuroscience, 26*, 5965–5969.

Volkmann, F. C., Riggs, L. A., White, K. D., & Moore, R. K. (1978). Contrast sensitivity during saccadic eye movements. *Vision Research, 18*, 1193–1199.

Von Holst, E., & Mittelstädt, H. (1954). Das Reafferenzprinzip. *Naturwissenschaften, 37*, 464–476.

Walker, M. F., Fitzgibbon, J., & Goldberg, M. E. (1995). Neurons of the monkey superior colliculus predict the visual result of impending saccadic eye movements. *Journal of Neurophysiology, 73*, 1988–2003.

Wang, X., Zhang, M., & Goldberg, M. E. (2008). *Perisaccadic elongation of receptive fields in the lateral intraparietal area (LIP).* Society for Neuroscience (Abstract), 855, 17/F23.

Watanabe, J., Noritake, A., Maeda, T., Tachi, S., & Nishida, S. (2005). Perisaccadic perception of continuous flickers. *Vision Research, 45*, 413–430.

Watson, T., & Krekelberg, B. (2011). An equivalent noise investigation of saccadic suppression. *Journal of Neuroscience, 31*, 6535–6541.

Watson, T. L., & Krekelberg, B. (2009). The relationship between saccadic suppression and perceptual stability. *Current Biology, 19*, 1040–1043.

Woodworth, R. S. (1906). Vision and localization during eye movements. *Psychological Bulletin, 3*, 68–70.

Wurtz, R. H. (2008). Neuronal mechanisms of visual stability. *Vision Research, 48*, 2070–2089. doi:10.1016/j.visres.2008.03.021.

Zhang, Z. L., Cantor, C., Ghose, T., & Schor, C. M. (2004). Temporal aspects of spatial interactions affecting stereo-matching solutions. *Vision Research, 44*, 3183–3192. doi:10.1016/j.visres.2004.07.024.

Zirnsak, M., Gerhards, R. G., Kiani, R., Lappe, M., & Hamker, F. H. (2011). Anticipatory saccade target processing and the presaccadic transfer of visual features. *Journal of Neuroscience, 31*, 17887–17891.

67 Plasticity of Eye Movement Control

PABLO M. BLAZQUEZ AND ANGEL M. PASTOR

When studying ocular motility, our eyes can be represented as spheres with zero mass whose angular position and movement is controlled by the action of three pairs of muscles. Importantly, because each muscle always moves the eyes in the same direction, eye movements can be readily translated into muscle activity, which is not common for most motor systems. For example, arm movements are controlled by the combined action of many muscles, have many degrees of freedom, and require different muscle activation patterns for the same arm movement depending on external factors such as arm weight, orientation with respect to gravity, posture, and so forth. Its apparent simplicity makes the oculomotor system an ideal proxy to study motor control and motor learning. Research on plasticity of eye movement control has focused over the last 35 years on four types of eye movements: the vestibulo-ocular reflex (VOR), the optokinetic reflex (OKR), the pursuit system, and the saccadic system.

MOTOR LEARNING IN THE VOR

The VOR is an exemplar system to study brain function because its behavior, neuroanatomy, and physiology are among the best understood. Moreover, its output is easily quantified in terms of gain and phase in relation to head movement during sinusoidal stimulation. The function of the VOR is to generate compensatory eye movements during head turns to maintain the visual image stable on the retina. Its sensory organs are located in the inner ear and include three pairs of semicircular canals, which detect angular accelerations of the head, and two pairs of otolith organs, which detect translational and gravitational forces (accelerations) on the head. In this chapter we deal only with angular VOR due to the scarcity of data on motor learning in the linear acceleration component of the VOR. The gain of the angular VOR is calculated as the ratio of eye to head velocity and is measured in the dark, when there is not visual feedback. In normal conditions the gain of this reflex is nearly one, but it can be modified following disease and trauma affecting the vestibular organs or the central nervous system. The ability of restoring the VOR gain to normal values after disease and trauma is a form of VOR motor learning called vestibular

compensation and is a key feature of VOR function. It is argued that this intrinsic adaptability of the reflex serves to continuously tune its operation to ongoing changes in function of semicircular canals, neurons, and muscles.

Research on VOR motor learning jumpstarted with the pioneering work carried out in the 1970s, when investigators designed experimental manipulations that engage plasticity in the VOR. Thus, Gonshor and Melvill Jones showed that human VOR gain can be short-term adapted using pursuit tracking that reversed the VOR (conflict visual–vestibular stimulation), and long-term adapted by the continuous (days–weeks) wearing of vision reverse prisms (Gonshor & Melvill Jones, 1973). Later, other investigators introduced the use of magnifying and minimizing goggles as experimental manipulations to increase and decrease the gain of the VOR (Gauthier & Robinson, 1975; Miles & Eighmy, 1980). Around the same time, Ito's group, using microstimulation and lesions, postulated that the cerebellum is essential for VOR motor learning and, later, that cerebellar LTD is the ultimate mechanism responsible for it (Ito, 1982; Ito & Miyashita, 1975; Ito, Sakurai, & Tongroach, 1982). Furthermore, single-unit recordings in the brainstem and cerebellar floccular complex (FL; flocculus and ventral paraflocculus; VPFL) as well as lesion studies support a role of these structures in VOR motor learning (Fukuda, Highstein, & Ito, 1972; Lisberger & Fuchs, 1978; Robinson, 1976). The above seminal papers introduced many of today's hot research topics in the VOR motor learning field: differences between high, low, and reversal VOR gain adaptation, the idea of VOR adaptation being frequency dependent, the differences between short-term versus long-term VOR adaptation, and the role of cerebellum and brainstem in memory formation and storage.

The Richness of VOR as a Model System to Study Plasticity and Learning

Researchers have at their disposal a wide range of experimental manipulations to induce VOR motor learning. For example, short-term VOR motor learning can be induced in just tens of minutes using conflicting visual–vestibular stimulation, long-term VOR motor learning can be induced in days, weeks, or months by wearing magnifying or minimizing goggles as well as reversing prisms, and VOR motor learning can be induced by passive or active head rotation. These experimental manipulations and their variations have unveiled the richness offered by the VOR system to study brain plasticity and motor learning.

The frequency of stimulation (head rotation) and the orientation of the subject with respect to gravity during VOR training are important factors that determine the extent of changes in the VOR gain as well as the VOR pathways undergoing plasticity. Lisberger, Miles, and Optican (1983) and later Pastor, de la Cruz, and Baker (1992) and Raymond and Lisberger (1996) showed that gain changes induced by single-frequency adaptation are larger at the training frequency, and that these changes can be modeled using parallel pathways containing controllers that are independently modifiable. Yakushin, Raphan, and Cohen (2003) showed that changes in VOR gain after short-term adaptation can be modeled using two independent components: gravity dependent and independent. Changes in the gravity independent component occur irrespective of the orientation used for testing, while changes in the gravity dependent component are larger for the head orientation used during training and smaller (null) at the orthogonal head orientations.

Other important factors determining the new behavior and the cellular mechanisms used to generate it are the direction of adaptation (VOR motor learning toward high or low gain) and the length of adaptation (long-term or short-term motor learning). Analysis of the VOR learning curves (gain and phase) of high- and low-gain VOR training following adaptation unveiled asymmetries in the rate of readaptation, thus suggesting different plastic mechanisms for VOR gain increases and decreases (Boyden & Raymond, 2003). In support, blockage of mGluR1 and GABAB receptors in the cerebellar flocculus prevent gain-up training but have little effect in gain-down training. Differential mechanisms are also suspected for short- and long-term VOR motor learning. VOR memories resulting from short-term VOR motor learning are labile while those resulting from long-term VOR motor learning are more stable. This process of consolidation that converts labile into consolidated VOR memories occurs soon after training. When animals are left in the dark without vestibular stimulation for more than 1 h after short-term VOR motor learning, VOR memories consolidate, especially after low-gain adaptation (Titley et al., 2007). The cerebellum plays a center role in this consolidation. Immediate cerebellar ablation after acute training demonstrated full retention of the modified response at the

shortest latency pathways of the reflex, indicating that memory transfer outside of the cerebellum had already occurred. Nonetheless, these short latency pathways require the cerebellum for adaptation as shown in chronic cerebellectomy studies (Pastor, de la Cruz, & Baker, 1994). Cerebellar lesion using muscimol immediately after training showed that short-term memories are erased, but long-term memories remain (Kassardjian et al., 2005). Yeo's laboratory, using eye blink conditioning, showed that inactivation of the cerebellar cortex with the GABAA receptor agonist muscimol immediately after training prevents consolidation while inactivation 90 min or more after training does not (Cooke, Attwell, & Yeo, 2004). What is still under debate is what are the cellular mechanisms used for VOR motor learning? For a long time, evidence pointed toward presynaptic and postsynaptic long-term potentiation (LTP) and postsynaptic long-term depression (LTD) in parallel fiber–Purkinje cell synapses, but recent studies have shown that mutant mice lacking cerebellar LTD can adapt their VOR gains, thus arguing against the necessity of cerebellar LTD for VOR motor learning (Schonewille et al., 2011).

The above features of VOR motor learning, that is, differences between high- and low-gain adaptation, short- and long-term gain adaptation, and the influence of frequency and gravity strengthened the notion of VOR as an ideal system to study motor learning because it allows us to investigate different forms of motor learning using the same simple behavior.

The VOR Circuitry and Neuronal Responses

The brainstem and the cerebellum are the main structures responsible for VOR behavior. The direct VOR pathway, also called the three-neuron reflex arc, includes the VIIIth nerve, secondary vestibular neurons in the brainstem, and motoneurons. This direct VOR pathway is responsible for the short latency response of the reflex. A major indirect VOR pathway integrates velocity signals coming from the VIIIth nerve into position signals which are used to command the eyes as well as to elongate the time constant of the VOR. This is the eye velocity to eye position integrator, which is mostly made up by a network of neurons located in the vestibular nuclei, the prepositus nuclei, the cerebellum and the interstitial nuclei of Cajal. A second major indirect VOR pathway involves also the cerebellum, specifically, the FL (flocculus and VPFL). The FL receives head velocity information through the vestibular nerve and secondary vestibular neurons in the vestibular nuclei, efferent copy signals from neurons in the vestibular nuclei, prepositus hypoglossi nuclei, and

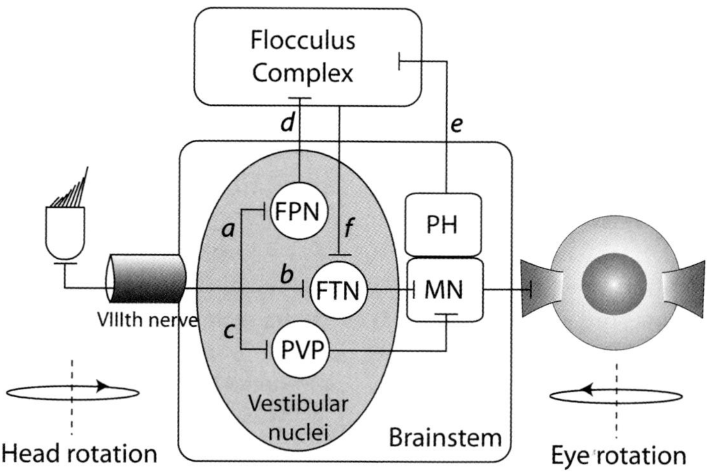

FIGURE 67.1   Main neuronal elements responsible for the vestibulo-ocular reflex (VOR). Vestibular input (head rotation) is detected by hair cells in the semicircular canals. Hair cells connect with vestibular afferents that send information to the brainstem through the vestibular nerve. Secondary vestibular neurons in the vestibular nuclei that participate in the VOR are separated into three groups: flocculus projecting neurons (FPNs), which send their axons to the flocculus complex; flocculus target neurons (FTNs), which receive inhibitory synapses from Purkinje cells in the floccular complex and send their axons to motoneurons (MNs); prepositus hypoglossi (PH), which send efferent copy information to the cerebellar flocculus complex; and position vestibular pause (PVP) neurons, which send their axons to MNs. Pathways indicated by labels a–f are explained in the text.

paramedian tract, and a mix of vestibular, eye movement, and visual signals from the dorsolateral pontine nuclei and the nucleus reticularis tegmenti pontis (NRTP). This second, indirect vestibular pathway has caught much interest because evidence from studies in other motor system and in VOR suggest that the cerebellum plays a major role in motor learning.

The direct VOR pathway includes three main types of oculomotor neurons: vestibular only (VO) neurons, which respond only to head movement; position vestibular pause neurons (PVP), which respond to changes in eye position and head velocity with a pause in firing rate during saccades; and flocculus target neurons (FTNs), which contain mostly eye velocity and head velocity information (see figure 67.1). Oculomotor VO neurons (many of them also called flocculus projecting neurons; FPNs), PVPs, and FTNs in the vestibular nuclei receive direct inputs from the ipsilateral VIIIth nerve and indirect projections from the contralateral VIIIth nerve. Lisberger's group and colleagues recorded the activity of these three classes of neurons before and after long-term VOR motor learning and hypothesized that vestibular pathways to FPNs and PVPs ("a" and "c" in figure 67.1) do not undergo plastic changes after long-term VOR adaptation, however, head velocity pathways to FTNs ("b") do (Lisberger et al., 1994a, 1994b). In support, Partsalis, Zhang, and Highstein (1995) examined the contribution of the flocculus and brainstem pathways ("a–d–f" and "b") to the response of

FTNs using reversible inactivation of the ipsilateral flocculus before and after long-term VOR motor learning (see figure 67.2). They showed that part of the changes in the response of vertical FTNs in the dorsal Y group occur because of changes in the direct vestibular pathway ("b").

More controversy exists about the role of the cerebellar pathways in VOR motor learning. Several laboratories have described the responses of FL gaze velocity Purkinje cells before and after VOR adaptation and showed that these neurons change their response during head turns such that the new response supports the new behavior (Lisberger et al., 1994a; Miles, Braitman, & Dow, 1980; Pastor, de la Cruz, & Baker, 1997). Specifically, Purkinje cell response helps slow down eye movements after low-gain adaptation and helps increase eye movements after high-gain adaptation. The intriguing finding however, is that, when the neuronal response is broken into its neuronal sensitivities to head and eye movement (sensitivity is understood as a measure of the strength of pathways "d" and "e" respectively in figure 67.1), the head velocity sensitivity of Purkinje cells changes in the wrong direction to support the new behavior after high-gain training. In fact, both eye and head velocity sensitivities ("d" and "e") increase after high- and low-gain VOR adaptation, with eye velocity signals dominating after low-gain adaptation and head velocity signals dominating after high-gain adaptation (Blazquez et al., 2003). Moreover, recordings of FTNs in the dorsal Y group before and after long-term adaptation identified a new modifiable element: the pathway from Purkinje cells to FTNs ("f"). This change is necessary to generate stable pursuit (Blazquez, Hirata, & Highstein, 2006). It is unclear what the advantages are of having widespread changes in the cerebellar pathways, but authors argue that changes in the cerebellar pathways are necessary to generate stable VOR gains (Hirata & Highstein, 2001; Lisberger et al., 1994a).

Model simulations built using the data mentioned above can reproduce the behavior in the adapted animal and the effect of FL lesions. These simulations suggest different plastic sites for the vertical and the horizontal VOR (Blazquez, Hirata, & Highstein, 2006; Lisberger, 1994). Specifically, in the vertical VOR, in addition to the plastic sites mentioned above, vestibular inputs to PVPs ("c") are modified (strengthen) after low-gain adaptation. Model simulations also explain why removal of the FL after long-term VOR motor learning does not restore the VOR gain to its preadaptation values. Namely, in the absence of a functioning cerebellum, the brainstem pathway alone can partially retain the new VOR gain values because its head velocity ("b") changes in the right direction to support the new behavior. However, lesion studies must be interpreted with caution because lesions change the network configuration; thus in actuality we are not evaluating the same circuit—before lesion the VOR circuit works as a combination of feedforward and feedback systems whereas after lesion the VOR circuit works exclusively as a feedforward system (Blazquez et al., 2004).

Short-term VOR adaptation also induces plastic changes in brainstem and cerebellar pathways. Hirata and colleagues recorded the activity of vertical gaze velocity Purkinje cells while subjects were trained to increase and decrease their VOR gain (Hirata & Highstein, 2001). They used model simulations and a system approach to dissect the circuit into its components and found that the head velocity component of the prefloccular–floccular and nonfloccular subsystems (pathway "a–d," and "b" in figure 67.1) changes with the VOR gain. No changes were found in the efferent copy or visual pathways (pathway "e" in figure 67.1). Further recording studies during short-term VOR training are necessary because these results cannot fully explain why removal of the FL after short-term adaptation would restore the VOR gain to its preadaptation value (Kassardjian et al., 2005; Rambold et al., 2002).

VOR motor learning is still at the forefront of learning and memory research. Today's leading questions center on the cellular mechanisms involved in this form of adaptation: How are LTP and LTD engaged during VOR motor learning? What is the role of climbing fibers? What is the role of cerebellar interneurons during VOR motor learning? What are the contributions of nonclassical forms of plasticity such as firing rate potentiation?

MOTOR LEARNING IN THE OKR

The VOR and OKR compensate for displacements of the visual image on the retina. The input signal driving these reflexes is a velocity signal: head velocity and retinal slip velocity, respectively. The OKR generates characteristic eye movements called nystagmus that consist of slow eye movements in the same direction as the movement of the visual image and saccades in the opposite direction. Here we consider the direction of the slow eye movements as the direction of the optokinetic nystagmus. These slow eye movements in primates and other foveated animals are generated by two neuronal pathways: a fast pathway that dominates during the initial part of the OKR and a slow building pathway (Waespe, Cohen, & Raphan, 1983) (figure 67.2). The fast pathway, also called direct visual pathway, is related to pursuit and can be abolished following lesions in the cerebellar FL (Büttner, 1989). The slow building pathway, also called indirect visual pathway, is related to the retinal reflex found in nonfoveated animals and is abolished following lesions in the vestibular nerve and the eye velocity to position integrator (e.g., prepositus hypoglossi nuclei) (Cannon & Robinson, 1987). The indirect pathway has storage capabilities (called eye velocity storage) that can be measured by quantifying the slow phase of nystagmus in the dark that follows OKR (optokinetic afternystagmus [OKAN]; figure 67.2B). In natural conditions, VOR and OKR work together to generate compensatory eye movements that cover the entire range of natural frequencies. At low frequencies the OKR takes a dominant role because, due to the mechanics of the semicircular canals, vestibular information is unreliable. At high frequencies the VOR dominates because the input of the OKR, the accessory optic system, saturates.

Behavioral studies showed that long-term VOR motor learning induced by the continuous wearing of magnifying and minimizing goggles can change the saturation velocity of the OKAN.

Lisberger and colleagues showed that increases and decreases in the gain of the horizontal VOR induce parallel changes in the saturation velocity of the horizontal OKAN and suggested that the modifiable elements (neurons) in the VOR carry head velocity information and indirect visual pathway information (Lisberger et al., 1981). These changes in the saturation velocity of the OKAN would help improve image

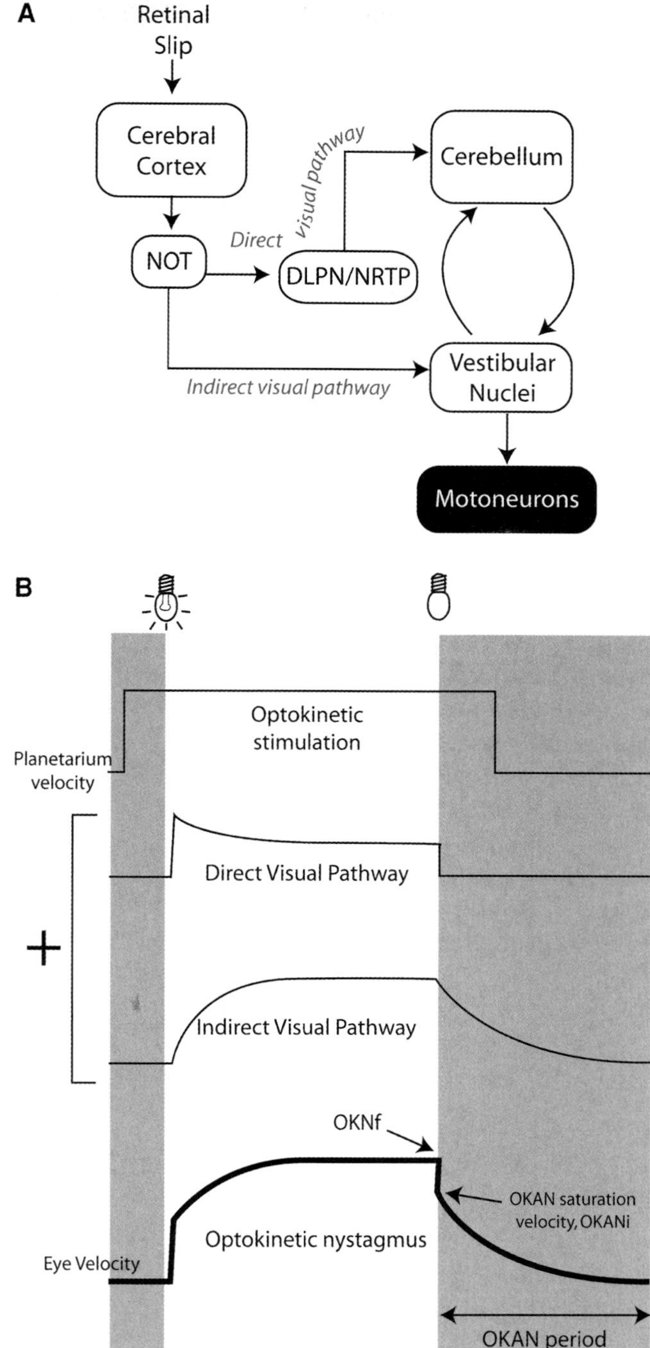

FIGURE 67.2   Schematic representation of the neuronal structures and behavior of the optokinetic response (OKR). (A) Structures involved in the generation of OKR and paths followed by the direct and indirect visual pathways. Downstream from the nucleus of the optic tract (NOT), the direct pathway travels to the cerebellar cortex while the indirect pathway travels to the vestibular nuclei. (B) Contribution of the direct and indirect visual pathways to OKR behavior. Direct and indirect visual pathways add up to generate the slow phase of OKR (eye velocity at the bottom). Symbols (lightbulbs) at the top indicate when optokinetic light was turned ON and OFF. Optokinetic stimulation began when the optokinetic light is turned ON. At the beginning of the optokinetic nystagmus, most of the eye velocity is generated by the direct visual pathway, but after several tens of seconds of optokinetic stimulation the optokinetic nystagmus final eye velocity (OKNf) is mostly generated by the indirect visual pathway. When the light is turned OFF (second gray area), optokinetic nystagmus shows residual eye movements called optokinetic after nystagmus (OKAN). High velocity optokinetic stimulation can charge maximally the indirect visual pathway. This maximal charge, called OKAN saturation velocity, can be measured as the eye velocity at the onset of OKAN (initial OKAN [OKANi]). DLPN, dorsolateral pontine nuclei; NRTP, nucleus reticularis tegmenti pontis.

stabilization at low frequencies of head turns after VOR adaptation. The vertical OKAN, however, is affected differently by long-term VOR motor learning; decreases in the gain of the vertical VOR do not induce decreases in the vertical OKAN saturation velocity, but increases in VOR gain do (Blazquez & Highstein, 2007).

NEURONAL REPRESENTATION OF OKR MOTOR LEARNING

Data obtained from neuronal recordings in the vestibular nuclei offer an explanation to the OKAN behavior observed in the normal subject as well as the OKAN behavior before and after long-term VOR adaptation. Thus, the tight correlation between the OKAN behavior and the VOR described by many laboratories (Blazquez & Highstein, 2007; Lisberger et al., 1981) is already observed at the level of secondary vestibular neurons. Second-order vestibular neurons with large sensitivity to head velocity also show large sensitivity to eye velocity during the OKAN and vice versa (Blazquez & Highstein, 2007; Waespe & Henn, 1977). Blazquez et al. (2007) suggested that changes in head velocity sensitivity of vestibular neurons through VOR adaptation will occur in parallel with changes in their sensitivity to eye velocity during OKAN. In support, vertical FTNs in the dorsal Y group increase their firing rate during upward OKAN after high-gain adaptation, and during downward OKAN after low-gain adaptation (see figure 67.3, Blazquez et al., 2007). These changes correlate with changes in the direct vestibular pathway to FTNs

---

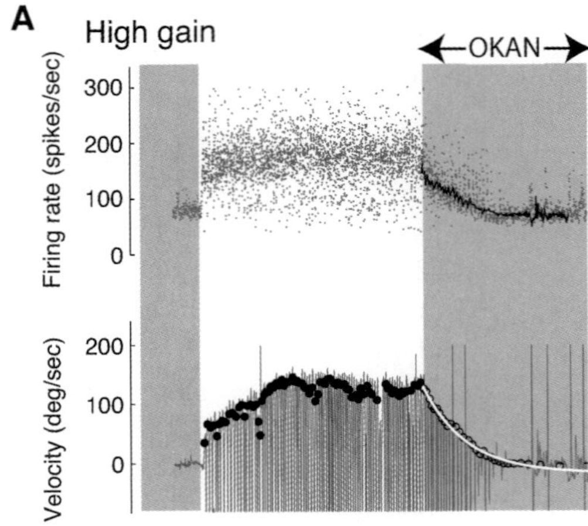

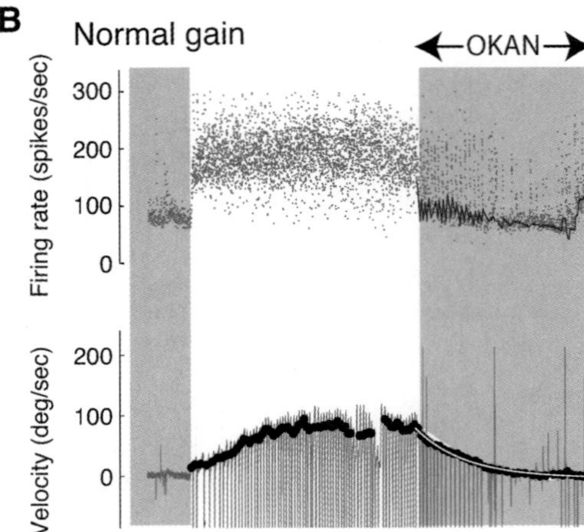

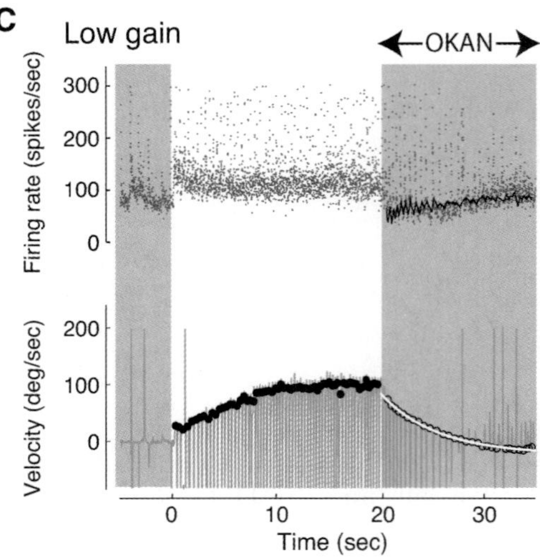

FIGURE 67.3 Changes in the response of representative vertical flocculus target neurons in the dorsal Y group to optokinetic stimulation following vestibulo-ocular reflex (VOR) adaptation. All panels (A–C) show: Top, response of a representative dorsal Y group neuron recorded from an animal in which VOR was adapted to high (A), normal (B), and low (C) gain. Bottom, corresponding behavioral response. Optokinetic stimulation consisted of a constant (160 deg/s) rotation of a planetarium system. Optokinetic stimulation was ON during the lighted portion of the recording and OFF during the shaded areas. If we look at the optokinetic afternystagmus (OKAN) period (second shaded area) we can observe a change in eye movement directional preference of dorsal Y group neurons. In the normal animal these neurons show little response to eye movements; in the high-gain animal they increase their response with upward eye velocity, and in the low-gain animal they decrease their response with upward eye velocity. White lines shown over the behavioral responses in the OKAN period represent the exponential fitting. Dark lines shown over the neuronal response during the OKAN period represent the estimated neuronal response after linear fitting using eye movement parameters. (From Blazquez et al., 2007.)

(pathway "b" in figure 67.1) and occur in the adequate direction to increase the OKAN saturation velocity after high-gain adaptation and decrease it after low-gain adaptation.

Lastly, results obtained from modeling and neuronal recording studies can explain the differential changes observed in the vertical and horizontal OKAN after VOR adaptation (Blazquez et al., 2007; Lisberger et al., 1994b). Lisberger and colleagues showed that horizontal PVPs do not change their head velocity sensitivity after high- and low-gain VOR adaptation (Lisberger et al., 1994b). Blazquez, Hirata, and Highstein (2006), using modeling studies, showed that vertical PVPs do not change their head velocity sensitivity after high-gain adaptation; however they do increase their head velocity sensitivity after low-gain adaptation. Blazquez et al. (2007) explained that if we assume that the relationship (ratio) between the OKAN eye velocity sensitivity and VOR head velocity sensitivity of PVPs is the same before and after VOR adaptation, as happens in FTNs, we can hypothesize that vertical OKAN saturation velocity does not change after low-gain VOR adaptation because the decrease in OKAN eye velocity sensitivity of vertical FTNs is canceled out by an increase in OKAN eye velocity sensitivity of vertical PVP neurons. This hypothesis fully explains the behavioral and neuronal data.

MOTOR LEARNING IN PURSUIT EYE MOVEMENTS

Pursuit eye movements are voluntary movements generated to keep moving visual objects stable on the fovea. These movements are highly developed in primates, which have well-developed foveas. The pursuit system is notably slow compared to the VOR: The latency of pursuit is about 100 ms in primates while the latency of the VOR can be as short as 14 ms. This delay between the visual input (target movement) and motor output (eye movement) provides a ~100-ms time window starting at the initiation of eye movement where the pursuit system works as an open-loop system—that is, eye movements generated exclusively by sensory information (moving visual stimulus). The pursuit gain is measured during this open-loop period as the eye velocity during the first 100 ms of pursuit divided by the target velocity during the previous 100 ms. The closed-loop period begins after the first 100 milliseconds of eye movement. In this period the eye movements generated by smooth pursuit are the result of combining the response to the sensory stimuli (the target displacement on the retina 100 ms earlier) and a feedback efferent copy of the ongoing eye movement. The step ramp pursuit task, developed in 1961 by Rashbass, is the task of choice today to study behavioral and neuronal aspects of pursuit (Rashbass, 1961). During this task the target is first stepped in the opposite direction to the subsequent pursuit eye movement such that the target passes by the center of fixation at the time pursuit begins (about 100 ms later). This task simplifies the study of pursuit eye movements because it reduces the need for catch-up saccades, thus effectively isolating the pursuit from the saccade system (see figure 67.4A), by reducing the target positional error (driving signal for saccades) while maintaining the target velocity error (driving signal for pursuit).

The pursuit system offers an opportunity to study learning and memory in the central nervous system because, like the VOR, its sensory input and behavioral output are easily controlled and measured by the experimenter. However, unlike VOR and OKR, which are reflexive behaviors, pursuit requires attention and motivation, resembling many motor behaviors we carry out throughout the day. Pursuit motor learning is commonly studied using two variations of the Rashbass task. In a first experimental manipulation, called gain adaptation paradigm, the target speed is increased or decreased at the time when the eyes start moving. This situation results in a mismatch between target and eye velocity during the open-loop pursuit period that generates an error signal (image slip). Upon repetition of the same task, the subject learns to generate the appropriate eye velocity in order to compensate for the change in target velocity, hence modifying the gain of pursuit (see figure 67.4). This experimental manipulation simulates a form of motor learning aimed to recalibrate the input/output relationship of the pursuit system, such as those that can occur to compensate for disease and trauma affecting the nervous system, and changes in the physical properties of the motor system. In a second variation of the pursuit task, the target direction changes at a fixed time after the initial target movement. This learning paradigm is useful to study how the brain learns to anticipate changes in the trajectory of moving targets. Kahlon and Lisberger (1996), using a gain adaptation paradigm, showed that pursuit learning at one speed generalizes to other target speeds but is direction specific. Because pursuit requires visuomotor transformations, where object features, speed, and direction are computed before generating the proper motor output, they suggested that the plastic site is located after the neuronal switch that channels information based on the direction of the target. Medina, Carey, and Lisberger (2005), using the pursuit directional change paradigm, found that the brain computes timing using a variety of clues such as movement time and distance to travel.

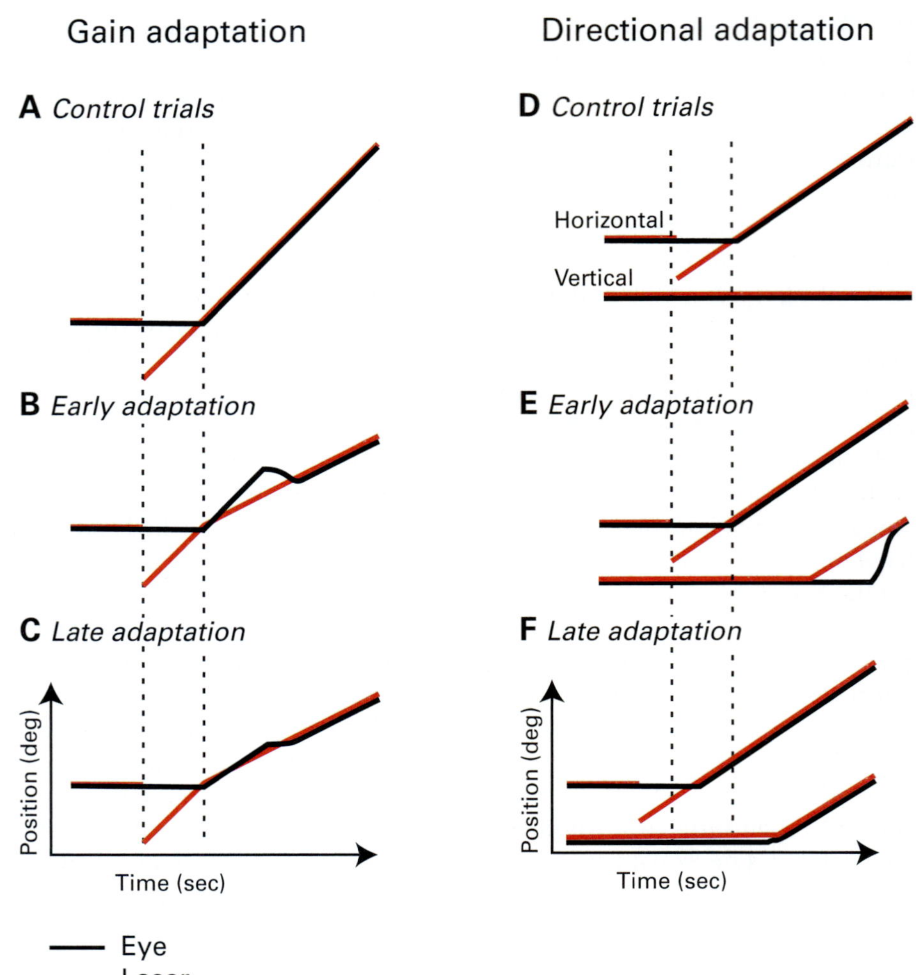

FIGURE 67.4 Behavioral tasks used to train the pursuit system to change its gain (A–C, low-gain adaptation) or anticipate a directional change in target motion (D–F). A and D show control pursuit behavior using the step ramp pursuit task first proposed by Rashbass (1961). The gain of pursuit is measured by dividing the eye velocity during the first 100 ms of pursuit (b) by the target velocity during the first 100 ms of target movement (a). In the gain adaptation task the target velocity is modified at the onset of pursuit (second vertical dashed line in A–C). At the beginning of training the subject generates pursuit eye movements of the same speed as the initial target movement; during the late phase of pursuit adaptation the subject modifies initial eye velocity to match the eye velocity of the target at the initiation of pursuit. In the directional adaptation pursuit task a directional change in the trajectory of the target is introduced at a fixed time from the onset of target movement (D–F). In this task the subject learns to generate anticipatory eye movements to correct the trajectory of pursuit. deg, degrees.

NEURONAL REPRESENTATION OF PURSUIT MOTOR LEARNING

The neuronal network responsible for pursuit behavior includes cortical and subcortical structures that transform image slip information into eye movement (Ilg & Thier, 2008). Neurons in primary visual cortex (V1) and middle temporal area (MT) carry information related with the properties of the visual stimulus while downstream cortical areas such as medial superior temporal area (MST) can carry combined sensory and eye movement information (Ilg, Schumann, & Thier, 2004; Ilg & Thier, 2008). The main subcortical structures involved in pursuit behavior include pontine nuclei, cerebellum, and brainstem.

Evidence suggests that there are multiple plastic sites responsible for pursuit learning and that these sites are located between MST and the output of the cerebellum. Experiments using neuronal recordings and electrical stimulation in the frontal pursuit area during pursuit gain learning suggest that this type of pursuit learning occurs downstream from the frontal pursuit area, with the dorsolateral pontine nuclei (DLPN) and the VPFL as the most likely candidates (Chou & Lisberger, 2004). Among these two structures, the VPFL has been the preferred one to study pursuit adaptation

     PABLO M. BLAZQUEZ AND ANGEL M. PASTOR

because there is strong evidence for a role of the cerebellar cortex in motor learning (Ito, 2006). However DLPN could also play an important role. For instance, NRTP neurons change their response to saccadic eye movements after saccade adaptation (Takeichi, Kaneko, & Fuchs, 2005); hence it would not be surprising if their pursuit homologues, DLPN neurons, change their response following pursuit adaptation.

Recent studies have provided important clues on the brain sites and mechanisms used for pursuit learning by examining the role of a brain structure throughout the course of adaptation. Thus, Carey, Medina, and Lisberger (2005), using microstimulation in MT during pursuit, proposed that neurons in MT provide the instructive signal for pursuit learning. Li et al. (2011) showed that individual frontal eye field (FEF) neurons do not change their response as predicted if they were responsible for the learned behavior; however, changes in the responses of VPFL neurons tend to follow behavioral learning. These results agree with the current notion of a role of cerebellar learning in pursuit adaptation. During training in the directional pursuit adaptation task, the presence of a complex spike in VPFL Purkinje cells in a learning trial is associated with a decrease in the simple spike activity of the same Purkinje cell in the next trial. This complex spike influence in simple spike activity suggests that complex spikes drive learning in simple spikes, which are partially responsible for behavioral learning (Medina & Lisberger, 2008).

The above neuronal recordings and stimulation studies suggest that although plasticity is distributed throughout the circuit (from MST to cerebellum), cerebellar learning is a key mechanism for pursuit adaptation, and that inferior olive neurons and cortical neurons provide the instructive signals for plasticity.

MOTOR LEARNING IN SACCADIC EYE MOVEMENTS

Saccades are fast eye movements, as short as 40 ms, that shift gaze toward new points on the visual scene. There are two types of saccadic eye movements: saccades associated with the vestibulo-ocular and the optokinetic reflexes and spontaneous saccades. The former move the eyes toward a more central position in the orbit and have a lesser role in shifting gaze toward points of interest because they also occur during the VOR in the dark and during the afternystagmus phase of the OKR. The latter could be generated during scanning of a stationary visual image or in response to sudden movements of a target of interest (catch-up saccades). In the laboratory, in order to control the parameters of saccade eye movements (e.g., saccade onset, initial and final eye position, saccade amplitude and direction, etc.), we use simple visually guided saccade tasks where subjects are instructed to fixate a center target and then to saccade to a new target location immediately after the fixation target is extinguished (see figure 67.5). The gain of the saccadic eye movement is defined as the amplitude of the saccade divided by the positional difference (degrees) between the initial fixation point and the new target location. Variations of this simple saccade task such as saccades toward previously remembered locations (memory-guided saccades) or saccades with delay initiation despite a new target having been presented (delayed saccades) are commonly used to study cognitive and sensorimotor processing by the central nervous system. To study plasticity in saccadic eye movements, we use the saccade adaptation task (Mclaughlin, 1967), which is thought to engage plastic mechanisms similar to those used during natural adaptation of saccades.

Saccade metrics can be adapted in the laboratory by taking advantage of the fact that visual information is suppressed during the performance of the saccade (reviewed in Hopp & Fuchs, 2004). In the saccade adaptation task the location of the target is modified during the execution of the saccade. The result is that at the time the saccade ends, and thus vision is enabled, the saccade final eye position does not coincide with the location of the target at the time the saccade ends, and the subject would need to perform a corrective saccade. The mismatch between the final eye position and the final target position at the end of the saccade signals that a motor error has been made, error that serves as the instruction to induce adaptation because proprioceptive signals do not seem to play a significant role in adaptation. It has been established that a visual error is more effective at driving adaptation when presented within 100 ms of saccade termination. Trial repetition leads to a progressive modification of the saccade amplitude until the saccade is performed to the secondary location without the need for a corrective saccade. If the secondary location of the target is closer to the fixation point than the initial location of the target, the saccadic gain (change in eye position divided by initial change in target position) should decrease to attain the target. Saccadic gain will increase when the secondary location of the target is aimed farther away from the fixation point than the initial target location (see figure 67.5). The rate of adaptation is an exponential function of trial repetition toward an asymptotic level in about 100 saccades in humans and about 1,000 saccades in primates that is called short-term adaptation since it

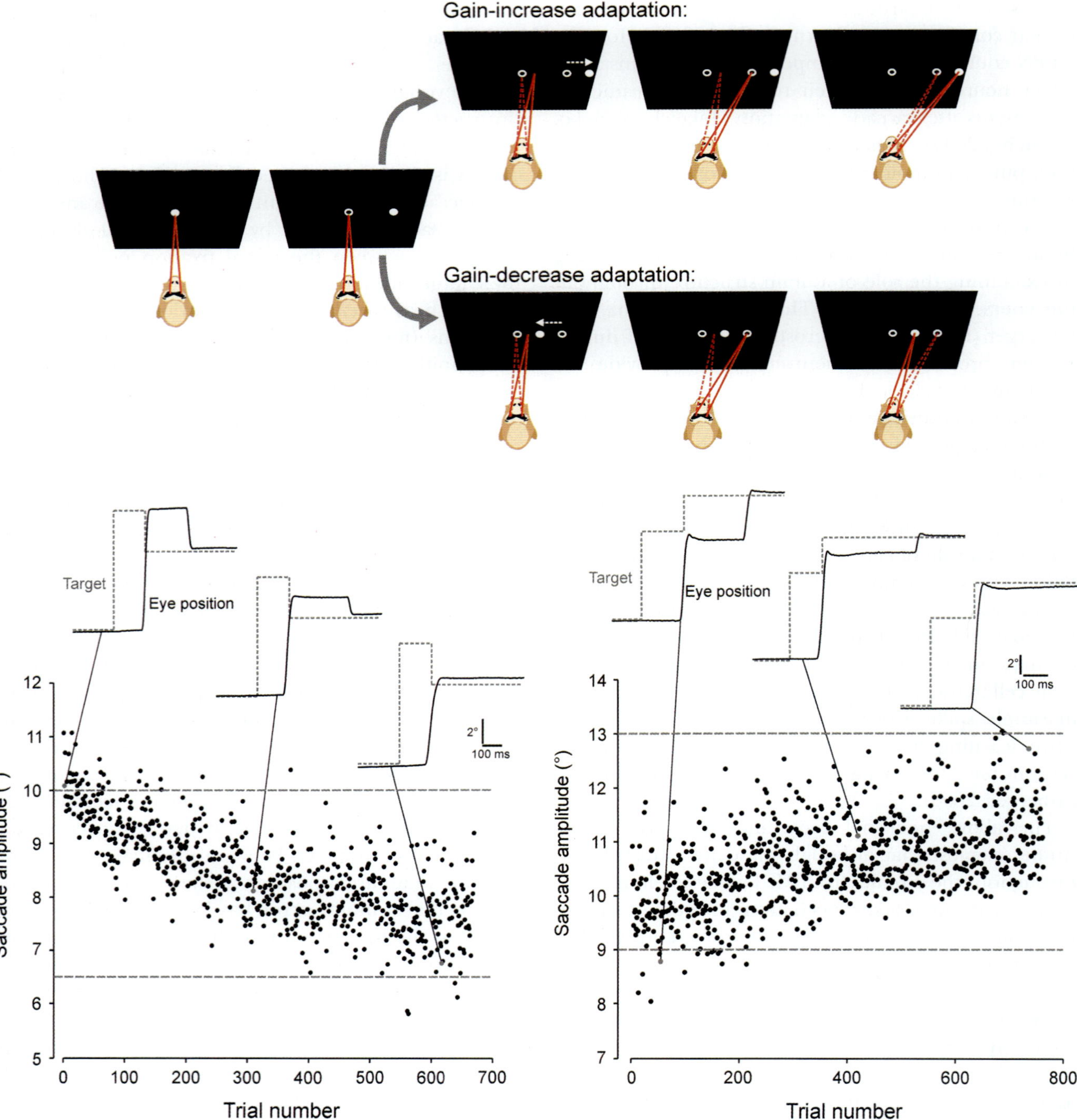

FIGURE 67.5 (Top) Depiction of successive target display configurations during the gain-increase/decrease saccadic adaptation paradigms. Filled and open white circles indicate current and previous target locations, respectively. Solid and dotted red traces indicate the current and previous eye positions, respectively. The horizontal dimension represents a chronological time line. (Bottom) Saccade amplitude as a function of trial number for examples of gain-decrease (left) and gain-increase (right) adaptation performed by a monkey. Insets provide eye position traces (solid) and information on target position (dotted) for three trials at different instances of the adaptation experiment. (From Prsa and Their, 2011.)

occurs within 1–2 h of training. These changes normally wane overnight, but long-term training endures the adaptive response.

Several properties arise from the saccade gain adaptation paradigm. Adaptation to gain increase, also called forward adaptation, is slightly slower than adaptation to low gain, also called backward adaptation. Firstly, it is argued that the saccadic system is designed to be hypometric because a hypometric saccade requires less muscle activation and it is easier to perform two consecutive saccades in the same direction as they do not require the recruitment of pools of neurons in both sides of the brain, and therefore it should be easier to decrease rather than increase saccadic gain (Hopp & Fuchs, 2004; Prsa & Thier, 2011). Secondly, the gain changes that occur during adaptation in one direction only partly transfer to saccades in other directions or even to saccades in the same direction but of different amplitude, and therefore adaptation is not parametric in the sense that a multiplicative factor will affect the saccade gain in all directions; thus it is more appropriate to speak of saccade amplitude adaptation. Finally, the amplitude adaptation mimics the natural condition that occurs after palsies of the oculomotor plant. For example, if the horizontal recti of an eye are weakened by tenetomy and the normal eye is patched, amplitude adaptation will occur in both eyes so that the weakened eye produces normal saccades and the patched eye becomes hypermetric. To the contrary, if the weakened eye is patched, then the normal eye will maintain a normal saccadic amplitude while the weakened eye becomes hypometric.

THE SACCADIC CIRCUITRY AND NEURONAL RESPONSES DURING ADAPTATION

Sensory signals related to target information reach visual cortex and then decision-making areas such as lateral intraparietal area (LIP), FEFs, and deep layers of the superior colliculus (SC), which receive projections heavily from the latter (Scudder, Kaneko, & Fuchs, 2002). The SC projects in parallel to the brainstem saccade generator and to the gate nucleus for the oculomotor vermis (OMV), the pontine NRTP. The cerebellar loop courses through the OMV, the fastigial oculomotor region (FOR), and back into the saccade generator area, which projects to oculomotor motoneurons.

Lesions to the OMV and FOR produce persistent hypermetric saccades (Optican & Robinson, 1980). Lesions to the OMV caused saccade hypermetria and permanent deficits in short-term saccade adaptation. But OMV lesions do not completely abolish long-term

saccade adaptation because saccade hypermetria is corrected over time. This suggests that short- and long-term mechanisms act independently (Barash et al., 1999).Consistent with lesion, muscimol inactivation, a GABA A receptor agonist, showed that bilateral inactivation of the FOR resulted in hypermetria of horizontal saccades that persisted irrespective of the visual error to reach the target. However when the animals were placed in the dark to wear off the effects of muscimol and tested again the next day, the saccades were hypometric (Robinson, Straube, & Fuchs, 1993). The explanation argues that the continuous hypermetric saccades during muscimol inactivation produced plastic changes upstream from the FOR. Unlike FOR or burst generator neurons, OMV neurons are not homogeneous, exhibiting idiosyncratic responses with burst for saccades of different sizes and directions. Therefore, to understand the role of OMV in saccade adaptation we need to look at the population response of Purkinje cells (Catz, Dicke, & Thier, 2008). Timing of populational burst of Purkinje cells is tightly adjusted to the termination of the saccades (Thier et al., 2000). Despite profound changes in the response of individual Purkinje cells (Kojima, Soetedjo, & Fuchs, 2010), the patterns of divergence of connections onto the target neurons in the FOR speak for a model of timing selection of mossy fiber inputs based on interactions with the climbing fiber responses. This interaction might determine the duration of the population response of Purkinje cells and thus of saccade duration which could serve as a mechanism for short-term changes of saccade amplitude (Catz, Dicke, & Thier, 2005). The deep layers of the SC receive saccade commands from the LIP and the FEF, and changes in the number of spikes in the burst of SC neurons correlate with saccade adaptation (reviewed in Iwamoto & Kaku, 2010). It is argued that changes in the spike activity of deep layers of the SC are used as part of the tecto–olivo–cerebellar pathway to transmit instructive signals to guide adaptation. In support, when pairing SC deep layer stimulation (subthreshold for movement production) with termination of the saccade over the course of hundreds of trials, saccades shift their direction and size; this shift is dependent on the stimulation point of the SC. However the NRTP receives innervation from the contralateral SC and projects bilaterally to the OMV as mossy fibers. Changes in the NRTP neurons during adaptation show that the number of spikes in the burst increases during amplitude decrease by adding spikes at the initiation of the burst and therefore cannot be regarded as secondary changes of FOR activity as the OMV increases activity toward the termination of the saccade. Activity changes in the FOR indicate that there is a gradual change during adaptation that courses in parallel with saccade adaptation. Neurons in the FOR increase their saccade-related discharge as the amplitude of the ipsiversive saccades adaptively decreases. Since neurons in the FOR project to the burst neurons, it is likely that the increase in FOR activity is causal to the change in saccade amplitude.

REFERENCES

Barash, S., Melikyan, A., Sivakov, A., Zhang, M., Glickstein, M., & Thier, P. (1999). Saccadic dysmetria and adaptation after lesions of the cerebellar cortex. *Journal of Neuroscience, 19,* 10931–10939.

Blazquez, M., Davis-Lopez de Carrizosa, M. A., Heiney, S. A., & Highstein, S. M. (2007). Neuronal substrates of motor learning in the velocity storage generated during optokinetic stimulation in the squirrel monkey. *Journal of Neurophysiology, 97,* 1114–1126. doi:10.1152/jn.00983.2006.

Blazquez, M., & Highstein, S. M. (2007). Visual–vestibular interaction in vertical vestibular only neurons. *Neuroreport, 18,* 1403–1406.

Blazquez, M., Hirata, Y., Heiney, S. A., Green, A. M., & Highstein, S. M. (2003). Cerebellar signatures of vestibulo-ocular reflex motor learning. *Journal of Neuroscience, 23,* 9742–9751.

Blazquez, M., Hirata, Y., & Highstein, S. M. (2004). The vestibulo-ocular reflex as a model system for motor learning: What is the role of the cerebellum? *Cerebellum (London, England), 3,* 188–192.

Blazquez, M., Hirata, Y., & Highstein, S. M. (2006). Chronic changes in inputs to dorsal Y neurons accompany VOR motor learning. *Journal of Neurophysiology, 95,* 1812–1825.

Boyden, E. S., & Raymond, J. L. (2003). Active reversal of motor memories reveals rules governing memory encoding. *Neuron, 39,* 1031–1042.

Büttner, U. (1989). The role of the cerebellum in smooth pursuit eye movements and optokinetic nystagmus in primates. *Revue Neurologique, 145,* 560–566.

Cannon, S. C., & Robinson, D. A. (1987). Loss of the neural integrator of the oculomotor system from brain stem lesions in monkey. *Journal of Neurophysiology, 57,* 1383–1409.

Carey, M. R., Medina, J. F., & Lisberger, S. G. (2005). Instructive signals for motor learning from visual cortical area MT. *Nature Neuroscience, 8,* 813–819.

Catz, N., Dicke, P. W., & Thier, P. (2005). Cerebellar complex spike firing is suitable to induce as well as to stabilize motor learning. *Current Biology, 15,* 2179–2189.

Catz, N., Dicke, P. W., & Thier, P. (2008). Cerebellar-dependent motor learning is based on pruning a Purkinje cell population response. *Proceedings of the National Academy of Sciences of the United States of America, 105,* 7309–7314.

Chou, I.-H., & Lisberger, S. G. (2004). The role of the frontal pursuit area in learning in smooth pursuit eye movements. *Journal of Neuroscience, 24,* 4124–4133.

Cooke, S. F., Attwell, P. J. E., & Yeo, C. H. (2004). Temporal properties of cerebellar-dependent memory consolidation. *Journal of Neuroscience, 24,* 2934–2941.

Fukuda, J., Highstein, S. M., & Ito, M. (1972). Cerebellar inhibitory control of the vestibulo-ocular reflex investigated in rabbit 3rd nucleus. *Experimental Brain Research, 14,* 511–526.

Gauthier, G. M., & Robinson, D. A. (1975). Adaptation of the human vestibuloocular reflex to magnifying lenses. *Brain Research, 92,* 331–335.

Gonshor, A., & Melvill Jones, G. (1973). Proceedings: Changes of human vestibulo-ocular response induced by vision-reversal during head rotation. *Journal of Physiology, 234,* 102P–103P.

Hirata, Y., & Highstein, S. M. (2001). Acute adaptation of the vestibuloocular reflex: Signal processing by floccular and ventral parafloccular Purkinje cells. *Journal of Neurophysiology, 85,* 2267–2288.

Hopp, J. J., & Fuchs, A. F. (2004). The characteristics and neuronal substrate of saccadic eye movement plasticity. *Progress in Neurobiology, 72,* 27–53.

Ilg, U. J., Schumann, S., & Thier, P. (2004). Posterior parietal cortex neurons encode target motion in world-centered coordinates. *Neuron, 43,* 145–151.

Ilg, U. J., & Thier, P. (2008). The neural basis of smooth pursuit eye movements in the rhesus monkey brain. *Brain and Cognition, 68,* 229–240.

Ito, M. (1982). Cerebellar control of the vestibulo-ocular reflex—Around the flocculus hypothesis. *Annual Review of Neuroscience, 5,* 275–296.

Ito, M. (2006). Cerebellar circuitry as a neuronal machine. *Progress in Neurobiology, 78,* 272–303.

Ito, M., & Miyashita, Y. (1975). The effect of chronic destruction of the inferior olive upon visual modification of the horizontal vestibulo-ocular reflex of rabbits. *Proceedings of the Japan Academy, 51,* 716–720.

Ito, M., Sakurai, M., & Tongroach, P. (1982). Climbing fibre induced depression of both mossy fibre responsiveness and glutamate sensitivity of cerebellar Purkinje cells. *Journal of Physiology, 324,* 113–134.

Iwamoto, Y., & Kaku, Y. (2010). Saccade adaptation as a model of learning in voluntary movements. *Experimental Brain Research, 204,* 145–162.

Kahlon, M., & Lisberger, S. G. (1996). Coordinate system for learning in the smooth pursuit eye movements of monkeys. *Journal of Neuroscience, 16,* 7270–7283.

Kassardjian, C. D., Tan, Y.-F., Chung, J.-Y. J., Heskin, R., Peterson, M. J., & Broussard, D. M. (2005). The site of a motor memory shifts with consolidation. *Journal of Neuroscience, 25,* 7979–7985.

Kojima, Y., Soetedjo, R., & Fuchs, A. F. (2010). Changes in simple spike activity of some Purkinje cells in the oculomotor vermis during saccade adaptation are appropriate to participate in motor learning. *Journal of Neuroscience, 30,* 3715–3727.

Li, J. X., Medina, J. F., Frank, L. M., & Lisberger, S. G. (2011). Acquisition of neural learning in cerebellum and cerebral cortex for smooth pursuit eye movements. *Journal of Neuroscience, 31,* 12716–12726.

Lisberger, S. G. (1994). Neural basis for motor learning in the vestibuloocular reflex of primates: III. Computational and behavioral analysis of the sites of learning. *Journal of Neurophysiology, 72,* 974–998.

Lisberger, S. G., & Fuchs, A. F. (1978). Role of primate flocculus during rapid behavioral modification of vestibuloocular reflex: I. Purkinje cell activity during visually guided horizontal smooth-pursuit eye movements and passive head rotation. *Journal of Neurophysiology, 41,* 733–763.

Lisberger, S. G., Miles, F. A., & Optican, L. M. (1983). Frequency-selective adaptation: Evidence for channels in the vestibulo-ocular reflex? *Journal of Neuroscience, 3*, 1234–1244.

Lisberger, S. G., Miles, F. A., Optican, L. M., & Eighmy, B. B. (1981). Optokinetic response in monkey: Underlying mechanisms and their sensitivity to long-term adaptive changes in vestibuloocular reflex. *Journal of Neurophysiology, 45*, 869–890.

Lisberger, S. G., Pavelko, T. A., Bronte-Stewart, H. M., & Stone, L. S. (1994a). Neural basis for motor learning in the vestibuloocular reflex of primates: II. Changes in the responses of horizontal gaze velocity Purkinje cells in the cerebellar flocculus and ventral paraflocculus. *Journal of Neurophysiology, 72*, 954–973.

Lisberger, S. G., Pavelko, T. A., & Broussard, D. M. (1994b). Neural basis for motor learning in the vestibuloocular reflex of primates: I. Changes in the responses of brain stem neurons. *Journal of Neurophysiology, 72*, 928–953.

Mclaughlin, S. (1967). Parametric adjustment in saccadic eye movements. *Perception & Psychophysics, 2*, 359–362.

Medina, J. F., Carey, M. R., & Lisberger, S. G. (2005). The representation of time for motor learning. *Neuron, 45*, 157–167.

Medina, J. F., & Lisberger, S. G. (2008). Links from complex spikes to local plasticity and motor learning in the cerebellum of awake-behaving monkeys. *Nature Neuroscience, 11*, 1185–1192.

Miles, F. A., Braitman, D. J., & Dow, B. M. (1980). Long-term adaptive changes in primate vestibuloocular reflex: IV. Electrophysiological observations in flocculus of adapted monkeys. *Journal of Neurophysiology, 43*, 1477–1493.

Miles, F. A., & Eighmy, B. B. (1980). Long-term adaptive changes in primate vestibuloocular reflex: I. Behavioral observations. *Journal of Neurophysiology, 43*, 1406–1425.

Optican, L. N., & Robinson, D. A. (1980). Cerebellar-dependent adaptive control of primate saccadic system. *Journal of Neurophysiology, 44*, 1058–1076.

Partsalis, A. M., Zhang, Y., & Highstein, S. M. (1995). Dorsal Y group in the squirrel monkey: II. Contribution of the cerebellar flocculus to neuronal responses in normal and adapted animals. *Journal of Neurophysiology, 73*, 632–650.

Pastor, A. M., de la Cruz, R. R., & Baker, R. (1992). Characterization and adaptive modification of the goldfish vestibuloocular reflex by sinusoidal and velocity step vestibular stimulation. *Journal of Neurophysiology, 68*, 2003–2015.

Pastor, A. M., de la Cruz, R. R., & Baker, R. (1994). Cerebellar role in adaptation of the goldfish vestibuloocular reflex. *Journal of Neurophysiology, 72*, 1383–1394.

Pastor, A. M., de la Cruz, R. R., & Baker, R. (1997). Characterization of Purkinje cells in the goldfish cerebellum during eye movement and adaptive modification of the vestibulo–ocular reflex. *Progress in Brain Research, 114*, 359–381.

Prsa, M., & Thier, P. (2011). The role of the cerebellum in saccadic adaptation as a window into neuronal mechanisms of motor learning. *European Journal of Neuroscience, 33*, 2114–2128.

Rambold, H., Churchland, A., Selig, Y., Jasmin, L., & Lisberger, S. G. (2002). Partial ablations of the flocculus and ventral paraflocculus in monkeys cause linked deficits in smooth pursuit eye movements and adaptive modification of the VOR. *Journal of Neurophysiology, 87*, 912–924.

Rashbass, C. (1961). The relationship between saccadic and smooth tracking eye movements. *Journal of Physiology, 159*, 326–338.

Raymond, J. L., & Lisberger, S. G. (1996). Behavioral analysis of signals that guide learned changes in the amplitude and dynamics of the vestibulo-ocular reflex. *Journal of Neuroscience, 16*, 7791–7802.

Robinson, D. A. (1976). Adaptive gain control of vestibuloocular reflex by the cerebellum. *Journal of Neurophysiology, 39*, 954–969.

Robinson, F. R., Straube, A., & Fuchs, A. F. (1993). Role of the caudal fastigial nucleus in saccade generation: II. Effects of muscimol inactivation. *Journal of Neurophysiology, 70*, 1741–1758.

Schonewille, M., Gao, Z., Boele, H.-J., Veloz, M. F. V., Amerika, W. E., Simek, A. A. M., et al. (2011). Reevaluating the role of LTD in cerebellar motor learning. *Neuron, 70*, 43–50.

Scudder, C. A., Kaneko, C. S., & Fuchs, A. F. (2002). The brainstem burst generator for saccadic eye movements: A modern synthesis. *Experimental Brain Research, 142*, 439–462.

Takeichi, N., Kaneko, C. R., & Fuchs, A. F. (2005). Discharge of monkey nucleus reticularis tegmenti pontis neurons changes during saccade adaptation. *Journal of Neurophysiology, 94*, 1938–1951.

Thier, P., Dicke, P. W., Haas, R., & Barash, S. (2000). Encoding of movement time by populations of cerebellar Purkinje cells. *Nature, 405*, 72–76.

Titley, H. K., Heskin-Sweezie, R., Chung, J.-Y. J., Kassardjian, C. D., Razik, F., & Broussard, D. M. (2007). Rapid consolidation of motor memory in the vestibuloocular reflex. *Journal of Neurophysiology, 98*, 3809–3812.

Waespe, W., Cohen, B., & Raphan, T. (1983). Role of the flocculus and paraflocculus in optokinetic nystagmus and visual-vestibular interactions: Effects of lesions. *Experimental Brain Research, 50*, 9–33.

Waespe, W., & Henn, V. (1977). Neuronal activity in the vestibular nuclei of the alert monkey during vestibular and optokinetic stimulation. *Experimental Brain Research, 27*, 523–538.

Yakushin, S. B., Raphan, T., & Cohen, B. (2003). Gravity-specific adaptation of the angular vestibuloocular reflex: Dependence on head orientation with regard to gravity. *Journal of Neurophysiology, 89*, 571–586.

68   The Neurology of Eye Movements: From Control Systems to Genetics to Ion Channels to Targeted Pharmacotherapy

DAVID S. ZEE AND AASEF G. SHAIKH

Eye movements are an elegant window to the function of the normal brain and to how it malfunctions when plagued by disease or trauma. Here we update our knowledge of two central aspects of ocular motor control using saccades as the archetypical eye movement. First, we will discuss recent developments about the neurophysiological underpinnings of saccades and recently based neuromimetic models that combine anatomical circuitry, the biomechanical properties of cellular membranes, ion channel kinetics, and the changes in neural activity that accompany normal and abnormal eye movements. A similar approach is now being applied to other subtypes of eye movements and gaze holding. Targeted pharmacotherapy of eye movement disorders is an important practical outcome of such advances. Secondly, we will focus on using internal feedback loops to understand the control of saccades, including applications to the adaptive control of saccades. Gaining new knowledge of ocular motor learning and compensation, a major focus of current scientific enquiry, will be translated into better ways to treat patients.

For five decades, control systems analysis, anatomical circuit diagrams, and the recording of neural activity within single neurons during eye movements from experimental animals have been the pillars of research into the control of eye movements. Bioengineers began modeling the ocular motor control system in the 1960s based upon recordings of eye movements in normal human subjects during various tracking and fixation tasks. However, key to validating these models was the ability to record eye movements simultaneously with activity from single neurons within the brainstem of alert behaving monkeys and so relate neural activity to the kinematic and dynamic characteristics of eye movements (e.g., Robinson, 1970). These physiological findings were used to test existing hypotheses that were based upon behavior alone and to develop more realistic models of the nature and flow of information within the neural circuits that generate premotor commands for eye movements. It followed naturally that one could use these newer, physiologically and anatomically based mathematical models to interpret disorders of eye movements in patients who were the unfortunate victims of disease and trauma due to the vicissitudes of nature. Quantification of abnormal eye movements in patients, in turn, was used to test further the concepts of how eye movement commands were generated in healthy humans (Ramat et al., 2007). These clinical studies led to refinement and sometimes refutation of long-held ideas about normal ocular motor control. Two striking examples, in that they had a major influence on how we view the brainstem circuits that generate saccade commands, were from patients who made abnormally slow saccades and from patients who had impaired fixation due to saccadic oscillations, unwanted back-to-back saccades, occurring one upon the other without any time between.

A PHYSIOLOGICAL AND CONTROL SYSTEMS APPROACH

Slow Saccades

The study of patients who make slow saccades, from which the idea of a local feedback loop controlling the generation of premotor saccade commands developed, typifies the ongoing reciprocal relationship between basic science and clinical observation. Each makes a contribution to our understanding of brain function, usually in small, but occasionally large, "paradigm changing" iterative steps. Initially saccades were conceived as "ballistic," preprogrammed movements that

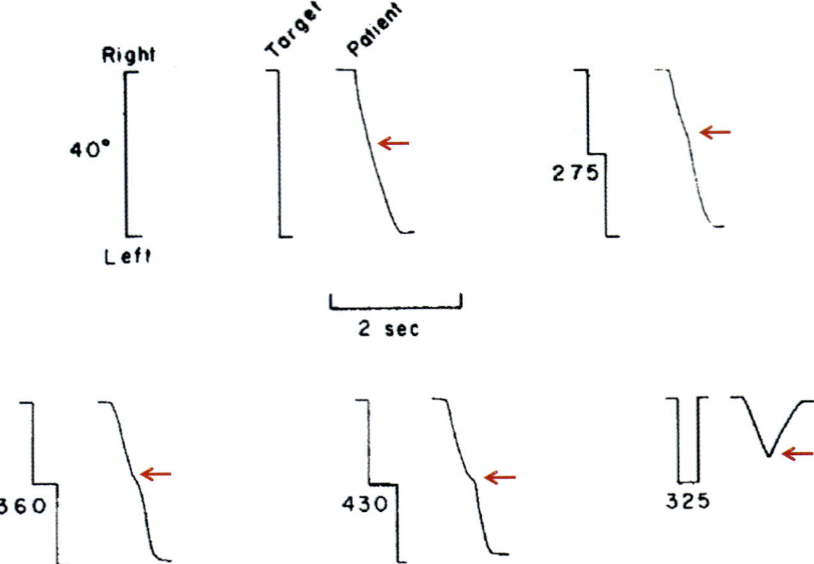

FIGURE 68.1   Response of a patient, who made slow saccades, to successive displacements of the target separated by different interstep intervals. Arrows indicate the inflection in the eye position trace which reflects the response to the second target jump. (Modified from Zee et al., 1976 with permission.)

could not be modified once launched. An obligatory refractory period was thought to follow each saccade during which time a new saccade could not be produced. The slow saccades made by patients, however, allow time to present new visual information that can be processed and potentially used to modify the saccade before it reaches its original intended location (Zee et al., 1976). Indeed we showed that the slow saccades of these patients could be modified and even turned around in midflight in response to a new desired target location (figure 68.1).

The behavior of patients who make slow saccades could not be explained by the prevailing ballistic model. In place of the ballistic model a continuous control model of saccade generation emerged (see figure 68.2). The continuous control model is based on the internal monitoring of the ongoing command to the eye during the saccade (figure 68.2, dashed brown path). This internal signal or "efference copy" is used to estimate where the eye actually is in the orbit as the saccade unfolds. The internal estimate of eye position then is continuously compared with the signal encoding the desired final position of the eye which is necessary for the saccade to deliver the fovea to its new target. In this construct the saccade is terminated automatically when the eye reaches its goal, that is, when the desired eye position and the internal estimate of eye position become the same.

It was natural to apply this new model based on the slow saccades made by a patient to the control of saccades in normal individuals with the caveat, of course, that the disease itself could have altered information processing for the control of saccades. Nevertheless, physiological findings based on neural recordings of the activity of premotor saccade-related "burst" neurons within the brainstem fit well with this new model of internal feedback control of saccades (Van Gisbergen, Robinson, & Gielen, 1981). Furthermore, pathological studies of the brain of a patient who had made slow saccades in life confirmed that the premotor, saccade-related, burst neurons within the pons had succumbed to the degenerative process (Geiner et al., 2008). For almost four decades, with a few, relatively small refinements, this local feedback model of saccade control has been the hypothetical construct driving most of the research into how the brainstem generates immediate premotor saccade commands (Jürgens, Becker, & Kornhuber, 1981).

Recent studies using transcranial magnetic stimulation (TMS) have revealed more about internal, online, feedback control of saccades (Xu-Wilson et al., 2011). TMS (or any "startling" stimulus—for example, a loud sound) delivered near the beginning of the saccade causes saccades to pause briefly beginning about 45 ms after the perturbation. After about 30 ms more the eye movement resumes, and its trajectory is corrected within the same saccade using compensatory motor commands that guide the eyes to the target. This within-saccade correction does not rely on visual input, suggesting that the brain monitors the ocular motor commands as the saccade unfolds, maintains a real-time estimate of the position of the eyes, and corrects for the

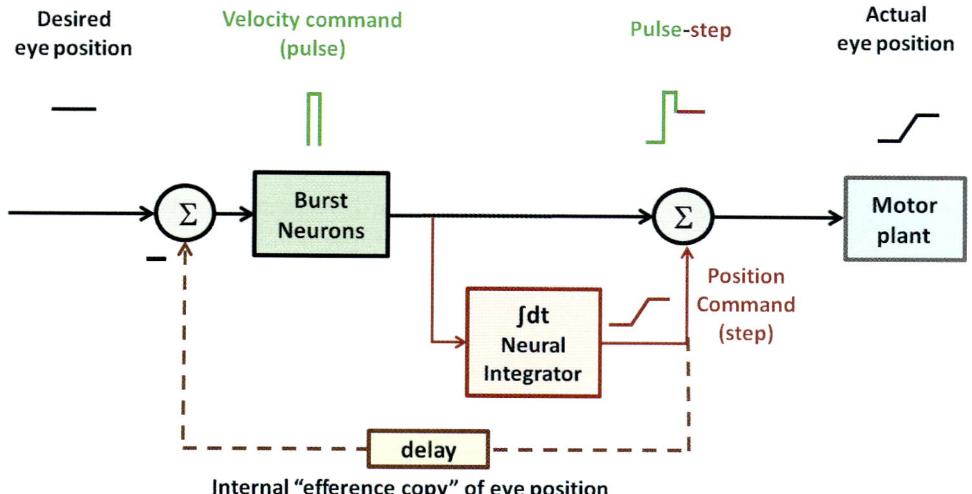

FIGURE 68.2   Simplified conceptual schematic for generating saccade commands using internal efference copy. Midbrain and pontine burst neurons produce a high-frequency discharge (pulse) encoding saccade velocity (green box). This signal is passed to the motor plant directly and also integrated mathematically by a neural integrator (red box) to produce a position command. This "pulse-step" of innervation drives the eye fast by overcoming orbital viscous forces and holds the eye in position at the end of the saccade by overcoming orbital elastic restoring forces. An efference copy of the position command (brown dashed line) is compared with the desired eye position, and when the two become equal, the neural signal driving the burst neurons becomes zero. The burst neurons then cease discharging and the saccade is over. Note the inherent delay in the efference copy position feedback signal that can make the system susceptible to oscillations (see the section on saccade oscillations).

perturbation. This corrective mechanism, perhaps elaborated by the cerebellum, affects ongoing motor commands upstream of the ocular motoneurons. Possible loci are at the level of the superior colliculus or more directly on the omnipause neurons within the pons (see the section on saccadic oscillations and figure 68.3 for the circuitry of generation of saccades, which includes pause neurons). Therefore, a TMS pulse (or equivalent startling stimulus) perturbs saccadic motor commands, but, because the progress of the eye is monitored internally, the saccade command can be adjusted, using immediate feedback, to ensure that the eye still gets to the target.

Continuous Control of Saccades Using Internal Models and the Consequences for Adaptation

The use of internal models that allow for continuous modification of the movements of the eye as the saccade unfolds is also an important aspect of the adaptive control of eye movements (see figure 68.4). Both natural development/aging and the vagaries of life with exposure to disease and trauma can alter both the central commands that move the eyes, as well as the effector organ, the eye muscles, and the other tissues within the orbit. These changes can lead to inaccuracy of saccades or dysmetria, which increases the time it takes to bring the image of interest onto the fovea and,

in turn, to process new visual information. Elimination of dysmetria is a fundamental requirement to ensure optimal visuomotor performance and, of course, increase the chance of an organism for survival. In the very short term, changes in motivation, attention, fatigue, or the prospect and timing of a reward for that particular motor behavior can alter saccade commands and, in turn, the speed and duration of the ensuing eye movement (e.g., Shadmehr et al., 2010; Xu-Wilson, Zee, & Shadmehr, 2009b). Regardless of these dynamic changes, saccades remain largely accurate during health, presumably due to internal models that can predict the sensory consequences of motor commands (see figure 68.4, top red path). These models can be used to make immediate adjustments when possible or, in the long term when faced with persistent errors, they can be recalibrated to restore the accuracy of saccades (Chen-Harris et al., 2008; Ethier, Zee, & Shadmehr, 2008; Xu-Wilson et al., 2009a). Such motor learning occurs over many time scales, from seconds to hours to days to weeks. The cerebellum (Golla et al., 2008; Iwamoto & Kaku, 2010; Jenkinson & Miall, 2010; Kojima, Soetedjo, & Fuchs, 2011; Ohki et al., 2009; Prsa & Thier, 2011; Schubert & Zee, 2010; Xu-Wilson et al., 2009a) and the superior colliculus (Kaku, Yoshida, & Iwamoto, 2009; Takeichi, Kaneko, & Fuchs, 2007) are important structures involved in adaptive control of saccades. The cerebral hemispheres, too, show changes in activity

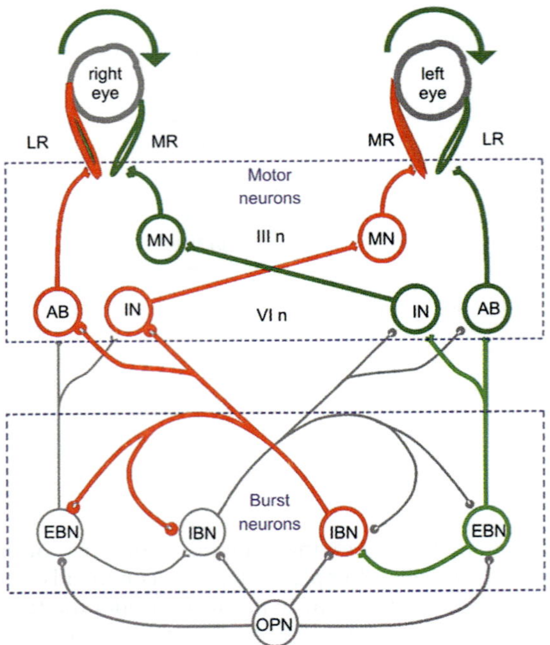

FIGURE 68.3 Brainstem circuit for controlling horizontal saccades. The premotor neurons are located in the pons and rostral medulla. The excitatory burst neurons (EBN) project to the ipsilateral abducens nucleus that contains interneurons (IN) and motor neurons (AB) which relay excitation to neurons innervating the agonist muscles and also to the region of the ipsilateral inhibitory burst neurons (IBN) (green). Axons of the IBN cross the midline and project to the contralateral interneuron (IN) and abducens motor neurons (MN) innervating antagonist muscles, and to the region of the contralateral EBN and IBN (red). When a rightward saccade is called for, the axons from right EBN carry impulses to excite motor neurons [or the relay IN] innervating the agonist muscles rotating the eyes to the right. At the same time, right IBN inhibit the antagonist muscles that would normally rotate the eyes to the left. The mutual inhibition between the burst neurons across the midline predisposes the neural circuit to instability and can lead to saccadic oscillations (uncalled for back-to-back saccades). OPN are pause neurons that inhibit burst neurons when fixation is required and cease discharging when a saccade is called for by releasing inhibition upon burst neurons. III n, oculomotor nerve to medial rectus (MR) and abducens nerve (VI n) to lateral rectus (LR). (Modified from Shaikh et al., 2007 with permission.)

during saccade adaptation (Blurton, Raabe, & Greenlee, 2012), though their role in driving adaptive changes is not firmly established (Gaymard et al., 2001).

Short-Term Saccade Adaptation: The Double-Step Paradigm

For relatively short-term learning, over minutes, saccade adaptation is commonly elicited using a double-step target paradigm that simulates dysmetria by displacing the target forward or backward to a new location once the saccade is launched toward the initial target location. When the saccade is completed, there is a new visual error signal requiring a corrective saccade. When repeated over many trials, subjects gradually increase or decrease the size of the saccade to the initial target location in response to the target jumping forward or backward, respectively. While widely used, the double-step paradigm introduces an isolated visual error signal, occurring with saccades only, to induce adaptation. This situation rarely occurs in natural behavior; more commonly adaptation is called for when there is a change in the strength of the eye muscles or in the properties of other tissues in the orbit. In this circumstance, unlike the isolated visual error signal provided in the double-step paradigm, there is another potential error signal, a mismatch between afferent (mediated by proprioception) and efferent signals about the position or movement of the globe. While proprioception does not play a role in the online modulation of saccades, it may be important for long-term adaptive changes in saccade accuracy, especially when the error is different between the two eyes calling for disconjugate adaptation (Shan et al., 2007a, 2007b). Indeed inverse models of the motor plant may make use of proprioceptive information for their adaptive calibration (see figure 68.4, middle path and legend). More central lesions that interrupt processing of premotor saccade commands by the brainstem or cerebellum would also lead to a different pattern of dysmetria that would have to be recognized and repaired by adaptive processes. With these caveats in mind, however, the double-step paradigm has been a useful and valid experimental tool to study motor learning in the saccade system (Scudder, Batourina, & Tunder, 1998).

Using the double-step paradigm, it has been shown that the brain appears to use two strategies to modify saccades, one for gain-up adaptation, which would be the response when saccades are too small, and the other for gain-down adaptation, which would be the response when saccades are too big (Ethier, Zee, & Shadmehr, 2008). This dichotomy was inferred from changes in the dynamic properties of saccades, that is, their peak-velocity amplitude relationships and their trajectories. For gain-down adaptation, the findings could be accounted for by a change in a forward model (see figure 68.4, top path). For gain-up adaptation, the finding could be accounted for by a remapping of visual (retinal) signals, with no change in the forward model (figure 68.4, left, input to motor command). A rationale for different adaptation strategies was that each is better suited, depending upon whether a reduction or an increase in saccade size is called for, to ensure the least expenditure of energy during the

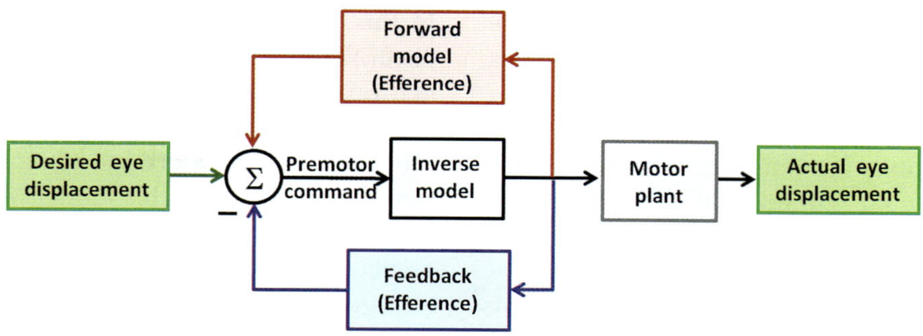

FIGURE 68.4   Simplified conceptual scheme for feedback control of saccades. Green box depicts the desired saccade. The output of the summing junction, the premotor command, reflects the combination of the activity of the midbrain and pontine burst neurons that produce the high-frequency pulse and the neurons that produce its integral, the step, from the neural integrator (see also figure 68.2). These signals will be processed by an "inverse model" (black box). The word "inverse" represents the inverse transformation of the action that will be performed by the motor plant (motor neurons, eye muscles, and other orbital tissues). In primates, a group of premotor neurons in the nucleus prepositus hypoglossi and medial vestibular nucleus (burst–tonic neurons), may represent the output of the inverse model (Ghasia, Meng, & Angelaki, 2008). The output of the inverse model is projected to the motor plant for the desired action. Simultaneously the efference copy of the output of the inverse model is projected back via a "forward" model (red box). Its function is to compare the motor commands with the expected sensory feedback and correct the motor command as required. The efference copy signal also provides the feedback regarding the actual motor action performed. This feedback signal is combined with the feedback signal from the forward model and compared with the desired displacement signal. The premotor command ends when the output of the summing junction becomes zero, that is, there is no longer an error signal to drive the burst neurons. Note that while we show the flow of information in this model through three separate paths, a priori they do not have to be independent. The cerebellum could represent the neural correlate of the forward and inverse models and, as a potential site of saccade adaptation, adjust their behaviors in the long term. Afferent (proprioceptive) information could also be used to calibrate these models. Another site of adaptation could be in the higher-level command to the saccade pulse generator, the desired eye displacement. This signal could be recalibrated directly or based on a remapping of sensory (visual) information.

movement associated with optimal speed and accuracy (i.e., to reduce the "motor costs" associated with the movement). For gain-up versus gain-down adaptation, Schnier and Lappe also showed differences in the changes in the dynamic properties of adapted saccades as well as in the transfer of adaptation to different types of saccades (Schnier & Lappe, 2011). Panouilleres and colleagues found evidence for separate mechanisms underlying gain increase and gain decrease adaptation by recording antisaccades (voluntary saccades made to the opposite, mirror location of a visually presented target) after conventional visually induced double-step adaptation (Panouilleres et al., 2009). They found transfer to antisaccades only after gain-down adaptation. There are caveats, however, in generalizing findings from short-term saccade adaptation, in response to artificial visual displacements of the target, to long-term and more enduring changes in saccade accuracy (Robinson, Soetedjo, & Noto, 2006). Context, prediction and anticipation, and perhaps the learning associated with developing skills rather than responding to errors in performance may show different optimization behaviors (Krakauer & Mazzoni, 2011).

Sources and Nature of Error Signals That Drive Adaptation

Much is unknown about adaptive control of saccades. What is the role of the amount of trial-to-trial variability (Srimal et al., 2008; Xu-Wilson et al., 2009a) or the size of the error signal, large or small, in determining the rate and completeness of adaptation (Wong & Shelhamer, 2011b)? What are the error signals (are they purely visual based on the discrepancy between the position of the eye at the end of the saccade and the location of the target, or are they based on some expectation of where the eye should end up relative to the target) (Wong & Shelhamer, 2011c)? How much of a role do prediction and context play in driving saccade adaptation (Herman, Harwood, & Wallman, 2009; Iwamoto & Kaku, 2010; Madelain et al., 2010; Pelisson et al., 2010; Tian & Zee, 2010; Wong & Shelhamer, 2011a)? How does the brain determine the source of the dysmetria (the "credit assignment problem") (Chen-Harris et al., 2008)? Does the dysmetria arise in the processing of error signals by the visual system, in the areas of the brain which produce motor commands, or in the final common pathway itself (ocular motoneurons, eye muscles, or orbit)? The adaptive mechanisms must know at what level within the nervous system the correction must be applied. For example, if the eye muscles are involved all the eye movement subsystems—saccades, pursuit, vestibular, and vergence—have to be recalibrated. However, if the problem is restricted to the pontine premotor saccade burst neurons, only the horizontal saccade circuits need to be readjusted. Similar considerations apply to attributing dysmetria to abnormalities in the circuits mediating the higher-level control of saccades. Is the dysmetria only for the more voluntary saccade types such as spontaneous or purposeful scanning of the environment or only for the more reactive, reflexive saccade types, such as the immediate response to the sudden appearance of a novel visual stimulus (Cotti et al., 2009; Hopp & Fuchs, 2010; Lavergne et al., 2011; Schnier & Lappe, 2011; Srimal & Curtis, 2010)? Even within the cerebellum the adaptive control of different types of saccades may be compartmentalized (Alahyane et al., 2008; Kojima, Soetedjo, & Fuchs, 2010). The problem of how to adapt eye movements becomes even more complicated as saccades are commonly combined with other types of eye movements, including pursuit (Schutz & Souto, 2011) and vergence. Naturally, especially for large changes in gaze, saccades occur with movements of the head; this adds the complexity of the interaction between a vestibular-induced slow phase and oppositely directed saccade. Adaptive mechanisms must decipher which part of these complicated motor synergies needs to be fixed and then parse the adjustments accordingly.

SACCADIC OSCILLATIONS

The local feedback model was also applied to understanding the origin of unintended saccades in patients with back-to-back, without an intersaccadic interval, saccadic oscillations (Ramat et al., 2007; Zee & Robinson, 1979). Two features of the local feedback model make it susceptible to oscillations, the inherent delay in the local feedback loop and the extremely high-frequency discharge of premotor saccadic burst neurons, which is needed to drive the eyes at high speed. If the delay becomes excessive, the system can become unstable (because it contains a negative feedback loop), and unwanted saccadic oscillations can emerge. The discovery of pause neurons within the pons, which project to and inhibit saccade burst neurons during steady fixation, was incorporated into the local feedback model to develop a hypothesis for saccade oscillations. Pause cells discharge steadily during fixation (and prevent any extraneous discharge of burst neurons) and, perforce, must cease discharging when a saccade is to be made.

In this way, saccadic oscillations could appear with a disturbance in which (1) pause cells no longer inhibited burst neurons during attempted fixation and (2) the delay in the internal feedback loop of eye position, which is used to determine when the saccade is over, was increased. While this model was a reasonable first approximation to understand pathological saccadic oscillations, it demanded physiologically unlikely changes in the delay within the feedback loop to account for the wide range of amplitudes and frequencies of saccadic oscillations shown by different patients or sometimes even within the same patient at different times.

Neuromimetic Models of Saccade Generation

A conceptual breakthrough to understanding the control of saccades and an understanding of saccadic oscillations in patients was based, in part, upon several recent observations of saccadic oscillations in normal subjects. Normal individuals show horizontal saccadic oscillations (1) when saccades are combined with vergence movements, (2) during purely large vertical saccades, and (3) with so-called "voluntary nystagmus," which actually is a series of small-amplitude, back-to-back saccades, one upon the other with no stopping in between, which some normal individuals can generate (Shaikh et al., 2008).

At the same time, models of the generation of saccades embraced two new ideas. First was the discovery of feedback loops between excitatory and inhibitory burst neurons within the brainstem with relatively brief delays between them (see figure 68.3). These delay loops were incorporated into the model and predispose the saccadic system to oscillations in a more physiological way than by simply increasing the delay in the (longer) local feedback loop carrying the efference copy of the eye (Ramat et al., 2005).

Secondly the role of membrane kinetics was incorporated into what was called a neuromimetic model for the generation of saccades (Miura & Optican, 2006). The behavior of specific ion channels and receptor neurotransmitters, along with the actual discharge rates of the different classes of neurons that participate during saccades, became the core features underlying the physiology of saccade behavior. The phenomenon of postinhibitory rebound (PIR), in which a neuron shows a spontaneous, brief burst of activity when it is suddenly released from inhibition, played a central role in this new model of saccade generation (Enderle, 2002). PIR provides the boost of activity that enables saccades to start promptly with high accelerations. Variations of this model have successfully explained slow

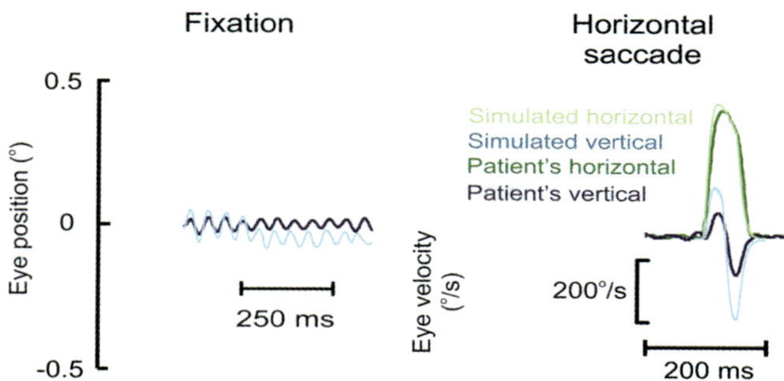

FIGURE 68.5   Simulation of saccade oscillations in a patient with microsaccade flutter. Continuous small-amplitude oscillations during fixation which become larger when associated with a saccade. Note the presence of vertical oscillations during the horizontal saccade, which reflects the fact that pause neurons project to and release inhibition from both horizontal and vertical burst neurons even when a pure horizontal saccade is called for. (From Shaikh et al., 2007 with permission.)

saccades after experimental lesions in the pause cell region in monkeys (Kaneko, 1996; Miura & Optican, 2006; Soetedjo, Kaneko, & Fuchs, 2002), slow saccades made under closed eyelids by normal human subjects (Shaikh et al., 2010b), saccadic oscillations made by normal human subjects during combined vergence and saccades (Ramat et al., 1999), the characteristics of the resumed saccades after interruption by TMS (Xu-Wilson et al., 2011), and pathological oscillations such as microsaccadic flutter in patients with familial saccadic oscillations (Shaikh et al., 2007) (see figure 68.5). One should remember, however, there is a downside to PIR because when coupled with the extremely high-frequency rate of discharge normally achieved by burst neurons, and the inherent delays in its feedback loops, the saccadic system lives on the edge of instability with the ever present risk of breaking out into saccadic oscillations. The model nicely captures these behavioral correlates of PIR and burst neuron discharge. The importance to the clinician of these ideas is that when oscillations are pathological or when uncalled for saccades occur, the activity within specific ion channels or the activity of specific neural transmitters can be targeted pharmacologically to reduce the excitability of the membrane and hence the propensity for intrusive extraneous saccades or saccadic oscillations (Serra et al., 2008; Shaikh et al., 2008; Shaikh et al., 2011b) (see figure 68.6).

One's genetic background now becomes an important consideration in understanding why some patients show pathological saccadic oscillations and others don't, when faced, for example, with the same metabolic insult. The degree to which one will be susceptible

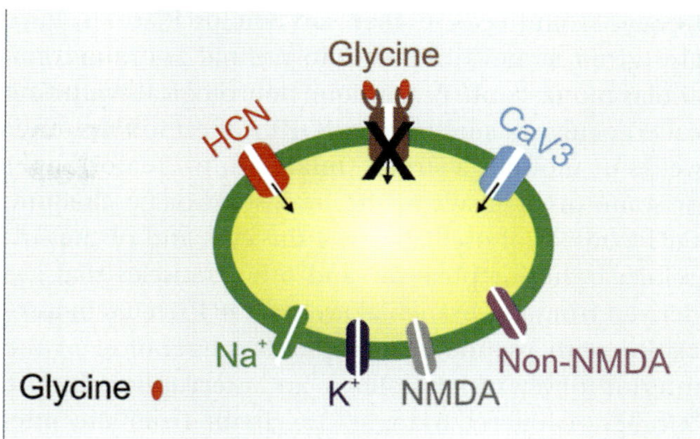

FIGURE 68.6   Membrane channels used in the conductance-based model of premotor burst neurons. Ion channels comprising the traditional Hodgkin–Huxley model of cell membranes were used to generate the action potential. In order to simulate the physiologically realistic neural behavior, pacemaker ion channels such as the hyperpolarization-activated cation current (HCN) and low-threshold calcium current (CaV3) were also included. Neural modulation as a response to neurotransmitter release was accomplished by simulated NMDA and non-NMDA excitatory glutamatergic channels as well as glycine or GABA sensitive inhibitory channels. A defect in glycinergic inhibition (black cross symbol) could cause saccadic oscillations. (Modified from Shaikh et al., 2007 with permission.)

to saccadic oscillations will depend on one's inherent complement of ion channel subtypes and their relative ability to increase membrane excitability. As an example, the strength of PIR that one develops and, consequently, the amplitude of the acceleration of the eye at saccade onset and ultimately the propensity for oscillations, could depend upon the genetically determined, relative proportion of fast- and slow-acting ion channel subtypes (Shaikh et al., 2007). In support of this idea is the report that the frequency of saccadic oscillations that comprise "voluntary nystagmus" that some normal subjects can produce runs true among members of the same family but differs between families (Neppert & Rambold, 2006). Such differences among individuals in saccade behavior based upon one's genetic background help to explain why the clinical expression of a particular disease can vary even when the nature of the neurological insult is the same.

PENDULAR OSCILLATIONS

Neuromimetic Models of Ocular Palatal Tremor

Neuromimetic models of other ocular motor abnormalities, such as pendular nystagmus in which the oscillations are quasi-sinusoidal and of a lower frequency and velocity than saccadic oscillations, have also given us new insights into normal and abnormal ocular motor control. A unique neurological syndrome called ocular palatal tremor (OPT) often follows, after weeks or months, a stroke (infarction or hemorrhage) or some other insult in the brainstem or cerebellum. OPT consists of oscillations of the eyes and of the soft palate, posterior pharynx, and other muscles that are derived from the branchial arches. OPT occurs in association with pseudohypertrophic degeneration of the inferior olive, which follows an interruption in the "Guillain–Mollaret triangle" (a circuit from the inferior olive to the deep cerebellar nuclei [and cerebellar cortex], and then through the superior cerebellar peduncle, passing through the red nucleus and then descending in the central tegmental tract back to the inferior olive). The vestibular nuclei, as displaced deep cerebellar nuclei, may also participate in this circuit. The oscillations of OPT are irregular, smooth, often disconjugate, and variable in shape from cycle to cycle (Shaikh et al., 2010a). An elegant animal model for the pathogenesis of OPT was developed by de Zeeuw et al. (de Zeeuw et al., 1990; Ruigrok, de Zeeuw, & Voogd, 1990). They produced hypertrophic degeneration of the inferior olive in the rat; this led to increases in connexin-mediated gap junctions between neighboring inferior olivary neurons as their somas ballooned in size, and they began increasingly to touch each other. In normal circumstances the gap junctions are restricted to the dendrites of the inferior olive. In OPT, because of the inferior olivary hypertrophy and the new somatosomatic gap junctions, local inferior olive patches become increasingly synchronized and, by virtue of their projections to the cerebellum, act as "pacemakers" that could both produce unwanted oscillations as well as drive maladaptive learning by the cerebellar cortex (Shaikh et al., 2010a). In this scheme the maladaptive cerebellar learning makes the oscillations arising from the abnormal inferior olive irregular. This hypothesis is supported by the effects of drugs given to patients with OPT (Shaikh et al., 2011a). Both gabapentin and memantine reduced the amplitude of OPT and affected the cycle-to-cycle variability of their frequency. While these drugs have many sites and modes of action, one can hypothesize they both reduce the excitability of cerebellar Purkinje neurons; gabapentin by decreasing calcium trafficking and memantine by blocking NMDA-glutamate receptors. Together these effects could reduce the amplitude and affect the irregularity of the frequency of OPT. While speculative, these findings encourage clinicians to develop more rational ways to use medications that target specific ion channels or transmitters.

Neuromimetic Models of Acquired Pendular Nystagmus with Multiple Sclerosis

Pendular nystagmus occurs frequently in patients who have multiple sclerosis and is severely disabling. Pendular nystagmus has been attributed to instability in the common ocular motor neural integrator, which is the network that mathematically integrates premotor conjugate velocity commands of all types—saccades, pursuit and vestibular—to produce position commands that hold the eyes steady in the orbit during fixation (Das et al., 2000). In support of this idea the phase of the ongoing oscillations of pendular nystagmus is reset when perturbed by a velocity signal such as a saccade (Das et al., 2000), suggesting that the integrator itself is directly involved in the generation of pendular nystagmus.

Aksay and colleagues formulated a scheme to account for the behavior of individual neurons within the integrator circuit (Aksay et al., 2001). The persistent tonic firing rate after the saccade is associated with the step-like changes in the interspike membrane potential of velocity–position integrator neurons. The interspike membrane potential and thus neuronal firing rate is directly proportional to the position of the eye. Within the integrator circuit there is a mutually excitatory feedback network among ipsilateral neurons and mutually inhibitory feedback between ipsi- and contralateral neurons. These connections determine yoked behavior of the neurons and the persistence of the firing rate (i.e., neural integration) (Aksay et al., 2007; Miri et al., 2011). This circuit architecture and the membrane properties of its neurons underscore the importance of strong network connections (as expected from neurons in proximity) in the efficiency of integration.

Based on these ideas, one can predict that a constant hyperpolarization of the membrane or disruption of integrator interconnections would prevent changes in the interspike membrane potential and subsequently impair the ability of the neural integrator to maintain a steady state change in the firing rate. Indeed injection of the hyperpolarizing agent, muscimol, at the putative site of the neural integrator in monkeys made it unstable while depolarization (with glutamate) reversed the effects (Arnold, Robinson, & Leigh, 1999). Additional pharmacological manipulations, however, make any simple interpretation imperfect since both agonists and antagonists of GABA, kainate, and glutamate cause abnormal neural integration, though in some cases the effects are opposite, either instability with runaway, velocity-increasing, slow phases or imperfect, "leaky" integration with velocity-decreasing slow phases

   DAVID S. ZEE AND AASEF G. SHAIKH

(Arnold, Robinson, & Leigh, 1999). Even so, these experiments point to a more rational approach to treating nystagmus in patients (Shaikh et al., 2011a). For example, it has been suggested that the severity of the instability of the neural integrator determines the amplitude of pendular nystagmus in patients with multiple sclerosis and that depolarization of the membrane would reduce the amplitude of nystagmus. Indeed, gabapentin and memantine, which could indirectly depolarize the cells of the nucleus prepositus hypoglossi, by blocking the alpha-2-delta subunit of calcium channels and antagonizing NMDA receptors at the cerebellar Purkinje neurons, reduce the amplitude of nystagmus in multiple sclerosis (Shaikh et al., 2011a; Thurtell et al., 2010b).

PATHOPHYSIOLOGY AND TARGETED PHARMACOTHERAPY IN OTHER OCULAR MOTOR DISORDERS

This novel approach to understanding and treating various forms of eye oscillations in the context of neurotransmitter and ion channel physiology is being expanded, especially in neurological conditions in which the ocular motor circuitry is not interrupted by a gross pathological lesion. Several examples include congenital nystagmus (or better yet called infantile nystagmus since it is not always obvious until several months of age), in which several drugs including acetazolamide can diminish the eye oscillations (Dell'osso et al., 2011; Thurtell et al., 2010a). Acetazolamide is effective in disorders with abnormal ion channels such as episodic ataxia type 2 in which the defect is in the calcium ion channel (Strupp et al., 2011). How acetazolamide might work in infantile nystagmus is unclear. Memantine and gabapentin are also effective in some patients with infantile nystagmus (McLean et al., 2007; McLean & Gottlob, 2009). Baclofen, with its action on GABA-B receptors, usually stops the oscillations of periodic alternating nystagmus (PAN) (Strupp et al., 2011). PAN emerges when inhibition on the velocity-storage circuit within the vestibular nuclei from the cerebellar nodulus is removed, allowing the vestibular system to become unstable. Memantine, too, may be used to treat PAN (Kumar et al., 2009). 4-aminopyridine, which acts on voltage-gated potassium channels, also may diminish nystagmus by replacing missing cerebellar inhibition and facilitating synchrony of discharge among Purkinje cells (Glasauer, Rossert, & Strupp, 2011; Strupp et al., 2011). Memantine, probably by acting upon excitatory neurotransmitters at the fastigial nuclei, diminishes unwanted saccadic intrusions that some cerebellar patients show (Serra et al., 2008).

FUTURE PERSPECTIVES

Based on these new studies and ideas, we now emphasize that the linkage between genetics and neural circuits helps us to understand the wide range of the phenotypic expression of genetic diseases as well as the variability in the response of otherwise normal individuals to external insults, and especially metabolic derangements. Genes determine one's particular complement of membrane ion channel subtypes, and so an abnormal complement of ion channels can lead to dysfunction in neural circuits without obvious structural abnormalities. In turn, regardless of cause, one can treat patients with molecules that target specific ion channels, or the release sites or targets of specific neurotransmitters, and, sometimes counterintuitively, alter the behavior of neural circuits to ameliorate ocular motor disorders. Any such path to understanding and treating ocular motor disorders will require a basic understanding of the anatomical connections of ocular motor circuits, of how neurons discharge during natural behavior, and how neural firing can be influenced using an understanding of ion channel and receptor physiology. At a behavioral level, another major challenge is to understand and learn to manipulate the multiple adaptive mechanisms that have evolved to compensate for disease and trauma and to optimize normal behavior in the face of natural development and aging. Choosing physical therapy programs that promote the quickest and most complete recovery after an insult and determining how to accelerate such recovery using medications are vital areas to be investigated.

REFERENCES

Aksay, E., Gamkrelidze, G., Seung, H. S., Baker, R., & Tank, D. W. (2001). In vivo intracellular recording and perturbation of persistent activity in a neural integrator. *Nature Neuroscience, 4*, 184–193.

Aksay, E., Olasagasti, I., Mensh, B. D., Baker, R., Goldman, M. S., & Tank, D. W. (2007). Functional dissection of circuitry in a neural integrator. *Nature Neuroscience, 10*, 494–504.

Alahyane, N., Fonteille, V., Urquizar, C., Salemme, R., Nighoghossian, N., Pelisson, D., et al. (2008). Separate neural substrates in the human cerebellum for sensory–motor adaptation of reactive and of scanning voluntary saccades. *Cerebellum, 7*, 595–601.

Arnold, D. B., Robinson, D. A., & Leigh, R. J. (1999). Nystagmus induced by pharmacological inactivation of the brainstem ocular motor integrator in monkey. *Vision Research, 39*, 4286–4295.

Blurton, S. P., Raabe, M., & Greenlee, M. W. (2012). Differential cortical activation during saccadic adaptation. *Journal of Neurophysiology, 107*, 1738–1747.

Chen-Harris, H., Joiner, W. M., Ethier, V., Zee, D. S., & Shadmehr, R. (2008). Adaptive control of saccades via internal feedback. *Journal of Neuroscience, 28,* 2804–2813.

Cotti, J., Panouilleres, M., Munoz, D. P., Vercher, J. L., Pelisson, D., & Guillaume, A. (2009). Adaptation of reactive and voluntary saccades: Different patterns of adaptation revealed in the antisaccade task. *Journal of Physiology, 587,* 127–138.

Das, V. E., Oruganti, P., Kramer, P. D., & Leigh, R. J. (2000). Experimental tests of a neural-network model for ocular oscillations caused by disease of central myelin. *Experiments in Brain Research, 133,* 189–197.

Dell'osso, L. F., Hertle, R. W., Leigh, R. J., Jacobs, J. B., King, S., & Yaniglos, S. (2011). Effects of topical brinzolamide on infantile nystagmus syndrome waveforms: Eyedrops for nystagmus. *Journal of Neuro-Ophthalmology, 31,* 228–233.

de Zeeuw, C. I., Ruigrok, T. J., Schalekamp, M. P., Boesten, A. J., & Voogd, J. (1990). Ultrastructural study of the cat hypertrophic inferior olive following anterograde tracing, immunocytochemistry, and intracellular labeling. *European Journal of Morphology, 28,* 240–255.

Enderle, J. D. (2002). Neural control of saccades. *Progress in Brain Research, 140,* 21–49.

Ethier, V., Zee, D. S., & Shadmehr, R. (2008). Changes in control of saccades during gain adaptation. *Journal of Neuroscience, 28,* 13929–13937.

Gaymard, B., Rivaud-Pechoux, S., Yelnik, J., Pidoux, B., & Ploner, C. J. (2001). Involvement of the cerebellar thalamus in human saccade adaptation. *European Journal of Neuroscience, 14,* 554–560.

Geiner, S., Horn, A. K., Wadia, N. H., Sakai, H., & Buttner-Ennever, J. A. (2008). The neuroanatomical basis of slow saccades in spinocerebellar ataxia type 2 (Wadia-subtype). *Progress in Brain Research, 171,* 575–581.

Ghasia, F. F., Meng, H., & Angelaki, D. E. (2008). Neural correlates of forward and inverse models for eye movements: Evidence from three-dimensional kinematics. *Journal of Neuroscience, 28,* 5082–5087.

Glasauer, S., Rossert, C., & Strupp, M. (2011). The role of regularity and synchrony of cerebellar Purkinje cells for pathological nystagmus. *Annals of the New York Academy of Sciences, 1233,* 162–167.

Golla, H., Tziridis, K., Haarmeier, T., Catz, N., Barash, S., & Thier, P. (2008). Reduced saccadic resilience and impaired saccadic adaptation due to cerebellar disease. *European Journal of Neuroscience, 27,* 132–144.

Herman, J. P., Harwood, M. R., & Wallman, J. (2009). Saccade adaptation specific to visual context. *Journal of Neurophysiology, 101,* 1713–1721.

Hopp, J. J., & Fuchs, A. F. (2010). Identifying sites of saccade amplitude plasticity in humans: Transfer of adaptation between different types of saccade. *Experimental Brain Research, 202,* 129–145.

Iwamoto, Y., & Kaku, Y. (2010). Saccade adaptation as a model of learning in voluntary movements. *Experimental Brain Research, 204,* 145–162.

Jenkinson, N., & Miall, R. C. (2010). Disruption of saccadic adaptation with repetitive transcranial magnetic stimulation of the posterior cerebellum in humans. *Cerebellum, 9,* 548–555.

Jürgens, R., Becker, W., & Kornhuber, H. H. (1981). Natural and drug-induced variations of velocity and duration of human saccadic eye movements: Evidence for a control of the neural pulse generator by local feedback. *Biological Cybernetics, 39,* 87–96.

Kaku, Y., Yoshida, K., & Iwamoto, Y. (2009). Learning signals from the superior colliculus for adaptation of saccadic eye movements in the monkey. *Journal of Neuroscience, 29,* 5266–5275.

Kaneko, C. R. S. (1996). Effect of ibotenic acid lesions of the omnipause neurons on saccadic eye movements in rhesus macaques. *Journal of Neurophysiology, 75,* 2229–2242.

Kojima, Y., Soetedjo, R., & Fuchs, A. F. (2010). Behavior of the oculomotor vermis for five different types of saccade. *Journal of Neurophysiology, 104,* 3667–3676.

Kojima, Y., Soetedjo, R., & Fuchs, A. F. (2011). Effect of inactivation and disinhibition of the oculomotor vermis on saccade adaptation. *Brain Research, 1401,* 30–39.

Krakauer, J. W., & Mazzoni, P. (2011). Human sensorimotor learning: Adaptation, skill, and beyond. *Current Opinion in Neurobiology, 21,* 636–644.

Kumar, A., Thomas, S., McLean, R., Proudlock, F. A., Roberts, E., Boggild, M., et al. (2009). Treatment of acquired periodic alternating nystagmus with memantine: A case report. *Clinical Neuropharmacology, 32,* 109–110.

Lavergne, L., Vergilino-Perez, D., Lemoine, C., Collins, T., & Dore-Mazars, K. (2011). Exploring and targeting saccades dissociated by saccadic adaptation. *Brain Research, 1415,* 47–55.

Madelain, L., Harwood, M. R., Herman, J. P., & Wallman, J. (2010). Saccade adaptation is unhampered by distractors. *Journal of Vision, 10,* 29.

McLean, R., Proudlock, F., Thomas, S., Degg, C., & Gottlob, I. (2007). Congenital nystagmus: Randomized, controlled, double-masked trial of memantine/gabapentin. *Annals of Neurology, 61,* 130–138.

McLean, R. J., & Gottlob, I. (2009). The pharmacological treatment of nystagmus: A review. *Expert Opinion on Pharmacotherapy, 10,* 1805–1816.

Miri, A., Daie, K., Arrenberg, A. B., Baier, H., Aksay, E., & Tank, D. W. (2011). Spatial gradients and multidimensional dynamics in a neural integrator circuit. *Nature Neuroscience, 14,* 1150–1159.

Miura, K., & Optican, L. M. (2006). Membrane channel properties of premotor excitatory burst neurons may underlie saccade slowing after lesions of omnipause neurons. *Journal of Computational Neuroscience, 20,* 25–41.

Neppert, B., & Rambold, H. (2006). Familial voluntary nystagmus. *Strabismus, 14,* 115–119.

Ohki, M., Kitazawa, H., Hiramatsu, T., Kaga, K., Kitamura, T., Yamada, J., et al. (2009). Role of primate cerebellar hemisphere in voluntary eye movement control revealed by lesion effects. *Journal of Neurophysiology, 101,* 934–947.

Panouilleres, M., Weiss, T., Urquizar, C., Salemme, R., Munoz, D. P., & Pelisson, D. (2009). Behavioral evidence of separate adaptation mechanisms controlling saccade amplitude lengthening and shortening. *Journal of Neurophysiology, 101,* 1550–1559.

Pelisson, D., Alahyane, N., Panouilleres, M., & Tilikete, C. (2010). Sensorimotor adaptation of saccadic eye movements. *Neuroscience and Biobehavioral Reviews, 34,* 1103–1120.

986   DAVID S. ZEE AND AASEF G. SHAIKH

Prsa, M., & Thier, P. (2011). The role of the cerebellum in saccadic adaptation as a window into neural mechanisms of motor learning. *European Journal of Neuroscience, 33,* 2114–2128.

Ramat, S., Leigh, R. J., Zee, D. S., & Optican, L. M. (2005). Ocular oscillations generated by coupling of brainstem excitatory and inhibitory saccadic burst neurons. *Experiments in Brain Research, 160,* 89–106.

Ramat, S., Leigh, R. J., Zee, D. S., & Optican, L. M. (2007). What clinical disorders tell us about the neural control of saccadic eye movements. *Brain, 130,* 10–35.

Ramat, S., Somers, J. T., Das, V. E., & Leigh, R. J. (1999). Conjugate ocular oscillations during shifts of the direction and depth of visual fixation. *Investigative Ophthalmology & Visual Science, 40,* 1681–1686.

Robinson, D. A. (1970). Oculomotor unit behavior in the monkey. *Journal of Neurophysiology, 33,* 393–403.

Robinson, F. R., Soetedjo, R., & Noto, C. (2006). Distinct short-term and long-term adaptation to reduce saccade size in monkey. *Journal of Neurophysiology, 96,* 1030–1041.

Ruigrok, T. J., de Zeeuw, C. I., & Voogd, J. (1990). Hypertrophy of inferior olivary neurons: A degenerative, regenerative or plasticity phenomenon. *European Journal of Morphology, 28,* 224–239.

Schnier, F., & Lappe, M. (2011). Differences in intersaccadic adaptation transfer between inward and outward adaptation. *Journal of Neurophysiology, 106,* 1399–1410.

Schubert, M. C., & Zee, D. S. (2010). Saccade and vestibular ocular motor adaptation. *Restorative Neurology and Neuroscience, 28,* 9–18.

Schutz, A. C., & Souto, D. (2011). Adaptation of catch-up saccades during the initiation of smooth pursuit eye movements. *Experiments in Brain Research, 209,* 537–549.

Scudder, C. A., Batourina, E. Y., & Tunder, G. S. (1998). Comparison of two methods of producing adaptation of saccade size and implications for the site of plasticity. *Journal of Neurophysiology, 79,* 704–715.

Serra, A., Liao, K., Martinez-Conde, S., Optican, L. M., & Leigh, R. J. (2008). Suppression of saccadic intrusions in hereditary ataxia by memantine. *Neurology, 70,* 810–812.

Shadmehr, R., Orban de Xivry, J. J., Xu-Wilson, M., & Shih, T. Y. (2010). Temporal discounting of reward and the cost of time in motor control. *Journal of Neuroscience, 30,* 10507–10516.

Shaikh, A. G., Hong, S., Liao, K., Tian, J., Solomon, D., Zee, D. S., et al. (2010a). Oculopalatal tremor explained by a model of inferior olivary hypertrophy and cerebellar plasticity. *Brain, 133,* 923–940.

Shaikh, A. G., Miura, K., Optican, L. M., Ramat, S., Leigh, R. J., & Zee, D. S. (2007). A new familial disease of saccadic oscillations and limb tremor provides clues to mechanisms of common tremor disorders. *Brain, 130,* 3020–3031.

Shaikh, A. G., Ramat, S., Optican, L. M., Miura, K., Leigh, R. J., & Zee, D. S. (2008). Saccadic burst cell membrane dysfunction is responsible for saccadic oscillations. *Journal of Neuro-Ophthalmology, 28,* 329–336.

Shaikh, A. G., Thurtell, M. J., Optican, L. M., & Leigh, R. J. (2011a). Pharmacological tests of hypotheses for acquired pendular nystagmus. *Annals of the New York Academy of Sciences, 1233,* 320–326. doi:10.1111/j.1749-6632.2011. 06118.x.

Shaikh, A. G., Wong, A. L., Optican, L. M., Miura, K., Solomon, D., & Zee, D. S. (2010b). Sustained eye closure slows saccades. *Vision Research, 50,* 1665–1675.

Shaikh, A. G., Zee, D. S., Optican, L. M., Miura, K., Ramat, S., & Leigh, R. J. (2011b). The effects of ion channel blockers validate the conductance-based model of saccadic oscillations. *Annals of the New York Academy of Sciences, 1233,* 58–63.

Shan, X., Tian, J., Ying, H. S., Quaia, C., Optican, L. M., Walker, M. F., et al. (2007a). Acute superior oblique palsy in monkeys: I. Changes in static eye alignment. *Investigative Ophthalmology & Visual Science, 48,* 2602–2611.

Shan, X., Ying, H. S., Tian, J., Quaia, C., Walker, M. F., Optican, L. M., et al. (2007b). Acute superior oblique palsy in monkeys: II. Changes in dynamic properties during vertical saccades. *Investigative Ophthalmology & Visual Science, 48,* 2612–2620.

Soetedjo, R., Kaneko, C. R., & Fuchs, A. F. (2002). Evidence that the superior colliculus participates in the feedback control of saccadic eye movements. *Journal of Neurophysiology, 87,* 679–695.

Srimal, R., & Curtis, C. E. (2010). Secondary adaptation of memory-guided saccades. *Experimental Brain Research, 206,* 35–46.

Srimal, R., Diedrichsen, J., Ryklin, E. B., & Curtis, C. E. (2008). Obligatory adaptation of saccade gains. *Journal of Neurophysiology, 99,* 1554–1558.

Strupp, M., Thurtell, M. J., Shaikh, A. G., Brandt, T., Zee, D. S., & Leigh, R. J. (2011). Pharmacotherapy of vestibular and ocular motor disorders, including nystagmus. *Journal of Neurology, 258,* 1207–1222.

Takeichi, N., Kaneko, C. R., & Fuchs, A. F. (2007). Activity changes in monkey superior colliculus during saccade adaptation. *Journal of Neurophysiology, 97,* 4096–4107.

Thurtell, M. J., Dell'osso, L. F., Leigh, R. J., Matta, M., Jacobs, J. B., & Tomsak, R. L. (2010a). Effects of acetazolamide on infantile nystagmus syndrome waveforms: Comparisons to contact lenses and convergence in a well-studied subject. *Open Ophthalmology Journal, 4,* 42–51.

Thurtell, M. J., Joshi, A. C., Leone, A. C., Tomsak, R. L., Kosmorsky, G. S., Stahl, J. S., et al. (2010b). Crossover trial of gabapentin and memantine as treatment for acquired nystagmus. *Annals of Neurology, 67,* 676–680.

Tian, J., & Zee, D. S. (2010). Context-specific saccadic adaptation in monkeys. *Vision Research, 50,* 2403–2410.

Van Gisbergen, J. A., Robinson, D. A., & Gielen, S. (1981). A quantitative analysis of generation of saccadic eye movements by burst neurons. *Journal of Neurophysiology, 45,* 417–442.

Wong, A. L., & Shelhamer, M. (2011a). Exploring the fundamental dynamics of error-based motor learning using a stationary predictive-saccade task. *PLoS ONE, 6,* e25225.

Wong, A. L., & Shelhamer, M. (2011b). Saccade adaptation improves in response to a gradually introduced stimulus perturbation. *Neuroscience Letters, 500,* 207–211.

Wong, A. L., & Shelhamer, M. (2011c). Sensorimotor adaptation error signals are derived from realistic predictions of movement outcomes. *Journal of Neurophysiology, 105,* 1130–1140.

Xu-Wilson, M., Chen-Harris, H., Zee, D. S., & Shadmehr, R. (2009a). Cerebellar contributions to adaptive control of saccades in humans. *Journal of Neuroscience, 29,* 12930–12939.

Xu-Wilson, M., Tian, J., Shadmehr, R., & Zee, D. S. (2011). TMS perturbs saccade trajectories and unmasks an internal feedback controller for saccades. *Journal of Neuroscience, 31,* 11537–11546.

Xu-Wilson, M., Zee, D. S., & Shadmehr, R. (2009b). The intrinsic value of visual information affects saccade velocities. *Experimental Brain Research, 196,* 475–481.

Zee, D. S., Optican, L. M., Cook, J. D., Robinson, D. A., & Engel, W. K. (1976). Slow saccades in spinocerebellar degeneration. *Archives of Neurology, 33,* 243–251.

Zee, D. S., & Robinson, D. A. (1979). A hypothetical explanation of saccadic oscillations. *Annals of Neurology, 5,* 405–414.

X CORTICAL MECHANISMS OF ATTENTION, COGNITION, AND MULTIMODAL INTEGRATION

69 Perceptual Learning

YUKA SASAKI AND TAKEO WATANABE

It is well known that the visual systems of the brain are most plastic during a specific "critical period" of one's early life, during which large-scale changes in visual processing can occur (Morishita & Hensch, 2008). It was once thought that hardly any change occurs in the visual system after the critical developmental period. However, it has since been found that repetitive training on a visual task can result in significant performance improvements on that task, even in adulthood. Moreover, these improvements persist for a long time. Such long-term improvement on a visual task is called perceptual learning (PL) (Fahle & Poggio, 2002; Gibson, 1963; Lu et al., 2011; Sagi, 2011; Sagi & Tanne, 1994; Sasaki, Nanez, & Watanabe, 2010; Seitz & Dinse, 2007) and may be a manifestation of ongoing plasticity in visual signal processing.

Recently PL has attracted a great deal of interest among researchers. In 2011 alone, as many as 40 papers with the words "perceptual learning" in the title were published. Though this metric is obviously crude, it illustrates the flurry of ongoing research on this topic. Though many PL researchers hope to find a general or unifying mechanism of PL, several key issues in the field are the subjects of increasingly controversial debates.

In this chapter, we will specifically discuss two of the most controversial issues: the brain region in which changes associated with PL occur and necessary factors that cause PL to occur during training.

If the reader is interested, there are other recent review papers on general PL issues (Lu et al., 2011; Sagi, 2011; Sasaki et al., 2010), clinical applications (Levi, 2005), and the role of sleep in PL (Walker & Stickgold, 2006), which we recommend.

NEURAL CHANGES ASSOCIATED WITH PL

The physiological site of PL remains a controversial point among researchers. PL could be associated with changes in early visual cortex, changes in the middle visual areas, or changes in readout as a result of connectivity within different visual areas or between visual areas and higher association areas. We will review research supporting each of these three hypotheses and then propose directions for future research.

Changes in Early Visual Cortex

Early studies found that PL was specific to the visual feature exposed during training. Karni and Sagi conducted the most influential series of studies in this line of research (Karni & Sagi, 1991, 1993; Karni et al., 1994). Their training stimulus consisted of a triad of texture elements whose orientation was different from that of an array of background texture elements. Subjects were instructed to first report whether a letter presented at the center of the display was "T" or "L" (to ensure the subjects' fixation) and then to indicate whether the orientation of the triad presented peripherally in one quadrant of the visual field was vertical or horizontal. This is called a texture-discrimination task (TDT).

Training resulted in a significantly lower threshold of stimulus onset asynchrony between the test stimulus and a mask. However, if the target was then presented in a different quadrant from the one in which the target triad had been trained, no PL effect was observed. In addition, after training was conducted with one eye, performance improvements observed with the trained eye were not found for the untrained eye. That is, PL of TDT was specific for location of the trained feature and trained eye. A number of other studies showed that PL is largely specific for orientation (Fiorentini & Berardi, 1980; Poggio, Fahle, & Edelman, 1992; Schoups, Vogels, & Orban, 1995), motion direction (Ball & Sekuler, 1981; Karni & Sagi, 1993; Koyama, Harner, & Watanabe, 2004; Vaina et al., 1998; Watanabe et al., 2002), contrast (Adini, Sagi, & Tsodyks, 2002; Shih et al., 2000), eye (Fahle & Edelman, 1993; Karni & Sagi, 1991; Seitz, Kim, & Watanabe, 2009), and location (Adini, Sagi, & Tsodyks, 2002; Ahissar & Hochstein, 1993, 1997; Crist et al., 1997; Fahle & Edelman, 1993; Fiorentini & Berardi, 1980; Karni & Sagi, 1993; Karni et al., 1994; McKee & Westheimer, 1978; Poggio, Fahle, & Edelman, 1992; Shiu & Pashler, 1992; Watanabe et al., 2002). These findings are in accord with a model in which repetitive training on a visual feature modifies early visual cortex, including the primary visual cortex (V1) of the adult's brain, in which neurons tend to respond more selectively to the features and retinotopic location of presented stimuli.

High specificity was also found for PL of task-irrelevant features, in which a visual feature is merely exposed in the periphery as task irrelevant, whereas subjects performed a central task on a different feature (for details of PL of task-irrelevant features, see the Role of Attention section below). While a rapid serial visual presentation task was performed in the center of the display, a motion stimulus, in which local dots moved in random directions within a certain range, was presented in the background. There are two types of perceived motion in this type of stimulus: the local motion signals of the individual dots and a global, averaged motion direction (Watamaniuk, Sekuler, & Williams, 1989). The range of local motion directions was varied across different subject groups. Performance increase was observed for all motion directions within the range of local motion, but not for the global motion percept. This indicates that PL occurred for task-irrelevant local motion directions, but not for the global motion direction (Watanabe et al., 2002). Once again, this result suggests that learning occurs in an early visual area where local motion signals are processed, but not global ones.

The hypothesis that PL results from changes in early visual cortex is supported by animal studies and human brain imaging research. It was found that training to discriminate orientations separated by a few degrees resulted in sharpening of the slope of the limbs of tuning curves in monkey V1 neurons that were tuned around ±15° from the discriminated orientations (Schoups et al., 2001). This result is consistent with the model for a fine-grained orientation discrimination task (Teich & Qian, 2003). In another study, training with monkeys failed to produce changes in the tuning properties of neurons, such as selectivity for location, size, and orientation selectivity. Nor did visual topography in V1 change. However, the influence of contextual stimuli presented outside the classical receptive field of trained neurons showed a change that was consistent with the trained discrimination (Crist, Li, & Gilbert, 2001). Recently, it was found that training in grating orientation identification resulted in increases in mean contrast sensitivity near the "trained" spatial frequency in V1 neurons of cats (Hua et al., 2010).

Although the results from animal studies are inconsistent (see below), data from brain imaging experiments strongly support the "early visual cortex" hypothesis. It has been found that spatiotemporal activation patterns with steep gradients of evoked potential field distributions over the primary visual cortex are correlated with PL as a result of exposure to vernier stimuli (Skrandies & Fahle, 1994) Changes in blood-oxygen-level-dependent (BOLD) signals in the region of the primary visual cortex corresponding to the retinotopic location of training have been reported in association with PL of TDT (Schwartz, Maquet, & Frith, 2002; Walker et al., 2005; Yotsumoto et al., 2009; Yotsumoto, Watanabe, & Sasaki, 2008) and orientation detection tasks (Furmanski, Schluppeck, & Engel, 2004). Enhancement of BOLD signals corresponding to the trained location in the primary visual cortex was also found while subjects slept following training (Yotsumoto et al., 2009). Recently Shibata and his colleagues (Shibata et al., 2011) developed a new online-feedback method that used decoded functional magnetic resonance imaging signals to induce activity patterns in early visual cortex (V1 and V2). These patterns corresponded to a preselected orientation and were induced without stimulus presentation or subjects' awareness of what was to be learned. Repeated induction of these activation patterns caused PL specific to the selected orientation. These results indicate that early visual cortex such as V1 and V2 are so plastic that mere inductions of activity patterns are sufficient to cause PL, although this finding does not rule out the possibility that some types of PL are associated with changes in higher areas or connectivity between different areas.

Changes in Middle Visual Areas

It has been found that in some experimental settings PL is not necessarily specific for the trained feature or its location. Training on a feature (e.g., contrast) at one location and additional training with an irrelevant feature/task (e.g., orientation) at a second location results in a complete transfer of learning of the feature (e.g., contrast) to the second location (Xiao et al., 2008). These results suggest that at least some types of PL are associated with changes in middle or higher stages of visual processing.

Some studies have found that PL is associated with tuning curve changes in middle-stage visual areas. John Maunsell and his colleagues conducted an electrophysiological study with monkeys. They did not find any pronounced changes in the response properties of neurons in V1 and V2 (Ghose, Yang, & Maunsell, 2002). Instead, they found that after training on a fine orientation discrimination task, neurons in V4 with receptive fields overlapping the trained location showed stronger responses and narrower orientation tuning properties than neurons with receptive fields in the untrained hemifield (Yang & Maunsell, 2004). Recently, it has been reported that training of coarse (around 90° apart) orientations discrimination resulted in robust increases in the capacity of V4 neurons to discriminate the trained orientations. This effect was observed during discrimination and also passive fixation (Adab & Vogels, 2011).

Changes in Connectivity within or beyond the Visual System

Although changes in visual areas associated with PL are generally thought to result from changes in the response properties of neurons, Barbara Dosher, Zhong-Lin Lu, and their colleagues have suggested that PL is largely a result of noise reduction in visual processing. In their model, PL is associated with connectivity changes between early visual cortex and higher visual areas, or perhaps areas beyond visual cortices (Davis et al., 2009; Dosher & Lu, 1998; Liu, Lu, & Dosher, 2010; Lu & Dosher, 2009).

Joshua Gold and his colleagues found that perceptual learning of coarse coherent motion direction discrimination was associated with tuning property changes in the lateral intraparietal area (LIP), an area known to be involved in perceptual decision making (Shadlen & Newsome, 2001), but not in the middle temporal area (MT), a visual area (Law & Gold, 2008), known to process visual motion. They further indicated that their data can be explained by changes in the weights of connections between MT and LIP (Law & Gold, 2009). In another monkey study (Chowdhury & DeAngelis, 2008), inactivation of MT by injection of the GABA agonist muscimol prior to coarse absolute depth discrimination training was shown to impair performance. However, fine relative depth discrimination training with MT inactivation neither impaired coarse depth discrimination nor led to changes in the disparity tuning of MT neurons. The authors concluded that when MT activation is inhibited, other areas in the ventral visual pathway are recruited and mediate coarse depth discrimination. These results suggest that PL occurs when decision units "learn" to more heavily weight disparity signals from the ventral pathway than those from MT.

Connectivity changes were also found in humans between the trained visual cortex and frontal–parietal areas (Lewis et al., 2009; Schwartz, Maquet, & Frith, 2002). After training on an identification task that employed a target stimulus presented in one visual quadrant, resting BOLD functional connectivity and directed mutual interaction between trained visual cortex and frontal–parietal areas were significantly modified. These changes were correlated with the degree of PL.

Possible Reasons for Variable Results

As the preceding discussion shows, different studies have used different experimental conditions with different parameter sets. Despite the variety of methods employed in this body of research, many studies overgeneralize their results and draw conclusions as if their specific results can be applied to all types of PL. However, it is neither certain nor even likely that all types of PL are subserved by the same and sole mechanism. The variance of experimental parameters outlined in the following list leaves the possibility that PL is subserved by multiple mechanisms.

TRAINED FEATURE The trained feature varies in PL studies. Commonly used features include orientation, motion direction, motion speed, and luminance contrast. These features are processed by different neural mechanisms and possibly by different brain areas or regions. Thus, it is not likely that PL of different features results from changes in the same mechanism.

TRAINED TASK Different tasks are used in PL research, even when the stimulus feature under study is held constant. For example, training in orientation or motion direction *discrimination* results in PL, as does training in orientation or motion direction *detection*. Just as the underlying mechanisms for performing these tasks may be different (Hol & Treue, 2001; Koyama, Harner, & Watanabe, 2004; Regan & Beverley, 1985; Vandenberghe et al., 1996), so too might be the mechanisms of PL for these tasks (Koyama, Harner, & Watanabe, 2004). Even within the seemingly narrow subdomain of discrimination tasks, coarse and fine discrimination may be subserved by significantly different mechanisms. In a coarse discrimination task, a stimulus is presented within noise and a subject is usually asked to indicate whether the presented orientation (or motion direction) is vertical (leftward) or horizontal (rightward) (Adab & Vogels, 2011; Law & Gold, 2008; Matarazzo et al., 2008). In a fine discrimination task, a subject is usually asked to discriminate between orientations (or motion directions) that are just a few degrees apart (Petrov & Van Horn, 2012; Schoups, Vogels, & Orban, 1995; Schoups et al., 2001). A most efficient way for the visual system to enhance performance in coarse discrimination would be to reduce the visual noise, whereas in fine discrimination, the visual system may sharpen the tuning curves of cells tuned around the discriminated stimuli.

TRAINING PHASE The time course or difficulty of training can also have profound effects on observed perceptual learning. For example, PL of TDT that demonstrated complete interocular transfer immediately after the offset of a training block showed incomplete transfer several hours later (Karni & Sagi, 1993). Another study that used a visual search task of variable difficulty found that an easier orientation discrimination was associated with higher location specificity of PL (Ahissar & Hochstein, 1997). These studies suggest that as learning proceeds, the involved mechanisms including the areas are changed.

Yotsumoto and their colleagues found that for the first few days of training of TDT, BOLD signals in the trained region of V1 in humans were enhanced, consistent with earlier studies (Schwartz, Maquet, & Frith, 2002). However, after several days BOLD signal amplitude in V1 decreased to pretraining levels, despite a sustained improvement of behavioral performance. This result indicates that different mechanisms may underlie the phase in which learning occurs and the phase in which learning is retained.

Conclusion: Neural Changes Associated with PL

The nature of the physiological changes that underlie PL—whether these occur in early visual cortex, in middle visual areas, or as a result of connectivity changes with nonvisual areas—remains a point of controversy. Each of the hypotheses is supported by reasonable evidence from psychophysics, brain imaging, and single-unit recordings. The most likely interpretation of these three competing hypotheses is that there is no single mechanism underlying all the types of PL. The area that undergoes change in association with PL seems to be dependent on a number of parameters. Thus, it would be fruitless to attack a competing hypothesis using experimental results obtained with a limited set of parameters. A more constructive approach may be to attempt to comprehensively explain different results obtained under different conditions. A recent interesting model that is in this line assumes that PL results from improved probabilistic inference induced by changes in early stages rather than in a decision unit, which could explain changes within and beyond visual areas (Bejjanki et al., 2011).

ROLES OF ATTENTION AND REWARD IN PL?

It has been found that at least two factors influence or are involved in the training of PL: attention and reward. These factors can exert powerful effects on PL, as in many other types of learning. How these factors influence PL is discussed in this section. As mentioned above, it is worth keeping in mind that the involvement of a particular factor in training does not necessarily mean that the brain area(s) related to that factor are changed in association with PL.

Role of Attention

Broadly speaking, the main role of attention is to select the most relevant and important aspects of information from the incoming environmental stimuli. This selection process may involve enhancing signals from relevant information while reducing irrelevant or unimportant signals.

Merrav Ahissar and Saul Hochstein suggested that attention significantly influences PL (Ahissar & Hochstein, 1993). In this study, subjects were first trained on a search task in which they had to indicate whether a stimulus array contained an element (target) of different orientation from the rest of the array's elements. Each presented array was either vertically or horizontally elongated (containing an extra row or column of elements), but this feature was not relevant to the target search task. After PL of the search task was complete, subjects were trained to indicate whether the same stimulus array as used in the search task was horizontally or vertically elongated. Little or no improvement was observed on this second task. The same tendency was observed when the order of the task trainings was switched. The authors concluded that attention plays an important role in PL since mere exposure to an unattended feature of the stimulus array (the elongation) failed to result in learning of that feature. Since then, a number of other studies have shown that PL of a visual feature that was merely exposed but not trained does not occur (Li, Piech, & Gilbert, 2004; Schoups et al., 2001; Shiu & Pashler, 1992). These results suggested that attention is necessary for PL to occur.

However, it has since been found that PL can occur as a result of exposure to a task-irrelevant and subthreshold coherent motion direction (Watanabe, Nanez, & Sasaki, 2001). This finding, called task-irrelevant perceptual learning (TIPL), indicates that while attention to a feature plays an important role in PL, it is not necessary for PL to occur. In other words, PL of a feature can occur without the involvement of attention to the feature during training. A number of subsequent studies have reported TIPL (Barbot, Landy, & Carrasco, 2011; Beste et al., 2011; Carrasco, Rosenbaum, & Giordano, 2008; Gutnisky et al., 2009; Rosenthal & Humphreys, 2010; Seitz & Watanabe, 2003; Seitz et al., 2005a; Tsushima, Seitz, & Watanabe, 2008; Watanabe et al., 2002; Xu, He, & Ooi, 2012; Zhang & Kourtzi, 2010). There now seems to be no doubt that TIPL occurs, though the existence of task-irrelevant learning does not diminish the important role of attention (Ahissar & Hochstein, 1993).

Why does TIPL occur in some cases but not others? To address this question, Tsushima and his colleagues conducted an experiment in which the coherence (and therefore detectability) of a task-irrelevant motion signal was varied during the performance of a central task (Tsushima et al., 2008). TIPL occurred only when the coherent motion signal was below or close to the observer's detection threshold. These results are consistent with a model in which the cognitive control system fails to detect weak or below-threshold signals and, therefore, fails to suppress these signals, thus allowing them to exert a stronger influence on information processing (Tsushima, Sasaki, & Watanabe, 2006; Yotsumoto et al., 2012). That is, when a task-irrelevant feature is very weak or under the threshold, these signals are not subject to cognitive control and, therefore, are not suppressed. As a result, task-irrelevant learning occurs only with these weak features. This result indicates that PL may fail in some studies (Ahissar & Hochstein, 1993; Li et al., 2004; Schoups et al., 2001; Shiu & Pashler, 1992) not because attention is necessary for PL of a feature to occur but because attention suppresses strong task-irrelevant signals, thus inhibiting PL of the feature (Tsushima et al., 2008). These results suggest a possible reinterpretation of studies showing that PL did not occur with features to which attention was not directed (Ahissar & Hochstein, 1993; Li et al., 2004; Schoups et al., 2001; Shiu & Pashler, 1992). It is possible that the exposed irrelevant features which failed to show learning, such as Ahissar and Hochstein's stimulus array, did so because these features were far above threshold and conspicuous and were suppressed by the cognitive control system.

However, other studies have shown that task-irrelevant PL can occur with a suprathreshold feature (Barbot et al., 2011; Beste et al., 2011; Gutnisky et al., 2009; Xu, He, & Ooi, 2012; Zhang & Kourtzi, 2010). Thus, there should be another mechanism that allows for TIPL. This idea will be discussed later.

Role of Reinforcement Signals

The examination of the role of reinforcement signals started with investigations of the effects of feedback (knowledge of performance accuracy provided to the observer) on PL. More recently, the effects of internal and external reward on PL have been studied. Feedback and reward have been largely thought to play the same or similar roles in reinforcement aspects of PL, as they are both different types of behavioral reinforcers, although this has not been empirically clarified.

Early studies found that PL occurs without feedback and that it is not a necessary condition for learning to occur. At the same, it has also been found that response feedback facilitates PL (Herzog & Fahle, 1997, 1999). Initially it was thought that response feedback served as a kind of supervisory learning signal. Michael Herzog and Manfred Fahle ultimately rejected this hypothesis because they found that feedback created a similar performance benefit regardless of whether feedback was provided trial-to-trial (as supervisory learning models would require) or in blocked format (as a report of "percent correct" after a group of trials is completed) (Herzog & Fahle, 1997). Shibata et al. had observers perform a PL task in which block feedback was provided to observers with a positive bias (reporting better than actual performance) or a negative bias (worse than actual performance). While PL was facilitated by positively biased feedback, it was not influenced by negatively biased feedback. These results suggest that PL is facilitated only by positive reinforcement signals (Shibata et al., 2009). Liu and his colleagues suggested that all of these feedback results can be predicted by an augmented Hebbian reweighting model of perceptual learning, in which feedback influences the effective rate of learning by serving as an additional input and not as a direct teaching signal (Liu et al., 2010; Petrov, Dosher, & Lu, 2005).

In the first study to investigate the role of reward in PL, subjects were asked to identify two white target letters in a sequence of black distractor letters (Seitz & Watanabe, 2003). Targets were consistently presented in synchrony with a preselected task-irrelevant subthreshold coherent motion direction. Distractor letters were synchronized with the presentation of other task-irrelevant subthreshold motion directions. Later tests of motion sensitivity revealed that PL occurred only with the directions paired with targets. In another study, two different subthreshold coherent motion directions were paired with the first and second targets in the letter sequence (Seitz et al., 2005a). When the first and second targets were presented within 400 ms of each other, the second target was difficult to identify, consistent with the known phenomenon of attentional blink (Raymond, Shapiro, & Arnell, 1992). In this condition, the coherent motion direction paired with the first target was learned while the direction paired with the second target was not. From these results, Seitz and Watanabe developed a model in which task-irrelevant PL occurs as a result of interactions between bottom-up signals from task-irrelevant stimuli and special diffusive reinforcement signals that are released as a result of successful task performance (Seitz & Watanabe, 2005). This model predicts that PL should occur for any visual feature presented contemporaneously with reward, even without subjects' involvement in an active task or a feeling of successful task performance. To test this hypothesis, subjects who had been deprived of water and food for 5 h prior to "training" were given water as reward when one preselected subthreshold orientation was presented. No water was provided during presentations of a second control orientation. PL occurred only with the orientation paired with the water reward (Seitz, Kim, & Watanabe, 2009). This result is in accord with the predictions of the model (Seitz & Watanabe, 2005).

Moreover, this model is consistent with the results of many other PL studies. For example, as discussed above, Law and Gold demonstrated that PL reflects changes in the LIP neurons of monkeys, but not MT neurons (Law & Gold, 2008). The authors proposed a model in which PL is associated with changes in connectivity between visual areas and a decision unit, essentially resulting in enhanced readout of visual signals (Dosher & Lu, 1998; Law & Gold, 2008). In a later study, Law and Gold suggested that their data fit well with a model in which connectivity is determined by reinforcement signals based on prediction error (Law & Gold, 2009). Along the same lines, Kahnt and his colleagues observed that changes in BOLD patterns in the anterior cingulate cortex of humans were correlated with performance improvements on an orientation discrimination task conducted with feedback. This is in accord with the hypothesis that PL is associated with changes in high-level decision areas in a fashion similar to reinforcement learning (Kahnt et al., 2011).

Differences between the Roles of Feedback and Reinforcement in PL

Although attention and reinforcement both significantly influence PL, their roles do not seem to be the same. Attention's general role is to enhance task-relevant signals while suppressing task-irrelevant signals. Thus, attention may facilitate task-relevant PL while reducing the amplitude of or eliminating task-irrelevant PL. Reinforcement processing is usually consistent with attention in PL (Roelfsema, van Ooyen, & Watanabe, 2010). However, the function of reinforcement processing may not include the selection of signals based on task relevancy, and, therefore, if experimentally separated from attention, a reinforcer may enhance both task-relevant and task-irrelevant signals (see figure 69.1). If true, a reinforcer may facilitate both task-relevant and task-irrelevant PL.

Conclusion: Roles of Attention and Reinforcement

As discussed above, attention and reinforcement processing may play important yet distinct roles in PL. Are attention and/or reward necessary for PL to occur?

PL of a feature can occur as a result of mere exposure to the feature, without directing attention to the feature. Therefore, attention to a feature is not necessary for PL to occur.

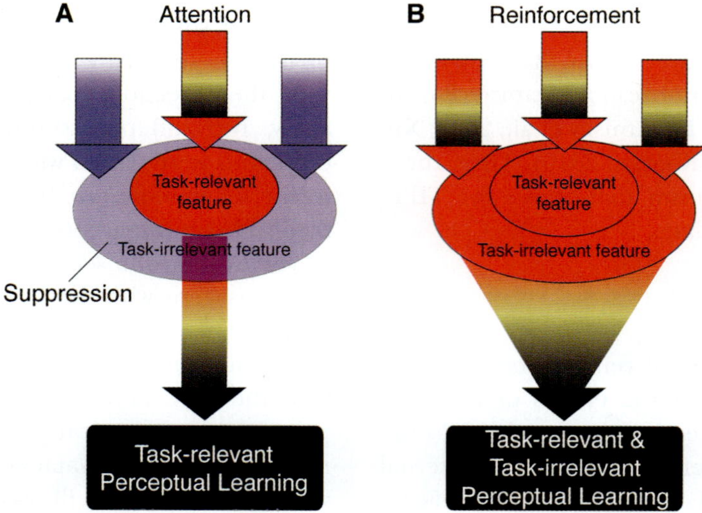

FIGURE 69.1 Two factors that influence perceptual learning (PL). (Modified from figure 1 in Sasaki, Nanez, & Watanabe, 2010.) (A) Attention, which enhances task-relevant signals and reduces task-irrelevant signals, leading mainly to task-relevant PL. (B) Reinforcement, which enhances both task-relevant and task-irrelevant signals, leading to both task-relevant and task-irrelevant PL.

Likewise, reinforcement may not be necessary for PL to occur. Although there are cases in which TIPL fails to occur without reinforcement signals driven by internal or external reward, there are also many reports in which PL occurs without explicit reward or feedback of any kind. The former cases could be due to the fact that a presented feature was subthreshold or very weak. Thus, it is probable that without enhancement of such signals via reinforcement processing, PL of the feature cannot occur. We conclude that although attention and reward can greatly influence PL, neither factor is necessary for PL to occur.

GENERAL CONCLUSION

Recently more and more studies have been published on PL, employing a variety of methods including psychophysics, brain imaging, and single-unit recording. Two scientific controversies have received increased attention as a result. One controversy is about the cortical areas that are changed in association with PL and its underlying neural mechanism. There are three different hypotheses that attempt to resolve this issue, implicating changes in early visual cortex, middle visual cortex, and connectivity between visual areas or visual and decision areas. As discussed above, each hypothesis is supported by reasonable empirical evidence. The second controversy is about the necessary conditions of PL. There are two questions embedded in this controversy. One is whether attention to a feature is necessary for learning of the feature to occur. The other is whether a reinforcer such as feedback or reward is necessary for PL to occur. As discussed above, it seems that neither attention nor a reinforcer is strictly necessary for PL to occur. Rather, these factors enhance weak feature signals, increasing the probability of PL for the feature.

Why have these become such serious issues in PL research? Generally speaking, experiments are designed to examine a certain type of PL using a certain set of parameters, yet researchers tend to generalize their results to all types of PL. This approach has misled PL research. In the future, it will be necessary to clarify how different sets of parameters lead to different types of PL.

REFERENCES

Adab, H. Z., & Vogels, R. (2011). Practicing coarse orientation discrimination improves orientation signals in macaque cortical area v4. *Current Biology, 21,* 1661–1666.

Adini, Y., Sagi, D., & Tsodyks, M. (2002). Context-enabled learning in the human visual system. *Nature, 415,* 790–793.

Ahissar, M., & Hochstein, S. (1993). Attentional control of early perceptual learning. *Proceedings of the National Academy of Sciences of the United States of America, 90,* 5718–5722.

Ahissar, M., & Hochstein, S. (1997). Task difficulty and the specificity of perceptual learning. *Nature, 387,* 401–406.

Ball, K., & Sekuler, R. (1981). Adaptive processing of visual motion. *Journal of Experimental Psychology. Human Perception and Performance, 7,* 780–794.

Barbot, A., Landy, M. S., & Carrasco, M. (2011). Exogenous attention enhances 2nd-order contrast sensitivity. *Vision Research, 51,* 1086–1098.

Bejjanki, V. R., Beck, J. M., Lu, Z. L., & Pouget, A. (2011). Perceptual learning as improved probabilistic inference in early sensory areas. *Nature Neuroscience, 14,* 642–648.

Beste, C., Wascher, E., Gunturkun, O., & Dinse, H. R. (2011). Improvement and impairment of visually guided behavior through LTP- and LTD-like exposure-based visual learning. *Current Biology, 21,* 876–882.

Carrasco, M., Rosenbaum, A., & Giordano, A. (2008). Exogenous attention: Less effort, more learning! *Journal of Vision, 8,* 1095a. doi:10.1167/8.6.1095.

Chowdhury, S. A., & DeAngelis, G. C. (2008). Fine discrimination training alters the causal contribution of macaque area MT to depth perception. *Neuron, 60,* 367–377.

Crist, R. E., Kapadia, M. K., Westheimer, G., & Gilbert, C. D. (1997). Perceptual learning of spatial localization: Specificity for orientation, position, and context. *Journal of Neurophysiology, 78,* 2889–2894.

Crist, R. E., Li, W., & Gilbert, C. D. (2001). Learning to see: Experience and attention in primary visual cortex. *Nature Neuroscience, 4,* 519–525.

Davis, C. M., Stevenson, G. W., Canadas, F., Ullrich, T., Rice, K. C., & Riley, A. L. (2009). Discriminative stimulus properties of naloxone in Long–Evans rats: Assessment with the conditioned taste aversion baseline of drug discrimination learning. *Psychopharmacology, 203,* 421–429. doi:10.1007/s00213-008-1233-5.

Dosher, B. A., & Lu, Z. L. (1998). Perceptual learning reflects external noise filtering and internal noise reduction through channel reweighting. *Proceedings of the National Academy of Sciences of the United States of America, 95,* 13988–13993. doi:10.1073/pnas.95.23.13988.

Fahle, M., & Edelman, S. (1993). Long-term learning in vernier acuity: Effects of stimulus orientation, range and of feedback. *Vision Research, 33,* 397–412.

Fahle, M., & Poggio, T. (2002). *Perceptual learning.* Cambridge, MA.: MIT Press.

Fiorentini, A., & Berardi, N. (1980). Perceptual learning specific for orientation and spatial frequency. *Nature, 287,* 43–44.

Furmanski, C. S., Schluppeck, D., & Engel, S. A. (2004). Learning strengthens the response of primary visual cortex to simple patterns. *Current Biology, 14,* 573–578.

Ghose, G. M., Yang, T., & Maunsell, J. H. (2002). Physiological correlates of perceptual learning in monkey V1 and V2. *Journal of Neurophysiology, 87,* 1867–1888.

Gibson, E. J. (1963). Perceptual learning. *Annual Review of Psychology, 14,* 29–56. doi:10.1146/annurev.ps.14.020163.000333.

Gutnisky, D. A., Hansen, B. J., Iliescu, B. F., & Dragoi, V. (2009). Attention alters visual plasticity during exposure-based learning. *Current Biology, 19,* 555–560.

Herzog, M. H., & Fahle, M. (1997). The role of feedback in learning a vernier discrimination task. *Vision Research, 37,* 2133–2141.

Herzog, M. H., & Fahle, M. (1999). Effects of biased feedback on learning and deciding in a vernier discrimination task. *Vision Research, 39,* 4232–4243.

Hol, K., & Treue, S. (2001). Different populations of neurons contribute to the detection and discrimination of visual motion. *Vision Research, 41,* 685–689.

Hua, T., Bao, P., Huang, C. B., Wang, Z., Xu, J., Zhou, Y., et al. (2010). Perceptual learning improves contrast sensitivity of V1 neurons in cats. *Current Biology, 20,* 887–894. doi:10.1016/j.cub.2010.03.066.

Kahnt, T., Grueschow, M., Speck, O., & Haynes, J. D. (2011). Perceptual learning and decision-making in human medial frontal cortex. *Neuron, 70,* 549–559.

Karni, A., & Sagi, D. (1991). Where practice makes perfect in texture discrimination: Evidence for primary visual cortex plasticity. *Proceedings of the National Academy of Sciences of the United States of America, 88,* 4966–4970. doi:10.1073/pnas.88.11.4966.

Karni, A., & Sagi, D. (1993). The time course of learning a visual skill. *Nature, 365,* 250–252.

Karni, A., Tanne, D., Rubenstein, B. S., Askenasy, J. J., & Sagi, D. (1994). Dependence on REM sleep of overnight improvement of a perceptual skill. *Science, 265,* 679–682.

Koyama, S., Harner, A., & Watanabe, T. (2004). Task-dependent changes of the psychophysical motion-tuning functions in the course of perceptual learning. *Perception, 33,* 1139–1147.

Law, C. T., & Gold, J. I. (2008). Neural correlates of perceptual learning in a sensory–motor, but not a sensory, cortical area. *Nature Neuroscience, 11,* 505–513.

Law, C. T., & Gold, J. I. (2009). Reinforcement learning can account for associative and perceptual learning on a visual-decision task. *Nature Neuroscience, 12,* 655–663.

Levi, D. M. (2005). Perceptual learning in adults with amblyopia: A reevaluation of critical periods in human vision. *Developmental Psychobiology, 46,* 222–232. doi:10.1002/dev.20050.

Lewis, C. M., Baldassarre, A., Committeri, G., Romani, G. L., & Corbetta, M. (2009). Learning sculpts the spontaneous activity of the resting human brain. *Proceedings of the National Academy of Sciences of the United States of America, 106,* 17558–17563. doi:10.1073/pnas.0902455106.

Li, W., Piech, V., & Gilbert, C. D. (2004). Perceptual learning and top-down influences in primary visual cortex. *Nature Neuroscience, 7,* 651–657.

Liu, J., Lu, Z. L., & Dosher, B. A. (2010). Augmented Hebbian reweighting: Interactions between feedback and training accuracy in perceptual learning. *Journal of Vision, 10*(10), 29. doi:10.1167/10.10.29.

Lu, Z. L., & Dosher, B. A. (2009). Mechanisms of perceptual learning. *Learning & Perception, 1,* 19–36.

Lu, Z. L., Hua, T., Huang, C. B., Zhou, Y., & Dosher, B. A. (2011). Visual perceptual learning. *Neurobiology of Learning and Memory, 95,* 145–151.

Matarazzo, L., Franko, E., Maquet, P., & Vogels, R. (2008). Offline processing of memories induced by perceptual visual learning during subsequent wakefulness and sleep: A behavioral study. *Journal of Vision, 8*(4), 7, 1–9. doi:10.1167/8.4.7.

McKee, S. P., & Westheimer, G. (1978). Improvement in vernier acuity with practice. *Perception & Psychophysics, 24,* 258–262.

Morishita, H., & Hensch, T. K. (2008). Critical period revisited: Impact on vision. *Current Opinion in Neurobiology, 18,* 101–107.

Petrov, A. A., Dosher, B. A., & Lu, Z. L. (2005). The dynamics of perceptual learning: An incremental reweighting model. *Psychological Review, 112,* 715–743.

Petrov, A. A., & Van Horn, N. M. (2012). Motion aftereffect duration is not changed by perceptual learning: Evidence against the representation modification hypothesis. *Vision Research, 61,* 4–14. doi:10.1016/j.visres.2011.08.005.

Poggio, T., Fahle, M., & Edelman, S. (1992). Fast perceptual learning in visual hyperacuity. *Science, 256,* 1018–1021.

Raymond, J. E., Shapiro, K. L., & Arnell, K. M. (1992). Temporary suppression of visual processing in an RSVP task: An attentional blink? *Journal of Experimental Psychology. Human Perception and Performance, 18,* 849–860.

Regan, D., & Beverley, K. I. (1985). Postadaptation orientation discrimination. *Journal of the Optical Society of America. A, Optics and Image Science, 2,* 147–155.

Roelfsema, P. R., van Ooyen, A., & Watanabe, T. (2010). Perceptual learning rules based on reinforcers and attention. *Trends in Cognitive Sciences, 14,* 64–71. doi:10.1016/j.tics.2009.11.005.

Rosenthal, O., & Humphreys, G. W. (2010). Perceptual organization without perception: The subliminal learning of global contour. *Psychological Science, 21,* 1751–1758. doi:10.1177/0956797610389188.

Sagi, D. (2011). Perceptual learning in *Vision Research. Vision Research, 51,* 1552–1566.

Sagi, D., & Tanne, D. (1994). Perceptual learning: Learning to see. *Current Opinion in Neurobiology, 4,* 195–199.

Sasaki, Y., Nanez, J. E., & Watanabe, T. (2010). Advances in visual perceptual learning and plasticity. *Nature Reviews. Neuroscience, 11,* 53–60.

Schoups, A. A., Vogels, R., & Orban, G. A. (1995). Human perceptual learning in identifying the oblique orientation: Retinotopy, orientation specificity and monocularity. *Journal of Physiology, 483,* 797–810.

Schoups, A. A., Vogels, R., Qian, N., & Orban, G. (2001). Practising orientation identification improves orientation coding in V1 neurons. *Nature, 412,* 549–553.

Schwartz, S., Maquet, P., & Frith, C. (2002). Neural correlates of perceptual learning: A functional MRI study of visual texture discrimination. *Proceedings of the National Academy of Sciences of the United States of America, 99,* 17137–17142. doi:10.1073/pnas.242414599.

Seitz, A., & Watanabe, T. (2005). A unified model for perceptual learning. *Trends in Cognitive Sciences, 9,* 329–334. doi:10.1016/j.tics.2005.05.010.

Seitz, A. R., & Dinse, H. R. (2007). A common framework for perceptual learning. *Current Opinion in Neurobiology, 17,* 148–153.

Seitz, A. R., Kim, D., & Watanabe, T. (2009). Rewards evoke learning of unconsciously processed visual stimuli in adult humans. *Neuron, 61,* 700–707.

Seitz, A. R., Lefebvre, C., Watanabe, T., & Jolicoeur, P. (2005a). Requirement for high-level processing in subliminal learning. *Current Biology, 15,* R753–R755. doi:10.1016/j.cub.2005.09.009.

Seitz, A. R., & Watanabe, T. (2003). Psychophysics: Is subliminal learning really passive? *Nature, 422,* 36.

Seitz, A. R., Yamagishi, N., Werner, B., Goda, N., Kawato, M., & Watanabe, T. (2005b). Task-specific disruption of perceptual learning. *Proceedings of the National Academy of Sciences of the United States of America, 102,* 14895–14900. doi:10.1073/pnas.0505765102.

Shadlen, M. N., & Newsome, W. T. (2001). Neural basis of a perceptual decision in the parietal cortex (area LIP) of the rhesus monkey. *Journal of Neurophysiology, 86,* 1916–1936.

Shibata, K., Watanabe, T., Sasaki, Y., & Kawato, M. (2011). Perceptual learning incepted by decoded fMRI neurofeedback without stimulus presentation. *Science, 334,* 1413–1415.

Shibata, K., Yamagishi, N., Ishii, S., & Kawato, M. (2009). Boosting perceptual learning by fake feedback. *Vision Research, 49,* 2574–2585.

Shih, J. J., Weisend, M. P., Davis, J. T., & Huang, M. (2000). Magnetoencephalographic characterization of sleep spindles in humans. *Journal of Clinical Neurophysiology, 17,* 224–231.

Shiu, L. P., & Pashler, H. (1992). Improvement in line orientation discrimination is retinally local but dependent on cognitive set. *Perception & Psychophysics, 52,* 582–588.

Skrandies, W., & Fahle, M. (1994). Neurophysiological correlates of perceptual learning in the human brain. *Brain Topography, 7,* 163–168.

Teich, A. F., & Qian, N. (2003). Learning and adaptation in a recurrent model of V1 orientation selectivity. *Journal of Neurophysiology, 89,* 2086–2100.

Tsushima, Y., Sasaki, Y., & Watanabe, T. (2006). Greater disruption due to failure of inhibitory control on an ambiguous distractor. *Science, 314,* 1786–1788.

Tsushima, Y., Seitz, A., & Watanabe, T. (2008). Task-irrelevant learning occurs only when the irrelevant feature is weak. *Current Biology, 16,* 516–517.

Vaina, L. M., Belliveau, J. W., des Roziers, E. B., & Zeffiro, T. A. (1998). Neural systems underlying learning and representation of global motion. *Proceedings of the National Academy of Sciences of the United States of America, 95,* 12657–12662. doi:10.1073/pnas.95.21.12657.

Vandenberghe, R., Dupont, P., De Bruyn, B., Bormans, G., Michiels, J., Mortelmans, L., et al. (1996). The influence of stimulus location on the brain activation pattern in detection and orientation discrimination. A PET study of visual attention. *Brain, 119,* 1263–1276. doi:10.1093/brain/119.4.1263.

Walker, M. P., & Stickgold, R. (2006). Sleep, memory, and plasticity. *Annual Review of Psychology, 57,* 139–166.

Walker, M. P., Stickgold, R., Jolesz, F. A., & Yoo, S. S. (2005). The functional anatomy of sleep-dependent visual skill learning. *Cerebral Cortex, 15,* 1666–1675.

Watamaniuk, S. N., Sekuler, R., & Williams, D. W. (1989). Direction perception in complex dynamic displays: The integration of direction information. *Vision Research, 29,* 47–59.

Watanabe, T., Nanez, J. E., & Sasaki, Y. (2001). Perceptual learning without perception. *Nature, 413,* 844–848.

Watanabe, T., Nanez, J. E., Sr., Koyama, S., Mukai, I., Liederman, J., & Sasaki, Y. (2002). Greater plasticity in lower-level than higher-level visual motion processing in a passive perceptual learning task. *Nature Neuroscience, 5,* 1003–1009.

Xiao, L. Q., Zhang, J. Y., Wang, R., Klein, S. A., Levi, D. M., & Yu, C. (2008). Complete transfer of perceptual learning across retinal locations enabled by double training. *Current Biology, 18,* 1922–1926.

Xu, J. P., He, Z. J., & Ooi, T. L. (2012). Perceptual learning to reduce sensory eye dominance beyond the focus of top-down visual attention. *Vision Research, 61,* 39–47. doi:10.1016/j.visres.2011.05.013.

Yang, T., & Maunsell, J. H. (2004). The effect of perceptual learning on neuronal responses in monkey visual area V4. *Journal of Neuroscience, 24,* 1617–1626.

Yotsumoto, Y., Sasaki, Y., Chan, P., Vasios, C. E., Bonmassar, G., Ito, N., et al. (2009). Location-specific cortical activation changes during sleep after training for perceptual learning. *Current Biology, 19,* 1278–1282. doi:10.1016/j.cub.2009.06.011.

Yotsumoto, Y., Seitz, A. R., Shimojo, S., Sakagami, M., Watanabe, T., & Sasaki, Y. (2012). Performance dip in motor response induced by task-irrelevant weaker coherent visual motion signals. *Cerebral Cortex, 22,* 1887–1893.

Yotsumoto, Y., Watanabe, T., & Sasaki, Y. (2008). Different dynamics of performance and brain activation in the time course of perceptual learning. *Neuron, 57,* 827–833.

Zhang, J., & Kourtzi, Z. (2010). Learning-dependent plasticity with and without training in the human brain. *Proceedings of the National Academy of Sciences of the United States of America, 107,* 13503–13508. doi:10.1073/pnas.1002506107.

70 Perceptual Learning and Plasticity in Primary Visual Cortex

WU LI AND CHARLES D. GILBERT

Processing of visual information in the brain engages a large number of cortical areas (Van Essen, Anderson, & Felleman, 1992). The traditional point of view on neural mechanisms of visual perception emphasizes a hierarchical order of the processing stream: Information about local simple image components such as line segments is first extracted in early visual cortex whereas more complex stimulus features such as global shapes are processed subsequently in higher-order cortical areas by integrating primitive components. The primary visual cortex (area V1) is anatomically situated at the lowest level of the hierarchically organized cortical pathways. It was initially thought that response properties of V1 neurons are rather simple, and that the neural circuitry in V1 of adults is hardwired, performing only stereotyped computations for detecting simple stimulus attributes such as the orientation of a line at a visual field location (Hubel & Wiesel, 1959).

Using more complex stimuli with center–surround or figure–ground structure rather than a patch of gratings or a single bar, subsequent studies consistently showed that visual neuronal responses to a stimulus placed in the receptive field (RF) are modified by the global context within which the stimulus is displayed (for reviews, see Albright & Stoner, 2002; Allman, Miezin, & McGuinness, 1985; Gilbert, 1998). This phenomenon, known as the contextual modulation, takes place throughout visual cortical areas, representing a general long-range integration mechanism that can account for many perceptual phenomena of contextual influences, including some rules of perceptual organization postulated by Gestalt psychology (Wertheimer, 1923). Contextual interactions widely seen in V1 indicate that V1 neurons can actually integrate stimuli across visual field areas much larger than the classically defined RF, and that they are selective for more complex features in visual scenes than simple stimulus attributes. The underlying neural substrates have been linked to the architecture of neural circuitry in V1. Neurons in V1 are interconnected by two major forms of connections (for reviews, see Callaway, 1998;

Gilbert, 1983), the local vertical connections that link neurons across different cortical layers for processing local stimulus at a retinotopic location, and the long-range horizontal connections that link neurons of like orientation preference for integrating information over a large visual field area (for review, see Gilbert, 1992). The latter is suited for mediating orientation-dependent contextual modulation in complex environments, such as linking contour elements arranged along a smooth path and segregating them from other scene components, as described by the Gestalt law of "good continuation."

In cluttered visual scenes, due to limited processing capacity at a given moment, the brain needs to employ attentional mechanism to selectively and efficiently process the part of visual information relevant to the immediate behavior. Early studies in awake monkeys, using simple stimuli and simple behavioral tasks, found that top-down attention significantly enhances neuronal responses in extrastriate cortex, but not in V1 (Moran & Desimone, 1985), leading to a misconception that V1 is simply driven by visual stimuli and virtually immune to top-down modulation. With accumulated evidence from functional magnetic resonance imaging (fMRI) and electrophysiological studies, it is now clear that attending to a spatial location or a stimulus in a cluttered environment can produce various modulatory effects in basically all visual cortical areas, including V1 (for reviews, see Carrasco, 2011; Kastner & Ungerleider, 2000). In fact, as we will see in this chapter, recent studies have shown that V1 neurons can dynamically adapt their response properties to the requirements of specific discrimination and detection tasks, indicating a task-dependent gating mechanism based on complex interactions between bottom-up, stimulus-driven and top-down, goal-directed processes.

The top-down modulation of response properties in V1, and the role of V1 in contour integration, is intimately linked to the contextual influences and the more integrative properties observed in V1. Though linking contour elements in natural scenes might seem to present a hopelessly large number of possible solutions, our visual system simplifies the problem greatly by taking into account the statistical properties of scene contours. The Gestalt rules of perceptual grouping, such as proximity, similarity, and good continuation, reflect the natural geometry of contours, which follow principles of collinearity and cocircularity (Geisler et al., 2001; Sigman et al., 2001). By the same token, the visual system contains an internal representation of these principles, as observed in psychophysical studies of contour saliency (Field, Hayes, & Hess, 1993; Li & Gilbert, 2002), in facilitatory contextual influences on neuronal responses in V1 (Kapadia, Westheimer, & Gilbert, 2000; Li, Piech, & Gilbert, 2006; McManus, Li, & Gilbert, 2011) and in the long range horizontal connections (Gilbert & Wiesel, 1989; Stettler et al., 2002). Together, these findings support the existence of an "association field," whereby lateral interactions in V1 mediate the linkage of scene elements into global contours.

The task-specific top-down modulation of V1 responses, when repeatedly exercised and thus gradually strengthened, plays an important role in perceptual learning (PL), which refers to improved perceptual ability with training in simple discrimination and detection tasks. Extensive practice in the same visual task on the same stimulus can repetitively invoke top-down influences specific to the task and, therefore, specifically potentiate the adaptive changes in V1 for more efficient processing of task-relevant features of the stimulus, resulting in long-lasting and specific improvement in discrimination and detection of the learned stimulus, and leading to automatization of the trained task. This chapter focuses on recent advances in understanding of learning-induced changes in V1, which could apply to the neural mechanisms of PL in general.

PLASTICITY IN V1 INDUCED BY ANOMALOUS VISUAL EXPERIENCE

The first form of experience-dependent plasticity in V1 was discovered by Hubel and Wiesel, which is well-known as the critical period for maturation of ocular dominance columns (Hubel & Wiesel, 1970; Hubel, Wiesel, & LeVay, 1977; Wiesel & Hubel, 1963). During early postnatal development, if visual input from one eye is deprived, the cortical territories in V1 that originally receive information from the deprived eye will be appropriated to represent the normal eye. Since the critical period is restricted to a limited time window during postnatal development, it was generally thought that in adults the neural circuitry in V1 becomes fixed by the end of the critical period.

Contrary to this idea, we now know from studies on cortical maps following retinal lesions that even in adult animals, V1 is capable of undergoing experience-dependent change (Calford et al., 2000; Chino et al., 1992; Eysel et al., 1999; Giannikopoulos & Eysel, 2006; Gilbert & Wiesel, 1992; Heinen & Skavenski, 1991; Kaas et al., 1990; Schmid et al., 1996). Focal binocular lesions initially silence the corresponding retinotopic region (the lesion projection zone; LPZ) in V1. During recovery following retinal lesions, neurons within the LPZ regain responsiveness to visual input from intact retinal regions surrounding the lesion area (Gilbert, 1992). This process is referred to as cortical reorganization and has been observed in nearly all sensory cortices associated with deafferentation (cutoff of sensory input) restricted to a region of the sensory surfaces (for review, see Weinberger, 1995). Cortical reorganization has been shown to be accompanied by rapid axonal sprouting and pruning in the LPZ (Yamahachi et al., 2009), an indication of remolding of the horizontal connections (see figure 70.1). The horizontal connections have been implicated in reorganization in other sensory systems, including the somatosensory system (Marik et al., 2010), with sprouting from nondeprived cortex to the LPZ. In addition to these changes of excitatory connections, reciprocal changes of inhibitory connections have been seen in both visual and somatosensory cortex (Marik, Yamahachi, & Gilbert, 2010; Marik et al, 2010).

This reorganization has adaptive value in mediating functional recovery following CNS lesions. Psychophysical experiments done in patients with macular degeneration show enhanced perceptual fill-in through parts of the visual field affected by the lesion (Zur & Ullman, 2003). It has been proposed that this is due to the connectivity mediating the cortical association field, where connections between columns of similar orientation preference allow perceptual completion of contours crossing the retinal scotomas (McManus, Ullman, & Gilbert, 2008). Consistent with this idea, the preexisting plexus of horizontal connections, with their property of iso-orientation connections, undergo sprouting, and the shifted RFs in the cortical LPZ maintain the same orientation preference as before the lesion (Das & Gilbert, 1995).

Reorganization of V1 retinotopic map was also observed using fMRI in macular degeneration patients with deteriorated central regions of the retina (Baker et al., 2005) and in stroke patients with partially damaged input fibers to V1 (Dilks et al., 2007). The perceptual consequence of cortical reorganization could be closely related to filling in or interpolation of missing information at the lesioned retinal locations (McManus, Ullman, & Gilbert, 2008). In such an

    WU LI AND CHARLES D. GILBERT

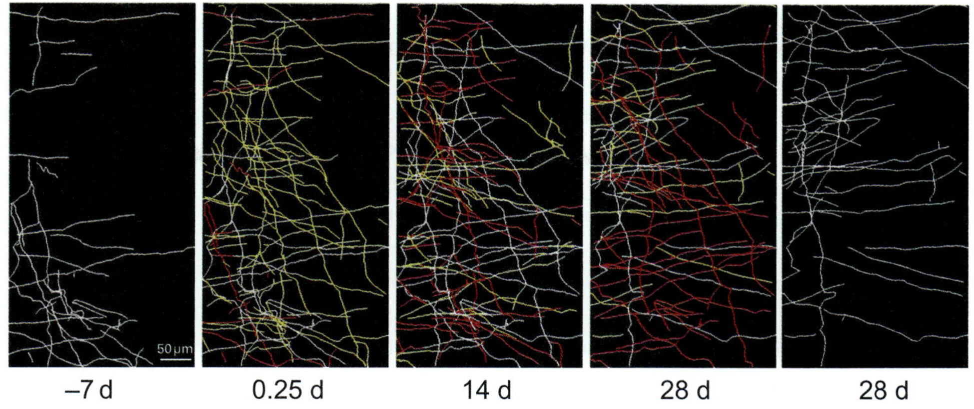

FIGURE 70.1   Axonal sprouting and pruning in monkey V1 induced by focal binocular lesions. Axonal tracing of Z stacks acquired through a depth of 200 microns in the lesion projection zone at baseline (−7 days) and various time points following the lesions (0.25, 14, and 28 days). Gray, axon segments that remained unchanged compared to the previous time point; yellow, segments that were added; red, segments eliminated. The last panel shows the axons present at the end of the imaging session. (From Yamahachi et al., 2009.)

extreme case as complete sight loss, visual cortical areas including V1 can be recruited to process auditory and somatosensory information, and even higher-order cognitive tasks (for reviews, see Burton, 2003; Pascual-Leone et al., 2005).

Plasticity in V1 has been observed not only during early postnatal development and in response to lesions but also in a dramatic change of visual experience. For example, a V1 neuron is normally driven by stimulation of its RF located in the contralateral visual field, but a prolonged reversal of the left and right visual field by wearing special spectacles causes some V1 cells to respond to a stimulus presented in either hemifield (Sugita, 1996).

The findings summarized above suggest that experience-dependent plasticity is an integral function of V1 throughout life. This leads to the speculation that repetitive experience with the same visual stimulus and task would result in plastic changes in V1 as well.

PERCEPTUAL LEARNING AT A GLANCE

Performance on various visual tasks can be remarkably improved with repetition, as is seen in a decrease of threshold for discrimination of the trained stimulus attributes such as orientation, or an increase in efficiency for detection of familiar shapes embedded in distracters (for review, see Sagi, 2011). Helmholtz described PL as follows: "The judgment of the senses may be modified by experience and by training derived under various circumstances, and may be adapted to the new conditions. Thus, persons may learn in some measure to utilize details of the sensation which otherwise would escape notice and not contribute to obtaining any idea of the object" (Helmholtz, 1866).

PL is a form of long-term learning. Learning can be divided into two major categories: explicit or declarative memory of objects, places, or events, which is mediated by a unified memory system—the medial temporal lobe; and the implicit or nondeclarative memory such as the formation of habits, and motor and perceptual skills, which is distributed over different brain regions (Squire, 2004). PL, as with other forms of implicit or unconscious memory, does not require the conventional memory system, since amnesic patients still show PL effects that are comparable to normal observers (Fahle & Daum, 2002).

The learning effect from perceptual training is usually highly restricted to the trained stimulus, including its simple attributes like orientation, and to the trained visual field location or stimulated retinal region (for review, see Sagi, 2011). As only early visual cortical neurons are selective for simple stimulus attributes, and their RFs are restricted to a small visual field area, the specificities of PL have been attributed to specific changes in early visual cortex like V1. However, some studies argue that the improved perceptual performance could simply be due to a refinement of executive control functions such as attention and decision making (Bartolucci & Smith, 2011; Dosher & Lu, 1998; Li, Levi, & Klein, 2004; Xiao et al., 2008; Yu, Klein, & Levi, 2004; Zhang et al., 2008, 2010). A dichotomy of learning mechanisms has also been proposed (Adini et al., 2004): Training under certain circumstances mainly enhances processing of sensory information in the visual cortex, which is less transferable between different stimuli, while training in other conditions mainly improves higher-order, cognitive functions such as decision making, which can be generalized to some untrained stimuli. When considering the unsettled issue on the exact cortical locus of the learning-induced changes, it is useful to take into account the fact that visual perception, as well as PL, is mediated by a chain of processes distributed across many cortical areas, including the visual cortex devoted to sensory processing, the frontal–parietal cortex responsible for attentional control, and the executive neural network involved in perceptual decisions. Different visual stimuli and behavioral tasks may mainly engage different cortical areas or functional modules, where the learning-induced changes are predominant. Therefore, the diverse visual stimuli, perceptual tasks, and learning paradigms used in the literature could catch different traits of the experience-dependent changes.

Despite a lack of consensus about the cortical loci where the plastic changes take place, recent physiological studies have shed light on the neural mechanisms of PL. An in-depth review of PL can be found in chapter 69 by Sasaki and Watanabe. This chapter focuses on PL-induced changes in early sensory cortex, especially in V1, which can provide a window for understanding the general mechanism underlying PL.

TASK-DEPENDENT CHANGES OF NEURONAL RESPONSE PROPERTIES IN V1

A prominent characteristic of PL, and of learning in general, is its task-specificity, which can be understood as a collection of cognitive settings in the brain, or a given brain state, that is specifically prepared for the execution of a particular task on a particular stimulus. As stated by Thorndike's law of identical elements, the transfer of any learning from one task to another cannot happen unless the two tasks share identical elements (Thorndike & Woodworth, 1901a, 1901b, 1901c). When applied to PL, identical elements encompass not only identical stimulus components but also the specific task performed on the stimulus.

PL rarely happens simply by repeated passive exposure to a stimulus; it requires the stimulus to be attended, and usually only the task-relevant feature is learned without generalization to other task-irrelevant features of the same stimulus (Ahissar & Hochstein, 1993; Saffell & Matthews, 2003; Shiu & Pashler, 1992). Task specificity of PL indicates that top-down influences pertaining to a specific task play an important role in encoding the learned information.

Task-dependent modulation of neuronal response properties was found in V1 (Li, Piech, & Gilbert, 2004). When monkeys were trained to perform either a bisection or vernier discrimination task on an identical set of stimulus patterns, responses of V1 neurons were strongly modulated by the stimulus components relevant to the task while task-irrelevant components exerted little contextual modulation (see figure 70.2). The task-dependent top-down control enables V1 neurons to carry more information about the stimulus features that are relevant to the immediate discrimination task.

Similar task-dependency of V1 responses was also observed in detection of global contours in a complex background. Discrete line segments that conform to the Gestalt laws of continuity and proximity are perceptually grouped, forming a global salient contour (see figure 70.3A). This contour integration process, which is critical for form perception, can be mediated by V1 neurons through the contextual modulation mechanism (Li, Piech, & Gilbert, 2006). A collinear contour running through and extending far beyond the classical RF of a V1 cell substantially facilitated neuronal responses when monkeys actively searched for the contour. The contour-related V1 responses were closely correlated with animal performance on the detection task and with the perceptual saliency of the contour: More salient contours, such as longer ones, evoked stronger neuronal responses (Li, Piech, & Gilbert, 2006). Strikingly, in a delayed match-to-sample task, when monkeys were cued to detect contours of different shapes (see figure 70.3B), the contextual influences on V1 responses were dynamically altered in such a way that visual contours with shapes resembling the cued one activated much stronger neuronal responses than the other shapes of contour (McManus, Li, & Gilbert, 2011).

The function-switching capability of V1 neurons among different tasks indicates that top-down signals specific to the task can dynamically modify neuronal selectivity for complex stimuli, increasing V1 signals useful for solving the perceptual task. This mechanism plays an important role in PL that is highly specific to the trained task. The way by which this can be implemented at the circuit level involves gating of the horizontal inputs, which provide information about stimulus context, by feedback connections, which carry information about behavioral context. In another sense, though the association field is a map of all the potential contextual interactions a neuron can express, input selection allows neurons to select components of the association field and thereby to express contextual influences that are appropriate for the task at hand.

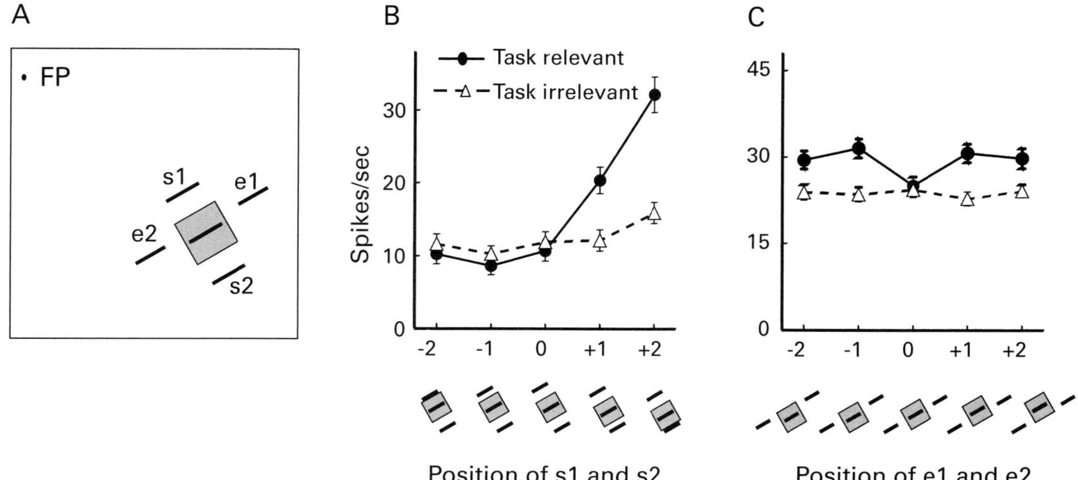

FIGURE 70.2   Task-dependent changes of neuronal response properties in V1. (A) Monkeys performed either of two discrimination tasks on an identical set of stimulus patterns, which consisted of five simultaneously presented lines: an optimally oriented line fixed in the receptive field (RF) center and flanked by four additional parallel lines surrounding the RF. (FP indicates fixation point.) In different trials, the positional arrangement of the two side flankers (s1, s2) was randomly chosen from a set of five different configurations (cartoons in B, labeled from –2 to +2). Each configuration differs from the others in the separation between the three side-by-side lines (in condition "0" the three lines were equidistant; in the other conditions either s1 or s2 was closer to the central line). In the same trials, the two end flankers (e1, e2) were also independently assigned a random configuration from a set of predefined arrangements, such that the two end flankers were collinear with each other but misaligned with the central line to either side (cartoons in C). The animal was cued to perform either a bisection task to discriminate the configurations of the three side-by-side lines or a vernier task to discriminate the configuration of the three end-to-end lines, using the same set of five-line stimuli. (B) Responses of a V1 cell were examined as a function of the position of the two side flankers s1 and s2 when the animal either performed the bisection task, in which s1 and s2 were task relevant, or performed the vernier task, in which the same s1 and s2 were task irrelevant. (C) Responses of a V1 cell were examined as a function of the position of the two end flankers e1 and e2 when the animal either performed the vernier task, in which e1 and e2 were task relevant, or performed the bisection task, in which the same e1 and e2 were task irrelevant. Note that V1 cells were much sensitive to the positions of the side or end flankers only if they were relevant to the task being performed by the animal. (From Li, Piech, & Gilbert, 2004.)

LEARNING-INDUCED CHANGES IN V1 AND THEIR TASK-DEPENDENCY

Early search for cortical plasticity induced by PL in the somatosensory and auditory systems showed an effect analogous to that caused by sensory deafferentation. After monkeys were trained to use a small skin area of a digit to discriminate the vibration frequency of tactile stimuli, remarkable reorganization of the primary somatosensory cortex (S1) was observed, resulting in a significant increase in the size and complexity of the territories representing the trained skin surface (Recanzone, Merzenich, & Jenkins, 1992; Recanzone, Merzenich, Jenkins, Grajski, & Dinse, 1992). Likewise, training monkeys on discrimination of sound frequencies greatly increased the cortical regions responding to the trained sound frequencies in the primary auditory cortex (A1) (Recanzone, Schreiner, & Merzenich, 1993). This reorganization process, referred to as cortical recruitment, deploys a larger cortical area and thus a greater number of neurons to encode the trained stimuli. Learning-induced cortical recruitment shows task specificity: Training in the tactile task while being passively stimulated by the auditory stimuli modifies only S1; training in the auditory task while being passively stimulated by the tactile stimuli modifies only A1.

In visual PL, an fMRI study reported that practicing a motion detection task caused a significant enlargement of the cortical territory representing the trained stimulus in the middle temporal visual cortex (area MT) (Vaina et al., 1998). However, similar recruitment has not been seen in early visual cortex. After extensive training of monkeys in a bisection discrimination task, neither the basic response properties of V1 neurons such as orientation tuning and RF size, nor the retinotopic map in the trained V1 area, are altered (Crist, Li, & Gilbert, 2001). The lack of transfer or interference of learning across visual field locations and between visual stimuli also argues against cortical recruitment as a potential mechanism of PL because taking up adjacent cortical regions would inevitably affect processing of some other stimuli. In fact, cortical recruitment is

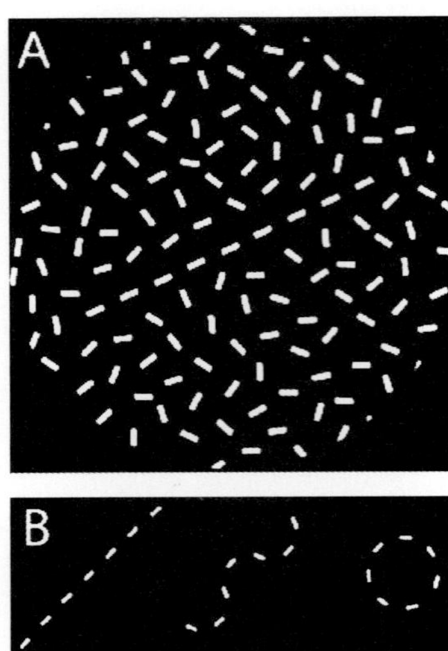

FIGURE 70.3 Stimuli used in studies on contour integration and its task dependency in monkey V1 (see the text for details). At the start of each trial the animal is given a shape cue (B, a straight line, wave shape or circle), and at the end of the trial the cued shape appears embedded in one of two patches of randomly oriented lines (such as the straight line shown in A). The other patch contains a false target. The animal then makes a saccade in the direction of the correct target.

not always associated with improved discrimination ability. For example, the recruitment within A1 seems unnecessary for enhanced performance on acoustic frequency discrimination (Brown, Irvine, & Park, 2004); overrepresentation of the familiar frequencies due to cortical recruitment in A1 is even detrimental to discrimination of the overrepresented frequencies (Han et al., 2007).

Other potential neural mechanisms to improve perceptual performance are to increase neuronal selectivity for the stimulus attribute that is relevant to the task, to enhance signal-to-noise ratio by selectively enhancing neuronal responsiveness to the task-relevant stimulus, and to achieve automatization and accelerated processing speed by shifting stimulus representation from higher to lower cortical areas. Learning-induced changes in these respects are described below.

Enhanced Stimulus Selectivity of V1 Neurons in Discrimination Learning

Some simple discrimination tasks, such as orientation discrimination, involve processing of basic stimulus attributes that are encoded by V1 neurons. Training monkeys on orientation discrimination sharpens orientation tuning functions of V1 neurons around the

trained orientation and at the trained retinotopic location, though this effect is rather weak (Schoups et al., 2001) (but see Ghose, Yang, & Maunsell, 2002). Similar and stronger effects were observed in higher visual cortical area V4 (Raiguel et al., 2006; Yang & Maunsell, 2004). While training on discrimination of simple stimuli can sharpen neuronal selectivity in early visual areas, learning to discriminate among complex objects was found to enhance object selectivity of neurons in the inferior temporal cortex (area IT) (Freedman et al., 2006; Kobatake, Wang, & Tanaka, 1998; Logothetis, Pauls, & Poggio, 1995). This enhancement also showed orientation dependency: Neuronal selectivity was stronger for stimuli presented at the trained orientation than for rotated versions of the same stimuli (Freedman et al., 2006; Logothetis, Pauls, & Poggio, 1995).

A cortical neuron is tuned to a bandwidth of a certain stimulus feature such as a range of orientations, and different neurons cover different but overlapped bandwidths. Therefore, a specific stimulus feature, such as the orientation of a line, can activate a population of neurons. If neuronal selectivity for the stimulus feature is sharpened by training, fewer cells would respond to the stimulus, leading to an overall decrease in population activity or an increase in sparseness of neural code. This speculation is consistent with some imaging studies: Training on orientation discrimination (Schiltz et al., 1999) and contrast discrimination (Mukai et al., 2007) reduces activation in visual cortical areas.

The theoretical interpretations of increased neuronal stimulus selectivity are mixed. While a modeling study was able to simulate improved ability in orientation discrimination based on a sharpening of orientation tuning curves (Teich & Qian, 2003), another modeling study predicted that a sharpening of tuning curves actually causes a general loss of information content conveyed by neuronal responses (Series, Latham, & Pouget, 2004).

Unlike discrimination of a simple stimulus attribute such as orientation, some discrimination tasks rely on information provided by contextual stimuli. For monkeys extensively trained in a three-line bisection discrimination task, the contextual modulatory effects seen in V1 can be remarkably changed (Crist, Li, & Gilbert, 2001): There is an overall increase in the modulatory strength, or even a reversal of the modulatory effect (see figure 70.4). That is, V1 neurons become more sensitive with training to positional displacement of parallel lines. This change was seen only in the trained retinotopic area when the monkey was doing the trained bisection task, suggesting top-down influences that were task-specific.

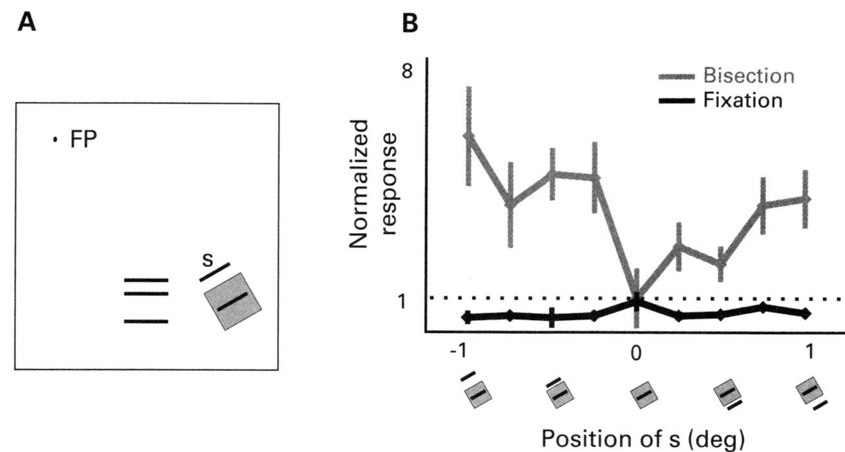

FIGURE 70.4   Task-specific modification of contextual influences in V1. (A) Monkeys were trained for months in the three-line bisection task (three horizontal lines indicate the task stimulus). Responses of single V1 neurons to another stimulus, the test stimulus, were recorded when the animal performed either the bisection task or a simple fixation task (FP indicates fixation point). The test stimulus consisted of two lines, an optimally oriented line fixed in the receptive field (RF) center and a second parallel line (indicated by s) placed at different locations on either side of the RF (cartoons in B). (B) The normalized responses of a typical V1 cell to the test stimulus as a function of the position of line s. When the animal was performing the fixation task, placing s on either side of the RF slightly suppressed neuronal responses relative to the responses at position 0°, where the two test lines were superimposed in the RF center. In contrast, when the animal was doing the trained bisection task, the weak contextual inhibition was changed into strong facilitation. (From Crist, Li, & Gilbert, 2001.)

Enhanced Responsiveness of V1 Neurons in Detection Learning

Different from discrimination of certain stimulus features, various detection tasks require identifying the presence or absence of low-saliency targets due to very low luminance contrast or strong interference from surrounding distracters and noise. Neuronal responsiveness to the trained, familiar target is usually specifically enhanced.

In naive monkeys that had never been trained on detection of the camouflaged contours (see figure 70.3A), V1 neurons carried little information about the global contours (Li, Piech, & Gilbert, 2008). Neuronal responses were found to be independent of the presence and length of the embedded contours when the animals did a simple fixation task (see figure 70.5A). The same was true when the animals attended to the target location but did a task irrelevant to contour detection—responding to a slight luminance change of the very central line of the contour pattern. These pre-training results indicate that contour integration in V1 is not a simple stimulus-driven process, and that spatial attention by itself does not suffice to enable and promote the integration mechanism. Over the course of training the animals to detect contours of different saliencies, however, striking contour-related responses

emerged in V1 (see figure 70.5B), which were parallel with the animals' detection performance. Consistent with many psychophysical studies on PL, the improvement in contour detection was specific to the trained retinotopic location in terms of both the behavioral and neuronal responses. These findings highlight the importance of interactions between bottom-up and top-down processes in generating a specific brain state and in inducing specific cortical changes. The same brain state is also important for retrieval of the implicit memory formed during PL, as doing a task that was irrelevant to contour detection significantly reduced V1 responses to the embedded contours in the trained animals (see figure 70.5C); a complete removal of all possible forms of top-down influences with anesthesia even completely abolished the contour-related responses in V1 (see figure 70.5D).

Similar to contour integration, training on detection of a difference in texture between center and surround stimuli significantly increases fMRI signals in early visual areas (Schwartz, Maquet, & Frith, 2002). Training on detection of an isolated target near contrast threshold can also selectively boost activity in early visual cortex (Furmanski, Schluppeck, & Engel, 2004). Enhancement of neuronal responsiveness with detection training has also been demonstrated in higher cortical areas along the visual processing streams. Training monkeys

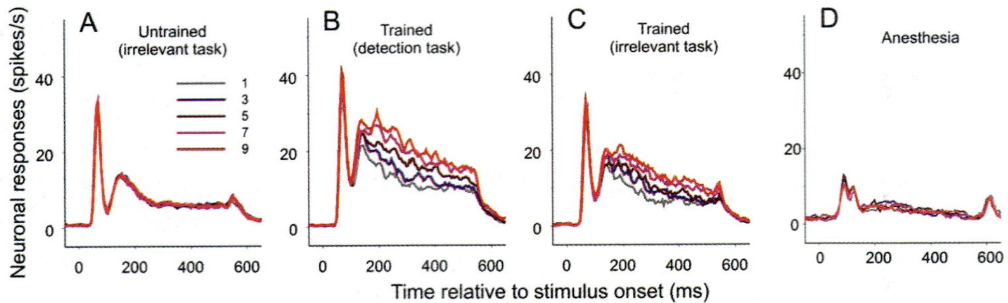

FIGURE 70.5   Learning-induced changes in monkey V1 with training on contour detection. Shown here are averaged neuronal responses to visual contours consisting of 1, 3, 5, 7, and 9 collinear lines embedded in an array of randomly oriented lines (e.g., see figure 70.3A). Time 0 indicates stimulus onset. (A) Before training, V1 responses were independent of contour lengths. The curves corresponding to different contour lengths are superimposed, indicating the absence of contour information in V1 responses. (B) Over the course of training the animals on contour detection, a late response component associated with contour saliency emerged—the longer the contours, the stronger the neuronal responses. (C) The contour-related V1 responses were much weakened when the trained animals performed a task irrelevant to contour detection. (D) Contour-related responses disappeared in the trained V1 region under anesthesia. (From Li, Piech, & Gilbert, 2008.)

to identify natural scene images that are degraded by adding noise specifically enhances V4 neuronal responses to those familiar and degraded pictures (Rainer, Lee, & Logothetis, 2004). In a motion detection task, an improvement in monkeys' performance is correlated with enhanced neuronal responses in area MT and the medial superior temporal area (Zohary et al., 1994) (but see Law & Gold, 2008).

Shift of Cortical Representation of the Learned Stimulus

Visual search is a kind of detection task in which a target is camouflaged within similar distracters. After monkeys were trained in an oddball detection task, increased neuronal responsiveness in V1 was observed in association with the animals' familiarity with the target (Lee et al., 2002). In addition to heightened activity in early visual areas, learning to search for a geometric shape within distracters causes a concomitant decrease in fMRI signals in higher visual areas responsible for shape processing (see figure 70.6) (Sigman et al., 2005). This finding suggests that extensive training can shift cortical representation of the learned shape from higher to lower visual areas for more efficient and less effortful processing. This idea is supported by the evidence that extensive training on a perceptual task significantly reduces activity in the frontoparietal attentional network (Mukai et al., 2007; Pollmann & Maertens, 2005; Sigman et al., 2005).

CONCLUDING REMARKS

Accumulated evidence indicates that V1—the earliest cortical stage along the processing hierarchy—is capable of running different computation "programs" tailored to different perceptual tasks and experiences. Task-dependent modification of neuronal response properties could be a general phenomenon in sensory processing, as it has also been reported in other sensory cortex like A1 (Fritz, Elhilali, & Shamma, 2005). The emerging view is that the information related to a given stimulus attribute is represented at the level of subsets of inputs to a cell, which are gated by the top-down signals via interactions between feedback connections from higher cortical areas and intrinsic connections within a sensory cortical area. This mechanism enables multiple attributes to be represented by the same cells without cross talk, greatly expanding information processing capability of neurons. This multiplexing mechanism can also account for PL that is stimulus and task specific: Repeated execution of the same perceptual task and, therefore, repetitive invoking of top-down influences specific to the task can facilitate and consolidate the dynamic changes useful for solving the perceptual tasks.

To understand further the neural mechanisms of PL, future studies need to dissociate the contributions of dynamic gating of neural connectivity through top-down modulatory mechanisms, from the contributions of plastic alteration of the neural circuitry per se. This differentiation will help to completely resolve the current debate on neural mechanisms of PL.

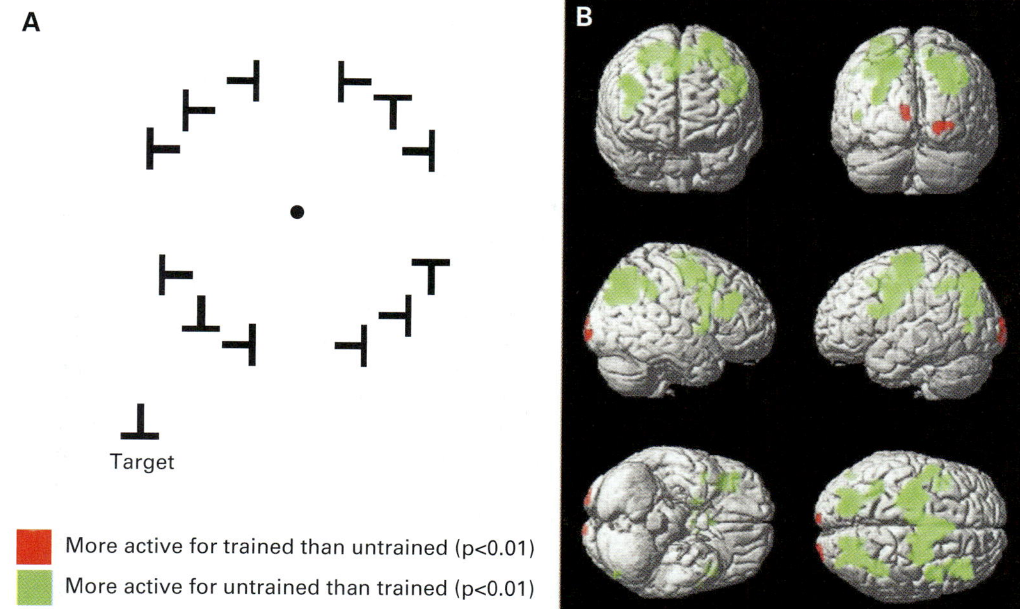

FIGURE 70.6   Shift of stimulus representation to early visual cortex after training human subjects in a visual search task. (A) The stimulus contained a number of Ts that were rotated by multiples of 90° and distributed in the four quadrants of the visual field. One rotated T was the only target. The task was to report the quadrant within which the target T appeared. (B) Differential activation between the trained and untrained conditions. The first row shows the front and back views of the brain; the second row, the left and right hemispheres; the third row, the ventral and dorsal (bottom and top) views. The untrained condition was more active (shown in green) than the trained condition over an extended network that mainly involved the parietal and frontal cortices and lateral occipital cortices. The trained condition was more active (shown in red) than the untrained condition in the middle occipital cortex, corresponding to early visual areas in retinotopic cortex including V1. (From Sigman et al., 2005.)

REFERENCES

Adini, Y., Wilkonsky, A., Haspel, R., Tsodyks, M., & Sagi, D. (2004). Perceptual learning in contrast discrimination: The effect of contrast uncertainty. *Journal of Vision, 4*(12), 993–1005.

Ahissar, M., & Hochstein, S. (1993). Attentional control of early perceptual learning. *Proceedings of the National Academy of Sciences of the United States of America, 90,* 5718–5722.

Albright, T. D., & Stoner, G. R. (2002). Contextual influences on visual processing. *Annual Review of Neuroscience, 25,* 339–379.

Allman, J., Miezin, F., & McGuinness, E. (1985). Stimulus specific responses from beyond the classical receptive field: Neurophysiological mechanisms for local–global comparisons in visual neurons. *Annual Review of Neuroscience, 8,* 407–430.

Baker, C. I., Peli, E., Knouf, N., & Kanwisher, N. G. (2005). Reorganization of visual processing in macular degeneration. *Journal of Neuroscience, 25,* 614–618.

Bartolucci, M., & Smith, A. T. (2011). Attentional modulation in visual cortex is modified during perceptual learning. *Neuropsychologia, 49,* 3898–3907.

Brown, M., Irvine, D. R. F., & Park, V. N. (2004). Perceptual learning on an auditory frequency discrimination task by cats: Association with changes in primary auditory cortex. *Cerebral Cortex, 14,* 952–965.

Burton, H. (2003). Visual cortex activity in early and late blind people. *Journal of Neuroscience, 23,* 4005–4011.

Calford, M. B., Wang, C., Taglianetti, V., Waleszczyk, W. J., Burke, W., & Dreher, B. (2000). Plasticity in adult cat visual cortex (area 17) following circumscribed monocular lesions of all retinal layers. *Journal of Physiology (London), 524,* 587–602.

Callaway, E. M. (1998). Local circuits in primary visual cortex of the macaque monkey. *Annual Review of Neuroscience, 21,* 47–74.

Carrasco, M. (2011). Visual attention: The past 25 years. *Vision Research, 51,* 1484–1525.

Chino, Y. M., Kaas, J. H., Smith, E. L., Langston, A. L., & Cheng, H. (1992). Rapid reorganization of cortical maps in adult cats following restricted deafferentation in retina. *Vision Research, 32,* 789–796.

Crist, R. E., Li, W., & Gilbert, C. D. (2001). Learning to see: Experience and attention in primary visual cortex. *Nature Neuroscience, 4,* 519–525.

Das, A., & Gilbert, C. D. (1995). Long-range horizontal connections and their role in cortical reorganization revealed by optical recording of cat primary visual cortex. *Nature, 375,* 780–784.

Dilks, D. D., Serences, J. T., Rosenau, B. J., Yantis, S., & McCloskey, M. (2007). Human adult cortical reorganization and consequent visual distortion. *Journal of Neuroscience, 27*, 9585–9594.

Dosher, B. A., & Lu, Z. L. (1998). Perceptual learning reflects external noise filtering and internal noise reduction through channel reweighting. *Proceedings of the National Academy of Sciences of the United States of America, 95*, 13988–13993.

Eysel, U. T., Schweigart, G., Mittmann, T., Eyding, D., Qu, Y., Vandesande, F., et al. (1999). Reorganization in the visual cortex after retinal and cortical damage. *Restorative Neurology and Neuroscience, 15*, 153–164.

Fahle, M., & Daum, I. (2002). Perceptual learning in amnesia. *Neuropsychologia, 40*, 1167–1172.

Field, D. J., Hayes, A., & Hess, R. F. (1993). Contour integration by the human visual system: Evidence for a local "association field." *Vision Research, 33*, 173–193.

Freedman, D. J., Riesenhuber, M., Poggio, T., & Miller, E. K. (2006). Experience-dependent sharpening of visual shape selectivity in inferior temporal cortex. *Cerebral Cortex, 16*, 1631–1644.

Fritz, J., Elhilali, M., & Shamma, S. (2005). Active listening: Task-dependent plasticity of spectrotemporal receptive fields in primary auditory cortex. *Hearing Research, 206*, 159–176.

Furmanski, C. S., Schluppeck, D., & Engel, S. A. (2004). Learning strengthens the response of primary visual cortex to simple patterns. *Current Biology, 14*, 573–578.

Geisler, W. S., Perry, J. S., Super, B. J., & Gallogly, D. P. (2001). Edge co-occurrence in natural images predicts contour grouping performance. *Vision Research, 41*, 711–724.

Ghose, G. M., Yang, T. M., & Maunsell, J. H. R. (2002). Physiological correlates of perceptual learning in monkey V1 and V2. *Journal of Neurophysiology, 87*, 1867–1888.

Giannikopoulos, D. V., & Eysel, U. T. (2006). Dynamics and specificity of cortical map reorganization after retinal lesions. *Proceedings of the National Academy of Sciences of the United States of America, 103*, 10805–10810.

Gilbert, C. D. (1983). Microcircuitry of the visual cortex. *Annual Review of Neuroscience, 6*, 217–247.

Gilbert, C. D. (1992). Horizontal integration and cortical dynamics. *Neuron, 9*, 1–13.

Gilbert, C. D. (1998). Adult cortical dynamics. *Physiological Reviews, 78*, 467–485.

Gilbert, C. D., & Wiesel, T. N. (1989). Columnar specificity of intrinsic horizontal and corticocortical connections in cat visual cortex. *Journal of Neuroscience, 9*, 2432–2442.

Gilbert, C. D., & Wiesel, T. N. (1992). Receptive field dynamics in adult primary visual cortex. *Nature, 356*, 150–152.

Han, Y. K., Kover, H., Insanally, M. N., Semerdjian, J. H., & Bao, S. (2007). Early experience impairs perceptual discrimination. *Nature Neuroscience, 10*, 1191–1197.

Heinen, S. J., & Skavenski, A. A. (1991). Recovery of visual responses in foveal V1 neurons following bilateral foveal lesions in adult monkey. *Experimental Brain Research, 83*, 670–674.

Helmholtz, H. (1866). *Treatise on physiological optics* (1962 ed., Vol. 3). New York: Dover Publications.

Hubel, D. H., & Wiesel, T. N. (1959). Receptive fields of single neurones in the cat's striate cortex. *Journal of Physiology (London), 148*, 574–591.

Hubel, D. H., & Wiesel, T. N. (1970). Period of susceptibility to physiological effects of unilateral eye closure in kittens. *Journal of Physiology (London), 206*, 419–436.

Hubel, D. H., Wiesel, T. N., & LeVay, S. (1977). Plasticity of ocular dominance columns in monkey striate cortex. *Philosophical Transactions of the Royal Society of London. Series B, Biological Sciences, 278*, 377–409.

Kaas, J. H., Krubitzer, L. A., Chino, Y. M., Langston, A. L., Polley, E. H., & Blair, N. (1990). Reorganization of retinotopic cortical maps in adult mammals after lesions of the retina. *Science, 248*, 229–231.

Kapadia, M. K., Westheimer, G., & Gilbert, C. D. (2000). Spatial distribution of contextual interactions in primary visual cortex and in visual perception. *Journal of Neurophysiology, 84*, 2048–2062.

Kastner, S., & Ungerleider, L. G. (2000). Mechanisms of visual attention in the human cortex. *Annual Review of Neuroscience, 23*, 315–341.

Kobatake, E., Wang, G., & Tanaka, K. (1998). Effects of shape-discrimination training on the selectivity of inferotemporal cells in adult monkeys. *Journal of Neurophysiology, 80*, 324–330.

Law, C.-T., & Gold, J. I. (2008). Neural correlates of perceptual learning in a sensory–motor, but not a sensory, cortical area. *Nature Neuroscience, 11*, 505–513.

Lee, T. S., Yang, C. F., Romero, R. D., & Mumford, D. (2002). Neural activity in early visual cortex reflects behavioral experience and higher-order perceptual saliency. *Nature Neuroscience, 5*, 589–597.

Li, R. W., Levi, D. M., & Klein, S. A. (2004). Perceptual learning improves efficiency by re-tuning the decision 'template' for position discrimination. *Nature Neuroscience, 7*, 178–183.

Li, W., & Gilbert, C. D. (2002). Global contour saliency and local colinear interactions. *Journal of Neurophysiology, 88*, 2846–2856.

Li, W., Piech, V., & Gilbert, C. D. (2004). Perceptual learning and top-down influences in primary visual cortex. *Nature Neuroscience, 7*, 651–657.

Li, W., Piech, V., & Gilbert, C. D. (2006). Contour saliency in primary visual cortex. *Neuron, 50*, 951–962.

Li, W., Piech, V., & Gilbert, C. D. (2008). Learning to link visual contours. *Neuron, 57*, 442–451.

Logothetis, N. K., Pauls, J., & Poggio, T. (1995). Shape representation in the inferior temporal cortex of monkeys. *Current Biology, 5*, 552–563.

Marik, S. A., Yamahachi, H., & Gilbert, C. D. (2010). Plasticity of inhibitory axonal arbors in visual cortex following retinal lesions. *Society for Neuroscience Abstracts*, 126.110.

Marik, S. A., Yamahachi, H., McManus, J. N., Szabo, G., & Gilbert, C. D. (2010). Axonal dynamics of excitatory and inhibitory neurons in somatosensory cortex. *PLoS Biology, 8*, e1000395. doi:10.1371/journal.pbio.1000395.

McManus, J. N., Li, W., & Gilbert, C. D. (2011). Adaptive shape processing in primary visual cortex. *Proceedings of the National Academy of Sciences of the United States of America, 108*, 9739–9746.

McManus, J. N. J., Ullman, S., & Gilbert, C. D. (2008). A computational model of perceptual fill-in following retinal degeneration. *Journal of Neurophysiology, 99*, 2086–2100.

Moran, J., & Desimone, R. (1985). Selective attention gates visual processing in the extrastriate cortex. *Science, 229*, 782–784.

Mukai, I., Kim, D., Fukunaga, M., Japee, S., Marrett, S., & Ungerleider, L. G. (2007). Activations in visual and attention-related areas predict and correlate with the degree of perceptual learning. *Journal of Neuroscience, 27*, 11401–11411.

Pascual-Leone, A., Amedi, A., Fregni, F., & Merabet, L. B. (2005). The plastic human brain cortex. *Annual Review of Neuroscience, 28*, 377–401.

Pollmann, S., & Maertens, M. (2005). Shift of activity from attention to motor-related brain areas during visual learning. *Nature Neuroscience, 8*, 1494–1496.

Raiguel, S., Vogels, R., Mysore, S. G., & Orban, G. A. (2006). Learning to see the difference specifically alters the most informative V4 neurons. *Journal of Neuroscience, 26*, 6589–6602.

Rainer, G., Lee, H., & Logothetis, N. K. (2004). The effect of learning on the function of monkey extrastriate visual cortex. *PLoS Biology, 2*, 275–283. doi:10.1371/journal.pbio.0020044.

Recanzone, G. H., Merzenich, M. M., & Jenkins, W. M. (1992). Frequency discrimination training engaging a restricted skin surface results in an emergence of a cutaneous response zone in cortical area 3A. *Journal of Neurophysiology, 67*, 1057–1070.

Recanzone, G. H., Merzenich, M. M., Jenkins, W. M., Grajski, K. A., & Dinse, H. R. (1992). Topographic reorganization of the hand representation in cortical area 3B of owl monkeys trained in a frequency-discrimination task. *Journal of Neurophysiology, 67*, 1031–1056.

Recanzone, G. H., Schreiner, C. E., & Merzenich, M. M. (1993). Plasticity in the frequency representation of primary auditory cortex following discrimination training in adult owl monkeys. *Journal of Neuroscience, 13*, 87–103.

Saffell, T., & Matthews, N. (2003). Task-specific perceptual learning on speed and direction discrimination. *Vision Research, 43*, 1365–1374.

Sagi, D. (2011). Perceptual learning in vision research. *Vision Research, 51*, 1552–1566.

Schiltz, C., Bodart, J. M., Dubois, S., Dejardin, S., Michel, C., Roucoux, A., et al. (1999). Neuronal mechanisms of perceptual learning: Changes in human brain activity with training in orientation discrimination. *NeuroImage, 9*, 46–62.

Schmid, L. M., Rosa, M. G. P., Calford, M. B., & Ambler, J. S. (1996). Visuotopic reorganization in the primary visual cortex of adult cats following monocular and binocular retinal lesions. *Cerebral Cortex, 6*, 388–405.

Schoups, A., Vogels, R., Qian, N., & Orban, G. (2001). Practising orientation identification improves orientation coding in V1 neurons. *Nature, 412*, 549–553.

Schwartz, S., Maquet, P., & Frith, C. (2002). Neural correlates of perceptual learning: A functional MRI study of visual texture discrimination. *Proceedings of the National Academy of Sciences of the United States of America, 99*, 17137–17142.

Series, P., Latham, P. E., & Pouget, A. (2004). Tuning curve sharpening for orientation selectivity: Coding efficiency and the impact of correlations. *Nature Neuroscience, 7*, 1129–1135.

Shiu, L. P., & Pashler, H. (1992). Improvement in line orientation discrimination is retinally local but dependent on cognitive set. *Perception & Psychophysics, 52*, 582–588.

Sigman, M., Cecchi, G. A., Gilbert, C. D., & Magnasco, M. O. (2001). On a common circle: Natural scenes and Gestalt rules. *Proceedings of the National Academy of Sciences of the United States of America, 98*, 1935–1940.

Sigman, M., Pan, H., Yang, Y. H., Stern, E., Silbersweig, D., & Gilbert, C. D. (2005). Top-down reorganization of activity in the visual pathway after learning a shape identification task. *Neuron, 46*, 823–835.

Squire, L. R. (2004). Memory systems of the brain: A brief history and current perspective. *Neurobiology of Learning and Memory, 82*, 171–177.

Stettler, D. D., Das, A., Bennett, J., & Gilbert, C. D. (2002). Lateral connectivity and contextual interactions in macaque primary visual cortex. *Neuron, 36*, 739–750.

Sugita, Y. (1996). Global plasticity in adult visual cortex following reversal of visual input. *Nature, 380*, 523–526.

Teich, A. F., & Qian, N. (2003). Learning and adaptation in a recurrent model of V1 orientation selectivity. *Journal of Neurophysiology, 89*, 2086–2100.

Thorndike, E. L., & Woodworth, R. S. (1901a). The influence of improvement in one mental function upon the efficiency of other functions (I). *Psychological Review, 8*, 247–261.

Thorndike, E. L., & Woodworth, R. S. (1901b). The influence of improvement in one mental function upon the efficiency of other functions: II. The estimation of magnitudes. *Psychological Review, 8*, 384–395.

Thorndike, E. L., & Woodworth, R. S. (1901c). The influence of improvement in one mental function upon the efficiency of other functions: III. Functions involving attention, observation and discrimination. *Psychological Review, 8*, 553–564.

Vaina, L. M., Belliveau, J. W., des Roziers, E. B., & Zeffiro, T. A. (1998). Neural systems underlying learning and representation of global motion. *Proceedings of the National Academy of Sciences of the United States of America, 95*, 12657–12662.

Van Essen, D. C., Anderson, C. H., & Felleman, D. J. (1992). Information processing in the primate visual system: An integrated systems perspective. *Science, 255*, 419–423.

Weinberger, N. M. (1995). Dynamic regulation of receptive fields and maps in the adult sensory cortex. *Annual Review of Neuroscience, 18*, 129–158.

Wertheimer, M. (1923). Untersuchungen zur Lehre von der Gestalt. *Psychologische Forschung, 4*, 301–350.

Wiesel, T. N., & Hubel, D. H. (1963). Single-cell responses in striate cortex of kittens deprived of vision in one eye. *Journal of Neurophysiology, 26*, 1003–1017.

Xiao, L.-Q., Zhang, J.-Y., Wang, R., Klein, S. A., Levi, D. M., & Yu, C. (2008). Complete transfer of perceptual learning across retinal locations enabled by double training. *Current Biology, 18*, 1922–1926.

Yamahachi, H., Marik, S. A., McManus, J. N. J., Denk, W., & Gilbert, C. D. (2009). Rapid axonal sprouting and pruning accompany functional reorganization in primary visual cortex. *Neuron, 64*, 719–729.

Yang, T. M., & Maunsell, J. H. R. (2004). The effect of perceptual learning on neuronal responses in monkey visual area V4. *Journal of Neuroscience, 24*, 1617–1626.

Yu, C., Klein, S. A., & Levi, D. M. (2004). Perceptual learning in contrast discrimination and the (minimal) role of context. *Journal of Vision, 4*(3), 169–182.

Zhang, J.-Y., Kuai, S.-G., Xiao, L.-Q., Klein, S. A., Levi, D. M., & Yu, C. (2008). Stimulus coding rules for perceptual learning. *PLoS Biology, 6*, e197. doi:10.1167/8.6.1136.

Zhang, J.-Y., Zhang, G.-L., Xiao, L.-Q., Klein, S. A., Levi, D. M., & Yu, C. (2010). Rule-based learning explains visual perceptual learning and its specificity and transfer. *Journal of Neuroscience, 30*, 12323–12328.

Zohary, E., Celebrini, S., Britten, K. H., & Newsome, W. T. (1994). Neuronal plasticity that underlies improvement in perceptual performance. *Science, 263*, 1289–1292.

Zur, D., & Ullman, S. (2003). Filling-in of retinal scotomas. *Vision Research, 43*, 971–982.

71 Selective Neuronal Synchronization and Attentional Stimulus Selection in Visual Cortex

THILO WOMELSDORF, CONRADO BOSMAN, AND PASCAL FRIES

Selective visual attention relies on dynamic restructuring of cortical information flow in order to ensure prioritized neuronal communication between those neuronal groups conveying information about behaviorally relevant information while reducing the influence from groups encoding irrelevant and distracting information. Electrophysiological evidence suggests that such selective neuronal communication is instantiated *and* sustained through selective neuronal synchronization of rhythmic activity at fast and slow temporal scales within and between neuronal groups processing the attentional-relevant visual information: Attentionally modulated synchronization patterns (1) evolve rapidly, (2) are evident even before sensory inputs arrive, (3) follow closely subjective readiness to process information in time, (4) can be sustained for prolonged time periods, and (5) convey specific information about perceptually selected sensory features as well as motor plans. The empirical survey of these functional signatures of selective synchronization patterns is complemented by recent insights about the mechanistic origins of rhythmic synchronization at micro- and macroscales of cortical neuronal processing, suggesting that selective attention is subserved by precise neuronal synchronization that is selective in space, time, and frequency.

INTRODUCTION

Top-down attention is the key mechanism to restructure cortical information flow in order to prioritize processing of behaviorally relevant over irrelevant and distracting information (Singer, 2011). The behavioral consequences of attentional restructuring of information flow are manifold (Gilbert & Sigman, 2007). Attended sensory inputs are processed more rapidly and accurately and with higher spatial resolution and sensitivity for fine changes while nonattended information appears lower in contrast and is sometimes not perceived at all (Simons & Rensink, 2005).

These functional consequences of attention require temporally dynamic and selective changes of neuronal interactions spanning multiple levels of neuronal information processing: Attentional selection (1) modulates interactions among single neurons within cortical microcircuits, (2) modulates the impact of selective local neuronal groups conveying relevant information within functionally specialized brain areas, and (3) imposes selective long-range interactions among neuronal groups from distant brain areas (Bosman et al., 2012; Gregoriou et al., 2009; Maunsell & Treue, 2006; Mitchell, Sundberg, & Reynolds, 2007; Womelsdorf & Fries, 2007).

For all these levels of neuronal interactions, converging evidence suggests that the selective modulation of interactions is critically built on selective synchronization (Fries, 2009; Siegel, Donner, & Engel, 2012). Neuronal synchronization is typically of oscillatory nature, that is, neurons fire and pause together in a common rhythm. When synchronization is rhythmic, it is often addressed as coherence, and we will use these terms interchangeably. This rhythmic synchronization can influence neuronal interactions in several ways: (1) Spikes that are synchronized will have a larger impact on a target neuron than spikes that are not synchronized (Azouz & Gray, 2003; Salinas & Sejnowski, 2001); (2) local inhibition that is rhythmically synchronized leaves periods without inhibition while nonsynchronized inhibition will prevent local network activity continuously (Knoblich et al., 2010; Tiesinga & Sejnowski, 2004); and (3) rhythmic synchronization of a local group of neurons will modulate the impact of input to that group, and therefore, the impact of rhythmic input will depend on the synchronization between input and target (Bosman et al., 2012; Womelsdorf et al., 2007). These mechanisms are at work on all levels of attentional selection: At the level of microcircuits, inhibitory interneuron networks have been shown to impose rhythmic synchronization capable of effectively controlling the gain of the neuronal spiking output (Bartos, Vida, & Jonas, 2007; Cardin et al., 2009; Tiesinga, Fellous, & Sejnowski, 2008; van Elswijk et al., 2010). At the level of local neuronal groups, attention selectively synchronizes the responses of those neurons conveying information about the attended feature or location (see Womelsdorf & Fries, 2007). And the coherent output from these local neuronal groups has been shown to selectively synchronize over long-range connections with task-relevant neuronal groups in distant brain regions (Bosman et al., 2012; Buschman & Miller, 2007; Gregoriou et al., 2009; Pesaran, Nelson, & Andersen, 2008; Saalmann, Pigarev, & Vidyasagar, 2007; Schoffelen, Oostenveld, & Fries, 2005; Siegel, Donner, & Engel, 2012).

These empirical insights suggest that mechanisms underlying neuronal synchronization are core mechanisms utilized by selective attention. In particular, top-down attention may act by biasing rhythmic synchronization to establish and sustain a selective neuronal communication structure (Fries, 2005, 2009). In the following, we begin by outlining this conceptual framework for selective neuronal communication through selective synchronization. We then survey basic insights from empirical and theoretical studies suggesting that rhythmic synchronization is particularly suited to serve as a gating mechanism (Bosman et al., 2012), effectively controlling the selective routing of neuronal information flow, and review how attention recruits these mechanisms across all levels of cortical processing.

ATTENTIONAL SELECTION AS A DYNAMIC INSTANTIATION OF A SELECTIVE NEURONAL COMMUNICATION STRUCTURE

During natural sensation, top-down control is dynamically established during ongoing processing. Experimentally, top-down signals are set by task instructions and by instructional cues defining relevant and irrelevant sensory features of the input stream during task performance. In typical paradigms of selective attention, the sensory input is kept identical across trials with variations only in covert attention to different aspects of that input, independent from any overt behavioral action selection. In such tasks, neuronal responses are modulated with rapid temporal dynamics and high spatial selectivity throughout the cerebral cortex (see figure 71.1A).

The temporal dynamics of attentional selection are illustrated by recent evidence of a rapid onset of selective neuronal response modulation in cortical areas as far apart as frontal cortex and primary visual cortices in the macaque brain (Gregoriou et al., 2009; Kaping et al., 2011; Khayat, Spekreijse, & Roelfsema, 2006; Monosov, Trageser, & Thompson, 2008). In these studies, monkeys were instructed to attentionally select a target stimulus in visual displays to guide saccadic eye movement. In prefrontal and parietal cortex, attentional selection occurred within the first 120 ms following the sensory onset of target and distractor stimulus information and allowed to predict the spatial focus of attention (Gottlieb, 2002; Kaping et al., 2011; Monosov, Trageser, & Thompson, 2008). Already about 30 ms later, top-down information changes neuronal responses at the earliest visual cortical processing stage in primary visual cortex (Khayat, Spekreijse, & Roelfsema, 2006; Roelfsema, Tolboom, & Khayat, 2007), evident in a response enhancement for neurons with receptive fields overlapping the attentional target stimulus. These findings demonstrate that top-down control restructures cortical activity to sensory inputs across distant cortical sites on a rapid time scale. Attention amplifies almost instantaneously (i.e., with the sensory response latency) the influence of local groups of neurons conveying behavioral relevant information and attenuates the influence of neuronal groups coding for irrelevant inputs. This finding suggests that those distributed groups processing "attended" inputs also interact effectively, establishing a selective neuronal communication structure on top of the existing infrastructure of anatomical connections (Fries, 2005) (figure 71.1A): Interactions among neurons conveying information about attended locations or features are rendered effective while anatomical connections between neuronal groups activated by distracting information are rendered ineffective.

Beyond the temporal dynamics of attentional selection, its spatial selectivity in restructuring cortical information flow is particularly evident across successive visual processing stages from primary visual cortex, extrastriate visual areas, and converging within inferotemporal (IT) cortex. Neurons at the highest visual processing stage in IT cortex have receptive fields that span much of a visual field and respond selectively to complex objects composed of simpler visual features. Part of this selectivity arises from their broad and convergent anatomical input from neurons at earlier processing stages having smaller receptive fields and simpler tuning properties. During natural vision, the large receptive field of an IT neuron will typically contain multiple objects. However, when attention is directed to only one of those objects, the IT neuron's response is biased toward the response that would be obtained if only the attended object were presented (Bosman et al., 2012; Chelazzi et al., 1993).

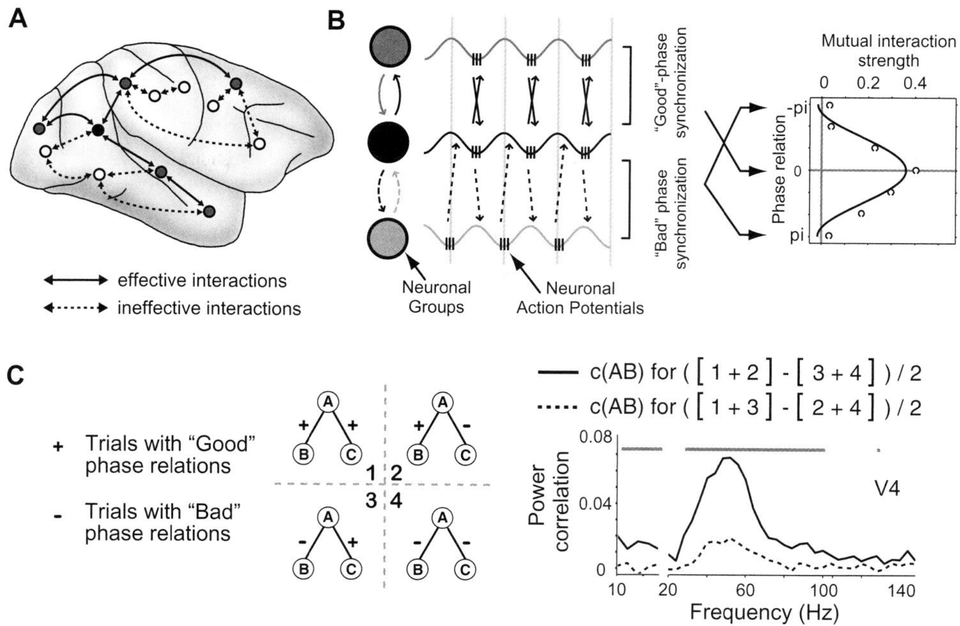

FIGURE 71.1   Selective synchronization renders neuronal interactions among subsets of neuronal groups effective. (A) Anatomical connectivity (sketched as lines) provides a rich infrastructure for neuronal communication among neuronal groups (circles) throughout the cortex. With selective attention, only a small subset of these connections are rendered effective (solid lines). Interactions among groups conveying irrelevant information (light gray circles) for the task at hand are rendered less effective (dashed lines). (B) Illustration of the hypothesized role of selective synchronization for selective communication among three neuronal groups (circles). Rhythmic activity (local field potential [LFP] oscillations with spikes in troughs) provide briefly reoccurring time windows of maximum excitability (LFP troughs), which are either in phase (black and dark gray groups), or in antiphase (black and light gray groups). The plot on the right shows that mutual interactions (upper axis, correlation of the power of the LFP and the neuronal spiking response between neuronal groups) are high during periods of in-phase synchronization and lower otherwise. (C) The trial-by-trial interaction pattern between neuronal groups (A to B and A to C) is predicted by the pattern of synchronization: If AB synchronizes at a good phase, their interaction is strongest, irrespective of whether A synchronizes with C at good or bad phase relations in the same trials. Thus, the spatial pattern of mutual interactions can be predicted by the phase of synchronization among rhythmically activated neuronal groups. (Panels in B and C adapted from Womelsdorf et al., 2007.)

Such dynamic biasing of responses in later visual processing stages could be achieved by selective enhancement (suppression) of the impact of those afferent inputs from neurons in earlier visual areas coding for the attended (nonattended) input (Reynolds, Chelazzi, & Desimone, 1999). However, the mechanisms underlying this up- and down-modulation of input gain for subsets of converging connections are only poorly understood at the level of single neuronal spiking responses. They likely entail a selective increase of temporally precise and coincident inputs from those neurons activated by an attended stimulus in earlier areas. This relevance of spike timing is suggested by fine-grained attentional modulation of precise neuronal synchronization within area V4 (Bichot, Rossi, & Desimone, 2005; Chalk et al., 2010; Fries et al., 2001; Taylor et al., 2005; Womelsdorf et al., 2006). Enhanced synchronization of the spiking output among those neuronal groups activated by attended sensory input (Fries et al., 2008) is

resulting in enhanced coincident arrival of their spikes at their postsynaptic target neurons at later stages of visual processing. Temporally coincident input is highly effective in driving neuronal activity (Azouz & Gray 2003; Salinas & Sejnowski 2001; Tiesinga, Fellous, & Sejnowski, 2008). It is therefore likely that selective synchronization within area V1 underlies attentional biasing in area V4, and that synchronization within area V4 underlies attentional biasing within IT cortex and could thus underlie effective spatial routing of information flow within visual cortex (Bosman et al., 2012).

Note that neuronal synchronization is in principle independent of firing rate, both in terms of metrics and physiology. The different metrics used for quantifying synchronization are typically normalized for firing rate. Physiologically, there are examples where enhanced firing rates are associated with strongly reduced synchronization, for example, the stimulus-induced alpha-band desynchronization in the superficial layers of monkey V4 (Fries et al., 2008; Buffalo et al., 2011). Neuronal gamma-band synchronization typically emerges when neuronal groups are activated, and therefore, it is in most cases associated with increased firing rates. However, firing rates and gamma-band synchronization can also be dissociated from each other, and this can be found primarily when firing rate changes are not driven by changes in bottom-up input (e.g., stimulus changes), but rather by changes in top-down input (e.g., attention or stimulus selection) (Bosman et al., 2012; Fries et al., 2002; Womelsdorf et al., 2006).

Synchronization is a neuronal population phenomenon, and it is often very difficult to assess it with recordings from isolated single units. Correspondingly, many studies of neuronal synchronization use recordings of multiunit activity and/or of the local field potential (LFP). The LFP reflects the summed transmembrane currents of neurons within a few hundred micrometers of tissue. Since synchronized currents sum up much more efficiently than unsynchronized currents, the LFP reflects primarily synchronized synaptic activity. Changes in LFP power typically correlate with changes in direct measures of neuronal synchronization.

Rhythmic synchronization within a neuronal group not only increases its impact on postsynaptic target neurons in a feedforward manner but also rhythmically modulates the group's ability to communicate, such that rhythmic synchronization between two neuronal groups likely subserves their interaction because rhythmic inhibition within the two groups is coordinated and mutual inputs are optimally timed. We capture these implications in the framework of selective neuronal communication through neuronal synchronization (Fries, 2005).

SELECTIVE NEURONAL COMMUNICATION THROUGH COHERENCE

Local neuronal groups frequently engage in periods of rhythmic synchronization. During activated states, rhythmic synchronization is typically evident in the gamma frequency band (30–90 Hz) (Fries, 2009). In vitro experiments and computational studies suggest that gamma-band synchronization emerges from the interplay of excitatory drive and rhythmic inhibition imposed by interneuron networks (Bartos, Vida, & Jonas, 2007; Börgers, Epstein, & Kopell, 2005). Interneurons impose synchronized inhibition onto the local network (Bartos, Vida, & Jonas, 2007; Cardin et al., 2009; Hasenstaub et al., 2005). The brief time periods between inhibition provide time windows for effective neuronal interactions with other neuronal groups because they reflect enhanced postsynaptic sensitivity to input from other neuronal groups as well as maximal excitability for generating spiking output to other neuronal groups (Azouz & Gray 2003; Fries, Nikolic, & Singer, 2007; Tiesinga, Fellous, & Sejnowski, 2008). As a consequence, when two neuronal groups open their temporal windows for interaction at the same time, they will be more likely to mutually influence each other (Womelsdorf et al., 2007). The consequences for selective neuronal communication are illustrated in figure 71.1B: If the rhythmic synchronization within neuronal groups is precisely synchronized between the two groups, they are maximally likely to interact. By the same token, if rhythmic activity within neuronal groups is uncorrelated between groups or synchronizes consistently out of phase, their interaction is curtailed (see figure 71.1B).

This scenario entails that the pattern of synchronization between neuronal groups flexibly structures the pattern of interactions between neuronal groups (see figure 71.1C). Consistent with this hypothesis, the interaction pattern of one neuronal group (group A) with two other groups (groups B and C) can be predicted by their pattern of precise synchronization (see figure 71.1C). This has recently been demonstrated for interactions of triplets of neuronal groups from within and between visual areas in awake cat and monkey visual cortex (Womelsdorf et al., 2007). This study measured the trial-by-trial changes in correlated amplitude fluctuation and changes in precise synchronization between pair AB and pair AC, using the spontaneous variation of neuronal activity during constant visual stimulation. The strength of amplitude covariation, that is, the covariation of power in the LFP and/or multiunit spiking responses, was considered the measure of mutual interaction strength. The results showed that the interaction strength of AB could be inferred from the phase of gamma-band synchronization between group A with group B, being rather unaffected by the phase of synchronization of group A with group C (see figure 71.1C). This finding was evident for triplets of neuronal groups spatially separated by as little as 650 µm, illustrating a high spatial resolution and specificity of the influence of precise phase synchronization between neuronal groups on the efficacy of neuronal interaction. Importantly, additional analysis supported a mechanistic role for the phase of synchronization between rhythmic activities to modulate the effective interaction strength (Womelsdorf et al., 2007). In particular, precise phase synchronization preceded higher-amplitude covariations in time by a few milliseconds, arguing for a causal influence of precise phase synchronization to trigger neuronal interactions. Taken together, these results provide strong evidence suggesting a critical mechanistic role of selective synchronization for neuronal interactions. They demonstrate that synchronization patterns can shape neuronal interactions with high specificity in time, space, and frequency.

Importantly, these same characteristics of selective neuronal interactions are the key elements underlying selective attention. Attentional selection dynamically evolves at a rapid time scale and with high spatial resolution by enhancing (reducing) the effective connectivity among neuronal groups conveying task-relevant (-irrelevant) information. Such dynamic restructuring of neuronal interactions could be accomplished through mechanisms evoking selective synchronization patterns within interneuron networks. Selective changes of precise synchronization in local neuronal groups are capable of modulating in a self-emergent manner with selective interaction patterns across neuronal groups (Börgers & Kopell, 2008; Mishra, Fellous, & Sejnowski, 2006; Tiesinga, Fellous, & Sejnowski, 2008).

The outlined scheme of selective attention implemented as selective neuronal synchronization comprises explicit assumptions that selective attention affects interneuron networks and synchronization patterns during task performance. The following surveys the available insights on interneuron networks and reviews the emerging signatures of attentional modulation through selective synchronization patterns in macaque cortex.

SYNCHRONIZATION IN INTERNEURON NETWORKS AND THEIR ATTENTIONAL MODULATION

Interneurons comprise about a fifth of the neuron population, but despite their ubiquitous presence, their functional roles underlying cortical computations or cognitive processes are far from understood (Markram et al., 2004). However, a central role for the control of local cortical network activity has been suggested for the large class of interneurons of the basket cell type (Wang, 2010). These neurons target perisomatic regions of principal cells and are thereby capable of determining the impact of synaptic inputs arriving at sites distal to a cell's soma. Such perisomatic connectivity critically controls the input gain of principal cells across a large population of principal cells (Buzsaki, Kaila, & Raichle, 2007; Cardin et al., 2009; Rudolph et al., 2007; Tiesinga & Sejnowski, 2004). As described above, the inhibitory synaptic influence of fast spiking basket cells is inherently rhythmic at high frequencies, carrying stronger gamma-band power than pyramidal cells (Bartos, Vida, & Jonas, 2007; Cardin et al., 2009; Hasenstaub et al., 2005).

The prominent role of these high-frequency inputs in shaping the spiking output of principal cells has recently been demonstrated directly in cat and rodent visual cortex. It was shown that the spiking of principal cells is indeed preceded by brief periods of reduced inhibition (Rudolph et al., 2007; see also figure 8 of Hasenstaub et al., 2005, and figure 4 of Cardin et al., 2009). Taken together, these findings suggest that interneurons are the source of rhythmic inhibition onto a local group of neurons synchronizing the discharge of pyramidal cells to the time windows between inhibition.

In the context of selective attention, interneuron networks could be activated by various possible sources. They may be activated by transient and spatially specific neuromodulatory inputs (Lin, Gervasoni, & Nicolelis, 2006; Rodriguez et al., 2004). Alternatively, selective attention could target local interneuron networks directly via top-down inputs from neurons in upstream areas (Buia & Tiesinga, 2008; Mishra, Fellous, & Sejnowski, 2006; Tiesinga, Fellous, & Sejnowski, 2008). In these models, selective synchronization emerges either by depolarizing selective subsets of interneurons (Buia & Tiesinga 2008; Tiesinga & Sejnowski, 2004), or by biasing the phase of rhythmic activity in a more global inhibitory interneuron pool (Mishra, Fellous, & Sejnowski, 2006). In either case, rhythmic inhibition controls the spiking responses of groups of excitatory neurons, enhancing the impact of those neurons spiking synchronously within the periods of disinhibition while actively reducing the impact of neurons spiking asynchronous to this rhythm. This suppressive influence on excitatory neurons, which are activated by distracting feedforward input, reflects the critical ingredient for the concept of selective attentional stimulus selection through selective synchronization: Attention not only enhances synchronization of already more coherent activity representing attended stimuli but actively suppresses the synchronization and impact of groups of neurons receiving strong, albeit distracting, inputs because these inputs arrive at nonoptimal phase relations to the noninhibited periods in the target group. The computational feasibility of both facilitatory and suppressive aspects, and the critical role of the timing of inhibitory circuits, have recently received direct support (Börgers & Kopell, 2008; Knoblich et al., 2010; Tiesinga & Sejnowski, 2010).

Despite the prominent computational role of interneuron activity for selective communication, there are only sparse insights into their implications in selective information processing during cognitive task performance. The basic prediction from the above models is that interneurons are attentionally modulated. Consistent with this presupposition, a recent study by

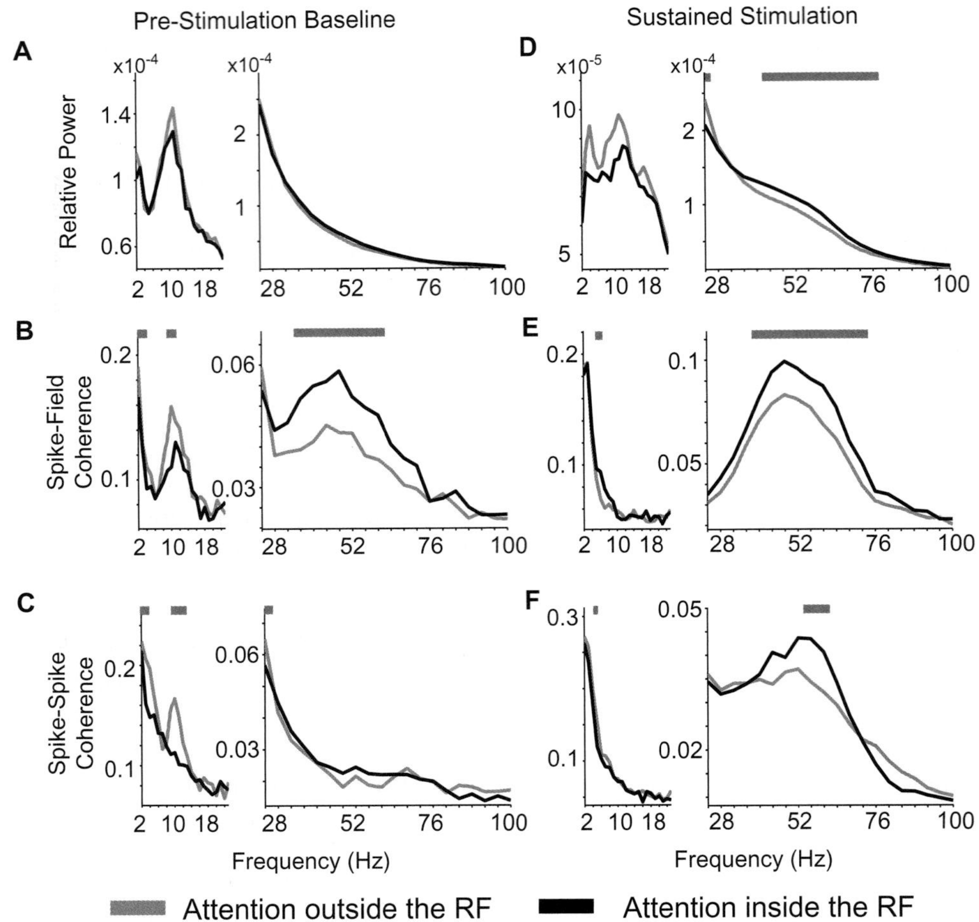

FIGURE 71.2   The pattern of attentional modulation of synchronization in macaque visual area V4 before and during sensory stimulation (A–C). Attentional modulation of relative local field potential (LFP) power (A), spike-to-LFP coherence (B), and spike-to-spike coherence (C) across low and high frequencies during the baseline period of a spatial attention task. Monkeys either attended (dark lines) or ignored (gray lines) the receptive field location of the recorded neuronal groups in blocks of trials. (D–F) Attentional modulation of the neuronal response during stimulation with an attended/ignored moving grating. Same format as in (A–C). Horizontal gray bars denote frequencies with significant attentional effects. RF, receptive field. (Adapted from Fries et al., 2008.)

Mitchell, Sundberg, and Reynolds (2007) reports a clear attentional modulation of putative interneurons in visual area V4 during a selective attention task requiring monkeys to track moving grating stimuli. Putative interneurons showed similar relative increases in firing rate and greater increases in reliability compared to putative pyramidal neurons. However, tests of more refined predictions about the relative modulation of synchronization and the phase relation of spiking responses of inhibitory and excitatory neuron types still have to be conducted (Buia & Tiesinga 2008).

SELECTIVE MODULATION OF SYNCHRONIZATION DURING ATTENTIONAL PROCESSING

Direct evidence for the functional significance of selective synchronization within and between local neuronal

groups for attentional selection has been obtained from recordings in macaque visual cortical areas V1, V2, and V4 (Bosman et al., 2012; Buffalo et al., 2011; Chalk et al., 2010; Fries et al., 2001; Gregoriou et al., 2009; Taylor et al., 2005; Womelsdorf et al., 2006). One consistent result across studies in V4 is that spatial attention enhances gamma-band synchronization within those neuronal groups with receptive fields overlapping the attended location. The attentionally enhanced rhythmic synchronization is strongly evident within the LFP signal and in more precise synchronization of neuronal spiking responses to the LFP (see figure 71.2) (Fries et al., 2008). Importantly, these effects on selective synchronization between spiking activity and LFP translate into enhanced spike-to-spike synchronization, demonstrating that those V4 neurons that convey attended stimulus information send more coherent spike output to their postsynaptic projection targets (figure 71.2)

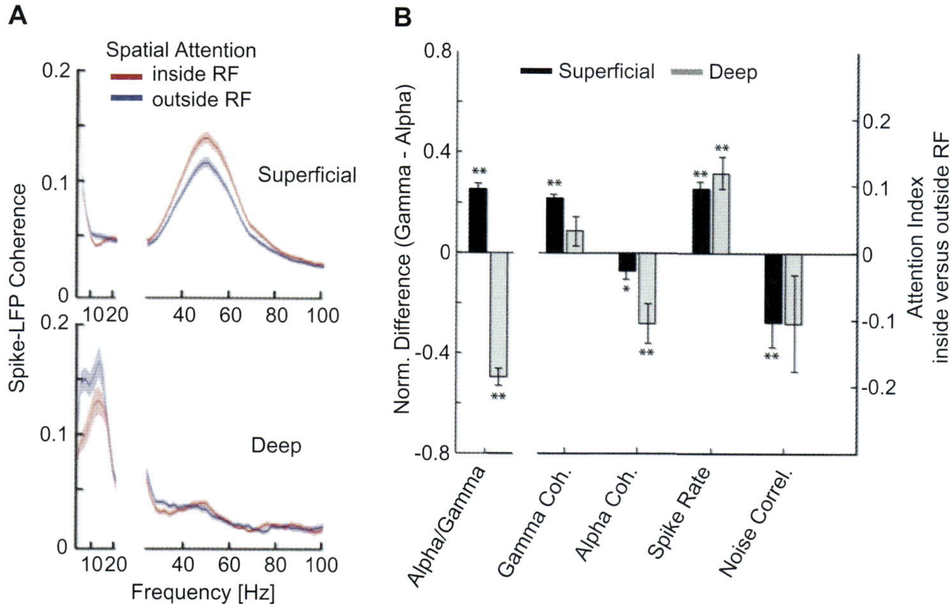

FIGURE 71.3   Laminar specificity of rhythmic synchronization in macaque visual cortex during stimulation with grating stimuli during selective attention. (A) Gamma rhythmic synchronization (spike– local field potential [LFP] coherence (Coh.), y-axis) in macaque visual area V4 during sensory stimulation with a single drifting grating is prominent in superficial cortical layers (top). Alpha-band synchronization is a signature of deep layers (bottom). Line colors indicate whether the monkeys attended (red) or ignored (blue) the stimulus inside the neuronal receptive field. (B) Quantification of the attention effect on the difference of gamma- and alpha-band synchronization in superficial (black bars) and deep (gray bars) layers (left, y-axis). The right axis illustrates that spatial attention (attention inside vs. outside the receptive field [RF] of the neurons) enhanced gamma coherence in superficial layers, reduced alpha coherence in deep layers, but had no apparent laminar specific effects on spike rate and noise correlations (Correl.). (Adapted from Buffalo et al., 2011.)

(Fries et al., 2008). A similar conclusion can be derived from recent findings that gamma-band spike–LFP synchronization is enhanced by attention particularly for neurons in the superficial layers, which host the majority of corticocortical feedforward projection neurons (Buffalo et al., 2011). Figure 71.3 illustrates that gamma-band synchronization is prominent in superficial V4 layers while deep-layer synchronization is characterized by a peak at lower, 6–16 Hz frequencies. The gamma-band synchronization in superficial layers is enhanced by attention into the respective receptive field. By contrast, the lower-frequency synchronization in deep layers is reduced by attention (Buffalo et al., 2011). A similar pattern of layer-specific attentional modulation is observed at the major input stages of area V4: Within areas V1 and V2, selective attention modulates gamma-band synchronization in superficial, that is, feedforward projecting, layers (Buffalo et al., 2011).

The strength of selective increases of spike–LFP gamma-band synchronization, as well as the attentional modulation of firing rates, is lower within V1 and V2 as compared to area V4 (Buffalo et al., 2010, 2011). This progression of attentional modulation from early to extrastriate visual areas may result from the increased demand of neurons with larger receptive fields in later

visual processing areas to select subsets of inputs from a broader set of converging inputs from neurons with smaller receptive fields. According to this suggestion, the larger attentional enhancement of selective gamma-band synchronization in area V4 is based on the requirement of V4 neurons to selectively gate the input from only those V1/V2 populations of neurons that convey attended, that is, relevant, stimulus information.

This selective gating of V1 to V4 connectivity has been demonstrated to be evident in selective interareal synchronization patterns in awake nonhuman primate cortex (Bosman et al., 2012). This study measured the neuronal coherence of electrocorticographic activity from one V4 neuronal population with activity of two spatially separate V1 populations (see figure 71.4A). The receptive fields of each V1 site converged within the larger receptive field extent of the V4 site and allowed the placement of two spatially separate stimuli into the two V1 receptive fields, and at the same time within the confines of the same V4 receptive field (figure 71.4B). With this spatial arrangement, the activity of the V4 recording site reflects the response to the combined, convergent input from the two V1 sites. When attention is selectively focused on one of the two stimuli, then only those connections should be

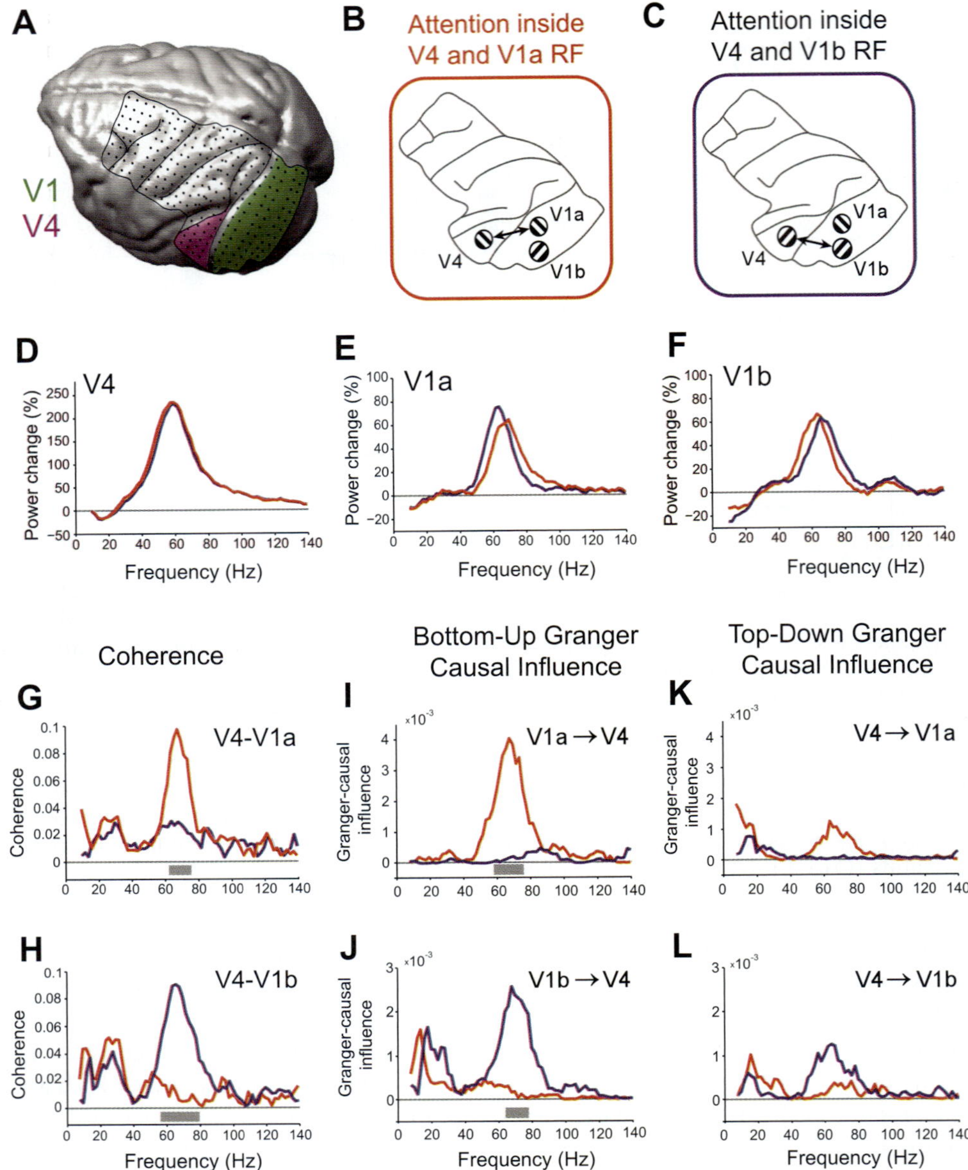

FIGURE 71.4 (A) The spatial coverage of electrocorticographic (ECoG) electrode locations (dots) projected on the rendered cortical surface of a macaque brain. Dots indicate the 252 electrodes of the high-density ECoG grid. Green and purple shaded regions indicate areas V1 and V4, respectively. (B, C) Illustration of the two attention conditions with contours showing the cortical surface and stimuli sketched on those locations that activated the V4 receptive field (RF) and two spatially separated receptive fields in V1. Arrows denote that the attentionally relevant stimulus was activating the V4 site and either the V1a site (B), or the V1b site (C). The color of the box contours (red, blue) is used in the remaining panels to denote the attention conditions. (D) Spectral power change relative to prestimulus baseline in V4. (E, F) Spectral power change relative to prestimulus baseline in V1a (E) and V1b (F) when the attended stimulus was overlaying V1a (red) or V1b (blue). (G, H) Coherence spectra of the V4 to V1a (G) and the V4 to V1b (H) sites when the attended stimulus overlaid V1a (red) and V1b (blue). (I, J) Bottom-up Granger causal influence for the V1a to V4 (I) and the V1b to V4 (J) connections when the attended stimulus overlaid with V1a (red) or V1b (blue). (K, L) Top-down Granger causal influence for the V4 to V1a (K) and the V4 to V1b (L) connection when the attended stimulus overlaid V4 and either V1a (red) or V1b (blue). Gray bars in (G–L) indicate the frequencies with a significant effect (p < 0.05, corrected for multiple comparisons across frequencies, nonparametric randomization across site pairs). (Adapted from Bosman et al., 2012.)

effective that link the V1 site processing the attended stimulus to V4 (figure 71.4C). Exactly this selective gating of effective connectivity was observed (Bosman et al., 2012). Figures 71.4D and 71.4E show the local effect of spatial attention, revealing a similar degree of gamma-band synchronization locally, that is, within each of the neuronal populations. When the animals focused spatial attention on one of the stimuli, gamma-band coherence emerged between V4 and the specific V1 site that conveyed the relevant, attended stimulus. At the same time, gamma-band synchronization was strongly reduced between V4 and the V1 site that conveyed the irrelevant, distracting stimulus (figure 71.4F). Thus, the top-down influence of selective attention determined which anatomically converging connections from primary visual cortex to extrastriate area V4 became functionally effective (Bosman et al., 2012). This emergence of selective functional connectivity was brought about by selective gamma-band coherence providing direct support for the concept of selective communication through coherence. Notably, communication though neuronal coherence could be based on a symmetrical mutual influence between neuronal groups, or it could arise from nonsymmetrical influences. The asymmetry or directedness of coherence can be estimated with the Granger causality metric that quantifies the influence of one neuronal group A (e.g., in V1) onto another group B (e.g., the V4 group) as the variance in B that is explained by the immediate past of A, rather than by the past of B (Dhamala, Rangarajan, & Ding, 2008). Applying the Granger causal metric, Bosman et al. (2012) revealed a predominance of Granger causal influences from the attended V1 neuronal group toward the neuronal group in V4 in the gamma frequency band (figure 71.4G–K) and only a subtle Granger causal feedback influence of V4 to the V1 site that overlaps the focus of attention (figure 71.4I,L). These findings resonate well with the predominance of gamma-band synchronization in superficial versus deep visual cortical areas reported above (Buffalo et al., 2011) because feedforward connectivity originates from superficial layers in primary visual cortex.

Functional Implications of Selective Gamma-Band Synchronization

In addition to the described selective attentional gating effect, various studies have demonstrated that the precision of local synchronization in visual area V4 is closely related to task performance, including behavioral accuracy and the reaction time to detect behaviorally relevant stimulus changes (Taylor et al., 2005; Womelsdorf et al., 2006). The relation to behavioral accuracy was derived from an error analysis of the pattern of synchronization in area V4 (Taylor et al., 2005). In this study, the spatial focus of attention could be inferred from the pattern of synchronization measured through epidural electrodes. Gamma-band synchronization was not only stronger for correct trials than for miss trials, but additionally, the degree of synchronization predicted whether the monkey was paying attention to the distractor. Thus, this study demonstrated that gamma-band synchronization reflects the actual allocation of attention rather than merely the attentional cueing itself. The link to reaction time was made in a study showing that the precision of stimulus-induced gamma-band synchronization within area V4 predicts how rapidly a stimulus change can be reported behaviorally (Womelsdorf et al., 2006) or, in a computational framework, how fast a local neuronal network transmits relevant information (Buehlmann & Deco, 2008). When monkeys were spatially cued to select one of two stimuli in order to detect a color change of the attended stimulus, the speed of change detection could be partly predicted by the strength of gamma-band synchronization shortly before the stimulus change actually occurred (Womelsdorf et al., 2006). Importantly, the reaction times to the stimulus change could not be predicted at times before the stimulus change by overall firing rates, or by synchronization outside of the gamma frequency band. Notably, the correlation of gamma-band synchronization with the speed of change detection showed high spatial selectivity: Neurons activated by an unattended stimulus engaged in lower synchronization when the monkeys were particularly fast in responding to the stimulus change at locations outside their receptive field. This finding rules out a possible influence of globally increased synchronization that may accompany states of enhanced nonspecific alertness and arousal (e.g., Rodriguez et al., 2004). And it argues for a fine-grained influence of synchronization to modulate the effective transmission of information about the stimulus change to postsynaptic target areas concerned with the planning and execution of responses.

These behavioral correlates of gamma-band synchronization during selective attention tasks are complemented by a variety of correlational results linking enhanced gamma-band synchronization to efficient task performance in various attention-demanding paradigms. For example, in memory-related structures, the strength of gamma-band synchronization has been linked to the successful encoding and retrieval of information (Jutras, Fries, & Buffalo, 2009; Sederberg et al., 2007).

Selective Gamma-Band Coherence beyond Visual Cortex

These results of selective gamma-band synchronization with selective spatial attention are supported by a growing number of converging findings from human EEG and magnetoencephalography (MEG) studies (Doesburg et al., 2007; Fan et al., 2007; Siegel, Donner, & Engel, 2012). Importantly, attention modulates gamma-band synchronization beyond sensory visual cortex. It has been reported for auditory cortex (Kaiser et al., 2006) and for somatosensory cortex (e.g., Bauer et al., 2006). Spatial attention for tactile discrimination at either the right or left index finger in humans enhanced stimulus-induced gamma-band synchronization in primary somatosensory cortex when measured with MEG. Similar topographies and dynamics of gamma-band synchronization correlate with the actual perception of somatosensory-induced pain (Gross et al., 2007). Importantly, enhanced oscillatory dynamics during tactile perception is not restricted to the somatosensory cortex (Ohara et al., 2006). In recent intracranial recordings in humans, synchronization was modulated across somatosensory cortex, medial prefrontal, and insular regions when subjects had to direct attention to painful tactile stimulation (Ohara et al., 2006).

Spatially Specific Synchronization Patterns during Preparatory Attentional States

The described gamma-band modulation of rhythmic activity is most prominent during activated states. However, attentional top-down control biases neuronal responses in sensory cortices already before sensory inputs impinge on the neuronal network (Fries et al., 2001, 2008; Schroeder & Lakatos, 2009; Siegel, Donner, & Engel, 2012). In many attention studies, the instructional cue period is followed by a temporal delay void of sensory stimulation. During these preparatory periods, top-down signals set the stage for efficient processing of expected stimulus information, rendering local neuronal groups ready to enhance the representation of attended sensory inputs. Intriguingly, the described preparatory bias is evident in selective synchronization patterns in the gamma band.

In macaque visual cortical area V4, neurons gamma synchronize their spiking responses to the LFP more precisely when monkeys expected a target stimulus at the receptive field location of the respective neuronal group (see figure 71.2B). This modulation was evident even though rhythmic activity proceeded at far lower levels in the absence of sensory stimulation compared

to synchronization strength during high contrast sensory drive. Lower overall strength, and correspondingly lower signal-to-noise ratio, may account for the lack of significant gamma-band modulation of LFP power or spike-to-spike synchronization during the prestimulus period when compared to attentional modulation during stimulation (see figure 71.2).

During preparatory periods, and thus in the absence of strong excitatory drive to the local network, rhythmic activity is dominated by frequencies lower than the gamma band (Womelsdorf et al., 2010b). In the described study from macaque V4, the prestimulus epochs were characterized by alpha-band peaks of local rhythmic synchronization when monkeys attended away from the receptive field of the neuronal group. Figure 71.2B and C demonstrates reduced locking of neuronal spiking in the alpha band to the LFP and to spiking output of nearby neurons (figure 71.2B, C). This finding agrees with various studies reporting reduced alpha-band activity during attentional processing (Bauer et al., 2006; Bollimunta et al., 2011; Pesaran et al., 2002; Rihs, Michel, & Thut, 2007; Siegel, Donner, & Engel, 2012; Worden et al., 2000). Notably, human EEG studies have extended this finding by showing that the degree of alpha frequency desynchronization during prestimulus intervals of visuospatial attention tasks indicates how fast a forthcoming target stimulus is processed (e.g., Thut et al., 2006). For example, reaction times to a peripherally cued target stimulus are partially predicted by the lateralization of alpha activity in a 1-s period before target appearance (Thut et al., 2006). While this predictive effect was based predominantly on reduced alpha-band responses over the hemisphere processing the attended position, recent studies suggest that alpha-band oscillations are selectively enhanced within local neuronal groups processing distracting information, that is, at unattended locations (Rihs, Michel, & Thut, 2007). These findings suggest that rhythmic alpha-band synchronization may play an active role in preventing the signaling of stimulus information (Jensen & Mazaheri, 2010). According to this hypothesis, attention up-regulates alpha-band activity of neuronal groups expected to process distracting stimulus information, rather than down-regulating local alpha-band synchronization for neuronal groups processing attended stimulus features and locations.

Synchronization Patterns Reflecting Temporal Expectancies of Target Processing

The previous paragraph surveyed evidence for an influence of spatially specific expectancy of target and distractor stimuli on synchronization patterns in visual cortex. In addition to spatially selective expectancy, the expectation of the occurrence of behaviorally relevant target events is known to influence neuronal synchronization patterns and firing rates in parietal, prefrontal, and cingulate cortex (Ghose & Maunsell, 2002; Gregoriou et al., 2009; Pesaran et al., 2008; Riehle, 2005; Schoffelen, Oostenveld, & Fries, 2005; Schroeder & Lakatos, 2009, Womelsdorf et al., 2010a). Attentional modulation of neuronal firing rates in the extrastriate middle temporal visual area is strongest around the time point at which the subjective anticipation for a target change, given that it had not occurred before in the trial (i.e., the hazard rate), is maximal (Ghose & Maunsell 2002). In premotor and motor cortex, the hazard rate is smoothly reflected in the strength of synchronization (Riehle, 2005; Schoffelen, Oostenveld, & Fries, 2005). Importantly, enhanced readiness to respond to attended sensory changes is thereby functionally closely linked to long-range synchronization of motor cortex with spinal motor units, suggesting a direct mechanistic influence of synchronization on the speed to respond to behaviorally relevant sensory events (Schoffelen, Oostenveld, & Fries, 2005).

An influence of temporal expectancy on synchronization in early sensory cortices has been demonstrated in recordings in primary visual cortex of macaques (Lakatos et al., 2008). In this study, monkeys were cued to detect deviant sensory stimuli in either an auditory or visual input stream to receive reward. Auditory and visual stimuli alternated, and both stimulus streams followed a noisy 1.55-Hz rhythm. This low-frequency rhythm of sensory inputs entrained neuronal responses in early visual cortex, such that responses to individual stimuli in the visual stream added to the entrained response. Attention to the visual stream amplified the entrainment (see figure 71.5A), but the most prominent attentional effect was evident in the phase of the 1.55-Hz entrainment in the superficial layers of visual cortex: This entrainment was always determined by the stimulus stream that was attended, that is, it switched by half a cycle when attention switched from the visual to the auditory stream (see figure 71.5B), which had a phase opposite to the visual stream. Importantly, low-frequency fluctuations in the LFP likely reflect fluctuations in neuronal excitability. With attention to the visual (auditory) input stream, the phase corresponding to maximal (minimal) neuronal excitability occurred around the average time when the target information was most likely to reach visual cortex. Consistent with a

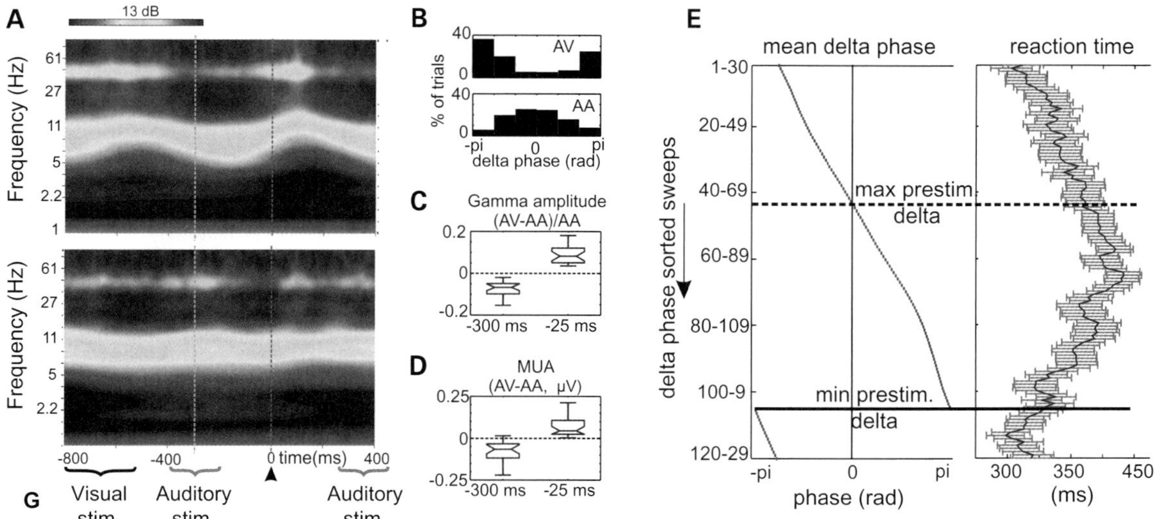

FIGURE 71.5 (A) Entrainment of synchronization from delta- to gamma-band frequencies in supragranular layers of the primary visual cortex during an auditory–visual change detection task. The panels show the time-frequency spectrograms during attention to the visual (upper panel; AV) and auditory (bottom panel; AA) input stream, aligned to the onset time of the visual stimulus. The task cued monkeys to detect infrequent deviant stimuli in either the auditory (white noise tones), or visual (red light flashes) input stream. Visual (auditory) stimuli were onset at a regular interval of 650 ms ± 150 ms, indicated below the time axis (mean stimulus rate = 1.55 Hz); stim., stimulus. (B) Entrainment of delta frequency phase in visual cortex measured as the delta (1.55 Hz) phase at the time of visual stimulus onset when attention was directed to the visual (upper panel) and auditory (lower panel) modality across recording sessions. (C, D) Modulation of gamma-band amplitude of the local field potential (C) and multiunit activity (MUA) (D) before and at visual stimulus onset. Positive values indicate enhancement with visual versus auditory attention. (E) Reaction times (x-axis) to the visual target stimulus sorted into groups of trials according to the prestimulus (prestim.) delta phase (at 0 ms to stimulus onset) (y-axis). Solid/dashed horizontal lines indicate the group of trials corresponding to maximum/minimum (max/min) delta amplitudes. (Adapted from Lakatos et al., 2008.)

functional role of the entrained delta phase, the authors reported the strongest attentional enhancement of gamma-band synchronization in the LFP and spiking activity around this time (see figure 71.5C and D) and showed that the detection of deviant visual stimuli was fastest (slowest) when the delta phase at stimulus onset corresponded to maximal (minimal) neuronal excitability (see figure 71.5E).

The described results suggest that top-down information selectively modulates excitability in early sensory cortices through changes in the phase of rhythmic entrainment in these areas (Lakatos et al., 2009). The exact frequency band underlying excitability modulations may extend from the low delta band, directly imposed by the stimulus structure in the described study, to the theta band around 4–8 Hz. This suggestion may be derived from the time-frequency evolution of LFP power in the theta band and its attentional modulation, shown in figure 71.5A. Intriguingly, similar to the effect of delta phase on the gamma-band response demonstrated directly in the discussed study (see figure 71.5B), previous studies have linked the phase of rhythmic activity in the theta band to the strength of high-frequency gamma-band synchronization in rodent hippocampus and over large regions in the human cortex (Bosman et al., 2012; Canolty et al., 2006). Additional hints suggesting a functional relevance of low-frequency phase synchronization for stimulus and action selection can be found in multiple interareal recording studies in the rodent (Womelsdorf et al., 2010b). In one of the earliest examples, Jones and Wilson (2005) reported that spiking responses in rodent prefrontal cortex phase lock to theta-band activity in the hippocampus during those task epochs requiring spatial decisions in a working memory context. In macaque visual cortex, the phase of theta-band synchronization has been directly linked to selective maintenance of task related information (Lee et al., 2005). Taken together, the emerging evidence demonstrates (1) that top-down, task-related information modulates low-frequency rhythmic activity, (2) that the phase of this rhythmic activity can be functionally related to task performance, and (3) that the phase of low-frequency activity shapes the strength of gamma-band synchronization in response to sensory inputs. As such, the pattern of selective synchronization in the gamma band described in the previous paragraphs could be tightly linked to underlying, selective low-frequency activity modulations. Whether both are coupled in an obligatory way, or whether the comodulation may be triggered by specific task demands, will be an interesting subject for future research (Schroeder & Lakatos, 2009).

Feature-Selective Modulation of Rhythmic Synchronization

The preceding sections discussed evidence for selective neuronal synchronization patterns evolving with space-based attentional selection of sensory inputs. However, in addition to spatial selection, attention frequently proceeds only on top-down information about the behaviorally relevant sensory feature and independent of the exact spatial location at which input impinges on sensory cortices. Such feature-based attention is known to modulate the responses of neurons tuned to the attended feature such as a particular motion direction or the color of a visual stimulus (Maunsell & Treue, 2006).

In visual area V4 feature-based attentional selection is apparent in selective spike-to-LFP synchronization by neurons tuned to the attended stimulus feature (Bichot, Rossi, & Desimone, 2005). In this study, spiking responses and LFPs were recorded in macaque visual area V4 while monkeys searched in multistimulus displays for a target stimulus defined either by color, shape, or both. When monkeys searched, for example, for a red stimulus by shifting their gaze across stimuli on the display, the nonfoveal receptive fields of the recorded neurons could either encompass nontarget stimuli (e.g., of blue color), or the (red) target stimulus prior to the time when the monkey detected the target. The authors found that neurons synchronized to the LFP stronger in response to their preferred stimulus feature when it was the attended search target feature rather than a distractor feature.

Thus, attention enhanced synchronization of the responses of those neurons sharing a preference for the attended target feature—and irrespective of the spatial location of attention (Bichot, Rossi, & Desimone, 2005). This feature-based modulation was also evident during a conjunction search task involving targets defined by two features: When monkeys searched for a target stimulus with a particular orientation and color (e.g., a red horizontal bar), neurons with preference to one of these features enhanced their neuronal synchronization (Bichot, Rossi, & Desimone, 2005). This enhancement was observed not only in response to the color–shape defined conjunction target but also in response to distractors sharing one feature with the target (e.g., red color). This latter finding corresponds well with the behavioral consequences of increased difficulty and search time needed for conjunction defined targets.

This study shows that feature salience is indexed not only by changes in firing rates (e.g., Martinez Trujillo & Treue, 2004) but also by selectively synchronizing neuronal responses depending on the similarity between neuronal feature preferences and the attended stimulus feature. The mechanisms behind this selective influence of featural top-down information could be based on a similar spatial weighting of interneuron network activity as implicated for spatial selection. Neuronal tuning to many basic sensory features is organized in regularly arranged local maps. Correspondingly, the tuning of groups of neurons measured with the LFP is locally highly selective. Importantly, neuronal stimulus preference is systematically related to the strength and precision of neuronal synchronization in the gamma frequency band. This has been demonstrated for stimulus orientation and spatial frequency (Vinck et al., 2010; Womelsdorf et al., 2012), the speed and direction of visual motion (Liu & Newsome, 2006), and the spatial motor intentions and movement directions (Scherberger, Jarvis, & Andersen, 2005). These findings show that rhythmic synchronization conveys feature-selective information. Feature-based attention appears to recruit this property with high spatial resolution by modulating which neurons synchronize to the local rhythmic activity.

Taken together, the previous subsections surveyed the accumulating evidence demonstrating selective neuronal synchronization patterns that evolve with selective spatial and feature-based attention within sensory cortices. Only a few studies have extended these insights to investigate how selective attention modulates selective neuronal interaction patterns between visual and higher-order cortical areas during task performance. These studies have begun to show that such dynamic interareal interaction patterns are evident in long-range synchronization patterns between cortical areas (for a comprehensive survey of the relevant human EEG and MEG studies, see Siegel, Donner, & Engel, 2012).

Selective Interareal Synchronization during Attentional Processing

In the preceding sections, selective synchronization patterns evolved for local neuronal groups in sensory cortices supporting a functional role for gamma-band synchronization for the selective restructuring of neuronal communication during attentional processing (see figure 71.1). However, attentional processing relies on effective interactions *between* local subsets of neuronal groups from distant cortical regions. So far, only a few studies have investigated these interareal interaction patterns during task epochs with selective attention (Bosman et al., 2012; Engel, Fries, & Singer, 2001; Singer, 2011; Womelsdorf & Fries, 2007). The emerging

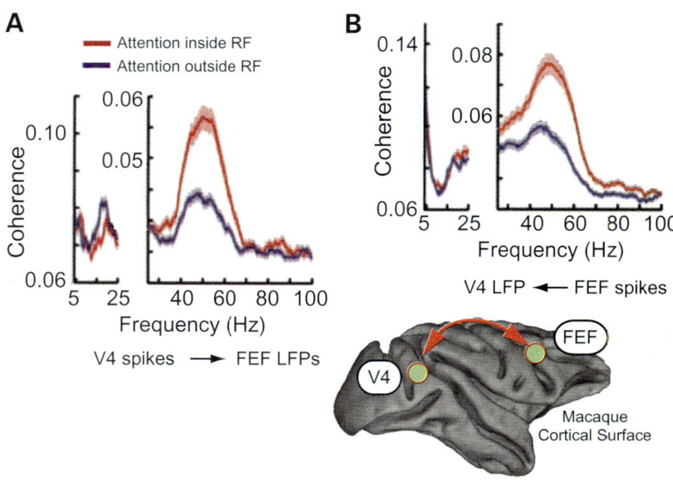

FIGURE 71.6   Selective modulation of long-range synchronization between the frontal eye field (FEF) and visual area V4 during states of selective visual attention (Gregoriou et al., 2009). (A) Coherence (y-axis) of spike trains in V4 to the local field potential (LFP) in FEF, across frequency (x-axis), when attention was directed to a stimulus inside (red) and outside (blue) the receptive field (RF) (V4) and response field (FEF). Recording areas are indicated in the macaque brain outline below the panel. (B) Same as in A, but depicting the coherence between spike trains in FEF and the LFP in V4. (Adapted from Gregoriou et al., 2009.)

evidence from these studies points toward a critical role of rhythmic long-range synchronization at gamma band (Gregoriou et al., 2009; see figure 71.6) as well as lower frequencies, most prominently at beta frequencies ranging from 15 Hz to 30 Hz (Siegel, Donner, & Engel, 2012).

Early studies in awake cats demonstrated transiently enhanced beta frequency synchronization among visual cortical and premotor regions, and between visual cortex and thalamus during nonselective states of expectancy of a behaviorally relevant stimulus (in, e.g., "go/no-go tasks") (Roelfsema et al., 1997; von Stein, Chiang, & König, 2000). Recent studies in the macaque monkey have extended these findings by showing that frontoparietal and intraparietal interactions between areas are accompanied by synchronization at beta frequencies (15–35 Hz) during task epochs requiring searching for and selecting behaviorally relevant visual stimuli (Buschman & Miller, 2007; Pesaran et al., 2008; Saalmann, Pigarev, & Vidyasagar, 2007). Figure 71.7 illustrates findings from a visual search task requiring monkeys to detect a search target that either is salient and pops out among distracting stimuli ("bottom-up search") or is nonsalient by sharing features with distracting stimuli (Buschman & Miller, 2007). In contrast to bottom-up salient targets, the nonsalient target

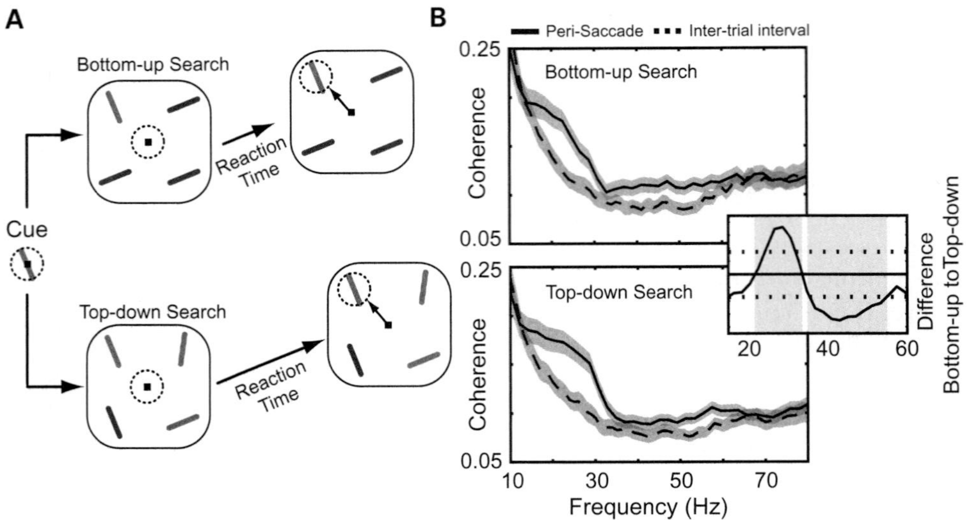

FIGURE 71.7   Selective modulation of long-range synchronization between frontal and parietal cortex during visual search. (A) Sketch of two visual search tasks used by Buschman and Miller (2007). A cue instructed monkeys about the orientation and color of a bar that was the later search target in a multistimulus display during a bottom-up search task (both target color and orientation was unique, upper panels) and during a top-down search task (target shared color or orientation with distracting stimuli, bottom panels). Monkeys covertly attended the multistimulus array and made a saccade to the target stimulus position as soon as they found it. The authors measured the coherence of the local field potential activity of neuronal groups in the frontal eye field and dorsolateral prefrontal cortex and lateral parietal area. (B) The line plots show the coherence (y-axis) for different frequency bands (x-axis) in the bottom-up and top-down tasks, along with the coherence difference across tasks (solid line in inset). The results show that attentional demand modulated long-range frontoparietal coherence at different frequency bands. (Adapted from Buschman & Miller, 2007.)

stimuli were detected more slowly, indicating that they require attentive search through the stimuli in the display before they are successfully detected ("top-down search"). Paralleling the difference in behavioral demands, the authors found a selective synchronization pattern among the LFPs in frontal and parietal cortex. While attentive "top-down search" enhanced specifically rhythmic synchronization at 20–35 Hz compared to the "bottom-up" search, the stimulus driven "bottom-up" search resulted in stronger interareal synchronization in the gamma frequency band (see figure 71.7B). The pattern of results is most likely due to relative differences in task demands in both search modes and was unaffected by differences in reaction times. Therefore these findings suggest that interareal communication during attentional top-down control is conveyed particularly through rhythmic synchronization in a high beta band, either in addition to, or separate from, the frequency of rhythmic interactions underlying bottom-up feedforward signaling (Engel & Fries, 2010).

Consistent with a functional role for top-down mediated long-range neuronal communication, various experimental paradigms demanding attentive processing have shown long-range synchronization in a broad beta band, although mostly at frequencies below 25 Hz.

The following provide a few examples of beta-band modulation in recent studies using very different task paradigms: Choosing freely a sequence of target stimuli for arm movements induces 15-Hz coherence among neurons in premotor cortex and the parietal reach region when compared to instructed searches (Pesaran, Nelson, & Andersen, 2008). Notably, during periods of enhanced interareal coherence, spiking activity in premotor cortex was more predictive of the direction of forthcoming arm movements compared to periods of lower coherence (Pesaran, Nelson, & Andersen, 2008). Variations in reaction times and readiness to respond to a sensory change event induced corresponding fine-grained variations of motor–spinal coherence in the beta band (Schoffelen, Oostenveld, & Fries, 2005). Somatosensory and motor cortex synchronize in the beta band during sensorimotor integration (Brovelli et al., 2004). Selective working memory maintenance in a delayed match-to-sample task results in stronger coherence in the beta band between higher visual areas in humans (Tallon-Baudry et al., 2001) and locally predicts performance in a similar task in the monkey (Tallon-Baudry et al., 2004). The failure to detect a target stimulus in a rapid stream of stimuli in the attentional blink paradigm is associated with reduced frontoparietal and frontotemporal beta-band synchronization (Gross et al., 2004). And as a last example for a potential functional role of beta-band activity, the perception of coherent objects from fragmented visual scenes goes along with transiently enhanced beta-band synchronization of the LFP among prefrontal, hippocampal, and lateral occipital sites (Sehatpour et al., 2008).

Taken together, these diverse findings agree to suggest that interareal synchronization critically subserves neuronal interactions during attentive processing. In the surveyed studies, synchronization in a broadly defined beta band occurred selectively during task epochs requiring effective neuronal integration of information across distributed cortical areas. However, further studies are needed to elucidate the properties of particular frequency bands and their characteristic recruitment during specific tasks (Kopell et al., 2000).

CONCLUDING REMARKS

Selective attention describes a central top-down process that restructures neuronal activity patterns to establish a selective representation of behavioral relevance. The surveyed evidence suggests that attention achieves this functional role by selectively synchronizing those neuronal groups conveying task-relevant information. Attentionally modulated synchronization patterns evolve rapidly, are evident even before sensory inputs arrive, follow closely subjective readiness to process information in time, can be sustained for prolonged time periods, and carry specific information about top-down selected sensory features and motor aspects.

In addition to these functional characteristics, insights into the physiological origins of synchronization have begun to shed light on the mechanistic underpinning of selective neuronal interaction patterns at all spatial scales of cortical processing: At the level of single neurons and local microcircuits, studies are deciphering the role of inhibitory interneuron networks, how precise timing information is conveyed and sustained even at high oscillation frequencies, and how rhythmic synchronization among interneurons is actively made robust against external influences (Bartos, Vida, & Jonas, 2007). These insights are integrated at the network level in models demonstrating how selective synchronization patterns evolve in a self-organized way (Börgers & Kopell, 2008; Tiesinga, Fellous, & Sejnowski, 2008). Acknowledging those basic physiological processes underlying the dynamic generation of selective synchronization seems to be pivotal to elucidate further the mechanistic working principles of selective attention in the brain.

REFERENCES

Azouz, R., & Gray, C. M. (2003). Adaptive coincidence detection and dynamic gain control in visual cortical neurons in vivo. *Neuron, 37*, 513–523.

Bartos, M., Vida, I., & Jonas, P. (2007). Synaptic mechanisms of synchronized gamma oscillations in inhibitory interneuron networks. *Nature Reviews. Neuroscience, 8*, 45–56.

Bauer, M., Oostenveld, R., Peeters, M., & Fries, P. (2006). Tactile spatial attention enhances gamma-band activity in somatosensory cortex and reduces low-frequency activity in parieto-occipital areas. *Journal of Neuroscience, 26*, 490–501.

Bichot, N. P., Rossi, A. F., & Desimone, R. (2005). Parallel and serial neural mechanisms for visual search in macaque area V4. *Science, 308*, 529–534.

Bollimunta, A., Mo, J., Schroeder, C. E., & Ding, M. (2011). Neuronal mechanisms and attentional modulation of corticothalamic alpha oscillations. *Journal of Neuroscience, 31*, 4935–4943.

Börgers, C., Epstein, S., & Kopell, N. J. (2005). Background gamma rhythmicity and attention in cortical local circuits: A computational study. *Proceedings of the National Academy of Sciences of the United States of America, 102*, 7002–7007. doi:10.1073/pnas.0502366102.

Börgers, C., & Kopell, N. J. (2008). Gamma oscillations and stimulus selection. *Neural Computation, 20*, 383–414.

Bosman, C. A., Schoffelen, J. M., Brunet, N., Oostenveld, R., Bastos, A. M., Womelsdorf, T., et al. (2012). Attentional stimulus selection through selective synchronization between monkey visual areas. *Neuron, 75*, 875–888.

Brovelli, A., Ding, M., Ledberg, A., Chen, Y., Nakamura, R., & Bressler, S. L. (2004). Beta oscillations in a large-scale sensorimotor cortical network: Directional influences revealed by Granger causality. *Proceedings of the National Academy of Sciences of the United States of America, 101*, 9849–9854. doi:10.1073/pnas.0308538101.

Buehlmann, A., & Deco, G. (2008). The neuronal basis of attention: Rate versus synchronization modulation. *Journal of Neuroscience, 28*, 7679–7686.

Buffalo, E. A., Fries, P., Landman, R., Buschman, T. J., & Desimone, R. (2011). Laminar differences in gamma and alpha coherence in the ventral stream. *Proceedings of the National Academy of Sciences of the United States of America, 108*, 11262–11267. doi:10.1073/pnas.1011284108.

Buffalo, E. A., Fries, P., Landman, R., Liang, H., & Desimone, R. (2010). A backward progression of attentional effects in the ventral stream. *Proceedings of the National Academy of Sciences of the United States of America, 107*, 361–365.

Buia, C. I., & Tiesinga, P. H. (2008). The role of interneuron diversity in the cortical microcircuit for attention. *Journal of Neurophysiology, 99*, 2158. doi:10.1152/jn.01004.2007.

Buschman, T. J., & Miller, E. K. (2007). Top-down versus bottom-up control of attention in the prefrontal and posterior parietal cortices. *Science, 315*, 1860–1862.

Buzsaki, G., Kaila, K., & Raichle, M. (2007). Inhibition and brain work. *Neuron, 56*, 771–783.

Canolty, R. T., Edwards, E., Dalal, S. S., Soltani, M., Nagarajan, S. S., Kirsch, H. E., et al. (2006). High gamma power is phase-locked to theta oscillations in human neocortex. *Science, 313*, 1626–1628.

Cardin, J. A., Carlen, M., Meletis, K., Knoblich, U., Zhang, F., Deisseroth, K., et al. (2009). Driving fast-spiking cells induces gamma rhythm and controls sensory responses. *Nature, 459,* 663–667.

Chalk, M., Herrero, J. L., Gieselmann, M. A., Delicato, L. S., Gotthardt, S., & Thiele, A. (2010). Attention reduces stimulus-driven gamma frequency oscillations and spike field coherence in V1. *Neuron, 66,* 114–125.

Chelazzi, L., Miller, E. K., Duncan, J., & Desimone, R. (1993). A neural basis for visual search in inferior temporal cortex. *Nature, 363,* 345–347.

Dhamala, M., Rangarajan, G., & Ding, M. (2008). Analyzing information flow in brain networks with nonparametric Granger causality. *NeuroImage, 41,* 354–362.

Doesburg, S. M., Roggeveen, A. B., Kitajo, K., & Ward, L. M. (2007). Large-scale gamma-band phase synchronization and selective attention. *Cerebral Cortex, 18,* 386–396. doi:10.1093/cercor/bhm073.

Engel, A. K., & Fries, P. (2010). Beta-band oscillations— Signalling the status quo? *Current Opinion in Neurobiology, 20,* 156–165.

Engel, A. K., Fries, P., & Singer, W. (2001). Dynamic predictions: Oscillations and synchrony in top-down processing. *Nature Reviews. Neuroscience, 2,* 704–716.

Fan, J., Byrne, J., Worden, M. S., Guise, K. G., McCandliss, B. D., Fossella, J., et al. (2007). The relation of brain oscillations to attentional networks. *Journal of Neuroscience, 27,* 6197–6206.

Fries, P. (2005). A mechanism for cognitive dynamics: Neuronal communication through neuronal coherence. *Trends in Cognitive Sciences, 9,* 474–480. doi:10.1016/j.tics.2005.08.011.

Fries, P. (2009). Neuronal gamma-band synchronization as a fundamental process in cortical computation. *Annual Review of Neuroscience, 32,* 209–224.

Fries, P., Nikolic, D., & Singer, W. (2007). The gamma cycle. *Trends in Neurosciences, 30,* 309–316. doi:10.1016/j.tins.2007.05.005.

Fries, P., Reynolds, J. H., Rorie, A. E., & Desimone, R. (2001). Modulation of oscillatory neuronal synchronization by selective visual attention. *Science, 291,* 1560–1563.

Fries, P., Schröder, J. H., Roelfsema, P. R., Singer, W., & Engel, A. K. (2002). Oscillatory neuronal synchronization in primary visual cortex as a correlate of stimulus selection. *Journal of Neuroscience, 22,* 3739–3754.

Fries, P., Womelsdorf, T., Oostenveld, R., & Desimone, R. (2008). The effects of visual stimulation and selective visual attention on rhythmic neuronal synchronization in macaque area V4. *Journal of Neuroscience, 28,* 4823–4835.

Ghose, G. M., & Maunsell, J. H. (2002). Attentional modulation in visual cortex depends on task timing. *Nature, 419,* 616–620.

Gilbert, C. D., & Sigman, M. (2007). Brain states: Top-down influences in sensory processing. *Neuron, 54,* 677–696.

Gottlieb, J. (2002). Parietal mechanisms of target representation. *Current Opinion in Neurobiology, 12,* 134–140.

Gregoriou, G. G., Gotts, S. J., Zhou, H., & Desimone, R. (2009). High-frequency, long-range coupling between prefrontal and visual cortex during attention. *Science, 324,* 1207–1210.

Gross, J., Schmitz, F., Schnitzler, I., Kessler, K., Shapiro, K., Hommel, B., et al. (2004). Modulation of long-range neural synchrony reflects temporal limitations of visual attention in humans. *Proceedings of the National Academy of Sciences of the United States of America, 101,* 13050–13055. doi:10.1073/pnas.0404944101.

Gross, J., Schnitzler, A., Timmermann, L., & Ploner, M. (2007). Gamma oscillations in human primary somatosensory cortex reflect pain perception. *PLoS Biology, 5,* e133. doi:10.1371/journal.pbio.0050133.

Hasenstaub, A., Shu, Y., Haider, B., Kraushaar, U., Duque, A., & McCormick, D. A. (2005). Inhibitory postsynaptic potentials carry synchronized frequency information in active cortical networks. *Neuron, 47,* 423–435.

Jensen, O., & Mazaheri, A. (2010). Shaping functional architecture by oscillatory alpha activity: Gating by inhibition. *Frontiers in Human Neuroscience, 4,* 186. doi:10.3389/fnhum.2010.00186.

Jones, M. W., & Wilson, M. A. (2005). Theta rhythms coordinate hippocampal–prefrontal interactions in a spatial memory task. *PLoS Biology, 3,* e402. doi:10.1371/journal.pbio.0030402.

Jutras, M. J., Fries, P., & Buffalo, E. A. (2009). Gamma-band synchronization in the macaque hippocampus and memory formation. *Journal of Neuroscience, 29,* 12521–12531.

Kaiser, J., Hertrich, I., Ackermann, H., & Lutzenberger, W. (2006). Gamma-band activity over early sensory areas predicts detection of changes in audiovisual speech stimuli. *NeuroImage, 30,* 1376–1382.

Kaping, D., Vinck, M., Hutchison, R. M., Everling, S., & Womelsdorf, T. (2011). Specific contributions of ventromedial, anterior cingulate, and lateral prefrontal cortex for attentional selection and stimulus valuation. *PLoS Biology, 9,* e1001224. doi:10.1371/journal.pbio.1001224.

Khayat, P. S., Spekreijse, H., & Roelfsema, P. R. (2006). Attention lights up new object representations before the old ones fade away. *Journal of Neuroscience, 26,* 138–142.

Kopell, N., Ermentrout, G. B., Whittington, M. A., & Traub, R. D. (2000). Gamma rhythms and beta rhythms have different synchronization properties. *Proceedings of the National Academy of Sciences of the United States of America, 97,* 1867–1872.

Knoblich, U., Siegle, J. H., Pritchett, D. L., & Moore, C. I. (2010). What do we gain from gamma? Local dynamic gain modulation drives enhanced efficacy and efficiency of signal transmission. *Frontiers in Human Neuroscience, 4,* 185. doi:10.3389/fnhum.2010.00185.

Lakatos, P., Karmos, G., Metha, A. D., Ulbert, I., & Schroeder, C. E. (2008). Entrainment of neuronal oscillations as a mechanism of attentional selection. *Science, 320,* 110–113.

Lakatos, P., O'Connell, M. N., Barczak, A., Mills, A., Javitt, D. C., & Schroeder, C. E. (2009). The leading sense: Supramodal control of neurophysiological context by attention. *Neuron, 64,* 419–430.

Lee, H., Simpson, G. V., Logothetis, N. K., & Rainer, G. (2005). Phase locking of single neuron activity to theta oscillations during working memory in monkey extrastriate visual cortex. *Neuron, 45,* 147–156.

Lin, S. C., Gervasoni, D., & Nicolelis, M. A. (2006). Fast modulation of prefrontal cortex activity by basal forebrain noncholinergic neuronal ensembles. *Journal of Neurophysiology, 96,* 3209–3219.

Liu, J., & Newsome, W. T. (2006). Local field potential in cortical area MT: Stimulus tuning and behavioral correlations. *Journal of Neuroscience, 26,* 7779–7790.

Markram, H., Toledo-Rodriguez, M., Wang, Y., Gupta, A., Silberberg, G., & Wu, C. (2004). Interneurons of the neocortical inhibitory system. *Nature Reviews. Neuroscience, 5,* 793–807.

Martinez Trujillo, J. C., & Treue, S. (2004). Feature-based attention increases the selectivity of population responses in primate visual cortex. *Current Biology, 14*, 744–751.

Maunsell, J. H., & Treue, S. (2006). Feature-based attention in visual cortex. *Trends in Neurosciences, 29*, 317–322. doi:10.1016/j.tins.2006.04.001.

Mishra, J., Fellous, J. M., & Sejnowski, T. J. (2006). Selective attention through phase relationship of excitatory and inhibitory input synchrony in a model cortical neuron. *Neural Networks, 19*, 1329–1346.

Mitchell, J. F., Sundberg, K. A., & Reynolds, J. H. (2007). Differential attention-dependent response modulation across cell classes in macaque visual area V4. *Neuron, 55*, 131–141.

Monosov, I. E., Trageser, J. C., & Thompson, K. G. (2008). Measurements of simultaneously recorded spiking activity and local field potentials suggest that spatial selection emerges in the frontal eye field. *Neuron, 57*, 614–625.

Ohara, S., Crone, N. E., Weiss, N., & Lenz, F. A. (2006). Analysis of synchrony demonstrates 'pain networks' defined by rapidly switching, task-specific, functional connectivity between pain-related cortical structures. *Pain, 123*, 244–253.

Pesaran, B., Nelson, M. J., & Andersen, R. A. (2008). Free choice activates a decision circuit between frontal and parietal cortex. *Nature, 453*, 406–409.

Pesaran, B., Pezaris, J. S., Sahani, M., Mitra, P. P., & Andersen, R. A. (2002). Temporal structure in neuronal activity during working memory in macaque parietal cortex. *Nature Neuroscience, 5*, 805–811.

Reynolds, J. H., Chelazzi, L., & Desimone, R. (1999). Competitive mechanisms subserve attention in macaque areas V2 and V4. *Journal of Neuroscience, 19*, 1736–1753.

Riehle, A. (2005). Preparation for action: One of the key functions of motor cortex. In A. Riehle & E. Vaadia (Eds.), *Motor cortex in voluntary movements: A dsitributed system for distributed functions* (Vol. 1, pp. 213–240). Boca Raton, FL: CRC Press.

Rihs, T. A., Michel, C. M., & Thut, G. (2007). Mechanisms of selective inhibition in visual spatial attention are indexed by alpha-band EEG synchronization. *European Journal of Neuroscience, 25*, 603–610.

Rodriguez, R., Kallenbach, U., Singer, W., & Munk, M. H. (2004). Short- and long-term effects of cholinergic modulation on gamma oscillations and response synchronization in the visual cortex. *Journal of Neuroscience, 24*, 10369–10378.

Roelfsema, P. R., Engel, A. K., König, P., & Singer, W. (1997). Visuomotor integration is associated with zero time-lag synchronization among cortical areas. *Nature, 385*, 157–161.

Roelfsema, P. R., Tolboom, M., & Khayat, P. S. (2007). Different processing phases for features, figures, and selective attention in the primary visual cortex. *Neuron, 56*, 785–792.

Rudolph, M., Pospischil, M., Timofeev, I., & Destexhe, A. (2007). Inhibition determines membrane potential dynamics and controls action potential generation in awake and sleeping cat cortex. *Journal of Neuroscience, 27*, 5280–5290.

Saalmann, Y. B., Pigarev, I. N., & Vidyasagar, T. R. (2007). Neural mechanisms of visual attention: How top-down feedback highlights relevant locations. *Science, 316*, 1612–1615.

Salinas, E., & Sejnowski, T. J. (2001). Correlated neuronal activity and the flow of neural information. *Nature Reviews. Neuroscience, 2*, 539–550.

Scherberger, H., Jarvis, M. R., & Andersen, R. A. (2005). Cortical local field potential encodes movement intentions in the posterior parietal cortex. *Neuron, 46*, 347–354.

Schoffelen, J. M., Oostenveld, R., & Fries, P. (2005). Neuronal coherence as a mechanism of effective corticospinal interaction. *Science, 308*, 111–113.

Schroeder, C. E., & Lakatos, P. (2009). Low-frequency neuronal oscillations as instruments of sensory selection. *Trends in Neurosciences, 32*, 9–18. doi:10.1016/j.tins.2008.09.012.

Sederberg, P. B., Schulze-Bonhage, A., Madsen, J. R., Bromfield, E. B., McCarthy, D. C., Brandt, A., et al. (2007). Hippocampal and neocortical gamma oscillations predict memory formation in humans. *Cerebral Cortex, 17*, 1190–1196. doi:10.1093/cercor/bhl030.

Sehatpour, P., Molholm, S., Schwartz, T. H., Mahoney, J. R., Mehta, A. D., Javitt, D. C., et al. (2008). A human intracranial study of long-range oscillatory coherence across a frontal–occipital–hippocampal brain network during visual object processing. *Proceedings of the National Academy of Sciences of the United States of America, 105*, 4399–4404. doi:10.1073/pnas.0708418105.

Siegel, M., Donner, T. H., & Engel, A. K. (2012). Spectral fingerprints of large-scale neuronal interactions. *Nature Reviews. Neuroscience, 13*, 121–134.

Simons, D. J., & Rensink, R. A. (2005). Change blindness: Past, present, and future. *Trends in Cognitive Sciences, 9*, 16–20.

Singer, W. (2011). Dynamic formation of functional networks by synchronization. *Neuron, 69*, 191–193.

Tallon-Baudry, C., Bertrand, O., & Fischer, C. (2001). Oscillatory synchrony between human extrastriate areas during visual short-term memory maintenance. *Journal of Neuroscience, 21*, RC177.

Tallon-Baudry, C., Mandon, S., Freiwald, W. A., & Kreiter, A. K. (2004). Oscillatory synchrony in the monkey temporal lobe correlates with performance in a visual short-term memory task. *Cerebral Cortex, 14*, 713–720.

Taylor, K., Mandon, S., Freiwald, W. A., & Kreiter, A. K. (2005). Coherent oscillatory activity in monkey area v4 predicts successful allocation of attention. *Cerebral Cortex, 15*, 1424–1437.

Thut, G., Nietzel, A., Brandt, S. A., & Pascual-Leone, A. (2006). Alpha-band electroencephalographic activity over occipital cortex indexes visuospatial attention bias and predicts visual target detection. *Journal of Neuroscience, 26*, 9494–9502.

Tiesinga, P., Fellous, J. M., & Sejnowski, T. J. (2008). Regulation of spike timing in visual cortical circuits. *Nature Reviews. Neuroscience, 9*, 97–107.

Tiesinga, P. H., & Sejnowski, T. J. (2004). Rapid temporal modulation of synchrony by competition in cortical interneuron networks. *Neural Computation, 16*, 251–275.

Tiesinga, P. H., & Sejnowski, T. J. (2010). Mechanisms for phase shifting in cortical networks and their role in communication through coherence. *Frontiers in Human Neuroscience, 4*, 196. doi:10.3389/fnhum.2010.00196.

van Elswijk, G., Maij, F., Schoffelen, J. M., Overeem, S., Stegeman, D. F., & Fries, P. (2010). Corticospinal beta-band synchronization entails rhythmic gain modulation. *Journal of Neuroscience, 30*, 4481–4488.

Vinck, M., Lima, B., Womelsdorf, T., Oostenveld, R., Singer, W., Neuenschwander, S., et al. (2010). Gamma-phase shifting in awake monkey visual cortex. *Journal of Neuroscience, 30*, 1250–1257.

von Stein, A., Chiang, C., & König, P. (2000). Top-down processing mediated by interareal synchronization. *Proceedings of the National Academy of Sciences of the United States of America, 97*, 14748–14753. doi:10.1073/pnas.97.26.14748.

Wang, X. J. (2010). Neurophysiological and computational principles of cortical rhythms in cognition. *Physiological Reviews, 90*, 1195–1268.

Womelsdorf, T., & Fries, P. (2007). The role of neuronal synchronization in selective attention. *Current Opinion in Neurobiology, 17*, 154–160.

Womelsdorf, T., Fries, P., Mitra, P. P., & Desimone, R. (2006). Gamma-band synchronization in visual cortex predicts speed of change detection. *Nature, 439*, 733–736.

Womelsdorf, T., Johnston, K., Vinck, M., & Everling, S. (2010a). Theta-activity in anterior cingulate cortex predicts task rules and their adjustments following errors. *Proceedings of the National Academy of Sciences of the United States of America, 107*, 5248–5253. doi:10.1073/pnas.0906194107.

Womelsdorf, T., Lima, B., Vinck, M., Neuenschwander, S., Oostenveld, R., Singer, W., et al. (2012). Orientation selectivity and noise correlation in awake monkey V1 are modulated by the gamma cycle. *Proceedings of the National Academy of Sciences of the United States of America, 109*, 4302–4307.

Womelsdorf, T., Schoffelen, J. M., Oostenveld, R., Singer, W., Desimone, R., Engel, A. K., et al. (2007). Modulation of neuronal interactions through neuronal synchronization. *Science, 316*, 1609–1612.

Womelsdorf, T., Vinck, M., Leung, L. S., & Everling, S. (2010b). Selective theta-synchronization of choice-relevant information subserves goal-directed behavior. *Frontiers in Human Neuroscience, 4*, 210. doi:10.3389/fnhum.2010.00210.

Worden, M. S., Foxe, J. J., Wang, N., & Simpson, G. V. (2000). Anticipatory biasing of visuospatial attention indexed by retinotopically specific alpha-band electroencephalography increases over occipital cortex. *Journal of Neuroscience, 20*, RC63.

72 Visuomotor Control

MELVYN A. GOODALE

The role of vision in the control of skilled movements has only recently attracted the attention of vision scientists. For most of the past century (and before), researchers have concentrated their efforts instead on working out how vision constructs our perception of the world (Goodale, 1983). Indeed, with the notable exception of eye movements, which have typically been regarded as an information-seeking adjunct to visual perception, little attention has been paid to the way in which vision is used to program and control our actions, particularly the movements of our hands and limbs. The last few decades, however, have seen considerable progress on this front. In this chapter, I will focus on the visual control of reach-to-grasp movements, a class of behavior that is exceptionally well-developed in human beings, and one that exemplifies the importance of vision in the control of action. I first review some of the research that has explored the role of different visual cues in the control of reach-to-grasp movements. I then discuss some of the research that has investigated the neural substrates of that control and how those substrates are related to those that mediate our perceptual representations of the world.

THE VISUAL CONTROL OF REACH-TO-GRASP MOVEMENTS

Over 30 years ago, Marc Jeannerod (1981, 1984, 1986, 1988) proposed that the reaching component of a grasping movement is relatively independent from the formation of the grip itself. According to his dual-channel account (Jeannerod, 1981), the kinematics of the reach component are largely determined by visual cues that are extrinsic to the goal object, such as its distance and location with respect to the grasping hand. In contrast, the kinematics of the grasp component reflect the size, shape, and other intrinsic properties of the goal object (see figure 72.1). Today, there is broad consensus that these two components show a good deal of functional independence and (as will be discussed later) are mediated by relatively independent neural circuitry.

Jeannerod's (1981) dual-channel hypothesis has not gone unchallenged however. Smeets and Brenner (1999, 2001; Smeets, Brenner, & Martin, 2009), for example, have proposed instead that the movements of each finger of the grasping hand are programmed and controlled independently. According to their account, when a person reaches out to grasp an object with a precision grip, the index finger is directed to one side of the object and the thumb to the other. The apparent scaling of grip aperture to object size is nothing more than an emergent property of the fact that the two digits are moving (independently) toward their respective end points. Simply put, it is location rather than size that drives grasping—and there is no need to separate grasping into transport and grip components, each sensitive to a different set of visual cues. Smeets and Brenner's (1999, 2001) "double-pointing" hypothesis has the virtue of being parsimonious. Nevertheless, it is not without its critics (e.g., Dubrowski et al., 2002; Mon-Williams & McIntosh, 2000; Mon-Williams & Tresilian, 2001; van de Kamp & Zaal, 2007). Moreover, (as we shall see later) the organization of the neural substrates of grasping as revealed by neuroimaging and neuropsychology can be more easily explained by the dual-channel than the double-pointing hypothesis.

When people reach out to grasp an object, their hand and arm rarely collide with other objects in the workspace. The ease with which this is accomplished belies the fact that a sophisticated obstacle avoidance system must be at work—a system that encodes possible obstructions to a goal-directed movement and incorporates this information into the motor plan. The few investigations that have examined obstacle avoidance have revealed an efficient system that is capable of altering the spatial and temporal trajectories of goal-directed reaching and grasping movements to avoid other objects in the workspace in a fluid manner (e.g., Castiello, 2001; Jackson, Jackson, & Rosicky, 1995; Tresilian, 1998; Vaughan, Rosenbaum, & Meulenbroek, 2001). There has been some debate as to whether the nongoal objects are always being treated as obstacles or instead as potential targets for action (e.g., Tipper, Howard, & Jackson, 1997) or even as frames of reference for the control of the movement (e.g., Diedrichsen et al., 2004; Obhi & Goodale, 2005). By positioning nongoal objects in different locations in the workspace with respect to the target, however, it is possible to show that most often

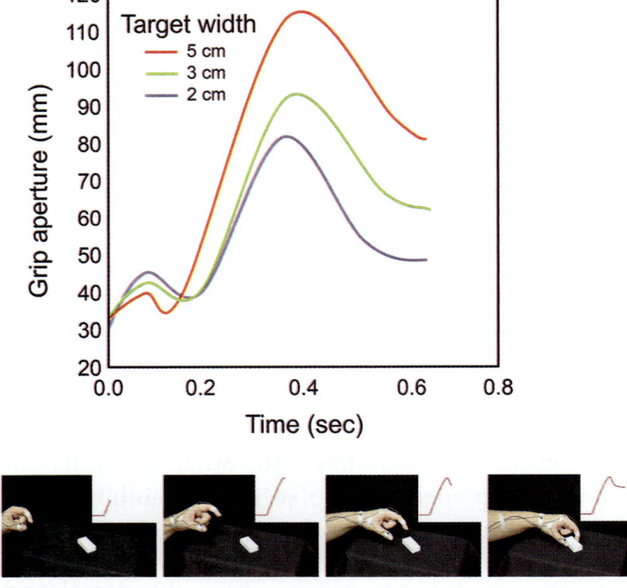

FIGURE 72.1 The sequence shown at the bottom of this figure is that of a hand reaching out to grasp a wooden block. Notice that the finger and thumb first open wider than the block and then close down as the hand approaches the block. The maximum extent of opening is referred to as the "maximum grip aperture," and this measure is strongly correlated with the width of the goal object (see graph at top).

these objects are being treated as obstacles and that when individuals reach out for the goal, the trial-to-trial adjustments of their trajectories are remarkably sensitive to the position of obstacles both in depth and in the horizontal plane, as well as to their height (Chapman & Goodale, 2008, 2010).

But what are the visual cues that are used to program and control grasping? As it turns out, some of the most powerful cues depend on the fact that we have two eyes, that is, we have binocular vision. Covering one eye has clear detrimental effects on grasping: People reach more slowly, show longer periods of deceleration, and execute more online adjustments of both their trajectory and their grip during the closing phase of the grasp (Keefe & Watt, 2009; Loftus et al., 2004; Melmoth & Grant, 2006; Servos, Goodale, & Jakobson, 1992; Watt & Bradshaw, 2000). In addition, it has been shown that adults with stereo-deficiencies exhibit slower and less accurate grasping movements (Melmoth et al., 2009). Nevertheless, individuals who have lost an eye are still able to grasp objects reasonably well. It turns out that they have learned to do this by making exaggerated head movements to generate retinal motion cues (Marotta et al., 1995a, 1995b).

Computing the required distance for the grasp appears to depend heavily on vergence whereas the scaling of the grasping movement and the final placement of the fingers depends more on retinal disparity (Melmoth et al., 2007; Mon-Williams & Dijkerman, 1999). Similarly, motion parallax contributes more to the computation of reach distance than it does to the formation of the grasp, although motion parallax becomes important only when binocular cues are no longer available (Marotta, Kruyer, & Goodale, 1998; Watt & Bradshaw, 2003). (It should be noted that the differential contributions that these cues make to the reach and grasp components, respectively, are much more consistent with Jeannerod's, 1981, dual-channel hypothesis than they are with Smeets & Brenner's, 1999, double-pointing account.) Static monocular cues can also be used to program and control grasping movements. Thus, Marotta and Goodale (1998, 2001) showed that pictorial cues, such as height in the visual scene and familiar size, can be exploited to program and control grasping—but the contributions from these cues are usually overshadowed by binocular information from vergence and/or retinal disparity.

The role of the shape and orientation of the goal object in determining the formation of the grasp is poorly understood. It is clear that the posture of the grasping hand is sensitive to these features (e.g., Cuijpers, Brenner, Smeets, 2006; Cuijpers, Smeets, & Brenner, 2004; Goodale et al., 1994b; van Bergen et al., 2007), but there have been only a few systematic investigations of how information about object shape and orientation is used to configure the hand during the planning and execution of grasping movements (e.g., Cuijpers, Smeets, & Brenner, 2004; Lee et al., 2008; Louw, Smeets, & Brenner, 2007; van Mierlo et al., 2009).

When attempting to pick up an object, people shift their gaze toward the object (e.g., Ballard et al., 1992; Johansson et al., 2001) and to those locations on the object where they intend to place their fingers, particularly the points where more visual feedback is required to position the fingers properly (e.g., Binsted et al., 2001; Brouwer, Franz, & Gegenfurtner, 2009). In cluttered workspaces, people also tend to direct their gaze toward obstacles that they might have to avoid (e.g., Johansson et al., 2001). Even when gaze is maintained elsewhere in the scene, there is evidence that attention is shifted covertly to the goal and is bound there until the movement is initiated (Deubel, Schneider, & Paprotta, 1998). In a persuasive account of the role of attention in reaching and grasping, Baldauf and Deubel (2010) have argued that the planning of a reach-to-grasp movement requires the formation of what they call an "attentional landscape" in which the locations of all the objects and features in the workspace that are relevant for the intended action are encoded.

Finally, it should be noted that the way in which different visual cues are weighted for the control of skilled movements is typically quite different from the way they are weighted for perceptual judgments. For example, Knill (2005) found that participants gave significantly more weight to binocular compared to monocular cues when they were asked to place objects on a slanted surface in a virtual display compared to when they were required to make explicit judgments about the slant. Similarly, Servos (2000) demonstrated that even though people relied much more on binocular than monocular cues when they grasped an object, their explicit judgments about the distance of the same object were no better under binocular than under monocular viewing conditions. These and other even more dramatic dissociations reviewed later underscore the fundamental differences between how vision is used for action and how it is used for perceptual report.

THE NEURAL SUBSTRATES OF VISION FOR ACTION

Despite the complexity of the interconnections between different visual areas in the cerebral cortex (Van Essen et al., 2001), two broad "streams" of projections from primary visual cortex have been identified in the macaque monkey brain: a ventral stream projecting eventually to the inferotemporal cortex and a dorsal stream projecting to the posterior parietal cortex (Ungerleider & Mishkin, 1982). Mounting evidence from neuroimaging suggests that the visual projections from early visual areas to the temporal and parietal lobes in the human brain also involve a separation into ventral and dorsal streams (Culham & Valyear, 2006; Grill-Spector & Malach, 2004; Tootell, Tsao, & Vanduffel, 2003; Van Essen et al., 2001; see figure 72.2).

Over 20 years ago, Goodale and Milner (1992) proposed that the dorsal stream, which is intimately connected with somatosensory and premotor cortex as well as motor centers in the brainstem, plays a critical role in the real-time control of action. In contrast, they argued, the ventral stream, which is closely linked with cognitive networks in the frontal and temporal lobes, helps to construct the rich and detailed representations of the world that allow us to identify objects and events, attach meaning and significance to them, and establish their causal relations. Thus, according to this two-visual-systems model, the dorsal stream mediates vision-for-action while the ventral stream mediates vision-for-perception (Goodale & Milner, 1992; Milner & Goodale, 2006, 2008). Of course, the dorsal and ventral streams have other roles to play as well. For example, the dorsal stream, together with areas in the ventral stream, plays a role in spatial navigation—and areas in the dorsal stream appear to be involved in some aspects of working memory (Kravitz et al., 2011). This review, however, will focus on the respective roles of the two streams in perception and action.

NEUROPSYCHOLOGICAL EVIDENCE FOR TWO STREAMS

It has been known for a long time that patients with lesions in the dorsal stream, particularly lesions of the

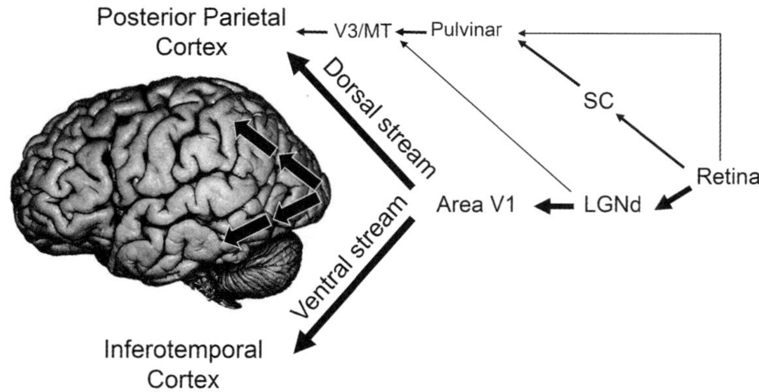

FIGURE 72.2   The two streams of visual processing in human cerebral cortex. The retina sends projections to the dorsal part of the lateral geniculate nucleus (LGNd), which projects in turn to primary visual cortex. Within the cerebral cortex, the ventral stream arises from early visual areas and projects to the inferotemporal cortex. The dorsal stream also arises from early visual areas but projects instead to the posterior parietal cortex. Recently, it has been shown that the posterior parietal cortex also receives visual input from the pulvinar via projections to MT and V3, as well as from the interlaminar layers of LGNd via projections to middle temporal area (MT) and V3. The pulvinar receives projections from both the retina and from the superior colliculus (SC). The approximate locations of the two streams are shown on a 3-D reconstruction of the pial surface of the brain. The two streams involve a series of complex interconnections that are not shown. (Adapted from Goodale, 2011.)

superior regions of the posterior parietal cortex that invade the territory of the intraparietal sulcus and/or the parieto-occipital sulcus, can have problems using vision to direct a grasp or aiming movement toward the correct location of a visual target placed in different positions in the visual field, particularly the peripheral visual field. This deficit is often described as *optic ataxia* (following Bálint, 1909; Bálint & Harvey, 1995). However, the failure to locate an object with the hand should not be construed as a problem in spatial vision; many of these patients, for example, can describe the relative position of the object in space quite accurately, even though they cannot direct their hand toward it (Perenin & Vighetto, 1988). (It should be pointed out, of course, that these patients typically have no difficulty using input from other sensory systems, such as proprioception or audition, to guide their movements.) Some of these patients are unable to use visual information to rotate their hand, scale their grip, or configure their fingers properly when reaching out to pick up an object (e.g., Jakobson et al., 1991), even though they have no difficulty describing the orientation, size, or shape of objects in that part of the visual field (see figure 72.3). Taken together this pattern of deficits and spared visual abilities suggests that the posterior parietal cortex plays a critical role in the visual control of skilled actions (for a more detailed discussion, see Milner & Goodale, 2006).

The opposite pattern of deficits and spared abilities can be seen in patients with visual agnosia. Take the case of patient DF, who developed a profound visual form agnosia following carbon monoxide poisoning (Goodale et al., 1991; Milner et al., 1991). Even though DF's "low-level" visual abilities are reasonably intact, she can no longer recognize everyday objects or the faces of her friends and relatives; nor can she identify even the simplest of geometric shapes. (If an object is placed in her hand, of course, she has no trouble identifying it by touch.) Remarkably, however, DF shows strikingly accurate guidance of her hand movements when she attempts to pick up the very objects she cannot identify. Thus, when she reaches out to grasp objects of different sizes, her hand opens wider midflight for larger objects than it does for smaller ones, just like it does in people with normal vision (see figure 72.3).[1] Similarly, she rotates her hand and wrist quite normally when she reaches out to grasp objects in different orientations, and she places her fingers correctly on the surface of objects with different shapes. At the same time, she is quite unable to distinguish between any of these objects when they are presented to her in simple discrimination tests. She even fails in manual "matching" tasks in which she is asked to show how wide an object is by opening her index finger and thumb a corresponding amount. This dissociation between compromised visual perception and spared visual control of action lends

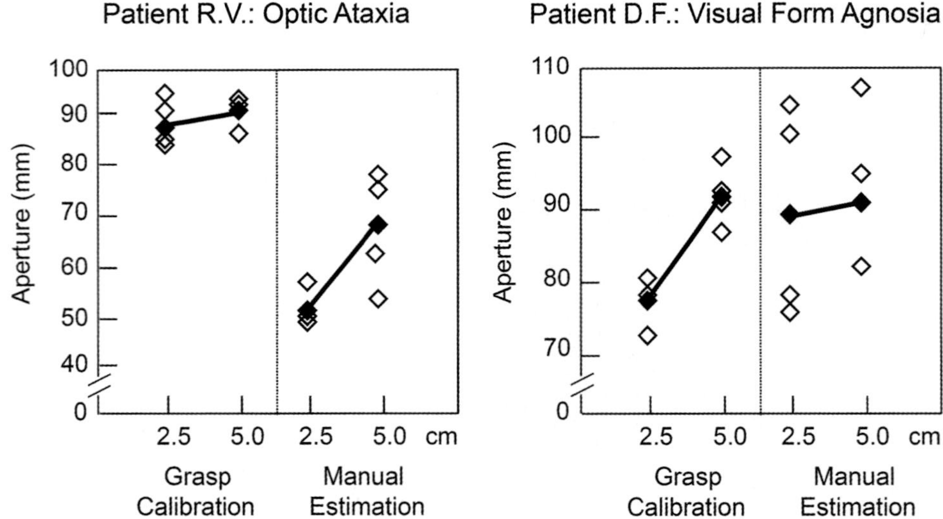

FIGURE 72.3 Graphs showing the size of the aperture between the index finger and thumb during object-directed grasping and manual estimates of object width for RV, a patient with optic ataxia, and DF, a patient with visual form agnosia. Left-hand panel shows that RV was able to indicate the size of the objects reasonably well (individual trials marked as open diamonds), but her maximum grip aperture in flight was not well tuned. She simply opened her hand as wide as possible on every trial. In contrast, right-hand panel shows that DF showed excellent grip scaling, opening her hand wider for the 50-mm-wide object than for the 25-mm-wide object. DF's manual estimates of the width of the two objects, however, were grossly inaccurate and showed enormous variability from trial to trial.

additional support to the idea that there are separate neural pathways for transforming incoming visual information for perception and action (for more details, see Goodale & Milner, 2004; Milner & Goodale 2006).

But where exactly is the damage in DF's brain? An early structural magnetic resonance imaging (MRI) scan showed evidence of extensive bilateral damage in ventrolateral occipital cortex. A more recent high-resolution MRI scan confirmed this damage but revealed that the lesions were focused in a region of the lateral occipital cortex (area LO) that we now know is involved in the visual recognition of objects, particularly their geometric structure (James et al., 2003; see figure 72.4). Even though this selective damage to area LO appears to have disrupted DF's ability to perceive the form of objects, the lesions have not interfered with her ability to use visual form information to shape her hand when she reaches out and grasp objects—presumably because the visuomotor networks in her dorsal stream are largely spared. Other patients with ventral-stream damage have also been identified who show strikingly similar dissociations between vision-for-perception and vision-for-action (Dijkerman et al., 2004; Karnath et al., 2009; Lê et al., 2002) Finally, it is worth noting that if one reads

the early clinical reports of patients with visual form agnosia, one can find a number of examples of what appears to be spared visuomotor skills in the face of massive deficits in form perception. Thus, Campion (1987), for example, reports that patient RC, who showed a profound visual form agnosia after carbon monoxide poisoning, "could negotiate obstacles in the room, reach out to shake hands and manipulate objects or [pick up] a cup of coffee."

Thus, the pattern of visual deficits and spared abilities in DF and other patients with visual form agnosia is in many ways the mirror image of that observed in the optic ataxia patients described earlier. In other words, patients with damage to the ventral stream can reach out and grasp objects whose form and orientation they cannot perceive, whereas patients with damage to their dorsal stream are unable to use vision to guide their reaching and/or grasping movements to objects whose form and orientation they perceive. Although this "double dissociation" provided the initial, and perhaps most compelling, evidence for the perception–action model proposed by Goodale and Milner (1992), the model is also supported by a wealth of anatomical, electrophysiological, and lesion studies in the monkey,

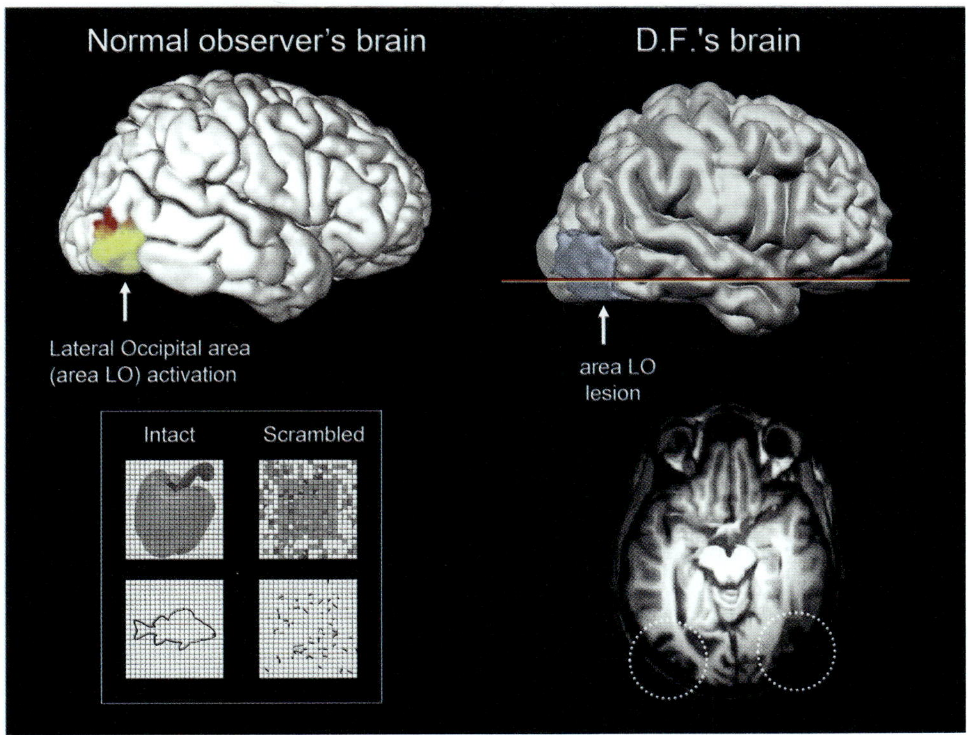

FIGURE 72.4   The lateral occipital area (LO), a ventral-stream area implicated in object recognition (particularly object form), has been localized on the brain of a healthy control subject by comparing fMRI activation to intact versus scrambled line drawings. Note that the lesion (marked in blue) on patient DF's right cerebral hemisphere encompasses all of area LO. Area LO in DF's left hemisphere is also completely damaged, as can be seen in the horizontal section shown on the bottom right. (Adapted with permission from James et al., 2003.)

which are too numerous to review here (for recent reviews, see Andersen & Buneo, 2003; Cohen & Andersen, 2002; Kravitz et al., 2011; Milner & Goodale, 2006; Tanaka, 2003). But perhaps some of the most convincing evidence for the perception–action proposal has come from functional magnetic resonance imaging (fMRI) studies of the dorsal and ventral streams in the human brain.

NEUROIMAGING EVIDENCE FOR TWO VISUAL STREAMS

In the ventral stream, regions have been identified that seem to be selectively responsive to different categories of visual stimuli. Early on, an area was isolated within the ventrolateral part of the occipital cortex (the lateral occipital area, or LO) that appears to be involved in object recognition (for review, see Grill-Spector, 2003). As mentioned earlier, DF has bilateral lesions in the ventral stream that include area LO in both hemispheres. Not surprisingly therefore, an fMRI investigation of activity in DF's brain revealed no differential activation for line drawings of common objects (vs. scrambled versions) anywhere in DF's remaining ventral stream, mirroring her poor performance in identifying the objects depicted in the drawings (James et al., 2003). Again, this strongly suggest that area LO is essential for form perception, generating the geometrical structure of objects by combining information about edges and surfaces that has already been extracted from the visual array by low-level visual feature detectors.

In addition to LO, other ventral-stream areas have been identified that code for faces, human body parts, and places or scenes (for review, see Milner & Goodale, 2006). Although there is a good deal of debate as to whether these areas are really category-specific or instead are particular nodes in a highly distributed system (Cant, Arnott, & Goodale, 2009; Cant & Goodale, 2007, 2011; Connolly et al., 2012; Downing et al., 2006; Haxby et al., 2001; Op de Beeck, Haushofer, & Kanwisher, 2008), the neuroimaging work continues to provide strong support for the idea that the ventral stream plays the major role in constructing our perceptual representation of the world (Kourtzi & Kanwisher, 2001).

fMRI has also revealed visuomotor areas in the dorsal stream that appear to be largely homologous with those in the monkey brain (for reviews, see Castiello, 2005; Culham & Kanwisher, 2001; Culham & Valyear, 2006). Early on, an area in the intraparietal sulcus of the posterior parietal cortex was identified that appeared to be activated when subjects shifted their gaze (or their covert attention) to visual targets. This area is thought by many investigators to be homologous with an area on the lateral bank of the intraparietal sulcus (area LIP) in the monkey that has been similarly associated with the visual control of eye movements and attention (for reviews, see Andersen & Buneo, 2003; Bisley & Goldberg, 2010), although in the human brain it is located more medially in the intraparietal sulcus (Culham, Cavina-Pratesi, & Singhal, 2006; Grefkes & Fink, 2005; Pierrot-Deseilligny, Milea, & Muri, 2004). A region in the anterior part of the intraparietal sulcus has been identified that is consistently activated when people reach out and grasp visible objects in the scanner (Binkofski et al., 1998; Culham, 2003; Culham et al., 2003). This area has been called human AIP (hAIP) because it is thought to be homologous with a region in the anterior intraparietal sulcus of the monkey that has also been implicated in the visual control of grasping (for review, see Sakata, 2003). Several more recent studies have also shown that hAIP is differentially activated during visually guided grasping (e.g., Cavina-Pratesi, Goodale, & Culham, 2007; Cavina-Pratesi et al., 2010b; Frey et al., 2005). Importantly, area LO in the ventral stream is not activated when subjects reach out and grasp objects (Cavina-Pratesi, Goodale, & Culham, 2007; Culham, 2004; Culham et al., 2003), suggesting that this object-recognition area in the ventral stream is not required for the programming and control of visually guided grasping and that hAIP and associated networks in the posterior parietal cortex (in association with premotor and motor areas) can do this independently. This conclusion is considerably strengthened by the fact that patient DF, who has large bilateral lesions of area LO, shows robust differential activation in area hAIP for grasping (compared to reaching) similar to that seen in healthy subjects (James et al., 2003; see figure 72.5).

But what about the visual control of reaching? As we saw earlier, the lesions associated with the misreaching that defines optic ataxia have been typically found in the posterior parietal cortex, including the intraparietal sulcus and sometimes extending into the inferior or superior parietal lobules (Perenin & Vighetto, 1988). More recent quantitative analyses of the lesion sites associated with misreaching have revealed several key foci in the parietal cortex, including the medial occipitoparietal junction, the superior occipital gyrus, the intraparietal sulcus, and the superior parietal lobule as well as parts of the inferior parietal lobule (Karnath & Perenin, 2005). As it turns out, these lesion sites map nicely onto the patterns of activation found in a recent fMRI study of visually guided reaching that showed reach-related activation both in a medial part of the intraparietal sulcus (near the intraparietal lesion site

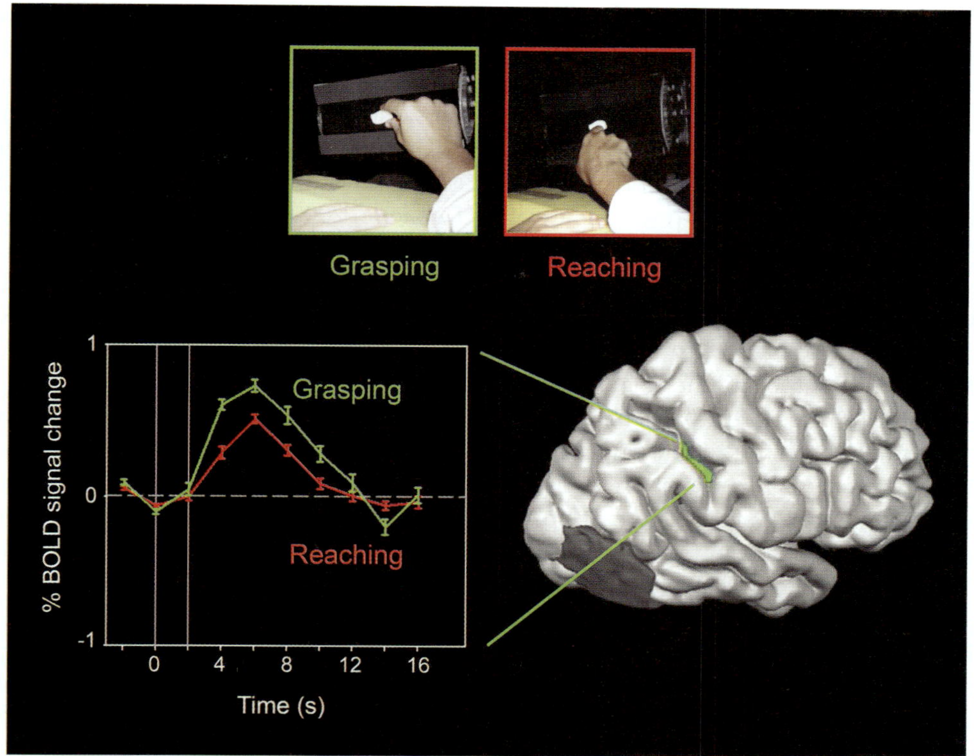

FIGURE 72.5   Brain activations in DF during reaching and grasping. Right: The activation corresponding to grasping versus reaching is shown in green in a region of DF's brain that is in the same location as hAIP (corresponding to anterior intraparietal sulcus) in the intact brain. Left: The averaged time course of activations for grasping and reaching trials in hAIP. Clearly, hAIP is functioning normally in spite of DF's brain damage. (The large lesion in DF's lateral occipital area is indicated in dark gray.) BOLD, blood-oxygen-level dependent. (Adapted with permission from James et al., 2003.)

identified by Karnath and Perenin) and in the medial occipitoparietal junction (Cavina-Pratesi et al., 2010b; Prado et al., 2005). In another study, it was found that the reach-related focus in the medial intraparietal sulcus was equally active for reaches with and without visual feedback, whereas an area the superior parietal occipital cortex (SPOC) was particularly active when visual feedback was available (Filimon et al., 2009). This suggests that the medial intraparietal region may reflect proprioceptive more than visual control of reaching whereas SPOC (see figure 72.6) may be more involved in visual control. Both these areas have also been implicated in the visual control of reaching in the monkey (Andersen & Buneo, 2003; Fattori et al., 2001; Snyder, Batista, & Andersen, 1997).

There is evidence to suggest that SPOC may play a role in some aspects of grasping, particularly wrist rotation (Grol et al., 2007; Monaco et al., 2011), a result that coincides with recent findings on a homologous area in the medial occipitoparietal cortex (area V6A) of the monkey (Fattori et al., 2004, 2010, 2012). But at the same time, it seems clear from the imaging data that

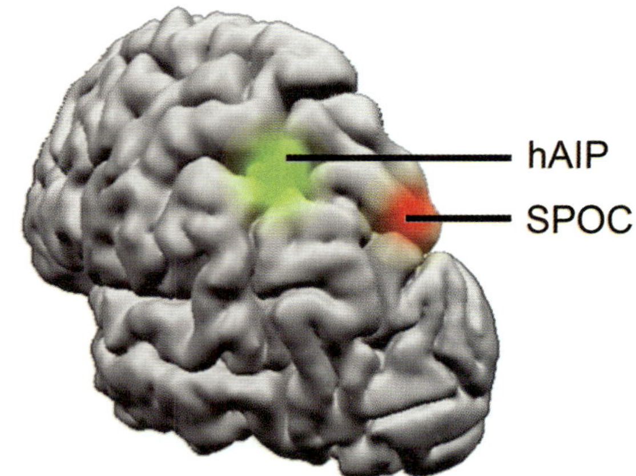

FIGURE 72.6   The location of the superior parieto-occipital cortex (SPOC), which is thought to correspond to V6A, and hAIP (corresponding to anterior intraparietal sulcus). These human areas have been shown to be activated during visually guided reaching and grasping, respectively. (Adapted with permission from Goodale, 2011.)

more anterior parts of the intraparietal sulcus, such as hAIP, play a unique role in visually guided grasping and appear not to be involved in the visual control of reaching movements. Moreover, patients with lesions of hAIP have deficits in grasping but retain the ability to reach toward objects (Binkofski et al., 1998), whereas other patients with lesions in more medial and posterior areas of the parietal lobe, including SPOC, show deficits in reaching but not grip scaling (Cavina-Pratesi et al., 2010a). The identification of areas in the human posterior parietal cortex for the visual control of reaching that are anatomically distinct from those implicated in the visual control of grasping, particularly the scaling of grip aperture, lends additional support to Jeannerod's (1981) proposal that the transport and grip components of reach-to-grasp movements are programmed and controlled relatively independently. None of these observations, however, can be easily accommodated within the double-pointing hypothesis of Smeets and Brenner (1999).

As mentioned earlier, not only are we adept at reaching out and grasping objects, but we are also able to avoid obstacles that might potentially interfere with our reach. There is persuasive neuropsychological evidence for a dorsal-stream locus for this coding. Thus, unlike healthy control subjects, patients with optic ataxia from dorsal-stream lesions do not automatically alter the trajectory of their grasp to avoid obstacles located to the left and right of the path of their hand as they reach out to touch a target beyond the obstacles—even though they certainly see the obstacles and can indicate the midpoint between them (Schindler et al., 2004). Conversely, patient DF shows normal avoidance of the obstacles in the same task, even though she is deficient at indicating the midpoint between the two obstacles (Rice et al., 2006). Although there are no neuroimaging studies of obstacle avoidance during reaching, there is evidence that an area in the intraparietal sulcus in the dorsal stream is activated when people avoid obstacles close to an object they are grasping—and that this dorsal-stream area modulates activity in early visual cortex (Chapman et al., 2011).

In summary, the neuropsychological and neuroimaging data that have been amassed over the last 2 decades suggest that vision-for-action and vision-for-perception depend on different and relatively independent visual pathways in the primate brain. Although I have focused almost entirely on the role of the dorsal stream in the control of action, it is important to emphasize that the posterior parietal cortex also plays a critical role in the deployment of attention, as well as in other high-level cognitive tasks, such as numeracy, working memory, and way-finding. Even so, a strong argument can be made that these functions of the dorsal stream (and associated networks in premotor cortex and more inferior parietal areas) grew out of the pivotal role that the dorsal stream plays in the control of eye movements and goal-directed limb movements (for more on these issues, see Kravitz et al., 2011; Moore, 2006; Nieder & Dehaene, 2009; Rizzolatti & Craighero, 1998; Rizzolatti et al., 1987).

DIFFERENT NEURAL COMPUTATIONS FOR PERCEPTION AND ACTION

Although the evidence from a broad range of empirical studies points to that fact that there are two relatively independent visual pathways in the primate cerebral cortex, the question remains as to why two separate systems evolved in the first place. Why couldn't one "general purpose" visual system handle both vision-for-perception and vision-for-action? The answer to this question lies in the differences in the computational requirements of vision-for-perception on the one hand and vision-for-action on the other. To be able to grasp an object successfully, the visuomotor system has to deal with the actual size of the object and its orientation and position with respect to the hand one intends to use to pick it up. These computations need to reflect the real metrics of the world, or at the very least, make use of learned "look-up tables" that link neurons coding a particular set of sensory inputs with neurons that code the desired state of the limb (Thaler & Goodale, 2010). The time at which these computations are performed is equally critical. Observers and goal objects rarely stay in a static relationship with one another, and, as a consequence, the egocentric location of a target object can often change radically from moment to moment. In other words, the required coordinates for action need to be computed at the very moment the movements are performed.

In contrast to vision-for-action, vision-for-perception does not need to deal with the absolute size of objects or their egocentric locations. In fact, very often such computations would be counterproductive because our viewpoint with respect to objects does not remain constant—even though our perceptual representations of those objects do show constancy. Indeed, one can argue that it would be better to encode the size, orientation, and location of objects relative to each other. Such a scene-based frame of reference permits a perceptual representation of objects that transcends particular viewpoints, while preserving information about spatial relationships (as well as relative size and orientation) as the observer moves around. The products of perception also need to be available over a much longer time scale than the visual information used in the control of action. To achieve this, the coding of the visual information has to be somewhat abstract—transcending particular viewpoint and viewing conditions. By working with perceptual representations that are object- or scene-based, we are able to maintain the constancies of size, shape, color, lightness, and relative location, over time and across different viewing conditions.

The differences in the metrics and frames of reference used by vision-for-perception and vision-for-action have been demonstrated in normal observers in experiments that have made use of pictorial illusions, particularly size-contrast illusions (see figure 72.7). Aglioti, DeSouza, and Goodale (1995), for example, showed that the scaling of grip aperture in flight is remarkably insensitive to the Ebbinghaus illusion, in which a target disk surrounded by smaller circles appears to be larger than the same disk surrounded by larger circles. They

Ebbinghaus Illusion

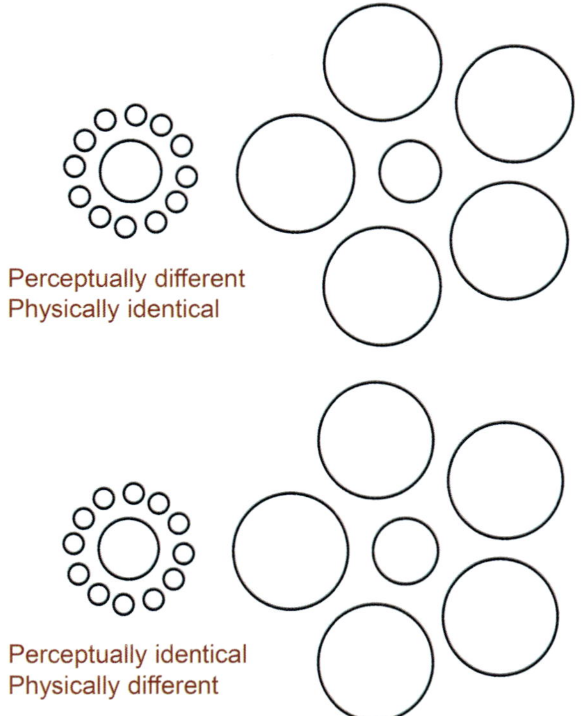

FIGURE 72.7   The effect of a size-contrast illusion on perception and action. Top: The traditional Ebbinghaus illusion in which the central circle in the annulus of larger circles is typically seen as smaller than the central circle in the annulus of smaller circles, even though both central circles are actually the same size. Bottom: The same display, except that the central circle in the annulus of larger circles has been made slightly larger. As a consequence, the two central circles now appear to be the same size.

found that maximum grip aperture was scaled to the real, not the apparent, size of the target disk (see figure 72.8). A similar dissociation between grip scaling and perceived size was reported by Haffenden and Goodale (1998) under conditions where participants had no visual feedback during the execution of grasping movements made to targets presented in the context of an Ebbinghaus illusion. Although grip scaling escaped the influence of the illusion, the illusion did affect performance in a manual matching task, a kind of perceptual report, in which participants were asked to open their index finger and thumb to indicate the perceived size of a disk. (This measure is akin to the typical magnitude estimation paradigms used in conventional psychophysics, but with the virtue that the manual estimation makes use of the same effector that is used in the grasping task.)

This dissociation between what people do and what they say they see underscores the differences between vision-for-action and vision-for-perception. The obligatory size-contrast effects that give rise to the illusion (in which different elements of the array are compared) presumably play a crucial role in scene interpretation, a central function of vision-for-perception. However, the execution of a goal-directed act, such as manual prehension, requires computations that are centered on the target itself, rather than on the relations between the target and other elements in the scene. In fact, the true size of the target for calibrating the grip can be computed from the retinal-image size of the object coupled with an accurate estimate of distance. Computations of this kind, which do not take into account the relative difference in size between different objects in the scene, would be expected to be quite insensitive to the kinds of pictorial cues that distort perception when familiar illusions are presented.

The initial demonstration by Aglioti, DeSouza, and Goodale (1995) that grasping is refractory to the Ebbinghaus illusion engendered a good deal of interest among researchers studying vision and motor control—and there have been numerous investigations of the effects (or not) of pictorial illusions on visuomotor control. Some investigators have replicated the original observations of Aglioti, DeSouza, and Goodale with the Ebbinghaus illusion (e.g., Amazeen & DaSilva, 2005; Fischer, 2001; Kwok & Braddick, 2003)—and others have observed a similar insensitivity of grip scaling to the Ponzo illusion (Brenner & Smeets, 1996; Jackson & Shaw, 2000), the horizontal–vertical illusion (Servos, Carnahan, & Fedwick, 2000), the Müller–Lyer illusion (Dewar & Carey, 2006), and the diagonal illusion (Stöttinger & Perner, 2006; Stöttinger et al., 2010, 2012). Others have reported that pictorial illusions

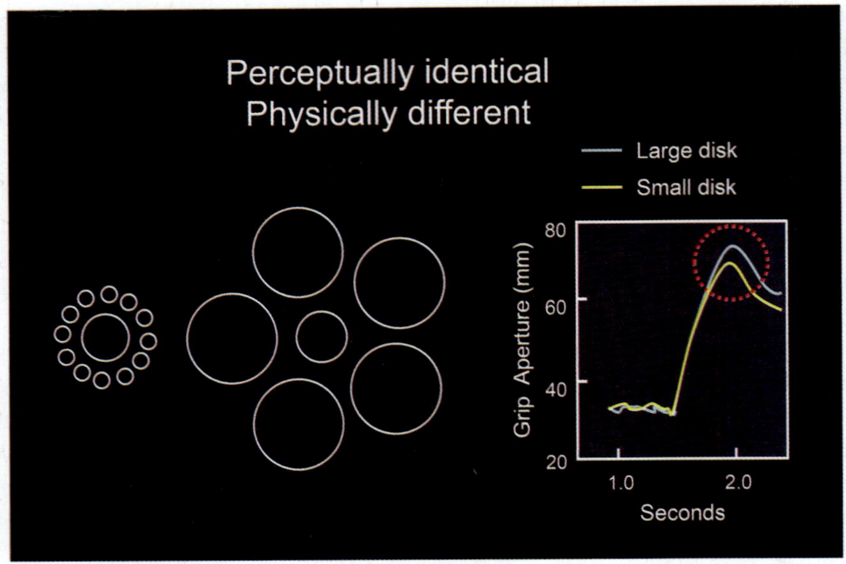

FIGURE 72.8   Performance on 3-D versions of the Ebbinghaus illusion. The graph on the right shows two trials in which the participant picked up the small disk on one trial and the large disk from the display shown on the left. Even though the two central disks were perceived as being the same size, the grip aperture in flight reflected the real, not the apparent, size of the disks. (Adapted with permission from Aglioti, DeSouza, & Goodale, 1995.)

affect some aspects of motor control but not others (e.g., Biegstraaten et al., 2007; Daprati & Gentilucci, 1997; Gentilucci et al., 1996; Glazebrook et al., 2005; van Donkelaar, 1999). And a few investigators have found no dissociation whatsoever between the effects of pictorial illusions on perceptual judgments and the scaling of grip aperture (e.g., Franz et al., 2000; Franz, Bülthoff, & Fahle, 2003).

Demonstrating that actions such as grasping are sometimes sensitive to illusory displays is not by itself a refutation of the idea of two visual systems. One should not be surprised that visual perception and visuomotor control can interact in the normal brain. The real surprise, at least for monolithic accounts of vision, is that there are clear instances where visually guided action is apparently unaffected by pictorial illusions, which, by definition, affect perception. However, from the standpoint of the duplex perception–action model such instances are to be expected (see Goodale, 2008; Milner & Goodale, 2006, 2008). Nevertheless, the fact that action has been found to be affected by pictorial illusions in some instances has led a number of authors to argue that the earlier studies demonstrating a dissociation had not adequately matched action and perception tasks for various input, attentional, and output demands (e.g., Smeets & Brenner, 2001; Vishton & Fabre, 2003)—and that when these factors are taken into account, the apparent differences between perceptual judgments and motor control could be resolved without invoking the idea of two visual systems.

As it turns out, however, there are several reasons why grip aperture might appear to be sensitive to illusions under certain testing conditions—even when it is not. In some cases, notably the Ebbinghaus illusion, the flanker elements can be treated as obstacles, influencing the posture of the fingers during the execution of the grasp (de Grave et al., 2005; Haffenden, Schiff & Goodale, 2001; Plodowski & Jackson, 2001). In other words, the apparent effect of the illusion on grip scaling in some experiments might simply reflect the operation of visuomotor mechanisms that treat the flanker elements of the visual arrays as obstacles to be avoided. Another critical variable is the timing of the grasp with respect to the presentation of the stimuli. When targets are visible during the programming of a grasping movement, maximum grip aperture is usually not affected by size-contrast illusions, whereas when vision is occluded before the command to initiate programming of the movement is presented, a reliable effect of the illusion on grip aperture is typically observed (Fischer, 2001; Hu & Goodale, 2000; Westwood & Goodale, 2003; Westwood, Heath, & Roy, 2000). As discussed earlier, vision-for-action is designed to operate in real time and is not normally engaged unless the target object is visible during the programming phase, when (bottom-up) visual information can be immediately converted into the appropriate motor commands. The observation that (top-down) memory-guided grasping is affected by the illusory display reflects the fact that the stored information about the target's dimensions was originally derived from the earlier operation of vision-for-perception (for a more detailed discussion of these and related issues, see Bruno, Bernardis, & Gentilucci, 2008; Goodale, Westwood, & Milner, 2004).

Nevertheless, some have argued that if the perceptual and grasping tasks are appropriately matched, then grasping can be shown to be as sensitive to size-contrast illusions as psychophysical judgments (Franz, 2001; Franz et al., 2000) Although this explanation, at least on the face of it, is a compelling one, it cannot explain the findings of a series of recent studies by Stöttinger and colleagues (2010, 2012), who showed that when one meets all the criteria that Franz argues must be satisfied before one can compare grip scaling with psychophysical judgments, grip scaling is still far less sensitive to the diagonal illusion than are perceptual judgments of target size.

Experiments by Ganel, Tanzer, and Goodale (2008) provide evidence that is even more difficult to explain away by appealing to a failure to match testing conditions and other task-related variables. In this experiment, which used a version of the Ponzo illusion, a real difference in size was pitted against a perceived difference in size in the opposite direction (see figure 72.9). The results were remarkably clear. Despite the fact that

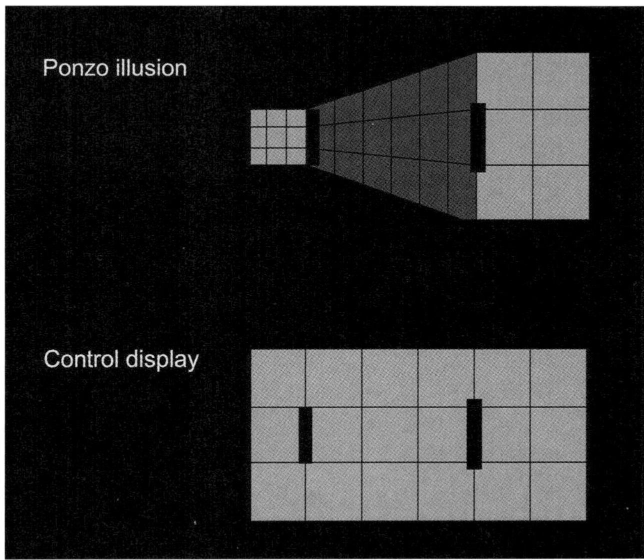

FIGURE 72.9   Stimuli used in the Ganel, Tanzer, and Goodale (2008) study. Top: The arrangement of the objects on incongruent trials in which real size and the illusory size were pitted against one another. In this example, the object on the right is perceived in most cases as shorter than the object on the left (due to the illusory context), although it is actually longer. Bottom: The real difference in size can be clearly here where the two objects are placed on the nonillusory control display. (Adapted with permission from Ganel, Tanzer, & Goodale, 2008.)

people believed that the shorter object was the longer one (or vice versa), their in-flight grip aperture reflected the real, not the illusory, size of the target objects. In other words, as can be seen in figure 72.10, on the same trials in which participants erroneously decided that one object was the longer (or shorter) of the two, the anticipatory opening between their fingers reflected the real direction and magnitude of size differences between the two objects. Moreover, the subjects in this experiment showed the same differential scaling to the real size of the objects whether the objects were shown on the illusory display or on the control display. Not surprisingly, as figure 72.10 also shows, when subjects were asked to use their finger and thumb to estimate the size of the target objects rather than pick them up, their manual estimates reflected the apparent, not the real, size of the targets. Overall, these results underscore once more the profound difference in the way visual information is transformed for action and perception. Importantly too, the results are difficult to reconcile with any argument that suggests that grip aperture is sensitive to illusions, and that the absence of an effect found in many studies is simply a consequence of differences in the task demands (Franz, 2001; Franz et al., 2000).

The fact that rapid skilled actions are often resistant to size-contrast illusions fits well with Smeets and Brenner's (1999, 2001) double-pointing hypothesis. According to their account, because grip scaling is simply an epiphenomenon of the independent finger trajectories, grip aperture would seem to be impervious to the effects of the illusion. Although, as discussed earlier, Smeets and Brenner's account has been challenged, it has to be acknowledged that their double-pointing model offers a convincing explanation of these findings (even if the neuropsychological and neuroimaging data are more consistent with Jeannerod's (1981) two-visuomotor channel account of reach-to-grasp movements).

Even so, there are some behavioral observations that cannot be accommodated within the Smeets and Brenner (1999, 2001) model. For example, as discussed earlier, if a delay is introduced between viewing the target and initiating the grasp, the scaling of the anticipatory grip aperture is much more likely to be sensitive to size-contrast illusions (Fischer, 2001; Goodale, Jakobson, Keillor, 1994a; Hu & Goodale, 2000; Westwood & Goodale, 2003; Westwood, Heath, & Roy, 2000). These results cannot be easily explained by the Smeets and Brenner model without conceding that—with delay—grip scaling is no longer a consequence of programming individual digit trajectories but instead reflects the perceived size of the target

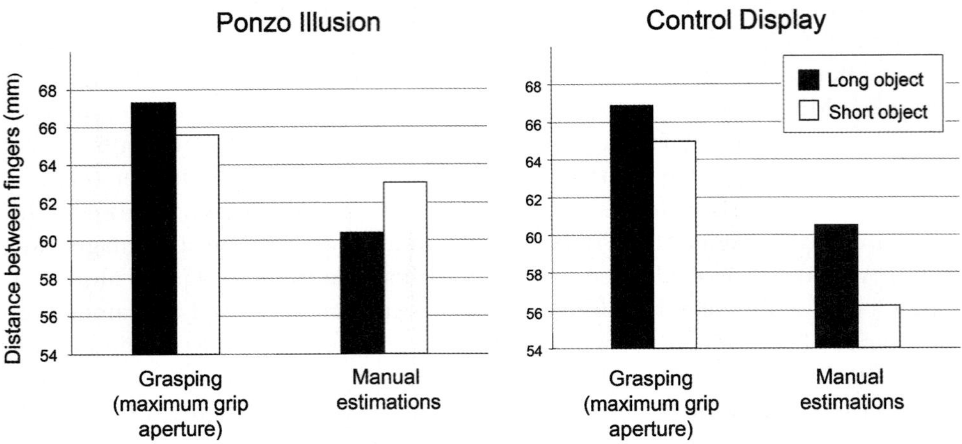

FIGURE 72.10   Maximum grip aperture and perceptual estimates of length for objects placed on the illusory display (left) and control display (right). Only incongruent trials in which participants made erroneous decisions about real size are shown for the grip aperture and estimates with the illusory display. As the left panel shows, despite the fact that participants erroneously perceived the physically longer object to be the shorter one (and vice versa), the opening between their finger and thumb during the grasping movements reflected the real difference in size between the objects. This pattern of results was completely reversed when they made perceptual estimates of the length of the objects. With the control display (right panel), both grip aperture and manual estimates went in the same direction. (Adapted with permission from Ganel, Tanzer, & Goodale, 2008.)

object. Nor can the Smeets and Brenner model explain what happens when unpracticed finger postures (e.g., the thumb and ring finger) are used to pick up objects in the context of a size-contrast illusion (see figure 72.11). In contrast to skilled grasping movements, grip scaling with unpracticed awkward grasping is quite sensitive to the illusory difference in size between the objects (Gonzalez et al., 2008). Only with practice does grip aperture begin to reflect the real size of the target objects. Smeets and Brenner's model cannot account for this result without positing that individual control over the digits occurs only after practice. Finally, recent neuropsychological findings with patient DF suggest that she could use action-related information about object size (presumably from her intact dorsal stream) to make explicit judgments regarding the length of an object that she was about to pick up (Schenk & Milner, 2006), suggesting that size rather than two separate locations is implicitly coded during grasping. At this point, the difference between the perception–action model (Goodale & Milner, 1992; Milner & Goodale, 2006) and the (modified) Smeets and Brenner account begins to blur. Both accounts posit that real-time control of skilled grasping depends on visuomotor transformations that are quite distinct from those involved in the control of delayed or unpracticed grasping movements. The difference in the two accounts turns on the nature of the control exercised over skilled movements performed in real time. But note that even if Smeets and Brenner are correct that

the trajectories of the individual digits are programmed individually on the basis of spatial information that ignores the size of the object, this would not obviate the idea of two visual systems, one for constructing our perception of the world and one for controlling our actions in that world. Indeed, the virtue of the perception–action model is that it accounts not only for the dissociations outlined above between the control of action and psychophysical report in normal observers in a number of different settings, but it also accounts for a broad range of neuropsychological, neurophysiological, and neuroimaging data (and is completely consistent with Jeannerod's dual-channel model of reach-to-grasp movements).

INTERACTIONS BETWEEN THE TWO STREAMS

When the idea of a separate vision-for-action system was first proposed over 20 years ago, the emphasis was on the independence of this system from vision-for-perception. But clearly the two systems must work closely together in the generation of purposive behavior. One way to think about the interaction between the two streams (an interaction that takes advantage of the complementary differences in their computational constraints) is in terms of a "teleassistance" model (Goodale & Humphrey, 1998). In teleassistance, a human operator, who has identified a goal object and decided what to do with it, communicates with a semi-autonomous robot that actually performs the required motor act on

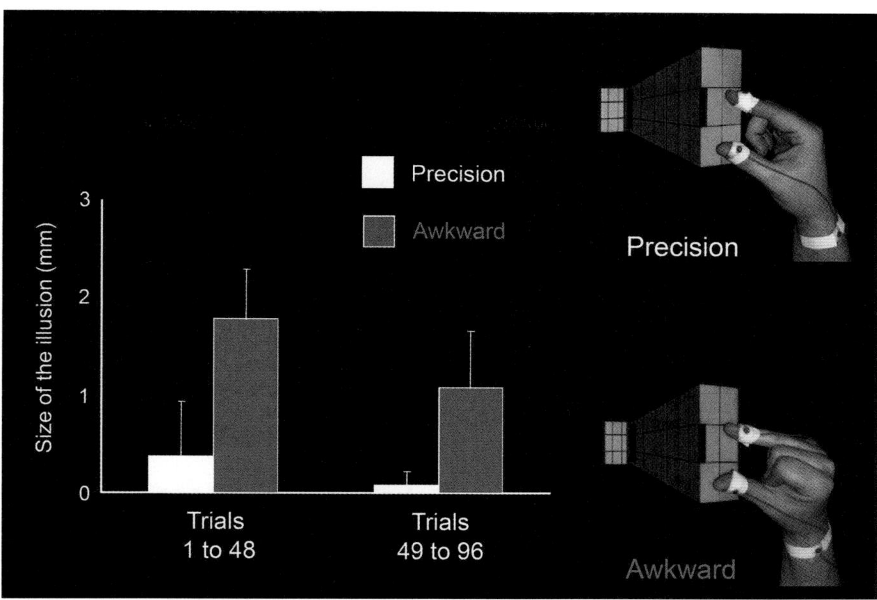

FIGURE 72.11   The effects of the Ponzo illusion on two different kinds of grasp posture: a normal and well-practiced precision grip, and a much more awkward and unfamiliar grip posture where the person uses the thumb and ring finger. The scaling of the precision grip is not influenced by the Ponzo illusion. In sharp contrast, the awkward grasp is—and this sensitivity decreases only slightly over the testing session. (Adapted with permission from Gonzalez et al., 2008.)

the flagged goal object (Pook & Ballard, 1996). In terms of this teleassistance metaphor, the perceptual–cognitive system in the ventral stream, with its rich and detailed representations of the visual scene (and links with cognitive systems), would be the human operator. Processes in the ventral stream participate in the identification of a particular goal and flag the relevant object in the scene, perhaps by means of an attention-like process. Once a particular goal object has been flagged, dedicated visuomotor networks in the dorsal stream (in conjunction with related circuits in premotor cortex, basal ganglia, and brainstem) are then activated to transform the visual information about the object into the appropriate coordinates for the desired motor act. This means that in many instances a flagged object in the scene will be processed in parallel by both ventral and dorsal stream mechanisms—each transforming the visual information in the array for different purposes. In other situations, where the visual stimuli are particularly salient, visuomotor mechanisms in the dorsal stream will operate without any immediate supervision by ventral stream perceptual mechanisms.

Of course, the teleassistance analogy is far too simplified. For one thing, the ventral stream by itself cannot be construed as an intelligent operator that can make assessments and plans. Clearly, there has to be some sort of top-down executive control—almost certainly engaging prefrontal mechanisms—that can initiate the

operation of attentional search and thus set the whole process of planning and goal selection in motion (for review, see Desimone & Duncan, 1995; Goodale & Haffenden, 1998). Reciprocal interactions between prefrontal/premotor areas and the areas in the posterior parietal cortex undoubtedly play a critical role in recruiting specialized dorsal-stream structures, such as LIP, which appear to be involved in the control of both voluntary eye movements and covert shifts of spatial attention in monkeys and humans (Bisley & Goldberg, 2010; Corbetta, Kincade, & Shulman, 2002). In terms of the teleassistance metaphor, area LIP can be seen as acting like a videocam on the robot, scanning the visual scene and thereby providing new inputs that the ventral stream can process and pass on to frontal systems that assess their potential importance. In practice, of course, the videocam/LIP system does not scan the environment randomly: It is constrained to a greater or lesser degree by top-down information about the nature of the potential targets and where those targets might be located, information that reflects the priorities of the operator/organism that are presumably elaborated in prefrontal systems.

What happens next goes beyond even these speculations. Before instructions can be transmitted to the visuomotor control systems in the dorsal stream, the nature of the action required needs to be determined. This means that praxis systems, perhaps located in the left hemisphere, need to "instruct" the relevant visuo-motor systems. After all, objects such as tools demand a particular kind of hand posture. Achieving this not only requires that the tool be identified, presumably using ventral-stream mechanisms (Valyear & Culham, 2010), but also that the required actions to achieve that posture be selected as well via a link to these praxis systems. At the same time, the ventral stream (and its related cognitive apparatus) has to communicate the locus of the goal object to these visuomotor systems in the dorsal stream. One way that this ventral–dorsal transmission could happen is via recurrent projections from foci of activity in the ventral stream back downstream to primary visual cortex and other adjacent visual areas. Once a target has been "highlighted" on these retinotopic maps, its location could then finally be forwarded to the dorsal stream for action (for a version of this idea, see Lamme & Roelfsema, 2000). Moreover, LIP itself, by virtue of the fact that it would be "pointing" at the goal object, could also provide the requisite coordinates, once it has been cued by recognition systems in the ventral stream.

After the particular disposition and location of the object with respect to the actor has been computed, that information has to be combined with the postural requirements of the appropriate functional grasp for the tool, that as I have already suggested are presumably provided by praxis systems that are in turn cued by recognition mechanisms in the ventral stream. At the same time, the initial fingertip forces that should be applied to the tool (or any object, for that matter) are based on estimations of its mass, surface friction, and compliance that are derived from visual information (e.g., Gordon et al., 1993). Once contact is made, somatosensory information can be used to fine-tune the applied forces—but the specification of the initial grip and lift forces must be derived from learned associations between the object's visual appearance and prior experience with similar objects or materials (Buckingham, Cant, & Goodale, 2009). This information presumably can be provided only by the ventral visual stream in conjunction with stored information about past interactions. Again, it must be emphasized that all of this is highly speculative. Nevertheless, whatever complex interactions might be involved, it is clear that goal-directed action is unlikely to be mediated by a simple serial processing system. Multiple iterative processing is almost certainly required, involving a constant interplay among different control systems at different levels of processing (for a more detailed discussion of these and related issues, see Milner & Goodale, 2006). One thing that is clear, however, is that the brain evolved not to enable us to think but to enable us to move.

Ultimately, all thinking (and by extension, all perception) is in the service of action.

NOTE

1. It has been recently suggested that DF's spared ability to scale her hand to the width of the target object depends more on haptic feedback than it does on visual processing (Schenk, 2012). This account, however, cannot explain why she continues to scale well even when objects of different sizes are randomly interleaved (Milner, Ganel, & Goodale, 2012).

REFERENCES

Aglioti, S., DeSouza, J., & Goodale, M. A. (1995). Size-contrast illusions deceive the eyes but not the hand. *Current Biology, 5,* 679–685.

Amazeen, E. L., & DaSilva, F. (2005). Psychophysical test for the independence of perception and action. *Journal of Experimental Psychology. Human Perception and Performance, 31,* 170–182.

Andersen, R. A., & Buneo, C. A. (2003). Sensorimotor integration in posterior parietal cortex. *Advances in Neurology, 93,* 159–177.

Baldauf, D., & Deubel, H. (2010). Attentional landscapes in reaching and grasping. *Vision Research, 50,* 999–1013.

Ballard, D. H., Hayhoe, M. M., Li, F., & Whitehead, S. D. (1992). Hand–eye coordination during sequential tasks. *Philosophical Transactions of the Royal Society of London. Series B, Biological Sciences, 337,* 331–338.

Bálint, R. (1909). Seelenlähmung des 'Schauens', optische Ataxie, räumliche Störung der Aufmerksamkeit. *Monatsschrift für Psychiatrie und Neurologie, 25,* 51–81.

Bálint, R., & Harvey, M. (1995). Psychic paralysis of gaze, optic ataxia, and spatial disorder of attention. *Cognitive Neuropsychology, 12,* 265–281.

Biegstraaten, M., de Grave, D. D. J., Brenner, E., & Smeets, J. B. J. (2007). Grasping the Müller–Lyer illusion: Not a change in perceived length. *Experimental Brain Research, 176,* 497–503.

Binkofski, F., Dohle, C., Posse, S., Stephan, K. M., Hefter, H., Seitz, R. J., et al. (1998). Human anterior intraparietal area subserves prehension: A combined lesion and functional MRI activation study. *Neurology, 50,* 1253–1259.

Binsted, G., Chua, R., Helsen, W., & Elliott, D. (2001). Eye–hand coordination in goal-directed aiming. *Human Movement Science, 20,* 563–585.

Bisley, J. W., & Goldberg, M. E. (2010). Attention, intention, and priority in the parietal lobe. *Annual Review of Neuroscience, 33,* 1-21. doi:10.1146/annurev-neuro-060909-152823.

Brenner, E., & Smeets, J. B. (1996). Size illusion influences how we lift but not how we grasp an object. *Experimental Brain Research, 111,* 473–476.

Brouwer, A. M., Franz, V. H., & Gegenfurtner, K. R. (2009). Differences in fixations between grasping and viewing objects. *Journal of Vision, 9,* 18, 1–24.

Bruno, N., Bernardis, P., & Gentilucci, M. (2008). Visually guided pointing, the Müller–Lyer illusion, and the functional interpretation of the dorsal–ventral split:

Conclusions from 33 independent studies. *Neuroscience and Biobehavioral Reviews, 32*, 423–437.

Buckingham, G., Cant, J. S., & Goodale, M. A. (2009). Living in a material world: How visual cues to material properties affect the way that we lift objects and perceive their weight. *Journal of Neurophysiology, 102*, 3111–3118.

Campion, J. (1987). Apperceptive agnosia: The specification and description of constructs. In G. W. Humphreys & M. J. Riddoch (Eds.), *Visual object processing: A cognitive neuropsychological approach* (pp. 197–232). London: Erlbaum.

Cant, J. S., Arnott, S. R., & Goodale, M. A. (2009). fMR-adaptation reveals separate processing regions for the perception of form and texture in the human ventral stream. *Experimental Brain Research, 192*, 391–405.

Cant, J. S., & Goodale, M. A. (2007). Attention to form or surface properties modulates different regions of human occipitotemporal cortex. *Cerebral Cortex, 17*, 713–731.

Cant, J. S., & Goodale, M. A. (2011). Scratching beneath the surface: New insights into the functional properties of the lateral occipital area and parahippocampal place area. *Journal of Neuroscience, 31*, 8248–8258.

Castiello, U. (2001). The effects of abrupt onset of 2-D and 3-D distractors on prehension movements. *Perception & Psychophysics, 63*, 1014–1025.

Castiello, U. (2005). The neuroscience of grasping. *Nature Reviews Neuroscience, 6*, 726–736.

Cavina-Pratesi, C., Goodale, M. A., & Culham, J. C. (2007). FMRI reveals a dissociation between grasping and perceiving the size of real 3D objects. *PLoS ONE, 2*, e424.

Cavina-Pratesi, C., Ietswaart, M., Humphreys, G. W., Lestou, V., & Milner, A. D. (2010a). Impaired grasping in a patient with optic ataxia: Primary visuomotor deficit or secondary consequence of misreaching? *Neuropsychologia, 48*, 226–234.

Cavina-Pratesi, C., Monaco, S., Fattori, P., Galletti, C., McAdam, T. D., Quinlan, D. J., et al. (2010b). Functional magnetic resonance imaging reveals the neural substrates of arm transport and grip formation in reach-to-grasp actions in humans. *Journal of Neuroscience, 30*, 10306–10323.

Chapman, C. S., Gallivan, J. P., Culham, J. C., & Goodale, M. A. (2011). Mental blocks: fMRI reveals top-down modulation of early visual cortex when obstacles interfere with grasp planning. *Neuropsychologia, 49*, 1703–1717.

Chapman, C. S., & Goodale, M. A. (2008). Missing in action: The effect of obstacle position and size on avoidance while reaching. *Experimental Brain Research, 191*, 83–97.

Chapman, C. S., & Goodale, M. A. (2010). Seeing all the obstacles in your way: The effect of visual feedback and visual feedback schedule on obstacle avoidance while reaching. *Experimental Brain Research, 202*, 363–375.

Cohen, Y. E., & Andersen, R. A. (2002). A common reference frame for movement plans in the posterior parietal cortex. *Nature Reviews Neuroscience, 3*, 553–562.

Connolly, A. C., Guntupalli, J. S., Gors, J., Hanke, M., Halchenko, Y. O., Wu, Y. C., et al. (2012). The representation of biological classes in the human brain. *Journal of Neuroscience, 32*, 2608–2618.

Corbetta, M., Kincade, M. J., & Shulman, G. L. (2002). Two neural systems for visual orienting and the pathophysiology of unilateral spatial neglect. In H.-O. Karnath, A. D. Milner, & G. Vallar (Eds.), *The cognitive and neural bases of spatial neglect* (pp. 259–273). Oxford, England: Oxford University Press.

Cuijpers, R. H., Brenner, E., & Smeets, J. B. (2006). Grasping reveals visual misjudgements of shape. *Experimental Brain Research, 175*, 32–44.

Cuijpers, R. H., Smeets, J. B., & Brenner, E. (2004). On the relation between object shape and grasping kinematics. *Journal of Neurophysiology, 91*, 2598–2606.

Culham, J. C. (2004). Human brain imaging reveals a parietal area specialized for grasping. In N. Kanwisher & J. Duncan (Eds.) *Attention and performance XX: Functional brain imaging of human cognition* (pp. 417–438). Oxford, England: Oxford University Press.

Culham, J. C., Cavina-Pratesi, C., & Singhal, A. (2006). The role of parietal cortex in visuomotor control: What have we learned from neuroimaging? *Neuropsychologia, 44*, 2668–2684.

Culham, J. C., Danckert, S. L., DeSouza, J. F. X., Gati, J. S., Menon, R. S., & Goodale, M. A. (2003). Visually-guided grasping produces activation in dorsal but not ventral stream brain areas. *Experimental Brain Research, 153*, 158–170.

Culham, J. C., & Kanwisher, N. G. (2001). Neuroimaging of cognitive functions in human parietal cortex. *Current Opinion in Neurobiology, 11*, 157–163.

Culham, J. C., & Valyear, K. F. (2006). Human parietal cortex in action. *Current Opinion in Neurobiology, 16*, 205–212.

Daprati, E., & Gentilucci, M. (1997). Grasping an illusion. *Neuropsychologia, 35*, 1577–1582.

de Grave, D. D., Biegstraaten, M., Smeets, J. B., & Brenner, E. (2005). Effects of the Ebbinghaus figure on grasping are not only due to misjudged size. *Experimental Brain Research, 163*, 58–64.

Desimone, R., & Duncan, J. (1995). Neural mechanisms of selective visual attention. *Annual Review of Neuroscience, 18*, 193–222.

Deubel, H., Schneider, W. X., & Paprotta, I. (1998). Selective dorsal and ventral processing: Evidence for a common attentional mechanism in reaching and perception. *Visual Cognition, 5*, 81–107.

Dewar, M. T., & Carey, D. P. (2006). Visuomotor 'immunity' to perceptual illusion: A mismatch of attentional demands cannot explain the perception–action dissociation. *Neuropsychologia, 44*, 1501–1508.

Diedrichsen, J., Werner, S., Schmidt, T., & Trommershauser, J. (2004). Immediate spatial distortions of pointing movements induced by visual landmarks. *Perception & Psychophysics, 66*, 89–103.

Dijkerman, H. C., Lê, S., Démonet, J. F., & Milner, A. D. (2004). Visuomotor performance in a patient with visual agnosia due to an early lesion. *Brain Research. Cognitive Brain Research, 20*, 12–25.

Downing, P. E., Chan, A. W., Peelen, M. V., Dodds, C. M., & Kanwisher, N. (2006). Domain specificity in visual cortex. *Cerebral Cortex, 16*, 1453–1461.

Dubrowski, A., Bock, O., Carnahan, H., & Jungling, S. (2002). The coordination of hand transport and grasp formation during single- and double-perturbed human prehension movements. *Experimental Brain Research, 145*, 365–371.

Fattori, P., Breveglieri, R., Amoroso, K., & Galletti, C. (2004). Evidence for both reaching and grasping activity in the medial parieto-occipital cortex of the macaque. *European Journal of Neuroscience, 20*, 2457–2466.

Fattori, P., Breveglieri, R., Raos, V., Bosco, A., & Galletti, C. (2012). Vision for action in the macaque medial posterior parietal cortex. *Journal of Neuroscience, 32*, 3221–3234.

Fattori, P., Gamberini, M., Kutz, D. F., & Galletti, C. (2001). 'Arm-reaching' neurons in the parietal area V6A of the macaque monkey. *European Journal of Neuroscience, 13,* 2309–2313.

Fattori, P., Raos, V., Breveglieri, R., Bosco, A., Marzocchi, N., & Galletti, C. (2010). The dorsomedial pathway is not just for reaching: Grasping neurons in the medial parieto-occipital cortex of the macaque monkey. *Journal of Neuroscience, 30,* 342–349.

Filimon, F., Nelson, J. D., Huang, R. S., & Sereno, M. I. (2009). Multiple parietal reach regions in humans: Cortical representations for visual and proprioceptive feedback during on-line reaching. *Journal of Neuroscience, 29,* 2961–2971.

Fischer, M. H. (2001). How sensitive is hand transport to illusory context effects? *Experimental Brain Research, 136,* 224–230.

Franz, V. H. (2001). Action does not resist visual illusions. *Trends in Cognitive Sciences, 5,* 457–459.

Franz, V. H., Bülthoff, H. H., & Fahle, M. (2003). Grasp effects of the Ebbinghaus illusion: Obstacle avoidance is not the explanation. *Experimental Brain Research, 149,* 470–477.

Franz, V. H., Gegenfurtner, K. R., Bülthoff, H. H., & Fahle, M. (2000). Grasping visual illusions: No evidence for a dissociation between perception and action. *Psychological Science, 11,* 20–25.

Frey, S. H., Vinton, D., Norlund, R., & Grafton, S. T. (2005). Cortical topography of human anterior intraparietal cortex active during visually guided grasping. *Brain Research. Cognitive Brain Research, 23,* 397–405.

Ganel, T., Tanzer, M., & Goodale, M. A. (2008). A double dissociation between action and perception in the context of visual illusions: Opposite effects of real and illusory size. *Psychological Science, 19,* 221–225.

Gentilucci, M., Chieffi, S., Daprati, E., Saetti, M. C., & Toni, I. (1996). Visual illusion and action. *Neuropsychologia, 34,* 369–376.

Glazebrook, C. M., Dhillon, V. P., Keetch, K. M., Lyons, J., Amazeen, E., Weeks, D. J., et al. (2005). Perception–action and the Müller–Lyer illusion: Amplitude or endpoint bias? *Experimental Brain Research, 160,* 71–78.

Gonzalez, C. L. R., Ganel, T., Whitwell, R. L., Morrissey, B., & Goodale, M. A. (2008). Practice makes perfect, but only with the right hand: Sensitivity to perceptual illusions with awkward grasps decreases with practice in the right but not the left hand. *Neuropsychologia, 46,* 624–631.

Goodale, M. A. (1983). Vision as a sensorimotor system. In T. E. Robinson (Ed.), *Behavioral approaches to brain research* (pp. 41–61). New York: Oxford University Press.

Goodale, M. A. (1995). The cortical organization of visual perception and visuomotor control. In S. Kosslyn & D. N. Oshershon (Eds), *An invitation to cognitive science: Vol. 2. Visual cognition and action* (2nd ed., pp. 167–214). Cambridge, MA: MIT Press.

Goodale, M. A. (2008). Action without perception in human vision. *Cognitive Neuropsychology, 25,* 891–919.

Goodale, M. A. (2011). Transforming vision into action. *Vision Research, 51,* 1567–1587.

Goodale, M. A., & Haffenden, A. M. (1998). Frames of reference for perception and action in the human visual system. *Neuroscience and Biobehavioral Reviews, 22,* 161–172.

Goodale, M. A., & Humphrey, G. K. (1998). The objects of action and perception. *Cognition, 67,* 179–205.

Goodale, M. A., Jakobson, L. S., & Keillor, J. M. (1994a). Differences in the visual control of pantomimed and natural grasping movements. *Neuropsychologia, 32,* 1159–1178.

Goodale, M. A., Meenan, J. P., Bülthoff, H. H., Nicolle, D. A., Murphy, K. S., & Racicot, C. I. (1994b). Separate neural pathways for the visual analysis of object shape in perception and prehension. *Current Biology, 4,* 604–610.

Goodale, M. A., & Milner, A. D. (1992). Separate visual pathways for perception and action. *Trends in Neurosciences, 15,* 20–25.

Goodale, M. A., & Milner, A. D. (2004). *Sight unseen: An exploration of conscious and unconscious vision.* Oxford, England: Oxford University Press.

Goodale, M. A., Milner, A. D., Jakobson, L. S., & Carey, D. P. (1991). A neurological dissociation between perceiving objects and grasping them. *Nature, 349,* 154–156.

Goodale, M. A., Westwood, D. A., & Milner, A. D. (2004). Two distinct modes of control for object-directed action. *Progress in Brain Research, 144,* 131–144.

Gordon, A. M., Westling, G., Cole, K. J., & Johansson, R. S. (1993). Memory representations underlying motor commands used during manipulation of common and novel objects. *Journal of Neurophysiology, 69,* 1789–1796.

Grefkes, C., & Fink, G. R. (2005). The functional organization of the intraparietal sulcus in humans and monkeys. *Journal of Anatomy, 207,* 3–17.

Grill-Spector, K. (2003). The neural basis of object perception. *Current Opinion in Neurobiology, 13,* 159–166.

Grill-Spector, K., & Malach, R. (2004). The human visual cortex. *Annual Review of Neuroscience, 27,* 649–677.

Grol, M. J., Majdandzić, J., Stephan, K. E., Verhagen, L., Dijkerman, H. C., Bekkering, H., et al. (2007). Parieto-frontal connectivity during visually guided grasping. *Journal of Neuroscience, 27,* 11877–11887.

Haffenden, A., & Goodale, M. A. (1998). The effect of pictorial illusion on prehension and perception. *Journal of Cognitive Neuroscience, 10,* 122–136.

Haffenden, A. M., Schiff, K. C., & Goodale, M. A. (2001). The dissociation between perception and action in the Ebbinghaus illusion: Nonillusory effects of pictorial cues on grasp. *Current Biology, 11,* 177–181.

Haxby, J. V., Gobbini, M. I., Furey, M. L., Ishai, A., Schouten, J. L., & Pietrini, P. (2001). Distributed and overlapping representations of faces and objects in ventral temporal cortex. *Science, 293,* 2425–2430.

Hu, Y., & Goodale, M. A. (2000). Grasping after a delay shifts size-scaling from absolute to relative metrics. *Journal of Cognitive Neuroscience, 12,* 856–868.

Jackson, S. R., Jackson, G. M., & Rosicky, J. (1995). Are non-relevant objects represented in working memory? The effect of non-target objects on reach and grasp kinematics. *Experimental Brain Research, 102,* 519–530.

Jackson, S. R., & Shaw, A. (2000). The Ponzo illusion affects grip-force but not grip-aperture scaling during prehension movements. *Journal of Experimental Psychology. Human Perception and Performance, 26,* 418–423.

Jakobson, L. S., Archibald, Y. M., Carey, D. P., & Goodale, M. A. (1991). A kinematic analysis of reaching and grasping movements in a patient recovering from optic ataxia. *Neuropsychologia, 29,* 803–809.

James, T. W., Culham, J., Humphrey, G. K., Milner, A. D., & Goodale, M. A. (2003). Ventral occipital lesions impair object recognition but not object-directed grasping: A fMRI study. *Brain, 126,* 2463–2475.

Jeannerod, M. (1981). Intersegmental coordination during reaching at natural visual objects. In J. Long & A. Baddeley (Eds.), *Attention and performance IX* (pp. 153–168). Hillsdale, NJ: Erlbaum.

Jeannerod, M. (1984). The timing of natural prehension movements. *Journal of Motor Behavior, 16,* 235–254.

Jeannerod, M. (1986). The formation of finger grip during prehension: A cortically mediated visuomotor pattern. *Behavioural Brain Research, 19,* 99–116.

Jeannerod, M. (1988). *The neural and behavioural organization of goal-directed movements.* Oxford, England: Clarendon Press.

Johansson, R., Westling, G., Bäckström, A., & Flanagan, J. R. (2001). Eye–hand coordination in object manipulation. *Journal of Neuroscience, 21,* 6917–6932.

Karnath, H. O., & Perenin, M.-T. (2005). Cortical control of visually guided reaching: Evidence from patients with optic ataxia. *Cerebral Cortex, 15,* 1561–1569.

Karnath, H. O., Rüter, J., Mandler, A., & Himmelbach, M. (2009). The anatomy of object recognition—Visual form agnosia caused by medial occipitotemporal stroke. *Journal of Neuroscience, 29,* 5854–5862.

Keefe, B. D., & Watt, S. J. (2009). The role of binocular vision in grasping: A small stimulus-set distorts results. *Experimental Brain Research, 194,* 435–444.

Knill, D. C. (2005). Reaching for visual cues to depth: The brain combines depth cues differently for motor control and perception. *Journal of Vision, 5*(16), 103–115.

Kourtzi, Z., & Kanwisher, N. (2001). Representation of perceived object shape by the human lateral occipital complex. *Science, 293,* 1506–1509.

Kravitz, D. J., Saleem, K. S., Baker, C. I., & Mishkin, M. (2011). A new neural framework for visuospatial processing. *Nature Reviews Neuroscience, 12,* 217–230.

Kwok, R. M., & Braddick, O. J. (2003). When does the Titchener circles illusion exert an effect on grasping? Two- and three-dimensional targets. *Neuropsychologia, 41,* 932–940.

Lamme, V. A. F., & Roelfsema, P. R. (2000). The distinct modes of vision offered by feedforward and recurrent processing. *Trends in Neurosciences, 23,* 571–579.

Lê, S., Cardebat, D., Boulanouar, K., Hénaff, M. A., Michel, F., Milner, D., et al. (2002). Seeing, since childhood, without ventral stream: A behavioural study. *Brain, 125,* 58–74.

Lee, Y. L., Crabtree, C. E., Norman, J. F., & Bingham, G. P. (2008). Poor shape perception is the reason reaches-to-grasp are visually guided online. *Perception & Psychophysics, 70,* 1032–1046.

Loftus, A., Servos, P., Goodale, M. A., Mendarozqueta, N., & Mon-Williams, M. (2004). When two eyes are better than one in prehension: Monocular viewing and end-point variance. *Experimental Brain Research, 158,* 317–327.

Louw, S., Smeets, J. B., & Brenner, E. (2007). Judging surface slant for placing objects: A role for motion parallax. *Experimental Brain Research, 183,* 149–158.

Marotta, J. J., & Goodale, M. A. (1998). The role of learned pictorial cues in the programming and control of grasping. *Experimental Brain Research, 121,* 465–470.

Marotta, J. J., & Goodale, M. A. (2001). The role of familiar size in the control of grasping. *Journal of Cognitive Neuroscience, 13,* 8–17.

Marotta, J. J., Kruyer, A., & Goodale, M. A. (1998). The role of head movements in the control of manual prehension. *Experimental Brain Research, 120,* 134–138.

Marotta, J. J., Perrot, T. S., Nicolle, D., & Goodale, M. A. (1995a). The development of adaptive head movements following enucleation. *Eye (London, England), 9,* 333–336.

Marotta, J. J., Perrot, T. S., Servos, P., Nicolle, D., & Goodale, M. A. (1995b). Adapting to monocular vision: Grasping with one eye. *Experimental Brain Research, 104,* 107–114.

Melmoth, D. R., Finlay, A. L., Morgan, M. J., & Grant, S. (2009). Grasping deficits and adaptations in adults with stereo vision losses. *Investigative Ophthalmology & Visual Science, 50,* 3711–3720.

Melmoth, D. R., & Grant, S. (2006). Advantages of binocular vision for the control of reaching and grasping. *Experimental Brain Research, 171,* 371–388.

Melmoth, D. R., Storoni, M., Todd, G., Finlay, A. L., & Grant, S. (2007). Dissociation between vergence and binocular disparity cues in the control of prehension. *Experimental Brain Research, 183,* 283–298.

Milner, A. D., Ganel, T., & Goodale, M. A. (2012). Does grasping in patient D.F. depend on vision? Trends in Cognitive Sciences, epub ahead of print. doi:10.1016/j.tics.2012.03.004.

Milner, A. D., & Goodale, M. A. (2006). *The visual brain in action* (2nd ed.). Oxford, England: Oxford University Press.

Milner, A. D., & Goodale, M. A. (2008). Two visual systems re-viewed. *Neuropsychologia, 46,* 774–785.

Milner, A. D., Perrett, D. I., Johnston, R. S., Benson, P. J., Jordan, T. R., Heeley, D. W., et al. (1991). Perception and action in "visual form agnosia." *Brain, 114,* 405–428.

Monaco, S., Cavina-Pratesi, C., Sedda, A., Fattori, P., Galletti, C., & Culham, J. C. (2011). Functional magnetic resonance adaptation reveals the involvement of the dorsomedial stream in wrist orientation for grasping. *Journal of Neurophysiology, 106,* 2248–2263.

Mon-Williams, M., & Dijkerman, H. C. (1999). The use of vergence information in the programming of prehension. *Experimental Brain Research, 128,* 578–582.

Mon-Williams, M., & McIntosh, R. D. (2000). A test between two hypotheses and a possible third way for the control of prehension. *Experimental Brain Research, 134,* 268–273.

Mon-Williams, M., & Tresilian, J. R. (2001). A simple rule of thumb for elegant prehension. *Current Biology, 11,* 1058–1061.

Moore, T. (2006). The neurobiology of visual attention: Finding sources. *Current Opinion in Neurobiology, 16,* 159–165.

Nieder, A., & Dehaene, S. (2009). Representation of number in the brain. *Annual Review of Neuroscience, 32,* 185–208.

Obhi, S. S., & Goodale, M. A. (2005). The effects of landmarks on the performance of delayed and real-time pointing movements. *Experimental Brain Research, 167,* 335–344.

Op de Beeck, H. P., Haushofer, J., & Kanwisher, N. G. (2008). Interpreting fMRI data: Maps, modules and dimensions. *Nature Reviews Neuroscience, 9,* 123–135.

Perenin, M.-T., & Vighetto, A. (1988). Optic ataxia: a specific disruption in visuomotor mechanisms. I. Different aspects of the deficit in reaching for objects. *Brain, 111,* 643–674.

Pierrot-Deseilligny, C. H., Milea, D., & Muri, R. M. (2004). Eye movement control by the cerebral cortex. *Current Opinion in Neurology, 17,* 17–25.

Plodowski, A., & Jackson, S. R. (2001). Vision: Getting to grips with the Ebbinghaus illusion. *Current Biology, 11*, R304–R306.

Pook, P. K., & Ballard, D. H. (1996). Deictic human/robot interaction. *Robotics and Autonomous Systems, 18*, 259–269.

Prado, J., Clavagnier, S., Otzenberger, H., Scheiber, C., & Perenin, M. T. (2005). Two cortical systems for reaching in central and peripheral vision. *Neuron, 48*, 849–858.

Rice, N. J., McIntosh, R. D., Schindler, I., Mon-Williams, M., Démonet, J. F., & Milner, A. D. (2006). Intact automatic avoidance of obstacles in patients with visual form agnosia. *Experimental Brain Research, 174*, 176–188.

Rizzolatti, G., & Craighero, L. (1998). Spatial attention: Mechanisms and theories. In M. Sabourin, F. Craik, & M. Robert (Eds.), *Advances in psychological science: Vol. 2. Biological and cognitive aspects* (pp. 171–198). East Sussex, England: Psychology Press.

Rizzolatti, G., Riggio, L., Dascola, I., & Umiltá, C. (1987). Reorienting attention across the horizontal and vertical meridians: Evidence in favor of a premotor theory of attention. *Neuropsychologia, 25*, 31–40.

Sakata, H. (2003). The role of the parietal cortex in grasping. *Advances in Neurology, 93*, 121–139.

Schenk, T. (2012). No dissociation between perception and action in patient D.F. when haptic feedback is withdrawn. *Journal of Neuroscience, 32*, 2013–2017.

Schenk, T., & Milner, A. D. (2006). Concurrent visuomotor behaviour improves form discrimination in a patient with visual form agnosia. *European Journal of Neuroscience, 24*, 1495–1503.

Schindler, I., Rice, N. J., McIntosh, R. D., Rossetti, Y., Vighetto, A., & Milner, A. D. (2004). Automatic avoidance of obstacles is a dorsal stream function: Evidence from optic ataxia. *Nature Neuroscience, 7*, 779–784.

Servos, P. (2000). Distance estimation in the visual and visuomotor systems. Experimental *Brain Research, 130*, 35–47.

Servos, P., Carnahan, H., & Fedwick, J. (2000). The visuomotor system resists the horizontal–vertical illusion. *Journal of Motor Behavior, 32*, 400–404.

Servos, P., Goodale, M. A., & Jakobson, L. S. (1992). The role of binocular vision in prehension: A kinematic analysis. *Vision Research, 32*, 1513–1521.

Smeets, J. B., & Brenner, E. (1999). A new view on grasping. *Motor Control, 3*, 237–271.

Smeets, J. B., & Brenner, E. (2001). Independent movements of the digits in grasping. *Experimental Brain Research, 139*, 92–100.

Smeets, J. B., Brenner, E., & Martin, J. (2009). Grasping Occam's razor. *Advances in Experimental Medicine and Biology, 629*, 499–522.

Snyder, L. H., Batista, A. P., & Andersen, R. A. (1997). Coding of intention in the posterior parietal cortex. *Nature, 386*, 167–170.

Stöttinger, E., & Perner, J. (2006). Dissociating size representation for action and for conscious judgment: Grasping visual illusions without apparent obstacles. *Consciousness and Cognition, 15*, 269–284.

Stöttinger, E., Pfusterschmied, J., Wagner, H., Danckert, J., Anderson, B., & Perner, J. (2012). Getting a grip on illusions: Replicating Stöttinger et al. [Exp Brain Res (2010) 202:79–88] results with 3-D objects. *Experimental Brain Research, 216*, 155–157.

Stöttinger, E., Soder, K., Pfusterschmied, J., Wagner, H., & Perner, J. (2010). Division of labour within the visual system: Fact or fiction? Which kind of evidence is appropriate to clarify this debate? *Experimental Brain Research, 202*, 79–88.

Tanaka, K. (2003). Columns for complex visual object features in the inferotemporal cortex: Clustering of cells with similar but slightly different stimulus selectivities. *Cerebral Cortex, 13*, 90–99.

Thaler, L., & Goodale, M. A. (2010). Beyond distance and direction: The brain represents target locations nonmetrically. *Journal of Vision, 10*(3), 1–27.

Tipper, S. P., Howard, L. A., & Jackson, S. R. (1997). Selective reaching to grasp: Evidence for distractor interference effects. *Visual Cognition, 4*, 1–38.

Tootell, R. B. H., Tsao, D., & Vanduffel, W. (2003). Neuroimaging weighs in: Humans meet macaques in "primate" visual cortex. *Journal of Neuroscience, 23*, 3981–3989.

Tresilian, J. R. (1998). Attention in action or obstruction of movement? A kinematic analysis of avoidance behavior in prehension. *Experimental Brain Research, 120*, 352–368.

Ungerleider, L. G., & Mishkin, M. (1982). Two cortical visual systems. In D. J. Ingle, M. A. Goodale, & R. J. W. Mansfield (Eds.), *Analysis of visual behavior* (pp. 549–586). Cambridge, MA: MIT Press.

Valyear, K. F., & Culham, J. C. (2010). Observing learned object-specific functional grasps preferentially activates the ventral stream. *Journal of Cognitive Neuroscience, 22*, 970–984.

van Bergen, E., van Swieten, L. M., Williams, J. H., & Mon-Williams, M. (2007). The effect of orientation on prehension movement time. *Experimental Brain Research, 178*, 180–193.

van de Kamp, C., & Zaal, F. T. J. M. (2007). Prehension is really reaching and grasping. *Experimental Brain Research, 182*, 27–34.

van Donkelaar, P. (1999). Pointing movements are affected by size-contrast illusions. *Experimental Brain Research, 125*, 517–520.

Van Essen, D. C., Lewis, J. W., Drury, H. A., Hadjikhani, N., Tootell, R. B., Bakircioglu, M., et al. (2001). Mapping visual cortex in monkeys and humans using surface-based atlases. *Vision Research, 41*, 1359–1378.

van Mierlo, C. M., Louw, S., Smeets, J. B., & Brenner, E. (2009). Slant cues are processed with different latencies for the online control of movement. *Journal of Vision, 9*(25), 1–8.

Vaughan, J., Rosenbaum, D. A., & Meulenbroek, R. G. (2001). Planning reaching and grasping movements: The problem of obstacle avoidance. *Motor Control, 5*, 116–135.

Vishton, P. M., & Fabre, E. (2003). Effects of the Ebbinghaus illusion on different behaviors: one- and two-handed grasping; one- and two-handed manual estimation; metric and comparative judgment. *Spatial Vision, 16*, 377–392.

Watt, S. J., & Bradshaw, M. F. (2000). Binocular cues are important in controlling the grasp but not the reach in natural prehension movements. *Neuropsychologia, 38*, 1473–1481.

Watt, S. J., & Bradshaw, M. F. (2003). The visual control of reaching and grasping: Binocular disparity and motion parallax. *Journal of Experimental Psychology. Human Perception and Performance, 29*, 404–415.

Westwood, D. A., & Goodale, M. A. (2003). Perceptual illusion and the real-time control of action. *Spatial Vision, 16*, 243–254.

Westwood, D. A., Heath, M., & Roy, E. A. (2000). The effect of a pictorial illusion on closed-loop and open-loop prehension. *Experimental Brain Research, 134*, 456–463.

73 The Evolution of Parietal Areas Associated with Visuomanual Behavior: From Grasping to Tool Use

DYLAN F. COOKE, ADAM GOLDRING, GREGG H. RECANZONE, AND LEAH KRUBITZER

One of the hallmarks of human evolution is the extraordinary degree to which we can manipulate the physical world with our hands or with tools that extend or amplify portions of our body. This manual dexterity coevolved with the expansion of areas of the brain associated with both visual processing and visuomanual coordination. Specifically, occipitotemporal cortex expanded and cortical areas emerged for the processing of motion, direction and heading (Britten, 2008), and the temporal lobe enlarged greatly to process information about object features such as faces. Cortical fields within posterior parietal cortex (PPC) also expanded and serve as an interface between perception and action. Areas in PPC combine the sensory information from multiple modalities with effector kinematics to compute and program visually guided reaching and grasping movements tailored to specific objects and contexts.

This link between manual dexterity and visual processing is best exemplified when one considers the importance of particular locations in extrapersonal space. Most manual activity takes place in front of the torso and face. This is the optimal space where the two hands meet, and from a visual and skeletal perspective, it is the most ergonomic bimanual work space. It is also a multimodal hot spot, the place where an object can easily be felt, seen, heard, smelled, and tasted. Perhaps unsurprisingly, monkeys' hands are found in this part of space more often than not, and nearly all grasping and manipulation occurs here (Graziano et al., 2004). In monkey motor cortex, electrical microstimulation causes arm movements that follow a similar pattern: overrepresentation of movements toward this central space, particularly when stimulation also evokes a grasp-like hand posture (Graziano et al., 2002, 2004), suggesting that object manipulation in this work space is a

fundamental feature of motor cortex organization. Human proprioceptive acuity for hand position in space is also nonuniform; that is, people have a more accurate kinesthetic sense of hand position at locations near the midline close to the body where this sense is continually being recalibrated with visual feedback (Rincon-Gonzalez, Buneo, & Helms Tillery, 2011; Tillery, Flanders, & Soechting, 1994). This space has been called a "motor fovea" (Tillery, Flanders, & Soechting, 1994) or "manual fovea" (Graziano et al., 2002). Like the overrepresentation of the fovea in the visual system, the brain regions contributing to this combination of manual behavior, sensory integration, and the representation of visual space may overrepresent this manual work space where food is prepared and eaten and tools are made and used.

This chapter focuses on some of the cortical areas that contribute to the visuomanual behaviors described above, which are located at the junction of sensory neocortex and PPC. These visuomotor behaviors are strongly dependent on tactile feedback, so we will begin with the anterior parietal area 2 and proceed to the more visually integrated regions in and around the banks of the intraparietal sulcus (IPS) such as area 5, the medial intraparietal area (MIP) and the anterior intraparietal area (AIP) (here and throughout, see table 73.1 for detailed explanations of abbreviations used), and lateral to this, 7b on the inferior parietal lobule. We discuss traditional views of these fields based on architectonic analysis and more modern concepts of how this region is subdivided based on multiple criteria including architecture, electrophysiological mapping data, and neuronal response properties. While most of our knowledge of these fields comes from studies in macaque monkeys, we broadly consider data from other nonhuman primates as well as humans to address

TABLE 73.1

Cortical areas, regions and somatosensory receptor locations

Abbreviation	Definition/description	Abbreviation	Definition/description
Parietal cortex (mostly figures 73.2, 73.3, 73.5)		IPS	Intraparietal sulcus
1	Area 1; cutaneous representation caudal to 3b	LIP	Lateral intraparietal area
		LIPd	LIP, dorsal division
1–2	Area 1–2 (galago and marmoset); of uncertain homology to areas 1 and 2 in other primates	LIPv	LIP, ventral division
		LS	Lateral sulcus
		MDP	Medial dorsal parietal area
2	Area 2; representation of deep receptors caudal to area 1	MIP	Medial intraparietal area
3a	Area 3a; somatosensory field rostral to 3b	Opt	Area Opt; overlaps caudomedial 7a
3b	Area 3b, primary somatosensory area, S1	PCS	Postcentral sulcus
5/BA5	Area 5/Brodmann's area 5; contemporary area 5 definitions encompass only part of BA5 which is used to refer only to Brodmann's original parcellation; on SPL	PE	Parietal area E; with other "PE" fields, mostly coextensive with BA5; on SPL of PPC
5D	5/BA5, dorsal division; on the rostral SPL	PEa	Parietal area E, anterior (not part of Seltzer and Pandya's (1986) PE); rostral/medial bank of IPS
5V	5/BA5, ventral division; in rostral bank of the IPS	PEc	Parietal area E, caudal (not part of Seltzer and Pandya's (1986) PE); between PE and PEa on the caudomedial SPL
5L	5/BA5, lateral division; from Seelke et al. (2012)	PEm	Parietal area Em, rostral division of PE from von Bonin and Bailey (1947)
7/BA7	Area 7/Brodmann's area 7; caudal bank of IPS and most or all of IPL	PEp	Parietal area Ep, posterior division of PE from von Bonin and Bailey (1947)
7a	Area 7a, caudal portion of 7; originally defined by Vogt and Vogt (1919) and subsequently further subdivided	PF	Parietal area F; rostral IPL; overlaps 7b
7a–l	Area 7a, lateral division	PFG	Parietal area FG; rostral IPL (transitional area between PF and PG) from Seltzer and Pandya (1986); may straddle 7a/7b border
7a–m	Area 7a, medial division		
7b	Area 7b, rostral portion of 7; originally defined by Vogt and Vogt (1919) and subsequently further subdivided	PG	Parietal area G; rostral IPL; overlaps 7a
		PGm	Parietal area Gm; on the medial wall portion of von Bonin's PE from Seltzer and Pandya (1986)
7op	Operculor area 7; lateral to 7b		
BA19	Area 19/Brodmann's area 19; overlaps several extrastriate visual areas	Pm	Parietal medial area (squirrel); may be homologous to primate PPC
aSMG	Anterior supramarginal gyrus (human); overlapping human PF	PO	Parietal occipital area (approximately V6 + V6a)
AIP	Anterior intraparietal area; overlaps BA7, Vogts' 7b (1919), POa	POa	Area POa (not part of PO); overlapping LIP and AIP; medial BA7
CIP	Caudal intraparietal area	PPc	Posterior parietal caudal area (galago, tree shrew)
CIP1	Caudal intralparietal 1, from Arcaro et al. (2011)		
		PPC	Posterior parietal cortex
CIP2	Caudal intralparietal 2, from Arcaro et al. (2011)	PPl	Posterior parietal lateral area (galago)
		Pm	Parietal medial area (squirrel)
CS	Central sulcus	PPr	Posterior parietal rostral area (tree shrew)
DZ	Dysgranular zone (rat); surrounded by S1	PR	Parietal rhinal area
Ig	Granular insular field; adjacent to S2 in LS, e.g., Friedman et al. (1986)	PRR	Parietal reach region
		PV	Parietal ventral area
IPd	Intraparietal depth area (in depth of IPS, adjacent to POa and PEa)	R	Rostral somatosensory area (striped possum, opossum)
IPL	Inferior parietal lobule		

 DYLAN F. COOKE, ADAM GOLDRING, GREGG H. RECANZONE, AND LEAH KRUBITZER

Abbreviation	Definition/description	Abbreviation	Definition/description
Ri	Retroinsular area; in the fundus of LS, adjacent to S2 and 7b, e.g., Friedman et al. (1986)	DP	Dorsal prelunate area
		M1	Primary motor cortex
S1	Primary somatosensory area	MST	Medial superior temporal area; visual
S2	Second somatosensory area	MT	Middle temporal area; visual
SC	Somatosensory caudal area (tree shrew, striped possum, opossum); perhaps homologous to primate area 1 or areas 1+2.	V1	Primary visual area
		V2	Second visual area
		V3	Third visual area
		V3d	V3, dorsal division
SPL	Superior parietal lobule	V3v	V3, ventral division
VIP	Ventral intraparietal area	V3A	Visual area V3A; lateral to V3; also called DM
VIPl	VIP, lateral division		
VIPm	VIP, medial division	V4	Fourth visual area; also called DL
V6	Visual area 6, medial to the caudal end of the IPS; with V6A, approximately overlapping PO	vis	Other visual areas
		Body parts and receptor types (figure 73.4)	
		cn	Chin
V6A	Visual area 6a, dorsal to (and not a division of) V6; with V6, approximately overlapping PO	cut	Cutaneous receptors on body part
		D1–5	Digits 1–5
		deep	Deep receptors on body part
VS	Ventral somatosensory area; part of the S2 complex from Krubitzer et al. (1995)	fa	Forearm
		gen	Genitals
Other cortex (mostly figure 73.2)		j	Jaw
A1	Primary auditory area	occ	Occiput
aud	Other auditory areas	sh	Shoulder
DL	Dorsolateral visual area; also called V4	sn	Snout
DLc	DL, caudal division	tr	Trunk
DLr	DL, rostral division	T1–5	Toes 1–5
DM	Dorsomedial visual area; visual; also called V3a	vis	Visual response

Note: All terms refer to macaque brains except where noted.

questions of how this boundary cortex evolved and how it covaries with sophisticated visuomanual behaviors that define primates.

SENSORY VERSUS NONSENSORY "ASSOCIATION" CORTEX

The cortical sheet is divided into multiple cortical fields defined by their architecture, function, and connectivity. Several cortical areas including primary sensory areas (V1, S1, A1) and secondary areas (e.g., V2, S2/PV, A2) have been described in all mammals investigated (see figure 73.1). In these primary areas there is a relatively simple, first-order transformation or representation of the entire sensory epithelium or array. For example, in the visual system, retinotopic order is maintained in V1 and V2, with no or few splits in the representation of the visual hemifield. In mammals in which the relative size of the neocortex is larger, more cortical fields are observed (see figure 73.2). Neurons in this expanded cortex generally do not respond to the kind of simple sensory stimulation that easily activates neurons in primary and secondary areas nor do they respond under most anesthetic conditions. Traditionally considered "association" areas, these regions contain neurons with much more complex response properties (e.g., neurons in inferotemporal cortex respond to faces). Such higher-order, often multisensory areas are located between traditional unimodal sensory areas, and also vary greatly across primates and across mammals in general.

One such region is PPC, situated between somatosensory and visual cortex. Although PPC has been described in primates such as New World monkeys and prosimians

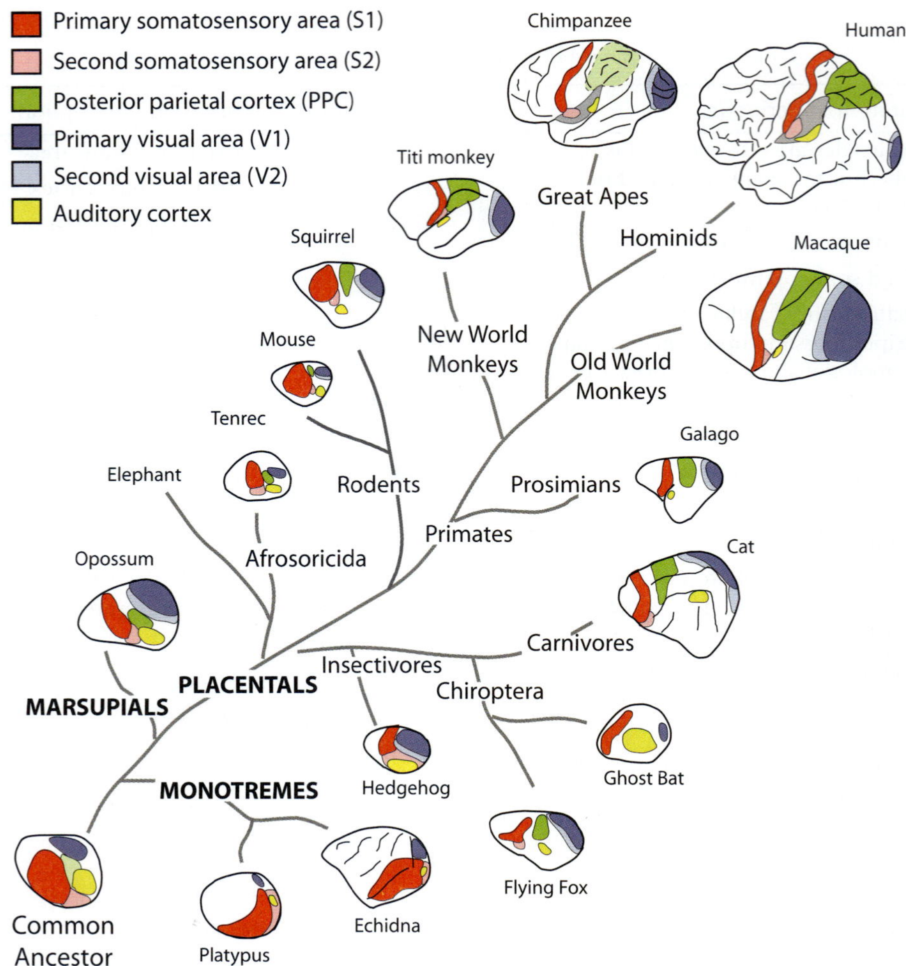

FIGURE 73.1   A cladogram illustrating the phylogenetic relationships for the major subclasses of mammals and some of the orders within each subclass. There is a constellation of cortical fields that is observed in all mammals. This constellation of fields is considered homologous and to have been present in the common ancestor. Posterior parietal cortex (green) or the presumptive posterior parietal cortex (light green) has also been described widely, but in species with small brains (e.g., mice and opossums), this generally encompasses a small zone of cortex between known visual (V1, dark blue; V2, light blue), somatosensory (S1, red; S2, pink), and auditory (yellow) fields in which neurons respond to stimulation of two or more modalities. See table 73.1 for abbreviations.

(e.g., Padberg, Disbrow, & Krubitzer, 2005; Padberg et al., 2007; Stepniewska, Fang, & Kaas, 2009) and other mammals such as carnivores and rodents (Krubitzer, Campi, & Cooke, 2011), most studies of PPC have used the macaque monkey as an animal model for humans. Studies in macaques and humans indicate that PPC contains complex maps that are not obviously topographic or retinotopic, and are involved in complex, multidimensional computations such as calculating optic flow (Merchant, Battaglia-Mayer, & Georgopoulos, 2001; Siegel & Read, 1997), generating an internal coordinate reference of the body (see Chang & Snyder, 2010, for review ; Grefkes & Fink, 2005), and even tool action (Peeters et al., 2009). The cortical magnification commonly seen in primary sensory fields, such as the

fovea in primary visual cortex, is more extreme in PPC. Cortical fields in PPC often do not have a complete representation of the body surface or retina (Arcaro et al., 2011; Patel et al., 2010; Seelke et al., 2012), but rather represent or are associated with special effector organs such as the eye or hand, or they represent very specific aspects of visual processing. For example, caudal regions of macaque PPC, including the caudal intraparietal area, the ventral intraparietal area (VIP), and the lateral intraparietal area (LIP), are generally associated with aspects of visual processing such as extracting three-dimensional features of objects, generating head-centered coordinates, or programming saccade endpoints, respectively (Grefkes & Fink, 2005). Rostral posterior parietal fields in and around the banks

   DYLAN F. COOKE, ADAM GOLDRING, GREGG H. RECANZONE, AND LEAH KRUBITZER

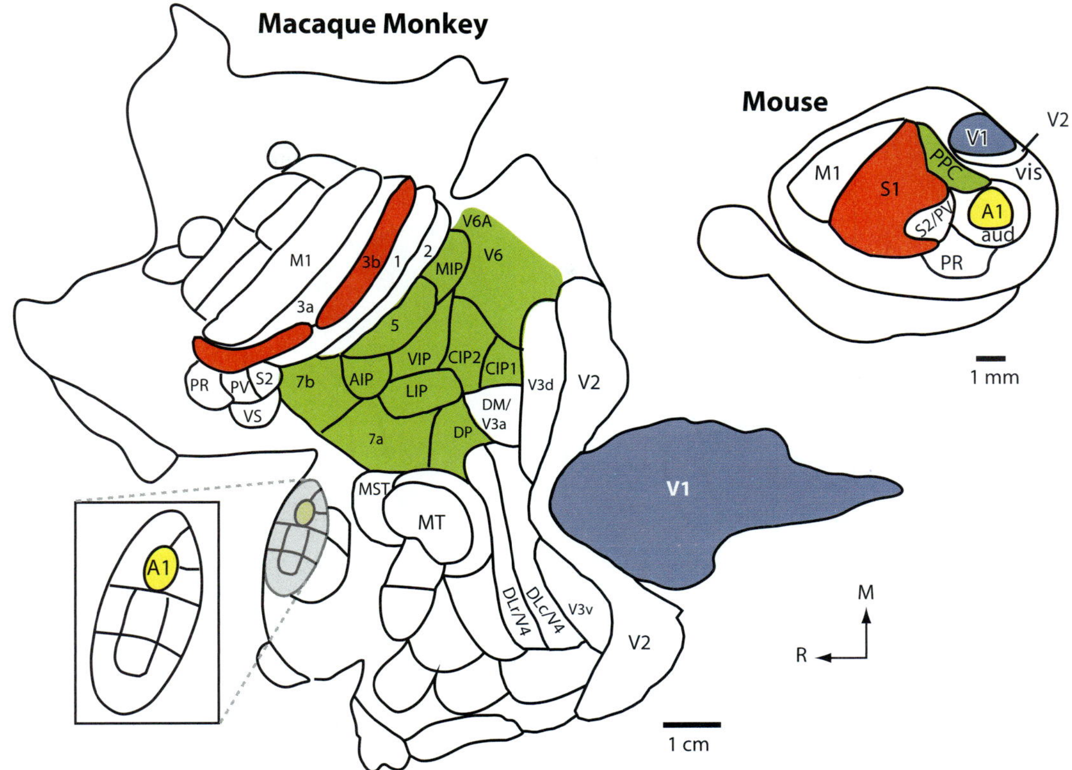

FIGURE 73.2   Drawings of a flattened cortex from macaque monkey and mouse illustrating the number and location of cortical fields. Some cortical fields such as the primary sensory areas (visual, blue; somatosensory, red; and auditory, yellow) have been described in all mammals investigated and are part of a homologous network. In animals with relatively large brains, such as macaque monkeys, the neocortex has greatly expanded and the number of visual and somatosensory areas has increased. Coincident with this increase in cortical sheet size and sensory cortical field number is also an increase in the relative size of posterior parietal cortex (PPC; green) and the number of subdivisions within this region. Divisions of PPC represent our interpretation of this region based on several studies including Stepniewska, Collins, and Kaas (2005), Orban (2008), Arcaro et al. (2011), and Seelke et al. (2012). See table 73.1 for abbreviations.

of the IPS, such as area 5, MIP, and AIP, are associated with initiating a reach, preshaping the hand, and matching object shape with grasp configuration, tasks that utilize both somatosensory and visual information (Debowy et al., 2001; Eskandar & Assad, 2002; Gallese et al., 1994; Murata et al., 2000). Although the focus of this chapter is PPC, in macaque monkeys we will describe this region in the context of adjacent parietal fields in 3 regions: area 5, area 7, and area 2.

Areas on the Rostral Bank of IPS

In macaque monkeys, there are several fields at the junction of S1 and PPC. Posterior parietal area 5 is one of these "higher-order" cortical areas. Historically, there has been a great deal of contention over the status of Brodmann's area 5 (BA5) and how it should be subdivided. BA5 was described architectonically in Old World monkeys as a large triangular-shaped field caudal to

area 2 (Brodmann, 1909; figure 73.3). BA5 encompassed the rostral/medial bank of the IPS, meeting Brodmann's area 7 (BA7) in the IPS. BA5 extends onto much of the caudal portion of the postcentral gyrus, especially in the medial portion where BA5 is widest. Medially, it includes a section of the medial wall extending just across the cingulate sulcus. Henceforth we will use "BA5" to mean this large field and "area 5" to refer to the more recent and variable definitions of a smaller version of this field which along with MIP, the medial division of VIP (VIPm), and the medial dorsal parietal area, probably overlap at least in part with BA5 (figure 73.3).

An alternative nomenclature grew out of von Economo's (1929) parcellation of the human brain in which the location of BA5 was labeled the parietal area E, or "PE." Von Bonin and Bailey (1947) adapted this nomenclature for macaques, splitting the field into rostral PEm and caudal PE/PEp divisions and including the

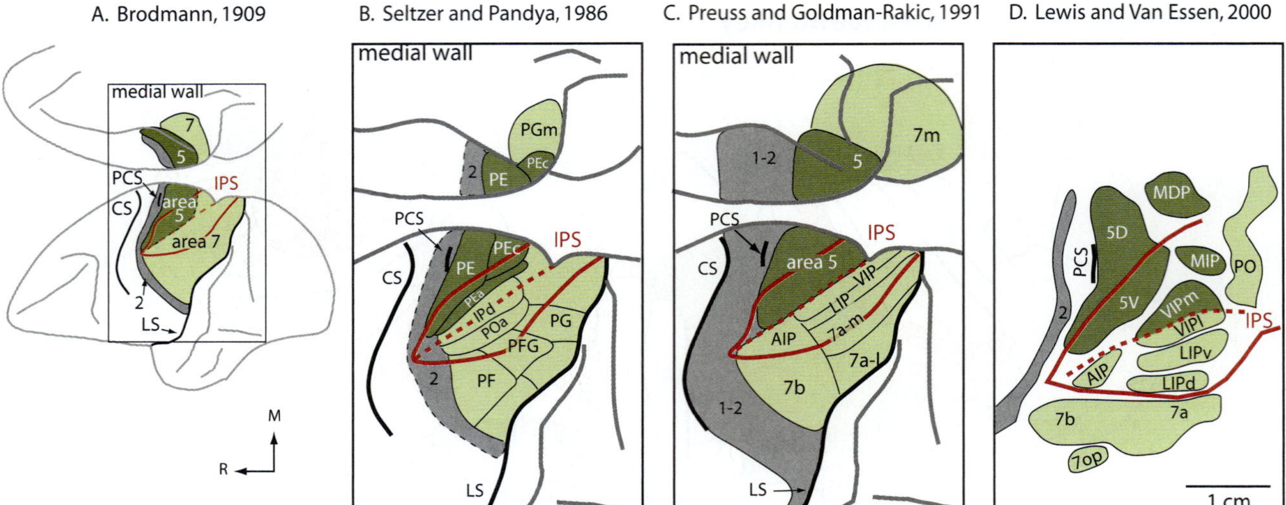

FIGURE 73.3 Architectonic parcellations of the posterior parietal cortex by different investigators. Brodmann (A) proposed the first comprehensive scheme of organization of posterior parietal cortex that included area 2 (gray), a large area 5 (dark green), and a large area 7 (light green). Subsequent schemes (B–D) have used different terminology and subdivided Brodmann's 5 and 7 divisions into multiple fields (B–D). Approximate correspondence between BA5 and BA7 and subsequent schemes are indicated by color (dark and light green, respectively). Note that as Brodmann (1909) generally did not illustrate cortical field boundaries within sulci, the borders of BA5 and BA7 (illustrated as a dashed line in panel A) as well as Brodmann's area 19 are not precisely known. In particular, correspondence of PO and adjacent fields to Brodmann's parcellation is uncertain. Intraparietal sulcus lips (solid line) and fundus (dashed line) are shown in red. Arrows next to A indicate rostral (R) and medial (M) anatomical directions. See table 73.1 for abbreviations.

entire portion of BA7 on the medial wall. Since these early publications, a number of histological techniques (e.g., Hof & Morrison, 1995) have been used to subdivide and refine the borders of area 5/PE. For example, Seltzer and Pandya (1980, 1986) subdivided PE into PE, PEc, and PEa (figure 73.3). More modern architectonic studies divided area 5 into two fields, dorsal area 5 (5D) and ventral area 5 (5V; Lewis & Van Essen, 2000b; figure 73.3B and D).

Functional studies of area 5 examining receptive field characteristics and neural response properties are variable in both their results and in the locations in which the recordings were made (Seelke et al., 2012). In early studies, most of the mediolateral extent of the rostral bank of the IPS and caudal portions of the postcentral gyrus was explored and considered to be "area 5" (e.g., Mountcastle et al., 1975; Sakata et al., 1973). The explored region did not include the lateral-most portion of the IPS or cortex on the medial wall. Subsequent studies of area 5 also differed greatly in the location of their recording sites (e.g., Gardner et al., 2007b; Iwamura, Iriki, & Tanaka, 1994; Kalaska, 1996; Taoka, Toda, & Iwamura, 1998; see Seelke et al., 2012, for review). Given these rather large differences in recording site location, it is not surprising that the results and interpretations regarding the function of area 5 varied as well.

That area 5 is unlike adjacent somatosensory fields on the postcentral gyrus first became apparent in early electrophysiological recording studies in awake monkeys by Duffy and Burchfiel (1971), Sakata and colleagues (1973), and Mountcastle and colleagues (1975; see figure 73.4). These investigators reported that receptive fields for neurons in area 5 are larger than in anterior parietal fields (3b, 1, and 2) and that neurons are most active during active arm movements. While individual neurons can be driven by a variety of stimuli, their optimal responses sometimes can involve highly specific and complex interactions of joint and skin stimulation (Sakata et al., 1973). Important for this review, a small proportion of neurons respond differentially during active reaching depending on the reward value of the reach target (Mountcastle et al., 1975).

Subsequent studies suggest that portions of BA5 play an important role in sophisticated functions such as the programming of intended movements (Debowy et al., 2001; Snyder, Batista, & Andersen, 1997), the coding of reach targets in body- or shoulder-centered coordinates (Ferraina & Bianchi, 1994; Lacquaniti et al., 1995), and the kinematics of object acquisition (e.g., Kalaska, 1996; Wise et al., 1997). Recent studies demonstrate that neurons in area 5 fire maximally during a reaching task before the target object is contacted with the hand (Gardner et al., 2007a, 2007b), and that neurons

 DYLAN F. COOKE, ADAM GOLDRING, GREGG H. RECANZONE, AND LEAH KRUBITZER

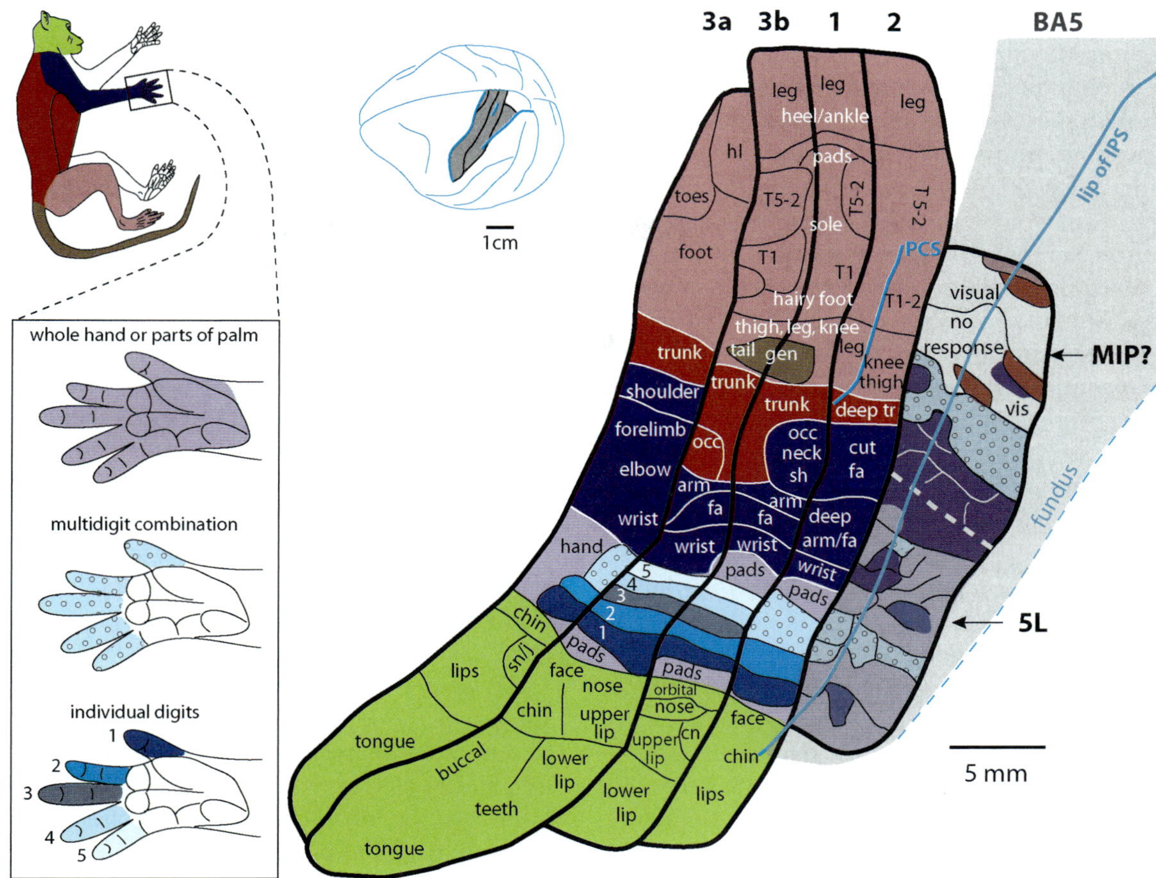

FIGURE 73.4   Summary of the functional subdivisions of anterior parietal areas 3a, 3b, 1, and 2 and posterior parietal areas 5L and the presumptive medial intraparietal area (MIP) from Seelke et al. (2012). Thick black lines are areal boundaries; sulci are shown in blue. Areas 3a, 3b, 1, and 2 have a clear parallel topographic organization not observed in posterior parietal areas. Areas 5L and the presumptive area MIP have fractured maps dominated by representations of the digits, hand and forelimb (thin black lines are body-part boundaries). Area 5L and MIP fall within Brodmann's area 5 (gray portion of enlarged view). In area 5L, digits are most often represented in various combinations (light blue stipple), or parts of the hand or the whole hand is represented (light purple). Schematic at top center shows location of these fields (gray) on the dorsolateral aspect of the brain. Schematic at left shows body-part color code: head, green; individual digits, various shades of blue; hand/forelimb, purple; trunk, red; hindlimb, pink; tail/genitals, brown. Topographic maps are redrawn from Krubitzer et al. (2004) (area 3a), Nelson et al. (1980) (areas 3b and 1), Pons et al. (1985) (area 2), and Rothemund et al. (2002) (genital/tail representation in areas 3b and 1). Conventions as in previous figures; see table 73.1 for abbreviations.

modulate their activity depending on how and when the hand is used in a grasp (Chen et al., 2009).

It is important to note that the architectonic divisions of area 5 do not correspond well to functional subdivisions, and some areas that may overlap BA5, such as the parietal reach region (PRR), have only been described using functional criteria. PRR has been proposed to overlap areas MIP and V6a (Snyder, Batista, & Andersen, 2000). While MIP is generally considered to be within BA5, V6a is probably not. Our own data and those of others indicate that BA5 contains a moderately sized functional area (5L) that is smaller than Brodmann's original architectonic description, and medial field (MIP). We use the term MIP because its location, architecture, and aspects of its organization are consistent with a previously explored area (termed MIP) defined using electrophysiological and/or architectonic criteria (Colby & Duhamel, 1991; Eskandar & Assad, 2002; Lewis & Van Essen, 2000b). Unlike anterior parietal areas, area 5L contains a fractured, nontopographically organized representation of muscles and joints of the hand and limb, and deep receptors of the skin (figure 73.4). This type of fractured representation is analogous to motor cortex and constitutes related groups of proprioceptors activated during behaviorally relevant movements.

Reports of neural response properties in this medial region vary greatly (Seelke et al., 2012); various locations of recordings and descriptions of these data have been reviewed by Seelke et al. (2012). Despite differences in recording location and behavioral tasks utilized, most studies indicate that cortical fields in the medial portions of the rostral bank of the IPS are involved in translating and combining multiple frames of reference (gaze centered, body centered, head centered) into a common coordinate system or integrated plan for reaching toward an intended target in immediate extrapersonal space (Buneo et al., 2002; Chang & Snyder, 2010). Laterally in the rostral bank of the IPS, area 5L appears to be involved in the kinematics of reaching, coordinating multiple limb parts for reaching and grasping actions, and matching visually determined object properties, such as size and shape, with grasping configurations (Chen et al., 2009).

Areas on the Inferior Parietal Lobule and Lateral Bank of the IPS

Brodmann's area 7 (BA7) is an enormous field that was architectonically defined in a variety of primates. As defined by Brodmann, it begins on the medial wall, it wraps around the caudal bank of the IPS, onto the inferior parietal lobule, and it continues around the upper bank of the lateral sulcus laterally (see figure 73.3). Subsequent studies have divided this region into 7a and 7b (Vogt & Vogt, 1919) or multiple cortical areas using various terminologies (see figure 73.3). More recently, using a combination of architecture, neuroanatomical data, and neurophysiological recording studies, Lewis and Van Essen (2000a, 2000b) divided BA7 into multiple regions, including 7b, 7op, 7a, AIP, divisions of LIP and VIP, MIP, and PO (figure 73.3D). As with BA5, architectonic divisions of area BA7 are often inconsistent with the location of functionally defined divisions of this region. We are most interested in the portions of area 7 that border known somatosensory fields, areas 7b and AIP as defined by Preuss and Goldman-Rakic (1991) and Lewis and Van Essen (2000b). These nomenclatures and divisions appear to be the most widely used in studies of this region of cortex.

Early electrophysiological investigations indicated that neurons in classically defined areas 7a and 7b of Vogt and Vogt (1919) responded primarily to visual fixation and eye movements (7a), or to somatosensory stimulation and passive movements of the arms and hands (7b) (Hyvarinen & Poranen, 1974; Leinonen et al., 1979; Mountcastle et al., 1975). Importantly, cells in both 7a and 7b were most active in awake animals when the monkey reached, grasped, and manipulated various visually targeted objects (Hyvarinen & Poranen, 1974).

Recent studies indicate that areas 7a and 7b overlap four architectonically distinct zones (Gregoriou et al., 2006; Pandya & Seltzer, 1982; Rozzi et al., 2006, 2008) that have distinct neuroanatomical connections: PG and Opt largely overlap 7a, PFG probably straddles the 7b/7a border, and PF lies within 7b. Collectively, these fields form a gradient of sensory and motor function that progresses from vision- and eye-related functions to somatosensation- and hand/face-related functions as one moves rostrolaterally along the IPS. Thus, PG contains visual fixation neurons and neurons responsive to eye and arm movements related to reaching and grasping while neurons in PFG and/or PF have visual responses to objects presented within manual space, are involved in goal-directed reaching, and appear to code motor acts predominantly related to hand use, orofacial movements, and hand–mouth coordination but are differentially active when these acts are embedded in different actions (e.g., Bonini et al., 2011; Fogassi et al., 2005; Yokochi et al., 2003).

AIP, located on the caudal (lateral) bank of the lateral portion of the IPS (figure 73.3C–D), was traditionally incorporated in area 7, and then 7b, but is now considered to be distinct from 7b (e.g., Gallese et al., 1994; Taira et al., 1990). Neurons in AIP are active during grasping as well as passive viewing of an object shaped to require the same grasp (Murata et al., 2000; Sakata et al., 1995) and code three-dimensional features of an object with shorter latency and less sensitivity to curves and edges than temporal cortical areas (Srivastava et al., 2009). AIP neurons also preferentially respond to visually targeted objects within reach (Srivastava et al., 2009), but many respond equally well to memory-guided object manipulations made in darkness (Murata et al., 1996). Importantly, neurons in AIP respond maximally during the early stages of prehension, increasing their firing just prior to contacting an object (Gardner et al., 2007b), consistent with AIP's role in preshaping the hand prior to a grasp (Debowy et al., 2001; Gallese et al., 1994).

Areas on the Postcentral Gyrus

Although area 2 is considered to be a proprioceptive area associated with somatosensory processing, it is important to establish its spatial relationship to area 5, as well as its contribution to reaching, grasping, bilateral coordination, and dexterity. Using multiunit electrophysiological recording techniques, Pons and colleagues (Pons & Kaas, 1985; Pons et al., 1985) determined that the organization of area 2 is parallel to that of 3b and 1 and contains an inverted representation of the body from medial (feet) to lateral (head; figure

   DYLAN F. COOKE, ADAM GOLDRING, GREGG H. RECANZONE, AND LEAH KRUBITZER

73.4). While some neurons in area 2 respond to stimulation of cutaneous receptors, the majority of neurons respond to stimulation of deep tissue (Taoka, Toda, & Iwamura, 1998) or noncutaneous stimulation of varying types (Hyvärinen & Poranen, 1978). When the area 2 representation of digits 1 and 2 are electrically stimulated using "long" duration pulse trains, the evoked grasp-like movement is the opposition of these digits. Interestingly, this area 2 "grasp zone" appears not to share direct connections with similar motor and premotor grasp zones, suggesting parallel pathways with slightly different functions (Gharbawie et al., 2011).

Recent studies indicate that neurons in the monkey S1 complex (including areas 3b and 1 as well as area 2) show changes in baseline firing rate when visual attention is directed to the tactile stimuli on specific digits of the hand (Meftah, Shenasa, & Chapman, 2002). This modulation is proposed to enhance salient features of the stimulus, such as texture, rather than simply improve overall tactile discrimination. In humans, studies using noninvasive imaging techniques demonstrate that viewing a stimulated body part modulated the response evoked by a tactile stimulus in somatosensory cortex (Forster & Eimer, 2005; Longo, Pernigo, & Haggard, 2011; Taylor-Clarke, Kennett, & Haggard, 2002), and these visual effects in somatosensory cortex improve tactile detection, tactile discrimination, and tactile acuity (Heller, 1982; Kennett, Taylor-Clarke, & Haggard, 2001; Ladavas et al., 1998). These data indicate that our concepts regarding unimodal sensory areas, their function, and their evolution should be reconsidered.

EVOLUTION OF PARIETAL CORTEX AND COEVOLUTION OF THE HAND IN PRIMATES

Thus far we have focused on the organization of portions of PPC in nonhuman primates, particularly macaque monkeys. However, because expansions of PPC are accompanied by expansions of sensory neocortex, it is important to discuss the relationship between sensory fields and the posterior parietal fields to which they provide inputs. The Kaas chapter in this volume provides an excellent overview on the evolution of the visual system of primates. Especially interesting is the expansion of the temporal lobe and the addition of new cortical fields associated with object and face identification and recognition. Areas at the boundary of occipital cortex and PPC have also expanded to include fields such as the middle temporal and medial superior temporal areas that are involved in object motion and self-motion (Britten, 2008). These additional areas in the temporal lobe do not appear to be present in nonprimate mammals, and if there are analogous areas, such as those that process motion, comparative studies suggest that they have evolved independently.

Similar expansions have occurred among somatosensory fields of the anterior parietal cortex and lateral sulcus and are clearly associated with and contribute to the unique processing that occurs in PPC. For example, while all mammals examined have a primary somatosensory area (S1, also termed 3b), cortex caudal to S1 appears to be differently organized in different mammals and even in different primates (see figure 73.5). Further, a region termed PPC, situated between S1 and V1, has been described in a variety of mammals including marsupials, rodents, and tree shrews (figure 73.5). While all primates examined have areas 3b/S1, S2, PV, and at least one other lateral sulcus area (Coq et al., 2004; Hinkley et al., 2007; Krubitzer & Kaas, 1990; Krubitzer et al., 1995; Qi, Lyon, & Kaas, 2002; Stepniewska, Preuss, & Kaas, 2006), cortex immediately caudal to 3b is different in different suborders. In galagos, cortex immediately caudal to area 3b contains neurons that are responsive to stimulation of deep receptors, muscles and joints or high-threshold cutaneous stimulation. This region is termed 1–2 because of uncertain homology with areas 1 and 2 in other primates. Cortex between area 1–2 and known extrastriate visual areas such as the dorsomedial visual area is termed PPC and is clearly a sensorimotor area like that described for subdivisions of PPC in macaque monkeys. Intracortical microstimulation of PPC in galagos reveals very gross topographic organization in which movements can be evoked from hind limb, then forelimb, then face in a mediolateral progression. Importantly, the evoked movements resemble ethologically relevant behaviors, and sites that evoke identifiable behavioral categories such as reaching, hand-to-mouth movements resembling feeding, and apparent defensive or aggressive movements are clustered (Stepniewska, Fang, & Kaas, 2005, 2009).

The organization of posterior portions of anterior parietal cortex and PPC itself is highly variable in New World monkeys. In the miniature marmoset monkey, cortex caudal to 3b is much like that in galagos and is also termed area 1–2 (Huffman & Krubitzer, 2001). However, there is a greater expansion of known visual cortex in marmosets (particularly in regions of the temporal lobe) and an accompanying increase in PPC. There has been some attempt to subdivide this rather large region of cortex to include areas VIP, LIP, and MIP based on connections, but no data on functional mapping or electrophysiological recordings in awake behaving marmosets have been published. Cortex just caudal to 3b in other New World Monkeys such as owl monkeys, titi monkeys, and squirrel monkeys have a

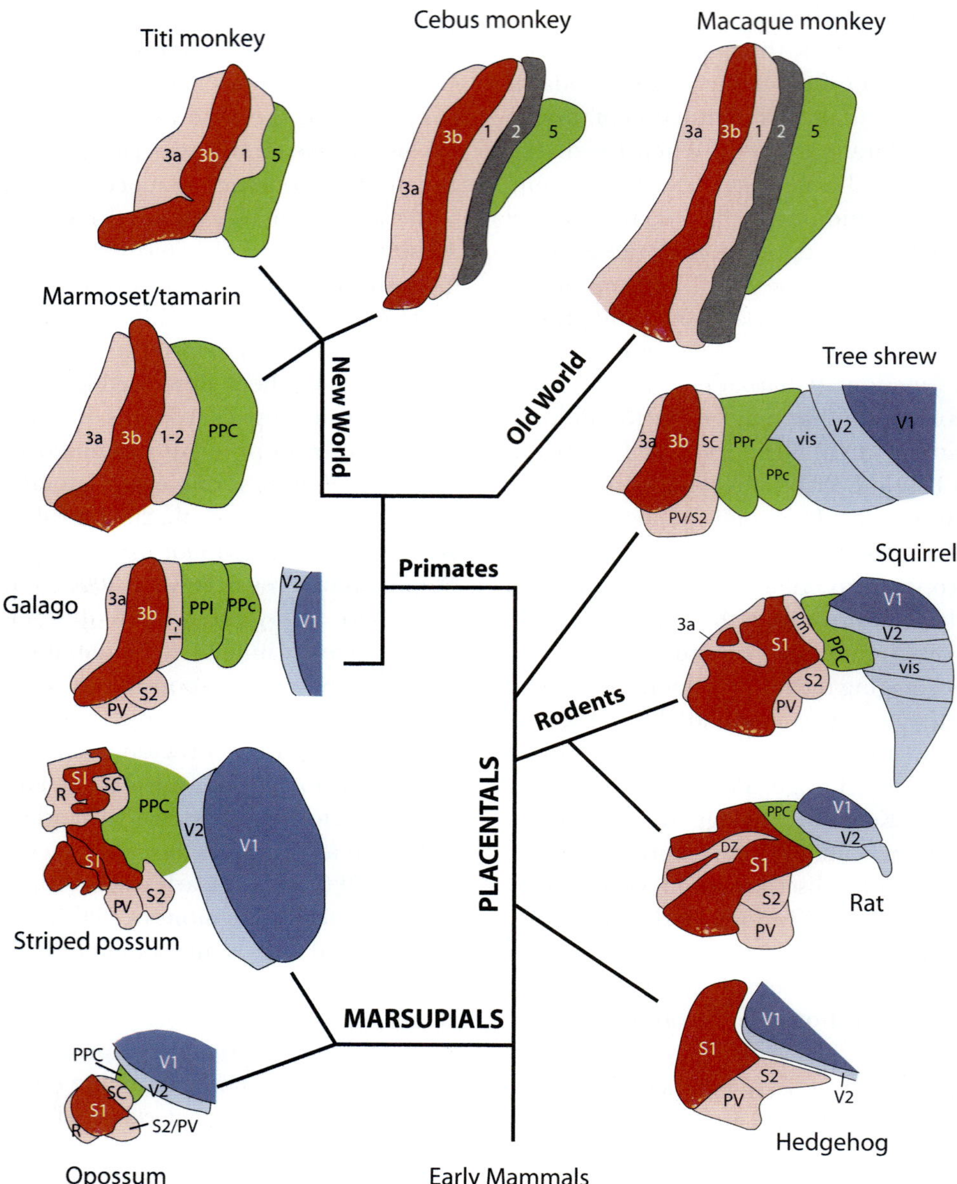

FIGURE 73.5 Cladogram showing the relationships of major mammalian lineages and changes that have taken place in the relationship between rostral somatosensory areas and posterior parietal cortex (PPC). In mammals whose ancestors branched off early from the primate lineage (e.g., opossums and hedgehogs) the amount of cortex considered PPC is small or nonexistent (hedgehogs) although in some marsupials such as striped possum there was an independent expansion of PPC. Rodents have a small PPC, but squirrels appear to have an expanded visual and somatosensory cortex and a relatively larger PPC than rats. PPC began to expand in the mammalian ancestor of tree shrews and primates and then greatly expanded in the primate lineage. For New and Old World monkeys, visual cortex and the caudal portion of PPC is not shown. However, these species illustrate the relationships between the expanding somatosensory cortex and PPC. Illustrations are modified from Huffman et al. (1999) (opossum, striped possum), Catania, Collins, and Kaas (2000) (hedgehog), Padberg, Disbrow, and Krubitzer (2005) and Padberg et al. (2007) (summary of titi, galago, macaque, and cebus monkey), Krubitzer, Campi, and Cooke (2011) (squirrel, rat), and Wong and Kaas (2009, 2010) (tree shrews, galagos). Conventions as in previous figures; see table 73.1 for abbreviations.

clearly defined area 1 that contains a complete representation of skin receptors of the contralateral body that mirrors that of area 3b (Padberg, Disbrow, & Krubitzer, 2005; Padberg et al., 2007). However, area 2 appears to be absent. Electrophysiological exploration of cortex caudal to area 1 in titi monkeys revealed a field that had many of the properties of area 5 in macaque monkeys, specifically an extreme magnification of the forelimb and hand and a fractured map. These studies indicate that although posterior parietal

     DYLAN F. COOKE, ADAM GOLDRING, GREGG H. RECANZONE, AND LEAH KRUBITZER

area 5 abuts area 2 in Old World monkeys, its relative position to somatosensory areas in most New World monkeys is different in that it is immediately adjacent to area 1 (or area 1–2). Based on its appearance in all monkeys that have been studied, the only certainty is that this posterior parietal region had already begun to expand in the common ancestor of New and Old World monkeys and likely contained multiple subdivisions.

New World cebus monkeys deviate from this general New World monkey plan of organization in which area 2 may be absent or combined with area 1. Rather, in cebus monkeys, a clear proprioceptive area 2 has emerged, and area 5L is a well-defined field with a distinct architecture and fractured topography (Padberg et al., 2007). We proposed that the emergence of area 2 in cebus monkeys coevolved with the large expansion of the neocortex, the evolution of the hand, and the emergence of opposable thumbs that allows for a precision grip. These differences also coevolved with changes in the corticospinal system in that there are direct and rich projections from motor cortex to alpha motor neurons of the ventral horn that control the muscles of the digits (Bortoff & Strick, 1993; Heffner & Masterton, 1983; Lemon & Griffiths, 2005). These alterations in both motor cortex and PPC are proposed to subserve the increased dexterity observed in cebus monkeys compared to their New World cousins and may form the basis for even greater cognitive, visuomanual abilities such as tool use.

Cebus monkeys selectively utilize tools under experimental conditions, and unlike macaques, also regularly use tools in the wild (e.g., Ferreira, Emidio, & Jerusalinsky, 2010; Schrauf, Huber, & Visalberghi, 2008; Visalberghi et al., 2007, 2009). They are arguably the best primate model for the study of tool use in humans. Cebus can also utilize the same tool for different tasks, they can use one tool to make another (Mannu & Ottoni, 2009), and they can learn to use a novel tool by observation (Fredman & Whiten, 2008). Thus, the implicit knowledge of object features and how they can be used to manipulate or transform other objects, once considered to be a unique feature of human brains, appears to be present to some degree in cebus monkeys as well.

Like cebus monkeys, these same abilities coevolved in humans with an expansion of the cortical sheet, an increase in the number of cortical areas, alterations in the connections within the brain, and changes in the morphology and use of the hand. The human hand–wrist consists of 27 bones and 39 intrinsic and extrinsic muscles (Hepp-Reymond, Huesler, & Maier, 1996) and has evolved a number of important changes including alterations in the size of the distal, middle, and proximal phalanges. The carpal and metacarpal joints—particularly the trapezium in the wrist, the articulation between the first and second carpals, and the metacarpophalangeal joints—have undergone significant alterations as did the size and position of associated ligaments (Lewis, 1977). The skin of the distal digit tips has evolved epidermal ridges and a high concentration of mechanosensory receptors such as Merkel disks and Meissner's corpuscles (Pare, Smith, & Rice, 2002). These transformations of the hand allow for an expanded repertoire of grips, including a precision grip in which the thumb can be opposed to digit 2 or other digits and independent use of individual digits. Many of these adaptations are proposed to have evolved for tool use (Marzke, 1997). Of course, in modern humans these adaptations of the hand and associated brain areas have been co-opted for tasks removed from those that would have exerted selective pressure on our hominid ancestors such as using a computer keyboard and intricate independent use of the fingers for playing a piano or violin. Interestingly, these are often learned as visuomanual skills but can progress to a stage (e.g., touch-typing) where vision is not necessary. This, in turn, allows for indirect but arguably more complex visuomotor integration such as sight-reading music. Thus, the most significant human adaptation for tool use may have been manual and cognitive flexibility.

While investigators have begun to explore the cortical areas associated with tool use in humans, the use of the traditionally used macaque model may not be appropriate for this specialization since they seldom use tools in the wild. Because of this dramatic ethological difference it is not surprising that although macaques and humans share homologous networks for hand use, including areas in PPC (e.g., 5 and AIP), ventral premotor cortex, and areas in temporal cortex (Gardner et al., 2007a, 2007b; Hinkley et al., 2009; Peeters et al., 2009; Valyear et al., 2007; Vingerhoets et al., 2009; Yalachkov, Kaiser, & Naumer, 2009), there are areas on the inferior parietal lobule associated with tool action that are unique to humans. Cebus monkeys, however, have independently evolved an opposable thumb, a precision grip, and tool selectivity, in addition to changes in motor cortex and PPC, so it is likely that they have evolved areas of the cortex analogous to those in humans (anterior supramarginal gyrus and superior parietal lobule) that represent tool action.

CONCLUSIONS

Our sense of three-dimensional space and our ability to move within and to physically manipulate objects in the world are, in addition to language, essential elements of human experience and culture. When considering how and why PPC evolved in humans, a broader perspective is needed regarding the array of changes that have necessarily coevolved with this region. For example, changes in effectors of the body, such as the frontal placement of the eyes and the ability to rapidly and accurately place the retinal fovea in a region of interest, arose along with a ballooning set of specialized visual cortical areas involved in perceiving complex motion and honing in specifically on faces and hands. Changes to the morphology of that other main effector, the hand, expanded the ability to generate grip types from forceful to fine. These too occurred in tandem with elaboration of somatosensory areas of the post central gyrus and lateral sulcus. These effectors, eyes and hands, meet in an optimal ergonomic extrapersonal work space that is constrained by the geometry of the skeletal system and the acuity of the visual system. PPC appears to have expanded in concert with the new visual and somatosensory areas of the neocortex, and these inputs generate networks in PPC devoted to the coordination of the hands and eyes through the transformation of sensory information into a coordinate system in which actions can be initiated. The further expansion of areas on the inferior parietal lobule in the human brain where tool action is encoded is a significant departure from the effector-based computations of other posterior parietal areas. These unique human cortical areas represent the ultimate, often indirect goals that tools can achieve.

REFERENCES

Arcaro, M. J., Pinsk, M. A., Li, X., & Kastner, S. (2011). Visuotopic organization of macaque posterior parietal cortex: A functional magnetic resonance imaging study. *Journal of Neuroscience, 31,* 2064–2078.

Bonini, L., Serventi, F. U., Simone, L., Rozzi, S., Ferrari, P. F., & Fogassi, L. (2011). Grasping neurons of monkey parietal and premotor cortices encode action goals at distinct levels of abstraction during complex action sequences. *Journal of Neuroscience, 31,* 5876–5886.

Bortoff, G. A., & Strick, P. L. (1993). Corticospinal terminations in two New-World primates: Further evidence that corticomotoneuronal connections provide part of the neural substrate for manual dexterity. *Journal of Neuroscience, 13,* 5105–5118.

Britten, K. H. (2008). Mechanisms of self-motion perception. *Annual Review of Neuroscience, 31,* 389–410.

Brodmann, K. (1909). *Vergleichende Lokalisationsiehre der Grosshirnrinde in ihren Prinzipien Dargestellt auf Grund des Zellenbaues.* Leipzig: Barth.

Buneo, C. A., Jarvis, M. R., Batista, A. P., & Andersen, R. A. (2002). Direct visuomotor transformations for reaching. *Nature, 416,* 632–636.

Catania, K. C., Collins, C. E., & Kaas, J. H. (2000). Organization of sensory cortex in the East African hedgehog (*Atelerix albiventris*). *Journal of Comparative Neurology, 421,* 256–274.

Chang, S. W., & Snyder, L. H. (2010). Idiosyncratic and systematic aspects of spatial representations in the macaque parietal cortex. *Proceedings of the National Academy of Sciences of the United States of America, 107,* 7951–7956. doi:10.1073/pnas.0913209107.

Chen, J., Reitzen, S. D., Kohlenstein, J. B., & Gardner, E. P. (2009). Neural representation of hand kinematics during prehension in posterior parietal cortex of the macaque monkey. *Journal of Neurophysiology, 102,* 3310–3328.

Colby, C. L., & Duhamel, J. R. (1991). Heterogeneity of extrastriate visual areas and multiple parietal areas in the macaque monkey. *Neuropsychologia, 29,* 517–537.

Coq, J. O., Qi, H. X., Collins, C. E., & Kaas, J. H. (2004). Anatomical and functional organization of somatosensory areas of the lateral fissure in the New World titi monkeys (*Callicebus moloch*). *Journal of Comparative Neurology, 476,* 363–387.

Debowy, D. J., Ghosh, S., Ro, J. Y., & Gardner, E. P. (2001). Comparison of neuronal firing rates in somatosensory and posterior parietal cortex during prehension. *Experimental Brain Research, 137,* 269–291.

Duffy, F. H., & Burchfiel, J. L. (1971). Somatosensory system—Organizational hierarchy from single units in monkey area-5. *Science, 172,* 273–275.

Eskandar, E. N., & Assad, J. A. (2002). Distinct nature of directional signals among parietal cortical areas during visual guidance. *Journal of Neurophysiology, 88,* 1777–1790.

Ferraina, S., & Bianchi, L. (1994). Posterior parietal cortex: Functional properties of neurons in area 5 during an instructed-delay reaching task within different parts of space. *Experimental Brain Research, 99,* 175–178.

Ferreira, R. G., Emidio, R. A., & Jerusalinsky, L. (2010). Three stones for three seeds: Natural occurrence of selective tool use by capuchins (*Cebus libidinosus*) based on an analysis of the weight of stones found at nutting sites. *American Journal of Primatology, 72,* 270–275.

Fogassi, L., Ferrari, P. F., Gesierich, B., Rozzi, S., Chersi, F., & Rizzolatti, G. (2005). Parietal lobe: From action organization to intention understanding. *Science, 308,* 662–667.

Forster, B., & Eimer, M. (2005). Vision and gaze direction modulate tactile processing in somatosensory cortex: Evidence from event-related brain potentials. *Experimental Brain Research, 165,* 8–18. doi:10.1007/s00221-005-2274-1.

Fredman, T., & Whiten, A. (2008). Observational learning from tool using models by human-reared and mother-reared capuchin monkeys (*Cebus apella*). *Animal Cognition, 11,* 295–309.

Friedman, D. P., Murray, E. A., O'Neill, J. B., & Mishkin, M. (1986). Cortical connections of the somatosensory fields of the lateral sulcus of macaques: Evidence for a corticolimbic pathway for touch. *Journal of Comparative Neurology, 252,* 323–347.

Gallese, V., Murata, A., Kaseda, M., Niki, N., & Sakata, H. (1994). Deficit of hand preshaping after muscimol injection in monkey parietal cortex. *Neuroreport, 5,* 1525–1529.

Gardner, E. P., Babu, K. S., Ghosh, S., Sherwood, A., & Chen, J. (2007a). Neurophysiology of prehension: III. Representation of object features in posterior parietal cortex of the macaque monkey. *Journal of Neurophysiology, 98,* 3708–3730.

Gardner, E. P., Babu, K. S., Reitzen, S. D., Ghosh, S., Brown, A. S., Chen, J., et al. (2007b). Neurophysiology of prehension: I. Posterior parietal cortex and object-oriented hand behaviors. *Journal of Neurophysiology, 97*, 387–406.

Gharbawie, O. A., Stepniewska, I., Qi, H., & Kaas, J. H. (2011). Multiple parietal–frontal pathways mediate grasping in macaque monkeys. *Journal of Neuroscience, 31*, 11660–11677.

Graziano, M. S., Cooke, D. F., Taylor, C. S., & Moore, T. (2004). Distribution of hand location in monkeys during spontaneous behavior. *Experimental Brain Research, 155*, 30–36.

Graziano, M. S., Taylor, C. S., Moore, T., & Cooke, D. F. (2002). The cortical control of movement revisited. *Neuron, 36*, 349–362.

Grefkes, C., & Fink, G. R. (2005). The functional organization of the intraparietal sulcus in humans and monkeys. *Journal of Anatomy, 207*, 3–17.

Gregoriou, G. G., Borra, E., Matelli, M., & Luppino, G. (2006). Architectonic organization of the inferior parietal convexity of the macaque monkey. *Journal of Comparative Neurology, 496*, 422–451.

Heffner, R. S., & Masterton, R. B. (1983). The role of the corticospinal tract in the evolution of human digital dexterity. *Brain, Behavior and Evolution, 23*, 165–183.

Heller, M. A. (1982). Visual and tactual texture perception: Intersensory cooperation. *Perception & Psychophysics, 31*, 339–344.

Hepp-Reymond, M.-C., Huesler, E. J., & Maier, M. A. (1996). Precision grip in humans: Temporal and spatial synergies. In A. M. Wing, P. Haggard, & J. R. Flanagan (Eds.), *Hand and brain* (pp. 37–68). San Diego, CA: Academic Press.

Hinkley, L., Krubitzer, L., Padberg, J., & Disbrow, E. (2009). Visual–manual exploration and posterior parietal cortex in humans. *Journal of Neurophysiology, 102*, 3433–3446.

Hinkley, L. B., Krubitzer, L. A., Nagarajan, S. S., & Disbrow, E. A. (2007). Sensorimotor integration in S2, PV, and parietal rostroventral areas of the human sylvian fissure. *Journal of Neurophysiology, 97*, 1288–1297.

Hof, P. R., & Morrison, J. H. (1995). Neurofilament protein defines regional patterns of cortical organization in the macaque monkey visual system: A quantitative immunohistochemical analysis. *Journal of Comparative Neurology, 352*, 161–186.

Huffman, K. J., & Krubitzer, L. A. (2001). Area 3a: Topographic organization and connections in marmoset monkeys. *Cerebral Cortex, 11*, 849–867.

Huffman, K. J., Nelson, J., Clarey, J., & Krubitzer, L. (1999). Organization of somatosensory cortex in three species of marsupials, *Dasyurus hallucatus, Dactylopsila trivirgata*, and *Monodelphis domestica*: Neural correlates of morphological specializations. *Journal of Comparative Neurology, 403*, 5–32.

Hyvarinen, J., & Poranen, A. (1974). Function of the parietal associative area 7 as revealed from cellular discharges in alert monkeys. *Brain, 97*, 673–692.

Hyvärinen, J., & Poranen, A. (1978). Receptive field integration and submodality convergence in the hand area of the post-central gyrus of the alert monkey. *Journal of Physiology, 283*, 539–556.

Iwamura, Y., Iriki, A., & Tanaka, M. (1994). Bilateral hand representation in the postcentral somatosensory cortex. *Nature, 369*, 554–556.

Kalaska, J. F. (1996). Parietal cortex area 5 and visuomotor behavior. *Canadian Journal of Physiology and Pharmacology, 74*, 483–498.

Kennett, S., Taylor-Clarke, M., & Haggard, P. (2001). Noninformative vision improves the spatial resolution of touch in humans. *Current Biology, 11*, 1188–1191. doi:10.1016/S0960-9822(01)00327-X.

Krubitzer, L., Campi, K. L., & Cooke, D. F. (2011). All rodents are not the same: A modern synthesis of cortical organization. *Brain, Behavior and Evolution, 78*, 51–93.

Krubitzer, L., Clarey, J., Tweedale, R., Elston, G., & Calford, M. (1995). A redefinition of somatosensory areas in the lateral sulcus of macaque monkeys. *Journal of Neuroscience, 15*, 3821–3839.

Krubitzer, L. A., Huffman, K. J., Disbrow, E., & Recanzone, G. H. (2004). Organization of area 3a in macaque monkeys: Contributions to the cortical phenotype. *Journal of Comparative Neurology, 471*, 97–111.

Krubitzer, L. A., & Kaas, J. H. (1990). The organization and connections of somatosensory cortex in marmosets. *Journal of Neuroscience, 10*, 952–974.

Lacquaniti, F., Guigon, E., Bianchi, L., Ferraina, S., & Caminiti, R. (1995). Representing spatial information for limb movement: The role of area 5 in monkey. *Cerebral Cortex, 5*, 391–409.

Ladavas, E., di Pellegrino, G., Farne, A., & Zeloni, G. (1998). Neuropsychological evidence of an integrated visuotactile representation of peripersonal space in humans. *Journal of Cognitive Neuroscience, 10*, 581–589.

Leinonen, L., Hyvarinen, J., Nyman, G., & Linnankoski, I. (1979). I. Functional properties of neurons in lateral part of associative area 7 in awake monkeys. *Experimental Brain Research, 34*, 299–320.

Lemon, R. N., & Griffiths, J. (2005). Comparing the function of the corticospinal system in different species: Organizational differences for motor specialization? *Muscle & Nerve, 32*, 261–279.

Lewis, J., & Van Essen, D. (2000a). Corticocortical connections of visual sensorimotor, multimodal processing areas in the parietal lobe of the macaque monkey. *Journal of Comparative Neurology, 428*, 112–137.

Lewis, J. W., & Van Essen, D. C. (2000b). Mapping of architectonic subdivisions in the macaque monkey, with emphasis on parieto–occipital cortex. *Journal of Comparative Neurology, 428*, 79–111.

Lewis, O. J. (1977). Joint remodelling and the evolution of the human hand. *Journal of Anatomy, 123*, 157–201.

Longo, M. R., Pernigo, S., & Haggard, P. (2011). Vision of the body modulates processing in primary somatosensory cortex. *Neuroscience Letters, 489*, 159–163.

Mannu, M., & Ottoni, E. B. (2009). The enhanced tool-kit of two groups of wild bearded capuchin monkeys in the Caatinga: Tool making, associative use, and secondary tools. *American Journal of Primatology, 71*, 242–251.

Marzke, M. W. (1997). Precision grips, hand morphology, and tools. *American Journal of Physical Anthropology, 102*, 91–110.

Meftah, E. M., Shenasa, J., & Chapman, C. E. (2002). Effects of a cross-modal manipulation of attention on somatosensory cortical neuronal responses to tactile stimuli in the monkey. *Journal of Neurophysiology, 88*, 3133–3149.

Merchant, H., Battaglia-Mayer, A., & Georgopoulos, A. P. (2001). Effects of optic flow in motor cortex and area 7a. *Journal of Neurophysiology, 86*, 1937–1954.

Mountcastle, V. B., Lynch, J. C., Georgopoulos, A., Sakata, H., & Acuña, C. (1975). Posterior parietal association cortex of the monkey: Command functions for operations within extrapersonal space. *Journal of Neurophysiology, 38*, 871–908.

Murata, A., Gallese, V., Kaseda, M., & Sakata, H. (1996). Parietal neurons related to memory-guided hand manipulation. *Journal of Neurophysiology, 75*, 2180–2186.

Murata, A., Gallese, V., Luppino, G., Kaseda, M., & Sakata, H. (2000). Selectivity for the shape, size, and orientation of objects for grasping in neurons of monkey parietal area AIP. *Journal of Neurophysiology, 83*, 2580–2601.

Nelson, R. J., Sur, M., Felleman, D. J., & Kaas, J. H. (1980). Representations of the body surface in postcentral parietal cortex of *Macaca fascicularis. Journal of Comparative Neurology, 192*, 611–643.

Orban, G. A. (2008). Higher order visual processing in macaque extrastriate cortex. *Physiological Reviews, 88*, 59–89.

Padberg, J., Disbrow, E., & Krubitzer, L. (2005). The organization and connections of anterior and posterior parietal cortex in titi monkeys: Do New World monkeys have an area 2? *Cerebral Cortex, 15*, 1938–1963. doi:10.1093/cercor/bhi071.

Padberg, J., Franca, J. G., Cooke, D. F., Soares, J. G., Rosa, M. G., Fiorani, M., Jr., et al. (2007). Parallel evolution of cortical areas involved in skilled hand use. *Journal of Neuroscience, 27*, 10106–10115.

Pandya, D. N., & Seltzer, B. (1982). Intrinsic connections and architectonics of posterior parietal cortex in the rhesus monkey. *Journal of Comparative Neurology, 204*, 196–210.

Pare, M., Smith, A. M., & Rice, F. L. (2002). Distribution and terminal arborizations of cutaneous mechanoreceptors in the glabrous finger pads of the monkey. *Journal of Comparative Neurology, 445*, 347–359.

Patel, G. H., Shulman, G. L., Baker, J. T., Akbudak, E., Snyder, A. Z., Snyder, L. H., et al. (2010). Topographic organization of macaque area LIP. *Proceedings of the National Academy of Sciences of the United States of America, 107*, 4728–4733. doi:10.1073/pnas.0908092107.

Peeters, R., Simone, L., Nelissen, K., Fabbri-Destro, M., Vanduffel, W., Rizzolatti, G., et al. (2009). The representation of tool use in humans and monkeys: Common and uniquely human features. *Journal of Neuroscience, 29*, 11523–11539.

Pons, T. P., Garraghty, P. E., Cusick, C. G., & Kaas, J. H. (1985). The somatotopic organization of area 2 in macaque monkeys. *Journal of Comparative Neurology, 241*, 445–466.

Pons, T. P., & Kaas, J. H. (1985). Connections of area 2 of somatosensory cortex with the anterior pulvinar and subdivisions of the ventroposterior complex in macaque monkeys. *Journal of Comparative Neurology, 240*, 16–36.

Preuss, T. M., & Goldman-Rakic, P. S. (1991). Architectonics of the parietal and temporal association cortex in the strepsirhine primate galago compared to the anthropoid primate Macaca. *Journal of Comparative Neurology, 310*, 475–506.

Qi, H. X., Lyon, D. C., & Kaas, J. H. (2002). Cortical and thalamic connections of the parietal ventral somatosensory area in marmoset monkeys (*Callithrix jacchus*). *Journal of Comparative Neurology, 443*, 168–182.

Rincon-Gonzalez, L., Buneo, C. A., & Helms Tillery, S. I. (2011). The proprioceptive map of the arm is systematic and stable, but idiosyncratic. *PLoS ONE, 6*, e25214. doi:10.1371/journal.pone.0025214.

Rothemund, Y., Qi, H. X., Collins, C. E., & Kaas, J. H. (2002). The genitals and gluteal skin are represented lateral to the foot in anterior parietal somatosensory cortex of macaques. *Somatosensory & Motor Research, 19*, 302–315.

Rozzi, S., Calzavara, R., Belmalih, A., Borra, E., Gregoriou, G. G., Matelli, M., et al. (2006). Cortical connections of the inferior parietal cortical convexity of the macaque monkey. *Cerebral Cortex, 16*, 1389–1417. doi:10.1093/cercor/bhj076.

Rozzi, S., Ferrari, P. F., Bonini, L., Rizzolatti, G., & Fogassi, L. (2008). Functional organization of inferior parietal lobule convexity in the macaque monkey: Electrophysiological characterization of motor, sensory and mirror responses and their correlation with cytoarchitectonic areas. *European Journal of Neuroscience, 28*, 1569–1588.

Sakata, H., Taira, M., Murata, A., & Mine, S. (1995). Neural mechanisms of visual guidance of hand action in the parietal cortex of the monkey. *Cerebral Cortex, 5*, 429–438.

Sakata, H., Takaoka, Y., Kawarasaki, A., & Shibutani, H. (1973). Somatosensory properties of neurons in the superior parietal cortex (area 5) of the rhesus monkey. *Brain Research, 64*, 85–102.

Schrauf, C., Huber, L., & Visalberghi, E. (2008). Do capuchin monkeys use weight to select hammer tools? *Animal Cognition, 11*, 413–422.

Seelke, A. M., Padberg, J. J., Disbrow, E., Purnell, S. M., Recanzone, G., & Krubitzer, L. (2012). Topographic maps within Brodmann's area 5 of macaque monkeys. *Cerebral Cortex, 22*, 1834–1850.

Seltzer, B., & Pandya, D. N. (1980). Converging visual and somatic sensory cortical input to the intraparietal sulcus of the rhesus monkey. *Brain Research, 192*, 339–351.

Seltzer, B., & Pandya, D. N. (1986). Posterior parietal projections to the intraparietal sulcus of the rhesus monkey. *Experimental Brain Research, 62*, 459–469.

Siegel, R. M., & Read, H. L. (1997). Analysis of optic flow in the monkey parietal area 7a. *Cerebral Cortex, 7*, 327–346.

Snyder, L. H., Batista, A. P., & Andersen, R. A. (1997). Coding of intention in the posterior parietal cortex. *Nature, 386*, 167–170.

Snyder, L. H., Batista, A. P., & Andersen, R. A. (2000). Intention-related activity in the posterior parietal cortex: A review. *Vision Research, 40*, 1433–1441.

Srivastava, S., Orban, G. A., De Maziere, P. A., & Janssen, P. (2009). A distinct representation of three-dimensional shape in macaque anterior intraparietal area: Fast, metric, and coarse. *Journal of Neuroscience, 29*, 10613–10626.

Stepniewska, I., Collins, C. E., & Kaas, J. H. (2005). Reappraisal of DL/V4 boundaries based on connectivity patterns of dorsolateral visual cortex in macaques. *Cerebral Cortex, 15*, 809–822.

Stepniewska, I., Fang, P. C., & Kaas, J. H. (2005). Microstimulation reveals specialized subregions for different complex movements in posterior parietal cortex of prosimian galagos. *Proceedings of the National Academy of Sciences of the United States of America, 102*, 4878–4883. doi:10.1073/pnas.0501048102.

Stepniewska, I., Fang, P. C., & Kaas, J. H. (2009). Organization of the posterior parietal cortex in galagos: I. Functional zones identified by microstimulation. *Journal of Comparative Neurology, 517*, 765–782.

Stepniewska, I., Preuss, T. M., & Kaas, J. H. (2006). Ipsilateral cortical connections of dorsal and ventral premotor areas in New World owl monkeys. *Journal of Comparative Neurology, 495*, 691–708.

Taira, M., Mine, S., Georgopoulos, A. P., Murata, A., & Sakata, H. (1990). Parietal cortex neurons of the monkey related to the visual guidance of hand movement. *Experimental Brain Research, 83*, 29–36.

Taoka, M., Toda, T., & Iwamura, Y. (1998). Representation of the midline trunk, bilateral arms, and shoulders in the monkey postcentral somatosensory cortex. *Experimental Brain Research, 123*, 315–322.

Taylor-Clarke, M., Kennett, S., & Haggard, P. (2002). Vision modulates somatosensory cortical processing. *Current Biology, 12*, 233–236.

Tillery, S. I., Flanders, M., & Soechting, J. F. (1994). Errors in kinesthetic transformations for hand apposition. *Neuroreport, 6*, 177–181.

Valyear, K. F., Cavina-Pratesi, C., Stiglick, A. J., & Culham, J. C. (2007). Does tool-related fMRI activity within the intraparietal sulcus reflect the plan to grasp? *NeuroImage, 36*(Suppl 2), T94–T108.

Vingerhoets, G., Acke, F., Vandemaele, P., & Achten, E. (2009). Tool responsive regions in the posterior parietal cortex: Effect of differences in motor goal and target object during imagined transitive movements. *NeuroImage, 47*, 1832–1843. doi:10.1016/j.neuroimage.2009.05.100.

Visalberghi, E., Addessi, E., Truppa, V., Spagnoletti, N., Ottoni, E., Izar, P., et al. (2009). Selection of effective stone tools by wild bearded capuchin monkeys. *Current Biology, 19*, 213–217.

Visalberghi, E., Fragaszy, D., Ottoni, E., Izar, P., de Oliveira, M. G., & Andrade, F. R. (2007). Characteristics of hammer stones and anvils used by wild bearded capuchin monkeys (*Cebus libidinosus*) to crack open palm nuts. *American Journal of Physical Anthropology, 132*, 426–444.

Vogt, C., & Vogt, O. (1919). Allgemeinere Ergebnisse unserer Hirnforschung. *Journal für Psychologie und Neurologie, 25*, 292–398.

Von Bonin, G., & Bailey, P. (1947). *The neocortex of Macaca mulatta. (Illinois Monographs in the Medical Sciences, 5)*. Champaign, IL: University of Illinois Press.

von Economo, C. (1929). *The cytoarchitectonics of the cerebral cortex*. London: Oxford University Press.

Wise, S. P., Boussaoud, D., Johnson, P. B., & Caminiti, R. (1997). Premotor and parietal cortex: Corticocortical connectivity and combinatorial computations. *Annual Review of Neuroscience, 20*, 25–42.

Wong, P., & Kaas, J. H. (2009). Architectonic subdivisions of neocortex in the tree shrew (*Tupaia belangeri*). *Anatomical Record (Hoboken, N.J.), 292*, 994–1027.

Wong, P., & Kaas, J. H. (2010). Architectonic subdivisions of neocortex in the galago (*Otolemur garnetti*). *Anatomical Record (Hoboken, N.J.), 293*, 1033–1069.

Yalachkov, Y., Kaiser, J., & Naumer, M. J. (2009). Brain regions related to tool use and action knowledge reflect nicotine dependence. *Journal of Neuroscience, 29*, 4922–4929.

Yokochi, H., Tanaka, M., Kumashiro, M., & Iriki, A. (2003). Inferior parietal somatosensory neurons coding face–hand coordination in Japanese macaques. *Somatosensory & Motor Research, 20*, 115–125.

74 Auditory–Visual Interactions

CHARLES SPENCE

The last few years have seen a rapid growth of interest in the study of multisensory perception, with the majority of research focusing on the nature of those interactions taking place between auditory and visual stimuli. In this chapter, the latest evidence concerning the key factors underlying the multisensory integration of, and interaction between, these senses in neurologically normal adult humans is reviewed. The large body of research concerning the role of spatiotemporal coincidence is briefly summarized. Additionally, the evidence that is now starting to emerge supporting the role of a number of other factors, including semantic congruency, crossmodal (or synesthetic) correspondence, the unity effect, the correlation between signals, Gestalt grouping, and attention in audiovisual integration, is highlighted. Our current understanding concerning the neural substrates and consequences of auditory–visual interactions is also reviewed, and, finally, a number of directions for future research signposted.

MULTISENSORY INTEGRATION

There has been a huge growth of interest in the nature of, and neural underpinnings underlying, crossmodal interactions and multisensory integration over the last 35 years or so (see Bremner, Lewkowicz, & Spence, 2012; Calvert, Spence, & Stein, 2004; Lewkowicz & Lickliter, 1994; Spence & Driver, 2004; Stein, 2012, for comprehensive reviews of the field). The majority of this research has tended to focus on those interactions taking place between auditory and visual stimuli. In part, this emphasis can be attributed to the rather mundane reason that it is simply much easier to present such signals under computer control in the psychophysics laboratory, not to mention in the scanner environment, than it is to present other combinations of sensory (e.g., involving tactile) stimuli. That said, the rules of multisensory integration that have now been documented for the audiovisual case likely also apply when it comes to the integration of other combinations of sensory inputs from the so-called spatial senses (e.g., comprising the modalities of vision, audition, and touch). This chapter is designed to provide an up-to-date review of the literature detailing the most important factors affecting the integration of auditory and visual signals.

At the outset, it is important to distinguish between two similar-sounding (and often interchangeably used) terms, namely *integration* and *interaction* (see also Spence & Bayne, forthcoming; Stein et al., 2010). The multisensory integration of auditory and visual signals carries with it (at least to certain researchers) the idea that the component unisensory signals are bound together in the neural representation of one and the same external object or event. By contrast, audiovisual interactions can be broadly defined in terms of those effects/phenomena whereby the presentation of a stimulus in one sensory modality (be it audition or vision in the context of the present chapter) has an impact on the processing of a stimulus in the other sensory modality (vision or audition, respectively). One can think of audiovisual integration, then, as constituting a subset of all audiovisual interactions. The focus of the present chapter will primarily be on audiovisual integration effects, which occur over a narrower range of stimulus onset asynchronies (SOAs) than do audiovisual interaction effects.

Prototypical examples of audiovisual integration include the binding of auditory and visual speech signals (e.g., Burnham, 1998; Dodd & Campbell, 1994; McGurk & MacDonald, 1976) and the integration of semantically related meaningful auditory and visual stimuli, such as the sound of a barking dog with a simultaneously presented image of a dog (e.g., Chen & Spence, 2010; Doehrmann & Naumer, 2008; Laurienti et al., 2004; Molholm et al., 2004; Naumer & Kaiser, 2010). By contrast, crossmodal alerting (Posner, 1978) and/or attentional capture effects (Spence & Driver, 2004; Wright & Ward, 2008) would appear to represent reasonably clear-cut examples of crossmodal interaction (if not necessarily of multisensory integration). Note, however, that in both of the latter two cases (e.g., crossmodal alerting and attentional orienting), the largest behavioral effects tend to occur when the auditory and visual signals are separated by a few hundred milliseconds (rather than being presented simultaneously). Crossmodal interactions between auditory and visual signals appear to be somewhat asymmetrical, with auditory signals giving rise to larger crossmodal alerting effects than visual signals (see Posner, 1978). Many researchers have also argued that auditory signals may give rise to somewhat more pronounced crossmodal attentional capture effects than do visual signals (e.g., Spence & Driver, 2004).

Part of the reason for the latter asymmetry may, in fact, be the increased localizability of the visual, as compared to auditory, cues used in many of the studies that have been published to date in this area (see Prime et al., 2008; Spence, McDonald, & Driver, 2004; Wright & Ward, 2008). Crossmodal attentional cuing effects resulting from the presentation of a spatially nonpredictive auditory cue have now been shown to influence early visual processing in a number of ways: They enhance the apparent contrast of visual targets, with changes in perceived contrast correlating positively with enlargements of the contralateral neural response (P1 and N1) starting within 100 ms of stimulus onset (Störmer, McDonald, & Hillyard, 2009); crossmodal spatial attentional cuing can also speed up the neural processing of visual target stimuli (a phenomenon known as prior entry; see Spence & Parise, 2010, for a review; though see also McDonald et al., 2005). That said, there is quite some debate about whether such effects are best conceptualized in terms of crossmodal spatial attention or in terms of multisensory integration (see Bolognini et al., 2005; McDonald, Teder-Sälejärvi, & Ward, 2001). It may well be that both accounts provide satisfactory explanations for the data that have been collected by various researchers at different SOAs and using somewhat different experimental paradigms.

CROSSMODAL FACILITATION

It is not always easy to determine unequivocally whether a particular behavioral effect, such as, for example, the frequently documented facilitatory effect resulting from the presentation of a task-irrelevant (and likely spatially nonpredictive) auditory cue on visual target detection/discrimination performance actually reflects an example of audiovisual multisensory integration rather than, say, some other kind of crossmodal interaction (see Spence & Ngo, 2012, for a review). So, for example, let's take Stein et al.'s (1996) early demonstration that the simultaneous presentation of a task-irrelevant tone led to a briefly presented visual stimulus being judged subjectively as brighter than when the visual stimulus was presented in silence. This result has often been taken to represent an example of multisensory integration in awake humans paralleling the neurophysiological enhancement effects documented in numerous studies of cellular responses (e.g., in the superior colliculus) in anesthetized animals (e.g., see Stein & Meredith, 1993; Stein & Stanford, 2008, for reviews). It is, however, important to note that there

are a number of alternative explanations for such results.

Marks and his colleagues, for example, have argued that many such examples of what may look at first like crossmodal facilitation may instead reflect nothing more than response bias (Arieh & Marks, 2008; Odgaard, Arieh, & Marks, 2003). That said, a number of more recent demonstrations of crossmodal facilitation have managed to rule out such an interpretation. Another plausible alternative explanation for such auditory–visual interactions is that the sound somehow increases the saliency, or perceived contrast, of the simultaneously presented visual event (e.g., Noesselt et al., 2008). Indeed, Noesselt et al. (2010) have shown that such crossmodal effects can affect the neural representation of visual stimuli in early visual areas in the occipital lobe, as well as in multisensory superior temporal sulcus, and may result, at least in part, from neural interactions taking place in the thalamic body (specifically, enhanced coupling between the lateral and medial geniculate bodies). Meanwhile other researchers have argued that the presentation of a task-irrelevant sound may reduce the temporal uncertainty concerning when a visual target has been presented (e.g., Lippert, Logothetis, & Kayser, 2007). Such an account may be particularly relevant under those conditions in which the visual target appears in among a stream of distractors (e.g., as in a rapid serial visual presentation stream; e.g., Chen & Yeh, 2009; Olivers & Van der Burg, 2008; Vroomen & de Gelder, 2000). One suggestion here is that the sound may provide a temporal marker that participants can use when allocating their attention across time. What appears increasingly clear from the latest research in this area is that transient signals may play a particularly important role in determining auditory–visual interactions (see Van der Burg et al., 2010; see also Andersen & Mamassian, 2008).

THE INTERSENSORY CONFLICT SITUATION

Researchers interested in assessing the relative contribution that auditory and visual stimuli make to perception have often used variants of the "intersensory conflict paradigm." In a typical study, participants are presented with incongruent, or discrepant, stimuli to eye and ear, and the question of interest is to try and determine how the human nervous system resolves that controversy (i.e., does the visual or auditory signal dominate the ensuing percept?). Traditionally, the stimuli presented in intersensory conflict research have been presented at a clearly suprathreshold level. Below, I will review the cognitive neuroscience evidence relevant to understanding three of the most frequently studied intersensory conflict situations. They are (1) the ventriloquism effect (Bertelson & de Gelder, 2004), (2) the McGurk effect (McGurk & MacDondald, 1976), and (3) the two-flash illusion (Shams, Kamitani, & Shimojo, 2000). These illusions provide examples of visual dominance, emergence, and auditory dominance, respectively. While all three illusions were originally documented in the psychophysics laboratory, cognitive neuroscientists have now started to uncover distinct neural underpinnings for these auditory–visual interactions.

THE VENTRILOQUISM EFFECT

One of the best-known examples of the dominance of vision over audition occurs when a spatial conflict is introduced between the location from which a pair of auditory and visual stimuli are presented. For example, in the cinema, we typically perceive the voices of the actors as originating from their lips on the screen (this despite the fact that the sounds are actually presented from loudspeakers situated elsewhere in the auditorium). In this illusion, known as the ventriloquism effect (see Bertelson & de Gelder, 2004, for a review), people tend to mislocalize sounds toward their apparent visual source. The majority of early studies of the ventriloquism effect tended to find that spatially discrepant sounds normally have very little, if any, effect on an observer's visual localization responses, thus leading to the widespread suggestion that vision tends to dominate our perception whenever audiovisual spatial information conflicts. Spatial ventriloquism turns out to be particularly strong in the case of audiovisual speech (presumably due to the highly correlated nature of the auditory and visual inputs resulting from the rich temporal structure inherent to speech stimuli; see below) although similar intersensory biasing effects have now been reported for a wide variety of spatially discrepant audiovisual stimuli, ranging from the sight and sound of whistling steaming kettles through to pure tone beeps paired with spatially incongruent flashes of light (e.g., Bertelson & Aschersleben, 1998; Jackson, 1953; Recanzone, 1998; Slutsky & Recanzone, 2001). What is more, repeated exposure to spatially discrepant auditory and visual stimuli can rapidly give rise to aftereffects in the localization of the component (usually auditory) stimuli, effects that can be seen in the primary auditory cortex (see also Recanzone, 2009).

It turns out that the spatial ventriloquism effect actually reflects the near-optimal statistical integration of the component auditory and visual signals. While vision normally dominates audition in the spatial domain, Alais and Burr (2004) elicited a shift from visual to

auditory dominance simply by blurring the visual stimulus, hence making its localization less reliable as compared to that of the sound. Such findings are consistent with the claim that the inputs from the different senses are combined by the human brain in order to minimize the variability associated with the final multisensory perceptual estimate (though see also Battaglia, Jacobs, & Aslin, 2003; Roach, Heron, & McGraw, 2006). Maximum-likelihood estimation (MLE) provides a mathematical formalization of the classic idea of "modality appropriateness" which has been debated in the field for decades (see Welch & Warren, 1980, 1986). Importantly, however, the MLE approach to multisensory integration also allows one to predict how a particular pair of auditory and visual signals will interact simply by estimating the variability associated with each of the individual signals (see also Heron, Whitaker, & McGraw, 2004).

Cognitive neuroscientists have studied the neural underpinnings underlying the ventriloquism effect (e.g., Bonath et al., 2007; Macaluso et al., 2004). For example, Bonath and colleagues conducted a combined event-related functional magnetic resonance imaging (fMRI) and event-related potential (ERP) study. They demonstrated, on a trial-by-trial basis, that a precisely timed biasing of the left–right balance of auditory cortex activity by spatially mismatching visual events (presented on either the left or the right side) correlated with the magnitude of the ventriloquism illusion. Specifically, on those trials in which the presentation of the visual stimulus elicited a spatial ventriloquism effect (i.e., when the perceived location of the sound was "pulled" toward the side on which the visual stimulus was presented), a relative reduction in the amplitude of the response, in both the blood-oxygen-level dependent (BOLD) and ERP signals, in the planum temporale (PT) of the hemisphere ipsilateral to the visual stimulus was observed. This gave rise to a relative enhancement in the response of the PT contralateral to the side of the shifted auditory percept. Note here that a similar asymmetrical neural response was also observed when sounds were actually presented on that side. That said, this audiovisual neural interaction was only observed in a relatively narrow time window (230–270 ms) after stimulus onset. The latter observation led to the further suggestion that this particular illusion-related asymmetry, attributable to vision's influence on auditory spatial processing, was likely mediated by pathways from visual cortex to multisensory association areas and from there back to auditory cortex.

It is interesting to note that while the majority of early research on the spatial ventriloquism effect tended to use stationary auditory and visual stimuli, recent studies have increasingly started to use dynamic (i.e., moving) stimuli instead (see Soto-Faraco et al., 2002). The results of numerous studies conducted with such stimuli over the last decade or so have demonstrated that visual motion (no matter whether it is apparent or real) also tends to dominate over (i.e., ventriloquize) the perceived direction of movement of auditory stimuli (e.g., Meyer et al., 2005; Sanabria, Spence, & Soto-Faraco, 2007; see also Freeman & Driver, 2008; Teramoto, Hidaka, & Sugita, 2010). As with many of the interactions that have been observed between auditory and visual stimuli, the largest behavioral effects tend to occur when the individual stimuli are individually only weakly effective (Wuerger, Hofbauer, & Meyer, 2003) and when they are presented from the same locations (Meyer et al., 2005; Soto-Faraco et al., 2002). However, that said, one should not necessarily jump to the conclusion that such results automatically provide support for the inverse effectiveness account of multisensory integration (see Holmes & Spence, 2005; Ross et al., 2007a; Stein & Meredith, 1993).

Researchers have recently started to investigate the neural correlates underlying crossmodal dynamic capture as well. Alink, Singer, and Muckli (2008), for example, observed a BOLD signal increase in the visual motion areas (hMT/V5+), together with a deactivation of auditory motion areas (the auditory motion complex) on those incongruent motion direction trials in which the participants experienced crossmodal dynamic capture (i.e., when the sound appeared to move in the direction of the visual stimulus even though it did not). This change in neural activity was seen bilaterally. The occurrence of crossmodal dynamic capture was associated with an increase of activation in the ventral intraparietal sulcus.

It should, of course, be borne in mind that the presentation of a single auditory and visual event in the majority of published studies of the spatial ventriloquism effect (not to mention in studies of the crossmodal dynamic capture effect) dramatically oversimplifies the multisensory binding (or pairing) problem that we typically face in the majority of real-world settings. For, outside of the average psychophysics laboratory, our senses tend to be bombarded by multiple sensory inputs at more or less the same time. As a result, trying to determine which of the many visual stimuli should be bound with which of the many auditory stimuli that may be available simultaneously soon turns into a nontrivial problem. Computational modelers have recently started to address this question by incorporating various coupling priors into their Bayesian decision theory models (cf. Beierholm, Quartz, & Shams, 2009; Körding et al., 2007).

THE MCGURK EFFECT

McGurk and MacDonald (1976) conducted a seminal study in which they assessed the relative contributions of auditory and visual inputs to speech perception. The participants in their study heard a voice uttering a particular syllable (e.g., "ba") while they simultaneously saw the synchronized lip movements associated with a speaker uttering another syllable (e.g., "ga"). Most of the participants reported hearing a syllable (e.g., "da") that was different from what had been presented in either sensory modality. In other words, the multisensory integration of nonredundant auditory and visual information resulted in the emergence of a qualitatively new percept. McGurk and McDonald's results demonstrate that what people *hear* can be modulated by the lip movements that they happen to *see*. The fact that the McGurk effect occurs even when people are perfectly aware that what they are hearing and seeing is incongruent suggests that this form of audiovisual integration may occur fairly automatically (though see Alsius et al., 2005; Walker, Bruce, & O'Malley, 1995).

In terms of the neural underpinnings of such audiovisual interactions, a number of neuroimaging studies have now demonstrated that the auditory cortex (even possibly A1; particularly on the left side) is activated not only by hearing speech sounds but also by watching lip movements and other facial articulatory gestures (e.g., Calvert et al., 1997; Pekkola et al., 2005). Taken together, such findings suggest the existence of modulatory signals on sound processing that originate, either directly or indirectly, from visual areas (see also Ghazanfar, Chandrasekaran, & Logothetis, 2008).

Similar audiovisual interactions have now also been reported in a variety of other nonspeech domains: so, for example, Saldaña and Rosenblum (1993) observed a similar visual biasing of auditory perception when participants, who all happened to be trained musicians, observed plucking and bowing movements on a string instrument (see also Schutz & Kubovy, 2009). Meanwhile, de Gelder and her colleagues have conducted a number of studies over the past decade or so demonstrating that people's perception of the emotional quality of a human voice can also be modified by affective visual information that happens to be presented at the same time (e.g., de Gelder & Vroomen, 2000a, b; Petrini, McAleer, & Pollick, 2010; though see also Massaro & Cohen, 2000).

In terms of audiovisual interactions in speech perception there is also an extensive body of research demonstrating the enhancement in signal-to-noise that can result when congruent lip movements are presented while participants are listening to speech in noise (e.g.,

Bernstein, Auer, & Takayanagi, 2004; Risberg & Lubker, 1978; Ross et al., 2007a; Sumby & Pollack, 1954). Interestingly, such crossmodal facilitation effects appear to be maximal at intermediate signal-to-noise levels (see also Manjarrez et al., 2007). Once again, it turns out that such behavioral results can be accounted for quite nicely by Bayesian models of optimal cue integration (see Ma et al., 2009).

THE TWO-FLASH ILLUSION

The two-flash illusion (Shams, Kamitani, & Shimojo, 2000) provides one of the rare examples of auditory dominance. In a typical study, participants have to report how many visual stimuli they see flashed in the periphery (normally somewhere between one and three). At the same time, an auditory stimulus, consisting of various numbers of beeps, is presented. The key result is that people often report seeing more flashes than were physically presented. So, for example, if a single flash is paired with two beeps, participants will likely report having seen two sequentially presented flashes of light (Innes-Brown & Crewther, 2009).

On those trials in which two sounds are presented together with a single peripheral visual event, participants sometimes report seeing a single flash (i.e., their report is veridical) while on other trials they report seeing an illusory double flash instead. The pattern of neural activation in early visual cortex has been shown to differ significantly between these two kinds of trial, thus suggesting once again that the audiovisual interaction underlying this example of auditory dominance can influence early visual processing (Watkins et al., 2007). Just as for many of the other auditory–visual interaction effects that have been documented thus far, the reliability of the individual sensory signals appears to determine which sense dominates under conditions of intersensory conflict (see Andersen, Tiippana, & Sams, 2004, 2005). Bayesian decision theory does a reasonably good job of accommodating the available psychophysical data that have been collected on this illusion (e.g., Andersen, Tiippana, & Sams, 2005; Wozny, Beierholm, & Shams, 2008).

INTERIM SUMMARY

The results of the research that has been reviewed thus far are consistent with the view that spatiotemporal coincidence constitutes an important factor underlying the multisensory integration of auditory and visual signals (e.g., Meienbrock et al., 2007; Meyer et al., 2005; Soto-Faraco et al., 2002; Van Wassenhove, Grant, & Poeppel, 2007). Our brain tries to integrate (or bind) the auditory and visual signals that it "believes" belong together (of course, determining what "belongs together" under stimulus presentation conditions that are complex, or crowded, requires the brain to make a number of further assumptions; see below). Both spatial and temporal ventriloquism effects appear to help realign sensory signals that, for whatever reason, have fallen out of spatiotemporal alignment (see Bertelson & de Gelder, 2004; Morein-Zamir, Soto-Faraco, & Kingstone, 2003; Parise & Spence, 2009; Recanzone, 2009; Slutsky & Recanzone, 2001; Van Wassenhove, Grant, & Poeppel, 2005).

That said, and contrary to what one often reads (e.g., see Meienbrock et al., 2007), it should be noted that for certain tasks, the relative spatial position from which the component auditory and visual signals are presented (i.e., same vs. different position) appears to have *no* impact whatsoever on the pattern of multisensory integration effects that are observed (see Innes-Brown & Crewther, 2009; Jones & Jarick, 2006; Recanzone, 2003; Regan & Spekreijse, 1977, for representative examples from the literature; see Spence, 2013, for a review). If one were to try and summarize the findings that have been published to date, it appears as though in the relatively uncluttered environments in which most psychophysical studies are normally conducted, spatial coincidence will enhance auditory–visual interactions under those conditions in which participants have to make a response that is in some sense spatial (see Spence, 2012; Spence & Driver, 2004). This might, for example, involve the participant making a speeded eye movement in the direction of the target (Frens, Van Opstal, & Van Der Willigen, 1995; Harrington & Peck, 1998) or perhaps discriminating the elevation of a target using a manual button-press response (Spence & Driver, 2004). By contrast, in those studies in which the participant is required to identify the target (e.g., when engaged in audiovisual speech perception), or when making some sort of response regarding the temporal properties of the stimulus (e.g., judging the rate or numerosity of stimuli presented), the brain appears to integrate the component auditory and visual signals just as effectively no matter whether the stimuli happen to be presented from the same spatial position or not (see Spence, 2012, 2013).

Now it is important to note that the story here might change under more realistic conditions whereby multiple signals (some relevant and others not) are presented simultaneously from different spatial positions. At present, though, the available evidence is most consistent with the claim that spatial coincidence (or relative spatial position) plays a surprisingly modest role, if any, in modulating audiovisual integration in nonspatial tasks. That said, different brain areas seem to be recruited as a function of whether auditory and visual stimuli are presented from the same location or not. For example, Meienbrock et al. (2007) conducted an fMRI study in which participants were presented with semantically meaning-congruent pairs of auditory and visual stimuli from either the same or different sides of fixation. The results suggested that whenever a spatial mismatch was detected between the location of the two stimuli in brain regions such as the intraparietal sulcus, neural processing resources may be redirected to low-level visual areas in order, presumably, to resolve the conflict.

Of course, such an apparent dissociation may lead to the suggestion that the idea of "what" (or "how") and "where" processing pathways, originally associated with the ventral and dorsal visual streams, may have some explanatory mileage when put in the context of multisensory integration (e.g., see Chan & Newell, 2008; Renier et al., 2009). One relevant trend that has started to emerge here in recent years is for researchers to try and determine whether audiovisual interactions may be preferentially related to either the magnocellular or parvocellular pathways (see Jaekl & Harris, 2009; Jaekl & Soto-Faraco, 2010; see also Maravita et al., 2008).

The mathematical approach embodied by Bayesian decision theory has been successfully used to model many of the auditory–visual interactions that have been documented to date (e.g., see Alais & Burr, 2004; Andersen, Tiippana, & Sams, 2005; Battaglia, Jacobs, & Aslin, 2003; Beierholm, Quartz, & Shams, 2009; Heron, Whitaker, & McGraw, 2004; Körding et al., 2007; Roach, Heron, & McGraw, 2006; Wozny, Beierholm, & Shams, 2008). It is currently unclear whether the Bayesian approach negates the need to invoke attention in terms of accounting for the range of auditory–visual interactions that have been documented to date. It may, of course, be that a small role for attention can be modeled in terms of a residual bias (or prior) that simply weights the information from one modality more highly (Andersen, Tiippana, & Sams, 2004; Battaglia, Jacobs, & Aslin, 2003; though see Witten & Knudsen, 2005).

SEMANTIC AND SYNESTHETIC CONGRUENCY

Over the last decade or so, a number of researchers have demonstrated that the degree of semantic congruency between simultaneously presented auditory and visual stimuli modulates multisensory integration (e.g., Chen & Spence, 2010, 2011; Molholm et al., 2004; Taylor et al., 2006). So, for example, several studies have now shown that people tend to respond to a visual target (e.g., the outline picture of a dog), say, more

rapidly and accurately when it happens to be presented together with a task-irrelevant and uninformative congruent sound (the sound of a dog barking) rather than an incongruent semantically meaningful sound (e.g., the sound of a cat meowing; e.g., Hein et al., 2007; Molholm et al., 2004). Furthermore, Chen and Spence (2010, 2011) have recently demonstrated that the presentation of a semantically congruent sound can also enhance the detectability of masked visual targets in an unspeeded signal detection study (i.e., the presentation of the task-irrelevant sound gave rise to a significant increase in d').

The last couple of years have also seen growing research interest in the role of crossmodal (or synesthetic) correspondence in auditory–visual integration (e.g., Evans & Treisman, 2010; Gallace & Spence, 2006; Parise & Spence, 2009; see Spence, 2011, for a review). So, for instance, Parise and Spence reported a series of experiments showing increased spatiotemporal integration (e.g., a larger spatial and temporal ventriloquism effect) when pairs of stimuli that matched crossmodally were presented than when pairs of mismatching stimuli were presented instead: Larger circles were preferentially bound with sounds having a lower pitch whereas smaller circles were preferentially bound with sounds having a higher pitch. Over the years, researchers have documented a number of different correspondences between auditory and visual stimuli, such as between auditory pitch and visual brightness, lightness, size, angularity, spatial frequency, and elevation (Spence, 2011). Building on Parise and Spence's work, Bien et al. (2012) have now demonstrated that the crossmodal correspondence between auditory pitch and visual size modulated the N1 and P1 ERP components. What is more, the magnitude of this modulation was well correlated with the size of the ventriloquism effect documented behaviorally. This study also showed that this crossmodal correspondence could be eliminated by applying transcranial magnetic stimulation (TMS) over the right intraparietal cortex.

It is, though, important to note that different crossmodal correspondences (e.g., statistical vs. semantic; see Spence, 2011) may be associated with different neural substrates. Hence further neuroimaging research will likely prove particularly helpful in disentangling these different kinds of correspondence that may have very similar effects on a participant's behavior (see Sadaghiani, Maier, & Noppeney, 2009). Such learned associations between auditory and visual stimuli may then subsequently facilitate the spread of attention across audiovisual sensory attributes (e.g., Busse et al., 2005; Fiebelkorn, Foxe, & Molholm, 2010). Crossmodal correspondences can be modeled in terms of Bayesian coupling priors. Such priors are needed in order to predict which sensory inputs will be bound together under conditions where multiple stimuli happen to be presented simultaneously and, hence, to help the brain solve the crossmodal binding problem (Beierholm, Quartz, & Shams, 2009).

THE UNITY EFFECT

It has long been argued that whenever two or more sensory inputs are perceived as being highly consistent (i.e., as being related in a way that they appear to "go together"), observers will be more likely to treat them as referring to a single audiovisual event (e.g., Welch & Warren, 1980). Consequently, they will be more likely to assume that they have a common spatiotemporal origin, and hence they will be more likely to bind them into a single multisensory perceptual object or event. The question of whether or not the unity assumption plays a significant role in audiovisual integration has proved to be one of the most contentious issues in multisensory research over the last 50 years (Jackson, 1953; see Vatakis & Spence, 2007, for a review). Some of the most robust psychophysical support for the influence of the unity assumption (or effect) on audiovisual integration comes from a series of audiovisual temporal order judgment (TOJ) experiments reported by Vatakis and Spence. The participants in these studies were presented with pairs of auditory and visual speech stimuli (either single syllables or words) at a range of different SOAs. The participants made unspeeded TOJs regarding whether the auditory or visual speech stream started first on each trial. On half of the trials, the auditory and visual speech stimuli were gender matched (i.e., a female face was presented together with the matching female voice) while on the remainder of the trials the auditory and visual speech stimuli were gender mismatched (i.e., the lip movements of a female face were presented together with a man's voice).

The participants found it significantly harder to judge which sensory modality had been presented first (suggesting enhanced multisensory integration) when evaluating the matched speech stimuli than when evaluating the mismatched speech stimuli. The presentation of the matched speech stimuli presumably resulted in more temporal ventriloquism taking place than was the case for the mismatched videos (Morein-Zamir, Soto-Faraco, & Kingstone, 2003). Vatakis and Spence's results therefore provide empirical support for the long-standing claim that the unity assumption can influence the multisensory integration of auditory and visual stimuli. Interestingly, however, such effects do not necessarily always seem to generalize beyond audiovisual speech stimuli (see Vatakis & Spence, 2008; see also Maier, Di Luca, & Noppeney, 2011; Petrini, Russell, & Pollick, 2009; Van Wassenhove, Grant, & Poeppel, 2007; Vroomen & Keetels, 2011). The importance and role of any correlation between the auditory and visual signals for this kind of auditory–visual interaction certainly deserves further consideration (see Parise, Spence, & Ernst, 2012; Van Wassenhove, Grant, & Poeppel, 2007). Again, the unity effect has now been modeled in terms of a prior in Bayesian decision theory (Sato, Toyoizumi, & Aihara, 2007).

GESTALT GROUPING

One factor that hasn't received as much neuroscientific interest as perhaps it might concerns the role of Gestalt grouping principles such as proximity, similarity, common fate, and good continuation on audiovisual interactions. The evidence that is currently available clearly demonstrates that the intramodal grouping of auditory and/or visual signals modulates the strength of audiovisual multisensory integration across a range of behavioral paradigms (e.g., O'Leary & Rhodes, 1984; Vroomen & de Gelder, 2000; see Spence & Chen, 2012, for a review). To give but one example here, the extent to which a sound presented at the moment that two circles coincide elicits the perception of bouncing (rather than streaming; this is referred to as the stream bounce illusion; Sekuler, Sekuler, & Lau, 1997) is itself modulated by whether the auditory stimulus segregates from a stream of background auditory stimuli (Watanabe & Shimojo, 2001). The interested reader is directed to a recent review article for a discussion of the various studies that have now documented the influence of intramodal auditory and/or visual grouping on the nature of auditory–visual interactions (Spence & Chen, 2011).

ATTENTION AND MULTISENSORY INTEGRATION

One of the most hotly debated areas in multisensory research on auditory–visual interactions in recent years has concerned the role of attention on multisensory integration (see Koelewijn, Bronkhorst, & Theeuwes, 2010; Navarra et al., 2009, for reviews). Performance on certain tasks, such as the ventriloquism effect observed in spatial localization tasks with discrepant auditory and visual stimuli, seems relatively immune to manipulations of a participant's attention (Vroomen, Bertelson, & de Gelder, 2001; see Spence, 2012; Eramudugolla et al., 2011). By contrast, performance in other tasks is clearly modulated by attentional manipulations (e.g., Alsius et al., 2005; Mozilic et al., 2008). Part of the answer here may turn out to relate to just how participants' attention was manipulated (e.g., via instruction vs. by manipulating the perceptual load of their task; see Lavie, 2005; Spence, 2009). It may also be the case that multisensory integration occurs more automatically for certain types of stimuli than for others. Certainly, audiovisual integration (e.g., for speech stimuli) has been shown to occur even when it is detrimental to a participant's performance (e.g., Driver, 1996). Currently, neuroscientists are using the various techniques at their disposal in order to try and figure out exactly what the relation between multisensory integration and attention really is (e.g., Busse et al., 2005; Driver & Noesselt, 2008; Senkowski et al., 2008; Talsma, Doty, & Woldorff, 2007; see Talsma et al., 2010, for a review).

AUDITORY–VISUAL INTERACTIONS: CONFLICT

Finally, it is worth noting that auditory–visual interactions can give rise to intersensory *competition* and not just *facilitation*. Such is the case with the Colavita visual dominance effect (Colavita, 1974). In this paradigm, participants are presented with a random sequence of auditory and visual target stimuli requiring a modality discrimination response (i.e., auditory or visual). On a small proportion of trials both the auditory and visual target are presented at the same time and participants are instructed to respond either by pressing both response keys or by pressing a third response key (indicating the presence of a bimodal target). Under such conditions, participants sometimes fail to respond to the auditory component of the bimodal trials (suggesting crossmodal extinction), this despite their responding near-perfectly on the unimodal target trials (see Spence, Parise, & Chen, 2011, for a review). Intriguingly, multisensory facilitation can be turned into crossmodal competition with exactly the same stimulus presentation profile but merely by altering the task demands slightly (e.g., see Sinnett, Soto-Farco, & Spence, 2008). Recently, neuroscientists have started to probe the neural circuits involved in such behavioral effects. Diaconescu, Alain, and McIntosh (2011), for example, have used magnetoencephalography in order to highlight the role of early (i.e., within 100 ms of stimulus onset) posterior parietal activity in multisensory facilitation effects. By contrast, multisensory conflict was found to result in the activation of the cingulate, temporal, and prefrontal cortices.

CONCLUSIONS

As the studies reviewed here have hopefully made clear, a number of factors modulate auditory–visual interactions. So, for example, visual signals are more likely to influence auditory perception if they happen to be presented from the same location (at least in spatial tasks) and at about the same time. Visual signals are also more likely to influence auditory perception if they happen to have a similar temporal structure (Parise, Spence, & Ernst, 2012) whereas intramodal perceptual grouping can reduce the magnitude of any auditory–visual interactions that are seen (Spence & Chen, 2012). Semantic congruency, crossmodal (or synesthetic) correspondence, and the unity effect, have been shown to influence the extent to which auditory and visual signals interact (e.g., Chen & Spence, 2010, 2011; Evans & Treisman, 2010; Parise & Spence, 2009; Sato, Toyoizumi, & Aihara, 2007; Vatakis & Spence, 2007). Research from the intersensory conflict situation has shown that vision tends to dominate over audition for spatial judgments (Alais & Burr, 2004) whereas audition tends to dominate over vision in temporal tasks (Welch, DuttonHurt, & Warren, 1986). Bayesian decision theory provides a satisfactory account for the majority of auditory–visual interaction effects that have been documented in this chapter (e.g., see Alais & Burr, 2004; Ma et al., 2009; Sato, Toyoizumi, & Aihara, 2007). It appears as though our brains normally weight the more reliable sensory signals more heavily.

While much of what is known concerning the constraints on crossmodal interactions between auditory and visual stimuli has emerged from carefully controlled psychophysical studies, the many techniques available to the neuroscientist, including TMS (e.g., Bien et al., 2012; Johnson, Strafella, & Zatorre, 2007) and transcranial direct current stimulation (Bolognini et al., 2011), are increasingly being used to probe the neural substrates and mechanisms underlying auditory–visual interaction effects. Researchers are now getting close to being able to link neuronal activity to the multisensory percepts experienced by the participants in human psychophysical studies (Ma & Pouget, 2008; Noesselt et al., 2010; Wener & Noppeney, 2010). Although there has not been space to discuss them here, the last couple of years have also seen the publication of a number of studies that have investigated how audiovisual interactions affect what it is that a person reports being aware of, often using ambiguous visual displays, such as those resulting from binocular rivalry (e.g., Chen, Yeh, & Spence, 2011; Conrad et al., 2010; van Ee et al., 2009; see also Sheth & Shimojo, 2004).

Furthermore, having started to understand the mechanisms underlying audiovisual interactions in neurologically normal adults, there is obviously a great deal of scope in trying to determine the developmental trajectory of audiovisual interactions/integration (see Bremner, Lewkowicz, & Spence, 2012; Laurienti et al., 2006; Lewkowicz & Ghazanfar, 2006; Nava, Scala, & Pavani, 2013; Neil et al., 2006; Tremblay et al., 2007). There has also been a recent growth of interest in trying to understand how those with a development absence of a sense, as in deaf individuals and/or those with congenital cataracts, integrate auditory and visual information following the restoration of their missing sense (e.g., Putzar et al., 2007; Rouger et al., 2007). The question of what may be going wrong in terms of multisensory integration in various special populations, such as schizophrenics (e.g., see De Gelder et al., 2005; Ross et al., 2007b) and those with autism spectrum disorder (Magnée et al., 2011) is also growing in popularity. And, finally, while beyond the scope of the present chapter, research documenting the results of specific brain damage also continues to attract scientific interest from those wishing to understand the neural underpinnings of multisensory integration (e.g., Phan et al., 2000; Van Vleet & Robertson, 2006).

REFERENCES

Alais, D., & Burr, D. (2004). The ventriloquist effect results from near-optimal bimodal integration. *Current Biology, 14,* 257–262.

Alink, A., Singer, W., & Muckli, L. (2008). Capture of auditory motion by vision is represented by an activation shift from auditory to visual motion cortex. *Journal of Neuroscience, 28,* 2690–2697.

Alsius, A., Navarra, J., Campbell, R., & Soto-Faraco, S. (2005). Audiovisual integration of speech falters under high attention demands. *Current Biology, 15,* 1–5. doi:10.1016/j.cub.2005.03.046.

Andersen, T. S., & Mamassian, P. (2008). Audiovisual integration of stimulus transients. *Vision Research, 48,* 2537–2544.

Andersen, T. S., Tiippana, K., & Sams, M. (2004). Factors influencing audiovisual fission and fusion illusions. *Brain Research. Cognitive Brain Research, 21,* 301–308.

Andersen, T. S., Tiippana, K., & Sams, M. (2005). Maximum likelihood integration of rapid flashes and beeps. *Neuroscience Letters, 380,* 155–160. doi:10.1016/j.neulet.2005.01.030.

Arieh, Y., & Marks, L. E. (2008). Cross-modal interaction between vision and hearing: A speed–accuracy analysis. *Perception & Psychophysics, 70,* 412–421.

Battaglia, P. W., Jacobs, R. A., & Aslin, R. N. (2003). Bayesian integration of visual and auditory signals for spatial localization. *Journal of the Optical Society of America. A, Optics, Image Science, and Vision, 20,* 1391–1397.

Beierholm, U. R., Quartz, S. R., & Shams, L. (2009). Bayesian priors are encoded independently from likelihoods in human multisensory perception. *Journal of Vision, 9*(5), 23, 1–9. doi:10.1167/9.5.23.

Bernstein, L. E., Auer, E. T., Jr., & Takayanagi, S. (2004). Auditory speech detection in noise enhanced by lipreading. *Speech Communication, 44,* 5–18. doi:10.1016/j.specom.2004.10.011.

Bertelson, P., & Aschersleben, G. (1998). Automatic visual bias of perceived auditory location. *Psychonomic Bulletin & Review, 5,* 482–489.

Bertelson, P., & de Gelder, B. (2004). The psychology of multimodal perception. In C. Spence & J. Driver (Eds.), *Crossmodal space and crossmodal attention* (pp. 141–177). Oxford, England: Oxford University Press.

Bien, N., ten Oever, S., Goebel, R., & Sack, A. T. (2012). The sound of size: Crossmodal binding in pitch-size synesthesia: A combined TMS, EEG, and psychophysics study. *NeuroImage, 59,* 663–672.

Bolognini, N., Frassinetti, F., Serino, A., & Làdavas, E. (2005). "Acoustical vision" of below threshold stimuli: Interaction among spatially converging audiovisual inputs. *Experimental Brain Research, 160,* 273–282.

Bolognini, N., Rossetti, A., Casati, C., Mancini, F., & Vallar, G. (2011). Neuromodulation of multisensory perception: A tDCS study of the sound-induced flash illusion. *Neuropsychologia, 49,* 231–237.

Bonath, B., Noesselt, T., Martinez, A., Mishra, J., Schwiecker, K., Heinze, H.-J., et al. (2007). Neural basis of the ventriloquist illusion. *Current Biology, 17,* 1697–1703.

Bremner, A., Lewkowicz, D., & Spence, C. (Eds.). (2012). *Multisensory development.* Oxford, England: Oxford University Press.

Burnham, D. (Ed.). (1998). *Hearing by eye: Vol. 2. Advances in the psychology of speechreading and auditory–visual speech.* Hove, England: Psychology Press.

Busse, L., Roberts, K. C., Crist, R. E., Weissman, D. H., & Woldorff, M. G. (2005). The spread of attention across modalities and space in a multisensory object. *Proceedings of the National Academy of Sciences of the United States of America, 102,* 18751–18756. doi:10.1073/pnas.0507704102.

Calvert, G. A., Bullmore, E. T., Brammer, M. J., Campbell, R., Williams, S. C., McGuire, P. K., et al. (1997). Activation of auditory cortex during silent lipreading. *Science, 276,* 593–596. doi:10.1126/science.276.5312.593.

Calvert, G., Spence, C., & Stein, B. E. (Eds.). (2004). *The handbook of multisensory processing.* Cambridge, MA: MIT Press.

Chan, J. S., & Newell, F. N. (2008). Behavioral evidence for task-dependent "what" versus "where" processing within and across modalities. *Perception & Psychophysics, 70,* 36–49.

Chen, Y.-C., & Spence, C. (2010). When hearing the bark helps to identify the dog: Semantically-congruent sounds modulate the identification of masked pictures. *Cognition, 114,* 389–404.

Chen, Y.-C., & Spence, C. (2011). Crossmodal semantic priming by naturalistic sounds and spoken words enhances visual sensitivity. *Journal of Experimental Psychology. Human Perception and Performance, 37,* 1554–1568.

Chen, Y.-C., & Yeh, S.-L. (2009). Catch the moment: Multisensory enhancement of rapid visual events by sound. *Experimental Brain Research, 198,* 209–219.

Chen, Y.-C., Yeh, S.-L., & Spence, C. (2011). Crossmodal constraints on human visual awareness: Can auditory semantic context modulate binocular rivalry? *Frontiers in Perception Science, 2,* 212. doi:10.3389/fpsyg.2011.00212.

Colavita, F. B. (1974). Human sensory dominance. *Perception & Psychophysics, 16,* 409–412.

Conrad, V., Bartels, A., Kleiner, M., & Noppeney, U. (2010). Audiovisual interactions in binocular rivalry. *Journal of Vision, 10*(10), 27, 1–15. doi:10.1167/10.10.27.

de Gelder, B., & Vroomen, J. (2000a). The perception of emotions by ear and eye. *Cognition and Emotion, 14*, 289–311.

de Gelder, B., & Vroomen, J. (2000b). Bimodal emotion perception: Integration across separate modalities, cross-modal perceptual grouping or perception of multimodal events? *Cognition and Emotion, 14*, 321–324.

de Gelder, B., Vroomen, J., de Jong, S. J., Masthoff, E. D., Trompenaars, F. J., & Hodimont, P. (2005). Multisensory integration of emotional faces and voices in schizophrenics. *Schizophrenia Research, 72*, 195–203.

Diaconescu, A. O., Alain, C., & McIntosh, A. R. (2011). The co-occurrence of multisensory facilitation and cross-modal conflict in the human brain. *Journal of Neurophysiology, 106*, 2896–2909.

Dodd, B., & Campbell, R. (Eds.). (1994). *Hearing by eye: The psychology of lip-reading*. Hillsdale, NJ: Erlbaum.

Doehrmann, O., & Naumer, M. J. (2008). Semantics and the multisensory brain: How meaning modulates processes of audio-visual integration. *Brain Research, 1242*, 136–150. doi:10.1016/j.brainres.2008.03.071.

Driver, J. (1996). Enhancement of selective listening by illusory mislocation of speech sounds due to lip-reading. *Nature, 381*, 66–68.

Driver, J., & Noesselt, T. (2008). Multisensory interplay reveals crossmodal influences on 'sensory-specific' brain regions, neural responses, and judgments. *Neuron, 57*, 11–23.

Eramudugolla, R., Kamke, M., Soto-Faraco, S., & Mattingley, J. B. (2011). Perceptual load influences auditory space perception in the ventriloquist aftereffect. *Cognition, 118*, 62–74.

Evans, K. K., & Treisman, A. (2010). Natural cross-modal mappings between visual and auditory features. *Journal of Vision, 10*(1), 6, 1–12. doi:10.1167/10.1.6.

Fiebelkorn, I. C., Foxe, J. J., & Molholm, S. (2010). Dual mechanisms for the cross-sensory spread of attention: How much do learned associations matter? *Cerebral Cortex, 20*, 109–120.

Freeman, E., & Driver, J. (2008). Direction of visual apparent motion driven by timing of a static sound. *Current Biology, 18*, 1262–1266.

Frens, M. A., Van Opstal, A. J., & Van Der Willigen, R. F. (1995). Spatial and temporal factors determine audio-visual interactions in human saccadic eye movements. *Perception & Psychophysics, 57*, 802–816.

Gallace, A., & Spence, C. (2006). Multisensory synesthetic interactions in the speeded classification of visual size. *Perception & Psychophysics, 68*, 1191–1203.

Ghazanfar, A. A., Chandrasekaran, C., & Logothetis, N. K. (2008). Interactions between the superior temporal sulcus and auditory cortex mediate dynamic face/voice integration in rhesus monkeys. *Journal of Neuroscience, 28*, 4457–4469.

Harrington, L. K., & Peck, C. K. (1998). Spatial disparity affects visual–auditory interactions in human sensorimotor processing. *Experimental Brain Research, 122*, 247–252.

Hein, G., Doehrmann, O., Müller, N. G., Kaiser, J., Muckli, L., & Naumer, M. J. (2007). Object familiarity and semantic congruency modulate responses in cortical audiovisual integration areas. *Journal of Neuroscience, 27*, 7881–7887.

Heron, J., Whitaker, D., & McGraw, P. V. (2004). Sensory uncertainty governs the extent of audio–visual interaction. *Vision Research, 44*, 2875–2884.

Holmes, N. P., & Spence, C. (2005). Multisensory integration: Space, time, and superadditivity. *Current Biology, 15*, R762–R764. doi:10.1016/j.cub.2005.08.058.

Innes-Brown, H., & Crewther, D. (2009). The impact of spatial incongruence on an auditory–visual illusion. *PLoS ONE, 4*, e6450. doi:10.1371/journal.pone.0006450.

Jackson, C. V. (1953). Visual factors in auditory localization. *Quarterly Journal of Experimental Psychology, 5*, 52–65.

Jaekl, P. M., & Harris, L. R. (2009). Sounds can affect visual perception mediated primarily by the parvocellular pathway. *Visual Neuroscience, 26*, 477–486.

Jaekl, P. M., & Soto-Faraco, S. (2010). Audiovisual contrast enhancement is articulated primarily via the M-pathway. *Brain Research, 10*, 1366–1385.

Johnson, J. A., Strafella, A. P., & Zatorre, R. J. (2007). The role of the dorsolateral prefrontal cortex in bimodal divided attention: Two transcranial magnetic stimulation studies. *Journal of Cognitive Neuroscience, 19*, 907–920.

Jones, J. A., & Jarick, M. (2006). Multisensory integration of speech signals: The relationship between space and time. *Experimental Brain Research, 174*, 588–594.

Koelewijn, T., Bronkhorst, A., & Theeuwes, J. (2010). Attention and the multiple stages of multisensory integration: A review of audiovisual studies. *Acta Psychologica, 134*, 372–384.

Körding, K. P., Beierholm, U., Ma, W. J., Tenenbaum, J. B., Quartz, S., & Shams, L. (2007). Causal inference in multisensory perception. *PLoS ONE, 2*, e943. doi:10.1371/journal.pone.0000943.

Laurienti, P. J., Burdette, J. H., Maldjian, J. A., & Wallace, M. T. (2006). Enhanced multisensory integration in older adults. *Neurobiology of Aging, 27*, 1155–1163.

Laurienti, P. J., Kraft, R. A., Maldjian, J. A., Burdette, J. H., & Wallace, M. T. (2004). Semantic congruence is a critical factor in multisensory behavioral performance. *Experimental Brain Research, 158*, 405–414.

Lavie, N. (2005). Distracted and confused? Selective attention under load. *Trends in Cognitive Sciences, 9*, 75–82.

Lewkowicz, D. J., & Ghazanfar, A. A. (2006). The decline of cross-species intermodal perception in human infants. *Proceedings of the National Academy of Sciences of the United States of America, 103*, 6771–6774. doi:10.1073/pnas.0602027103.

Lewkowicz, D. J., & Lickliter, R. (Eds.). (1994). *The development of intersensory perception: Comparative perspectives*. Hillsdale, NJ: Erlbaum.

Lippert, M., Logothetis, N. K., & Kayser, C. (2007). Improvement of visual contrast detection by a simultaneous sound. *Brain Research, 1173*, 102–109.

Ma, W. J., & Pouget, A. (2008). Linking neurons to behavior in multisensory perception: A computational review. *Brain Research, 1242*, 4–12. doi:10.1016/j.brainres.2008.04.082.

Ma, W. J., Zhou, X., Ross, L. A., Foxe, J. J., & Parra, L. C. (2009). Lip-reading aids word recognition most in moderate noise: A Bayesian explanation using high-dimensional feature space. *PLoS ONE, 4*, e4638. doi:10.1371/journal.pone.0004638.

Macaluso, E., George, N., Dolan, R., Spence, C., & Driver, J. (2004). Spatial and temporal factors during processing of audiovisual speech perception: A PET study. *NeuroImage, 21*, 725–732.

Magnée, M. J. C. M., de Gelder, B., van Engeland, H., & Kemner, C. (2011). Multisensory integration in autism spectrum disorder: Critical influence of attention and noise. *PLoS ONE, 6,* e24196. doi:10.1371/journal.pone. 0024196.

Maier, J. X., Di Luca, M., & Noppeney, U. (2011). Audiovisual asynchrony detection in human speech. *Journal of Experimental Psychology. Human Perception and Performance, 37,* 245–256.

Manjarrez, E., Mendez, I., Martinez, L., Flores, A., & Mirasso, C. R. (2007). Effects of auditory noise on the psychophysical detection of visual signals: Cross-modal stochastic resonance. *Neuroscience Letters, 415,* 231–236. doi:10.1016/ j.neulet.2007.01.030.

Maravita, A., Bolognini, N., Bricolo, E., Marzi, C. A., & Savazzi, S. (2008). Is audio-visual integration subserved by the superior colliculus in humans? Clues from the redundant signal effect. *Neuroreport, 19,* 271–275.

Massaro, D. W., & Cohen, M. M. (2000). Fuzzy logical model of bimodal emotion perception: Comment on "The perception of emotions by ear and by eye" by de Gelder and Vroomen. *Cognition and Emotion, 14,* 313–320.

McDonald, J. J., Teder-Sälejärvi, W. A., Di Russo, F., & Hillyard, S. A. (2005). Neural basis of auditory-induced shifts in visual time-order perception. *Nature Neuroscience, 8,* 1197–1202.

McDonald, J. J., Teder-Sälejärvi, W. A., & Ward, L. M. (2001). Multisensory integration and crossmodal attention effects in the human brain. *Science, 292,* 1791.

McGurk, H., & MacDonald, J. (1976). Hearing lips and seeing voices. *Nature, 264,* 746–748.

Meienbrock, A., Naumer, M. J., Doehrmann, O., Singer, W., & Muckli, L. (2007). Retinotopic effects during spatial audio–visual integration. *Neuropsychologia, 45,* 531–539.

Meyer, G. F., Wuerger, S. M., Röhrbein, F., & Zetzsche, C. (2005). Low-level integration of auditory and visual motion signals requires spatial co-localisation. *Experimental Brain Research, 166,* 538–547.

Molholm, S., Ritter, W., Javitt, D. C., & Foxe, J. J. (2004). Multisensory visual–auditory object recognition in humans: A high-density electrical mapping study. *Cerebral Cortex, 14,* 452–465.

Morein-Zamir, S., Soto-Faraco, S., & Kingstone, A. (2003). Auditory capture of vision: Examining temporal ventriloquism. *Brain Research. Cognitive Brain Research, 17,* 154–163.

Mozolic, J. L., Hugenschmidt, C. E., Peiffer, A. M., & Laurienti, P. J. (2008). Modality-specific selective attention attenuates multisensory integration. *Experimental Brain Research, 184,* 39–52.

Naumer, M. J., & Kaiser, J. (Eds.). (2010). *Multisensory object perception in the primate brain.* New York: Springer.

Nava, E., Scala, G. G., & Pavani, F. (2013). Changes in sensory preference during childhood. Converging evidence from the Colavita effect and the sound-induced flash illusion. *Child Development, 84,* 604–616. doi:10.1111/j.1467-8642. 2012.01856.x.

Navarra, J., Alsius, A., Soto-Faraco, S., & Spence, C. (2009). Assessing the role of attention in the audiovisual integration of speech. *Information Fusion, 11,* 4–11.

Neil, P. A., Chee-Ruiter, C., Scheier, C., Lewkowicz, D. J., & Shimojo, S. (2006). Development of multisensory spatial integration and perception in humans. *Developmental Psychology, 9,* 454–464.

Noesselt, T., Bergmann, D., Hake, M., Heinze, H. J., & Fendrich, R. (2008). Sound increases the saliency of visual events. *Brain Research, 1220,* 157–163.

Noesselt, T., Tyll, S., Boehler, C. N., Budinger, E., Heinze, H.-J., & Driver, J. (2010). Sound-induced enhancement of low-intensity vision: Multisensory influences on human sensory-specific cortices and thalamic bodies relate to perceptual enhancement of visual detection sensitivity. *Journal of Neuroscience, 30,* 13609–13623.

Odgaard, E. C., Arieh, Y., & Marks, L. E. (2003). Cross-modal enhancement of perceived brightness: Sensory interaction versus response bias. *Perception & Psychophysics, 65,* 123–132.

O'Leary, A., & Rhodes, G. (1984). Cross-modal effects on visual and auditory object perception. *Perception & Psychophysics, 35,* 565–569.

Olivers, C. N. L., & Van der Burg, E. (2008). Bleeping you out of the blink: Sound saves vision from oblivion. *Brain Research, 1242,* 191–199.

Parise, C., & Spence, C. (2009). 'When birds of a feather flock together': Synesthetic correspondences modulate audiovisual integration in non-synesthetes. *PLoS ONE, 4,* e5664. doi:10.1371/journal.pone.0005664.

Parise, C. V., Spence, C., & Ernst, M. (2012). When correlation implies causation in multisensory integration. *Current Biology, 22,* 1–4. doi:10.1016/j.cub.2011.11.039.

Pekkola, J., Ojanen, V., Autti, T., Jääskeläinen, J. P., Möttönen, R., Tarkiainen, A., et al. (2005). Primary auditory cortex activation by visual speech: An fMRI study at 3T. *Neuroreport, 16,* 125–128.

Petrini, K., McAleer, P., & Pollick, F. (2010). Audiovisual integration of emotional signals from music improvisation does not depend on temporal correspondence. *Brain Research, 1323,* 139–148.

Petrini, K., Russell, M., & Pollick, F. (2009). When knowing can replace seeing in audiovisual integration of actions. *Cognition, 110,* 432–439.

Phan, M. L., Schendel, K. L., Recanzone, G. H., & Robertson, L. C. (2000). Auditory and visual spatial localization deficits following bilateral parietal lobe lesions in a patient with Balint's syndrome. *Journal of Cognitive Neuroscience, 12,* 583–600.

Posner, M. I. (1978). *Chronometric explorations of mind.* Hillsdale, NJ: Erlbaum.

Prime, D. J., McDonald, J. J., Green, J., & Ward, L. M. (2008). When crossmodal attention fails: A controversy resolved? *Canadian Journal of Experimental Psychology, 62,* 192–197.

Putzar, L., Goerendt, I., Lange, K., Rösler, F., & Röder, B. (2007). Early visual deprivation impairs multisensory interactions in humans. *Nature Neuroscience, 10,* 1243–1245.

Recanzone, G. H. (1998). Rapidly induced auditory plasticity: The ventriloquism aftereffect. *Proceedings of the National Academy of Sciences of the United States of America, 95,* 869–875.

Recanzone, G. H. (2003). Auditory influences on visual temporal rate perception. *Journal of Neurophysiology, 89,* 1078–1093.

Recanzone, G. H. (2009). Interactions of auditory and visual stimuli in space and time. *Hearing Research, 258,* 89–99.

Regan, D., & Spekreijse, H. (1977). Auditory–visual interactions and the correspondence between perceived auditory space and perceived visual space. *Perception, 6,* 133–138.

Renier, L. A., Anurova, I., De Volder, A. G., Carlson, S., VanMeter, J., & Rauschecker, J. P. (2009). Multisensory integration of sounds and vibrotactile stimuli in processing streams for "what" and "where." *Journal of Neuroscience, 29,* 10950–10960.

Risberg, A., & Lubker, J. (1978). Prosody and speech-reading. *STL-Quarterly Progress and Status Report, 19*, 1–16.

Roach, N. W., Heron, J., & McGraw, P. V. (2006). Resolving multisensory conflict: A strategy for balancing the costs and benefits of audio–visual integration. *Proceedings. Biological Sciences, 273*, 2159–2168. doi:10.1098/rspb.2006.3578.

Ross, L. A., Saint-Amour, D., Leavitt, V. M., Javitt, D. C., & Foxe, J. J. (2007a). Do you see what I am saying? Exploring visual enhancement of speech comprehension in noisy environments. *Cerebral Cortex, 17*, 1147–1153. doi:10.1093/cercor/bhl024.

Ross, L. A., Saint-Amour, D., Leavitt, V. M., Molholm, S., Javitt, D. C., & Foxe, J. J. (2007b). Impaired multisensory processing in schizophrenia: Deficits in the visual enhancement of speech comprehension under noisy environmental conditions. *Schizophrenia Research, 97*, 173–183.

Rouger, J., Lagleyre, S., Fraysse, B., Deneve, S., Deguine, O., & Barone, P. (2007). Evidence that cochlear-implanted deaf patients are better multisensory integrators. *Proceedings of the National Academy of Sciences of the United States of America, 104*, 7295–7300. doi:10.1073/pnas.0609419104.

Sadaghiani, S., Maier, J. X., & Noppeney, U. (2009). Natural, metaphoric, and linguistic auditory direction signals have distinct influences on visual motion processing. *Journal of Neuroscience, 29*, 6490–6499.

Saldaña, H. M., & Rosenblum, L. D. (1993). Visual influences on auditory pluck and bow judgments. *Perception & Psychophysics, 54*, 406–416.

Sanabria, D., Spence, C., & Soto-Faraco, S. (2007). Perceptual and decisional contributions to audiovisual interactions in the perception of apparent motion: A signal detection study. *Cognition, 102*, 299–310.

Sato, Y., Toyoizumi, T., & Aihara, K. (2007). Bayesian inference explains perception of unity and ventriloquism aftereffect: Identification of common sources of audiovisual stimuli. *Neural Computation, 19*, 3335–3355.

Schutz, M., & Kubovy, M. (2009). Causality and cross-modal integration. *Journal of Experimental Psychology. Human Perception and Performance, 35*, 1791–1810.

Sekuler, R., Sekuler, A. B., & Lau, R. (1997). Sound alters visual motion perception. *Nature, 385*, 308.

Senkowski, D., Schneider, T. R., Foxe, J. J., & Engel, A. K. (2008). Crossmodal binding through neural coherence: Implications for multisensory processing. *Trends in Neurosciences, 31*, 401–409.

Shams, L., Kamitani, Y., & Shimojo, S. (2000). What you see is what you hear: Sound induced visual flashing. *Nature, 408*, 788.

Sheth, B. R., & Shimojo, S. (2004). Sound-aided recovery from and persistence against visual filling-in. *Vision Research, 44*, 1907–1917.

Sinnett, S., Soto-Faraco, S., & Spence, C. (2008). The co-occurrence of multisensory competition and facilitation. *Acta Psychologica, 128*, 153–161.

Slutsky, D. A., & Recanzone, G. H. (2001). Temporal and spatial dependency of the ventriloquism effect. *Neuroreport, 12*, 7–10.

Soto-Faraco, S., Lyons, J., Gazzaniga, M., Spence, C., & Kingstone, A. (2002). The ventriloquist in motion: Illusory capture of dynamic information across sensory modalities. *Brain Research. Cognitive Brain Research, 14*, 139–146.

Spence, C. (2011). Crossmodal correspondences: A tutorial review. *Attention, Perception & Psychophysics, 73*, 971–995.

Spence, C. (2012). Multisensory perception, cognition, and behavior: Evaluating the factors modulating multisensory integration. In B. E. Stein (Ed.), *The new handbook of multisensory processing* (pp. 241–264). Cambridge, MA: MIT Press.

Spence, C. (2013). Just how important is spatial coincidence to multisensory integration? Evaluating the spatial rule. *Annals of the New York Academy of Sciences.*

Spence, C., & Bayne, T. (in press). Is consciousness multisensory? In M. Matthen & D. Stokes (Eds.), *The senses.* Oxford, England: Oxford University Press.

Spence, C., & Chen, Y.-C. (2012). Intramodal and crossmodal perceptual grouping. In B. E. Stein (Ed.), *The new handbook of multisensory processing* (pp. 265–282). Cambridge, MA: MIT Press.

Spence, C., & Driver, J. (Eds.). (2004). *Crossmodal space and crossmodal attention.* Oxford, England: Oxford University Press.

Spence, C., McDonald, J., & Driver, J. (2004). Exogenous spatial cuing studies of human crossmodal attention and multisensory integration. In C. Spence & J. Driver (Eds.), *Crossmodal space and crossmodal attention* (pp. 277–320). Oxford, England: Oxford University Press.

Spence, C., & Ngo, M.-C. (2012). Does attention or multisensory integration explain the crossmodal facilitation of masked visual target identification? In B. E. Stein (Ed.), *The new handbook of multisensory processing* (pp. 345–358). Cambridge, MA: MIT Press.

Spence, C., & Parise, C. (2010). Prior entry. *Consciousness and Cognition, 19*, 364–379.

Spence, C., Parise, C., & Chen, Y.-C. (2011). The Colavita visual dominance effect. In M. M. Murray & M. Wallace (Eds.), *Frontiers in the neural bases of multisensory processes* (pp. 523–550). Boca Raton, FL: CRC Press.

Stein, B. E. (Ed.). (2012). *The new handbook of multisensory processing.* Cambridge, MA: MIT Press.

Stein, B. E., Burr, D., Costantinides, C., Laurienti, P. J., Meredith, A. M., Perrault, T. J., Jr., et al. (2010). Semantic confusion regarding the development of multisensory integration: A practical solution. *European Journal of Neuroscience, 31*, 1713–1720. doi:10.1111/j.1460-9568.2010.07206.x.

Stein, B. E., London, N., Wilkinson, L. K., & Price, D. P. (1996). Enhancement of perceived visual intensity by auditory stimuli: A psychophysical analysis. *Journal of Cognitive Neuroscience, 8*, 497–506.

Stein, B. E., & Meredith, M. A. (1993). *The merging of the senses.* Cambridge, MA: MIT Press.

Stein, B. E., & Stanford, T. R. (2008). Multisensory integration: Current issues from the perspective of the single neuron. *Nature Reviews. Neuroscience, 9*, 255–267.

Störmer, V. S., McDonald, J. J., & Hillyard, S. A. (2009). Crossmodal cueing of attention alters appearance and early cortical processing of visual stimuli. *Proceedings of the National Academy of Sciences of the United States of America, 106*, 22456–22461. doi:10.1073/pnas.0907573106.

Sumby, W. H., & Pollack, I. (1954). Visual contribution to speech intelligibility in noise. *Journal of the Acoustical Society of America, 26*, 212–215.

Talsma, D., Doty, T. J., & Woldorff, M. G. (2007). Selective attention and audiovisual integration: Is attending to both modalities a prerequisite for early integration? *Cerebral Cortex, 17*, 691–701.

Talsma, D., Senkowski, D., Soto-Faraco, S., & Woldorff, M. G. (2010). The multifaceted interplay between attention and multisensory integration. *Trends in Cognitive Sciences, 14,* 400–410.

Taylor, K. I., Moss, H. E., Stamatakis, E. A., & Tyler, L. K. (2006). Binding crossmodal object features in perirhinal cortex. *Proceedings of the National Academy of Sciences of the United States of America, 103,* 8239–8244. doi:10.1073/pnas.0509704103.

Teramoto, W., Hidaka, S., & Sugita, Y. (2010). Sounds move a static visual object. *PLoS ONE, 5,* e12255. doi:10.1371/journal.pone.0012255.

Tremblay, C., Champoux, F., Voss, P., Bacon, B. A., Lepore, F., & Théoret, H. (2007). Speech and non-speech audio-visual illusions: A developmental study. *PLoS ONE, 8,* e742. doi:10.1371/journal.pone.0000742.

Van der Burg, E., Cass, J., Olivers, C. N. L., Theeuwes, J., & Alais, D. (2010). Efficient visual search from synchronized auditory signals requires transient audiovisual events. *PLoS ONE, 5,* e10664. doi:10.1371/journal.pone.0010664.

van Ee, R., van Boxtel, J. J. A., Parker, A. L., & Alais, D. (2009). Multimodal congruency as a mechanism for willful control over perceptual awareness. *Journal of Neuroscience, 29,* 11641–11649.

Van Vleet, T., & Robertson, L. C. (2006). Cross-modal interactions in time and space: Auditory influences on visual attention in patients with hemispatial neglect. *Journal of Cognitive Neuroscience, 18,* 1368–1379.

Van Wassenhove, V., Grant, K. W., & Poeppel, D. (2005). Visual speech speeds up the neural processing of auditory speech. *Proceedings of the National Academy of Sciences of the United States of America, 102,* 1181–1186.

Van Wassenhove, V., Grant, K. W., & Poeppel, D. (2007). Temporal window of integration in audio–visual speech perception. *Neuropsychologia, 45,* 598–607.

Vatakis, A., Ghazanfar, A., & Spence, C. (2008). Facilitation of multisensory integration by the 'unity assumption': Is speech special? *Journal of Vision, 8*(9), 14, 1–11. doi:10.1167/8.9.14.

Vatakis, A., & Spence, C. (2007). Crossmodal binding: Evaluating the "unity assumption" using audiovisual speech stimuli. *Perception & Psychophysics, 69,* 744–756.

Vatakis, A., & Spence, C. (2008). Evaluating the influence of the 'unity assumption' on the temporal perception of realistic audiovisual stimuli. *Acta Psychologica, 127,* 12–23.

Vroomen, J., Bertelson, P., & de Gelder, B. (2001). The ventriloquist effect does not depend on the direction of automatic visual attention. *Perception & Psychophysics, 63,* 651–659.

Vroomen, J., & de Gelder, B. (2000). Sound enhances visual perception: Cross-modal effects of auditory organization on vision. *Journal of Experimental Psychology: Human Perception and Performance, 26,* 1583–1590.

Vroomen, J., & Keetels, J. J. (2011). Perception of intersensory synchrony in audiovisual speech: Not that special. *Cognition, 118,* 78–86.

Walker, S., Bruce, V., & O'Malley, C. (1995). Facial identity and facial speech processing: Familiar faces and voices in the McGurk effect. *Perception & Psychophysics, 57,* 1124–1133.

Watanabe, K., & Shimojo, S. (2001). When sound affects vision: Effects of auditory grouping on visual motion perception. *Psychological Science, 12,* 109–116.

Watkins, S., Shams, L., Josephs, O., & Rees, G. (2007). Activity in human V1 follows multisensory perception. *NeuroImage, 37,* 572–578.

Welch, R. B., DuttonHurt, L. D., & Warren, D. H. (1986). Contributions of audition and vision to temporal rate perception. *Perception & Psychophysics, 39,* 294–300.

Welch, R. B., & Warren, D. H. (1980). Immediate perceptual response to intersensory discrepancy. *Psychological Bulletin, 3,* 638–667.

Welch, R. B., & Warren, D. H. (1986). Intersensory interactions. In K. R. Boff, L. Kaufman, & J. P. Thomas (Eds.), *Handbook of perception and performance: Vol. 1. Sensory processes and perception* (pp. 25–1–25–36). New York: Wiley.

Wener, S., & Noppeney, U. (2010). Superadditive responses in superior temporal sulcus predict audiovisual benefits in object categorization. *Cerebral Cortex, 20,* 1829–1842.

Witten, I. B., & Knudsen, E. I. (2005). Why seeing is believing: Merging auditory and visual worlds. *Neuron, 48,* 489–496.

Wozny, D. R., Beierholm, U. R., & Shams, L. (2008). Human trimodal perception follows optimal statistical inference. *Journal of Vision, 8*(3), 24, 1–11. doi:10.1167/8.3.24.

Wright, R. D., & Ward, L. M. (2008). *Orienting of attention.* Oxford, England: Oxford University Press.

Wuerger, S. M., Hofbauer, M., & Meyer, G. F. (2003). The integration of auditory and visual motion signals at threshold. *Perception & Psychophysics, 65,* 1188–1196.

75 Neuroimaging Studies on Human Attention Networks in Visual and Frontoparietal Cortex

GEOFFREY M. BOYNTON AND SABINE KASTNER

We all know from our subjective visual experience that we can only pay attention to a few things at any moment in time. This is not due to a limitation of our sensory systems; as long as our eyes are open, the array of photoreceptors in the human retina responds faithfully to the incoming stream of visual information. Somewhere between the retinal input and our conscious awareness the vast majority of this visual information is discarded, presumably due to capacity limitations, or "bottlenecks," inherent with the serial, focused nature of conscious awareness.

The term "visual attention" has a variety of definitions, but all have in common the process of selecting a subset of the incoming stream of visual information for further processing. Until fairly recently it was believed that attentional filtering of visual information operated by higher brain areas selecting information from the early visual areas that respond much like a feed-forward image processing machine. However, advances in neuroscience such as animal electrophysiological and human neuroimaging techniques have led to the discovery that higher brain areas are actually modulating neural responses at all stages of processing that receive feedback inputs including the visual thalamus.

This chapter focuses on what is known about influences of selective attention on responses in the human visual system, focusing on what has been learned from neuroimaging studies in humans. The chapter is divided into two sections. The first section reviews the effects of attention on the neuronal representation of visual stimuli, that is, at the site of the stimulus representation in the visual cortex. The second section discusses neuroimaging studies on higher brain areas beyond visual cortex that are involved with generating modulatory signals, thereby constituting the presumed "sources" of neuronal modulations that can be observed within visual cortex.

THE "SITE" OF ATTENTION: EFFECTS OF ATTENTION ON NEUROIMAGING SIGNALS IN THE HUMAN VISUAL CORTEX

In a very simple experiment for studying attention, a variety of stimuli are presented, each having a number of different features (e.g., color, direction of motion), and the observer is instructed to select one of the stimuli to detect a change in one property of that stimulus. If the experiment is designed properly, changes can occur both within the attended property and in other properties so that the physical changes in the stimulus do not vary systematically with the task. Any corresponding changes in behavioral performance or neuronal responses can therefore be attributed to the act of attentional selection, and not to physical changes in the stimulus. This first section focuses on how these instructions to attend modulate brain imaging signals in the early visual areas of humans, and how these modulations may lead to changes in behavioral performance.

Parallel Studies Using Human Functional Magnetic Resonance Imaging and Monkey Electrophysiology

Not long after the development of methods for recording electrophysiological signals from individual neurons in the awake and behaving macaque, it was discovered that neural signals in extrastriate visual cortex could be strongly modulated by directed attention (Moran & Desimone, 1985). Over the next decade, the first human neuroimaging studies appeared using positron emission tomography (PET) and functional magnetic resonance imaging (fMRI) to show effects of task-based attention in visual cortex. Since then, there has been a growing parallel literature on the attentional effects observed with monkey electrophysiology and human neuroimaging approaches. Both fields have benefited from the results provided by the other.

It is important to keep in mind that brain imaging signals (e.g., fMRI and EEG) reflect pooled responses across large populations of neurons that are selective to and modulated by a range of visual stimuli and attentional tasks. The fMRI signal reflects changes in the ratio of oxygenated and deoxygenated hemoglobin that serves hundreds of thousands of neurons within a given voxel whereas electrophysiological signals typically reflect local responses from single or multiple neurons, as in single- or multi-unit recordings, or small local neuronal populations, as in local field potential recordings. The two methods have advantages and disadvantages and should be regarded as largely complementary. It is likely that there are certain conditions in which effects of attention that are measurable in individual neurons will not be measurable with fMRI. For example, the fMRI signal could be unaffected if attention happened to both increase and decrease firing rates within subpopulations of neurons equally within a given fMRI voxel. On the other hand, the fMRI signal could actually be more powerful for detecting small changes than the single-unit measurements due to the massive pooling across the neuronal population by the hemodynamic coupling process (see Boynton, 2011, for a discussion). Therefore, it is important to consider the electrophysiology literature when discussing the effects of attention on the human neuroimaging signal.

A Simple Model of Attentional Effects in Visual Cortex

A simple model of how attention affects the responses of visual neurons is that it increases the responsivity of neurons selective to the location and features of an attended stimulus. A minor variant to this model is to assume that attention also suppresses the responses within neuronal populations tuned for different features and appearing at different locations than the attended stimulus. In this section, we will consider the effects of attention in the human visual cortex in the context of this simple model, which accounts for a surprisingly large fraction of what is known about the effects of attention in both the monkey and human visual cortex.

An illustration of this simple model is shown in figure 75.1. Within a visual area there is a population of neurons showing maximal sensitivity to a specific spatial location (its receptive field) and feature (e.g., direction of motion, orientation or color). The neuronal response across this population can be represented as a "neural image" in which the brightness of the image represents

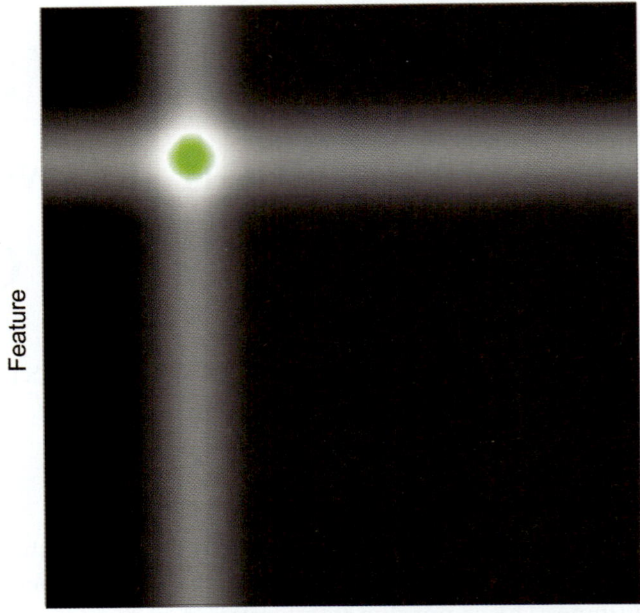

FIGURE 75.1 Simple model of how spatial and feature based attention affects responses in visual cortex. Each location in the image represents a neuronal population selective to a specific spatial position and location.

the responsiveness of neurons with the corresponding spatial and feature selectivity. If attention is directed to a specific location (spatial attention) and feature (feature-based attention), as shown by the green dot in the figure, then the simple model predicts that all neurons selective to the attended location and attended feature will have an enhanced response. This corresponds to a "plus" shaped pattern of activity in the neural image. The plus shape predicts that attention affects neurons selective to either the attended location *or* the attended feature. Also, the pattern of modulation depends on the attended location and feature, and not the physical stimulus itself.

A model of this simplicity is undoubtedly limited, but it does provide a convenient framework for interpreting the results of neuroimaging studies of selective attention in humans. More detailed models have been recently proposed that explain a range of attentional phenomena including how the effects of attention interact with the physical stimulus, but most are extensions of this basic simple model (Boynton, 2009; Reynolds & Heeger, 2009). The "normalization model of attention" by Reynolds and Heeger is a particularly elegant model that incorporates attention into an established feed-forward model and predicts a wide range of phenomena, including how the spatial spread of attention affects behavioral performance as a function of stimulus contrast (Herrmann et al., 2010).

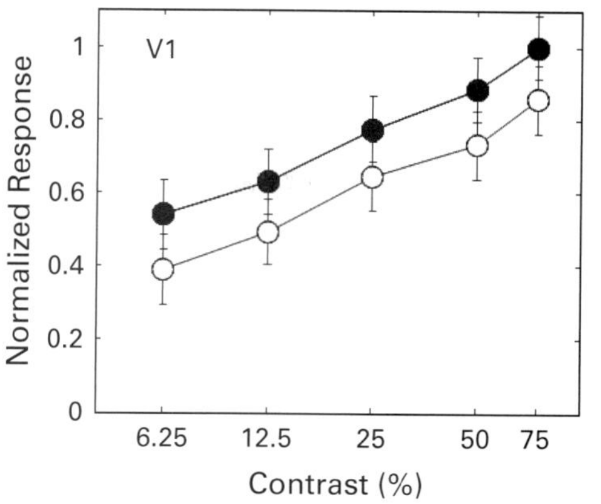

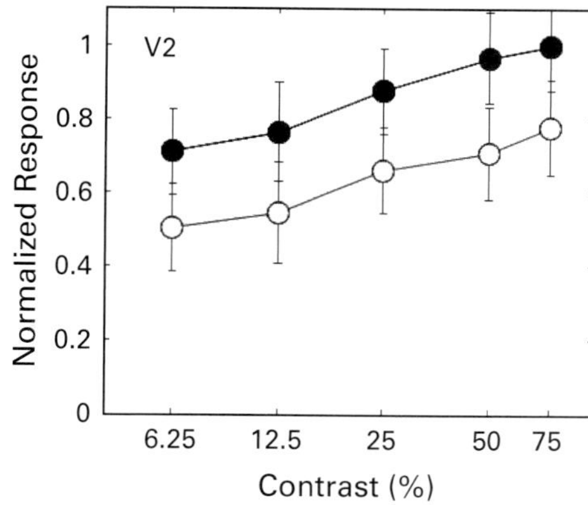

FIGURE 75.2   Effects of spatial attention on the functional magnetic resonance imaging responses averaged across four human subjects in area V1 and V2 (data from Buracas & Boynton, 2007). Filled symbols and dark lines represent the response to the attended stimulus versus an unattended blank field, and open symbols with gray lines for the estimated response to the unattended stimulus versus an unattended blank field. Error bars are standard errors of the mean.

Effects of Spatial Attention

The cortical visual areas in humans and other primates are organized retinotopically, whereby nearby neurons within a cortical area have nearby receptive fields so that the receptive field centers form an organized (though distorted) map of the visual field. Using fMRI, retinotopic organization has been revealed with a wedge stimulus rotating around fixation or a ring stimulus expanding from fixation, thereby producing characteristic waves in the blood-oxygen-level-dependent (BOLD) signal that can be easily visualized on a flattened representation of the visual cortex. The size of retinotopic maps are on the spatial scale of centimeters, and the spatial resolution of fMRI is on the order of millimeters, so such a mechanism of spatial attention is ideally suited for measurement with fMRI.

The effects of spatial attention on the BOLD signal can be reliably found within the retinotopic map of primary visual cortex (V1) (Gandhi, Heeger, & Boynton, 1999; Martinez et al., 1999; Tootell et al., 1998). Shifting attention around the visual field shifts the fMRI signal within V1 much like a shifting physical stimulus (Brefczynski & DeYoe, 1999).

While the effects of spatial attention on the fMRI response in V1 are robust, they are even stronger in extrastriate visual areas. The strength of the effects increase along the visual hierarchy to more advanced ventral and dorsal processing stages (Kastner, De Weerd, Desimone, & Ungerleider, 1998; Martinez et al., 1999). This is consistent with the general finding that early processing stages maintain a more veridical representation of a visual stimulus while higher stages reflect more strongly the behavioral relevance of a stimulus (Boynton, 2004).

An exception has been observed in the lateral geniculate nucleus (LGN), where even though it lies early in the processing stream, the effects of spatial attention are roughly equal in strength as compared to higher visual areas such as V4, TEO, and middle temporal visual cortex (MT) (O'Connor et al., 2002). This finding suggests that attentional modulation in the LGN is driven by feedback from higher visual areas.

The basic model poses that the effects of attention on neuronal responses depend on the attended location and feature, but not necessarily on the actual stimulus. This predicts that attention should increase the firing rates of neurons even in the absence of a stimulus. This is indeed the case for fMRI signals in early visual areas. A variety of studies have reported "anticipatory" fMRI responses either before a stimulus is presented or on trials in which a stimulus is simply not present (Kastner et al., 1999; Murray, 2008; Ress, Backus, & Heeger, 2000; Sestieri et al., 2008; Silver, Ress, & Heeger, 2007).

Most surprisingly, these anticipatory or baseline changes in the fMRI signal with spatial attention dominate the attentional effects in the human visual cortex. In fact, the effects of spatial attention are largely independent of the strength (contrast) of the physical stimulus (Buracas & Boynton, 2007; Murray, 2008) (see figure 75.2). This is in apparent contradiction to the results from macaque studies showing that spatial attention appears either to increase the effective contrast of the stimulus (Martinez Trujillo & Treue, 2002; Reynolds, Pasternak, & Desimone, 2000) or to act as a multiplicative gain factor (Williford & Maunsell, 2006). This discrepancy might be due to a number of issues including interspecies differences and differences between the electrophysiology and fMRI signals.

Further, it is possible that spatial attention has both a baseline and a stimulus-dependent effect on neuronal responses, but the pooling nature of the fMRI signal favors the baseline effect. Indeed, a close inspection of the electrophysiology literature shows that there is a small, but consistent, baseline effect with spatial attention (see Boynton, 2009, for review), which may dominate the fMRI measurement (see Boynton, 2011, for a discussion).

Task-Specific Attention

Some areas of the human visual cortex exhibit functional specialization, that is, specific brain areas respond better to specific classes of stimuli, and tasks. For example, area MT+ (or V5), believed to be the homologue of macaque area MT, is highly responsive to moving stimuli while the fusiform face area, or FFA, a region in the ventral surface of the temporal lobe on the fusiform gyrus, responds strongly to face stimuli. The basic model of attention is easily compatible with the concept of modular organization of the visual cortex. Just like for spatial attention, a simple mechanism for improving performance on a given task is to boost the responses of all neurons within the modular areas of the brain associated with that task. Like spatial attention, this sort of modulation should occur at a spatial scale that is also well-suited for measurement with fMRI.

Corbetta and colleagues demonstrated this basic mechanism of task-specific attention in an early PET study in humans (Corbetta et al., 1990). While keeping the stimulus constant, regions in the extrastriate cortex specialized for shape, color, and speed showed enhanced regional cerebral blood flow while subjects discriminated attributes along the corresponding stimulus dimensions.

The first fMRI study showing attentional effects in the human visual cortex demonstrated such a task-specific effect by showing that responses in human MT+ increased when subjects attended to a field of moving dots, compared to when they attended to overlapping stationary dots. Similarly, attention to faces and houses has been shown to modulate the corresponding stimulus-selective ventral brain areas (Wojciulik, Kanwisher, & Driver, 1998). During viewing of moving bar stimuli and performing a speed discrimination task, fMRI responses increased in area MT+ as compared with performance of a contrast discrimination task (Huk & Heeger, 2000; but see Buracas, Fine, & Boynton, 2005).

Sensory cortex is organized in a modular fashion not only within a sensory modality but also across modalities, with areas of the cortex specifically processing different sensory inputs. It is therefore perhaps not surprising that attention to one multimodal signal modulates the BOLD signal in cortical areas associated with the attended sensory input. For example, when subjects are alternating between a speed discrimination task on a visual stimulus and a frequency discrimination task on an auditory stimulus, the corresponding BOLD signals are modulated in the visual and auditory cortices, respectively (Ciaramitaro, Buracas, & Boynton, 2007), and attention to a tactile stimulus increases the BOLD signal in the primary sensory cortex.

Feature-Based Attention

The simple model outlined above predicts that attention to a specific feature, such as a direction of motion or a color, should enhance the response in neurons selectively tuned to that feature and possibly suppress responses in neurons tuned to other features. Stimulus features are represented in the visual cortex at a finer spatial scale than spatial position and stimulus class. For example, V1 neurons in the primate cortex selective to a given orientation are clustered within vertical columns of cortex at a spatial scale on the order of about a millimeter. Similarly in macaque MT, direction selectivity can be found within homogenous columns on a similar spatial scale. Implementing the simple attention model in the feature domain is therefore not as straightforward as for space or stimulus class. A feature-specific gain modulation on similarly tuned neurons would have to operate at the spatial scale of the column, or perhaps even smaller for features such as color that may be organized at an even finer scale. Nevertheless, such feature-specific modulation does seem to occur for motion and has been proposed formally in the "feature similarity gain model" by Treue and Martinez-Trujillo (1999).

The tight spatial scale of modulation within the cortex is also problematic for studying feature-based attention with fMRI. A given fMRI voxel is typically around 3 mm cubed, which would typically cover populations of neurons selective to a range of features. Any feature-specific modulation at a small spatial scale may therefore be buried in noise originating from other signals within an individual voxel. Recent developments in applying multivariate statistics to fMRI data have attempted to tackle and overcome some of these issues. The so-called "multivoxel pattern analysis" on fMRI data takes advantage of small but systematic differences in the pattern of responses across voxels within a visual area (Kamitani & Tong, 2005). Using this technique, Kamitani and Tong were able to find systematic variability in the pattern of responses in V1 for different stimulus orientations, but also systematic variability when attention was directed to one of two orientations of the same plaid stimulus. Presumably, attention to one of the two orientations enhanced responses within V1 neurons selective for the attended orientation (and possibly suppressed other V1 neurons), leading to the systematic effects on the pattern of responses across voxels. Similar effects for attention to direction of motion on the pattern of responses were found across voxels in area MT+ and other early visual areas.

Like the simple model applied to spatial attention, the effect of feature-based attention depends on the attended feature, not the physical stimulus. This means that feature-based attention should affect neuronal responses even outside the spatial focus of attention. This is indeed what was found in area MT of the macaque. A neuron selective to upward motion with a receptive field outside the spatial focus of attention increased its firing rate when attention was directed to an upward moving stimulus appearing elsewhere in the visual field, and it was suppressed when attention was directed to a downward stimulus outside the receptive field (Treue & Martinez Trujillo, 1999).

A similar spatially global effect of feature-based attention has been found in the human visual cortex with fMRI. The first evidence came from a study that had subjects attend alternately to overlapping upward and downward moving dots in one visual hemifield while ignoring an upward field of dots in the opposite hemifield. The fMRI response in areas V1, V2, V3, V4, and MT+ to the unattended upward moving stimulus was greater when attention was directed away to upward motion compared to downward motion (Saenz, Buracas, & Boynton, 2002). These results are consistent with the simple model where attention in one location to upward motion increased the response to all neurons selective to upward motion, regardless of their receptive fields. This increase was measurable in the population-based fMRI response because neurons responsive to the unattended upward motion were more strongly modulated by attention than the rest of the neuronal population. Though this result is exactly what is expected from the single-unit studies on attention to motion in the monkey, the same result was also found for color. Attention to one of two colors in one visual hemifield increased the fMRI response in areas V1, V2, V3, V4, and MT+ to a stimulus of the attended color in the unattended hemifield. This is interesting because not much is known about how the feature of color is represented in the visual cortex, especially in extrastriate areas like MT+.

Recall that the simple attention model does not require information about the physical stimulus. This correctly predicted that there is a significant modulation of the fMRI signal with spatial attention even in the absence of a stimulus, and that the attentional effects are largely independent of stimulus contrast. Similarly, an effect of feature-based attention has been found in the absence of visual stimulation (Serences & Boynton, 2007). Subjects performed a motion discrimination task on stimuli presented to only one visual hemifield. Using a multivoxel pattern classification method, Serences and Boynton (2007) found that the pattern of responses across voxels in areas V1, V2, V3, V4, V3a, and MT+ that did not represent the attended spatial location was systematically affected by the attended feature, a particular direction of motion. Presumably, direction-selective neurons with receptive fields outside the focus of attention were increasing and decreasing their baseline responses with the attended direction of motion, as predicted by the simple model.

Object-Based Attention

Any study of visual attention requires instructions to an observer to make a decision about some physical stimulus, or object. It therefore makes sense to think of objects as being the fundamental unit for selection rather than spatial location or feature. The basic idea behind "object-based" attention is that attention to an object enhances the representations of all features associated with that object (Duncan, 1984). For an object at a fixed location with fixed features, this is consistent with the basic model of attention described here. The primary difference is that if objects are the fundamental unit for selection, then spatial and feature-based attention are not specific attentional mechanisms themselves but instead are a consequence of an automatic selection or binding process associated with object perception.

There is fMRI evidence supporting a mechanism of object-based attention. O'Craven and colleagues (O'Craven, Downing, & Kanwisher, 1999) took advantage of the modular organization of the extrastriate visual areas and measured fMRI responses to areas specifically responsive to faces (FFA), houses (parahippocampal place area), or motion (area MT+). Subjects viewed transparently superimposed houses and faces in which one of the two stimuli was moving. Subjects attended to the houses, the faces, or the motion of the object. It was found that fMRI responses increased not only in the area associated with the task but also in areas associated with task-irrelevant attributes of the attended stimulus. For example, if subjects conducted a decision about a face that happened to be moving, fMRI responses increased not only in the FFA but also in area MT+ (O'Craven, Downing, & Kanwisher, 1999). This result is not consistent with the simple model because the motion of the face was not an attended feature. Instead, it suggests that attention automatically enhances neurons selective to task-irrelevant features associated with an attended object.

Linking Attentional Effects in the Brain with Behavioral Performance

Consider a task where a subject is presented with two stimuli simultaneously in two locations. A target could appear in either location. On some trials, a cue is provided that indicates reliably which location will contain the target. No cue is provided on other trials. This classic paradigm consistently shows that the cue leads to improved performance (e.g., Posner, Snyder, & Davidson, 1980). It may seem obvious that the mechanism behind the spatial cueing paradigm is to increase the brain's response to stimuli representing the cued location. However, there are alternate explanations for the behavioral effect. The link between the effects of attention on neuronal responses and behavioral performance falls into three categories: *response enhancement, noise reduction,* and *sensory selection* (Serences, 2011).

The general fact that attention increases the response rate of some neurons and reduces that of others makes sense if attention operates to facilitate decisions about an attended stimulus. Common sense dictates that a greater response should lead to better performance for tasks that rely on these boosted neurons, and reducing the response to ignored stimuli should help exclude interference from these stimuli. Response enhancement can lead to improved behavioral performance if the noise, or standard deviation of neuronal firing rates, increases less quickly than the signal, or mean (Shadlen & Newsome, 1998). Increased signal-to-noise by response enhancement should then allow the observer to compute a more reliable estimate of the stimulus location and properties and, therefore, better (and faster) performance on the attended stimulus. As discussed above, response enhancement is a ubiquitous effect on neuronal responses, and it seems likely that this is an important factor in the link between attentional effects on neuronal responses and behavioral performance.

A second way in which attention can increase the signal-to-noise ratio of the response to an attended stimulus is to reduce the noise. This is not easily studied with the slow hemodynamic response because it does not provide a measure of the temporal variability of the neuronal response. However, there is evidence that attending to an object reduces the variance of the firing rates to attended stimuli in area V4 of the monkey (Mitchell, Sundberg, & Reynolds, 2007). Recent studies

have also shown that spatial attention decreases the pair-wise covariance of neuronal responses in areas MT and V4 of the monkey, which is another way of effectively reducing the noise in a population-based response (Cohen & Maunsell, 2011; Mitchell, Sundberg, & Reynolds, 2009).

A third way for attention to improve behavioral performance is for it to determine which subpopulations of neurons to monitor for a given task. In the extreme, this *response selection* mechanism would not require attention to modify the neuronal response to a stimulus at all. In the cueing paradigm, allowing the brain to monitor only neurons representing the cued location should reduce the probability that a response at the uncued location should interfere with the decision-making process. This idea has been formally implemented in the context of signal detection theory and can explain a wide range of behavioral effects, even without the need to enhance or alter the sensory representation (Palmer, Ames, & Lindsey, 1993). Recently, a study applied a cueing paradigm to compare the effects of spatial attention on the fMRI signal in V1 to behavioral performance on a contrast discrimination task (Pestilli et al., 2011). The study replicated previous results showing that spatial attention increased the fMRI response in V1 by a constant amount across stimulus contrasts. However, the effects of attention on behavioral performance varied with stimulus contrast. Therefore, the increased neuronal response with attention was not sufficient to predict behavioral performance. Instead, the authors made a compelling argument that response selection also played an important role in the link between neuronal responses and behavioral performance.

Of course, area V1 may not be the site holding the representation of the stimulus for making the perceptual decision. More likely, the response enhancements in V1 with attention serve to help bias competition in later stages of visual processing toward attended stimuli.

Filtering of Unwanted Information

Most of the attentional mechanisms that we discussed thus far applied to situations where only a single visual stimulus was present in the visual field. However, in real life situations, there is always a multitude of visual information present among which we need to select the most relevant to ongoing behavior. A large body of evidence from both single-cell physiology and neuroimaging suggests that multiple stimuli present at the same time are not processed independently but interact with each other in a mutually suppressive way. In physiology studies (e.g., Reynolds, Chelazzi, & Desimone, 1999),

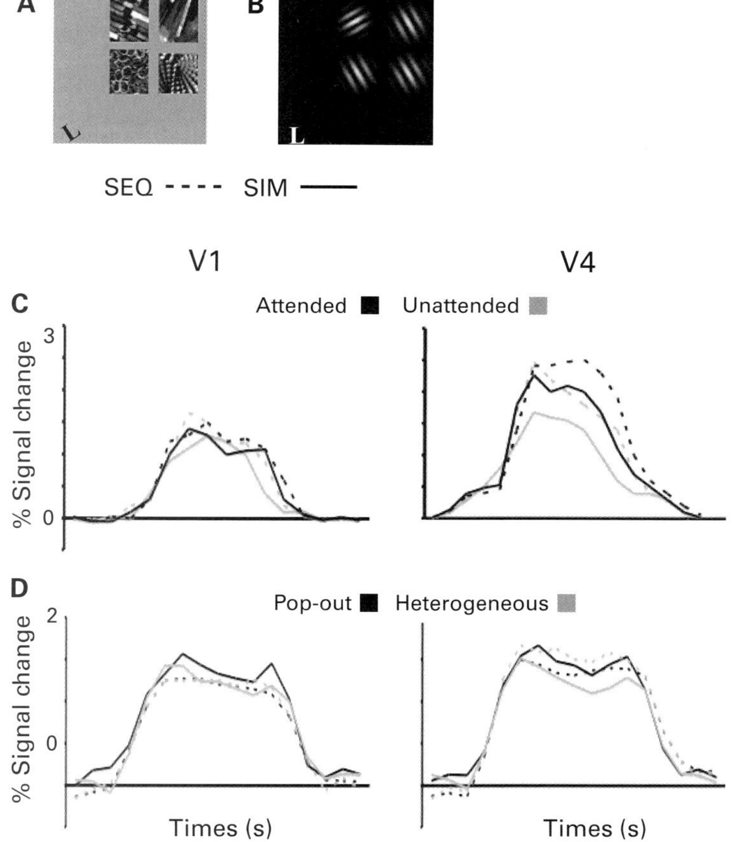

FIGURE 75.3   Top-down and bottom-up biases in resolving neural competition in visual cortex. In two experiments, competitive interactions among stimuli were assessed in the presence of top-down and bottom-up biases. Competition was measured by comparing functional magnetic resonance imaging responses to four stimuli presented either simultaneously (SIM; potentially competing) or sequentially (SEQ; noncompeting). (A) In the study investigating top-down attention biases (Kastner et al., 1998), the four stimuli were complex colored images. In the attended conditions, subjects monitored the lower left location for the appearance of a target stimulus whereas in the unattended condition they ignored the colorful stimuli and instead detected targets at fixation. (B) In the bottom-up pop-out study (Beck & Kastner, 2005), the stimuli were four gabors; either all differed in color and orientation (heterogeneous condition; not shown) or one gabor differed in color and orientation from the rest (pop-out condition; shown here without color). (C–D) Dashed curves indicate activity evoked by sequential presentations, and solid curves indicate activity evoked by simultaneous presentations for attention (C) and pop-out (D) studies. Black lines represent the conditions in which a top-down or bottom-up bias was probed while gray lines represent conditions during which stimuli were competing without any bias. For both studies, competition was partially overcome in V4 when a bias was present (black curves), resulting in smaller differences in responses evoked by sequential and simultaneous presentations (dashed and solid curves).

the responses to paired stimuli were found to be smaller than the sum of the responses evoked by each stimulus individually. In particular, the response evoked by a pair of stimuli was a weighted average of the individual responses (Reynolds, Chelazzi, & Desimone, 1999). Suppressive interactions, such as these, among multiple stimuli have been found in several visual areas in the monkey brain, including V2, V4, MT, MST, and IT, and have been interpreted as the neural substrate of an ongoing competition among multiple simultaneously present stimuli for representation in visual cortex.

In the human brain, evidence for neural competition has been found using an fMRI paradigm (Beck & Kastner, 2005; Kastner et al., 1998, 2001), in which four colorful and patterned visual stimuli that optimally activate ventral visual cortex were presented in four nearby locations to the periphery of the visual field while subjects maintained fixation (see figure 75.3A). Critically, these stimuli were presented under two different presentation conditions, *sequential* and *simultaneous*. In the sequential presentation condition, each stimulus was presented alone in one of the four locations, one after the other. In the simultaneous presentation condition, the same four stimuli appeared simultaneously in the four locations. Thus, integrated over time, the physical stimulation parameters were identical in each of the four locations in the two presentation conditions. However, competitive interactions among stimuli could take place only in the simultaneous, not in the sequential, presentation condition. Consistent with the physiology literature, simultaneous presentations evoked weaker responses than sequential presentations in areas V1, V2/VP, V4, TEO, V3A, and MT. The response differences were smallest in V1 and increased in magnitude toward ventral extrastriate areas V4 (figure 75.3C) and TEO, and dorsal extrastriate areas V3A and MT. In other words, the suppressive interactions among the four stimuli scaled with the increasing receptive field (RF) sizes across visual cortex, consistent with the idea that the stimuli were competing for representation at the level of the RF. The idea that multiple objects are represented as a weighted average of individual responses in extrastriate cortex has been corroborated by recent studies targeted at human object-selective cortex (MacEvoy & Epstein, 2009; Reddy, Kanwisher, & VanRullen, 2009). It is important to note that the suppressive (competitive) interactions across visual cortex discussed thus far occurred automatically and in the absence of attentional allocation to the stimuli. In fact, in these paradigms, participants were engaged in an attention-demanding task at fixation. Thus, neural competition would appear to be pervasive in the representation of cluttered visual scenes. In order to overcome this less than optimal representation of objects within a visual scene, there need to be mechanisms by which this ongoing competition among multiple stimuli can be resolved. There is evidence that both top-down and bottom-up processes can contribute to resolve competitive interactions among multiple stimuli.

Single-cell recording studies have shown that spatially directed attention represents one top-down mechanism that can bias the competition among multiple stimuli in favor of the attended stimulus by modulating competitive interactions. When a monkey directed attention to one of two competing stimuli within an RF, the responses in extrastriate areas V2, V4, and MT to the pair of stimuli were as large as those to that stimulus presented alone (e.g., Reynolds, Chelazzi, & Desimone, 1999). A similar mechanism appears to operate in the human visual cortex. Kastner et al. (1998) studied the effects of spatially directed attention on multiple competing visual stimuli in a variation of the paradigm described above. In addition to the two different presentation conditions, sequential and simultaneous, two different attentional conditions were tested, attended and unattended. During the unattended condition, attention was directed away from the peripheral visual display by having subjects count letters at fixation. In the attended condition, subjects were instructed to attend covertly to the peripheral stimulus location in the display closest to fixation and to count the occurrences of a target stimulus. Directing attention to this location led to greater activity for simultaneously presented stimuli than to sequentially presented stimuli in areas V4 and TEO, but not in early visual areas such as V1 (see figure 75.3C). Like the competition effects, the magnitude of the attentional effects scaled with RF size, with the strongest reduction of suppression occurring in ventral extrastriate areas V4 and TEO (see also Bles et al., 2006). It is important to note that the stage at which these attention mechanisms will operate in the visual system is flexibly determined by the spatial scale of the displays, similar to the spatial scaling of competition mechanisms to RF size. For example, targets in displays of large spatial scale will engage areas with larger RFs and will shift the locus of attentional selection to more anterior extrastriate areas relative to targets in displays of smaller spatial scale (Buffalo et al., 2005; Hopf et al., 2006). Together, these findings support the idea that directed attention enhances information processing of stimuli at the attended location by counteracting suppression induced by nearby stimuli. This may be an important mechanism by which unwanted information is filtered out from nearby distractors.

Properties of the stimulus, that is, bottom-up stimulus-driven mechanisms, have also been shown to modulate competition (Beck & Kastner, 2005; McMains & Kastner, 2010; Reynolds & Desimone, 2003). For example, Beck and Kastner (2005) found that competition could be influenced by visual salience, a bottom-up stimulus-driven mechanism that does not depend on the focus of attention. Instead of the complex images used in the previous studies, in which the sequential/simultaneous paradigm was investigated, they used four Gabor patches of different colors and orientations. These Gabor patches could then be presented in two display contexts: pop-out displays, in which a single item differed from the others in color and orientation (see figure 75.3B), and heterogeneous displays, in which all items differed from each other in both dimensions. As has been shown with the heterogeneous displays used in previous experiments, the heterogeneous displays produced robust suppressive interactions among multiple stimuli in areas V2/VP and V4 (see figure 75.3D). However, this suppression was eliminated when the same stimuli were presented in the context of pop-out displays, consistent with the prediction that visual salience can bias competitive interactions among multiple stimuli in intermediate processing areas (see figure 75.3D). As in previous studies, no evidence of competitive interactions was found in area V1, presumably due to the small RF sizes in that area. However, an effect of display context was evident in this early visual area: Simultaneously presented pop-out displays evoked more activity than any of the other three conditions, consistent with both physiology studies and computational models suggesting that pop-out may be computed as early as in area V1 (e.g., Kastner, Nothdurft, & Pigarev, 1997; Knierim & van Essen, 1992).

Taken together, these data suggest that V1 may be the source of the signal that biases neural competition in extrastriate cortex when salient stimuli are present in the visual scene, which further distinguishes this bottom-up bias from top-down biases that are thought to have their source in frontoparietal cortex (Kastner et al., 1999; Moore & Armstrong, 2003), as will be discussed in the next section. Other bottom-up factors that may determine the degree of competition among multiple stimuli are grouping factors that operate according to Gestalt rules. For example, stimuli that group by similarity or form an illusory object evoke less competition than nongrouped items (McMains & Kastner, 2010).

THE "SOURCE" OF ATTENTION: EFFECTS OF ATTENTION ON FRONTAL AND PARIETAL CORTEX

Evidence for Involvement of Higher-Order Cortex in Attentional Selection

There is a long history indicating that unilateral brain lesions, especially of higher-order cortex, may cause impairment in spatially directing attention to the contralateral hemifield, a syndrome known as visuospatial hemineglect. In severe cases, patients suffering from neglect will completely disregard the visual hemifield contralateral to the side of the lesion (e.g., Bisiach & Vallar, 1988). Typical examples include that these patients will read from only one side of a book, apply make-up to only one half of their face, or eat from only one side of a plate. In less severe cases, the deficit is more subtle and becomes apparent only if the patient is confronted with competing stimuli, as in the case of visual extinction. In visual extinction, patients are able to orient attention to a single visual object presented to their impaired visual hemifield; however, if two stimuli are presented simultaneously, one in the impaired and the other in the intact hemifield, the patients will only detect the one presented to the intact side. These findings suggest that visual extinction reflects an attentional bias toward the intact hemifield in the presence of competing objects (Kinsbourne, 1993).

Visuospatial neglect may follow unilateral lesions at different sites, including the parietal lobe and superior temporal cortex (e.g., Karnath, Ferber, & Himmelbach, 2001; Vallar & Perani, 1987), regions of the frontal lobe (e.g., Damasio, Damasio, & Chui, 1980), the anterior cingulate cortex (e.g., Janer & Pardo, 1991), the basal ganglia (e.g., Damasio, Damasio, & Chui, 1980), and the thalamus, in particular the pulvinar (e.g., Karnath, Himmelbach, & Rorden, 2002). The lesion sites that have been observed most frequently in visuospatial hemineglect are the parietal cortex, especially its inferior part as well as the temporoparietal junction (Mort et al., 2003), and the superior temporal cortex (Karnath, Ferber, & Himmelbach, 2001). The issue which of the two most prominent sites is critical for causing the syndrome is still heavily contested. Neglect occurs more often with right-sided lesions than with left-sided lesions, which has been taken as evidence for a specialized role for the right hemisphere (RH) in space-based attentional selection.

Models of Space-Based Attention Control

To account for the hemispheric asymmetry observed with the neglect syndrome, a "hemispatial" theory has been proposed by Heilman and colleagues (Heilman & Van Den Abell, 1980). In this account, the RH directs attention to both visual hemifields whereas the left hemisphere (LH) directs attention to the right visual field (RVF) only (see also Mesulam, 1981). Thus, while LH damage can be compensated for by the RH, such compensation will not be possible with RH damage, thereby resulting in neglect of the left visual field (LVF). An alternative account, Kinsbourne's "interhemispheric competition" theory, has proposed an opponent processor control system, in which each hemisphere directs attention toward the contralateral visual field with the LH processor being more powerful than the RH processor. In an intact system, the two hemispheric processors are balanced through mutual reciprocal inhibition, presumably through direct callosal connections. Alternatively, such inhibitory control could be achieved through corticosubcortical interactions between parietal cortex and the superior colliculus (SC) (Sprague, 1966). According to the interhemispheric competition account, neglect occurs as a consequence of an imbalanced system following damage to one of the processors, thereby resulting in a bias toward the ipsilesional visual field that is controlled by the remaining intact, unopposed processor (Kinsbourne, 1977). Despite their differences, these accounts make at least two predictions that we will further discuss in the following sections: (1) Higher-order cortex appears to be involved in controlling attentional operations throughout the visual field, and (2) higher-order cortex may contain discrete representations of visual space.

Attention-Related Activations in Higher-Order Cortex

In meeting the first prediction, functional brain imaging studies have identified areas in parietal and frontal cortex that are involved in the allocation of spatial attention. When subjects attend to a location in space in anticipation of the appearance of a stimulus, neural signals increase in a frontoparietal network consisting of the superior parietal lobule (SPL), the frontal eye field (FEF), and the supplementary eye field (SEF) extending into the anterior cingulate cortex (see figure 75.4B; for a meta-analysis, see Kastner & Ungerleider, 2000). This so-called dorsal frontoparietal attention network has been implicated in many visuospatial tasks, regardless of the specific task requirements, that is, whether target stimuli were detected, discriminated, or tracked in visual space, to give just a few examples. Further, the frontoparietal attention network is not limited to controlling spatial attention. It is also activated when subjects select nonspatial information. In a typical nonspatial attention experiment, subjects are instructed to attend to a particular feature of a multi-feature display (e.g., detect a color change in a red moving dot field) or to direct attention to one of two overlapping objects (Liu et al., 2003; Serences et al., 2004). Since the feature or object information that is selected is presented in the same portion of visual space, the selection mechanism is considered "nonspatial" and purely based on the featural and object information that guides behavior. Such feature- and object-based selections activate similar regions of the frontoparietal network that largely overlap with those observed during space-based selection. However, it is not yet clear whether the same neural populations are engaged in these different selection processes or whether the selection of spatial and nonspatial information occurs at different processing times within shared neural populations. Together, these findings suggest that object- or feature-based attention engages the frontoparietal attention network, possibly indicating a general domain-independent mechanism of attentional control (Yantis & Serences, 2003).

In addition to activations within a dorsal frontoparietal attention network, selective attention has also been shown to engage a ventral attention network consisting of regions in temporoparietal cortex and inferior frontal cortex (Corbetta & Shulman, 2002). This network is largely lateralized to the RH and does not appear to contribute to the maintenance of goal-directed behavior in selecting relevant stimuli but rather appears to be specialized for detecting behaviorally relevant information from the environment that is salient or occurs unexpectedly, thereby requiring a reorienting of spatial attention and a resetting of the attentional task set to a new location in visual space. We will discuss the functions of the ventral attention network in more detail in a later section.

In the human brain, evidence that the frontoparietal attention network is the source of the modulatory signals found in the visual system has been provided by imaging studies in which subjects were cued to direct attention to a location in the visual field, where a target stimulus would appear with a delay (Hopfinger, Buonocore, & Mangun, 2000; Kastner et al., 1999). In one study, it was shown that during directed attention to the target location in the delay period, thus in the absence of visual stimuli, there was a stronger increase in baseline activity in frontal and parietal areas of the attention network than in visual cortex. Importantly, there was no further increase in activity evoked by the attended stimulus presentations in these higher-order areas (see figure 75.4C; Kastner et al., 1999). Rather, there was sustained activity throughout the expectation period and the attended presentations, indicating that the activity reflected the attentional operations of the task and not visual processing. This evidence has been elegantly corroborated by recent studies that combined transcranial magnetic stimulation (TMS) concurrently with fMRI, or EEG measurements (Ruff et al., 2006; Taylor, Nobre, & Rushworth, 2007). With TMS, a brief magnetic pulse is applied through the skull to an underlying brain region, which may cause a variety of phenomena depending on the specific brain region that is targeted, but importantly often includes interference with ongoing neural processing for a short period of time. Thus, this method provides a causal measure of brain–behavior relationships. Stimulating the human FEF evoked a characteristic and topographic pattern of activity in early and intermediate visual cortex that was accompanied by alterations in visual perception (Ruff et al., 2006). These effects on neural activity occurred in the absence and in the presence of visual stimuli. TMS applied over parietal cortex also influenced neural activity in downstream visual areas (Ruff et al., 2008), and the pattern of activity changes was different from frontal stimulation, suggesting different roles for parietal and frontal cortex in cognitive control. Together, these findings have provided strong evidence in support of the notion that these parietal and frontal areas are the sources of feedback that generate the top-down biasing signals seen in visual cortex.

   GEOFFREY M. BOYNTON AND SABINE KASTNER

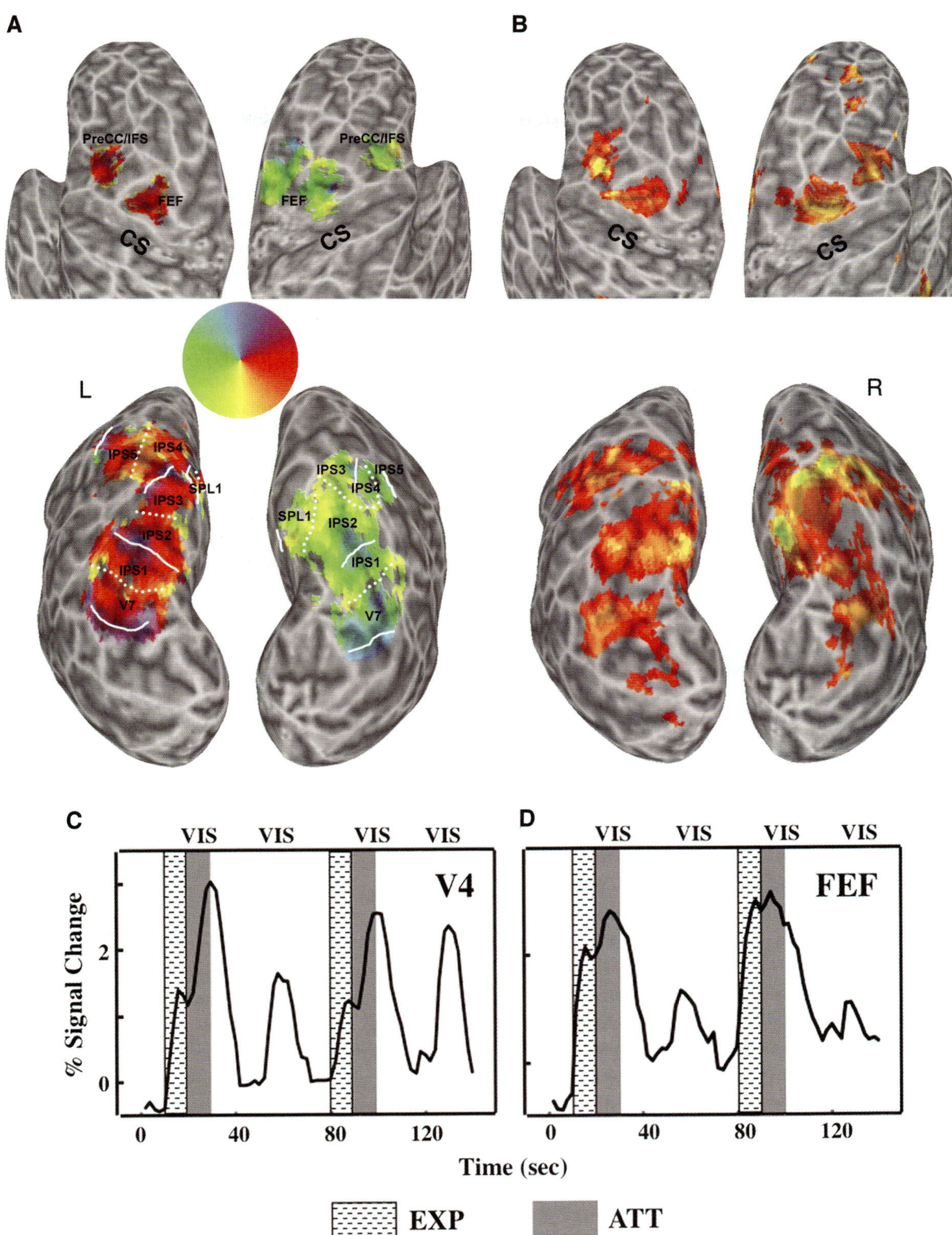

FIGURE 75.4   The frontoparietal attention network. (A) Topographic organization of areas in frontal and parietal cortex. Using a memory-guided saccade task, several areas with a systematic representation of the contralateral visual field were identified along the intraparietal sulcus (IPS0–IPS5), adjacent superior parietal cortex (SPL1), and in superior (FEF) and inferior aspects of precentral cortex (preCC/IFS). (B) Attention-related activations within parietal and frontal cortex in a spatial attention task. There was significant overlap between attention-related activations and topographic representations in higher-order cortex. (C) Time series of functional magnetic resonance imaging (fMRI) signals in V4. Directing attention to a peripheral target location in the absence of visual (VIS) stimulation led to an increase of baseline activity (textured blocks; EXP), which was followed by a further increase after the onset of the stimuli (gray shaded blocks; ATT). Baseline increases were found in both striate and extrastriate visual cortex. (Bottom row) Time series of fMRI signals in FEF. Directing attention to the peripheral target location in the absence of visual stimulation led to a stronger increase in baseline activity than in visual cortex; the further increase of activity after the onset of the stimuli was not significant.

As mentioned earlier, the existence of topographic representations in sensory systems has greatly facilitated the study of functional specialization of cortical areas. The methods of measuring cortical fMRI responses under passive conditions in order to reveal retinotopic organization has recently been extended to a variety of more complex tasks and stimuli. Such "cognitive mapping" approaches have revealed topographic organization in parietal and frontal cortex, thereby meeting the second prediction made by models of space-based attention control.

The first evidence of topographic organization within human parietal cortex was provided by Sereno and colleagues (Sereno, Pitzalis, & Martinez, 2001), employing a memory-guided saccade task in which the location of a target stimulus was remembered during a delay period, followed by a saccadic eye movement to the remembered location. The location of the target stimulus and saccade endpoint systematically traversed the visual field, and analysis of fMRI responses revealed a topographic map in posterior parietal cortex (PPC). Subsequent investigations used a variety of experimental paradigms including a visual spatial attention task (Silver, Ress, & Heeger, 2005), presentation of a colorful and dynamic periodic mapping stimulus (Swisher et al., 2007), and a variation of the memory-guided saccade task originally used by Sereno and colleagues (Konen & Kastner, 2008; Schluppeck, Glimcher, & Heeger, 2005) to characterize topographic organization of responses in human PPC. Thus far, seven topographically organized parietal areas have been described: Six of these areas form a contiguous band along the intraparietal sulcus (IPS0–IPS5), and one area branches off into the superior parietal lobule (SPL1) (see figure 75.4A). Each of these topographic areas contains a continuous representation of the contralateral visual field and is separated from neighboring areas by reversals in the orientation of the visual field representation. Boundaries between these areas correspond to alternating representations of the upper (denoted in blue in figure 75.4A) and lower (denoted in yellow) vertical meridian.

Topographic maps have also been discovered in frontal cortex in a variety of periodic tasks (Hagler & Sereno, 2006; Kastner et al., 2007). One such map is located in the superior branch of precentral cortex (PreCC), in the approximate location of the human FEF, and a second one in the inferior branch of PreCC (see figure 75.4A). Both of these areas are also activated by visually guided saccadic eye movements (Kastner

et al., 2007), and their topographic representations have several characteristic features. First, there is a bias toward a contralateral representation of both saccade directions and memorized locations in each area. Second, similar saccade directions and memorized locations are commonly represented in neighboring locations of each topographic map. Finally, for each area, particular saccade directions or memorized locations are represented redundantly in several parts of the topographic map. Thus, the representation of visual space in these frontal maps appears to be different from the organization of occipital and parietal maps, which typically exhibit a one-to-one mapping between locations in visual space and locations on the cortical sheet. In contrast, in the frontal maps, particular saccade directions and memorized locations were found to be sometimes represented in multiple locations in each topographic area. Remarkably, topographic representations in frontal cortex showed significant variability across subjects but were highly reproducible within subjects.

Importantly, there is a great amount of overlap between the attention-related activations discussed in the last section with topographically organized frontal and parietal areas (see figure 75.4A, B), which permits the systematic study of attention control systems, as represented in topographically defined higher-order cortical areas in individual subjects. Such an approach promises to yield a mechanistic understanding of the neural underpinnings of cognitive control processes related to selective attention as well as others.

Defining Control Properties of the Frontoparietal Attention Network

Following such a research strategy, attention signals in topographic frontal and parietal areas were investigated in individual subjects in a study in which subjects were instructed to covertly direct attention to a peripheral location in either the left or right visual hemifield and to detect targets embedded in a stream of visual stimuli (Szczepanski, Konen, & Kastner, 2010). Activity in most of the topographic PPC areas was found to be spatially specific, with stronger responses associated with directing attention to the contralateral as compared to the ipsilateral visual field, thereby generating a contralateral spatial biasing signal (see figure 75.5A). With the exception of left SPL1, each topographic area in frontal and parietal cortex generated an individual contralateral spatial bias that was on average balanced between the two hemispheres (see figure 75.5B). Importantly, two hemispheric asymmetries were noted (Szczepanski, Konen, & Kastner, 2010). First, right, but not left, SPL1

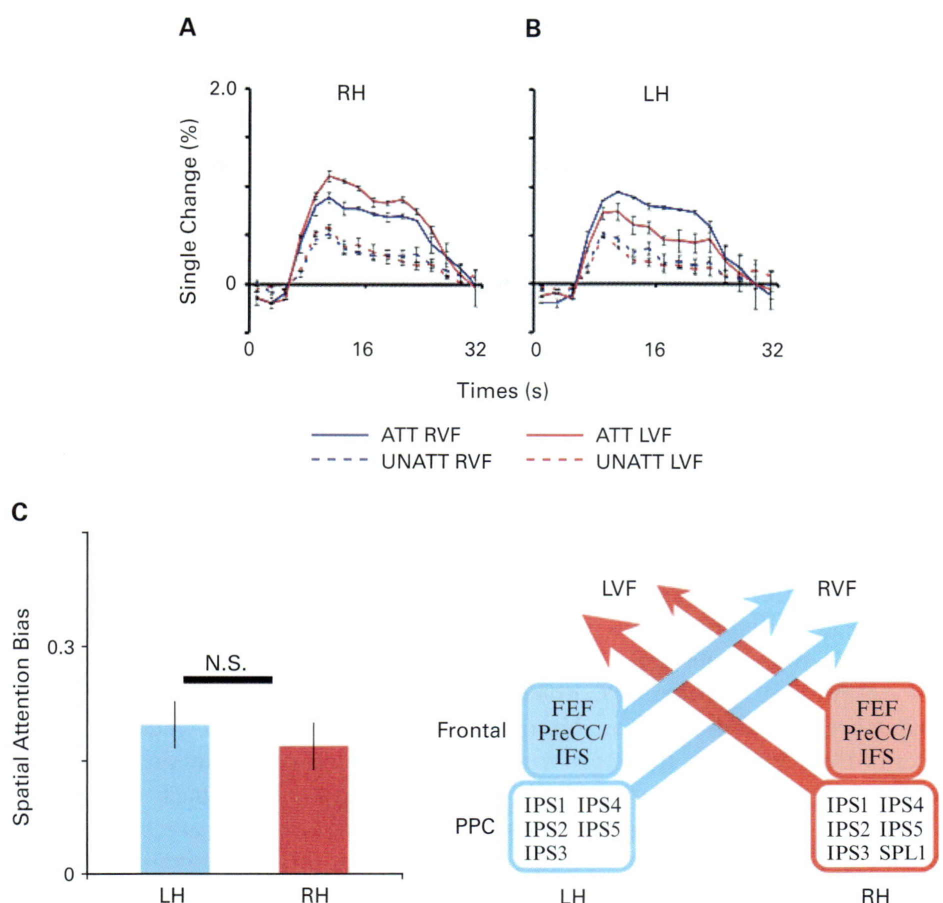

FIGURE 75.5   An interhemispheric competition model of spatial attention control. (A) Spatial attention signals averaged across subjects in topographic regions of frontal cortex ($n = 9$) are shown separately for each hemisphere. Solid curves indicate activity evoked by the attended conditions (ATT), and dashed curves indicate activity evoked by the unattended (UNATT) conditions. Red curves correspond to directed attention to the left visual field (LVF) while blue curves correspond to directed attention to the right visual field (RVF). Error bars indicate standard errors of the mean across subjects. (B) Attention bias signals toward the RVF or LVF were computed as index measures for each topographic area and averaged across topographic frontal and parietal areas within a hemisphere. The contralateral biases exerted by each hemisphere were similar, suggesting that the control of spatial attention across the visual field is exerted by a balanced competition of spatial biasing signals between the two hemispheres. (C) A model of spatial attention control is proposed, in which a number of topographically organized areas in frontoparietal cortex bias attention to the contralateral visual field. Arrows toward the RVF or LVF denote the direction of attentional bias for each lobe (frontal cortex or PPC) while the thickness of each arrow refers to the strength of the attentional bias, or "attentional weight." LH, left hemisphere; RH, right hemisphere; FEF, frontal eye field; preCC/IFS, inferior aspects of precentral cortex; IPS0–IPS5, areas with a systematic representation of the contralateral visual field identified along the intraparietal sulcus and (SPL1) adjacent superior parietal cortex.

carried spatial attention signals, as measured with fMRI. Second, left FEF and left IPS1/2 generated stronger contralateral biasing signals than their counterparts in the RH.

These results, based on data from the normal adult brain, potentially provide a neural basis for interhemispheric competition accounts of space-based attention control (Kinsbourne, 1977; Smania et al., 1998). Specifically, with this account (see figure 75.5C), each of the LH and RH areas contributes to the control of spatial attention across the visual field by generating a spatial bias, or "attentional weight" toward the contralateral hemifield. The sum of the weights contributed by each area within a hemisphere constitutes the overall spatial bias that can be exerted over contralateral space. The net output of attentional weights generated by the two hemispheres is similar, indicating that the control system is normally balanced. This balance of attentional weights across the hemispheres may be achieved through reciprocal interhemispheric inhibition of corresponding areas. However, the higher-order control system appears to be complicated by right SPL1's unique role in spatial attention. The attentional weight generated by this area was not counteracted by left SPL1. Instead, left FEF and left IPS1/2 generated stronger attentional weights than their right-hemispheric counterparts, possibly to maintain a balanced distribution of attentional resources across the visual field. Thus, the LH control system requires the cooperation of several distributed subcomponents in order to balance the spatial biasing signals generated by the two hemispheres (see figure 75.5C).

This empirical model of space-based attention control makes a number of predictions that can account for key findings from behavioral studies in neglect patients. First, the model predicts that dysfunction of right PPC (including right SPL1) will result in a spatial bias toward the RVF and a neglect of the LVF due to the shift of the residual attentional weights toward contralateral space. This is in agreement with data from visuospatial neglect patients with damage to right parietal cortex (Mesulam, 1999). Second, the model predicts that if a portion of either PPC or FEF and surrounding frontal regions is dysfunctional in the LH, a strong neglect of the RVF is less likely to result, since the intact portion of left frontal or parietal cortex may be able to compensate. Therefore, isolated dysfunction of left PPC or left FEF is *less* likely to cause a strong shift in the weights toward the LVF. This is in agreement with observations that RVF neglect occurs much less frequently with isolated frontal or parietal lesions. Therefore, together, the first two predictions are consistent with the notion of right-hemispheric dominance of attention function (Heilman & Van Den Abell, 1980; Mesulam, 1981).

Third, the model predicts that both left frontal cortex and left PPC need to be dysfunctional before right neglect is strongly manifested. In general, too few studies have examined patients in great enough detail to determine the exact neural correlates of right neglect. However, interestingly and in accordance with the model predictions, right neglect has been reported to occur more frequently in patients with extensive LH damage covering frontal, parietal, and temporal cortex. Several studies have reported extensive lesions (Bartolomeo, Chokron, & Gainotti, 2001) or tissue dysfunction (Kleinman et al., 2007) of frontal, parietal, and temporal cortex in patients exhibiting right neglect. Beis et al. (2004) compared a large number of patients with left and right neglect and found that those patients with large anteroposterior strokes, covering frontal and parietal cortex in the RH or the LH, respectively, performed relatively similarly on a comprehensive battery of tests for unilateral neglect. Thus, these studies appear to provide at least some preliminary evidence in support of the model predictions regarding the occurrence of right neglect.

Importantly, studies in patients suffering from hemineglect following a stroke to the RH show reduced activity in the right relative to the left PPC, even though these brain regions are structurally intact (Corbetta et al., 2005). Thus, the attentional deficits observed in these patients may be explained by a distal impact of the lesion, which is typically including inferior parts of the right parietal lobe and adjacent superior temporal cortex, on structurally intact brain issue in the right dorsal attention network, thereby resulting in an imbalance of attentional weights generated by each hemisphere to control contralateral space. This imbalance is also accompanied by a breakdown of functional connectivity within the dorsal network between the two hemispheres.

Modeling the Information Flow through the Attention Network

While there is good evidence that areas in frontal and parietal areas are the sources that generate modulatory signals for feedback to sensory processing areas, neuroimaging data have limitations in testing hypotheses about the directionality of the flow of information within the network and to derive information in the time domain about the ongoing computations due to the sluggishness and poor temporal resolution of hemodynamic signals. Functional connectivity models have recently begun to tackle some of these issues and provide first evidence for a future connectivity model of selective attention that will include the entire large-scale network that has been discussed in this chapter. Thus far, measures of Granger causality have confirmed that greater influences in the "top-down" direction, that is, from FEF and regions within the IPS to intermediate and lower-level tiers of visual cortex than in the reverse direction (Bressler et al., 2008). Moreover, these studies suggest greater causal influences from FEF to IPS than from IPS to FEF, suggesting together with corroborating evidence from monkey physiology (Buschman & Miller, 2007; Gregoriou et al., 2009) that areas in frontal cortex are key nodes for attentional control within the rest of the network. Attention signals in parietal cortex appear to precede those in visual cortex by about 200 ms (Lauritzen et al., 2009). Future network modeling studies will be invaluable sources of information about the involvement of certain nodes within a given task epoch as well as across different tasks and about the directionality of the information flow within a network. Linking such models to individual behavior will be a

    GEOFFREY M. BOYNTON AND SABINE KASTNER

particularly powerful means of obtaining information about brain–behavior relationships at the network level.

The Ventral Attention Network and Its Many Functions

As mentioned earlier, selective attention appears to involve two large-scale networks that are spatially and functionally segregated, a dorsal frontoparietal network that we have discussed thus far and a ventral frontoparietal network, respectively. While the dorsal network engages both hemispheres with the few asymmetries that were noted above, the ventral network is largely lateralized to the RH and consists of areas in the region of the temporoparietal junction (TPJ) and inferior frontal cortex (Corbetta & Shulman, 2002). Studies investigating activity patterns during a resting state confirmed that these networks constitute separate entities that were identifiable by characteristic coherence of signal fluctuations during the resting state that can be used as neural signatures of functional connectivity within large-scale networks (He et al., 2007).

Initial studies suggested that the ventral network reflected a reorienting of spatial attention to a new location in the visual field, where a stimulus had unexpectedly occurred (Corbetta et al., 2000). Particularly, it was shown that the dorsal network, but not the ventral network, became activated when spatial attention was directed to and maintained at a cued location, thus when a specific attentional set was instantiated and maintained. In contrast, the ventral network was activated when spatial attention needed to be redirected to a location that was not previously cued, where a target had appeared unexpectedly. Subsequently it was shown that such activation did not require a spatial redirection of attention but occurred also in so-called oddball tasks performed at a fixed spatial location, in which infrequently presented target stimuli had to be detected within a series of visual stimuli. Importantly, the ventral network is engaged during the detection of infrequently occurring target stimuli independent of their sensory modality or the specific response demands of the task (Downar et al., 2000). Together, these studies indicated that the ventral attention network constituted a general domain mechanism for the detection of salient and infrequently occurring events in the environment. It should be noted that the ventral attention network is considered a "nonspatial" network, that is, a network that does not represent spatial information systematically. Indeed, topographic organization has not yet been identified in the brain regions that constitute the ventral attention network.

Importantly, it has been demonstrated that activation of the ventral attention network does not occur with any salient or infrequent sensory stimulus, but only with stimuli that are relevant to the current task set. Further, the TPJ has been shown to be deactivated depending on task relevance and subjects' focus (Shulman et al., 2007), which has been interpreted as a filter mechanism that prevents reorienting to unimportant sensory stimuli. Such deactivation of the ventral network may be controlled by the dorsal attention network to maintain and optimize its attentional task set.

The ventral attention network, and in particular the TPJ, have also been shown to be activated in a number of tasks that involve self-referencing and are important in social cognition. For example, the TPJ is involved in "theory of mind" cognition; these are tasks during which subjects reason about other people's mental states (Fletcher et al., 1995; Gallagher & Frith, 2003) or during which they feel empathy for others (Decety & Lamm, 2007). While it is not clear whether the same populations of neurons in the region of the TPJ are involved with tasks that involve redirecting attention versus those that involve referencing oneself to another person's state of mind, it is conceivable that the reorienting function of the TPJ is not restricted to sensory events in the environment but can also be applied to changing a self-referential context (Graziano & Kastner, 2011). Thus, the ventral network may be a general purpose network for rereferencing social context that could involve events driven by the environment as well as by cognitive states.

CONCLUSION

Evidence from functional brain imaging reveals that attention operates at various processing levels within the human visual system and beyond. These attention mechanisms appear to be controlled by a distributed network of higher-order areas in frontal and parietal cortex, which generate top-down signals that are transmitted via feedback connections to the visual system. Together, these widely distributed brain systems cooperate to mediate the selection of behaviorally relevant information that can be further utilized in other cognitive networks to ultimately guide goal-directed action.

ACKNOWLEDGMENTS

We would like to thank the National Eye Institute, the National Institute of Mental Health, and the National Science Foundation for supporting our research.

REFERENCES

Bartolomeo, P., Chokron, S., & Gainotti, G. (2001). Laterally directed arm movements and right unilateral neglect after left hemisphere damage. *Neuropsychologia, 39*, 1013–1021.

Beck, D. M., & Kastner, S. (2005). Stimulus context modulates competition in human extrastriate cortex. *Nature Neuroscience, 8*, 1110–1116.

Beis, J. M., Keller, C., Morin, N., Bartolomeo, P., Bernati, T., Chokron, S., et al. (2004). Right spatial neglect after left hemisphere stroke: Qualitative and quantitative study. *Neurology 63*, 1600–1605.

Bisiach, E., & Vallar, G. (1988). Hemineglect in humans. In J. G. F. Boller (Ed.), *Handbook of neuropsychology* (Vol. 1, pp. 195–222). Amsterdam: Elsevier.

Bles, M., Schwarzbach, J., De Weerd, P., Goebel, R., & Jansma, B. M. (2006). Receptive field size-dependent attention effects in simultaneously presented stimulus displays. *Neuroimage, 30*, 506–511, doi:10.1016/j.neuroimage.2005.09.042.

Boynton, G. M. (2004). Adaptation and attentional selection. *Nature Neuroscience, 7*, 8–10.

Boynton, G. M. (2009). A framework for describing the effects of attention on visual responses. *Vision Research, 49*, 1129–1143.

Boynton, G. M. (2011). Spikes, BOLD, attention, and awareness: A comparison of electrophysiological and fMRI signals in V1. *Journal of Vision, 11*(5), 12, doi:10.1167/11.5.12.

Brefczynski, J. A., & DeYoe, E. A. (1999). A physiological correlate of the 'spotlight' of visual attention. *Nature Neuroscience, 2*, 370–374.

Bressler, S. L., Tang, W., Sylvester, C. M., Shulman, G. L., & Corbetta, M. (2008). Top-down control of human visual cortex by frontal and parietal cortex in anticipatory visual spatial attention. *Journal of Neuroscience, 28*, 10056–10061.

Buffalo, E. A., Bertini, G., Ungerleider, L. G., & Desimone, R. (2005). Impaired filtering of distracter stimuli by TE neurons following V4 and TEO lesions in macaques. *Cerebral Cortex, 15*, 141–151.

Buracas, G. T., & Boynton, G. M. (2007). The effect of spatial attention on contrast response functions in human visual cortex. *Journal of Neuroscience, 27*, 93–97.

Buracas, G. T., Fine, I., & Boynton, G. M. (2005). The relationship between task performance and functional magnetic resonance imaging response. *Journal of Neuroscience, 25*, 3023–3031.

Buschman, T. J., & Miller, E. K. (2007). Top-down versus bottom-up control of attention in the prefrontal and posterior parietal cortices. *Science, 315*, 1860–1862.

Ciaramitaro, V. M., Buracas, G. T., & Boynton, G. M. (2007). Spatial and cross-modal attention alter responses to unattended sensory information in early visual and auditory human cortex. *Journal of Neurophysiology, 98*, 2399–2413.

Cohen, M. R., & Maunsell, J. H. (2011). When attention wanders: How uncontrolled fluctuations in attention affect performance. *Journal of Neuroscience, 31*, 15802–15806.

Corbetta, M., Kincade, J. M., Ollinger, J. M., McAvoy, M. P., & Shulman, G. L. (2000). Voluntary orienting is dissociated from target detection in human posterior parietal cortex. *Nature Neuroscience, 3*, 292–297.

Corbetta, M., Kincade, M. J., Lewis, C., Snyder, A. Z., & Sapir, A. (2005). Neural basis and recovery of spatial attention deficits in spatial neglect. *Nature Neuroscience, 8*, 1603–1610.

Corbetta, M., Miezin, F. M., Dobmeyer, S., Shulman, G. L., & Petersen, S. E. (1990). Attentional modulation of neural processing of shape, color, and velocity in humans. *Science, 248*, 1556–1559.

Corbetta, M., & Shulman, G. L. (2002). Control of goal-directed and stimulus-driven attention in the brain. *Nature Reviews. Neuroscience, 3*, 201–215.

Damasio, A. R., Damasio, H., & Chui, H. C. (1980). Neglect following damage to frontal lobe or basal ganglia. *Neuropsychologia, 18*, 123–132.

Decety, J., & Lamm, C. (2007). The role of the right temporoparietal junction in social interaction: How low-level computational processes contribute to meta-cognition. *Neuroscientist, 13*, 580–593.

Downar, J., Crawley, A. P., Mikulis, D. J., & Davis, K. D. (2000). A multimodal cortical network for the detection of changes in the sensory environment. *Nature Neuroscience, 3*, 277–283.

Duncan, J. (1984). Selective attention and the organization of visual information. *Journal of Experimental Psychology. General, 113*, 501–517.

Fletcher, P. C., Happe, F., Frith, U., Baker, S. C., Dolan, R. J., Frackowiak, R. S., & Frith, C. D. (1995). Other minds in the brain: A functional imaging study of "theory of mind" in story comprehension. *Cognition 57*, 109–128.

Gallagher, H. L., & Frith, C. D. (2003). Functional imaging of 'theory of mind.' *Trends in Cognitive Sciences, 7*, 77–83.

Gandhi, S. P., Heeger, D. J., & Boynton, G. M. (1999). Spatial attention affects brain activity in human primary visual cortex. *Proceedings of the National Academy of Sciences of the United States of America, 96*, 3314–3319.

Graziano, M. S., & Kastner, S. (2011). Human consciousness and its relationship to social neuroscience: A novel hypothesis. *Cognitive Neuroscience, 2*, 98–113.

Gregoriou, G. G., Gotts, S. J., Zhou, H., & Desimone, R. (2009). High-frequency, long-range coupling between prefrontal and visual cortex during attention. *Science, 324*, 1207–1210.

Hagler, D. J., Jr., & Sereno, M. I. (2006). Spatial maps in frontal and prefrontal cortex. *Neuroimage, 29*, 567–577.

He, B. J., Snyder, A. Z., Vincent, J. L., Epstein, A., Shulman, G. L., & Corbetta, M. (2007). Breakdown of functional connectivity in frontoparietal networks underlies behavioral deficits in spatial neglect. *Neuron, 53*, 905–918.

Heilman, K. M., & Van Den Abell, T. (1980). Right hemisphere dominance for attention: The mechanism underlying hemispheric asymmetries of inattention (neglect). *Neurology 30*, 327–330.

Herrmann, K., Montaser-Kouhsari, L., Carrasco, M., & Heeger, D. J. (2010). When size matters: Attention affects performance by contrast or response gain. *Nature Neuroscience, 13*, 1554–1559.

Hopf, J. M., Boehler, C. N., Luck, S. J., Tsotsos, J. K., Heinze, H. J., & Schoenfeld, M. A. (2006). Direct neurophysiological evidence for spatial suppression surrounding the focus of attention in vision. *Proceedings of the National Academy of Sciences of the United States of America, 103*, 1053–1058.

Hopfinger, J. B., Buonocore, M. H., & Mangun, G. R. (2000). The neural mechanisms of top-down attentional control. *Nature Neuroscience, 3*, 284–291.

Huk, A. C., & Heeger, D. J. (2000). Task-related modulation of visual cortex. *Journal of Neurophysiology, 83*, 3525–3536.

Janer, K. W., & Pardo, J. V. (1991). Deficits in selective attention following bilateral anterior cingulotomy. *Journal of Cognitive Neuroscience, 3*, 231–241.

Kamitani, Y., & Tong, F. (2005). Decoding the visual and subjective contents of the human brain. *Nature Neuroscience, 8*, 679–685.

Karnath, H. O., Ferber, S., & Himmelbach, M. (2001). Spatial awareness is a function of the temporal not the posterior parietal lobe. *Nature, 411*, 950–953.

Karnath, H. O., Himmelbach, M., & Rorden, C. (2002). The subcortical anatomy of human spatial neglect: Putamen, caudate nucleus and pulvinar. *Brain, 125*, 350–360.

Kastner, S., DeSimone, K., Konen, C. S., Szczepanski, S. M., Weiner, K. S., & Schneider, K. A. (2007). Topographic maps in human frontal cortex revealed in memory-guided saccade and spatial working-memory tasks. *Journal of Neurophysiology, 97*, 3494–3507.

Kastner, S., De Weerd, P., Desimone, R., & Ungerleider, L. G. (1998). Mechanisms of directed attention in the human extrastriate cortex as revealed by functional MRI. *Science, 282*, 108–111.

Kastner, S., De Weerd, P., Pinsk, M. A., Elizondo, M. I., Desimone, R., & Ungerleider, L. G. (2001). Modulation of sensory suppression: Implications for receptive field sizes in the human visual cortex. *Journal of Neurophysiology, 86*, 1398–1411.

Kastner, S., Nothdurft, H. C., & Pigarev, I. (1997). Neuronal correlates of pop-out in cat striate cortex. *Vision Research, 37*, 371–376.

Kastner, S., Pinsk, M. A., De Weerd, P., Desimone, R., & Ungerleider, L. G. (1999). Increased activity in human visual cortex during directed attention in the absence of visual stimulation. *Neuron, 22*, 751–761.

Kastner, S., & Ungerleider, L. G. (2000). Mechanisms of visual attention in the human cortex. *Annual Review of Neuroscience, 23*, 315–341.

Kinsbourne, M. (1977). Hemi-neglect and hemisphere rivalry. *Advances in Neurology, 18*, 41–49.

Kinsbourne, M. (1993). Orientation bias model of unilateral neglect: Evidence from attentional gradients within hemispace. In I. H. Robertson & J. C. Marshall (Eds.), *Unilateral neglect: Clinical and experimental studies* (pp. 63–86). Hillsdale, NJ: Erlbaum.

Kleinman, J. T., Newhart, M., Davis, C., Heidler-Gary, J., Gottesman, R. F., & Hillis, A. E. (2007). Right hemispatial neglect: Frequency and characterization following acute left hemisphere stroke. *Brain and Cognition, 64*, 50–59.

Knierim, J. J., & van Essen, D. C. (1992). Neuronal responses to static texture patterns in area V1 of the alert macaque monkey. *Journal of Neurophysiology, 67*, 961–980.

Konen, C. S., & Kastner, S. (2008). Representation of eye movements and stimulus motion in topographically organized areas of human posterior parietal cortex. *Journal of Neuroscience, 28*, 8361–8375.

Lauritzen, T. Z., D'Esposito, M., Heeger, D. J., & Silver, M. A. (2009). Top-down flow of visual spatial attention signals from parietal to occipital cortex. *Journal of Vision, 9*(13), 8, 11–14. doi:10.1167/9.13.18.

Liu, T., Slotnick, S. D., Serences, J. T., & Yantis, S. (2003). Cortical mechanisms of feature-based attentional control. *Cerebral Cortex, 13*, 1334–1343.

MacEvoy, S. P., & Epstein, R. A. (2009). Decoding the representation of multiple simultaneous objects in human occipitotemporal cortex. *Current Biology, 19*, 943–947.

Martinez, A., Anllo-Vento, L., Sereno, M. I., Frank, L. R., Buxton, R. B., Dubowitz, D. J., et al. (1999). Involvement of striate and extrastriate visual cortical areas in spatial attention. *Nature Neuroscience, 2*, 364–369.

Martinez Trujillo, J., & Treue, S. (2002). Attentional modulation strength in cortical area MT depends on stimulus contrast. *Neuron, 35*, 365–370.

McMains, S. A., & Kastner, S. (2010). Defining the units of competition: Influences of perceptual organization on competitive interactions in human visual cortex. *Journal of Cognitive Neuroscience, 22*, 2417–2426.

Mesulam, M. M. (1981). A cortical network for directed attention and unilateral neglect. *Annals of Neurology, 10*, 309–325.

Mesulam, M. M. (1999). Spatial attention and neglect: Parietal, frontal and cingulate contributions to the mental representation and attentional targeting of salient extrapersonal events. *Philosophical Transactions of the Royal Society of London. Series B, Biological Sciences, 354*, 1325–1346.

Mitchell, J. F., Sundberg, K. A., & Reynolds, J. H. (2007). Differential attention-dependent response modulation across cell classes in macaque visual area V4. *Neuron, 55*, 131–141.

Mitchell, J. F., Sundberg, K. A., & Reynolds, J. H. (2009). Spatial attention decorrelates intrinsic activity fluctuations in macaque area V4. *Neuron, 63*, 879–888.

Moore, T., & Armstrong, K. M. (2003). Selective gating of visual signals by microstimulation of frontal cortex. *Nature, 421*, 370–373.

Moran, J., & Desimone, R. (1985). Selective attention gates visual processing in the extrastriate cortex. *Science, 229*, 782–784.

Mort, D. J., Malhotra, P., Mannan, S. K., Rorden, C., Pambakian, A., Kennard, C., & Husain, M. (2003). The anatomy of visual neglect. *Brain, 126*, 1986–1997, doi:10.1093/brain/awg200.

Murray, S. O. (2008). The effects of spatial attention in early human visual cortex are stimulus independent. *Journal of Vision, 8*(10), 1–11, doi:10.1167/8.10.2.

O'Connor, D. H., Fukui, M. M., Pinsk, M. A., & Kastner, S. (2002). Attention modulates responses in the human lateral geniculate nucleus. *Nature Neuroscience, 5*, 1203–1209.

O'Craven, K. M., Downing, P. E., & Kanwisher, N. (1999). fMRI evidence for objects as the units of attentional selection. *Nature, 401*, 584–587.

Palmer, J., Ames, C. T., & Lindsey, D. T. (1993). Measuring the effect of attention on simple visual search. *Journal of Experimental Psychology. Human Perception and Performance, 19*, 108–130.

Pestilli, F., Carrasco, M., Heeger, D. J., & Gardner, J. L. (2011). Attentional enhancement via selection and pooling of early sensory responses in human visual cortex. *Neuron, 72*, 832–846.

Posner, M. I., Snyder, C. R., & Davidson, B. J. (1980). Attention and the detection of signals. *Journal of Experimental Psychology, 109*, 160–174.

Reddy, L., Kanwisher, N. G., & VanRullen, R. (2009). Attention and biased competition in multi-voxel object representations. *Proceedings of the National Academy of Sciences of the United States of America, 106*, 21447–21452.

Ress, D., Backus, B. T., & Heeger, D. J. (2000). Activity in primary visual cortex predicts performance in a visual detection task. *Nature Neuroscience, 3*, 940–945.

Reynolds, J. H., Chelazzi, L., & Desimone, R. (1999). Competitive mechanisms subserve attention in macaque areas V2 and V4. *Journal of Neuroscience, 19,* 1736–1753.

Reynolds, J. H., & Desimone, R. (2003). Interacting roles of attention and visual salience in V4. *Neuron, 37,* 853–863.

Reynolds, J. H., & Heeger, D. J. (2009). The normalization model of attention. *Neuron, 61,* 168–185.

Reynolds, J. H., Pasternak, T., & Desimone, R. (2000). Attention increases sensitivity of V4 neurons. *Neuron, 26,* 703–714.

Ruff, C. C., Bestmann, S., Blankenburg, F., Bjoertomt, O., Josephs, O., Weiskopf, N., et al. (2008). Distinct causal influences of parietal versus frontal areas on human visual cortex: Evidence from concurrent TMS–fMRI. *Cerebral Cortex, 18,* 817–827.

Ruff, C. C., Blankenburg, F., Bjoertomt, O., Bestmann, S., Freeman, E., Haynes, J. D., et al. (2006). Concurrent TMS–fMRI and psychophysics reveal frontal influences on human retinotopic visual cortex. *Current Biology, 16,* 1479–1488.

Saenz, M., Buracas, G. T., & Boynton, G. M. (2002). Global effects of feature-based attention in human visual cortex. *Nature Neuroscience, 5,* 631–632.

Schluppeck, D., Glimcher, P., & Heeger, D. J. (2005). Topographic organization for delayed saccades in human posterior parietal cortex. *Journal of Neurophysiology, 94,* 1372–1384.

Serences, J. T. (2011). Mechanisms of selective attention: Response enhancement, noise reduction, and efficient pooling of sensory responses. *Neuron, 72,* 685–687.

Serences, J. T., & Boynton, G. M. (2007). Feature-based attentional modulations in the absence of direct visual stimulation. *Neuron, 55,* 301–312.

Serences, J. T., Yantis, S., Culberson, A., & Awh, E. (2004). Preparatory activity in visual cortex indexes distractor suppression during covert spatial orienting. *Journal of Neurophysiology, 92,* 3538–3545.

Sereno, M. I., Pitzalis, S., & Martinez, A. (2001). Mapping of contralateral space in retinotopic coordinates by a parietal cortical area in humans. *Science, 294,* 1350–1354.

Sestieri, C., Sylvester, C. M., Jack, A. I., d'Avossa, G., Shulman, G. L., & Corbetta, M. (2008). Independence of anticipatory signals for spatial attention from number of nontarget stimuli in the visual field. *Journal of Neurophysiology, 100,* 829–838.

Shadlen, M. N., & Newsome, W. T. (1998). The variable discharge of cortical neurons: Implications for connectivity, computation, and information coding. *Journal of Neuroscience, 18,* 3870–3896.

Shulman, G. L., Astafiev, S. V., McAvoy, M. P., d'Avossa, G., & Corbetta, M. (2007). Right TPJ deactivation during visual search: Functional significance and support for a filter hypothesis. *Cerebral Cortex, 17,* 2625–2633.

Silver, M. A., Ress, D., & Heeger, D. J. (2005). Topographic maps of visual spatial attention in human parietal cortex. *Journal of Neurophysiology, 94,* 1358–1371.

Silver, M. A., Ress, D., & Heeger, D. J. (2007). Neural correlates of sustained spatial attention in human early visual cortex. *Journal of Neurophysiology, 97,* 229–237.

Smania, N., Martini, M. C., Gambina, G., Tomelleri, G., Palamara, A., Natale, E., et al. (1998). The spatial distribution of visual attention in hemineglect and extinction patients. *Brain, 121,* 1759–1770.

Sprague, J. M. (1966). Interaction of cortex and superior colliculus in mediation of visually guided behavior in the cat. *Science, 153,* 1544–1547.

Swisher, J. D., Halko, M. A., Merabet, L. B., McMains, S. A., & Somers, D. C. (2007). Visual topography of human intraparietal sulcus. *Journal of Neuroscience, 27,* 5326–5337.

Szczepanski, S. M., Konen, C. S., & Kastner, S. (2010). Mechanisms of spatial attention control in frontal and parietal cortex. *Journal of Neuroscience, 30,* 148–160.

Taylor, P. C., Nobre, A. C., & Rushworth, M. F. (2007). FEF TMS affects visual cortical activity. *Cerebral Cortex, 17,* 391–399.

Tootell, R. B., Hadjikhani, N., Hall, E. K., Marrett, S., Vanduffel, W., Vaughan, J. T., & Dale, A. M. (1998). The retinotopy of visual spatial attention. *Neuron, 21,* 1409–1422.

Treue, S., & Martinez Trujillo, J. C. (1999). Feature-based attention influences motion processing gain in macaque visual cortex. *Nature, 399,* 575–579.

Vallar, G., & Perani, D. (1987). The anatomy of unilateral neglect after right-hemisphere stroke lesions: A clinical/CT-scan correlation study in man. *Neuropsychologia, 24,* 609–622.

Williford, T., & Maunsell, J. H. (2006). Effects of spatial attention on contrast response functions in macaque area V4. *Journal of Neurophysiology, 96,* 40–54.

Wojciulik, E., Kanwisher, N., & Driver, J. (1998). Covert visual attention modulates face-specific activity in the human fusiform gyrus: fMRI study. *Journal of Neurophysiology, 79,* 1574–1578.

Yantis, S., & Serences, J. T. (2003). Cortical mechanisms of space-based and object-based attentional control. *Current Opinion in Neurobiology, 13,* 187–193.

76 Attentional "Spotlight" in Early Visual Cortex

DAVID C. SOMERS

The "spotlight" metaphor for visual attentional selection suggests that (1) there is a single focus of attention; (2) it may be spatially separated from the center of eye gaze; (3) items within the focus of attention receive enhanced processing; and (4) items outside of the focus of attention receive little, if any, processing. Human neuroimaging studies have demonstrated that a neural correlate of the attentional spotlight operates in early visual processing areas, including primary visual cortex and the lateral geniculate nucleus (LGN) of the thalamus. As predicted by the spotlight metaphor, attention enhances processing of stimuli that lie within the retinotopic representation of the attentional focus. However, neuroimaging studies also reveal attentional selection mechanisms that go beyond the simple predictions of the spotlight metaphor. Outside the focus of attention, visual cortical processing is not simply unmodulated but rather is suppressed relative to baseline conditions. Furthermore, compelling evidence has also been provided that attentional selection may be split into multiple spatially distinct spotlights.

BACKGROUND

Although visual perception often feels effortless, in many complex situations there is far more information reaching our eyes than our cognitive systems can act on at one time. Although the retina performs massively parallel processing, visual cognition operates on no more than a few items at once. Given these limitations, what we do or do not perceive is largely determined by attentional mechanisms that select information for enhanced cognitive processing. William James (1890/1950), commenting on experiments by von Helmholtz (1910/1924; experiments were circa 1866), proposed that visual attention selects a single object or region of visual space while largely neglecting the rest of the visual field. This idea is at the core of the notion of an attentional spotlight. This spotlight model was further refined by Eriksen and St. James (1986) as a "zoom lens" mechanism that selects one contiguous, convex spatial window. It has successfully addressed a wealth of psychophysical data; however, more recent work (e.g., Alvarez & Franconeri, 2007; Awh & Pashler, 2000; Bettencourt & Somers, 2009; McMains & Somers, 2004, 2005; Pylyshyn & Storm, 1988; Somers & McMains, 2005) indicates that visual attention may simultaneously select a small number of objects at once while ignoring intervening regions. This article summarizes functional magnetic resonance imaging (fMRI) based research explorations of the mechanisms of spatial attention in human occipital cortex, including primary visual cortex.

SPATIAL MAPS IN EARLY VISUAL CORTICAL AREAS

In order to investigate the mechanisms of visual attention it is first important to review the organization of spatial maps in human visual cortex. The cerebral cortical representations of the visual field were first studied over 100 years ago by Inouye, an ophthalmologist working in the Russo–Japanese War (Inouye, 1909). He observed a correlation between the location of shrapnel damage to the back of the skull and the extent and location of visual field loss experienced by the injured soldiers. Not only did he observe evidence for contralateral representations, in which damage to the left occipital lobe resulted in right visual field loss and damage to the right occipital lobe resulted in left visual field loss, but he also observed that superior or dorsal occipital lobe damage correlated with loss of the lower visual field while inferior or ventral damage correlated with loss of the upper visual field. From these observations, he correctly inferred that the occipital lobe contains a map-like representation of the visual field that is roughly rotated 180° from the orientation of the visual field (see figure 76.1). These observations have been confirmed many times over the last century in humans (e.g., Holmes, 1918; Horton & Hoyt, 1991) and in nonhuman primates (e.g., Daniel & Whitteridge, 1961; Talbot & Marshall, 1942; Van Essen, Newsome, & Maunsell, 1984). This cortical map is distorted in a way that reflects the density of photoreceptors in the retina; the fovea, or center of the visual field, has a large amount of cortex dedicated to it while relatively smaller amounts of cortex are dedicated to the peripheral visual field.

With the advent of fMRI, noninvasive methods using phase-encoded stimuli were developed for rapidly and accurately mapping visual field representations within human occipital cortex (DeYoe et al., 1996; Engel, Glover, & Wandell, 1997; Sereno et al., 1995). The standard retinotopy method indexes the visual field into polar coordinates, mapping out the polar angle and eccentricity dimensions in two separate scan runs using flickering checkerboard stimuli that slowly circulate across one spatial dimension for several cycles. The moving checkerboard stimuli induce traveling waves of activity across visual cortex. Phase-encoded analysis first identifies voxels that are significantly modulated at the temporal frequency of the stimulus, then uses the phase of the peak traveling wave signal in each voxel to infer the voxel's preferred visual field location. The resulting phase information is typically displayed on inflated and flattened cortical surface representations for the occipital lobes (see figure 76.1). These fMRI methods have revealed no fewer than 20 distinct visual field maps tiling the occipital lobe and extending into posterior portions of parietal and temporal cortex (Arcaro et al., 2009; DeYoe et al., 1996; Engel, Glover, & Wandell, 1997; Sereno et al., 1995; Silver & Kastner, 2009; Swisher et al., 2007; Tootell et al., 1997; Wandell, Dumoulin, & Brewer, 2007). In early visual cortex, areas V1, V2, V3/VP, the representation of the fovea, or center of the visual field, converges at the occipital pole while more peripheral representations lie anterior to the occipital

pole. The dorsal portions of V1, V2, and V3 lie above the fundus of the calcarine sulcus, and each represents the lower contralateral quadrant of the visual field, while the ventral regions lying below the calcarine fundus represent the upper contralateral quadrant. In area V1, the map is a mirror image representation of the visual field; however, this flips to a nonmirror image representation in area V2 and flips back to a mirror-image representation to in V3/VP. The boundary between areas V1 and V2 lies at their representations of the vertical meridian of the visual field while the boundary between areas V2 and V3 lies at their representations of the horizontal meridian. Note that in this nomenclature area V3 contains only a lower field representation and area VP contains an upper field representation. Together V3 and VP constitute a full representation of the visual field. Although functional differences between these regions in monkeys have been demonstrated, such differences have not been observed in humans, and thus most authors refer to the human areas together as V3/VP or simplify this nomenclature to refer to the combined region as just V3 (e.g., Tootell et al., 1996; Wandell, Dumoulin, & Brewer, 2007). Anterior to V3, visual areas contain hemifield representations and their boundaries occur at vertical meridian representations. Area V3A lies anterior to dorsal V3 while hV4 lies anterior to ventral V3 (or VP).

These fMRI retinotopic mapping methods are an important initial step in the analysis of spatial attentional manipulations in early visual cortex. These maps can be obtained for each hemisphere of each individual subject. This permits the researcher to identify the visual cortical representations of each of multiple visual stimuli placed in the visual field. These functionally

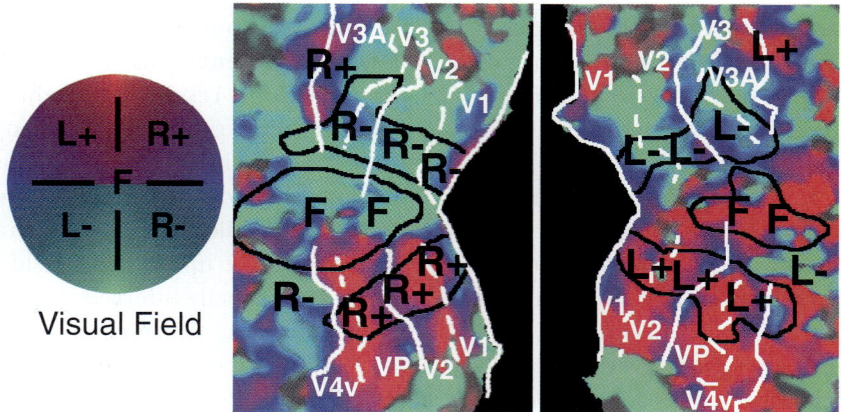

FIGURE 76.1   Retinotopic mapping of polar angle representations in early visual cortex. Left and right cortical hemispheric surfaces were reconstructed and inflated by computer; the occipital lobe representations were cut off and flattened to create these occipital patches. The color code reflects the preferred polar angle representations of each voxel obtained via functional mapping of retinotopy, using flashing checkerboard stimuli that slowly and periodically sweep through polar angles. (Reprinted from McMains & Somers, 2004.)

defined representations then can be used as regions of interest to study the cortical effects of directing attention toward or away from a particular stimulus.

ATTENTIONAL MODULATION IN V1 AND OTHER EARLY VISUAL CORTEX

Our initial fMRI investigations of attention began with the question of how visual cortical processing of a moving stimulus is influenced by attention (Somers et al., 1999). These experiments depended critically on the ability to distinguish the visual cortical representations of different parts of the visual field. The cortical representation of the center of the visual field (or fovea) occupies a large swath of cortex in the center of the flattened occipital cortex patches (see figure 76.2b, c). More peripheral eccentricities are represented as horizontal bands above and below the central region. The stimulus display was configured to exploit the eccentricity bias of the cortical map. A two-part stimulus display was utilized (see figure 76.2a). It consisted of an annular region containing a grating pattern, which rotated either clockwise or counterclockwise on a given trial, and a central disk in which letters were displayed in a rapid-serial-visual-presentation (RSVP) format. In this annulus-disk configuration, the RSVP letters would drive the central eccentricity band while the motion annulus would drive the horizontal bands above and below the central region.

Subjects held central fixation on this two-part display and directed their attention to one portion or the other. Eye position measurements performed in the scanner confirmed that subjects could hold central fixation during these experiments. Subjects performed alternating blocks of trials in which they either judged the rotation direction of the motion annulus or identified five consecutive letters appearing in the central RSVP stream. The central task was designed to be highly demanding in order to strongly draw attention away from the motion annulus. Comparison of fMRI activation between the two conditions revealed spatially specific attentional modulations across all early visual cortical areas, including V1, V2, V3, VP, V3A, V4v. When attention was directed to the motion annulus, fMRI activation increased in the iso-eccentricity bands corresponding to the cortical representation of the annulus (see figure 76.2d, e, figure 76.3). When attention was directed to the fovea, fMRI activation increased in the cortical representation of the fovea (see figure 76.3). These findings were also confirmed with a comparison of equivalent tasks, in which the RSVP stream was replaced by a motion disk stimulus that rotated independently of the annulus.

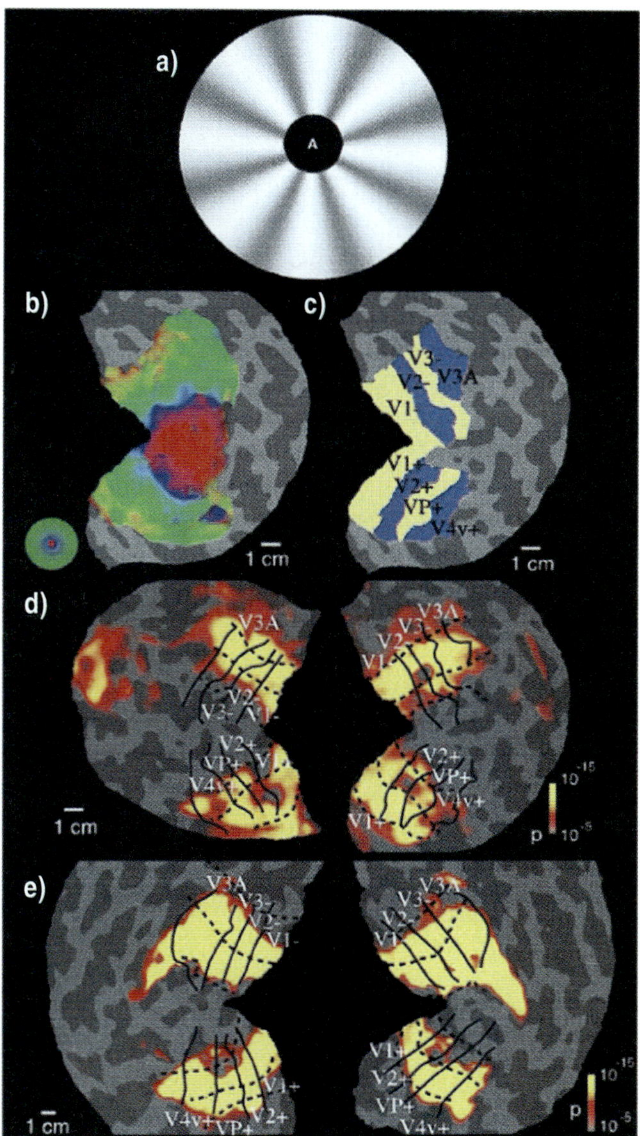

FIGURE 76.2 Attentional modulation of striate and extrastriate visual cortex. (a) Visual stimuli were composed of an annulus with rotating radial wedge patterns and a central target that was either a fixation point or single letters in a rapid-serial-visual-presentation (RSVP) stream. (b) Functional mapping of visual eccentricity, with the foveal representation in the center of the flattened patch. (c) Functionally defined visual cortical areas. (d, e) Patterns of statistically significant increased activation for attend extrafoveal motion versus attend foveal letters for both hemispheres for two subjects, extending across all labeled retinotopic areas. (Reprinted from Somers, Dale, Seiffert, & Tootell, 1999.)

At the time, the robust attentional modulations in area V1 were surprising. These attentional modulations in area V1 were as large as 1% signal change in some subjects. Overall, the attentional modulations alone comprised 60% of the combined stimulus plus attention activation (compared to passive viewing of a blank

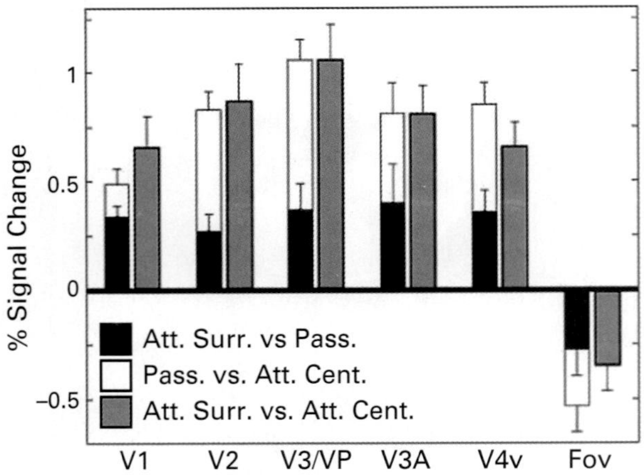

FIGURE 76.3 Facilitation and suppression in visual cortex. Three different attentional contrasts were run: Attend Surround versus Attend Center (gray), Attend Surround versus Passive Viewing (black), and Passive Viewing versus Attend Center (white). In the nonfoveal regions of interest (ROIs), any modulation in the Attend Surround versus Passive Viewing contrast reflects attentional facilitation while any modulation in the Passive Viewing versus Attend Center contrast reflects attentional suppression. In the foveal ROI, the negative values reflect greater activation for attending to the center of the stimulus than to the surround; facilitation and suppression effects of attention are also observed in this ROI. These findings reveal both facilitation and suppression effects of visual spatial attention in early visual cortex, demonstrating the push–pull manner of spatial attention. Att., attend; Surr., surround; Pass., passive; Cent., center; Fov, foveal. (Reprinted from Somers, Dale, Seiffert, & Tootell, 1999.)

fixation display). This implies that the attentional modulation amplitude was of the same order as the pure stimulus drive activation and was perhaps even greater than the stimulus drive. The conventional wisdom had held that primary visual cortex was a preprocessing stage that was "cognitively impenetrable." The seminal primate electrophysiology studies of Moran and Desimone in 1985 had observed attentional modulations in area V4, but not V1. As late as 1998, a leading research group predicted that attention could not operate in area V1. Our findings flew in the face of this conventional wisdom and were supported by nearly simultaneous reports of V1 attention from other fMRI laboratories using widely different stimuli and task paradigms (Brefczynski & DeYoe, 1999; Gandhi, Heeger, & Boynton, 1999; Martinez et al., 1999; Tootell et al., 1998; Watanabe et al., 1998). This wave of studies resulted in the quick acceptance of the V1 attention finding (Posner & Gilbert, 1999). Thus, the V1 attention result is an example of a finding that was widely accepted on the basis of human fMRI data before it was embraced on the basis of primate physiology. Motter

(1993) had reported V1 attentional effects in monkey, but this finding was questioned due to concerns that the effects might be due to small eye movement artifacts. The large areas of cortex modulated in the fMRI studies could not be explained away as an eye movement effect. Ironically, the relatively coarse spatial resolution of fMRI was an important asset in establishing this finding. Significantly, Kastner and colleagues (O'Connor et al., 2002) have shown robust spatial attention modulations in the LGN of the thalamus.

FACILITATION AND SUPPRESSION BY VISUAL SPATIAL ATTENTION

Further experiments were aimed at determining whether the observed attentional modulations reflected an enhancement of activity at the attended locations or suppression of activity at the unattended locations or both (Somers et al., 1999). In the initial experiments this issue could not be resolved since there was a direct comparison of attention directed to two spatially complementary regions. This issue was addressed by performing experiments in which each attentional state (attend center, attend periphery) was paired with a condition in which the stimulus was passively viewed. The passive viewing condition represents the attentional baseline condition. Each experiment makes complementary predictions about the modulations to be observed in the two spatial regions. In the comparison of Attend Periphery versus Passive Viewing, modulations observed in the periphery would be interpreted as attentional enhancements while modulations observed in the fovea region would be interpreted as attentional suppressions. Conversely, in the Attend Center versus Passive Viewing comparison, peripheral modulations would reflect suppression and central modulations would reflect enhancement. In both sets of experiments, attentional modulations were observed in both spatial regions (in antiphase relationship), implying that spatial attention acts in a "push–pull" manner (e.g., Posner, Snyder, & Davidson, 1980), increasing responses at the cortical representations of the attended locations and diminishing responses at the nonattended cortical representations (see figure 76.3).

MULTIFOCAL VISUAL ATTENTION

In real-world situations, such as driving, team sports, and videogame playing, we often have a need to attend to multiple objects while ignoring irrelevant distractors. There has been a long-standing debate as to whether spatial attention could be split to simultaneously attend to multiple distinct objects or regions of space. William

James (1890) famously defined attention as "the taking possession by the mind, in clear and vivid form, of *one* out of what seem several simultaneous objects or trains of thought." This unifocal view of spatial attention dominated psychology until the turn of the millennium. However, numerous experiments, spanning behavioral, fMRI, EEG, and primate electrophysiology methods, demonstrate that multifocal attention is not only possible but offers behavioral advantages (Awh & Pashler, 2000; Cavanagh & Alvarez, 2005; Cave & Bichot, 1999; McMains & Somers, 2004, 2005; Muller et al., 2003a; Niebergall et al., 2011; Pylyshyn & Storm, 1988). Historically three different mechanisms have been proposed to account for attention to multiple objects: (1) The "multiple spotlight" model suggests that spatial attention may be split to simultaneously attend to a small number of objects; (2) the "zoom lens" model (Castiello & Umilta, 1990; Eriksen & St. James, 1986; Muller et al., 2003b) suggests that unifocal attention spreads out to select multiple objects; the limitations to this model are that attentional resources are diluted as the overall spatial extent of the zoom lens increases and that it also requires selection of any distractors that lie in between targets of interests; thus additional processing would be required to ignore these selected distractors; and (3) the "rapidly moving spotlight" model (Shulman, Remington, & McLean, 1979; Tsal, 1983) suggests that a unifocal attentional spotlight rapidly moves between multiple locations. Variants of this model suggest either (a) that the spotlight remains on while switching, thus briefly selecting regions and objects that lie between targets (much like the zoom lens model predicts), or (b) that the spotlight rapidly turns off when it moves and turns back on when it reaches a target of interest. This model requires that the attentional spotlight switch between targets multiple times per second. While there is little doubt that we have the ability to switch attention between multiple objects, the speed limit of the attentional switching is a major point of debate in this literature. When subjects intentionally move their attentional spotlight, it takes a minimum of 200–250 ms to shift from one target to another, thus implying that a full cycle of attention to two distinct objects should take 400 ms or longer.

Our position is that all three mechanisms may account for attentional selection in different contexts. Of the three proposals, only the multiple spotlight mechanism is controversial; thus the remainder of this section will focus on evaluation of this model. We performed a series of experiments that demonstrate that spatial attention can be split into multiple distinct spotlights of activation in human occipital cortical regions (McMains & Somers, 2004, 2005). Since fMRI studies are limited by their temporal resolution and since spatial attention is capable of moving as fast as every 200–500 ms (e.g., Peterson & Juola, 2000; Reeves & Sperling, 1986; Weichselgartner & Sperling, 1987), it is extremely difficult to rule out the movement of spatial attention as a factor in the pattern of activation observed if one relies purely on the fMRI data. A central feature of these experiments was to employ a psychophysical task that excluded the possibility that spatial attention was rapidly switching between locations of interest.

Subjects were required to compare the identity of targets simultaneously displayed in two separated locations (see figure 76.4a, b). The targets appeared in RSVP streams and thus were only briefly visible and were immediately masked by another stimulus. The targets were digits appearing in RSVP streams of letters (use of letter targets did not change the psychophysical findings) and there were no other overt cues to the appearance of a target. Subjects were required to report whether the target digits were the same or different. This task can only be performed at above chance levels if both targets are identified. These two RSVP streams were imbedded in a display consisting of five RSVP streams, a central one surrounded by four equidistant peripheral streams, one per visual field quadrant (see figure 76.4a). The central RSVP stream lay directly in between the two attended RSVP streams (which were placed in opposing visual field quadrants). In order to make the central stream highly distracting, only digits appeared in this stream. If information appearing in the central stream were selected along with the information in the two peripheral streams of interest, it would likely interfere with performance of the digit comparison task. Thus, for this task, it was advantageous to avoid selecting the central region while selecting the two peripheral streams. Psychophysical performance in this task was investigated by parametrically varying the letter presentation duration of the RSVP streams. Threshold level performance ($d' = 1$) was observed at letter durations of 67 ms. This rate is much faster than the minimum estimates for how quickly spatial attention can select a target, deploy to a new location, and select a second target (200–500 ms). In addition to the Attend2 task, an Attend1 condition was included in which attention was covertly directed only to a (third) peripheral RSVP stream while viewing the same display as that used in the Attend2 condition. In the Attend1 task subjects had to identify digits appearing in the stream and report whether they matched a predefined target digit. Performance in the Attend1 task exhibited a small but statistically significant enhancement relative to Attend2 task performance. Although the Attend1 task was somewhat easier than the Attend2 task,

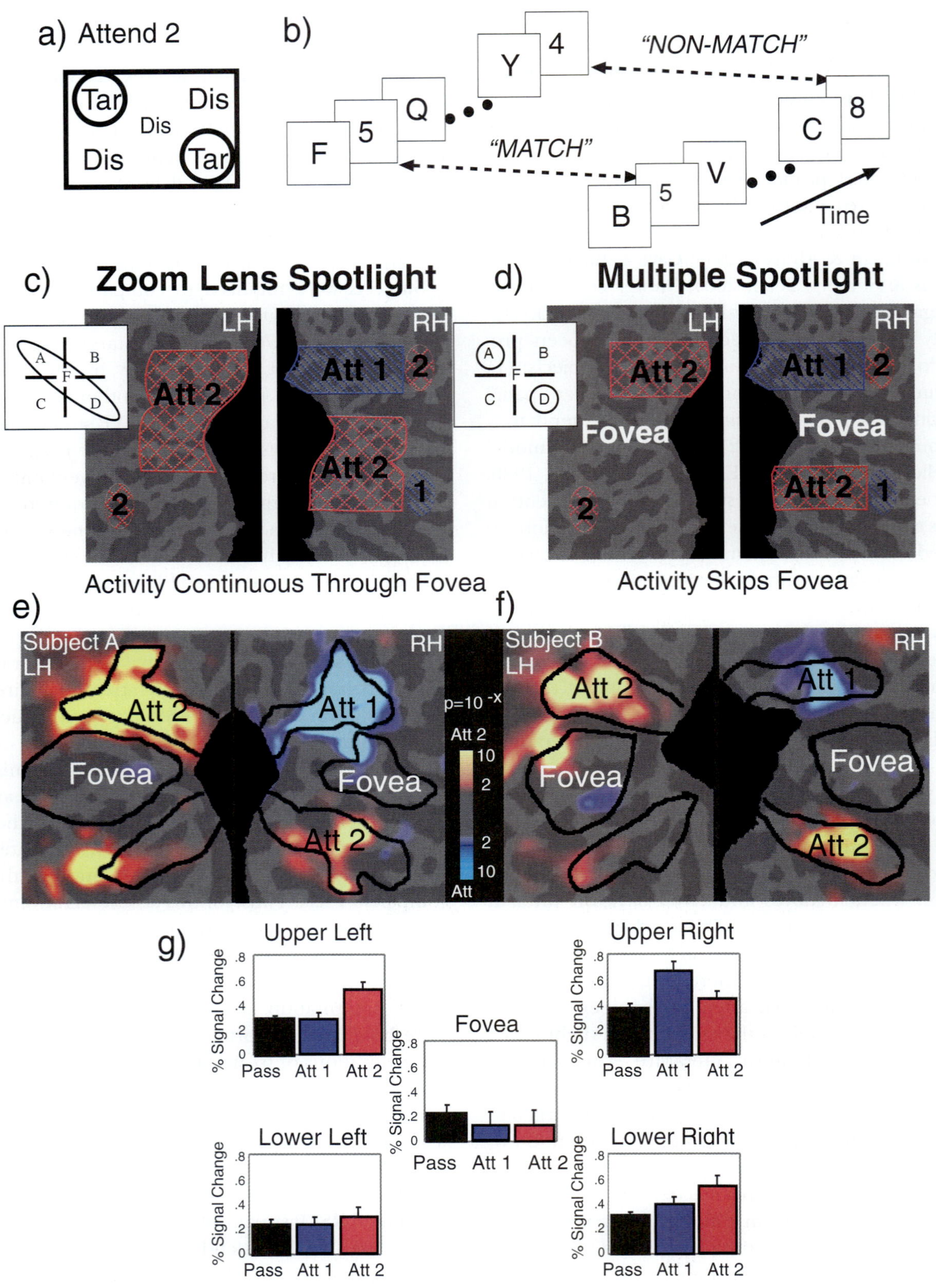

FIGURE 76.4   Multiple spotlights of spatial attention. (a, b) Multiple RSVP letter streams were presented. In the Attend2 (Att 2) condition, subjects were instructed to attend to two letter streams while ignoring an intermediately positioned stream and two others. Digits would appear simultaneously in both streams, and subjects had to report whether they were the same or different. (c, d) Predicted activation patterns for Attend2 versus Attend1 (Att 1) for the two competing hypotheses. (e, f) Activation patterns for Attend2 versus Attend1, for two subjects, supporting the multiple spotlight hypothesis. (g) Average activations in each of the five regions of interest confirm that subjects split their attentional spotlight between two peripheral regions and spared the fovea in the Attend2 task. Tar, target; Dis, distractor; LH, left hemisphere; RH, right hemisphere. (Reprinted from McMains & Somers, 2004.)

performance differences were far less than would be predicted if the Attend2 task were done by serial processing—a serial model would predict for a given performance level of sufficient difficulty that the Attend2 task should require twice the time that the Attend1 task required. However, the slope of the duration versus performance curves for the two tasks differed only by 13%, not by the 100% predicted by the serial model. Moreover, at threshold (d' = 1) in the Attend1 task was 59 ms, only 8 ms shorter than for the Attend2 task. Since no experiments have ever revealed evidence for 8-ms attentional switching (or 125 times per second), we reject the hypothesis that subjects were rapidly moving a single attentional spotlight. Even recent work that argues for rapid switching (e.g., Buschman & Miller, 2010; Hogendoorn, Carlson, & Verstraten, 2007) suggests substantially slower switch rates than could account for our findings. Thus the psychophysical results demonstrate that spatial attention was simultaneously deployed to the two RSVP streams of interest.

Having thus ruled out the rapid switching hypothesis for this task, two hypotheses then remained that could explain selection of the two peripheral streams. The zoom lens model predicts that attention stretches to select the two RSVP streams and necessarily also selects the distracting central RSVP stream (see figure 76.4c). The alternate hypothesis suggests that the spatial window of attention can be split into two distinct spotlights that select the two RSVP streams of interest while filtering out the central stream (see figure 76.4d). This question was addressed using fMRI of occipital lobe activation during task performance. Comparison of Attend2 versus Attend1 activation (see figure 76.4e, f) revealed two hot spots of activation at the visual cortical representations of the two Attend2 streams. Critically, the cortical representation of the central RSVP stream was not activated (nor was it activated in comparisons with the passive viewing condition). This pattern was observed for all subjects. This result demonstrates that the window of spatial attention may be split to select multiple distinct regions. This pattern was observed for all subjects. Group analysis revealed that this split attentional spotlight effect occurred in all "early" visual cortical areas, including primary visual cortex. As expected, the Attend1 condition (vs. Attend2) revealed a single hot spot of activation at the cortical representation of the attended RSVP stream (see figure 76.4e, f, g).

Our findings have been erroneously critiqued (Jans, Peters, & DeWeerd, 2010) and rightly defended (Cave, Bush, & Taylor, 2010) in recent review papers. The Jans et al. critique neglects the fact that performance of our task requires selection of both briefly presented stimuli and neglects the fact that subjects could perform this difficult task much faster than any serial model could predict given the Attend1 task data.

We repeated these experiments using a different spatial configuration of RSVP streams in which all streams were placed in one hemifield and the two RSVP streams of interest in the Attend2 task were placed in one visual field quadrant—one at the fovea and one in the periphery. The critical distractor was placed in between at a mid-eccentricity. The Attend1 RSVP stream was placed in the other visual quadrant. The fMRI results again revealed two hot spots of activation corresponding to the locations of the two targets, with a sparing of the intermediate region. This demonstrates that the attentional spotlight could be split not only within a single visual hemifield but also within a single quadrant of the visual field. The finding that the foveal representation may either be included or excluded as one focus of attention demonstrates that attention may be divided between overt and covert targets or divided between two covertly monitored targets. In these regards, the ability to divide spatial attention exhibits remarkable flexibility.

In a follow-up experiment (McMains & Somers, 2005), we directly compared zoom lens and split spotlight attention deployed across 1, 2, or 3 stimuli. As expected, behavioral performance declined as the number of attended RSVP streams increased. There was no difference in performance between attending to two adjacent RSVP streams (Zoom2 condition) and attending to two RSVP streams separated by a distractor RSVP stream (Split2 condition). Moreover, performance in the split spotlight (Split2) condition was superior to performance when zoom lens attention selected all three RSVP streams (Zoom3) in which both conditions covered the same spatial extent. These behavioral patterns were mirrored in the fMRI BOLD signal attentional modulation observed in occipital cortex. Thus, multifocal attentional deployment to two targets exhibited no overhead costs, in terms of behavior or occipital lobe activation, relative to zoom lens selection of the same number of targets, and multifocal attention exhibited a resource conservation benefit relative to zoom lens attentional selection over the same spatial extent.

Other paradigms have also revealed behavioral evidence for multifocal visual attention selection (Awh & Pashler, 2000; Cavanagh & Alvarez, 2005; Pylyshyn & Storm, 1988). Steady state visual evoked potential studies also provide strong evidence for the ability to simultaneously select two target locations while ignoring an intervening distractor region (Muller et al., 2003a). Our findings have also been replicated in fMRI and EEG experiments using different paradigms (Drew et al., 2009; Morawetz et al., 2007). Significantly, a recently published set of nonhuman primate electro-physiological experiments observed multifocal visual attention mechanisms at work within the middle temporal (MT) visual cortical area of the macaque (Niebergall et al., 2011).

In hindsight, the requirement that spatial attention be a serial process seems unnecessary. The capacity for parallel processing is an implicit feature of spatial representations. Once spatial attention was demonstrated to operate in early visual cortical areas (Brefczynski & DeYoe, 1999; Gandhi, Heeger, & Boynton, 1999; Martinez et al., 1999; Somers et al., 1999; Tootell et al., 1998; Watanabe et al., 1998), the general requirement for a single spotlight of spatial attention was ready to fall. These cortical areas primarily contain neurons with spatially limited receptive fields (RFs), so multiple targets need not compete for resources at these stages, provided that targets are far enough apart. Even in higher visual cortical areas, where attention shrinks down the normally large RFs to isolate single targets (e.g., Desimone, 1998; Desimone & Duncan, 1995), attentional selection may divide the pool of relevant neurons between multiple targets rather than assigning them in a "winner-take-all" fashion to one target location. This strategy might result in some cost for attending to multiple targets but would not necessarily result in a complete failure to select some locations. A modestly parallel spatial attention system of this form appears much simpler to implement physiologically than does a rapidly switching serial spatial attention system.

Control of spatial attention in early visual cortex is likely directed by regions of the posterior parietal cortex and lateral prefrontal cortex. Neuroimaging studies have revealed the existence of multiple visuotopic maps in parietal and frontal cortex (Schluppeck, Glimcher, & Heeger, 2005; Silver & Kastner, 2009; Silver, Ress, & Heeger, 2005; Swisher et al., 2007; Wandell, Dumoulin, & Brewer, 2007). Neuroimaging studies also demonstrate that activation in these regions is proportional to the number of tracked targets, suggesting that frontal and parietal lobe regions contain spatial attention maps that simultaneously index multiple locations (Culham et al., 1998; Culham, Cavanagh, & Kanwisher, 2001; Jovicich et al., 2001; Schluppeck, Glimcher, & Heeger, 2005; Shim et al., 2010; Silver et al., 2005; Swisher et al., 2007).

SUMMARY

Visuospatial attention is directed by frontoparietal cortex, but the influences are observed in all visual cortical areas, including V1, and even the LGN. Attention boosts BOLD signal activation at the retinotopic cortical representations of attended targets while suppressing activation for ignored distractors. In fMRI experiments, these spatial attention effects are additive/subtractive in nature. The visual system is limited in its attentional capacity, but this limit is not a single item, as implied by the spotlight model, but rather extends to a small number of objects. Multifocal spatial attention has been found to modulate early visual cortical representations of multiple targets while treating intervening stimuli as distractors.

REFERENCES

Alvarez, G. A., & Franconeri, S. L. (2007). How many objects can you track? Evidence for a resource-limited attentive tracking mechanism. *Journal of Vision, 7*(13), 14.1–14.10. doi:10.1167/7.13.14.

Arcaro, M., McMains, S., Singer, B., & Kastner, S. (2009). Retinotopic organization of human ventral visual cortex. *Journal of Neuroscience, 29,* 10638–10652.

Awh, E., & Pashler, H. (2000). Evidence for split attentional foci. *Journal of Experimental Psychology. Human Perception and Performance, 26,* 834–846.

Bettencourt, K. C., & Somers, D. C. (2009). Effects of target enhancement and distractor suppression on multiple object tracking capacity. *Journal of Vision, 9*(7), 1–11. doi:10.1167/9.7.9.

Brefczynski, J. A., & DeYoe, E. A. (1999). A physiological correlate of the 'spotlight' of visual attention. *Nature Neuroscience, 2,* 370–374.

Buschman, T., & Miller, E. K. (2010). Shifting the spotlight of attention: Evidence for discrete computations in cognition. *Frontiers in Human Neuroscience, 4,* 194.1–194.9. doi:10.3389/fnhum.2010.00194.

Castiello, U., & Umilta, C. (1990). Size of the attentional focus and efficiency of processing. *Acta Psychologica, 73,* 195–209.

Cavanagh, P., & Alvarez, G. (2005). Tracking multiple targets with multifocal attention. *Trends in Cognitive Sciences, 9,* 349–354.

Cave, K., & Bichot, N. (1999). Visuospatial attention: Beyond a spotlight model. *Psychonomic Bulletin & Review, 6,* 204–223.

Cave, K., Bush, W., & Taylor, T. (2010). Split attention as part of a flexible attentional system for complex scenes: Comment on Jans, Peters, and De Weerd (2010). *Psychological Review, 117,* 685–696.

Culham, J. C., Brandt, S. A., Cavanagh, P., Kanwisher, N. G., Dale, A. M., & Tootell, R. B. (1998). Cortical fMRI activation produced by attentive tracking of moving targets. *Journal of Neurophysiology, 80,* 2657–2670.

Culham, J. C., Cavanagh, P., & Kanwisher, N. G. (2001). Attention response functions: Characterizing brain areas using fMRI activation during parametric variations of attentional load. *Neuron, 32,* 737–745.

Daniel, P. M., & Whitteridge, D. (1961). The representation of the visual field on the cerebral cortex in monkeys. *Journal of Physiology, 159,* 203–221.

Desimone, R. (1998). Visual attention mediated by biased competition in extrastriate visual cortex. *Philosophical Transactions of the Royal Society of London. Series B, Biological Sciences, 353,* 1245–1255.

Desimone, R., & Duncan, J. (1995). Neural mechanisms of selective visual attention. *Annual Review of Neuroscience, 18,* 193–222.

DeYoe, E. A., Carman, G. J., Bandettini, P., Glickman, S., Wieser, J., Cox, R., et al. (1996). Mapping striate and extrastriate visual areas in human cerebral cortex. *Proceedings of the National Academy of Sciences of the United States of America, 93,* 2382–2386.

Drew, T., McCollough, A. W., Horowitz, T. S., & Vogel, E. K. (2009). Attentional enhancement during multiple-object tracking. *Psychonomic Bulletin & Review, 16,* 411–417.

Engel, S. A., Glover, G. H., & Wandell, B. A. (1997). Retinotopic organization in human visual cortex and the spatial precision of functional MRI. *Cerebral Cortex, 7,* 181–192.

Eriksen, C. W., & St. James, J. D. (1986). Visual attention within and around the field of focal attention: A zoom lens model. *Perception & Psychophysics, 40,* 225–240.

Gandhi, S. P., Heeger, D. J., & Boynton, G. M. (1999). Spatial attention affects brain activity in human primary visual cortex. *Proceedings of the National Academy of Sciences of the United States of America, 96,* 3314–3319. doi:10.1073/pnas.96.6.3314.

Hogendoorn, H., Carlson, T., & Verstraten, F. (2007). The time course of attentive tracking. *Journal of Vision, 7,* 2.1–2.10. doi:10.1167/7.14.2.

Holmes, G. (1918). Disturbances of vision by cerebral lesions. *British Journal of Ophthalmology, 2,* 353–384.

Horton, J. C., & Hoyt, W. F. (1991). The representation of the visual field in human striate cortex. *Archives of Ophthalmology, 109,* 816–824.

Inouye, T. (1909). *Die Sehstroungen bei Schussverietzungen der korti- kalen Sehsphare.* Leipzig, Germany: Engelmann.

James, W. (1890/1950). *The principles of psychology.* New York: Dover.

Jans, B., Peters, J. C., & DeWeerd, P. (2010). Visual spatial attention to multiple locations at once: The jury is still out. *Psychological Review, 117,* 637–684.

Jovicich, J., Peters, R. J., Koch, C., Braun, J., Chang, L., & Ernst, T. (2001). Brain areas specific for attentional load in a motion-tracking task. *Journal of Cognitive Neuroscience, 13,* 1048–1058.

Martinez, A., Anllo-Vento, L., Sereno, M. I., Frank, L. R., Buxton, R. B., Dubowitz, D. J., et al. (1999). Involvement of striate and extrastriate visual cortical areas in spatial attention. *Nature Neuroscience, 2,* 364–369.

McMains, S. A., & Somers, D. C. (2004). Multiple spotlights of attentional selection in human visual cortex. *Neuron, 42,* 677–686.

McMains, S. A., & Somers, D. C. (2005). Processing efficiency of divided spatial attention mechanisms in human visual cortex. *Journal of Neuroscience, 25,* 9444–9448.

Moran, J., & Desimone, R. (1985). Selective attention gates visual processing in the extrastriate cortex. *Science, 229,* 782–784.

Morawetz, C., Holz, P., Baudewig, J., Treue, S., & Dechent, P. (2007). Split of attentional resources in human visual cortex. *Visual Neuroscience, 24,* 817–826.

Motter, B. C. (1993). Focal attention produces spatially selective processing in visual cortical areas V1, V2, and V4 in the presence of competing stimuli. *Journal of Neurophysiology, 70,* 909–919.

Muller, M. M., Malinowski, P., Gruber, T., & Hillyard, S. A. (2003a). Sustained division of the attentional spotlight. *Nature, 424,* 309–312.

Muller, N. G., Bartelt, O. A., Donner, T. H., Villringer, A., & Brandt, S. A. (2003b). A physiological correlate of the "zoom lens" of visual attention. *Journal of Neuroscience, 23,* 3561–3565.

Niebergall, R., Khayat, P., Treue, S., & Martinez-Trujillo, J. (2011). Multifocal attention filters targets from distracters within and beyond primate MT neurons' receptive field boundaries. *Neuron, 72,* 1067–1079.

O'Connor, D. H., Fukui, M. M., Pinsk, M. A., & Kastner, S. (2002). Attention modulates responses in the human lateral geniculate nucleus. *Nature Neuroscience, 5,* 1203–1209.

Peterson, M. S., & Juola, J. F. (2000). Evidence for distinct attentional bottlenecks in attentional blink tasks. *Journal of General Psychology, 127,* 6–26.

Posner, M., & Gilbert, C. (1999). Attention and primary visual cortex. *Proceedings of the National Academy of Sciences of the United States of America, 96,* 2585–2587.

Posner, M. I., Snyder, C. R. R., & Davidson, B. J. (1980). Attention and the detection of signals. *Journal of Experimental Psychology, 109,* 160–174.

Pylyshyn, Z. W., & Storm, R. W. (1988). Tracking multiple independent targets: Evidence for a parallel tracking mechanism. *Spatial Vision, 3,* 179–197. doi:10.1163/156856888X00122.

Reeves, A., & Sperling, G. (1986). Attention gating in short-term visual memory. *Psychological Review, 93,* 180–206.

Schluppeck, D., Glimcher, P., & Heeger, D. J. (2005). Topographic organization for delayed saccades in human posterior parietal cortex. *Journal of Neurophysiology, 94,* 1372–1384.

Sereno, M. I., Dale, A. M., Reppas, J. B., Kwong, K. K., Belliveau, J. W., Brady, T. J., et al. (1995). Borders of multiple visual areas in humans revealed by functional magnetic resonance imaging. *Science, 268,* 889–893.

Shim, W. M., Alvarez, G. A., Vickery, T. J., & Jiang, Y. V. (2010). The number of attentional foci and their precision are dissociated in the posterior parietal cortex. *Cerebral Cortex, 19,* 916–925.

Silver, M. A., & Kastner, S. (2009). Topographic maps in human frontal and parietal cortex. *Trends in Cognitive Sciences, 13,* 488–495.

Silver, M. A., Ress, D., & Heeger, D. J. (2005). Topographic maps of visual spatial attention in human parietal cortex. *Journal of Neurophysiology, 94,* 1358–1371.

Shulman, G. L., Remington, R. W., & McLean, J. P. (1979). Moving attention through visual space. *Journal of Experimental Psychology. Human Perception and Performance, 5,* 522–526.

Somers, D. C., Dale, A. M., Seiffert, A. E., & Tootell, R. B. (1999). Functional MRI reveals spatially specific attentional modulation in human primary visual cortex. *Proceedings of the National Academy of Sciences of the United States of America, 96,* 1663–1668.

Somers, D. C., & McMains, S. A. (2005). Spatially-specific attentional modulation revealed by fMRI. In L. Itti, G. Rees, & J. Tsotsos (Eds.), *Neurobiology of attention* (pp. 377–382). New York: Academic Press.

Swisher, J. D., Halko, M. A., Merabet, L. B., McMains, S. A., & Somers, D. C. (2007). Visual topography of human intraparietal sulcus. *Journal of Neuroscience, 27,* 5326–5337.

Talbot, S. A., & Marshall, W. H. (1942). Physiological studies on the neural mechanisms of visual location and discrimination. *American Journal of Ophthalmology, 24*, 1255–1263.

Tootell, R. B., Dale, A. M., Sereno, M. I., & Malach, R. (1996). New images from human visual cortex. *Trends in Neurosciences, 19*, 481–489.

Tootell, R. B., Hadjikhani, N., Hall, E. K., Marrett, S., Vanduffel, W., Vaughan, J. T., et al. (1998). The retinotopy of visual spatial attention. *Neuron, 21*, 1409–1422.

Tootell, R. B., Mendola, J. D., Hadjikhani, N. K., Ledden, P. J., Liu, A. K., Reppas, J. B., et al. (1997). Functional analysis of V3A and related areas in human visual cortex. *Journal of Neuroscience, 17*, 7060–7078.

Tsal, Y. (1983). Movements of attention across the visual field. *Journal of Experimental Psychology. Human Perception and Performance, 9*, 523–530.

Van Essen, D. C., Newsome, W. T., & Maunsell, J. R. (1984). The visual field representation in striate cortex of the macaque mokey: Asymmetries, anisotropies, and individual variability. *Vision Research, 24*, 429–448.

von Helmholtz, H. (1910/1924 trans.). *Treatise on physiological optics.* New York: Dover.

Wandell, B., Dumoulin, S., & Brewer, A. (2007). Visual field maps in human cortex. *Neuron, 56*, 366–383.

Watanabe, T., Sasaki, Y., Miyauchi, S., Putz, B., Fujimaki, N., Nielsen, M., et al. (1998). Attention-regulated activity in human primary visual cortex. *Journal of Neurophysiology, 79*, 2218–2221.

Weichselgartner, E., & Sperling, G. (1987). Dynamics of automatic and controlled visual. *Science, 238*, 778–780.

77 Feature-Based Attention in Primates: Mechanisms and Theoretical Considerations

JULIO C. MARTINEZ-TRUJILLO AND PAUL S. KHAYAT

We can easily recognize that the Canadian flag has red and white thick stripes oriented vertically while the American flag has white and red thin stripes oriented horizontally. To make the distinction, our visual system has recognized the presence of the different colors, of the vertically and horizontally oriented stripes, and of the stripes' low or high spatial frequency. If we examine the properties of neurons in areas such as V1 and V4 of the primate visual pathways, we find individual neurons that are selective for each one of the listed features, that is, a given neuron fires strongly when a red stripe is shown inside its receptive field (RF) and weakly when a white stripe is shown at the same location. Because different neurons with overlapping RFs are selective for different features, we can infer that the primate visual cortex contains spatial/retina-centered maps of stimulus features. It is thought that the activity evoked by signals entering the retina within these cortical maps underlies visual perception.

A consensus among visual neuroscientists is that during behavior, neuronal activity within cortical maps does not faithfully reflect all signals entering the retinas. Behavioral and physiological studies have demonstrated a mechanism that allows stimuli sharing behaviorally relevant features to be preferentially represented over other stimuli. For example, when searching for the Canadian flag among those of other countries, flags that are red and white, such as the Austrian flag, will be better represented in visual cortical maps than flags that do not share these colors, such as the Brazilian flag, which is composed of green and yellow colors. This mechanism, known as feature-based attention (FBA), allows the brain to focus processing resources on the potentially behaviorally relevant signals. It protects the visual and other sensory systems from information overload, determining our performance in many tasks (Posner & Rothbart, 1980; Posner, Snyder, & Davidson, 1980; Van Essen, Anderson, & Felleman, 1992).

Decades of experimental work have led to the identification of key structures and mechanisms related to FBA. The deepest insight into such mechanisms has been provided by single-cell studies in macaque monkeys, the closest animal model to humans in terms of brain structure and function that is used by neuroscientists to measure neuronal activity during behavioral tasks. In this chapter, we will review important studies that have measured the effects of FBA on the responses of single neurons in different areas of the macaque brain, and that have revealed two different mechanisms that modulate neuronal responses: feature matching and feature-similarity gain. Feature matching operates by increasing the responses of neurons whose RFs include stimuli that share the attended feature(s). For example, searching for a red-colored object in a scene will enhance the responses of neurons that contain red objects inside their RFs. This mechanism operates across the entire visual field and is not dependent on the neuronal feature-selectivity (e.g., whether the neuron is selective for red). We will argue that it may serve as a global process that enhances the neuronal representation of the attended feature to guide visual search. The second mechanism, feature-similarity gain, modulates responses based on the similarity between the attended feature and the neuron's selectivity for that feature (i.e., the modulation could be an enhancement or a suppression of responses depending on the neuron's selectivity for the attended feature). We will argue that this mechanism provides a sharpening of the neuronal representation of attended features for perception. In the following sections, we will first focus on studies in visual area V4 and middle temporal visual cortex (MT) that have provided evidence in favor of feature matching or feature-similarity gain. Second we will highlight studies in areas of the frontal lobe. Finally, we will elaborate on a unified theory of FBA, in which these two mechanisms may operate concurrently.

One of the most targeted brain areas by studies of attention in macaques is extrastriate area V4 of the ventral visual pathway. V4 receives direct projections from areas V1 and V2 and contains neurons with RFs predominantly representing the contralateral visual hemifield that are selective for basic features such as orientation, color, and contrast (Desimone & Schein, 1987). Importantly, V4 neurons project toward object-selective neurons in areas of the temporal lobe. Thus, one can infer that if FBA affects the responses of V4 neurons, this will affect visual processing at subsequent stages.

One of the first demonstrations of the effects of FBA on the responses of V4 neurons was reported in 1988 by Haenny and colleagues (Haenny, Maunsell, & Schiller, 1988). They recorded the responses of V4 units to gratings of different orientations during a match-to-sample task and found that more than half of neurons responded more strongly when the animals were instructed to match a particular orientation. This result was later confirmed by another V4 study in which monkeys had to attend, within an array composed of red and green stimuli, to stimuli of one color (e.g., red) while ignoring stimuli of the other color (e.g., green) (Motter, 1994). When the animal attended to red and a red stimulus was inside the neuron's RF, the response was stronger than when the animal attended to green and the same red stimulus was inside the RF. This FBA modulation of responses will be referred to as feature matching since neuronal responses to stimuli matching the attended feature are enhanced. This mechanism could be highly relevant to tasks such as visual search in which the searched target is defined by one or several features.

A demonstration of the effect of feature matching on the response of V4 neurons during visual search was given by Bichot, Rossi, and Desimone (2005). They trained macaques to search within a multielement array composed of colored shapes for a target with a particular color or shape by freely making saccades to any of the stimuli. When the saccade landed on the target, the animals were rewarded with drops of juice, which motivated them to find the target with the fewest saccades possible (6.3 saccades on average). Because the authors precisely measured eye movements, they could map the RF location of a recorded V4 neuron during periods when the animals kept gaze stationary (see figure 77.1a).

The most striking result of the study was that even when the animals did not plan a saccade toward the target, which happened to be inside the RF of the recorded neuron, responses were increased by ~30% relative to when the same identical stimulus was not the search-for target (see figure 77.1b, red solid line). Moreover the authors showed, during a more complex, conjunction search task, that when a stimulus sharing one feature with the target was inside the RF the responses were also increased relative to when the stimulus in the RF did not share any feature with the target. This result, which has been recently replicated by Zhou and Desimone (2011), suggests that FBA guides visual search by highlighting neuronal representations of stimuli sharing the attended features and thus resembling the target.

Other studies in area V4 have isolated a modulatory effect of FBA different from feature matching. McAdams and Maunsell (2000) recorded single-unit responses when macaque monkeys attended either to the color or the orientation of a given stimulus. They found that responses to an ignored stimulus inside the neurons' RFs were modulated depending on whether attention was allocated to one or the other attribute. Importantly, unlike the visual search task employed by Bichot, Rossi, and Desimone (2005), the stimulus inside the neurons' RFs was never behaviorally relevant for the animals. This discards a possible role of multifocal attention that would have enhanced the spatial representations of potentially relevant stimuli (Niebergall, Khayat, Treue, & Martinez-Trujillo, 2011).

Using a similar design, Cohen and Maunsell (2011) recorded simultaneously from many V4 neurons using multielectrode arrays implanted in both hemispheres of two macaques. During the task, animals covertly attended to a grating at a cued fixed position in the visual field and detected a change in one of the stimulus features (orientation or spatial frequency). By positioning a second irrelevant grating inside the neurons' RF and alternating blocks in which the animals detected orientation or spatial frequency changes, they could compare responses when one or the other feature dimension was attended or when different values of the same feature (e.g., different orientations or spatial frequencies) were attended (see figure 77.1c). They found that when the animals switched attention between different feature values of a grating, neuronal responses varied. This FBA modulation was dependent on the relationship between the attended stimulus feature and the cell's preferred feature. For example, the response of a neuron when animals attended to a particular orientation or spatial frequency was enhanced if the unit preferred that orientation/spatial frequency, but it was suppressed if the neuron was poorly selective for the attended feature.

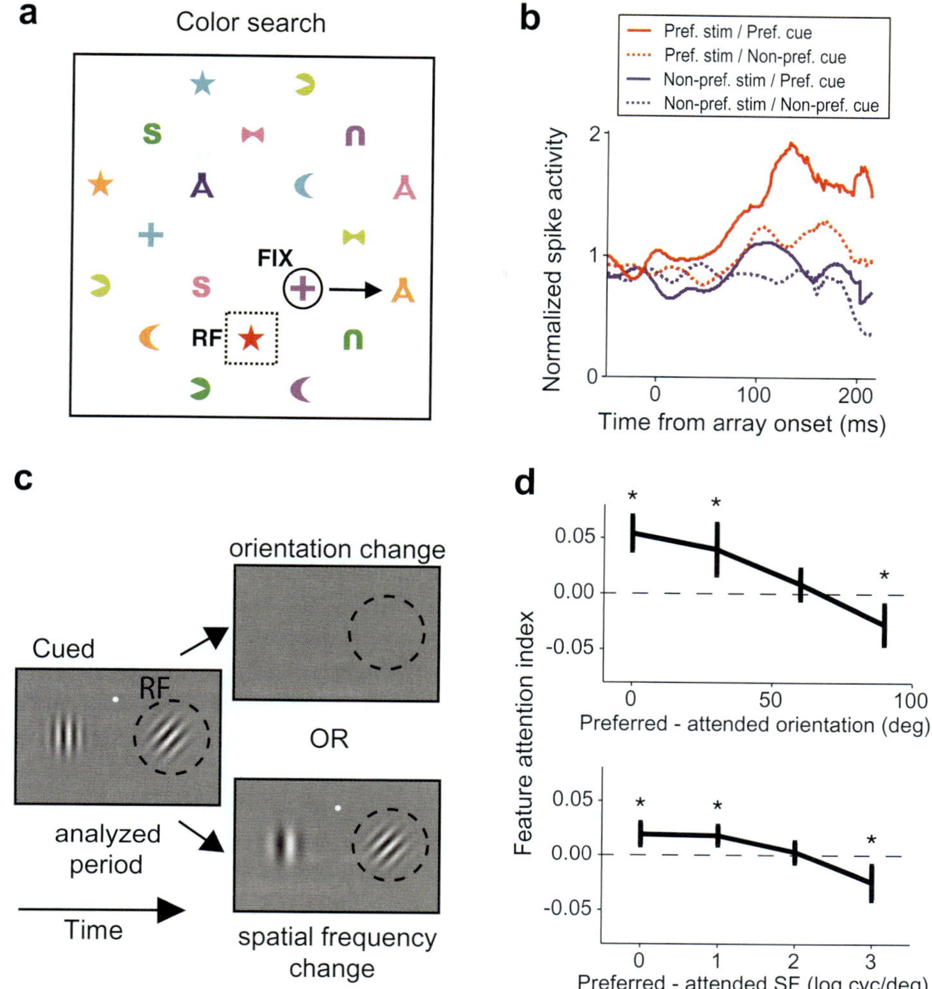

FIGURE 77.1 Effects of feature-based attention (FBA) on the responses of V4 neurons. (a) Monkeys searched for a precued color target in an array of distracters. Neuronal responses were examined during fixation periods (circle) when the preferred or nonpreferred stimulus in the receptive field (RF) (dotted square) was not the goal of the impending saccade (arrow) but was either the search-for target or a distractor. (b) Responses of V4 neurons during color search trials, when the RF stimulus (Stim) was of the neurons' preferred (Pref.; red line) or nonpreferred (blue lines) color, and when the search-for-target was of the preferred (solid lines) or nonpreferred (dotted lines) color (see legend and text). (c) Monkeys were presented with two Gabor patches at both sides of a fixation spot. In some blocks they had to detect a change in spatial frequency while in other blocks a change in orientation of the patch outside the recorded neuron RF (dashed circle). (d) A feature attention index between responses when the animals attended to the stimulus outside the RF relative to when the animal ignored that stimulus as a function of the similarity between the attended orientation (top panel) or spatial frequency (SF; lower panel) and the neuron's preferred feature. Positive values indicate response enhancement, and negative response suppression. cyc, cycles. (a and b adapted from Bichot, Rossi, & Desimone, 2005; c and d from Cohen & Maunsell, 2011.)

In general, the degree to which a neuron's firing rate was modulated depended on the distance in feature space between the neuron's preferred and attended feature (see figure 77.1d). This type of FBA modulation (feature-similarity gain) has been also described in area MT of macaques (Treue & Martinez-Trujillo, 1999).

One question that arises from the results discussed thus far is whether FBA could change the tuning properties of V4 neurons. A study by David and colleagues provided an answer to this question (David et al., 2008). They trained macaques in a match-to-sample task using complex images and found that tuning of neurons for spatial frequency is altered by FBA. Basically, neurons shift their preferred spatial frequency to match the frequency of the attended stimuli. So far, this effect has not been replicated in other studies, and it therefore remains a challenge for models of FBA to reveal the mechanisms of this change in the tuning of V4 neurons for basic features.

In sum, two global FBA mechanisms, feature matching and feature-similarity gain, have been shown to modulate the responses of neurons in extrastriate visual area V4. The former seems to operate during visual search and the latter during tasks in which subjects detect a change in the feature of a behaviorally relevant stimulus without changes in the focus of spatial attention.

EFFECTS OF FBA ON THE RESPONSES OF MT NEURONS

Together with V4, another highly targeted area by studies of FBA in macaques has been area MT. MT neurons are tuned for the direction, speed, and visual disparity of moving stimuli. They receive projections from V1 neurons and send projections toward higher-order areas, such as the medial superior temporal (MST), where cells are selective for complex optical flow patterns (Born & Bradley, 2005). Hence, any effects of attention on the firing rates of MT neurons will likely influence functions such as the perception of object's and self-motion.

One of the first demonstrations of FBA in area MT was provided by Treue and Martinez-Trujillo (1999), who recorded the responses of MT direction-selective neurons during a motion detection task. During these experiments a moving random dot pattern (RDP) was positioned inside a neuron's RF and a second RDP was positioned farther away in the opposite hemifield. In some trials the pattern inside the RF moved in the neuron's preferred direction (direction of motion evoking the strongest response) while in others in the neuron's antipreferred direction (evoking the weakest response). The pattern outside the RF could also move in either direction, yielding four different stimulus combinations (inside-preferred and outside-preferred, inside-preferred and outside-antipreferred, inside-antipreferred and outside-preferred, inside-antipreferred and out-side-antipreferred) (see figure 77.2a). The animal's task was to covertly attend to the RDP positioned outside the RF in order to detect a change in its motion direction or speed. In such a task, two different outcomes could be expected according to the FBA effects described thus far in area V4. First, according to feature matching, responses will be enhanced when both the RF pattern and the attended pattern move in the same direction, compared to when they move in opposite directions. Second, according to feature-similarity gain, whereby responses depend on the relationship between the attended feature and the neuron's selectivity for that feature, responses will be enhanced when attention is directed to the pattern moving in the neuron's preferred direction, compared to when it is directed to the antipreferred pattern. The main finding of the study was that the responses of the recorded neurons to either of the "ignored" stimuli in the RF (i.e., preferred or antipreferred) increased when the attended RDP moved in the preferred direction relative to when it moved in the antipreferred direction (see figure 77.2b). This is consistent with feature-similarity gain. Notably, this effect was not caused by differences in stimulus configuration between the different conditions since the response modulation disappeared when the animals attended to the fixation cross and ignored both RDPs.

Based on these results, the authors proposed a feature-similarity gain model of attention in which responses of visual neurons are modulated depending on the relationship between the attended feature and the cells' preferred feature. The model predicts that attending to a given feature will increase the response of neurons encoding that feature as preferred while it will decrease the response of neurons encoding that feature as antipreferred. Note that this is very different from the feature matching effect described by visual search studies (see figure 77.1a), which would have predicted a completely opposite result to the one shown in figure 77.2b (right panel). Indeed, according to the feature matching hypothesis, the response to a stimulus that matches the features of the attended stimulus (shown in blue, i.e., configuration where both stimuli move in the antipreferred direction) would always be enhanced compared to the response in the configuration where the two stimuli did not match (shown in red).

In order to further test the feature-similarity gain model, the authors expanded the experimental design illustrated in figure 77.2a by recording the responses of MT neurons to a RDP moving in one of many different directions while monkeys attended to an RDP outside the RF that moved in the same direction (Martinez-Trujillo & Treue, 2004) (see figure 77.3a).

By comparing the responses when the animals detected a direction change in the RDP located outside the neurons' RF against the ones when the animals ignored both RDPs and detected a small color change of the fixation square (fixation condition), they isolated the effects of attending to different motion directions on the neurons' response. They found that relative to the fixation condition the responses of neurons were modulated as a function of the relationship between the attended and the neuron's preferred direction (see figure 77.3b). The modulation varied from response enhancement, when the preferred direction was attended, to response suppression, when the

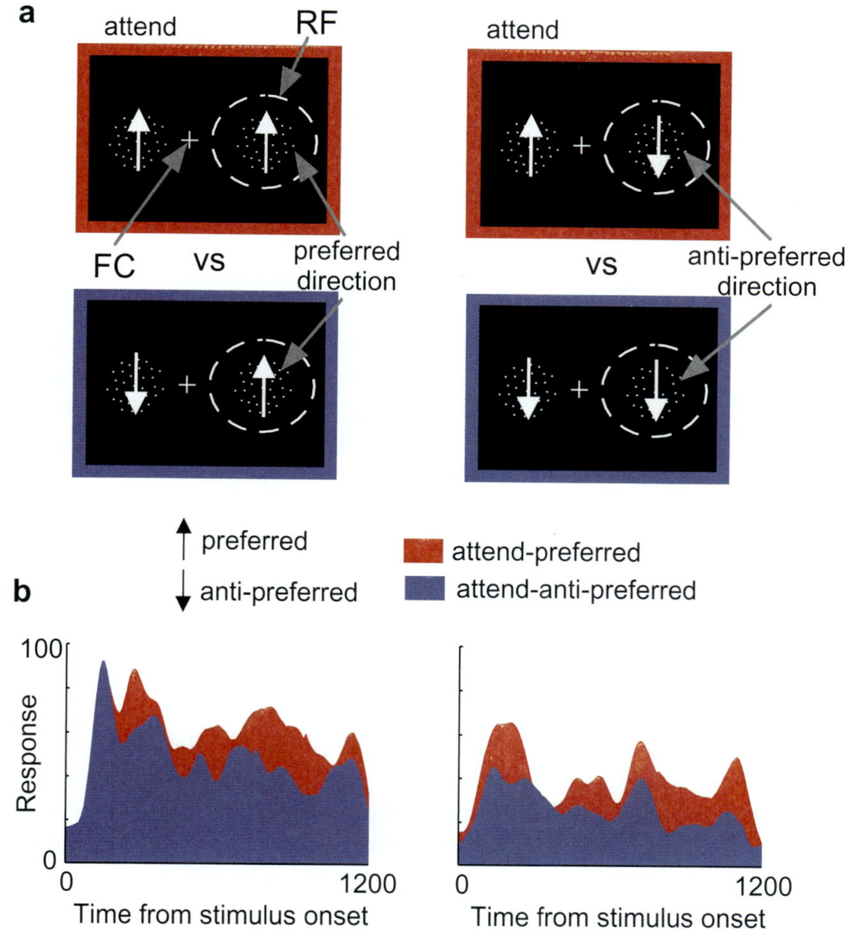

FIGURE 77.2   Effects of feature-based attention (FBA) on the responses of middle temporal visual cortex (MT) neurons. (a) Monkeys were trained to detect a change in the direction or speed of the pattern outside the receptive field (RF) (dashed circle) while maintaining gaze on the fixation cross (FC). Across trials, the attended pattern moved in the neuron's preferred (red frames) or antipreferred (blue frames) direction. The pattern inside the RF could also move in either motion direction. See text for details. (b) Responses of an MT neuron to the preferred (left) and antipreferred (right) directions as a function of stimulus onset time (msec) when the animals attended to the preferred (red) or antipreferred (blue) stimulus outside the RF. (From Treue & Martinez-Trujillo, 1999.)

antipreferred direction was attended. Again, this effect could not be explained by feature matching, which would have yielded a constant response modulation across attended RDP directions since on every trial the RF stimulus always matched the attended stimulus.

The shape of the modulation was effectively described by a linear regression model whose slope represents the increase in response, or gain, relative to fixation (right panel in figure 77.3b). For the population of neurons in MT the relationship between the attended motion direction and the modulation was described by the following equation:

$$FSG = 0.00083*\text{angDist} +1.06 \qquad (77.1)$$

FSG (feature-similarity gain) is the factor that multiplies the sensory (during fixation) response of a neuron, and

angDist is the distance in degrees between the neuron's preferred direction and the attended direction. When the distance is 0°, FBA multiplies the response by a factor of 1.06, producing a 6% increase in firing rate. When the distance is 180°, FBA multiplies the response by 0.91, producing a 9% decrease in response (see figure 77.3c). When this FSG equation is applied to a population of direction-selective neurons activated by a moving stimulus, we can easily predict the change in the activity profile. To illustrate this point, we have simulated the activity profile of MT direction-selective neurons (*pop*Res) evoked by an upward moving stimulus using tuning curve parameters from a population of 64 MT neurons recorded during fixation (the blue curve in the right panel of figure 77.3c). We used the following equation:

$$pop\,\mathrm{Res} = baseline + height * e^{-\frac{1}{2}*\left(\frac{center-prefdir}{width}\right)^2} \qquad (77.2)$$

The parameter *baseline* represents the responses of the neurons encoding upward motion as antipreferred, *height* is the difference between *baseline* and the activation of the neuronal population encoding upward motion as preferred, *center* indicates the direction of the stimulus (upward in this case), and *width* represents the spread of the activation across the population (Martinez-Trujillo & Treue, 2004). The effects of FBA are illustrated by the red curve. The curve was obtained by multiplying the vectors *FSG* and *pop*Res. Basically, FBA decreases the *baseline* of the curve and increases the *height*. If one considers the latter as a measurement of how direction selective the response of the population is, then FBA has increased the selectivity of the population to represent the direction of the moving stimulus. This effect may underlie the behavioral effects of FBA in motion perception (White & Carrasco, 2011).

FBA with Multiple Stimuli in the RF

One question regarding the mechanisms of FBA is whether during a given task FBA preferentially acts on feature-selective neurons within a particular area or it does so across different areas in the hierarchy of visual processing. For example, many neurons in areas V1, MT, and MST are selective for the direction of moving stimuli (Born & Bradley, 2005). When in a task such as the one illustrated in figure 77.3a a monkey attends to a particular motion direction, is FBA acting solely on the responses of MT direction-selective neurons? An answer to this question may be provided by a recent experiment conducted by Khayat, Niebergall, and Martinez-Trujillo (2010a). They positioned two moving RDPs inside an MT neuron RF; one RDP always moved in the neurons antipreferred direction (AP pattern) and the other changed direction from trial to trial (test pattern). The range of possible directions of the test pattern varied between the recorded neuron's preferred direction (i.e., 180° away from the antipreferred) in steps of 15° to 90° away from the preferred direction (i.e., orthogonal to both the preferred and antipreferred directions) (see figure 77.4a). They also positioned a similar pair of RDPs outside the recorded neuron RF, in the opposite hemifield. The animals were cued in some trials to covertly detect a direction change in the antipreferred pattern positioned outside the RF (attend-AP out condition), and in other trials to ignore all patterns and detect a subtle contrast change in a central fixation spot (fixation condition). The key prediction of feature-similarity gain acting at the level of

MT neurons is a decrease in response when the animals attended to the antipreferred pattern relative to the fixation condition that is constant across all the directions of the test pattern. This is because during this task, the relationship between the attended direction and the neuron's preferred direction never changes.

Contrary to that prediction, the study found that the attentional modulation changed as a function of the test pattern direction (see figure 77.4b, upper panels). The authors proposed that in these circumstances the feature-similarity gain modulation would result from a direction-dependent attentional mechanism that occurs before the visual signals arrive into MT neurons (see also Khayat, Niebergall, & Martinez-Trujillo, 2010b), that is, in an upstream direction-selective area that projects to MT and where neurons would have smaller RFs that include only a single stimulus. For example, in this particular experimental design, the size of the RDPs approximated the size of V1 or V3 neurons RFs. Feature-similarity gain would act at this stage.

To illustrate this idea, let us consider two distinct populations of V1/V3 neurons activated by the antipreferred and the test pattern, respectively, and providing inputs into the recorded MT neuron. If the responses of the neuronal population having the antipreferred stimulus in their RF are always enhanced when this feature is attended relative to the fixation condition, but the responses of the neuronal population having the test stimulus in the RF are differentially modulated depending on what the most active neurons are, one could explain the observed effects through feature-similarity gain.

To make this clear, we use a computer simulation in which two putative populations of direction-selective neurons in area V1 are activated by the antipreferred and test stimuli, respectively. A critical detail in the simulation is that while the population activated by the antipreferred stimulus is always composed of the same units (solid curve in figure 77.4c, bottom panel), the population of neurons activated by the test stimulus changes its relative composition depending on the direction of that stimulus (dashed curves). For example, when the test stimulus moves in the MT neuron preferred direction, the most active V1 units are the ones encoding the same direction as preferred, but when the test stimulus moves in a direction orthogonal to the preferred, the same V1 units become less active, compared to a different population that encodes the orthogonal direction as preferred. This explains why we have represented several of the dashed curves along the abscissa in the plot, one corresponding to each direction. We have used the same abscissa for the population

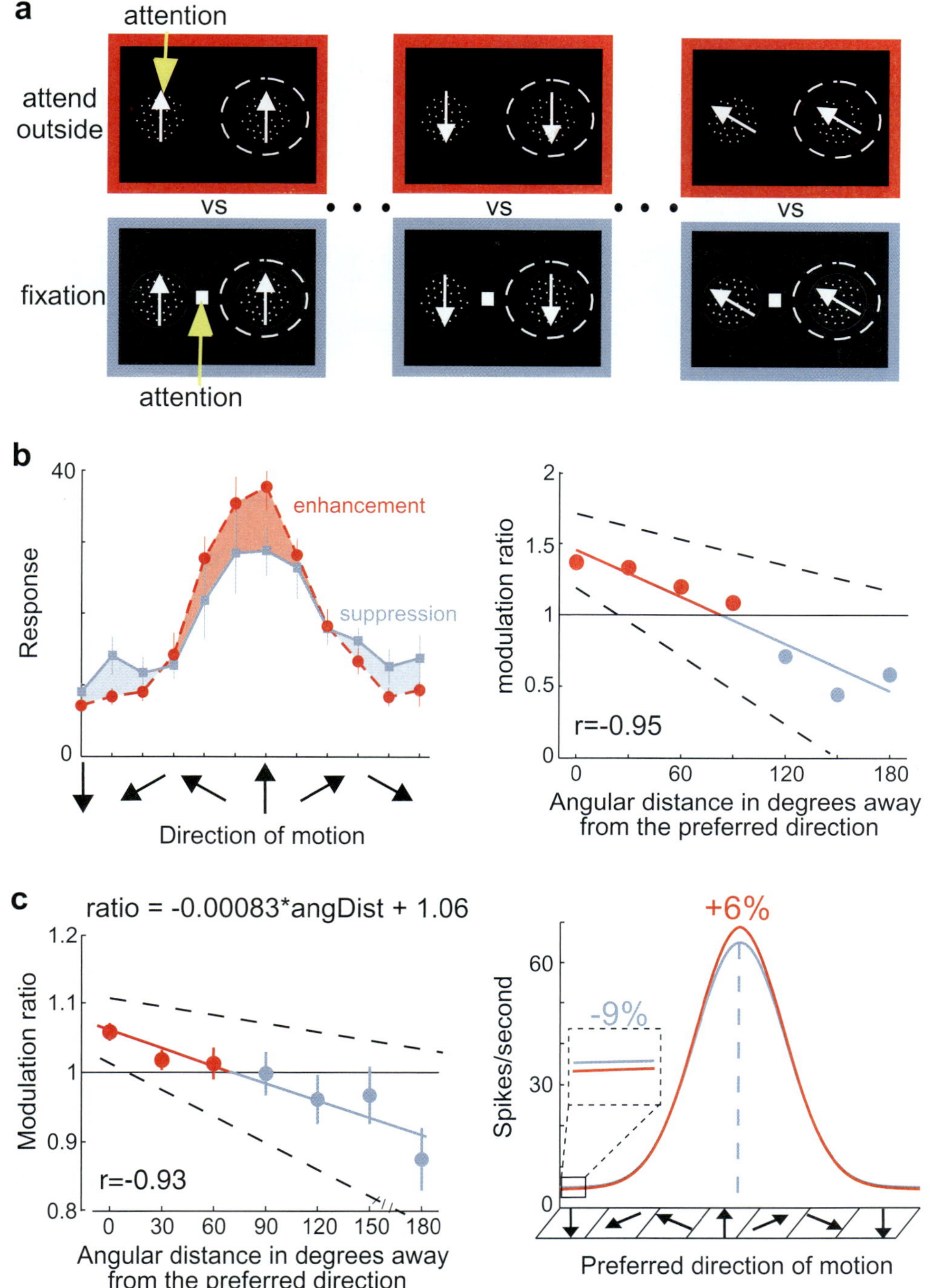

FIGURE 77.3 Effects of feature-similarity gain on the responses of middle temporal visual cortex (MT) neurons. (Adapted from Martinez-Trujillo & Treue, 2004). (a) Experimental design. The stimulus configuration was similar to the one in figure 77.2; however the stimuli inside and outside the RF always moved in one of 12 possible directions. See text for details. (b) Tuning curves of a direction-selective neuron during the attend RDP outside condition (red) and the attend fixation condition (blue). Observe that the attend RDP curve is higher than the attend fixation curve at the neuron's preferred direction and lower at the neuron's antipreferred direction. This is better illustrated in the right panel that plots the modulation ratios between responses in the two conditions as a function of the similarity between the attended direction and the neuron's preferred direction. The line represents the linear regression. The red dots illustrate response enhancement relative to fixation, and blue dots response suppression. (c) Left panel, average ratios across a sample of 135 MT neurons. Right panel, feature-similarity gain effect on a simulated response (ordinate) of a population of MT neurons preferring different directions (abscissa) to a stimulus moving upward (red is attended and blue unattended). The insert shows a magnification of the effect at the nonpreferred direction.

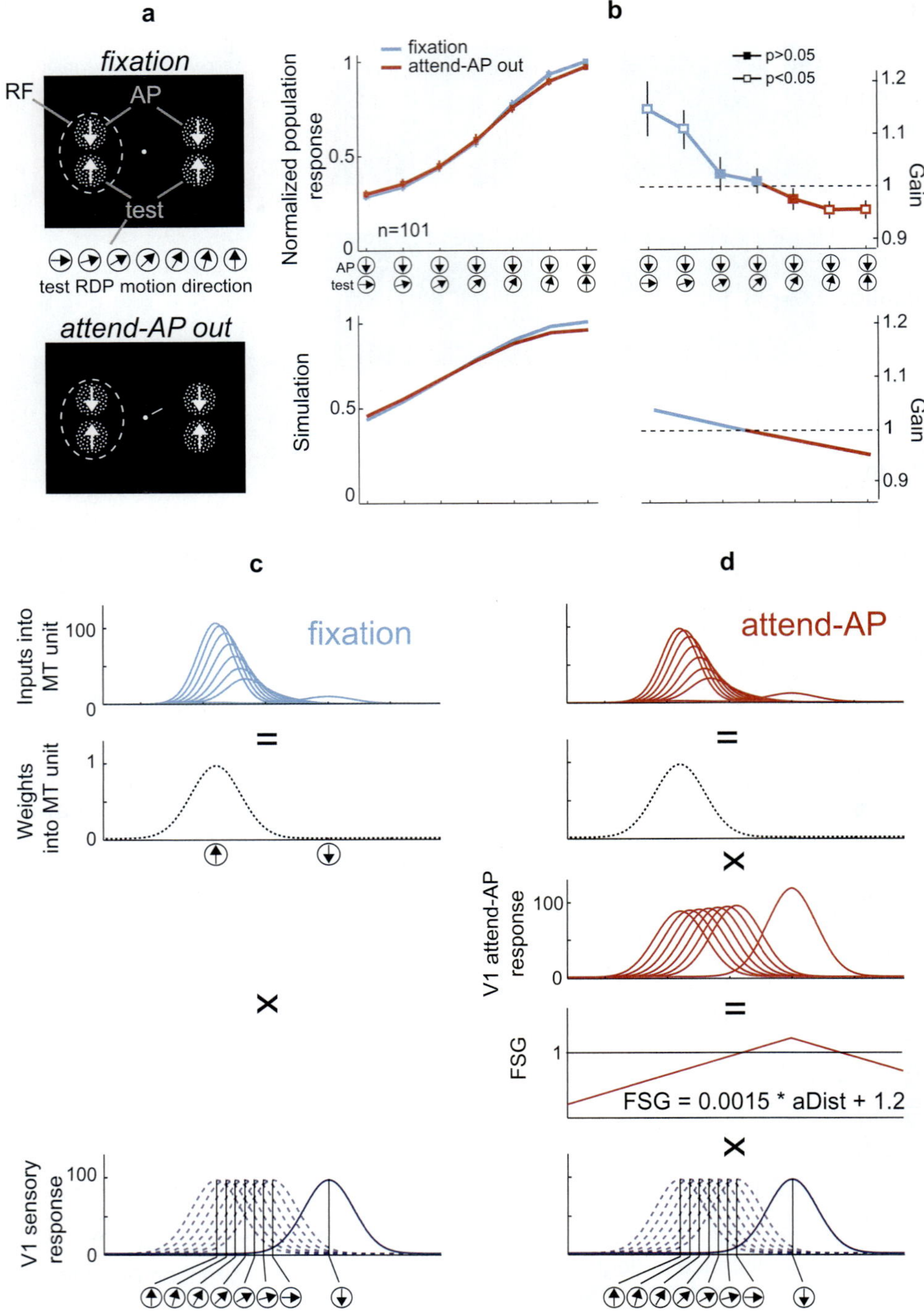

FIGURE 77.4 Effects of feature-based attention on the responses of middle temporal visual cortex (MT) neurons to two stimuli inside the receptive field (RF). (a) Experimental design by Khayat, Niebergall, and Martinez-Trujillo (2010a). The animals were presented with two random dot patterns (RDPs) inside the RF of the recorded MT neurons and an identical pair outside. One pattern always moved in the neuron's antipreferred direction (AP) and the second, test pattern could change direction from trial to trial, from preferred to 90° away (see sketches). In the attend-AP out condition the animals attended to the antipreferred pattern outside the RF and detected a change in its motion direction, and in the fixation condition they detected a contrast change on the fixation spot. (b) The upper panels show the average population responses of MT neurons in both conditions (left) and the ratio of the responses in the attend-AP relative to fixation (right). The lower panels illustrate results of a computer simulation bearing similar effects (see text for details). (c, d) Simulation illustrating the responses of V1 neurons to each separate RDP in the fixation (c) and the attend-AP condition (d). See text for details. FSG, feature-similarity gain.

responses corresponding to both patterns to make the plot easy to interpret.

Given that V1 provides direction-selective inputs to area MT, one can model the weights of these inputs into a given MT neuron as a Gaussian function in which the smallest weights (w = 0.1) correspond to inputs from V1 neurons tuned for the MT cell antipreferred direction and the largest weights (w = 1) correspond to V1 units selective for the MT cell preferred direction (second row from the top in figure 77.4c) (Rust et al., 2006). The contributions of the weighted inputs to the MT neuron response (product of second and third row) appear in the first row (blue curves) with the area under each curve representing the contribution of inputs activated by one stimulus. For a given stimulus combination (e.g., antipreferred pattern + test pattern moving in the preferred direction) we can simply add the contribution of the corresponding sets of inputs and obtain an estimate of the direction-selective inputs activating the MT neuron. This input linearity assumption agrees with models of how MT neurons transform inputs into spikes (Simoncelli & Heeger, 1998). This is assuming that all stimuli have the same contrast and thus that the subsequent normalization step applied to the inputs described by the same model is constant across different stimulus combinations. Then one can describe the firing rate of the MT neuron as the sum of the inputs divided by a constant. The lower panel in figure 77.4b shows the sum of inputs for the different combinations normalized to the largest input (blue line). The curve closely resembles the MT neurons' response to the different stimulus combinations during the fixation condition (blue line in top panel).

In order to simulate the response with feature-similarity gain applied to the responses of the population of MT neurons (i.e., when the animal attended to the antipreferred pattern outside the MT neuron RF) we have multiplied the response of each neuron in the corresponding population by a feature-similarity gain factor depending on the distance between the neuron's preferred direction and the attended direction (i.e., antipreferred direction of the MT neuron, the second panel from the bottom in figure 77.4d). The resulting activity profile (red curves in the third panel) is then multiplied by the weights (second panel), and the inputs into the MT neuron are computed (top panel). Then a similar procedure as in figure 77.4c is applied to compute the population response (red curve in lower left panel, figure 77.4b). One detail in the simulation is that for obtaining the red curve that crosses the blue curve we had to increase the slope and intercept of the feature-similarity gain equation previously described in MT neurons (Martinez-Trujillo & Treue, 2004).

Basically, we have made the effects of FBA stronger in V1 neurons (see the equation in figure 77.4d). An estimate of the overall effects is obtained by computing the ratio between the simulated responses (figure 77.4b, lower right panel). As we can appreciate the simulated effects follow the same profile as the effects isolated by the authors in MT neurons (upper right panel).

From the recorded data and the simulation we can draw two main conclusions. First, it is possible to obtain the observed FBA effects in area MT by applying feature-similarity gain on a population of direction-selective neurons with small RFs including single stimuli and projecting toward MT. Second, the magnitude of FBA effects and therefore feature-similarity gain seems to vary depending on the circumstances. For example, our simulation predicts stronger FBA effects in the response of V1 neurons under the experimental conditions used by Khayat, Niebergall, and Martinez-Trujillo (2010a) than in MT neurons under the experimental conditions used by Martinez-Trujillo and Treue (2004). One may here consider that the degree of difficulty of the task used in the former study, where the stimuli were small and close together, was likely higher than the degree of difficulty used in the latter study, where stimuli were large and located in opposite hemifields. Because task difficulty increases the modulatory effects of attention (Spitzer, Desimone, & Moran, 1988), this may provide an interpretation to the different values of feature-similarity gain. However, further experimental work is needed to fully clarify this issue.

Another study in area MT has examined the effect of FBA in the responses of neurons when monkeys attend to a feature that the cells are selective for (e.g., motion direction) relative to when the animals attend to a feature that neurons encode poorly (e.g., color) (Chen et al., 2011). The authors found that about 25% of the neurons show differences in firing rate between the two conditions. However, this effect did not show a particular bias (e.g., stronger responses when attending to direction than to color). One possible interpretation of these results is that the modulation observed in direction-selective neurons was shaped by feature-similarity gain and thus included response enhancements when the cell's preferred and similar directions were attended, and response suppression when the antipreferred and similar directions were attended (Martinez-Trujillo & Treue, 2004). It is also possible that attention to color somewhat modulates the responses of MT neurons through a mechanism similar to feature matching. For example, once the target feature has been identified, spatial attention may enhance the representations of stimuli sharing that feature in retinal maps across different visual areas.

In sum, studies in area MT have mainly described the effect of feature-similarity gain. Some of this effect cannot be explained by a model in which FBA acts on the responses of neurons within the area, but they are better accounted for by a model in which FBA acts at different stages in the hierarchy of motion processing. Futures studies using visual search tasks must clarify whether feature matching also modulates the responses of MT neurons.

FBA BEYOND VISUAL CORTEX

One issue that has been poorly investigated is how and where FBA signals originate in the brain. For specific tasks such as visual search it has been proposed that neurons in areas of the frontal and parietal lobe holding working memory representations of attended features modulate the responses of neurons with similar feature preferences in visual cortex (Desimone & Duncan, 1995). In order to investigate this issue Zhou and Desimone (2011) conducted simultaneous recordings

from areas V4 and the frontal eye fields (FEFs) of macaques during a visual search task. The animals were required to remember a visual cue presented at the beginning of a trial and then search, in a display composed of an array of different objects, for the one that matches the cue by directing gaze to single items (see figure 77.5a). The FEF is located in the prefrontal cortex and contains neurons that encode the position of a visual stimulus, as well as the intended gaze position (Tehovnik et al., 2000). Some degree of shape selectivity has also been reported in FEF neurons (Peng et al., 2008). Several studies have supported the role of the FEF as a source of top-down spatial attention signals that reach neurons in area V4 and modulate their sensitivity to visual inputs (Gregoriou et al., 2009; Moore & Armstrong, 2003).

Zhou and Desimone (2011) found that during the visual search task, neurons in V4 and the FEF respond more strongly to the target stimulus or to stimuli sharing the target features than to other stimuli. The authors discarded the possible role of spatial attention

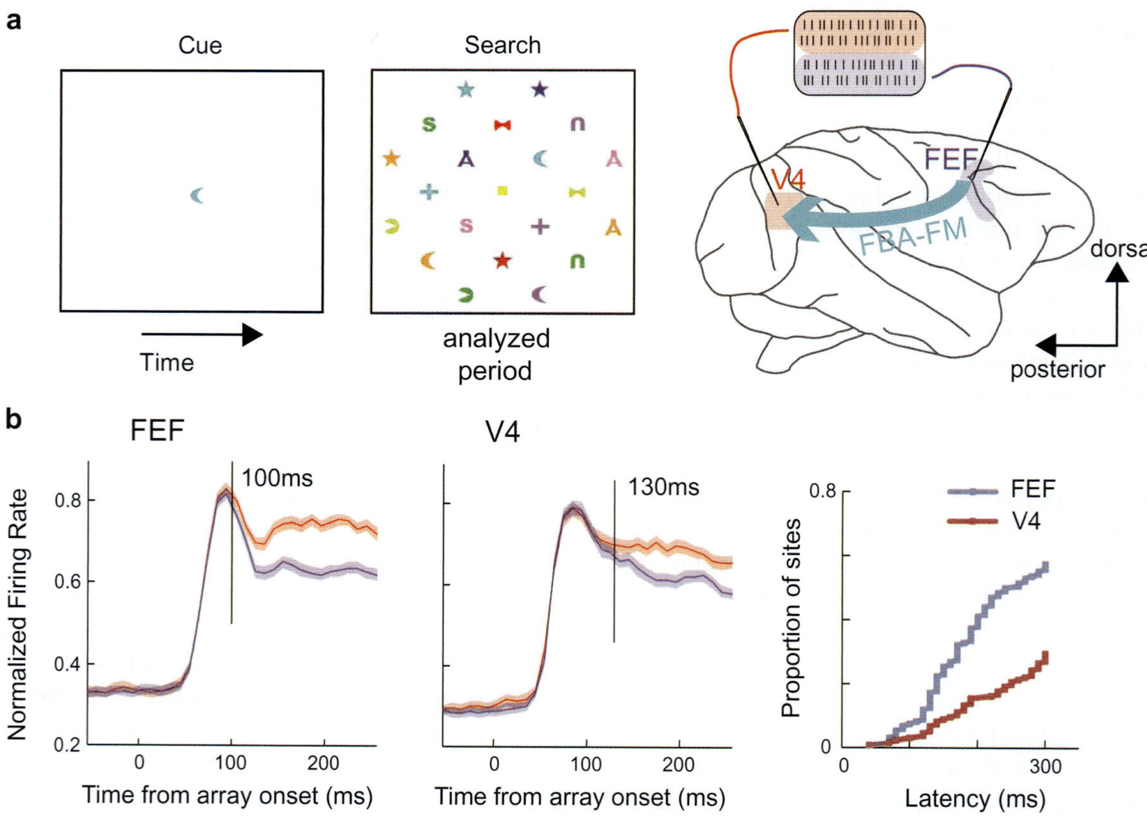

FIGURE 77.5   Effects of feature-based attention (FBA) on the responses of frontal eye field (FEF) and V4 neurons. (a) Stimulus display used by Zhou and Desimone (2011) and illustration of the simultaneous recordings from the FEF and V4 (right panel). (b) Population responses of FEF (left) and V4 neurons (right) as a function of stimulus array onset for stimuli that shared the target features (red) and stimuli that did not (blue). The latency of the FBA effects is indicated for each area. The panel on the right displays the cumulative distribution of latencies for each area (also see text). FM, feature matching.

by analyzing trials in which saccades were made to a stimulus away from the RF of the recorded neurons. Because in these trials the focus of spatial attention was not on the stimulus inside the neurons' RF but elsewhere at the position of the future saccadic eye movement, the increase in response to stimuli matching the attended features was likely due to FBA. Importantly, the latency of this effect was shorter in FEF than in V4 neurons (figure 77.5b), and the intensity of the response modulation was predictive of the efficiency of the visual search—as quantified by the number of saccades needed to find the target. This demonstrates that the FEF is a potential source of top-down signals during visual-search tasks.

One interesting finding of the study is that feature-selectivity arose earlier in V4 than in FEF neurons. Thus, the FEF may combine incoming feature information from V4 with working memory signals from other areas carrying information about the relevant features to compute a saliency map that highlights the locations of potential targets. This map not only guides gaze but also provides feedback signals to V4 in order to enhance the processing of stimuli sharing the target features. Note that according to this hypothesis, although the trigger signal for the FEF saliency computation is a stimulus feature, the nature of the top-down signal is spatial since it highlights locations of potential targets, that is, it enhances responses of neurons with RFs that include stimuli resembling the target. Because the attentional enhancement occurred in any neuron containing a stimulus that matches the target feature within its RF, independently of the unit's selectivity, the reported effects in the FEF are compatible with feature matching rather than with feature-similarity gain. The origins of feature-similarity gain effects remain unknown.

Other studies in monkeys have described effects of attention in the dorsolateral prefrontal cortex (dlPFC), areas 8/46 of macaque monkeys during match-to-sample tasks. They have described a strong modulation of responses to complex objects without changes in the spatial allocation of attention. For example, Everling et al. (2002) used complex objects (e.g., fish and hamburger) presented serially at the same location on a computer screen while monkeys matched one of the objects to a predefined sample object. They found that dlPFC neurons strongly respond to objects similar to the target while suppressing their responses to other objects. Although the authors did not directly test whether sharing the target feature had an influence in the cell response, one would assume that the representation of the attended object and therefore of their features was enhanced. Here one must consider another

form of attention known as object-based attention (OBA) in which objects' representations and not individual features are the units of selection. Evidence in favor of this type of attention has been found in area V1 (Roelfsema, Lamme, & Spekreijse, 1998). To what extent OBA and FBA are interrelated is still a matter of debate.

In sum, FBA has not been extensively explored outside the visual cortex. Areas related to working memory coding in the frontal and possibly parietal lobe seem to carry FBA signals that could be the source or the trigger for the modulation observed in visual cortex. More research in this direction is needed to fully understand the origins of FBA.

TOWARD A UNIFIED THEORY OF FBA

In sum, the accumulated evidence by studies of FBA across different areas of the macaque brain has shown that in tasks requiring attending to stimulus features, two different mechanisms are responsible for the modulation of responses across cortical maps. Feature matching increases the responses of neurons whose RFs include stimuli sharing the attended feature(s). If the stimulus in the RF shares more than one feature with the target, the response will be more strongly enhanced relative to when the stimulus shares fewer or no features with the target. This type of effect has been mainly described during visual search tasks. The general consensus is that by highlighting the potential target locations within retinotopic or spatiotopic maps in cortical areas feature matching could make search more efficient.

The second effect described by FBA studies in macaques is feature-similarity gain. This effect is portrayed by the feature-similarity gain model of attention. The model postulates that attending to a particular feature enhances the responses of neurons encoding that feature as preferred, independently of their spatial location. This phenomenon has been mainly described in tasks where the focus of attention remains fixed at the same position in space and subjects detect a change in one of the target features. How feature-similarity gain would favor the detection process is not clear. One possibility is that in these circumstances an enhanced representation of the attended feature across the whole cortical map of space contributes to sharpen the representation of the relevant feature by the neurons that contribute the most to the task.

The physiological mechanisms underlying each one of these effects remain obscure. However, for the case of feature matching, two recent reports have revealed important details. First, the FEF, an area classically involved in top-down signals related to spatial attention, exhibits attentional enhancement of targets matching the attended feature, an effect independent of the neurons' feature preference (Zhou & Desimone, 2011). Second, the spotlight of spatial attention is not necessarily unitary but can be split into multiple foci (Niebergall et al., 2011). Putting together these two findings, one can elaborate on a mechanism where the attentional system "locates" potentially relevant targets based on their features. This could occur through convergence of top-down working memory signals from the neighboring dlPFC area 46 and bottom-up visual information about the stimulus in an area such as the FEF. The result of this signal integration may produce a saliency map in which the spatial positions of stimuli resembling the target the most are "highlighted" while the positions of dissimilar stimuli are "suppressed." This process does not need to yield to a binary map, but the degree of enhancement/suppression could depend on the similarity between the stimuli and the task-relevant target. Supporting this hypothesis, one study has shown that the similarity between a target and a distracter is encoded in the responses of dlPFC neurons (Lennert & Martinez-Trujillo, 2011). This explanation integrates feature matching and multifocal spatial attention under the same umbrella.

The mechanisms underlying feature-similarity gain are unknown. So far no study has reported the effects of feature-similarity gain in areas of the parietal and frontal lobe. However, if feature-similarity gain is implemented through feedback projections from these areas into the visual cortex, one may predict that (a) neurons in areas of the frontal and parietal lobe should show selectivity for visual features such as orientation and direction of motion, (b) the effects of feature-similarity gain should arise first in the frontal and/or parietal lobe then in the visual cortex, and (c) neurons carrying these effects should send "diffuse" direct or indirect projections toward visual areas reaching visual neurons with similar feature selectivity independently of their RF location. Another possibility is that feature-similarity gain is implemented locally at the level of feature maps in visual cortex through interactions between different units. This hypothesis would predict direct or indirect connectivity between neurons encoding the same feature as preferred likely through excitatory synapses, and between neurons encoding the opposite features through inhibitory synapses. Moreover, the proportion of excitatory/inhibitory connections between two neurons would be determined by the similarity between their preferred feature(s). This point is actually supported by studies on motion adaptation; adapting to a given motion direction produces a motion aftereffect in the opposite direction, which reflects imbalances in the activity of specific populations of direction-selective neurons mutually interconnected (Huk, Ress, & Heeger, 2001). However, further studies are needed to corroborate these hypotheses.

In conclusion, we have learned that depending on the task, FBA can distinctively shape the responses of neurons across sensory cortices, and that the key to the control of this process may be in executive areas of the frontal lobe. FBA can therefore be seen as a flexible mechanism used by brain systems to modulate the responses of neurons during different behavioral tasks to ultimately protect sensory systems from information overload and improve perception.

REFERENCES

Bichot, N. P., Rossi, A. F., & Desimone, R. (2005). Parallel and serial neural mechanisms for visual search in macaque area V4. *Science, 308*, 529–534.

Born, R. T., & Bradley, D. C. (2005). Structure and function of visual area MT. *Annual Review of Neuroscience, 28*, 157–189.

Chen, X., Hoffmann, K. P., Albright, T. D., & Thiele, A. (2011). Effect of feature-selective attention on neuronal responses in macaque area MT. *Journal of Neurophysiology, 107*, 1530–1543.

Cohen, M. R., & Maunsell, J. H. (2011). Using neuronal populations to study the mechanisms underlying spatial and feature attention. *Neuron, 70*, 1192–1204.

David, S. V., Hayden, B. Y., Mazer, J. A., & Gallant, J. L. (2008). Attention to stimulus features shifts spectral tuning of V4 neurons during natural vision. *Neuron, 59*, 509–521.

Desimone, R., & Duncan, J. (1995). Neural mechanisms of selective visual attention. *Annual Review of Neuroscience, 18*, 193–222.

Desimone, R., & Schein, S. J. (1987). Visual properties of neurons in area V4 of the macaque: Sensitivity to stimulus form. *Journal of Neurophysiology, 57*, 835–868.

Everling, S., Tinsley, C. J., Gaffan, D., & Duncan, J. (2002). Filtering of neural signals by focused attention in the monkey prefrontal cortex. *Nature Neuroscience, 5*, 671–676.

Gregoriou, G. G., Gotts, S. J., Zhou, H., & Desimone, R. (2009). High-frequency, long-range coupling between prefrontal and visual cortex during attention. *Science, 324*, 1207–1210.

Haenny, P. E., Maunsell, J. H., & Schiller, P. H. (1988). State dependent activity in monkey visual cortex: II. Retinal and extraretinal factors in V4. *Experimental Brain Research, 69*, 245–259.

Huk, A. C., Ress, D., & Heeger, D. J. (2001). Neuronal basis of the motion aftereffect reconsidered. *Neuron, 32*, 161–172.

Khayat, P. S., Niebergall, R., & Martinez-Trujillo, J. C. (2010a). Attention differentially modulates similar neuronal responses evoked by varying contrast and direction stimuli in area MT. *Journal of Neuroscience, 30*, 2188–2197.

Khayat, P. S., Niebergall, R., & Martinez-Trujillo, J. C. (2010b). Frequency-dependent attentional modulation of local field potential signals in macaque area MT. *Journal of Neuroscience, 30*, 7037–7048.

Lennert, T., & Martinez-Trujillo, J. (2011). Strength of response suppression to distracter stimuli determines attentional-filtering performance in primate prefrontal neurons. *Neuron, 70,* 141–152.

Martinez-Trujillo, J. C., & Treue, S. (2004). Feature-based attention increases the selectivity of population responses in primate visual cortex. *Current Biology, 14,* 744–751.

McAdams, C. J., & Maunsell, J. H. (2000). Attention to both space and feature modulates neuronal responses in macaque area V4. *Journal of Neurophysiology, 83,* 1751–1755.

Moore, T., & Armstrong, K. M. (2003). Selective gating of visual signals by microstimulation of frontal cortex. *Nature, 421,* 370–373.

Motter, B. C. (1994). Neural correlates of attentive selection for color or luminance in extrastriate area V4. *Journal of Neuroscience, 14,* 2178–2189.

Niebergall, R., Khayat, P. S., Treue, S., & Martinez-Trujillo, J. C. (2011). Multifocal attention filters targets from distracters within and beyond primate MT neurons' receptive field boundaries. *Neuron, 72,* 1067–1079.

Peng, X., Sereno, M. E., Silva, A. K., Lehky, S. R., & Sereno, A. B. (2008). Shape selectivity in primate frontal eye field. *Journal of Neurophysiology, 100,* 796–814.

Posner, M. I., & Rothbart, M. K. (1980). The development of attentional mechanisms. *Nebraska Symposium on Motivation, 28,* 1–52.

Posner, M. I., Snyder, C. R., & Davidson, B. J. (1980). Attention and the detection of signals. *Journal of Experimental Psychology, 109,* 160–174.

Roelfsema, P. R., Lamme, V. A., & Spekreijse, H. (1998). Object-based attention in the primary visual cortex of the macaque monkey. *Nature, 395,* 376–381.

Rust, N. C., Mante, V., Simoncelli, E. P., & Movshon, J. A. (2006). How MT cells analyze the motion of visual patterns. *Nature Neuroscience, 9,* 1421–1431.

Simoncelli, E. P., & Heeger, D. J. (1998). A model of neuronal responses in visual area MT. *Vision Research, 38,* 743–761. doi:10.1016/S0042-6989(97)00183-1.

Spitzer, H., Desimone, R., & Moran, J. (1988). Increased attention enhances both behavioral and neuronal performance. *Science, 240,* 338–340.

Tehovnik, E. J., Sommer, M. A., Chou, I. H., Slocum, W. M., & Schiller, P. H. (2000). Eye fields in the frontal lobes of primates. *Brain Research. Brain Research Reviews, 32,* 413–448.

Treue, S., & Martinez Trujillo, J. C. (1999). Feature-based attention influences motion processing gain in macaque visual cortex. *Nature, 399,* 575–579.

Van Essen, D. C., Anderson, C. H., & Felleman, D. J. (1992). Information processing in the primate visual system: An integrated systems perspective. *Science, 255,* 419–423.

White, A. L., & Carrasco, M. (2011). Feature-based attention involuntarily and simultaneously improves visual performance across locations. *Journal of Vision, 11*(6), 5.1–5.10. doi:10.1167/11.6.15.

Zhou, H., & Desimone, R. (2011). Feature-based attention in the frontal eye field and area V4 during visual search. *Neuron, 70,* 1205–1217.

78 Parietal Mechanisms of Attentional Guidance: The Role of Learning and Cognition

JACQUELINE GOTTLIEB

Since the nineteenth century, neuropsychological evidence has suggested that the parietal cortex is important in spatial attention. Patients with damage to the parietal lobe can develop the syndrome of neglect, a specific inability to notice sensory stimuli in the contralateral space. A milder form of the disease is that of extinction or perceptual rivalry, where patients can perceive a contralateral object when it is presented alone but, if presented with two objects simultaneously, tend to choose the ipsilateral one. Accompanying their attentional deficits, patients with parietal lesions can also have optic ataxia, difficulty looking at or reaching contralateral targets. These syndromes suggest that the parietal lobes are important for linking perception and action and, specifically, in assigning importance (or "interest") to the contralateral space.

The view of the parietal lobe emerging from neurological studies is upheld in neurophysiological work in monkeys. Based on anatomical and physiological evidence, the monkey parietal lobe has been subdivided into two broad sectors, the superior and inferior parietal lobes (SPL and IPL), which lie, respectively, dorsal and ventral to the intraparietal sulcus (IPS) (see figure 78.1). Physiologically, these sectors have different responses to sensory and motor stimulation. The SPL is responsive primarily to touch (somatosensation) and skeletal (limb) movements and represents the near peripersonal space (Colby & Goldberg, 1999). The IPL by contrast responds to visual inputs and movements of the eyes and head and seems to represent the farther visual space (Colby & Goldberg, 1999).

The IPL, which has been most intensively studied in relation to attention, is further subdivided into three areas. The ventral intraparietal area (VIP) lies in the fundus of the IPS and is a transitional area that responds to both visual and somatosensory stimulation. VIP neurons have bimodal sensory and visual receptive fields (RFs) that represent the space near the upper body and face (Colby, Duhamel, & Goldberg, 1993).

Area 7a is a visual area that covers a large expanse on the cortical convexity lateral to the IPL and conveys information about full-field visual motion, visual salience, and eye position (Constantinidis & Steinmetz, 2005; Raffi & Siegel, 2005). Finally the lateral intraparietal area (LIP) is a small area in the lateral bank of the intraparietal sulcus that has strong ties to the oculomotor system. LIP neurons have visual RF located mostly in the contralateral hemifield and respond to small stimuli that drive attention and shifts of gaze (Blatt, Andersen, & Stoner, 1990). This area receives visual inputs from the lateral pulvinar thalamic nucleus and from cortical areas in the dorsal and ventral visual streams and has strong anatomical connections to two oculomotor areas—the frontal eye fields (FEF) and the superior colliculus (SC) (Blatt, Andersen, & Stoner, 1990; Lewis & Van Essen, 2000).

As will become clear in this chapter, area LIP has been intensively investigated and is believed to reflect internal processes related to attention, learning, and decision formation. This chapter will focus on these studies and their implications for attention and eye movement control.

Adaptive Spatial Representation

Although LIP contains several types of cells, the vast majority of studies have focused on neurons that have visual RFs—that is, respond to visual inputs in a limited range of retinal locations. In terms of its layout in the IPS the retinotopic map in LIP has only a coarse organization as adjacent neurons can respond to disparate retinal locations (Ben Hamed & Duhamel, 2002; Ben Hamed et al., 2002). Nevertheless by virtue of their restricted RFs, the neurons convey spatial information— that is, encode the location of objects in the external world.

The spatiotopic map in LIP is distinguished from maps in earlier visual areas by several characteristics.

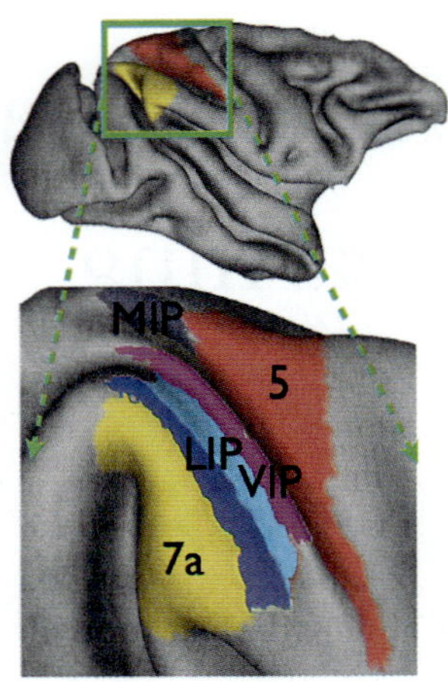

FIGURE 78.1   The parietal lobe in the monkey. Lateral view of the rhesus monkey brain indicating the approximate location of individual parietal areas. The inferior parietal lobe (IPL) is located ventral and posterior to the intraparietal sulcus and includes areas 7a, LIP (the lateral intraparietal area, ventral and dorsal divisions), and VIP (ventral intraparietal area). The superior parietal lobe is located dorsal and medial to the sulcus and includes area 5 and the medial intraparietal area (MIP). (Modified with permission from Gottlieb & Snyder, 2010.)

First, LIP neurons encode not the mere presence but the salience or relevance of a visual input. Second, even though the RF of LIP neurons are retinotopic (meaning that they move relative to the external world each time the eye moves), neurons maintain an accurate memory of salient locations across shifts of gaze. And third, in addition to their visual responses LIP neurons are sensitive to a wide array of behavioral ("extraretinal") factors.

A good illustration of these response properties comes from an experiment that emulated natural viewing conditions by training monkeys to make eye movements across a stable visual scene (Gottlieb, Kusunoki, & Goldberg, 1998). A circular array containing several objects remained stable on the screen (see figure 78.2, top cartoon). On each trial the monkey looked at (fixated) a peripheral location where no array stimulus was in a neuron's RF and then made a saccade to a central location that brought one of the stimuli into the RF (labeled RF1 and RF2). This mode of presentation closely emulates natural viewing conditions, where stimuli are stable and inconspicuous in their environment but move on the retina by virtue of the observer's motion.

Using two variants of this stable array task, investigators tested whether LIP neurons encode the mere presence of a stimulus or the special salience of an abrupt onset. In one condition all stimuli remained stable and unchanging for a long block of trials, and one of these stimuli entered the RF by virtue of the monkey's saccade. In a second condition the visual stimulation to the neurons' RF was identical, but the stimulus entering the RF was rendered salient by abruptly appearing and disappearing on each trial.

Even though the visual stimulation to their RF was identical in both cases, LIP neurons distinguished between the stable and recent-onset contexts. Neurons had barely any response when a stable stimulus entered their RF (figure 78.2, left) but responded exuberantly to the same object if it had recently flashed on (figure 78.2, right). Thus neurons have a remarkable ability to filter out most objects in rich visual scenes, selectively respond to salient locations, and automatically track these locations across shifts of gaze.

To see if neurons also encode top-down, deliberate selection, experimenters used a modified version of the stable array task where the RF stimulus was inconspicuous and stable but could be behaviorally relevant or irrelevant on each trial (see figure 78.3). On each trial after achieving fixation, monkeys saw a cue that matched one of the stable stimuli and instructed the monkey to make a saccade to it. The cue was flashed at the center of gaze outside the neurons' RF, so that it did not directly activate the cell. Nevertheless, neurons began responding selectively if the stable stimulus in their RF became the designated target (e.g., upward saccade in the left column). In a separate condition the cue itself appeared in the RF (figure 78.3, right column). In this condition neurons had two responses—the first to the flashed cue and the second to the selected stable target (which could be at any location in the array). The cue response occurred early and the target response arose later during the delay, but the two responses overlapped in time in the neural population (Gottlieb, Kusunoki, & Goldberg, 1998a, 2005).

Thus LIP neurons encode a spatially organized selective visual representation, which integrates information about visual conspicuity and task demands and tracks salient locations across shifts of gaze. The spatial accuracy of the LIP response can be best appreciated in figure 78.2, where neurons responded to a visual event (abrupt onset) that had occurred outside the RF (before the shift of gaze), after the location of the past event entered the RF because of the monkey's saccade. This spatially accurate remapping is thought to reflect the integration of retinotopic visual information with extraretinal responses—in particular, information about eye position and corollary discharge signals specifying the direction of the upcoming saccade (Bisley & Goldberg, 2010). Such retinal–extraretinal integration may be the basis of spatial encoding in different reference frames

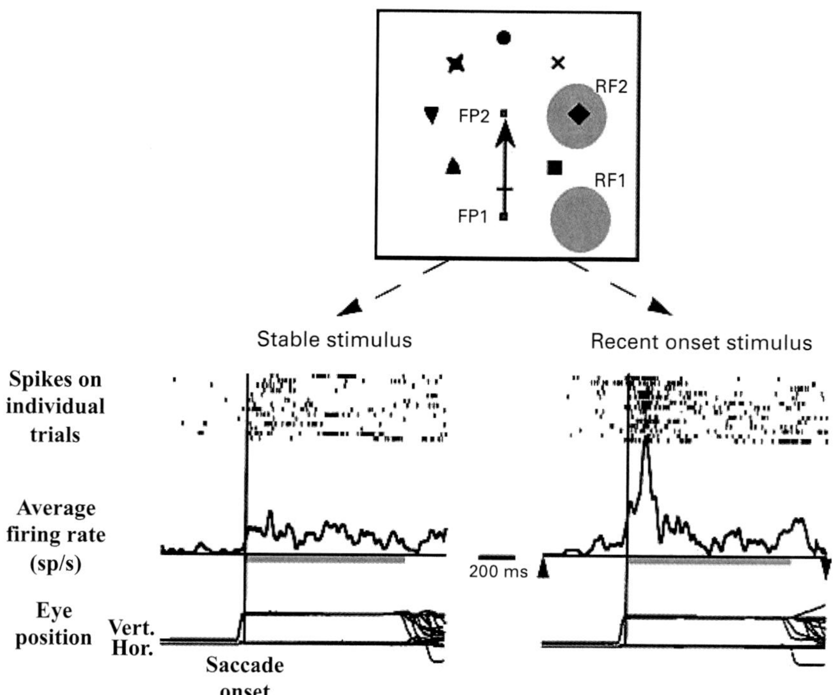

FIGURE 78.2   LIP neurons respond selectively to salient events. The top cartoon shows the experiment design. A circular object array was continuously present on the screen. The monkey began each trial by looking at an initial fixation point (FP1), chosen so that the receptive field (RF) of the neuron under study was on a blank visual location (RF1). The fixation point then jumped to the center of the array (FP2), and the monkey made a saccade to it (arrow). This saccade brought the neuron's RF onto one of the array objects (RF2). The bottom panels show the activity of one LIP neuron. Raster plots (top) show the times of individual action potentials relative to the saccade that brings the stimulus in the RF. The traces underneath the rasters (middle) show the average firing rate in spikes per second (sp/s). Bottom traces show the horizontal (Hor.) and vertical (Vert.) eye position. The neuron had only weak activation if the stimulus entering its RF was stable for a long period of time (left). In contrast the neuron responded strongly if the same stimulus had been rendered salient by virtue of an abrupt onset (right). (Modified with permission from Gottlieb, Kusunoki, & Goldberg, 1998.)

(Pouget, Deneve, & Duhamel, 2002). It may also account for the fact that spatial deficits in neglect can occur in multiple reference frames as patients ignore stimuli that are contralateral with respect to the center of gaze, with respect to their head or body, or with respect to external landmarks (Behrmann & Geng, 2002).

Neurons Encode Visual Selection

As described above, the selective responses in LIP signal a vector, or pointer, from the center of gaze to an interesting object or location. Such a retinotopic vector is in principle suitable for both influencing visual processing and guiding shifts of gaze. For the visual system this vector could act as a top-down bias facilitating sensory processing at the selected locations. For the oculomotor system it could act as a motor command, specifying the direction and amplitude of a desired saccade. Indeed, if the same signal were broadcast simultaneously to the visual and oculomotor systems, this could

explain the close natural coordination between attention and gaze.

Psychophysical studies, however, show that despite their close relationship, eye movements and attention are not identical but have partly dissociable mechanisms. The links between eye movements and attention are asymmetric. On one hand when an overt saccade is made, there does seem to be an obligatory link since perceptual performance is by default improved at the saccade goal and attending to nontarget locations necessarily entails a cost in saccade performance (Kowler et al., 1995). On the other hand, attention can be allocated without shifts of gaze, as if by an internal— "covert"—movement of the mind's eye (Carrasco & Yeshurun, 2009). Covert attention contributes to efficient visual analysis as it allows the brain to extract information from multiple locations before a shift of gaze. It is also beneficial in social situations. For example, it is often desirable to attend to a dominant individual without shifting gaze, as direct eye contact can signify a threat.

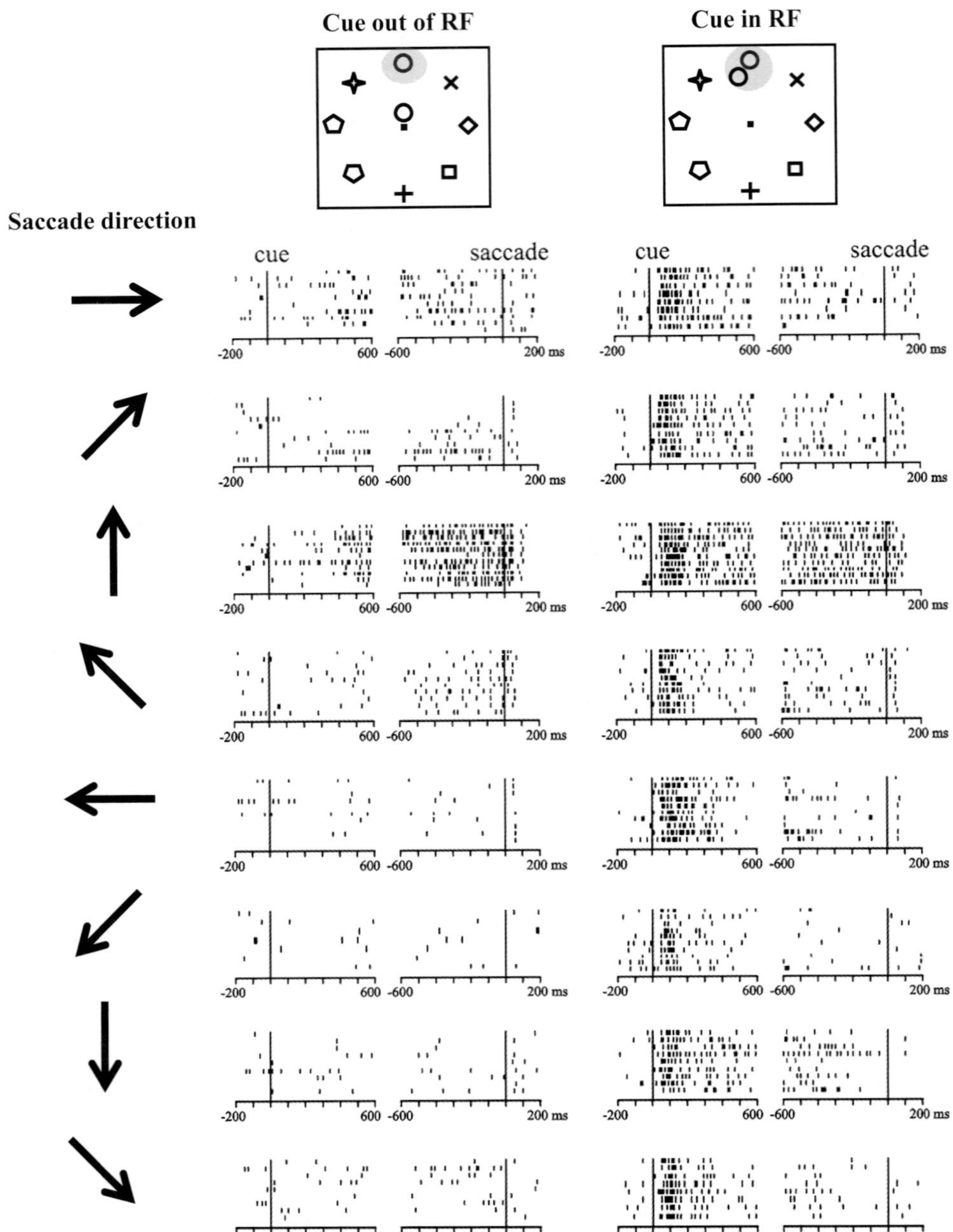

FIGURE 78.3   Lateral intraparietal neurons respond selectively to relevant targets. The top panels show the task design. Monkeys viewed a circular array containing eight peripheral saccade targets, of which one was constantly in the neuron's receptive field (RF) (shaded oval). On each trial a cue was flashed for 200 ms, and after a 600- to 800-ms delay monkeys were rewarded for making a saccade to the target matching the cue. The rasters show responses of a representative neuron sorted according to the direction of the saccade. If the cue was out of the RF (left column) the neuron had no response to the cue; however, gradually during the delay period it developed a response to saccade direction—for example, became active only if the saccade was directed to its RF (upward saccades). If the cue was in the RF (right column) the neuron responded first to the cue and later encoded saccade direction. (Modified with permission from Gottlieb, Kusunoki, & Goldberg, 2005.)

The capacity for covert attention therefore allows animals important flexibility as to whether and how they translate mental processes into action. From the point of view of neural processing, this flexibility implies that the brain must have partially dissociable mechanisms of visual selection on one hand, and eye movement planning on the other. Multiple studies have addressed the question of which process is more closely encoded in LIP, and the bulk of the evidence from these studies favors an interpretation in terms of visual selection.

Perhaps the simplest and strongest evidence in this direction is the fact that LIP neurons respond strongly to salient stimuli even when these do not evoke saccades. This can be appreciated in the experiment shown in figure 78.2, where monkeys were not rewarded for making saccades to the salient object or to the briefly flashed cue. And yet in this study (and many like it) LIP neurons emit some of their strongest responses to salient stimuli even as monkeys deliberately avoid looking at these stimuli (Gottlieb, Kusunoki, & Goldberg, 2005; Powell & Goldberg, 2000).

A follow-up experiment further expanded on this observation by asking how neurons respond in a condition of visuomotor conflict—when monkeys make saccades away from a salient cue (Gottlieb & Goldberg, 1999). On each trial in the task monkeys were first shown a cue that flashed briefly at one of two possible locations—either inside or opposite the neurons' RF. Depending on the color of the fixation point, monkeys had to make either a saccade toward the cue (an easy or habitual "prosaccade") or a saccade to an unmarked location opposite the cue (a more difficult, controlled "antisaccade"). By randomly interleaving the two cue locations and the two mapping rules the task achieved a full spatial dissociation, such that on any given trial the cue, the saccade goal, neither, or both could be in the RF.

As expected from the earlier results, LIP neurons had strong transient responses when the cue flashed in their RF. These responses were highly consistent regardless of whether the cue instructed a pro- or an antisaccade. Also consistent with prior observations, the neurons responded before saccades to their RF; this presaccadic response, however, was inconsistent and dependent on visual stimulation. A neuron showing this visual dependence is shown in figure 78.4A. On prosaccade trials when the movement was in its preferred direction the neuron had strong responses (top-row) and, if a delay was imposed between the cue and the saccade, it had a dip in activity followed by a reactivation just before the movement itself (top right). Despite this apparently movement-related profile, however, all of the cell's responses vanished in the antisaccade condition even though the monkey's movements remained the same (bottom row). Thus, even though the cell's response seemed to be movement-related, it was strongly dependent on prior visual stimulation. This visual dependence was the rule for most cells, with most responding weakly before antisaccades (figure 78.4B). Therefore, while neurons conveyed reliable information about cue location (reflecting the consistency of their visual response), they transmitted hardly any information about saccade direction even as monkeys were making that saccade (figure 78.4C). Thus, LIP neurons have robust responses to visual stimuli and visually guided

saccades but only weak responses for internally generated movements that are not congruent with a visual cue.

What do these findings imply about the relation between LIP activity and overt saccades? Converging evidence shows that saccade production depends on a gradual visuomotor transformation that is implemented in neural populations distributed throughout LIP, FEF, and the SC. Many neurons in the FEF and SC show visual dependence similar to LIP (Schall et al., 2011). These cells seem to encode an internal layer of visual selection. However, the FEF and SC contain a distinct population of "movement" neurons that are scarce or absent in LIP, and which seem to encode a distinct motor stage. In contrast with the neurons I described so far, movement cells respond invariably before saccades but not to simple visual stimulation and, in FEF, seem to reach a fixed threshold before an actual saccade (Hanes & Schall, 1996). These cells therefore are reliable reporters of an actual saccade plan and of the integration of evidence toward such a plan (Schall et al., 2011; Stanford et al., 2010).

Given their projections to FEF and the SC, LIP neurons are likely to influence, directly or indirectly, these movement mechanisms. Clearly, however, the integration of the visual selection signal into a motor plan is not automatic but is gated in a task-dependent fashion. One gating mechanism is thought to involve inhibition of movement neurons by the substantia nigra and collicular "fixation cells" (Lo & Wang, 2006; Pouget et al., 2011). These cells are tonically active during sustained fixation and may prevent an automatic visual saccade if a movement has to be withheld (e.g., as in figure 78.2). Thus, LIP neurons can contribute to visual selection and suggest possible eye movement targets. However additional mechanisms that proceed in different areas are required to generate saccades.

A final noteworthy point is that rather than reflecting a feedforward motor effect, much of the saccade-related response in LIP may reflect feedback from downstream motor mechanisms. Information about saccade direction continues to accumulate in LIP after the saccade itself (Powell & Goldberg, 2000) and, for antisaccades, arises at delays longer than the natural latency of the saccade (Zhang & Barash, 2004). It is therefore possible that the saccade response reflects, at least in part, corollary information about a saccade plan that is computed in FEF or the SC and is fed back to LIP, where it modulates a visual response (Sommer & Wurtz, 2002). This hypothesis can explain both the visual dependence of the LIP response and the psychophysical finding that attention is automatically drawn to a saccade goal.

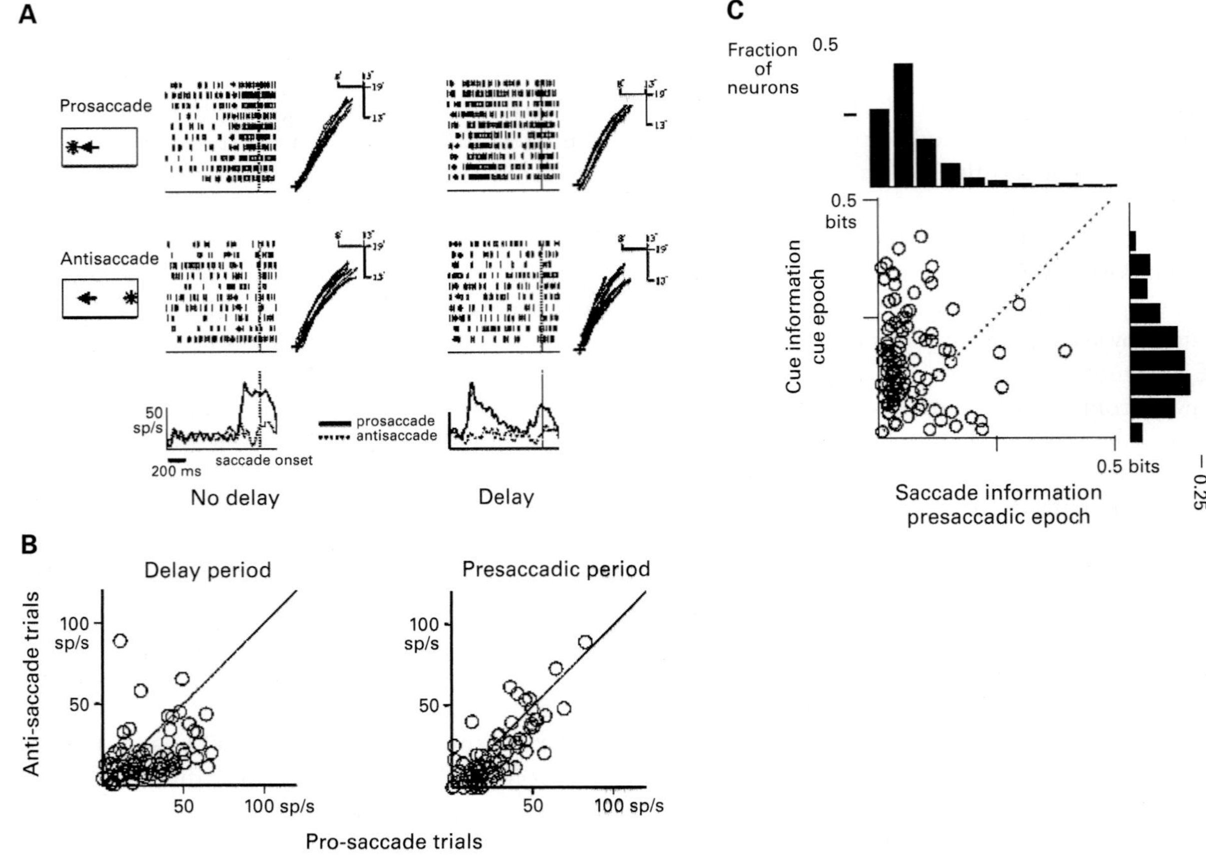

FIGURE 78.4   Lateral intraparietal neurons convey more reliable visual than saccadic information. (A) The responses of a representative neuron on trials in which the monkey made a prosaccade (top) or an antisaccade (bottom) toward its RF. The raster plots show action potentials aligned on saccade onset, and the time of cue onset is indicated by a dot. The left column shows no-delay trials, in which the saccade was made immediately upon cue onset, and the right column shows delay trials, in which a variable delay was imposed after cue presentation. The bottom panels show average firing rates in spikes per second (sp/s) for pro- and antisaccades. The traces next to each raster show the two-dimensional trajectory (horizontal and vertical position) for the corresponding saccades. (B) The vast majority of cells showed stronger responses on delayed prosaccades relative to antisaccade trials, both during the delay period and immediately before the saccade. (C) The vast majority of cells transmitted more information about the location of the cue in their visual response than about the location of the saccade goal in their presaccadic response. Each dot shows the information transmitted by one cell, and the histograms show the marginal distributions of cue and saccade direction information. (Modified with permission from Gottlieb & Goldberg, 1999.)

LIP Activity and Visual Attention

As described above, the selective visual responses in LIP seem to have a direct correspondence with visual attention. However, unlike activity in earlier visual areas, LIP neurons are not feature selective and do not describe the sensory world. Rather they convey a more abstract measure of the "interestingness" or attention worthiness of an object or location.

According to the biased competition hypothesis, selective responses such as the one encoded in LIP can modulate perception through top-down feedback to earlier sensory representations (Desimone & Duncan, 1995). Such feedback signals may selectively increase the gain of neurons tuned to the selected features or locations (Reynolds & Chelazzi, 2004; Reynolds & Heeger, 2009; Reynolds, Pasternak, & Desimone, 2000). In addition, top-down signals can enhance competitive weight, allowing the selected item them to suppress competition from irrelevant distractors (Reynolds, Chelazzi, & Desimone, 1999; Reynolds & Heeger, 2009). To see whether LIP neurons are indeed a possible source of attentional bias, several studies have tested the relation between LIP responses and perceptual selection, as assessed psychophysical measures of contrast sensitivity and visual search.

Bisley and Goldberg (2003) used a dual-task design that measured the monkeys' locus of attention based on contrast thresholds on a visual discrimination. Monkeys were first shown a saccade target whose

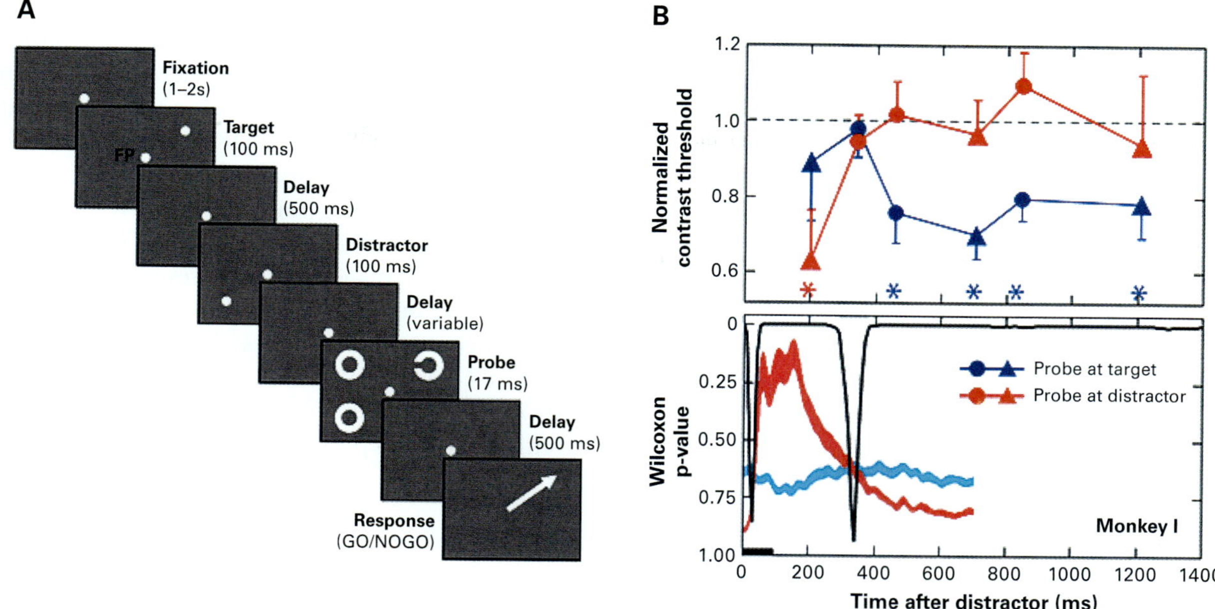

FIGURE 78.5   Lateral intraparietal (LIP) activity predicts the locus of perceptual attention. (A) The task design, showing the parameters of target and distractor presentation, followed by the presentation of the no-go instruction (the Landoldt-C) that also served as an attentional probe. (B) Data from monkey I. The top panel shows normalized contrast thresholds for discriminating the probe when it appeared at the target versus distractor locations. Thresholds are normalized to a value of 1 that represents the threshold at the remaining two locations that contained neither the target nor the distractor. Stars indicate thresholds that are significantly lower than the reference value. The bottom panel shows the population response of LIP neurons to the target and distractor in their receptive field. The black trace shows the p value from a moving window paired test, showing that these responses were reliably different for most of the delay period. The brief window of ambiguity when the two responses were equivalent (shaded window, downward deflection in the p-value trace) corresponds to times when attention shifted from the distractor to the target location. (Modified with permission from Bisley & Goldberg, 2010.)

location they had to memorize throughout a memory (delay) period (see figure 78.5A). On some trials a distractor was flashed during this period, which monkeys were instructed to ignore. While engaged in this task monkeys received a secondary task designed to test covert attention. At some point after target presentation monkeys were shown a visual cue—a c-like stimulus that could have two possible orientations. The cue's orientation instructed the monkey whether to proceed with the planned saccade or to cancel it and maintain fixation. By presenting this cue very briefly at several randomized locations and levels of contrast, experimenters could measure the monkeys' contrast sensitivity at different locations. The locus of attention was defined as the location that showed the lowest contrast threshold for cue discrimination.

Using this approach, Bisley and Goldberg showed that while holding stable fixation, monkeys shifted their attention from the target to the distractor and back again, and these shifts correlated with the balance between target and distractor activity in LIP. As shown in figure 78.5B, LIP neurons responded to the target with a transient response followed by a lower-level sustained activation (blue trace), and also had a strong response to the distractor (red trace). This dual responsiveness confirms the earlier finding that although selective, the population of LIP neurons can encode more than one stimulus at a time (Gottlieb, Kusunoki, & Goldberg, 2005). In contrast with this dual response, however, perceptual attention (the locus of lowered thresholds) was only found at a single location at a time, and this was the location that elicited the higher response in LIP. Attention was allocated to the distractor for as long as the distractor response exceeded that to the target, but it shifted back to the target location when the balance of activity reversed in LIP.

These findings suggest that, just as LIP plays an indirect role in driving saccades, it may have an indirect role in controlling attention. While the priority map in LIP may select several locations as candidates for enhanced discrimination, subsequent processing may implement this enhancement at only one location as if by a winner-take-all mechanism. This additional selection may be implemented in a discrete stage subsequent to LIP, or simply reflect the properties of sensory processing itself. It is also important to note that the experiment of Bisley and Goldberg used very brief (16-ms) masked visual presentations and shows that attention has a unitary locus on these very brief time scales. It remains possible, however, that attention rapidly shifts between locations and, on longer time scales, it is effectively distributed across multiple locations.

While Bisley and Goldberg focused on saccade-related and bottom-up attention, a second set of studies focused on top-down attention during covert visual search. These studies capitalized on the classic observation that the time required for finding an inconspicuous target increases as a function of the number of distractors, a set-size effect indicative of inefficient or attentionally demanding search (Haslam, Porter, & Rothschild, 2001). An involvement of LIP in difficult search was suggested by two experiments using reversible inactivation with the GABA agonist muscimol. These studies showed that inactivation impairs visual selection during difficult search, whether the search occurs covertly or in conjunction with overt saccades (Wardak, Olivier, & Duhamel, 2002, 2004).

To determine the neural responses underlying this function, Balan et al. (2008) trained monkeys to discriminate a visual cue—an E-like shape—that appeared in the visual periphery in an array of distractors (see figure 78.6A). Monkeys reported cue orientation by releasing a bar and performed the task using central

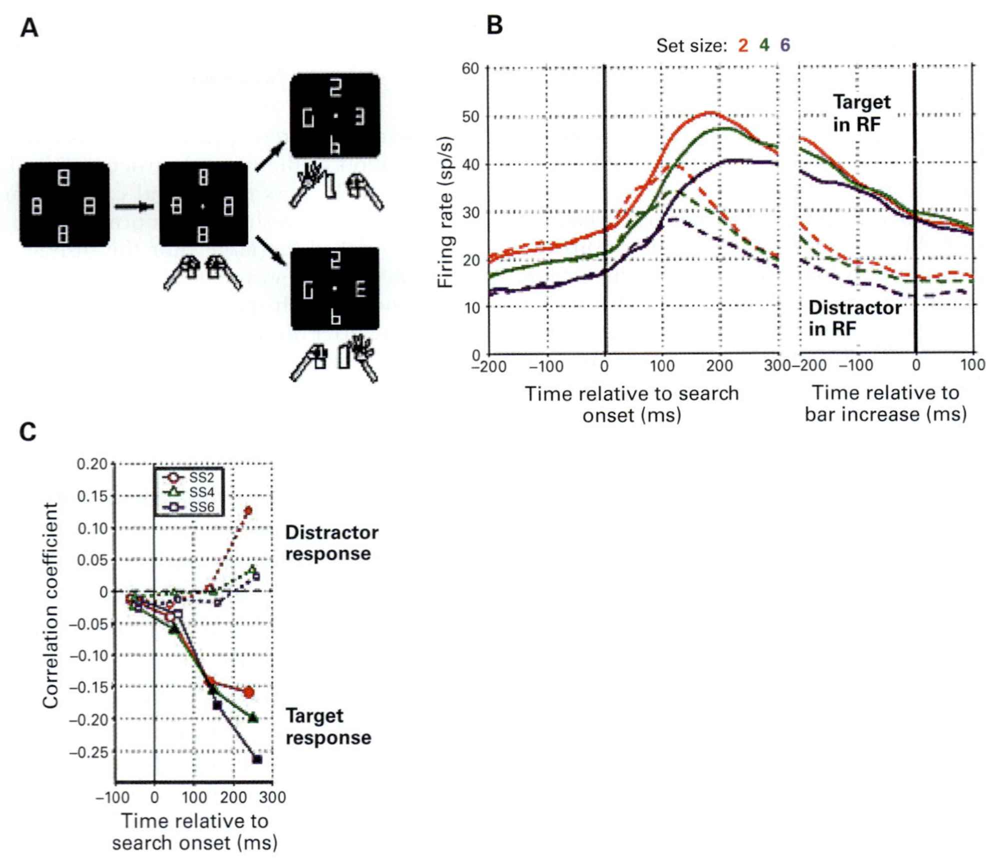

FIGURE 78.6   Lateral intraparietal (LIP) activity encodes covert top-down attention and the set-size effect. (A) The task design. An array of figure-8 placeholders remained stably on the screen. In interleaved blocks of trials the array contained two, four, or six placeholders. Each trial began when the monkey achieved fixation, bringing a placeholder into the receptive field (RF). Two line segments were then removed from each placeholder, revealing a display with one target and several distractors. The target (a right- or left-facing letter "E") appeared at a random location. Without shifting gaze, monkeys had to find the target and report its orientation by releasing a bar. (B) During the time between search onset (stimulus presentation) and the manual release, LIP neurons encoded target location, responding more strongly when the "E" relative to a distractor appeared in their RF. These responses declined as a function of set size. The set-size effect was evident even before search onset, when a different number of placeholders was present on the screen. (C) Correlation coefficients between neuronal firing rates and reaction times for trials in which the target (solid lines) or a distractor (dotted lines) were in the RF. Significant correlations (filled symbols) were found for the target response, showing that the larger this response, the shorter the reaction time. In contrast the distractor response was more weakly related to performance. sp/s, spikes per second; SS, set size. (Modified with permission from Balan et al., 2008.)

fixation without shifts of gaze, thus eliminating any motor orienting toward the attended cue. Performance showed a set-size effect in both reaction time and accuracy, indicating effortful, attentionally demanding search.

LIP neurons had robust responses encoding the cue location, responding more strongly if the cue rather than a distractor was in their RF (figure 78.6B). Consistent with the behavioral set-size effect, these responses diminished as a function of the number of distractors. The set-size-related decline in activity was seen in the fixation period, when the monkeys viewed a variable number of placeholders but could not yet begin their search (figure 78.6B, −200 to 0 ms). Behavioral testing showed that monkeys were insensitive to changes in location probability per se, suggesting that the set-size effect was not due to changes in probability estimates. Most likely, therefore, the neural set-size effect reflected competitive visual interactions triggered by surrounding distractors (Falkner, Krisha, & Goldberg, 2006), suggesting that these interactions explain capacity limitations during visual search.

The robust responses on this task also show that LIP encodes visual selection in order to guide skeletal actions, that is, is independent of a specific motor output. Notably, however, while in saccade tasks neural responses typically increase up to the time of a saccade (e.g., figure 78.4A), in this task activity peaked in the middle of the reaction time and declined well before the motor response. This suggests that LIP accumulates visual information before passing control to skeletomotor mechanisms.

The results also provided data regarding the specific behavioral correlates of the response. If a distractor was in the RF, neurons showed an enduring set-size effect that persisted until the bar release (figure 78.6B, dashed traces). By contrast if the target was in the RF, the set-size effect diminished gradually over time, so that the response reached a comparable peak level at all set sizes (solid traces). In addition the monkeys' reaction times were better correlated with the target rather than the distractor response (figure 78.6C). These findings are consistent with a biased competition account whereby the influence of distractors is filtered through top-down control. In addition they suggest that the top-down signal may primarily influence performance through its actions on target representations.

These studies therefore establish correlations between LIP activity and perceptually defined attention. These relationships arise whether attention is driven in bottom-up or top-down fashion, whether it is allocated during saccade planning or maintained fixation, and whether it guides ocular or skeletal actions.

COMBINING BOTTOM-UP AND TOP-DOWN INFORMATION

Psychophysical studies show that attention is controlled jointly by top-down and bottom-up information. Importantly the effects of salient distractors are not immutable but depend strongly on task context and the observers' search sets (Burnham, 2007). Given that LIP neurons integrate these two types of information, they are a possible candidate for mediating these behavioral effects.

To address this question, Balan and Gottlieb (2006) used the "E"-search task described in figure 78.6 but, in addition to the target, introduced a visually salient perturbation. On each trial the appearance of the search display (target and distractors) was preceded by a 50-ms visual perturbation—which could be a brief change in the position, luminance, or color of a stable placeholder or the brief appearance of a frame around a placeholder. The significance of the perturbation was varied in randomly interleaved blocks of trials. In some blocks the perturbation appeared at the same location as the search target—and was thus a valid cue to the target location. In other blocks the perturbation appeared at a different, unrelated location and was irrelevant to the search. These statistical associations remained constant during a block, and monkeys could exploit them to adjust their attentional strategy. Indeed, in perturbation-relevant blocks the monkey's reaction time was shorter than in perturbation-irrelevant blocks, showing that they learned to use the perturbation as a valid cue in relevant blocks but block its distracting effects in irrelevant blocks.

LIP neurons also showed contextual sensitivity, emitting enhanced responses to the perturbation in the relevant relative to the irrelevant context. Moreover, the relative size of the perturbation response correlated with the behavioral effect. In the perturbation-relevant block responses were higher overall (see figure 78.7A), and an increase in the visual response correlated with a decrease in reaction time (figure 78.7B), consistent with the fact that the perturbation had a facilitatory effect. In the perturbation-irrelevant block, by contrast, responses were lower (figure 78.7B) and a higher visual response was correlated with longer search reaction times, indicating a distracting effect. Interestingly, the increment in the visual response was accompanied by an increase in the neurons' baseline activity even before the perturbation occurred, and the fractional increase in baseline activity was not different from the fractional increase in the visual response (figure 78.7A). Multiplicative gain was shown previously to enhance responses to weak stimuli when these are the focus of attention

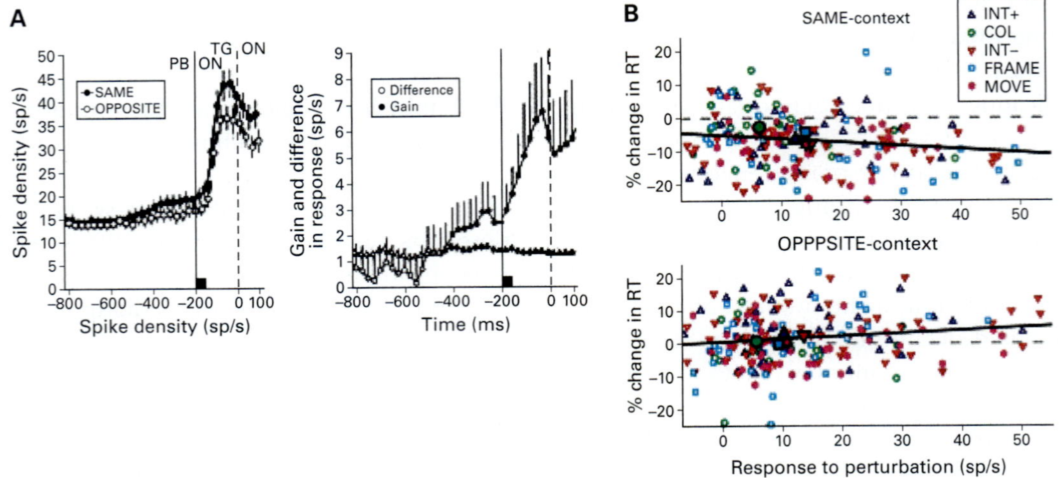

FIGURE 78.7   Lateral intraparietal (LIP) neurons integrate bottom-up and top-down information. (A) Left panel shows the population responses on trials in which a salient perturbation was relevant ("same") or irrelevant ("opposite") for the location of the search target. Responses are aligned on the time of the perturbation (PB ON) and truncated after the appearance of the target (TG ON). An enhancement is seen in the relevant blocks starting 200–300 ms before onset of the perturbation. The right panel shows the difference and ratio (gain) between firing rates (in spikes per second, sp/s) in the two conditions. While the difference increases sharply at the time of the visual response, the gain remains constant throughout the trial. (B) The perturbation response in LIP correlates with the effect of the perturbation on search reaction times. In the "same" context a larger perturbation response is associated with a large fractional decrease in reaction time (RT), while in the "opposite" context a larger response produces a larger increase in reaction time. Each point shows the data from one neuron and one perturbation type (increase or decrease in luminance, INT+ and INT–), isoluminant color change (COL), appearance of a frame (FRAME), or movement of a placeholder (MOVE). (Modified with permission from Balan & Gottlieb, 2006.)

(McAdams & Maunsell, 1999; Williford & Maunsell, 2006). In this task, however, response gain increased in a nonspatial fashion before the animal could direct attention to a specific location. Thus, abstract knowledge of task context can also produce multiplicative effects that influence sensitivity to sensory stimulation.

A task-dependent encoding of salient stimuli was also shown in a visual search task in which a pop-out (color-contrasting) distractor was present along with an inconspicuous target (Ipata et al., 2006). Neurons showed variable responses to this distractor, which correlated with the monkeys' ability to suppress saccades to it, supporting the idea that LIP is a possible substrate for integrating top-down and bottom-up information.

Determinants of Visual Selection, I: Semantic Associations

The studies described above show that the selective response in LIP can be a source of top-down bias that selectively guides or channels visual processing. However this raises an important question. How does the brain generate this bias, and how do LIP neurons know the "relevance" of a visual cue?

Psychophysical studies suggest that three key factors establish the significance of visual cues. These include the learned contextual and action associations of the cues, the usefulness of a cue in providing information, and the Pavlovian (reward) associations of the cues. In the following sections I discuss the behavioral factors that modulate the selective response in LIP and their possible contribution to a relevance computation.

A fundamental step in computing relevance is learning the associations between stimuli and the broader task, including the context, actions, or goals of that task. Imagine, for example, that you want to drive to a remote location. By virtue of information stored in long-term memory, you have learned to associate the task of driving with the concept of a "car." This can in turn activate a visual template (of a "big object with wheels") which can produce an attentional bias and ultimately a focus on the car, as implemented in computational models (Huang & Grossberg, 2010; Navalpakkam & Itti, 2005). The idea of associative top-down control is consistent with psychophysical evidence that attention can be guided by context or gist (Oppermann et al., 2012) or by the motor associations (affordances) of a visual cue (Roberts & Humphreys, 2011a, 2011b).

Consistent with the importance of associative learning, a number of studies have shown that the target selection response in LIP is not stereotyped but is shaped by the motor, visual, and categorical associations of visual cues. Evidence regarding motor associations comes from the covert visual search task described in figure 78.6 (Oristaglio et al., 2006). As described above, in that task monkeys were required find a visual cue while maintaining central fixation and report its orientation with a manual release. Specifically, monkeys learned to release a bar held in the right paw if the cue was a right-facing "E" but to release a bar held in the left paw if the cue was a left-facing "E" (a "3"). This manual response occurred outside the monkeys' field of view and it was nontargeting—that is, independent of the location of the cue. Nevertheless a given motor response had a nonspatial (semantic) association of a given cue, and this association was encoded in LIP.

In about half of the cells encoding target location, the cue-evoked response was not constant but depended on the manual release. Of these cells, some had stronger responses if the cue appeared in the RF and instructed a left-bar release (see figure 78.8A, left column, red vs. blue). Other cells had the complementary preference, responding best for the cue signaling right release. Control experiments showed that the effect was not linked to the cue's orientation or to the location of the limb. Thus, when monkeys were trained to indicate the orientation of a new set of cues (an upright or inverted-U-like shape), a majority of neurons showed the same limb preference as they did for the E-like cues. Second, when monkeys were trained to perform the same task with their limbs crossed across the body midline, limb effects remained unchanged: Neurons that preferred the right or left limb in the standard arm position continued to have the same preference in the crossed position (figure 78.8B). Thus, neurons encoding the cue location were also sensitive to the associated limb.

An important aspect of these limb effects is that, while they occurred reliably in over half the cells, they were not the neurons' primary response. Rather they modulated a primary response to visual selection. The cell in figure 78.8A, for example, had robust limb selectivity if the cue appeared in its RF. However the neuron remained silent if a distractor was in the RF even though monkeys performed the same manual action (right column). The primacy of the visual response was confirmed by a further experiment using reversible inactivation, which impaired only visual but not motor selection (Balan & Gottlieb, 2009). Infusion of muscimol into LIP in one hemisphere impaired performance in a spatially selective manner—that is, if the cue was contralateral but not if it was ipsilateral to the inactivation site (see figure 78.8B). However inactivation caused no deficits in the manual release—neither a global

impairment in the manual action nor a limb-specific deficit. These findings therefore suggest that the limb responses in LIP do not indicate a primary involvement of this area in the execution or release of a grasp. Rather they are an indirect influence of a motor association on visual selection.

Other experiments have shown that LIP neurons can also reflect more abstract properties such as the categorical membership or visuovisual associations of an instructive cue (Freedman & Assad, 2011). In a task where monkeys were trained to report the category of a motion stimulus using a delayed match to sample task, neural responses to the stimulus were category selective, with different neurons responding more for one or the other category (Freedman & Assad, 2006). Similar differentiation was found in different tasks testing visuovisual associations (Fitzgerald, Freedman, & Assad, 2011) or visual search based on feature conjunctions. Although using very different tasks, these studies converge on a similar conclusion—that LIP neurons reflect not only the location but the learned visual, action, and semantic associations of task-relevant cues.

These findings were proposed to reflect the encoding of abstract, motor-independent decisions (Freedman & Assad, 2011). In addition, given the role of LIP in visual selection, they may also contribute to the learning of relevance and top-down control. This hypothesis raises new questions and expands our view about this control. In the traditional view of attention, top-down signals for features or locations are thought to arise from different neurons and recruit different attentional mechanisms (Desimone & Duncan, 1995; Maunsell & Treue, 2006). The multifaceted nature of the LIP response, however, suggests that top-down signals may be selective for both space and high-level features (Gottlieb & Snyder, 2010). For example, if neurons such as the limb-selective cells in figure 78.7A provide the top-down bias, this implies that a different population of neurons provides the bias for cues signaling right or left release. In other words, directing attention to a given location may be accomplished by different neuronal populations depending on the significance of the attended object. While counterintuitive at first glance such a differentiated (combinatorial) signal may provide an efficient interface with visual and motor mechanisms, which are also selective for feature and/or categories. An alternative possibility is that neurons with dual spatial/nonspatial selectivity represent only the hidden layer in a network computing relevance, and the output of the network (the attentional bias itself) may be conveyed by the neurons that lack associative effects. Future experiments are needed to clarify this question.

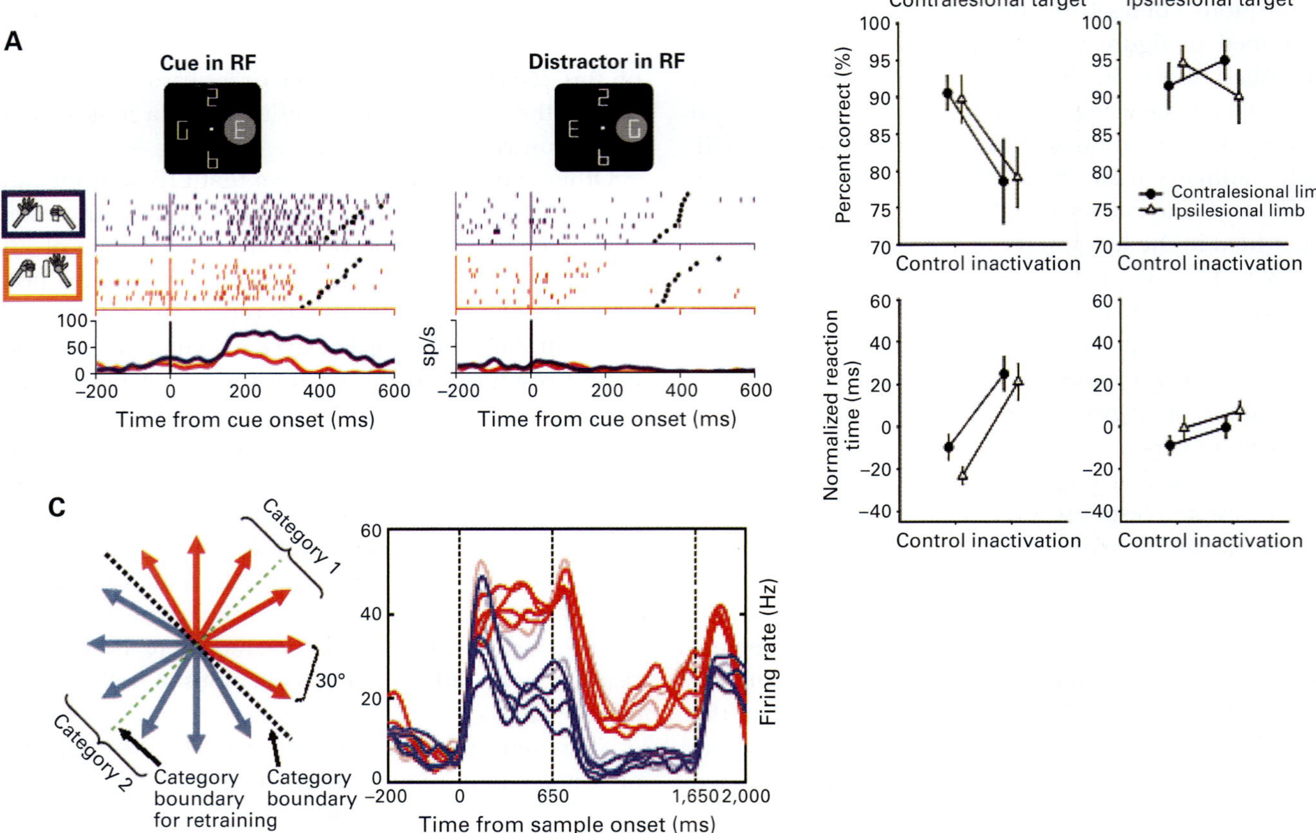

FIGURE 78.8 Lateral intraparietal (LIP) neurons encode motor and categorical associations. (A) Modulation by manual release. Responses of a representative neuron in the "E" search task shown in figure 78.6. Monkeys were rewarded for maintaining fixation and reporting the orientation of the "E"—right or left facing—by releasing a bar held, respectively, in the right or left paw. The bars themselves were outside of the field of view. Rightward-facing cues could appear on the left and vice versa, so the laterality of the motor response was independent of the laterality of the visual cue. The neuron responded only if the "E" appeared in the receptive field (RF) but was silent if a distractor did (left vs. right column). In addition, when an "E" appeared in its RF, the cell was more active if the monkey released the left bar than the right bar (blue vs. red traces). Raster plots in the top panels show individual trials. Each dot represents the time of an action potential aligned on cue onset, and the black dots show the time of manual release. Trials are sorted offline in order of manual reaction time. The bottom panel shows the corresponding averaged spike density histograms (smoothed with a Gaussian kernel, sigma 10 ms). (Adapted with permission from Oristaglio et al., 2006.) (B) Muscimol inactivation impairs visual but not motor selection Performance of the "E" search task in control conditions and after muscimol inactivation of LIP in one hemisphere. The top row shows discrimination accuracy and the bottom row, reaction times. Symbols show average and standard errors. Data are segregated according to the hemifield of the target and the side of the active limb (contralateral or ipsilateral to the inactivation site). (Modified with permission from Balan & Gottlieb, 2009.) (C) Modulation by stimulus category. The left panel illustrates the behavioral task. Monkeys viewed a sample stimulus containing random dot motion in one of eight possible directions. After a delay period (650 to 1,650 ms) a test motion stimulus appeared and monkeys had to release a bar if the test stimulus matched the category of motion of the sample but continue to hold the bar otherwise. Monkeys were initially trained to categorize directions according to one category boundary (black dotted line) and then retrained to use a different boundary (green dashed line). Top right panel shows representative LIP neuron that had visual and delay period activity following presentation of a sample inside its RF as well as sensitivity to sample category. Firing rates were much more strongly modulated by changes in direction across, relative to within, a category boundary, dissociating this modulation from simple selectivity for motion direction. (Modified with permission from Freedman & Assad, 2006.)

A second set of questions concerns the mechanism generating associative effects. One possibility is that the effects reflect online feedback from structures that accumulate evidence regarding categories or actions. Alternatively, however, it is possible that through prolonged training, visual associations produce plasticity in visual areas themselves or in the connections between these areas and LIP. Such plasticity can be produced in simple reinforcement models (Ferrera & Grinband, 2006) and is consistent with the presence of

FIGURE 78.9   Eye movements during active tasks. (A) Eye movements of a subject filling water in a teakettle. (B) Eye movements of a subject preparing a sandwich. (Reproduced with permission from Land, 2009.)

semantic effects in earlier visual areas such as V4 (Mirabella et al., 2007).

Determinants of Visual Selection, II: Utility and Information

While semantic associations are necessary, they are not sufficient for determining relevance: To efficiently guide attention, the brain must also estimate the usefulness of a visual cue. To appreciate this distinction, consider the pattern of eye movements made by a subject during natural behavior—a simple task such as filling a kettle for making tea (see figure 78.9A). As described by the seminal work of Yarbus in the early twentieth century, the vast majority of eye movements made in such conditions are directed to task-relevant cues. However these movements do not target all the stimuli that have task associations, which in this example would include the kitchen walls, cabinets, and floor. Rather, they are selective for stimuli that bring new information—that is, the faucet and sink that guide the manual actions. This suggests that additional mechanisms must operate within the set of task-associated cues to further select those stimuli that inform—and thus increase the likelihood of success of—a future action.

Consistent with this idea, multiple studies have shown that oculomotor decisions are sensitive to expected reward and have proposed that the target selection response in LIP (and possibly other areas) encodes the utility of alternative actions (Kable & Glimcher, 2009; Sugrue, Corrado, & Newsome, 2004, 2005). Monkeys are easily trained to direct gaze for liquid rewards, and human subjects make optimal trade-offs between expected gain and salience during visual search (Navalpakkam et al., 2010). Reward effects have been described in multiple structures including the FEF (Ding & Hikosaka, 2006) and SC (Ikeda & Hikosaka, 2003). Similarly in LIP, the target selection response increases monotonically with expected reward, whether this comes about through variations in reward magnitude, probability, or timing of a reward, and whether animals perform instructed or free-choice tasks (Dorris & Glimcher, 2004; Louie & Glimcher, 2010; Louie, Grattan, & Glimcher, 2011; Platt & Glimcher, 1999; Sugrue, Corrado, & Newsome, 2004). These findings suggest the powerful hypothesis that the target selection response in LIP encodes a common currency of visual utility regardless of the specific factors through which utility is conveyed (Kable & Glimcher, 2009; Sugrue, Corrado, & Newsome, 2005).

An important question, however, is if neurons encode expected reward per se or the specific form of utility that is important for attention—the utility of information. Rather than being solely determined by expected reward, the usefulness of information depends on the observer's uncertainty. A sensitivity to uncertainty (or the flip side, new information) is critical in order for attention to filter out redundant signals and arbitrate between competing motor demands (Sprague & Ballard 2005; Tatler et al., 2011). Consider again the eye movements of the subject in figure 78.9A. To complete his task, the subject is simultaneously manipulating the kettle and walking from the stove to the sink. Both his hand and leg actions have high value and are necessary for the task. However the eye movements exclusively guide the hand actions and are not directed to the floor, the target of the leg action. While valuable in the task, this action has little uncertainty or need for information.

While the specific utility of information has not been examined in the oculomotor domain, insights into this question come from reinforcement learning studies of attention in humans and rats (Pearce & Mackintosh, 2011). A central idea in these studies is that attentional weight is assigned not based on expected reward but based on the prediction errors associated with a cue. A prediction error is a measure of the difference between the agent's expectations and the actual experienced outcome. Outcomes that are better or worse than expected generate, respectively, positive and negative errors, and these errors drive learning in iterative algorithms. Reinforcement learning theories of attention postulate that the "informativeness" of a sensory signal can be estimated through the absolute value of the prediction errors associated with that signal. Cues that are reliable predictors will reduce uncertainty and have low associated errors while cues that make uncertain predictions will have higher prediction errors.

Consistent with this framework, psychophysical studies show that human attention is guided by the predictive qualities of visual cues (Hogarth, Dickinson, & Duka, 2010). Furthermore, studies in rats suggest that this guidance may involve two distinct mechanisms (Pearce & Mackintosh, 2011). One mechanism, dubbed "attention for action," is thought to depend on the frontal lobe and assign priority to stimuli that have low prediction errors—that is, are reliable predictors that can guide future actions (e.g., the faucet or sink in figure 78.9A). A second mechanism, dubbed "attention for learning," involves the amygdala and parietal lobe and assigns value to novel or uncertain cues in order to learn *about* such cues.

So far little is known about the importance of prediction errors in the visual and oculomotor system as most studies have focused on the role of expected reward. A key question for future work is to integrate this work with a reinforcement learning framework and understand how parietal and frontal neurons assign priority based on predictive or informational properties of visual cues.

Determinants of Attention, III: Pavlovian (Emotional) Associations

A central assumption of reinforcement theories of attention is that attention is guided by the absolute magnitude of prediction errors regardless of their sign. In this way attention can be directed to an informative cue whether that stimulus brings "good" or "bad" news. This seems to correspond to our intuition of how attention should proceed: While seeing a tiger on your hiking trail is an unpleasant prospect, there should not be a problem in attending to confirm that there is, in fact, a tiger in the bush.

In contrast with this purely informational view, however, converging psychophysical evidence shows that attention does have strong emotional components and is sensitive to the social or affective valence of external cues (Vuilleumier, 2005). Stimuli that are initially neutral but gain positive reward associations can automatically attract attention (Anderson, Laurent, & Yantis, 2011; Libera & Chelazzi, 2009) while stimuli with negative associations can repel gaze and reduce fixation time at their locations (Hogarth, Dickinson, & Duka, 2010). Unlike the goal-directed attention discussed above, these effects arise automatically even when the stimuli are irrelevant to the task. Affective biases in attention are strong in psychiatric disease (Williams, Mathews, & MacLeod, 1996). For example, patients with drug addiction have automatic biases toward drug-related cues that may reinstate drug craving and cause relapse even after periods of abstinence (Flagel, Akil, & Robinson, 2009).

Despite the possible importance of emotional attention, little is known about its neural mechanisms. However a recent experiment shows that conditioned cues also produce valence-specific biases in the neural responses in LIP (Peck et al., 2009). Monkeys performed a probabilistic reward task where each trial had a 50% prior probability of ending in a reward or no reward. At the outset of a trial monkeys viewed a peripheral conditioned stimulus (CS) that provided full information about the trial's reward. Some cues brought "good news," signaling that the trial will end in a reward (CS+); others brought "bad news," signaling that, even if correctly completed, the trial will end in no reward (CS–). If attention depends solely on predictive validity, it should be directed in a similar fashion to the positive and negative cues as both conveyed reliable information. By contrast, if attention depends on reward associations, it may be differently allocated to the positive and negative cues.

Consistent with the latter possibility, attentional biases differed according to affective value—such that "good news" cues (CS+) attracted attention while "bad news" (CS–) cues repelled attention from their location. As shown in figure 78.10A, upon presentation of a CS in the RF, LIP neurons had a fast transient visual response followed by sustained activity that lasted after disappearance of the cue. The sustained response that followed a CS+ was excitatory, resulting in a higher response when the cue appeared in the RF relative to when it appeared at the opposite location (see figure 78.10B, top). By contrast, the response evoked by a CS– was suppressive, producing lower activity at the cue location relative to the opposite location (see figure 78.10B, bottom).

To see whether these neural biases had behavioral correlates, monkeys were trained to maintain fixation for a brief delay period and then make a saccade to a second target that appeared unpredictably either at the opposite or same location as the cue. By comparing saccades that were directed toward versus opposite the CS location, investigators could detect whether the CS had attractive or repulsive effects.

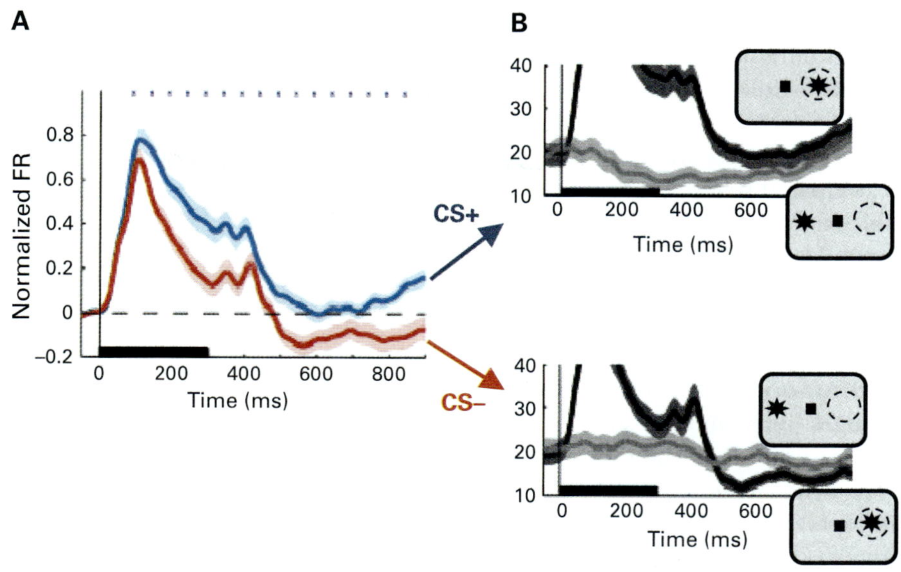

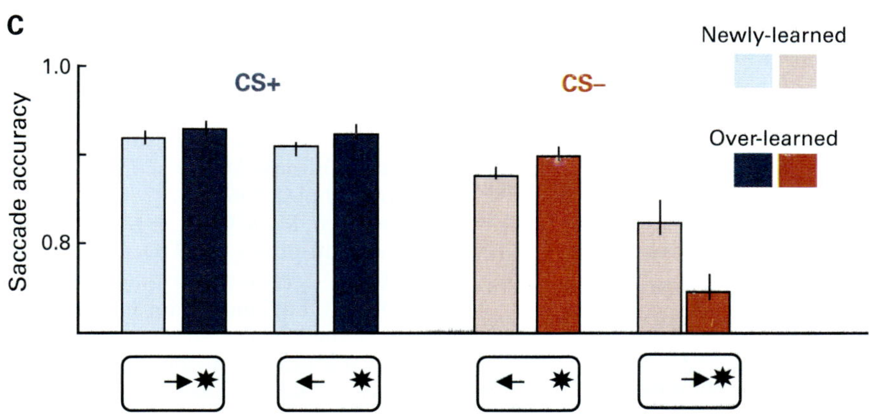

FIGURE 78.10   Pavlovian cues bias attention and lateral intraparietal (LIP) activity. (A) Population responses to Pavlovian cues. A cue flashes in the receptive field (RF) for 300 ms (dark bar) and is followed by a delay period during which monkeys maintain fixation. LIP neurons have transient and sustained responses that are selective for cue value, being stronger for a positive conditioned stimulus (CS) predicting a reward (CS+, blue) relative to a negative CS predicting no reward (CS–, red). The dots at the top show time bins with a significant difference between the two conditions. The bottom dashed line shows the precue level of activity. Shading shows the standard error of the mean. (B) Cue-evoked biases are spatially specific. The dark traces in each panel are the same as, respectively, the blue and red traces in panel A but are shown on an expanded vertical axis (with the peak of the traces intentionally truncated). The gray traces show responses when the cues appeared opposite the RF. Responses evoked by an RF cue are higher than (CS+) or lower than (CS–) those evoked by a cue at the opposite location. (C) Attentional biases are spatially specific. Panels show mean and standard error of the mean for saccade accuracy across all sessions. Opaque colors show trials with overlearned CS, which had consistent reward associations for tens of thousands of trials. Pale colors show trials with novel CS, which were introduced and trained within a single session. Measurement of anticipatory licking showed that monkeys learned the value of the novel CS within the first 5 to 10 trials. Data collection began after this learning was complete. On CS– trials (red) accuracy is somewhat lower than on CS+ trials, indicating a modest effect of motivation. The strongest effect, however, is on CS– congruent trials, when the saccade is directed toward the CS– location (right-most bars). Moreover, the accuracy on these trials decreases further with training, being lower after overlearned relative to newly learned cues. (Modified with permission from Peck et al., 2010.)

Consistent with the neural biases in LIP, saccades were slightly facilitated if they were directed toward a CS+ location but strongly impaired if directed toward a CS– location (see figure 78.10C). Performance on a CS– trial was much better if the saccade was directed away from the CS–, showing that the impairment was spatially specific and distinct from a global decrease in motivation on unrewarded trials (see figure 78.10C, right-most two bars). Moreover, the impairment on CS– congruent trials affected both reaction times and accuracy. The dysmetria (inaccuracy) produced by a CS– is particularly important because it distinguishes this repulsion from inhibition of return, which has only been reported to prolong latency (Fecteau & Munoz, 2006). In addition, by reducing accuracy, the CS– produced many targeting errors and lowered the monkeys' rate of reward. Despite its detrimental consequences, this CS– induced repulsion increased rather than decreasing with training, becoming stronger for over-learned relative to newly learned cues (see figure 78.10C, solid vs. pale colors).

These effects therefore are bona fide correlates of the emotional biases documented in human subjects. Like these biases, the effects produced by the CS depended on the valence of the cue, arose automatically, and persisted despite their maladaptive effects.

These findings indicate the need to distinguish between two classes of reward-based mechanisms in attention. One is a goal-directed mechanism that estimates the operant value of a cue in informing future actions. This mechanism may be guided by the same mechanisms that drive action valuation, or by unsigned prediction errors that can measure "new information" or surprise, and may be encoded in cortical areas (den Ouden et al., 2010; Itti & Baldi, 2009). A second mechanism, however, seems guided by the intrinsic (conditioned) Pavlovian associations of a sensory cue. This mechanism may be guided by signed prediction errors that signal both reliability and affective valence, and may be encoded by dopamine neurons of the midbrain (Flagel et al., 2011). Affective biases may be very useful in detecting biologically relevant information. As described above, however, this form of attention is a double-edge sword, which tends to operate automatically even when it has maladaptive effects.

CONCLUSIONS

Converging evidence from anatomical, single neuron recordings and inactivation studies implicates area LIP in the allocation of visual attention. Neurons in this area encode a sparse visual representation that selects candidate stimuli for shifts of attention and gaze and integrates sensory and behavioral information.

The complex behavioral influences in LIP raise important questions about the nature of top-down control. Understanding these influences requires that we consider the multiple mechanisms by which the brain assigns relevance, including the task associations, information content, and emotional value of cues (Gottlieb, 2012). Thus a critical goal for future research will be to integrate evidence from psychophysical and behavioral research toward a better understanding of the neurophysiological mechanisms.

ACKNOWLEDGMENTS

This research was supported by the National Alliance for Research on Schizophrenia and Depression, the Gatsby Charitable Foundation, and the National Eye Institute.

REFERENCES

Anderson, B. A., Laurent, P. A., & Yantis, S. (2011). Value-driven attentional capture. *Proceedings of the National Academy of Sciences of the United States of America, 108,* 10367–10371. doi:10.1073/pnas.1104047108.

Balan, P. F., & Gottlieb, J. (2006). Integration of exogenous input into a dynamic salience map revealed by perturbing attention. *Journal of Neuroscience, 26,* 9239–9249.

Balan, P. F., & Gottlieb, J. (2009). Functional significance of nonspatial information in monkey lateral intraparietal area. *Journal of Neuroscience, 29,* 8166–8176.

Balan, P. F., Oristaglio, J., Schneider, D. M., & Gottlieb, J. (2008). Neuronal correlates of the set-size effect in monkey lateral intraparietal area. *PLoS Biology, 6,* e158. doi:10.1371/journal.pbio.0060158.

Behrmann, M., & Geng, J. J. (2002). What is "left" when all is said and done? In H.-O. Karnath, D. Milner, & G. Vallar (Eds.), *Spatial coding in hemispatial neglect* (pp. 85–100). Oxford, England: Oxford University Press.

Ben Hamed, S., & Duhamel, J. R. (2002). Ocular fixation and visual activity in the monkey lateral intraparietal area. *Experimental Brain Research, 142,* 512–528.

Ben Hamed, S. B., Duhamel, J. R., Bremmer, F., & Graf, W. (2002). Visual receptive field modulation in the lateral intraparietal area during attentive fixation and free gaze. *Cerebral Cortex, 12,* 234–245.

Bisley, J., & Goldberg, M. (2010). Attention, intention, and priority in the parietal lobe. *Annual Review of Neuroscience, 33,* 1–21. doi:10.1146/annurev-neuro-060909-152823.

Bisley, J. W., & Goldberg, M. E. (2003). Neuronal activity in the lateral intraparietal area and spatial attention. *Science, 299,* 81–86.

Blatt, G. J., Andersen, R. A., & Stoner, G. R. (1990). Visual receptive field organization and cortico–cortical connections of the lateral intraparietal area (area LIP) in the macaque. *Journal of Comparative Neurology, 299,* 421–445.

Burnham, B. R. (2007). Displaywide visual features associated with a search display's appearance can mediate attentional capture. *Psychonomic Bulletin & Review, 14,* 392–422.

Carrasco, M., & Yeshurun, Y. (2009). Covert attention effects on spatial resolution. *Progress in Brain Research, 176,* 65–86.

Colby, C. L., Duhamel, J.-R., & Goldberg, M. E. (1993). Ventral intraparietal area of the macaque: Anatomic location and visual response properties. *Journal of Neurophysiology, 69,* 902–914.

Colby, C. L., & Goldberg, M. E. (1999). Space and attention in parietal cortex. *Annual Review of Neuroscience, 23,* 319–349.

Constantinidis, C., & Steinmetz, M. A. (2005). Posterior parietal cortex automatically encodes the location of salient stimuli. *Journal of Neuroscience, 25*, 233–238.

den Ouden, H. E., Daunizeau, J., Roiser, J., Friston, K. J., & Stephan, K. E. (2010). Striatal prediction error modulates cortical coupling. *Journal of Neuroscience, 30*, 3210–3219.

Desimone, R., & Duncan, J. (1995). Neural mechanisms of selective visual attention. *Annual Review of Neuroscience, 18*, 183–222.

Ding, L., & Hikosaka, O. (2006). Comparison of reward modulation in the frontal eye field and caudate of the macaque. *Journal of Neuroscience, 26*, 6695–6703.

Dorris, M. C., & Glimcher, P. W. (2004). Activity in posterior parietal cortex is correlated with the relative subjective desirability of action. *Neuron, 44*, 365–378.

Falkner, A. L., Krisha, B. S., & Goldberg, M. E. (2006). Lateral inhibitory interactions in the lateral intraparietal area (LIP) of the monkey. *Abstracts of the Society for Neuroscience, 37*, 606.608.

Fecteau, J. H., & Munoz, D. P. (2006). Salience, relevance, and firing: A priority map for target selection. *Trends in Cognitive Sciences, 10*, 382–390. doi:10.1016/j.tics.2006.06.011.

Ferrera, V. P., & Grinband, J. (2006). Walk the line: Parietal neurons respect category boundaries. *Nature Neuroscience, 9*, 1207–1208.

Fitzgerald, J. K., Freedman, D. J., & Assad, J. A. (2011). Generalized associative representations in parietal cortex. *Nature Neuroscience, 14*, 1075–1079.

Flagel, S. B., Akil, H., & Robinson, T. E. (2009). Individual differences in the attribution of incentive salience to reward-related cues: Implications for addiction. *Neuropharmacology, 56*, 139–148.

Flagel, S. B., Clark, J. J., Robinson, T. E., Mayo, L., Czul, A., Willuhn, I., et al. (2011). A selective role for dopamine in stimulus–reward learning. *Nature, 469*, 53–57. doi:10.1038/nature09588.

Freedman, D. J., & Assad, J. A. (2006). Experience-dependent representation of visual categories in parietal cortex. *Nature, 443*, 85–88.

Freedman, D. J., & Assad, J. A. (2011). A proposed common neural mechanism for categorization and perceptual decisions. *Nature Neuroscience, 14*, 143–146.

Gottlieb, J. (2012). Attention, learning and the value of information. *Neuron, 76*, 281–295.

Gottlieb, J., & Goldberg, M. E. (1999). Activity of neurons in the lateral intraparietal area of the monkey during an antisaccade task. *Nature Neuroscience, 2*, 906–912.

Gottlieb, J., Kusunoki, M., & Goldberg, M. E. (1998). The representation of visual salience in monkey parietal cortex. *Nature, 391*, 481–484.

Gottlieb, J., Kusunoki, M., & Goldberg, M. E. (2005). Simultaneous representation of saccade targets and visual onsets in monkey lateral intraparietal area. *Cereb Cortex, 15*, 1198–1206.

Gottlieb, J., & Snyder, L. H. (2010). Spatial and non-spatial functions of the parietal cortex. *Current Opinion in Neurobiology, 20*, 731–740.

Hanes, D. P., & Schall, J. D. (1996). Neural control of voluntary movement initiation. *Science, 274*, 427–430.

Haslam, N., Porter, M., & Rothschild, L. (2001). Visual search: Efficiency continuum or distinct processes? *Psychonomic Bulletin & Review, 8*, 742–746.

Hogarth, L., Dickinson, A., & Duka, T. (2010). Selective attention to conditioned stimuli in human discrimination learning: Untangling the effects of outcome prediction, valence, arousal and uncertainty. In C. J. Mitchell & M. E. LePelley (Eds.), *Attention and associative learning: From brain to behavior* (pp. 71–97). Oxford, England: Oxford University Press.

Huang, T. R., & Grossberg, S. (2010). Cortical dynamics of contextually cued attentive visual learning and search: Spatial and object evidence accumulation. *Psychological Review, 117*, 1080–1112.

Ikeda, T., & Hikosaka, O. (2003). Reward-dependent gain and bias of visual responses in primate superior colliculus. *Neuron, 39*, 693–700.

Ipata, A. E., Gee, A. L., Gottlieb, J., Bisley, J. W., & Goldberg, M. E. (2006). LIP responses to a popout stimulus are reduced if it is overtly ignored. *Nature Neuroscience, 9*, 1071–1076.

Itti, L., & Baldi, P. (2009). Bayesian surprise attracts human attention. *Vision Research, 49*, 1295–1306. doi:10.1016/j.visres.2008.09.007.

Kable, J. W., & Glimcher, P. W. (2009). The neurobiology of decision: Consensus and controversy. *Neuron, 63*, 733–745.

Kowler, E., Anderson, E., Dosher, B., & Blaser, E. (1995). The role of attention in the programming of saccades. *Vision Research, 35*, 1897–1916. doi:10.1016/004-6989(94)00279-U.

Land, M. F. (2009). Vision, eye movements, and natural behavior. *Visual Neuroscience, 26*, 51–62.

Lewis, J. W., & Van Essen, D. C. (2000). Corticocortical connections of visual, sensorimotor, and multimodal processing areas in the parietal lobe of the macaque monkey. *Journal of Comparative Neurology, 428*, 112–137.

Libera, C. D., & Chelazzi, L. (2009). Learning to attend and to ignore is a matter of gains and losses. *Psychological Science, 20*, 778–784.

Lo, C. C., & Wang, X. J. (2006). Cortico–basal ganglia circuit mechanism for a decision threshold in reaction time tasks. *Nature Neuroscience, 9*, 956–963.

Louie, K., & Glimcher, P. W. (2010). Separating value from choice: Delay discounting activity in the lateral parietal area. *Journal of Neuroscience, 30*, 5498–5507.

Louie, K., Grattan, L. E., & Glimcher, P. W. (2011). Reward value-based gain control: Divisive normalization in parietal cortex. *Journal of Neuroscience, 31*, 10627–10639.

Maunsell, J. H., & Treue, S. (2006). Feature-based attention in visual cortex. *Trends in Neurosciences, 29*, 317–322. doi:10.1016/j.tins.2006.04.001.

McAdams, C. J., & Maunsell, J. H. R. (1999). Effects of attention on orientation-tuning functions of single neurons in macaque cortical area V4. *Journal of Neuroscience, 19*, 431–441.

Mirabella, G., Bertini, G., Samengo, I., Kilavik, B. E., Frilli, D., Della Libera, C., et al. (2007). Neurons in area V4 of the macaque translate attended visual features into behaviorally relevant categories. *Neuron, 54*, 303–318. doi:10.1016/j.neuron.2007.04.007.

Navalpakkam, V., & Itti, L. (2005). Modeling the influence of task on attention. *Vision Research, 45*, 205–231.

Navalpakkam, V., Koch, C., Rangel, A., & Perona, P. (2010). Optimal reward harvesting in complex perceptual environments. *Proceedings of the National Academy of Sciences of the United States of America, 107*, 5232–5237. doi:10.1073/pnas.0911972107.

Oppermann, F., Hassler, U., Jescheniak, J. D., & Gruber, T. (2012). The rapid extraction of gist-early neural correlates of high-level visual processing. *Journal of Cognitive Neuroscience, 24*, 521–529.

Oristaglio, J., Schneider, D. M., Balan, P. F., & Gottlieb, J. (2006). Integration of visuospatial and effector information during symbolically cued limb movements in monkey lateral intraparietal area. *Journal of Neuroscience, 26*, 8310–8319.

Pearce, J. M., & Mackintosh, N. J. (2011). Two theories of attention: A review and a possible integration. In C. J. Mitchell & M. E. LePelley (Eds.), *Attention and associative learning: From brain to behavior* (pp. 11–39). Oxford, England: Oxford University Press.

Peck, C. J., Jangraw, D. C., Suzuki, M., Efem, R., & Gottlieb, J. (2010). Reward modulates attention independently of action value in posterior parietal cortex. *Journal of Neuroscience, 29*, 11182–11191.

Platt, M. L., & Glimcher, P. W. (1999). Neural correlates of decision variables in parietal cortex. *Nature, 400*, 233–238.

Pouget, A., Deneve, S., & Duhamel, J. R. (2002). A computational perspective on the neural basis of multisensory spatial representations. *Nature Reviews Neuroscience, 3*, 741–747.

Pouget, P., Logan, G. D., Palmeri, T. J., Boucher, L., Pare, M., & Schall, J. D. (2011). Neural basis of adaptive response time adjustment during saccade countermanding. *Journal of Neuroscience, 31*, 12604–12612.

Powell, K. D., & Goldberg, M. E. (2000). Response of neurons in the lateral intraparietal area to a distractor flashed during the delay period of a memory-guided saccade. *Journal of Neurophysiology, 84*, 301–310.

Raffi, M., & Siegel, R. M. (2005). Functional architecture of spatial attention in the parietal cortex of the behaving monkey. *Journal of Neuroscience, 25*, 5171–5186.

Reynolds, J. H., & Chelazzi, L. (2004). Attentional modulation of visual processing. *Annual Review of Neuroscience, 27*, 611–647.

Reynolds, J. H., Chelazzi, L., & Desimone, R. (1999). Competitive mechanisms subserve attention in macaque areas V2 and V4. *Journal of Neuroscience, 19*, 1736–1753.

Reynolds, J. H., & Heeger, D. J. (2009). The normalization model of attention. *Neuron, 61*, 168–185.

Reynolds, J. H., Pasternak, T., & Desimone, R. (2000). Attention increases sensitivity of V4 neurons. *Neuron, 26*, 703–714.

Roberts, K. L., & Humphreys, G. W. (2011a). Action-related objects influence the distribution of visuospatial attention. *Quarterly Journal of Experimental Psychology, 64*, 669–688.

Roberts, K. L., & Humphreys, G. W. (2011b). Action relations facilitate the identification of briefly-presented objects. *Attention, Perception & Psychophysics, 73*, 597–612.

Schall, J. D., Purcell, B. A., Heitz, R. P., Logan, G. D., & Palmeri, T. J. (2011). Neural mechanisms of saccade target selection: Gated accumulator model of the visual–motor cascade. *European Journal of Neuroscience, 33*, 1991–2002.

Sommer, M. A., & Wurtz, R. H. (2002). A pathway in primate brain for internal monitoring of movements. *Science, 296*, 1480–1482.

Sprague, N., & Ballard, D. H. (2005). Modeling embodied visual behaviors. *ACM Transactions on Applied Perception, 1*, 1–26. doi:10.1145/1265957.1265960.

Stanford, T. R., Shankar, S., Massoglia, D. P., Costello, M. G., & Salinas, E. (2010). Perceptual decision making in less than 30 milliseconds. *Nature Neuroscience, 13*, 379–385.

Sugrue, L. P., Corrado, G. S., & Newsome, W. T. (2004). Matching behavior and the representation of value in the parietal cortex. *Science, 304*, 1782–1787.

Sugrue, L. P., Corrado, G. S., & Newsome, W. T. (2005). Choosing the greater of two goods: Neural currencies for valuation and decision making. *Nature Reviews Neuroscience, 6*, 363–375.

Tatler, B. W., Hayhoe, M. N., Land, M. F., & Ballard, D. H. (2011). Eye guidance in natural vision: Reinterpreting salience. *Journal of Vision, 11*(5), 5–25. doi:10.1167/11.5.5.

Vuilleumier, P. (2005). How brains beware: Neural mechanisms of emotional attention. *Trends in Cognitive Sciences, 9*, 585–594. doi:10.1016/j.tics.2005.10.011.

Wardak, C., Olivier, E., & Duhamel, J. R. (2002). Saccadic target selection deficits after lateral intraparietal area inactivation in monkeys. *Journal of Neuroscience, 22*, 9877–9884.

Wardak, C., Olivier, E., & Duhamel, J. R. (2004). A deficit in covert attention after parietal cortex inactivation in the monkey. *Neuron, 42*, 501–508.

Williams, J. M., Mathews, A., & MacLeod, C. (1996). The emotional Stroop task and psychopathology. *Psychological Bulletin, 120*, 3–24.

Williford, T., & Maunsell, J. H. (2006). Effects of spatial attention on contrast response functions in macaque area V4. *Journal of Neurophysiology, 96*, 40–54.

Zhang, M., & Barash, S. (2004). Persistent LIP activity in memory antisaccades: Working memory for a sensorimotor transformation. *Journal of Neurophysiology, 91*, 1424–1441.

XI INVERTEBRATE VISION

79 Invertebrate Vision: Optics and Behavior

MICHAEL F. LAND

Early Evolution

Almost all metazoan animals show some sort of light-controlled behavior, and in nearly all cases the key molecules involved are G-coupled opsin receptor proteins. These molecules are unique to animals and, with the possible exception of sponges, are found in all animal phyla. Opsins thus presumably evolved at the very beginning of animal evolution. Two main forms of opsin, each associated with its own transduction biochemistry, have been in existence since before the splitting of bilaterian animals into two main divisions, the protostomes (flatworms, molluscs, annelids, and arthropods) and deuterostomes (echinoderms and chordates). It was thought until recently that the protostome line had only r-opsins (rhabdomeric) and the deuterostome line c-opsins (ciliary), but this is only true of their principal receptors for sight. Some vertebrate retinal ganglion cells, for example, contain r-opsins complete with the appropriate transduction machinery, and photoreceptors in the brains of annelid worms contain c-opsins (Arendt et al., 2004).

The protostome–deuterostome split presumably occurred deep in the Precambrian, well before anything resembling what we usually think of as eyes evolved. The sudden appearance of recognizable eyes in the fossil record occurred in the early Cambrian, 530 million years ago (mya) and was associated with an increase in the size of animals, their mobility, and crucially for palaeontologists the development of hard external skeletons. All of this implies an increase in carnivory, with both predators and prey needing to be able to see better, either for hunting or defense. The Cambrian was the period when most of the eye types that resolve well, of both compound and single-chambered designs, evolved into more or less their modern forms (Land & Nilsson, 2012).

Light-Controlled Behaviors

One way of categorizing the sophistication of visual behavior is in terms of the amount of information needed to control it. Some behaviors, visual predation for example, require well-resolved spatial vision; others such as simple phototaxis toward or away from light require much less. The scheme here follows Nilsson's (2009) classification of light-controlled behaviors, which allows us to trace the evolutionary stages that have led from basic light sensitivity to the kind of multipurpose vision possessed by a bee, an octopus, or a fish. This scheme has four levels, each characterized by a structural change in the way light is manipulated (see figure 79.1).

LEVEL 1: NONDIRECTIONAL PHOTORECEPTION Many behaviors require no photoreceptor specialization other than photopigment and a transduction system. Examples are circadian rhythm control, protection against high light levels, especially short wavelengths, and the use of light as a depth gauge. These responses involve large slow-intensity changes and so do not require densely stacked photoreceptor membranes, and the receptors involved are typically inconspicuous. Defensive shadow responses, mediated, for example, by skin receptors in molluscs, also require no specializations although there is a greater requirement here for speed of response.

LEVEL 2: DIRECTIONAL PHOTORECEPTION Animals whose habitat needs require approach to or avoidance of light are often equipped with receptors backed by screening pigment so that they respond preferentially to light from only part of the surroundings. Then, either by scanning their eyespots from side to side (klinotaxis) or simultaneously comparing the output from two eyespots pointing in different directions (tropotaxis), an appropriate course can be maintained

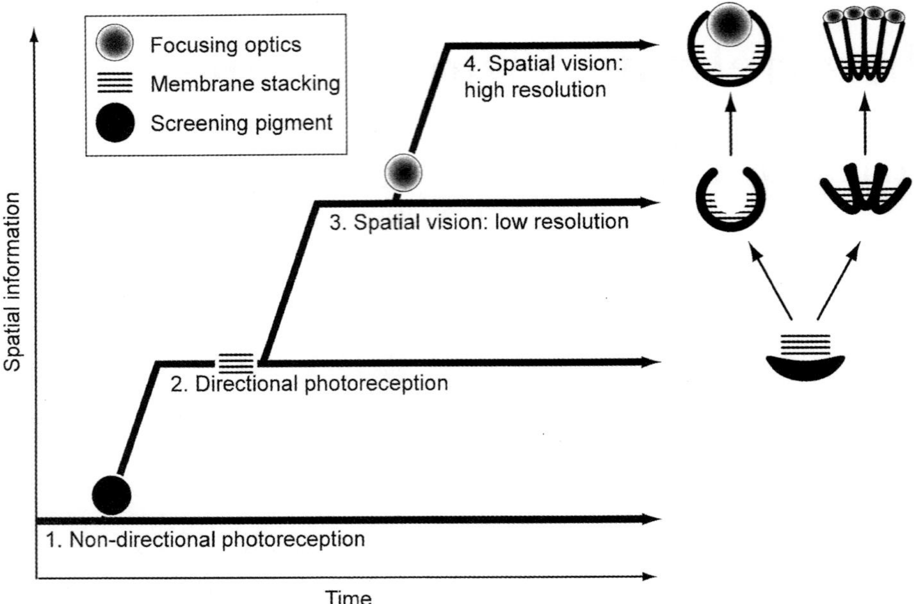

FIGURE 79.1   Four levels of light-controlled behavior. Level 1 requires only photopigment and a transduction pathway. Level 2 requires screening pigment to make receptors directional, and membrane stacking to increase photopigment density. In Level 3, basic optical mechanisms are needed to provide modest spatial resolution. Level 4 vision, typical of cephalopod molluscs, arthropods, and vertebrates, requires greatly improved optics and large brains. (Modified from Nilsson, 2009.)

(Fraenkel & Gunn, 1940). Such animals include some flatworms, nematodes, many larvae of marine invertebrates, and the larvae of insects such as flies. The field of view of such receptors may be as great as 180°. However, they need to be able to respond to relatively small changes in intensity over a time scale of seconds or less, so they will require higher sensitivities than level 1 receptors, and so an increase in the number of photoreceptor molecules will be needed. This may involve specialized membrane folding, especially when the eyespots are to be used in dim conditions.

LEVEL 3: LOW RESOLUTION SPATIAL VISION   Tasks such as object avoidance, tracking large landmarks or celestial objects, monitoring self-motion, or detecting a predator before it casts a shadow all require an eye to be able to measure the intensity of light coming from different directions: That is, they must have true spatial vision. The resolution need not be high: 5°–30° would be adequate for all these tasks. Pigment cup eyes, or compound structures without optical specializations, would be adequate (see figure 79.2a and b). Such eyes are found in cubozoan jellyfish, flatworms, annelid worms, gastropod and bivalve molluscs, and many larval arthropods. Because the acceptance angle of each receptor is now much reduced compared with level 2 photoreception, there is a greater need for increased numbers of photopigment molecules, and at this level

photoreceptors all have highly folded membranes. Typically these take the form of tightly packed microvillous extensions of the cell membrane (rhabdomeres) although lamellar structures are also found.

LEVEL 4: HIGH-RESOLUTION SPATIAL VISION   To be able to spot prey or predators at a distance, recognize the landmarks around a home or feeding site, distinguish the signals made by conspecifics, or determine the visual features of a food plant are all tasks that require much higher resolution than any we have encountered so far. The eyes of adult insects resolve in the range of 5° (*Drosophila*) to 0.25° (dragonfly). Crustaceans and spiders cover a similar range. Octopus and squid, with large single-chambered eyes, are much better, with a resolution of about 1 arcmin, comparable with the best vertebrates. All this requires the evolution of a competent optical system, and this will be the subject of the next section. It also requires mechanisms for stabilizing the eye against movements of the animal's body, and in the larger eyes, mechanisms for focusing on objects at different distances. Nearly all highly resolving eyes have the capacity for color vision, or at least some ability to resolve wavelength, which means that they must have multiple visual pigments (at least two and, in some mantis shrimps, up to twelve, including four in the ultraviolet). Many can also resolve the pattern of polarized light in the sky or surroundings. Such eyes can

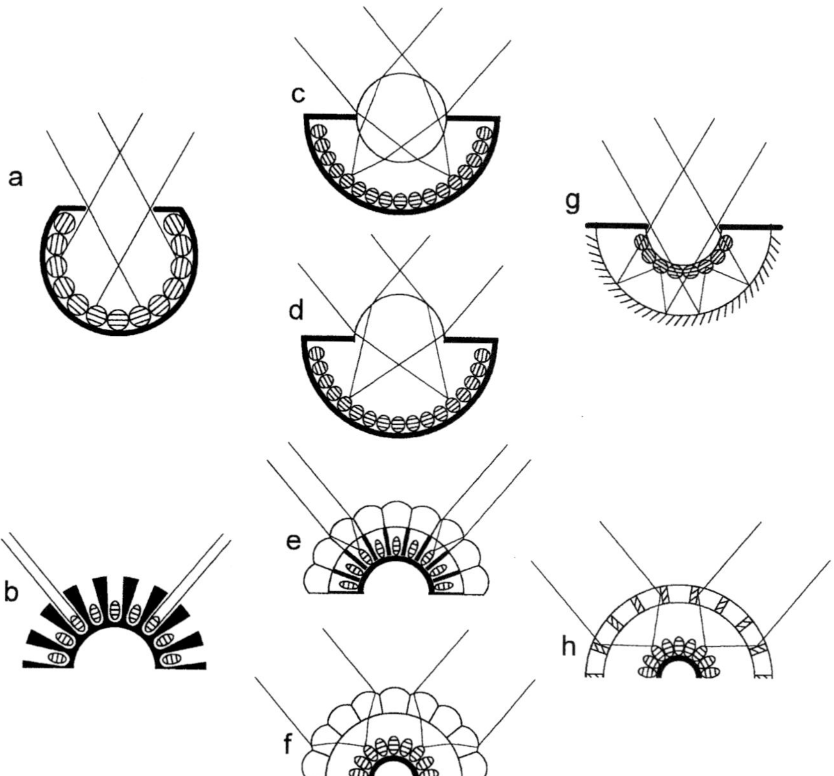

FIGURE 79.2   Mechanisms of image formation in animals. Single-chambered eyes: (a) Pit/pinhole eye, (c) spherical lens eye, (d) corneal eye, (g) concave mirror eye. Compound eyes. (b) Lensless compound eye, (e) apposition eye with lenses, (f) refracting superposition compound eye, (h) reflecting superposition compound eye. Details in the text. (From Land & Nilsson, 2012.)

support a wide variety of behaviors and, not surprisingly, need large brains to cope with the information they supply (Land & Nilsson, 2006).

TYPES OF IMAGE FORMATION

An eye capable of spatial vision (at level 3 or 4 above) requires a structure that will split up the incoming light according to its direction of origin, and in nature, as in technology, there are three main ways of doing this: light restricting apertures, refractive structures such as curved surfaces and lenses, and mirrors. By the end of the Cambrian, about 490 mya, all three were in use. All three come in two basic forms depending on whether the light receptors are contained in single concave pits (single-chambered eyes, often misnamed "simple" eyes) or on protruding convex structures with the receptors in outwardly directed tubes (compound eyes). The major classes of eye are shown in figure 79.2.

Single-Chambered Eyes

SHADOWED PITS   See figure 79.2a. Many simple animals, from flatworms to annelids and molluscs, some crustaceans, and many larval forms, rely on pit eyes with a pigmented aperture that restricts light reaching single receptors to a broad cone, typically 25° or more wide. Such eyes are certainly capable of directing the animal to brighter or darker regions of the surroundings, but not a great deal more.

PINHOLES   See figure 79.2a. A logical way of improving the performance of a pit eye is to reduce the aperture to something approaching a pinhole. In giant clams (*Tridacna*) the mantle contains many hundred such eyes, each with an aperture that restricts the field of view of single receptors to about 15° (Land, 2002). The receptors give OFF responses, and as a result the clam can respond, by shutting, to large moving objects such as fish, before they come close enough to do damage. The most famous pinhole eye is that of *Nautilus* (see figure 79.3a), a cephalopod mollusc with large (1-cm) eyes. The eyes have mobile irises and eye muscles that keep the eye stabilized in the vertical plane. Other cephalopods have lenses, but *Nautilus* does not, and its visual behavior seems to be mainly restricted to stabilizing the animal's course against involuntary rotation while *Nautilus* grazes along the edge of the reef (Muntz & Raj, 1984). Pinhole eyes have the severe drawback that restricting the aperture to improve resolution reduces the amount of light reaching the retina, and opening the aperture destroys resolution. The real answer is a lens.

SPHERICAL LENSES See figure 79.2c. Many molluscs, particularly cephalopods such as *Octopus*, cuttlefish, and squid, some annelids, and famously fish, have eyes with spherical lenses. Presumably these evolved by increasing the concentration of refractile material in the chamber of a pit-like eye and then confining this material into an enclosed structure. These lenses are hardly ever optically homogeneous, but all have an approximately parabolic refractive index gradient rising from not much denser than seawater (1.34) at the periphery to about 1.52 at the center, where the material is almost dry and crystalline (Jagger, 1992). The effect of this gradient is to greatly reduce the spherical aberration inherent in a homogeneous structure by weakening the ray-bending power of the regions furthest from the axis. The fact that rays bend continuously within the lens also reduces the focal length to about 2.5 lens radii (Matthiessen's ratio). Such lenses can provide an excellent image, giving fish and cephalopods a resolution of a few arc minutes. Subtle variations in the gradient in the lenses of fish also go some way to overcoming chromatic aberration by providing separate zones with different focal lengths, which permit different wavelengths to come to a common focus (Kröger et al., 1999). From different evolutionary origins, the parallel "discovery" of this type of lens has enabled fish and cephalopods to develop the full range of behaviors associated with acute vision.

CORNEAL EYES See figure 79.2d. With the colonization of the land, by invertebrates and vertebrates, another method of image formation became available: a curved surface separating air from fluid. Corneal eyes also occur in arthropods, notably the spiders (figure 79.3b), where they can provide excellent images. In the jumping spider *Portia* the principal eyes have interreceptor angles of 2.4 arcmin, only five times greater than in the human eye (Williams & McIntyre, 1980), and in others such as *Dinopis*, the huge (1.3-mm) eyes provide bright enough images for the spider to catch cockroaches in forests at night (Blest & Land 1977). In web-building spiders vision is much less acute, and the eyes play other roles such as sun- and polarized light compass navigation. In flying insects corneal eyes are found as the three dorsal ocelli, concerned with stabilizing flight relative to the horizon (Stange, 1981). Corneal eyes are also the main organs of sight in many insect larvae.

MIRROR EYES See figure 79.2g. Concave mirrors have been used to produce images since the days of Newton's telescope, but they are rare in nature. Scallops (*Pecten* spp., figure 79.3d) provide probably the best example. Scallops have up to 100 1-mm eyes around the mantle, serving the same "burglar alarm" function as the eyes of giant clams. They are, however, much more acute, responding to movements of as little as 2°. The image-forming surface is a hemispherical concave reflector, the argentea, at the back of the eye, which throws an image onto a retina of OFF-responding receptors in the focal plane near the center of the eye (Land, 1965). When crossed by the image of a dark object, these cells respond, and the animal closes its shell. (There is a second, out-of-focus, ON-responding retina between the OFF-responding cells and the mirror, which seems to be involved in orientation while the animal is swimming.) The mirror itself consists of about 10 layers of 1 μm-square guanine crystals, each with an optical thickness of a quarter of a wavelength of green light, and separated from the next layer by a film of cytoplasm of the same optical thickness. The whole multilayer produces constructive interference for green light, and the reflectance approaches 100%. Similar reflectors are found in fish scales and butterfly wings. A reason why eyes with this mirror design are uncommon is that light has passed through the retina unfocused before it returns to form the image, so the image is inevitably of reduced contrast. A recently discovered mirror eye in a vertebrate, where light reaches the retina directly, is found in the spookfish *Dolichopteryx longipes* (Wagner et al., 2009). This is the only other image-forming mirror eye of any size.

Compound Eyes

SIMPLE TUBES See figure 79.2b. Conceptually, the simplest type of compound eye is a series of pigmented tubes oriented radially around the surface of a hemisphere, with each tube containing a receptor. Something very like this is found in the eyes of the bivalve mollusc *Arca* and its relatives (see figure 79.3c), which have a large number of such eyes around the mantle. They have no lenses and rely on the shadowing effect of each tube to provide each receptor with an acceptance angle of 10°–20°. As in giant clams and scallops, the function of these eyes is movement detection, and their role is defense (Nilsson, 1994). Somewhat similar eyes are found in sabellid tube worms and some starfish.

APPOSITION EYES See figure 79.2e. The addition of a lens to each element of a tube eye of the type just mentioned has the effect of greatly increasing the light that reaches each receptor and also allows each element to

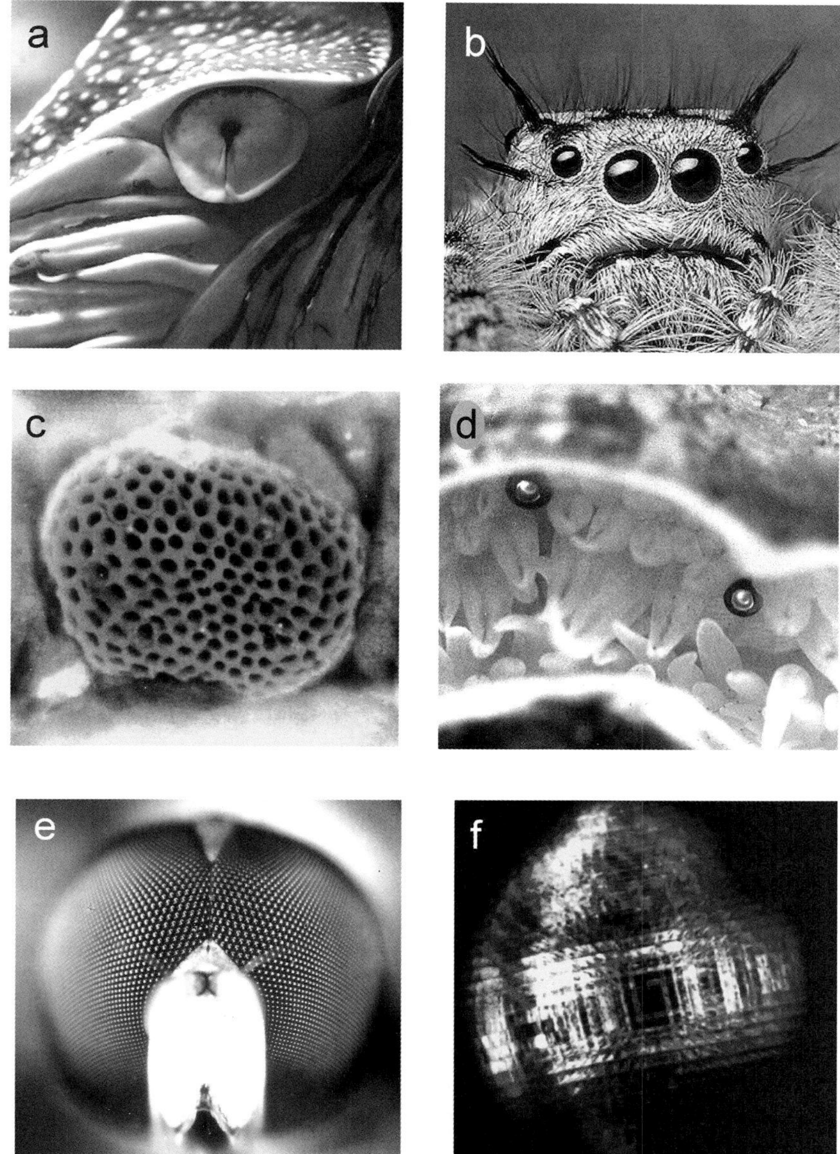

FIGURE 79.3   A selection of different optical eye types. (a) Pinhole eye of *Nautilus*. (b) Corneal eyes of the jumping spider *Phidippus mystaceus*. (c) Lensless compound eye of the clam *Barbatia*. (d) Concave mirror eyes of the scallop *Pecten*. (e) Apposition eye of the male hoverfly *Syritta*. (f) Reflecting superposition eye of the shrimp *Palaemonetes*. (Photos: a, Hans Hillewart; b, Opotserer; c–f, from Land & Nilsson, 2012.)

have a much smaller acceptance angle. Such eyes are known as apposition eyes because the fields of view of each element (or ommatidium) are apposed to each other to form a single, erect, composite image. It is this image that the animal sees rather than the tiny inverted images produced by each lens. True apposition eyes are found in the tubeworm *Branchiomma*, but it is in the arthropods that these eyes have become the typical organs of sight. Indeed, some of the first eyes for which we have reliable fossil evidence are the apposition eyes of trilobites from the Cambrian about 525 mya. These eyes have lenses made of the mineral calcite, so they

came prefossilized (Levi-Setti, 1993). Apposition compound eyes occur in the myriapods, xiphosurans (*Limulus*), many crustaceans, and most adult insects.

In the honeybee, an insect with a typical apposition eye, each ommatidium consists of a lens which throws an image onto the distal tip of a long rod (the rhabdom) made up of photopigment-containing microvilli contributed by eight receptors. The tip of the rhabdom picks up light from the central region of the image and views about 1° of outside space. The rhabdom has a higher refractive index than the surrounding tissue and acts as a light guide within which the light becomes scrambled, so that there is no further spatial resolution. However, the photopigments contained in the microvilli from the different receptors have different spectral sensitivities, to green, blue, and ultraviolet light, so although there is no further spatial resolution within an ommatidium, there is spectral resolution. In the dorsal region of the eye, pointing at the sky, there is a special region where the photopigment molecules in the microvilli are aligned, and these are sensitive to the polarization pattern in skylight (for reviews, see Nilsson, 1989; Wehner, 1981).

Apposition eyes, and compound eyes in general, have a serious drawback. Because the lenses are necessarily small, their resolution is limited by diffraction. A 25-µm lens can only resolve 1°. To double the resolution of a compound eye requires the diameter of each facet to double, but to benefit from this, twice the number of ommatidia are required, and so the size of a compound eye goes up as the square of its resolution (Kirschfeld, 1976). As early as 1894 Mallock had estimated that a compound eye with human (1 arcmin) resolution would be more than 20 m high. Some insects, such as dragonflies and some dipteran flies, do squeeze in extra resolution into parts of the eye (e.g., see figure 79.3e), but they pay for it in increased size and reduced resolution elsewhere in the field.

It should be mentioned that dipteran flies (figure 79.3e) have a variant of the apposition eye in which the image in each ommatidium *is* resolved. The eight receptors (6 + 2 inline) have separate, unfused, microvillar structures (rhabdomeres). However, this arrangement is not used to increase the resolution of the eye but its sensitivity. Each rhabdomere in one ommatidium shares a field of view with one in each of the six surrounding ommatidia. In the next neural layer, the lamina, the axons of six of the eight receptors with the same field of view join up with each other, so that each lamina cell has a single, shared, field of view; however, the signal is six times stronger than that of a single receptor. This arrangement has been described as *neural superposition*, and by improving sensitivity its effect is to buy time at dawn and dusk when many dipterans flies swarm and the eyesight of their predators is failing (Kirschfeld, 1972).

REFRACTING SUPERPOSITION   See figure 79.2f. This version of the compound eye provides much more light for the receptors than the apposition type and is found in nocturnal insects (moths, fireflies) and some deeper water marine crustaceans (krill). Instead of having lenses that are private to each ommatidium, in these eyes many optical elements contribute (superpose) their ray paths jointly to give a single (erect!) image

lying about halfway out from the center of curvature of the eye, which is where the receptors are located. The optics of this arrangement were described by Exner (1891/1989). Basically, each optical element behaves as a two-lens telescope with each member of the pair having roughly the same power, so that the combination behaves as a simple inverter. A ray incident on the outer surface of each element, making some angle alpha with the structure's axis, is bent through an angle 2 alpha, to emerge at an angle −alpha on the same side of the axis as the incident ray. As we shall see below, this is essentially the same as the path of a ray deflected by a mirror. When such elements are arranged radially around a spherical surface, the central ray is not deflected, but at increasing distances from this axis rays become deflected at increasing angles, and as a result they all arrive at approximately the same focal point within the eye. The beauty of this arrangement is that up to 100 or even 1,000 optical elements contribute to each image point, and the amount of light reaching the receptors is correspondingly augmented. Butterflies have the same double-lens optical arrangement as their moth relatives but have restricted the light reaching each rhabdom to a single facet, so that for most purposes they operate as apposition eyes (Nilsson, Land, & Howard, 1988).

As Exner realized, the superposition elements do not actually consist of two lenses, but, in beetles at least, are often a single continuous structure. He worked out that the same ray path could be achieved by having a cylinder with a refractive index gradient (in a similar way to fish lenses) with the highest index along the central axis. This gradient was finally demonstrated by Hausen in 1973, using interference microscopy. In such a structure (a lens cylinder) rays are bent continuously, and depending on its length it can behave as a single lens (as in the apposition eye of *Limulus*) or as a two lens inverter in eyes of the refracting superposition type.

REFLECTING SUPERPOSITION   See figure 79.2h. The long-bodied decapod crustaceans (shrimp, crayfish, and lobsters) have a unique version of the superposition eye, based on mirrors rather than lens cylinders (see figure 79.3f). The mirrors are arranged radially, and when seen in an idealized cross-section, they produce much the same ray paths as in refracting superposition eyes, inverting the paths of rays and arriving at a single focus within the eye. However, rays not in this ideal plane are potentially not so well behaved. This is the reason for the unusual geometry of these eyes. The mirrors are arranged not in a hexagonal lattice but in an array of square tapered boxes. This ensures that most of the rays entering the eye are reflected from two sides of a box, which thus behaves as a corner reflector. Such a reflector ensures that when viewed along the axis of the box, such rays are reflected through 180°, ensuring that they travel along paths parallel to their original direction (as happens in a refracting superposition eye) and are thereby brought to a common focus (Vogt, 1980). These eyes seem to be comparable in performance to their refracting counterparts.

Interestingly, the larvae of most of these animals, which tend to have planktonic lives in the well-lit upper waters, have apposition eyes with hexagonal facets, and these only "square off" when the animals become adult and sink to deeper, darker water. The related brachyuran crabs, which live mostly around the shoreline or even on land, have largely retained the apposition eye structure. In one group of swimming crabs, the portunids, and in hermit crabs, there is an interesting hybrid arrangement in which axial rays are focused by a lens into a light guide, to reach a deep-lying rhabdom. Oblique rays, however, encounter the silvered parabolic sides of the crystalline cone and are made parallel again, to form a superposition image, much as in other superposition eyes. This most recently discovered variant has been described as *parabolic superposition* (Nilsson, 1988).

EYES, BRAINS, AND BEHAVIOR

Simple and Complex Vision

In general, the amount of information an eye can supply matches the size and complexity of the brain it serves. There are only three animal groups—the vertebrates, the cephalopod molluscs, and the insects and higher crustaceans—in which the visual systems are fully multipurpose (one might also add the salticid spiders). To take one example, there is not much that a bee cannot do visually. It can use vision to guide and stabilize flight, it can locate home and food sources using landmarks, it can navigate using the sun and sky polarization, it can recognize form and color in flowers, and males can locate and catch females on the wing. The information from the same eyes is used in a host of different ways, depending on task and circumstance. In each of the groups mentioned the brains are relatively large, and the proportion devoted to vision is greater than 50% (Land & Nilsson, 2006).

In some other animals, still with level 4 vision, the range of visual tasks is much more restricted. For example, the alciopid polychaete worms and the heteropod sea snails have evolved excellent lens eyes, which they use just to detect and capture planktonic prey. The mirror eyes of scallops and compound eyes of sabellid worms are concerned with detecting predators—and little else. The eyes of some pontellid copepods, with good optics but remarkably few receptors, are concerned principally with detecting mates (only the males have large eyes). In all these cases the brains are small, similar to those of animals with vision at levels 1–3.

It is far from clear what evolutionary routes visual systems took to become multipurpose. The animals mentioned in the last paragraph, with specialized single-purpose vision, went no further; they certainly did not evolve into fish, cephalopods, or crabs. It seems more likely that the origins of good vision were humbler. It may have begun in animals with poorly resolved level 3 vision, which they used to keep track of their own locomotion and to not bump into things. This would have involved the beginnings of motion detection and of rudimentary pattern vision, the two basic ingredients of more advanced sight. Simply tuning up these processes, and applying them in new circumstances, would have provided a good way for vision to progress.

Eye Movements in Invertebrates

For animals with high-resolution vision (level 4), movements of the eyes are essential to maintain a clear image. Photoreceptors are slow, typically taking 10–20 ms to respond fully to a change, and this means that motion blur will set in if the eye moves relative to the surroundings at more than a minimum speed of about one receptor acceptance angle per response time. For humans this is not much above 1° per second. This means that movements of the eyes that would result from locomotion must be compensated for by stabilizing reflexes (vestibular–ocular reflex and optokinetic reflex in vertebrates). However, gaze must also relocate, ultimately because otherwise the eyes will become trapped at a backstop when the animal turns. These relocations are always fast, to minimize visual "downtime," and these fast saccades alternate with stabilized fixations (Carpenter, 1988). To what extent is this characteristic "saccade and fixate" pattern shared by animals with different types of eye and from different lineages?

Among insects and crustaceans this pattern seems to be common if not universal. Insects have eyes that are part of the head, so head and eye movements are the same thing. In walking and in flying flies (see figure 79.4) rapid movements of the head on the thorax ensure that gaze changes are much more rapid than the more sluggish body movements, and that gaze is stabilized against body rotation between turns (Schilstra & van Hateren, 1998). Stabilization is brought about in part by signals from the halteres—oscillating gyroscopes

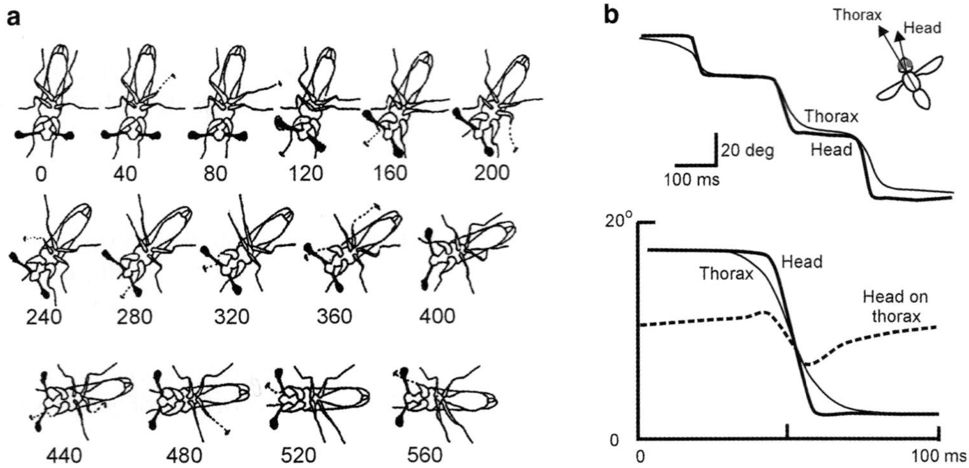

FIGURE 79.4   Head movements of walking and flying flies. (a) Stalk-eyed fly turning on a glass plate. Note large head saccades at 120 and 400 ms with head stabilization between these as the body continues to turn. (b) Head and thorax movements of a flying blowfly. The head turns faster than the thorax, with the neck (head on thorax) first turning away from the turn, then rapidly into the turn, then away from the turn again. In this way the head makes the saccade while compensating for the slower thorax motion. (Redrawn from Schilstra & van Hateren, 1998.)

that take the place of the second wings. Bees and wasps show similar behavior. Hoverflies have such mastery of flight that they do not have to move their heads but exhibit a similar pattern of saccades and fixations in the flight behavior itself. Some male hoverflies (*Syritta*) actually track females during flight with a smooth pursuit system reminiscent of that of primates (Collett & Land, 1975). Dragonflies can track prey with their heads, independent of the flight maneuvers of the body. Praying mantids also track prey while waiting in ambush, but in general this is achieved with a succession of small saccades.

Shore crabs do have eyes that move separately from the head, and when the animal turns, the eyes stay locked to surroundings using a combination of optokinetic and statocyst-mediated stabilizing reflexes (Paul, Nalbach, & Varjú, 1990). Sand and mudflat crabs, such as fiddler crabs (*Uca*) and ghost crabs (*Ocypode*), also keep their eyes aligned with the horizon. This provides them with a mechanism for distinguishing predators (above the horizon) from conspecifics (below). Cephalopods (*Sepia*) also show optokinetic (nystagmus-like) eye movements when turning.

It does seem that the fundamental problem imposed by the slow speed of the photoreceptor process has been solved in much the same way by animals in all the three phyla that have achieved Level 4 vision. The ability to keep gaze still in spite of the animals' own locomotory movements is as crucial a part of advanced vision as the provision of a well-resolved image. Indeed, the higher the quality of the image, the better gaze stabilization needs to be.

One- and Two-Dimensional Vision

The majority of eyes, whether single chambered or compound, produce a two-dimensional image in which intensities are resolved according to their direction of origin. Often wavelength distribution and polarization are analyzed in the same two-dimensional frame. There are, however, eyes which provide only a single-dimensional image from a linear retina. A consequence of such a design is that these eyes must scan to provide a second dimension. This arrangement has evolved once in the molluscs (the heteropod sea snails), once in the spiders (the salticid jumping spiders), twice in the crustaceans (the pontellid copepods and stomatopod mantis shrimps), and once in the insects (diving beetle larvae).

The predatory planktonic sea snail *Oxygyrus* has an eye whose optical construction is rather like that of a fish, with a spherical lens (see figure 79.5a). The retina, however, is a strip about 410 receptors long but only 3 wide, with a field of view of 180 × 3 degrees. Each eye pivots about a horizontal axis, so that the retina scans the sea below the animal, moving rapidly downward and more slowly upward. Presumably what the animal is doing is scanning for potential prey objects that glisten in sunlight against the dark of the abyss below (Land, 1982). The arrangement in male pontellid copepods of the genus *Labidocera* is rather different. The paired lens eyes are directed upward and share a single retina containing only 10 receptors arranged as a transverse line subtending about 40°. The retina swings below the lenses through about 40°, so that the

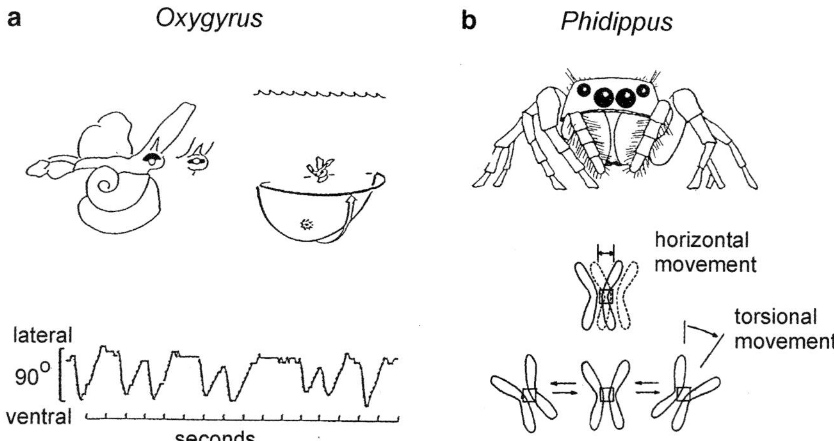

FIGURE 79.5   Two examples of eyes with one-dimensional retinas. (a) the heteropod sea snail *Oxygyrus*. Upper left: Drawing of the animal, which swims shell down. The eye is shown in down-pointing and (inset) side-pointing orientations. Right: The field of view of one eye as it sweeps upward through the dark water below the animal. *Oxygyrus* is planktonic, living close below the ocean surface (wavy line). Below: Record from a video of eight scans. (After Land, 1982.) (b) The jumping spider *Phidippus*. Top: Head-on view of a female, showing the principal and anterolateral eyes, and the predominance of leg outlines making 30° angles with the vertical. Bottom: Horizontal and torsional scanning movements of the vertically elongated retinas of the principal eyes. Horizontal scans have a period of 1–2 s, torsional scans 5–8 s. (After Land, 1969.)

field covered by the scan is about 40° × 40° above the animal. Since only the males possess these enlarged scanning eyes, it is a reasonable assumption that this remarkable system has evolved to detect females, which have long blue bodies. Tropical mantis shrimps have conventional two-dimensional apposition compound eyes, but running through the center of each eye is a linear band six ommatidial rows wide which contains a tiered color vision system containing 12 visual pigments (in four ommatidial rows) plus two rows that analyze both linear and circular polarized light (Marshall & Oberwinkler, 1999). Eye movements are of two kinds: saccades and optokinetic movements associated with the main parts of the eyes, and smaller, slower scanning movements about three axes that maneuver the midband around in space. How the one- and two-dimensional systems cooperate with each other remains a mystery. In jumping spiders, which have eight eyes of the corneal type, the situation is clearer (see figure 79.5b). Six of the eyes are fixed, and their main function is to detect motion, and then direct the animal to face the source of movement with the larger forward-pointing principal eyes. These eyes have vertically elongated retinas, about five receptors wide by 200 high. These retinas (but not the lenses) are moved by six muscles, which can shift the retinal fields of view by up to 50° horizontally and vertically (Land, 1969). When presented with a small target, the principal eyes typically scan it for a few seconds, with both horizontal and torsional eye movements, apparently looking for appropriate contours that signify the presence and disposition of legs in the object under scrutiny (figure 79.5b). The logic seems to be that if there are enough legs making the correct angles, this is likely to be another jumping spider; if not, it is probably prey. Appropriate behaviors then ensue. Diving beetle larvae of the genera *Thermonectes* and *Acilius* have two pairs of enlarged ocelli, each with a horizontal linear retina subtending 30–50° by 2°. They are used to align the body with prey such as mosquito larvae before a strike. Unlike the animals mentioned above, there are no eye muscles, and the vertical scanning required by the four horizontal retinas is brought about by press-up-like movements of the neck and thorax–abdomen joints (Buschbeck, Sbita, & Morgan, 2007).

One other animal deserves mention. Females of the copepod *Copilia quadrata* also have scanning eyes, but here the retina is little more than a single spot detector, containing 5–7 receptors, and with a field of view of only about 3°. The optical system of the two eyes resembles a pair of telescopes, in which the rear elements, with the receptors attached, scan laterally over about 15° (Gregory, Ross, & Moray, 1964). Thus a zero-dimension retina scans a one-dimensional line. On its own this will not be of great use for predation, and presumably a second dimension is provided either by the undulations of the swimming body, or even the vertical migration of the prey itself.

CONCLUSIONS

The physical ways that light can be manipulated to form images are the same in biology as they are in optical technology, and among the invertebrates almost every method has been tried out (Fresnel and zoom lenses seem to be exceptions). It is particularly interesting that many of these have evolved repeatedly and independently in different phyla. Molluscs, annelids, and some crustaceans have all evolved spherical lenses which use the same refractive index gradient to overcome spherical aberration that vertebrates arrived at with the lenses of early fish. Similarly, compound eyes of the apposition type have evolved independently in molluscs, annelids, and at least twice in the arthropods. Convergent evolution of optical structures appears to be common in the invertebrates, and much of the experimentation that led to this occurred in the 40-million-year period of the Cambrian. Good designs do tend to be retained within phyla (e.g., the lens eyes of cephalopods and the apposition and superposition eyes of insects and crustaceans), but even then there are puzzling exceptions. How did *Nautilus* manage to retain a large lensless eye when the evolution of lens eyes appears quite straightforward? Why, in the bivalves molluscs, are three types of optical system (pinholes, concave mirrors, and compound eyes) all used in different families do the same job of predator detection? Why is it that insects have single-lens eyes (larval and dorsal ocelli) which would have been capable, if exploited, of producing much better resolved images than compound eyes, which are severely hampered by diffraction? The only conclusion that makes sense is that, if a structure is adequate for the behavioral needs of an animal, it does not get "improved" further.

This same profligacy seen in optical mechanisms is not apparent in the biochemistry of vision, where the machinery involved is both more ancient and more conservative. The two main opsin types and their associated transduction cascades date from the earliest years of the bilateria, perhaps 700–800 mya, and were only later equipped with the optical structures that would enable them to be used in spatial vision. Some aspects of eye genetics are susceptible to "tweaking," the spectral sensitivities of receptors for example, but others, it seems, are not.

Vision, in the sense that we understand it as humans, has had much more to do with advances in the handling by the brain of the information provided by the image than advances in optical resolution. The image in the eye of a bee or wasp is roughly 100 times less well resolved than that in the human eye, and yet the diversity of visually informed activities undertaken by a bee is comparable with that of most vertebrates. It even extends, in some wasps, to the recognition of the faces of other individuals (Sheehan & Tibbetts, 2011). The eyes are just the starting point.

REFERENCES

Arendt, D., Tessmar-Raible, K., Snyman, H., Dorresteijn, A. W., & Wittbrodt, J. (2004). Ciliary photoreceptors with a vertebrate-type opsin in an invertebrate brain. *Science, 306,* 869–871.

Blest, A. D., & Land, M. F. (1977). Physiological optics of *Dinopis subrufus* L Koch: A fish-lens in a spider. *Proceedings of the Royal Society of London. Series B, Biological Sciences, 196,* 197–222.

Buschbeck, E., Sbita, S. J., & Morgan, R. C. (2007). Scanning behavior by larvae of the predaceous diving beetle, *Thermonectes marmoratus* (Coleoptera: Dytiscidae) enlarges visual field prior to prey capture. *Journal of Comparative Physiology. A, Neuroethology, Sensory, Neural, and Behavioral Physiology, 193,* 973–982.

Carpenter, R. H. S. (1988). *Movements of the eyes* (2nd ed.). London: Pion.

Collett, T. S., & Land, M. F. (1975). Visual control of flight behaviour in the hoverfly, *Syritta pipiens* L. *Journal of Comparative Physiology, 99,* 1–66.

Exner, S. (1989). *The physiology of the compound eyes of insects and crustaceans* (R. C. Hardie, Trans.). Berlin: Springer-Verlag. (Original work published 1891)

Fraenkel, G. S., & Gunn, D. L. (1940). *The orientation of animals. Oxford University Press: Oxford. Reprinted (1961).* New York: Dover.

Gregory, R. L., Ross, H. S., & Moray, N. (1964). The curious eye of *Copilia. Nature, 201,* 1166.

Hausen, K. (1973). Die Brechungsindex im Kristallkegel der Mehlmotte *Ephestia kühniella. Journal of Comparative Physiology, 82,* 365–378.

Jagger, W. S. (1992). The optics of the spherical fish lens. *Vision Research, 32,* 1271–1284. doi:10.1016/0042-6989(92)90222-5.

Kirschfeld, K. (1972). The visual system of *Musca:* Studies on optics structure and function. In R. Wehner (Ed.), *Information processing in the visual systems of arthropods* (pp. 61–74). Berlin: Springer.

Kirschfeld, K. (1976). The resolution of lens and compound eyes. In F. Zettler & R. Weiler (Eds.), *Neural principles in vision* (pp. 354–370). Berlin: Springer.

Kröger, R. H. H., Campbell, M. C. W., Fernald, R. D., & Wagner, H. J. (1999). Multifocal lenses compensate for chromatic defocus in vertebrate eyes. *Journal of Comparative Physiology. A, Neuroethology, Sensory, Neural, and Behavioral Physiology, 184,* 361–369.

Land, M. F. (1965). Image formation by a concave reflector in the eye of the scallop, *Pecten maximus. Journal of Physiology, 179,* 138–153.

Land, M. F. (1969). Movements of the retinae of jumping spiders in response to visual stimuli. *Journal of Experimental Biology, 51,* 471–493.

Land, M. F. (1982). Scanning eye movements in a heteropod mollusc. *Journal of Experimental Biology, 96,* 427–430.

Land, M. F. (2002). The spatial resolution of the pinhole eyes of the giant clam (*Tridacna maxima*). *Proceedings of the*

Royal Society of London. Series B, Biological Sciences, 270, 185–188.

Land, M. F., & Nilsson, D.-E. (2006). General purpose and special purpose visual systems. In E. J. Warrant & D.-E. Nilsson (Eds.), *Invertebrate vision* (pp. 167–210). Cambridge, England: Cambridge University Press.

Land, M. F., & Nilsson, D.-E. (2012). *Animal eyes*. Oxford, England: Oxford University Press.

Levi-Setti, R. (1993). *Trilobites*. Chicago: Chicago University Press.

Mallock, A. (1894). Insect sight and the defining power of composite eyes. *Proceedings of the Royal Society of London. Series B, Biological Sciences, 55*, 85–90. doi:10.1098/rspl.1894.0016.

Marshall, N. J., & Oberwinkler, J. (1999). The colourful world of the mantis shrimp. *Nature, 401*, 873–874.

Muntz, W. R. A., & Raj, U. (1984). On the visual system of *Nautilus pompilius*. *Journal of Experimental Biology, 109*, 253–263.

Nilsson, D.-E. (1988). A new type of imaging optics in compound eyes. *Nature, 332*, 76–78.

Nilsson, D.-E. (1989). Optics and evolution of the compound eye. In D. G. Stavenga & R. C. Hardie (Eds.), *Facets of vision* (pp. 30–73). Berlin: Springer.

Nilsson, D.-E. (1994). Eyes as optical alarm systems in fan worms and ark clams. *Philosophical Transactions of the Royal Society of London. B, 346*, 195–212.

Nilsson, D.-E. (2009). The evolution of eyes and visually guided behaviour. *Philosophical Transactions of the Royal Society of London. B, 364*, 2833–2847.

Nilsson, D.-E., Land, M. F., & Howard, J. (1988). Optics of the butterfly eye. *Journal of Comparative Physiology. A, Neuroethology, Sensory, Neural, and Behavioral Physiology, 162*, 341–366.

Paul, H., Nalbach, H.-O., & Varjú, D. (1990). Eye movements in the rock crab *Pachygrapsus marmoratus* walking along straight and curved paths. *Journal of Experimental Biology, 154*, 81–97.

Schilstra, C., & van Hateren, J. H. (1998). Stabilizing gaze in flying blowflies. *Nature, 395*, 654.

Sheehan, M. J., & Tibbetts, E. A. (2011). Specialized face learning is associated with individual recognition in paper wasps. *Science, 334*, 1272–1275.

Stange, G. (1981). The ocellar component of flight equilibrium control in dragonflies. *Journal of Comparative Physiology. A, Neuroethology, Sensory, Neural, and Behavioral Physiology, 141*, 335–347.

Vogt, K. (1980). Die Spiegeloptik des Flusskrebsauges. The optical system of the crayfish eye. *Journal of Comparative Physiology, 135*, 1–19.

Wagner, H.-J., Douglas, R. H., Frank, T. M., Roberts, N. W., & Partridge, J. C. (2009). *Dolichopteryx longipes*, a deep-sea fish with a bipartite eye using both refractive and reflective optics. *Current Biology, 19*, 108–114.

Wehner, R. (1981). Spatial vision in arthropods. In H. Autrum (Ed.), *Handbook of sensory physiology* (Vol. VII/6C, pp. 287–616). Berlin: Springer.

Williams, D. S., & McIntyre, P. (1980). The principal eyes of a jumping spider have a telephoto component. *Nature, 288*, 578–580.

80 Visual Navigation Strategies in Insects: Lessons from Desert Ants

RÜDIGER WEHNER, KEN CHENG, AND HOLK CRUSE

Desert ants are among the most fascinating insect navigators. Oddly enough, it is the lifestyle of a thermophilic scavenger that has rendered them one of the champions in the world of animal navigation. The workers of these heat-tolerant species—be they members of the Saharan *Cataglyphis*, the Namibian *Ocymyrmex*, or the Australian *Melophorus* tribes—perform wide-ranging foraging journeys during which they search for dead arthropod matter. In concentrating their outdoor activities to the hottest times of day and year, they are strictly diurnal, visually guided central-place foragers. As food densities are rather low in their arid-land habitats, they must cover large foraging distances and thus expose themselves to increasingly stressful environmental conditions. Hence their maxim must be: Venture out as far as necessary, and return home as quickly as possible. These conflicting demands have selected for high running speeds and sophisticated means of spatial orientation. The latter are the topic of this chapter.

HOLOPTIC VISION

The visual world of the high-speed desert ants is of bewildering diversity. For some species such as the salt-pan species *Cataglyphis fortis* it is flat and largely featureless, but for others such as the sand-dune species *Cataglyphis bombycina* it is three-dimensional, and for most species it is cluttered to various degrees with stony gravel, grass tussocks, low shrubs, and loosely scattered trees. Due to their holoptic eyes and panoramic vision, *Cataglyphis* ants can perceive these diverse visual surroundings simultaneously. Taken together, the visual fields of the two eyes cover 93% of the surface of the unit sphere (4π steradians). Even though eye size varies in relation to body size from 600 ommatidia in the smallest workers of *Cataglyphis bicolor* to nearly 1,300 ommatidia in the largest ones, the following considerations on the allometric scaling of morphological and optical parameters clearly show that the size of the total visual field remains constant (Zollikofer, Wehner, & Fukushi, 1995).

The number of ommatidia N_o per eye scales with head width W_h (a proxy for body size, with which it is correlated isometrically):

$$N_o \sim W_h^{0.74}. \tag{80.1}$$

Furthermore, the interommatidial angle $\Delta\varphi$ (degrees) and the area A_o (steradians) covered by one ommatidium on the unit sphere become smaller as head width increases:

$$\Delta\varphi \sim W_h^{-0.39}. \tag{80.2}$$

$$A_o \sim W_h^{-0.70}. \tag{80.3}$$

By combining equations 80.1 and 80.3, we arrive at the total visual area A_{total} of one compound eye:

$$A_{total} = N_o A_o \sim W_h^{0.74-0.70} \sim W_h^{0.04} \approx \text{constant}.$$

Hence, while nearly all anatomical and optical parameters of the *Cataglyphis* compound eye—the numbers and corneal sizes of the ommatidia, the ommatidial divergence angles and fields of view of the individual ommatidia, as well as the size and curvature of the eye—vary allometrically with body size, the total field of view of the eye stays constant. Consequently, smaller individuals have wider interommatidial angles and hence lower spatial resolution than their larger conspecifics, meaning that they forgo visual acuity for a whole-field view of the world through which they move. Hence holoptic vision must have been strongly selected for. Moreover, *Cataglyphis* ants possess a marked visual streak, an equatorial zone of increased spatial resolution ($\Delta\varphi = 3.5°$). *Melophorus bagoti*, which covers about the same range of body sizes as *Cataglyphis bicolor* does, but which has a much smaller number of ommatidia (420–590) per compound eye, lacks such a high-acuity belt as well as a fully holoptic visual field (Schwarz, Narendra, & Zeil, 2011). Instead it seems to have opted for high-contrast vision in the frontal field of view. Whether these intergeneric differences in the optics of the eye correlate with different behavioral needs—navigation in visually more or less cluttered environments—remains to be investigated.

In any case, how do navigating ants exploit the enormous amount of spatial information impinging on their wide-angle visual systems at any one time? What features do they extract, and what spatial information do they finally acquire about the lay of their foraging land? Consider that all this information must be handled and used by a 0.1-mg brain, and moreover, that only particular parts of this miniature brain might be involved in accomplishing spatial orientation tasks. Further note that at least some centers of the *Cataglyphis* brain exhibit high synaptic plasticity during the ant's life history (Stieb et al., 2010). For example, those parts of the mushroom bodies—the insect's higher integration and learning centers—that are processing visual information (the collar regions) undergo massive dendritic expansions of the postsynaptic Kenyon cells and rearrangements of the microglomerular synaptic complexes once the animals are exposed to light, either normally during their first appearances at the nest entrance or precociously by artificial illumination at earlier stages of their lives. Obviously, the full function of these centers is made available only when it is needed.

In the following we first treat some particular navigational routines that have been well studied in desert ants. Then we discuss how these multiple routines might interact and whether these interactions would finally result in a unified spatial representation of the ant's foraging grounds.

VECTOR NAVIGATION

Path integration (vector navigation) is the ant's predominant mode of navigation (for a review, see Wehner & Srinivasan, 2003). In featureless or unfamiliar terrain it is the only means of acquiring positional information, but even in cluttered environments when landmark-guidance routines dominate the ant's behavior, the path integrator keeps running in the background and provides the animal with an internal safety line readily available should other navigational routines fail. Reversed in sign, it can subsequently be used as a food vector to guide the ant to a previously visited food site. This home/food vector is acquired by a continuous integration of information provided by compass and odometer systems.

Skylight Compass

The compass—a visual compass depending largely on the pattern of polarized light in the sky—is a well-studied part of the ant's path integrator (for reviews, see Wehner, 1994; Wehner & Labhart, 2006), and it is already here that some common design features of

the insect's navigational toolkit come to the fore. The module-specific design of the polarization compass is characterized by some special means of peripheral sensory coding. A set of specialized photoreceptors confined to a small region of the eye, the dorsal rim area (DRA) first discovered in *Cataglyphis* ants (Herrling, 1975), picks up polarized light information from the sky. Among the four optical parameters that characterize the light emanating from each pixel of the celestial hemisphere—angle of polarization, degree of polarization, spectral content, and radiant intensity—the first one is most robust against atmospheric disturbances and hence provides the most reliable compass cue. In the ant's visual system it is freed from the confounding effects of the other three parameters by specific adaptations of the DRA photoreceptors and their underlying interneurons: Interference by spectral cues is eliminated by the homochromacy of the DRA compass system (exclusive involvement of only one—the ultraviolet—type of photoreceptor); interference by skylight intensity is eliminated by the crossed-analyzer arrangement and antagonistic interaction of pairs of DRA receptors within individual detector units, the DRA ommatidia (polarization opponency and contrast enhancement); finally, interference by the degree of polarization is eliminated most likely by some gain-control mechanism operating at an interneuronal level (as proposed by Sakura, Lambrinos, & Labhart, 2008). Furthermore, the outputs of the DRA ommatidia are further processed by a limited number of second-order interneurons located in the medulla and sampling information across large parts of the celestial hemisphere. These neurobiological data and various kinds of behavioral results (e.g., Wehner & Müller, 2006) support the hypothesis that a bilaterally symmetrical array of a few spatial low-pass skylight filters extracts from the sky what turns out to be the decisive compass information: the angular position of the symmetry plane of the pattern (the solar–antisolar meridian) relative to the animal's longitudinal body axis.

This hypothesis can be readily reconciled with neurobiological findings obtained in the central complex of locusts (Homberg & el Jundi, chapter 84, this volume). Within this midline system of neuropiles—a system that "supervises walking" (Strausfeld, 1999)—the 16 columns of the protocerebral bridge (PB) provide a topographical representation of zenithal angles of polarization (0°–180° on either side of the midline) integrated over the insect's right and left visual field. As the architectural design of the central complex is rather similar in all insects examined so far, we can assume that this representation holds true for ants as well. However, how the PB polarization system is finally used for one or another kind of spatial orientation might well differ between locusts (migratory insects) and ants and bees (central place foragers). Moreover, one should emphasize that the array of polarization detecting PB neurons does not imply that the ants are able to detect and analyze individual angles of polarization in whatever part of the sky. Our behavioral experiments really rule out this possibility. Rather it will be the wide-field, binocularly integrated response modulation occurring across the PB as the animal rotates about its vertical body axis that provides the proper compass information. In conclusion, due to the peripheral coding mechanisms outlined above, the skylight information finally reaching the central complex has been transformed in a way that renders it compatible with bilateral information arriving at the central complex from other sensory channels, either visual ones (e.g., about landmark skylines) or nonvisual (e.g., proprioceptive) ones.

As to the operation of the ant's polarization compass it is important to note that *Cataglyphis* does not refer to the zenithal region of the celestial hemisphere as assumed for crickets and locusts but to areas seen by frontolateral parts of its eyes (centered about 50°–55° above the horizontal and 45° laterally from the forward direction). This can be deduced from experiments in which the ants, while walking underneath a movable trolley, had their fields of view restricted in certain ways, and in which the resulting compensatory head movements and changes in navigation accuracy were recorded (Duelli, 1975). For example, as shown by cinematographic analyses, *Cataglyphis* tilts its head upward and downward when the boundary of a light-tight screen extends for elevations of greater than 45° upward from the horizon and greater than 30° downward from the zenith, respectively, into the ant's field of view (see figure 80.1B). These observations had been made shortly before we had established the necessity and sufficiency of the DRA as the input region of the ant's polarization compass. With the latter information at hand, we now can conclude that when the ants are provided with only limited parts of the skylight patterns, they prefer to employ the frontal and spatially most widely extended parts of their DRAs (see figure 80.1A). This preference is so strong that when presented with a zenithal patch of sky, *Cataglyphis* straightens up, that is, raises its body as far as it can, until it finally somersaults backward. Inquiries into the functional significance of this preference will certainly shed light on the mechanism of the ant's skylight compass. Most likely, in expressing the behavior described above, the ants try to determine the orientation of the solar or antisolar meridian relative to their walking (forward) direction. In accomplishing this task, the additional use of the spectral gradients in the sky will help (Wehner, 1997). Indeed, each pixel of the celestial hemisphere is viewed by a polarization-sensitive but spectrally blind visual unit of the contralateral DRA, but also by a spectrally competent but polarization blind unit of the ipsilateral dorsal retina outside the DRA.

Odometer

While the compass of the ant's path integrator is predominantly visual (and only complemented by a mechanoreceptive wind compass (Müller & Wehner, 2007), the odometer is a stride integrator, which most likely relies on proprioceptive signals associated with the movements of the legs (Wittlinger, Wehner, & Wolf, 2006). Vision—the use of self-induced image flow as employed by flying insects such as honeybees (Srinivasan et al., chapter 85, this volume)—plays a minor role (Ronacher & Wehner, 1995). The stride integrator is sufficient for gauging distances traveled because ants performing their inbound runs by walking on stilts or stumps, and with the ventral halves of their eyes occluded, overestimate and underestimate their travel distances, respectively, exactly as predicted by the changes in stride length alone (M. Wittlinger, H. Wolf, personal communication). For obvious reasons, the question whether the use of the stride integrator is not only sufficient but also necessary for odometry cannot be answered straightforwardly. However, the fact that ants carried by nest-mates from one nest of a colony to another nest are also able to record, at least to a certain degree, the distances traveled while being carried (as shown when later "decoupled" from the transporter; Duelli, 1976) offers an opportunity to inquire into this question (M. Wittlinger, work in progress). Note, however, that immobilized ants, which are moved above the ground in a small sledge, do not record the distances covered passively by the experimental device (Seidl, Knaden, & Wehner, 2006), but while being transported by a nest-mate the transportee is in another motivational state.

At present, little is known about where in the ant's brain compass and odometer information is combined and, hence, where the updating of the path integrator occurs. Most likely this happens in the central complex or in the large-synapse region of the lateral accessory lobe, which receives inputs from columnar neurons of the central complex. But we do know that confluence of both types of information is necessary for the updating process. Whenever the proprioceptive odometer information is not associated with visual compass

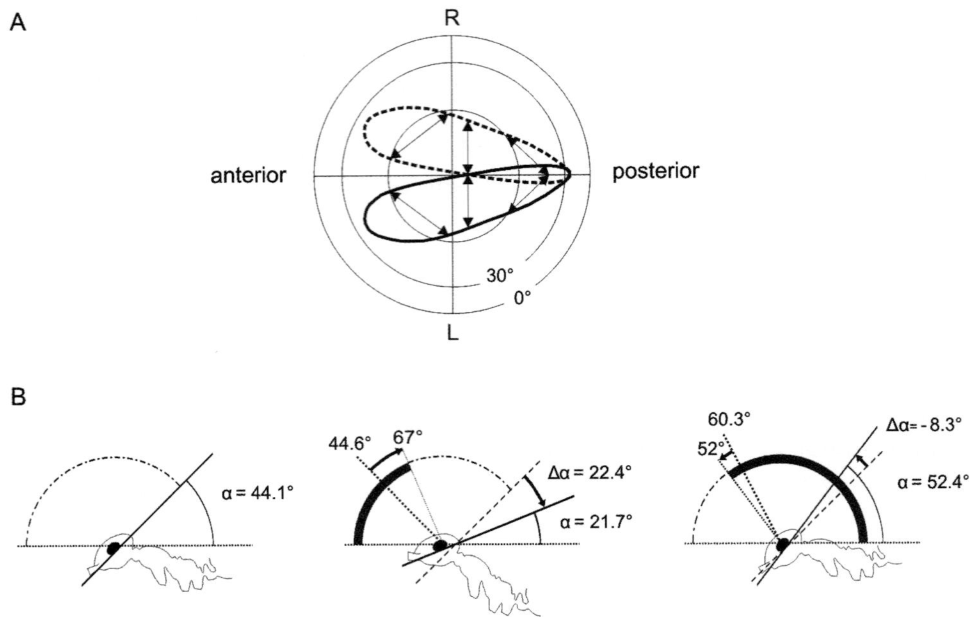

FIGURE 80.1   (A) Visual fields of the contralaterally looking left and right dorsal rim area (DRA) (surrounded by dotted and continuous lines, respectively) of *Cataglyphis bicolor.* L and R indicate the left and right visual field; zenith projection. The double-headed arrows mark the mean polarization tuning axes of the ommatidial units in the frontal, middle, and caudal parts of the DRAs (Pol areas). (B) Postural changes in navigating ants, *Cataglyphis bicolor,* which are subjected to optical restrictions in their lower and upper fields of view. The part of the visual field marked by a gray boundary is shielded by a light-tight screen. Left-hand figure: Normal tilt angle of head relative to the horizontal. Middle figure: A screen extending up to 67° from the horizontal causes the ants to tilt upward by Δα = 22.4°, meaning that the lower margin of the preferred Pol compass region in the ant's eye is 44.6°. The same rationale applies to the upper margin (right-hand figure). The postural changes are achieved mainly by raising and lowering the alitrunk ("thorax") rather than tilting the head relative to the alitrunk. (A adapted from Wehner, 1992; Wehner & Labhart, 2006; B from Duelli, 1975.)

information, the path integrator stops running (Ronacher, Westwig, & Wehner, 2006; Sommer & Wehner, 2005).

Employing the Path Integrator

Recently, path integration as performed by desert ants has inspired theoreticians to inquire about several aspects of this behavioral routine, for example, the pros and cons of egocentric and geocentric, continuous and discontinuous, idiothetic and allothetic mechanisms (Collett & Collett, 2000; Vickerstaff & Cheung, 2010). Rather than fully delving into this subject proper, let us illustrate the power of path integration in the ant's overall system of navigation by presenting some recent behavioral data from *Cataglyphis.* We start by assuming that when an ant moves from home to a familiar place, it continually compares an internally stored reference vector (R) acquired during previous foraging journeys with the current state of its path integrator, the current vector (C). Once the latter matches the former, the ant has arrived at the known place. By employing a particular experimental paradigm we succeeded in training the ants to move from home alternatively to two food sites, to either A or B, so that the ants acquired and stored two reference food vectors, R_A and R_B (see figure 80.2). If then deprived of food, say, at A, the ants would directly move along a novel route from A to B. In this case the choice of a novel route, usually considered as evidence for map-based behavior, can be completely explained by path integration: At site A an ant's current vector C matches R_A; when finding no food there, the ant loads down R_B, and by again trying to match its current vector with its then retrieved reference vector, that is, by following the output (travel) vector $T = R_B - C$, moves to site B just by path integration. On the other hand, preliminary experiments indicate that zero-vector ants, that is, ants that have not walked from home to the familiar place A, but have been passively displaced to it, do not move to place B.

LANDMARK GUIDANCE

Place Recognition

Due to the inevitable accumulation of errors just mentioned, the path integrator will usually guide the animal to a location that is close to but not exactly at the goal.

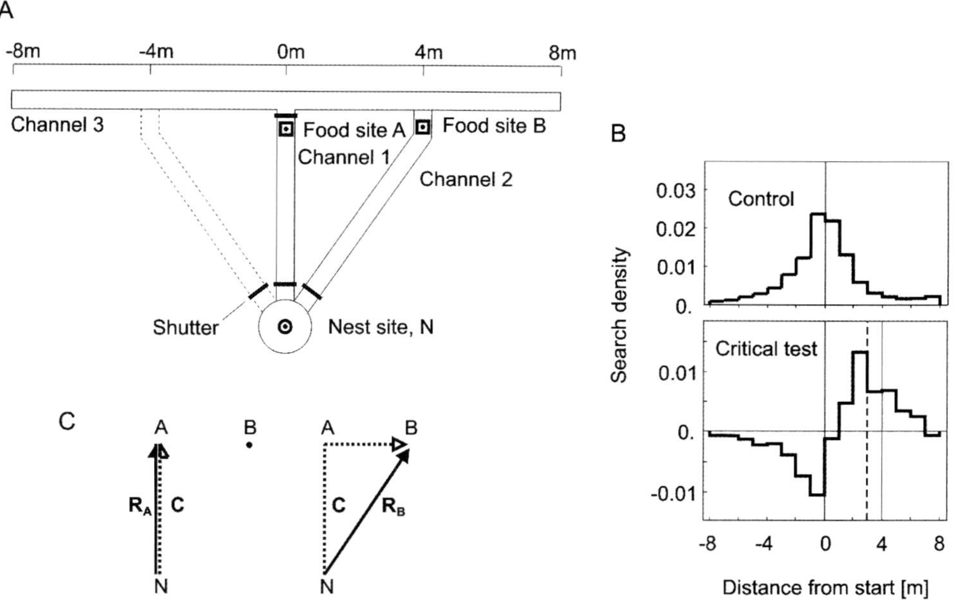

FIGURE 80.2   Novel shortcuts performed by path integration. (A) Experimental paradigm: array of open-topped channels installed at a desert field site. Channels 1 and 2: training channels leading to food sites A and B, respectively. In a separate group of ants channel 2 was used in a mirror-imaged position (dotted lines). Channel 3: test channel. (B) Results: search distribution of ants, *Cataglyphis fortis*, in test channel 3. The upper panel depicts the search distribution of ants that had been trained exclusively to site A (control). The lower panel refers to ants that had been trained from the nest to sites A and B. If in the critical test they were deprived of food, they traveled along a novel route from A to B. The graph presents the difference of the ants' search distribution and the control distribution. Dashed vertical line: ant-subjective location of site B (true position of site B is marked by the solid vertical line). As in the channel device the ants had only a partial view of the sky, their polarization compass induced systematic errors (Wehner & Müller, 2006). If these errors are taken into account, the position of site B as indicated by the ant's path integrator shifts slightly toward site A. (C) Interpretation. C, current vector; R_A and R_B, reference vectors. (Data from B. Vögeli, M. Knaden, & R. Wehner.)

In this situation previously acquired information about terrestrial signposts can assist in finally pinpointing the goal. Experiments performed with sets of artificial landmarks on desert ants as well as other ants and bees have impressively shown that in using visual landmarks for defining goal locations (*Cataglyphis*: Wehner & Räber, 1979; Wehner, Michel, & Antonsen, 1996; *Melophorus*: Cheng et al., 2009) the insects rely on retinotopic views as seen from these locations. Based on such studies, several computational models have been designed to account for this type of view-based local visual homing. According to the classical snapshot (template matching; TM) model, the insect acquires and stores a rather unprocessed retinotopic two-dimensional (snapshot) image, or template, of the world around the goal and later moves so as to gradually improve the correspondence between the current view and the snapshot (Cartwright & Collett, 1983; Möller, 2001; Mangan & Webb, 2009; see figure 80.3A). In this TM model the ant is supposed to determine the differences in bearing and retinal size between any landmark image in the snapshot and its closest landmark image in the current view and to use these differences to compute tangential and radial vectors that reduce the differences in bearing and size, respectively. By moving in the direction of the sum of all tangential and radial component vectors, the ant as well as the robot Sahabot-2, in which this algorithm has been implemented (Lambrinos et al., 2000; Möller et al., 2001), improve the correspondence between template and current view.

A different matching procedure relies on gradient descent. In natural scenes, the image difference between template and current view rises gradually and monotonically as the spatial distance between current location and goal increases (Zeil, Hoffmann, & Chahl, 2003; Möller & Vardy, 2006; Wystrach, Beugnon, & Cheng, 2011), defining an image difference function (IDF) with the global minimum at the goal. An unprocessed array of pixels, each assigned a gray-scale value, can be used for this pixel-by-pixel matching process. The agent samples different directions of movement and then moves in a direction that reduces image difference between target and current views, thus descending an image-difference gradient. Crucial to all these

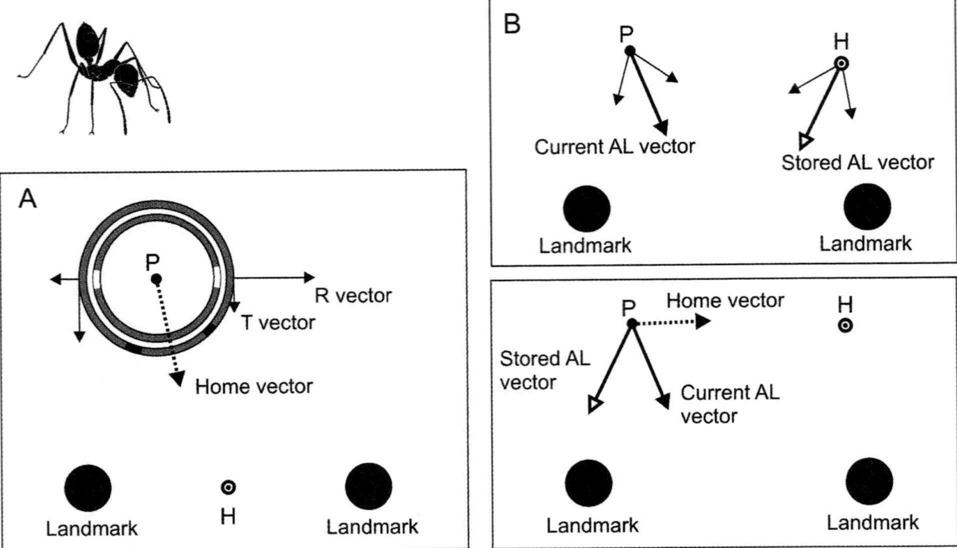

FIGURE 80.3   Two classical models of place recognition. The models describe how an ant—or, in general, an agent—can local-ize a familiar place (e.g., home, H), by comparing the currently obtained information about the surrounding visual scene (at the current position, P) with the stored information previously acquired at the goal location (H). (A) Template matching model. At the current position (P) the ant relates the stored view of the two landmarks as seen from the nest (template: inner circle) with the current view (outer circle). It does so by computing tangential and radial vectors (T and R vectors), the lengths of which are proportional to the differences in bearing and size between the image of a landmark in the template and its closest neighbor in the current view. Moving in the direction of the home vector (the sum of all component vectors) improves the match between template and current view in the best possible way. (B) Average landmark vector model. Upper panel: At each location an average landmark (AL) vector is computed as the sum of unit landmark (UL) vectors. Lower panel: The difference between the current and stored AL vectors yields the home vector. Rather than referring to the "center of mass" of the landmark images (as done here), ALV models can also use edges of objects. The figure of a *Cataglyphis fortis* in the upper left is to indicate that in both models a compass reference, for example, a constant orientation of the animal, is needed to align template and current view. (For references to the theoretical formulations of these models, see text.)

matching processes is to align the target view and the current view in the same compass direction. Short of mental rotation of images (thought to be computation-ally intensive), multiple views associated with different compass directions might have to be encoded, so that the ant could later select the one pertaining to the current compass bearing. Various modifications of TM models have been proposed. Moreover, as navigating animals move, optic-flow information resulting from movement near the goal may also be exploited (Dittmar et al., 2010), so that static cues might be complemented or replaced by dynamic ones.

Image matching procedures, the essence of any TM model, are not required in what has been called the average landmark vector (ALV) model (Lambrinos et al., 2000; Möller, 2001; Mangan & Webb, 2009; Yu & Kim, 2011; see figure 80.3B). This model assumes that at the goal the ant computes and stores an average landmark vector, that is, the average (vector sum) of unit vectors pointing toward individual landmarks. By subtracting the stored ALV from the ALV pertaining to the ant's current position, a home vector is computed.

As in this case the entire panoramic image is condensed, so to speak, into one parameter vector, the ALV model is on the one hand holistic and on the other hand a parameter-based rather than template-based model (Möller & Vardy, 2006). Computationally it is much cheaper than the latter, but it still needs a compass to align either images or vectors, and the visual scene around the animal must be dissected into individual landmarks or landmark features such as edges. Other methods such as the image warping method (Franz et al., 1998) do not depend on either of these precondi-tions but due to their high computational complexity might not fit into an insect's brain. Furthermore, it might well be that in the ant's panoramic field of view a segregation into individual landmarks and surround-ing background does not occur at all. In recent studies, in which full 360° pictures of the ant's environment have been recorded and analyzed, the ants' behavior could indeed be explained by just comparing stored and current panoramic images without the need for isolating "landmarks" from the remainder of a scene (Wystrach, Beugnon, & Cheng, 2011).

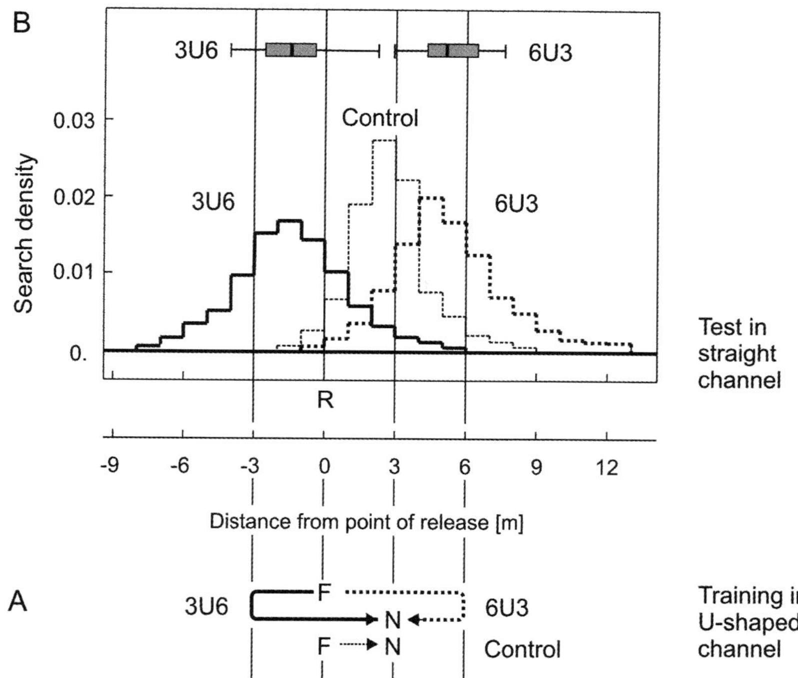

FIGURE 80.4   Local vectors. Test paradigm in which site-based (local) vectors outcompete path-integration (global) vectors. (A) Training paradigms: In a linear channel array desert ants, *Cataglyphis fortis*, had to walk along U-turn detour paths from a feeder (F) to the nest (N). For example, the solid-arrow paradigm (3U6) means that the ants first walked 3 m in the opposite direction to the nest, performed a U-turn, and then walked 6 m to the nest. (B) Test: search distributions of ants that had been transferred from F to a straight channel and released at R. The bold solid and dotted lines depict the search density distributions exhibited by the two groups of ants in a linear test channel. Thin dashed line: control, that is, search density distribution of ants that had been trained directly (without detour) from F to N. The box plots in the upper part of the figure indicate the ants' first turning points after departure from the feeder. (Adapted from Knaden, Lange, & Wehner, 2006.)

Site-Based Steering Commands (Local Vectors)

Stored views are used not only to pinpoint familiar places but also to guide the ants away from these places and forward on their paths. Ants can learn to steer a particular course once a particular waypoint is reached and hence attach site-based (local) steering commands to these points. Such associations of local views with motor commands are acquired during path integration but can later be employed completely independently. As cue-competition experiments show, the direction of such site-based vectors can be related either allothetically to (1) a global, for example, skylight reference, system (Bisch & Wehner, 1998; Collett et al., 1998; Knaden, Lange, & Wehner, 2006) or to (2) the local landmark scenery (Collett, Collett, & Wehner, 2001) or even idiothetically to (3) the preceding section of the animal's path (Bisch-Knaden & Wehner, 2001). Under natural conditions these systems will interact in concert, but in particularly designed experiments they can be disentangled and studied in isolation. A striking example of such a local steering command is shown in figure 80.4. In this experimental paradigm the ants are forced to travel between nest and an artificial feeder along a U-turn detour path and are later transferred from the feeder into a straight test channel. For example, if the detour path is designed such that the starting direction of the homebound ants (the direction of the local vector anchored at the food site) points in the opposite direction to that of the nest (indicated by the global path integration vector), the ants highly significantly follow the local vector.

Route Navigation

One of the most thought-provoking recent achievements in navigation research is the discovery of remarkably large sets of landmark memories acquired and used by desert ants. These animals can successively learn a number of complex landmark-based routes in an idiosyncratic way, learn different routes for outbound and subsequent inbound runs, and remember these routes for extended periods of time (Kohler & Wehner, 2005; Sommer, von Beeren, & Wehner, 2008). In principle, they could travel along routes by combining a series of snapshot memories, each associated with a local vector.

After one local vector is run off, the view and its associated snapshot would prime the retrieval of the next local vector, and so on (sequential priming followed by serial recall). More economically, the ants could employ gradient descent of the IDF as the matching procedure. However, the ants' behavior on a route does not comply with such a gradient descent routine. Instead, the ants occasionally scan the environment by pirouetting on the spot (Baddeley et al., 2012; Wehner, 1994) as described for the learning walks of *Cataglyphis* species (Stieb et al., 2010; Wehner, Meier, & Zollikofer, 2004) and *Ocymyrmex robustior* (Müller & Wehner, 2010). In this case some kind of rotational matching might occur, in which the ant uses the panorama as a visual compass (Philippides et al., 2011; Zeil, Hoffmann, & Chahl, 2003). A rotational image difference function (rIDF) may be defined, in which mismatches are calculated as a function of different orientations of the current panoramic view with respect to the target image until the best match is found and the rIDF reaches a marked minimum. (Note that the IDF introduced above can now be defined more specifically as a translational image difference function, tIDF; Zeil, 2011.) Recently, Wystrach, Beugnon, and Cheng (2012) found that a model of rotational image matching worked well for ants traveling on familiar routes even if in the model only approximate skyline heights measured every 15° in azimuthal distance were used. In sum, the rIDF model provides the animal with a local "visual panorama compass" in addition to its global "visual skylight compass."

Rather than dissecting the route into a sequence of waypoints, the ants could encode routes in a more holistic way (Baddeley et al., 2011; Philippides et al., 2011). In this approach, views are characterized by a set of feature detectors as to whether or not they belong to a learned route, and facing the ant's correct direction of travel. A scanning routine would then allow the animal to move in the direction that is most likely part of the route. Hence, views are used to define steering commands, that is, actions, rather than places. Recently Baddeley et al. (2012) have further refined the idea of holistic route representation by introducing the concept of familiarity discrimination based on the information-maximization (Infomax) neural network approach of Lulham et al. (2011). When simulated ants walking through an artificial cluttered environment employed such an Infomax system, and used silhouettes of objects against the sky as the only visual input, they learned and recalled routes nearly as rapidly and accurately as real ants had done. They did so without relying on global compass or odometer information, without storing multiple views, even any view, and thus without having to burden their brains with unrealistically high memory capacities. Furthermore, by including information obtained during learning walks around the goal, the ants could use the same navigational routine for being guided on a route and searching for the final destination, that is, for accomplishing tasks that so far have been treated separately.

In conclusion, a lot of theoretical development has taken place since the early TM and ALV models had been designed. Skyline matching, rIDFs, and holistic encoding of routes are recent proposals, all needing much more research to see how they apply to insect navigation. Experiments using manipulations to test the models directly, as contrasted with running the models on ants' navigational behaviors in natural habitats, are especially pertinent.

THE NAVINET

How can the various navigational modules described in the preceding sections be combined to accomplish particular behavioral tasks under particular environmental conditions? For example, path integration serves as a scaffold in acquiring landmark information, but later landmark guidance can be freed from path integration constraints. Furthermore, depending on salience, local (site-based) vectors can outcompete global (path integration) vectors, which nevertheless continue to be updated in the background and come to the fore again when during the course of a home run landmark information fades away or becomes unreliable. On the other hand, there are constraints which at first glance might be irritating. Desert ants can learn visual landmark routes quickly and reliably but cannot retrace them in the opposite direction (Wehner et al., 2006). Obviously, view-based landmark guidance routines provide the ant with *procedural* rather than *positional* information (Graham, 2010; Wehner, 2008). For example, landmark memories are not coupled to path integration coordinates. Following procedures rather than reading positions from a unitary representation of foraging space seems to be the essence of the insect's navigational abilities.

A decentralized procedural memory architecture indeed shows that information acquired and stored in various navigational routines can be retrieved and used in ways that are fully compatible with what we observe in ants. This "Navinet" model (see figure 80.5) consists of four parts. The first part comprises a controller based on path integration and including the ability to perform area concentrated search (random generator). The latter behavior is performed by ants when the home vector has reached its zero state but the nest entrance has not been found yet.

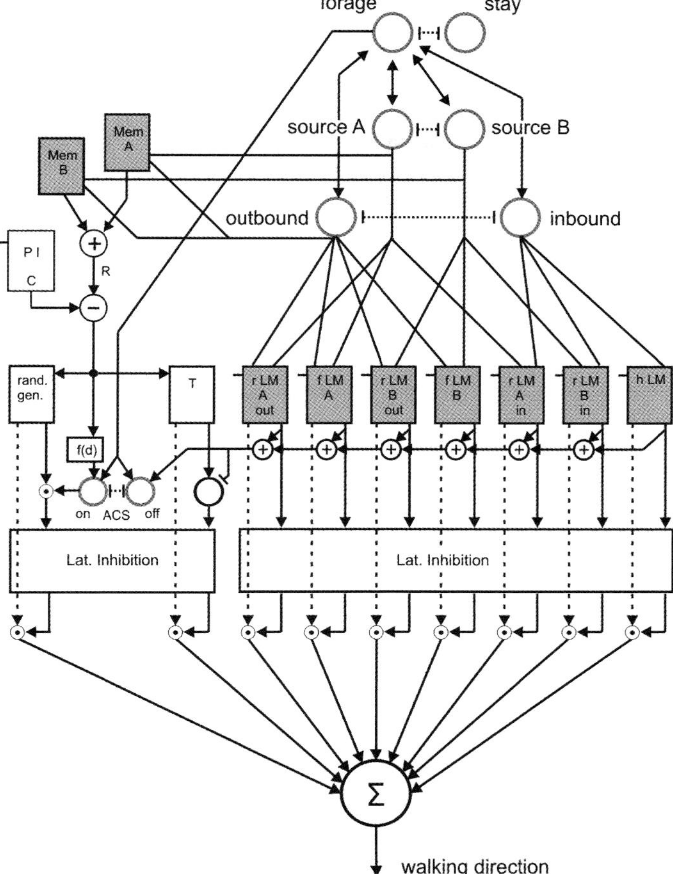

FIGURE 80.5   A decentralized memory architecture allowing for navigation exploiting landmarks and path integration. It consists of eight motivation units (gray lines), a bank of memory elements (gray rectangles), MemA, MemB, and seven further elements (two food landmark elements, fLMA, fLMB, one home landmark element, hLM, and four route landmark elements, rLMAout, rLMBout, rLMAin, rLMBin). The path integrator (box marked PI) provides the current vector, C. Food place memories (MemA, MemB) provide the reference vector, R. Computation (R − C) provides the travel vector, see box marked T. The box "rand. gen." and the motivation units ACSon, and ACSoff control "area-concentrated search" walks. The box marked f (d) represents a function that activates unit ACSon the less, the larger the difference d = R − C increases. Boxes Lat. (lateral) Inhibition provide weighting factors (multiplication symbols) to decide which memory elements control the motor output (walking direction). For further details see text. (Adapted from Cruse & Wehner, 2011.)

The second part contains a bank of memory elements storing either (1) vectors that point from home to food places (food place vectors memA and MemB), (2) information that represents particular landmark views plus the local vectors associated with them (route landmarks, rLMs), or (3) landmark elements that represent food places (food landmarks; fLMs) or the home place (home landmark; hLM) plus the vectors leading to the goal.

The third part consists of a network of so-called motivation units. In figure 80.5 these units are depicted in gray. They form a recurrently connected artificial neural network that can relax to various attractors, each of which determines a specific internal state, which the whole system can adopt. At the highest level, these are the two states Sleep and Forage. Within the state Forage there are the substates Outbound and Inbound. In addition, these states are characterized by the food source that has currently been selected (A or B). The activation of the motivation network forms the context determining which of the memory elements may possibly be retrieved and which can currently not be attended. For example, a possible set of activated motivation units might consist of units Forage, Outbound, and sourceA. A memory element can only be retrieved if all its motivation units are active.

The fourth part of the model concerns the treatment of the outputs of the active memory elements to determine the walking direction. In essence, this decision is conducted by a weighted lateral inhibition network (boxes Lat. Inhibition) as detailed by Cruse and Wehner (2011). Hoinville, Wehner, and Cruse (2012) provide an extension of Navinet that includes learning the food source positions and qualities.

The memory architecture of the Navinet allows one to simulate all navigational performances described in the previous sections (Cruse & Wehner, 2011). In addition, and as also outlined in the latter account, the modular structure of the Navinet and its inherent recurrent network characteristics find their neuroanatomical and neurophysiological counterparts in the circuitry properties of the ant's and bee's mushroom bodies (e.g., Rybak & Menzel, 2010; Strausfeld, 2002; Strausfeld et al., 2009). The neural substrate to which the output channels might finally, though indirectly (e.g., via the lateral protocerebrum), project is hypothesized to be the central complex (e.g., Strausfeld, 1999; Strauss & Berg, 2010), which is also the major input station of the ant's compass system. Most likely it is at this stage that the directions of movement are finally determined. To unravel these circuitries in the light of properly designed behavioral paradigms is one of the most fascinating, promising, and pressing enterprises of future neuroethological research on insect navigation, and one that seems to be within reach.

REFERENCES

Baddeley, B., Graham, P., Husbands, P., & Philippides, A. (2012). Model of ant route navigation driven by scene familiarity. *PLoS Computational Biology*, *8*, e1002336. doi:10.1371/journal.pcbi.1002336.

Baddeley, B., Graham, P., Philippides, A., & Husbands, P. (2011). Holistic visual encoding of ant-like routes: Navigation without waypoints. *Adaptive Behavior*, *19*, 3–15. doi:10.1177/1059712310395410.

Bisch, S., & Wehner, R. (1998). Visual navigation in ants: Evidence for site-based vectors. *Proceedings of the Neurobiology Conference Göttingen*, *26*, 417.

Bisch-Knaden, S., & Wehner, R. (2001). Egocentric information helps desert ants to navigate around familiar objects. *Journal of Experimental Biology*, *204*, 4177–4184.

Cartwright, B. A., & Collett, T. S. (1983). Landmark learning in bees: Experiments and models. *Journal of Comparative Physiology. A, Neuroethology, Sensory, Neural, and Behavioral Physiology*, *151*, 521–543.

Cheng, K., Narendra, A., Sommer, S., & Wehner, R. (2009). Travelling in clutter: Navigation in the central Australian desert ant *Melophorus bagoti*. *Behavioural Processes*, *80*, 261–268.

Collett, M., & Collett, T. S. (2000). How do insects use path integration for their navigation? *Biological Cybernetics*, *83*, 245–259.

Collett, M., Collett, T. S., Bisch, S., & Wehner, R. (1998). Local and global vectors in desert ant navigation. *Nature*, *394*, 269–272.

Collett, M., Collett, T. S., & Wehner, R. (2001). The guidance of desert ants by extended landmarks. *Journal of Experimental Biology*, *204*, 1635–1639.

Cruse, H., & Wehner, R. (2011). No need for a cognitive map: Decentralized memory for insect navigation. *PLoS Computational Biology*, *7*, e1002009. doi:10.1371/journal.pcbi.1002009.

Dittmar, L., Stürzl, W., Baird, E., Boeddeker, N., & Egelhaaf, M. (2010). Goal seeking in honey bees: Matching of optic flow snapshots? *Journal of Experimental Biology*, *213*, 2913–2923.

Duelli, P. (1975). A fovea for e-vector orientation in the eye of *Cataglyphis bicolor* (Formicidae, Hymenoptera). *Journal of Comparative Physiology*, *102*, 43–56.

Duelli, P. (1976). Distanzdressuren von getragenen Ameisen (*Cataglyphis bicolor*). *Revue Suisse de Zoologie*, *83*, 413–418.

Franz, M. O., Schölkopf, B., Mallot, H. P., & Bülthoff, H. H. (1998). Where did I take that snapshot? Scene based homing by image matching. *Biological Cybernetics*, *79*, 191–202.

Graham, P. (2010). Insect navigation. In M. D. Breed & J. Moore (Eds.), *Encyclopedia of animal behavior* (Vol. 2, pp. 167–175). London: Academic Press.

Herrling, P. L. (1975). Topographische Untersuchungen zur funktionellen Anatomie der Retina von Cataglyphis bicolor (*Formicidae, Hymenoptera*). Ph.D. Thesis, University of Zurich.

Hoinville, T., Wehner, R., & Cruse, H. (2012). Learning and retrieval of memory elements in a navigational task. In T. T. Prescott, N. F. Lepora, A. Mura, & P. F. M. J. Verschure (Eds.) *Biomimetic and Biohybrid Systems. Proceedings of the First International Conference, Living Machines* (pp. 120–131). Lecture Notes in Artificial Intelligence. Berlin: Springer.

Knaden, M., Lange, C., & Wehner, R. (2006). The importance of procedural knowledge in desert-ant navigation. *Current Biology*, *16*, R916–R917. doi:10.1016/j.cub.2006.09.059.

Kohler, M., & Wehner, R. (2005). Idiosyncratic route-based memories in desert ants, *Melophorus bagoti*: How do they interact with path-integration vectors? *Neurobiology of Learning and Memory*, *83*, 1–12. doi:10.1016/j.nlm.2004.05.011.

Lambrinos, D., Möller, R., Labhart, T., Pfeifer, R., & Wehner, R. (2000). A mobile robot employing insect strategies for navigation. *Robotics and Autonomous Systems*, *30*, 39–64.

Lulham, A., Bogacz, R., Vogt, S., & Brown, M. W. (2011). An Infomax algorithm can perform both familiarity discrimination and feature extraction in a single network. *Neural Computation*, *23*, 909–926.

Mangan, M., & Webb, B. (2009). Modelling place memory in crickets. *Biological Cybernetics*, *101*, 307–323.

Möller, R. (2001). Do insects use templates or parameters for landmark navigation? *Journal of Theoretical Biology*, *210*, 33–45.

Möller, R., Lambrinos, D., Roggendorf, T., Pfeifer, R., & Wehner, R. (2001). Insect strategies of visual homing in mobile robots. In B. Webb & T. R. Consi (Eds.), *Biorobotics: Methods and applications* (pp. 37–66). Menlo Park, CA: AAAI Press/MIT Press.

Möller, R., & Vardy, A. (2006). Local visual homing by matched-filter descent in image distances. *Biological Cybernetics*, *95*, 413–430.

Müller, M., & Wehner, R. (2007). Wind and sky as compass cues in desert ant navigation. *Naturwissenschaften*, *94*, 589–594.

Müller, M., & Wehner, R. (2010). Path integration provides a scaffold for landmark learning in desert ants. *Current Biology*, *20*, 1368–1371. doi:10.1016/j.cub.2010.06.035.

Philippides, A., Baddeley, B., Cheng, K., & Graham, P. (2011). How might ants use panoramic views for route navigation? *Journal of Experimental Biology*, *214*, 445–451.

Ronacher, B., & Wehner, R. (1995). Desert ants, *Cataglyphis fortis*, use self-induced optic flow to measure distances travelled. *Journal of Comparative Physiology. A, Neuroethology, Sensory, Neural, and Behavioral Physiology*, *177*, 21–27.

Ronacher, B., Westwig, E., & Wehner, R. (2006). Integrating two-dimensional paths: Do desert ants process distance information in the absence of celestial compass cues? *Journal of Experimental Biology*, *209*, 3301–3308.

Rybak, J., & Menzel, R. (2010). Mushroom body of the honey bee. In G. M. Shepherd & S. Grillner (Eds.), *Handbook of brain microcircuits* (pp. 433–440). Oxford, England: Oxford University Press.

Sakura, M., Lambrinos, D., & Labhart, T. (2008). Polarized skylight navigation in insects: Model and electrophysiology of e-vector coding by neurons in the central complex. *Journal of Neurophysiology*, *99*, 667–682.

Schwarz, S., Narendra, A., & Zeil, J. (2011). The properties of the visual system in the Australian desert ant *Melophorus bagoti*. *Arthropod Structure & Development*, *40*, 128–134.

Seidl, T., Knaden, M., & Wehner, R. (2006). Desert ants: Is active locomotion a prerequisite for path integration? *Journal of Comparative Physiology. A, Neuroethology, Sensory, Neural, and Behavioral Physiology*, *192*, 1125–1131.

Sommer, S., von Beeren, C., & Wehner, R. (2008). Multiroute memories in desert ants. *Proceedings of the National Academy of Sciences of the United States of America*, *105*, 317–322.

Sommer, S., & Wehner, R. (2005). Vector navigation in desert ants, *Cataglyphis fortis*: Celestial compass cues are essential for the proper use of distance information. *Naturwissenschaften*, *92*, 468–471.

Stieb, S. M., Münz, T. S., Wehner, R., & Rössler, W. (2010). Visual experience and age affect synaptic organization in the mushroom bodies of the desert ant *Cataglyphis fortis*. *Developmental Neurobiology, 70*, 408–423.

Strausfeld, N. J. (1999). A brain region in the insect that supervises walking. *Progress in Brain Research, 123*, 273–284.

Strausfeld, N. J. (2002). Organization of the honey bee mushroom bodies: Representation of the calyx within the vertical and gamma lobes. *Journal of Comparative Neurology, 450*, 4–33.

Strausfeld, N. J., Sinakevitch, I., Brown, S. M., & Farris, S. M. (2009). Ground plan of the insect mushroom body: Functional and evolutionary implications. *Journal of Comparative Neurology, 513*, 265–291.

Strauss, R., & Berg, C. (2010). The central control of oriented locomotion in insects—Towards a neurobiological model. *IEEE World Congress on Computational Intelligence 2010*, 3919–3926.

Vickerstaff, R. J., & Cheung, A. (2010). Which coordinate system for modelling path integration? *Journal of Theoretical Biology, 263*, 242–261.

Wehner, R. (1982). Himmelsnavigation bei Insekten. Neurophysiologie und Verhalten. *Neujahrsblatt der Naturforschenden Gesellschaft Zürich, 184*, 1–132.

Wehner, R. (1994). The polarization-vision project: Championing organismic biology. In K. Schildberger & N. Elsner (Eds.), *Neural basis of behavioural adaptations* (pp. 103–143). Stuttgart, Germany: G. Fischer.

Wehner, R. (1997). The ant's celestial compass system: Spectral and polarization channels. In M. Lehrer (Ed.), *Orientation and communication in arthropods* (pp. 145–185). Basel: Birkhäuser.

Wehner, R. (2008). The desert ant's navigational toolkit: Procedural rather than positional knowledge. *Journal of Insect Navigation, 55*, 101–114.

Wehner, R., Boyer, M., Loertscher, F., Sommer, S., & Menzi, U. (2006). Desert ant navigation: One-way routes rather than maps. *Current Biology, 16*, 75–79.

Wehner, R., & Labhart, T. (2006). Polarization vision. In E. Warrant & D.-E. Nilsson (Eds.), *Invertebrate vision* (pp. 291–348). Cambridge, England: Cambridge University Press.

Wehner, R., Meier, C., & Zollikofer, C. (2004). The ontogeny of foraging behaviour in desert ants, *Cataglyphis bicolor*. *Ecological Entomology, 29*, 240–250.

Wehner, R., Michel, B., & Antonsen, P. (1996). Visual navigation in insects: Coupling of egocentric and geocentric information. *Journal of Experimental Biology, 199*, 129–140.

Wehner, R., & Müller, M. (2006). The significance of direct sunlight and polarized skylight in the ant's celestial system of navigation. *Proceedings of the National Academy of Sciences of the United States of America, 103*, 12575–12579. doi:10.1073/pnas.0604430103.

Wehner, R., & Räber, F. (1979). Visual spatial memory in desert ants, *Cataglyphis bicolor* (Hymenoptera: Formicidae). *Experientia 35*, 1569–1571.

Wehner, R., & Srinivasan, M. V. (2003). Path integration in insects. In K. J. Jeffery (Ed.), *The neurobiology of spatial behaviour* (pp. 9–30). Oxford, England: Oxford University Press.

Wittlinger, M., Wehner, R., & Wolf, H. (2006). The ant odometer: Stepping on stilts and stumps. *Science, 312*, 1965–1967.

Wystrach, A., Beugnon, G., & Cheng, K. (2011). Landmarks or panoramas: What do navigating ants attend for guidance? *Frontiers in Zoology, 8*, 21. doi:10.1186/1742-9994-8-21.

Wystrach, A., Beugnon, G., & Cheng, K. (2012). Ants might use different view-matching strategies on and off the route. *Journal of Experimental Biology, 215*, 44–55.

Yu, S. E., & Kim, D. E. (2011). Landmark vectors with quantized distance information for homing navigation. *Adaptive Behavior, 19*, 121–141. doi:10.1177/1059712311398669.

Zeil, J. (2011). Visual homing: An insect perspective. *Current Opinion in Neurobiology, 22*, 1–9. doi:10.1016/j.conb.2011.12.008.

Zeil, J., Hoffmann, M. I., & Chahl, J. S. (2003). The catchment areas of panoramic snapshots in outdoor scenes. *Journal of the Optical Society of America. A, Optics, Image Science, and Vision, 20*, 450–469.

Zollikofer, C. P. E., Wehner, R., & Fukushi, T. (1995). Optical scaling in conspecific *Cataglyphis* ants. *Journal of Experimental Biology, 198*, 1637–1646.

81 Vision and Body Coloration in Marine Invertebrates

JUSTIN MARSHALL AND KAREN L. CHENEY

COLORS OF MARINE INVERTEBRATES AND THEIR BACKGROUNDS, AN INTRODUCTION TO THE SYSTEM

The thought of colorful marine invertebrates conjures up images of gaudy nudibranchs or brilliantly colored sponges. Perhaps a colorful flatworm gliding through a coral reef studded with starfish of various hues and seashells adorned with multicolor stripes and spots (see figure 81.1, Wicksten, 1989). These chromatic festivals of our imagination do exist, but generally only in a limited number of places in shallow seas. This chapter starts with an examination of selected marine invertebrates, their colors, and where they live but then returns to the neuroscience of seeing, processing, and behaving relative to these colors. It is the physics and chemistry of light and color underwater that sets the limits of function and performance of the visual systems described (Lythgoe, 1979).

The Ocean Background—A Critical Factor in Determining Colors

Terrestrial life is dominated by green and brown with a backdrop of blue or gray sky, and against this the occasional colorful animal such as a parrot or poison dart frog may appear (Lythgoe & Partridge, 1989). Much of the ocean is also mostly green and brown with a backdrop of blue or gray, depending on the sea or ocean clarity (Jerlov, 1976). As with the land, green underwater is largely the result of reflection from chlorophyll, but along with the green of growing objects such as seaweed (algae), water itself may appear green due to suspended algal cells. Even rocks or coral are largely green or at least greenish brown due to their algal content or coating. Water may also contain other dissolved organic matter and inorganic sediments, contributing to its variable coloration and making it much more opaque and murky than air (Jerlov, 1976). As a result, for visual systems operating underwater, the water itself may be a local backdrop to camouflage or display against.

Corals, and indeed other marine invertebrates such as anemones, jellyfish, nudibranchs, and even giant clams, also contain within their tissues symbiotic algae, sometimes termed zooxanthellae. This tightly linked algal–animal relationship is critical for life underwater, and the colors of many aquatic animals are brown, gold, or green, resulting from their algal inhabitants' chlorophyll content or its porphyrin derivatives. Algae provide carbohydrate-based nutrients and receive protection and proteins in return.

Many marine invertebrates color themselves to be camouflaged against their green and brown background, to avoid detection for defense or surprise attack (Cott, 1940; Stevens & Merilata, 2011). In the relatively clear open ocean, matching a brown/green background is replaced by transparency for camouflage, or, for a few near the ocean surface—such as the colonial hydrozoan *Porpita* and its predatory nudibranch, *Glaucus*—matching the brilliant violet–blue of the open ocean surface (Marshall & Johnsen, 2011). Transparency in particular is a common strategy in midwater oceanic invertebrates as there is rarely anything to hide behind or in front of.

CAMOUFLAGE OR COMMUNICATION AND SOME EXAMPLES OF INVERTEBRATE COLOR The reef scene in figure 81.1e is at around 10-m depth and shows apparently colorful invertebrates such as soft and hard corals, crinoids, and ascidians, but the colors here are not accurate, being enhanced with flash photography. In reality, whether viewing over depth or looking at objects horizontally through a body of water, the combined absorbance and scattering of light strips away either end of the spectrum (see figure 81.2a), making red objects dull and removing some of the short wavelengths also (Lythgoe, 1979).

Objects in very shallow waters viewed at close range may display the full spectrum of vibrant colors provided by encrusting invertebrates such as sponges. It is also notable that many motile marine invertebrates are strikingly colored, some taking on the brilliant colors of the background sessile invertebrates upon which they live

FIGURE 81.1   The marine substrate and colorful marine invertebrates. (a) Sacoglossan sea slug *Cyerce nigricans*. (b) Nudibranch, *Chromodoris kuniei*. (Photograph: Luke Gordon.) (c) Blue-ringed octopus, *Haplochlaena lunulata*. (Photograph: Roy Caldwell.) (d) *Octopus cyanea*. (e) A coral reef scene at around 5-m depth. (f) A less colorful area of reef with a scorpion fish (*scorpaenidae*) camouflaged as rock and encrusting invertebrates. As if to complete its own disguise, the fish allows a conspicuous nudibranch (*Nembrotha* sp. arrowed) to glide over it as if it were a rock. (g) Polyclad tiger flatworms (*Pseudobiceros dimidiatus*). (h) Another scorpaenid fish (see f) apparently camouflaged as red encrusting sponges. (Photographs: Steve Parish unless otherwise stated.)

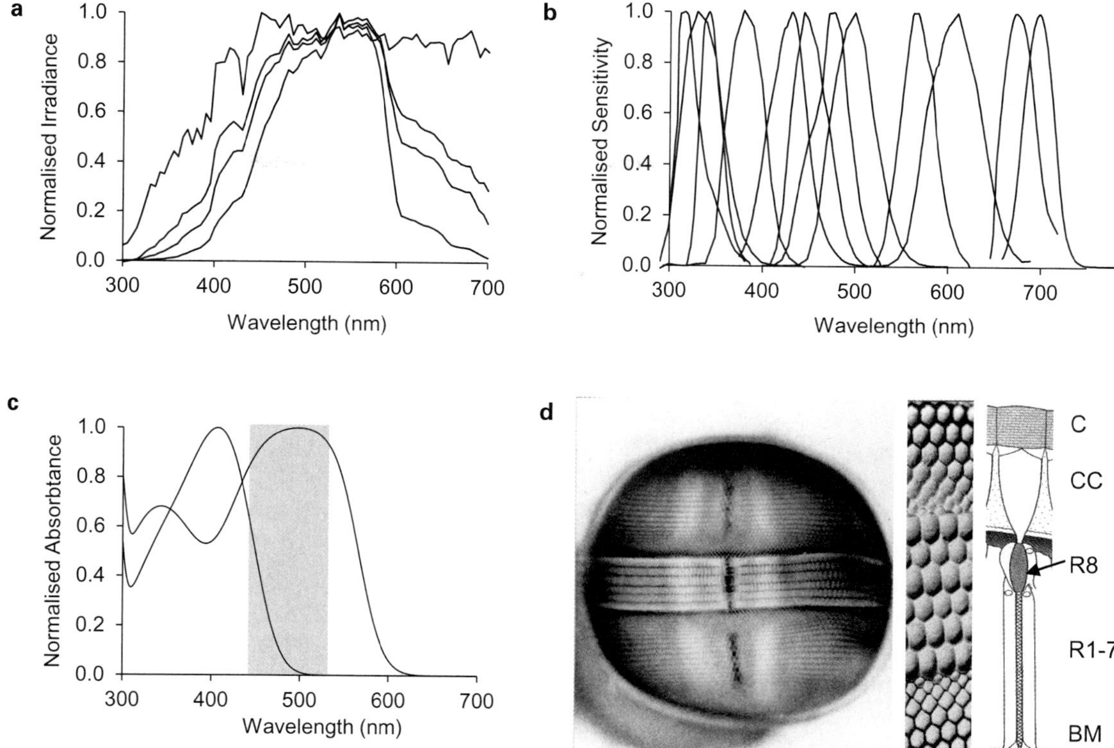

FIGURE 81.2 Spectral light distribution and spectral sensitivities in marine invertebrates. (a) Down-welling irradiance in waters close to tropical reefs at 0.1-, 3-, 6-, and 10-m depth, the curves narrowing around 500 nm (Marshall et al., 2003). (b) The twelve electrophysiologically determined spectral sensitivities of the stomatopod crustacean, *Neogonodactylus oerstedii*, most likely used for color vision. (c) The two spectral sensitivities of mesopelagic shrimp, *Systelaspis debilis*, a typical red crustacean from these depths. The shaded box encloses the region of spectral emission peaks of most bioluminescence in the ocean. (d) Left: the stomatopod apposition compound eye of *Odontodactylus scyllarus* (figure 81.4c). Note the central midband and the flanking hemispheres. Middle: expanded section of cornea of stomatopod eye showing six midband rows of ommatidia. Right: diagram of typical crustacean ommatidium in longitudinal section: C, cornea; CC, crystalline cones; R8, photoreceptor retinular cell 8; R1–7, retinular cells 1–7 comprising the main rhabdom; BM, basement membrane separating the retina from the lamina, the first optic neuropil.

(see figure 81.1). In this instance it is tempting to assume that these vibrant colors match the vibrantly colored background for camouflage, but colors may have several functions including those with no relevance to vision.

The cephalopods, including octopus and cuttlefish (see figures 81.1 and 81.3) are famous for their camouflage ability, but they may also flash intense colors and shades such as red, black, or white to communicate with each other. The blue-ringed octopus, *Haplochlaena lunulata*, flashes its blue rings in warning defensive mode (see figure 81.1c), but the blue rings may be "turned off" when crypsis is required, similar to *Octopus cyanea*, master of camouflage against the reef, employing both colors and texture to match its background (see figure 81.1d). The blue rings in several *Haplochlaena* species are highlighted against yellow chromatophores, and this striking yellow and blue combination warns of their toxic, tetrodotoxin-laden bite. Such intentionally conspicuous coloration that warns of noxiousness of one form or other is known as aposematic coloration (Cott, 1940; Ruxton, Sherratt, & Speed, 2005).

Yellow and blue is a color combination shared by several nudibranch, flatworm, crustacean, and other invertebrate species as well as marine fish (see figures 81.1 and 81.4; Marshall 2000a, b). It is specifically effective underwater as these are the colors that transmit or stand out best in the ocean (Lythgoe, 1979; Marshall & Vorobyev, 2003). It is notable that both the blue-ringed octopus and many other nudibranchs, including *Chromodoris elisabethina* (figure 81.4a) and *C. kuniei* (figure 81.1b) also utilize black and white to enhance the edges of such presumably conspicuous displays. Yellow and blue are complementary colors in the spectrum and are powerfully contrasting to many and indeed to probably the first color vision systems (Hurlbert, 1997). The yellow–blue psychophysical axis is evolutionarily basal in primates (Jacobs, 1993) and is similar in its constituent parts to almost any dichromatic color vision system that possesses a short-wavelength and a long-wavelength photoreceptor. It seems that many marine (and terrestrial animals) exploit this relatively ubiquitous behavioral flag and neural pathway either for aposematism or other conspicuous communications.

The sacoglossan sea slug *Cyerce nigricans* (figure 81.1a) employs conspicuous aposematic colors, to warn of noxious chemicals (in this case pyrone toxins) or unpalatability. If attacked, *C. nigricans* will drop its "leaves" or cerata, and these wiggle like a lizard's tail as distracters. Another mollusc, the nudibranch, *Chromodoris kuniei* (figure 81.1b), is covered in apparently conspicuous spots, ringed in white and black possibly for extra contrast. This strategy of edge accentuation is used in many animal patterns to increase conspicuousness (Cott, 1940, and see figures 81.1c and 81.4a).

Relatively drab areas of green and brown ocean bed are more common per unit area than the apparently colorful reef areas we see in books (see figure 81.1e). Even what appears to be rock is often "living rock" with dull-colored encrusting sponges and coralline algae and typically brown hard corals. In the photograph of the ocean floor in figure 81.1f, a scorpion fish imitates the colors and patterns of these dull invertebrates, matching the background perfectly. So apparently convinced of its camouflage, the fish allows a conspicuous, aposematically colored nudibranch (*Nembrotha* sp., arrowed) to glide over its surface. Some scorpaenid fish and other sit and wait predators, such as the shallow water anglerfish (Marshall & Johnsen, 2011), also camouflage themselves as encrusting marine invertebrates but have instead selected, through evolution, to imitate conspicuous red or yellow sponges (see figure 81.1h). The effectiveness of this disappearing trick argues strongly for the necessity to be a conspicuous color in a colorful background rather than just ordinary brown or green benthic colors. By being the color of a chosen host sponge, camouflage is maintained at any depth (see figure 81.1).

The functions of colors may also have nothing to do with camouflage or communication but may be a byproduct of diet, food storage, and other metabolic processes, related to temperature control or shading from the sun (Fox, 1976). As humans, we are easily distracted by pretty things and like to assign purpose to the colors; however, a bright orange sponge that lives under a rock or a bright red shrimp that lives in the inky depths of the ocean may have evolved that color for nonvisual reasons.

Colors may be used for different tasks in different circumstances, and this makes determining their function complex. The polyclad tiger flatworms (*Pseudobiceros dimidiatus*) in figure 81.1g move from one background against which they may be conspicuous to a potentially camouflaging substrate and illustrate these uncertainties well. Understanding such phenomena requires the care of observational biology—which is no longer fashionable—and a case-by-case consideration, rather than taking a flash photograph and attempting to fit this observation into an ill-defined overall trend.

Color Change

Many organisms do not just use color passively but may change over several different time scales, ranging from subsecond through diurnal, seasonal, or longer-term ontogenetic changes that may be linked to a change in habitat. The different circumstances inducing color change may be associated with hiding or sexual display or functions that are not visually related, such as heat regulation (Fox, 1976). Fiddler crabs are known to regulate their color and conspicuousness against the background, apparently in response to increased predatory threat and/or temperature control on the hot mudflats (Hemmi et al., 2006). Both decapod prawns and stomatopods can regulate the colors of their carapace to match different local backgrounds, a trick also shared by many crustaceans that inhabit specifically colored encrusting invertebrates (Wade et al., 2012). These color changes may be performed over hours by hormonally controlled chromatophore expansion and contraction or over several moult cycles as in stomatopods. There are several examples of colors changing on a seasonal basis before migrations in the lobster, for example (Wade et al., 2008) or apparently as part of mating displays. Examples of this in crabs are discussed below.

The most remarkable control of color change in marine invertebrates is among the cephalopods that can perform almost instantaneous background matches using color, intensity pattern, and texture (see figures 81.1d and 81.3g). Notably, this is achieved despite almost all cephalopods being color blind as they possess only a single visual pigment resulting in a single spectral sensitivity throughout the eye (Marshall & Messenger, 1996). As with obvious cephalopod communications (see figure 81.1c), the chromatic content here is directed toward other animals and likely predators, such as fish and crustaceans, that may be able to see or be fooled by the camouflaged body coloration.

Of course, luminous intensity, or black and white and pattern components of the rapid color changes in cephalopods, may be directed at other cephalopods, and it is notable that many cephalopod displays that are used

    JUSTIN MARSHALL AND KAREN L. CHENEY

in a social and sexual context are largely conducted in black and white (Hanlon et al., 2007). The "passing cloud" agonistic display of *Sepia apama*, the giant Australian cuttlefish, is a good example (figure 81.3h), while the dazzling pulsating display of a hunting broad-club cuttlefish is actually designed to confuse its crustacean prey (see "The Brainy Bunch," ABC TV documentary, 1997).

VISION AND COLOR VISION UNDERWATER

Compared to terrestrial systems, vision, and specifically color vision, is less well developed in the marine invertebrates. In this environment, invertebrates are spread over more phyla than on land, and many possess rudimentary eyes, often without lenses (Land, 1981). Of those investigated, the majority possess a single spectral sensitivity (e.g., 60 out of 86 crustaceans detailed in Marshall, Kent, & Cronin, 1999) although this sample may be skewed by the large number of deeper-living species examined. As a result, color vision is less important for marine invertebrates than for their terrestrial counterparts. It is therefore likely that much of the diversity seen in marine invertebrate colors has evolved for communication to other animals such as the fish that share or look into their watery habitat and that do possess color vision.

Two examples of crustaceans that do have a color sense are discussed below, the spectrally obsessed stomatopods with 12 color channels and sexual communication with two color channels in crabs. It is the crustaceans, among the marine invertebrates, that have gone some way toward color vision, several possessing a green (around 500 nm) and a UV/violet spectral sensitivity (around 400 nm) that could allow for dichromacy (12 out of 86 in Marshall, Kent, & Cronin, 1999). One beach-dwelling isopod, *Ligia exotica*, has been shown to possess three spectral sensitivities (340 nm, 460 nm, and 520 nm; Hariyama, Tsukahara, & Meyer-rochow, 1993), perhaps providing trichromacy in this largely out-of-water, spectrally broad world.

Color Patterns, Spatial Resolution, and the Perception of Too Much Detail

Light never travels far underwater due to being scattered from particles and water molecules. This is one reason for the relatively poor spatial resolution in most aquatic animals, even among the fish with eyes essentially identical to land vertebrates, as there are no distant objects to be resolved. Fish viewing marine invertebrates possess spatial resolution usually around 10 times worse than humans (with some exceptions;

Marshall, 2000a). Our human-biased, large terrestrial visually oriented animal approach probably leads to overinterpretation of many of the patterns and colors of marine animals, and a nonsubjective view is essential. Among the invertebrates, resolving power is highly variable, ranging from those that just see large shadows (5°–30°) due to deliberate low-pass-filtering defocus or indeed no lens element in the eye at all—to the cephalopods, some of which are comparable to the best vertebrates (see Land, this volume, and Land, 1988).

Crustaceans possess compound eyes with a spatial resolution that is necessarily lower than the "simple" single-chamber eyes of cephalopods. Only very few crustaceans, the stomatopods (Marshall & Land, 1993) and some land-based crabs (Zeil, Nalbach, & Nalbach, 1986), approach the 0.25° resolving capability of the best insects such as the dragonfly. In general, many crustaceans are also at least 10 times worse than this (Land, 1984). What this means for colors and color vision underwater is that, even viewed close up, the fine spots and stripes of color exhibited by several marine creatures are blurred, and this may result in a combined color that is in fact cryptic rather than conspicuous (Marshall, 2000a).

The Spectral Envelope and Visual Ecology Underwater

The colors yellow through red are conspicuous, and, due to their similarity to ripening fruit, visually significant to human and primate vision. This relevance is reduced in the ocean realm, partly because of the attenuation of long wavelength photons below a few meters and partly because, as a corollary of this, many (but not all) marine visual systems lack long-wavelength sensitivities (see figure 81.2; Marshall, 2000b). As the spectral envelope closes in around 475 nm, the best transmitted wavelength in clear ocean water, or other transmission maxima, depending on the water color (figure 81.2a is for in shore reef water, which transmits best around 550 nm), visual systems respond by modifying their spectral sensitivities. This is one of the best-known examples of visual ecology (Lythgoe, 1979). Essentially, unless there is other light provided through, for example, bioluminescence, there is no point in being sensitive to wavelengths at which there are few or no photons. Light environments on land are much less variable, although local background may play a part in defining visual systems (Endler, 1993), but there is not the same ecological restriction on spectral sensitivity there.

Not surprisingly, therefore, the spectral sensitivities in most marine invertebrates (just as in fish; Lythgoe,

FIGURE 81.3 Signals and camouflage in cephalopods and crustaceans. (a–d) Stomatopod crustaceans all exhibiting a "meral spread" display that exposes the species-specifically colored meral spot—white, orange, magenta, and blue from a–d. Species are, respectively, *Neogonodactylus wennerae, Gonodactylus glabrous, Gonodactylus smithii* (deep phase), and *G. smithii* (shallow phase), the color differences between these two matching the local habitat colors. (Photographs: Roy Caldwell.) (e) Fiddler crab (males) *Uca elegans* fighting. (Photograph: Martin How.) (f) The seasonally red cheliped of the female blue crab *Callinectes exasperatus.* (Photograph: Sonke Johnsen.) (g) The giant Australian cuttlefish *S. apama* in camouflage mode. (h) *S. apama* showing an agonistic display to another male, largely in black and white. (Photographs: Chris Talbot.)

1979) generally match the light available, whether this be from light in water or bioluminescence (Lythgoe, 1980). Crustaceans that possess a single spectral sensitivity have a single rhabdom, or photoreceptor, built from a number of cells containing blue- or green-sensitive visual pigment. Several crustacean eyes also possess a second class of photoreceptor, the R8 (rhabdomere number 8) photoreceptors, and it is these cells that carry UV or violet visual pigments (see figure 81.2). R8s are distally placed relative to the main rhabdom, which is usually constructed from seven rhabdomeres, the R1–7 cells (Eguchi & Waterman, 1973). It is likely that color discrimination in crustaceans, where present, is built around this potentially dichromatic UV/violet—blue/green comparison.

The Function of Colors in Marine Invertebrates: Three Examples

Few well-substantiated examples of color vision are known in marine invertebrates outside of the crustaceans. As we noted initially, nonanthropomorphic interpretations of marine invertebrate colors are rare, and it seems that many invertebrate colors have evolved to be visible to fish and other vertebrates. Here three examples of color communication are detailed briefly. The first two, sexual and agonistic displays in crustaceans, are colors produced by invertebrates for their own species, the last a hypothetical model of communication from nudibranchs to fish color vision.

SEXUAL COMMUNICATION IN CRABS Fiddler crabs are semi-terrestrial species, the males of which wave large colorful chelipeds both in male–male competition and to attract females (see figure 81.4e). Because these displays are in air, there is no spectral restriction to potential displays, and it is notable that many claw colors are yellow, orange, or red, potentially taking advantage of this and also providing strong contrast against sky or vegetation background (Hemmi et al., 2006). Color vision in these engaging flag-wavers has been proposed several times but only recently demonstrated behaviorally with some of the appropriate controls. *Uca mjoebergi* and *Uca capricornis* identify gender and neighboring individuals based on learned color patterns (Detto, Hemmi, & Backwell, 2008). Mate choice in *U. mjoebergi* is influenced by claw color (Detto, 2007), and as with other crustaceans (aside from stomatopods) this color discrimination is most likely mediated by R8/R1–7 dichromatic mechanisms. Spectral sensitivity in fiddler crabs is difficult to measure, but Horch and colleagues (2002) found evidence for a two-visual-pigment system (one at 430 nm and another between 500 and 540 nm)

in *Uca thayeri*, using extracellular recordings. Jordão, Cronin, and Oliveira (2007) found main rhabdom (R1–7) sensitivities varying from 508 and 530 nm in different species using microspectrophotometry but could not measure the small R8 cells that presumably would contain the violet sensitivity found in many crustaceans. These authors also note that the actual spectral sensitivity of the rhabdom will be influenced by different screening pigments in a manner similar to some of the stomatopod systems (Marshall, 1988).

What remains lacking are clear intracellular spectral sensitivity measurements in any fiddler crab species. Recent molecular evidence suggests three different opsin genes in the sand fiddler crab, *Uca pugilator*, with some coexpression of two genes in one cell (Rajkumar et al., 2010). While this study points toward the potential for trichromacy in this species and clear identification of an R8 population and two different opsins in the R1–7 cells, the spectral sensitivities of this system are yet to be confirmed.

Fiddler crabs are also able to change color relatively rapidly and do so in response to increased predation pressure and other stressors (Hemmi et al., 2006). However, it is not clear if these color changes are significant to the crab's color vision system or more directed at the visual systems of potential predators such as birds. The hypothesis put forward by Detto, Hemmi, and Backwell (2008) is that pattern changes are related to individual recognition, but this may rely on the pattern and not the color component of the signal.

Blue crabs are also known for their colorful agonistic displays. The bright blue, very sharp claws (chelipeds) of males are used to frighten would-be attackers in both inter- and intraspecific specific displays. In clear marine waters, blue is a good color for long-distance transmission, but it is the red claws of females that have received recent attention. Mate choice behavior, in which male *Callinectes sapidus* show a preference for females that exhibit red claws, has recently been documented (Baldwin & Johnsen, 2009). Controls demonstrate that this choice is based on chromatic cues and demonstrate color vision in this species, most likely based on opponent interaction between their known R8–440-nm, R1–7–508-nm spectral sensitivities. Further experiments (Baldwin & Johnsen, 2012) suggest that luminance is also important for male choice of females in this species, indicating that multiple cues guide mate choice as is known to occur in some insect species (Kelber & Osorio, 2010). Presumably, both redness and luminance are somehow correlated with fecundity in females, but the details of this are not yet known except for the fact that the colors change with sexual maturity. Blue crab claw color may also be used in species recognition, as there are around 16 species showing different color patterns (Baldwin & Johnsen, 2012).

The use of a red signal for close-range sexual interaction is interesting in this instance as red will be an effective signal close up but will rapidly attenuate in marine waters over distance, preventing potentially unwanted eavesdropping. Furthermore, red will contrast strongly against a blue or green water background, adding to the effectiveness of the signal. A significant problem, however, is that the information content in such a red signal will degrade rapidly with depth, and this may also explain the concomitant reliance on luminance as well.

INTRASPECIFIC COMMUNICATION AND AGGRESSION IN STOMATOPODS Stomatopod crustaceans from several families possess the potential for complex color vision with 12 spectral sensitivities, all situated in the top four rows of the six-row midband region of the eye (figure 81.2d; Cronin & Marshall, 1989; Marshall, 1988). Behavioral evidence shows that stomatopods are indeed capable of color vision (Marshall, Jones, & Cronin, 1996) as well as both linear and circular polarization vision using other eye regions (Chiou et al., 2008). These behavioral tests, however, do not indicate what stomatopods use their color vision for, and while it is tempting to suggest that they could use it for whatever they like, a look at their color signaling systems suggests some concrete possibilities.

Stomatopods are sometimes colorful animals (see figures 81.3 and 81.4). The majority of their external markings, when viewed in context, are better classified as camouflage. However, and in common with fiddler crabs, stomatopods also employ species-specific markings, the meral spots (see figure 81.3) These are uncovered and shown off in stereotyped displays known as "meral spreads" (Caldwell & Dingle, 1976). Meral spots are named for their position on the inner surface of the meral segment of the large raptorial appendages that stomatopods use to spear and batter conspecifics, prey, and potential predators (Caldwell & Dingle, 1976). Stomatopods are highly aggressive animals, and meral spread displays are often given prior to combat, apparently as a warning behavior. Meral spots may be colored, white, yellow, red, magenta, or purple and are often accentuated with a white surround (see figure 81.3). The two meral spots have been suggested to resemble eyes, so that the meral spread is seen as a revealing of startling eyes to scare would-be attackers. Intriguingly, different species show different levels of aggression, and it has also been suggested that different colored meral spots are used by stomatopods as a label of which species to avoid and which to potentially combat (Caldwell & Dingle, 1976).

Recent studies of stomatopod colors also suggest that they may be adapted under different light conditions and in different modalities for interspecific and sexual selection (Cheroske & Cronin, 2005). However, given the complexity of the colors and the visual system of stomatopods (see figure 81.2) along with the color complexity of the reef-top habitat that many species inhabit, there are many questions remaining. Certainly the nature of the spectral reflection from even the most colorful of stomatopods (see figures 81.3 and 81.4) still does not explain the complexity of their potential color vision system.

COMMUNICATION TO THE VERTEBRATES, APOSEMATISM IN NUDIBRANCHS Many colorful marine invertebrates use their bright and conspicuous coloration to warn predators that they are toxic or unpalatable. This phenomenon is known as aposematism (Ruxton, Sherratt, & Speed, 2005). Sponges are toxic to most fish and contain sharp spicules, ascidians can release sulphuric or tunic acid, and cnidarians (anemones and corals) possess stinging cells. Nudibranchs and flatworms can be toxic, and some can even borrow the stinging cells from the cnidarians they eat as a means of defense (Cortesi & Cheney, 2010). Nudibranchs are a well-known example of presumed aposematic coloration, but much of this hypothesis is based on supposition. We do not know how predators learn the colors of aposematic animals although some color themes such as black and yellow are said to be an innate "beware" signal for many animals (Ruxton, Sherratt, & Speed, 2005). Very few nudibranchs have been quantified for their toxicity or inedibility (but see Cortesi & Cheney, 2010), and indeed many are very well camouflaged (see figure 81.4e and f). Recent studies have begun to quantify the colors and relative toxicity of different nudibranch species in order to examine whether colorful and clearly visible nudibranchs are more toxic than camouflaged species. One evolutionary hypothesis predicts that colorful nudibranchs evolved from an ancestral stock of camouflaged species that achieved defense through crypsis rather than aposematism. By increasing their levels of toxicity at the same time as increasing the colorfulness of their warning signals, nudibranchs were able to avoid being predated on as they became more noticeable to potential predators.

In attempting to understand aposematism, it is important to determine which eyes the colors of nudibranchs evolved for, and one simplifying factor is that they do not see their own colors as their vision is rudimentary or nonexistent. As a result, we can exclude

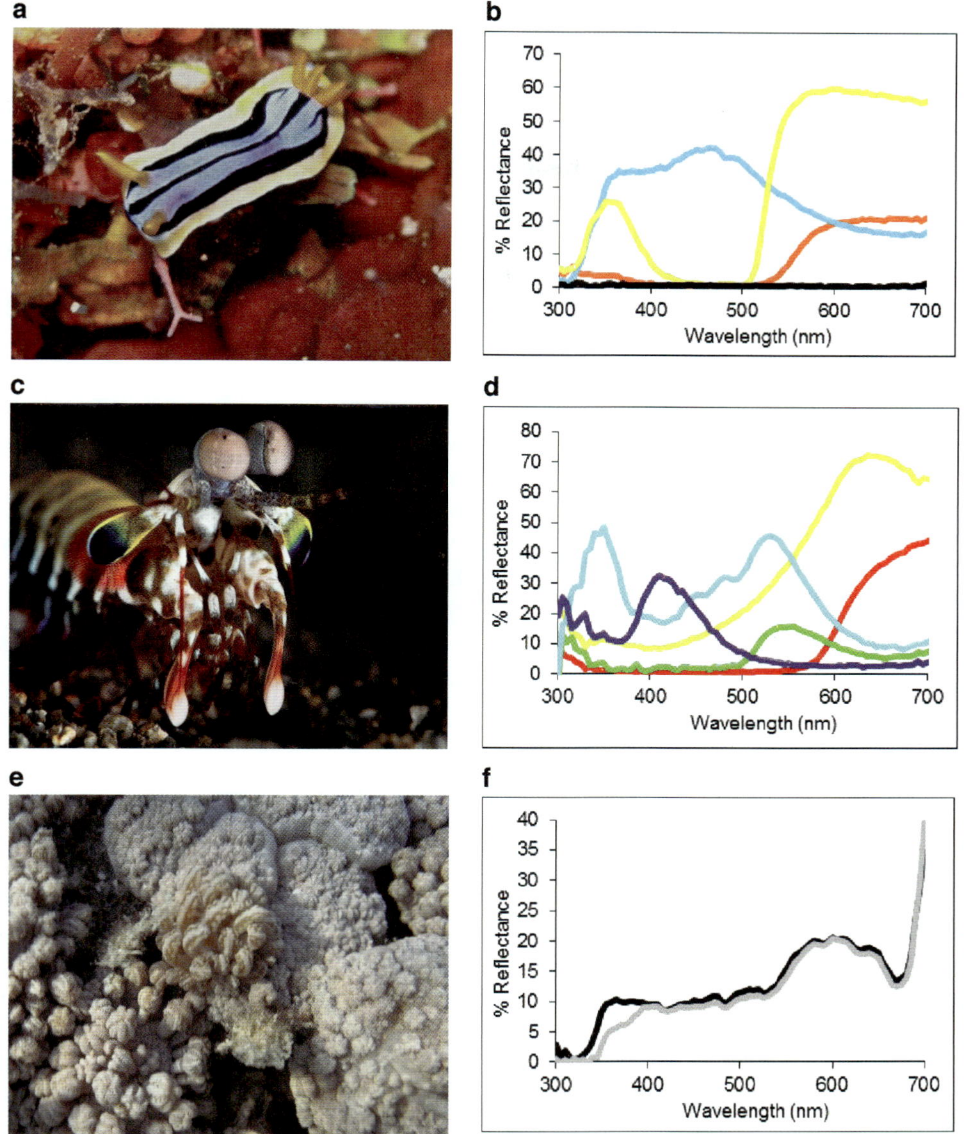

FIGURE 81.4 Spectral reflection of conspicuous and camouflaged invertebrates. (a) The nudibranch mollusc, *Chromodoris elisabethina*. (Photograph: Luke Gordon.) (b) Spectral reflectance of the colors of *C. elisabethina*. The colors of the lines match the image in (a), for example, the orange gills and horns. (c) Stomatopod crustacean, *Odontodactylus scyllarus*. (Photograph: Roy Caldwell.) (d) Spectral reflectances of the colors of *O. scyllarus*. Line colors approximately match body colors. (e) The cryptic nudibranch, *Phyllodesmium lizae*, on a soft coral. (f) Spectral reflectances of the colors of *P. lizae* (black line) and soft coral on which it is found (gray line). (Photograph: Karen Cheney.)

sexual selection or any other conspecific visual process as an evolutionary driver for their colors. As slow moving, benthic animals that lack the protective shells of their close gastropod neighbors, nudibranchs might make an easy meal for fish. Here we model what the nudibranch *C. elisabethina* would look like to the color vision system of a known trichromatic triggerfish, *Rhinecanthus aculeatus*, and for comparison, a colorful stomatopod, *O. scyllarus*, seen by the same fish figure 81.5 (Pignatelli et al., 2010). *R. aculeatus* is an inquisitive, benthic omnivorous feeder on coral reefs and

associated shallow areas and could encounter both nudibranch and stomatopod and identify either as a source of food.

The model we use is similar to others previously described (Kelber, Vorobyev, & Osorio, 2003; Vorobyev & Osorio, 1998). In brief, having measured spectral reflectance from different areas of an animal (see figure 81.4), the resultant spectra are then plotted in a color vision space (see figure 81.5b and c). In order to do this, the spectral sensitivities of the observing animal, in this case a trichromat with short- (S), medium- (M), and long- (L) cone sensitivities (see figure 81.5a), the illuminant or irradiance at the object (here assumed surface waters, figure 81.2a, 0.1-m irradiance), and anatomical features of the retina, such as comparative cone number, are needed. The spectra are plotted as dots in the weighted color space in which each corner essentially represents maximum excitation for its (S, M, or L) spectral sensitivity. Therefore as we see here, in simple terms, red spectra tend to plot near the L corner and blue spectra near the S corner (Kelber, Vorobyev, & Osorio, 2003).

The distance between dots representing spectra can then be correlated, up to a point, with the possibility of distinguishing the spectra from one another. Importantly, as this distance increases past discrimination threshold, larger dot separation is not necessarily correlated to increased discrimination so such models operate best where colors match or differ slightly. Notable from this exercise are the following points:

1. Not surprisingly, the obviously cryptic color of nudibranch *Phyllodesmium lizae* (see figure 81.4e and f) is a good match to its local background. As the spectra of animal and background are so congruent (see figure 81.4e and f), this would be similar for any color vision space.

2. The pale blue of both the nudibranch and stomatopod, while conspicuous to our visual system, lies close to the achromatic point (triangle center) for *R. aculeatus*, about which many of the background spectra also cluster. This color is therefore effectively more like "gray" to the fish and well camouflaged in the background colors.

3. The yellow and orange nudibranch colors are chromatically highly conspicuous to the fish color vision system against the background as they lie far from the background dots and toward the edge of the triangle (figure 81.5b). Red and yellow sponge colors (such as those in figure 81.4a) are included in the background spectra plotted so their colors are significantly less saturated to the visual system of the fish than the nudibranch colors.

4. Aside from one of the blues and red, the stomatopod colors are close to the cluster of background colors also plotted, suggesting effective camouflage against the colors of the bottom (figure 81.5c).

5. Particularly notable and unusual for human vision, the yellow of the stomatopod *O. scyllarus* is an excellent match to the average reef background (brown circle in figure 81.5c), suggesting this yellow is well camouflaged against this background (see similar outcomes for reef fish yellows in Marshall, 2000a, 2000b; Marshall & Vorobyev, 2003).It is quite different from the highly saturated yellow of *C. elisabethina* (figure 81.4b and figure 81.5b).

These conclusions demonstrate the importance of describing the potential conspicuousness or crypticity of animals in the eyes of potential predators and not through human vision. Other reef fish have different visual systems (Losey et al., 2003; Lythgoe et al., 1994) and may see these spectra differently. Also, not just the colors but the color patterns should be taken into account in such visualizations through other animal eyes (Endler et al., 2005). For example an estimate of what a fish such as *R. aculeatus* would actually see when viewing a benthic scene at 15 m, including a nudibranch, is illustrated in figure 81.5d and e. Here we take into account not specifically the color vision of the fish but the illumination weighting at this depth and spatial resolution of the fish (around 10 times worse than human main retina). Although a subjective estimate, this clearly shows that red objects at this depth are likely to appear dark or blue and that, at least from this distance, the spatial detail that our human-biased camera systems provide is unrealistically high.

CONCLUSIONS

With a few exceptions, most marine invertebrates have rudimentary vision, and few have color vision. Crustaceans are the clearest exception, but even here, monochromacy is the norm, and dichromacy is found in some. Stomatopods, however, stand out as an extreme counterexample. These crustaceans may achieve color perception using entirely novel principles that have not been evolved by any other living creature known so far, despite 400 million years of evolution (Manning, Schiff, & Abbott, 1984).

It has been suggested that color vision evolved underwater to remove intensity fluctuations that might result from wave flicker, which might cause confusion between luminance changes and object movements in a simple nervous system (Maximov, 2000). Two opponent channels are all that is required for eliminating susceptibility to flicker. In this context, it is notable that many crustacean and fish species seem to have converged on two spectral sensitivities, even among species that live close to the surface. Stomatopods and several species of reef fish, or indeed other freshwater fish, clearly provide exceptions, but two spectral sensitivities may be the basal norm for color sense in the sea.

The perhaps surprising lack of color vision in the ocean may reflect the nature of attenuation of light in water, and, specifically, the information content in signals. Animals that signal with color may convey

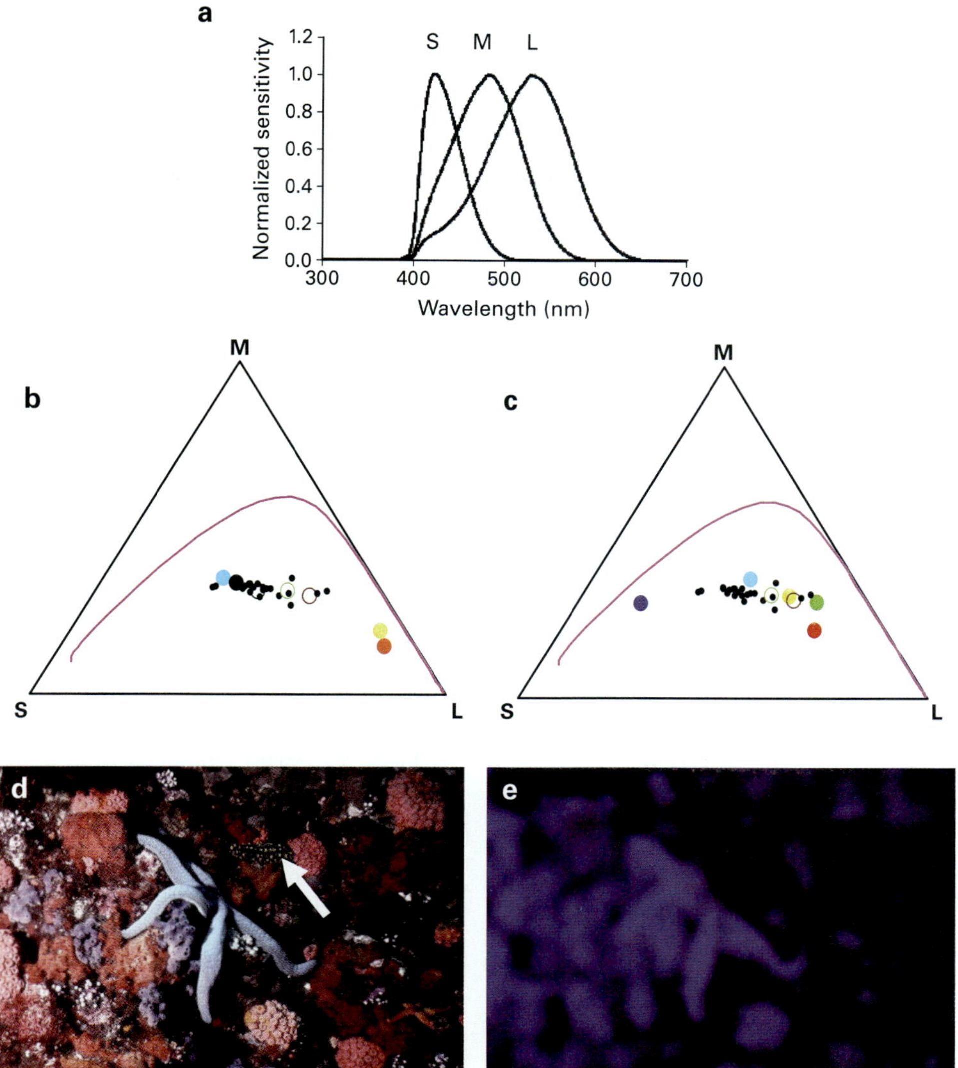

FIGURE 81.5 Fish-eye views of colorful invertebrates. (a) The spectral sensitivities of the three cone photoreceptors in the triggerfish *Rhinecanthus aculeatus* (S, M, L: short-, middle-, and long-wavelength sensitivities), a shallow water benthic invertebrate feeder around reefs. Note that these are also filtered by the ocular media (lens and cornea) transmission (Marshall et al., 2003). (b) Color vision space of *R. aculeatus*. Pink line is the monochromatic locus for the visual space. Black dots: the positions of 21 background benthic features such as corals and sponges (some would be called colorful) in the visual space. Solid colored dots: the visual space positions of the spectra of nudibranch *C. elisabethina*, color coded as in figure 81.4 a, b. Brown circle: position of an average of 255 coral spectra (Marshall, 2000a). Green circle: position of an average of 30 algal spectra. Black circle: position of the color spectrum of the cryptic nudibranch, *P. lizae*, figure 81.4 e, f. The spectrum of its background plots as the dot in the lower right of the black circle. See text for further details. (c) As (b) but for colors of the stomatopod, *O. scyllarus* (figure 81.4 c, d). (d) Photograph of a benthic scene of the sort that *R. aculeatus* might hunt in using flash illumination. Many invertebrates are present including a nudibranch, *Phyllidiella* sp., arrowed. (e) Same photograph as in (d) but a more accurate representation of the scene seen at 10 m where little red light remains and through the 10 times worse than human resolution of the eyes of *R. aculeatus*. (Photograph: Steve Parish.)

precise information about fecundity, territory, or strength with colors, and if this message is degraded, as is the case over even relatively short distances or depths in water, the signaling system breaks down. This simple fact may have prevented the proliferation of the colors and color vision systems that we see on land, such as the flowers and their various pollinators such as bees. Both the cephalopods and many of the crustaceans seem to have opted for polarization vision in place of color (Marshall, Kent, & Cronin, 1999), and this is a communication system we are only just beginning to discover (Pignatelli et al., 2011; Temple et al., 2012).

The polarization content of objects underwater is more reliable at different depths, and perhaps this is the prevalent information currency in the ocean?

Finally, motion has hardly been mentioned in this chapter, but it is often linked to color signals, such as the claws of crabs, meral spread of stomatopods, or the wave of nudibranch fringe (see figure 81.1b). Many visual systems are exquisitely tuned to motion and indeed motion differences (see chapters 54 and 56), and in any full system description of invertebrate colors and color vision, it is also important to consider how the color is presented in time.

REFERENCES

Baldwin, J., & Johnsen, S. (2009). The importance of color in mate choice of the blue crab *Callinectes sapidus*. *Journal of Experimental Biology, 212*, 3762–3768. doi:10.1242/jeb. 028027.

Baldwin, J., & Johnsen, S. (2012). The male blue crab, *Callinectes sapidus*, uses both chromatic and achromatic cues during mate choice. *Journal of Experimental Biology, 215*, 1184–1191. doi:10.1242/jeb.067512.

Caldwell, R. L., & Dingle, H. (1976). Stomatopods. *Scientific American, 234*, 80–89.

Cheroske, A. G., & Cronin, T. W. (2005). Variation in stomatopod (*Gonodactylus smithii*) color signal design associated with organismal condition and depth. *Brain, Behavior and Evolution, 66*, 99–113.

Chiou, T. H., Kleinlogel, S., Cronin, T., Caldwell, R., Loeffler, B., Siddiqi, A., et al. (2008). Circular polarization vision in a stomatopod crustacean. *Current Biology, 18*, 429–434. doi:10.1016/j.cub.2008.02.066.

Cortesi, F., & Cheney, K. L. (2010). Conspicuousness is correlated with toxicity in marine opisthobranchs. *Journal of Evolutionary Biology, 23*, 1509–1518. doi:10.1111/j.1420-9101.2010.02018.x.

Cronin, T. W., & Marshall, N. J. (1989). Multiple spectral classes of photoreceptors in the retinas of gonodactyloid stomatopod crustaceans. *Journal of Comparative Physiology. A, Neuroethology, Sensory, Neural, and Behavioral Physiology, 166*, 261–275.

Cott, H. B. (1940). *Adaptive colouration in animals*. London: Methuen.

Detto, T. (2007). The fiddler crab *Uca mjoebergi* uses colour vision in mate choice. *Proceedings. Biological Sciences, 274*, 2785–2790. doi:10.1098/rspb.2007.1059.

Detto, T., Hemmi, J. M., & Backwell, P. R. Y. (2008). Colouration and colour changes of the fiddler crab, *Uca capricornis*: A descriptive study. *Plos One, 3*, e1629. doi:10.1371/journal. pone.0001629.

Eguchi, E., & Waterman, T. H. (1973). Orthogonal microvillus pattern in the eighth rhabdomere of the rock crab *Grapsus*. *Zeitschrift fur Zellforschung und Mikroskopische Anatomie, 137*, 145–157.

Endler, J. A. (1993). The color of light in forests and its implications. *Ecological Monographs, 63*, 1–27. doi:10.2307/ 2937121.

Endler, J. A., Westcott, D. A., Madden, J. R., & Robson, T. (2005). Animal visual systems and the evolution of color patterns: Sensory processing illuminates signal evolution. *Evolution; International Journal of Organic Evolution, 59*, 1795–1818. doi:10.1554/04-669.1.

Fox, D. L. (1976). *Animal biochromes and structural colours: Physical, chemical, distributional and physiological features* (2nd ed.). Berkeley: University of California Press.

Hanlon, R. T., Naud, M. J., Forsythe, J. W., Hall, K., Watson, A. C., & McKechnie, J. (2007). Adaptable night camouflage by cuttlefish. *American Naturalist, 169*, 543–551.

Hariyama, T., Tsukahara, Y., & Meyerrochow, V. B. (1993). Spectral responses, including a UV-sensitive cell type, in the eye of the isopod *Ligia exotica*. *Naturwissenschaften, 80*, 233–235.

Hemmi, J. M., Marshall, J., Pix, W., Vorobyev, M., & Zeil, J. (2006). The variable colours of the fiddler crab *Uca vomeris* and their relation to background and predation. *Journal of Experimental Biology, 209*, 4140–4153. doi:10.1242/jeb. 02483.

Horch, K., Salmon, M., & Forward, R. (2002). Evidence for a two pigment visual system in the fiddler crab, *Uca thayeri*. *Journal of Comparative Physiology. A, Neuroethology, Sensory, Neural, and Behavioral Physiology, 188*, 493–499.

Hurlbert, A. (1997). Primer—Colour vision. *Current Biology, 7*, R400–R402.

Jacobs, G. H. (1993). The distribution and nature of color vision among the mammals. *Biological Reviews of the Cambridge Philosophical Society, 68*, 413–471.

Jerlov, N. G. (1976). *Marine optics*. Amsterdam: Elsevier.

Jordão, J. M., Cronin, T. W., & Oliveira, R. F. (2007). Spectral sensitivity of four species of fiddler crabs (*Uca pugnax, Uca pugilator, Uca vomeris* and *Uca tangeri*) measured by in situ microspectrophotometry. *Journal of Experimental Biology, 210*, 447–453.

Kelber, A., & Osorio, D. (2010). From spectral information to animal colour vision: Experiments and concepts. *Proceedings. Biological Sciences, 277*, 1617–1625. doi:10.1098/ rspb.2009.2118.

Kelber, A., Vorobyev, M., & Osorio, D. (2003). Animal colour vision—Behavioural tests and physiological concepts. *Biological Reviews of the Cambridge Philosophical Society, 78*, 81–118.

Land, M. F. (1981). Optics and vision in invertebrates. In H. Autrum (Ed.), *Handbook of sensory physiology* (Vol. 7/B6, pp. 471–592). Berlin: Springer-Verlag.

Land, M. F. (1984). Crustacea. In M. A. Ali (Ed.), *Photoreception and vision in invertebrates* (pp. 401–438). New York: Plenum Publishing.

Land, M. F. (1988). The optics of animal eyes. *Contemporary Physics, 29*, 435–455.

Losey, G. S., McFarland, W. N., Loew, E. R., Zamzow, J. P., Nelson, P. A., & Marshall, N. J. (2003). Visual biology of Hawaiian coral reef fishes: I. Ocular transmission and visual pigments. *Copeia*, 433–454. doi:10.1643/01-053.

Lythgoe, J. N. (1979). *The ecology of vision*. Oxford, England: Clarendon Press.

Lythgoe, J. N. (1980). Vision in fishes: Ecological adaptations. In M. A. Ali (Ed.), *Environmental physiology of fishes* (Vol. 35, pp. 431–445). London: Plenum Publishing.

Lythgoe, J. N., Muntz, W. R. A., Partridge, J. C., Shand, J., & Williams, D. M. (1994). The ecology of the visual pigments of snappers (*Lutjanidae*) on the Great-Barrier-Reef. *Journal of Comparative Physiology. A, Neuroethology, Sensory, Neural, and Behavioral Physiology, 174*, 461–467.

Lythgoe, J. N., & Partridge, J. C. (1989). Visual pigments and the acquisition of visual information. *Journal of Experimental Biology, 146*, 1–20.

Manning, R. B., Schiff, H., & Abbott, B. C. (1984). Eye structure and the classification of stomatopod crustacea. *Zoologica Scripta, 13*, 41–44. doi:10.1111/j.1463-6409.1984.tb00021.x.

Marshall, N. J. (1988). A unique color and polarization vision system in mantis shrimps. *Nature, 333*, 557–560.

Marshall, N. J. (2000a). Communication and camouflage with the same 'bright' colours in reef fishes. *Philosophical Transactions of the Royal Society of London. Series B, Biological Sciences, 355*, 1243–1248.

Marshall, N. J. (2000b). The visual ecology of reef fish colours. In Y. Espmark, T. Amundsen, & G. Rosenquist (Eds.), *Animal signals: Signalling and signal design in animal communication* (pp. 83–120). Trondheim, Norway: Tapier.

Marshall, N. J., Jennings, K., McFarland, W. N., Loew, E. R., & Losey, G. S. (2003). Visual biology of Hawaiian coral reef fishes: III. Environmental light and an integrated approach to the ecology of reef fish vision. *Copeia*, 467–480.

Marshall, N. J., & Johnsen, S. (2011). Camouflage in the marine environment. In M. Stevens & S. Merilaita (Eds.), *Animal camouflage: Mechanisms and function* (pp. 186–211). New York: Cambridge University Press.

Marshall, N. J., Jones, J. P., & Cronin, T. W. (1996). Behavioural evidence for colour vision in stomatopod crustaceans. *Journal of Comparative Physiology. A, Neuroethology, Sensory, Neural, and Behavioral Physiology, 179*, 473–481.

Marshall, N. J., Kent, J., & Cronin, T. W. (1999). Visual adaptations in crustaceans. In S. N. Archer, M. B. A. Djamgoz, E. Lowe, J. C. Partridge, & S. Vallerga (Eds.), *Adaptive mechanisms in the ecology of vision* (pp. 285–328). London: Kluwer Academic Publishers.

Marshall, N. J., & Land, M. F. (1993). Some optical features of the eyes of Stomatopods: I. Eye shape, optical zxes and resolution. *Journal of Comparative Physiology. A, Neuroethology, Sensory, Neural, and Behavioral Physiology, 173*, 565–582.

Marshall, N. J., & Messenger, J. B. (1996). Colour-blind camouflage. *Nature, 382*, 408–409.

Marshall, N. J., & Vorobyev, M. (2003). The design of color signals and color vision in fishes. In S. P. Collin & J. N. Marshall (Eds.), *Sensory processing in aquatic environments* (pp. 194–222). New York: Springer.

Maximov, V. V. (2000). Environmental factors which may have led to the appearance of colour vision. *Philosophical Transactions of the Royal Society of London. Series B, Biological Sciences, 355*, 1239–1242. doi:10.1098/rstb.2000.0675.

Pignatelli, V., Champ, C., Marshall, J., & Vorobyev, M. (2010). Double cones are used for colour discrimination in the reef fish, *Rhinecanthus aculeatus*. *Biology Letters, 6*, 537–539. doi:10.1098/rsbl.2009.1010.

Pignatelli, V., Temple, S. E., Chiou, T.-H., Roberts, N. W., Collin, S. P., & Marshall, N. J. (2011). Behavioural relevance of polarization sensitivity as a target detection mechanism in cephalopods and fishes. *Philosophical Transactions of the Royal Society of London. Series B, Biological Sciences, 366*, 734–741. doi:10.1098/rstb.2010.0204.

Rajkumar, P., Rollmann, S. M., Cook, T. A., & Layne, J. E. (2010). Molecular evidence for color discrimination in the Atlantic sand fiddler crab, *Uca pugilator*. *Journal of Experimental Biology, 213*, 4240–4248. doi:10.1242/jeb.051011.

Ruxton, G. D., Sherratt, T. N., & Speed, M. P. (2005). *Avoiding attack. The evolutionary ecology of crypsis, warning signals and mimicry.* Oxford, England: Oxford University Press.

Stevens, M., & Merilata, S. (2011). *Animal camouflage: Mechanisms and function.* Cambridge, England: Cambridge University Press.

Temple, S. E., Pignatelli, V., Cook, T., How, M. J., Chiou, T. H., Roberts, N. W., et al. (2012). High-resolution polarisation vision in a cuttlefish. *Current Biology, 22*, R121–R122. doi:10.1016/j.cub.2012.01.010.

Vorobyev, M., & Osorio, D. (1998). Receptor noise as a determinant of colour thresholds. *Proceedings. Biological Sciences, 265*, 351–358.

Wade, N. M., Anderson, M., Sellars, M. J., Tume, R. K., Preston, N. P., & Glencross, B. D. (2012). Mechanisms of colour adaptation in the prawn *Penaeus monodon*. *Journal of Experimental Biology, 215*, 343–350. doi:10.1242/jeb.064592.

Wade, N. M., Melville-Smith, R., Degnan, B. M., & Hall, M. R. (2008). Control of shell colour changes in the lobster, *Panulirus cygnus*. *Journal of Experimental Biology, 211*, 1512–1519. doi:10.1242/jeb.012930.

Wicksten, M. K. (1989). Why are there bright colors in sessile marine invertebrates? *Bulletin of Marine Science, 45*, 519–530.

Zeil, J., Nalbach, G., & Nalbach, H. O. (1986). Eyes, eye stalks and the visual world of semi-terrestial crabs. *Journal of Comparative Physiology. A, Neuroethology, Sensory, Neural, and Behavioral Physiology, 159*, 801–811.

82 The Cognitive Structure of Visual Navigation in Honeybees

RANDOLF MENZEL

Animals use all their senses to explore the world, return to safe places, discover locations of importance, and travel between them. The phylogenetic history of the respective species equips it with information about the physics of the world, but locations need to be learned according to their relations to physical and chemical signals. Memory established by exploration and successful outcome of navigation links environmental cues and may lead to complex forms of neural representation of space. Cognitive levels of navigation range from guidance by predominantly innate stimulus–response connections to goal-directed planning based on highly integrated combinations of multisensory inputs. In this sense the study of navigation and its neural underpinnings presents a paradigmatic case of cognitive neuroscience.

Having accumulated a wide range of observations from "foraging" *Caenorhabditis* worms to human goal finding, behaviorists noticed that animals may apply rather simple strategies to head toward a goal or to return to a place. In an attempt to apply parsimony arguments strictly, behavioral biologists hesitate to assume any more complex forms of neural integration. Thus the assumption of a memory structure like a cognitive map is hotly debated and may even be entirely rejected (Bennett, 1996; Shettleworth, 2010). Neuroscientists, in contrast, refer to a cognitive map even for experimental settings in the lab where it cannot be excluded that the animal may have simply steered toward the goal by following a stimulus gradient, headed toward a beacon at the goal, or performed sequential matching procedures, gradually reducing the mismatch between an image learned close to the goal and that at the current position (O'Keefe & Nadel, 1978). Tolman's (1948) criterion of a novel shortcut as an indication for a cognitive map is indeed rather weak if other explanations like dead reckoning, beacon orientation, and image matching are not ruled out. Here I explore the cognitive levels of visual navigation in a flying insect, the honeybee, and present data and concepts that go beyond elementary forms of navigation. In so doing, I refer to the sophisticated form of social communication in honeybees, the waggle dance.

THE SPATIAL PRIMITIVES OF NAVIGATION

Elements of Spatial Primitives

The elements of spatial primitives are discussed at length in this volume, with many examples applying also to the honeybee. They belong to two groups: a basic visual recognition group and a more advanced group that integrates such elements in a performance-related way. Elements of the first group are, for example, various forms of taxes; object segmentation, discrimination, and learning; segmentation of celestial cues (sun, blue sky); odometry; detection of movement direction; associating flight vectors, objects, and circadian time to meaning (e.g., reward, expected outcome of own performance); and many more. Compositions of these basic elements control elements of navigation—for example, learning patterns of objects for the use of image matching (steering toward a goal by reducing the mismatch between the current and the learned pattern), learning the local time-compensated sun compass, relating the pattern of polarized light of the sky to the great circle of the sun, deriving sun compass directions from extended landmarks, estimating and learning sequences of objects, estimating a direct path by integrating partial vectors (dead reckoning and path integration), and many more. Here I dwell first on two of these components, path integration and image matching, because they are frequently thought to fully explain navigation performance in honeybees (Collett & Collett, 2002; Cruse & Wehner, 2011).

Path Integration

Path integration in its basic form provides the information for an animal to return to a starting point—for example, the nest—by a memory of its own movements. From spiders to mammals, animals possess an accurate system for keeping track of relative spatial locations by integrating linear and angular motion even without access to external cues (Mittelstaedt & Mittelstaedt, 1982). Angular motion is extracted under daylight conditions from visual inputs like landmarks and celestial cues (see chapter 85, this volume) and from traveled distance from an odometer. The outcome of the path integration process provides a running estimate, leading the animal back to the starting point at any time during its excursion. Thus such home vectors can be computed solely on egocentric information and may include landmark information as an allocentric reference for the two components of path integration, namely, angular and translatory movement.

Formal descriptions distinguish between four cases (egocentric with polar or Cartesian coordinate systems, geocentric with polar or Cartesian coordinate systems) and conclude that the geocentric Cartesian system provides the most robust home vector information (Vickerstaff & Cheung, 2010). This latter finding is particularly relevant for the buildup of metric spatial representation by path integration, an idea first put forward by O'Keefe (1976). Location-dependent information may be used for corrections due to error accumulation in path integration and for estimates of spatial relations via single or multiple visits to these locations (Biegler, 2000; Gallistel, 1990; McNaughton et al., 2006).

Two fundamentally different learning situations need to be distinguished in path integration: the use of the running estimate of the home vector during exploratory movements in an unknown landscape and the movement along frequently traveled routes. I shall distinguish nonassociative dead reckoning, the basic form of path integration during exploration, from associative dead reckoning that requires traveling along a route multiple times. The latter strategy is most important for embedding egocentric into allocentric reference systems and is of particular interest here. Bees, for example, learn the sequence of objects experienced along a multiply flown path and use the sequence for distance estimation (Chittka & Geiger, 1995; Menzel et al., 2010). After training bees to two feeders and analyzing their straight flight components (SFCs) during homing behavior, we found multiple SFCs that resemble multiple vector memories (Menzel et al., 2012) (see figure 82.1). The vector memories belong to two forms: the experienced flight vectors reflecting the routes between the hive and each feeder (see figure 82.1a) and the vectors derived from vector integration. Two of these derived vectors connect the two feeders (see figure 82.1b); other flight vectors connect the release site or any location after some search flights and the hive (see figure 82.1c). We observed that all SFCs resembled the direct vectors between the feeders and

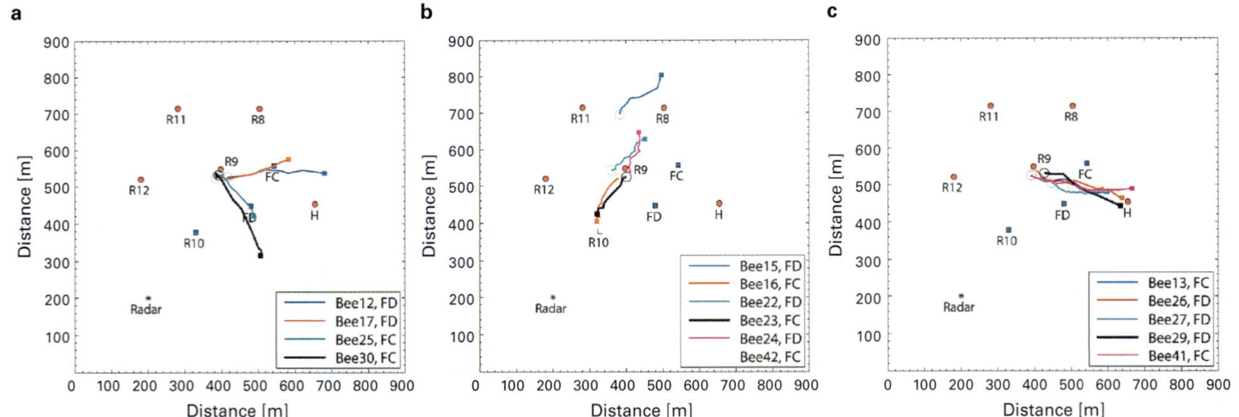

FIGURE 82.1   Representative examples of straight flight components (SFCs) of test bees released at the same release site (R9). The bees were trained from the hive (H) to two feeders (FC, FD). They did not experience the straight flight between FC and FD during training. The number of the test bee and the feeder from which it was collected is given in the respective inset. The two feeders are called FC and FD and form an approximately equal-sided triangle with the hive (length of side = 160 m). The circle gives the starting point of the respective SFC and the filled square its end point. The landmarks characterizing the feeding places and the release site are rather inconspicuous so that animals may not be able to notice at the release site that they have been moved to an unexpected location. (a) Animals flew along a path resembling homing flight vectors from the feeder from which they were collected. Thus the animals switched motivation from outbound to inbound flights. (b) Animals followed the vector components of the bee line between the two feeders. Two of the animals (bee 15 and bee 42) first performed extended search flights before the first SFC shown here. (c) Animals performed novel shortcut flights back to the hive. (After Menzel et al., 2012.)

were not performed at the feeder sites (feeders and all material were removed from the respective site during tests) but further away from the hive than the bee line between the feeders and the hive, suggesting that the animals related their choices to the overall spatial relations of the three sites—the hive and the two feeders.

The two kinds of derived vectors represent novel shortcut flights and may result from the integration of at least two associative dead reckoning vectors or from activation of site-specific memory. We observed that half of the first SFCs at release site R9 were directed toward the hive whereas none of the first SFCs from animals released at R7, R8, R10, and R12 belonged to this flight direction category, possibly indicating that the area around R9 more closely resembled the landmarks characterizing the two feeding sites. This would favor the activation of both home vectors from the two feeders leading to either a compromise flight vector or to vector integration of both vectors, an interpretation suggested by Menzel et al. (1998) for a similar constellation. An alternative interpretation assumes that the geometric relations between R9 and the hive favored the novel shortcut toward the hive, a behavior that would require knowledge of the geometric relations between the respective locations. As in the study by Menzel et al. (2005) these data do not allow one to distinguish between these two interpretations.

It has often been argued that the process of retrieving a vector memory on the basis of landmarks, switching motivation between outbound and inbound flights, and vector subtraction concepts are more parsimonious neural procedures than a geometric representation in spatial memory in ants and bees (Collett & Collett, 2002; see chapter 80, this volume). Indeed, a simple model of memory retrieval, motivational switch and path integration (Cruse & Wehner, 2011), formally meets the requirements for predicting the navigational performance of bees as tested in the study presented here and in the Menzel et al. (2005) study. Whether the model captures a more parsimonious neural implementation is a different question and must be kept open as long as we do not have any data on the neural processes in the insect brain allowing the animal to navigate over long distances in a highly flexible way. Running and flying insects may differ substantially in their capacity to relate the egocentric measures to an allocentric reference since the bird's eye view of flying insects offers a geometric layout as a primary visual source of information, whereas running insects are bound to multiple sequential views of cluttered objects whose geometric relations are only indirectly accessible and difficult to extract.

Beacon Orientation and Image Matching

When leaving the hive or a food patch, honeybees perform a characteristic scanning behavior and learn the immediate surroundings in spatial relation to the hive entrance (see chapter 85, this volume). Whether bees use the same kind of image learning for close and far distance landmarks (panorama) is not clear as image matching experiments have only tested bees for close landmarks (Cartwright & Collett, 1987). The panorama is learned during exploratory orientation flights and route flights. Such forms of learning differ substantially from image learning at a vantage point. Furthermore, a flying insect like the bee will have access to the geometric layout around the hive and at further distances soon after it is in air. It is therefore questionable whether bees follow homing strategies similar to those of ants that travel in a visually cluttered environment and appear to follow the image-matching strategy even along multiply traveled routes further away from the nest (Philippides et al., 2011; see chapter 80, this volume). It is more likely that bees use different spatial learning strategies than do ants, and an extension of concepts developed for close-up image matching may not be adequate.

Mapping to Compass Values

Animals are innately prepared to relate their movements to one or several compass systems. Evidence in favor of spatial mapping comes from free-ranging animals under natural conditions both for far distance navigation and navigation within the home range. Analysis of a large database on bird navigation has led to the concept that intersecting stimulus gradients form a multicoordinate system by which any point in space is characterized by a unique combination of coordinates (Wallraff, 2005; Wiltschko & Wiltschko, 2003). These coordinate values appear to provide a global allocentric reference frame with respect to which items in spatial long-term memory (places, landmarks, home) could be represented. Pigeon homing has been conceptualized by a "mosaic map" that stores gradients associated with compass directions (Wallraff, 1974, 2005; Wiltschko & Wiltschko, 2003). Lipp et al. (2004) tracked many homing flights of pigeons carrying global positioning system devices and found that they followed highways, performed turns on intersections, and accepted detours if they were marked by such gradients.

Bees associate gradients (e.g., forest edges, roads) with sun compass directions and read the sun compass direction from these gradients when the sky is overcast (Dyer & Gould, 1981; von Frisch &

Lindauer, 1954). In their waggle dances bees report distance and direction of a feeding site or a nest site referring to celestial cues or cues derived from landmarks. They then transform the directional information into a code relative to gravity and encode the distance estimated by the outbound flight via their visual odometer. Thus waggle dance communication can be used to read the structure of their spatial memory (see below).

Beyond the Primitives of Navigation

The cognitive building blocks of navigation can be considered as letters and words arranged by rules to become a meaningful text. It is the search for the rules that characterizes the cognitive approach. Does the honeybee solve a navigational task that cannot be explained by a set of primitives and requires a higher level of integration? The basic design of experiments performed to address this question is the catch-and-release paradigm under natural conditions. An animal whose knowledge about the environment is known as much as possible by prior training is caught in a defined motivational state (e.g., when leaving a feeding place to return to the hive, when leaving the hive after following a waggle dance) and transported to an unexpected release site within its explored area. After being released its full flight path is recorded by harmonic radar. The test conditions require excluding beacon orientation and image matching as a navigation strategy. If the animal is able to return to the hive or steer toward any other important place (feeding place or dance-directed place) directly along a novel path (novel shortcut), one needs to conclude that it uses a strategy beyond navigation primitives. However, the structure of the spatial reference is not elucidated by such a result. One still needs to ask which landmark features guided the animal and what the structure of its navigation memory is. Experiments under natural conditions make it difficult to address these questions, but resorting to the lab or to simpler test conditions is not an option because reducing the environment may not allow the animal to apply its cognitive capacities. Furthermore, the question about the cognitive dimensions of navigation is not limited to path finding. It also includes motivational components, decision making, planning, and in the case of the honeybee, social communication. Do bees communicate primitives of flight vectors or places that are associated with meaning and create expectancy about a particular place both in the transmitting and the receiving bee?

Experience from Route Flights Is Not Required for Novel Shortcuts

The method of training individually marked bees to a feeding site has been a major source of discoveries since its introduction into behavioral biology by the Nobel Laureate Karl von Frisch more than 100 years ago (von Frisch, 1967). Bees learn the distance and direction of their route flights between hive and feeder and report the outbound flight vector in the waggle dance. This vector is stored in memory and dominates the behavior of bees in catch-and-release experiments (Menzel et al., 2005). As long as only the vanishing bearings of released bees could be recorded, the stereotypic perpetuation of the sun compass–related direction of this vector gave the impression that bees' navigation is bound to an egocentric frame of reference and relies solely on the information gathered during route training (Wehner & Menzel, 1990). If this were true, bees would be lost if they were trained such that they did not learn a route vector. But bees are not lost. Figure 82.2 shows the flight time of two groups of bees under similar test conditions. One group was trained along a route, the other to a feeder close to the hive that was rotated around the hive. It took animals without route training no longer to return to the hive from five release sites around the hive than it took route-trained animals when they were released at their training site. Thus bees must be able to refer to a different spatial memory than that formed during route training, and this kind of memory cannot come from earlier foraging activities since the areas around the release sites differed considerably with respect to potential forage. Furthermore, beacon orientation toward the hive and image matching with the panorama were not possible because the view toward the hive was blocked either from R2 or R5.

Route-trained animals and animals without route training were also compared with respect to their homing flights using harmonic radar for tracking (Menzel et al., 2005) (see figure 82.3). The test area did not provide any panorama cues, and the animals relied on local ground structures for navigation. The initiation points of direct homing flights (homing points marked with a red star in figure 82.3 a and b) lie outside the visual catchment area around the hive excluding the possibility of beacon orientation. The distribution of homing points does not differ between V- and C-bees, indicating that they refer to a spatial memory not derived from route training. Accumulation of homing points in both groups of bees south of the hive overlaps with a long-ranging landmark (a border line between

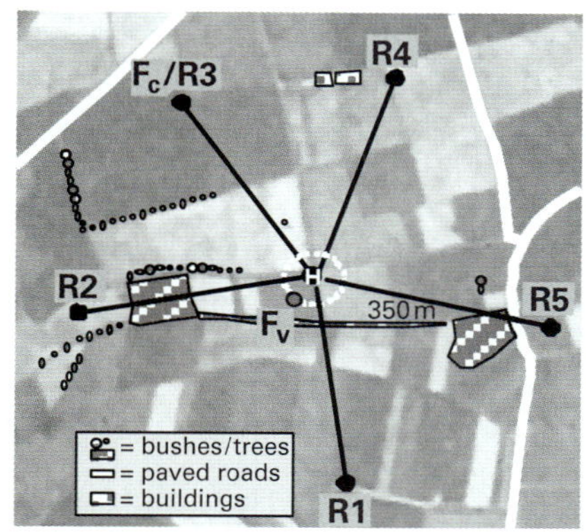

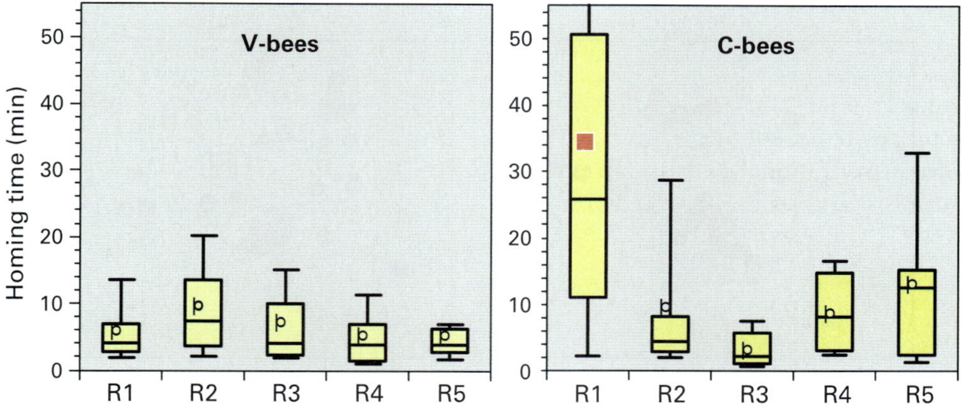

FIGURE 82.2   Catch-and-release experiment with two groups of bees. V-bees were trained to a variable feeder close to the hive (H) that rotated around the hive (white dotted line around the hive in the upper figure). C-bees were trained along a route to a stationary feeder (Fc/R3 in the upper figure). Fc and the 5 release sites (R1–R5) had a distance to the hive of 350 m. The two lower figures give the flight times between the respective release sites and the hive for both groups of bees. In both cases the bees were collected individually at one of the two feeders once they had sucked their fill and prepared for departure to the hive. Only the red marked bar (R1 of C-bees) is statistically significantly different from that of all other bars. C-bees released at R1 take longer to return to the hive because they first fly further away from the hive as they first apply their home bound vector memory from Fc/R3 to the hive (Menzel et al., 2000).

two differently cut pastures stretching north–east to south–west) and a patch of local landmarks (tents). Obviously these two landmarks have been learned by the bees in spatial relation to the hive, independently of route flights. Since the experiments were performed when no natural forage was available, the novel short-cuts to the hive from more or less all directions could not have been learned during foraging flights.

Shortcutting and Learning during Orientation Flights

A social animal and central place forager like the bee needs to return safely to its colony. Bees also need to learn a range of properties of the environment before initiating foraging flights. These properties relate to the sun compass, the time of the day and the local ephemeris function, and possibly also to the calibration of the visual odometer. In one of the most fascinating series of experiments Karl von Frisch and Martin Lindauer (1954) showed that bees use extended landmarks (such as straight forest boundaries) as guides for sun compass orientation. Later Dyer and Gould (1981) called the same phenomenon a backup system for cloudy days and related the connection between sun compass orientation and landmark orientation to a safety system. However, it is more likely that the tight connections between extended landmarks and sun compass need to be seen in the context of calibrating the properties of the sun compass. In any case, extended landmarks are obviously of special importance for the bee.

Exploratory orientation flights bring the bee in narrow loops into the surrounding environment (Capaldi & Dyer, 1999; Capaldi et al., 2000). One component of these orientation flights is nonassociative dead reckoning (see above). The other component is associative dead reckoning during which bees learn about the spatial relations of extended and local

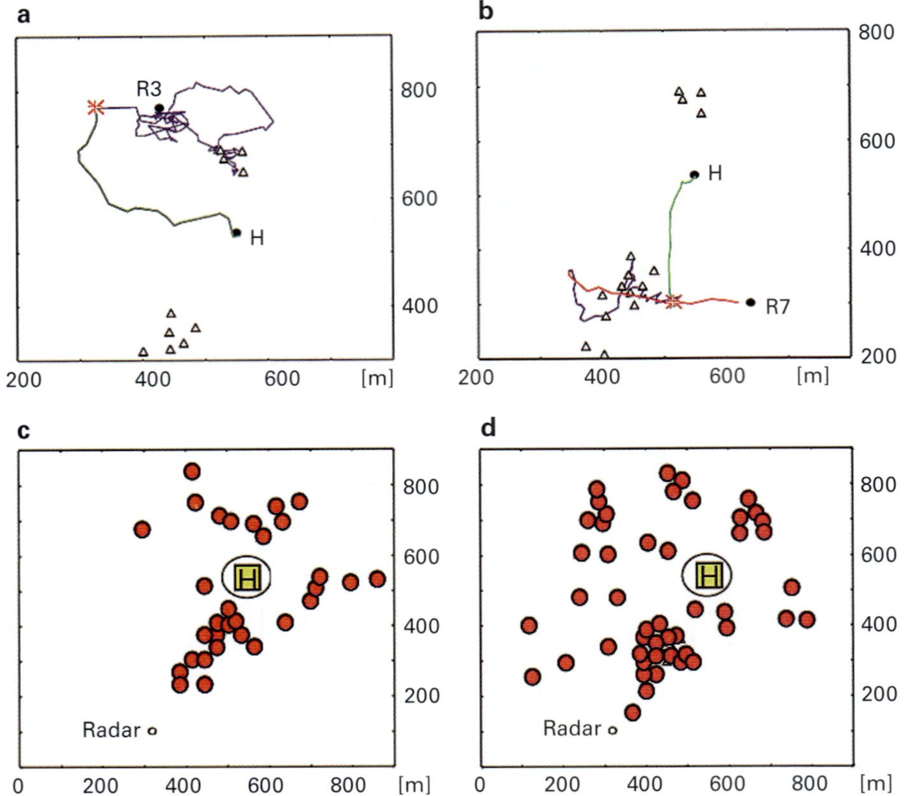

FIGURE 82.3   Homing behavior of V-bees and C-bees in a catch-and-release experiment where flight paths were recorded by harmonic radar (Menzel et al., 2005). The two upper figures (a, b) show representative examples of full flight paths of a V-bee (a) and a C-bee (b). (a) V-bees lack the vector flight, and start with a search flight at the release site (here at R3). The homing flight is shown in green starting at the red star (homing point) (b) The red line shows the vector flight of a route-trained bee whose feeder was 200 m east of the hive after it was released at R7. The vector flight is followed by a search flight component (in blue). Frequently bees return to the release site before commencing the homing flight. (c and d) The homing points for V-bees (c) and C-bees (d). The triangles in a and b mark tents in an otherwise rather homogeneous landscape. The circle around the hive (H) in c and d indicates the visual catchment area of the hive. Note the different scales in the upper and lower figures.

landmarks. Recently we found that bees return home faster after a single orientation flight when released in the explored sector as compared to releases in the unexplored sector. We also saw that multiple sequential orientation flights of the same animal are directed into different sectors with increasing range of exploration, suggesting that the surrounding environment is systematically explored. These data indicate that allocentric relations are learned during orientation flights.

Decision Making in Novel Shortcuts

Most interestingly, bees trained to a distant feeder returned home not only by direct flights to the hive but also via the feeder (see figure 82.4). The ability to decide between the hive and the feeder as the destination for a homing flight requires some form of relational representation of the two locations. Given that neither of these two locations could be approached with the help of a beacon or the panorama, it is tempting to conclude that bees made decisions between potential goals by referring to a map-like structure of their spatial memory. However, one can also argue they may have learned to associate home-directed vectors with local landmarks. This would explain the direct home flights, but an additional process would be required to explain the results shown in figure 82.4. This additional process may be based on the integration of memory of far-ranging vectors, one that leads to the hive from a particular location and one that was learned during multiple route flights from the hive to the feeder. Two motivations would have to be active at the same time, homing toward the hive and outbound flight from the hive to the feeder. A single motivation as claimed by Cruse and Wehner (2011) would not suffice. All these vector operations would have to be made on the level of a form of working memory in

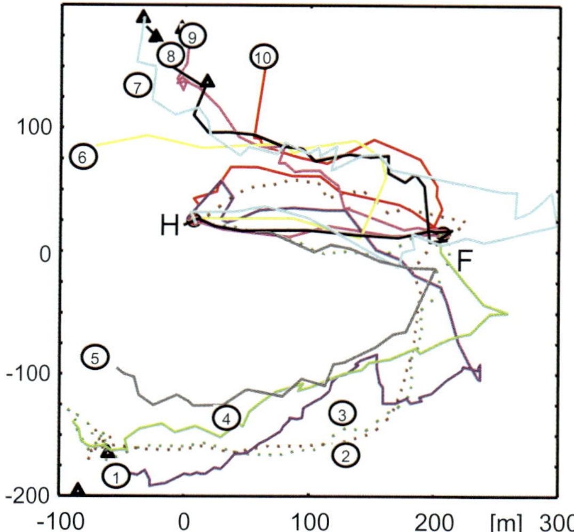

FIGURE 82.4   Final part of nine flight tracks of bees that flew back to the hive (H) via an area close to the feeder (F) (Menzel et al., 2005). One of these bees landed at the feeder.

which representations of these vectors are available for integration.

In the context of the data shown in figure 82.1 it may be argued that such operations on the level of working memory are basically not different from a map-like representation.

SOCIAL COMMUNICATION AND SPATIAL LOCATION

Social Communication in Honeybees

Honeybees use various kinds of stereotyped motion patterns for social communication (Seeley, 1995; von Frisch, 1967). The round and waggle dance communicates spatial relations to the hive. In the waggle dance, a dancing bee executes fast and short forward movements straight ahead on the comb surface, returns in a semicircle in the opposite direction, and starts the cycle again in regular alternation (each waggle dance involves several of these cycles). The straight portion of this course, called the waggle-run, is emphasized by lateral waggling motions of the abdomen. The length of single waggle-runs and the number of sound pulses increase with the distance flown to reach the source, and their angles relative to gravity correlate with the direction of the foraging flights relative to the sun's azimuth in the field and sun-linked patterns of polarized skylight. Thus by encoding the visually measured distance and the direction toward the goal, the waggle dance provides vector information toward a desirable goal. But what does the dancer really indicate? This will depend on both the transmitter (dancer) and the receiver (recruit).

Early detour experiments by von Frisch and colleagues (reviewed in von Frisch, 1967) indicated that the bees' odometer is primarily decoupled from directional information processing, indicating that no global flight vector is reported in the context of the waggle dance. These early findings were recently confirmed by manipulating the navigational information provided to a dancing bee (De Marco & Menzel, 2005). Thus one might ask whether the waggle dance encodes spatial information provided only by the actual flight path. The detour experiments by von Frisch and the results of von Frisch and Lindauer (1954) cited above suggest that the directional component reported in the waggle dance may also be derived from landmarks. This idea is not without precursors. Early experiments showed that with increasing experience of the terrain, directional information available during the inbound flight (and not only the outbound flight) may be computed for the purpose of directional indication in the waggle dance (Otto, 1959). It thus appears that bees may rely on some form of geocentric reference system.

Is there a symbolic component in the bee dance? To answer this question, we need to know what kind of neural or mental state the dancing bee refers to when it communicates a location of particular properties. Does she transmit only the motor performances to be applied by the recruit, or does she express her memory of the location of the site in the same geometric reference frame as the recruit? Does she read out the memory of the experience made with the site, or does she just convert a stereotypical measure of quality (of the food source, of the potential nest site) into a dance parameter? We do not know (yet?).

A Common Frame of Spatial Memory in Navigation and Communication

How do recruits deal with the information they receive from the dancer? Do they treat this information for their sensorimotor performance during the outbound flight, or do they integrate the spatial components of this information into their memory about the landscape? We addressed these questions in experiments in which a group of bees foraged at a feeding site (the trained food site FT) and later experienced that FT did not provide any food anymore (Menzel et al., 2011). As a consequence they gave up foraging at FT and became recruits to two other bees performing dances for a food site (the dance-indicated food site FD) at the same distance as FT but at either 30° or 60° to FT (see figure 82.5). As in all other experiments with the harmonic radar we did not use any odor at the food site and the two locations could not be seen by the animals over

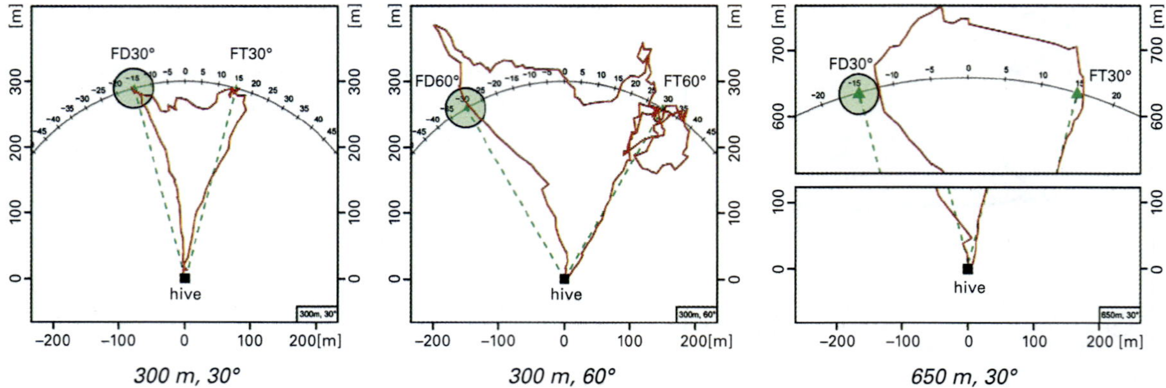

FIGURE 82.5   The figure shows three conditions under which recruits performed shortcuts between the dance-indicated location (FD 60°, FD 30°) and a location the animal had learned before (FT 60°, FT 30°). The distance between hive and FD or FT was either 300 m or 650 m, and the angle between the directions from hive to FD and that to FT was either 30° (FD 30°, FT 30°) or 60° (FD 60°, FT 60°). Test bees did not fly shortcuts between FD and FT (or FT and FD) for distances of 650 m and an angle of 60° (not shown). These findings indicate a common frame of reference for experienced and communicated locations. Recruits perform shortcuts according to the absolute and relative distance between the respective locations (Menzel et al., 2011).

distances greater than 50 m and without the help of the panorama. We found that recruits performed differently depending on the difference between their own foraging experience and the information transmitted in dance communication. The number of outbound flights to either FT or FD depended on the angular difference between FT and FD. Furthermore, recruits performed a range of novel flight behaviors. In the 30° arrangement some of them deviated from the course toward FD during their outbound flights and crossed over to FT. Most importantly, after arriving at either FD or FT some of them performed cross flights to the respective other location (see figure 82.5). From these observations we conclude that locations FD and FT are both stored in spatial memory in such a way that bees are able to fly directly from one location to the other following a novel shortcut.

We asked whether the decision for FD or FT depends on the number of waggle-runs followed by the recruited bee and found that more information is needed by recruits to fly to FD, the dance-indicated location. Bees that followed fewer waggle-runs either flew to their experienced feeding site, returned to the hive after a short excursion, or did not leave the hive. Following more waggle-runs (in our experiment on average 25 runs) resulted in FD flights indicating that the motivation to apply the information collected about FD is enhanced after longer dance following. However, the information about FD has been learned also during shorter dance following since animals that flew first to FT performed shortcut flights from FT to FD (see figure 82.5). Obviously dance communication involves two separate components, a motivational and an instructive component, the former requiring less information transfer. The motivational component appears to remind a recruit about its own foraging experience.

Given the bees' rich navigational memory, one may ask what exactly is communicated by the waggle dance: just the outbound vector or the location of the goal? In the first case the amount of vector information accumulated by the recruit may have to pass a certain threshold before new vector information can be applied. In the latter case the recruit would compare the expected properties of the indicated location with its own knowledge of this location and other potential foraging options from its own experience before reaching a decision about where to fly. Since we interpret our radar tracking data to document a rich form of a common allocentric memory as the structure of navigational working memory, it is tempting to conclude that vector information from the waggle dance is incorporated into such a common memory, and thus it too has an allocentric structure.

HOW COGNITIVE IS THE COGNITIVE MAP?

Arguments against a Cognitive Map in Bees

The structure of a cognitive map should allow the animal to localize itself within the explored environment irrespective of how it reached the current location and to perform novel paths to an intended goal along a short route (novel shortcut). Such a behavior requires the capacity to spot the current location and to estimate the direction and distance of the intended goal. Additional properties of a cognitive map can be assumed. The animal may be able to decide between two or more goals on the basis of the expected outcome when arriving at the goal, it may qualify these goals (nest, feeding sites, higher or lower ranking feeding or nest sites), and it may make its decision dependent on its own motivational state.

Five arguments have been put forward against the hypothesis that bees navigate with reference to a memory structure best described as a "cognitive map" as introduced by Tolman (1948) for rats and humans:

1. *The cognitive map is not the most parsimonious explanation* Parsimony is a strong argument in the interpretation of experimental data (Bennett, 1996). Although it should not be overlooked that radical forms of parsimony as applied to behavioral science were and may still, at least partially, be a historical burden (Menzel & Fischer, 2011), it is essential that "simpler" explanations be excluded in the experimental design. Parsimony can be understood as a formal criterion and, in the case of behavioral biology, as an argument for the simplest implementation in neural structures. Both aspects depend on what needs to be explained. Even if we ignore the evidence that bees make decisions according to the expected outcome (see below) and do not take into account any qualitative evaluation of the intended goal, we are left with the conclusion that bees either perform some sort of long-distance vector integration with at least three vectors to be considered or they refer to geometric relations of landmarks, best conceptualized as a cognitive map. The former is a geometric map, too, so the difference lies only in the procedure by which spatial relations are established and used.

2. *Small brains like that of the bee do not support a memory structure like that of a cognitive map* The parsimony argument is often combined with the statement that brains as small as those of bees cannot support such a memory structure. Furthermore, it is argued (Collett & Collett, 2002) that small brains need to solve their tasks with less "cognition," meaning with a toolbox of loosely interrelated elementary functions rather than an integrated, allocentric level of spatial representation. It should be recognized that we simply do not know whether the integration of multiple and complex sensory and procedural neural processes into a common spatial memory with geometric organization (a map) may not be a more economical and thus simpler way of representing sequential experiences during navigation (Griffin, 1984). It is likely that the mushroom bodies with their 360,000 neurons, each equipped with thousands of synapses, receive highly processed visual information (Gronenberg, 2001), freeing it from low-level processing. Are these neurons too few to encode geometric relations between identified objects? We simply do not know.

3. *Bees should fly directly home from the release site* Indeed they don't. They first perform a flight according to the active state of their working memory. This behavior is not an argument against a cognitive map because the spatial memory bees need to refer to has been learned during orientation flights, and this memory is obviously not active when they follow their route flights or fly according to the dance information; it has to be recruited from remote memory.

4. *Bees should not fly into a region they have not explored— for example, out on a lake* Gould and Gould (1982) reported that bees reject dance information which would have brought them out on a lake. Wray et al. (2008) interpreted their data as showing that bees have no problem flying out on a lake after following a dance; however they had to use odor marking of the feeding station on a boat, and although they tried to downgrade an odor effect, they still could not eliminate the possibility that bees flew out on the lake because they were attracted by the odor. In our view, the topic of bees not accepting dance information into a white spot of their navigational memory is not yet resolved and requires testing with the harmonic radar. Let's assume for the moment that bees do not hesitate to fly into an unknown area. Does this mean they do not have a cognitive map? Certainly not, because white spots are surrounded by known area, and why should bees (like humans) not explore the unknown?

5. *As long as you cannot rotate the landmarks used by bees relative to their sun compass one cannot believe in the cognitive map concept* We all know from our discussions in science that sometimes abstruse arguments are put forward that ask for something impossible. This is such an argument. It has its roots in a tradition of experimentation in which the experimenter controls for all possible parameters, varies just one, and finds that the animal performs according to the hypothesis behind just that variable. It is then concluded that the animal can do only what was just tested. Navigation does not deal with close-up object recognition. Navigation in bees cannot be tested in a white 1×1-m box with three black stripes on the wall. Navigation occurs in the natural environment when bees fly over many hundreds of meters.

The Structure of a Cognitive Map in Bees

What could be the structure of such an overarching memory? Vectors are formally the most efficient way of specifying a location (Biegler, 2000; Gallistel & Cramer, 1996; Vickerstaff & Cheung, 2010). If these vectors are anchored to landmarks, they provide a geocentric reference frame. Vectors are reported in the waggle dance. Thus bees appear to take advantage of the formal applicability of such a spatial measure and need to encode only two parameters. This does not have to mean that all they are communicating is the vector. The directional component of the communicated vector may be retrieved from the memory of spatial relations to extended landmarks (gradients) because these are also defined by their relations to compass directions. Such gradients could compose a memory for a rather simple "bearing map" as proposed by Jacobs and Schenk (2003). Such a rough bearing map does not require a large amount of neural encoding and storage but would provide a geometric representation of the whole experienced environment at a coarse resolution. Picture memories ("sketch maps," in the terminology of Jacobs and Schenk) could exist loosely distributed and only partially connected to each other, leaving white spots in between. Way finding (and possibly communication about ways) could therefore consist first in identifying the sketch map of current location, the spotting of that sketch map in the bearing map, and then the creation of a novel shortcut flight according to the compass direction to the goal as derived from the sketch map. The bee would travel through "unknown" territory (white spots) whenever she leaves a sketch map memory and has not yet reached another one, but she would not be lost because at any point she has access to the bearing map. If such a scenario applies, bees would dance for a location in a bearing map, and recruits would interpret the message according to such a map. The finding by von Frisch and Lindauer (1954) that extended landmarks can replace access to the sun compass on overcast days could in fact indicate the use of such "gradients" providing a primary source of information for navigation. The link to the compass may just be a side effect of learning about such gradients. Since there is no vocabulary for particular gradients in the dance, the flight direction has to be encoded into a compass direction.

What We Need to Ask Next

The kinds of questions to be asked in the future in navigation and communication studies in honeybees differ from those addressed so far. The sensorimotor routines involved are well understood, and they have been analyzed by asking "What can the animal do?" Now we need to ask what kind and how the information is stored in their working memory, how this information

is processed, and how decisions are made. We will thus have to analyze the structure of internal representations. Dance communication provides us with a window into these processes, and carefully designed experiments will allow access to processes beyond behavioral acts. These operations are far from simple and transcend elemental forms of associations (Menzel & Giurfa, 2001; Menzel, 2012). The richness of these operations is accessible only in animals behaving in their natural environment, and the methods for collecting the relevant data are now available. Ultimately we want to know how and where the bee's small brain performs these operations; the answers lie in the future.

REFERENCES

Bennett, A. T. D. (1996). Do animals have cognitive maps? *Journal of Experimental Biology, 199,* 219–224.

Biegler, R. (2000). Possible uses of path integration in animal navigation. *Animal Learning & Behavior, 28,* 257–277. doi:10.3758/BF03200260.

Capaldi, E. A., & Dyer, F. C. (1999). The role of orientation flights on homing performance in honeybees. *Journal of Experimental Biology, 202,* 1655–1666.

Capaldi, E. A., Smith, A. D., Osborne, J. L., Fahrbach, S. E., Farris, S. M., Reynolds, D. R., et al. (2000). Ontogeny of orientation flight in the honeybee revealed by harmonic radar. *Nature, 403,* 537–540.

Cartwright, B. A., & Collett, T. S. (1987). Landmark maps for honeybees. *Biological Cybernetics, 57,* 85–93.

Chittka, L., & Geiger, K. (1995). Can honeybees count landmarks? *Animal Behaviour, 49,* 159–164.

Collett, T. S., & Collett, M. (2002). Memory use in insect visual navigation. *Nature Reviews. Neuroscience, 3,* 542–552.

Cruse, H., & Wehner, R. (2011). No need for a cognitive map: Decentralized memory for insect navigation. *PLoS Computational Biology, 7,* e1002009. doi:10.1371/journal.pcbi.1002009.

De Marco, R. J., & Menzel, R. (2005). Encoding spatial information in the waggle dance. *Journal of Experimental Biology, 208,* 3885–3894. doi:10.1242/jeb.01832.

Dyer, F. C., & Gould, J. L. (1981). Honey bee orientation: A backup system for cloudy days. *Science, 214,* 1041–1042.

Gallistel, C. R. (1990). *The organization of learning.* Cambridge, MA: MIT Press.

Gallistel, C. R., & Cramer, A. E. (1996). Computations on metric maps in mammals: Getting oriented and choosing a multi-destination route. *Journal of Experimental Biology, 199,* 211–217.

Gould, J. L., & Gould, C. G. (1982). The insect mind: Physics or metaphysics? In D. R. Griffin (Ed.), *Animal mind–human mind* (pp. 269–298). New York: Springer.

Griffin, D. R. (1984). *Animal thinking.* Cambridge, MA: Harvard University Press.

Gronenberg, W. (2001). Subdivisions of hymenopteran mushroom body calyces by their afferent supply. *Journal of Comparative Neurology, 436,* 474–489.

Jacobs, L. F., & Schenk, F. (2003). Unpacking the cognitive map: The parallel map theory of hippocampal function. *Psychological Review, 110,* 285–315.

Lipp, H. P., Vyssotski, A. L., Wolfer, D. P., Renaudineau, S., Savini, M., Troster, G., et al. (2004). Pigeon homing along highways and exits. *Current Biology, 14,* 1239–1249. doi:10.1016/j.cub.2004.07.024.

McNaughton, B. L., Battaglia, F. P., Jensen, O., Moser, E. I., & Moser, M. B. (2006). Path integration and the neural basis of the "cognitive map." *Nature Reviews. Neuroscience, 7,* 663–678.

Menzel, R. (2012). The honeybee as a model for understanding the basis of cognition. *Nature Reviews Neuroscience, 13,* 758–768. doi:10.1038/nrn3357.

Menzel, R., Brandt, R., Gumbert, A., Komischke, B., & Kunze, J. (2000). Two spatial memories for honeybee navigation. *Proceedings. Biological Sciences, 267,* 961–968.

Menzel, R., & Fischer, J. (Eds.). (2011). *Animal thinking: Contemporary issues in comparative cognition* (pp. 1–416). Cambridge, MA: MIT Press.

Menzel, R., Fuchs, J., Nadler, L., Weiss, B., Kumbischinski, N., Adebiyi, D., et al. (2010). Dominance of the odometer over serial landmark learning in honeybee navigation. *Naturwissenschaften, 97,* 763–767.

Menzel, R., Geiger, K., Müller, U., Joerges, J., & Chittka, L. (1998). Bees travel novel homeward routes by integrating separately acquired vector memories. *Animal Behaviour, 55,* 139–152.

Menzel, R., & Giurfa, M. (2001). Cognitive architecture of a mini-brain: The honeybee. *Trends in Cognitive Sciences, 5,* 62–71. doi:10.1016/S1364-6613(00)01601-6.

Menzel, R., Greggers, U., Smith, A., Berger, S., Brandt, R., Brunke, S., et al. (2005). Honeybees navigate according to a map-like spatial memory. *Proceedings of the National Academy of Sciences of the United States of America, 102,* 3040–3045. doi:10.1073/pnas.0408550102.

Menzel, R., Kirbach, A., Haass, W. D., Fischer, B., Fuchs, J., Koblofsky, M., et al. (2011). A common frame of reference for learned and communicated vectors in honeybee navigation. *Current Biology, 21,* 645–650.

Menzel, R., Lehmann, K., Manz, G., Fuchs, J., Koblofsky, M., & Greggers, U. (2012). Vector integration and novel shortcutting in honeybee navigation. *Apidologie, 43,* 229–243. doi:10.1007/s13592-012-0127-z.

Mittelstaedt, H., & Mittelstaedt, M. L. (1982). Homing by path integration. In W. Papi (Ed.), *Avian navigation* (pp. 290–297). Berlin: Springer-Verlag.

O'Keefe, J. (1976). Place units in the hippocampus of the freely moving rat. *Experimental Neurology, 51,* 78–109.

O'Keefe, J., & Nadel, J. (1978). *The hippocampus as a cognitive map.* New York: Oxford University Press.

Otto, F. (1959). Die Bedeutung des Rückfluges für die Richtungs- und Entfernungsangabe der Bienen. *Zeitschrift fur Vergleichende Physiologie, 42,* 303–333.

Philippides, A., Baddeley, B., Cheng, K., & Graham, P. (2011). How might ants use panoramic views for route navigation? *Journal of Experimental Biology, 214,* 445–451.

Seeley, T. D. (1995). *The wisdom of the hive: The social physiology of honey bee colonies.* Cambridge, MA: Harvard University Press.

Shettleworth, S. J. (2010). *Cognition, evolution, and behavior* (2nd ed.). New York: Oxford University Press.

Tolman, E. C. (1948). Cognitive maps in rats and men. *Psychological Review, 55,* 189–208.

Vickerstaff, R. J., & Cheung, A. (2010). Which coordinate system for modeling path integration? *Journal of Theoretical Biology, 263,* 242–261.

von Frisch, K. (1967). *The dance language and orientation of bees.* Cambridge, MA: Harvard University Press.

von Frisch, K., & Lindauer, M. (1954). Himmel und Erde in Konkurrenz bei der Orientierung der Bienen. *Naturwiss, 41,* 245–253.

Wallraff, H. G. (1974). *Das Navigationssystem der Vögel.* Munich, Germany: R. Oldenbourg Verlag.

Wallraff, H. G. (2005). *Avian navigation: Pigeon homing as a paradigm.* Berlin: Springer.

Wehner, R., & Menzel, R. (1990). Do insects have cognitive maps? *Annual Review of Neuroscience, 13,* 403–414.

Wiltschko, R., & Wiltschko, W. (2003). Avian navigation: From historical to modern concepts. *Animal Behaviour, 65,* 257–272.

Wray, M. K., Klein, B. A., Mattila, H. R., & Seeley, T. D. (2008). Honeybees do not reject dances for "implausible" locations: Reconsidering the evidence for cognitive maps in insects. *Animal Behaviour, 76,* 261–279.

83 Neurobiology of Movement-Sensitive Behavior in Flies

ALEXANDER BORST

Visual motion provides the animal with a rich source of information about the environment. When the animal is at rest, a moving object can indicate a potential prey, a conspecific, or a predator—in each case something important for survival. Here, motion increases the saliency of an object and attracts the attention of the observing animal to the patch of the image where motion occurred. In addition, visual motion cues also occur when the observer itself is actively moving. Then, the whole image is moving across the observer's retinas, causing a distribution of motion vectors called the "optic flow" (Gibson, 1950). Since the optic flow is characteristic for each kind of ego-motion (translation along or rotation around particular body axes), it is widely used for visual guidance and stabilization in space. Furthermore, self-motion allows stationary objects to be separated from a more distant background via relative motion cues. All effects relying on self-motion should be more pronounced, the faster the animal moves. It is not surprising, therefore, that flies have been found to often respond vigorously and often in a highly stereotyped way to particular types of visual motion stimuli and, accordingly to devote a large part of their nervous system to the analysis of visual motion. In the following, I will review our current understanding of the different movement-sensitive behaviors that have been studied in flies, as well as what is known about the neural control circuits underlying these behaviors.

OPTOMOTOR RESPONSE

The optomotor response undoubtedly represents the behavioral paradigm that has been most influential in the study of insect motion vision. When a fly is tethered in the center of a striped drum that is rotating clockwise, the insect tries to turn clockwise, too (see figure 83.1a). When the drum is moving in the opposite direction, the insect turns counterclockwise. This is true for rotations around the vertical ("Yaw"), the transverse ("Pitch"), as well as the longitudinal ("Roll") body axis (Blondeau & Heisenberg, 1982). Thus, the optomotor

response consists of a following reaction around any body axis, syndirectional with the motion of the surround, that builds up slowly over several seconds. Measuring it in a tethered animal with its head fixed to its thorax offers two advantages. First, it isolates the visual response component from the proprioceptive, vestibular one. Second, it allows the visual motion stimulus to be presented on the animal's retina with ultimate precision. This approach was pioneered by Hassenstein and Reichardt (1956), who analyzed the turning tendency of the beetle *Chlorophanus viridis* walking on a spherical Y-maze. Following these seminal studies, more sophisticated devices to measure the insect's turning tendency in flight or during walking were introduced, such as the torque meter (Goetz, 1964), the wing beat analyzer (Goetz, 1987), or a patterned Styrofoam ball, the movement of which was detected optoelectronically (Buchner, 1976).

Studying the dependency of the optomotor response on the wavelength, contrast, and velocity of moving gratings led to the proposal of the "Reichardt model" for elementary motion detection (Hassenstein & Reichardt, 1956; Reichardt, 1987). The idea is that a two-dimensional array of such elementary detectors, each having only a small receptive field, covers the whole visual field of the animal. The observed optomotor response then is produced by spatially pooling the output of all these local motion detectors. To provide an estimate of the local image velocity, such a detector unit cross-correlates the luminance values of adjacent image pixels after one of them has been delayed or low-pass filtered by a certain time constant. This model leads to a number of rather counterintuitive predictions: (1) The response strength should increase with increasing pattern contrast; (2) unlike a speedometer, the response should reveal a velocity-optimum: For velocities beyond this optimum, the response should decrease with increasing velocity; and (3) grating patterns with different spatial wavelengths should have different velocity optima, such that the ratio of the optimal velocity and the spatial wavelength, which has the

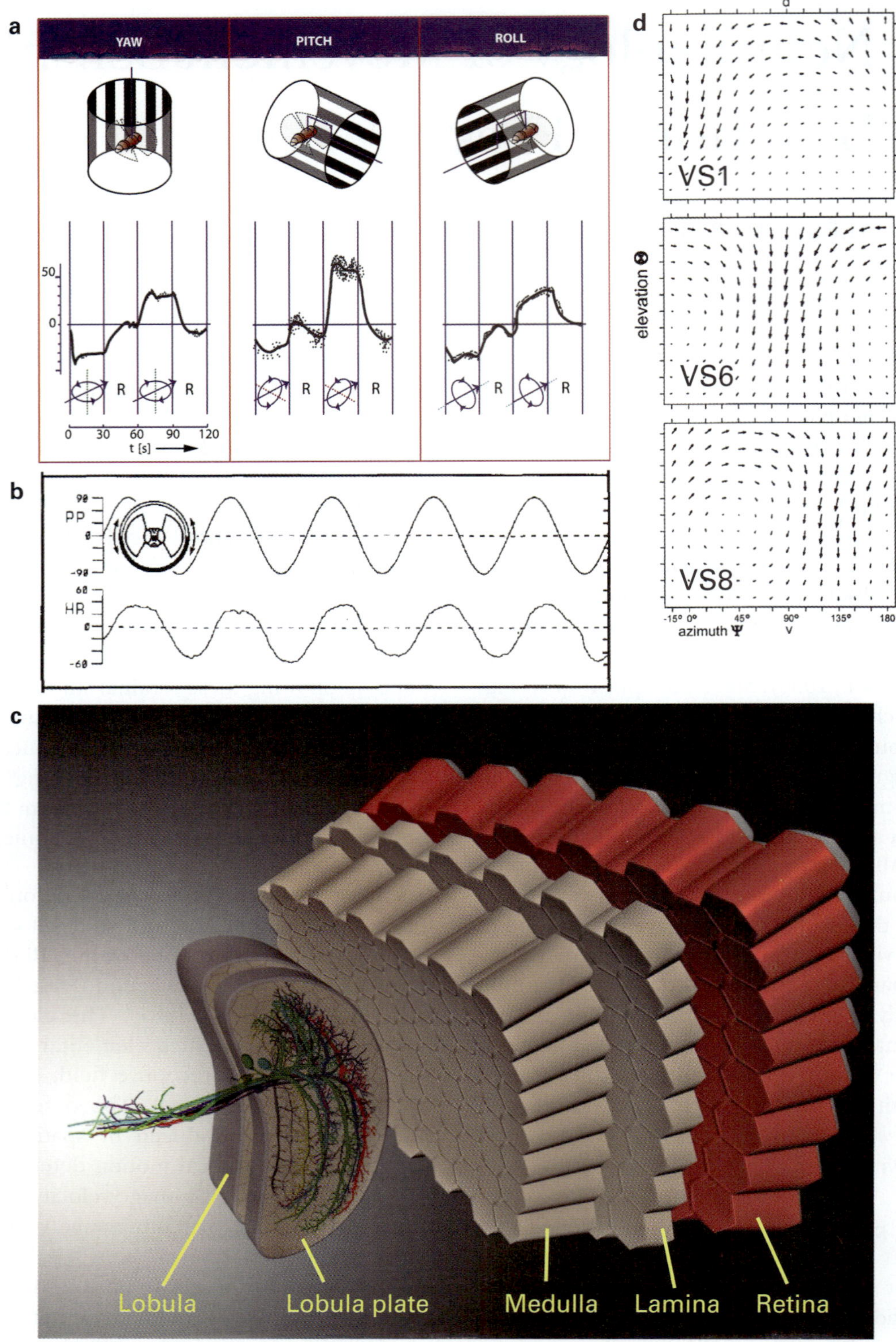

FIGURE 83.1   Optomotor response and lobula plate tangential cells. (a) A tethered flying *Drosophila* performs an optomotor response around all three principal body axes (modified from Blondeau & Heisenberg, 1982). Between 30 and 60, as well as 90 and 120 s, the pattern is at rest (R). (b) Tethered flying blowflies perform head roll (upper trace: pattern position, PP; lower trace: head roll, HR) in response to a visual surround rotating around the longitudinal body axis (upper trace; from Hengstenberg, 1988). (c) The fly optic lobe consists of four neuropile layers, each revealing a columnar structure that repeats the facets of the eye. In the lobula plate, a set of about 50, individually identifiable large tangential cells integrate the signals of local, motion-sensitive elements. Shown are the cells of the vertical system (VS-cells; modified from Borst, Haag, & Reiff, 2010). (d) Receptive fields of three VS-cells as a function of the azimuth and elevation. At each location, the orientation of the arrow indicates the local preferred direction of the cell, and the length of the arrow represents the response strength (from Krapp, Hengstenberg, & Hengstenberg, 1998).

dimension of a temporal frequency, remains constant. Surprisingly, all these predictions could be corroborated in the optomotor response of various fly species (for review, see Borst, Haag, & Reiff, 2010).

The torque exerted by the fly is not the only behavioral response component that can be influenced by large-field motion. When the head of the fly is not fixed to the thorax but free to move, head movements can as well be measured reliably in tethered flies, as shown for the head role response in figure 83.1b (Hengstenberg, 1988). Again, as for the torque response, head movements are syndirectional with the pattern motion, obviously compensating for the motion of the environment. Of course, such head movements have a mechanically limited range and can only cover +/−20°–30°, depending on the axis of rotation. Taken together, body torque as well as head movements can be reliably elicited in tethered flying flies by rotating the whole visual environment around the animal. Obviously, this "whole world" rotation is best interpreted by the nervous system as a self-rotation in the opposite direction. Compensating for it results in a following reaction. For this response to occur in a precisely timed way around the precise body axis requires (1) in a first stage, measuring the optic flow by a two-dimensional array of local motion detectors, and (2) in a second stage, integrating this information across the visual field of the fly. Such a control circuit has been shown to indeed mimic the fly's optomotor response under open-loop as well as closed-loop conditions (Warzecha & Egelhaaf, 1996).

As neural control elements for this optomotor response, the large-field motion-sensitive tangential cells of the lobula plate have been proposed (see figure 83.1c; for review, see Borst, Haag, & Reiff, 2010). These cells are located in the third layer of the optic lobe and comprise a group of about 50 different, individually identifiable neurons per hemisphere. With their large dendrites, they spatially pool the output signals of local elementary motion detectors. Furthermore, tangential cells also interact with each other synaptically, resulting in a rather complex network of lobula plate circuitry. All tangential cells have large receptive fields with the local preferred direction varying within the visual field in a systematic way (see figure 83.1d; Krapp & Hengstenberg, 1996; Krapp, Hengstenberg, & Hengstenberg, 1998). As in the case of the optomotor response, these neurons have been demonstrated to receive input from local motion detectors of the Reichardt type (Haag, Denk, & Borst, 2004; Joesch et al., 2008; Single & Borst, 1998). A large-scale modeling study revealed that the direct dendritic input from local motion detectors together with the lateral input from other tangential cells can indeed account for such particular receptive field structures (Borst & Weber, 2011). Interestingly, the receptive field as described by a vector field resembles an optic flow as occurring during certain types of egomotion (see figure 83.1d). Since each tangential cell is characterized by a different receptive field, the idea has been put forward that the tangential cells can act as a matched filter (Franz & Krapp, 2000): For rotation around each body axis, the resulting optic flow stimulates a specific tangential cell maximally, which could then initiate the respective corrective steering maneuver, counteracting the involuntary deviation from a straight course.

All the above properties make tangential cells ideal candidates for the neural control of the optomotor response. Indeed, when tangential cells were ablated surgically or via gene mutation, the corresponding optomotor response was found to be severely impaired (Heisenberg, Wonneberger, & Wolf, 1978; Geiger & Nässel, 1981; Hausen & Wehrhahn, 1983). However, for technical reasons, the above ablation studies always affected a large and ill-defined group of tangential cells: So far, the effect of ablating or silencing a single tangential cell on the torque response of tethered animals has never been measured. One point casting serious doubt on the role of tangential cells in the control of the optomotor response was the fact that optomotor responses in blowflies and fruit flies had a significantly higher temporal frequency optimum than was found in tangential cells: Since information about the temporal frequency of the stimulus is lost after spatial pooling, this could not be remedied by some sort of postprocessing. However, recent studies revealed that the tangential cells when measured in tethered walking or flying animals undergo a pronounced change in their response properties: Not only do they increase their spontaneous activity and overall response amplitude (Maimon, Straw, & Dickinson, 2010), they also shift their temporal frequency optimum toward higher values (Chiappe et al., 2010; Jung, Borst, & Haag, 2011). Mechanistically, this can be explained by release of octopamine shortening the time constant of the delay filter in the elementary motion detector (Jung, Borst, & Haag, 2011).

A further complication arises from the fact that the anatomical connections from the tangential cells onto the thoracic neurons of the flight motor are unknown in detail. This situation is different when head movements are considered instead of the torque response. Here, tangential cells are known to synapse either directly, or indirectly via a group of descending neurons, onto motor neurons of the various neck muscles, responsible for head rotations around different axes (Gronenberg, Milde, & Strausfeld, 1995; Haag,

Wertz, & Borst, 2007, 2010; Strausfeld & Bassemir, 1985; Strausfeld & Seyan, 1985; Wertz, Borst, & Haag, 2008). Apart from the optic flow input from the lobula plate neurons, these neck motor neurons receive additional input from a variety of other sensory modalities including the halteres, wind receptors on the antennae, as well as a central input reflecting the behavioral state of the animal (Haag, Wertz, & Borst, 2010).

ESCAPE AND LANDING RESPONSE

Optomotor following reactions are elicited by optic flow stimuli corresponding to a rotation of the animal around various body axes. Each rotational optic flow is independent of the distance between the observer and any objects or walls confining the physical space. In contrast, the translational optic flow depends, in addition to the flight direction and velocity, on the inverse of the distance between the observer and an object at each location of visual space (Koenderink & van Doorn, 1987). This makes translational optic flows ideal indicators of impending collisions, leading to corrective steering maneuvers or landing responses during flight and initiating escape responses while at rest.

Tammero, Frye, and Dickinson (2004) investigated the influence of expanding flow-fields at various locations within the visual field on the torque response of tethered flying *Drosophila*. They found that, under open-loop conditions, flies robustly turn away from the focus of expansion (see figure 83.2a, bottom graph). Interestingly, this response can be understood as the sum of the turning responses to the individual components of the expanding flow-field: Due to a reversal of the turning response to stimuli presented in the rear visual field of the fly, the resulting torque response to an expanding stimulus exceeds the response to a rotational stimulus. A similar effect can be seen when the stimulus is again split into a frontal and a posterior part and the latter is gradually switched from syndirectional motion, corresponding to a rotational stimulus, to antidirectional motion, corresponding to a lateral expansion: The torque amplitude in response to image expansion is almost three times as large as the response to an image rotation (figure 83.2b). When considering the response as a function of the azimuth position of the focus of expansion, the strongest responses are found for positions straight to the left and to the right of the fly (figure 83.2c). Under closed-loop conditions where the flies are given full control over the azimuth position of the focus of expansion, flies consequently choose to turn away from the focus of expansion. This brings the focus of expansion stably behind the fly, and the contracting pole in front of it (figure 83.2d). These

findings together result in a consistent description of the fly's turning tendency with respect to both rotational and translational optic flow: Considering the response to each stimulus component alone, the overall resulting response can be understood as a linear superposition of the response to each individual stimulus segment (Tammero, Frye, & Dickinson, 2004).

For all these reasons, the optomotor and the avoidance response would be expected to be fed by the same set of elementary motion detectors. Whether this is indeed the case was investigated explicitly in another study where the dependency of both response types was investigated as a function of the spatial, temporal, and contrast properties of expansion and rotation stimuli, quantifying the time course and amplitude of the turning responses during tethered flight (Duistermars et al., 2007). As an example, the response amplitude is shown as a function of the temporal frequency of the stimulus using grating patterns with different spatial wavelengths both for the rotation and the expansion stimulus (see figure 83.2f): Clearly, the responses to both types of stimuli reveal an almost identical dependency, peaking between 2 and 10 Hz, independent on the spatial wavelength of the grating pattern. These results support the conclusion that expansion and rotation responses draw from the same system of elementary motion detectors.

Apart from the avoidance response, many studies have shown that an expanding optic flow with a frontal point of expansion can lead to the initiation of a landing response in tethered flying flies (Braitenberg & Taddei Ferretti, 1966; Wehrhahn, Hausen, & Zanker, 1981). This landing response is a highly stereotyped sequence of leg extensions (see figure 83.3a), accompanied by a shift of the wing-beat plane and a reduction of the thrust force (Borst, 1986; Borst & Bahde, 1988). Under these conditions, the only free parameter in the behavioral response is the latency, which turned out to depend on the spatial wavelength, velocity, and contrast of the stimulus grating in a systematic way. Exploiting this relationship revealed that the landing response is driven by the same type of elementary motion detectors as is the optomotor response (Borst & Bahde, 1986). However, in contrast to the optomotor response, which usually is measured under steady-state conditions, that is, when pattern motion is constant for several seconds, the landing response usually occurs within less than 100 ms after the pattern has started to expand, requiring taking the transient response properties of motion detectors into account. Integrating the responses of appropriately aligned local motion detectors over space and time followed by a constant threshold leads to response latencies which match the behavioral observations in

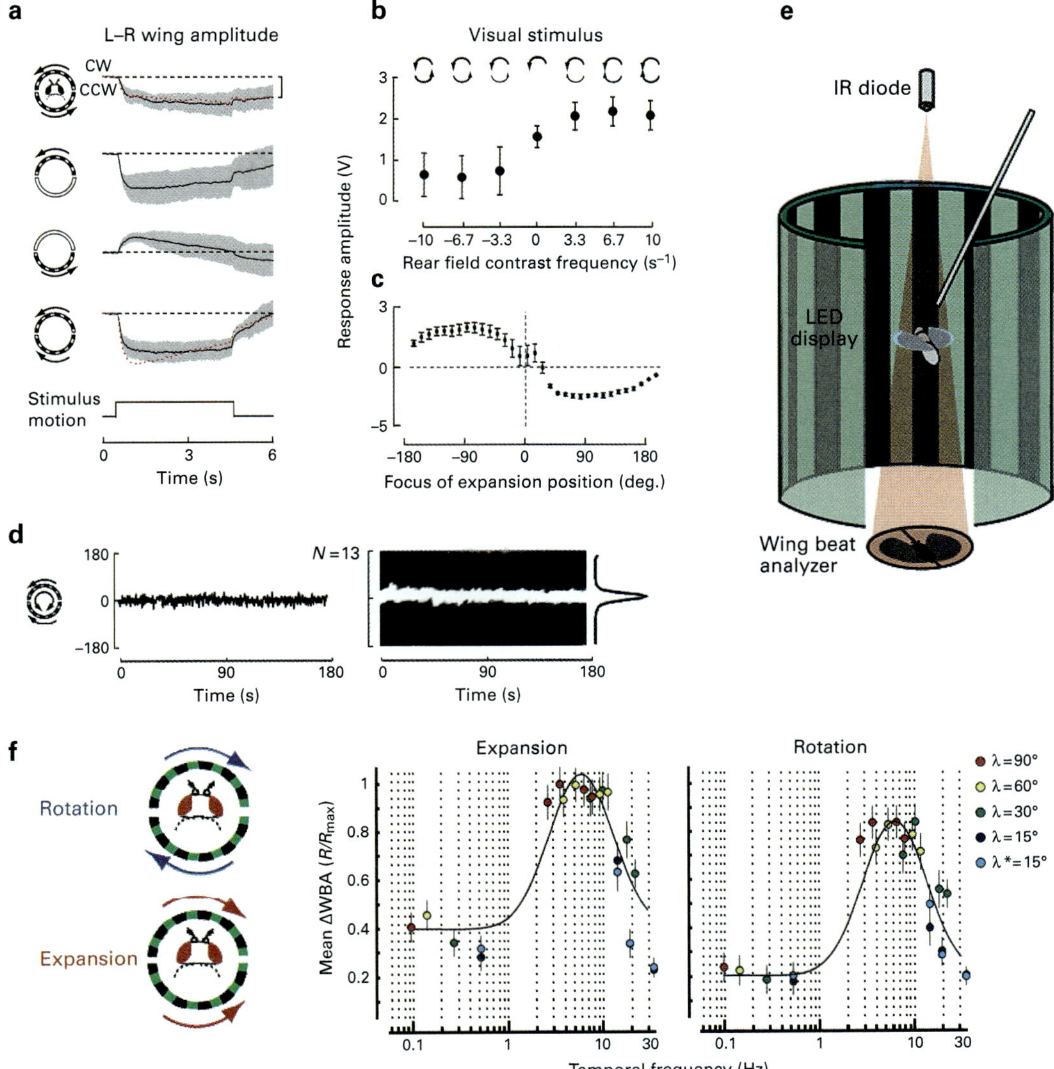

FIGURE 83.2   Avoidance response. (a) Open-loop torque responses of tethered flying *Drosophilae* (stimulus conditions shown to the left; from Tammero, Frye, & Dickinson, 2004). (b) Mean torque as a function of the grating's temporal frequency in the rear visual field. The temporal frequency of the grating in the frontal visual field is constant (from Tammero, Frye, & Dickinson, 2004). (c) Mean torque response to an expanding optic flow as a function of the focus of expansion (from Tammero, Frye, & Dickinson, 2004). A focus of expansion at 90° means an expansion stimulus delivered to the left of the animal. In both c and d, positive responses represent a counterclockwise turning tendency. (d) Closed-loop response of a tethered flying *Drosophila* to an expanding optic flow. The fly keeps the focus of expansion permanently at 0° position, that is, behind itself. Left: single trace. Right: Average of 13 flies (from Tammero, Frye, & Dickinson, 2004). (e) Experimental setup. An infrared (IR) diode casts a shadow of the wing-beat envelope on two photodiodes. The difference of the wing-beat amplitudes is a measure of the torque generated by the fly (from Duistermars et al., 2007). (f) Mean open-loop torque response to an expanding and a rotating optic flow as a function of the grating's temporal frequency (from Duistermars et al., 2007). Here, a positive response denotes a clockwise turning tendency.

amazing detail (see figure 83.3b and c; Borst & Bahde, 1986, 1988). Differences in the latency of both the collision-avoidance and the landing response with the expansion rate of the grating support the hypothesis that the two behaviors are mediated by separate pathways (Tammero & Dickinson, 2002a).

When flies are at rest, an expanding visual stimulus signals the approach of a predator or another threat as, for example, a human equipped with a fly swatter. It is common knowledge that flies are extremely hard to catch under these conditions. The reason why is that image expansion elicits a fast and reliable escape start

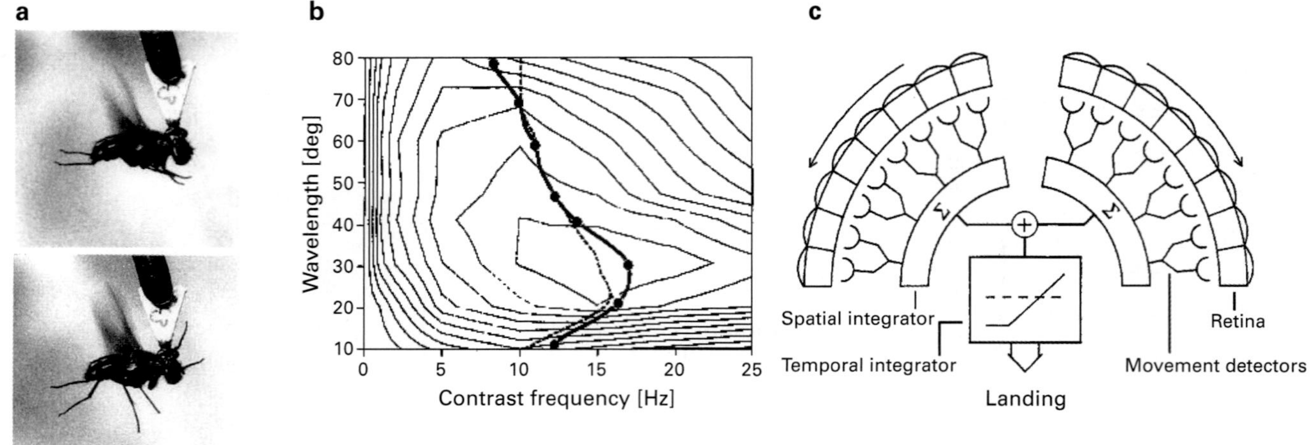

FIGURE 83.3   Landing and escape response. (a) A tethered flying housefly during flight (top) and during landing (bottom) (from Borst, 1990). (b) Contour plot of the response latency as a function of the grating's temporal frequency (on the x-axis) and spatial wavelength (on the y-axis). Overlaid lines represent, for each wavelength, the temporal frequency at which minimum response latency is observed. Solid line: experimental data. Broken line: model predictions (from Borst & Bahde, 1988). (c) Model for the release of the landing response (from Borst & Bahde, 1988).

in flies. Such an escape start was thought for long to be a highly stereotyped fixed action pattern. However, a more recent study by Card and Dickinson (2008) of the escape behavior of the fruit fly, *Drosophila*, revealed that this is not the case. In contrast, flies use visual information to plan a jump directly away from a looming threat. Using high-speed videography, they found that approximately 200 ms before takeoff, flies begin a series of postural adjustments that determine the direction of their escape. These movements position their center of mass so that leg extension will push them away from the expanding visual stimulus.

As for the neural control elements of the various responses to expanding stimuli, surprisingly little is known. Apart from unidentified neurons recorded in the cervical connective of blowflies that respond to an expanding optic flow (Borst, 1991), only a single most recent study reports about looming-sensitive neurons found within the lobula and the lobula plate of *Drosophila* (deVries & Clandinin, 2012). As do all other tangential cells, these cells have large dendrites (see figure 83.4a). They respond most strongly when stimulated by looming disks with a time course reminiscent of what has been found in looming sensitive neurons of the locust (Hatsopoulos, Gabbiani, & Laurent, 1995). Intriguingly however, their response is almost identical when varying the position of the focus of expansion over almost 90° within the visual field of the fly (see figure 83.4b and c). This rules out a mechanism involving local motion detectors with a fixed orientation in visual space. While the general mechanism underlying

their response selectivity remains unsolved, the role of these looming-sensitive neurons for behavior could be established in a most convincing way by two types of experiments. First, the authors genetically blocked the synaptic output from these neurons and demonstrated that the escape response, elicited by a looming object, was strongly reduced. This establishes the necessity of these cells for the escape response. Sufficiency of the looming cells to elicit the response was established by optogenetic stimulation in blind flies: When activating these cells in blind flies, flies readily take off and start lifting themselves up in the air, exactly as is observed in intact flies in response to an approaching object (see figure 83.4d and e).

OBJECT RESPONSE

The reflexes discussed above are all elicited by large-field moving stimuli. However, flies also respond to small objects that move relative to a background either because these objects are other animals that actively move in physical space or because relative motion of the object on the fly's retina is caused by ego-motion of the fly. As was the case of the optomotor response, seminal studies were performed on tethered walking or flying flies.

When a fly is tethered to a torque meter inside a transparent cylinder containing a single vertical black stripe and given control over the angular velocity of the stripe by feeding back the fly's torque onto the motor controlling the arena ("closed-loop"), Reichardt and

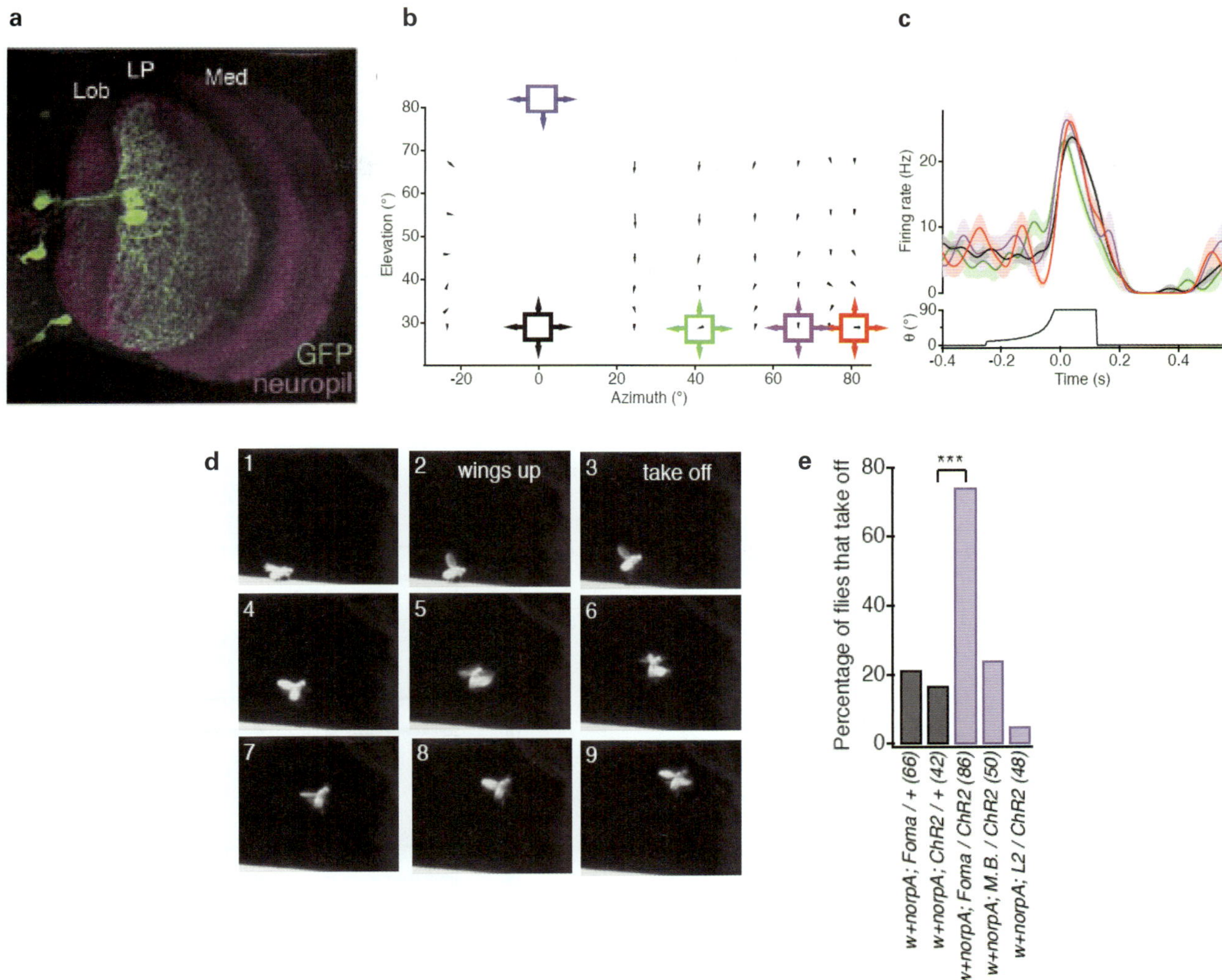

FIGURE 83.4   (a) A looming-sensitive neuron in *Drosophila*. (b) Five different foci of expansion used in the experiments. Arrows represent the local preferred directions of the neurons as determined by a moving dot. (c) Responses as recorded from the looming-sensitive neuron. The color corresponds to the specific focus of expansion tested (see b). The neuron responds the same way, no matter where in the visual field the disk started to expand. (d) A blind fly takes off in response to illumination when channelrhodopsin is expressed in looming-sensitive neurons. (e) Summary results of such experiments, including various control lines (from left to right): Only driver gene, but no effector (Foma/+), only effector, but no driver (ChR2/+), driver and effector (Foma/ChR2), mushroom body driver and effector (M.B./ChR2), L2 driver and effector (L2/ChR2). All figures from DeVries and Clandinin (2012).

colleagues were the first to describe that houseflies tend to keep the stripe in front of them, a behavior they called "object fixation" (see figure 83.5a; Reichardt & Wenking, 1969; Poggio & Reichardt, 1973). Searching for the underlying mechanism, they determined the fly's torque, in an "open-loop" experiment, as a function of the stripe's position. They found a particular function, termed D(Ψ), showing a rather linear relationship within the frontal visual field from about −45° to +45° (figure 83.5b; see also figure 3 in Reiser & Dickinson, 2010). Put in simple words, flies turn to the left when the stripe is to their left side and turn right

when the stripe is positioned to their right. Obviously, such a relationship will bring the stripe stably in the frontal position under closed-loop conditions. In the next step of analysis, the visual surround was made more complex by introducing a background and an equally textured stripe that either moved coherently with the background or relative to it with a given phase relation (Egelhaaf, 1985a; Reichardt & Poggio, 1979). These experiments revealed that, as soon as the stripe starts moving relative to the background, flies shift their mean torque toward the side where the stripe is positioned with marked large-amplitude responses

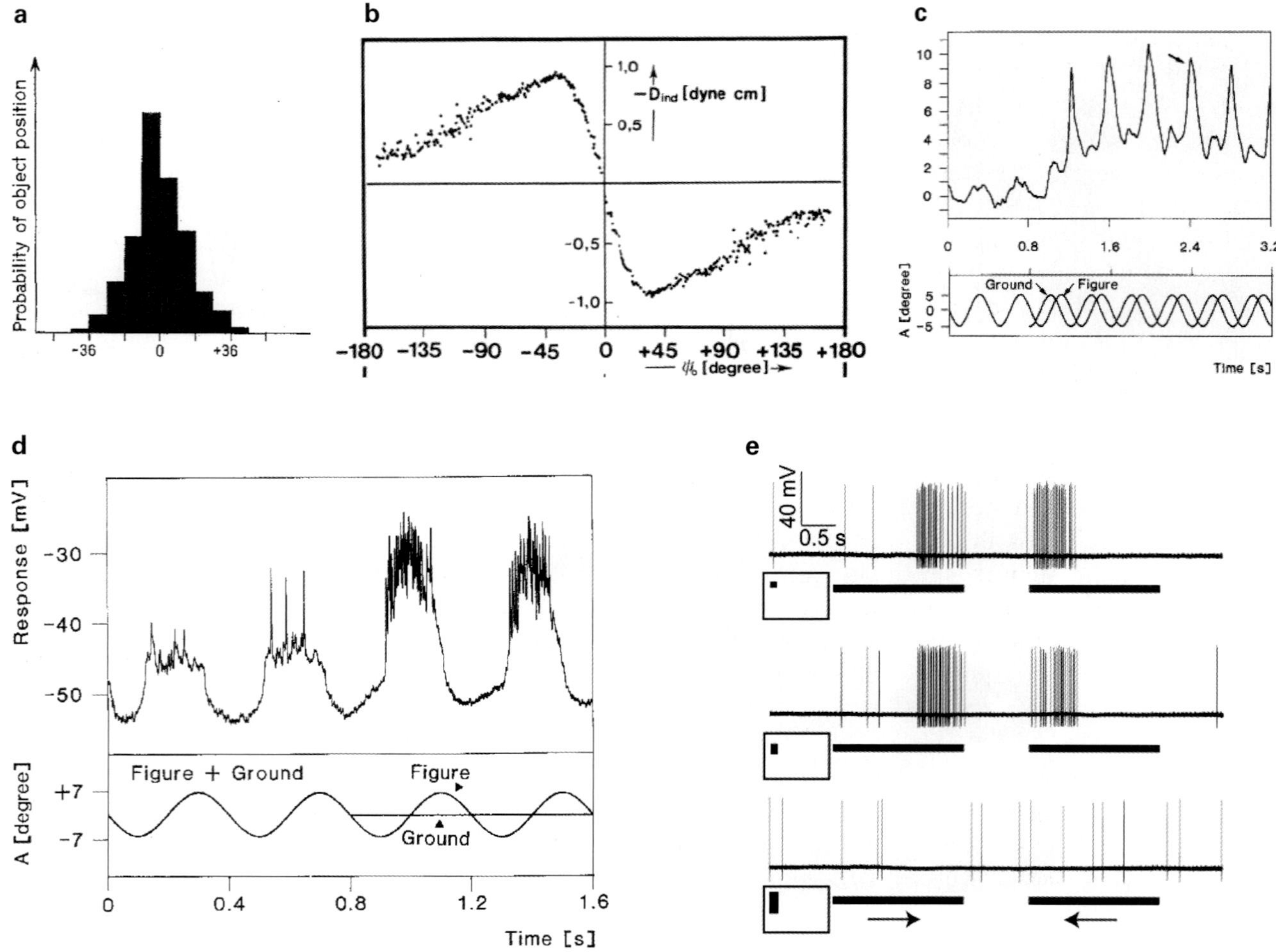

FIGURE 83.5   (a) Fixation of objects by tethered flying houseflies under closed-loop conditions (from Reichardt & Wenking, 1969). (b) Average torque response as a function of the stripe position under open-loop conditions (from Poggio & Reichardt, 1973). (c) Time course of the torque response (upper trace) during synchronous motion of figure and background (lower trace, first two cycles, A = angle) and after introduction of a phase shift (from Egelhaaf, 1985a.) (d) Example recording from a figure-detection neuron in a blowfly (from Egelhaaf, 1985b). (e) Example recording from a visual neuron of the hoverfly *Eristalis tenax* responding selectively to the motion of a small target (from Nordstroem, Barnett, & O'Carroll, 2006).

superimposed during front-to-back motion of the stripe (figure 83.5c; Egelhaaf, 1985a). Again, under closed-loop conditions, this would bring the stripe into the frontal position of the fly. Similar studies on fruit flies revealed that *Drosophilae*, as their large cousins, robustly fixate a single object in front of them (Goetz, 1987). There is, however, one observation indicating that fixation behavior is not a hardwired reflex but is better described as the result of operant conditioning: When the coupling between the fly's motor output and the motor controlling the velocity of the stripe is inverted such that the stripe is moving further away from the frontal position whenever the fly is producing a torque toward the stripe, flies, after some time, manage to stabilize the stripe in front of them nevertheless (Heisenberg & Wolf, 1984).

In the lobula plate of blowflies, neurons have been found responding selectively to small-object motion. In contrast to other tangential cells held responsible for the optomotor response, these neurons respond with a small amplitude depolarization only when the whole visual background is moving in their preferred direction. In contrast, when the background stops and only a small stripe is moving, these responses become enlarged in size, with a burst of action potentials riding on top of it (see figure 83.5d; Egelhaaf, 1985b). These cells, named "figure-detection" or "FD-cells," represent a heterogeneous group of neurons with different morphologies but similar response properties: As a common feature, they seem to be inhibited by large-field motion and exhibit a somewhat variable optimal stripe width of around 5°–30° (Gauck & Borst, 1999). Selective laser ablation could demonstrate that another tangential cell, the so-called "CH-cell" is responsible for this large-field inhibition of FD-cells (Warzecha, Borst, & Egelhaaf, 1992).

In hoverflies, similar neurons were described with an even more pronounced selectivity for small targets (Nordstroem, Barnett, & O'Carroll, 2006; Nordstroem & O'Carroll, 2006). Although many of these neurons, as the FD-cells mentioned above, are inhibited by the motion of a background pattern, some of them respond to target motion within the receptive field under a large range of background motion stimuli, even when there is no velocity difference between target and background. Analysis of responses to targets smaller than the size of the visual field of single photoreceptors or to targets with reduced contrast shows that these neurons have extraordinarily high contrast sensitivity. As an example, the neuron shown in figure 83.5e, responds vigorously to motion of a square of only 0.8° by 0.8°. In contrast to FD-cells, this neuron is not directionally selective, that is, responds whether the target is moving from the left to the right or in the opposite direction. As a further distinction, the response of this neuron is completely at base level when the moving target is vertically elongated to a stripe of 15° height (see figure 83.5e): Therefore, such neurons could not mediate the fixation behavior discussed above.

PUTTING IT ALL TOGETHER: ANALYSIS OF FREELY MOVING FLIES

In contrast to tethered flight, experimental parameters are less well defined under free flight conditions where multiple sensory inputs are integrated. Besides imbalances caused by turbulent air that lead to rotations of the animal, image expansions might occur in different parts of the visual field when the animal is getting close to the wall of the flight chamber. Moreover, in addition to the visual input, the animal is informed about its ego-motion by numerous proprioreceptors, first of all its halteres (see figure 83.6a, upper left; Chan, Prete, & Dickinson, 1998; Mayer et al., 1988; Nalbach & Hengstenberg, 1994). Like the visual optomotor system, halteres are known to respond to rotational velocities around all body axes, leading to compensatory head and body responses. However, while the visual input predominates for low oscillation velocities, haltere-mediated responses grow in the high-velocity range (figure 83.6a; Sherman & Dickinson, 2003). An additional complication arises from the imprecision of the behavioral readout: While the position and orientation of the body of the fly might be recordable quite easily, determining its precise head position via the same tracking system is usually not possible because of its limited spatial resolution given the large field of view. This leaves the visual input often ill-defined and makes it hard to estimate the contribution of each of the response components discussed above to the overall motor output.

Despite all these qualifications, plain observation of free-flying flies revealed a particular flight characteristic: During cruising flights, flies typically fly in rather straight lines for several hundreds of milliseconds before turning rather abruptly (Wagner, 1986a; Wehrhahn, Poggio, & Buelthoff, 1982). As revealed by a coil system attached to the thorax and the head, these so-called "saccades" are short time episodes that consist of coordinated turns of the body and the head (see figure 83.6b; Schilstra & van Hateren, 1998, 1999): A banked turn is initiated by a combined roll and yaw movement of the body while the head remains fixed for about 10 ms. This then is followed by a rapid turn of the head into the new flight direction, while the body again follows with a somewhat slower time course. Overall, a body saccade lasts for about 30 ms while the saccade movement of the head is completed within 20 ms or less.

From experiments utilizing a magnetic tether in which flies are fixed in space but free to rotate about their yaw axis, it is known that body saccades in *Drosophila* are largely controlled by the halteres: Altering visual feedback (by rotating the visual panorama whenever the fly initiated a saccade) had no significant effect on the dynamics of saccades whereas increasing and decreasing the amount of haltere-mediated feedback (by putting additional mass on the end knob or ablating one haltere completely) decreased and increased saccade amplitude, respectively (Bender & Dickinson, 2006). This, however, does not rule out visual feedback to be used for course control between the saccades. Indeed, high-speed video analysis of fruit flies flying inside a transparent cylinder (see figure 83.6c) showed that free-flight behavior is dramatically influenced by rotation of a surrounding textured drum (Mronz & Lehmann, 2008). When the drum is stationary, flies display their typical saccade-like flight structure with rather straight episodes interspersed by rapid changes of their flight direction (see figure 83.6d). In contrast, when the drum is rotating, the flight path becomes much more curved, syndirectional to the rotation of the drum, with straight flight episodes and saccades almost absent (see figure 83.6e). In addition, Tammero and Dickinson (2002b) demonstrated visual input to have a strong influence on saccade initiation: Flies surrounded by a high-contrast textured drum have much shorter straight flight episodes and make saccades much more frequently than when surrounded by a milky plexiglass

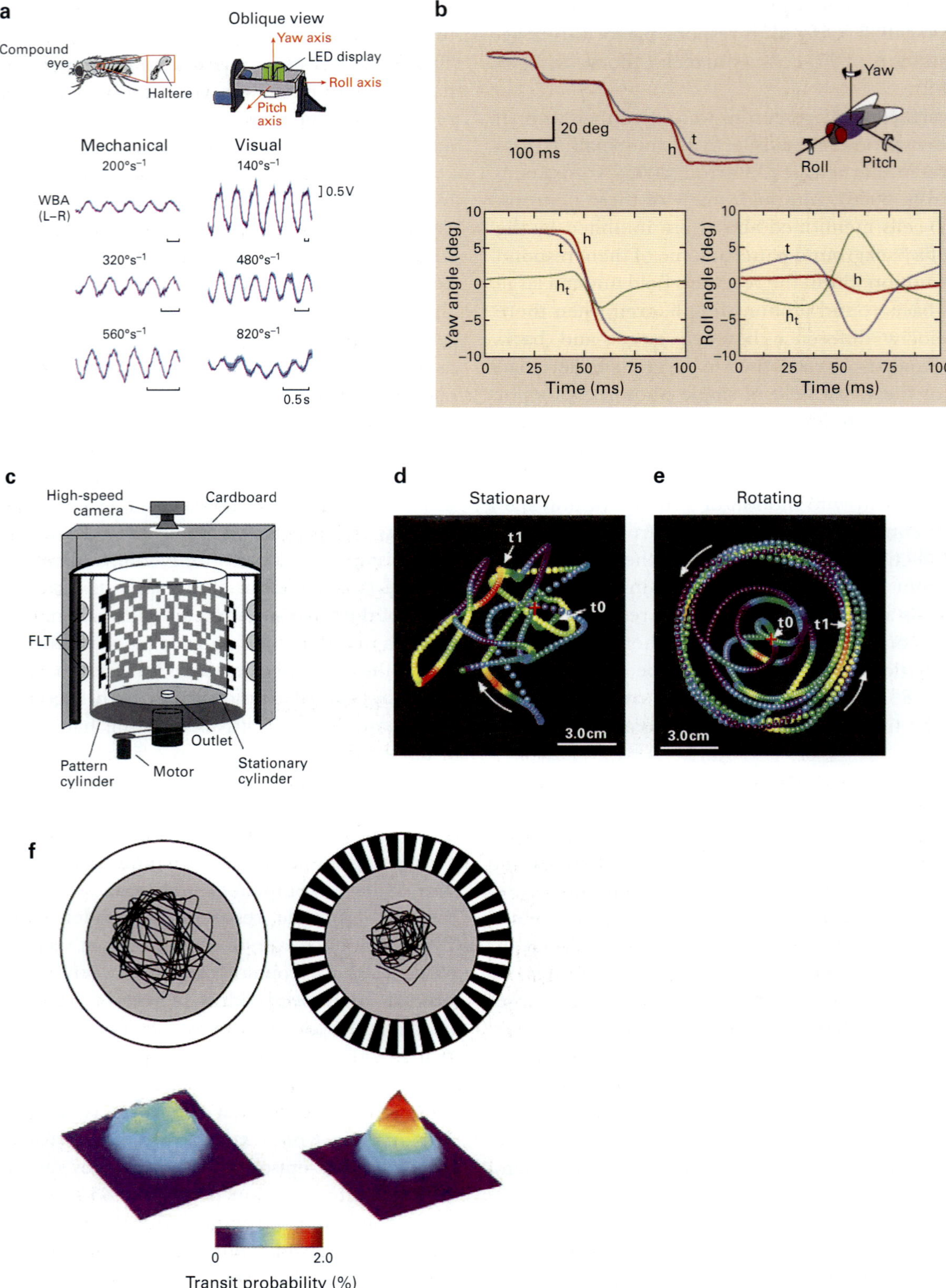

Figure 83.6 (a) Top left: The mechanosensory haltere is a small club shaped modified hind wing. Top right: Apparatus for delivering visual and mechanosensory stimuli. A wrap-around light-emitting diode (LED) display is mounted within a 3°-of-freedom rotational gimbal. The fly is mounted in the center of the visual display, above a sensor that measures the left and right wingbeat amplitudes (WBA). Below: Torque reactions of tethered flying *Drosophilae* in response to mechanical (left) and visual (right) stimulation, using three different peak velocities of the oscillation (from Sherman & Dickinson, 2003). (b) Head (h), thorax (t) and head-to-thorax (h_t) angle of free-flying blowflies (from Schilstra & van Hateren, 1998). (c–e) Flight arena (c) with tracks from individual fruit flies, with a stationary panorama (d) and while the panorama is rotating at a constant speed (e) (from Mronz & Lehmann, 2008). (f) Top: Top view of tracks of a freely flying *Drosophila* surrounded by a milky glass (left) and by a patterned wall (right). Bottom: Distribution histograms of the same fly under both conditions (from Tammero & Dickinson, 2002b).

cylinder (see figure 83.6f). The experiments can be summarized in the following picture of flight control in flies: During straight flight, flies use their visual opto-motor control system, while they rely on proprioceptive feedback from the halteres when performing their rapid saccades. To what extend the head saccade is visually controlled remains to be seen.

While the above experiments unambiguously demonstrate the importance of the optomotor response for the free-flight behavior, the situation is less clear in case of the responses of flies to an expanding optic flow. Here, analysis on tethered flies revealed an optic flow based mechanism with subsequent temporal integration and thresholding (Borst & Bahde, 1988). In contrast, analysis of freely moving flies landing on a sphere inside a cage led to the proposal that flies use a different control variable, namely, the change of the angular size of the landing site (Wagner, 1982). One possibility to resolve this contradiction is that spatiotemporal integration of an expansion flow, applied to the physical situation of a black sphere in front of a distant background, leads to similar predictions as does the relative retinal expansion velocity: In other words, flies might indeed use the latter control variable but approximate it by a spatiotemporal integration of the optic flow.

As was the case for tethered flying flies, vigorous fixation behavior is also observed in freely walking flies (but with wings cut!) on a circular platform surrounded by a water-filled moat, with two vertical stripes located on opposite sides. Under these conditions, flies walk back and forth on an almost straight line between the two inaccessible landmarks, as if they couldn't decide which one to choose, a set-up called "Buridan's paradigm" in the literature (see figure 83.7a; Goetz, 1980; Strauss & Pichler, 1998). A similar fixation behavior is observed in free flight when *Drosophilae* are video-tracked inside a cage with a bar suspended in the midst of it (see figure 83.7b; Maimon, Straw, & Dickinson, 2008): Here, flies clearly tend to circle around the bar. However, when the bar is substantially shortened, flies no longer become attracted by it but rather avoid the space surrounding it. Thus, shortening a bar reverses the attractiveness of a visible object, switching the behavior from a fixation to an avoidance response.

But what visual cues are flies using during fixation behavior? Is it stereopsis or peering, is it the expansion of the object on the retina, or is the relative motion against the background, that is, motion parallax during walking? To answer these questions, Goetz and colleagues (Schuster, Strauss, & Goetz, 2002; Strauss, Schuster, & Goetz, 1997) applied virtual environment techniques in freely walking fruit flies where a tracker-controlled 360° panorama converts a fruit fly's changing coordinates into object illusions that require the perception of specific cues to appear at preselected distances up to infinity. These studies revealed that en route sampling of retinal-image changes accounts for distance discrimination within a surprising range of at least 8–80 body lengths (20–200 mm). Stereopsis and peering are not involved. Rather, distance from image translation in the expected direction (motion parallax) outweighs distance from image expansion. The ability to discriminate distances is robust to artificially delayed updating of image translation within a time window of about 2 s (Schuster, Strauss, & Goetz, 2002; Strauss, Schuster, & Goetz, 1997).

Species-specific differences become obvious when considering the responses to small objects: Whereas fruit flies tend to avoid small objects (see above), freely flying male houseflies rather chase a small moving object, as represented by a conspecific female, by quite aerobatic flight maneuvers reaching rotational velocities of several thousand degrees per second (see figure 83.7c; Land & Collett, 1974; Wagner, 1986b; Wehrhahn, Poggio, & Buelthoff, 1982). Cross-correlating the angular orientation of the leading fly and the angular velocity of the chasing fly revealed a rather linear relationship between them, with a peak at a 30-ms time lag, which represents the total delay in the chasing fly.

CONCLUSIONS

Within the last 50 years, the processing of visual motion information in flies has been analyzed in unprecedented detail. Here, the optomotor response explored in open-loop experiments revealed the principal mechanism of elementary motion detection. Its contribution to free flight control and interplay with proprioceptive feedback from halteres and other visually mediated reflexes turned out to be much more complex and is still under investigation. At the neuronal level, several new developments hold great promises for the future.

First of all, single-cell recording in behaving animals has now been demonstrated to be feasible in several studies (Chiappe et al., 2010; Jung, Borst, & Haag, 2011; Maimon, Straw, & Dickinson, 2010). The results obtained so far already reveal that the neuronal response obtained in fixed animals can significantly differ from the behavioral response—not because the recorded neuron does not participate in the behavioral control but because its response properties depend on the behavioral state of the animal. In extreme cases, the neuron might simply be shut down when the animal is at rest and only becomes suprathreshold during walking or flight.

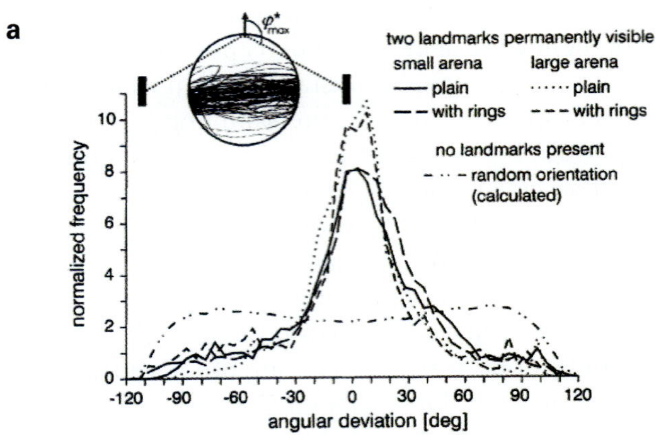

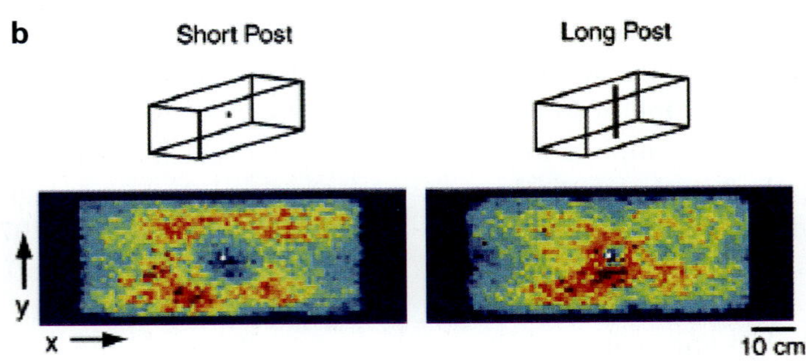

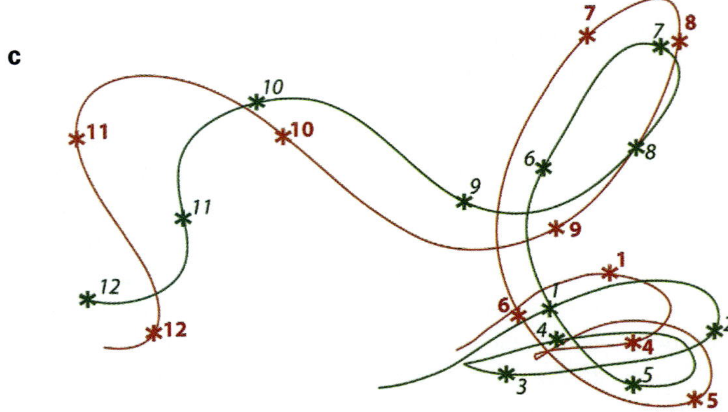

FIGURE 83.7   (a) In "Buridan's paradigm," fruit flies walk continuously back and forth between two distant stripes (from Strauss & Pichler, 1998). (b) X-Y position distribution of freely flying *Drosophilae* in a cage with a short (left) and a long (right) post suspended in the center (from Maimon, Straw, & Dickinson, 2008). (c) Flight tracks (viewed from above) of two male houseflies one chasing the other (leading fly in green, chasing fly in red) (modified from Wehrhahn, Poggio, & Buelthoff, 1982).

Second, with respect to another way of probing the participation of individual neurons in certain behaviors, *Drosophila* allows for silencing a genetically targeted neuron while observing the behavior, even in freely moving animals (for review, see Borst, 2009). This, together with a growing sophistication in manipulating the visual feedback in virtual environments, will certainly lead to a rapidly growing knowledge about the neural circuitry controlling the various flight maneuvers that have fascinated researchers for so long.

ACKNOWLEDGMENTS

I am grateful to Juergen Haag and Franz Weber for critically reading the manuscript.

REFERENCES

Bender, J. A., & Dickinson, M. H. (2006). A comparison of visual and haltere-mediated feedback in the control of body saccades in *Drosophila melanogaster*. *Journal of Experimental Biology, 209*, 4597–4606. doi:10.1242/jeb.02583.

Blondeau, J., & Heisenberg, M. (1982). The three-dimensional optomotor torque system of *Drosophila melanogaster*. Studies on wildtype and the mutant optomotor blind H31. *Journal of Comparative Physiology A, 145*, 321–329.

Borst, A. (1986). Time course of the houseflies' landing response. *Biological Cybernetics, 54*, 379–383.

Borst, A. (1990). How do flies land? From behavior to neuronal circuits. *Bioscience, 40*, 292–299.

Borst, A. (1991). Fly visual interneurons responsive to image expansion. *Zool. Jb. Physiol., 95*, 305–313.

Borst, A. (2009). *Drosophila*'s view on insect vision. *Current Biology, 19*, R36–R47.

Borst, A., & Bahde, S. (1986). What kind of movement detector is triggering the landing response of the housefly? *Biological Cybernetics, 55*, 59–69.

Borst, A., & Bahde, S. (1988). Spatio-temporal integration of motion: A simple strategy for safe landing in flies. *Naturwissenschaften, 75*, 265–267.

Borst, A., Haag, J., & Reiff, D. F. (2010). Fly motion vision. *Annual Review of Neuroscience, 33*, 49–70.

Borst, A., & Weber, F. (2011). Neural action fields for optic flow based navigation: A simulation study of the fly lobula plate network. *PLoS ONE, 6*, e16303. doi:10.1371/journal.pone.0016303.

Braitenberg, V., & Taddei Ferretti, C. (1966). Landing reaction of *Musca domestica* induced by visual stimuli. *Naturwissenschaften, 6*, 155. doi:10.1007/BF00591892.

Buchner, E. (1976). Elementary movement detectors in an insect visual system. *Biological Cybernetics, 24*, 86–101.

Card, G., & Dickinson, M. H. (2008). Visually mediated motor planning on the escape response of *Drosophila*. *Current Biology, 18*, 1300–1307. doi:10.1016/j.cub.2008.07.094.

Chan, W. P., Prete, F., & Dickinson, M. H. (1998). Visual input to the efferent control system of a fly's "gyroscope." *Science, 280*, 289–292.

Chiappe, M. E., Seelig, J. D., Reiser, M. B., & Jayaraman, V. (2010). Walking modulates speed sensitivity in *Drosophila* motion vision. *Current Biology, 20*, 1470–1475. doi:10.1016/j.cub.2010.06.072.

DeVries, S. E. J., & Clandinin, T. R. (2012). Loom sensitive neurons link computation to action in the *Drosophila* visual system. *Current Biology, 22*, 353–362.

Duistermars, B. J., Chow, D. M., Condro, M., & Frye, M. A. (2007). The spatial, temporal and contrast properties of expansion and rotation flight optomotor response in *Drosophila*. *Journal of Experimental Biology, 210*, 3218–3227.

Egelhaaf, M. (1985a). On the neuronal basis of figure–ground discrimination by relative motion in the visual system of the fly: I. Behavioral constraints imposed on the neuronal network and the role of the optomotor system. *Biological Cybernetics, 52*, 123–140.

Egelhaaf, M. (1985b). On the neuronal basis of figure–ground discrimination by relative motion in the visual system of the fly: II. Figure-detection cells, a new class of visual interneurones. *Biological Cybernetics, 52*, 195–209.

Franz, M. O., & Krapp, H. G. (2000). Wide-field, motion-sensitive neurons and matched filters for optic flow fields. *Biological Cybernetics, 83*, 185–197.

Gauck, V., & Borst, A. (1999). Spatial response properties of contralateral inhibited lobula plate tangential cells in the fly visual system. *Journal of Comparative Neurology, 406*, 51–71.

Geiger, G., & Nässel, D. R. (1981). Visual orientation behaviour of flies after selective laser beam ablation of interneurons. *Nature, 293*, 398–399.

Gibson, J. J. (1950). *Perception of the visual world*. Boston: Houghton Mifflin.

Goetz, K. G. (1964). Optomotorische Untersuchung des visuellen Systems einiger Augenmutanten der Fruchtfliege Drosophila. *Kybernetik, 2*, 77–92. doi:10.1007/BF00288561.

Goetz, K. G. (1980). Visual guidance in *Drosophila*. In O. Siddiqi, P. Babu, L. M. Hall, & J. C. Hall (Eds.), *Development and neurobiology of Drosophila* (pp. 391–407). New York: Plenum Press.

Goetz, K. G. (1987). Course-control, metabolism and wing interference during ultralong tethered flight in *Drosophila melanogaster*. *Journal of Experimental Biology, 128*, 35–46.

Gronenberg, W., Milde, J. J., & Strausfeld, N. J. (1995). Oculomotor control in Calliphorid flies—Organization of descending neurons to neck motor-neurons responding to visual-stimuli. *Journal of Comparative Neurology, 361*, 267–284.

Haag, J., Denk, W., & Borst, A. (2004). Fly motion vision is based on Reichardt detectors regardless of the signal-to-noise ratio. *Proceedings of the National Academy of Sciences of the United States of America, 101*, 16333–16338.

Haag, J., Wertz, A., & Borst, A. (2007). Integration of lobula plate output signals by DNOVS1, an identified premotor descending neuron. *Journal of Neuroscience, 27*, 1992–2000.

Haag, J., Wertz, A., & Borst, A. (2010). Central gating of fly optomotor response. *Proceedings of the National Academy of Sciences of the United States of America, 107*, 20104–20109.

Hatsopoulos, N., Gabbiani, F., & Laurent, G. (1995). Elementary computation of object approach by a wide-field neuron. *Science, 270*, 1000–1003.

Hassenstein, B., & Reichardt, W. (1956). Systemtheoretische Analyse der Zeit-, Reihenfolgen- und Vorzeichenauswertung bei der Bewegungsperzeption des Rüsselkäfers *Chlorophanus*. *Zeitschrift fur Naturforschung. Teil B. Anorganische Chemie, Organische Chemie, Biochemie, Biophysik, Biologie, 11b*, 513–524.

Hatsopoulos, N., Gabbiani, F., & Laurent, G. (1995). Elementary computation of object approach by a wide-field neuron. *Science, 270*, 1000–1003.

Hausen, K., & Wehrhahn, C. (1983). Microsurgical lesion of horizontal cells changes optomotor yaw response in the blowfly *Calliphora erythrocephala*. *Proceedings of the Royal Society of London. Series B, Biological Sciences, 219*, 211–216.

Heisenberg, M., & Wolf, R. (1984). *Vision in Drosophila*. Berlin: Springer.

Heisenberg, M., Wonneberger, R., & Wolf, R. (1978). Optomotor-blind (H31): A *Drosophila* mutant of the lobula plate giant neurons. *Journal of Comparative Physiology A, 124*, 287–296.

Hengstenberg, R. (1988). Mechanosensory control of compensatory head roll during flight in the blowfly *Calliphora erythrocephala* Meig. *Journal of Comparative Physiology, A, 163,* 151–165.

Joesch, M., Plett, J., Borst, A., & Reiff, D. F. (2008). Response properties of motion-sensitive visual interneurons in the lobula plate of *Drosophila melanogaster. Current Biology, 18,* 368–374.

Jung, S. N., Borst, A., & Haag, J. (2011). Flight activity alters velocity tuning of fly motion-sensitive neurons. *Journal of Neuroscience, 31,* 9231–9237.

Koenderink, J. J., & van Doorn, A. J. (1987). Facts on optic flow. *Biological Cybernetics, 56,* 247–254.

Krapp, H. G., & Hengstenberg, R. (1996). Estimation of self-motion by optic flow processing in single visual interneurons. *Nature, 384,* 463–466.

Krapp, H. G., Hengstenberg, B., & Hengstenberg, R. (1998). Dendritic structure and receptive-field organization of optic flow processing interneurons in the fly. *Journal of Neurophysiology, 79,* 1902–1917.

Land, M. F., & Collett, T. S. (1974). Chasing behaviour of houseflies (*Fannia canicularis*): A description and analysis. *Journal of Comparative Physiology, 89,* 331–357.

Maimon, G., Straw, A. D., & Dickinson, M. H. (2008). A simple vision-based algorithm for decision making in flying *Drosophila. Current Biology, 18,* 464–470.

Maimon, G., Straw, A. D., & Dickinson, M. H. (2010). Active flight increases the gain of visual motion processing in *Drosophila. Nature Neuroscience, 13,* 393–399.

Mayer, M., Vogtmann, K., Bausenwein, B., Wolf, R., & Heisenberg, M. (1988). Flight control during 'free yaw turns' in *Drosophila melanogaster. Journal of Comparative Physiology, A, 163,* 389–399.

Mronz, M., & Lehmann, F. O. (2008). The free-flight response of *Drosophila* to motion of the visual environment. *Journal of Experimental Biology, 211,* 2026–2045. doi:10.1242/jeb.008268.

Nalbach, G., & Hengstenberg, R. (1994). The halteres of the blowfly *Calliphora*: II. Three-dimensional organization of compensatory reactions to real and simulated rotations. *Journal of Comparative Physiology, A, 175,* 695–708.

Nordstroem, K., & O'Carroll, D. C. (2006). Small object detection neurons in female hoverflies. *Proceedings of the Royal Society B, Biological Sciences, 273,* 1211–1216.

Nordstroem, K., Barnett, P. D., & O'Carroll, D. C. (2006). Insect detection of small targets moving in visual clutter. *PLoS Biology, 4,* e54. doi:10.1371/journal.pbio.0040054.

Poggio, T., & Reichardt, W. (1973). A theory of the pattern induced flight orientation of the fly *Musca domestica. Kybernetik, 12,* 185–203. doi:10.1007/BF00270572.

Reichardt, W. (1987). Evaluation of optical motion information by movement detectors. *Journal of Comparative Physiology, A, 161,* 533–547.

Reichardt, W., & Poggio, T. (1979). Figure–ground discrimination by relative movement in the visual system of the fly: I. Experimental results. *Biological Cybernetics, 35,* 81–100.

Reichardt, W., & Wenking, H. (1969). Optical detection and fixation of objects in fixed flying flies. *Naturwissenschaften, 56,* 424–425. doi:10.1007/BF00593644.

Reiser, M. B., & Dickinson, M. H. (2010). *Drosophila* fly straight by fixating objects in the face of expanding optic flow. *Journal of Experimental Biology, 213,* 1771–1781. doi:10.1242/jeb.035147.

Schilstra, C., & van Hateren, J. H. (1998). Stabilizing gaze in flying blowflies. *Nature, 395,* 654.

Schilstra, C., & van Hateren, J. H. (1999). Blowfly flight and optic flow: II. Head movement during flight. *Journal of Experimental Biology, 202,* 1491–1500.

Schuster, S., Strauss, R., & Goetz, K. G. (2002). Virtual-reality techniques resolve the visual cues used by fruit flies to evaluate object distances. *Current Biology, 12,* 1591–1594. doi:10.1016/S0960-9822(02)01141-7.

Sherman, A., & Dickinson, M. H. (2003). A comparison of visual and haltere-mediated equilibrium reflexes in the fruit fly *Drosophila melanogaster. Journal of Experimental Biology, 206,* 295–302. doi:10.1242/jeb.00075.

Single, S., & Borst, A. (1998). Dendritic integration and its role in computing image velocity. *Science, 281,* 1848–1850.

Strausfeld, N. J., & Bassemir, U. K. (1985). Lobula plate and ocellar interneurons converge onto a cluster of descending neurons leading to neck and leg motor neuropil in Calliphora erythrocephala. *Cell and Tissue Research, 240,* 617–640.

Strausfeld, N. J., & Seyan, H. S. (1985). Convergence of visual, haltere and prosternal inputs at neck motor neurons of *Calliphora erythrocephala. Cell and Tissue Research, 240,* 601–615.

Strauss, R., & Pichler, J. (1998). Persistence of orientation toward a temporarily invisible landmark in *Drosophila melanogaster. Journal of Comparative Physiology, A, 182,* 411–423.

Strauss, R., Schuster, S., & Goetz, K. G. (1997). Processing of artificial visual feedback in the walking fruit fly *Drosophila melanogaster. Journal of Experimental Biology, 200,* 1281–1296.

Tammero, L. F., & Dickinson, M. H. (2002a). Collision-avoidance and landing responses are mediated by separate pathways in the fruit fly, *Drosophila melanogaster. Journal of Experimental Biology, 205,* 2785–2798.

Tammero, L. F., & Dickinson, M. H. (2002b). The influence of visual landscape on the free flight behavior of the fruit fly *Drosophila melanogaster. Journal of Experimental Biology, 205,* 327–343.

Tammero, L. F., Frye, M. A., & Dickinson, M. H. (2004). Spatial organization of visuomotor reflexes in *Drosophila. Journal of Experimental Biology, 207,* 113–122.

Wagner, H. (1982). Flow-field variables trigger landing in flies. *Nature, 297,* 147–148.

Wagner, H. (1986a). Flight performance and visual control of flight of the free-flying housefly (*Musca domestica* L.): I. Organization of the flight motor. *Philosophical Transactions of the Royal Society of London. Series B, Biological Sciences, 312,* 527–551.

Wagner, H. (1986b). Flight performance and visual control of flight of the free-flying housefly (*Musca domestica* L.): II. Pursuit of targets. *Philosophical Transactions of the Royal Society of London. Series B, Biological Sciences, 312,* 553–579.

Warzecha, A. K., Borst, A., & Egelhaaf, M. (1992). Neural circuit for the detection of small-field motion as revealed by laser ablation of single neurons. *Neuroscience Letters, 141,* 119–122.

Warzecha, A. K., & Egelhaaf, M. (1996). Intrinsic properties of biological motion detectors prevent the optomotor control system from getting unstable. *Philosophical Transactions of the Royal Society of London. Series B, Biological Sciences, 351*, 1579–1591.

Wehrhahn, C., Hausen, K., & Zanker, J. M. (1981). Is the landing response of the housefly (*Musca*) driven my motion of a flow field? *Biological Cybernetics, 41*, 91–99.

Wehrhahn, C., Poggio, T., & Buelthoff, H. (1982). Tracking and chasing in houseflies (*Musca*). *Biological Cybernetics, 45*, 123–130.

Wertz, A., Borst, A., & Haag, J. (2008). Nonlinear integration of binocular optic flow by DNOVS2, a descending neuron of the fly. *Journal of Neuroscience, 28*, 3131–3140.

84 Polarization Vision in Arthropods

UWE HOMBERG AND BASIL EL JUNDI

Electromagnetic radiation perceived as light has three properties: It can differ in spectral composition, perceived as color, in intensity, perceived as brightness, and in the plane of oscillation, termed polarization. While most visual systems exploit differences in brightness and color for visual discrimination, the third attribute, its polarization, is analyzed more rarely. Nevertheless, polarization sensitivity (PS) is present in all major animal taxa, including most vertebrate groups. Evidence from various species points to three major biological contexts of PS: (1) spatial orientation and navigation, (2) enhancement of contrast, especially at the water surface and underwater, and (3) intraspecific communication (Horváth & Varjú, 2004). While PS in vertebrates is still largely considered as a marginal visual sense, perhaps restricted to certain species, data in arthropods, especially insects, crustaceans, and spiders, have established PS as a major visual quality. Recent progress suggesting a role for circularly polarized light in arthropods and cephalopods and increasing insights into neuronal mechanisms of polarization vision in crustaceans and insects show that central nervous analysis of polarization reaches a complexity that could be comparable to the analysis of chromatic contrast.

This chapter provides an overview of behavioral evidence for polarized-light detection and its biological significance in arthropods. In addition we summarize current understanding of photoreceptor mechanisms for polarization vision and focus on recent advances in neural processing of polarization signals, particularly in insects.

SOURCES OF POLARIZED LIGHT

Light consists of electromagnetic waves that are characterized by orthogonal electric and magnetic field oscillations perpendicular to the direction of propagation. In natural light sources such as the sun, the orientations of the electric field vectors (*E*-vectors) of the molecular light emitters are not correlated with each other, resulting in unpolarized light. Light is linearly polarized when the electric field is oriented in a single plane (see figure 84.1A). In analogy to wavelength, saturation, and intensity in color vision, polarized light is characterized by the angle/orientation of the *E*-vector, by the relative percentage of polarized light (the degree of polarization), and by its intensity (Bernard & Wehner, 1977). In contrast to linearly polarized light, elliptically and circularly polarized light are characterized by a rotating *E*-vector orientation in a light beam owing to the overlap of two perpendicular oscillations of linearly polarized light of identical periodicity but differing in phase by 90° (leading to circularly polarized light; see figure 84.1A) or by a value between 0° and 90° (leading to elliptically polarized light). Depending on the rotational direction of the *E*-vector, circular/elliptical polarized light can be characterized as right-handed (clockwise) or left-handed (counterclockwise).

Linearly polarized light originates from light scattering or reflections (Waterman, 1981). Scattering of light from the sun leads to a characteristic pattern of polarization in the sky and underwater (see figure 84.1), and light reflected from shiny dielectric surfaces such as leaves, rocks, or bodies of water is likewise linearly polarized. Skylight polarization originates from scattering of sunlight in the atmosphere (Strutt, 1871). As a result, *E*-vectors in the blue sky are arranged along concentric circles around the sun (see figure 84.1B). The degree of polarization strongly depends on clouding conditions. It is highest at a scattering angle of 90° from the sun but even under the most favorable conditions (cloudless blue sky) does not exceed 75% (Brines & Gould, 1982) and gradually decreases to 0% toward the sun. A similar pattern, but with much lower intensity, exists in the night sky around the moon. Because the sky polarization pattern is fixed relative to the position of the sun, it can serve as an external compass, especially when the sun is not directly visible. This aspect of polarized light perception is in fact regarded as the most prevalent biological role of PS in insects.

In water, the refraction of light at the air–water boundary compresses the 180° celestial hemisphere and its polarization pattern into an angle of 97.2° (Snell's window). Outside this critical angle, scattering of sunlight in water itself leads to polarization. At moderate depth, the overall pattern of *E*-vector orientations is similar to that observed in the sky (Cronin & Shashar, 2001; see figure 84.1C) and might be used to locate the sun's azimuth down to 200 m or more in clear waters (Waterman, 2006). Scattering of light in the water itself

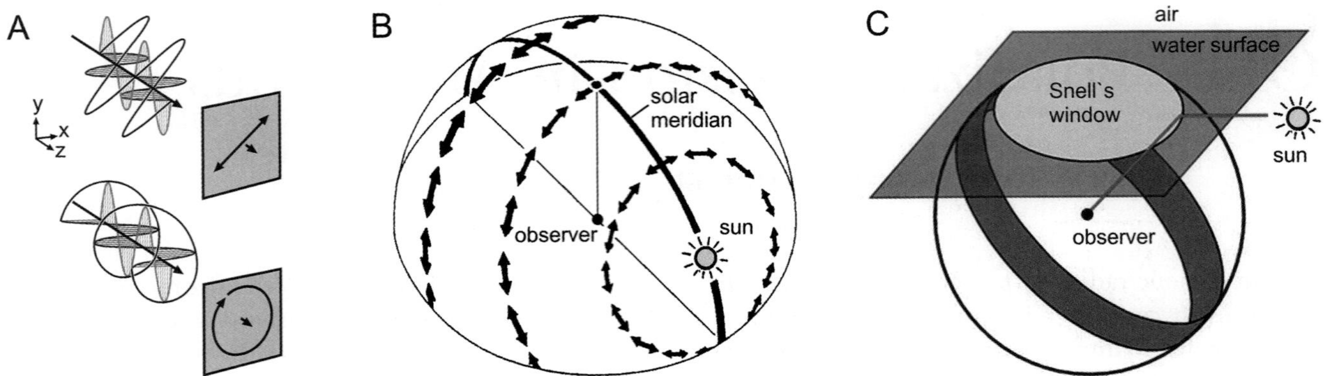

FIGURE 84.1   A polarized light stimulus. (A) Physical basis of polarized light. Linearly polarized light can be regarded as being composed of horizontally and vertically oscillating electric field components, oscillating in phase with each other (upper panel). Circularly polarized light is obtained when the two components have identical amplitudes, but a relative phase shift of 90° (lower panel). Depending on whether the phase shift is negative or positive, the changing *E*-vector will rotate in the clockwise or the counterclockwise direction. (B) Pattern of polarization in the sky. *E*-vectors (double arrows) are oriented tangentially to concentric circles around the sun. The thickness of the double arrows indicates the degree of polarization. (C) Underwater, light is polarized in all directions. The highest degree of polarization occurs—as in the sky—in a band perpendicular to the refracted sunlight, shown here for a solar position at the horizon. *E*-vectors are oriented perpendicularly to the plane of scattering. (A modified from Wehner & Labhart, 2006; B from Rossel & Wehner, 1987 with permission from Springer; C from Hawryshyn, 1992.)

leads to horizontally polarized light. It increasingly dominates underwater polarization with increasing depth inside and outside the critical angle (Waterman, 1981, 2006). The degree of polarization depends on the volume of water scattering the light but rarely exceeds 50% (Waterman, 2006). Light coming from the open sea is more strongly polarized than light coming from the shore, which might explain the mechanism underlying shore flight, the movement of small crustaceans away from the shore into deeper water to avoid shallow water predators (Schwind, 1999).

PHOTORECEPTOR MECHANISMS

The molecular basis of PS lies in the dichroic nature of retinal, and hence of rhodopsin, which absorbs light maximally if it oscillates parallel to the excitable double bonds of retinal. Photoreceptor neurons are, therefore, polarization sensitive to the degree to which rhodopsin molecules are aligned in parallel with respect to the incoming light. In arthropod photoreceptors, rhodopsin is incorporated into microvillar membranes, finger-like extrusions from the cells, which together form the rhabdomeric microvillar seam (see figure 84.2). Even if the chromophores are randomly distributed in the plane of the photoreceptor membrane, a single microvillus should be more sensitive to light oscillating parallel to the micovillar axis than to light oscillating in a perpendicular plane, resulting in a theoretical PS of 2 (Moody & Parriss, 1961). However, measurements have

revealed much higher PS values of up to 10 or more, suggesting that additional mechanisms enhance the alignment of rhodopsins. Candidate factors increasing alignment are anchoring of rhodopsin molecules to cytoskeletal elements, to scaffolding proteins of the phototransduction signaling complex (Xu et al., 1998), or alignment across adjacent microvillar membranes, but all of these mechanisms are still hypothetical (Roberts, Porter, & Cronin, 2011). In addition to parallel orientation of rhodopsin in the microvillar membrane, the microvilli themselves must be aligned in parallel to generate high PS of the photoreceptor cell. This is in fact the case for many crustacean photoreceptors (see figure 84.2B) and for insect photoreceptors of the polarization-sensitive dorsal rim area (DRA) of the eye (see below). Outside the DRA, microvilli are poorly aligned or even twisted along the rhabdomere, thereby greatly reducing PS (Wehner & Labhart, 2006).

BIOLOGICAL SIGNIFICANCE AND BEHAVIORAL EVIDENCE

Spatial Orientation and Navigation

Just over 60 years ago, Karl von Frisch first described polarized light perception in an animal by showing that the dances of honeybees could be manipulated by changing the *E*-vector orientation of a patch of blue sky on a horizontal dance platform in front of the beehive (von Frisch, 1949). Meanwhile compass orientation by

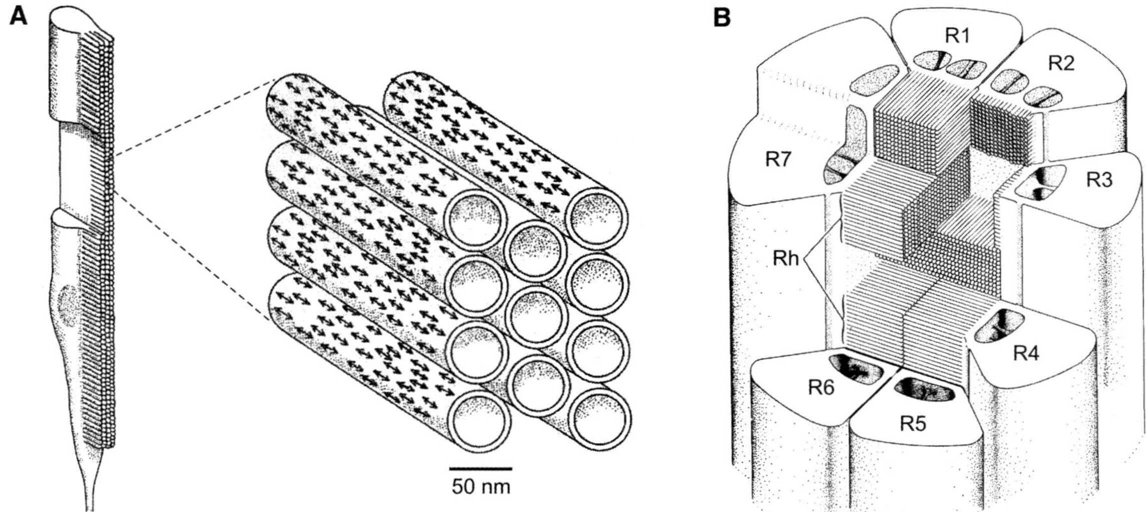

FIGURE 84.2   Rhabdomeric photoreceptors of an insect (A) and a crustacean (B). (A) In insect compound eye photoreceptors the rhabdomeric microvilli are oriented perpendicularly to the incident light. Enlargement illustrates parallel orientation of the chromophore dipoles in the microvillar membranes, favoring polarization sensitivity. (B) Orthogonal orientation of microvilli in a crab ommatidium showing the 7 large photoreceptor cells (R1–R7). Rh, rhabdomere. (A modified from Wehner, 1976; B from Stowe, 1980 with permission from Springer.)

skylight polarization has been demonstrated in several insect species, including honeybees, desert ants, dung beetles, fruit flies, and monarch butterflies, and more indirect evidence suggests that a wide range of insects from all major taxa detect the sky polarization pattern (Horváth & Varjú, 2004; Labhart & Meyer, 1999; Weir & Dickinson, 2012). In all species studied, a small DRA of the eye is specifically adapted for polarization vision. In the DRA, photoreceptors are homochromatic, usually in the UV or blue range, and microvilli in each DRA ommatidium form two blocks of orthogonal orientation (see figure 84.3). Central place foragers like honeybees and desert ants largely rely on the sky polarization pattern to perform path integration during their homeward orientation, and in the case of honeybees, even use it to communicate the direction of a food source to nest mates (Rossel, 1993; Wehner, 2003). In seasonal migrants like the monarch butterfly, the preferential orientation to the angle of skylight polarization is innate and allows the animals to find their overwintering grounds in central mountain ranges in Mexico (Reppert, Zhu, & White, 2004). Dung beetles, on the other hand, just need to take a straight path away from a dung pile, and some species even rely on the weak polarization pattern around the moon to perform this task (Dacke, Nordström, & Scholtz, 2003).

Evidence for navigation by sky polarization also exists for three families of spiders (Dacke, Doan, & O'Carroll, 2001; Görner, 1962; Ortega-Escobar & Muñoz-Cuevas, 1999). While the funnel web spider

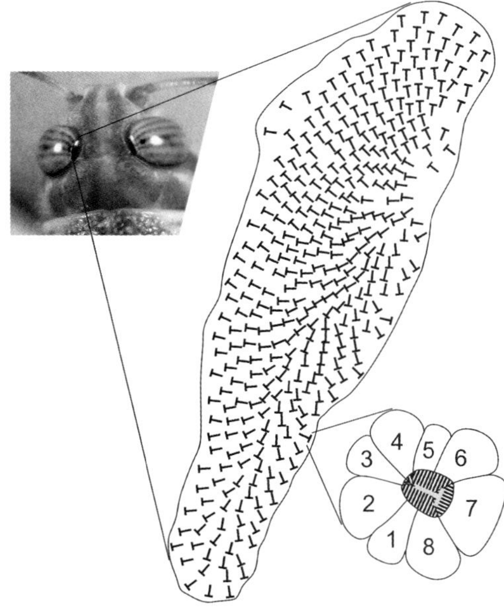

FIGURE 84.3   Polarization-sensitive dorsal rim area (DRA) in the compound eye of the desert locust *Schistocerca gregaria*. T-shaped symbols illustrate the fan-shaped arrangement of microvillar orientations throughout the DRA. In each DRA ommatidium, the microvilli of receptor cell 7 are oriented perpendicularly to those of receptor cells 1, 2, 5, 6, and 8. Microvilli of photoreceptor cells 3 and 4 are irregular. (Adapted from Homberg, 2004.)

Agelena labyrinthica and the wolf spiders *Arctose variana* and *Lycosa tarantula* use their anterior median eyes for polarization vision to navigate back to their retreat, Dacke et al. (1999) discovered a unique role for the posterior median eyes of the gnaphosid spider *Drassodes cupreus* as a sky compass organ. These eyes apparently act as two *E*-vector analyzers with perpendicular *E*-vector tuning. The concurrent receptive fields of both eyes in the zenith and a unique canoe-shaped tapetum boost PS and make this dual-eye system an ideal sky polarization compass, particularly during sunset when the spider is active.

Both inside and outside Snell's window, the polarization pattern underwater provides orientational information and might be used for sun compass navigation (Waterman, 2006). Although this is assumed to be an important function for polarization vision in crustaceans, behavioral evidence is scarce. A notable exception is the grass shrimp, *Palaemonetes vulgaris*, which uses the polarization pattern for compass orientation in off-shore escape responses (Goddard & Forward, 1991). Schwind (1999) provided evidence that the stronger polarized light from the open ocean than from the shore might underlie the mechanism by which planctonic crustaceans move away from the shore to avoid shallow water predators.

Perception of Water Surfaces

Depending on the reflections of the sky and vegetation, water surfaces such as lakes are difficult to detect reliably based on intensity and color cues alone (Horváth & Varjú, 2004). In contrast, light reflected from water surfaces is always polarized horizontally. Therefore, many insects associated with water detect water surfaces by polarotaxis (Schwind, 1989, 1995). These include water bugs, nonbiting midges, dragonflies, mayflies, tabanid flies, and aquatic beetles (reviewed by Horváth & Varjú, 2004; Kriska et al., 2009). Schwind (1984, 1985) showed that a plunge reaction of flying backswimmers, *Notonecta glauca*, into water is elicited specifically by the strong horizontal polarization of the water surface. Desert locusts, likewise, apparently detect water surfaces by polarized light reflections and might use this to avoid flying over large bodies of water (Shashar, Sabbah, & Aharoni, 2005). Accordingly, a specialized polarization-sensitive ventral eye region has been demonstrated in *Notonecta* (Schwind, 1985) and the water strider, *Gerris lacustris* (Schneider & Langer, 1969). Ventral sensitivity to polarized light was recently also found in the fruit fly, *Drosophila*, but here, unlike the situation in other eyes, multiple spectral classes of photoreceptors are involved and a distinct polarization-sensitive ventral eye region is missing (Wernet et al., 2012).

Contrast Enhancement

Besides a role in navigation, the biological use for underwater polarization vision is generally assumed to lie in contrast enhancement (Johnsen, Marshall, & Widder, 2011; Lythgoe & Hemmings, 1967; Wehner, 2001). Light scattering in water creates a veil of light which strongly reduces the range of visibility. The amount of scattering can be reduced by analysis through vertical or crossed polarizers, thereby greatly increasing visibility (Cronin & Marshall, 2011; Wehner 2001). In addition, polarization analysis improves the visibility of prey by the detection of scattered radiance from zooplankton against horizontally polarized backgrounds (Johnsen, Marshall, & Widder, 2011). Contrast enhancement may also play a role in terrestrial environments but has hardly been studied. In *Papilio* butterflies, reflected polarized light is detected throughout the eye and appears to be used to enhance color or intensity contrast for recognition of leaves for egg deposition or flowers for foraging (Kelber, Thunell, & Arikawa, 2001; Kinoshita, Yamazato, & Arikawa, 2011).

Intraspecific Communication—A Role for Circularly Polarized Light?

Iridescent scales of *Heliconius* butterflies, which are adapted to deep forest habitats, strongly reflect linearly polarized light. This is attractive to male conspecifics, indicating that polarized light serves a role in mate recognition (Sweeney, Jiggins, & Johnsen, 2003). In contrast, the metallic iridescent appearances of reflections from jewel beetles and related species are circularly polarized owing to complex cuticular surface specializations (Sharma et al., 2009; Stavenga et al., 2011). Because circularly polarized light occurs rarely in nature, it might serve as a specific communication channel in these insects (Cronin et al., 2003). Recent behavioral experiments on this topic gave mixed results. While the scarab beetle, *Chrysina gloriosa*, showed a differential orientation response to circularly polarized light and, thus, might use it for communication with conspecifics (Brady & Cummings 2010), no responses were found in four different scarab species (Blahó et al., 2012). Stomatopod crustaceans (mantis shrimps) may likewise use circularly polarized light for intraspecific communication (Chiou et al., 2008; Marshall et al., 1999). These animals have an extraordinary visual system that allows them not only to detect linearly polarized light but also to distinguish between clockwise and counterclockwise rotating circularly polarized light. In addition to linear polarization analyzers present in the dorsal and ventral hemispheres of the eye, two rows of ommatidia in a specific midband region appear to work as circularly polarized light analyzers. To achieve this, the short rhabdoms of distal photoreceptors R8 work as quarter wave plates. They convert circularly to linearly polarized light, which is then analyzed by the underlying orthogonal rhabdoms of photoreceptors R1–7 (Chiou et al., 2008; Kleinlogel & White, 2008). It is proposed that besides a role in contrast enhancement, the system may be used to play a role in sexual signaling and mate choice through circularly polarized light reflections from the telson and other body parts that exist in males but not in females (Chiou et al., 2008).

POLARIZATION SENSITIVITY VERSUS POLARIZATION VISION

In its most simple form, for example, to enhance visual contrast, sensitivity to polarized light may not need to be disentangled from other visual qualities such as brightness or color (Kelber, Thunell, & Arikawa, 2001). "True polarization vision," on the other hand, requires determining the E-vector angle independently from variations in color, intensity, and degree of polarization (Bernard & Wehner, 1977; Wehner, 2001). Wavelength independence is achieved by using homochromatic photoreceptors for polarized light detection, and intensity invariance results from orthogonal E-vector analyzers and polarization opponency (see below). Photoreceptors with orthogonal microvillar orientations, and hence orthogonal E-vector tuning, are common in ommatidia throughout the eyes of decapods and in the DRA of insects. A polarization system of crossed E-vector analyzers would, however, still have confusion states when both analyzers are excited equally. This can be overcome by scanning movements of the animal, which is frequently observed in navigating ants (sequential method; Wehner, 1989). On the other hand, comparison of signals from adjacent ommatidia with differently oriented cross analyzers would also solve this problem (simultaneous method). A recent study demonstrating learning of E-vector orientations in fixed honeybees suggests an underlying instantaneous algorithm for measuring E-vector orientation, but in that study independence from light intensity was not studied (Sakura, Okada, & Aonuma, 2012). The ultimate polarization vision system is clearly present in mantis shrimps. In these animals, the combination of input signals from ommatidia of the dorsal and ventral eye hemispheres (signaling linearly polarized light) and of rows 5/6 of the midband of the eye (signaling circularly polarized

light) provides an impressive basis for true polarization vision (Marshall et al., 1999), allowing these animals to measure all essential parameters (three Stokes's parameters) of any state of polarization of light (Kleinlogel & White, 2008).

NEURAL MECHANISMS

Neural processing of polarized-light stimuli, studied in a few species of insects and decapod crustaceans, has provided first insights into central mechanisms underlying sky compass orientation in insects and the contribution of polarized light to object and movement detection in crabs and crayfish.

Insect Sky Compass Orientation

POLARIZATION VISION PATHWAYS Tracing studies in locusts, supplemented by work in crickets, bees, and butterflies, support a specific visual pathway underlying polarized-light-dependent sky compass orientation in insects (see figure 84.4D). Photoreceptors from the DRA of the eye project axons to distinct DRAs of the lamina and the medulla (Homberg & Paech, 2002; Pfeiffer & Kinoshita, 2012). In locusts, neurons of a specific medulla layer connect to its dorsal rim and form an early neural network of polarized-light signaling throughout all medulla cartridges (el Jundi, Pfeiffer, & Homberg, 2011). Transmedulla neurons send axonal fibers from the dorsal rim of the medulla to a small neuropil in the central brain, the lower unit of the anterior optic tubercle (AOTu) (Homberg et al., 2003a; Pfeiffer & Kinoshita, 2012). From the AOTus of both brain hemispheres, projection neurons make contact with tangential neurons of the central complex, an unpaired midline neuropil that appears to act as an internal sky compass (Heinze & Reppert, 2011; Homberg et al., 2011; Pfeiffer & Kinoshita, 2012; Sakura, Lambrinos, & Labhart, 2007). Finally, outputs from the central complex, demonstrated in locusts, project to posterior regions of the lateral accessory lobes, and polarization information is transferred via descending neurons to the motor control centers in the thoracic ganglia (Träger & Homberg, 2011). In addition to convergence of polarization signals from the left and right eye in the central complex, bilateral connections are also made at two more peripheral levels, via heterolateral neurons, termed POL1 neurons in crickets, connecting the right and left medulla (Labhart & Meyer, 2002; Loesel & Homberg 2001; el Jundi, Pfeiffer, & Homberg, 2011), and via neurons connecting the right and left AOTus (Pfeiffer & Kinoshita, 2012; Pfeiffer, Kinoshita, & Homberg, 2005).

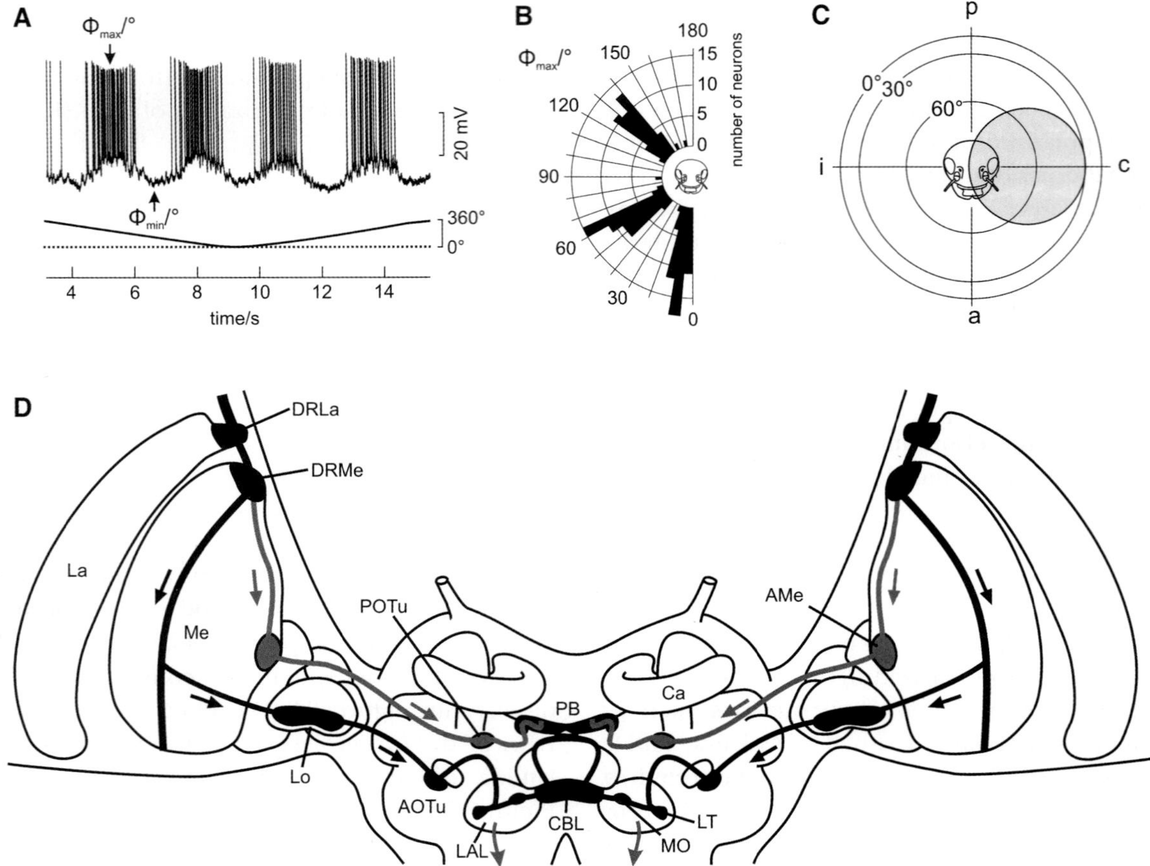

FIGURE 84.4   Polarization-sensitive (POL) neurons in the insect brain. (A) Intracellular recording from a POL1 neuron in the brain of the cricket *Gryllus campestris*. The neuron shows polarization opponency during stimulation with a rotating polarizer from dorsal direction (0°–360° clockwise and counterclockwise rotation). (B) Distribution of angular tunings to polarized light (Φ_{max}) from 142 recordings from POL1 neurons. (C) Receptive field structure of POL1 neurons. The receptive field is centered at an elevation of 60° in the contralateral hemisphere. a, anterior; i, ipsilateral; p, posterior; c, contralateral. (D) Polarization vision pathways in the brain of the desert locust *Schistocerca gregaria*. Lines and areas in black illustrate the polarization vision pathway from the dorsal rim area of the medulla (DRMe) via the medulla (Me), the lobula complex (Lo), the anterior optic tubercle (AOTu), and the lateral accessory lobe (LAL) to the lower division of the central body (CBL). Lines and areas in gray show a proposed second polarization vision pathway from the DRMe via the accessory medulla (AMe) and posterior optic tubercle (POTu) to the protocerebral bridge (PB) of the central complex. Ca, calyx; DRLa, dorsal rim area of the lamina; La, lamina; LT, lateral triangle; MO, median olive. (A adapted from Wehner & Labhart, 2006; B from Labhart & Meyer, 2002 with permission from Elsevier; C from Labhart, Petzold, & Helbling, 2001, with permission from *Journal of Experimental Biology*.)

POLARIZATION OPPONENCY   POL neurons in insects show sinusoidal modulations of spiking activity with a 180° periodicity during stimulation of the animal with a rotating polarizer, and many of these neurons show polarization opponency (cricket: Labhart, 1988; Labhart, Petzold, & Helbling, 2001; cockroach: Loesel & Homberg, 2001; monarch butterfly: Heinze & Reppert, 2011; locust: Pfeiffer, Kinoshita, & Homberg, 2005; Vitzthum, Müller, & Homberg, 2002). Particularly well-studied examples are POL1 neurons connecting the right and left medulla in crickets (see figure 84.4). The neurons are maximally activated at a particular *E*-vector orientation (Φ_{max}) and are maximally inhibited at an *E*-vector orientation orthogonal to Φ_{max} (Φ_{min}) (see figure 84.4A; Labhart, 1988). It is assumed that antagonistic inputs from the crossed *E*-vector analyzers in DRA ommatidia give rise to polarization opponency in POL1. POL1 neurons occur as three physiological subtypes with Φ_{max} orientations at 0°, 60°, and 120° (see figure 84.4B). Their relative activities, however, do not qualify for an internal compass signal, because their receptive fields are not centered in the zenith but at an elevation of about 60° contralaterally, which renders their responses dependent on solar elevation (see figure 84.4C; Labhart, Petzold, & Helbling, 2001). In general, polarization opponency has two important consequences: (1) It increases the sensitivity to polarized light, and (2) it leads to an intensity-independent response (Labhart, 1988; Labhart & Petzold, 1993), which is essential for signaling solar azimuth independent of variations in clouding conditions.

Several neurons in locusts and monarch butterflies were found to receive only excitatory input at Φ_{max} or inhibitory input at Φ_{min} without showing opponency (el Jundi, Pfeiffer, & Homberg, 2011; Heinze & Reppert, 2011; Pfeiffer, Kinoshita, & Homberg, 2005). These neurons are expected to confuse E-vector orientation with brightness, but many of these also respond to chromatic stimuli in an azimuth-dependent way (see below), thereby differing considerably from the "pure" POL1 neurons in crickets.

INTEGRATION WITH OTHER CELESTIAL CUES   The plane of skylight polarization alone does not allow distinguishing between the solar and the antisolar sky hemispheres (see figure 84.1B). A possible way to overcome this ambiguity in directional signaling is to combine information on sky polarization with information on the chromatic and intensity gradient of the sky. Behavioral evidence for an integration of these signals in navigation indeed exists for the honeybee and desert ant (Wehner, 1997; Wehner & Rossel, 1985). In an orientation assay, honeybees interpret a green light spot as being the sun whereas a UV light spot is interpreted as lying in the antisolar hemisphere of the sky (Wehner & Rossel, 1985). Accordingly, POL neurons in the medulla and AOTu of locusts and central complex of monarchs were found to respond additionally to unpolarized chromatic stimuli in an azimuth-dependent way (el Jundi, Pfeiffer, & Homberg, 2011; Heinze & Reppert, 2011; Pfeiffer & Homberg, 2007). In the experiments, the animals were presented with an unpolarized green or a UV light spot moving around the head at an elevation of 45° while recording intracellularly from a POL neuron. Whereas POL neurons of the locust medulla were tuned to ipsilateral unpolarized light spots largely independently of wavelength and time of day, responses of POL neurons in the following stage of processing, the AOTu, were more complex. AOTu neurons showed color opponency to UV and green light, which in some cell types was combined with spatial opponency, for example, excitation versus inhibition at opposite positions of the UV light stimulus (Pfeiffer & Homberg, 2007). Moreover, the neurons changed their E-vector tuning characteristics during the day in a way that suggests that they compensate for changes in solar elevation in their combined response to sky chromatic contrast and E-vector orientation. These properties make them suited to distinguish between the solar and

antisolar sky hemispheres (Kinoshita, Pfeiffer, & Homberg, 2007; Pfeiffer & Homberg, 2007). Likewise, POL neurons in the central brain of monarch butterflies respond to spectral and polarization signals consistent with a compensatory mechanism for changes in solar elevation (Heinze & Reppert, 2011). In contrast to the locust AOTu neurons, however, neurons in the monarch butterfly showed a wavelength-independent response, suggesting that they use the solar azimuth itself as a navigational cue.

REPRESENTATION OF AZIMUTHAL SPACE   The central complex is a final key processing center for skylight polarization (locust: Heinze & Homberg, 2007, 2009; Vitzthum, Müller, & Homberg, 2002; cricket: Sakura, Lambrinos, & Labhart, 2007; monarch butterfly: Heinze & Reppert, 2011) and appears to serve a role as an internal sky compass. The central complex is a group of midline brain areas including the protocerebral bridge (PB) and the upper (CBU) and lower divisions of the central body (CBL) (see figure 84.5). The three subdivisions of the central complex are arranged in linear arrays of 16 columns, and the CBU and CBL are additionally subdivided into horizontal layers. Information flow through the locust central complex has been divided into three stages (Heinze, Gotthardt, & Homberg, 2009; Homberg et al., 2011). At the input stage, tangential neurons of the CBL (TL2/3 neurons, figure 84.5) receive polarization input from AOTu neurons. A set of intermediate columnar neurons (CL1 neurons, figure 84.5) ramifies in the CBL and likely transfers polarized light signals to tangential neurons of the PB (TB1 neurons, figure 84.5). TB1 neurons connect columns of the right and left PB hemisphere that are precisely eight columns apart (column r1–18; r2–17, etc.). These topographic projections correspond linearly to E-vector tuning angles in TB1 neurons and create a compass-like representation of zenithal E-vectors covering 180° in each PB hemisphere (Heinze & Homberg, 2007; figure 84.5). At the output stage, several types of columnar neurons (CP1, CP2, CPU1 neurons, figure 84.5) provide sky compass information to postsynaptic neurons of the lateral accessory lobes. E-vector tuning in these neurons is, as in TB1 neurons, linearly dependent on innervated column, but is perpendicular to E-vector tuning in the TB1 neurons suggesting that tangential and columnar neurons are connected via inhibitory synapses (Heinze & Homberg, 2007). A striking decrease in the amplitude of E-vector-dependent spike frequency modulation from the input to the output stage of the central-complex network suggests that nonpolarization inputs such as landmark information or behavioral context increasingly

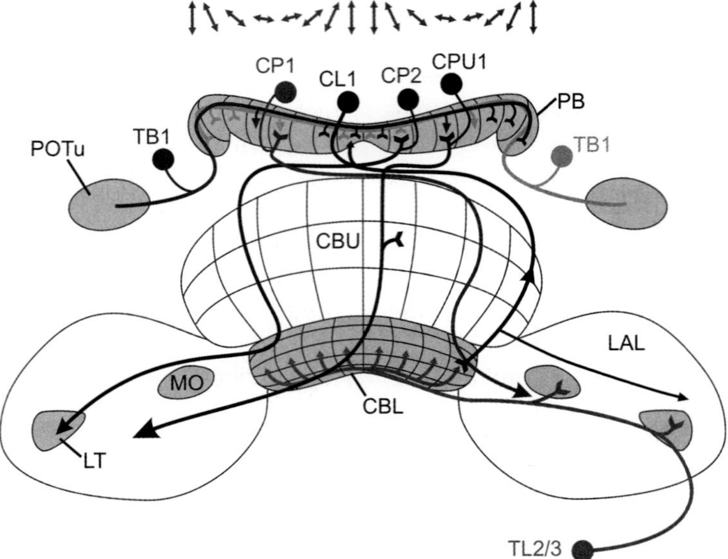

FIGURE 84.5 Hypothetical neural network underlying polarized-light processing in the central complex of the desert locust. At the input stage, TL2/3 neurons transfer polarized light signals from the lateral triangle (LT) and the median olive (MO) to the lower division of the central body (CBL). CL1 neurons are intercalated between the CBL and the protocerebral bridge (PB) and transfer polarization signals to TB1 tangential cells, which have additional ramifications in the posterior optic tubercle (POTu). Columnar output neurons of the central complex (CP1, CP2, CPU1) provide a compass signal to different subdivisions of the lateral accessory lobe (LAL). TB1, CP1, CP2, and CPU1 cells contribute to a compass-like representation of *E*-vectors in the 16 columns of the PB. Double arrows above the PB illustrate *E*-vector tunings of CP1, CP2, and CPU neurons. CBU, upper division of the central body. Based on data from Heinze and Homberg (2007, 2009); Heinze, Gotthardt, & Homberg (2009); and Vitzthum, Müller, & Homberg (2002).

contribute to signaling at later stages of the polarization vision network.

TIME COMPENSATED CELESTIAL ORIENTATION Maintaining constant directions during long-distance migration is only feasible if the internal sky compass network receives daytime-dependent input compensating for the diurnal shift in solar azimuth. Time compensation has, indeed, been demonstrated for homing behavior in bees and ants and during migration of monarch butterflies (Lindauer, 1960; Mouritsen & Frost, 2002; Wehner, 1992). In locusts, POL neurons descending from the brain to thoracic ganglia show a shift of 21.5°/h in *E*-vector tuning throughout the day, which is in the range of diurnal changes of solar azimuth (15°/h) and, thus, send time-compensated directional information to motor control centers in the thoracic ganglia (Träger & Homberg, 2011). How circadian signals are integrated with the sky compass network is still poorly understood. Experiments on monarch butterflies suggest that circadian inputs from the antennae are required for time-compensated compass navigation (Merlin, Gegear, & Reppert, 2009). In locusts, the accessory medulla, a small neuropil in the optic lobe, is discussed as a candidate to transfer daytime signals to the polarization vision network.

Studies in cockroaches and flies have established the accessory medulla as internal master clock controlling circadian changes in locomotor activity (Helfrich-Förster, Stengl, & Homberg, 1998; Homberg, Reischig, & Stengl, 2003b). In locusts, large tangential POL neurons of the medulla provide connections to the accessory medulla, which in turn is linked via the posterior optic tubercle to the PB of the central complex (see figure 84.4D; el Jundi & Homberg, 2010). Although physiological studies are still lacking, this second hypothetical polarization vision pathway might deliver time-compensated polarization information to the internal sky compass in the central complex.

Crustaceans

In contrast to those of insects, photoreceptors of crabs and crayfish show high PS throughout their eyes. With the exception of stomatopods, ommatidia through most parts of the eye have identical microvillar orientations (vertical and horizontal) supporting a two-channel polarization system (Glantz, 2007; Shaw, 1966). Studies in the optic lobe of crabs and crayfish support a contribution of PS to contrast enhancement and motion detection (Glantz, 1996a, 1996b, 2001, 2008; Leggett, 1976; Waterman, 1981). None of the studied

     UWE HOMBERG AND BASIL EL JUNDI

types of POL neuron responded exclusively to changes in *E*-vector orientation, but all were additionally sensitive to contrast vision signals. In the medulla externa of crayfish, two types of neuron, one tuned to vertical *E*-vectors (sustaining fibers) and one tuned to horizontal *E*-vectors (dimming fibers) could be distinguished (Glantz, 2001). In the more central medulla interna (corresponding to the insect lobula), movement-sensitive neurons also responded to motion of pure polarization contrast with corresponding directional sensitivities (Glantz, 2008). These data support earlier findings in the swimming crab, *Scylla serrata*, showing that neurons were particularly sensitive to a moving (rotating) polarizer and one class of units, directionally sensitive with striking difference in response to clockwise and counterclockwise rotation of the polarizer (Leggett, 1976). Taken together, these data suggest that *E*-vector angle in crabs and crayfish is not analyzed and processed as a separate visual modality but is specifically used to enhance spatial and temporal contrast at low contrast environments, particularly for motion vision.

CONCLUSIONS

Sensitivity to polarized light in arthropods encompasses a wide range of behavioral functions, highly dependent on the ecological situation of the animal. While the receptor mechanisms for polarized light detection may largely be conserved, mechanisms of central neural processing apparently differ considerably and are tailored to the sensory discrimination task to be solved. Advances in the analysis of neural mechanisms underlying the detection of polarized skylight in insects have opened the door to a number of key principles in sky polarization vision systems including polarization opponency, multisensory integration, and compass-like internal representation of celestial *E*-vectors. The recent addition of *Drosophila* as a research object in polarization-dependent orientation behavior (Hardie, 2012) promises to further stimulate this line of research with focus on central mechanisms of celestial navigation and space representation as well as detection of objects and surfaces based on polarized-light reflections.

REFERENCES

Bernard, G. D., & Wehner, R. (1977). Functional similarities between polarization vision and color vision. *Vision Research, 17*, 1019–1028. doi:10.1016/0042-6989(77)90005-0.

Blahó, M., Egri, A., Hegedüs, R., Jósvai, J., Tóth, M., Kertész, K., et al. (2012). No evidence for behavioural responses to circularly polarized light in four scarab beetle species with circularly polarizing exocuticle. *Physiology & Behavior, 105*, 1067–1075.

Brady, P., & Cummings, M. (2010). Differential response to circularly polarized light by the jewel scarab beetle *Chrysina gloriosa*. *American Naturalist, 175*, 614–620.

Brines, M. L., & Gould, J. L. (1982). Skylight polarization patterns and animal orientation. *Journal of Experimental Biology, 96*, 69–91.

Chiou, T. H., Kleinlogel, S., Cronin, T., Caldwell, R., Loeffler, B., Siddiqi, A., et al. (2008). Circular polarization vision in a stomatopod crustacean. *Current Biology, 18*, 429–434.

Cronin, T. W., & Marshall, J. (2011). Patterns and properties of polarized light in air and water. *Philosophical Transactions of the Royal Society of London. Series B, Biological Sciences, 366*, 619–626.

Cronin, T. W., & Shashar, N. (2001). The linearly polarized light field in clear tropical marine waters: Spatial and temporal variation of light intensity, degree of polarization and e-vector angle. *Journal of Experimental Biology, 204*, 2461–2467.

Cronin, T. W., Shashar, N., Caldwell, R. L., Marshall, J., Cheroske, A. G., & Chiou, T. H. (2003). Polarization vision and its role in biological signaling. *Integrative and Comparative Biology, 43*, 549–558.

Dacke, M., Doan, T. A., & O'Carroll, D. C. (2001). Polarized light detection in spiders. *Journal of Experimental Biology, 204*, 2481–2490.

Dacke, M., Nilsson, D. E., Warrant, E. J., Blest, A. D., Land, M. F., & O'Carroll, D. C. (1999). Built-in polarizers form part of a compass organ in spiders. *Nature, 401*, 470–473.

Dacke, M., Nordström, P., & Scholtz, C. H. (2003). Twilight orientation to polarized light in the crepuscular dung beetle *Scarabaeus zambesianus*. *Journal of Experimental Biology, 206*, 1535–1543.

el Jundi, B., & Homberg, U. (2010). Evidence for the possible existence of a second polarization-vision pathway in the locust brain. *Journal of Insect Physiology, 56*, 971–979.

el Jundi, B., Pfeiffer, K., & Homberg, U. (2011). A distinct layer of the medulla integrates sky compass signals in the brain of an insect. *PLoS ONE, 6*, e27855. doi:10.1371/journal.pone.0027855.

Frisch, K. v. (1949). Die Polarisation des Himmelslichtes als orientierender Faktor bei den Tänzen der Bienen. *Experientia, 5*, 142–148.

Glantz, R. M. (1996a). Polarization sensitivity in crayfish lamina monopolar neurons. *Journal of Comparative Physiology. A, Neuroethology, Sensory, Neural, and Behavioral Physiology, 178*, 413–425.

Glantz, R. M. (1996b). Polarization sensitivity in the crayfish optic lobe: Peripheral contributions to opponency and directionally selective motion detection. *Journal of Neurophysiology, 76*, 3404–3414.

Glantz, R. M. (2001). Polarization analysis in the crayfish visual system. *Journal of Experimental Biology, 204*, 2383–2390.

Glantz, R. M. (2007). The distribution of polarization sensitivity in the crayfish retinula. *Journal of Comparative Physiology. A, Neuroethology, Sensory, Neural, and Behavioral Physiology, 193*, 893–901.

Glantz, R. M. (2008). Polarization vision in crayfish motion detectors. *Journal of Comparative Physiology. A, Neuroethology, Sensory, Neural, and Behavioral Physiology, 194*, 565–575.

Goddard, S. M., & Forward, R. B. (1991). The role of underwater polarized light pattern in sun compass navigation of the grass shrimp, *Palaemonetes vulgaris*. *Journal of Comparative*

Physiology. A, Neuroethology, Sensory, Neural, and Behavioral Physiology, 169, 479–491.

Görner, P. (1962). Die Orientierung der Trichterspinne nach polarisiertem Licht. *Zeitschrift für Vergleichende Physiologie, 45,* 307–314.

Hardie, R. (2012). Polarization vision: *Drosophila* enters the arena. *Current Biology, 22,* R12–R14. doi:10.1016/j.cub.2011.11.016.

Hawryshyn, C. W. (1992). Polarization vision in fish. *American Naturalist, 80,* 164–175.

Heinze, S., Gotthardt, S., & Homberg, U. (2009). Transformation of polarized light information in the central complex of the locust. *Journal of Neuroscience, 29,* 11783–11793.

Heinze, S., & Homberg, U. (2007). Maplike representation of celestial *E*-vector orientations in the brain of an insect. *Science, 315,* 995–997.

Heinze, S., & Homberg, U. (2009). Linking the input to the output: New sets of neurons complement the polarization vision network in the locust central complex. *Journal of Neuroscience, 29,* 4911–4921.

Heinze, S., & Reppert, S. M. (2011). Sun compass integration of skylight cues in migratory monarch butterflies. *Neuron, 69,* 345–358.

Helfrich-Förster, C., Stengl, M., & Homberg, U. (1998). Organization of the circadian system in insects. *Chronobiology International, 15,* 567–594.

Homberg, U. (2004). In search of the sky compass in the insect brain. *Naturwissenschaften, 91,* 199–208.

Homberg, U., Heinze, S., Pfeiffer, K., Kinoshita, M., & el Jundi, B. (2011). Central neural coding of sky polarization in insects. *Philosophical Transactions of the Royal Society of London. Series B, Biological Sciences, 366,* 680–687.

Homberg, U., Hofer, S., Pfeiffer, K., & Gebhardt, S. (2003a). Organization and neural connections of the anterior optic tubercle in the brain of the locust, *Schistocerca gregaria. Journal of Comparative Neurology, 462,* 415–430.

Homberg, U., & Paech, A. (2002). Ultrastructure and orientation of ommatidia in the dorsal rim area of the locust compound eye. *Arthropod Structure & Development, 30,* 271–280.

Homberg, U., Reischig, T., & Stengl, M. (2003b). Neural organization of the circadian system of the cockroach *Leucophaea maderae. Chronobiology International, 20,* 577–591.

Horváth, G., & Varjú, D. (2004). *Polarized light in animal vision—Polarization patterns in nature.* Berlin: Springer.

Johnsen, S., Marshall, N. J., & Widder, E. A. (2011). Polarization sensitivity as a contrast enhancer in pelagic predators: Lessons from *in situ* polarization imaging of transparent zooplankton. *Philosophical Transactions of the Royal Society of London. Series B, Biological Sciences, 366,* 655–670.

Kelber, A., Thunell, C., & Arikawa, K. (2001). Polarisation-dependent colour vision in *Papilio* butterflies. *Journal of Experimental Biology, 204,* 2469–2480.

Kinoshita, M., Pfeiffer, K., & Homberg, U. (2007). Spectral properties of identified polarized-light sensitive interneurons in the brain of the desert locust *Schistocerca gregaria. Journal of Experimental Biology, 210,* 1350–1361.

Kinoshita, M., Yamazato, K., & Arikawa, K. (2011). Polarization-based brightness discrimination in the foraging butterfly, *Papilio xuthus. Philosophical Transactions of the Royal Society of London. Series B, Biological Sciences, 366,* 688–696.

Kleinlogel, S., & White, A. G. (2008). The secret world of shrimps: Polarisation vision at its best. *PLoS ONE, 3,* e2190. doi:10.1371/journal.pone.0002190.

Kriska, G., Bernáth, B., Farkas, R., & Horváth, G. (2009). Degrees of polarization of reflected light eliciting polarotaxis in dragonflies (*Odonata*), mayflies (*Ephemeroptera*) and tabanid flies (*Tabanidae*). *Journal of Insect Physiology, 55,* 1167–1173.

Labhart, T. (1988). Polarization-opponent interneurons in the insect visual system. *Nature, 331,* 435–437.

Labhart, T., & Meyer, E. P. (1999). Detectors for polarized skylight in insects: A survey of ommatidial specializations in the dorsal rim area of the compound eye. *Microscopy Research and Technique, 47,* 368–379.

Labhart, T., & Meyer, E. P. (2002). Neural mechanisms in insect navigation: Polarization compass and odometer. *Current Opinion in Neurobiology, 12,* 707–714.

Labhart, T., & Petzold, J. (1993). Processing of polarized light information in the visual system of crickets. In K. Wiese, F. G. Gribakin, A. V. Popov, & G. Renninger (Eds.), *Sensory systems of arthropods* (pp. 158–169). Basel , Switzerland: Birkhäuser.

Labhart, T., Petzold, J., & Helbling, H. (2001). Spatial integration in polarization-sensitive interneurones of crickets: A survey of evidence, mechanisms and benefits. *Journal of Experimental Biology, 204,* 2423–2430.

Leggett, L. M. W. (1976). Polarised light-sensitive interneurones in a swimming crab. *Nature, 262,* 709–711.

Lindauer, M. (1960). Time-compensated sun orientation in bees. *Cold Spring Harbor Symposia on Quantitative Biology, 25,* 371–377.

Loesel, R., & Homberg, U. (2001). Anatomy and physiology of neurons with processes in the accessory medulla of the cockroach *Leucophaea maderae. Journal of Comparative Neurology, 439,* 193–207.

Lythgoe, J. N., & Hemmings, C. C. (1967). Polarized light and underwater vision. *Nature, 213,* 893–894.

Marshall, J., Cronin, T. W., Shashar, N., & Land, M. (1999). Behavioural evidence for polarisation vision in stomatopods reveals a potential channel for communication. *Current Biology, 9,* 755–758.

Merlin, C., Gegear, R. J., & Reppert, S. M. (2009). Antennal circadian clocks coordinate sun compass orientation in migratory monarch butterflies. *Science, 325,* 1700–1704.

Moody, M. F., & Parriss, J. R. (1961). The discrimination of polarized light by *Octopus*: A behavioural and morphological study. *Zeitschrift für Vergleichende Physiologie, 44,* 268–291.

Mouritsen, H., & Frost, B. J. (2002). Virtual migration in tethered flying monarch butterflies reveals their orientation mechanisms. *Proceedings of the National Academy of Sciences of the United States of America, 99,* 10162–10166. doi:10.1073/pnas.152137299.

Ortega-Escobar, J., & Muñoz-Cuevas, A. (1999). Anterior median eyes of *Lycosa tarentula* (Araneae, Lycosidae) detect polarized light: Behavioral experiments and electroretinographic analysis. *Journal of Arachnology, 27,* 663–671.

Pfeiffer, K., & Homberg, U. (2007). Coding of azimuthal directions via time-compensated combination of celestial compass cues. *Current Biology, 17,* 960–965.

Pfeiffer, K., & Kinoshita, M. (2012). Segregation of visual inputs from different regions of the compound eye in two parallel pathways through the anterior optic tubercle of the bumblebee (*Bombus ignitus*). *Journal of Comparative Neurology*, *520*, 212–229.

Pfeiffer, K., Kinoshita, M., & Homberg, U. (2005). Polarization-sensitive and light-sensitive neurons in two parallel pathways passing through the anterior optic tubercle in the locust brain. *Journal of Neurophysiology*, *94*, 3903–3915.

Reppert, S. M., Zhu, H., & White, R. H. (2004). Polarized light helps monarch butterflies navigate. *Current Biology*, *14*, 155–158.

Roberts, N. W., Porter, M. L., & Cronin, T. W. (2011). The molecular basis of mechanisms underlying polarization vision. *Philosophical Transactions of the Royal Society of London. Series B, Biological Sciences*, *366*, 627–637.

Rossel, S. (1993). Navigation by bees using polarized skylight. *Comparative Biochemistry and Physiology. Part A, Physiology*, *104*, 695–708.

Rossel, S., & Wehner, R. (1987). The bee's e-vector compass. In R. Menzel & A. Mercer (Eds.), *Neurobiology and behaviour of honeybees* (pp. 76–93). Berlin: Springer.

Sakura, M., Lambrinos, D., & Labhart, T. (2007). Polarized skylight navigation in insects: Model and electrophysiology of e-vector coding by neurons in the central complex. *Journal of Neurophysiology*, *99*, 667–682.

Sakura, M., Okada, R., & Aonuma, H. (2012). Evidence for instantaneous e-vector detection in the honeybee using an associative learning paradigm. *Proceedings of the Royal Society of London. Series B, Biological Sciences*, *279*, 535–542.

Schneider, L., & Langer, H. (1969). Die Struktur des Rhabdoms im "Doppelauge" des Wasserläufers *Gerris lacustris*. *Zeitschrift für Zellforschung*, *99*, 538–559.

Schwind, R. (1984). The plunge reaction of the backswimmer *Notonecta glauca*. *Journal of Comparative Physiology. A, Neuroethology, Sensory, Neural, and Behavioral Physiology*, *155*, 319–321.

Schwind, R. (1985). Sehen unter und über Wasser, Sehen von Wasser. *Naturwissenschaften*, *72*, 343–352.

Schwind, R. (1989). A variety of insects are attracted to water by reflected polarized light. *Naturwissenschaften*, *76*, 377–378.

Schwind, R. (1995). Spectral regions in which aquatic insects see reflected polarized light. *Journal of Comparative Physiology. A, Neuroethology, Sensory, Neural, and Behavioral Physiology*, *177*, 439–448.

Schwind, R. (1999). *Daphnia pulex* swims towards the most strongly polarized light - A response that leads to 'shore flight.' *Journal of Experimental Biology*, *202*, 3631–3635.

Sharma, V., Crne, M., Park, J. O., & Srinivasarao, M. (2009). Structural origin of circularly polarized iridescence in jeweled beetles. *Science*, *325*, 449–451.

Shashar, N., Sabbah, S., & Aharoni, N. (2005). Migrating locusts can detect polarized reflections to avoid flying over the sea. *Biology Letters*, *1*, 472–475.

Shaw, S. (1966). Polarized light responses from crab retinula cells. *Nature*, *211*, 92–93.

Stavenga, D. G., Wilts, B. D., Leertouwer, H. L., & Hariyama, T. (2011). Polarized iridescence of the multilayered elytra of the Japanese jewel beetle *Chrysochroa fulgidissima*.

Philosophical Transactions of the Royal Society of London. Series B, Biological Sciences, *366*, 709–723.

Stowe, S. (1980). Rapid synthesis of photoreceptor membrane and assembly of new microvilli in a crab at dusk. *Cell and Tissue Research*, *211*, 419–440.

Strutt, J. W. (1871). On the light from the sky, its polarization and color. *Philosophical Magazine*, *41*, 274–279.

Sweeney, A., Jiggins, C., & Johnsen, S. (2003). Polarized light as a butterfly mating signal. *Nature*, *423*, 31–32.

Träger, U., & Homberg, U. (2011). Polarization-sensitive descending neurons in the locust: Connecting the brain to thoracic ganglia. *Journal of Neuroscience*, *31*, 2238–2247.

Vitzthum, H., Müller, M., & Homberg, U. (2002). Neurons of the central complex of the locust *Schistocerca gregaria* are sensitive to polarized light. *Journal of Neuroscience*, *22*, 1114–1125.

Waterman, T. H. (1981). Polarization sensitivity. In H. Autrum (Ed.), *Handbook of sensory physiology: Vol. 7/6B. Vision in invertebrates* (pp. 281–469). Berlin: Springer.

Waterman, T. H. (2006). Reviving a neglected celestial underwater polarization compass for aquatic animals. *Biological Reviews of the Cambridge Philosophical Society*, *81*, 111–115.

Wehner, R. (1976). Polarized-light navigation by insects. *Scientific American*, *235*, 106–115.

Wehner, R. (1989). Neurobiology of polarization vision. *Trends in Neurosciences*, *12*, 353–359.

Wehner, R. (1992). Arthropods. In F. Papi (Ed.), *Animal homing* (pp. 45–144). London: Chapman and Hall.

Wehner, R. (1997). The ant's celestial compass system: Spectral and polarization channels. In M. Lehrer (Ed.), *Orientation and communication in arthropods* (pp. 145–185). Basel, Switzerland: Birkhäuser.

Wehner, R. (2001). Polarization-vision—A uniform sensory capacity? *Journal of Experimental Biology*, *204*, 2589–2596.

Wehner, R. (2003). Desert ant navigation: How miniature brains solve complex tasks. *Journal of Comparative Physiology. A, Neuroethology, Sensory, Neural, and Behavioral Physiology*, *189*, 579–588.

Wehner, R., & Labhart, T. (2006). Polarization vision. In E. Warrant & D. E. Nilsson (Eds.), *Invertebrate vision* (pp. 291–348). Cambridge, England: Cambridge University Press.

Wehner, R., & Rossel, S. (1985). The bee's celestial compass——A case study in behavioural neurobiology. In B. Hölldobler & M. Lindauer (Eds.), *Fortschritte der Zoologie* (Vol. 31, pp. 11–53). Stuttgart, Germany: Fischer.

Weir, P. T., & Dickinson, M. H. (2012). Flying *Drosophila* orient to sky polarization. *Current Biology*, *22*, 1–7.

Wernet, M. F., Velez, M. M., Clark, D. A., Baumann-Klausener, F., Brown, J. R., Klovstad, M., et al. (2012). Genetic dissection reveals two separate retinal substrates for polarization vision in *Drosophila*. *Current Biology*, *22*, 12–20. doi:10.1016/j.cub.2011.11.028.

Xu, X. Z. S., Choudhury, A., Li, X., & Montrall, C. (1998). Coordination of an array of signaling proteins through homo- and heteromeric interactions between PDF domains and target proteins. *Journal of Cell Biology*, *142*, 545–555.

85 Vision and Navigation in Insects, and Applications to Aircraft Guidance

MANDYAM V. SRINIVASAN, RICHARD J. D. MOORE, SAUL THURROWGOOD, DEAN SOCCOL, DANIEL BLAND, AND MICHAEL KNIGHT

Flying insects provide living proof that small eyes and brains do not necessarily compromise sensorimotor performance. This review outlines our growing understanding of how these creatures use visual information to stabilize flight, avoid collisions with objects, regulate flight speed, navigate to distant food sources, and perform smooth landings. We also sketch how these insights are being used to develop novel, biologically inspired strategies for the guidance of autonomous, airborne vehicles.

INTRODUCTION

A glance at a fly landing flawlessly on a teacup, at a dragonfly catching an insect on the wing, or at a honeybee returning home unerringly after a 10-km search for food, reveals that insects possess exquisitely sensitive vision, fine motor control, and reliable navigation. Clearly, in such creatures, nature has evolved vision and navigation systems that are not compromised by diminutive size. Indeed, we hope this chapter will illustrate that the constraints on size and weight have promoted the evolution of eye designs and nervous systems that perform the required computations using strategies that are simple, elegant, and often unexpectedly novel.

The compound eyes of insects have a very different appearance from the so-called "simple" eyes of vertebrates. Typically, a vertebrate eye is composed of a single lens that focuses light from the outside world on to a sheet of photoreceptive neurons in the retina—rather like a camera. The compound eye of an insect, on the other hand, is composed of a large number of "little eyes," so-called "ommatidia," arranged on a curved surface (e.g., Borst, 2009; Goodman, 2003). Each ommatidium carries a small lens that captures light from a small patch of the environment, and focuses it onto a small group of eight or nine photoreceptors. In a honeybee, the two compound eyes together constitute an array of 11,000 ommatidia, providing the insect with a nearly all-round view of the environment. Each ommatidium contributes to this view by collecting information about the intensity and color (and possibly other properties) of light arriving from a small patch of the environment that is about $2°$ in diameter (Srinivasan, 2011).

While compound eyes possess distinct advantages, they also pose challenges for their owner. Unlike vertebrates, insects have immobile eyes with fixed-focus optics. Therefore, they cannot infer the distance of an object from the extent to which the directions of gaze must converge to view the object or by monitoring the refractive power that is required to bring the image of the object into focus on the retina. Furthermore, compared with human eyes, the eyes of insects are positioned much closer together and possess inferior spatial acuity. Therefore, even if an insect possessed the neural apparatus required for binocular stereopsis, such a mechanism would be relatively imprecise and would be restricted to measuring ranges of only a few centimeters (Rossel, 1983; Srinivasan, 1993). It comes as no surprise, then, that insects have evolved alternative visual strategies for "seeing" the world in three dimensions and guiding their flight in it. Many of these strategies rely on using cues derived from the image motion that the animal experiences in its eyes when it moves in its environment. In the following sections we will outline the ways in which insects exploit these motion cues to extract information about the environment and use this information to fly safely through it.

USING VISION TO CONTROL FLIGHT

Stabilizing Flight Direction and Attitude

When an insect, intending to fly in a straight line, is blown to the left by a gust of wind, the image on its frontal retina moves to the right. This causes the flight motor system to generate a corrective turning response—the so-called "optomotor response"—that steers the insect back on course. By responding in this way to the motion of the panoramic image of the world in the eye, the insect stabilizes unwanted rotations not only in yaw but also in roll and pitch. These rotations are sensed by neurons in the lobula plate, which represents the fourth stage of processing in the visual pathway of the fly. This region of the brain carries a number of motion sensitive neurons, with large visual fields, that are selectively responsive to rotations of the head about the yaw, pitch, and roll axes, as well as other axes (Borst, 2009; Egelhaaf, 2008; Joesch et al., 2008; Krapp, 2000; Nordstrom et al., 2008) (see also chapter 83, this volume). The ensemble of responses across these neurons encodes information about the direction of head rotation and drives downstream motorneurons, involved in flight control, to generate appropriate corrective rotations of the head and body.

Avoiding Obstacles and Negotiating Narrow Gaps

When an insect flies in a straight line, the image of a point in the scene, along a particular direction of view, will move in the eye at a speed (angular speed, measured in degrees per second) that is proportional to the speed of flight and is inversely proportional to the distance of the point from the eye. Thus, distant objects in the scene will induce low image speeds, whereas nearby objects will generate rapid image motion. Indeed, experiments using moving stimuli reveal that honeybees tend to avoid, or steer away from, regions of the visual field that experience rapid image motion (Srinivasan, 2011), suggesting that this is a strategy for sensing dangerously close objects and avoiding collisions with them. A bee flying through a narrow passage, such as a tunnel, positions itself such that the two eyes experience approximately the same image velocity (see figure 85.1a). This guidance strategy ensures that the two walls are at the same distance from the bee, enabling a safe, collision-free flight through the middle of the tunnel (Dyhr & Higgins, 2010; Serres et al., 2008; Srinivasan, 2011). That bees indeed use such a strategy has been confirmed by moving the visual texture on one of the walls, either in the direction of the bee's flight or against it. The bees then no longer fly through the middle of the tunnel but along a new axis, positioned such that the two eyes again experience equal image velocities (Srinivasan, 2011). When an object is approached, its image expands in the eye at a rate that increases with the object's proximity. Fruit flies appear to use image expansion as a cue for detecting nearby objects and avoiding collisions with them (Tammero & Dickinson, 2002).

Regulating Flight Speed

Fruit flies and honeybees appear to monitor and regulate their flight speed by measuring the average image velocity that is experienced by the two eyes. Bees passing through a tunnel (as in figure 85.1b) fly at a speed such that the speeds of the images in the lateral fields of view are held constant, at about 300°/s (Srinivasan, 2011). However, they increase their flight speed when the patterns on the two walls are moved in the same direction as the bee's flight, and they decrease it when the patterns are moved in the opposite direction, always holding the lateral image velocity at about 300°/s (Baird et al., 2005). Fruit flies display a similar behavior (Fry et al., 2009). An important consequence of this response is that the flight speed is automatically reduced when the insect negotiates a narrow passage, and increased when it flies in a wide-open environment. This has been confirmed by filming bees flying through a tapered tunnel (see figure 85.1c), as well as in tunnels of different widths (Baird, Kornfeldt, & Dacke, 2010; Barron & Srinivasan, 2006). Thus, insects appear to have a simple, elegant strategy for automatically reducing their flight speed to lower and safer levels when they enter a densely cluttered environment. Recent work (Baird, Kornfeldt, & Dacke, 2010) suggests that image velocity is measured in the lateral as well as the frontolateral parts of the visual field, which introduces a component of "anticipation" into the regulation of flight speed.

Controlling Flight Height

When an insect flies at a fixed speed, the speed of image motion in the ventral field of view provides an indication of its height above the ground—the higher the image speed, the lower the altitude. In principle, this information can be used to control the altitude of flight. Observations of bees flying in a tunnel in which the image speed generated by the floor is artificially manipulated suggest that they indeed use this cue (Baird et al., 2006). Movement of the floor pattern in the direction of flight causes the bees to fly at a lower height while maintaining the same air speed (Portelli, Ruffier, & Franceschini, 2010a). In this study the bees maintained a ventral image velocity of about 265°/s, a value not very different from that used to regulate flight speed (ca. 300°/s, see above). Therefore, it is possible that both the speed and the height of flight are regulated by the same kind of movement-detecting system. Under certain conditions, flight speed may be regulated by maintaining a prescribed image velocity in the lateral fields of view, and height by maintaining a

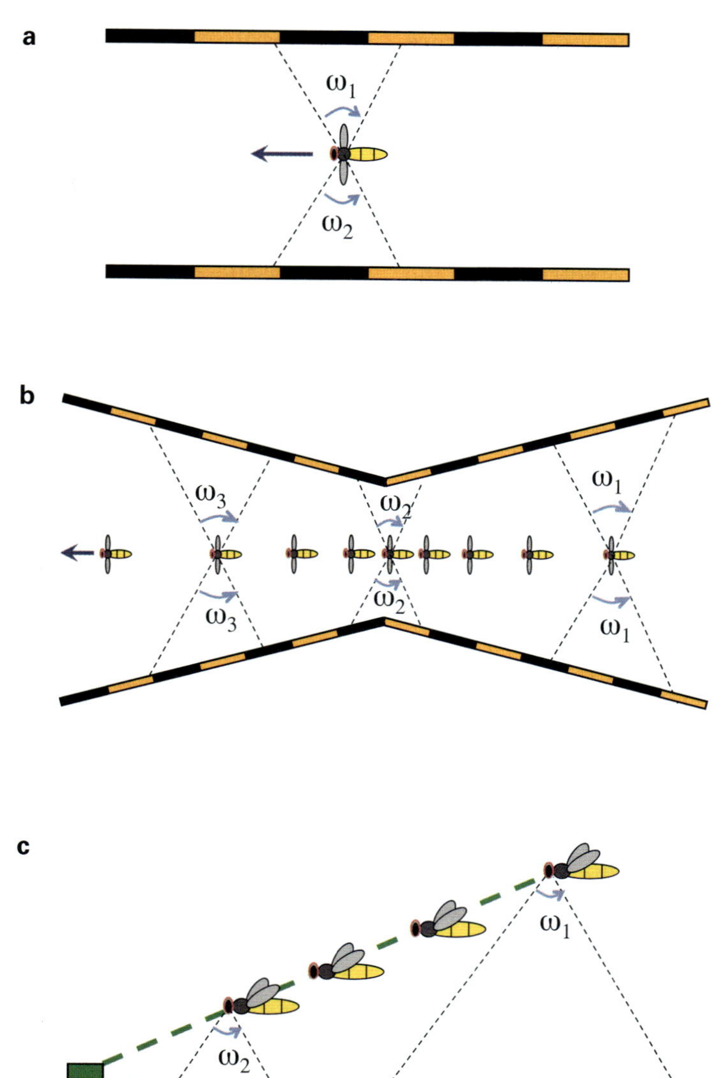

FIGURE 85.1   Illustration of some ways in which flying insects use optic flow for visual guidance. (a) Narrow passages are negotiated safely by flying a trajectory such that both eyes experience the same image velocity in the lateral fields of view ($\omega_1 = \omega_2$). (b) Flight speed is controlled by holding the image velocities in the lateral fields of view constant throughout the flight ($\omega_1 = \omega_2 = \omega_3$). (c) Smooth landings are executed by decreasing flight speed progressively such that the image velocity in the ventral field of view is held constant ($\omega_1 = \omega_2$). (Adapted with permission from Srinivasan, 2011b.)

prescribed image velocity in the ventral field (Portelli, Ruffier, & Franceschini, 2010a). When flying through wide tunnels, honeybees no longer fly down the middle, but closer to one wall, at a constant distance from it (Serres et al., 2008). This "wall-following" strategy appears to be achieved by holding constant the speed of the image of the closer wall as when regulating height. In *Drosphila*, there is evidence that height regulation is achieved by sensing expansions or contractions of the image of the ground, which correspond to decreases or increases of height, respectively (Straw, Lee, & Dickinson, 2010).

Achieving Smooth Landings

Orchestrating a safe landing requires considerable skill. How do insects achieve this? Investigations of bees performing grazing landings on horizontal surfaces have revealed that the speed of flight is progressively reduced as the surface is approached by holding constant the angular speed of the image of the ground in the eye (Srinivasan et al., 2000). This automatically ensures that flight speed decreases as the ground is approached, leading to a smooth touchdown (see figure 85.1c). The simplicity and elegance of this landing strategy lies in the fact that it does not require knowledge of the instantaneous height above the ground or the instantaneous speed of flight—all that is required is the measurement (and regulation) of the speed of the image in the eye. During landing, bees maintain image velocities of about 500°/s in the ventral field of view (Srinivasan et al., 2000). A mathematical model of this strategy successfully predicts all of the observed characteristics of the landing trajectories (Srinivasan et al., 2000).

Gauging Distance Traveled

Insects that make repeated foraging excursions and return home safely each time—the so-called "central place foragers"—must possess competent navigation systems that include a reliable method of estimating how far they have traveled. A reliable "odometer" would not only help them navigate back home but also return to a food source if it merits repeated visits. Behavioral experiments have revealed that bees estimate distance flown by summing over time (i.e., integrating) the optic flow that is generated by the movement of the image of the environment in their eyes during their journey (Srinivasan, 2011). Desert ants, on the other hand, gauge travel distance by monitoring the motions of the legs, in a process akin to counting footsteps (Wittlinger, Wehner, & Wolf, 2007).

Using the Horizon to Sense Attitude

In addition to the compound eyes, many insects possess three small, rudimentary, "simple" eyes, known as ocelli, located at the top of the head. One, termed the median ocellus, is positioned dorsofrontally and views the horizon in front of the flying insect. Two others, known as the lateral ocelli, are situated dorsolaterally and view the horizon to either side. A clockwise roll of the head causes the right lateral ocellus to see more of the sky and less of the ground and, therefore, receive more light than the left lateral ocellus, which would experience the opposite effect. The difference between the optical stimulations received by the left and right ocelli can be used to sense and correct deviations in roll attitude. Similarly, the median ocellus receives more or less light according to whether the insect pitches up or down. Therefore, the signals from the three ocelli, when processed appropriately, can provide information on the orientation of the insect's head relative to the horizon and, thus, an accurate estimate of roll and pitch. Indeed, behavioral studies with tethered, flying dragonflies have shown that compensatory rolling and pitching movements of the head can be elicited by

appropriate optical stimulation of the three ocelli, using light guides (Stange, 1981). Attitude can also be gauged by sensing the actual position of the horizon in the visual fields of the left, right, and median ocelli by detecting the changes in color that occur across the boundary between the ground and the sky. There is recent evidence that the second-order visual interneurons in the ocellar pathways are indeed capable of using spectral information to segregate the ground from the sky (Berry, van Kleef, & Stange, 2007; van Kleef, James, & Stange, 2005). Estimating attitude in this way, although computationally more complex, is likely to be more accurate than a strategy that simply compares the total numbers of photons received by the three ocelli because the latter approach would be more susceptible to errors caused by bright regions of the sky (such as sun or the clouds) or by bright patches on the ground.

Extracting Compass Information from the Sky

If a foraging insect is to find its way back home safely, it needs to know not only how far it has traveled from home but also its direction of travel. Bees and desert ants monitor their course by using the sun as a compass (Wehner, 1992). When the sun is hidden by a cloud, but a patch of open sky is still visible, then the pattern of polarized light that is created in the sky by the sun, and is visible in this patch, is used as a compass in lieu of the sun (Kraft et al., 2011; von Frisch, 1993). As the sun moves across the sky during the course of the day, the polarization pattern moves with it. The sun and the pattern of polarized light that it creates in the sky constitute a "celestial compass" (Wehner & Labhart, 2006). In many insects, including bees, ants, crickets, and locusts, the ommatidia in the uppermost dorsal parts of the compound eyes (the so-called dorsal rim area) carry photoreceptors that are sensitive to the orientation of polarized light (Homberg, 2004; Labhart & Meyer, 2002; Wehner & Labhart, 2006) (see also chapter 84, this volume). Each photoreceptor responds maximally when the light is polarized in a particular orientation (the "preferred" orientation), and minimally when it is polarized in the orthogonal orientation. Information from these polarization-sensitive photoreceptors is fed to interneurons in the medulla and eventually to the central-complex region of the brain. Recordings from the medulla reveal three classes of polarization-sensitive neurons, with broad orientation tuning curves and preferred orientations separated by about 120° (Labhart & Meyer, 2002). Recordings from the central complex, on the other hand, reveal polarization-sensitive neurons that are individually much more sharply tuned. In this population of neurons the preferred directions are distributed more or less uniformly in all compass directions (Heinze & Homberg, 2007). The preferred orientations of these neurons change during the course of the day, indicating that they are being continually recalibrated by signals from an internal circadian clock to compensate for the movement of the polarization pattern across the sky (Pfeiffer & Homberg, 2007; see also chapter 84, this volume).

APPLICATIONS TO ROBOTICS

Over the past 2 decades there has been considerable interest in implementing some of the insights gained from the study of vision and navigation in insects to the guidance of terrestrial and aerial vehicles. The reasons for doing this are twofold. First, robotic platforms offer a means of testing, under rigorous, real-world conditions, our concepts of how insects see and steer their way through the environment. Second, biological inspiration may provide novel solutions to difficult and persistent problems in the design of navigation systems for autonomous vehicles.

We have seen above that insects rely heavily on cues derived from optic flow to gauge the distances to various obstacles and surfaces, and to maneuver safely in their three-dimensional environment. From the standpoint of machine vision, the use of optic flow information for navigation is less demanding, computationally, than is the more classical approach of using stereo vision for this purpose. The makes the optic-flow-based approach attractive for use in small, compact, and lightweight aerial vehicles (Floreano et al., 2009). Here we outline some of the ways in which the principles of insect vision and navigation have been used for the guidance of autonomously navigating vehicles, on land and in the air.

Steering Robots along Corridors

The principle by which bees fly safely through narrow passages offers a simple strategy for guiding a robot along a corridor. A robot can progress along the middle of the corridor, without colliding with the walls, by balancing the speeds of the images of the two sidewalls. In addition, the speed of the robot can be adjusted to a safe value by holding constant the average velocity of the images of the two walls. The feasibility of this technique has been demonstrated in simulations (e.g., Portelli et al., 2010b) as well as in real robots (e.g., Humbert & Hyslop, 2010; Srinivasan, 2011; Srinivasan, Thurrowgood, & Soccol, 2009).

Hardware for Emulating Insect Vision

Figure 85.2 shows one way in which insect-like vision can be realized using standard off-the-shelf hardware (Moore at al., 2011a, 2011b; Thurrowgood et al., 2009, 2010). This system ("iEye") comprises two miniature video cameras, each equipped with a wide-angle lens that conveys a field of view of 190° (slightly larger than a hemisphere). The cameras are directed approximately laterally—one facing the left and the other the right—with a small frontal convergence of 10° for each camera. This arrangement provides nearly panoramic vision, with a small blind zone at the rear, and a frontal visual field in which there is binocular overlap—as in bees—in a thin, vertically elongated section of the frontal visual field, shaped like a slice of an orange, measuring 130° vertically, and 30° horizontally at the equator.

Terrain-Following Guidance for Aircraft

Flying at a constant, low height above the ground is important when there is a need to prevent detection by enemy radar or to perform close-up exploration of terrain. If the ground speed of the aircraft is known (e.g., from measurements of airspeed or from global positioning system information), then, following the strategy of the honeybee, the height above the ground can be computed from the optic flow that is generated by the aircraft's motion above the ground. This magnitude of the optic flow can be used to control the aircraft's height above the ground, and to achieve terrain following. This approach is attractive because optic flow can be measured by incorporating a small, inexpensive, low-resolution video camera in the aircraft. This vision-based approach is cheaper and requires a smaller payload compared to other strategies for height measurement, such as radar or ultrasound. This technique of terrain following has been implemented successfully in fixed-wing (Barrows, Chahl, & Srinivasan, 2003) as well as rotary-wing aircraft (Garratt & Chahl, 2008; Srinivasan, 2011; Srinivasan, Thurrowgood, & Soccol, 2009).

Using the Horizon to Sense and Control Attitude

As we discussed above, one way of determining attitude is to sense the position and shape of the horizon profile in a panoramic image (see figure 85.3) acquired by the vision system (see figure 85.2) during flight. The horizon (red profile, figure 85.3) is detected by an algorithm that uses intensity and spectral information to classify pixels as belonging to the ground or the sky

FIGURE 85.2 (a) View of near-panoramic iEye aircraft vision system, featuring two miniature video cameras, each equipped with a wide-angle lens that provides a 190° field of view. (b) Upper panel: Raw images captured by the two cameras. Lower panel: Remapping of the raw images onto a single, near-panoramic, equirectangular image in which the horizontal and vertical axes represent azimuth and elevation, respectively. The vision system captures approximately 88% of the total view sphere. The red area indicates the region of binocular overlap. (Reproduced with permission from R. J. D. Moore, 2012.)

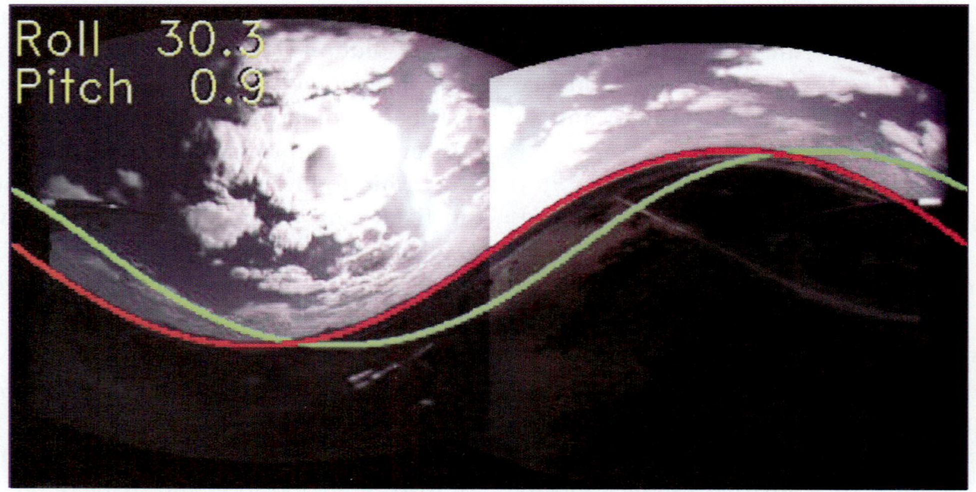

FIGURE 85.3 Determination of aircraft attitude (roll and pitch) from the horizon profile in a panoramic view of the environment. Details in text.

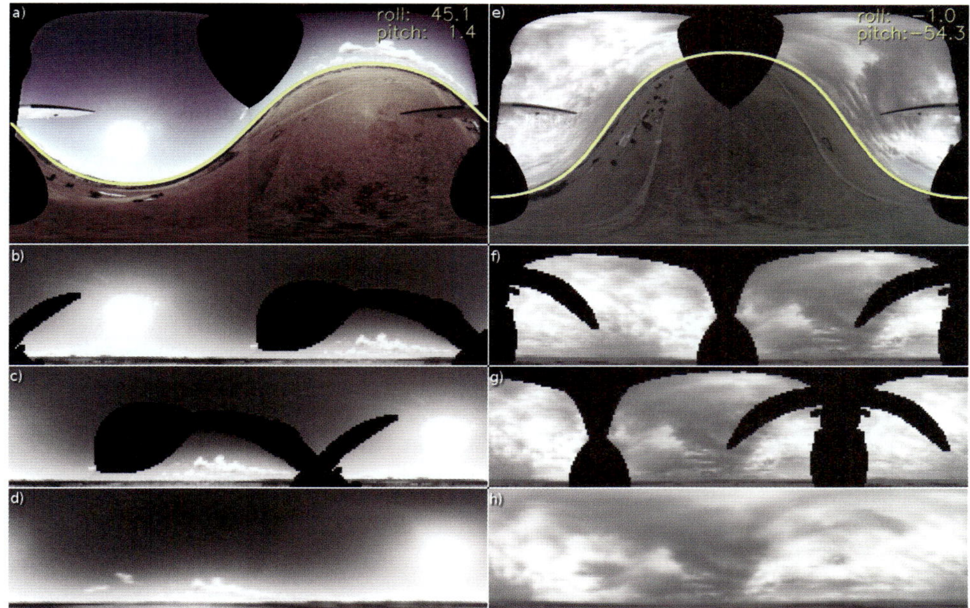

FIGURE 85.4 Illustration of a scheme for estimating and controlling heading direction. Two examples are shown, one for a clear day in which the sun was visible (left-hand column) and another for a cloudy day (right-hand column). Panels (a, e) show the raw panoramic image, with the detected horizon profile (yellow curve) and the instantaneous roll and pitch of the aircraft, computed from the horizon profile as described in the text. Panels (b, f) show remapped versions of the panoramic image, using the horizon as the reference. Panels (c, g) show the remapped image registered with the reference image. Panels (d, h) show the fully generated reference image. (Reproduced with permission from Moore et al., 2011b.) Details in text.

(Moore et al., 2011a, 2011b; Thurrowgood et al., 2009, 2010). The numbers indicate the estimates of the instantaneous roll and pitch of the aircraft, based on the shape of the horizon profile. The green curve represents the horizon profile corresponding to the roll and yaw as sensed by the aircraft's onboard inertial measurement unit. It is clear that the horizon-based estimate is very close to the ground truth. The inertial estimate is considerably different and is less accurate. This is to be expected because inertial measurements have the disadvantage that the instantaneous attitude is determined by integrating rate signals from the gyroscopes over time and combining these with accelerometer readings, which are presumed to indicate the direction of gravity. However, the accelerometer readings are influenced not only by gravity but also by accelerations of the aircraft. Furthermore, the error in the estimate of attitude from the gyro signals will increase with time because they are noisy rate signals that are being integrated over time. These problems do not arise if the horizon is used to estimate attitude because the horizon provides an absolute external reference. This is precisely what insect ocelli seem to be organized to achieve. Recent work has demonstrated that this technique for monitoring and controlling

aircraft attitude can also be used to automate the execution of a variety of aerobatic maneuvers such as loops and Immelmann turns (Thurrowgood et al., 2010).

Using the Sky to Set and Maintain Course Direction

In addition to using the horizon to monitor and control pitch and roll (as described above), visual information from the sky can be used to estimate and control heading direction. Information on heading direction can be obtained by monitoring the spatial distribution of intensity across the sky, due to the presence of the sun and clouds (Moore et al., 2011b). The strategy is to capture a panoramic view of the sky (top panels, figure 85.4) and remap it digitally to obtain a horizon-referenced image (second panels, figure 85.4). In the remapped image the horizontal axis represents the azimuth (varying from 0° to 360°), and the vertical axis represents the elevation (varying from 10° below the horizon to 80° at the top). In these images, some parts of the sky are occluded by aircraft structures. This problem is eliminated by combining a series of partially remapped images of the sky images, acquired during the initial phase of the flight, to obtain a complete,

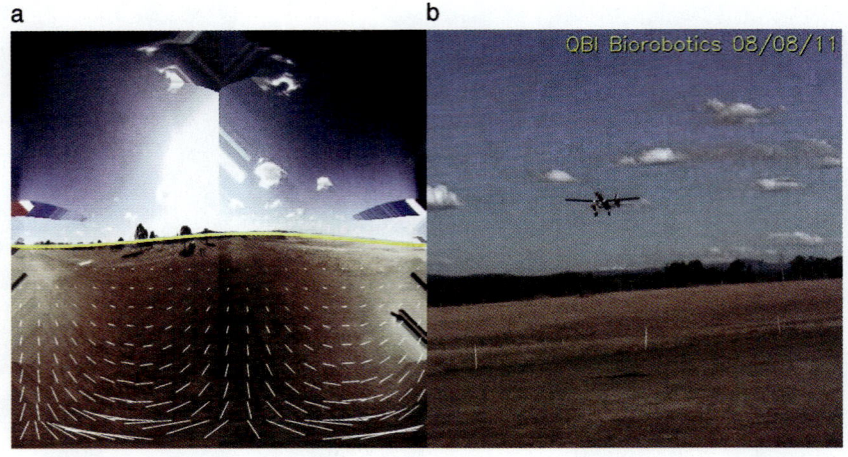

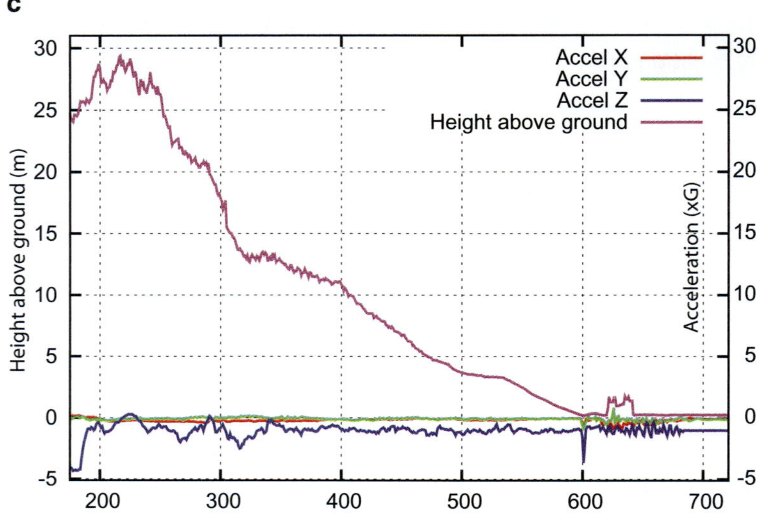

FIGURE 85.5   Automatic aircraft landing. (a) The left-hand panel shows the detected horizon profile (yellow curve), which is used to estimate and control the aircraft's roll and pitch. It also shows the optic flow induced by the translational component of the aircraft's motion, which is monitored to control the aircraft's descent. (b) A view of the landing aircraft. (c) Performance of the automatic landing system, showing the variation of height with time (expressed in video frame numbers at a frame rate of 25/s) and the readings of the onboard accelerometers. The negative offset in the Z acceleration trace represents the acceleration due to gravity (−1G). Accel, Acceleration.

occlusion-free image of the sky panorama (bottom panels, figure 85.4). The unoccluded image of the sky is then used as a reference image, against which the image acquired at any instant, such as that shown in the third panels of figure 85.4, is compared to compute the angular rotation (in yaw) of the aircraft with respect to the direction defined by the reference image. The heading direction at any instant, relative to that defined by the reference image, is computed by determining the horizontal shift that should be applied to the current image (second panels) to best match it to the reference image (bottom panels), as indicated by a "sum of absolute differences" algorithm (Wikipedia, 2012). The match between the two images is computed only in the unoccluded region of the sky.

Robot Navigation Using a Polarization Compass

The efficacy of using the polarized light pattern in the sky for navigation has been tested by incorporating an insect-inspired polarization compass into autonomous robots (Lambrinos et al., 2000). "Sahabot," a terrestrial robot steered by a polarization compass, was put through its paces in the arid, featureless landscapes of the Saharan desert, where only the polarization pattern of the sky provided cues as to the robot's heading. The robot's compass used three polarization-sensitive channels, with preferred directions separated by 120°. Each channel consisted of a polarizing filter positioned over a photodetector. The signals from the three photodetectors were analyzed to compute the orientation of the polarization vector in the overhead sky and to steer the robot in any prescribed direction. The heading information was also combined with information from wheel-based odometry to determine, with impressive accuracy, exactly where the robot was located at any point along its journey.

Robot Odometry Based on Honeybee Vision

Several laboratories have used the honeybee-based principle of integrating optic flow to implement visual odometry successfully. In one study a robot was able to traverse a corridor repeatedly, always stopping after it had traveled a fixed distance as reported by its visually driven odometer. The error in the stopping position was less than 2% of the traverse distance (Weber, Venkatesh, & Srinivasan, 1997). In another study (Chahl & Srinivasan, 1996) a robot used optic flow to compute all of its translations and rotations along any arbitrary route that it took from a starting point, and it was able to use this information to (1) compute where it was in relation to the starting point, at all times, and (2) return successfully to the starting location—again relying solely on optic flow information and without knowledge of or reference to any landmarks in the environment. Visual odometry based on optic flow has also been implemented successfully in large automotive vehicles, by using a downward-facing video camera to capture the optic flow generated by the road surface (Nourani-Vatani, Roberts, & Srinivasan, 2009).

Visual Guidance of Aircraft Landing

We have seen that honeybees perform grazing landings by adjusting their flight speed so as to hold the image velocity of the ground constant during the approach. Autonomous landing approaches using approximations of this strategy have been achieved with fixed-wing aircraft (Beyeler, 2009; Chahl, Srinivasan, & Zhang, 2004). It is difficult to implement the honeybee landing strategy exactly on a fixed-wing aircraft because this requires the ground speed to approach arbitrarily small values (below stall speed) as the aircraft nears the ground. Consequently, a modified landing strategy has been implemented and tested. In this strategy, the aircraft approaches the ground at a constant, known speed in the initial phase of the landing. This is achieved by using a prescribed throttle setting and pitch attitude. During this period, the optic flow generated by the ground is monitored and used to control the rate of descent of the aircraft (see figure 85.5). The desired rate of descent is achieved by controlling the height (via the elevators) in such a way that the magnitude of the optic flow increases (ramps up) at the appropriate rate. When close to touchdown, the optic flow information tends to become unreliable and is eschewed in favor of stereo-based range information, for guidance during the final moments of descent. There is some evidence that honeybees rely on stereo vision during their hover phase just prior to touchdown (Evangelista et al., 2010) although this remains to be explored further. An illustration of the efficacy of this landing strategy is shown in figure 85.5c, which shows the descent profiles for an automatic landing and the forces of impact at touchdown as measured by onboard accelerometers. The results indicate that automatic landings performed using this strategy result in touchdowns that are as smooth as in manually (remotely) controlled landings, and are more predictable and reproducible—especially under windy conditions.

CONCLUSIONS AND OUTLOOK

We have seen that, with regard to visual guidance and navigation, insects are scarcely handicapped by their miniscule brains and limited processing capacity. Instead, the demand for neural parsimony appears to have driven the evolution of simple, elegant, and effective solutions to deal with the challenges. Because of their small interocular separations, many insects appear not to use stereo vision, except for objects that are at very close range (Evangelista et al., 2010; Rossel, 1983). Instead, they rely heavily on cues derived from optic flow. We have seen that many important behaviors, such as collision avoidance, control of flight speed, landing, and odometry, are mediated by cues derived from optic flow. The continuing challenges in this field are, first, to better understand the neural processing mechanisms that underlie these remarkable faculties, and second, to explore whether these biological insights can provide further and better inspiration for endeavors in engineering.

ACKNOWLEDGMENTS

Some of the research described here was supported partly by the Australian Research Council Centre of Excellence in Vision Science (CE0561903), by U.S. Army Research Office Grant MURI ARMY-W911NF041076, and by a Queensland Premier's Fellowship.

REFERENCES

Baird, E., Kornfeldt, T., & Dacke, M. (2010). Minimum viewing angle for visually guided ground speed control in bumblebees. *Journal of Experimental Biology, 213,* 1625–1632.

Baird, E., Srinivasan, M. V., Zhang, S. W., & Cowling, A. (2005). Visual control of flight speed in honeybees. *Journal of Experimental Biology, 208,* 3895–3905.

Baird, E., Srinivasan, M. V., Zhang, S. W., Lamont, R., & Cowling, A. (2006). Visual control of flight speed and height in the honeybee. From Animals to Animats 9, Proceedings of the 9th International Conference on Simulations of Adaptive Behavior, 4095, 40–51.

Barron, A., & Srinivasan, M. V. (2006). Visual regulation of ground speed and headwind compensation in freely flying honey bees (*Apis mellifera L.*). *Journal of Experimental Biology, 209,* 978–984.

Barrows, G. L., Chahl, J. S., & Srinivasan, M. V. (2003). Biologically inspired visual sensing and flight control. *Aeronautical Journal, 107,* 159–168.

Berry, R., van Kleef, J., & Stange, G. (2007). The mapping of visual space by dragonfly lateral ocelli. *Journal of Comparative Physiology. A, Neuroethology, Sensory, Neural, and Behavioral Physiology, 193,* 495–513.

Beyeler, A. (2009). Vision-based control of near-obstacle flight. Thesis No. 4456, École Polytechnique Fédérale de Lausanne, Lausanne, Switzerland.

Borst, A. (2009). Drosophila's view on insect vision. *Current Biology, 19,* R36–R47.

Chahl, J. S., & Srinivasan, M. V. (1996). Visual computation of egomotion using an image interpolation technique. *Biological Cybernetics, 74,* 405–411.

Chahl, J. S., Srinivasan, M. V., & Zhang, S. W. (2004). Landing strategies in honeybees and applications to uninhabited airborne vehicles. *International Journal of Robotics Research, 23,* 101–110.

Dyhr, J. P., & Higgins, C. M. (2010). The spatial-frequency tuning of optic-flow-dependent behaviors in the bumblebee *Bombus impatiens. Journal of Experimental Biology, 213,* 1643–1650.

Egelhaaf, M. (2008). Fly vision: Neural mechanisms of motion computation. *Current Biology, 18,* R339–R341. doi:10.1016/j.cub.2008.02.046.

Evangelista, C., Kraft, P., Dacke, M., Reinhard, J., & Srinivasan, M. V. (2010). The moment before touchdown: Landing manoeuvres of the honeybee *Apis mellifera. Journal of Experimental Biology, 213,* 262–270.

Floreano, D., Zufferey, J.-C., Srinivasan, M. V., & Ellington, C. (2009). *Flying insects and robots.* Berlin: Springer.

Fry, S. N., Rohrseitz, N., Straw, A. D., & Dickinson, M. H. (2009). Visual control of flight speed in *Drosophila melanogaster. Journal of Experimental Biology, 212,* 1120–1130.

Garratt, M. A., & Chahl, J. S. (2008). Vision-based terrain following for an unmanned aircraft. *Journal of Field Robotics, 25,* 284–301.

Goodman, L. (2003). (Ed.). Form and function in the honeybee. Cardiff, England: International Bee Research Association.

Heinze, S., & Homberg, U. (2007). Maplike representation of celestial E-vector orientations in the brain of an insect. *Science, 315,* 995–997.

Homberg, U. (2004). In search of the sky compass in the insect brain. *Naturwissenschaften, 91,* 199–208. doi:10.1007/s00114-004-0525-9.

Humbert, J. S., & Hyslop, A. M. (2010). Bioinspired visuomotor convergence. *IEEE Transactions on Robotics, 26,* 121–130.

Joesch, M., Plett, J., Borst, A., & Reiff, D. F. (2008). Response properties of motion-sensitive visual interneurons in the lobula plate of *Drosophila melanogaster. Current Biology, 18,* 368–374.

Kraft, P., Evangelista, C., Dacke, M., Labhart, T., & Srinivasan, M. (2011). Honeybee navigation: Following routes using polarized-light cues. *Philosophical Transactions of the Royal Society of London. Series B, Biological Sciences, 366,* 703–708.

Krapp, H. G. (2000). Neuronal matched filters for optic flow processing in flying insects. *International Review of Neurobiology, 44,* 93–120.

Labhart, T., & Meyer, E. P. (2002). Neural mechanisms in insect navigation: Polarization compass and odometer. *Current Opinion in Neurobiology, 12,* 707–714.

Lambrinos, D., Moller, R., Labhart, T., Pfeifer, R., & Wehner, R. (2000). A mobile robot employing insect strategies for navigation. *Robotics and Autonomous Systems, 30,* 39–64.

Moore, R. J. D. (2012). Vision systems for autonomous aircraft guidance. Ph.D. thesis, University of Queensland, Brisbane.

Moore, R. J. D., Thurrowgood, S., Bland, D., Soccol, D., & Srinivasan, M. V. (2011a, 25–30 September 2011). A fast and adaptive method for estimating UAV attitude from the visual horizon. Paper presented at the Proceedings, IEEE / RSJ International Conference on Intelligent Robots and Systems, San Francisco, CA.

Moore, R. J. D., Thurrowgood, S., Bland, D., Soccol, D., & Srinivasan, M. V. (2011b, 7–9 December 2011). A method for the visual estimation and control of 3-DOF attitude for UAVs. Paper presented at the Proceedings, Thirteenth Australasian Conference on Robotics and Automation (ACRA 2011), Melbourne, Australia.

Nordstrom, K., Barnett, P. D., Moyer de Miguel, I., & O'Carroll, D. C. (2008). Sexual dimorphism in the hoverfly motion vision pathway. *Current Biology, 18,* 661–667.

Nourani-Vatani, N., Roberts, J., & Srinivasan, M. V. (2009). Practical visual odometry for car-like vehicles. Paper presented at the IEEE International Conference on Robotics and Automation, Kobe, Japan.

Pfeiffer, K., & Homberg, U. (2007). Coding of azimuthal directions via time-compensated combination of celestial compass cues. *Current Biology, 17,* 960–965.

Portelli, G., Ruffier, F., & Franceschini, N. (2010a). Honeybees change their height to restore their optic flow. *Journal of Comparative Physiology. A, Neuroethology, Sensory, Neural, and Behavioral Physiology, 196,* 307–313.

Portelli, G., Serres, J., Ruffier, F., & Franceschini, N. (2010b). Modelling honeybee visual guidance in a 3-D environment. *Journal of Physiology, Paris, 104,* 27–39.

Rossel, S. (1983). Binocular stereopsis in a insect. *Nature, 302,* 821–822.

Serres, J. R., Masson, G. P., Ruffier, F., & Franceschini, N. (2008). A bee in the corridor: Centering and wall-following. *Naturwissenschaften, 95,* 1181–1187. doi:10.1007/s00114-008-0440-6.

Srinivasan, M. (1993). How insects infer range from visual motion. In F. Miles & J. Wallman (Eds.), *Visual motion and its role in the stabilization of gaze* (pp. 139–156). Amsterdam: Elsevier.

Srinivasan, M., Thurrowgood, S., & Soccol, D. (2009). From flying insects to autonomously navigating robots. *IEEE Robotics & Automation Magazine, 16,* 59–71.

Srinivasan, M. V. (2011a). Honeybees as a model for the study of visually guided flight, navigation, and biologically inspired robotics. *Physiological Reviews, 91,* 389–411.

Srinivasan, M. V. (2011b). Visual control of navigation in insects and its relevance for robotics. *Current Opinion in Neurobiology, 21,* 535–543.

Srinivasan, M. V., Zhang, S. W., Chahl, J. S., Barth, E., & Venkatesh, S. (2000). How honeybees make grazing landings on flat surfaces. *Biological Cybernetics, 83,* 171–183.

Stange, G. (1981). The ocellar component of flight equilibrium control in dragonflies. *Journal of Comparative Physiology, 141,* 335–347.

Straw, A. D., Lee, S., & Dickinson, M. H. (2010). Visual control of altitude in flying *Drosophila. Current Biology, 20,* 1–7.

Tammero, L. F., & Dickinson, M. H. (2002). The influence of visual landscape on the free flight behavior of the fruit fly *Drosphila melanogaster. Journal of Experimental Biology, 205,* 327–343.

Thurrowgood, S., Moore, R. J. D., Bland, D., Soccol, D., & Srinivasan, M. V. (2010, 1–3 December 2010). UAV attitude control using the visual horizon. Paper presented at the Proceedings, Twelfth Australasian Conference on Robotics and Automation, Brisbane, Australia.

Thurrowgood, S., Soccol, D., Moore, R. J. D., Bland, D., & Srinivasan, M. V. (2009). A vision based system for attitude estimation of UAVs. Paper presented at the IEEE/RSJ International Conference on Intelligent Robots and Systems.

van Kleef, J., James, A. C., & Stange, G. (2005). A spatiotemporal white noise analysis of photoreceptor responses to UV and green light in the dragonfly median ocellus. *Journal of General Physiology, 126,* 481–497.

von Frisch, K. (1993). *The dance language and orientation of bees.* Cambridge, MA: Harvard University Press.

Weber, K., Venkatesh, S., & Srinivasan, M. V. (1997). *Insect inspired behaviours for the autonomous control of mobile robots: From living eyes to seeing machines* (pp. 226–248). Oxford, England: Oxford University Press.

Wehner, R. (1992). Arthropods. In F. Papi (Ed.), *Animal homing* (pp. 45–144). London: Chapman and Hall.

Wehner, R., & Labhart, T. (2006). Polarization vision. In E. Warrant & D.-E. Nilsson (Eds.), *Invertebrate vision* (pp. 291–348). Cambridge, England: Cambridge University Press.

Wikipedia. (2012). Sum of absolute differences.

Wittlinger, M., Wehner, R., & Wolf, H. (2007). The desert ant odometer: A stride integrator that accounts for stride length and walking speed. *Journal of Experimental Biology, 210,* 198–207.

XII   THEORETICAL

PERSPECTIVES

86 The Evolution of the Visual System in Primates

JON H. KAAS

Compared to most mammals, primates are highly visual. While the over 350 species of primates vary in their behavioral adaptations and thus brain organizations, primates in general share a number of distinctive features of brain organization and of visual system organization. Thus, an experienced investigator, given any primate brain, can look at appropriately prepared brain sections and say, yes, this is a primate. Here we review the features of the visual system of primates that distinguish them from those of other mammals, consider the ways that visual systems of primates are known to vary, and suggest how the distinctive features of the visual systems of primates emerged in evolution and came to vary. To do this, we first outline the evolutionary history of primates, starting with early mammals, mention some overall features of primate brains that distinguish them from those of other mammals, and proceed to discuss the major components of the visual systems (the retina, visual thalamus, superior colliculus, and visual cortex) of primates in terms of what primates share and how primates vary.

WHAT IS SPECIAL ABOUT PRIMATE BRAINS?

Primate brains are different than the brains of other mammalian taxa, and many of these differences relate to the visual system and the emphasis on vision that characterize primates. An experienced comparative neuroanatomist can recognize a brain from being from one of the 350 or so primate species by its fissure pattern (always a lateral fissure and a calcarine fissure) and its expansive occipital and temporal regions (containing visual areas). In a series of brain sections, a distinctive pattern of lamination in the dorsal lateral geniculate (visual) nucleus most clearly identifies the brains as from a primate, but other features are also characteristic of primates.

A fundamental difference in primate brains compared to the brains of other mammals is the number of neurons they have, given the size of their brain (Herculano-Houzel et al., 2007). In general, species with larger brains have more neurons, and by implication, more computational power. However, the increase in neuron numbers with progressively larger brains in nonprimate mammals is less than that for primates with a similarly sized brain. For example, a primate brain that is roughly the same size as that of a rodent has twice the number of neurons that are more densely packed than other mammals with progressively larger brains. This is partly a result of an increase in the average size of neurons in the brains of these nonprimate mammals, but also because the ratio of glia and other cells per neuron increases in nonprimates. As an example of this difference, the brain of a small monkey of about 16 g has 1.5 billion neurons while a rodent brain of about the same size (18 g) has only 0.9 billion neurons (see figure 86.1). Even the largest rodent that ever lived, the now extinct South American rodent, phoberomus, at 700 kg (1,540 lb) had a brain with only an estimated 7.5 billion neurons, while a human of 81 kg (180 lb) has a brain of about 86 billion neurons (Avzevedo et al., 2009). Elephants with brains three times the size of a human brain nevertheless fall short with about half the number of at least cortical neurons as the human brain (Hart & Hart, 2007). Neuron densities are higher in primate brains because neurons on average are smaller although primates also have a great range of neuron sizes, shapes, and functions (Sherwood & Hof, 2007). This seems to be especially the case in neocortex, where primary visual cortex, V1, has very small neurons in layer 4, hence the name, koniocortex. Yet, there are also a few very large pyramidal neurons (Meynert neurons). Because differences in neuron sizes and packing densities vary across visual areas and primate taxa (Collins et al., 2010), we return to this issue in the section on visual cortex.

In addition to differences in cellular composition, size, and density, primate brains differ from the brains of other mammals in a number of other ways. First, although variable across the primate taxa, in general primates have more cortical areas compared to nonprimate mammals, and this difference is particularly noted

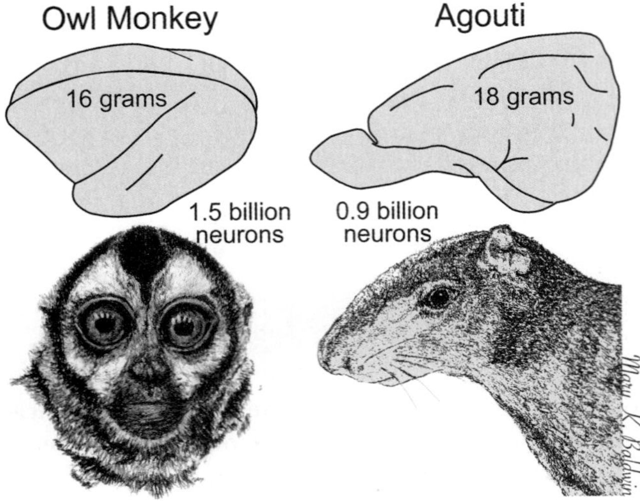

FIGURE 86.1   Differences in the number of neurons in the brains of about the same size from a small monkey and a large rodent. Primates' brains of comparable sizes have more neurons than rodent brains, and this difference increases with greater brain size. This difference is even greater when neuron numbers are related to body size. The largest rodent that ever existed (now extinct) had a body size of 700 kg (1540 lb) while having an estimated 7.5 billion neurons, while the brain of an 81-kg (180-lb) human would have roughly 100 billion neurons. These estimates are based on neuron scaling rules from Herculano-Houzel et al. (2007).

in the visual cortex, where well-studied macaque monkeys are proposed to have more than 30 cortical fields (Felleman & Van Essen, 1991; Kaas, 1989). A second difference is that matched cortical areas in the two cerebral hemispheres tend to specialize in different ways, at least in the anthropoid primates with larger brains (Corballis, 2007). Obvious hemispheric differences in humans include left-hemisphere specialization for language and handedness, but right-hemisphere specializations for face recognition and other visual functions have been reported as well. Another important difference is in the size of the peripheral olfactory system, which has gotten relatively smaller and less emphasized in primates (Bhatnagar & Smith, 2007) while frontal and parietal networks for sensorimotor abilities have expanded greatly and diversified for a variety of primate lifestyles.

PRIMATE EVOLUTION

Before considering the evolution of visual systems of primates in detail, it is useful to briefly outline the phylogenetic relationships of present-day primates and their close relatives, and how they evolved. Early mammals gave rise to the six major branches of present-day mammals including the early branches of the

egg-laying monotremes, and the marsupials, which rear their young in a pouch of skin or a pseudo-pouch where they are attached to the nipples and hang like a cluster of grapes. The four main branches of placental (eutherian) mammals diverged over 90 million years ago (see figure 86.2). While carnivores such as domestic cats have long been used in studies of the visual system, and ferrets more recently, carnivores are not closely related to primates. As members of the Euarchontoglire superclade, primates are most closely related to tree shrews (*Scandentia*) and flying lemurs (*Deromoptera*). Despite this common name, flying lemurs glide, rather than fly, and they are not lemurs. Rodents and Lagomorphs are the other members of the Euarchontoglire superclade and thus more closely related to primates than carnivores. This relationship implies that if we are interested in inferring what the brains were like of the most recent nonprimate ancestors of primates, we should look at the brains of tree shrews, flying lemurs (largely unavailable for study), rodents, and rabbits.

Early primates emerged as a branch of the Euarchontoglire superclade over 80 million years ago (Meredith et al., 2011). They diversified into several lines of archaic primates that became extinct, and the euprimates, which led to the present-day lemurs, lorises, and galagos, the visually specialized tarsiers, and the great radiation of anthropoid primates (New World monkeys, Old World monkeys, apes, and humans). Early primates were small and resembled present-day galagos and some of the lemurs in brain size relative to body size, although their brains were somewhat smaller than those of present-day prosimians. Their fossil remains indicate that they were likely nocturnal, and fed in the thin branches of bushes and trees on insects, small vertebrates, fruits, and buds (Ross & Martin, 2007). Feeding while positioned on thin, often moving branches and reaching for food implies that early primates had already evolved neural networks for exceptional visuomotor control of hand movements (Block & Boyer, 2002), one of the most distinguishing features of the behaviors of extant primates (Whishaw, 2003). While present-day prosimians are varied, and some are now diurnal, galagos and a few other prosimians, such as mouse lemurs, may have changed the least from the general body form and brain organization of early primates. Galagos are African primates that are the size of cats, or smaller, arboreal, nocturnal, and feed on fruits, gums, and insects. One line of early prosimian primate evolution led to present-day tarsiers and anthropoids. Tarsiers are highly specialized, small, nocturnal, visual predators of invertebrates and small vertebrates, and they apparently eat no plants. The earliest anthropoids, as well as the shared ancestors with tarsiers, were small,

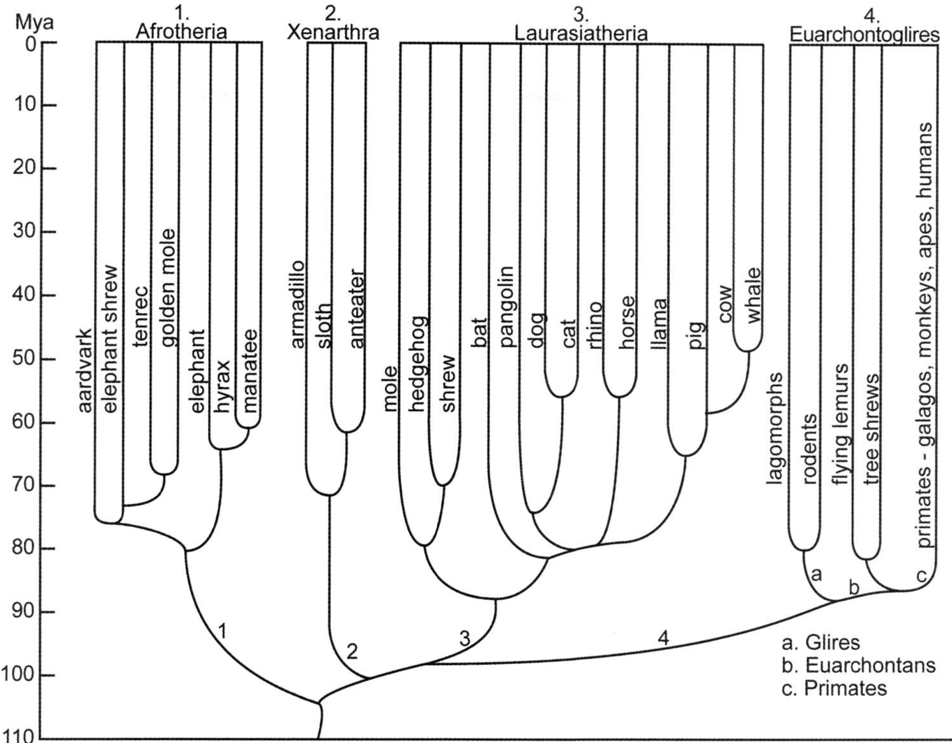

FIGURE 86.2   The emergence and radiation of the four major clades of placental (eutherian) mammals. In each clade, a few representative present-day species and other groups are listed. Primates are part of the Euarchontoglire radiation. The illustrated phylogenetic relationships suggest that inferences about the visual systems of early primates would be most productive when the visual systems comparisons are made with other members of the Euarchontoglire clade. Time on the left is in millions of years ago (mya).

diurnal visual predators, with later stem anthropoids (monkeys) evolving larger body sizes, and diversified diets that included fruit, leaves, and other parts of plants. Many of these primates formed social groups, possibly as diurnal life made them less cryptic and more dependent on others for protection and warnings. A line of anthropoid monkeys some 25–30 million years ago led to apes, which formed a successful radiation, occupying rainforest environments. One line of apes diverged 6–8 million years ago into a branch that led to modern chimpanzees and bonobos, and to bipedal primates—the hominins, including modern humans (de Sousa & Wood, 2007). Hominins probably emerged as one line of apes adapted to a drier climate with grassland and savannah. The early hominins, australopiths, roamed the earth about 5 million years ago. They were bipedal, but they retained skeletal features for climbing trees. Their brains were about the size of the brains of apes of similar body size. The genus *Homo* emerged about 2 million years ago with *Homo habilis*, followed 1.7 million years ago by *Homo erectus*, which spread out of Africa, and *Homo sapiens*, some 250,000 years ago. Other *Homo* species included Neanderthals, which died out only 35,000 years ago. There is evidence for an archaic

hominin population from Siberia, the Denisovans, which also coexisted with modern humans (Skoglund & Jakobsson, 2011).

Extant primates constitute one of the most diversified taxonomic groups. It includes 14 families, over 350 species, and a range of body sizes from the 40-g mouse lemur to the 200-kg male gorilla, a 5,000-fold difference that is important as body size relates to brain size (Jerison, 1973). This variability in body and brain size makes a comparative study of primate brains a challenging task, especially since all members of the remarkable and rapid hominin radiation are now extinct except for us. Yet a wealth of recent studies has led to much progress, and a lot that can be inferred about the brains of unstudied species.

THE EYE AND THE RETINA

The eyes and retinas of modern primates reflect their ancestral condition as well as subsequent adaptations to new visual environments (Preuss, 2007). Ancestral primates were thought to be nocturnal (crepuscular), being most active during the dim light just after sunset and just before sunrise (Ross & Martin, 2007). Early primates, as well as extant primates, had large forward-facing eyes, which favored stereoscopic vision and produced a sharper image for central vision. Most extant strepsirrhine primates have retained a nocturnal lifestyle, and their eyes have retained visual specializations for functioning in dim light. Thus, the retina lacks a fovea, and even in the central retina, there is a convergence of a number of receptors on bipolar cells to promote the detection of light. In addition, the back of the retina is lined with a tapetum, a reflecting layer absent in some modern prosimians (diurnal lemurs): This enables receptors to be stimulated by reflected as well as direct light. Also, the retinas of nocturnal prosimian primates are dominated by rods, which are more sensitive than cones and responsive to a broader range of wavelengths. As for most mammals, strepsirrhine primates have two types of cones, the short wavelength-sensitive or blue (S) cones, and the medium-to-long wavelength (green/red or M/L cones). The S-cone pigment has been lost in some nocturnal primates (lorises, galagos, and owl monkeys) (Wikler & Rakic, 1990).

Tarsiers are haplorhine primates, as are anthropoid primates. The ancestors of tarsiers, in common with nearly all anthropoid primates, were probably diurnal, resulting in the loss of the reflecting tapetum, an increase in the proportion of cones to rods, and a cone-dominated fovea, all adaptations for higher visual acuity in bright light. When the more immediate ancestors of tarsiers reverted to nocturnal life, they compensated for the lack of a tapetum by evolving very large eyes and by adding rods to a degenerate fovea. Despite their nocturnal lifestyle, present-day tarsiers retain a distribution of functional S cones (Hendrickson et al., 2000).

Early anthropoid primates (monkeys) retained the M/L and the S cones. The M/L gene is on the X chromosome, and different alleles for this gene emerged to produce pigments with slightly different sensitivities to wavelength. Thus, for some species of New World monkeys, females may have different alleles on each of their two X chromosomes, which produce two slightly different M/L. Having two classes of M/L cones as well as S cones allows some females to function somewhat as trichromats. The males, with only one X chromosome, remain dichromats. Whether this enables the females to better detect ripe fruit and other colored food is unknown. However, one group of New World monkeys, the howler monkeys (Jacobs et al., 1996), evolved a three-cone retina by duplicating the ancestral M/L pigment gene. Evolved differences in the wavelength sensitivities of the pigments of photoreceptors, which were produced by the duplicated genes on the same X chromosome, allowed both males and females

to have trichromatic vision. A common ancestor of all Old World monkeys independently duplicated the M/L gene to produce (green) M and L (yellow–green) pigment genes, and all species of catarrhine primates (Old World monkeys, apes, and humans) are trichromats. In some humans, one of these genes may become nonfunctional, so that individuals are functionally dichromats. As these genes are on the X chromosome, this condition is more likely to exist in males (with only one X chromosome).

The intrinsic connections of the retina produce three main types of retinal ganglion cells. The midget ganglion cells for the parvocellular pathway (P), the much larger parasol ganglion cells for the magnocellular pathway (M), and the small ganglion cells of the koniocellular pathway (K). These neuronal streams from the retina are named after their targets in the parvocellular, magnocellular, and koniocellular layers of the dorsal lateral geniculate nucleus. The three types of ganglion cells appear to have homologues in other mammals, but their relative numbers and targets differ. In many species of the K and M classes (also called W and Y) dominate (Casagrande, Khaytin, & Boyd, 2007) while the P class (also called X) is small. In primates, the P class includes about 80% of the ganglion cells, and the K and M classes are each about 10% (Weller & Kaas, 1989). In many mammals, most ganglion cells of the retina project to the superior colliculus, but in primates only about 20% of the ganglion cells, the K and M cells, project to the superior colliculus. However, nearly all ganglion cells project to the lateral geniculate nucleus, and most or all P cells appear to project exclusively to the lateral geniculate nucleus (Weller & Kaas, 1989). Thus, the role of the P (midget) ganglion cells has vastly changed in primate evolution compared to their nonprimate ancestors, and the primary target of the retina has moved from the superior colliculus to the lateral geniculate nucleus as cortical processing of visual information became more important.

THE SUPERIOR COLLICULUS

The superior colliculus is a laminated visual structure in the dorsal midbrain of all mammals. In other vertebrates, the superior colliculus is called the optic tectum, and it is the major structure for generating visual behavior. In primates, the superior colliculus is generally considered a visuomotor structure, although it provides visual inputs to the lateral geniculate nucleus (Harting et al., 1991) and pulvinar (see figure 86.5) of the dorsal thalamus. In many mammals, the projections of the superior colliculus activate pulvinar neurons that provide an important second source of visual information to temporal visual cortex, thereby preserving some visual abilities after damage to primary visual cortex. Primates appear to be much more dependent on primary visual cortex for the activation of extrastriate cortex, suggesting that the role of the superior colliculus in perception has diminished. Other outputs of the superior colliculus are to groups of motor neurons in the brainstem that mediate eye and head movements, so that the eyes and central vision are directed toward objects of potential interest. Visual and other information reaches the superior colliculus via direct retinal projections (Kaas, 1978) and projections from visual and visuomotor areas of cortex (Baldwin & Kaas, 2012; Collins, Lyon, & Kaas, 2005).

The projections of the retina to the superior colliculus in all examined primates (e.g., Lane et al., 1973) differ from all nonprimate patterns in that only the "hemiretinas" that are devoted to the contralateral visual hemifield project to each superior colliculus (see figure 86.3). This pattern must have been present when the first primates evolved, but not much before, because tree shrews, rats, and rabbits (close relatives of primates) all have the more generalized condition that includes inputs from the complete retina of the

contralateral eye (Kaas, Harting, & Guillery, 1974). It was once thought that megachiropteran (fruit) bats also had the primate pattern, and this was considered evidence that these bats might actually be primates (Pettigrew, 1986), but further study did not support this conclusion (Thiele, Vogelsang, & Hoffman, 1991). For other reasons, bats are now placed on the Laurasiatheria clade, and are no longer considered closely related to primates (see figure 86.2). This primate specialization of only representing the contralateral hemifield in each superior colliculus likely relates to the emphasis on binocular and central vision that occurred in early primates.

The other major change that took place in the superior colliculus with the emergence of primates was a dramatic increase in the inputs from the ipsilateral eye, so that the major portion of the superior colliculus was responsive to both eyes. As the eyes rotated forward to create a large binocular visual field in the evolution of early primates, an increased input from the ipsilateral eye likely spread over most of the superior colliculus to terminate with a visuotopic pattern that matched that of the inputs from the contralateral eye. In present-day galagos, the contralateral retinal inputs to the superior

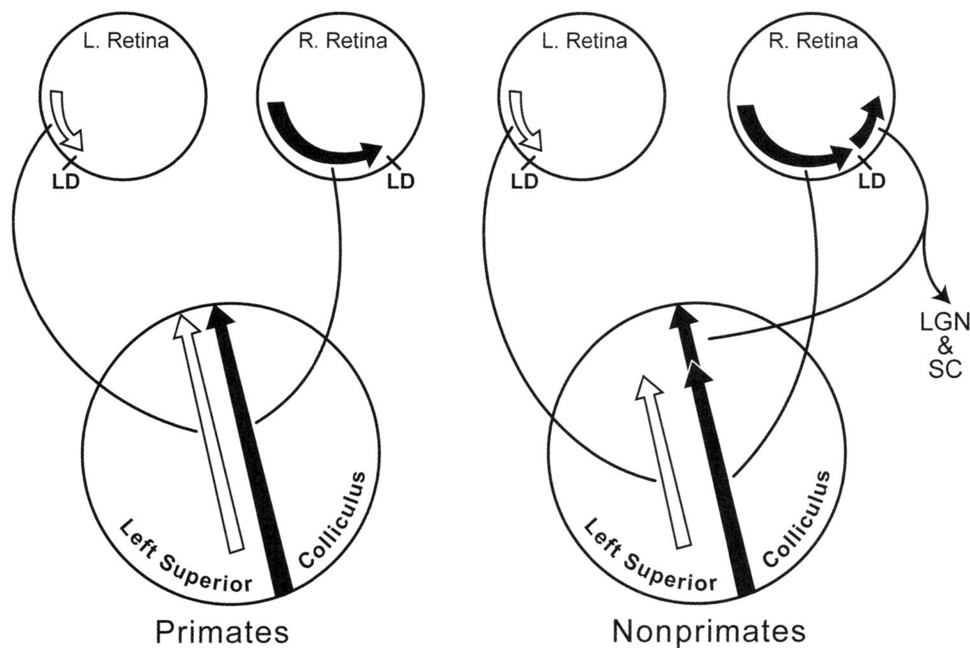

FIGURE 86.3   All primates share a unique type of retinotopic organization in the superior colliculus. In all primates, the left superior colliculus receives superimposed inputs from the nasal hemiretina of the right (R) eye and the temporal hemiretina of the left (L) eye, resulting in a representation of the right hemifield that is largely binocular. The right superior colliculus (SC) has a comparable pattern of inputs from the left and right eyes. In other studied mammals, the complete retina of the contralateral eye projects to the SC, even though the temporal retina also projects ipsilaterally to the lateral geniculate nucleus (LGN), as in primates. The temporal retina may also project weakly to the ipsilateral SC in a species-variable manner. In nonprimate mammals, the SC represents the complete visual field of the contralateral eye, including the binocular part of the ipsilateral hemifield. LD, line of decussation of retinal projections to the two hemispheres. See Kaas and Preuss (1993).

colliculus terminate more superficially in the superficial gray layer of the superior colliculus than the inputs from the contralateral eye (Tigges & Tigges, 1970). In contrast, the retinal inputs from both eyes terminate at the same superficial level in New World and Old World monkeys, but in a patchy pattern so that terminations from the ipsilateral eye may alternate with those from the contralateral eye (Weller & Kaas, 1989). With either pattern, terminations from the ipsi and contra eyes are at least partially segregated.

Another variable feature of the superior colliculus is in the input from cortex. All mammals with functional visual cortex have projections from visual cortex to the superficial layers of the superior colliculus, but most nonprimate mammals do not have the vast array of cortical visual areas present in primates (Felleman & Van Essen, 1991). Thus, the number of visual areas contributing inputs to the superior colliculus increased with the expansion of the visual cortex and an increase in the number of visual areas in the ancestors of early primates. Primates also differ from most nonprimates in having a large region of posterior parietal cortex that is involved in visuomotor functions (Kaas, Gharbawie, & Stepniewska, 2011), and these visuomotor areas project to the superior colliculus (Baldwin & Kaas, 2012; Collins, Lyon, & Kaas, 2005). Although all primates have a frontal eye field where electrical microstimulation evokes eye movements, this field does not project strongly to the superior colliculus in prosimian galagos as it does in anthropoid primates. This observation suggests that significant visuomotor projections from frontal cortex to the superior colliculus were not present in early primates, but they emerged with the evolution of the diurnal ancestors of anthropoid primates.

THE VISUAL THALAMUS: THE DORSAL LATERAL GENICULATE NUCLEUS AND THE PULVINAR

The main visual structures of the dorsal thalamus of all mammals are the dorsal lateral geniculate nucleus, LGN, and the pulvinar complex. In primates, the target of almost all retinal ganglion cells is the LGN (Weller & Kaas, 1989). All primates have a laminated LGN that reflects a basic pattern of four main layers (see figure 86.4). The two thinner ventral layers of large cells (the magnocellular layers) next to the optic tract receive inputs from the M cells of the retina while the two thicker dorsal layers of smaller cells (the parvocellular layers) receive inputs from the much more numerous P ganglion cells. The outer (external) layers of each pair receive inputs from the contralateral eye, and the inner (internal) layers receive inputs from the ipsilateral eye. However, there is some evidence for a modification of this pattern in tarsiers so that the external magnocellular layer receives inputs from the ipsilateral eye (Pettigrew et al., 1989). The K ganglion cells of the retina project to a distinct pair of koniocellular layers of small cells in the LGN of prosimian primates, and in other primates, project to a scattering of small cells between layers. The P-cell layers divide to form additional sublayers in the LGN of some of the New World monkeys, all of the Old World monkeys, and the great apes and humans.

In cats, the results of early studies of the response properties of retinal ganglion cells led investigators to define X, Y, and W cell classes, which are very similar to the P, M, and K classes of primates and likely represent homologous classes of cells in perhaps all mammals (Casagrande, Khaytin, & Boyd, 2007). Allowing for this premise, we can use comparative evidence to reconstruct the probable evolution of the laminar organization of the LGN from early mammals to present-day primates. Early mammals, as for many species of present-day mammals, likely had only weakly differentiated layers in the LGN (Kaas, 2007), with cryptic lamination revealed as a central pocket of inputs from the ipsilateral eye bordered on each side by tissue with inputs from the contralateral eye (see figure 86.4). The contra layer most distant from the optic tract likely had a mixture of P (X) and M (Y) retinal inputs, as did the central ipsi layer. The larger layer near the optic tract had K (W) retinal cell inputs, or a mixture of M (Y) and K (W) inputs, and was dominated by inputs from the contralateral eye, with some inputs from the ipsilateral eye, perhaps forming an ipsilateral region central in this layer, as in some rodents.

The closest relatives of primates, the tree shrews and flying lemurs, also have a laminated LGN, but they differ in the arrangement of inputs from the two eyes (see figure 86.4). In addition, LGN layers in tree shrews segregate inputs from classes of ganglion cells in a different way than primates. Most notably, the neurons with P (X) and M (Y) retinal inputs are mixed in the same layers, while those P and M neurons that respond to light onset (ON cells) are in different layers than the layers that have neurons that respond to light offset (OFF cells). Thus, distinctly laminated LGNs evolved independently and differently in tree shrews, flying lemurs, and primates. The early primates likely evolved a basic four-layer pattern from a more generalized LGN, with some segregation of inputs by eye of origin and some segregation by type of input from ganglion cell classes. Specialized layers for K-cell inputs were probably part of this early pattern, but these layers were diminished or lost as diurnal primates emerged, so that

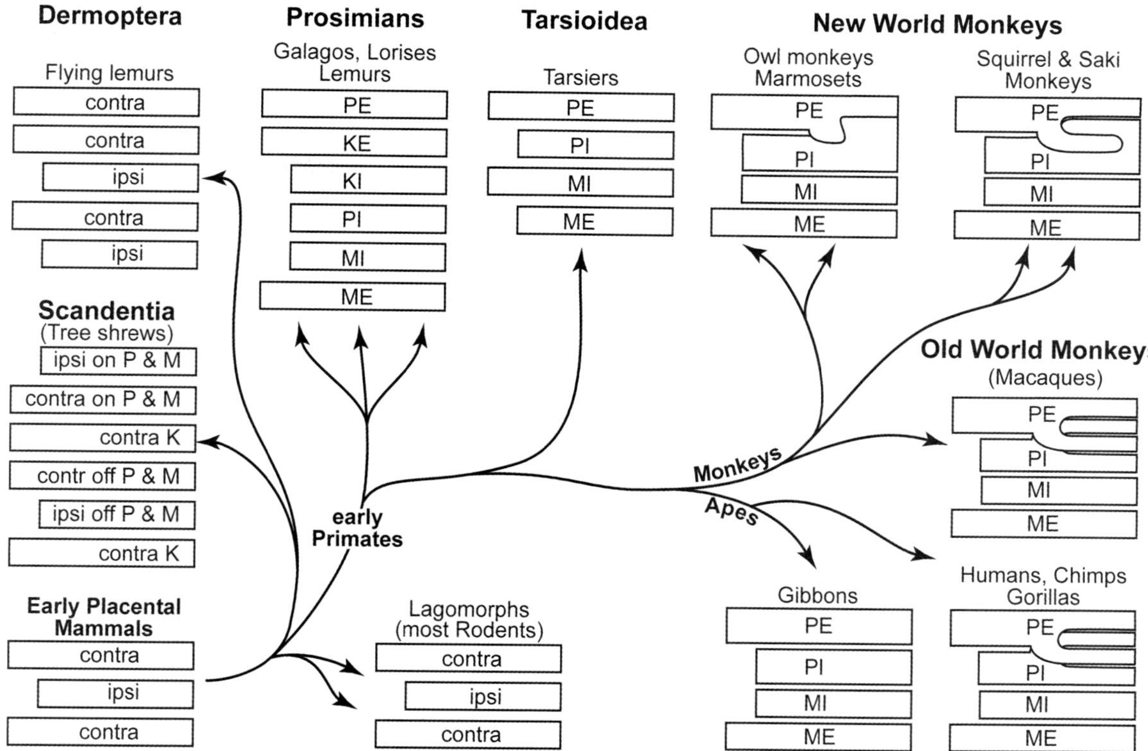

FIGURE 86.4   Laminar patterns of the dorsal lateral geniculate nucleus (LGN) in primates. The schematics are of the laminar patterns in brain sections cut across layers that are stacked like slices of bread from ventral (or along the optic tract) to dorsal (near the pulvinar). Early mammals had a simple pattern of LGN lamination that segregated inputs from the ipsilateral retina in a central region between outer regions with inputs from the contralateral retina. All primates have a more elaborate pattern of four basic layers that evolved with the first primates. Close relatives of primates, tree shrews and flying lemurs, evolved different patterns of lamination. In flying lemurs, layers have been identified histologically and by having inputs from the contralateral (contra) or ipsilateral (ipsi) eye (Pettigrew et al., 1989). In tree shrews microelectrode recordings suggest that M- and P-cell (termed Y and X) inputs are mixed in the same layers, unlike primates, while ON and OFF ganglion cell inputs are segregated in different layers, unlike in primates. ON retinal ganglion cells respond to the onset of light in the receptive field center while OFF cells respond to dimming or offset of light. Other layers appear to have K-cell inputs. The basic primate pattern of lamination includes two parvocellular (P) and two magnocellular (M) layers, with the longer (for the monocular field) outer (external) layers receiving from the contralateral retina (PE, ME) and the shorter inner or internal layers (PI, MI) from the ipsilateral eye. Present-day tarsiers may have a modified pattern, as there is evidence that the two M layers have a reversed pattern of input in regard to eye of origin, so that ME receives input from the ipsilateral retina (Pettigrew et al., 1989). All prosimians have an extra pair of layers, the koniocellular layers (KE and KI). In other primates, K cells are scattered between layers where they do not form distinct layers. In many anthropoid primates, the parvocellular layers divide to form sublayers and leaflets of sublayers. In the diagram, the layers with input from the ipsilateral eye are shorter as they represent less of a contralateral visual hemifield (0°–90°).

LGN K cells were scattered between layers. Tarsiers did not regain K layers with a return to nocturnal life, but tarsiers and the only nocturnal monkey, the owl or night monkey, evolved a thick K-cell region, but not distinct K-cell layers, between the P and M layers. In anthropoid primates, the emphasis on diurnal vision resulted in a great increase in the proportion of the LGN devoted to P-cell inputs. These layers became thicker, especially in parts representing central vision, and they subdivided to form four or more P "layers" (actually partial layers or leaflets of layers). The appearance of four leaflets of P layers over the part of the LGN representing central vision, plus the two M layers, led to the description of the LGN as having six layers. However, in apes and humans, even subleaflets of leaflets can be found, and this would lead to an even greater number if counted as full layers. A more parsimonious interpretation is that the LGN of primates has two M cell layers, two P cell layers that sometimes subdivide, and various distributions of K cells that are more pronounced in nocturnal primates, where the distributions of K cells are sometimes recognized as forming layers.

The comparative evidence on the relative sizes and types of LGN layers in primates is consistent with the view that the P system is critically involved in detailed vision and in the use of cone mediated sensitivities to color differences (Casagrande, Khaytin, & Boyd, 2007). Thus, P-cell information from the retina goes exclusively (or nearly so) to the LGN in anthropoid primates (Weller & Kaas, 1989). The smaller M system is used to detect change in the visual field, usually produced by object or self-motion. This retinal information goes to both the LGN and superior colliculus, likely via collateral branches of the same axons. The small K-cell retinal output goes to both the LGN and superior colliculus via collaterals. Some K cells carry information from the short S-wavelength cones of the retina (White et al., 1998). K-cell layers and neuron distributions in the LGN also receive projections from the superior colliculus (Harting et al., 1991), and they project both to layer 3 of primary visual cortex, but also to other extrastriate visual areas such as the middle temporal area, MT (Sincich et al., 2004; Stepniewska, Qi, & Kaas, 1999). These features suggest that the functions of the K-cell system are quite distinct from those of the P and M cell systems. While the proportionately larger K-cell layers and regions in nocturnal primates indicate that K-cell inputs are likely important in vision in dim light, the roles of the K-cell system in vision remain uncertain.

The other major visual structure of the dorsal thalamus is the pulvinar complex. It consists of several subdivisions or nuclei that are considered visual because they receive inputs from the superior colliculus or visual cortex and project to visual cortex (see figure 86.5). The pulvinar complex of primates has been traditionally divided into the inferior pulvinar, the lateral pulvinar, and the medial pulvinar. The medial pulvinar is connected with multisensory and other regions of cortex and is not considered to be a strictly visual structure. A proposed dorsomedial division of the lateral pulvinar is also not solely a visual structure. The pulvinar was once thought to be a part of the thalamus specific to primates, and this incorrect assumption is responsible for parts, or all, of the pulvinar complex being called the lateral posterior nucleus or nuclei in nonprimate mammals.

Subdivisions or nuclei of the visual pulvinar in anthropoid primates have been revealed by differences in the expression of different substances (acetylcholinesterase, calbindin, glutamate transporters, cytochrome oxidase, etc.) and differences in connections, as well as by having different representations of the contralateral visual hemifield (Kaas & Lyon, 2007). All anthropoid primates appear to have the divisions of the visual pulvinar illustrated in figure 86.5, but the relative sizes and positions of the nuclei differ. For example, nuclei projecting to primary visual cortex are proportionately

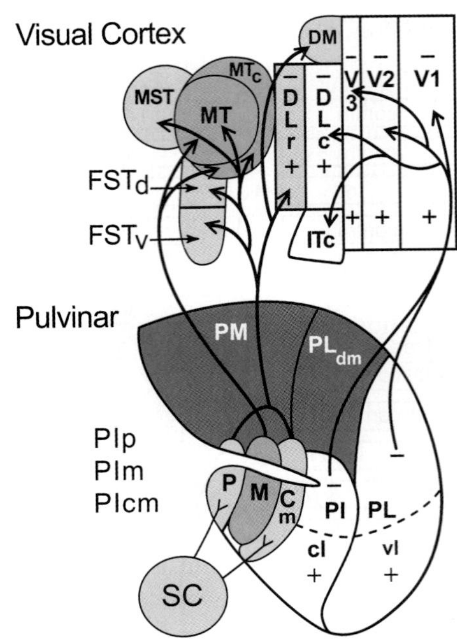

FIGURE 86.5   Proposed subdivisions (nuclei) of the pulvinar complex which are likely common to most or all primates. Two large, retinotopically organized nuclei, the central lateral nucleus of the inferior pulvinar (PIcl) and the ventral lateral nucleus of the lateral pulvinar (PLvl) project to visual areas early in cortical processing (V1–V3) and areas in the ventral processing stream (caudal DL or V4 and caudal inferior temporal cortex), and are activated by inputs from early cortical areas. Two nuclei with dense inputs from the superior colliculus (SC), the posterior and central medial nuclei (PIp and PIcm) of the inferior pulvinar project to a collection of visual areas in the dorsal stream of visual processing (MT, dorsal and ventral divisions of FST, MST, MTc, DM, (V3a), and rostral DL or rostral V4). Comparative evidence suggests that PIp and PIcm differentiated from a single caudal pulvinar nucleus, with SC inputs and projections to temporal cortex, in the nonprimate ancestors of primates. The dorsal medial nucleus of the lateral pulvinar (PLdm) has frontal and parietal lobe connections while the medial pulvinar (PM) has connections with frontal, cingulate, insular, and temporal cortex. (Based on Kaas & Lyon, 2007.)

larger in primates that are diurnal and have emphasized foveal vision (i.e., nuclei more related to the ventral stream cortical processing). Unfortunately, the histological preparations, which have been so useful in revealing subdivisions of the visual pulvinar in anthropoid primates, have been less useful in prosimian primates, such that clear homologues of nuclei are less apparent (Wong et al., 2009). However, two large regions of the pulvinar in prosimian galagos that project to primary visual cortex (V1) appear to be homologous with the ventrolateral division of the lateral pulvinar and the central lateral division of the inferior pulvinar of anthropoids, while a caudal division of the pulvinar complex of galagos with inputs from the superior colliculus appears to be homologous to the posterior and central medial divisions of the inferior pulvinar of anthropoids. If so, the mediodorsal position of the caudal pulvinar in galagos is positioned much as the dorsal pulvinar nucleus is in tree shrews (Lyon, Jain, & Kaas, 2003) and the caudal pulvinar nucleus in squirrels (Baldwin et al., 2011). All these nuclei receive inputs from the superior colliculus and project to temporal visual cortex. Thus, we are beginning to be able to relate divisions of the pulvinar complex in primates to those of rodents and tree shrews and infer how differences might have evolved. For now, it seems that a caudal nucleus in rodents, tree shrews, and galagos, and possibly many other mammals, is homologous to parts of the inferior pulvinar in anthropoid primates.

VISUAL CORTEX

Early primates were characterized by a large expanse of visual cortex that was divided into a number of visual areas. Because the areal organization of visual cortex is not fully understood in any primate, much less in a range of primate species, only limited comparisons across members of the major branches of primate evolution are possible. Thus, the arrangement and number of cortical visual areas of early primates can only be inferred in part.

All primates have a large primary visual area, V1, which is rotated by the expansion of extrastriate visual cortex and posterior parietal cortex from its common caudal–dorsal position in other mammals to occupy the caudal pole of the hemisphere. As a result, much of V1 is on the ventral surface and medial wall of the hemisphere, and in the calcarine sulcus, a fissure found only in primates. Primary visual cortex in extant primates is approximately two to three times larger than would be expected for mammals of a similar body size, but V1 is smaller in prosimian primates than in simian primates of similar size. Primates tend to devote one-third to one-half of V1 to the first 10° of central vision, with this expansion of central vision being more pronounced in diurnal primates (Rosa et al., 1997). All primates have a similar projection pattern from LGN layers to V1 to that inputs from M layers terminate in a sublayer of layer 4 that is external to the sublayer with P layer inputs (Casagrande & Kaas, 1994). All primates appear to have a modular subdivisions of layer 3 such that a dot-like pattern of cytochrome oxidase (CO) dense "blobs," which receive afferents from the K cells of the LGN (Casagrande & Kaas, 1994), and nonblob surrounds (Preuss & Kaas, 1996) are visible. Tree shrews do not have the CO blob pattern, although myelin stains reveal a pattern of myelin-poor regions that are blob-like

(Lyon, Jain, & Kaas, 1998). Both primates and tree shrews have orderly arrangements of orientation-selective neurons in V1 (Bosking et al., 1997). As this arrangement is not found in rodents, systematic arrays of orientation-selective "columns" of neurons likely evolved in the common ancestors of tree shrews and primates. However, carnivores appear to have independently evolved an orderly arrangement of orientation-selective "columns" or modules in V1 (Kaschube et al., 2010).

Primates are unusual in that LGN inputs related to the contralateral and ipsilateral eyes tend to terminate in layer 4 of V1 in separate bands or clusters, forming the basis for the ocular dominance columns. However, the patterns created by the segregation of these inputs are quite variable across primate species, and even within a species (Adams & Horton, 2003). Such variability likely reflects the interplay of changing neural activity patterns, the proportions of axon terminals related to each eye, and the chemical signals that guide termination patterns (Kaas & Catania, 2002). Ocular dominance columns, or bands, are not found in the closest relatives of primates (tree shrews, rodents, and lagomorphs).

The major cortical outputs from V1 in all studied primates (Lyon & Kaas, 2001, 2002a, 2002b, 2002c) are in decreasing magnitudes to V2, V3, MT, and DM (see figure 86.5). Most of the projections to MT are from deep layer 3 (3C, incorrectly identified as layer 4B in most studies; see Casagrande & Kaas, 1994) while the middle of layer 3 (3B) provides most of the projections to the other cortical targets.

The laminar and sublaminar specializations of V1 vary somewhat across primates, with layers being least distinct in prosimian primates and layers and sublayers being most distinct in tarsiers, where V1 occupies proportionately more neocortex than in any other primate (Collins, Lyon, & Kaas, 2005). Humans have a specialization of a sublayer of layer 3 that suggests a role in the magnocellular processing stream involved in motion analysis (Preuss, Qi, & Kaas, 1999). Across primate taxa, V1 gets bigger with brain size, but only up to the great apes, as the much larger human brain has a V1 about the same size as in chimpanzees. Thus, the size of V1 relative to the rest of neocortex actually decreases in humans compared to other primates. Over a range of primate species from small prosimian brains to the large baboon brain, the ratio of V1 neurons to LGN neurons increases, so that as V1 increases in size, each LGN relay neuron relates to a larger number of V1 neurons. In addition, V1 of primates differs from V1 of other mammals by having more neurons for its size. Thus, V1 of primates has three to four times more densely packed neurons than other cortical areas. This difference is the least pronounced in prosimian galagos, more so in New World monkeys, even more in macaques and baboons (Collins et al., 2010). Such differences in neuron packing densities are not marked in nonprimates. As high densities of small neurons provide a framework for preserving the details of visual scenes, V1 is specialized for detailed vision in all primates, with this feature much enhanced in Old World monkeys, apes, and humans. In these anthropoids, even V2 has more densely packed neurons than most cortical areas.

Most of the outer border of V1 is formed by a narrow belt-like second visual area, V2, in all extant primates. This outer border of V1 represents the line of decussation of the retina (the zero vertical meridian), which corresponds to the full extent of the V1/V2 border. The rest of V1, corresponding to the temporal margin of monocular vision, is bordered by a small, poorly understood area, the prostriata (Rosa et al., 1997). Early primates, as in present-day nocturnal prosimians, devoted less of V1 to the representation of central vision. As a result, less of the border of V1 corresponded to the vertical meridian and a shared border with V2. Thus, V2 was less elongated than in extant diurnal primates.

In monkeys, V2 is characterized by a modular organization that is revealed by stains for CO as they mark alternating CO-dense and CO-light bands crossing the rostrocaudal extent of V2. The CO-dense bands appear to be of two types, "thick" and "thin." Each of three types of bands is thought to have different connections with V1 modules and sublayers and with other visual areas (Hubel & Livingstone, 1987; Krubitzer & Kaas, 1990; Roe, 2004). Such a pattern of modular organization is less apparent in prosimian primates although CO-dense bands may be weakly apparent in favorable material (Preuss, Beck, & Kaas, 1993; Preuss & Kaas, 1996). However, modular differences in connections of V2 of galagos suggest that an anatomical framework for bands exists in V2 (Collins, Stepniewska, & Kaas, 2001). Tree shrews also have a banding pattern of V1 connections in V2 (Lyon, Jain, & Kaas, 1998; Sesma, Casagrande, & Kaas, 1984), but the functional significance of this segregation is not clear. These observations suggest that V2 had modular subdivisions in early primates, but the modules may have been less regular and distinct than in present-day simians. Nevertheless, a specific pattern of modular processing in V2 was part of the visual specializations of the primate branch of mammalian evolution.

The outer border of V2 is bordered by V3 in all primates. The extent of V3 with its shared boundary with V2 was previously contentious in that some investigators considered ventral V3 as separate visual area (VP), differing from dorsal V3 by its lack of connections with V1 (Lyon & Connolly, 2012). However there is now compelling evidence for dorsal V3 and ventral VP being parts of the same visual area, with evidence of matching V1 connections to ventral and dorsal V3 in galagos, several species of New World monkeys, and macaques (Lyon & Kaas, 2001, 2002a, 2002b, 2002c; Lyon, Jain, & Kaas, 2002). Given this broad distribution, early primates undoubtedly had a V3. However, there is no clear evidence yet for a V3 in tree shrews (Lyon, Jain, & Kaas, 1998) or in rodents and rabbits (Rosa & Krubitzer, 1999), so V3 may not have been present in the common ancestors of primates, tree shrews, and rodents. This conclusion should be held cautiously, however, as further study could provide evidence for V3 in tree shrews and other mammals. As a primate-like V3 has been described in cats, this difference in distribution suggests that V3 in primates and cats evolved independently.

A similar conclusion relates to the evolution of the middle temporal visual area (MT). MT is easily identified in primates by its projection pattern from V1, dense myelination, retinotopic organization, and neurons that are dominated by a relay of the M-cell pathway though V1 (Casagrande & Kaas, 1994). This distinctive visual area has been found in galagos (Allman, Kaas, & Lane, 1973) and in all other studied primates, including humans, but not in tree shrews, rodents, or rabbits (Kaas, 1997). Because of this distribution, MT appears to be a visual area that first evolved with the early primates. Thus, cats are unlikely to have a homologue of MT, as is sometimes postulated.

Finally, in all primates, much of posterior parietal cortex is involved in processing visual information so that separate domains within this region can help plan, initiate, and guide adaptive movements of eyes, hands, and other parts of the body (Kaas, Gharbawie, & Stepniewska, 2011). Some of these domains, including those for grasping, reaching, defense of the face, and eye movements (see figure 86.5) have been defined by microstimulation in galagos, New World monkeys, and macaques, and they likely exist in all primates. Such movement domains do not appear to be present in posterior parietal cortex of tree shrews and rodents, which have very little posterior parietal cortex. Thus, this large region of cortex with movement-specific domains must have emerged in the immediate ancestors of the first primates. However, the region differs in organization and complexity, as primates differ in the skilled use of their hands, and more functional domains likely exist in humans than in prosimians and monkeys (Orban et al., 2006).

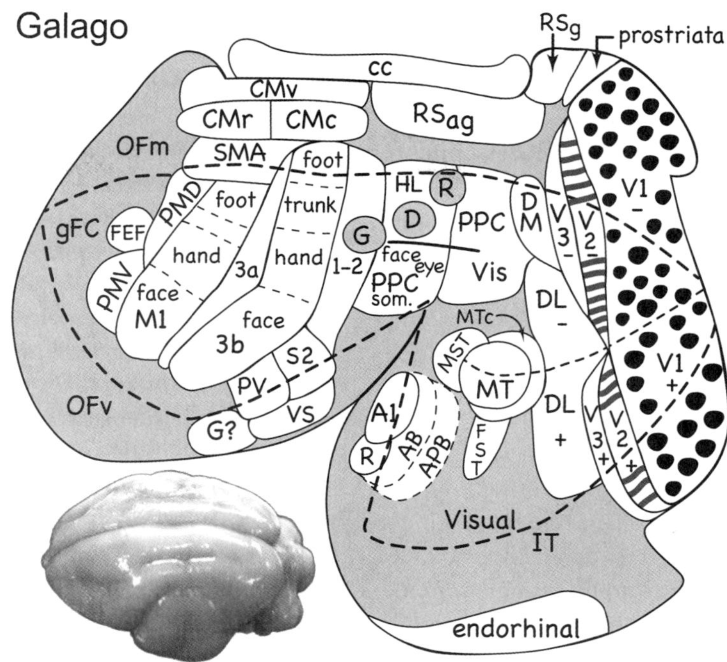

FIGURE 86.6   Proposed subdivisions of visual cortex in a prosimian primate (galago). Other areas are included for reference. A dorsolateral view of a galago brain is on the lower left. On the upper right, the cortex from the left hemisphere has been flattened so that subdivisions on hidden surfaces are visible. The dashed line on the flattened cortex outlines cortex that is visible on a dorsolateral view of the intact hemisphere. The areas shown here are common to all studied primates, and likely existed in the first primates. The primary visual area, V1, is characterized by a distribution of cytochrome oxidase (CO) dense patches of cortex known as blobs. These appear to exist in all primates, but not the close relatives of primates tree shrews and rodents. The second visual area, V2, has a sequence of myelin-dense bands crossing the width of the area. In anthropoid primates, these bands are of two functional types, thin and thick; bands are only weakly apparent in prosimian primates. V2 is bordered by a narrow V3. Other areas common to primates include the dorsolateral visual area, DL, also known as V4, the middle temporal visual area, MT, the MT crescent (MTc), the fundal area of the superior temporal sulcus (FST), the medial superior temporal area (MST), and the dorsal medial visual area, DM. The caudal half of posterior parietal cortex (PPC) is primarily visual, but functional subdivisions are unknown. The rostral half is visual and somatosensory, with subdivisions (domains) for guiding grasping (G), defense (D), reaching (R), and other movement sequences. The frontal eye field (FEF) is involved in eye movements. Inferior temporal (IT) cortex is visual, but functional subdivisions are unknown. Prostriata is a visual area common to mammals. Motor areas include primary motor cortex (M1), ventral (PMV) and dorsal (PMD) premotor cortex, the supplementary motor area (SMA), and ventral (CMv), rostral (CMr), and caudal (CMc) cingulate motor areas. Somatosensory areas include areas 3a, 3b, 1 (or 1 plus 2) of Brodmann, the second area (S2), the parietal ventral area (PV), a ventral somatosensory area (VS), and a proposed gustatory (G?) area. Auditory areas include primary (A1) and rostral (R) auditory areas, and the auditory belt (AB) and parabelt (PB). Retrosplenial cortex includes granular (RSg) and agranular (RSag) areas. Frontal cortex includes granular frontal cortex (gFC) and ventral (OFv) and medial (OFm) orbital frontal areas.

TOWARD A FURTHER UNDERSTANDING OF HOW THE HUMAN VISUAL SYSTEM EVOLVED

Despite much progress, we are at the early stages of fully describing the organization of the human visual system, and we know little about the visual systems of apes. Members of the major branches of primate evolution cover a range of body and brain sizes, and they are specialized in various ways, so we can expect variations in the visual system within as well as across taxonomic groups. In Old World monkeys, over 30 visual areas have been proposed (Felleman & Van Essen, 1991), and these primates have greatly expanded regions of visual cortex in the temporal and parietal lobes compared to most New World monkeys and all prosimians. Quite possibly, Old World monkeys have more visual areas than most or all New World monkeys. Further increases in overall brain size and the total extent of visual cortex undoubtedly occurred in apes and in our early hominid ancestors. Possibly there was a further increase in the number of visual areas, but this is uncertain. What is clear from the fossil record is that our ancestors of only 3 million years ago had brains only slightly larger (600–800 cc) than those of present-day African apes (Kaas & Preuss, 2012), and these brains may have resembled those of African apes in general organization.

Over the last 2 million years, the brains of our ancestors increased rapidly in size to the present volume of about 1,400 cc, but we know little about the structural changes in brain organization that occurred during that time. What is known from comparing human brains with those of apes is that the easily identified primary visual area increased only slightly in absolute size while occupying proportionately much less of the total neocortex. This observation is consistent with the conclusion that the number of cortical areas, including visual areas, has increased with the evolution of modern humans.

Changes in the numbers and sizes of cortical areas in primate evolution have functional consequences. The functions of cortical areas are altered as they change in size (Kaas, 2000). In smaller cortical areas, individual neurons are activated by a larger proportion of inputs. Thus, neurons in small cortical areas represent proportionally more of the inputs, and smaller cortical areas have intrinsic connections that cover more of the cortical area. These differences in connections make neurons in smaller areas more suitable for global comparisons as they have larger receptive field centers and surrounds. In contrast, large areas, such as V1, have mainly neurons that receive proportionally few of the inputs, as these neurons have smaller dendritic arbors (Elston, 2007). Thus, larger areas are good for processing the local detail of a stimulus, and smaller areas more broadly integrate information. This may be why the large human brain is not simply a larger version of a monkey brain. V1 and at least several other areas did not increase proportionally in size as brains became larger and new areas were likely added. Only a few visual areas would need to be large and preserve detailed information about the visual scene while smaller visual areas could be specialized in various ways. The evolution of the complex human brain may have depended on the addition of gene copies that subsequently differentiated and acquired new functions (Ohno, 1970), possibly including those that increased the number of visual areas. The addition of new cortical areas allowed both added and original visual areas to differentiate, specialize, and mediate new capacities (Kaas, 1989).

Presently, we know that human brains are about three times as large as those of the African apes, but we still know very little about how human and ape brains differ structurally. Histochemical and traditional architectonic approaches can help identify nuclei and cortical areas that apes and humans have in common (Hackett, Preuss, & Kaas, 2001; Qi, Preuss, & Kaas, 2007), specialized features of those areas (Preuss, Qi, & Kaas, 1999), and possibly nuclei and areas unique to humans. The goal of determining the similarities and differences in cortical organization across primates has been expressed from the time of Brodmann (1909), but now we have powerful modern methods to add force to such studies.

Finally, much can be learned about the evolution of complex visual systems from theoretical approaches that examine scaling issues related to brain size (Herculano-Houzel, 2011), as well as from our growing understanding of the modes of brain development and how development can be altered in evolution (Molnár, 2011). Comparative studies of areal differences and similarities in the histochemistry and patterns of gene expression of visual cortex of humans, apes, and other primates are also of great value (Preuss et al., 2004; Takahata et al., 2012).

REFERENCES

Adams, D. L., & Horton, J. C. (2003). Capricious expression of cortical columns in the primate brain. *Nature Neuroscience, 6,* 113–114.

Allman, J. M., Kaas, J. H., & Lane, R. H. (1973). The middle temporal visual area (MT) in the bushbaby (*Galago senegalensis*). *Brain Research, 57,* 197–202.

Avzevedo, F. A. C., Carvalho, L. R. B., Grinberg, L. T., Farfel, J. M., Ferretti, R. E. J., Leite, R. E. P., et al. (2009). Equal numbers of neuronal and nonneuronal cells make the human brain an isometrically scaled-up primate brain. *Journal of Comparative Neurology, 513,* 532–541. doi:10.1002/cne.21974.

Baldwin, M. K. L., & Kaas, J. H. (2012). Cortical projections to the superior colliculus in prosimian galagos (*Otolemur garnetti*). *Journal of Comparative Neurology, 520,* 2002–2020. doi:10.1002/cne.23025.

Baldwin, M. K. L., Wong, P., Reed, J. L., & Kaas, J. H. (2011). Superior colliculus connections with visual thalamus in grey squirrels (Sciurus carolinensis): Evidence for four subdivisions within the pulvinar complex. *Journal of Comparative Neurology, 519,* 1071–1094.

Bhatnagar, K. P., & Smith, T. D. (2007). The vomeronasal organ and its evolutionary loss in catarrhine primates. In J. H. Kaas & T. M. Preuss (Eds.), *Evolution of nervous systems* (Vol. 4, pp. 142–152). London: Elsevier.

Block, J. I., & Boyer, D. M. (2002). Grasping primate origins. *Science, 298,* 1606–1610.

Bosking, W. H., Zhang, Y. M., Schofield, B., & Fitzpatrick, D. (1997). Orientation selectivity and the arrangement of horizontal connections in tree shrew striate cortex. *Journal of Neuroscience, 17,* 2112–2127.

Brodmann, K. (1909). *Vergleichende Lokalisationslehre der Grosshirnrhinde.* Leipzig, Germany: Barth.

Casagrande, V. A., & Kaas, J. H. (1994). The afferent, intrinsic, and efferent connections of primary visual cortex in primates. In A. Peters & K. Rockland (Eds.), *Cerebral cortex* (Vol. 10, pp. 201–259). New York: Plenum Press.

Casagrande, V. A., Khaytin, I., & Boyd, J. (2007). The evolution of parallel visual pathways in the brains of primates. In J. H. Kaas & T. M. Preuss (Eds.), *Evolution of nervous systems* (Vol. 4, pp. 87–108). London: Elsevier.

Collins, C. E., Airey, D. C., Young, N. A., Leitch, D. B., & Kaas, J. H. (2010). Neuron densities vary across and within cortical areas in primates. *Proceedings of the National Academy of Sciences of the United States of America, 107,* 15927–15932. doi:10.1073/pnas.1010356107.

Collins, C. E., Lyon, D. C., & Kaas, J. H. (2005). Distribution across cortical areas of neurons projecting to the superior colliculus in New World monkeys. *Anatomical Record, 285A,* 619–627.

Collins, C. E., Stepniewska, I., & Kaas, J. H. (2001). Topographic patterns of V2 cortical connections in a prosimian primate (*Galago garnetti*). *Journal of Comparative Neurology, 431,* 155–167.

Corballis, M. C. (2007). Evolution of hemispheric specialization of the human brain. In J. H. Kaas & T. M. Preuss (Eds.), *Evolution of nervous systems* (Vol. 4, pp. 379–396). London: Elsevier.

de Sousa, A., & Wood, B. (2007). The hominin fossil record and the emergence of the modern human central nervous system. In J. H. Kaas & T. M. Preuss (Eds.), *Evolution of nervous systems* (Vol. 4, pp. 291–336). London: Elsevier.

Elston, G. N. (2007). Specialization of the neocortical pyramidal cell during primate evolution. In J. H. Kaas & T. M. Preuss (Eds.), *Evolution of nervous systems* (Vol. 4, pp. 191–242). London: Elsevier.

Felleman, D. J., & Van Essen, D. C. (1991). Distributed hierarchical processing in the primate cerebral cortex. *Cerebral Cortex, 1,* 1–47. doi:10.1093/cercor/1.1.1-a.

Hackett, T. A., Preuss, T. M., & Kaas, J. H. (2001). Architectonic identification of the core region in auditory cortex of macaques, chimpanzees, and humans. *Journal of Comparative Neurology, 441,* 197–222.

Hart, B. L., & Hart, L. A. (2007). Evolution of the elephant brains: A paradox between brain size and cognitive behavior. In J. H. Kaas & L. A. Krubitzer (Eds.), *Evolution of nervous systems* (Vol. 3, pp. 491–497). London: Elsevier.

Harting, J. K., Huerta, M. F., Hashikawa, T., & Van Lieshout, D. P. (1991). Projections of the mammalian superior colliculus upon the dorsal lateral geniculate nucleus: Organization of tectogeniculate pathways in nineteen species. *Journal of Comparative Neurology, 304,* 275–306.

Hendrickson, A., Djajadi, H. R., Nakamura, L., Possin, D. E., & Sajuthi, D. (2000). Nocturnal tarsier retina has both short and long/medium-wavelength cones in an unusual topography. *Journal of Comparative Neurology, 424,* 718–730.

Herculano-Houzel, S. (2011). Not all brains are made the same: New views on brain scaling in evolution. *Brain, Behavior and Evolution, 78,* 22–36.

Herculano-Houzel, S., Collins, C. E., Wong, P., & Kaas, J. H. (2007). Cellular scaling rules for primate brains. *Proceedings of the National Academy of Sciences of the United States of America, 104,* 3562–3567. doi:10.1073/pnas.0611396104.

Hubel, D. H., & Livingstone, M. S. (1987). Segregation of form, color, and stereopsis in primate area 18. *Journal of Neuroscience, 7,* 3378–3415.

Jacobs, G. H., Neitz, M., Deegan, J. F., & Neitz, J. (1996). Trichromatic colour vision in New World monkeys. *Nature, 382,* 156–158.

Jerison, H. J. (1973). *Evolution of the brain and intelligence.* New York: Academic Press.

Kaas, J. H. (1978). Organization of visual cortex in primates. In C. R. Noback (Ed.), *Sensory systems of primates* (pp. 151–179). New York: Plenum Press.

Kaas, J. H. (1989). Why does the brain have so many visual areas? *Journal of Cognitive Neuroscience, 1,* 121–135.

Kaas, J. H. (1997). Theories of visual cortex organization in primates. In K. S. Rockland, J. H. Kaas, & A. Peters (Eds.), *Cerebral cortex: Extrastriate cortex in primates* (Vol. 12, pp. 91–125). New York: Plenum Press.

Kaas, J. H. (2000). Why is brain size so important: Design problems and solutions as neocortex gets bigger or smaller. *Brain and Mind, 1,* 7–23.

Kaas, J. H. (2007). The evolution of the dorsal thalamus in mammals. In J. H. Kaas & L. A. Krubitzer (Eds.), *Evolution of nervous systems* (Vol. 3, pp. 500–516). London: Elsevier.

Kaas, J. H., & Catania, K. C. (2002). How do features of sensory representations develop? *BioEssays, 24,* 334–343.

Kaas, J. H., Gharbawie, O. A., & Stepniewska, I. (2011). The organization and evolution of dorsal stream multisensory motor pathways in primates. *Frontiers in Neuroanatomy, 5,* 1–7. doi:10.3389/fnana.2011.00034.

Kaas, J. H., Harting, J. K., & Guillery, R. W. (1974). Representation of the complete retina in the contralateral superior colliculus of some mammals. *Brain Research, 65,* 343–346.

Kaas, J. H., & Lyon, D. C. (2007). Pulvinar contributions to the dorsal and ventral streams of visual processing in primates. *Brain Research. Brain Research Reviews, 55,* 285–296.

Kaas, J. H., & Preuss, T. M. (2012). Human brain evolution. In L. R. Squire (Ed.), *Fundamental neuroscience* (4th ed.). San Diego, CA: Academic Press.

Kaschube, M., Schnabel, M., Löwel, S., Coppola, D. M., White, L. E., & Wolf, F. (2010). Universality in the evolution of orientation columns in the visual cortex. *Science, 330,* 1113–1116.

Krubitzer, L. A., & Kaas, J. H. (1990). Convergence of processing channels in the extrastriate cortex of monkeys. *Visual Neuroscience, 5,* 609–613.

Lane, R. H., Allman, J. M., Kaas, J. H., & Miezin, F. M. (1973). The visuotopic organization of the superior colliculus of the owl monkey (*Aotus trivirgatus*) and the bush baby (*Galago senegalensis*). *Brain Research, 60,* 335–349.

Lyon, D. C., & Connolly, J. D. (2012). The case for primate V3. *Proceedings. Biological Sciences, 279,* 625–633.

Lyon, D. C., Jain, N., & Kaas, J. H. (1998). Cortical connections of striate and extrastriate visual areas in tree shrews. *Journal of Comparative Neurology, 401,* 109–128.

Lyon, D. C., Jain, N., & Kaas, J. H. (2003). The visual pulvinar in tree shrews: I. Multiple subdivisions revealed through acetylcholinesterase and Cat-301 chemoarchitecture. *Journal of Comparative Neurology, 467,* 593–606.

Lyon, D. C., & Kaas, J. H. (2001). Connectional and architectonic evidence for dorsal and ventral V3, and dorsomedial area in marmoset monkeys. *Journal of Neuroscience, 21,* 249–261.

Lyon, D. C., & Kaas, J. H. (2002a). Connectional evidence for dorsal and ventral V3, and other extrastriate areas in the prosimian primate, *Galago garnetti. Brain, Behavior and Evolution, 59,* 114–129.

Lyon, D. C., & Kaas, J. H. (2002b). Evidence for a modified V3 with dorsal and ventral halves in macaque monkeys. *Neuron, 33,* 453–461.

Lyon, D. C., & Kaas, J. H. (2002c). Evidence from V1 connections for both dorsal and ventral subdivisions of V3 in three species of New World monkeys. *Journal of Comparative Neurology, 449,* 281–297.

Lyon, D. C., Xu, X., Casagrande, V. A., Stefansic, J. D., Shima, D., & Kaas, J. H. (2002). Optical imaging reveals retinotopic organization of dorsal V3 in New World owl monkeys.

Proceedings of the National Academy of Sciences of the United States of America, 99, 15735–15742. doi:10.1073/pnas. 242600699.

Meredith, R. W., Janecka, J. E., Gatesy, J., Ryder, O. A., & Fisher, C. A. (2011). Impacts of the cretaceous terrestrial revolution and KPg extinction on mammal diversification. *Science, 334,* 521–524.

Molnár, Z. (2011). Evolution of cerebral cortical development. *Brain, Behavior and Evolution, 78,* 94–107.

Ohno, S. (1970). *Evolution of gene duplication.* New York: Springer.

Orban, G. A., Claeys, K., Nelissen, K., Smans, R., Sunaert, S., Todd, J. T., et al. (2006). Mapping the parietal cortex of human and non-human primates. *Neuropsychologia, 44,* 2647–2667. doi:10.1016/j.neuropsychologia.2005.11.001.

Pettigrew, J. P. (1986). Flying primates? Megabats have the advanced pathway from eye to midbrain. *Science, 231,* 1304–1306.

Pettigrew, J. P., Jamieson, B. G. M., Robson, S. K., Hall, L. S., McNally, K. I., & Cooper, H. M. (1989). Phylogenetic relations between microbats, megabats and primates. *Philosophical Transactions of the Royal Society of London. Series B, Biological Sciences, 325,* 489–559.

Preuss, T. M. (2007). Primate brain evolution in phylogenetic context. In J. H. Kaas & T. M. Preuss (Eds.), *Evolution of nervous systems* (Vol. 4, pp. 1–34). Oxford: Elsevier.

Preuss, T. M., Beck, P. D., & Kaas, J. H. (1993). Areal, modular, and connectional organization of visual cortex in a prosimian primate, the slow loris (*Nycticebus coucang*). *Brain, Behavior and Evolution, 42,* 321–335.

Preuss, T. M., Cáceres, M., Oldham, M. C., & Geschwind, D. H. (2004). Human brain evolution: Insights from microarrays. *Nature Reviews. Genetics, 5,* 850–860.

Preuss, T. M., & Kaas, J. H. (1996). Cytochrome oxidase 'blobs' and other characteristics of primary visual cortex in a lemuroid primate, *Cheirogaleus medius. Brain, Behavior and Evolution, 47,* 103–112.

Preuss, T. M., Qi, H., & Kaas, J. H. (1999). Distinctive compartmental organization of human primary visual cortex. *Proceedings of the National Academy of Sciences of the United States of America, 96,* 11601–11606. doi:10.1073/pnas. 96.20.11601.

Qi, H. X., Preuss, T. M., & Kaas, J. H. (2007). Somatosensory areas of the cerebral cortex: Architectonic characteristics and modular organization. In E. Gardner & J. H. Kaas (Eds.), *The senses: A comprehensive reference* (Vol. 6, pp. 143–169). London: Elsevier.

Roe, A. W. (2004). Modular complexity of area V2 in the macaque monkey. In J. H. Kaas & C. E. Collins (Eds.), *The primate visual system* (pp. 109–138). Boca Raton, FL: CRC Press.

Rosa, M. G. P., Casagrande, V. A., Preuss, T., & Kaas, J. H. (1997). Visual field representation in striate and prostriate cortices of a prosimian primate (*Galago garnetti*). *Journal of Neurophysiology, 77,* 3193–3217.

Rosa, M. G. P., & Krubitzer, L. A. (1999). The evolution of visual cortex: Where is V2? *Trends in Neurosciences, 22,* 242–248.

Ross, C. F., & Martin, R. D. (2007). The role of vision in the origin and evolution of primates. In J. H. Kaas & T. M. Preuss (Eds.), *Evolution of nervous systems* (Vol. 4, pp. 59–78). London: Elsevier.

Sesma, M. A., Casagrande, V. A., & Kaas, J. H. (1984). Cortical connections of area 17 in tree shrews. *Journal of Comparative Neurology, 230,* 337–351.

Sherwood, C. C., & Hof, P. R. (2007). The evolution of neuron types and cortical histology in apes and humans. In J. H. Kaas & T. M. Preuss (Eds.), *Evolution of nervous systems* (Vol. 4, pp. 355–378). London: Elsevier.

Sincich, L. C., Park, K. F., Wohlgemuth, M. J., & Horton, J. C. (2004). Bypassing V1: A direct geniculate input to area MT. *Nature Neuroscience, 7,* 1123–1128.

Skoglund, P., & Jakobsson, M. (2011). Archaic human ancestry in East Asia. *Proceedings of the National Academy of Sciences of the United States of America, 108,* 18301–18306. doi:10.1073/ pnas.1108181108.

Stepniewska, I., Qi, H. X., & Kaas, J. H. (1999). Do superior colliculus projection zones in the inferior pulvinar project to MT in primates? *European Journal of Neuroscience, 11,* 469–480.

Takahata, T., Shukla, R., Yamamori, T., & Kaas, J. H. (2012). Differential expression patterns of striate cortex enriched genes among Old World, New World, and prosimian primates. Cerebral Cortex 22, 2313-2321.

Thiele, A., Vogelsang, M., & Hoffman, K. P. (1991). Pattern of retinotectal projection in the megachiropteran bat (*Rousettus aegyptiacus*). *Journal of Comparative Neurology, 314,* 671–683.

Tigges, M., & Tigges, J. (1970). The retinofugal fibers and their terminal nuclei in *Galago crassicaudatus. Journal of Comparative Neurology, 138,* 87–102.

Weller, R. E., & Kaas, J. H. (1989). Parameters affecting the loss of ganglion cells of the retina following ablations of striate cortex in primates. *Visual Neuroscience, 3,* 327–349.

Whishaw, I. Q. (2003). Did a change in sensory control of skilled movements stimulate the evolution of the primate frontal cortex? *Behavioural Brain Research, 146,* 31–41.

White, A. J., Wilder, H. D., Goodchild, A. K., Secfton, A. J., & Martin, P. R. (1998). Segregation of receptive field properties in the lateral geniculate nucleus of a New World monkey, the marmoset *Callithrix jacchus. Journal of Neurophysiology, 80,* 2063–2076.

Wikler, K. C., & Rakic, P. (1990). Distribution of photoreceptor subtypes in the retina of diurnal and nocturnal primates. *Journal of Neuroscience, 10,* 3390–3401.

Wong, P., Collins, C. E., Baldwin, M. K., & Kaas, J. H. (2009). Cortical connections of the visual pulvinar complex in prosimian galagos (*Otolemur garnetti*). *Journal of Comparative Neurology, 517,* 493–511.

87 What Natural Scene Statistics Can Tell Us about Cortical Representation

BRUNO A. OLSHAUSEN AND MICHAEL S. LEWICKI

Over the past 50 years, visual neuroscience has sought to characterize how neurons respond to specific stimulus properties such as shape, texture, color, and motion. While this approach has revealed many interesting and important aspects of neural coding in the visual system, we still remain largely ignorant of how neural populations represent these properties as they appear within the context of dynamic, natural scenes. The problem is that what constitutes "the stimulus" in a natural scene is far from obvious, whereas much of visual neuroscience has proceeded by assuming it is given to begin with. Neuroscience has taken a reductionist approach, while vision is largely a holistic process.

Consider for example the simple scene of a log against a background of rocks, as in figure 87.1. It takes little conscious effort to comprehend what is going on in this scene—the boundary of the log appears obvious to most observers. However, if we put ourselves in the position of a patch of neurons in V1 getting input from a local patch of this image, things are far less clear. The right panel of figure 87.1 shows the response of an array of model V1, orientation-selective units analyzing a local patch of the image, with the boundary of the log superimposed as a faint gray line. As one can see, almost nowhere along this boundary are there neurons firing indicating the position and orientation of the boundary. Instead, one finds neurons firing at many different positions and orientations that signal structure in the background and foreground, but with little relation to the boundary itself. Thus, simply measuring oriented contrast in an image does not give us a direct measure of the shape of things in the visual world. Extracting those properties involves something much more complicated.

Unfortunately, our introspections about how we see the world are a poor guide for how to go about studying it. Indeed, engineers attempting to build artificial vision systems came to the same conclusion decades ago: The definition of a feature as elementary as an edge or contour is essentially an ill-posed problem as it depends heavily on context and high-level knowledge. Even the definition of contrast, a seemingly fundamental stimulus property, is difficult as it is a relative measure that depends on specifying a region over which to measure the local luminance, and there is not one right answer for how to do this in natural scenes.

As Helmholtz wisely observed more than 100 years ago, perception is a process of "unconscious inferences." Inferring properties of the world depends upon combining data (the image) together with *prior knowledge* about the world. (See also chapter 88.) The obvious questions, then, are how is this knowledge learned, and how is it instantiated in the neural circuits of the visual cortex? In order to answer these questions it is necessary to have mathematical models that can describe the wealth of variability and structure in natural scenes. The visual system has to describe not only simple features like edges but many forms of structure in natural scenes, much of which may not correspond to things we are usually aware of when we look at a scene. In other words, there is a need to educate ourselves about the structure of natural scenes and the different ways of modeling them before we can understand how neurons represent and exploit this information. The problem of characterizing the wide range of structure, from edges to textures to subtle patterns of shading, is what we will refer to as *natural scene statistics*.

This chapter reviews work over the past several decades on modeling natural scene statistics and their relation to cortical representation. It should be mentioned from the outset that most of these models are of *images* of natural scenes, not of the physical scenes themselves. The underlying hypothesis is that the nervous system can eventually learn models of the world working from the statistics of the input stream, which are sensed as images. We also focus on models that may be directly related to neural mechanisms and ultimately tested in neurophysiological experiments. There is much work exploring the links between natural scene statistics and psychophysical measurements such as contrast sensitivity (Bex, Solomon, & Dakin, 2009), color sensitivity (Yoonessi & Kingdom, 2008), contour

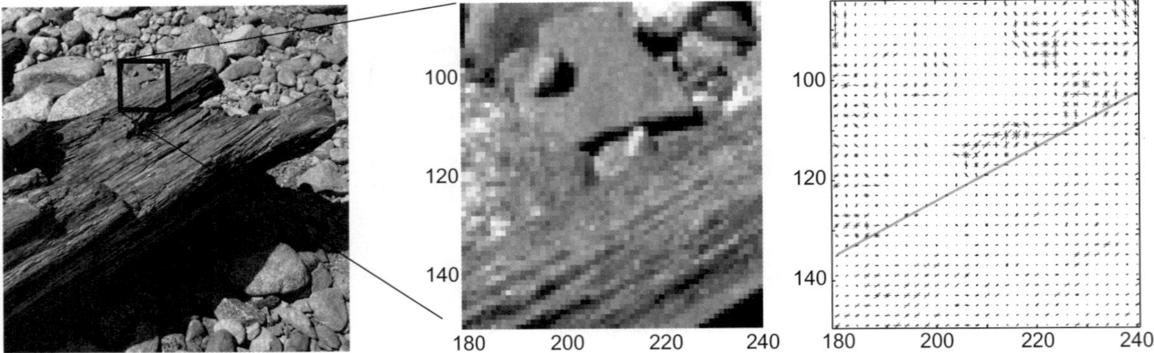

FIGURE 87.1   What constitutes a feature, or "the stimulus," within a natural scene is far from obvious. The right panel shows how a hypothetical array of model V1 neurons (Gabor filters at four different orientations) would respond to the image subregion shown at left. The length of each line segment indicates the magnitude of response of a neuron whose receptive field is situated at that position and orientation. Simply measuring oriented contrast tells one very little about the structures of interest in natural scenes.

detection (Geisler et al., 2001), and depth perception (Burge et al., 2010) which we do not include here. The reader is referred to Geisler (2008) for an excellent review of much of the work in this area. Another excellent source for work on natural scene statistics and image coding is the text by Hyvärinen, Hurri, and Hoyer (2009).

We begin here by introducing the theory of efficient coding, which forms the foundation for much of the work on natural scene statistics. We then discuss theories of sparse representation and hierarchical representation and their relation to the response properties of cortical neurons.

EFFICIENT CODING

Just as eyes have evolved to form an image, so too has the circuitry required to process it. In the case of eyes, there are a large number of factors that determine the quality of a focused image and its adaptability to environmental conditions (Land & Nilsson, 2012). Each animal will have evolved toward the optimal trade-off between various constraints and the optical performance that is required for it to thrive in its environmental niche. This point represents a local optimum in adaptive space and so tends to be relatively stable. If the animal's environment and behavioral requirements are known, we can speak of a theoretical explanation from the principles of optics and the physical constraints on the system. This basic theoretical approach carries over to visual information processing: We can provide a theoretical explanation of visual processing and representation if we understand the natural visual environment, an animal's behavioral requirements, and the principles of visual information processing.

The goal of *efficient coding* is to represent the most relevant visual information with the fewest physical and metabolic resources. Clearly, determining what constitutes relevant information for a mammal is plagued with its own problems since most animals perform a multitude of tasks, from the simple pupillary reflex all the way to visual scene analysis. Nevertheless, we can make progress by choosing computational goals for which we can derive an optimal solution, given appropriate constraints, and relevant visual stimuli. The solution to this problem constitutes a theoretical prediction of the neural system and thus gives a falsifiable model. In the approaches described below, it is largely assumed that the early visual system is forming a generic representation that is useful for myriad tasks, and so the goal is to preserve all information about the scene that is captured in the image. Later, we elaborate this theoretical approach beyond coding to the idea of recovering abstract properties of scenes.

Theory of Redundancy Reduction and Whitening

Attneave (1954) was the first to point out that there could be a formal relationship between the statistical properties of images and certain aspects of visual perception. This notion was then put into concrete mathematical and neurobiological terms by Barlow (1961, 1989), who proposed a self-organizing strategy for sensory nervous systems based on the principle of *redundancy reduction*—that is, the idea that neurons should encode information in such a way as to minimize statistical dependencies among their activities. Barlow reasoned that such representations make more efficient use of neural resources in transmitting information since they do not duplicate information in different neurons.

The first strides in quantitatively testing the theory of redundancy reduction came from the work of Simon Laughlin and M.V. Srinivasan. They measured both the histograms and spatial correlations of image pixels in the natural visual environment of flies and then used this knowledge to make quantitative predictions about the response properties of neurons in early stages of the visual system (Laughlin, 1981; Srinivasan, Laughlin, & Dubs, 1982). They showed that the contrast response function of bipolar cells in the fly's eye performs histogram equalization (so that all output values are equally likely), and that lateral inhibition among these neurons serves to decorrelate their responses for natural scenes, confirming two predictions of the redundancy reduction hypothesis.

Another advance was made 10 years later by Atick and Redlich (1992), and van Hateren (1992, 1993), who formulated a theory of coding in the retina based on whitening the power spectrum of natural images in space and time. Since it had been shown by Field (1987) that natural scenes possess a characteristic $1/f^2$ spatial power spectrum, they reasoned that the optimal decorrelating filter should attempt to whiten the power spectrum—that is, make it flat, or uniform. Since the signal amplitude (square root of power) falls as $1/f$, then the optimal whitening filter has a transfer function that simply rises linearly with spatial frequency in order to produce a flat power spectrum in the output signals of the retina. The whitening filter is combined with a low-pass filter that cuts out noise at the highest spatial frequencies. Taking the inverse Fourier transform of the combined filter, assuming zero phase, results in a spatial filter that is qualitatively similar to the center–surround antagonistic receptive fields of retinal ganglion cells and neurons in the lateral geniculate nucleus (LGN). The spatiotemporal extension of this theory was tested in the LGN of cats, where it was shown that the temporal power spectrum is whitened in response to natural movies (Dan, Atick, & Reid, 1996). Importantly, testing this theory requires using natural scenes or other stimuli with the same spatiotemporal correlations, not simply white noise.

Robust Coding

Although redundancy reduction plays a central role in sensory codes, it is not the whole story because redundancy itself is essential for robustness to noise (Atick & Redlich, 1990; Barlow, 2001; Ruderman, 1994). In the peripheral visual system there are many sources of noise and uncertainty. Blurring due to the optics and photoreceptor transduction noise are two, but the more limiting factor is that neurons can only transmit information with finite precision, which has been estimated to be around 1–2 bits per spike (Borst & Theunissen, 1999). Without redundancy in the neural code itself, information about scene structure will be lost.

Doi and Lewicki (2007) used the framework of *robust coding* (Doi, Balcan, & Lewicki, 2007; Doi & Lewicki, 2005) to develop a model of retinal coding using noisy units, sensory noise, and optical blur. The model optimizes the trade-off between redundancy and efficiency to learn a code that minimizes the mean squared reconstruction error of the stimulus. Unlike earlier methods based on power spectra, the model can have an arbitrary number of coding units and can accurately predict receptive field structures and their adaptation to noise in both the fovea, where the photoreceptor to (midget) ganglion cell ratio is close to 1:1 (with combined ON and OFF channels), and the periphery, where it exceeds 20:1. The optimal robust code can be decomposed into the traditional Wiener filter, which optimally compensates for noise and distortion in the input, and an optimal code for a noisy Gaussian channel (Doi & Lewicki, 2011), which approximates a population of limited capacity neurons.

Optimal codes generated by redundancy reduction and robustness approaches are not necessarily unique and can generate a family of solutions with equivalent performance (e.g., information transmission or mean squared error reconstruction) (e.g., Atick & Redlich, 1992; Doi & Lewicki, 2007). To explain the center–surround structure of retinal ganglion cells, for example, it is necessary to impose additional constraints, such as a cost on the weights (Vincent & Baddeley, 2003). This can be viewed as a more general statement of the redundancy reduction principle because there are many other factors that influence the structure and overall metabolic efficiency of the neural population code (Balasubramanian, Kimber, & Berry, 2001; Laughlin, 2001; Tkacik et al., 2010). Recently, more biologically accurate models that minimize the metabolic cost of spiking and adapt non-linear neural response functions to maximize information transmission account for rectification and predict the ON and OFF center–surround structure of retinal ganglion cells (Karklin & Simoncelli, 2011).

Beyond Efficient Coding

While the principles of redundancy reduction and robust coding have made some inroads in accounting for response properties of neurons in early vision, it would seem that other considerations come into play in the cortex. An important difference between the retina and the cortex is that the retina is faced with a severe structural constraint, the optic nerve, which limits the number of axon fibers leaving the eye. Given the net convergence of approximately 7 million cones (and at least 10 times as many rods) onto 1.5 million ganglion cells, redundancy reduction would appear to constitute a sensible coding strategy for making the most use of the limited resources of the optic nerve. V1, by contrast, expands the image representation coming from the LGN by having far more neurons for representation than it has inputs. In layer 4 of macaque V1 alone the ratio of stellate cells to geniculate input fibers is on the order of 100:1, and it is even higher in the fovea (Barlow, 1981). So what is being gained by spending extra neural resources in this way?

First, it must be recognized that the real goal of sensory representation is to model the *causes* of the redundancy in images, not necessarily to reduce it (Barlow, 2001). What we really want is a meaningful representation—something that captures the causal properties of images, or what's "out there" in the environment. Second, redundancy reduction provides a valid probabilistic model of images only to the extent that the world can meaningfully be described in terms of statistically independent factors. While some aspects of the visual world do seem well described in terms of independent factors (e.g., surface reflectance is independent of illumination), most seem awkward to describe in this framework (e.g., body parts can move fairly independently yet are also oftentimes coordinated to accomplish certain tasks). Thus, in order to understand how the cortex forms useful representations of scene structure we must appeal to principles other than redundancy reduction that are beyond the basic framework of efficient coding and help us move toward inferring underlying causes.

SPARSE, DISTRIBUTED REPRESENTATION

In 1972, Horace Barlow put forth a second theoretical proposal—dubbed the "neuron doctrine" of perception—which proposed that neural representations have been organized to describe sensory stimuli using the fewest possible number of active neurons, and furthermore that such representations are matched to the statistics of natural stimuli (Barlow, 1972). Since then a number of investigators have developed quantitative models based on this idea that attempt to account for the response properties of cortical neurons. Here we describe the theoretical motivations for this approach, models that have been developed and their various elaborations, and the relation to V1 response properties.

Theory of Sparse Representation

One way of potentially achieving a meaningful representation of sensory information is by finding a way to group inputs together so that the world can be described in terms of a small number of events at any given moment. In terms of a neural representation, this means that activity is distributed among a small fraction of neurons, forming a *sparse, distributed representation.* Such a representation converts the higher-order redundancy present in images (i.e., the complex dependencies among pixel values) into a simple first-order redundancy in each neuron (Field, 1994). Each neuron's activity is redundant since it is highly predictable—it spends most of its time at zero—but so long as this can provide a meaningful description of images, then it is potentially more useful than a dense representation in which all redundancy has been reduced.

Consider for example a local region of the image containing an edge at a particular orientation (see figure 87.2). In the retina or LGN, many neurons with circularly symmetric receptive fields will need to be active to completely represent the change in luminance along the length of the edge. However a set of cortical neurons with elongated receptive fields can represent this structure with many fewer active units—just those whose orientation and position is aligned with the edge. Since the receptive fields of these neurons are better matched to the edge, fewer are needed to describe it. Note that nothing has been gained in the ability to describe the edge element per se—that is, there is no gain in information—but the description is now in a more *explicit* format.

All information about the image is present in the photoreceptors, but it is not in a form that is easily accessible or useful for driving behavior (except perhaps for controlling the pupillary reflex or accommodation). For a representation to drive behavior it needs to make explicit the properties of the environment that are relevant to behavior. In this particular example, the activity of a single cortical unit conveys more meaning about what is going on in the scene—the presence of the edge and its orientation—than does a single neuron in retina or LGN. An edge is admittedly still far removed from what we need to drive useful behavior, but nevertheless it is a first step in pulling out structure about the scene.

Sparse representations are sometimes (mistakenly) put in the same category as "grandmother cells" or other winner-take-all representation schemes. Such schemes would require an enormous number of neurons to represent the large variety of input patterns that occur, since a population of N neurons can only represent N input patterns. Here, we are concerned

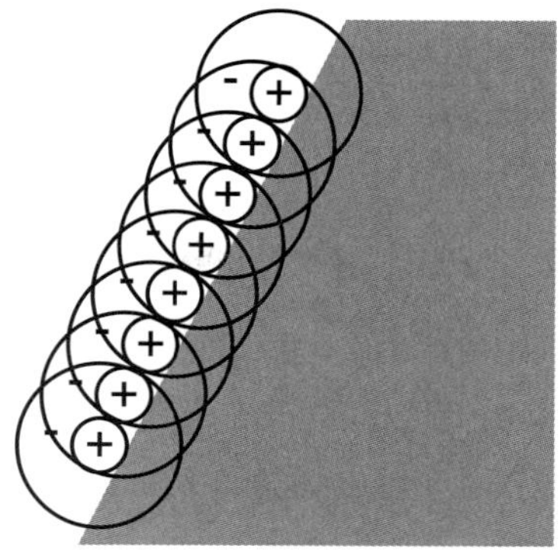 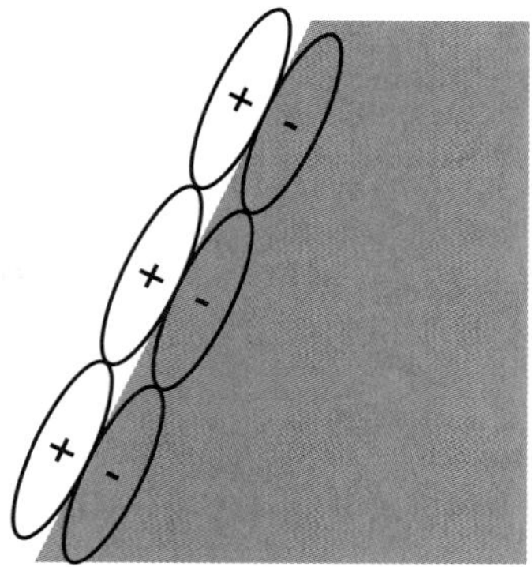

FIGURE 87.2   Sparse representation of an edge. A population containing only neurons with circularly symmetric receptive fields would require many active neurons to completely represent the structure along the edge (left). By contrast, a population of neurons with elongated receptive fields at different orientations will require many fewer active units because each of the active units is better matched to the structure in the image (right).

with sparse *distributed* representations in which multiple active units are still used to encode any given stimulus. This dramatically increases the representational capacity, since a population of N neurons constrained to have only k active units can in theory represent "N choose k" or $N!/k!(N-k)!$ patterns (Foldiak & Young, 1995).

Sparse Coding Model of V1

The first quantitative link between the principle of sparse representation and the oriented receptive fields of neurons in visual cortex was provided by Field (1987). He modeled the oriented receptive fields of V1 neurons with Gabor functions (as was proposed previously by Marcelja, 1980, and Daugman, 1985) and examined the histogram of their responses to a diverse set of natural images. By exploring different settings of the Gabor function parameters (spatial frequency bandwidth and aspect ratio), he was able to show that the setting which maximizes the concentration of activity into the fewest number of units is roughly the same as those found for many cortical neurons—that is, around one octave in bandwidth and 1.3 in aspect ratio (length to width). In other words, the particular shapes of V1 simple-cell receptive fields appear well suited for achieving a sparse representation of natural images.

Olshausen and Field (1996) took this a step further by using a nonparametric model that makes no specific assumptions about the functional form of the receptive fields and attempts to adapt a population of units to the statistics of natural images so as to maximize sparseness. Instead of considering the representation as an array of filter outputs, they formulated the problem in terms of a linear generative model:

$$I(\vec{x}) = \sum_i a_i \phi_i(\vec{x}) + \varepsilon(\vec{x}), \qquad (87.1)$$

where $I(\vec{x})$ denotes the spatial distribution of intensity within an image (typically a local image patch, on the order 16×16 pixels), $\phi_i(\vec{x})$ is a basis function, or "dictionary element," defined over the same spatial domain as the image, a_i describes how much of function $\phi_i(\vec{x})$ is needed to describe the image, and $\varepsilon(\vec{x})$ is a residual term that accounts for structure not well described by the model. ($\vec{x}$ represents the two-dimensional position within the image.) The coefficient values a_i are taken to represent the activities of neurons within a patch of cortex representing the image region $I(\vec{x})$. They are computed by minimizing an energy function that consists of the squared reconstruction error plus a penalty on the coefficients:

$$E = \sum_{\vec{x}} \left[I(\vec{x}) - \sum_i a_i \phi_i(\vec{x}) \right]^2 + \lambda \sum_i C(a_i), \qquad (87.2)$$

where C is a cost function appropriate for encouraging sparsity (often chosen to be absolute value) and λ controls the trade-off between sparsity and reconstruction error. When the number of basis functions is equal to the number of image pixels and there is no noise ($\varepsilon(\vec{x}) = 0$), the model is equivalent to so-called "independent components analysis" (ICA) (Bell & Sejnowski, 1995; Olshausen & Field, 1997).

Solutions to the energy minimization (87.2) may be computed efficiently by a neural circuit consisting of leaky integrators, threshold units, and lateral inhibition (similar to a Hopfield network) (Hopfield, 1984; Rozell et al., 2008). Learning of the basis functions is accomplished by minimizing the same energy function, typically via stochastic gradient descent, which leads to a Hebb-like rule between the inputs and outputs of the circuit (see also Foldiak, 1990, Rehn & Sommer, 2007, and Zylberberg, Murphy, & DeWeese, 2011, for alternative formulations).

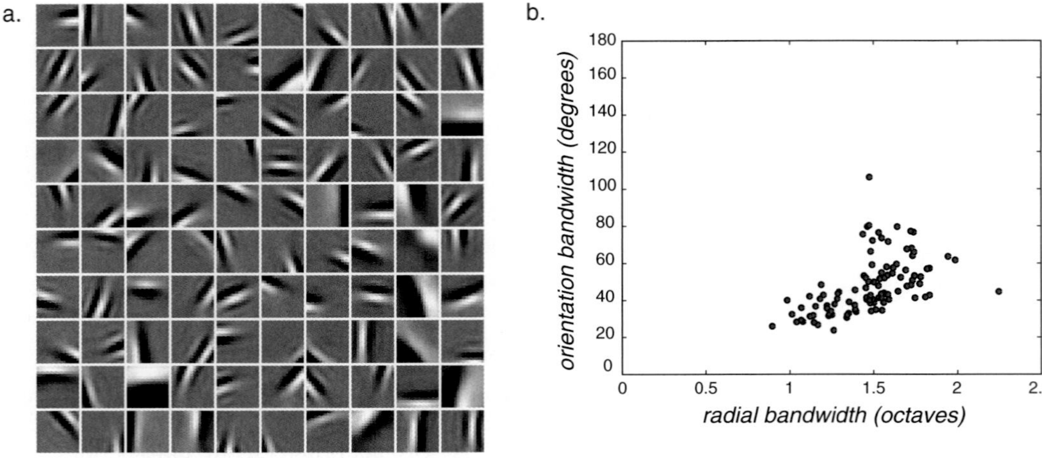

FIGURE 87.3 (a) Basis functions $\phi_i(\vec{x})$ learned from training on natural images. (b) Scatterplot of spatial frequency bandwidths and orientation bandwidths of the learned functions. (From Olshausen, Cadieu, & Warland, 2009.)

Adapting this model to millions of image patches extracted from natural scenes results in basis functions that are spatially localized, oriented, and bandpass (selective to structure at different scales) as shown in figure 87.3. When characterized in terms of their Fourier spectrum, the learned functions fall around 1.5 octaves spatial frequency bandwidth and 30°–40° orientation bandwidth, again similar to the spatial properties of V1 simple-cell receptive fields. Note however that making a direct comparison between the basis functions and receptive fields is complicated because the activity of a neuron in the model indicates how much of its basis function is present in the image whereas the receptive field usually refers to the spatial weighting of the input that a neuron uses to compute its output. These two things are equivalent only when the basis is orthogonal, which it is not. However, in the neural circuit implementation of the model, the feedforward drive to each unit is given by the inner product of its basis function with the image, so in that sense it acts similarly to a receptive field. In addition, when one maps out the receptive field of a unit in the model using single-pixel stimuli, the result resembles the basis function (Olshausen & Field, 1996, figure 4b). Other models that are formulated directly in terms of receptive fields or filters achieve qualitatively similar results (Bell & Sejnowski, 1997; Osindero, Welling, & Hinton, 2006; van Hateren & van der Schaaf, 1998).

Ideally, one would like to compare not just the form of individual basis functions but also how the population as a whole tiles the joint space of position, orientation, and spatial frequency. However, to do such a comparison properly would require exhaustively recording from all neurons within a hypercolumn of V1. From the partial assays of parafoveal neurons currently available, it would seem there is an overabundance of neurons tuned to low spatial frequencies as compared to the model (DeValois, Albrecht, & Thorell, 1982; Parker & Hawken, 1988; van Hateren & van der Schaaf, 1998). However real neurons have a certain level of precision with which they can code information, whereas in the model there is no limit in precision imposed upon the coefficient amplitudes (i.e., they have essentially infinite precision in amplitude). It seems likely that when such biophysical constraints are taken into account, the bias toward low spatial frequencies could be explained since the low spatial frequencies in natural scenes have a higher signal-to-noise ratio than high spatial frequencies (Doi & Lewicki, 2005).

It should be noted that such a linear model of images cannot possibly hope to capture the full richness of the structures contained in natural scenes. One important reason for this is that the true causes of images—light reflecting off the surfaces of objects—combine by the rules of occlusion, which are highly nonlinear (Ruderman, 1997; see also chapter 88). It remains to be seen how the solutions are affected in a model incorporating these nonlinearities and whether it is still consistent with V1 response properties, though see Lücke et al. (2009) and Le Roux et al. (2011) for promising work in this direction.

Overcomplete Representation

In general, a complete code where there are as many outputs as inputs can represent information without loss. However if the goal is to discover and represent structure in the input, there is no reason to impose this limitation. Indeed, much work in signal analysis and image processing has demonstrated the importance of

1252   BRUNO A. OLSHAUSEN AND MICHAEL S. LEWICKI

using *overcomplete representations*—where the number of outputs is greater than the number of inputs—when one wishes to ascribe meaning to the code outputs in terms of structures they represent in the input (Chen, Donoho, & Saunders, 2001; Lewicki & Olshausen, 1999; Mallat & Zhang, 1993; Simoncelli et al., 1992).

Rehn and Sommer (2007) developed a sparse coding model that enforces "hard sparsity"—where coefficients are forced to be either active or exactly zero—and showed that when the representation is made three times overcomplete a greater diversity emerges in the learned basis functions. Besides localized, oriented functions, one also finds unoriented, circularly symmetric functions in addition to grating-like functions with more oscillations. Interestingly, these results are better matched to the actual diversity seen among V1 receptive fields (Ringach, 2002). Another study systematically explored the effect of increasing either sparsity or overcompleteness (up to 10 times) and obtained similar diverse families of functions as either of these parameters is increased (Olshausen, Cadieu, & Warland, 2009).

As noted above, layer 4 of V1 contains on the order of 100 times as many neurons as there are input fibers from the LGN. Thus, the models explored to date are still far below the neurobiological regime. Why is V1 so overcomplete, and what are the extra dimensions being used for? At least part of the answer must have to do with the fact that we are still missing many other stimulus dimensions such as time, color, and disparity.

Sparse Coding in Time, Color, and Stereo

Retinal images are not static but change continuously over time as an observer moves through the world. The sparse coding model may be extended to describe this structure as well. Van Hateren and Ruderman (1998) approached the problem simply by considering time as another dimension and then learning a basis over x,y,t image cubes. The learned basis functions resemble the previous solution in terms of their spatial characteristics, but they also translate over time in a direction that is orthogonal to the orientation, and at different rates to represent structures moving at different speeds in the image. The disadvantage of a blocked coding scheme of this type, however, is that it results in many copies of the same space–time function centered at different points in time within a block. Olshausen (2002, 2003) used a convolution model to cover the time domain so that any given basis function can be applied at any point in time. The resulting functions are qualitatively similar to those learned in a blocked ICA scheme but require many fewer code elements. The learned basis functions resemble at least qualitatively the inseparable space–time receptive fields of V1 simple cells (DeAngelis, Ohzawa, & Freeman, 1995). Making a quantitative comparison or prediction of neural response properties, however, will demand training on movie ensembles that are representative of the spatiotemporal structure falling on the retina.

The sparse coding model has also been extended to describe spatiochromatic structure and disparities arising from stereo image pairs of natural scenes. Wachtler, Lee, and Sejnowski (2001) trained an ICA model on hyperspectral images and showed that color opponent receptive fields emerge for the low spatial frequency basis functions while high spatial frequency basis functions remain in luminance only. Hoyer and Hyvärinen (2000) obtained similar results training on RGB images, as did Doi et al. (2003) using a realistic cone mosaic. Johnson, Kingdom, and Baker (2005) analyzed the spatiochromatic structure of natural scenes by examining correlations among luminance and color-opponent channels; their results suggest that changes in coarse-scale color information correlate with changes in fine-scale texture information, but how this higher-order structure could be captured in a neural coding scheme has not been explored.

Hoyer and Hyvärinen (2000) trained an ICA model on stereo images of natural scenes and showed that a population of binocular neurons emerges spanning a range of ocular dominance and mimicking observed disparity tuning properties such as "tuned excitatory," "tuned inhibitory," "near," and "far." However making a meaningful quantitative comparison or prediction of neural response properties will require collecting and training on stereo image pairs that are representative of the range of fixations that occur during natural viewing of the 3-D environment.

Nonclassical Receptive Fields as an Emergent Property of Sparse Coding

Beyond accounting for known receptive field properties, the sparse coding model also makes predictions about the types of nonlinearities and interactions among neurons expected in response to natural images. As mentioned previously, each neuron's output is computed by minimizing the energy function in equation 87.2. The result of this minimization is a nonlinear mapping from the input image $I(\vec{x})$ to neuron activities a_i. Thus, although the image model itself is linear, the encoding of images is nonlinear. The nature of this nonlinearity is such that each output unit is modified by a suppressive interaction with its neighbors (those units with overlapping receptive fields) (Olshausen & Field, 1997; Rozell et al., 2008). Specifically, the response of a neuron is sparser than expected from simply computing the inner product of its basis function with the image. In other words, responses are pruned out, or *sparsified*, so that only those units that best describe the image structure are active. In the probabilistic version of the model, this is known as "explaining away" (Hinton & Ghahramani, 1997; see also chapter 88).

Simulations by Zhu and Rozell (2010), in addition to Lee et al. (2007), show that these sparsifying nonlinearities can account for many of the nonclassical receptive properties observed in V1 neurons such as end stopping, contrast-invariant orientation tuning, and orientation-specific surround suppression. Importantly, these effects were not built into the model but rather emerge as a consequence of inference in a generative model that has been adapted to the structure of natural images. (Other accounts based on natural scene statistics have been proposed by Schwartz & Simoncelli, 2001, and Karklin & Lewicki, 2009, as described below, and in terms of "predictive coding" by Rao & Ballard, 1999, and Spratling, 2010.)

The experiments of Vinje and Gallant (2000, 2002) also lend support to the idea of sparsification. They recorded from V1 neurons in an awake behaving monkey while natural image sequences obtained from free-viewing were played both within and surrounding a neuron's receptive field. They showed that when neurons are exposed to progressively more context around their classical receptive field, their responses become sparser. In the model, this happens because units are effectively competing to describe the image at any given moment. With little or no context, there is more ambiguity about which basis functions are best suited to describe structure within the image, and so the response resembles what is predicted from a linear weighting of the image. This effect could also be the result of top-down influences from higher cortical areas, as would be expected from explaining away in a hierarchical graphical model (Lewicki & Sejnowski, 1997; see also chapter 88).

Modeling Group Dependencies

When the sparse coding model is reformulated in terms of a probabilistic, graphical model, the cost function on the coefficients corresponds to a "prior" that encourages both sparseness and statistical independence (Olshausen & Field, 1997). After adapting the model to natural images the coefficients are indeed sparse, but not statistically independent (Bethge, 2006). There are a variety of reasons for the remaining dependencies. One is due to covariations in contrast, leading to correlations among magnitudes of neighboring basis

function coefficients (Simoncelli & Buccigrossi, 1997). Another is due to the existence of contours and other more extended forms of structure in images which can not be captured by a simple basis function model. For example, Geisler et al. (2001) and Sigman et al. (2001) have shown that edge co-occurrence statistics in natural scenes follow a cocircular pattern that extends far beyond the receptive field of any given oriented neuron.

Given that such dependencies exist, what should the cortex do about it? According to redundancy reduction, dependencies should be removed. This approach was taken by Schwartz and Simoncelli (2001) who proposed removing dependencies among oriented units via a specific divisive normalization operation derived from natural scene statistics. The resulting model provides a good account for contextual effects measured in V1 neurons using spatial frequency gratings. An alternative approach, taken by Garrigues and Olshausen (2008) and Lyu and Simoncelli (2007), is to use an Ising or Markov random field model to capture the dependencies. These models use a nonfactorial prior that includes a pairwise coupling term between units that condition coefficient magnitudes. Neurons that exhibit dependencies in their responses to natural images thus learn facilitatory connections between them. Such facilitatory interactions are consistent with a substantial body of psychophysics (Field, Hayes, & Hess, 1993; Polat & Sagi, 1993) and physiology (Kapadia, Westheimer, & Gilbert, 2000). A number of computational models of contour integration employ similar mechanisms (Parent & Zucker, 1989; Ullman & Sha'ashua, 1988; Yen & Finkel, 1998).

Another factor contributing to statistical dependencies among coefficients is that the basis functions in a sparse code need to cooperate in order to interpolate structures that vary along a continuum. For example, edges will occur at different positions along a continuum, but there is only a discrete set of basis functions available—not enough to match each and every position of an edge. Thus, two or more basis functions must add together in order to describe things that occur in between them (figure 87.4a,b). The signature of this dependency is that groups of basis functions related in position, orientation, or scale will exhibit a circularly symmetric distribution in their coefficient values, as opposed to a star-like distribution that would result from sparse variables that are statistically independent (Zetzsche, Krieger, & Wegmann, 1999) (fig. 87.4c). In the time domain, these dependencies give rise to temporal correlations in coefficient magnitude, or "bubble-like" behavior, in the temporal envelope of activity (Hyvärinen, Hurri, & Väyrynen, 2003). Similar forms of

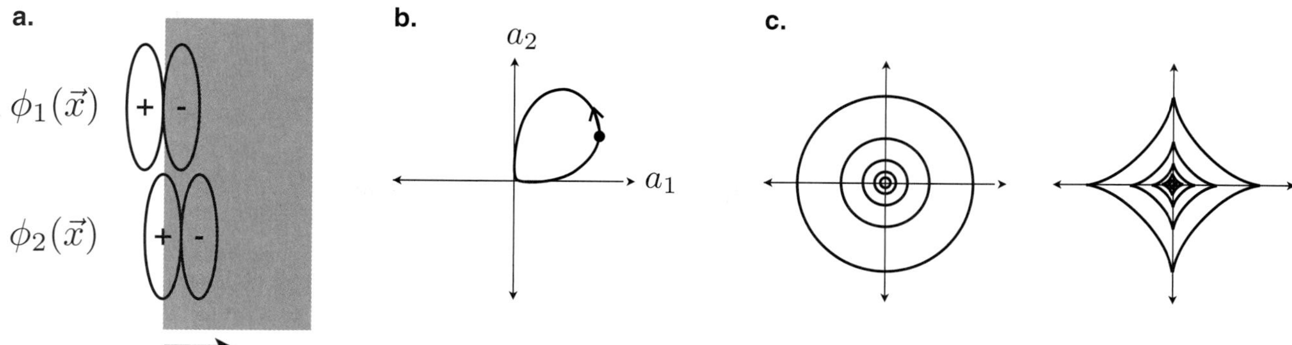

FIGURE 87.4   Statistical dependencies among coefficients arise due to interpolation of image features occurring at different positions along a continuum. (a) An edge moving continuously over two vertically oriented basis functions centered at different positions in the image (functions are shown displaced vertically to avoid clutter). (b) In the joint space of the coefficients, the movement of the edge traces out an arc. Dot indicates the current position of the edge. (c) When averaged over many different contrasts and polarities as well as edge types (lines vs. edges), the resulting joint distribution of the coefficients is circularly symmetric yet sparse (peaked at zero with heavy tails). This is in contrast to the star-shaped distribution (right) that would result from two sparse, statistically independent variables.

dependency can also arise simply from common contrast fluctuations (Eichhorn, Sinz, & Bethge, 2009; Lyu, 2011).

One approach toward modeling this form of dependency is to group together related basis functions so that their coefficients share a common amplitude component:

$$a_{ij} = \sigma_i u_{ij},$$

where i denotes the group, and j indexes the elements within a group. Group amplitudes $\sigma_i \geq 0$ are constrained to be sparse and independent whereas the normalized coefficients within a group u_{ij} have no sparseness constraint. This approach was used by Wainwright, Simoncelli, and Willsky (2001) to model statistical dependencies among wavelet coefficients in terms of Gaussian scale mixtures (where the u_{ij} are Gaussian and σ_i has a heavy-tailed distribution). Hyvärinen and Hoyer (2000) used this approach in a modified ICA model, termed "subspace ICA," in which basis functions are separated into nonoverlapping groups, or subspaces, of size 2, 4, or 8. After training on natural images, each group learns a set of basis functions having similar orientations but with shifted positions or phases, and the group amplitude σ_i exhibits phase invariance and shift invariance similar to complex cells. Other investigators have since elaborated on this idea by having overlapping groups that are organized topographically into a 2-D map, allowing one to visualize the dependencies among an entire population of learned basis functions (Garrigues & Olshausen, 2010; Gregor & LeCun, 2010; Hyvärinen, Hoyer, & Inki, 2001; Osindero, Welling, & Hinton, 2006). The resulting maps bear a striking resemblance to orientation maps and nonoriented

"blobs" in V1. Note however that in all of these models the group structure is fixed, not learned—only the basis functions are learned to fit this structure. Hyvärinen and Köster (2007) explored a range of different group sizes and found that groups of 16–32 basis functions yield the best fit (in terms of log-likelihood) for natural images using 24×24 pixel image patches.

Temporal dependencies may be modeled in a similar fashion by imposing a "slowness prior" (via a penalty on the temporal derivative) on the σ_i so that they vary smoothly or persist over time in response to a video sequence. The general idea of imposing slowness was initially proposed by Foldiak (1991) and Wiskott and Sejnowski (2002) as a way to learn invariant representations of visual input—also known as "slow feature analysis." Berkes, Turner, and Sahani (2009) employ this approach to learn local invariances from natural video sequences. The learned group structure is similar to that obtained with subspace ICA, and the amplitude responses also resemble those of complex cells, but in this case the group size is also learned, yielding around 2–4 bases per group. Cadieu and Olshausen (2009, 2012) employ the idea of temporal persistence in a complex-valued basis function model. The coefficients are split into amplitude and phase, with slowness imposed upon the amplitudes. They show that the resulting phase variables tend to precess in a linear fashion that is suitable for learning and representing transformations (i.e., motion) in natural video sequences.

HIERARCHICAL MODELS

The approaches and models discussed above have either been concerned with forming an efficient code of the visual image or a sparse representation based on a linear combination of features. The goal of vision, however, is not merely to describe the image but to understand it—that is, to deduce, from the raw 2-D image, properties of the 3-D scene and the surfaces and objects within it. In other words, the problems of coding and representation lead to deeper problems of computation and inference. These problems become especially relevant as one moves beyond V1 to consider the functions of higher stages of cortical processing.

One approach to understanding higher levels of processing is to develop hierarchical models composed of multiple layers, in which a higher level captures structure in the lower level, which is itself a nonlinear transform of earlier levels (Fukushima, 1980; George & Hawkins, 2005; Hinton, 2007, 2010; LeCun et al., 1989; Lee, Ekanadham, & Ng, 2008; Serre et al., 2007; see also chapter 88). (Note that nonlinearity between levels is key because, without it, a concatenation of two linear models could be trivially reduced to a single-layer linear model.) Such models are loosely based on the hierarchical structure of visual cortex and can be viewed as an extension of the methods discussed above for learning group dependencies. Similar approaches are currently being pursued in machine learning and computer vision (Hinton & Salakhutdinov, 2006; Jarrett et al., 2009; Le et al., 2012; Lee et al., 2009; Ranzato & Hinton, 2010). Although these models show promise in capturing higher-order structure and exhibit impressive performance in classification tasks, the connections to biological systems are less clear.

In order to provide insights into the functions of visual cortex, it is necessary to define computational objectives that are biologically relevant—that is, objectives that capture aspects of the problems that natural vision systems have evolved to solve. Our own view is that the problem of "object recognition" as it has been defined in computer vision—assigning labels to images—is too narrow to capture the range of tasks we use our visual systems for. Tasks such as navigation, locomotion (e.g., foot placement), grasping, foraging, and social interaction are more than labeling problems. They demand rich and dynamic representations of 3-D shape and scene layout that are suitable for planning and driving actions, or subtle details of reflectance that provide cues regarding material properties. A formal, rigorous specification of these tasks remains an open problem however. In the meantime we focus on two subproblems which we believe are essential components of many tasks: learning abstract properties of images and factorizing form and motion.

Learning Abstract Properties of Images

An important aspect of vision is *abstraction*—that is, the ability to generalize from specific instances to general categories or more abstract visual features. In the context of natural scenes, Karklin and Lewicki (2009) addressed the fact that contours are typically composed of a great variety of image edges due to the changing textures of the foreground and background surfaces. Therefore, to encode (or "recognize") a contour, the visual system must at some point generalize from these specific instances of edges to an abstract representation of a contour that is invariant to the specific form. Linear image models such as ICA or sparse coding are not able to capture this structure because they encode the image literally or exactly. In terms of a probabilistic model, they assume a single distribution for all natural images.

To define a computational objective for the generalization problem, Karklin and Lewicki use a hierarchical generative model whose density is conditionally dependent on a higher-level, distributed representation. When adapted to natural images, the first level of the model learns a standard linear image representation composed of Gabor-like functions. The second level of the model learns an efficient representation of image distributions, in terms of the first-level functions, that are typical in natural scenes. The model learns abstract representations of a variety of image structures such as contours, junctions, and textures (see figure 87.5). More pertinent to biological visual systems is that the model also predicts a number of nonlinear properties of complex cells, including insensitivity to phase. Moreover, the higher-level units also exhibit functional subunits similar to those obtained with spike-triggered covariance analysis of V1 complex cells. Thus, the model can predict the dimensions in stimulus space to which V1 complex cells are either sensitive or insensitive.

Other models along similar lines have also been proposed to learn abstract properties of images. Schwartz, Sejnowski, and Dayan (2006) and Coen-Cagli, Dayan, and Schwartz (2012) formulate the problem of modeling densities in terms of Gaussian scale mixtures and show that the second-layer variables also show invariances resembling those of complex cells, in addition to accounting for perceptual salience. Ranzato and Hinton (2010) use the second-layer variables to model the mean and covariance in an RBM (restricted Boltzmann machine) model and show that these variables can be used for image classification. Zoran and Weiss (2011) propose a discrete Gaussian mixture model, which may be viewed as a special case of Karklin and Lewicki's model when the second layer is restricted to having only

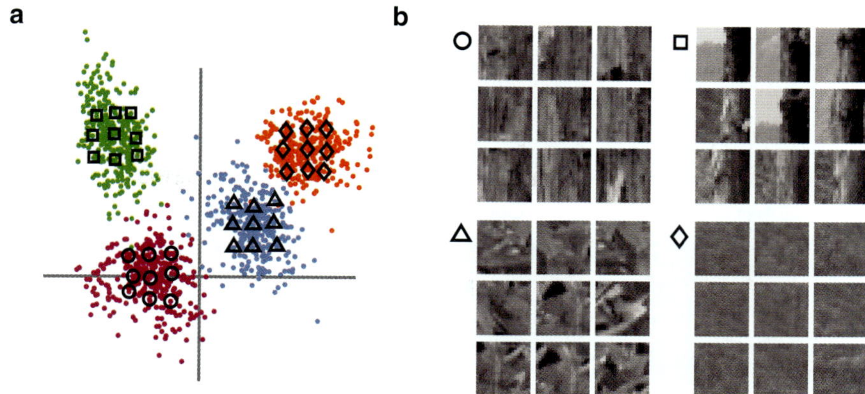

FIGURE 87.5   Generalization across variability in natural scenes. (a) A two-dimensional projection of the model's second-layer representation reveals well-separated clusters corresponding to abstract image properties. (b) Each 3×3 group of images shows a sample of the images in each cluster shown at left. Despite the variability in the appearance of edges and textures, the model's representation of natural images generalizes within each region while still distinguishing among them. (From Karklin & Lewicki, 2009.)

one unit active, and the learned features in the first layer are subdivided into nonoverlapping groups. Their model achieves high log-likelihood and good performance at image denoising and deblurring tasks. (Similar denoising results were obtained using a distributed representation of the image covariance [Karklin, 2007].) Shan, Zhang, and Cottrell (2007) developed a "recursive ICA" model that stacks layers of ICA with a compressive nonlinearity in between each stage. This simpler, extendible model learns representations that capture similar structures in higher layers but without explicit causal inference.

Learning to Separate Form and Motion

Time-varying natural images contain highly complex and dynamic structure. This is due to many factors, but a major component is due to the projection of the 3-D environment onto the image plane as an observer moves about the world. These two factors—the observer motion and the structure of the world—are entangled together in the time-varying pixel intensities. A reasonable goal for a perceptual system then is to disentangle these two factors from the raw data because doing so would recover the motion of the observer and the 3-D structure of the scene.

Cadieu and Olshausen (2012) have proposed a hierarchical model that factorizes time-varying images into components appropriate for learning form and motion (see figure 87.6a). The first layer forms a sparse, linear decomposition of image content in terms of a set of spatial features as described above. The resulting sparse representation is then factorized into two sets of

variables that disentangle the different contributions due to form and motion: a set of amplitudes $a(t)$ that represent the contrast of each feature, and a set of phases $\alpha(t)$ that represent the local shift of each feature. The second layer is then able to learn the patterns contained in the amplitude and phase variables in response to time-varying images. These learned patterns, expressed in the second layer weights B and D, act as a prior over the form and motion during the inference of these variables (see figure 87.6b,c). Importantly, the learned motion patterns provide a rich basis for regularizing optic flow beyond conventional "smoothing" priors (see also Sun et al., 2008, for related work).

The inferred second-layer variables v and w provide a distributed representation of the patterns of form and motion, respectively, that occur in natural movies. This separation mirrors the separation of form and motion found in the ventral and dorsal streams in visual cortex. The v variables are selective to texture boundaries, contours, and differential orientation structure, which are hypothesized to be represented in ventral regions such as areas V2 and V4 (Gallant, Braun, & Van Essen, 1993; von der Heydt & Peterhans, 1989; Willmore, Prenger & Gallant, 2010). Likewise, the w variables are selective to motion patterns such as local translation or differential motion which are hypothesized to be represented in dorsal regions such as areas MT and MST (Andersen, 1997). The manner in which these properties are computed, however, is markedly different than previous hierarchical models that have been proposed for form and motion processing (Serre et al., 2007; Simoncelli & Heeger, 1998) in two important respects: (1) the

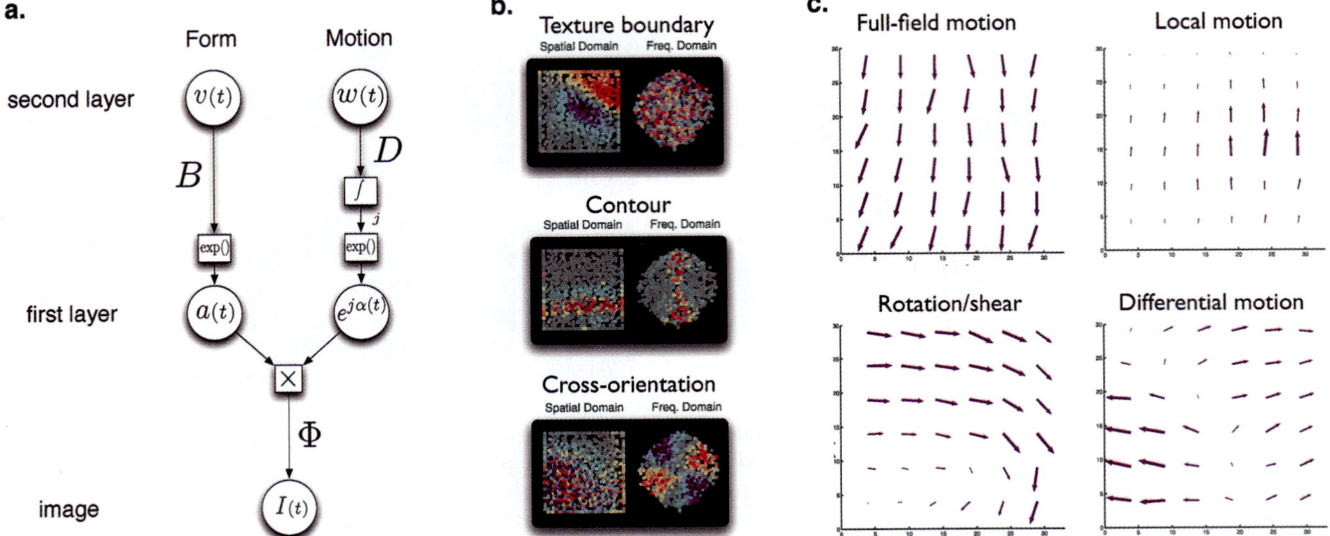

FIGURE 87.6 (a) Hierarchical model for factorizing form and motion. (b) Examples of learned form components, *B*. Weights are visualized in terms of the spatial position (left) or the peak spatial frequency and orientation (right) of the first-layer features they connect to. The learned form patterns capture texture boundaries, contours, and differential orientation structure. (c) Examples of learned motion components, *D*. Each is visualized according to the optic flow pattern it captures in the input. The learned motion patterns include full-field motion, local motion, rotation and shear, and differential motion. (From Cadieu & Olshausen, 2012.)

representations are *learned* from the statistics of time-varying natural scenes, rather than being designed by hand to match physiological data or to optimize performance for a pre-specified task, and (2) form and motion are inferred by a *factorization* process, which means that the computation of these quantities is intimately intertwined (not unlike factorizing a number), rather than being computed in separate, independent streams. Testing the specific predictions arising from these aspects of the model is the subject of ongoing investigation.

CONCLUSIONS

Over the past 30 years, work on natural scene statistics has steadily progressed from characterizing simple image pixel statistics to more abstract properties of form and motion. From these mathematical models arise predictions about neural coding and representation that may be compared to neurophysiological data. In many cases, such as in the retina and area V1, these models give us a new way to think about neural response properties that are already known and have been well characterized. What has been gained here is that we have a *linking principle* between these response properties and the statistics of natural scenes. To the extent that these principles may be generalized, then, we can make predictions about response properties that are

heretofore unknown. The hierarchical models described above, for example, provide new hypotheses about what to look for in the representations of higher level areas such as V2 and MT.

As we noted at the outset, we are referring to this body of work as "natural *scene* statistics" even though most of the models are actually of *images*, which are simply 2-D projections of scenes. The hope is that explicit representations of scene properties will eventually emerge from models of images. In the meantime though it would also make sense to make direct measurements of scene properties—for example, surface shape, material properties, and dynamics—and build these into models to infer properties of the world. Indeed there is promising work in this direction (Barron & Malik, 2012), and it remains a rich area for future exploration.

Although we are claiming to make predictions from these models about response properties of cortical neurons, there is much more that needs to be done in formulating these models in terms of specific neural mechanisms. In terms of Marr's levels of analysis, our models are formulated at the level of computational theory, yet we are tying them to phenomena at the level of implementation. Nothing has been said for example about what layers are implementing these models or what cell types are involved. In addition, the "units" in these models are far removed from neurons in that they represent continuous, graded values rather than spikes, and they integrate their inputs through simple linear summation as opposed to the types of nonlinear integration that occurs in dendrites (Polsky, Mel, & Schiller, 2004). In order to make concrete predictions that may be directly and seriously compared to real neural substrates, it will be necessary to rethink how these models are implemented at the biophysical level.

Perhaps the most immediate application of these models in neuroscience and psychophysics is to use them to parameterize the properties of natural scenes in a way that may be manipulated and controlled in experiments. In the same way that these models form internal representations of scene structure, they may be used in the opposite direction to *generate* scenes from stochastic perturbations to the latent variables. In this way, it is possible to generate classes of complex visual stimuli that adhere to natural scene statistics to varying degrees, and then ask what aspects observers, or different populations of neurons, are most sensitive to. In this way, models of natural scene statistics may offer us a more ecologically valid set of stimuli for probing the visual system.

ACKNOWLEDGMENTS

Supported by the National Science Foundation (IIS-1111654), National Institutes of Health (EY019965), National Geospatial-Intelligence Agency, and the Canadian Institute for Advanced Research.

REFERENCES

Andersen, R. A. (1997). Neural mechanisms of visual motion perception in primates. *Neuron, 18*, 865–872.

Atick, J. J., & Redlich, A. N. (1990). Towards a theory of early visual processing. *Neural Computation, 2*, 308–320.

Atick, J. J., & Redlich, A. N. (1992). What does the retina know about natural scenes? *Neural Computation, 4*, 196–210.

Attneave, F. (1954). Some informational aspects of visual perception. *Psychological Review, 61*, 183–193.

Balasubramanian, V., Kimber, D., & Berry, M. J. (2001). Metabolically efficient information processing. *Neural Computation, 13*, 799–815.

Barlow, H. B. (1961). Possible principles underlying the transformation of sensory messages. In W. Roserblith (Ed.), *Sensory communication* (pp. 217–234). Cambridge, MA: MIT Press.

Barlow, H. B. (1972). Single units and sensation: A neuron doctrine for perceptual psychology? *Perception, 1*, 371–394.

Barlow, H. B. (1981). The Ferrier Lecture, 1980: Critical limiting factors in the design of the eye and visual cortex. *Proceedings of the Royal Society of London. Series B, Biological Sciences, 212*, 1–34.

Barlow, H. B. (1989). Unsupervised learning. *Neural Computation, 1*, 295–311.

Barlow, H. B. (2001). Redundancy reduction revisited. *Network, 12*, 241–253.

Barron, J., & Malik, J. (2012). Shape, albedo, and illumination from a single image of an unknown object. *Conference on Computer Vision and Pattern Recognition*, 16–21 June 2012 (pp. 334–341).

Bell, A. J., & Sejnowski, T. J. (1995). An information-maximization approach to blind separation and blind deconvolution. *Neural Computation, 7*, 1129–1159.

Bell, A. J., & Sejnowski, T. J. (1997). The "independent components" of natural scenes are edge filters. *Vision Research, 37*, 3327–3338.

Berkes, P., Turner, R. E., & Sahani, M. (2009). A structured model of video reproduces primary visual cortical organisation. *PLoS Computational Biology, 5*, e1000495. doi:10.1371/journal.pcbi.1000495.g010.

Bethge, M. (2006). Factorial coding of natural images: How effective are linear models in removing higher-order dependencies? *Journal of the Optical Society of America. A, Optics, Image Science, and Vision, 23*, 1253–1268.

Bex, P. J., Solomon, S. G., & Dakin, S. C. (2009). Contrast sensitivity in natural scenes depends on edge as well as spatial frequency structure. *Journal of Vision, 9*(10), 1.1–1.19. doi:10.1167/9.10.1.

Borst, A., & Theunissen, F. E. (1999). Information theory and neural coding. *Nature Neuroscience, 2*, 947–957.

Burge, J., Fowlkes, C. C., & Banks, M. S. (2010). Natural-scene statistics predict how the figure–ground cue of convexity affects human depth perception. *Journal of Neuroscience, 30*, 7269–7280. doi:10.1523/JNEUROSCI.5551-09.2010.

Cadieu, C. F., & Olshausen, B. A. (2009). Learning transformational invariants from natural movies. *Advances in Neural Information Processing Systems, 21*, 209–216.

Cadieu, C. F., & Olshausen, B. A. (2012). Learning intermediate-level representations of form and motion from natural movies. *Neural Computation, 24*, 827–866. doi:10.1162/NECO_a_00247.

Chen, S. S., Donoho, D. L., & Saunders, M. A. (2001). Atomic decomposition by basis pursuit. *SIAM Review, 43*, 129–159.

Coen-Cagli, R., Dayan, P., & Schwartz, O. (2012). Cortical surround interactions and perceptual salience via natural scene statistics. *PLoS Computational Biology, 8*, e1002405.

Dan, Y., Atick, J. J., & Reid, R. C. (1996). Efficient coding of natural scenes in the lateral geniculate nucleus: Experimental test of a computational theory. *Journal of Neuroscience, 16*, 3351–3362.

Daugman, J. G. (1985). Uncertainty relation for resolution in space, spatial frequency, and orientation optimized by two-dimensional visual cortical filters. *Journal of the Optical Society of America. A, Optics and Image Science, 2*, 1160–1169.

DeAngelis, G. C., Ohzawa, I., & Freeman, R. D. (1995). Receptive-field dynamics in the central visual pathways. *Trends in Neurosciences, 18*, 451–458.

De Valois, R. L., Albrecht, D. G., & Thorell, L. G. (1982). Spatial frequency selectivity of cells in macaque visual cortex. *Vision Research, 22*, 545–559.

Doi, E., Balcan, D. C., & Lewicki, M. S. (2007). Robust coding over noisy overcomplete channels. *IEEE Transactions on Image Processing, 16*, 442–452.

Doi, E., Inue, T., Lee, T.-W., Wachtler, T., & Sejnowski, T. J. (2003). Spatiochromatic receptive field properties derived from information-theoretic analyses of cone mosaic responses to natural scenes. *Neural Computation, 15,* 397–417.

Doi, E., & Lewicki, M. S. (2005). Sparse coding of natural images using an overcomplete set of limited capacity units. *Advances in Neural Information Processing Systems, 17,* 377–384.

Doi, E., & Lewicki, M. S. (2007). A theory of retinal population coding. *Advances in Neural Information Processing Systems, 19,* 353–368.

Doi, E., & Lewicki, M. S. (2011). Characterization of minimum error linear coding with sensory and neural noise. *Neural Computation, 23,* 2498–2510. doi:10.1162/NECO_a_00181.

Eichhorn, J., Sinz, F., & Bethge, M. (2009). Natural image coding in V1: How much use is orientation selectivity? *PLoS Computational Biology, 5,* e1000336. doi:10.1371/journal.pcbi.1000336.t003.

Field, D. J. (1987). Relations between the statistics of natural images and the response properties of cortical cells. *Journal of the Optical Society of America. A, Optics and Image Science, 4,* 2379–2394.

Field, D. J. (1994). What is the goal of sensory coding? *Neural Computation, 6,* 559–601.

Field, D. J., Hayes, A., & Hess, R. F. (1993). Contour integration by the human visual system: Evidence for a local "association field." *Vision Research, 33,* 173–193.

Foldiak, P. (1990). Forming sparse representations by local anti-Hebbian learning. *Biological Cybernetics, 64,* 165–170. doi:10.1007/BF02331346.

Foldiak, P. (1991). Learning invariance from transformation sequences. *Neural Computation, 3,* 194–200.

Foldiak, P., & Young, M. P. (1995). Sparse coding in the primate cortex. In M. A. Arbib (Ed.), *The handbook of brain theory and neural networks* (pp. 895–898). Cambridge, MA: MIT Press,.

Fukushima, K. (1980). Neocognitron: A self organizing neural network model for a mechanism of pattern recognition unaffected by shift in position. *Biological Cybernetics, 36,* 193–202.

Gallant, J. L., Braun, J., & Van Essen, D. C. (1993). Selectivity for polar, hyperbolic, and Cartesian gratings in macaque visual cortex. *Science, 259,* 100–103.

Garrigues, P. J., & Olshausen, B. A. (2008). Learning horizontal connections in a sparse coding model of natural images. *Advances in Neural Information Processing Systems, 20,* 505–512.

Garrigues, P. J., & Olshausen, B. A. (2010). Group sparse coding with a Laplacian scale mixture prior. *Advances in Neural Information Processing Systems, 24,* 676–684.

Geisler, W. S. (2008). Visual perception and the statistical properties of natural scenes. *Annual Review of Psychology, 59,* 167–192. doi:10.1146/annurev.psych.58.110405.085632.

Geisler, W. S., Perry, J. S., Super, B. J., & Gallogly, D. P. (2001). Edge co-occurrence in natural images predicts contour grouping performance. *Vision Research, 41,* 711–724.

George, D., & Hawkins, J. (2005). A hierarchical Bayesian model of invariant pattern recognition in the visual cortex. *IEEE International Joint Conference on Neural Networks,* 1812–1817.

Gregor, K., & LeCun, Y. (2010). Emergence of complex-like cells in a temporal product network with local receptive fields. *CoRR, abs/1006.0448,* arXiv:1006.0448v1.

Hinton, G. E. (2007). Learning multiple layers of representation. *Trends in Cognitive Sciences, 11,* 428–434. doi:10.1016/j.tics.2007.09.004.

Hinton, G. E. (2010). Learning to represent visual input. [Review]. *Philosophical Transactions of the Royal Society of London. Series B, Biological Sciences, 365,* 177–184. doi:10.1098/rstb.2009.0200.

Hinton, G. E., & Ghahramani, Z. (1997). Generative models for discovering sparse distributed representations. *Philosophical Transactions of the Royal Society of London. Series B, Biological Sciences, 352,* 1177–1190. doi:10.1098/rstb.1997.0101.

Hinton, G. E., & Salakhutdinov, R. R. (2006). Reducing the dimensionality of data with neural networks. *Science, 313,* 504–507. doi:10.1126/science.1127647.

Hopfield, J. J. (1984). Neurons with graded response have collective computational properties like those of two-state neurons. *Proceedings of the National Academy of Sciences of the United States of America, 81,* 3088–3092.

Hoyer, P. O., & Hyvärinen, A. (2000). Independent component analysis applied to feature extraction from colour and stereo images. *Network (Bristol, England), 11,* 191–210.

Hyvärinen, A., & Hoyer, P. O. (2000). Emergence of phase- and shift-invariant features by decomposition of natural images into independent feature subspaces. *Neural Computation, 12,* 1705–1720.

Hyvärinen, A., Hoyer, P. O., & Inki, M. (2001). Topographic independent component analysis. *Neural Computation, 13,* 1527–1558. doi:10.1162/089976601750264992.

Hyvärinen, A., Hurri, J., & Hoyer, P. O. (2009). *Natural image statistics.* New York: Springer-Verlag.

Hyvärinen, A., Hurri, J., & Väyrynen, J. (2003). Bubbles: A unifying framework for low-level statistical properties of natural image sequences. *Journal of the Optical Society of America. A, Optics, Image Science, and Vision, 20,* 1237–1252.

Hyvärinen, A., & Köster, U. (2007). Complex cell pooling and the statistics of natural images. *Network (Bristol, England), 18,* 81–100. doi:10.1080/09548980701418942.

Jarrett, K., Kavukcuoglu, K., Ranzato, M. A., & LeCun, Y. (2009). What is the best multi-stage architecture for object recognition? *2009 IEEE 12th International Conference on Computer Vision (ICCV)* (pp. 2146–2153).

Johnson, A., Kingdom, F. A. A., & Baker, C. L. (2005). Spatiochromatic statistics of natural scenes: First- and second-order information and their correlational structure. *Journal of the Optical Society of America. A, Optics, Image Science, and Vision, 22,* 2050–2059.

Kapadia, M. K., Westheimer, G., & Gilbert, C. D. (2000). Spatial distribution of contextual interactions in primary visual cortex and in visual perception. *Journal of Neurophysiology, 84,* 2048–2062.

Karklin, Y. (2007). Hierarchical statistical models of computation in the visual cortex. Ph.D. Thesis, Department of Computer Science, Carnegie Mellon University. Pittsburgh, PA.

Karklin, Y., & Lewicki, M. S. (2009). Emergence of complex cell properties by learning to generalize in natural scenes. *Nature, 457,* 83–86. doi:10.1038/nature07481.

Karklin, Y., & Simoncelli, E. P. (2011). Efficient coding of natural images with a population of noisy linear–nonlinear neurons. *Advances in Neural Information Processing Systems, 12,* 999–1007.

Land, M. F., & Nilsson, D. E. (2012). *Animal eyes* (2nd ed.). New York: Oxford University Press.

Laughlin, S. B. (1981). A simple coding procedure enhances a neuron's information capacity. *Zeitschrift fur Naturforschung. Section C. Biosciences, 36,* 910–912.

Laughlin, S. B. (2001). Energy as a constraint on the coding and processing of sensory information. *Current Opinion in Neurobiology, 11*, 475–480.

Le, Q. Y., Ranzato, M. A., Monga, R., Matthieu, D., Chen, K., Corrado, G. S., Dean, J., & Ng, A. Y. (2012). Building high-level features using large scale unsupervised learning. arXiv:1112.6209.

LeCun, Y., Boser, B., Denker, J. S., Henderson, D., Howard, R. E., Hubbard, W., et al. (1989). Backpropagation applied to handwritten zip code recognition. *Neural Computation, 1*, 541–551.

Lee, H., Battle, A., Raina, R., & Ng, A. Y. (2007). Efficient sparse coding algorithms. *Advances in Neural Information Processing Systems, 19*, 801–808.

Lee, H., Ekanadham, C., & Ng, A. Y. (2008). Sparse deep belief net model for visual area V2. *Advances in Neural Information Processing Systems, 20*, 873–880.

Lee, H., Grosse, R., Ranganath, R., & Ng, A. Y. (2009). Convolutional deep belief networks for scalable unsupervised learning of hierarchical representations. *ICML 26th Annual International Conference*, New York.

Le Roux, N., Heess, N., Shotton, J., & Winn, J. (2011). Learning a generative model of images by factoring appearance and shape. *Neural Computation, 23*, 593–650. doi:10.1162/NECO_a_00086.

Lewicki, M. S., & Olshausen, B. A. (1999). Probabilistic framework for the adaptation and comparison of image codes. *Journal of the Optical Society of America. A, Optics, Image Science, and Vision, 16*, 1587–1601.

Lewicki, M. S., & Sejnowski, T. J. (1997). Bayesian unsupervised learning of higher order structure. *Advances in Neural Information Processing Systems, 9*, 529–535.

Lücke, J., Turner, R. E., Sahani, M., & Henniges, M. (2009). Occlusive components analysis. *Advances in Neural Information Processing Systems, 22*, 1069–1077.

Lyu, S. (2011). Dependency reduction with divisive normalization: Justification and effectiveness. *Neural Computation, 23*, 2942–2973. doi:10.1162/NECO_a_00197.

Lyu, S., & Simoncelli, E. P. (2007). Statistical modeling of images with fields of Gaussian scale mixtures. *Advances in Neural Information Processing Systems, 19*, 945–952.

Mallat, S., & Zhang, Z. (1993). Matching pursuits with time-frequency dictionaries. *IEEE Transactions on Signal Processing, 41*, 3397–3415.

Marcelja, S. (1980). Mathematical description of the responses of simple cortical cells. *Journal of the Optical Society of America, 70*, 1297–1300.

Olshausen, B. A. (2002). Sparse codes and spikes. In R. P. N. Rao, B. A. Olshausen, & M. S. Lewicki (Eds.), *Probabilistic models of the brain* (pp. 235–248). Cambridge, MA: MIT Press.

Olshausen, B. A. (2003). Learning sparse, overcomplete representations of time-varying natural images. *International Conference on Image Processing, 41*, I-41–44.

Olshausen, B. A., Cadieu, C. F., & Warland, D. K. (2009). Learning real and complex overcomplete representations from the statistics of natural images. *Proceedings of SPIE-Society for International Optics & Photonics, 7446*.

Olshausen, B. A., & Field, D. J. (1996). Emergence of simple-cell receptive field properties by learning a sparse code for natural images. *Nature, 381*, 607–609. doi:10.1038/381607a0.

Olshausen, B. A., & Field, D. J. (1997). Sparse coding with an overcomplete basis set: A strategy employed by V1? *Vision Research, 37*, 3311–3325.

Osindero, S., Welling, M., & Hinton, G. E. (2006). Topographic product models applied to natural scene statistics. *Neural Computation, 18*, 381–414. doi:10.1162/089976606775093936.

Parent, P., & Zucker, S. W. (1989). Trace inference, curvature consistency, and curve detection. *IEEE Transactions on Pattern Analysis and Machine Intelligence, 11*, 823–839. doi:10.1109/34.31445.

Parker, A. J., & Hawken, M. J. (1988). Two-dimensional spatial structure of receptive fields in monkey striate cortex. *Journal of the Optical Society of America. A, Optics and Image Science, 5*, 598–605.

Polat, U., & Sagi, D. (1993). Lateral interactions between spatial channels: Suppression and facilitation revealed by lateral masking experiments. *Vision Research, 33*, 993–999.

Polsky, A., Mel, B. W., & Schiller, J. (2004). Computational subunits in thin dendrites of pyramidal cells. *Nature Neuroscience, 7*, 621–627. doi:10.1038/nn1253.

Ranzato, M. A., & Hinton, G. E. (2010). Modeling pixel means and covariances using factorized third-order Boltzmann machines. *IEEE Conference on Computer Vision and Pattern Recognition (CVPR)*, 2551–2558.

Rao, R. P., & Ballard, D. H. (1999). Predictive coding in the visual cortex: A functional interpretation of some extra-classical receptive-field effects. *Nature Neuroscience, 2*, 79–87.

Rehn, M., & Sommer, F. T. (2007). A network that uses few active neurones to code visual input predicts the diverse shapes of cortical receptive fields. *Journal of Computational Neuroscience, 22*, 135–146. doi:10.1007/s10827-006-0003-9.

Ringach, D. L. (2002). Spatial structure and symmetry of simple-cell receptive fields in macaque primary visual cortex. *Journal of Neurophysiology, 88*, 455–463.

Rozell, C. J., Johnson, D. H., Baraniuk, R. G., & Olshausen, B. A. (2008). Sparse coding via thresholding and local competition in neural circuits. *Neural Computation, 20*, 2526–2563. doi:10.1162/neco.2008.03-07-486.

Ruderman, D. L. (1994). Designing receptive fields for highest fidelity. *Network (Bristol, England), 5*, 147–155.

Ruderman, D. L. (1997). Origins of scaling in natural images. *Vision Research, 37*, 3385–3398.

Schwartz, O., Sejnowski, T. J., & Dayan, P. (2006). Soft mixer assignment in a hierarchical generative model of natural scene statistics. *Neural Computation, 18*, 2680–2718. doi:10.1162/neco.2006.18.11.2680.

Schwartz, O., & Simoncelli, E. P. (2001). Natural signal statistics and sensory gain control. *Nature Neuroscience, 4*, 819–825.

Serre, T., Wolf, L., Bileschi, S., Riesenhuber, M., & Poggio, T. (2007). Robust object recognition with cortex-like mechanisms. *IEEE Transactions on Pattern Analysis and Machine Intelligence, 29*, 411–426. doi:10.1109/TPAMI.2007.56.

Shan, H., Zhang, L., & Cottrell, G. W. (2007). Recursive ICA. *Advances in Neural Information Processing Systems, 19*, 1273–1280.

Sigman, M., Cecchi, G. A., Gilbert, C. D., & Magnasco, M. O. (2001). On a common circle: Natural scenes and Gestalt rules. *Proceedings of the National Academy of Sciences of the United States of America, 98*, 1935–1940. doi:10.1073/pnas.031571498.

Simoncelli, E. P., & Buccigrossi, R. T. (1997). Embedded Wavelet Image Compression Based on a Joint Probability Model. *International Conference on Image Processing.*

Simoncelli, E. P., Freeman, W. T., Adelson, E. H., & Heeger, D. J. (1992). Shiftable multiscale transforms. *IEEE Transactions on Information Theory, 38*, 587–607.

Simoncelli, E. P., & Heeger, D. J. (1998). A model of neuronal responses in visual area MT. *Vision Research, 38*, 743–761.

Srinivasan, M. V., Laughlin, S. B., & Dubs, A. (1982). Predictive coding: A fresh view of inhibition in the retina. *Proceedings of the Royal Society of London. Series B, Biological Sciences, 216*, 427–459.

Spratling, M. W. (2010). Predictive coding as a model of response properties in cortical area V1. *Journal of Neuroscience, 30*, 3531–3543.

Sun, D., Roth, S., Lewis, J., & Black, M. J. (2008). Learning optical flow. *Proceedings of the 10th European Conference on Computer Vision: Part III*, 83–97.

Tkacik, G., Prentice, J. S., Balasubramanian, V., & Schneidman, E. (2010). Optimal population coding by noisy spiking neurons. *Proceedings of the National Academy of Sciences of the United States of America, 107*, 14419–14424. doi:10.1073/pnas.1004906107.

Ullman, S., & Sha'ashua, A. (1988). Structural saliency: The detection of globally salient structures using a locally connected network. *A.I. Memo No. 1061*: Massachusetts Institute of Technology, Artificial Intelligence Laboratory.

van Hateren, J. H. (1992). A theory of maximizing sensory information. *Biological Cybernetics, 68*, 23–29.

van Hateren, J. H. (1993). Spatiotemporal contrast sensitivity of early vision. *Vision Research, 33*, 257–267.

van Hateren, J. H., & Ruderman, D. L. (1998). Independent component analysis of natural image sequences yields spatio-temporal filters similar to simple cells in primary visual cortex. *Proceedings. Biological Sciences, 265*, 2315–2320. doi:10.1098/rspb.1998.0577.

van Hateren, J. H., & van der Schaaf, A. (1998). Independent component filters of natural images compared with simple cells in primary visual cortex. *Proceedings. Biological Sciences, 265*, 359–366. doi:10.1098/rspb.1998.0303.

Vincent, B., & Baddeley, R. (2003). Synaptic energy efficiency in retinal processing. *Vision Research, 43*, 1283–1290.

Vinje, W. E., & Gallant, J. L. (2000). Sparse coding and decorrelation in primary visual cortex during natural vision. *Science, 287*, 1273–1276.

Vinje, W. E., & Gallant, J. L. (2002). Natural stimulation of the nonclassical receptive field increases information transmission efficiency in V1. *Journal of Neuroscience, 22*, 2904–2915.

von der Heydt, R., & Peterhans, E. (1989). Mechanisms of contour perception in monkey visual cortex. I. Lines of pattern discontinuity. *Journal of Neuroscience, 9*, 1731–1748.

Wachtler, T., Lee, T.-W., & Sejnowski, T. J. (2001). Chromatic structure of natural scenes. *Journal of the Optical Society of America. A, Optics, Image Science, and Vision, 18*, 65–77.

Wainwright, M. J., Simoncelli, E. P., & Willsky, A. S. (2001). Random cascades on wavelet trees and their use in analyzing and modeling natural images. *Applied and Computational Harmonic Analysis, 11*, 89–123. doi:10.1006/acha.2000.0350.

Willmore, D. B., Prenger, R. J., & Gallant, J. L. (2010). Neural representation of natural images in visual area V2. *Journal of Neuroscience, 30*, 2102–2114.

Wiskott, L., & Sejnowski, T. J. (2002). Slow feature analysis: Unsupervised learning of invariances. *Neural Computation, 14*, 715–770.

Yen, S.-C., & Finkel, L. H. (1998). Extraction of perceptually salient contours by striate cortical networks. *Vision Research, 38*, 719–741.

Yoonessi, A., & Kingdom, F. A. A. (2008). Comparison of sensitivity to color changes in natural and phase-scrambled scenes. *Journal of the Optical Society of America. A, Optics, Image Science, and Vision, 25*, 676–684.

Zetzsche, C., Krieger, G., & Wegmann, B. (1999). The atoms of vision: Cartesian or polar? *Journal of the Optical Society of America. A, Optics, Image Science, and Vision, 16*, 1554–1565.

Zhu, M., & Rozell, C. J. (2010). Sparse coding models demonstrate some non-classical receptive field effects. *BMC Neuroscience, 11*(Suppl 1), O21. doi:10.1186/1471-2202-11-S1-O21.

Zoran, D., & Weiss, Y. (2011). From learning models of natural image patches to whole image restoration. *IEEE International Conference on Computer Vision (ICCV)*.

Zylberberg, J., Murphy, J. T., & Deweese, M. R. (2011). A sparse coding model with synaptically local plasticity and spiking neurons can account for the diverse shapes of V1 simple cell receptive fields. *PLoS Computational Biology, 7*, e1002250. doi:10.1371/journal.pcbi.1002250.

88 Vision: Bayesian Inference and Beyond

DANIEL KERSTEN AND ALAN YUILLE

Although research has provided an enormous amount of knowledge about visual brain anatomy, physiology, and neural mechanisms, this knowledge is insufficient to quantitatively describe neural dynamics of large systems of neurons (except in certain restricted regimes). Moreover, even precise knowledge of neural dynamics would only yield partial understanding of brain function (knowing the electronic dynamics of a computer does not give much insight into the algorithm that the computer is running). This suggests that direct approaches based on studying the neurobiology of visual circuitry must be augmented by understanding visual processing at a more abstract level (see figure 88.1). Such understanding should take into account behavioral functions or tasks and the computational problems the organism must solve. We get insight into tasks through the study of how animals, such as ourselves, use vision (Milner & Goodale, 2006) as well as their relationships to the development of computer vision systems that are required to achieve specific goals (Ullman, 2000).

This chapter describes how visual processing can be modeled and understood in terms of *probabilistic inference*, or, equivalently, as a decoding problem where the goal is to determine information about the world from *image patterns* reaching the eyes. Information gathering, however, is not a passive process and depends on the goals and abilities of the *organism* performing the decoding. From this perspective, three important questions are as follows: (1) what types of image patterns occur? (2) what information can be extracted from these patterns for a given task? and (3) how does the organism's inferences compare with optimal inference given answers to (1) and (2)? (See figure 88.2.)

At this abstract level, the primary concern is with how the required computations and algorithms for inference and learning constrain neural processing. This priority is arguably a consequence of the inherent complexity of natural image patterns. Marr wrote in 1982: "... the nature of the computations that underlie perception depends more upon the computational problems that have to be solved than upon the particular hardware in which their solutions are implemented" (27). Theories of human perceptual inference require an understanding of the limits of perceptual inference through optimal decoding theories. These theories, in turn, require an understanding of the transformations and variations introduced in pattern formation. Advances in applied mathematics, probability theory, information theory, artificial neural networks, and artificial intelligence have helped to develop a *Bayesian* framework for approaching vision as well as other problems, including sensorimotor control (Battaglia & Schrater, 2007; Franklin & Wolpert, 2011; Körding & Wolpert, 2006; Schlicht & Schrater, 2007) and cognition (Griffiths et al., 2010). This framework includes a common language for describing vision problems, mathematical techniques for modeling them, and algorithms that can be applied to solve them. These advances have been facilitated by the enormous increase in computer power which has made it possible to explore increasingly complicated probability models.

The Bayesian framework reduces to standard signal detection theory as a special case (Green & Swets, 1966) but goes beyond it in its range of applicability, the power of its techniques, and the kinds of questions asked (Chater, Tenenbaum, & Yuille, 2006; Kersten & Schrater, 2002). The development of signal detection theory in the 1950s was strongly motivated by the idea that errors in psychophysical decisions varied because of noise and bias inside the observer. However, not long after computer vision began attempts to mimic human perception in the late 1960s, it became apparent that the fundamental limit to accurate and reliable perceptual decisions (by machine or organism) was not noise in the internal processing or sensory measurements but rather the ambiguity in what local image measurements implied about the external world. This problem was exacerbated with the realization that useful information, that is, signals like "curved line," "round shape," or "baseball player," require substantial computation to be decrypted from input image patterns. While understanding how the brain deals with noisy input and neurons is important, this chapter focuses on elements of the Bayesian program that are important for studying vision when internal noise can be neglected, and where the challenge is to discover and extract important information given the inherent ambiguities and complexities of images.

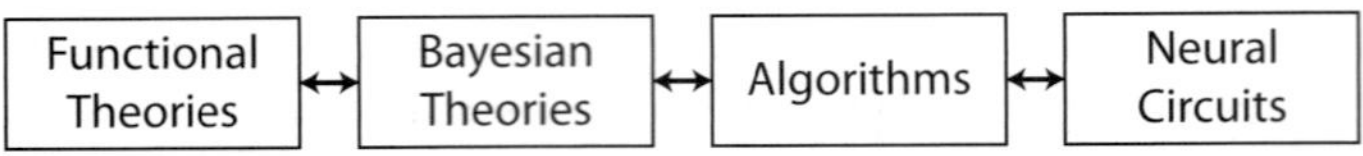

FIGURE 88.1   Probabilistic/Bayesian theories fall between the functional ("computational") and algorithmic levels of analysis in Marr's categorization of three levels of analysis for the study of a complex information processing system such as vision (Marr, 2010).

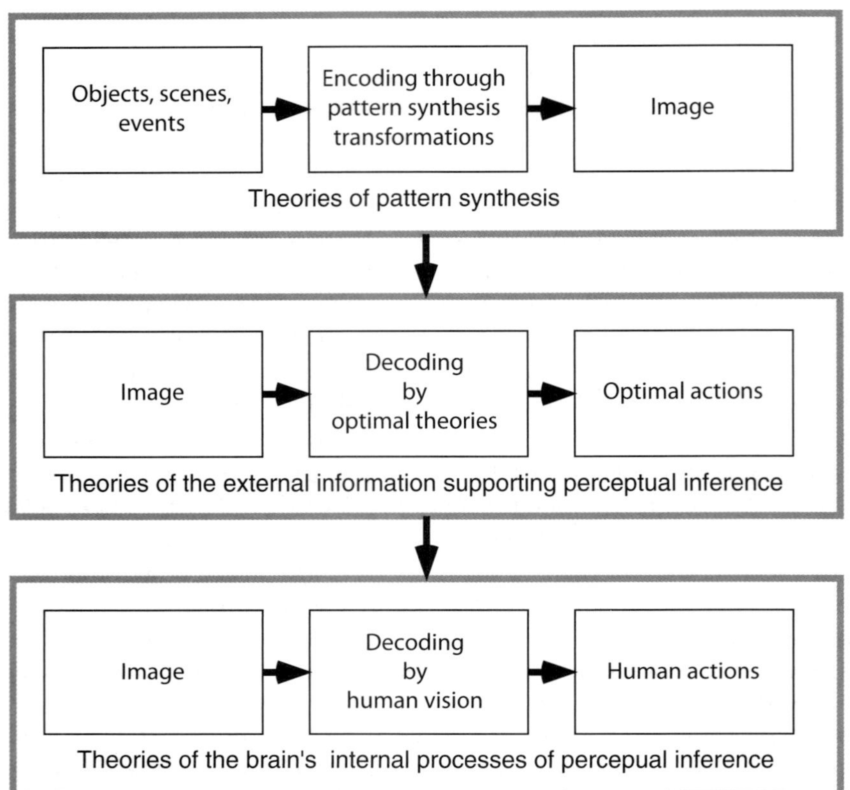

FIGURE 88.2   Three levels of study in perceptual inference.

The past decade has produced a substantial body of research interpreting human behavior and neural coding from a Bayesian framework.[1] For practical experimental reasons, this work has focused on modular problems, where the tasks are limited and the stimulus descriptions simple. Yet, human vision deals with images that are enormously complex and for a wide range of tasks. Our aim in this chapter is to show how recent developments in Bayesian models can provide insight into how the visual system deals with the complexities of natural images, and its relation to the hierarchical organization of the visual cortex.

The rest of this chapter is divided into three main sections: (1) "Bayesian Modeling" describes basic probabilistic concepts and operations; (2) "Dealing with Image Complexity and Task Flexibility" applies ideas from computer vision to questions of ventral stream feature hierarchy; (3) "Behavioral and Neural Evidence for Bayesian Computations" describes results of several experiments consistent with hierarchical, probabilistic computations in the visual system.

BAYESIAN MODELING

Basic Ingredients

We assume that the knowledge required for perception is represented in terms of a probability distribution, $p(I,s)$, over random variables that are measurable, I, and hidden variables, s, some or all of which need to be estimated.[2] At the most basic level, we think of measurable variables as the intensity pattern at the eye, I. However measurables are often assumed to be features that are "easily" computed from the image early in processing. Deciding what is a measurable feature and what is a state to be inferred is a modeling assumption. For example, estimating the distance of an object from an image requires a "measurement" of angular size; however, determining the angular size of an object is itself a nontrivial inference in a natural image with background clutter.

Hidden variables can represent interpretable, external states of the world such as objects and events or more abstract causes or "latent variables" that capture regularities in the image.

Using the product rule to condition the joint distribution on a measurement I reduces the uncertainty about s:

$$p(s \mid I) = p(s, I) / p(I).$$

This distribution is called the posterior distribution of s on I and is the basis for optimal inference.

Typically more than one cause influences a measurable image variable (see figure 88.3B). One component of cause s in $s = (s_1, s_2)$ may be important to estimate, whereas another is a confounding or "nuisance" variable to be discounted (see figure 88.3C). For example, suppose the image intensity is $I = s_1.s_2$, with s_1 and s_2 representing reflectance ("grayness" of surface) and illumination (level of light falling on a surface), respectively. The pattern of light, s_2, falling on an object is usually considered a nuisance variable whereas s_1 is important because it is an invariant surface property useful for object recognition (e.g., "Is it a white or black piece of paper?"). Like conditioning, discounting can reduce uncertainty about the remaining variables through application of the "sum rule," called marginalization:

$$p(s_1 \mid I) = \sum_{s_2} p(s_1, s_2 \mid I).$$

Because the problem of confounding variables is the rule rather than the exception, marginalization can be viewed as a basic operation for inference that underlies useful decisions. Its effects could be built in, as in distinct visual pathways specialized for different functions (Fang & He, 2005; Milner & Goodale, 2006), or it could involve dynamic neural processes that adapt to the task at hand. Freeman (1994) showed how marginalizing with respect to viewpoint and illumination could substantially reduce uncertainty about shape and material. Integrating out illumination direction can also disambiguate depth from cast shadows (Kersten, 1999). Freeman (1994) further showed that marginalization could be viewed as choosing the estimate least sensitive to variations in the confounding variable, providing a Bayesian interpretation of the generic viewpoint principle (see Lowe, 1987; Nakayama & Shimojo, 1992). Beck, Latham, and Pouget (2011) show that given certain assumptions about spike train statistics, some kinds of marginalization can be achieved through divisive normalization (usually thought of in terms of gain control and attentional processes; Reynolds & Heeger, 2009; Carandini & Heeger, 2011).

Generative Models

The posterior probability, $p(s \mid I)$, represents the information required for reliable estimates of states of the world from image input. The posterior could be as simple as a lookup table assigning probabilities to scenes for a given input, without an explicit model of input variation.[3] However visual images are complex, inferences are computationally hard, and there are big advantages to expressing the posterior in terms of

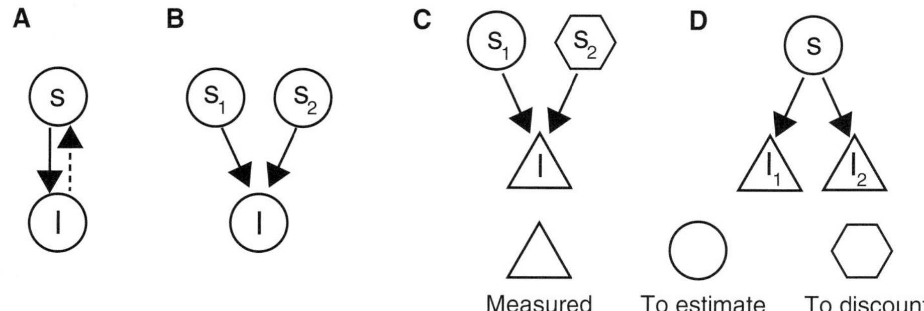

FIGURE 88.3   Simple directed graphs. The solid arrows represent conditional probability relationships between parents and descendants. (A) "Basic Bayes," where the joint is factorized as $p(s, I) = p(I \mid s)p(s)$. The dashed arrow indicates the bottom-up direction of inference. (B) More than one cause contributes to measurements, $p(I, s_1, s_2) = p(I \mid s_1, s_2)p(s_1)p(s_2)$. This kind of graph implies conditional dependence of the causes given a measurement, and marginal independence over measurements. Causes can effectively compete to explain a measurement, a Bayesian phenomenon called "explaining away" (Pearl, 1988). (C) If only one parent variable is important to estimate, the other one needs to be integrated out, that is, discounted. (D) One cause leads to two effects, that is, image measurements. The graph implies that the two measurements are conditionally independent given s. This generative model is the basis for tests of optimal cue integration.

generative components, that is, a description of the regularities in the causes (state variables describing objects, scenes, events, independent of the resulting images) and how those causes create images.[4] To provide some intuition as to why generative knowledge is useful, consider the many images that could result from a pair of gardening shears. Its appearance will vary with lighting, position, 3-D orientation, and the opening angle. In practice, it is impossible to have a lookup table that can anticipate all possible appearances. While some variations may be efficiently handled bottom-up, others such as discounting 3-D orientation or opening angle can be much more efficiently handled with object-specific knowledge—that is, that a shears can rotate in depth, and open and close (Ullman, 2000). Bayes's rule describes the relationship between the posterior and information about how images could be formed. It says

$$p(s \mid I) = \frac{p(I \mid s)p(s)}{p(I)} = \frac{p(I \mid s)p(s)}{\sum_{s'} p(I \mid s')p(s')}.$$

This theorem reexpresses $p(s \mid I)$, the probability of the state given the measurement, in terms of $p(I \mid s)$, the probability of the measurement given the state (called the likelihood of s), and $p(s)$, the probability of the state variables (called the prior). It tells us that we can look to models of how image patterns are formed by states of the world (e.g., shape, material, lighting, and viewpoint) as well as models of regularities of those states (e.g., surface smoothness, pigmentation, etc.).

Elaborating Graphical Structure

Bayes's formula makes generative knowledge explicit in terms of two basic factors represented as "Basic Bayes" in the graph of figure 88.3A; however, more structure is needed to cope with the range of tasks and the complex interdependency of image intensities. It is impossible to do Bayesian inference on high dimensional joint distributions, $p(I_1, I_1, ..., s_1, s_2, ...)$ without appropriate knowledge structuring. A straightforward next step is to consider slightly more complex graphs, for example involving more than one cause (figures 88.3B and 88.3C), one cause leading to more than one measurement (figure 88.3D), and simple three-level hierarchies which are discussed in a later section (see *Bayesian Coarse-to-Fine*, figure 88.7).

Consider the case where one cause produces more than one measurement, providing the basis for cue integration (see figure 88.3D). A distinctive prediction of Bayesian observers is that decisions should be based on full knowledge of the posterior distribution, that is, including higher-order moments. Following the work of Clark and Yuille (1990), Jacobs (1999), Ernst and Banks (2002), and Ernst and Bülthoff (2004), numerous studies have tested whether humans combine sensory information weighted by reliability (the reciprocal of the second moment, variance). The majority of these studies confirm optimality, with possible exceptions (cf. Cheng et al., 2007; Gori et al., 2008). Most studies of cue integration have been restricted to continuous-valued measurements whose precision is modeled using standard Gaussian and independence assumptions, and thus a linear weighting function. Stevenson and Körding (2009) devised a Bayesian ideal observer that uses occlusion, a nonmetric cue, and showed that their model accurately predicts human depth judgments.

While small graphs are useful for low-dimensional, modular analyses of visual behavior, substantially more structure is needed to understand recognition with natural images, and we return to this later when we discuss a possible relationship of cortical visual architecture to graphical representations.

Making Decisions and Estimates

To make a decision, one can choose values of the state variables that maximize the appropriately conditioned and marginalized distribution—this is the maximum a posterior (MAP) estimate, which is optimal in the sense of minimizing the average error rate. However, human perception is flexible, and some tasks may need several estimates, each requiring different levels of precision, and perhaps at different times. To pick up a ball, you need to know how far away and how big it is. Given a measurement of angular size, I, and a constraint $I \approx s_2 / s_1$, planning the reach requires a good estimate of distance (s_1) while the actual grasping requires a good estimate of physical size (s_2). Bayesian decision theory generalizes the operation of marginalization and defines optimality for an action as minimizing the risk:

$$R(\alpha; I) = \sum_s L(\alpha, s)p(s \mid I),$$

where the loss function $L(\alpha, s)$ is the cost of deciding to take action α when the true world state is s (see Geisler & Kersten, 2002; Maloney & Mamassian, 2009). To keep things simple, the rest of this chapter assumes that conditioning and marginalization choices are sufficient to characterize different task requirements.

As discussed below, Bayesian models for object recognition represent knowledge as probabilities over complex, but structured graphs. Inference on complex graphs requires message passing algorithms that update local conditional distributions at all nodes by passing information between neighbors, given that the values of some nodes are fixed and others integrated out. Belief propagation is one such method (MacKay, 2003; Pearl, 1988). Messages that update probabilities rather than passing on decisions respect the Bayesian version of Marr's principle of least commitment (Marr, 2010). The question of whether communication between neural populations observes this principle (cf. Ma et al., 2006) or instead involves decisions that get passed from one population to the next (cf. Lennie, 1998) is a fundamental unanswered question in brain science.

So far we have treated decisions at an abstract level without distinguishing deliberate from automatic decisions, or those at the level of subprocesses. An intriguing proposal is that perception does not actually commit to a decision in the usual sense. Instead a conscious percept reflects a stochastic sample drawn from a posterior distribution. The posterior is represented by a collection of exemplars that approximate the frequencies of occurrence of the perceptual hypothesis conditioned on the measurements (Fiser et al., 2010; Lee & Mumford, 2003). There are a number of behavioral results, including perceptual bistability, consistent with this idea (Battaglia, Kersten, & Schrater, 2011; Gershman, Vul, & Tenenbaum, 2012; Moreno-Bote, Knill, & Pouget, 2011; Sundareswara & Schrater, 2008).

In summary, visual knowledge can be conceptualized as a joint distribution over possible images and interpretations. The operations of conditioning and marginalizing narrow the space of possibilities through image measurements and task assumptions, respectively. Optimal decisions require computations that take full account of the knowledge available, including knowledge of uncertainty.

DEALING WITH IMAGE COMPLEXITY AND TASK FLEXIBILITY

While Bayesian models have provided compelling accounts of how human visual behavior manages uncertainty for low-dimensional problems, a central problem is to understand how the visual system is organized to deal with the high dimensionality of raw image data. In the next sections, we compare what is known about the neural basis of object processing and recognition in the ventral stream with Bayesian computer vision systems for recognizing objects in natural images.

The basic assumption is that the organization of cortical areas reflects the structure of natural images for efficient representation, learning, and inference.

Object Recognition and Ventral Stream Processing

A large body of results from primate neurophysiology and human neuroimaging is consistent with the idea that visual object processing is based on a hierarchical organization of stages through which image information is successively transformed from a high-dimensional set of local feature measurements with a small number of types (e.g., edges at many locations) to increasingly lower-dimensional representations of many types (e.g., dog, baseball player, . . .), and that this increase in selectivity is accompanied by increased invariance to illumination, translation, and scale (cf. DiCarlo, Zoccolan, & Rust, 2012; Grill-Spector & Malach, 2004; Kourtzi & Connor, 2011; Roe et al., 2012).

Further, activity in these cortical stages is modulated by context and task, suggesting computations that operate within and between these areas. However, from a computational perspective, there is no clear consensus as to exactly what information visual cortical areas represent, why they have the form they do, or what operations act on these representations. (For various theories and analyses, see DiCarlo, Zoccolan, & Rust, 2012; Epshtein, Lifshitz, & Ullman, 2008; Friston, 2005; Marr, 2010; Poggio, 2011; Ullman, 1995, 2007.)

Bayesian Systems for Object Recognition

From a Bayesian perspective, the structure of images can be formulated in terms of probability distributions over graphs. To do this, we assume the following:

1. The visual system's structure should rely on the compositional structure of natural images (Geman, Potter, & Chi, 2002; Jin & Geman, 2006; Yuille, 2011). Compositionality refers to the ability to construct hierarchical representations, whereby features/parts are used and shared to describe a virtually unlimited number of relational compositions. One argument is that without such a structure, we could not account for the speed with which humans can acquire and generalize visual knowledge.

2. Visual knowledge can be represented as a graphical structure in which nodes are random variables that represent hypotheses about features, parts, objects, and their relations, and the links capture the statistical dependencies between nodes. It is important to note that this assumption involves explicit representations with accessible semantic interpretations, not just sequential banks of spatial filters (see below). This seems to be necessary to perform a range of visual tasks at different levels of analysis and for the ability to transfer learning (e.g., not just detect objects like cats, but also locate their parts and spatial relations, and even

learn to recognize hybrid objects, e.g., which have cat torsos and dog legs).

3. The system should be organized to be able to detect conjunctions of features that belong together as part of an object while at the same time discounting, through disjunction, sources of variation, such as scale, illumination, position, and articulation. An illustration of this principle are AND/OR graphs (see below).

4. Inference and task flexibility is achieved by fixing values of nodes based on image measurements, priming, or attention, together with integrating out variables that are unimportant for a given task.

Motivated by stiff competitions within the computer vision community to build recognition systems that can handle the enormous variation of natural images (e.g., Everingham et al., 2006), algorithms have been developed that can do flexible inference as well as learn a generative structure. For a review of several Bayesian hierarchical composition models for recognition, see Zhu, Chen, and Yuille (2011). Next we describe an example to illustrate how knowledge is represented at different levels, how it performs different tasks, and how it can perform bottom-up and top-down inference. As with the primate ventral stream, higher level nodes represent hypotheses more coarsely, with increased specificity and invariance from bottom to top.

An Example of a Computer Model for Recognition

The task is to detect and parse a baseball player, given the naturally occurring wide range of variation in lighting, poses, and background clutter. Figure 88.4 illustrates a model of a baseball player formulated as a probabilistic model defined over an AND/OR graph (Zhu et al., 2010a). Selectivity and invariance are achieved through AND and OR operations. Although the logical operations are reminiscent of the "simple" and "complex" cell units in Riesenhuber and Poggio (1999), here the nodes represent random variables with semantic content and with associated distributions, and the links involve two-way interactions. The high-level nodes represent body parts such as head, torso, and feet—while the lower-level nodes represent the subparts. The high-level nodes/parts are either compositions of lower nodes/subparts or are choices between different alternative subparts. The compositions and choices correspond to logical AND and OR operations, respectively.

The representation of a baseball player requires specifying the state variables of all the nodes in the hierarchy. The high-level nodes only contain executive-level descriptions—for example, the center position, size, and orientation of the head—leaving more precise

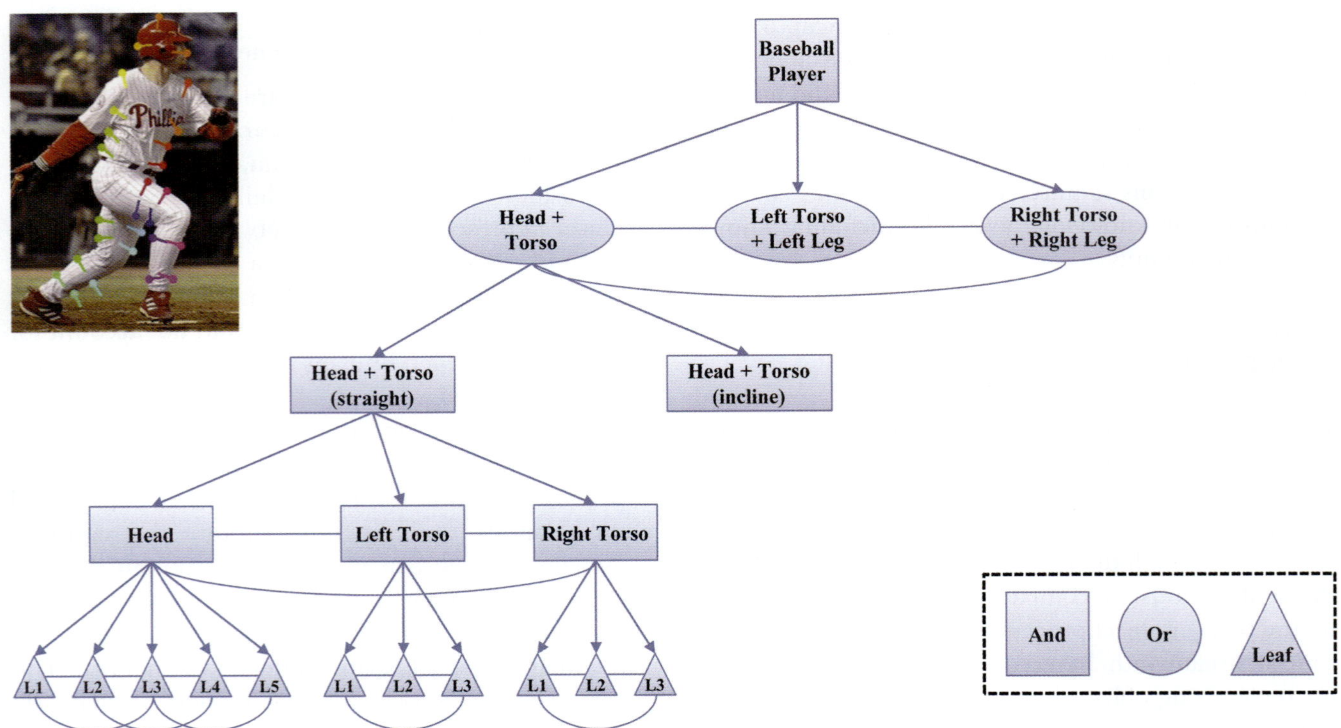

FIGURE 88.4   The AND/OR graph model (Zhu et al., 2010a). The baseball player is an AND of the head and torso, and left and right legs, but the head is an OR of straight head and torso or an inclined head and torso (top left). L, level.

information—for example, the precise position of the boundary of the head—to be represented by the states of the lower-level nodes. This is similar to the gradual loss of positional specificity as one moves up the ventral stream. The existence of some representation of spatial relations even at higher levels is consistent with behavioral and with functional magnetic resonance imaging (fMRI) studies (Kravitz, Kriegeskorte, & Baker, 2010).

The edges between the different graph nodes, and the local conditional probabilities defined over them (see description of Markov random fields [MRFs] in the section *From Edges to Surfaces*, figure 88.6), capture the statistical relations about the spatial configurations of the baseball player (i.e., the likely relative positions of the body parts, and which body parts are likely to occur taking into account viewpoint and pose changes).

Because the AND/OR graph is a probability model defined over a graph, one can perform inference to estimate the states of the unknown graph nodes, conditioned on the values of a subset of the nodes by message passing algorithms. This leads to a bottom-up and top-down strategy which is driven by input from the images (i.e., the model is conditioned on the states of the bottom levels). Hypotheses are computed at the lower levels and propagated up the hierarchy to form hypotheses for larger parts of the object, and top-down processing is used to remove the "false" hypotheses at lower levels. Intuitively there will be many possible hypotheses for the small subparts of the object since small features and parts of the object are easy to confuse with the background clutter. Compositions of subparts, however, are less likely to occur by chance in the background. Thus as the algorithm passes messages up the hierarchy it will tend to converge to the correct solution. Convergence speed depends on the reliability of the initial measurements.

One can also run the model in an attentional/priming mode where some of the top-level nodes are fixed (or "conditioned on"—that is, the system is primed to see a baseball player but does not know exactly where it is—while the bottom level nodes are also specified by the data (if we condition only on the top-level nodes, allowing no input from the image, then this is like imagining, or dreaming, a baseball player). This requires passing messages both top-down (from the primed nodes) and bottom-up (from the input nodes).

Learning Hierarchical Structure in Natural Images

While the brain clearly needs to be adaptive to image structures relevant for successful behaviors, the complexity of natural images suggests that part of the brain's solution involves the discovery of hierarchical structure in images themselves. A property of natural images is that intensities are to a first approximation, piece-wise smooth, so that one can predict pixel intensities from nearby pixels. Barlow argued that if image content is recoded to remove these and other higher-order statistical dependencies (through sparse coding), it becomes easier to compute useful information with probabilities (Barlow, 2001). Then one can detect "suspicious coincidences" ("Is $p(s_1, s_2) \gg p(s_1)p(s_2)$?") and learn to predict them so they become unsuspicious.

The principle of detecting suspicious coincidences provides the means to build up a hierarchical model of features, parts, objects, and scenes (Zhu et al., 2008).

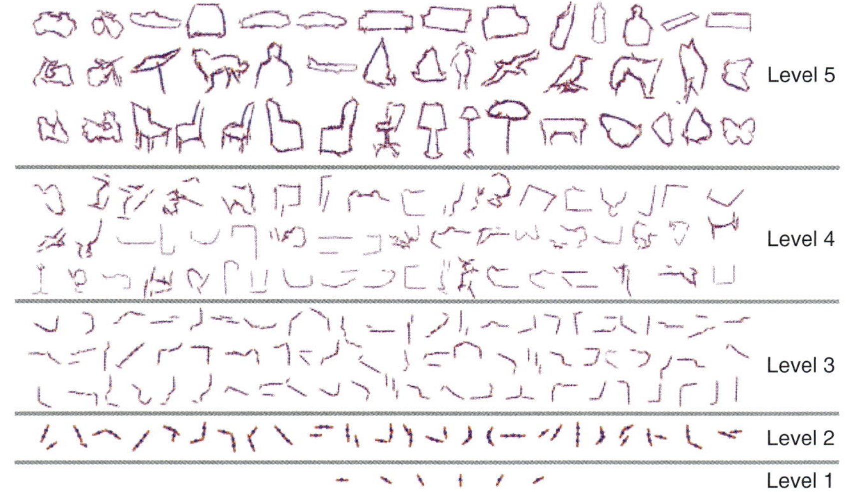

FIGURE 88.5 Examples of the mean shapes of visual concepts automatically learned for multiple objects with part sharing between objects (Zhu et al., 2010b). The first few levels (1–3) are generic, and higher levels (4 and 5) contain more specific concepts, or object parts, which occur for a limited number of objects. (Adapted from Zhu et al., 2010b.)

Learning relies on a compositional prior for hierarchical world structure in which an image can be composed from meaningful, "reusable" parts as in language (Geman, Potter, & Chi, 2002). A number of learning algorithms have been developed to learn hierarchical models from data (cf. Hinton, 2007; Zhu et al., 2010b; Zeiler, Taylor, & Fergus, 2011). We describe the results of one by Zhu et al. (2010b) that explicitly relies on the compositional nature of collections of natural objects.

Figure 88.5 shows the hierarchy of features, parts, and shapes that result from learning models for the small parts first and then combining them to learn models for the larger parts. Unlike other hierarchical models (e.g., deep-belief networks), the graph structure of the probability model is learned, not just the weights. This allows the model to adapt and transfer to novel stimuli. The algorithm can rapidly learn to recognize a new object if it can be constructed by combining parts that have already been learned for other objects.

What aspects of the above Bayesian models might be shared by the primate visual system? The large proportion of our knowledge of primate vision is early level, and the next section focuses on evidence that the visual system is well adapted to the generative structure of images, and that these early representations provide the basis for lateral and hierarchical probabilistic computations that begin at the bottom level of the visual hierarchy. This section is followed by a description of several behavioral and neuroimaging studies that support interactions between lower- and higher-level processing stages.

BEHAVIORAL AND NEURAL EVIDENCE FOR BAYESIAN COMPUTATIONS

Early Representation and Processing of Images

A common theme underlying explanations of neural population architecture in V1 has been in terms of efficient codes that exploit the redundancy or regularities in natural images. For example it has been shown that neural response properties, such as orientation and spatial frequency tuning in V1 neurons, may be accounted for in terms of a sparse coding strategy adapted to the statistics of natural images (Hyvärinen, 2010; Olshausen, 1996). Neurons in primary visual cortex also show nonlinear divisive-normalization behavior in which responses are inhibited by contrast variation outside the classical receptive field. Divisive normalization results in a reduction of statistical dependencies (Schwartz & Simoncelli, 2001), providing an efficient representation potentially useful for discovering suspicious coincidences (Barlow, 1990). Both sparse coding and contrast normalization have been explained in terms of an underlying generative model of natural images called a Gaussian scale mixture model (Wainwright & Simoncelli, 2000).

How do orientation- and spatial-frequency-selective spatial filter descriptions relate to visual behavior, to object perception, and to information and computations needed for higher-level areas? There are many answers to these questions depending on task, but the key step is to interpret neural population activity as representing a space of hypotheses about the causes of the stimulus (e.g., object boundary) rather than the stimulus itself (e.g., change in intensity). Here we focus on edges and surface regions, with a Bayesian view to understanding V1's role in object recognition.

From Local Spatial Filters to Edges

Hubel and Wiesel provided one clear functional interpretation of primary visual cortex (V1) – that a population of orientation-tuned simple and complex cells could represent an edge or line (Hubel, 1982). While there is wide consensus that information about orientation, as represented in neural populations of orientation-selective neurons, is a fundamental aspect of early visual spatial processing, exactly how such information is used to reliably determine object properties, such as contour boundaries, or internal textures is still a challenging problem.

Given certain assumptions (Poisson-like spiking behavior), the pattern of activity across a population of orientation-tuned neurons can be assumed to represent the likelihood of orientation and thus provide the basis for Bayesian decoding of edge orientation (Pouget, Dayan, & Zemel, 2000). However, determining whether such an edge is part of an object boundary requires additional inference. Natural images are full of local edges, many due to accidents of illumination, occlusion, and background clutter. Distinguishing those which belong together as part of an object of interest likely requires a combination of lateral (within a cortical area) and top-down (between cortical areas) interactions. Local computation could be based on natural, local smoothness constraints (i.e., priors) on object boundaries (Geisler & Perry, 2009) to link nearby edges of similar orientation (see figure 88.6) (for relationship to independent coding see Garrigues & Olshausen, 2008). However, several decades of computer vision studies have shown that local, lateral constraints are insufficient, and top-down processes that incorporate intermediate-level, gestalt constraints (parallels, symmetry), and object-specific information are required for robust segmentation (see the intermediate-level generic

     DANIEL KERSTEN AND ALAN YUILLE

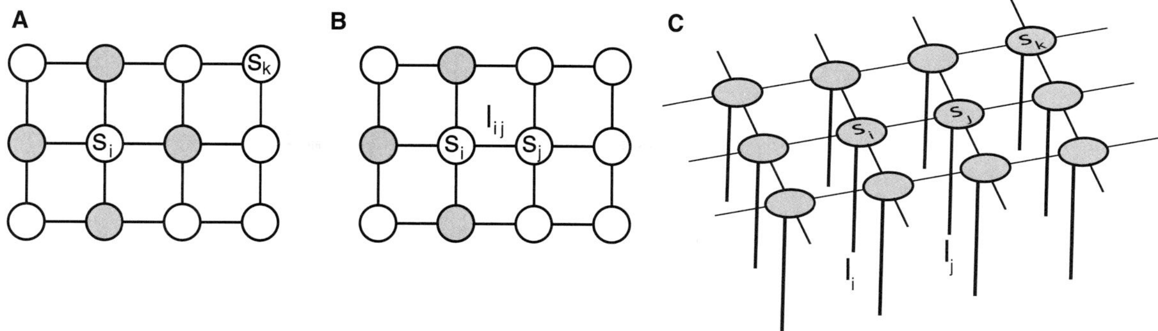

FIGURE 88.6   Markov random fields (MRFs) can be used to model smoothness and piece-wise smoothness in image intensities or intrinsic images. The nodes represent random variables and the links the relationships. (A) An undirected graph representing an MRF. The probability distribution of s_i depends only on variables in a predefined neighborhood (e.g., in this example, the four nearest neighbors, dashed nodes). Prior probabilities on smoothness and piece-wise smoothness constraints can be represented with this kind of graph. (B) An MRF with explicit line processes that when switched on break the neighborhood relations. (C) An MRF where the values to be estimated are linked to the image measurements, I_i.

and later shape-specific visual concepts in figure 88.5). Neurophysiological results are increasingly consistent with this view (cf. McManus, Li, & Gilbert, 2011). We return to the question of hierarchical representations of object knowledge, and how feedback may interact with edge representations, later.

FROM EDGES TO SURFACE REGIONS   The visual system is sensitive to shape as well as to surface properties of color, reflectance, and texture. An open question has been whether regional surface properties are represented in a dense retinotopic format. An image-like representation of surface properties, shape, and reflectance could support more robust matching of input to higher-level templates. Could such a representation exist in early topographic visual areas, such as V1 or V2?

Neuroimaging experiments have shown that human V1 responds to lightness change, a correlate of reflectance, almost as strongly as it does to luminance change, and in V2 just as strongly (Boyaci et al., 2007). Using the same kind of stimulus, work in anesthetized monkeys has found luminance responses to temporal changes in spatially uniform regions to neurons in cytochrome oxidase blobs of primary visual cortex (V1), and responses to illusory lightness in cells in the color-activated regions (thin stripes) of V2 (Roe et al., 2005).

To represent reflectance, however, the visual system would have to solve a nontrivial inference problem. The first is that reflectance, shape, and illumination (three "intrinsic images") are all confounded in the (luminance) image. Solving this problem requires prior knowledge of the spatial regularities of the intrinsic images. For example, reflectance could be roughly approximated as piece-wise constant, shape as piece-wise smooth, and illumination as smooth. One way to

characterize these properties is through measurements of statistics on intrinsic images (Grosse et al., 2009). One recent model factors out all three intrinsic images by using priors from objectively measured intrinsic images (Barron & Malik, 2012).

There is evidence for mechanisms that smoothly "fill-in" surface color between boundaries (Komatsu, 2006; Lee & Yuille, 2007), suggesting neural processes that assume smoothness to construct a map of surface color. Graph structures such as those illustrated in figure 88.6A can be used to describe smoothness priors on reflectance. Given some nodes initialized with reliable measurements (e.g., at edges), missing or noisy values can be smoothed out by interpolation. There are a number of algorithms to do this (cf. Gibbs sampling and Markov Chain Monte Carlo [MCMC] methods in MacKay, 2003). Edges can be made explicit as random variables, called "line-processes," which have their own 1-D smoothness priors, and when coupled with the regional surface process, serve to break neighborhood relations (see figure 88.6B and Koch, Marroquin, & Yuille, 1986; Kersten, 1991). Whether and how such algorithms are neurally implemented is an open question. Distinct populations for boundaries and regions together with linking mechanisms have been proposed (Grossberg & Hong, 2006; Roe et al., 2005); however further experiments are needed to determine if and how such neural populations interact within and across visual areas. MRFs can also be used to model the strength of spatial grouping of features and parts at more abstract levels in a visual hierarchy. This is discussed in a later section.

A final problem is that the perception of lightness is sensitive to 3-D layout of surfaces, (Gilchrist, 1977), suggesting that reflectance computations require nonlocal computations. For example, occlusion can interpose one surface over another, spatially separating image regions with the same underlying reflectance. There is evidence that, as early as V1, regions spatially disconnected by occlusion, with similar but different luminances, are grouped together as a surface with a common lightness (Boyaci et al., 2010). It is possible that both lateral MRF-like and top-down mechanisms are involved in early lightness computations, suggesting interactions in the cortical hierarchy.

Top-Down, Bottom-Up Interactions

What is the evidence for top-down/bottom-up Bayesian computations over hierarchical representations in human perception? Let's consider what this question might mean. An advantage of generative models is that they allow for flexible inference. Consider the three-level hierarchical graphical model shown in figure 88.7A. Even for this simple model, there are a large number of possible inferences depending on which variables are known and which are integrated out. For example, suppose the task requires estimating the top-level value m representing an object category. The MAP solution would be to find that value of m which maximizes $p(m \mid I_1, I_2)$. This could be achieved with a purely bottom-up algorithm with the middle variables, s_1, s_2, integrated out from the posterior, $p(s_1, s_2, m \mid I_1, I_2)$. Further, as noted earlier in the discussion of cue integration, a distinctive aspect of a Bayesian solution is to integrate the image measurements based on full knowledge of the uncertainty represented by the posterior. On the other hand, suppose the category m is known (from some other source of knowledge), and one wants to estimate the middle parameters, s_1, s_2, representing possible shapes within a specific category m'. The MAP solution would be to find s_1 and s_2 that maximize $p(s_1, s_2 \mid I_1, I_2, m')$. An algorithm could use generative knowledge, that $p(s_1, s_2 \mid I_1, I_2, m') \propto p(I_1, I_2 \mid s_1, s_2) p(s_1, s_2 \mid m')$. Note that the factors on the right-hand side use knowledge both of how the category influences shape, and of how shape influences the image measurements. Below, we discuss a related third possibility, "Bayesian coarse-to-fine," in which the first priority is to make an accurate high-level decision, followed by estimation of lower-level parameters. "Explaining-away" is another kind of inference which uses top-down generative knowledge, also discussed below (figure 88.3B).

Currently there is no direct evidence for neural populations representing hypotheses rather than decisions, or for probabilistic computations (as in message passing). However, there are behavioral and neuroimaging results that are suggestive of Bayesian computations at different levels of abstraction, and between cortical areas. We briefly mention some of them.

BAYESIAN COARSE-TO-FINE A hierarchical architecture allows for Bayesian coarse-to-fine processing in which an initial high-level inference, at a coarse level of abstraction, constrains a subsequent finer-grained analysis of lower-level features. It is well-established that

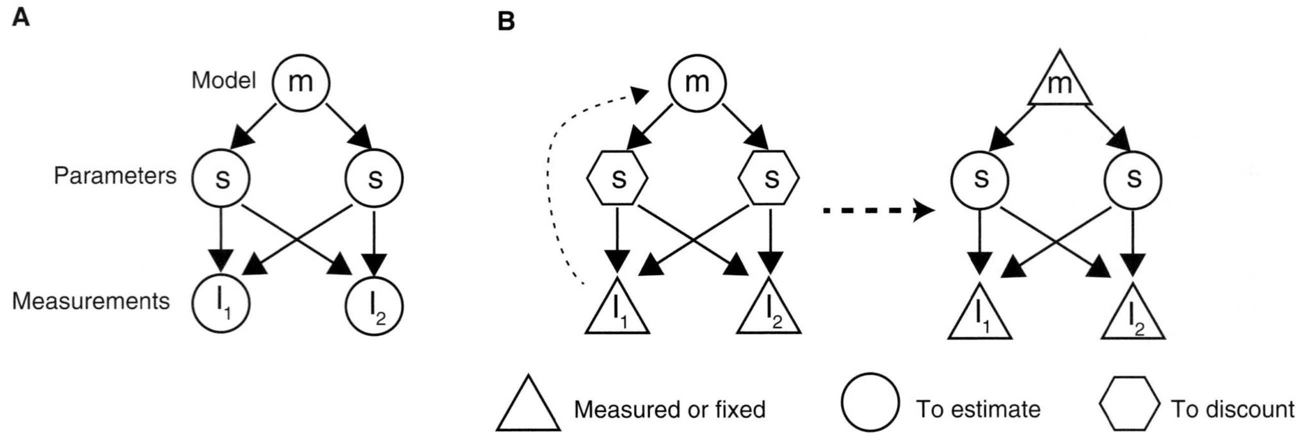

FIGURE 88.7   (A) A simple hierarchical graphical model. Values of a discrete random variable at the top level of the hierarchy represent "model" choices. Various models represent various categories of lower-level parameters. Depending on the task, some nodes are fixed (either through measurement at the lowest level or through prior decisions, attention, or priming at the highest, model level), others get integrated out, and others are estimated. For example, the network can be run in a generative fashion where samples, I, are drawn conditional on model choices, m. (B) "Conditioned perception" (Stocker & Simoncelli, 2008), is the particular sequence of marginalization where (1) intermediate-level parameters are integrated out and the high-level model values estimated (the dashed arrow shows the direction of inference), and then (2) the value of the inferred model is held fixed and the parameters are estimated.

certain visual decisions ("animal present or not?") can be made extremely fast (Rousselet, Thorpe, & Fabre-Thorpe, 2004). This is consistent with the above flexible Bayesian-style architecture, in which decisions are effectively confidence driven. For example in the AND/OR baseball player models, hypotheses can be rapidly propagated bottom-up, and if there is little uncertainty at the lower levels, there may be no need for top-down removal of false hypotheses. However, in the process the top-down information can help to specify the precise positions of the baseball player's boundaries (and subparts). More generally, Bayesian hierarchical models can allow reductions in uncertainty at a high level of abstraction to later affect decisions at lower levels.

Perceptual decisions can be automatic, driven by strong, reliable context evidence (either spatial information elsewhere in the image or temporal as in priming) or consciously task driven and specified by some even higher-level "executive." Both strategies are consistent with a coarse-to-fine computation in which a high-level decision "fixes" the value in the upper level of a hierarchical model, constraining or biasing subsequent lower-level decisions. An optimal decision restricted to a high level requires marginalization over intermediate-level parameters. We briefly describe several behavioral results which are consistent with Bayesian coarse-to-fine computations over a simple hierarchical graph structure (see figure 88.7).

Knill (2003) showed that, when estimating surface orientation from texture cues, the visual system uses different texture models depending on evidence in the image or nonlinearity weighting of cues. The model types, texture orientation parameters, and image measurements (cues) can be represented at the top, middle, and bottom levels of the hierarchy, respectively. The system usually interprets a 2-D texture as caused by an underlying isotropic 3-D texture. However, textures may also be anisotropic. Human surface judgments were well modeled by a Bayesian observer that uses a texture cue to make a coarse inference to decide the texture model type and then applies this model to interpret the texture cues. Other work has shown that humans infer causal structure (i.e., "Common or independent sources?") when integrating sound and visual stimuli to estimate direction, consistent with Bayesian estimation (Körding et al., 2007; Shams, 2010). Human perception of complex motion fields is also consistent with Bayesian selection of motion type (rotation, expansion, translation) influencing discrimination performance (Wu, Lu, & Yuille, 2008).

Human biases in motion direction estimation following a conscious classification decision (Jazayeri & Movshon, 2007) have been also been explained in terms of model selection (Stocker & Simoncelli, 2008), figure 88.7B.

TOP-DOWN PREDICTION Top-down hypotheses can compete to explain data. The Bayesian graphical interpretation is shown in figure 88.3B. A number of perceptual phenomena, in which contextual changes drastically change local visual interpretations, are consistent with "explaining away" (Kersten, Masmassian, & Yuille, 2004). There are several ways in which the representation of the probability distributions could change. In the earlier computer vision baseball example, top-down processes suppressed false hypotheses at a lower level, keeping those consistent with the larger picture emerging as a consequence of message passing. Another strategy is "predictive coding" in which s^{n+1} is a provisional guess at stage $n + 1$ that can use generative knowledge, f, to predict the input, s^n, of an earlier stage n. If the prediction error, $|s^n - f(s^{n+1})|$ is small (represented by low neural activity), the likelihood is high and the hypothesis is a good fit; if not, another hypothesis needs to be tried (Friston & Kiebel, 2009; Lee & Mumford, 2003; Rao & Ballard, 1999; Yuille & Kersten, 2006).

There is growing evidence from human fMRI studies for context-dependent suppression of neural activity in an earlier area in certain cases (Alink et al., 2010; Cardin, Friston, & Zeki, 2011; Fang, Kersten, & Murray, 2008; Murray et al., 2002; Rauss, Schwartz, & Pourtois, 2011; Summerfield et al., 2006). However, the spatial resolution of fMRI is too coarse to tell whether reduced activity may be due to suppression of false (Grossberg, 1999) or true hypotheses (Rao & Ballard, 1999). In fact, because of different functional requirements, the suppression of lower-level false and true hypotheses could both be occurring. A top-down attentional task might suppress competing lower-level ambiguous hypotheses in order to select true hypotheses relevant to the task (Grossberg, 2007). But the system may also rely on automatic processes to detect new information that does not fit with the current interpretation. The first strategy corresponds to executive, top-down ("endogenous") attention, and the second to stimulus-driven, bottom-up ("exogenous") attention (Desimone & Duncan, 1995; Petersen & Posner, 2011). Thus there may be two types of neural populations, those whose activity codes errors and those whose increased activity represents increased confidence in local feature hypotheses coded in that area (Friston, 2005). Enhancement of consistent features could be accompanied by suppression of false positives in one population but consistent features suppressed in another population. The laminar pattern of feedback connections raises the possibility that these representations may lie in distinct cortical layers. A recent study used ultrahigh field fMRI with submillimeter resolution and found stronger fMRI response in middle cortical layers of V1 during the presentation of scrambled objects as compared with intact objects (Olman et al., 2012), similar to what one might expect from error units.

Egner, Monti, and Summerfield (2010) proposed that distinct populations of "representational" units (feature detectors) and "error" units could be activated independently by behavioral manipulations of expectation and surprise, respectively. They manipulated expectation by training with cues that were diagnostic of the appearance of either a face or a house on a given trial. The subject's task was to detect an occasional and thus surprising, upside-down face or house. The authors were able to account for the pattern of results in human fusiform face area in terms of a sum of activity due to separate populations of units whose activity reflected expectation and surprise rather than by face-specific responses predicted by bottom-up selectivity alone.

CONCLUSIONS AND FUTURE DIRECTIONS

Although it was observed by Helmholtz more than 100 years ago that perception is a process of "unconscious inference," developing and testing quantitative models that embrace this idea has occurred only recently. From this body of work we have gained new insights about inferential processes occurring in perception and new ways of thinking about neural computation in the brain. However, achieving a full understanding of perception and its neural basis will remain an important and challenging problem for the future. A major aspect of the problem is that there is no consensus on what the overall neural or psychological constraints on a generative or inferential model should look like. We have presented the case that hierarchical, compositional models are necessary (Feldman, 2009; Geman, Potter, & Chi, 2002); however, it is unclear what the levels should represent (truth values vs. distributions on hypotheses) and what kinds of computations are done over them (decision sequences vs. propagating hypotheses and updating distributions). Variables in a level may represent higher-order image statistics, or they could (also) have accessible semantic interpretations, such as "edge," "curved line," "limbs," and "baseball player."

Models of image generation must eventually capture the flexibility of humans to interpret images that deviate substantially from natural, real-world experience. Human perception is almost never stumped—it always comes up with an interpretation, even if bizarre, wrong, or self-contradictory (consider a Dali painting, a cartoon, or a video game). While developing probabilistic models on graphical representations is an important direction, the need for an even richer structure has been recognized in proposals for image grammars (Mumford & Desolneux, 2010) and probabilistic programs (Goodman et al., 2008).

ACKNOWLEDGMENTS

Comments may be sent to the author at kersten@umn.edu. D.K. and A.Y. were supported by the WCU (World Class University) program funded by the Ministry of Education, Science and Technology through the National Research Foundation of Korea (R31–10008).

NOTES

1. For a general introduction to Bayesian inference in cognitive science, see Griffiths and Yuille (2008). For Bayesian models applied to ideal observer psychophysics and natural systems analysis, see Geisler (2008, 2011). For an overview and critique applied to cognitive neuroscience, see Colombo and Seriès (2012). For Bayesian decision theory, see Körding (2007) and Maloney and Mamassian (2009), and for optimal learning, Fiser et al. (2010). For an earlier review of object perception, see Kersten, Masmassian, and Yuille (2004). For examples of applications of ideal observers to learning, attention, and letter detection, see Trenti, Barraza, and Eckstein (2010), Eckstein, Drescher, and Shimozaki (2006), Pelli et al. (2006), to saliency and eye movements, Chikkerur et al. (2010), Itti and Baldi (2009), Torralba et al. (2006), Zhang et al. (2008), to contour and shape, Feldman (2001), Feldman and Singh (2005), Wilder, Feldman, and Singh (2011), to filling-in, Zhaoping and Jingling (2008), and to cross-modal interactions, Battaglia, Kersten, and Schrater (2011), Körding et al. (2007), Shams (2010). For neural coding, see Pouget, Dayan, and Zemel (2000), Knill and Pouget (2004), Fiser et al. (2010), Ma (2010), Berkes et al. (2011).
2. The term "hidden" emphasizes the fact that the values of states of the world are effectively encrypted and need to be decoded. This is the central mystery of perception. The shape of an object may appear to be a direct measurement (and at some higher level could be treated as such), but computational theory and experiments have shown that inferring shape from image patterns requires inferential processes.
3. Decisions based on $p(s \mid I)$ in the absence of a generative model are called discriminative. The distinction between discriminative and generative models is related to the distinction made between policy- and model-based learning in Bayesian reinforcement learning theory (Strens, 2000).
4. Generative means that, in principle, one can generate observables I by first drawing a sample from $p(s)$, call it s', and then drawing a sample of I from $p(I \mid s')$.

REFERENCES

Alink, A., Schwiedrzik, C. M., Kohler, A., Singer, W., & Muckli, L. (2010). Stimulus predictability reduces responses in primary visual cortex. *Journal of Neuroscience, 30*, 2960–2966.

Barlow, H. (1990). Conditions for versatile learning, Helmholtz's unconscious inference, and the task of perception. *Vision Research, 30,* 1561–1571.

Barlow, H. (2001). Redundancy reduction revisited. *Network: Computation in Neural Systems, 12,* 241–253. doi:10.1080/net.12.3.241.253.

Barron, J. T., & Malik, J. (2012). Shape, albedo, and illumination from a single image of an unknown object. *Computer Vision and Pattern Recognition (CVPR), 2012 IEEE Conference on,* 334–341.

Battaglia, P. W., Kersten, D., & Schrater, P. R. (2011). How haptic size sensations improve distance perception. *PLoS Computational Biology, 7,* e1002080. doi:10.1371/journal.pcbi.1002080.

Battaglia, P. W., & Schrater, P. (2007). Humans trade off viewing time and movement duration to improve visuomotor accuracy in a fast reaching task. *Journal of Neuroscience, 27,* 6984–6994.

Beck, J. M., Latham, P. E., & Pouget, A. (2011). Marginalization in neural circuits with divisive normalization. *Journal of Neuroscience, 31,* 15310–15319.

Berkes, P., Orb'an, G., Lengyel, M., & Fiser, J. (2011). Spontaneous cortical activity reveals hallmarks of an optimal internal model of the environment. *Science, 331,* 83–87.

Boyaci, H., Fang, F., Murray, S. O., & Kersten, D. (2007). Responses to lightness variations in early human visual cortex. *Current Biology, 17,* 989–993.

Boyaci, H., Fang, F., Murray, S. O., & Kersten, D. (2010). Perceptual grouping-dependent lightness processing in human early visual cortex. *Journal of Vision, 10*(9), 1–12. doi:10.1167/10.9.4.

Carandini, M., & Heeger, D. J. (2011). Normalization as a canonical neural computation. *Nature Reviews Neuroscience, 13,* 51–62. doi:10.1038/nrn3136.

Cardin, V., Friston, K. J., & Zeki, S. (2011). Top-down modulations in the visual form pathway revealed with dynamic causal modeling. *Cerebral Cortex, 21,* 550–562.

Chater, N., Tenenbaum, J. B., & Yuille, A. (2006). Probabilistic models of cognition: Conceptual foundations. *Trends in Cognitive Sciences, 10,* 287–291. doi:10.1016/j.tics.2006.05.007.

Cheng, K., Shettleworth, S. J., Huttenlocher, J., & Rieser, J. J. (2007). Bayesian integration of spatial information. *Psychological Bulletin, 133,* 625–637.

Chikkerur, S., Serre, T., Tan, C., & Poggio, T. (2010). What and where: A Bayesian inference theory of attention. *Vision Research, 50,* 2233–2247.

Clark, J. J., & Yuille, A. L. (1990). *Data fusion for sensory information processing systems. Springer.* Norwell, MA: Kluwer Academic Publishers.

Colombo, M., & Seriès, P. (2012). Bayes in the brain—on Bayesian modelling in neuroscience. *British Journal for the Philosophy of Science,* 1–27. doi:10.1093/bjps/axr043.

Desimone, R., & Duncan, J. (1995). Neural mechanisms of selective visual attention. *Annual Review of Neuroscience, 18,* 193–222.

DiCarlo, J. J., Zoccolan, D., & Rust, N. C. (2012). How does the brain solve visual object recognition? *Neuron, 73,* 415–434.

Eckstein, M. P., Drescher, B., & Shimozaki, S. S. (2006). Attentional cues in real scenes, saccadic targeting, and Bayesian priors. *Psychological Science, 17,* 973.

Egner, T., Monti, J. M., & Summerfield, C. (2010). Expectation and surprise determine neural population responses in the ventral visual stream. *Journal of Neuroscience, 30,* 16601–16608.

Epshtein, B., Lifshitz, I., & Ullman, S. (2008). Image interpretation by a single bottom-up top-down cycle. *Proceedings of the National Academy of Sciences of the United States of America, 105,* 14298. doi:10.1073/pnas.0800968105.

Ernst, M. O., & Banks, M. S. (2002). Humans integrate visual and haptic information in a statistically optimal fashion. *Nature, 415,* 429–433.

Ernst, M. O., & Bülthoff, H. H. (2004). Merging the senses into a robust percept. *Trends in Cognitive Sciences, 8,* 162–169. doi:10.1016/j.tics.2004.02.002.

Everingham, M., Zisserman, A., Williams, C., Van Gool, L., Allan, M., Bishop, C. M., et al. (2006). The 2005 PASCAL visual object classes challenge. Machine learning challenges: Evaluating predictive uncertainty, visual object classification, and recognising textual entailment. *Lecture Notes in Computer Science, 3944,* 117–176. doi:10.1007/11736790_8.

Fang, F., & He, S. (2005). Cortical responses to invisible objects in the human dorsal and ventral pathways. *Nature Neuroscience, 8,* 1380–1385.

Fang, F., Kersten, D., & Murray, S. O. (2008). Perceptual grouping and inverse fMRI activity patterns in human visual cortex. *Journal of Vision, 8* (7), 2, 1–9. doi:10.1167/8.7.2.

Feldman, J. (2001). Bayesian contour integration. *Attention, Perception & Psychophysics, 63,* 1171–1182.

Feldman, J. (2009). Bayes and the simplicity principle in perception. *Psychological Review, 116,* 875–887.

Feldman, J., & Singh, M. (2005). Information along contours and object boundaries. *Psychological Review, 112,* 243.

Fiser, J., Berkes, P., Orbán, G., & Lengyel, M. (2010). Statistically optimal perception and learning: From behavior to neural representations. *Trends in Cognitive Sciences, 14,* 119–130. doi:10.1016/j.tics.2010.01.003.

Franklin, D. W., & Wolpert, D. M. (2011). Computational mechanisms of sensorimotor control. *Neuron, 72,* 425–442.

Freeman, W. (1994). The generic viewpoint assumption in a framework for visual perception. *Nature, 368,* 542–545.

Friston, K. (2005). A theory of cortical responses. *Philosophical Transactions of the Royal Society B. Biological Sciences, 360,* 815–836. doi:10.1098/rstb.2005.1622.

Friston, K., & Kiebel, S. (2009). Predictive coding under the free-energy principle. *Philosophical Transactions of the Royal Society B. Biological Sciences, 364,* 1211–1221. doi:10.1098/rstb.2008.0300.

Garrigues, P., & Olshausen, B. A. (2008). Learning horizontal connections in a sparse coding model of natural images. *Advances in Neural Information Processing Systems, 20,* 505–512.

Geisler, W. S. (2008). Visual perception and the statistical properties of natural scenes. *Annual Review of Psychology, 59,* 167–192.

Geisler, W. S. (2011). Contributions of ideal observer theory to vision research. *Vision Research, 51,* 771–781.

Geisler, W. S., & Kersten, D. (2002). Illusions, perception and Bayes. *Nature Neuroscience, 5,* 508–510.

Geisler, W. S., & Perry, J. (2009). Contour statistics in natural images: Grouping across occlusions. *Visual Neuroscience, 26,* 109–121.

Geman, S., Potter, D., & Chi, Z. (2002). Composition systems. *Quarterly of Applied Mathematics, 60,* 707–736.

Gershman, S., Vul, E., & Tenenbaum, J. B. (2012). Multistability and perceptual inference. *Neural Computation, 24*, 1–24. doi:10.1162/NECO_a_00226.

Gilchrist, A. L. (1977). Perceived lightness depends on perceived spatial arrangement. *Science, 195*, 185–187.

Goodman, N., Mansinghka, V., Roy, D., Bonawitz, K., & Tenenbaum, J. B. (2008). Church: A language for generative models. *Uncertainty in Artificial Intelligence, 22*, 23.

Gori, M., Del Viva, M., Sandini, G., & Burr, D. C. (2008). Young children do not integrate visual and haptic form information. *Current Biology, 18*, 694–698.

Green, D. M., & Swets, J. A. (1966). *Signal detection theory and psychophysics.* New York: Wiley.

Griffiths, T., Chater, N., Kemp, C., Perfors, A., & Tenenbaum, J. B. (2010). Probabilistic models of cognition: Exploring representations and inductive biases. *Trends in Cognitive Sciences, 14*, 357–364. doi:10.1016/j.tics.2010.05.004.

Griffiths, T., & Yuille, A. (2008). A primer on probabilistic inference. In N. Chater & M. Oaksford (Eds.), *The probabilistic mind: Prospects for Bayesian cognitive science* (pp. 33–57). Oxford: Oxford University Press. doi:10.1093/acprof:oso/9780199216093.003.0002.

Grill-Spector, K., & Malach, R. (2004). The human visual cortex. *Annual Review of Neuroscience, 27*, 649–677.

Grossberg, S. (1999). The link between brain learning, attention, and consciousness. *Consciousness and Cognition, 8*, 1–44. doi:10.1006/ccog.1998.0372.

Grossberg, S. (2007). Towards a unified theory of neocortex: Laminar cortical circuits for vision and cognition. *Progress in Brain Research, 165*, 79–104.

Grossberg, S., & Hong, S. (2006). A neural model of surface perception: Lightness, anchoring, and filling-in. *Spatial Vision, 19*, 263–321.

Grosse, R., Johnson, M. K., Adelson, E. H., & Freeman, W. T. (2009). Ground-truth dataset and baseline evaluations for intrinsic image algorithms. *International Conference on Computer Vision*, 2335–2342.

Hinton, G. E. (2007). Learning multiple layers of representation. *Trends in Cognitive Sciences, 11*, 428–434.

Hubel, D. (1982). Evolution of ideas on the primary visual cortex, 1955–1978: A biased historical account. *Bioscience Reports, 2*, 435–469.

Hyvärinen, A. (2010). Statistical models of natural images and cortical visual representation. *Topics in Cognitive Science, 2*, 251–264. doi:10.1111/j.1756-8765.2009.01057.x.

Itti, L., & Baldi, P. (2009). Bayesian surprise attracts human attention. *Vision Research, 49*, 1295–1306.

Jacobs, R. (1999). Optimal integration of texture and motion cues to depth. *Vision Research, 39*, 3621–3629.

Jazayeri, M., & Movshon, J. A. (2007). A new perceptual illusion reveals mechanisms of sensory decoding. *Nature, 446*, 912–915.

Jin, Y., & Geman, S. (2006). Context and hierarchy in a probabilistic image model. *IEEE Computer Society Conference on Computer Vision and Pattern Recognition, 2*, 2145–2152.

Kersten, D. (1991). Transparency and the cooperative computation of scene attributes. In M. S. Landy (Ed.), *Computational models of visual processing* (pp. 209–228). Cambridge, MA: MIT Press.

Kersten, D. (1999). High-level vision as statistical inference. In M. Gazzaniga (Ed.), *The new cognitive neurosciences* (pp. 353–364). Cambridge, MA: MIT Press.

Kersten, D., Masmassian, P., & Yuille, A. (2004). Object perception as Bayesian inference. *Annual Review of Psychology, 55*, 271–304.

Kersten, D., & Schrater, P. (2002). Pattern inference theory: A probabilistic approach to vision. In R. Mausfeld & D. Heyer (Eds.), *Perception and the physical world* (pp. 191–228). Chichester, England: Wiley.

Knill, D. C. (2003). Mixture models and the probabilistic structure of depth cues. *Vision Research, 43*, 831–854.

Knill, D. C., & Pouget, A. (2004). The Bayesian brain: The role of uncertainty in neural coding and computation. *Trends in Neurosciences, 27*, 712–719. doi:10.1016/j.tins.2004.10.007.

Koch, C., Marroquin, J., & Yuille, A. (1986). Analog "neuronal" networks in early vision. *Proceedings of the National Academy of Sciences of the United States of America, 83*, 4263–4267. doi:10.1073/pnas.83.12.4263.

Komatsu, H. (2006). The neural mechanisms of perceptual filling-in. *Nature Reviews Neuroscience, 7*, 220–231.

Körding, K. P. (2007). Decision theory: What "should" the nervous system do? *Science, 318*, 606–610.

Körding, K. P., Beierholm, U., Ma, W. J., Quartz, S., Tenenbaum, J. B., & Shams, L. (2007). Causal inference in multisensory perception. *PLoS ONE, 2*, e943. doi:10.1371/journal.pone.0000943.

Körding, K. P., & Wolpert, D. M. (2006). Bayesian decision theory in sensorimotor control. *Trends in Cognitive Sciences, 10*, 319–326. doi:10.1016/j.tics.2006.05.003.

Kourtzi, Z., & Connor, C. E. (2011). Neural representations for object perception: Structure, category, and adaptive coding. *Annual Review of Neuroscience, 34*, 45–67.

Kravitz, D. J., Kriegeskorte, N., & Baker, C. I. (2010). High-level visual object representations are constrained by position. *Cerebral Cortex, 20*, 2916–2925.

Lee, T., & Mumford, D. (2003). Hierarchical Bayesian inference in the visual cortex. *Journal of the Optical Society of America. A, Optics, Image Science, and Vision, 20*, 1434–1448.

Lee, T., & Yuille, A. (2007). *Efficient coding of visual scenes by grouping and segmentation.* In D. Kenji, S. Ishii, A. Pouget, & R. P. N. Rao (Eds.), *Bayesian brain: Probabilistic approaches to neural coding* (pp. 145–188). Cambridge, MA: MIT Press.

Lennie, P. (1998). Single units and visual cortical organization. *Perception, 27*, 889–936.

Lowe, D. (1987). The viewpoint consistency constraint. *International Journal of Computer Vision, 1*, 57–72.

Ma, W. J. (2010). Signal detection theory, uncertainty, and Poisson-like population codes. *Vision Research, 50*, 2308–2319.

Ma, W. J., Beck, J. M., Latham, P. E., & Pouget, A. (2006). Bayesian inference with probabilistic population codes. *Nature Neuroscience, 9*, 1432–1438.

MacKay, D. J. C. (2003). *Information theory, inference, and learning algorithms.* Cambridge, England: Cambridge University Press.

Maloney, L. T., & Mamassian, P. (2009). Bayesian decision theory as a model of human visual perception: Testing Bayesian transfer. *Visual Neuroscience, 26*, 147.

Marr, D. (2010). *Vision: A computational investigation into the human representation and processing of visual information.* Cambridge, MA: MIT Press.

McManus, J. N. J., Li, W., & Gilbert, C. D. (2011). Adaptive shape processing in primary visual cortex. *Proceedings of the*

National Academy of Sciences of the United States of America, 108, 9739–9746. doi:10.1073/pnas.1105855108.

Milner, D., & Goodale, M. (2006). *The visual brain in action* (2nd ed.). New York: Oxford University Press.

Moreno-Bote, R., Knill, D. C., & Pouget, A. (2011). Bayesian sampling in visual perception. *Proceedings of the National Academy of Sciences of the United States of America, 108,* 12491–12496. doi:10.1073/pnas.1101430108.

Mumford, D., & Desolneux, A. (2010). *Pattern theory: The stochastic analysis of real-world signals.* Natick, MA: Peters.

Murray, S. O., Kersten, D., Olshausen, B. A., Schrater, P., & Woods, D. L. (2002). Shape perception reduces activity in human primary visual cortex. *Proceedings of the National Academy of Sciences of the United States of America, 99,* 15164–15169. doi:10.1073/pnas.192579399.

Nakayama, K., & Shimojo, S. (1992). Experiencing and perceiving visual surfaces. *Science, 257,* 1357–1363.

Olman, C. A., Harel, N., Feinberg, D. A., He, S., Zhang, P., Ugurbil, K., et al. (2012). Layer-specific fMRI reflects different neuronal computations at different depths in human V1. *PLoS ONE, 7,* e32536. doi:10.1371/journal.pone.0032536.

Olshausen, B. A. (1996). Emergence of simple-cell receptive field properties by learning a sparse code for natural images. *Nature, 381,* 607–609.

Pearl, J. (1988). *Probabilistic reasoning in intelligent systems: networks of plausible inference* (1st ed.). Morgan Kaufmann.

Pelli, D. G., Burns, C. W., Farell, B., & Moore-Page, D. C. (2006). Feature detection and letter identification. *Vision Research, 46,* 4646–4674.

Petersen, S. E., & Posner, M. I. (2011). The attention system of the human brain: 20 years after. *Annual Review of Neuroscience, 35,* 73-89. doi:10.1146/annurev-neuro-062111-150525.

Poggio, T. (2011). *The computational magic of the ventral stream: Towards a theory.* Nature Proceedings.

Pouget, A., Dayan, P., & Zemel, R. (2000). Information processing with population codes. *Nature Reviews Neuroscience, 1,* 125. doi:10.1038/35039062.

Rao, R. P. N., & Ballard, D. (1999). Predictive coding in the visual cortex: A functional interpretation of some extra-classical receptive-field effects. *Nature Neuroscience, 2,* 79–87.

Rauss, K., Schwartz, S., & Pourtois, G. (2011). Top-down effects on early visual processing in humans: A predictive coding framework. *Neuroscience and Biobehavioral Reviews, 35,* 1237–1253.

Reynolds, J. H., & Heeger, D. J. (2009). The normalization model of attention. *Neuron, 61,* 168–185.

Riesenhuber, M., & Poggio, T. (1999). Hierarchical models of object recognition in cortex. *Nature Neuroscience, 2,* 1019–1025.

Roe, A. W., Chelazzi, L., Connor, C. E., Conway, B. R., Fujita, I., Gallant, J. L., et al. (2012). Toward a unified theory of visual area V4. *Neuron, 74,* 12–29. doi:10.1016/j.neuron.2012.03.011.

Roe, A. W., Lu, H. D., Hung, C. P., & Kaas, J. H. (2005). Cortical processing of a brightness illusion. *Proceedings of the National Academy of Sciences of the United States of America, 102,* 3869–3874. doi:10.1073/pnas.0500097102.

Rousselet, G. A., Thorpe, S. J., & Fabre-Thorpe, M. (2004). How parallel is visual processing in the ventral pathway? *Trends in Cognitive Sciences, 8,* 363–370. doi:10.1016/j.tics.2004.06.003.

Schlicht, E. J., & Schrater, P. R. (2007). Impact of coordinate transformation uncertainty on human sensorimotor control. *Journal of Neurophysiology, 97,* 4203–4214.

Schwartz, O., & Simoncelli, E. (2001). Natural signal statistics and sensory gain control. *Nature Neuroscience, 4,* 819–825.

Shams, L. (2010). Probability matching as a computational strategy used in perception. *PLoS Computational Biology, 6,* e1000871. doi:10.1371/journal.pcbi.1000871.

Stevenson, I., & Körding, K. P. (2009). Structural inference affects depth perception in the context of potential occlusion. In Y. Bengio, D. Schuurmans, J. Lafferty, C. K. I. Williams and A. Culotta (Eds.), *Advances in neural informational processing systems* (Vol. 22, pp. 1777–1784). La Jolla, CA: NIPS Foundation.

Stocker, A. A., & Simoncelli, E. (2008). A Bayesian model of conditioned perception. In J.C. Platt, D. Koller, Y. Singer and S. Roweis (Eds.), *Advances in neural informational processing systems* (Vol. 20, pp. 1409–1416). Cambridge, MA: MIT Press.

Strens, M. (2000). A Bayesian framework for reinforcement learning. *Proceedings of the Seventeenth International Conference on Machine Learning (ICML-2000),* 943–950.

Summerfield, C., Egner, T., Greene, M., Koechlin, E., Mangels, J., & Hirsch, J. (2006). Predictive codes for forthcoming perception in the frontal cortex. *Science, 314,* 1311–1314.

Sundareswara, R., & Schrater, P. (2008). Perceptual multistability predicted by search model for Bayesian decisions. *Journal of Vision, 8*(5), 12. doi:10.1167/8.5.12.

Torralba, A., Oliva, A., Castelhano, M. S., & Henderson, J. M. (2006). Contextual guidance of eye movements and attention in real-world scenes: The role of global features in object search. *Psychological Review, 113,* 766–786.

Trenti, E. J., Barraza, J. F., & Eckstein, M. P. (2010). Learning motion: Human vs. optimal Bayesian learner. *Vision Research, 50,* 460–472.

Ullman, S. (1995). Sequence seeking and counter streams: A computational model for bidirectional information flow in the visual cortex. *Cerebral Cortex, 5,* 1–11. doi:10.1093/cercor/5.1.1.

Ullman, S. (2000). *High-level vision: Object recognition and visual cognition.* Cambridge, MA: MIT Press.

Ullman, S. (2007). Object recognition and segmentation by a fragment-based hierarchy. *Trends in Cognitive Sciences, 11,* 58–64. doi:10.1016/j.tics.2006.11.009.

Wainwright, M. J., & Simoncelli, E. (2000). Scale mixtures of Gaussians and the statistics of natural images. In D. Koller, D. Schuurmans, & Y. B. L. Bottou (Eds.), *Advances in neural information processing systems* (Vol. 12, pp. 855–861). Cambridge, MA: MIT Press.

Wilder, J., Feldman, J., & Singh, M. (2011). Superordinate shape classification using natural shape statistics. *Cognition, 119,* 325–340.

Wu, S., Lu, H., & Yuille, A. (2009). Model selection and velocity estimation using novel priors for motion patterns. In D. Koller, D. Schuurmans, & Y. B. L. Bottou (Eds.), *Advances in neural information processing systems* (Vol. 21, pp. 1793–1800). Cambridge, MA: MIT Press.

Yuille, A. (2011). Towards a theory of compositional learning and encoding of objects. *IEEE International Conference on Computer Vision Workshops (ICCV Workshops),* 1448–1455.

Yuille, A., & Kersten, D. (2006). Vision as Bayesian inference: Analysis by synthesis? *Trends in Cognitive Sciences, 10,* 301–308. doi:10.1016/j.tics.2006.05.002.

Zeiler, M., Taylor, G., & Fergus, R. (2011). Adaptive deconvolutional networks for mid and high level feature learning. *IEEE International Conference on Computer Vision (ICCV)*, 2018–2025.

Zhang, L., Tong, M. H., Marks, T. K., Shan, H., & Cottrell, G. W. (2008). SUN: A Bayesian framework for saliency using natural statistics. *Journal of Vision, 8*(7), 32. doi:10.1167/8.7.32.

Zhaoping, L., & Jingling, L. (2008). Filling-in and suppression of visual perception from context: A Bayesian account of perceptual biases by contextual influences. *PLoS Computational Biology, 4*, e14. doi:10.1371/journal.pcbi.0040014.

Zhu, L., Chen, Y., Lin, C., & Yuille, A. (2010a). Max margin learning of hierarchical configural deformable templates (HCDTs) for efficient object parsing and pose estimation. *International Journal of Computer Vision, 93*, 1–21. doi:10.1007/s11263-010-0375-1.

Zhu, L., Chen, Y., Torralba, A., Freeman, W., & Yuille, A. (2010b). Part and appearance sharing: Recursive compositional models for multi-view multi-object detection. *IEEE Computer Society Conference on Computer Vision and Pattern Recognition*, 1919–1926.

Zhu, L., Chen, Y., & Yuille, A. (2011). Recursive compositional models for vision: Description and review of recent work. *Journal of Mathematical Imaging and Vision, 41*, 122–146.

Zhu, L., Lin, C., Huang, H., Chen, Y., & Yuille, A. (2008). *Unsupervised structure learning: Hierarchical recursive composition, suspicious coincidence and competitive exclusion.* Los Angeles: Department of Statistics, UCLA.

89 Neural Oscillations and Synchrony as Mechanisms for Coding, Communication, and Computation in the Visual System

FRIEDRICH T. SOMMER

Oscillatory structure in the mass activity of neurons is prevalent throughout the nervous system and across a wide variety of species. The phenomenon was already discovered in the nineteenth century by neurophysiologists who recorded from the exposed brain with mirror galvanometers. The recorded rhythms were unrelated to structure in the stimuli, as well as to heart or breathing rhythms, and thus correctly identified as an intrinsic feature of brain activity (Caton, 1875). Beck (1890) investigated the visual system by recording in occipital areas of rabbits and dogs during visual stimulation. He reported visually evoked potentials as well as ongoing oscillatory signals that could be suppressed by the arrival of stimuli, a phenomenon that is now referred to as stimulus-dependent desynchronization (Zayachkivska, Gzhegotsky, & Coenen, 2011). While the functional interpretation was straightforward for the stimulus-evoked potentials, it remained elusive for the oscillatory signals.

In the 1950s and 1960s, single-cell physiology in visual areas provided a more detailed picture of stimulus-evoked and rhythmic activity. Stimulus-dependent spike rate changes revealed the specific responses to localized visual stimuli (Hubel & Wiesel, 1962; Kuffler, 1953) in retina, lateral geniculate nucleus, and primary visual cortex. The experimental findings in primary visual cortex led to a powerful conceptual model of the two main types of excitatory cells involved in visual coding, simple and complex cells. In this model, a simple cell receives direct thalamic input and responds selectively to a conjunction of active inputs, thereby detecting specific features, such as localized, oriented edges. In contrast, a complex cell pools over several simple cells that coincide in certain features, such as a specific orientation, but differ in other features, such as location. The response of a complex cell is a disjunctive combination of its inputs. Specifically, the cell fires if any of the simple cells it is connected to is active, thereby signaling a specific orientation, somewhat independent of its exact location (or spatial phase) (Hubel & Wiesel, 1962). This conceptual model of consecutive layers of simple and complex cells within a cortical area has led to a canonical hierarchical model of visual processing (Fukushima, 1980; Riesenhuber & Poggio, 1999b; Serre, Oliva, & Poggio, 2007a) that will be important in what follows. The model describes the visual system as a feedforward cascade of processing modules, each consisting of simple-cell-like units that feed into complex-cell-like units. The idea is that cascading conjunctive and disjunctive feature combinations could produce specific yet invariant visual representations suited for object recognition and other functions performed by biological visual systems.

Single-cell physiology in early stages of the visual system also revealed that spike trains often exhibit peaks in the autocorrelation function, in the interspike-interval histogram, or in the Fourier transform, even without stimulation. These peaks reflect a dominant periodicity in the spiking—a hallmark of neural oscillations. For example, in ganglion cells, oscillatory activity is found in both anesthetized (Laufer & Verzeano, 1967; Ogawa, Bishop, & Levick, 1966; Rodieck, 1967) and unanesthetized preparations (Heiss & Bornschein, 1966; Steinberg, 1966). While anesthesia can increase neural oscillations, the presence of oscillations in the awake brain suggests that they could serve a function in vision. This intrinsic organization discovered in neural activity raised the question of its purpose and function. However, it was not until the mid-seventies that concrete hypotheses about the function of intrinsically paced periodic firing of neurons were proposed.

CORRELATION THEORY OF BRAIN FUNCTION

Based on earlier ideas of Milner (1974) and Grossberg (1976), Christoph von der Malsburg (1981) published a technical report entitled "Correlation Theory of Brain Function," possibly one of the highest impact technical reports in scientific history. The report started by identifying potential problems with the canonical hierarchical model of visual processing. Two problems were highlighted that hamper the model to reproduce the capabilities of biological visual systems. First, it cannot explain how the brain solves the problem of identifying objects in images. If one stage in the canonical model extracts some set of features, the next higher stage loses access to the relative spatial relationships or context between features that would be critical to identify objects. For example, if the lower stage extracts local edge and color features in a picture of a red triangle and a green square, the higher stage cannot access the information that the triangle was red and not the square. This problem of visual feature binding is one instance of the more general neural binding problem (Feldman, 2013).

A second problem of the canonical hierarchical model lies in supporting invariant object recognition; that is, the inability to produce representations unique to an object but independent of incidental variables, such as position, pose, or scale. As a consequence of the pooling in complex cells, their response is somewhat independent from the exact position of the edge structure. For instance, if the stimulus is the character L, its representation, produced by complex cells with vertical and horizontal orientation, is invariant under small shifts of the L. However, the set of complex cells that represent L's will also be activated by other shapes with vertical and horizontal edges, such as a letter T and other combinations of a vertical and a horizontal edge. Thus, representations in the canonical model are invariant but not unique because, like above, context is discarded, in this case the exact relative positions of the edge features.

Von der Malsburg's report proceeded by sketching how the correlation theory offers a solution to these problems. The pivotal proposition of the theory is that the intrinsic structure of neural signals is not noise but essential for capturing the otherwise lost contextual information between features. This theory, like the canonical model, is only conceptual, not a full-fledged computational model. By proposing a set of coherent hypotheses, it became a manifesto about potential roles of intrinsic rhythmic activity in the visual system and the brain, spurring, guiding, and sometimes also biasing varied experimental and modeling work over the ensuing decades. For this reason, I will use four essential elements in the correlation theory as a scaffold to structure the material in the remainder of this chapter.

Is Feature Binding a Problem in Vision?

The first hypothesis of the correlation theory is that there is a binding problem in vision. It is conjectured that in addition to visual features the (contextual) relationships between the features have to be encoded for modeling the functions of the visual stream of humans and animals, such as producing invariant object recognition or actions. It is further conjectured that it is practically impossible to capture context by just adding contextual features in the canonical hierarchical model because of the combinatorial explosion of such features.

Some researchers argue that there is no binding problem in vision. Barlow (1985) has postulated that the visual system might be able to operate if it contained a manageable number of "cardinal cells" that represent context between visual features. Since this claim is hard to address directly in vision, it has first been investigated in another domain, the representation of text documents. It would be quite ambiguous to encode words just by the sets of their characters without representing feature context, in this case, the order of letters. For example, word pairs like "stare" and "tears" could not be distinguished. However, the use of features that represent some limited order information, such as n-tuples of subsequent characters, can decrease the ambiguity of the representations drastically and still result in a manageable number of features (Wickelgren, 1969). The representation of the above example words by letter pairs is 'st,' 'ta,' 'ar,' 're,' and 'te,' 'ea,' 'ar,' 'rs,' respectively. This representation is already easy to disambiguate since there is only one common feature. This result has been used to propose that the binding problem of vision could be fully solved by adding a manageable number of disambiguating midlevel features in the canonical model (Mel & Fiser, 2000).

Another argument in support of this view comes from modeling studies. Riesenhuber and Poggio (1999a) demonstrated that a simulation of the canonical standard model for vision combined with state-of-the-art classifiers can reach high performance in a classification task even if the images contain background clutter (i.e., paperclip stimuli, similar to those used in Missal, Vogels, and Orban, 1997). Some recent studies further amplified this view by demonstrating that the canonical model of vision can reach human-level performance in image classification, for example, the task of determining whether or not there is an animal in the scene (Serre et al., 2007b). Criticism has been raised against these demonstrations. For example, a method was developed that can trace back which features in an individual image were strongly indicative for the presence of animals. In some instances, these features were located in the background and not part of the animal (Landecker et al., 2010). Furthermore, it has been argued that although classification might be solvable without the full contextual information, other behaviorally relevant tasks of the visual system, such as interacting with objects in arbitrary poses or producing actions, may not.

From the perspective of perceptual psychology, Treisman (1999) and Wolfe and Cave (1999) argue that a hard binding problem exists in human vision. Illusory conjunctions are exquisite examples of this. When subjects must report on the identity of items in briefly presented arrays of colored shapes, they often report seeing a stimulus made up of the color from one array element and the shape from a different array element (e.g., Prinzmetal, 1981; Treisman & Schmidt, 1982). These experiments demonstrate that perceptual features can become unbound from their original objects and can be spuriously recombined to form a new object representation.

Is There Feature Binding by Synchrony in the Brain?

The second hypothesis in the correlation theory proposes a specific neural coding scheme whereby the information about feature binding is represented in the brain. It is postulated that the synchronous structure of intrinsic fluctuations in neural signals encodes the relationship between features. This statement of the correlation theory, often referred to as binding by synchrony, was the first to gain strong traction in the field. Starting in the late eighties and continuing for about a decade, numerous studies tested this hypothesis in primary visual cortex. Some of the different positions in the field about the validity and usefulness of the binding-by-synchrony hypothesis were captured in a series of review articles appearing in a special issue of *Neuron* (Roskies, 1999).

A first wave of experiments reported evidence for the binding hypothesis in primary visual cortex of anesthetized cats (Eckhorn et al., 1988; Engel et al., 1991; Gray et al., 1989). During visual stimulation these studies reported prominent gamma-band activity (30–60 Hz). The coherence of these oscillations in simultaneously measured cells was larger if the cells represented features that were part of a common object than if the cells

represented features of two independent objects. A second wave of experiments investigated the existence of stimulus-evoked gamma-band oscillations in visual areas of monkeys, with somewhat mixed results (for reviews see Gray, 1999; Shadlen and Movshon, 1999). Oscillatory activity was not evident either in inferotemporal cortex in alert monkeys (Tovee & Rolls, 1992) or in striate cortex or middle temporal visual cortex of anesthetized monkey (Young, Tanaka, & Yamane, 1992). On the other hand, it was demonstrated that gamma-band activity is a robust property of neural responses in V1 and V2 of alert and also anesthetized monkey (Eckhorn et al., 1993; Friedman-Hill, Robertson, & Treisman, 1995; Frien et al., 1994).

Another line of experiments argued against the idea that binding takes place specifically in the cortex because high gamma-band activity is often not stimulus dependent and can be driven by gamma-band activity in the lateral geniculate nucleus (LGN) (Ghose & Freeman, 1992, 1997).

Reynolds and Desimone (1999) acknowledge that the binding problem exists for illusory conjunctions. However, they argue that most experimental evidence suggests that the problem is solved by top-down mechanisms of attention rather than by a bottom-up binding-by-synchrony mechanism.

A number of theoretical studies have modeled binding by synchrony in spiking networks (e.g., Knoblauch & Palm, 2001; Ritz et al., 1994; Wennekers & Palm, 2000). Other studies have proposed alternative memory-based models of how the binding problem in vision could be solved without resorting to neuronal synchrony. One is the shifter circuit or routing circuit model (Anderson & Van Essen, 1987; Olshausen, Anderson, & Van Essen, 1993); another related model is the map-seeking circuit (Arathorn, 2002). The map-seeking circuit is able to solve challenging invariant recognition tasks in real images. However, there has been no direct experimental evidence for either of these models.

Do Neural Oscillations Enable Signal Communication?

The correlation theory makes an important statement about signal communication between different sets of neurons. It suggests that fast intrinsic signal fluctuations carry contextual information in a frequency band that is separated from the frequency band corresponding to feature changes in stimuli which occur at a slower, behaviorally relevant time scale. It also describes how correlations of signal fluctuations can route the contextual information specifically to downstream targets that receive convergent input from features grouped within the same context. Von der Malsburg hypothesized different signal propagation paths that could benefit from such multiplexing of information: bottom-up, conveying sensory information from lower to higher visual areas, and top-down, such as in visual attention. In the last 15 years, this communication aspect of the correlation theory has probably received the most attention from neuroscience.

MULTIPLEXING Various multiplexing schemes have been developed in engineering to communicate multiple messages separately in a single information channel. They fall under two broad classes. In time division multiplexing, the time axis is divided into interleaved nonoverlapping time windows, each exclusively reserved for one of the messages. This scheme works if the sampling rate of the time window is above the Nyquist limits of the signals to be communicated. In frequency division multiplexing, the frequency domain is divided into nonoverlapping frequency bands, each carrying one of the messages. If the transmitted signals occupy overlapping frequency bands, frequency-division multiplexing relies on methods for shifting signals to nonoverlapping bands in the frequency domain.

It is instructive to apply these definitions from engineering to the communication with spike trains in the brain. Clearly, the correlation theory proposes a form of frequency-division multiplexing, as it assumes that the intrinsic fluctuations used for coding context are in a higher-frequency regime than the signal changes directly reflecting sensory inputs. At the same time, the mechanisms postulated for synchronizing periodic fluctuations between neurons introduce a scheme of time-division multiplexing within the high-frequency band. If a group of neurons represents features with a common context, their rhythmic activity synchronizes and confines the neurons' firing to narrow time windows within the oscillation cycle. If the integration window of downstream neurons is small enough, this temporal patterning allows neurons with a common context to recruit downstream neurons preferentially. This selection effect has been called feedforward coincidence detection (Fries, 2009) and has been observed experimentally in cortical neurons (Bruno & Sakmann, 2006). Furthermore, it has been revealed that the activity of inhibitory interneurons in cortical areas can exhibit strong power in the gamma range. Thus, the alignment or de-alignment of the phases of inhibition with the time windows of synchronized excitatory input provides a mechanism for how neurons can actively select which inputs they are sensitive to (Fries, Nikolic, & Singer, 2007). For simulation experiments exploring

feedforward coincidence detection and input selection by inhibition, see Tiesinga, Fellous, and Sejnowski (2008).

There is evidence for multiplexing in various sensory systems. In the vibrissae system, frequency division multiplexing using periodic intrinsic signals has been proposed for various computational functions (Ahissar et al., 1997; Ahrens et al., 2002; Kleinfeld & Mehta, 2006). In the olfactory system, the meaning of a spike can depend on the phase of a reference signal (Friedrich, Habermann, & Laurent, 2004). A theoretical model showed how the relative phase of a reference oscillation can be used to multiplex multiple visual signals in one spike train (Nadasdy, 2009). For general techniques on how to estimate the information rate that spikes carry about intrinsic periodic signals, see (Agarwal & Sommer, 2013).

GAMMA ENHANCEMENT DURING VISUAL ATTENTION Strong evidence has been presented supporting the idea that coherence in neuronal fluctuations might be crucial in mediating top-down effects of attention (Engel, Fries, & Singer, 2001). For example, in a study where alert monkeys attended to behaviorally relevant stimuli while ignoring distractors, it was shown that V4 neurons activated by attended stimuli exhibited increased gamma activity compared to neurons nearby in V4 that were activated by distractors (Fries et al., 2001). Another study reported a direct correlation between gamma-band synchrony and visually triggered behavior. The response time to a stimulus change can be predicted by the degree of gamma-band synchronization among those neurons in monkey visual area V4 that are activated by the behaviorally relevant stimulus (Womelsdorf et al., 2006).

BOTTOM-UP COMMUNICATION IN THE VISUAL SYSTEM In addition to endogenous rhythms, the cortex also seems to inherit oscillations that emerge in retina and LGN (Castelo-Branco, Neuenschwander, & Singer, 1998; Ghose & Freeman, 1992, 1997; Neuenschwander & Singer, 1996) and which are present both with and without anesthesia (Heiss & Bornschein, 1965, 1966). In spike trains from retina and LGN, the gamma oscillations and the stimulus-evoked changes are well separated in the frequency domain. The visual information encoded by spike rate occupies only the lower 25 Hz of the frequency spectrum, reflecting the fact that the spectral power of natural visual signals decays as the inverse of the frequency (Dong & Atick, 1995).

A recent study investigated how oscillations in the retina might be used by the thalamus to transmit information downstream (Koepsell et al., 2009). It was shown that the spike trains of a single thalamic relay cell can transmit two separate streams of information, one encoded by firing rate and the other in gamma oscillations (Koepsell et al., 2009). The study combined computational methods (Koepsell & Sommer, 2008) with the technique of whole-cell recording in vivo (Wang et al., 2007), which allowed the detection of both retinothalamic synaptic potentials and the action potentials they evoke from single relay cells. In other words, it was possible to reconstruct the spike trains of the inputs and outputs of single relay cells. In many cells, it was found that both spike trains had an oscillatory component. To explore whether or not these oscillations were transmitted by the thalamic cell, the phase of the oscillation of the retinal inputs was used to dejitter the timing of thalamic spikes across repeated trials of the stimulus. The result of the realignment was dramatic, as illustrated in figure 89.1A. Although the oscillation was not visible in the raw peristimulus histogram (PSTH), it generated a pronounced modulation in the amplitude of the PSTH made from the dejittered signal (see figure 89.1B).

By estimating the amount of information conveyed by the dejittered spike train, it was shown that most relay cells receiving periodic synaptic inputs transmitted a significant amount of information in the gamma frequency band. For some cells, the amount of information in the oscillation-based (high-frequency) channel was several fold higher than that conveyed by the rate-coded channel (1.2 vs. 0.4 bits/spike in the example shown in figures 89.1C and 89.1D). Thus, gamma oscillations in retina and thalamus provide a channel for conveying information through LGN to the cortex.

There are various possibilities for how this channel could contribute to visual function. One is the case in which the retinal oscillations do not contain information about the visual stimulus. Even if the oscillations are an uninformative carrier, they might increase the amount of information about local retinal features transmitted by the thalamic rate code. They would do so by a process akin to amplitude modulation, in which information about the retinal feature is reproduced in the frequency band of the oscillations. This redundant information could be read out and decoded in the cortex by mechanisms such as feedforward coincidence detection. A specific role for the oscillation-based channel could be denoising (see also Kenyon et al., 2004). Further, the modulation of the afferent spike train with a carrier might enable cortical oscillations to route the incoming sensory information or to direct attention to a particular feature.

A second possibility is that retinal oscillations are influenced by the stimulus, specifically, by displacements of the retinal image caused by eye movements. Thus, periodic activity in the retina might encode spatial information in the temporal domain, similar to the whisker system (Ahissar & Arieli, 2001; Rucci, 2008). This idea is motivated by the similarity between the dominant frequency bands in the local field potential recorded from primary visual cortex and fixational eye movements (note also that oscillatory eye movements are found in species ranging from turtle to humans; Greschner et al., 2002; Martinez-Conde, Macknik, & Hubel, 2004).

A third potential role for retinal oscillations involves computational analysis of visual stimuli. Since retinal oscillations are formed by distributed networks, they might be sensitive to spatially extensive features and/or context. In fact, there are many models of oscillatory neural networks that are able to transform spatial structure from visual input into temporal structure in neural activity. These models, which were originally developed to simulate cortical computations, are built with phase-coupled oscillatory neurons, for example, Baldi and Meir (1990), Schillen and Koenig (1994), Sompolinsky, Golomb, and Kleinfeld (1991), Sporns, Tonioni, and Edelman (1991), Ursino et al. (2006), von der Malsburg and Buhmann (1992), and Wang and Terman (1997). It would be worthwhile to further develop such models for describing and exploring possible roles of oscillations in retinal and thalamic function. What needs to be tested experimentally is whether the oscillation-based channel might transmit large-scale information such as segments in the retinal image, conveying the gist of a scene (Navon, 1977). Through feedforward coincidence detection the oscillations could preferentially activate cells in V1 whose features are most consistent with the image segments. Thus, retinal and thalamic oscillations could help select cortical visual representations that not only carry fine-grained image information but are also helpful for guiding behaviors like object recognition or the interaction with objects (Koepsell et al., 2010).

A behavioral role for retinal gamma oscillations along those lines has been clearly established in the frog. Specifically, looming stimuli designed to simulate shadows cast by predators evoke synchronous oscillatory discharges in neural "dimming detectors." By contrast, small dark spots that mimic prey fail to induce such activity (Ishikane, Kawana, & Tachibana, 1999). The consequence of the synchronous oscillations among retinal dimming detectors is important for an animal's survival since it triggers escape behavior (Arai et al., 2004). Further strengthening the link between synchronous retinal activity and behavior, it was shown that pharmacological suppression of gamma

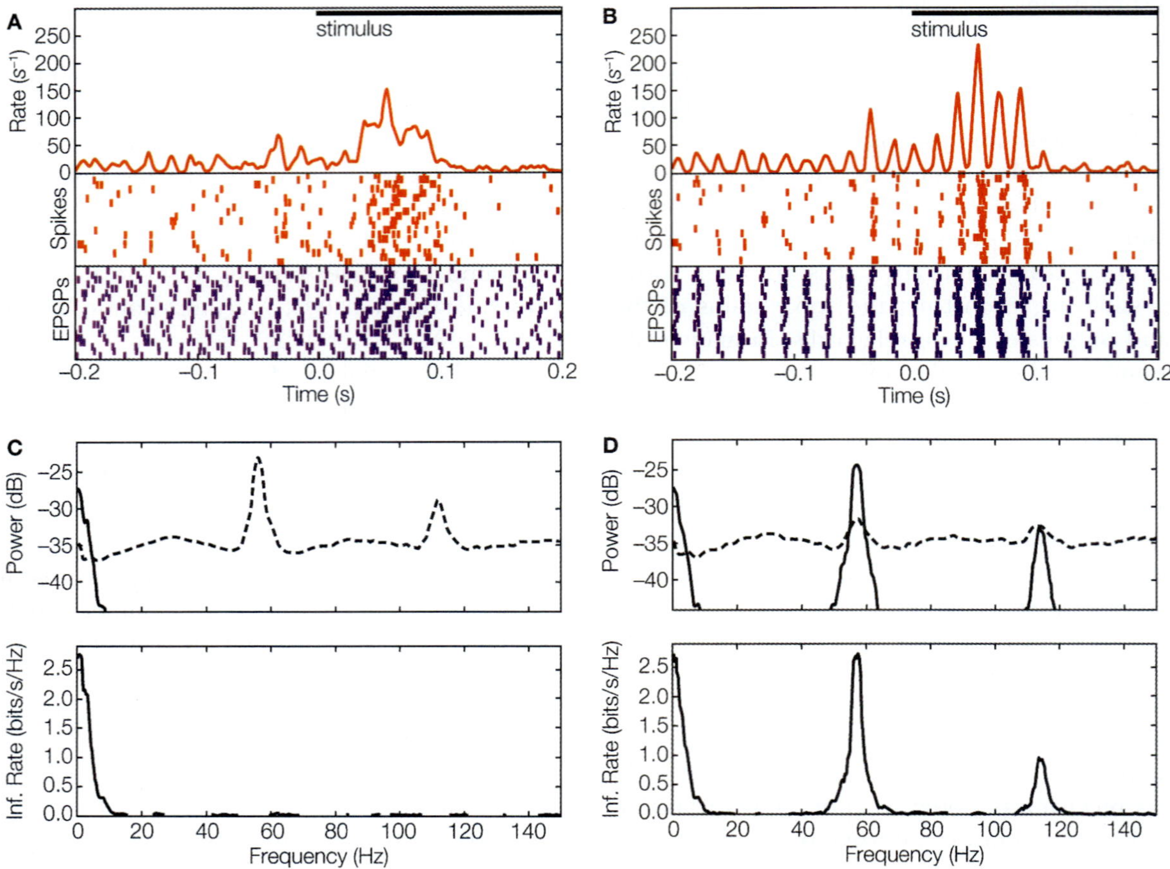

FIGURE 89.1   Multiplexed information (Inf.) in the lateral geniculate nucleus. (A) Event times aligned to stimulus onset displayed as averaged spike rate (red curve) and rasters for spikes (red) and excitatory postsynaptic potentials (EPSPs) (blue) for 20 trials of a movie clip; spike rasters were smoothed with a Gaussian window (2 ms) before averaging. (B) Responses corrected for small variation in latency (< 10 ms) by aligning the phase of the periodicity in the ongoing (retinal) activity that preceded stimulus onset; conventions as in A. (C) Top, power spectrum of thalamic spike trains decomposed into signal (solid line) and noise (dashed line). Bottom, estimate for spectral information rate, taken from the area under the curve, is 12.7 bit/s; the mean spike rate of 29 spikes/s yields a value of 0.4 bit/spike. (D) Power spectrum (top) of dejittered spike train decomposed into signal (solid line) and noise (dashed line); spectral information rate (bottom). Dejittering increased the total information from 0.4 bit/spike (C) to 1.2 bit/spike (Koepsell et al., 2009). The movie stimulus was presented with 19–50 frames/s on a monitor with a high refresh rate (140 Hz). The neural response did not lock to the frame update or monitor refresh. (Reprinted from Koepsell et al., 2010.)

oscillations abolishes escape responses but spares the slower modulation of spike rate evoked by small objects (Ishikane et al., 2005). Thus, in the frog, information about different types of visual signals seems to be multiplexed in different frequency bands of neural spike trains.

CROSS-FREQUENCY COUPLING   Studies in a variety of sensory systems have shown that the power of gamma oscillations is modulated by the phase of lower-frequency intrinsic brain rhythms, such as theta waves (Canolty et al., 2006; Lakatos et al., 2005) and alpha waves. It is believed that this modulation of the gamma power could shape the brain activity into cycles for selection and processing of a particular aspect of sensory input (Fries, 2009; Schroeder & Lakatos, 2009) (see also Freeman, 2000).

Biological Mechanisms Supporting Visual Processing with Neural Oscillations

The correlation theory makes a very specific hypothesis about the basic computational mechanism involved in visual processing. It postulates that a fast form of synaptic plasticity (or learning) is crucial for encoding of context information and for forming invariant visual representations (Bienenstock & von der Malsburg, 1987; Wiskott & von der Malsburg, 1996). Such fast synaptic plasticity could easily interact with neuronal oscillations to introduce correlations between neurons for representing related items. Interestingly, this postulate predated the discovery of fast types of synaptic plasticity such as spike-timing dependent plasticity (STDP) (Bi & Poo, 1998; Markram et al., 1997). However, although some studies have reported response changes of visual neurons induced by STDP during vision (Yao & Dan, 2001), currently there seems to be little evidence that the interaction between oscillations and STDP is a crucial mechanism for visual perception.

There is a large body of literature studying the mechanisms for the production and synchronization of oscillations in cortical circuits (Bartos, Vida, & Jonas, 2007; Tiesinga & Sejnowski, 2009). Specifically, three mechanisms have been proposed for producing synchrony in a cortical region (Tiesinga & Sejnowski, 2009). First, by inheritance of synchrony from upstream areas via their feedforward projections (Ghose & Freeman, 1997; Koepsell et al., 2009; Neuenschwander & Singer, 1996; Tiesinga, Fellous, & Sejnowski, 2008); second, by activation of inhibitory networks via the interneuron gamma (ING) mechanism (Whittington, Traub, & Jeffreys, 1995); and third, by activation of reciprocally connected networks of excitatory and inhibitory neurons via the pyramidal-interneuron gamma (PING) mechanism (Börgers & Kopell, 2005) as reviewed in Whittington et al. (2000). In the ING mechanism, only small effects are expected from activating the excitatory cells whereas activating inhibitory cells will increase the inhibitory cell firing rate and synchrony. Recent optogenetic methods allow for testing these proposed mechanisms quite directly. Studies that selectively modulated the activity in interneurons with optogenetic methods favored the PING mechanism (Cardin et al., 2009; Sohal et al., 2009). However, there is also experimental support for ING (Whittington, Traub, & Jeffreys, 1995), and the current evidence for PING is not strong enough to rule out ING entirely (Tiesinga & Sejnowski, 2009).

CONCLUSIONS: THE RISE AND FALL, AND RISE AGAIN, OF OSCILLATIONS

More than a generation after the appearance of von der Malsburg's technical report, opinions and viewpoints regarding the significance of oscillatory activity for visual processing in the brain have undergone several fundamental shifts. In the late nineties, the discussion was quite narrowly focused on two aspects of gamma oscillations, evidence for the binding-by-synchrony hypothesis and assessments of how reliably stimulus-evoked gamma oscillations occur during visual perception. Following this first wave of experiments and modeling, it appeared that the evidence for binding by synchrony, though existent, was not conclusive. Likewise, stimulus-evoked gamma oscillations were reported in some experimental configurations but not in others.

In the face of these quite inconclusive results, it might appear surprising that studies of oscillatory neural activity would increase throughout the 2000s rather than die out. These newer studies differ from those of the first generation by a shift in perspective and also by taking a broader outlook. For example, Pascal Fries and colleagues (see chapter 71) have studied how the coherence of gamma oscillations in higher visual areas is correlated with focused attention necessary to solve a visual task in the presence of distractors (Fries et al., 2001). Rather than considering attention as a competitor to oscillation-based computations, the question here is how oscillatory mechanisms might be involved in creating attention-dependent biased competition between different sensory inputs. The involvement of oscillatory activity in focused attention and its impact on behavior has now become well established in the field.

Renewed interest in gamma oscillations has also come about as the result of new methods and findings in neuroscience. First, new optogenetic techniques allow for dissecting the mechanisms for how pyramidal cells and interneurons are involved in the generation of gamma oscillations and their synchronization (Sohal et al., 2009). Second, the coupling of gamma oscillations to other, slower and more global brain waves has become an active field of research (Canolty et al., 2006). Third, there is now increased awareness that gamma oscillations in primary visual cortex have multiple origins, and so there is probably not a unique functional interpretation, such as binding by synchrony. To disentangle the puzzle, the reexamination of oscillations in the early visual pathway (Ghose & Freeman, 1997; Neuenschwander & Singer, 1996) may be crucial.

The strong impact of correlation theory in driving the investigation of oscillatory neuronal activity in vision is a striking example of the lasting power that a computational theory can have. However, since its inception, many new experimental methodologies and observations have emerged, and our appreciation of the challenges of visual processing has matured. Thus we may seek to extend the original theory to address the current incarnations of some fundamental open questions: How can oscillatory structure as observed in brain activity contribute to the powerful parallel and recurrent computations that neural circuits seem to perform? Can oscillation-driven schemes close the performance gap between brains and computer algorithms? To approach these questions, theorists should design models of how oscillations, as observed in brain activity, can produce, organize, and drive distributed computation. Such models can be tested on technical benchmark problems, for example in image recognition. In tasks that biological visual systems can solve, these models should favorably compare to state-of-the-art computer algorithms and clearly outperform the canonical feedforward model of vision. In addition to influencing technology, such computational models might motivate and guide future experiments to yield a deeper understanding of the periodic structure of brain activity.

ACKNOWLEDGMENTS

The author thanks Gautam Agarwal, Kilian Koepsell, Bruno Olshausen, Ryan Canolty, Joe Goldbeck, and the Redwood Center for Theoretical Neuroscience for many helpful discussions. Funding was provided by National Science Foundation grant (IIS-0713657) to F.T.S.

REFERENCES

Agarwal, G., & Sommer, F. T. (2013). Measuring information in spike trains about intrinsic brain signals. In P. M. DiLorenzo & J. D. Victor (Eds.), *Spike timing: Mechanisms and functions* (pp. 137–152). Boca Raton, FL: CRC Press. doi:10.1201/b14859-8.

Ahissar, E., & Arieli, A. (2001). Figuring space by time. *Neuron, 32*, 185–201.

Ahissar, E., Haidarliu, S., & Zacksenhouse, M. (1997). Decoding temporally encoded sensory input by cortical oscillations and thalamic phase comparators. *Proceedings of the National Academy of Sciences of the United States of America, 94*, 11633–11638. doi:10.1073/pnas.94.21.11633.

Ahrens, K. F., Levine, H., Suhl, H., & Kleinfeld, D. (2002). Spectral mixing of rhythmic neuronal signals in sensory cortex. *Proceedings of the National Academy of Sciences of the United States of America, 99*, 15176–15181. doi:10.1073/pnas.222547199.

Anderson, C. H., & Van Essen, D. C. (1987). Shifter circuits: A computational strategy for dynamic aspects of visual processing. *Proceedings of the National Academy of Sciences of the United States of America, 84*, 6297–6302.

Arai, I., Yamada, Y., Asaka, T., & Tachibana, M. (2004). Light-evoked oscillatory discharges in retinal ganglion cells are generated by rhythmic synaptic inputs. *Journal of Neurophysiology, 92*, 715–725.

Arathorn, D. (2002). *Map-seeking circuits in visual cognition: A computational mechanism for biological and machine vision.* Stanford, CA: Stanford University Press.

Baldi, P., & Meir, R. (1990). Computing with arrays of coupled oscillators: An application to preattentive texture discrimination. *Neural Computation, 2*, 458–471.

Barlow, H. B. (1985). The Twelfth Bartlett Memorial Lecture: The role of single neurons in the psychology of perception. *Quarterly Journal of Experimental Psychology, 37*, 121–145.

Bartos, M., Vida, I., & Jonas, P. (2007). Synaptic mechanisms of synchronized gamma oscillations in inhibitory interneuron networks. *Nature Reviews Neuroscience, 8*, 45–56. doi:10.1038/nrn2044.

Beck, A. (1890). Die Stroeme der Nervencentren. *Centralbibliothek Physiologie, 4*, 572–573.

Bi, G. Q., & Poo, M. M. (1998). Synaptic modifications in cultured hippocampal neurons: Dependence on spike timing, synaptic strength, and postsynaptic cell type. *Journal of Neuroscience, 18*, 10464–10472.

Bienenstock, E., & von der Malsburg, C. (1987). A neural network for invariant pattern recognition. *Europhysics Letters, 4*, 121–126.

Börgers, C., & Kopell, N. (2005). Effects of noisy drive on rhythms in networks of excitatory and inhibitory neurons. *Neural Computation, 17*, 557–608.

Bruno, R. M., & Sakmann, B. (2006). Cortex is driven by weak but synchronously active thalamocortical synapses. *Science, 312*, 1622–1627.

Canolty, R. T., Edwards, E., Dalal, S. S., Soltani, M., Nagarajan, S. S., Kirsch, H. E., et al. (2006). High gamma power is phase-locked to theta oscillations in human neocortex. *Science, 313*, 1626–1628. doi:10.1126/science.1128115.

Cardin, J. A., Carlen, M., Meletis, K., Knoblich, U., Zhang, F., Deisseroth, K., et al. (2009). Driving fast-spiking cells induces gamma rhythm and controls sensory responses. *Nature, 459*, 663–667. doi:10.1038/nature08002.

Castelo-Branco, M., Neuenschwander, S., & Singer, W. (1998). Synchronization of visual responses between the cortex, lateral geniculate nucleus, and retina in the anesthetized cat. *Journal of Neuroscience, 18*, 6395–6410.

Caton, R. (1875). The electric currents of the brain. *British Medical Journal, 2*, 278.

Dong, D. W., & Atick, J. J. (1995). Statistics of natural time-varying images. *Network, 6*, 345–358.

Eckhorn, R., Bauer, R., Jordan, W., Brosch, M., Kruse, W., Munk, M., et al. (1988). Coherent oscillations: A mechanism of feature linking in the visual cortex? Multiple electrode and correlation analyses in the cat. *Biological Cybernetics, 60*, 121–130.

Eckhorn, R., Frien, A., Bauer, R., Woelbern, T., & Kehr, H. (1993). High frequency (60–90 Hz) oscillations in primary visual cortex of awake monkey. *Neuroreport, 5*, 2273–2277.

Engel, A. K., Fries, P., & Singer, W. (2001). Dynamic predictions: Oscillations and synchrony in top-down processing. *Nature Reviews, 2*, 704–716.

Engel, A. K., Koenig, P., Kreiter, A. K., & Singer, W. (1991). Interhemispheric synchronization of oscillatory neuronal responses in cat visual cortex. *Science, 252*, 1177–1178.

Feldman, J. (2013). The neural binding problems. *Cognitive Neurodynamics, to appear,* Online First publication: DOI 10.1007/s11571-012-9219-8.

Freeman, W. J. (2000). Mesoscopic neurodynamics: From neuron to brain. *Journal of Physiology, Paris, 94*, 303–322.

Friedman-Hill, S. R., Robertson, L. C., & Treisman, A. (1995). Parietal contributions to visual feature binding: Evidence from a patient with bilateral lesions. *Science, 269*, 853–855.

Friedrich, R. W., Habermann, C. J., & Laurent, G. (2004). Multiplexing using synchrony in the zebrafish olfactory bulb. *Nature Neuroscience, 7*, 862–871.

Frien, A., Eckhorn, R., Bauer, R., Woelbern, T., & Kehr, H. (1994). Stimulus-specific fast oscillations at zero phase between visual areas V1 and V2 of awake monkey. *Neuroreport, 5,* 2273–2277.

Fries, P. (2009). Neuronal gamma-band synchronization as a fundamental process in cortical computation. *Annual Review of Neuroscience, 32,* 209–224. doi:10.1146/annurev.neuro.051508.135603.

Fries, P., Nikolic, D., & Singer, W. (2007). The gamma cycle. *Trends in Neurosciences, 30,* 309–316. doi:10.1016/j.tins.2007.05.005.

Fries, P., Reynolds, J. H., Rorie, A. E., & Desimone, R. (2001). Modulation of oscillatory neuronal synchronization by selective visual attention. *Science, 291,* 1560–1563. doi:10.1126/science.291.5508.1560.

Fukushima, K. (1980). Neocognitron: A hierarchical neural network capable of visual pattern recognition. *Neural Networks, 1,* 119–130.

Ghose, G. M., & Freeman, R. (1997). Intracortical connections are not required for oscillatory activity in the visual cortex. *Visual Neuroscience, 14,* 963–979.

Ghose, G. M., & Freeman, R. D. (1992). Oscillatory discharge in the visual system: Does it have a functional role? *Journal of Neurophysiology, 68,* 1558–1574.

Gray, C. M. (1999). The temporal correlation hypothesis review of visual feature integration: Still alive and well. *Neuron, 24,* 31–47.

Gray, C. M., Konig, P., Engel, A. K., & Singer, W. (1989). Oscillatory responses in cat visual cortex exhibit inter-columnar synchronization which reflects global stimulus properties. *Nature, 338,* 334–337. doi:10.1038/338334a0.

Greschner, M., Bongard, M., Rujan, P., & Ammermuller, J. (2002). Retinal ganglion cell synchronization by fixational eye movements improves feature estimation. *Nature Neuroscience, 5,* 341–347.

Grossberg, S. (1976). Adaptive pattern classification and universal recoding, II: Feedback, expectations, olfaction, and illusions. *Biological Cybernetics, 23,* 187–202.

Heiss, W. D., & Bornschein, H. (1965). Distribution of impulse of continuous activity of single optic nerve fibers: Effects of light, ischemia, strychnine and barbiturate. *Pflügers Archiv für die Gesamte Physiologie des Menschen und der Tiere, 286,* 1–18.

Heiss, W. D., & Bornschein, H. (1966). Multimodal interval histograms of the continuous activity of retinal cat neurons. *Kybernetik, 3,* 187–191.

Hubel, D. H., & Wiesel, T. N. (1962). Receptive fields, binocular interaction and functional architecture in the cat's visual cortex. *Journal of Physiology, 160,* 106–154.

Ishikane, H., Kawana, A., & Tachibana, M. (1999). Short- and long-range synchronous activities in dimming detectors of the frog retina. *Visual Neuroscience, 16,* 1001–1014.

Ishikane, H., Gangi, H., Honda, S., & Tachibana, M. (2005). Synchronized retinal oscillations encode essential information for escape behavior in frogs. *Nature Neuroscience, 8,* 1087–1095.

Kenyon, G., Theiler, J., George, J., Travis, B., & Marshak, D. (2004). Correlated firing improves stimulus discrimination in a retinal model. *Neural Computation, 16,* 2261–2291. doi:10.1162/0899766041941916.

Kleinfeld, D., & Mehta, S. B. (2006). Spectral mixing in nervous systems: Experimental evidence and biologically plausible circuits. *Progress of Theoretical Physics Supplement, 161,* 86–98. doi:10.1143/PTPS.161.86.

Knoblauch, A. & Palm, G. (2001). Pattern separation and synchronization in spiking associative memories and visual areas. *Neural Networks, 14,* 763–780. doi:10.1016/S0893-6080(01)00084-3.

Koepsell, K., & Sommer, F. T. (2008). Information transmission in oscillatory neural activity. *Biological Cybernetics, 99,* 403–416.

Koepsell, K., Wang, X., Hirsch, J. A., & Sommer, F. T. (2010). Exploring the function of neural oscillations in early sensory systems. *Frontiers in Neuroscience 4,* 53. doi:10.3389/neuro.01.010.2010.

Koepsell, K., Wang, X., Vaingankar, V., Wei, Y., Wang, Q., Rathbun, D. L., et al. (2009). Retinal oscillations carry visual information to cortex. *Frontiers in Systems Neuroscience, 3,* 4.

Kuffler, S. (1953). Discharge patterns and functional organization of the mammalian retina. *Journal of Neurophysiology, 16,* 37–68.

Lakatos, P., Shah, A. S., Knuth, K. H., Ulbert, I., Karmos, G., & Schroeder, C. E. (2005). An oscillatory hierarchy controlling neuronal excitability and stimulus processing in the auditory cortex. *Journal of Neurophysiology, 94,* 1904–1911. doi:10.1152/jn.00263.2005.

Landecker, W., Brumby, S., Thomure, M., Kenyon, G., Bettencourt, L., & Mitchell, M. (2010). Visualizing classification decisions of hierarchical models of cortex. Paper presented at the Computational and Systems Neuroscience Conference.

Laufer, M., & Verzeano, M. (1967). Periodic activity in the visual system of the cat. *Vision Research, 7,* 215–229.

Markram, H., Lubke, J., Frotscher, M., & Sakmann, B. (1997). Regulation of synaptic efficacy by coincidence of postsynaptic APs and EPSPs. *Science, 275,* 213–215.

Martinez-Conde, S., Macknik, S. L., & Hubel, D. H. (2004). The role of fixational eye movements in visual perception. *Nature Reviews Neuroscience, 5,* 229–240. doi:10.1038/nrn1348.

Mel, B. W., & Fiser, J. (2000). Minimizing binding errors using learned conjunctive features. *Neural Computation, 12,* 731–762.

Milner, P. M. (1974). A model for visual shape recognition. *Psychological Review, 81,* 521–535.

Missal, M., Vogels, R., & Orban, G. (1997). Responses of macaque inferior temporal neurons to overlapping shapes. *Cerebral Cortex, 7,* 758–767.

Nadasdy, Z. (2009). Information encoding and reconstruction from the phase of action potentials. *Frontiers in Systems Neuroscience, 3,* 6.

Navon, D. (1977). Forest before trees: The precedence of global features in visual perception. *Cognitive Psychology, 9,* 353–383.

Neuenschwander, S., & Singer, W. (1996). Long-range synchronization of oscillatory light responses in the cat retina and lateral geniculate nucleus. *Nature, 379,* 728–732.

Ogawa, T., Bishop, P. O., & Levick, W. R. (1966). Temporal characteristics of responses to photic stimulation by single ganglion cells in the unopened eye of the cat. *Journal of Neurophysiology, 29,* 1–30.

Olshausen, B. A., Anderson, C. H., & Van Essen, D. C. (1993). A neurobiological model of visual attention and invariant pattern recognition based on dynamic routing of information. *Journal of Neuroscience, 13,* 4700–4719.

Prinzmetal, W. (1981). Principles of feature integration in visual perception. *Attention, Perception & Psychophysics, 30*, 330–340.

Reynolds, J. H., & Desimone, R. (1999). The role of neural mechanisms of attention in solving the binding problem. *Neuron, 24*, 19–29, 111–125.

Riesenhuber, M., & Poggio, T. (1999a). Are cortical models really bound by the "binding problem"? *Neuron, 24*, 87–93, 111–125.

Riesenhuber, M., & Poggio, T. (1999b). Hierarchical models of object recognition in cortex. *Nature Neuroscience, 2*, 1019–1025. doi:10.1038/14819.

Ritz, R., Gerstner, W., Fuentes, U., & van Hemmen, J. L. (1994). A biologically motivated and analytically soluble model of collective oscillations in the cortex. *Biological Cybernetics, 71*, 349–358. doi:10.1007/BF00239622.

Rodieck, R. W. (1967). Maintained activity of cat retinal ganglion cells. *Journal of Neurophysiology, 5*, 1043–1071.

Roskies, A. L. (1999). The binding problem. *Neuron, 24*, 7–9, 111–125.

Rucci, M. (2008). Fixational eye movements, natural image statistics, and fine spatial vision. *Network, 19*, 253–285. doi:10.1080/09548980802520992.

Schillen, T., & Koenig, P. (1994). Binding by temporal structure in multiple feature domains of an oscillatory neuronal network. *Biological Cybernetics, 70*, 397–405.

Schroeder, C. E., & Lakatos, P. (2009). The gamma oscillation: Master or slave? *Brain Topography, 22*, 24–26. doi:10.1007/s10548-009-0080-y.

Serre, T., Oliva, A., & Poggio, T. (2007a). A feedforward architecture accounts for rapid categorization. *Proceedings of the National Academy of Sciences of the United States of America, 104*, 6424–6429. doi:10.1073/pnas.0700622104.

Serre, T., Wolf, L., Bileschi, S., Riesenhuber, M., & Poggio, T. (2007b). Robust object recognition with cortex-like mechanisms. *IEEE Transactions on Pattern Analysis and Machine Intelligence, 29*, 411–426. doi:10.1109/TPAMI.2007.56.

Shadlen, M. N., & Movshon, J. A. (1999). Synchrony unbound: A critical evaluation of the temporal binding hypothesis. *Neuron, 24*, 67–77, 111–125.

Sohal, V. S., Zhang, F., Yizhar, O., & Deisseroth, K. (2009). Parvalbumin neurons and gamma rhythms enhance cortical circuit performance. *Nature, 459*, 698–702. doi:10.1038/nature07991.

Sompolinsky, H., Golomb, D., & Kleinfeld, D. (1991). Cooperative dynamics in visual processing. *Physical Review A., 43*, 6990–7011.

Sporns, O., Tonioni, G., & Edelman, G. (1991). Modeling perceptual grouping in figure–ground segregation by means of active reentrant connections. *Proceedings of the National Academy of Sciences of the United States of America, 88*, 129–133.

Steinberg, R. H. (1966). Oscillatory activity in the optic tract of cat and light adaptation. *Journal of Neurophysiology, 29*, 139–156.

Tiesinga, P., Fellous, J. M., & Sejnowski, T. J. (2008). Regulation of spike timing in visual cortical circuits. *Nature Reviews Neuroscience, 9*, 97–107.

Tiesinga, P., & Sejnowski, T. J. (2009). Cortical enlightenment: Are attentional gamma oscillations driven by ING or PING? *Neuron, 63*, 727–732. doi:10.1016/j.neuron.2009.09.009.

Tovee, M. J., & Rolls, E. T. (1992). Oscillatory activity is not evident in the primate temporal visual cortex with static stimuli. *Neuroreport, 3*, 369–372.

Treisman, A. (1999). Solutions to the binding problem: Review progress through controversy summary and convergence. *Neuron, 24*, 105–110.

Treisman, A., & Schmidt, H. (1982). Illusory conjunctions in the perception of objects. *Cognitive Psychology, 14*, 107–141.

Ursino, M., Magosso, E., La Cara, G. E., & Cuppini, C. (2006). Object segmentation and recovery via neural oscillators implementing the similarity and prior knowledge gestalt rules. *Bio Systems, 85*, 201–218.

von der Malsburg, C. (1981). The correlation theory of brain function. MPI Biophysical Chemistry, Internal Report 81-2. Reprinted in E. Domany, J. L. van Hemmen, & K. Schulten (Eds.), *Models of neural networks II* (1985, pp. 95–106). Berlin: Springer.

von der Malsburg, C., & Buhmann, J. (1992). Sensory segmentation with coupled neural oscillators. *Biological Cybernetics, 67*, 233–242.

Wang, D., & Terman, D. (1997). Image segmentation based on oscillatory correlation. *Neural Computation, 9*, 805–836.

Wang, X., Wei, Y., Vaingankar, V., Wang, Q., Koepsell, K., Sommer, F. T., et al. (2007). Feedforward excitation and inhibition evoke dual modes of firing in the cat's visual thalamus during naturalistic viewing. *Neuron, 55*, 465–478.

Wennekers, T., & Palm, G. (2000). Cell Assemblies, associative memory and temporal structure in brain signals. In R. Miller (Ed.), *Time and the Brain* (pp. 251–273). Amsterdam: Harwood Academic Publishers. doi:10.4324/9780203304570_chapter_10.

Whittington, M. A., Traub, R. D., & Jeffreys, J. G. R. (1995). Synchronized oscillations in interneuron networks driven by metabotropic glutamate receptor activation. *Nature, 373*, 612–615.

Whittington, M. A., Traub, R. D., Kopell, N., Ermentrout, B., & Buhl, E. H. (2000). Inhibition-based rhythms: Experimental and mathematical observations on network dynamics. *International Journal of Psychophysiology, 38*, 315–336.

Wickelgren, W. A. (1969). Context-sensitive coding, associative memory, and serial order in (speech) behavior. *Psychological Review, 76*, 1–39.

Wiskott, L., & von der Malsburg, C. (1996). Face recognition by dynamic link matching. In J. Sirosh, R. Miikkulainen, & Y. Choe (Eds.), *Lateral interactions in the cortex: Structure and function*. Austin: University of Texas. Hypertextbook: http://www.cs.utexas.edu/users/nn/web-pubs/htmlbook96/.

Wolfe, J. M., & Cave, K. R. (1999). The psychophysical evidence for a binding problem in human vision. *Neuron, 24*, 11–17, 111–125.

Womelsdorf, T., Fries, P., Mitra, P. P., & Desimone, R. (2006). Gamma-band synchronization in visual cortex predicts speed of change detection. *Nature, 439*, 733–736. doi:10.1038/nature04258.

Yao, H., & Dan, Y. (2001). Stimulus timing-dependent plasticity in cortical processing of orientation. *Neuron, 32*, 315–323.

Young, M. P., Tanaka, K., & Yamane, S. (1992). On oscillating neuronal responses in the visual cortex of the monkey. *Journal of Neurophysiology, 67*, 1464–1474.

Zayachkivska, O., Gzhegotsky, M., & Coenen, A. (2011). Impact on electroencephalography of Adolf Beck, a prominent Polish scientist and founder of the Lviv School of Physiology. *International Journal of Psychology, 85*, 3–6.

XIII   MOLECULAR AND DEVELOPMENTAL PROCESSES

90 Development of Retinal Arbors and Synapses

JEREMY N. KAY AND JOSHUA R. SANES

The retina is a parallel processor of visual input in two distinct ways. First, light information transduced by photoreceptors is split into multiple parallel channels, each of which is devoted to extracting specific information about the visual scene. Second, these computations are performed simultaneously for each portion of the visual world. In this chapter, we will consider the neurodevelopmental mechanisms that establish the parallel circuits of the retina. How does the developing retina allocate different neurons to different circuits? What are the cellular and molecular mechanisms that ensure the circuits remain discrete? And how are the circuits patterned so that they observe appropriate-sized swaths of visual space?

Parallel processing of the first type—multiple representations of the world—depends on the tremendous neuronal diversity found in the retina. Photoreceptors pass information to at least 12 different bipolar cell (BC) subtypes, which pass information to more than 20 retinal ganglion cell (RGC) subtypes. On its way to RGCs, this information is filtered by horizontal cells (1 to 4 subtypes, depending on species) and up to 50 different amacrine cell (AC) subtypes. These interactions shape the firing patterns of RGCs, which in turn provide information about distinct visual features—for example, motion in specific directions—to retinorecipient regions of the brain (Gollisch & Meister, 2010; Sanes & Zipursky, 2010; Roska and Meister, chapter 13 in this volume).

To understand how this type of processing arises during development, we first need to know which neuronal subtypes comprise the various circuits. Given the extreme cell type diversity, this would seem to be a difficult problem—but luckily the anatomy of the inner retina reveals many details of the retina's complex wiring diagram. The inner plexiform layer (IPL), the neuropil that separates the inner nuclear layer (INL) from the ganglion cell layer (GCL; see figure 90.1) contains many parallel sublaminae. These are the sites where BCs, ACs, and RGCs that comprise individual circuits come together to make synapses—each neuron projects to specific IPL strata, and within those strata it

encounters its synaptic partners. Thus, in order to ask how the parallel circuits of the retina are assembled, we can investigate the developmental mechanisms that influence the lamina-specific projection patterns of AC, BC, and RGC subtypes (Sanes & Yamagata, 2009). This is the subject of the first part of the chapter.

Parallel processing of the second type—full coverage of the visual field—imposes additional requirements on the developing retinal circuitry. Each portion of the retina must contain a full set of RGC subtypes and an adequate complement of ACs and BCs to supply them. Further, the spread of pre- and postsynaptic arbors must be coordinated to generate receptive fields with appropriate sizes and overlaps. The mechanisms underlying this kind of patterning in the plane of the visual field have long been mysterious but are now beginning to yield their secrets. In the second part of this chapter we describe recent advances in our understanding of lateral patterning during retinal development, with an emphasis on the cellular and molecular mechanisms that act during development to wire the retina for parallel processing.

In describing recent progress, we draw from recent work on mice, chick, and zebrafish. For a discussion of earlier steps in vertebrate retinal development, such as eye morphogenesis, patterning, and neurogenesis, we refer readers to the first edition of this book (Wong & Godinho, 2003) as well as more recent reviews (Agathocleous & Harris, 2009; Fuhrmann, 2010; Reese, 2011).

LAMINAR SPECIFICITY AND SYNAPTIC CHOICES

The Phenomenon

The IPL first appears early in retinal development as a small space—smaller than a cell diameter—separating the GCL from the rest of the neural retina (Godinho et al., 2005). Over the course of development, as ACs, BCs, and RGCs elaborate their axonal or dendritic arbors within it, the IPL grows substantially. By maturity it has differentiated into a highly organized structure with at least 10 anatomically and physiologically distinct sublaminae (Roska & Werblin, 2001; Siegert et al., 2009). The function of each individual substratum has yet to be worked out, but the sublayers can be divided broadly into two groups: The ON sublaminae are located in the inner half (approximately) of the IPL, and they contain projections from neurons that depolarize when light levels increase. The OFF sublaminae, containing projections from neurons that depolarize when light levels decrease, are found in the outer half of the IPL, closer to the INL (see figure 90.1). Each type of AC, BC, and RGC confines its arbors to a specific subset of these IPL substrata, often a single sublayer. What are the developmental mechanisms that impart this specificity?

There are three phases to IPL development, each of which is critical to the final laminar pattern. The first step is formation of the IPL at the interface of the INL and GCL, defining the site where axons and dendrites will arborize (see figure 90.1A, B). The second phase of IPL development is a period of rapid expansion of the neuropil, accompanied by the appearance of anatomically distinct sublaminae (see figure 90.1B–E). During this stage, most neurons make laminar-specific projections, but in some cases they target different sublaminae than they do in mature retina. The third stage of IPL development, which overlaps with the second, is the refinement of these immature projections to generate the final pattern of laminar-specific connectivity (see figure 90.1D, E). The cellular and molecular mechanisms underlying these three phases of IPL development are beginning to be understood, and we describe key findings here.

Cellular Mechanisms

IPL FORMATION   Early in retinal histogenesis—at time points corresponding to midgestation in mice, or ~25 h postfertilization (hpf) in zebrafish—the only postmitotic inner retinal neurons are RGCs. At this stage there is no clear neuropil separating RGCs in the inner retina from neuroblasts in the outer retina. The first appearance of IPL neuropil is correlated with the differentiation of ACs (Godinho et al., 2005; Kay et al., 2004), suggesting that RGC dendrites are not sufficient to form the IPL on their own. Prior to AC neurogenesis, RGCs send axons toward brain targets, and their dendrites are rudimentary (see figure 90.1). But as early-born ACs appear, RGCs switch from axon-growth to dendrite-growth mode, perhaps in response to AC-derived factors (Goldberg et al., 2002). These dendrites,

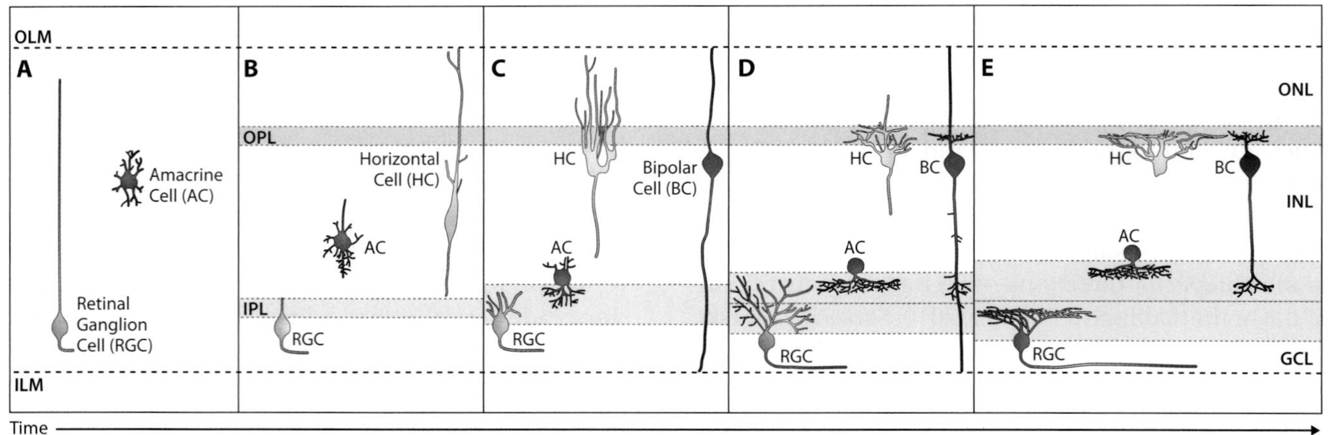

FIGURE 90.1   Neuronal differentiation and plexiform layer development. Schematic drawing showing the development of neuronal arbors for four classes of retinal neuron. Panels A–E depict five successive time points. Amacrine cells (ACs) and retinal ganglion cells (RGCs) are the first neurons to differentiate (A), and the inner plexiform layer (IPL) forms between these two populations (B). ACs initially project short processes in all directions before eventually stabilizing those oriented toward the IPL (C). As they elaborate their arbors, many ACs, such as the one shown here, exhibit "initially precise" targeting of a specific IPL sublayer (D, E). By contrast, the RGC shown here exhibits an "initially diffuse" projection pattern, elaborating dendrites throughout the IPL (D) before pruning to target a specific layer (E). Newly postmitotic bipolar cells (BCs) initially project across the entire width of the retina, from the outer to the inner limiting membrane (OLM; ILM). Subsequently they elaborate arbors in the outer plexiform layer (OPL) and in a specific IPL sublayer (C–E). The IPL grows over time, as neurons elaborate their processes within it. Horizontal cells exhibit transient vertically oriented dendrites that are important for mosaic formation (B, C). These are pruned and replaced by mature dendrites oriented horizontally in the OPL (D, E). GCL, GCL: ganglion cell layer. (Adapted from Morgan & Wong, 2007; Sanes & Zipursky, 2010.)

together with the nascent processes of early-born ACs, form a rudimentary IPL that is approximately 1 cell diameter wide at birth in mice.

How do the interactions among RGCs and ACs generate the IPL? Newborn ACs send out transient processes in all directions, which presumably sample the environment and concentrate neurite growth in the nascent IPL (Deans et al., 2011; Godinho et al., 2005; Hinds & Hinds, 1978; Prada et al., 1987). Indeed, AC processes directed toward the GCL are selectively stabilized during this period, suggesting the GCL-derived signals help ACs determine where the IPL should form (see figure 90.1; Godinho et al., 2005). When RGC-derived signals are removed, via genetic deletion of RGCs in the zebrafish *lakritz* (*atoh7*) mutant, the earliest-born ACs show a striking dendritic disorganization, which leads to a delay in IPL formation (Kay et al., 2004). Eventually the ACs do form a neuropil in these mutants, however, suggesting that ACs can form an IPL in the absence of RGCs. The AC-driven IPL that forms in *atoh7* mutants is not perfect—some ACs get trapped on the wrong side of the IPL, and some of the early disorganization is not corrected, leading to local perturbations in sublaminar targeting (Kay et al., 2004). Still, RGC-derived signals seem to play a fairly minor role in IPL formation, and those signals (or ones that can substitute for them) appear to be provided by ACs as well. The overall picture that emerges is that ACs are key to determining when and where the IPL will develop.

How do ACs accomplish this task? There may be a particular subtype of AC with a privileged role in forming the IPL. In zebrafish, prior to IPL formation, a subset of ACs delaminates from the INL and moves inward to the border of the GCL, lying on top of the RGCs (Godinho et al., 2005). These "displaced" ACs send processes toward the INL, contacting the processes of INL ACs. Eventually the intermingled processes of these two populations form a plexus that becomes the nascent IPL. In this way, the IPL forms as a consequence of interactions among ACs, rather than between ACs and RGCs. In several key respects, these zebrafish ACs bear an intriguing resemblance to the cholinergic "starburst" ACs, a subtype that is found in most vertebrate retinas and, at least in mice, is the earliest-born AC subset (Voinescu, Kay, & Sanes, 2009). In mature retina, starburst ACs exist as two separate populations, one in the INL and one displaced to the GCL. During development, starbursts start out as a single population prior to IPL formation; a subset then migrates to the forming GCL, and the IPL appears as a gap between these two starburst populations (Knabe et al., 2007; Nguyen et al., 2000; Prada et al., 1999). We do not yet know if the ACs studied by Godinho et al.

(2005) are cholinergic, but either way, the mechanism of IPL formation driven by specific subsets of early-born ACs appears to be quite similar from fish to mammals.

IPL STRATIFICATION   Once formed, the IPL expands rapidly, and anatomically distinct sublaminae appear. The exact details of where and how sublayers appear are species specific—zebrafish sublayers, for instance, appear more or less in a vitreal-to-scleral order, whereas chick sublayers appear in the opposite order (Drenhaus, Morino, & Veh, 2003; Mumm et al., 2006). Universally, however, the increases in IPL sublaminar complexity are temporally coupled to differentiation, arbor growth, and synapse formation by the RGCs, ACs, and BCs that project to the IPL. Based on anatomical and time-lapse studies, the most likely cellular mechanism for sublayer formation is a process that has been termed "self-assembly" (Sanes & Zipursky, 2010): As differentiating neurons ramify their arbors in the IPL, cells fated to project to the same layer begin to cofasciculate with each other. Each successive wave of differentiation and arbor growth both increases the IPL volume and adds new fascicles of stratified neurites. Although self-assembly is an attractive hypothesis, a systematic understanding of the link between neuronal differentiation, IPL expansion, and sublayer formation is still lacking.

Having helped to form the IPL, ACs in general and starburst ACs in particular also appear to play critical roles in its sublamination (see figure 90.1). Many ACs are stratified before BCs have even begun to differentiate (Drenhaus, Morino, & Veh, 2003; Mumm et al., 2006; Stacy & Wong, 2003), and AC-directed IPL sublamination is normal enough to guide the laminar projections of BCs in retinas that lack RGCs (Kay et al., 2004). In mouse retina, starbursts already show stratification in two IPL sublamina at P0, an age where the IPL is scarcely one cell diameter wide. Because starbursts stratify so early, it is tempting to speculate that their two sublaminae divide the young IPL into three zones that are targeted by later-differentiating cells. Such a mechanism could lead to the progressive addition of new layers above, below, or between the two starburst AC strata, thereby explaining how the "self-assembled" IPL strata become stacked in a particular order.

IPL SUBLAMINAR TARGETING AND REFINEMENT   How do neurons target the correct layers, especially given that the IPL is constantly growing and adding new sublayers during the period in which this targeting takes place? In principle, neurons might use three basic strategies to find their layers. In an "initially precise" strategy, neurons project directly to the appropriate sublayer as the IPL expands. Alternatively, neurons might use an opposite strategy in which "initially diffuse" projections are pruned, thereby localizing to specific sublayers once they have formed. The third strategy is an intermediate one in which neurons show initially precise projections but then add to, subtract from, or shift their arbors as development proceeds. Which of these strategies predominates has been a subject of much debate over the last decade, but we now know that all three occur. RGCs, for example, were for many years thought to use exclusively a strategy involving exuberant growth followed by pruning (see figure 90.1) (Bodnarenko, Jeyarasasingam, & Chalupa, 1995; Tian & Copenhagen, 2003). However, with the identification of genetically modified zebrafish and mouse strains in which subtypes of RGCs are labeled with fluorescent proteins, it has become clear that different RGC subtypes use all three of these targeting strategies (Kim et al., 2010; Mumm et al., 2006; Stacy & Wong, 2003). The ON–OFF direction-selective RGCs, like their starburst AC presynaptic partners, show early and specific targeting of the correct layers (Kim et al., 2010). Other RGC subtypes, by contrast, refine initially diffuse arbors, and still others project to intermediate targets or add specific sublayers as they mature. Although we have less information about how specific subtypes of ACs and BCs stratify, there do appear to be subtype-intrinsic temporal patterns of layer restriction for these neurons as well (Godinho et al., 2005; Morgan et al., 2006). The task going forward will be to understand how interactions within and across subtypes coordinate the histogenesis of IPL sublayers with the targeting of neuronal processes to those sublayers.

For those subtypes that remodel their projections, one important question is whether neural activity is involved in this remodeling. Hyperpolarization of BCs using a drug that targets metabotropic glutamate receptors delays pruning of excess RGC dendrites, as does deprivation of visual experience in the period following eye opening (Bodnarenko, Jeyarasasingam, & Chalupa, 1995; Tian & Copenhagen, 2003). On the other hand, genetic silencing of rod- and ON-BCs by expression of tetanus toxin light chain does not prevent ON-RGCs from choosing their correct BC synaptic partners (Kerschensteiner et al., 2009). These results are consistent with the idea that RGC dendrite pruning is not sensitive to the balance of activity between ON and OFF circuits, but rather to global activity levels.

Molecular Mechanisms

IPL FORMATION   Newborn ACs send out transient dendrites and filopodia that sample the space around them (see figure 90.1); eventually those dendrites oriented toward the GCL become stable while others are pruned away (Godinho et al., 2005; Hinds & Hinds, 1978; Prada et al., 1987). This cellular behavior suggests the existence of a molecular signal present in the GCL that can inform newborn ACs where to begin elaborating their processes for IPL formation. A recent study implicates *Fat3* in this signaling process. *Fat3* encodes a cell-surface molecule of the atypical cadherin family that can act as a receptor for signals that alter cell shape (Deans et al., 2011). FAT3 is expressed in the GCL and in ACs at late embryonic stages, and the protein localizes strongly to the nascent IPL. In *Fat3* mutants, newborn migrating ACs make two developmental mistakes that implicate FAT3 signaling in recognition of the GCL. First, some ACs fail to stop migrating at the edge of the GCL, leading to an excess of displaced ACs in these mutants. Second, newborn ACs often fail to prune processes oriented toward the outer retina, suggesting that their ability to distinguish the location of the GCL is impaired. Accompanying these AC behaviors are striking errors in IPL formation: There are two ectopic IPLs, one in the middle of the INL, and the other on the vitreal side of the GCL (Deans et al., 2011). These results suggest that ACs use FAT3 as a receptor for GCL-derived signals, and in its absence, when ACs are unable to properly locate the GCL, they can begin forming the IPL in mistaken locations. At present the ligand that is detected by the FAT3 receptor is unknown.

Signals coming from the opposite direction, informing ACs about the location of the outer retina, also have a critical role in IPL formation. Repulsive ligands of the Semaphorin family, Sema5A and Sema5B, are expressed in the outer neuroblast layer of the mouse retina starting around birth while their receptors, PlexinA1 and PlexinA3, are broadly expressed by RGCs and ACs. This expression pattern suggests that the Sema5/PlexinA system is well positioned to discourage developing ACs and RGCs from forming the IPL in the wrong place. Abrogation of this signaling system, in either *PlxnA1; PlxnA3* double mutants or *Sema5a; Sema5b* double mutants, leads to formation of ectopic IPL neuropil within the INL, suggesting that some newborn neurons have mistakenly oriented the wrong way during IPL formation (Matsuoka et al., 2011a). Despite the presence of ectopic IPLs in these mutants, as well as the *Fat3* mutant, a "standard" IPL does form and shows relatively normal histology, indicating that newborn ACs and RGCs are not completely lacking positional information and that additional orienting cues are likely to exist. Regardless, these two studies are the first to define molecular signals that constrain the IPL to form between the INL and GCL.

SUBLAMINAR TARGETING   The idea that neurons self-assemble IPL sublaminae through selective cofasciculation is supported by molecular studies of sublaminar targeting. In chick retina, four closely related homophilic cell adhesion molecules of the immunoglobulin (Ig) superfamily, Sidekick (Sdk) 1, Sdk2, Dscam, and DscamL, are selectively concentrated in distinct IPL sublaminae (Yamagata & Sanes, 2008; Yamagata, Weiner, & Sanes, 2002). Each of the four molecules is expressed by a unique, nonoverlapping subset of RGCs and interneurons that project to those specific sublaminae (see figure 90.2). This expression pattern suggests that costratifying neurons might use Ig superfamily molecules for homophilic adhesion in order to locate their correct laminar partners during IPL growth and stratification. Indeed, misexpression of these four molecules is sufficient to redirect AC and RGC dendrites to the sublayer appropriate to the misexpressed protein, while neurons deprived of their normal Sdk/Dscam molecule via RNAi-mediated knockdown make sublaminar targeting errors. These results are consistent with a model in which, as growing dendrites build the IPL, homophilic adhesion among dendrites destined to costratify facilitates the emergence of distinct sublaminae. If this is a general principle for IPL stratification, then there must be other adhesion molecules besides Sdks and Dscams—after all, there are more than four IPL sublayers. Recent work indicates that a related set of Ig superfamily molecules, the contactins, may have important roles in development of the non-Sdk, non-Dscam sublaminae (Yamagata & Sanes, 2012).

While homophilic attraction may promote proper connectivity within sublaminae, it is unclear how such attractive signals could arrange the sublaminae in a particular order. It is appealing to imagine that longer-range signals are involved in this process, and recently one was found. In mouse, during the first postnatal week, the repulsive guidance molecule Sema6A and its receptor, PlexinA4, are localized to complementary zones of the IPL—Sema6A is in the ON sublaminae of the IPL, while PlexinA4 is in the OFF zone (see figure 90.2). PlexinA4 is expressed by a relatively small subset of ACs, while Sema6A is more broadly expressed by ACs and RGCs. Loss of either component of this receptor-ligand complex in mutant mice causes sublaminar targeting defects by AC subsets that express PlexinA4— instead of exclusively targeting OFF sublayers, they project to ON sublayers as well (Matsuoka et al., 2011b). This study demonstrates a critical role for repulsive cues in IPL layer selection—the mutant mouse phenotypes are consistent with the notion that Sema6A-mediated repulsion is required to keep properly targeted dendrites from straying out of their designated layer. Furthermore, the study reveals a surprising interplay between attractive and repulsive signals in the matching of synaptic partners within the IPL. For one PlexinA4-positive AC subtype, the dopaminergic ACs, mislocalization of projections in an ectopic ON IPL sublamina

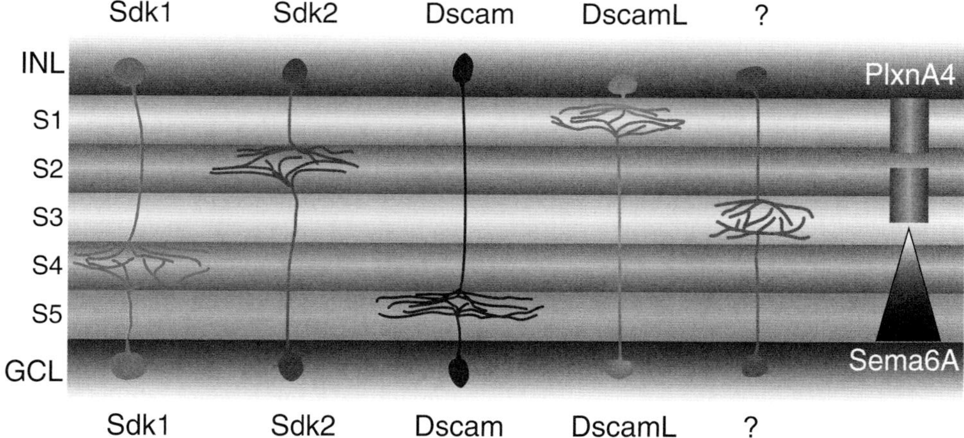

FIGURE 90.2   Molecular cues for inner plexiform layer (IPL) sublaminar targeting. A schematic of the IPL, divided into five sets of sublaminae labeled S1–S5. Neurons in the inner nuclear layer (INL) and ganglion cell layer (GCL) that project to the same sublayer coexpress the adhesion molecules Sdk1, Sdk2, Dscam, and DscamL. Diagram shows one sublayer for each gene, but the real situation is more complex. The homophilic attraction mediated by these proteins brings together synaptic partners in the same sublayer as demonstrated by gain- and loss-of-function studies (Yamagata & Sanes, 2008; Yamagata, Weiner, & Sanes, 2002). Some cells do not express any of these four molecules, but many of them express additional attractive adhesion cues such as contactins (Yamagata and Sanes, 2012). Long-range repulsive signals, mediated by the ligand Sema6A and its receptor PlxnA4, are also present in the IPL and appear to have a critical role in positioning sublaminae in the correct part of the IPL (Matsuoka et al., 2011b).

induces the same projection error in a synaptic partner RGC subtype, the M1 melanopsin RGCs (Matsuoka et al., 2011b). Because M1 RGCs do not express PlexinA4, the mistargeting of their projections to ON IPL must be secondary to the errors made by dopaminergic ACs, thereby demonstrating a strong attractive signal between these two cell types that remains intact even in the absence of the Sema6A/PlexinA4 repulsive signal. This finding suggests an intriguing hypothesis about the differing roles of attraction and repulsion during IPL development: that attractive (often homophilic) signals bring synaptic partners together in sublaminae while repulsive cues help place the sublaminae properly within the IPL.

PATTERNING RETINAL CIRCUITS IN THE PLANE OF THE VISUAL FIELD

The Phenomenon

As we have seen, the retina is a parallel processor of light stimuli, with the parallel circuits formed in IPL sublaminae. Matching of synaptic partners in proper sublaminae—the z axis of the retina—wires these parallel circuits correctly. In addition, the retina needs to be patterned in the x–y plane, to ensure that each portion of the retinal surface, and hence the entire visual scene, can be analyzed for each visual feature. Neurons of each subtype must be evenly spaced across the retina, which ensures complete coverage; axonal and dendritic arbors must grow to an appropriate size, which determines the size of their projective and receptive fields (Peichl & Wässle, 1983); and arbor shapes must be appropriate to facilitate the synaptic arrangements that allow the neuron to perform its computations. Defects in the developmental mechanisms that pattern circuits in the x–y plane would presumably lead both to "blind spots," where the absence of specific circuits impairs detection of specific visual features, and to alterations in receptive field properties.

Cellular Mechanisms of Patterning in the Plane of the Visual Field

What are the patterning mechanisms that determine a retinal neuron's receptive field size, shape, and spacing? Although the particulars vary from subtype to subtype, we can identify four phenomena that are broadly important: tiling, selective fasciculation, mosaicism, and self-avoidance (see figure 90.3).

TILING Tiling refers to an arrangement of axons or dendrites in which the arbors of neighboring cells belonging to a single subtype show minimal overlap (see figure 90.3A). The degree of tiling can be quantified as a "coverage factor," which describes the number of cells whose arbors cover any given point on the retinal surface. If the coverage factor is 1, the subtype shows perfect tiling; if it is 10, any point in space is served by processes of 10 cells of that subtype, and so on. Tiling can in principle solve two of the critical x–y patterning problems faced by retinal neurons—establishing receptive field size and ensuring even coverage of the visual field by each subtype. The hypothesized cellular mechanism is quite straightforward: If neurons simply grow their dendrites until they encounter dendrites of a homotypic neighbor, then each cell will have a roughly uniform arbor size, and together those arbors will cover the entire retinal surface. Indeed, few retinal neuron subtypes have coverage factors less than 1, which would leave retinal regions unserved by the subtype in question, creating potential blind spots. However, coverage factors of ~1 are also rare—for instance, the dendrites of certain RGC subtypes exhibit near-perfect tiling, as do the axon terminals of most cone BC subtypes (Dacey, 1993; Field et al., 2010; Vaney, 1994; Wässle, Peichl, & Boycott, 1981; Wässle et al., 2009). For these cell types, arbor growth arrest via homotypic contacts could be an important x–y plane patterning mechanism. The rest of the subtypes have coverage factors substantially greater than 1, meaning that they do not tile (Devries & Baylor, 1997; Peichl & Wässle, 1979; Tauchi & Masland, 1984). However, even in these cases, the proposed contact-mediated mechanism could still work to limit arbor size and overlap, if contact slows growth rather than halting it immediately.

Does such a mechanism in fact operate in retinal neurons, and if so, does it really influence receptive field size and the distribution of arbors across the retina? There are several lines of evidence supporting this mechanism. First, RGC subtypes that are present at high densities usually have smaller dendritic arbors than those present at low densities. This relationship is consistent with a contact-inhibition model. Second, experimentally induced or natural variations in RGC density influence arbor size in a manner consistent with the model—for a given subtype, manipulations that increase cell density produce smaller arbors whereas lowering RGC density increases arbor size (reviewed in Wingate, 1996). Third, RGCs of the same subtype make selective contacts with each other during the developmental period when their dendrites are growing, whereas they avoid contacts with the dendrites of other RGC subtypes. Moreover, these contacts appear to stifle additional dendritic growth beyond the site of contact. Such contact inhibition is not limited to cells that show

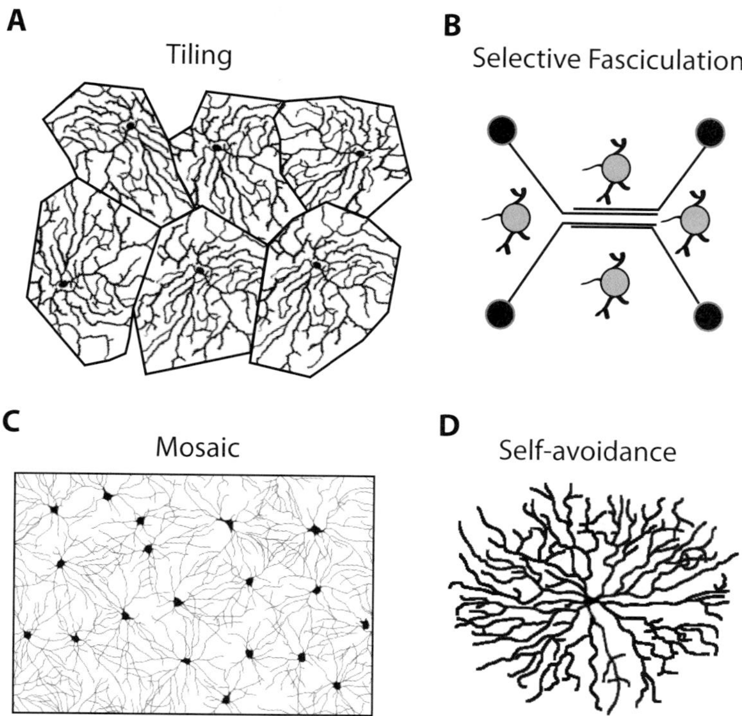

FIGURE 90.3   Mechanisms for patterning in the plane of the visual field. (A) Tiling: Neurons of the same subtype grow their dendrites until they touch, leading to an arrangement whereby each point on the retinal surface is covered by the dendrites of just one cell of that subtype. Thick lines show borders of dendritic territories of each cell. (B) Selective fasciculation: Neurons of types that do not tile (black cells) sometimes form fascicles with each other as they project through the retinal x–y plane while avoiding contact with dendrites originating from other cell types (gray cells). (C) Mosaicism: Cells of a single subtype are found near each other less often than would be predicted by chance. This arrangement, called a mosaic, occurs both for cell types that tile and for those with overlapping arbors. It leads to an even spacing of each cell type across the retina. (D) Self-avoidance: Dendrites of a single cell avoid crossing each other.

perfect tiling, suggesting that it could be a general mechanism for establishing arbor size and coverage factor (Lohmann & Wong, 2001). Fourth, and perhaps most strikingly, ablation of a small patch of RGCs in neonatal retina causes dendrites of surrounding RGCs to orient toward and grow into the barren site, as if the inhibitory contacts with homotypic neighbors had been removed (Eysel, Peichl, & Wässle, 1985; Perry & Linden, 1982).

In contrast to these results, some other findings are not consistent with the simplest form of the tiling hypothesis. First, in ablation experiments, the arbors of RGCs bordering the ablated area are not capable of growing to fill it entirely (Eysel, Peichl, & Wässle, 1985). This stands in contrast to results in other systems where neuronal arbors tile, such as the somatosensory neurons of *Drosophila* or zebrafish larvae—in these systems, ablation is followed by near-complete reinnervation of empty territory by remaining cells (Grueber & Sagasti, 2010). Furthermore, in mouse mutants where a large fraction (more than 80%) of RGCs either degenerate or are never born, the arbors of spared RGCs are relatively normal (Lin, Wang, & Masland, 2004). These results suggest that RGCs have intrinsic limits on their growth, and that tiling is unlikely to be the sole mechanism that determines receptive field size. Instead, tiling may modulate intrinsically determined arbors to adjust the extent of dendritic overlap. The intrinsic cellular and molecular mechanisms that determine arbor size remain a mystery.

SELECTIVE FASCICULATION   Like tiling, selective fasciculation involves interactions among dendrites or axons in the retinal x–y plane. Unlike tiling, however, these interactions are attractive rather than repulsive (see figure 90.3B). As neuronal processes course across the retinal surface through a particular IPL sublamina, they sometimes fasciculate with neurites of particular neuron subclasses that also project to the same sublayer. These are often homotypic interactions, although there are examples of heterotypic fasciculation—for example, dendrites of starburst ACs (which have a coverage factor of >20) form synapses with each other, and dendrites of starburst ACs also fasciculate with ON–OFF direction-selective RGCs. Since fasciculation presumably facilitates the formation of synapses between closely apposed neurites, its role may be to impart synapse specificity in the x–y plane, controlling both the number of synapses received along a dendrite and which presynaptic cells are making those synapses. Selective fasciculation can also influence dendritic arbor shape. The mechanisms underlying retinal selective fasciculation have not been studied in much detail, so we understand very little of how it is controlled during development.

MOSAICISM Neuronal somata of most subtypes are more evenly spaced across the retinal surface than would be expected by chance alone. This arrangement has been termed a "mosaic." While not as precise as the mosaics found in the insect compound eye, vertebrate retinal mosaics are generally strikingly regular, as can be appreciated in whole-mounts stained for a single neuronal subtype (see figures 90.3C and 90.4A). As far as is known, every functionally distinct retinal neuron subclass makes a mosaic. Moreover, these mosaics are independent of each other—cells of each subtype arrange themselves nonrandomly without regard for the location of cells belonging to other subtypes (Rockhill, Euler, & Masland, 2000; Wässle & Boycott, 1991). Similar to tiling, these overlapping mosaics could serve as a mechanism to ensure even coverage of visual space by each functionally specialized circuit.

The defining feature of a retinal mosaic is an "exclusion zone"—a region around each cell in which another neuron of the same type is rarely found. Despite the fact that different subtypes are present at different densities, and that some mosaics are more regular than others, nearly all mosaics can be modeled accurately if the exclusion zone size is known. In fact, the rule for modeling mosaics is a remarkably simple one, whereby cells are randomly arranged except for the exclusion zone around each cell in the array (Eglen, 2006; Galli-Resta et al., 1997, 1999). This result demonstrates that global, nonrandom geometrical patterning of neurons across the retina can arise from mechanisms that act locally around each cell—the mechanisms that create exclusion zones. The search for a cellular mechanism for mosaic patterning, therefore, boils down to the question of how exclusion zones arise during development.

There are at least three cellular mechanisms that could create exclusion zones. First, regularity could arise at or before neurogenesis: Specified progenitors could be arranged mosaically in the ventricular zone, or newborn neurons could inhibit nearby progenitors from generating another neuron of the same subtype.

A second possible mechanism is cell death—when two neurons are present in the same exclusion zone, one might kill the other. Finally, homotypic repulsion could lead neurons initially positioned too close together to move away from each other.

The first mechanism operates in the *Drosophila* eye to generate a regular arrangement of photoreceptors; however, unlike in the insect compound eye, developing cells of the vertebrate retina are motile in the x–y plane, meaning that the site of a neuron's birth is a poor indicator of its final retinal location (Reese, Harvey, & Tan, 1995). Local inhibition of cell fate decisions, therefore, would not necessarily impose exclusion zones on the mature mosaic. Furthermore, some cell types show less regular or even random spacing shortly after their birth, only to sharpen their mosaics later (Galli-Resta et al., 1997; Jeyarasasingam et al., 1998). On balance, the evidence indicates that cell fate mechanisms alone cannot explain exclusion zone formation.

The two remaining possible mechanisms for creating exclusion zones, cell death and cell movement, both involve interactions between postmitotic developing neurons that allow such cells to determine the location of their neighbors. These interactions must be homotypic because, as noted above, neurons only keep away from cells of the same type; they are perfectly happy to sit next to cells of different types (Rockhill, Euler, & Masland, 2000). Because every subtype makes a mosaic, every subtype must have a mechanism for figuring out whether its neighbor is the same or different in order to establish its exclusion zone. With as many as 100 different subtypes making overlapping, independent mosaics, this is a daunting molecular problem. There is great interest, therefore, in figuring out both the molecular basis of homotypic recognition and the cellular outcome of these signals—do homotypic neighbors have the ability to kill each other, or do they simply move apart?

Of the many mosaics that have been described, only a few have been studied from a developmental viewpoint. These are the horizontal cells of the mouse; two mouse AC subtypes, the starbursts and the dopaminergic ACs; and two RGC subtypes, alpha cells in cat and intrinsically photosensitive (ip) RGCs in mouse. These studies indicate that the cellular mechanisms for establishing exclusion zones are cell type specific. For dopaminergic ACs, ipRGCs, and cat alpha RGCs, cell death plays an important role in creating exclusion zones (Jeyarasasingam et al., 1998; Keeley et al., 2011; Raven et al., 2003). These cell types, like many others throughout the nervous system, have as part of their normal development a brief period of naturally occurring cell death that eliminates up to 50% of the original population. In mouse mutants where naturally occurring cell death is blocked, dopaminergic ACs and ipRGCs are found near each other much more frequently than in control animals. As a result, mosaics are disrupted; in fact, the effect on proximity is so strong that cell arrays show evidence of homotypic attraction rather than repulsion (Keeley et al., 2011; Raven et al., 2003). The model for exclusion zone formation in these cell types is that, during the period of naturally occurring cell death, homotypic interactions within the exclusion zone radius increase the probability of apoptosis. Because dopaminergic ACs do not move laterally as much as some other cell types, they may be particularly dependent on cell death mechanisms for development of mosaic spacing.

By contrast, horizontal cells and starburst ACs undergo limited naturally occurring cell death, so they must use a different mechanism for exclusion zone formation. Indeed, substantial evidence suggests that they use homotypic interactions to establish exclusion zones through cell–cell repulsion. As with all retinal neurons, starbursts and horizontal cells exit the cell cycle adjacent to the outer limiting membrane and then migrate in the z axis to their final laminar position (see figure 90.1). During this period of their development, the cells are randomly positioned in the x–y plane (Galli-Resta et al., 1997). Once they arrive in their correct z position (GCL for ON starbursts, inner INL for OFF starbursts, outer INL for horizontal cells), they begin elaborating dendrites. The cells still show random spacing at this time. It is only when their nascent dendrites grow to reach the dendrites of their homotypic neighbors that regular spacing is established (Galli-Resta, Novelli, & Viegi, 2002; Huckfeldt et al., 2009; Poché et al., 2008). The idea, then, is that homotypic dendritic contacts allow each cell in the array to know the location of its neighbors and to move accordingly in order to establish exclusion zones. Through a mechanism perhaps related to tiling, repulsive interactions between the dendrites of neighboring cells carve out each cell's territory. During this earliest phase of arbor growth, exclusion zone size is equal to a cell's dendritic arbor size (Huckfeldt et al., 2009). When a single horizontal cell is laser ablated, neighboring cells rapidly send dendrites into the vacated area; within a matter of hours these neighboring survivors have adjusted the location of their cell bodies within their now-expanded arbors to once again produce regular soma spacing (Huckfeldt et al., 2009).

Mature starbursts and horizontal cells have coverage factors greater than 20, so this phase of dendritic tiling must be strictly limited to the period of development when it is needed for establishing exclusion zones.

Horizontal cells accomplish this temporal regulation by growing transient, vertically oriented dendrites with a morphology clearly distinct from their mature, horizontal dendritic arbor (see figure 90.1). In mice, the vertical dendrites exist for about four days, starting shortly before the animal's birth. As the mature horizontal arbors begin to grow, the vertical dendrites are pruned away (see figure 90.1). Whereas the vertical dendrites tile, the mature dendrites overlap extensively, suggesting that the two types of dendrites are cell-biologically specialized to accomplish their unique functions (Huckfeldt et al., 2009). Specialized dendrites have not been documented in starburst ACs, but it is possible that, at the phase when their dendrites are first extending, they might have a special transient function in exclusion zone formation. Consistent with this idea, starburst arbors undergo extensive remodeling and shape changes during the perinatal period in mice (Lefebvre et al., 2012). Thus, as dendrites transition from a role in exclusion zone establishment to behaving as true dendrites, there might be structural changes that accompany these functional changes.

SELF-AVOIDANCE   Self-avoidance is a pattern of axon or dendrite development in which individual processes can recognize and avoid neurites from the same cell while interacting freely with the processes of their homotypic neighbors. A neuron that adopts this patterning strategy has dendritic arbors that fill its receptive field without any instances of dendrites crossing each other (see figure 90.3D). At the same time, however, such a neuron will freely cross the receptive fields, and thus the dendrites, of neighboring cells. Self-avoidance was first described in leech mechanosensory neurons, but has since been found in a wide variety of cell types throughout invertebrate and vertebrate nervous systems (reviewed in Grueber & Sagasti, 2010). Examples of retinal neurons that exhibit self-avoidance are the starburst ACs and several subclasses of RGCs (Montague & Friedlander, 1991; Tauchi & Masland, 1984). For subclasses with coverage factors substantially greater than 1, self-avoidance is a particularly useful patterning mechanism—in these cases, interactions between dendrites of different cells are inevitable and extensive. Self-avoidance allows individual neurons to adopt maximal and uniform dendritic coverage of their own dendritic fields without interfering with their neighbors' dendrite patterning. While self-avoidance is not universal in the retina, every neuron does develop a subtype-specific pattern of axonal or dendritic morphology, and self-avoidance is a useful model for understanding how these morphologies arise during development.

Molecular Mechanisms of Patterning in the Plane of the Visual Field

The cell biological studies discussed so far suggest that homotypic recognition events are key to each of the major patterning phenomena—repellent interactions among neighbors' dendrites for tiling and mosaicism; adhesion among dendrites for selective fasciculation; and repulsion between dendrites of a single cell for self-avoidance. A key task, then, is to identify the recognition molecules involved. Here we describe recently identified candidate mediators of mosaic formation and dendritic self-avoidance.

MOSAICS   Studies of the cellular mechanisms for mosaic patterning make predictions about the molecular nature of the signals underlying creation of exclusion zones during development. First, because each cell subtype makes an independent mosaic, the molecules for homotypic recognition must be highly cell type specific. Second, at least in the case of starbursts and horizontal cells, the repulsive signal for exclusion zone formation is likely to act at short range, and perhaps be membrane-tethered (Galli-Resta, 2000). Finally, expression of the repulsive signal (or its receptor) is likely to be temporally restricted—not present too early, because newborn neurons do not generally show mosaic spacing, nor too late, because then the repulsive system might interfere with attractive homotypic interactions such as selective fasciculation.

In a screen for genes enriched in specific neuron subsets during mouse retinal development, *Multiple EGF-like domains 10* (*Megf10*) was identified as a gene with these characteristics (Kay, Chu, & Sanes, 2012). This gene encodes a single-pass transmembrane protein with a large extracellular domain, containing 17 EGF-like repeats, and an intracellular domain that can couple to well-defined downstream signaling pathways (Reddien & Horvitz, 2004; Suzuki & Nakayama, 2007; Wu et al., 2009). It is selectively expressed by starburst ACs and horizontal cells, and its expression in starbursts is restricted to the developmental interval during which starburst mosaics form (Kay, Chu, & Sanes, 2012). When MEGF10 is misexpressed in vivo in neurons that do not normally have it, starbursts begin avoiding those cells as if they were homotypic neighbors, providing striking evidence that MEGF10 is a homotypic recognition cue. Most important, starbursts are morphologically normal but randomly positioned in mutant mice lacking MEGF10, suggesting a complete failure of homotypic recognition in the absence of this one molecule (see figure 90.4).

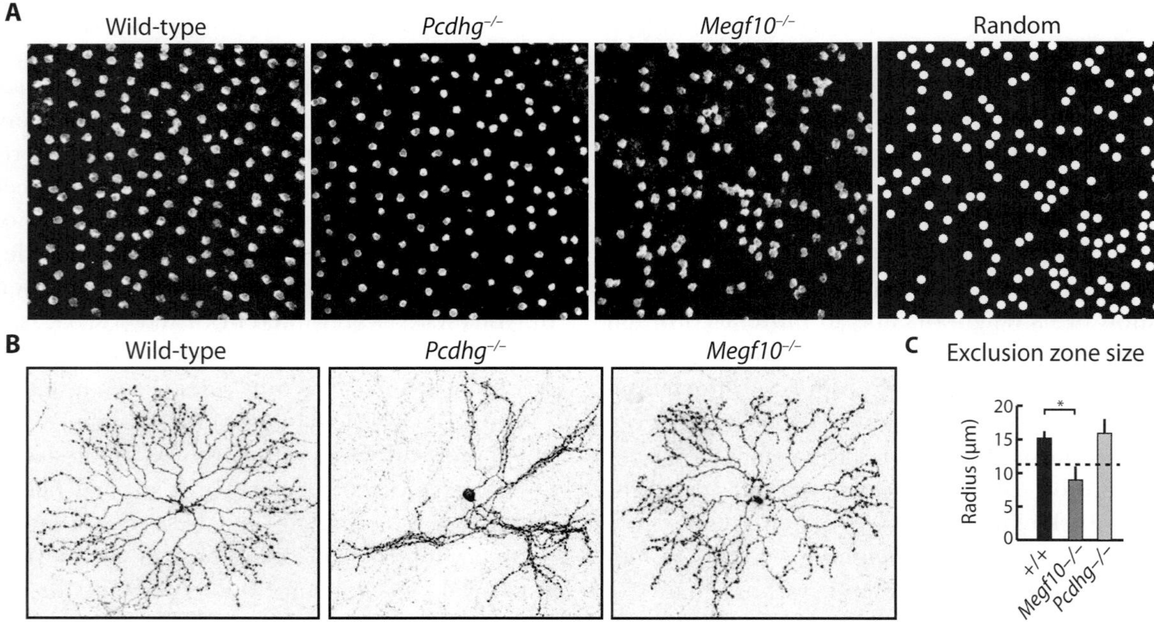

FIGURE 90.4   Molecular mechanisms for homotypic and isoneuronal recognition in mosaic spacing and self-avoidance. (A) The starburst AC mosaic. Wild-type and *Protocadherin-γ* (*Pcdhg*) mutant mice have evenly spaced starburst mosaics, but spacing is strongly disrupted in *Megf10* mutant mice. A random cellular array, matched in cell size and density to the real data, is shown for comparison (right panel). The arrangement of starbursts in Megf10 mutants was indistinguishable from that predicted for a random distribution. (B) Starburst single-cell morphology. Cells from wild-type and *Megf10⁻/⁻* mice show self-avoidance—the dendrites of individual cells rarely cross. In *Pcdhg⁻/⁻* mice, however, starbursts show striking self-avoidance defects. (C) Quantification of exclusion zone size in starburst mosaic formation. Dashed line shows what would be expected for a random array. (From Kay, Chu, & Sanes, 2012, and Lefebvre et al., 2012.)

Horizontal cells also express MEGF10, but mosaic defects in this cell type are modest in the absence of MEGF10. It turns out that horizontal cells also express and use the closely related (and also specifically expressed) molecule MEGF11. In MEGF10; MEGF11 double mutants, defects in the horizontal cell mosaic are profound (Kay, Chu, & Sanes, 2012).

The model that emerges from the cellular and molecular studies is the following: Newborn starbursts and horizontal cells display MEGF10 (and, in the case of horizontal cells, MEGF11 too) on their cell surface as they begin growing dendrites. Because MEGF10 acts as its own receptor (Kay, Chu, & Sanes, 2012), only a homotypic neighbor—another starburst or horizontal cell—will be able to see this molecular cue on the dendrite's surface. When the dendrites of adjacent starbursts interact, MEGF10–MEGF10 signaling triggers a repulsive response in both cells, leading to the establishment of unique dendritic territories and, eventually, exclusion zones. Once exclusion zones are established, MEGF10 is downregulated, allowing starburst dendrites to overlap extensively. Clearly, homotypic recognition systems must exist for numerous other neuronal subtypes, and these remain to be discovered. There is only a single additional gene in the MEGF10/11 family, so genes of other families are likely to be involved Nevertheless, the molecular logic of the MEGF10/11 system, and its signal transduction mechanism (currently under investigation) may provide clues to identification of genes that mediate formation of other mosaics.

One other pair of molecules that has been implicated in patterning of mosaics are the Dscams (Fuerst et al., 2008, 2009). As we saw in the section on laminar specificity, *Dscam* and its close relative *DscamL* encode homophilic adhesion molecules of the Ig superfamily. In mouse *Dscam* mutants, the homotypic cell–cell recognition underlying mosaic formation is intact, as evidenced by the fact that mosaics initially form normally only to degrade over time. As they degrade, cells aggregate into subtype-specific clumps, indicating that cells can still identify their homotypic neighbors. Instead of a direct effect on exclusion zone formation, these mosaic defects are most likely secondary to other *x*–*y* plane patterning defects in these mutants that we will discuss below. Dscam also affects mosaic regularity via an influence over cell death (Keeley et al., 2011). Because the mechanism through which Dscam promotes cell death is not yet understood, we do not yet know exactly how this Dscam phenotype is related to the cellular clumping phenotype.

self-avoidance For cells that use self-avoidance to pattern their dendritic fields, each individual dendrite must have a way to recognize and avoid other dendrites of the same cell. In some cases, dendrites must also distinguish self from nonself—that is, whether neighboring dendrites arise from the same cell or a different cell. Such recognition is required to allow isoneuronal avoidance without affecting heteroneuronal interactions such as selective fasciculation. Subtype-specific signals such as those that underlie mosaicism endow every cell of the same subtype with the same cues and therefore cannot explain the distinction between isoneuronal and heteroneuronal interactions. Instead, one must postulate cell-specific cues—each cell must have its own molecular signature. What kind of molecular mechanism could provide such a signature? In *Drosophila* several (but not all) cases of self-avoidance are mediated by the *Dscam1* gene, the fly ortholog of the vertebrate *Dscam* and *DscamL* genes that we have discussed above. Drosophila *Dscam1* is alternatively spliced to generate thousands of related homophilic adhesion molecules, which, quite remarkably, each shows homophilic binding specificity. This homophilic binding leads to dendritic repulsion. Because each neuron stochastically expresses a slightly different collection of Dscam1 isoforms, the homophilic binding activity of Dscam1 allows the dendrites of a single neuron to specifically recognize and avoid each other while failing to interact with the dendrites from other cells (reviewed in Zipursky & Sanes, 2010).

Vertebrate Dscam1 homologues, Dscam and DscamL, have also been implicated in dendritic self-avoidance (Fuerst et al., 2008, 2009; Tadenev et al., chapter 91 in this volume). Each of these closely related adhesion molecules is expressed in a nonoverlapping subset of ACs and RGCs. When either molecule is knocked out in mice, the cells that would normally express it show strikingly abnormal dendrite patterning that is consistent with dendritic aggregation. In fact, it is not just dendrites of a single cell that aggregate in these mutants—somata also aggregate into subtype-specific clumps (leading to the secondary alterations in mosaics mentioned above). However, the mechanism by which vertebrate Dscam influences self-avoidance must be different than in flies, because the vertebrate genes do not undergo the kind of alternative splicing seen in fly *Dscam1*. Consistent with this idea, vertebrate Dscams do not appear to affect isoneuronal and heteroneuronal interactions differentially.

There are, however, vertebrate candidates genes that might act in a manner analogous to fly *Dscam1*. The clustered *Protocadherin* locus encodes 58 cadherin-related transmembrane signaling molecules, arranged in three subgroups called alpha, beta, and gamma, comprising 22, 14, and 22 isoforms, respectively. Small subsets of *protocadherin* genes are expressed stochastically in neurons throughout the mouse brain, and at least the Protocadherin-γ (Pcdhg) isoforms interact homophilically in an isoform-specific manner (Kaneko et al., 2006; Schreiner & Weiner, 2010; Wu & Maniatis, 1999; Zipursky & Sanes, 2010). Initial studies of mutant mice lacking all 22 *Pcdhg* genes found cell-type-specific defects in neuronal survival and synaptic function in the spinal cord and retina (Lefebvre et al., 2008; Wang et al., 2002; Weiner et al., 2005). More recent analysis, using genetic tools to specifically ablate the *Pcdhg* gene cluster in the retina, revealed that *Pcdhg* deletion also causes dramatic self-avoidance defects in starburst ACs (see figure 90.4). These defects are rescued when *Pcdhg* mutants are crossed to transgenic mice expressing single *Pcdhg* isoforms, indicating that as long as a starburst has a single *Pcdhg* isoform, its dendrites are able to self-avoid (Lefebvre et al., 2012). However, in these mice where diversity of *Pcdhg* expression is severely reduced, the dendrites of neighboring starbursts show abnormal repulsion (Lefebvre et al., 2012). These findings strongly suggest that *Pcdhg* diversity is essential for starburst dendrites to distinguish their isoneuronal dendritic neighbors from heteroneuronal neighbors. Thus, stochastic expression of a few *Pcdh* isoforms could endow each starburst with the unique molecular identity required for self-avoidance.

CONCLUSION

Once retinal neurons have been generated, they assemble complex circuits. Many of the steps required for this assembly depend on the ability of neurons to recognize and discriminate among other neurons. For example, recognition and discrimination are required for neurons to identify their synaptic partners, to locate homotypic neighbors, and to distinguish isoneuronal from heteroneuronal processes. Similar recognition events are undoubtedly critical for patterning circuits throughout the brain. An important future direction, therefore, will be to extend studies of retinal circuit patterning deeper into the visual system—for example to the superior colliculus, the lateral geniculate nucleus, and the visual cortex. We are confident that methods and insights gleaned from retinal studies will facilitate efforts to understand cellular and molecular mechanisms governing wiring specificity in the visual centers of the brain.

REFERENCES

Agathocleous, M., & Harris, W. A. (2009). From progenitors to differentiated cells in the vertebrate retina. *Annual Review of Cell and Developmental Biology, 25*, 45–69.

Bodnarenko, S. R., Jeyarasasingam, G., & Chalupa, L. M. (1995). Development and regulation of dendritic stratification in retinal ganglion cells by glutamate-mediated afferent activity. *Journal of Neuroscience, 15*, 7037–7045.

Dacey, D. M. (1993). The mosaic of midget ganglion cells in the human retina. *Journal of Neuroscience, 13*, 5334–5355.

Deans, M. R., Krol, A., Abraira, V. E., Copley, C. O., Tucker, A. F., & Goodrich, L. V. (2011). Control of neuronal morphology by the atypical cadherin Fat3. *Neuron, 71*, 820–832.

Devries, S. H., & Baylor, D. A. (1997). Mosaic arrangement of ganglion cell receptive fields in rabbit retina. *Journal of Neurophysiology, 78*, 2048–2060.

Drenhaus, U., Morino, P., & Veh, R. W. (2003). On the development of the stratification of the inner plexiform layer in the chick retina. *Journal of Comparative Neurology, 460*, 1–12.

Eglen, S. J. (2006). Development of regular cellular spacing in the retina: Theoretical models. *Mathematical Medicine and Biology, 23*, 79–99.

Eysel, U. T., Peichl, L., & Wässle, H. (1985). Dendritic plasticity in the early postnatal feline retina: Quantitative characteristics and sensitive period. *Journal of Comparative Neurology, 242*, 134–145.

Field, G. D., Gauthier, J. L., Sher, A., Greschner, M., Machado, T. A., Jepson, L. H., et al. (2010). Functional connectivity in the retina at the resolution of photoreceptors. *Nature, 467*, 673–677. doi:10.1038/nature09424.

Fuerst, P. G., Bruce, F., Tian, M., Wei, W., Elstrott, J., Feller, M. B., et al. (2009). DSCAM and DSCAML1 function in self-avoidance in multiple cell types in the developing mouse retina. *Neuron, 64*, 484–497. doi:10.1016/j.neuron.2009.09.027.

Fuerst, P. G., Koizumi, A., Masland, R. H., & Burgess, R. W. (2008). Neurite arborization and mosaic spacing in the mouse retina require DSCAM. *Nature, 451*, 470–474.

Fuhrmann, S. (2010). Eye morphogenesis and patterning of the optic vesicle. *Current Topics in Developmental Biology, 93*, 61–84.

Galli-Resta, L. (2000). Local, possibly contact-mediated signalling restricted to homotypic neurons controls the regular spacing of cells within the cholinergic arrays in the developing rodent retina. *Development (Cambridge, England), 127*, 1509–1516.

Galli-Resta, L., Novelli, E., Kryger, Z., Jacobs, G. H., & Reese, B. E. (1999). Modelling the mosaic organization of rod and cone photoreceptors with a minimal-spacing rule. *European Journal of Neuroscience, 11*, 1461–1469.

Galli-Resta, L., Novelli, E., & Viegi, A. (2002). Dynamic microtubule-dependent interactions position homotypic neurones in regular monolayered arrays during retinal development. *Development (Cambridge, England), 129*, 3803–3814.

Galli-Resta, L., Resta, G., Tan, S. S., & Reese, B. E. (1997). Mosaics of islet-1-expressing amacrine cells assembled by short-range cellular interactions. *Journal of Neuroscience, 17*, 7831–7838.

Godinho, L., Mumm, J. S., Williams, P. R., Schroeter, E. H., Koerber, A., Park, S. W., et al. (2005). Targeting of amacrine cell neurites to appropriate synaptic laminae in the developing zebrafish retina. *Development (Cambridge, England), 132*, 5069–5079. doi:10.1242/dev.02075.

Goldberg, J. L., Klassen, M. P., Hua, Y., & Barres, B. A. (2002). Amacrine-signaled loss of intrinsic axon growth ability by retinal ganglion cells. *Science, 296*, 1860–1864.

Gollisch, T., & Meister, M. (2010). Eye smarter than scientists believed: Neural computations in circuits of the retina. *Neuron, 65*, 150–164.

Grueber, W. B., & Sagasti, A. (2010). Self-avoidance and tiling: Mechanisms of dendrite and axon spacing. *Cold Spring Harbor Perspectives in Biology, 2*, a001750. doi:10.1101/cshperspect.a001750.

Hinds, J. W., & Hinds, P. L. (1978). Early development of amacrine cells in the mouse retina: An electron microscopic, serial section analysis. *Journal of Comparative Neurology, 179*, 277–300.

Huckfeldt, R. M., Schubert, T., Morgan, J. L., Godinho, L., Di Cristo, G., Huang, Z. J., et al. (2009). Transient neurites of retinal horizontal cells exhibit columnar tiling via homotypic interactions. *Nature Neuroscience, 12*, 35–43.

Jeyarasasingam, G., Snider, C. J., Ratto, G. M., & Chalupa, L. M. (1998). Activity-regulated cell death contributes to the formation of ON and OFF alpha ganglion cell mosaics. *Journal of Comparative Neurology, 394*, 335–343.

Kaneko, R., Kato, H., Kawamura, Y., Esumi, S., Hirayama, T., Hirabayashi, T., et al. (2006). Allelic gene regulation of Pcdh-alpha and Pcdh-gamma clusters involving both monoallelic and biallelic expression in single Purkinje cells. *Journal of Biological Chemistry, 281*, 30551–30560. doi:10.1074/jbc.M605677200.

Kay, J. N., Chu, M. W., & Sanes, J. R. (2012). MEGF10 and MEGF11 mediate homotypic interactions required for mosaic spacing of retinal neurons. *Nature, 483*, 465–469.

Kay, J. N., Roeser, T., Mumm, J. S., Godinho, L., Mrejeru, A., Wong, R. O. L., et al. (2004). Transient requirement for ganglion cells during assembly of retinal synaptic layers. *Development, 131*, 1331–1342.

Keeley, P. W., Sliff, B., Lee, S. C. S., Fuerst, P. G., Burgess, R. W., Eglen, S. J., et al. (2011). Neuronal clustering and fasciculation phenotype in dscam- and bax-deficient mouse retinas. *Journal of Comparative Neurology, 520*, 1349–1364.

Kerschensteiner, D., Morgan, J. L., Parker, E. D., Lewis, R. M., & Wong, R. O. L. (2009). Neurotransmission selectively regulates synapse formation in parallel circuits in vivo. *Nature, 460*, 1016–1020.

Kim, I.-J., Zhang, Y., Meister, M., & Sanes, J. R. (2010). Laminar restriction of retinal ganglion cell dendrites and axons: Subtype-specific developmental patterns revealed with transgenic markers. *Journal of Neuroscience, 30*, 1452–1462.

Knabe, W., Washausen, S., Happel, N., & Kuhn, H. J. (2007). Development of starburst cholinergic amacrine cells in the retina of *Tupaia belangeri*. *Journal of Comparative Neurology, 502*, 584–597.

Lefebvre, J. L., Kostadinov, D., Chen W. V., Maniatis, T., & Sanes, J. R. (2012). Protocadherins mediate dendritic self-avoidance in the mammalian nervous system. *Nature, 488*, 517–521.

Lefebvre, J. L., Zhang, Y., Meister, M., Wang, X., & Sanes, J. R. (2008). gamma-Protocadherins regulate neuronal survival but are dispensable for circuit formation in retina. *Development (Cambridge, England), 135*, 4141–4151.

Lin, B., Wang, S. W., & Masland, R. H. (2004). Retinal ganglion cell type, size, and spacing can be specified independent of homotypic dendritic contacts. *Neuron, 43*, 475–485.

Lohmann, C., & Wong, R. O. (2001). Cell-type specific dendritic contacts between retinal ganglion cells during development. *Journal of Neurobiology, 48*, 150–162.

Matsuoka, R. L., Chivatakarn, O., Badea, T. C., Samuels, I. S., Cahill, H., Katayama, K.-I., et al. (2011a). Class 5 transmembrane semaphorins control selective mammalian retinal lamination and function. *Neuron, 71*, 460–473. doi:10.1016/j.neuron.2011.06.009.

Matsuoka, R. L., Nguyen-Ba-Charvet, K. T., Parray, A., Badea, T. C., Chédotal, A., & Kolodkin, A. L. (2011b). Transmembrane semaphorin signalling controls laminar stratification in the mammalian retina. *Nature, 470*, 259–263.

Montague, P. R., & Friedlander, M. J. (1991). Morphogenesis and territorial coverage by isolated mammalian retinal ganglion cells. *Journal of Neuroscience, 11*, 1440–1457.

Morgan, J. L., Dhingra, A., Vardi, N., & Wong, R. O. L. (2006). Axons and dendrites originate from neuroepithelial-like processes of retinal bipolar cells. *Nature Neuroscience, 9*, 85–92.

Morgan, J. L., & Wong, R. O. L. (2007). Development of cell types and synaptic connections in the retina. In H. Kolb, R. Nelson, E. Fernandez, & B. Jones (Eds.), *Webvision*. http://webvision.med.utah.edu/.

Mumm, J. S., Williams, P. R., Godinho, L., Koerber, A., Pittman, A. J., Roeser, T., et al. (2006). In vivo imaging reveals dendritic targeting of laminated afferents by zebrafish retinal ganglion cells. *Neuron, 52*, 609–621. doi:10.1016/j.neuron.2006.10.004.

Nguyen, L. T., De Juan, J., Mejia, M., & Grzywacz, N. M. (2000). Localization of choline acetyltransferase in the developing and adult turtle retinas. *Journal of Comparative Neurology, 420*, 512–526.

Peichl, L., & Wässle, H. (1979). Size, scatter and coverage of ganglion cell receptive field centres in the cat retina. *Journal of Physiology, 291*, 117–141.

Peichl, L., & Wässle, H. (1983). The structural correlate of the receptive field centre of alpha ganglion cells in the cat retina. *Journal of Physiology, 341*, 309–324.

Perry, V. H., & Linden, R. (1982). Evidence for dendritic competition in the developing retina. *Nature, 297*, 683–685.

Poché, R. A. Raven, M. A., Kwan, K. M., Furuta, Y., Behringer, R. R., & Reese, B. E. (2008). Somal positioning and dendritic growth of horizontal cells are regulated by interactions with homotypic neighbors. *European Journal of Neuroscience, 27*, 1607–1614.

Prada, C., Puelles, L., Genis-Gálvez, J. M., & Ramírez, G. (1987). Two modes of free migration of amacrine cell neuroblasts in the chick retina. *Anatomy and Embryology, 175*, 281–287.

Prada, F., Medina, J. I., López-Gallardo, M., López, R., Quesada, A., Spira, A., et al. (1999). Spatiotemporal gradients of differentiation of chick retina types I and II cholinergic cells: Identification of a common postmitotic cell population. *Journal of Comparative Neurology, 410*, 457–466.

Raven, M. A., Eglen, S. J., Ohab, J. J., & Reese, B. E. (2003). Determinants of the exclusion zone in dopaminergic amacrine cell mosaics. *Journal of Comparative Neurology, 461*, 123–136.

Reddien, P. W., & Horvitz, H. R. (2004). The engulfment process of programmed cell death in caenorhabditis elegans. *Annual Review of Cell and Developmental Biology, 20*, 193–221.

Reese, B. E. (2011). Development of the retina and optic pathway. *Vision Research, 51*, 613–632. doi:10.1016/j.visres.2010.07.010.

Reese, B. E., Harvey, A. R., & Tan, S. S. (1995). Radial and tangential dispersion patterns in the mouse retina are cell-class specific. *Proceedings of the National Academy of Sciences of the United States of America, 92,* 2494–2498. doi:10.1073/pnas.92.7.2494.

Rockhill, R. L., Euler, T., & Masland, R. H. (2000). Spatial order within but not between types of retinal neurons. *Proceedings of the National Academy of Sciences of the United States of America, 97,* 2303–2307. doi:10.1073/pnas.030413497.

Roska, B., & Werblin, F. (2001). Vertical interactions across ten parallel, stacked representations in the mammalian retina. *Nature, 410,* 583–587.

Sanes, J. R., & Yamagata, M. (2009). Many paths to synaptic specificity. *Annual Review of Cell and Developmental Biology, 25,* 161–195.

Sanes, J. R., & Zipursky, S. L. (2010). Design principles of insect and vertebrate visual systems. *Neuron, 66,* 15–36.

Schreiner, D., & Weiner, J. A. (2010). Combinatorial homophilic interaction between gamma-protocadherin multimers greatly expands the molecular diversity of cell adhesion. *Proceedings of the National Academy of Sciences of the United States of America, 107,* 14893–14898. doi:10.1073/pnas. 1004526107.

Siegert, S., Scherf, B. G., Del Punta, K., Didkovsky, N., Heintz, N., & Roska, B. (2009). Genetic address book for retinal cell types. *Nature Neuroscience, 12,* 1197–1204.

Stacy, R. C., & Wong, R. O. L. (2003). Developmental relationship between cholinergic amacrine cell processes and ganglion cell dendrites of the mouse retina. *Journal of Comparative Neurology, 456,* 154–166.

Suzuki, E., & Nakayama, M. (2007). MEGF10 is a mammalian ortholog of CED-1 that interacts with clathrin assembly protein complex 2 medium chain and induces large vacuole formation. *Experimental Cell Research, 313,* 3729–3742. doi:10.1016/j.yexcr.2007.06.015.

Tauchi, M., & Masland, R. H. (1984). The shape and arrangement of the cholinergic neurons in the rabbit retina. *Proceedings of the Royal Society of London. Series B, Biological Sciences, 223,* 101–119.

Tian, N., & Copenhagen, D. R. (2003). Visual stimulation is required for refinement of ON and OFF pathways in postnatal retina. *Neuron, 39,* 85–96.

Vaney, D. I. (1994). Territorial organization of direction-selective ganglion cells in rabbit retina. *Journal of Neuroscience, 14,* 6301–6316.

Voinescu, P. E., Kay, J. N., & Sanes, J. R. (2009). Birthdays of retinal amacrine cell subtypes are systematically related to their molecular identity and soma position. *Journal of Comparative Neurology, 517,* 737–750.

Wang, X., Weiner, J. A., Levi, S., Craig, A. M., Bradley, A., & Sanes, J. R. (2002). Gamma protocadherins are required for survival of spinal interneurons. *Neuron, 36,* 843–854.

Weiner, J. A., Wang, X., Tapia, J. C., & Sanes, J. R. (2005). Gamma protocadherins are required for synaptic development in the spinal cord. *Proceedings of the National Academy of Sciences of the United States of America, 102,* 8–14. doi:10.1073/pnas.0407931101.

Wingate, R. J. (1996). Retinal ganglion cell dendritic development and its control: Filling the gaps. *Molecular Neurobiology, 12,* 133–144.

Wong, R. O., & Godinho, L. (2003). Development of the vertebrate retina. In L. M. Chalupa & J. S. Werner (Eds.), *The visual neurosciences* (pp. 77–93). Cambridge, MA: MIT Press.

Wu, H., Bellmunt, E., Scheib, J. L., Venegas, V., Burkert, C., Reichardt, L. F., et al. (2009). Glial precursors clear sensory neuron corpses during development via Jedi-1, an engulfment receptor. *Nature Neuroscience, 12,* 1534–1541. doi:10.1038/nn.2446.

Wu, Q., & Maniatis, T. (1999). A striking organization of a large family of human neural cadherin-like cell adhesion genes. *Cell, 97,* 779–790.

Wässle, H., & Boycott, B. B. (1991). Functional architecture of the mammalian retina. *Physiological Reviews, 71,* 447–480.

Wässle, H., Peichl, L., & Boycott, B. B. (1981). Dendritic territories of cat retinal ganglion cells. *Nature, 292,* 344–345.

Wässle, H., Puller, C., Müller, F., & Haverkamp, S. (2009). Cone contacts, mosaics, and territories of bipolar cells in the mouse retina. *Journal of Neuroscience, 29,* 106–117.

Yamagata, M., & Sanes, J. R. (2008). Dscam and Sidekick proteins direct lamina-specific synaptic connections in vertebrate retina. *Nature, 451,* 465–469.

Yamagata, M., & Sanes, J. R. (2012).Expanding the immunoglobulin superfamily code for laminar specificity in retina: Expression and role of contactins. *Journal of Neuroscience, 32,* 14402–14414.

Yamagata, M., Weiner, J. A., & Sanes, J. R. (2002). Sidekicks: Synaptic adhesion molecules that promote lamina-specific connectivity in the retina. *Cell, 110,* 649–660.

Zipursky, S. L., & Sanes, J. R. (2010). Chemoaffinity revisited: Dscams, protocadherins, and neural circuit assembly. *Cell, 143,* 343–353.

91 The Role of DSCAMs in the Neural Development of the Retina and Visual System

ABIGAIL L. D. TADENEV, ANDREW M. GARRETT, AND
ROBERT W. BURGESS

In this chapter, we will discuss some of the general anatomical features of the vertebrate retina, including the diversity of neuronal cell types, the positioning of the cell bodies in both vertical and horizontal space, the arborization of the processes of specific cell types, and the relationship of this anatomy to circuitry in the retina. We will then discuss Down syndrome cell adhesion molecules (DSCAMs), proteins that have been extensively studied in *Drosophila*, which are now known to be important factors in vertebrate retinal development. DSCAMs are implicated in self-recognition and self-avoidance, cell body positioning, laminar specificity and dendrite arborization, controlling cell number, and axon guidance and targeting in the visual system. We will describe the data supporting each of these functions, as well as emerging cellular and molecular mechanisms of DSCAM function.

ANATOMICAL CONSIDERATIONS FOR VISUAL SYSTEMS

The primary function of the eye is to create a neural representation of the external visual field. This is true regardless of the type of eye being studied, from the compound eye of *Drosophila* to the "simple" vertebrate eye. While the simplest eyes in the animal kingdom merely detect the amount of ambient light, here we will focus on visual systems that detect not only light levels but also shapes, motion, and often color. In order to create a cellular representation of their environment, animals have developed eyes that contain arrays of photosensitive cells, much like a digital camera's image sensor. The paths by which light reaches the photoreceptors of the retinal array dictate the visual field of the eye, and each cell communicates information from a given point in that field. Unlike the image sensor, however, the retina also uses interneurons to perform multiple processing functions, such as motion detection, on the signal transmitted by its photoreceptive element. The circuitry by which the photoreceptors' information transits the retina determines the receptive field and response properties of the ganglion cells, the output neurons of the system. Ultimately, this convergent processing produces a faithful neural representation of the visual world. The spatial mapping of this array must be maintained from the photoreceptors to the brain, across multiple synaptic junctions, and each region of the visual field must be processed by the appropriate subtypes of interneurons. Thus, it has long been assumed that the spacing and morphology of each neuron in the chain is of critical importance for the fidelity of visual processing. Indeed, the highly stereotyped nature of the anatomy of eyes, from the number and spacing of cells of each subtype to the size, shape, and position of their dendritic and axonal arbors, speaks to its importance.

Certain recurring themes are seen in the architecture of retinal neurons, as they often have the common task of transmitting information from a given physical area with maximal efficiency and minimal redundancy needed for fidelity. One of these themes is self-avoidance, where neurites from a given cell actively avoid contacting one another. This helps a cell to develop even and economical coverage within its receptive field. Another theme is that of tiling, and is exhibited by certain retinal cell types. Tiling refers to the complete coverage of an area (typically the entire retina) by a given cell type, where the dendritic arbors of the cells abut, but do not overlap, each other. This allows for the nonredundant transmission of information, helping to maintain the precise mapping of the visual field. Other cell types also show even coverage of the retina, but with overlap, allowing for some redundancy and a probable increase in sensitivity. The extent to which cells of a given subtype overlap is termed the coverage factor, which is defined as the number of neurons of that subtype that overlap a particular point. In this way, the coverage factor equals one for a cell type that tiles, and increases as overlap increases.

Many variables exist for the anatomical arrangement of a cell type, including the number of cells, the spacing of those cells, and the size and shape of each cell's dendritic arbor. How, then, does a cell know whether it is in the proper location and has the correct receptive field? Especially in more complex animals where the number of cells and their precise spacing vary slightly between individuals, this information is not developmentally hardwired. Rather, cells must obtain information from their local environment in order to determine whether they should move this way or that or remain stationary and whether to retract or extend their neurites and, if so, in which direction. Given the vast number of cell types that intermingle within the retina (at least 50 for the mouse retina) (Coombs et al., 2006; MacNeil & Masland, 1998; Rockhill et al., 2002; Wassle et al., 2009), it is tempting to postulate that there is a complex recognition code (Zipursky & Sanes, 2010) that allows a developing neurite to determine whether it has encountered "self" (a sister branch from the same neuron, also called isoneuronal), "self-type" (a branch from a different neuron of the identical subtype, also called homotypic), or "nonself" (another cell type or substrate such as extracellular matrix, also called heterotypic). As an additional layer of complexity, it is likely that the recognition code also conveys information about subcellular localization, distinguishing the cell body, axon, and dendrite, and perhaps even distinct domains along a neurite.

The Down syndrome cell adhesion molecules (DSCAMs) are a family of proteins that have been implicated in both isoneuronal and homotypic recognition. As we will discuss, while there are significant differences between insect and vertebrate DSCAMs, they have similar functions in neuronal morphogenesis. In *Drosophila*, DSCAMs mediate both self-avoidance and tiling, in part by an extraordinary diversity of Dscam1 imparted by a large set of alternatively spliced exons. In mice, no such broad molecular diversity exists, but DSCAMs are nevertheless critical for generating appropriate cell numbers, spacing, and arborization within the retina.

IDENTIFICATION AND INITIAL
CHARACTERIZATION OF DSCAMs

Vertebrate DSCAMs

DSCAM is an immunoglobulin (Ig) superfamily member first identified as a gene in the Down syndrome critical interval 21q22.2–22.3 that was thought likely to be involved in the neurological phenotypes of the disease given its widespread expression in the nervous system and similarity to neural cell adhesion molecules, including Contactins, and Deleted in colorectal carcinoma (DCC) (Yamakawa et al., 1998). It was classified as the defining member of a new class of neural cell adhesion molecules due to its unique structure (see figure 91.1B) (Yamakawa et al., 1998). Subsequent studies confirmed its expression in the adult and developing mouse nervous system (Agarwala et al., 2001a; Barlow et al., 2001, 2002) and identified a highly related gene, *Dscam-like 1* (*DSCAML1*), which is also widely expressed in the nervous system but in a largely nonoverlapping subset of neurons (Agarwala et al., 2001b; Barlow et al., 2002). Both forms exhibit homophilic binding in cell aggregation assays (Agarwala et al., 2000, 2001b; Barlow et al., 2002).

Drosophila *Dscams*

Drosophila Dscam1 was first identified in a screen of proteins that interact with Dock (Schmucker et al., 2000), an SH2–SH3 adaptor protein implicated in neural circuit formation (Desai et al., 1999; Garrity et al., 1996). While its extracellular region has the same domain structure as mammalian DSCAM, their intracellular domains are divergent. Consistent with the similarity of their extracellular domains, *Drosophila* Dscam1 also shows homophilic binding (Wojtowicz et al., 2004). In stark contrast to mammalian DSCAMs, insect Dscam1 shows an almost unparalleled level of diversity generated by alternative splicing (see figure 91.1A). With 12 alternate exon 4s, 48 exon 6s, and 33 exon 9s, plus two alternate transmembrane domains (exon 17), mutually exclusive splicing can generate over 38,000 different isoforms (Schmucker et al., 2000). In addition, the inclusion or exclusion of exons 19 and 23 allows for structural and functional diversity of the Dscam1 endodomain (Yu et al., 2009). Surprisingly, more than 95% of isoforms exhibit preferential isoform-specific homophilic binding, and this occurs in a modular fashion where homophilic interactions between all three variable regions (corresponding roughly to Ig domains 2, 3, and 7) are required (Wojtowicz et al., 2007). Crystal structures support the hypothesis that each variable Ig domain binds to its counterpart on a sister Dscam1 molecule in an antiparallel fashion (Meijers et al., 2007; Sawaya et al., 2008) (see figure 91.1C). A stochastic method of alternative exon usage determines the expression of *Dscam1* isoforms within a cell, and each cell expresses approximately 14–50 isoforms (Neves et al., 2004); however, isoform choices are

  ABIGAIL L. D. TADENEV, ANDREW M. GARRETT, AND ROBERT W. BURGESS

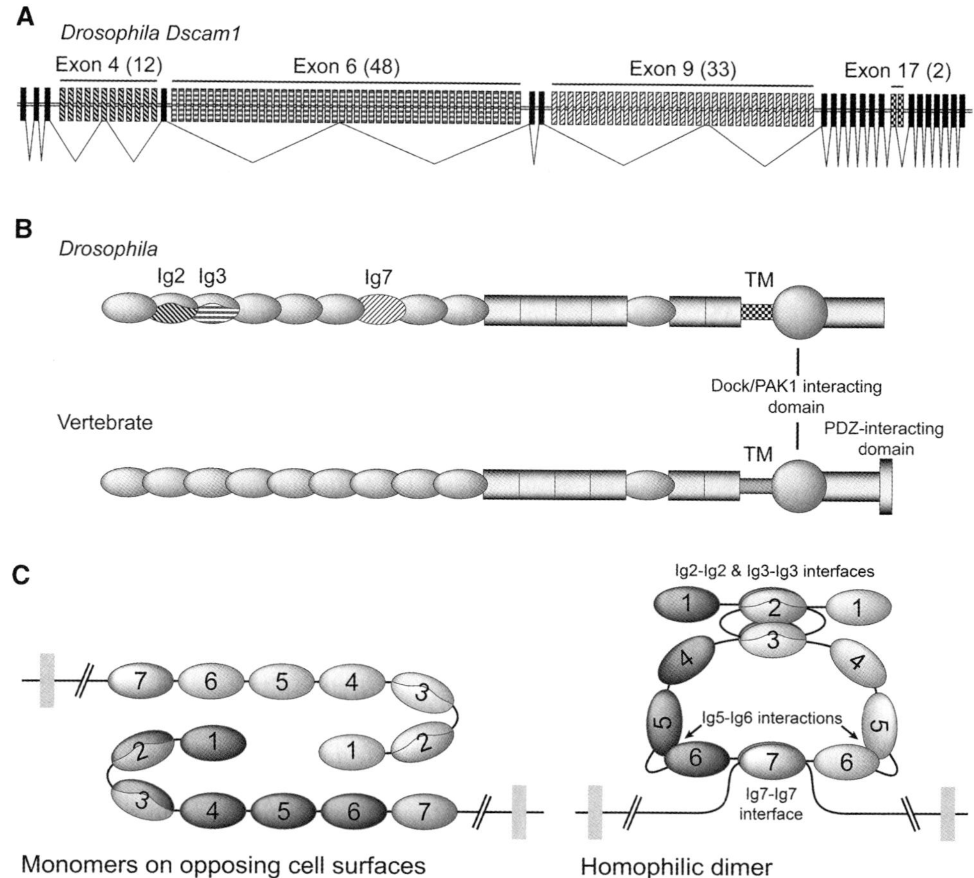

FIGURE 91.1　Fly and vertebrate Down syndrome cell adhesion molecules (*DSCAMs*). (A) *Drosophila Dscam1* contains four sets of alternative exons. Random choice of variable exons 4, 6, and 9 determines the adhesive specificities of immunoglobulin (Ig) domains 2, 3, and 7 respectively, while variable exon 17 encodes the transmembrane domain (B). Unlike in *Drosophila*, vertebrate *Dscam* is not extensively alternatively spliced, but the protein domain structures are nevertheless highly similar. DSCAMs contain 10 Ig domains, the 10th of which is located between the fourth and fifth of six fibronectin repeats. The intracellular domain of both fly and vertebrate DSCAMs can activate P21-activated kinase (Pak1): vertebrate DSCAM by directly binding Pak1 via its juxtamembrane domain and *Drosophila* Dscam1 through the adaptor protein Dock. Vertebrate DSCAMs have canonical PDZ interacting domains at the c-terminus, while *Drosophila* Dscams do not. (C) *Drosophila* Dscam1 binds in trans by forming an S-shape, allowing homophilic interactions between Ig domains 2, 3, and 7. (Modified from Hattori, Millard, Wojtowicz, & Zipursky, 2008.)

also biased, influenced by both spatial and temporal regulation (Celotto & Graveley, 2001; Zhan et al., 2004). Thus, Dscam1 can provide each cell that expresses it with a unique molecular signature on its cell surface.

EARLY STUDIES ON DSCAM FUNCTION: NEURONAL MORPHOGENESIS IN *DROSOPHILA*

Dscam1 Mediates Self-Avoidance

Much of our understanding of the function of DSCAMs in the visual system comes from studies in *Drosophila* in other regions of the body. In the initial description of *Drosophila* Dscam1, Schmucker et al. (2000) sought to identify axon guidance receptors via an association with Dock, and indeed they demonstrated a role for Dscam1 in axon pathfinding within Bolwig's nerve and the ventral nerve cord. Subsequent studies revealed axonal defects in other cell types. *Dscam1* mutant flies have altered branching and impaired divergence of axonal branches in mushroom body neurons (Wang et al., 2002) and defects in the fasciculation of mushroom body axons within the peduncle (Zhan et al., 2004). Mechanosensory neurons have reduced axonal arborization (Chen et al., 2006), and olfactory receptor neurons from both the antennae and maxillary palp exhibit defects in axonal targeting and elaboration (Hummel et al., 2003).

More recently, defects in the morphogenesis of dendritic arbors have also been reported. Projection neurons that normally innervate only a single glomerulus within the antennal lobe still target to the correct glomerulus, but the dendritic arbor is significantly reduced, and dendritic branches of a single projection neuron are often clumped together (Zhu et al., 2006). These data support a previously proposed role for Dscam1 in self-avoidance (Zhan et al., 2004). Once the importance of Dscam1 for dendritic morphogenesis was understood, attention shifted from the *Drosophila* olfactory system to a set of neurons better known for their elaborate dendrites, the dendritic arborization (da) neurons of the larval body wall. There are four classes of da neurons, I–IV, which are categorized based on increasing size and complexity of their dendritic arbors. A trio of independent studies in 2007 revealed dramatic defects in the dendritic arbors of da neurons (Hughes et al., 2007; Matthews et al., 2007; Soba et al., 2007). *Dscam1* mutant da neurons display defective self-avoidance as evidenced by an increased number of self-crossings and fasciculation within the arbor of an individual cell; however, the tiling of class III (Hughes et al., 2007) and IV (Matthews et al., 2007) neurons is unaffected. Expression of single Dscam1 isoforms within mutant cells is sufficient to rescue self-avoidance, and intracellular signaling is likely required, as Dscam1 lacking its cytoplasmic domain is unable to rescue. The implication of these results is that Dscam1 is required cell-autonomously in order to confer individual identity, and homophilic interaction between identical Dscam1 isoforms on sister neurites induces repulsion. When *Dscam1* is ablated, neurites can no longer recognize and avoid each other.

How many Dscam1 isoforms are needed to confer individual cell identity and maintain the fidelity of neurite arborization? Several studies have addressed this issue by genetically reducing the total possible number of splice variants, and these experiments generally agree that, while some reduction in diversity can be tolerated, thousands of isoforms are needed to construct the wild-type (WT) anatomy (Chen et al., 2006; Hattori et al., 2007, 2009; Wang et al., 2004). Thus, the vast diversity of Dscam1 is key to its function.

Drosophila *Dscam2 Promotes Tiling in the Visual System*

In addition to *Dscam1*, the *Drosophila* genome contains three other DSCAM family members, *Dscam2*, *Dscam3*, and *Dscam4*; however, these genes have not attracted the same attention that *Dscam1* has because, although they share the same protein domain structure, they lack *Dscam1*'s extensive isoform diversity. Dscam3 functions redundantly with Dscam1 to promote CNS midline crossing of axons (Andrews et al., 2008). Like Dscam1, Dscam2 has also been shown to be important in neuronal morphogenesis. However, it has taken on a distinct role in this process: Dscam2 helps mediate axonal tiling within the medulla of the *Drosophila* visual system (Millard et al., 2007). The *Drosophila* compound eye is made up of 750 units called ommatidia, each of which contains eight photoreceptor neurons, termed R cells. Cells R1–R6 terminate within columnar "cartridges" in a region directly below the retina called the lamina, where they synapse on lamina neurons L1–L5. The axons of R7 and R8 photoreceptors do not terminate in the lamina, but instead, along with the L1–L5 neurons, project to the medulla of the optic lobe. Here the projections terminate in columns, and each column contains one axon of each cell type. Like Dscam1, Dscam2 binds homophilically to mediate repulsion between neurites. Ablation of *Dscam2* causes axonal arbors to extend beyond the boundaries of their resident columns in the medulla. However, while Dscam1 isoforms differ from cell to cell and confer self-avoidance, Dscam2 does not differ between sister cells and instead confers "self-type" avoidance. In this way, Dscam2 is more similar to vertebrate DSCAMs, which are not extensively alternatively spliced.

Can the knowledge that has been gained about the function of Dscams in *Drosophila* be applied to vertebrate visual systems? As we will discuss, while the enormous diversity of Dscam1 is restricted to insect lineages, many of the basic principles, including self-avoidance, are conserved.

THE ROLE OF DSCAMs IN THE MOUSE VISUAL SYSTEM

Not long after the importance of *Dscams* in neuronal morphogenesis was demonstrated in *Drosophila*, a spontaneously occurring mouse *Dscam* mutant was described (Fuerst et al., 2008). The *Dscam*[del17] mouse carries a 38-bp deletion in exon 17, leading to premature termination and dramatic loss of messenger RNA and protein. *Dscam* mutant mice have reduced viability, which is particularly prominent on some inbred backgrounds, and adults display an overt neurological phenotype, including incoordination, spontaneous seizures, and kyphosis (increased dorsal/ventral curvature of the thoracic spine).

DSCAMs Mediate Self-Avoidance in the Retina

Given its elaborate, stereotyped structure and the breadth of knowledge and tools available, the retina makes a highly tractable system to study questions of neuronal morphogenesis and intercellular recognition in mammals. As *Dscam* is robustly expressed in the mouse retina, Fuerst et al. (2008) examined the effects of *Dscam* mutation on retinal anatomy. In cross-section, hematoxylin/eosin-stained mutant retinas reveal disorganization of the inner nuclear and ganglion cell layers. Molecular markers exist for many retinal cell types, and profound defects were found in dopaminergic and bNOS-positive amacrine cells. Double in situ hybridization confirmed that affected cell types were, in fact, those that express *Dscam*. These cell types were then labeled and imaged in flat, whole-mount retinal preparations, revealing dramatic morphological defects. In the absence of DSCAM, these cells are abnormally spaced, and their dendrites are no longer evenly elaborated but instead fasciculate in a cell type–specific manner (see figure 91.2). In the WT retina, cell bodies of most neuronal subtypes are nonrandomly spaced, such that fewer cells are located in close proximity to each other than would be expected if they were randomly distributed. This phenomenon can be quantified using a statistical test, density recovery profiling (DRP), which measures the density of sister cells located at incremental distances from the reference cell (Rodieck, 1991). DRP analyses demonstrated that the "exclusion zone" around WT cells is not observed in the *Dscam* mutant retina, indicating a defect in the cell spacing mechanism. A similar spacing defect is seen in the dendrites of these cells. As opposed to the spreading arborization observed in the WT retina, the dendrites of *Dscam* mutant amacrine cells fasciculate, often forming large, rope-like entanglements. Interestingly, dendrites from neighboring dopaminergic and bNOS-positive mutant amacrine cells do not aggregate across cell types but instead clump only with their sister dendrites.

Subsequent studies revealed similar phenotypes in several subpopulations of retinal ganglion cells (RGCs) (Fuerst et al., 2009). These populations were labeled with antibodies against melanopsin and SMI-32 and a YFP-transgenic reporter line (Mito-Y). Each labeled population is itself heterogeneous, with anti-melanopsin labeling multiple classes of intrinsically

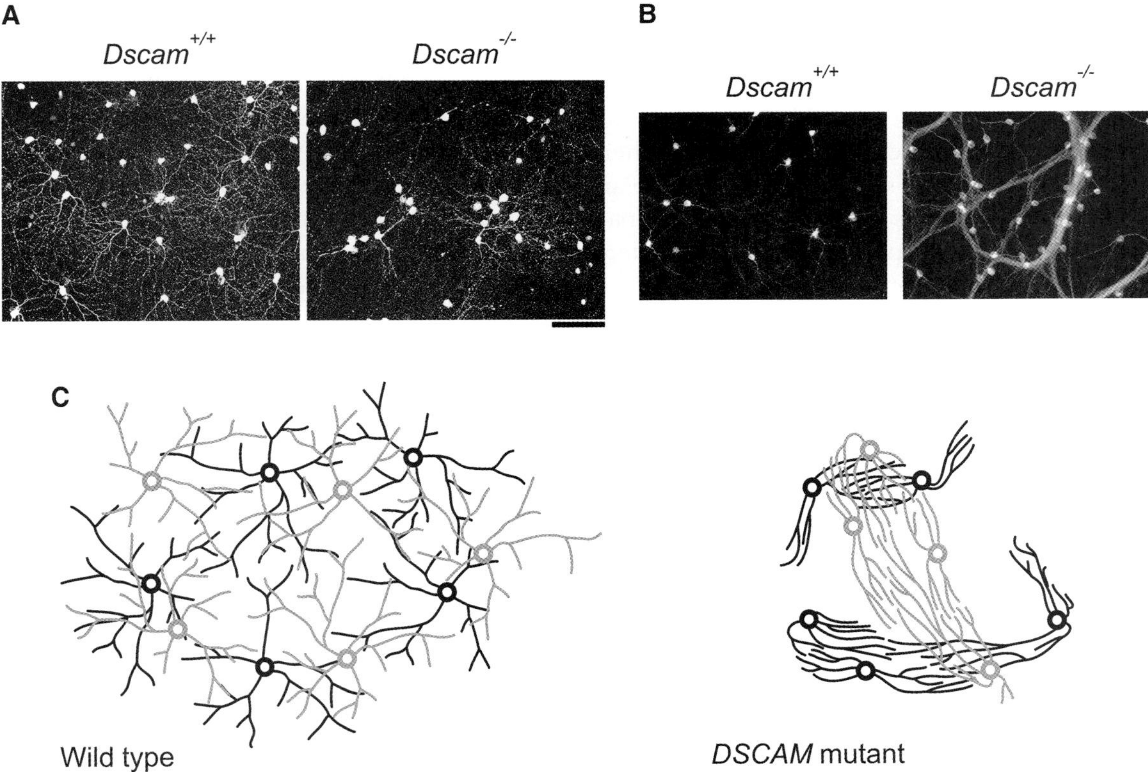

FIGURE 91.2 Down syndrome cell adhesion molecule (DSCAM) prevents dendrite fasciculation and cell body clumping between homotypic cells. Normal cell body spacing and dendrite arborization is disrupted in *Dscam⁻/⁻* retinas. The phenotype is limited to cells that normally express *Dscam*, including retinal ganglion cell populations (A) and amacrine cell populations (B). Clumping and fasciculation is cell type specific: Clumps containing multiple cell types are not observed (illustrated in C).

photosensitive retinal ganglion cells (ipRGCs), anti-SMI-32 labeling at least four subtypes of RGCs (Coombs et al., 2006), and Mito-Y labeling at least three subtypes of RGCs. DRP analyses demonstrated loss of the normal cellular "exclusion zone" in these *Dscam* mutant RGCs, as described above for amacrine cells. Each labeled group of mutant RGCs also had extensive fasciculation of their dendrites, and although markers did not exist to examine whether the constituent subtypes of these heterogeneous groups aggregated independently, analyses using the depth of dendritic stratification within the inner plexiform layer (IPL) as an indicator of ipRGC subtype suggest that the aggregates are indeed cell subtype-specific. Studies using a conditional allele of *Dscam* indicate that DSCAM mediates self-avoidance in a cell type–specific manner: Cells of a given type clump and fasciculate both when all members of the group are *Dscam* mutant and when chimerism allows WT and mutant cells to intermingle (Fuerst et al., 2012). These results indicate that DSCAM indeed functions by binding homophilically, and this is consistent with contact-dependent self-avoidance in a "cell type–autonomous" manner.

The closely related gene *Dscaml1* appears to function in much the same way as *Dscam*, although in different cell populations, dependent upon its expression pattern (Fuerst et al., 2009). DSCAML1 is found on AII amacrine and rod bipolar cells, and in a *Dscaml1*-null mouse line generated via gene trapping (*Dscaml1^{GT}*), these cell types display defects reminiscent of *Dscam* mutant cells. The AII amacrine cell bodies are clumped together instead of being mosaically spaced, and rod bipolar cells, which are not mosaically spaced in the WT retina, display fasciculation of their proximal dendrites. Interestingly, there are also defects in the distribution of dopaminergic amacrine (DA) cell neurites, but not in spacing of their cell bodies. *Dscaml1* is not expressed in this population; the likely explanation is that the DA cell phenotype is secondary to the clumping of AII amacrine cells, which are a synaptic target of the DA cells.

Together, these studies reveal a dramatic defect in self-avoidance within mutant cell populations that would normally express *Dscams*. Other populations that do not normally express *Dscam*, such as cholinergic starburst amacrines, are unaffected in the *Dscam* mutant mouse (Fuerst et al., 2008). While these data implicate DSCAMs in self-avoidance in a subset of cells, they are not consistent with a simple repulsive mechanism as seems to be the case in *Drosophila*. In the mouse retina, there is little to no *Dscam* isoform diversity, and cell types that express the gene overlap extensively, with coverage factors averaging between two and nine

(Fuerst et al., 2009). Thus, it has been hypothesized that DSCAM normally passively mediates self-avoidance by acting as a "nonstick coating," countering an underlying predilection for adhesion among sister cells (Fuerst & Burgess, 2009). In support of this, a recent study examined the normal morphological development of DA cells and inferred that these neurons extend their neurites with indifference to, and not active avoidance of, their neighboring sister cells (Keeley & Reese, 2010).

DSCAMs and Cell Number

In addition to the clumping and fasciculation phenotypes described above, both *Dscam* and *Dscaml1* mutant mice display increases in cell number for populations that would normally express the genes; In general, other cell types are unaffected. The molecular reasons behind this increase in cell number are unclear, but the proximal cause appears to be a lack of developmental cell death. There is no evidence for an increase in proliferation and no change in RGC cell number at birth; the increase instead manifests around P5, after the normal window of programmed cell death. Indeed, a decrease in TUNEL-positive DA cells was observed during the normal developmental wave of cell death, and the clumped DA cells were not clonal, indicating that the clumping is not due to overproliferation of progenitor cells (Fuerst et al., 2008). It is possible that the clustering of homotypic cells induces some level of resistance to programmed cell death. In order to examine whether the clumping and fasciculation phenotypes are a by-product of the increase in cell number, *Bax* knockout mice, in which programmed cell death is inhibited, were studied (Fuerst et al., 2009; Keeley et al., 2011). These mice show an increase in cell number that is variable between cell types but is of comparable magnitude to the increase seen in *Dscam* mutants. In studying DA and ipRGC cell populations in the *Bax* mutant, some defects were seen in cell spacing and fasciculation, but these varied in severity between cell types and were milder than those seen in the *Dscam* mutant retina. Indeed, although DA cells are much more numerous in the *Bax* knockout, they exhibit significantly less clumping than in the *Dscam* mutant (Keeley et al., 2011). Thus, while the increase in cell number may contribute to the morphological changes observed in *Dscam* and *Dscaml1* mutant mice, especially in some cell types such as ipRGCs, it cannot account entirely for the clumping and fasciculation observed in other cell types, such as dopaminergic amacrine cells (Fuerst et al., 2012; Keeley et al., 2011).

DSCAMs have also been implicated in laminar specificity of neuronal processes in the plexiform layers of the retina (Fuerst et al., 2010; Yamagata & Sanes, 2008). In the chick IPL, Dscam, Dscam-like (DscamL), and the related proteins Sidekick-1 and -2 each label a characteristic group of sublaminae. By identifying other markers for RGCs expressing these proteins, Yamagata and Sanes (2008) observed diffusion of dendrites away from this tight stratification when the adhesion molecules were targeted by RNA interference. In addition, ectopic expression of Dscams or Sidekicks redirects neurites to the sublaminae normally occupied by the ectopically expressed adhesion molecule. While the underlying mechanism remains unclear, the authors suggest that adhesion is important for establishing IPL laminar specificity and demonstrate that a human neural cell line transfected with plasmids encoding either Dscam or Sidekick induces aggregation and accumulation of synaptic vesicles at sites of cell–cell contact, suggesting a homophilic attraction mechanism.

In mice, the case for involvement of DSCAMs in laminar specificity is more complicated and probably depends on the genetic background of the mice used. Both the $Dscam^{del17}$ and $Dscaml1^{GT}$ mice described above were maintained on a mixed background to circumvent the significant perinatal lethality that occurs on some inbred backgrounds (Fuerst et al., 2010). Within these strains, IPL laminar specificity defects are mild and inconsistent and difficult to separate from general retinal disorganization (Fuerst et al., 2009). However, a different mutant allele of $Dscam$, $Dscam^{2J}$, which arose on a different inbred genetic background, does display some defects in laminar specificity within some cell populations in which $Dscam$ regulates cell spacing, such as bNOS-positive amacrines, but not in other such populations, such as DA cells or melanopsin-positive RGCs (Fuerst et al., 2012, 2010). Indeed, calbindin-positive amacrine cells are tightly laminated in the $Dscam^{del17}$, but not the $Dscam^{2J}$, retina (Fuerst et al., 2010). Notably, laminar specificity of cholinergic amacrine cells, which do not express $Dscam$, is also impaired in the $Dscam^{2J}$ mutant retina; this suggests some non-cell-autonomy in $Dscam$'s involvement in this phenotype (Fuerst et al., 2010, 2012). Both the $Dscam^{del17}$ and the $Dscam^{2J}$ mutations are severe loss-of-function mutations with truncating frame shifts, so the phenotypic differences between the two may result from differences in the genetic background rather than from differences in allele severity. These results suggest that while $Dscam$ is involved in neurite laminar specificity, this function is heavily influenced by other genetic factors in mammals.

Closely related to the issue of laminar specificity is that of synaptic specificity, as IPL laminae are defined by the stratified synaptic contacts between various groups of cellular partners. If laminar specificity is perturbed, will the synaptic specificity of the resident cell types then be disturbed? Recent evidence suggests that the answer may be no. Mutation of either Semaphorin6A or its receptor, PlexinA4, disrupts the laminar specificity of dopaminergic amacrine cells and M1-type ipRGCs, but the colocalization of these fibers within the IPL remains largely intact (Matsuoka et al., 2011). Synaptic specificity has also been studied in $Dscam$ mutant settings, albeit not in the $Dscam^{2J}$ mice. Fuerst et al. (2009) examined the coupling of two pairs of cell types whose spacing and arborization are affected by the $Dscaml1^{GT}$ mutation: rod photoreceptors and rod bipolar cells (RBCs), and RBCs and AII amacrine cells. Electron microscopy revealed that synapses between these cell types still form in the $Dscaml1$-null retina, and electrophysiological recordings demonstrated functionality of these connections; however, synapses appeared somewhat abnormal in both of these assays. Thus, although these cells and their neurites are no longer in their correct positions, they are still able to recognize and form synapses with their partners, but whether the appropriate number of connections form, and whether promiscuous connections form with inappropriate targets, remains to be determined.

Data from other species also provide evidence for DSCAM function in synapses. In *Aplysia*, DSCAM is involved in both de novo and learning-induced synaptogenesis via reorganization of glutamate receptors (Li et al., 2009). Studies in *Drosophila* suggest that Dscams may play a role in determining the synaptic composition of multiple-contact synapses (Millard et al., 2010), which are those that include multiple presynaptic elements, postsynaptic elements, or both. In the lamina of the *Drosophila* visual system, photoreceptors form tetrad synapses containing four postsynaptic elements. Two of these are invariant and come from an L1- and an L2-type cell, respectively, while the remaining two elements are contributed by other cell types. Defects in composition of tetrad synapses occur in two cases: (1) when "self-type" avoidance and the fidelity of cartridge (column containing one cell of each type) composition are lost via $Dscam2$ mutation and (2) when both "self" and "self-type" avoidance are lost via loss of both $Dscam1$ and $Dscam2$. In the first case, there is a modest percentage of "fused cartridges" containing two cells of each type, and loss of Dscam2-mediated "self-type" avoidance allows two L2 cells to connect at a single synapse. In the remaining single cartridges, Dscam1 is able to compensate for Dscam2, yielding WT synapses. In the second case, even within single cartridges, two processes from a single L1 or L2 cell can connect at a single synapse because both "self" and "self-type" avoidance mechanisms are nonfunctional. This suggests a redundant role for both Dscam1 and Dscam2 in ensuring appropriate L1 and L2 contributions to tetrad synapses. It will be fascinating to see in future studies whether vertebrate DSCAMs play any such role in multiple-contact synapses; if so, this could suggest that DSCAMs are capable of mediating repulsion for the purpose of synaptic specificity in these species as well as in *Drosophila*.

Beyond the Retina: DSCAMs and Circuit Refinement

DSCAMs also appear to play a role in the vertebrate visual system outside of the retina. It is easy to imagine that the visual maps that appear in various forms in the brain, like the retina, rely heavily on a highly stereotyped anatomy, and that DSCAMs may help to organize cells within these maps. A recent study by Blank et al. (2011) suggests that DSCAMs do indeed play a role in organizing higher visual centers. RGC axons from both eyes project out of the retina and terminate in the lateral geniculate nucleus (LGN), a bilateral structure within the thalamus. Here, fibers from the ipsilateral eye occupy a region within the center of the LGN while fibers from the contralateral eye fill the remainder. There is normally a modest degree of overlap of these two regions. Using a mouse model of Down syndrome, Ts65Dn, which carries three copies of *Dscam* (along with many other genes) and the *Dscam^{del17}* mutant mouse line, it was shown that *Dscam* regulates refinement of this segregation in a dose-dependent manner: In *Dscam^{+/-}* mice, segregation of inputs is defective, resulting in a larger region of overlap; in *Dscam^{-/-}* mice, the segregation pattern is disrupted, with the ipsilateral fibers forming multiple foci; and in Ts65Dn;*Dscam^{+/+}* mice carrying three copies of *Dscam*, segregation is enhanced, resulting in a smaller region of overlap. Notably, reducing the number of copies of *Dscam* to two in the context of trisomy (Ts65Dn;*Dscam^{+/-}* mice) does not fully restore the WT phenotype, suggesting that although *Dscam* is an important player in the segregation process, there are additional genes located within the Down syndrome interval that are also involved (Blank et al., 2011). Consistent with the dose-dependent effects of *Dscam* in segregation, the cell number and spacing of dopaminergic amacrine cells, and the cell number, spacing, and dendritic arborization

of melanopsin-positive RGCs, are also sensitive to *Dscam* dosage (Blank et al., 2011; Fuerst et al., 2012; Keeley et al., 2011). The extent to which the defects in other visual areas of the brain are secondary to phenotypes in the retina remains to be determined, but evidence that DSCAMs function outside the retina, and indeed outside of the visual system, is provided by the overt phenotype of the mice and by defects in other CNS regions such as the brain stem rhythmicity center (Amano et al., 2009).

MECHANISM OF DSCAM FUNCTION

Extracellular Domain

Much of the function of DSCAMs has been ascribed to their homophilic binding on adjacent sister neurites or processes from different cells. The structural basis for this interaction, and indeed for the highly specific homophilic binding shown by *Drosophila* Dscam1 isoforms, is well understood (see figure 91.1). There is, however, also evidence for heterophilic binding of DSCAMs to other ligands. In *Drosophila*, Dscam1 and Dscam3 function redundantly in midline crossing of CNS axons; this is partly mediated by binding the secreted protein Netrin but also seems to have a Netrin-independent component (Andrews et al., 2008). Vertebrate DSCAM has also been suggested to have the ability to bind Netrin and mediate midline guidance and Netrin-induced turning of commissural axons (Liu et al., 2009; Ly et al., 2008). However, recent genetic evidence using *Dscam* knockout mice suggests that, in vertebrates, DSCAM is not required for the outgrowth and guidance of commissural axons directed by Netrin (Palmesino et al., 2012). Apparent discrepancies between these studies may reflect the compensation that can occur in cells in which a gene has been totally ablated, and which may not be observed during transient suppression by RNA interference. DSCAM may also have an indirect effect on Netrin-induced phenotypes; specifically, *Drosophila* Dscam1's repulsive effects may counterbalance Netrin-mediated attraction (Matthews & Grueber, 2011). DSCAM physically interacts with DCC, a canonical Netrin receptor (Ly et al., 2008); while the consequences of this interaction are unclear, it is possible that this allows DSCAM to modulate Netrin-mediated signaling.

Draxin, another secreted protein, is proposed to bind DSCAM and could inhibit axon extension (Ahmed et al., 2011). While the heterophilic binding of DSCAMs to such molecules is intriguing, it is unlikely to be a major mechanism mediating the function of DSCAMs in retinal development given the short distances involved; data also suggest that DSCAMs' effects in the retina are contact-dependent (Fuerst et al., 2012).

Intracellular Signaling

Drosophila Dscam1 was first identified in a screen of proteins that bind to Dock, an SH2/SH3 protein (Schmucker et al., 2000), in an interaction that may be dependent upon Dscam1 phosphorylation (Muda et al., 2002). Dscam1/Dock recruit Pak (p21-activated serine/threonine kinase); these three proteins act together to affect development of Bolwig's nerve (Schmucker et al., 2000). While the intracellular domain of Dscam1 is crucial for its effects on the arborization of da neurons (Hughes et al., 2007; Matthews et al., 2007; Soba et al., 2007), it appears that Dock and Pak are dispensable for this function as loss-of-function mutations in these genes do not phenocopy loss of *Dscam1* (Hughes et al., 2007; Matthews et al., 2007). DSH3PX1, a putative *Drosophila* sorting nexin that interacts with the actin cytoskeleton, was also identified as a component of the Dscam1/Dock complex (Worby et al., 2001); it remains to be seen if this protein plays any role in the Dscam1 phenotype in *Drosophila*.

The intracellular domains of DSCAMs differ greatly between insect and mammalian homologues. Nevertheless, human DSCAM also activates Pak1, in this case directly instead of via an intermediary as in *Drosophila* (Li & Guan, 2004). Mammalian DSCAM is also capable, directly or indirectly, of activating Fyn, a Src family kinase (Liu et al., 2009), and the MAP kinases Jnk and p38 (Li & Guan, 2004). Jnk activation plays a role in mediating morphological changes induced by transfection of heterologous cells with the DSCAM intracellular domain (Li & Guan, 2004). It will be interesting to learn whether Jnk is involved in DSCAM functions within the visual system. Vertebrate DSCAMs are predicted to have a PDZ-interacting C terminus, and bind to multi-PDZ domain-containing proteins, including PSD-95, Chapsyn110, SAP102, Pist/Cal, and MAGI-1 and-2 (Yamagata & Sanes, 2010). MAGI-1 can recruit Sidekick-2, a protein closely related to DSCAMs, to photoreceptor synapses (Yamagata & Sanes, 2010). Multi-PDZ domain proteins such as the MAGIs are attractive candidates for mediators of DSCAM function in the retina as they can interact with multiple transmembrane proteins simultaneously. This could help to explain DSCAMs' apparent function in counteracting an intrinsic attraction between cells of the same subtype; perhaps both DSCAM and the adhesive molecules are present in the same complex. Thus, the potential role of MAGI proteins in development of the microanatomy of the visual system is likely to be an exciting line of investigation. Moreover, MAGIs, like the related PSD-95 protein, have been proposed to be important for synaptic function, in part by their association with synaptic proteins such as the *N*-methyl-D-aspartate (NMDA) receptor, glutamate receptor, and Stargazin, a transmembrane AMPA receptor regulating protein (Deng et al., 2006; Hirao et al., 1998; Stetak et al., 2009). This along with the persistent expression of DSCAM in the adult and the observation of abnormal ribbon synapses in *Dscaml1*-null mice (Fuerst et al., 2009) suggest that DSCAMs may be involved in maintenance and/or plasticity of synapses as well as in their development.

SUMMARY

In the fly visual system, DSCAMs are involved in tiling and synaptic specificity, while in the mouse, they have been shown to regulate cell number and spacing, dendritic arborization, laminar specificity, and segregation of axonal projections into the thalamus (see figure 91.3). Whereas much of the function of DSCAMs originates from homophilic binding, the downstream effects of that binding also differ among species: In *Drosophila*, DSCAMs mainly induce repulsion whereas they mediate indifference in mouse retinal neurons and perhaps even adhesion in the chick IPL. The divergence of intracellular domains between insect and vertebrate DSCAMs is likely to underlie many of these differences, but it is also possible that the expression, distribution, and regulation of various intracellular signaling pathways endows DSCAMs with versatile effects at different developmental stages. One can imagine how this could be exquisitely regulated so that DSCAM could mediate repulsion, adhesion, or indifference in particular subcellular compartments, at different time points, or in response to various external or internal cues. These attractive and repulsive forces are likely important to counterbalance such forces generated by other transmembrane proteins, including those suggested by the cell type–specific adhesion of *Dscam* mutant mouse retinal neurons. It will be fascinating to uncover the identities of these molecules and those of proteins that may have a similar function in cells lacking expression of DSCAMs. This knowledge will help us to understand the mechanisms of DSCAM function in the regulation of cell number, spacing, and dendrite arborization and may also elucidate the role of DSCAM in axon guidance. The retina and visual system provide an excellent model for examining the neurodevelopmental role of DSCAMs, but the principles learned will certainly apply to other regions of the central nervous system as well.

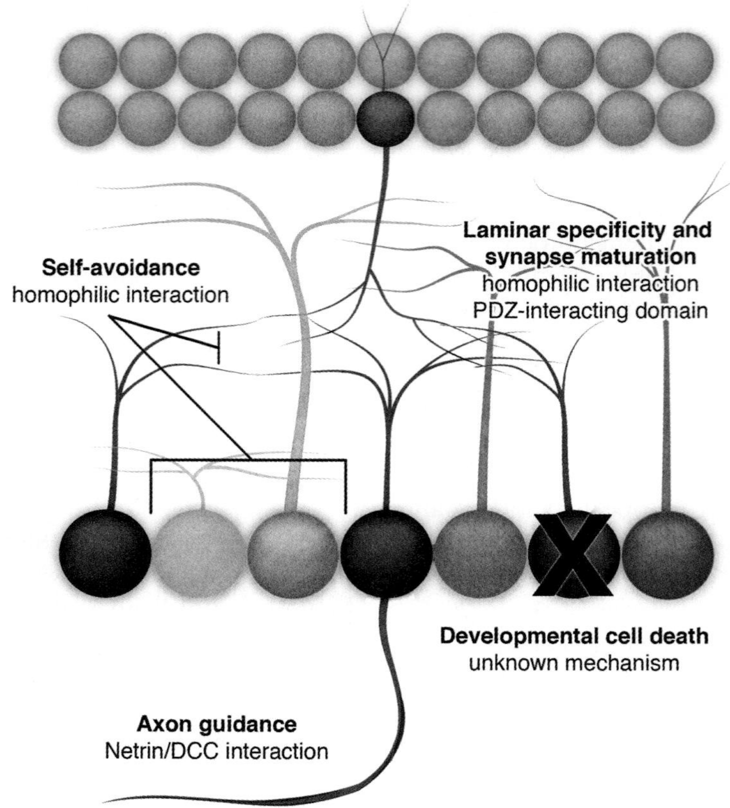

FIGURE 91.3 Down syndrome cell adhesion molecules (DSCAMs) influence multiple aspects of vertebrate visual system development. DSCAMs are implicated in self-avoidance functions analogous to *Drosophila* Dscam1. In mice, this serves to prevent dendrite fasciculation and to maintain mosaic cell body spacing between cells of a specific subtype. DSCAMs are also implicated in laminar specificity, in which synaptic partners costratify in the inner plexiform layer of the retina, and in the formation of fully matured ribbon synapses. DSCAMs control cell number, and in their absence, those cells that would normally express them are overly abundant due to an apparent decrease in developmental cell death. DSCAM also interacts with netrin and its receptor Deleted in colorectal carcinoma (DCC), possibly influencing axon outgrowth and guidance. PDZ, PSD-95/disks large/ ZO-1.

REFERENCES

Agarwala, K. L., Ganesh, S., Amano, K., Suzuki, T., & Yamakawa, K. (2001a). DSCAM, a highly conserved gene in mammals, expressed in differentiating mouse brain. *Biochemical and Biophysical Research Communications, 281*, 697– 705. doi: 10.1006/bbrc.2001.4420 S0006–291X(01)94420–1.

Agarwala, K. L., Ganesh, S., Tsutsumi, Y., Suzuki, T., Amano, K., & Yamakawa, K. (2001b). Cloning and functional characterization of DSCAML1, a novel DSCAM-like cell adhesion molecule that mediates homophilic intercellular adhesion. *Biochemical and Biophysical Research Communications, 285*, 760–772. doi: 10.1006/bbrc.2001.5214 S0006291X01952143.

Agarwala, K. L., Nakamura, S., Tsutsumi, Y., & Yamakawa, K. (2000). Down syndrome cell adhesion molecule DSCAM mediates homophilic intercellular adhesion. *Brain Research. Molecular Brain Research, 79*, 118–126. doi: S0169328X00 00108X.

Ahmed, G., Shinmyo, Y., Ohta, K., Islam, S. M., Hossain, M., Naser, I. B., et al. (2011). Draxin inhibits axonal outgrowth through the netrin receptor DCC. *Journal of Neuroscience, 31*, 14018–14023. doi:10.1523/JNEUROSCI.0943-11.2011.

Amano, K., Fujii, M., Arata, S., Tojima, T., Ogawa, M., Morita, N., et al. (2009). DSCAM deficiency causes loss of preinspiratory neuron synchroneity and perinatal death. *Journal of Neuroscience, 29*, 2984–2996. doi:10.1523/ jneurosci.3624-08.2009.

Andrews, G. L., Tanglao, S., Farmer, W. T., Morin, S., Brotman, S., Berberoglu, M. A., et al. (2008). Dscam guides embryonic axons by Netrin-dependent and -independent functions. *Development, 135*, 3839–3848. doi:10.1242/dev.023739.

Barlow, G. M., Micales, B., Chen, X. N., Lyons, G. E., & Korenberg, J. R. (2002). Mammalian DSCAMs: Roles in the development of the spinal cord, cortex, and cerebellum? *Biochemical and Biophysical Research Communications, 293*, 881–891. doi: 10.1016/S0006-291X(02)00307-8 S0006- 291X(02)00307-8.

Barlow, G. M., Micales, B., Lyons, G. E., & Korenberg, J. R. (2001). Down syndrome cell adhesion molecule is conserved in mouse and highly expressed in the adult mouse brain. *Cytogenetics and Cell Genetics, 94*, 155–162.

Blank, M., Fuerst, P. G., Stevens, B., Nouri, N., Kirkby, L., Warrier, D., et al. (2011). The Down syndrome critical region regulates retinogeniculate refinement. *Journal of Neuroscience, 31*, 5764–5776. doi:10.1523/JNEUROSCI.6015-10.2011.

Celotto, A. M., & Graveley, B. R. (2001). Alternative splicing of the *Drosophila* Dscam pre-mRNA is both temporally and spatially regulated. *Genetics, 159*, 599–608.

Chen, B. E., Kondo, M., Garnier, A., Watson, F. L., Puettmann-Holgado, R., Lamar, D. R., et al. (2006). The molecular diversity of Dscam is functionally required for neuronal wiring specificity in *Drosophila. Cell, 125*, 607–620. doi:10.1016/j.cell.2006.03.034.

Coombs, J., van der List, D., Wang, G.-Y., & Chalupa, L. M. (2006). Morphological properties of mouse retinal ganglion cells. *Neuroscience, 140*, 123–136. doi:10.1016/j.neuroscience.2006.02.079.

Deng, F., Price, M. G., Davis, C. F., Mori, M., & Burgess, D. L. (2006). Stargazin and other transmembrane AMPA receptor regulating proteins interact with synaptic scaffolding protein MAGI-2 in brain. *Journal of Neuroscience, 26*, 7875–7884. doi:10.1523/jneurosci.1851-06.2006.

Desai, C. J., Garrity, P. A., Keshishian, H., Zipursky, S. L., & Zinn, K. (1999). The *Drosophila* SH2–SH3 adapter protein Dock is expressed in embryonic axons and facilitates synapse formation by the RP3 motoneuron. *Development, 126*, 1527–1535.

Fuerst, P. G., Bruce, F., Rounds, R. P., Erskine, L., & Burgess, R. W. (2012). Cell autonomy of DSCAM function in retinal development. *Developmental Biology, 361*, 326–337. doi:10.1016/j.ydbio.2011.10.028.

Fuerst, P. G., Bruce, F., Tian, M., Wei, W., Elstrott, J., Feller, M. B., et al. (2009). DSCAM and DSCAML1 function in self-avoidance in multiple cell types in the developing mouse retina. *Neuron, 64*, 484–497. doi:10.1016/j.neuron.2009.09.027.

Fuerst, P. G., & Burgess, R. W. (2009). Adhesion molecules in establishing retinal circuitry. *Current Opinion in Neurobiology, 19*, 389–394. doi:10.1016/j.conb.2009.07.013.

Fuerst, P. G., Harris, B. S., Johnson, K. R., & Burgess, R. W. (2010). A novel null allele of mouse DSCAM survives to adulthood on an inbred C3H background with reduced phenotypic variability. *Genesis (New York, N.Y.), 48*, 578–584. doi:10.1002/dvg.20662.

Fuerst, P. G., Koizumi, A., Masland, R. H., & Burgess, R. W. (2008). Neurite arborization and mosaic spacing in the mouse retina require DSCAM. *Nature, 451*, 470–474. doi:10.1038/nature06514.

Garrity, P. A., Rao, Y., Salecker, I., McGlade, J., Pawson, T., & Zipursky, S. L. (1996). *Drosophila* photoreceptor axon guidance and targeting requires the dreadlocks SH2/SH3 adapter protein. *Cell, 85*, 639–650.

Hattori, D., Chen, Y., Matthews, B. J., Salwinski, L., Sabatti, C., Grueber, W. B., et al. (2009). Robust discrimination between self and non-self neurites requires thousands of Dscam1 isoforms. *Nature, 461*, 644–648. doi:10.1038/nature08431.

Hattori, D., Demir, E., Kim, H. W., Viragh, E., Zipursky, S. L., & Dickson, B. J. (2007). Dscam diversity is essential for neuronal wiring and self-recognition. *Nature, 449*, 223–227. doi:10.1038/nature06099.

Hattori, D., Millard, S. S., Wojtowicz, W. M., & Zipursky, S. L. (2008). Dscam-mediated cell recognition regulates neural circuit formation. *Annual Review of Cell and Developmental Biology, 24*, 597–620. doi:10.1146/annurev.cellbio.24.110707.175250.

Hirao, K., Hata, Y., Ide, N., Takeuchi, M., Irie, M., Yao, I., et al. (1998). A novel multiple PDZ domain-containing molecule interacting with N-methyl-D-aspartate receptors and neuronal cell adhesion proteins. *Journal of Biological Chemistry, 273*, 21105–21110.

Hughes, M. E., Bortnick, R., Tsubouchi, A., Baumer, P., Kondo, M., Uemura, T., et al. (2007). Homophilic Dscam interactions control complex dendrite morphogenesis. *Neuron, 54*, 417–427. doi:10.1016/j.neuron.2007.04.013.

Hummel, T., Vasconcelos, M. L., Clemens, J. C., Fishilevich, Y., Vosshall, L. B., & Zipursky, S. L. (2003). Axonal targeting of olfactory receptor neurons in *Drosophila* is controlled by Dscam. *Neuron, 37*, 221–231. doi:10.1016/S0896-6273(02)01183-2.

Keeley, P. W., & Reese, B. E. (2010). Morphology of dopaminergic amacrine cells in the mouse retina: Independence from homotypic interactions. *Journal of Comparative Neurology, 518*, 1220–1231. doi:10.1002/cne.22270.

Keeley, P. W., Sliff, B., Lee, S. C. S., Fuerst, P. G., Burgess, R. W., Eglen, S. J., et al. (2011). Neuronal clustering and fasciculation phenotype in Dscam- and Bax-deficient mouse retinas. *Journal of Comparative Neurology*. doi:10.1002/cne.23027.

Li, H.-L., Huang, B. S., Vishwasrao, H., Sutedja, N., Chen, W., Jin, I., et al. (2009). Dscam mediates remodeling of glutamate receptors in *Aplysia* during de novo and learning-related synapse formation. *Neuron, 61*, 527–540. doi:10.1016/j.neuron.2009.01.010.

Li, W., & Guan, K. L. (2004). The Down syndrome cell adhesion molecule (DSCAM) interacts with and activates Pak. *Journal of Biological Chemistry, 279*, 32824–32831. doi:10.1074/jbc.M401878200.

Liu, G., Li, W., Wang, L., Kar, A., Guan, K.-L., Rao, Y., et al. (2009). DSCAM functions as a netrin receptor in commissural axon pathfinding. *Proceedings of the National Academy of Sciences of the United States of America, 106*, 2951–2956. doi:10.1073/pnas.0811083106.

Ly, A., Nikolaev, A., Suresh, G., Zheng, Y., Tessier-Lavigne, M., & Stein, E. (2008). DSCAM is a netrin receptor that collaborates with DCC in mediating turning responses to netrin-1. *Cell, 133*, 1241–1254. doi:10.1016/j.cell.2008.05.030.

MacNeil, M. A., & Masland, R. H. (1998). Extreme diversity among amacrine cells: Implications for function. *Neuron, 20*, 971–982.

Matsuoka, R. L., Nguyen-Ba-Charvet, K. T., Parray, A., Badea, T. C., Chédotal, A., & Kolodkin, A. L. (2011). Transmembrane semaphorin signalling controls laminar stratification in the mammalian retina. *Nature, 470*, 259–263. doi:10.1038/nature09675.

Matthews, B. J., & Grueber, W. B. (2011). Dscam1-mediated self-avoidance counters netrin-dependent targeting of dendrites in *Drosophila. Current Biology, 21*, 1480–1487. doi:10.1016/j.cub.2011.07.040.

Matthews, B. J., Kim, M. E., Flanagan, J. J., Hattori, D., Clemens, J. C., Zipursky, S. L., et al. (2007). Dendrite self-avoidance is controlled by Dscam. *Cell, 129*, 593–604. doi:10.1016/j.cell.2007.04.013.

Meijers, R., Puettmann-Holgado, R., Skiniotis, G., Liu, J. H., Walz, T., Wang, J. H., et al. (2007). Structural basis of Dscam isoform specificity. *Nature, 449*, 487–491. doi:10.1038/nature06147.

Millard, S. S., Flanagan, J. J., Pappu, K. S., Wu, W., & Zipursky, S. L. (2007). Dscam2 mediates axonal tiling in the *Drosophila* visual system. *Nature, 447*, 720–724. doi:10.1038/nature05855.

Millard, S. S., Lu, Z., Zipursky, S. L., & Meinertzhagen, I. A. (2010). *Drosophila* dscam proteins regulate postsynaptic specificity at multiple-contact synapses. *Neuron, 67*, 761–768. doi:10.1016/j.neuron.2010.08.030.

Muda, M., Worby, C. A., Simonson-Leff, N., Clemens, J. C., & Dixon, J. E. (2002). Use of double-stranded RNA-mediated interference to determine the substrates of protein tyrosine kinases and phosphatases. *Biochem J, 366*, 73–77. doi:10.1042/BJ20020298.

Neves, G., Zucker, J., Daly, M., & Chess, A. (2004). Stochastic yet biased expression of multiple Dscam splice variants by individual cells. *Nature Genetics, 36*, 240–246. doi:10.1038/ng1299.

Palmesino, E., Haddick, P. C., Tessier-Lavigne, M., & Kania, A. (2012). Genetic analysis of DSCAM's role as a netrin-1 receptor in vertebrates. *Journal of Neuroscience, 32*, 411–416. doi:10.1523/JNEUROSCI.3563-11.2012.

Rockhill, R. L., Daly, F. J., MacNeil, M. A., Brown, S. P., & Masland, R. H. (2002). The diversity of ganglion cells in a mammalian retina. *Journal of Neuroscience, 22*, 3831–3843. doi: 20026369.

Rodieck, R. W. (1991). The density recovery profile: A method for the analysis of points in the plane applicable to retinal studies. *Visual Neuroscience, 6*, 95–111.

Sawaya, M. R., Wojtowicz, W. M., Andre, I., Qian, B., Wu, W., Baker, D., et al. (2008). A double S shape provides the structural basis for the extraordinary binding specificity of Dscam isoforms. *Cell, 134*, 1007–1018. doi:10.1016/j.cell.2008.07.042.

Schmucker, D., Clemens, J. C., Shu, H., Worby, C. A., Xiao, J., Muda, M., et al. (2000). *Drosophila* Dscam is an axon guidance receptor exhibiting extraordinary molecular diversity. *Cell, 101*, 671–684. doi:10.1016/S0092-8674(00)80878-8.

Soba, P., Zhu, S., Emoto, K., Younger, S., Yang, S. J., Yu, H. H., et al. (2007). *Drosophila* sensory neurons require Dscam for dendritic self-avoidance and proper dendritic field organization. *Neuron, 54*, 403–416. doi:10.1016/j.neuron.2007.03.029.

Stetak, A., Horndli, F., Maricq, A. V., van den Heuvel, S., & Hajnal, A. (2009). Neuron-specific regulation of associative learning and memory by MAGI-1 in *C. elegans*. *PLoS ONE, 4*, e6019. doi:10.1371/journal.pone.0006019.

Wang, J., Ma, X., Yang, J. S., Zheng, X., Zugates, C. T., Lee, C. H., et al. (2004). Transmembrane/juxtamembrane domain-dependent Dscam distribution and function during mushroom body neuronal morphogenesis. *Neuron, 43*, 663–672. doi:10.1016/j.neuron.2004.06.033.

Wang, J., Zugates, C. T., Liang, I. H., Lee, C. H., & Lee, T. (2002). *Drosophila* Dscam is required for divergent segregation of sister branches and suppresses ectopic bifurcation of axons. *Neuron, 33*, 559–571. doi:10.1016/S0896-6273(02)00570-6.

Wassle, H., Puller, C., Muller, F., & Haverkamp, S. (2009). Cone contacts, mosaics, and territories of bipolar cells in the mouse retina. *Journal of Neuroscience, 29*, 106–117. doi:10.1523/jneurosci.4442-08.2009.

Wojtowicz, W. M., Flanagan, J. J., Millard, S. S., Zipursky, S. L., & Clemens, J. C. (2004). Alternative splicing of *Drosophila* Dscam generates axon guidance receptors that exhibit isoform-specific homophilic binding. *Cell, 118*, 619–633. doi:10.1016/j.cell.2004.08.021.

Wojtowicz, W. M., Wu, W., Andre, I., Qian, B., Baker, D., & Zipursky, S. L. (2007). A vast repertoire of Dscam binding specificities arises from modular interactions of variable Ig domains. *Cell, 130*, 1134–1145. doi:10.1016/j.cell.2007.08.026.

Worby, C. A., Simonson-Leff, N., Clemens, J. C., Kruger, R. P., Muda, M., & Dixon, J. E. (2001). The sorting nexin, DSH3PX1, connects the axonal guidance receptor, Dscam, to the actin cytoskeleton. *Journal of Biological Chemistry, 276*, 41782–41789. doi: 10.1074/jbc.M107080200 M107080200.

Yamagata, M., & Sanes, J. R. (2008). Dscam and Sidekick proteins direct lamina-specific synaptic connections in vertebrate retina. *Nature, 451*, 465–469. doi:10.1038/nature06469.

Yamagata, M., & Sanes, J. R. (2010). Synaptic localization and function of Sidekick recognition molecules require MAGI scaffolding proteins. *Journal of Neuroscience, 30*, 3579–3588. doi:10.1523/JNEUROSCI.6319-09.2010.

Yamakawa, K., Huot, Y. K., Haendelt, M. A., Hubert, R., Chen, X. N., Lyons, G. E., et al. (1998). DSCAM: A novel member of the immunoglobulin superfamily maps in a Down syndrome region and is involved in the development of the nervous system. *Human Molecular Genetics, 7*, 227–237. doi:10.1093/hmg/7.2.227.

Yu, H.-H., Yang, J. S., Wang, J., Huang, Y., & Lee, T. (2009). Endodomain diversity in the *Drosophila* Dscam and its roles in neuronal morphogenesis. *Journal of Neuroscience, 29*, 1904–1914. doi:10.1523/JNEUROSCI.5743-08.2009.

Zhan, X.-L., Clemens, J. C., Neves, G., Hattori, D., Flanagan, J. J., Hummel, T., et al. (2004). Analysis of Dscam diversity in regulating axon guidance in *Drosophila* mushroom bodies. *Neuron, 43*, 673–686. doi:10.1016/j.neuron.2004.07.020.

Zhu, H., Hummel, T., Clemens, J. C., Berdnik, D., Zipursky, S. L., & Luo, L. (2006). Dendritic patterning by Dscam and synaptic partner matching in the *Drosophila* antennal lobe. *Nature Neuroscience, 9*, 349–355. doi:10.1038/nn1652.

Zipursky, S. L., & Sanes, J. R. (2010). Chemoaffinity revisited: Dscams, protocadherins, and neural circuit assembly. *Cell, 143*, 343–353. doi:10.1016/j.cell.2010.10.009.

92 The Development of Retinal Decussations

CAROL MASON, TAKAAKI KUWAJIMA, AND QING WANG

In the vertebrate visual system, retinal ganglion cell (RGC) axons from each eye grow toward the brain and meet at the midline in the ventral diencephalon, where they establish an X-shaped pathway, the optic chiasm. In higher vertebrates, RGC axons from each eye diverge from one another at the optic chiasm, enter the optic tract on the same (ipsilateral) and opposite (contralateral) sides of the brain, and then extend to targets in the thalamus (dorsal lateral geniculate nucleus; dLGN) and the midbrain (superior colliculus; SC; or optic tectum in lower vertebrates). The divergence of ipsi- and contralateral axons at the chiasm is critical to binocular vision.

Here we will review optic chiasm development, including the axon paths and behaviors of RGCs during their growth as they form the chiasm. We will present recent progress in the understanding of molecular mechanisms that guide axons during avoidance or passage through the midline at the chiasm, transcription factors that specify ipsilateral and contralateral programs of cell identity, and genes that pattern the retina into two domains giving rise to these two projections. We will then discuss the relationship of axon divergence at the chiasm to eye-specific innervation of the dLGN. Finally, we will highlight the visual system defects of albinos and other mutants in which the balance of ipsi- and contralateral retinal projections is perturbed. We also direct the reader to several reviews on retinal axon divergence that have appeared since the previous publication of these volumes (Erskine & Herrera, 2007; Herrera & Mason, 2006; Petros & Mason, 2008; Petros, Rebsam, & Mason, 2008; Williams, Mason, & Herrera, 2004).

PRINCIPLES OF OPTIC CHIASM ORGANIZATION

The patterns of RGC axon projection at the optic chiasm in different species range from complete crossing, with segregation of fibers from each eye, to partial decussation, with complex intermingling of the fibers from the two eyes. The presence and relative size of the ipsilateral projection depends on the degree of binocular overlap in the visual field (see figure 92.1). In lower vertebrates that lack binocular vision, such as fish, the fibers from each eye form an entirely crossed projection. The tadpoles of some amphibian species, such as *Xenopus laevis*, also have an entirely crossed projection, but an ipsilateral projection develops at metamorphosis when the eyes change position and binocularity develops (Hoskins & Grobstein, 1985; Mann & Holt, 2001). Adult chickens have an entirely crossed projection, but in early development there is a transient ipsilateral projection (O'Leary, Gerfen, & Cowan, 1983; Thanos & Bonhoeffer, 1984).

In mammals, partial decussation occurs, and the degree of crossing varies widely among species (see figure 92.1). In humans and primates, all RGC axons originating from the nasal retina cross the midline and project into the contralateral optic tract, and all RGC axons from the temporal retina project into the ipsilateral optic tract, resulting in ~40% of RGC axons projecting ipsilaterally (Chalupa & Lia, 1991; Polyak, 1957). In ferrets and cats ~12%–15% of RGCs project ipsilaterally. By contrast, in mouse, only a small proportion of RGCs (<5%) located in the peripheral ventro-temporal (VT) retina project ipsilaterally (see figure 92.2), and these cells are intermingled with contralaterally projecting RGCs (Guillery, Mason, & Taylor, 1995; Jeffery, 2001). Eye position on the head is correlated with degree of binocularity: The more frontal the eyes, the higher the proportion of ipsilaterally projecting RGCs.

OPTIC CHIASM FORMATION: RETINAL AXON DIVERGENCE

Perhaps the most complex change in RGC axon arrangement along the visual pathway is the divergence of crossed and uncrossed axons at the optic chiasm. An early view of RGC axon behavior during divergent growth was made by Ramón y Cajal. His drawing of the kitten chiasm revealed that some uncrossed axons grow

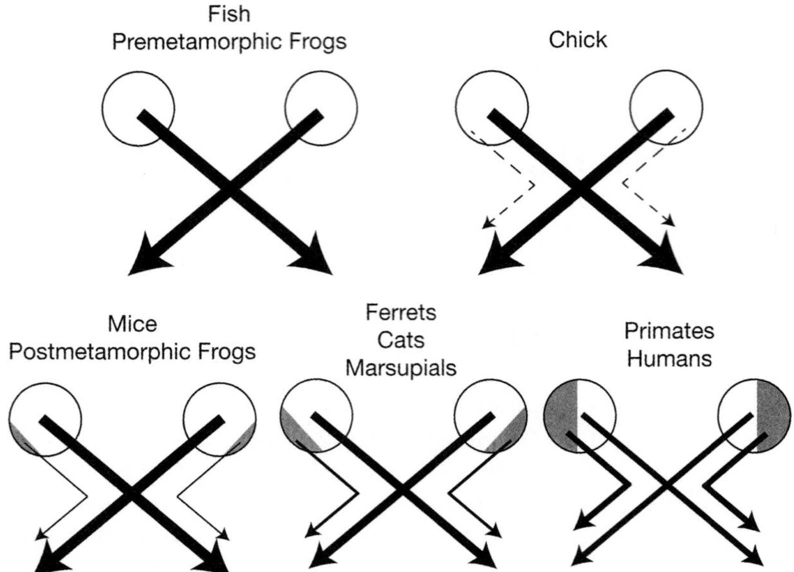

FIGURE 92.1   Retinal axon decussation at the optic chiasm of different species. Topographic origin and relative proportion of crossed and uncrossed retinal projections varies across species.

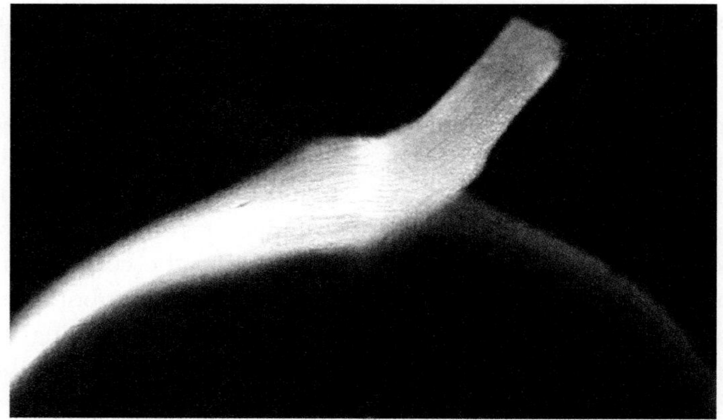

FIGURE 92.2   The mouse optic chiasm. The projection of retinal ganglion cell axons at the optic chiasm of an E17 mouse embryo, viewed in a whole mount from the ventral surface of the brain. A crystal of DiI was applied to the left optic disk. The majority of fibers cross the midline in the optic chiasm, and a small subset of fibers (<5%) extend ipsilateral to the labeled optic nerve. Some retinal ganglion cell axons course into the other optic nerve during development and into adulthood of diverse species (Plump et al., 2002).

directly into the ipsilateral optic tract while others project toward the midline and then curve back to project ipsilaterally (Ramon y Cajal, 1911). By contrast, in marsupials, the ipsilaterally projecting axons diverge laterally and never approach the midline (Jeffery & Harman, 1992; Taylor & Guillery, 1995). Direct growth of RGC axons into the ipsilateral optic tract occurs also during early stages of mouse chiasm development (Colello & Guillery, 1990; Marcus & Mason, 1995), the formation of the transient ipsilateral projection in the chick (Thanos & Bonhoeffer, 1984), and possibly at the primate and human chiasm (Hoyt & Luis, 1963; Meissirel & Chalupa, 1994). In later stages of mouse development, however, the axons that give rise to the permanent ipsilateral projection separate from axons projecting contralaterally close to the midline (Godement, Salaun, & Mason, 1990; Godement, Wang, & Mason, 1994). Although the routes taken by RGC axons through the chiasm may vary, the mechanisms that regulate axon divergence at the chiasm, in all species, include distinct properties encoded by the ipsilaterally and contralaterally projecting RGCs such that their axons respond differentially to guidance cues arrayed at the chiasm midline (see below).

     CAROL MASON, TAKAAKI KUWAJIMA, AND QING WANG

Time Course of Optic Chiasm Formation

Studies in the past 30 years have characterized the birth-dates of ipsilateral and contralateral RGCs and the timing of axon growth through the optic chiasm (Colello & Guillery, 1990; Dräger, 1985; Sretavan, 1990). RGC axon decussation in the mouse takes place at three discrete stages of optic chiasm development (see figure 92.3). From embryonic day (E) 12 to E13, the earliest-born RGCs originating from dorsocentral (DC) retina extend through the optic stalk and along the pia of the ventral diencephalon both contralaterally and ipsilaterally. However, the fate of these early-born ipsilaterally projecting RGCs is unknown as they cannot be retrogradely labeled in adulthood. During the peak phase of retinal axon divergence from E14 to E17, RGC differentiation expands from the DC region to the peripheral circumference of the retina. The permanent ipsilateral projection emerges from RGCs in the most peripheral VT retinal crescent, and the contralateral projection forms from RGCs outside of the VT crescent. During the late phase of RGC axon extension from E17 to postnatal day (P) 0, newborn RGCs in the VT crescent project contralaterally rather than ipsilaterally. Thus, from E14 to E17, RGCs from the ventrotemporal retina project primarily to ipsilateral targets, but after E17, the last-born RGCs within VT retina project contralaterally. Questions remain on whether the earliest born DC RGCs pioneer the retinal axon pathway as seen in zebrafish (Pittman, Law, & Chien, 2008), whether the DC RGCs that project ipsilaterally have the same molecular profile as the permanent ipsilateral VT RGCs, and whether they die, retract their ipsilateral projections, or migrate to the VT retina.

Cellular Composition of the Optic Chiasm Region

Studies characterizing the cellular and molecular specializations in the optic chiasm have shown that radial glial cells are positioned at the base of third ventricle, and their processes form a palisade on both sides of the chiasm midline. The chiasmatic radial glial cells express glial markers such RC2, brain lipid-binding protein (BLBP), and glutamate-aspartate transporter (GLAST) during the formation of the retinal decussation. The other major chiasmatic cell population is a phalanx of early-born neurons caudal to the chiasm that extends a raphe into the midline. These cells are morphologically distinct from the radial glia and express β-tubulin, SSEA-1, and CD44 (Marcus et al., 1995; Marcus &

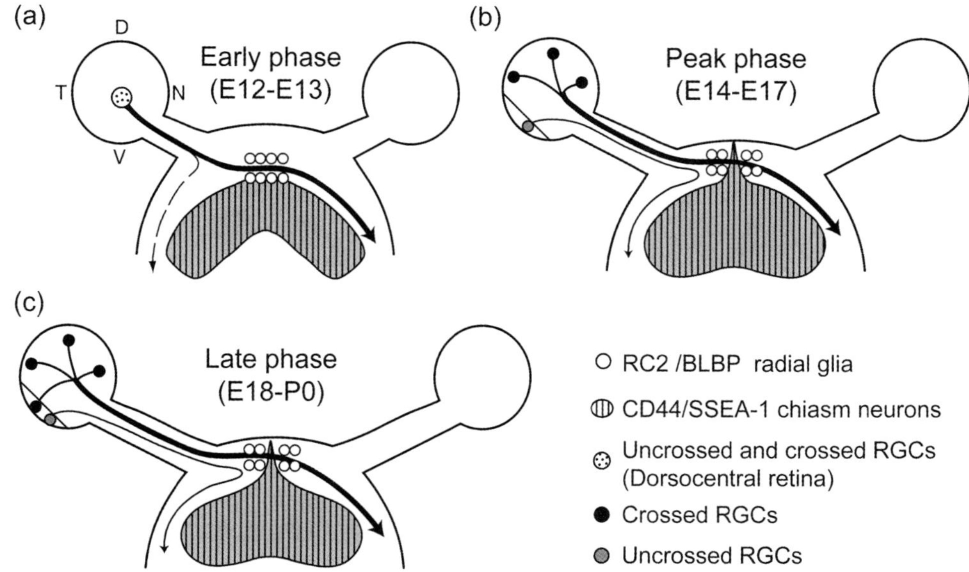

FIGURE 92.3   Three phases of retinal ganglion cell (RGC) axon outgrowth during mouse optic chiasm formation. Horizontal view of the retinal axon projection through the optic chiasm. Chiasmatic radial glia form a palisade on either side of the midline and express glial markers such as RC2 and BLBP. The early born neurons expressing CD44 and SSEA-1 are positioned caudal to the chiasm and form a raphe rostrally through the midline. (a) During the early phase (E12–13), the first RGCs from dorsocentral retina project both contralaterally and ipsilaterally. Crossed axons traverse the chiasm midline while transient uncrossed axons do not reach the chiasm palisade and turn laterally. (b) During the peak phase (E14–17), the permanent ipsilateral projection emerges from RGCs in ventrotemporal (VT) retina and the contralateral projection from RGCs in extra-VT retina. Both the ipsilateral and contralateral RGC axons approach the midline, where they diverge. (c) During the late phase (E18–P0), RGCs in VT retina project contralaterally, as do RGCs in extra-VT retina. E, embryonic day.

Mason, 1995; Mason & Sretavan, 1997; Petros, Rebsam, & Mason, 2008). Recent studies show that each cell population expresses distinct guidance and transcription factors (see below).

Growth Cone Behaviors at the Chiasm Midline

During optic chiasm formation, the growing tips, or growth cones, of both crossed and uncrossed RGC axons enter the chiasm region. Uncrossed axons diverge from crossed axons in a zone 100–200 microns proximal to the midline (Dingwell, Holt, & Harris, 2000; Mason & Erskine, 2000). While uncrossed RGC axons form highly branched and complex growth cones that turn away from the chiasm midline, crossed axons have simple growth cones and traverse the radial glial palisade and midline raphe of the SSEA-1/CD44 neurons (Godement, Salaun, & Mason, 1990; Marcus et al., 1995; Mason & Sretavan, 1997). Live imaging of growth cone behaviors has revealed that although both uncrossed and crossed growth cones pause at the midline and undergo extension–retraction behavior, uncrossed axons pause for longer periods and form Y-shaped or multiple growth cones as compared to crossed axons, which bear individual, more simple growth cones (Godement, Wang, & Mason, 1994). RGC axons cross the midline at more dorsal levels through the glial palisades while the uncrossed axons turn at more ventral levels of the midline (Colello & Coleman, 1997). These observations suggested that guidance cues for retinal axon divergence are located within the cell populations around the midline of the optic chiasm.

Analysis of retina–chiasm cocultures showed that the chiasmatic neurons and glia provide such cues for retinal axon growth and divergence. When cocultured with chiasm explants in collagen gel cultures, RGCs from all retinal regions display reduced neurite outgrowth (Wang et al., 1996), indicating that chiasm cells express diffusible inhibitory cues to retinal axons. However, when grown directly on dissociated chiasm cells in two-dimensional cultures, VT explants show a 60% reduction in neurite outgrowth whereas extra-VT (e.g., dorsotemporal; DT) retinal explants show only a 20% reduction (Wang et al., 1995). Thus, while the chiasm appears to be somewhat inhibitory to all RGC axon growth, it inhibits ipsilateral RGC axon growth to a greater extent, reflecting the differential behavior of axons of these two populations of RGCs at the midline in vivo. These results suggest that multiple guidance cues presented by chiasm cells underlie the differential response of the axons of crossed and uncrossed RGCs through both diffusible and contact-mediated interactions.

GUIDANCE MOLECULES THAT DIRECT RETINAL GANGLION CELL AXON DIVERGENCE

Although early studies hinted at extracellular cues that guide axons along their projection pathway and regulate growth cone behavior, the molecular identity of these cues was unknown until the last decade. Several studies have led to the identification of conserved families of guidance molecules such as netrins, slits, semaphorins, and ephrins, as well as noncanonical guidance factors such as morphogens (Kolodkin & Tessier-Lavigne, 2011). In addition, cell adhesion molecules (CAMs) such as Ng-CAM-related cell adhesion molecule (NrCAM), L1, and Down syndrome cell adhesion molecule (DSCAM) regulate axon guidance through their interactions with other CAMs, semaphorins, or other guidance receptors (Ly et al., 2008; Nawabi & Castellani, 2011; Sakurai, 2012). Many of these molecules function at the midline of the vertebrate spinal cord and in the CNS of the fly (Dickson & Gilestro, 2006) and are now known to also participate in retinal decussation (see figure 92.4).

Avoiding the Midline

The first guidance cue that was implicated in retinal axon divergence at the optic chiasm, Ephrin-B, is expressed at the chiasm of metamorphic but not pre-metamorphic *Xenopus* and repels EphB-expressing RGCs during the formation of the ipsilateral projection. Precocious expression of ephrin-B in the tadpole chiasm is sufficient for inducing an ipsilateral projection from EphB⁺ RGCs. Ephrin-Bs are also expressed in the chiasm of mammals with binocular vision but not in the chiasm of fish and birds with monocular vision (Nakagawa et al., 2000).

Ephrin-B2 is expressed by radial glial cells of the mouse optic chiasm during the peak phase of RGC axon outgrowth when the permanent ipsilateral projection develops. Moreover, ephrinB2 specifically inhibits outgrowth of uncrossed RGC axons, and blocking ephrinB2 relieves this inhibition in vitro. EphB1, a receptor for ephrinB2, is expressed in regions of retina that give rise to the ipsilateral projection during the same window of expression as that of ephrinB2 at the chiasm midline. *EphB1⁻/⁻* mutants display a reduced ipsilateral projection (Williams et al., 2003). Ectopic EphB1 expression in extra-VT retina redirects RGCs ipsilaterally, an effect that requires both the extracellular and juxtamembrane domains of EphB1 (Petros, Shrestha, & Mason, 2009). Together, these data suggest that the function of B-class Ephs and ephrins in patterning binocular vision is conserved between species, and

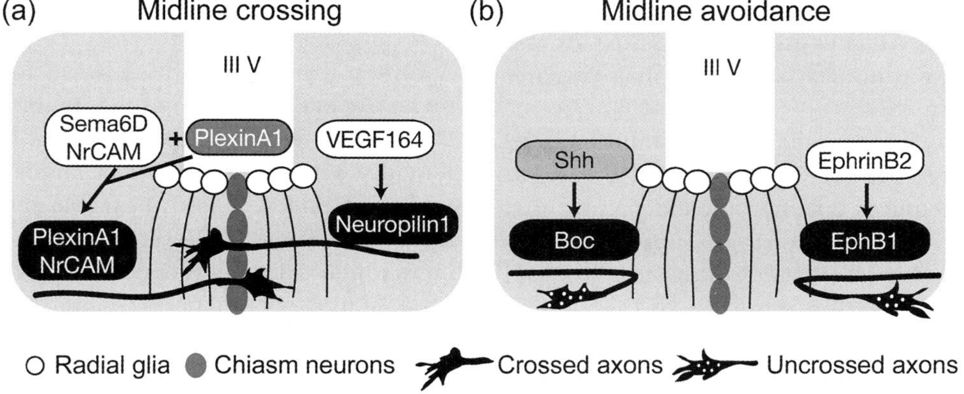

FIGURE 92.4   Molecular interactions for crossing and avoidance at the optic chiasm midline. Frontal view of the optic chiasm at the peak phase of retinal axon outgrowth E14–17). (a) Midline crossing: VEGF164 is expressed at the optic chiasm and acts as a diffusible cue to attract growth cones of neuropilin1[+] RGCs. Sema6D, expressed by radial glial cells, is inhibitory to outgrowth of Plexin-A1[+]/NrCAM[+] contralateral RGC axons. Plexin-A1 on chiasm neurons, together with NrCAM on radial glia, converts growth-inhibition to growth-promotion. (b) Midline avoidance: Ephrin-B2 is expressed by radial glial cells and inhibits outgrowth of EphB1-expressing ipsilateral axons. Shh at the optic chiasm repels ipsilateral RGCs, which also express Boc. E, embryonic day.

the inhibitory EphB1-ephrinB2 interaction drives VT axons to project ipsilaterally at the mouse optic chiasm midline.

Sonic hedgehog signaling has also been implicated in chiasm crossing (Sanchez-Camacho & Bovolenta, 2008; Trousse et al., 2001). Boc, a receptor to the morphogen sonic hedgehog (Shh), is expressed in ipsilateral RGCs of the developing retina. Boc[+] ipsilateral RGC axons retract in the presence of Shh, while $Boc^{-/-}$ axons show no response to Shh in vitro. Moreover, $Boc^{-/-}$ mice have a decreased ipsilateral projection, and ectopic expression of Boc in contralateral RGCs causes axons to project ipsilaterally (Fabre, Shimogori, & Charron, 2010). Although these results implicate an inhibitory interaction between Boc and Shh in the formation of the ipsilateral projection, whether Boc directs RGC axons as a guidance receptor for chiasm Shh or instead influences ipsilateral RGC specification within the retina remains to be determined.

Crossing the Midline

After the discovery of ipsilateral guidance programs, efforts turned to determining whether midline crossing at the optic chiasm involves attractive or growth-supporting cues, similar to the role of netrin in the ventral midline of the mouse spinal cord (Dickson & Gilestro, 2006). However, such signals have not been identified at the optic chiasm until recently (Erskine et al., 2011). Netrin-1 is expressed at the optic disk rather than at the chiasm midline (Deiner et al., 1997), and accordingly, in mutants of netrin-1 or its receptor, deleted in colon cancer (DCC), axons fail to enter the optic nerve while retinal decussation at the optic chiasm remains normal (Deiner & Sretavan, 1999).

The semaphorins, a large family of axon guidance cues, have also been reported to mediate midline crossing of RGC axons at the optic chiasm and commissural axons in the spinal cord (Nawabi & Castellani, 2011). Neuropilin-1, a receptor for the Sema3 subfamily, is expressed in crossed RGCs during development (Erskine et al., 2011). VEGF164, the neuropilin-binding isoform of classical VEGF, is expressed at the chiasm midline from E12 to E17. Intriguingly, VEGF164 acts as an attractive diffusible cue that promotes axon outgrowth and attracts growth cones of neuropilin-1[+] RGCs in vitro. Mutants lacking Neuropilin1 or VEGF164 display an increased ipsilateral projection and highly defasciculated RGC axons in the chiasm. Thus, neuropilin and VEGF interactions are thought to facilitate the formation of the contralateral projection by providing permissive signals to retinal axons.

NrCAM, a member of the immunoglobulin (Ig) superfamily of cell adhesion molecules, is also expressed in crossed RGCs throughout the retina during all phases of axon outgrowth (Lustig et al., 2001; Williams et al., 2006). Blocking NrCAM function leads to decreased axon outgrowth of crossed RGCs cocultured with dissociated chiasm cells and to an increased ipsilateral projection in semi-intact visual system preparations. $NrCAM^{-/-}$ mice similarly display an increased ipsilateral projection in vivo, but these fibers arise only from the late-born VT RGCs (Williams et al., 2006). Thus, NrCAM is necessary for the contralateral projection at the later age of chiasm formation.

As midline zones contain both inhibitory and growth-supporting cues (Nawabi et al., 2010; Parra & Zou, 2010), attraction is not the only mechanism that can implement midline crossing. Supporting this hypothesis, contralateral RGC growth cones spend several hours advancing and retracting at the midline before rapidly crossing the midline in vivo, as though abrogation of an inhibitory cue occurs. In zebrafish, which have a completely crossed retinal projection, the repulsive guidance cue Sema3d is expressed at the midline of the ventral diencephalon (Sakai & Halloran, 2006). Both ubiquitous Sema3d overexpression and Sema3d knockdown exclusively at the chiasm midline drive axons to project ipsilaterally. These data imply that inhibitory signals, such as Sema3d, are released by the midline. Two hypotheses can explain how repulsive guidance cues mediate midline crossing: (1) Inhibitory midline signals are converted to attractive signals; (2) inhibitory signals are diminished by degradation or cleavage of their receptors or are offset by other attractive signals.

In a recent study, transmembrane protein Sema6D was shown to be expressed by radial glial cells in the mouse optic chiasm along with NrCAM (Kuwajima et al., 2012). Plexin-A1, a known receptor for Sema6D, is expressed by SSEA-1$^+$ chiasm neurons. NrCAM and Plexin-A1 are coexpressed by RGCs that cross the midline. In vitro, Sema6D alone is inhibitory only to crossed RGCs, but the presence of both NrCAM and Plexin-A1 on chiasm cells can convert the repulsive effects of Sema6D to growth promotion. Binding studies explain this result by demonstrating that NrCAM is a receptor for Sema6D and interacts with the Sema6D-Plexin-A1 complex. Finally, in vivo analysis of *Sema6D*$^{-/-}$ and *Plexin-A1*$^{-/-}$;*NrCAM*$^{-/-}$ mutants indicates that NrCAM and Plexin-A1 expression is required in both RGCs and chiasm cells, along with chiasm Sema6D for efficient decussation and fasciculation of retinal axons at the optic chiasm.

Although the molecular programs described above appear to be important for axon pathfinding at the chiasm, as shown by severe chiasm perturbations in mice with mutations in the guidance factors discussed above, the majority of crossed axons still cross the midline (Erskine et al., 2011; Kuwajima et al., 2012). Thus, multiple signals from different guidance families, including many not yet identified, must act together to mediate midline crossing and axonal organization at the optic chiasm.

Axon Fasciculation and Midline Crossing

Axon–axon fasciculation is critical to axon guidance, including at the midline of the neuraxis (Bak & Fraser, 2003; Moon & Gomez, 2005; Myers & Bastiani, 1993; Raper & Mason, 2010). In fish, fibers from each optic nerve are bundled, and these two bundles overlap to form the chiasm. However, in higher vertebrates, increasingly finer degrees of eye-specific bundles interleave at the chiasmatic crossing (Guillery Mason, & Taylor, 1995).

Several molecules have been identified that might function in fasciculation generally and in retinal axon decussation at the chiasm (figure 92.5). For example, Sema3d modulates levels of the cell adhesion molecule L1 to mediate axon–axon interactions in the zebrafish spinal cord (Wolman et al., 2004). Members of the Slit family of guidance cues, Slit1 and Slit2, are expressed in the domains surrounding the path of growing retinal axons and around the chiasm midline; their receptor, Robo2, is expressed in RGCs (Erskine et al., 2000). Both Slit1 and Slit2 inhibit outgrowth of all RGC axons in vitro. While neither the *Slit1*$^{-/-}$ nor *Slit2*$^{-/-}$ single mutant displays severe axon guidance defects, retinal axons of *Slit1*$^{-/-}$;*Slit2*$^{-/-}$ double mutant mice project into the contralateral optic nerve and extend ectopically dorsal or lateral to the chiasm (Plump et al., 2002). Similar to *Slit1*$^{-/-}$;*Slit2*$^{-/-}$ double mutants, *Robo2*$^{-/-}$ mice also show an expansion of the chiasm along the rostrocaudal axis (Plachez et al., 2008). Slits are thus thought to establish a repulsive corridor to channel retinal axons and encourage fasciculation but are not critical role for axon divergence at the chiasm.

Other molecules that mediate axon fasciculation at the chiasm include heparan sulfate (HSPGs) and chondroitin sulfate proteoglycans, which are important for cell surface interactions (Leung, Taylor, & Chan, 2003; Reese et al., 1997). The HSPG modifying enzymes Hs2st and Hs6st1 are expressed by RGCs and cells within the ventral diencephalon (Pratt et al., 2006). *Hs2st*$^{-/-}$ and *Hs6st1*$^{-/-}$ mutants display axon misrouting and defasciculation at the chiasm; however, like Slit mutants, neither mutant displays defects in midline decussation. Thus, mechanisms regulating axonal fasciculation and organization at the optic chiasm may be independent of those directing RGC divergence.

TRANSCRIPTIONAL CONTROL OF RETINAL AXON DECUSSATION

A major question in the field of axon guidance is how transcription factors that specify cell identity relate to the guidance programs that these cell types employ.

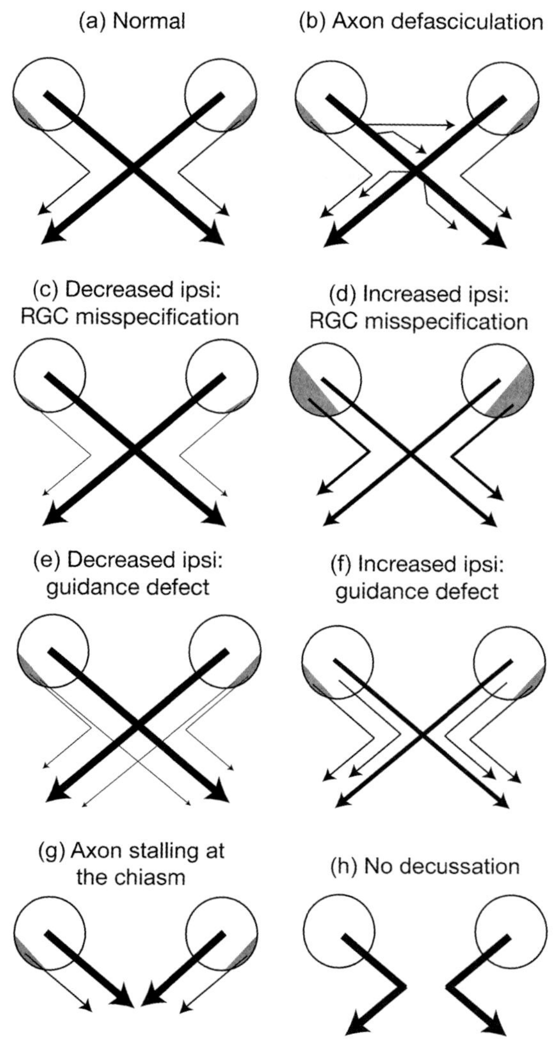

FIGURE 92.5 Aberrant retinal decussation at the optic chiasm. Representative abnormal retinal decussations at the mouse optic chiasm. (a) Normal pattern of mouse retinal ganglion cell (RGC) axon decussation. Mutants displaying defects include: (b) *Slit1⁻/⁻;Slit2⁻/⁻*, (c) albino (*Tyr^(c-2J/c-2J)*) and *Foxd1⁻/⁻*, (d) *Foxg1⁻/⁻*, (e) *EphB1⁻/⁻*, Zic2 hypomorph, (f) *Plexin-A1⁻/⁻; NrCAM⁻/⁻* and *VEGF164⁻/⁻*, (g) *Vax1⁻/⁻*, (h) *belladonna* (zebrafish), and achiasmatic Belgian sheepdog and humans. Ipsi, ipsilateral.

Studies in cortex, spinal cord, and retina have revealed transcription factor codes that regulate neuronal subtype identity and their corresponding axon projections (Butler & Tear, 2007; Polleux, Ince-Dunn, & Ghosh, 2007). In the spinal cord, combinatorial codes of LIM homeobox transcription factors within spinal levels and Hox transcription factors along the rostro-caudal axis encode the identity of distinct motor neuron pools and their projections to different muscle groups (Briscoe et al., 2000; Dasen et al., 2005). LIM homeodomain proteins have been further shown to pattern these projections by regulating the topographic expression of

Eph receptors and ephrins in motor neurons and limb targets (Kania & Jessell, 2003; Luria et al., 2008; Palmesino et al., 2010). In the cortex, a network of transcription factors regulates fate determination in subpopulations of projection neurons (Leone et al., 2008; Molyneaux et al., 2007). Whether these transcription factors also regulate the expression of guidance receptors as they do in motor neurons remains to be determined.

In the visual system, a number of transcription factors regulate the topographic patterning of the retina and RGC axon targets, such as chick Vax, mouse Vax2 (Schulte et al., 1999), and *Xenopus* Tbx5 (Koshiba-Takeuchi et al., 2000), which are involved in dorsoventral patterning of the retina. More specifically, Zic4 expression in the dLGN plays a role in the retinotopic targeting of ipsilateral RGC fibers (Horng et al., 2009) but not in the segregation of ipsilateral and contralateral RGC afferents. Although transcription factor Brn3b regulates retinal axon pathfinding at multiple points along the projection pathway (Erkman et al., 2000), none of these transcription factors appear to direct decussation or targeting of ipsilateral and contralateral RGCs.

Transcriptional Control of the Ipsilateral Guidance Program

The first discovery of a transcriptional regulator of midline decussation in the vertebrate nervous system was Zic2. The Zic family of zinc finger transcription factors (Zic1–5) is critical for early neural patterning and midline formation in the developing embryo (Merzdorf, 2007; Nagai et al., 1997), and Zic2 mutations have been implicated in human developmental defects such as holoprosencephaly (Brown et al., 1998). Zic2 is expressed early in the optic vesicle and stalk during eyecup formation but later downregulated (Nagai et al., 1997). During the peak phase of retinal outgrowth (E14 to E17), Zic2 is again upregulated but only in RGCs of the peripheral VT crescent from which the ipsilateral retinal axons arise, indicating a tight spatiotemporal restriction of Zic2 to ipsilaterally projecting RGCs. In fact, retinal Zic2 expression tightly correlates with the size of the ipsilateral projection within different vertebrate species, as it is absent from species lacking binocular vision such as chick and zebrafish and, like EphB1, is upregulated during metamorphosis in *Xenopus* when the ipsilateral projection forms. Zic2 hypomorph mice have a severely reduced ipsilateral projection, and cultured DT retinal axons overexpressing Zic2 respond to the growth inhibitory cues of cocultured chiasm cells, similar to VT axons (Herrera et al.,

2003). Together, these experiments demonstrate that a single transcription factor is necessary and sufficient for determining laterality of RGC projections.

Interestingly, both Zic2 and EphB1 are important for RGC axon divergence at the chiasm midline and share a similarly restricted pattern of expression to ipsilaterally projecting RGCs in the VT retina. Further studies then showed that EphB1 expression is lost in Zic2 hypomorph mice, and Zic2 overexpression in extra-VT RGCs is sufficient for eliciting midline avoidance and repulsion by Ephrin-B2, mostly through EphB1-dependent mechanisms (Garcia-Frigola et al., 2008; Lee, Petros, & Mason, 2008). However, Zic2 overexpression is more efficient than EphB1 in inducing this ectopic ipsilateral projection (Petros, Shrestha, & Mason, 2009), and Zic2 overexpression can induce a small ipsilateral projection even in $EphB1^{-/-}$ mice (Garcia-Frigola et al., 2008). Moreover, Zic2 has not been shown to directly regulate EphB1 transcription. These findings on Zic2 and EphB1 point to EphB1-independent ipsilateral guidance programs operating in the retina and also to additional transcription factors that regulate these programs.

Transcriptional Control of the Contralateral Guidance Program

The studies characterizing the role of Zic2 in VT retina lead to the question of whether a parallel transcriptional regulator exists for the contralateral projection. The LIM homeodomain transcription factor Islet 2 (Isl2) is expressed in the developing retina exclusively in contralaterally projecting RGCs. In the VT retina, Isl2 is expressed in a nonoverlapping pattern with Zic2 and is upregulated in late-born RGCs that project contralaterally. As the ipsilateral projection is increased in $Isl2^{-/-}$ mice, Isl2 seemed a good candidate for a Zic2-equivalent in specifying contralaterality. However, the additional ipsilaterally projecting axons in the $Isl2^{-/-}$ mice originate only from VT retina; moreover, Zic2 expression is upregulated in additional cells within the VT but does not expand into the extra-VT domain (Pak et al., 2004). Thus, similar to the $NrCAM^{-/-}$ phenotype, loss of Isl2 appears to only convert the laterality of late-born, contralaterally projecting RGCs within the VT retina.

The above studies support a model in which Isl2 specifies RGC laterality by repressing an ipsilateral guidance program involving Zic2 and EphB1 that is unique to the VT retina. However, it remains to be determined whether Isl2 indeed represses Zic2 and/or EphB1, whether this genetic interaction is through direct transcriptional regulation, and whether Zic2 similarly represses contralateral identity through repressing Isl2

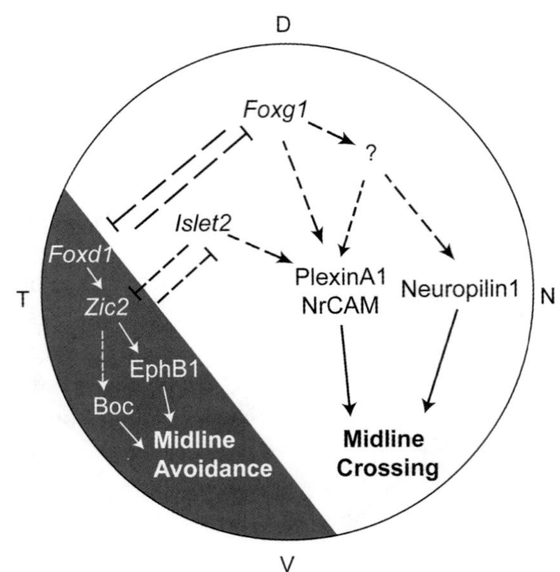

FIGURE 92.6 Genetic network regulating decussation of retinal ganglion cell axons. Genetic pathways within the ipsilateral (gray) and contralateral (white) retinal ganglion cell domains, including hypothesized interactions (dotted arrows). D, dorsal; V, ventral; N, nasal; T, temporal.

and downstream genes (see figure 92.6). Studies in the spinal cord have demonstrated that cell specification can be regulated by cross-repression between spatially restricted transcription factors (Dasen, 2009; Surmeli et al., 2011). Such cross-regulatory networks may similarly act in the retina to restrict uncrossed RGCs to the VT crescent.

As Isl2 appears to be necessary for laterality decisions only within late-born VT RGCs and is expressed in only one-third of all contralaterally projecting RGCs (Pak et al., 2004), other transcription factors are likely the important determinants of laterality in the extra-VT retinal domain. The POU-domain transcription factor Brn3a (Pou4f1) is also expressed in only contralaterally projecting RGCs; however, unlike Isl2, $Brn3a^{-/-}$ mutants show no defects in RGC midline crossing or in expression of genes controlling laterality (Quina et al., 2005).

Thus far, a genetic program directing the major contralateral RGC projection from transcription factor to guidance receptor that parallels Zic2 and EphB1 for the ipsilateral projection has not been identified. Although recent efforts have revealed candidate genes that directly regulate retinal expression of Plexin-A1 and NrCAM (T. Kuwajima & C. Mason, unpublished), whether they can indeed specify laterality in RGCs residing outside of the VT retina is not known. Alternatively, midline crossing may be an inherent property of RGCs, given that the contralateral projection appears to be the "ground state" in lower vertebrates, and may be difficult to eliminate when guidance or transcription factors are mutated.

Genes That Define Ipsilateral and Contralateral RGC Identity

While transcription factors can direct axon trajectory through regulation of guidance factor expression, they can also indirectly participate in these decisions by specifying cell identity. Although Zic2 has been shown to be both necessary and sufficient for inducing RGC axons to avoid the optic chiasm midline, a recent study demonstrated that Zic2 also plays a role in axon refinement at RGC targets via the serotonin transporter (SERT) (Garcia-Frigola & Herrera, 2010). In the retina, SERT expression is restricted to ipsilaterally projecting RGCs of the VT and dependent on direct transcriptional activation by Zic2. Through regulating serotonin availability, SERT has been shown to participate in axon arbor remodeling, which is important in the segregation of ipsilateral and contralateral fibers projecting to dLGN and SC (Gaspar, Cases, & Maroteaux, 2003). As Zic2 has now been shown to control multiple stages of ipsilateral RGC growth from axon pathfinding (embryonic) to axon refinement (postnatal), it is likely to be a more upstream regulator of ipsilateral RGC identity.

Despite our growing understanding of the genetic network that specifies the ipsilateral and contralateral RGC projections, many gaps remain where unidentified molecules are likely to play key functions (see figure 92.6). In overexpression studies, Zic2 is more potent than EphB1 in converting laterality (Petros, Shrestha, & Mason, 2009), suggesting that Zic2 accomplishes this through regulating additional downstream factors important for midline guidance. Moreover, knockout mouse models for retinal guidance receptors all show only partial diverting of RGC projections to the opposite side. Finally, the transcriptional control of the contralateral projection remains unclear. Thus, further identification of transcriptional regulators, guidance factors, and other genes important for ipsilateral and contralateral RGC subtype identity is needed. Microarray screens of these two RGC subtypes, purified based on their anatomical projections, have revealed a number of genes that may be involved in axon guidance, cell differentiation, or specification of ipsilateral versus contralateral RGCs (Q. Wang & C. Mason, unpublished).

Patterning the Ipsilateral and Contralateral Retinal Domains

Given the distinct identities of RGCs that project ipsilaterally versus contralaterally, the sectors in which they reside must be patterned to have similarly distinct properties. In vertebrates, the designation of nasal versus temporal retina is reflected in the complementary expression of winged helix transcription factors Foxg1 (BF-1) and Foxd1 (BF-2), respectively (Hatini, Tao, & Lai, 1994). Studies in humans point to signals from outside the eye in establishing the nasotemporal axis (Lambot et al., 2005), while studies on early zebrafish eye development pinpointed FGF as this signal that, in turn, regulates Foxg1 expression (Picker et al., 2009). In the mouse, Foxd1 is broadly expressed in the temporal retina when the first ganglion cells are born (E11) and acts upstream of Zic2 and EphB1 in the genetic program associated with the uncrossed RGC projection (Herrera et al., 2004). Foxd1 has been further shown to imprint temporal identity within the mouse retina by directing the expression of EphA6 and ephrinA5, which play a role in rostrocaudal topographic mapping within RGC targets (Carreres et al., 2011).

Mice lacking Foxg1 have an increased ipsilateral projection (Pratt et al., 2004; Tian, Pratt, & Price, 2008), suggesting a role for Foxg1 in repressing the ipsilateral and promoting the contralateral guidance pathways. Foxd1 is expressed in a region broader than the ipsilateral RGC domain, and both Fox genes are expressed in a gradient rather than in clearly delineated sectors. Thus, Foxd1 and Foxg1 could act as patterning genes, establishing general ipsilateral and contralateral domains that are further refined by downstream transcription factors such as Zic2. However, it is unclear whether the Fox genes indeed hold upstream positions within the transcription factor networks that specify ipsilateral and contralateral RGC identity or whether they control fate specification through gating proliferation and differentiation as they do in the cortex (Hanashima et al., 2002) (see figure 92.6). Moreover, how Foxd1 and Foxg1 interact to establish the crossed and uncrossed domains of the retina remains to be determined. Both Foxd1 and Foxg1, as well as Zic2, are additionally expressed in the ventral diencephalon (Herrera et al., 2004; Marcus et al., 1999) and help to pattern the resident cell ensembles in and around the optic chiasm. Thus, any loss-of-function studies involving global knockouts that produce apparent disruptions in retinal decussation cannot distinguish between the role these genes play in patterning the retina versus the optic chiasm.

PROJECTION OF IPSILATERAL AND CONTRALATERAL RGCs TO THE DORSAL LATERAL GENICULATE NUCLEUS

The partial decusation of RGC axons at the optic chiasm midline ensures that targets on each side of the brain

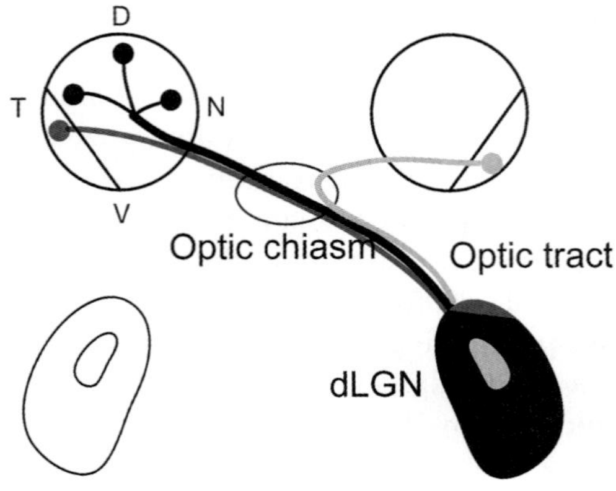

FIGURE 92.7   The mouse retinogeniculate projection. Ipsilateral (light gray) and contralateral (black and dark gray) retinal ganglion cell (RGC) afferents segregate from each other within the dorsal lateral geniculate nucleus (dLGN). Late-born contralateral RGCs in ventrotemporal (VT) retina project into the dorsal tip of the dLGN (dark gray) whereas contralateral RGCs from extra-VT retina project to the outer shell of the dLGN (black). D, dorsal; V, ventral; N, nasal; T, temporal.

receive inputs from both eyes. Within the dLGN, RGC afferents from each eye are segregated into distinct territories. In mice, the retina maps topographically onto the dLGN, directed by guidance factors and receptors in the Eph/ephrin A family (Huberman et al., 2005; Pfeiffenberger et al., 2005). In addition, retinal inputs from both eyes segregate into distinct subregions within the dLGN (see figure 92.7). Rather than adopting the laminar organization found in higher vertebrates, mouse ipsilateral fibers from VT retina form a core surrounded by contralateral inputs (Godement, Salaun, & Imbert, 1984; Jaubert-Miazza et al., 2005). In contrast, the late-projecting contralateral inputs from VT retina innervate the dorsal tip of the dLGN (Pfeiffenberger et al., 2005). This precise organization of eye-specific inputs is dependent on both molecular factors and neural activity (Huberman, Feller, & Chapman, 2008).

Although the mouse dLGN has been widely used as a model for activity-dependent refinement, we are only beginning to understand the developmental progression of RGC afferent targeting in the dLGN and the molecular mechanisms directing these processes. The overall time course of projection and refinement has been outlined (Dhande et al., 2011; Jaubert-Miazza et al., 2005), yet the precise differences between timing and point of entry for ipsilateral versus contralateral inputs, as well as the extent to which these two cohorts are segregated in the optic tract, are not well understood.

Molecular Control of Eye-Specific Targeting

The identification of eye-specific markers within the corresponding recipient zones of the dLGN has lagged behind an understanding of the molecules that mediate topographic connections (Huberman, Feller, & Chapman, 2008). In ferret, genes have been annotated in eye-specific layers within the dLGN (Kawasaki et al., 2004), but only after innervation is established. The best candidate thus far for ipsilateral retinogeniculate targeting is Ten_m3, an adhesion molecule in the teneurin family. Ten_m3 is expressed in the ventral retina, and *Ten_m3$^{-/-}$* mice show targeting defects of only ipsilateral retinal axons in the dLGN, producing deficits in binocular vision without affecting midline guidance (Leamey et al., 2007). Ongoing gene profiling efforts by multiple groups will likely identify molecules in both retina and dLGN that mediate eye-specific innervation of visual targets.

Activity-Dependent Refinement of Eye-Specific Inputs

The segregation of ipsilateral and contralateral inputs from an initially overlapping pattern is an activity-dependent process (Huberman, Feller, & Chapman, 2008; see also Chapman, 2004). Neural activity in the form of "waves" generated by amacrine and retinal ganglion cells has been implicated in the patterning of dLGN connectivity (Pfeiffenberger et al., 2005; Shatz, 1996; Torborg, Hansen, & Feller, 2005), but the precise aspects of neural activity that act on eye-specific innervation and refinement have long been under debate (Chalupa, 2009; Feller, 2009). This issue has been explored more precisely in two recent studies. In the first, Xu et al. demonstrated that wave size, rather than the presence or pattern of activity, drives eye-specific segregation (Xu et al., 2011). In another study, Ullian et al. selectively reduced glutamate release from ipsilateral-projecting RGCs and showed that glutamatergic transmission plays a role in excluding competing axons from inappropriate target regions during visual circuit refinement but not in consolidation or maintenance of ipsilateral axonal territory, implicating other molecular mechanisms for such processes (Koch et al., 2011). However, the role neural activity plays in expression of molecules important for eye-specific targeting and segregation has not been examined.

   CAROL MASON, TAKAAKI KUWAJIMA, AND QING WANG

*Relationship of Chiasm Decussation to Eye-Specific
Targeting*

Genetic models such as the Siamese cat (Kaas, 2005)
and *Drosophila* robo mutants (Wolf & Chiba, 2000) have
raised the issue of how laterality decisions relate to
targeting, a question that can be studied using mouse
mutants with RGC midline guidance defects. In
$EphB1^{-/-}$ mice, the RGCs from VT retina are misrouted
contralaterally but terminate in the correct general
ipsilateral-recipient region within the contralateral
dLGN as a small segregated patch. Moreover, when
EphB1 is ectopically expressed in retinal regions giving
rise to contralateral projections, misrouted EphB1-
expressing RGCs still innervate the contralateral-
recipient zones of the ipsilateral dLGN (Rebsam, Petros,
& Mason, 2009). These results indicate that RGC axons
possess the ability to respond to instructional cues for
targeting that are distinct from those used in chiasmatic
decussation. Semaphorins and plexins are candidates
for implementing retinogeniculate targeting in addition to chiasm midline crossing, as they are expressed
in the dLGN (A. Rebsam, personal communication)
and have been shown to function in both axon guidance and targeting in other regions (Pecho-Vrieseling
et al., 2009; Yoshida et al., 2006). With defects in both
fasciculation and in RGC axon divergence at the optic
chiasm, mutants of these molecules are likely to have
even more complex aberrations in retinogeniculate targeting.

*Relationship of Pretarget Axon Sorting to Eye-Specific
Targeting*

While significant strides have been made in the last two
decades in identifying and characterizing molecules
involved in axon guidance and targeting during neural
circuit formation, questions arise regarding the organization of axons between intermediate and final targets.
In the olfactory system, pretarget sorting of olfactory
sensory neuron axons en route to the olfactory bulb is
required for correct topographic organization of target
innervation (Imai et al., 2009; Sakano, 2010; Satoda
et al., 1995). Likewise, chick motor axons innervating
fast and slow muscle fibers are molecularly distinct from
each other and bundle in separate cohorts prior to
target innervation (Xue & Honig, 1999).

In the developing cat, mouse, and ferret retinal axon
pathway, the extent to which axons are organized by
retinal origin, chronotopy, or eye-specificity and
whether such organization relates to proper targeting
have been in some dispute (Horton, Greenwood,
& Hubel, 1979; Simon & O'Leary, 1991). When

fasciculation or bundling of axons is disrupted by antibodies to the glycoprotein neural cell adhesion molecule (NCAM), the youngest retinal axons can target to
proper poles of the optic tectum, but older axons
misroute (Thanos, Bonhoeffer, & Rutishauser, 1984).
Each of these types of organization changes within different segments of the pathway (Colello & Guillery,
1992; Guillery & Walsh, 1987). In chronotopic order,
the youngest axons extend along radial glial endfeet at
the pial surface in the chiasm but cross the midline in
eye-specific bundles deep to the pia (Colello & Guillery,
1998; Walsh & Guillery, 1985). Chronotopic order is
reestablished as the axons enter the optic tract (Reese,
1996). In the mouse optic stalk, RGC axons are roughly
topographically ordered after exiting the eye but
become progressively disorganized as they near the
optic chiasm. After decussating at the chiasm, axons
regain topographic order in the optic tract, corresponding to the dorsal–ventral but not nasal–temporal retinal
axis (Chan & Chung, 1999). At the base of the optic
tract, axons from the dorsal and ventral retina are
biased medially and laterally, respectively; the tract then
twists just ventral to the LGN so that axons from dorsal
and ventral retina switch positions (Plas, Lopez, & Crair,
2005).

How ipsilateral and contralateral RGC axons are
grouped in the optic tract and whether such putative
segregation influences eye-specific targeting are not
clear. Early studies using anterograde labeling of one
eye indicated that ipsilateral RGC axons are segregated
to the lateral edge of the optic tract while contralateral
fibers run throughout (Godement, Salaun, & Imbert,
1984). Whole and partial eye labeling with modern
tracers and refined microscopy have revealed that eye-
specific organization in the optic tract does in fact exist
(A. Sitko & C. Mason, unpublished). Further exploration of the molecular mechanisms of optic tract development and organization using mutant mice lacking
guidance factors important for retinal axon decussation, fasciculation, and/or targeting should illuminate
whether segregation of fibers from each eye is indeed
necessary for proper eye-specific targeting.

DEFECTS IN CHIASM FORMATION

Lack of Ocular Pigment and Retinal Decussation

Albinos have a hypopigmentation disorder that can
result from different genetic mutations (King et al.,
2003; Oetting et al., 2003). The universal phenotype of
all forms of albinism is a decreased ipsilateral projection, resulting in perturbed binocular vision (Kaas,
2005). Recent studies have described an additional

defect in $Tyr^{c-2J/c-2J}$ mice, which harbor a mutation in a key enzyme in melanin synthesis, tyrosinase. Anterograde labeling revealed that similar to the $EphB1^{-/-}$ dLGN, a segregated patch forms within the albino dLGN; however, the position of this patch differs, in that it is adjacent to the target zone of the late-born contralateral VT RGCs (Rebsam, Bhansali, & Mason, 2012). Interestingly, the abnormal patches in both $Tyr^{c-2J/c-2J}$ and $EphB1^{-/-}$ mice derive from RGCs in the VT retina, and their segregation is dependent on normal retinal activity (Rebsam, Bhansali, & Mason, 2012).

The picture emerging from the perturbations along the ipsilateral RGC projection pathway in the albino suggests a broader defect in ipsilateral RGC specification rather than a defect in midline axon guidance as seen in $EphB1^{-/-}$ mice. Indeed, the number of cells that express both Zic2 and EphB1 in VT retina is reduced in albino mice, in accordance with the reduction of the ipsilateral projection (Herrera et al., 2003; Rebsam, Bhansali, & Mason, 2012). Moreover, RGC neurogenesis is disrupted in the albino retina with respect to the timing of cell differentiation and regulation of cell cycle parameters (Rachel et al., 2002; Tibber, Whitmore, & Jeffery, 2006; Webster & Rowe, 1991). Preliminary evidence suggests that aberrant RGC production and cell cycle timing occurs specifically in the VT retinal domain in $Tyr^{c-2J/c-2J}$ embryos (Bhansali et al., submitted). Ongoing analyses will determine whether cells that are born later are specified to project contralaterally because they have missed the temporal window during which Zic2 specifies ipsilateral RGC identity.

The defects manifested in albinism are due to perturbed biogenesis or packaging of melanin in the retinal pigment epithelium (RPE), the monolayer of epithelial cells surrounding the neural retina. How these RPE defects alter retinal axon decussation at the optic chiasm is not understood. One hypothesis is that factors in the melanogenic pathway in the developing RPE influence precursors in the neural retina to express genes that specify RGC laterality. As tyrosinase also mediates the production of L-DOPA, an intermediate product during melanin synthesis, L-DOPA has been proposed as such a key factor (Kralj-Hans et al., 2006; Lopez et al., 2008). Unanswered questions include how regulatory factors from RPE are transferred to neural retina and whether developmental processes influenced by these factors include both neurogenesis and cell type specification.

Failure of Optic Chiasm Development

In a breed of Belgian sheepdog with an autosomal recessive mutation in an unknown gene, the optic chiasm fails to develop (Hogan & Williams, 1995). Instead, all RGCs extend ipsilaterally but terminate appropriately in their targets (see figure 92.5). Consequently, noncongruent mirror-image maps of visual space are formed in adjacent layers in the dLGN. Although no other gross brain malformations have been identified in these dogs (Hogan & Williams, 1995), the zebrafish *belladonna* mutant, in which the optic chiasm also fails to form, shows perturbations in cellular organization and gene expression within the ventral diencephalon (Seth et al., 2006) (see figure 92.7). Interestingly, a small abnormal chiasm does develop in a few of these Belgian sheepdogs. The cells that give rise to these residual crossed axons are distributed throughout the retina (Hogan & Williams, 1995), as in some mouse mutants in which guidance factors within the chiasm region are perturbed (Kuwajima et al., 2012).

Similar defects in optic chiasm development have been described in a small number of humans with non-decussating retinofugal fiber syndrome (Apkarian, Bour, & Barth, 1994). With the exception of abnormal eye movements, the affected individuals are otherwise mentally and physically normal and have no other gross brain abnormalities. Their vision is also remarkably normal, possibly as a result of transfer of visual information between the visual hemispheres via the corpus callosum (Victor et al., 2000).

SUMMARY AND PERSPECTIVES

The studies discussed above shed light on how the circuit from retina to brain is patterned to establish the optic chiasm. Studies of retinal guidance at the midline have illuminated how multiple signals interact to control retinal axon divergence at the optic chiasm. The interactions between the genes that control the specification of the ipsilateral versus contralateral RGCs should reveal mechanisms that designate the line of decussation within the human retina. Understanding how multiple guidance factors cooperate during development of the retinogeniculate pathway should help resolve the current debate on the relative roles of molecular factors and neural activity in eye-specific targeting. Additionally, how fasciculation and bundling of like axonal cohorts contribute to targeting will provide insight into how the binocular circuit and other neural pathways form.

REFERENCES

Apkarian, P., Bour, L., & Barth, P. G. (1994). A unique achiasmatic anomaly detected in non-albinos with misrouted retinal-fugal projections. *European Journal of Neuroscience, 6,* 501–507.

Bak, M., & Fraser, S. E. (2003). Axon fasciculation and differences in midline kinetics between pioneer and follower axons within commissural fascicles. *Development, 130,* 4999–5008.

Briscoe, J., Pierani, A., Jessell, T. M., & Ericson, J. (2000). A homeodomain protein code specifies progenitor cell identity and neuronal fate in the ventral neural tube. *Cell, 101,* 435–445.

Brown, S. A., Warburton, D., Brown, L. Y., Yu, C. Y., Roeder, E. R., Stengel-Rutkowski, S., et al. (1998). Holoprosencephaly due to mutations in ZIC2, a homologue of *Drosophila* odd-paired. *Nature Genetics, 20,* 180–183.

Butler, S. J., & Tear, G. (2007). Getting axons onto the right path: The role of transcription factors in axon guidance. *Development, 134,* 439–448.

Carreres, M. I., Escalante, A., Murillo, B., Chauvin, G., Gaspar, P., Vegar, C., et al. (2011). Transcription factor Foxd1 is required for the specification of the temporal retina in mammals. *Journal of Neuroscience, 31,* 5673–5681.

Chalupa, L. M. (2009). Retinal waves are unlikely to instruct the formation of eye-specific retinogeniculate projections. *Neural Development, 4,* 25. doi:10.1186/1749.8104.4.25.

Chalupa, L. M., & Lia, B. (1991). The nasotemporal division of retinal ganglion cells with crossed and uncrossed projections in the fetal rhesus monkey. *Journal of Neuroscience, 11,* 191–202.

Chan, S. O., & Chung, K. Y. (1999). Changes in axon arrangement in the retinofugal [correction of retinofungal] pathway of mouse embryos: Confocal microscopy study using single- and double-dye label. *Journal of Comparative Neurology, 406,* 251–262.

Chapman, B. (2004). The development of eye-specific segregation in the retino-geniculo-striate pathway. In L. M. Chalupa & J. S. Werner (Eds.), *The Visual Neurosciences.* Cambridge, MA: MIT Press.

Colello, R. J., & Guillery, R. W. (1990). The early development of retinal ganglion cells with uncrossed axons in the mouse: Retinal position and axonal course. *Development, 108,* 515–523.

Colello, R. J., & Guillery, R. W. (1992). Observations on the early development of the optic nerve and tract of the mouse. *Journal of Comparative Neurology, 317,* 357–378.

Colello, S. J., & Coleman, L. A. (1997). Changing course of growing axons in the optic chiasm of the mouse. *Journal of Comparative Neurology, 379,* 495–514.

Colello, S. J., & Guillery, R. W. (1998). The changing pattern of fibre bundles that pass through the optic chiasm of mice. *European Journal of Neuroscience, 10,* 3653–3663.

Dasen, J. S. (2009). Transcriptional networks in the early development of sensory–motor circuits. *Current Topics in Developmental Biology, 87,* 119–148.

Dasen, J. S., Tice, B. C., Brenner-Morton, S., & Jessell, T. M. (2005). A Hox regulatory network establishes motor neuron pool identity and target-muscle connectivity. *Cell, 123,* 477–491.

Deiner, M. S., Kennedy, T. E., Fazeli, A., Serafini, T., Tessier-Lavigne, M., & Sretavan, D. W. (1997). Netrin-1 and DCC mediate axon guidance locally at the optic disk: Loss of function leads to optic nerve hypoplasia. *Neuron, 19,* 575–589.

Deiner, M. S., & Sretavan, D. W. (1999). Altered midline axon pathways and ectopic neurons in the developing hypothalamus of netrin-1- and DCC-deficient mice. *Journal of Neuroscience, 19,* 9900–9912.

Dhande, O. S., Hua, E. W., Guh, E., Yeh, J., Bhatt, S., Zhang, Y., et al. (2011). Development of single retinofugal axon arbors in normal and beta2 knock-out mice. *Journal of Neuroscience, 31,* 3384–3399.

Dickson, B. J., & Gilestro, G. F. (2006). Regulation of commissural axon pathfinding by slit and its Robo receptors. *Annual Review of Cell and Developmental Biology, 22,* 651–675.

Dingwell, K. S., Holt, C. E., & Harris, W. A. (2000). The multiple decisions made by growth cones of RGCs as they navigate from the retina to the tectum in Xenopus embryos. *Journal of Neurobiology, 44,* 246–259.

Dräger, U. C. (1985). Birth dates of retinal ganglion cells giving rise to the crossed and uncrossed optic projections in the mouse. *Proceedings of the Royal Society of London. Series B, Biological Sciences, 224,* 57–77. doi:10.1098/rspb.1985.0021.

Erkman, L., Yates, P. A., McLaughlin, T., McEvilly, R. J., Whisenhunt, T., O'Connell, S. M., et al. (2000). A POU domain transcription factor-dependent program regulates axon pathfinding in the vertebrate visual system. *Neuron, 28,* 779–792.

Erskine, L., & Herrera, E. (2007). The retinal ganglion cell axon's journey: Insights into molecular mechanisms of axon guidance. *Developmental Biology, 308,* 1–14. doi:10.1016/j.ydbio.2007.05.013.

Erskine, L., Reijntjes, S., Pratt, T., Denti, L., Schwarz, Q., Vieira, J. M., et al. (2011). VEGF signaling through neuropilin 1 guides commissural axon crossing at the optic chiasm. *Neuron, 70,* 951–965.

Erskine, L., Williams, S. E., Brose, K., Kidd, T., Rachel, R. A., Goodman, C. S., et al. (2000). Retinal ganglion cell axon guidance in the mouse optic chiasm: Expression and function of robos and slits. *Journal of Neuroscience, 20,* 4975–4982.

Fabre, P. J., Shimogori, T., & Charron, F. (2010). Segregation of ipsilateral retinal ganglion cell axons at the optic chiasm requires the Shh receptor Boc. *Journal of Neuroscience, 30,* 266–275.

Feller, M. B. (2009). Retinal waves are likely to instruct the formation of eye-specific retinogeniculate projections. *Neural Development, 4,* 24. doi:10.1186/1749.8104.4.24.

Garcia-Frigola, C., Carreres, M. I., Vegar, C., Mason, C., & Herrera, E. (2008). Zic2 promotes axonal divergence at the optic chiasm midline by EphB1-dependent and -independent mechanisms. *Development, 135,* 1833–1841.

Garcia-Frigola, C., & Herrera, E. (2010). Zic2 regulates the expression of Sert to modulate eye-specific refinement at the visual targets. *European Molecular Biology Organization Journal, 29,* 3170–3183. doi:10.1038/emboj.2010.172.

Gaspar, P., Cases, O., & Maroteaux, L. (2003). The developmental role of serotonin: News from mouse molecular genetics. *Nature Reviews Neuroscience, 4,* 1002–1012.

Godement, P., Salaun, J., & Imbert, M. (1984). Prenatal and postnatal development of retinogeniculate and retinocollicular projections in the mouse. *Journal of Comparative Neurology, 230,* 552–575.

Godement, P., Salaun, J., & Mason, C. A. (1990). Retinal axon pathfinding in the optic chiasm: Divergence of crossed and uncrossed fibers. *Neuron, 5,* 173–186.

Godement, P., Wang, L. C., & Mason, C. A. (1994). Retinal axon divergence in the optic chiasm: Dynamics of growth cone behavior at the midline. *Journal of Neuroscience, 14,* 7024–7039.

Guillery, R. W., Mason, C. A., & Taylor, J. S. (1995). Developmental determinants at the mammalian optic chiasm. *Journal of Neuroscience, 15*, 4727–4737.

Guillery, R. W., & Walsh, C. (1987). Changing glial organization relates to changing fiber order in the developing optic nerve of ferrets. *Journal of Comparative Neurology, 265*, 203–217.

Hanashima, C., Shen, L., Li, S. C., & Lai, E. (2002). Brain factor-1 controls the proliferation and differentiation of neocortical progenitor cells through independent mechanisms. *Journal of Neuroscience, 22*, 6526–6536.

Hatini, V., Tao, W., & Lai, E. (1994). Expression of winged helix genes, BF-1 and BF-2, define adjacent domains within the developing forebrain and retina. *Journal of Neurobiology, 25*, 1293–1309.

Herrera, E., Brown, L., Aruga, J., Rachel, R. A., Dolen, G., Mikoshiba, K., et al. (2003). Zic2 patterns binocular vision by specifying the uncrossed retinal projection. *Cell, 114*, 545–557.

Herrera, E., Marcus, R., Li, S., Williams, S. E., Erskine, L., Lai, E., et al. (2004). Foxd1 is required for proper formation of the optic chiasm. *Development, 131*, 5727–5739.

Herrera, E., & Mason, C. A. (2006). The evolution of crossed and uncrossed retinal pathways in mammals. In J. H. Kaas (Ed.), *Evolution of nervous systems* (Vol. 3, pp. 307–318). Amsterdam: Elsevier.

Hogan, D., & Williams, R. W. (1995). Analysis of the retinas and optic nerves of achiasmatic Belgian sheepdogs. *Journal of Comparative Neurology, 352*, 367–380.

Horng, S., Kreiman, G., Ellsworth, C., Page, D., Blank, M., Millen, K., et al. (2009). Differential gene expression in the developing lateral geniculate nucleus and medial geniculate nucleus reveals novel roles for Zic4 and Foxp2 in visual and auditory pathway development. *Journal of Neuroscience, 29*, 13672–13683.

Horton, J. C., Greenwood, M. M., & Hubel, D. H. (1979). Non-retinotopic arrangement of fibres in cat optic nerve. *Nature, 282*, 720–722.

Hoskins, S. G., & Grobstein, P. (1985). Development of the ipsilateral retinothalamic projection in the frog Xenopus laevis: II. Ingrowth of optic nerve fibers and production of ipsilaterally projecting retinal ganglion cells. *Journal of Neuroscience, 5*, 920–929.

Hoyt, W. F., & Luis, O. (1963). The primate chiasm: Details of visual fiber organization studied by silver impregnation techniques. *Archives of Ophthalmology, 70*, 69–85.

Huberman, A. D., Feller, M. B., & Chapman, B. (2008). Mechanisms underlying development of visual maps and receptive fields. *Annual Review of Neuroscience, 31*, 479–509.

Huberman, A. D., Murray, K. D., Warland, D. K., Feldheim, D. A., & Chapman, B. (2005). Ephrin-As mediate targeting of eye-specific projections to the lateral geniculate nucleus. *Nature Neuroscience, 8*, 1013–1021.

Imai, T., Yamazaki, T., Kobayakawa, R., Kobayakawa, K., Abe, T., Suzuki, M., et al. (2009). Pre-target axon sorting establishes the neural map topography. *Science, 325*, 585–590.

Jaubert-Miazza, L., Green, E., Lo, F. S., Bui, K., Mills, J., & Guido, W. (2005). Structural and functional composition of the developing retinogeniculate pathway in the mouse. *Visual Neuroscience, 22*, 661–676.

Jeffery, G. (2001). Architecture of the optic chiasm and the mechanisms that sculpt its development. *Physiological Reviews, 81*, 1393–1414.

Jeffery, G., & Harman, A. M. (1992). Distinctive pattern of organisation in the retinofugal pathway of a marsupial: II. Optic chiasm. *Journal of Comparative Neurology, 325*, 57–67.

Kaas, J. H. (2005). Serendipity and the Siamese cat: The discovery that genes for coat and eye pigment affect the brain. *ILAR Journal, 46*, 357–363.

Kania, A., & Jessell, T. M. (2003). Topographic motor projections in the limb imposed by LIM homeodomain protein regulation of ephrin-A:EphA interactions. *Neuron, 38*, 581–596.

Kawasaki, H., Crowley, J. C., Livesey, F. J., & Katz, L. C. (2004). Molecular organization of the ferret visual thalamus. *Journal of Neuroscience, 24*, 9962–9970.

King, R. A., Pietsch, J., Fryer, J. P., Savage, S., Brott, M. J., Russell-Eggitt, I., et al. (2003). Tyrosinase gene mutations in oculocutaneous albinism 1 (OCA1): Definition of the phenotype. *Human Genetics, 113*, 502–513.

Koch, S. M., Dela Cruz, C. G., Hnasko, T. S., Edwards, R. H., Huberman, A. D., & Ullian, E. M. (2011). Pathway-specific genetic attenuation of glutamate release alters select features of competition-based visual circuit refinement. *Neuron, 71*, 235–242.

Kolodkin, A. L., & Tessier-Lavigne, M. (2011). Mechanisms and molecules of neuronal wiring: A primer. *Cold Spring Harbor Perspectives in Biology, 3*. doi:10.1101/cshperspect.a001727.

Koshiba-Takeuchi, K., Takeuchi, J. K., Matsumoto, K., Momose, T., Uno, K., Hoepker, V., et al. (2000). Tbx5 and the retinotectum projection. *Science, 287*, 134–137.

Kralj-Hans, I., Tibber, M., Jeffery, G., & Mobbs, P. (2006). Differential effect of dopamine on mitosis in early postnatal albino and pigmented rat retinae. *Journal of Neurobiology, 66*, 47–55.

Kuwajima, T., Yoshida, Y., Takegahara, N., Petros, T. J., Kumanogoh, A., Jessell, T. M., et al. (2012). Optic chiasm presentation of Semaphorin6D in the context of Plexin-A1 and Nr-CAM promotes retinal axon midline crossing. *Neuron, 74*, 676–690.

Lambot, M. A., Depasse, F., Noel, J. C., & Vanderhaeghen, P. (2005). Mapping labels in the human developing visual system and the evolution of binocular vision. *Journal of Neuroscience, 25*, 7232–7237.

Leamey, C. A., Merlin, S., Lattouf, P., Sawatari, A., Zhou, X., Demel, N., et al. (2007). Ten_m3 regulates eye-specific patterning in the mammalian visual pathway and is required for binocular vision. *PLoS Biology, 5*, e241. doi:10.1371/journal.pbio.0050241.

Lee, R., Petros, T. J., & Mason, C. A. (2008). Zic2 regulates retinal ganglion cell axon avoidance of ephrinB2 through inducing expression of the guidance receptor EphB1. *Journal of Neuroscience, 28*, 5910–5919.

Leone, D. P., Srinivasan, K., Chen, B., Alcamo, E., & McConnell, S. K. (2008). The determination of projection neuron identity in the developing cerebral cortex. *Current Opinion in Neurobiology, 18*, 28–35.

Leung, K. M., Taylor, J. S., & Chan, S. O. (2003). Enzymatic removal of chondroitin sulphates abolishes the age-related axon order in the optic tract of mouse embryos. *European Journal of Neuroscience, 17*, 1755–1767.

Lopez, V. M., Decatur, C. L., Stamer, W. D., Lynch, R. M., & McKay, B. S. (2008). L-DOPA is an endogenous ligand for OA1. *PLoS Biology, 6*, e236. doi:10.1371/journal.pbio.0060236.

Luria, V., Krawchuk, D., Jessell, T. M., Laufer, E., & Kania, A. (2008). Specification of motor axon trajectory by ephrin-B:EphB signaling: Symmetrical control of axonal patterning in the developing limb. *Neuron, 60,* 1039–1053.

Lustig, M., Erskine, L., Mason, C. A., Grumet, M., & Sakurai, T. (2001). Nr-CAM expression in the developing mouse nervous system: Ventral midline structures, specific fiber tracts, and neuropilar regions. *Journal of Comparative Neurology, 434,* 13–28.

Ly, A., Nikolaev, A., Suresh, G., Zheng, Y., Tessier-Lavigne, M., & Stein, E. (2008). DSCAM is a netrin receptor that collaborates with DCC in mediating turning responses to netrin-1. *Cell, 133,* 1241–1254. doi:10.1016/j.cell. 2008.05.030.

Mann, F., & Holt, C. E. (2001). Control of retinal growth and axon divergence at the chiasm: Lessons from Xenopus. *BioEssays, 23,* 319–326.

Marcus, R. C., Blazeski, R., Godement, P., & Mason, C. A. (1995). Retinal axon divergence in the optic chiasm: Uncrossed axons diverge from crossed axons within a midline glial specialization. *Journal of Neuroscience, 15,* 3716–3729.

Marcus, R. C., & Mason, C. A. (1995). The first retinal axon growth in the mouse optic chiasm: Axon patterning and the cellular environment. *Journal of Neuroscience, 15,* 6389–6402.

Marcus, R. C., Shimamura, K., Sretavan, D., Lai, E., Rubenstein, J. L., & Mason, C. A. (1999). Domains of regulatory gene expression and the developing optic chiasm: Correspondence with retinal axon paths and candidate signaling cells. *Journal of Comparative Neurology, 403,* 346–358.

Mason, C., & Erskine, L. (2000). Growth cone form, behavior, and interactions in vivo: Retinal axon pathfinding as a model. *Journal of Neurobiology, 44,* 260–270.

Mason, C. A., & Sretavan, D. W. (1997). Glia, neurons, and axon pathfinding during optic chiasm development. *Current Opinion in Neurobiology, 7,* 647–653.

Meissirel, C., & Chalupa, L. M. (1994). Organization of pioneer retinal axons within the optic tract of the rhesus monkey. *Proceedings of the National Academy of Sciences of the United States of America, 91,* 3906–3910. doi:10.1073/ pnas.91.9.3906.

Merzdorf, C. S. (2007). Emerging roles for zic genes in early development. *Developmental Dynamics, 236,* 922–940.

Molyneaux, B. J., Arlotta, P., Menezes, J. R., & Macklis, J. D. (2007). Neuronal subtype specification in the cerebral cortex. *Nature Reviews Neuroscience, 8,* 427–437.

Moon, M. S., & Gomez, T. M. (2005). Adjacent pioneer commissural interneuron growth cones switch from contact avoidance to axon fasciculation after midline crossing. *Developmental Biology, 288,* 474–486.

Myers, P. Z., & Bastiani, M. J. (1993). Cell–cell interactions during the migration of an identified commissural growth cone in the embryonic grasshopper. *Journal of Neuroscience, 13,* 115–126.

Nagai, T., Aruga, J., Takada, S., Gunther, T., Sporle, R., Schughart, K., et al. (1997). The expression of the mouse Zic1, Zic2, and Zic3 gene suggests an essential role for Zic genes in body pattern formation. *Developmental Biology, 182,* 299–313.

Nakagawa, S., Brennan, C., Johnson, K. G., Shewan, D., Harris, W. A., & Holt, C. E. (2000). Ephrin-B regulates the ipsilateral routing of retinal axons at the optic chiasm. *Neuron, 25,* 599–610.

Nawabi, H., Briancon-Marjollet, A., Clark, C., Sanyas, I., Takamatsu, H., Okuno, T., et al. (2010). A midline switch of receptor processing regulates commissural axon guidance in vertebrates. *Genes & Development, 24,* 396–410.

Nawabi, H., & Castellani, V. (2011). Axonal commissures in the central nervous system: How to cross the midline? *Cellular and Molecular Life Sciences, 68,* 2539–2553. doi:10.1007/ s00018-011-0691-9.

Oetting, W. S., Fryer, J. P., Shriram, S., & King, R. A. (2003). Oculocutaneous albinism type 1: The last 100 years. *Pigment Cell Research, 16,* 307–311.

O'Leary, D. M., Gerfen, C. R., & Cowan, W. M. (1983). The development and restriction of the ipsilateral retinofugal projection in the chick. *Brain Research, 312,* 93–109.

Pak, W., Hindges, R., Lim, Y. S., Pfaff, S. L., & O'Leary, D. D. (2004). Magnitude of binocular vision controlled by islet-2 repression of a genetic program that specifies laterality of retinal axon pathfinding. *Cell, 119,* 567–578.

Palmesino, E., Rousso, D. L., Kao, T. J., Klar, A., Laufer, E., Uemura, O., et al. (2010). Foxp1 and lhx1 coordinate motor neuron migration with axon trajectory choice by gating Reelin signalling. *PLoS Biology, 8,* e1000446. doi:10.1371/journal.pbio.1000446.

Parra, L. M., & Zou, Y. (2010). Sonic hedgehog induces response of commissural axons to Semaphorin repulsion during midline crossing. *Nature Neuroscience, 13,* 29–35.

Pecho-Vrieseling, E., Sigrist, M., Yoshida, Y., Jessell, T. M., & Arber, S. (2009). Specificity of sensory–motor connections encoded by Sema3e-Plxnd1 recognition. *Nature, 459,* 842–846.

Petros, T. J., & Mason, C. A. (2008). Early development of the optic stalk, chiasm and astrocytes. In L. M. Chalupa & R. W. Williams (Eds.), *Eye, retina, and visual system of the mouse* (pp. 381–400). Cambridge, MA: MIT Press.

Petros, T. J., Rebsam, A., & Mason, C. A. (2008). Retinal axon growth at the optic chiasm: To cross or not to cross. *Annual Review of Neuroscience, 31,* 295–315.

Petros, T. J., Shrestha, B. R., & Mason, C. (2009). Specificity and sufficiency of EphB1 in driving the ipsilateral retinal projection. *Journal of Neuroscience, 29,* 3463–3474.

Pfeiffenberger, C., Cutforth, T., Woods, G., Yamada, J., Renteria, R. C., Copenhagen, D. R., et al. (2005). Ephrin-As and neural activity are required for eye-specific patterning during retinogeniculate mapping. *Nature Neuroscience, 8,* 1022–1027.

Picker, A., Cavodeassi, F., Machate, A., Bernauer, S., Hans, S., Abe, G., et al. (2009). Dynamic coupling of pattern formation and morphogenesis in the developing vertebrate retina. *PLoS Biology, 7,* e1000214. doi:10.1371/journal. pbio.1000214.

Pittman, A. J., Law, M. Y., & Chien, C. B. (2008). Pathfinding in a large vertebrate axon tract: Isotypic interactions guide retinotectal axons at multiple choice points. *Development, 135,* 2865–2871. doi:10.1242/dev.025049.

Plachez, C., Andrews, W., Liapi, A., Knoell, B., Drescher, U., Mankoo, B., et al. (2008). Robos are required for the correct targeting of retinal ganglion cell axons in the visual pathway of the brain. *Molecular and Cellular Neurosciences, 37,* 719–730.

Plas, D. T., Lopez, J. E., & Crair, M. C. (2005). Pretarget sorting of retinocollicular axons in the mouse. *Journal of Comparative Neurology, 491,* 305–319.

Plump, A. S., Erskine, L., Sabatier, C., Brose, K., Epstein, C. J., Goodman, C. S., et al. (2002). Slit1 and Slit2 cooperate to prevent premature midline crossing of retinal axons in the mouse visual system. *Neuron, 33,* 219–232.

Polleux, F., Ince-Dunn, G., & Ghosh, A. (2007). Transcriptional regulation of vertebrate axon guidance and synapse formation. *Nature Reviews Neuroscience, 8,* 331–340.

Polyak, S. L. (1957). *The vertebrate visual system.* Chicago: University of Chicago Press.

Pratt, T., Conway, C. D., Tian, N. M., Price, D. J., & Mason, J. O. (2006). Heparan sulphation patterns generated by specific heparan sulfotransferase enzymes direct distinct aspects of retinal axon guidance at the optic chiasm. *Journal of Neuroscience, 26,* 6911–6923.

Pratt, T., Tian, N. M., Simpson, T. I., Mason, J. O., & Price, D. J. (2004). The winged helix transcription factor Foxg1 facilitates retinal ganglion cell axon crossing of the ventral midline in the mouse. *Development, 131,* 3773–3784. doi:10.1242/dev.01246.

Quina, L. A., Pak, W., Lanier, J., Banwait, P., Gratwick, K., Liu, Y., et al. (2005). Brn3a-expressing retinal ganglion cells project specifically to thalamocortical and collicular visual pathways. *Journal of Neuroscience, 25,* 11595–11604.

Rachel, R. A., Dolen, G., Hayes, N. L., Lu, A., Erskine, L., Nowakowski, R. S., et al. (2002). Spatiotemporal features of early neuronogenesis differ in wild-type and albino mouse retina. *Journal of Neuroscience, 22,* 4249–4263.

Ramon y Cajal, S. (1911). *Histologie du système nerveux de l'homme et des vertébrés* (Vol. II). Paris: Maloine.

Raper, J., & Mason, C. (2010). Cellular strategies of axonal pathfinding. *Cold Spring Harbor Perspectives in Biology, 2,* a001933. doi:10.1101/cshperspect.a001933.

Rebsam, A., Bhansali, P., & Mason, C. A. (2012). Eye-specific projections of retinogeniculate axons are altered in albino mice. *Journal of Neuroscience, 32,* 4821–4826.

Rebsam, A., Petros, T. J., & Mason, C. A. (2009). Switching retinogeniculate axon laterality leads to normal targeting but abnormal eye-specific segregation that is activity dependent. *Journal of Neuroscience, 29,* 14855–14863.

Reese, B. E. (1996). The chronotopic reordering of optic axons. *Perspectives on Developmental Neurobiology, 3,* 233–242.

Reese, B. E., Johnson, P. T., Hocking, D. R., & Bolles, A. B. (1997). Chronotopic fiber reordering and the distribution of cell adhesion and extracellular matrix molecules in the optic pathway of fetal ferrets. *Journal of Comparative Neurology, 380,* 355–372.

Sakai, J. A., & Halloran, M. C. (2006). Semaphorin 3d guides laterality of retinal ganglion cell projections in zebrafish. *Development, 133,* 1035–1044. doi:10.1242/dev.02272.

Sakano, H. (2010). Neural map formation in the mouse olfactory system. *Neuron, 67,* 530–542.

Sakurai, T. (2012). The role of NrCAM in neural development and disorders—Beyond a simple glue in the brain. *Molecular and Cellular Neurosciences, 49,* 351–363.

Sanchez-Camacho, C., & Bovolenta, P. (2008). Autonomous and non-autonomous Shh signalling mediate the in vivo growth and guidance of mouse retinal ganglion cell axons. *Development, 135,* 3531–3541. doi:10.1242/dev.023663.

Satoda, M., Takagi, S., Ohta, K., Hirata, T., & Fujisawa, H. (1995). Differential expression of two cell surface proteins, neuropilin and plexin, in Xenopus olfactory axon subclasses. *Journal of Neuroscience, 15,* 942–955.

Schulte, D., Furukawa, T., Peters, M. A., Kozak, C. A., & Cepko, C. L. (1999). Misexpression of the Emx-related homeobox genes cVax and mVax2 ventralizes the retina and perturbs the retinotectal map. *Neuron, 24,* 541–553.

Seth, A., Culverwell, J., Walkowicz, M., Toro, S., Rick, J. M., Neuhauss, S. C., et al. (2006). Belladonna/(Ihx2) is required for neural patterning and midline axon guidance in the zebrafish forebrain. *Development, 133,* 725–735.

Shatz, C. J. (1996). Emergence of order in visual system development. *Proceedings of the National Academy of Sciences of the United States of America, 93,* 602–608.

Simon, D. K., & O'Leary, D. D. (1991). Relationship of retinotopic ordering of axons in the optic pathway to the formation of visual maps in central targets. *Journal of Comparative Neurology, 307,* 393–404.

Sretavan, D. W. (1990). Specific routing of retinal ganglion cell axons at the mammalian optic chiasm during embryonic development. *Journal of Neuroscience, 10,* 1995–2007.

Surmeli, G., Akay, T., Ippolito, G. C., Tucker, P. W., & Jessell, T. M. (2011). Patterns of spinal sensory–motor connectivity prescribed by a dorsoventral positional template. *Cell, 147,* 653–665.

Taylor, J. S., & Guillery, R. W. (1995). Does early monocular enucleation in a marsupial affect the surviving uncrossed retinofugal pathway? *Journal of Anatomy, 186,* 335–342.

Thanos, S., & Bonhoeffer, F. (1984). Development of the transient ipsilateral retinotectal projection in the chick embryo: A numerical fluorescence-microscopic analysis. *Journal of Comparative Neurology, 224,* 407–414.

Thanos, S., Bonhoeffer, F., & Rutishauser, U. (1984). Fiber–fiber interaction and tectal cues influence the development of the chicken retinotectal projection. *Proceedings of the National Academy of Sciences of the United States of America, 81,* 1906–1910. doi:10.1073/pnas.81.6.1906.

Tian, N. M., Pratt, T., & Price, D. J. (2008). Foxg1 regulates retinal axon pathfinding by repressing an ipsilateral program in nasal retina and by causing optic chiasm cells to exert a net axonal growth-promoting activity. *Development, 135,* 4081–4089. doi:10.1242/dev.023572.

Tibber, M. S., Whitmore, A. V., & Jeffery, G. (2006). Cell division and cleavage orientation in the developing retina are regulated by L-DOPA. *Journal of Comparative Neurology, 496,* 369–381.

Torborg, C. L., Hansen, K. A., & Feller, M. B. (2005). High frequency, synchronized bursting drives eye-specific segregation of retinogeniculate projections. *Nature Neuroscience, 8,* 72–78.

Trousse, F., Marti, E., Gruss, P., Torres, M., & Bovolenta, P. (2001). Control of retinal ganglion cell axon growth: A new role for Sonic hedgehog. *Development, 128,* 3927–3936.

Victor, J. D., Apkarian, P., Hirsch, J., Conte, M. M., Packard, M., Relkin, N. R., et al. (2000). Visual function and brain organization in non-decussating retinal-fugal fibre syndrome. *Cereb Cortex, 10,* 2–22. doi:10.1093/cercor/10.1.2.

Walsh, C., & Guillery, R. W. (1985). Age-related fiber order in the optic tract of the ferret. *Journal of Neuroscience, 5,* 3061–3069.

Wang, L. C., Dani, J., Godement, P., Marcus, R. C., & Mason, C. A. (1995). Crossed and uncrossed retinal axons respond differently to cells of the optic chiasm midline in vitro. *Neuron, 15,* 1349–1364.

Wang, L. C., Rachel, R. A., Marcus, R. C., & Mason, C. A. (1996). Chemosuppression of retinal axon growth by the mouse optic chiasm. *Neuron, 17*, 849–862.

Webster, M. J., & Rowe, M. H. (1991). Disruption of developmental timing in the albino rat retina. *Journal of Comparative Neurology, 307*, 460–474.

Williams, S. E., Grumet, M., Colman, D. R., Henkemeyer, M., Mason, C. A., & Sakurai, T. (2006). A role for Nr-CAM in the patterning of binocular visual pathways. *Neuron, 50*, 535–547.

Williams, S. E., Mann, F., Erskine, L., Sakurai, T., Wei, S., Rossi, D. J., et al. (2003). Ephrin-B2 and EphB1 mediate retinal axon divergence at the optic chiasm. *Neuron, 39*, 919–935.

Williams, S. E., Mason, C. A., & Herrera, E. (2004). The optic chiasm as a midline choice point. *Current Opinion in Neurobiology, 14*, 51–60.

Wolf, B. D., & Chiba, A. (2000). Axon pathfinding proceeds normally despite disrupted growth cone decisions at CNS midline. *Development, 127*, 2001–2009.

Wolman, M. A., Liu, Y., Tawarayama, H., Shoji, W., & Halloran, M. C. (2004). Repulsion and attraction of axons by semaphorin3D are mediated by different neuropilins in vivo. *Journal of Neuroscience, 24*, 8428–8435.

Xu, H. P., Furman, M., Mineur, Y. S., Chen, H., King, S. L., Zenisek, D., et al. (2011). An instructive role for patterned spontaneous retinal activity in mouse visual map development. *Neuron, 70*, 1115–1127.

Xue, Y., & Honig, M. G. (1999). Ultrastructural observations on the expression of axonin-1: Implications for the fasciculation of sensory axons during axonal outgrowth into the chick hindlimb. *Journal of Comparative Neurology, 408*, 299–317.

Yoshida, Y., Han, B., Mendelsohn, M., & Jessell, T. M. (2006). PlexinA1 signaling directs the segregation of proprioceptive sensory axons in the developing spinal cord. *Neuron, 52*, 775–788.

93 Mechanisms of Axon Guidance and Adhesion Signaling in Thalamocortical Axon Targeting

PATRICIA F. MANESS

Correct synaptic targeting of axons from distinct nuclei of the thalamus to different primary cortical areas is essential for vision, touch, and hearing, and motor function. Elucidation of the mechanisms governing the development of thalamocortical (TC) connectivity is challenging because of its precise topographic organization along both rostrocaudal and mediolateral axes of the thalamus and neocortex. In the TC pathway, thalamic axons from motor (ventroanterior/ventrolateral; VA/VL), somatosensory (ventrobasal complex; VB), and visual thalamic nuclei (dorsal lateral geniculate; dLGN) project topographically to the primary cortical areas (motor, M1; somatosensory, S1; visual, V1) (see figure 93.1). Current evidence obtained largely from studies in mice reveals that TC topography arises from a progressive series of axon guidance decisions at critical choice points from the dorsal thalamus (DT), through a corridor in the ventral telencephalon (VTe), and into the neocortex (see figure 93.1). Within the cortex, TC axons are further influenced by intrinsic cues that direct them to area- and lamina-specific regions. Initially diffuse projections are refined by activity-dependent and activity-independent mechanisms enabling them to synapse with appropriate synaptic partners. Recent progress has been made in identifying the molecular cues that guide TC axons from the thalamus to the cortex although important gaps in the knowledge of mechanisms involved remain to be filled. In this review we will focus on specific roles identified for Ephrins, Semaphorins, Netrins, Slits, Neuregulin1 (Nrg1), and in particular, neural cell adhesion molecules (CAMs) in controlling TC axon guidance at pivotal choice points along the TC projection but will not consider transcriptional regulation influencing TC development.

ENTRANCE OF THALAMOCORTICAL AXONS INTO THE VENTRAL TELENCEPHALON

Thalamic axons in the DT exit through the ventral thalamus and proceed ventrally around embryonic day (E) 12.5 in the mouse, orienting in the direction of the hypothalamus (see figure 93.2; for general reference see Price et al., 2006). TC axons from the rostral and caudal DT are well segregated at the thalamic eminence as they leave the thalamus. As axons cross the boundary between the diencephalon and telencephalon, they turn acutely toward the telencephalon. At approximately E14.5, TC axons project through a permissive corridor in the region of the medial ganglionic eminence (MGE) toward the cortex. By E16.5 TC axons have crossed the palllial–subpallial boundary and entered the cortical intermediate zone. From E16.5–16 TC axons wait in the cortical subplate and later reach layer 4 where they arborize and form synapses postnatally.

During the early stages of TC axon growth from the DT, the hypothalamus expresses high levels of secreted guidance molecules of the Slit family while TC axons express their immunoglobulin-class Robo receptors. Slits1–3 function typically as repulsive guidance cues by signaling through Robo1–4 receptors. Consistent with a repellent function of Slit1/2 on TC axon guidance, TC axon contingents in Slit 2 null, Slit1/2 double null, and Robo1/2 null mice do not turn toward the telencephalon but instead misproject ventrally to the hypothalamus, and thalamic axons are repelled from Slit2 in vitro (Braistead et al., 2009; Lopez-Bendito et al., 2007). These findings support the conclusion that Slit1/2 repels TC axons from the hypothalamus, thus promoting axon turning toward the telencephalon.

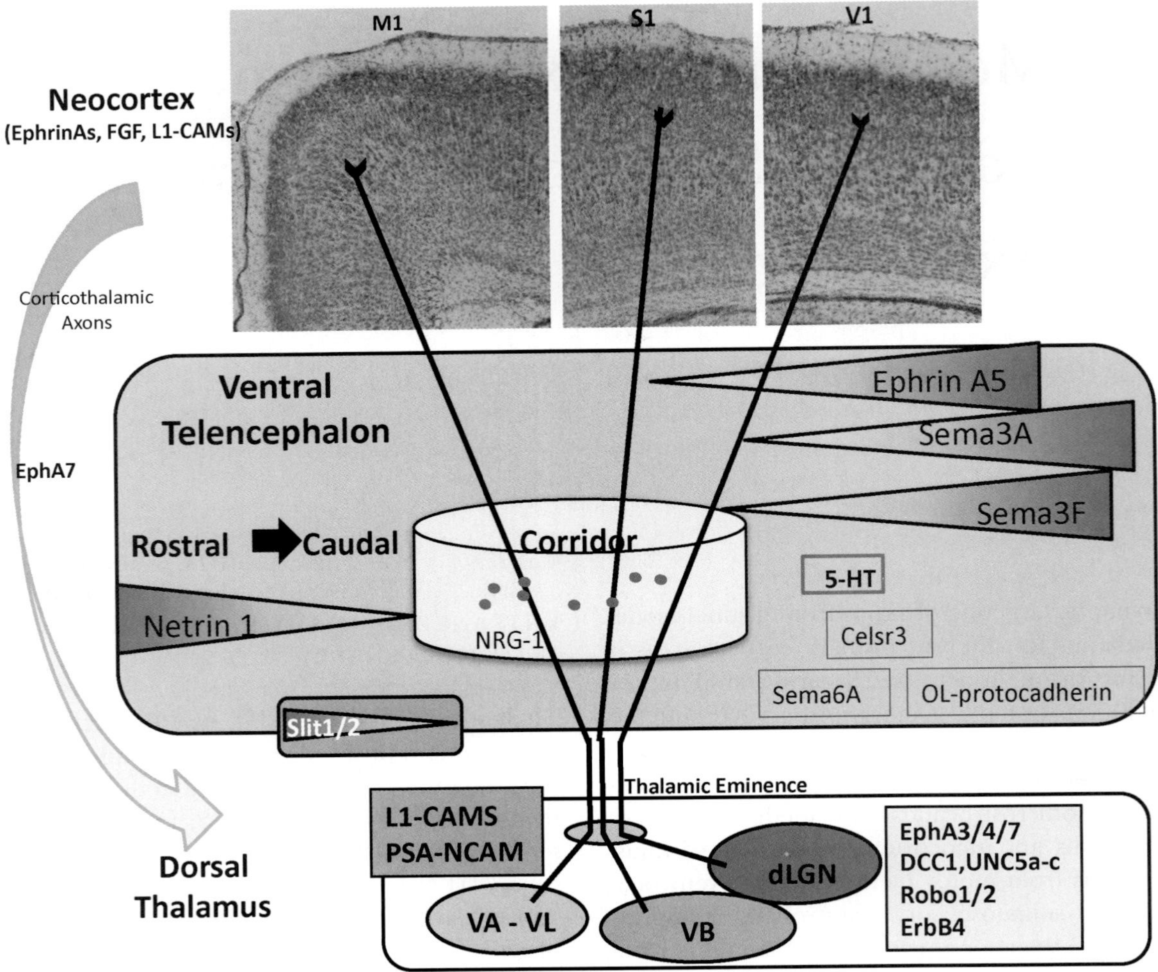

FIGURE 93.1 Scheme of thalamocortical axon (TCA) trajectories from dorsal thalamic nuclei in the subpallium/ventral telencephalon to distinct neocortical areas. TCAs from different nuclei in the dorsal thalamus emerge at the thalamic eminence en route to the neocortex and are sorted within a corridor of Islet1-positive cells along the rostral to caudal axis of the ventral telencephalon (embryonic day 13.5–15.5 in the mouse). Within this region, TCAs expressing different combinations of axon guidance cue receptors (listed in the box within the dorsal thalamus) are guided by gradients of repellent and attractant cues (EphrinA5, Semaphorin3A, Semaphorin3F, Semaphorin6A, Netrin1, Slit1/2) and influenced by Neuregulin-1(NRG1) and serotonin (5-HT, 5-hydroxytryptamine). L1 family cell adhesion molecules (L1-CAMs), neural cell adhesion molecule NCAM (polysialylated), and cadherin family members Celsr3 and OL-protocadherin contribute to thalamic axon guidance. Reciprocal corticothalamic axon projections are controlled in part by EphA7. TCAs from thalamic motor nuclei (VA/VL, ventroanterior/ventrolateral), somatosensory nuclei (VB, ventrobasal complex), and visual nucleus (dLGN, dorsal lateral geniculate) finally target primary motor cortex (M1), somatosensory cortex (S1), and visual cortex (V1). Within the neocortex, a plethora of growth cone guidance cues and cell adhesion molecules as well as activity-dependent mechanisms promote the final synaptic targeting of TCAs.

Polysialylated NCAM, an immunoglobulin-class adhesion molecule expressed on TC axons, promotes TC pathfinding from the DT through the ventral thalamus and into the VTe (Schiff et al., 2011). Polysialylation, an embryonic modification of NCAM due to sialyltransferase activity, reduces affinity of homophilic NCAM interactions, increasing neuronal responses to other guidance cues. Removal of polysialic acid prevents TC axons from turning sharply into the VTe, suggesting that this modification enables axons to respond to Slits. NCAM deletion in mice results in normal TC axon sorting in the VTe at embryonic stages; however, rostral DT axons show pathfinding defects as they establish cortical connections, resulting in an altered somatosensory map (Enriquez-Barreto et al., 2012).

A transmembrane Semaphorin, Sema6A, is also required for dLGN axons in the VTe to turn and enter the cortex (Leighton et al., 2001; Little et al., 2009). In the absence of Sema6A, dLGN axons go the amygdalar region. As a possible consequence, some VB axons

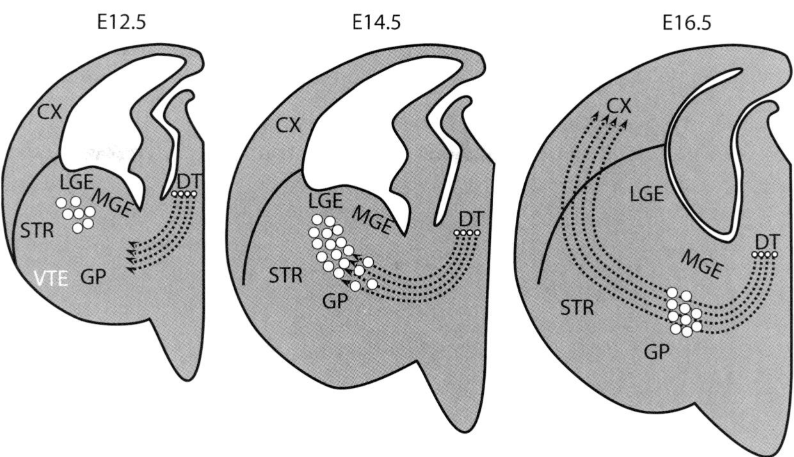

FIGURE 93.2   Projection of thalamocortical axons during mouse embryonic development. Thalamic axons project from the DT to cross the DTB (embryonic day [E] 12.5) and enter the ventral telencephalon (VTe) through a corridor of Islet1[+] cells (E14.5), then pass the pallial–subpallial boundary into the neocortex (E16.5). Dorsal thalamus (DT), cortex (CX), lateral and medial ganglionic eminence (LGE, MGE), striatum (STR), globus pallidus (GP), diencephalic-telencephalic boundary (DTB). Dashed lines, thalamic axons; large white circles, corridor cells.

misproject to V1, although in adults this phenotype is not observed, suggesting that cortical cues may correct aberrant projections. Members of the cadherin family of calcium-dependent CAMs also contribute to TC axon pathfinding in the VTe. Celsr3, an atypical cadherin expressed in the VTe, mediates entry of both TC and corticothalamic (CT) axons into this subdomain (Zhou et al., 2009). OL-Protocadherin also influences early pathfinding of TC and CT axons in the VTe, where it may regulate growth of striatal axons and increase the permissive environment of the region (Uemura et al., 2007).

Within the VTe lies a corridor of cells that function as a substrate to allow TC axons to grow through relatively nonpermissive territory within the MGE. Islet1+ corridor cells are GABAergic neurons that migrate ventrally from the lateral ganglionic eminence from E11.5 to E14 and form a permissive substrate that pioneers the entrance of TC axons into the VTe (Bielle et al., 2011; Lopez-Bendito et al., 2006). Slit2 expression in the ventral MGE serves as a repellent to appropriately position the migrating corridor cells (Bielle et al., 2011). Corridor cells express NRG-1, a membrane-associated attractant that signals through the ErbB4 tyrosine kinase receptor to activate multiple intracellular signaling pathways. Axon tracing in NRG-1 and ErbB4 mutant mice and explant coculture experiments collectively demonstrate that TC axon navigation through the corridor is achieved via NRG-1/ErbB4 signaling, although other cues likely contribute. Factors that control TC axon segregation and early guidance within the thalamus are an important area for further research.

INFLUENCE OF EPHRINAS AND SLIT1 ON THALAMOCORTICAL AXON SORTING WITHIN THE VENTRAL TELENCEPHALON

The VTe is a crucial intermediate target where TC axons are sorted rostromedially to caudolaterally toward appropriate neocortical areas under the influence of multiple guidance signals. This sorting is generated by a diversity of molecular cues expressed within the VTe independently of cortex-derived signals. Both repellent and attractant guidance cues exhibit graded expression patterns in the striatal area of the VTe (Ephrins, Semaphorins, Netrins). Countergradients of EphrinA ligands and their EphA tyrosine kinase receptors control sorting of rostral TC axon contingents within the VTe (Dufour et al., 2003; Egea et al., 2005). Axons from the rostral DT express high levels of EphA4 and EphA7 receptors and are repelled from the high-caudolateral to low-rostromedial gradient of EphrinA5 in the VTe. This mechanism enables correct targeting of motor thalamic axons from the ventrolateral (VL) nucleus to M1. In mice deficient in EphA4, EphA7, or both EphrinA5 and EphA4, contingents of VL axons are caudally shifted in the VTe and misproject to S1, where they form stable connections (Dufour et al., 2003) (see figure 93.3). EphrinA5/EphA4 double mutants are the most severely impacted and display intra-areal as well as interareal mistargeting of TC axons, as shown by altered terminations of VPM axons within S1 (Dufour et al., 2003). Intra-areal mistargeting of VB axons to S1 depends on EphA4 kinase activity (Dufour et al., 2006; Egea et al., 2005). In addition a fraction of afferents from the

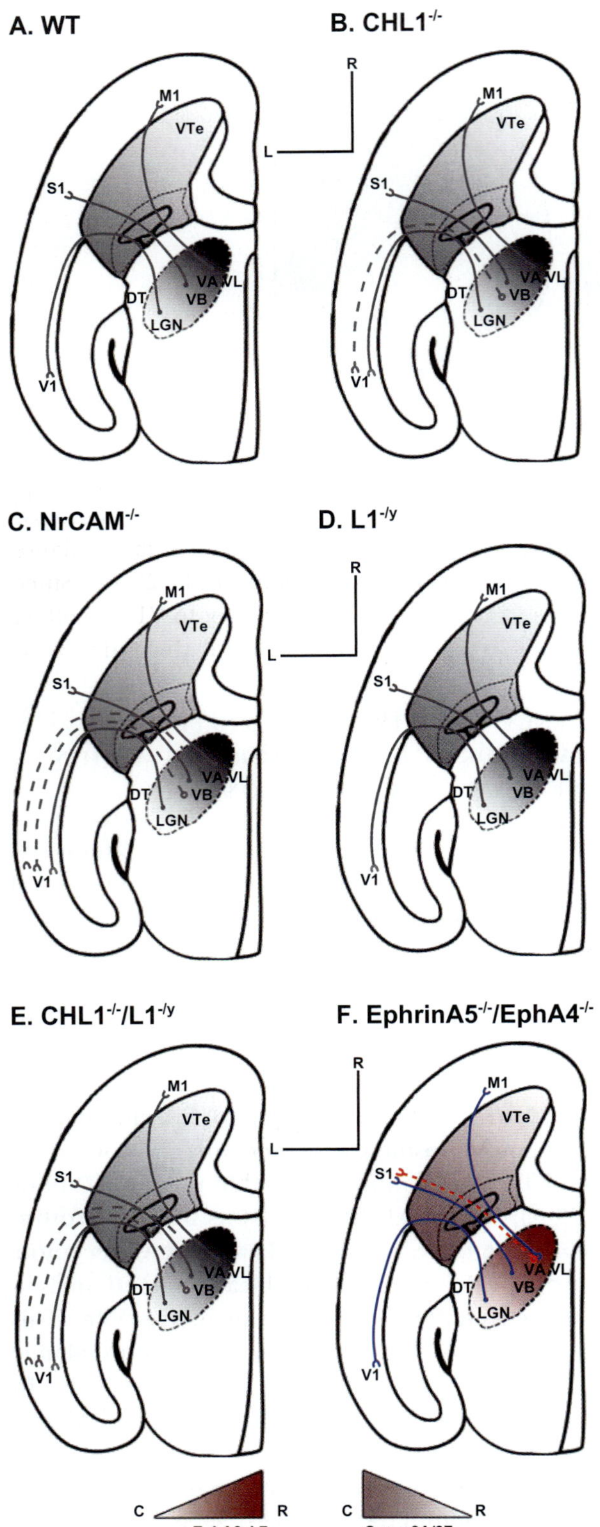

FIGURE 93.3 Scheme of thalamocortical trajectories in mouse null mutants. Based on axon tracing, normal projections are depicted as solid lines; misprojections, as dashed lines. (A) wild type (WT), (B) CHL1⁻/⁻, (C), NrCAM⁻/⁻, (D) L1⁻/ʸ (null mutant; L1 gene is on X chromosome), (E) CHL1⁻/⁻ L1⁻/ʸ double mutant, (F) EphrinA5⁻/⁻/EphA4⁻/⁻ double mutant. The rostrocaudal expression gradients of Sema3A, Sema3F, and EphrinA5 in the ventral telencephalon (VTe) is shown, and rostrocaudal gradients of EphA3,4,7 in the dorsal thalamus are indicated by shading. R, right; L, left; M1, primary motor cortical area; S1, primary somatosensory cortical area; V1, primary visual cortical area; VA VL, ventroanterior/ventrolateral; VB, ventrobasal complex; LGN, lateral geniculate nucleus; DT, dorsal thalamus.

laterodorsal thalamic nucleus of EphrinA5 null mice caudally misproject to S1 (Uziel et al., 2002). EphrinA5 appears to act at several levels in the TC pathway. In vitro studies implicate ephrinA5 as both a repellent (Gao et al., 1998) and attractant for thalamic and cortical axons (Castellani et al., 1998; Mann et al., 2002). Within the cortex, loss of EphrinA5 restricts the arborization of thalamic axon terminals (Uziel, Muhlfriedel, & Bolz, 2008) and may promote compensatory branching of spiny stellate cell dendrites, a target of VB projections in layer 4 of S1 (Guellmar, Rudolph, & Bolz, 2009).

The axon guidance molecule Netrin1 provides a counterforce to EphrinA5-induced thalamic axon repulsion in the VTe (Bonnin et al., 2007; Braistead et al., 2000; Powell et al., 2008). Netrin1 is expressed in a rostral-high gradient opposite to the gradient of EphrinA5 in the VTe. Netrin1 plays a dual role in attracting rostral thalamic axons and repelling caudal thalamic axons (Bonnin et al., 2007; Powell et al., 2008). These opposing effects are mediated by different levels of the Netrin1 receptors DCC (deleted in colorectal carcinoma) and Unc5a,c expressed on thalamic axons. TC axon responses in the VTe can be modulated by serotonin (5-hydroxytryptamine; 5HT) (Bonnin et al., 2007). Recent findings also demonstrate a combinatorial role for Netrin1 and Slit1 within the corridor, in which parallel rostrocaudal gradients of Netrin1 and Slit1 coordinate to guide rostral TC axons (Bielle et al., 2011).

L1 CELL ADHESION MOLECULES MODULATE GUIDANCE OF THALAMOCORTICAL AXONS TO GRADIENTS OF SEMAPHORINS AND EPHRINS

L1 family CAMs (L1, Close Homolog of L1 [CHL1], Neuron-glial related cell adhesion molecule [NrCAM], and Neurofascin) are structurally related immunoglob-ulin-class axon guidance molecules (see figure 93.4). In situ hybridization has revealed distinct and overlapping patterns of expression of different L1-CAMs within the embryonic DT but not smooth gradients. During growth of TC axons through the VTe, L1-CAMs are localized to TC axons and growth cones but not to cells within the VTe substrate (Demyanenko et al., 2011a; Wright et al., 2007). L1-CAMs promote or inhibit axon growth depending on whether they engage in homophilic *trans*-interactions with L1-CAMs on opposing cells, or heterophilic *cis*-interactions with repellent receptors on the axon (Castellani et al., 2000, 2002).

L1-CAMs regulate pathfinding of different ensembles of TC axons by mediating repellent responses to gradients of class III Semaphorins (Sema3s), which are expressed by cells in the VTe, rather than in thalamic neurons. Sema3A-G are secreted axon guidance cues exhibiting graded expression in the VTe during TC axon pathfinding (Tran, Kolodkin, & Bharadwaj, 2007). Sema3s are well known for their ability to induce growth cone collapse and axon retraction, but they can also act as attractants. Sema3s bind Npn1/2 receptors, which recruit PlexinA signaling subunits (PlexA1–4) to form a complex in the plasma membrane. PlexinAs have intrinsic Rac1-GTPase activity (Pasterkamp & Giger, 2009; Tran, Kolodkin, & Bharadwaj, 2007), which has the potential to bring about cytoskeletal rearrangements that influence growth cone collapse and axon guidance. L1 was first shown to regulate Sema3A-induced axon repulsion by associating with Npn1 (Castellani et al., 2000) and also mediates repulsion to attraction. Interestingly, Sema3A-mediated repulsion can be converted to attraction (Castellani et al., 2002).

CHL1 specifies targeting of VB axon contingents to S1 by binding Npn-1 and mediating Sema3A-induced repellent guidance in the VTe (Wright et al., 2007). Genetic lesion of CHL1 in mice causes caudal shifting of VB axon contingents within the VTe, resulting in mistargeting to V1 rather than to S1 (Wright et al., 2007) (see figure 93.3). Projection of dLGN axons to V1 appears unaffected in the mutants. Moreover, Npn-1$^{\text{Sema3A-/-}}$ knock-in mice, which express a mutated Npn-1 that does not bind Sema3A (Gu et al., 2003), display the same phenotype of caudal shifting of VB axons in the VTe (Wright et al., 2007). CHL1 associates with Npn-1 through a sequence (FASNRL) in the CHL1 Ig1 domain, which is required for growth cone collapse of thalamic neurons in culture. In vivo, Sema3A is expressed in a high-caudolateral to low-rostromedial gradient in the VTe, where it can act as a repellent to direct VB axons toward S1.

Unlike CHL1, deletion of L1 in mice does not alter area-specific topographic targeting of TC axons from VA/VL, VB, or dLGN (Wiencken-Barger et al., 2004) (see figure 93.3). But notably, when both L1 and CHL1 are deleted in mice, a more severe phenotype is observed than in mice with single gene deletions (Demyanenko et al., 2011b). Thalamic projections from rostral (VA/VL) as well as VB nuclei mistarget to V1 in double mutants, whereas dLGN axons project normally to V1 (see figure 93.3). This phenotype supports a cooperative role for L1 and CHL1 in TC axons, possibly for responding to Sema3A or other repellent cues in the VTe. L1, like CHL1 (Wright et al., 2007), binds Npn-1 through a homologous motif (FASNKL) in its Ig1 domain, which is required for growth cone collapse to Sema3A (Castellani et al., 2000). L1-induced collapse is selective for Sema3A, as L1 does not mediate collapse to Sema3B or Sema3E (Castellani et al., 2000). Since CHL1 loss causes VB axon to mistarget to V1, the effect of dual deletion likely impacts rostral thalamic axons selectively.

In addition to transducing Sema3A responses, L1 and CHL1 mediate EphrinA5-induced growth cone collapse. Both cortical and thalamic neuron cultures from L1 or CHL1 null mutant mice are impaired in their growth cone collapse responses to EphrinA5 (Demyanenko et al., 2011b). L1 and CHL1 coimmunoprecipitate with the principal EphrinA5 receptors expressed in the DT (EphA3, EphA4, and EphA7) but display preferences for different EphA isoforms (Demyanenko et al., 2011b). L1 associates with EphA3, -A4, and -A7 whereas CHL1 associates only with EphA7 (Demyanenko et al., 2011b). In the mouse TC pathway different patterns of expression have been described for EphrinAs (EphrinA3, -A4, and -A5) and their EphA receptors (EphA3, -A4, and -A7) (Mackarehtschian et al., 1999; Torii & Levitt, 2005; Uziel et al., 2006). Moreover, EphrinAs show selective binding affinities for different EphA receptors, as shown for EphA4 and EphA7 (Janis, Cassidy, & Kromer, 1999). These results suggest that differences in receptor and ligand expression may confer differential responsiveness on thalamic neuron subpopulations, which are modulated by L1 and CHL1.

NrCAM plays an analogous role to CHL1 in guiding TC axon populations from VB nuclei to S1, but also in guiding axons from more rostral nuclei (VA/VL) to M1 (see figure 93.3). Deletion of NrCAM in null mutant mice causes contingents of axons from VA/VL and VB to become caudally shifted in the VTe and to mistarget to V1, whereas dLGN axons target V1 normally (Demyanenko et al., 2011a). Npn-2 null mutant mice display the same phenotype of axon misguidance in the VTe. Sema3F is expressed in a high-caudolateral

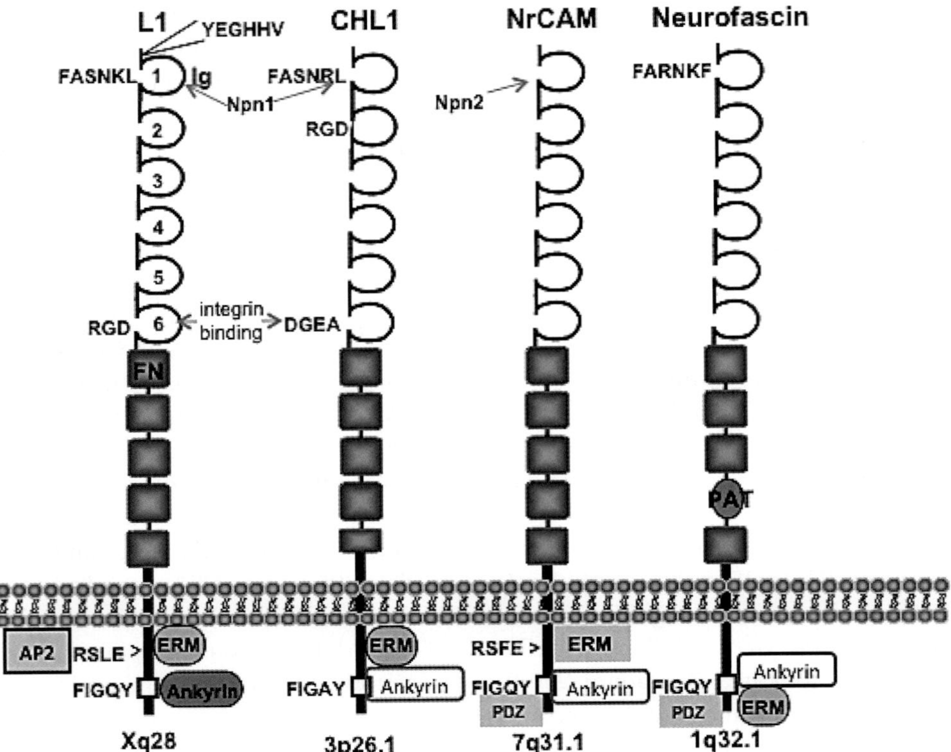

FIGURE 93.4   Structure and interactions of L1 family cell adhesion molecules. Domain structures of L1, Close Homolog of L1 (CHL1), Neuron-glial related cell adhesion molecule (NrCAM) and Neurofascin and sequences that bind interaction partners are indicated. Conserved cytoplasmic domains bind the clathrin adaptor AP2, ERMs (ezrin, radixin, moesin), ankyrin, and PDZ domain containing proteins. Neurofascin has an extracellular PAT domain not present in the other L1-CAMs. The human chromosomal location of each L1-CAM is indicated below.

to low-rostromedial gradient in the VTe, where it is in position to repel TC axon contingents toward M1 and S1. Thalamic neurons from NrCAM null mice are defective in growth cone collapse to Sema3F. NrCAM physically associates with Npn-2 in E15.5 brain (Demyanenko et al., 2011a) and is selective for inducing growth cone collapse to Sema3F but not Sema3A (Falk et al., 2005). Neurofascin is expressed in subsets of neurons in the DT at E14.5–15.5, but deletion of this gene in null mutant mice did not affect final targeting of TC axons to M1, S1, or V1 (Demyanenko et al., 2011b). It remains open whether thalamic projections to other areas depend on this L1-CAM.

An alternative consideration for topographic mapping described in the olfactory pathway involves selective fasciculation of axon subsets. Olfactory map topography depends on pretarget sorting of olfactory sensory axon contingents within the nerve, which is achieved in part by Sema3A- and Npn1-repulsion among axons within the fascicle (Dickstein et al., 2011). Topographic order in the olfactory map is also influenced by Sema3F, which is secreted by early arriving axons and repels axons that arrive later (Takeuchi et al., 2010). In

contrast, Sema3A and Sema3F in the TC pathway are expressed on the VTe substrate rather than on TC axons (Demyanenko et al., 2011a; Wright et al., 2007). Moreover, rostral and caudal DT axons remain well segregated upon leaving the thalamic eminence in L1, NrCAM, Npn1/2, and Sema3A null mutant embryos (Demyanenko et al., 2011a; Wright et al., 2007). However, Sema3 gradients in the VTe might induce subtle defasciculation of TC axons, which could increase responses to other factors, as TC axons in L1 null mutant mice appear hyperfasciculated (Ohyama et al., 2004; Wiencken-Barger et al., 2004).

EFFECT OF NRCAM DELETION ON
VISUAL FUNCTION

Loss of NrCAM has functional consequences on visual cortical responses in the mouse, suggesting that visual wiring critically depends on NrCAM expression. NrCAM null mice exhibit impaired visual acuity, measured by visually evoked potentials recorded in the binocular zone of V1 at P2 (Demyanenko et al., 2011a). Deletion of NrCAM preferentially reduced the ipsilaterally evoked eye responses, suggesting that ipsilateral inputs are weak or specifically mistargeted. NrCAM loss did not alter retinogeniculate axon targeting to the dLGN, eye-specific segregation of ipsilateral and contralateral inputs within the dLGN, or geniculocortical targeting (Demyanenko et al., 2011a). More work is needed to determine if altered visual responses are due to mistargeted TC inputs from VA/VL or VM nuclei or to other alterations in retinotopic pathway. For example, a subpopulation of contralaterally projecting retinal axons misprojects ipsilaterally at the optic chiasm (E17-P0) (Falk et al., 2005; Williams et al., 2006). However, in view of the normal retinogeniculate map observed in NrCAM null mice (postnatal day [P] 10 to P12) (Demyanenko et al., 2011a), these retinal axons may not reach the dLGN or are eliminated postnatally.

It is noteworthy that genetic deletion in NrCAM (Demyanenko et al., 2011a), CHL1 (Wright et al., 2007), L1/CHL1 (Demyanenko et al., 2011b), EphA4/ EphrinA5 (Dufour et al., 2003), and Sema6A (Little et al., 2009) all result in a common pattern of caudal mistargeting of TC axons from inappropriate nuclei to V1. Mouse mutants in Npn1/2, Sema3A/3F, and Netrin1 (Demyanenko et al., 2011a; Powell et al., 2008; Wright et al., 2007) also display caudal deviation in pathfinding within the VTe, although they were not examined for final targeting to V1 postnatally. The reason for this common default projection to V1 is not understood but might result from the action of strong gradients of rostral repellents or caudal attractants in the VTe. The extent to which loss of these molecules impairs visual acuity and ocularity can be addressed by analyzing visual cortical responses in the other null mutant mice.

CYTOSKELETAL MECHANISMS OF L1-CAM-MEDIATED AXON REPULSION

The highly conserved cytoplasmic domains of L1-CAMs reversibly engage cytoskeletal adaptors ankyrin and ezrin, radixin, moesin (ERMs), which link L1-CAMs to the submembranous actin cytoskeleton (see figure 93.4). Alternative splice variants of L1 and NrCAM harbor an insert in the cytoplasmic domain (RSL/FE) that binds the clathrin adaptor AP2, which participates in receptor endocytosis. Dynamic interactions of adaptor proteins with L1-CAMs can modulate functions important for TC axon pathfinding, including (a) growth cone/substrate adhesion, (b) motility, (c) fasciculation, and (d) trafficking of surface receptors.

L1-ankyrin binding is regulated by phosphorylation of the FIGQ/AY binding motif on tyrosine residue Y1229, which releases ankyrin (Garver et al., 1997).

Ankyrin binding promotes adhesion and stationary behavior by reducing retrograde flow of actin (Gil et al., 2003). The serine/threonine kinase ERK indirectly mediates FIGQ/AY phosphorylation and ankyrin dissociation from L1 (Whittard et al., 2006). EphB receptor tyrosine kinases can phosphorylate L1 at FIGQY, mediated by the nonreceptor tyrosine kinase Src, to cause ankyrin dissociation and promotion of neurite outgrowth (Dai et al., 2012; Zisch et al., 1998). Because growth cone collapse requires detachment from the substrate, L1-Y1229 phosphorylation and ankyrin dissociation may foster TC axon repulsion from high concentrations of ephrins.

L1-ankyrin binding has been studied in vivo in a mouse mutant line with a point mutation (L1-Y1229H) that prevents ankyrin binding to L1 (Buhusi et al., 2008). Retinocollicular mapping performed in these mutants demonstrated a critical role for L1-ankyrin binding in mapping of retinal ganglion call axons to correct topographic termination zones along the mediolateral axis of the superior colliculus, which is governed by an EphrinB gradient (Hindges et al., 2002; McLaughlin et al., 2003). Mediolateral axon targeting is also controlled by L1-dependent and EphrinB-modulated adhesion to ALCAM, an Ig-CAM expressed on cells of the superior colliculus (Buhusi et al., 2009).

L1–ERM interactions underlie both repellent and attractive responses in growth cones. There are two ERM binding sites containing critical tyrosine residues (Y1151, Y1176) in the L1 cytoplasmic domain, one of which (Y1176) overlaps with the AP2/clathrin binding site (Cheng, Itoh, & Lemmon, 2005; Dickson et al., 2002; Mintz et al., 2008). Mutation of the more N-terminal L1–ERM sequence disrupts L1 association and inhibits Sema3A-induced growth cone collapse in cortical neuron cultures (Mintz et al., 2008; Schlatter et al., 2008). Receptor endocytosis is a crucial component of the growth cone collapse mechanism and involves actin rearrangements that may depend on ERM interactions. Indeed, Sema3A induces internalization of Npn1 together with L1 during the collapse response (Castellani, Falk, & Rougon, 2004). Sema3A-induced repulsion might also play a role in segregating TC and corticothalamic axons in an L1-dependent manner during pathfinding by promoting homotypic axon fasciculation. ERM function is also required for neurotrophin-induced attractive responses in growth cones of sensory neurons (Marsick, San Miguel-Ruiz, & Letourneau, 2012).

Finally, NrCAM and Neurofascin are the only L1-CAMs with PDZ binding domains at their carboxyl terminus (see figure 93.4). The NrCAM PDZ binding domain has been shown to engage the PDZ-domain scaffold protein PSD95, located at postsynaptic densities (Davey et al., 2005; Dirks, Thomas, & Montag, 2006). Little is known about the role of NrCAM–PDZ interactions, but they might function in the transition to synaptogenesis or in stabilizing transient synaptic contacts. All L1-CAMs are expressed in cortical neurons in complex patterns in the developing brain, in addition to thalamic neurons, and thus have the potential to perform postsynaptic roles.

In summary, interactions of L1 family molecules with cytoskeletal and scaffold proteins may modulate diverse cellular functions, including adhesion, endocytosis, actomyosin contraction, growth cone collapse, and synaptogenesis. A current challenge is to identify which of these interactions regulate responses in migrating TC growth cones and how they coordinate to guide TC axons to their cortical targets.

CORTICAL CUES REGULATING THALAMOCORTICAL AXON TARGETING

Some of the same molecules and mechanisms acting in the VTe may operate at multiple levels in the TC pathway to establish topographic patterning. Within the cortex, EphrinA/EphA gradients regulate the precision and reciprocity of TC and CT projections between and within neocortical areas (Bolz et al., 2004; Cang et al., 2005; Torii & Levitt, 2005). EphA7/ephrinA interactions also influence intra-areal targeting of CT axons from neurons in the auditory cortex to the medial geniculate body (Torii et al., 2013). In particular, EphA7 expressed in cortical neurons regulates reciprocal CT projections from S1 and V1 enabling specific terminations within their respective target nuclei in the VB complex and dLGN (Torii & Levitt, 2005). Gradients of EphrinAs and their receptors also function as intracortical cues for mapping dLGN axons along the mediolateral axis of V1 and for creation of an appropriate functional map (Cang et al., 2005). High expression of EphrinA2,3,5 lateral and medial to V1 may direct its location, probably in cooperation with other molecules. Class 3 Semaphorins also act as dual attractive and repellent guidance cues for cortical axons in vitro, and their expression patterns are consistent with a guidance function for corticofugal projections (Bagnard et al., 1998). Diffusible growth factors with graded expression in the embryonic cortex also function as intracortical determinants of TC patterning. In particular, fibroblast growth factor FGF8 is a morphogen that patterns the rostral cortical region, and when ectopically expressed in posterior cortex, the barrel field is duplicated,

suggesting that thalamic axons use cortical positional information for targeting (Fukuchi-Shimogori & Grove, 2001).

It is anticipated that more axon guidance cues and receptors will cooperate at subcortical as well as cortical levels, and together with activity-dependent mechanisms, will ultimately direct TC axons to their precise interareal and intra-areal cortical targets. Elucidation of new guidance cues for TC topographic mapping is a timely avenue for future research. For molecules that are expressed at multiple levels along the TC axon trajectory, deciphering their individual roles will require development of conditional strains of mice that target gene deletions to thalamic or cortical neurons selectively.

COMBINATORIAL MODEL OF ADHESION AND RECEPTOR SIGNALING

Distinct and overlapping expression of Semaphorins, Ephrins, Netrins, and other cues in the VTe may ensure precision in guiding TC axon subpopulations bearing diverse complements of receptors to correct neocortical areas. L1-CAMs with distinctive patterns of expression among thalamic neuron subpopulations (Demyanenko et al., 2011a, 2011b; Wright et al., 2007) and in the cortex add another level of regulation. Guidance receptor complexes in thalamic neurons may serve to integrate signals from multiple cues within subdomains of the growth cone membrane (Marquardt et al., 2005) as presented in a speculative model (see figure 93.5). Downstream signaling from activated receptors within growth cone subdomains may impinge asymmetrically on actin filaments localized in the growth cone central region (Burnette et al., 2008; Zhang et al., 2003) resulting in localized retraction of filopodial and lamellipodia (Schaefer et al., 2008). Such asymmetry could provide a means for directional navigation of the growth cone essential for correct pathfinding.

Important goals for the future will be to identify the intracellular signaling networks activated by each guidance receptor and CAM complex at the crucial choice points along the TC pathway. Molecular mechanisms of TC pathfinding within the cortex and guidance of CT axons to the thalamus also remain frontier areas for research. Several key questions remain to be understood. Specifically, how do TC axon subpopulations navigating within the cortical intermediate zone recognize appropriate sites of entry into the cortical plate? Are there distinct determinants of mediolateral and rostrocaudal targeting of TC axons within the neocortex? What signals cause TC axons to wait in the cortical

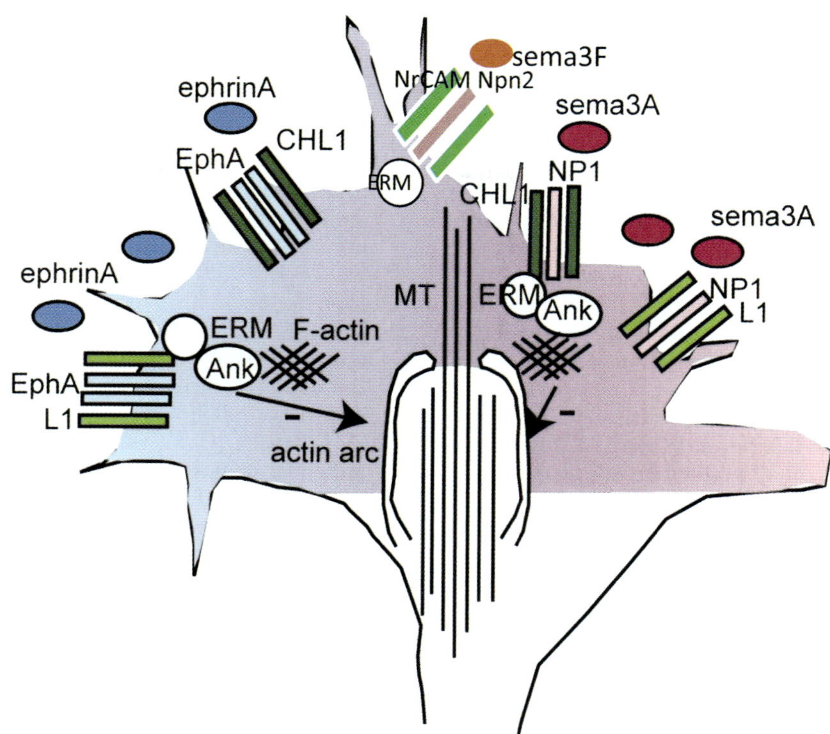

FIGURE 93.5   Speculative model for signaling by receptor and adhesion complexes in growth cone subdomains. Different guidance receptors and L1 cell adhesion molecule (L1-CAM) complexes are suggested to signal from the growth cone membrane in response to extracellular ligands (ephrinA, sema3A, sema3F). Dynamic interactions of L1-CAM cytoplasmic domains with cytoskeletal adaptors (ERM, ankyrin [ANK]) are suggested to ultimately impinges on the actin cytoskeletal structures (F-actin, actin arcs) in the central region of the growth cone adjacent to microtubules (MT) to induce localized growth cone collapse. Shading in the growth cone represents intracellular signaling induced by different gradients of sema3A /sema3F (red) or ephrinA (blue), and convergent signaling (pink) inside the growth cone. NrCAM, neuron-glial related cell adhesion molecule; CHL1, Close Homolog of L1.

subplate, arborize in layer 4, and form synapses? How are activity-dependent and activity-independent mechanisms coordinated for precise connectivity? Are the same molecules and mechanisms utilized for modulating TC connectivity during adult plasticity and aging?

ACKNOWLEDGMENTS

This work was supported by grants from the National Science Foundation (0822969), Autism Speaks (1847), and National Institutes of Health (MH 098138) funding of the University of North Carolina Silvio O. Conte Center for the Neuroscience of Mental Disorders (MH064065).

REFERENCES

Bagnard, D., Lohrum, M., Uziel, D., Puschel, A. W., & Bolz, J. (1998). Semaphorins act as attractive and repulsive guidance signals during the development of cortical projections. *Development, 125*, 5043–5053.

Bielle, F., Marcos-Mondejar, P., Keita, M., Mailhes, C., Verney, C., Nguyen Ba-Charvet, K., et al. (2011). Slit2 activity in the migration of guidepost neurons shapes thalamic projections during development and evolution. *Neuron, 69*, 1085–1098. doi:10.1016/j.neuron.2011.02.026.

Bolz, J., Uziel, D., Muhlfriedel, S., Gullmar, A., Peuckert, C., Zarbalis, K., et al. (2004). Multiple roles of ephrins during the formation of thalamocortical projections: Maps and more. *Journal of Neurobiology, 59*, 82–94. doi:10.1002/neu.10346.

Bonnin, A., Torii, M., Wang, L., Rakic, P., & Levitt, P. (2007). Serotonin modulates the response of embryonic thalamocortical axons to netrin-1. *Nature Neuroscience, 10*, 588–597.

Braisted, J. E., Catalano, S. M., Stimac, R., Kennedy, T. E., Tessier-Lavigne, M., Shatz, C. J., et al. (2000). Netrin-1 promotes thalamic axon growth and is required for proper development of the thalamocortical projection. *Journal of Neuroscience, 20*, 5792–5801.

Braisted, J. E., Ringstedt, T., & O'Leary, D. D. (2009). Slits are chemorepellents endogenous to hypothalamus and steer thalamocortical axons into ventral telencephalon. *Cerebral Cortex, 19*(Suppl 1), i144–i151.

Buhusi, M., Demyanenko, G. P., Jannie, K. M., Dalal, J., Darnell, E. P., Weiner, J. A., et al. (2009). ALCAM regulates mediolateral retinotopic mapping in the superior colliculus. *Journal of Neuroscience, 29,* 15630–15641. doi:10.1523/JNEUROSCI.2215-09.2009.

Buhusi, M., Schlatter, M. C., Demyanenko, G. P., Thresher, R., & Maness, P. F. (2008). L1 interaction with ankyrin regulates mediolateral topography in the retinocollicular projection. *Journal of Neuroscience, 28,* 177–188.

Burnette, D. T., Ji, L., Schaefer, A. W., Medeiros, N. A., Danuser, G., & Forscher, P. (2008). Myosin II activity facilitates microtubule bundling in the neuronal growth cone neck. *Developmental Cell, 15,* 163–169.

Cang, J., Kaneko, M., Yamada, J., Woods, G., Stryker, M. P., & Feldheim, D. A. (2005). Ephrin-as guide the formation of functional maps in the visual cortex. *Neuron, 48,* 577–589.

Castellani, V., Chedotal, A., Schachner, M., Faivre-Sarrailh, C., & Rougon, G. (2000). Analysis of the L1-deficient mouse phenotype reveals cross-talk between Sema3A and L1 signaling pathways in axonal guidance [see comments]. *Neuron, 27,* 237–249.

Castellani, V., De Angelis, E., Kenwrick, S., & Rougon, G. (2002). Cis and trans interactions of L1 with neuropilin-1 control axonal responses to semaphorin 3A. *European Molecular Biology Organization Journal, 21,* 6348–6357.

Castellani, V., Falk, J., & Rougon, G. (2004). Semaphorin3A-induced receptor endocytosis during axon guidance responses is mediated by L1 CAM. *Molecular and Cellular Neurosciences, 26,* 89–100.

Castellani, V., Yue, Y., Gao, P. P., Zhou, R., & Bolz, J. (1998). Dual action of a ligand for Eph receptor tyrosine kinases on specific populations of axons during the development of cortical circuits. *Journal of Neuroscience, 18,* 4663–4672.

Cheng, L., Itoh, K., & Lemmon, V. (2005). L1-mediated branching is regulated by two ezrin-radixin-moesin (ERM)-binding sites, the RSLE region and a novel juxtamembrane ERM-binding region. *Journal of Neuroscience, 25,* 395–403.

Dai, J., Dalal, J. S., Thakar, S., Henkemeyer, M., Lemmon, V. P., Harunaga, J. S., et al. (2012). EphB regulates L1 phosphorylation during retinocollicular mapping. *Molecular and Cellular Neurosciences, 50,* 201–210. doi:10.1016/j.mcn.2012.05.001.

Davey, F., Hill, M., Falk, J., Sans, N., & Gunn-Moore, F. J. (2005). Synapse associated protein 102 is a novel binding partner to the cytoplasmic terminus of neurone-glial related cell adhesion molecule. *Journal of Neurochemistry, 94,* 1243–1253.

Demyanenko, G. P., Riday, T. T., Tran, T. S., Dalal, J., Darnell, E. P., Brennaman, L. H., et al. (2011a). NrCAM deletion causes topographic mistargeting of thalamocortical axons to the visual cortex and disrupts visual acuity. *Journal of Neuroscience, 31,* 1545–1558. doi:10.1523/JNEUROSCI.4467-10.2011.

Demyanenko, G. P., Siesser, P. F., Wright, A. G., Brennaman, L. H., Bartsch, U., Schachner, M., et al. (2011b). L1 and CHL1 cooperate in thalamocortical axon targeting. *Cerebral Cortex, 21,* 401–412. doi:10.1093/cercor/bhq115.

Dickson, T. C., Mintz, C. D., Benson, D. L., & Salton, S. R. (2002). Functional binding interaction identified between the axonal CAM L1 and members of the ERM family. *Journal of Cell Biology, 157,* 1105–1112.

Dickstein, L. P., Zoghbi, S. S., Fujimura, Y., Imaizumi, M., Zhang, Y., Pike, V. W., et al. (2011). Comparison of 18F- and 11C-labeled aryloxyanilide analogs to measure translocator protein in human brain using positron emission tomography. *European Journal of Nuclear Medicine and Molecular Imaging, 38,* 352–357. doi:10.1007/s00259-010-1622-y.

Dirks, P., Thomas, U., & Montag, D. (2006). The cytoplasmic domain of NrCAM binds to PDZ domains of synapse-associated proteins SAP90/PSD95 and SAP97. *European Journal of Neuroscience, 24,* 25–31.

Dufour, A., Egea, J., Kullander, K., Klein, R., & Vanderhaeghen, P. (2006). Genetic analysis of EphA-dependent signaling mechanisms controlling topographic mapping in vivo. *Development, 133,* 4415–4420.

Dufour, A., Seibt, J., Passante, L., Depaepe, V., Ciossek, T., Frisen, J., et al. (2003). Area specificity and topography of thalamocortical projections are controlled by ephrin/Eph genes. *Neuron, 39,* 453–465. doi:10.1016/S0896-6273(03)00440-9.

Egea, J., Nissen, U. V., Dufour, A., Sahin, M., Greer, P., Kullander, K., et al. (2005). Regulation of EphA 4 kinase activity is required for a subset of axon guidance decisions suggesting a key role for receptor clustering in Eph function. *Neuron, 47,* 515–528. doi:10.1016/j.neuron.2005.06.029.

Enriquez-Barreto, L., Palazzetti, C., Brennaman, L. H., Maness, P. F., & Fairén, A. (2012). NCAM regulates thalamocortical axon pathfinding and the organization of the cortical somatosensory representation in mouse. *Frontiers in Neuroscience, 5,* 1–13. doi:10.3389/fnmol.2012.00076.

Falk, J., Bechara, A., Fiore, R., Nawabi, H., Zhou, H., Hoyo-Becerra, C., et al. (2005). Dual functional activity of semaphorin 3B is required for positioning the anterior commissure. *Neuron, 48,* 63–75. doi:10.1016/j.neuron.2005.10.024.

Fukuchi-Shimogori, T., & Grove, E. A. (2001). Neocortex patterning by the secreted signaling molecule FGF8. *Science, 294,* 1071–1074.

Gao, P. P., Yue, Y., Zhang, J. H., Cerretti, D. P., Levitt, P., & Zhou, R. (1998). Regulation of thalamic neurite outgrowth by the Eph ligand ephrin-A5: Implications in the development of thalamocortical projections. *Proceedings of the National Academy of Sciences of the United States of America, 95,* 5329–5334.

Garver, T. D., Ren, Q., Tuvia, S., & Bennett, V. (1997). Tyrosine phosphorylation at a site highly conserved in the L1 family of cell adhesion molecules abolishes ankyrin binding and increases lateral mobility of neurofascin. *Journal of Cell Biology, 137,* 703–714.

Gil, O. D., Sakurai, T., Bradley, A. E., Fink, M. Y., Cassella, M. R., Kuo, J. A., et al. (2003). Ankyrin binding mediates L1CAM interactions with static components of the cytoskeleton and inhibits retrograde movement of L1CAM on the cell surface. *Journal of Cell Biology, 162,* 719–730. doi:10.1083/jcb.200211011.

Gu, C., Rodriguez, E. R., Reimert, D. V., Shu, T., Fritzsch, B., Richards, L. J., et al. (2003). Neuropilin-1 conveys semaphorin and VEGF signaling during neural and cardiovascular development. *Developmental Cell, 5,* 45–57. doi:10.1016/S1534-5807(03)00169-2.

Guellmar, A., Rudolph, J., & Bolz, J. (2009). Structural alterations of spiny stellate cells in the somatosensory cortex in ephrin-A5-deficient mice. *Journal of Comparative Neurology, 517,* 645–654.

Hindges, R., McLaughlin, T., Genoud, N., Henkemeyer, M., & O'Leary, D. D. (2002). EphB forward signaling controls directional branch extension and arborization required for dorsal–ventral retinotopic mapping. *Neuron, 35,* 475–487.

Janis, L. S., Cassidy, R. M., & Kromer, L. F. (1999). Ephrin-A binding and EphA receptor expression delineate the matrix compartment of the striatum. *Journal of Neuroscience, 19,* 4962–4971.

Leighton, P. A., Mitchell, K. J., Goodrich, L. V., Lu, X., Pinson, K., Scherz, P., et al. (2001). Defining brain wiring patterns and mechanisms through gene trapping in mice. *Nature, 410,* 174–179. doi:10.1038/35065539.

Little, G. E., Lopez-Bendito, G., Runker, A. E., Garcia, N., Pinon, M. C., Chedotal, A., et al. (2009). Specificity and plasticity of thalamocortical connections in Sema6A mutant mice. *PLoS Biology, 7,* e98. doi:10.1371/journal.pbio.1000098.

Lopez-Bendito, G., Cautinat, A., Sanchez, J. A., Bielle, F., Flames, N., Garratt, A. N., et al. (2006). Tangential neuronal migration controls axon guidance: A role for neuregulin-1 in thalamocortical axon navigation. *Cell, 125,* 127–142. doi:10.1016/j.cell.2006.01.042.

Lopez-Bendito, G., Flames, N., Ma, L., Fouquet, C., Di Meglio, T., Chedotal, A., et al. (2007). Robo1 and Robo2 cooperate to control the guidance of major axonal tracts in the mammalian forebrain. *Journal of Neuroscience, 27,* 3395–3407. doi:10.1523/JNEUROSCI.4605-06.2007.

Mackarehtschian, K., Lau, C. K., Caras, I., & McConnell, S. K. (1999). Regional differences in the developing cerebral cortex revealed by ephrin-A5 expression. *Cerebral Cortex, 9,* 601–610.

Mann, F., Peuckert, C., Dehner, F., Zhou, R., & Bolz, J. (2002). Ephrins regulate the formation of terminal axonal arbors during the development of thalamocortical projections. *Development, 129,* 3945–3955.

Marquardt, T., Shirasaki, R., Ghosh, S., Andrews, S. E., Carter, N., Hunter, T., et al. (2005). Coexpressed EphA receptors and ephrin-A ligands mediate opposing actions on growth cone navigation from distinct membrane domains. *Cell, 121,* 127–139. doi:10.1016/j.cell.2005.01.020.

Marsick, B. M., San Miguel-Ruiz, J. E., & Letourneau, P. C. (2012). Activation of ezrin/radixin/moesin mediates attractive growth cone guidance through regulation of growth cone actin and adhesion receptors. *Journal of Neuroscience, 32,* 282–296.

McLaughlin, T., Hindges, R., Yates, P. A., & O'Leary, D. D. (2003). Bifunctional action of ephrin-B1 as a repellent and attractant to control bidirectional branch extension in dorsal–ventral retinotopic mapping. *Development, 130,* 2407–2418.

Mintz, C. D., Carcea, I., McNickle, D. G., Dickson, T. C., Ge, Y., Salton, S. R., et al. (2008). ERM proteins regulate growth cone responses to Sema3A. *Journal of Comparative Neurology, 510,* 351–366. doi:10.1002/cne.21799.

Ohyama, K., Tan-Takeuchi, K., Kutsche, M., Schachner, M., Uyemura, K., & Kawamura, K. (2004). Neural cell adhesion molecule L1 is required for fasciculation and routing of thalamocortical fibres and corticothalamic fibres. *Neuroscience Research, 48,* 471–475.

Pasterkamp, R. J., & Giger, R. J. (2009). Semaphorin function in neural plasticity and disease. *Current Opinion in Neurobiology, 19,* 263–274.

Powell, A. W., Sassa, T., Wu, Y., Tessier-Lavigne, M., & Polleux, F. (2008). Topography of thalamic projections requires attractive and repulsive functions of Netrin-1 in the ventral telencephalon. *PLoS Biology, 6,* e116. doi:10.1371/journal.pbio.0060116.

Price, D. J., Kennedy, H., Dehay, C., Zhou, L., Mercier, M., Jossin, Y., et al. (2006). The development of cortical connections. *European Journal of Neuroscience, 23,* 910–920. doi:10.1111/j.1460-9568.2006.04620.x.

Schaefer, A. W., Schoonderwoert, V. T., Ji, L., Mederios, N., Danuser, G., & Forscher, P. (2008). Coordination of actin filament and microtubule dynamics during neurite outgrowth. *Developmental Cell, 15,* 146–162.

Schiff, M., Rockle, I., Burkhardt, H., Weinhold, B., & Hildebrandt, H. (2011). Thalamocortical pathfinding defects precede degeneration of the reticular thalamic nucleus in polysialic acid-deficient mice. *Journal of Neuroscience, 31,* 1302–1312.

Schlatter, M. C., Buhusi, M., Wright, A. G., & Maness, P. F. (2008). CHL1 promotes Sema3A-induced growth cone collapse and neurite elaboration through a motif required for recruitment of ERM proteins to the plasma membrane. *Journal of Neurochemistry, 104,* 731–744.

Takeuchi, H., Inokuchi, K., Aoki, M., Suto, F., Tsuboi, A., Matsuda, I., et al. (2010). Sequential arrival and graded secretion of Sema3F by olfactory neuron axons specify map topography at the bulb. *Cell, 141,* 1056–1067. doi:10.1016/j.cell.2010.04.041.

Torii, M., Hackett, T. A., Rakic, P., Levitt, P., & Polley, D. B. (2013). EphA signaling impacts development of topographic connectivity in auditory corticofugal systems. *Cerebral Cortex, 23,* 775–785. doi:10.1093/cercor/bhs066.

Torii, M., & Levitt, P. (2005). Dissociation of corticothalamic and thalamocortical axon targeting by an EphA7-mediated mechanism. *Neuron, 48,* 563–575.

Tran, T. S., Kolodkin, A. L., & Bharadwaj, R. (2007). Semaphorin regulation of cellular morphology. *Annual Review of Cell and Developmental Biology, 23,* 263–292.

Uemura, M., Nakao, S., Suzuki, S. T., Takeichi, M., & Hirano, S. (2007). OL-Protocadherin is essential for growth of striatal axons and thalamocortical projections. *Nature Neuroscience, 10,* 1151–1159.

Uziel, D., Garcez, P., Lent, R., Peuckert, C., Niehage, R., Weth, F., et al. (2006). Connecting thalamus and cortex: The role of ephrins. *Anatomical Record. Part A, Discoveries in Molecular, Cellular, and Evolutionary Biology, 288,* 135–142. doi:10.1002/ar.a.20286.

Uziel, D., Muhlfriedel, S., & Bolz, J. (2008). Ephrin-A5 promotes the formation of terminal thalamocortical arbors. *Neuroreport, 19,* 877–881.

Uziel, D., Muhlfriedel, S., Zarbalis, K., Wurst, W., Levitt, P., & Bolz, J. (2002). Miswiring of limbic thalamocortical projections in the absence of ephrin-A5. *Journal of Neuroscience, 22,* 9352–9357.

Whittard, J. D., Sakurai, T., Cassella, M. R., Gazdoiu, M., & Felsenfeld, D. P. (2006). MAP kinase pathway-dependent phosphorylation of the L1-CAM ankyrin binding site regulates neuronal growth. *Molecular Biology of the Cell, 17,* 2696–2706.

Wiencken-Barger, A. E., Mavity-Hudson, J., Bartsch, U., Schachner, M., & Casagrande, V. A. (2004). The role of L1 in axon pathfinding and fasciculation. *Cerebral Cortex, 14,* 121–131.

Williams, S. E., Grumet, M., Colman, D. R., Henkemeyer, M., Mason, C. A., & Sakurai, T. (2006). A role for Nr-CAM in the patterning of binocular visual pathways. *Neuron, 50,* 535–547.

Wright, A. G., Demyanenko, G. P., Powell, A., Schachner, M., Enriquez-Barreto, L., Tran, T. S., et al. (2007). Close homolog of L1 and neuropilin 1 mediate guidance of thalamocortical axons at the ventral telencephalon. *Journal of Neuroscience, 27,* 13667–13679. doi:10.1523/JNEUROSCI.2888-07.2007.

Zhang, X. F., Schaefer, A. W., Burnette, D. T., Schoonderwoert, V. T., & Forscher, P. (2003). Rho-dependent contractile responses in the neuronal growth cone are independent of classical peripheral retrograde actin flow. *Neuron, 40,* 931–944.

Zhou, L., Qu, Y., Tissir, F., & Goffinet, A. M. (2009). Role of the atypical cadherin Celsr3 during development of the internal capsule. *Cerebral Cortex, 19*(Suppl 1), i114–i119.

Zisch, A. H., Kalo, M. S., Chong, L. D., & Pasquale, E. B. (1998). Complex formation between EphB2 and Src requires phosphorylation of tyrosine 611 in the EphB2 juxtamembrane region. *Oncogene, 16,* 2657–2670.

94 Development of Direction Selectivity

AARON M. HAMBY AND MARLA B. FELLER

Motion constitutes one of the most salient cues for both vertebrate and invertebrate visual systems. Indeed, multiple distinct circuits at various levels of the visual system are sensitive to the presence of moving objects. Perhaps the best understood circuit that addresses the extraction of motion is that of the ON–OFF direction-selective ganglion cell (DSGC) in the mammalian retina. DSGCs respond strongly to an image moving in the preferred direction and weakly to an image moving in the opposite, or null, direction. Though the cellular and synaptic bases of this computation are reasonably well understood, the wiring up of this circuit during development is mostly a mystery. In this chapter, we describe how electrophysiological, imaging, genetic, and optogenetic techniques have allowed us to begin addressing questions regarding the development of direction-selective circuits in the retina. This topic has also been the subject of several recent reviews (Borst & Euler, 2011; Elstrott & Feller, 2009; Wei & Feller, 2011).

OVERVIEW OF THE DIRECTION-SELECTIVE CIRCUIT

Several distinct populations of DSGCs have been identified. Transgenic mice that label single subtypes of DSGCs with green fluorescent protein (GFP) indicate that the various subtypes (see figure 94.1b and discussion below) may display intrinsic genetic differences. However, we are far from a complete genetic description. Therefore, the present classification of DSGCs is based on their response properties, morphologies, and target areas in the brain.

DSGCs fall into roughly three populations (reviewed in Berson, 2008; Borst & Euler, 2011; Vaney & Taylor, 2002; Wei & Feller, 2011): (1) ON–OFF DSGCs that project to the lateral geniculate nucleus and superior colliculus, (2) ON-DSGCs that project to the medial terminal nucleus of the accessory optic system, and (3) OFF-DSGCs cells that project to the superior colliculus.

ON–OFF DSGCs comprise four subtypes, each preferring motion along one of the cardinal directions: nasal, temporal, dorsal, or ventral (see figure 94.2A). Each subtype forms an independent mosaic that tiles the retina with little intrasubtype dendritic overlap. Thus, each point on the retina falls within the receptive fields of DSGCs that are sensitive to motion in all four cardinal directions.

ON-DSGCs comprise three subtypes, each preferring motion along one of the axes of rotation imposed by the extraocular muscles. Each subtype of this DSGC class is also thought to form a mosaic although these neurons are less populous than ON–OFF DSGCs and have larger dendritic trees (Rockhill et al., 2002; Sun, Li & He, 2002). In addition, ON-DSGCs are more narrowly tuned to slower velocities (Sun et al., 2006). Despite the different projection patterns and tuning properties, ON–OFF DSGCs and ON-DSGCs share many circuit properties in the adult and appear to follow a similar developmental program as described below.

OFF-DSGCs comprise only one subtype, known as J-RGC, and it prefers motion in the superior direction. These cells also tile the retina in a regular mosaic, and they are marked by an immunoglobin family protein, junctional adhesion molecule b (called JAM-B). The majority of JAM-B expressing neurons found in the central, nasal, and temporal retina, roughly 85% of the entire population, have asymmetrical dendritic trees that point down in retinal coordinates, parallel to the preferred direction. The remaining 15% are located in the ventral retina, have symmetrical dendritic trees, and are not direction selective. Thus, the asymmetrical structure of OFF-DSGCs seems to play a role in generating their direction selectivity.

The direction-selective circuit of ON–OFF DSGCs is schematized in figure 94.1A. In brief, photoreceptors activate ON and OFF bipolar cells that provide excitatory glutamatergic input to the ON and OFF sublaminae of DSGCs. Bipolar cells also provide glutamatergic input to two mirror-symmetrical populations of the GABAergic interneurons called starburst amacrine cells. One population of these amacrine cells has somas residing in the ganglion cell layer with processes costratifying with the ON-sublamina processes of the ON–OFF DSGCs; the other population has somas residing in the inner nuclear layer with processes costratifying with the OFF sublamina processes of these DSGCs.

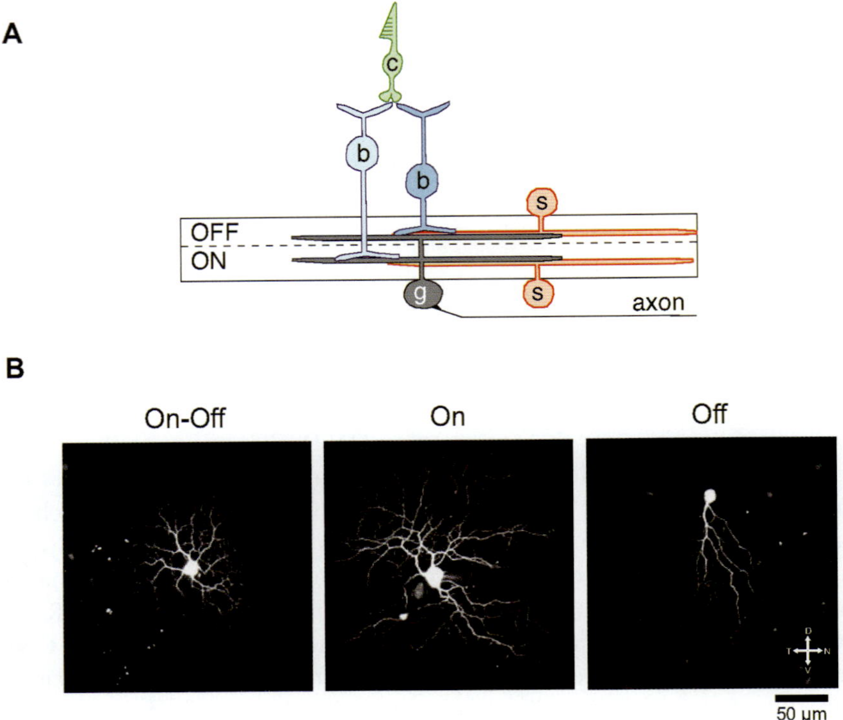

FIGURE 94.1   Direction-selective retinal circuit and ganglion cell types. (A) Schematic of the retinal circuit that mediates direction selectivity. The direction-selective ganglion cell (DSGC) (g) projects dendrites into the ON and OFF halves (or sublaminae) of the retina's inner synaptic layer. The dendrites in each sublamina receive synapses from the corresponding type of bipolar cell (b) that is relaying signals from cone photoreceptors (c), and starburst amacrine cells (s). (From Demb, 2007.) (B) Live two-photon images of a postnatal day (P) 10 ON–OFF DSGC (left), a P9 ON-DSGC (middle), and a P11 OFF-DSGC (right), all before eye opening. The ON–OFF- and ON-DSGCs were filled with Alexa 594 before imaging. (From Elstrott & Feller, 2010.)

The primary circuit model for generating direction selectivity in the retina states that stimulation in the null direction leads to stronger inhibition than stimulation in the preferred direction. Consistent with this hypothesis, paired recordings from starburst amacrine cells and DSGCs in mouse retina reveal that depolarization of a starburst amacrine cell on the null side induces significantly larger inhibitory currents in the DSGC than depolarization of this amacrine cell on the preferred side (Fried, Munch, & Werblin, 2002; Lee, Kim, & Zhou, 2010; Wei et al., 2011). These findings suggest that starburst amacrine cells create the asymmetrical inhibitory fields that underlie direction selectivity in DSGCs via an asymmetrical release of GABA that depends on the direction of motion of the light stimulus.

How does light moving in one direction lead to more GABA release than light moving in the other direction? A stimulus that leads to directional responses in DSGCs evokes symmetrical responses from the starburst amacrine cells when measured at the level of soma depolarization. However, at the distal tips of these cells, two-photon calcium imaging reveals that calcium transients are larger for motion outward from the soma than for motion toward the soma (Euler, Detwiler, & Denk, 2002). Although this increase in intracellular calcium has not yet been shown to drive more GABA release, it has led to the hypothesis that the directionality of the starburst amacrine cell process underlies the asymmetrical release of GABA.

Support for this hypothesis that directionality of a single starburst amacrine cell process contributes to direction selectivity in the retina was recently provided by a serial electron microscopic reconstruction of this circuit (Briggman, Helmstaedter, & Denk, 2011). This reconstruction showed that individual starburst amacrine cell processes—rather than individual starburst cells—selectively wire to particular subtypes of DSGCs. In particular, the local orientation of the process dictates the size of the synapse: Processes oriented parallel to the null direction of the postsynaptic DSGC form a large (and therefore presumably strong) synapse while processes oriented in other directions form smaller synapses or no synapses at all.

It is important to note that several other aspects of the retinal circuit contribute to the computation of direction selectivity (see chapters 8 and 13, Borst &

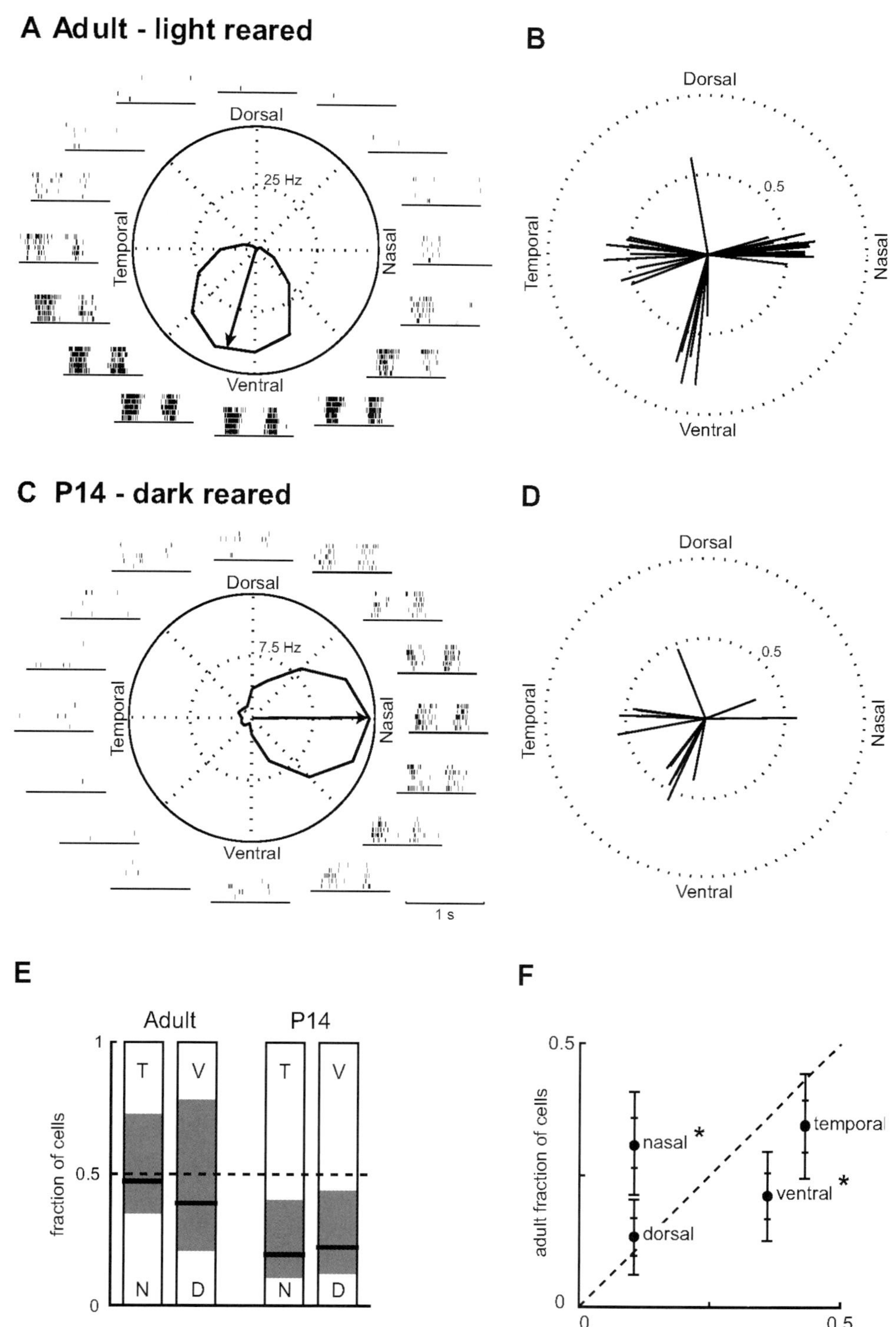

FIGURE 94.2   Responses of ON–OFF direction-selective ganglion cells (DSGCs) to moving gratings reveal strong direction selectivity in the adult and P14 dark reared mouse. (A, C) Polar plot of the mean spike rate across five repetitions in response to motion in 16 directions. Spike traces for all directions are shown for one period of the moving grating. The arrow indicates the mean preferred direction of the cell. Firing rate is calculated as the total number of spikes divided by the stimulus duration (10 s). Adult: n = 31 cells; P14 dark reared: n = 15 cells. (B, D) The preferred directions of all ON–OFF DSGCs within one animal preparation. These directions fall clearly into four groups along the cardinal directions for the adult (B) but are more variable at P14. (D). The preferred direction for each cell is shown as the vector sum of the response, so the length of each line indicates the normalized response magnitude. (E) The symmetry of preferred directions along a given axis (temporal–nasal or ventral–dorsal) is determined by calculating the fraction of cells that fall in one quadrant along the axis (e.g., temporal) versus the fraction of cells that fall in the opposite quadrant (e.g., nasal). A perfectly symmetrical axis would show 50% of the cells in opposite quadrants. The thick black bars show the actual distribution of cells along the axes for adult and P14. The shaded gray regions show the 95% confidence interval around each value based on resampling of the adult and P14 distributions. These data indicate that there is a nonuniform distribution of cells among the four cardinal directions. In addition, the representation of opposite directions along the same axis was strongly biased, with nasal and dorsal quadrants significantly underrepresented compared to their opposites (T, temporal; N, nasal; D, doral; V, ventrral.) (F) A comparison of adult and P14 distributions of preferred directions on a quadrant-by-quadrant basis. These data indicate that the relative representation of the nasal and ventral directions was significantly changed between P14 and adult. With a uniform distribution, each quadrant would encompass 25% of the cells. (From Elstrott et al., 2008.)

Euler, 2011; Wei & Feller, 2011). However, the aspect of asymmetrical starburst amacrine cell wiring to DSGCs is the primary focus of the studies on the development of direction selectivity that are described below.

DIRECTION SELECTIVITY IS PRESENT AT EYE OPENING AND FORMS INDEPENDENT OF VISUAL EXPERIENCE

Several studies have concluded that direction selectivity is wired up at eye opening and is independent of visual experience. Using large-scale multi-electrode array recordings (Elstrott et al., 2008) or single cell-attached recordings (Chan & Chiao, 2008; Chen et al., 2009; Masland, 1977), researchers have detected robust ON–OFF directional responses before or at eye opening in mice and rabbits. This also holds true for mice that are dark reared prior to eye opening (Chan & Chiao, 2008; Elstrott et al., 2008). Similar results have been found for ON-DSGCs (Yonehara et al., 2009). At this young age, both the excitatory inputs (Tian & Copenhagen, 2001) and the intrinsic excitability of retinal ganglion cells (Chen et al., 2009; Qu & Myhr, 2008; Sun et al., 2008) lack maturity, so the firing rate in the preferred direction is lower than that in adult animals. However, the firing rate in the null direction is as low as the rate in adult, indicating that the circuits mediating null-side inhibition are established early in development.

Support for this early establishment of direction selectivity was provided by a recent study in which direction-selective responses were detected in the visual cortex of mice at eye opening (Rochefort et al., 2011). In stark contrast, ferrets exhibit a direction selectivity in the cortex that occurs after eye opening and does require visual experience (Li, Fitzpatrick, & White, 2006; Li et al., 2008). Interestingly, in mouse visual cortex, the distribution of directional preference is not uniform at eye opening, with a larger percentage of cells preferring anterior and dorsal motion direction over the opposite directions (Rochefort et al., 2011). This anisotropy mimics the distribution that has been reported in retinas where a larger percentage of cells prefer motion toward either the temporal or the ventral part of the retina (Elstrott et al., 2008). In adult rabbits, the ultimate distribution of DSGC preferred directions is unchanged even if the rabbits are raised in a visual environment consisting entirely of vertical stripes rotating in one direction (Daw & Wyatt, 1974). Together, these results show that the initial establishment of retinal direction selectivity in mice and rabbits occurs independent of visual experience.

TRANSGENIC MOUSE LINES ARE USED TO STUDY DIRECTION SELECTIVITY PRIOR TO MATURATION OF THE LIGHT RESPONSE

Addressing the developmental events in the establishment of direction selectivity has been significantly hindered in the past because there was no reliable way to identify a DSGC without first measuring a light response. Since the synaptic connections fundamental to the response become established before eye opening and the emergence of visual function, identification of nascent DSGCs needs to be based on something other than the tuning curve.

The field was recently transformed by two technological breakthroughs. First, genetically altered mouse lines have been identified that express GFP selectively in certain DSGC subtypes, specifically ON-DSGCs (Yonehara et al., 2008, 2009), ON–OFF DSGCs (Huberman et al., 2009; Kay et al., 2011; Rivlin-Etzion et al., 2011; Trenholm et al., 2011), and OFF-DSGCs (Kay et al., 2011; Kim et al., 2008, 2010). These mice have allowed the morphological characterization and identification of projection patterns from identified DSGCs (see chapter 90 by Kay and Sanes). Second, the technique of two-photon targeted recording has been applied to the retina. Typically, recording light responses from GFP-expressing retinal neurons requires illuminating the retina with light at a wavelength that is appropriate to excite GFP. However, this light also stimulates the photopigment, so it leads to rapid and irreversible bleaching of the light response. To avoid this, two-photon excitation based on infrared light has been used to visualize the neurons since infrared light spares the photopigment and therefore the light response (Euler et al., 2009; Wei, Elstrott, & Feller, 2010). Using two-photon targeted recording on these lines of transgenic mice has enabled clear characterization of the light responses of identified subtypes of DSGCs.

One intriguing outcome of the above approaches is that some markers label specific subsets of DSGC subtypes. For example, SPIG1-GFP is expressed solely in ON-DSGCs that prefer motion in the upward direction. Similarly, the BAC-transgenic mouse, where GFP is expressed under the type 4 dopamine receptor, expresses GFP in a subset of ON–OFF DSGCs that prefer motion in the posterior direction. These lines enable us to address two important questions: How do DSGCs divide into their respective subtypes during development, and is this process partially or fully under the control of genetic specification? By isolating GFP (+) populations of neurons by fluorescence-activated cell sorting (FACS), researchers can perform a molecular characterization of individual DSGC subtypes.

Indeed, this method has identified several cell-surface markers that uniquely mark subtypes of ON–OFF DSGCs (Kay et al., 2011). Since these distinguishing markers are found prior to eye opening, the different subtypes of ON–OFF DSGCs are thought to exist independent of visual experience (Kay et al., 2011). Studies such as these have just begun, but similar methods are likely to lead to insights on how individual subtypes are specified (see below).

SPECIFIC INHIBITORY CIRCUITS ARE ESTABLISHED PRIOR TO EYE OPENING

A fundamental component of the direction-selective circuit is the asymmetrical wiring of starburst amacrine cells to DSGCs. In the adult retina, this asymmetry has been determined by paired recordings between starburst amacrine cells and DSGCs—if a starburst amacrine cell is located on the null side, then there is stronger overall synaptic input than if the starburst amacrine cell is located on the preferred side. Recordings from DSGCs at eye opening indicate that the null-side inhibition is in place at this early stage of development. Questions then arise—when and how does this asymmetry evolve during development? There are two options. On one hand, the asymmetry could reflect a

very specific synaptic connection between the null-side starburst amacrine cells and DSGCs at the point of synaptogenesis. On the other hand, the starburst amacrine cells could initially make GABAergic contacts with both preferred- and null-side DSGCs and through the course of development could either keep or strengthen only the null-side synapses.

This issue has been addressed using two methods. First, targeted paired recordings have been performed on transgenic mice in which both posterior-preferring DSGCs and starburst amacrine cells express GFP (Wei et al., 2011; see figure 94.3). These mice are double transgenics that are a cross between the DRD4-GFP line and mGluRII-GFP line (which labels starburst amacrine cells). Though both starburst amacrine cells and DSGCs express GFP, they can be distinguished by their morphologies and their intracellular patterns of GFP expression. The recordings show that, as early as postnatal day (P) 4, inhibitory synapses between starburst amacrine cells and DSGCs are detected, and, until P7, DSGCs receive symmetrical GABAergic input from starburst amacrine cells on both the preferred and null sides. However, by P14, the strength of the GABAergic conductance between DSGCs and null-side starburst amacrine cells increases several fold while that between DSGCs and preferred-side starburst amacrine cells

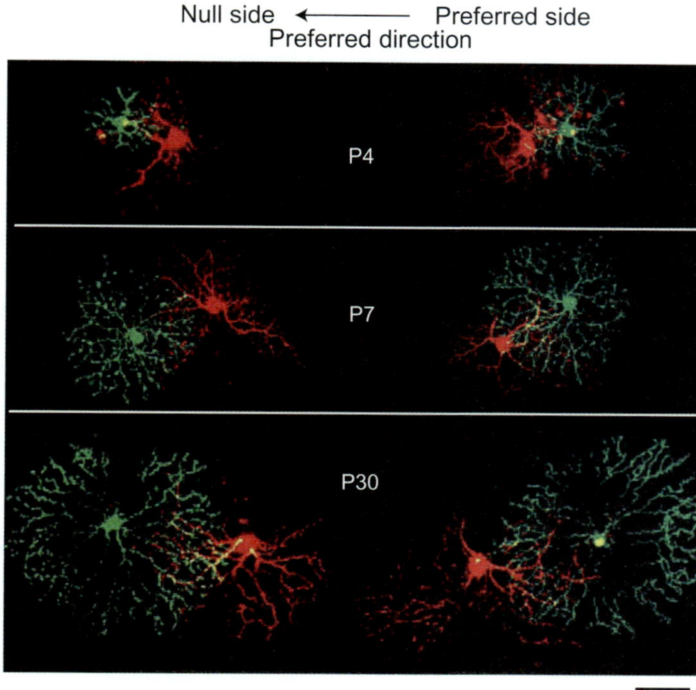

FIGURE 94.3   Synaptically connected, dye-filled starburst amacrine cell–DSGC (direction-selective ganglion cell) pairs at P4, P7 and P30. On the left are null-side pairs where the starburst amacrine cells (filled with green fluorescent dye) are located on the null side of the DSGCs (filled with red fluorescent dye). On the right are preferred-side pairs. Scale bar: 50 μm. (From Wei et al., 2011.)

remains weak. Hence, the asymmetry of connectivity is thought to arise between P7 and P14.

The second method used to address the question of asymmetry in development is that of using optogenetics to characterize the pattern of inhibitory synapses during development (Yonehara et al., 2011). In these studies, channelrhodopsin (ChR) is expressed in a large percentage of the starburst amacrine cells in mice where ON-DSGCs express GFP. In adult, inhibitory currents in ON-DSGCs that are evoked by photostimulating the null side are larger than those evoked by photostimulating the preferred side. This ChR-evoked response in the ON DSGC is symmetrical at P6 and sharply transitions to an asymmetrical pattern at P8, indicating that this synaptic strengthening occurs over only a 2-day period.

Neither of these methods was able to determine whether the developmental change in overall synaptic strength is due to changes in the number of synapses or to changes in the strength of individual synapses. However, they do both indicate that a subset of GAB-Aergic synapses made between the two cell types are destined to be formed or strengthened during a relatively short time in development. The cellular mechanisms that underlie this synapse-specific formation and/or strengthening remain to be determined.

MOLECULAR SPECIFICITY AND DENDRITIC ARBORIZATION

One likely set of mechanisms that underlies the specificity of direction-selective circuit architecture involves the interplay of signaling molecules that dictate the connections between specific cell types. In this scenario, each subtype of DSGC expresses a particular molecule on its dendrites, and this molecule interacts with a cognate molecular partner expressed selectively on starburst amacrine cell processes that are oriented in the null direction of the partnering DSGC. Such hypotheses have been offered to explain how a circuit comprised of cells with symmetrical morphology can wire in an asymmetrical way. While the hunt for molecular players that might mediate the specific wiring of direction-selective circuits is under way, the key molecular players have not yet been identified. However, recent advances have hinted that the retina uses both cell adhesion molecules and transmembrane guidance cues to establish these circuits.

The evidence that several cell adhesion molecules are involved in the development of retinal circuits is reviewed in Fuerst and Burgess (2009). One particular family of cell adhesion molecules, the Down syndrome cell adhesion molecules (DSCAMs), has been shown to play a role in the laminar stratification (Yamagata & Sanes, 2008) of the retina, as well as in its mosaic spacing (Fuerst et al., 2009; Keeley et al., 2012). DSCAMs are members of the immunoglobulin superfamily and mediate both attractive and repulsive intercellular events, depending on the setting and species. In *Drosophila*, DSCAMs are known to undergo extensive alternative splicing, giving rise to up to ~38,000 isoforms, and to control mosaic spacing, axonal tiling and branching, and synaptic specificity. The vertebrate homologues are far less diverse, consisting of only a handful of genes that do not undergo alternative splicing. Nevertheless, in chicken—a species that is not known to have DSGCs—deletion or misexpression of DSCAMs and the closely related DSCAM-like, Sidekick, and Contactin molecules reveals that they collectively specify stratification and synaptic partner choice in the inner plexiform layer (IPL) (Yamagata & Sanes, 2008, 2012; Yamagata, Weiner, & Sanes, 2002). This contrasts with the case in mouse. There, deletion of DSCAMs or DSCAM-L does not alter lamination, but these molecules do act as repulsive cues that prevent the adhesion of neurons of the same type, essentially acting to enforce proper mosaic spacing (Fuerst et al., 2009). Thus, in DSCAM knockout mice, many neurons form aggregated clumps where previously they were evenly spaced. However, their dendrites still seek out and terminate in the proper sublaminae and associate with their normal synaptic partners. These findings indicate that lamination and mosaic spacing in mice are likely mediated by different factors. While this study did not specifically investigate the direction-selective circuit, the preservation of dendritic targeting despite the gross disorganization of neuronal somas implies that DSCAMs do not help organize mammalian retinal circuits at the level of synapse specificity and therefore they are likely not responsible for wiring up specific direction-selective connections.

Repulsive transmembrane guidance cues have also been implicated in playing a role in retinal lamination by a few very recent studies (Matsuoka et al., 2011a, b). In these works, transmembrane semaphorins signal through the Plexin receptors and direct the processes of select subtypes of retinal neurons to specific sublaminae within the IPL via a process of repulsion. In particular, a subset of semaphorins (Sema5A and Sema5b) signals through its receptors (PlexinA1 and PlexinA3) and prevents dendrites from residing in the OFF sublamina. Though knockouts of these signaling molecules impair OFF visual responses, direction selectivity is not affected.

These recent studies implicate a variety of attractive and repulsive cues in establishing synaptic lamination patterns in the inner retina. At the time of these studies, markers for DSGCs were not easily available, having only been described in the last couple of years, so a phenotype for DSGCs has not yet been found. However, growing evidence indicates that individual direction-selective subtypes do in fact express unique molecular profiles, including some transmembane adhesion molecules (Kay et al., 2011). Genetic profiling based on FACS sorted neurons has identified factors that are differentially expressed across various DSGCs neurons. Specifically, microarray analysis has obtained expression profiles for three of the four directional subtypes of ON-Off DSGCs (nasal, ventral, and dorsal), starburst amacrine cells. and non-direction-selective RCGs. Some of these differentially expressed markers are molecules with an extracellular site of action and therefore may play a role in mediating intercellular interactions. These include collagen 25a1 (Col25a1), cadherin 6 (Cdh6), and matrix-metalloprotease 17 (Mmp17). Col25a1 and Cdh6 are expressed only in subtypes that prefer vertical motion (ventral and dorsal), while Mmp17 is expressed in subtypes that prefer nasal motion. Another transcript, cocaine and amphetamine related transcript (CART), is expressed by all ON-OFF DSGCs subtypes, but not by any other retinal neurons. Of all the remaining retinal neurons, only starburst amacrine cells express both Col25a1 and Cdh6, consistent with the idea that they mediate starburst amacrine cell-DSGC interactions. Furthermore, when cdh-6 positive cells are marked as early as E10 using an inducible CRE system, the majority of labeled RGCs (~80%) are ON-OFF DSGCs at P14, indicating that the fate of these cells are specified shortly after birth (De la Huerta et al., 2012). Subsequent studies will be required to test whether these candidates play a role in the development of direction-selective circuits, but these findings provide compelling evidence that genetic molecular factors likely play a large role in specifying certain elements of these circuits.

Insights into whether or not the circuits that mediate direction selectivity are established by molecular interactions can be found by examining the development of dendritic stratification and pruning. An accepted hypothesis is that if, during development, dendrites overarborize and then prune, this process is probably influenced by neural activity; in contrast, if dendrites simply grow in well-defined strata, their lamination is probably determined primarily by molecular interactions. The overall size of dendrites is thought to be a consequence of activity and of molecular interactions among dendrites from the same cell type (for reviews, see Reese et al., 2011; Tian, 2011). Kim et al. (2010) found that one subtype of DSGC, the OFF-DSGC, overarborizes within the IPL so that its dendrites enter and remain in the ON-sublamina during the first postnatal

week but are absent at eye opening. In contrast, dendrites of bistratified RGCs are narrow even at the earliest ages that they are observed (Kim et al., 2010; Stacy & Wong, 2003). However, other non-direction-selective subtypes show a far more dynamic process of innervation, with original overshooting or diffuse branching that is pruned or rearranged during the period between P5 and P13. Indeed, there appears to be no general rule for which subtypes undergo pruning and which do not (Tian, 2008, 2011), and future studies will be required to determine the role of molecular interactions in establishing directional circuits.

A ROLE FOR RETINAL WAVES IN THE ESTABLISHMENT OF DIRECTION SELECTIVITY?

Does neural activity play a role in the development of direction selectivity? The presence, at eye opening, of direction-selective responses and asymmetrical wiring of inhibitory synapses, even in dark-reared animals, indicates that sensory experience does not play a major role. However, prior to eye opening, retinal waves provide a robust source of highly structured activity and could potentially serve as a "proxy" for a moving stimulus in the previsual retina. These retinal waves are instructive for patterning the projections of retinal ganglion cells to their retinorecipient targets (Huberman, Feller, & Chapman, 2008; Xu et al., 2010). Thus, they are attractive candidates for investigation in direction selectivity for three key reasons: (1) Waves are present during the maturation of synapses between many retinal cell types, including the period of null-side synapse strengthening described above; (2) waves correlate the depolarization of neighboring cell types in the inner retina, including starburst amacrine cells and DSGCs (Elstrott & Feller, 2010; see figure 94.4), and therefore could possibly drive activity-dependent synaptic strengthening or synaptic refinement; and (3) waves have been shown to have a directional bias (Elstrott & Feller, 2010; Stafford et al., 2009).

To determine the role of retinal waves in establishing direction selectivity, investigators have employed a number of approaches. Intraocular injections of a voltage-gated sodium channel blocker (tetrodotoxin), a GABAA receptor agonist (muscimol), GABAA receptor antagonists (bicululline methiodide, SR95531/gabazine), or a cholinergic receptor agonist (epibatidine) (Sun, Han, & He, 2011; Wei et al., 2011; see figure 94.5) all fail to prevent the development of retinal direction selectivity. In addition, knockout mice that lack normal retinal waves during the first postnatal week maintain their direction-selective responses. However, these are all negative results, meaning they show activity

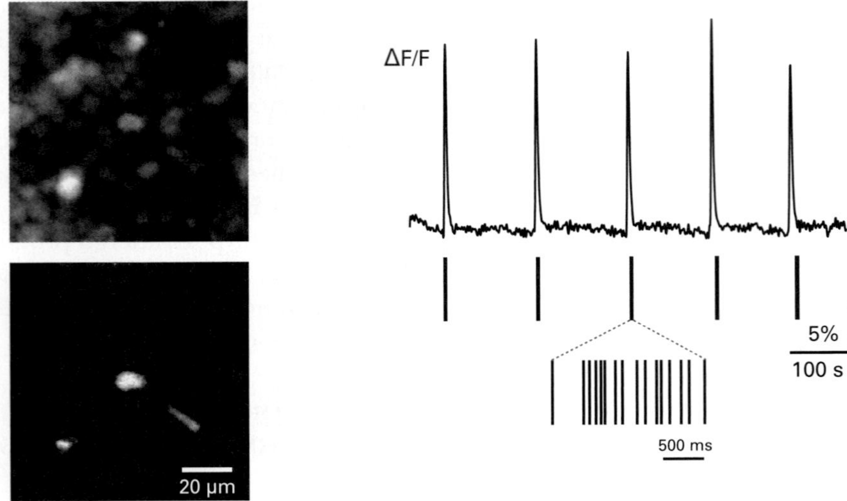

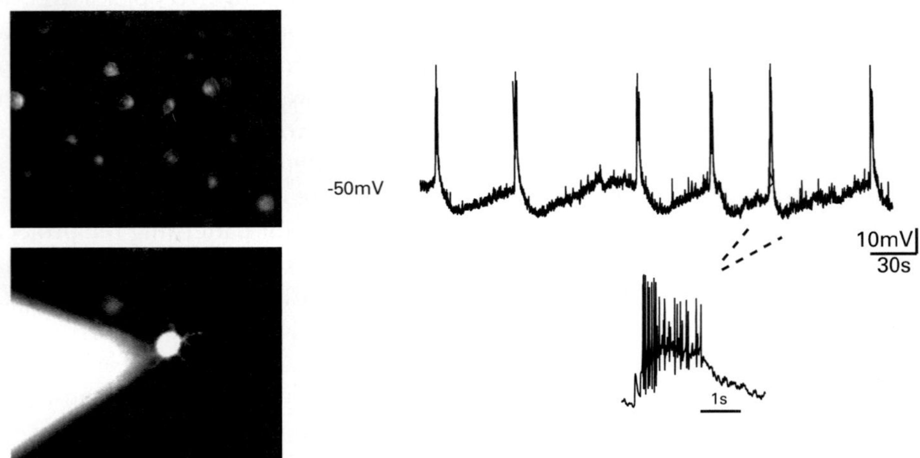

FIGURE 94.4   Direction-selective ganglion cells (DSGCs) and starburst amacrine cells are depolarized by retinal waves. (A) Left: Live fluorescence imaging (top) of P5 Spig1 retina (where green fluorescent protein [GFP] is expressed in ON-DSGCs) bolus loaded with the calcium indicator OGB-1 AM and the same retina with the ON-DSGC targeted for cell-attached recording configuration (bottom). Right: ΔF/F trace reveals periodic retinal waves that pass over a 200×200 µm region centered on the ON-DSGC cell, and the spike bursts below are elicited from the cell during these waves. Inset shows a magnified view of the spikes evoked by a single wave. (From Elstrott & Feller, 2010.) (B) Left: Live fluorescence image of a mGluR2-GFP retina (top) and the same retina with the starburst amacrine cell targeted for whole-cell recording and filled with Alexa 568 (bottom). Scale bar: 10 µm. Right: Current-clamp perforated-patch recording from the starburst amacrine cell shows wave-induced depolarizations followed by slow afterhyperpolarizations. Inset shows a magnified view of the calcium spikes during a wave-induced depolarization. (From Ford, Felix, & Feller, 2012.)

manipulations that do not alter the development of direction selectivity. Thus, future experiments based on targeted genetic manipulations will be needed to address more specific hypotheses on the role of neural activity in establishing various aspects of the circuit that mediates direction selectivity.

SUMMARY

The development of direction selectivity is a new and growing field with many unanswered questions. Although several studies indicate that direction selectivity is present at eye opening, it is not yet known if this

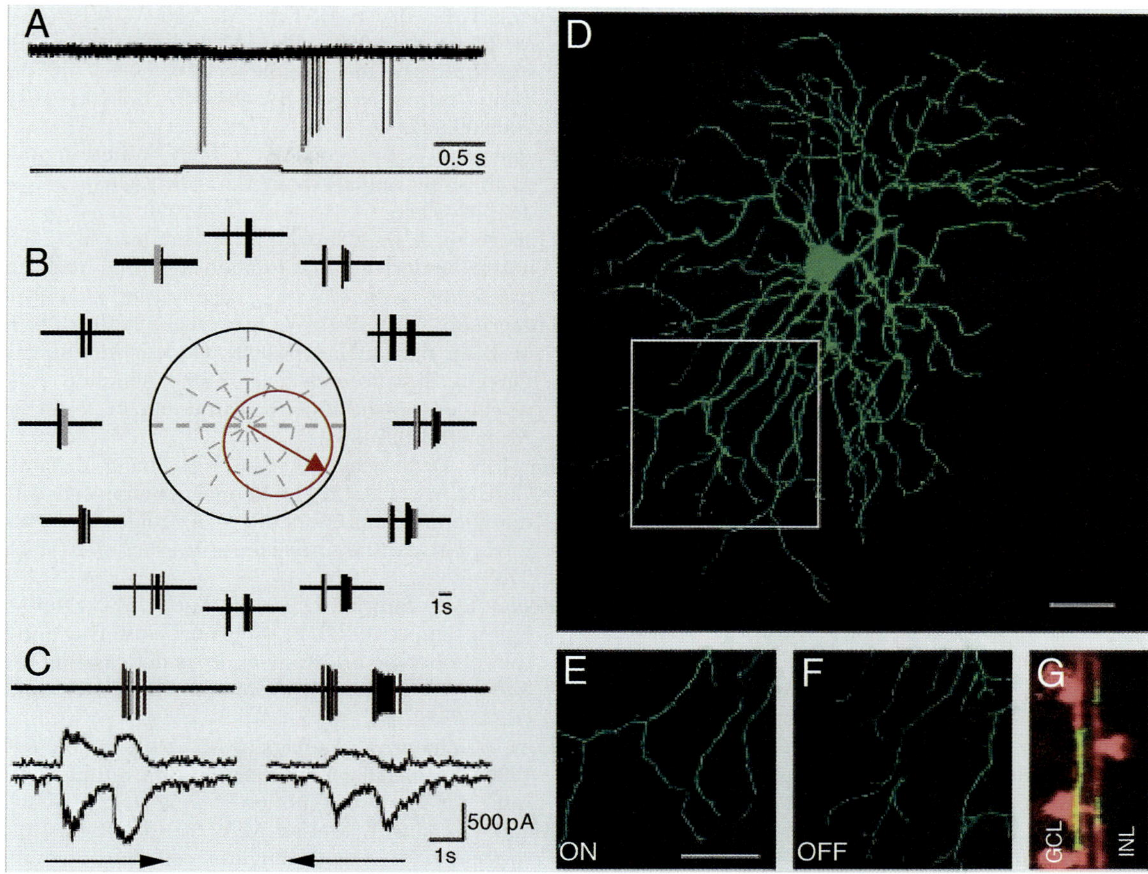

FIGURE 94.5   Retinas that receive daily intraocular injections of GABA-A receptor antagonists display normal direction-selective ganglion cells (DSGCs). (A) DSGC response to the onset and termination of a stationary flashing spot. (B) Directional responses of this same DSGC to a rectangle moving in 12 directions. The preferred direction is represented by the arrow in the middle of the polar plot. The traces of action potentials that surround this plot are the responses evoked by motion in each direction. (C) Asymmetrical synaptic inputs are recorded when the membrane potential is held at −65 mV and 0 mV. Much larger inhibitory inputs are observed when the visual stimulus moves in the null direction (left) as opposed to preferred direction (right). (D) Dendritic morphology of the recorded cell. A section of the dendritic field is enlarged to show the clear segregation between ON (E) and OFF (F) layers. (G) These dendrites tightly costratify with the cholinergic processes of starburst amacrine cells (red). Scale bar: 30 μm. GCL, ganglion cell layer; INL, inner nuclear layer. (From Sun, Han, & He, 2011.)

represents a mature state of the system. For example, the optokinetic reflex, a visually guided behavior mediated by ON-direction-selective cells (Yoshida et al., 2001), is not mature for 2 weeks after birth in rodents (Faulstich, Onori, & du Lac, 2004) and for several weeks in primates (Distler & Hoffmann, 2011). Whether this slow maturation reflects maturation within the retina or of circuits downstream is not yet known. Moreover, though some functional retinal mosaics are established at eye opening (Anishchenko et al., 2010), whether or not this is the case for direction-selective cells is also not known. Finally, the relative role of activity and molecular interactions in all of these maturational processes has yet to be established. With the advent of an ever-growing number of genetic and

imaging tools, we anticipate that the years to come will reveal tremendous progress in this field.

ACKNOWLEDGMENTS

Support contributed by National Institutes of Health (NIH) Grant RO1EY013528 (MBF), RO1EY019498 (MBF), National Science Foundation Grant IOS-0818983 (MBF), and NIH Grant F31NS071784 (AMH).

REFERENCES

Anishchenko, A., Greschner, M., Elstrott, J., Sher, A., Litke, A. M., Feller, M. B., et al. (2010). Receptive field mosaics of retinal ganglion cells are established without visual experience. *Journal of Neurophysiology, 103,* 1856–1864. doi:10.1152/jn.00896.2009.

Berson, D. M. (2008). Retinal ganglion-cell types and their central projections. In R. H. Masland & T. D. Albright (Eds.), *The senses: A comprehensive reference* (Vol. 1, pp. 491–520). San Diego, CA: Academic Press.

Borst, A., & Euler, T. (2011). Seeing things in motion: Models, circuits, and mechanisms. *Neuron, 71,* 974–994.

Briggman, K. L., Helmstaedter, M., & Denk, W. (2011). Wiring specificity in the direction-selectivity circuit of the retina. *Nature, 471,* 183–188.

Chan, Y.-C., & Chiao, C.-C. (2008). Effect of visual experience on the maturation of ON–OFF direction selective ganglion cells in the rabbit retina. *Vision Research, 48,* 2466–2475. doi:10.1016/j.visres.2008.08.010.

Chen, M., Weng, S., Deng, Q., Xu, Z., & He, S. (2009). Physiological properties of direction-selective ganglion cells in early postnatal and adult mouse retina. *Journal of Physiology, 587,* 819–828.

Daw, N. W., & Wyatt, H. J. (1974). Raising rabbits in a moving visual environment: An attempt to modify directional sensitivity in the retina. *Journal of Physiology, 240,* 309–330.

De la Huerta, I., Kim, I. J., Voinescu, P. E., & Sanes, J. R. (2012). Direction-selective retinal ganglion cells arise from molecularly specified multipotential progenitors. *Proceedings of the National Academy of Sciences of the United States of America, 109,* 17663–17668. doi:10.1073/pnas.1215806109.

Demb, J. B. (2007). Cellular mechanisms for direction selectivity in the retina. *Neuron, 55,* 179–186.

Distler, C., & Hoffmann, K. P. (2011). Visual pathway for the optokinetic reflex in infant macaque monkeys. *Journal of Neuroscience, 31,* 17659–17668.

Elstrott, J., Anishchenko, A., Greschner, M., Sher, A., Litke, A. M., Chichilnisky, E. J., et al. (2008). Direction selectivity in the retina is established independent of visual experience and cholinergic retinal waves. *Neuron, 58,* 499–506. doi:10.1016/j.neuron.2008.03.013.

Elstrott, J., & Feller, M. B. (2009). Vision and the establishment of direction-selectivity: A tale of two circuits. *Current Opinion in Neurobiology, 19,* 293–297.

Elstrott, J., & Feller, M. B. (2010). Direction-selective ganglion cells show symmetric participation in retinal waves during development. *Journal of Neuroscience, 30,* 11197–11201.

Euler, T., Detwiler, P. B., & Denk, W. (2002). Directionally selective calcium signals in dendrites of starburst amacrine cells. *Nature, 418,* 845–852.

Euler, T., Hausselt, S. E., Margolis, D. J., Breuninger, T., Castell, X., Detwiler, P. B., et al. (2009). Eyecup scope—Optical recordings of light stimulus-evoked fluorescence signals in the retina. *Pflugers Arch- European. Journal of Physiology, 457,* 1393–1414.

Faulstich, B. M., Onori, K. A., & du Lac, S. (2004). Comparison of plasticity and development of mouse optokinetic and vestibulo-ocular reflexes suggests differential gain control mechanisms. *Vision Research, 44,* 3419–3427. doi:10.1016/j.visres.2004.09.006.

Ford, K. J., Felix, A. L., & Feller, M. B. (2012). Cellular mechanisms underlying spatiotemporal features of cholinergic retinal waves. *Journal of Neuroscience, 32,* 850–863.

Fried, S. I., Munch, T. A., & Werblin, F. S. (2002). Mechanisms and circuitry underlying direction selectivity in the retina. *Nature, 420,* 411–414.

Fuerst, P. G., Bruce, F., Tian, M., Wei, W., Elstrott, J., Feller, M. B., et al. (2009). DSCAM and DSCAML1 function in self-avoidance in multiple cell types in the developing mouse retina. *Neuron, 64,* 484–497. doi:10.1016/j.neuron.2009.09.027.

Fuerst, P. G., & Burgess, R. W. (2009). Adhesion molecules in establishing retinal circuitry. *Current Opinion in Neurobiology, 19,* 389–394.

Huberman, A. D., Feller, M. B., & Chapman, B. (2008). Mechanisms underlying development of visual maps and receptive fields. *Annual Review of Neuroscience, 31,* 479–509.

Huberman, A. D., Wei, W., Elstrott, J., Stafford, B. K., Feller, M. B., & Barres, B. A. (2009). Genetic identification of an On–Off direction-selective retinal ganglion cell subtype reveals a layer-specific subcortical map of posterior motion. *Neuron, 62,* 327–334.

Kay, J. N., De la Huerta, I., Kim, I. J., Zhang, Y., Yamagata, M., Chu, M. W., et al. (2011). Retinal ganglion cells with distinct directional preferences differ in molecular identity, structure, and central projections. *Journal of Neuroscience, 31,* 7753–7762.

Keeley, P. W., Sliff, B., Lee, S. C., Fuerst, P. G., Burgess, R. W., Eglen, S. J., et al. (2012). Neuronal clustering and fasciculation phenotype in dscam- and bax-deficient mouse retinas. *Journal of Comparative Neurology, 520,* 1349. doi:10.1002/cne.23033.

Kim, I., Zhang, Y., Yamagata, M., Meister, M., & Sanes, J. (2008). Molecular identification of a retinal cell type that responds to upward motion. *Nature, 452,* 478–482.

Kim, I.-J., Zhang, Y., Meister, M., & Sanes, J. R. (2010). Laminar restriction of retinal ganglion cell dendrites and axons: Subtype-specific developmental patterns revealed with transgenic markers. *Journal of Neuroscience, 30,* 1452–1462.

Lee, S., Kim, K., & Zhou, Z. J. (2010). Role of ACh-GABA cotransmission in detecting image motion and motion direction. *Neuron, 68,* 1159–1172.

Li, Y., Fitzpatrick, D., & White, L. E. (2006). The development of direction selectivity in ferret visual cortex requires early visual experience. *Nature Neuroscience, 9,* 676–681.

Li, Y., Van Hooser, S., Mazurek, M., White, L. E., & Fitzpatrick, D. (2008). Experience with moving visual stimuli drives the early development of cortical direction selectivity. *Nature, 456,* 952–956.

Masland, R. H. (1977). Maturation of function in the developing rabbit retina. *Journal of Comparative Neurology, 175,* 275–286.

Matsuoka, R. L., Chivatakarn, O., Badea, T. C., Samuels, I. S., Cahill, H., Katayama, K., et al. (2011a). Class 5 transmembrane semaphorins control selective mammalian retinal lamination and function. *Neuron, 71,* 460–473. doi:10.1016/j.neuron.2011.06.009.

Matsuoka, R. L., Nguyen-Ba-Charvet, K. T., Parray, A., Badea, T. C., Chedotal, A., & Kolodkin, A. L. (2011b). Transmembrane semaphorin signalling controls laminar stratification in the mammalian retina. *Nature, 470,* 259–263.

Qu, J., & Myhr, K. L. (2008). The development of intrinsic excitability in mouse retinal ganglion cells. *Developmental Neurobiology, 68,* 1196–1212.

Reese, B. E., Keeley, P. W., Lee, S. C., & Whitney, I. E. (2011). Developmental plasticity of dendritic morphology and the establishment of coverage and connectivity in the outer retina. *Developmental Neurobiology, 71,* 1273–1285.

Rivlin-Etzion, M., Zhou, K., Wei, W., Elstrott, J., Nguyen, P. L., Barres, B. A., et al. (2011). Transgenic mice reveal unexpected diversity of on–off direction-selective retinal ganglion cell subtypes and brain structures involved in motion processing. *Journal of Neuroscience, 31,* 8760–8769. doi:10.1523/JNEUROSCI.0564-11.2011.

Rochefort, N. L., Narushima, M., Grienberger, C., Marandi, N., Hill, D. N., & Konnerth, A. (2011). Development of direction selectivity in mouse cortical neurons. *Neuron, 71,* 425–432.

Rockhill, R. L., Daly, F. J., MacNeil, M. A., Brown, S. P., & Masland, R. H. (2002). The diversity of ganglion cells in a mammalian retina. *Journal of Neuroscience, 22,* 3831–3843.

Stacy, R. C., & Wong, R. O. L. (2003). Developmental relationship between cholinergic amacrine cell processes and ganglion cell dendrites of the mouse retina. *Journal of Comparative Neurology, 456,* 154–166.

Stafford, B. K., Sher, A., Litke, A. M., & Feldheim, D. A. (2009). Spatial–temporal patterns of retinal waves underlying activity-dependent refinement of retinofugal projections. *Neuron, 64,* 200–212.

Sun, C., Speer, C. M., Wang, G. Y., Chapman, B., & Chalupa, L. M. (2008). Epibatadine application in vitro blocks retinal waves without silencing all retinal ganglion cell action potentials in the developing retina of the mouse and ferret. *Journal of Neurophysiology, 100,* 3253–3263.

Sun, L., Han, X., & He, S. (2011). Direction-selective circuitry in rat retina develops independently of GABAergic, cholinergic and action potential activity. *PLoS ONE, 6,* e19477. doi:10.1371/journal.pone.0019477.

Sun, W., Deng, Q., Levick, W. R., & He, S. (2006). ON direction-selective ganglion cells in the mouse retina. *Journal of Physiology, 576,* 197–202.

Sun, W., Li, N., & He, S. (2002). Large-scale morphological survey of mouse retinal ganglion cells. *Journal of Comparative Neurology, 451,* 115–126.

Tian, N. (2008). Synaptic activity, visual experience and the maturation of retinal synaptic circuitry. *Journal of Physiology, 586,* 4347–4355.

Tian, N. (2011). Developmental mechanisms that regulate retinal ganglion cell dendritic morphology. *Developmental Neurobiology, 71,* 1297–1309.

Tian, N., & Copenhagen, D. R. (2001). Visual deprivation alters development of synaptic function in inner retina after eye opening. *Neuron, 32,* 439–449.

Trenholm, S., Johnson, K., Li, X., Smith, R. G., & Awatramani, G. B. (2011). Parallel mechanisms encode direction in the retina. *Neuron, 71,* 683–694.

Vaney, D. I., & Taylor, W. R. (2002). Direction selectivity in the retina. *Current Opinion in Neurobiology, 12,* 405–410.

Wei, W., Elstrott, J., & Feller, M. B. (2010). Two-photon targeted recording of GFP-expressing neurons for light responses and live-cell imaging in the mouse retina. *Nature Protocols, 5,* 1347–1352. doi:10.1038/nprot.2010.106.

Wei, W., & Feller, M. B. (2011). Organization and development of direction-selective circuits in the retina. *Trends in Neurosciences, 34,* 638–645. doi:10.1016/j.tins. 2011.08.002.

Wei, W., Hamby, A. M., Zhou, K., & Feller, M. B. (2011). Development of asymmetric inhibition underlying direction selectivity in the retina. *Nature, 469,* 402–406.

Xu, H. P., Furman, M., Mineur, Y. S., Chen, H., King, S. L., Zenisek, D., et al. (2010). An instructive role for patterned spontaneous retinal activity in mouse visual map development. *Neuron, 70,* 1115–1127. doi:10.1016/j.neuron. 2011.04.028.

Yamagata, M., & Sanes, J. R. (2008). Dscam and Sidekick proteins direct lamina-specific connections in vertebrate retina. *Nature, 451,* 465–469.

Yamagata, M., & Sanes, J. R. (2012). Expanding the Ig superfamily code for laminar specificity in retina: Expression and role of contactins. *Journal of Neuroscience, 41,* 14402–14414. doi:10.1523/JNEUROSCI.3193-12.2012.

Yamagata, M., Weiner, J. A., & Sanes, J. R. (2002). Sidekicks: Synaptic adhesion molecules that promote lamina-specific connectivity in the retina. *Cell, 110,* 649.

Yonehara, K., Balint, K., Noda, M., Nagel, G., Bamberg, E., & Roska, B. (2011). Spatially asymmetric reorganization of inhibition establishes a motion-sensitive circuit. *Nature, 469,* 407–420.

Yonehara, K., Ishikane, H., Sakuta, H., Shintani, T., Nakamura-Yonehara, K., Kamiji, N., et al. (2009). Identification of retinal ganglion cells and their projections involved in central transmission of information about upward and downward image motion. *PLoS ONE, 4,* e4320. doi:10.1371/journal.pone.0004320.

Yonehara, K., Shintani, T., Suzuki, R., Sakuta, H., Takeuchi, Y., Nakamura-Yonehara, K., & Noda, M. (2008). Expression of SPIG1 reveals development of a retinal ganglion cell subtype projecting to the medial terminal nucleus in the mouse. *PLoS ONE, 3,* e1533. doi:10.1371/journal.pone. 0001533.

Yoshida, K., Watanabe, D., Ishikane, H., Tachibana, M., Pastan, I., & Nakanishi, S. (2001). A key role of starburst amacrine cells in originating retinal directional selectivity and optokinetic eye movement. *Neuron, 30,* 771–780.

95 Mechanisms of Visual Cortex Plasticity during Development

IKUE NAGAKURA, NIKOLAOS MELLIOS, AND MRIGANKA SUR

A prominent feature of the mammalian visual cortex is that its proper development requires visual experience, and that development of its synapses, neurons, and circuits is highly influenced by electrical activity. This makes the visual cortex an attractive model system for studying the role of sensory experience in refining cortical circuits and function. Ocular dominance (OD) plasticity is a classic form of experience-dependent plasticity that refers to changes in visual cortical circuitry due to unbalanced inputs from the two eyes. Hubel and Wiesel first demonstrated almost 50 years ago that altering visual inputs from one eye by eyelid closure during development changes responses of cells located in the primary visual cortex (V1) of cats; responses to the deprived (closed) eye are weakened while responses to the nondeprived (open) eye are strengthened (Wiesel & Hubel, 1963). This paradigm, called monocular deprivation (MD), induces OD plasticity during a specific developmental time window termed the critical period (Gordon & Stryker, 1996; Hubel & Wiesel, 1970). In rodents, however, the definition of the critical period is controversial due to the observation that plasticity still exists outside a classically defined developmental window, leading to a consensus view of the critical period as a particularly sensitive period of development during which even a brief alteration in visual experience induces significant cortical plasticity. In recent decades, many discoveries have revealed a surprisingly rich variety of molecular and cellular mechanisms that underlie OD plasticity, which have, somewhat surprisingly, propelled an understanding of mechanisms underlying a variety of developmental brain disorders. The purpose of this review is to organize these findings into a coherent conceptual framework and suggest areas of further research.

TWO ASPECTS OF PLASTICITY: FEEDFORWARD (HEBBIAN) AND FEEDBACK (HOMEOSTATIC)

The V1 of rodents contains a binocular region that receives inputs from both eyes, with the contralateral eye providing a much greater number and larger extent of projections compared to the ipsilateral eye, leading to a contralateral bias in visual responses. Various recording techniques, such as visually evoked potentials (VEPs), intrinsic signal optical imaging, single-unit recording, and two-photon calcium imaging, have been employed to measure OD plasticity of single neurons and neuronal populations—importantly from the contralateral toward the ipsilateral eye following MD of the contralateral eye. Strengthening and weakening of responses from each eye are measured according to the technique used; for example, VEP recordings measure summed synaptic currents of thalamocortical input to layer 4 while optical and calcium imaging are limited to layer 2/3 neurons located less than 500 μm of the cortical surface and measure intrinsic activities of a population of cells or calcium transients in single cells, respectively (Smith, Heynen & Bear, 2009). Using these techniques, it is now clear that a shift in OD is observed after brief (3 days of) MD, which is mediated by weakening of deprived-eye responses whereas a longer duration (5–7 days) of MD induces an additional OD shift due to strengthening of open-eye responses (Frenkel & Bear, 2004; Kaneko et al., 2008). The framework of feedforward and feedback regulation was developed to understand these two components of plasticity, namely the initial weakening of deprived-eye responses and later strengthening of open-eye responses, respectively (Tropea, Van Wart, & Sur, 2009b). It is suggested that feedforward regulation is mediated by synapse-specific Hebbian forms of plasticity while feedback regulation is mediated by cell-wide global homeostatic mechanisms, which together comprise mechanisms underlying OD plasticity (see figure 95.1).

MECHANISMS OF FEEDFORWARD PLASTICITY

The idea that the initial weakening of deprived-eye responses is mediated by Hebbian, feedforward plasticity comes from the observation that molecular and cellular changes of the deprived eye follow mechanisms similar to homosynaptic long-term depression (LTD), which is induced by decorrelated firing of pre- and

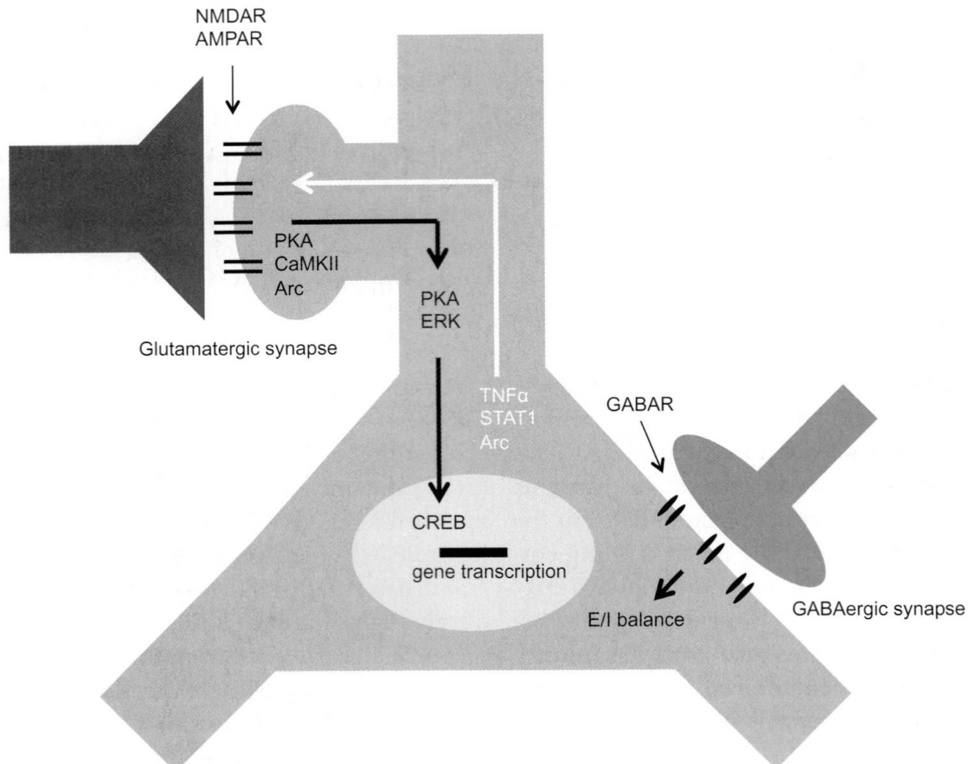

FIGURE 95.1   Schematic of key intracellular molecules that mediate feedforward and feedback plasticity in visual cortex. A pyramidal neuron (middle) receives inputs from an excitatory neuron through a glutamatergic synapse (top left) and from an inhibitory neuron through a gamma-aminobutyric acid-releasing (GABAergic) synapse (bottom right). Feedforward plasticity is shown with thick black arrows; following brief monocular deprivation (MD), *N*-methyl-D-aspartate receptor (NMDAR) activation leads to activation of intracellular kinases such as calcium/calmodulin-dependent protein kinase II (CaMKII), protein kinase A (PKA), and extracellular signal-regulated kinase 1,2 (ERK). PKA may regulate glutamate receptor function at synaptic sites and, together with ERK, also leads to cyclic AMP response element (CRE) mediated gene transcription through CRE-binding protein (CREB). An immediate-early gene, Arc, mediates weakening of deprived-eye responses through alpha-amino-3-hydroxy-5-methyl-4-isoxazole propionic acid receptor (AMPAR) internalization at synaptic sites. A balance between excitation (E) and inhibition (I) is maintained at a proper level through modulation of inhibition at GABAergic synaptic sites. Feedback plasticity is shown with a white arrow; following longer duration of MD, molecules of the classic immune system such as tumor necrosis factor alpha (TNF-alpha) and STAT1 regulate strengthening of open-eye responses, possibly through synaptic scaling and increased AMPAR trafficking. Arc may enhance open-eye responses through NMDAR-dependent long-term potentiation and/or synaptic scaling.

postsynaptic neurons leading to synaptic weakening. Brief MD has been reported to induce homosynaptic LTD in visual cortex (Heynen et al., 2003). Several studies have demonstrated involvement of similar molecular machinery as LTD in feedforward plasticity, including requirement for *N*-methyl-D-aspartate receptors (NMDARs) and activation of intracellular calcium signaling. Maintaining a proper balance between excitation and inhibition is a crucial factor in initiating and terminating critical period plasticity, which importantly involves modulation of inhibition. Finally, functional changes in deprived-eye responses are accompanied by structural changes in dendritic spines.

Changes in NMDA Receptor Subunit

Modulation of excitatory synaptic drive takes place partly through changes in glutamate receptors such as NMDARs. Several lines of evidence suggest that changes in the composition of NR2A and NR2B subunits of NMDARs are influenced by visual experience during postnatal development. For example, rearing rodents in darkness decreases the level of NR2A and/or increases the level of NR2B depending on the duration of dark rearing, leading to a reduction in the ratio of NR2A/NR2B subunits (Chen & Bear, 2007). This reduction in the NR2A/2B ratio, which normally transitions from low to high during development (Quinlan, Olstein, & Bear, 1999), may serve as a condition for OD plasticity to occur even during adulthood (He, Hodos, & Quinlan, 2006). NR2A subunits have a reduced calcium influx and shorter NMDA-mediated synaptic currents, while NR2B subunits have high calcium permeability and thus appear to enhance plasticity (Flint et al., 1997). The reduction in the NR2A/2B ratio may be permissive for strengthening open-eye responses since there is a significant reduction in the ratio following 5–7 days (later, homeostatic component) of MD (Chen & Bear, 2007). Taken together, these studies suggest that developmental and experience-dependent changes in NMDA-mediated excitatory transmission can regulate the capacity for OD plasticity.

Intracellular Signaling: PKA, ERK, CREB

NMDAR-dependent LTD in the hippocampus is associated with dephosphorylation of the GluA1 (GluR1) subunit of alpha-amino-3-hydroxy-5-methyl-4-isoxazole propionic acid receptors (AMPARs) at Ser-845, which is the phosphorylation site for the cyclic AMP (cAMP)-dependent protein kinase (protein kinase A; PKA) (Kameyama et al., 1998). Similarly in the visual cortex, MD induces synaptic depression and dephosphorylation of GluA1 at Ser-845 (Heynen et al., 2003). Moreover, pharmacological blockade of PKA inhibits OD shift in cats (Beaver et al., 2001). These results thus suggest the involvement of PKA and GluA1 in OD plasticity. Extracellular signal-regulated kinase 1,2 (ERK) is another intracellular kinase that is required for OD plasticity (Di Cristo et al., 2001). Following visual stimulation after a brief period of dark rearing, ERK, together with PKA, enhances downstream cAMP response element (CRE)-mediated gene expression (Cancedda et al., 2003). Blocking the function of CRE-binding protein (CREB), which controls CRE-mediated gene expression, prevents an OD shift (Mower et al., 2002). Interestingly, the role of CRE-mediated gene expression following MD seems unique to the open eye; CRE-mediated *lacZ* expression is mostly observed in the areas receiving inputs from the open eye but not the deprived eye in a time course of 1–3 days (Pham et al., 1999), suggesting the involvement of CRE-mediated gene expression in rapid feedback plasticity for the open eye.

Layer-Specific Mechanisms

Metabotropic glutamate receptors (mGluRs) are another type of glutamatergic receptors implicated in visual cortical plasticity (Daw, Reid, & Beaver, 1999). However, the requirement of mGluRs in LTD in visual cortex and OD plasticity depends on specific layers (Daw et al., 2004). It appears that LTD in layers 2/3 and 5 depends on NMDARs but not mGluRs while LTD in layer 6 requires mGluRs but not NMDARs (Rao & Daw, 2004). It had been long thought that the feedforward pathway into and through the V1 is mainly comprised as retina → thalamus → layer 4 → layer 2/3 (superficial layers) → layer 5/6 (deep layers) → extrastriate and subcortical structures. This notion is brought into a question by findings that support the idea that there is a parallel pathway into layer 2/3 directly from thalamus with distinct molecular requirements. For example, LTD induced in layer 2/3 requires an endocannabinoid CB1 receptor whereas LTD induced in layer 4 is independent of CB1 but instead dependent on PKA and AMPAR endocytosis (Crozier et al., 2007).

Modulation by Inhibition

Although many of the earlier findings emphasized the role of excitatory transmission mediated through glutamate receptors, there is growing evidence for the role of inhibition in OD plasticity. A minimal level of inhibition is necessary for initiation of OD plasticity during the critical period, and maturation of cortical inhibitory circuits limits plasticity in adulthood. Mice with genetic deletion of a gamma-aminobutyric acid (GABA)-synthetic enzyme (GAD65 knockout) show lack of OD shift after MD (Hensch et al., 1998). Conversely, manipulations that accelerate GABA circuit function trigger premature plasticity before the critical period (Di Cristo et al., 2007; Sugiyama et al., 2008). Brain-derived neurotrophic factor (BDNF) is a key neurotrophin that triggers maturation of inhibitory circuits, and overexpression of BDNF leads to precocious termination of the critical period for OD plasticity (Huang et al., 1999). Another molecule, Lynx1, which binds to the nicotinic acetylcholine receptor, was discovered as a crucial protein maintaining the balance between excitation and inhibition through cholinergic inhibition, without which plasticity is extended into adulthood (Morishita et al., 2010). It has become increasingly clear that a GABA-releasing (GABAergic) cell type, the parvalbumin (PV)-positive basket cell, is a key player in critical period plasticity. PV cells are the largest class of inhibitory interneuron in the cortex and comprise about 40% of the GABAergic cell population in the mouse visual cortex (Gonchar, Wang, & Burkhalter, 2007). PV cells regulate OD plasticity by sending inputs to GABA$_A$ receptor-α1 subunits in excitatory neurons (Fagiolini et al., 2004). Lynx1 colocalizes with PV cells (Morishita et al., 2010), and molecules such as BDNF and the embryonic homeoprotein Otx2 are suggested to regulate OD plasticity by controlling the maturation of PV cells (Huang et al., 1999; Sugiyama et al., 2008).

Structural Modification

Functional changes in OD plasticity are tightly linked to structural changes at dendritic spines, which can be observed even on a time scale of hours (Yu, Majewska, & Sur, 2011). Degradation of the extracellular matrix (ECM) and increased spine motility are key components of such structural changes (Majewska & Sur, 2003). Brief MD during the critical period increases spine motility and decreases dendritic spines in layer 2/3, which are dependent on degradation of the ECM due to tissue-type plasminogen activator (tPA)/plasmin proteolytic cascade (Mataga, Mizuguchi, & Hensch, 2004; Oray, Majewska, & Sur, 2004), and an OD shift is prevented without tPA (Mataga, Nagai, & Hensch, 2002). Conversely, inducing ECM degradation with chondroitinase-ABC, which selectively degrades chondroitin sulfate proteoglycans (CSPGs) in extracellular perineuronal nets (PNNs, which are components of the ECM), restores OD plasticity in adult animals (Pizzorusso et al., 2002). Proper sulfation of CSPGs in PNNs is particularly required for the accumulation of Otx2 and maturation of PV cells, which controls the termination of the critical period (Miyata et al., 2012).

One important question is whether these structural changes occur at spines receiving input from the deprived or open eye. A recent study addressed this issue using a genetically engineered Förster resonance energy transfer probe for the detection of calcium/calmodulin-dependent protein kinase II (CaMKII) activity in identified spines in vivo (Mower et al., 2011). Brief MD specifically activates CaMKII in spines within deprived-eye regions, and spines that are eliminated receive input from the deprived eye and have low basal CaMKII activity while spines that are preserved show increased CaMKII activation following MD (Mower et al., 2011). These results not only demonstrate the deprived-eye specificity of synapse elimination but also imply a protective role for activated CaMKII against synapse loss following reduction of input drive.

MECHANISMS OF FEEDBACK PLASTICITY

The studies described above strongly suggest that Hebbian-based, feedforward plasticity and its underlying cellular and molecular mechanisms are key for one component of OD plasticity, specifically the rapid weakening of deprived-eye responses. However, several lines of evidence indicate that OD plasticity also includes homeostatic feedback regulation that leads to the later strengthening of open-eye responses; first, strengthening of open-eye responses in the mouse V1 is observed 2–3 days after weakening of deprived-eye responses (Frenkel & Bear, 2004; Kaneko et al., 2008), suggesting a homeostatic-like feedback mechanism in response to initial weakening of deprived-eye responses; and second, even cells responding to the deprived eye can strengthen their responses after longer duration of MD (Mrsic-Flogel et al., 2007), suggesting non-cell-specific, global feedback regulation. Homeostatic feedback regulation may exist to preserve a net visual drive for each neuron through mechanisms such as synaptic scaling and intrinsic excitability.

Synaptic Scaling

Homeostatic regulation is necessary to prevent neural circuits from becoming hyper- or hypoactive, without which Hebbian forms of plasticity such as long-term potentiation (LTP) and LTD could drive neuronal activity toward runaway excitation or total quiescence (Turrigiano & Nelson, 2004). The most studied mechanism for homeostatic plasticity is synaptic scaling, which was first discovered in rat visual cortical culture, where blockade of neuronal activity with tetrodotoxin (TTX) or increasing activity with bicuculline results in an increase or decrease in miniature excitatory postsynaptic currents (mEPSCs) amplitude, respectively (Turrigiano et al., 1998). Synaptic scaling induces a multiplicative change in synaptic weights across the entire neuron, such that the entire distribution of mEPSC amplitudes is scaled up or down proportionally (Turrigiano et al., 1998). Thus, synaptic scaling appears to be a global process involving all synapses in a postsynaptic neuron, which might account for feedback regulation in OD plasticity.

Indeed, 2 days of MD scale up the amplitude of mEPSCs onto monocular pyramidal neurons in a layer- and age-dependent manner (Desai et al., 2002). Scaling up of mEPSC amplitude by MD or dark rearing starts to take place at the opening of the critical period and persists throughout adulthood in layer 2/3 while mEPSCs in layer 4 pyramidal neurons can be scaled up at an earlier postnatal age P14–16 but is unaffected during the critical period (Desai et al., 2002; Goel & Lee, 2007), suggesting developmental regulation of synaptic scaling in different layers. It is also interesting to note that synaptic scaling in adult mice in layer 2/3 is not multiplicative in nature, that is, not all the synapses are scaled up (Goel & Lee, 2007). This raises an important question of whether a given postsynaptic neuron can preferentially scale one type of synapse while leaving others unaffected.

Another line of evidence that synaptic scaling underlies the homeostatic component of OD plasticity comes from studies that use genetically modified mice for a molecule that is important for synaptic scaling. Tumor necrosis factor alpha (TNF-alpha) is a proinflammatory cytokine that is released by glial cells and acts on neurons through its receptor, TNFR1 (Stellwagen et al., 2005; Stellwagen & Malenka, 2006). In optical imaging recording, mice lacking TNF-alpha show impairment in open-eye responses (i.e., no increase in the responses following 5–6 days of MD) while deprived-eye responses are left intact (i.e., normal decrease in the responses) (Kaneko et al., 2008). These mice show normal LTP in the visual cortex but lack synaptic scaling induced by activity blockade (Kaneko et al., 2008). While the downstream molecular mechanism of TNF-alpha regulation of the homeostatic component of OD plasticity is not clear, one possibility is that it occurs through increased surface insertion of the GluA1 subunit of AMPARs (Stellwagen et al., 2005).

An interesting interaction was suggested between TNF-alpha and signal transducer and activator of transcription1 (STAT1). Using optical imaging, STAT1 was found to negatively regulate the homeostatic component of OD plasticity by inhibiting TNF-alpha signaling; STAT1 knockout mice show accelerated increase of open-eye responses following 4 days of MD, which is reversed with cortical infusion of a TNF-alpha inhibitor (Nagakura et al., 2012). STAT1 might regulate OD plasticity through trafficking of AMPARs since accelerated increase of open-eye responses in STAT1 knockout mice is accompanied by increased surface levels of GluA1 AMPARs (Nagakura et al., 2012).

Scaling of Intrinsic Excitability

Relatively unexplored but potentially just as important as synaptic scaling for homeostatic OD plasticity is scaling of intrinsic excitability. Changes in intrinsic excitability affect neuronal circuit plasticity through alterations of a neuron's input–output function without changing synaptic weights (Turrigiano, 2011). In visual cortical cultures, activity blockade by TTX enhances intrinsic excitability by increasing sodium currents and lowering the threshold for action potentials, so that neurons fire more to the same synaptic input (Desai, Rutherford, & Turrigiano, 1999). Interestingly, MD by lid suture does not scale up mEPSC amplitude in layer 2/3 pyramidal neurons while intraocular activity blockade by TTX does (Desai et al., 2002; Maffei & Turrigiano, 2008). Instead, lid suture increases homeostatic plasticity through an increase in intrinsic excitability in layer 2/3 neurons (Maffei & Turrigiano, 2008). It is

suggested that lid suture induces stronger decorrelation of sensory drive to cortex (Linden et al., 2009), which drives strong synaptic depression in layer 2/3 (Maffei & Turrigiano, 2008), and this depression exceeds the ability of synaptic scaling to compensate and instead triggers intrinsic homeostatic plasticity (Turrigiano, 2011). Consistently, when synaptic scaling normally takes place in layer 4, intrinsic excitability is not affected (Maffei, Nelson, & Turrigiano, 2004).

MOLECULES AND PATHWAYS THAT LINK FEEDFORWARD AND FEEDBACK PLASTICITY

Although mechanisms for feedforward and feedback plasticity appear to be distinct and seemingly independent processes, they are not totally separable. Signaling cascades inside cells and even individual molecules are likely to be involved in the coordinated regulation of feedforward and feedback plasticity. Gene expression analyses in V1 during the critical period, and following variable periods of visual deprivation, provide a comprehensive picture of cortical changes during normal development and experience-dependent plasticity. The critical period in mice is accompanied by a distinct transcriptional profile that includes a relative abundance of genes involving the actin cytoskeleton, G protein signaling, transcription, and myelination, and MD until around P28, the peak of the critical period, reverses the expression pattern of the majority of these genes (Lyckman et al., 2008), suggesting that a surprisingly large number of genes and molecules have a role in feedback regulation. Another microarray study revealed that dark rearing and MD both activate feedforward and feedback mechanisms, and MD specifically activates sets of genes that comprise molecular pathways related to growth factors and neuronal degeneration (Tropea et al., 2006). In particular, expression of a binding protein of insulin-like growth factor-1 (IGF1) is highly up-regulated after MD whereas IGF1 and components of its downstream phosphatidylinositol 3-kinase (PI3K)/Akt signaling pathway are down-regulated. Exogenous application of IGF1 up-regulates these signals and prevents OD plasticity in vivo (Tropea et al., 2006).

A good example of a single molecule that appears to coregulate feedforward and feedback plasticity is an immediate-early gene, Arc (Arg3.1). Arc has diverse roles in activity-dependent plasticity, both in Hebbian and homeostatic forms of plasticity, through regulation of AMPAR trafficking (Shepherd & Bear, 2011). A study using Arc knockout mice demonstrated that OD plasticity is completely blocked, where both deprived- and open-eye responses are unaffected following MD

(McCurry et al., 2010). Weakening of deprived-eye responses, measured as decreased intrinsic signal in optical imaging and synaptic depression in VEP recordings, is suggested to be blocked by the lack of a LTD-like mechanism that requires internalization of AMPARs since AMPAR internalization is impaired in these mice. The lack of subsequent strengthening of open-eye responses, that is, increased intrinsic signal and synaptic potentiation, could be due to one of two reasons, which are not mutually exclusive. Visual deprivation may cause a metaplastic adjustment of the properties of NMDAR-dependent LTP (the threshold for LTP lowered by induction of NMDAR-dependent LTD) that enables open-eye potentiation, which is impaired without Arc (McCurry et al., 2010). Alternatively, Arc may be required for strengthening of open-eye responses through mechanisms of synaptic scaling. While these possibilities are difficult to test because Arc has a role in both LTP and synaptic scaling, they point to crucial mechanisms that link multiple forms of plasticity.

Despite the plethora of protein-coding genes shown to influence OD plasticity, very little is known about the potential role of noncoding RNAs in modulating experience-dependent plasticity in the visual cortex. Noncoding RNAs are abundantly expressed in the brain and have recently emerged as novel molecular regulators of various cellular processes inside the nervous system (Qureshi & Mehler, 2011). Moreover, a type of evolutionarily conserved small noncoding RNAs, known as microRNAs (miRNAs), have already been shown to be a critical component of brain development, maturation, and synaptic plasticity (Mellios & Sur, 2012). Two pivotal studies have recently demonstrated the involvement of an activity-dependent miRNA, miR-132, in the modulation of the critical period and OD plasticity. In one study, virus-mediated inhibition of miR-132 function in the visual cortex during the critical period resulted in increases in GTPase p250GAP (Mellios et al., 2011), which is a bona fide target of miR-132 capable of inhibiting dendritic spine growth by blocking Rac activation (Vo et al., 2005; Wayman et al., 2008) (see also figure 95.2). As a result, dendritic spine number is reduced and synaptic maturation is delayed (Mellios et al., 2011). In the second study, infusion of a miR-132 mimic through an osmotic minipump resulted in an opposite shift in dendritic spine morphology toward increased maturation (Tognini et al., 2011). Intriguingly, both in vivo manipulations of miR-132 expression abrogate OD plasticity, suggesting that an optimum level of miR-132 is required to maintain the appropriate timing of synaptic maturation necessary for visual cortical plasticity. Furthermore, these findings suggest the possibility that miR-132 is important for the onset and termination of the critical period. It is known that miR-132 can interact with NMDA and neurotrophin signaling (Miller et al., 2012; Remenyi et al., 2010), and its expression can be robustly activated by CREB (Remenyi et al., 2010; Vo et al., 2005) (see also figure 95.2). Moreover, miR-132 can indirectly increase the expression of BDNF (Klein et al., 2007), a known enhancer of miR-132 transcription (Remenyi et al., 2010). Such a complex feedback loop may allow the regulation of both Hebbian and homeostatic parameters of cortical plasticity through miR-132.

PLASTICITY MECHANISMS THAT REVEAL MECHANISMS OF DEVELOPMENTAL BRAIN DISORDERS

Disorders of human brain development such as autism appear to be related significantly to abnormal development of synapses and, potentially, to abnormal plasticity of synapses and circuits. OD plasticity in mice, a classic model of how experience modifies synapses and circuits in cortex throughout development, can thus be a useful window into mechanisms of developmental brain disorders. It is particularly useful for exploring the role of genes in cortical plasticity, and mice with a deletion of a particular gene that is nonfunctional in these disorders give us an important insight into the underlying mechanisms causing developmental deficits. Mechanisms and even therapeutics emerging from understanding OD plasticity have already made a clinical impact on disorders such as Rett syndrome (RTT) and fragile X syndrome (FXS).

Rett Syndrome

RTT is a subset of autism and an X-linked neurodevelopmental disorder affecting approximately 1 in 10,000 girls, mostly caused by mutations in the gene *MECP2* encoding X-linked methyl-CpG-binding protein 2 (MeCP2) (Amir et al., 1999). MeCP2 is a transcriptional repressor involved in chromatin remodeling and modulation of RNA splicing, aberrations of which result in neuropathophysiology underlying RTT symptoms such as loss of speech, repetitive hand movements, and cognitive symptoms (Chahrour & Zoghbi, 2007). Expression of MeCP2 is regulated in a manner that correlates with neuronal maturation (Cohen et al., 2003), and mice lacking MeCP2 show RTT-like symptoms, which are rescued by postnatal expression of MeCP2 (Giacometti et al., 2007; Guy et al., 2007). This reversible feature of RTT supports the idea that neuronal circuits are not atrophic but rather remain in a labile, immature state, and activation of MeCp2 even at a later stage of the

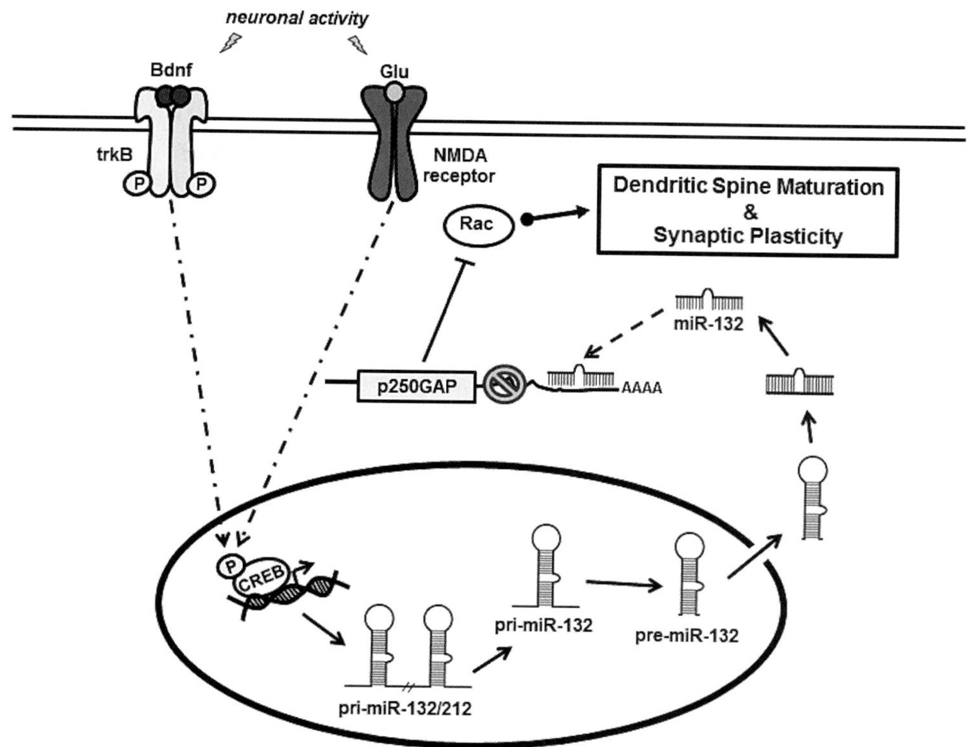

FIGURE 95.2   Neuronal activity-induced microRNA miR-132 expression regulates synaptic plasticity. Following neuronal activation, brain-derived neurotrophic factor (Bdnf) and Glutamate (Glu) activate trkB and *N*-methyl-D-aspartate (NMDA) receptors, respectively, which phosphorylate cyclic AMP response element binding protein (CREB). Phosphorylated (P) CREB then binds to the promoter of miR-132/miR-212, leading to the transcription of a multicistronic primary transcript containing both miRNAs (pri-miR-132/212). Pri-miR-132 is then cleaved to produce pre-miR-132, which translocates to the cytoplasm, where it is further cleaved to generate a miRNA duplex intermediate, which is finally unwound so as to produce mature miR-132. Mature miR-132 then inhibits its target p250GAP, by binding to its 3′ untranslated region, and p250GAP in turn inhibits Rac, a molecule known to promote dendritic spine growth and synaptic plasticity.

disorder can repair the symptoms through subsequent maturation (Tropea et al., 2009a).

MeCP2 null mice show extended OD plasticity into adulthood, presumably because of immature cortical circuits. As mentioned above, decreased IGF1 signaling is correlated with OD plasticity following MD (Tropea et al., 2006), and IGF1 application prevents the effects of MD, suggesting that IGF1 treatment might reverse the synaptic and circuit effects of MeCP2 loss in MeCP2 null mice. Indeed, systemic administration of IGF1 causes a partial to complete reversal of a wide range of symptoms in MeCP2 null mice, including improvements in organismal measures such as lifespan, breathing and heart rate, levels of signaling and synaptic molecules, and closure of the abnormally long critical period for OD plasticity in these mice (Castro et al., 2011; Tropea et al., 2009a). These studies using OD plasticity as a model for cortical maturation identify IGF1 and its molecular pathway as an attractive therapeutic target for RTT, and clinical trials treating RTT patients with IGF1 are currently ongoing.

Fragile X Syndrome

FXS is the most common form of inherited mental retardation and the leading identified cause of autism, accounting for around 4% in the autism population. FXS is caused by silencing of the gene *FMR1* that codes for the fragile X mental retardation protein (FMRP), an RNA-binding protein normally produced in response to activation of group-1 mGluRs (Belmonte & Bourgeron, 2006). *Fmr1* mutant mice exhibit acceleration of OD plasticity following 3 days of MD, suggesting excessive plasticity without FMRP (Dolen et al., 2007). This phenotype is reversed with 50% reduction in group-1 mGluR5 in these mice, indicating mGluR5 as a significant contributor for the pathogenesis of FXS (Dolen et al., 2007). In the visual cortex, group-1 mGluR signaling is highest during the critical period (Dudek & Bear, 1989), suggesting that developmental down-regulation of mGluR signaling may be important for normal synaptic and circuit maturation (Dolen & Bear, 2008). It is proposed that both mGluR5 and FMRP regulate protein synthesis but in opposite directions; mGluR5 activation initiates protein synthesis while FMRP suppresses it, and without FMRP, mGluR5 activation leads to excessive protein synthesis that might underlie clinical features of FXS (Dolen & Bear, 2008). Treatments for FXS using antagonists to mGluR5 are also undergoing clinical trials.

CONCLUSION

Significant advances have been made in the past decades toward understanding molecular and cellular mechanisms of OD plasticity, in particular, for revealing feedforward components underlying weakening of deprived-eye responses. However, the list of potential mechanisms that impact feedback components of OD plasticity is growing rapidly as well. The importance of understanding mechanisms behind a simple but very robust form of cortical plasticity extends far beyond just OD plasticity alone, as already demonstrated by the fascinating insights these mechanisms have provided into pathological conditions that afflict normal brain development.

REFERENCES

Amir, R. E., Van den Veyver, I. B., Wan, M., Tran, C. Q., Francke, U., & Zoghbi, H. Y. (1999). Rett syndrome is caused by mutations in X-linked MECP2, encoding methyl-CpG-binding protein 2. *Nature Genetics, 23*, 185–188.

Beaver, C. J., Ji, Q., Fischer, Q. S., & Daw, N. W. (2001). Cyclic AMP-dependent protein kinase mediates ocular dominance shifts in cat visual cortex. *Nature Neuroscience, 4*, 159–163.

Belmonte, M. K., & Bourgeron, T. (2006). Fragile X syndrome and autism at the intersection of genetic and neural networks. *Nature Neuroscience, 9*, 1221–1225.

Cancedda, L., Putignano, E., Impey, S., Maffei, L., Ratto, G. M., & Pizzorusso, T. (2003). Patterned vision causes CRE-mediated gene expression in the visual cortex through PKA and ERK. *Journal of Neuroscience, 23*, 7012–7020.

Castro, J., Kwok, S., Garcia, R., & Sur, M. (2011). Effects of recombinant human IGF1 treatment in a mouse model of Rett syndrome. *Society for Neuroscience Meeting Washington*, 59.17/DD26.

Chahrour, M., & Zoghbi, H. Y. (2007). The story of Rett syndrome: From clinic to neurobiology. *Neuron, 56*, 422–437.

Chen, W. S., & Bear, M. F. (2007). Activity-dependent regulation of NR2B translation contributes to metaplasticity in mouse visual cortex. *Neuropharmacology, 52*, 200–214.

Cohen, D. R., Matarazzo, V., Palmer, A. M., Tu, Y., Jeon, O. H., Pevsner, J., et al. (2003). Expression of MeCP2 in olfactory receptor neurons is developmentally regulated and occurs before synaptogenesis. *Molecular and Cellular Neurosciences, 22*, 417–429.

Crozier, R. A., Wang, Y., Liu, C. H., & Bear, M. F. (2007). Deprivation-induced synaptic depression by distinct mechanisms in different layers of mouse visual cortex. *Proceedings of the National Academy of Sciences of the United States of America, 104*, 1383–1388. doi:10.1073/pnas.0609596104.

Daw, N., Rao, Y., Wang, X. F., Fischer, Q., & Yang, Y. (2004). LTP and LTD vary with layer in rodent visual cortex. *Vision Research, 44*, 3377–3380. doi:10.1016/j.visres.2004.09.004.

Daw, N. W., Reid, S. N., & Beaver, C. J. (1999). Development and function of metabotropic glutamate receptors in cat visual cortex. *Journal of Neurobiology, 41*, 102–107.

Desai, N. S., Cudmore, R. H., Nelson, S. B., & Turrigiano, G. G. (2002). Critical periods for experience-dependent synaptic scaling in visual cortex. *Nature Neuroscience, 5*, 783–789.

Desai, N. S., Rutherford, L. C., & Turrigiano, G. G. (1999). Plasticity in the intrinsic excitability of cortical pyramidal neurons. *Nature Neuroscience, 2*, 515–520.

Di Cristo, G., Berardi, N., Cancedda, L., Pizzorusso, T., Putignano, E., Ratto, G. M., et al. (2001). Requirement of ERK activation for visual cortical plasticity. *Science, 292*, 2337–2340.

Di Cristo, G., Chattopadhyaya, B., Kuhlman, S. J., Fu, Y., Belanger, M. C., Wu, C. Z., et al. (2007). Activity-dependent PSA expression regulates inhibitory maturation and onset of critical period plasticity. *Nature Neuroscience, 10*, 1569–1577.

Dolen, G., & Bear, M. F. (2008). Role for metabotropic glutamate receptor 5 (mGluR5) in the pathogenesis of fragile X syndrome. *Journal of Physiology, 586*, 1503–1508.

Dolen, G., Osterweil, E., Rao, B. S., Smith, G. B., Auerbach, B. D., Chattarji, S., et al. (2007). Correction of fragile X syndrome in mice. *Neuron, 56*, 955–962.

Dudek, S. M., & Bear, M. F. (1989). A biochemical correlate of the critical period for synaptic modification in the visual cortex. *Science, 246*, 673–675.

Fagiolini, M., Fritschy, J. M., Low, K., Mohler, H., Rudolph, U., & Hensch, T. K. (2004). Specific GABAA circuits for visual cortical plasticity. *Science, 303*, 1681–1683.

Flint, A. C., Maisch, U. S., Weishaupt, J. H., Kriegstein, A. R., & Monyer, H. (1997). NR2A subunit expression shortens NMDA receptor synaptic currents in developing neocortex. *Journal of Neuroscience, 17*, 2469–2476.

Frenkel, M. Y., & Bear, M. F. (2004). How monocular deprivation shifts ocular dominance in visual cortex of young mice. *Neuron, 44*, 917–923.

Giacometti, E., Luikenhuis, S., Beard, C., & Jaenisch, R. (2007). Partial rescue of MeCP2 deficiency by postnatal activation of MeCP2. *Proceedings of the National Academy of Sciences of the United States of America, 104*, 1931–1936. doi:10.1073/pnas.0610593104.

Goel, A., & Lee, H. K. (2007). Persistence of experience-induced homeostatic synaptic plasticity through adulthood in superficial layers of mouse visual cortex. *Journal of Neuroscience, 27*, 6692–6700.

Gonchar, Y., Wang, Q., & Burkhalter, A. (2007). Multiple distinct subtypes of GABAergic neurons in mouse visual cortex identified by triple immunostaining. *Frontiers in Neuroanatomy, 1*, 3. doi:10.3389/neuro.05.003.2007.

Gordon, J. A., & Stryker, M. P. (1996). Experience-dependent plasticity of binocular responses in the primary visual cortex of the mouse. *Journal of Neuroscience, 16*, 3274–3286.

Guy, J., Gan, J., Selfridge, J., Cobb, S., & Bird, A. (2007). Reversal of neurological defects in a mouse model of Rett syndrome. *Science, 315*, 1143–1147.

He, H. Y., Hodos, W., & Quinlan, E. M. (2006). Visual deprivation reactivates rapid ocular dominance plasticity in adult visual cortex. *Journal of Neuroscience, 26*, 2951–2955.

Hensch, T. K., Fagiolini, M., Mataga, N., Stryker, M. P., Baekkeskov, S., & Kash, S. F. (1998). Local GABA circuit control of experience-dependent plasticity in developing visual cortex. *Science, 282*, 1504–1508.

Heynen, A. J., Yoon, B. J., Liu, C. H., Chung, H. J., Huganir, R. L., & Bear, M. F. (2003). Molecular mechanism for loss of visual cortical responsiveness following brief monocular deprivation. *Nature Neuroscience, 6*, 854–862.

Huang, Z. J., Kirkwood, A., Pizzorusso, T., Porciatti, V., Morales, B., Bear, M. F., et al. (1999). BDNF regulates the maturation of inhibition and the critical period of plasticity in mouse visual cortex. *Cell, 98*, 739–755.

Hubel, D. H., & Wiesel, T. N. (1970). The period of susceptibility to the physiological effects of unilateral eye closure in kittens. *Journal of Physiology, 206*, 419–436.

Kameyama, K., Lee, H. K., Bear, M. F., & Huganir, R. L. (1998). Involvement of a postsynaptic protein kinase A substrate in the expression of homosynaptic long-term depression. *Neuron, 21*, 1163–1175. doi:10.1016/S0896-6273(00)80633-9.

Kaneko, M., Stellwagen, D., Malenka, R. C., & Stryker, M. P. (2008). Tumor necrosis factor-alpha mediates one component of competitive, experience-dependent plasticity in developing visual cortex. *Neuron, 58*, 673–680.

Klein, M. E., Lioy, D. T., Ma, L., Impey, S., Mandel, G., & Goodman, R. H. (2007). Homeostatic regulation of MeCP2 expression by a CREB-induced microRNA. *Nature Neuroscience, 10*, 1513–1514.

Linden, M. L., Heynen, A. J., Haslinger, R. H., & Bear, M. F. (2009). Thalamic activity that drives visual cortical plasticity. *Nature Neuroscience, 12*, 390–392.

Lyckman, A. W., Horng, S., Leamey, C. A., Tropea, D., Watakabe, A., Van Wart, A., et al. (2008). Gene expression patterns in visual cortex during the critical period: Synaptic stabilization and reversal by visual deprivation. *Proceedings of the National Academy of Sciences of the United States of America, 105*, 9409–9414. doi:10.1073/pnas.0710172105.

Maffei, A., Nelson, S. B., & Turrigiano, G. G. (2004). Selective reconfiguration of layer 4 visual cortical circuitry by visual deprivation. *Nature Neuroscience, 7*, 1353–1359.

Maffei, A., & Turrigiano, G. G. (2008). Multiple modes of network homeostasis in visual cortical layer 2/3. *Journal of Neuroscience, 28*, 4377–4384.

Majewska, A., & Sur, M. (2003). Motility of dendritic spines in visual cortex in vivo: Changes during the critical period and effects of visual deprivation. *Proceedings of the National Academy of Sciences of the United States of America, 100*, 16024–16029. doi:10.1073/pnas.2636949100.

Mataga, N., Mizuguchi, Y., & Hensch, T. K. (2004). Experience-dependent pruning of dendritic spines in visual cortex by tissue plasminogen activator. *Neuron, 44*, 1031–1041.

Mataga, N., Nagai, N., & Hensch, T. K. (2002). Permissive proteolytic activity for visual cortical plasticity. *Proceedings of the National Academy of Sciences of the United States of America, 99*, 7717–7721. doi:10.1073/pnas.102088899.

McCurry, C. L., Shepherd, J. D., Tropea, D., Wang, K. H., Bear, M. F., & Sur, M. (2010). Loss of Arc renders the visual cortex impervious to the effects of sensory experience or deprivation. *Nature Neuroscience, 13*, 450–457.

Mellios, N., Sugihara, H., Castro, J., Banerjee, A., Le, C., Kumar, A., et al. (2011). miR-132, an experience-dependent microRNA, is essential for visual cortex plasticity. *Nature Neuroscience, 14*, 1240–1242.

Mellios, N., & Sur, M. (2012). The emerging role of microRNAs in schizophrenia and autism specrum disorders. *Frontiers in Psychiatry, 3*. doi:10.3389/fpsyt.2012.00039.

Miller, B. H., Zeier, Z., Xi, L., Lanz, T. A., Deng, S., Strathmann, J., et al. (2012). MicroRNA-132 dysregulation in schizophrenia has implications for both neurodevelopment and adult brain function. *Proceedings of the National Academy of Sciences of the United States of America, 109*, 3125–3130. doi:10.1073/pnas.1113793109.

Miyata, S., Komatsu, Y., Yoshimura, Y., Taya, C., & Kitagawa, H. (2012). Persistent cortical plasticity by upregulation of chondroitin 6-sulfation. *Nature Neuroscience, 15*, 414–422.

Morishita, H., Miwa, J. M., Heintz, N., & Hensch, T. K. (2010). Lynx1, a cholinergic brake, limits plasticity in adult visual cortex. *Science, 330*, 1238–1240.

Mower, A. F., Kwok, S., Yu, H., Majewska, A. K., Okamoto, K., Hayashi, Y., et al. (2011). Experience-dependent regulation of CaMKII activity within single visual cortex synapses in vivo. *Proceedings of the National Academy of Sciences of the United States of America, 108*, 21241–21246. doi:10.1073/pnas.1108261109.

Mower, A. F., Liao, D. S., Nestler, E. J., Neve, R. L., & Ramoa, A. S. (2002). cAMP/Ca2+ response element-binding protein function is essential for ocular dominance plasticity. *Journal of Neuroscience, 22*, 2237–2245.

Mrsic-Flogel, T. D., Hofer, S. B., Ohki, K., Reid, R. C., Bonhoeffer, T., & Hubener, M. (2007). Homeostatic regulation of eye-specific responses in visual cortex during ocular dominance plasticity. *Neuron, 54*, 961–972.

Nagakura, I., Van Wart, A., Tropea, D., Crawford, B., & Sur, M. (2012). Novel role of STAT1 immune signaling in cortical plasticity and autism. *Society for Neuroscience Meeting New Orleans*, 443.20/F20.

Oray, S., Majewska, A., & Sur, M. (2004). Dendritic spine dynamics are regulated by monocular deprivation and extracellular matrix degradation. *Neuron, 44*, 1021–1030.

Pham, T. A., Impey, S., Storm, D. R., & Stryker, M. P. (1999). CRE-mediated gene transcription in neocortical neuronal plasticity during the developmental critical period. *Neuron, 22*, 63–72.

Pizzorusso, T., Medini, P., Berardi, N., Chierzi, S., Fawcett, J. W., & Maffei, L. (2002). Reactivation of ocular dominance plasticity in the adult visual cortex. *Science, 298*, 1248–1251.

Quinlan, E. M., Olstein, D. H., & Bear, M. F. (1999). Bidirectional, experience-dependent regulation of N-methyl-D-aspartate receptor subunit composition in the rat visual cortex during postnatal development. *Proceedings of the National Academy of Sciences of the United States of America, 96*, 12876–12880. doi:10.1073/pnas.96.22.12876.

Qureshi, I. A., & Mehler, M. F. (2011). Non-coding RNA networks underlying cognitive disorders across the lifespan. *Proceedings of the National Academy of Sciences of the United States of America, 17*, 337–346.

Rao, Y., & Daw, N. W. (2004). Layer variations of long-term depression in rat visual cortex. *Journal of Neurophysiology, 92*, 2652–2658.

Remenyi, J., Hunter, C. J., Cole, C., Ando, H., Impey, S., Monk, C. E., et al. (2010). Regulation of the miR-212/132 locus by MSK1 and CREB in response to neurotrophins. *The Biochemical Journal, 428*, 281–291.

Shepherd, J. D., & Bear, M. F. (2011). New views of Arc, a master regulator of synaptic plasticity. *Nature Neuroscience, 14*, 279–284.

Smith, G. B., Heynen, A. J., & Bear, M. F. (2009). Bidirectional synaptic mechanisms of ocular dominance plasticity in visual cortex. *Philosophical Transactions of the Royal Society of London. Series B, Biological Sciences, 364*, 357–367.

Stellwagen, D., Beattie, E. C., Seo, J. Y., & Malenka, R. C. (2005). Differential regulation of AMPA receptor and GABA receptor trafficking by tumor necrosis factor-alpha. *Journal of Neuroscience, 25*, 3219–3228.

Stellwagen, D., & Malenka, R. C. (2006). Synaptic scaling mediated by glial TNF-alpha. *Nature, 440*, 1054–1059.

Sugiyama, S., Di Nardo, A. A., Aizawa, S., Matsuo, I., Volovitch, M., Prochiantz, A., et al. (2008). Experience-dependent transfer of Otx2 homeoprotein into the visual cortex activates postnatal plasticity. *Cell, 134*, 508–520.

Tognini, P., Putignano, E., Coatti, A., & Pizzorusso, T. (2011). Experience-dependent expression of miR-132 regulates ocular dominance plasticity. *Nature Neuroscience, 14*, 1237–1239.

Tropea, D., Giacometti, E., Wilson, N. R., Beard, C., McCurry, C., Fu, D. D., et al. (2009a). Partial reversal of Rett syndrome-like symptoms in MeCP2 mutant mice. *Proceedings of the National Academy of Sciences of the United States of America, 106*, 2029–2034. doi:10.1073/pnas.0812394106.

Tropea, D., Kreiman, G., Lyckman, A., Mukherjee, S., Yu, H., Horng, S., et al. (2006). Gene expression changes and molecular pathways mediating activity-dependent plasticity in visual cortex. *Nature Neuroscience, 9*, 660–668.

Tropea, D., Van Wart, A., & Sur, M. (2009b). Molecular mechanisms of experience-dependent plasticity in visual cortex. *Philosophical Transactions of the Royal Society of London. Series B, Biological Sciences, 364*, 341–355.

Turrigiano, G. (2011). Too many cooks? Intrinsic and synaptic homeostatic mechanisms in cortical circuit refinement. *Annual Review of Neuroscience, 34*, 89–103.

Turrigiano, G. G., Leslie, K. R., Desai, N. S., Rutherford, L. C., & Nelson, S. B. (1998). Activity-dependent scaling of quantal amplitude in neocortical neurons. *Nature, 391*, 892–896.

Turrigiano, G. G., & Nelson, S. B. (2004). Homeostatic plasticity in the developing nervous system. *Nature Reviews. Neuroscience, 5*, 97–107.

Vo, N., Klein, M. E., Varlamova, O., Keller, D. M., Yamamoto, T., Goodman, R. H., et al. (2005). A cAMP-response element binding protein-induced microRNA regulates neuronal morphogenesis. *Proceedings of the National Academy of Sciences of the United States of America, 102*, 16426–16431. doi:10.1073/pnas.0508448102.

Wayman, G. A., Davare, M., Ando, H., Fortin, D., Varlamova, O., Cheng, H. Y., et al. (2008). An activity-regulated microRNA controls dendritic plasticity by down-regulating p250GAP. *Proceedings of the National Academy of Sciences of the United States of America, 105*, 9093–9098. doi:10.1073/pnas.0803072105.

Wiesel, T. N., & Hubel, D. H. (1963). Single-cell responses in striate cortex of kittens deprived of vision in one eye. *Journal of Neurophysiology, 26*, 1003–1017.

Yu, H., Majewska, A. K., & Sur, M. (2011). Rapid experience-dependent plasticity of synapse function and structure in ferret visual cortex in vivo. *Proceedings of the National Academy of Sciences of the United States of America, 108*, 21235–21240. doi:10.1073/pnas.1108270109.

96 Role of Glial Cells and Immune Molecules in Visual Development

ALLISON R. BIALAS AND BETH STEVENS

The proper development of the visual system is dependent on communication between remarkably diverse cell types with specialized roles at each developmental stage. Until recently, visual development had been viewed largely through a "neurocentric" lens. However, new research is revealing that glia, the "other" cells of the visual system, actively communicate with neurons and one another to influence nervous system development and brain wiring.

Much of our general knowledge of glial function has come from seminal studies carried out in the visual system. In particular, the mammalian retina and optic nerve have been critical model systems for studying neuron–glia communication during development for many reasons. First, all major glial cell types are represented (see figure 96.1), and their anatomical interactions and biological functions can be studied in vitro and in vivo. Second, the accessibility of the eye is particularly amenable to pharmacological and molecular manipulations of neuron–glia interactions in vivo. Importantly, development of pioneering methods to purify and culture retinal neurons and each of the main glial cell types has been a major advance in the field (B. Barres et al., 1988; Foo et al., 2011; Mi & Barres, 1999; Shi, Marinovich, & Barres, 1998). This approach allows for investigation of specific questions about how neurons and glia influence each other and the mechanisms involved, which are often difficult to address in vivo. Isolation of highly purified visual system neurons and glia has led to the elucidation of the astrocyte, oligodendrocyte, neuron, and Muller glial transcriptomes (Cahoy et al., 2008; Lovatt et al., 2007; Roesch et al., 2008), rich resources that continue to provide important insight into molecular mechanisms governing glial roles in the visual system. Together, these and other studies have revealed a prominent role for glia in wiring the visual system by regulating diverse functions including neuronal survival, axon guidance, myelination, synapse development, and plasticity.

The goal of this chapter is to introduce glia, their functional roles in visual system development, and critical neuron–glia signals involved, including several immune-related signaling pathways that help refine immature synaptic circuits. Here, we will highlight some of the recent discoveries that brought these long-overlooked cells into the spotlight.

GLIA OF THE VISUAL SYSTEM

From the specialized Muller glia of the retina that act as living optic fibers to direct light to the photoreceptors (Franze et al., 2007) to the astrocytes of the visual cortex that help define receptive fields (Schummers, Yu, & Sur, 2008), virtually every level of visual processing is in some way dependent on glia. Unlike neurons, glia are electrically inexcitable cells; however, they still actively communicate with neurons and one another by chemical signals, secreted proteins, and contact-dependent molecular interactions. Before exploring glial function during visual development, it is important to define the types of glia in the visual system, which exhibit remarkable diversity with respect to their origin, morphology, and function.

In general, glia fall into three main classes (see figure 96.2): (1) Astrocytes, which extend processes that intimately associate with synapses, blood vessels, and axons; (2) oligodendrocytes, which ensheath and myelinate axons; and (3) microglia, the resident "immune" cells in the brain that rapidly respond to changes in the extracellular milieu. These general classes and functions of glia have been observed and characterized in many model systems, from flies and worms to vertebrates, suggesting that many glial functions are widely conserved (Freeman et al., 2003; Freeman & Doherty, 2006). Although it is convenient to place glia into morphological groups, it is becoming increasingly clear that the current categories fail to represent the extraordinary diversity and heterogeneity of glial morphology and function. This issue will be discussed in the sections below.

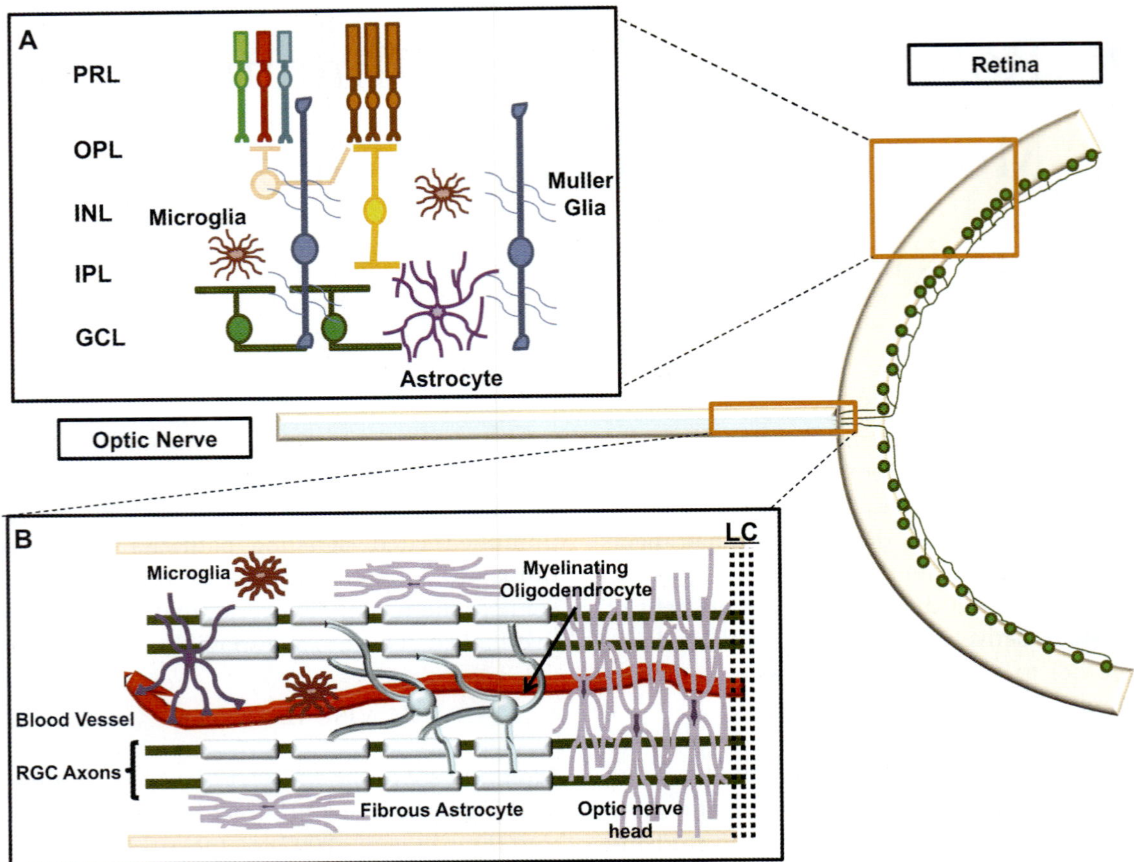

FIGURE 96.1   Glia of the visual system. (A) Distribution of types of glia found in the layers of the retina. PRL, photoreceptor layer; OPL, outer plexiform layer; INL, outer nuclear layer; IPL, inner plexiform layer; GCL, retinal ganglion cell layer. (B) Many glial subtypes can be found in the optic nerve, including specialized optic nerve head astrocytes with unique functions and morphologies. RGC, retinal ganglion cell. LC, lamina cribrosa.

Astrocytes

Astrocytes, named for their star-like appearance, are a numerous and diverse class of glia. In addition to performing "neural support roles," such as buffering the ionic environment and removing excess glutamate via glutamate transporters, astrocytes are active participants at each major stage of neural development. They maintain neuronal survival, contact vasculature to modulate blood flow, stimulate neurite outgrowth, guide axons, and promote the formation and plasticity of synapses.

Astrocytes come in two general classes: fibrous (white matter/type 2) astrocytes and protoplasmic (gray matter) astrocytes. Fibrous astrocytes associate longitudinally with white matter tracts, such as the optic nerve (see figure 96.1B), whereas protoplasmic astrocytes are uniformly distributed throughout gray matter and synaptic regions (see figures 96.1 and 96.2B) (Oberheim, Goldman, & Nedergaard, 2012). Until recently, our understanding of astrocyte morphology was based largely on immunostaining for a protein highly expressed by astrocytes, glial fibrillary acidic protein (GFAP). While this protein is found in most astrocytes in the brain, its localization is hardly representative of astrocytic morphology, as previously assumed. Recent in vivo imaging and three-dimensional reconstructions of astrocytes revealed that, despite their name, both fibrous and protoplasmic astrocytes are not "star-shaped" after all and have complex branching patterns (see figure 96.2B).

Protoplasmic astrocytes are heterogeneous and diverse in their functions, many of which have yet to be discovered (Oberheim, Goldman, & Nedergaard, 2012; Zhang & Barres, 2010). They classically buffer ions and neurotransmitters in the extracellular space, but astrocytes also express receptors for most neurotransmitters and can release a variety of neuroactive, synaptogenic, and trophic factors at synapses. The term "tripartite synapse" has become a well-known phrase to denote the

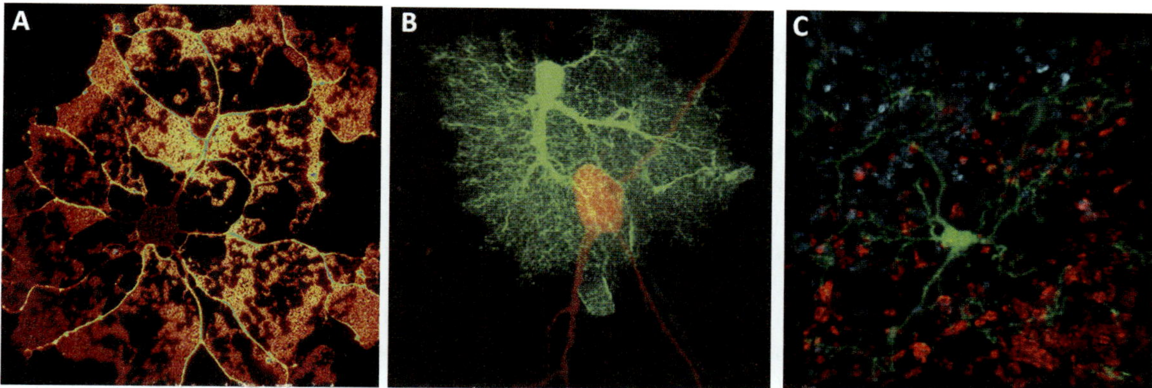

FIGURE 96.2   Main classes of glia. (A) Oligodendrocytes (red) are the myelinating cells of the brain. They produce large sheets of fatty myelin (yellow) that wrap axons. (B) Astrocytes (green) have many diverse functions including ensheathing synapses and interacting with neurons (red) to modulate activity (Bushong et al., 2002). (C) Microglia (green) are the immune cells of the brain and recently have been shown to help prune synapses (red and blue) during synaptic refinement (Schafer et al., 2012).

important contribution of an synapse-ensheathing astrocytic process to synapse function (Haydon, 2001). A single protoplasmic astrocyte can extend thousands of processes that intimately associate with synapses. In fact, it has been estimated that a single astrocyte can associate with an average of four neurons (Halassa et al., 2007) and contact over 100,000 synapses (Bushong et al., 2002)! Moreover, the densely packed processes of adjacent astrocytes are organized into large, nonoverlapping anatomical domains, the functional significance of which is unknown. It has been hypothesized that astrocytic tiling helps connect neurons into microcircuits, especially as astrocytes are tightly coupled via gap junctions (Nagy & Rash, 2000). Astrocytic processes from a single protoplasmic astrocyte can contact blood vessels as well as synapses, placing them in the perfect position to regulate blood brain barrier development and neurovascular coupling (Haydon & Carmignoto, 2006; Rubin & Staddon, 1999).

Fibrous astrocytes can be distinguished by morphology and molecular markers, including ganglioside and GFAP expression, and they are also quite heterogeneous (Raff et al., 1983; Zamanian et al., 2012). Even in the optic nerve, one can observe substantial morphological diversity (see figure 96.1B). Recent imaging studies using GFAP reporter mice have identified that individual optic nerve head (ONH) astrocytes are more elaborate and extensive than most fibrous astrocytes. In fact, the processes of ONH astrocytes span most of the width of the nerve and overlap domains of other astrocytes (see figure 96.1B). The function of these specialized astrocytes has yet to be characterized during development, but they are particularly of interest in glaucoma research, where abnormalities in the ONH are thought to be early signs of pathology (Sun et al., 2010).

Oligodendrocytes

Oligodendrocytes, the myelinating glia of the CNS, are active participants in visual system development (see figures 96.1B and 96.2A). The myelin sheath regulates action potential propagation, axon diameter, as well as the expression and distribution of voltage-gated sodium channels (Rasband & Shrager, 2000; Sánchez et al., 1996; Schafer et al., 2006). Myelinating oligodendrocytes also provide trophic support for axons (Nave, 2010) and are potent regulators of neurite outgrowth and structure (Bandtlow, Zachleder, & Schwab, 1990; McKerracher et al., 1994). Moreover, myelin has been proposed to play a role in limiting regeneration and visual system plasticity (McGee et al., 2005; Schwab, 1990).

During postnatal development, different types of oligodendrocytes associate with axons in discrete waves of myelination throughout the CNS. Myelinating oligodendrocytes are derived from oligodendrocyte precursor cells (OPCs), which are born during late embryonic and early postnatal development. OPCs undergo a precise developmental program driven by environmental and intrinsic cues described later in this chapter that regulate their development into myelinating oligodendrocytes (Dugas et al., 2006). Interestingly, early studies suggest that OPC proliferation may be regulated by neuronal activity (Barres & Raff, 1993), among other molecular cues.

Recent studies also demonstrate that a subset of OPCs expressing the chondroitin sulfate proteoglycan, NG2, have unique characteristics. These "NG2" cells are multipotent and can give rise to neurons, astrocytes, and oligodendrocytes (Nishiyama et al., 2009; Zhu, Bergles, & Nishiyama, 2008), but perhaps their most unique property is the fact that they can receive synaptic input from neurons and may be integrated into the neural circuitry of the brain (Bergles et al., 2000; Lin & Bergles, 2003). These synaptic junctions can be associated with either glutamatergic or GABAergic neurons and exhibit all the characteristics of a typical neuron–neuron synapse. The significance of these synapses is still unknown, but it has been suggested that these connections enable rapid communication between neurons and these glial progenitors to alter the behavior of these cells (Bergles, Jabs, & Steinhäuser, 2010).

Mature myelinating oligodendrocytes are also remarkably heterogeneous. They are generally classified into subtypes based on their branching pattern (del Río-Hortega, 1928), the type of axon they myelinate, and the thickness of myelin formed (Butt et al., 1995). Oligodendrocytes are excluded from the retina; therefore, retinal ganglion cell (RGC) axons within the eye are unmyelinated (Boiko et al., 2001). There are also nonmyelinating oligodendrocytes throughout the CNS, but their role in visual system development is unclear, although the close proximity of these cells to neurons and synapses suggests they may help to regulate the microenvironment around neurons.

Microglia

Microglia are the primary immune and phagocytic cells of the CNS (see figures 96.1 and 96.2C). Microglia are myeloid-derived cells and, until recently, they were thought to develop from peripheral macrophages that entered and colonized the brain throughout development and into adulthood. We now know that most microglia arise from a unique pool of yolk-sac derived myeloid progenitors and migrate into the CNS by embryonic day E10.5 (Ginhoux et al., 2010). Following migration into the CNS, microglia undergo a slow maturation and differentiation process that persists through late postnatal development (Ransohoff & Perry, 2009). Microglia start as highly phagocytic, amoeboid cells and maintain this morphology through the period of programmed cell death when they actively remove debris and dead cells (Ferrer et al., 1990). By the first postnatal week in mouse, these cells develop branches, which are used to continually survey the environment and down-regulate their phagocytic receptors (see figure 96.2C) (Ling & Wong, 1993). In the event of injury or infection, microglia quickly adopt a more classically "reactive" state and migrate to the site of injury where they can shield the injury site, engulf pathogenic material or debris, and release many cytokines and chemokines (Davalos et al., 2005; Nimmerjahn, Kirchhoff, & Helmchen, 2005). While past work had focused on the role of microglia during disease, recent studies have revealed

dynamic interactions between microglia and synapses in the healthy brain, particularly during developmental synaptic pruning (Schafer, Lehrman, & Stevens, 2013; Tremblay et al., 2011).

The preponderance of glia throughout the visual system and their seemingly endless array of functions have led developmental neurobiologists to ask how these diverse cells influence visual system development. Understanding how glia communicate with neurons and each other to wire up the visual system has evolved into a broad and prolific area of research in developmental neurobiology. In the following sections, we will explore some of the major neuron–glia signaling interactions that help guide each major stage of visual system development.

HOW ARE THE DIVERSE NEURONS AND GLIA OF THE VISUAL SYSTEM GENERATED?

One of the earliest roles for glia in the developing brain is the generation of neurons and other glial cells. In fact, it is well established that radial glia in the ventricular zone serve as precursors to neurons and some glia, including astrocytes and oligodendrocytes (Alvarez-Buylla & García-Verdugo, 2002; Doetsch, 2003). Radial glia, as their name suggests, each extend a long process toward the pial surface that guides the outward migration of newly born neurons.

In the retina, specialized radial glia called Muller cells give rise to many of the cells of the retina, including amacrine cells, horizontal cells, photoreceptors, Muller glia, and bipolar cells (Prada et al., 1991) (see figure 96.1A). Unlike radial glia of the ventricular zone which are transient, Muller glia persist throughout life and have unique functions in the retina. Their radial processes extend throughout the layers of the retina and act as "living optic fibers," allowing light to be transmitted to the photoreceptors without distortion (Franze et al., 2007) (see figure 96.1). In addition to their important role in light transmission, Muller glia, a type of specialized astrocyte, perform many of the functions carried out by astrocytes in the rest of the brain, including modulation of blood flow and neuronal activity (de Melo Reis et al., 2008; Newman & Zahs, 1998). Interestingly, recent work has also uncovered neurogenic potential for adult Muller cells from many organisms including humans, suggesting novel therapeutic potential for these cells for eye diseases (Bhairavi et al., 2011; Fischer & Reh, 2001; Ramírez & Lamas, 2009; Takeda et al., 2008).

Where do the rest of visual system glia come from and how are they guided to their appropriate territories? Early in development, pools of glial precursor cells residing in the ventricular zones follow one of several migratory streams to their appropriate location. During gliogenesis, six individual glial precursor cell types have been identified (Lee, Mayer-Proschel, & Rao, 2000); however, rigorous fate-mapping studies following glial development have yet to be performed. Glial precursor cells are derived from the same radial glia that generate neurons. The birth of oligodendrocytes and astrocytes begins as neurogenesis ends; however, astrocytes and oligodendrocytes remain in an immature state in the first few weeks of life.

The arrival and maturation of the various glial subtypes differ significantly. The majority of microglia seem to migrate into the CNS by embryonic day E10.5 (Ginhoux et al., 2010; Saederup et al., 2010); however, they do not reach their mature branched state until the second postnatal week. OPCs are the first glial cells born from radial glia after neurogenesis. OPCs undergo a maturation process that is not fully complete in all areas of the brain until adulthood (see myelination section). Astrocytes are the last cells born in the brain, and they are derived from radial glia. They migrate along the optic nerve and arrive in the optic nerve fiber and RGC layers of the retina, along with the rest of the visual system, toward the end of embryonic and early postnatal development (Watanabe & Raff, 1988). These cells also have an immature phenotype with a distinct gene expression profile including increased expression of intermediate filaments such as nestin and vimentin, some synaptogenic cues, and other neuroactive substances (Cahoy et al., 2008; Christopherson et al., 2005).

As glia populate the nervous system, they profoundly affect neuronal function; however, newborn neurons play an active role in regulating gliogenesis and glial function as well. In the retina, for example, RGC axons secrete Shh and additional growth factors which promote expansion of astrocyte precursors in the embryonic and postnatal optic nerve (Burne & Raff, 1997; Dakubo et al., 2003). Another progenitor cell pool that is modulated by neuronal signaling is the NG2 cell pool, discussed earlier in this chapter. These cells retain the potential to generate neurons and astrocytes into adulthood, and it is postulated that special synaptic connections between neurons and NG2 cells may be a way in which neurons can rapidly trigger proliferation of this population (Nishiyama et al., 2009; Zhu, Bergles, & Nishiyama, 2008).

HOW ARE NEURONS AND AXONS GUIDED TO THEIR PROPER LOCATIONS?

It has long been thought that glia play a key role in neuronal migration and guidance. Indeed, not only do radial glia give rise to neurons but also their processes serve as roadmaps guiding newborn neurons to their appropriate locations. Once neurons populate the visual system, they next must specify axons and dendrites to connect with appropriate targets. Glia play an important role in guiding axons along their path by secreting guidance cues and providing structural support for axons.

Glia-derived cues have been shown to influence axon structure and guidance in vivo. In *Drosophila*, for example, photoreceptor axons are guided along their appropriate path by a stream of glial cells, known as retinal basal glia (RBG), migrating from the optic stalk into the optic disk (Rangarajan, Gong, & Gaul, 1999). When migration of RBG cells into the optic disk is blocked, as in the *gish* loss-of-function mutant, photoreceptor cells extend axons away from instead of into the optic stalk (Hummel et al., 2002). Once photoreceptor cell axons reach their targets, another *Drosophila* glial cell type, the lamina glia, help guide these axons to the appropriate lamina of the optic lobe, reminiscent of classic guidepost cells (Bentley & Caudy, 1983; Poeck et al., 2001; Suh et al., 2002). Interestingly, however, lamina glia initially require axonal cues to migrate to their stereotyped laminar positions. They migrate along optic lobe axons to their correct position. Absent or aberrant projections from optic lobe neurons result in a failure of glial migration (Dearborn & Kunes, 2004).

In the mammalian visual system, glia also are critical for RGC axon guidance. Radial glia localized to a region of the ventral midline known as the "glial palisade" secrete molecular cues which guide RGC axons to form the optic chiasm at this position (Mason & Sretavan, 1997). Glial Ephrin-B2 and EphB1 signaling have been identified as critical guidance cues at the chiasm. During development of RGC ipsilateral projections, ephrin-B2 is expressed by radial glial cells at the midline. Ipsilaterally projecting RGCs, which express EphB1, are repelled by ephrin-B2-producing radial glia of the midline. Consistent with a role for ephrin-B2 and EphB1 signaling in axon guidance at the chiasm, blockade of ephrin-B2 prevents the formation of ipsilateral projections in mice and mice deficient in EphB1 show a dramatic reduction in the number of ipsilateral projections (Williams et al., 2003).

The target neurons for RGCs in the lateral geniculate nucleus (LGN) also appear to be guided into their appropriate positions by glia. In most mammals, the LGN is divided into several cell layers including magnocellular and parvocellular layers that receive input from distinct types of RGCs. The cues that regulate the formation of these layers are not well understood; however, laminations in the distribution of glial proteins such as

GFAP and vimentin were observed prior to the development of neuronal layers, suggesting that a glial-derived cue may direct neurons to their locations (Hutchins & Casagrande, 1988).

HOW DOES MYELIN FORM AND FUNCTION IN THE DEVELOPING VISUAL SYSTEM?

Myelination is one of the most complex cellular interactions in biology. The process requires recognition, association, ensheathment, and elaboration of glial membrane around an appropriate axon at precisely the right time in development. The end result is the formation of compact myelin sheath, which is essential for rapid propagation of nerve impulses along axons.

Much of our understanding of the formation and function of myelin comes from studies carried out in the visual system. In particular, the vertebrate optic nerve (see figure 96.1B) has served as an excellent model system in which the stages of myelination have been extensively studied both in vitro and in vivo. These and other studies have uncovered critical molecular pathways regulating oligodendrocyte maturation and myelination (Dugas et al., 2006, 2010; Emery, 2010a; Pogoda et al., 2006). The process begins with the differentiation of an OPC into a postmitotic, premyelinating oligodendrocyte (see figure 96.3). Once oligodendrocytes mature, environmental cues drive the progression of myelination. The factors promoting myelination, therefore, include both intrinsic oligodendrocyte factors that promote maturation and extrinsic factors from the axon or other cell types (Emery, 2010a,

2010b). A recently developed coculture system that enables rapid myelination of CNS axons offers new opportunities for molecular dissection of multiple stages of myelination (Watkins et al., 2008).

Oligodendrocytes are critically dependent on continual communication with axons during embryonic and early postnatal development. A number of molecular pathways have been identified that regulate myelination either directly or indirectly. For example, Wnt/β-catenin signaling is critical to drive terminal oligodendrocyte differentiation. Wnt signaling is specifically activated as terminal oligodendrocyte differentiation begins, but in mature oligodendrocytes, down-regulation of Wnt signaling appears to be necessary for myelination as mutations preventing Wnt down-regulation lead to myelination defects (Fancy et al., 2009; Fu et al., 2009). Conversely, cues such as Jagged (Wang et al., 1998), polysialyated-neural cell adhesion molecule (PSA-NCAM) (Grinspan & Franceschini, 1995), and LINGO-1 (Mi et al., 2005), all inhibit either OPC differentiation or myelination.

In addition to these molecular cues, neural activity is thought to regulate myelination. Early studies in the optic nerve demonstrated that mice reared in the dark developed fewer myelinated axons compared with control mice (Gyllensten & Malmfors, 1963). Hypomyelination was also observed in optic nerve of the naturally blind cape-mole rat (Omlin, 1997) whereas prematurely opening the eyes in rabbit accelerated myelination (Tauber, Waehneldt, & Neuhoff, 1980). A number of recent in vitro studies have provided additional information on activity-dependent myelination.

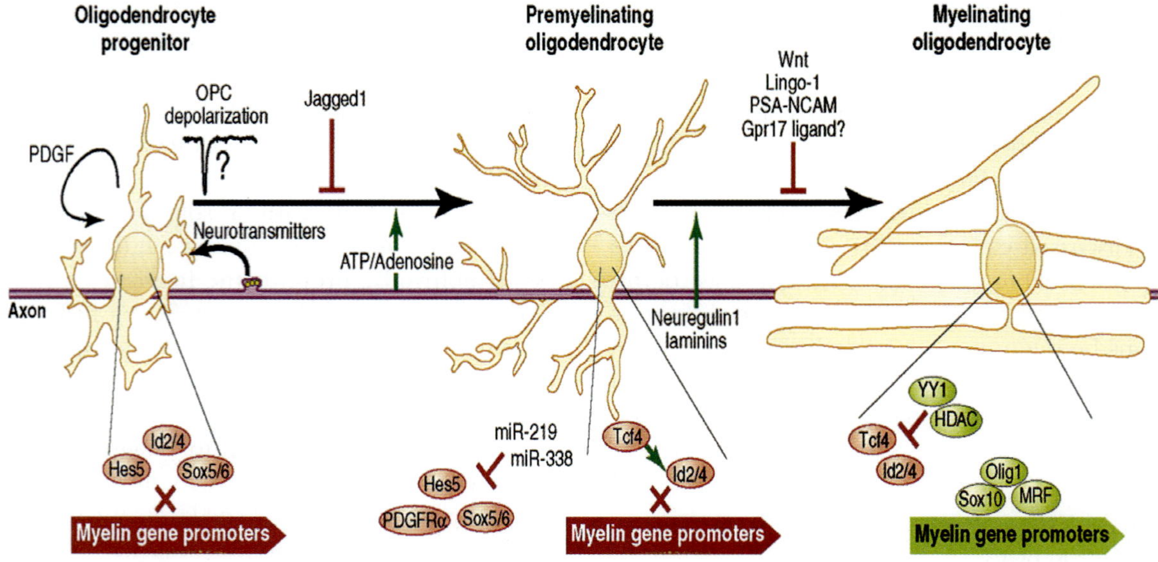

FIGURE 96.3   The stages of oligodendrocyte maturation and myelination. Critical intrinsic and extrinsic factors regulate the complex interactions that guide an immature oligodendrocyte precursor cell (OPC) to a mature, myelinating oligodendrocyte. (From Emery, 2010a.)

Activity-dependent regulation of axon-derived cues is one proposed mechanism (Itoh et al., 1995; Makinodan et al., 2012; Wake, Lee, & Fields, 2011). For example, axonal release of ATP and adenosine is thought to stimulate purinergic receptors on OPCs promoting oligodendrocyte maturation and myelination (Stevens et al., 2002). More recent experiments suggest that astrocytes influence activity-dependent regulation of myelination via ATP-dependent production of leukemia inhibitory factor (LIF). LIF could then promote OPC maturation and myelination (Ishibashi et al., 2006, 2009). The precise molecular mechanisms underlying activity-dependent effects on myelination remain an active area of investigation (Emery, 2010a).

What are the intrinsic signals controlling myelination? The oligodendrocyte transcriptome has revealed many highly expressed, oligodendrocyte-specific molecules whose roles are mostly unknown (Cahoy et al., 2008; Nielsen et al., 2006). One such key factor is myelin regulatory factor (MRF). MRF is a nuclear protein developmentally expressed specifically in postmitotic oligodendrocytes. RNA interference knockdown of MRF in oligodendrocytes suppresses expression of CNS myelin genes while overexpression of MRF promotes myelin gene expression. Mice in which MRF has been knocked out specifically in oligodendrocytes fail to generate myelin, display severe neurological abnormalities, and die in the first few postnatal weeks (Emery et al., 2009). Moreover, oligodendrocytes in these mice are arrested in the premyelinating stage, suggesting that MRF is a master switch for myelination. Several other transcription factors have been identified that either promote or inhibit maturation of OPCs (Dugas et al., 2010). The transcription factor Olig2 is initially responsible for specifying the oligodendrocyte lineage (Lu et al., 2002; Zhou, Choi, & Anderson, 2001). Downstream transcription factors including Olig1, Ascl1, Nkx2.2, Sox10, YY1, and Tcf4 as well as MRF then promote maturation of OPCs into myelinating oligodendrocytes while cells expressing Id2, Id4, Hes5, and Sox6 maintain an OPC phenotype (Wegner, 2008) (see figure 96.3).

Other Functions of Myelin

In addition to increasing the efficiency and speed of action potentials, myelinating oligodendrocytes wear many other hats. They appear to regulate axon structure, neurite outgrowth, and the expression and localization of ion channels along axons (Sánchez et al., 1996; Starr et al., 1996). For example, recent work has shown that contact between RGCs and myelinating oligodendrocytes enhances axon function by altering voltage-gated sodium channel (Na_v) expression and distribution. In mature RGCs, the voltage-gated sodium channel, $Na_v1.2$, exclusively localizes to unmyelinated portions of RGC axons within the eye while a different type, $Na_v1.6$, is localized to nodes of Ranvier along the myelinated axon. Interestingly, during development, $Na_v1.2$ is first expressed throughout the axon, including at nodes of Ranvier. As myelination progresses, however, $Na_v1.2$ channels are replaced at mature nodes of Ranvier with $Na_v1.6$ channels, suggesting that myelin-derived cues guide channel expression and localization (Boiko et al., 2001).

Myelin is also a potent inhibitor of neurite outgrowth. As myelination proceeds in waves in the CNS, regions that become myelinated early may act as repulsive guidance cues for CNS tracts that form later in development (Schwab & Schnell, 1991). Candidate myelin-derived proteins that inhibit axon growth include Nogo, NI-35, NI-250, and myelin-associated glycoprotein (MAG) (Caroni & Schwab, 1988a, 1988b; McGee et al., 2005; McKerracher et al., 1994), and these molecules have been extensively studied in vitro and in vivo. Interestingly, peripheral nervous system myelin does not inhibit regeneration or growth, suggesting that unique components of CNS myelin underlie growth inhibition (Mukhopadhyay et al., 1994). Understanding the mechanisms of growth inhibition in the CNS is an area of active research in the regeneration field and may also be relevant to critical period plasticity, which we will discuss.

WHAT DO ASTROCYTES DO AT DEVELOPING SYNAPSES?

Establishment of the correct numbers and types of synapses is crucial for proper visual system development and function. We know a great deal about the neuronal proteins required for the assembly and maturation of CNS synapses; however, pioneering studies in the mammalian visual system have identified astrocytes as critical mediators of CNS synaptogenesis through several secreted signals. The spatiotemporal correlation between the appearance of immature astrocytes at CNS synapses and the onset of synaptogenesis suggested that astrocytes might provide important synaptogenic cues. For example, in the rodent visual system, RGC axons innervate the superior colliculus by birth (Lund, 1972). However, there is a 1-week delay before the majority of synapses are formed. This delay coincides with the birth and proliferation of astrocytes.

Do astrocytes promote synaptogenesis? The establishment of methods to purify and culture rodent RGCs allowed Pfrieger and Barres to ask whether synapses could form in the absence of astrocytes (Pfrieger & Barres, 1997). Purified RGCs were healthy, elaborated dendrites and axons but formed few synapses when cultured without astrocytes. In contrast, addition of a feeding layer of astrocytes or astrocyte-conditioned medium significantly increased RGC synaptic activity and the number of structural synapses that formed (Pfrieger & Barres, 1997; Ullian et al., 2001). These and other findings indicated that secreted factors from astrocytes strongly enhance pre- and postsynaptic function in developing RGCs (Ullian et al., 2001) and prompted further exploration into the specific molecules that control this process.

What Are the Astrocyte-Secreted Synaptogenic Factors?

One class of astrocyte-secreted molecules that was recently identified are thrombospondins (TSPs) (Christopherson et al., 2005), a large multimeric family of extracellular matrix proteins (see table 96.1). In vivo, TSP-1 and -2 localize to astrocytic processes and synapses during development. Using purified RGC cultures, it was later found that astrocyte-derived TSPs are necessary and sufficient for the assembly of structural synapses. The synapses induced by TSPs are ultrastructurally normal, presynaptically active, but postsynaptically silent. Mice deficient for both TSP-1 and -2 exhibited fewer synaptic puncta by immunostaining, demonstrating the importance of this molecule in normal development in vivo. Further investigation revealed that a calcium channel subunit, alpha2delta-1, is the receptor for TSP required for synaptogenesis (Eroglu et al., 2009). The molecular mechanism by which TSP-alpha2delta-1 interaction mediates synaptogenesis has yet to be fully elucidated. Interestingly, alpha2delta-1 is also the receptor for an anti-epileptic drug, gabapentin. Treating cultured RGCs or mice with gabapentin during synaptogenesis significantly reduced the number of synapses formed, identifying gabapentin as a potent inhibitor of synaptogenesis (Eroglu et al., 2009).

Are There Other Astrocyte-Derived Synaptogenic Signals?

Recent work has identified additional astrocyte-secreted proteins, hevin and SPARC, that regulate synaptogenesis (Kucukdereli et al., 2011). Both of these proteins are enriched in the superior colliculus coinciding with robust synaptogenesis during the postnatal period. Hevin, like TSP, promotes the formation of ultrastructurally normal synapses that are postsynaptically silent. Conversely, SPARC inhibits synaptogenesis by specifically antagonizing hevin. Although SPARC does

TABLE 96.1

Astrocyte-derived molecules regulate synaptogenesis

Molecule	Effect on Synapses	Citation
Thrombospondins	+ Structural synapses + Presynaptic function	Christopherson et al. (2005)
Hevin	+ Structural synapses + Presynaptic function	Kucukdereli et al. (2011)
SPARC	− Structural synapses	Kucukdereli et al. (2011)
ApoE/cholesterol	+ Synapse number and efficacy	Mauch et al. (2001)
Glypican	+ Structural and functional synapses +Pre- and postsynaptic function	Allen et al. (2012)

not directly interact with hevin, it appears to compete with hevin for a common binding partner. Consistent with in vitro results, mice deficient in SPARC show an increase in synapse number in the superior colliculus while hevin-deficient mice show a reduction (see table 96.1). The discovery of these molecules and their interaction demonstrates that astrocytes can both positively and negatively regulate synaptogenesis to influence synapse number.

What is the elusive factor that converts silent synapses into fully functional connections? Recent work has identified glypicans as a novel class of astrocyte-derived proteins that regulate the maturation of postsynaptic machinery, specifically by inserting GluA1 subunits into the postsynaptic membrane (Allen et al., 2012). Glypican-4 and -6 (see table 96.1) were identified as molecules sufficient to induce fully functional synapses in RGC cultures and in vivo (Allen et al., 2012). Glypicans are the first astrocyte-derived molecules to be identified that can induce fully functional synapses, raising many questions about the underlying mechanisms.

Since the initial finding that astrocytes regulate synaptogenesis, the list of astrocyte-derived synaptogenic factors continues to grow (see table 96.1). For example, cholesterol complexed with apoE enhances presynaptic function in RGC cultures by regulating presynaptic release machinery and potentially enhancing dendritic maturation (Goritz, Mauch, & Pfrieger, 2005; Mauch et al., 2001). Understanding how these astrocyte-derived factors cooperate to regulate synapse formation is an area of active investigation. One hypothesis is that astrocytes regulate distinct phases of synaptogenesis (i.e. initial adhesion vs. maturation) via different factors and signaling pathways. Alternatively, many of these signals could act in a complex to initiate a common downstream pathway required to assemble functional CNS synapses.

HOW IS CONNECTIVITY REFINED?

Glia and Immune Molecules Sculpt Developing Circuits in the Visual System

The circuitry established during synaptogenesis is initially imprecise and includes many weak synapses which compete with each other for postsynaptic territory. This activity-dependent competition leads to synaptic refinement—the selective pruning of inappropriate synapses and strengthening of appropriate synaptic connections (reviewed in Hua & Smith, 2004; Huberman, Feller, & Chapman, 2008; Katz & Shatz, 1996). While the role of neuronal activity in developmental synaptic pruning is well established (Hua & Smith, 2004; Katz & Shatz, 1996; Sanes & Lichtman, 1999) (and covered in more detail in other chapters), the molecular mechanisms linking neuronal activity with synaptic pruning are less clear. Recent studies implicate glia and immune molecules in sculpting synaptic circuits in the visual system.

The previous section discussed a role for astrocytes in forming new synapses. Could glia also influence synapse elimination? A screen to determine how astrocytes influence neuronal gene expression first identified the innate immune protein, C1q, as one of the few genes that was highly up-regulated in purified retinal ganglion neurons (RGCs) in response to astrocyte-derived secreted factors (Stevens et al., 2007). This was a surprising finding since C1q was not thought to be expressed by neurons in the healthy brain and microglia were presumed to be the only source of C1q in the brain. Perhaps even more unexpected was the finding that C1q and downstream complement protein C3 localize to subsets of synapses throughout the postnatal brain and retina (Stevens et al., 2007). In the immune system, C1q is the initiating protein in the classical complement cascade. The complement cascade opsonizes or "tags" pathogenic microbes and cellular debris for rapid elimination by phagocytic macrophages or complement-mediated cell lysis. Could these classic immune molecules be similarly opsonizing or tagging immature synapses for elimination?

This idea was tested in the mouse retinogeniculate system—a classical model for studying activity-dependent developmental synapse elimination and refinement (reviewed in (Guido, 2008; Hong & Chen, 2011; Huberman, 2007; Sretavan & Shatz, 1986). Early in development, RGCs form transient synaptic connections with relay neurons in the dorsal LGN (dLGN) of the thalamus (see figure 96.4). During the first 2 postnatal weeks, many of these transient retinogeniculate synapses are permanently eliminated (Campbell & Shatz, 1992; Hooks & Chen, 2006; Sretavan & Shatz, 1984). Consistent with a role for the classical complement cascade in synaptic pruning, mice deficient in C1q and downstream C3 exhibit sustained defects in synapse elimination, as shown by the failure to segregate into eye-specific territories and the retention of multi-innervated LGN relay neurons (Stevens et al., 2007) (see figure 96.4). These mice still undergo a substantial degree of synaptic refinement (Stevens et al., 2007), suggesting that complement proteins cooperate with other pathways and neuronal activity to refine developing retinogeniculate circuits. Together, these findings raise many questions regarding the underlying cellular and molecular mechanisms. For example, if complement is tagging synapses, how then are complement-tagged synapses eliminated?

A NOVEL ROLE FOR MICROGLIA IN DEVELOPMENTAL SYNAPTIC PRUNING

Emerging evidence implicates microglia as key players in developmental synaptic pruning (Ransohoff & Stevens, 2011; Schafer et al., 2012; Schafer, Lehrman, & Stevens, 2013). As discussed earlier, process-bearing phagocytic microglia have been observed in the dLGN and several other postnatal brain regions including hippocampus, cerebellum, and olfactory bulb (Dalmau et al., 1998; Perry, Hume, & Gordon, 1985), but until recently, the function of microglia in normal brain had remained a mystery. Microglia express an array of phagocytic receptors, including complement receptor 3 (CR3/CD11b/CD18) that could mediate the engulfment of synaptic elements during development. In the brain, microglia are the only resident cells expressing CR3 (Bobak et al., 1987; Graeber, 2010; Guillemin & Brew, 2004; Ransohoff & Perry, 2009). In the immune system, the activated C3 fragment C3b (iC3b) opsonizes the surface of cells/debris and "tags" them for elimination by phagocytic macrophages that express C3 receptors (CR3/CD11b) (Carroll, 2004; Gasque, 2004; van Lookeren Campagne, Wiesmann, & Brown, 2007).

Using the mouse retinogeniculate system as a model, microglia were found to engulf RGC presynaptic inputs during a peak pruning period in the developing dLGN (Schafer et al., 2012). Moreover, genetic or pharmacological disruptions in microglia-mediated engulfment during the postnatal period resulted in sustained

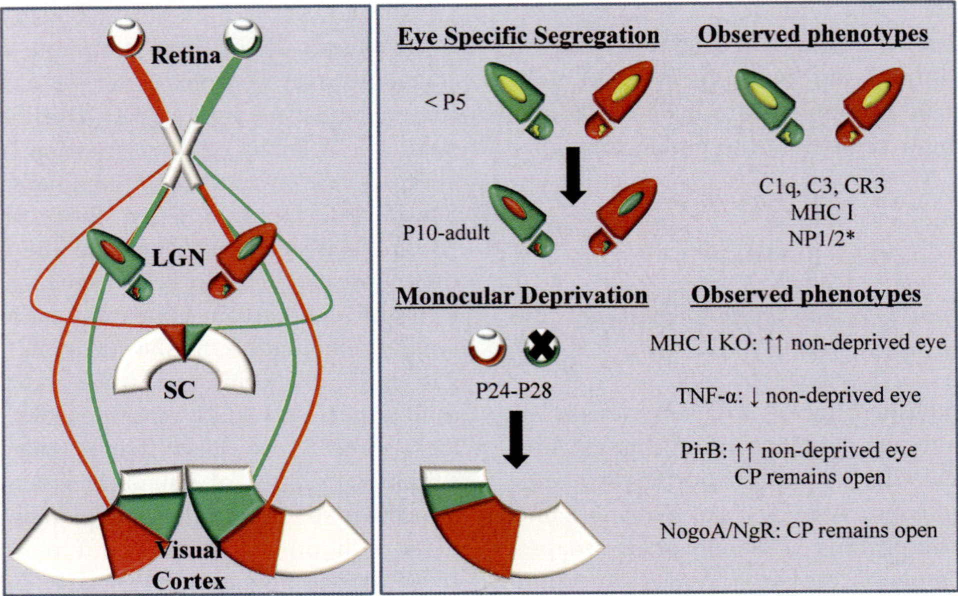

FIGURE 96.4   Immune molecules regulate refinement and plasticity of the visual system. Immune system molecules have been implicated in synaptic refinement of the retinogeniculate system. When NP1/2, complement (C1q or C3), or major histocompatibility complex (MHC) I signaling are disrupted, eye specific territories fail to form. Ocular dominance plasticity can be assessed by monocular deprivation during the critical period for ocular dominance (~postnatal day 24–28) and seeing a shift in cortical responses toward the nondeprived eye. Mice deficient in MHC I, tumor necrosis factor alpha (TNF-alpha), PirB, or Nogo show various phenotypes in ocular dominance plasticity, showing that immune molecules regulate multiple levels of this process. *The phenotype in NP1/2 knockout (KO) mice is transient. LGN, lateral geniculate nucleus; SC, superior colliculus; CP, critical period.

functional deficits in eye-specific segregation. Furthermore, microglia-mediated engulfment of synaptic inputs was dependent upon signaling between CR3, expressed specifically by microglia, and complement component C3, which is highly expressed in the postnatal dLGN (see figure 96.4) (Schafer et al., 2012). Interestingly, microglia-mediated engulfment was found to be regulated by neuronal activity. When competition between inputs from the two eyes was enhanced, microglia preferentially engulfed inputs from the eye with reduced neuronal activity. Although it is not yet known whether or how microglia target specific "weaker" synapses, these data are consistent with previous work demonstrating a decreased synaptic territory of the "weaker" inputs and increased territory of "stronger" inputs within the dLGN (see figure 96.5) (Cook, Prusky, & Ramoa, 1999; Del Rio & Feller, 2006; Huberman, Feller, & Chapman, 2008; Penn et al., 1998; Shatz, 1990; Shatz & Stryker, 1988; Stellwagen & Shatz, 2002).

Recent studies also suggest that microglia associate with postsynaptic elements during synaptic remodeling in the hippocampus and juvenile visual cortex, raising the question of whether microglia-dependent pruning is a global mechanism of synaptic remodeling in the CNS (Paolicelli et al., 2011; Tremblay, Lowery, & Majewska, 2010). Paolicelli et al. (2011) demonstrated a role for the fractalkine receptor (CX3CR1), expressed on the surface of microglia, in hippocampal synapse development and maturation. $Cx3cr1^{KO}$ mice have a transient reduction in the number of microglia in the postnatal brain; thus, fractalkine signaling could interact with complement and other signals to regulate microglia-mediated developmental pruning by influencing microglia number or, possibly, recruitment to synaptic sites in the postnatal brain (Ransohoff & Stevens, 2011). Together these new findings raise several fundamental questions related to the underlying mechanisms of microglia- and complement-mediated pruning including how neural activity, complement, and microglia may interact to sculpt developing visual circuits.

Microglia may not be the only cells in the CNS phagocytosing material in the healthy, normal brain—astrocytes also show evidence of phagocytic capabilities (Al-Ali & Al-Hussain, 1996; Bechmann & Nitsch, 1997; Cahoy et al., 2008). Although it is not yet clear whether astrocyte refine circuitry like microglia, a specialized class of astrocytes in the ONH myelin transition zone (MTZ) has been shown to normally express a common phagocytic marker, the galactose-specific lectin Mac-2 (also known as Lgals3 or galectin-3) (Nguyen et al., 2011; Sun et al., 2009) (see figure 96.1). In normal, healthy mice, large inclusions of axonal material have been observed in these astrocytes, suggesting that MTZ astrocytes may phagocytose axonal components as a

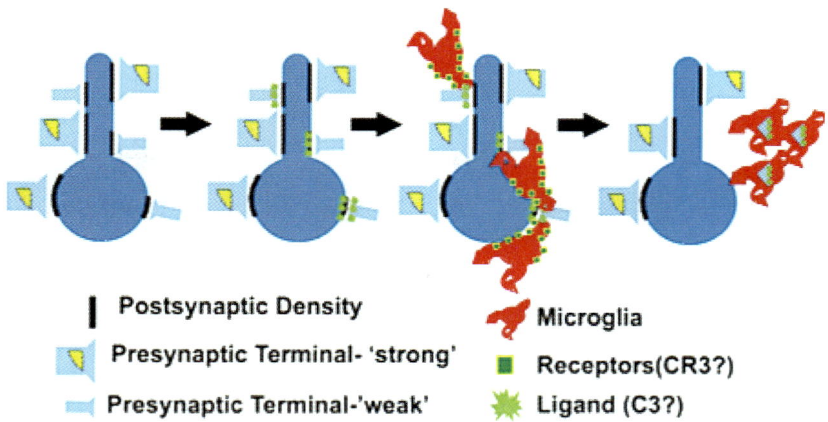

FIGURE 96.5   Model for microglia- and complement-dependent synapse elimination. Recent work demonstrated that microglia engulf less active inputs. The model for complement-dependent synapse elimination suggests that complement (C1q or C3) specifically tags these synapses for elimination by microglial engulfment via complement receptor 3(CR3)–C3 signaling (Schafer et al., 2012).

form of axonal maintenance in the normal animal (Nguyen et al., 2011).

Other Immune Molecules

Complement is among several immune-related molecules that have been identified as mediators of synaptic refinement and plasticity in the visual system (reviewed in Boulanger, 2009; Shatz, 2009). These include neuronal pentraxins (e.g., NP1/2, NARP) and components of the adaptive immune system (e.g., class I major histocompatibility complex [MHC I] family of proteins and receptors) (Bjartmar et al., 2006; Corriveau, Huh, & Shatz, 1998; Datwani et al., 2009; Huh et al., 2000). It is not clear at this point if these pathways interact with glia and/or complement or represent parallel pathways for refinement.

Class I MHC molecules were the first to be identified as mediators of synaptic refinement. These molecules were initially shown to be expressed in an activity-dependent manner in the LGN. Infusion of tetrodotoxin to block all action potential activity in the LGN significantly reduced expression of class I MHC molecules. Furthermore, increasing neuronal firing with kainic acid induced seizure resulted in dramatic increases in class I MHC expression (Corriveau, Huh, & Shatz, 1998). The activity-dependent regulation of this gene suggested a potential role for the class I MHC signaling pathway in synaptic refinement. The role for class I MHC in activity-driven structural remodeling and synaptic plasticity was later demonstrated by examining three mouse strains deficient in class I MHC signaling: beta 2–microglobulin knockout mice, CD3 zeta-deficient mice, and a strain deficient in TAP1 and beta 2–microglobulin (Goddard, Butts, & Shatz, 2007; Huh

et al., 2000). All three mouse mutants showed similar defects in synaptic refinement in the retinogeniculate system, assayed by neuroanatomical tracing of eye specific territories (see figure 96.4). Mice deficient in class I MHC also showed enhanced long-term potentiation (LTP) and absent long-term depression in the adult hippocampus (Huh et al., 2000), implicating MHCs in plasticity in other brain regions as well. In the hippocampus, class I MHC was shown to modulate plasticity by inhibiting NMDA receptor-regulated alpha-amino-3-hydroxy-5-methyl-4-isoxazole propionic acid (AMPA) receptor (AMPAR) trafficking (Fourgeaud et al., 2010), but it is not clear if this same mechanism plays a role in visual system plasticity.

Understanding the mechanism by which class I MHC signaling promotes synaptic refinement and plasticity are important questions and areas of active investigation; however, recent work demonstrated that MHC I affects even earlier stages of development. MHC I is present at synapses in visual cortex during the peak of synaptogenesis and works as a negative regulator of excitatory and inhibitory synaptogenesis. In vitro, knocking down MHC I signaling increased excitatory and inhibitory synapse density, while beta 2-microglobulin deficient mice showed an increase in synapse density throughout the brain in vivo (Glynn et al., 2011). The strength of excitatory and inhibitory synapses was also modulated by MHC I, suggesting that this molecule has diverse and important functions throughout development. Interestingly, recent work demonstrated a colocalization of C1q and MHC I proteins at RGC synapses in the postnatal LGN (Datwani et al., 2009), hinting that these proteins may interact during development. Recent work also has revealed enhanced ocular dominance plasticity in mice class I MHC signaling supporting a role for this pathway in experience-dependent plasticity, as discussed in the following section (Syken et al., 2006).

Neuronal pentraxins (NP1 and 2) were found to play a role in retinogeniculate refinement. Neuronal pentraxins are synaptic proteins with homology to pentraxins of the peripheral immune system, which are traditionally involved in opsonization and phagocytosis of dead cells in the immune system (Nauta et al., 2003). Mice deficient in neuronal pentraxins, NP1 and NP2, as well as the receptor, NPR, have transient defects in eye-specific segregation in the dLGN (see figure 96.4) (Bjartmar et al., 2006). These defects were not maintained into adulthood, and the eye-specific retinogeniculate inputs became segregated by postnatal day 30 (P30). A similar age-dependent phenotype results when spontaneous activity in the retina is blocked from P1–P10 and then allowed to recover. When retinal activity was measured in mice deficient in neuronal pentraxins, levels of RGC spiking activity were dramatically increased, which may explain the phenotype in eye specific segregation (Bjartmar et al., 2006). Interestingly, PTX3, a long pentraxin, which has homology to neuronal pentraxins, can enhance microglial phagocytic activity (Jeon et al., 2010). In addition, neuronal pentraxins are significantly homologous to short pentraxins such as C-reactive protein, which is a well-described binding partner of C1q. Thus, neuronal pentraxins could potentially serve as synaptic binding partners for C1q during synapse development.

Neuronal pentraxins also have other functions in the developing visual system. Neuronal pentraxins are required for normal acetylcholine-mediated retinal wave activity during development. NP1 can also interact with another neuronal pentraxin, Narp, to influence synaptogenesis and synaptic plasticity. Narp was identified as an activity-dependent immediate early gene that can promote synaptogenesis and the clustering of AMPARs in other systems (O'Brien et al., 1999; Tsui et al., 1996). When complexed with NP1, Narp has an enhanced synaptogenic effect, mediating both activity-dependent and activity-independent clustering of AMPARs (Bjartmar et al., 2006; Xu et al., 2003). These interesting acute phase proteins seem to have critical roles in both synaptogenesis and synaptic refinement.

ROLE OF GLIA AND IMMUNE MOLECULES IN VISUAL EXPERIENCE-DEPENDENT PLASTICITY

After eye opening, sensory experience helps to shape developing visual circuitry and tune visual responses. As the visual cortex matures, it develops exquisite organization, and visual receptive fields, along with ocular dominance and orientation selectivity columns, are mapped to the cortex (Hensch, 2004; Sur & Rubenstein, 2005). Recent work has demonstrated that astrocytes not only develop finely tuned responses to visual stimuli that mimic the mapping of neuronal responses but also influence the magnitude and duration of adjacent visually driven neuronal responses via glutamate transporters (Schummers, Yu, & Sur, 2008). The effects of sensory experience on glia and the role of glial cells in opening and closing the critical periods for visual plasticity have yet to be fully explored, although there have been clues that astrocytes, oligodendrocytes, and microglia influence cortical plasticity.

One of the first clues that astrocytes might play an instructive role in ocular dominance plasticity came from a study in the late 1980s by Müller and Best. They showed that introducing kitten astrocytes into the mature cat visual cortex could reopen the critical period for ocular dominance (Müller & Best, 1989). Dark-rearing kittens, which delays maturation of the visual cortex, has also been shown to delay astrocyte maturation in visual cortex, linking neuronal activity with astrocyte maturation in this area (Müller, 1990). The neuron–astrocyte signaling pathways required for critical period plasticity have yet to be identified, however, and are an open area of active research.

Microglia have also been recently implicated in experience-dependent plasticity. Recent in vivo imaging experiments have shown that microglia processes are highly dynamic and often associate with dendritic spines during the critical period for ocular dominance (Tremblay, Lowery, & Majewska, 2010; Tremblay et al., 2011; Wake et al., 2009). In experiments in which retinal activity was reduced via tetrodotoxin or enucleation of both eyes, microglia retracted their processes away from spines and showed reduced motility overall (Wake et al., 2009). Microglia were also shown to participate in the experience-dependent structural remodeling that occurs after dark adapting mice for 1 week and exposing them to light. Dark exposure decreases synaptic strength, and microglia processes were often found associated with subsets of spines. The presence of phagocytic compartments in microglial processes also increased with dark adapting. Upon exposure to light, microglia remained phagocytic, however were less often associated with spines (Tremblay, Lowery, & Majewska, 2010). These experience-dependent changes in microglial dynamics support a role for microglia in experience-dependent structural remodeling.

Myelinating oligodendrocytes, as well as specific myelin proteins, have also been implicated in limiting experience-dependent plasticity in the adult visual cortex.

　　ALLISON R. BIALAS AND BETH STEVENS

Myelin-related proteins such as Nogo, MAG, and OMgp have been extensively studied as potent inhibitors of axonal growth and regeneration (Filbin, 2003). Nogo, which is expressed on the surface of oligodendrocytes, interacts with the Nogo receptor (NgR) expressed on axons to inhibit axonal growth. Interestingly, these proteins also seem to limit cortical plasticity (GrandPré, Li, & Strittmatter, 2002). When myelination is disrupted as in the NgR−/− mouse, mice enter the critical period normally; however, experience-dependent plasticity persists into adulthood such that even at P60 or beyond, monocular deprivation can trigger an increase in cortical responses to the nondeprived eye (see figure 96.4) (McGee et al., 2005). Myelin-derived Nogo, MAG, and OMgp, which have a broad distribution throughout the layers of visual cortex in adult versus juvenile mice, are thought to consolidate the neural circuitry established during experience-dependent plasticity. In addition, recent work has demonstrated that social experience regulates oligodendrocyte maturation and myelination in prefrontal cortex during a temporally restricted "critical period" (Makinodan et al., 2012). It will be interesting to see if experience-dependent changes in oligodendrocytes drive myelination during critical periods for visual development as well.

A critical role for immune molecules has also been discovered in ocular dominance plasticity. In this system, monocular visual deprivation results in the weakening and pruning of inputs projecting from regions innervated by the deprived eye and an expansion of inputs from the nondeprived eye (figure 96.4). Class I MHC molecules as well as proteins that interact with MHC I, such as paired-immunoglobulin–like receptor B (PirB), have been implicated in limiting ocular dominance plasticity. In mice that lack the transmembrane portion of PirB, thus preventing PirB signaling, ocular dominance plasticity, particularly the expansion of nondeprived inputs, is more robust and can be induced at any age (Syken et al., 2006). Interestingly, tumor necrosis factor alpha (TNF-alpha), a cytokine produced by astrocyte cultures and microglia, is also required for the strengthening of nondeprived inputs. Mice deficient in TNF-alpha exhibit the expected loss of deprived-eye responses (figure 96.4), but the increase in cortical response to stimulation of the open eye is absent (Kaneko et al., 2008). This result supports the hypothesis that the weakening and strengthening of inputs in response to monocular deprivation are two distinct processes.

The mechanism by which TNF-alpha is acting in ocular dominance plasticity has not been fully elucidated. Beattie and colleagues showed that the pro-inflammatory cytokine TNF-alpha enhanced synaptic efficacy by increasing surface expression of AMPARs (Beattie et al., 2002), which could explain the failure of synaptic strengthening during ocular dominance plasticity. In addition, TNF-alpha recently has been implicated in homeostatic synaptic scaling, a form of synaptic plasticity that involves uniform adjustments in the strength of all synapses on a cell in response to prolonged changes in the cell's synaptic activity such as blockade of synaptic function. Stellwagen and Malenka showed that TNF-alpha mediates synaptic scaling in response to prolonged activity blockade (Stellwagen & Malenka, 2006). Another possibility is that TNF-alpha affects plasticity as a result of its ability to induce MHC I expression (Israel et al., 1989). In fact, many proinflammatory cytokines including interleukin (IL-) 6, IL-1 beta, IL-2, IL-18, IL-8, and interferon-alpha and -beta all can inhibit hippocampal LTP, suggesting convergence on some common pathway (Boulanger, 2009). A major missing piece of the puzzle, however, is the identity of the molecules that scale down activity in the projections from the deprived eye.

DISCUSSION

This chapter has highlighted the diversity of glial morphology and function in the visual system. The normal progression of visual development, originally viewed through a "neurocentric" lens, is now understood to be dramatically influenced by the diverse types of visual system glia. Conversely, "activation" and dysfunction of glial cells can be damaging to the CNS and are hallmarks of many CNS diseases. Thus, understanding the role of glia and immune molecules during normal development could provide important insight into the mechanisms underlying neurodevelopmental disorders and neurodegenerative diseases affecting visual system function.

The recent work reviewed here has provided a strong foundation for understanding the many roles that glia play in visual system development and function and has paved the way for many future studies. Many important and open questions remain, including the following: What are the cues that promote the terminal differentiation of glial precursors into their diverse morphological subtypes? What is the significance of "NG2" cells in visual function? Do astrocytes and other glial cells directly regulate development of inhibitory synapses? How do glia help control experience-dependent remodeling of synaptic circuits? How do microglia sense and read out activity-dependent cues to help sculpt developing synaptic circuits? Do immune molecules, such as class I MHC and complement, have other functions in the developing visual system?

Despite recent advances in understanding the molecules and pathways that underlie neuron–glia and neural–immune communication during visual development, there is still a need for better molecular tools to manipulate subsets of glial cells at the right place and time in vivo. Regardless, the exciting new findings highlighted here lay a strong foundation in a new area of neuroscience, in which there is much to explore and discover.

REFERENCES

Al-Ali, S., & Al-Hussain, S. (1996). An ultrastructural study of the phagocytic activity of astrocytes in adult rat brain. *Journal of Anatomy, 188*, 257–262.

Allen, N. J., Bennett, M. L., Foo, L. C., Wang, G. X., Chakraborty, C., Smith, S. J., & Barnes, B. A. (2012). Astrocyte glypicans 4 and 6 promote formation of excitatory synapses via GluA1 AMPA receptors. *Nature, 486*, 410–414. doi:10.1038/nature11059.

Alvarez-Buylla, A., & García-Verdugo, J. M. (2002). Neurogenesis in adult subventricular zone. *Journal of Neuroscience, 22*, 629–634.

Bandtlow, C., Zachleder, T., & Schwab, M. E. (1990). Oligodendrocytes arrest neurite growth by contact inhibition. *Journal of Neuroscience, 10*, 3837–3848.

Barres, B., & Raff, M. (1993). Proliferation of oligodendrocyte precursor cells depends on electrical activity in axons. *Nature, 361*, 258–260.

Barres, B. A., Silverstein, B. E., Corey, D. R., & Chun, L. L. Y. (1988). Immunological, morphological, and electrophysiological variation among retinal ganglion cells purified by panning. *Neuron, 1*, 791–803.

Beattie, E. C., Stellwagen, D., Morishita, W., Bresnahan, J. C., Ha, B. K., Von Zastrow, M., et al. (2002). Control of synaptic strength by glial TNFalpha. *Science, 295*, 2282–2285. doi:10.1126/science.1067859.

Bechmann, I., & Nitsch, R. (1997). Astrocytes and microglial cells incorporate degenerating fibers following entorhinal lesion: A light, confocal, and electron microscopical study using a phagocytosis-dependent labeling technique. *Glia, 20*, 145–154.

Bentley, D., & Caudy, M. (1983). Pioneer axons lose directed growth after selective killing of guidepost cells. *Nature, 304*, 62–65.

Bergles, D. E., Jabs, R., & Steinhäuser, C. (2010). Neuron–glia synapses in the brain. *Brain Research. Brain Research Reviews, 63*, 130–137.

Bergles, D. E., Roberts, J. D. B., Somogyi, P., & Jahr, C. E. (2000). Glutamatergic synapses on oligodendrocyte precursor cells in the hippocampus. *Nature, 405*, 187–191.

Bhairavi, B., Hari, J., Shweta, S., Jones, M. F., & Astrid, L. G. (2011). Differences between the neurogenic and proliferative abilities of Müller glia with stem cell characteristics and the ciliary epithelium from the adult human eye. *Experimental Eye Research, 93*, 852–861.

Bjartmar, L., Huberman, A. D., Ullian, E. M., Renteria, R. C., Liu, X., Xu, W., et al. (2006). Neuronal pentraxins mediate synaptic refinement in the developing visual system. *Journal of Neuroscience, 26*, 6269–6281. doi:10.1523/JNEUROSCI.4212-05.2006.

Bobak, D. A., Gaither, T., Frank, M., & Tenner, A. (1987). Modulation of FcR function by complement: Subcomponent C1q enhances the phagocytosis of IgG-opsonized targets by human monocytes and culture-derived macrophages. *Journal of Immunology, 138*, 1150–1156.

Boiko, T., Rasband, M. N., Levinson, S. R., Caldwell, J. H., Mandel, G., Trimmer, J. S., et al. (2001). Compact myelin dictates the differential targeting of two sodium channel isoforms in the same axon. *Neuron, 30*, 91–104. doi:10.1016/S0896-6273(01)00265-3.

Boulanger, L. M. (2009). Immune proteins in brain development and synaptic plasticity. *Neuron, 64*, 93–109.

Burne, J. F., & Raff, M. C. (1997). Retinal ganglion cell axons drive the proliferation of astrocytes in the developing rodent optic nerve. *Neuron, 18*, 223–230.

Bushong, E. A., Martone, M. E., Jones, Y. Z., & Ellisman, M. H. (2002). Protoplasmic astrocytes in CA1 stratum radiatum occupy separate anatomical domains. *Journal of Neuroscience, 22*, 183–192.

Butt, A., Ibrahim, M., Ruge, F., & Berry, M. (1995). Biochemical subtypes of oligodendrocyte in the anterior medullary velum of the rat as revealed by the monoclonal antibody Rip. *Glia, 14*, 185–197. doi:10.1002/glia.440140304.

Cahoy, J. D., Emery, B., Kaushal, A., Foo, L. C., Zamanian, J. L., Christopherson, K. S., et al. (2008). A transcriptome database for astrocytes, neurons, and oligodendrocytes: A new resource for understanding brain development and function. *Journal of Neuroscience, 28*, 264–278. doi:10.1523/JNEUROSCI.4178-07.2008.

Campbell, G., & Shatz, C. J. (1992). Synapses formed by identified retinogeniculate axons during the segregation of eye input. *Journal of Neuroscience, 12*, 1847–1858.

Caroni, P., & Schwab, M. E. (1988a). Antibody against myelin associated inhibitor of neurite growth neutralizes nonpermissive substrate properties of CNS white matter. *Neuron, 1*, 85–96.

Caroni, P., & Schwab, M. E. (1988b). Two membrane protein fractions from rat central myelin with inhibitory properties for neurite growth and fibroblast spreading. *Journal of Cell Biology, 106*, 1281–1288.

Carroll, M. C. (2004). The complement system in regulation of adaptive immunity. *Nature Immunology, 5*, 981–986.

Christopherson, K. S., Ullian, E. M., Stokes, C. C. A., Mullowney, C. E., Hell, J. W., Agah, A., et al. (2005). Thrombospondins are astrocyte-secreted proteins that promote CNS synaptogenesis. *Cell, 120*, 421–433. doi:10.1016/j.cell.2004.12.020.

Cook, P. M., Prusky, G., & Ramoa, A. S. (1999). The role of spontaneous retinal activity before eye opening in the maturation of form and function in the retinogeniculate pathway of the ferret. *Visual Neuroscience, 16*, 491–501.

Corriveau, R. A., Huh, G. S., & Shatz, C. J. (1998). Regulation of class I MHC gene expression in the developing and mature CNS by neural activity. *Neuron, 21*, 505–520.

Dakubo, G. D., Wang, Y. P., Mazerolle, C., Campsall, K., McMahon, A. P., & Wallace, V. A. (2003). Retinal ganglion cell-derived sonic hedgehog signaling is required for optic disk and stalk neuroepithelial cell development. *Development, 130*, 2967–2980.

Dalmau, I., Finsen, B., Zimmer, J., Gonzalez, B., & Castellano, B. (1998). Development of microglia in the postnatal rat hippocampus. *Hippocampus, 8*, 458–474.

Datwani, A., McConnell, M. J., Kanold, P. O., Micheva, K. D., Busse, B., Shamloo, M., et al. (2009). Classical MHCI molecules regulate retinogeniculate refinement and limit ocular dominance plasticity. *Neuron, 64*, 463–470. doi:10.1016/j.neuron.2009.10.015.

Davalos, D., Grutzendler, J., Yang, G., Kim, J. V., Zuo, Y., Jung, S., et al. (2005). ATP mediates rapid microglial response to local brain injury in vivo. *Nature Neuroscience, 8*, 752–758. doi:10.1038/nn1472.

Dearborn, R., & Kunes, S. (2004). An axon scaffold induced by retinal axons directs glia to destinations in the *Drosophila* optic lobe. *Development, 131*, 2291–2303.

de Melo Reis, R. A., Ventura, A. L. M., Schitine, C. S., de Mello, M. C. F., & de Mello, F. G. (2008). Müller glia as an active compartment modulating nervous activity in the vertebrate retina: Neurotransmitters and trophic factors. *Neurochemical Research, 33*, 1466–1474.

Del Rio, T., & Feller, M. B. (2006). Early retinal activity and visual circuit development. *Neuron, 52*, 221–222.

del Río-Hortega, P. (1928). Tercera aportación al conocimiento morfológico e interpretación funcional de la oligodendroglía. *Memorias de la Real Sociedad Española de Historia Natural, 14*, 40–122.

Doetsch, F. (2003). The glial identity of neural stem cells. *Nature Neuroscience, 6*, 1127–1134.

Dugas, J. C., Cuellar, T. L., Scholze, A., Ason, B., Ibrahim, A., Emery, B., et al. (2010). Dicer1 and miR-219 are required for normal oligodendrocyte differentiation and myelination. *Neuron, 65*, 597–611. doi:10.1016/j.neuron.2010.01.027.

Dugas, J. C., Tai, Y. C., Speed, T. P., Ngai, J., & Barres, B. A. (2006). Functional genomic analysis of oligodendrocyte differentiation. *Journal of Neuroscience, 26*, 10967–10983.

Emery, B. (2010a). Regulation of oligodendrocyte differentiation and myelination. *Science, 330*, 779–782.

Emery, B. (2010b). Transcriptional and post-transcriptional control of CNS myelination. *Current Opinion in Neurobiology, 20*, 601–607.

Emery, B., Agalliu, D., Cahoy, J. D., Watkins, T. A., Dugas, J. C., Mulinyawe, S. B., et al. (2009). Myelin gene regulatory factor is a critical transcriptional regulator required for CNS myelination. *Cell, 138*, 172–185. doi:10.1016/j.cell.2009.04.031.

Eroglu, C., Allen, N. J., Susman, M. W., O'Rourke, N. A., Park, C. Y., Özkan, E., et al. (2009). Gabapentin receptor [alpha] 2 [delta]-1 is a neuronal thrombospondin receptor responsible for excitatory CNS synaptogenesis. *Cell, 139*, 380–392. doi:10.1016/j.cell.2009.09.025.

Fancy, S. P. J., Baranzini, S. E., Zhao, C., Yuk, D. I., Irvine, K. A., Kaing, S., et al. (2009). Dysregulation of the Wnt pathway inhibits timely myelination and remyelination in the mammalian CNS. *Genes & Development, 23*, 1571–1585. doi:10.1101/gad.1806309.

Ferrer, I., Bernet, E., Soriano, E., Del Rio, T., & Fonseca, M. (1990). Naturally occurring cell death in the cerebral cortex of the rat and removal of dead cells by transitory phagocytes. *Neuroscience, 39*, 451–458.

Filbin, M. T. (2003). Myelin-associated inhibitors of axonal regeneration in the adult mammalian CNS. *Nature Reviews Neuroscience, 4*, 703–713.

Fischer, A. J., & Reh, T. A. (2001). Muller glia are a potential source of neural regeneration in the postnatal chicken retina. *Nature Neuroscience, 4*, 247–252.

Foo, L. C., Allen, N. J., Bushong, E. A., Chung, W. S., Zhou, L., Cahoy, J. D., et al. (2011). Development of a method for the purification and culture of rodent astrocytes. *Neuron, 71*, 799–811. doi:10.1016/j.neuron.2011.07.022.

Fourgeaud, L., Davenport, C. M., Tyler, C. M., Cheng, T. T., Spencer, M. B., & Boulanger, L. M. (2010). MHC class I modulates NMDA receptor function and AMPA receptor trafficking. *Proceedings of the National Academy of Sciences of the United States of America, 107*, 22278–22283. doi:10.1073/pnas.0914064107.

Franze, K., Grosche, J., Skatchkov, S. N., Schinkinger, S., Foja, C., Schild, D., et al. (2007). Müller cells are living optical fibers in the vertebrate retina. *Proceedings of the National Academy of Sciences of the United States of America, 104*, 8287–8292. doi:10.1073/pnas.0611180104.

Freeman, M. R., Delrow, J., Kim, J., Johnson, E., & Doe, C. Q. (2003). Unwrapping glial biology: Gcm target genes regulating glial development, diversification, and function. *Neuron, 38*, 567–580.

Freeman, M. R., & Doherty, J. (2006). Glial cell biology in *Drosophila* and vertebrates. *Trends in Neurosciences, 29*, 82–90.

Fu, Q., Li, X., Shi, J., Xu, G., Wen, W., Lee, D. H. S., et al. (2009). Synaptic degeneration of retinal ganglion cells in a rat ocular hypertension glaucoma model. *Cellular and Molecular Neurobiology, 29*, 575–581. doi:10.1007/s10571-009-9349-7.

Gasque, P. (2004). Complement: A unique innate immune sensor for danger signals. *Molecular Immunology, 41*, 1089–1098.

Ginhoux, F., Greter, M., Leboeuf, M., Nandi, S., See, P., Gokhan, S., et al. (2010). Fate mapping analysis reveals that adult microglia derive from primitive macrophages. *Science, 330*, 841–845. doi:10.1126/science.1194637.

Glynn, M. W., Elmer, B. M., Garay, P. A., Liu, X. B., Needleman, L. A., El-Sabeawy, F., et al. (2011). MHCI negatively regulates synapse density during the establishment of cortical connections. *Nature Neuroscience, 14*, 442–451. doi:10.1038/nn.2764.

Goddard, C. A., Butts, D. A., & Shatz, C. J. (2007). Regulation of CNS synapses by neuronal MHC class I. *Proceedings of the National Academy of Sciences of the United States of America, 104*, 6828–6833.

Goritz, C., Mauch, D. H., & Pfrieger, F. W. (2005). Multiple mechanisms mediate cholesterol-induced synaptogenesis in a CNS neuron. *Molecular and Cellular Neurosciences, 29*, 190–201.

Graeber, M. B. (2010). Changing face of microglia. *Science's STKE, 330*, 783–788.

GrandPré, T., Li, S., & Strittmatter, S. M. (2002). Nogo-66 receptor antagonist peptide promotes axonal regeneration. *Nature, 417*, 547–551.

Grinspan, J., & Franceschini, B. (1995). Platelet-derived growth factor is a survival factor for PSA-NCAM+ oligodendrocyte pre-progenitor cells. *Journal of Neuroscience Research, 41*, 540–551.

Guido, W. (2008). Refinement of the retinogeniculate pathway. *Journal of Physiology, 586*, 4357–4362.

Guillemin, G. J., & Brew, B. J. (2004). Microglia, macrophages, perivascular macrophages, and pericytes: A review of function and identification. *Journal of Leukocyte Biology, 75*, 388–397.

Gyllensten, L., & Malmfors, T. (1963). Myelinization of the optic nerve and its dependence on visual function—A quantitative investigation in mice. *Journal of Embryology and Experimental Morphology, 11,* 255–266.

Halassa, M. M., Fellin, T., Takano, H., Dong, J. H., & Haydon, P. G. (2007). Synaptic islands defined by the territory of a single astrocyte. *Journal of Neuroscience, 27,* 6473–6477.

Haydon, P. G. (2001). Glia: Listening and talking to the synapse. *Nature Reviews Neuroscience, 2,* 185–193.

Haydon, P. G., & Carmignoto, G. (2006). Astrocyte control of synaptic transmission and neurovascular coupling. *Physiological Reviews, 86,* 1009–1031.

Hensch, T. K. (2004). Critical period regulation. *Annual Review of Neuroscience, 27,* 549–579.

Hong, Y. K., & Chen, C. (2011). Wiring and rewiring of the retinogeniculate synapse. *Current Opinion in Neurobiology, 21,* 228–237.

Hooks, B. M., & Chen, C. (2006). Distinct roles for spontaneous and visual activity in remodeling of the retinogeniculate synapse. *Neuron, 52,* 281–291.

Hua, J. Y., & Smith, S. J. (2004). Neural activity and the dynamics of central nervous system development. *Nature Neuroscience, 7,* 327–332.

Huberman, A. D. (2007). Mechanisms of eye-specific visual circuit development. *Current Opinion in Neurobiology, 17,* 73–80.

Huberman, A. D., Feller, M. B., & Chapman, B. (2008). Mechanisms underlying development of visual maps and receptive fields. *Annual Review of Neuroscience, 31,* 479–509.

Huh, G. S., Boulanger, L. M., Du, H., Riquelme, P. A., Brotz, T. M., & Shatz, C. J. (2000). Functional requirement for class I MHC in CNS development and plasticity. *Science, 290,* 2155–2159.

Hummel, T., Attix, S., Gunning, D., & Zipursky, S. L. (2002). Temporal control of glial cell migration in the *Drosophila* eye requires gilgamesh, hedgehog, and eye specification genes. *Neuron, 33,* 193–203.

Hutchins, J. B., & Casagrande, V. A. (1988). Glial cells develop a laminar pattern before neuronal cells in the lateral geniculate nucleus. *Proceedings of the National Academy of Sciences of the United States of America, 85,* 8316–8320.

Ishibashi, T., Dakin, K. A., Stevens, B., Lee, P. R., Kozlov, S. V., Stewart, C. L., et al. (2006). Astrocytes promote myelination in response to electrical impulses. *Neuron, 49,* 823–832. doi:10.1016/j.neuron.2006.02.006.

Ishibashi, T., Lee, P. R., Baba, H., & Fields, R. D. (2009). Leukemia inhibitory factor regulates the timing of oligodendrocyte development and myelination in the postnatal optic nerve. *Journal of Neuroscience Research, 87,* 3343–3355.

Israel, A., Le Bail, O., Hatat, D., Piette, J., Kieran, M., Logeat, F., et al. (1989). TNF stimulates expression of mouse MHC class I genes by inducing an NF kappa B-like enhancer binding activity which displaces constitutive factors. *European Molecular Biology Organization Journal, 8,* 3793–3800.

Itoh, K., Stevens, B., Schachner, M., & Fields, R. D. (1995). Regulated expression of the neural cell adhesion molecule L1 by specific patterns of neural impulses. *Science, 270,* 1369–1372.

Jeon, H., Lee, S., Lee, W. H., & Suk, K. (2010). Analysis of glial secretome: The long pentraxin PTX3 modulates phagocytic activity of microglia. *Journal of Neuroimmunology, 229,* 63–72.

Kaneko, M., Stellwagen, D., Malenka, R. C., & Stryker, M. P. (2008). Tumor necrosis factor-α mediates one component of competitive, experience-dependent plasticity in developing visual cortex. *Neuron, 58,* 673–680.

Katz, L. C., & Shatz, C. J. (1996). Synaptic activity and the construction of cortical circuits. *Science, 274,* 1133–1138.

Kucukdereli, H., Allen, N. J., Lee, A. T., Feng, A., Ozlu, M. I., Conatser, L. M., et al. (2011). Control of excitatory CNS synaptogenesis by astrocyte-secreted proteins Hevin and SPARC. *Proceedings of the National Academy of Sciences of the United States of America, 108,* E440–E449. doi:10.1073/pnas.1104977108.

Lee, J., Mayer-Proschel, M., & Rao, M. (2000). Gliogenesis in the central nervous system. *Glia, 30,* 105–121.

Lin, S., & Bergles, D. E. (2003). Synaptic signaling between GABAergic interneurons and oligodendrocyte precursor cells in the hippocampus. *Nature Neuroscience, 7,* 24–32.

Ling, E. A., & Wong, W. C. (1993). The origin and nature of ramified and amoeboid microglia: A historical review and current concepts. *Glia, 7,* 9–18.

Lovatt, D., Sonnewald, U., Waagepetersen, H. S., Schousboe, A., He, W., Lin, J. H. C., et al. (2007). The transcriptome and metabolic gene signature of protoplasmic astrocytes in the adult murine cortex. *Journal of Neuroscience, 27,* 12255–12266. doi:10.1523/JNEUROSCI.3404-07.2007.

Lu, Q. R., Sun, T., Zhu, Z., Ma, N., Garcia, M., Stiles, C. D., et al. (2002). Common developmental requirement for Olig function indicates a motor neuron/oligodendrocyte connection. *Cell, 109,* 75–86. doi:10.1016/S0092-8674(02)00678-5.

Lund, R. D. (1972). Synaptic patterns in the superficial layers of the superior colliculus of the monkey, *Macaca mulatta. Experimental Brain Research, 15,* 194–211.

Makinodan, M., Rosen, K. M., Ito, S., & Corfas, G. (2012). A critical period for social experience—Dependent oligodendrocyte maturation and myelination. *Science, 337,* 1357–1360. doi:10.1126/science.1220845.

Mason, C. A., & Sretavan, D. W. (1997). Glia, neurons, and axon pathfinding during optic chiasm development. *Current Opinion in Neurobiology, 7,* 647–653.

Mauch, D. H., Nagler, K., Schumacher, S., Goritz, C., Muller, E. C., Otto, A., et al. (2001). CNS synaptogenesis promoted by glia-derived cholesterol. *Science, 294,* 1354–1357. doi:10.1126/science.294.5545.1354.

McGee, A. W., Yang, Y., Fischer, Q. S., Daw, N. W., & Strittmatter, S. M. (2005). Experience-driven plasticity of visual cortex limited by myelin and Nogo receptor. *Science, 309,* 2222–2226.

McKerracher, L., David, S., Jackson, D., Kottis, V., Dunn, R., & Braun, P. (1994). Identification of myelin-associated glycoprotein as a major myelin-derived inhibitor of neurite growth. *Neuron, 13,* 805–811.

Mi, H., & Barres, B. A. (1999). Purification and characterization of astrocyte precursor cells in the developing rat optic nerve. *Journal of Neuroscience, 19,* 1049–1061.

Mi, S., Miller, R. H., Lee, X., Scott, M. L., Shulag-Morskaya, S., Shao, Z., et al. (2005). LINGO-1 negatively regulates myelination by oligodendrocytes. *Nature Neuroscience, 8,* 745–751. doi:10.1038/nn1460.

Mukhopadhyay, G., Doherty, P., Walsh, F. S., Crocker, P. R., & Filbin, M. T. (1994). A novel role for myelin-associated glycoprotein as an inhibitor of axonal regeneration. *Neuron, 13,* 757–767.

Müller, C. M. (1990). Dark-rearing retards the maturation of astrocytes in restricted layers of cat visual cortex. *Glia, 3*, 487–494.

Müller, C. M., & Best, J. (1989). Ocular dominance plasticity in adult cat visual cortex after transplantation of cultured astrocytes. *Nature, 342*, 427–430.

Nagy, J. I., & Rash, J. E. (2000). Connexins and gap junctions of astrocytes and oligodendrocytes in the CNS. *Brain Research. Brain Research Reviews, 32*, 29–44.

Nauta, A. J., Daha, M. R., Kooten, C., & Roos, A. (2003). Recognition and clearance of apoptotic cells: A role for complement and pentraxins. *Trends in Immunology, 24*, 148–154.

Nave, K. A. (2010). Myelination and the trophic support of long axons. *Nature Reviews Neuroscience, 11*, 275–283.

Newman, E. A., & Zahs, K. R. (1998). Modulation of neuronal activity by glial cells in the retina. *Journal of Neuroscience, 18*, 4022–4028.

Nguyen, J. V., Soto, I., Kim, K. Y., Bushong, E. A., Oglesby, E., Valiente-Soriano, F. J., et al. (2011). Myelination transition zone astrocytes are constitutively phagocytic and have synuclein dependent reactivity in glaucoma. *Proceedings of the National Academy of Sciences of the United States of America, 108*, 1176–1181. doi:10.1073/pnas.1013965108.

Nielsen, J. A., Maric, D., Lau, P., Barker, J. L., & Hudson, L. D. (2006). Identification of a novel oligodendrocyte cell adhesion protein using gene expression profiling. *Journal of Neuroscience, 26*, 9881–9891.

Nimmerjahn, A., Kirchhoff, F., & Helmchen, F. (2005). Resting microglial cells are highly dynamic surveillants of brain parenchyma in vivo. *Science, 308*, 1314–1318.

Nishiyama, A., Komitova, M., Suzuki, R., & Zhu, X. (2009). Polydendrocytes (NG2 cells): Multifunctional cells with lineage plasticity. *Nature Reviews Neuroscience, 10*, 9–22.

Oberheim, N. A., Goldman, S. A., & Nedergaard, M. (2012). Heterogeneity of astrocytic form and function. *Methods in Molecular Biology, 814*, 23–45.

O'Brien, R. J., Xu, D., Petralia, R. S., Steward, O., Huganir, R. L., & Worley, P. (1999). Synaptic clustering of AMPA receptors by the extracellular immediate-early gene product Narp. *Neuron, 23*, 309–323.

Omlin, F. X. (1997). Optic disk and optic nerve of the blind cape mole-rat (*Georychus capensis*): A proposed model for naturally occurring reactive gliosis. *Brain Research Bulletin, 44*, 627–632.

Paolicelli, R. C., Bolasco, G., Pagani, F., Maggi, L., Scianni, M., Panzanelli, P., et al. (2011). Synaptic pruning by microglia is necessary for normal brain development. *Science, 333*, 1456–1458. doi:10.1126/science.1202529.

Penn, A. A., Riquelme, P. A., Feller, M. B., & Shatz, C. J. (1998). Competition in retinogeniculate patterning driven by spontaneous activity. *Science, 279*, 2108–2112.

Perry, V., Hume, D. A., & Gordon, S. (1985). Immunohistochemical localization of macrophages and microglia in the adult and developing mouse brain. *Neuroscience, 15*, 313–326.

Pfrieger, F. W., & Barres, B. A. (1997). Synaptic efficacy enhanced by glial cells in vitro. *Science, 277*, 1684–1687.

Poeck, B., Fischer, S., Gunning, D., Zipursky, S. L., & Salecker, I. (2001). Glial cells mediate target layer selection of retinal axons in the developing visual system of *Drosophila*. *Neuron, 29*, 99–113.

Pogoda, H. M., Sternheim, N., Lyons, D. A., Diamond, B., Hawkins, T. A., Woods, I. G., et al. (2006). A genetic screen identifies genes essential for development of myelinated axons in zebrafish. *Developmental Biology, 298*, 118–131. doi:10.1016/j.ydbio.2006.06.021.

Prada, C., Puga, J., Pérez-Méndez, L., López, R., & Ramírez, G. (1991). Spatial and temporal patterns of neurogenesis in the chick retina. *European Journal of Neuroscience, 3*, 559–569.

Raff, M., Abney, E., Cohen, J., Lindsay, R., & Noble, M. (1983). Two types of astrocytes in cultures of developing rat white matter: Differences in morphology, surface gangliosides, and growth characteristics. *Journal of Neuroscience, 3*, 1289–1300.

Ramírez, M., & Lamas, M. (2009). NMDA receptor mediates proliferation and CREB phosphorylation in postnatal Müller glia-derived retinal progenitors. *Molecular Vision, 15*, 713–721.

Rangarajan, R., Gong, Q., & Gaul, U. (1999). Migration and function of glia in the developing *Drosophila* eye. *Development, 126*, 3285–3292.

Ransohoff, R. M., & Perry, V. H. (2009). Microglial physiology: Unique stimuli, specialized responses. *Annual Review of Immunology, 27*, 119–145.

Ransohoff, R. M., & Stevens, B. (2011). How many cell types does it take to wire a brain? *Science, 333*, 1391–1392.

Rasband, M. N., & Shrager, P. (2000). Ion channel sequestration in central nervous system axons. *Journal of Physiology, 525*, 63–73.

Roesch, K., Jadhav, A. P., Trimarchi, J. M., Stadler, M. B., Roska, B., Sun, B. B., et al. (2008). The transcriptome of retinal Müller glial cells. *Journal of Comparative Neurology, 509*, 225–238. doi:10.1002/cne.21730.

Rubin, L., & Staddon, J. (1999). The cell biology of the blood–brain barrier. *Annual Review of Neuroscience, 22*, 11–28.

Saederup, N., Cardona, A. E., Croft, K., Mizutani, M., Cotleur, A. C., Tsou, C. L., et al. (2010). Selective chemokine receptor usage by central nervous system myeloid cells in CCR2-red fluorescent protein knock-in mice. *PLoS ONE, 5*, e13693. doi:10.1371/journal.pone.0013693.

Sánchez, I., Hassinger, L., Paskevich, P. A., Shine, H. D., & Nixon, R. A. (1996). Oligodendroglia regulate the regional expansion of axon caliber and local accumulation of neurofilaments during development independently of myelin formation. *Journal of Neuroscience, 16*, 5095–5105.

Sanes, J. R., & Lichtman, J. W. (1999). Development of the vertebrate neuromuscular junction. *Annual Review of Neuroscience, 22*, 389–442.

Schafer, D. P., Custer, A. W., Shrager, P., & Rasband, M. N. (2006). Early events in node of Ranvier formation during myelination and remyelination in the PNS. *Neuron Glia Biology, 2*, 69–79.

Schafer, D. P., Lehrman, E. K., Kautzman, A. G., Koyama, R., Mardinly, A. R., Yamasaki, R., et al. (2012). Microglia sculpt postnatal neural circuits in an activity and complement-dependent manner. *Neuron, 74*, 691–705. doi:10.1016/j.neuron.2012.03.026.

Schafer, D. P., Lehrman, E. K., & Stevens, B. (2013). The "quad-partite" synapse: Microglia–synapse interactions in the developing and mature CNS. *Glia, 61*, 24–36.

Schummers, J., Yu, H., & Sur, M. (2008). Tuned responses of astrocytes and their influence on hemodynamic signals in the visual cortex. *Science, 320*, 1638–1643.

Schwab, M., & Schnell, L. (1991). Channeling of developing rat corticospinal tract axons by myelin-associated neurite growth inhibitors. *Journal of Neuroscience, 11*, 709–721.

Schwab, M. E. (1990). Myelin-associated inhibitors of neurite growth. *Experimental Neurology, 109*, 2–5.

Shatz, C. J. (1990). Competitive interactions between retinal ganglion cells during prenatal development. *Journal of Neurobiology, 21*, 197–211.

Shatz, C. J. (2009). MHC class I: An unexpected role in neuronal plasticity. *Neuron, 64*, 40–45.

Shatz, C. J., & Stryker, M. P. (1988). Prenatal tetrodotoxin infusion blocks segregation of retinogeniculate afferents. *Science, 242*, 87–89.

Shi, J., Marinovich, A., & Barres, B. A. (1998). Purification and characterization of adult oligodendrocyte precursor cells from the rat optic nerve. *Journal of Neuroscience, 18*, 4627–4636.

Sretavan, D., & Shatz, C. J. (1984). Prenatal development of individual retinogeniculate axons during the period of segregation. *Nature, 308*, 845–848.

Sretavan, D. W., & Shatz, C. J. (1986). Prenatal development of retinal ganglion cell axons: Segregation into eye-specific layers within the cat's lateral geniculate nucleus. *Journal of Neuroscience, 6*, 234–251.

Starr, R., Attema, B., DeVries, G., & Monteiro, M. (1996). Neurofilament phosphorylation is modulated by myelination. *Journal of Neuroscience Research, 44*, 328–337.

Stellwagen, D., & Malenka, R. C. (2006). Synaptic scaling mediated by glial TNF-alpha. *Nature, 440*, 1054–1059.

Stellwagen, D., & Shatz, C. J. (2002). An instructive role for retinal waves in the development of retinogeniculate connectivity. *Neuron, 33*, 357–367.

Stevens, B., Allen, N. J., Vazquez, L. E., Howell, G. R., Christopherson, K. S., Nouri, N., et al. (2007). The classical complement cascade mediates CNS synapse elimination. *Cell, 131*, 1164–1178. doi:10.1016/j.cell.2007.10.036.

Stevens, B., Porta, S., Haak, L. L., Gallo, V., & Fields, R. D. (2002). Adenosine: A neuron–glial transmitter promoting myelination in the CNS in response to action potentials. *Neuron, 36*, 855–868.

Suh, G. S. B., Poeck, B., Chouard, T., Oron, E., Segal, D., Chamovitz, D. A., et al. (2002). *Drosophila* JAB1/CSN5 acts in photoreceptor cells to induce glial cells. *Neuron, 33*, 35–46. doi:10.1016/S0896-6273(01)00576-1.

Sun, D., Lye-Barthel, M., Masland, R. H., & Jakobs, T. C. (2009). The morphology and spatial arrangement of astrocytes in the optic nerve head of the mouse. *Journal of Comparative Neurology, 516*, 1–19.

Sun, D., Lye-Barthel, M., Masland, R. H., & Jakobs, T. C. (2010). Structural remodeling of fibrous astrocytes after axonal injury. *Journal of Neuroscience, 30*, 14008–14019.

Sur, M., & Rubenstein, J. L. R. (2005). Patterning and plasticity of the cerebral cortex. *Science's STKE, 310*, 805–810.

Syken, J., Grandpre, T., Kanold, P. O., & Shatz, C. J. (2006). PirB restricts ocular-dominance plasticity in visual cortex. *Science, 313*, 1795–1800.

Takeda, M., Takamiya, A., Jiao, J., Cho, K. S., Trevino, S. G., Matsuda, T., et al. (2008). α-Aminoadipate induces progenitor cell properties of Müller glia in adult mice. *Investigative Ophthalmology & Visual Science, 49*, 1142–1150. doi:10.1167/iovs.07-0434.

Tauber, H., Waehneldt, T., & Neuhoff, V. (1980). Myelination in rabbit optic nerves is accelerated by artificial eye opening. *Neuroscience Letters, 16*, 235–238.

Tremblay, M. È., Lowery, R. L., & Majewska, A. K. (2010). Microglial interactions with synapses are modulated by visual experience. *PLoS Biology, 8*, e1000527. doi:10.1371/journal.pbio.1000527.

Tremblay, M. È., Stevens, B., Sierra, A., Wake, H., Bessis, A., & Nimmerjahn, A. (2011). The role of microglia in the healthy brain. *Journal of Neuroscience, 31*, 16064–16069.

Tsui, C. C., Copeland, N. G., Gilbert, D. J., Jenkins, N. A., Barnes, C., & Worley, P. F. (1996). Narp, a novel member of the pentraxin family, promotes neurite outgrowth and is dynamically regulated by neuronal activity. *Journal of Neuroscience, 16*, 2463–2478.

Ullian, E. M., Sapperstein, S. K., Christopherson, K. S., & Barres, B. A. (2001). Control of synapse number by glia. *Science, 291*, 657–661.

van Lookeren Campagne, M., Wiesmann, C., & Brown, E. J. (2007). Macrophage complement receptors and pathogen clearance. *Cellular Microbiology, 9*, 2095–2102.

Wake, H., Lee, P. R., & Fields, R. D. (2011). Control of local protein synthesis and initial events in myelination by action potentials. *Science, 333*, 1647. doi:10.1126/science.1206998.

Wake, H., Moorhouse, A. J., Jinno, S., Kohsaka, S., & Nabekura, J. (2009). Resting microglia directly monitor the functional state of synapses in vivo and determine the fate of ischemic terminals. *Journal of Neuroscience, 29*, 3974–3980.

Wang, S., Sdrulla, A. D., diSibio, G., Bush, G., Nofziger, D., Hicks, C., et al. (1998). Notch receptor activation inhibits oligodendrocyte differentiation. *Neuron, 21*, 63–75. doi:10.1016/S0896-6273(00)80515-2.

Watanabe, T., & Raff, M. C. (1988). Retinal astrocytes are immigrants from the optic nerve. *Nature, 332*, 834–837.

Watkins, T. A., Emery, B., Mulinyawe, S., & Barres, B. A. (2008). Distinct stages of myelination regulated by [gamma]-secretase and astrocytes in a rapidly myelinating CNS coculture system. *Neuron, 60*, 555–569.

Wegner, M. (2008). A matter of identity: Transcriptional control in oligodendrocytes. *Journal of Molecular Neuroscience, 35*, 3–12.

Williams, S. E., Mann, F., Erskine, L., Sakurai, T., Wei, S., Rossi, D. J., et al. (2003). Ephrin-B2 and EphB1 mediate retinal axon divergence at the optic chiasm. *Neuron, 39*, 919–935. doi:10.1016/j.neuron.2003.08.017.

Xu, D., Hopf, C., Reddy, R., Cho, R. W., Guo, L., Lanahan, A., et al. (2003). Narp and NP1 form heterocomplexes that function in developmental and activity-dependent synaptic plasticity. *Neuron, 39*, 513–528. doi:10.1016/S0896-6273(03)00463-X.

Zamanian, J., Xu, L., Foo, L., Nouri, N., Zhou, L., Giffard, R., et al. (2012). Genomic analysis of reactive astrogliosis. *Journal of Neuroscience, 32*, 6391–6410. doi:10.1523/JNEUROSCI.6221-11.2012.

Zhang, Y., & Barres, B. A. (2010). Astrocyte heterogeneity: An underappreciated topic in neurobiology. *Current Opinion in Neurobiology, 20*, 588–594.

Zhou, Q., Choi, G., & Anderson, D. J. (2001). The bHLH transcription factor Olig2 promotes oligodendrocyte differentiation in collaboration with Nkx2.2. *Neuron, 31*, 791–807.

Zhu, X., Bergles, D. E., & Nishiyama, A. (2008). NG2 cells generate both oligodendrocytes and gray matter astrocytes. *Development, 135*, 145–157.

97 Optic Nerve Regeneration

LARRY I. BENOWITZ AND SILMARA DE LIMA

Unlike most pathways that convey sensory information to the brain, the optic nerve is an integral part of the central nervous system (CNS) itself. The retina develops from the diencephalon and the axons that extend from retinal ganglion cells (RGCs) to the brain are ensheathed by oligodendrocytes. Consequently, like most CNS pathways in mature mammals, the optic nerve cannot regenerate if injured, leaving victims of traumatic nerve injury, glaucoma, and other types of nerve damage with irreversible losses in vision. Investigators as far back as Cajal and even earlier observed that, following optic nerve injury in mature mammals, damaged axons regress from the injury site and transiently grow back as far as the glial scar but do not advance any further (Ramon y Cajal, 1991). This process is followed by a near-complete loss of RGCs over the next month or so (Grafstein & Ingoglia, 1982; Villegas-Perez et al., 1988), making the possibility of visual recovery seem remote. Over the past few decades, however, we have learned a great deal about factors that limit or promote the regeneration of injured axons through the optic nerves of lower vertebrates and immature mammals, through peripheral nerve grafts in mature mammals, and more recently, through the mature mammalian optic nerve itself. As a result of this research, the optic nerve has gone from being a prime example of regenerative failure in the mature CNS to a system in which some degree of functional recovery is starting to look feasible.

GENERAL CONSIDERATIONS

Several conditions must be met in order to achieve successful regeneration. RGCs begin to die several days after nerve injury, and this loss clearly has to be arrested for regeneration to succeed. Although the mechanisms underlying axotomy-induced death are not fully understood, several components of the cell death pathway have been identified, and treatments aimed at blocking these prolong RGC survival. Some neuroprotective treatments also promote axon regeneration, but many do not. Thus, although regeneration is obviously predicated upon RGC survival, the molecular mechanisms underlying the two processes are distinct. Another requirement for regeneration to succeed is the need to activate the genetic program for long-distance axon growth. In lower vertebrates in which regeneration occurs spontaneously, and in mammals in which regeneration can be induced experimentally, regeneration involves dramatic changes in RGCs' program of gene expression. These changes reflect regulation at the levels of gene transcription and protein translation, and are marked by dramatic shifts in the expression of some genes that play a role in the initial development of the visual system and others that are specific to the regenerative state. Another issue concerns cell-extrinsic inhibitors of growth. Several proteins associated with myelin and the glial scar that forms at the site of injury suppress axon regeneration. Although counteracting these inhibitory signals allows for a moderate amount of regeneration, it is generally insufficient for long-distance growth. However, once RGCs' intrinsic growth state has been activated, strategies to overcome cell-extrinsic inhibitors of growth strongly amplify the level of regeneration. A fourth issue is axon guidance. A great deal has been learned about the molecular signals that enable RGC axons to navigate to their appropriate destinations during development. However, our information about the persistence of guidance signals in the adult mammalian brain is limited, as is our knowledge of whether mature RGC axons express the receptors and transduction mechanisms required to respond to these signals. Finally, for efficient signaling to take place, it is essential that regenerating axons become myelinated and form physiologically functional synapses with appropriate target neurons.

OPTIC NERVE REGENERATION IN LOWER VERTEBRATES

Fish and amphibia can regenerate their optic nerves and recover vision throughout life, providing a vertebrate model in which to study cellular and molecular phenomena associated with successful regeneration. Regeneration begins a few days after nerve injury, and within a few weeks, axons have grown back to the brain and begin to sort themselves out, culminating in full visual recovery within 2–4 months (Fawcett & Gaze, 1981; Sperry, 1948, 1963; Yoon, 1971).

Cell-Intrinsic Changes

Following nerve injury, RGCs undergo massive changes in RNA, protein, and lipid biosynthesis (Grafstein & Murray, 1969; Heacock et al., 1984; Ingoglia et al., 1973; Murray, 1973; Murray & Grafstein, 1969; Sbaschnig-Agler et al., 1985), marked by increased expression of transcription factors (Herdegen et al., 1993; Takeda et al., 2000; Veldman et al., 2007), regulators of RNA and protein metabolism (Veldman et al., 2007), trophic factors and cell-survival proteins (Koriyama et al., 2007; Nagashima et al., 2011), cytoskeletal proteins (Asch et al., 1998; Hall & Schechter, 1991; Heacock & Agranoff, 1976; Hieber et al., 1998), cell-adhesion molecules (Bernhardt et al., 1996; Blaugrund et al., 1990; Jung, Petrausch, & Stuermer, 1997), scaffold proteins associated with lipid rafts (Langhorst, Reuter, & Stuermer, 2005), multiple enzymes (Ballestero et al., 1997; Eitan & Schwartz, 1993; Koriyama et al., 2009; Sugitani et al., 2006), and regulators of growth cone dynamics, including GAP-43 (Benowitz, Shashoua, & Yoon, 1981; Benowitz, Yoon, & Lewis, 1983; Skene & Willard, 1981b). At the same time, the expression of genes related to ion channels and neurotransmission decreases (Veldman et al., 2007).

Role of Specific Proteins

One of the earliest changes seen after nerve injury in goldfish is an increase in expression of heat shock factor-1 (HSF-1). HSF-1 regulates the expression of genes that promote RGC survival, including genes encoding heat-shock protein (HSP)-70 and Bcl-2, while suppressing expression of the pro-apoptotic protein Bax (Nagashima et al., 2011). Insulin-like growth factor-1 (IGF)-1 is up-regulated around day 4 and activates the phosphatidyl inositol-3 kinase (PI3 kinase) pathway, thereby promoting both cell-survival and axon growth (see figure 97.1) (Koriyama et al., 2007). Because of their role in orchestrating many downstream genes, transcription factors are of particular interest. One prominent change that occurs during optic nerve regeneration in zebrafish is an increase in expression of the transcription factors Krüppel-like family (KLF) 6a and -7a (Veldman et al., 2007), whose downstream targets include T alpha1, a tubulin isoform that is essential for axon regeneration (Veldman, Bemben, & Goldman, 2010). Several other gene products that are up-regulated in goldfish or zebrafish have also been shown to directly influence the regenerative process, including the enzyme transglutaminase (Eitan & Schwartz, 1993; Sugitani et al., 2006), neuronal nitric oxide synthase (nNOS: figure 97.1G,H) (Koriyama

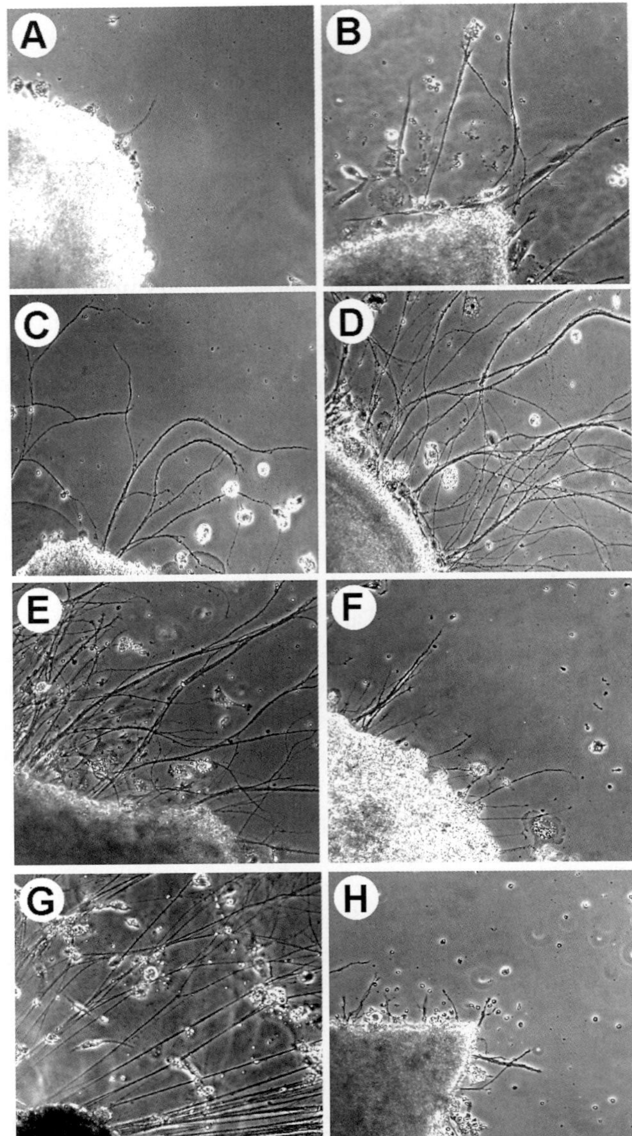

FIGURE 97.1   Regulators of optic nerve regeneration in goldfish. Neurite outgrowth from adult goldfish retinal explants after 5 days in culture. (A) Normal unprimed retina. (B) Primed retina explanted 5 days after optic nerve injury. (C, D) Insulin-like growth factor-1 (100 nM) added to unprimed (C) and primed (D) retinas augments outgrowth. (E, F) Role of nitric oxide signaling. Addition of dbcGMP (dibutyrylguanosine 3:5′-cyclic monophosphate; 400 μM) promotes neurite outgrowth from unprimed retina (E) whereas a short hairpin RNA directed against neuronal nitric oxide synthase inhibits outgrowth from a primed retina (F). (G, H) Role of the retinol-binding protein purpurin. Addition of purpurin to unprimed retina (G) augments growth, whereas a small interfering RNA directed against purpurin suppresses the ability of primed retina to grow (H) (Matsukawa et al., 2004; Koriyama et al., 2007, 2009; Nagashima et al., 2009; Koriyama et al., 2012).

et al., 2009), and the lipid raft-associated reggie proteins (flotillins; Munderloh et al., 2009).

Cell-Extrinsic Activators of Regeneration

Cell-extrinsic factors appear to play a role in transforming goldfish RGCs into a regenerative state. Goldfish retinal explants exhibit much stronger outgrowth when placed in culture if they had first been "primed" by leaving the retina in situ for several days after optic nerve injury rather than being explanted without prior surgery (see figure 97.1A, B) (Landreth & Agranoff, 1976, 1979). Although "naive" retinal explants will eventually begin to extend axons, they never show as much growth as primed explants. One interpretation of these results is that the priming exposes RGCs to cell-extrinsic factors that are not available in culture. Complementation (add-back) studies show that factors secreted by nonneuronal cells of the optic nerve enable naive RGCs to extend lengthy axons in culture (see figure 97.2A, B) (Schwalb et al., 1995, 1996). The most potent of these was identified as a simple carbohydrate (see figure 97.2C). At low micromolar concentrations, mannose or glucose (see figure 97.2D), but not other carbohydrates, stimulates extensive outgrowth, and this effect is independent of their role in energy metabolism (Li et al., 2003). The retinol-binding protein purpurin also promotes regeneration in a non-cell-autonomous fashion. Within a few days of optic nerve injury, photoreceptors increase expression of purpurin, which transports retinol to RGCs. Retinol-bound purpurin stimulates RGCs to up-regulate expression of proteins involved in retinoic acid (RA) metabolism, RA-dependent transcription factors, and presumably, downstream RA-dependent genes, and culminates in extensive axon outgrowth (see figure 97.1G, H) (Matsukawa et al., 2004; Nagashima et al., 2009).

Cell-Extrinsic Barriers to Regeneration

Along with an ability to reactivate RGCs' intrinsic growth state, lower vertebrates seem to have fewer cell-extrinsic barriers to growth than mammals. In goldfish, phagocytic cells enter the injury site within a few days of nerve damage and clear myelin debris more rapidly than occurs in mammals (Battisti et al., 1995; Colavincenzo & Levine, 2000; Nona, Thomlinson, & Stafford, 1998). In addition, unlike mammals, oligodendrocytes of the goldfish optic nerve have been reported to be permissive to axon growth (Bastmeyer et al., 1991; Sivron, Schwab, & Schwartz, 1994) and the goldfish variant of the protein Nogo, unlike its mammalian homolog, enhances the growth of RGC axons (Abdesselem et al., 2009). An astrocytic scar does not form at the injury site in goldfish, and instead there is an influx of fibroblasts that support growth (Hirsch, Cahill, & Stuermer, 1995).

Axon Guidance Signals

The guidance of regenerating axons to their appropriate target sites involves many of the same signals that govern the initial development of the visual system. Slit-Robo interactions play a critical role in repelling axons from inappropriate areas of the di- and mesencephalon during development and play a similar role during optic nerve regeneration. In zebrafish, RGCs undergoing axon regeneration up-regulate expression of Robo2, a receptor for guidance signals of the Slit family, at the time their axons enter the brainstem, whereas expression of the complementary Slit ligands persists at high levels in appropriate regions of the brainstem throughout life (Wyatt et al., 2010). Regenerating goldfish RGCs also increase expression of EphA receptors along an ascending nasal-to-temporal gradient whereas the optic tectum (superior colliculus) displays a marked gradient of the complementary ligands, Ephrin A2 and -A5. As in development, these molecules are critical for reestablishing a topographically organized map of RGC axons onto their targets (Becker, Meyer, & Becker, 2000; Rodger et al., 2000, 2004). In *Xenopus*, Eph and Ephrin gradients appear to remain unchanged throughout life (Higenell et al., 2012). Goldfish RGCs also express deleted in colon cancer (DCC), a receptor for netrin ligands that is involved in guiding axons to the optic nerve head and through the ventral hypothalamus (Petrausch et al., 2000).

DEVELOPMENTAL LOSS OF REGENERATIVE ABILITY IN MAMMALS

Change in Intrinsic Growth State

In mammals, the ability of RGCs to regenerate injured axons declines sharply over the course of development. In newborn hamsters, RGCs can regrow severed axons back into the superior colliculus only until postnatal day 3 (So, Schneider, & Ayres, 1981) but retain the ability to extend short terminal branches a few days more (Jhaveri, Edwards, & Schneider, 1991; So & Schneider, 1978). Correspondingly, studies in explant culture show a marked decline in RGCs' intrinsic ability to extend lengthy axons around postnatal day 2 (Chen, Jhaveri, & Schneider, 1995), as do studies using RGCs isolated from newborn rats (see figure 97.3A) (Goldberg et al., 2002). This developmental decline is

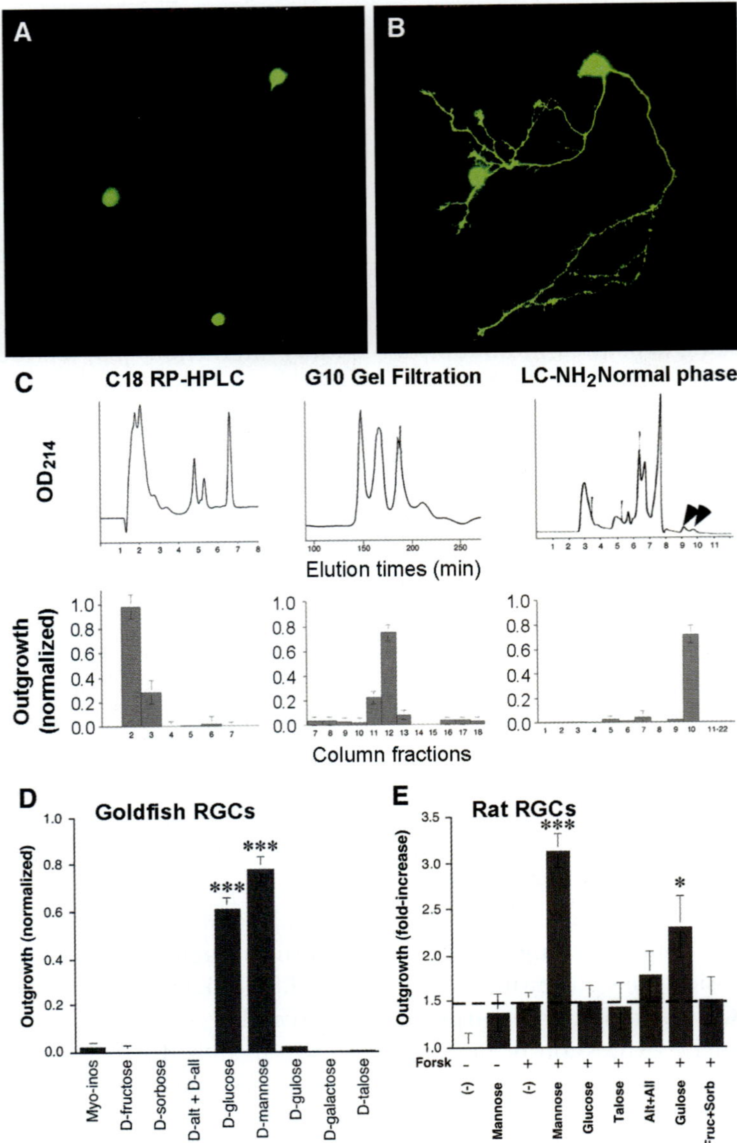

FIGURE 97.2 Small carbohydrates are cell-extrinsic activators of regeneration. (A, B) Dissociated neurons from normal goldfish retina show little outgrowth in defined, serum-free media (A) but show extensive growth when treated with factors secreted by optic nerve glia or fluid from the posterior chamber of the eye (B). (C) Sequential separation of molecules from the vitreous by C18 reversed-phase (RP) HPLC, G10 Gel-filtration, and LC-NH₂ normal-phase liquid chromatography leads to the purification of a molecule that stimulates outgrowth from unprimed goldfish RGCs. Top row: chromatograms, visualized by optical density at 214 nm (OD_{214}). Bottom: Axon outgrowth (percentage of retinal ganglion cells [RGCs] with axons ≥ 75 μm in length) induced by column fractions. The active component was identified by mass spectroscopy as a simple carbohydrate ($M_r = 180.16$ Daltons). (D) Of the carbohydrates tested, only mannose and glucose induce outgrowth from RGCs. (E) In rat RGCs, only mannose (and to a lesser extent, gulose) induces outgrowth and also requires elevation of [cyclic AMP]$_i$. Data in C–E: mean ±S.E.M., n $\geq$ 4 replicate wells (Li et al., 2003; Schwalb et al., 1995).

triggered by a cell-surface interaction between RGCs and amacrine cells, which transforms RGCs from an axon-elongation state to a dendritogenic state (see figure 97.3B) (Goldberg et al., 2002). This switch is marked by a dramatic shift in RGCs' program of gene expression (Wang et al., 2007), axonally transported proteins (Moya et al., 1988), and ability to up-regulate growth-associated proteins after injury (Skene & Willard, 1981a). At the same time that RGCs' intrinsic growth potential is declining, levels of cell-extrinsic inhibitors of axon growth associated with myelin, glial scarring, and the perineuronal net increase, along

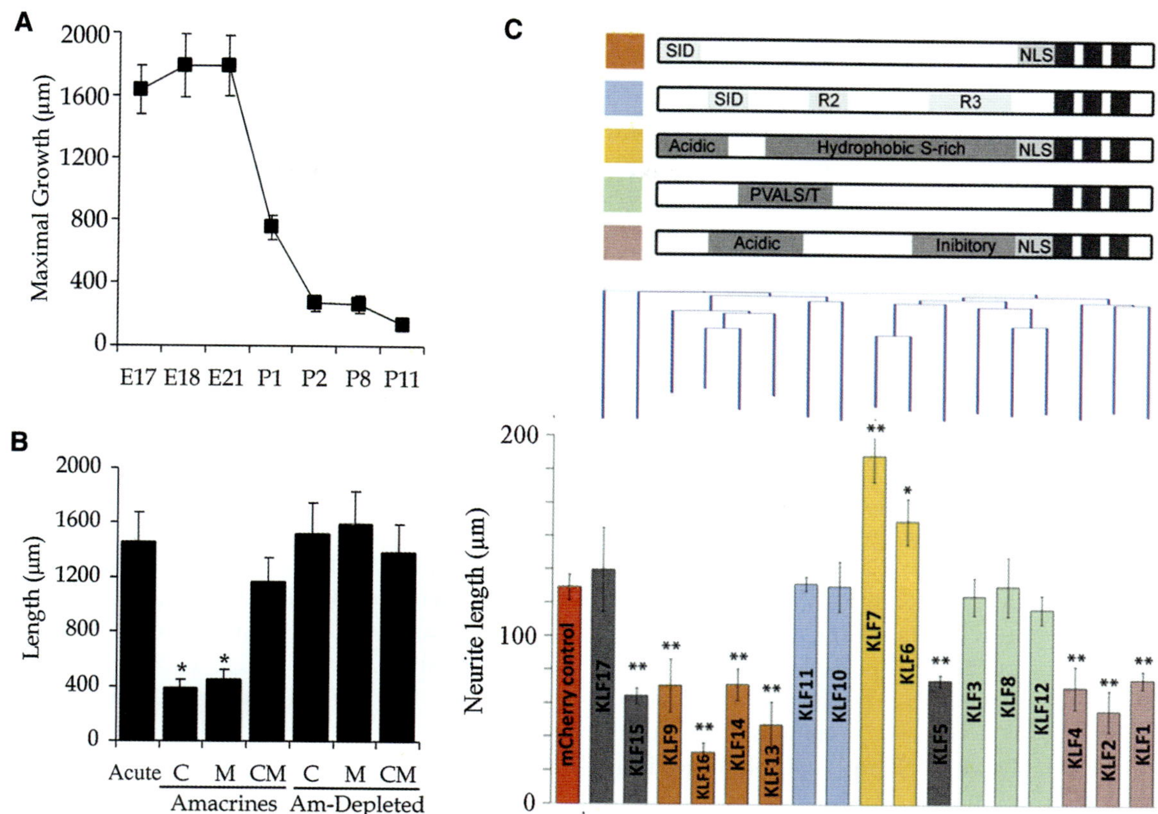

FIGURE 97.3 Developmental decline in the intrinsic growth potential of mammalian retinal ganglion cells (RGCs): role of amacrine cells and Krüppel-like family (KLF) transcription factors. (A) Decline in axon outgrowth between embryonic day 17 (E17) and postnatal day 11 (P11). RGCs were isolated by immunopanning and cultured under highly permissive conditions. (B) Amacrine cells regulate RGCs' intrinsic growth state. E20 RGCs were either acutely purified (left bar, positive control) or aged 3–4 days with components of amacrine cells or of retinal cells depleted of amacrine cells (Am-depleted). Incubation occurred in the presence of whole cells, C, cell membranes, M, or conditioned medium, CM, from the two cell populations. RGCs were then dissociated, replated, and evaluated for axon outgrowth after 3 days. *$P < 0.05$ compared to positive controls. (C) KLF transcription factors regulate axon growth. Top: KLF family members grouped according to structural domains. Bottom: P5 cortical neurons were transfected with individual KLFs and grown for 3 days on laminin. Graph shows average neurite length. *$P < 0.05$, **$P < 0.01$ compared to transfection control (red) (Goldberg et al., 2002; Moore et al., 2009.)

with the response of RGCs to these signals (Shewan, Berry, & Cohen, 1995). Developing RGCs also show a decline in their ability to respond positively to the growth-permissive protein laminin (Cohen et al., 1986).

Transcription factors belonging to the KLF family may contribute to the decline in RGCs' intrinsic growth potential (Moore et al., 2009). KLF4 is a suppressor of axon growth that is up-regulated at the time RGCs lose their intrinsic growth capacity (see figure 97.3C), and deleting the *klf4* gene in RGCs allows for moderate levels of axon regeneration following optic nerve injury in adult mice (Moore et al., 2009). At the same time that KLF4 levels increase, the expression of KLF6 and -7, two enhancers of axon growth (see figure 97.3C),

declines sharply. As noted earlier, KLF6 and -7 are important for optic nerve regeneration in zebrafish (Veldman et al., 2007; Veldman, Bemben, & Goldmanm, 2010).

Another cell-intrinsic change that may contribute to RGCs' loss of regenerative potential is a decline in expression of the transcriptional coactivator P300. P300 is a histone acetyltransferase that is involved in chromatin remodeling and transcriptional regulation. In rat RGCs, P300 expression declines after postnatal day 21 and decreases further after optic nerve damage. Over-expression of P300 in RGCs increases expression of GAP-43 and SPRR1A and is sufficient to promote moderate levels of regeneration after optic nerve damage (Gaub et al., 2010).

AXON REGENERATION THROUGH A PERIPHERAL NERVE GRAFT

Early last century, Tello, a student of Cajal, cut the optic nerve of a mature rabbit and replaced it with a peripheral nerve (PN) graft, enabling some axons to regenerate into the graft (Ramon y Cajal, 1991). This discovery represented the first evidence that neurons of the mature mammalian CNS have not entirely lost their intrinsic capacity to regrow axons and suggested that regenerative failure might be due mostly to environmental factors. More systematic studies by Aguayo and colleagues later showed that RGCs could regenerate axons long distances through a PN graft and form functional synapses if the far end of the graft were inserted into the superior colliculus (Aguayo et al., 1991; Bray, Vidal-Sanz, & Aguayo, 1987; Carter, Bray, & Aguayo, 1989; Keirstead et al., 1989; Sauve, Sawai, & Rasminsky, 1995; So & Aguayo, 1985; Vidal-Sanz et al., 1987). Sauve, Sawai, and Rasminsky (2001) reported a slight tendency for regenerating RGC axons to reestablish a topographic map upon the superior colliculus. One group reported that animals with PN grafts exhibit some visual discrimination, though this finding was inconsistent and was based on a small number of animals (Thanos, Naskar, & Heiduschka, 1997). Another study reported some recovery of the pupillary light reflex (PLR) when the distal end of a PN graft was directed to the olivary pretectal nucleus (Vidal-Sanz et al., 2002). However, the interpretation of these findings is uncertain in view of recent studies showing that a PLR can occur even without neural input due to the intrinsic photosensitivity of the iris (Xue et al., 2011).

CELL-EXTRINSIC BARRIERS TO OPTIC NERVE REGENERATION

Multiple proteins associated with myelin, the perineuronal net, and the scar that forms at an injury site inhibit CNS regeneration. The discovery that oligodendrocytes and CNS myelin are inhibitory to axon growth (Schwab & Caroni, 1988; Schwab & Thoenen, 1985) led to the discovery of several potent inhibitory molecules, including Nogo-A, myelin-associated glycoprotein (MAG), oligodendrocyte-myelin glycoprotein (OMgp), ephrin B3, versican V2, Semaphorins 4D and 5A, and lipid sulfatides (Benson et al., 2005; Chen et al., 2000; Goldberg et al., 2004; GrandPre et al., 2000; McKerracher et al., 1994; Moreau-Fauvarque et al., 2003; Mukhopadhyay et al., 1994; Wang et al., 2002b; Winzeler et al., 2011). NogoA, MAG, and OMgp exert their effects on growing axons via two shared receptors, the glycosylphosphatidylinositol-anchored Nogo receptor (NgR) and the paired immunoglobulin-like receptor PirB (Atwal et al., 2008; Filbin, 2003; Fournier, GrandPre, & Strittmatter, 2001; Fournier et al., 2002; Wang et al., 2002b), as well as ligand-specific receptors (Goh et al., 2008; Schweigreiter et al., 2004). NgR forms a multiprotein receptor complex with LINGO1 and either the low-affinity neurotrophin receptor P75 or the protein TROY (Mi et al., 2004; Park et al., 2005; Wang et al., 2002a). The binding of ligands to this receptor complex activates the small GTPase RhoA, which, through its downstream effectors, leads to cytoskeletal destabilization and growth cone collapse (Lehmann et al., 1999; Niederost et al., 2002; Perrot, Vazquez-Prado, & Gutkind, 2002; Yamashita & Tohyama, 2003).

In addition to the myelin-associated inhibitors described above, axon growth is suppressed by chondroitin- and keratin sulfate proteoglycans (CSPGs, KSPGs) that are associated with the perineuronal net and the astrocytic scar that forms at the site of injury (Fitch & Silver, 2008; Rudge & Silver, 1990; Silver & Miller, 2004). The effects of proteoglycans on axon growth are mediated through several receptors that include protein tyrosine phosphatase receptor sigma (Shen et al., 2009), transmembrane leukocyte common antigen-related phosphatase (Fisher et al., 2011), and two NgR isotypes (Dickendesher et al., 2012). Like myelin-associated inhibitors of growth, the inhibitory effects of CSPGs are mediated largely via activation of RhoA (Monnier et al., 2003; Schweigreiter et al., 2004).

Counteracting cell-extrinsic inhibitors of growth allows for a moderate level of optic nerve regeneration in mature animals, and these effects are strongly enhanced when RGCs' intrinsic growth state becomes activated (see below). An antibody to NogoA or virally mediated expression of a "decoy" Nogo receptor in RGCs does not induce appreciable regeneration in mature rats (Chierzi et al., 1999; Fischer, He, & Benowitz, 2004a). However, a loss of all three *ngr* genes has a more comprehensive effect and leads to appreciable regeneration in mice (Dickendesher et al., 2012; Wang et al., 2012).

The downstream signaling pathway activated by many inhibitory molecules activates RhoA; inactivating RhoA or its downstream effector, Rho kinase, represents a comprehensive way to overcome multiple inhibitory signals at once. However, this strategy has only modest effects on optic nerve regeneration in rats (Fischer et al., 2004b; Lehmann et al., 1999; Lingor et al., 2007).

RGC SURVIVAL

As noted earlier, RGCs begin to die within a few days of axotomy, precluding any possibility for extensive regeneration. We have learned a great deal about the mechanisms that underlie RGC death after axonal injury, but there are critical steps in this process that are not well understood. This research has led to the development of several neuroprotective strategies, but the effects of most of these are transient. In addition, whereas several neuroprotective strategies induce some level of regeneration, most do not, emphasizing again the separation between cell-signaling pathways underlying cell survival and axon outgrowth.

Trophic Factors

Among the neurotrophins (NTs), brain-derived neurotrophic factor (BDNF) and NT-4/5 have the clearest effects on RGC survival, attenuating, though not ultimately preventing, RGC death after optic nerve damage (Di Polo et al., 1998; Mansour-Robaey et al., 1994; Mey & Thanos, 1993; Peinado-Ramon et al., 1996). The effect of BDNF is enhanced by overexpressing its receptor, trkB, in RGCs (Cheng et al., 2002), continuously infusing BDNF into the cortex (Weber et al., 2010), or inducing a limited inflammatory response in the eye (Pernet & Di Polo, 2006). The TGF-beta family members glial-derived neurotrophic factor (GDNF) and neurturin delay RGC death after optic nerve damage and have additive effects with BDNF (Klocker et al., 1997; Koeberle & Ball, 1998; Schmeer et al., 2002; Yan et al., 1999). IGF-1 also partially improves RGC survival (Homma et al., 2007). The role of CTNF is discussed in a later section.

Voltage-Gated Potassium Channels

The activation of voltage-gated potassium channels (K_vs) is an early event in apoptosis, leading to an efflux of K^+ and cell shrinkage. These events precede DNA degradation, apoptosome formation, and mitochondrial depolarization (Bortner & Cidlowski, 2007). RGCs express several Type 1 Kv's (Kv1), and either pharmacological blockade or small interfering RNA-mediated depletion of Kv1.1 or Kv1.3 attenuates RGC loss after axotomy (Koeberle, Wang, & Schlichter, 2010). As with many other treatments, however, the protection that is afforded is transitory.

Bcl-2 Family Members

Proteins of the Bcl-2 (B-cell lymphoma-2) family play a central role in controlling the life and death of cells by regulating the mitochondrial transition pore. Mature rat RGCs expresses low levels of the anti-apoptotic proteins Bcl-2 and Bcl-x$_L$, and levels decline further after optic nerve injury (Chaudhary et al., 1999; Isenmann et al., 1997; Levin et al., 1997). Of the many neuroprotective strategies that have been tested, overexpression of Bcl-2 or Bcl-x$_L$ provides the strongest protection of axotomized RGCs described to date (Bonfanti et al., 1996; Malik et al., 2005). Bcl-2 overexpression has also been reported to prolong RGCs' ability to regenerate axons into the optic tectum for a short while after the elongation phase normally ends (Chen et al., 1997), but it does not enable mature RGCs to regenerate injured axons through the mature optic nerve (Chierzi et al., 1999) or through a peripheral nerve graft (Inoue et al., 2002). Likewise, overexpression of Bcl-x$_L$ only slightly enhances regeneration (Malik et al., 2005). In addition to causing a down-regulation of anti-apoptotic Bcl family members, optic nerve injury leads to a transient elevation of the pro-apoptotic proteins Bax (Isenmann et al., 1997) and Bim (Napankangas et al., 2003). Suppressing expression or activity of either of these proteins transiently protects RGCs (Isenmann et al., 1999; McKernan & Cotter, 2007; Qin, Patil, & Sharma, 2004).

Caspase Inhibition

Caspases are cysteine proteases that are involved in apoptotic cell death and other cellular processes. Caspase-3 is activated by upstream caspases and is part of a final common pathway that leads to cell death by both ligand-gated and intrinsic pathways. Multiple intraocular injections of a caspase-3 inhibitor attenuate RGC death after axotomy, but RGCs nevertheless continue to die (Kermer et al., 1999). Likewise, multiple intraocular injections of an inhibitor of Caspase-9, an upstream activator of cell death pathways, have transitory effects in attenuating RGC death (Kermer et al., 2000). These findings suggest that axotomized RGCs may die by both apoptotic and nonapoptotic mechanisms. Inhibition of caspases-6 and -8 promotes RGC survival to a similar extent as inhibition of caspases-3 and -9 and induces a modest amount of axon regeneration (see figure 97.4: Monnier et al., 2011).

Nitric Oxide (NO) and Reactive Oxygen Species (ROS)

NO levels increase in many retinal cells within a few days of optic nerve damage. NO increases in microglia and Muller cells as a result of increased expression of inducible nitric oxide synthase (iNOS), and at the same time, levels of NO increase in RGCs due to increased expression of constitutive NOS (cNOS, neuronal NOS, nNOS) (Koeberle & Ball, 1999). Blocking the activity of either iNOS or cNOS increases RGC survival 2–3 fold

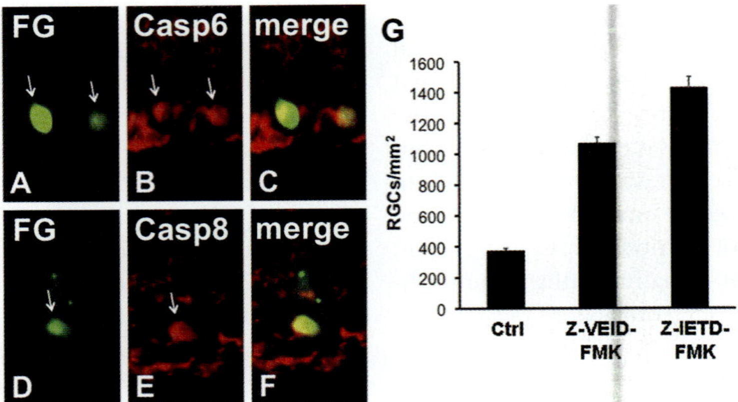

FIGURE 97.4 Caspases-6 and -8 (CASP6, CASP8) contribute to retinal ganglion cell (RGC) loss after optic nerve injury. (A–F) Localization of CASP6 (A–C) and CASP8 (D–F) in RGCs 7 days after optic nerve damage. (A, D) RGCs retrogradely labeled with Fluorogold (FG). (B, E) Immunostaining for CASP6 and CASP8, respectively. (C, F) Merged images show that CASP6 and -8 are in RGCs. (G) Intraocular injection of inhibitors of CASP6 (Z-VEID-FMK) or -8 (Z-IETD-FMK) promotes RGC survival 14 days after nerve injury.

after 2 weeks, but RGCs still continue to die, and multiple injections are no more effective than one (Koeberle & Ball, 1999). Thus, although NO contributes to the death of RGCs after optic nerve injury, it is unlikely to be an early trigger of death or sole mediator of this loss. The anti-inflammatory cytokines IL-4 and IL-10 attenuate RGC death by suppressing NO production and protein nitrotyrosylation (Koeberle, Gauldie, & Ball, 2004).

Optic nerve injury leads to the generation of superoxide in a small fraction of RGCs and peaks at 4–5 days (Kanamori et al., 2010). Inactivation of superoxide with the enzyme superoxide dismutase (SOD) decreases the number of RGCs undergoing apoptosis but has only a modest effect on overall RGC death (Kanamori et al., 2010). Thus, superoxide production appears to be a relatively early event in RGC loss, but again, not an obligatory pathway toward RGC death. Superoxide can also be attenuated using tris (2-carboxyethyl) phosphine, a sulfhydryl reducing agent, which may act by preventing oxidative damage to proteins (Swanson et al., 2005).

AXON REGENERATION IN THE MATURE MAMMALIAN OPTIC NERVE

Intraocular Inflammation as a Stimulus for Regeneration

Although axon regeneration through the mature optic nerve was long considered to be impossible, several studies have now obtained extensive regeneration by stimulating an inflammatory reaction in the eye. In 1996, Berry, Carlile, and Hunter reported that implanting a fragment of peripheral nerve into the posterior chamber of the eye caused RGCs to extend lengthy axons down the rat optic nerve, and they ascribed this effect to trophic factors secreted by Schwann cells. However, these implants also contained numerous inflammatory cells (Berry, Carlile, & Hunter, 1996), and subsequent studies found that other methods that induce intraocular inflammation evoke at least as much regeneration as the implants. Some of these alternative manipulations include injuring the lens (Fischer, Heiduschka, & Thanos, 2001; Fischer et al., 2004b; Leon et al., 2000; Lorber, Berry, & Logan, 2005) injecting Zymosan, a yeast cell wall extract (Leon et al., 2000; Yin et al., 2003), or Pam3Cys, a toll-like receptor-2 agonist (Hauk et al., 2010), or the chemokine ciliary neurotrophic factor (CNTF) (Cen et al., 2007). Intraocular inflammation causes RGCs to shift their program of gene expression and extend lengthy axons beyond the site of nerve injury (see figure 97.5A, B).

Oncomodulin, cAMP, and Mannose

The proregenerative effect of inflammation is mediated in large part by Oncomodulin (Ocm), an atypical growth factor. As noted earlier, studies in lower vertebrates had indicated that optic nerve regeneration is facilitated by non-cell-autonomous factors that include a small carbohydrate. Parallel studies using rat retinal neurons show that low micromolar concentrations of the carbohydrate mannose similarly cause rat RGCs to extend axons in culture provided intracellular cyclic AMP levels ($[cAMP]_i$) are elevated (see figure 97.2E) (Li et al., 2003). However, because mannose is abundant in the mammalian vitreous and cerebrospinal

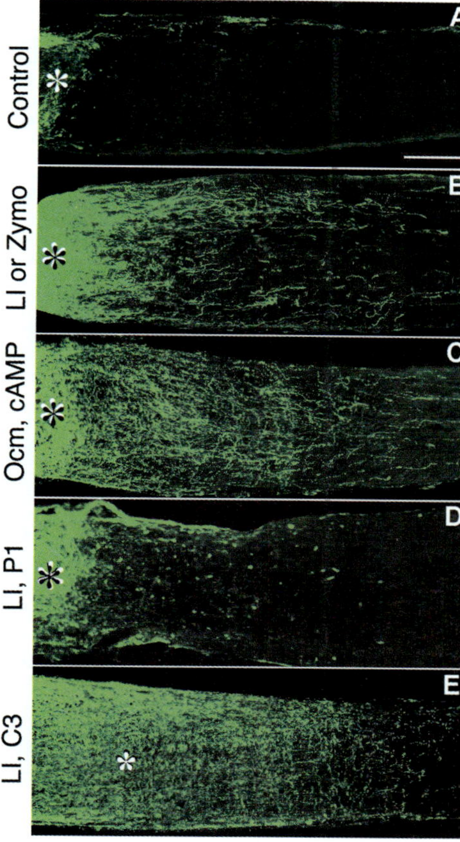

FIGURE 97.5 Axon regeneration in the mature rat optic nerve: role of oncomodulin (Ocm) and cell-extrinsic inhibitors of growth. Longitudinal sections through the mature rat optic nerve 2 weeks after nerve crush. Axons regenerating beyond the injury site (asterisk) are visualized by immunostaining for GAP-43 (green fluorescence). (A) Little or no regeneration is seen beyond the injury site under normal circumstances. (B) Intraocular inflammation, induced by either lens injury (LI) or Zymosan (Zymo), leads to appreciable regeneration (Yin et al., 2003). (C) Gain of function studies: slow-release of oncmodulin (Ocm) from polymeric beads, when combined with a cyclic AMP (cAMP) analogue, induces appreciation regeneration. Translocation of the Ocm receptor to the cell surface is cAMP-dependent, and thus neither Ocm or cAMP is effective alone (Yin et al., 2006). (D) Loss-of-function: Regeneration induced by LI is suppressed by P1, a peptide that prevents Ocm from binding to its receptor (Yin et al., 2009). (E) Combinatorial treatment: Ectopic expression of C3 ribosyltransferase in RGCs inactivates RhoA and increases the effects of intraocular inflammation. (cf. panel B) (Fischer et al., 2004b). Scale for all images: 200 μm.

fluid, its role in regeneration is likely to be permissive rather than regulatory. Elevation of [cAMP]$_i$ increases the effects of mannose and other growth factors but does not induce extensive regeneration by itself (Cui et al., 2003; Kurimoto et al., 2010; Monsul et al., 2004; Yin et al., 2006). The principal factor from inflammatory cells that enables RGCs to switch into an active growth state is the small Ca^{2+}-binding protein Ocm (Yin et al., 2006). Upon inducing an inflammatory response, neutrophils and macrophages enter the eye and secrete high levels of Ocm, which binds to a high-affinity receptor on RGCs in a cAMP-dependent manner (Kurimoto et al., 2010; Yin et al., 2006). Slow-release of Ocm plus a cAMP analogue from a biopolymer mimics the proregenerative effects of intraocular inflammation (see figure 97.5C), and conversely, a peptide antagonist of Ocm, or immune-depletion of Ocm, suppresses inflammation-induced regeneration (see figure 97.5D: Kurimoto et al., 2010; Yin et al., 2006, 2009). Administering a cAMP analogue at the time of inflammation prolongs the binding of Ocm to its receptor and doubles the amount of regeneration induced by intraocular inflammation. Conversely, an antagonist of the cAMP-dependent protein kinase diminishes Ocm binding to RGCs and suppresses inflammation-induced regeneration (Kurimoto et al., 2010). One study failed to detect Ocm in the eye after intraocular inflammation and reported that neither immune-depletion of Ocm nor a macrophage toxin suppressed regeneration (Hauk et al., 2008). The likely explanations for these discrepant results have been discussed elsewhere (Yin et al., 2009).

Role of Other Trophic Factors in Regeneration

Several observations suggest that, in addition to Ocm, other trophic factors are likely to contribute to inflammation-induced regeneration. One indication that other factors are involved is that, whereas inflammation enhances both axon growth and RGC survival, Ocm antagonists only block regeneration while not affecting the prosurvival effects of inflammation (Kurimoto et al., 2010; Yin et al., 2009). In addition, the ability of Ocm to bind to its receptor requires factors that increase [cAMP]$_i$ in RGCs. One factor that normally contributes to RGC survival in the rat and augments the effects of Ocm following intraocular inflammation is stromal-derived factor-1 (Yin et al., 2012).

CNTF and leukemia inhibitory factor (LIF) may also contribute to maintaining RGC survival but are unlikely to be major effectors of inflammation-induced axon growth *per se*. One group reported that multiple intraocular injections of high concentrations of CNTF enable RGCs to regenerate axons when explanted into culture (Leibinger et al., 2009; Muller, Hauk, & Fischer, 2007), but many laboratories have either failed to find appreciable effects of CNTF in vivo or in culture (Cohen, Bray, & Aguayo, 1994; Leon et al., 2000; Lingor et al., 2007; Pernet & Di Polo, 2006; Smith et al., 2009; Yin et al., 2009) or have found that CNTF can stimulate growth via an indirect mechanism that involves macrophage chemotaxis (Cen et al., 2007). The absence of appreciable regeneration when the effects of intraocular inflammation are blocked with Ocm inhibitors (Kurimoto et al., 2010) provides further evidence that neither CNTF nor any other factors are sufficient to stimulate regeneration when Ocm is blocked. Another group reported positive effects of virally expressed CNTF (Leaver et al., 2006), but this was not found in an earlier study (Weise et al., 2000). The normally low response of mature RGCs to CNTF can be attributed to the high levels of SOCS3, a suppressor of intracellular signaling through the Jak-STAT pathway (Park et al., 2009), in mature RGCs, and the further elevation of SOCS3 that occurs following axonal injury and intraocular inflammation (Fischer et al., 2004b). On the other hand, deletion of the *cntf* and *lif* genes causes RGCs to die rapidly after axonal injury (Leibinger et al., 2009), suggesting that these latter two factors may contribute to the normal survival of RGCs.

Several other growth factors can affect RGC survival, outgrowth, or both but have not been implicated in inflammation-induced regeneration. Fibroblast growth factor-2 has a modest effect on optic nerve regeneration (Sapieha et al., 2003). As noted above, BDNF enhances RGC survival after optic nerve damage but reverses the effects of intraocular inflammation on regeneration (Pernet & Di Polo, 2006). GDNF, neurturin, and IGF-1 all enhance RGC survival, but their effects on regeneration have not been reported (Koeberle & Ball, 2002). In cats, a combination of CNTF, BDNF, and cAMP augments the ability of RGCs to regenerate axons through a peripheral nerve graft, but this may be secondary to the effect of these agents on cell survival (Watanabe et al., 2003).

Changes in Neuronal Gene Expression

Optic nerve regeneration involves a dramatic shift in RGCs' program of gene expression. Whereas earlier studies used in situ hybridization or immunohistochemistry to investigate changes in the expression of particular molecules of interest, recent studies have used more comprehensive approaches, for example, transcriptional profiling of RGCs isolated at various stages of regeneration (Fischer et al., 2004b; Sun et al., 2012). Many of the changes seen during regeneration resemble those seen in regenerating RGCs of lower vertebrates (Veldman et al., 2007) and/or sensory neurons of the dorsal root ganglia regenerating their axons after peripheral nerve injury (Fischer et al., 2004b). For example, the phosphoprotein GAP-43 is up-regulated in RGCs of lower vertebrates during optic nerve regeneration, DRG neurons of higher vertebrates during

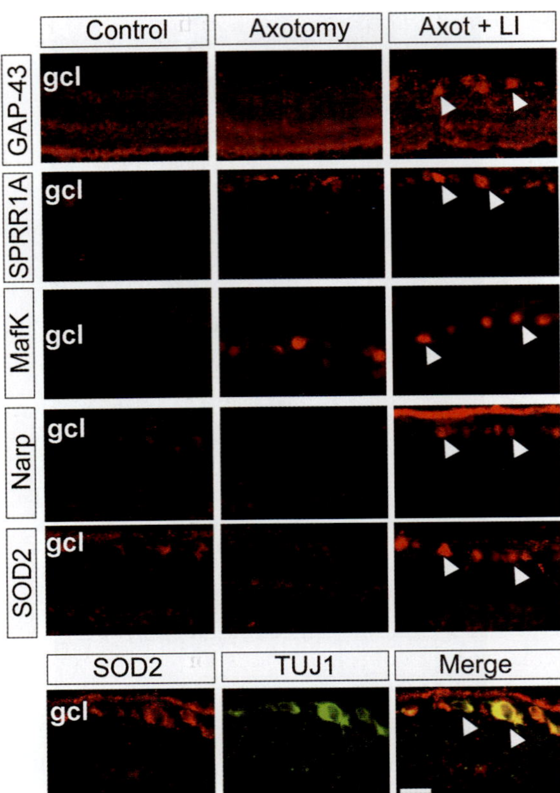

FIGURE 97.6 Changes in protein expression during optic nerve regeneration. Regeneration induced by intraocular inflammation leads to up-regulation of proteins associated with the cytoskeleton and growth cone, including GAP-43 and SPRR1A, the transcription factor MafK, the secreted synaptic protein Narp, and the neuroprotective protein SOD2. Shown are levels of proteins detected by immunofluorescence in retinal ganglion cells (RGCs) of normal control rat or 4 days after axotomy alone, or 4 days after nerve damage combined with intraocular inflammation induced by lens injury (Axot + LI). All of these changes occur within RGCs per se, as demonstrated by the example in the bottom row (double-immunostaining for SOD2 and the RGC marker βIII-tubulin). Scale bar: 20 μm (Leon et al., 2000; Fischer et al., 2004b).

peripheral nerve regeneration, and mature mammalian RGCs undergoing axon regeneration through a peripheral nerve graft or, when stimulated, through the optic nerve itself (see figure 97.6) (Benowitz & Lewis, 1983; Berry, Carlile, & Hunter, 1996; Doster et al., 1991; Leon et al., 2000; Skene & Willard, 1981a, 1981b). GAP-43 is important for axon pathfinding and subsequent synaptic plasticity (Benowitz & Routtenberg, 1997). Other components of RGCs' growth program include increased expression of additional cytoskeletal components (e.g., MAP1B, Sprr1a, Fn14), cell adhesion molecules, transcription factors (c-Jun, MafK, C/EBPs, ATF3), prosurvival factors (IGF-1, Hsp27, SOD2), regulators of growth through the extracellular matrix (TIMP1, metalloproteinases) and many other genes whose contributions to regeneration remain to be determined (see figure 97.6) (Ahmed et al., 2006; Becker, Becker, & Meyer, 2001; Dieterich et al., 2002; Fischer et al., 2004b; Hull & Bahr, 1994; Sun et al., 2012).

Intracellular Signaling Pathways

The ability of Ocm to stimulate regeneration involves multiple signal transduction pathways. Pharmacological blockers of the MAP kinase, PI3 kinase, or Jak-STAT pathways fail to suppress the effects of Ocm when tested individually, but blocking any two of these blocks the effects of Ocm completely (Yin et al., 2006). This finding suggests that the effects of Ocm involve several signal-transduction pathways acting in parallel but leaves open the question of whether one or more of these pathways is activated by Ocm or needs to be active constitutively, and whether this activation is complete or remains limited. These questions are answered in part by the discovery that deleting a suppressor of intracellular signaling strongly amplifies the effects of intraocular inflammation.

The protein phosphatase and tensin homologue (PTEN) suppresses signaling through the PI3 kinase pathway, and deleting the *pten* gene in RGCs is sufficient to promote appreciable optic nerve regeneration in mice (Park et al., 2008). This effect is due in part to activation of the mTOR pathway, which controls protein translation (Sun et al., 2012). Combining *pten* deletion with intraocular inflammation and cAMP increases long-distance regeneration 10-fold over the level induced by any of these treatments alone (see figure 97.7) and enables RGCs to regenerate axons into the basal diencephalon within 6 weeks (Kurimoto et al., 2010). These results imply that Ocm signaling only partially activates the PI3 kinase pathway when PTEN is present, and that deleting the *pten* gene greatly augments the effects of Ocm. Conversely, these results show that activation of PI3 kinase alone does not produce maximal levels of regeneration, and that further activation of intracellular signaling by Ocm (and perhaps other factors associated with inflammation) is required for greater levels of regeneration.

Suppressor of cytokine signaling-3 (SOCS3) suppresses signal transduction through the Jak-STAT pathway, which is activated by trophic factors of the CNTF family. CNTF by itself has little effect on regeneration when SOCS3 is present but becomes active when SOCS3 is deleted (Smith et al., 2009). *socs3* deletion combined with CNTF synergistically enhances regeneration induced by deleting the *pten* gene although the number of regenerating axons declines with distance along the optic nerve and few axons cross the optic chiasm (Sun et al., 2012).

The protein kinase Mst3b (STK24) also plays an important role in optic nerve regeneration. Mst3b is a neuron-specific homolog of Sterile20, an upstream

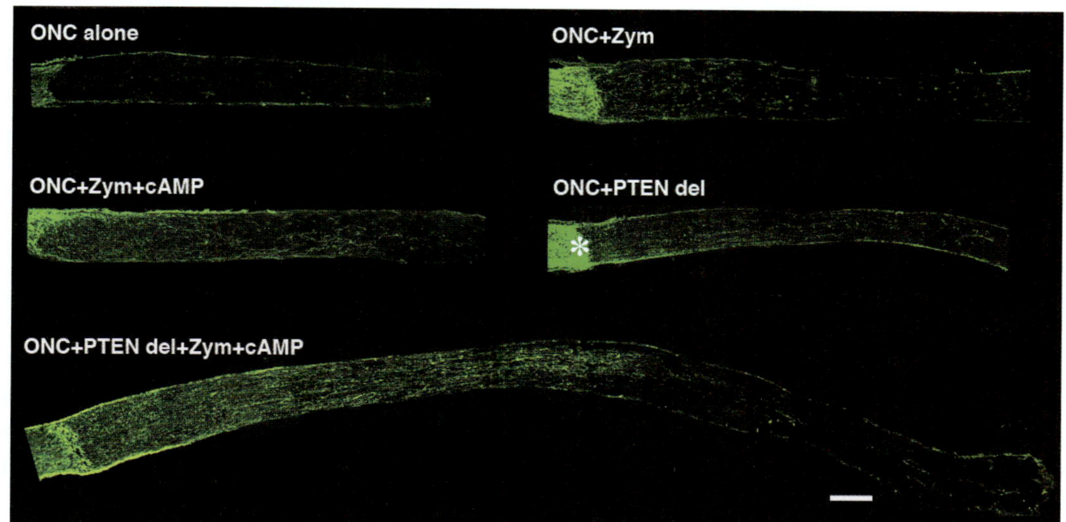

FIGURE 97.7   Intraocular inflammation, [cAMP]ᵢ elevation, and *pten* gene deletion have synergistic effects on regeneration. Longitudinal sections through the adult mouse optic nerve 2 weeks after optic nerve crush (ONC) coupled with various treatments as shown. Regeneration increases progressively with injections of Zymosan (Zym), a cyclic AMP (cAMP) analogue, and deletion of the *pten* gene in RGCs of *pten*^(flx/flx) mice (via intraocular injection of adeno-associated viruses, AAV2, expressing Cre recombinase). Controls received AAV2 expressing green fluorescent protein. Optic nerves were crushed 1 mm behind the eye 2 weeks after virus injection (Kurimoto et al., 2010). Scale bar for all images: 200 µm.

activator of a MAP kinase signaling pathway that controls budding in yeast (Dan, Watanabe, & Kusumi, 2001; Huang et al., 2002; Schinkmann & Blenis, 1997). Expression of a constitutively active Mst3b mutant enables RGCs to extend axons in the absence of Ocm or other trophic factors, and conversely, blocking either the expression or activity of Mst3b blocks the axon-promoting effects of Ocm in culture and the effects of inflammation on optic nerve regeneration in vivo (Lorber et al., 2009).

Activation of RGCs' Intrinsic Growth State Combined with Counteracting Inhibition

As noted above, overcoming cell-extrinsic inhibitors of axon growth results in only limited optic nerve regeneration. However, once RGCs' intrinsic growth state has been activated, overcoming inhibitory signals has a dramatic effect. Expressing a dominant-negative form of the Nogo receptor (NgR) in RGCs or genetically deleting one or more forms of the *NgR* gene strongly enhances the amount of regeneration stimulated by intraocular inflammation (Dickendesher et al., 2012; Fischer, He, & Benowitz, 2004a; Wang et al., 2012).

The binding of most cell-extrinsic inhibitors of axon growth to their receptors activates the small GTPase RhoA, which leads to growth cone collapse and/or arrest of axon elongation (Lehmann et al., 1999; Niederost et al., 2002; Perrot, Vazquez-Prado, & Gutkind, 2002; Yamashita & Tohyama, 2003). RhoA can be inactivated irreversibly by a bacterial ADP-ribosyltransferase, which, when introduced into RGCs, enables these cells to regenerate a modest number of axons into the optic nerve (Lehmann et al., 1999). However, when combined with trophic factors released from inflammatory cells, RhoA inactivation increases regeneration several-fold over the level induced by inflammation alone (see figure 97.5E: Fischer et al., 2004b). Inflammation-induced regeneration can also be augmented with taxol, a microtubule stabilizing agent (Sengottuvel et al., 2011).

Axon Guidance

Interactions between EphA receptors on growing RGC axons and EphrinA ligands in visual target areas play a central role in the establishment of a topographic map of visual space onto the superior colliculus and lateral geniculate nucleus (Feldheim & O'Leary, 2010). Rat RGCs continue to express a gradient of EphA5 receptors ascending from nasal to temporal retina into adulthood whereas the gradient of the cognate ligands EphrinA2 and -A5 was reported by one group to decrease as rats mature but become reestablished after axonal injury (Symonds et al., 2007) and by another group to persist unabated from the postnatal period on (Knoll et al., 2001). DCC and Unc5h2, the receptors for the guidance signal netrin, are expressed in mature RGCs but decrease after nerve injury, even when RGCs are stimulated to regenerate axons through a peripheral nerve graft (Ellezam et al., 2001; Petrausch et al., 2000).

Restoration of Central Visual Circuits

With adequate stimulation, RGCs will regenerate their axons the entire length of the visual pathway and reestablish visual circuits in adult mice.

As noted earlier, combining intraocular inflammation with cAMP and *pten* gene deletion enables RGCs to regrow damaged axons the full length of the optic nerve, across the optic chiasm, and into the diencephalon within 6 weeks (Kurimoto et al., 2010). By 10 weeks, axons navigate to their appropriate target sites and terminate in the suprachiasmatic nucleus, dorsal lateral geniculate nucleus, olivary pretectal nucleus, medial terminal nucleus, and superior colliculus, with little evidence of miswiring. At the electron microscopic level, many regenerating axons become remyelinated (see figure 97.8) (de Lima et al., 2012).

At a behavior level, mice with central reinnervation show some evidence of visual recovery. Although their behavior remains well below that of normal animals, mice with central reinnervation show evidence of depth avoidance on a visual cliff apparatus, an optomotor response in tracking a pattern of rotating vertical lines, and circadian entrainment, albeit with a considerable time lag (de Lima et al., 2012). These results indicate that there may not be any insurmountable barriers to regeneration when the intrinsic growth potential of RGCs is adequately stimulated.

CONCLUSIONS AND FUTURE PROSPECTS

The foregoing studies show that optic nerve regeneration, long known to occur in lower vertebrates and through peripheral nerve grafts in mature mammals, can also occur in mature mammals despite the existence of many obstacles. In a period of 15 years or so, we have progressed from the point of thinking that optic nerve regeneration in mature mammals might be impossible to the point where, with adequate stimulation, RGCs can be induced to regenerate injured axons the full length of the optic nerve and on into the brain, reinnervating their appropriate central target areas and

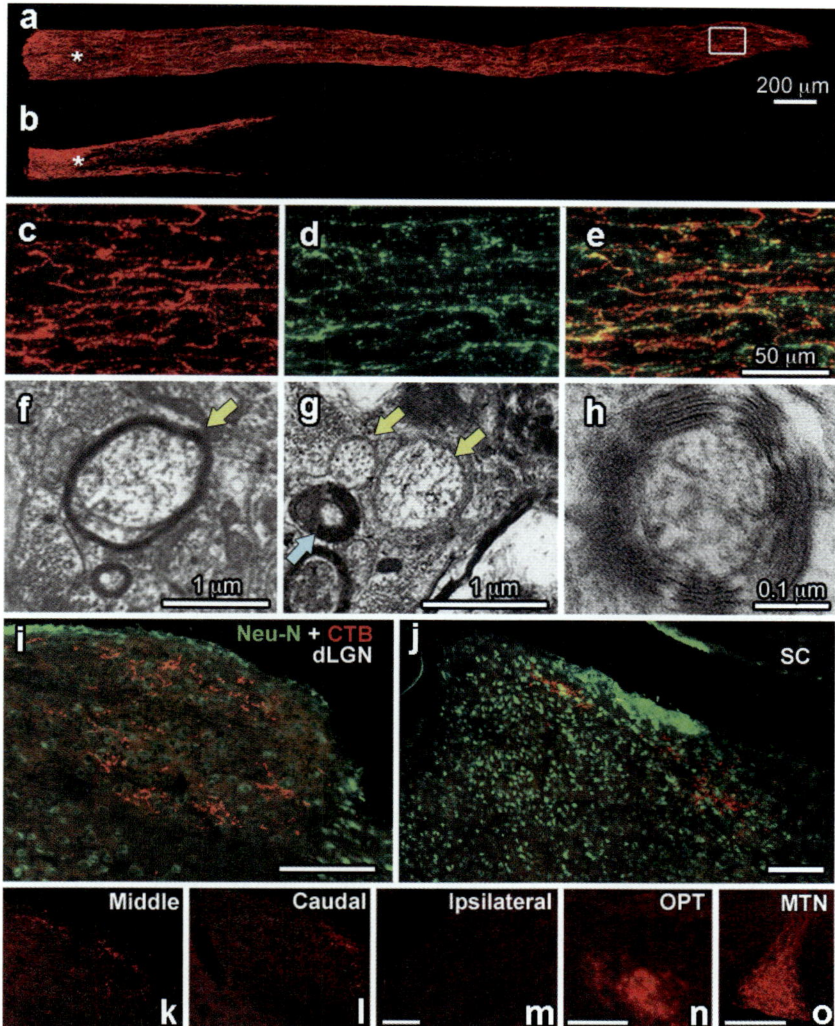

FIGURE 97.8   Restoration of central visual projections. (a, b) Axon regeneration 10 weeks after optic nerve injury in adult *pten*^{flx/flx} mice with either the *pten* gene deleted (via intraocular injection of AAV2-Cre as above, a) or present (b) in RGCs. All mice received three injections of Zymosan and CPT-cyclic AMP. Axons were traced using the anterograde tracer cholera toxin B (CTB). Asterisks denote the lesion site. (c–e) Double-staining of inset area in *a* with antibodies to GAP-43 and CTB. Some axons are still growing but are not labeled with CTB, some are labeled with CTB but are not still growing, and some show both labels. (f–h) Electron micrographs show myelinated axons halfway down the optic nerves of mice with successful axon regeneration. Some axons show thin myelin (f), some are unmyelinated (g, yellow arrows), and some show thick myelin (g, blue arrow) with multiple lamellae (h). (i, j) Regenerating axons in the dorsal lateral geniculate nucleus (dLGN, i) and superior colliculus (SC, j) contralateral to the regenerating optic nerve. Sections are immunostained for CTB (red) to visualize regenerating axons and Neu-N (green) to visualize neurons. (k, l) Other levels of the dLGN showing reinnervation at the middle (K) and caudal (L) levels. (m) dLGN ipsilateral to the lesioned side does not show any regenerating fibers. (n, o) Reinnervation of the olivary pretectal nucleus (OPT) and the medial terminal nucleus (MTN) (DeLima et al., 2012). Scale bar: *i-o* 100 µm.

enabling some functional recovery to occur. The extent to which this regeneration can be improved upon remains to be determined. Among the many questions remaining to be answered are the nature of the guidance signals that direct axons to their proper destinations, whether there exist treatments that can increase the number of RGCs that regenerate their axons, the extent to which central target areas retain their connections with secondary projection areas, whether specific types of ganglion cells are more likely to regenerate than others, whether regenerating axons can reestablish a topographic map of visual space upon central target areas, and whether visual relearning is necessary or even possible to improve functional outcome. At the same time that we explore these questions in animal models, we can begin to think how preclinical findings can be moved in a translational direction. Although the methods that have thus far induced extensive regeneration in animal models are not directly applicable to humans, they point to methods that might be. Gene therapy has been shown to be safe and effective for other ocular diseases (Bainbridge et al., 2008; Chung, Lee, & Maguire, 2009; Maguire et al., 2008) and can be improved upon for the treatment of acute disorders (Yokoi et al., 2007). Other methods of intraocular drug delivery are also available (Lavik, Kuehn, & Kwon, 2011; Thrimawithana et al., 2011), lending encouragement to the possibility of developing clinically feasible ways to provide trophic factors such as Ocm while at the same time diminishing the expression of genes that encode suppressors of growth. All in all, the prospects for optic nerve regeneration and functional recovery in animal models and eventually in the clinic are looking more promising than ever.

REFERENCES

Abdesselem, H., Shypitsyna, A., Solis, G. P., Bodrikov, V., & Stuermer, C. A. (2009). No Nogo66- and NgR-mediated inhibition of regenerating axons in the zebrafish optic nerve. *Journal of Neuroscience, 29*, 15489–15498.

Aguayo, A. J., Rasminsky, M., Bray, G. M., Carbonetto, S., McKerracher, L., Villegas-Perez, M. P., et al. (1991). Degenerative and regenerative responses of injured neurons in the central nervous system of adult mammals. *Philosophical Transactions of the Royal Society of London. Series B, Biological Sciences, 331*, 337–343.

Ahmed, Z., Suggate, E. L., Brown, E. R., Dent, R. G., Armstrong, S. J., Barrett, L. B., et al. (2006). Schwann cell-derived factor-induced modulation of the NgR/p75NTR/EGFR axis disinhibits axon growth through CNS myelin in vivo and in vitro. *Brain, 129*, 1517–1533.

Asch, W. S., Leake, D., Canger, A. K., Passini, M. A., Argenton, F., & Schechter, N. (1998). Cloning of zebrafish neurofilament cDNAs for plasticin and gefiltin: Increased mRNA expression in ganglion cells after optic nerve injury. *Journal of Neurochemistry, 71*, 20–32.

Atwal, J. K., Pinkston-Gosse, J., Syken, J., Stawicki, S., Wu, Y., Shatz, C., et al. (2008). PirB is a functional receptor for myelin inhibitors of axonal regeneration. *Science, 322*, 967–970.

Bainbridge, J. W., Smith, A. J., Barker, S. S., Robbie, S., Henderson, R., Balaggan, K., et al. (2008). Effect of gene therapy on visual function in Leber's congenital amaurosis. *New England Journal of Medicine, 358*, 2231–2239.

Ballestero, R. P., Wilmot, G. R., Agranoff, B. W., & Uhler, M. D. (1997). gRICH68 and gRICH70 are 2′,3′-cyclic-nucleotide 3′-phosphodiesterases induced during goldfish optic nerve regeneration. *Journal of Biological Chemistry, 272*, 11479–11486. doi:10.1074/jbc.272.17.11479.

Bastmeyer, M., Beckmann, M., Schwab, M. E., & Stuermer, C. A. (1991). Growth of regenerating goldfish axons is inhibited by rat oligodendrocytes and CNS myelin but not but not by goldfish optic nerve tract oligodendrocytelike

cells and fish CNS myelin. *Journal of Neuroscience, 11*, 626–640.

Battisti, W. P., Wang, J., Bozek, K., & Murray, M. (1995). Macrophages, microglia, and astrocytes are rapidly activated after crush injury of the goldfish optic nerve: A light and electron microscopic analysis. *Journal of Comparative Neurology, 354*, 306–320.

Becker, C. G., Becker, T., & Meyer, R. L. (2001). Increased NCAM-180 immunoreactivity and maintenance of L1 immunoreactivity in injured optic fibers of adult mice. *Experimental Neurology, 169*, 438–448.

Becker, C. G., Meyer, R. L., & Becker, T. (2000). Gradients of ephrin-A2 and ephrin-A5b mRNA during retinotopic regeneration of the optic projection in adult zebrafish. *Journal of Comparative Neurology, 427*, 469–483.

Benowitz, L. I., & Lewis, E. R. (1983). Increased transport of 44,000- to 49,000-dalton acidic proteins during regeneration of the goldfish optic nerve: A two-dimensional gel analysis. *Journal of Neuroscience, 3*, 2153–2163.

Benowitz, L. I., & Routtenberg, A. (1997). GAP-43: An intrinsic determinant of neuronal development and plasticity. *Trends in Neurosciences, 20*, 84–91.

Benowitz, L. I., Shashoua, V. E., & Yoon, M. G. (1981). Specific changes in rapidly transported proteins during regeneration of the goldfish optic nerve. *Journal of Neuroscience, 1*, 300–307.

Benowitz, L. I., Yoon, M. G., & Lewis, E. R. (1983). Transported proteins in the regenerating optic nerve: Regulation by interactions with the optic tectum. *Science, 222*, 185–188.

Benson, M. D., Romero, M. I., Lush, M. E., Lu, Q. R., Henkemeyer, M., & Parada, L. F. (2005). Ephrin-B3 is a myelin-based inhibitor of neurite outgrowth. *Proceedings of the National Academy of Sciences of the United States of America, 102*, 10694–10699.

Bernhardt, R. R., Tongiorgi, E., Anzini, P., & Schachner, M. (1996). Increased expression of specific recognition molecules by retinal ganglion cells and by optic pathway glia accompanies the successful regeneration of retinal axons in adult zebrafish. *Journal of Comparative Neurology, 376*, 253–264.

Berry, M., Carlile, J., & Hunter, A. (1996). Peripheral nerve explants grafted into the vitreous body of the eye promote the regeneration of retinal ganglion cell axons severed in the optic nerve. *Journal of Neurocytology, 25*, 147–170.

Blaugrund, E., Bartsch, U., Martini, R., Schachner, M., & Schwartz, M. (1990). Immunological evidence that the neural adhesion molecule L1 is expressed in fish brain and optic nerve: Possible association with optic nerve regeneration. *Brain Research, 530*, 239–244.

Bonfanti, L., Strettoi, E., Chierzi, S., Cenni, M. C., Liu, X. H., Martinou, J. C., et al. (1996). Protection of retinal ganglion cells from natural and axotomy-induced cell death in neonatal transgenic mice overexpressing bcl-2. *Journal of Neuroscience, 16*, 4186–4194.

Bortner, C. D., & Cidlowski, J. A. (2007). Cell shrinkage and monovalent cation fluxes: Role in apoptosis. *Archives of Biochemistry and Biophysics, 462*, 176–188.

Bray, G. M., Vidal-Sanz, M., & Aguayo, A. J. (1987). Regeneration of axons from the central nervous system of adult rats. *Progress in Brain Research, 71*, 373–379.

Carter, D. A., Bray, G. M., & Aguayo, A. J. (1989). Regenerated retinal ganglion cell axons can form well-differentiated synapses in the superior colliculus of adult hamsters. *Journal of Neuroscience, 9*, 4042–4050.

Cen, L. P., Luo, J. M., Zhang, C. W., Fan, Y. M., Song, Y., So, K. F., et al. (2007). Chemotactic effect of ciliary neurotrophic factor on macrophages in retinal ganglion cell survival and axonal regeneration. *Investigative Ophthalmology & Visual Science, 48*, 4257–4266.

Chaudhary, P., Ahmed, F., Quebada, P., & Sharma, S. C. (1999). Caspase inhibitors block the retinal ganglion cell death following optic nerve transection. *Brain Research. Molecular Brain Research, 67*, 36–45.

Chen, D. F., Jhaveri, S., & Schneider, G. E. (1995). Intrinsic changes in developing retinal neurons result in regenerative failure of their axons. *Proceedings of the National Academy of Sciences of the United States of America, 92*, 7287–7291.

Chen, D. F., Schneider, G. E., Martinou, J. C., & Tonegawa, S. (1997). Bcl-2 promotes regeneration of severed axons in mammalian CNS. *Nature, 385*, 434–439.

Chen, M. S., Huber, A. B., van der Haar, M. E., Frank, M., Schnell, L., Spillmann, A. A., et al. (2000). Nogo-A is a myelin-associated neurite outgrowth inhibitor and an antigen for monoclonal antibody IN-1. *Nature, 403*, 434–439.

Cheng, L., Sapieha, P., Kittlerova, P., Hauswirth, W. W., & Di Polo, A. (2002). TrkB gene transfer protects retinal ganglion cells from axotomy-induced death in vivo. *Journal of Neuroscience, 22*, 3977–3986.

Chierzi, S., Strettoi, E., Cenni, M. C., & Maffei, L. (1999). Optic nerve crush: Axonal responses in wild-type and bcl-2 transgenic mice. *Journal of Neuroscience, 19*, 8367–8376.

Chung, D. C., Lee, V., & Maguire, A. M. (2009). Recent advances in ocular gene therapy. *Current Opinion in Ophthalmology, 20*, 377–381.

Cohen, A., Bray, G. M., & Aguayo, A. J. (1994). Neurotrophin-4/5 (NT-4/5) increases adult rat retinal ganglion cell survival and neurite outgrowth in vitro. *Journal of Neurobiology, 25*, 953–959.

Cohen, J., Burne, J. F., Winter, J., & Bartlett, P. (1986). Retinal ganglion cells lose response to laminin with maturation. *Nature, 322*, 465–467.

Colavincenzo, J., & Levine, R. L. (2000). Myelin debris clearance during Wallerian degeneration in the goldfish visual system. *Journal of Neuroscience Research, 59*, 47–62.

Cui, Q., Yip, H. K., Zhao, R. C., So, K. F., & Harvey, A. R. (2003). Intraocular elevation of cyclic AMP potentiates ciliary neurotrophic factor-induced regeneration of adult rat retinal ganglion cell axons. *Molecular and Cellular Neurosciences, 22*, 49–61.

Dan, I., Watanabe, N. M., & Kusumi, A. (2001). The Ste20 group kinases as regulators of MAP kinase cascades. *Trends in Cell Biology, 11*, 220–230.

De Lima, S., Koriyama, Y., Kurimoto, T., Oliveira, J. T., Yin, Y., Li, Y., Gilbert, H.-Y., Fagiolini, M., Martinez, A. M. B., & Benowitz, L. I. (2012). Full-length axon regeneration in the adult mouse optic nerve and partial recovery of simple visual behaviors. *Proceedings of the National Academy of Sciences of the United States of America, 109*, 9149–9154.

Dickendesher, T. L., Baldwin, K. T., Mironova, Y. A., Koriyama, Y., Raiker, S. J., Askew, K. L., et al. (2012). NgR1 and NgR3 are receptors for chondroitin sulfate proteoglycans. *Nature Neuroscience, 15*, 703–712. doi:10.1038/nn.3070.

Dieterich, D. C., Trivedi, N., Engelmann, R., Gundelfinger, E. D., Gordon-Weeks, P. R., & Kreutz, M. R. (2002).

Partial regeneration and long-term survival of rat retinal ganglion cells after optic nerve crush is accompanied by altered expression, phosphorylation and distribution of cytoskeletal proteins. *European Journal of Neuroscience, 15*, 1433–1443.

Di Polo, A., Aigner, L. J., Dunn, R. J., Bray, G. M., & Aguayo, A. J. (1998). Prolonged delivery of brain-derived neurotrophic factor by adenovirus-infected Muller cells temporarily rescues injured retinal ganglion cells. *Proceedings of the National Academy of Sciences of the United States of America, 95*, 3978–3983.

Doster, S. K., Lozano, A. M., Aguayo, A. J., & Willard, M. B. (1991). Expression of the growth-associated protein GAP-43 in adult rat retinal ganglion cells following axon injury. *Neuron, 6*, 635–647.

Eitan, S., & Schwartz, M. (1993). A transglutaminase that converts interleukin-2 into a factor cytotoxic to oligodendrocytes. *Science, 261*, 106–108.

Ellezam, B., Selles-Navarro, I., Manitt, C., Kennedy, T. E., & McKerracher, L. (2001). Expression of netrin-1 and its receptors DCC and UNC-5H2 after axotomy and during regeneration of adult rat retinal ganglion cells. *Experimental Neurology, 168*, 105–115.

Fawcett, J. W., & Gaze, R. M. (1981). The organization of regenerating axons in the *Xenopus* optic nerve. *Brain Research, 229*, 487–490.

Feldheim, D. A., & O'Leary, D. D. (2010). Visual map development: Bidirectional signaling, bifunctional guidance molecules, and competition. *Cold Spring Harbor Perspectives in Biology, 2*, a001768. doi:10.1101/cshperspect.a001768.

Filbin, M. T. (2003). Myelin-associated inhibitors of axonal regeneration in the adult mammalian CNS. *Nature Reviews Neuroscience, 4*, 703–713.

Fischer, D., He, Z., & Benowitz, L. I. (2004a). Counteracting the Nogo receptor enhances optic nerve regeneration if retinal ganglion cells are in an active growth state. *Journal of Neuroscience, 24*, 1646–1651.

Fischer, D., Heiduschka, P., & Thanos, S. (2001). Lens-injury-stimulated axonal regeneration throughout the optic pathway of adult rats. *Experimental Neurology, 172*, 257–272.

Fischer, D., Petkova, V., Thanos, S., & Benowitz, L. I. (2004b). Switching mature retinal ganglion cells to a robust growth state in vivo: Gene expression and synergy with RhoA inactivation. *Journal of Neuroscience, 24*, 8726–8740.

Fisher, D., Xing, B., Dill, J., Li, H., Hoang, H. H., Zhao, Z., et al. (2011). Leukocyte common antigen-related phosphatase is a functional receptor for chondroitin sulfate proteoglycan axon growth inhibitors. *Journal of Neuroscience, 31*, 14051–14066.

Fitch, M. T., & Silver, J. (2008). CNS injury, glial scars, and inflammation: Inhibitory extracellular matrices and regeneration failure. *Experimental Neurology, 209*, 294–301.

Fournier, A. E., Gould, G. C., Liu, B. P., & Strittmatter, S. M. (2002). Truncated soluble Nogo receptor binds Nogo-66 and blocks inhibition of axon growth by myelin. *Journal of Neuroscience, 22*, 8876–8883.

Fournier, A. E., GrandPre, T., & Strittmatter, S. M. (2001). Identification of a receptor mediating Nogo-66 inhibition of axonal regeneration. *Nature, 409*, 341–346.

Gaub, P., Tedeschi, A., Puttagunta, R., Nguyen, T., Schmandke, A., & Di Giovanni, S. (2010). HDAC inhibition promotes neuronal outgrowth and counteracts growth cone collapse through CBP/p300 and P/CAF-dependent p53 acetylation. *Cell Death and Differentiation, 17,* 1392–1408.

Goh, E. L., Young, J. K., Kuwako, K., Tessier-Lavigne, M., He, Z., Griffin, J. W., et al. (2008). beta1-integrin mediates myelin-associated glycoprotein signaling in neuronal growth cones. *Molecular Brain, 1,* 10. doi:10.1186/1756-6606-1-10.

Goldberg, J. L., Klassen, M. P., Hua, Y., & Barres, B. A. (2002). Amacrine-signaled loss of intrinsic axon growth ability by retinal ganglion cells. *Science, 296,* 1860–1864.

Goldberg, J. L., Vargas, M. E., Wang, J. T., Mandemakers, W., Oster, S. F., Sretavan, D. W., et al. (2004). An oligodendrocyte lineage-specific semaphorin, Sema5A, inhibits axon growth by retinal ganglion cells. *Journal of Neuroscience, 24,* 4989–4999.

Grafstein, B., & Ingoglia, N. A. (1982). Intracranial transection of the optic nerve in adult mice: Preliminary observations. *Experimental Neurology, 76,* 318–330.

Grafstein, B., & Murray, M. (1969). Transport of protein in goldfish optic nerve during regeneration. *Experimental Neurology, 25,* 494–508.

GrandPre, T., Nakamura, F., Vartanian, T., & Strittmatter, S. M. (2000). Identification of the Nogo inhibitor of axon regeneration as a Reticulon protein. *Nature, 403,* 439–444.

Hall, C. M., & Schechter, N. (1991). Expression of neuronal intermediate filament proteins ON1 and ON2 during goldfish optic nerve regeneration: Effect of tectal ablation. *Neuroscience, 41,* 695–701.

Hauk, T. G., Leibinger, M., Muller, A., Andreadaki, N., Knippschild, U., & Fischer, D. (2010). Stimulation of axon regeneration in the mature optic nerve by intravitreal application of the Toll-like receptor 2 agonist Pam3Cys. *Investigative Ophthalmology & Visual Science, 51,* 459–464.

Hauk, T. G., Muller, A., Lee, J., Schwendener, R., & Fischer, D. (2008). Neuroprotective and axon growth promoting effects of intraocular inflammation do not depend on oncomodulin or the presence of large numbers of activated macrophages. *Experimental Neurology, 209,* 469–482.

Heacock, A. M., & Agranoff, B. W. (1976). Enhanced labeling of a retinal protein during regeneration of optic nerve in goldfish. *Proceedings of the National Academy of Sciences of the United States of America, 73,* 828–832.

Heacock, A. M., Klinger, P. D., Seguin, E. B., & Agranoff, B. W. (1984). Cholesterol synthesis and nerve regeneration. *Journal of Neurochemistry, 42,* 987–993.

Herdegen, T., Bastmeyer, M., Bahr, M., Stuermer, C., Bravo, R., & Zimmermann, M. (1993). Expression of JUN, KROX, and CREB transcription factors in goldfish and rat retinal ganglion cells following optic nerve lesion is related to axonal sprouting. *Journal of Neurobiology, 24,* 528–543.

Hieber, V., Dai, X., Foreman, M., & Goldman, D. (1998). Induction of alpha1-tubulin gene expression during development and regeneration of the fish central nervous system. *Journal of Neurobiology, 37,* 429–440.

Higenell, V., Han, S. M., Feldheim, D. A., Scalia, F., & Ruthazer, E. S. (2012). Expression patterns of Ephs and ephrins throughout retinotectal development in *Xenopus laevis. Developmental Neurobiology, 72,* 547–563.

Hirsch, S., Cahill, M. A., & Stuermer, C. A. (1995). Fibroblasts at the transection site of the injured goldfish optic nerve and their potential role during retinal axonal regeneration. *Journal of Comparative Neurology, 360,* 599–611.

Homma, K., Koriyama, Y., Mawatara, K., Yoshihiro, H., Kosaka, J., & Kato, S. (2007). Caspase activation of mammalian sterile 20-like kinase 3 (Mst3): Nuclear translocation and induction of apoptosis. *Neurochemistry International, 50,* 741–748.

Huang, C. Y., Wu, Y. M., Hsu, C. Y., Lee, W. S., Lai, M. D., Lu, T. J., et al. (2002). Caspase activation of mammalian sterile 20-like kinase 3 (Mst3): Nuclear translocation and induction of apoptosis. *Journal of Biological Chemistry, 277,* 34367–34374. doi:10.1074/jbc.M202468200.

Hull, M., & Bahr, M. (1994). Regulation of immediate-early gene expression in rat retinal ganglion cells after axotomy and during regeneration through a peripheral nerve graft. *Journal of Neurobiology, 25,* 92–105.

Ingoglia, N. A., Grafstein, B., McEwen, B. S., & McQuarrie, I. G. (1973). Axonal transport of radioactivity in the goldfish optic system following intraocular injection of labelled RNA precursors. *Journal of Neurochemistry, 20,* 1605–1615.

Inoue, T., Hosokawa, M., Morigiwa, K., Ohashi, Y., & Fukuda, Y. (2002). Bcl-2 overexpression does not enhance in vivo axonal regeneration of retinal ganglion cells after peripheral nerve transplantation in adult mice. *Journal of Neuroscience, 22,* 4468–4477.

Isenmann, S., Engel, S., Gillardon, F., & Bahr, M. (1999). Bax antisense oligonucleotides reduce axotomy-induced retinal ganglion cell death in vivo by reduction of Bax protein expression. *Cell Death and Differentiation, 6,* 673–682.

Isenmann, S., Wahl, C., Krajewski, S., Reed, J. C., & Bahr, M. (1997). Up-regulation of Bax protein in degenerating retinal ganglion cells precedes apoptotic cell death after optic nerve lesion in the rat. *European Journal of Neuroscience, 9,* 1763–1772.

Jhaveri, S., Edwards, M. A., & Schneider, G. E. (1991). Initial stages of retinofugal axon development in the hamster: Evidence for two distinct modes of growth. *Experimental Brain Research, 87,* 371–382.

Jung, M., Petrausch, B., & Stuermer, C. A. (1997). Axon-regenerating retinal ganglion cells in adult rats synthesize the cell adhesion molecule L1 but not TAG-1 or SC-1. *Molecular and Cellular Neurosciences, 9,* 116–131.

Kanamori, A., Catrinescu, M. M., Kanamori, N., Mears, K. A., Beaubien, R., & Levin, L. A. (2010). Superoxide is an associated signal for apoptosis in axonal injury. *Brain, 133,* 2612–2625.

Keirstead, S. A., Rasminsky, M., Fukuda, Y., Carter, D. A., Aguayo, A. J., & Vidal-Sanz, M. (1989). Electrophysiologic responses in hamster superior colliculus evoked by regenerating retinal axons. *Science, 246,* 255–257.

Kermer, P., Ankerhold, R., Klocker, N., Krajewski, S., Reed, J. C., & Bahr, M. (2000). Caspase-9: Involvement in secondary death of axotomized rat retinal ganglion cells in vivo. *Brain Research. Molecular Brain Research, 85,* 144–150.

Kermer, P., Klocker, N., & Bahr, M. (1999). Long-term effect of inhibition of ced 3-like caspases on the survival of axotomized retinal ganglion cells in vivo. *Experimental Neurology, 158,* 202–205.

Klocker, N., Braunling, F., Isenmann, S., & Bahr, M. (1997). In vivo neurotrophic effects of GDNF on axotomized retinal ganglion cells. *Neuroreport, 8,* 3439–3442.

Knoll, B., Isenmann, S., Kilic, E., Walkenhorst, J., Engel, S., Wehinger, J., et al. (2001). Graded expression patterns of ephrin-As in the superior colliculus after lesion of the adult mouse optic nerve. *Mechanisms of Development, 106,* 119–127.

Koeberle, P. D., & Ball, A. K. (1998). Effects of GDNF on retinal ganglion cell survival following axotomy. *Vision Research, 38,* 1505–1515. doi:10.1016/S0042-6989(02)00364-7.

Koeberle, P. D., & Ball, A. K. (1999). Nitric oxide synthase inhibition delays axonal degeneration and promotes the survival of axotomized retinal ganglion cells. *Experimental Neurology, 158,* 366–381.

Koeberle, P. D., & Ball, A. K. (2002). Neurturin enhances the survival of axotomized retinal ganglion cells in vivo: Combined effects with glial cell line-derived neurotrophic factor and brain-derived neurotrophic factor. *Neuroscience, 110,* 555–567.

Koeberle, P. D., Gauldie, J., & Ball, A. K. (2004). Effects of adenoviral-mediated gene transfer of interleukin-10, interleukin-4, and transforming growth factor-beta on the survival of axotomized retinal ganglion cells. *Neuroscience, 125,* 903–920.

Koeberle, P. D., Wang, Y., & Schlichter, L. C. (2010). Kv1.1 and Kv1.3 channels contribute to the degeneration of retinal ganglion cells after optic nerve transection in vivo. *Cell Death and Differentiation, 17,* 134–144.

Koriyama, Y., Homma, K., Sugitani, K., Higuchi, Y., Matsukawa, T., Murayama, D., et al. (2007). Upregulation of IGF-I in the goldfish retinal ganglion cells during the early stage of optic nerve regeneration. *Neurochemistry International, 50,* 749–756.

Koriyama, Y., Sugitani, K., Matsukawa, T., & Kato, S. (2012). An application for mammalian optic nerve repair by fish regeneration-associated genes. *Advances in Experimental Medicine and Biology, 723,* 161–166.

Koriyama, Y., Yasuda, R., Homma, K., Mawatari, K., Nagashima, M., Sugitani, K., et al. (2009). Nitric oxide-cGMP signaling regulates axonal elongation during optic nerve regeneration in the goldfish in vitro and in vivo. *Journal of Neurochemistry, 110,* 890–901.

Kurimoto, T., Yin, Y., Omura, K., Gilbert, H. Y., Kim, D., Cen, L. P., et al. (2010). Long-distance axon regeneration in the mature optic nerve: Contributions of oncomodulin, cAMP, and pten gene deletion. *Journal of Neuroscience, 30,* 15654–15663.

Landreth, G. E., & Agranoff, B. W. (1976). Explant culture of adult goldfish retina: Effect of prior optic nerve crush. *Brain Research, 118,* 299–303.

Landreth, G. E., & Agranoff, B. W. (1979). Explant culture of adult goldfish retina: A model for the study of CNS regeneration. *Brain Research, 161,* 39–55.

Langhorst, M. F., Reuter, A., & Stuermer, C. A. (2005). Scaffolding microdomains and beyond: The function of reggie/flotillin proteins. *Cellular and Molecular Life Sciences, 62,* 2228–2240. doi:10.1007/s00018-005-5166-4.

Lavik, E., Kuehn, M. H., & Kwon, Y. H. (2011). Novel drug delivery systems for glaucoma. *Eye (London, England), 25,* 578–586.

Leaver, S. G., Cui, Q., Plant, G. W., Arulpragasam, A., Hisheh, S., Verhaagen, J., et al. (2006). AAV-mediated expression of CNTF promotes long-term survival and regeneration of adult rat retinal ganglion cells. *Gene Therapy, 13,* 1328–1341. doi:10.1038/sj.gt.3302791.

Lehmann, M., Fournier, A., Selles-Navarro, I., Dergham, P., Sebok, A., Leclerc, N., et al. (1999). Inactivation of Rho signaling pathway promotes CNS axon regeneration. *Journal of Neuroscience, 19,* 7537–7547.

Leibinger, M., Muller, A., Andreadaki, A., Hauk, T. G., Kirsch, M., & Fischer, D. (2009). Neuroprotective and axon growth-promoting effects following inflammatory stimulation on mature retinal ganglion cells in mice depend on ciliary neurotrophic factor and leukemia inhibitory factor. *Journal of Neuroscience, 29,* 14334–14341.

Leon, S., Yin, Y., Nguyen, J., Irwin, N., & Benowitz, L. I. (2000). Lens injury stimulates axon regeneration in the mature rat optic nerve. *Journal of Neuroscience, 20,* 4615–4626.

Levin, L. A., Schlamp, C. L., Spieldoch, R. L., Geszvain, K. M., & Nickells, R. W. (1997). Identification of the bcl-2 family of genes in the rat retina. *Investigative Ophthalmology & Visual Science, 38,* 2545–2553.

Li, Y., Irwin, N., Yin, Y., Lanser, M., & Benowitz, L. I. (2003). Axon regeneration in goldfish and rat retinal ganglion cells: Differential responsiveness to carbohydrates and cAMP. *Journal of Neuroscience, 23,* 7830–7838.

Lingor, P., Teusch, N., Schwarz, K., Mueller, R., Mack, H., Bahr, M., et al. (2007). Inhibition of Rho kinase (ROCK) increases neurite outgrowth on chondroitin sulphate proteoglycan in vitro and axonal regeneration in the adult optic nerve in vivo. *Journal of Neurochemistry, 103,* 181–189.

Lorber, B., Berry, M., & Logan, A. (2005). Lens injury stimulates adult mouse retinal ganglion cell axon regeneration via both macrophage- and lens-derived factors. *European Journal of Neuroscience, 21,* 2029–2034.

Lorber, B., Howe, M. L., Benowitz, L. I., & Irwin, N. (2009). Mst3b, an Ste20-like kinase, regulates axon regeneration in mature CNS and PNS pathways. *Nature Neuroscience, 12,* 1407–1414.

Maguire, A. M., Simonelli, F., Pierce, E. A., Pugh, E. N., Jr., Mingozzi, F., Bennicelli, J., et al. (2008). Safety and efficacy of gene transfer for Leber's congenital amaurosis. *New England Journal of Medicine, 358,* 2240–2248. doi:10.1056/NEJMoa0802315.

Malik, J. M., Shevtsova, Z., Bahr, M., & Kugler, S. (2005). Long-term in vivo inhibition of CNS neurodegeneration by Bcl-XL gene transfer. *Molecular Therapy, 11,* 373–381.

Mansour-Robaey, S., Clarke, D. B., Wang, Y. C., Bray, G. M., & Aguayo, A. J. (1994). Effects of ocular injury and administration of brain-derived neurotrophic factor on survival and regrowth of axotomized retinal ganglion cells. *Proceedings of the National Academy of Sciences of the United States of America, 91,* 1632–1636.

Matsukawa, T., Sugitani, K., Mawatari, K., Koriyama, Y., Liu, Z., Tanaka, M., et al. (2004). Role of purpurin as a retinol-binding protein in goldfish retina during the early stage of optic nerve regeneration: Its priming action on neurite outgrowth. *Journal of Neuroscience, 24,* 8346–8353.

McKernan, D. P., & Cotter, T. G. (2007). A critical role for Bim in retinal ganglion cell death. *Journal of Neurochemistry, 102,* 922–930.

McKerracher, L., David, S., Jackson, D. L., Kottis, V., Dunn, R. J., & Braun, P. E. (1994). Identification of myelin-associated glycoprotein as a major myelin-derived inhibitor of neurite growth. *Neuron, 13,* 805–811.

Mey, J., & Thanos, S. (1993). Intravitreal injections of neurotrophic factors support the survival of axotomized retinal ganglion cells in adult rats in vivo. *Brain Research, 602,* 304–317. doi:10.1016/0006-8993(93)90695-J.

Mi, S., Lee, X., Shao, Z., Thill, G., Ji, B., Relton, J., et al. (2004). LINGO-1 is a component of the Nogo-66 receptor/p75 signaling complex. *Nature Neuroscience, 7*, 221–228.

Monnier, P. P., D'Onofrio, P. M., Magharious, M., Hollander, A. C., Tassew, N., Szydlowska, K., et al. (2011). Involvement of caspase-6 and caspase-8 in neuronal apoptosis and the regenerative failure of injured retinal ganglion cells. *Journal of Neuroscience, 31*, 10494–10505.

Monnier, P. P., Sierra, A., Schwab, J. M., Henke-Fahle, S., & Mueller, B. K. (2003). The Rho/ROCK pathway mediates neurite growth-inhibitory activity associated with the chondroitin sulfate proteoglycans of the CNS glial scar. *Molecular and Cellular Neurosciences, 22*, 319–330.

Monsul, N. T., Geisendorfer, A. R., Han, P. J., Banik, R., Pease, M. E., Skolasky, R. L., Jr., et al. (2004). Intraocular injection of dibutyryl cyclic AMP promotes axon regeneration in rat optic nerve. *Experimental Neurology, 186*, 124–133.

Moore, D. L., Blackmore, M. G., Hu, Y., Kaestner, K. H., Bixby, J. L., Lemmon, V. P., et al. (2009). KLF family members regulate intrinsic axon regeneration ability. *Science, 326*, 298–301.

Moreau-Fauvarque, C., Kumanogoh, A., Camand, E., Jaillard, C., Barbin, G., Boquet, I., et al. (2003). The transmembrane semaphorin Sema4D/CD100, an inhibitor of axonal growth, is expressed on oligodendrocytes and upregulated after CNS lesion. *Journal of Neuroscience, 23*, 9229–9239.

Moya, K. L., Benowitz, L. I., Jhaveri, S., & Schneider, G. E. (1988). Changes in rapidly transported proteins in developing hamster retinofugal axons. *Journal of Neuroscience, 8*, 4445–4454.

Mukhopadhyay, G., Doherty, P., Walsh, F. S., Crocker, P. R., & Filbin, M. T. (1994). A novel role for myelin-associated glycoprotein as an inhibitor of axonal regeneration. *Neuron, 13*, 757–767.

Muller, A., Hauk, T. G., & Fischer, D. (2007). Astrocyte-derived CNTF switches mature RGCs to a regenerative state following inflammatory stimulation. *Brain, 130*, 3308–3320. doi:10.1093/brain/awm257.

Munderloh, C., Solis, G. P., Bodrikov, V., Jaeger, F. A., Wiechers, M., Malaga-Trillo, E., & Stuermer, C. A. (2009). Reggies/flotillins regulate retinal axon regeneration in the zebrafish optic nerve and differentiation of hippocampal and N2a neurons. *Journal of Neuroscience, 29*, 6607–6615.

Murray, M. (1973). 3 H-uridine incorporation by regenerating retinal ganglion cells of goldfish. *Experimental Neurology, 39*, 489–497.

Murray, M., & Grafstein, B. (1969). Changes in the morphology and amino acid incorporation of regenerating goldfish optic neurons. *Experimental Neurology, 23*, 544–560.

Nagashima, M., Fujikawa, C., Mawatari, K., Mori, Y., & Kato, S. (2011). HSP70, the earliest-induced gene in the zebrafish retina during optic nerve regeneration: Its role in cell survival. *Neurochemistry International, 58*, 888–895.

Nagashima, M., Sakurai, H., Mawatari, K., Koriyama, Y., Matsukawa, T., & Kato, S. (2009). Involvement of retinoic acid signaling in goldfish optic nerve regeneration. *Neurochemistry International, 54*, 229–236.

Napankangas, U., Lindqvist, N., Lindholm, D., & Hallbook, F. (2003). Rat retinal ganglion cells upregulate the pro-apoptotic BH3-only protein Bim after optic nerve transection. *Brain Research. Molecular Brain Research, 120*, 30–37.

Niederost, B., Oertle, T., Fritsche, J., McKinney, R. A., & Bandtlow, C. E. (2002). Nogo-A and myelin-associated glycoprotein mediate neurite growth inhibition by antagonistic regulation of RhoA and Rac1. *Journal of Neuroscience, 22*, 10368–10376.

Nona, S. N., Thomlinson, A. M., & Stafford, C. A. (1998). Temporary colonization of the site of lesion by macrophages is a prelude to the arrival of regenerated axons in injured goldfish optic nerve. *Journal of Neurocytology, 27*, 791–803.

Park, J. B., Yiu, G., Kaneko, S., Wang, J., Chang, J., He, X. L., et al. (2005). A TNF receptor family member, TROY, is a coreceptor with Nogo receptor in mediating the inhibitory activity of myelin inhibitors. *Neuron, 45*, 345–351.

Park, K. K., Hu, Y., Muhling, J., Pollett, M. A., Dallimore, E. J., Turnley, A. M., et al. (2009). Cytokine-induced SOCS expression is inhibited by cAMP analogue: Impact on regeneration in injured retina. *Molecular and Cellular Neurosciences, 41*, 313–324.

Park, K. K., Liu, K., Hu, Y., Smith, P. D., Wang, C., Cai, B., et al. (2008). Promoting axon regeneration in the adult CNS by modulation of the PTEN/mTOR pathway. *Science, 322*, 963–966.

Peinado-Ramon, P., Salvador, M., Villegas-Perez, M. P., & Vidal-Sanz, M. (1996). Effects of axotomy and intraocular administration of NT-4, NT-3, and brain-derived neurotrophic factor on the survival of adult rat retinal ganglion cells: A quantitative in vivo study. *Investigative Ophthalmology & Visual Science, 37*, 489–500.

Pernet, V., & Di Polo, A. (2006). Synergistic action of brain-derived neurotrophic factor and lens injury promotes retinal ganglion cell survival, but leads to optic nerve dystrophy in vivo. *Brain, 129*, 1014–1026. doi:10.1093/brain/awl015.

Perrot, V., Vazquez-Prado, J., & Gutkind, J. S. (2002). Plexin B regulates Rho through the guanine nucleotide exchange factors leukemia-associated Rho GEF (LARG) and PDZ-RhoGEF. *Journal of Biological Chemistry, 277*, 43115–43120. doi:10.1074/jbc.M206005200.

Petrausch, B., Jung, M., Leppert, C. A., & Stuermer, C. A. (2000). Lesion-induced regulation of netrin receptors and modification of netrin-1 expression in the retina of fish and grafted rats. *Molecular and Cellular Neurosciences, 16*, 350–364.

Qin, Q., Patil, K., & Sharma, S. C. (2004). The role of Bax-inhibiting peptide in retinal ganglion cell apoptosis after optic nerve transection. *Neuroscience Letters, 372*, 17–21. doi:10.1016/j.neulet.2004.08.075.

Ramon y Cajal, S. (1991). *Degeneration and regeneration of the nervous system.* New York: Oxford University Press.

Rodger, J., Bartlett, C. A., Beazley, L. D., & Dunlop, S. A. (2000). Transient up-regulation of the rostrocaudal gradient of ephrin A2 in the tectum coincides with reestablishment of orderly projections during optic nerve regeneration in goldfish. *Experimental Neurology, 166*, 196–200.

Rodger, J., Vitale, P. N., Tee, L. B., King, C. E., Bartlett, C. A., Fall, A., et al. (2004). EphA/ephrin-A interactions during optic nerve regeneration: Restoration of topography and regulation of ephrin-A2 expression. *Molecular and Cellular Neurosciences, 25*, 56–68.

Rudge, J. S., & Silver, J. (1990). Inhibition of neurite outgrowth on astroglial scars in vitro. *Journal of Neuroscience, 10*, 3594–3603.

Sapieha, P. S., Peltier, M., Rendahl, K. G., Manning, W. C., & Di Polo, A. (2003). Fibroblast growth factor-2 gene delivery stimulates axon growth by adult retinal ganglion cells after acute optic nerve injury. *Molecular and Cellular Neurosciences, 24*, 656–672.

Sauve, Y., Sawai, H., & Rasminsky, M. (1995). Functional synaptic connections made by regenerated retinal ganglion cell axons in the superior colliculus of adult hamsters. *Journal of Neuroscience, 15*, 665–675.

Sauve, Y., Sawai, H., & Rasminsky, M. (2001). Topological specificity in reinnervation of the superior colliculus by regenerated retinal ganglion cell axons in adult hamsters. *Journal of Neuroscience, 21*, 951–960.

Sbaschnig-Agler, M., Ledeen, R. W., Alpert, R. M., & Grafstein, B. (1985). Changes in axonal transport of phospholipids in the regenerating goldfish optic system. *Neurochemical Research, 10*, 1499–1509.

Schinkmann, K., & Blenis, J. (1997). Cloning and characterization of a human STE20-like protein kinase with unusual cofactor requirements. *Journal of Biological Chemistry, 272*, 28695–28703.

Schmeer, C., Straten, G., Kugler, S., Gravel, C., Bahr, M., & Isenmann, S. (2002). Dose-dependent rescue of axotomized rat retinal ganglion cells by adenovirus-mediated expression of glial cell-line derived neurotrophic factor in vivo. *European Journal of Neuroscience, 15*, 637–643.

Schwab, M. E., & Caroni, P. (1988). Oligodendrocytes and CNS myelin are nonpermissive substrates for neurite growth and fibroblast spreading in vitro. *Journal of Neuroscience, 8*, 2381–2393.

Schwab, M. E., & Thoenen, H. (1985). Dissociated neurons regenerate into sciatic but not optic nerve explants in culture irrespective of neurotrophic factors. *Journal of Neuroscience, 5*, 2415–2423.

Schwalb, J. M., Boulis, N. M., Gu, M. F., Winickoff, J., Jackson, P. S., Irwin, N., et al. (1995). Two factors secreted by the goldfish optic nerve induce retinal ganglion cells to regenerate axons in culture. *Journal of Neuroscience, 15*, 5514–5525.

Schwalb, J. M., Gu, M. F., Stuermer, C., Bastmeyer, M., Hu, G. F., Boulis, N., et al. (1996). Optic nerve glia secrete a low-molecular-weight factor that stimulates retinal ganglion cells to regenerate axons in goldfish. *Neuroscience, 72*, 901–910.

Schweigreiter, R., Walmsley, A. R., Niederost, B., Zimmermann, D. R., Oertle, T., Casademunt, E., et al. (2004). Versican V2 and the central inhibitory domain of Nogo-A inhibit neurite growth via p75NTR/NgR-independent pathways that converge at RhoA. *Molecular and Cellular Neurosciences, 27*, 163–174.

Sengottuvel, V., Leibinger, M., Pfreimer, M., Andreadaki, A., & Fischer, D. (2011). Taxol facilitates axon regeneration in the mature CNS. *Journal of Neuroscience, 31*, 2688–2699.

Shen, Y., Tenney, A. P., Busch, S. A., Horn, K. P., Cuascut, F. X., Liu, K., et al. (2009). PTPsigma is a receptor for chondroitin sulfate proteoglycan, an inhibitor of neural regeneration. *Science, 326*, 592–596.

Shewan, D., Berry, M., & Cohen, J. (1995). Extensive regeneration in vitro by early embryonic neurons on immature and adult CNS tissue. *Journal of Neuroscience, 15*, 2057–2062.

Silver, J., & Miller, J. H. (2004). Regeneration beyond the glial scar. *Nature Reviews Neuroscience, 5*, 146–156.

Sivron, T., Schwab, M. E., & Schwartz, M. (1994). Presence of growth inhibitors in fish optic nerve myelin: Postinjury changes. *Journal of Comparative Neurology, 343*, 237–246.

Skene, J. H., & Willard, M. (1981a). Axonally transported proteins associated with axon growth in rabbit central and peripheral nervous systems. *Journal of Cell Biology, 89*, 96–103.

Skene, J. H., & Willard, M. (1981b). Changes in axonally transported proteins during axon regeneration in toad retinal ganglion cells. *Journal of Cell Biology, 89*, 86–95.

Smith, P. D., Sun, F., Park, K. K., Cai, B., Wang, C., Kuwako, K., et al. (2009). SOCS3 deletion promotes optic nerve regeneration in vivo. *Neuron, 64*, 617–623.

So, K. F., & Aguayo, A. J. (1985). Lengthy regrowth of cut axons from ganglion cells after peripheral nerve transplantation into the retina of adult rats. *Brain Research, 328*, 349–354.

So, K. F., & Schneider, G. E. (1978). Abnormal recrossing retinotectal projections after early lesions in Syrian hamsters: Age-related effects. *Brain Research, 147*, 277–295.

So, K. F., Schneider, G. E., & Ayres, S. (1981). Lesions of the brachium of the superior colliculus in neonate hamsters: Correlation of anatomy with behavior. *Experimental Neurology, 72*, 379–400.

Sperry, R. W. (1948). Patterning of central synapses in regeneration of the optic nerve in teleosts. *Physiological Zoology, 21*, 351–361.

Sperry, R. W. (1963). Chemoaffinity in the orderly growth of nerve fiber patterns and connections. *Proceedings of the National Academy of Sciences of the United States of America, 50*, 703–710.

Sugitani, K., Matsukawa, T., Koriyama, Y., Shintani, T., Nakamura, T., Noda, M., et al. (2006). Upregulation of retinal transglutaminase during the axonal elongation stage of goldfish optic nerve regeneration. *Neuroscience, 142*, 1081–1092.

Sun, F., Park, K. K., Belin, S., Wang, D., Lu, T., Chen, G., et al. (2012). Sustained axon regeneration induced by co-deletion of PTEN and SOCS3. *Nature, 480*, 372–375.

Swanson, K. I., Schlieve, C. R., Lieven, C. J., & Levin, L. A. (2005). Neuroprotective effect of sulfhydryl reduction in a rat optic nerve crush model. *Investigative Ophthalmology & Visual Science, 46*, 3737–3741.

Symonds, A. C., King, C. E., Bartlett, C. A., Sauve, Y., Lund, R. D., Beazley, L. D., et al. (2007). EphA5 and ephrin-A2 expression during optic nerve regeneration: A 'two-edged sword.' *European Journal of Neuroscience, 25*, 744–752.

Takeda, M., Kato, H., Takamiya, A., Yoshida, A., & Kiyama, H. (2000). Injury-specific expression of activating transcription factor-3 in retinal ganglion cells and its colocalized expression with phosphorylated c-Jun. *Investigative Ophthalmology & Visual Science, 41*, 2412–2421.

Thanos, S., Naskar, R., & Heiduschka, P. (1997). Regenerating ganglion cell axons in the adult rat establish retinofugal topography and restore visual function. *Experimental Brain Research, 114*, 483–491.

Thrimawithana, T. R., Young, S., Bunt, C. R., Green, C., & Alany, R. G. (2011). Drug delivery to the posterior segment of the eye. *Drug Discovery Today, 16*, 270–277.

Veldman, M. B., Bemben, M. A., & Goldman, D. (2010). Tuba1a gene expression is regulated by KLF6/7 and is necessary for CNS development and regeneration in zebrafish. *Molecular and Cellular Neurosciences, 43*, 370–383.

Veldman, M. B., Bemben, M. A., Thompson, R. C., & Goldman, D. (2007). Gene expression analysis of zebrafish retinal ganglion cells during optic nerve regeneration identifies KLF6a and KLF7a as important regulators of axon regeneration. *Developmental Biology, 312*, 596–612.

Vidal-Sanz, M., Aviles-Trigueros, M., Whiteley, S. J., Sauve, Y., & Lund, R. D. (2002). Reinnervation of the pretectum in adult rats by regenerated retinal ganglion cell axons: Anatomical and functional studies. *Progress in Brain Research, 137*, 443–452.

Vidal-Sanz, M., Bray, G. M., Villegas-Perez, M. P., Thanos, S., & Aguayo, A. J. (1987). Axonal regeneration and synapse formation in the superior colliculus by retinal ganglion cells in the adult rat. *Journal of Neuroscience, 7*, 2894–2909.

Villegas-Perez, M. P., Vidal-Sanz, M., Bray, G. M., & Aguayo, A. J. (1988). Influences of peripheral nerve grafts on the survival and regrowth of axotomized retinal ganglion cells in adult rats. *Journal of Neuroscience, 8*, 265–280.

Wang, J. T., Kunzevitzky, N. J., Dugas, J. C., Cameron, M., Barres, B. A., & Goldberg, J. L. (2007). Disease gene candidates revealed by expression profiling of retinal ganglion cell development. *Journal of Neuroscience, 27*, 8593–8603.

Wang, K. C., Kim, J. A., Sivasankaran, R., Segal, R., & He, Z. (2002a). p75 interacts with the Nogo receptor as a co-receptor for Nogo, MAG and OMgp. *Nature, 420*, 74–78.

Wang, K. C., Koprivica, V., Kim, J. A., Sivasankaran, R., Guo, Y., Neve, R. L., et al. (2002b). Oligodendrocyte-myelin glycoprotein is a Nogo receptor ligand that inhibits neurite outgrowth. *Nature, 417*, 941–944.

Wang, X., Hasan, O., Arzeno, A., Benowitz, L. I., Cafferty, W. B., & Strittmatter, S. M. (2012). Axonal regeneration induced by blockdge of glial inhibitors coupled with activation of intrinsic growth pathways. *Experimental Neurology, 237*, 55–69.

Watanabe, M., Tokita, Y., Kato, M., & Fukuda, Y. (2003). Intravitreal injections of neurotrophic factors and forskolin enhance survival and axonal regeneration of axotomized beta ganglion cells in cat retina. *Neuroscience, 116*, 733–742.

Weber, A. J., Viswanathan, S., Ramanathan, C., & Harman, C. D. (2010). Combined application of BDNF to the eye and brain enhances ganglion cell survival and function in the cat after optic nerve injury. *Investigative Ophthalmology & Visual Science, 51*, 327–334.

Weise, J., Isenmann, S., Klocker, N., Kugler, S., Hirsch, S., Gravel, C., et al. (2000). Adenovirus-mediated expression of ciliary neurotrophic factor (CNTF) rescues axotomized rat retinal ganglion cells but does not support axonal regeneration in vivo. *Neurobiology of Disease, 7*, 212–223.

Winzeler, A. M., Mandemakers, W. J., Sun, M. Z., Stafford, M., Phillips, C. T., & Barres, B. A. (2011). The lipid sulfatide is a novel myelin-associated inhibitor of CNS axon outgrowth. *Journal of Neuroscience, 31*, 6481–6492.

Wyatt, C., Ebert, A., Reimer, M. M., Rasband, K., Hardy, M., Chien, C. B., et al. (2010). Analysis of the astray/robo2 zebrafish mutant reveals that degenerating tracts do not provide strong guidance cues for regenerating optic axons. *Journal of Neuroscience, 30*, 13838–13849.

Xue, T., Do, M. T., Riccio, A., Jiang, Z., Hsieh, J., Wang, H. C., et al. (2011). Melanopsin signalling in mammalian iris and retina. *Nature, 479*, 67–73.

Yamashita, T., & Tohyama, M. (2003). The p75 receptor acts as a displacement factor that releases Rho from Rho-GDI. *Nature Neuroscience, 6*, 461–467.

Yan, Q., Wang, J., Matheson, C. R., & Urich, J. L. (1999). Glial cell line-derived neurotrophic factor (GDNF) promotes the survival of axotomized retinal ganglion cells in adult rats: Comparison to and combination with brain-derived neurotrophic factor (BDNF). *Journal of Neurobiology, 38*, 382–390.

Yin, Y., Cui, Q., Gilbert, H. Y., Yang, Y., Yang, Z., Berlinicke, C., et al. (2009). Oncomodulin links inflammation to optic nerve regeneration. *Proceedings of the National Academy of Sciences of the United States of America, 106*, 19587–19592. doi:10.1073/pnas.0907085106.

Yin, Y., Cui, Q., Li, Y., Irwin, N., Fischer, D., Harvey, A. R., et al. (2003). Macrophage-derived factors stimulate optic nerve regeneration. *Journal of Neuroscience, 23*, 2284–2293.

Yin, Y., Li, Y., Cen, L.-P., Gilbert, H.-Y., Cui, Q., & Benowitz, L. (2012). Stromal cell-derived factor-1 (SDF-1) contributes to inflammation-induced optic nerve regeneration and retinal ganglion cell survival. Program No. 531.04. New Orleans, LA, Society for Neuroscience.

Yin, Y., Henzl, M. T., Lorber, B., Nakazawa, T., Thomas, T. T., Jiang, F., et al. (2006). Oncomodulin is a macrophage-derived signal for axon regeneration in retinal ganglion cells. *Nature Neuroscience, 9*, 843–852.

Yokoi, K., Kachi, S., Zhang, H. S., Gregory, P. D., Spratt, S. K., Samulski, R. J., et al. (2007). Ocular gene transfer with self-complementary AAV vectors. *Investigative Ophthalmology & Visual Science, 48*, 3324–3328.

Yoon, M. (1971). Reorganization of retinotectal projection following surgical operations on the optic tectum in goldfish. *Experimental Neurology, 33*, 395–411.

Conformity and Specificity of Primate Corticogenesis

MARION BETIZEAU, COLETTE DEHAY, AND HENRY KENNEDY

The cortex detects statistical regularities in the environment by sensing local and global correlations of neuronal activity. In this way, invariant characteristics of the world can be inferred such as color constancy and object segmentation. With regards to cortical development, this notion of cortical function tempers thinking of brain development as the sum of its parts. Instead it favors taking into account the role of the environment and ultimately argues for the combined consideration of both the development and function of the cortex. It suggests that for understanding function, it is necessary to consider the statistical features of the environment that the brain is able to detect, and to directly link the structural/functional properties of the brain to its perceptual capacities (Shepard, 2001). This approach suggests we need to take into account the environmental factors that have molded the phylogenetic history of the cortex, and that self-organization and natural selection are two facets of the evolutionary process (Halley & Winkler, 2008; Kauffman, 2000). The developing sensory apparatus produces environmental information from which the brain needs to extract behaviorally relevant patterns. By rewiring or reweighting connections, it tunes itself to or learns about coherent (and presumably relevant) patterns in its input. This unsupervised classification procedure is used to generate self-organized maps. One can speculate that the neuronal mechanisms of ontogenetic self-organization actually persist into adulthood where they mediate adaptive changes in learning and memory.

The species as a whole is subject to environmental patterns that exert pressure through natural selection, thereby promoting the development of circuits and processing modules that are tuned to the exigencies required for the survival of the current generation (Geisler & Diehl, 2002). The proposed process carries the prediction that corticogenesis even at very early stages of development will be influenced by extrinsic factors, echoing earlier stages of phylogeny. During the early 1970s there were considerable efforts headed by

Van der Loos to show that corticogenesis is indeed regulated by extrinsic factors related to the sensory periphery (Van der Loos, 1977). This work was also supported by the observation that visual experience plays an important role in the elaboration of the functional architecture of the primary visual cortex (LeVay, Wiesel, & Hubel, 1980; Thompson, Kossut, & Blakemore, 1983). This understanding of corticogenesis was later referred to as protocortex theory. Taking its roots in the work of the British empiricist John Locke, it considered the immature cortical plate as a uniform and unspecified structure that would take on its regional features in response to specific sets of ascending inputs (O'Leary, 1989; Sur & Rubenstein, 2005). The extreme version of protocortex concept has been largely displaced by the equally radical protomap theory. Here it is postulated that corticogenesis is driven uniquely by intrinsic molecular mechanisms. While in recent years there has been overwhelming evidence in favor of a genetic specification of cortical areas, there is a real need to reconsider the integration of intrinsic and extrinsic specification (Kennedy & Dehay, 1993; Kennedy et al., 2007; Rakic, 1988b).

A hallmark of the primate neocortex is the selective enlargement of the supragranular layers (SGL, layers II and III) compartment (see figure 98.1). Interspecies comparisons of the number and organization of cortical neurons imply evolutionary changes in the developmental program that generates these primate neurons. During mammalian evolution, the subventricular zone (SVZ) enlarges along with cortical expansion (Cheung et al., 2010; Molnar et al., 2006; Reillo et al., 2011). Work on the monkey cortical germinal zones has shown that the primate has evolved a specialized and greatly expanded SVZ that is split into cytoarchitecturally distinct compartments: the inner SVZ (ISVZ) and the outer SVZ (i.e., OSVZ) (Smart et al., 2002). A number of studies have since confirmed in both human and non-human primates, that the cellular architecture and composition of the OSVZ differ strikingly from that of

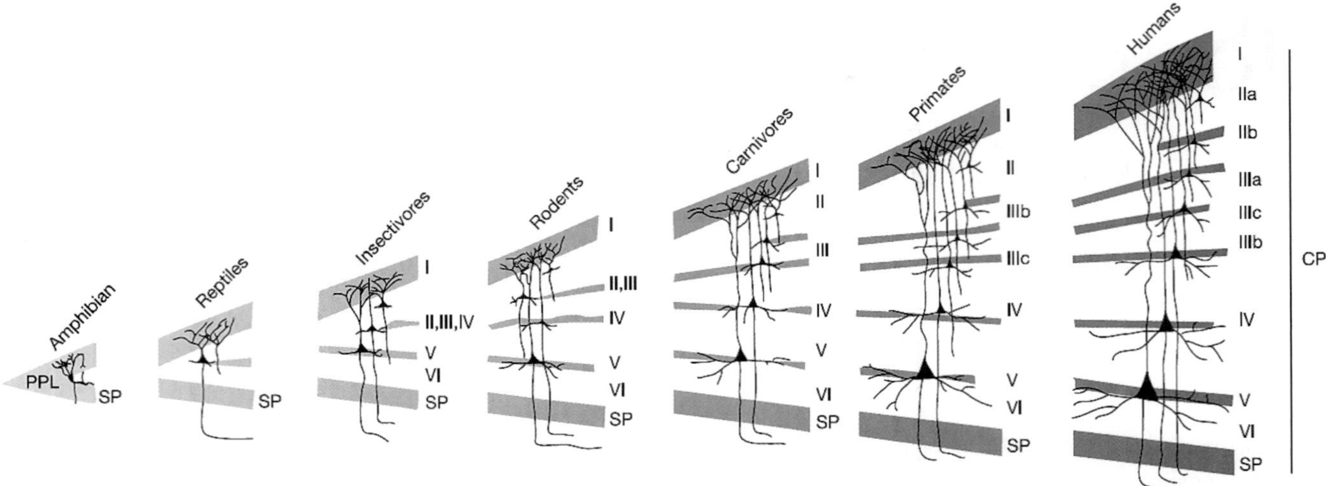

FIGURE 98.1   Evolution of the supragranular layers (SGL) of the cortex. Supragranular cortical layers (layers II and III), generated late in neurogenesis, are greatly expanded in the primate cerebral cortex, especially in humans. In primates SGL neurons form local patchy connections and feedforward long-distance corticocortical connections. CP, cortical plate; PPL, primordium plexiform layer; SP, subplate. (Reprinted with permission from Hill & Walsh, 2005.)

the rodent SVZ (Fietz et al., 2010; Hansen et al., 2010; Lukaszewicz et al., 2005; Smart et al., 2002). Further, in the human and macaque monkey cortex, the OSVZ reaches its maximum thickness during peak production of SGL neurons (Fietz et al., 2010; Hansen et al., 2010; Smart et al., 2002). The link between the OSVZ and the SGL has been formally established by birthdating experiments in the monkey (Lukaszewicz et al., 2005).

The discovery of the OSVZ in the primate has prompted numerous teams to reexamine the structure of the cortical germinal zones in a wide range of mammalian species. These studies have led to a revision of the classical compartmentalization of the germinal zones into ventricular zone (VZ) and SVZ and have uncovered a diversification of the germinal zones in carnivores (ferret, Fietz et al., 2010; Reillo et al., 2011) and larger brained mammals (pig, agouti, Garcia-Moreno et al., 2011). It is now considered that the SVZ of primates and of gyrencephalic mammals is subdivided into an inner SVZ and an outer SVZ while the SVZ of the laboratory rodent remains a thin and more homogeneous structure (Martinez-Cerdeno et al., 2012; Smart, 1973). However, it is only in the primate order that the outer region of the SVZ takes on the full morphological features of a distinctive OSVZ. Recent data however show that some precursor cell types akin to those observed in the primate OSVZ are also found in the laboratory rodent SVZ, albeit at much lower frequencies (Martinez-Cerdeno et al., 2012; Shitamukai, Konno, & Matsuzaki, 2011; Wang et al., 2011).

Compared to the other orders, the primate cortex is characterized by an overrepresented SGL compartment where neurons are dedicated to the transfer of information between cortical areas. A number of features of the SGL suggest that these layers play an important role in the evolution of the cortex. For instance they express high levels of genes associated with long-term potentiation and calcium signaling, reflecting the considerable plasticity found in these layers (see figure 98.1) (Bernard et al., 2012). They also contain high numbers of double-bouquet cells, which are particularly developed in primates and may have a cortical origin during development (Letinic, Zoncu, & Rakic, 2002; Yanez et al., 2005). In the monkey cortex, the SGL are generated by a primate-specific germinal zone during in utero development (Lukaszewicz et al., 2005). In the present chapter we shall examine how environmental factors contribute to shaping the function of these cortical layers and how this may be a highly conserved feature of brain development. Here we shall focus on the SGL of the cortex and the mechanisms that ensure their expansion in primates (Dehay & Kennedy, 2007).

DEVELOPMENTAL ORIGINS OF SGL INTERNEURONS

Local recurrent and long range projecting cortical excitatory neurons include stellate and pyramidal glutamatergic neurons. Activity is finely tuned by locally projecting GABAergic neurons, also called inhibitory interneurons, which serve to synchronize excitatory neurons, thereby facilitating synaptic potentiation and the generation of potentiated cell ensembles (Bartos, Vida, & Jonas, 2007; Fishell & Rudy, 2011).

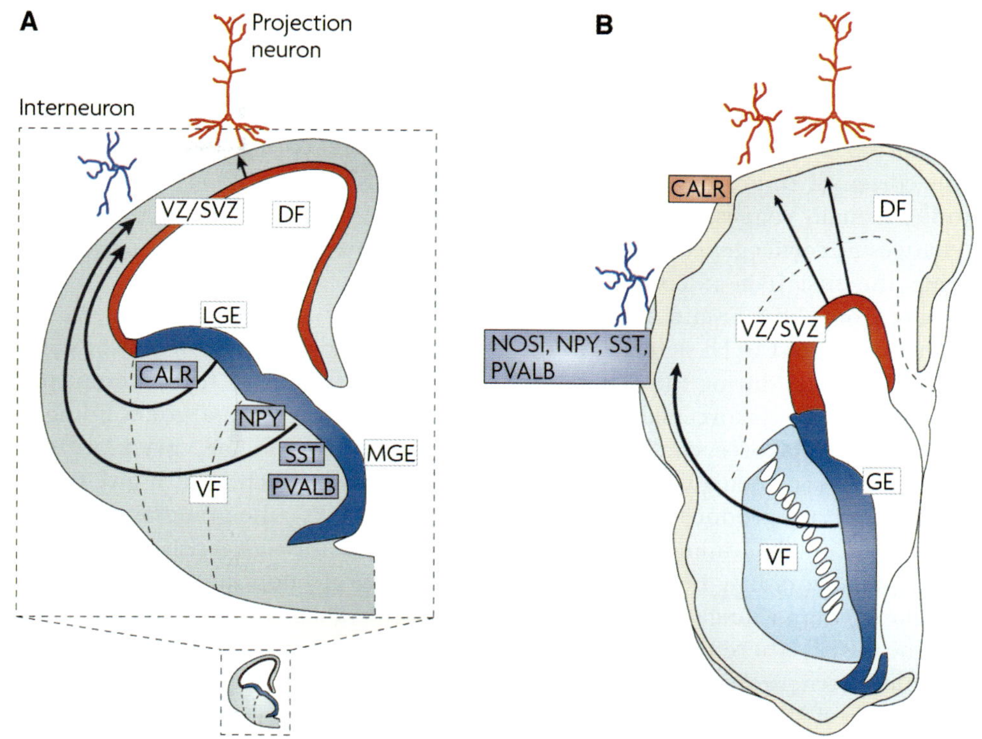

FIGURE 98.2   Origins of cortical neurons in rodents and primates. These schemes depict cross-sections of a hemisphere of a rodent (A) and a human (B) fetal brain. They are drawn to scale, but a zoom of the rodent section is displayed for better readability. Excitatory neurons in both rodents and primates are generated in the cortical ventricular zone/subventricular zone (VZ/SVZ) and migrate radially to settle in the cortical plate. In rodents the majority of cortical interneurons originate from the ganglionic eminence (GE), subdivided into a lateral and medial part (LGE and MGE, respectively), and migrate tangentially into the cortex. Some primate cortical interneurons originate also from the GE, especially at early stages, but a new source appears in the cortical VZ/SVZ at midgestation during the generation of the SGL. Different interneuron types are produced by the two sources in primates. Nitric oxide synthase 1 (NOS1), neuropeptide Y (NPY), somatostatin (SST), and parvalbumin (PVALB) positive interneurons originate from the GE, and calretinin (CALR) positive cells from the cortical VZ/SVZ. DF: dorsal forebrain; VF: ventral forebrain. (Reprinted by permission from Macmillan Publishers Ltd: [*Nature Reviews Neuroscience*] [Rakic, 2009], copyright 2009.)

Primate and nonprimate interneurons differ in two major respects: their relative abundance compared to excitatory cells and their increased diversity in primates (Defelipe, 2002; Gabbott & Bacon, 1996). Interneurons account for 25% to 30% of all cortical neurons in primates and exhibit areal and laminar differences (Hendry et al., 1987) whereas they are thought to compose only 15% to 20% in rodents (Beaulieu, 1993). Interneurons constitute a heterogeneous population differing in their morphologies, connectivity, electrophysiological properties, and the expression of the calcium binding proteins calbindin, somatostatin, calretinin, and parvalbumin (Ascoli et al., 2008). Interneurons achieve their highest level of differentiation in primates (Defelipe, 2002), and at least one type seems to be unique to this order: the double-bouquet cells (Defelipe, 2011; Jones et al., 1988; Yanez et al., 2005).

The increased number and diversity of primate interneurons is likely to arise through a more complex developmental process that is still not entirely resolved. Glutamatergic neurons are generated exclusively by the cortical germinal zones in the dorsal forebrain and migrate radially to the overlying cortical plate. In contrast, interneurons in rodent and ferret originate subcortically from the ganglionic eminence (GE) in the ventral forebrain (see figure 98.2) (Anderson et al., 1997, 2002). They reach the cortical plate via two migration streams in the marginal and intermediate zones (Marin & Rubenstein, 2001; Metin et al., 2006; Tamamaki, Fujimori, & Takauji, 1997). Genetic fate mapping studies in mouse showed that parvalbumin positive cells are generated in the medial GE (MGE), somatostatin positive cells from the MGE and the caudal GE (CGE), and calretinin positive cells from the CGE (for review see Wonders & Anderson, 2006).

The ventral origin of some interneurons has been confirmed in primates (Jakovcevski, Mayer, & Zecevic, 2011; Letinic, Zoncu, & Rakic, 2002; Petanjek, Berger,

& Esclapez, 2009; Zecevic, Hu, & Jakovcevski, 2011). However in primates an important fraction of cortical interneurons are produced locally in the cortical germinal zones of the dorsal forebrain. Rakic's group estimated that in human 65% of cortical interneurons originate from the cortex (Letinic, Zoncu, & Rakic, 2002). Several studies actually suggest that the source of interneurons changes as corticogenesis proceeds in the primate. During the first half of corticogenesis, interneurons originate from the GE, and at midgestation interneuron precursors start to be detected in the primate cortical germinal zones during the generation of the SGL (see figure 98.2) (Jakovcevski Mayer, & Zecevic, 2011; Zecevic, Hu, & Jakovcevski, 2011).

Although these studies convincingly showed that many primate interneurons are produced in the cortical germinal zones, they did not determine their lineage. Cortical interneuron precursors may be generated by ventral precursors that undergo tangential migration and settle in the cortical SVZ, where they resume proliferation and generate interneurons. Such precursors have recently been described in late mouse corticogenesis (Wu et al., 2011). On the other hand, they could be generated by cortical precursors in the VZ or ISVZ/OSVZ, as suggested in humans by in vitro genetic fate mapping (Yu & Zecevic, 2011) and ex vivo time-lapse experiments (Hansen et al., 2010). These two lineages are not mutually exclusive. The diversity of cortical interneuron precursors identified by the expression of various ventral markers is in favor of multiple lineages (Jakovcevski, Mayer, & Zecevic, 2011).

It is clear that the primate cortical germinal zones generate many more interneurons compared to nonprimates, where this process seems to be marginal. Only 5% of the interneurons are produced dorsally in the mouse, and they seem to be destined mainly to the olfactory bulb (Kohwi et al., 2007), although one study reported a dorsal origin of some cortical interneurons produced postnatally (Inta et al., 2008).

The higher diversity of interneuron sources in primates might be the path followed by evolution to provide a bigger cortex with more numerous and more diverse types of interneurons. The SGL of the cortex are the most developed and complex layers of the primate brain. The timing of interneuron production in primates suggests that interneurons destined for the SGL have mainly a cortical origin. It is also interesting to note that the primate-specific double-bouquet interneurons are mainly found in the SGL (Defelipe, 2011; Defelipe et al., 1999) and thus are likely to derive from cortical interneuron precursors. At midgestation, when cortical interneuron precursors are detected, the OSVZ is the main proliferative compartment (Lukaszewicz et al., 2005). These findings highlight a possible link between cortical production of interneurons, OSVZ precursor increase in number and in potentialities, and primate-specific expansion and specialization of the SGL of the cortex (Petanjek, Kostovic, & Esclapez, 2009).

DEVELOPMENTAL ORIGIN OF SGL EXCITATORY NEURONS

The pyramidal projection neurons of the neocortex are generated by at least five different precursor cell types located in the germinal zones lining the ventricle. The first precursors in the cortex are the neuroepithelial cells (NECs), which in turn give rise to radial glial precursors (RGPs), the grandmother cell of cortical development as they give rise directly or indirectly to all cortical projection neurons (Heins et al., 2002; Malatesta et al., 2003; Miyata et al., 2004; Noctor et al., 2004). Both NECs and RGPs divide at the apical surface and constitute the VZ, which is the first formed germinal zone. Iain Smart observed, as far back as 1973, that early in corticogenesis, crowding of precursors leading to queuing at the apical surface is relieved by some precursors abandoning apical mitosis and shifting to undergo mitosis at the top of the ventricular zone (Smart, 1973). These basal dividing precursors have been named intermediate progenitors (IPs) and shown to express the transcription factor Tbr2 (Englund et al., 2005), which distinguishes them from the NECs and RGPs, which express Prominin 1 and the transcription factor Pax6 (Götz, Stoykova, & Gruss, 1998; Hartfuss et al., 2001; Malatesta, Hartfuss, & Götz, 2000; Miyata et al., 2004; Noctor et al., 2004). Finally a short neural precursor (SNP) was identified in the early VZ, which like the RGPs undergoes apical or near apical division (Gal et al., 2006; Stancik et al., 2010).

A major research breakthrough occurred when it was shown that the RGPs are not only part of a glial scaffolding but also constitute multipotent cortical precursors (Malatesta, Hartfuss, & Götz, 2000; Noctor et al., 2001, 2002). There is further heterogeneity of RGPs in the embryonic primate, where a fraction cease to proliferate and instead function only as a migration scaffolding for several months before reinitiating proliferation and generating astrocytes (Rakic, 2003; Schmechel & Rakic, 1979).

SPECIFICITIES OF THE PRIMATE OSVZ

Primate corticogenesis is characterized by an important expansion of the SVZ to form an imposing outer SVZ (OSVZ), which is not observed in the rodent (Smart et al., 2002) (see figure 98.3). The OSVZ is a distinct histological structure that is bounded by two clearly defined fiber pathways (the outer and inner fiber layers) that have no parallel in the developing rodent or carnivore. The homologous structure of the rodent SVZ in the primate could be the ISVZ, which in these species is in contact with the VZ. Contrarily to what is observed in the rodent, where the VZ is the major germinal compartment throughout corticogenesis, the primate VZ declines rapidly during the course of corticogenesis. This decline is associated with an early appearance of the SVZ quickly followed by the OSVZ. (Smart et al., 2002).

The differences in the dimensions of developmental compartments in the rodent and the primate shown in figure 98.3 reflect differences in mitotic activity. Hence, the distribution of mitoses in the developing primate cortical germinal zones is dramatically different from that observed in the rodent (Fietz et al., 2010; Hansen et al., 2010; Martinez-Cerdeno et al., 2012; Smart et al., 2002). In the rodent, the majority of mitotic figures are observed in the VZ during the whole period of corticogenesis (Martinez-Cerdeno et al., 2012; Smart, 1973). In primates, basally dividing precursors appear at very early stages. Once formed, the OSVZ remains the principal proliferative compartment throughout corticogenesis (Hansen et al., 2010; Lukaszewicz et al., 2005; Martinez-Cerdeno et al., 2012; Smart et al., 2002). Basally located mitotic figures appear at comparatively later stages in the rodent. For instance it has been noted that during the production of infragranular layers at E65 in the primate, more than 43% of dividing precursors are located basally in the OSVZ (Lukaszewicz et al., 2005; Martinez-Cerdeno et al., 2012) whereas only 3% of mitotic precursors are reported in a basal SVZ at a comparable stage in the rodent (Martinez-Cerdeno et al., 2012).

This primate-specific organization of the SVZ in an ISVZ and OSVZ, first described in the monkey (Smart et al., 2002), has also been subsequently observed in the developing human cortex (Zecevic, Chen, & Filipovic, 2005). Recent studies report a compartmentalization of the germinal zones in ferret, agouti, and marmoset based on the density of Tbr2 expressing cells and the presence of Pax6+ cells in a basal location (Garcia-Moreno et al., 2011; Kelava et al., 2012; Martinez-Cerdeno et al., 2012; Reillo et al., 2011). These authors found a basal proliferative compartment above the SVZ with a low density of Tbr2+ cells, which they refer to as OSVZ. However, this proliferative zone appears radically different from the primate OSVZ: (1) It appears later during corticogenesis compared to the primate OSVZ, (2) the contribution of

precursors from this zone is low compared to the VZ, (3) it lacks the distinctive cytoarchitecture of the primate OSVZ (absence of the tightly packed radial orientated cells), and (4) it is not bounded by fiber pathways. The dense, radially orientated precursors of an imposing OSVZ therefore constitute a unique primate feature.

There is also a major difference in the cellular composition of the primate OSVZ with respect to the rodent SVZ. Whereas in the rodent all RGP nuclei are restricted to the VZ, RGPs' somata are morphologically identified in the OSVZ of the primate (Levitt, Cooper, & Rakic, 1981; Lukaszewicz et al., 2005). There is evidence that precursors of the primate OSVZ express Pax6, which characterizes RGP identity in the rodent (Fish et al., 2008).

The Pax6 positive precursors in the OSVZ have been further characterized in human (Fietz et al., 2010; Hansen et al., 2010) and in ferret (Fietz et al., 2010; Reillo et al., 2011). A subset of Pax6 expressing cells divide basally and display features reminiscent of the RGPs located in the VZ and have thus been called basal radial glial-like cells (bRG) (Kelava et al., 2012). Other studies refer to them as OSVZ radial glial-like cells (Fietz et al., 2010; Hansen et al., 2010) or intermediate radial glial-like cells (Reillo et al., 2011). This precursor type accounts for around 40% of the OSVZ precursor pool in human (Hansen et al., 2010). bRG express the RGP markers Pax6 and Sox2 as well as GFAP, Nestin, and Glast but not the IP marker Tbr2. They possess a long basal process that reaches the pia, and they retain it during mitosis. They rarely exhibit an apical process, and if present, it does not contact the VZ surface. Despite those epithelial features, they fail to express apical domain proteins, and their centrosome is not located at the VZ border but instead close to the nucleus in the OSVZ (Fietz et al., 2010; Hansen et al., 2010). The bRG show another major difference from the RGPs. The latter exhibit stereotypic migration of their nucleus in phase with the cell cycle, which ensures that the RGP division occurs at the ventricular border subsequent to an apical directed movement. In contrast, prior to division, bRG undergo a basally directed mitotic translocation (Hansen et al., 2010). This basal movement is thought to contribute to the radial expansion of the OSVZ.

There is an ongoing debate as to whether the newly discovered bRG are characteristic of gyrencephalic brains. First found in human and ferret, several studies have later described this precursor type in lissencephalic rodents (Martinez-Cerdeno et al., 2012; Shitamukai, Konno, & Matsuzaki, 2011; Wang et al., 2011), in the agouti, a gyrencephalic rodent (Garcia-Moreno

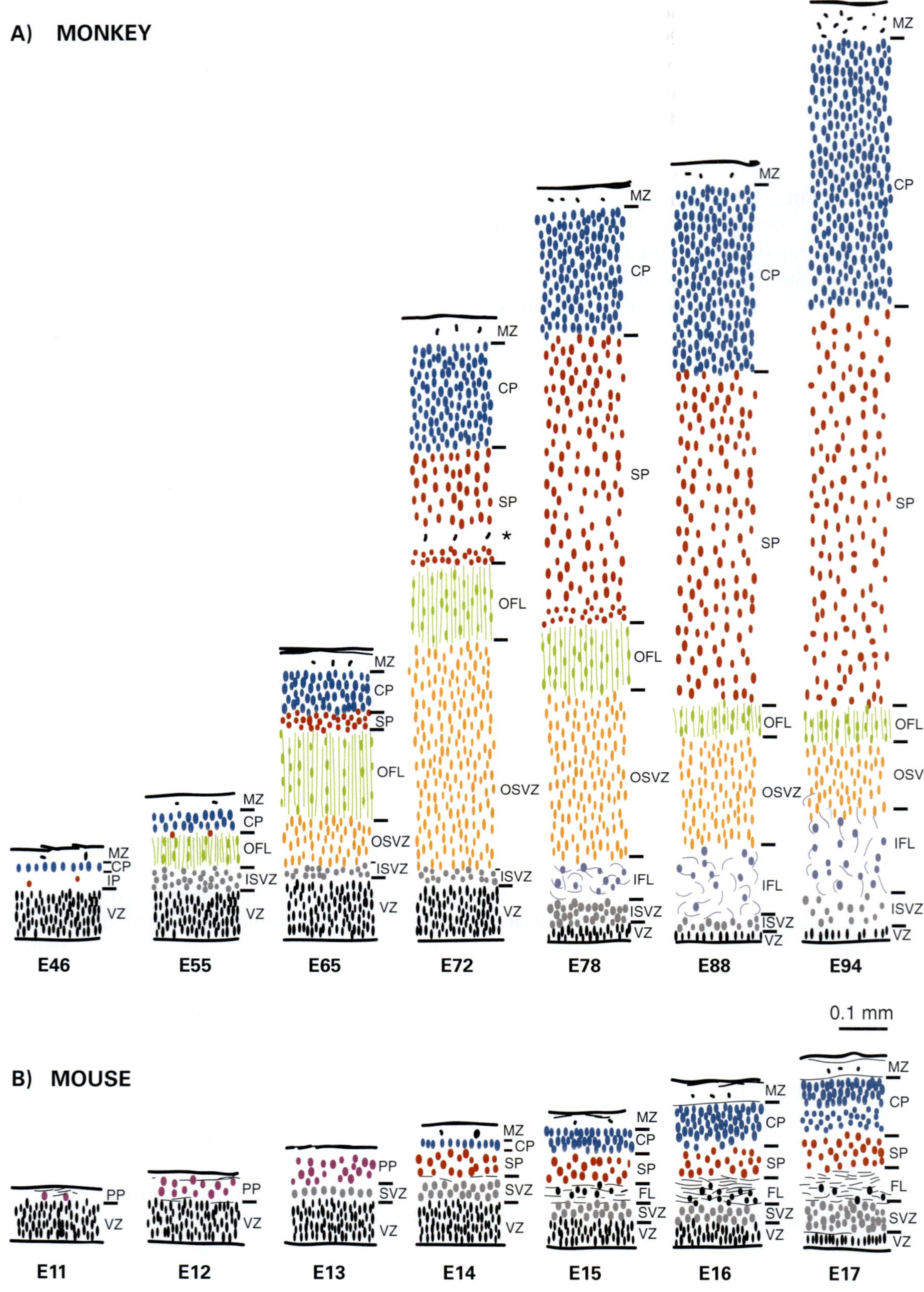

FIGURE 98.3 Comparison of primate (A) and mouse (B) germinal zones at equivalent developmental stages. These drawings are transects through presumptive area 17 in monkey and mouse dorsal cortex at comparable developmental stages. The depth of each layer is drawn to a common scale. In the primate, an early appearing outer fiber layer (OFL) forms a major landmark from embryonic day (E) 55 onwards. The ventricular zone (VZ) declines progressively after E65. The subventricular zone, by contrast, increases progressively in depth and by E72 is divided into an inner subventricular zone (ISVZ) and outer subventricular zone (OSVZ) by an intruding inner fiber layer (IFL). The increase in the OSVZ is particularly important between E65 and E72 and occurs as the VZ declines. CP, cortical plate; SP, subplate; IP, intermediate plate; FL, fiber layer, MZ, marginal zone; PP, preplate. (Reprinted with permission from Smart et al., 2002.)

et al., 2011), and in the marmoset, a lissencephalic primate (Garcia-Moreno et al., 2011; Kelava et al., 2012). In contrast to small rodents where bRG are rare (5% of precursors), the relative proportions of bRG are claimed to be similar in human and ferret (40% of SVZ progenitors) (Kelava et al., 2012). However, it seems that the potential of bRG might be different in primates and nonprimates. Whereas in primates bRG generate neurons via an IP step (Hansen et al., 2010; Lui, Hansen, & Kriegstein, 2011), ferret bRG principally give rise to astrocytes (Reillo et al., 2011), suggesting that only primate bRG contribute to the increased rates of neuron production.

DOES THE OSVZ HAVE A SPECIAL RELATIONSHIP TO SGL?

Numerous observations link production of infragranular layers to VZ precursors and SVZ precursors to production of SGL. Although SVZ precursors are thought to be derived from VZ precursors, there are clear differences in gene expression between the two precursor pools and these differences correlate with distinct neuronal progeny. For instance, Otx1 and Fez1 are both expressed in VZ precursors and are then subsequently up-regulated in subsets of deep-layer neurons and down-regulated in the SVZ (Arlotta et al., 2005; B. Chen, Schaevitz, & McConnell, 2005; J. G. Chen et al., 2005; Frantz et al., 1994; Molyneaux et al., 2005). Furthermore, both Otx1 and Fez1 play a crucial role in specifying the axonal projections of subsets of lower-layer neurons. Recent studies in mice showed that several transcription factors (Cux2, Tbr2, Satb2, and Nex) (Britanova et al., 2005; Nieto et al., 2004; Pinto et al., 2009; Wu et al., 2005; Zimmer et al., 2004), as well as the noncoding RNA Svet-1 (Tarabykin et al., 2001), are selectively expressed in both the SVZ and SGL neurons. This congruency of gene expression first in SVZ precursors and subsequently in SGL together with time-lapse microscopy observations and birthdating experiments suggest that the SGL is generated by the SVZ (Lukaszewicz et al., 2005; Noctor et al., 2001, 2004; Tarabykin et al., 2001; Zimmer et al., 2004).

Consistent with these findings, distinct molecular mechanisms have been identified for the specification of infra- and supragranular neuronal lineages. Studies from mutant mice show that Ngn1 and Ngn2 activity is required for the specification of a subset of infragranular neurons but not for the specification of SGL neurons. In contrast, Pax6 and Tlx, two genes required for the formation of the SVZ (Nieto et al., 2004; Roy et al., 2004; Zimmer et al., 2004), are synergistically involved in the specification of SGL (Schuurmans et al., 2004). It therefore can be hypothesized that the selective expansion of the SGL compartment in the primate cortex results from modifications of the Pax6/Tlx related specification independently of the Ngn specification mechanisms (Schuurmans et al., 2004).

In apparent contrast to the above-mentioned links between the SVZ and the SGL is the observation that rodent IPs contribute neurons to all cortical layers (Haubensak et al., 2004; Kowalczyk et al., 2009; Miyata et al., 2004; Noctor et al., 2004; Shen et al., 2006). This has led to the suggestion that rodent IPs are universal generators of neurons, and it has been argued the separation of the germinal zones into VZ and SVZ is therefore arbitrary, and instead we should distinguish between apically (NECs and RGPs) and basally dividing progenitors (IPs) (Kowalczyk et al., 2009). However, while it is undoubtedly true that IPs contribute neurons to all cortical layers, there is also evidence that IPs in the VZ and SVZ display important differences, thereby potentially constituting two distinct pools of precursors.

In rodent a subset of Tbr2+ precursors undergo apical or subapical mitosis and constitute as much as 5%–20% of the total Tbr2+ population (Kowalczyk et al., 2009). They are reported to resemble SNPs (Gal et al., 2006). However, some findings suggest important differences between IPs in the VZ and SVZ in rodents. Firstly, whereas the IPs in the VZ have radial morphologies, those in the SVZ and at the interface between SVZ and VZ are multipolar as first described in 1973 by Smart (Kowalczyk et al., 2009; Smart, 1973). Secondly, time lapse observations show that IPs in the VZ undergo only one or two proliferative divisions. Given the decrease of the neurogenic fraction in the SVZ (Kowalczyk et al., 2009), it is likely that basal IPs undergo more frequent proliferative divisions than VZ IPs. Thirdly, IPs from the VZ produce infragranular neurons whereas SVZ IPs generate supragranular neurons (Kowalczyk et al., 2009; Stancik et al., 2010). Further investigations are needed to determine if the different categories of IP cells have distinct cell-cycle kinetics to confirm the distinction between these two sets of precursors (Arai et al., 2011). However, the fact that IPs contribute neurons to all layers fails to invalidate the observed links between SVZ and SGL. This is particularly important in the primate visual cortex, where more than 75% of cortical neurons destined for the upper layers originate from the OSVZ precursors (Lukaszewicz et al., 2005). This contrasts with the situation in rodent, where we predict that only a fraction of SGL neurons are derived from the SVZ (Kowalczyk et al., 2009).

A number of studies suggest that IPs play a role in regulating the amplification of neuron production, thereby contributing to determining laminar thickness and areal extent of the cortex (Pontious et al., 2008). The increased expansion of the occipital pole of the fetal monkey is thought to be related to the extensive size of the OSVZ in this region. This suggests a link between gyrification and IPs (Lukaszewicz et al., 2006; Smart et al., 2002), which has been further explored in the ferret (Kriegstein, Noctor, & Martinez-Cerdeno, 2006; Martinez-Cerdeno, Noctor, & Kriegstein, 2006; Reillo et al., 2011). Because IPs respond to extrinsic signals released by the thalamic fibers (Dehay et al., 2001) and to genes that control specification of areas and growth (Bedford et al., 2005; Cappello et al., 2006; Chen et al., 2006; Holm et al., 2007; Land & Monaghan, 2003; Quinn et al., 2007; Roy et al., 2004; Schuurmans et al., 2004; Yun et al., 2004; Zhou et al., 2006), these results could mean that IP amplification plays a role in determining cortical dimensions (Cheung et al., 2007; Molnar et al., 2006).

In rodent, the lack of evidence of IP amplification influencing expansion of cortical surface area is the major argument put forward by Pontious and colleagues (Pontious et al., 2008) in favor of an increased importance of a RGP pool regulation in this species. In contrast to IPs, factors influencing RGPs in rodents have been shown to influence expansion of the cortical surface (Chenn & Walsh, 2002; Hevner, 2005; Inglis-Broadgate et al., 2005; Kuida et al., 1998). However, the objection that induced changes in the abundance of IPs in rodents fails to impact on surface area could be uniquely relevant to rodents. In primates removal of thalamic input to the cortex via bilateral in utero enucleation drastically reduces the dimensions of cortical area 17 (Dehay et al., 1989, 1991; Rakic, 1988a). Because the removal of inputs to the cortex occurred after the early phase of expansion of the RGPs, it would seem to be a direct consequence of alteration of the proliferation of IPs. This supports the importance of IP amplification in primates. The argument that cortical expansion is either based on RGPs or IPs is largely attributable to the erroneous notion that RGPs undergo early symmetrical division to establish the pool of RGPs prior to the onset of neuron production. In fact it is now established that RGP proliferation occurs continuously throughout corticogenesis (Kowalczyk et al., 2009; Miyata et al., 2004; Noctor et al., 2004). Hence, factors that influence IPs could *also* be influencing RGPs either directly or indirectly. That such influences might exist is illustrated in recent experiments aimed at accelerating the cell cycle by reducing selectively the G1 phase via transfection of Cyclin E in rodents (Pilaz et al., 2009). The cell-cycle acceleration of Pilaz et al. (2009) confirmed the hypothesis of the control of mode of division by cell-cycle duration (Calegari & Huttner, 2003; Dehay et al., 2001; Götz & Huttner, 2005; Lukaszewicz et al., 2002, 2005) and is pertinent to theories of cortical expansion.

Pilaz et al. (2009) carried out a cell-cycle-acceleration study in mouse. This showed that reduction of the G1 phase of the cell cycle led to an increase in the frequency of proliferative divisions where both daughter cells reenter the cell cycle accompanied by a decrease in differentiative neurogenic divisions where daughter cells quit the cell cycle to become neurons (Pilaz et al., 2009). This increase in cell-cycle reentry led to a transient expansion of the precursor pool followed by a subsequent surge in neuron production. Interestingly, the increase in cell-cycle reentry of the Pax6+ RGPs was very short-lived and did not appear to have any consequence on the dimensions of the RGPs pool nor on the cortical architecture. This contrasted with the effects of cell-cycle acceleration on the Tbr2+ population, which showed a much bigger amplification leading to a marked and relatively long-lived increase in the dimensions of the SVZ. This IP pool amplification was followed by an increase in the production of SGL neurons and subsequently an increased thickness of the SGL. This "primatization" of the rodent cortex provides a conceptual exploration of the potential contribution of IP amplification to the expansion of the cortex.

The numerical values of the different projections neuron phenotypes determine the hierarchy and network features of the cortex (Kennedy & Dehay, 2012). There is considerable evidence that the phenotypic fate is sealed during the final round of mitosis in the germinal zones and that the timing of this event is highly significant (B. Chen, Schaevitz, & McConnell, 2005; J. G. Chen et al., 2005; Molyneaux et al., 2005; Polleux et al., 2001; Shen et al., 2006). It seems that fate specification is entirely premigratory as neurons destined for a given cortical layer that end up in an inappropriate layer do not acquire the connectivity of the inappropriate layer (Polleux et al., 2001).

The timing of the differentiative division that generates a given neuron type is not only directly involved in the specification but also in determining the number of neurons generated. Delaying the final differentiative division of a precursor pool leads to increasing the number of proliferative divisions, which in turn leads to its amplification and hence expansion of the numbers of neurons generated (Pilaz et al., 2009; Polleux et al., 1997). Mathematical modeling of these events shows a high predictability of the numbers of neurons generated confirming the relationship between the kinetics of the cycling precursor and the cytoarchitecture of the cortex (Lukaszewicz et al., 2005; Pilaz et al., 2009).

Cell-cycle acceleration experiments by Calegari's group revealed a 300% increase in cortical surface area (Lange, Huttner, & Calegari, 2009), which could suggests that the increased IP population has a feedback control on proliferation in the RGPs pool in turn leading to the increase in cortical size. Such feedback influences of the IPs on the RGPs are certainly required to support the claims of the role of IPs in gyrification (Kriegstein, Noctor, & Martinez-Cerdeno, 2006; Lukaszewicz et al., 2006; Martinez-Cerdeno, Noctor, & Kriegstein, 2006; Smart et al., 2002). For instance in the E80 fetal monkey at the onset of SGL production there is a very prominent OSVZ which decreases in size concomitantly with the increased rates of neuron production and the very rapid growth of the occipital pole and the formation of the lunate gyrus. During this period of ballooning out of the occipital pole the cortical plate of the incipient area 17 is actually thinner than the presumptive area 18 despite the fact that later in the adult it will house a greater number of neurons in its thickness (Lukaszewicz et al., 2006). During this period of rapid growth of the occipital pole the maintenance of the VZ will require an increase in rates of proliferation and implies that there is a concerted mitogenic effect relayed from the OSVZ down to the VZ.

Species differences in the RGP and IP pools in terms of the self-renewal potential of their respective precursors and their mutual influences can explain the primate cortical expansion. The rodent SVZ is only partially self-sustaining and instead has to receive a constant supply of precursors from the VZ (Haubensak et al., 2004; Miyata et al., 2004; Noctor et al., 2004; Reznikov, Acklin, & van der Kooy, 1997; Wu et al., 2005). Enlargement of the SVZ precursor pool in the primate with progenitors possessing a higher self-renewal potential and possibly acting back on the RGP pool might correspond to an evolutionary adaptive mechanism ensuring the increased neuronal output necessary to build a more highly developed neocortex involving both an expanded and pronounced cytological complexification of the SGL and an increase in cortical surface related to gyrification (Dehay et al., 1993; Lukaszewicz et al., 2005; Smart et al., 2002).

MODULATION OF CELL CYCLE AND
DETERMINATION OF NEURON NUMBER

The determination of neuron numbers composing either a cortical area or a cortical layer has been shown—by a technique referred to as the mitotic history of the neuron (Dehay & Kennedy, 2007)—to crucially depend on the modulation of the mode of division where high rates of proliferative division lead to increases in the production of neurons (Polleux, Dehay, & Kennedy, 1998; Polleux et al., 1997). Because cytoarchitecture is characterized by differences in the numbers of neurons in individual layers, arealization and lamination are two sides of the same coin. Hence the spatial and temporal modulation of the frequency of proliferative divisions of cortical precursors determines the cytoarchitecture of the cortex (Dehay & Kennedy, 2007).

The coordinated regulation of two cardinal cell-cycle parameters of cortical precursors determines neuronal production via the regulation of the size of the precursor pool: the duration of the cell cycle and the relative frequency of cell-cycle reentry compared with cell-cycle exit. In vivo and ex vivo analyses of the cell-cycle regulation of VZ and OSVZ precursors of the primate visual cortex have shed light on the molecular correlates of area-specific differences in proliferation that underlie areal differences in the thickness of the SGL (Dehay et al., 1993; Lukaszewicz et al., 2005).

Importantly, the lengthening of the cell-cycle duration in primates could have consequences on fate determination, for instance by allowing more time for a precursor to interact with the microenvironment, or by loosening temporal constraints on gene transcription during the cell cycle. The prolonged duration of cell cycle in monkey cortical precursors has been hypothesized to be an adaptive feature related to the evolutionary expansion of neocortex in primates (Rakic, 1995). Because environmental signals are known to determine cortical precursor fate during the cell cycle as well as to regulate rates of proliferation, the prolonged duration of the primate cell cycle may serve to ensure a fine adjustment of rates of production of phenotypically defined neurons (Dehay et al., 2001; McConnell & Kaznowski, 1991; Polleux et al., 2001).

Compared to area 18 OSVZ precursors, area 17 OSVZ precursors are characterized by both a shorter cell-cycle duration, due to a reduction of the G1 phase, and an increased relative frequency of cell-cycle reentry. These areal differences on OSVZ precursor cell-cycle regulation are associated with significant differences in the level of expression of molecular regulators of the G1/S transition namely p27kip1 and Cyclin E. The ex vivo up and down modulation of the level of expression of these regulators significantly affects cell-cycle reentry and the rate of cell-cycle progression. These experiments highlight the role of the G1 phase regulation in corticogenesis (Lukaszewicz et al., 2005, 2006). Mathematical modeling of the observed differences in both rates of cell-cycle reentry and in G1 phase duration shows that the combined variation of these two parameters is sufficient to generate the enlarged SGL that distinguishes area 17 from area 18. Further, the modeling shows that a hypothetical shortening of the cell cycle without a change in the mode of division would have little effect on the numbers of neurons produced. These results show that variations of G1 phase duration and the coordinated change in mode of division contribute directly to regulate neuron number (Götz & Huttner, 2005; Lukaszewicz et al., 2002).

EXTRINSIC FACTORS INFLUENCING CORTICOGENESIS

The timetable of thalamic axons' ingrowth into the presumptive visual cortex (Rakic, 1977) coincides with the increase of thickness of the OSVZ (Smart et al., 2002). Using both fetal intraocular tritiated proline injections followed by 4-day survival periods to allow in vivo transynaptic transport, and DiI tracing in fixed embryonic brain, we have characterized the location of thalamic fibers in the developing primate neuroepithelium. The bulk of ingrowing geniculostriate axons reside in the outer fiber layer that lies immediately above the OSVZ (Smart et al., 2002) (see figure 98.3). Interestingly this constitutes an inverted pattern with respect to what is found in nonprimates where the corticopetal pathway lies above and not below the corticofugal pathway (Bicknese et al., 1994; Miller, Chou, & Finlay, 1993; Woodward, & Coull, 1984). A substantial number of individual embryonic geniculostriate axons are also observed within the depth of the OSVZ (Dehay & Kennedy, 2007).

In vitro work on mouse cortical precursors (Dehay et al., 2001) indicates that thalamic afferents control corticogenesis by modulating rates of proliferation. Embryonic thalamic axons release a mitogenic factor that increases the proliferative capacity of mouse cortical precursors during generation of the SGL by decreasing the G1 duration and by promoting cell-cycle reentry (Dehay et al., 2001). In the monkey, the geniculostriate axons that selectively project onto the OSVZ of area 17 could be responsible for the temporal and spatially restricted stimulation of proliferation that results in the transient upsurge of the size of SGL precursor pool in area 17 (Dehay et al., 1993; Lukaszewicz et al., 2005; Smart et al., 2002). There are also in vivo indications in primates that embryonic thalamic axons impact on areal size and specification via an early influence on neuron production during cortical neurogenesis (Dehay et al., 1989, 1991). Because thalamic axons are precisely targeted on to distinct cortical areas (Lopez-Bendito & Molnar, 2003), they will be able to differentially affect rates of precursor proliferation and of neuron production across the germinal zones and therefore determine local areal cytoarchitectonic features.

The observation that thalamic afferents are modulating proliferation of cortical precursors is in agreement with extensive findings in lower vertebrates and invertebrates suggesting that afferents in development determine neuron number via a modulation of neuronal proliferation (Baptista, Gershon, & Macagno, 1990; Gong & Shipley, 1995; Kollros, 1953, 1982; Selleck & Steller, 1991; Williams & Herrup, 1988). In this way, it would seem that the modulation of cortical neuron number by thalamic afferents is a widely phylogenetically conserved mechanisms for the orchestration of development.

CONCLUDING REMARKS

We have seen that there are a number of highly characteristic features in the development of the primate cortex such as the expansion of the OSVZ, the relatively high proportion of interneurons generated in the cortex, and some important changes in the relative timing of developmental events. In contrast there are highly conserved aspects of primate corticogenesis such as the regulation of proliferation by the ascending afferents to the cortex.

These primate-specific processes are likely to explain the evolutionary expansion and complexification of the primate SGL. The expansion of the OSVZ enables an amplification of the production of both excitatory and inhibitory neurons. The primate OSVZ also contains precursors exhibiting unique behaviors that are able to generate neuronal types specific to primates with the example of the double-bouquet cells. These unique developmental features provide the material necessary for the elaboration of the primate-specific cortical networks that constitute the best cognitive system known so far: the human brain.

The attractor or epigenetic landscape considers that the cell phenotype is defined in a state space in which the cell has a specific trajectory (Kauffman, 1993). This leads to phenotypic variability, which allows complex cellular interactions that underlie the emergence of biocomplexity. The peaks and saddles separating the valleys in the attractor landscape can be transversed, via forced gene expression. We have explored this concept in the above-mentioned experiments where we have caused an acceleration of the cell cycle of cortical precursors during the production of SGL (Pilaz et al., 2009). This gain-of-function experiment led to an expansion of the mouse precursor pool, resulting in an increased production of the SGL of the mouse. The increased size of the SVZ and the SGL of the mouse appears as a form of "primatization" of the rodent brain. Interestingly, with the expansion of the SVZ precursor pool there is also an important increase in the fraction of SVZ precursors that coexpress Pax6 and Tbr2. In mouse very few precursors coexpress the two transcription factors (Englund et al., 2005) while this is a characteristic feature of the primate OSVZ and to a lesser extent of the ferret SVZ (Hansen et al., 2010; Reillo et al., 2011).

One can interpret these experimental results in terms of a modification of the normal regulatory control of mouse corticogenesis. One can hypothesize that the gain-of-function experiment described above is essentially revealing what would be the consequence of an increase in the thalamic regulation of corticogenesis, and that this could be mimicking the expansion of cortex during evolution. The finding that interests us here is that the experimental gain-of-function experiment resulting from cell-cycle acceleration induces the expression of Pax6 in precursors that would normally be uniquely expressing Tbr2. Understanding this experimental result in the framework of the Kauffman hypothesis suggests that the attractor landscape of the mouse cortical precursors has been modified via the G1 phase reduction. This mimics what is happening during normal development under the influence of the thalamic control of corticogenesis, suggesting that the regulatory loops are shaping the attractor landscape of the cortical precursor cells via modification of the gene regulatory network (Kauffman, 1993). These results explore the idea that cortical development is tightly linked to its evolution, that it involves long-distance cell–cell interactions with far-reaching consequences, and that phenotypic variability plays a crucial role.

The changing of the relative timing of developmental events, referred to as heterochrony, is thought to be a fundamental process throughout phylogeny that allows large phenotypic variation with very little genetic change (Kennedy & Dehay, 1993; King & Wilson, 1975). If heterochrony via the relative early arrival of thalamic axons in primates is leading to the expansion of the OSVZ in this order, it illustrates the importance of conservation of afferent specification in the emergence of unique features in phylogeny.

ACKNOWLEDGMENTS

This work was supported by FP6–2005 IST-1583 (HK); FP7–2007 ICT-216593 (HK-CD); ANR-05-NEUR-088 (HK); ANR-06-NEUR-CMMCS (CD), and LabEx CORTEX (CD-HK).

REFERENCES

Anderson, S. A., Eisenstat, D. D., Shi, L., & Rubenstein, L. R. (1997). Interneuron migration from basal forebrain to neocortex: Dependence on *Dlx* genes. *Science, 278*, 474–476.

Anderson, S. A., Kaznowski, C. E., Horn, C., Rubenstein, J. L., & McConnell, S. K. (2002). Distinct origins of neocortical projection neurons and interneurons in vivo. *Cerebral Cortex, 12*, 702–709.

Arai, Y., Pulvers, J. N., Haffner, C., Schilling, B., Nusslein, I., Calegari, F., et al. (2011). Neural stem and progenitor cells shorten S-phase on commitment to neuron production. *Nature Communications, 2*, 154. doi:10.1038/ncomms1155.

Arlotta, P., Molyneaux, B. J., Chen, J., Inoue, J., Kominami, R., & Macklis, J. D. (2005). Neuronal subtype-specific genes that control corticospinal motor neuron development in vivo. *Neuron, 45*, 207–221.

Ascoli, G. A., Alonso-Nanclares, L., Anderson, S. A., Barrionuevo, G., Benavides-Piccione, R., Burkhalter, A., et al. (2008). Petilla terminology: Nomenclature of features of GABAergic interneurons of the cerebral cortex. *Nature Reviews. Neuroscience, 9*, 557–568. doi:10.1038/nrn2402.

Baptista, C. A., Gershon, T. R., & Macagno, E. R. (1990). Peripheral organs control central neurogenesis in the leech. *Nature, 346*, 855–858.

Bartos, M., Vida, I., & Jonas, P. (2007). Synaptic mechanisms of synchronized gamma oscillations in inhibitory interneuron networks. *Nature Reviews. Neuroscience, 8*, 45–56.

Beaulieu, C. (1993). Numerical data on neocortical neurons in adult rat, with special reference to the GABA population. *Brain Research, 609*, 284–292.

Bedford, L., Walker, R., Kondo, T., van Cruchten, I., King, E. R., & Sablitzky, F. (2005). Id4 is required for the correct timing of neural differentiation. *Developmental Biology, 280*, 386–395.

Bernard, A., Lubbers, L. S., Tanis, K. Q., Luo, R., Podtelezhnikov, A. A., Finney, E. M., et al. (2012). Transcriptional architecture of the primate neocortex. *Neuron, 73*, 1083–1099. doi:10.1016/j.neuron.2012.03.002.

Bicknese, A. R., Sheppard, A. M., O'Leary, D. D. M., & Pearlman, A. L. (1994). Thalamocortical axons extend along a chondroitin sulfate proteoglycan-enriched pathway coincident with the neocortical subplate and distinct from the efferent path. *Journal of Neuroscience, 14*, 3500–3510.

Britanova, O., Akopov, S., Lukyanov, S., Gruss, P., & Tarabykin, V. (2005). Novel transcription factor Satb2 interacts with matrix attachment region DNA elements in a tissue-specific manner and demonstrates cell-type-dependent expression in the developing mouse CNS. *European Journal of Neuroscience, 21*, 658–668.

Calegari, F., & Huttner, W. B. (2003). An inhibition of cyclin-dependent kinases that lengthens, but does not arrest, neuroepithelial cell cycle induces premature neurogenesis. *Journal of Cell Science, 116*, 4947–4955.

Cappello, S., Attardo, A., Wu, X., Iwasato, T., Itohara, S., Wilsch-Brauninger, M., et al. (2006). The Rho-GTPase cdc42 regulates neural progenitor fate at the apical surface. *Nature Neuroscience, 9*, 1099–1107. doi:10.1038/nn1744.

Chen, B., Schaevitz, L. R., & McConnell, S. K. (2005). Fezl regulates the differentiation and axon targeting of layer 5 subcortical projection neurons in cerebral cortex. *Proceedings of the National Academy of Sciences of the United States of America, 102,* 17184–17189. doi:10.1073/pnas.0508732102.

Chen, J. G., Rasin, M. R., Kwan, K. Y., & Sestan, N. (2005). Zfp312 is required for subcortical axonal projections and dendritic morphology of deep-layer pyramidal neurons of the cerebral cortex. *Proceedings of the National Academy of Sciences of the United States of America, 102,* 17792–17797. doi:10.1073/pnas.0509032102.

Chen, L., Liao, G., Yang, L., Campbell, K., Nakafuku, M., Kuan, C. Y., et al. (2006). Cdc42 deficiency causes Sonic hedgehog-independent holoprosencephaly. *Proceedings of the National Academy of Sciences of the United States of America, 103,* 16520–16525. doi:10.1073/pnas.0603533103.

Chenn, A., & Walsh, C. A. (2002). Regulation of cerebral cortical size by control of cell cycle exit in neural precursors. *Science, 297,* 365–369.

Cheung, A. F., Kondo, S., Abdel-Mannan, O., Chodroff, R. A., Sirey, T. M., Bluy, L. E., et al. (2010). The subventricular zone is the developmental milestone of a 6-layered neocortex: Comparisons in metatherian and eutherian mammals. *Cerebral Cortex, 20,* 1071–1081. doi:10.1093/cercor/bhp168.

Cheung, A. F., Pollen, A. A., Tavare, A., DeProto, J., & Molnar, Z. (2007). Comparative aspects of cortical neurogenesis in vertebrates. *Journal of Anatomy, 211,* 164–176.

Defelipe, J. (2002). Cortical interneurons: From Cajal to 2001. *Progress in Brain Research, 136,* 215–238.

Defelipe, J. (2011). The evolution of the brain, the human nature of cortical circuits, and intellectual creativity. *Frontiers in Neuroanatomy, 5,* 29. doi:10.3389/fnana.2011.00029.

Defelipe, J., Gonzalez-Albo, M. C., Del Rio, M. R., & Elston, G. N. (1999). Distribution and patterns of connectivity of interneurons containing calbindin, calretinin, and parvalbumin in visual areas of the occipital and temporal lobes of the macaque monkey. *Journal of Comparative Neurology, 412,* 515–526.

Dehay, C., Giroud, P., Berland, M., Smart, I., & Kennedy, H. (1993). Modulation of the cell cycle contributes to the parcellation of the primate visual cortex. *Nature, 366,* 464–466.

Dehay, C., Horsburgh, G., Berland, M., Killackey, H., & Kennedy, H. (1989). Maturation and connectivity of the visual cortex in monkey is altered by prenatal removal of retinal input. *Nature, 337,* 265–267.

Dehay, C., Horsburgh, G., Berland, M., Killackey, H., & Kennedy, H. (1991). The effects of bilateral enucleation in the primate fetus on the parcellation of visual cortex. *Brain Research. Developmental Brain Research, 62,* 137–141.

Dehay, C., & Kennedy, H. (2007). Cell-cycle control and cortical development. *Nature Reviews. Neuroscience, 8,* 438–450.

Dehay, C., Savatier, P., Cortay, V., & Kennedy, H. (2001). Cell-cycle kinetics of neocortical precursors are influenced by embryonic thalamic axons. *Journal of Neuroscience, 21,* 201–214.

Englund, C., Fink, A., Lau, C., Pham, D., Daza, R. A., Bulfone, A., et al. (2005). Pax6, Tbr2, and Tbr1 are expressed sequentially by radial glia, intermediate progenitor cells, and postmitotic neurons in developing neocortex. *Journal of Neuroscience, 25,* 247–251. doi:10.1523/JNEUROSCI.2899-04.2005.

Fietz, S. A., Kelava, I., Vogt, J., Wilsch-Brauninger, M., Stenzel, D., Fish, J. L., et al. (2010). OSVZ progenitors of human and ferret neocortex are epithelial-like and expand by integrin signaling. *Nature Neuroscience, 13,* 690–699. doi:10.1038/nn.2553.

Fish, J. L., Dehay, C., Kennedy, H., & Huttner, W. B. (2008). Making bigger brains—The evolution of neural-progenitor-cell division. *Journal of Cell Science, 121,* 2783–2793.

Fishell, G., & Rudy, B. (2011). Mechanisms of inhibition within the telencephalon: "Where the wild things are." *Annual Review of Neuroscience, 34,* 535–567.

Frantz, G. D., Weimann, J. M., Levin, M. E., & McConnell, S. K. (1994). Otx1 and Otx2 define layers and regions in developing cerebral cortex and cerebellum. *Journal of Neuroscience, 14,* 5725–5740.

Gabbott, P. L. A., & Bacon, S. J. (1996). Local circuit neurons in the medial prefrontal cortex (areas 24a, b, c, 25 and 32) in the monkey: II. Quantitative areal and laminar distribution. *Journal of Comparative Neurology, 364,* 609–636.

Gal, J. S., Morozov, Y. M., Ayoub, A. E., Chatterjee, M., Rakic, P., & Haydar, T. F. (2006). Molecular and morphological heterogeneity of neural precursors in the mouse neocortical proliferative zones. *Journal of Neuroscience, 26,* 1045–1056.

Garcia-Moreno, F., Vasistha, N. A., Trevia, N., Bourne, J. A., & Molnar, Z. (2011). Compartmentalization of cerebral cortical germinal zones in a lissencephalic primate and gyrencephalic rodent. *Cerebral Cortex, 22,* 482–492.

Geisler, W. S., & Diehl, R. L. (2002). Bayesian natural selection and the evolution of perceptual systems. *Philosophical Transactions of the Royal Society of London. Series B, Biological Sciences, 357,* 419–448. doi:10.1098/rstb.2001.1055.

Gong, Q. H., & Shipley, M. T. (1995). Evidence that pioneer olfactory axons regulate telencephalon cell cycle kinetics to induce the formation of the olfactory bulb. *Neuron, 14,* 91–101.

Götz, M., & Huttner, W. B. (2005). The cell biology of neurogenesis. *Nature Reviews. Molecular Cell Biology, 6,* 777–788.

Götz, M., Stoykova, A., & Gruss, P. (1998). Pax6 controls radial glia differentiation in the cerebral cortex. *Neuron, 21,* 1031–1044.

Halley, J. D., & Winkler, D. A. (2008). Critical-like self-organization and natural selection: Two facets of a single evolutionary process? *Bio Systems, 92,* 148–158.

Hansen, D. V., Lui, J. H., Parker, P. R., & Kriegstein, A. R. (2010). Neurogenic radial glia in the outer subventricular zone of human neocortex. *Nature, 464,* 554–561.

Hartfuss, E., Galli, R., Heins, N., & Götz, M. (2001). Characterization of CNS precursor subtypes and radial glia. *Developmental Biology, 229,* 15–30.

Haubensak, W., Attardo, A., Denk, W., & Huttner, W. B. (2004). Neurons arise in the basal neuroepithelium of the early mammalian telencephalon: A major site of neurogenesis. *Proceedings of the National Academy of Sciences of the United States of America, 101,* 3196–3201.

Heins, N., Malatesta, P., Cecconi, F., Nakafuku, M., Tucker, K. L., Hack, M. A., et al. (2002). Glial cells generate neurons: The role of the transcription factor Pax6. *Nature Neuroscience, 5,* 308–315. doi:10.1038/nn828.

Hendry, S. H. C., Schwark, H. D., Jones, E. G., & Yan, J. (1987). Numbers and proportions of GABA-immunoreactive neurons in different areas of monkey cerebral cortex. *Journal of Neuroscience, 7,* 1503–1519.

Hevner, R. F. (2005). The cerebral cortex malformation in thanatophoric dysplasia: Neuropathology and pathogenesis. *Acta Neuropathologica, 110,* 208–221.

Hill, R. S., & Walsh, C. A. (2005). Molecular insights into human brain evolution. *Nature, 437,* 64–67.

Holm, P. C., Mader, M. T., Haubst, N., Wizenmann, A., Sigvardsson, M., & Götz, M. (2007). Loss- and gain-of-function analyses reveal targets of Pax6 in the developing mouse telencephalon. *Molecular and Cellular Neurosciences, 34,* 99–119.

Inglis-Broadgate, S. L., Thomson, R. E., Pellicano, F., Tartaglia, M. A., Pontikis, C. C., Cooper, J. D., et al. (2005). FGFR3 regulates brain size by controlling progenitor cell proliferation and apoptosis during embryonic development. *Developmental Biology, 279,* 73–85. doi:10.1016/j.ydbio.2004.11.035.

Inta, D., Alfonso, J., von Engelhardt, J., Kreuzberg, M. M., Meyer, A. H., van Hooft, J. A., et al. (2008). Neurogenesis and widespread forebrain migration of distinct GABAergic neurons from the postnatal subventricular zone. *Proceedings of the National Academy of Sciences of the United States of America, 105,* 20994–20999. doi:10.1073/pnas.0807059105.

Jakovcevski, I., Mayer, N., & Zecevic, N. (2011). Multiple origins of human neocortical interneurons are supported by distinct expression of transcription factors. *Cerebral Cortex, 21,* 1771–1782.

Jones, E. G., DeFelipe, J., Hendry, S. H. C., & Maggio, J. E. (1988). A study of tachykinin-immunoreactive neurons in monkey cerebral cortex. *Journal of Neuroscience, 8,* 1206–1224.

Kauffman, S. A. (1993). *The origins of order.* New York: Oxford University Press.

Kauffman, S. A. (2000). *Investigations.* New York: Oxford University Press.

Kelava, I., Reillo, I., Murayama, A. Y., Kalinka, A. T., Stenzel, D., Tomancak, P., et al. (2012). Abundant occurrence of basal radial glia in the subventricular zone of embryonic neocortex of a lissencephalic primate, the common marmoset *Callithrix jacchus. Cerebral Cortex, 22,* 469–481. doi:10.1093/cercor/bhr301.

Kennedy, H., & Dehay, C. (1993). Cortical specification of mice and men. *Cerebral Cortex, 3,* 27–35.

Kennedy, H., & Dehay, C. (2012). Self-organization and inter-areal networks in the primate cortex. *Progress in Brain Research, 195,* 341–360.

Kennedy, H., Douglas, R., Knoblauch, K., & Dehay, C. (2007). Self-organization and pattern formation in primate cortical networks. *Novartis Foundation Symposium, 288,* 178–194 discussion 195–198, 276–281.

King, M. C., & Wilson, A. C. (1975). Evolution at two levels in humans and chimpanzees. *Science, 188,* 107–116.

Kohwi, M., Petryniak, M. A., Long, J. E., Ekker, M., Obata, K., Yanagawa, Y., et al. (2007). A subpopulation of olfactory bulb GABAergic interneurons is derived from Emx1- and Dlx5/6-expressing progenitors. *Journal of Neuroscience, 27,* 6878–6891. doi:10.1523/JNEUROSCI.0254-07.2007.

Kollros, J. J. (1953). The development of the optic lobes in the frog: I. The effects of unilateral enucleation in embryonic stages. *Journal of Experimental Zoology, 123,* 153–187.

Kollros, J. J. (1982). Peripheral control of midbrain mitotic activity in the frog. *Journal of Comparative Neurology, 205,* 171–178.

Kowalczyk, T., Pontious, A., Englund, C., Daza, R. A., Bedogni, F., Hodge, R., et al. (2009). Intermediate neuronal progenitors (basal progenitors) produce pyramidal-projection neurons for all layers of cerebral cortex. *Cerebral Cortex, 19,* 2439–2450. doi:10.1093/cercor/bhn260.

Kriegstein, A., Noctor, S., & Martinez-Cerdeno, V. (2006). Patterns of neural stem and progenitor cell division may underlie evolutionary cortical expansion. *Nature Reviews. Neuroscience, 7,* 883–890.

Kuida, K., Haydar, T. F., Kuan, C. Y., Gu, Y., Taya, C., Karasuyama, H., et al. (1998). Reduced apoptosis and cytochrome c-mediated caspase activation in mice lacking caspase 9. *Cell, 94,* 325–337. doi:10.1016/S0092-8674(00)81476-2.

Land, P. W., & Monaghan, A. P. (2003). Expression of the transcription factor, tailless, is required for formation of superficial cortical layers. *Cerebral Cortex, 13,* 921–931.

Lange, C., Huttner, W. B., & Calegari, F. (2009). Cdk4/cyclinD1 overexpression in neural stem cells shortens G1, delays neurogenesis, and promotes the generation and expansion of basal progenitors. *Cell Stem Cell, 5,* 320–331.

Letinic, K., Zoncu, R., & Rakic, P. (2002). Origin of GABAergic neurons in the human neocortex. *Nature, 417,* 645–649.

LeVay, S., Wiesel, T. N., & Hubel, D. H. (1980). The development of ocular dominance columns in normal and visually deprived monkeys. *Journal of Comparative Neurology, 191,* 1–51.

Levitt, P., Cooper, M. L., & Rakic, P. (1981). Coexistence of neuronal and glial precursor cells in the cerebral ventricular zone of the fetal monkey: An ultrastructural immunoperoxidase analysis. *Journal of Neuroscience, 1,* 27–39.

Lopez-Bendito, G., & Molnar, Z. (2003). Thalamocortical development: How are we going to get there? *Nature Reviews. Neuroscience, 4,* 276–289.

Lui, J. H., Hansen, D. V., & Kriegstein, A. R. (2011). Development and evolution of the human neocortex. *Cell, 146,* 18–36.

Lukaszewicz, A., Cortay, V., Giroud, P., Berland, M., Smart, I., Kennedy, H., et al. (2006). The concerted modulation of proliferation and migration contributes to the specification of the cytoarchitecture and dimensions of cortical areas. *Cerebral Cortex, 16*(Suppl. 1), i26–i34. doi:10.1093/cercor/bhk011.

Lukaszewicz, A., Savatier, P., Cortay, V., Giroud, P., Huissoud, C., Berland, M., et al. (2005). G1 phase regulation, area-specific cell cycle control, and cytoarchitectonics in the primate cortex. *Neuron, 47,* 353–364. doi:10.1016/j.neuron.2005.06.032.

Lukaszewicz, A., Savatier, P., Cortay, V., Kennedy, H., & Dehay, C. (2002). Contrasting effects of basic fibroblast growth factor and neurotrophin 3 on cell cycle kinetics of mouse cortical stem cells. *Journal of Neuroscience, 22,* 6610–6622.

Malatesta, P., Hack, M. A., Hartfuss, E., Kettenmann, H., Klinkert, W., Kirchhoff, F., et al. (2003). Neuronal or glial progeny: Regional differences in radial glia fate. *Neuron, 37,* 751–764.

Malatesta, P., Hartfuss, E., & Götz, M. (2000). Isolation of radial glial cells by fluorescent-activated cell sorting reveals a neuronal lineage. *Development, 127,* 5253–5263.

Marin, O., & Rubenstein, J. L. (2001). A long, remarkable journey: Tangential migration in the telencephalon. *Nature Reviews. Neuroscience, 2,* 780–790.

Martinez-Cerdeno, V., Cunningham, C. L., Camacho, J., Antczak, J. L., Prakash, A. N., Cziep, M. E., et al. (2012). Comparative analysis of the subventricular zone in rat, ferret and macaque: Evidence for an outer subventricular zone in rodents. *PLoS ONE, 7*, e30178. doi:10.1371/journal.pone.0030178.

Martinez-Cerdeno, V., Noctor, S. C., & Kriegstein, A. R. (2006). The role of intermediate progenitor cells in the evolutionary expansion of the cerebral cortex. *Cerebral Cortex, 16*(Suppl. 1), i152–i161.

McConnell, S. K., & Kaznowski, C. E. (1991). Cell cycle dependence of laminar determination in developing neocortex. *Science, 254*, 282–285.

Metin, C., Baudoin, J. P., Rakic, S., & Parnavelas, J. G. (2006). Cell and molecular mechanisms involved in the migration of cortical interneurons. *European Journal of Neuroscience, 23*, 894–900.

Miller, B., Chou, L., & Finlay, B. L. (1993). The early development of thalamocortical and corticothalamic projections. *Journal of Comparative Neurology, 335*, 16–41.

Miyata, T., Kawaguchi, A., Saito, K., Kawano, M., Muto, T., & Ogawa, M. (2004). Asymmetric production of surface-dividing and non-surface-dividing cortical progenitor cells. *Development, 131*, 3133–3145.

Molnar, Z., Metin, C., Stoykova, A., Tarabykin, V., Price, D. J., Francis, F., et al. (2006). Comparative aspects of cerebral cortical development. *European Journal of Neuroscience, 23*, 921–934. doi:10.1111/j.1460-9568.2006.04611.x.

Molyneaux, B. J., Arlotta, P., Hirata, T., Hibi, M., & Macklis, J. D. (2005). Fezl is required for the birth and specification of corticospinal motor neurons. *Neuron, 47*, 817–831.

Nieto, M., Monuki, E. S., Tang, H., Imitola, J., Haubst, N., Khoury, S. J., et al. (2004). Expression of Cux-1 and Cux-2 in the subventricular zone and upper layers II–IV of the cerebral cortex. *Journal of Comparative Neurology, 479*, 168–180. doi:10.1002/cne.20322.

Noctor, S. C., Flint, A. C., Weissman, T. A., Dammerman, R. S., & Kriegstein, A. R. (2001). Neurons derived from radial glial cells establish radial units in neocortex. *Nature, 409*, 714–720.

Noctor, S. C., Flint, A. C., Weissman, T. A., Wong, W. S., Clinton, B. K., & Kriegstein, A. R. (2002). Dividing precursor cells of the embryonic cortical ventricular zone have morphological and molecular characteristics of radial glia. *Journal of Neuroscience, 22*, 3161–3173.

Noctor, S. C., Martinez-Cerdeno, V., Ivic, L., & Kriegstein, A. R. (2004). Cortical neurons arise in symmetric and asymmetric division zones and migrate through specific phases. *Nature Neuroscience, 7*, 136–144.

O'Leary, D. D. M. (1989). Do cortical areas emerge from a protocortex? *Trends in Neurosciences, 12*, 400–406. doi:10.1016/0166-2236(89)90080-5.

Petanjek, Z., Berger, B., & Esclapez, M. (2009). Origins of cortical GABAergic neurons in the cynomolgus monkey. *Cerebral Cortex, 19*, 249–262.

Petanjek, Z., Kostovic, I., & Esclapez, M. (2009). Primate-specific origins and migration of cortical GABAergic neurons. *Frontiers in Neuroanatomy, 3*, 26. doi:10.3389/neuro.05.026.2009.

Pilaz, L. J., Patti, D., Marcy, G., Ollier, E., Pfister, S., Douglas, R. J., et al. (2009). Forced G1-phase reduction alters mode of division, neuron number, and laminar phenotype in the cerebral cortex. *Proceedings of the National Academy of Sciences of the United States of America, 106*, 21924–21929. doi:10.1073/pnas.0909894106.

Pinto, L., Drechsel, D., Schmid, M. T., Ninkovic, J., Irmler, M., Brill, M. S., et al. (2009). AP2gamma regulates basal progenitor fate in a region- and layer-specific manner in the developing cortex. *Nature Neuroscience, 12*, 1229–1237. doi:10.1038/nn.2399.

Polleux, F., Dehay, C., Goffinet, A., & Kennedy, H. (2001). Pre- and post-mitotic events contribute to the progressive acquisition of area-specific connectional fate in the neocortex. *Cerebral Cortex, 11*, 1027–1039.

Polleux, F., Dehay, C., & Kennedy, H. (1998). Neurogenesis and commitment of corticospinal neurons in reeler. *Journal of Neuroscience, 18*, 9910–9923.

Polleux, F., Dehay, C., Moraillon, B., & Kennedy, H. (1997). Regulation of neuroblast cell-cycle kinetics plays a crucial role in the generation of unique features of neocortical areas. *Journal of Neuroscience, 17*, 7763–7783.

Pontious, A., Kowalczyk, T., Englund, C., & Hevner, R. F. (2008). Role of intermediate progenitor cells in cerebral cortex development. *Developmental Neuroscience, 30*, 24–32.

Quinn, J. C., Molinek, M., Martynoga, B. S., Zaki, P. A., Faedo, A., Bulfone, A., et al. (2007). Pax6 controls cerebral cortical cell number by regulating exit from the cell cycle and specifies cortical cell identity by a cell autonomous mechanism. *Developmental Biology, 302*, 50–65. doi:10.1016/j.ydbio.2006.08.035.

Rakic, P. (1977). Genesis of the dorsal lateral geniculate nucleus in the rhesus monkey: Site and time of origin, kinetics of proliferation, routes of migration and pattern of distribution of neurons. *Journal of Comparative Neurology, 176*, 23–52.

Rakic, P. (1988a). Defects of neuronal migration and the pathogenesis of cortical malformations. *Progress in Brain Research, 73*, 15–37.

Rakic, P. (1988b). Specification of cerebral cortical areas. *Science, 241*, 170–176.

Rakic, P. (1995). A small step for the cell, a giant leap for mankind: A hypothesis of neocortical expansion during evolution. *Trends in Neurosciences, 18*, 383–388. doi:10.1016/0166-2236(95)93934-P.

Rakic, P. (2003). Developmental and evolutionary adaptations of cortical radial glia. *Cerebral Cortex, 13*, 541–549.

Rakic, P. (2009). Evolution of the neocortex: A perspective from developmental biology. *Nature Reviews. Neuroscience, 10*, 724–735.

Reillo, I., de Juan Romero, C., Garcia-Cabezas, M. A., & Borrell, V. (2011). A role for intermediate radial glia in the tangential expansion of the mammalian cerebral cortex. *Cerebral Cortex, 21*, 1674–1694.

Reznikov, K., Acklin, S. E., & van der Kooy, D. (1997). Clonal heterogeneity in the early embryonic rodent cortical germinal zone and the separation of subventricular from ventricular zone lineages. *Developmental Dynamics, 210*, 328–343.

Roy, K., Kuznicki, K., Wu, Q., Sun, Z., Bock, D., Schutz, G., et al. (2004). The Tlx gene regulates the timing of neurogenesis in the cortex. *Journal of Neuroscience, 24*, 8333–8345. doi:10.1523/JNEUROSCI.1148-04.2004.

Schmechel, D. E., & Rakic, P. (1979). Arrested proliferation of radial glial cells during midgestation in rhesus monkey. *Nature, 277*, 303–305.

Schuurmans, C., Armant, O., Nieto, M., Stenman, J. M., Britz, O., Klenin, N., et al. (2004). Sequential phases of cortical specification involve Neurogenin-dependent and -independent pathways. *European Molecular Biology Organization Journal, 23,* 2892–2902. doi:10.1038/sj.emboj.7600278.

Selleck, S. B., & Steller, H. (1991). The influence of retinal innervation on neurogenesis in the 1st optic ganglion of drosophila. *Neuron, 6,* 83–99.

Shen, Q., Wang, Y., Dimos, J. T., Fasano, C. A., Phoenix, T. N., Lemischka, I. R., et al. (2006). The timing of cortical neurogenesis is encoded within lineages of individual progenitor cells. *Nature Neuroscience, 9,* 743–751. doi:10.1038/nn1694.

Shepard, R. N. (2001). Perceptual-cognitive universals as reflections of the world. *Behavioral and Brain Sciences, 24,* 581–601, discussion 652–671.

Shitamukai, A., Konno, D., & Matsuzaki, F. (2011). Oblique radial glial divisions in the developing mouse neocortex induce self-renewing progenitors outside the germinal zone that resemble primate outer subventricular zone progenitors. *Journal of Neuroscience, 31,* 3683–3695.

Smart, I. H. (1973). Proliferative characteristics of the ependymal layer during the early development of the mouse neocortex: A pilot study based on recording the number, location and plane of cleavage of mitotic figures. *Journal of Anatomy, 116,* 67–91.

Smart, I. H., Dehay, C., Giroud, P., Berland, M., & Kennedy, H. (2002). Unique morphological features of the proliferative zones and postmitotic compartments of the neural epithelium giving rise to striate and extrastriate cortex in the monkey. *Cerebral Cortex, 12,* 37–53.

Stancik, E. K., Navarro-Quiroga, I., Sellke, R., & Haydar, T. F. (2010). Heterogeneity in ventricular zone neural precursors contributes to neuronal fate diversity in the postnatal neocortex. *Journal of Neuroscience, 30,* 7028–7036.

Sur, M., & Rubenstein, J. L. (2005). Patterning and plasticity of the cerebral cortex. *Science, 310,* 805–810.

Tamamaki, N., Fujimori, K. E., & Takauji, R. (1997). Origin and route of tangentially migrating neurons in the developing neocortical intermediate zone. *Journal of Neuroscience, 17,* 8313–8323.

Tarabykin, V., Stoykova, A., Usman, N., & Gruss, P. (2001). Cortical upper layer neurons derive from the subventricular zone as indicated by Svet1 gene expression. *Development, 128,* 1983–1993.

Thompson, I. D., Kossut, M., & Blakemore, C. (1983). Development of orientation columns in cat striate cortex revealed by 2-deoxyglucose autoradiography. *Nature, 301,* 712–715.

Van der Loos, H. (1977). Structural changes in the cerebral cortex upon modification of the periphery: Barrels in somatosensory cortex. *Philosophical Transactions of the Royal Society of London. Series B, Biological Sciences, 278,* 373–376.

Wang, X., Tsai, J. W., LaMonica, B., & Kriegstein, A. R. (2011). A new subtype of progenitor cell in the mouse embryonic neocortex. *Nature Neuroscience, 14,* 555–561.

Williams, R. W., & Herrup, K. (1988). The control of neuron number. *Annual Review of Neuroscience, 11,* 423–454.

Wonders, C. P., & Anderson, S. A. (2006). The origin and specification of cortical interneurons. *Nature Reviews. Neuroscience, 7,* 687–696.

Woodward, W. R., & Coull, B. M. (1984). Localization and organization of geniculocortical and corticofugal fiber tracts within the subcortical white matter. *Neuroscience, 12,* 1089–1099.

Wu, S., Esumi, S., Watanabe, K., Chen, J., Nakamura, K. C., Nakamura, K., et al. (2011). Tangential migration and proliferation of intermediate progenitors of GABAergic neurons in the mouse telencephalon. *Development, 138,* 2499–2509. doi:10.1242/dev.063032.

Wu, S. X., Goebbels, S., Nakamura, K., Nakamura, K., Kometani, K., Minato, N., et al. (2005). Pyramidal neurons of upper cortical layers generated by NEX-positive progenitor cells in the subventricular zone. *Proceedings of the National Academy of Sciences of the United States of America, 102,* 17172–17177. doi:10.1073/pnas.0508560102.

Yanez, I. B., Munoz, A., Contreras, J., Gonzalez, J., Rodriguez-Veiga, E., & DeFelipe, J. (2005). Double bouquet cell in the human cerebral cortex and a comparison with other mammals. *Journal of Comparative Neurology, 486,* 344–360.

Yu, X., & Zecevic, N. (2011). Dorsal radial glial cells have the potential to generate cortical interneurons in human but not in mouse brain. *Journal of Neuroscience, 31,* 2413–2420.

Yun, K., Mantani, A., Garel, S., Rubenstein, J., & Israel, M. A. (2004). Id4 regulates neural progenitor proliferation and differentiation in vivo. *Development, 131,* 5441–5448.

Zecevic, N., Chen, Y., & Filipovic, R. (2005). Contributions of cortical subventricular zone to the development of the human cerebral cortex. *Journal of Comparative Neurology, 491,* 109–122.

Zecevic, N., Hu, F., & Jakovcevski, I. (2011). Interneurons in the developing human neocortex. *Developmental Neurobiology, 71,* 18–33.

Zhou, C. J., Borello, U., Rubenstein, J. L., & Pleasure, S. J. (2006). Neuronal production and precursor proliferation defects in the neocortex of mice with loss of function in the canonical Wnt signaling pathway. *Neuroscience, 142,* 1119–1131.

Zimmer, C., Tiveron, M. C., Bodmer, R., & Cremer, H. (2004). Dynamics of Cux2 expression suggests that an early pool of SVZ precursors is fated to become upper cortical layer neurons. *Cerebral Cortex, 14,* 1408–1420.

99 Neural Limitations on Visual Development in Primates: Beyond Striate Cortex

LYNNE KIORPES AND J. ANTHONY MOVSHON

Since the pioneering work of Wiesel and Hubel in the 1960s and 1970s, we have known that the primary visual cortex shows evidence of postnatal development and is susceptible to effects of visual experience during an early critical period. It was therefore widely assumed that the maturation of striate cortex presented an important limit on the development of visual function (Wiesel, 1982). However, despite substantial investigation over the past 50 years into the anatomical and physiological processes that take place postnatally in the primate visual system, it remains unclear what neural mechanisms limit the development of vision. In nonhuman primates, where we have the strongest data across levels of analysis—from cellular to behavioral—it is clear that early postnatal changes at the level of the retina pose no important limit on performance (Kiorpes et al., 2003). Similarly, for spatial vision, development of neuronal response properties at the level of the lateral geniculate nucleus (LGN) and striate cortex are insufficient to account for behavioral development of spatial resolution and contrast sensitivity (Kiorpes & Movshon, 2004a; Movshon et al., 2005; Zheng et al., 2007). Conversely, temporal resolution, as represented by the high temporal frequency cutoff of the temporal modulation transfer function, is well matched to the temporal frequency response properties of magnocellular LGN neurons (Stavros & Kiorpes, 2008). Thus, although the striate cortex has long been thought to be the likely site of limitations on vision in infants, development of neuronal response properties at this level of the visual system does not seem to account for behaviorally measured visual function (Kiorpes & Movshon, 2004a). This chapter will evaluate the idea that later, hierarchical development of extrastriate visual areas is permitting the progress of visual development and will address additional hypotheses regarding the quality of information available to the behavioral output in infants.

It is well accepted that the cortical visual system is organized in a hierarchical fashion, with dominant extrastriate pathways emerging from primary visual cortex (Felleman & Van Essen, 1991; see chapter 17, this volume) and with higher-level visual areas involved in ever more complex visual processing. Although there is good evidence to suggest that the brain develops hierarchically as well (Guillery, 2005), with subcortical development preceding cortical, and primary sensory cortices preceding secondary and higher areas, there has to date been no full analysis of this process at the functional level. The basic organization of striate cortex is laid down prenatally, with ocular dominance and orientation domains essentially adult-like at birth in nonhuman primates (Blasdel, Obermayer, & Kiorpes, 1995; Horton & Hocking, 1996; Wiesel & Hubel, 1974). Cytochrome oxidase organization—blobs, patches, and stripes—is also quite mature in areas V1 and V2 in neonatal macaques (Baldwin et al., 2012; Horton & Hocking, 1996) and humans (Burkhalter, Bernardo, & Charles, 1993). Visual experience is not required for this organization to develop although abnormal visual experience can disrupt its maintenance (Horton & Hocking, 1996, 1997; Wiesel & Hubel, 1974). As reviewed by Kennedy and Burkhalter (2004), the organization of feedforward and feedback interareal connections is also largely laid down prenatally, linking areas that are hierarchically related in the adult monkey. Topographic specificity of feedforward projections is relatively adult-like near birth and undergoes relatively little refinement postnatally. Feedback projections are in general more diffuse and exuberant initially and require substantial postnatal refinement. Nevertheless, they are apparently fully mature by 2–4 months after birth (Baldwin et al., 2012; Kennedy & Burkhalter, 2004). A similar pattern of cortical maturation is found in human visual cortex although in humans refinement of feedback projections continues until 18–24 months

(Burkhalter, 1993). The types of changes described at the anatomical level could support a modest decrease in the size of receptive field centers and more substantial development or reorganization of receptive field surround mechanisms, similar to what has been reported previously based on electrophysiology in V1 and V2 of macaques (Baldwin et al., 2012; Chino et al., 1997; Kiorpes & Movshon, 2004a; Zheng et al., 2007). However, even basic visual functions measured behaviorally are still quite immature in nonhuman primates at 16 weeks when these structural and electrophysiological processes are found to be completely mature. This suggests that the important limitations lie downstream from V1 and V2. Therefore it is of interest to evaluate the development of visual functions that may depend on higher-level visual areas.

A number of higher-order visual functions have been studied in young macaques. These more complex visual functions are generally integrative in nature, requiring combination of information across space or space-time to yield a coherent or global percept. Some integrative functions are thought to reflect the processing of particular extrastriate visual areas. For example, the middle temporal area MT/V5 in the dorsal stream is well-known to be involved in the processing of global motion stimuli and provides signals that support the perception of motion (Britten et al., 1992; Newsome & Pare, 1988). On the other hand, perception of static global form stimuli, such as Glass patterns, and figure–ground segmentation tasks are thought to depend on processing in ventral stream areas like V4 or lateral occipital complex (Altmann, Bülthoff, & Kourtzi, 2003; Ostwald et al., 2008). Thus, examination of the development of sensitivity to integrative stimuli should provide a window into the role of extrastriate areas in the vision of young primates.

One of the best-documented correlations between neuronal activity and perception is that of the electrophysiological properties of neurons in area MT and the perception of motion. In adult macaques and human patients, damage to area MT results in changes in perception of global motion (Newsome & Pare, 1988; Rudolph & Pasternak, 1999; Shipp et al., 1994; Vaina et al., 2005). Also, electrical stimulation of selected populations of MT neurons biases perceived motion direction in a predictable fashion (Salzman et al., 1992). Naturally, then, one would expect a close correlation between the development of neural function in this area and development of motion sensitivity. Behavioral development of motion sensitivity has been described in infant macaques (Hall-Haro & Kiorpes, 2008; Kiorpes & Movshon, 2004b; Kiorpes et al., 2012; Mikami & Fujita, 1992). With either 1-D grating stimuli or

integrative motion stimuli based on random dot kinematograms (RDKs), infant macaques are able to detect visual motion and discriminate its direction shortly after birth. However, motion sensitivity in infants is substantially reduced compared with that in adults and improves over a long, slow developmental time course that comprises the first 2–3 years after birth (see figure 99.1A). Interestingly, the perception of 2-D pattern motion diverges from this profile. Perception of coherent motion of a plaid pattern develops comparatively late. Infants under the age of 12 weeks failed to identify the direction of plaid motion while they had no difficulty discriminating the direction of 1-D grating motion on a comparable task (Hall-Haro & Kiorpes, 2008). This ability was evident in all animals by 18 weeks after birth and, although individuals varied greatly in absolute sensitivity, pattern motion sensitivity reached asymptotic levels around the end of the first postnatal year (see figure 99.1B). Clearly the developmental time course for motion perception depends on the type of motion under test, and the neural processes supporting perception of motion in RDKs and plaids are not the same.

To investigate the relative maturity of neurons in area MT of infant macaques, our group recorded single-neuron response properties in 1-, 4-, and 16-week-old infants and in adults (Movshon et al., 2003). We found infant MT neurons to be well tuned for speed and direction of 1-D motion similar to those of adults. However, there were some immaturities. As noted in earlier cortical areas (V1 and V2) (see Kiorpes & Movshon, 2004a; Zheng et al., 2007), infant MT neurons were comparatively weakly responsive with longer latencies and lower spike rates than found in those of adults. Unlike earlier cortical areas, these properties were not fully adult in MT at 16 weeks. One striking difference between infant and adult MT was a dearth of pattern direction selective (PDS) neurons, those that signal the direction of motion of complex patterns, such as plaid patterns (Movshon et al., 1985; Rodman & Albright, 1989). PDS responses emerge from the elaboration of specific feedforward projections from cells in V1 to MT, and the late development of these cells may reflect the delayed refinement of that pattern of connections (Rust et al., 2006). The proportion of PDS cells in 16-week-olds was only one-half that of adults and was substantially smaller at younger ages (see figure 99.1C). While it is not straightforward to make a direct comparison between behavioral and electrophysiological results, these neural data suggest that infant macaques have the neural machinery to detect and discriminate simple motion but, based on the dearth of PDS cells, the youngest infants may not have the ability to appreciate complex pattern motion. These predictions are borne out by the

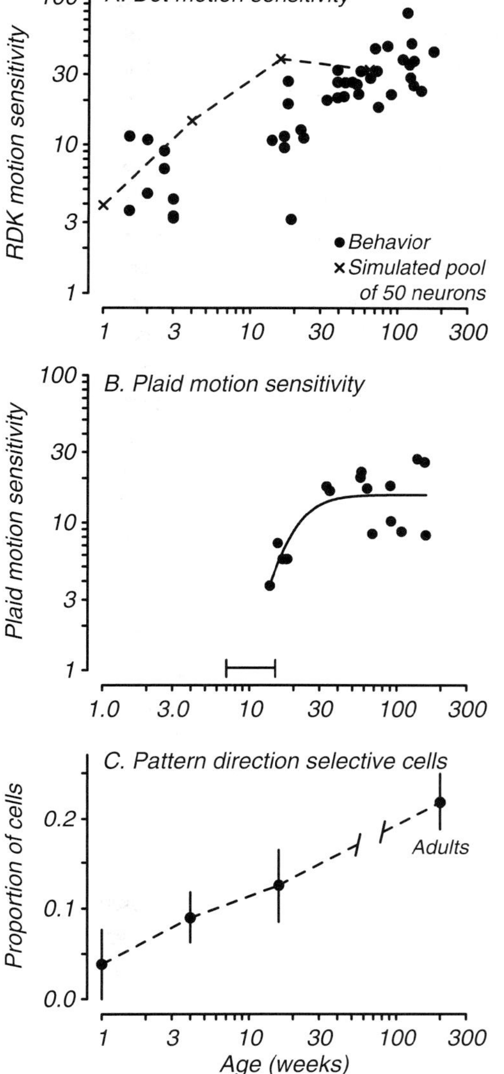

FIGURE 99.1 Development of "dorsal stream" motion mechanisms. (A) Development of sensitivity to random dot motion. Coherence represents the strength of the motion signal in the random dot kinematogram (RDK) stimulus. Filled symbols are data from individual infant monkeys plotted as a function of age (data from Kiorpes & Movshon, 2004b). The Xs and dashed line show the estimated sensitivity of the population of middle temporal visual cortex (MT) neurons recorded at each age, based on a pool size of 50 neurons. (B) Development of sensitivity to the direction plaid pattern motion. The filled symbols are data from individual monkeys plotted as a function of age. The horizontal bar along the abscissa delimits the range of ages over which infants failed to demonstrate the perception of plaid motion. The line fit to the data is a Michaelis–Menton function which captures the rate of development and the age at which sensitivity reaches 50% of adults levels (after Hall-Haro & Kiorpes, 2008). (C) Proportion of pattern direction cells in MT recorded at each age (mean ± SE).

behavioral data (see figure 99.1A, B) showing early onset of RDK motion discrimination but delayed development of pattern motion discrimination. What then explains the comparatively poor performance of infants shown in figure 99.1A and the extended developmental profile? As mentioned above, although MT neurons are well tuned in the youngest infants, they are only weakly responsive, and overall responsivity changes with age. To capture neuronal coherence thresholds as a function of age, we performed a pooling analysis similar to that used by Shadlen et al. (1996). An estimate of population-level coherence sensitivity at each age revealed parallel improvement between neuronal sensitivity and behavior up to 16 weeks when adult-level neuronal sensitivity was reached (Xs and dashed line in figure 99.1A). The correlation between neural and behavioral development at the early ages is appealing although this leaves the remaining extended developmental period for RDK motion sensitivity unexplained.

A variety of visual abilities that are thought to rely on ventral stream function have been studied in young macaques. Ventral stream function is characterized by perceptual challenges such as organization of static elements into a coherent form, figure–ground segmentation, object recognition, and object invariance, all of which rely on global form perception. To segment elements of a scene into objects and background, it is necessary to distinguish contours and textures that define their boundaries. Orientation and texture perception are apparent quite early in monkeys and humans (Atkinson & Braddick, 1992; El-Shamayleh, Movshon, & Kiorpes, 2010; Sireteanu & Rieth, 1992). Orientation and texture discrimination ability are evident in macaques as early as 5–6 weeks after birth (El-Shamayleh, Movshon, & Kiorpes, 2010). Interestingly, the ability to discriminate patterns based solely on second-order textures cues matured *earlier* than first-order spatial contrast sensitivity (see figure 99.2A, B). Thus the building blocks for global form perception develop quite early.

A canonical measure of global form perception is the ability to appreciate global structure in Glass patterns (Glass, 1969). Glass patterns are composed of oriented dipoles, the global structure of which is perceptible only by integration of dipole patterns across space. We studied the development of sensitivity to Glass pattern structure and found comparatively late onset of this ability, which was followed by an extended developmental time course that was reminiscent of that for global motion perception (see figure 99.2C) (Kiorpes et al., 2012). Few monkeys were able to discriminate Glass pattern structure from incoherent control patterns before about 15 weeks; adult levels of sensitivity were approached 2–3 years after birth. This profile seems to be fairly general for global form perception. Contour integration requires both perceptual organization and figure–ground segmentation skills. It is typically measured by the ability to link Gabor patches or other discrete elements together to extract a global shape from a background of randomly arrayed noise elements (Field, Hayes, & Hess, 1993). This ability also develops quite late compared to spatial contrast sensitivity, which is measureable near birth in macaques and approaches adult levels near the end of the first postnatal year (see figure 99.2A) (El-Shamayleh, Movshon, & Kiorpes, 2010; Stavros & Kiorpes, 2008). Kiorpes and Bassin (2003) reported a failure of contour integration ability prior to about 16 weeks in young macaques followed by development to adult levels over at least the succeeding year (see figure 99.2D). Late development of contour integration ability has been reported in human children as well, with onset after 3 years of age and improvement continuing into the teenage years (Kovács et al., 1999). Consistent with psychophysical measures of global form development, several studies of object concepts in young macaques show development over the

first 4 or more months after birth (Diamond, 1990; Ha, Kimpo, & Sackett, 1997; Hall-Haro et al., 2008).

The neural processes underlying global form perception in adults are a matter of ongoing debate. Second-order orientation-defined texture discrimination of the kind tested by El-Shamayleh, Movshon, and Kiorpes (2010) is correlated with activity both in and downstream of early visual areas V1 and V2 in human visual cortex (Larsson, Landy, & Heeger, 2006). Consistent with that report, El-Shamayleh and Movshon (2011) failed to find evidence for second-order texture coding by neurons in macaque V2. Other studies have identified neural correlates of shape and texture processing at the level of V4 (De Weerd, Desimone, & Ungerleider, 1996; Huxlin et al., 2000; Schiller, 1995). The ability to extract a coherent percept from integration of elements imbedded in noise, as required for contour integration, appears to involve higher occipitotemporal areas as well as contextual interactions in early visual cortex (Altmann, Bülthoff, & Kourtzi, 2003; Kourtzi et al., 2003; Li, Piëch, & Gilbert, 2006, 2008). The onset of contour integration ability in macaques coincides with maturation of feedback projections to V1 and V2

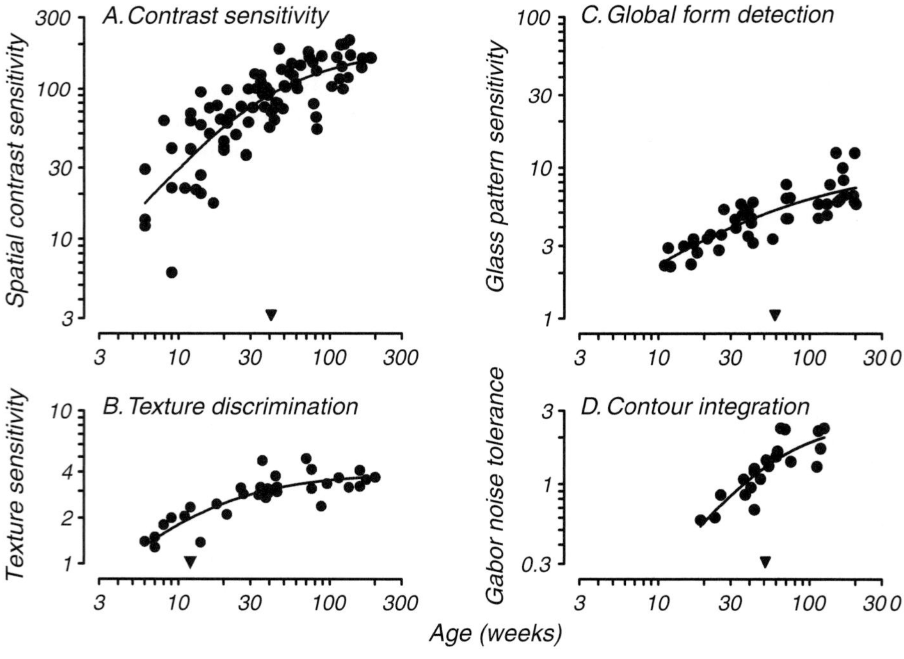

FIGURE 99.2   Development of "ventral stream" form mechanisms. (A) Development of peak spatial contrast sensitivity (data from Stavros & Kiorpes, 2008). (B) Development of sensitivity to second-order orientation defined texture (data from El-Shamayleh, Movshon, & Kiorpes, 2010). (C) Development of sensitivity to global structure in concentric Glass patterns (data from Kiorpes et al., 2012). (D) Development of contour integration ability (data from Kiorpes & Bassin, 2003). Filled symbols in each panel are data from individual infant monkeys plotted as a function of age. The line fit to the data in each panel is a Michaelis–Menton function which captures the rate of development and the age at which sensitivity reaches 50% of adults' levels (indicated by the filled triangles along the abscissa).

and the subsequent reorganization of receptive field surrounds as described by Baldwin et al. (2012). Extraction of global form in Glass patterns seems to rely on activity in higher-order extrastriate areas (Ostwald et al., 2008); activity in early visual cortex in the presence of Glass patterns can be attributed to local orientation signals (Smith, Bair, & Movshon, 2002; Smith, Kohn, & Movshon, 2007). Together with a preponderance of evidence for shape processing in inferotemporal visual areas, it is likely that global form perception depends on the function of areas beyond V2. Unfortunately little is known about the development of neuronal response properties in areas beyond V2. Rodman and colleagues recorded from inferior temporal (IT) neurons in macaques ranging in age from 5 weeks to adult (Rodman, Scalaidhe, & Gross, 1993). They found dramatically reduced responsiveness of IT neurons compared with MT neurons recorded under the same conditions in the same animals. However, IT neurons that were responsive in young monkeys (6–8 weeks of age) showed typical tuning for shapes, patterns, and faces. It is important to note that they encountered a substantially higher proportion of responsive neurons in awake infant recordings compared with anesthetized preparations, a distinction that was not found for MT. As reported by Movshon et al. (2003) for MT neurons, infant IT neurons were less responsive overall and responded with longer latency compared with those of adults regardless of state. These results suggest that ventral stream areas may be somewhat less mature in young monkeys than dorsal stream areas, which is consistent with metabolic assays of maturation (Distler et al., 1996). However they leave the late onset of global form perception and its long developmental time course unexplained.

Taken together, electrophysiological data on neuronal response properties and anatomical studies of cortical organization show that visual brain areas linked with higher-order perception in adults are substantially more mature than would be expected based on behavioral assessment of visual function in infants. Therefore, stimulus information is available to the infant—encoded with reasonable fidelity—that is not being used or is not sufficient to support appropriate behavior. A number of possibilities exist to explain the comparatively poor performance of infants. One obvious candidate is a limitation on the behavior itself or on the ability of the animal to understand the task. These explanations are unlikely since behavioral ability has been clearly demonstrated by success on tests of basic visual function at very young ages (e.g., Kiorpes & Bassin, 2003; Stavros & Kiorpes, 2008) or under comparable task demands (Hall-Haro & Kiorpes, 2008). Another possibility is the weak responsiveness of infant neurons: Although the tuning properties of most infant neurons appear to be adult-like at very young ages, they respond at quite low rates, with long latency. The weaker responses could translate to lower signal-to-noise ratio in infants or greater response variability. Curiously, an analysis of the reliability of infant V1 neurons concluded that in fact their responses are *more* reliable, not less so (Rust, Schultz, & Movshon, 2002). However, there is good evidence that the infant visual system is "noisier" than that of the adult and that developmental changes in contrast sensitivity parallel a decrease in signal-to-noise ratio which may reflect improved signaling by cortical neurons (Brown, 1994; Kiorpes & Movshon, 1998; Norcia, 2004).

Up to this point, we have primarily considered the maturation of single-neuron properties as a substrate for visual development. As we have seen, those properties provide a partial account of the changes during visual development, but not a complete one. Of course, it is unlikely that behavioral performance depends on the output of individual neurons; rather it must reflect the activity of a large number of neurons active together. Thus, it is instructive to consider how the properties of infant neurons might impact the response of the population. The rather weak responses of infant neurons combined with their broader latency distribution (as noted in MT and IT) might result in a less coherent population output from extrastriate areas to their downstream targets, even though individual neuron responses reflect stimulus attributes quite well. As responsivity increases and the latency distribution narrows with age, the population of neurons in a given area would generate a stronger and more coherent readout pattern of activity for downstream areas to decode, allowing performance on visual tasks to improve disproportionately more than the properties of individual neurons. Equally plausible, it might be the rules for how neuronal responses are combined to formulate a population response—such as pool size or identity of neurons in the pool—that change with development. Changes in pooling rules could result in a population response that becomes stronger with maturation, driving an improvement in the quality of information available to the behavioral output.

We know that the tuning properties of infant neurons are quite mature, so let's suppose that the population output from midlevel extrastriate areas in fact faithfully represents the encoded visual stimulus. If so, then it may be that the neural "decoder"—the downstream process responsible for translating this encoded information into a perceptual decision—is untuned or inefficient in young infants. For instance, an accurate readout from an upstream visual area may be incorrectly interpreted or inappropriately weighted by the decision process, with the result that behavioral performance is poor. In this case, the developmental time course reflected by behavioral assay might be defined by a process of "tuning" or learning by the decoder to correctly interpret the sensory readout rather than by a refinement of neuronal response properties. One can imagine a developmental process that is analogous to perceptual learning in the adult, but on a more dramatic scale. Law and Gold (2008) recorded simultaneously from MT and lateral intraparietal area (LIP) while adult monkeys practiced a motion direction discrimination, with an eye movement as the behavioral output. They found substantial modification of responses in LIP with extended practice but observed no concurrent changes in MT. Their interpretation was that improvement in the animals' performance was related to more efficient decoding of the upstream information. A similar change might take place during normal development as the animal interacted with the visual world.

In reality, the processes by which downstream areas decode their inputs to form perceptual decisions may not in fact differ from those used to combine inputs for the creation of more elaborate receptive field properties. The construction of linear likelihood-based decoders for the output of MT as used in motion discrimination tasks reveals that a feedforward pattern of connections involving summation of signals from nearby directions and suppression by signals from remote directions is optimal for coarse discrimination tasks like the ones we study (Graf et al., 2011; Jazayeri & Movshon, 2006). This decoding theory also provides a very accurate account of the perceptual learning results of Law and Gold (2008). Consider figure 99.3. On the left is the deduced pattern of output connections from a population of V1 neurons inferred by a likelihood decoding model from the recordings of Graf et al. (2011). This is an optimal connection pattern for identifying orientation on the basis of V1 activity. Note the excitatory feedforward output from neurons tuned near the direction of choice and the suppressive output from neurons tuned to more remote directions. On the right is shown the connectional architecture deduced by Rust et al. (2006) for the computation of pattern motion by MT neurons. According to this account, pattern selective responses emerge in large part from the particular pattern of feedforward connections. The similarity of that pattern to the architecture required for decoding is obvious. Thus it may be that the connections that are responsible for the computation of a more elaborate response property—for example, one that permits the analysis of higher-order motion signals—are the same ones needed for the computation of sensory decision variables. Perhaps these can be achieved in the same type of cortical circuits using the same algorithms. Note also that,

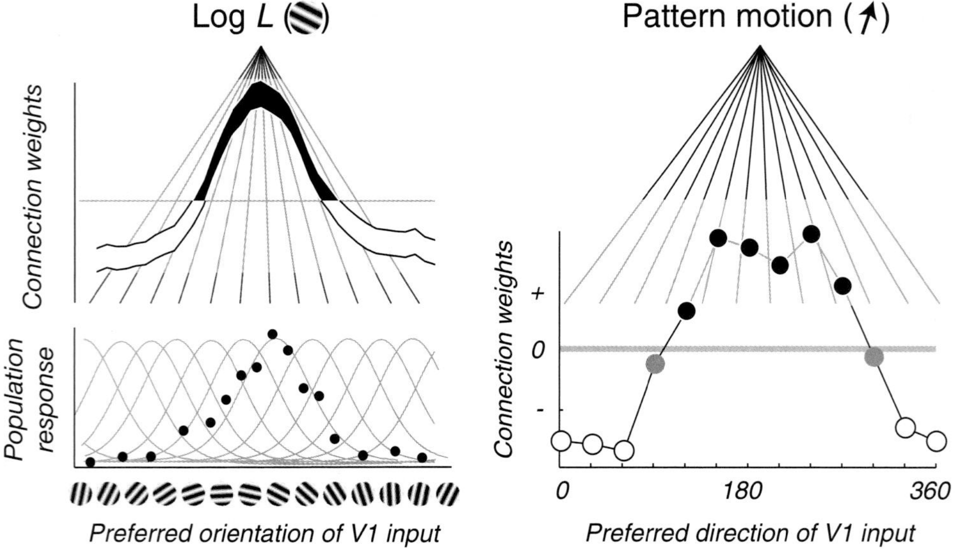

FIGURE 99.3   Computing an optimal linear decoder may involve the same circuits and computations as higher-order receptive field properties. The left panel shows results from Graf et al. (2011), showing the average connection weights (top) for the optimal estimation of orientation from a population response of V1 neurons (schematically illustrated at the bottom); mean positive weights for one set of data (± 1 SEM) are shown in the solid black band and negative weights as an open band. The right panel shows the feedforward connectional architecture inferred by Rust et al. (2006) for the V1 inputs to an example pattern selective neuron in MT. Positive weights are again filled, negative ones open. The pattern of connections for decoding orientation (left) is essentially identical to that for encoding pattern motion (right) (x axes in degrees).

as shown above, the computation of pattern motion by MT neurons develops relatively late (see figure 99.1); similarly the delayed onset of performance at integrative visual tasks suggests that the precision of V1 decoding might also develop late.

A parsimonious account of a large amount of data is that visual development depends on two separate processes: The age of onset of a particular visual function may reflect the maturation of a limiting underlying neural substrate while the subsequent development to adult levels may reflect the learning or tuning of a perceptual template by the decoder. These two processes are intertwined and can be achieved by the same architecture. The delayed development of higher-order stimulus selectivity in downstream cortical areas could be the result of the late establishment of this pattern of functional connections, and as a consequence performance—which depends on optimal decoding—might be similarly delayed even though the signals carried by input neurons are quite adult-like. Substantiation of these ideas will require a combination of population-level assessment of neural activity at multiple stages of the visual system hierarchy and direct evaluation of decoding principles in play during development.

ACKNOWLEDGMENTS

The research from our laboratories described in this chapter was supported primarily by National Institutes of Health grants EY05864, EY04440, and EY2017 and Center grant RR00166 to the Washington National Primate Research Center. We are grateful to many individual students and colleagues for their contributions as well as excellent animal care and oversight by Michael Gorman and the NYU Office of Veterinary Resources.

REFERENCES

Altmann, C. F., Bülthoff, H. H., & Kourtzi, Z. (2003). Perceptual organization of local elements into global shapes in the human visual cortex. *Current Biology, 13,* 342–349.

Atkinson, J., & Braddick, O. (1992). Visual segmentation of oriented textures by infants. *Behavioural Brain Research, 49,* 123–131.

Baldwin, M. K., Kaskan, P. M., Zhang, B., Chino, Y. M., & Kaas, J. H. (2012). Cortical and subcortical connections of V1 and V2 in early postnatal macaque monkeys. *Journal of Comparative Neurology, 520,* 544–569.

Blasdel, G., Obermayer, K., & Kiorpes, L. (1995). Organization of ocular dominance and orientation columns in the striate cortex of neonatal macaque monkeys. *Visual Neuroscience, 12,* 589–603.

Britten, K. H., Shadlen, M. N., Newsome, W. T., & Movshon, J. A. (1992). The analysis of visual motion: A comparison of neuronal and psychophysical performance. *Journal of Neuroscience, 12,* 4745–4765.

Brown, A. M. (1994). Intrinsic contrast noise and infant visual contrast discrimination. *Vision Research, 34,* 1947–1964. doi:10.1016/0042-6989(94)90025-6.

Burkhalter, A. (1993). Development of forward and feedback connections between areas V1 and V2 of human visual cortex. *Cerebral Cortex, 3,* 476–487.

Burkhalter, A., Bernardo, K. L., & Charles, V. (1993). Development of local circuits in human visual cortex. *Journal of Neuroscience, 13,* 1916–1931.

Chino, Y. M., Smith, E. L., III, Hatta, S., & Cheng, H. (1997). Postnatal development of binocular disparity sensitivity in neurons of the primate visual cortex. *Journal of Neuroscience, 17,* 296–307.

De Weerd, P., Desimone, R., & Ungerleider, L. G. (1996). Cue-dependent deficits in grating orientation discrimination after V4 lesions in macaques. *Visual Neuroscience, 13,* 529–538.

Diamond, A. (1990). The development and neural bases of memory functions as indexed by the AB and delayed response tasks in human infants and infant monkeys. *Annals of the New York Academy of Sciences, 608,* 267–317.

Distler, C., Bachevalier, J., Kennedy, C., Mishkin, M., & Ungerleider, L. G. (1996). Functional development of the corticocortical pathway for motion analysis in the macaque monkey, a 14C–2-deoxyglucose study. *Cerebral Cortex, 6,* 184–195.

El-Shamayleh, Y., & Movshon, J. A. (2011). Neuronal responses to texture-defined form in macaque visual area V2. *Journal of Neuroscience, 31,* 8543–8555.

El-Shamayleh, Y., Movshon, J. A., & Kiorpes, L. (2010). Development of sensitivity to visual texture modulation in macaque monkeys. *Journal of Vision, 10,* 1–12. doi:10.1167/10.11.11.

Felleman, D. J., & Van Essen, D. C. (1991). Distributed hierarchical processing in the primate cerebral cortex. *Cerebral Cortex, 1,* 1–47. doi:10.1093/cercor/1.1.1-a.

Field, D. J., Hayes, A., & Hess, R. F. (1993). Contour integration by the human visual system: Evidence for a local "association field." *Vision Research, 33,* 173–193.

Glass, L. (1969). Moiré effect from random dots. *Nature, 223,* 578–580.

Graf, A. B., Kohn, A., Jazayeri, M., & Movshon, J. A. (2011). Decoding the activity of neuronal populations in macaque primary visual cortex. *Nature Neuroscience, 14,* 239–245.

Guillery, R. W. (2005). Is postnatal neocortical maturation hierarchical? *Trends in Neurosciences, 28,* 512–517. doi:10.1016/j.tins.2005.08.006.

Ha, J. C., Kimpo, C. L., & Sackett, G. P. (1997). Multiple-spell, discrete-time survival analysis of developmental data: Object concept in pigtailed macaques. *Developmental Psychology, 33,* 1054–1059.

Hall-Haro, C., Johnson, S. P., Price, T. A., Vance, J. A., & Kiorpes, L. (2008). Development of object concepts in macaque monkeys. *Developmental Psychobiology, 50,* 278–287.

Hall-Haro, C., & Kiorpes, L. (2008). Normal development of pattern motion sensitivity in macaque monkeys. *Visual Neuroscience, 25,* 675–684.

Horton, J. C., & Hocking, D. R. (1996). An adult-like pattern of ocular dominance columns in striate cortex of newborn monkeys prior to visual experience. *Journal of Neuroscience, 16,* 1791–1807.

Horton, J. C., & Hocking, D. R. (1997). Timing of the critical period for plasticity of ocular dominance columns in macaque striate cortex. *Journal of Neuroscience, 17*, 3684–3709.

Huxlin, K. R., Saunders, R. C., Marchionini, D., Pham, H. A., & Merigan, W. H. (2000). Perceptual deficits after lesions of inferotemporal cortex in macaques. *Cerebral Cortex, 10*, 671–683.

Jazayeri, M., & Movshon, J. A. (2006). Optimal representation of sensory information by neural populations. *Nature Neuroscience, 9*, 690–696.

Kennedy, H., & Burkhalter, A. (2004). Ontogenesis of cortical connectivity. In L. M. Chalupa & J. S. Werner (Eds.), *The visual neurosciences* (pp. 146–158). Cambridge, MA: MIT Press.

Kiorpes, L., & Bassin, S. A. (2003). Development of contour integration in macaque monkeys. *Visual Neuroscience, 20*, 567–575.

Kiorpes, L., & Movshon, J. A. (1998). Peripheral and central factors limiting the development of contrast sensitivity in macaque monkeys. *Vision Research, 38*, 61–70.

Kiorpes, L., & Movshon, J. A. (2004a). Neural limitations on visual development in primates. In L. M. Chalupa & J. S. Werner (Eds.), *The visual neurosciences* (pp. 159–173). Cambridge, MA: MIT Press.

Kiorpes, L., & Movshon, J. A. (2004b). Development of sensitivity to visual motion in macaque monkeys. *Visual Neuroscience, 21*, 851–859.

Kiorpes, L., Price, T., Hall-Haro, C., & Movshon, J. A. (2012). Development of sensitivity to global form and motion in macaque monkeys (*Macaca nemestrina*). *Vision Research, 63*, 34. doi:10.1016/j.visres.2012.04.018.

Kiorpes, L., Tang, C., Hawken, M. J., & Movshon, J. A. (2003). Ideal observer analysis of the development of spatial contrast sensitivity in macaque monkeys. *Journal of Vision, 3*(10), 630–641. doi:10.1167/3.10.6.

Kourtzi, Z., Tolias, A. S., Altmann, C. F., Augath, M., & Logothetis, N. K. (2003). Integration of local features into global shapes: Monkey and human fMRI studies. *Neuron, 37*, 333–346.

Kovács, I., Kozma, P., Fehér, A., & Benedek, G. (1999). Late maturation of visual spatial integration in humans. *Proceedings of the National Academy of Sciences of the United States of America, 96*, 12204–12209. doi:10.1073/pnas.96.21.12204.

Larsson, J., Landy, M. S., & Heeger, D. J. (2006). Orientation-selective adaptation to first- and second-order patterns in human visual cortex. *Journal of Neurophysiology, 95*, 862–881.

Law, C., & Gold, J. I. (2008). Neural correlates of perceptual learning in a sensory-motor, but not a sensory, cortical area. *Nature Neuroscience, 11*, 505–513.

Li, W., Piëch, V., & Gilbert, C. D. (2006). Contour saliency in primary visual cortex. *Neuron, 50*, 951–962.

Li, W., Piëch, V., & Gilbert, C. D. (2008). Learning to link visual contours. *Neuron, 57*, 442–451.

Mikami, A., & Fujita, K. (1992). Development of the ability to detect visual motion in infant macaque monkeys. *Developmental Psychobiology, 25*, 345–354.

Movshon, J. A., Adelson, E. H., Gizzi, M. S., & Newsome, W. T. (1985). The analysis of moving visual patterns. In C. Chagas, R. Gattass, & C. Gross (Eds.), *Pattern recognition mechanisms* (pp. 117–151). Rome: Vatican Press.

Movshon, J. A., Kiorpes, L., Hawken, J. A., & Cavanaugh, J. R. (2005). Functional maturation of the macaque's lateral geniculate nucleus. *Journal of Neuroscience, 25*, 2712–2722.

Movshon, J. A., Rust, N. C., Kohn, A., Kiorpes, L., & Hawken, M. J. (2003). Receptive field properties of MT neurons in infant macaques. *Perception, 33*, ECVP Abstract Supplement. doi:10.1068/v040593.

Newsome, W. T., & Pare, E. B. (1988). A selective impairment of motion perception following lesions of the middle temporal visual area (MT). *Journal of Neuroscience, 8*, 2201–2211.

Norcia, A. M. (2004). Development of spatial selectivity and response timing in humans. In L. M. Chalupa & J. S. Werner (Eds.), *The visual neurosciences* (pp. 174–188). Cambridge, MA: MIT Press.

Ostwald, D., Lam, J. M., Li, S., & Kourtzi, Z. (2008). Neural coding of global form in the human visual cortex. *Journal of Neurophysiology, 99*, 2456–2469.

Rodman, H. R., & Albright, T. D. (1989). Single-unit analysis of pattern-motion selective properties in the middle temporal visual area (MT). *Experimental Brain Research, 75*, 53–64.

Rodman, H. R., Scalaidhe, S. P., & Gross, C. G. (1993). Response properties of neurons in temporal cortical visual areas of infant monkeys. *Journal of Neurophysiology, 70*, 1115–1136.

Rudolph, K., & Pasternak, T. (1999). Transient and permanent deficits in motion perception after lesions of cortical areas MT and MST in the macaque monkey. *Cerebral Cortex, 9*, 90–100.

Rust, N. C., Mante, V., Simoncelli, E. P., & Movshon, J. A. (2006). How MT cells analyze the motion of visual patterns. *Nature Neuroscience, 9*, 1421–1431.

Rust, N. C., Schultz, S. R., & Movshon, J. A. (2002). A reciprocal relationship between reliability and responsiveness in developing visual cortical neurons. *Journal of Neuroscience, 22*, 10519–10523.

Salzman, C. D., Murasugi, C. M., Britten, K. H., & Newsome, W. T. (1992). Microstimulation in visual area MT: Effects on direction discrimination performance. *Journal of Neuroscience, 12*, 2331–2355.

Shadlen, M. N., Britten, K. H., Newsome, W. T., & Movshon, J. A. (1996). A computational analysis of the relationship between neuronal and behavioral responses to visual motion. *Journal of Neuroscience, 16*, 1486–1510.

Schiller, P. H. (1995). Effect of lesions in visual cortical area V4 on the recognition of transformed objects. *Nature, 376*, 342–344.

Shipp, S., de Jong, B. M., Zihl, J., Frackowiak, R. S., & Zeki, S. (1994). The brain activity related to residual motion vision in a patient with bilateral lesions of V5. *Brain, 117*, 1023–1038.

Sireteanu, R., & Rieth, C. (1992). Texture segregation in infants and children. *Behavioural Brain Research, 49*, 133–139.

Smith, M. A., Bair, W., & Movshon, J. A. (2002). Signals in macaque striate cortical neurons that support the perception of Glass patterns. *Journal of Neuroscience, 22*, 8334–8345.

Smith, M. A., Kohn, A., & Movshon, J. A. (2007). Glass pattern responses in macaque V2 neurons. *Journal of Vision, 7*(3), 5. doi:10.1167/7.3.5.

Stavros, K. A., & Kiorpes, L. (2008). Behavioral measurements of temporal contrast sensitivity development in macaque monkeys (*Macaca nemestrina*). *Vision Research, 48,* 1335–1344. doi:10.1016/j.visres.2008.01.031.

Vaina, L. M., Cowey, A., Jakab, M., & Kikinis, R. (2005). Deficits of motion integration and segregation in patients with unilateral extrastriate lesions. *Brain, 128,* 2134–2145.

Wiesel, T. N. (1982). Postnatal development of the visual cortex and the influence of environment. *Nature, 299,* 583–591.

Wiesel, T. N., & Hubel, D. H. (1974). Ordered arrangement of orientation columns in monkeys lacking visual experience. *Journal of Comparative Neurology, 158,* 307–318.

Zheng, J., Zhang, B., Bi, H., Maruko, I., Watanabe, I., Nakatsuka, C., et al. (2007). Development of temporal response properties and contrast sensitivity of V1 and V2 neurons in macaque monkeys. *Journal of Neurophysiology, 97,* 3905–3916.

The Molecular and Structural Basis of Amblyopia

JASON E. COLEMAN, ARNOLD J. HEYNEN, AND MARK F. BEAR

Amblyopia is a common cause of vision loss during infancy and childhood (Holmes & Clarke, 2006). Visual dysfunction in amblyopia stems from conditions that degrade image formation or eye alignment prior to adolescence; these include strabismus, uncorrected refractive errors (e.g., anisometropia), and cataracts (Doshi & Rodriguez, 2007). The abnormal development of synaptic connections in the primary visual cortex (V1) is responsible for the loss of vision.

The substrate for binocular vision in mammals is the convergence of visuotopically matched inputs onto common postsynaptic targets in V1. Early in development, the wiring of visual connections from dorsal lateral geniculate nucleus (LGN) to V1 is largely guided by a genetic program and intrinsic neural activity (Crair et al., 2001; Crowley & Katz, 2000; LeVay, Stryker, & Shatz, 1978). Beyond this period, the refinement and maintenance of synaptic connections for encoding behaviorally relevant information depends upon the quality of visual experience from the outside world. In amblyopia, the disruption of normal binocular visual experience leads to the weakening and reorganization of visual circuitry in V1.

The pioneering work of David Hubel and Torsten Wiesel, who received the 1981 Nobel Prize in Physiology or Medicine for their studies on visual cortical function and plasticity, has served as the platform for modern mechanistic studies of amblyopia. They found that a prolonged period of monocular lid suture begun shortly after birth caused a shift in the ocular dominance (OD) of cortical neurons so that the majority failed to respond to stimulation of the deprived eye (Hubel & Wiesel, 1964). Using this monocular deprivation (MD) paradigm, they also discovered anatomical correlates of visual impairment (Wiesel & Hubel, 1963b). The most robust correlate of the OD shift was the demonstration that in binocular cortex the territory innervated by open-eye thalamocortical afferents in layer 4 expands after MD whereas the territory innervated by deprived-eye afferents shrinks (LeVay, Wiesel, & Hubel, 1980). Together, these studies provided the first descriptions of what is now known as OD plasticity, and this paradigm has proven to be invaluable for studying mechanisms underlying experience-dependent plasticity in cortex and amblyopia.

In this chapter, we provide an overview of work that has led to a greater understanding of the molecular and structural mechanisms of amblyopia.

INVESTIGATING THE MECHANISMS OF AMBLYOPIA IN THE MOUSE

As research efforts have increasingly focused on elucidating the molecular and cellular bases of amblyopia, the mouse visual system (see figure 100.1A) has become the model of choice. The fine organization of primary visual input to V1 (i.e., thalamocortical axons) differs between rodents and carnivores or primates, but the basic circuitry underlying vision and the transmission of visual information from retina to thalamus to V1 is conserved. It is important at this point to draw a distinction between OD plasticity and OD *column* plasticity. OD columns reflect the segregation of LGN afferents in layer 4 of visual cortex and are arranged as regularly spaced bands or stripes. Present in some carnivores (e.g., cat and ferret) and some primates (e.g., macaque and human), they provide a convenient pattern for visualizing eye-specific thalamocortical input, which can be illuminated by injecting a transsynaptic tracer into one eye (Hubel & Wiesel, 1968). However, they are not a consistent feature of binocular visual cortex and are not a predictor of the capacity for OD plasticity, even among primates (Horton & Adams, 2005). OD plasticity can be measured physiologically by recording from single neurons or populations of neurons that receive input from both eyes, or anatomically by monitoring synaptic and axonal structures, without regard to whether they are in eye-specific columns (Hubel & Wiesel, 1968) or innervate cortex in a salt-and-pepper arrangement (Mrsic-Flogel et al., 2007).

In addition to the fact that mice are inbred and can be genetically manipulated, mouse visual cortex

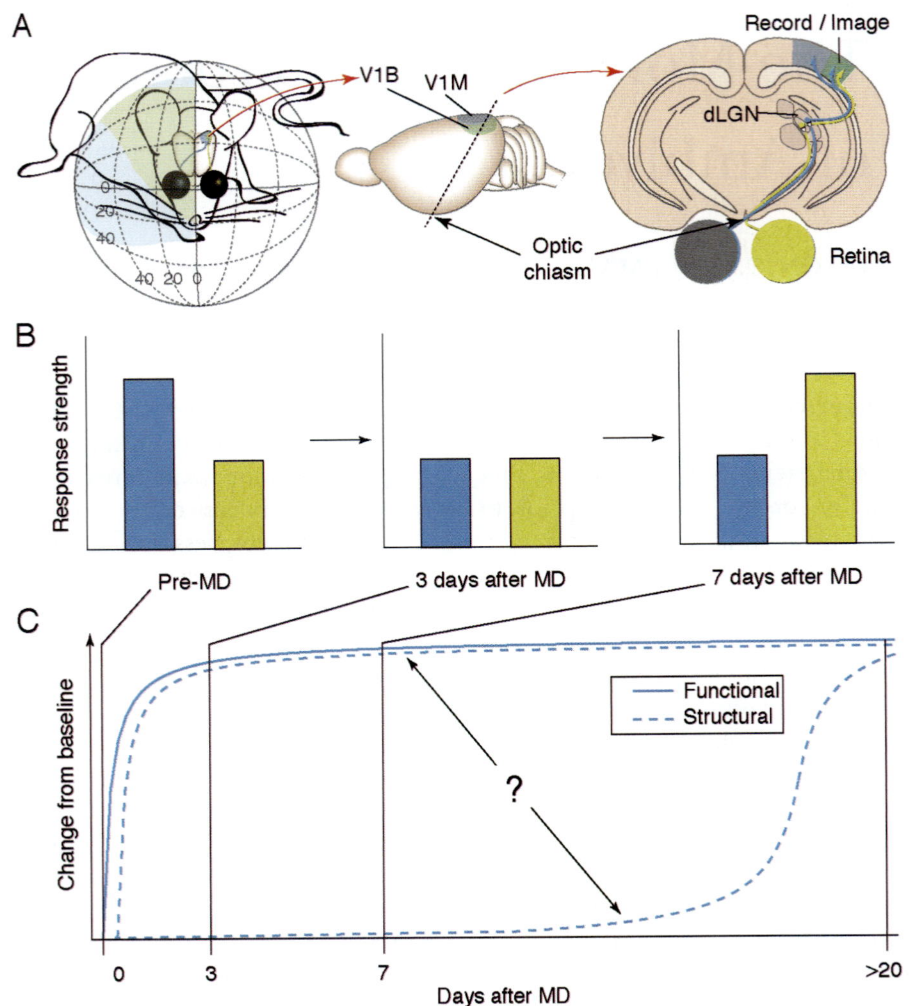

FIGURE 100.1   Binocular vision and ocular dominance (OD) plasticity in the mouse. (A) The representation of visual space in the cortex and the site in cortex where OD plasticity is monitored. The left and middle panels show how information for one visual hemifield viewed by the ipsilateral eye (blue region, monocular visual field) and by both eyes (green region, binocular visual field) is represented in V1. The eye ipsilateral to the hemifield (blue) projects to the contralateral hemisphere whereas the contralateral eye viewing the same space (yellow) projects to the ipsilateral hemisphere (numbers indicate visual degrees; globe not drawn to scale). The right panel shows a schematic of the mouse visual pathway (V1B, binocular segment; V1M, monocular segment; dLGN, dorsal lateral geniculate nucleus). Binocular responses (and structures) are recorded (and imaged) in V1B contralateral to the deprived eye. (B) Physiological measurements of plasticity at thalamocortical synapses (e.g., layer 4 visually evoked potentials) of the mouse reveal two sequential changes—deprived-eye depression after 3 days of monocular deprivation (MD) and open-eye potentiation after 7 days of MD (blue = contralateral, deprived eye; yellow = ipsilateral, open eye). (C) Proposed model for how molecular and structural changes coincide with and contribute to pathway-specific functional plasticity in V1B (e.g., deprived-eye depression and thalamocortical axon retraction). The graph highlights two key points: (1) The functional consequences of short-term and long-term MD are the same, and (2) the kinetics (i.e., rapid or slow as indicated by arrows) of structural plasticity relative to functional plasticity have yet to be fully elucidated.

possesses several characteristics that have made it advantageous as a model for studying OD plasticity. First, cortical processing of binocularity begins with the convergence of thalamic inputs onto layer 4 neurons in V1 rather than with the convergence of layer 4 inputs onto layer 2/3 neurons in V1 that occurs in primates, a pattern of connectivity that potentially simplifies the analyses of the synaptic changes that accompany OD plasticity. Second, the absence of a columnar organization makes feasible the use of chronic field potential recordings from awake animals since cells with varying degrees of eye dominance are intermingled in the binocular segment of V1 (Mrsic-Flogel et al., 2007). Third, the fact that the mouse visual cortex is relatively undifferentiated (e.g., compared to monkey V1) suggests that insights gained here might apply broadly across

1434   JASON E. COLEMAN, ARNOLD J. HEYNEN, AND MARK F. BEAR

species and cortical areas (Coleman, Law, & Bear, 2009).

Using chronic visually evoked potential recordings in layer 4 of mouse V1, Frenkel and Bear (2004) showed that the OD shift toward the nondeprived eye occurs by two distinct phases: a weakening of deprived-eye responses after 3 days of MD and a strengthening of open-eye responses after 7 days of MD (see figure 100.1B). This sequence of rapid deprived-eye depression and delayed open-eye potentiation has also been noted using other measures of plasticity (Hofer et al., 2006; Kaneko et al., 2008; Mrsic-Flogel et al., 2007) and in other species (Mioche & Singer, 1989) and reflects functionally meaningful changes in sensory processing. In rats, a dramatic reduction in visual acuity assessed through visually guided behavior occurs through the deprived eye following MD (Iny et al., 2006; Prusky, West, & Douglas, 2000). In the same visually guided task, open-eye performance was enhanced following MD (Iny et al., 2006).

HOMOSYNAPTIC LONG-TERM DEPRESSION AND OCULAR DOMINANCE PLASTICITY

An important question when considering the mechanism of deprived-eye depression is whether it is simply a consequence of the absence of activity in the deprived eye or whether it is triggered by activity in the deprived eye that no longer correlates with a strong cortical response. The question is critical because it cleanly divides potential mechanisms. In the former model, deprived-eye depression is simply explained by disuse, the silencing of afferent synapses. In the latter model, the plasticity is triggered instead by the untimely release of neurotransmitter by afferent synapses. A number of lines of evidence suggest that the deprivation effects are triggered by activity rather than inactivity. For example, the depression of input from one eye after strabismus cannot be explained by disuse but is readily explained by a loss of pre- and postsynaptic correlations. Moreover, simple blurring of images on the retina with an overcorrecting contact lens is as effective as monocular lid closure in inducing an OD shift (Rittenhouse et al., 2006). Finally, intraocular injection of tetrodotoxin, which silences all output from the retina, fails to induce robust deprived-eye depression in the cortex (Frenkel & Bear, 2004; Rittenhouse et al., 1999). Together, the data suggest that synaptic depression in cortex is driven to occur by poorly correlated afferent activity. Because deprived-eye synapses apparently trigger their own demise via untimely release of neurotransmitter, the mechanisms responsible are said to be "homosynaptic" (Smith, Heynen, & Bear, 2009).

To study the mechanisms of homosynaptic depression, a paradigm was introduced by Dudek and Bear (1992) in which tetanic electrical stimulation of synapses was used to induce long-term depression (LTD) of synaptic transmission in brain slices (reviewed by Bear, 2003). Although it is now appreciated that there are many mechanisms for LTD in different brain regions, some of these are well conserved (Malenka & Bear, 2004). The study of LTD in hippocampus and visual cortex has led to a detailed understanding of how activity triggers a loss of synaptic strength.

Intracortical circuitry is complex, and it is reasonable to question where the primary site of plasticity is. Although the work of Hubel and Wiesel clearly implicated plasticity of geniculocortical synapses, this could be a consequence rather than a primary cause of deprived-eye depression. Recent studies in mouse have indicated, however, that depression of thalamocortical synaptic transmission after MD is very rapid, occurring at the same rate as the OD shift (Khibnik, Cho, & Bear, 2010). Thus, although modifications of other synapses clearly occur after MD, the changes in thalamocortical synaptic transmission alone are sufficient to account for the OD shift. These findings justify a focus on plasticity of excitatory synaptic transmission as the primary cause of deprived-eye depression. As mentioned, LTD of excitatory synaptic transmission has been studied in slices of visual cortex to gain insight into the mechanisms of the OD shift. To assess the relevance of LTD to naturally occurring plasticity, two approaches have been taken: (1) tests of the hypothesis that LTD is induced in visual cortex by MD and (2) investigation of shared requirements for induction or expression of LTD and deprived-eye depression. Although the contribution of LTD mechanisms to OD plasticity was once considered to be controversial, it is now clear that these same molecular mechanisms are engaged by MD and necessary for the loss of visual responses. Some of this evidence is summarized below.

LTD Is Induced by MD and Necessary for the Loss of Visual Responsiveness

The "canonical" LTD mechanism in the CA1 region of hippocampus was used to guide early mechanistic studies in the visual cortex. In CA1, weak activation of N-methyl-D-aspartate (NMDA) receptors and a rise in postsynaptic Ca^{2+} activate a postsynaptic protein phosphatase cascade that alters the phosphorylation state of AMPA receptors, which are subsequently internalized by clathrin-dependent endocytosis (Malenka & Bear, 2004). These changes can be detected biochemically using phosphorylation site-specific antibodies and assays of receptor surface expression. The biochemical signature of LTD has been used as a "molecular fingerprint" to test whether similar changes occur in visual cortex following a period of MD. This has been directly examined in both rat and mouse visual cortex. The results support the hypothesis that MD sets in motion changes that are identical to LTD in visual cortex (Heynen et al., 2003; McCurry et al., 2010).

To find out if LTD and deprived-eye depression utilize the same molecular cascades, one can ask whether LTD in slices is occluded by prior MD. Indeed, brief periods of MD (i.e., the minimum time required to cause maximal deprived-eye depression) cause a significant reduction in the amount of LTD elicited in rat (Heynen et al., 2003) and mouse visual cortical slices (Crozier et al., 2007). Mimicry and occlusion are two criteria that must be satisfied to conclude that two methods to induce synaptic depression converge on common mechanisms. By these criteria, both MD and LTD engage the same molecular mechanisms.

Although the evidence summarized above indicates that MD induces LTD, these studies do not reveal the relative contribution of LTD mechanisms to the OD shift after MD. There are two ways one can address this question of whether LTD mechanisms are *necessary* for deprived-eye depression after MD: (1) Determine whether both processes are affected by the same *modulators* and (2) determine whether both processes depend on the same *mediators*. The distinction between modulators and mediators is important. Modulators are factors that may alter the induction requirements under specific experimental conditions. Mediators are the molecular events that directly couple synaptic activity to the change in synaptic strength (Malenka & Bear, 2004).

An example of a common modulator is the state of inhibition. Because LTD induction requires an appropriate level of voltage-dependent NMDA receptor activation, the stimulation requirements for LTD will vary depending on the status of postsynaptic excitability (Steele & Mauk, 1999). Excitability is finely controlled by GABAergic inhibition. The reduction in visual cortical inhibition caused by genetic deletion of glutamic acid decarboxylase 67 both impairs the ocular dominance shift (Hensch et al., 1998) and prevents LTD with standard stimulation protocols (Choi et al., 2002). Although this correlation supports the general notion that OD plasticity and LTD have similar requirements, shared modulation is not particularly strong evidence because it is to be expected that electrical stimulation protocols used to study LTD and patterns of LGN and cortical activity in vivo may be differentially susceptible to genetic or pharmacological manipulations.

Much stronger evidence has come from the study of mediators of LTD. The most well-understood forms of LTD in visual cortex require for induction activation of NMDA receptors (Crozier et al., 2007; Kirkwood & Bear, 1994), and it is now very well established that NMDA receptor activation is also required for OD plasticity in visual cortex (Bear et al., 1990; Ramoa et al., 2001). Even subtle manipulations of NMDA receptor subunit composition similarly affect LTD and deprived-eye depression (Cho et al., 2009). A limitation on the interpretation of these findings, however, is that NMDA receptors trigger multiple forms of synaptic plasticity, not just LTD.

As mentioned previously, one form of LTD is mediated by clathrin-dependent endocytosis of AMPA receptors, and loss of surface AMPA receptors has been observed after MD. NMDA receptor–dependent AMPA receptor endocytosis is selectively blocked by peptides that mimic regions of the GluR2 subunit C-terminus required for binding of clathrin adaptor proteins. Recent studies have shown that expression of these peptides in cortical neurons selectively prevents the depression of deprived-eye responses after MD (Yang et al., 2011; Yoon et al., 2009). Another requirement for AMPA receptor endocytosis is expression of the immediate early gene Arc (Chowdhury et al., 2006), and deprived-eye depression is absent in the Arc knockout mouse (McCurry et al., 2010). These data provide strong evidence that NMDA receptor–mediated AMPA receptor endocytosis is a critical mediator of OD plasticity.

Most of the aforementioned studies were focused on plasticity in layer 4 of mouse visual cortex, because this layer receives the bulk of the LGN input. However, studies of plasticity in layer 3 (which also receives direct LGN input in the mouse) have revealed an interesting variation in the mechanism of LTD and OD plasticity. Although LTD is still NMDA receptor–dependent in layer 3, it is not mediated by AMPA receptor endocytosis. Rather, layer 3 LTD is mediated by a mechanism that involves activation of presynaptic cannabinoid (CB) receptors (Crozier et al., 2007). An inhibitor of CB1 receptors, AM251, reliably blocks layer 3 LTD, whereas the GluR2 C-terminus peptides do not. In excellent agreement with the "LTD hypothesis" of visual cortical plasticity, OD plasticity in layer 3 (but not layer 4) is also blocked by AM251. Together, these data show that the earliest event in the development of amblyopia is LTD of synaptic transmission caused by weak NMDA receptor activation in response to poorly structured activity arising from the deprived eye.

    JASON E. COLEMAN, ARNOLD J. HEYNEN, AND MARK F. BEAR

Structural Plasticity after MD

The following facts have been established: (1) Early molecular changes modify the strength of synaptic transmission soon after onset of MD, and (2) a long-term consequence of MD is structural modification of geniculocortical synapses and axon arbors. The important questions remain as to how rapidly structural plasticity occurs and how this is related to early synaptic plasticity.

Early anatomical studies of OD columns in kittens seemed to indicate that OD plasticity is a consequence of the modification of synapses in layer 4, where geniculocortical axons first exert their influence in visual cortex. First visualized as eye-specific stripes in V1, LeVay, Wiesel, and Hubel (1980) showed that the shrinkage and expansion of deprived and nondeprived thalamic afferents in layer 4 appear to faithfully and clearly reflect the loss and gain of function of nondeprived and deprived inputs, respectively. It was also noted that the degree of OD column plasticity matched well with the age-related decrease in physiological plasticity (LeVay, Wiesel, & Hubel, 1980; Wiesel & Hubel, 1963a). These observations were sufficient to support the view that the thalamocortical synapse was the site for the initial plasticity. However, in searching for the limits of the critical period, it was also noted that the OD shift was only detected physiologically in superficial layers when MD was initiated in greater than 1-year-old monkeys (Blakemore, Garey, & Vital-Durand, 1978; Hubel, Wiesel, & LeVay, 1977). Thus, structural plasticity of thalamocortical inputs is not necessary for the OD shift under all conditions.

Nearly two decades later, Antonini and Stryker (1993) demonstrated at the single-axon level that there was a dramatic pruning of deprived-eye thalamocortical axons after only 7 days of MD in kitten V1. However, although more rapid than previously suspected, this pruning was incomplete by 4 days post-MD, long after saturation of the physiological OD shift within 2 days of MD (Trachtenberg, Trepel, & Stryker, 2000). Thus, even though early anatomical studies of OD columns in kittens suggested that OD plasticity is indeed a consequence of the modification of synapses in layer 4, they concluded that this was still too slow to account for or contribute to the earliest functional consequences of MD: a loss of visual function in the deprived-eye pathway. Comparable studies in the mouse yielded a similar conclusion (Antonini, Fagiolini, & Stryker, 1999).

In a search for a faster anatomical correlate of OD plasticity, Trachtenberg and Stryker (2001) showed that anatomical rearrangements of long-range horizontal connections in superficial layers were concurrent with a 2-day OD shift in strabismic kittens, which was in line with earlier studies showing that physiological measures of OD plasticity revealed rapid OD shifts in layer 2/3 (Trachtenberg, Trepel, & Stryker, 2000). Whether similar anatomical plasticity of intracortical axons occurs in mouse visual cortex after 3 days of MD is unknown. Again, because the mouse visual cortex lacks OD columns, monitoring horizontal connectivity in a meaningful way with regard to eye-specific function is technically difficult. However, we can garner clues from 2-photon imaging studies of cortical axons in V1 during functional reorganization induced by discrete retinal lesions. These studies reveal a high capacity for the rapid anatomical plasticity of intracortical axons, demonstrating the potential for such changes to contribute to rapid functional reorganization in mouse (Keck et al., 2011).

RELATING FINE-SCALE STRUCTURAL PLASTICITY AND RAPID FUNCTIONAL PLASTICITY In order to establish whether functional and structural plasticity are distinct processes or part of a continuum, it is first necessary to identify where synaptic rearrangements are concurrent with deprived-eye depression. One approach to monitor changes in connectivity in cortex is to examine dendritic spines, which serve as a proxy for observing excitatory synapses (Holtmaat & Svoboda, 2009). The majority of these structures receive axonal input from other cortical neurons whereas a much smaller subset (<10%) receive thalamocortical input (Ahmed et al., 1994, 1997; da Costa & Martin, 2011). To date, there are relatively few studies that have been designed to examine spine turnover during brief MD in vivo, but the results so far are consistent with a role for rapid structural plasticity in deprived-eye depression. In ferret, spine loss was observed in deprived-eye cortical domains that underwent functional depression within hours of MD (Yu, Majewska, & Sur, 2012). Accordingly, in this same study, rapid spine growth was also observed in deprived-eye cortical domains that showed functional recovery when normal binocular vision was restored after MD. In mouse, 4 days of MD increases the motility of dendritic spines located in superficial layers belonging to layer 5 pyramidal neurons in mice (Oray, Majewska, & Sur, 2004). In studies of fixed tissue, Mataga, Mizuguchi, and Hensch (2004) reported a transient decrease in spine density in layer 2/3 pyramidal cells in a confined 25-μm-long region of their proximal apical dendrites.

The rapid, MD-induced changes in dendritic spine dynamics found in mouse V1 are dependent on the tissue-type plasminogen activator (tPA)/plasmin proteolytic cascade. Brief MD elevates tPA activity in the cortex, and genetic deletion of tPA both reduces the magnitude of the OD shift assayed through single-unit recordings and prevents the loss of dendritic spines during MD (Mataga, Mizuguchi, & Hensch, 2004; Mataga, Nagai, & Hensch, 2002). Furthermore, spine motility can be increased with plasmin treatment, an effect which is occluded by prior MD (Oray, Majewska, & Sur, 2004). Together, these findings suggest that degradation of the extracellular matrix by the tPA/plasmin cascade is an essential component of OD plasticity, possibly due to its role in promoting dendritic spine dynamics and structural plasticity. However, whether the modified synapses are of thalamocortical or intracortical origin remains unknown, and their contribution to functional plasticity is unclear. It should be noted that in mouse V1 thalamocortical axons richly innervate the region examined in these studies, cortical layers 1–3 (see figure 100.2A–B).

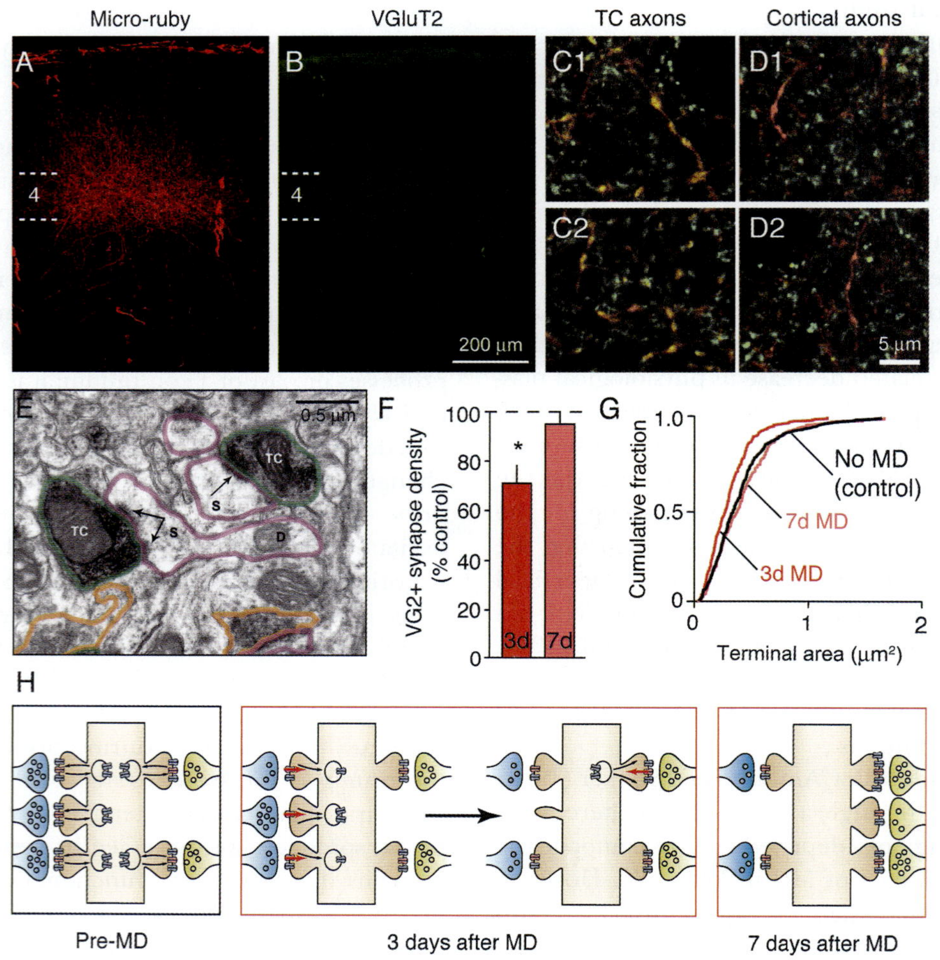

FIGURE 100.2   Evidence for rapid structural plasticity of thalamocortical (TC) synaptic terminals in mouse V1B. (A, B) Confocal images showing labeled TC axons (red, A) and VGluT2 puncta (green, B) in mouse V1B. (C, D) Close-up images demonstrate that VGluT2 puncta colocalize (yellow) with TC axonal varicosities (C1–2), but not varicosities along corticocortical axons (D1–2). (E) Transmission electron micrograph of VGluT2-labeled synaptic terminals (green), which predominately contact dendritic spines (purple). Unlabeled corticocortical terminals are highlighted orange. (F, G) Summary of results after 3 or 7 days of monocular deprivation (MD). Note that both the density and size of thalamocortical synaptic terminals are reduced by 3 days of MD relative to non-MD control animals (dashed line in F), which is concurrent with maximal deprived-eye depression (* = p < 0.05, *t* test). In addition, there is a net recovery of both density and size after 7 days of MD. (H) Model showing how MD-induced homosynaptic long-term depression relates to the structural remodeling of TC synapses. Upon MD (middle panel), Ca^{2+} entry (red arrows) through weakly activated *N*-methyl-D-aspartate (NMDA) receptors (red subunits) driven by poorly correlated input from the deprived eye (blue terminals) initiate the cascade responsible for internalizing AMPA receptors (blue subunits). This "homosynaptic" depression of synaptic transmission is rapidly followed by a full retraction and loss of corresponding TC synaptic terminals. Long-term potentiation-like mechanisms, such as Ca^{2+}-triggered insertion of AMPA receptors, may drive open-eye (yellow terminals) potentiation, and the apparent recovery of TC terminals after 7 days of MD (right panel) suggests that axonal/synaptic growth may contribute to strengthening of the open-eye input. (Images and data in A–G adapted from Coleman et al., 2010).

Only recently has it been possible to examine the state of thalamocortical synapses within the time frame of maximal deprived-eye depression (3 days post-MD in mouse). Capitalizing on the discovery that the VGluT2 vesicular glutamate transporter is selectively localized to thalamocortical terminals (Nahmani & Erisir, 2005), we quantified the density of thalamocortical synaptic terminals in layer 4 of mouse V1 after 3 and 7 days of MD (Coleman et al., 2010) (see figure 100.2A–E). Both the density and size of synaptic terminals in layer 4 were decreased within 3 days of MD (see figure 100.2F–G). These findings are consistent with previous work showing that thalamocortical synaptic terminals exhibit morphological signatures of retraction and weakening after long-term MD (Tieman, 1984), but on a time scale that matches maximal deprived-eye depression (see figure 100.2H). It is also interesting to note that, like physiological OD plasticity, these anatomical correlates of modified thalamocortical transmission are a consequence of uncorrelated ascending activity rather than an absence of retinal activity, suggesting that common cellular-level mechanisms are at work (Coleman et al., 2010).

These synapse-scale structural responses to MD are entirely consistent with the hypothesis that deprived-eye depression (and loss of visual function in amblyopia) occurs through LTD mechanisms. It has been shown that LTD is associated with structural reorganization of presynaptic axonal boutons and a retraction of dendritic spines (Bastrikova et al., 2008; Nagerl et al., 2004; Zhou, Homma, & Poo, 2004).

How can fine-scale structural changes in mice be reconciled with the comparatively slower retraction of entire deprived-eye thalamocortical axons observed in kittens? The earliest time anyone has assayed for thalamocortical axon changes after MD in the mouse visual cortex is 20 days, wherein no significant difference was found between the complexity of deprived-eye and open-eye arbors reconstructed from both hemispheres of the same animal (Antonini, Fagiolini, & Stryker, 1999). This observation does not preclude more discrete changes along individual branches of thalamocortical arbors in response to MD however, which could be obscured by the heterogeneous structure of contralateral-eye axons or the intermingling of OD within layer 4 of mouse V1 (Bence & Levelt, 2005). Although smaller scale remodeling of axonal branches may be difficult to detect, it is reasonable to expect such changes based on the synaptic loss found in the ultrastructural studies described above.

What is the anatomical scale of the physical reorganization, if any, of axons after 3 days of MD? We have begun to test the hypothesis that some portion of thalamocortical axonal branches reorganize fast enough to contribute to deprived-eye depression using 2-photon laser-scanning microscopy (see figure 100.3A). Nearly all 2-photon imaging studies in sensory cortex have been limited to superficial layers because the resolution is relatively poor in deeper layers of V1 (i.e., ~400–500 µm to reach layer 4). However, an advantage of the mouse visual system is that the same thalamocortical axons projecting to layer 4 send significant collaterals to layers 1–3 as well (see figure 100.3B; Antonini, Fagiolini, & Stryker, 1999; Rubio-Garrido et al., 2009). As shown in figure 100.3B–D, thalamocortical branches of deprived-eye axons can be labeled with GFP and tracked in layer 1 before and after MD. In the example images, two branch tips of an axon show significant retractions during the first 3 days of MD with no further changes during an additional 4 days of MD.

While more work is clearly needed, the observations described here reveal that branch remodeling on the order of tens to hundreds of micrometers, as well as changes in dendritic spine structure and number, rapidly occur in visual cortex at an age when deprived-eye depression is robust. In addition to dendritic spines, terminal and en passant bouton dynamics can also be monitored and provide a reliable proxy for visualizing synaptic turnover in vivo (De Paola et al., 2006; Gogolla, Galimberti, & Caroni, 2007). Recent advances in the tools used for labeling and imaging cells in vivo have given us a greater appreciation for how even relatively subtle changes in synaptic structures can significantly alter cortical function (Gogolla, Galimberti, & Caroni, 2007; Holtmaat & Svoboda, 2009; Hubener & Bonhoeffer, 2010). With these tools, we are now poised to investigate the relationship between "functional" and "structural" plasticity in the context of OD plasticity with greater temporal and spatial precision than in the past.

FUTURE CHALLENGES

While much progress has been made toward unveiling the molecular and anatomical substrates of OD plasticity, several questions remain unanswered. What is the functional impact of relatively small-scale changes to thalamocortical synapses? What is the threshold for the amount of synapse loss required to alter the receptive field properties of an individual visually driven neuron? Are LTD mechanisms required for rapid structural plasticity of corticocortical and/or thalamocortical connections? Does the disruption of structural plasticity alone interfere with or diminish functional plasticity or visually guided behavior? What lessons can be learned from

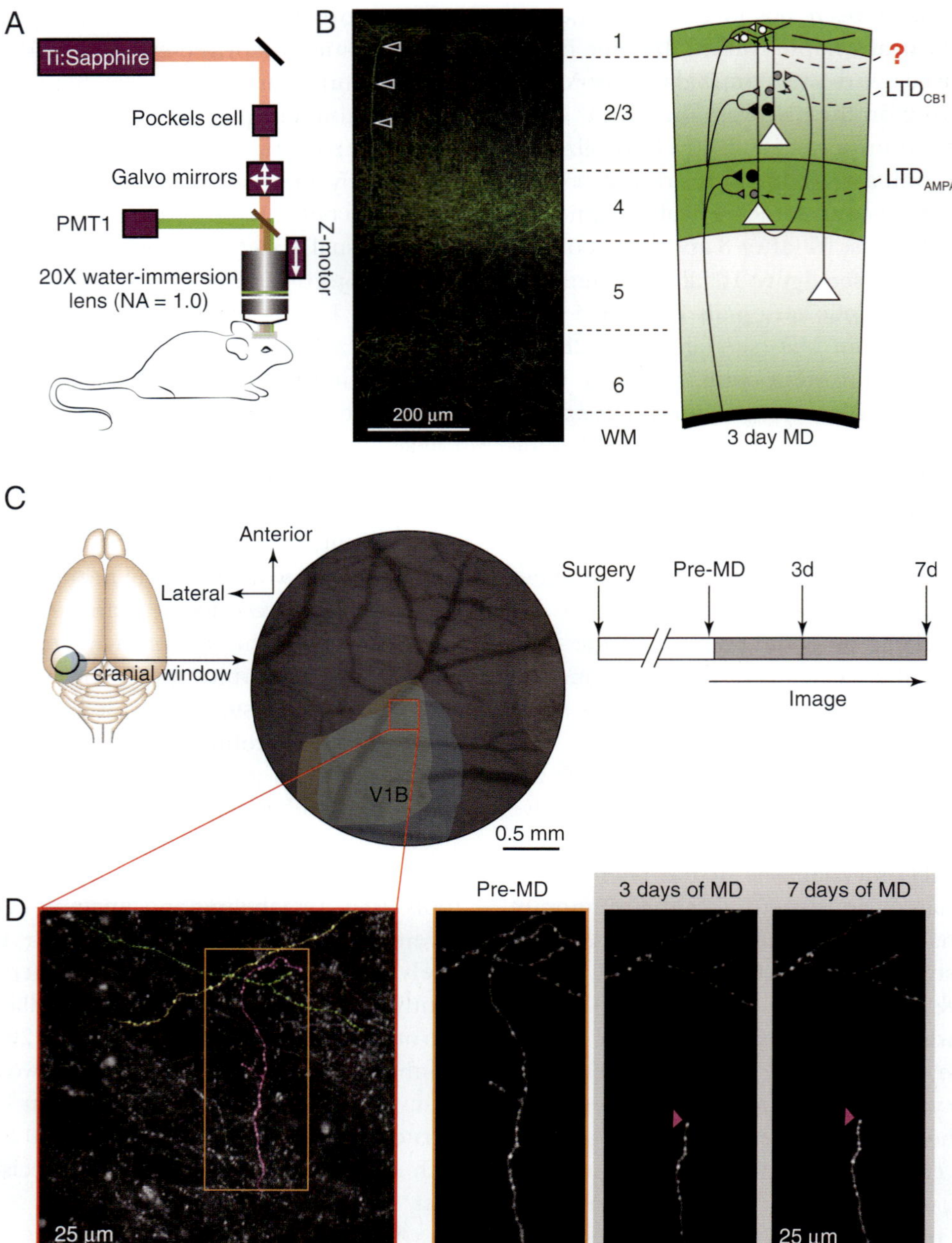

FIGURE 100.3   Chronic 2-photon imaging of thalamocortical axons in mouse V1. (A) Schematic of a basic 2-photon laser-scanning microscope setup. The near-infrared light source is a Ti:sapphire pulsed laser delivered to the sample through an optically sensitive wide-field water-immersion lens. Laser power is controlled by a Pockels cell and allows the user to incrementally increase in power as the focal plane (controlled by the Z-motor) penetrates deeper into the tissue. The beam is scanned in the X and Y dimensions by a galvanometer-controlled mirror to excite the fluorescent label (e.g., green fluorescent protein; GFP). Emitted photons are detected by a sensitive photomultiplier tube (PMT). NA, numerical aperture. (B) Left panel: Confocal image of thalamocortical axon labeling with GFP by injection of a lentiviral vector into the dorsal lateral geniculate nucleus (LGN). Afferents from the LGN form a dense plexus in layer 4 that extends into layer 2/3 and layer 1 (arrowhead marks ascending collateral) in binocular V1 (V1B). Right panel: Schematic showing sites of thalamocortical synapses and sites where long-term depression (LTD) is expressed through AMPA receptor–dependent (LTD$_{AMPAR}$) and cannabinoid (CB) receptor–dependent (LTD$_{CB1}$) mechanisms. (Cartoon adapted from Smith, Heynen, & Bear, 2009). While it is not precisely known how thalamocortical synapses in layer 1 contribute to LTD and deprived-eye depression, the major targets of these projections are most likely spines located along the apical dendrites of pyramidal neurons in layers 2/3, 4, and 5. Thus, structural changes in layer 1 may also contribute to deprived-eye depression in deeper layers and can be readily imaged in vivo. V1B, binocular V1; WM, white matter. (C) Schematic showing the placement of the cranial window relative to V1, which is functionally identified by intrinsic signal imaging (yellow = signal from ipsilateral eye stimulation, blue = signal from binocular stimulation). Vasculature patterns seen through the cranial window (photograph) aid in pinpointing areas for repeated imaging as shown in the experimental timeline to the right. (D) Far left image: Best maximum-intensity Z-projection of GFP-labeled axons within a volume of V1B (XYZ = 240 × 240 × 137 µm). The three axonal branches highlighted green, yellow, and magenta were manually traced in 3-D. The panels to the right show the thresholded images of these branches before and after monocular deprivation (MD). This particular example shows the retraction of two branches (indicated by arrowheads) from one axon that occurs during 3 days of MD. Portions of stable branches are visible in the upper part of each image.

our current understanding of OD plasticity to facilitate recovery of visual function in amblyopia?

A long-standing challenge in the field has been to clearly label eye-specific inputs into the cortex in living tissue. Without this information it is difficult to interpret many of the results in the literature. For example, MD produces changes in dendritic spines (Mataga, Mizuguchi, & Hensch, 2004; Oray, Majewska, & Sur, 2004), but it is unclear whether these changes are restricted to spines receiving input from a particular eye. Similarly, studies of thalamocortical synaptic terminals in mice have been limited to examining net changes to these inputs because the eye-specificity of these afferents cannot be determined. In contrast, one can ascertain precisely what synapses are modified if specific populations of axons are selectively labeled and monitored. New virus-based tools for anterograde transsynaptic labeling of axons and genetic manipulation of eye-specific pathways may provide a long sought after solution to this challenge (Beier et al., 2011; Gradinaru et al., 2010). With this technology, one could test the prediction that occlusion of LTD by prior MD (Crozier et al., 2007) is restricted to deprived-eye, but not open-eye, inputs into layer 4. Likewise, one could monitor open-eye thalamocortical axons to determine whether growth of these axons and synapses contributes to open-eye potentiation, as suggested from previous studies (Coleman et al., 2010). Furthermore, eye-specific synapses could be manipulated and/or labeled using trans-synaptic viruses that allow for perturbations of connected cells during the OD shift and/or expression of genetically encoded molecules for visualizing synapse loss and gain in situ with high fidelity (Kim et al., 2012).

To better understand how synapse remodeling affects cellular function, the development of combinatorial approaches for simultaneously monitoring neural anatomy and function during MD are needed. Recent studies suggest it may soon be feasible to map functionally defined afferent inputs with single-synapse resolution (Chen et al., 2011; Jia et al., 2010). In addition, advances in genetically encoded calcium sensors may permit large numbers of neurons (along with their axons, dendrites, and spines) to be observed repeatedly over the course of MD (Looger & Griesbeck, 2012; Tian et al., 2012). Specifically, the addition of improved intensio- and ratiometric calcium sensors to the imager's toolbox may make these experiments feasible (Andermann, Kerlin, & Reid, 2010; Andermann et al., 2012; Lutcke et al., 2010; Tian et al., 2009). Together, these advances could open the door to investigating the contribution of specific synapses to OD plasticity and whether individual neurons in a cortical network respond to MD in a cell-autonomous manner.

Further work is needed to determine whether any form of structural plasticity during MD requires LTD mechanisms. For example, as highlighted in figure 100.3B, a simple experiment would be to determine whether structural changes to axon branches in layer 1 require CB receptor activation or clathrin-mediated AMPA receptor trafficking in a manner similar to deprived-eye depression. Reagents that disrupt structural changes, but not physiological processes, would help establish a causal role of anatomical plasticity in experience-dependent plasticity. While it is not known whether open-eye potentiation mirrors deprived-eye depression by using LTP-like mechanisms, there is some indication that the former may be supported by different molecular players (Kaneko et al., 2008; Ranson et al., 2012). Given that MD affects both deprived-eye as well as open-eye responses, it is critical to determine if alterations in spine dynamics associated with MD relate to the depression or potentiation of visual responses. There is already convincing evidence that the rapid growth of new spines provides a substrate for strengthening open-eye connections (Yu, Majewska, & Sur, 2012). Likewise, if the model we propose in figure 100.2H is correct, the growth of new open-eye thalamocortical synapses may provide a means for increasing the strength of this input (Coleman et al., 2010). It will be important to follow up these findings by simultaneously imaging axons and spines in in vivo longitudinal studies of structure. Techniques that would allow one to ablate new synaptic growth or selectively block structural plasticity altogether while monitoring cellular function would provide valuable insights into the contribution of anatomical plasticity to functional changes.

Finally, previous studies using the so-called "reverse-suture" paradigm show that weakening of previously open-eye synapses occurs readily whereas strengthening of previously deprived-eye synapses is modest, with little to no sign of recovery at the level of thalamocortical axons (Antonini et al., 1998). Thus, rejuvenating structural plasticity may lead to a more complete functional recovery in amblyopia. It is important to understand whether structural plasticity is causal to other forms of experience-dependent plasticity known to enhance cortical function, as these paradigms may offer opportunities to investigate ways to restore structural connectivity. For example, functional and structural recovery from deprivation amblyopia can be dramatically improved when reverse suture is preceded by brief periods of dark adaptation (He et al., 2007; Montey & Quinlan, 2011). Interestingly, both dark adaptation and treatment with fluoxetine show a capacity for reactivating structural plasticity in the adult visual cortex following MD (Chen et al., 2011; Montey & Quinlan, 2011). However, the kinetics of thalamocortical axon plasticity or eye-specific postsynaptic morphological plasticity have not been examined during open-eye potentiation and warrant investigation. Therapeutic strategies that capitalize on knowledge of experience-dependent plasticity during development and/or perceptual learning have high potential. Increasing visual function through mechanisms of perceptual learning (Cooke & Bear, 2010; Sale et al., 2010) and pharmacological interventions that facilitate deprived-eye recovery (Baroncelli, Maffei, & Sale, 2011; Maya Vetencourt et al., 2008; Yang et al., 2011) are beginning to show promise in humans (Levi & Li, 2009; Maurer & Hensch, 2012).

In conclusion, studies of OD plasticity in mouse visual cortex have improved our understanding of the detailed sequence of events that culminate in the physical disconnection of visual inputs from their targets in V1 after MD. The knowledge of how experience-dependent structural changes overlap or connect with molecular cascades associated with LTD will aid in the development of novel interventions for amblyopia that either strengthen remaining inputs or restore those connections that were lost (Cho & Bear, 2010; Smith, Heynen, & Bear, 2009).

ACKNOWLEDGMENTS

We thank Sam Cooke, Lena Khibnik, Rachel Schecter, and Sue Semple-Rowland for helpful discussions and comments and Lauren Herring, Suzanne Meagher, and Erik Sklar for excellent technical assistance.

REFERENCES

Ahmed, B., Anderson, J. C., Douglas, R. J., Martin, K. A., & Nelson, J. C. (1994). Polyneuronal innervation of spiny stellate neurons in cat visual cortex. *Journal of Comparative Neurology, 341*, 39–49.

Ahmed, B., Anderson, J. C., Martin, K. A., & Nelson, J. C. (1997). Map of the synapses onto layer 4 basket cells of the primary visual cortex of the cat. *Journal of Comparative Neurology, 380*, 230–242.

Andermann, M. L., Kerlin, A. M., & Reid, R. C. (2010). Chronic cellular imaging of mouse visual cortex during operant behavior and passive viewing. *Frontiers in Cellular Neuroscience, 4*, 3.

Andermann, M. L., Kerlin, A. M., Roumis, D. K., Glickfeld, L. L., & Reid, R. C. (2012). Functional specialization of mouse higher visual cortical areas. *Neuron, 72*, 1025–1039.

Antonini, A., & Stryker, M. P. (1993). Rapid remodeling of axonal arbors in the visual cortex. *Science, 260*, 1819–1821.

Antonini, A., Fagiolini, M., & Stryker, M. P. (1999). Anatomical correlates of functional plasticity in mouse visual cortex. *Journal of Neuroscience, 19*, 4388–4406.

Antonini, A., Gillespie, D. C., Crair, M. C., & Stryker, M. P. (1998). Morphology of single geniculocortical afferents and functional recovery of the visual cortex after reverse monocular deprivation in the kitten. *Journal of Neuroscience, 18*, 9896–9909.

Baroncelli, L., Maffei, L., & Sale, A. (2011). New perspectives in amblyopia therapy on adults: A critical role for the excitatory/inhibitory balance. *Frontiers in Cellular Neuroscience, 5*, 1–6.

Bastrikova, N., Gardner, G. A., Reece, J. M., Jeromin, A., & Dudek, S. M. (2008). Synapse elimination accompanies functional plasticity in hippocampal neurons. *Proceedings of the National Academy of Sciences of the United States of America, 105*, 3123–3127. doi:10.1073/pnas.0800027105.

Bear, M. F. (2003). Bidirectional synaptic plasticity: From theory to reality. *Philosophical Transactions of the Royal Society of London. Series B, Biological Sciences, 358*, 649–655.

Bear, M. F., Kleinschmidt, A., Gu, Q. A., & Singer, W. (1990). Disruption of experience-dependent synaptic modifications in striate cortex by infusion of an NMDA receptor antagonist. *Journal of Neuroscience, 10*, 909–925.

Beier, K. T., Saunders, A., Oldenburg, I. A., Miyamichi, K., Akhtar, N., Luo, L., et al. (2011). Anterograde or retrograde transsynaptic labeling of CNS neurons with vesicular stomatitis virus vectors. *Proceedings of the National Academy of Sciences of the United States of America, 108*, 15414–15419. doi:10.1073/pnas.1110854108.

Bence, M., & Levelt, C. N. (2005). Structural plasticity in the developing visual system. *Progress in Brain Research, 147*, 125–139.

Blakemore, C., Garey, L. J., & Vital-Durand, F. (1978). The physiological effects of monocular deprivation and their reversal in the monkey's visual cortex. *Journal of Physiology, 283*, 223–262.

Chen, J. L., Lin, W. C., Cha, J. W., So, P. T., Kubota, Y., & Nedivi, E. (2011). Structural basis for the role of inhibition in facilitating adult brain plasticity. *Nature Neuroscience, 14*, 587–594.

Chen, X., Leischner, U., Rochefort, N. L., Nelken, I., & Konnerth, A. (2011). Functional mapping of single spines in cortical neurons in vivo. *Nature, 475*, 501–505.

Cho, K. K., & Bear, M. F. (2010). Promoting neurological recovery of function via metaplasticity. *Future Neurology, 5*, 21–26.

Cho, K. K., Khibnik, L., Philpot, B. D., & Bear, M. F. (2009). The ratio of NR2A/B NMDA receptor subunits determines the qualities of ocular dominance plasticity in visual cortex. *Proceedings of the National Academy of Sciences of the United States of America, 106*, 5377–5382. doi:10.1073/pnas.0808104 106.

Choi, S. Y., Morales, B., Lee, H. K., & Kirkwood, A. (2002). Absence of long-term depression in the visual cortex of glutamic acid decarboxylase-65 knock-out mice. *Journal of Neuroscience, 22*, 5271–5276.

Chowdhury, S., Shepherd, J. D., Okuno, H., Lyford, G., Petralia, R. S., Plath, N., et al. (2006). Arc/Arg3.1 interacts with the endocytic machinery to regulate AMPA receptor trafficking. *Neuron, 52*, 445–459. doi:10.1016/j.neuron.2006.08. 033.

Coleman, J. E., Law, K., & Bear, M. F. (2009). Anatomical origins of ocular dominance in mouse primary visual cortex. *Neuroscience, 161*, 561–571.

Coleman, J. E., Nahmani, M., Gavornik, J. P., Haslinger, R., Heynen, A. J., Erisir, A., et al. (2010). Rapid structural remodeling of thalamocortical synapses parallels experience-dependent functional plasticity in mouse primary visual cortex. *Journal of Neuroscience, 30,* 9670–9682. doi:10.1523/JNEUROSCI.1248-10.2010.

Cooke, S. F., & Bear, M. F. (2010). Visual experience induces long-term potentiation in the primary visual cortex. *Journal of Neuroscience, 30,* 16304–16313.

Crair, M. C., Horton, J. C., Antonini, A., & Stryker, M. P. (2001). Emergence of ocular dominance columns in cat visual cortex by 2 weeks of age. *Journal of Comparative Neurology, 430,* 235–249.

Crowley, J. C., & Katz, L. C. (2000). Early development of ocular dominance columns. *Science, 290,* 1321–1324.

Crozier, R. A., Wang, Y., Liu, C. H., & Bear, M. F. (2007). Deprivation-induced synaptic depression by distinct mechanisms in different layers of mouse visual cortex. *Proceedings of the National Academy of Sciences of the United States of America, 104,* 1383–1388.

da Costa, N. M., & Martin, K. A. (2011). How thalamus connects to spiny stellate cells in the cat's visual cortex. *Journal of Neuroscience, 31,* 2925–2937.

De Paola, V., Holtmaat, A., Knott, G., Song, S., Wilbrecht, L., Caroni, P., et al. (2006). Cell type–specific structural plasticity of axonal branches and boutons in the adult neocortex. *Neuron, 49,* 861–875. doi:10.1016/j.neuron.2006.02.017.

Doshi, N. R., & Rodriguez, M. L. (2007). Amblyopia. *American Family Physician, 75,* 361–367.

Dudek, S. M., & Bear, M. F. (1992). Homosynaptic long-term depression in area CA1 of hippocampus and effects of N-methyl-D-aspartate receptor blockade. *Proceedings of the National Academy of Sciences of the United States of America, 89,* 4363-4367.

Frenkel, M. Y., & Bear, M. F. (2004). How monocular deprivation shifts ocular dominance in visual cortex of young mice. *Neuron, 44,* 917–923.

Gogolla, N., Galimberti, I., & Caroni, P. (2007). Structural plasticity of axon terminals in the adult. *Current Opinion in Neurobiology, 17,* 516–524.

Gradinaru, V., Zhang, F., Ramakrishnan, C., Mattis, J., Prakash, R., Diester, I., et al. (2010). Molecular and cellular approaches for diversifying and extending optogenetics. *Cell, 141,* 154–165. doi:10.1016/j.cell.2010.02.037.

He, H. Y., Ray, B., Dennis, K., & Quinlan, E. M. (2007). Experience-dependent recovery of vision following chronic deprivation amblyopia. *Nature Neuroscience, 10,* 1134–1136.

Hensch, T. K., Fagiolini, M., Mataga, N., Stryker, M. P., Baekkeskov, S., & Kash, S. F. (1998). Local GABA circuit control of experience-dependent plasticity in developing visual cortex. *Science, 282,* 1504–1508.

Heynen, A. J., Yoon, B. J., Liu, C. H., Chung, H. J., Huganir, R. L., & Bear, M. F. (2003). Molecular mechanism for loss of visual cortical responsiveness following brief monocular deprivation. *Nature Neuroscience, 6,* 854–862.

Hofer, S. B., Mrsic-Flogel, T. D., Bonhoeffer, T., & Hubener, M. (2006). Prior experience enhances plasticity in adult visual cortex. *Nature Neuroscience, 9,* 127–132.

Holmes, J. M., & Clarke, M. P. (2006). Amblyopia. *Lancet, 367,* 1343–1351. doi: 101016/S0140–6736(06)68581–4.

Holtmaat, A., & Svoboda, K. (2009). Experience-dependent structural synaptic plasticity in the mammalian brain. *Nature Reviews. Neuroscience, 10,* 647–658.

Horton, J. C., & Adams, D. L. (2005). The cortical column: A structure without a function. *Philosophical Transactions of the Royal Society of London. Series B, Biological Sciences, 360,* 837–862.

Hubel, D. H., & Wiesel, T. N. (1964). Effects of monocular deprivation in kittens. *Naunyn-Schmiedebergs Archiv fur Experimentelle Pathologie und Pharmakologie, 248,* 492–497.

Hubel, D. H., & Wiesel, T. N. (1968). Receptive fields and functional architecture of monkey striate cortex. *Journal of Physiology, 195,* 215–243.

Hubel, D. H., Wiesel, T. N., & LeVay, S. (1977). Plasticity of ocular dominance columns in monkey striate cortex. *Philosophical Transactions of the Royal Society of London. Series B, Biological Sciences, 278,* 377–409.

Hubener, M., & Bonhoeffer, T. (2010). Searching for engrams. *Neuron, 67,* 363–371.

Iny, K., Heynen, A. J., Sklar, E., & Bear, M. F. (2006). Bidirectional modifications of visual acuity induced by monocular deprivation in juvenile and adult rats. *Journal of Neuroscience, 26,* 7368–7374.

Jia, H., Rochefort, N. L., Chen, X., & Konnerth, A. (2010). Dendritic organization of sensory input to cortical neurons in vivo. *Nature, 464,* 1307–1312.

Kaneko, M., Stellwagen, D., Malenka, R. C., & Stryker, M. P. (2008). Tumor necrosis factor-alpha mediates one component of competitive, experience-dependent plasticity in developing visual cortex. *Neuron, 58,* 673–680.

Keck, T., Scheuss, V., Jacobsen, R. I., Wierenga, C. J., Eysel, U. T., Bonhoeffer, T., et al. (2011). Loss of sensory input causes rapid structural changes of inhibitory neurons in adult mouse visual cortex. *Neuron, 71,* 869–882. doi:10.1016/j.neuron.2011.06.034.

Khibnik, L. A., Cho, K. K., & Bear, M. F. (2010). Relative contribution of feedforward excitatory connections to expression of ocular dominance plasticity in layer 4 of visual cortex. *Neuron, 66,* 493–500.

Kim, J., Zhao, T., Petralia, R. S., Yu, Y., Peng, H., Myers, E., et al. (2012). mGRASP enables mapping mammalian synaptic connectivity with light microscopy. *Nature Methods, 9,* 96–102. doi:10.1038/nmeth.1784.

Kirkwood, A., & Bear, M. F. (1994). Homosynaptic long-term depression in the visual cortex. *Journal of Neuroscience, 14,* 3404–3412.

LeVay, S., Stryker, M. P., & Shatz, C. J. (1978). Ocular dominance columns and their development in layer IV of the cat's visual cortex: A quantitative study. *Journal of Comparative Neurology, 179,* 223–244.

LeVay, S., Wiesel, T. N., & Hubel, D. H. (1980). The development of ocular dominance columns in normal and visually deprived monkeys. *Journal of Comparative Neurology, 191,* 1–51.

Levi, D. M., & Li, R. W. (2009). Improving the performance of the amblyopic visual system. *Philosophical Transactions of the Royal Society of London. Series B, Biological Sciences, 364,* 399–407.

Looger, L. L., & Griesbeck, O. (2012). Genetically encoded neural activity indicators. *Current Opinion in Neurobiology, 22,* 18–23.

Lutcke, H., Murayama, M., Hahn, T., Margolis, D. J., Astori, S., Zum Alten Borgloh, S. M., et al. (2010). Optical recording of neuronal activity with a genetically-encoded calcium indicator in anesthetized and freely moving mice. *Frontiers in Neural Circuits, 4,* 9.1–12. doi:10.3389/fncir.2010.00009.

Malenka, R. C., & Bear, M. F. (2004). LTP and LTD: An embarrassment of riches. *Neuron, 44*, 5–21.

Mataga, N., Mizuguchi, Y., & Hensch, T. K. (2004). Experience-dependent pruning of dendritic spines in visual cortex by tissue plasminogen activator. *Neuron, 44*, 1031–1041.

Mataga, N., Nagai, N., & Hensch, T. K. (2002). Permissive proteolytic activity for visual cortical plasticity. *Proceedings of the National Academy of Sciences of the United States of America, 99*, 7717–7721. doi:10.1073/pnas.102088899.

Maurer, D., & Hensch, T. K. (2012). Amblyopia: Background to the special issue on stroke recovery. *Developmental Psychobiology, 54*, 224–238.

Maya Vetencourt, J. F., Sale, A., Viegi, A., Baroncelli, L., De Pasquale, R., O'Leary, O. F., et al. (2008). The antidepressant fluoxetine restores plasticity in the adult visual cortex. *Science, 320*, 385–388. doi:10.1126/science.1150516.

McCurry, C. L., Shepherd, J. D., Tropea, D., Wang, K. H., Bear, M. F., & Sur, M. (2010). Loss of Arc renders the visual cortex impervious to the effects of sensory experience or deprivation. *Nature Neuroscience, 13*, 450–457.

Mioche, L., & Singer, W. (1989). Chronic recordings from single sites of kitten striate cortex during experience-dependent modifications of receptive-field properties. *Journal of Neurophysiology, 62*, 185–197.

Montey, K. L., & Quinlan, E. M. (2011). Recovery from chronic monocular deprivation following reactivation of thalamocortical plasticity by dark exposure. *Nature Communications, 2*, 317.

Mrsic-Flogel, T. D., Hofer, S. B., Ohki, K., Reid, R. C., Bonhoeffer, T., & Hubener, M. (2007). Homeostatic regulation of eye-specific responses in visual cortex during ocular dominance plasticity. *Neuron, 54*, 961–972.

Nagerl, U. V., Eberhorn, N., Cambridge, S. B., & Bonhoeffer, T. (2004). Bidirectional activity-dependent morphological plasticity in hippocampal neurons. *Neuron, 44*, 759–767.

Nahmani, M., & Erisir, A. (2005). VGluT2 immunochemistry identifies thalamocortical terminals in layer 4 of adult and developing visual cortex. *Journal of Comparative Neurology, 484*, 458–473.

Oray, S., Majewska, A., & Sur, M. (2004). Dendritic spine dynamics are regulated by monocular deprivation and extracellular matrix degradation. *Neuron, 44*, 1021–1030.

Prusky, G. T., West, P. W., & Douglas, R. M. (2000). Experience-dependent plasticity of visual acuity in rats. *European Journal of Neuroscience, 12*, 3781–3786.

Ramoa, A. S., Mower, A. F., Liao, D., & Jafri, S. I. (2001). Suppression of cortical NMDA receptor function prevents development of orientation selectivity in the primary visual cortex. *Journal of Neuroscience, 21*, 4299–4309.

Ranson, A., Cheetham, C. E., Fox, K., & Sengpiel, F. (2012). Homeostatic plasticity mechanisms are required for juvenile, but not adult, ocular dominance plasticity. *Proceedings of the National Academy of Sciences of the United States of America, 109*, 1311–1316.

Rittenhouse, C. D., Shouval, H. Z., Paradiso, M. A., & Bear, M. F. (1999). Monocular deprivation induces homosynaptic long-term depression in visual cortex. *Nature, 397*, 347–350.

Rittenhouse, C. D., Siegler, B. A., Voelker, C. C., Shouval, H. Z., Paradiso, M. A., & Bear, M. F. (2006). Stimulus for rapid ocular dominance plasticity in visual cortex. *Journal of Neurophysiology, 95*, 2947–2950.

Rubio-Garrido, P., Perez-de-Manzo, F., Porrero, C., Galazo, M. J., & Clasca, F. (2009). Thalamic input to distal apical dendrites in neocortical layer 1 is massive and highly convergent. *Cerebral Cortex, 19*, 2380–2395.

Sale, A., De Pasquale, R., Bonaccorsi, J., Pietra, G., Olivieri, D., Berardi, N., et al. (2010). Visual perceptual learning induces long-term potentiation in the visual cortex. *Neuroscience, 172*, 219–225. doi:10.1016/j.neuroscience.2010.10.078.

Smith, G. B., Heynen, A. J., & Bear, M. F. (2009). Bidirectional synaptic mechanisms of ocular dominance plasticity in visual cortex. *Philosophical Transactions of the Royal Society of London. Series B, Biological Sciences, 364*, 357–367.

Steele, P. M., & Mauk, M. D. (1999). Inhibitory control of LTP and LTD: Stability of synapse strength. *Journal of Neurophysiology, 81*, 1559–1566.

Tian, L., Akerboom, J., Schreiter, E. R., & Looger, L. L. (2012). Neural activity imaging with genetically encoded calcium indicators. *Progress in Brain Research, 196*, 79–94.

Tian, L., Hires, S. A., Mao, T., Huber, D., Chiappe, M. E., Chalasani, S. H., et al. (2009). Imaging neural activity in worms, flies and mice with improved GCaMP calcium indicators. *Nature Methods, 6*, 875–881. doi:10.1038/nmeth.1398.

Tieman, S. B. (1984). Effects of monocular deprivation on geniculocortical synapses in the cat. *Journal of Comparative Neurology, 222*, 166–176.

Trachtenberg, J. T., & Stryker, M. P. (2001). Rapid anatomical plasticity of horizontal connections in the developing visual cortex. *Journal of Neuroscience, 21*, 3476–3482.

Trachtenberg, J. T., Trepel, C., & Stryker, M. P. (2000). Rapid extragranular plasticity in the absence of thalamocortical plasticity in the developing primary visual cortex. *Science, 287*, 2029–2032.

Wiesel, T. N., & Hubel, D. H. (1963a). Effects of visual deprivation on morphology and physiology of cells in the cat's lateral geniculate body. *Journal of Neurophysiology, 26*, 978–993.

Wiesel, T. N., & Hubel, D. H. (1963b). Single-cell responses in striate cortex of kittens deprived of vision in one eye. *Journal of Neurophysiology, 26*, 1003–1017.

Yang, K., Xiong, W., Yang, G., Kojic, L., Wang, Y. T., & Cynader, M. (2011). The regulatory role of long-term depression in juvenile and adult mouse ocular dominance plasticity. *Scientific Reports, 1*, 203.

Yoon, B. J., Smith, G. B., Heynen, A. J., Neve, R. L., & Bear, M. F. (2009). Essential role for a long-term depression mechanism in ocular dominance plasticity. *Proceedings of the National Academy of Sciences of the United States of America, 106*, 9860–9865. doi:10.1073/pnas.0901305106.

Yu, H., Majewska, A. K., & Sur, M. (2012). Rapid experience-dependent plasticity of synapse function and structure in ferret visual cortex in vivo. *Proceedings of the National Academy of Sciences of the United States of America, 108*, 21235–21240. doi:10.1073/pnas.1108270109.

Zhou, Q., Homma, K. J., & Poo, M. M. (2004). Shrinkage of dendritic spines associated with long-term depression of hippocampal synapses. *Neuron, 44*, 749–757.

XIV TRANSLATIONAL VISUAL NEUROSCIENCE

101 Translational Research for Optic Nerve Disorders: Overview

NEIL R. MILLER

The fundamental role of the ophthalmologist is to preserve and restore vision. Although ophthalmologists have been successful in a number of areas, including treatment of cataracts and diseases of the cornea and retina, the treatment of optic nerve (ON) disease is much less successful. Processes that damage the ON, including increased intracranial pressure (ICP), inflammation, ischemia, trauma, and hereditary metabolic dysfunction often cause visual loss for which there is no current treatment. In this chapter, I will review the clinical manifestations and current translational research regarding the pathogenesis, treatment, and prevention of three important nonglaucomatous ON disorders for which the most translational research is being performed: papilledema, nonarteritic anterior ischemic optic neuropathy (NAION), and Leber hereditary optic neuropathy (LHON). Other nonglaucomatous optic neuropathies either are associated with spontaneous recovery and thus do not invite much translational research (e.g., optic neuritis) or are so heterogeneous that translational research has been relatively minimal (e.g., traumatic optic neuropathy).

PAPILLEDEMA

Papilledema is defined as swelling of the optic disk associated with increased ICP (see figure 101.1).

This potentially vision-threatening ophthalmological condition results from transmission of the increased ICP to the subarachnoid space (SAS) of the ON. The associated increase in cerebrospinal fluid (CSF) volume in the perioptic space results in unfolding of the ON sheath, and it is believed that this, in turn, compresses the nerve, causing stasis of axonal transport, thus producing swelling of the ON axons (Hayreh, 1968; Hayreh & Hayreh, 1977; Hayreh, March, & Anderson, 1979; Tso & Hayreh, 1977a, 1977b). Pressure measurements in the SAS surrounding the ON have not been performed in this setting, but it has been assumed that the pressure in the SAS is the same as that measured during lumbar puncture and that there is free bidirectional flow of CSF between the intracranial SAS and the ON SAS. If this were the case, one would expect that reduction of pressure on the ON by, for example, a CSF diversion procedure such as a lumboperitoneal or ventriculoperitoneal shunt or an ON sheath fenestration (ONSF) would result in resolution of papilledema (Brazis, 2009; Feldon, 2007; Garton, 2004). In fact, it is not uncommon for papilledema to persist despite an apparently functioning shunt or a successful ONSF (Kelman et al., 1991; Wilkes & Siatkowski, 2009). In such cases, permanent visual loss may result (Friedman & Jacobson, 2004). Recent evidence suggests that the flow of CSF between intracranial SAS and ON SAS is neither continuous nor bidirectional (Killer et al., 2007). In patients with increased ICP and papilledema, this may lead to CSF segregation and the development of a biologically unfavorable environment from a reduced CSF recycling time within the perioptic space. Indeed, Jaggi et al. (2010) placed a silicone band around one ON in seven sheep. The band was tightened just enough to compress the SAS surrounding the nerve, thus blocking the flow of CSF without compressing the ON itself. After 4 to 21 days, both the ligated and untouched ONs were removed and evaluated histologically with both light and electron microscopy.

All treated ONs showed marked loss of axons, destruction of myelin and swelling of meningoepithelial cells, most pronounced in the proximal ON adjacent to the globe at the location most distant to the ligature (see figure 101.2). There was no significant difference in histological findings between the ONs that were ligated for 4 days and those with 21 days of ligature. These results indicate that CSF segregation in the ON by blocking the SAS leads within 4 days to severe nerve damage. The increasing severity of these changes with increasing distance from the site of the ligature argues against simple pressure- or microperfusion-dependent effects and supports the hypothesis that interruption of CSF flow in the SAS of the ON, such as that which may occur in patients with papilledema, can produce damage due to a change of CSF flow and content. This

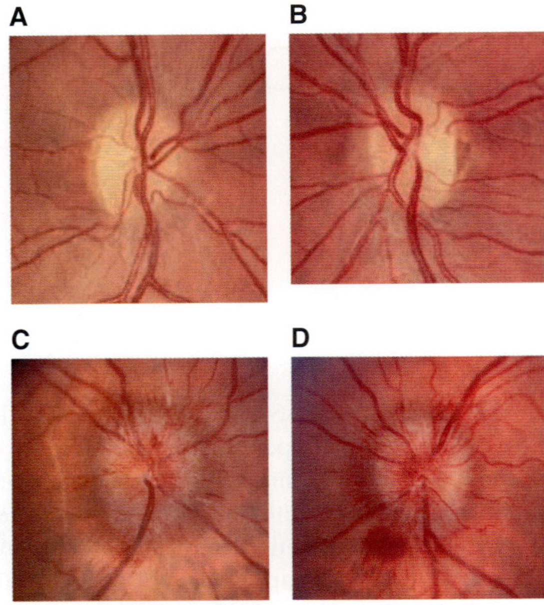

FIGURE 101.1 Normal optic disks compared with bilateral papilledema. (A and B) Appearance of normal right (A) and left (B) optic disks. Note clear disk margins and lack of disk swelling or peripapillary hemorrhages. (C and D) Appearance of bilateral papilledema involving right (C) and left (D) optic disks of a patient with increased intracranial pressure. Note blurred disk margins, elevation and hyperemia of the disks, and presence of peripapillary hemorrhages.

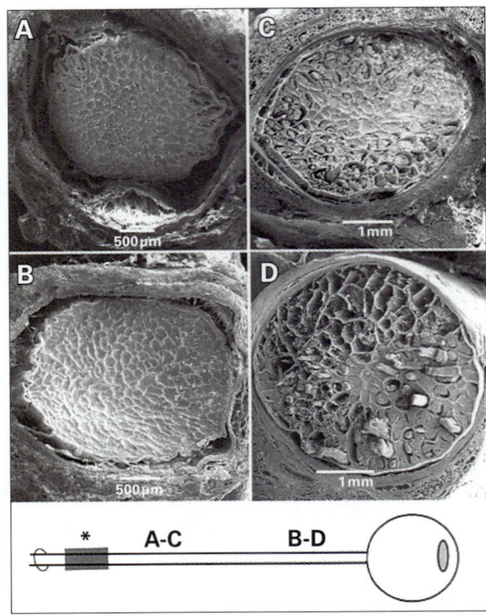

FIGURE 101.2 Scanning electron microscopy of cross-sections of normal (untreated side as control) (A, B) and ligatured (C, D) optic nerves (ONs) along with a drawing showing the region of the ligature (asterisk) and the general areas that were examined. (A, C) Posterior (distal) portion. (B, D) Anterior (proximal, adjacent to globe) portion. In (A), note the normally wide subarachnoid space (SAS) in the posterior portion of a control ON. In anterior part of a control ON (B), note normally narrow SAS (this contrasts to human ON). In posterior part of an ON that was ligatured for 4 days (C), note extensive myelin damage in the periphery of the nerve with few defects visible in its center. In anterior part of an ON that was ligatured for 4 days (D), note massive peripheral and central axon and myelin destruction. (From Jaggi et al., 2010.)

phenomenon may account for the failure of CSF diversion procedures and ON decompression surgery in some patients (Killer, Jaggi, & Miller, 2009). To further test this concept, Killer et al. (2011) performed CT cisternography in 10 patients with increased ICP and compared the findings with those from two subjects with normal ICP who were enrolled in an ongoing prospective study unrelated to papilledema. In addition, they determined the concentration gradient of lipocalin-like prostaglandin D synthase (L-PGDS) between CSF from the ON SAS obtained during ONSF and lumbar CSF in seven of their patients with increased ICP. In increased concentrations, this protein has been shown to be toxic to astrocytes in vitro (Xin et al., 2009) (see figure 101.3), and its gradient has been reported to be a biomarker of CSF segregation in the perioptic space (Killer et al., 2006).

CT cisternography showed a progressively reduced influx of contrast-loaded CSF from intracranial CSF spaces into the SAS. The lowest concentration of contrast-loaded CSF was found in the region of the ON immediately behind the globe, where the ON sheath was widened (possibly by unfolding) in all patients compared with normal subjects. The concentration of L-PGDS differed between the spinal CSF and the CSF in the SAS, with a markedly higher concentration in the SAS (see figure 101.4).

The results of this study suggest that CSF turnover in the SAS of the ON is reduced in patients with papilledema from various causes and that the composition of CSF differs between spinal CSF and that surrounding the ON. All of the studies described above suggest that although increased ICP plays a major role in the development of papilledema, alterations in the content of the stagnant CSF surrounding the ON in patients with papilledema also may play a role and that treatment of these alterations may be critical in preventing visual loss in patients with this condition.

NONARTERITIC ANTERIOR ISCHEMIC OPTIC NEUROPATHY

Anterior ischemic optic neuropathy (AION) results from a sudden ischemic insult to the proximal portion of the ON. There are two main forms of AION:

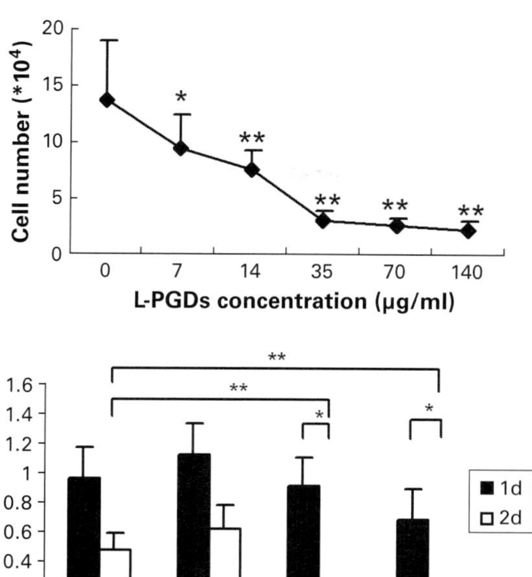

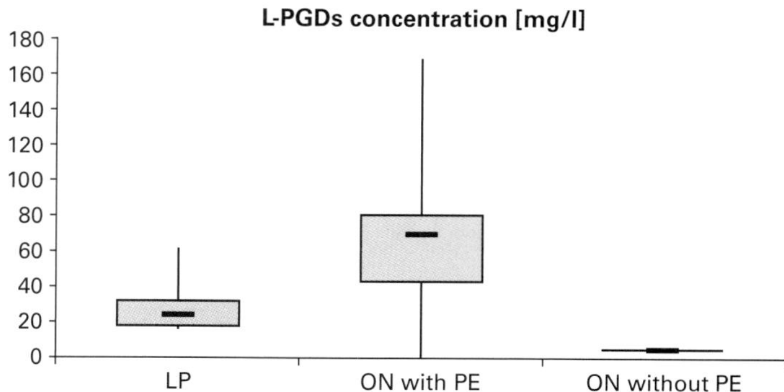

FIGURE 101.4   Concentration of lipocalin-like prostaglandin D synthase (L-PGDS) in cerebrospinal fluid taken from the lumbar subarachnoid space, lumbar puncture (LP), and from the subarachnoid space surrounding the optic nerves with papilloedema (ON with PE) and in control subjects (ON without PE). Box plot indicating median value ± quartile, maximum, and minimum values (mg/l). (From Killer et al., 2011.)

FIGURE 101.3   Toxicity of lipocalin-like prostaglandin D synthase (L-PGDS) to astrocytes in vitro. Top: Effect of L-PGDS on astrocyte proliferation. Proliferation of astrocytes decreased with an increase of the concentration of L-PGDS. Evident decrease of cell number was detected at L-PGDS concentration ranging from 14 to 140 µg/mL. Results are mean ± standard deviation of six experiments. Statistically different data (p < 0.005 compared with controls) are indicated by double asterisks (**). Bottom: Number of astrocytes (LN 229) incubated with different concentrations of L-PGDS as determined by a cell analyzer system (CASY1). (From Xin et al., 2009.)

nonarteritic (NAION) and arteritic (AAION). The two forms are etiologically distinct. NAION comprises 85% of all cases of AION, with AAION the remainder. NAION is the most common cause of sudden ON-related vision loss and typically affects individuals over 55 years of age (Arnold, 2005) whereas AAION is less common, usually is caused by giant cell (temporal) arteritis, and most often affects persons over the age of 70.

Controversy clouds the actual etiology of NAION. A well-recognized risk factor is a small ON passage through the sclera, presumably causing crowding of axons within a tight dural sheath (the "crowded disk" or "disk at risk"); however, a host of other factors, including age, nocturnal or situational (e.g., perioperative) hypotension, diabetes mellitus, hypercholesterolemia, hypertension, obstructive sleep apnea, use of certain medications or drugs, and possibly a number of genetic sequence polymorphisms involving either mitochondria or genes associated with vascular function, may contribute (Bosley, Abu-Amero, & Ozand, 2004; Ischemic Optic Neuropathy Decompression Trial Study

Group, 1996; Palombi et al., 2006; Pomeranz & Bhavsar, 2005; Sakai et al., 2007; Salomon et al., 1999, 2004). In addition, although it initially was believed that the ON in pre-insult individuals was functionally normal, some individuals predisposed to NAION can have subtle vascular differences from control (non-NAION predisposed) eyes (Collignon-Robe, Feke, & Rizzo, 2004; Leiba et al., 2000), suggesting that there are subclinical changes that may increase susceptibility. These include reduced ON head blood flow as measured by laser Doppler flowmetry, in addition to the small ON outlet (Leiba et al., 2000). The presence of electrophysiological abnormalities in the clinically normal, fellow eye of patients with NAION supports this notion (Janaky et al., 2006). The combined factors of hypertension and age have led a number of investigators to suggest that NAION may be triggered by temporary loss of vascular homeostasis, with initially subtle tissue edema accumulating in a confined area, resulting in a tissue compartment syndrome (Levin & Danesh-Meyer, 2008). Levin and Danesh-Meyer (2008) even suggested that transient functional or anatomic occlusion of central retinal vein tributaries within the ON may initiate NAION. Regardless of the initiating event, the subsequent edema and further compression of ON capillaries in the restricted space of the ON, with its closely confining, thick sheath, results in vascular compression and ischemia. This, in turn, leads to an ischemic axonopathy, with resultant stasis of axonal flow, disruption of electrical and growth factor-associated signaling, and disconnection of action potential-related communication between retinal and higher structures in the central nervous system (CNS). In this respect, NAION is similar to other sudden

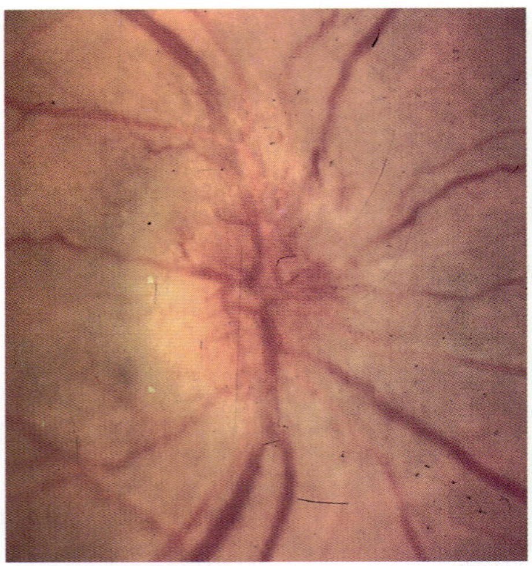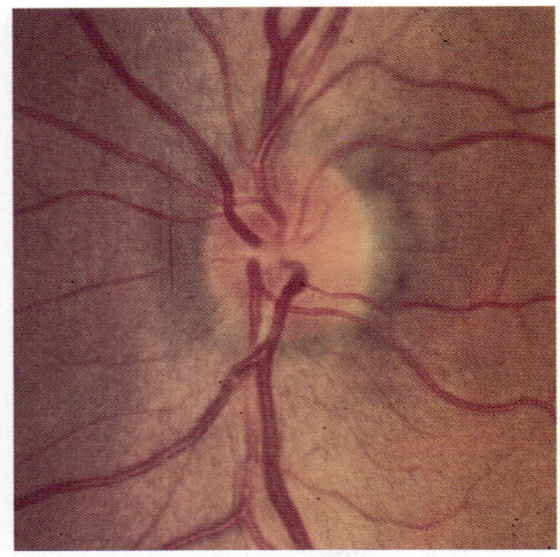

FIGURE 101.5   Nonarteritic anterior ischemic optic neuropathy. Right: The right optic disk is hyperemic and swollen. Several peripapillary retinal hemorrhages are present. Left: The left optic disk is small and has no cup (the "disk-at-risk").

axonopathic conditions, such as ON stretch, crush, and axotomy (Nadal-Nicolas et al., 2009; Schlamp et al., 2001).

NAION is the single leading cause of sudden ON-related vision loss around the world. Although typically unilateral, 15–20% of individuals with unilateral NAION will experience NAION in the contralateral eye over the subsequent 5 years, and there is no consistently effective treatment to date, either to improve vision in an eye affected by NAION or to prevent visual loss from NAION in the fellow eye.

The most obvious finding in patients with NAION is edema of the ON head (i.e., the optic disk) usually associated with flame-shaped hemorrhages on or adjacent to the optic disk (see figure 101.5). These hemorrhages may be related to vascular compression at the disk or to ischemia-reperfusion injury, with rebleeding from the damaged vessels.

A perfect (i.e., clinically identical) animal model of human NAION not only would take into account the associated pathophysiological contributing factors (e.g., the "crowded disk," underlying vascular risk factors, age), but the ON itself would be physiologically similar to the human nerve with respect to structure, vasculature, size, and control mechanisms. Unfortunately, such a model does not exist.

In all species, the ON is a white-matter CNS tract formed by the long-axon neurons of retinal ganglion cells (RGCs) and myelinated by oligodendrocytes rather than Schwann cells. The ON sheath is continuous with the sclera. The mouse ON possesses roughly 50,000 axons (Williams et al., 1996); the rat, 100,000 (Perry, Henderson, & Linden, 1983); the rhesus monkey and human, 1.1–1.2 million (Morrison et al., 1990). Axon

fiber diameters are similar among all species, and diffusion coefficients are identical among all mammalian species (Fukuda et al., 1988). Thus, there is a 20-fold factor in axon numbers between rhesus and mouse ON, with a corresponding increase in intraneural vascularization. The differences in number of axons in the ONs of these species results in a substantial difference in their ON size. The mouse ON has a maximum diameter of 150–200 microns, whereas the rat ON has, at maximum, a 500-micron diameter, and the ON of the rhesus monkey and the human ON have similar diameters of 1.5–3 mm, depending on whether the portion measured is myelinated or unmyelinated (Fukuda et al., 1988; Hayreh & Vrabec, 1966; Morrison et al., 1990). Additionally, the primate ON sheath is thicker than that of mice and rats. This may predispose primates to more damage from intraneural edema via a compartment syndrome than can occur in rodents (Tesser, Niendorf, & Levin, 2003).

Another major difference in ON-related anatomy between rodents and primates is in the structure of the lamina cribrosa (LC) (Albrecht, 2008). The LC is a complex region that permits the RGC axons to penetrate the globe and emerge as the ON.

The primate LC consists of both a glial laminar element and a connective tissue laminar element. The latter consists of connective tissue sheets that form the pores through which the RGC axons pass (Rizzo, 2005) (see figure 101.6, left).

Considerable research has focused on the primate LC, and an extensive description is beyond the scope of this review; however, briefly, the LC subdivides the intraocular portion of the ON into three zones: prelaminar (visible by fundus examination), intralaminar, and

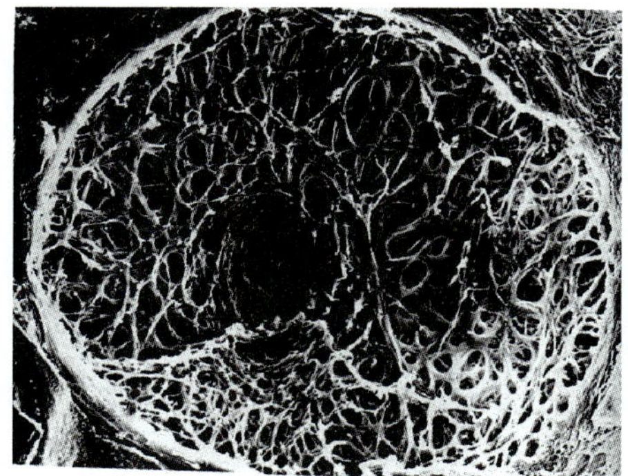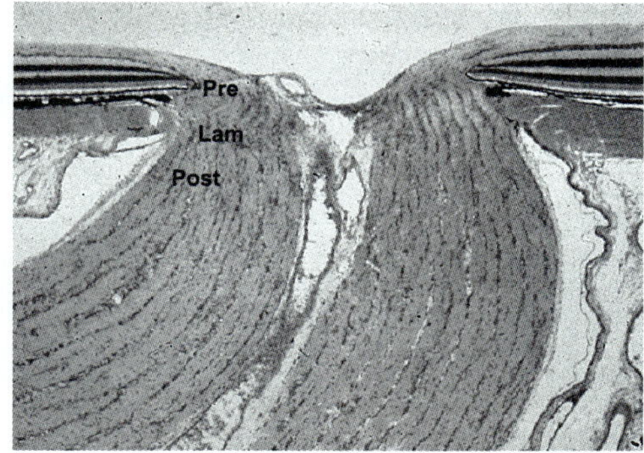

FIGURE 101.6   Left: The human lamina cribrosa. Right: Longitudinal section through a human optic nerve with its prelaminar (Pre), laminar (Lam), and postlaminar (Post) segments indicated.

postlaminar (Rizzo, 2005) (see figure 101.6, right). Within this junctional zone, axons pass from a region of higher pressure (intraocular) to lower pressure (intraorbital–intracranial), as they form the ON. There are also changes in composition of the glial components, with few astrocytes and unmyelinated axons in the prelaminar region and many more astrocytes and myelinating oligodendrocytes in the postlaminar region.

In rodents, the glial component of the lamina is similar to that of primates, but there are marked differences in the connective tissue component between rodent and primate (Johansson, 1987; Morrison et al., 1995). Rather than tissue sheets, the connective tissue component in rats is comprised of fiber bundles (Albrecht, 2008). In mice, no LC connective tissue component exists. This undoubtedly accounts for some of the visible differences in edema seen after AION induction in the mouse versus the rat (see below). To summarize, LC collagen sheets seen in primates are replaced by collagen bundles in rats and are largely absent in mice (May & Lutjen-Drecoll, 2002). These differences likely alter the impact of compartmentation associated with ON ischemia, which is believed to be integral to the development of NAION. In particular, the absence or minimization of the collagenous component of the LC in rodents and the thinness of the sclera in most small animals are likely to change the response of their eyes to the pressure exerted by tissue edema in AION, resulting in marked differences in the appearance of ON head edema.

Vascularization of the anterior ON region is complex and, once again, is very different among species. In primates and rats, both the retina and ON are supplied by vessels that penetrate the sclera around the ON (see figure 101.7).

Inner retinal circulation is supplied via the central retinal vessels that enter (central retinal artery) or leave

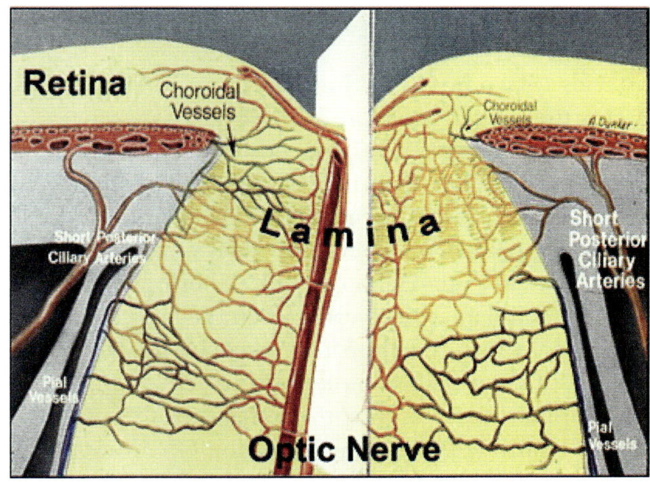

FIGURE 101.7   Optic nerve (ON) vasculature. Short posterior ciliary arteries perforate through the sclera (intrascleral region) around the optic nerve and supply the choroid, as well as forming an anastomotic vascular circle (the circle of Zinn–Haller [ZH]), which contributes to the vascular supply of the anterior ON. ON capillaries are supplied from the ZH and anastomose with choroidal vessels, as well as intraretinal capillaries in the region of the optic nerve. ON capillaries posterior to the lamina are also supplied by pial vessels near the ON sheath and also receive contributions from the central retinal artery and central retinal vein. These large central vessels ultimately supply the inner retina.

(central retinal vein) the eye through the ON. The intraretinal portion of the ON is supplied by capillaries derived from the same central retinal vessels supplying the inner retina, although this can be quite variable in rats (Bernstein et al., 2003; Morrison et al., 1999). In distinction, choroidal circulation supplying outer retinal (photoreceptor) function and the perineural vascular arcade supplying the intrascleral portion of the ON (the perioptic nerve arteriolar anastomoses forming the "circle of Zinn–Haller") in both primates (humans) and rats are derived from the short posterior ciliary

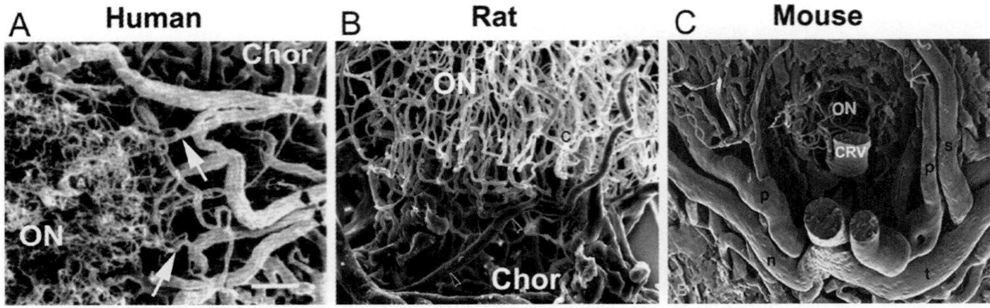

FIGURE 101.8   Comparison of scanning electron micrographs of optic nerve (ON) capillary supply in human (A), rat (B), and mouse (C). In humans, ON capillaries are largely derived from feeder branches (arrows) derived from the choroidal vasculature (Chor) surrounding the optic nerve (ON). Rats have a similar ON capillary structure largely derived from the choroid (Chor). In mice, the ON capillary structure is much sparser, with few capillaries draining into the central retinal vein (V). In figures A and C: A, artery. In figure B: c, capillaries. (From Bernstein, Johnson, & Miller, 2011.)

arteries (Hayreh, 1974; Morrison et al., 1999; Olver & McCartney, 1989) (see figure 101.8).

The vascularization of the posterior (postlaminar) portion of the ON in primates and rats is supplied by vessels originating from the pial plexus surrounding the ON and, possibly, recurrent arterioles emanating from the circle of Zinn–Haller as well. There also may be variations and anastomoses between the two circulatory supplies. In rats, the choroidal vascular ring surrounding the ON is usually complete (Bernstein et al., 2003), with capillaries perforating the intrascleral ON. Because of the relatively large diameter of the ON in rats compared with mice, the vascular supply of the rat ON is similar to that of primates in the relative abundance of ON capillaries (see figure 101.8). Mice, on the other hand, typically have an incomplete vascular ring surrounding the nerve at the level of the choroid (Goldenberg-Cohen et al., 2005; May & Lutjen-Drecoll, 2002). The lack of a complete ring may be an important contributor to the overall variability of the extent of ischemic neuropathy following AION induction. Mice also have a recurrent arterial supply derived from the retinal circulation that, due to its larger diameter, may be more resistant to photothrombotic ischemic induction. There is also a generally sparse vascular supply to the mouse ON, due to its small ON diameter. Primates also have more exuberant choroidal/choriocapillaris vasculature than rats or mice (Zhang, 1994).

Not only are the differences in ON structure and vasculature among rats, mice, and primates significant, their retinae vary greatly in the extent of intraretinal stereotopic organization. For example, primates have a fovea/foveolar complex for improved central visual acuity whereas rats and mice do not (Albrecht, 2008; Jeon, Strettoi, & Masland, 1998; Ogden, 1994). All of these differences must be considered when determin-

ing the ideal animal model for the study of ON disorders, particularly AION.

CNS stroke models utilizing laser-induced photothrombotic induction of CNS ischemia have been produced by a number of investigators (Frontczak-Baniewicz & Gajkowska, 2001; Watson et al., 1985; Zhao et al., 2002). This approach also has been used to produce retinal ischemia (Mosinger & Olney, 1989). Bernstein et al. (2003) induced NAION using Rose Bengal (RB), an iodinated fluorescein derivative that fluoresces brightly over a broad range of visible wavelengths, including argon-green laser light (514 nm) (Fluhler, Hurley, & Kochevar, 1989) and frequency-doubled neodymium-YAG (Fd-YAG) laser light (532 nm) (Rodgers, 1981).

Previous studies in CNS tissue and in vitro have determined that it is the generation of singlet oxygen, and not light-derived heat energy, that is responsible for vascular damage in the majority of these models (Lambert et al., 1996; Rodgers, 1981; Wilson & Hatchell, 1991). It also appears that it is damage to the capillary endothelium and not direct singlet-oxygen derived platelet activation that is the primary pathophysiological event (Inamo, Belougne, & Doutremepuich, 1996). RB is excreted hepatically, with first-order kinetics. Thus, circulating RB levels quickly drop after intravenous administration, similar to that seen after intravenous injection of fluorescein. Other singlet dyes have been used in modified AION models in which the intraorbital portion of the ON is exposed. These dyes have a slower rate of excretion than RB, enabling longer exposure times or time till induction (Duan et al., 2010); however, these approaches are limited by the need for microsurgical skill by the investigator and postoperative recovery by the experimental animal.

1452   NEIL R. MILLER

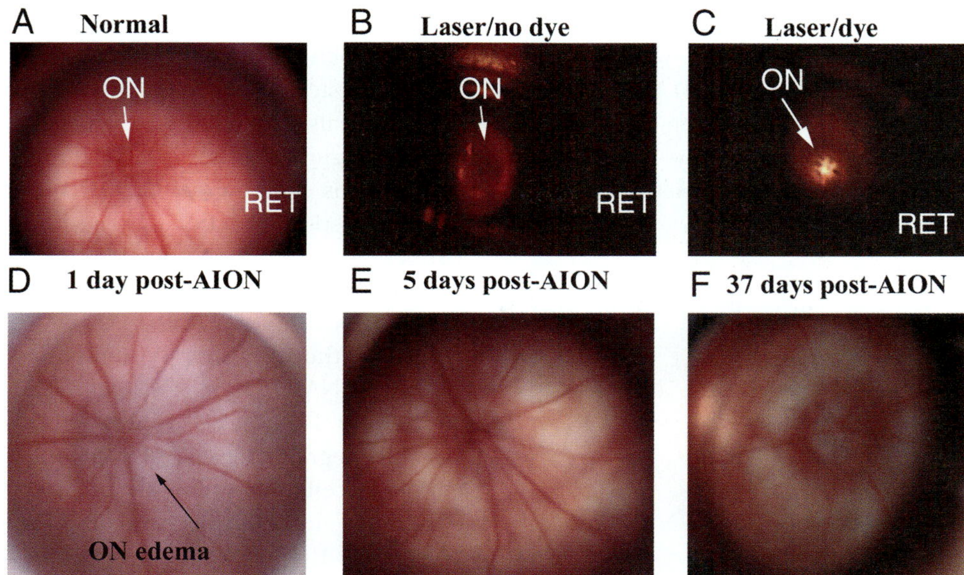

FIGURE 101.9   Rat anterior ischemic optic neuropathy (AION). Comparison of optic disk (ON) and retina (RET) before, during, and after induction. Slit-lamp biomicroscopic view (high magnification) seen with rat fundus contact lens. (A) Normal fundus. The ON is flat. Radial retinal vessels emerge from the ON to supply the inner retina. (B) Laser exposure without Rose Bengal (RB) dye administration. The ON is dark. (C) Laser exposure after RB dye administration. The central vessels glow with a golden color, indicating dye activation. (D) ON 1 day postinduction. ON edema is present, and there is obscuration of the disk margin. (E) ON appearance 5 days postinduction. ON edema has resolved. (F) ON appearance 37 days postinduction. The ON disk is pale. (From Bernstein Johnson, & Miller, 2011.)

A major problem in producing consistent laser treatment of small animals is the difficulty of focusing and fixing the rodent eye for stable treatments. This was resolved by Bernstein et al. (2003) who used a small fundus contact lens that fits on the eye. Once the lens is in place, these investigators injected intravenously RB and then used a standard Fd-YAG laser or argon ophthalmic laser coupled to a slit-lamp within 1 min following injection to induce rodent AION (rAION) (Bernstein et al., 2003, Bernstein et al., 2011; Goldenberg-Cohen et al., 2005). They determined the focal laser spot size by the size of the ON (about 500 microns in the rat), with the laser light centered on the ON. Laser-induced RB activation results in capillary thrombosis at the level of the anterior ON, with the larger central retinal vessels being relatively resistant to this treatment. Thus, the rAION model spares the inner retinal vasculature and selectively damages circulation to the ON. One day postinduction, the rat ON disk shows swelling typical of ON head ischemia (see figure 101.9).

In humans, NAION-associated optic disk swelling usually takes at least 2 weeks to resolve, whereas in rats, rAION-induced ON edema resolves more rapidly, usually within 5 days, being replaced by pallor of the optic disk, suggesting decreased function. The mouse model of AION shows similar changes (Goldenberg-Cohen et al., 2005). Because of the major differences in the anatomy and vascularization of primate versus murine retinas and ONs discussed above, a primate model of AION is preferable for the study of human responses to ON ischemia, and the Bernstein lab chose the rhesus monkey to do so.

Chen and colleagues (2008) generated a model of primate NAION (pAION) using a technique similar to that used to produce rAION. Following IV administration of 2.5mg/kg RB (identical with that used in rats and mice), the intraocular portion of the rhesus ON was irradiated with a 200 mW laser spot of 1.06 mm diameter positioned over the center of the disk, using a human pediatric (Glasser) laser fundus contact lens. The power was chosen because the 1.06-mm spot diameter is approximately 4 times larger in surface area than a 500-micron laser spot; thus, the fluence (energy distribution across the area of irradiation) decreases by fourfold if the power is not increased. Initial laser exposure time is 7–10 s, depending on the age of the animal and the severity of pAION to be produced.

In this model, 1 day postinduction, the induced eyes showed a relative afferent pupillary defect indicative of ON dysfunction, and there is a variable amount of optic disk swelling, often with peripapillary retinal hemorrhage, similar to that seen in human NAION (see figure 101.10).

Intravenous fluorescein angiography performed in animals following induction of pAION reveals breakdown of the blood–retinal barrier (BRB), with fluorescein leakage at the disk. Reconstitution of the BRB occurs by 14 days postinduction. These data suggest that, like the breakdown of the BBB seen following

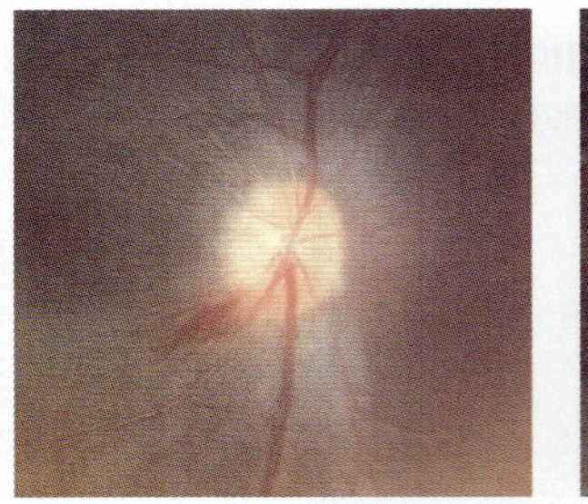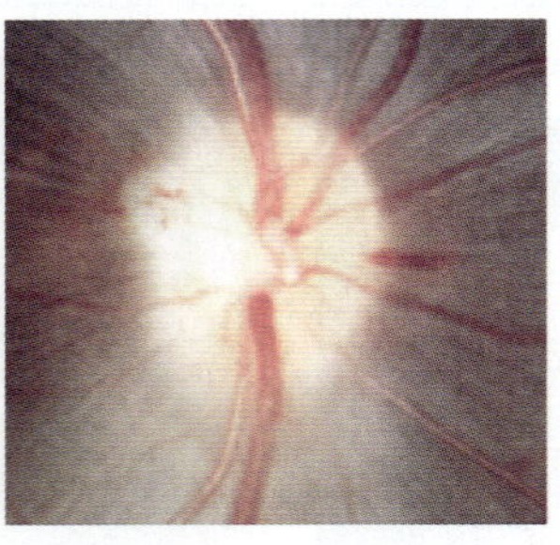

FIGURE 101.10   Primate anterior ischemic optic neuropathy (AION). Two examples showing swelling of the optic disks associated with blurring of the disk margins and flame-shaped peripapillary retinal hemorrhages.

human CNS infarcts, a considerable time period ensues during which the affected ON is exposed to the systemic immune system. High-speed indocyanine green angiography reveals that in the early stages of pAION, there can be regional restriction of central venous return, with venous sludging at the ON head, which resolves by 1 week postinduction. These findings support the concept that acute ON edema results in intraretinal venous restriction, venous dilatation, and intraretinal vascular extravasation but that this restriction rapidly resolves without causing permanent retinal damage. As in human NAION, the disk swelling of pAION resolves over about 1–2 weeks, eventually being replaced by a variable degree of optic disk pallor, with evidence of ganglion cell loss and optic atrophy by both optical coherence tomography and histological assessment (see figure 101.11). At the same time, electrophysiological testing (visual evoked potentials [VEP], pattern electroretinography [PERG], full-field electroretinography [ffERG]) reveals progressive reduction in VEP

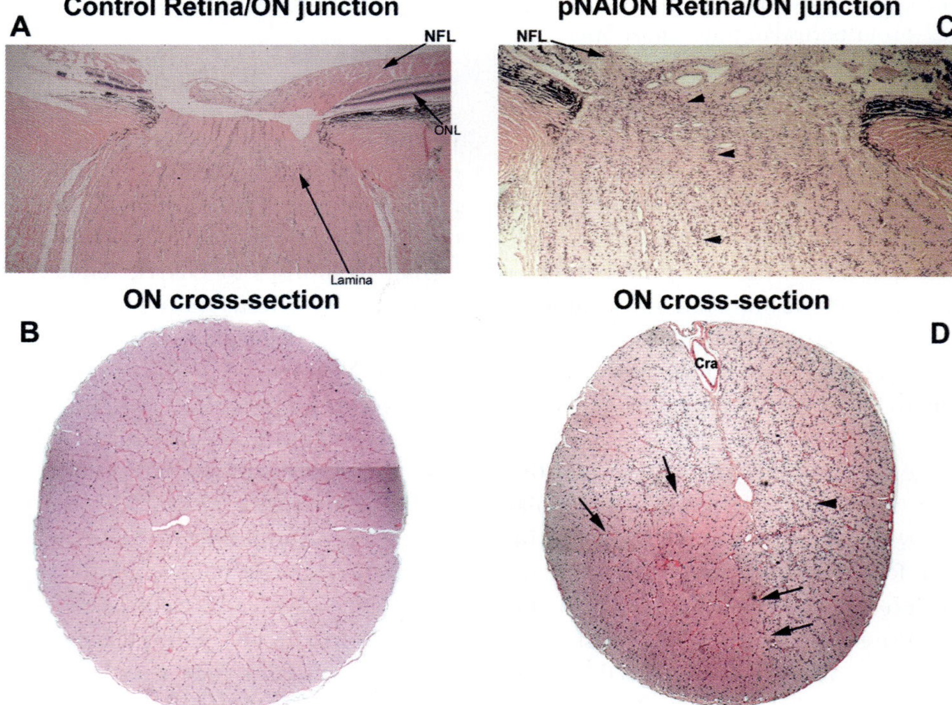

FIGURE 101.11   Primate nonarteritic anterior ischemic optic neuropathy (pNAION) associated histological changes in the rhesus macaque lamina and optic nerve (ON). (A) Normal (control) retina: ON junction. The laminar region appears slightly pale (long arrow); the nerve fiber layer (NFL) is thick. There is a regular columnar organization of the retinal ganglion cell axon bundles from the lamina through the distal ON regions. (B) ON cross-section. The axon bundles are regular, and there is a suggestion of the septae surrounding the axons. (C) Same regions in a rhesus monkey 75 days postinduction of pAION. The NFL is thinned (panel C, arrow). There is disruption of the normal columnar structures, with increased cellularity. Columnar organization is regionally maintained (arrowheads). (D) 75 days pNAION-induced, ON cross-section. There is a collapse of the normal septal bundles in many areas, and increased cellularity (arrowhead). There is a normal appearing region, with reduced cellularity and preservation of the normal septated bundles (arrows). Cra, central retinal artery. ONL, outer nuclear layer. (From Bernstein, Johnson, & Miller, 2011.)

amplitudes, delayed reduction in PERG amplitudes, and normal results of ffERG, consistent with primary axonal damage with eventual loss of RGC function.

Because inflammation can both initiate and exacerbate CNS infarct-related conditions (Nakase et al., 2008; Pineau et al., 2010), questions regarding NAION-associated inflammation are important from the point of view of both initiation and treatment. Postinfarct inflammation is a well-defined phenomenon in other regions of the CNS, and both the murine and primate models of AION developed by the Bernstein laboratory show definite evidence of an inflammatory response that mirrors the time course of optic disk swelling and resolution (Bernstein, Johnson, & Miller, 2011; Salgado et al., 2011; Zhang et al., 2009) (see figures 101.12 and 101.13).

In addition, there is a single histological report of ON findings 28 days post-NAION onset (Tesser, Niendorf, & Levin, 2003), in which it was suggested that there was no obvious inflammation; however, this study did not utilize specialized antibodies to identify immune cells. In fact, the authors of the single human clinical study were kind enough to provide the Bernstein laboratory with unstained sections from the affected ON, which, when stained for inflammatory cells, did, in fact, show evidence of acute inflammation in the region of the infarct as well as in the surrounding regions (Salgado et al., 2011). The Bernstein laboratory currently is using both their murine and primate models of AION to assess the relative importance of acute inflammation in NAION and the ability of certain substances such as prostaglandin (PGJ2) to decrease ON damage and improve visual function. The relative ease of the application of these and other models of AION, combined with the ability to evaluate responses to acute isolated ON ischemia without direct retinal compromise,

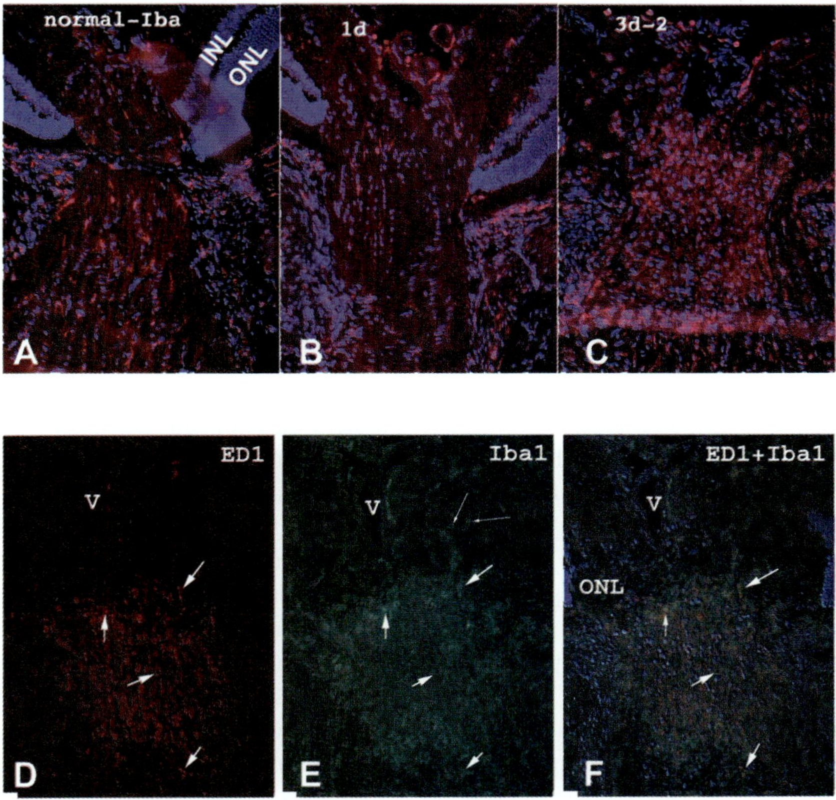

Figure 101.12 Early appearance of inflammatory cells in rat optic nerve (ON) following rodent anterior ischemic optic neuropathy (rAION) induction. (A–C) IBA1 (ionized calcium channel protein) immunostaining. (A) Control retina/optic nerve. IBA1+ microglia are scattered randomly throughout the ON, with occasional cells in the retina. (B) One day postinduction. IBA-1+ cells are apparent in the retinal vessels, as well as the choroid, surrounding the area of the primary infarct. (C) Three days postinduction. There is infiltration of IBA-1+ cells in the ON region of the infarct. (D–F) Early invasion and localization of extrinsic macrophages following rAION. (D) ED1 immunoreactivity. ED1+ cells are detectable in the center of the rAION lesion at 3 days postinduction. (E) IBA1 immunoreactivity. (F) ED1/IBA1 colocalization. The center of the rAION lesion is infiltrated by extrinsic (blood-borne) macrophages early postinduction. INL, inner nuclear later; ONL, outer nuclear layer; v, vein. (From Bernstein, Johnson, & Miller, 2011.)

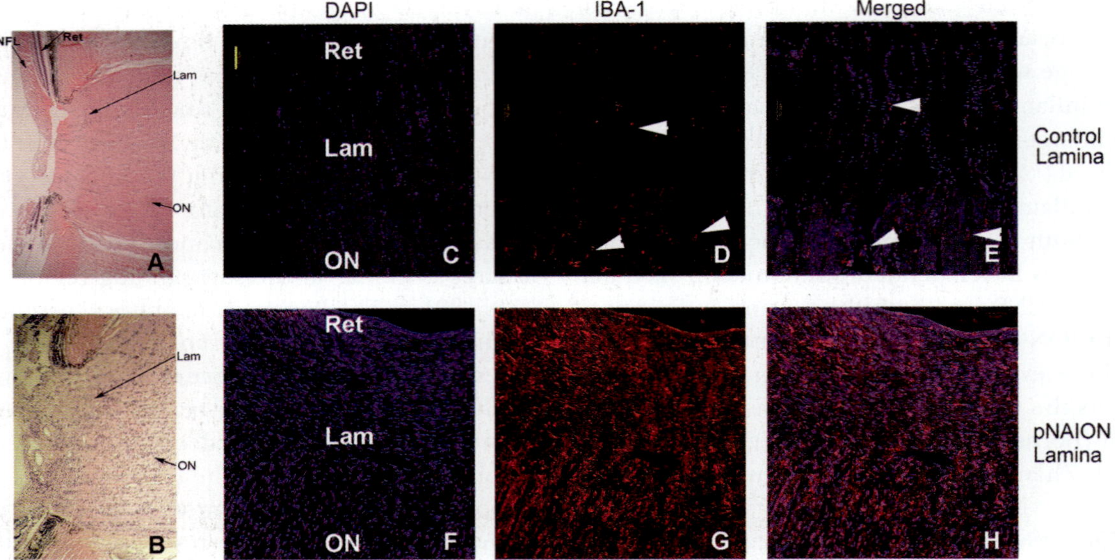

FIGURE 101.13   Post primate nonarteritic anterior ischemic optic neuropathy (pNAION) changes in inflammatory cell infiltration in the laminar region. (A and B) H&E (hematoxylin and eosin) stained cross-sections of nonhuman primate laminar regions. (A) Control lamina. The retina is visible to the left, with the optic nerve on the right. There is a columnar organization in the normal lamina, with cellularity distributed between the eosinophilic columns. (B) Lamina of a pNAION-induced eye (75 days postinduction). There is disruption of the normal columnar organization, and increased cellularity. (C–E) Confocal analysis of inflammatory cells in the normal lamina. (C) DAPI (4',6-diamidino-2-phenylindole) staining. Normal columnar structure of nuclei. (D) Inflammatory (IBA1) cells. There are a few scattered IBA1+ cells across the lamina. (E) Merged image. (F–H) Confocal analysis of inflammatory cells in the laminar region of a pNAION-induced (75d) eye. There is disruption of the normal columnar structure, with infiltration of many IBA1+ cells across the lamina. NFL, nerve fiber layer; Ret, retina; Lam, lamina; ON, optic nerve. Scale bar in panel C: 50 microns. (From Bernstein, Johnson, & Miller, 2011.)

improves the understanding of the mechanisms responsible for human NAION-associated dysfunction. Although differences remain among all animal models of human disease, even nonhuman primate models, and the clinical disorder they mimic, these models nevertheless can provide valuable insight into the physiological mechanisms associated with the response to isolated, sudden ON ischemia.

LEBER HEREDITARY OPTIC NEUROPATHY

LHON is a maternally inherited optic neuropathy that occurs primarily but not exclusively in men. The onset of visual loss typically occurs between the ages of 15 and 35 years, but otherwise classic LHON has been reported in many individuals both younger and older (Koilkonda & Guy, 2011; Newman, 2005), with a range in molecularly confirmed cases of 2–80 years. This age variability occurs even among members of the same pedigree. Visual loss in patients with LHON typically is painless. In about 50% of cases, both eyes are affected simultaneously. In the other 50%, one eye is affected days, weeks, months, or even years before the second eye. The course is acute or subacute, with deterioration of visual

function stabilizing within 3–6 months. The vision in most patients with LHON deteriorates to worse than 20/200, associated with severe color vision loss and central or cecocentral scotomas. The optic disks in the acute phase of the disorder may appear completely normal or there may be a characteristic, albeit often subtle, triad of circumpapillary telangiectatic microangiopathy, swelling of the nerve fiber layer around the disk (pseudoedema), and absence of leakage from the disk or papillary region on fluorescein angiography (distinguishing the LHON disk from a disk that is truly swollen) (see figure 101.14A and B). As the disease progresses, the telangiectatic vessels disappear, the pseudoedema of the disk resolves, and, eventually, optic disk pallor develops, with nerve fiber layer dropout that usually is most pronounced in the papillomacular bundle (see figure 101.14C and D).

Most patients who develop LHON have permanent bilateral visual loss; however, in some cases, recovery of excellent central vision occurs years after visual deterioration (Newman, 2005). The recovery may occur gradually over 6 months to 1 year after initial visual loss or may suddenly occur up to 10 years after onset (Brunette & Bernier, 1969). It usually is characterized by a breakup

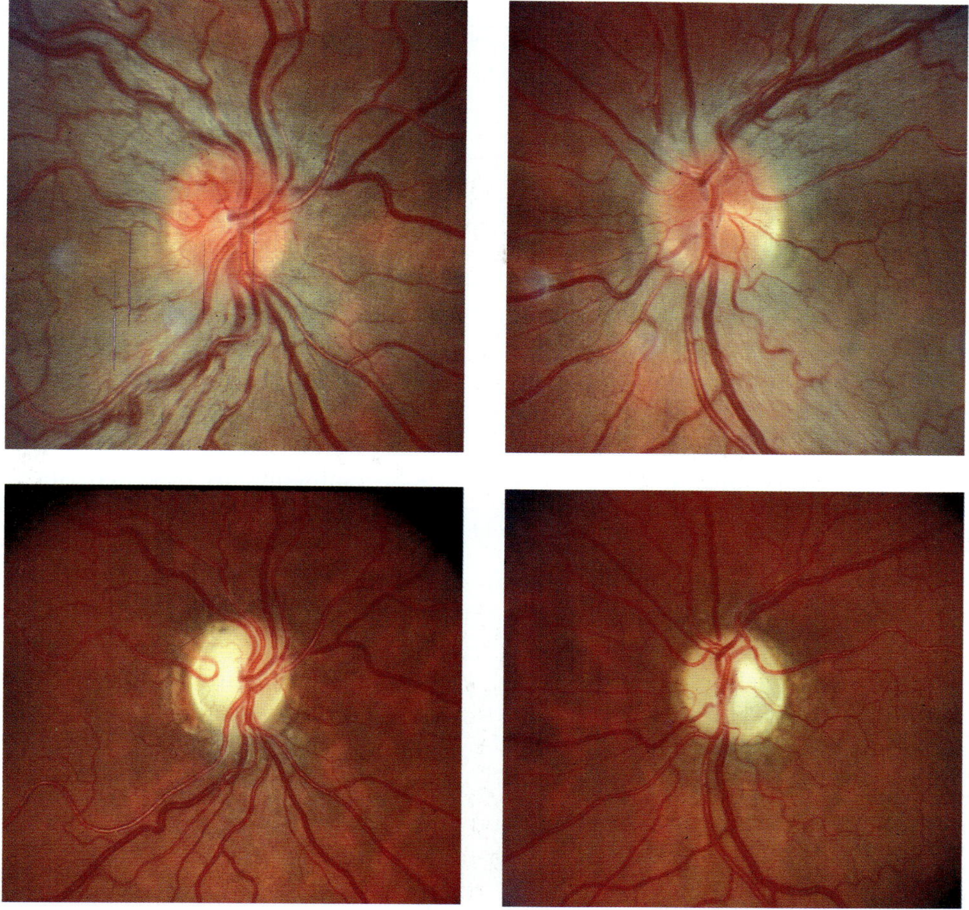

FIGURE 101.14   Leber hereditary optic neuropathy in the acute (upper left and right) and late (lower left and right) phases. Upper left and right: Note hyperemic optic disks with faint telangiectatic vessels on the disk surface and in the peripapillary region. Lower left and right: The optic disks are diffusely pale. Loss of nerve fibers is most pronounced in the papillomacular bundle.

of the central scotoma, such that small islands of clear vision develop within the larger scotoma. Recovery usually is bilateral but may be unilateral. Patients whose vision improves most substantially appear to have a lower mean age at the time of initial visual loss (Riordan-Eva et al., 1995). Furthermore, the particular mitochondrial DNA (mtDNA) mutation also influences prognosis, with the 11778 mutation carrying the worst prognosis for vision and the 14484 mutation the best.

Theories on the pathogenesis of LHON usually begin with oxidative phosphorylation and the generation of adenosine triphosphate (ATP). The three primary LHON mutations are found in genes that code for protein subunits of complex I of the respiratory chain. As complex I is the site of NADH reoxidation, it may be that maintenance of the proper NAD/NADH redox balance is crucial for normal maintenance and function of RGCs or their axons, especially for axonal transport (Howell, 1998; Levin & Schalow, 1998). Transporter defects involving channels and pumps particularly reliant on mitochondrial ATP production could have specific tissue effects (Schon, Bonilla, & DiMauro, 1997). Others propose that free radical damage is crucial in LHON pathogenesis (Kirkinezos & Moraes, 2001; Qi et al., 2003a; Schapira, 1997). Inhibition of the respiratory chain causes increased generation of free radicals, and the respiratory chain in turn may be inhibited by oxidative damage. This self-amplifying cycle of oxidative damage and respiratory chain dysfunction could cause direct or indirect damage to vulnerable tissues such as the RGC or its axon. The different mtDNA mutations may cause different amounts of free radical production at the level of complex I, and this increased production of reactive oxygen species may occur particularly in a neuronal cell environment (Qi et al., 2003b).

Even less clear are the subsequent mechanisms by which respiratory chain dysfunction or free radical production result in generally irreversible damage to the RGCs and their axons. Mitochondria play a major role in the induction of apoptosis (Danielson et al., 2002). The collapse of the mitochondrial transmembrane potential and opening of the so-called permeability transition pore leads to the release of proapoptotic proteins, including cytochrome c, and resultant cell death (Ghelli et al., 2003; Schapira, 1997). Many physiologic and pathologic stimuli can directly trigger opening of the mitochondrial transition pore. It is also probable that multiple consequences of mitochondrial dysfunction, including collapse of the transmembrane potential, uncoupling of the respiratory chain, hyperproduction of superoxide anions, disruption of mitochondrial biogenesis, outflow of matrix calcium and glutathione, and release of soluble intermembrane proteins, can indirectly activate apoptosis or necrosis (Zamzami et al., 1997). Alternatively, excitotoxic death can be initiated directly by reduced mitochondrial energy production, which leads to activation of the NMDA-type glutamate receptors (Schapira, 1997). Additionally, free radicals are a key component in some excitotoxicity pathways and may also trigger apoptotic cell death directly (Newman, 2005).

Various mechanisms have been proposed on how any of these theorized pathophysiologic processes result in selective damage to the ON, but supportive evidence is scarce. Based on the abnormal appearance of the papillary and peripapillary vasculature in many LHON patients, a primary vascular process with subsequent ischemia has been proposed (Nikoskelainen et al., 1996); however, these vascular changes are likely secondary, especially given the intact vascular endothelium as demonstrated by lack of fluorescein leakage, and the absence of these fundus findings in many patients. A mechanical mechanism also has been suggested in which the severe angle turn that the ON axons need to make in the prelaminar area results in a "chokepoint" region where there is impaired or labile axoplasmic transport (Howell, 1998). Small-caliber, constantly firing P-cells may be particularly vulnerable (Sadun et al., 2000). Histochemical studies of the ON in animals have shown a high degree of mitochondrial respiratory activity within this unmyelinated, prelaminar portion of the ON, suggesting a particularly high requirement for mitochondrial function in this region (Andrews et al., 1999; Bristow et al., 2002; Lessell & Horovitz, 1972). The RGCs and their axons appear particularly vulnerable to defects in mitochondrial function (Carelli, Ross-Cisneros, & Sadun, 2004; Newman, 2002). The results of treatment of LHON generally have been disappointing. Systemic steroids, hydroxycobalamin, cyanide antagonists, and topical brimonidine tartrate—an alpha-2 agonist with anti-apoptotic properties—all have proved ineffective (Newman, 2002; Newman et al., 2005). Idebenone, a quinol that has significant effects on mitochondrial bioenergetics including the stimulation of ATP formation (Giorgio et al., 2012), appears to result in an improved visual outcome in some patients, but its overall efficacy is less than impressive (Carelli et al., 2011; Klopstock et al., 2011; Newman, 2011).

Until recently, the treatment of LHON was hampered by the lack of a bona fide animal model of the disorder for LHON; however, this is no longer the case. The first animal model for LHON was made by Zhang, Jones, and Gonzalez-Lima (2002), who administered rotenone, an irreversible complex I inhibitor, to mice. Histological analysis showed 43% thinning of the RGC layer by 1 day after the rotenone injections. Qi et al. (2003b) subsequently designed ribozymes to degrade the messenger RNA (mRNA) encoding a critical nuclear-encoded subunit gene of complex I (NDUFA1), resulting in markedly reduced complex I activity in murine cells. Using an adeno-associated viral (AAV) vector to deliver the ribozymes into the mouse vitreous cavity, the authors found loss of RGCs and axons that produced an optic neuropathy, the histopathology of which resembled that of LHON (Qi et al., 2003c) (see figure 101.15).

This model system also implicated oxidative stress in the pathogenesis of the degenerative process. These investigators then performed intraocular injections of AAV-expressing ribozymes designed to degrade mitochondrial superoxide dismutase (*SOD2*) mRNA and found that the injections induced further loss of axons and myelin in the ON and RGCs (Qi et al., 2003b). When the investigators injected an AAV that overexpressed SOD2 into eyes that also had received the NDUFA1 ribozymes, they found that RGC and axonal loss were substantially decreased (Qi et al., 2004). Qi et al. (2007a) subsequently attempted to rescue cybrid cells with the G11778A mutation in mtDNA from galactose-induced apoptotic cell death by infecting them with AAV-*SOD2*. The control cells were treated with AAV-GFP (green fluorescent protein). Within 3 days of growth in galactose media, LHON cell survival increased by 89%. These findings suggested that antioxidant genes may offer a therapeutic strategy directed at the pathophysiologic mechanisms of LHON. Still, it remained unclear whether or not the findings in these mouse models are representative of the pathogenic processes that occur in patients with LHON.

To generate a more representative animal model of LHON, Qi et al. (2007b) constructed a mutant ND4 subunit gene that was designed to express the arginine-to-histidine substitution at amino acid 340 characteristic of the mutant human LHON ND4 protein. Delivery of this construct with the AAV vector injected into the

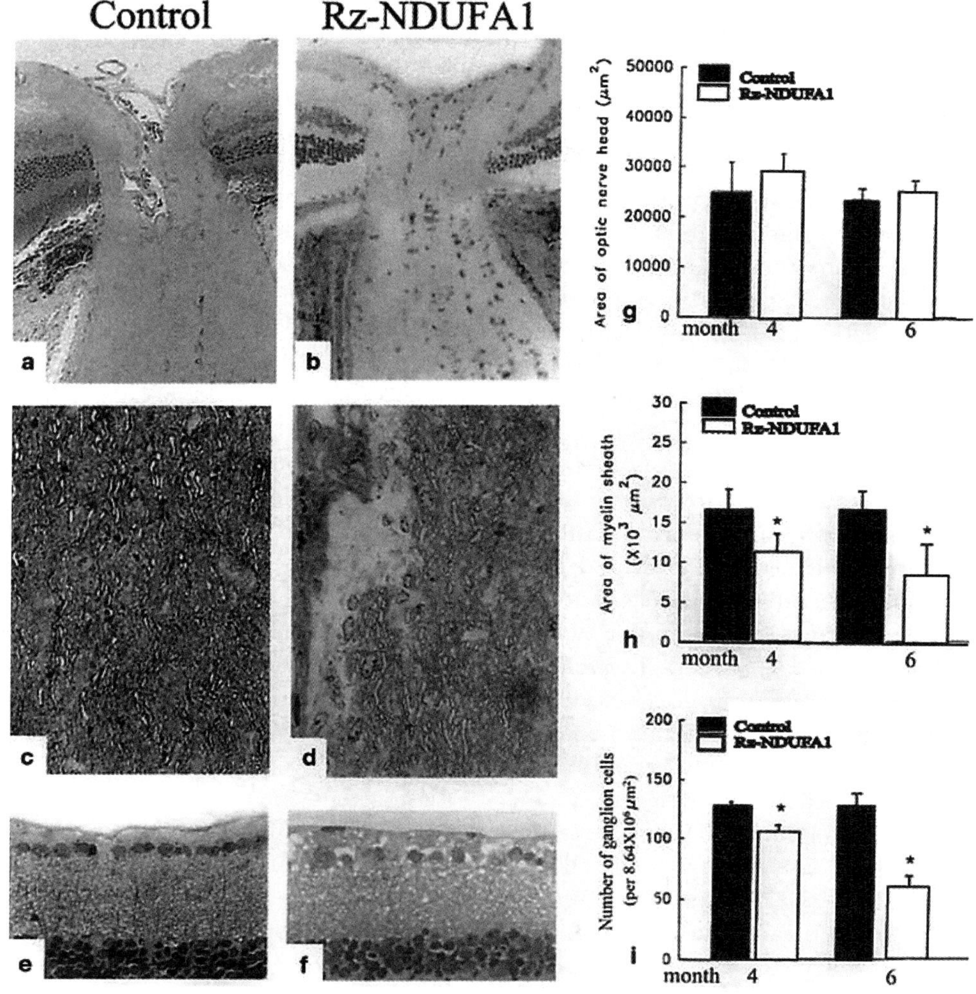

FIGURE 101.15   Light microscopy of the effects of recombinant adeno-associated virus (rAAV) ribozymes in the mouse eye. Relative to mock-infected eyes (a, c, e), some eyes infected with Rz-NDUFA1 (b, d, f) exhibit mild optic disk edema with filling of the central cup and lateral displacement of the peripapillary retina (b) 4 months after AAV infection. Quantitative analysis showed that the optic nerve head area was 14% greater for ribozyme-infected eyes, but this was not significantly different from mock-infected eyes (g). In contrast, the retrobulbar optic nerves of Rz-NDUFA1-infected eyes were demyelinated with loss of toluidine blue staining (c, d) by 33% at 4 months and nearly 50% at 6 months (h). Relative to mock-infected eyes, retinal ganglion cell numbers in Rz-NDUFA1-infected eyes showed an 18% loss at 4 months (e, f) and 47% at 6 months (both $p < 0.01$) (i). Asterisks indicate statistical significance ($p < 0.01$). (From Qi et al., 2003c.)

mouse vitreous cavity resulted in optic disk swelling. Several months later, the ON became atrophic, RGCs were lost, and ultrastructural analysis revealed disruption of mitochondrial cytoarchitecture. Mouse eyes injected with AAV containing the normal human ND4 showed no evidence of pathology. As the mutant and human ND4 constructs differed only in the arginine-to-histidine transition at amino acid 340 (mutant ND4), the results of these studies indicate that pathogenicity of this mutation is the cause of LHON. Ellouze et al. (2008) subsequently introduced the mutant human ND4 gene into rat eyes and found that it caused RGC degeneration and a decline in visual performance in these animals; however, an important difference between human LHON and these rodent models is that disease in rodents occurred even in the presence of endogenous mouse (or rat) ND4. In human LHON, wild-type ND4 usually is absent. That being the case, rescue of the LHON rodent model by the addition of more wild-type ND4 may not be possible.

One of the most promising emerging technologies that may result in successful treatment of patients with LHON is "allotopic expression," wherein a nuclear version of the mitochondrial gene is constructed by partially recoding the mtDNA gene in the nuclear genetic code. It was through allotopic expression of the mutant human ND4 subunit gene that an LHON-like phenotype was induced in rodent models as discussed in the previous paragraph. Changing the ATA codon to ATG is necessary to achieve allotopic expression, as the ATA encodes for methionine in mitochondria but isoleucine in the nucleus. In addition, the TGA codon that specifies tryptophan in mitochondria is a stop codon in the nucleus. Therefore, this codon also must be corrected for the full-length ND4 protein to be translated on cytoplasmic ribosomes. Protein import into mitochondria is then directed by the addition of a mitochondrial targeting sequence (MTS) to the amino terminus (Guy et al., 2002; Larsson, 2002). Protein expression can then be monitored with an epitope tag appended to the carboxy terminus. Guy et al. (2002) were the first to use this approach with a human ND4 gene to rescue the defects of oxidative phosphorylation in G11778A LHON cells. They constructed a synthetic ND4 subunit from overlapping 80-mer oligonucleotides, packaged it in an AAV vector, and used it to transduce cells harboring 100% G11778A-mutated mtDNA. One of the constructs successfully increased ATP synthesis by threefold in LHON cell lines relative to controls treated with GFP or transduced with the same ND4 gene that had a different MTS and epitope tag that was not imported into the mitochondria.

Guy et al. (2009) also have demonstrated that allotopic delivery of the normal human ND4 subunit gene into the vitreous cavity of the murine eye is safe. There was no difference in total RGC counts, measured as Thy1.2 positive cells, between these experimental eyes and controls injected with AAV-GFP. Moreover, the pattern and flash ERG amplitudes after the injections remained unchanged from their baseline values before the injections. This important finding indicates that injection of the allotopic human ND4 did not compromise murine RGC function. Using immunoprecipitation of the 45 subunit complex I, these investigators demonstrated that the FLAG-tagged human ND4 incorporated into the holoenzyme of infected murine retinal and ON tissues. To validate the technique, they submitted nine bands pulled down by complex I immunoprecipitation of murine mitochondria isolated from the ON, brain, spinal cord, or retina for identification by mass spectroscopy. They were positively identified as subunits of the NADH-ubiquinone oxidoreductase. No other respiratory complexes (II–V) were detected. Therefore, the FLAG-tagged human ND4 detected by this assay proves that it effectively integrated into the murine complex I.

In 2002, Guy et al. showed that LHON cells expressing the same allotopic human ND4, but fused to a different MTS (aldehyde dehydrogenase) or epitope tag (GFP), did import into the mitochondria. Consistent with this finding, the latter construct did not rescue

LHON cells from glucose-free galactose media-induced cell death or improve their ATP synthesis.

In support of a paramount role for the MTS in directing mitochondrial trafficking, Supekova et al. (2010) showed that the allotopic import of a mutant COX2 was dependent on the MTS, but not the mitochondrial targeting 3′UTR. Nevertheless, the studies of Bonnet et al. (2007) clearly demonstrated the benefits of the *COX10* 3′UTR when used in conjunction with the CIS acting elements of the *COX10* MTS. Relative to controls, the *COX10-ND4* or *COX10-ND4* 3′UTR constructs each increased G11778A LHON cell survival in galactose media and improved their ATP synthesis. Their findings support the earlier studies of Guy and coworkers' successful allotopic ND4 import into mitochondria (Guy et al., 2002). Clearly, the testing of constructs for allotopic expression that include the MTS, protein, epitope tag, or 3′UTR is largely a trial-and-error endeavor (Shokolenko et al., 2010). As cell culture studies can sometimes be misleading (Perales-Clemente et al., 2011), confirmation in appropriate animal models is vital to demonstrating the safety and effectiveness of allotopic ND4 expression before it can be applied to LHON patients.

Using their MTS and 3′UTR model system, Ellouze et al. (2008) introduced the mutant human ND4 subunit gene harboring the G11778A mutation into rat eyes, by in vivo electroporation. This led to loss of vision and degeneration of almost half the RGCs, as previously described by Qi et al. (2007b); however, when Ellouze et al. (2008) introduced a normal copy of the human ND4 gene, visual and RGC loss did not occur, at least over a period of 75 days when some of the animals were sacrificed. Thus, the data accumulated to date provide overwhelming evidence that allotopic expression of a mutant ND4 causes RGC degeneration and a wild-type version does not. More importantly, they show that the wild-type ND4 can rescue an LHON animal model, at least in the short term. Clearly, allotopic delivery of a normal ND4 is a promising approach in the quest for an effective remedy for LHON caused by mutated G11778A mtDNA. For this to occur, an effective and safe delivery system is necessary for ND4 gene therapy. In this regard, the single-stranded (ss) AAV2 used in allotopic mouse experiments has proven safe in several phase I human ocular gene therapy trials (Hauswirth et al., 2008), and, thus, the retinal layer exclusively affected in LHON can be targeted by optimizing this vector and by choosing intravitreal injection as the route of administration (Koilkonda & Guy, 2011).

Newer generation vectors include the self-complementary (sc) AAV that contains both positive and negative complementary strands. As second-strand

  NEIL R. MILLER

synthesis is believed to be the rate-limiting step for expression of single-stranded vectors, it is not surprising that scAAV vectors increase the speed and efficiency of transgene expression (McCarty, 2008). Other AAVs with mutations in the capsid proteins also increase the efficiency of transgene expression (Zhong et al., 2008a, 2008b). They were designed to reduce cellular degradation of AAV, thus increasing cellular levels of AAV virions. By taking advantage of scAAV to deliver the allotopic ND4 into the mouse eye, Koilkonda et al. (2010) doubled RGC expression relative to the single-stranded AAV that is the current standard vehicle for gene delivery. With scAAV, FLAG-tagged ND4 was seen in almost all murine RGCs (90%). Such newer generation vectors may be highly advantageous for LHON gene therapy. They can be used at lower doses, thus minimizing immunologic responses against the viral capsid that could prevent expression of ND4, without sacrificing efficiency (Zaiss & Muruve, 2008). This could have important implications for treatment of LHON patients, in whom a prior injection into the first eye, should it generate an immune response, might limit expression with later injections of AAV-ND4 into the second eye.

As discussed in the section on NAION above, great care must be taken in extrapolating the results achieved in rodents to the human disorder, particularly under pathological conditions. For example, the studies of Oca-Cossio et al. (2003) suggest that if ND4 is not correctly processed into mitochondria, it may be harmful. In addition, if LHON is primarily an axonopathy, the allotopic ND4 may not get to the target tissue (i.e., the mitochondria in the axons of the ON) for rescue in patients with acute LHON. On the other hand, if LHON is primarily a disorder of RGCs, then rescue may indeed be possible. Further studies in lower vertebrates are needed to delineate the best window for intervention.

Many other experimental techniques have been proposed to address disorders caused by mutated mtDNA. One particular approach that might be applicable to LHON is that of trans-kingdom allotopic expression, also called "xenotopic expression." This technique was pioneered by Ojaimi et al. (2002) who experimentally restored defects in complex V of the electron transport chain. Xenotopic technology has the advantage whereby a single construct, NDI1, can treat *all* LHON cases caused by mutated ND4, ND1, or ND6 complex I subunits. Allotopic expression of human complex I subunits requires three separate constructs, one for each of the three mutated subunit genes. It remains unclear, however, whether introduction of a gene from an entirely different species is acceptable for human therapy. Moreover, it has not been demonstrated how NDI1 interacts with complexes II–V of the human respiratory chain. A recent publication showing the extension of the fly life span with the NDI1 gene suggests that the mechanism of benefit was not achieved by improving oxidative phosphorylation but rather by decreased production of ROS (Bahadorani et al., 2010). Nevertheless, this technique shows promise. In the meantime, based on the work of Guy and others, plans are being made to treat patients with long-standing (and, ultimately, acute) LHON with gene therapy (Lam et al., 2010).

REFERENCES

Albrecht, M. C. (2008). Comparative anatomy of the optic nerve head and inner retina in non-primate animal models used for glaucoma research. *Open Ophthalmology Journal, 2,* 94–101.

Andrews, R. M., Griffiths, P. G., Johnson, M. A., & Turnbull, D. M. (1999). Histochemical localization of mitochondrial enzyme activity in human optic nerve and retina. *British Journal of Ophthalmology, 83,* 231–235.

Arnold, A. C. (2005). Ischemic optic neuropathy. In N. R. Miller, N. J. Newman, V. Biousse, & J. B. Kerrison (Eds.), *Walsh and Hoyt's clinical neuro-ophthalmology* (6th ed., Vol. 1, pp. 349–384). Baltimore: Lippincott-Williams & Wilkins.

Bahadorani, S., Cho, J., Lo, T., Contreras, H., Lawal, H. O., Krantz, D. E., et al. (2010). Neuronal expression of a single-subunit yeast NADH-ubiquinone oxidoreductase (Ndi1) extends *Drosophila* lifespan. *Aging Cell, 9,* 191–202.

Bernstein, S. L., Guo, Y., Kelman, S. E., Flower, R. W., & Johnson, M. A. (2003). Functional and cellular responses in a novel rodent model of anterior ischemic optic neuropathy. *Investigative Ophthalmology & Visual Science, 44,* 4153–4162. doi:10.1167/iovs.03-0274.

Bernstein, S. L., Johnson, M. A., & Miller, N. R. (2011). Non-arteritic anterior; ischemic optic neuropathy (NAION) and its experimental models. *Progress in Retinal and Eye Research, 30,* 167–187.

Bonnet, C., Kaltimbacher, V., Ellouze, S., Augustin, S., Bénit, P., Forster, V., et al. (2007). Allotopic mRNA localization to the mitochondrial surface rescues respiratory chain defects in fibroblasts harboring mitochondrial DNA mutations affecting complex I or V subunits. *Rejuvenation Research, 10,* 127–143.

Bosley, T. M., Abu-Amero, K. K., & Ozand, P. T. (2004). Mitochondrial DNA nucleotide changes in non-arteritic ischemic optic neuropathy. *Neurology, 63,* 1305–1308.

Brazis, P. W. (2009). Surgery for idiopathic intracranial hypertension. *Journal of Neuro-Ophthalmology, 29,* 271–274.

Bristow, E. A., Griffiths, P. G., Andrews, R. M., Johnson, M. A., & Turnbull, D. M. (2002). The distribution of mitochondrial activity in relation to optic nerve structure. *Archives of Ophthalmology, 120,* 791–796.

Brunette, J. R., & Bernier, G. (1969). Study of a family of Leber's optic atrophy with recuperation. In J. R. Brunette & A. Barbeau (Eds.), *Progress in neuro-ophthalmology* (pp. 91–97). Amsterdam: Excerpta Medica.

Carelli, V., La Morgia, C., Valentino, M. L., Rizzo, G., Carbonelli, M., De Negri, A. M., et al. (2011). Idebenone treatment in Leber's hereditary optic neuropathy. *Brain, 134,* 1–5. doi:10.1093/brain/awr180.

Carelli, V., Ross-Cisneros, F. N., & Sadun, A. A. (2004). Mitochondrial dysfunction as a cause of optic neuropathies. *Progress in Retinal and Eye Research, 23*, 53–89. doi:10.1016/j.preteyeres.2003.10.003.

Chen, C. S., Johnson, M. A., Flower, R. A., Slater, B. J., Miller, N. R., & Bernstein, S. L. (2008). A primate model of nonarteritic anterior ischemic optic neuropathy (pNAION). *Investigative Ophthalmology & Visual Science, 49*, 2985–2992. doi:10.1167/iovs.07-1651.

Collignon-Robe, N. J., Feke, G. T., & Rizzo, J. F. (2004). Optic nerve head circulation in nonarteritic anterior ischemic optic neuropathy and optic neuritis. *Ophthalmology, 111*, 1663–1672. doi:10.1016/j.ophtha.2004.05.020.

Danielson, S. R., Wong, A., Carelli, V., Martinuzzi, A., Schapira, A. H., & Cortopassi, G. A. (2002). Cells bearing mutations causing Leber's hereditary optic neuropathy are sensitized to fas-induced apoptosis. *Journal of Biological Chemistry, 277*, 5810–5815. doi:10.1074/jbc.M110119200.

Duan, Y., Thurston, J., Watson, B., Hernandez, E., Fern, R., & Goldberg, J. L. (2010). Retinal ganglion cell survival and optic nerve glial response after rat optic nerve ischemic injury. *Investigative Ophthalmology & Visual Science, 48*, 644.

Ellouze, S., Augustin, S., Bouaita, A., Bonnet, C., Simonutti, M., Forster, V., et al. (2008). Optimized allotropic expression of the human mitochondrial ND4 prevents blindness in a rat model of mitochondrial dysfunction. *American Journal of Human Genetics, 83*, 373–387.

Feldon, S. (2007). Visual outcomes comparing surgical techniques for management of severe idiopathic intracranial hypertension. *Neurosurgical Focus, 23*, 1–7. doi:10.3171/foc.2007.23.5.7.

Fluhler, E. N., Hurley, J. K., & Kochevar, I. E. (1989). Laser intensity and wavelength dependence of Rose Bengal-photosensitized inhibition of red blood cell acetylcholinesterase. *Biochimica et Biophysica Acta, 990*, 269–275.

Friedman, D. I., & Jacobson, D. M. (2004). Idiopathic intracranial hypertension. *Journal of Neuro-Ophthalmology, 24*, 138–145.

Frontczak-Baniewicz, M., & Gajkowska, B. (2001). Focal ischemia in the cerebral cortex has an effect on the neurohypophysis: II. Angiogenesis in the neurohypophysis is a consequence of the focal ischemia in the cerebral cortex. *Neurology Endocrinology Letters, 22*, 87–92.

Fukuda, Y., Watanabe, M., Wakakuwa, K., Sawai, H., & Morigiwa, K. (1988). Intraretinal axons of ganglion cells in the Japanese monkey (*Macaca fuscata*): Conduction velocity and diameter distribution. *Neuroscience Research, 6*, 53–71.

Garton, J. H. L. (2004). Cerebrospinal diversion procedures. *Journal of Neuro-Ophthalmology, 24*, 146–155.

Ghelli, A., Zanna, C., Porcelli, A. M., Schapira, A. H., Martinuzzi, A., Carelli, V., et al. (2003). Leber's hereditary optic neuropathy (LHON) pathogenic mutations induce mitochondrial-dependent apoptotic death in transmitochondrial cells incubated with galactose medium. *Journal of Biological Chemistry, 278*, 4145–4150. doi:10.1074/jbc.M210285200.

Giorgio, V., Petronilli, V., Ghelli, A., Carelli, V., Rugolo, M., Lenaz, G., & Bernardi, P. (2012). The effects of idebenone on mitochondrial bioenergetics. *Biochimica et Biophysica Acta (BBA)—Bioenergetics, 1817*, 363–369. doi:10.1016/j.bbabio.2011.10.012.

Goldenberg-Cohen, N., Guo, Y., Margolis, F. L., Miller, N. R., Cohen, Y., & Bernstein, S. L. (2005). Oligodendrocyte dysfunction following induction of experimental anterior optic nerve ischemia. *Investigative Ophthalmology & Visual Science, 46*, 2716–2725. doi:10.1167/iovs.04-0547.

Guy, J., Qi, X., Koilkonda, R. D., Arguello, T., Chou, T. H., Ruggeri, M., et al. (2009). Efficiency and safety of AAV-mediated gene delivery of the human ND4 complex I subunit in the mouse visual system. *Investigative Ophthalmology & Visual Science, 50*, 4205–4214. doi:10.1167/iovs.08-3214.

Guy, J., Qi, X., Pallotti, F., Schon, E. A., Manfredi, G., Carelli, V., et al. (2002). Rescue of a mitochondrial deficiency causing Leber hereditary optic neuropathy. *Annals of Neurology, 52*, 534–542. doi:10.1002/ana.10354.

Hauswirth, W. W., Aleman, T. S., Kaushal, S., Cideciyan, A. V., Schwartz, S. B., Wang, L., et al. (2008). Treatment of Leber congenital amaurosis due to RPE65 mutations by ocular subretinal injection of adeno-associated virus gene vector: Short-term results of a phase I trial. *Human Gene Therapy, 19*, 979–990.

Hayreh, M. S., & Hayreh, S. S. (1977). Optic disk edema in raised intracranial pressure: I. Evolution and resolution. *Archives of Ophthalmology, 95*, 1237–1244. doi:10.1001/archopht.1977.04450070135013.

Hayreh, S. S. (1968). Pathogenesis of oedema of the optic disk. *Documenta Ophthalmologica, 24*, 289–411. doi:10.1007/BF02550944.

Hayreh, S. S. (1974). Anterior ischemic optic neuropathy: I. Terminology and pathogenesis. *British Journal of Ophthalmology, 58*, 955–963.

Hayreh, S. S., March, W., & Anderson, D. R. (1979). Pathogenesis of block of rapid orthograde axonal transport by elevated intraocular pressure. *Experimental Eye Research, 28*, 515–523.

Hayreh, S. S., & Vrabec, F. (1966). The structure of the head of the optic nerve in rhesus monkey. *American Journal of Ophthalmology, 61*, 136–150.

Howell, N. (1998). Leber hereditary optic neuropathy: Respiratory chain dysfunction and degeneration of the optic nerve. *Vision Research, 38*, 1495–1504. doi:10.1016/S0042-6989(97)00444-6.

Inamo, J., Belougne, E., & Doutremepuich, C. (1996). Importance of photo activation of rose bengal for platelet activation in experimental models of photochemically induced thrombosis. *Thrombosis Research, 83*, 229–235. doi:10.1016/0049-3848(96)00131-4.

Ischemic Optic Neuropathy Decompression Trial Study Group. (1996). Characteristics of patients with nonarteritic anterior ischemic optic neuropathy eligible for the Ischemic Optic Neuropathy Decompression Trial. *Archives of Ophthalmology, 114*, 1366–1374.

Jaggi, G. P., Harlev, M., Ziegler, U., Dotan, S., Miller, N. R., & Killer, H. E. (2010). Cerebrospinal fluid segregation optic neuropathy: An experimental model and a hypothesis. *British Journal of Ophthalmology, 94*, 1088–1093.

Janaky, M., Fulop, Z., Palffy, A., Benedek, K., & Benedek, G. (2006). Electrophysiological findings in patients with nonarteritic anterior ischemic optic neuropathy. *Clinical Neurophysiology, 117*, 1158–1166.

Jeon, C. J., Strettoi, E., & Masland, R. H. (1998). The major cell populations of the mouse retina. *Journal of Neuroscience, 18*, 8936–8946.

Johansson, J. O. (1987). The lamina cribrosa in the eyes of rats, hamsters, gerbils and guinea pigs. *Acta Anatomica, 128*, 55–62.

Kelman, S. E., Sergott, R. C., Cioffi, G. A., Savino, P. J., Bosley, T. M., & Elman, M. J. (1991). Modified optic nerve decompression in patients with functioning lumboperitoneal shunts and progressive visual loss. *Ophthalmology, 98,* 1449–1453.

Killer, H. E., Jaggi, G. P., Flammer, J., Miller, N. R., & Huber, A. R. (2006). The optic nerve: A new window into cerebrospinal fluid composition? *Brain, 129,* 1027–1030.

Killer, H. E., Jaggi, G. P., Flammer, J., Miller, N. R., Huber, A. R., & Mironov, A. (2007). Cerebrospinal fluid dynamics between the intracranial and the subarachnoid space of the optic nerve. Is it always bidirectional? *Brain, 130,* 514–520.

Killer, H. E., Jaggi, G. P., & Miller, N. R. (2009). Papilledema revisited: Is its pathophysiology really understood? *Clinical & Experimental Ophthalmology, 37,* 444–447.

Killer, H. E., Jaggi, G. P., Miller, N. R., Landolt, H., Mironov, A., Meyer, P., et al. (2011). Cerebrospinal fluid dynamics between the basal cisterns and the subarachnoid space of the optic nerve in patients with papilloedema. *British Journal of Ophthalmology, 95,* 822–827.

Kirkinezos, I. G., & Moraes, C. T. (2001). Reactive oxygen species and mitochondrial diseases. *Seminars in Cell & Developmental Biology, 12,* 449–457.

Klopstock, T., Yu-Wai-Man, P., Dimitriadis, K., Rouleau, J., Heck, S., Bailie, M., et al. (2011). A randomized placebo-controlled trial of idebenone in Leber's hereditary optic neuropathy. *Brain, 134,* 2677–2686. doi:10.1093/brain/awr170.

Koilkonda, R. D., Chou, T.-H., Porciatti, V., Hauswirth, W. W., & Guy, J. (2010). Induction of rapid and highly efficient expression of the human ND4 complex I subunit in the mouse visual system by self-complementary adenoassociated virus. *Archives of Ophthalmology, 128,* 876–883.

Koilkonda, R.D., & Guy, J. (2011). Leber's hereditary optic neuropathy-gene therapy: From benchtop to bedside. *Journal of Ophthalmology, Epub,* 1–16. doi:10.1155/2011/179412.

Lam, B. L., Feuer, W. J., Abukhalil, F., Porciatti, V., Hauswirth, W. W., & Guy, J. (2010). Leber hereditary optic neuropathy gene therapy clinical trial recruitment: Year 1. *Archives of Ophthalmology, 128,* 1129–1135.

Lambert, C. R., Stiel, H., Leupold, D., Lynch, M. C., & Kochevar, I. E. (1996). Intensity-dependent enzyme photosensitization using 532 nm nanosecond laser pulses. *Photochemistry and Photobiology, 63,* 154–160.

Larsson, N. G. (2002). Leber hereditary optic neuropathy: A nuclear solution of a mitochondrial problem. *Annals of Neurology, 52,* 529–530.

Leiba, H., Rachmiel, R., Harris, A., Kagemann, L., Pollack, A., & Zalish, M. (2000). Optic nerve head blood flow measurements in non-arteritic anterior ischaemic optic neuropathy. *Eye (London, England), 14,* 828–833.

Lessell, S., & Horovitz, B. (1972). Histochemical study of enzymes of optic nerve of monkey and rat. *American Journal of Ophthalmology, 74,* 118–126.

Levin, L. A., & Danesh-Meyer, H. V. (2008). Hypothesis: A venous etiology for nonarteritic anterior ischemic optic neuropathy. *Archives of Ophthalmology, 126,* 1582–1585. doi:10.1001/archopht.126.11.1582.

Levin, L. A., & Schalow, K. (1998). Differential susceptibility of retinal ganglion cells to redox modulation: A possible link to mitochondrial optic neuropathy [abstract]. *Investigative Ophthalmology & Visual Science, 39,* S808.

May, C. A., & Lutjen-Drecoll, E. (2002). Morphology of the murine optic nerve. *Investigative Ophthalmology & Visual Science, 43,* 2206–2212.

McCarty, D. M. (2008). Self-complementary AAV vectors: Advances and applications. *Molecular Therapy, 16,* 1648–1656.

Morrison, J. C., Cork, L. C., Dunkelberger, G. R., Brown, A., & Quigley, H. A. (1990). Aging changes in the rhesus monkey optic nerve. *Investigative Ophthalmology & Visual Science, 31,* 1623–1627.

Morrison, J. C., Farrell, S., Johnson, E., Deppmeier, L., Moore, C. G., & Grossmann, E. (1995). Structure and composition of the rodent lamina cribrosa. *Experimental Eye Research, 60,* 127–135.

Morrison, J. C., Johnson, E. C., Cepurna, W. O., & Funk, R. H. (1999). Microvasculature of the rat optic nerve head. *Investigative Ophthalmology & Visual Science, 40,* 1702–1709.

Mosinger, J. L., & Olney, J. W. (1989). Photothrombosis-induced ischemic neuronal degeneration in the rat retina. *Experimental Neurology, 105,* 110–113.

Nadal-Nicolas, F. M., Jimenez-Lopez, M., Sobrado-Calvo, P., Nieto-Lopez, L., Canovas-Martinez, I., Salinas-Navarro, M., et al. (2009). Brn3a as a marker of retinal ganglion cells: Qualitative and quantitative time course studies in naive and optic nerve-injured retinas. *Investigative Ophthalmology & Visual Science, 50,* 3860–3868.

Nakase, T., Yamazaki, T., Ogura, N., Suzuki, A., & Nagata, K. (2008). The impact of inflammation on the pathogenesis and prognosis of ischemic stroke. *Journal of the Neurological Sciences, 271,* 104–109.

Newman, N. J. (2002). From genotype to phenotype in Leber hereditary optic neuropathy: Still more questions than answers. *Journal of Neuro-Ophthalmology, 22,* 257–261.

Newman, N. J. (2005). Hereditary optic neuropathies. In N. R. Miller, N. J. Newman, V. Biouse, & J. B. Kerrison (Eds.), *Walsh and Hoyt's clinical neuro-ophthalmology* (6th ed., Vol. 1, pp. 465–501). Baltimore: Lippincott-Williams & Wilkins.

Newman, N. J. (2011). Treatment of Leber hereditary optic neuropathy. *Brain, 134,* 2447–2450.

Newman, N. J., Biousse, V., David, R., Bhatti, M. T., Hamilton, S. R., Farris, B. K., et al. (2005). Prophylaxis for second eye involvement in Leber hereditary optic neuropathy: An open-label, nonrandomized multicenter trial of topical brimonidine purite. *American Journal of Ophthalmology, 140,* 407–415.

Nikoskelainen, E. K., Huoponen, K., Juvonen, V., Lamminen, T., Nummelin, K., & Savontaus, M. L. (1996). Ophthalmologic findings in Leber hereditary optic neuropathy, with special reference to mtDNA mutations. *Ophthalmology, 103,* 504–514.

Oca-Cossio, J., Kenyon, L., Hao, H., & Moraes, C. T. (2003). Limitations of allotopic expression of mitochondrial genes in mammalian cells. *Genetics, 165,* 707–720.

Ogden, T. E. (1994). Topography of the retina. In S. J. Ryan (Ed.), *The retina* (pp. 32–36). St. Louis, MO: Mosby.

Ojaimi, J., Pan, J., Santra, S., Snell, W. J., & Schon, E. J. (2002). An algal nucleus-encoded subunit of mitochondrial ATP synthase rescues a defect in the analogous human mitochondrial-encoded subunit. *Molecular Biology of the Cell, 13,* 3836–3844.

Olver, J. M., & McCartney, A. C. (1989). Orbital and ocular micro-vascular corrosion casting in man. *Eye (London, England), 3,* 588–596.

Palombi, K., Renard, E., Levy, P., Chiquet, C., Deschaux, C., Romanet, J. P., et al. (2006). Non-arteritic anterior ischaemic optic neuropathy is nearly systematically associated with obstructive sleep apnoea. *British Journal of Ophthalmology, 90*, 879–882.

Perales-Clemente, E., Fernández-Silva, P., Acin-Pérez, R., Pérez-Martos, A., & Enriquez, J. A. (2011). Allotopic expression of mitochondrial-encoded genes in mammals: Achieved goal, undemonstrated mechanism or impossible task? *Nucleic Acids Research, 39*, 229–234.

Perry, V. H., Henderson, Z., & Linden, R. (1983). Postnatal changes in retinal ganglion cell and optic axon populations in the pigmented rat. *Journal of Comparative Neurology, 219*, 356–368.

Pineau, I., Sun, L., Bastien, D., & Lacroix, S. (2010). Astrocytes initiate inflammation in the injured mouse spinal cord by promoting the entry of neutrophils and inflammatory monocytes in an IL-1 receptor/MyD88-dependent fashion. *Brain, Behavior, and Immunity, 24*, 540–553.

Pomeranz, H. D., & Bhavsar, A. R. (2005). Nonarteritic ischemic optic neuropathy developing soon after use of sildenafil (Viagra): A report of seven new cases. *Journal of Neuro-Ophthalmology, 25*, 9–13.

Qi, X., Lewin, A. S., Hauswirth, W. W., & Guy, J. (2003a). Oxygen toxicity induced by complex I deficiency is rescued with AAV mediated gene transfer of human SOD2. *Neurology, 60*(Suppl. 1), A26.

Qi, X., Lewin, A. S., Hauswirth, W. W., & Guy, J. (2003b). Optic neuropathy induced by reductions in mitochondrial superoxide dismutase. *Investigative Ophthalmology & Visual Science, 44*, 1088–1096.

Qi, X., Lewin, A. S., Hauswirth, W. W., & Guy, J. (2003c). Suppression of complex I gene expression induces optic neuropathy. *Annals of Neurology, 53*, 198–205.

Qi, X., Lewin, A. S., Sun, L., Hauswirth, W. W., & Guy, J. (2004). SOD2 gene transfer protects against optic neuropathy induced by deficiency of complex I. *Annals of Neurology, 56*, 182–191.

Qi, X., Sun, L., Hauswirth, W. W., Lewin, A. S., & Guy, J. (2007a). Use of mitochondrial antioxidant defenses for rescue of cells with a Leber hereditary optic neuropathy-causing mutation. *Archives of Ophthalmology, 125*, 268–272.

Qi, X., Sun, L., Lewin, A. S., Hauswirth, W. W., & Guy, J. (2007b). The mutant human ND4 subunit of complex I induces optic neuropathy in the mouse. *Investigative Ophthalmology & Visual Science, 48*, 1–10. doi:10.1167/iovs.06-0789.

Riordan-Eva, P., Sanders, M. D., Govan, G. G., Sweeney, M. G., Da Costa, J., & Harding, A. E. (1995). The clinical features of Leber's hereditary optic neuropathy defined by the presence of a pathogenic mitochondrial DNA mutation. *Brain, 118*, 319–337.

Rizzo, J. F., III. (2005). Embryology, anatomy and physiology of the anterior visual pathway. In N. R. Miller, N. J. Newman, V. Biouse, & J. B. Kerrison (Eds.), *Walsh and Hoyt's clinical neuro-ophthalmology* (6th ed., Vol. 1, pp. 3–92). Baltimore: Lippincott-Williams & Wilkins.

Rodgers, M. A. J. (1981). Light-induced generation of singlet oxygen in solutions of rose bengal. *Chemical & Physics Letters, 78*, 509–514.

Sadun, A. A., Win, P., Ross-Cisneros, F., Walker, S. O., & Carelli, V. (2000). Leber's hereditary optic neuropathy differentially affects smaller axons in the optic nerve. *Transactions of the American Ophthalmological Society, 98*, 1–13.

Sakai, T., Shikishima, K., Matsushima, M., & Kitahara, K. (2007). Endothelial nitric oxide synthase gene polymorphisms in non-arteritic anterior ischemic optic neuropathy. *Graefes Archive for Clinical and Experimental Ophthalmology, 245*, 288–292.

Salgado, C., Vilson, F., Miller, N. R., & Bernstein, S. L. (2011). Cellular inflammation in nonarteritic anterior ischemic optic neuropathy and its primate model. *Archives of Ophthalmology, 129*, 1583–1591.

Salomon, O., Huna-Baron, R., Kurtz, S., Steinberg, D. M., Moisseiev, J., Rosenberg, N., et al. (1999). Analysis of prothrombotic and vascular risk factors in patients with nonarteritic anterior ischemic optic neuropathy. *Ophthalmology, 106*, 739–742.

Salomon, O., Rosenberg, N., Steinberg, D. M., Huna-Baron, R., Moisseiev, J., Dardik, R., et al. (2004). Nonarteritic anterior ischemic optic neuropathy is associated with a specific platelet polymorphism located on the glycoprotein Ibalpha gene. *Ophthalmology, 111*, 184–188.

Schapira, A. H. (1997). Mitochondrial disorders: An overview. *Journal of Bioenergetics and Biomembranes, 29*, 105–107.

Schlamp, C. L., Johnson, E. C., Li, Y., Morrison, J. C., & Nickells, R. W. (2001). Changes in Thy1 gene expression associated with damaged retinal ganglion cells. *Molecular Vision, 7*, 192–201.

Schon, E. A., Bonilla, E., & DiMauro, S. (1997). Mitochondrial DNA mutations and pathogenesis. *Journal of Bioenergetics and Biomembranes, 29*, 131–149.

Shokolenko, I. N., Alexeyev, M. F., Ledoux, S. P., & Wilson, G. L. (2010). The approaches for manipulating mitochondrial proteome. *Environmental and Molecular Mutagenesis, 51*, 451–461.

Supekova, L., Supek, F., Greer, J. E., & Schultz, P. G. (2010). A single mutation in the first transmembrane domain of yeast COX2 enables its allotopic expression. *Proceedings of the National Academy of Sciences of the United States of America, 107*, 5047–5052. doi:10.1073/pnas.1000735107.

Tesser, R. A., Niendorf, E. R., & Levin, L. A. (2003). The morphology of an infarct in nonarteritic anterior ischemic optic neuropathy. *Ophthalmology, 110*, 2031–2035.

Tso, M. O., & Hayreh, S. S. (1977a). Optic disk edema in raised intracranial pressure: III. A pathologic study of experimental papilledema. *Archives of Ophthalmology, 95*, 1448–1457. doi:10.1001/archopht.1977.04450080158022.

Tso, M. O., & Hayreh, S. S. (1977b). Optic disk edema in raised intracranial pressure: IV. Axoplasmic transport in experimental papilledema. *Archives of Ophthalmology, 95*, 1458–1462. doi:10.1001/archopht.1977.04450080168023.

Watson, B. D., Dietrich, W. D., Busto, R., Wachtel, M. S., & Ginsberg, M. D. (1985). Induction of reproducible brain infarction by photochemically initiated thrombosis. *Annals of Neurology, 17*, 497–504.

Wilkes, B. N., & Siatkowski, R. M. (2009). Progressive optic neuropathy in idiopathic intracranial hypertension after optic nerve sheath fenestration. *Journal of Neuro-Ophthalmology, 29*, 281–283.

Williams, R. W., Strom, R. C., Rice, D. S., & Goldowitz, D. (1996). Genetic and environmental control of variation in retinal ganglion cell number in mice. *Journal of Neuroscience, 16*, 7193–7205.

Wilson, C. A., & Hatchell, D. L. (1991). Photodynamic retinal vascular thrombosis. *Investigative Ophthalmology & Visual Science, 32*, 2357–2365.

Xin, X., Huber, A., Meyer, P., Flammer, J., Neutzner, A., Miller, N. R., et al. (2009). L-PGDS (Betatrace protein) inhibits astrocyte proliferation and mitochondrial ATP production in vitro. *Journal of Molecular Neuroscience, 39*, 366–371.

Zaiss, A. K., & Muruve, D. A. (2008). Immunity to adeno-associated virus vectors in animals and humans: A continued challenge. *Gene Therapy, 15*, 808–816.

Zamzami, N., Hirsch, T., Dallaporta, B., Petit, P. X., & Kroemer, G. (1997). Mitochondrial implication in accidental and programmed cell death: Apoptosis and necrosis. *Journal of Bioenergetics and Biomembranes, 29*, 185–193.

Zhang, C., Guo, Y., Miller, N. R., & Bernstein, S. L. (2009). Optic nerve infarction and post-ischemic inflammation in the rodent model of anterior ischemic optic neuropathy (rAION). *Brain Research, 1264*, 67–75. doi:10.1016/j.brainres.2008.12.075.

Zhang, H. R. (1994). Scanning electron-microscopic study of corrosion casts on retinal and choroidal angioarchitecture in man and mammals. *Progress in Retinal and Eye Research, 13*, 243–270.

Zhang, X., Jones, D., & Gonzalez-Lima, F. (2002). Mouse model of optic neuropathy caused by mitochondrial complex I dysfunction. *Neuroscience Letters, 326*, 97–100.

Zhao, B. Q., Suzuki, Y., Kondo, K., Kawano, K., Ikeda, Y., & Umemura, K. (2002). A novel MCA occlusion model of photothrombotic ischemia with cyclic flow reductions: Development of cerebral hemorrhage induced by heparin. *Brain Research. Brain Research Protocols, 9*, 85–92.

Zhong, L., Li, B., Jayandharan, G., Mas, C. S., Govindasamy, L., Agbandje-McKenna, M., et al. (2008a). Tyrosine phosphorylation of AAV2 vectors and its consequences on viral intracellular trafficking and transgene expression. *Virology, 381*, 194–202.

Zhong, L., Li, B., Mah, C. S., Govindasamy, L., Agbandje-McKenna, M., Cooper, M., et al. (2008b). Next generation of adenoassociated virus 2 vectors: Point mutations in tyrosines lead to high-efficiency transduction at lower doses. *Proceedings of the National Academy of Sciences of the United States of America, 105*, 7827–7832. doi:10.1073/pnas.0802866105.

102 Transcriptional Regulation of Photoreceptor Development

VINOD RANGANATHAN AND DONALD J. ZACK

One of the most fascinating questions of developmental biology is how vastly different cell types originate from a single cell. While many intriguing questions remain unanswered, the patterning and transcriptional regulation of complex neuronal tissues is perhaps best understood in the retina (Quan, Ramaekers, & Hassan, 2012). The vertebrate retina is spatially arranged in a laminar configuration consisting of six major types of neurons: photoreceptors (rods and cones), horizontal cells, bipolar cells, ganglion cells, amacrine cells, and Müller glia cells. Development arises from multipotent retinal progenitor cells (RPCs) that divide and differentiate into the cells of the retina through the interplay of cell-intrinsic and cell-extrinsic factors. Intrinsic regulation, the focus of this chapter, is mediated through a complex network consisting of *trans*-acting transcription factors and their complementary DNA *cis*-regulatory elements, other regulatory proteins, and also epigenetic mechanisms that are only now beginning to be understood in the retina. At the molecular level, differential expression of transcription factors is the key determinant regulating cell-type specific gene expression and the determination of cell fate; proper orchestration of this ultimately leads from a single cell to the complexity that has been appreciated in the retina for over 100 years (Cajal, 1892). The *trans*-acting factors, which can act as transcriptional activators or repressors (or sometimes both, depending on the setting), bind to *cis*-regulatory DNA elements that are generally located in regions adjacent to the genes they regulate (Davidson, 2006). However, regulatory elements such as enhancers can sometimes be tens of thousands, and even hundreds of thousands, of base pairs from their target genes (Bulger & Groudine, 2011). In this chapter, since the whole field of retinal development and transcriptional regulation has become too large to cover in a short review, we will try to briefly summarize and highlight the key transcription factors that induce progenitors cells to differentiate into rod and cone photoreceptor cells.

PHOTORECEPTOR DEVELOPMENT

Vision begins with visual pigment-mediated light quanta capture by photoreceptors, which through the photo-transduction process generates a neural chemical signal, which is then integrated and processed by bipolar, horizontal, retinal ganglion, and amacrine cells and ultimately transmitted to the brain via the optic nerve. In mammals, photoreceptors constitute approximately 50% of the neural retinal. Photoreceptor cells belong to one of two classes: rods, which function under dim light conditions, and cones, which function in bright light and mediate color and high acuity vision. In humans, rods comprise over 95% of the photoreceptor cells, and slightly more in mice (Carter-Dawson & LaVail, 1979a; Roorda & Williams, 1999). The rod visual pigment, rhodopsin, has a peak spectral sensitivity of approximately 500 nm and is sensitive enough to respond to a single quanta of light. In contrast, cone cells, which differ in their spectral sensitivities, mediate color vision. In most mammals there are two types of cone opsins that confer dichromatic color vision: S-opsin (short-wavelength or blue-sensitive) and M-opsin (medium-long wavelength or green-sensitive). Humans and some nonhuman primates possess an additional opsin, L-opsin (long-wavelength or red-sensitive), which confers trichromatic color vision; their peak spectral absorbances are 426 nm (S-opsin), 530 nm (M-opsin), and ~555 nm (L-opsin) (Merbs & Nathans, 1992; Nathans et al., 1992).

Retinal ganglion cells, horizontal cells, rod and cone photoreceptors, amacrine cells, bipolar cells, and Müller glial cells all arise from a common multipotent RPC (Turner & Cepko, 1987; Wetts & Fraser, 1988). Decades ago, classical birthdating studies in mice found a highly conserved, yet overlapping, temporal order to the birth date of retinal cells: retinal ganglion cells are produced first, then cones, rods, bipolar interneurons, and finally Müller glial cells (Carter-Dawson & LaVail, 1979b; Sidman, 1961; Young, 1985). The presence of overlapping time window suggests that these precursor cells retain some cell fate plasticity; however, because there is a specific order to the birth of cell types, plasticity appears to become progressively limited over time (Cepko et al., 1996). The pathway from a multipotent retinal precursor cell to a fully functional photoreceptor cell can be divided into five developmental phases: (1) division and proliferation of the RPC; (2) lineage restriction; (3) exiting the cell cycle and formation of postmitotic precursors; (4) immature photoreceptors forming rod and cone precursors; and finally, (5) terminal differentiation, which includes proper opsin expression, outer segment maturation, and synapse formation.

The coordinated process of proliferation, specification, and differentiation that generates the cellular diversity and three-dimensional architecture of the retina is ultimately controlled through the temporally regulated developmental process of gene expression. In vertebrates, the key transcription factors in the most uncommitted retinal stem cells are *Pax6*, *Rx1*, *Six3*, *Six6*, *Vsx2*, and *Lhx2* (Marquardt et al., 2001). The paired-box 6 gene (*Pax6*) belongs to a family of transcription factors characterized by the presence of a paired box DNA-binding domain and a homeodomain (Treisman et al., 1989). The Pax family genes play a wide role in the nervous system and the eye, and in particular, *Pax6* is thought to be one of the master regulators for visual system development (Shaham et al., 2012). Widely conserved across evolution, *Pax6* expression is also found in all RPCs. Ectopic expression of Pax6 can induce eye formation in *Drosophila* and in vertebrates (Chow, et al., 1999; Halder, Callaerts, & Gehring, 1995) while *PAX6* mutations are associated with diseases such as aniridia, Peter's anomaly, and others (Hanson et al., 1994; Kokotas & Petersen, 2010).

Early in development, RPCs divide and proliferate in response to external signaling molecules such as Notch, Wnt, Sonic hedgehog, retinoic acid, and others (Agathocleous & Harris, 2009). The Notch pathway links external signals to intrinsic regulation by modulating transcription in retinal precursor cells. Inhibition of Notch signaling is one of the earliest steps in rod and cone development, driving RPCs toward a photoreceptor precursor fate; on the other hand, maintenance of Notch signaling leads to the expression of two repressive basic helix-loop-helix (bHLH) transcription factors, Hes1 and Hes5, which act to maintain the pool of cycling progenitor cells (Jadhav, Mason, & Cepko, 2006; Yaron et al., 2006). Downstream of Notch inhibition are proneuronal bHLH transcription factors such as NeuroD1, Mash1, neurogenin 2, Math3, and Math5—the interplay of which specify cell fate differentiation

(Hatakeyama & Kageyama, 2004). The proneural bHLH protein NeuroD and Mash1 appear to play an important role in the genesis of photoreceptors. NeuroD is an early expression marker for photoreceptors and is also expressed in adulthood (Morrow et al., 1999).

In humans, photoreceptor development occurs prenatally with cones first appearing around fetal week 8 and rods at fetal week 10 (Cornish, Hendrickson, & Provis, 2004; Hendrickson et al., 2008). While cone formation is largely complete at birth, rod formation continues through a short postnatal period. Although both photoreceptors are morphologically differentiated, the expression of opsins, formation of a functional outer segment, and synapse formation, takes place later. In primates, opsin expression follows about 1 month after photoreceptor formation with rhodopsin expression occurring first, even though cones are formed first (Dorn, Hendrickson, & Hendrickson, 1995; La Vail, Rapaport, & Rakic, 1991). In mice, cones are also born first, between embryonic day (E) 11.5 to E18.5, while rods first begin to appear at E12.5, peak at birth, and continue forming until postnatal day 7 (Carter-Dawson & LaVail, 1979b; Young, 1985). The first detectable opsin during development is S-opsin, which in mice is not detected until around E18, indicating that there is a lag in time between the birth of a photoreceptor and the expression of its opsin (Morrow, Belliveau, & Cepko, 1998).

TRANSCRIPTION FACTORS IN ROD PHOTORECEPTOR DEVELOPMENT AND DIFFERENTIATION

The expression and interplay between several key transcription factors play critical roles in the development of rod and cones from photoreceptor precursors. Rods and cones for the most part employ specific and mutually exclusive transcriptions factors although some factors are common to both cell types. A common feature in the transcriptional networks that regulate photoreceptor development is complementary antagonism, in which a gene network that promotes expression of genes specific to a particular cell fate simultaneously represses genes characteristic of another, mutually exclusive cell-type.

OTX2

OTX2 is paired-type homeobox transcription factor whose expression is critical for the cell fate determination of both rods and cones. Originally identified in *Drosophila*, *orthodenticle* (*otd*) is required for developmental formation of the brain and is expressed in photoreceptors (Finkelstein et al., 1990; Vandendries, Johnson, & Reinke, 1996). *Otd* regulates the expression of rhodopsin genes in *Drosophila*, where it plays a dual role of activating specific rhodopsin expression in a subset of photoreceptors and repressing expression in others. *Drosophila* contains six rhodopsin genes (rh1–6), and promoter analysis identified the highly conserved Otd-binding sequence, 5'-TAATCC-3', upstream three of the genes (rh3, rh5, rh6); Otd is required for the expression of two of the three Otd-binding site containing rhodopsin genes (rh3, rh5), and it is required for repressing the other (rh6) (Tahayato et al., 2003).

There are three Otd homologues in mammals, OTX1, OTX2, and CRX, all of which play important roles in ocular development. All three contain the *bicoid*-like homeodomain from *Drosophila*. This 60-amino-acid domain contains a critical lysine at position 50 which is crucial for both recognition and binding to the core 5'-TAAT-3' consensus (Hanes & Brent, 1989, 1991). All three Otd orthologs contain a high degree of sequence homology, particularly in the homeodomain found near the N-terminus, as well as in segments in the C-terminal region known as the Otx-tail (Chen et al., 1997; Furukawa, Morrow, & Cepko, 1997). *Otx1* and *Otx2* are involved with brain and head development while the more divergent *Crx* (otx5 ortholog in fish, amphibians, and chicken) plays a critical role in the development of photoreceptors (see below) and pinealocytes in the pineal gland (Plouhinec et al., 2003; Vandendries, Johnson, & Reinke, 1996). Cross and interspecies experiments have shown that these proteins are remarkably conserved across evolution, not only structurally but also functionally; the *Drosophila otd* can functionally rescue Otx1 and Otx2 in mouse development, and the three human orthologs (*OTX1, OTX2,* and *CRX*) rescue *otd* mutants in *Drosophila* (Acampora et al., 1998; Adachi et al., 2001; Nagao et al., 1998; Terrell et al., 2012).

The homozygous *Otx2* knockout mouse is embryonic lethal, however the heterozygous mouse is viable and exhibits multiple ocular defects (Matsuo et al., 1995). The conditional knockout mouse model has shown that acute loss of *Otx2* in developing photoreceptors results in the complete absence of both rods and cones. Interestingly, these mice exhibit a compensatory increase in amacrine-like neurons, suggesting that loss of *Otx2* alters the postmitotic precursor cell fate (Nishida et al., 2003). Furthermore, retroviral-mediated overexpression of *Otx2* in RPCs promotes the photoreceptor cell fate while decreasing numbers of bipolar, amacrine, and Müller cells (Koike et al., 2007; Nishida et al., 2003). Transcriptional profiling of the *Otx2*-deficient

retina identified several other photoreceptor transcription factors that are strongly reduced: *Crx, Nrl, Nr2e3, Esrrb, Isl1, Blimp1, Pias3,* and *NeuroD* (Omori et al., 2011). This data places *Otx2* as the most upstream gene in the photoreceptor transcriptional network, playing a key role in early photoreceptor development and in specifying cell fate.

Through development, *Otx2* expression is found in the forebrain, midbrain, and eye; however its expression in the mature animal is largely restricted to the neural retina and to the retinal pigment epithelium (RPE), where it regulates the expression of other RPE-specific transcription factors (Martinez-Morales et al., 2001, 2003). In the neural retina, *Otx2* expression is found in a subset of postmitotic cells undergoing terminal differentiation: ganglion cells, bipolar cells, and photoreceptors (Baas et al., 2000).

While little is known about the regulatory expression of *Otx2* itself in the retina, recent advances in retinal electroporation techniques identified an approximately 600 bp evolutionarily conserved element (ECR2) located ~5 kb from the start site. This region is sufficient to drive expression in early postmitotic cells and is primarily confined to photoreceptor cells; however it is also detected in horizontal cells. As reporter expression from the construct was undetected in bipolar cells, this data suggest that the ECR2 region reflects a molecular switch directing cell fate that is functionally restricted to photoreceptors and horizontal cells (Emerson & Cepko, 2011).

As photoreceptors and bipolar cells both express *Otx2*, it is clear that expression *Otx2* expression itself is not sufficient to determine photoreceptor cell fate. An insight into this issue was provided by two recent studies that found that *Blimp1*, a transcriptional target of *Otx2*, plays a key role in promoting photoreceptor cell fate by directly binding and repressing the key bipolar marker *Chx10* (Brzezinski, Lamba, & Reh, 2010; Katoh et al., 2010). An interesting question that remains to be answered is what limits *Blimp1* expression in *Otx2* expressing cells that are destined to become bipolar cells.

Located on 14q22.3, *OTX2* (MIM #600037) mutations have been identified in patients with Leber's congenital amaurosis (LCA), a group of early-onset retinal dystrophies leading to severe retinal dysfunction and vision loss (Chatelain et al., 2006). Consistent with embryonic lethality in homozygous null mice, mutations in *Otx2* are likely hypomorphic, leading to haplo-insufficiency. Additionally, *OTX2* mutations have been identified in syndromic microphthalmia, a disease whose spectrum covers ocular malformations, as well as pituitary insufficiency (Ragge et al., 2005).

CRX—Cone–Rod Homeobox Protein

Another important *otd* homologue in retinal development is the cone–rod homeobox (CRX) protein, a key transcriptional regulator of many photoreceptor genes, including the opsins (Chen et al., 1997; Furukawa, Morrow, & Cepko, 1997). OTX2 binds and directly regulates expression of *Crx*, through *cis*-regulatory elements in the *Crx* promoter region (Nishida et al., 2003). The sequential expression of *Otx2* and *Crx* is required for the differentiation of both rods and cones from photoreceptor precursors; both OTX2 and CRX bind to the consensus sequence 5'-TTAATC-3', a commonly found DNA motif in the promoter or enhancer regions of many photoreceptor genes (Chen et al., 1997; Chen & Zack, 1996; Furukawa, Morrow, & Cepko, 1997).

Homozygous knockout of *Crx* does not significantly affect photoreceptor cell number (i.e., cell lineage is largely intact), but does significantly affect photoreceptor development and function. In the *Crx*$^{-/-}$ mouse, photoreceptors appear normal in early development; however, photoreceptors fail to form outer segments, fail to properly express phototransduction and other photoreceptor-specific genes genes, and subsequently degenerate (Furukawa, Morrow, & Cepko, 1997; Koike et al., 2007; Morrow et al., 2005). Developmental analysis found that *Crx* is expressed in early postmitotic precursors that have exited the cell cycle, at a time when Pax-6 expression appears to be down-regulated (Garelli, Rotstein, & Politi, 2006). In mice, *Crx* transcripts are first detectable in the neural retina at E12.5 during the time of cone genesis, with peak expression occurring at postnatal day (P) 3, coincident with rhodopsin expression. This expression continues to be detected in mature photoreceptors throughout adulthood (Chen et al., 1997; Furukawa, Morrow, & Cepko, 1997). *Crx* also appears to be expressed in bipolar cells (Liu et al., 2001). In the nonneural retina, *Crx* is expressed in the RPE, although at lower levels (Esumi et al., 2009).

CRX (MIM #602225), which maps to 19q13.33, has been associated with several diseases (Rivolta, Berson, & Dryja, 2001). *CRX* is located within the region mapped for the Cone–rod dystrophy 2 (CORD2) locus, and *CRX* mutations are associated with *CORD2* (Freund et al., 1997). Cone–rod dystrophy (CORD) is a disease with early cone dysfuntion (i.e., vision loss beginning with decreased color vision and acuity), which is then followed by rod-mediated loss of night blindness. Later studies identified *CRX* mutations in additional diseases such as autosomal dominant retinitis pigmentosa (adRP) and LCA (Freund et al., 1998; Rivolta, Berson, & Dryja, 2001; Sohocki et al., 1998; Swaroop et al., 1999). Not surprisingly, many of the mutations that have been identified are found in the homeodomain and affect DNA binding (Chen et al., 2002).

The central role of *Crx* in regulating photoreceptor genes came from several studies. Gene expression profiling by microarray and serial analysis of gene expression (SAGE) on *Crx*$^{-/-}$ mice found that hundreds of rod photoreceptor genes were misregulated (Blackshaw et al., 2001; Livesey et al., 2000). Furthermore, retroviral overexpression of *Crx* resulted in a similar, but reduced, *Otx2* overexpression phenotype: an increase in rod differentiation with a loss of amacrine and Müller cells, and no change in bipolar cells (Furukawa, Morrow, & Cepko, 1997). And more recently, ChIP experiments found that CRX directly binds the promoter or enhancer regions of many photoreceptor genes (Peng & Chen, 2005).

By coupling high-throughput sequencing technology with chromatin immunoprecipitation, ChIP-seq, the CRX-bound regions (CBR) were validated as the consensus sequence, 5'-TTAATC-3' (Corbo et al., 2010). The widespread presence of this *cis*-regulatory DNA motif in retinal promoters allows expression of a single gene, *Crx*, to play a key role in orchestrating the transcriptional network of photoreceptor specification (Hsiau et al., 2007; Qian et al., 2005). Due to the hundreds, to potentially thousands, of genes regulated by *Crx*, it is best thought of as a "terminal selector gene" for photoreceptors (Hobert, 2008). This term refers to the concept that a single gene can regulate a multitude of other genes, the result of which is to specify a terminal differentiation phenotype. These terminal selector genes, like *Crx*, often regulate their own expression requisite for maintenance of the terminal phenotype and can also provide feed-forward loops to modulate expression with other coregulators (i.e., NRL-enhanced expression of rhodopsin; see below).

NRL—Requirement for Rod Photoreceptor Fate

The chief determinant for rod photoreceptor cell fate is *Nrl* (neural retina leucine zipper protein), which acts as a molecular switch between rod and cone cell fates: the expression of *Nrl*, perhaps the earliest marker of rod cell identity, drives a photoreceptor precursor to become a rod by activating expression of rhodopsin and other rod-specific genes. NRL is a basic motif-leucine zipper (bZIP) transcription factor belonging to the Maf family of transcription factors (Swaroop et al., 1992). *Nrl* is highly expressed in the retina, where transcripts are detectable at E14.5 and continue into adulthood (Liu et al., 1996). The regulation of *Nrl* occurs through the binding of OTX2, CRX, and RORβ, three retinal transcription factors that bind to a highly conserved

30-nucleotide *cis*-regulatory region in the *Nrl* promoter (Jia et al., 2009; Kautzmann et al., 2011; Montana et al., 2011).

bZIP proteins are well known to interact with other proteins by forming homo- or heterodimers. One such interaction is between the leucine zipper domain of NRL and the homeodomain of CRX (Mitton et al., 2000). In combination with CRX, NRL synergistically enhances the expression of rhodopsin (Chen et al., 1997). The activation of rhodopsin expression is additionally enhanced through the binding of another zinc-finger protein, FIZ1, when in complex with CRX and NRL (Mali et al., 2007). The interaction of NRL with the rhodopsin promoter occurs through the *Nrl*-response elements (NRE), an AP-1 like sequence that bZIP proteins bind (Kumar et al., 1996; Rehemtulla et al., 1996). NRL has been shown to associate with the TATA-binding protein (TBP), a function consistent with recruitment of the transcriptional machinery (Friedman et al., 2004). In combination with CRX, NRL promotes the expression of other rod-specific genes. Microarray studies comparing wildtype and the *Nrl* knockout mouse indicate that NRL targets a many genes involved with phototransduction, transcriptional regulation, signaling pathways, structural proteins, and others (Akimoto et al., 2006; Yoshida et al., 2004; Yu et al., 2004). One recently identified target, *Mef2c*, was also shown to promote the expression of rhodopsin (Hao et al., 2011).

Consistent with a role in promoting rod cell formation, loss of NRL in mice leads to the complete absence of rods and an increase in S-cones (see below, *rd7* mouse and enhanced S-cone syndrome) (Daniele et al., 2005; Mears et al., 2001). Conversely, studies in mice that placed *Nrl* under control of a *Crx* promoter found that all photoreceptor precursors were driven to a rod-cell fate. Interestingly, transgenic mice engineered to drive *Nrl* from the S-opsin promoter indicated the presence of rod–cone hybrid cells. NRL was also found to be associated with S-opsin and the *Trβ2* promoters, suggesting that NRL might play a directly role in suppressing cone genes (Oh et al., 2007). NRL may repress cone cell genes directly (Peng & Chen, 2005), or indirectly, by activating repression of cone genes through the expression of the nuclear orphan receptor *Nr2e3* (Chen, Rattner, & Nathans, 2005; Cheng et al., 2004, 2006; Corbo & Cepko, 2005; Oh et al., 2007; Peng & Chen, 2011). These studies concluded that *Nrl* is required for rod cell fate determination.

NRL (MIM #162080) maps to 14q11.2 and is also associated with retinal diseases. Missense mutations in *NRL* are found in adRP while recessive mutations are associated with clumped pigmentary retinal degeneration that results in the loss of rods with normal cone function (Bessant et al., 1999; Nishiguchi et al., 2004).

NR2E3 Nr2e3 (Nuclear receptor, class 2, subfamily E, member 3) or PNR (photoreceptor-specific nuclear receptor) belongs to a superfamily of ligand-modulated nuclear hormone receptor transcription factors (Kobayashi et al., 1999). The core binding site for nonsteroidal nuclear hormone receptors known as the DR1 response site is AG(G/T)TCA, and NR2E3 was found to bind to the consensus DNA sequence 5' (A/G) AG(A/G)TCAAA(A/G)(A/G)TCA-3', indicating that it likely binds as a dimer (Chen, Rattner, & Nathans, 2005). Expression of Nr2e3 is localized to the retina and was found by in situ hybridization studies to be restricted to the outer nuclear layer (Bookout et al., 2006; Haider et al., 2000). Consistent with its proposed role in repressing cone-specific gene expression, immunostaining experiments confirmed that *Nr2e3* is expressed exclusively in rod cells. Temporal analysis of *Nr2e3* found that expression occurred in rod precursor cells, even prior to rhodopsin expression, one of the earliest markers of rod cells. Furthermore, *Nr2e3* expression is excluded from cone cells (Bumsted O'Brien et al., 2004; Chen, Rattner, & Nathans, 2005).

Several important insights into the function of NR2E3 came from animal studies. First, it was shown that coexpression of *Nr2e3* and *Nrl* in photoreceptor precursors of the *Nrl*[−/−] mouse inhibited cone formation, and that functional rod formation was absolutely dependent on *Nrl* expression, as *Nr2e3* alone was insufficient (Cheng et al., 2006; Oh et al., 2007). Secondly, the *retinal degeneration* 7 mouse (rd7/rd7), a spontaneously occurring recessive mutation in the *Nr2e3* gene caused by an antisense L1 retrotransposon insertion, is characterized by progressive photoreceptor degeneration, a 1.5–2 fold increase in S-cones, and retinal folds (Akhmedov et al., 2000; Chen, Rattner, & Nathans, 2006; Haider et al., 2000). The presence of retinal folds in these mice, a phenotype also seen in the *Nrl*[−/−] mouse, is caused by abnormally high S-cones, which can be rescued by partial ablation of cone cells (Chen & Nathans, 2007). This data suggests that although cones contribute to only 3% of the photoreceptors, mutations that affect transcription factors and result in only a moderate changes in cell type and number can have a significant effect on the global structure of the retina. Finally, the exclusive expression of *Nrl* and *Nr2e3* in rods, the similar retinal phenotype observed between the *Nrl*[−/−] and rd7 mouse, the lack of *Nr2e3* expression in the *Nrl*[−/−] mouse, and gene profiling experiments all suggest that NR2E3 acts downstream of NRL in rod cell development (Mears et al., 2001; Swain et al., 2001; Yoshida et al., 2004).

Coimmunoprecipitation studies from bovine extract found the NR2E3 interacts in a complex with NRL, CRX, and another orphan nuclear receptor NR1D1 (Rev-erbα). Furthermore, evidence supporting complex formation and transcriptional enhancer function came from transient transfection promoter studies examining the expression of rhodopsin, which found a synergistic increase when NR2E3 was present with NRL, CRX, and NR1D1. Conversely, this complex was found to repress S- or M-cone opsin expression (Cheng et al., 2004).

As a transcriptional activator of rod-specific genes, such as rhodopsin, NR2E3 forms a complex with NRL/CRX/NR1D1 through interactions with the CRX DNA-binding domain. Chromatin immunoprecipitation experiments (ChIP) also found this complex located at cone-specific promoters, further confirming that NR2E3 has a dual function in promoting and maintaining rod cell fate: through inducing rod-specific genes and by inhibiting cone-specific genes (Cheng et al., 2004; Peng et al., 2005).

Mutations in *NR2E3* (MIM #604485), located in 15q23, result in enhanced S-cone syndrome (ESCS), an autosomal recessive retinopathy (Haider et al., 2000). ECSC patients are characterized by hypersensitivity to blue light, and progressive night blindness, often leading to complete vision loss. Histopathology has demonstrated an excess of S-cones and a lack of rods, recapitulating the phenotype seen in the rd7 mouse retina. Though S-cones normally represent approximately 7% of the cones in the human retina, in ESCS patients, S-cones represent 92% of the total cones, and there is a twofold overall increase in cones. Expression analysis found that in addition to the skewed S-cone-to-M/L-cone ratio, some cones expressed S-opsin together with either M- or L-opsin (Milam et al., 2002). Studies in mice and humans have shown that mutations in *NR2E3* are critical for rod cell differentiation and S-opsin repression.

Pias3

As many proteins in the cell need to be precisely regulated in a temporal fashion, one common strategy is through posttranslational modifications, such as phosphorylation, ubiquitination, and SUMOylation; more recently, the occurrence and importance of posttranslational modification of transcription factors in photoreceptor development has become apparent. One such family of proteins that regulate general transcription factors is PIAS (protein inhibitor of activated STAT), a member of which is *Pias3*, an E3-like SUMO ligase (Kotaja et al., 2002). SUMOylation, which is closely

related to ubiquitination, is a reversible modification that covalently links a SUMO molecule to lysine residues on a target protein through an isopeptide bond (Meulmeester & Melchior, 2008). Although *Pias3* is expressed in a broad range of tissues, it is specifically expressed in early postmitotic and immature photoreceptors, coincident with the temporal expression pattern of *Crx*, *Nrl*, and *Nr2e3* (Blackshaw et al., 2004).

PIAS3 was found to physically associate with CRX and NR2E3 through immunoprecipitation studies, and ChiP experiments confirmed these proteins form a complex at both rod and cone promoters. Furthermore, NR2E3 was indeed found to be SUMOylated by PIAS3 in the developing retina. This SUMOylation was required for the repression of S-cone-specific genes, as *Pias3* knockdown experiments resulted in increased cone-like photoreceptors. Overexpression of *Pias3* results in increased rod-like cells, and rhodopsin promoter expression is increased when PIAS3 is present with CRX, NRL, and NR2E3. Together, this data indicates that Pias3 is required for NR2E3 SUMOylation, and that this mechanism is critical for the repression of S-cone genes during rod differentiation (Onishi et al., 2009). Further implicating the role of SUMO regulation in photoreceptor development, NRL was also recently found to be SUMOylated on two lysine residues, the presence of which was shown to enhance expression from the rhodopsin promoter (Roger et al., 2010).

Another important photoreceptor transcription factor is the estrogen-related receptor (ERRβ), an orphan nuclear hormone receptor that was identified through gene expression analysis as being enriched in rod photoreceptors (Blackshaw et al., 2001, 2004). Recent studies have shown that ERRβ regulates a number of rod photoreceptor genes, including rhodopsin, and that loss of ERRβ leads to loss of outer segments followed by rod degeneration overtime. Compared to *Nrl* and *Nr2e3* knockouts, there is no increase in S-opsin expression, suggesting that ERRβ does not play a significant role in repressing cone opsin expression (Onishi et al., 2010b).

THE *CIS*-REGULATORY ELEMENTS AND *TRANS*-ACTING FACTORS THAT CONTROL RHODOPSIN EXPRESSION

Rhodopsin is a 40-kDa transmembrane apoprotein belonging to an ancient class of receptors that transduces light into a G-protein mediated signal cascade. The mammalian opsin genes are present in single copy in the genome and found on chromosome 3 in humans and on chromosome 6 in mice (Elliott et al., 1990;

Martin et al., 1986; Nathans, Thomas, & Hogness, 1986b). Evolutionary studies indicate that rhodopsin evolved from an ancestral cone opsin gene from which it diverged in both sequence and function (Arendt et al., 2004; Okano et al., 1992).

Rhodopsin regulation is primarily regulated at the level of transcription, and numerous studies have contributed to a detailed understanding of the *cis*-regulatory elements and *trans*-acting factors that control its expression. Based on appearance of Otx2 at E11.5, Crx transcripts at E12.5, Nrl at E14.5, and Nr2E3 at E16.5, a temporal relationship for rod cell fate can be deduced (Cheng et al., 2004; Furukawa, Morrow, & Cepko, 1997; Liu et al., 1996; Nishida et al., 2003).

Promoter mapping studies in transgenic mice found that proper temporal and spatial expression of rhodopsin is contained within a 1.5–2.0 kb of DNA sequence found upstream of a canonical TATA box and transcriptional start site (Zack et al., 1991). Within this sequence, there are two critically important regions that function to specify photoreceptor expression and proper levels. The first, known as the rhodopsin proximal promoter region (RPPR), directs expression in photoreceptors, and the second, a more upstream region known as the rhodopsin enhancer region (RER), is required for high levels of transcription (Nie et al., 1996).

The RPPR, a region that spans −176 to +70 in the rhodopsin promoter, itself contains several conserved DNA motifs (bovine sequence and coordinates are given due to historical reasons): (1) Ret-4, found at position −54 to −40, contains the sequence 5'- GGAGCT TAGGGAGGG -3' (Chen & Zack, 1996); (2) immediately adjacent to Ret-4 at position −68 to −56 is the Nrl response element, or NRE, which contains the sequence 5'- TGCTGATTCAGCC-3' (Kumar et al., 1996; Rehemtulla et al., 1996; Swaroop et al., 1992); (3) the BAT-1 sequence, 5'- ATTAATATGATTAATAACGCC -3', is located at −103 to −83 (DesJardin & Hauswirth, 1996); (4) the Ret-1/PCE-1 sequence 5'-GGCCCCACCT-GGGAAGCCAATTAAGCCC-3', which is bound by the QRX transcription factor, is located further upstream at positions −148 to −122 bp and is a common binding site found in other rod photoreceptor-expressed genes (Kikuchi et al., 1993; Morabito, Yu, & Barnstable, 1991; Wang et al., 2004); and (5) Eopsin, the most distal element in the RPPR, is located at nucleotides −152 to −147 and contains the short motif 5'- CACCTG -3' (Ahmad, 1995). The RER region is an approximately 100 nt sequence; in the bovine gene it is located at −2044 to −1943 bp, and in the human gene it is at −1906 to −1805 bp. Exhibiting a high degree of homology to 37 bp of the green and red locus control regions (LCR), this region contains a highly conserved core sequence,

5'- CGATGG -3', and a Ret-3 sequence, 5'- ccttgagcagct-gcgccccacccacccg -3' located at −1970 to −1943 (Nie et al., 1996). Additionally, although this region is highly conserved across vertebrates, the relative distance from the start site varies in different species. Recent studies using chromosomal capture assays (3C) have shown that, in rods, intrachromosomal loops occur at the rhodopsin locus, juxtaposing the RER and RPPR regions. Consistent with the likely functional importance of these loops, this organization does not occur for untranscribed loci—for example, for the cone opsin loci in rod photoreceptors (Peng & Chen, 2011).

TRANSCRIPTION FACTORS IN CONE PHOTORECEPTOR DEVELOPMENT AND DIFFERENTIATION

Trβ2 and Rxrγ

Thyroid hormone receptors (TRs) belong to a superfamily of nuclear receptors that act as ligand-inducible transcription factors in the presence of 3,5,3'-triiodothyronine (T3)—thyroid hormone. When ligand bound, these receptors form homodimers or heterodimers with retinoid X receptors (RXR) and bind to thyroid hormone response elements (TRE) [5'-(A/G) GGT(C/A/G)A], where they act as *cis*-acting transcription factors. In humans, there are two thyroid receptors genes, encoded by two TRα and TRβ (THRA or THRB), both of which produce three isoforms (Song, Yao, & Ying, 2011). The transcription factor TRβ2 is a thyroid hormone nuclear receptor formed from a splice variant of the thyroid receptor B gene. While most isoforms are widely expressed, TRβ2 expression is uniquely found in immature cone photoreceptors as well as the cochlea, hypothalamus, and pituitary (Nunez et al., 2008).

Transient transfection studies have shown that TRβ2 can activate M-opsin transcription while inhibiting S-opsin transcription (Yanagi, Takezawa, & Kato, 2002). Knockout mice that lack TRβ2 result in the complete loss of M-cones with the compensatory increase in S-cones, indicating that TRβ2 plays a role in promoting M-opsin expression (Ng et al., 2001). Additionally, treatment of newborn mice with (T3) reduces S-opsin expression, and TSH receptor-deficient mice, which exhibit severe hypothyroidism, fail to express M-opsin at normal levels (Lu et al., 2009; Roberts et al., 2006). A rise in circulating thyroid hormone levels in development could also provide a temporal cue for M-opsin expression (Lu et al., 2009).

As mentioned above, TRs mediate transcriptional activation through either homodimerization or heterodimerization, and previous studies provided one notable candidate for TRβ2 heterodimerization formation—Rxrγ. The retinoid-related receptor family (Rxr) also contains three members, RXRα, RXRβ, RXRγ, with various expression patterns in the retina. Rxrγ was of particular interest in cone function as it is specifically expressed in developing cones (Janssen et al., 1999; Mori et al., 2001). The Rxrγ knockout mouse indicated that S-opsin was produced in all cones, suggesting a role in repressing S-opsin expression, similar to the TRβ2 knockout. However, one difference between the Rxrγ knockout and TRβ2 knockout is that cones in the Rxrγ knockout display the normal M-opsin gradient (Roberts et al., 2005). This data suggested two things: (1) TRβ2 and Rxrγ form heterodimers which are responsible for suppressing S-opsin expression, and (2) TRβ2 plays a role in promoting M-opsin expression, likely through homodimerization which is modulated in response to T3 (Roberts et al., 2006).

Rorβ2—A Positive Regulator of S-opsin

The retinoid-related orphan receptor family (Ror) consists of three receptors, RORα, RORβ, RORγ, each with multiple isoforms that are typically expressed in a tissue-specific manner. The protein structure consists of four major domains: an N-terminal domain (A/B), a DNA-binding domain, a hinge domain, and a C-terminal ligand-binding domain. RORs bind to the DNA consensus sequence 5'-RGGTCA- 3' which follows a short A/T rich sequence (Jetten, 2009). The RORβ isoform is highly expressed in the retina and pineal gland, and knockout mice are blind (Andre et al., 1998a; Andre, Gawlas, & Becker-Andre, 1998b).

In the presence of CRX, RORβ has been shown to synergistically activate S-opsin, which contains two conserved and adjacent consensus binding sites (RORE1 and RORE2): 5'- AATATAGGTCA -3' and 5'- GATTGAG GTCA -3'. Furthermore, knockout mice fail to express S-opsin, without affecting M-opsin, and photoreceptors fail to form proper outer segments (Srinivas et al., 2006). Interestingly, studies that looked at rod cell formation in *Rorβ* knockout mice found the absence of rods, and gene expression analysis showed that *Nrl* and *Nr2e3* were not expressed; reintroduction of Nrl converted the cells to a rod-like state, indicating that *Rorβ* also plays a role in rod cell formation, perhaps even upstream of *Nrl* (Jia et al., 2009). The related RORα, in combination with CRX, enhances S-opsin and M-opsin (Fujieda et al., 2009).

Recently, a role for PIAS3 has been reported in regulating cone opsin expression, similarly to the SUMOylation of NR2E3 that occurs in rod cells. *Pias3* expression was noted in cone cells, and promoter analysis identified a highly conserved thyroid hormone element, 5'- TGGGCCTGA -3', that was responsive to Trβ2 and RXRγ, and subsequently confirmed by ChIP analysis. Overexpression of *Pias3* found that M-opsin expression was enhanced, while S-opsin expression was downregulated, and conversely, knockdown experiments showed an increase in S-opsin expression. *Pias3* was found to play two complementary roles: (1) It activated M-opsin expression through a T3/Trβ-dependent function, and (2) it suppressed S-opsin expression through SUMOylation of RORα (Onishi et al., 2010a). Taken together, these suggest that in cones *Pias3* functions to promote the M-cone cell fate.

CONE OPSIN EXPRESSION

S-Opsin versus M-Opsin

Humans and old-world primates evolved trichromacy through a recent gene-duplication event on the X chromosome followed by sequence divergence and the acquisition of different spectral absorption properties (Jacobs, 1993; Nathans, 1999). In order to discriminate between visual stimuli and prevent sensory input overlap, any given cone cell expresses one—and only one—type of opsin (although in some species there are exceptions to this general rule); this arrangement, analogous to olfactory receptors, occurs through the mutually exclusive expression of a single opsin type per photoreceptor. Committed photoreceptor cells that do not receive signals to become rods—namely through *Nrl*—will become cones. The density, spatial distribution, type, and peak absorption of visual pigments varies greatly in vertebrates. In most mammals there are three types of pigments: rhodopsin and two cone pigments that confer dichromatic vision (S-opsin and M-opsin).

Unlike rhodopsin and L/M-opsin genes, which have defined upstream enhancer regions, the S-opsin gene does not. Instead, the three critical regions in the S-opsin promoter are the RORE1, RORE2, and CRX response element (CRXE) that are found within 300 nucleotides upstream of the TATA box (Srinivas et al., 2006). This region is likely analogous to the RER of rhodopsin and the LCR of the M/L-opsin locus, as chromosomal looping between this upstream region and the promoter correlates with S-opsin expression (Peng & Chen, 2011).

Though beyond the scope of this chapter, in addition to cell intrinsic factors, rod cells also secrete factors—critical to cone survival—known as rod-derived cone viability factors (RdCVFs) (Leveillard & Sahel, 2010). The genes NXNL1 and NXNL2 encode two of these factors, RdCVF1 and RdCVF2. NXNL1 contains binding sites for CHX10/VSX2, VSX1 and PAX4, and SP3, while NXNL2 contains a CRX binding site located in its promoter region (Lambard et al., 2010; Reichman et al., 2010).

L-Opsin versus M-Opsin

Molecular cloning of the human visual pigments indicated that the red and green pigments share 96% protein identify and are found on the X chromosome in a head-to-tail tandem array (Nathans et al., 1986a; Nathans, Thomas, & Hogness, 1986b; Vollrath, Nathans, & Davis, 1988). The L-opsin gene lies 5' to the M-opsin gene, which is often duplicated or found in higher copy numbers. The mutually exclusive expression of either the L or M alleles occurs through a highly conserved upstream LCR located ~3 kb from the start site of the L-opsin gene (Nathans et al., 1989; Smallwood, Wang, & Nathans, 2002; Wang et al., 1992). Within this region is a highly conserved 37-nucleotide sequence. Similar to the situation in rods, where intrachromosomal loops occur at the rhodopsin locus, 3C assay data indicates that the cone opsins form intrachromosomal loops with the LCR and either the L- or M-opsin promoter (Peng & Chen, 2011). Furthermore, the ratio of loop formation to the L-opsin versus M-opsin promoter was approximately 3:1, the L/M cone ratio seen in Caucasian men (Carroll, Neitz, & Neitz, 2002).

In the Caucasian population, approximately 25% of the males carry a single M gene, 50% carry two, and the remaining carry three or more. Studies have shown that only the first two genes in the tandem array have the propensity to be expressed (Hayashi, Motulsky, & Deeb, 1999; Yamaguchi, Motulsky, & Deeb, 1997). The presence of a single LCR ensures the expression of either the red or green pigment, which occurs through a stochastic yet stable *cis*-acting association between the LCR and L- or M-opsin promoter (Wang et al., 1999). There are two consequences due to the X chromosomal location of the red and green pigment genes: (1) The decision to express either the L- or M-opsin only has to be made on a single allele, given X-inactivation in females and X chromosomal hemizygosity in males, and (2) the prevalence of the X-linked inherited form of red–green color blindness, which occurs in ~8% of males.

Blue cone monochromacy (BCM) is an X-linked recessive syndrome resulting in monochromatism from loss of both L- and M-cone function. One cause of BCM is from deletions that map to the visual pigment cluster (Xq28), approximately 4 kb upstream of the red pigment gene (Nathans et al., 1989). Further studies identified these deletions as encompassing the LCR indicating the critical function this *cis*-regulatory region plays in M- and L-opsin (Wang et al., 1992). Another form of BCM can occur through inactivating point mutations found in patients that only carry a single pigment gene at the L/M locus (Nathans et al., 1993).

SUMMARY

Several lines of evidence have led to the hypothesis that photoreceptor precursors follow a default pathway leading to an S-cone fate unless transcriptional regulatory signals direct another cell fate (Swaroop, Kim, & Forrest, 2010; Szel, van Veen, & Rohlich, 1994; Yanagi, Takezawa, & Kato, 2002). First, molecular phylogenic analysis of the opsin pigments indicates that S-opsin is the more ancient of the present day opsin genes, which is consistent with a default pathway leading to S-cones (Bowmaker, 2008). Secondly, S-opsin is expressed before the other opsins in the developing retina. Studies in rat suggested that S-cones are produced first, appearing at P5, while M-cones appeared at P9, the timing of which correlated with a precipitous drop in S-cone formation. Furthermore, antibody detection to S-opsin and M-opsin indicated that early in development cones were only reactive to the S-opsin antibody, and late in development antibodies reacted with either S-opsin or M-opsin. However during the intermediate period between P9 and P20, many cones were labeled with both antibodies. This suggested the presence of both opsins in developing cones, which could result from an intermediary state reflecting an S-opsin to M-opsin transition (Szel, van Veen, & Rohlich, 1994). In other studies, transgenic mice engineered to express *β-galactosidase* from the human L-opsin promoter and LCR indicated that the transgene was properly expressed in cone cells. Analysis by antibody detection found that transgene expression colocalized with both L-cones and S-cones. Cone distribution across the mouse retina is not random, with M-cones predominating the superior half and S-cones dominating the inferior half. Spatial–temporal analysis indicated that the gradient of transgene expression, which began in the inferior retina at P6, spreads to the superior half by P10 (Wang et al., 1992). Similarly in humans, S-opsin mRNA and protein can be detected earlier than L/M opsin; S-opsin protein can be detected at fetal week 10.9 while L/M opsin can only be detected more than 1 month later at fetal week 14–15. At fetal week 19, S-cones comprise nearly 90% of the retina, and at this stage the developing retina contains nearly three times the number of S-cones as the adult retina. Additionally, coexpression of S-opsin and M/L-opsin can be detected in humans as well (Xiao & Hendrickson, 2000).

Studies from mice and human disease indicate that the loss of function of particular transcription factors can lead to the increased expression of S-opsin. In the retina of the *Nrl* knockout mouse, an excess of S-cones compensates for the lack of rods cells. Likewise, in the *r7* mouse or in ESCS, mutations in *NR2E3* also lead to excessive S-cones. While these findings indicate that rod factors are involved with suppressing the S-cone cell fate, M-cones similarly repress S-opsin expression; *Trβ2* knockout mice lack M-opsin and express S-opsin in all of their cone cells. Finally, PIAS3-dependent SUMOylation is also required in both rods and cones, and though its targets are different in each type of photoreceptor, its role is similar with regard to repressing S-opsin production. More recently, knock-in mice engineered to express Trβ2 from the Nrl locus were found to lack rods, as expected; however, an excess of M-cones was also observed (Ng et al., 2011). Taken together, these data suggest that S-cones are formed through a default pathway, and that when specific transcription factors reach a threshold, cell fate can be altered. First, the expression of *Nrl* determines the rod versus cone cell fate, and if the precursor cell does not become a rod, then expression of *Trβ2* determines whether the cell becomes an M-cone. Thus, the expression and levels of these two factors, *Nrl* and *Trβ2*, are critical determinants that play key roles in altering the default S-cone pathway toward either rods or M-cones, respectively.

Though the notion of a default S-cone pathway is attractive for the reasons given, it may represent an oversimplified paradigm for photoreceptor development. For one, RORβ2 synergizes with CRX to promote S-opsin expression. Rorβ2 knockout mice do not express S-opsin during the appropriate postnatal time, indicating the requirement of an additional positive regulator of S-opsin for proper S-cone development.

In this chapter, we have tried to summarize much of what is known about the transcriptional network that leads a multipotent retinal precursor cell to become a functional photoreceptor cell. First, division and proliferation of the RPC is mediated by homeodomain-containing transcription factors such as *Pax6*, *Rx1*, *Six1*, *Six3*, *Vsx2*, *Lhx2*, as well as soluble and extracellular factors such as Notch and Sonic hedgehog. Second, lineage, or competence, restriction occurs over time in response to both cell-intrinsic and cell-extrinsic factors. The cell-intrinsic factors largely belong to either activating or repressive bHLH transcription factors such as Hes1 and Hes5 or Math 3 and Math 5, respectively. Third, cells exit the cell cycle, and the formation of postmitotic precursors occurs in response to levels of the cell fate specification transcription factors Otx2,

Crx, Nrl, Nr2e3, Trβ2, Rxrγ, or RORβ. Fourth, immature photoreceptors begin to express rod- and cone-specific genes like rhodopsin or cone opsins. Fifth, terminal differentiation and maintenance follows, which includes proper opsin expression, outer segment morphogenesis, and synapse formation.

Numerous discoveries over the last 10 years have contributed to a detailed grasp of the transcriptional mechanisms underlying retinal cell fate. Coupled with recent advances in human embryonic stem cell or induced pluripotent stem cell–based therapies, understanding the conditions for lineage-restricted differentiation will be paramount to ensuring safe and effective therapeutic strategies. As we witness the emergence of clinical therapeutics derived from detailed mechanistic studies, like those described here, it is an exciting time for translational vision research.

REFERENCES

Acampora, D., Avantaggiato, V., Tuorto, F., Barone, P., Reichert, H., Finkelstein, R., et al. (1998). Murine Otx1 and *Drosophila* otd genes share conserved genetic functions required in invertebrate and vertebrate brain development. *Development, 125,* 1691–1702.

Adachi, Y., Nagao, T., Saiga, H., & Furukubo-Tokunaga, K. (2001). Cross-phylum regulatory potential of the ascidian Otx gene in brain development in *Drosophila* melanogaster. *Development Genes and Evolution, 211,* 269–280.

Agathocleous, M., & Harris, W. A. (2009). From progenitors to differentiated cells in the vertebrate retina. *Annual Review of Cell and Developmental Biology, 25,* 45–69. doi:10.1146/annurev.cellbio.042308.113259.

Ahmad, I. (1995). Mash-1 is expressed during ROD photoreceptor differentiation and binds an E-box, E(opsin)-1 in the rat opsin gene. *Brain Research. Developmental Brain Research, 90,* 184–189.

Akhmedov, N. B., Piriev, N. I., Chang, B., Rapoport, A. L., Hawes, N. L., Nishina, P. M., et al. (2000). A deletion in a photoreceptor-specific nuclear receptor mRNA causes retinal degeneration in the rd7 mouse. *Proceedings of the National Academy of Sciences of the United States of America, 97,* 5551–5556.

Akimoto, M., Cheng, H., Zhu, D., Brzezinski, J. A., Khanna, R., Filippova, E., et al. (2006). Targeting of GFP to newborn rods by Nrl promoter and temporal expression profiling of flow-sorted photoreceptors. *Proceedings of the National Academy of Sciences of the United States of America, 103,* 3890–3895. doi:10.1073/pnas.0508214103.

Andre, E., Conquet, F., Steinmayr, M., Stratton, S. C., Porciatti, V., & Becker-Andre, M. (1998a). Disruption of retinoid-related orphan receptor beta changes circadian behavior, causes retinal degeneration and leads to vacillans phenotype in mice. *EMBO Journal, 17,* 3867–3877. doi:10.1093/emboj/17.14.3867.

Andre, E., Gawlas, K., & Becker-Andre, M. (1998b). A novel isoform of the orphan nuclear receptor RORbeta is specifically expressed in pineal gland and retina. *Gene, 216,* 277–283.

Arendt, D., Tessmar-Raible, K., Snyman, H., Dorresteijn, A. W., & Wittbrodt, J. (2004). Ciliary photoreceptors with a vertebrate-type opsin in an invertebrate brain. *Science, 306*, 869–871. doi:10.1126/science.1099955.

Baas, D., Bumsted, K. M., Martinez, J. A., Vaccarino, F. M., Wikler, K. C., & Barnstable, C. J. (2000). The subcellular localization of Otx2 is cell-type specific and developmentally regulated in the mouse retina. *Brain Research. Molecular Brain Research, 78*, 26–37.

Bessant, D. A., Payne, A. M., Mitton, K. P., Wang, Q. L., Swain, P. K., Plant, C., et al. (1999). A mutation in NRL is associated with autosomal dominant retinitis pigmentosa. *Nature Genetics, 21*, 355–356. doi:10.1038/7678.

Blackshaw, S., Fraioli, R. E., Furukawa, T., & Cepko, C. L. (2001). Comprehensive analysis of photoreceptor gene expression and the identification of candidate retinal disease genes. *Cell, 107*, 579–589.

Blackshaw, S., Harpavat, S., Trimarchi, J., Cai, L., Huang, H., Kuo, W. P., et al. (2004). Genomic analysis of mouse retinal development. *PLoS Biology, 2*, E247. doi:10.1371/journal.pbio.0020247.

Bookout, A. L., Jeong, Y., Downes, M., Yu, R. T., Evans, R. M., & Mangelsdorf, D. J. (2006). Anatomical profiling of nuclear receptor expression reveals a hierarchical transcriptional network. *Cell, 126*, 789–799. doi:10.1016/j.cell.2006.06.049.

Bowmaker, J. K. (2008). Evolution of vertebrate visual pigments. *Vision Research, 48*, 2022–2041. doi:10.1016/j.visres.2008.03.025.

Brzezinski, J. A., Lamba, D. A., & Reh, T. A. (2010). Blimp1 controls photoreceptor versus bipolar cell fate choice during retinal development. *Development, 137*, 619–629. doi:10.1242/dev.043968.

Bulger, M., & Groudine, M. (2011). Functional and mechanistic diversity of distal transcription enhancers. *Cell, 144*, 327–339.

Bumsted O'Brien, K. M., Cheng, H., Jiang, Y., Schulte, D., Swaroop, A., & Hendrickson, A. E. (2004). Expression of photoreceptor-specific nuclear receptor NR2E3 in rod photoreceptors of fetal human retina. *Investigative Ophthalmology & Visual Science, 45*, 2807–2812. doi:10.1167/iovs.03-1317.

Cajal, S. R. Y. (1892). *La rétine des vértebrés* (La Cellule, English trans.; S. Thorpe & M. Glickstein, 1972). Springfield, IL: Charles C. Thomas.

Carroll, J., Neitz, J., & Neitz, M. (2002). Estimates of L:M cone ratio from ERG flicker photometry and genetics. *Journal of Vision, 2*(8), 531–542. doi: 10:1167/2.8.1.

Carter-Dawson, L. D., & LaVail, M. M. (1979a). Rods and cones in the mouse retina: I. Structural analysis using light and electron microscopy. *Journal of Comparative Neurology, 188*, 245–262. doi:10.1002/cne.901880204.

Carter-Dawson, L. D., & LaVail, M. M. (1979b). Rods and cones in the mouse retina: II. Autoradiographic analysis of cell generation using tritiated thymidine. *Journal of Comparative Neurology, 188*, 263–272. doi:10.1002/cne.901880205.

Cepko, C. L., Austin, C. P., Yang, X., Alexiades, M., & Ezzeddine, D. (1996). Cell fate determination in the vertebrate retina. *Proceedings of the National Academy of Sciences of the United States of America, 93*, 589–595.

Chatelain, G., Fossat, N., Brun, G., & Lamonerie, T. (2006). Molecular dissection reveals decreased activity and not dominant negative effect in human OTX2 mutants. *Journal of Molecular Medicine, 84*, 604–615. doi:10.1007/s00109-006-0048-2.

Chen, J., & Nathans, J. (2007). Genetic ablation of cone photoreceptors eliminates retinal folds in the retinal degeneration 7 (rd7) mouse. *Investigative Ophthalmology & Visual Science, 48*, 2799–2805. doi:10.1167/iovs.06-0922.

Chen, J., Rattner, A., & Nathans, J. (2005). The rod photoreceptor-specific nuclear receptor Nr2e3 represses transcription of multiple cone-specific genes. *Journal of Neuroscience, 25*, 118–129. doi:10.1523/JNEUROSCI.3571-04.2005.

Chen, J., Rattner, A., & Nathans, J. (2006). Effects of L1 retrotransposon insertion on transcript processing, localization and accumulation: Lessons from the retinal degeneration 7 mouse and implications for the genomic ecology of L1 elements. *Human Molecular Genetics, 15*, 2146–2156. doi:10.1093/hmg/ddl138.

Chen, S., Wang, Q. L., Nie, Z., Sun, H., Lennon, G., Copeland, N. G., et al. (1997). Crx, a novel Otx-like paired-homeodomain protein, binds to and transactivates photoreceptor cell-specific genes. *Neuron, 19*, 1017–1030.

Chen, S., Wang, Q. L., Xu, S., Liu, I., Li, L. Y., Wang, Y., et al. (2002). Functional analysis of cone–rod homeobox (CRX) mutations associated with retinal dystrophy. *Human Molecular Genetics, 11*, 873–884.

Chen, S., & Zack, D. J. (1996). Ret 4, a positive acting rhodopsin regulatory element identified using a bovine retina in vitro transcription system. *Journal of Biological Chemistry, 271*, 28549–28557.

Cheng, H., Aleman, T. S., Cideciyan, A. V., Khanna, R., Jacobson, S. G., & Swaroop, A. (2006). In vivo function of the orphan nuclear receptor NR2E3 in establishing photoreceptor identity during mammalian retinal development. *Human Molecular Genetics, 15*, 2588–2602. doi:10.1093/hmg/ddl185.

Cheng, H., Khanna, H., Oh, E. C., Hicks, D., Mitton, K. P., & Swaroop, A. (2004). Photoreceptor-specific nuclear receptor NR2E3 functions as a transcriptional activator in rod photoreceptors. *Human Molecular Genetics, 13*, 1563–1575. doi:10.1093/hmg/ddh173.

Chow, R. L., Altmann, C. R., Lang, R. A., & Hemmati-Brivanlou, A. (1999). Pax6 induces ectopic eyes in a vertebrate. *Development, 126*, 4213–4222.

Corbo, J. C., & Cepko, C. L. (2005). A hybrid photoreceptor expressing both rod and cone genes in a mouse model of enhanced S-cone syndrome. *PLOS Genetics, 1*, e11. doi:10.1371/journal.pgen.0010011.

Corbo, J. C., Lawrence, K. A., Karlstetter, M., Myers, C. A., Abdelaziz, M., Dirkes, W., et al. (2010). CRX ChIP-seq reveals the cis-regulatory architecture of mouse photoreceptors. *Genome Research, 20*, 1512–1525. doi:10.1101/gr.109405.110.

Cornish, E. E., Hendrickson, A. E., & Provis, J. M. (2004). Distribution of short-wavelength-sensitive cones in human fetal and postnatal retina: Early development of spatial order and density profiles. *Vision Research, 44*, 2019–2026. doi:10.1016/j.visres.2004.03.030.

Daniele, L. L., Lillo, C., Lyubarsky, A. L., Nikonov, S. S., Philp, N., Mears, A. J., et al. (2005). Cone-like morphological, molecular, and electrophysiological features of the photoreceptors of the Nrl knockout mouse. *Investigative Ophthalmology & Visual Science, 46*, 2156–2167. doi:10.1167/iovs.04-1427.

Davidson, E. H. (2006). *The regulatory genome: Gene regulatory networks in development and evolution.* Boston: Academic Press.

DesJardin, L. E., & Hauswirth, W. W. (1996). Developmentally important DNA elements within the bovine opsin upstream region. *Investigative Ophthalmology & Visual Science, 37,* 154–165.

Dorn, E. M., Hendrickson, L., & Hendrickson, A. E. (1995). The appearance of rod opsin during monkey retinal development. *Investigative Ophthalmology & Visual Science, 36,* 2634–2651.

Elliott, R. W., Sparkes, R. S., Mohandas, T., Grant, S. G., & McGinnis, J. F. (1990). Localization of the rhodopsin gene to the distal half of mouse chromosome 6. *Genomics, 6,* 635–644.

Emerson, M. M., & Cepko, C. L. (2011). Identification of a retina-specific Otx2 enhancer element active in immature developing photoreceptors. *Developmental Biology, 360,* 241–255. doi:10.1016/j.ydbio.2011.09.012.

Esumi, N., Kachi, S., Hackler, L., Jr., Masuda, T., Yang, Z., Campochiaro, P. A., et al. (2009). BEST1 expression in the retinal pigment epithelium is modulated by OTX family members. *Human Molecular Genetics, 18,* 128–141. doi:10.1093/hmg/ddn323.

Finkelstein, R., Smouse, D., Capaci, T. M., Spradling, A. C., & Perrimon, N. (1990). The orthodenticle gene encodes a novel homeo domain protein involved in the development of the *Drosophila* nervous system and ocellar visual structures. *Genes & Development, 4,* 1516–1527.

Freund, C. L., Gregory-Evans, C. Y., Furukawa, T., Papaioannou, M., Looser, J., Ploder, L., et al. (1997). Cone-rod dystrophy due to mutations in a novel photoreceptor-specific homeobox gene (CRX) essential for maintenance of the photoreceptor. *Cell, 91,* 543–553.

Freund, C. L., Wang, Q. L., Chen, S., Muskat, B. L., Wiles, C. D., Sheffield, V. C., et al. (1998). De novo mutations in the CRX homeobox gene associated with Leber congenital amaurosis. *Nature Genetics, 18,* 311–312.

Friedman, J. S., Khanna, H., Swain, P. K., Denicola, R., Cheng, H., Mitton, K. P., et al. (2004). The minimal trans-activation domain of the basic motif-leucine zipper transcription factor NRL interacts with TATA-binding protein. *Journal of Biological Chemistry, 279,* 47233–47241. doi:10.1074/jbc.M408298200.

Fujieda, H., Bremner, R., Mears, A. J., & Sasaki, H. (2009). Retinoic acid receptor-related orphan receptor alpha regulates a subset of cone genes during mouse retinal development. *Journal of Neurochemistry, 108,* 91–101. doi:10.1111/j.1471-4159.2008.05739.x.

Furukawa, T., Morrow, E. M., & Cepko, C. L. (1997). Crx, a novel otx-like homeobox gene, shows photoreceptor-specific expression and regulates photoreceptor differentiation. *Cell, 91,* 531–541.

Garelli, A., Rotstein, N. P., & Politi, L. E. (2006). Docosahexaenoic acid promotes photoreceptor differentiation without altering Crx expression. *Investigative Ophthalmology & Visual Science, 47,* 3017–3027. doi:10.1167/iovs.05-1659.

Haider, N. B., Jacobson, S. G., Cideciyan, A. V., Swiderski, R., Streb, L. M., Searby, C., et al. (2000). Mutation of a nuclear receptor gene, NR2E3, causes enhanced S cone syndrome, a disorder of retinal cell fate. *Nature Genetics, 24,* 127–131. doi:10.1038/72777.

Halder, G., Callaerts, P., & Gehring, W. J. (1995). Induction of ectopic eyes by targeted expression of the eyeless gene in *Drosophila. Science, 267,* 1788–1792.

Hanes, S. D., & Brent, R. (1989). DNA specificity of the bicoid activator protein is determined by homeodomain recognition helix residue 9. *Cell, 57,* 1275–1283.

Hanes, S. D., & Brent, R. (1991). A genetic model for interaction of the homeodomain recognition helix with DNA. *Science, 251,* 426–430.

Hanson, I. M., Fletcher, J. M., Jordan, T., Brown, A., Taylor, D., Adams, R. J., et al. (1994). Mutations at the PAX6 locus are found in heterogeneous anterior segment malformations including Peters' anomaly. *Nature Genetics, 6,* 168–173. doi:10.1038/ng0294-168.

Hao, H., Tummala, P., Guzman, E., Mali, R. S., Gregorski, J., Swaroop, A., et al. (2011). The transcription factor neural retina leucine zipper (NRL) controls photoreceptor-specific expression of myocyte enhancer factor Mef2c from an alternative promoter. *Journal of Biological Chemistry, 286,* 34893–34902. doi:10.1074/jbc.M111.271072.

Hatakeyama, J., & Kageyama, R. (2004). Retinal cell fate determination and bHLH factors. *Seminars in Cell & Developmental Biology, 15,* 83–89. doi:10.1016/j.semcdb.2003.09.005.

Hayashi, T., Motulsky, A. G., & Deeb, S. S. (1999). Position of a 'green-red' hybrid gene in the visual pigment array determines colour-vision phenotype. *Nature Genetics, 22,* 90–93. doi:10.1038/8798.

Hendrickson, A., Bumsted-O'Brien, K., Natoli, R., Ramamurthy, V., Possin, D., & Provis, J. (2008). Rod photoreceptor differentiation in fetal and infant human retina. *Experimental Eye Research, 87,* 415–426. doi:10.1016/j.exer.2008.07.016.

Hobert, O. (2008). Regulatory logic of neuronal diversity: Terminal selector genes and selector motifs. *Proceedings of the National Academy of Sciences of the United States of America, 105,* 20067–20071. doi:10.1073/pnas.0806070105.

Hsiau, T. H., Diaconu, C., Myers, C. A., Lee, J., Cepko, C. L., & Corbo, J. C. (2007). The cis-regulatory logic of the mammalian photoreceptor transcriptional network. *PLoS ONE, 2,* e643. doi:10.1371/journal.pone.0000643.

Jacobs, G. H. (1993). The distribution and nature of colour vision among the mammals. *Biological Reviews of the Cambridge Philosophical Society, 68,* 413–471.

Jadhav, A. P., Mason, H. A., & Cepko, C. L. (2006). Notch 1 inhibits photoreceptor production in the developing mammalian retina. *Development, 133,* 913–923. doi:10.1242/dev.02245.

Janssen, J. J., Kuhlmann, E. D., van Vugt, A. H., Winkens, H. J., Janssen, B. P., Deutman, A. F., et al. (1999). Retinoic acid receptors and retinoid X receptors in the mature retina: Subtype determination and cellular distribution. *Current Eye Research, 19,* 338–347.

Jetten, A. M. (2009). Retinoid-related orphan receptors (RORs): Critical roles in development, immunity, circadian rhythm, and cellular metabolism. *Nuclear Receptor Cignaling, 7,* e003. doi:10.1621/nrs.07003.

Jia, L., Oh, E. C., Ng, L., Srinivas, M., Brooks, M., Swaroop, A., et al. (2009). Retinoid-related orphan nuclear receptor RORbeta is an early-acting factor in rod photoreceptor development. *Proceedings of the National Academy of Sciences of the United States of America, 106,* 17534–17539. doi:10.1073/pnas.0902425106.

Katoh, K., Omori, Y., Onishi, A., Sato, S., Kondo, M., & Furukawa, T. (2010). Blimp1 suppresses Chx10 expression in differentiating retinal photoreceptor precursors to ensure proper photoreceptor development. *Journal of Neuroscience, 30,* 6515–6526. doi:10.1523/JNEURO-SCI.0771-10.2010.

Kautzmann, M. A., Kim, D. S., Felder-Schmittbuhl, M. P., & Swaroop, A. (2011). Combinatorial regulation of photoreceptor differentiation factor, neural retina leucine zipper gene NRL, revealed by in vivo promoter analysis. *Journal of Biological Chemistry, 286,* 28247–28255. doi:10.1074/jbc.M111.257246.

Kikuchi, T., Raju, K., Breitman, M. L., & Shinohara, T. (1993). The proximal promoter of the mouse arrestin gene directs gene expression in photoreceptor cells and contains an evolutionarily conserved retinal factor-binding site. *Molecular and Cellular Biology, 13,* 4400–4408.

Kobayashi, M., Takezawa, S., Hara, K., Yu, R. T., Umesono, Y., Agata, K., et al. (1999). Identification of a photoreceptor cell-specific nuclear receptor. *Proceedings of the National Academy of Sciences of the United States of America, 96,* 4814–4819.

Koike, C., Nishida, A., Ueno, S., Saito, H., Sanuki, R., Sato, S., et al. (2007). Functional roles of Otx2 transcription factor in postnatal mouse retinal development. *Molecular and Cellular Biology, 27,* 8318–8329. doi:10.1128/MCB.01209-07.

Kokotas, H., & Petersen, M. B. (2010). Clinical and molecular aspects of aniridia. *Clinical Genetics, 77,* 409–420. doi:10.1111/j.1399-0004.2010.01372.x.

Kotaja, N., Karvonen, U., Janne, O. A., & Palvimo, J. J. (2002). PIAS proteins modulate transcription factors by functioning as SUMO-1 ligases. *Molecular and Cellular Biology, 22,* 5222–5234.

Kumar, R., Chen, S., Scheurer, D., Wang, Q. L., Duh, E., Sung, C. H., et al. (1996). The bZIP transcription factor Nrl stimulates rhodopsin promoter activity in primary retinal cell cultures. *Journal of Biological Chemistry, 271,* 29612–29618.

Lambard, S., Reichman, S., Berlinicke, C., Niepon, M. L., Goureau, O., Sahel, J. A., et al. (2010). Expression of rod-derived cone viability factor: Dual role of CRX in regulating promoter activity and cell-type specificity. *PLoS ONE, 5,* e13075. doi:10.1371/journal.pone.0013075.

La Vail, M. M., Rapaport, D. H., & Rakic, P. (1991). Cytogenesis in the monkey retina. *Journal of Comparative Neurology, 309,* 86–114. doi:10.1002/cne.903090107.

Leveillard, T., & Sahel, J. A. (2010). Rod-derived cone viability factor for treating blinding diseases: From clinic to redox signaling. *Science Translational Medicine, 2,* 26ps16. doi:10.1126/scitranslmed.3000866.

Liu, Q., Ji, X., Breitman, M. L., Hitchcock, P. F., & Swaroop, A. (1996). Expression of the bZIP transcription factor gene Nrl in the developing nervous system. *Oncogene, 12,* 207–211.

Liu, Y., Shen, Y., Rest, J. S., Raymond, P. A., & Zack, D. J. (2001). Isolation and characterization of a zebrafish homologue of the cone rod homeobox gene. *Investigative Ophthalmology & Visual Science, 42,* 481–487.

Livesey, F. J., Furukawa, T., Steffen, M. A., Church, G. M., & Cepko, C. L. (2000). Microarray analysis of the transcriptional network controlled by the photoreceptor homeobox gene Crx. *Current Biology, 10,* 301–310.

Lu, A., Ng, L., Ma, M., Kefas, B., Davies, T. F., Hernandez, A., et al. (2009). Retarded developmental expression and pat-terning of retinal cone opsins in hypothyroid mice. *Endocrinology, 150,* 1536–1544. doi:10.1210/en.2008-1092.

Mali, R. S., Zhang, X., Hoerauf, W., Doyle, D., Devitt, J., Loffreda-Wren, J., et al. (2007). FIZ1 is expressed during photoreceptor maturation, and synergizes with NRL and CRX at rod-specific promoters in vitro. *Experimental Eye Research, 84,* 349–360. doi:10.1016/j.exer.2006.10.009.

Marquardt, T., Ashery-Padan, R., Andrejewski, N., Scardigli, R., Guillemot, F., & Gruss, P. (2001). Pax6 is required for the multipotent state of retinal progenitor cells. *Cell, 105,* 43–55.

Martin, R. L., Wood, C., Baehr, W., & Applebury, M. L. (1986). Visual pigment homologies revealed by DNA hybridization. *Science, 232,* 1266–1269.

Martinez-Morales, J. R., Dolez, V., Rodrigo, I., Zaccarini, R., Leconte, L., Bovolenta, P., et al. (2003). OTX2 activates the molecular network underlying retina pigment epithelium differentiation. *Journal of Biological Chemistry, 278,* 21721–21731. doi:10.1074/jbc.M301708200.

Martinez-Morales, J. R., Signore, M., Acampora, D., Simeone, A., & Bovolenta, P. (2001). Otx genes are required for tissue specification in the developing eye. *Development, 128,* 2019–2030.

Matsuo, I., Kuratani, S., Kimura, C., Takeda, N., & Aizawa, S. (1995). Mouse Otx2 functions in the formation and patterning of rostral head. *Genes & Development, 9,* 2646–2658.

Mears, A. J., Kondo, M., Swain, P. K., Takada, Y., Bush, R. A., Saunders, T. L., et al. (2001). Nrl is required for rod photoreceptor development. *Nature Genetics, 29,* 447–452. doi:10.1038/ng774.

Merbs, S. L., & Nathans, J. (1992). Absorption spectra of human cone pigments. *Nature, 356,* 433–435. doi:10.1038/356433a0.

Meulmeester, E., & Melchior, F. (2008). Cell biology: SUMO. *Nature, 452,* 709–711. doi:10.1038/452709a.

Milam, A. H., Rose, L., Cideciyan, A. V., Barakat, M. R., Tang, W. X., Gupta, N., et al. (2002). The nuclear receptor NR2E3 plays a role in human retinal photoreceptor differentiation and degeneration. *Proceedings of the National Academy of Sciences of the United States of America, 99,* 473–478. doi:10.1073/pnas.022533099.

Mitton, K. P., Swain, P. K., Chen, S., Xu, S., Zack, D. J., & Swaroop, A. (2000). The leucine zipper of NRL interacts with the CRX homeodomain: A possible mechanism of transcriptional synergy in rhodopsin regulation. *Journal of Biological Chemistry, 275,* 29794–29799. doi:10.1074/jbc.M003658200.

Montana, C. L., Lawrence, K. A., Williams, N. L., Tran, N. M., Peng, G. H., Chen, S., et al. (2011). Transcriptional regulation of neural retina leucine zipper (nrl), a photoreceptor cell fate determinant. *Journal of Biological Chemistry, 286,* 36921–36931. doi:10.1074/jbc.M111.279026.

Morabito, M. A., Yu, X., & Barnstable, C. J. (1991). Characterization of developmentally regulated and retina-specific nuclear protein binding to a site in the upstream region of the rat opsin gene. *Journal of Biological Chemistry, 266,* 9667–9672.

Mori, M., Ghyselinck, N. B., Chambon, P., & Mark, M. (2001). Systematic immunolocalization of retinoid receptors in developing and adult mouse eyes. *Investigative Ophthalmology & Visual Science, 42,* 1312–1318.

Morrow, E. M., Belliveau, M. J., & Cepko, C. L. (1998). Two phases of rod photoreceptor differentiation during rat retinal development. *Journal of Neuroscience, 18*, 3738–3748.

Morrow, E. M., Furukawa, T., Lee, J. E., & Cepko, C. L. (1999). NeuroD regulates multiple functions in the developing neural retina in rodent. *Development, 126*, 23–36.

Morrow, E. M., Furukawa, T., Raviola, E., & Cepko, C. L. (2005). Synaptogenesis and outer segment formation are perturbed in the neural retina of Crx mutant mice. *BMC Neuroscience, 6*, 5. doi:10.1186/1471-2202-6-5.

Nagao, T., Leuzinger, S., Acampora, D., Simeone, A., Finkelstein, R., Reichert, H., et al. (1998). Developmental rescue of *Drosophila* cephalic defects by the human Otx genes. *Proceedings of the National Academy of Sciences of the United States of America, 95*, 3737–3742.

Nathans, J. (1999). The evolution and physiology of human color vision: Insights from molecular genetic studies of visual pigments. *Neuron, 24*, 299–312.

Nathans, J., Davenport, C. M., Maumenee, I. H., Lewis, R. A., Hejtmancik, J. F., Litt, M., et al. (1989). Molecular genetics of human blue cone monochromacy. *Science, 245*, 831–838.

Nathans, J., Maumenee, I. H., Zrenner, E., Sadowski, B., Sharpe, L. T., Lewis, R. A., et al. (1993). Genetic heterogeneity among blue-cone monochromats. *American Journal of Human Genetics, 53*, 987–1000.

Nathans, J., Merbs, S. L., Sung, C. H., Weitz, C. J., & Wang, Y. (1992). Molecular genetics of human visual pigments. *Annual Review of Genetics, 26*, 403–424. doi:10.1146/annurev.ge.26.120192.002155.

Nathans, J., Piantanida, T. P., Eddy, R. L., Shows, T. B., & Hogness, D. S. (1986a). Molecular genetics of inherited variation in human color vision. *Science, 232*, 203–210.

Nathans, J., Thomas, D., & Hogness, D. S. (1986b). Molecular genetics of human color vision: The genes encoding blue, green, and red pigments. *Science, 232*, 193–202.

Ng, L., Hurley, J. B., Dierks, B., Srinivas, M., Salto, C., Vennstrom, B., et al. (2001). A thyroid hormone receptor that is required for the development of green cone photoreceptors. *Nature Genetics, 27*, 94–98. doi:10.1038/83829.

Ng, L., Lu, A., Swaroop, A., Sharlin, D. S., & Forrest, D. (2011). Two transcription factors can direct three photoreceptor outcomes from rod precursor cells in mouse retinal development. *Journal of Neuroscience, 31*, 11118–11125. doi:10.1523/JNEUROSCI.1709-11.2011.

Nie, Z., Chen, S., Kumar, R., & Zack, D. J. (1996). RER, an evolutionarily conserved sequence upstream of the rhodopsin gene, has enhancer activity. *Journal of Biological Chemistry, 271*, 2667–2675.

Nishida, A., Furukawa, A., Koike, C., Tano, Y., Aizawa, S., Matsuo, I., et al. (2003). Otx2 homeobox gene controls retinal photoreceptor cell fate and pineal gland development. *Nature Neuroscience, 6*, 1255–1263. doi:10.1038/nn1155.

Nishiguchi, K. M., Friedman, J. S., Sandberg, M. A., Swaroop, A., Berson, E. L., & Dryja, T. P. (2004). Recessive NRL mutations in patients with clumped pigmentary retinal degeneration and relative preservation of blue cone function. *Proceedings of the National Academy of Sciences of the United States of America, 101*, 17819–17824. doi:10.1073/pnas.0408183101.

Nunez, J., Celi, F. S., Ng, L., & Forrest, D. (2008). Multigenic control of thyroid hormone functions in the nervous system. *Molecular and Cellular Endocrinology, 287*, 1–12. doi:10.1016/j.mce.2008.03.006.

Oh, E. C., Khan, N., Novelli, E., Khanna, H., Strettoi, E., & Swaroop, A. (2007). Transformation of cone precursors to functional rod photoreceptors by bZIP transcription factor NRL. *Proceedings of the National Academy of Sciences of the United States of America, 104*, 1679–1684. doi:10.1073/pnas.0605934104.

Okano, T., Kojima, D., Fukada, Y., Shichida, Y., & Yoshizawa, T. (1992). Primary structures of chicken cone visual pigments: Vertebrate rhodopsins have evolved out of cone visual pigments. *Proceedings of the National Academy of Sciences of the United States of America, 89*, 5932–5936.

Omori, Y., Katoh, K., Sato, S., Muranishi, Y., Chaya, T., Onishi, A., et al. (2011). Analysis of transcriptional regulatory pathways of photoreceptor genes by expression profiling of the Otx2-deficient retina. *PLoS ONE, 6*, e19685. doi:10.1371/journal.pone.0019685.

Onishi, A., Peng, G. H., Chen, S., & Blackshaw, S. (2010a). Pias3-dependent SUMOylation controls mammalian cone photoreceptor differentiation. *Nature Neuroscience, 13*, 1059–1065. doi:10.1038/nn.2618.

Onishi, A., Peng, G. H., Hsu, C., Alexis, U., Chen, S., & Blackshaw, S. (2009). Pias3-dependent SUMOylation directs rod photoreceptor development. *Neuron, 61*, 234–246. doi:10.1016/j.neuron.2008.12.006.

Onishi, A., Peng, G. H., Poth, E. M., Lee, D. A., Chen, J., Alexis, U., et al. (2010b). The orphan nuclear hormone receptor ERRbeta controls rod photoreceptor survival. *Proceedings of the National Academy of Sciences of the United States of America, 107*, 11579–11584. doi:10.1073/pnas.1000102107.

Peng, G. H., Ahmad, O., Ahmad, F., Liu, J., & Chen, S. (2005). The photoreceptor-specific nuclear receptor Nr2e3 interacts with Crx and exerts opposing effects on the transcription of rod versus cone genes. *Human Molecular Genetics, 14*, 747–764. doi:10.1093/hmg/ddi070.

Peng, G. H., & Chen, S. (2005). Chromatin immunoprecipitation identifies photoreceptor transcription factor targets in mouse models of retinal degeneration: New findings and challenges. *Visual Neuroscience, 22*, 575–586. doi:10.1017/S0952523805225063.

Peng, G. H., & Chen, S. (2011). Active opsin loci adopt intrachromosomal loops that depend on the photoreceptor transcription factor network. *Proceedings of the National Academy of Sciences of the United States of America, 108*, 17821–17826. doi:10.1073/pnas.1109209108.

Plouhinec, J. L., Sauka-Spengler, T., Germot, A., Le Mentec, C., Cabana, T., Harrison, G., et al. (2003). The mammalian Crx genes are highly divergent representatives of the Otx5 gene family, a gnathostome orthology class of orthodenticle-related homeogenes involved in the differentiation of retinal photoreceptors and circadian entrainment. *Molecular Biology and Evolution, 20*, 513–521. doi:10.1093/molbev/msg085.

Qian, J., Esumi, N., Chen, Y., Wang, Q., Chowers, I., & Zack, D. J. (2005). Identification of regulatory targets of tissue-specific transcription factors: Application to retina-specific gene. *Nucleic Acids Research, 33*, 3479–3491. doi:10.1093/nar/gki658.

Quan, X. J., Ramaekers, A., & Hassan, B. A. (2012). Transcriptional control of cell fate specification: Lessons from the fly retina. *Current Topics in Developmental Biology, 98*, 259–276.

Ragge, N. K., Brown, A. G., Poloschek, C. M., Lorenz, B., Henderson, R. A., Clarke, M. P., et al. (2005). Heterozygous mutations of OTX2 cause severe ocular malformations. *American Journal of Human Genetics, 76*, 1008–1022. doi:10.1086/430721.

Rehemtulla, A., Warwar, R., Kumar, R., Ji, X., Zack, D. J., & Swaroop, A. (1996). The basic motif-leucine zipper transcription factor Nrl can positively regulate rhodopsin gene expression. *Proceedings of the National Academy of Sciences of the United States of America, 93*, 191–195.

Reichman, S., Kalathur, R. K., Lambard, S., Ait-Ali, N., Yang, Y., Lardenois, A., et al. (2010). The homeobox gene CHX10/VSX2 regulates RdCVF promoter activity in the inner retina. *Human Molecular Genetics, 19*, 250–261. doi:10.1093/hmg/ddp484.

Rivolta, C., Berson, E. L., & Dryja, T. P. (2001). Dominant Leber congenital amaurosis, cone–rod degeneration, and retinitis pigmentosa caused by mutant versions of the transcription factor CRX. *Human Mutation, 18*, 488–498. doi:10.1002/humu.1226.

Roberts, M. R., Hendrickson, A., McGuire, C. R., & Reh, T. A. (2005). Retinoid X receptor (gamma) is necessary to establish the S-opsin gradient in cone photoreceptors of the developing mouse retina. *Investigative Ophthalmology & Visual Science, 46*, 2897–2904. doi:10.1167/iovs.05-0093.

Roberts, M. R., Srinivas, M., Forrest, D., Morreale de Escobar, G., & Reh, T. A. (2006). Making the gradient: Thyroid hormone regulates cone opsin expression in the developing mouse retina. *Proceedings of the National Academy of Sciences of the United States of America, 103*, 6218–6223. doi:10.1073/pnas.0509981103.

Roger, J. E., Nellissery, J., Kim, D. S., & Swaroop, A. (2010). Sumoylation of bZIP transcription factor NRL modulates target gene expression during photoreceptor differentiation. *Journal of Biological Chemistry, 285*, 25637–25644. doi:10.1074/jbc.M110.142810.

Roorda, A., & Williams, D. R. (1999). The arrangement of the three cone classes in the living human eye. *Nature, 397*, 520–522. doi:10.1038/17383.

Shaham, O., Menuchin, Y., Farhy, C., & Ashery-Padan, R. (2012). Pax6: A multi-level regulator of ocular development. Progress in Retinal and Eye Research [Epub ahead of print, May 3]. doi:10.1016/j.preteyeres.2012.04.002.

Sidman, R. (1961). *Histologenesis of the mouse retina studied with thymidine-3H.* New York: Academic Press.

Smallwood, P. M., Wang, Y., & Nathans, J. (2002). Role of a locus control region in the mutually exclusive expression of human red and green cone pigment genes. *Proceedings of the National Academy of Sciences of the United States of America, 99*, 1008–1011. doi:10.1073/pnas.022629799.

Sohocki, M. M., Sullivan, L. S., Mintz-Hittner, H. A., Birch, D., Heckenlively, J. R., Freund, C., et al. (1998). A range of clinical phenotypes associated with mutations in CRX, a photoreceptor transcription-factor gene. *American Journal of Human Genetics, 63*, 1307–1315. doi:10.1086/302101.

Song, Y., Yao, X., & Ying, H. (2011). Thyroid hormone action in metabolic regulation. *Protein & Cell, 2*, 358–368. doi:10.1007/s13238-011-1046-x.

Srinivas, M., Ng, L., Liu, H., Jia, L., & Forrest, D. (2006). Activation of the blue opsin gene in cone photoreceptor development by retinoid-related orphan receptor beta. *Molecular Endocrinology (Baltimore, Md.), 20*, 1728–1741. doi:10.1210/me.2005-0505.

Swain, P. K., Hicks, D., Mears, A. J., Apel, I. J., Smith, J. E., John, S. K., et al. (2001). Multiple phosphorylated isoforms of NRL are expressed in rod photoreceptors. *Journal of Biological Chemistry, 276*, 36824–36830. doi:10.1074/jbc.M105855200.

Swaroop, A., Xu, J. Z., Pawar, H., Jackson, A., Skolnick, C., & Agarwal, N. (1992). A conserved retina-specific gene encodes a basic motif/leucine zipper domain. *Proceedings of the National Academy of Sciences of the United States of America, 89*, 266–270.

Swaroop, A., Wang, Q. L., Wu, W., Cook, J., Coats, C., Xu, S., et al. (1999). Leber congenital amaurosis caused by a homozygous mutation (R90W) in the homeodomain of the retinal transcription factor CRX: Direct evidence for the involvement of CRX in the development of photoreceptor function. *Human Molecular Genetics, 8*, 299–305.

Swaroop, A., Kim, D., & Forrest, D. (2010). Transcriptional regulation of photoreceptor development and homeostasis in the mammalian retina. *Nature Reviews. Neuroscience, 11*, 563–576. doi:10.1038/nrn2880.

Szel, A., van Veen, T., & Rohlich, P. (1994). Retinal cone differentiation. *Nature, 370*, 336. doi:10.1038/370336a0.

Tahayato, A., Sonneville, R., Pichaud, F., Wernet, M. F., Papatsenko, D., Beaufils, P., et al. (2003). Otd/Crx, a dual regulator for the specification of ommatidia subtypes in the *Drosophila* retina. *Developmental Cell, 5*, 391–402.

Terrell, D., Xie, B., Workman, M., Mahato, S., Zelhof, A., Gebelein, B., et al. (2012). OTX2 and CRX rescue overlapping and photoreceptor-specific functions in the *Drosophila* eye. *Developmental Dynamics, 241*, 215–228. doi:10.1002/dvdy.22782.

Treisman, J., Gonczy, P., Vashishtha, M., Harris, E., & Desplan, C. (1989). A single amino acid can determine the DNA binding specificity of homeodomain proteins. *Cell, 59*, 553–562.

Turner, D. L., & Cepko, C. L. (1987). A common progenitor for neurons and glia persists in rat retina late in development. *Nature, 328*, 131–136. doi:10.1038/328131a0.

Vandendries, E. R., Johnson, D., & Reinke, R. (1996). Orthodenticle is required for photoreceptor cell development in the *Drosophila* eye. *Developmental Biology, 173*, 243–255. doi:10.1006/dbio.1996.0020.

Vollrath, D., Nathans, J., & Davis, R. W. (1988). Tandem array of human visual pigment genes at Xq28. *Science, 240*, 1669–1672.

Wang, Q. L., Chen, S., Esumi, N., Swain, P. K., Haines, H. S., Peng, G., et al. (2004). QRX, a novel homeobox gene, modulates photoreceptor gene expression. *Human Molecular Genetics, 13*, 1025–1040. doi:10.1093/hmg/ddh117.

Wang, Y., Macke, J. P., Merbs, S. L., Zack, D. J., Klaunberg, B., Bennett, J., et al. (1992). A locus control region adjacent to the human red and green visual pigment genes. *Neuron, 9*, 429–440.

Wang, Y., Smallwood, P. M., Cowan, M., Blesh, D., Lawler, A., & Nathans, J. (1999). Mutually exclusive expression of human red and green visual pigment-reporter transgenes occurs at high frequency in murine cone photoreceptors. *Proceedings of the National Academy of Sciences of the United States of America, 96*, 5251–5256.

Wetts, R., & Fraser, S. E. (1988). Multipotent precursors can give rise to all major cell types of the frog retina. *Science, 239*, 1142–1145.

Xiao, M., & Hendrickson, A. (2000). Spatial and temporal expression of short, long/medium, or both opsins in human fetal cones. *Journal of Comparative Neurology, 425,* 545–559.

Yamaguchi, T., Motulsky, A. G., & Deeb, S. S. (1997). Visual pigment gene structure and expression in human retinae. *Human Molecular Genetics, 6,* 981–990.

Yanagi, Y., Takezawa, S., & Kato, S. (2002). Distinct functions of photoreceptor cell-specific nuclear receptor, thyroid hormone receptor beta2 and CRX in one photoreceptor development. *Investigative Ophthalmology & Visual Science, 43,* 3489–3494.

Yaron, O., Farhy, C., Marquardt, T., Applebury, M., & Ashery-Padan, R. (2006). Notch1 functions to suppress cone-photoreceptor fate specification in the developing mouse retina. *Development, 133,* 1367–1378. doi:10.1242/dev.02311.

Yoshida, S., Mears, A. J., Friedman, J. S., Carter, T., He, S., Oh, E., et al. (2004). Expression profiling of the developing and mature Nrl–/– mouse retina: Identification of retinal disease candidates and transcriptional regulatory targets of Nrl. *Human Molecular Genetics, 13,* 1487–1503. doi:10.1093/hmg/ddh160.

Young, R. W. (1985). Cell differentiation in the retina of the mouse. *Anatomical Record, 212,* 199–205. doi:10.1002/ar.1092120215.

Yu, J., He, S., Friedman, J. S., Akimoto, M., Ghosh, D., Mears, A. J., et al. (2004). Altered expression of genes of the Bmp/Smad and Wnt/calcium signaling pathways in the cone-only Nrl–/– mouse retina, revealed by gene profiling using custom cDNA microarrays. *Journal of Biological Chemistry, 279,* 42211–42220. doi:10.1074/jbc.M408223200.

Zack, D. J., Bennett, J., Wang, Y., Davenport, C., Klaunberg, B., Gearhart, J., et al. (1991). Unusual topography of bovine rhodopsin promoter-lacZ fusion gene expression in transgenic mouse retinas. *Neuron, 6,* 187–199.

103 Retinopathy of Prematurity: A Template for Studying Retinal Vascular Disease

MARY ELIZABETH HARTNETT

Abnormal blood vessel growth into the vitreous gel, that is, aberrant intravitreal angiogenesis, can have blinding consequences in a number of retinovascular conditions, including proliferative diabetic retinopathy, retinal vein occlusion, retinopathy of prematurity (ROP), and inflammatory diseases of the retina. Although there are differences in the early causative mechanisms of these diseases, there are similarities in the events involving the development of aberrant intravitreal angiogenesis. The pathogenic mechanisms of aberrant intravitreal angiogenesis cannot be safely studied in the human eye. Therefore, other methods must be used, including models involving animals. The most commonly used models of aberrant intravitreal angiogenesis are those that expose newborn animals to oxygen stresses and are called oxygen-induced retinopathy (OIR) models. Models of OIR use species of animals that complete retinal vascular development after birth unlike the human, in which retinal vascular development is complete at full-term birth (i.e., 40 weeks' gestation). In both human preterm infants and oxygen-susceptible newborn animal species, newly developed retinal capillaries are sensitive to oxygen stresses and can constrict in high oxygen, leaving areas of nonperfused retina that influence later development of aberrant intravitreal angiogenesis. Some OIR models expose animals to repeated oxygen fluctuations that delay retinal vascular development and mimic ROP. Through genetic or pharmacological manipulation, OIR models can be used to study the effects of various oxygen exposures or other exogenous stresses on signaling pathways that mediate abnormalities in retinal vascular development or aberrant intravitreal angiogenesis (Hartnett & Penn, 2012). A clear question, with an understanding of the limitations of the OIR model as related to the human disease of interest, is important when setting out to study the mechanisms of a disease or a pathway involving aberrant intravitreal angiogenesis. This chapter will use ROP as an example of how OIR models are used to develop and explore hypotheses related to human disease. The chapter will discuss the temporal relationships between neural retinal development and that of vascular beds in the eye and how defects in the coordination of timing of these can be associated with pathology in the human preterm infant with ROP. It will also provide evidence from models of OIR that have been translated into clinical studies in ROP.

RETINOPATHY OF PREMATURITY

Epidemiological and Clinical Characteristics

ROP is a retinovascular disease that only affects premature infants and is a leading cause of childhood blindness worldwide (Chen & Smith, 2007). ROP is complex, in that it is affected by genetic (Bizzarro et al., 2006) and environment factors (Hartnett, 2010a,b). It has been reported that 70% of the variance in ROP is due to genetic factors (Bizzarro et al., 2006), but most studies presenting candidate genes are small or have not been replicated among different populations (Hutcheson et al., 2005). Environmental factors associated with ROP have included oxygen stresses such as high oxygen at birth (Ashton, Ward, & Serpell, 1954; Michaelson, 1948; Patz, 1954, 1985) and fluctuations in oxygen during an infant's neonatal course (Cunningham et al., 1995; York et al., 2004), oxidative stress, nutritional factors, maternal factors, and reduced postnatal weight gain (Cunningham et al., 1995; Darlow et al., 2005; Hellstrom et al., 2009).

Human ROP is characterized by retinal zones, stages of disease severity, and the presence or absence of plus disease (The Committee for the Classification of Retinopathy of Prematurity, 1984). The zones correspond to the area of retina with clinically apparent vascular development of the inner retinal vascular plexus (see figure 103.1).

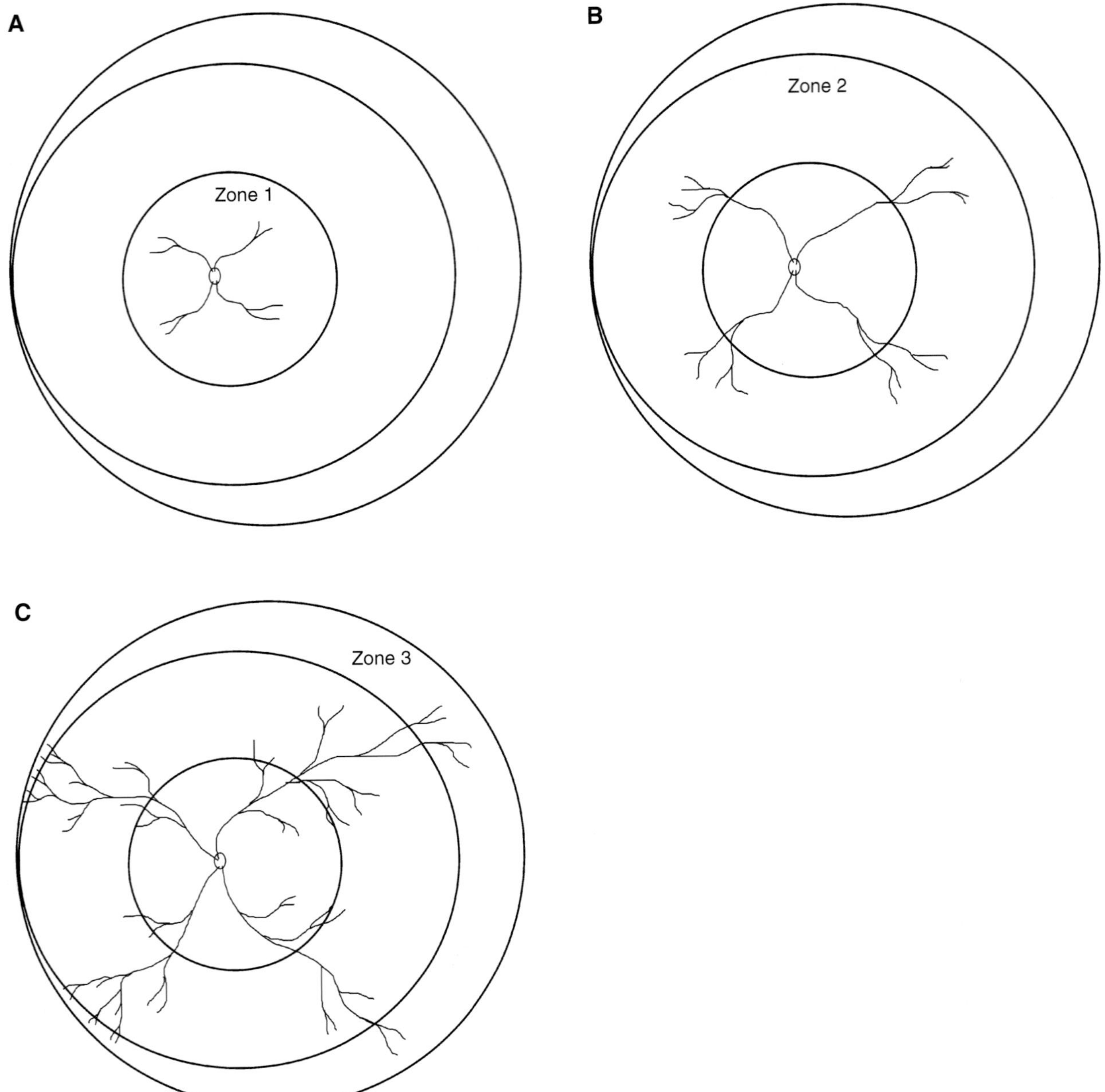

FIGURE 103.1   Diagram of the zones in retinopathy of prematurity (ROP). (A) Zone I is a circular area with the center at the optic nerve and the radius equal to twice the distance between the optic nerve and where the fovea will develop. Eyes with zone I vascularization have retinal vessels that extend almost to the circumference of zone I. Eyes that have ROP in zone I have worse prognoses than eyes with ROP in zones II or III. (B) Zone II is a circular area with the center at the optic nerve and the radius equal to the distance between the optic nerve and the nasal ora serrata. Eyes with zone II vascularization have retinal vessels that do not extend to the 2 clock h on either side of the horizontal meridian of the nasal ora serrata and which fall within the circumference of zone II. (C) Zone III represents vascularization outside zones I and II. Eyes with or without ROP in zone III have the best prognoses.

The stages of ROP describe the appearance of the retina at the junction between vascular and avascular retina in the vascular proliferative stages of ROP, stages 1 through 3 (see figure 103.2A and 103.2C), or the extent of retinal detachment in the fibrovascular stages, stage 4 (partial, figure 103.2B) or stage 5 (total) ROP.

Plus disease is the presence of tortuosity and dilation of the retinal arterioles and veins and is seen in different manifestations of severe ROP (figure 103.3).

Severe ROP is diagnosed by the combination of several parameters (see table 103.1) (Cryotherapy for Retinopathy of Prematurity Cooperative Group, 2002;

1484   MARY ELIZABETH HARTNETT

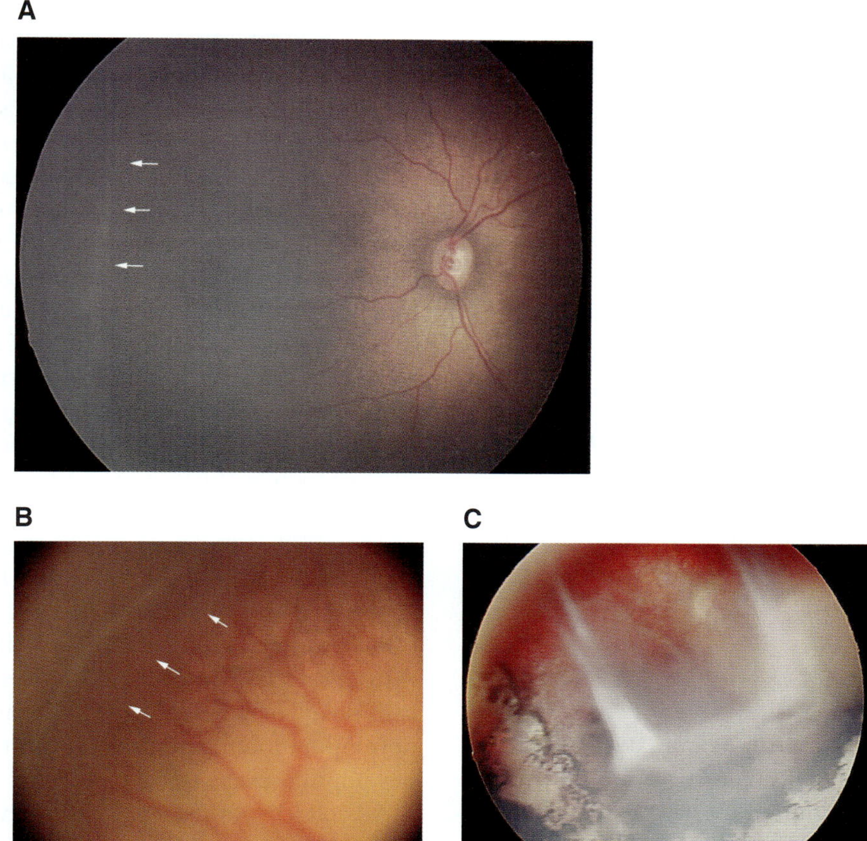

FIGURE 103.2   Images of retinopathy of prematurity (ROP) stages. (A) Image of human infant retina of a right eye with stage 1–2 ROP in zone II. The peripheral retina (left side of image) lacks retinal vessels whereas the posterior retina (right side of image) is vascularized. A line (stage 1) and, in some areas, ridge (stage 2) are apparent at the junction of the vascularized and avascularized retina (arrows). The optic nerve is apparent at the right side of the image. The fovea is immature. Image taken with wide-angle photography (Retcam, Clarity). (B) Magnified image of human infant retina with stage 3 ROP showing aberrant intravitreal angiogenesis posterior to the ridge (arrow) and at the junction of vascular (right side of image) and avascular retina (left side of image). Image taken with wide-angle photography (Retcam, Clarity). (C) Image of human infant retina of a right eye with stage 4 ROP. In focus is the attached retina seen at the top of the image and at the left. The pigmented areas are where previous laser had been performed. The white areas represent fibrous scar tissue that formed in the regions of previous aberrant intravitreal angiogenesis that had only partially regressed following laser. Retina between the fibrous tissue and the optic nerve that appears out of focus represents detached retina. Image taken with wide-angle photography (Retcam, Clarity).

Early Treatment for Retinopathy of Prematurity Cooperative Group, 2003), including extension of the retinal vasculature within zone I or zone II, the presence of plus disease, and stage 2 or stage 3 ROP (aberrant intravitreal angiogenesis). In threshold ROP (see table 103.1), the risk of a bad outcome (e.g., stage 5 ROP) without treatment approaches 50%, whereas in type 1 ROP ("prethreshold"), the risk approaches 15%. Currently, management involves examining infants at risk of severe ROP and then prompt treatment of severe ROP to prevent stages of retinal detachment, that is, stage 4 or 5 ROP, because even with successful surgical reattachment of the detached retina in these infants, visual acuity outcomes are often poor (Repka et al., 2011).

ROP develops 1 or more months after birth and usually in infants no younger than 32 weeks' postgestational age, defined as the sum of gestational age and chronologic age in weeks. ROP often progresses from stage 1 to 2, or less often to stage 3, but then regresses, clinically disappears, and vascularization of the retina proceeds. However, in about 10% of preterm infants, ROP can develop severe features (stage 3, that is, aberrant intravitreal angiogenesis, and plus disease; see figure 103.3), at which point there is increased risk of developing stage 4 or 5 ROP with retinal detachment. Preterm infants are screened with retinal examinations based on gestational age and birth weight; cutoffs vary throughout the world partly based on resources available to monitor and regulate oxygen (Gilbert et al.,

Threshold ROP (all three criteria must be met; unfavorable outcome—50%)*

- Zones I or II
- Five contiguous or 8 clock h of stage 3 ROP
- Four quadrants of plus disease (CRYO-ROP definition of plus)

Type 1 ROP (any one of these constitutes severe ROP; unfavorable outcome—15%)**

- Zone I, any stage ROP with plus disease*
- Zone I, stage 3 ROP without plus disease*
- Zone II, stage 2 or 3 with plus disease* (ETROP definition of plus is 2 quadrants)

*CRYO-ROP—Cryotherapy for Retinopathy of Prematurity Study (Cryotherapy for Retinopathy of Prematurity Cooperative Group, 1990). **ETROP—Early Treatment for Retinopathy of Prematurity Study (Early Treatment for Retinopathy of Prematurity Cooperative Group, 2003).

1997). The timing of repeat examinations is determined by the severity of ROP using zone and stage. Longitudinal retinal examinations are performed until retinal vascularization has extended to the ora serrata or severe ROP is diagnosed. If severe ROP develops, the peripheral avascular retina is treated with laser therapy (see figures 103.3B [recent laser] and 103.4 [healed laser]).

Alternative, but still experimental approaches include anti-angiogenic agents (Mintz-Hittner & Kuffel, 2008; Mintz-Hittner, Kennedy, & Chuang, 2011), such as neutralizing monoclonal antibodies to vascular endothelial growth factor (VEGF). In stage 4 or 5 ROP, surgery is performed to reattach the retina. In all infants, methods are implemented to maximize visual development through the use of corrective lenses or other treatment, as needed.

Role of Oxygen

ROP has different manifestations throughout the world and since the 1940s (Terry, 1942) when it was first described in the United States. In countries that lack resources and equipment to implement oxygen monitoring and regulation, ROP may develop in preterm infants born at larger birth weights and older gestational ages than in countries in which oxygen is routinely regulated (Gilbert et al., 1997). In addition, regional differences in prenatal care, maternal health and nutrition, and the incidence of teen or multiple births (Sapieha et al., 2010) may account for differences in outcomes among neonatal intensive care units within the United States or in individual countries.

In the United States in the 1940s, high oxygen at birth was identified as a major risk factor for developing ROP (Ashton, Ward, & Serpell, 1954; Michaelson, 1948; Patz, 1954, 1985). Since then, technological advancements in oxygen monitoring and regulation in the United States reduced the incidence of ROP, but with further improvements in neonatal care and greater ability to save infants of low birth weights and young gestational ages, ROP has reemerged and additional risk factors are now recognized (Allen, Donohue, & Dusman, 1993). Besides high oxygen at birth, which may constrict and attenuate newly developed blood vessels and capillaries, other stresses such as fluctuations in oxygen, oxidative stress, or supplemental oxygen during the later course in the neonatal intensive care units have been associated with severe ROP (Cunningham et al., 1995; Darlow et al., 2005; Hellstrom et al., 2009; York et al., 2004). Also, poor postnatal weight gain and intrauterine growth restriction (Darlow et al., 2005; Hellstrom et al., 2009) can delay or interfere with normal retinal vascular development and increase the preterm infant's risk of developing severe ROP.

A broad hypothesis has been developed to describe biological events in ROP given different oxygen exposures and appearances throughout the world. First, an infant born prematurely has incomplete retinal vascular development, and, therefore, peripheral avascular retina exists. Second, events during the neonatal course of the infant slow ongoing retinal vascular development or in some cases cause constriction or injury to newly developed capillaries. At birth, there is also a relatively high oxygen concentration compared to in utero, in which the partial pressure of oxygen in the uterine vein is often < 30 mm Hg (Brodsky & Martin, 2010). This change in oxygen exposure can cause newly formed capillaries to constrict and retract (Smith, 2008) (Sapieha et al., 2010). It is now rare to have high constant oxygen at birth in the United States, and other stresses may be more relevant, such as repeated fluctuations in oxygen (Cunningham et al., 1995), which have been found to delay developmental angiogenesis in animal models (Niesman, Johnson, & Penn, 1997; Penn, 1992; Saito et al., 2007). In addition, the preterm infant has a limited capacity to reduce oxidative compounds, which are more common in the preterm than full-term infant's circulation. These compounds can trigger a number of signaling events that release inflammatory cytokines (Hardy et al., 2005; Lambert, Pedersen, & Poulsen, 2006) that, when increased, constrict vessels. All these processes can increase avascular retina. As the infant is weaned from oxygen support,

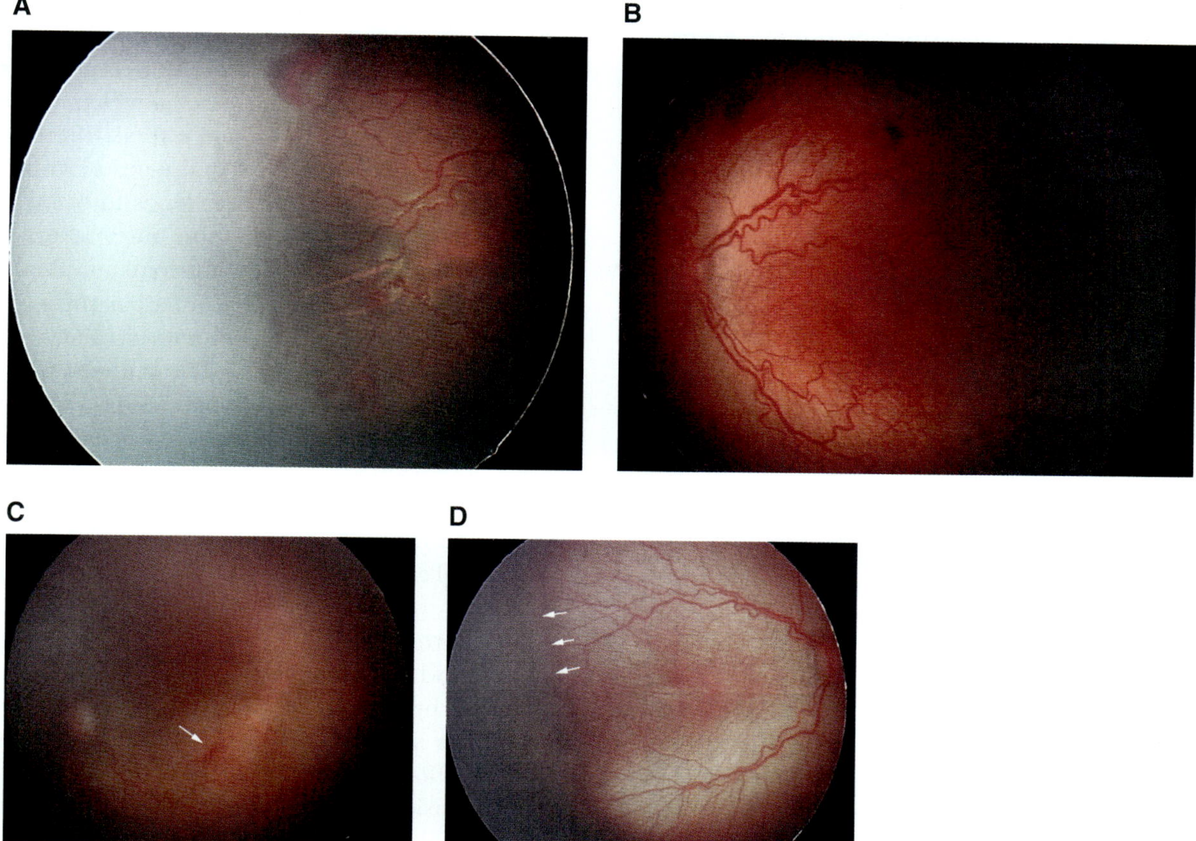

FIGURE 103.3   Images of severe retinopathy of prematurity (ROP). (A) Image of human infant retina of a right eye with moderate plus disease of the retinal vessels in zone I to posterior zone II. The optic nerve is not visible in the image but would be further to the right. A ridge and hemorrhage are seen at the junction of vascular (right of image) and avascular retina (left of image). Image taken with wide-angle photography (Retcam, Clarity). (B) Image of human infant retina of a left eye with aggressive posterior ROP (APROP) showing plus disease in zone I. The ROP, diagnosed at a young postgestational age (33 weeks), developed rapidly and did not progress through stages. This would be classified as type 1 ROP and had recently been treated with laser (yellow spots seen in the right of the image). Image taken with wide-angle photography (Retcam, Clarity). (C) Image of human infant retina of a left eye with APROP showing flat neovascularization (arrow) and difficulty in discerning the avascular from vascular retina. This would be classified as type 1 ROP. Image taken with wide-angle photography (Retcam, Clarity). (D) Image of human infant retina of a right eye with peripheral severe ROP, showing optic nerve at right of image, plus disease, and aberrant intravitreal angiogenesis (stage 3 ROP—arrows) at the junction of the vascular and avascular retina. This would be classified as type 1 ROP. Image taken with wide-angle photography (Retcam, Clarity).

avascular retina becomes hypoxic, and hypoxia triggers signaling pathways within glial and neural cells to up-regulate angiogenic factors (Akula et al., 2007; Hartnett & Penn, 2012). These factors interact with receptors on endothelial cells, but instead of causing vascular growth within the retina, they induce aberrant intravitreal angiogenesis. A third event, therefore, seems necessary, of disordering the physiological retinal angiogenesis into aberrant intravitreal angiogenesis where vessels grow into the vitreous (Hartnett, 2010b) (see also section IV).

Severe Forms of ROP

It might seem logical if severe ROP manifested at different postgestational ages, depending upon the gestational age of the infant. However, most severe ROP clusters at certain postgestational ages, regardless of birth weight or gestational age. The more common form of severe ROP (peripheral severe ROP; PSROP; see figure 103.3D) occurs about 37 weeks postgestational age (Cryotherapy for Retinopathy of Prematurity Cooperative Group., 2002). Infants who develop a more

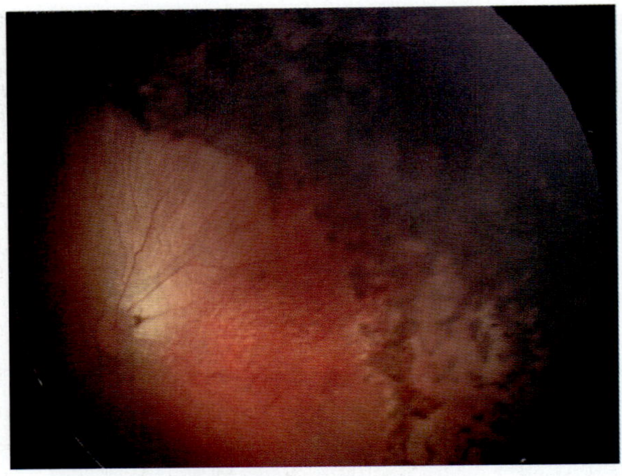

FIGURE 103.4   Laser treatment for severe retinopathy of prematurity (ROP). Image of human infant retina of a left eye that had been treated for aggressive posterior ROP with two sessions of laser treatment followed by a surgical lens-sparing vitrectomy to reattach stage 4 ROP. Healed laser scars are apparent in the right of the image. Image taken with wide-angle photography (Retcam, Clarity).

severe form of ROP, aggressive posterior ROP (APROP; see figure 103.3A–C), manifest severe ROP at younger postgestational ages than PSROP, often 31 through 34 weeks. Treatment with laser photocoagulation is performed for both types of severe ROP, but outcomes are not as good in APROP as in PSROP. Also, in APROP new aberrant intravitreal angiogenesis can develop at about 37 weeks' corrected age. The commonality in postgestational age between a recurrence in APROP and in PSROP raises the question of the potential role of other developmental and metabolic events occurring concomitantly in the retina. We will review the events of development of the neural retinal and of the ocular vasculatures as they relate to the development of severe forms of ROP. First, we will review the development of the ocular vasculatures in the human eye, recognizing that the ability to study this in human eyes is limited by the lack of intact human ocular tissue at many gestational ages, especially beyond 22 weeks' gestation (Chan-Ling, 1997).

VASCULAR DEVELOPMENT OF THE EYE AND THE ASSOCIATION OF NEURAL AND VASCULAR DEVELOPMENT IN THE HUMAN

The proliferation and differentiation of neural cells, and the development of signal transduction, metabolic processing, and synaptic connections within the retina can be associated temporally with the development and regression of ocular vasculatures. Migration and proliferation of progenitor retinal cells begin in the posterior

eye and extend to the periphery, and from the outer layers of the common neuroblastic layer toward the inner retina, whereas differentiation tends to occur within the inner retina first and is followed later by specialization of cells in the outer layers. Differentiation of progenitor cells also begins in the posterior pole and extends to the periphery, but it tends to occur first in the inner retina and later in the outer layers. These processes require oxygen and nutrients supplied by the developing and regressing ocular vasculatures.

The three vascular compartments in the eye include the choroid, the hyaloid, and the retinal vasculature. All develop by one or more processes, including vasculogenesis (the assembly of vascular precursors or angioblasts), angiogenesis (migration and proliferation of endothelial cells from existing blood vessels), and hemovasculogenesis (differentiation of blood cells and blood vessel cells from a common precursor, the hemangioblast).

The choroidal and retinal circulations persist into adulthood whereas the hyaloidal circulation undergoes regression by term birth. Both retinal and choroidal circulations have the ability to autoregulate, or maintain blood flow over a range of perfusion pressures and oxygen tensions while adjusting to meet metabolic needs (see review, Sapieha et al., 2010). Autoregulation can involve the constriction or dilation of vessels in response to oxygen and metabolites to maintain blood flow. In retinal vessels, high oxygen causes vasoconstriction in the adult. However, the preterm and newborn infant cannot constrict the vasculature in response to high oxygen. One hypothesis is that preterm infants have increased levels of prostaglandins, nitric oxide, and potentially carbon dioxide that maintain the vessels in a dilated state (Kermorvant-Duchemin et al., 2010). It is believed that this dilation permits excessive exposure of oxygen to the newly formed vessels, leading to endothelial cell death and altered cell determination of progenitor endothelial cells (Uno et al., 2007). These processes lead to the production of regions of avascular hypoxic retina in animal models studied.

Choriocapillaris/Choroid

The adult choroid is made up of three layers: a fenestrated choriocapillaris and layers of medium and large sized choroidal vessels that are not fenestrated. The choriocapillaris arises from secondary mesoderm and is one of the first vasculatures to develop (Hasegawa et al., 2007; Lutty et al., 2010; see chapter 104). The initial choriocapillaris develops by hemovasculogenesis at about 6 to 8 weeks' gestation (Hasegawa et al., 2007; Lutty et al., 2010) with the presence of nucleated

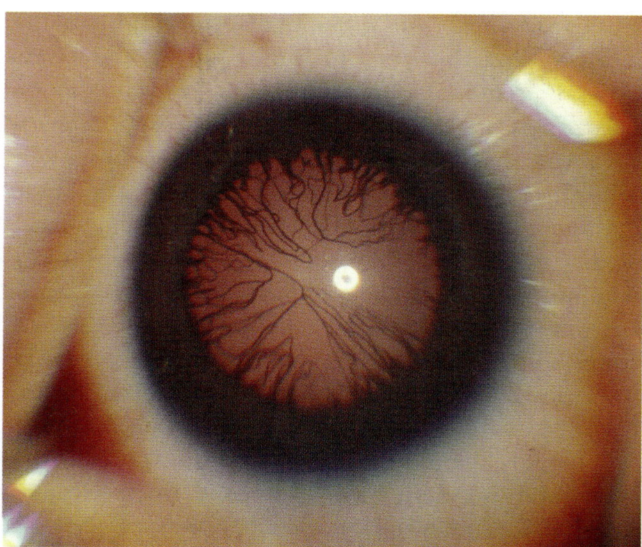

FIGURE 103.5   Periestent pupillary membrane. Image of left eye of infant showing anterior segment with dilated vessels of pupillary membrane on the surface of the iris and lens, most apparent in the pupil. Image taken with Zeiss FF3, vertically mounted.

erythrocytes often coexpressing epsilon hemoglobin and endothelial cell markers, such as CD31, CD34, and CD39. This early vasculature coincides with the appearance of the RPE. At 11 to 12 weeks, the choriocapillaris develops a more definitive structure and the deeper choroidal vessels develop in the posterior pole through the process of angiogenesis, extending the choroidal vasculature to the periphery through about 14 weeks' gestation (coincidently, the retinal vasculature begins to form through vasculogenesis). Early, but scarce, choriocapillaris fenestrations are seen at 16 weeks' gestation and are more prominent in the posterior pole. The choroid has developed three layers of blood vessels by about 21–22 weeks' gestation, and early inner segments. These early inner segments of the photoreceptors have mitochondria and are likely metabolically active. At 24–26 weeks, when the inner segments are more mature (Hendrickson & Yuodelis, 1984), the choriocapillaris is fenestrated mostly on the sides facing the RPE, presumably to provide passive transit of fluids, macromolecules, nutrients and oxygen to the RPE and outer retina or to remove waste from the RPE (Lutty et al., 2010).

Hyaloidal Circulation

The hyaloidal circulation, which includes the hyaloid artery, the vasa hyaloidea propria, the tunica vasculosa lentis, and the pupillary membrane (see figure 103.5), develops by hemovasculogenesis and supports the entire developing eye (Barishak, 1992; McLeod et al., 2012). As the RPE differentiates, and the inner and outer neuroblastic layers develop, the hyaloid artery invades the optic fissure at about 4 weeks and becomes more prominent by the 5th week of development. At 7 weeks' gestation, ganglion cells and Mueller cells start to differentiate and migrate inwards. The hyaloidal circulation becomes most prominent between 8 and 12 weeks. Regression of the hyaloidal circulation begins at 12 weeks (Mann, 1964) with the vasa hyaloidea propria, followed by the tunica vasculosa lentis, that is, the pupillary membrane, and finishes at about 35 to 36 weeks' gestation with the loss of the hyaloidal artery. A proteomic analysis of the vitreous was conducted in the developing mouse eye. Unlike in the human, the mouse hyaloidal circulation develops between embryonic day 10.5 and is completed by embryonic day 13.5 (Mitchell, Risau, & Drexler, 1998). Regression begins at postnatal day (P) 1 and is mostly complete at P16–30. The investigators found that groups of proteins tended to shift from those known important in energy metabolism, cell proliferation, and development at P1 to those involved in signal transduction at P16. Throughout the time course, apoptotic and anti-angiogenic proteins were also present (Albe et al., 2008). Evidence supports the notion that the hyaloid regresses from high oxygen and decreased concentration of the angiogenic transcription factor, HIF 1 alpha (Kurihara et al., 2010). Other mechanisms of regression proposed include the expression of wnt 7b from perivascular macrophages and the role of angiopoietin 2 in suppressing endothelial survival factors and in stimulating wnt 7b expression (Lobov et al., 2005; Lobov, Brooks, & Lang, 2002).

Retinal Vasculature

A pool of precursor cells arises from the outer retina to form CD39+ angioblasts before retinal vessels become apparent (Hasegawa et al., 2008; McLeod et al., 2006). At 7 weeks' gestation, precursor CXCR4+, CD39+ cells from the inner neuroblastic layer are attracted to migrate inwardly toward a gradient of stromal derived factor (SDF-1) in the posterior region of the eye near the inner limiting membrane (ILM). SDF-1 is the ligand for receptor CXCR4. SDF-1 can be up-regulated by hypoxia in other situations, and this may be the stimulus for its presence in the fetal human retina, but the cell source is unknown. SDF-1 is produced by stem cells in other organs (Tachibana et al., 1998). The CXCR4+ cells that migrate from the inner neuroblastic layer are believed to be multipotent and may develop into ganglion cells, neurons, astrocytes, or microglia. The origin of the precursor cells remains unclear, but the greatest evidence supports that they arise from the optic nerve rather than from adjacent mesoderm. At 12 weeks, CD39+ angioblasts appear at the ILM (Hasegawa et al., 2008). The CD39+ cells at the ILM develop cords and give rise to endothelial cells that make up initial retinal vessels manifesting at 14 weeks' gestation. This initial vasculature thus occurs by vasculogenesis from preexisting CD39+ angioblasts (McLeod et al., 2006). Vessels are found to associate with Mueller cell end feet in the inner retina at 17 weeks' gestation. There is evidence from histopathologic studies that ganglion cell axons may provide guidance for developing blood vessels. Neurofilament labeled ganglion cell processes associate with angioblasts, and the angioblasts form blood vessels through the help from neuropilin-1 on ganglion cell axons. Also, impaired ganglion cell development in transgenic mice is associated with abnormalities in retinal vascular development and structure as early as P1 (Edwards et al., 2012).

Thus, the retinal circulation appears at about 14 weeks' gestation (Hasegawa et al., 2008), coinciding with the migration of cells from the outer neuroblastic layer to the inner retina to form bipolar and later horizontal cells while those from the inner neuroblastic layer form amacrine cells. At this developmental time point, ganglion cells are also developing dendrites. Through 22 weeks' gestation, the inner plexus of the posterior retina becomes vascularized by vasculogenesis only to about a region within zone I (McLeod et al., 2006). Because of the inability to experiment on humans and/or obtain tissue from fetuses older than 22 weeks' gestation, it is unclear how the rest of the retinal vasculature develops in humans although, generally, expansion of the inner plexus and the formation of the outer plexus are believed to occur through angiogenesis. Preceding the angioblasts that become endothelial cells are Pax2+/GFAP– astrocyte precursors at about 14 weeks' gestation (Chan-Ling et al., 2004). In other species, these cells are postulated to migrate in front of endothelial cells, sense physiological hypoxia, and up-regulate VEGF, which creates a gradient for proliferating and migrating endothelial cells (Chan-Ling, Gock, & Stone, 1995; Stone et al., 1995), extending the inner plexus from the posterior retina to the ora serrata by a process of angiogenesis. Deeper plexus vessels are also believed to occur by angiogenesis in association with the maturation of the photoreceptors. At the 4th month, the outermost cells of the neuroblastic layer, which later become photoreceptors, remain attached to the RPE through adherens junctions and begin to differentiate in the posterior pole and extend toward the periphery. At the same time, the ganglion cells become differentiated. As the hyaloid regresses, there is less oxygenation for the developing and differentiating neurons, and this need may be compensated by the expanding retinal vasculature both peripherally and outwardly toward the inner nuclear layer (for review, see Barishak & Spierer, 2005).

During the 5th and 6th months, photoreceptor differentiation continues. At the same time the surface area of the retina and eyecup increase. Cones differentiate in the posterior pole in the 6th month and manifest mitochondria. At 7 months, the fovea develops as cones shift toward the fovea. The rods differentiate from posterior to anterior retina and are believed to become fully differentiated at the ora serrata after birth. The nasal retina is vascularized by about 36 weeks' gestation and the temporal by 40 weeks' (for review, see Barishak & Spierer, 2005).

Association between Developmental Age of Neural Retina and Severe ROP

Oxygen concentration appears to be associated with the developing retina and with human preterm ROP. Two forms of severe ROP, APROP (see figure 103.3A–C) and PSROP (see figure 103.3D) occur at different postgestational ages. Several decades ago, most cases of severe ROP manifested as PSROP (Cryotherapy for Retinopathy of Prematurity Cooperative Group., 2002). However, APROP has become more common, possibly due to advances in neonatal care and the ability to save preterm infants of very young gestational age (23 to 24 weeks' gestation). APROP manifests at an earlier age than PSROP due to the incomplete retinal development at this young gestational age. Based on human studies, preterm infants born at age 22 or 23 weeks' gestation have retinal vascularization limited to zone I or posterior zone II (McLeod et al., 2006). At this time, the photoreceptors are just starting to differentiate in the posterior pole. By 24 weeks, photoreceptors have differentiated into cones and, by 28 weeks, into rods. Ganglion cells start to mature by about 16 weeks' gestation, also beginning in the posterior pole and extending peripherally. Neuronal proliferation, differentiation, synapse formation, and metabolism expend energy and therefore require oxygen. With incomplete retinal vascularization, there is little oxygen to support the developing and differentiating retinal neurons. In addition, hypoxia triggers angiogenic events, and excessive signaling of angiogenic pathways can lead to disordered angiogenesis as seen in cancer development (Zeng et al., 2007) and animal models of OIR (Hartnett et al., 2008). This disordered angiogenesis may manifest as intravitreal aberrant angiogenesis and interfere with the normal ordered intraretinal angiogenesis, further maintaining peripheral zones of avascular retina.

Laser treatment to hypoxic, avascular retina can be successful initially, but new aberrant intravitreal angiogenesis often forms several weeks later. At least four events occur that may lead to the new development of severe ROP. First, rod photoreceptor differentiation has extended to the equatorial and peripheral retina where retinal vascularization has not yet occurred, causing new metabolic demand to outweigh available oxygen. Second, flat neovascularization, common in APROP (see figure 103.3C), has regressed, leaving new regions of avascular retina and a new junction between vascular and avascular retina. Third, the retinal surface expands with transverse shifting of the retinal cells, causing areas devoid of laser to enlarge. Fourth, postnatal stresses slow retinal vascularization, such that vascular supply cannot keep up with the oxygen demand associated with the increase in retinal area. New aberrant intravitreal angiogenesis develops at about 37 weeks' postgestational age, similar to the peak time when PSROP develops in infants of older gestational ages (see table 103.2).

Infants born at older gestational ages have vascularized retinas beyond zone I and may have also begun vascularization of the deeper, outer plexus. Like the intravitreal angiogenesis that develops following laser photocoagulation for APROP, PSROP develops in this more peripheral location due to the region of developing retina. Oxygen stresses delay ongoing angiogenesis leading into peripheral avascular retina. As photoreceptors differentiate in the peripheral retina and synaptic connections develop, rods elongate, initiate phototransduction, and have their outer segments turn over. These events increase oxygen demand that stimulates the development of PSROP (see figure 103.3D).

Clearly, oxygen is not the only factor in ROP. Infants born at older gestational ages may develop APROP, and infants born at young gestational ages do not develop severe ROP. Intrauterine growth restriction is associated with ROP (Darlow et al., 2005), and poor postnatal weight gain also appears to play a role in infants of older gestational age (11). In some cases, the premature condition appears to result in defects in growth factor regulation important for systemic growth and in growth factor production or regulation locally in the retina (Wang et al., 2012). There is evidence that oxidative stress is associated with features of severe ROP (Raju et al., 1997). In infants of young gestational ages who do not develop severe ROP, protective factors may be present. In animal models, these factors include antioxidants, nutritional factors such as peptides (Neu

et al., 2006), and omega-3 fatty acids (Connor et al., 2007) and growth factors such as insulin-like growth factor 1 (IGF-1) (Hellstrom et al., 2003; Vanhaeserbrouck et al., 2009).

Animal studies have provided evidence for an association between metabolic activity and ROP. Berkowitz showed with high-resolution manganese-enhanced magnetic resonance imaging that ion demand affected retinal circulation in models of OIR (Berkowitz et al., 2007), supporting changes in metabolic activity. Also, deficits in rod response parameters have been identified in OIR models (Fulton et al., 1999; Reynaud, Hansen, & Fulton, 1995) that are related to insufficient oxygen supply. The hypoxia may lead to long-term functional deficits. Theories for the abnormalities in rod response with electroretinograms have included reduced rhodopsin content, shorter length of photoreceptor outer segments, or reduced number of rods. However, Fulton et al. failed to find evidence of this in animal models of ROP compared to controls, but did find that the outer segments of rods were mildly disorganized in the ROP rat compared to controls, using electron micrographs and microspectrophotometry of single photoreceptors (Fulton et al., 1999). The subtle alteration in rod outer segment structure was associated with the proximal cause of the rod dysfunction. Besides dysfunction associated with vascular abnormalities, early rod dysfunction appeared to predict later vascular abnormalities found in rat models of OIR (Akula et al., 2007), and these vascular changes could be partly prevented with a visual cycle modulator (Akula et al., 2010) that reduced metabolic demand. Significant deficits in rod photoreceptor sensitivity are also present in human infants and persist into childhood (Harris et al., 2011). The deficit in the rod response parameters varied significantly with severity of antecedent acute vascular ROP (Fulton et al., 2001).

Repeated oxygen fluctuations in the rat ROP model increase oxidative signaling that contributes to persistent avascular retina. Treatment with analogues of vitamins C and E (Penn, Tolman, & Bullard, 1997), manganese superoxide dismutase (Niesman, Johnson, & Penn, 1997), or apocynin, an inhibitor of NADPH oxidase, reduced avascular retinal area (Saito et al., 2007) or aberrant intravitreal angiogenesis in the rat ROP model with oxygen fluctuations and in the mouse OIR model (Al Shabrawey et al., 2005; Byfield, Budd, & Hartnett, 2009; Saito et al., 2008).

Thus, in the preterm infant retina, a mismatch in oxygen demand appears to affect the function of retinal neurons, and the metabolism of differentiating neurons can drive aberrant intravitreal angiogenesis in severe ROP.

TABLE 103.2

Hypothesis as to development of severe retinopathy of prematurity (ROP) in association with vascular and neural development

Developmental age (weeks)	Ocular vascular development			Neural development	Eye growth	Gestational age at birth (weeks)		
	Hyaloid	Choroid	Retina			23–25	26–29	31+
4	Artery invades			RPE begins to develop	Eye growth occurs—cells proliferate and migrate	Retinal vessels in zone I by vasculogenesis. Hyaloid vessels still present	Retinal vessels form by vasculogenesis in zone I	
6–8		Begins	Precursor angioblasts migrate to inner retina	Ganglion and Mueller cells migrate and develop				
12	Prominent, begins regression		Angioblasts present at inner retina					
14		Covers extent of eye cup	First vasculature seen and develops by vasculogenesis	Bipolar horizontal and amacrine cells migrate from neuroblastic layers				
16				RPE and photoreceptors remain attached through adherens juctions; ganglion cells differentiate				
21–22		3 layers, with fenestrations in cc	Vasculature in zone I vasculogenesis	Photoreceptor differentiation continues				
24–26		Inner segments and metabolism starts		Cones differentiate		Postnatal stresses delay developing vessels and/or cause vessel constriction	Outer plexus and peripheral retina vascularize by angiogenesis into zone II	
28				Fovea begins development; rods differentiate; Synapses forming				
31–33						APROP seen	Postnatal stresses delay developing vessels	
36–38				Rods continue to differentiate in peripheral retina; synapses	Retinal surface increases by shifting of cells	New areas of avascular retina appear with regression of flat NV and increased retinal surface area; new severe ROP seen	PSROP seen	No ROP seen
40+				Fovea matures				

Note. Vascular and neural events are associated with the formation of the three ocular vasculatures and regression of the hyaloid. RPE, retinal pigment epithelium; APROP, aggressive posterior ROP; PSROP, peripheral severe ROP; NV, neovascularization.

MODELS OF OIR: RELEVANCE TO ROP AND ADULT RETINOVASCULAR DISEASES

High Constant Oxygen Followed by Room Air Exposure

When ROP was first observed as retrolental fibroplasia in the 1940s (Terry, 1942), technology had not been developed to monitor and regulate oxygen levels. However, it was assumed that high oxygen was needed to resuscitate infants at birth. Animal models were then developed to test the role of high oxygen at birth on the development of ROP. Using animals that vascularize their retinas after birth, cat (Ernest & Goldstick, 1984), mouse (Smith et al., 1994), rat (Penn, Tolman, & Henry, 1994), and beagle (McLeod, Brownstein, & Lutty, 1996), high oxygen exposure has been found to cause avascular retinal regions within the already vascularized retina to form, which is then followed by aberrant intravitreal angiogenesis from existing retinal vessels once the animals have been removed from high oxygen. (The result of these early studies was that preterm infants were exposed to lower oxygen, and ROP was reduced for a time until neonatal advances permitted smaller preterm infants of younger gestational ages to survive.) These models are used today to study the effects of high oxygen exposure and relative retinal hypoxia on angiogenesis. Typically, animals are exposed to constant oxygen (75%) at about postnatal day 7. High oxygen can change the fate of endothelial precursors (Uno et al., 2007) and causes newly developed capillaries to constrict, in part, from apoptotic death of endothelial cells (see figure 103.6), which leads to avascular retina. In these animal models, this process has been termed vaso-obliteration (McLeod, Brownstein, & Lutty, 1996; Smith et al., 1994). Following return to room air, endothelial budding into the vitreous is driven by the relative hypoxia in the avascular retinal regions. While the mouse OIR model is particularly valuable for studying genetic mechanisms and potentially for studying early ROP of the 1940s, it remains unknown if vaso-obliteration actually occurred or occurs in human infants. However, today, high oxygen at birth is avoided in neonatal units that have implemented technology to monitor and regulate oxygen. It remains a concern in units in which technology to save preterm infants is available, but resources are insufficient to implement oxygen monitoring and regulation (Gilbert et al., 1997). However, these models can be useful to study diseases associated with loss of already developed retinal capillary support, such as in proliferative diabetic retinopathy or retinal vein occlusions.

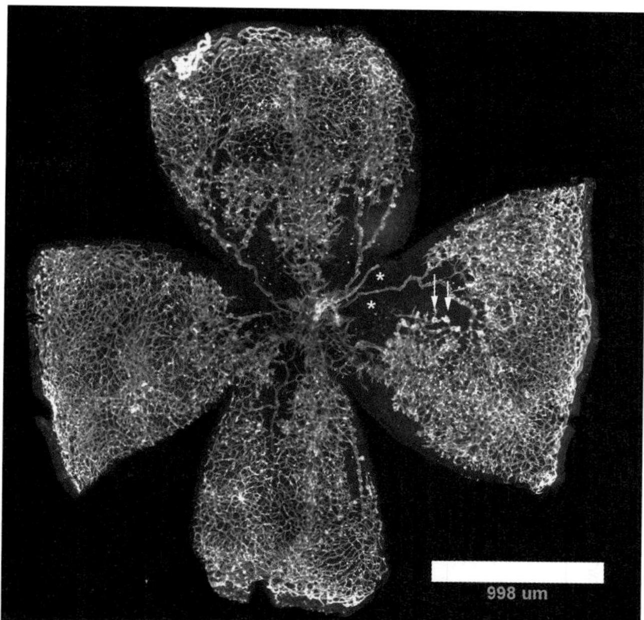

FIGURE 103.6   Mouse oxygen-induced retinopathy. Retinal flat mount from a postnatal day 17 mouse pup that had been placed into 75% oxygen for 5 days starting at postnatal day 7 and then placed into room air. The flat mount is stained with lectin to show the retinal vasculature. Central avascular retina exists as dark spaces adjacent to the region of the optic nerve (asterisk), and budding endothelial cells extend into the vitreous (arrows).

Variable Oxygen in Animal Models

There are a number of models of variable oxygen stresses, mostly involving rat (Berkowitz & Penn, 1998; Dorey et al., 1996; Gao et al., 2002). The most characterized is the rat 50/10 OIR model (referred to as the ROP model), in which newborn rat pups are exposed to 50% oxygen for 24 h, followed by 10% oxygen for 24 h (Penn, Henry, & Tolman, 1994). The 24-h cycles are repeated until day 14, at which time animals are returned to room air. At postnatal day 14, there is peripheral avascular retina, and by day 18, aberrant intravitreal angiogenesis at the junction of vascular and avascular retina (see figure 103.7).

The rat 50/10 OIR model may simulate the pattern of oxygen exposure experienced by preterm infants who develop PSROP in the United States today. The oxygen extremes cause arterial oxygen levels similar to transcutaneous oxygen levels measured in human preterm infants (Cunningham et al., 1995). Fluctuations in oxygen are believed to be commonly associated with severe ROP since constant oxygen levels are avoided, if possible, and the preterm infant has a number of abnormalities in ventilation and perfusion

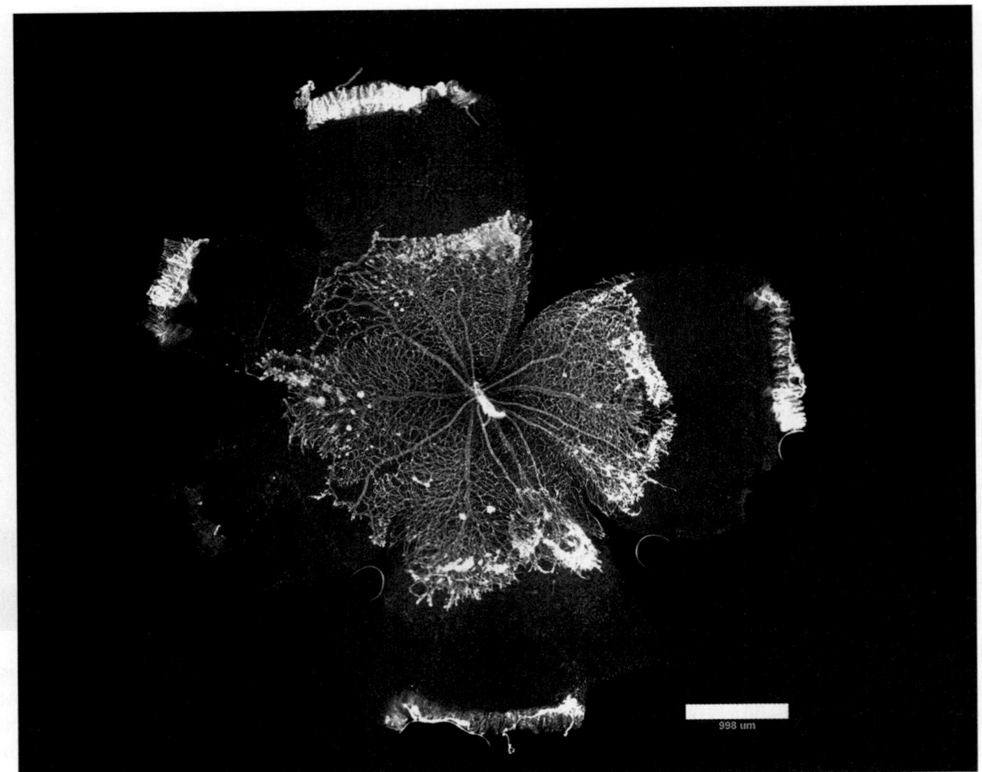

FIGURE 103.7   Rat retinopathy of prematurity (ROP) model of variable oxygen (50/10 oxygen-induced retinopathy model or ROP model). Retinal flat mount from postnatal day 18 rat pup that had been placed into an oxygen environment that fluctuated inspired oxygen between 50% and 10% every 24 h for 14 days and then brought into room air environment. Peripheral avascular retina is noted as well as aberrant intravitreal angiogenesis at the junction of vascular and avascular retina.

that lead to minute-to-minute fluctuations in arterial oxygen concentration (the oxygenation status in the retina during changes in systemic oxygen concentrations remains unknown). The appearance of the retina is similar to PSROP with a central vascularized retina, but a peripheral avascular retina (figure 103.7). Hypoxic retina is present in the avascular peripheral retina as well as in the vascularized regions of retinal vessels (Saito et al., 2008). Aberrant intravitreal angiogenesis later develops at the junction of vascular and avascular retina.

OIR to Study Other Retinovascular Diseases

The mouse OIR model has been used as a model of adult retinovascular diseases with aberrant intravitreal angiogenesis. The loss of capillary support with subsequent retinal hypoxia is similar to what is believed to occur in human proliferative diabetic retinopathy and ischemic retinal vein occlusion. The rat 50/10 OIR model also simulates diabetic retinopathy because of shared molecular mechanisms with those noted in hyperglycemia-induced alteration in integrin associated protein and IGF-1 induced endothelial cell permeability (Maile et al., 2012).

VASCULAR ENDOTHELIAL GROWTH FACTOR IN DEVELOPMENT AND PATHOLOGIC ANGIOGENESIS

VEGF is a 45 kD protein that is important in vascular permeability and angiogenesis (Ferrara, Gerber, & Lecouter, 2003). VEGF is also one of the most important angiogenic factors that causes pathologic choroidal neovascular diseases such as in age-related macular degeneration, or retinovascular diseases such as in diabetic retinopathy and retinal vein occlusion, which are characterized by aberrant intravitreal angiogenesis (Penn et al., 2008). Injections of large doses of VEGF into the vitreous of mice caused endothelial cell filopodia to point into the vitreous (Gerhardt et al., 2003) and to cause retinal vascular microaneurysms and permeability (Tolentino et al., 2002). Inhibition of VEGF bioactivity reduced pathology in animal models and in human age-related macular degeneration (Aiello et al., 1995; Pierce et al., 1995; Rosenfeld et al., 2006).

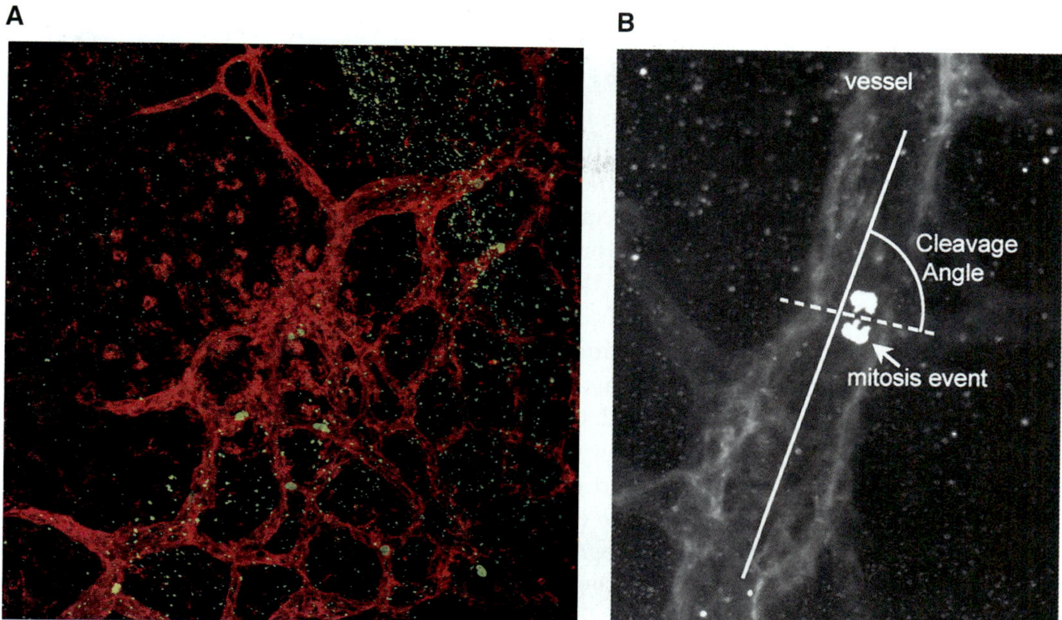

FIGURE 103.8   Figure and flat mount of phosphohistone stained anaphase in lectin-stained flat mount in rat 50/10 oxygen-induced retinopathy model. (A) Lectin-stained (red) vessels in a retinal flat mount taken from postnatal day 14 rat pup and labeled with antibody to phosphohistone H3 to visualize the cleavage planes (green) during anaphase. (B) Diagram of mitotic figure overlying a lectin-stained vessel from a retinal flat mount from a postnatal day 14 rat to demonstrate how cleavage angles between the cleavage plane and long axis of the vessel were determined and measured. (Adapted from Hartnett et al., 2008.)

Much evidence in the literature supports the thinking that reducing VEGF bioactivity by different methods will inhibit intravitreal endothelial cell growth during relative hypoxia in animal models (Robinson et al., 1996; Smith et al., 1994). Some investigators have found that inner glial VEGF was necessary for the development of aberrant angiogenesis (Weidemann et al., 2010) whereas evidence from models with conditional VEGF knockout in Mueller cells supported the notion that Mueller cell derived VEGF was important for increasing permeability of vessels (Wang et al., 2010) and extraretinal neovascularization (Bai et al., 2009).

However, VEGF has been found essential for physiological retinal vascular development, which is ongoing in the preterm infant (Gerhardt et al., 2003; Hiratsuka et al., 1998; Kearney et al., 2004; Taomoto et al., 2000). Therefore, there is the potential that inhibiting VEGF to reduce aberrant intravitreal angiogenesis will also prevent ongoing retinal vascularization. Other pathways (IGF-1, stromal derived factor, pigment epithelium derived factor, as examples) are also important, but VEGF has been found to be an important mediator of many of these pathways or of their biological outcomes (Bentley, Gerhardt, & Bates, 2008; Budd & Hartnett, 2010; Hellstrom et al., 2007; Jones et al., 2008; Sonmez, Drenser, Capone, & Trese, 2008). There is also the concern of potential harm to retinal neurons since

VEGF is a survival factor. For example, ganglion cells and RPE produce VEGF, and the question arises whether inhibition of VEGF would be detrimental to the health of these cells (Ford et al., 2011). Long-term clinical evidence for benefit from anti-VEGF treatment in retinal vascular diseases exists for adults, but not for the preterm infant. In the developing infant, there is concern about inhibiting a factor that is important in other vascular beds that are developing, such as the lung, brain, and kidney (Sorenson & Sheibani, 2011). It is also important to take into consideration that preterm infants have lower blood volumes than adults, which could result in a higher circulatory drug concentration than that in adults after giving intravitreal doses.

Evidence of VEGF Disordering Developmental Angiogenesis

In the rat 50/10 model of variable oxygen, VEGF signaling through VEGF receptor (R)2 caused venous dilation and arteriolar tortuosity, a condition similar in appearance to human plus disease in severe ROP (see figure 103.3). Lectin-stained flat mounts colabeled with antiphosphohistone H3 antibody demonstrated that increased VEGFR2 signaling altered cleavage angles of dividing vascular cells (see figure 103.8).

Vessel tortuosity and dilation were reduced after an intravitreal injection of a neutralizing antibody to VEGF (Hartnett et al., 2008). In development, VEGFR1 is believed to trap VEGF and regulate its signaling through VEGFR2. The question was posed if increased signaling of VEGF through VEGFR2 would disorder developmental angiogenesis. A single allele of VEGF, its splice variants, or receptors is lethal. Therefore, an embryonic stem cell model, which allows study of genes that produce nonviable knockouts, that is, VEGF or receptors, VEGF(R) 1 or R2, was used to enable viable knockout of VEGFR1 (murine equivalent is *flt-1$^{-/-}$*). In the *flt-1$^{-/-}$* embryonic stem cell model, VEGF–VEGFR2 signaling was increased. Dividing endothelial cells were found to form disordered vascular networks based on measurement of cleavage angles during anaphase in the *flt-1$^{-/-}$* model compared to wild type. Ordered angiogenesis was restored with a PECAM-sflt-1 transgene, providing evidence that excessive VEGF–VEGFR2 signaling caused disordered angiogenesis (Zeng et al., 2007).

There is evidence suggesting that exceeding a threshold of VEGF signaling may be detrimental to the retinal vasculature and affect the orderliness of developmental angiogenesis (Zeng et al., 2007), such that restoring physiological VEGF signaling may facilitate retinal vascular development by reducing disordered angiogenesis. Studies support the idea that too much VEGF signaling through VEGFR2 may cause disordered developmental angiogenesis such as that seen in stage 3 severe ROP. In the 50/10 OIR model, neutralizing VEGF with antibody or with a VEGFR2 tyrosine kinase inhibitor reduced aberrant intravitreal angiogenesis without interfering with ongoing retinal vascular development (Budd et al., 2009; Geisen et al., 2008) at certain doses. In a beagle model of ROP, Lutty et al. reported that a low dose of a "VEGFtrap," a fusion protein, composed of IgG Fc fragments fused to VEGFR1 domain 2 and VEGFR2 domain 3, stopped aberrant intravitreal angiogenesis without interfering with intraretinal angiogenesis. However, a high dose of VEGFtrap virtually stopped both aberrant intravitreal angiogenesis and physiological angiogenesis (Lutty et al., 2011). These studies show the importance of dose, but it is not yet possible to establish the correct dose in the individual infant eye. The finding that increased VEGFR2 activation caused venous dilation and arteriolar tortuosity in the rat 50/10 variable oxygen model suggests that anti-VEGF treatment could be administered when human plus disease becomes clinically apparent (Hartnett et al., 2008). Clinical observations indicate that treatment should be implemented prior to the onset of fibrovascular changes (Trese & Capone, 2004). Thus, a window of treatment with anti-VEGF therapy may be at the time of plus disease but prior to the onset of fibrovascular activity. This timing in humans is usually prior to 37 weeks' postgestational age. How early treatment can be safely given to infants remains unknown, but based on one clinical study, 33 weeks' corrected age in the presence of plus disease in APROP may be safe for at least some infants (Mintz-Hittner, Kennedy, & Chuang, 2011).

Inhibiting VEGF Alters the Natural History of ROP

Clinical studies using anti-VEGF agents, such as bevacizumab, suggest that inhibition of the bioactivity of vascular endothelial growth factor (VEGF) in some subsets of infants, such as those with severe ROP advanced beyond the recommended level of severity for treatment, was more effective than laser treatment, but sufficient follow-up was absent to assess the effect on physiological developmental angiogenesis (Mintz-Hittner, Kennedy, & Chuang, 2011) or on safety. Another clinical study reported progression to stage 4 ROP after anti-VEGF treatment for (severity) ROP (Drenser, 2009). The BEAT-ROP (Bevacizumab Eliminates the Angiogenic Threat of ROP) study enrolled 150 infants with advanced severe ROP. The laser and bevacizumab groups each had 75 infants. Although eyes treated with bevacizumab had fewer recurrences than those treated with laser (P = 0.002), the recurrence rate for the laser treated group was 26%, much higher than the 9% reported in a multicenter clinical trial testing laser for severe ROP (Early Treatment Retinopathy of Prematurity Study) (Early Treatment for Retinopathy of Prematurity Cooperative Group, 2003). This observation may reflect the more advanced nature of severe ROP in the sample population enrolled in BEAT-ROP. Further study is warranted regarding dose and long-term safety before recommending bevacizumab for broad treatment of ROP.

Erythropoietin

Because of safety concerns and the fact that data from OIR models or from human clinical studies failed to show complete vascularization in the retina of some infants with severe ROP (Mintz-Hittner, Kennedy, & Chuang, 2011), the role of other growth factors and signaling pathways have been considered. One candidate is erythropoietin, which stimulates erythropoiesis. Erythropoeitin has been used in preterm infants for anemia of prematurity. However, retrospective studies reported an association between severe ROP, and therefore poor outcomes in preterm infants, with the use of erythropoietin (Brown et al., 2006; Ohlsson & Aher,

2006). Retrospective studies by design have limitations. There has been new interest in studying the effect of erythropoietin on neuroprotection and cognitive development in preterm infants (Brown et al., 2009). We found that local erythropoietin production in the retina was reduced in the rat 50/10 variable oxygen model in part, mediated through the Janus kinase/signaling transducer and activator of transcription 3 pathway that was triggered by excessive VEGF signaling through VEGFR2 (Wang et al., 2012). This evidence suggests that inhibiting VEGF in an ROP model can lead to compensatory signaling through other angiogenic pathways and may not be sufficient to reduce the stimulus for aberrant intravitreal angiogenesis. This provides additional evidence for the importance of long-term follow up of preterm infants treated with anti-VEGF agents.

SUMMARY

ROP is a complex disease involving genetic and environmental factors. Early hypotheses of its development have required refinement as neonatal care has advanced and infants of younger gestational ages and larger birth weights are surviving. There are limitations to studying ROP because of the inability to experiment on preterm human infant eyes or to obtain human tissue at all gestational ages. The association between retinal neural development (migration, proliferation, synapse formation, and signal transduction) with development of the ocular vasculatures and regression of the hyaloid provides support that the two broad processes are linked as to function and disease. OIR models can be helpful in studying ROP or other retinovascular diseases associated with aberrant intravitreal angiogenesis, but there are differences between human and animal species including type and role of precursor cells, times of vascular development, and differences in retinal structure. Taking all these into consideration and using several approaches is helpful: cell culture and transgenic animals to test mechanisms, appropriate animal models to test environmental factors, and human tissue or cells to corroborate proteins or cells of interest. Once sufficient evidence of efficacy and safety is assessed, clinical trials are needed before interventions are recommended in humans.

ACKNOWLEDGMENTS

Sarah Moyer, CRA, FOPS, and James Gilman CRA, FOPS, are acknowledged for their assistance in obtaining and formatting images and creating figures. Y. Robert Barishak, M.D., is thanked for clarifying embryological details in human retinal development.

REFERENCES

Aiello, L. P., Pierce, E. A., Foley, E. D., Takagi, H., Chen, H., Riddle, L., et al. (1995). Suppression of retinal neovascularization in vivo by inhibition of vascular endothelial growth factor (VEGF) using soluble VEGF-receptor chimeric proteins. *Proceedings of the National Academy of Sciences of the United States of America, 92*, 10457–10461. doi:10.1073/pnas.92.23.10457.

Akula, J. D., Hansen, R. M., Martinez-Perez, M. E., & Fulton, A. B. (2007). Rod photoreceptor function predicts blood vessel abnormality in retinopathy of prematurity. *Investigative Ophthalmology & Visual Science, 48*, 4351–4359. doi:10.1167/iovs.07-0204.

Akula, J. D., Hansen, R. M., Tzekov, R., Favazza, T. L., Vyhovsky, T. C., Benador, I. Y., et al. (2010). Visual cycle modulation in neurovascular retinopathy. *Experimental Eye Research, 91*, 153–161. doi:10.1016/j.exer.2010.04.008.

Albe, C. E., Chang, J. H., Azar, N. F., Ivanov, A. R., & Azar, D. T. (2008). Proteomic analysis of the hyaloid vascular system regression during ocular development. *Journal of Proteome Research, 7*, 4904–4913. doi:10.1021/pr800551m.

Allen, M. B., Donohue, P. K., & Dusman, A. E. (1993). The limit of viability—Neonatal outcome of infants born at 22 to 25 weeks' gestation. *New England Journal of Medicine, 329*, 1597–1601.

Al Shabrawey, M., Bartoli, M., El Remessy, A. B., Platt, D. H., Matragoon, S., Behzadian, M. A., et al. (2005). Inhibition of NAD(P)H oxidase activity blocks vascular endothelial growth factor overexpression and neovascularization during ischemic retinopathy. *American Journal of Pathology, 167*, 599–607. doi:10.1016/S0002-9440(10)63001-5.

Ashton, N., Ward, B., & Serpell, G. (1954). Effect of oxygen on developing retinal vessels with particular reference to the problem of retrolental fibroplasia. *British Journal of Ophthalmology, 38*, 397–430.

Bai, Y., Ma, J. X., Guo, J., Wang, J., Zhu, M., Chen, Y., et al. (2009). Muller cell-derived VEGF is a significant contributor to retinal neovascularization. *Journal of Pathology, 219*, 446–454. doi:10.1002/path.2611.

Barishak, R., & Spierer, A. (2005). Embryology of the posterior segment and developmental disorders: A. In M. E. Hartnett, M. Trese, A. Capone Jr., B. K. Keats, & S. M. Steidl (Eds.), *Pediatric retina* (1st ed., pp. 3–12). Philadelphia: Lippincott Williams & Wilkins.

Barishak, Y. R. (1992). Embryology of the eye and its adnexae. *Developments in Ophthalmology, 24*, 1–142.

Bentley, K., Gerhardt, H., & Bates, P. A. (2008). Agent-based simulation of notch-mediated tip cell selection in angiogenic sprout initialisation. *Journal of Theoretical Biology, 250*, 25–36.

Berkowitz, B. A., & Penn, J. S. (1998). Abnormal panretinal response pattern to carbogen inhalation in experimental retinopathy of prematurity. *Investigative Ophthalmology & Visual Science, 39*, 840–845.

Berkowitz, B. A., Roberts, R., Penn, J. S., & Gradianu, M. (2007). High-resolution manganese-enhanced MRI of experimental retinopathy of prematurity. *Investigative Ophthalmology & Visual Science, 48*, 4733–4740. doi:10.1167/iovs.06-1516.

Bizzarro, M. J., Hussain, N., Jonsson, B., Feng, R., Ment, L. R., Gruen, J. R., et al. (2006). Genetic susceptibility to retinopathy of prematurity. *Pediatrics, 118*, 1858–1863. doi:10.1542/peds.2006-1088.

Brodsky, D., & Martin, C. (2010). Cardiology. In *Neonatology review* (2nd ed., pp. 109–111). Hanley & Belfus.

Brown, M. S., Baron, A. E., France, E. K., & Hamman, R. F. (2006). Association between higher cumulative doses of recombinant erythropoietin and risk for retinopathy of prematurity. *Journal of American Association for Pediatric Ophthalmology and Strabismus, 10,* 143–149.

Brown, M. S., Eichorst, D., LaLa-Black, B., & Gonzalez, R. (2009). Higher cumulative doses of erythropoietin and developmental outcomes in preterm infants. *Pediatrics, 124,* e681–e687.

Budd, S. J., & Hartnett, M. E. (2010). Increased angiogenic factors associated with peripheral avascular retina and intravitreous neovascularization: A model of retinopathy of prematurity. *Archives of Ophthalmology, 128,* 589–595.

Budd, S., Byfield, G., Martiniuk, D., Geisen, P., & Hartnett, M. E. (2009). Reduction in endothelial tip cell filopodia corresponds to reduced intravitreous but not intraretinal vascularization in a model of ROP. *Experimental Eye Research, 89,* 718–727.

Byfield, G. E., Budd, S., & Hartnett, M. E. (2009). Supplemental oxygen can cause intravitreous neovascularization through JAK/STAT pathways in a model of retinopathy of prematurity. *Investigative Ophthalmology & Visual Science.* Epub[Mar 5], PMID: 19264880.

Chan-Ling, T. (1997). Glial, vascular and neuronal cytogenesis in whole-mounted cat retina. *Microscopy Research and Technique, 36,* 1–16. doi:10.1002/(SICI)1097-0029(19970101) 36:1<1:AID-JEMT1>3.0.CO;2-V.

Chan-Ling, T., Gock, B., & Stone, J. (1995). The effect of oxygen on vasoformative cell division: Evidence that "physiological hypoxia" is the stimulus for normal retinal vasculogenesis. *Investigative Ophthalmology & Visual Science, 36,* 1201–1214.

Chan-Ling, T., McLeod, D. S., Hughes, S., Baxter, L., Chu, Y., Hasegawa, T., et al. (2004). Astrocyte–endothelial cell relationships during human retinal vascular development. *Investigative Ophthalmology & Visual Science, 45,* 2020–2032. doi:10.1167/iovs.03-1169.

Chen, J., & Smith, L. E. (2007). Retinopathy of prematurity. *Angiogenesis, 10,* 133–140.

Connor, K. M., SanGiovanni, J. P., Lofqvist, C., Aderman, C. M., Chen, J., Higuchi, A., et al. (2007). Increased dietary intake of omega-3-polyunsaturated fatty acids reduces pathological retinal angiogenesis. *Nature Medicine, 13,* 868–873. doi:10.1038/nm1591.

Cryotherapy for Retinopathy of Prematurity Cooperative Group. (1990). Multicenter trial of cryotherapy for retinopathy of prematurity: Three month outcome. *Archives of Ophthalmology, 108,* 195–204.

Cryotherapy for Retinopathy of Prematurity Cooperative Group. (2002). Multicenter trial of cryotherapy for retinopathy of prematurity: Natural history ROP: Ocular outcome at 5(1/2) years in premature infants with birth weights less than 1251g. *Archives of Ophthalmology, 120,* 595–599.

Cunningham, S., Fleck, B. W., Elton, R. A., & McIntosh, N. (1995). Transcutaneous oxygen levels in retinopathy of prematurity. *Lancet, 346,* 1464–1465. doi:10.1016/S0140-6736(95)92475-2.

Darlow, B. A., Hutchinson, J. L., Henderson-Smart, D. J., Donoghue, D. A., Simpson, J. M., & Evans, N. J. (2005). Prenatal risk factors for severe retinopathy of prematurity among very preterm infants of the Australian and New Zealand neonatal network. *Pediatrics, 115,* 990–996. doi:10.1542/peds.2004-1309.

Dorey, C., Aouididi, S., Reynaud, X., Dvorak, H. F., & Brown, L. F. (1996). Correlation of vascular permeability factor/vascular endothelial growth factor with extraretinal neovascularisation in the rat. *Archives of Ophthalmology, 114,* 1210–1217.

Drenser, K. A. (2009). Anti-angiogenic therapy in the management of retinopathy of prematurity. *Developments in Ophthalmology, 44,* 89–97.

Early Treatment for Retinopathy of Prematurity Cooperative Group. (2003). Revised indications for the treatment of retinopathy of prematurity: Results of the early treatment for retinopathy of prematurity randomized trial. *Archives of Ophthalmology, 121,* 1684–1694.

Edwards, M. M., McLeod, D. S., Renzhong, L., Grebe, R., Bhutto, I., Mu, X., et al. (2012). The deletion of Math5 disrupts retinal blood vessel and glial development in mice. *Experimental Eye Research, 96,* 147–156. doi:10.1016/j. exer.2011.12.005.

Ernest, J. T., & Goldstick, T. K. (1984). Retinal oxygen tension and oxygen reactivity in retinopathy of prematurity in kittens. *Investigative Ophthalmology & Visual Science, 25,* 1129–1134.

Ferrara, N., Gerber, H. P., & Lecouter, J. (2003). The biology of VEGF and its receptors. *Nature Medicine, 9,* 669–676.

Ford, K. M., Saint-Geniez, M., Walshe, T., Zahr, A., & D'Amore, P. A. (2011). Expression and role of VEGF in the adult retinal pigment epithelium. *Investigative Ophthalmology & Visual Science, 52,* 9478–9487.

Fulton, A. B., Hansen, R. M., Peterson, R. A., & Vanderveen, D. K. (2001). The rod photoreceptors in retinopathy of prematurity: An electroretinographic study. *Archives of Ophthalmology, 119,* 499–505.

Fulton, A. B., Reynaud, X., Hansen, R. M., Lemere, C. A., Parker, C., & Williams, T. P. (1999). Rod photoreceptors in infant rats with a history of oxygen exposure. *Investigative Ophthalmology & Visual Science, 40,* 168–174.

Gao, G., Li, Y., Fant, J., Crosson, C. E., Becerra, S. P., & Ma, J. (2002). Difference in ischemic regulation of vascular endothelial growth factor and pigment epithelium-derived factor in Brown Norway and Sprague Dawley rats contributing to different susceptibilities to retinal neovascularization. *Diabetes, 51,* 1218–1226. doi:10.2337/diabetes.51.4. 1218.

Geisen, P., Peterson, L., Martiniuk, D., Uppal, A., Saito, Y., & Hartnett, M. (2008). Neutralizing antibody to VEGF reduces intravitreous neovascularization and does not interfere with vascularization of avascular retina in an ROP model. *Molecular Vision, 14,* 345–357.

Gerhardt, H., Golding, M., Fruttiger, M., Ruhrberg, C., Lundkvist, A., Abramsson, A., et al. (2003). VEGF guides angiogenic sprouting utilizing endothelial tip cell filopodia. *Journal of Cell Biology, 161,* 1163–1177. doi:10.1083/ jcb.200302047.

Gilbert, C., Rahi, J., Eckstein, M., O'Sullivan, J., & Foster, A. (1997). Retinopathy of prematurity in middle-income countries. *Lancet, 350,* 12–14. doi:10.1016/S0140-6736(97)01107-0.

Hardy, P., Beauchamp, M., Sennlaub, F., Gobeil, J., Tremblay, L., Mwaikambo, B., et al. (2005). New insights into the retinal circulation: Inflammatory lipid mediators in ischemic retinopathy. *Prostaglandins, Leukotrienes, and Essential Fatty Acids, 72*, 301–325.

Harris, M., Moskowitz, A., Fulton, A., & Hansen, R. (2011). Long-term effects of retinopathy of prematurity (ROP) on rod and rod-driven function. *Documenta Ophthalmologica, 122*, 19–27.

Hartnett, M. E. (2010a). The effects of oxygen stresses on the development of features of severe retinopathy of prematurity: Knowledge from the 50/10 OIR model. *Documenta Ophthalmologica. Advances in Ophthalmology, 120*, 25–39.

Hartnett, M. E. (2010b). Studies on the pathogenesis of avascular retina and neovascularization into the vitreous in peripheral severe retinopathy of prematurity (an american ophthalmological society thesis). Transactions of the American Ophthalmological Society, 108, 96–119.

Hartnett M. E., Martiniuk D. J., Byfield G. E., Geisen P., Zeng G., & Bautch V. L. (2008). Neutralizing VEGF decreases tortuosity and alters endothelial cell division orientation in arterioles and veins in rat model of ROP: Relevance to plus disease. *Investigative Ophthalmology & Visual Science*, Mar 31 epub[49], 7, 3107–3114. doi:10.1167/iovs.08-1780.

Hartnett M. E., & Penn, J. S. (2012). Mechanisms and management of retinopathy of prematurity. *New England Journal of Medicine, 367*, 2515–2526.

Hasegawa, T., McLeod, D. S., Bhutto, I. A., Prow, T., Merges, C. A., Grebe, R., et al. (2007). The embryonic human choriocapillaris develops by hemo-vasculogenesis. *Developmental Dynamics, 236*, 2089–2100.

Hasegawa, T., McLeod, D. S., Prow, T., Merges, C., Grebe, R., & Lutty, G. A. (2008). Vascular precursors in developing human retina. *Investigative Ophthalmology & Visual Science, 49*, 2178–2192. doi:10.1167/iovs.07-0632.

Hellstrom, A., Engstrom, E., Hard, A. L., Albertsson-Wikland, K., Carlsson, B., Niklasson, A., et al. (2003). Postnatal serum insulin-like growth factor I deficiency is associated with retinopathy of prematurity and other complications of premature birth. *Pediatrics, 112*, 1016–1020.

Hellstrom, A., Hard, A. L., Engstrom, E., Niklasson, A., Andersson, E., Smith, L., et al. (2009). Early weight gain predicts retinopathy in preterm infants: New, simple, efficient approach to screening. *Pediatrics, 123*, e638–e645. doi:10.1542/peds.2008-2697.

Hellstrom, M., Phng, L. K., Hofmann, J. J., Wallgard, E., Coultas, L., Lindblom, P., et al. (2007). Dll4 signalling through Notch1 regulates formation of tip cells during angiogenesis. *Nature, 445*, 776–780. doi:10.1038/nature05571.

Hendrickson, A. E., & Yuodelis, C. (1984). The morphological development of the human fovea. *Ophthalmology, 91*, 603–612.

Hiratsuka, S., Minowa, O., Kuno, J., Noda, T., & Shibuya, M. (1998). Flt-1 lacking the tyrosine kinase domain is sufficient for normal development and angiogenesis in mice. *Proceedings of the National Academy of Sciences of the United States of America, 95*, 9349–9354. doi:10.1073/pnas.95.16.9349.

Hutcheson, K. A., Paluru, P. C., Bernstein, S. L., Koh, J., Rappaport, E. F., Leach, R. A., et al. (2005). Norrie disease gene sequence variants in an ethnically diverse population with retinopathy of prematurity. *Molecular Vision, 11*, 501–508.

Jones, C. A., London, N. R., Chen, H., Park, K. W., Sauvaget, D., Stockton, R. A., et al. (2008). Robo4 stabilizes the vascular network by inhibiting pathologic angiogenesis and endothelial hyperpermeability. *Nature Medicine, 14*, 448–453. doi:10.1038/nm1742.

Kearney, J. B., Kappas, N. C., Ellerstrom, C., DiPaola, F. W., & Bautch, V. L. (2004). The VEGF receptor flt-1 (VEGFR-1) is a positive modulator of vascular sprout formation and branching morphogenesis. *Blood, 103*, 4527–4535. doi:10.1182/blood-2003-07-2315.

Kermorvant-Duchemin, E., Sapieha, P., Sirinyan, M., Beauchamp, M., Checchin, D., Hardy, P., et al. (2010). Understanding ischemic retinopathies: Emerging concepts from oxygen-induced retinopathy. *Documenta Ophthalmologica, 120*, 51–60.

Kurihara, T., Kubota, Y., Ozawa, Y., Takubo, K., Noda, K., Simon, M. C., et al. (2010). von Hippel-Lindau protein regulates transition from the fetal to the adult circulatory system in retina. *Development, 137*, 1563–1571. doi:10.1242/dev.049015.

Lambert, I. H., Pedersen, S. F., & Poulsen, K. A. (2006). Activation of PLA2 isoforms by cell swelling and ischaemia/hypoxia. *Acta Physiologica (Oxford, England), 187*, 75–85. doi:10.1111/j.1748-1716.2006.01557.x.

Lobov, I. B., Brooks, P. C., & Lang, R. A. (2002). Angiopoietin-2 displays VEGF-dependent modulation of capillary structure and endothelial cell survival in vivo. *Proceedings of the National Academy of Sciences of the United States of America, 99*, 11205–11210. doi:10.1073/pnas.172161899.

Lobov, I. B., Rao, S., Carroll, T. J., Vallance, J. E., Ito, M., Ondr, J. K., et al. (2005). WNT7b mediates macrophage-induced programmed cell death in patterning of the vasculature. *Nature, 437*, 417–421. doi:10.1038/nature03928.

Lutty, G. A., Hasegawa, T., Baba, T., Grebe, R., Bhutto, I., & McLeod, D. S. (2010). Development of the human choriocapillaris. *Eye (London, England), 24*, 408–415.

Lutty, G. A., McLeod, D. S., Bhutto, I., & Wiegand, S. J. (2011). Effect of VEGF trap on normal retinal vascular development and oxygen-induced retinopathy in the dog. *Investigative Ophthalmology & Visual Science, 52*, 4039–4047. doi:10.1167/iovs.10-6798.

Maile, L. A., Gollahon, K., Wai, C., Byfield, G., Hartnett, M. E., & Clemmons, D. (2012). Disruption of IAP/SHPS-1 association inhibits pathophysiologic changes in retinal endothelial function in diabetic rats. *Diabetologia, 55*, 835–844. doi:10.1007/s00125-011-2416-x.

Mann, I. (1964). *The development of the human eye* (1st ed.). New York: Grune & Stratton.

McLeod, D. S., Brownstein, R., & Lutty, G. A. (1996). Vaso-obliteration in the canine model of oxygen-induced retinopathy. *Investigative Ophthalmology & Visual Science, 37*, 300–311.

McLeod, D. S., Hasegawa, T., Baba, T., Grebe, R., Galtier d'Auriac, I., Merges, C., Edwards, M., & Lutty, G. A. (2012). From blood islands to blood vessels: morphologic observations and expression of key molecules during hyaloid vascular system development. *Investigative Ophthalmology & Visual Science, 53*, 7912–7927. doi:10.1167/iovs.12-10140.

McLeod, D. S., Hasegawa, T., Prow, T., Merges, C., & Lutty, G. (2006). The initial fetal human retinal vasculature develops by vasculogenesis. *Developmental Dynamics, 235*, 3336–3347.

Michaelson, I. C. (1948). The mode of development of the vascular system of the retina: With some observations on its significance for certain retinal diseases. *Transactions of the Ophthalmological Societies of the United Kingdom, 68*, 137–180.

Mintz-Hittner, H. A., & Kuffel, R. R., Jr. (2008). Intravitreal injection of bevacizumab (Avastin) for treatment of stage 3 retinopathy of prematurity in zone I or posterior zone II. *Retina (Philadelphia, Pa.), 28,* 831–838.

Mintz-Hittner, H. A., Kennedy, K. A., & Chuang, A. Z. (2011). Efficacy of intravitreal bevacizumab for stage 3+ retinopathy of prematurity. *New England Journal of Medicine, 364,* 603–615.

Mitchell, C. A., Risau, W., & Drexler, H. A. (1998). Regression of vessels in the tunica vasculosa lentis is initiated by coordinated endothelial apoptosis: A role for vascular endothelial growth factor as a survival factor for endothelium. *Developmental Dynamics, 213,* 322–333.

Neu, J., Afzal, A., Pan, H., Gallego, E., Li, N., Calzi, S. L., et al. (2006). The dipeptide Arg-Gln inhibits retinal neovascularization in the mouse model of oxygen-induced retinopathy. *Investigative Ophthalmology & Visual Science, 47,* 3151–3155. doi:10.1167/iovs.05-1473.

Niesman, M. R., Johnson, K. A., & Penn, J. S. (1997). Therapeutic effect of liposomal superoxide dismutase in an animal model of retinopathy of prematurity. *Neurochemical Research, 22,* 597–605.

Ohlsson, A., & Aher, S. M. (2006). Early erythropoietin for preventing red blood cell transfusion in preterm and/or low birth weight infants. *Cochrane Database of Systematic Reviews, 3,* CD004863.

Patz, A. (1954). Oxygen studies in retrolental fibroplasia. *American Journal of Ophthalmology, 38,* 291–308.

Patz, A. (1985). Observations on the retinopathy of prematurity. *American Journal of Ophthalmology, 100,* 164–168.

Penn, J. S. (1992). Oxygen-induced retinopathy in the rat: Vitamins C and E as potential therapies. *Investigative Ophthalmology & Visual Science, 33,* 1836–1845.

Penn, J. S., Henry, M. M., & Tolman, B. L. (1994). Exposure to alternating hypoxia and hyperoxia causes severe proliferative retinopathy in the newborn rat. *Pediatric Research, 36,* 724–731.

Penn, J. S., Madan, A., Caldwell, R. B., Bartoli, M., Caldwell, R. W., & Hartnett, M. E. (2008). Vascular endothelial growth factor in eye disease. *Progress in Retinal and Eye Research, 27,* 331–371.

Penn, J. S., Tolman, B. L., & Bullard, L. E. (1997). Effect of a water-soluble vitamin E analog, Trolox C, on retinal vascular development in an animal model of retinopathy of prematurity. *Free Radical Biology & Medicine, 22,* 977–984.

Penn, J. S., Tolman, B. L., & Henry, M. M. (1994). Oxygen-induced retinopathy in the rat: Relationship of retinal nonperfusion to subsequent neovascularization. *Investigative Ophthalmology & Visual Science, 35,* 3429–3435.

Pierce, E. A., Avery, R. L., Foley, E. D., Aiello, L. P., & Smith, L. E. H. (1995). Vascular endothelial growth factor/vascular permeability factor expression in a mouse model of retinal neovascularization. *Proceedings of the National Academy of Sciences of the United States of America, 92,* 905–909.

Raju, T. N. K., Langenberg, P., Bhutani, V., & Quinn, G. E. (1997). Vitamin E prophylaxis to reduce retinopathy of prematurity: A reappraisal of published trials. *Journal of Pediatrics, 131,* 844–850.

Repka, M. X., Tung, B., Good, W. V., Capone, A., Jr., & Shapiro, M. J. (2011). Outcome of eyes developing retinal detachment during the early treatment for retinopathy of prematurity study. *Archives of Ophthalmology, 129,* 1175–1179.

Reynaud, X., Hansen, R. M., & Fulton, A. B. (1995). Effect of prior oxygen exposure on the electroretinographic responses of infant rats. *Investigative Ophthalmology & Visual Science, 36,* 2071–2079.

Robinson, G. S., Pierce, E. A., Rook, S. L., Foley, E., Webb, R., & Smith, L. E. H. (1996). Oligodeoxynucleotides inhibit retinal neovascularization in a murine model of proliferative retinopathy. *Proceedings of the National Academy of Sciences of the United States of America, 93,* 4851–4856. doi:10.1073/pnas.93.10.4851.

Rosenfeld, P. J., Brown, D. M., Heier, J. S., Boyer, D. S., Kaiser, P. K., Chung, C. Y., et al. (2006). Ranibizumab for neovascular age-related macular degeneration. *New England Journal of Medicine, 355,* 1419–1431. doi:10.1056/NEJMoa054481.

Saito, Y., Geisen, P., Uppal, A., & Hartnett, M. E. (2007). Inhibition of NAD(P)H oxidase reduces apoptosis and avascular retina in an animal model of retinopathy of prematurity. *Molecular Vision, 13,* 840–853.

Saito, Y., Uppal, A., Byfield, G., Budd, S., & Hartnett, M. E. (2008). Activated NAD(P)H oxidase from supplemental oxygen induces neovascularization independent of VEGF in retinopathy of prematurity model. *Investigative Ophthalmology & Visual Science, 49,* 1591–1598.

Sapieha, P., Joyal, J. S., Rivera, J. C., Kermorvant-Duchemin, E., Sennlaub, F., Hardy, P., et al. (2010). Retinopathy of prematurity: Understanding ischemic retinal vasculopathies at an extreme of life. *Journal of Clinical Investigation, 120,* 3022–3032. doi:10.1172/JCI42142.

Smith, L. E. (2008). Through the eyes of a child: Understanding retinopathy through ROP. The Friedenwald Lecture. *Investigative Ophthalmology & Visual Science, 49,* 5177–5182. doi:10.1167/iovs.08-2584.

Smith, L. E. H., Wesolowski, E., McLellan, A., Kostyk, S. K., D'Amato, R., Sullivan, R., et al. (1994). Oxygen induced retinopathy in the mouse. *Investigative Ophthalmology & Visual Science, 35,* 101–111.

Sonmez, K., Drenser, K. A., Capone, A. Jr., & Trese, M. T. (2008). Vitreous levels of stromal cell-derived factor 1 and vascular endothelial growth factor in patients with retinopathy of prematurity. *Ophthalmology, 115,* 1065–1070.

Sorenson, C. M., & Sheibani, N. (2011). Anti-vascular endothelial growth factor therapy and renal thrombotic microangiopathy. *Archives of Ophthalmology, 129,* 1082.

Stone, J., Itin, A., Alon, T., Peer, J., Gnessin, H., Chan-Ling, T., et al. (1995). Development of retinal vasculature is mediated by hypoxia-induced vascular endothelial growth factor (VEGF) expression by neuroglia. *Journal of Neuroscience, 15,* 4738–4747.

Tachibana, K., Hirota, S., Iizasa, H., Yoshida, H., Kawabata, K., Kataoka, Y., et al. (1998). The chemokine receptor CXCR4 is essential for vascularization of the gastrointestinal tract. *Nature, 393,* 591–594.

Taomoto, M., McLeod, D. S., Merges, C., & Lutty, G. A. (2000). Localization of adenosine A2a receptor in retinal development and oxygen-induced retinopathy. *Investigative Ophthalmology & Visual Science, 41,* 230–243.

Terry, T. L. (1942). Extreme prematurity and fibroblastic overgrowth of persistent vascular sheath behind each crystalline lens: I. Preliminary report. *American Journal of Ophthalmology, 25,* 203–204.

The Committee for the Classification of Retinopathy of Prematurity. (1984). An international classification of

     MARY ELIZABETH HARTNETT

retinopathy of prematurity. *Archives of Ophthalmology, 102*, 1130–1134.

Tolentino, M. J., McLeod, D. S., Taomoto, M., Otsuji, T., Adamis, A. P., & Lutty, G. A. (2002). Pathologic features of vascular endothelial growth factor-induced retinopathy in the nonhuman primate. *American Journal of Ophthalmology, 133*, 373–385.

Trese, M. T., & Capone, A. (2004). Retinopathy of prematurity: Evolution of stages 4 and 5 ROP and management: A. Evolution to retinal detachment and physiologically based management. In M. E. Hartnett, M. T. Trese, A. Capone, B. J. K. Keats, & S. M. Steidl (Eds.), *Pediatric retina* (pp. 411–416). Philadelphia: Lippincott Williams & Wilkins.

Uno, K., Merges, C. A., Grebe, R., Lutty, G. A., & Prow, T. W. (2007). Hyperoxia inhibits several critical aspects of vascular development. *Developmental Dynamics, 236*, 981–990.

Vanhaesebrouck, S., Daniels, H., Moons, L., Vanhole, C., Carmeliet, P., & De Zegher, F. (2009). Oxygen-induced retinopathy in mice: Amplification by neonatal IGF-I deficit and attenuation by IGF-I administration. *Pediatric Research, 65*, 307–310.

Wang, H., Byfield, G., Jiang, Y., Smith, G. W., McCloskey, M., & Hartnett, M. E. (2012). VEGF-mediated STAT3 activation inhibits retinal vascularization by downregulating erythropoietin expression. *American Journal of Pathology, 180*, 1243.

Wang, J., Xu, X., Elliott, M. H., Zhu, M., & Le, Y. Z. (2010). Müller cell-derived VEGF is essential for diabetes-induced retinal inflammation and vascular leakage. *Diabetes, 59*, 2297–2305. doi:10.2337/db09-1420.

Weidemann, A., Krohne, T. U., Aguilar, E., Kurihara, T., Takeda, N., Dorrell, M. I., et al. (2010). Astrocyte hypoxic response is essential for pathological but not developmental angiogenesis of the retina. *Glia, 58*, 1177–1185. doi:10.1002/glia.20997.

York, J. R., Landers, S., Kirby, R. S., Arbogast, P. G., & Penn, J. S. (2004). Arterial oxygen fluctuation and retinopathy of prematurity in very-low-birth-weight infants. *Journal of Perinatology, 24*, 82–87.

Zeng, G., Taylor, S. M., McColm, J. R., Kappas, N. C., Kearney, J. B., Williams, L. H., et al. (2007). Orientation of endothelial cell division is regulated by VEGF signaling during blood vessel formation. *Blood, 109*, 1345–1352. doi:10.1182/blood-2006-07-037952.

D. SCOTT McLEOD AND GERARD A. LUTTY

The human choroid is a thin, pigmented, and highly vascularized tissue that lies beneath the sensory retina and forms the posterior portion of the uveal tract (the iris, ciliary body, and choroid). The inner boundary of the choroid is Bruch's membrane (BrMb), a pentalaminar structure on which a monolayer of cuboidal retinal pigment epithelium (RPE) sits. The adult choroidal vasculature has three layers: the inner choriocapillaris (CC) with broad, flat lumens (12 to 20 mm diameter); Sattler's layer of intermediate vessels in the middle; and the outermost Haller's layer with large vessels (Bhutto & Lutty, 2006). The CC, the capillary component of the choroidal vasculature, lies adjacent and posterior to BrMb. The choroidal vasculature has a lobular pattern and is arranged in a single layer restricted to the inner portion of the choroid. Feeding arterioles and draining venules enter the capillary plexus from Sattler's layer at right angles in the posterior pole. In the peripheral and equatorial areas of choroid, arterioles and venules are located in the plane of the CC, and the CC has a ladder-like arrangement. Inner choroidal vessels (CC and medium-sized vessels) are sandwiched between two pigmented cell types, apical RPE of neuroepithelial origin and outer choroidal melanocytes of neural crest origin. The adult CC is unique in that its basement membrane comprises the most posterior layer of BrMb, and the endothelial cells are fenestrated mostly on the retinal side. Fenestrated endothelium is a characteristic of tissues that are involved in secretion and/or filtration. The sidedness of the CC is also true in terms of receptor expression: Vascular endothelial cell growth factor (VEGF) receptor-1, -2, and -3 (VEGF-R1, R2, R3) are expressed mainly on the retinal side of the vasculature (Blaauwgeers et al., 1999). VEGF is secreted from the basal side of the RPE, which is thought to be necessary in maintaining the CC fenestrations (Blaauwgeers et al., 1999) and its survival (Marneros et al., 2005; Saint-Geniez et al., 2009). CC is also one of the few capillary systems in which the endothelial cells constitutively express ICAM-1 (McLeod et al., 1995).

The major roles of the adult CC are to transport nutrients and oxygen to the RPE and photoreceptors (PR) and remove waste from the RPE. The anatomical arrangement of CC permits this vasculature to provide all of the metabolic requirements for the PR from serum components to 90% of the O_2 consumed by the PR in darkness (Wangsa-Wirawan & Linsenmeier, 2003). The PRs consume more oxygen per gram of tissue weight than any cell in the body. The tissue oxygen level at the inner segments is near zero in the dark (Wangsa-Wirawan & Linsenmeier, 2003). This suggests that any disruption in choroidal blood flow would be detrimental to PR function and/or survival. Abnormalities in the choroidal vasculature result in many kinds of congenital and adult diseases, from choroidal coloboma to age-related macular degeneration (AMD) (Daufenbach et al., 1998; Lutty et al., 1999; McLeod et al., 2009).

The intent of this review is to summarize our studies on the development and maturation of the human CC from 5.5 until 22 weeks' gestation (WG). The CC is a primitive vascular system at 5.5 WG and will mature to have most of the adult morphological characteristics by 22 WG.

THE INITIAL CC DEVELOPS BY HEMO-VASCULOGENESIS (5.5–8 WEEKS' GESTATION)

There are three mechanisms by which a vasculature develops: (1) angiogenesis, the development by migration and proliferation of endothelial cells from an existing blood vessel; (2) vasculogenesis, coalescence, and differentiation of vascular progenitors or angioblasts; and (3) hemo-vasculogenesis, differentiation of vascular and blood cells (hematopoiesis [expression of CD34$^+$] and erythropoiesis [presence of epsilon chain of hemoglobin or Hb-ε$^+$]) from a common progenitor, the hemangioblast. We have recently demonstrated that the initial human CC develops by hemo-vasculogenesis between 5.5 and 8 WG (Hasegawa et al., 2007). At 5.5–7 WG, erythroblasts (nucleated erythrocytes expressing epsilon hemoglobin [Hbε$^+$]) were observed within blood islands in the CC layer and were scattered within the adjacent choroidal stroma (see figure 104.1).

The apparent erythroblasts were within primitive lumens, in the walls of primitive lumens and free within the choroidal stroma (see figure 104.1). Using the endothelial cell markers (CD31, CD34, CD39) in conjunction with Hbε labeling, we often found that the

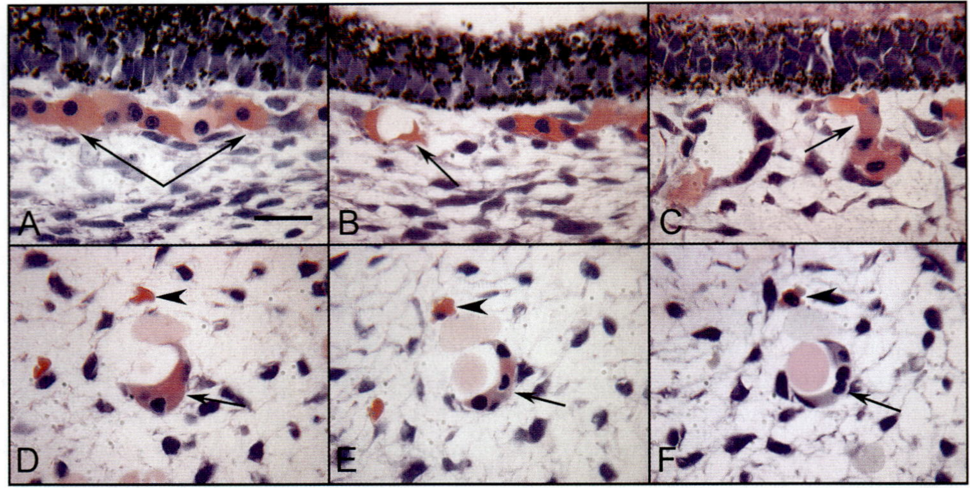

FIGURE 104.1   Hemo-vasculogenesis is demonstrated in cross-sections of 6.5 weeks' gestation fetal choroid (A–F). In the choriocapillaris layer, erythroblasts (bright pink cytoplasm) can form solid cord-like structures (double arrow) without a lumen (A). Erythroblast, hematopoetic, and vascular cells develop in situ with erythroblasts sometimes forming a lumen (arrow in D). Eventually, the outer cells become primarily endothelial cells and the inner cells become primarily erythroblasts (F). Free erythroblasts (arrowheads in D–F) are present in stroma of choroid. The monolayer of pigmented retinal pigment epithelium cells is visible at the top in (A–C). (Scale bar: 10 μm; Giemsa stained JB4 sections.)

same cells coexpressed Hbε as well as endothelial cell (CD31), hematopoietic (CD34), and hemangioblast (VEGFR-2) markers (see figure 104.2). This suggested that these cells were hemangioblasts or progenitors derived from hemangioblasts and that their aggregation and subsequent differentiation was the mechanism by which the initial CC develops (Hasegawa et al., 2007). When flat mount choroids were labeled for CD39, a marker for endothelial cells and angioblasts (McLeod et al., 2006), the blood island appearance of the vascular structures were apparent, and erythroblasts were both inside and outside of the islands (see figure 104.3A). Only a single layer of vasculature (CD31[+], CD39[+], CD34[+]) is observed in choroid at this time, suggesting that the CC formed independent of a functional blood supply (see figure 104.4A). There were very few erythroblasts associated with CC by 8 WG, and vascular lumens were apparent. The erythroblasts that remain are only within formed lumens by 9 WG in flat mounts, when Hbε production declines (see figure 104.3B). Ultrastructurally, the association of erythroblasts with endothelial-like progenitors is apparent in the early stages of hemo-vasculogenesis (see figure 104.5A). In some structures, the endothelial cell progenitor almost encloses an erythroblast in completing early lumen formation (see figure 104.5B), but in other examples only slit-like empty lumens form (Baba et al., 2009; Sellheyer, 1990; Sellheyer & Spitznas, 1988). Development of the CC by hemo-vasculogenesis from islands of progenitors explains how this vasculature forms without a source of blood (Hasegawa et al., 2007). Formation of blood vessels by hemo-vasculogenesis has been observed in embryonic mouse in several organ systems (Sequeira Lopez et al., 2003).

The expression of three enzymes known to be important in adult CC function was evaluated as an index of the functional properties of the CC: endothelial nitric oxide synthase (eNOS), carbonic anhydrase IV (CA IV), and alkaline phosphatase (APase). CA IV is an ecto-enzyme that controls local pH when coupled to the electrogenic sodium bicarbonate cotransporter (NBC1) (Srere, 1987; Yang et al., 2005). As early as 8 WG, CA IV and eNOS were expressed at low levels during hemo-vasculogenesis (5.5–9 WG) (Baba et al., 2009). Even though eNOS's presumed function is vasodilation of blood vessels, only slit-like lumens were present at this time (Baba et al., 2009). Curiously, neuronal NOS (nNOS) was present in the nuclei of vascular progenitors in CC as well as in retina and vitreous and also in the nucleus of RPE cells and progenitors throughout choroid (McLeod et al., 2012). We have used APase enzyme histochemistry as a marker for choroidal vessels in adult human (McLeod & Lutty, 1994), and forming CC already had APase activity at 7 WG, even though the primitive ECs were just starting to differentiate from hemangioblasts. However, APase is used as a marker for human embryonic stem cells and other progenitors extensively, so the observation of high APase activity in the blood islands is not unexpected (Saito, Liu, & Yokoyama, 2004; Saito et al., 2005).

   D. SCOTT MCLEOD AND GERARD A. LUTTY

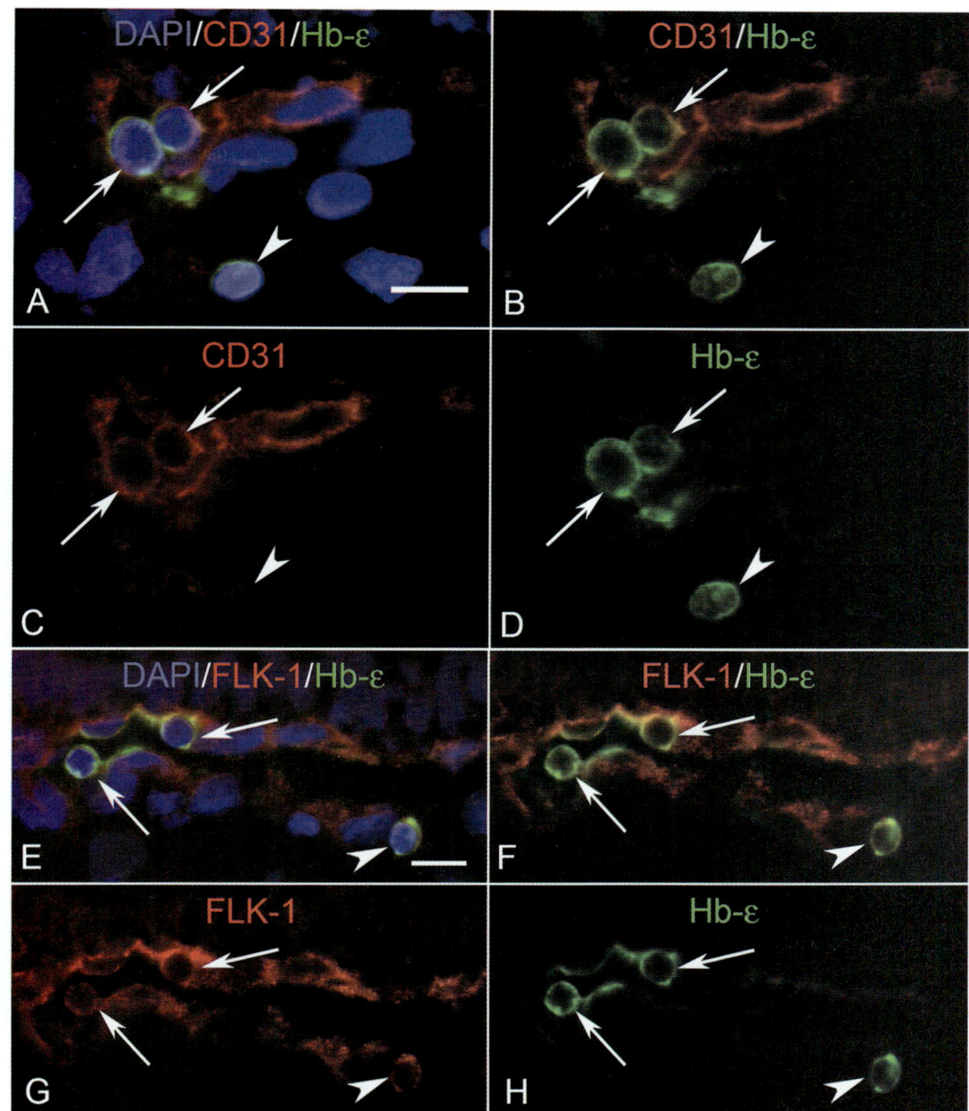

FIGURE 104.2   Colocalization of embryonic hemoglobin (Hb-ε) and endothelial markers (CD31 and VEGFR-2 or FLK-1) in developing choriocapillaris (CC). (A–D) CD31 (red) and Hb-ε (green) are colocalized in cells of the developing CC (arrows) and single cells within the choroidal stroma (arrowhead). (E–H) FLK-1 and Hb-ε coexpression in cells lining a developing lumen (arrows) and in cells located outside of the structure (arrowhead). (Scale bars: 10 µm; counterstained with DAPI [4',6-diamidino-2-phenylindole], blue.) (From Lutty et al., 2010.)

9–12 WEEKS' GESTATION

Development of intermediate and deeper choroidal vessels was observed between 9 and 12 WG. The developmental stages are centripedal in that this development was more advanced in the posterior pole than in the equatorial choroid (see figure 104.4). The forming vessels expressed EC markers including CD31, CD34, and CD39. Some ECs were proliferating and budding from the scleral side of the CC as shown with CD34 and Ki67 double labeling, suggesting that intermediate vessels form by angiogenesis (Hasegawa et al., 2007).

Flat CD39 labeled preparations of choroid (see figure 104.3B) demonstrated a rather linear pattern at 9 WG and then a chicken wire-like pattern of blood vessels with a few free CD39 positive cells between the vascular segments at 12 WG (figure 104.4C) (Hasegawa et al., 2007). This suggests that some angioblasts were still present in the choroid (McLeod et al., 2006) at 12 WG, but the pattern of CC was approaching an adult-like lobular pattern.

As mentioned, the adult CC is fenestrated, so we investigated the presence of fenestrations and their components at different ages. At and before 12 WG, all

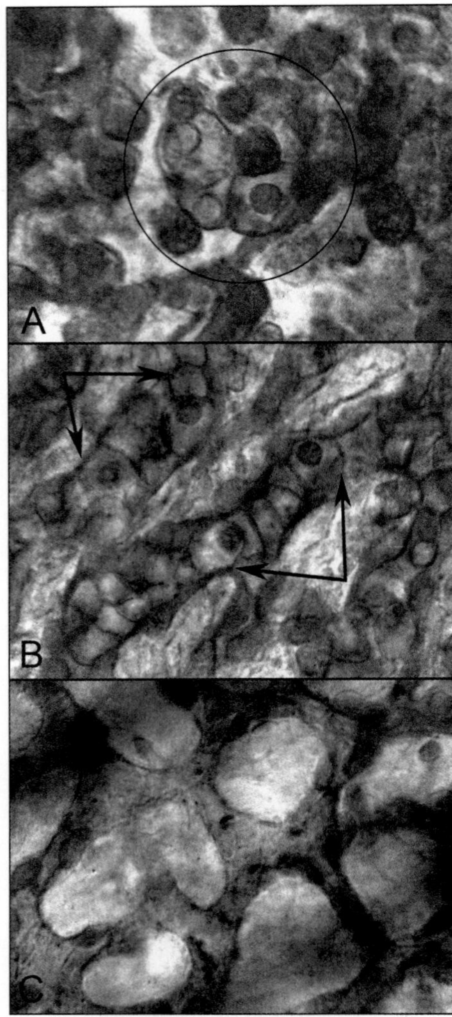

FIGURE 104.3   CD39 immunolabeled flat choroids showing the pattern of the developing choriocapillaris in embryonic and fetal human eyes. At 5.5 weeks' gestation (WG) (A), CD39-positive cells and erythroblasts are organized into blood islands (circle). At 9 WG (B), highly cellular linear cord-like structures without apparent lumen have formed. By 12 WG (C), the capillaries are narrower, lumens have formed, and cellularity has decreased markedly.

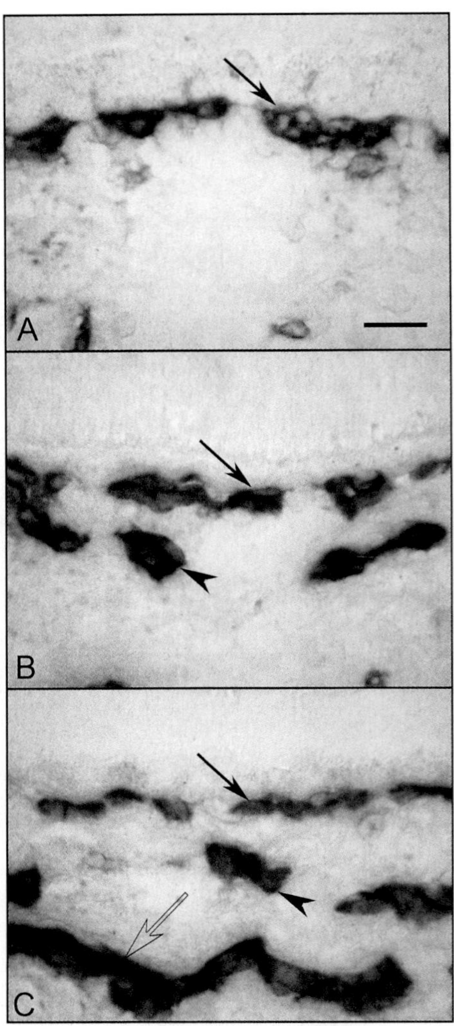

FIGURE 104.4   CD31 immunolabeling of choroidal sections from embryonic eyes at 6 weeks' gestation (WG) (A), and fetal eyes at 12 WG (B) and 20 WG (C). At 6 WG (A), only a highly cellular rudimentary choriocapillaris (CC) with poorly defined lumen (arrow) is present. At 12 WG (B), vessels are budding from the CC into the deeper choroid (arrowhead). At 22 WG (C), the CC (arrow), medium-size vessels of the Sattler's layer (arrowhead), and the larger outer blood vessels (open arrow) are all present. (Scale bar: 30 μm.)

vessels were negative for PV-1, PLVAP (plasmalemmal vesicle associated protein), which is an integral membrane glycoprotein in the diaphragms of fenestrations (Stan et al., 1999a; Stan, Kubitza, & Palade, 1999b). Transmission electron microscopy (TEM) demonstrated occasional fenestrations associated with the filipodial-like structures both in and around the lumen. These fenestrations were probably not functional because of their unusual position. The CC at this time period was composed of aggregates of progenitor cells with only slit-like lumens (Sellheyer & Spitznas,1988). Some cells had assumed an adventitial position relative to the cells lining the primitive lumens, but the two cell types were indistinguishable based on ultrastructural appearance in peripheral CC (Baba et al., 2009). There were some tight junctions present between the plump progenitor cells that bordered the lumens. Some Wieble–Palade bodies, organelles, found only in vascular endothelial cells, were present. More definitive pericyte-like cells were found on the more developed vessels in central choroid (area from disk to equator) (Baba et al., 2009). In the more mature central blood vessels, the pericytes had a nucleus that appeared more distinct from the endothelium. The slit-like lumens were often

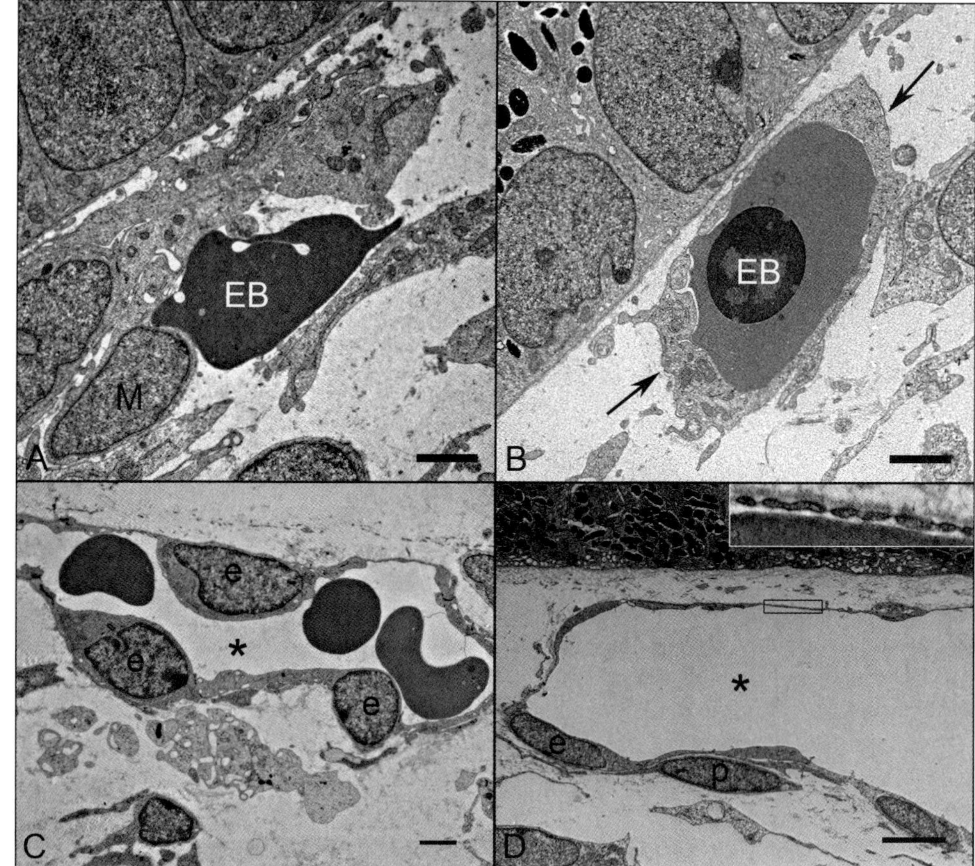

FIGURE 104.5   Ultrastructure of the developing embryonic and fetal human choriocapillaris (CC) at 6.5 weeks' gestation (WG) (A, B), 16 WG (C), and 22 WG (D). In ultrathin sections from 6.5 WG choroid (A and B), free progenitor cells (M) make contact with free erythroblasts (EB). In other areas, immature endothelial cells (ECs) envelop the erythroblasts to form blood islands (B). The retinal pigment epithelium (RPE) is present in the upper left hand corner in (A) and (B). By 16 WG, lumens are apparent (asterisk); the EC nuclei (e) with more organized chromatin were reduced in volume and had decreased cytoplasmic projections. By 22 WG, lumens were broad and flat, ECs were thin and fusiform, and definitive pericytes (p) were present on the outer surface of the capillaries. At this stage of development, fenestrations were present along inner aspect of the CC (inset). Fenestrations (inset) were present on the retinal side of CC lumens under RPE (top) at 22 WG. (Scale bars: 20 μm [A, B]; 2 μm [C]; 4 μm [D].)

filled with complex membranous infoldings that resembled filopodial processes from the luminal cells. In some lumens at the equator that were more open, the filopodia appeared to touch erythrocytes in the lumen, and the plasma membranes of the two cells could not be discerned (Baba et al., 2009). Basal lamina was not observed around these developing vessels (Baba et al., 2009).

14–16 WEEKS' GESTATION

By 14 WG in peripheral CC, progenitors in the ablumenal position or pericyte-like cells formed peg-in-socket-like contacts with endothelial cells lining the lumen, a characteristic of normal adult microvasculature (Baba et al., 2009). Antibodies for two pericyte markers were used to evaluate the maturation of these ablumenal cells: alpha smooth muscle actin (aSMA), present in mature pericytes and smooth muscle cells, and NG-2, a glycosaminoglycan present on the surface of pericytes. There was limited aSMA immunoreactivity at 14 WG while NG2 immunoreactivity was very prominent. Some areas of CC were weakly positive with PV-1 antibody at 16 WG, suggesting the presence of some fenestrations. TEM confirmed that there were a few fenestrations in the CC at this age, but they were not continuous in the thin endothelial cell processes on the retinal side of the lumens. The CC in the posterior pole was most mature morphologically and had the most fenestrations compared to peripheral regions. The number of filopodia in the broader lumens at 16 WG (see figure 104.5C). appeared greatly reduced compared to 11 WG. EC

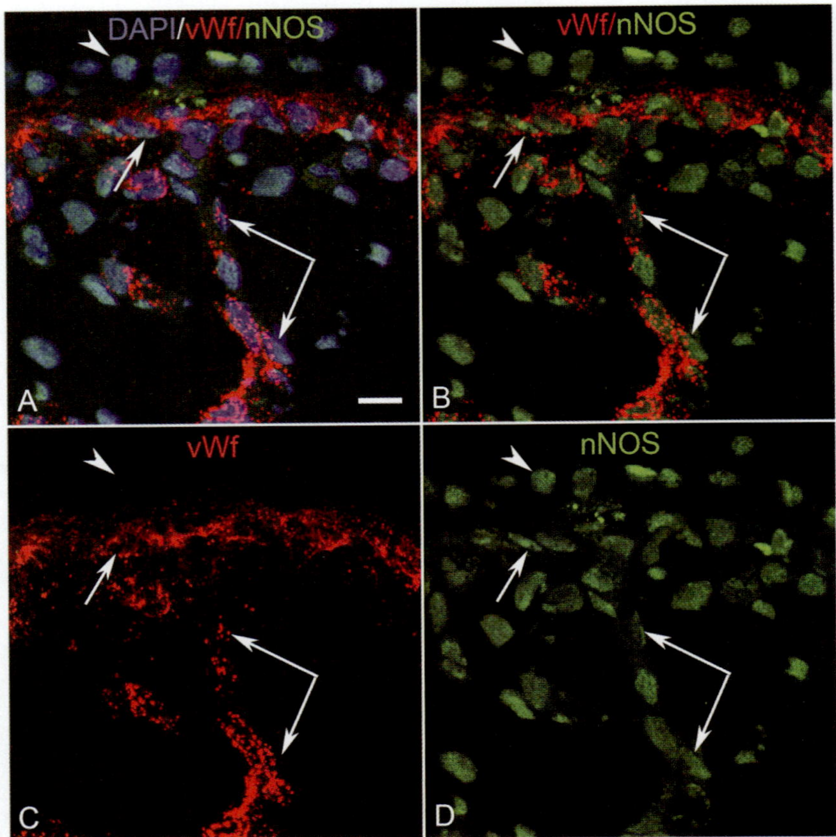

FIGURE 104.6   Choroid of a 21 weeks' gestation (WG) fetal eye double-labeled for von Willebrand's factor (vWf) (red) and neuronal nitric oxide synthase (nNOS) (green). Nuclear nNOS expression is seen in retinal pigment epithelium (arrowhead), endothelial cells of choriocapillaris (arrow), and large choroidal vessels (paired arrow). Nuclei were counterstained with DAPI (4',6-diamidino-2-phenylindole), blue. (Scale bar in A: 10 μm.)

nuclei were more oval and uniform in shape with less dense chromatin, and BrMb appeared more organized. Intermediate choroidal vessels and CC had prominent CA IV and eNOS immunoreactivity and intense APase at 14 WG.

21–22 WEEKS' GESTATION

At 21 WG, three layers of blood vessels were apparent within the posterior pole as demonstrated with EC markers (see figure 104.4C). This is the first time point at which there is evidence of PR maturation. Short rudimentary inner segments were present at the outermost portion of the neuroblastic layer (Baba et al., 2009). PV-1 immunoreactivity was present in most of the CC, but again, it was more intense in the posterior pole than in the periphery. However, in the adult human eye used as a positive control, the PV-1 was uniformly intense and more apparent on the retinal side of the CC lumens (Baba et al., 2009). Endothelial NOS was prominent in the endothelial cell cytoplasm of all choroidal blood vessels (see figure 104.6). Neuronal NOS was mostly nuclear in both endothelial cells labeled with von Willebrand's factor (vWf) (see figure 104.6) as well as pericytes and smooth muscle cells labeled with aSMA (McLeod et al., 2012).

TEM at 22 WG demonstrated that the CC endothelial cells were thin and fusiform. The CC had broad open lumens and contiguous areas of fenestrations in the narrow endothelial processes on the retinal side of the lumen (see figure 104.5D). Well-formed tight junctions were present between endothelial cells, Wieble–Palade bodies had increased in the cytoplasm, and a continuous basement membrane was present. BrMb was also more developed with collagen and elastin dispersed under the RPE basement membrane.

aSMA increased continuously with age until 22 WG when aSMA+ cells were present in the CC, and also around intermediate and large choroidal blood vessels. NG2 was most prominent at 22 WG when pericytes were apparent by TEM at this age (see figure 104.5D), and they were located primarily on the scleral side of the CC.

VEGF is known to play a key role in vascular development in many tissues including the retina (Ferrara, Gerber, & LeCouter, 2003; Gerhardt, 2008). The $VEGF_{165}$ isoform is thought to be the most critical isoform of VEGF-A in pathological angiogenesis (Carmeliet & Jain, 2000). Recently, it has been reported that VEGF-A has two splice variants VEGFxxx and VEGFxxxb from the VEGF-A gene product in addition to the known isoforms with varying heparin binding affinities and molecular sizes: $VEGF_{121'\ 145'\ 165'\ 189'\ 206}$ (Bates et al., 2002; Ladomery, Harper, & Bates, 2007; Qiu et al., 2009). The VEGFxxxb family members are anti-angiogenic and the splicing is stimulated by TGF beta through splicing factor SRp55 (serine/arginine protein 55) (Nowak et al., 2008). On the contrary, VEGFxxx is pro-angiogenic, and insulin-like growth factor 1 and tumor necrosis factor alpha upregulate splicing through ASF/SF2 (alternative splicing factor/splicing factor 2) (Nowak et al., 2008, 2010). We recently investigated the presence of these two antagonistic splice variants in developing choroid. During hemovasculogenesis, $VEGF_{165}$ is prominent in the forming CC only, while $VEGF_{165}b$ is not present (Baba et al., 2012). $VEGF_{165}$ expression increased with time in the forming vasculature and became very prominent in the basal portion of RPE and in CC at 12 WG. At 17 WG, $VEGF_{165}$ was very prominent in the basal RPE and in developing choroidal vessels; $VEGF_{165}b$ expression is present in some choriocapillaris (Baba et al., 2012). By 21 WG, the level of the two splice variants was comparable, but both appear to be present in the same cells (see figure 104.7). VEGF was present mostly in the cytoplasm of

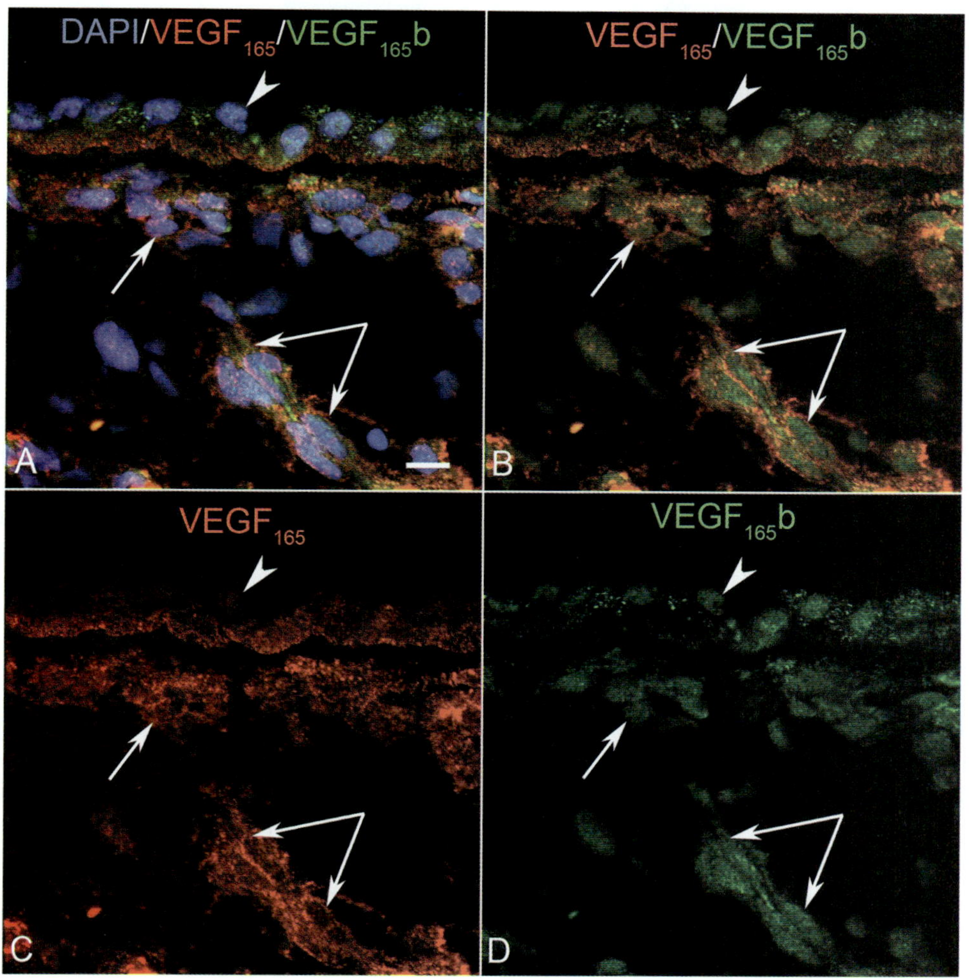

FIGURE 104.7   Vascular endothelial cell growth factor $(VEGF)_{165}$ (red) and $VEGF_{165}b$ (green) coexpression in the choriocapillaris (arrows), deeper choroidal vessels (paired arrows) and the retinal pigment epithelium (RPE) (arrowhead). $VEGF_{165}b$ has an apparent nuclear localization in the endothelium of blood vessels. In the RPE, it is associated with the nuclei and in the apical portion of cells. $VEGF_{165}$ is diffuse in endothelium and mostly localized to the basal portion of the RPE. Nuclei were counter stained with DAPI (4',6-diamidino-2-phenylindole), blue. (Scale bar in A: 10 μm.)

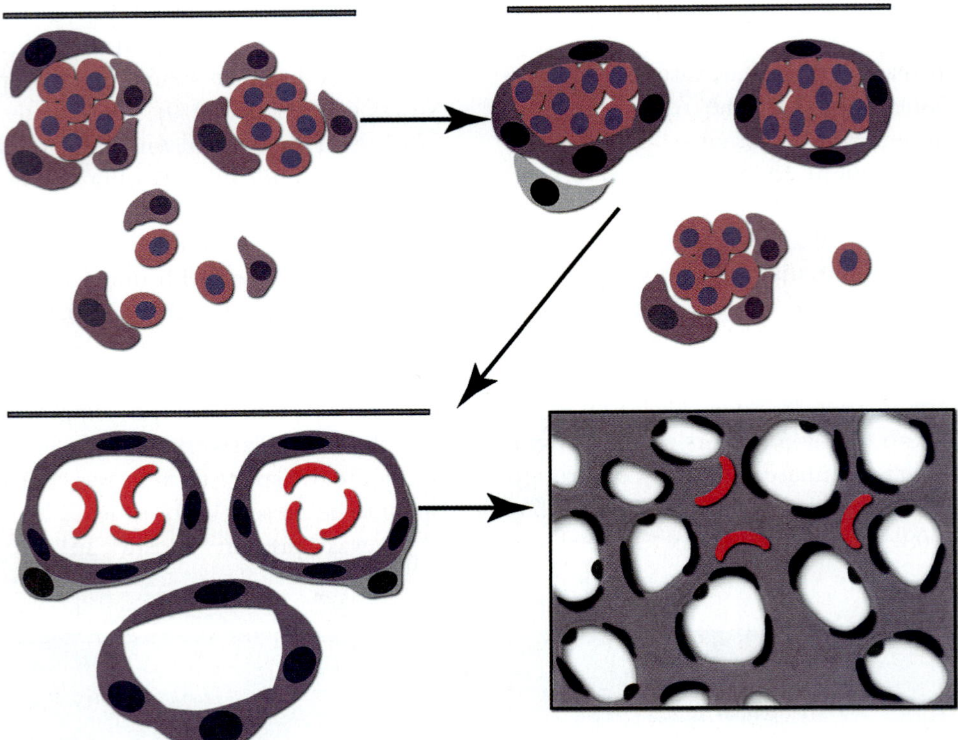

FIGURE 104.8   A schematic representation of embryonic and fetal human choriocapillaris (CC) development by the initial process of hemo-vasculogenesis. Loose progenitor cells and free erythroblasts are aggregated to form blood islands. Cells in the blood islands differentiate and organize into primitive blood vessels: Endothelial cell-like cells line a lumen filled with erythroblasts and other blood cells, while pericyte-like cells occupy an ablumenal position. The new radially oriented vessels become united to each other by other progenitors eventually yielding a chicken-wire pattern that will become the lobular adult CC.

endothelial cells and basal cytoplasm of the RPE while VEGF$_{165}$b localized to the nuclei of these cells (see figure 104.7). Therefore, VEGF$_{165}$ is prominent during hemo-vasculogenesis and angiogenesis while VEGF$_{165}$b is only present as vascular development nears completion.

DISCUSSION AND SUMMARY

The adult human CC is a fenestrated lobular vasculature that is similar to kidney glomeruli and unlike the terminal vascular bed of the retina. Therefore, it is reasonable that its mode of development would be unique from the retinal vasculature. Ida Mann provided the first elegant documentation of the development of human CC by paraffin histology and light microscopy (Mann, 1928). This was followed by the meticulous ultrastructural study of Sellheyer and Spitznas which demonstrated the primitive appearance of the early CC (Sellheyer & Spitznas, 1988). Our studies demonstrate that human choroidal vasculature development involves several processes. The initial human CC forms by

hemo-vasculogenesis: the progenitors having a hemato-endothelial phenotype (Hbε as well as CD31, CD34, VEGFR-2) differentiate into endothelial cells (CD31[+], CD39[+], vWf[+]), pericytes (NG2[+], aSMA[+]), erythroblasts (Hbε[+]), and hematopoietic cells (CD34[+]) (figure 104.8). This process, as occurs in blood islands of the yolk sac, has recently been described to occur throughout the mouse embryo (Sequeira Lopez et al., 2003). In the fetal period (9 WG and older), hemo-vasculogenesis was complete, and new blood vessels appeared to bud from the scleral side of the newly formed CC by angiogenesis since ECs were proliferating (Ki67[+]) (see figure 104.9) (Hasegawa et al., 2007). The intermediate blood vessels continue to develop by angiogenesis and eventually anastomose with the capillaries and larger vessels as observed by Drake and associates in mouse embryo (Drake & Fleming, 2000). The CC never reached the lobular pattern in the posterior pole or ladder pattern in periphery or vascular density of the adult (McLeod & Lutty, 1994) in the ages included in our studies (5.5–22 WG), suggesting that significant expansion and remodeling of the system will occur after 22 WG. Free

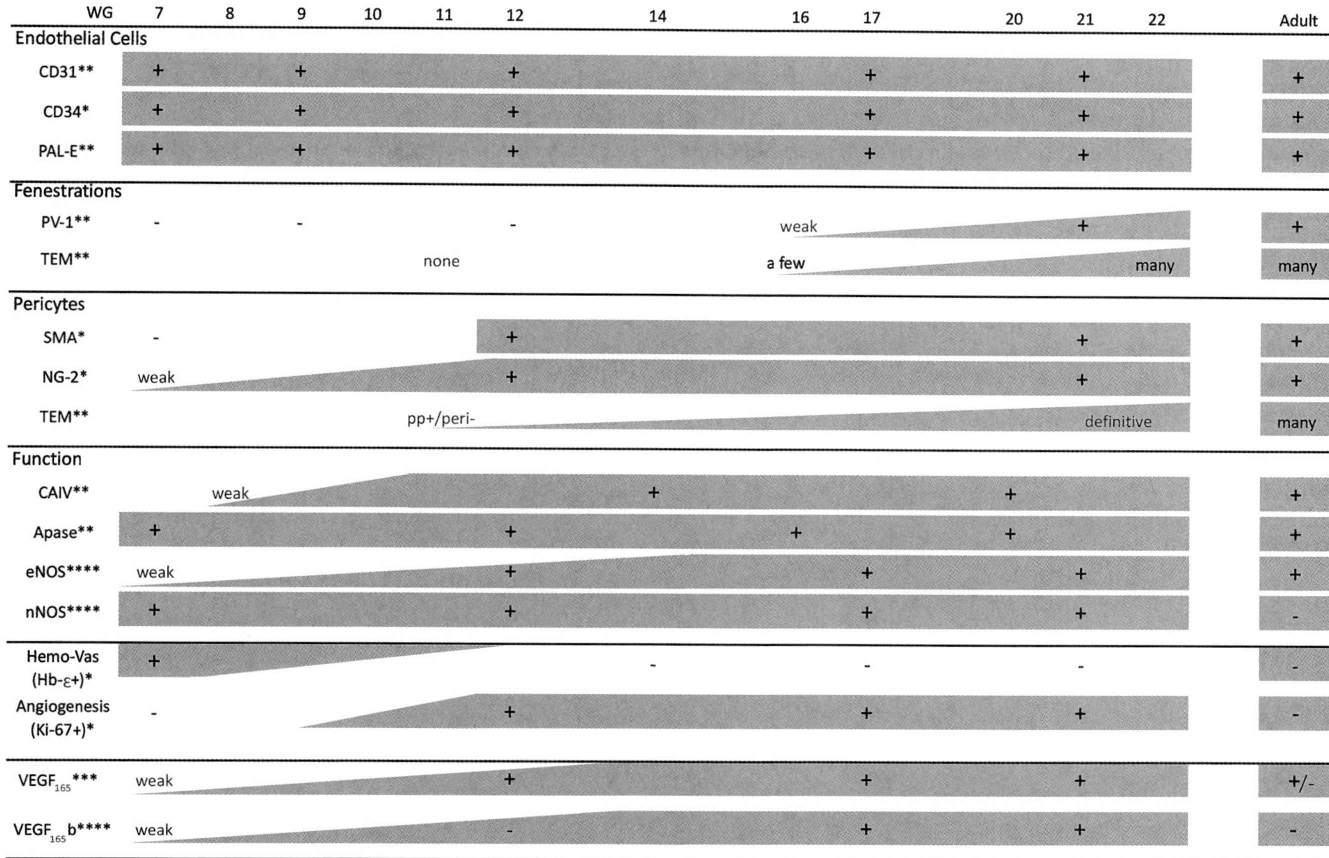

*Data from reference Hasegawa, 2007
** Data from reference Baba, 2009
*** Data from reference Baba, 2012
**** Data from reference McLeod, 2012
Hemo-vas + Hemo-vasculogenesis; TEM = Transmission election microscopy

Figure 104.9   Time line of choriocapillaris development. WG, weeks' gestation; Hemo-Vas, hemovasculogenesis; TEM, transmission electron microscopy; SMA, smooth muscle actin; APase, alkaline phosphatase; eNOS, endothelial nitric oxide synthase; nNOS, neuronal nitric oxide synthase; VEGF, vascular endothelial cell growth factor.

CD39+ angioblasts were still present in between formed segments of CC, so additional vessels may form by the process of vasculogenesis. Alternatively, the additional segments may form by angiogenesis because proliferation is still occurring in CC at 22 WG.

Transport of solutes and small molecules to and from CC occurs at least in part through fenestrations. Contiguous fenestrations were present at 22 WG, but the relationship and interactions between CC and RPE is incomplete in that BrMb is very immature. The highly metabolically active PRs are also dependent on the CC for removal of the end products after PR outer segment disk shedding and subsequent RPE phagocytosis. Only inner segments have formed at 22 WG, so outer segment disk shedding has not commenced. However, PR metabolism has increased since inner segments are rich in mitochondria at this time. Thus, it appears CC maturation is concomitant with both PR differentiation and RPE maturation.

Vascular maturation is not complete until blood vessels are invested with adventitial cells: pericytes around capillaries and venules, and smooth muscle cells (SMC) in the walls of arterioles and arteries. Little is known regarding the origins, differentiation, and appearance of contractile, adventitial cells associated with the choroidal vasculature during embryonic and fetal development. Our observations demonstrated that luminal (presumably endothelial cells) and perivascular cells (presumably pericytes) were indistinguishable ultrastructurally in the early developing human CC, suggesting that choroidal pericytes and ECs are probably derived from a common progenitor. Although NG2 was present in low levels at 7 WG and more prominently in vascular structures at 12 WG (see figure 104.9), aSMA, the predominant actin isoform found in mature SMC and pericytes (Herman, 1993), was not present until 22 WG. Our in vitro studies of retinal angioblasts demonstrate that the same progenitor may differentiate into either an endothelial cell or pericyte depending on the culture conditions (Lutty et al., 2006). TEM demonstrated that apparent mature pericytes were present at 22 WG and were located predominantly on the scleral side of the capillary wall (see figure 104.5D).

A key event in blood vessel development is lumen formation. The initial lumenal spaces of the early CC were slit-like lumenal structures filled with filopodial extensions of the primitive endothelial cells, as observed by Sellheyer as early as 6.5 WG (Sellheyer & Spitznas, 1988). Our studies show that even the erythroblasts can participate in lumen formation during hemo-vasculo-genesis (figure 104.1D, E) (Hasegawa et al., 2007). Even at this stage, the luminal cells made recognizable junctional complexes (see figure 104.5B) that are needed for a mature vasculature, suggesting that these lumenal cells were committed to being endothelial cells. Cyto-plasmic extensions from developing endothelial cells were observed in chick brain by Roy et al., and the number of cytoplasmic extensions decreased as the lumens became broader (Roy et al., 1974), similar to our observations in developing CC. Angioblasts and ECs use these processes to touch and interlock with each other (Maina, 2004) and erythroblasts. ECs finally become fusiform at 22 WG, when the wall of the vessels thin and lumen become broad and flat. This probably reflects changes associated with blood flow. Finally, peri-cytes have assumed a flatter morphology, and their pro-cesses ensheath the vessel wall (see figure104.5D).

The CC is one of the few fenestrated capillary beds in the body. The fenestrations of CC are unique pore-like structures that have a diaphragm, which has a protein recognized by the PV-1 antibody. Fenestrations allow passive transport of some fluids and macromole-cules of the correct charge, which is critical in their providing PR and RPE with nutrients, ions, and small molecules as well as transport of the waste from the RPE. Contiguous fenestrations were not observed until 22 WG when PV-1 immunoreactivity was greatly increased in CC.

The development of the CC by hemo-vasculogenesis probably bestows certain unique developmental charac-teristics. Formation of capillaries by hemovasculogene-sis may explain the development of the lobular pattern of CC in that the first blood vessels are islands. The islands eventually connect to each other without any contribution from blood flow since intermediate and large blood vessels are not yet connected (see figure 104.8), similar to the sequence of events in kidney

development (Ballermann, 2005). Eventually, the origi-nal islands and the interconnecting segments become the lobular system of flat broad capillaries whose lumens are separated by the intercapillary septa or pillars of matrix, seen in adult CC (McLeod & Lutty, 1994). The final mature CC is very similar to the capillaries of kidney glomeruli: large flat, fenestrated capillaries that are lobular in pattern (Ballermann, 2005). Functional maturity may occur in advance of structural maturity (see figure 104.9). Fenestrations form late in matura-tion (21–22 WG), which is coordinated with the differ-entiation of PRs that Hendrickson and coworkers have shown begins around 24–26 WG when inner segments form (Hendrickson & Yuodelis, 1984). After this time, the CC will be fenestrated mostly on the RPE side, which is critical for its adult function in supporting the viability and survival of PRs and RPE cells.

The final product of this unique developmental process forebodes susceptibility to choroidal vascular diseases in the adult. Although the volume of blood in the choroidal vasculature is large compared to the retinal vasculature, the velocity of red blood cells (RBCs) in the CC is 4 times slower than in retinal capil-laries (Braun, Dewhirst, & Hatchell, 1997; Wajer et al., 2000). The lobular pattern of the CC means that there are no long linear segments for RBC transit; each time the RBC moves, it meets a vascular wall adjacent to an intercapillary septa. Because ICAM-1 is constitutively expressed in CC (McLeod et al., 1995), activated leuko-cytes with CD11/CD18 can bind to the endothelial cell's lumenal surfaces, further slowing blood cell veloc-ity. In diabetic CC, we found significantly more neutro-phils than in control CC, and they were often associated with nonviable capillary segments (Cao et al., 1998; Lutty, Cao, & McLeod, 1997). Loss of capillary segments in diabetic choroid was 5 times greater than in CC of age-comparable nondiabetics (Cao et al., 1998). This may be further compounded by atherosclerosis, which can result in plaque formation in intermediate or large choroidal blood vessels, or stenosis as we have often observed in AMD choroids (McLeod et al., 2009). The reduction in supply vessel blood volume or velocity may contribute to reduced flow in CC that Grunwald and associates have observed in AMD (Grunwald et al., 2005; Metelitsina et al., 2008) or even loss in capillary segments as we observed in AMD (McLeod et al., 2009). Since outer retina is dependent on CC, reduction in flow can be detrimental to PRs, which consume most of the oxygen they are provided from CC, resulting in an almost anoxic environment in the dark (Wangsa-Wirawan & Linsenmeier, 2003). The choroidal vascula-ture was once thought to have a vast capillary system with great redundancy, so loss of some capillary segments would not be detrimental. However, the loss of capillaries that has been seen in AMD and diabetes, and the knowledge that PRs deplete most of the oxygen provided in the dark, suggests that PRs would be at risk with loss in CC. Furthermore, RPE may become hypoxic with adjacent loss in CC upregulating hypoxia-inducible growth factors like VEGF (Adamis et al., 1993), which would stimulate the choroidal neovascularization formation from CC that occurs in wet AMD and diabetic choroidopathy (Cao et al., 1998; McLeod et al., 2009).

ACKNOWLEDGMENTS

Grant support: NIH grants EY016151 (GL), EY01765 (Wilmer); RPB unrestricted funds (Wilmer), the Altsheler-Durell Foundation; and a gift from the Himmelfarb Family Foundation in the name of Morton F. Goldberg. Gerard Lutty is an RPB Senior Investigator. The authors acknowledge the excellent electron microscopy of Rhonda Grebe, and confocal microscopy of Takuya Hasegawa and Takayuki Baba. Takuya Hasegawa and Takayuki Baba were Bausch and Lomb Japan Vitreoretinal Research Fellows, and Takayuki Baba was also a Uehara Memorial Foundation Research Fellow.

REFERENCES

Adamis, A. P., Shima, D. T., Yeo, K. T., Yeo, T. K., Brown, L. F., Berse, B., et al. (1993). Synthesis and secretion of vascular permeability factor/vascular endothelial growth factor by human retinal pigment epithelial cells. *Biochemical and Biophysical Research Communications, 193*, 631–638. doi:10.1006/bbrc.1993.1671.

Baba, T., Grebe, R., Hasegawa, T., Bhutto, I., Merges, C., McLeod, D. S., et al. (2009). Maturation of the fetal human choriocapillaris. *Investigative Ophthalmology & Visual Science, 50*, 3503–3511. doi:10.1167/iovs.08-2614.

Baba, T., McLeod, D. M., Edwards, M. M., Merges, C., Sen, T., Sinha, D., et al. (2012). VEGF165b in the developing vasculatures of the fetal human eye. *Developmental Dynamics, 241*, 595. doi:10.1002/dvdy.23743.

Ballermann, B. J. (2005). Glomerular endothelial cell differentiation. *Kidney International, 67*, 1668–1671.

Bates, D. O., Cui, T. G., Doughty, J. M., Winkler, M., Sugiono, M., Shields, J. D., et al. (2002). VEGF165b, an inhibitory splice variant of vascular endothelial growth factor, is downregulated in renal cell carcinoma. *Cancer Research, 62*, 4123–4131.

Bhutto, I. A., & Lutty, G. A. (2006). The vasculature of choroid. In D. Schepro & P. A. D'Amore (Eds.), *Encyclopedia of microvasculatures*. San Diego: Elsevier.

Blaauwgeers, H. G., Holtkamp, G. M., Rutten, H., Witmer, A. N., Koolwijk, P., Partanen, T. A., et al. (1999). Polarized vascular endothelial growth factor secretion by human retinal pigment epithelium and localization of vascular endothelial growth factor receptors on the inner choriocapillaris: Evidence for a trophic paracrine relation.

American Journal of Pathology, 155, 421–428. doi:10.1016/S0002-9440(10)65138-3.

Braun, R. D., Dewhirst, M. W., & Hatchell, D. L. (1997). Quantification of erythrocyte flow in the choroid of the albino rat. *American Journal of Physiology, 272*, H1444–H1453.

Cao, J., McLeod, S., Merges, C. A., & Lutty, G. A. (1998). Choriocapillaris degeneration and related pathologic changes in human diabetic eyes. *Archives of Ophthalmology, 116*, 589–597.

Carmeliet, P., & Jain, R. K. (2000). Angiogenesis in cancer and other diseases. *Nature, 407*, 249–257.

Daufenbach, D. R., Ruttum, M. S., Pulido, J. S., & Keech, R. V. (1998). Chorioretinal colobomas in a pediatric population. *Ophthalmology, 105*, 1455–1458.

Drake, C. J., & Fleming, P. A. (2000). Vasculogenesis in the day 6.5 to 9.5 mouse embryo. *Blood, 95*, 1571–1579.

Ferrara, N., Gerber, H. P., & LeCouter, J. (2003). The biology of VEGF and its receptors. *Nature Medicine, 9*, 669–676.

Gerhardt, H. (2008). VEGF and endothelial guidance in angiogenic sprouting. *Organogenesis, 4*, 241–246.

Grunwald, J. E., Metelitsina, T. I., Dupont, J. C., Ying, G. S., & Maguire, M. G. (2005). Reduced foveolar choroidal blood flow in eyes with increasing AMD severity. *Investigative Ophthalmology & Visual Science, 46*, 1033–1038. doi:10.1167/iovs.04-1050.

Hasegawa, T., McLeod, D. S., Bhutto, I. A., Prow, T., Merges, C. A., Grebe, R., et al. (2007). The embryonic human choriocapillaris develops by hemo-vasculogenesis. *Developmental Dynamics, 236*, 2089–2100. doi:10.1002/dvdy.21231.

Hendrickson, A. E., & Yuodelis, C. (1984). The morphological development of the human fovea. *Ophthalmology, 91*, 603–612.

Herman, I. M. (1993). Actin isoforms. *Current Opinion in Cell Biology, 5*, 48–55.

Ladomery, M. R., Harper, S. J., & Bates, D. O. (2007). Alternative splicing in angiogenesis: The vascular endothelial growth factor paradigm. *Cancer Letters, 249*, 133–142.

Lutty, G. A., Cao, J., & McLeod, D. S. (1997). Relationship of polymorphonuclear leukocytes (PMNs) to capillary dropout in the human diabetic choroid. *American Journal of Pathology, 151*, 707–714.

Lutty, G., Grunwald, J., Majji, A. B., Uyama, M., & Yoneya, S. (1999). Changes in choriocapillaris and retinal pigment epithelium in age-related macular degeneration. *Molecular Vision, 5*, 35.

Lutty, G. A., Hasegawa, T., Baba, T., Grebe, R., Bhutto, I. A., & McLeod, D. S. (2010). Development of the human choriocapillaris. *Eye (London), 24*, 408–415.

Lutty, G. A., Merges, C., Grebe, R., Prow, T., & McLeod, D. S. (2006). Canine retinal angioblasts are multipotent. *Experimental Eye Research, 83*, 183–193.

Maina, J. N. (2004). Systematic analysis of hematopoietic, vasculogenic, and angiogenic phases in the developing embryonic avian lung, *Gallus gallus* variant *domesticus*. *Tissue & Cell, 36*, 307–322.

Mann, I. C. (1928). *The development of the human eye.* Cambridge, England: Cambridge University Press.

Marneros, A. G., Fan, J., Yokoyama, Y., Gerber, H. P., Ferrara, N., Crouch, R. K., et al. (2005). Vascular endothelial growth factor expression in the retinal pigment epithelium is essential for choriocapillaris development and visual function.

American Journal of Pathology, 176, 1451–1459. doi:10.1016/S0002-9440(10)61231-X.

McLeod, D., Baba, T., Bhutto, I., & Lutty, G. A. (2012). Co-expression of endothelial and neuronal nitric oxide synthases in the developing vasculatures of the human fetal eye. *Graefes Archive for Clinical and Experimental Ophthalmology, 250,* 839. doi:10.1007/s00417-012-1969-9.

McLeod, D. S., Grebe, R., Bhutto, I., Merges, C., Baba, T., & Lutty, G. A. (2009). Relationship between RPE and choriocapillaris in age-related macular degeneration. *Investigative Ophthalmology & Visual Science, 50,* 4982–4991. doi:10.1167/iovs.09-3639.

McLeod, D. S., Hasegawa, T., Prow, T., Merges, C., & Lutty, G. (2006). The initial fetal human retinal vasculature develops by vasculogenesis. *Developmental Dynamics, 235,* 3336–3347. doi:10.1002/dvdy.20988.

McLeod, D. S., Lefer, D. J., Merges, C., & Lutty, G. A. (1995). Enhanced expression of intracellular adhesion molecule-1 and P-selectin in the diabetic human retina and choroid. *American Journal of Pathology, 147,* 642–653.

McLeod, D. S., & Lutty, G. A. (1994). High resolution histologic analysis of the human choroidal vasculature. *Investigative Ophthalmology & Visual Science, 35,* 3799–3811.

Metelitsina, T. I., Grunwald, J. E., DuPont, J. C., Ying, G. S., Brucker, A. J., & Dunaief, J. L. (2008). Foveolar choroidal circulation and choroidal neovascularization in age-related macular degeneration. *Investigative Ophthalmology & Visual Science, 49,* 358–363.

Nowak, D. G., Amin, E. M., Rennel, E. S., Hoareau-Aveilla, C., Gammons, M., Damodoran, G., et al. (2010). Regulation of vascular endothelial growth factor (VEGF) splicing from pro- angiogenic to anti-angiogenic isoforms: A novel therapeutic strategy for angiogenesis. *Journal of Biological Chemistry, 285,* 5532–5540. doi:10.1074/jbc.M109.074930.

Nowak, D. G., Woolard, J., Amin, E. M., Konopatskaya, O., Saleem, M. A., Churchill, A. J., et al. (2008). Expression of pro- and anti-angiogenic isoforms of VEGF is differentially regulated by splicing and growth factors. *Journal of Cell Science, 121,* 3487–3495. doi:10.1242/jcs.016410.

Qiu, Y., Hoareau-Aveilla, C., Oltean, S., Harper, S. J., & Bates, D. O. (2009). The anti- angiogenic isoforms of VEGF in health and disease. *Biochemical Society Transactions, 37,* 1207–1213. doi:10.1042/BST0371207.

Roy, S., Hirano, A., Kochen, J. A., & Zimmerman, H. M. (1974). The fine structure of cerebral blood vessels in chick embryo. *Acta Neuropathologica, 30,* 277–285.

Saint-Geniez, M., Kurihara, T., Sekiyama, E., Maldonado, A. E., & D'Amore, P. A. (2009). An essential role for RPE-derived soluble VEGF in the maintenance of the choriocapillaris. *Proceedings of the National Academy of Sciences of the United States of America, 106,* 18751–18756. doi:10.1073/pnas.0905010106.

Saito, S., Liu, B., & Yokoyama, K. (2004). Animal embryonic stem (ES) cells: Self-renewal, pluripotency, transgenesis and nuclear transfer. *Human Cell, 17,* 107–115.

Saito, S., Yokoyama, K., Tamagawa, T., & Ishiwata, I. (2005). Derivation and induction of the differentiation of animal ES cells as well as human pluripotent stem cells derived from fetal membrane. *Human Cell, 18,* 135–141.

Sellheyer, K. (1990). Development of the choroid and related structures. *Eye (London), 4,* 255–261.

Sellheyer, K., & Spitznas, M. (1988). The fine structure of the developing human choriocapillaris during the first trimester. *Graefes Archive for Clinical and Experimental Ophthalmology, 226,* 65–74.

Sequeira Lopez, M. L., Chernavvsky, D. R., Nomasa, T., Wall, L., Yanagisawa, M., & Gomez, R. A. (2003). The embryo makes red blood cell progenitors in every tissue simultaneously with blood vessel morphogenesis. *American Journal of Physiology. Regulatory, Integrative and Comparative Physiology, 284,* R1126–R1137.

Srere, P. A. (1987). Complexes of sequential metabolic enzymes. *Annual Review of Biochemistry, 56,* 89–124.

Stan, R. V., Ghitescu, L., Jacobson, B. S., & Palade, G. E. (1999a). Isolation, cloning, and localization of rat PV-1, a novel endothelial caveolar protein. *Journal of Cell Biology, 145,* 1189–1198. doi:10.1083/jcb.145.6.1189.

Stan, R. V., Kubitza, M., & Palade, G. E. (1999b). PV-1 is a component of the fenestral and stomatal diaphragms in fenestrated endothelia. *Proceedings of the National Academy of Sciences of the United States of America, 96,* 13203–13207. doi:10.1073/pnas.96.23.13203.

Wajer, S. D., Taomoto, M., McLeod, M., McCally, R. L., Nishiwaki, H., Fabry, M. E., et al. (2000). Velocity measurements of normal and sickle red blood cells in the rat retinal and choroidal vasculatures. *Microvascular Research, 60,* 281–293. doi:10.1006/mvre.2000.2270.

Wangsa-Wirawan, N. D., & Linsenmeier, R. A. (2003). Retinal oxygen: Fundamental and clinical aspects. *Archives of Ophthalmology, 121,* 547–557.

Yang, Z., Alvarez, B. V., Chakarova, C., Jiang, L., Karan, G., Frederick, J. M., et al. (2005). Mutant carbonic anhydrase 4 impairs pH regulation and causes retinal photoreceptor degeneration. *Human Molecular Genetics, 14,* 255–265.

105 Treatment of Neovascular Age-Related Macular Degeneration

RAJENDRA S. APTE AND RITHWICK RAJAGOPAL

Age-related macular degeneration (AMD) is the leading cause of blindness in developed countries. The disease is divided into two clinical entities, nonexudative, or "dry," AMD and neovascular, or "wet," AMD. Dry AMD accounts for over 85% of the incidence of AMD and is characterized by the presence of drusen in the macula. Although only 10% of patients develop the neovascular type, also referred to as exudative AMD, it accounts for greater than 90% of advanced visual loss from AMD. Identified risk factors for AMD include advancing age, white race, and history of tobacco smoking. In addition to these, several genetic loci that predispose individuals to AMD have been identified. These include complement factor H (*CFH*), complement factor H-related genes 1–5 (*CFHR1–5*), complement factor C2 and C3, age-related maculopathy susceptibility 2 (*ARMS2*), HTRA Serine Peptidase 1 (*HTRA1*), and more recently immune regulatory genes such as Chemokine (C-X3-C motif) Receptor 1 (*CX3CR1*), Toll-Like Receptor 3 (*TLR3*), and Toll-like Receptor 4 (*TLR4*) (Deangelis et al., 2011).

DIAGNOSIS

The key to establishing a diagnosis of wet AMD is the combined use of biomicroscopy, fundus photography, fluorescein angiography and optical coherence tomography (OCT). Patients often complain of decreasing vision, metamorphopsia, or central scotoma when symptoms related to wet AMD first arise. The hallmark of wet AMD is the presence of a choroidal neovascular membrane (CNV) within the macula. A CNV may exhibit several features, such as subretinal fluid, an elevated greenish-gray lesion deep in the retina, a solid retinal pigment epithelial detachment, subretinal hemorrhage, or even intraretinal hemorrhage. OCT is a valuable adjunct to ophthalmoscopy when trying to determine the presence of these features, especially in the cases of subtle subretinal fluid or small retinal pigment epithelial detachments. Recently, enhanced depth imaging of the choroid using OCT has been utilized to image the subretinal pigment epithelial space and monitor changes in that space with treatment (Spaide, 2009). The use of fluorescein angiography is crucial in determining the presence of an active CNV, especially when the size of the lesion is small. Leakage of fluorescein dye from a suspicious lesion is a strong indicator that disease activity is present. In addition to identifying the presence of CNV, fluorescein angiography is used to classify it into several subtypes, which may impact treatment choice. Furthermore, fluorescein angiography allows the clinician to classify the location of the CNV as subfoveal, juxtafoveal, extrafoveal, or peripapillary, a distinction that dictates which treatment modalities may be used. Both OCT and fluorescein angiography also serve as useful tools in monitoring progress of disease or response to therapy.

PATTERNS OF NEOVASCULARIZATION IN AMD

The two originally described patterns of neovascularization in AMD are classic and occult CNV. The macular photocoagulation studies defined classic CNV as that which demonstrates clearly visible and well-demarcated hyperfluorescence in the early phase, with increasing leakage in the late phase (Macular Photocoagulation Study Group, 1990) (see figure 105.1). Occult CNV occurs in two categories, fibrovascular pigment epithelial detachments and "late leakage of undetermined source." Fibrovascular pigment epithelial detachments display early heterogeneous or "stippled" hyperfluorescence patterns that later display mild or even minimal leakage (see figure 105.2). These are distinguished from a serous pigment epithelial detachment (which is not a finding specific to AMD) by the absence of uniform homogenous hyperfluorescence that pools into a contained space. "Late leakage from undetermined source" refers to stippled or nonuniform mild or minimal leaking hyperfluorescence patterns that occur only late in the angiogram. Gass originally used the terms type 1 and type 2 neovascularization to refer to the histological correlates of these angiographic

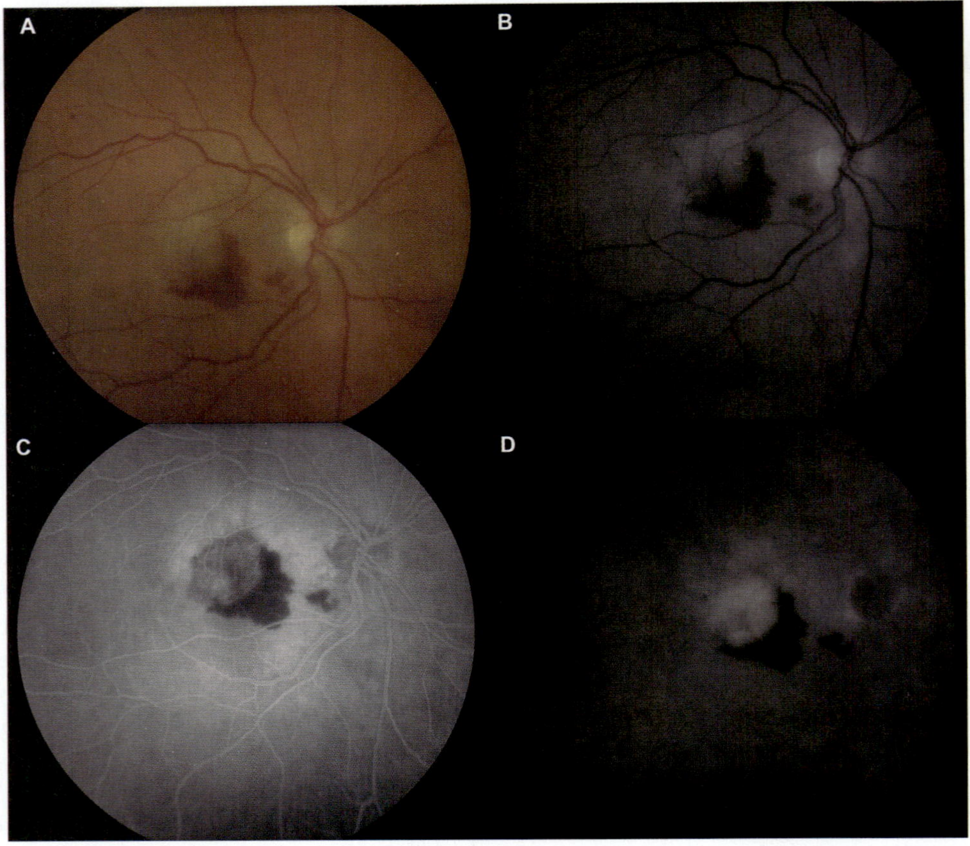

FIGURE 105.1   Classic choroidal neovascularization. Color fundus photograph showing an elevated macular lesion with associated subretinal hemorrhage (A), corresponding red-free image (B), early frame angiogram demonstrating uniform lesion hyperfluoresence (C), and late frame angiogram showing uniform leakage (D).

patterns. He defined type 1 CNV as that which occurs inferior to the retinal pigment epithelium and type 2 CNV as that which occurs superior to it (Gass, 1997). Type 2 CNV is associated with a classic CNV pattern whereas type 1 CNV demonstrates an occult CNV pattern. More recently Freund and colleagues have proposed a type 3 CNV, which angiographically behaves as occult CNV (Freund et al., 2008; Yannuzzi, Freund, & Takahashi, 2008). This form of neovascularization is also referred to as retinal angiomatous proliferation (RAP). The key difference between this and other types of neovascularization in AMD is the presence of intraretinal vascular nets. Clinically these lesions present with intraretinal hemorrhages and levels of macular edema that are out of proportion to what is typically observed in AMD. Intraretinal cystic changes can also occur. Serous retinal pigment epithelial detachments may accompany RAP lesions, distinguishing them from macular telangiectasia, another condition complicated by intraretinal neovascularization not associated with AMD (Freund et al., 2008; Scott & Bressler, 2010; Yannuzzi, Freund, & Takahashi, 2008).

TREATMENT

The treatment of neovascular AMD has seen a rapid evolution in the past 30 years. Physicians initially treated AMD with thermal laser and then later adopted surgical interventions and photoselective laser therapies and finally moved on to highly targeted pharmacotherapy. Ophthalmic surgery to remove CNVs by submacular dissection has been described for neovascular complexes from AMD, histoplasmosis, high myopia, and idiopathic causes. The submacular surgery trials reported no benefit to the surgical removal of neovascular membranes in AMD (Bressler et al., 2000). This type of surgery is now very rarely employed for AMD but still continues to be used in select cases of CNV from histoplasmosis or idiopathic causes. Macular translocation surgery, which involves repositioning the macula away from diseased retinal pigment epithelium and choroid and onto an uninvolved site, has been used successfully to treat wet AMD but is only rarely employed nowadays.

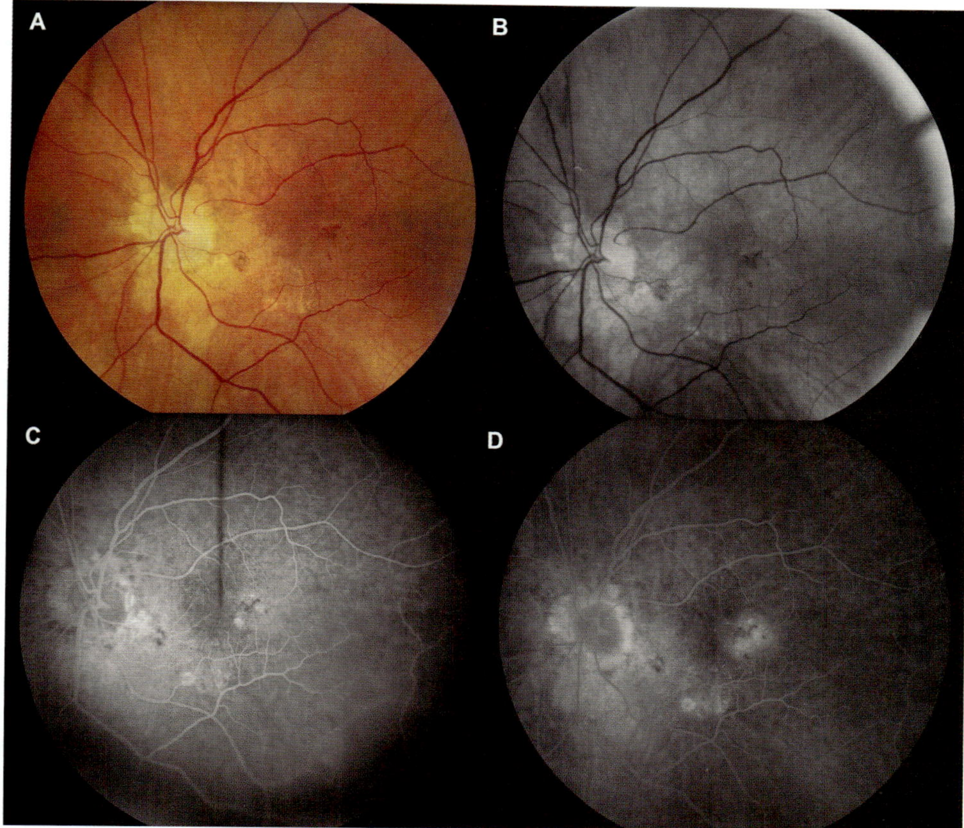

FIGURE 105.2   Occult choroidal neovascularization. Color fundus photograph showing a poorly defined macular lesion with hemorrhage (A), corresponding red-free image (B), early frame angiogram demonstrating stippled hyperfluoresence (C), and late frame angiogram showing leakage (D).

Thermal Laser Photocoagulation

Laser photocoagulation for neovascular AMD (NVAMD) has been used by the retinal specialist for almost 2 decades. Ablation of CNV using this technique requires focal application of thermal laser energy to the lesion under direct visualization through contact lens biomicroscopy. The goal of this type of therapy is to reduce the risk of severe vision loss.

The macular photocoagulation studies were a series of several clinical trials that tested the efficacy of thermal laser in treating CNV from AMD (Macular Photocoagulation Study Group, 1990, 1991a, 1991b, 1991c, 1994). Both argon laser and krypton laser photocoagulation were used to treat lesions that were categorized as either subfoveal (within 200 microns of the foveal center) or extrafoveal (greater than or equal to 200 microns from the foveal center). In the study of argon laser treatment of extrafoveal lesions from AMD, the relative risk of untreated eyes to lose six or more lines of vision was 1.5 compared to laser-treated eyes at the 5-year time point. Mean vision loss at 5 years was 7.1 lines in the untreated group compared to 5.2 lines in the treated group.

Recurrent CNV was observed in 54% of the laser-treated population at 5 years. While this study concluded that argon laser photocoagulation could be employed for well-defined extrafoveal lesions, the authors observed a high rate of persistence of CNV as well as recurrence into the foveal center (Macular Photocoagulation Study Group, 1991a).

The krypton laser study for CNV from AMD also demonstrated a benefit for laser, although in this study both subfoveal and extrafoveal lesions were included. The theoretical advantage of the red laser produced by krypton, which is not absorbed by macular xanthophyll pigment, was to reduce collateral damage to the foveal avascular zone during treatment. At 3 years, 49% of laser treated eyes versus 58% of untreated eyes had lost six or more lines of visual acuity. However, this study also demonstrated that laser-treated CNV has a high rate of recurrence (Macular Photocoagulation Study Group, 1990).

In the subfoveal study of argon and krypton photocoagulation, 20% of treated eyes lost six or more lines of vision compared with only 11% of untreated eyes at the 3-month time point. However, at the 2-year follow-up period, 20% of laser-treated eyes had lost six or more lines of vision compared with 37% of untreated eyes. Most patients who were treated suffered permanent central scotomas after laser but retained better long-term contrast discrimination than their untreated counterparts. Persistent or recurrent CNV was observed in 51% of laser-treated eyes at 2 years. The authors concluded that thermal laser was effective in reducing long-term loss of vision from subfoveal CNV due to AMD, but it came at the cost of creating an immediate and permanent central scotoma (Macular Photocoagulation Study Group, 1990, 1991b, 1991c).

At the time, these studies supported the use of thermal laser, either produced by krypton or argon, for treatments of lesions in AMD not directly under the fovea. In all studies, laser coagulation produced a permanent scotoma. For this reason, thermal laser has now largely been replaced by the newer therapies discussed in the upcoming sections and is only rarely used as salvage therapy of lesions that are extrafoveal with well-defined borders.

Photodynamic Therapy with Verteporfin

Photodynamic therapy (PDT) utilizes an intravenously administered agent, verteporfin or Visudyne, which is activated by application of 689-nm-wavelength laser light. The theoretical benefit of this therapy is that only new vessels bind verteporfin and will be treated by the laser, as opposed to thermal laser which will damage all tissues in its treatment area. The drug is administered by intravenous infusion over 10 min at a recommended dose of 6 mg/m^2 of body surface area. Fifteen minutes after infusion, nonthermal laser light at 689 nm, with an intensity of 600 mW/cm^2, is delivered onto the retinal surface for 83 s for a total energy of 50 J/cm^2. This process results in the activation of cytotoxic free radicals that damage the endothelial cells of the target tissue (Michels & Schmidt-Erfurth, 2001).

The Treatment of Age-Related Macular Degeneration with Photodynamic Therapy study (TAP) included 609 people with classic CNV or those with mixed classic and occult components (TAP Study Group, 1999). Lesions were no larger than 5,400 microns, and entry visual acuity had to be between 20/40–20/200. Patients were randomized to treatment with PDT with verteporfin or sham treatment. Retreatment was permitted after 3 months if there was activity demonstrated on fluorescein angiogram. The primary outcome measure was the proportion of patients with less than 15 letter loss of visual acuity. TAP demonstrated a clear benefit for PDT over sham, with 61% of treated patients losing less than 15 letters after 1 year compared to 46% of placebo-treated patients. The 2-year results were similar. Subgroup analysis revealed that those with predominantly classic lesions, defined as having a classic component greater than 50% of the lesion area, showed the greatest benefit with PDT. On average, patients were treated with PDT 3.4 times during the first year and 2.2 times during the second year (TAP Study Group, 1999).

The Verteporfin in Photodynamic therapy (VIP) study included patients with occult CNV who were excluded from the TAP study (Verteporfin in Photodynamic Therapy Study Group, 2001). Entry visual acuity had to be better than 20/100, and lesions could not be larger than 5,400 microns. A total of 339 people were randomized to treatment with PDT or sham. The primary endpoint was the same as in TAP (loss of fewer than 15 letters). In contrast to TAP, the VIP study found no statistically significant difference between groups with regard to the primary outcome at 12 months. Subgroup analysis demonstrated that there was a treatment benefit for occult lesions which were less than four disk areas or if visual acuity was less than 20/50. Patients in this study received an average of five treatments over the 2-year study period (Verteporfin in Photodynamic Study Group, 2001).

Although the observed visual gains were modest, these studies were pivotal in the treatment of subfoveal CNV from AMD by establishing a role for PDT for a subset of patients with AMD. Prior to this study, the majority of patients with AMD were destined to lose vision, either from natural progression of disease or from the treatments for it. Now, a select subset of AMD patients could be potentially spared further loss of vision from their disease. Adverse effects were few, but the major shortcoming of this treatment was that not all patients with active CNV benefited from therapy.

Vasogenic Antagonism

The contribution of vasogenic growth factor signaling to pathologic angiogenesis in the eye has been known for over 15 years. The vascular endothelial growth factor (VEGF) family consists of homodimeric glycoproteins involved in various growth promoting signaling cascades. There are at least six known genes that belong to the VEGF family, including *vegf-a*, *vegf-b*, *vegf-c*, *vegf-d*, *vegf-e*, and placental growth factor (*pgf*). The *vegf-a* gene products are primarily implicated in regulation of normal and pathogenic angiogenesis and are the primary target of therapies to combat neovascularization (from here onward, we will refer to the *vegf-a* protein product as VEGF). The *vegf-a* gene is composed of eight axons, which when alternatively spliced, results in several isoforms including VEGF121, VEGF165,

VEGF189, and VEGF206. There was initial evidence that the VEGF165 isoform was the critical mediator in the development of pathogenic ocular vascularization and permeability, but subsequent data suggest that multiple *vegf-a* isoforms are causative of pathologic neovascularization in the eye (Ferrara, 2004).

In 1971, Folkman proposed that certain cancers could be treated by antagonizing their vascular supply (Folkman et al., 1971). Since then, not only has VEGF signaling been shown to be elevated in several human cancers, but it has also been demonstrated that tumor growth and metastases were dependent on it. VEGF levels are also markedly elevated in the eye, as demonstrated in several different model systems. Intraocular levels of VEGF have been shown to be markedly elevated in diabetic retinopathy, iris neovascularization, and retinopathy of prematurity (Adamis et al., 1994; Aiello et al., 1994; Malecaze et al., 1994). Antagonism of VEGF signaling was shown in several in vivo and preclinical studies to be effective in treating choroidal neovascularization (Adamis et al., 1996; Krzystolik et al., 2002; Rosenfeld et al., 2005b).

These findings heralded the development of therapeutic anti-VEGF agents. The first of these to be Food and Drug Administration approved for intraocular use was pegaptanib, a pegylated RNA aptamer which binds VEGF165. The pegaptanib molecule is a synthesized RNA oligonucleotide with a polyethylene glycol moiety attached to the five prime end to improve bioavailability and is the first such drug of this class to be approved for treatment of any human disease (Ng et al., 2006). Its use in wet AMD was validated in the VEGF Inhibition Study in Ocular Neovascularization (VISION) studies (Chakravarthy et al., 2006; D'Amico et al., 2006; Gonzales, 2005). Patients with all angiographic types of CNV up to 12 disk diameters in size were included for the study. Patients received either pegaptanib by intravitreal injection every 6 weeks for 1 year or sham injections. The primary outcome was percentage loss of 15 letters or fewer on Early Treatment of Diabetic Retinopathy Study (ETDRS) visual acuity examination (Klein et al., 1983). Rescue therapy was permitted with PDT if indicated. At 1 year, 70% of patients treated with pegaptanib had lost less than 15 letters of acuity compared to 55% for control subjects. In addition, pegaptanib-treated patients were more likely to gain visual acuity compared to sham treated patients (33% compared to 23%). In the 2-year outcome data, 59% of treated subjects compared to 40% of control subjects had lost fewer than 15 letters. In terms of safety, pegaptanib injection into the eye was well tolerated without any observable morphologic or functional toxicity to the retina or retinal pigment epithelium. Intravitreal injection of this drug was associated with a less than 0.1% risk of endophthalmitis. No serious systemic adverse events such as bleeding or cerebrovascular accident were observed in the VISION trials (Apte, 2008; Chakravarthy et al., 2006; D'Amico et al., 2006; Gonzales, 2005).

The next generation of anti-VEGF agents came in the form of antibodies that recognize and neutralize VEGF. Kim and Ferrara first developed antibodies to VEGF by immunizing mice with human VEGF165 (Kim et al., 1992). They describe the isolation of four IgG1 molecules which had varying degrees of affinity to the immunogen. Bevacizumab is a full-length humanized murine IgG1 that is derived from the most avid of these original antibodies (Lien & Lowman, 2008). Bevacizumab demonstrated equal in vitro antagonism of VEGF as the parent molecule in endothelial proliferation assays and tumor growth assays (Presta et al., 1997). The half-life of bevacizumab is between 1 and 2 weeks (Lin et al., 1999). Ranibizumab is a humanized murine Fab derived from the same parent molecule as bevacizumab (Lien & Lowman, 2008). Experiments using a 150-kD full-length antibody precursor to bevacizumab showed that it was not able to penetrate through inner limiting membrane of the retina in a primate model (Mordenti et al., 1999). It should be pointed out that subsequent to these initial studies, bevacizumab has been shown to penetrate all layers of the rabbit retina (Shahar et al., 2006). Nevertheless, ranibizumab was developed for better ocular penetration than the full-length antibody. The smaller ranibizumab molecule, only 48 kD, was easily able to penetrate all layers of the retina (Gaudreault et al., 2007). Furthermore, ranibizumab demonstrated a 22-fold greater binding affinity for VEGF as measured by competition assays and a greater than 100-fold affinity as measured by surface plasmon resonance. It also demonstrates between a 30- and 100-fold increase in anti-proliferative potency in vitro compared to bevacizumab (Chen et al., 1999; Ferrara et al., 2006).

Several clinical studies have demonstrated a benefit for both ranibizumab and bevacizumab in the treatment of wet AMD (Bashshur et al., 2006; Rosenfeld, Fung, & Puliafito, 2005a; Schmid-Kubista et al., 2009; Tufail et al., 2010). However, the first large phase 3 clinical study to demonstrate efficacy and safety of anti-VEGF antibodies in the treatment of AMD was the Anti-VEGF Antibody for the Treatment of Predominantly Classic Choroidal Neovascularization in Age-Related Macular Degeneration (ANCHOR) trial (Brown et al., 2006). This was a double-blinded, randomized, standard-of-care-controlled study in which 423 patients either received 0.3 mg or 0.5 mg of ranibizumab by intravitreal injection every month for 1 year or received

PDT with verteporfin. Patients in this trial had classic choroidal neovascularization as assessed by fluorescein angiography and lesions that were no larger than 5,400 microns, with no evidence of permanent structural damage to the central retina; 96.4% of those patients receiving 0.5 mg of ranibizumab compared to 64.3% of verteporfin-treated patients had lost less than 15 letters on visual acuity examination. In fact, vision improved by greater than 15 letters in 35.7% of the 0.3-mg group and 40.3% of the 0.5-mg group, compared to only 5.6% of the verteporfin group. These visual gains had persisted through the 2-year follow-up period. In addition, patients treated with ranibizumab had demonstrably less activity on fluorescein angiogram compared to baseline, indicating that the drug was effectively targeting the CNV (Brown et al., 2006).

The sister study to ANCHOR, called the Minimally Classic/Occult Trial of the Anti-VEGF Antibody Ranibizumab in the Treatment of Neovascular Age-Related Macular Degeneration (MARINA) trial, randomized 716 patients to treatment with one of two doses of ranibizumab given monthly versus sham injections (Rosenfeld et al., 2006). As with ANCHOR, this was a multi-center double blinded study, but only patients who had CNV that behaved angiographically as occult lesions or had minimally classic components were included n he MARINA trial. This trial produced very similar results, with 94.5% of the 0.3-mg group and 94.6% of the 0.5-mg group losing fewer than 15 letters on the visual acuity chart at 12 months, compared with 62.2% in the control group. Improvement of 15 letters or more was seen in 24.8% in the 0.3-mg group and 33.8% in the 0.5-mg group compared with 5% of controls. Again, these gains in visual acuity continued on to the 2-year follow-up period. As in ANCHOR, patients in the MARINA study who were treated with ranibizumab showed less leakage on fluorescein angiography at the conclusion of the study as compared to baseline. Adverse affects were rare, with endophthalmitis in 1% of patients being the most frequent serious adverse outcome (Rosenfeld et al., 2006).

These two studies represent landmark achievements in the treatment of AMD. As outlined above, prior therapies could only offer patients *less loss* of vision. ANCHOR and MARINA demonstrated for the first time in phase 3 clinical trials that a therapeutic agent could allow patients to *gain* back vision previously lost to AMD. While this resulted in much optimism among ophthalmologists treating AMD, there were several concerns. First, ANCHOR and MARINA did not include all patients with wet AMD. Entry criteria in these studies included visual acuities that ranged from 20/30–20/320 (measured by ETDRS standards), lesion size

limitations, no prior treatment, and no evidence of permanent physical damage to the central macula (Brown et al., 2006; Klein et al., 1983; Rosenfeld et al., 2006). Patients with vision worse than 20/320 (ETDRS), large lesions, or with signs of scarring or atrophy were not included. This begs the question as to whether VEGF antagonism would benefit these subsets of patients similarly. Secondly, patients in the ANCHOR and MARINA trial received intravitreal injections once every month for 2 years for a total of 24 injections. Intravitreal injection of medication is associated with introduction of infection into the eye (albeit rarely) as well as the more commonly associated side effects of ocular irritation, injection site pain, subconjunctival hemorrhage, and patient anxiety. Injection of anti-VEGF antibodies has also been loosely linked to a slight increase in the risk of cerebrovascular accident (discussed in more detail in the upcoming sections). In light of these issues, another important question that arose from these studies is whether monthly administration is necessary and whether less frequent dosing regimens are equally effective.

To address the issue of monthly dosing, several follow-up studies to ANCHOR and MARINA were conducted. The PIER study randomized patients with both classic and occult lesions to receive either ranibizumab in two different doses or control (sham injection). Patients treated with ranibizumab were given monthly injections for the first 3 months and then switched to quarterly dosing (i.e., every 3 months). The primary outcome measure in this study was mean change in visual acuity as measured by ETDRS standards. At 1 year, patients in the 0.3-mg and 0.5-mg ranibizumab groups lost an average of 1.6 letters and 0.2 letters, respectively, compared to a loss of 16.3 letters in the controls. Importantly, whereas these patients fared better in terms of visual acuity change compared to their controls, visual gains with quarterly dosing was much less than observed with mandatory monthly dosing in the ANCHOR and MARINA trials. Patients in the PIER study reached a peak of visual gain at the 3-month time interval (+2.9 letters and +4.3 letters in the 0.3-mg and 0.5-mg groups, respectively) but then lost those visual gains during the quarterly dosing interval. Interestingly, the PIER study demonstrated that quarterly dosing had lasting effects on reduction of leakage from the CNV on fluorescein angiogram as well as lasting effects on reduction in macular thickness as measured by OCT. However, this study demonstrated that fixed quarterly dosing in wet AMD was not sufficient to control progression of the disease (Regillo et al., 2008).

In the Prospective Optical Coherence Tomography Imaging of Patients with Neovascular AMD Treated with

Intraocular Ranibizumab (PrONTO) study, the effect of a variable *pro re nata* dosing schedule was investigated (Fung et al., 2007; Lalwani et al., 2009). This was a single-center, nonrandomized, uncontrolled prospective study that enrolled 40 patients with wet AMD to treatment with intravitreal ranibizumab. Patients were not screened based on the behavior of their lesions on fluorescein angiogram. Importantly, unlike in ANCHOR, MARINA, and PIER, patients were eligible for the study even if they had received prior treatment. Patients received three initial monthly injections and were then reassessed monthly for (1) loss of at least five letters in conjunction with thickening on OCT, (2) thickening of the macula on OCT by at least 100 microns, (3) new onset classic CNV, (4) new macular hemorrhage, and (5) persistent macular fluid on OCT detected at least 1 month after the previous injection. During the second year of the study, the retreatment criteria were amended such that any qualitative increase in macular fluid on OCT was eligible for retreatment (Lalwani et al., 2009). At 1 year, patients showed a mean improvement of 9.3 letters and a mean decrease in OCT thickness of 178 microns. After the second year, patients gained an average of 11.1 letters with a mean reduction in macular thickness of 212 microns. These results were achieved with an average of 5.6 injections over the first year and an additional 4.3 injections over the second year for a total of 9.9 injections over 24 months. In PrONTO, patients were initially placed under close surveillance following the first injection. Interestingly, OCT measurements demonstrated that the effect of intravitreal ranibizumab injection could be observed within 1 day. While these results were a promising indicator that *pro re nata* dosing could be effective for treatment of AMD, the study was limited by its lack of a control group, lack of masking, and of course its smaller sample size compared to the prior phase 3 studies (Fung, et al., 2007; Lalwani, et al., 2009).

The European-conducted SUSTAIN study also investigated the role of *pro re nata* variable dosing for treatment of NVAMD but enrolled a far greater number of patients than PrONTO (Holz et al., 2011). In SUSTAIN, 513 patients who were naive to treatment with ranibizumab, but may have had prior PDT, were enrolled to received monthly intravitreal doses of 0.3 mg ranibizumab for the first three treatments and were then treated as needed on the basis of retreatment criteria that were very similar to PrONTO. Unlike PrONTO, treating physicians could defer an injection if a patient's visual acuity was better than 79 ETDRS letters or if the patient's OCT thickness was less than 225 microns. Although this study differed from PrONTO in that it was a multicenter trial, it also did not include a control

group which received mandatory injections. Mean visual acuity gain peaked at month 3, with +5.8 letters, but then dropped by 12 months to +3.6 letters, which was lower than the gains observed in the smaller PrONTO study (Fung et al., 2007; Holz et al., 2011; Lalwani et al., 2009). OCT thickness was reduced by 101.1 microns by month 3 and 91.5 microns by month 12. The average number of injections was 5.7 over 12 months, similar to the PrONTO findings (Holz et al., 2011). The major shortcoming of this study was the lack of a control group, which therefore prevents a head-to-head comparison with a mandatory dosing schedule. With that said, the mean visual gains in SUSTAIN were far less compared to that observed at 12 months in the ANCHOR and MARINA trials (Brown et al., 2006; Holz et al., 2011; Rosenfeld et al., 2006).

In the recently reported HARBOR trial, patients with subfoveal CNV from AMD were randomized in a 1:1:1:1 manner to receive either 0.5 mg ranibizumab, given either monthly or *pro re nata* after three initial loading doses, or 2.0 mg ranibizumab, also given either monthly or *pro re nata*. This study found that there was no difference in visual acuity outcomes between the higher and lower dosing groups and also found no difference between monthly or variable *pro re nata* dosing (Busbee et al., 2011).

While the above studies shed some light on the efficacy of less frequent dosing schedules, they were all performed with ranibizumab. Another major question that lingered was whether the use of the bevacizumab would be as efficacious as ranibizumab. Even prior to the release of ANCHOR and MARINA data in 2005, ophthalmologists were widely using bevacizumab to treat wet AMD as a cheaper off-label alternative to ranibizumab and were achieving good results widely reported in the literature. The difficulty in comparing these two drugs is that based on several years of experience with them, the expected difference between the two is small. The sample size needed to demonstrate this effect in a head-to-head superiority trial would be enormous, raising feasibility issues. Instead, the National Eye Institute fortunately spearheaded a publicly funded multicenter, double-masked, randomized, noninferiority trial comparing intravitreal injection of ranibizumab with bevacizumab that was completed in 2011 (Martin et al., 2011). The Comparison of Age-Related Macular Degeneration Treatments Trials (CATT) randomized 1,208 patients with wet AMD to receive 0.5 mg ranibizumab or 1.25 mg bevacizumab, both given in either a monthly dosing schedule or a *pro re nata* regimen. The primary outcome measure was mean change in visual acuity at 1 year, and the noninferiority limit was 5 letters. Inclusion criteria were similar to those in the ANCHOR and MARINA trials, with the exception that patients with visual acuities of 20/25–20/320 could be included (ANCHOR and MARINA enrolled only patients with 20/40 or worse acuity). Although all patients received entry fluorescein angiograms and OCT, patients were included regardless of the angiographic appearance of their lesions. Retreatment criteria were more stringent for the *pro re nata* groups compared with the PrONTO and SUSTAIN trials. Any loss of vision, increase in fluid, activity on fluorescein angiogram, or presence of new hemorrhage on exam qualified a patient to receive an intravitreal injection. The study found that monthly dosing of bevacizumab was noninferior to monthly ranibizumab, with mean gains of +8.0 and +8.5 letters at 12 months in each group, respectively. The as-needed regimens of both these drugs were also equivalent, with a mean gain of +5.9 letters in the bevacizumab group and +6.8 letters in the ranibizumab group. Monthly ranibizumab was also equivalent to as-needed dosing. Comparison of monthly bevacizumab and *pro re nata* bevacizumab was inconclusive. In the as-needed-dosing groups, patients received an average of 6.9 injections of ranibizumab and 7.7 injections of bevacizumab over the first year. Monthly dosing of ranibizumab yielded the most effective decrease in central macular thickness as measured by OCT (mean decrease of 196 microns in the monthly ranibizumab group, compared to 152–168 microns in the other groups). The findings of CATT validated the widespread off-label use of bevacizumab for wet AMD, which has important ramifications for health economics in ophthalmology (Sears, Bena, & Singh, 2011). This study also supported the use of a less frequent variable dosing schedule, as long as close surveillance is maintained, confirming the notion put forth by the PrONTO and SUSTAIN studies (Davis et al., 2011; Martin et al., 2011).

Safety of Anti-VEGF Antibody Injections

Intravitreal injections of anti-VEGF agents have some potential systemic adverse effects. With the exception of pegaptanib, all marketed anti-VEGF agents are detectable in the serum after intravitreal injection. Serum levels of bevacizumab are detectable 1 day after intravitreal injection with a concomitant decrease in serum VEGF levels (Matsuyama et al., 2010). Furthermore, effects of intravitreal injection can be detected in the contralateral eye (Yoon et al., 2009). These findings indicate that even after intravitreal injection, bevacizumab can enter the systemic circulation. Ranibizumab has also been detected in the serum after intravitreal injection and also displays a contralateral eye effect (Rouvas et al., 2009). Systemic toxicity of anti-VEGF

agents has been documented in clinical trials that used bevacizumab for the treatment of cancers, but the systemic dose is greater than 250 times the dose delivered into the eye. The most serious of these unwanted side effects were arterial and venous thrombotic events, gastrointestinal perforation and hemorrhage, poor wound healing, and hypertension. Indeed, in the phase 3 clinical trials using ranibizumab and bevacizumab by *intravitreal* administration, these side effects were also observed. Antiplatelet Trialists Collaboration (APTC)-defined arteriothrombotic events (such as cerebrovascular accident and myocardial infarction) were observed in 3.8% of the higher dose ranibizumab group (compared to 2.5% in controls) in the MARINA trial, and in 2.8% of the higher dose ranibizumab group (compared to 1.4% in controls) in ANCHOR. Nonocular hemorrhages were noted in 8.8% of ranibizumab-treated patients versus 5.5% of controls in MARINA, and in 6.4% of treated patients versus 2.1% in controls in ANCHOR (Brown et al., 2006; Rosenfeld, et al., 2006). In CATT, arteriothrombotic events occurred in 2.4% of patients receiving ranibizumab monthly and in 2.1% of patients receiving bevacizumab monthly. In the as-needed groups, these events occurred in 2.0% of the ranibizumab group and in 2.7% of the bevacizumab group. There was also a higher incidence of patients with "serious systemic adverse events," defined as that needing primary hospitalization, in the bevacizumab group compared to ranibizumab (24.1% vs. 19.0%) (Martin et al., 2011).

The Safety Assessment of Intravitreous Lucentis for AMD (SAILOR) study was a phase 3 trial in which the primary outcome measures were those concerning safety. Ocular and nonocular adverse events were recorded along with serious adverse events. Patients had subfoveal CNV due to AMD and were treated with intravitreal injection of ranibizumab. This study enrolled 4,300 patients into two cohorts with different dosing schedules. Cohort 1 received three initial mandatory monthly doses followed by variable dosing guided by vision and OCT results. Patients in cohort 1 received one of two doses of ranibizumab, 0.3 mg or 0.5 mg. Cohort 2 received variable dosing with 0.5 mg after the first injection, with subsequent doses determined at the treating physician's discretion. The average number of injections was 4.9 in cohort 1 and 3.6 in cohort 2. Incidence of vascular and nonvascular deaths in the study was comparable across all group and doses and ranged between 0.7 and 1.5%. Stroke rates were 0.7% in the 0.3-mg group, 1.2% in the 0.5-mg group, and 0.6% in cohort 2. In the subgroup analysis of patients who had a history of stroke, there was a 9.6% incidence of stroke in the 0.5-mg group. This compared to 2.7%

of the 0.3-mg group with prior stroke and none of cohort 2 patients with prior stroke. However, the number of patients in this subgroup analysis was small, and the differences between groups were not statistically significant. Moreover, epidemiological studies of patients with AMD suggest that the annual rates of stroke in patients with newly diagnosed neovascular AMD is 3.8% and is 56.4% for those who had experienced an ischemic stroke during the prior year (Alexander et al., 2007). Although comparison across studies is limited, the number observed in the epidemiological trials is higher than those observed in SAILOR (Boyer et al., 2009).

Future Therapies

The newest drug to target VEGF pathways in the treatment of wet AMD is VEGF-Trap Eye, also called aflibercept. VEGF-Trap Eye is a recombinant protein that fuses one extracellular domain of VEGFR2 and one extracellular domain of VEGFR3 with the Fc portion of human IgG. This fusion protein has between 10- and 100-fold greater binding affinity for all VEGF-A isoforms compared to the native receptors. It also binds and inhibits PGF, but it is unclear whether this has therapeutic potential in AMD. Two large multicenter, phase 3 trials conducted in either Europe or North America, termed VEGF Trap-Eye: Investigation of Efficacy and Safety in Wet Age-Related Macular Degeneration (VIEW1 and VIEW2), demonstrated both safety and efficacy of this drug for wet AMD. VIEW1 and VIEW2 recruited approximately 1,200 patients in each study, and each of these patients was randomized 1:1:1:1 to receive either 2 mg aflibercept every 4 weeks, 0.5 mg aflibercept every 4 weeks, 2 mg aflibercept every 8 weeks, or 0.5 mg ranibizumab every 4 weeks. The primary endpoint was maintenance of vision, that is, percentage of patients losing less than 15 letters. The secondary endpoint was percentage of patients who experienced greater than 15 letter gains in visual acuity. The average number of injections for the q4 dosing groups was between 12 and 13 over the first year, compared to 7.6–7.7 in the q8 dosing group for aflibercept. Across all groups, between 94% and 96% of patients demonstrated fewer than 15 letters of loss over the first year. The aflibercept groups were statistically noninferior to ranibizumab in terms of the primary outcome. Among patients receiving treatment every 4 weeks, the integrated results showed a mean visual acuity gain of 8.7 letters in the ranibizumab group, compared to 8.3 letters in the 0.5 mg aflibercept group and 9.3 letters in the 2 mg aflibercept group. Patients receiving 2 mg of aflibercept every 8 weeks demonstrated a mean visual gain of 8.4 letters. All groups demonstrated similar proportions of patients who gained greater than 15 letters at 1 year, ranging from 30–33%. All groups also demonstrated equivalent decreases in mean macular thickness as measured by OCT. In terms of safety there were no differences between aflibercept and ranibizumab in the incidence of ocular adverse effects, including endophthalmitis, or nonocular adverse events, including hypertension, myocardial infarction, and stroke (Schmidt-Erfurth et al., 2011).

Formulations of small molecule inhibitors of VEGF receptors delivered as topical eye drops or intravitreal solutions have also been investigated. Pazopanib, a small molecule tyrosine kinase inhibitor of VEGFR1 and VEGFR2, is a drug used to treat renal cell carcinoma. It is currently being tested in a phase 2 trial for the treatment of NVAMD. The drug is formulated as an eye drop given daily to patients with active NVAMD and either minimally classic or occult lesions. Patients with classic CNV or RAP lesions and those who have had prior treatment failures with either bevacizumab or ranibizumab were excluded. The primary outcome measure is OCT thickness at 1 month of treatment, with secondary outcomes being OCT thickness at 3 months and safety and tolerability. This study is ongoing and actively recruiting (ClinicalTrials.gov, 2011a). AL-39324 is a small molecule inhibitor of VEGFR tyrosine kinase activity that is delivered into the eye as an intravitreal suspension. It has been studied in a phase 2 trial using ranibizumab treatment as a control arm. The primary endpoints were those concerning safety and tolerability of the drug, but change in OCT thickness was also reviewed. This study is completed, and results are pending (ClinicalTrials.gov, 2011e).

Small interfering ribonucleic acid (siRNA)-based therapies that target VEGF pathways have also been successfully reported in treating CNV from AMD. Sirna-027 is a siRNA directed against VEGFR-1. In a phase 1 study of this treatment, patients with NVAMD were treated with a single injection of one of six doses of Sirna-027 with primary outcome measures being safety related. Change in OCT thickness was also studied. The drug was well tolerated without any serious adverse effects. Preliminary analysis suggests that this drug is effective in decreasing mean OCT thickness in as few as 2 weeks postinjection (Kaiser et al., 2010).

Pathways independent of the canonical VEGF signaling cascade have also been studied. The complement cascade has been extensively linked to the pathogenesis of AMD, both neovascular and nonneovascular, and has been the target of therapeutic approaches. Several members of the complement cascade have been linked to either promoting or protecting against the development of AMD, including complement component 2, complement component 3, and complement factor H (Bradley, Zipfel, & Hughes, 2011; Deangelis, et al., 2011; Telander, 2011). ARC 1905 is an anti-C5 aptamer that is currently in phase 1 testing for treatment of NVAMD by intravitreal injection (ClinicalTrials.gov, 2011b). POT-4, a cyclic-peptide inhibitor of complement factor 3, has completed phase 1 testing for intravitreal injection therapy in NVAMD, and phase 2 testing is currently under way (ClinicalTrials.gov, 2011d).

Mediators of the cellular autophagy response have been linked to the pathogenesis of AMD. The increasing inability to prevent accumulation of cytotoxic metabolic by-products in the RPE and photoreceptors during the aging process may trigger cellular responses such as autophagy, which may lead to NVAMD (Kaarniranta, 2010; Ryhanen et al., 2009; Wang et al., 2009). The mammalian target of rapamycin (mTOR) pathway has been implicated in the regulation of autophagy in this setting (Hyttinen et al., 2011). Palomid 529 is an inhibitor of the mTOR pathway, a regulator of cellular proliferation. A clinical trial sponsored by the National Eye Institutes is currently in the recruiting stage for testing efficacy and safety of intravitreal Palomid 529 in NVAMD, against ranibizumab-treated patients as control (ClinicalTrials.gov, 2011c). Sirolimus (rapamycin) itself has also been tested in the eye for treatment of AMD. In phase 2 trials, Sirolimus was delivered as either a subconjunctival injection or an intravitreal injection in conjunction with intravitreal ranibizumab for the treatment of patients with NVAMD. Preliminary data suggests that subconjunctival injections are as effective as intravitreal injections, and both were well tolerated in the eye. This drug is currently in phase 3 testing (ClinicalTrials.gov).

SUMMARY

The hallmark of neovascular AMD is the presence of a CNV which may present as several angiographically distinct lesions. Neovascular disease occurs in a small minority of patients with AMD but accounts for the majority of vision loss from it. Treatment has seen a remarkable evolution in the past 3 decades, which began with surgical removal and laser ablation of lesions, progressed to selective ablation of lesions with the use of photo-activatable dyes, and finally progressed to highly targeted therapy against VEGF. Therapies currently employed for AMD heavily focus on treatment of an active CNV; however, future therapies should target factors that promote the genesis of CNV from the AMD-stricken retina.

REFERENCES

Adamis, A. P., Miller, J. W., Bernal, M. T., D'Amico, D. J., Folkman, J., Yeo, T. K., et al. (1994). Increased vascular endothelial growth factor levels in the vitreous of eyes with proliferative diabetic retinopathy. *American Journal of Ophthalmology, 118,* 445–450.

Adamis, A. P., Shima, D. T., Tolentino, M. J., Gragoudas, E. S., Ferrara, N., Folkman, J., et al. (1996). Inhibition of vascular endothelial growth factor prevents retinal ischemia-associated iris neovascularization in a nonhuman primate. *Archives of Ophthalmology, 114,* 66–71.

Aiello, L. P., Avery, R. L., Arrigg, P. G., Keyt, B. A., Jampel, H. D., Shah, S. T., et al. (1994). Vascular endothelial growth factor in ocular fluid of patients with diabetic retinopathy and other retinal disorders. *New England Journal of Medicine, 331,* 1480–1487.

Alexander, S. L., Linde-Zwirble, W. T., Werther, W., Depperschmidt, E. E., Wilson, L. J., Palanki, R., et al. (2007). Annual rates of arterial thromboembolic events in Medicare neovascular age-related macular degeneration patients. *Ophthalmology, 114,* 2174–2178.

Apte, R. S. (2008). Pegaptanib sodium for the treatment of age-related macular degeneration. *Expert Opinion on Pharmacotherapy, 9,* 499–508.

Bashshur, Z. F., Bazarbachi, A., Schakal, A., Haddad, Z. A., El Haibi, C. P., & Noureddin, B. N. (2006). Intravitreal bevacizumab for the management of choroidal neovascularization in age-related macular degeneration. *American Journal of Ophthalmology, 142,* 1–9. doi:10.1016/j.ajo.2006.02.037.

Boyer, D. S., Heier, J. S., Brown, D. M., Francom, S. F., Ianchulev, T., & Rubio, R. G. (2009). A phase IIIb study to evaluate the safety of ranibizumab in subjects with neovascular age-related macular degeneration. *Ophthalmology, 116,* 1731–1739.

Bradley, D. T., Zipfel, P. F., & Hughes, A. E. (2011). Complement in age-related macular degeneration: A focus on function. *Eye (London), 25,* 683–693.

Bressler, N. M., Bressler, S. B., Hawkins, B. S., Marsh, M. J., Sternberg, P., Jr., & Thomas, M. A. (2000). Submacular surgery trials randomized pilot trial of laser photocoagulation versus surgery for recurrent choroidal neovascularization secondary to age-related macular degeneration: I. Ophthalmic outcomes submacular surgery trials pilot study report number 1. *American Journal of Ophthalmology, 130,* 387–407.

Brown, D. M., Kaiser, P. K., Michels, M., Soubrane, G., Heier, J. S., Kim, R. Y., et al. (2006). Ranibizumab versus verteporfin for neovascular age-related macular degeneration. *New England Journal of Medicine, 355,* 1432–1444.

Busbee, B. G., Murahashi, W. Y., Li, Z., & Rubio, R. G. (2011). *Efficacy and Safety of 2.0-mg or 0.5-mg Ranibizumab in Patients with Subfoveal Neovascular AMD: HARBOR Study.* Paper presented at the American Academy of Ophthalmology Annual Meeting 2011.

Chakravarthy, U., Adamis, A. P., Cunningham, E. T., Jr., Goldbaum, M., Guyer, D. R., Katz, B., & Patel, M. (2006). Year 2 efficacy results of 2 randomized controlled clinical trials of pegaptanib for neovascular age-related macular degeneration. *Ophthalmology, 113,* 1508 e1501–1525.

Chen, Y., Wiesmann, C., Fuh, G., Li, B., Christinger, H. W., McKay, P., et al. (1999). Selection and analysis of an optimized anti-VEGF antibody: Crystal structure of an affinity-matured Fab in complex with antigen. *Journal of Molecular Biology, 293*, 865–881.

ClinicalTrials.gov. Phase 2 Study of an ocular sirolimus (rapamycin) formulation in combination with Lucentis® in patients with age-related macular degeneration (EMERALD). From http://clinicaltrials.gov/ct2/show/NCT00766337?term=sirolimus+amd&rank=5

ClinicalTrials.gov. (2011a). 12 week patient study in neovascular age-related macular degeneration (AMD). From http://clinicaltrials.gov/ct2/show/NCT01362348?term=pazopanib+amd&rank=4

ClinicalTrials.gov. (2011b). ARC1905 (ANTI-C5 APTAMER) given either in combination therapy with Lucentis® 0.5 mg/eye in subjects with neovascular age-related macular degeneration. From http://clinicaltrials.gov/ct2/show/NCT00709527?term=arc+1905+amd&rank=2

ClinicalTrials.gov. (2011c). Palomid 529 in patients with neovascular age-related macular degeneration. From http://clinicaltrials.gov/ct2/show/NCT01271270?term=palomid+amd&rank=1

ClinicalTrials.gov. (2011d). Safety of intravitreal POT-4 therapy for patients with neovascular age-related macular degeneration (AMD) (ASaP). From http://clinicaltrials.gov/ct2/show/NCT00473928?term=pot-4+amd&rank=1

ClinicalTrials.gov. (2011e). WALTZ—Wet age-related macular degeneration (AMD) AL-39324 treatment examination. From http://clinicaltrials.gov/ct2/show/NCT00992563?term=AL-39324+amd&rank=1

D'Amico, D. J., Masonson, H. N., Patel, M., Adamis, A. P., Cunningham, E. T., Jr., Guyer, D. R., & Katz, B. (2006). Pegaptanib sodium for neovascular age-related macular degeneration. *Ophthalmology, 113*, 992–1001 e1006. doi:10.1016/j.ophtha.2006.02.027.

Davis, J., Olsen, T. W., Stewart, M., & Sternberg, P., Jr. (2011). How the comparison of age-related macular degeneration treatments trial results will impact clinical care. *American Journal of Ophthalmology, 152*, 509–514.

Deangelis, M. M., Silveira, A. C., Carr, E. A., & Kim, I. K. (2011). Genetics of age-related macular degeneration: Current concepts, future directions. *Seminars in Ophthalmology, 26*, 77–93.

Ferrara, N. (2004). Vascular endothelial growth factor: Basic science and clinical progress. *Endocrine Reviews, 25*, 581–611.

Ferrara, N., Damico, L., Shams, N., Lowman, H., & Kim, R. (2006). Development of ranibizumab, an anti-vascular endothelial growth factor antigen binding fragment, as therapy for neovascular age-related macular degeneration. *Retina, 26*, 859–870.

Folkman, J., Merler, E., Abernathy, C., & Williams, G. (1971). Isolation of a tumor factor responsible for angiogenesis. *Journal of Experimental Medicine, 133*, 275–288.

Freund, K. B., Ho, I. V., Barbazetto, I. A., Koizumi, H., Laud, K., Ferrara, D., et al. (2008). Type 3 neovascularization: The expanded spectrum of retinal angiomatous proliferation. *Retina, 28*, 201–211.

Fung, A. E., Lalwani, G. A., Rosenfeld, P. J., Dubovy, S. R., Michels, S., Feuer, W. J., et al. (2007). An optical coherence tomography-guided, variable dosing regimen with intravitreal ranibizumab (Lucentis) for neovascular age-related macular degeneration. *American Journal of Ophthalmology, 143*, 566–583.

Gass, J. D. (1997). Stereoscopic atlas of macular diseases (4th ed.). St. Louis, MO: Mosby.

Gaudreault, J., Fei, D., Beyer, J. C., Ryan, A., Rangell, L., Shiu, V., et al. (2007). Pharmacokinetics and retinal distribution of ranibizumab, a humanized antibody fragment directed against VEGF-A, following intravitreal administration in rabbits. *Retina, 27*, 1260–1266. doi:10.1097/IAE.0b013e318134eecd.

Gonzales, C. R. (2005). Enhanced efficacy associated with early treatment of neovascular age-related macular degeneration with pegaptanib sodium: An exploratory analysis. *Retina, 25*, 815–827.

Holz, F. G., Amoaku, W., Donate, J., Guymer, R. H., Kellner, U., Schlingemann, R. O., et al. (2011). Safety and efficacy of a flexible dosing regimen of ranibizumab in neovascular age-related macular degeneration: The SUSTAIN study. *Ophthalmology, 118*, 663–671.

Hyttinen, J. M., Petrovski, G., Salminen, A., & Kaarniranta, K. (2011). 5'-adenosine monophosphate-activated protein kinase-Mammalian target of rapamycin axis as therapeutic target for age-related macular degeneration. *Rejuvenation Research, 14*, 651–660.

Kaarniranta, K. (2010). Autophagy—Hot topic in AMD. *Acta Ophthalmologica, 88*, 387–388.

Kaiser, P. K., Symons, R. C., Shah, S. M., Quinlan, E. J., Tabandeh, H., Do, D. V., et al. (2010). RNAi-based treatment for neovascular age-related macular degeneration by Sirna-027. *American Journal of Ophthalmology, 150*, 33–39 e32. doi:10.1016/j.ajo.2010.02.006.

Kim, K. J., Li, B., Houck, K., Winer, J., & Ferrara, N. (1992). The vascular endothelial growth factor proteins: Identification of biologically relevant regions by neutralizing monoclonal antibodies. *Growth Factors (Chur, Switzerland), 7*, 53–64.

Klein, R., Klein, B. E., Moss, S. E., & DeMets, D. (1983). Interobserver variation in refraction and visual acuity measurement using a standardized protocol. *Ophthalmology, 90*, 1357–1359.

Krzystolik, M. G., Afshari, M. A., Adamis, A. P., Gaudreault, J., Gragoudas, E. S., Michaud, N. A., et al. (2002). Prevention of experimental choroidal neovascularization with intravitreal anti-vascular endothelial growth factor antibody fragment. *Archives of Ophthalmology, 120*, 338–346.

Lalwani, G. A., Rosenfeld, P. J., Fung, A. E., Dubovy, S. R., Michels, S., Feuer, W., et al. (2009). A variable-dosing regimen with intravitreal ranibizumab for neovascular age-related macular degeneration: Year 2 of the PrONTO Study. *American Journal of Ophthalmology, 148*, 43–58 e41. doi:10.1016/j.ajo.2009.01.024.

Lien, S., & Lowman, H. B. (2008). Therapeutic anti-VEGF antibodies. *Handbook of Experimental Pharmacology, 181*, 131–150.

Lin, Y. S., Nguyen, C., Mendoza, J. L., Escandon, E., Fei, D., Meng, Y. G., et al. (1999). Preclinical pharmacokinetics, interspecies scaling, and tissue distribution of a humanized monoclonal antibody against vascular endothelial growth factor. *Journal of Pharmacology and Experimental Therapeutics, 288*, 371–378.

Macular Photocoagulation Study Group. (1990). Krypton laser photocoagulation for neovascular lesions of age-related macular degeneration: Results of a randomized clinical trial. *Archives of Ophthalmology, 108*, 816–824.

Macular Photocoagulation Study Group. (1991a). Argon laser photocoagulation for neovascular maculopathy: Five-year results from randomized clinical trials. *Archives of Ophthalmology, 109,* 1109–1114.

Macular Photocoagulation Study Group. (1991b). Laser photocoagulation of subfoveal neovascular lesions in age-related macular degeneration: Results of a randomized clinical trial. *Archives of Ophthalmology, 109,* 1220–1231.

Macular Photocoagulation Study Group. (1991c). Subfoveal neovascular lesions in age-related macular degeneration: Guidelines for evaluation and treatment in the macular photocoagulation study. Archives of Ophthalmology, 109, 1242–1257.

Macular Photocoagulation Study Group. (1994). Laser photocoagulation for juxtafoveal choroidal neovascularization: Five-year results from randomized clinical trials. *Archives of Ophthalmology, 112,* 500–509.

Malecaze, F., Clamens, S., Simorre-Pinatel, V., Mathis, A., Chollet, P., Favard, C., et al. (1994). Detection of vascular endothelial growth factor messenger RNA and vascular endothelial growth factor-like activity in proliferative diabetic retinopathy. *Archives of Ophthalmology, 112,* 1476–1482.

Martin, D. F., Maguire, M. G., Ying, G. S., Grunwald, J. E., Fine, S. L., & Jaffe, G. J. (2011). Ranibizumab and bevacizumab for neovascular age-related macular degeneration. *New England Journal of Medicine, 364,* 1897–1908.

Matsuyama, K., Ogata, N., Matsuoka, M., Wada, M., Takahashi, K., & Nishimura, T. (2010). Plasma levels of vascular endothelial growth factor and pigment epithelium-derived factor before and after intravitreal injection of bevacizumab. *British Journal of Ophthalmology, 94,* 1215–1218.

Michels, S., & Schmidt-Erfurth, U. (2001). Photodynamic therapy with verteporfin: A new treatment in ophthalmology. *Seminars in Ophthalmology, 16,* 201–206.

Mordenti, J., Cuthbertson, R. A., Ferrara, N., Thomsen, K., Berleau, L., Licko, V., et al. (1999). Comparisons of the intraocular tissue distribution, pharmacokinetics, and safety of 125I-labeled full-length and Fab antibodies in rhesus monkeys following intravitreal administration. *Toxicologic Pathology, 27,* 536–544.

Ng, E. W., Shima, D. T., Calias, P., Cunningham, E. T., Jr., Guyer, D. R., & Adamis, A. P. (2006). Pegaptanib, a targeted anti-VEGF aptamer for ocular vascular disease. *Nature Reviews. Drug Discovery, 5,* 123–132.

Presta, L. G., Chen, H., O'Connor, S. J., Chisholm, V., Meng, Y. G., Krummen, L., et al. (1997). Humanization of an anti-vascular endothelial growth factor monoclonal antibody for the therapy of solid tumors and other disorders. *Cancer Research, 57,* 4593–4599.

Regillo, C. D., Brown, D. M., Abraham, P., Yue, H., Ianchulev, T., Schneider, S., et al. (2008). Randomized, double-masked, sham-controlled trial of ranibizumab for neovascular age-related macular degeneration: PIER study year 1. *American Journal of Ophthalmology, 145,* 239–248.

Rosenfeld, P. J., Brown, D. M., Heier, J. S., Boyer, D. S., Kaiser, P. K., Chung, C. Y., et al. (2006). Ranibizumab for neovascular age-related macular degeneration. *New England Journal of Medicine, 355,* 1419–1431. doi:10.1056/NEJMoa054481.

Rosenfeld, P. J., Fung, A. E., & Puliafito, C. A. (2005a). Optical coherence tomography findings after an intravitreal injection of bevacizumab (Avastin) for macular edema from central retinal vein occlusion. *Ophthalmic Surgery, Lasers & Imaging, 36,* 336–339.

Rosenfeld, P. J., Schwartz, S. D., Blumenkranz, M. S., Miller, J. W., Haller, J. A., Reimann, J. D., et al. (2005b). Maximum tolerated dose of a humanized anti-vascular endothelial growth factor antibody fragment for treating neovascular age-related macular degeneration. *Ophthalmology, 112,* 1048–1053. doi:10.1016/j.optha.2005.01.043.

Rouvas, A., Liarakos, V. S., Theodossiadis, P., Papathanassiou, M., Petrou, P., Ladas, I., et al. (2009). The effect of intravitreal ranibizumab on the fellow untreated eye with subfoveal scarring due to exudative age-related macular degeneration. *Ophthalmologica, 223,* 383–389. doi:10.1159/000228590.

Ryhanen, T., Hyttinen, J. M., Kopitz, J., Rilla, K., Kuusisto, E., Mannermaa, E., et al. (2009). Crosstalk between Hsp70 molecular chaperone, lysosomes and proteasomes in autophagy-mediated proteolysis in human retinal pigment epithelial cells. *Journal of Cellular and Molecular Medicine, 13*(9B), 3616–3631. doi:10.1111/j.1582-4934.2008.00577.x.

Schmid-Kubista, K. E., Krebs, I., Gruenberger, B., Zeiler, F., Schueller, J., & Binder, S. (2009). Systemic bevacizumab (Avastin) therapy for exudative neovascular age-related macular degeneration. The BEAT-AMD-Study. *British Journal of Ophthalmology, 93,* 914–919.

Schmidt-Erfurth, U. M., Heier, J. S., Anderesi, M., Vitti, R., Sandbrink, R., Norenberg, C., et al. (2011). *VEGF Trap-Eye vs. Ranibizumab in Wet AMD: Integrated Analysis of the VIEW 1 and VIEW 2 Studies.* Paper presented at the American Academy of Ophthalmology Annual Meeting 2011.

Scott, A. W., & Bressler, S. B. (2010). Retinal angiomatous proliferation or retinal anastomosis to the lesion. *Eye (London), 24,* 491–496.

Sears, J., Bena, J., & Singh, A. D. (2011). CATT: Into the eye of a tiger. *British Journal of Ophthalmology, 95,* 761.

Shahar, J., Avery, R. L., Heilweil, G., Barak, A., Zemel, E., Lewis, G. P., et al. (2006). Electrophysiologic and retinal penetration studies following intravitreal injection of bevacizumab (Avastin). *Retina (Philadelphia, Pa.), 26,* 262–269. doi:10.1097/00006982-200603000-00002.

Spaide, R. F. (2009). Enhanced depth imaging optical coherence tomography of retinal pigment epithelial detachment in age-related macular degeneration. *American Journal of Ophthalmology, 147,* 644–652.

Telander, D. G. (2011). Inflammation and age-related macular degeneration (AMD). *Seminars in Ophthalmology, 26,* 192–197.

Treatment of age-related macular degeneration with photodynamic therapy (TAP) Study Group. (1999). Photodynamic therapy of subfoveal choroidal neovascularization in age-related macular degeneration with verteporfin: One-year results of 2 randomized clinical trials—TAP report. *Archives of Ophthalmology, 117,* 1329–1345.

Tufail, A., Patel, P. J., Egan, C., Hykin, P., da Cruz, L., Gregor, Z., et al. (2010). Bevacizumab for neovascular age related macular degeneration (ABC Trial): Multicentre randomised double masked study. *BMJ (Clinical Research Ed.), 340,* c2459. doi:10.1136/bmj.c2459.

Verteporfin in Photodynamic Therapy Study Group. (2001). Verteporfin therapy of subfoveal choroidal neovascularization in age-related macular degeneration: Two-year results of a randomized clinical trial including lesions with occult with no classic choroidal neovascularization—Verteporfin in photodynamic therapy report 2. *American Journal of Ophthalmology, 131*, 541–560.

Wang, A. L., Lukas, T. J., Yuan, M., Du, N., Tso, M. O., & Neufeld, A. H. (2009). Autophagy and exosomes in the aged retinal pigment epithelium: Possible relevance to drusen formation and age-related macular degeneration. *PLoS ONE, 4*, e4160. doi:10.1371/journal.pone.0004160.

Yannuzzi, L. A., Freund, K. B., & Takahashi, B. S. (2008). Review of retinal angiomatous proliferation or type 3 neovascularization. *Retina, 28*, 375–384.

Yoon, Y. H., Kim, J. G., Chung, H., & Lee, S. Y. (2009). Rapid progression of subclinical age-related macular degeneration in the untreated fellow eye after intravitreal bevacizumab. *Acta Ophthalmologica, 87*, 685–687.

106 Bruch's Membrane in Outer Retinal Health and Disease

CHRISTINE A. CURCIO

First described by Karl Bruch in 1854, Bruch's membrane (BrM) is a thin (2–4 μm), acellular, 5-layered extracellular matrix located between the metabolically active retinal pigment epithelium (RPE) (Strauss, 2005) and a remarkable vascular bed, the choroid. BrM serves two major functions: a support element/attachment site for RPE and a vessel wall. Unhindered translocation of nutrients to and metabolites from the RPE across BrM, a semipermeable filtration barrier, is essential for normal vision. BrM has outsized clinical significance because of its involvement in age-related macular degeneration (AMD) and other chorioretinal diseases, a significance that is destined to grow. AMD is prevalent in the fastest growing population segment, the elderly. The principal extracellular lesions of AMD, drusen and basal linear deposits (BlinD), form outside the blood–retina barrier on BrM's inner surface, as does basal laminar deposit, an extracellular matrix secreted by stressed RPE. Optical coherence tomography imaging has made the choroid newly accessible for routine observation in clinics and laboratories (Margolis & Spaide, 2009). BrM will thus become better understood in its appropriate context.

Readers are directed to comprehensive (Curcio & Johnson, 2012; Curcio et al., 2009) and concise (Curcio et al., 2011) summaries by the author and colleagues of most topics covered herein, with complete references. This chapter provides only those references not cited in those previous articles. Additional information on BrM's early history, inherited disorders with distinctive BrM pathology, and detailed derivation of quantitative transport properties is available (Curcio & Johnson, 2012).

PHOTORECEPTORS AND THEIR SUPPORT SYSTEM: A NEUROVASCULAR UNIT

Photoreceptors, RPE, and BrM–choroid together can be viewed as a neurovascular unit, a concept first applied to brain (Chen et al., 2011). A neurovascular unit consists of microvessels, neurons, glial cells, pericytes, and extracellular matrix. Vascular abnormalities are accompanied by damage to neural or glial cells and vice versa.

Photoreceptors rely on the support of the RPE and choroid. The RPE is unique among polarized epithelia in that it faces neurosensory retina on its apical surface, forming the blood–retina barrier, and it faces the systemic circulation, through the choroid on its basal surface. This monolayer performs diverse functions essential for photoreceptor and vascular health, including maintaining the outer blood–retina barrier, controlling metabolite exchange, daily phagocytosis and recycling of photoreceptor outer segment tips, vitamin A metabolism, coordination of cytokine-mediated immune protection, and maintenance of the choriocapillaris, a dense capillary bed. The choroid is unique among vasculatures. With the iris and ciliary body, the choroid forms the eye's uveal tract, a continuous layer of blood vessels, connective tissue, and melanocytes. Ophthalmic artery branches penetrate the sclera with the optic nerve and ramify to form the choroid. Up to 300 μm thick, the choroid has the highest blood flow per unit tissue perfused in the body, with flow sevenfold higher in the primate macula than the periphery (Alm & Bill, 1973). The choroid is essential to photoreceptor metabolism, contributing 90% of the oxygen consumed in darkness and all of the oxygen consumed during light adaptation. Conspicuous structural features of the choriocapillary endothelium are fenestrations that are permeable to macromolecules.

STRUCTURE OF BRUCH'S MEMBRANE IN THE YOUNG ADULT EYE

Hogan's five-layer nomenclature for BrM is widely used (for layers, see figure 106.1; for molecular constituents, see table 106.1). Bounded externally by choriocapillary endothelium, BrM functions as a distinctive subendothelial space within the outer retinal neurovascular unit. All vessel walls have an endothelial cell lining and an associated basement membrane. Although adjacent to choriocapillaries, BrM contains an inner collagenous

TABLE 106.1

Structural and molecular components of Bruch's membrane

Layer (common abbreviation)	Component; age change
Basal laminar deposit (BlamD)	**+ Fibronectin, laminin, IV α4-5, VI, endostatin, EFEMP1**
RPE–basal lamina (RPE-BL)	IV α1-5, V, laminins 1, 5, 10, and 11, nidogen-1, heparan sulfate, chondroitin sulfate
Lipid Wall/basal linear deposit (BlinD)	**+ Lipoproteins**, apoE, apoB
Inner collagenous layer (ICL)	I, III, V, fibronectin, chondroitin sulfate, dermatan sulfate, **lipoproteins** ↑, apoE, heme, clusterin, vitronectin
Elastic layer (EL)	**Elastin** ↑, **calcium phosphate** ↑
Outer collagenous layer (OCL)	I, III, V, fibulin-5, fibronectin, chondroitin sulfate, dermatan sulfate, **lipoproteins** ↑, apoE, clusterin, complement factor H
ChC–basal lamina	IV α1,2, V, VI, laminin, heparan sulfate, chondroitin sulfate, endostatin
Bruch's, throughout or layer not specified	**I**↑, **collagen solubility**↓, perlecan, fibrillin, **MMP-2**↑, **MMP-9**↑, **TIMP-2; TIMP-3**↑, **pentosidine** ↑, **CML**↑, **GA-AGE**↑, RGR-d, apoB, oxidized apoB-100, 7-KCh, MDA, LHP, **HHE**↑, **DHP-lys**↑, **C3d**↑, **C5b-9**↑, **pentraxin-3**↑, **endostatin**↓, thrombospondin-1, zinc

Note. Table shows definitely localized components. Complete reference list is in Curcio and Johnson (2012). Most determinations were made in macula. Studies showing histochemical/immunohistochemical verification of biochemistry and ultrastructural validation of structures identified by light microscopy techniques were given greater weight. Known changes with advancing age are bold with an arrow indicating direction of change. New additions with age are shown with a plus (+). Plain text means no change or not tested. MDA, malondialdehyde; 7-KCh, 7 keto-cholesterol; CML, carboxymethyl-lysine; DHP-Lys, dihydropyridine lysine; GA-AGE, glycolaldehyde derived advanced glycation end products; HHE, 4-hydroxyhexenal.

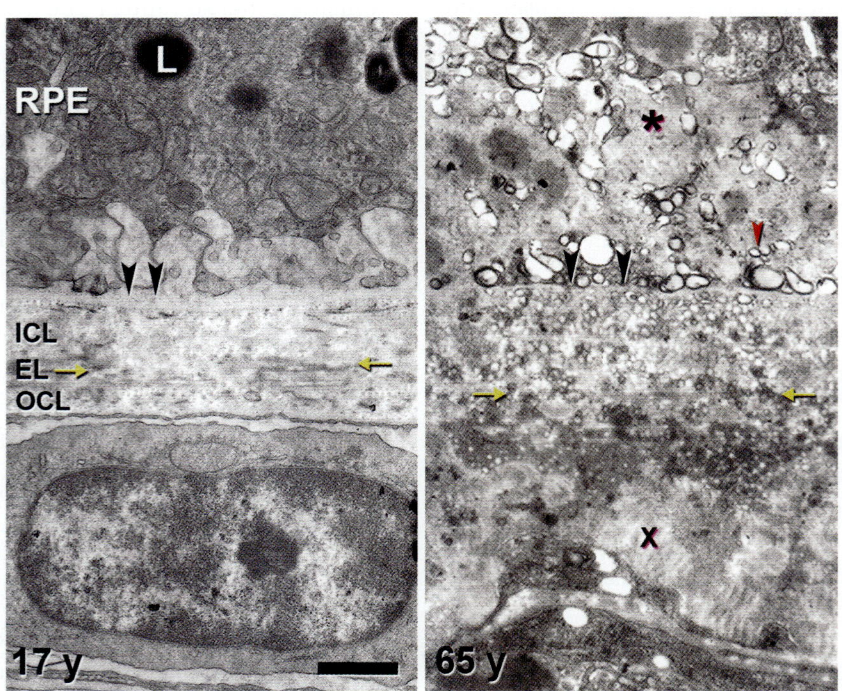

FIGURE 106.1 Bruch's membrane structure in youth and maturity. Arrowheads indicate retinal pigment epithelium (RPE) basal lamina. ICL, inner collagenous layer; EL, elastic layer (between yellow arrows); OCL, outer collagenous layer; L, lipofuscin. Basal lamina of the choriocapillary endothelium is not clear in this specimen. Age 17: Electron-dense amorphous debris and lipoproteins are absent. Leukocyte is in choriocapillaris lumen. Scale bar: 1 μm. Age 65: Electron-dense amorphous debris and lipoproteins are abundant. Membranous debris, also called lipoprotein-derived debris (red arrow), has electron-dense exteriors within basal laminar deposit (BlamD, *). X, type VI collagen in OCL, more typically found in BlamD.

layer (ICL) and an outer collagenous layer (OCL) and is thus thicker than a capillary wall. The innermost wall of large arteries, the tunica intima, is supported by a network of elastic fibers in a fenestrated plate (internal elastic lamina). BrM architecture is thus similar to vascular intima, with the EL corresponding to the internal elastic lamina. BrM shares compositional characteristics with intima, including extracellular matrix components typical of a subendothelial space (type I, III, IV, VI collagens, elastin, fibrillin, laminin, fibronectin, and glycosaminoglycans [GAGs]) (Raghow, Seyer, & Kang, 2006). BrM's abluminal aspect differs from other vessel walls in that it abuts a basal lamina of an epithelium, the RPE.

The RPE basal lamina (RPE-BL) is a ~ 0.15-μm-thick meshwork of fine fibers like other basal laminas and resembling the basement lamina of the choriocapillaris endothelium excepting collagen VI. The RPE-BL contains collagen IV α3–5, like that of kidney glomerulus, another organ with specialized filtration and transport functions. The RPE synthesizes specific laminins that preferentially adhere BrM to the RPE through interaction with integrins.

The ICL is ~1.4 μm thick and contains 70-nm-diameter fibers of collagens I, III, and V in a multilayered crisscross, parallel to the plane of BrM. The collagen grid is associated with interacting molecules, particularly the negatively charged proteoglycans, chondroitin sulfate and dermatan sulfate.

The elastic layer (EL) consists of stacked layers of linear elastin fibers, crisscrossing to form a 0.8-μm-thick sheet with interfibrillary spaces of ~1 μm. This sheet extends from the edge of the optic nerve to the ciliary body pars plana. The EL confers on BrM its biomechanical properties, vascular compliance during pulsatile blood flow, and anti-angiogenic barrier functions.

The OCL is molecularly similar to the ICL, and the collagen fibrils running parallel to the choriocapillaris additionally form prominent bundles. This layer features periodic outward extensions between individual choriocapillary lumens called intercapillary pillars that form an indistinct border with choroidal stroma. Between pillars, OCL thickness can range from 1 to 5 μm.

The 0.07-μm-thick choriocapillaris basal lamina (ChC-BL) is discontinuous with respect to BrM due to interruptions by the intercapillary pillars. It is continuous with respect to the complex network of spaces (lake of blood) defined by the choriocapillary lumens, because the basal lamina envelops the complete circumference of the endothelium. This basal lamina may inhibit endothelial cell migration into BrM, as do basal laminas associated with retinal capillaries.

Comparative Anatomy: Bruch's Membrane in Other Species

Braekevelt investigated BrM ultrastructure in 30 species and found a pentalaminate structure in all nontapetal vertebrates, except for the teleost (bony fishes) infraclass (Braekevelt, Smith, & Smith, 1998), which has a trilaminar BrM lacking an EL. In birds, the EL is displaced toward the choriocapillaris.

DEVELOPMENT OF BRUCH'S MEMBRANE

The bipartite character of BrM arises from embryology. When the optic cup invaginates and folds, its inner layer forms the neural retina, and its outer layer, the RPE, which contacts mesenchyme. At this apposition, by 6–7 weeks' gestation, inner BrM forms from ectodermal tissue and outer BrM, from mesodermal, with the EL becoming visible later, at 11–12 weeks. By week 13, fenestrations are apparent in the subjacent choriocapillary endothelium, suggesting functional transport at this stage. Choroidal endothelial cells originate from para-ocular mesenchyme.

Collagen and elastin are apparently synthesized by invading fibroblasts and the filopodia of endothelial cells lining the adjacent choriocapillaris. The two basal laminas are produced by their associated cell layers. In addition to collagen IV subunits specific to specialized basal lamina, RPE expresses genes for structural collagen III and angiostatic collagen XVIII in a developmentally regulated manner linked to photoreceptor maturation. Development of the choroid, and BrM with it, depends on differentiated RPE and its production of inductive signals, including basic fibroblast growth factor and vascular endothelial growth factor (VEGF).

BRUCH'S MEMBRANE IN AN AGED EYE

Aging is the largest risk factor for developing AMD. Identification of factors predisposing to disease progression is a priority. Current opinion holds that RPE and BrM age in concert, and normal BrM aging transforms insidiously into AMD pathology. Early electron microscopists described aged BrM as filled with debris, including amorphous electron-dense material, membrane fragments, vesicles, and calcification, beginning in macular ICL and OCL in early adulthood and later in equatorial regions. Identifying this material has been a fruitful approach to understanding early pathways involved in disease. The profound accumulation of lipids in BrM, ongoing throughout life yet first revealed by aging, is the antecedent of AMD's signature lipid-rich lesions (drusen), with implications for intraocular transport, RPE physiology, nutrition of outer retina, and maintenance of photoreceptor health.

Lipid Accumulation: Bruch's Membrane Lipoproteins

Clinical observations on fluid-filled RPE detachments in older adults led to Bird and Marshall's hypothesis that a lipophilic barrier in BrM blocked a normal, outwardly directed fluid efflux from the RPE. This hypothesis motivated Pauleikhoff et al.'s seminal histochemical study, which demonstrated oil red O-binding material (EC, esterified cholesterol; TG, triglyceride; FA, fatty acid) localized exclusively to BrM, unlike other stains, focusing attention on its composition. This lipid was absent at <30 years of age, variably present at 31–60 years of age, and abundant at ≥61 years of age. EC is now known as the major component of the oil red O binding material (see figure 106.2A–C), by histochemistry and physicochemical studies. EC in macular BrM, revealed by filipin histochemistry in enzymatically hydrolyzed tissue, rose linearly from near zero in early adulthood to high and variable levels in aged eyes. EC was detectable in periphery at ~1/7 macular levels and increased significantly with age. Hot stage polarizing microscopy also demonstrated BrM EC, as liquid crystals when examined through a polarizing filter. Few birefringent crystals signifying the neutral lipid TG were found.

Perifibrous Lipid, the Antecedent to Atherosclerosis

An understanding of perifibrous lipid, a systemic process of cholesterol deposition in extracellular matrix, is required to contextualize BrM lipid deposition. Essential for membranes, cholesterol assumes two chemical forms, unesterified (or free, UC) and esterified to fatty acids (EC). Three physical forms of cholesterol (oily droplets, membranes, and crystals) have different proportions of UC, EC, and phospholipid. Lipids binding the histochemical stain oil red O increase with age in normal human connective tissues, including sclera, cornea (arcus), tendons, and intima of large arteries. Although perifibrous lipid is a normal age change, excessive lipid deposition in these locations is pathologic (cornea—lipid keratopathy; tendon—xanthoma; intima—atherosclerosis). The oil red O-positive material comprises small (60–200 nm) extracellular droplets highly enriched in EC relative to UC (69% EC, 22% UC, and 9% phospholipid), closely associated with elastin and collagen fibers. The source of EC in sclera, cornea, and intima is low-density lipoproteins (LDLs) of hepatic origin that insudate from plasma into connective tissues, lose their apolipoproteins, and fuse. LDL-derived cholesterol is present in the earliest forms of the lipid-rich core of an atherosclerotic plaque. Ultrastructural, histochemical, direct assay of cholesterol forms, and profiling of fatty acids esterified to cholesterol established that the earliest-appearing intimal lipids came from plasma, as an extension of perifibrous lipid, and not from dying foam cells (cholesterol-engorged macrophages). This work formed a key piece of rationale behind plasma lipid-lowering therapies (i.e., statins) for coronary artery disease. Knowledge of this literature thus raised the question of whether the oil red O-positive material that increases with age in human BrM is EC-rich particles and whether lipid deposition in BrM is an ocular manifestation of systemic perifibrous lipid or a phenomenon unique to the eye. Remarkably, it appears unique to the eye involving deposition of an intraocularly produced lipoprotein in an unusual vessel wall, BrM, while sharing common molecules and mechanisms from these other tissues.

Converging histochemical, ultrastructural, biochemical, gene expression, and cell biological evidence now indicate that the EC-rich material accumulating with age in BrM is a lipoprotein containing apolipoprotein B, assembled by the RPE. The numerous small (<100 nm), round electron-lucent vesicular profile, suggesting aqueous interiors, as described by decades of electron microscopists, are actually solid, lipid-containing particles when viewed with lipid-preserving preparation techniques (see figure 106.3) and combined with extraction studies. These methods include postfixation in osmium tannic acid paraphenylenediamine and, most strikingly, quick-freeze/deep-etch, a freeze fracture method with an etching step to remove frozen water. The solid particles are now considered lipoproteins, displaying a characteristic surface-and-core appearance and varying in size from 60 to 100 nm.

Lipoprotein particles first appear among EL fibrils in early adulthood, extending inward to ultimately fill most of the ICL by the seventh decade. Most fatefully, a new layer, the Lipid Wall, then forms with solid particles stacked 3–4 deep occupying nearly 100% of a space between RPE basal lamina and OCL of many older eyes. The Lipid Wall displaces ICL collagen fibrils that anchor the RPE basal lamina (see figure 106.3) and sets the stage for BlinD, a specific AMD lesion (see below). Of further note is coated membrane-bounded bodies, which are bags of apparent lipoprotein particles in aged BrM in situ and in homogenates along with fragmented membranes and disbursed contents.

Lipoprotein composition can provide clues to sources of its components. Isolated BrM lipoproteins (see figure 106.2D–F) are EC enriched and TG poor (EC/total

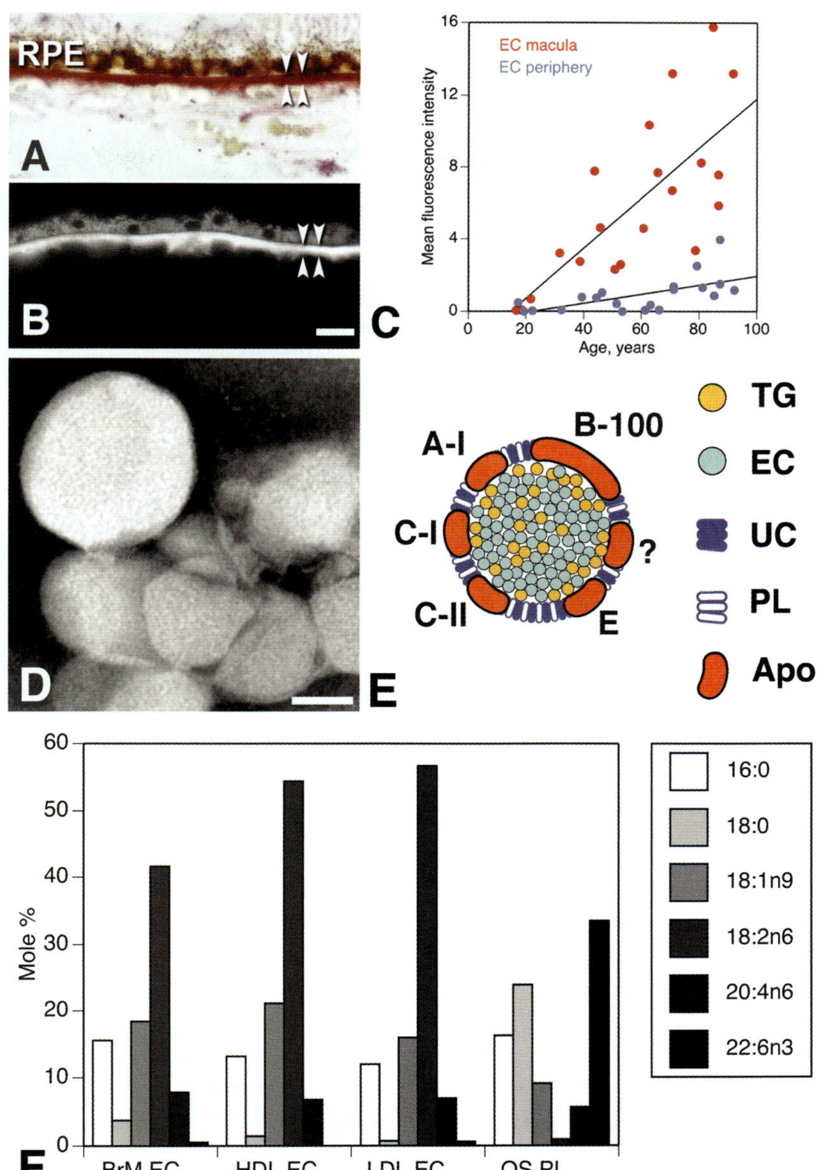

FIGURE 106.2   Lipoproteins rich in cholesterol accumulate with age in Bruch's membrane (BrM). (A) Oil red O stains BrM and retinal pigment epithelium (RPE) lipofuscin in an elderly donor eye. RPE is at the top and choroid at the bottom. BrM is bracketed by arrowheads. Scale bar: 20 μm. (B) Filipin for esterified cholesterol (EC) stains BrM. RPE lipofuscin is autofluorescent. (C) Filipin fluorescence due to EC increases markedly with age in normal eyes, more in the macula than in the periphery (Curcio et al., 2001). (D) Isolated lipoprotein particles are large and spherical; negative stain electron microscopy. Scale bar: 50 nm. (E) Bruch's membrane lipoprotein composition inferred from direct assay, druse composition, and RPE gene expression. TG, triglyceride; EC, esterified cholesterol; UC, unesterified cholesterol; PL, phospholipid; Apo, apolipoproteins; ?, unknown additional apolipoproteins. (F) Fatty acids in BrM/EC are enriched in linoleate, like plasma lipoproteins HDL (high-density lipoprotein) and LDL (low-density lipoprotein), and poor in docosahexaenoate, unlike photoreceptor membrane phospholipids (OS PL). 16:0, cholesteryl palmitate; 18:0, cholesteryl stearate; 18:1n9, cholesteryl oleate; 18:2n6, cholesteryl linoleate; 20:4n6, cholesteryl arachidonate; 22:6n3, cholesteryl docosahexaenoate. (Panel F was originally published in Curcio et al., 2010, © the American Society for Biochemistry and Molecular Biology.)

cholesterol = 0.56; EC/TG = 4–11). For comparison, hepatic very-low-density lipoprotein (VLDL), of similar diameter, is TG rich. Abundant EC points to the only mechanism by which neutral lipids are released directly from cells, an apoB-containing lipoprotein, like hepatic VLDL or intestinal chylomicrons. Significantly, RPE expresses the apoB gene and protein, along with microsomal triglyceride transfer protein (MTP), required for apoB lipidation and secretion. Lack of functional MTP is the basis of abetalipoproteinemia, a rare inherited

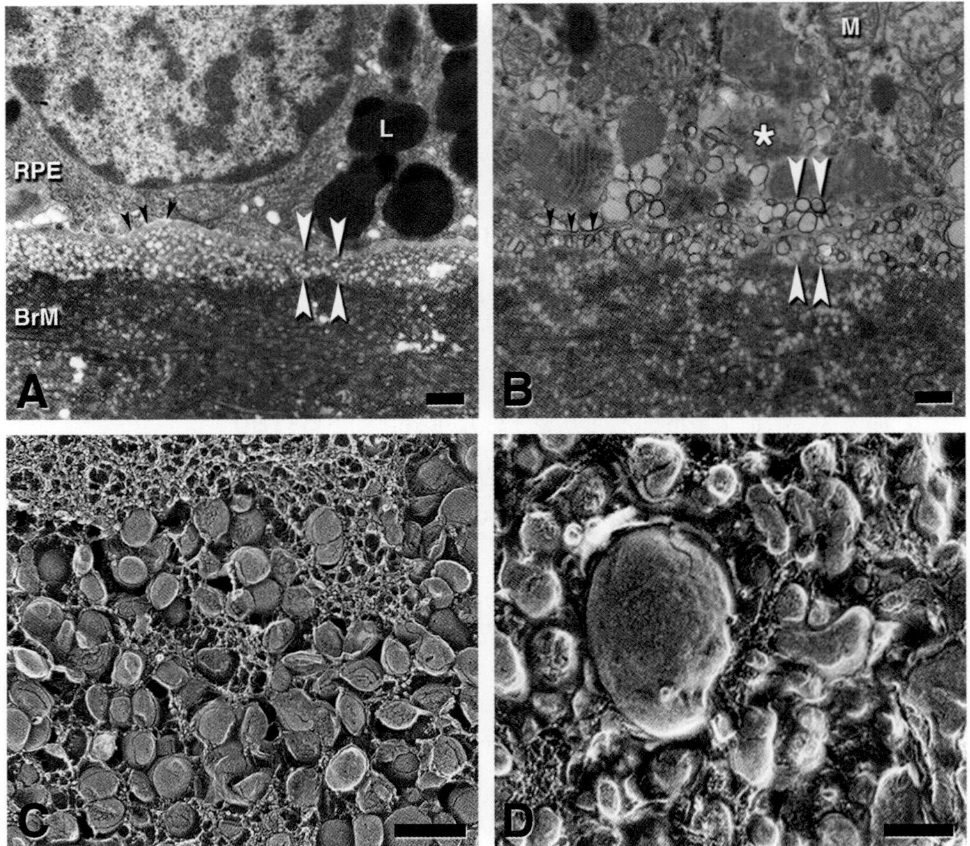

FIGURE 106.3   The Lipid Wall is a precursor to basal linear deposits (BlinD), a specific lesion of AMD. (A, B) Thin section transmission electron micrograph shows lipoproteins as vesicles; osmium postfixation, vertical plane, scale bars: 1 μm. (C, D) Quick freeze deep etch shows modified lipoproteins as solid particles; oblique plane, scale bars: 200 nm. (A) Lipoproteins (spherical vesicles of uniform diameter) accumulate 3–4 deep between the retinal pigment epithelium basal lamina (RPE-BL) (black arrowheads) and inner collagenous layer (Lipid Wall, white arrowheads). L, lipofuscin. (B) BlinD (white arrowheads) appears as membrane-bounded vesicles in the same plane as the Lipid Wall in A. High electron density at vesicular borders is associated with increased UC content. Black arrowheads, RPE-BL. M, mitochondrion. *, basal laminar deposit. (C) Tightly packed Bruch's membrane (BrM) lipoproteins in the Lipid Wall display classic core and surface of lipoproteins (Huang et al., 2007). (D) In BlinD from a 78-year-old donor with geographic atrophy, lipoproteins have more heterogeneous sizes and shapes. Pooled lipid is also apparent, consistent with a model of surface degradation and particle fusion. (From Curcio et al., 2011 with permission.)

syndromic disorder with a pigmentary retinopathy, signifying that lipoprotein assembly is essential for retinal health. The combination of apoB and MTP within native RPE marks these cells as constitutive lipoprotein secretors. Full-length apoB can be secreted by RPE cell lines of human and rat origin. Consistent with an RPE origin, BrM particles in situ first appear in the EL and fill in toward the RPE.

Indirect evidence that BrM lipoproteins are of intraocular origin also emerges from epidemiology. If the EC deposition in BrM and AMD-associated lesions were a manifestation of systemic perifibrous lipid and atherosclerosis, then a strong positive correlation between disease status and plasma lipoprotein levels, like that documented for coronary artery disease, might be expected. Such has not emerged (Smith et al., 2001).

Identifying the upstream sources of BrM lipoprotein constituents is essential for understanding the biological purpose of this pathway and the prospects for its eventual clinical exploitation. Studies using isolated lipoproteins from BrM and BrM–choroid EC report a high mole percentage of linoleate (>40%) and low docosahexaenoate (<1%) for all lipid classes (see figure 106.2F). This composition strongly points away from photoreceptor outer segments (35% docosahexaenoate in membrane phospholipids) as an upstream source, as long postulated, and toward plasma lipoproteins (45–55% linoleate in all lipid classes). These data have been interpreted to signify that plasma lipoproteins are major contributors *upstream* to an apoB lipoprotein of RPE origin. In contrast, source(s) of UC and cholesterol esterified to fatty acids in BrM lipoproteins are not

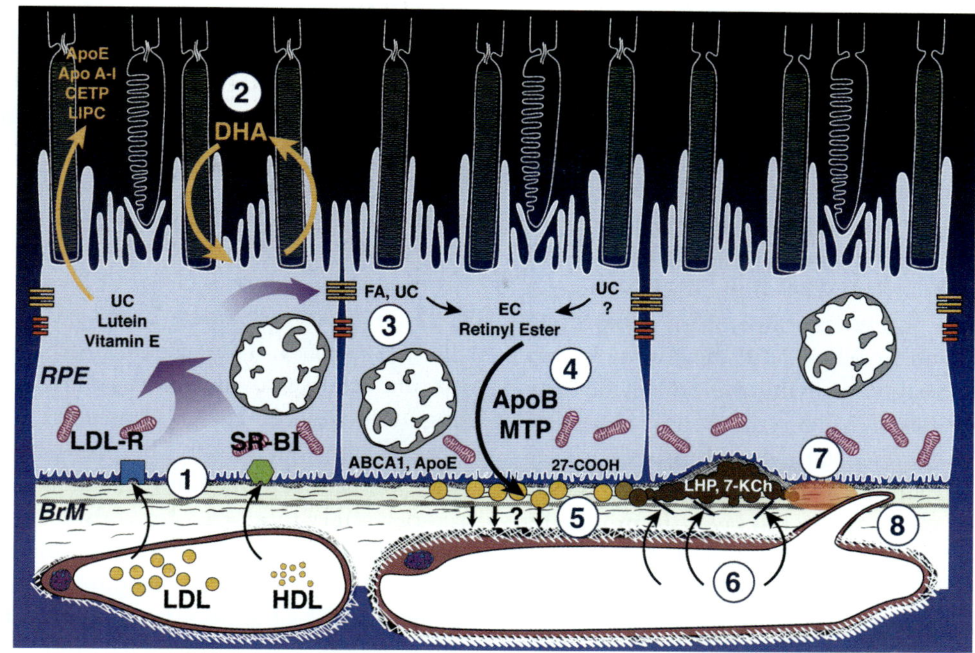

FIGURE 106.4 Lipoproteins in Bruch's membrane (BrM) in age-related macular degeneration pathobiology. From top to bottom, photoreceptor outer segments, retinal pigment epithelium (RPE), BrM, and choriocapillaris are depicted. RPE cells contain nuclei and mitochondria (magenta). Lipofuscin is omitted for clarity. (1) Plasma LDL (low-density lipoprotein) and HDL (high-density lipoprotein) delivering lipophilic nutrients are taken up at LDL and SRB-I receptors on the basolateral RPE. (2) Proteins classically associated with plasma HDL metabolism are expressed in the subretinal space and may be involved in rapid turnover of UC in the neurosensory retina. DHA is cycled between RPE and retina. (3) A basolaterally secreted apoB lipoprotein is assembled from multiple sources, including uptake of plasma lipoproteins, endogenously synthesized lipids, and/or photoreceptor degradation products. Fatty acids, especially linoleate (18:2n6), come largely from taken-up plasma lipoproteins. UC, from sources not yet determined, is reesterified (not shown). (4) RPE expresses both apoB and microsomal triglyceride transfer protein (MTP) and secretes esterified cholesterol (EC) rich particles into BrM (gold circles), where they are retained and eventually cleared through the choriocapillary endothelium. (5) Lipoprotein particles begin to accumulate during adulthood for unknown reasons, building up a layer 3–4 deep external to the RPE basal lamina (the Lipid Wall). Other outflow pathways for UC include apoE, ABCA1-mediated transfer to circulating HDL, apoE secretion, and enzymatic conversion to oxysterols (27-COOH). (6) Over time, pro-inflammatory/toxic species like linoleate hydroperoxide and 7-ketocholesterol appear. Particles could fuse to form lipoprotein-derived debris, the principal component of BlinD and soft drusen. Translocation of necessary compounds from plasma is impeded (blocked arrows). (7) Inflammation is elicited in inner BrM. (8) Choroidal neovascularization (type I) ensues. (From Curcio et al., 2011 with permission.)

yet known and could be outer segments, plasma lipoproteins, endogenous synthesis, or a combination.

Lipoproteins may thus be assembled from several sources, including outer segments, remnants from the photoreceptor nutrient supply system, and endogenous synthesis. According to this model (see figure 106.4), plasma lipoproteins are vehicles for delivering lipophilic nutrients (carotenoids, vitamin E, and cholesterol) to photoreceptors by RPE, which has functional LDL and high-density lipoprotein (HDL) receptors. Photoreceptor-destined nutrients are stripped from lipoproteins by the RPE, and the remnants are repackaged for secretion into BrM within apoB-containing lipoproteins, like a curbside recycling program. There they accumulate and become toxically modified to instigate inflammation in AMD. Studies showing associations of AMD with genes historically part of plasma HDL metabolism, without consistent effects on plasma HDL itself, may be revealing the activity of upstream pathways within the eye (Curcio, 2011; Ebrahimi & Handa, 2011; Rodriguez & Larrayoz, 2010).

Other Aging Changes

BrM thickens throughout adulthood (20–100 years of age) by two- to threefold under the macula. Thickening is highly variable between older individuals. Equatorial BrM changes little while BrM near the ora serrata thickens twofold. A large ultrastructural study demonstrated an age-related increase in EL thickness and stable EL integrity. RPE-BL and ChC-BL thicknesses are also stable with age.

Unbalanced regulation of extracellular matrix molecules and their modulators are thought to result in ICL and OCL thickening. Increased histochemical reactivity for glycoconjugates, GAGs, collagen, and elastin is seen in the macula relative to equator and near the ora serrata. Collagen solubility declines with age. Metalloproteinases MMP-2 and MMP-3 increase with age, as does a potent inhibitor of metalloproteinases, TIMP-3, which reaches adult levels in several systems at age 30 years, signifying the end of developmental organogenesis. TIMP-3 reduction or absence is pro-angiogenic, via homeostatic metalloproteinase regulation and direct binding to pro-angiogenic VEG-F (Ebrahem et al., 2011). BrM thickness is modifiable by dietary zinc, of interest because zinc is both part of the Age-Related Eye Disease Study (2001) antioxidant formulation given to AMD patients and a cofactor for complement factor H (CFH). The latter is a fluid phase regulator in the innate immunity system strongly implicated in AMD pathogenesis by genetics and immunohistochemistry.

At the molecular level, aging BrM contains evidence of many biological activities including lipoprotein accumulation, extracellular matrix remodeling, oxidative damage, and inflammation, consistent with its role as a multifunctional subendothelial space. Long-lived proteins like collagens are modified in vivo by nonenzymatic Maillard and free radical reactions to yield advanced glycation end products (AGEs) and the formation of lipid-derived reactive carbonyl species like malondialdehyde (MDA), and 4-hydroxyhexenal, collectively called age-related lipoperoxidation end products (ALEs). Accumulation of AGEs and ALEs, characteristic of diabetes and atherosclerosis, also occurs in aging BrM (see table 106.1). MDA binds CFH, suggesting a role in sequestering oxidized lipids for this protein. Immunoreactivity but not Raman spectroscopy signals for MDA have been detected in human BrM to date (Beattie et al., 2010; Weismann et al., 2011). A common BrM age change is calcification and ensuing brittleness, which involves fine deposition of electron-dense particulate calcium phosphate on individual elastin fibrils. That lipid deposition might precede BrM basophilia and fragmentation of calcified EL was hypothesized early, perhaps because similar events occur in lipid-rich atherosclerotic plaques (Duer et al., 2008). Other components prominent in aged BrM include complement components C3d, C5b-9, and pentraxin-3, a homologue of the acute phase respondent C-reactive protein. Importantly, CFH localizes particularly along intercapillary pillars of the OCL.

Regional Differences

BrM structure and age changes vary with retinal region. Relative to periphery, macular BrM has more debris deposition, OCL thickening, cholesterol-rich lipoprotein accumulaton, and predilection for AMD's sight-threatening lesions. In normal eyes of all ages, the EL has significant regional differences. Macular EL is thinner (134 nm) by a factor of 3 relative to periphery (392 nm) and covers less BrM surface than periphery (37% vs 92.3%). The sevenfold greater blood flow in macula relative to periphery (Alm & Bill, 1973) is a huge effect whose basis within outer retinal metabolic needs is not understood. This gradient outpaces that of both photoreceptor/mm^2 (macula/temporal equatorial periphery ratio = 2.7 for rods, 1.9 for cones; Curcio et al., 1990) and retinal thickness (macula/periphery = 2.6 for whole retina, 2.3 for outer retina only; unpublished data). Explaining topographic differences within the outer retinal neurovascular unit will likely provide major insight into AMD's macular predilection and antecedents in normal ocular physiology.

FUNCTION OF BRUCH'S MEMBRANE

Structural Role of Bruch's Membrane

As a vessel wall, BrM's primary function is structural, and it requires elasticity. Encircling more than half the eye, BrM stretches with the corneoscleral envelope as intraocular pressure fluctuates, stretches to accommodate choroidal blood volume, and may join the choroid as a spring pulling the lens during accommodation. The modulus of elasticity in human BrM–choroid is estimated as 7–19 MPa. After early adulthood, this parameter increases at a rate of ~1% per year (indicating stiffening), with no further difference noted in a small number of AMD eyes. The modulus of BrM–choroid is similar to sclera, consistent with a load-bearing role.

Transport Role of Bruch's Membrane

A second important function is transport. BrM's luminal aspect faces a fenestrated vascular endothelium and basal lamina, making it structurally analogous to the renal glomerulus. RPE and glomerular basement laminas contain similar specialized collagens, providing a basis for commonality between retinal and kidney disease and between fluid and macromolecular transport in the renal glomerulus and BrM. The transport processes served by BrM are numerous. Oxygen, electrolytes, nutrients, cytokines, signaling factors, and plasma-lipoprotein-borne vitamins destined for RPE and photoreceptors pass through BrM from the choriocapillaris. Waste products travel back in the opposite direction for elimination in the circulation, including RPE-produced lipoproteins and water from the subretinal space pumped outward by the RPE.

HYDRAULIC CONDUCTIVITY OF BRUCH'S MEMBRANE GAGs are concentrated in the interphotoreceptor matrix and corneal stroma, where these highly charged macromolecules maintain geometric fidelity essential for vision (collagen spacing for corneal transparency, photoreceptor spacing for visual sampling). GAGs generate significant swelling pressure, and without a mechanism to maintain tissue deturgescence, GAGs would imbibe fluid, swell, destroy tissue geometry, and degrade visual function. Corneal endothelium forestalls swelling by continuously pumping fluid out, as does RPE, and RPE failure can lead to retinal detachment. A driving force adequate to overcome the collective flow resistance of RPE, BrM, and choriocapillaris endothelium is provided by a gradient in fluid pressure and oncotic pressure (osmotic pressure generated by plasma proteins). This balance is embodied by Starling's Law, which characterizes the relationship between fluid flux (q: flow per unit area; positive when flow is out of the blood vessel) across a capillary vessel wall and driving forces:

$$q = L_p * (\Delta P - \sigma \Delta \Pi). \qquad (106.1)$$

L_p is hydraulic conductivity, a measure of how easily fluids flow cross a vessel wall. If blood vessel surface area is A, then $1/(L_p A)$ is flow resistance of the vessel wall. ΔP is the difference between fluid pressure within the vessel (P_{cc}) and at the basal RPE (P_{RPE}). $\Delta \Pi$ is the difference between oncotic pressure within the vessel (Π_{cc}) and at the basal RPE (Π_{RPE}). σ is a reflection coefficient that characterizes the extent to which a vessel wall rejects a molecular species generating $\Delta \Pi$, ranging from 0 (freely permeable) to 1 (completely rejected).

The magnitude of $\Delta P - \sigma \Delta \Pi$ in the normal eye can be estimated as approximately –5.5 mm Hg, pulling fluid into the choriocapillaris. Since L_p of macular BrM–choroid of healthy young humans ranges from 20 to 100×10^{-10} m/s/Pa, use of equation 106.1 indicates that, with a healthy BrM, the choriocapillaris is capable of adsorbing a flow rate per unit area of 500–2,500 μL/hr/cm². As the RPE pumping rate has been measured as approximately $q = 11$ μL/hr/cm², it is clear that in the normal adult eye, this fluid can be easily adsorbed into the choriocapillaris.

AGE-RELATED CHANGES IN HYDRAULIC CONDUCTIVITY AND DISEASE Fisher first measured L_p of human BrM, finding that L_p decreased significantly with age. Marshall and Hussain revisited these measurements using BrM–choroid with RPE removed, showing with laser ablation that flow resistance was entirely due to BrM. They also found that flow rate increased linearly with driving pressure, indicating that BrM L_p is relatively insensitive to pressure <25 mm Hg. Importantly, L_p of macular BrM exhibited a dramatic, exponential decline throughout life (see figure 106.5), dropping from 130×10^{-10} m/s/Pa in young children to 0.52×10^{-10} m/s/Pa in old age. Macular L_p dropped more rapidly with age than peripheral L_p. The lowest value measured for BrM L_p in normal eyes (0.5×10^{-10} m/s/Pa), just barely sufficient to allow fluid resorption by the RPE (see equation 106.1).

Determining BrM L_p in isolated AMD macula is difficult due to advanced pathology including scars. In the periphery of AMD eyes, BrM L_p is lower than in age-matched normal eyes (see figure 106.5) (Hussain, Starita, & Marshall, 2004). Assuming that similar processes occur in the macula due to its profound lipid accumulation, then in diseased eyes, the RPE must generate even higher pressures to drive fluid into the choriocapillaris, with potentially deleterious consequences. Above some threshold, higher pressure will cause the

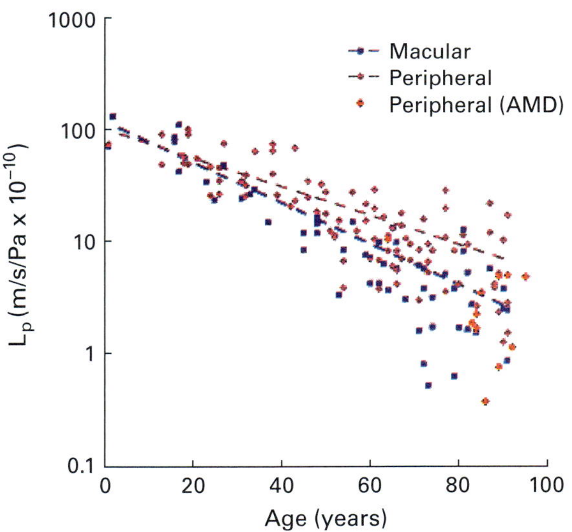

FIGURE 106.5 Hydraulic conductivity (L_p) of Bruch's membrane as a function of age. Dotted lines are exponential fits to data from macular and peripheral regions, respectively. All data from eyes with age-related macular degeneration (AMD) (peripheral region only) have lower values of L_p than the best fit to data taken from peripheral Bruch's membrane of non-diseased eyes (Hussain, Starita, & Marshall, 2004). (From Curcio and Johnson, 2012 with permission.)

RPE-BL to separate from the ICL. Along with direct plasma leakage from choroidal neovascular membranes (Spaide, 2009), this nonresorbed fluid accumulates and leads to RPE detachment, as seen in 12–20% of AMD patients (Pauleikhoff et al., 2002).

The dramatic age-related decrease in BrM L_p can be plausibly ascribed to age-related lipoprotein accumulation. In fact, lipid particles trapped in an extracellular matrix can generate higher flow resistance than would be expected based simply on particle size and number. However, Marshall and Hussain observed that the most marked decline in L_p occurred before age 40 (see figure 106.6A) while overall BrM lipid content increased after age 40. They concluded that other age-related phenomena must account for diminished L_p and focused subsequent studies on regulation of extracellular matrix turnover.

A very different conclusion is reached from examining age effects on flow *resistivity*, the *inverse* of L_p and a parameter more intuitively correlated with blockage of fluid by a hydrophobic barrier. Resistivity increases from a low of $R = 10^8$ Pa/m/s for young individuals to $R = 10^{10}$ Pa/m/s for aged persons. BrM resistivity R and histochemically detected EC plotted against age agree almost perfectly (see figure 106.6B). This is strong evidence that BrM's increasing lipid content and progressive hydrophobicity can indeed impair fluid transfer with age. Laser ablation studies localizing flow resistance to the ICL further support this conclusion because lipoproteins accumulate prominently in the ICL. Further, more laser pulses were required to abolish flow resistance in the oldest eyes, consistent with presence of a Lipid Wall. Thus, decreased L_p and increased resistivity R of aging BrM is closely related to the age-related accumulation of lipids, primarily EC. Macular L_p decreases more markedly with age than peripheral L_p, consistent with macula's greater lipid content.

PERMEABILITY OF BRUCH'S MEMBRANE TO SOLUTE TRANSPORT Along with bulk fluid flow, individual molecular species, including dissolved gases, nutrients, cytokines, and waste products, are transported across BrM driven by passive diffusion. Flow across BrM is too slow to influence this process under physiological conditions.

Diffusion follows Fick's law, whereby diffusive flux per unit area (j) is proportional to the diffusion coefficient (D) of that species in the medium through which it passes and to the concentration difference across the medium (ΔC). It is inversely proportional to the diffusion length:

$$j = D\, \Delta C\, /L. \qquad (106.2)$$

Tissue *permeability* to a given species is defined as $P = j/\Delta C$, and $P = D/L$. For example, the permeability of BrM to oxygen is ~ 0.067 cm/s. Since diffusion moves down a concentration gradient, one species (e.g., oxygen) might diffuse toward the RPE while another (e.g., carbon dioxide) diffuses simultaneously in the other direction. With high diffusion coefficients and little interaction with extracellular matrix, small molecules like oxygen diffuse quickly across BrM. However, macromolecules have much smaller free solution diffusion coefficients due to their size. Coefficients are further reduced by interactions with extracellular

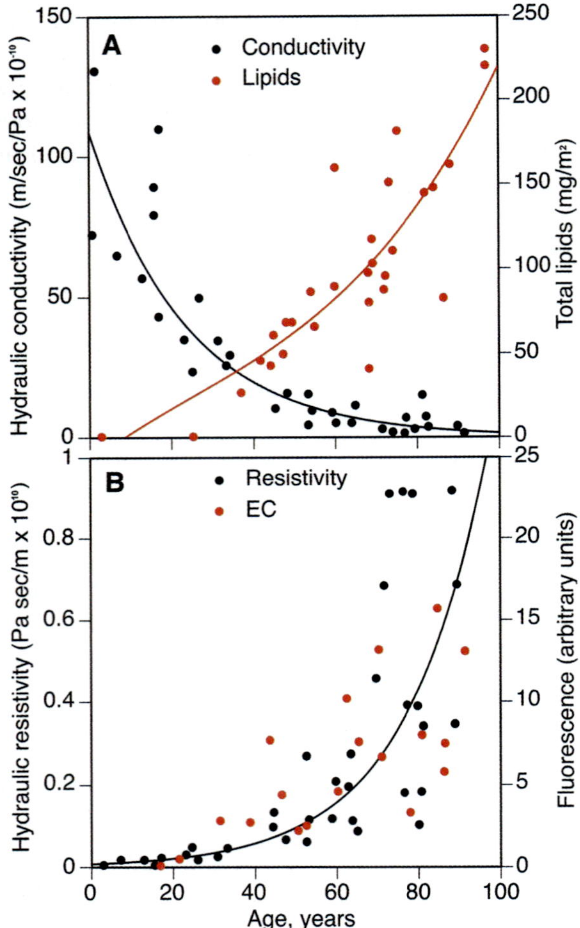

FIGURE 106.6 Conductivity, resistivity, and Bruch's membrane (BrM) lipid content. (A) Hydraulic conductivity L_p of human macular BrM–choroid as a function of age and total lipid accumulation in human macular BrM; lines are exponential fits to the data (Marshall et al., 1998). (B) Hydraulic resistivity (R) of human macular BrM–choroid as a function of age (Marshall et al., 1998) as compared to esterified cholesterol accumulation in human macular BrM (Curcio et al., 2001); lines are exponential fits to the data and nearly overlie one another. (From Curcio and Johnson, 2012 with permission.)

matrix and lipoproteins, impeding macromolecular transport.

The transport rate across human BrM declines linearly with age for all molecular species measured, despite difficulties in determining the diffusion coefficient of BrM independent of choroid. Amino acids exhibited permeabilities of 0.6×10^{-4} cm/s (phenylalanine) to 1.2×10^{-4} cm/s (glycine) for young BrM and a modest ≤ 2-fold decline with aging. Serum proteins decrease more markedly, dropping by an order of magnitude from first to ninth decade of life (3.5×10^{-6} cm/s to 0.2×10^{-6} cm/s), especially for proteins larger than 100 kD. Macular BrM decreased more with age than did peripheral BrM (Hussain et al., 2010).

Decreased permeability of BrM to transport is thus likely due to a decrease in diffusion coefficients, especially for larger molecular species affected by interaction with extracellular matrix and lipoproteins. Increased path length due to age-related BrM thickening could also have a significant effect (equation 106.2).

Because RPE cells internalize plasma LDL from the choroid, and because lipoproteins accumulate with age in BrM, Cankova et al. examined the reflection coefficient of bovine BrM to plasma LDL (diameter, 22 nm; molecular weight, 2.28×10^{3} kD; Crouse et al., 1985). They measured a reflection coefficient of 0.58, lower than coefficients of 0.998 and 0.827 for arterial endothelium and intima, respectively. Thus, while LDL did not pass freely through BrM, it could nonetheless pass. This work indicates that a molecular weight exclusion limit to BrM macromolecule transport, if one exists, must be much higher than 66–200 kD, as originally proposed.

These considerations are relevant not only to understanding the mass transfer between the choriocapillaris and RPE that underlies outer retinal nutrition but also for trans-scleral drug delivery strategies including anti-angiogenic agents for AMD and steroids for diabetic retinopathy. The transport of lipophilic solutes is impacted by BrM–choroid while the transport of hydrophilic moieties is impacted by the RPE.

FUNCTIONAL IMPLICATIONS BrM's physiological roles are structural and facilitating transport. Transport is increasingly hindered with age, due at least partly to the marked age-related accumulation of EC-rich lipoproteins in this tissue, impeding outward pumping of fluid from RPE. A $\geq 90\%$ decrease in transport of some species from the choroid may include lipophilic essentials delivered by lipoproteins, with functional consequences for photoreceptors. A well-characterized change occurring through the life span of individuals with healthy maculas is slowed dark adaptation, attributed to impaired translocation of retinoids across the RPE–BrM interface. This slowing, worse in AMD patients, can be partly ameliorated by short-term administration of high-dose vitamin A, presumably overcoming this translocation deficit via mass action. Another well-characterized age change is a 30% loss of perifoveal rods in apparently normal human maculas. It will be important to determine if this distinctive pattern reflects poor transport in the underlying BrM.

PATHOLOGY OF BRUCH'S MEMBRANE

AMD's Specific Lesions

In aging and AMD, characteristic extracellular lesions accumulate anterior to the ICL, in the sub-RPE space (see figure 106.7). Known as drusen and BlinD, these lipid-containing aggregations ultimately impact RPE and photoreceptor health by impairing transport, causing inflammation, and predisposing to choroidal neovascularization (CNV). Another lesion consisting primarily of extracellular matrix material, basal laminar deposit (BlamD), accumulates in a patchy or a continuous manner, between the RPE and the RPE-BL. Investigation of lesion morphology and composition has revealed points of potential therapeutic intervention and provided glimpses of previously unrecognized complexities in the cellular and molecular physiology required for optimal photoreceptor homeostasis and excellent vision.

DRUSEN AND BlinD Drusen are the major intraocular risk factors for AMD progression and dynamic lesions whose composition signifies major pathways perturbed by disease. In a fundus view, drusen are 30–300 μm-diameter yellow–white deposits posterior to the RPE. By optical coherence tomography, they appear as variably hyporeflective spaces in the same location. Histologically, drusen are focal, domed lesions between the RPE-BL and the ICL, that is, in the same sub-RPE tissue compartment as the Lipid Wall and BlinD. Found in most older adults, drusen are more numerous in larger peripheral retina than in smaller macula. Drusen are typically classified as "hard" and "soft" by the appearance of their borders. Soft drusen confer high risk of advanced disease and, importantly, are found only in the macula. Other rare druse types exist and are being characterized via differential light absorption and reflection in the ocular fundus enabled by high-resolution multimodal clinical imaging (Guigui et al., 2011).

BlinD is a thin (0.4–2 μm) layer located in the same sub-RPE compartment as soft drusen but not yet visible clinically. BlinD is rich in solid lipoprotein particles and

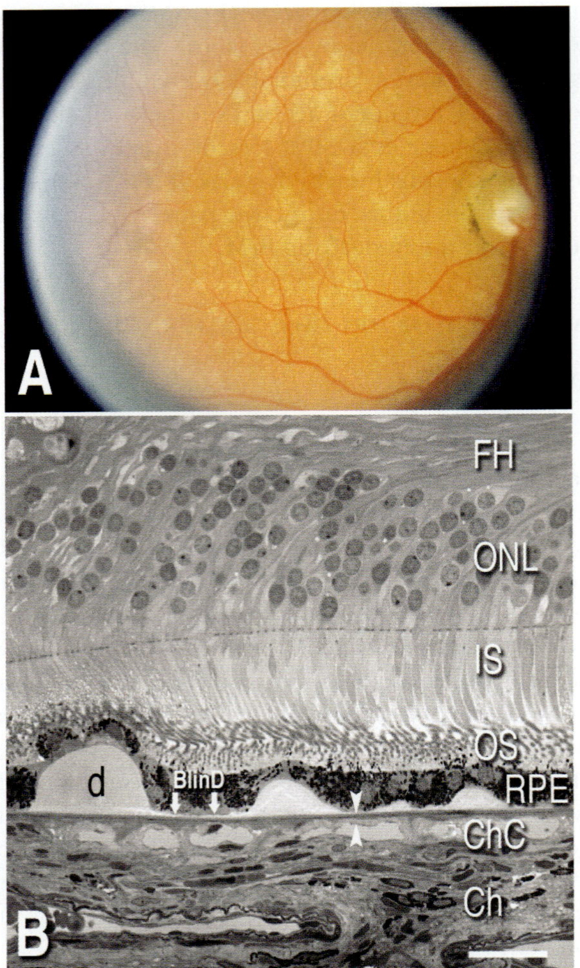

FIGURE 106.7   Clinical and histological views of drusen. (A) Fundus of a patient with nonneovascular age-related macular degeneration (AMD). Numerous yellowish spots are drusen on the anterior surface of Bruch's membrane (BrM). (B) Histological section (1 μm-thick, toluidine-O-blue) of a macula of a different patient, cut in the vertical plane. Drusen (d) are focal deposits of debris between the retinal pigment epithelium basal lamina (RPE-BL) and BrM (arrowheads). BlinD, basal linear deposit, diffusely distributed soft druse material ("the Oil Spill in BrM"); FH, Henle fibers; ONL, outer nuclear layer (cones only); IS, inner segments; OS, outer segments; ChC, choriocapillaris; Ch, choroid. Scale bar: 40 μm.

lipid pools (see figure 106.3). BlinD and soft drusen are considered alternate forms (layer and lump) of the same entity and may interchange over time. Soft drusen are oily, are difficult to isolate individually, and are biomechanically more fragile than hard drusen, properties attributable to BlinD by inference. ApoE and apoB are present in BlinD and its precursor, the Lipid Wall (see table 106.1).

Early detection is a surrogate for abundance, so it significant that Wedl noted that drusen looked like fat

globules in 1854. All drusen contain histochemically detectable lipid. Extractable lipids account for ≥40% of hard druse volume, making them the largest single component on a per druse basis. Lipid content is likely higher for macular soft drusen, which include large EC-rich lakes, reminiscent of atherosclerotic plaque cores. Immunoreactivity for apolipoproteins appears frequently in drusen (100%, apoE; >80% apoB; 60%, A-I). Plasma-abundant apoC-III is present in fewer drusen than plasma-sparse apoC-I, indicating a specific retention mechanism for plasma-derived apolipoproteins or an intraocular source. Importantly, hard drusen contain many solid, extractable electron-dense particles of the same diameter as the lipoproteins that accumulate with age in BrM. These observations together with the appearance of membranous debris in soft drusen (below) make an RPE-secreted lipoprotein particle an efficient mechanism to place multiple lipids and apolipoproteins within lesion compartments.

Discrete nonlipid components in some drusen include amyloid assemblies and granules of lipofuscin or melanin indicating cellular origin (see table 106.2). Other constituents present in all drusen include vitronectin, TIMP-3, complement factor H, complement components C3 and C8, crystallins, and zinc (see table 106.2).

The principal lipid-containing component of soft drusen and BlinD was called membranous debris by Sarks, Sarks, and Killingsworth (1994) and lipoprotein-derived debris by Curcio et al. (2009). The revised nomenclature was proposed because these lesions are richer in histochemically detectable UC than surrounding cellular membranes and because they have abundant neutral lipid not accounted for by UC-containing membranes. It has been argued the building blocks of membranous debris are more plausibly the UC-rich exteriors of lipoproteins (native and fused) whose neutral lipid interiors are not well preserved in postmortem tissue.

Long-standing theories for druse formation evoked transformation of overlying RPE and deposition of materials onto BrM, with deposition now accepted (Hageman & Mullins, 1999). The RPE has been implicated as a source of many druse components, via budding of membrane bounded cytoplasmic packets and secretion. Recently a culture system of highly differentiated RPE cells was shown to constitutively secrete to the basolateral aspect apoE-immunoreactive material in amorphous and vesicular aggregations (Curcio, 2011; Johnson et al., 2011), the latter closely resembling coated membrane bounded bodies. Plasma contributors to drusen have been difficult to study, but in this culture system, complement cascade

Table 106.2

Localized constituents of drusen

Component	Direct assay	Abundance (per druse if available or per eye)
Lipoproteins (EC, UC, phospholipid)	✓	All drusen; >40% of hard druse volume; EC pools in soft drusen
Apolipoproteins (apoB, A-I, C-I, E)	apoE ✓	60–100% of hard drusen; higher rates in periphery than macula
Melanin/lipofuscin granules		6% of hard and soft drusen
Cells (dendritic, others)		3–6% of hard drusen only
Amyloid vesicles (0.25–10 μm)		2% of hard drusen, 40% of compound drusen, frequent in eyes with many drusen, some AMD eyes
Calcification	✓	43% of macular hard drusen, 1.6% of soft drusen, 2% of peripheral hard drusen
Clusterin, TIMP3, vitronectin, apolipoprotein E, complement factor H, complement components 8, 9, C-reactive protein	✓	Reliably detected by multiple studies; high abundance inferred
Components of classic, lectin, alternative, terminal complement pathways; C3 fragments indicating activation	Some ✓	Many pathway components evidence key role of complement
RGR-d		All drusen
αA- and αB-crystallin	✓	N.A.; higher in BrM, more in AMD drusen
Ubiquitin		Most drusen in most eyes
Exosome markers CD63, CD81, and LAMP2		N.A.
Bestrophin, membrane-bound		N.A.
Carbohydrates		All drusen
Zinc	✓	Many drusen

Note: Complete reference list is in Curcio and Johnson (2012). EC, esterified cholesterol; UC, unesterified cholesterol; BrM, Bruch's membrane; AMD: age-related macular disease; N.A., not available. Localization methods: immunohistochemistry, histochemistry, immunogold transmission electron microscopy. Direct assays: proteomics, western blot, microprobe synchrotron X-ray fluorescence for zinc. Varying estimates of particulate druse components are due to differences in location of samples and druse types examined.

components, added exogenously via serum, bound RPE-secreted material (Johnson et al., 2011). These experimental findings support the idea that the RPE secretes large lipoproteins containing apoB and apoE, like hepatic VLDL, that can evoke systemic inflammatory response, including complement activation regulated by intraocular mechanisms such as CD46 and CD59.

Significance ascribed to molecules sequestered in drusen (table 106.2), and by inference, BlinD, include toxicity to overlying RPE and evidence of formative processes such as extrusion of cellular materials, secretion, extracellular enzymatic processing, and cellular activity. Additional significance can be ascribed to lesions as physical objects that increase path length between choriocapillaries and retina, raising ischemia risk (Stefansson, Geirsdottir, & Sigurdsson, 2011), and provide a biomechanically unstable cleavage plane between RPE-BL and ICL. The dynamism of drusen (Sarks, Sarks, & Killingsworth, 1994), which can appear, coalesce, and disappear over a time scale of months, has been dramatically demonstrated by natural history studies utilizing automated reconstruction of optical coherence tomograms (Yehoshua et al., 2011).

BASAL LAMINAR DEPOSIT Basal laminar deposit (BlamD) forms small pockets between the RPE and the RPE-BL in many older normal eyes or a continuous layer as thick as 15 μm in AMD eyes. It is not formally part of BrM, but it is abundant in AMD eyes. Ultrastructurally, BlamD resembles basement membrane material, containing laminin, fibronectin, type IV, and type VI collagen. Thick BlamD, associated with advanced AMD risk, contains histochemically detectable lipid including UC and EC and is a classically described site for membranous debris. By lipid-preserving methods, solid particles are seen in BlamD. Especially enriched in basal mounds, lipoprotein-derived debris in BlamD may be considered as retained in transit from the RPE to BlinD and/or drusen. BlamD also contains vitronectin, MMP-7, TIMP-3, C3, C5b-9, EC, and UC. Present in numerous mouse models of aging, stress, and genetic manipulation, BlamD is a reliable marker of RPE stress.

Response-to-Retention Hypothesis of AMD

The parallels between pathology of arterial intima and that of BrM are striking. Both diseases feature cholesterol-rich lesions in subendothelial compartments within the systemic circulation, involving many of the same molecules and biological processes at multiple steps, as anticipated for decades. According to the response-to-retention theory of atherosclerosis, plasma lipoproteins cross the vascular endothelium of large arteries and bind to extracellular matrix. By itself, this process (perifibrous lipid) is not pathologic. However, lipoprotein components become modified via oxidative and nonoxidative processes and launch numerous downstream deleterious events, including inflammation, macrophage recruitment, and neovascularization leading to disease. Analogous with apoB-lipoprotein-instigated disease in arterial intima, an intraocular response-to-retention involving the RPE and BrM in aging and AMD would begin with age-related accumulation of lipoproteins *of local, intraocular origin*. Oxidation, perhaps driven by reactive oxygen species from nearby mitochondria in basolateral RPE, would initiate a pathological process resembling that in the vascular system with inflammation-driven downstream events including complement activation and structurally unstable lesions.

Neovascular AMD

CNV, the major sight-threatening complication of AMD, involves macular angiogenesis along several vectors: vertically across BrM, and either laterally external to the RPE (type 1 CNV), laterally within the subretinal space (type 2 CNV), or further anteriorly into the retina (type 3 CNV) (Klein & Wilson, 2011). CNV is a multifactorial nonspecific wound healing response to various specific stimuli, involving VEGF stimulation of choriocapillaris endothelium, compromise to BrM, and participation of macrophages. Impaired transport across BrM in AMD increasingly isolates the RPE from its metabolic source in the choriocapillaris and hinders waste product disposal. VEGF released by RPE as a stress signal initiates an angiogenic response by the endothelium.

BrM compromise enables CNV progression. In paired donor eyes with and without CNV secondary to AMD, progressed eyes are distinguished by calcification and breaks in BrM. The EL, an important anti-angiogenic barrier, is thinner and more discontinuous in eyes with neovascular AMD, as well as in aging macula relative to the periphery. Mouse models provide supporting evidence. In a mouse with VEGF overexpressing and an intact BrM, intrachoroidal neovascularization occurs without CNV (Schwesinger et al., 2001). Laser-induced neovascularization in mice deficient in lysyl oxidase-like 1, required for elastin polymerization, is extensive. Mice overexpressing the AMD susceptibility gene HTRA1 (encoding a serine protease) have a severely disrupted EL and polypoidal choroidal vasculopathy, an aneurismal CNV (Jones et al., 2011).

Soft drusen and BlinD further this process by acting as a horizontal cleavage plane that is permissive to invading capillaries. The lipid-rich composition, relative lack of structural elements like collagen fibrils, lesion biomechanical instability (Rudolf et al., 2008), and pro-inflammatory, pro-angiogenic compounds like 7-ketocholesterol and linoleate hydroperoxide likely promote vessel growth in the sub-RPE compartment. Surgically excised CNV membranes have little associated BlinD but frequent BlamD (Grossniklaus et al., 2005). Since BlinD's clinical correlates are soft drusen and soft drusen are CNV risk factors, this finding is interpreted as indicating that CNV grows into the compartment previously occupied by BlinD and either replaces or removes the deposit (Spaide, Armstrong, & Browne, 2003).

MODEL SYSTEMS FOR BRUCH'S MEMBRANE EXPLORATION

In Vivo: Monkey and Mouse

Macaque monkeys have maculas, life spans up to 30 years in captivity, and many similarities to human retinal aging and AMD, with evidence for genetically susceptible lineages, predilection for the ARMS2 susceptibility locus (chromosome 10q26), and a low level of neovascular complications (Francis et al., 2008; Stafford, Anness, & Fine, 1984). Like humans, macaque EL is discontinuous in macula and thick and continuous in periphery (Gouras et al., 2010), and BrM contains oil red O-binding lipids and oxysterols. Clinically evident drusen are angiographically similar to human (Chang et al., 1998) and contain many proteins familiar from human drusen (Umeda et al., 2005); the lipid content is not known. Heterogeneous material, comprised of membranous, granular, and cellular components, are present in the ICL and OCL (Ulshafer et al., 1987b). Some micrographs appear to show extensive thick BlinD and lipid pools (Ulshafer et al., 1987a).

Creation of mouse models replicating signature lesions has been a major milestone for progress on other age-related diseases (Breslow, 1996; Price et al., 1998). Studies of established atherosclerotic mouse models with hyperlipidemia have reported BrM changes comparable to alterations in humans, include thickening, loss of lamination, and accumulation of apparent lipid vesicles. The best-described models are based on the C57BL/6 line and include deletions of apoE or LDL-receptor or transgenes of human apoB-100 or apoE. Several generalizations are possible: (1) Mice are naturally resistant to atherosclerosis and differ from humans in key aspects of lipoprotein metabolism; (2) interpretation of studies using transgenic mice expressing lipoprotein-related genes are complicated by RPE expression of the same genes, preventing definitive attribution of effects to plasma lipoproteins of hepatic/intestinal origin or to lipoproteins of RPE origin; (3) ocular changes seen in mouse models generally occur in the setting of hyperlipidemia, which is not associated with AMD; (4) murine BrM, only 250 to ~600 nm thick, is challenging to study but nevertheless amenable to demonstration of neutral lipid by specialized histochemical and ultrastructural methods, as well immunohistochemical detection of druse-abundant apoE; and (5) mouse models created to test other AMD-relevant hypotheses (oxidative stress, immune factors) have BrM changes and sub-RPE deposits (e.g., Imamura et al., 2006) that have not been assessed for lipid or apolipoprotein content and therefore may yet have this added value.

Two recent studies tested the intraocular response-to-retention hypothesis with systemically injected exogenous LDL. In a mouse model of BrM aging (D-galactose evoked AGE formation), exogenous LDL particles are retained (Cano et al., 2011). Lipoprotein lipase in these animals, also shown in AMD BrM and choroid, may promote retention as a bridging molecule as it does in atherosclerosis. LDL injections alone in a different mouse evoked BrM thickening and early photoreceptor apoptosis without BlinD or BlamD, unlike human AMD (Yin et al., 2012). Although a plasma source cannot be definitively excluded by this work, and BrM structure can indeed be altered in some hypercholesterolemia models (e.g., Rojas et al., 2011), evidence so far indicates that LDL of systemic origin is insufficient to initiate BrM pathology (Yin et al., 2012).

Ex Vivo: Bruch's Membrane–Choroid Explants

Human and bovine full-thickness BrM–choroid explants have been utilized for experimental studies of transport across pentalaminate BrM in a large eye. Investigation of transport properties will further benefit from a newly scaled down Ussing chamber system for mouse eyes (or portions of large eyes) and methods for following multiple tracers in single samples via gel exclusion chromatography (Zayas-Santiago, Marmorstein, & Marmorstein, 2011). Initial studies indicate differential diffusion rates for tracers of different sizes, with globular proteins as large as ferritin (MW = 450 kD) able to pass in either direction across BrM.

Human and porcine BrM–choroid organ culture systems have also been used extensively to develop means for refurbishing BrM for RPE transplants (Del Priore, Tezel, & Kaplan, 2006; Gullapalli et al., 2005; Sugino et al., 2011). A surgical approach to AMD includes replacing or repairing damaged RPE (via transplantation, translocation, or autologous cell proliferation), immune suppression (for allografts), and reconstruction or replacement of BrM to restore substrate integrity for proper cell attachment (Del Priore, Tezel, & Kaplan, 2006). Experiments by Del Priore and colleagues (2006) designed to make BrM a more hospitable substrate for cell replacement also documented that RPE β1 integrin was an important ligand of BrM extracellular matrix. Further, survival was enhanced by secure attachment of RPE, choice of ICL as the transplant bed, and cleaning following by coating with exogenous extracellular matrix. Zarbin and colleagues found that transplanted RPE cells survive better if BrM is treated with extracts that include collagen XVI, a fibril-associated collagen with interrupted helices, not thought to be secreted from RPE (Sugino et al., 2011). These studies did not test the effect of removing accumulated lipoproteins, so it is remains open whether this treatment could additionally improve survival of RPE transplanted into aged eyes.

SUMMARY

BrM serves essential functions as substrate to the RPE and vessel wall of the outer retina. Its layers and constituent proteins collectively represent a barrier that keeps choroidal vessels at bay, providing a route for water, solutes, and macromolecules that transfer between RPE and choroid while supporting the structural integrity of both. It is unusual among human tissues in accumulating a high content of EC-rich neutral lipid over the life span. A natural history and biochemical model now suggests this lipid is due to apoB lipoprotein secretion by RPE, which may be part of an outer retinal nutrition system. This deposition can account for the impaired outward movement of fluid from RPE, increasing risk for RPE detachments more common in older persons, and impaired macromolecular transport also leading to RPE stress. Oxidation of these lipid deposits in BrM likely initiates an inflammatory process that leads to formation of pathognomonic drusen and BlinD, and choroidal neovascularization in AMD. Model systems exist for testing these hypotheses in a rigorous manner.

REFERENCES

Age-Related Eye Disease Study Research Group. (2001). A randomized, placebo-controlled, clinical trial of high-dose supplementation with vitamins c and e, beta carotene, and zinc for age-related macular degeneration and vision loss: AREDS Report No. 8. *Archives of Ophthalmology, 119*, 1417–1436. PMCID: 1462955. doi:10.1001/archopht.119.10.1417.

Alm, A., & Bill, A. (1973). Ocular and optic nerve blood flow at normal and increased intraocular pressures in monkeys (*Macaca irus*): A study with radioactively labelled microspheres including flow determinations in brain and some other tissues. *Experimental Eye Research, 15*, 15–29.

Beattie, J. R., Pawlak, A. M., Boulton, M. E., Zhang, J., Monnier, V. M., McGarvey, J. J., et al. (2010). Multiplex analysis of age-related protein and lipid modifications in human Bruch's membrane. *FASEB Journal, 24*, 4816–4824. doi:10.1096/fj.10-166090.

Braekevelt, C. R., Smith, S. A., & Smith, B. J. (1998). Fine structure of the retinal pigment epithelium of *Oreochromis niloticus* L. (Cichlidae; Teleostei) in light- and dark-adaptation. *Anatomical Record, 252*, 444–452.

Breslow, J. L. (1996). Mouse models of atherosclerosis. *Science, 272*, 685–688.

Cano, M., Fijalkowski, N., Kondo, N., Dike, S., & Handa, J. (2011). Advanced glycation endproduct changes to Bruch's membrane promotes lipoprotein retention by lipoprotein lipase. *American Journal of Pathology, 179*, 850–859. doi:10.1016/j.ajpath.2011.04.010.

Chang, A. A., Morse, L. S., Handa, J. T., Morales, R. B., Tucker, R., Hjelmeland, L., et al. (1998). Histologic localization of indocyanin green dye in aging primate and human ocular tissues with clinical angiographic correlation. *Ophthalmology, 105*, 1060–1068.

Chen, W., Wang, Z., Zhou, X., Li, B., & Zhang, H. (2011). Choroidal and photoreceptor layer thickness in myopic population. *European Journal of Ophthalmology, 22*, 590–597. doi:10.5301/ejo.5000092.

Crouse, J. R., Parks, J. S., Schey, H. M., & Kahl, F. R. (1985). Studies of low density lipoprotein molecular weight in human beings with coronary artery disease. *Journal of Lipid Research, 26*, 566–574.

Curcio, C. A. (2011). Complementing apolipoprotein secretion by retinal pigment epithelium. *Proceedings of the National Academy of Sciences of the United States of America, 108*, 18569–18570.

Curcio, C. A., & Johnson, M. (2012). Structure, function, and pathology of Bruch's membrane. In S. J. Ryan, A. P. Schachat, C. P. Wilkinson, D. R. Hinton, S. Sadda, & P. Wiedemann (Eds.), *Retina* (5th ed., Vol. 1, pp. 466–481). London: Elsevier.

Curcio, C. A., Johnson, M., Huang, J.-D., & Rudolf, M. (2009). Aging, age-related macular degeneration, and the response-to-retention of apolipoprotein B-containing lipoproteins. *Progress in Retinal and Eye Research, 28*, 393–422.

Curcio, C. A., Johnson, M., Huang, J.-D., & Rudolf, M. (2010). Apolipoprotein B-containing lipoproteins in retinal aging and age-related maculopathy. *Journal of Lipid Research, 51*, 451–467.

Curcio, C. A., Johnson, M., Rudolf, M., & Huang, J.-D. (2011). The oil spill in ageing Bruch's membrane. *British Journal of Ophthalmology, 95*, 1638–1645.

Curcio, C. A., Millican, C. L., Bailey, T., & Kruth, H. S. (2001). Accumulation of cholesterol with age in human Bruch's membrane. *Investigative Ophthalmology & Visual Science, 42*, 265–274.

Curcio, C. A., Sloan, K. R., Kalina, R. E., & Hendrickson, A. E. (1990). Human photoreceptor topography. *Journal of Comparative Neurology, 292*, 497–523.

Del Priore, L. V., Tezel, T. H., & Kaplan, H. J. (2006). Maculoplasty for age-related macular degeneration: Reengineering Bruch's membrane and the human macula. *Progress in Retinal and Eye Research, 25*, 539–562. doi:10.1016/j.preteyeres.2006.08.001.

Duer, M. J., Friscic, T., Proudfoot, D., Reid, D. G., Schoppet, M., Shanahan, C. M., et al. (2008). Mineral surface in calcified plaque is like that of bone: Further evidence for regulated mineralization. *Arteriosclerosis, Thrombosis, and Vascular Biology, 28*, 2030–2034. doi:10.1161/ATVBAHA.108.172387.

Ebrahem, Q., Qi, J. H., Sugimoto, M., Ali, M., Sears, J., Cutler, A., et al. (2011). Increased neovascularization in mice lacking tissue inhibitor of metalloproteinases-3. *Investigative Ophthalmology & Visual Science, 52*, 6117–6123. doi:10.1167/iovs.10-5899.

Ebrahimi, K. B., & Handa, J. T. (2011). Lipids, lipoproteins, and age-related macular degeneration. *Journal of Lipids, 802059*. doi:10.1155/2011/802059.

Francis, P. J., Appukuttan, B., Simmons, E., Landauer, N., Stoddard, J., Hamon, S., et al. (2008). Rhesus monkeys and humans share common susceptibility genes for age-related macular disease. *Human Molecular Genetics, 17*, 2673–2680. doi:10.1093/hmg/ddn167.

Gouras, P., Ivert, L., Neuringer, M., & Mattison, J. A. (2010). Topographic and age-related changes of the retinal epithelium and Bruch's membrane of rhesus monkeys. *Graefe's Archive for Clinical and Experimental Ophthalmology, 248*, 973–984. doi:10.1007/s00417-010-1325-x.

Grossniklaus, H. E., Miskala, P. H., Green, W. R., Bressler, S. B., Hawkins, B. S., Toth, C., et al. (2005). Histopathologic and ultrastructural features of surgically excised subfoveal choroidal neovascular lesions: Submacular surgery trials report no. 7. *Archives of Ophthalmology, 123*, 914–921. doi:10.1001/archopht.123.7.914.

Guigui, B., Leveziel, N., Martinet, V., Massamba, N., Sterkers, M., Coscas, G., et al. (2011). Angiography features of early onset drusen. *British Journal of Ophthalmology, 95*, 238–244. doi:10.1136/bjo.2009.178400.

Gullapalli, V. K., Sugino, I. K., Van Patten, Y., Shah, S., & Zarbin, M. A. (2005). Impaired RPE survival on aged submacular human Bruch's membrane. *Experimental Eye Research, 80*, 235–248.

Hageman, G. S., & Mullins, R. F. (1999). Molecular composition of drusen as related to substructural phenotype. *Molecular Vision, 5,* 28–37.

Huang, J.-D., Presley, J. B., Chimento, M. F., Curcio, C. A., & Johnson, M. (2007). Age-related changes in human macular Bruch's membrane as seen by quick-freeze/deep-etch. *Experimental Eye Research, 85,* 202–218.

Hussain, A. A., Starita, C., Hodgetts, A., & Marshall, J. (2010). Macromolecular diffusion characteristics of ageing human Bruch's membrane: Implications for age-related macular degeneration (AMD). *Experimental Eye Research, 90,* 703–710.

Hussain, A. A., Starita, C., & Marshall, J. (2004). Transport characteristics of ageing human Bruch's membrane: Implications for age-related macular degeneration (AMD). In O. R. Ioseliani (Ed.), *Focus on macular degeneration research.* Hauppauge, NY: Nova Biomedical Books.

Imamura, Y., Noda, S., Hashizume, K., Shinoda, K., Yamaguchi, M., Uchiyama, S., et al. (2006). Drusen, choroidal neovascularization, and retinal pigment epithelium dysfunction in SOD1-deficient mice: A model of age-related macular degeneration. *Proceedings of the National Academy of Sciences of the United States of America, 103,* 11282–11287.

Johnson, L. V., Forest, D. L., Banna, C. D., Radeke, C. M., Maloney, M. A., Hu, J., et al. (2011). Cell culture model that mimics drusen formation and triggers complement activation associated with age-related macular degeneration. *Proceedings of the National Academy of Sciences of the United States of America, 108,* 18277–18282. doi:10.1073/pnas.1109703108.

Jones, A., Kumar, S., Zhang, N., Tong, Z., Yang, J. H., Watt, C., et al. (2011). Increased expression of multifunctional serine protease, HTRA1, in retinal pigment epithelium induces polypoidal choroidal vasculopathy in mice. *Proceedings of the National Academy of Sciences of the United States of America, 108,* 14578–14583. doi:10.1073/pnas.1102853108.

Klein, M. L., & Wilson, D. J. (2011). Clinicopathologic correlation of choroidal and retinal neovascular lesions in age-related macular degeneration. *American Journal of Ophthalmology, 151,* 161–169. doi:10.1016/j.ajo.2010.07.020.

Margolis, R., & Spaide, R. F. (2009). A pilot study of enhanced depth imaging optical coherence tomography of the choroid in normal eyes. *American Journal of Ophthalmology, 147,* 811–815. doi:10.1016/j.ajo.2008.12.008.

Marshall, J., Hussain, A. A., Starita, C., Moore, D. J., & Patmore, A. L. (1998). Aging and Bruch's membrane. In M. F. Marmor & T. J. Wolfensberger (Eds.), *The retinal pigment epithelium: Function and disease* (pp. 669–692). New York: Oxford University Press.

Pauleikhoff, D., Loffert, D., Spital, G., Radermacher, M., Dohrmann, J., Lommatzsch, A., et al. (2002). Pigment epithelial detachment in the elderly: Clinical differentiation, natural course and pathogenetic implications. *Graefe's Archive for Clinical and Experimental Ophthalmology, 240,* 533–538.

Price, D. L., Tanzi, R. E., Borchelt, D. R., & Sisodia, S. S. (1998). Alzheimer's disease: Genetic studies and transgenic models. *Annual Review of Genetics, 32,* 461–493. doi:10.1146/annurev.genet.32.1.461.

Raghow, R., Seyer, J., & Kang, A. (2006). Connective tissues of the subendothelium. In M. A. Creager, V. J. Dzau, & J. Loscalzo (Eds.), *Vascular medicine: A companion to Braunwald's heart disease* (pp. 31–60). Philadelphia: Saunders.

Rodriguez, I. R., & Larrayoz, I. M. (2010). Cholesterol oxidation in the retina: Implications of 7KCh formation in chronic inflammation and age-related macular degeneration. *Journal of Lipid Research, 51,* 2847–2862. doi:10.1194/jlr.R004820.

Rojas, B., Ramirez, A. I., Salazar, J. J., de Hoz, R., Redondo, A., Raposo, R., et al. (2011). Low-dosage statins reduce choroidal damage in hypercholesterolemic rabbits. *Acta Ophthalmologica, 89,* 660–669. doi:10.1111/j.1755-3768.2009.01829.x.

Rudolf, M., Clark, M. E., Chimento, M., Li, C.-M., Medeiros, N. E., & Curcio, C. A. (2008). Prevalence and morphology of druse types in the macula and periphery of eyes with age-related maculopathy. *Investigative Ophthalmology & Visual Science, 49,* 1200–1209.

Sarks, J. P., Sarks, S. H., & Killingsworth, M. C. (1994). Evolution of soft drusen in age-related macular degeneration. *Eye (London, England), 8,* 269–283.

Schwesinger, C., Yee, C., Rohan, R. M., Joussen, A. M., Fernandez, A., Meyer, T. N., et al. (2001). Intrachoroidal neovascularization in transgenic mice overexpressing vascular endothelial growth factor in the retinal pigment epithelium. *American Journal of Pathology, 158,* 1161–1172.

Smith, W., Assink, J., Klein, R., Mitchell, P., Klaver, C. C. W., Klein, B. E. K., et al. (2001). Risk factors for age-related macular degeneration: Pooled findings from three continents. *Ophthalmology, 108,* 697–704.

Spaide, R. F. (2009). Enhanced depth imaging optical coherence tomography of retinal pigment epithelial detachment in age-related macular degeneration. *American Journal of Ophthalmology, 147,* 644–652. doi:10.1016/j.ajo.2008.10.005.

Spaide, R. F., Armstrong, D., & Browne, R. (2003). Continuing medical education review: Choroidal neovascularization in age-related macular degeneration—What is the cause? *Retina (Philadelphia, Pa.), 23,* 595–614.

Stafford, T. J., Anness, S. H., & Fine, B. S. (1984). Spontaneous degenerative maculopathy in the monkey. *Ophthalmology, 91,* 513–521.

Stefansson, E., Geirsdottir, A., & Sigurdsson, H. (2011). Metabolic physiology in age related macular degeneration. *Progress in Retinal and Eye Research, 30,* 72–80. doi:10.1016/j.preteyeres.2010.09.003.

Strauss, O. (2005). The retinal pigment epithelium in visual function. *Physiological Reviews, 85,* 845–881.

Sugino, I. K., Rapista, A., Sun, Q., Wang, J., Nunes, C. F., Cheewatrakoolpong, N., et al. (2011). A method to enhance cell survival on Bruch's membrane in eyes affected by age and age-related macular degeneration. *Investigative Ophthalmology & Visual Science, 52,* 9598–9609. doi:10.1167/iovs.11-8400.

Ulshafer, R. J., Allen, C. B., Nicolaissen, B., Jr., & Rubin, M. L. (1987a). Scanning electron microscopy of human drusen. *Investigative Ophthalmology & Visual Science, 28,* 683–689.

Ulshafer, R. J., Engel, H. M., Dawson, W. W., Allen, C. B., & Kessler, M. J. (1987b). Macular degeneration in a community of rhesus monkeys: Ultrastructural observations. *Retina (Philadelphia, Pa.), 7,* 198–203.

Umeda, S., Suzuki, M. T., Okamoto, H., Ono, F., Mizota, A., Terao, K., et al. (2005). Molecular composition of drusen and possible involvement of anti-retinal autoimmunity in two different forms of macular degeneration in cynomolgus monkey (*Macaca fascicularis*). *FASEB Journal, 19,* 1683–1685.

Weismann, D., Hartvigsen, K., Lauer, N., Bennett, K. L., Scholl, H. P., Charbel Issa, P., et al. (2011). Complement factor H binds malondialdehyde epitopes and protects from oxidative stress. *Nature, 478*, 76–81. doi:10.1038/nature10449.

Yehoshua, Z., Wang, F., Rosenfeld, P. J., Penha, F. M., Feuer, W. J., & Gregori, G. (2011). Natural history of drusen morphology in age-related macular degeneration using spectral domain optical coherence tomography. *Ophthalmology, 118*, 2434–2441. doi:10.1016/j.ophtha.2011.05.008.

Yin, L., Shi, Y., Liu, X., Zhang, H., Gong, Y., Gu, Q., et al. (2012). A rat model for studying the biological effects of circulating LDL in the choriocapillaris-BrM-RPE complex. *American Journal of Pathology, 180*, 541–549. doi:10.1016/j.ajpath.2011.10.015.

Zayas-Santiago, A., Marmorstein, A. D., & Marmorstein, L. Y. (2011). Relationship of Stokes radius to the rate of diffusion across Bruch's membrane. *Investigative Ophthalmology & Visual Science, 52*, 4907–4913. doi:10.1167/iovs. 10-6595.

107 Macular Pigment: Characteristics and Role in the Older Eye

IAN J. MURRAY

PREAMBLE

There is a conspicuous yellow-pigmented spot concentrated in the central 5 to 6 mm of the retina of humans and higher primates called the macula lutea or macular pigment (MP). Under normal ophthalmoscopic viewing, MP is seen as a dark region illustrated in figure 107.1. The presence and possible role of MP has preoccupied scientists and ophthalmic researchers for more than 200 years. It was identified as being derived from the naturally occurring carotenoids by George Wald (1945) and described as being composed of lutein (L) and zeaxanthin (Z) by Bone, Landrum, and Tarsis (1985). MP cannot be synthesized de novo in animals; it is obtained from dark green leafy vegetables and is absorbed directly from the bloodstream. It is therefore a sensitive index of the health of the diet and of the retina. What is of particular fascination is that its deposition is far from passive; rather, it is controlled by active local retinal mechanisms. Hence, in those with an adequate diet, it occurs in the central retina at remarkably high concentrations.

MP has been described as a sun blocker for the retina because it absorbs light in the blue region of the visible spectrum, corresponding to the wavelengths that are most effective at damaging photoreceptors. This prompted the suggestion that it may serve to protect the photoreceptors which are extremely sensitive to damage from short-wavelength light. However, interest in MP was greatly increased in the early 1990s following reports that it also has strong antioxidant properties. At around this time there was an increase in the awareness of age-related macular degeneration (AMD), a debilitating disease of older eyes in which, due to photoxidative effects, the central retina is damaged, leading to profound loss of vision. The discovery of its antioxidant properties raised the possibility of a second role for MP; the repair of tissue damaged by oxidative stress in age-related neurodegenerative disease. The incidence of AMD has been increasing steadily in the last 2 decades, and it is now by far the major cause of blindness in older

people in the developed world. Apart from being consistent with the biology of carotenoids in the natural world, the neuroprotective hypothesis is supported by substantial epidemiological evidence of an inverse relationship between MP and the risk for AMD. There is, in addition, more direct evidence of the benefits of MP in the form of placebo-controlled intervention trials in which participants' levels of MP are enhanced by absorbing food supplements containing L and Z.

Although by no means universal, the view that MP may act to mitigate the effects of AMD is widely acknowledged in the ophthalmological fraternity. AMD affects mainly central vision and therefore leads to severe visual impairment having a devastating effect on everyday life. For demographic and environmental reasons, it will continue to increase in the coming decades. If augmenting MP is widely accepted as part of a management strategy for AMD, this will be a prime example of how careful, detailed applied visual neuroscience can have an impact on the health and well-being of the community at large.

This chapter provides a contemporary overview of the role MP might play in maintaining the healthiness of the macular region in the older human eye. Some aspects of the biology, measurement, and putative ocular benefits of MP are discussed.

THE MACULA

The macula is a specialized region of the retina of humans and higher primates. It is approximately 5–6 mm (15–18°) in diameter. See figure 107.1. At its center lies the fovea, where the cone photoreceptors reach maximum density, ~ 200/mm^2, declining rapidly within the central 3°. The fovea is therefore the site of maximum visual acuity (VA). Rods are absent in the central 1.25° but rapidly increase their density to reach a peak (~140,000/mm^2), the celebrated rod ring (Packer, Hendrickson, & Curcio, 1989) at around 15° eccentricity. Individuals vary, but the typical foveola (central 1°25'), where cones are at maximum density,

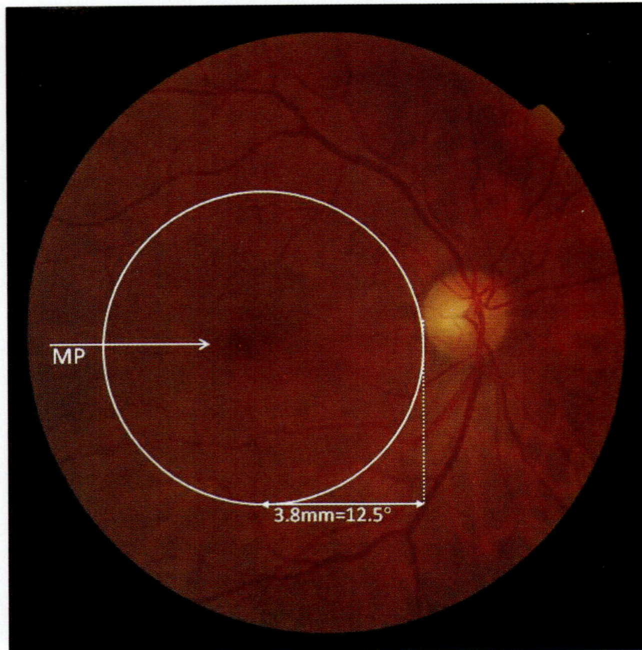

FIGURE 107.1 The human fundus. The white circle denotes the macula. The macular pigment (MP) is the darkened region at the center of the macula. (Note these dimensions assume a simple model eye. The transposition between retinal distance R and visual angle V is given by tan V = R/η, where η is the distance from nodal point to cornea, here 17 mm.)

contains between 7,000 and 10,000 cones, and there are of the order of 90,000 in the central 5.5°, emphasizing the curious fact that our central vision depends on only a small percentage of the total cone count of around 5 million (Curcio et al., 1990). The ratio of rods to cones is around 1:1 at a surprisingly central location of 0.4–0.5 mm. Individuals vary, but this means that rods may outnumber cones in the macula by about 12:1. As a result of the large numbers of photoreceptors the macula is metabolically highly active. The outer segments of photoreceptors are phagocytosed by the retinal pigment epithelium (RPE) and this results in the generation of so-called reactive oxygen intermediaries such as hydrogen peroxide and singlet oxygen (Beatty et al., 1999). The eye is designed to collect light which is accurately focused at the fovea, and there is therefore a powerful mix of ingredients for phototoxicity. The macula is highly sensitive to anoxia, partly because photoreceptors, and particularly rods, are very energy demanding because of their phototransduction properties.

After the age of 55 many individuals develop accumulations of abnormal material between Bruch's membrane and the RPE called drusen. At the same time the RPE becomes loaded with lipofuscin granules which arise from incomplete digestion of rod outer segments (Nguyen-Legros & Hicks, 2000). Drusen are composed of highly oxidized lipids (Adler et al., 1999), indicating they are formed from RPE cells that are no longer able to support phagocytosis of photoreceptor outer segments. Hence there is a direct link between the very high metabolic demands of vast numbers of photoreceptors, their topographical distribution, and the accumulation of drusen. The formation of drusen is a sign that the renewal/disposal cycle of photoreceptor outer segments is compromised. There is evidence that they cause oxidative stress because they act as an obstructive barrier between the choroid and the RPE, hindering the diffusion of oxygen from the choroid. Hence aging changes enter a cause-and-effect cycle, the formation of drusen promoting thickening and subsequent reduced permeability of Bruch's membrane.

MP accumulates in the axons of the photoreceptors. It was first described in compelling detail by Max Schultze in 1866 (translated in the Appendix of Werner, Donnelly, & Kliegl, 1987). Schultze describes removing three small segments of the fovea. He notes that these did not change on exposure to colored lights and that the colored substance, located in the inner retinal layers but not in cones, was of an intense lemon–orange color and was not soluble in water. Although minute in quantity, MP reaches remarkably high concentrations in the human retina, between 0.3 and 13 mM, which is the highest concentration of any carotenoid in the body (Landrum & Bone, 2001) and of the order of 10,000 times higher than in serum. It is composed primarily of three isomeric carotenoids, L, Z, and meso zeaxanthin (meso Z); these are described in the next section.

CAROTENOIDS

Carotenoids are defined by a common $C_{40}H_{56}$ backbone structure. Around 600 have been identified in nature although humans can ingest only 40 (Ahmed et al., 2005). These divide into the more numerous xanthophylls, which have an oxygen-containing functional group, and carotenes that do not contain oxygen. Carotenoids represent a remarkably diverse class of natural pigments, accounting for the red, orange, and yellow in higher plants such as daffodils and for the striking coloration in insects, fish, and birds. Familiar examples are the pigments in egg yolk, the plumage of flamingos, and the coloration of tomatoes, which contain lycopene.

Carotenoids are synthesized de novo by plants and are obtained only through the diet in animals. Universally, their role is coloration, acting as a biological signaling mechanism, and in photoprotection, serving to

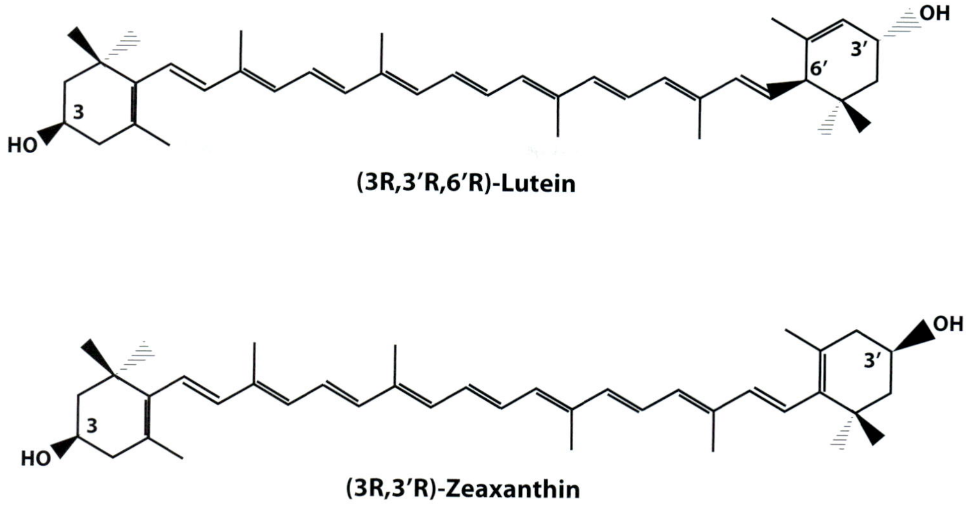

(3R,3′R,6′R)-Lutein

(3R,3′R)-Zeaxanthin

FIGURE 107.2   The chemical structure of the main retinal carotenoids.

absorb blue light in order to protect chlorophyll from harmful shortwave radiation. Hence, all carotenoids absorb shortwave visible light, are insoluble in water, and are bleached on exposure to oxygen or light. Many of the carotenoids and their isomers are found in human serum and are concentrated in the ovaries, testes, liver, brain, and eyes (Ahmed et al., 2005).

In the human and higher primate eye, the xanthophylls, L, Z, and meso Z, are concentrated in the central, macular region. These are now commonly referred to as the retinal carotenoids. All are the same molecular weight and are stereoisomers of each other. Their chemical structures, identified by Bone, Landrum, and Tarsis (1985), are illustrated in figure 107.2.

There can be little doubt that, as with all carotenoids, the biological function of L, Z, and meso Z is intimately linked to their structure. It is no coincidence that these compounds, rather than, say, β-carotene, are present in the macula. The complexities of the stereochemistry, structure, and physiology are described in a review by Landrum and Bone (2001). They discuss how the slight differences in chemical structure may account for different functional properties. While Z occupies a site that is perpendicular to the membrane surface, L may be deposited in a nonorthogonal orientation. These chemical properties almost certainly account for the location of MP deposition in the retinal layers and the fact that L and Z have different spatial distributions. Landrum and Bone (2001) also discuss the possible biological and chemical pathways for the photoprotective function of MP and in particular the different protective properties of L and Z based on their chemistry. Unlike L and Z, meso Z is not normally obtained from the diet, and there is strong evidence that there is a

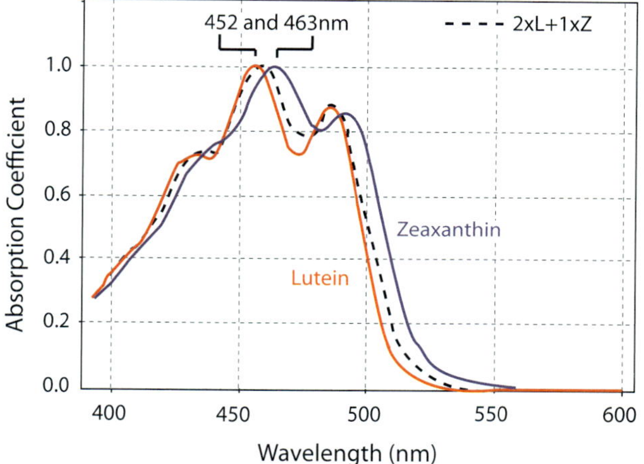

FIGURE 107.3   Absorption spectra of lutein and zeaxanthin in olive oil. The dashed line is a mix of the two pigments (Bernstein et al., 2009).

local specific biochemical pathway which is responsible for conversion of L into meso Z.

ABSORPTION SPECTRUM OF MP

All carotenoids have a characteristic absorption spectrum with a central maximum peaking at around 450 nm and a distinct shoulder at shorter wavelengths. They have evolved to screen out higher-energy shortwave light and allow the absorption of other wavelengths vital for photosynthesis. L and Z represent slight variations on this theme. They are different in color, with zeaxanthin having a rosy appearance consistent with having a slightly broader absorption spectrum with a shift of around 6 nm to 9 nm toward longer wavelengths,

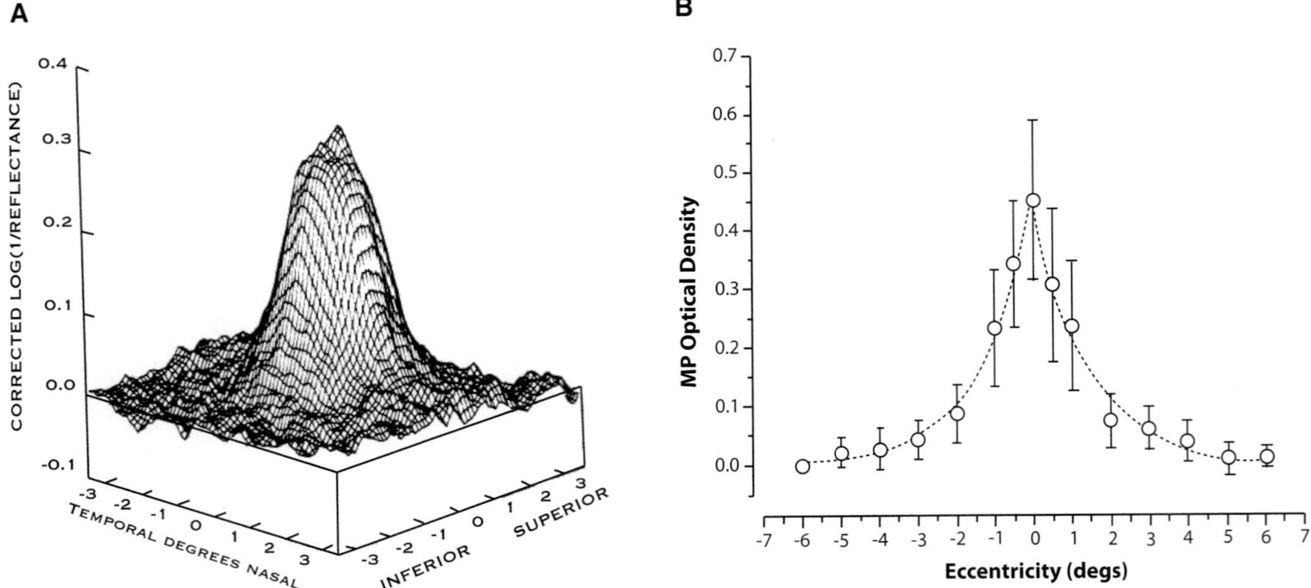

FIGURE 107.4   (A) Macular pigment (MP) optical density topography of the seven observers in Kilbride et al. (1989). (B) Spatial distribution of MP for seven observers (van der Veen et al., 2009a).

depending on the solvent. The absorption spectra of L and Z illustrated in figure 107.3 is for olive oil. Note these characteristics are slightly different in ethanol.

As discussed by Davies and Morland (2004), there will be inevitable individual differences in the absorption of MP at longer wavelengths. This is partly due to the chemical environment of the L and Z and partly because of their likely idiosyncratic distribution in individual retinas, linked to such factors as local topography and diet. Note that it is now known that L and Z are obtained from dark green leafy vegetables and different orange and yellow fruits and vegetables (Sommerburg et al., 1998). Eggs are a particularly effective source, because of their high fat content. Meso Z is nondietary and is thought to be synthesized in the retina from L (Bone et al., 1997) as stated above.

MP RETINAL DISTRIBUTION

It has been known for many years that the MP is localized mainly in the outer and inner plexiform layers (Snodderly, Auran, & Delori, 1984). It peaks at the center of the macular and declines approximately exponentially (Hammond et al., 1997). This characteristic distribution was described in vivo in a seminal study by Kilbride et al. (1989). They used a television-based imaging fundus reflectometer to obtain digital images at a range of wavelengths between 462 and 697 nm to derive the data illustrated in figure 107.4A. This characteristic distribution was confirmed by many others.

For example, van der Veen et al. (2009b) exploited the phenomenon to compare two techniques for assessing MP (discussed below). Their data, based on heterochromatic flicker photometry (HFP), are illustrated in figure 107.4B. Note that the standard error increases toward the center, illustrating the wide interindividual variation, which mainly reflects individual differences in MP depending on factors such as diet, foveal topography, and the absolute numbers of photoreceptor axons as discussed in Snodderly, Auran, and Delori (1984). Beyond around 5° there are only small traces of MP, so there is very little difference between individuals, and this is reflected in the small between-subjects errors in figure 107.4B.

To a large extent the distribution of MP is unimodal and circularly symmetrical, but it has been shown that many individuals show a more complex change with eccentricity (Robson et al., 2003). This was described as a distinct ring-like pattern by Delori et al. (2006) and Berendschot and van Norren (2006). Intriguingly, the former study found women were more likely than men to exhibit this phenomenon. Bernstein et al. (2009) have described different categories of MP distribution based on autofluorescence imaging, which, as described below, also provides a topographical plot of MP distribution across the central macula. In general the nonuniform spatial distribution is not seen in HFP-based methods.

Of particular interest is the different spatial distribution of L and Z. This has been known for some time

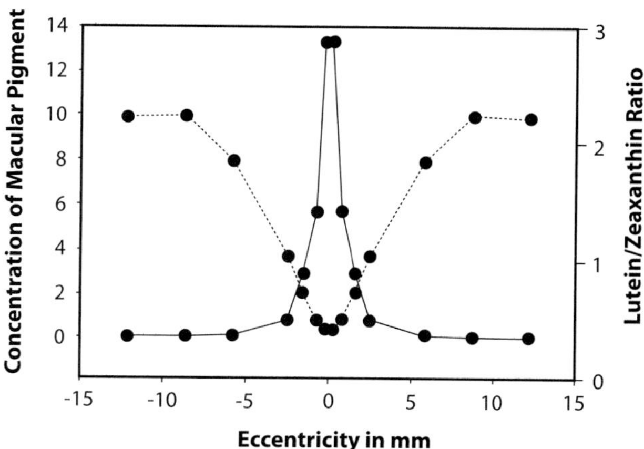

FIGURE 107.5   The relative distribution of lutein and zeaxanthin based on high-pressure liquid chromatography (Landrum & Bone, 2001).

from high-pressure liquid chromatography (HPLC) (Bone et al., 1988, 2001). It seems that Z is predominant in the central retinal tissue and L in the more peripheral structures, as shown for HPLC studies in figure 107.5. Interestingly, this distribution matches that of the rods and cones, but it is unlikely that this is significant because of the existence of L and Z in the largely rod-free fovea (Bone et al., 1988).

The differential distributions of L and Z was confirmed in a novel, noninvasive study by van de Kraats et al. (2008). They used an optical technique (described below) to measure the reflectance of a 1° spot projected on to the macula over a range of 400 nm to 800 nm. The levels of L and Z were extracted using a model based on their different extinction spectra. They tested two groups, sampling at 5 eccentricities. One group supplemented their diet with a Z-only capsule for 6 months. The two subjects who took the Z-only supplement showed a decrease in L and a substantial increase in Z. In all observers, they found the spatial profiles of L and Z to match those obtained with the in vitro methods illustrated in figure 107.5.

MP MEASUREMENT

The optical density of MP is a measure of the attenuation of blue light incident on the retina. Consistent with Beer's Law, it is linearly related to the concentration integrated over the region sampled. Typical values range between 0 and 1.0 OD units. Laboratory methods such HPLC cannot be used on living eyes, so the MP optical density (MPOD) serves as an indirect estimate of the amount of MP in the retinal tissue.

The most frequently used, noninvasive method is based on HFP. This psychophysical procedure has been known since the early 1900s for equalizing the luminance of two lights of different wavelengths (Lennie, Pokorny, & Smith, 1993). When the lights alternate at around 15–20 Hz and are of different luminance, they appear to flicker. In HFP, the intensity of one of the lights is adjusted until the flicker is eliminated, and the lights are then formally defined as of equal luminance. The intensity ratio of the two lights provides an index of the relative visibility of the lights. The technique provides the basis for the field of photometry in that it is used to determine the luminous efficiency of the eye, known as the Vλ function (Wyszecki & Stiles, 1982).

HFP has been extensively used in a Maxwellian-view setup (Pease, Adams, & Nuccio, 1987; Werner, Donnelly, & Kliegl, 1987) to measure MPOD. Here, the two lights are typically blue (around 460 nm) and green (around 540 nm), and the observer adjusts their intensity ratio until flicker is eliminated while viewing a 1° target presented at the central fovea where MP is maximal. A relatively large amount of blue is required to reach the isoluminant point because of the presence of the blue-absorbing MP. The intensity of the blue light Bc in the isoluminant ratio is then a measure of its perceived luminance. The blue–green flickering target is then presented in the peripheral retina, where MP is assumed to be absent, and the task of determining the minimum flicker point is repeated. The intensity of the blue light Bp in the new peripheral blue–green ratio then serves as a reference to be compared with that obtained centrally in the equation

$$\text{MPOD} = \log_{10}[Bc/Bp]. \qquad (107.1)$$

Variations in the technology were described by Hammond and Fuld (1992), and Wooten, Hammond, Land, and Snodderly (1999) showed that free-view and Maxwellian optics methods are compatible. The introduction of LEDs, which allow accurate intensity control, and virtually instantaneous switching, were essential in the development of free-view systems. Beatty et al. (2000) showed that the absorption spectrum of the MP can be estimated in a free-view optical system. They used fiber optics to superimpose the output of a monochromator and a cluster of blue LEDs to form a 15-Hz flickering target. The observer then minimized flicker as described above. This approach of minimizing flicker was used in many other instruments (e.g., Bone & Landrum, 2004; Hammond & Wooten, 2005; Mellerio et al., 2002; Snodderly, Auran, & Delori, 2004; van de Kraats et al., 2006).

One of the limitations of HFP is that it is rather a demanding task to perform, especially for older observers. This problem was addressed by van der Veen et al. (2009a), who used an alternative approach to obtain the isoluminant point. In their method, the MPS9000, the temporal frequency of the two lights is gradually ramped down (6Hz/s) from above the critical fusion frequency. The task for observers is to detect, rather than eliminate, flicker. A series of different blue–green ratios are tested until a v-shaped curve is obtained, and the minimum of this function then provides the isoluminant point for the two lights. The test is repeated in the peripheral field, and the log ratio of the luminance is calculated from equation 107.1. The technique is regarded as easier to perform than the conventional method. Another advantage is that it takes account of the observers' sensitivity to flicker. In addition, the peripheral isoluminant setting can be estimated from the observer's age. This strategy, described for 5,616 observers in Makridaki, Carden, and Murray (2009), means the measurement can be completed in less than 5 minutes even with older observers. The issue of observer flicker sensitivity is important. It was recognized by Snodderly, Auran, and Delori (2004), who optimized the flicker frequency for each observer.

The veracity of the data derived from flicker techniques is frequently questioned, and one complication is almost certainly the individual differences in spatial profiles of MP and the fact that different observers use a region at, or close to, the edge of the stimulus to set their flicker threshold. Note also that, as reported by Robson et al. (2003), the total amount of MP present is not correlated with the peak measurement according to HFP. In order to address this issue quantitatively, van der Veen et al. (2009b) conducted a comparison of fundus reflectometry, using the macular pigment reflectometer (MPR; described below) and a flicker method (MPS 9000), measuring 19 healthy individuals at eccentricities of 0, 0.5, 1, 2, 3, 4, 5, 6, and 7°. They used a modeling technique based on the spatial profiles of L and Z derived from the MPR. They found a correction factor of 0.05 OD units at all eccentricities was required to take account of the assumption of nonzero values in the peripheral retina. A second correction factor, dependent on the steepness of the L and Z profiles, was also derived. The data from the two techniques are virtually identical when these correcting factors are used. It turned out that, as has been speculated previously, observers, on average, use a point close to the edge of the (1°) stimulus when setting flicker thresholds in the HFP method. Crucially, however, different observers are inclined to use slightly different locations. This point is discussed further below.

In motion photometry, two lights, one absorbed by MP and the other not, are equalized for luminance using the well-established techniques of minimum motion photometry and apparent motion photometry

(Moreland, 2004). Typically a moving square wave grating composed of 450 nm and 580 nm is presented to the observer, who is required to either minimize or reverse the direction of motion by adjusting the intensity of the two lights. The technique yields MP maps of OD units because the test field is typically divided in to annular fields at eccentricities 0.8 to 11. It was found to be highly compatible with other techniques as reported by Robson et al. (2003).

The visual evoked potential (VEP) is extremely sensitive to luminance differences between lights of different wavelengths. In a unique approach to MP measurement Robson and Parry (2008) used a steady state contrast reversing blue–green grating to identify a null point which represents isoluminance. By presenting the stimulus in several concentric annuli, they were able to estimate the retinal distribution of MP; this was compatible with HFP and minimum motion photometry. In a more recent method, also based on VEPs, Parry and Robson (2012) have devised a so-called pan-isoluminant blue–yellow grating whose luminance ratio varies with eccentricity, effectively allowing for individual variation in the spatial profile of the MP. They present psychophysical and electrophysiological evidence that the Parry–Robson grating is highly selective for the S-cone mediated visual pathway and minimizes luminance intrusion, even when large (18°) stimuli are employed.

Fundus reflectometry has been used for many years to obtain MP measurements. It was recognized in the early developments of the technique that oxygenated hemoglobin, melanin, and lens pigments could be confounded with the reflectance spectrum of the MP and a model of their known absorption spectra should be incorporated into estimates of MP (Van Norren & Tiemeijer, 1986). By subtracting two digitized images, one at 558 nm and the other at 462 nm, Kilbride et al. (1989) were able to derive the "double density" of the MP comparing reflectances from bleached foveal and peripheral retinal regions. One problem, common to all techniques involving reflectance spectra, is that the two-way absorption spectrum obtained for MP is influenced by light scatter which has not passed twice through the reflecting structures of the retina.

In contemporary methods a CCD is combined with a scanning laser ophthalmoscope (SLO), which minimizes light scatter by virtue of its confocal optics (e.g., Berendschot & van Norren, 2004; Wustemeyer et al., 2003), and blue (488 nm) and green (514 nm) images are obtained. These are subtracted after careful alignment and logarithmic transformation. In some implementations, an average of the peripheral values is taken as a reference. Light incident at the fovea will traverse the MP twice. As described above it is also absorbed and reflected by other retinal structures (van de Kraats et al., 2006). Unlike HFP, this method provides both the optical density and topographic distribution of MP. A crucial parameter is the area of the retina sampled. Of course it is assumed that the value obtained represents the integrated MPOD over this area, but the steepness of the spatial profile will influence this. It is for this reason that maps are invariably generated and why different authors use different sampling areas. One minor disadvantage, if the technique is to be used in large-scale investigations, is that observers' pupils must be dilated. SLO-based fundus reflectometry is regarded by some as the nearest available technique to a gold standard for MP measurement.

A more recent development, which avoids the complications of imaging and is based on spectral reflectometry and an optical model of the fundus layers, has been described by van de Kraats et al. (2006). The device, called a macular pigment reflectometer (MPR) is small and compact and uses a CCD device to obtain a reflectance spectrum at multiple wavelengths. The advantages are that it is objective, takes only a few minutes to collect the data, and, apart from good fixation, does not require any action on the part of the observer. Crucially, the method can be used without dilating the pupil. It has been used in a comparison with a flicker technique as described above (van Der Veen et al., 2009). By taking account of the assumptions of the two methods, the data are highly compatible. It would appear that provided the issue of sampling region is taken in to account, reflectance and HFP methods have more in common than at first thought. As described above, it can be used to separate L and Z and thereby measure their optical density. This is the only known method of obtaining this information in vivo.

Autofluorescence imaging can also be used to measure MP. First described by Delori (1994), and subsequently in imaging form (Delori, 2004), this method determines the MPOD by the fact that it attenuates the fluorescence of lipofuscin in the RPE. It is possible to excite lipofuscin emission at two different wavelengths, one of which (e.g., 488-nm Argon laser) is absorbed by MP and the other (e.g., Yag laser 532 nm) is not. Incident light of both wavelengths causes RPE lipofuscin to fluoresce, and both of these are imaged. The MP attenuates the shorter wavelength image, and this can be transformed into a topographical map. Problems associated with the intrinsic fluorescence of the crystalline lens can be avoided by restricting the fluorescence wavelengths to longer wavelengths (Sharifzadeh, Bernstein, & Gellermann, 2006). The technique has been combined with fundus reflectometry in the same SLO as described in Berendschot and van Norren (2005), for example.

The technique of resonance Raman spectroscopy measures the excitation of bond vibrations in molecules in MP which is proportional to the concentration of carotenoids in the retinal tissue (Bernstein et al., 1998). The observer fixates a 1-mm argon laser spot which resonates with the MP. The Raman scattered light provides a unique signature of a particular molecule in that a characteristic shift in scattered light occurs. Carotenoids exhibit a strong Raman signal because the incident light overlaps with their absorption spectrum and therefore produce three characteristic Raman peaks. MPOD is calculated after subtraction of the background fluorescence. The intensity of the Raman scatter has been shown to be linearly related to the total carotenoid content using HPLC-based measurement of excised retinal tissue samples. Its main advantage is that, unlike all the other methods, it is highly chemically specific to retinal carotenoids.

FACTORS AFFECTING MP DEPOSITION

Typical Values

In general, the mean value for MPOD measured centrally varies markedly around a midrange of 0.3 OD units. According to a consensus panel (Bernstein et al., 2009), a value below 0.2 should be regarded as low and a value greater than 0.5 as high. This of course ignores such factors as iris pigmentation and ethnicity which covary and, as described below, systematically influence MPOD. Data collected from different studies in the United States from a total of 846 eyes show that 46% had a value of 0.2 and 16%, a value less than 0.1 (Wooten & Hammond, 2002). This is similar to a population of 828 in Ireland, the mean of which was 0.3 (Nolan et al., 2007b). Perhaps one of the largest samples was that reported by van der Veen et al. (2009a), who found a mean of 0.33 ± 0.18 from a population of 5,581 taken from 48 optometric practices across the Midwest of the United States. Their data are illustrated in figure 107.6, and the details are tabulated in their paper (see below). Overall, there was a slight increase in MP with age in this population, but it is clear many other factors apart from age are at work. There is overwhelming evidence that age is a secondary factor as reviewed in Davies and Morland (2004). This was reinforced by Berendschot and van Norren (2005), who conducted a rigorous study using different techniques without finding an age effect. In van der Veen et al. (2009a), there were also male–female differences, with younger males having higher MP than younger females, whereas in the older

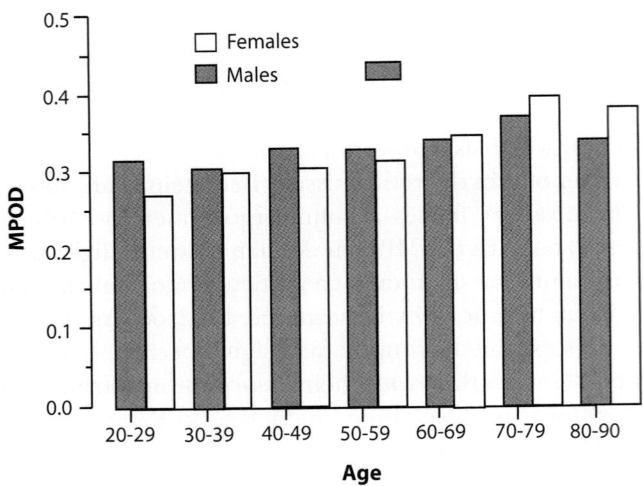

FIGURE 107.6   Macular pigment optical density (MPOD) distribution in 5,881 observers (van der Veen et al., 2009a).

groups this pattern was reversed. For confidentiality reasons, only the gender and age are known in this population. As discussed by van der Veen et al. (2009a) there is a strong suggestion that the main factors influencing MP in this particular population are diet and lifestyle.

GENDER   The discussion on whether females have slightly less MP than males continues but is closely linked to the role of adiposity and MP deposition. In one study (Hammond et al., 1996a), after correcting for age and caloric intake, males were found to have 38% greater MP than females. A study performed in the Netherlands using fundus reflectometry on 199 men and 236 women did not find any difference (Berendschot et al., 2002). Similarly Ciulla et al. (2001) did not detect any systematic differences between men and women in 280 participants in a study based in the U.S. Midwest. In their work, dietary effects were the strongest correlate of MPOD. However other surveys have suggested that adipose tissue acts as a sink for carotenoids, thereby compromising their deposition in the retina. Johnson et al. (2000) reported a significant inverse relationship between adipose tissue concentration and MP optical density in females. Broekmans et al. (2002) showed that MP was correlated with lutein adipose tissue in men but not women. It would seem that when dietary factors are removed, the mechanisms controlling the deposition of L and Z in adipose tissues are different in men and women.

IRIS PIGMENTATION   A further not well understood factor is the influence of iris color. In general it has been known for many years (Hammond, Fuld, & Snodderly, 1996b) that individuals with darker irides have higher MP than those with light irides. They found that when observers' eyes were categorized in to blue/gray, green/hazel or brown/black, there was a strong correlation between MPOD and eye color, despite similarities in diet and blood concentrations of carotenoids. The link with eye color is not technique specific. It was also seen in 118 normal observers who were divided according to ethnicity in to either white non-Hispanics or African. Autofluorescence maps were used in this study (Wolf-Schnurrbusch et al., 2007) which revealed a difference of 0.23 OD units between the two groups, which was highly statistically significant. There are no satisfactory explanations for this well-established phenomenon. It has been speculated that a shared mechanism for depositing both melanin and MP may have evolved, or it may be linked to the intensity of light to which eyes are exposed. The retinas of highly pigmented eyes are exposed to much less light than pale eyes, and it may be that MP is more rapidly depleted due to an increase in oxidative stress in those eyes.

SMOKING   One well-established factor which is universally accepted as having a deleterious affect on MP accumulation is tobacco smoking. This was reported by Hammond, Wooten, and Snodderly (1996c), who found that mean MP was 0.16 OD units lower in 34 smokers compared with 34 controls. These effects were highly significant, and factors such as age, weight, iris color, and dietary intake of carotenoids were controlled. Similar findings have been reported; Delori et al. (2001) used autofluorescence mapping to compare 27 smokers with 102 nonsmokers and reported values of 0.40 ± 0.15 and 0.49 ± 0.16, respectively. However, when the two groups were compared with fundus reflectance, there was no difference between them. Other studies have found a link between smoking and reduced MP (e.g., Mellerio et al., 2002; Nolan et al., 2007a). Taken with the observation that smoking is associated with significantly reduced serum carotenoid levels, and, as reported by Klein et al. (1993), is strongly associated with macular disease, the opinion that tobacco has adverse effects on eye health is more or less universal.

REPRODUCIBILITY AND ACCURACY   Measurements of MPOD are somewhat variable, regardless of whether the technique relies on a psychophysical method or reflectance or absorption spectra. All techniques are subject to uncontrollable variables. In HFP there is the issue of exactly which part of the (usually 1°) target the observer uses to make their setting. Some have argued that the luminance contrast edge between the stimulus and its background is used, the so-called edge hypothesis. However, this is a gross simplification. As pointed out in van der Veen et al. (2009b) the contrast of this edge will vary with the MPOD and its spatial distribution. They showed, using a modeling technique, and by comparing an HFP method (target size 1°) with optical reflectance (integration area 1°), that for a population of 19 observers, the fixation point in the HFP method averaged at 0.4° for 0° fixation and 0.9° for 0.5° fixation. The paper illustrates well the fact that the two techniques yield almost the same values despite their different underlying assumptions and methodology, which are explored in detail. It is clear that in all the psychophysical methods, individuals having slightly different criteria for flicker threshold, different fixation strategies, and perhaps most significant of all, different MP spatial profiles, there are some variables that are difficult to control. The net result is that unless the operator is familiar with the idiosyncrasies of a particular instrument/technique, measures of repeatability will be rather varied across studies as discussed in Howells, Eperjesi, and Bartlett (2011), who summarized the findings of 26 surveys.

It is important to recognize that so-called objective techniques, whereby observer responses (apart from steady fixation) are not required, cannot be regarded as a panacea of MP measurement. All in vivo techniques are also vulnerable to a range of only partly controllable variables. One of the main complications in fundus reflectometry, for example, is light scatter within the retina, because an unknown proportion of the light detected may be scattered at intraretinal surfaces or reflected specularly from the internal limiting membrane. Light from the measuring beam may also be backscattered from the cornea and/or lens, or forward scattered from the surface of the crystalline lens.

These factors are evident in the repeatability data; Berendschot et al. (2000) reported within subject variability of 10% for MP maps obtained with a scanning laser ophthalmoscope, and Delori et al. (2001) discuss the variability in three techniques, autofluorescence, reflectance, and HFP. They also emphasize the role of the shape of the absorbance profile of the MP (see their figure 8) and the impact this might have on the different estimates of MPOD. They found an absolute difference between separate measurements of 0.042 OD units and 0.039 OD units for fluorometry and reflectometry, respectively. The technique produced test–retest differences of 9% of the mean density estimates for the autofluorescence and 22% for reflectance, that is, smaller variances for autofluorescence. Wustemeyer et al. (2003) found the coefficient of variation to be 6.2% using an SLO to obtain reflectance maps. In common with others they reported substantially higher values for autofluorescence than for reflectance maps. They tested 109 observers, and found a highly significant decline with age for the reflectance technique but no effect of age in the autofluorescence maps.

Berendschot and van Norren (2005) provide an excellent perspective of the different techniques in that they used two forms of multispectral fundus reflectometry, an SLO to generate reflectance and autofluorescence maps and a flicker photometry technique; in all, seven estimates of MPOD were obtained for each of 53 observers. Each technique has its own assumptions and limitations in terms of reproducibility, ease of use, and absolute accuracy. When these methods are compared, for the same population, the data may have high degree of association, but they rarely, if ever, produce the same absolute values.

In summary it is worth emphasizing that given the many different techniques, the effects of age and gender are probably second order. It seems likely that diet can override these factors in most individuals. There is a consensus regarding iris color however. It is likely that in the future, when assessing whether or not an individual's MP is normal or low, there should be an allowance for the color of their eyes, so that a value of, say, 0.2 may be regarded as acceptable in a pale-colored eye but lower than average in a dark eye. Similarly, there is agreement on smoking being a factor; the only difference in opinion appears to be size of the effect. This again will be influenced by other confounders such as eye color and diet. The final point to make here is that when judged overall, the factors tending toward low MP are also those that are risk factors for AMD, and this is discussed below.

AGE-RELATED MACULAR DEGENERATION

As discussed above the macular region is particularly susceptible to damage in later years. There can be no doubt that AMD represents a major clinical problem in ophthalmology (see Ambati et al., 2003, and Neelam et al., 2009, for reviews). It is predicted that the number of cases will increase almost twofold in the United States between 2010 and 2050 (Rein et al., 2006). AMD is by far the leading reason for entry to the blind registers in the United States, Europe, and Australasia. Approximately 23% of the Caucasian populations over 75 in the developed world are thought to be affected by early stage disease (Klein, Klein, & Linton, 1992a). As the numbers in this age group are set to rise in the next 2 decades faster than any others, the ophthalmic community can expect a dramatic increase in numbers of apparently intractable cases of severe vision loss. Although AMD will pose a significant public health problem in the near future, a preventative strategy based on modifying lifestyle, including enhancing MP, is quite realistic as discussed in Bernstein et al. (2009). Much depends on the extent to which awareness of the main issues can be raised in at-risk individuals before the onset of symptoms. Hence early detection of signs and genetic screening might form the cornerstone of management in the future.

The early stage of the disease is referred to as age-related maculopathy (ARM) (Bird et al., 1995). Patients in this category have normal or close to normal VA. The diagnosis is based on the appearance of large (>125 μm), confluent drusen and hypo- or hyperpigmentation on the fundus. These findings are pathognomic for the subsequent development of geographic atrophy and extensive loss of vision. Typically, in the early stages, patients are asymptomatic or may report mild visual disturbance such as occasional distortion of straight edges. Subsequently there is a steady loss of central vision over a period of 2 to 10 years. As they have not been studied in detail (but see Hogg & Chakravarthy, 2006, for an excellent review), other manifestations of the vision impairment are not well understood, but there is histological (Curcio, Medeiros, & Millican, 1996) and psychophysical (Owsley et al., 2000) evidence that rod-mediated vision is particularly vulnerable in the early stages of the disease. More recent studies, involving robust psychophysical procedures, indicate that dynamic aspects of photopic and scotopic sensitivity recovery following a bleach are sensitive assays of the visual pathology in early stage AMD (Dimitrov et al., 2011).

MP LINK WITH AMD; EPIDEMIOLOGICAL AND OBSERVATIONAL FINDINGS

AMD is a classic example of a multifactorial disease. Although there is a genetic component, the understanding of the heritance is hindered by interaction with the many environmental factors. The one undisputed risk factor, apart from age, is family history. Other hypothesized risk factors include tobacco smoking, obesity, cardiovascular disease/raised blood pressure, and pale irides, and in some studies, females are said to be more susceptible than males. Certainly all forms of AMD are more common in Caucasians than in darkly pigmented races, but there is no consensus on the reasons for this (Ambati et al., 2003). In sum, it appears that certain individuals inherit a predisposition for AMD and if they are exposed to the correct cocktail of harmful environmental factors, then the incidence is high, particularly among whites; in the Beaver Dam study (4,926 whites between 43 and 86 years old) overall

prevalence of early stage disease was 15.6%, varying from 8.45% of those between 43 and 54 years to 29.7% of those over 75 years. For late-stage disease this ranged from 0.1% to 7.1% in the same age groups (Klein et al., 1992b).

The enormous human and economic impact of AMD easily justifies the extensive investigation for the prospect of preventing the onset of disease by reducing the modifiable risk factors and changes in diet. One of the first observational studies of the link between retinal carotenoids and the risk for AMD was a follow-up to the Eye Disease Case–Control Study (EDCCS) (Seddon et al., 1994). They showed that those with the highest dietary intake and serum levels of L and Z had a 43% risk reduction for AMD. Subsequently there have been many such studies, the majority of which have been consistent with this finding (Delcourt et al., 2006; Gale et al., 2003; Mares-Perlman et al., 1996). For reviews of these issues see Loane et al. (2008) and Bernstein et al. (2009). In a recent analysis of data from the Age-Related Eye Disease Study (AREDS; described below), a high dietary intake of L and Z was associated with reduced likelihood of AMD (SanGiovanni et al., 2007), and an arm of the Blue Mountain Study reached the same conclusion (Tan et al., 2008).

Those most likely to request guidance based on firm evidence linking MP with AMD are ophthalmic clinicians who are regularly faced with patients who find they are at high risk. Should they recommend dietary supplementation or simply an improvement in lifestyle and a healthy diet? Although the aggregate of data from epidemiological studies is compelling, drawing conclusions from these reports must be approached with caution. There will be inevitable sampling bias arising from such factors as country, race, region from which participants are recruited, genetic profile, lifestyle, and many other difficult-to-control factors. Volunteers, by definition, are rarely a representative sample of a population. In fact, as observed by Evans (2007), volunteers are frequently affluent and already health conscious. Invariably, the individuals who are most likely to benefit from changing diet and lifestyle are the ones who do not volunteer for health-related studies.

In an attempt to gain a wider perspective on the effect of dietary supplements, Chong et al. (2007) employed a meta-analysis of the effectiveness of dietary oxidants in the primary prevention of AMD, and their literature search included only studies showing no sign of AMD at baseline. They looked at studies which used a 1-year follow-up and primary ARM as primary outcome. Twelve studies met their inclusion criteria, and only five of these reported on the effects of L and Z. Unfortunately none of the studies measured MP, and a dietary questionnaire was administered only at the start of the studies cited. Perhaps unsurprisingly, they concluded that there is inadequate evidence that L and Z prevent primary onset of AMD.

In general it would seem that healthy individuals have higher MP than those with disease. For example some studies have shown a correlation between patients with AMD and low MP (Beatty et al., 2001). A slightly different approach is to investigate healthy individuals. A study of 828 healthy observers revealed a positive correlation between dietary intake of L and Z, serum levels, and MPOD, and an inverse relationship was seen between MPOD and risk factors for AMD (Nolan et al., 2007a).

MP LINK WITH AMD; INTERVENTION TRIALS

The study that represents the current official position, as far as the direct benefits of dietary supplements are concerned, is AREDS, which is perhaps the largest and most frequently cited. In this 5-year prospective trial over 4,700 participants between 55 and 80 years old were given either a placebo or a high dose antioxidant cocktail of vitamin C, vitamin E, beta-carotene, and zinc. It was concluded that patients with moderate to advanced disease had a 25% risk reduction compared with those in the placebo group (SanGiovanni et al., 2007). Patients in the AREDS trial did not however receive L or Z because these compounds were not commercially available at the start of the study. The trial had other limitations; zinc levels were rather high, and the formulation contained beta-carotene, which, in high doses, is linked to lung cancer in smokers. A new AREDS trial, AREDS 2, aimed at addressing some of the problems with the earlier study, is now in progress. Enrollment to AREDS 2 concluded in June 2008. Patients will be followed for 5 to 6 years, and initial data are expected to be available in 2013. Participants will receive L and Z in one of the active formulations, but MP will be measured in only a small cohort of patients.

There are now many studies that have shown that an increase in dietary intake of retinal carotenoids gives rise to a substantial increase in MPOD (e.g., Berendschot et al., 2000; Bernstein & Gellermann, 2002; Bone et al., 2003; Koh et al., 2004; Landrum, Bone, & Kilburn, 1997; Richer et al., 2004; Rodriguez-Carmona et al., 2006; Schalch et al., 2007; Trieschmann et al., 2007). One crucial fact to emerge is that the increase in MP following a given level of carotenoid supplementation varies markedly between individuals from zero, so-called nonresponders, to increases of around 60–80%. It seems that those with low MP are most likely to show a substantial improvement in their MP following

supplementation, and this emerged in the LUtein Nutrition Effects Measured by Autofluorescence (LUNA) study (Trieschmann et al., 2007). One important observation was made by Koh et al. (2004), who showed that the MP in patients with early AMD can be augmented just as effectively as in healthy volunteers. Typical values of supplementation are a daily tablet containing around 5–10 mg of L. As might be expected, the time course for increasing MPOD is totally different to that for the increase in serum L and Z. Serum levels increase substantially within a few weeks of the start of the supplementation while the increase in MP may take between 1 and 3 months.

Although it is relatively easy to show MP can be augmented by supplementation, it is a different matter to establish unequivocally that there are genuine functional and/or morphological benefits in AMD patients. In the Lutein Antioxidant Supplementation Trial study, Richer et al. (2004) demonstrated improvements in VA and in contrast sensitivity compared with placebo following 1-year supplementation of 90 AMD patients. It was found that the VA changes were seen regardless of whether supplementation was composed of L alone or with L in combination with other supplements. Although the effects were modest, they were statistically significant. More recently, the CLEAR (Combination of Lutein Effects in the Aging Retina) study (Berendschot et al., 2011) obtained similar findings. In this two-center investigation based in The Netherlands and the United Kingdom, 80 early stage AMD patients were tested, with half being given a daily placebo and the other half a 10-mg L-only capsule. The supplementation lasted 1 year with patients being seen at 0 (baseline) 4, 8, and 12 months. A subgroup of patients who had abnormal VA at baseline showed significant improvement in VA compared with placebo, and these changes were correlated with the change in MP over the supplementation period. Uniquely, this investigation also revealed a significant improvement in dark adaptation in the intervention group compared with the placebo group.

Similar effects of improvement in VA were reported by Weigert et al. (2011) for 84 patients who received a L-only supplement for 6 months. They also reported a weak, but statistically significant, association between MP changes and VA changes. It may be that the duration of their study was too short. Berendschot et al. (2011) found that overall, VA improvements were not evident for at least 3 months after the MP had shown a substantial increase. In the CLEAR study, genuinely enhanced levels of MP were not seen within around 2–3 months after the onset of the supplementation. In another intervention trial Richer et al. (2011) used two interventions; one contained Z only, and the other, L and Z. The overall MP increased in the intervention groups, and there was an improvement of VA by 1.5 lines in the group who received the Z-only supplement. Although the relationship between VA and MP change is rather weak in all these investigations, there appears to be accumulating evidence of a genuine functional benefit of the supplementation in some but not all participants. These effects are all the more convincing when compared to the slight deterioration in VA experienced by the patients in the corresponding placebo groups, an observation particularly evident in the CLEAR study (Murray et al 2013; Berendschot et al., 2011).

CONCLUDING COMMENTS

There appears to be a biologically plausible pathway for how MP might play a role in the protection of the older retina from degenerative disease. It is now universally accepted that patients who show signs of early stage AMD should be advised to optimize their diet and lifestyle. Although the AREDS formulation is most commonly cited as the evidence base for supplementation, it does not comply with European Union regulations. It is known that a diet rich in L and Z will not enhance the MP as effectively as taking supplements containing these carotenoids. On the other hand supplements containing L and Z are expensive and not generally available through health care schemes. The most obvious strategy for the high-risk patient, particularly if he or she has already lost vision in one eye, is to be expedient and take the supplements despite the absence of unequivocal evidence of their benefits. It can be said with confidence that these patients are reducing their risk, but accurate prediction of benefit in individual patients awaits the results of more intervention trials where pretrial risk calculations can be compared with actual improvement according to recognized clinical measures.

ACKNOWLEDGMENTS

I am grateful for the help of Maria Makridaki and David Carden in the preparation of this chapter.

REFERENCES

Adler, R., Curcio, C., Hicks, D., Price, D., & Wong, F. (1999). Cell death in age-related macular degeneration. *Molecular Vision, 5,* 31.

Ahmed, S. S., McGregor, N., Lott, M. D., & Marcus, D. M. (2005). The macular xanthophylls. *Survey of Ophthalmology, 50,* 183–193.

Ambati, J., Ambati, B. K., Yoo, S. H., Ianchulev, S., & Adamis, A. P. (2003). Age-related macular degeneration: Etiology, pathogenesis, and therapeutic strategies. *Survey of Ophthalmology, 48,* 257–293.

Beatty, S., Boulton, M., Henson, D., Koh, H. H., & Murray, I. J. (1999). Macular pigment and age related macular degeneration. *British Journal of Ophthalmology, 83,* 867–877.

Beatty, S., Koh, H. H., Carden, D., & Murray, I. J. (2000). Macular pigment optical density measurement: A novel compact instrument. *Ophthalmic & Physiological Optics, 20,* 105–111.

Beatty, S., Murray, I. J., Henson, D. B., Carden, D., Koh, H., & Boulton, M. E. (2001). Macular pigment and risk for age-related macular degeneration in subjects from a Northern European population. *Investigative Ophthalmology & Visual Science, 42,* 439–446.

Berendschot, T. T., Broekmans, W. M., Klopping-Ketelaars, I. A., Kardinaal, A. F., Van Poppel, G., & Van Norren, D. (2002). Lens aging in relation to nutritional determinants and possible risk factors for age-related cataract. *Archives of Ophthalmology, 120,* 1732–1737.

Berendschot, T. T., Goldbohm, R. A., Klopping, W. A., van de Kraats, J., van Norel, J., & van Norren, D. (2000). Influence of lutein supplementation on macular pigment, assessed with two objective techniques. *Investigative Ophthalmology & Visual Science, 41,* 3322–3326.

Berendschot, T. T. J. M., Makridaki, M., van der Veen, R. L. P., Parry, N. R. A. Carden, D., & Murray, I. J. (2011). The Clear (combination (of) Lutein Effects (on) Aging Retina) Study; Lutein supplementation improves visual acuity and night vision in early Amd; A two-centre, placebo-controlled study. *Investigative Ophthalmology & Visual Science,* e-abstract 3631.

Berendschot, T. T., & van Norren, D. (2004). Objective determination of the macular pigment optical density using fundus reflectance spectroscopy. *Archives of Biochemistry and Biophysics, 430,* 149–155.

Berendschot, T. T., & van Norren, D. (2005). On the age dependency of the macular pigment optical density. *Experimental Eye Research, 81,* 602–609.

Berendschot, T. T., & van Norren, D. (2006). Macular pigment shows ringlike structures. *Investigative Ophthalmology & Visual Science, 47,* 709–714.

Bernstein, P. S., Delori, F. C., Richer, S., van Kuijk, F. J., & Wenzel, A. J. (2009). The value of measurement of macular carotenoid pigment optical densities and distributions in age-related macular degeneration and other retinal disorders. *Vision Research, 50,* 716–728.

Bernstein, P. S., & Gellermann, W. (2002). Measurement of carotenoids in the living primate eye using resonance Raman spectroscopy. *Methods in Molecular Biology (Clifton, N.J.), 196,* 321–329.

Bernstein, P. S., Yoshida, M. D., Katz, N. B., McClane, R. W., & Gellermann, W. (1998). Raman detection of macular carotenoid pigments in intact human retina. *Investigative Ophthalmology & Visual Science, 39,* 2003–2011.

Bird, A. C., Bressler, N. M., Bressler, S. B., Chisholm, I. H., Coscas, G., Davis, M. D., et al. (1995). An international classification and grading system for age-related maculopathy and age-related macular degeneration. The International ARM Epidemiological Study Group. *Survey of Ophthalmology, 39,* 367–374. doi:10.1016/S0039-6257(05) 80092-X.

Bone, R. A., & Landrum, J. T. (2004). Heterochromatic flicker photometry. *Archives of Biochemistry and Biophysics, 430*, 137–142.

Bone, R. A., Landrum, J. T., Fernandez, L., & Tarsis, S. L. (1988). Analysis of the macular pigment by HPLC: Retinal distribution and age study. *Investigative Ophthalmology & Visual Science, 29*, 843–849.

Bone, R. A., Landrum, J. T., Friedes, L. M., Gomez, C. M., Kilburn, M. D., Menendez, E., et al. (1997). Distribution of lutein and zeaxanthin stereoisomers in the human retina. *Experimental Eye Research, 64*, 211–218.

Bone, R. A., Landrum, J. T., Guerra, L. H., & Ruiz, C. A. (2003). Lutein and zeaxanthin dietary supplements raise macular pigment density and serum concentrations of these carotenoids in humans. *Journal of Nutrition, 133*, 992–998.

Bone, R. A., Landrum, J. T., Mayne, S. T., Gomez, C. M., Tibor, S. E., & Twaroska, E. E. (2001). Macular pigment in donor eyes with and without AMD: A case–control study. *Investigative Ophthalmology & Visual Science, 42*, 235–240.

Bone, R. A., Landrum, J. T., & Tarsis, S. L. (1985). Preliminary identification of the human macular pigment. *Vision Research, 25*, 1531–1535.

Broekmans, W. M., Berendschot, T. T., Klopping-Ketelaars, I. A., de Vries, A. J., Goldbohm, R. A., Tijburg, L. B., et al. (2002). Macular pigment density in relation to serum and adipose tissue concentrations of lutein and serum concentrations of zeaxanthin. *American Journal of Clinical Nutrition, 76*, 595–603.

Chong, E. W., Wong, T. Y., Kreis, A. J., Simpson, J. A., & Guymer, R. H. (2007). Dietary antioxidants and primary prevention of age related macular degeneration: Systematic review and meta-analysis. *BMJ (Clinical Research Ed.), 335*, 755.

Ciulla, T. A., Curran-Celantano, J., Cooper, D. A., Hammond, B. R., Jr., Danis, R. P., Pratt, L. M., et al. (2001). Macular pigment optical density in a midwestern sample. *Ophthalmology, 108*, 730–737.

Curcio, C. A., Medeiros, N. E., & Millican, C. L. (1996). Photoreceptor loss in age-related macular degeneration. *Investigative Ophthalmology & Visual Science, 37*, 1236–1249.

Curcio, C. A., Sloan, K. R., Kalina, R. E., & Hendrickson, A. E. (1990). Human photoreceptor topography. *Journal of Comparative Neurology, 292*, 497–523.

Davies, N. P., & Morland, A. B. (2004). Macular pigments: Their characteristics and putative role. *Progress in Retinal and Eye Research, 23*, 533–559. doi:10.1016/j.preteyeres.2004.05.004.

Delcourt, C., Carriere, I., Delage, M., Barberger-Gateau, P., & Schalch, W. (2006). Plasma lutein and zeaxanthin and other carotenoids as modifiable risk factors for age-related maculopathy and cataract: The POLA Study. *Investigative Ophthalmology & Visual Science, 47*, 2329–2335. doi:10.1167/iovs.05-1235.

Delori, F. C. (1994). Spectrophotometer for noninvasive measurement of intrinsic fluorescence and reflectance of the ocular fundus. *Applied Optics, 33*, 7439–7452. doi:10.1364/AO.33.007439.

Delori, F. C. (2004). Autofluorescence method to measure macular pigment optical densities fluorometry and autofluorescence imaging. *Archives of Biochemistry and Biophysics, 430*, 156–162.

Delori, F. C., Goger, D. G., Hammond, B. R., Snodderly, D. M., & Burns, S. A. (2001). Macular pigment density measured by autofluorescence spectrometry: Comparison with reflectometry and heterochromatic flicker photometry. *Journal of the Optical Society of America. A, Optics, Image Science, and Vision, 18*, 1212–1230. doi:10.1364/JOSAA.18.001212.

Delori, F. C., Goger, D. G., Keilhauer, C., Salvetti, P., & Staurenghi, G. (2006). Bimodal spatial distribution of macular pigment: Evidence of a gender relationship. *Journal of the Optical Society of America. A, Optics, Image Science, and Vision, 23*, 521–538.

Dimitrov, P. N., Robman, L. D., Varsamidis, M., Aung, K. Z., Makeyeva, G. A., Guymer, R. H., et al. (2011). Visual function tests as potential biomarkers in age-related macular degeneration. *Investigative Ophthalmology & Visual Science, 52*, 9457–9469. doi:10.1167/iovs.10-7043.

Evans, J. (2007). Primary prevention of age related macular degeneration. *BMJ (Clinical Research Ed.), 335*, 729.

Gale, C. R., Hall, N. F., Phillips, D. I., & Martyn, C. N. (2003). Lutein and zeaxanthin status and risk of age-related macular degeneration. *Investigative Ophthalmology & Visual Science, 44*, 2461–2465. doi:10.1167/iovs.02-0929.

Hammond, B. R., Jr., Curran-Celentano, J., Judd, S., Fuld, K., Krinsky, N. I., Wooten, B. R., et al. (1996a). Sex differences in macular pigment optical density: Relation to plasma carotenoid concentrations and dietary patterns. *Vision Research, 36*, 2001–2012.

Hammond, B. R., Jr., & Fuld, K. (1992). Interocular differences in macular pigment density. *Investigative Ophthalmology & Visual Science, 33*, 350–355.

Hammond, B. R., Jr., Fuld, K., & Snodderly, D. M. (1996b). Iris color and macular pigment optical density. *Experimental Eye Research, 62*, 293–297.

Hammond, B. R., Jr., Johnson, E. J., Russell, R. M., Krinsky, N. I., Yeum, K. J., Edwards, R. B., et al. (1997). Dietary modification of human macular pigment density. *Investigative Ophthalmology & Visual Science, 38*, 1795–1801.

Hammond, B. R., Jr., & Wooten, B. R. (2005). CFF thresholds: Relation to macular pigment optical density. *Ophthalmic & Physiological Optics, 25*, 315–319.

Hammond, B. R., Jr., Wooten, B. R., & Snodderly, D. M. (1996c). Cigarette smoking and retinal carotenoids: Implications for age-related macular degeneration. *Vision Research, 36*, 3003–3009.

Hogg, R. E., & Chakravarthy, U. (2006). Visual function and dysfunction in early and late age-related maculopathy. *Progress in Retinal and Eye Research, 25*, 249–276.

Howells, O., Eperjesi, F., & Bartlett, H. (2011). Measuring macular pigment optical density in vivo: A review of techniques. *Graefe's Archive for Clinical and Experimental Ophthalmology, 249*, 315–347.

Johnson, E. J., Hammond, B. R., Yeum, K. J., Qin, J., Wang, X. D., Castaneda, C., et al. (2000). Relation among serum and tissue concentrations of lutein and zeaxanthin and macular pigment density. *American Journal of Clinical Nutrition, 71*, 1555–1562.

Kilbride, P. E., Alexander, K. R., Fishman, M., & Fishman, G. A. (1989). Human macular pigment assessed by imaging fundus reflectometry. *Vision Research, 29*, 663–674.

Klein, R., Klein, B. E., & Linton, K. L. (1992a). Prevalence of age-related maculopathy. The Beaver Dam Eye Study. *Ophthalmology, 99*, 933–943.

Klein, R., Klein, B. E., Linton, K. L., & DeMets, D. L. (1993). The Beaver Dam Eye Study: The relation of age-related maculopathy to smoking. *American Journal of Epidemiology, 137*, 190–200.

Klein, R., Meuer, S. M., Moss, S. E., & Klein, B. E. (1992b). Detection of drusen and early signs of age-related maculopathy using a nonmydriatic camera and a standard fundus camera. *Ophthalmology, 99*, 1686–1692.

Koh, H. H., Murray, I. J., Nolan, D., Carden, D., Feather, J., & Beatty, S. (2004). Plasma and macular responses to lutein supplement in subjects with and without age-related maculopathy: A pilot study. *Experimental Eye Research, 79*, 21–27.

Landrum, J. T., & Bone, R. A. (2001). Lutein, zeaxanthin, and the macular pigment. *Archives of Biochemistry and Biophysics, 385*, 28–40.

Landrum, J. T., Bone, R. A., & Kilburn, M. D. (1997). The macular pigment: A possible role in protection from age-related macular degeneration. *Advances in Pharmacology (San Diego, Calif.), 38*, 537–556.

Lennie, P., Pokorny, J., & Smith, V. C. (1993). Luminance. *Journal of the Optical Society of America. A, Optics and Image Science, 10*, 1283–1293. doi:10.1364/JOSAA.10.001283.

Loane, E., Kelliher, C., Beatty, S., & Nolan, J. M. (2008). The rationale and evidence base for a protective role of macular pigment in age-related maculopathy. *British Journal of Ophthalmology, 92*, 1163–1168.

Makridaki, M., Carden, D., & Murray, I. J. (2009). Macular pigment measurement in clinics: Controlling the effect of the ageing media. *Ophthalmic & Physiological Optics, 29*, 338–344.

Mares-Perlman, J. A., Klein, R., Klein, B. E., Greger, J. L., Brady, W. E., Palta, M., et al. (1996). Association of zinc and antioxidant nutrients with age-related maculopathy. *Archives of Ophthalmology, 114*, 991–997.

Mellerio, J., Ahmadi-Lari, S., van Kuijk, F., Pauleikhoff, D., Bird, A., & Marshall, J. (2002). A portable instrument for measuring macular pigment with central fixation. *Current Eye Research, 25*, 37–47.

Moreland, J. D. (2004). Macular pigment assessment by motion photometry. *Archives of Biochemistry and Biophysics, 430*, 143–148.

Murray, I. J., Makridaki, M., van der Veen, R., Carden, D., Parry, N. R. A., & Berendschot, T. T. (under review). Lutein supplementation over a one year period in early AMD might have a mild beneficial effect on visual acuity; the CLEAR study. *Investigative Ophthalmology & Visual Science.*

Neelam, K., Nolan, J., Chakravarthy, U., & Beatty, S. (2009). Psychophysical function in age-related maculopathy. *Survey of Ophthalmology, 54*, 167–210.

Nguyen-Legros, J., & Hicks, D. (2000). Renewal of photoreceptor outer segments and their phagocytosis by the retinal pigment epithelium. *International Review of Cytology, 196*, 245–313.

Nolan, J. M., Stack, J., O'Connell, E., & Beatty, S. (2007a). The relationships between macular pigment optical density and its constituent carotenoids in diet and serum. *Investigative Ophthalmology & Visual Science, 48*, 571–582.

Nolan, J. M., Stack, J., O'Donovan, O. D., Loane, E., & Beatty, S. (2007b). Risk factors for age-related maculopathy are associated with a relative lack of macular pigment. *Experimental Eye Research, 84*, 61–74.

Owsley, C., Jackson, G. R., Cideciyan, A. V., Huang, Y., Fine, S. L., Ho, A. C., et al. (2000). Psychophysical evidence for rod vulnerability in age-related macular degeneration. *Investigative Ophthalmology & Visual Science, 41*, 267–273.

Packer, O., Hendrickson, A. E., & Curcio, C. A. (1989). Photoreceptor topography of the retina in the adult pigtail macaque (*Macaca nemestrina*). *Journal of Comparative Neurology, 288*, 165–183.

Parry, N. R., & Robson, A. G. (2012). Optimization of large field tritan stimuli using concentric isoluminant annuli. *Journal of Vision, 12.*1–13

Pease, P. L., Adams, A. J., & Nuccio, E. (1987). Optical density of human macular pigment. *Vision Research, 27*, 705–710.

Rein, D. B., Zhang, P., Wirth, K. E., Lee, P. P., Hoerger, T. J., McCall, N., et al. (2006). The economic burden of major adult visual disorders in the United States. *Archives of Ophthalmology, 124*, 1754–1760.

Richer, S., Stiles, W., Graham-Hoffman, K., Levin, M., Ruskin, D., Wrobel, J., et al. (2011). Randomized, double-blind, placebo-controlled study of zeaxanthin and visual function in patients with atrophic age-related macular degeneration. *Optometry, 82*, 667–680.

Richer, S., Stiles, W., Statkute, L., Pulido, J., Frankowski, J., Rudy, D., et al. (2004). Double-masked, placebo-controlled, randomized trial of lutein and antioxidant supplementation in the intervention of atrophic age-related macular degeneration: The Veterans LAST study (Lutein Antioxidant Supplementation Trial). *Optometry (St. Louis, Mo.), 75*, 216–230.

Robson, A. G., Moreland, J. D., Pauleikhoff, D., Morrissey, T., Holder, G. E., Fitzke, F. W., et al. (2003). Macular pigment density and distribution: Comparison of fundus autofluorescence with minimum motion photometry. *Vision Research, 43*, 1765–1775. doi:10.1016/S0042-6989(03)00280-3.

Robson, A. G., & Parry, N. R. (2008). Measurement of macular pigment optical density and distribution using the steady-state visual evoked potential. *Visual Neuroscience, 25*, 575–583.

Rodriguez-Carmona, M., Kvansakul, J., Harlow, J. A., Kopcke, W., Schalch, W., & Barbur, J. L. (2006). The effects of supplementation with lutein and/or zeaxanthin on human macular pigment density and colour vision. *Ophthalmic & Physiological Optics, 26*, 137–147.

SanGiovanni, J. P., Chew, E. Y., Clemons, T. E., Ferris, F. L., III, Gensler, G., Lindblad, A. S., et al. (2007). The relationship of dietary carotenoid and vitamin A, E, and C intake with age-related macular degeneration in a case–control study: AREDS Report No. 22. *Archives of Ophthalmology, 125*, 1225–1232.

Schalch, W., Cohn, W., Barker, F. M., Kopcke, W., Mellerio, J., Bird, A. C., et al. (2007). Xanthophyll accumulation in the human retina during supplementation with lutein or zeaxanthin—the LUXEA (LUtein Xanthophyll Eye Accumulation) study. *Archives of Biochemistry and Biophysics, 458*, 128–135.

Seddon, J. M., Ajani, U. A., Sperduto, R. D., Hiller, R., Blair, N., Burton, T. C., et al. (1994). Dietary carotenoids, vitamins A, C, and E, and advanced age-related macular degeneration. Eye Disease Case–Control Study Group. *Journal of the American Medical Association, 272*, 1413–1420. doi:10.1001/jama.1994.03520180037032.

Sharifzadeh, M., Bernstein, P. S., & Gellermann, W. (2006). Nonmydriatic fluorescence-based quantitative imaging of human macular pigment distributions. *Journal of the Optical*

Society of America. A, Optics, Image Science, and Vision, 23, 2373–2387. doi:10.1364/JOSAA.23.002373.

Snodderly, D. M., Auran, J. D., & Delori, F. C. (1984). The macular pigment: II. Spatial distribution in primate retinas. *Investigative Ophthalmology & Visual Science, 25,* 674–685.

Snodderly, D. M., Mares, J. A., Wooten, B. R., Oxton, L., Gruber, M., & Ficek, T. (2004). Macular pigment measurement by heterochromatic flicker photometry in older subjects: The carotenoids and age-related eye disease study. *Investigative Ophthalmology & Visual Science, 45,* 531–538.

Sommerburg, O., Keunen, J., Bird, A., & van Kuijk, F. J. G. M. (1998). Fruits and vegetables that are sources for lutein and zeaxanthin: The macular pigment in human eyes. *British Journal of Ophthalmology, 82,* 907–910.

Tan, J. S., Wang, J. J., Flood, V., Rochtchina, E., Smith, W., & Mitchell, P. (2008). Dietary antioxidants and the long-term incidence of age-related macular degeneration: The Blue Mountains Eye Study. *Ophthalmology, 115,* 334–341.

Trieschmann, M., Beatty, S., Nolan, J. M., Hense, H. W., Heimes, B., Austermann, U., et al. (2007). Changes in macular pigment optical density and serum concentrations of its constituent carotenoids following supplemental lutein and zeaxanthin: The LUNA study. *Experimental Eye Research, 84,* 718–728.

van de Kraats, J., Berendschot, T. T., Valen, S., & van Norren, D. (2006). Fast assessment of the central macular pigment density with natural pupil using the macular pigment reflectometer. *Journal of Biomedical Optics, 11,* 064031. doi:10.1117/1.2398925.

van de Kraats, J., Kanis, M. J., Genders, S. W., & van Norren, D. (2008). Lutein and zeaxanthin measured separately in the living human retina with fundus reflectometry. *Investigative Ophthalmology & Visual Science, 49,* 5568–5573. doi:10.1167/iovs.08-1939.

van der Veen, R. L., Berendschot, T. T., Hendrikse, F., Carden, D., Makridaki, M., & Murray, I. J. (2009a). A new desktop instrument for measuring macular pigment optical density based on a novel technique for setting flicker thresholds. *Ophthalmic & Physiological Optics, 29,* 127–137.

van der Veen, R. L., Berendschot, T. T., Makridaki, M., Hendrikse, F., Carden, D., & Murray, I. J. (2009b). Correspondence between retinal reflectometry and a flicker-based technique in the measurement of macular pigment spatial profiles. *Journal of Biomedical Optics, 14,* 064046.

Van Norren, D., & Tiemeijer, L. F. (1986). Spectral reflectance of the human eye. *Vision Research, 26,* 313–320.

Wald, G. (1945). Human vision and the spectrum. *Science, 101,* 653–658.

Weigert, G., Kaya, S., Pemp, B., Sacu, S., Lasta, M., Werkmeister, R. M., et al. (2011). Effects of lutein supplementation on macular pigment optical density and visual acuity in patients with age-related macular degeneration. *Investigative Ophthalmology & Visual Science, 52,* 8174–8178. doi:10.1167/iovs.11-7522.

Werner, J. S., Donnelly, S. K., & Kliegl, R. (1987). Aging and human macular pigment density. Appended with translations from the work of Max Schultze and Ewald Hering. *Vision Research, 27,* 257–268.

Wolf-Schnurrbusch, U. E., Roosli, N., Weyermann, E., Heldner, M. R., Hohne, K., & Wolf, S. (2007). Ethnic differences in macular pigment density and distribution. *Investigative Ophthalmology & Visual Science, 48,* 3783–3787. doi:10.1167/iovs.06-1218.

Wooten, B. R., & Hammond, B. R. (2002). Macular pigment: Influences on visual acuity and visibility. *Progress in Retinal and Eye Research, 21,* 225–240.

Wooten, B. R., Hammond, B. R., Jr., Land, R. I., & Snodderly, D. M. (1999). A practical method for measuring macular pigment optical density. *Investigative Ophthalmology & Visual Science, 40,* 2481–2489.

Wustemeyer, H., Moessner, A., Jahn, C., & Wolf, S. (2003). Macular pigment density in healthy subjects quantified with a modified confocal scanning laser ophthalmoscope. *Graefe's Archive for Clinical and Experimental Ophthalmology, 241,* 647–651.

Wyszecki, G., & Stiles, W. S. (1982). *Color science: Concept and methods, quantitative data and formulae* (2nd ed.). Wiley.

108 Molecular Mechanisms Underlying Non-Neovascular Age-Related Macular Degeneration

MARK E. KLEINMAN AND JAYAKRISHNA AMBATI

Age-related macular degeneration (AMD) is the leading cause of permanent vision loss in people over 50 years old in the United States and many other industrialized nations (Smith et al., 2001). The first clinical record of AMD was published in 1884 by Nettleship, who described it as a degenerative disease of the retina and used the term central senile areolar choroidal atrophy (Penfold et al., 2001). Several reports of atrophic AMD followed in the literature and are well documented in mid-twentieth-century ophthalmic atlases and texts, including Duke-Elder's *System of Ophthalmology*. In the modern era, with increasing life span, expanding populations, and improved diagnostics, AMD has become an epidemic in the developed world and is now estimated to affect 25% of the population over 75 years old with significantly higher prevalence with further aging (Klein, Klein, & Linton, 1992; Leibowitz et al., 1980).

The natural history of AMD is uniquely characterized by advancement to either geographic atrophy (GA) or choroidal neovascularization (CNV) in the macula. Advanced dry AMD is diagnosed by the subretinal accumulation of extracellular debris called drusen followed by GA of the retinal pigment epithelium (RPE) and degeneration of the overlying photoreceptors in the neural retina. The remaining affected population manifests an alternate disease phenotype termed exudative, or "wet," AMD, which is due to the formation of CNV and subsequent detachment of the neurosensory retina and/or RPE (Allikmets et al., 1997). While the pathogenesis of CNV has been well studied in a surrogate laser injury animal model leading to the development of clinically proven therapeutics, there is a paucity of data on the cellular events leading to GA due to the lack of a suitable animal model. Meanwhile, with the aging population in the United States, blindness due to GA is becoming a common problem in eye clinics. 90% of patients with AMD have the dry phenotype, and the prevalence of advanced atrophic AMD is nearly fourfold higher than neovascular AMD in patients over the age of 85 (Klein et al., 2007). Dry atrophic AMD is characterized by the focal loss of RPE, overlying photoreceptors, and the underlying choriocapillaris. Over several years, these areas may coalesce to form a large central zone of macular atrophy and a visually debilitating central scotoma (see figure 108.1). While there are over 40 clinical trials ongoing for the treatment of GA, there is currently no proven pharmacological approach to slow the progression of atrophy and preserve vision. Yet, as we begin to improve the fundamental knowledge in retinal cell survival and death, many significant scientific advances are being applied to translational aims in the diagnosis and treatment of dry AMD.

CLINICAL FEATURES OF ATROPHIC DRY MACULAR DEGENERATION

Currently, the precise etiology of dry AMD remains unknown. The hallmark of this disease is the formation of drusen which are sub-RPE deposits within Bruch's membrane containing lipoproteinaceous debris and oxidized pigment. Drusen phenotypes vary from hard fine pinpoint lesions to large basal laminar deposits with specific features that increase risk of disease progression. The Age-Related Eye Disease Study (AREDS) classified components of specific clinical features of AMD including drusen size, pigment deposition, RPE atrophy, and CNV formation to devise a simple scheme that allowed for improved risk assessment for disease progression (AREDS Study Group, 2001). Dry AMD advances over a period of years to decades depending on the number of risk factors and phenotypic features, whereas the switch to neovascular AMD appears to be more rapid, evolving over a period of days to months. Progression studies report that the enlargement of atrophic areas is between 1 and 3 mm^2/year (Holz et al., 2007; Klein et al., 2010; Yehoshua et al., 2011). The

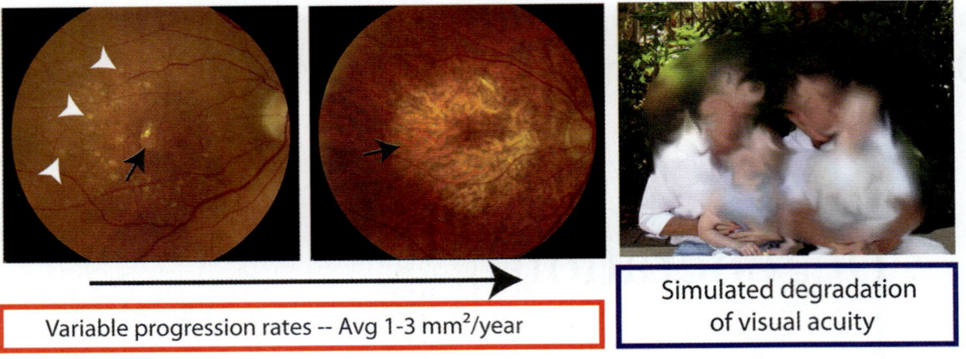

FIGURE 108.1   Geographic atrophy occurs in late stage age-related macular degeneration (AMD) and results in significant vision loss. Intermediate AMD (left) characterized by medium to large size drusen develop large areas of retinal pigment epithelium (RPE) loss (middle) overlying these lesions that may progress to complete wipeout of the macula over a period of several years. The loss of visual acuity is often severe with significant degradation of the central field (right). There is currently no effective treatment, symptomatic or otherwise, for the treatment of this disease. White arrowheads, drusen; black arrows, areas of RPE atrophy; Avg, average.

pattern of atrophy is very particular in that it will often begin in areas of large drusen, pigment clumping, or pigment epithelial detachments with eventual focal loss of the overlying RPE (Elman et al., 1986; Sarks, Sarks, & Killingsworth, 1988). The atrophic zone then frequently enlarges into a horseshoe shape, most commonly in the temporal macula, sparing the fovea (Maguire & Vine, 1986). This fortunate sparing phenomenon is hypothesized to be a result of increased levels of antioxidant pigments lutein and zeaxanthine in the fovea (Weiter, Delori, & Dorey, 1988) and decreased lipofuscin levels due to the absence of rod photoreceptors and cone predominance (Curcio et al., 1993). The atrophic zone will eventually coalesce to form a continuous band of atrophy encircling the fovea which may then become involved at the end stages of the disease. Additionally, other types of drusen have been now characterized including reticular drusen (pseudodrusen or subretinal drusenoid deposits), which are located within the outer retina, in contrast to the sub-RPE space, and are highly associated with GA (Schmitz-Valckenberg et al., 2011). With the rapid ascent of multimodal imaging including optical coherence tomography (OCT) and fundus autofluorescence (FAF), retinal visualization in vivo has allowed for the additional classification of dry AMD clinical features. FAF is a surrogate measure of RPE cell function and correlates with levels of lipofuscin accumulation (Delori et al., 1995), a by-product derived from RPE mediated phagocytosis of photoreceptor outer segments. As dry AMD progresses, large soft drusen may form from existing basal linear deposits and lead to atrophy of the overlying RPE and photoreceptors. This can be visualized in vivo as a hypopigmented patch on ophthalmoscopy, by OCT as absence of a distinct RPE layer, or by

dark areas on FAF (see figure 108.2). While the technique and application of FAF has been previously described by others (Delori et al., 1995; von Ruckmann, Fitzke, & Bird, 1997), the identification of specific patterns of FAF in AMD has been well described by the contemporary Fundus Autofluorescence in AMD (FAM) and Geographic Atrophy Progression (GAP) study groups (Holz et al., 2007). In these studies, several different FAF phenotypes were identified in GA including focal, patchy, and diffuse, which is the most common. The importance of FAF imaging was initially realized by studies that correlated specific FAF signatures with higher rates of GA progression. For example, while patchy or focal FAF may exist in patients with average GA progression rates of approximately 2 mm²/yr, the diffuse trickling subtype has been associated with more rapid loss of RPE approaching 3 mm²/yr. Another feature of FAF patterning is that it is symmetrical in up to 80% of patients with bilateral lesions (Bellmann et al., 2002), which aligns with previous studies of disease concordance in fellow eyes with color fundus photography (Sarks, Sarks, & Killingsworth, 1988). Further research on FAF and high-risk patterning is anticipated to generate important insights into the pathogenetic mechanisms of dry AMD.

Another important aspect of improving our understanding of FAF and lipofuscin accumulation is that numerous visual cycle modifiers are currently in clinical trials for the treatment of dry AMD. The hypothesis relies on the foundation that aged and dysfunctional RPE cells are unable to efficiently process visual cycle molecules, and thus decreasing the available pool of retinoid derivatives may alleviate this metabolic demand and subsequent accumulation of oxidized and pro-inflammatory by-products such as

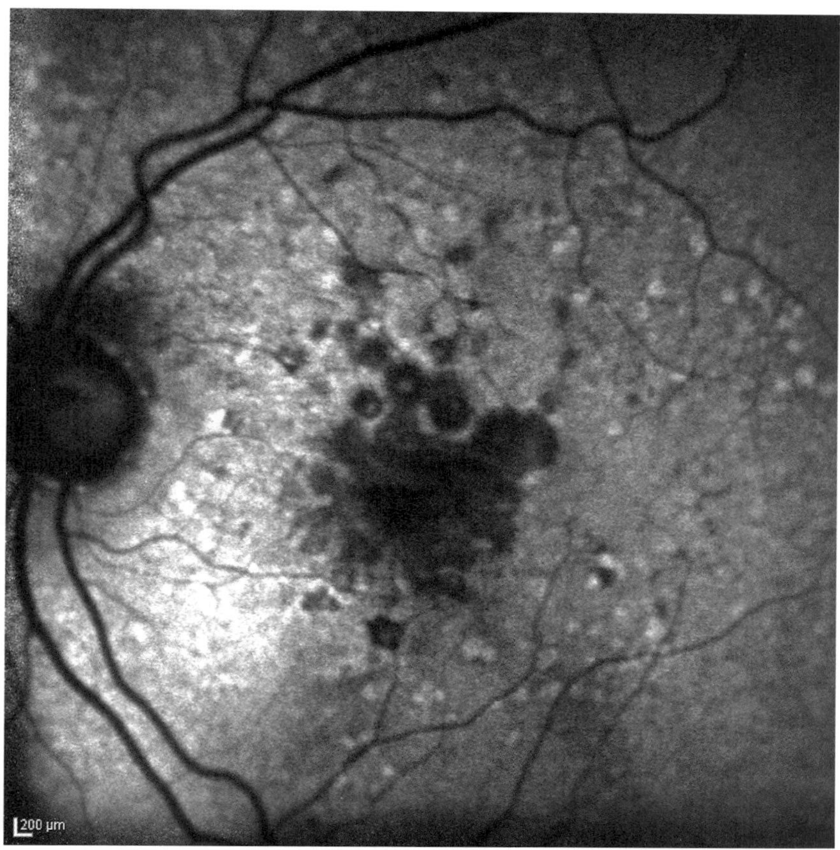
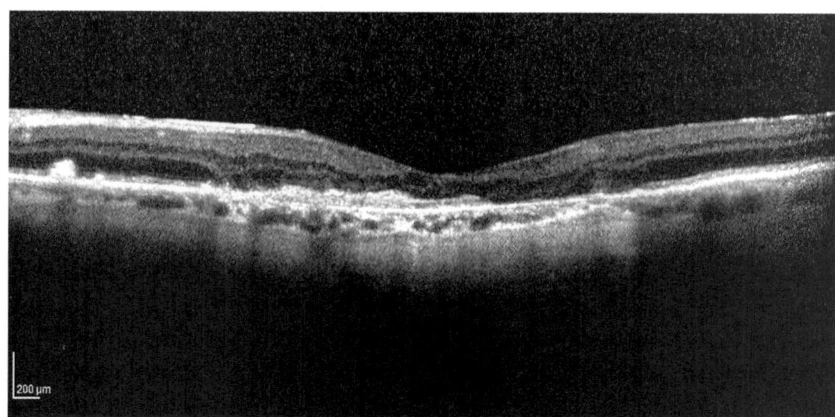

FIGURE 108.2   Multimodal imaging of non-neovascular age-related macular degeneration (AMD). Blue light (488 nm) fundus autofluorescence (top) in intermediate AMD demonstrating abnormal punctate hyperfluorescence surrounding dark areas of focal retinal pigment epithelium (RPE) atrophy. Spectral-domain optical coherence tomography imaging (bottom) reveals laminar sub-RPE deposits with areas of complete loss of RPE layer and overlying photoreceptor outer segments and disruption of the external limiting membrane.

N-retinylidene-N-retinylethanolamine (A2E). Fenretinide (4-hydroxy(phenyl)retinamide) is an orally administered retinoid derivative that displaces endogenous retinol from its chaperone to decrease RPE concentrations. This compound was studied in a phase II trial and showed promising interim results but no significant protective benefit at the 2-year readout. Another visual cycle modifier, ACU-4429 (Acucela®), targets *trans-cis* conversion of retinol by the protein RPE65, is effective at reducing scotopic b-wave amplitude (Kubota et al., 2012), and is in a phase II study for dry AMD.

MICROSCOPIC AND STRUCTURAL FEATURES OF DRY AMD

Detailed gross and microscopic analyses have succeeded in charting the natural history of this disease but have failed to reveal the underlying molecular events that lead to its initiation and progression. Drusen, the hallmark of dry AMD, are aggregates of extracellular debris containing glycoprotein and lipid classically localized between the RPE and Bruch's membrane. Several studies have focused exclusively on compositional analyses of drusen with the hope that an elemental understanding of these pathologic deposits may lead to a strategy to induce resorption or prevent further accumulation (Dentchev et al., 2003; Johnson et al., 2002). The results revealed numerous proteins within subretinal deposits including amyloid-β, complement factors, clusterin, apolipoprotein E, serum amyloid P, and ubiquitin (Anderson et al., 2004; Mullins et al., 2000). Many of these constituents are believed to be biologically active and involved in the progression of early AMD to advanced forms of the disease. On microscopic evaluation in explanted tissues, GA can be identified by in situ immunofluorescence of the intercellular RPE junctional protein, zona-occludens-1 (ZO-1), in macular flat mounts from affected human eyes (see figure 108.3). Light microscopy reveals areas of absent RPE bordered by hypertrophy and hyperplasia with associated pigment extrusion and basal laminar deposits.

Ultrastructural studies of dry AMD eyes with electron microscopy have demonstrated drusen as an amorphous granular substance and associated significant RPE changes including membrane blebbing and loss of mitochondria (Ding, Patel, & Chan, 2009; Ishibashi et al., 1986). An excellent overview of additional ultrastructural features of dry AMD is included in chapter 106.

MACROPHAGE RESPONSES IN DRUSEN BIOGENESIS AND RPE ATROPHY

In the past 20 years, there has been a major paradigm shift toward inflammatory and immunologic mediators playing a significant role in the development of AMD. Monocytes (Mo) are released from the bone marrow and recruited to sites of inflammation and reticuloendothelial networks where they differentiate into mature macrophages (MΦ) capable of scavenging pathogens, foreign particles, and cellular debris (Crocker & Gordon, 1985; Kennedy & Abkowitz, 1998; Volkman & Gowans, 1965). Resident tissue MΦ within the central nervous system, retina, and choroid are also referred to as microglia. With the high metabolic state of the macula and constant production of discarded photoreceptor disks, efficient removal of this debris is critical to the maintenance of retinal health and visually optimal microenvironment of the photoreceptor–RPE interface. MΦ are essential for this process and have

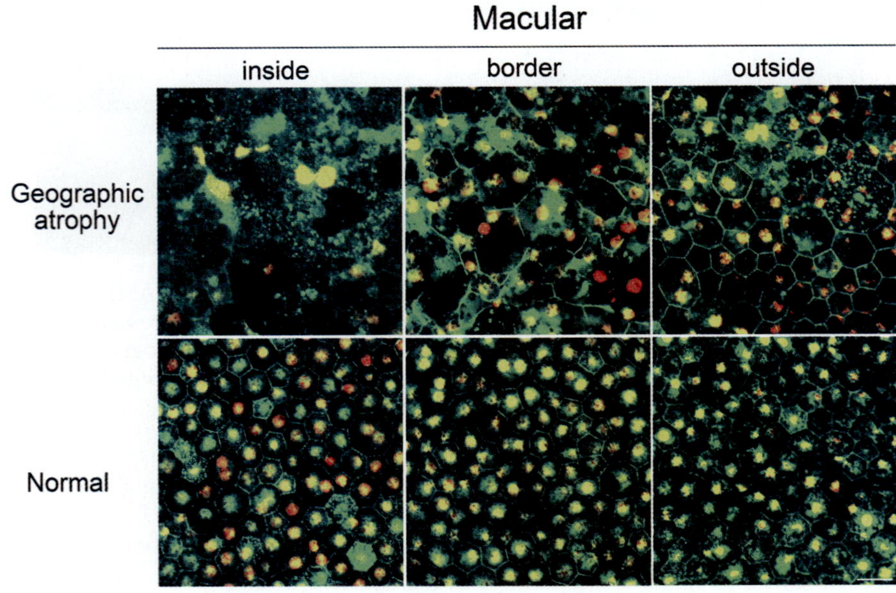

FIGURE 108.3   Retinal pigment epithelium (RPE) dysmorphology in geographic atrophy. Zona-occludens-1 and propidium iodide stained immunofluorescent RPE/choroid flat mounts from human eyes with geographic atrophy and normal age-matched controls demonstrating significant cell loss, hypertrophy, and disorganization of cell junctions. (Image courtesy of Hiroki Kaneko, M.D., Ph.D.)

been visualized on the choriocapillaris aspect of Bruch's membrane below sites of drusen deposition with processes extending into the lesions in AMD eyes (Killingsworth, Sarks, & Sarks, 1990) suggesting their key involvement in drusen clearance and potential dysfunction is drusen biogenesis.

In 2003, the first mouse model of AMD was described which offered significant support for the purported role for MΦ trafficking and function in drusen biogenesis and clearance. Two transgenic mouse models deficient in CCL2 (also known as monocyte chemotactic protein-1; MCP1) and/or CCR2 (a receptor/ligand pair critical to the peripheral recruitment of Mo/MΦ cells) were found to develop drusen-like deposits resembling those found in patients with AMD (see figure 108.4) (Ambati et al., 2003). Ultrastructural electron microscopic analyses of these spots revealed basal laminar deposits, diffuse drusen, and hard excrescences such as those found in the human condition. Similar findings in mice deficient in either MCP1 and/or CCR2 were subsequently reported by other investigators (Chowers et al., 2006; Shen et al., 2006). Using MCP1 migration assays, wild-type peritoneal MΦ were found to be immobilized by C5a and, to a lesser degree, IgG, both of which are known components of drusen. These complement factors were also colocalized to the drusen-like deposits found in 9-month-old MCP1- and CCR2-deficient mice.

Flow cytometry of RPE and choroid samples showed an age-dependent increase in the percentage of F4/80+ MΦ in wild-type mice which was significantly decreased compared to MCP1 and CCR2 knockout mice. These data suggest that MΦ infiltration into aged choroid tissue mediated by the MCP1/CCR2 chemotactic axis may play a role in the resorption of immune complexes that are found in drusen, and that a deficiency in MΦ trafficking and function may hasten the progression of dry AMD.

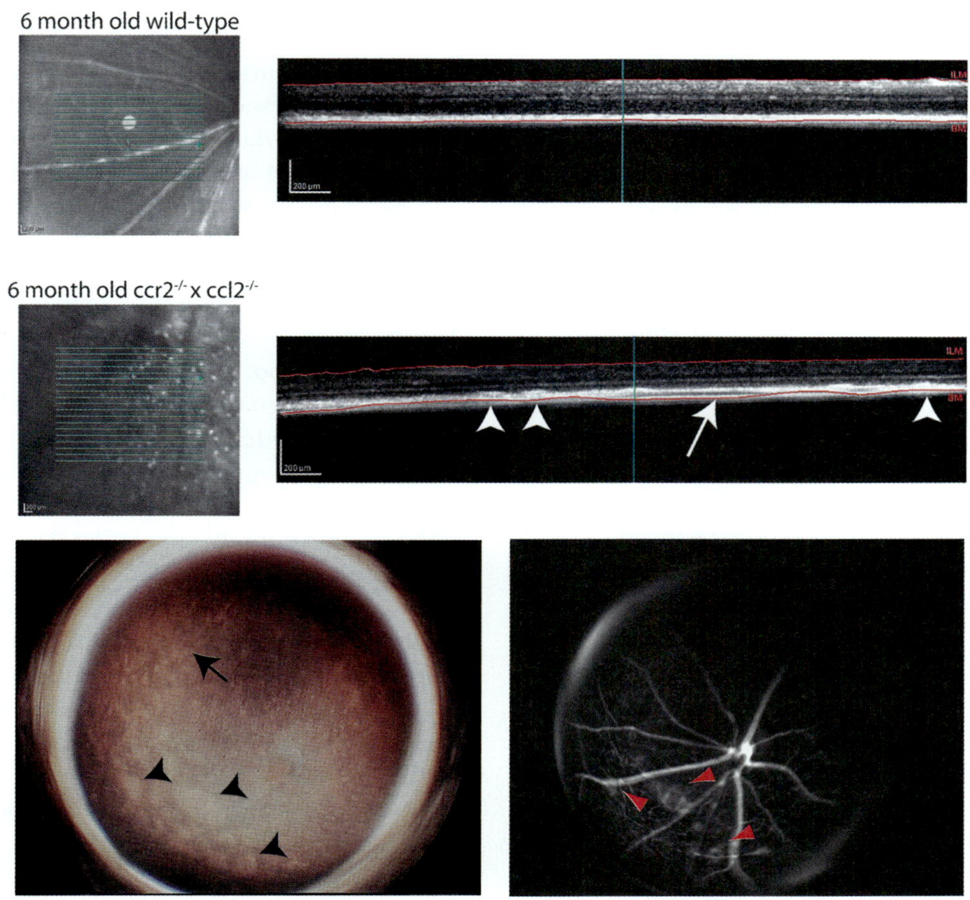

FIGURE 108.4   Mouse model of drusen-like retinal lesions and retinal pigment epithelium (RPE) atrophy. Infrared reflectance and spectral-domain optical coherence tomography of age-matched wild-type C57/Bl6 mouse and double-knockout *Ccr2/Ccl2* mouse demonstrating large deposits (~ 100 μm) in RPE layer (arrowheads) and area of RPE detachment from Bruch's membrane (BM) (arrow). Red line designates internal limiting membrane (ILM) and BM. Color fundoscopy (bottom left) and fluorescein angiography (bottom right) in 12-month-old double-knockout mice reveal RPE atrophy (black arrowheads) and window defects (red arrowheads) consistent with advanced dry AMD.

More evidence to support a role for this chemokine axis is that young mouse retina and RPE express low levels of MCP1, which increases during aging and during acute inflammation (Chen et al., 2008). Thus, there may be a protective effect of increased MΦ signaling with elevated MCP1 concentrations in aged retina. Alternatively, increased MCP1 may regulate a low level of chronic inflammation to assist with debris removal and drusen clearance. A recently defined cellular process, termed para-inflammation, was introduced by Medzhitov to describe the cellular state in which stressed tissues are exposed to chronic low-grade inflammation, resulting in activation of tissue resident MΦs (Medzhitov, 2008). This mechanism may actually be protective in the aging retina and assist in the processing and clearance of cellular debris while buffering the threshold for acute inflammation. This balance is hypothesized to be disrupted in the transgenic mouse model of AMD due to impaired phagocytic function and migration of resident and infiltrating MΦ leading to increased cytokine production and pro-apoptotic factors (Chen, Forrester, & Xu, 2011).

There have been several follow-up studies to this initial report of the mouse model of AMD which have generated controversy about the cause of subretinal deposits in the MCP1-deficient animals. FAF studies and immunofluorescence revealed lipofuscin and pigment bloated F4/80+ CD68+ MΦ coregistering with highly AF drusen-like deposits in these mice (Luhmann et al., 2009). Some have proposed that MCP1/CCR2 deficiency only results in severely impaired MΦ trafficking whereas associated retinal degeneration in this model may be dependent on aging alone. There are now reports that aged C57BL/6N mice from commercial vendors may exhibit features of chorioretinal thinning, drusen formation, and degeneration due to genetic lines harboring an rd8 mutation (Mattapallil et al., 2012). Such contamination of commercial mouse strains needs to be seriously addressed, as the related and unanticipated genetic effects of a homozygous rd8 mutation on the wild-type mouse background would severely confound the results of many murine retinal degeneration models. However, the rd8 mutation is not present in the reported model, as it was developed on a C57BL/6J background, not C57BL/6N. Additionally, no pan-retinal drusen such as those found in MCP1/CCR2 deficient mice were observed in age-matched control animals.

Within the field of immunology, there is a cadre of investigators focused on the heterogeneous function of MΦ as scavenger and/or pro-inflammatory cells that induce their effects through the expression of scavenger receptors and cytokines. The existence of such cellular subsets is now an area of intense investigation, as the surface markers that define resident versus inflammatory MΦ and Mo populations has significantly improved with advanced techniques in multicolor flow cytometry. The surface expression profile to differentiate these cells was dissected in elegantly designed adoptive transfer experiments that established the migratory patterns of circulating populations of Mo/MΦ progenitors which are driven into either resident or pro-inflammatory MΦ lineages (Geissmann, Jung, & Littman, 2003). These experiments demonstrated that CCR2 and CX3CR1, the fractalkine receptor, were expressed on MΦ progenitors that migrated to sites of inflammation and exhibited a shorter half-life, whereas CX3CR1, but not CCR2, was expressed on progenitor cells that eventually became resident MΦ. The discovery that mice deficient in CCR2 or MCP-1 spontaneously develop drusen-like accumulations with fundoscopic and ultrastructural features similar to human AMD suggests that infiltrating MΦ are required for maintenance of the RPE interface and clearance of drusen. Further work has confirmed the existence of a CX3CR1 single nucleotide polymorphism (SNP), T280M, which leads to decreased expression and increased AMD risk (Tuo et al., 2004). These data support the concept that a balance between resident and pro-inflammatory MΦ may regulate AMD progression. More studies have now characterized the ocular phenotypes of mouse strains deficient in CX3CR1 or both CX3CR1 and CCR2, which also exhibit AMD-like features (Combadiere et al., 2007; Tuo et al., 2007).

Scattered reports have emerged describing pro-inflammatory signatures in circulating MΦ progenitors from animal models of retinal degeneration and acute retinal injury (London et al., 2011; Xu et al., 2005). An early report analyzed tumor necrosis factor (TNF) alpha production in circulating Mo harvested from AMD patients and stimulated in MΦ growth conditions (Cousins, Espinosa-Heidmann, & Csaky, 2004). Currently, a number of groups have analyzed human peripheral blood from patients with dry and neovascular AMD to determine specific imbalances in Mo/MΦ subsets. Still, the modern understanding of trafficking in this immune cell axis and the precise function in AMD is crudely charted. Future studies with the updated subset surface markers will deepen our insight into the multifunctional roles of MΦ in dry AMD progression.

MOLECULAR REPONSES OF THE INNATE IMMUNE SYSTEM IN DRY AMD

Activation of the immune system is regulated by two integrated pathways: the innate and adaptive (or acquired) systems. Innate immunity is defined as the immediate nonspecific host response that is initiated upon pathogen invasion. To benefit host survival, it is essential that this activation be extremely swift and potent to safeguard the remaining cells from infection. Such an acute and robust response is capable through the complement cascade which recognizes foreign antigens and autolysis of affected cells as well as through a broad array of toll-like receptors (TLRs) that react to specific pathogen-associated molecular patterns. Interestingly, both of these pathways have been implicated in the pathogenesis of dry AMD and will be discussed further here.

Complement Activation in Dry AMD

Immune complement factors bind directly to pathogens to enhance host-mediated phagocytosis, a process termed opsonization, via the classical, alternative, and lectin pathways. All of these complement pathways channel into a common terminus. There is sparse evidence of antibody mediated complement fixation or localization of specific classical pathway components in AMD (Anderson et al., 2002; Johnson et al., 2001). It is important to recognize that in contrast to the classical pathway, alternative and lectin pathways do not require specific antibody–antigen interaction for complement mediated cell lysis. Some of the more significant gene associations with advanced AMD have been linked to components of the alternative complement pathway system (Gehrs et al., 2006). Many of these components are expressed in the RPE and choroid and colocalize to drusen and surrounding areas in AMD (Hageman et al., 2005; Johnson et al., 2000). One of the stronger associations is with an endogenous soluble complement inhibitor, complement factor H (CFH), capable of suppressing alternative pathway activation via binding C3B and factor I (Jozsi et al., 2007) thus preventing complement mediated host cell death in tissues. In 2006, a group of independent studies reported SNPs within the human *cfh* gene that significantly increase AMD risk (Edwards et al., 2005; Hageman et al., 2005; Haines et al., 2005; Klein et al., 2005) with the Y402H variant elevating risk of advanced AMD by approximately 2.7-fold in patients of European descent. The physiological effects of the Y402H mutation are thought to be imparted through multiple pathways. Firstly, the reduction of heparin binding affinity of the Y402H CFH variant to host cell membranes has been demonstrated, which could lead to decreased inhibition of C3b and increased downstream alternative complement activation (Clark et al., 2006). More recently, CFH was found to bind to and neutralize the pro-inflammatory

lipid peroxidation product, malondialdehyde (MDA), while the Y402H mutation significantly decreased CFH binding affinity for MDA (Weismann et al., 2011). Without CFH blockade, MDA epitopes are capable of increasing expression of IL-8 and TNF-alpha in MΦ and RPE, which may promote inflammation and disease progression in AMD. Interestingly, mice deficient in CFH exhibited some visual changes, but only subtle fundoscopic and microscopic features were consistent with human AMD (Coffey et al., 2007). There is no clear explanation for the unmatched phenotype, but biological differences between total gene deletions versus expression of the mutant *cfh* gene or, more simply, differences in species may be responsible. Still, with the obvious importance of the CFH discovery in AMD pathogenesis, several translational avenues are currently in preclinical phases with approaches including repletion via recombinant CFH and a fusion protein technology.

Several other SNPs in C2, C3, and Factor B genes were found to be in disequilibrium in affected patients (Gold et al., 2006; Yates et al., 2007). No association with C5 SNPs has been demonstrated to date (Baas et al., 2010). In research designs to translate complement inhibition for the treatment of dry AMD, most industry efforts have focused on C3 and C5 as these components are responsible for activating the terminal cascade. There are several pioneering clinical studies of complement inhibitors for dry AMD that will have final data readouts this year. In 2007, the first Food and Drug Administration–approved complement inhibitor was studied, an anti-C5 humanized monoclonal antibody (eculizumab, Solaris) from Alexion for the treatment of another complement related disease, paroxysmal nocturnal hemoglobinura. A phase II study is currently active for the study of eculizumab in the treatment of dry AMD with interim 6-month data showing no improvement in treated groups (Garcia Filho et al., 2012) and a final 12-month readout scheduled for 2012. Ophthotech is in phase I trials of its anti-C5 aptamer, ARC1905, and is scheduled for completion in 2012. Jerini Ophthalmics has focused on the C5a receptor with small molecule peptidomimetic antagonists and is currently investigating two lead candidates (JPE-1375 and JSM-7717) in preclinical studies. A small molecule inhibitor of C3, compstatin (POT-4, AL-78898A), had little benefit in exudative AMD and is awaiting testing in advanced dry AMD. In 2011, a report surfaced of another alternative pathway component, complement factor D (CFD), demonstrating a SNP association in patients with AMD. Importantly, this finding was confined to females and could not be independently replicated in a subsequent investigation (Zeng et al., 2010).

Regardless, an anti-CFD antibody (FCFD4514S, Genentech) is currently being studied in a phase Ib/II clinical trial for the treatment of advanced dry AMD.

Alternate Innate Immune Activation in Dry AMD

In order for the innate immune system to rapidly respond to specific antigenic elements found in virulent pathogens, an array of pattern recognition receptors has evolved which is most well represented by the highly conserved type I transmembrane proteins called toll-like receptors (Medzhitov & Janeway, 2000). Ten functional TLRs have been identified in humans (12 in mice), several of which have been found to be involved in various ocular inflammatory processes (Chang, McCluskey, & Wakefield, 2006). Some of the well-characterized responses include TLR2, which binds peptidoglycan motifs from gram-positive bacteria cell walls, TLR3 for double-stranded RNA (dsRNA) from viral genomes, replicate intermediates, and endogenous RNA processing, and TLR4 for lipopolysaccharide or endotoxin, a highly virulent molecule located in the membrane of gram-negative bacteria (Oda & Kitano, 2006). Upon binding and dimerization, TLRs initiate signaling cascades and nuclear translocation of various transcription factors involved in antiviral defense. Downstream effects result in the expression of interferons and cytokines with a robust pro-apoptotic signature (Kawai & Akira, 2006).

RPE cells express most TLRs (1–7, 9, 10) (Kumar et al., 2004), and, in addition to their role in immune surveillance, data have emerged that they are involved in physiological homeostasis in the retina. For instance, TLR4 may aid RPE phagocytosis of photoreceptor outer segments (Kindzelskii et al., 2004), a process thought to be deficient in AMD (Rakoczy et al., 2002). The fundamental understanding of TLRs in AMD biology remains woefully inadequate, and data from genome-wide association studies are controversial. The TLR4 SNP, D299G, which dampens signal activation (Arbour et al., 2000), was associated with increased risk of AMD in a moderately sized cohort (Zareparsi et al., 2005), but follow-up studies with European, United States, and Indian AMD cohorts did not concur (Despriet et al., 2008; Kaur et al., 2006). However, more studies of extended pedigrees identified an association between the 9q33 locus, where TLR4 resides, and AMD (Abecasis et al., 2004; Seddon et al., 2003), suggesting that it may be involved in some patients. Another genetic mapping analysis revealed a link between AMD and the 4q32 locus (Majewski et al., 2003) where the TLR2 gene maps to, but there are currently no reports on TLR2 SNPs.

Translational investigations of TLR3, the dsRNA sensor, have also created controversy. In a worldwide collaboration, a genetic association between advanced dry AMD and a hypomorphic SNP in the TLR3 gene called L412F was reported with correlate data from a mouse model of dsRNA-induced RPE cell death that recapitulated the human condition (Yang et al., 2008). The study concluded that carriers of the mutation are protected from the development of RPE atrophy and that TLR3 activation may be deleterious to the macular RPE. In this study, experiments revealed that intraocular injections of a potent TLR3 agonist, poly (I:C) resulted in focal RPE atrophy with a lack of frank vitreous or anterior segment inflammation. Yet, as had occurred with several of the other SNP studies with TLRs, additional investigations could not confirm the protective effect of the TLR3 mutation (Allikmets et al., 2009; Edwards et al., 2008). Regardless, further molecular characterization of the animal model of advanced dry AMD with dsRNA stimulation was reported. Interrogations of the downstream signaling apparatus in RPE cell loss was found to channel through interferon regulatory factor 3 (IRF3), resulting in caspase-3 activation in apoptosis (Kleinman & Ambati, 2012; Kleinman et al., 2012). More startling insight into the importance of dsRNA biology in AMD was revealed in experiments demonstrating the increased expression of Alu RNA and its cytotoxic effects on the RPE (Kaneko et al., 2011). *Alu* is a transposon with highly repetitive elements that comprises up to 10% of the human genome. Here, it was found that the dsRNA processing enzyme, DICER1, an important component in the micro-RNA (miRNA) pathway, was down-regulated in the RPE from GA eyes compared to age-matched controls. Mouse models of *Dicer1* ablation or enforced *Alu* expression likewise formed large zones of GA-like lesions in RPE similar to poly I:C or synthetic dsRNA treated mouse eyes (see figure 108.5). This RNA induced toxicity was not due to perturbations in miRNA sequences or transcriptional regulation but via a direct effect on the RPE. In fact, it does not appear that Alu RNA induces apoptosis via TLR3 but through activation of the inflammasome and terminal caspase activation through a MyD88 and IL-18 dependent mechanism (Tarallo et al., 2012).

AUTOIMMUNE-RELATED FINDINGS IN ADVANCED DRY AMD
Patients with AMD may harbor increased circulating anti-retinal autoantibodies that are thought to be active against neural components of the retina based on serum staining patterns in eye sections (Penfold et al., 2001). Certainly, it is possible that these antibodies may also accelerate the photoreceptor and RPE degeneration that occurs during the progression of dry AMD. In

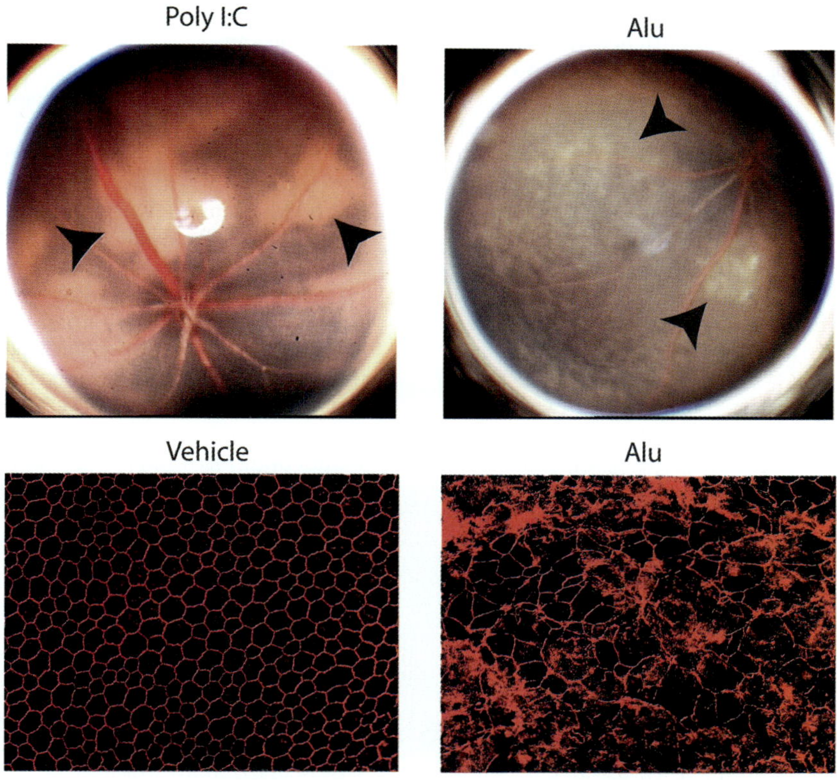

FIGURE 108.5   Mouse models of geographic atrophy–like retinal pigment epithelium (RPE) atrophy. The long synthetic double-stranded RNA (dsRNA), poly I:C, and expression of Alu induce focal loss of RPE as evident on color fundoscopy (top panels) 7 days after treatment. Zona-occludens-1 immunofluorescence (bottom panels) reveals significant disruption of the RPE inter-cellular junctions with hypertrophy in Alu treated eyes compared to the normal hexagonal tessylated pattern in vehicle treated animals. Similar findings to Alu were present with synthetic dsRNA.

view of these data, there are advocates for the utilization of these autoantibodies as biomarkers of early AMD (Cherepanoff et al., 2006; Patel et al., 2005). Recent data have identified a protein adduct, carboxyethylpyr-role (CEP), generated during lipid oxidation in the retina leading to the formation of harmful cross-reactive anti-CEP antibodies that bind to the retina and induce degeneration (Gu et al., 2003; Hollyfield et al., 2008). This may be another mechanism by which advanced dry AMD progresses and would explain the presence of antiretinal antibodies in affected patients. Moreover, similar to the interaction of CFH with MDA by-products, CEP could also be sequestered via protein binding but released for antigen presentation in indi-viduals with specific risk alleles.

CONCLUSION

With an aging population, the toll of vision loss from AMD has reached staggering proportions with some-where between 30 and 50 million affected worldwide. Critical insights into the morphological, cellular, and molecular events that occur during the progression of dry AMD have revealed the multifactorial nature of this disease. Although there is currently no effective therapeutic for dry AMD, a recent burst in molecular signaling data have yielded some very promising phar-macological targets. Technical advances in tomographic and fluorescent imaging of the retina and choroid have aided this effort greatly with the ability to visualize pathology in vivo. Generation of reactive oxygen species, phagocyte dysfunction, lipofuscin and visual cycle inter-mediate overload, immunogenic protein modifications, and cytotoxic RNA molecules all have been proposed as pathologic mechanisms for this disease. While these molecular constituents are further studied and the scope of clinical trial data expands, the threat of this blinding disease will hopefully fade with new diagnostic modalities and treatment approaches.

REFERENCES

Abecasis, G. R., Yashar, B. M., Zhao, Y., Ghiasvand, N. M., Zareparsi, S., Branham, K. E., et al. (2004). Age-related macular degeneration: A high-resolution genome scan for susceptibility loci in a population enriched for late-stage disease. *American Journal of Human Genetics, 74*, 482–494. doi:10.1086/382786.

Allikmets, R., Bergen, A. A., Dean, M., Guymer, R. H., Hageman, G. S., Klaver, C. C., et al. (2009). Geographic atrophy in age-related macular degeneration and TLR3. *New England Journal of Medicine, 360*, 2251–2254; Author Reply, 2255–2256.

Allikmets, R., Shroyer, N. F., Singh, N., Seddon, J. M., Lewis, R. A., Bernstein, P. S., et al. (1997). Mutation of the Stargardt disease gene (ABCR) in age-related macular degeneration. *Science, 277*, 1805–1807. doi:10.1126/science.277.5333.1805.

Ambati, J., Anand, A., Fernandez, S., Sakurai, E., Lynn, B. C., Kuziel, W. A., et al. (2003). An animal model of age-related macular degeneration in senescent Ccl-2- or Ccr-2-deficient mice. *Nature Medicine, 9*, 1390–1397.

Anderson, D. H., Mullins, R. F., Hageman, G. S., & Johnson, L. V. (2002). A role for local inflammation in the formation of drusen in the aging eye. *American Journal of Ophthalmology, 134*, 411–431.

Anderson, D. H., Talaga, K. C., Rivest, A. J., Barron, E., Hageman, G. S., & Johnson, L. V. (2004). Characterization of beta amyloid assemblies in drusen: The deposits associated with aging and age-related macular degeneration. *Experimental Eye Research, 78*, 243–256.

Arbour, N. C., Lorenz, E., Schutte, B. C., Zabner, J., Kline, J. N., Jones, M., et al. (2000). TLR4 mutations are associated with endotoxin hyporesponsiveness in humans. *Nature Genetics, 25*, 187–191.

AREDS Study Group. (2001). A randomized, placebo-controlled, clinical trial of high-dose supplementation with vitamins C and E, beta carotene, and zinc for age-related macular degeneration and vision loss: AREDS report no. 8. *Archives of Ophthalmology, 119*, 1417–1436.

Baas, D. C., Ho, L., Ennis, S., Merriam, J. E., Tanck, M. W., Uitterlinden, A. G., et al. (2010). The complement component 5 gene and age-related macular degeneration. *Ophthalmology, 117*, 500–511. doi:10.1016/j.ophtha.2009.08.032.

Bellmann, C., Jorzik, J., Spital, G., Unnebrink, K., Pauleikhoff, D., & Holz, F. G. (2002). Symmetry of bilateral lesions in geographic atrophy in patients with age-related macular degeneration. *Archives of Ophthalmology, 120*, 579–584.

Chang, J. H., McCluskey, P. J., & Wakefield, D. (2006). Toll-like receptors in ocular immunity and the immunopathogenesis of inflammatory eye disease. *British Journal of Ophthalmology, 90*, 103–108.

Chen, H., Liu, B., Lukas, T. J., & Neufeld, A. H. (2008). The aged retinal pigment epithelium/choroid: A potential substratum for the pathogenesis of age-related macular degeneration. *PLoS ONE, 3*, e2339. doi:10.1371/journal.pone.0002339.

Chen, M., Forrester, J. V., & Xu, H. (2011). Dysregulation in retinal para-inflammation and age-related retinal degeneration in CCL2 or CCR2 deficient mice. *PLoS ONE, 6*, e22818. doi:10.1371/journal.pone.0022818.

Cherepanoff, S., Mitchell, P., Wang, J. J., & Gillies, M. C. (2006). Retinal autoantibody profile in early age-related macular degeneration: Preliminary findings from the Blue Mountains Eye Study. *Clinical & Experimental Ophthalmology, 34*, 590–595.

Chowers, I., Deleon, E., & Obolensky, A., Lederman, Alper, R., Berenstein, E., Chevion, M., & Banin, E. (2006). *Iron-Associated Oxidative Injury as a Novel Therapeutic Target in Age Related Macular Degeneration. Toward the Prevention of Age Related Macular Degeneration.* Paper presented at the ARVO 2006 Summer Eye Research Conference (SERC), Fort Myers, Florida.

Clark, S. J., Higman, V. A., Mulloy, B., Perkins, S. J., Lea, S. M., Sim, R. B., et al. (2006). H384 allotypic variant of factor H associated with age-related macular degeneration has different heparin-binding properties from the non-disease-associated form. *Journal of Biological Chemistry, 281*, 24713–24720. doi:10.1084/jbc.M605083200.

Coffey, P. J., Gias, C., McDermott, C. J., Lundh, P., Pickering, M. C., Sethi, C., et al. (2007). Complement factor H deficiency in aged mice causes retinal abnormalities and visual dysfunction. *Proceedings of the National Academy of Sciences of the United States of America, 104*, 16651–16656. doi:10.1083/pnas.0705079104.

Combadiere, C., Feumi, C., Raoul, W., Keller, N., Rodero, M., Pezard, A., et al. (2007). CX3CR1-dependent subretinal microglia cell accumulation is associated with cardinal features of age-related macular degeneration. *Journal of Clinical Investigation, 117*, 2920–2928.

Cousins, S. W., Espinosa-Heidmann, D. G., & Csaky, K. G. (2004). Monocyte activation in patients with age-related macular degeneration: A biomarker of risk for choroidal neovascularization? *Archives of Ophthalmology, 122*, 1013–1018. doi:10.1001/archopht.122.7.1013.

Crocker, P. R., & Gordon, S. (1985). Isolation and characterization of resident stromal macrophages and hematopoietic cell clusters from mouse bone marrow. *Journal of Experimental Medicine, 162*, 993–1014.

Curcio, C. A., Millican, C. L., Allen, K. A., & Kalina, R. E. (1993). Aging of the human photoreceptor mosaic: Evidence for selective vulnerability of rods in central retina. *Investigative Ophthalmology & Visual Science, 34*, 3278–3296.

Delori, F. C., Dorey, C. K., Staurenghi, G., Arend, O., Goger, D. G., & Weiter, J. J. (1995). In vivo fluorescence of the ocular fundus exhibits retinal pigment epithelium lipofuscin characteristics. *Investigative Ophthalmology & Visual Science, 36*, 718–729.

Dentchev, T., Milam, A. H., Lee, V. M., Trojanowski, J. Q., & Dunaief, J. L. (2003). Amyloid-beta is found in drusen from some age-related macular degeneration retinas, but not in drusen from normal retinas. *Molecular Vision, 9*, 184–190.

Despriet, D. D., Bergen, A. A., Merriam, J. E., Zernant, J., Barile, G. R., Smith, R. T., et al. (2008). Comprehensive analysis of the candidate genes CCL2, CCR2, and TLR4 in age-related macular degeneration. *Investigative Ophthalmology & Visual Science, 49*, 364–371. doi:10.1167/iovs.07-0656.

Ding, X., Patel, M., & Chan, C. C. (2009). Molecular pathology of age-related macular degeneration. *Progress in Retinal and Eye Research, 28*, 1–18. doi:10.1016/j.preteyeres.2008.10.001.

Edwards, A. O., Chen, D., Fridley, B. L., James, K. M., Wu, Y., Abecasis, G., et al. (2008). Toll-like receptor polymorphisms and age-related macular degeneration. *Investigative Ophthalmology & Visual Science, 49*, 1652–1659.

Edwards, A. O., Ritter, R., III, Abel, K. J., Manning, A., Panhuysen, C., & Farrer, L. A. (2005). Complement factor H polymorphism and age-related macular degeneration. *Science, 308*, 421–424.

Elman, M. J., Fine, S. L., Murphy, R. P., Patz, A., & Auer, C. (1986). The natural history of serous retinal pigment epithelium detachment in patients with age-related macular degeneration. *Ophthalmology, 93*, 224–230.

Garcia Filho, C. A. A., Yehoshua, Z., Gregori, G., Li, Y., Feuer, W., Penha, F. M., et al. (2012). Efficacy of the systemic complement inhibitor Eculizumab in AMD patients with drusen: The COMPLETE Study. *Investigative Ophthalmology & Visual Science, 53*, 2045.

Gehrs, K. M., Anderson, D. H., Johnson, L. V., & Hageman, G. S. (2006). Age-related macular degeneration—Emerging pathogenetic and therapeutic concepts. *Annals of Medicine, 38*, 450–471.

Geissmann, F., Jung, S., & Littman, D. R. (2003). Blood monocytes consist of two principal subsets with distinct migratory properties. *Immunity, 19*, 71–82.

Gold, B., Merriam, J. E., Zernant, J., Hancox, L. S., Taiber, A. J., Gehrs, K., et al. (2006). Variation in factor B (BF) and complement component 2 (C2) genes is associated with age-related macular degeneration. *Nature Genetics, 38*, 458–462.

Gu, X., Meer, S. G., Miyagi, M., Rayborn, M. E., Hollyfield, J. G., Crabb, J. W., et al. (2003). Carboxyethylpyrrole protein adducts and autoantibodies, biomarkers for age-related macular degeneration. *Journal of Biological Chemistry, 278*, 42027–42035.

Hageman, G. S., Anderson, D. H., Johnson, L. V., Hancox, L. S., Taiber, A. J., Hardisty, L. I., et al. (2005). A common haplotype in the complement regulatory gene factor H (HF1/CFH) predisposes individuals to age-related macular degeneration. *Proceedings of the National Academy of Sciences of the United States of America, 102*, 7227–7232.

Haines, J. L., Hauser, M. A., Schmidt, S., Scott, W. K., Olson, L. M., Gallins, P., et al. (2005). Complement factor H variant increases the risk of age-related macular degeneration. *Science, 308*, 419–421.

Hollyfield, J. G., Bonilha, V. L., Rayborn, M. E., Yang, X., Shadrach, K. G., Lu, L., et al. (2008). Oxidative damage-induced inflammation initiates age-related macular degeneration. *Nature Medicine, 14*, 194–198.

Holz, F. G., Bindewald-Wittich, A., Fleckenstein, M., Dreyhaupt, J., Scholl, H. P., & Schmitz-Valckenberg, S. (2007). Progression of geographic atrophy and impact of fundus autofluorescence patterns in age-related macular degeneration. *American Journal of Ophthalmology, 143*, 463–472. doi:S0002–9394(06)01327–4.

Ishibashi, T., Patterson, R., Ohnishi, Y., Inomata, H., & Ryan, S. J. (1986). Formation of drusen in the human eye. *American Journal of Ophthalmology, 101*, 342–353.

Johnson, L. V., Leitner, W. P., Rivest, A. J., Staples, M. K., Radeke, M. J., & Anderson, D. H. (2002). The Alzheimer's A beta-peptide is deposited at sites of complement activation in pathologic deposits associated with aging and age-related macular degeneration. *Proceedings of the National Academy of Sciences of the United States of America, 99*, 11830–11835.

Johnson, L. V., Leitner, W. P., Staples, M. K., & Anderson, D. H. (2001). Complement activation and inflammatory processes in drusen formation and age related macular degeneration. *Experimental Eye Research, 73*, 887–896.

Johnson, L. V., Ozaki, S., Staples, M. K., Erickson, P. A., & Anderson, D. H. (2000). A potential role for immune complex pathogenesis in drusen formation. *Experimental Eye Research, 70*, 441–449.

Jozsi, M., Oppermann, M., Lambris, J. D., & Zipfel, P. F. (2007). The C-terminus of complement factor H is essential for host cell protection. *Molecular Immunology, 44*, 2697–2706.

Kaneko, H., Dridi, S., Tarallo, V., Gelfand, B. D., Fowler, B. J., Cho, W. G., et al. (2011). DICER1 deficit induces Alu RNA toxicity in age-related macular degeneration. *Nature, 471*, 325–330.

Kaur, I., Hussain, A., Hussain, N., Das, T., Pathangay, A., Mathai, A., et al. (2006). Analysis of CFH, TLR4, and APOE polymorphism in India suggests the Tyr402His variant of CFH to be a global marker for age-related macular degeneration. *Investigative Ophthalmology & Visual Science, 47*, 3729–3735.

Kawai, T., & Akira, S. (2006). TLR signaling. *Cell Death and Differentiation, 13*, 816–825. doi:10.1038/sj.cdd.4401850.

Kennedy, D. W., & Abkowitz, J. L. (1998). Mature monocytic cells enter tissues and engraft. *Proceedings of the National Academy of Sciences of the United States of America, 95*, 14944–14949.

Killingsworth, M. C., Sarks, J. P., & Sarks, S. H. (1990). Macrophages related to Bruch's membrane in age-related macular degeneration. *Eye (London, England), 4*, 613–621.

Kindzelskii, A. L., Elner, V. M., Elner, S. G., Yang, D., Hughes, B. A., & Petty, H. R. (2004). Toll-like receptor 4 (TLR4) of retinal pigment epithelial cells participates in transmembrane signaling in response to photoreceptor outer segments. *Journal of General Physiology, 124*, 139–149.

Klein, M. L., Ferris, F. L., III, Francis, P. J., Lindblad, A. S., Chew, E. Y., Hamon, S. C., et al. (2010). Progression of geographic atrophy and genotype in age-related macular degeneration. *Ophthalmology, 117*, 1554–1559.

Klein, R., Klein, B. E., Knudtson, M. D., Meuer, S. M., Swift, M., & Gangnon, R. E. (2007). Fifteen-year cumulative incidence of age-related macular degeneration: The Beaver Dam Eye Study. *Ophthalmology, 114*, 253–262. doi:S0161–6420(06)01478–3.

Klein, R., Klein, B. E., & Linton, K. L. (1992). Prevalence of age-related maculopathy. The Beaver Dam Eye Study. *Ophthalmology, 99*, 933–943.

Klein, R. J., Zeiss, C., Chew, E. Y., Tsai, J. Y., Sackler, R. S., Haynes, C., et al. (2005). Complement factor H polymorphism in age-related macular degeneration. *Science, 308*, 385–389.

Kleinman, M. E., & Ambati, J. (2012). A window to innate neuroimmunity: Toll-like receptor-mediated cell responses in the retina. *Advances in Experimental Medicine and Biology, 723*, 3–9.

Kleinman, M. E., Kaneko, H., Cho, W. G., Dridi, S., Fowler, B. J., Blandford, A. D., et al. (2012). Short-interfering RNAs induce retinal degeneration via TLR3 and IRF3. *Molecular Therapy, 20*, 101–108.

Kubota, R., Boman, N. L., David, R., Mallikaarjun, S., Patil, S., & Birch, D. (2012). Safety and effect on rod function of ACU-4429, a novel small-molecule visual cycle modulator. *Retina (Philadelphia, Pa.), 32*, 183–188.

Kumar, M. V., Nagineni, C. N., Chin, M. S., Hooks, J. J., & Detrick, B. (2004). Innate immunity in the retina: Toll-like receptor (TLR) signaling in human retinal pigment epithelial cells. *Journal of Neuroimmunology, 153*, 7–15.

Leibowitz, H. M., Krueger, D. E., Maunder, L. R., Milton, R. C., Kini, M. M., Kahn, H. A., et al. (1980). The Framingham Eye Study monograph: An ophthalmological and epidemiological study of cataract, glaucoma, diabetic retinopathy, macular degeneration, and visual acuity in a general population of 2631 adults, 1973–1975. *Survey of Ophthalmology, 24*(Suppl.), 335–610.

London, A., Itskovich, E., Benhar, I., Kalchenko, V., Mack, M., Jung, S., et al. (2011). Neuroprotection and progenitor cell renewal in the injured adult murine retina requires healing monocyte-derived macrophages. *Journal of Experimental Medicine, 208*, 23–39.

Luhmann, U. F., Robbie, S., Munro, P. M., Barker, S. E., Duran, Y., Luong, V., et al. (2009). The drusenlike phenotype in aging Ccl2-knockout mice is caused by an accelerated accumulation of swollen autofluorescent subretinal macrophages. *Investigative Ophthalmology & Visual Science, 50*, 5934–5943.

Maguire, P., & Vine, A. K. (1986). Geographic atrophy of the retinal pigment epithelium. *American Journal of Ophthalmology, 102*, 621–625.

Majewski, J., Schultz, D. W., Weleber, R. G., Schain, M. B., Edwards, A. O., Matise, T. C., et al. (2003). Age-related macular degeneration—A genome scan in extended families. *American Journal of Human Genetics, 73*, 540–550.

Mattapallil, M. J., Wawrousek, E. F., Chan, C. C., Zhao, H., Roychoudhury, J., Ferguson, T. A., et al. (2012). The rd8 mutation of the Crb1 gene is present in vendor lines of C57BL/6N mice and embryonic stem cells, and confounds ocular induced mutant phenotypes. *Investigative Ophthalmology & Visual Science, 53*, 2921–2927. doi:10.1167/iovs.12-9662.

Medzhitov, R. (2008). Origin and physiological roles of inflammation. *Nature, 454*, 428–435.

Medzhitov, R., & Janeway, C., Jr. (2000). The toll receptor family and microbial recognition. *Trends in Microbiology, 8*, 452–456.

Mullins, R. F., Russell, S. R., Anderson, D. H., & Hageman, G. S. (2000). Drusen associated with aging and age-related macular degeneration contain proteins common to extracellular deposits associated with atherosclerosis, elastosis, amyloidosis, and dense deposit disease. *Federation of American Societies for Experimental Biology Journal, 14*, 835–846.

Oda, K., & Kitano, H. (2006). A comprehensive map of the toll-like receptor signaling network. *Molecular Systems Biology, 2*, 2006.0015. doi:10.1038/msb4100057.

Patel, N., Ohbayashi, M., Nugent, A. K., Ramchand, K., Toda, M., Chau, K. Y., et al. (2005). Circulating anti-retinal antibodies as immune markers in age-related macular degeneration. *Immunology, 115*, 422–430.

Penfold, P. L., Madigan, M. C., Gillies, M. C., & Provis, J. M. (2001). Immunological and aetiological aspects of macular degeneration. *Progress in Retinal and Eye Research, 20*, 385–414.

Rakoczy, P. E., Zhang, D., Robertson, T., Barnett, N. L., Papadimitriou, J., Constable, I. J., et al. (2002). Progressive age-related changes similar to age-related macular degeneration in a transgenic mouse model. *American Journal of Pathology, 161*, 1515–1524.

Sarks, J. P., Sarks, S. H., & Killingsworth, M. C. (1988). Evolution of geographic atrophy of the retinal pigment epithelium. *Eye (London, England), 2*, 552–577. doi:10.1038/eye.1988.106.

Schmitz-Valckenberg, S., Alten, F., Steinberg, J. S., Jaffe, G. J., Fleckenstein, M., Mukesh, B. N., et al. (2011). Reticular drusen associated with geographic atrophy in age-related macular degeneration. *Investigative Ophthalmology & Visual Science, 52*, 5009–5015.

Seddon, J. M., Santangelo, S. L., Book, K., Chong, S., & Cote, J. (2003). A genomewide scan for age-related macular degeneration provides evidence for linkage to several chromosomal regions. *American Journal of Human Genetics, 73*, 780–790.

Shen, D., Wen, R., Tuo, J., Bojanowski, C. M., & Chan, C. C. (2006). Exacerbation of retinal degeneration and choroidal neovascularization induced by subretinal injection of Matrigel in CCL2/MCP-1-deficient mice. *Ophthalmic Research, 38*, 71–73.

Smith, W., Assink, J., Klein, R., Mitchell, P., Klaver, C. C., Klein, B. E., et al. (2001). Risk factors for age-related macular degeneration: Pooled findings from three continents. *Ophthalmology, 108*, 697–704.

Tarallo, V., Hirano, Y., Gelfand, B. D., Dridi, S., Kerur, N., Kim, Y., et al. (2012). DICER1 loss and Alu RNA induce age-related macular degeneration via the NLRP3 inflammasome and MyD88. *Cell, 149*, 847–859.

Tuo, J., Bojanowski, C. M., Zhou, M., Shen, D., Ross, R. J., Rosenberg, K. I., et al. (2007). Murine ccl2/cx3cr1 deficiency results in retinal lesions mimicking human age-related macular degeneration. *Investigative Ophthalmology & Visual Science, 48*, 3827–3836.

Tuo, J., Smith, B. C., Bojanowski, C. M., Meleth, A. D., Gery, I., Csaky, K. G., et al. (2004). The involvement of sequence variation and expression of CX3CR1 in the pathogenesis of age-related macular degeneration. *Federation of American Societies for Experimental Biology Journal, 18*, 1297–1299.

Volkman, A., & Gowans, J. L. (1965). The origin of macrophages from bone marrow in the rat. *British Journal of Experimental Pathology, 46*, 62–70.

von Ruckmann, A., Fitzke, F. W., & Bird, A. C. (1997). Fundus autofluorescence in age-related macular disease imaged with a laser scanning ophthalmoscope. *Investigative Ophthalmology & Visual Science, 38*, 478–486.

Weismann, D., Hartvigsen, K., Lauer, N., Bennett, K. L., Scholl, H. P., Charbel Issa, P., et al. (2011). Complement factor H binds malondialdehyde epitopes and protects from oxidative stress. *Nature, 478*, 76–81. doi:10.1038/nature10449.

Weiter, J. J., Delori, F., & Dorey, C. K. (1988). Central sparing in annular macular degeneration. *American Journal of Ophthalmology, 106*, 286–292.

Xu, H., Manivannan, A., Dawson, R., Crane, I. J., Mack, M., Sharp, P., et al. (2005). Differentiation to the CCR2+ inflammatory phenotype in vivo is a constitutive, time-limited property of blood monocytes and is independent of local inflammatory mediators. *Journal of Immunology (Baltimore, MD.: 1950), 175*, 6915–6923.

Yang, Z., Stratton, C., Francis, P. J., Kleinman, M. E., Tan, P. L., Gibbs, D., et al. (2008). Toll-like receptor 3 and geographic atrophy in age-related macular degeneration. *New England Journal of Medicine, 359*, 1456–1463.

Yates, J. R., Sepp, T., Matharu, B. K., Khan, J. C., Thurlby, D. A., Shahid, H., et al. (2007). Complement C3 variant and the risk of age-related macular degeneration. *New England Journal of Medicine, 357*, 553–561.

Yehoshua, Z., Rosenfeld, P. J., Gregori, G., Feuer, W. J., Falcao, M., Lujan, B. J., et al. (2011). Progression of geographic atrophy in age-related macular degeneration imaged with spectral domain optical coherence tomography. *Ophthalmology, 118*, 679–686.

Zareparsi, S., Buraczynska, M., Branham, K. E., Shah, S., Eng, D., Li, M., et al. (2005). Toll-like receptor 4 variant D299G is associated with susceptibility to age-related macular degeneration. *Human Molecular Genetics, 14*, 1449–1455.

Zeng, J., Chen, Y., Tong, Z., Zhou, X., Zhao, C., Wang, K., et al. (2010). Lack of association of CFD polymorphisms with advanced age-related macular degeneration. *Molecular Vision, 16*, 2273–2278.

109 Gene Therapy for Retinal Degeneration

CURTIS R. BRANDT

Diseases that involve degeneration of the retina are a significant cause of inherited vision loss in the world. Traditionally, diseases that have been referred to as retinal dystrophies involved loss of photoreceptor cells, either through the inheritance of mutations that directly affected such cells or indirectly through mutations that negatively affected the function of retinal pigment epithelial cells. The scope of retinal degeneration can be expanded, however, to include glaucoma, which involves the death of retinal ganglion cells, and therefore is a retinal degeneration. Although these diseases have a wide diversity of initiating mechanisms, they share one common feature; the loss of vision ultimately occurs through neuronal apoptosis. In this chapter, we will focus on diseases that affect photoreceptor viability, but some of the concepts can be applied broadly to other diseases that involve neuronal apoptosis as the final pathological step.

Retinal degenerations present a number of significant challenges for the development of therapeutic strategies. Mutations in a large number of genes can cause degenerative disease. Even within a single gene there can be a large number of causative mutations. This genetic heterogeneity presents significant difficulties in developing therapies. For small molecule and gene therapy approaches, there are numerous potential targets, and for many gene therapy approaches targeted to a specific gene, the causative gene must be identified for each patient. Genetic diagnostics are improving but remain expensive. Given that numerous mutations in a single gene can be causative, targeted therapies might even require the identification of the specific mutation. Deciding on a therapeutic approach based on the clinical phenotype is not possible because mutations in any of several genes can lead to a common phenotype. In fact, mutations in the same gene can also lead to very different clinical phenotypes (den Hollander et al., 2010). In addition, the nature of the mutation also influences the potential therapeutic approaches. Retinal degenerations can be inherited in an autosomal recessive, autosomal dominant, or X-linked manner.

Autosomal dominant mutations present a significant challenge in that simply delivering a normal copy of the gene to a cell will not correct the defect. Haploinsufficiency phenotypes are a challenge as treatment might involve finely adjusting the level of gene expression (Frio et al., 2008). In addition to these issues, there are challenges involved in the actual delivery of the gene, the choice of vectors, the availability of suitable promoters, and innate and acquired host immune responses that need to be considered.

Although this list of challenges seems daunting, there is hope. Recently completed and ongoing clinical trials of gene therapy for Leber's congenital amaurosis (LCA) have resulted in some gain of function for individuals treated with an adeno-associated virus (AAV) vector expressing the RPE 65 gene.

THE NATURE OF INHERITED RETINAL DYSTROPHIES

Currently, there are 27 defined categories of retinal degenerative disease, and as of December 27, 2012, 231 genes or loci have been mapped for the different categories (https://sph.uth.tmc.edu/RetNet/). A total of 191 genes have been identified. This diversity creates significant issues for gene delivery strategies because one needs to know what gene to deliver for treatment. Looking at a single category such as autosomal recessive retinitis pigmentosa (arRP), we find that 23 specific genes have been linked to the disease. For some other forms of retinal degeneration, fewer numbers of genes have been identified. The gene encoding rhodopsin is mutated in a number of patients with RP, but just considering rhodopsin, over 100 specific mutations have been identified leading to the phenotypes mentioned above (Buch, Bainbridge, & Ali, 2008; den Hollander et al., 2010). For some gene delivery strategies, it would be important to identify the specific mutation. In addition to identifying the gene in a particular patient, one needs to have access to vectors to deliver each of the genes, and obtaining final approval for therapeutic use of this many vectors would be prohibitive at the present time. As we will discuss below, there are some strategies that would reduce the numbers of vectors needed.

Even considering a single gene such as rhodopsin, individual patients can have unique mutations that lead to different clinical phenotypes involving appearance of the retina, the timing of the onset of vision loss, and the progression of the disease (den Hollander et al., 2010). These factors also influence gene therapy strategies. For example, it would be important to identify, diagnose, and treat patients before significant numbers of photoreceptor cells had been lost in order to preserve as much function as possible. Very early onset of cell loss would require treatment during childhood, and diseases with rapid cell loss would require treatment as soon as possible after diagnosis.

STRATEGIES FOR OCULAR GENE DELIVERY

To illustrate some of the issues involved in gene therapy, we will look at examples representing an autosomal recessive disease (LCA), an autosomal dominant disease (autosomal dominant RP), and an X-linked disease (X-linked retinoschisis; XLR). Each of these requires different approaches due to the nature of the mutation.

LEBER'S CONGENITAL AMAUROSIS—AN AUTOSOMAL RECESSIVE GENE THERAPY SUCCESS STORY

The RPE65 protein (553 amino acids) encodes an isomerase that is involved in regenerating 11-*cis*-retinal from the all-*trans*-retinol generated during photoisomerization (Cai, Conley, & Naash, 2009). The gene for *RPE65* was mapped to human chromosome 1 in 1994. The gene is large, containing 14 exons spanning 20 kb in the genome (Cai, Conley, & Naash, 2009). Currently, there are few vectors that have a carrying capacity this large and can efficiently deliver a transgene. Fortunately, the spliced complementary DNA (cDNA) can be used, which substantially reduces the size of the transgene so that transfer can be achieved with vectors having small carrying capacities.

To date, over 60 different mutations have been associated with inherited retinal dystrophy and have been identified in multiple diseases including LCA and autosomal recessive RP. The observation that mutations in the same gene can cause different diseases is common for the retinal dystrophies. Approximately 16% of the cases of LCA are due to mutations in *RPE65* (Cideciyan, 2010). Within those mutations in *RPE65*, there is heterogeneity in the severity of the disease with some forms progressing much faster than others. The heterogeneity is related to the effect of the mutation on the protein. In some forms, RPE65 is unstable (mutations Y144D, P363T, and Y368H) and is rapidly degraded, resulting in faster progression. Mutations in the RPE 65 protein that result in the retention of some function, such as R91W, P25L, and L22P, have slower progressing phenotypes (Cai, Conley, & Naash, 2009).

A number of animal models are available, and strategies for using gene therapy were initially developed using these models. The models include the Briard dog, a naturally occurring mouse model ($Rpe65^{rd12}$), a genetically engineered deletion mutant ($Rpe65^{-/-}$), the $Rpe65^{R91W/R91W}$ mouse, and the *rd12 mouse* (Bemelmans et al., 2006). Most of these studies have used either adeno-associated virus vectors or lentiviral-based vectors (vectors are described below).

Given that LCA is autosomal recessive, it requires that the patient carry two copies of the mutant allele. It should therefore be relatively straightforward to correct the defect by delivering a normal copy of the gene as long as haploinsufficiency is not an issue. Haploinsufficiency occurs when the phenotype is dependent on the expression of normal levels of a protein and is more common with structural proteins. Since RPE65 is an enzyme, lower levels of protein appear to be sufficient because people who are heterozygous do not appear to have retinal dysfunction.

Initial studies in animal models, most notably the Briard dog, demonstrated that delivery of a wild-type copy of the *RPE65* gene using an adeno-associated virus vector (see below) could restore function as measured by electroretinograms (ERGs) and, in the case of the dog model, the ability to navigate an obstacle course (Acland et al., 2001). In addition to improvement in visual function, functional magnetic resonance imaging (fMRI) revealed that gene delivery resulted in increased cortical activity in the lateral gyrus (Aguirre et al., 2007). This was somewhat surprising because of the photoreceptor loss and loss of vision at an early age but was very encouraging because it raises the possibility that significant improvements in visual function might be obtainable even after a substantial loss of photoreceptors.

The success in the Briard dog model led to the initiation of phase I clinical trials at multiple sites with a total of 30 patients to date (Bainbridge et al., 2008; Cideciyan et al., 2008; Maguire et al., 2008, 2009; Simonelli et al., 2010; Colella and Auricchio, 2012). In all of the trials, AAV2 vectors expressing the *RPE65* gene were delivered by subretinal injection, but each vector used different promoters (Bainbridge et al., 2008; Cideciyan et al., 2009; Hauswirth et al., 2008). Two studies used the chicken β-actin promoter, but one was modified to include a Kozak sequence to improve translation. The third trial used a short RPE65 promoter. The results were somewhat variable in terms of objective measures of visual function but improvements in light sensitivity, reduction of nystagmus, increased visual fields, and improvement in navigation tasks were reported (Colella & Auricchio, 2012). Thus, these early results are very encouraging.

The reason for the variability in the response is not clear but could result from the use of different promoters or differences in the delivery of the vector. The sites of subretinal injection were slightly different in each patient, and the fovea was not targeted in all of the patients (Cideciyan, 2010). Individual patient differences might also be involved. These and other factors that might contribute to variable outcomes following gene delivery will be discussed in more detail below.

One critical question arises from the severity of the photoreceptor loss which results in symptomatic presentation at birth or in the first few months after birth. Since gene therapy does not currently result in neuronal regeneration, it will only protect the remaining photoreceptors. Since the patients in the trials had advanced disease, it was not clear if any cortical functionality had been retained. Cortical function in the trial subjects was evaluated using fMRI, and it was found that the patients demonstrated increased responses to a range of contrast stimuli (Ashtari et al., 2011) and that the improved responses were seen in the areas where the transgene was delivered. These results and similar results in dogs (Aguirre et al., 2007) suggest that severe photoreceptor loss does not eliminate visual cortex function and that more significant therapeutic outcomes might be achieved if therapy were to be initiated earlier.

AUTOSOMAL DOMINANT RETINITIS PIGMENTOSA—A DIFFERENT GENE THERAPY CHALLENGE

RP is a heterogeneous group of retinal dystrophies involving gradual loss of photoreceptors and vision. Overall, the frequency is 1:4000. Of the inheritance patterns, 30% of patients will have the autosomal dominant (adRP) form. Commonly, rod photoreceptors are lost first, resulting in night blindness followed by the loss of cone photoreceptors. Currently 19 different genes have been mapped in adRP patients (https://sph.uth.tmc.edu/RetNet/). Thus, the situation is similar to other retinal dystrophies where multiple genes can be involved and successful gene therapy may require that the mutant genes be indentified in each patient. In addition, mutations in some of the adRP genes have also been linked to other forms of retinal degeneration (den Hollander et al., 2010). In the autosomal dominant form, the patient need only inherit one copy of the mutant allele. Expression of the mutant form of the protein overrides the function of the remaining normal copy; thus the strategy of delivering a normal copy of the gene, as was done for LCA, will not be therapeutic.

One of the genes that causes adRP is rhodopsin, the photopigment in rod photoreceptors (den Hollander et al., 2010). Many of the mutations in rhodopsin are thought to result in improper folding or posttranslational processing of the protein resulting in trapping of the protein in the endoplasmic reticulum, activation of the misfolded protein stress response, and ultimately activation of apoptosis and rod cell death. Gene therapy strategies targeting apoptosis are discussed later. The presence of normal rhodopsin expressed from a wild-type allele will not prevent the pathology, so other strategies need to be developed. These include correcting the mutation and restoring the genome to wild type, eliminating expression of the mutant allele, or correcting the altered folding or processing of the mutant rhodopsin.

Targeted mutagenesis of rhodopsin in the retinas of mice has been described, suggesting the strategy is potentially viable (Chan et al., 2011). This required the introduction of a specific nuclease site in a transgenic rhodopsin-GFP (green fluorescent protein) mouse followed by the delivery of the nuclease in an AAV vector to the photoreceptor cells. The nuclease cleaved the DNA, triggering cellular DNA repair pathways that restored rhodopsin-GFP expression either by end-joining or single-strand annealing homologous recombination. It appeared that 100% of the transduced rod cells were affected. In this study, a mutation was induced rather than a reversion of a mutated gene. The reliance on the introduction of a specific nuclease site would not be feasible in patients.

Protein folding in cells is guided, in many cases, by interaction with proteins called chaperones. Since mutations in rhodopsin result in improper folding, one potential strategy to correct the defect would be to restore correct folding by delivering a gene that improves chaperone activity. Heat shock proteins or other chaperones with general activity might be worth exploring. Some proteins have specific chaperones, so delivery of these genes might have a therapeutic effect. The Hsp70 related protein BiP/Grp78 is a key endoplasmic reticulum stress sensor that plays a role in modulating intracellular calcium levels by increasing storage. The end result is to inhibit Ca^{++}-mediated apoptosis. In the P23H rat model, overexpression of BiP/Grp78, through AAV-mediated gene delivery, protected photoreceptors and restored visual function (Gorbatyuk et al., 2010). Rhodopsin is glycosylated, and some adRP mutations are known to affect glycosylation. The calnexin protein is a chaperone for glycosylated proteins, and it has been shown that calnexin can improve rhodopsin folding in the presence of 11-*cis*-retinal (Noorwez, Sama, & Kaushal, 2009). In calnexin null (*cnx*$^{-/-}$) mice, there are subtle changes in the structure of the retina but rhodopsin expression and outer segment structure appear normal (Kraus et al., 2010), so it is not clear whether calnexin delivery would have a protective effect, but there are other chaperones that might. Increasing the efficiency of the phototransduction cascade through delivery of an enhanced arrestin has been shown to have a protective effect (Song et al., 2009). It is possible that other mechanisms to enhance phototransduction in mutated photoreceptors might have protective effects.

Another strategy would be to eliminate or substantially reduce expression of the mutant allele. This could be achieved by introducing a stop codon very near the amino-terminus of the protein, but these types of strategies are currently inefficient (Andrieu-Soler et al., 2007). A more feasible strategy is to use short hairpin RNA (shRNA) or short interfering RNA (siRNA) (Wood, Yin, & McClorey, 2007) targeted to the mutant allele to reduce expression (see figure 109.1). Continued long-term expression of the shRNA or siRNA gene would suppress synthesis of the mutant protein, and specific targeting would result from the presence of the sequence change in the mutant allele compared to the wild-type copy of the gene. This strategy has been successfully applied in mouse and rat models of adRP (Gorbatyuk et al., 2007; Jiang et al., 2011).

Although targeting of the mutant allele with shRNA or siRNA seems straightforward, there are at least two significant problems. To date, over 200 mutations leading to retinal degeneration have been identified in rhodopsin. The shRNA and siRNA strategies both rely on specific targeting of the ribozyme to the mutated sequence in the genome. Therefore, not only would gene therapy utilizing delivery of a specific ribozyme require that rhodopsin be identified as the causative gene in a specific patient, but the exact sequence of the mutated allele would also need to be identified. Testing of each patient would be prohibitively expensive. Furthermore, over 200 different sh or si RNAs would need to be developed, tested, approved by the Food and Drug Administration, and stockpiled for use. The effort and expense needed to achieve this is not feasible given the number of people that are affected.

An alternative gene delivery strategy for treating an autosomal dominant disease would be to use an siRNA or shRNA strategy to suppress expression of both the mutant and wild-type alleles coupled with delivery of a replacement wild-type copy of the gene that lacked the si/shRNA targeting site (Millington-Ward et al., 2011), illustrated by figure 109.1. This approach has been successfully demonstrated in several animal models including Rho$^{-/-}$, Rho$^{+/-Pro/347Ser}$, Rho^{P23H}, Rho$\Delta^{I-255/256}$, Rho^{G909D} mice and the Rho^{P23H} rat (Chadderton et al., 2009; Millington-Ward et al., 2011; O'Reilly et al., 2007). Both hammerhead and hairpin ribozymes have been used. Long-term (up to 1 year after delivery) therapeutic effects, using this approach, have been demonstrated in a mouse model of the autosomal dominant cone dystrophy caused by mutations in the Ca^{++}-binding protein GCAP1 (Jiang et al., 2011). This approach holds great promise because it would require that the mutation be mapped to a specific gene in a patient but the actual mutation would not need to be identified. This would reduce diagnostic costs. In addition, only two vectors would need to be validated, the ribozyme that cleaves messenger RNA (mRNA) expressed from both alleles and the wild-type allele with the mutated si/shRNA target site.

Because rhodopsin is a structural protein, delivery of one functional copy of the gene might result in a haploinsufficiency phenotype. Although this has not apparently been an issue in the animal studies to date, haploinsufficiency is a potential problem. This could be addressed through the choice of promoter driving expression of the transgene or by including multiple copies of the transgene in the vector. The expression might also be altered by the inclusion of *cis*-acting sequences that regulate mRNA turnover or translation.

X-LINKED RETINOSCHISIS (XLR)

Several types of X-linked retinal dystrophies have been identified, including cone or rod dystrophies, congenital stationary night blindness, optic atrophy, RP, and combined syndromic/systemic diseases with retinopathy. To date, 18 genes and loci have been identified as having an X-linked inheritance pattern, and 11 of the genes have been identified (https://sph.uth.tmc.edu/RetNet/). Although X-linked diseases are separated categorically because of the inheritance pattern, for gene therapy, the approaches that can be used are similar to autosomal recessive diseases, particularly when heterozygous females are asymptomatic carriers. Delivery of a wild-type allele to the retina should result in a positive therapeutic outcome.

X-linked retinoschisis is caused by mutations in the RS1 gene (retinoschisin) which was identified in 1997

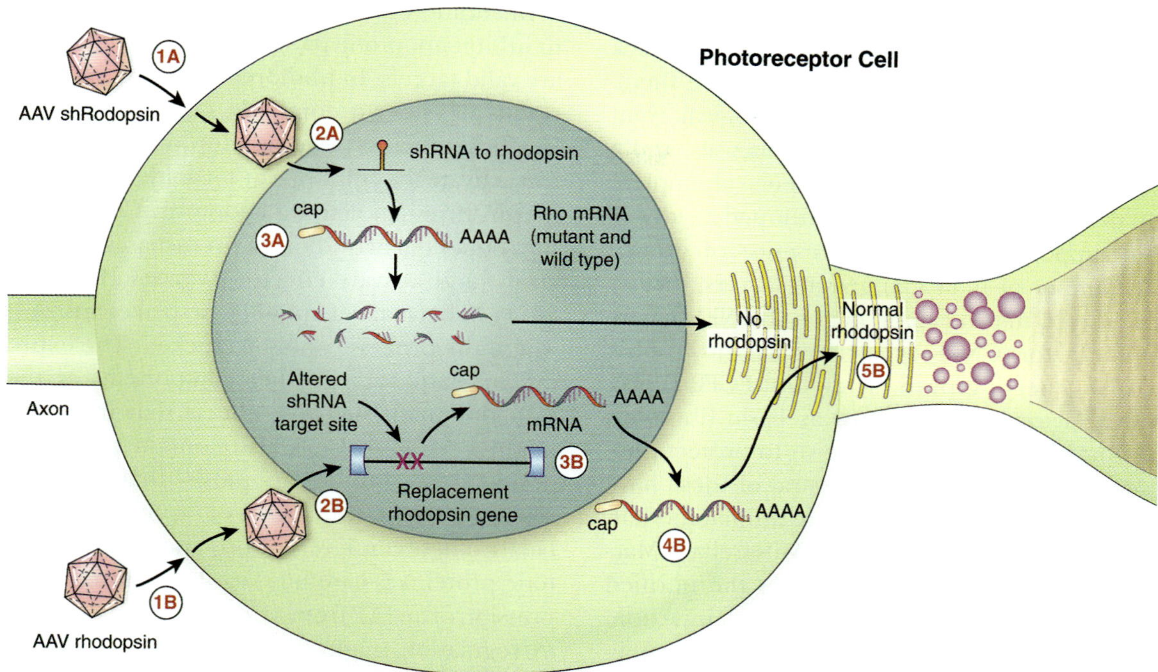

FIGURE 109.1   RNA interference strategies for retinal gene therapy in autosomal dominant diseases. The vector virus expressing a short hairpin RNA (shRNA), short interfering RNA (siRNA), or a ribozyme attaches to the cell (1A) and enters the nucleus (2A). The shRNA targeting the rhodopsin (both wild type and mutant) RNA is expressed, binds to the target messenger RNA (mRNA), and triggers degradation of the mRNA (3A), resulting in reduced levels of rhodopsin protein. At the same time, another vector that expresses a wild-type rhodopsin carrying mutations in the shRNA target site is delivered (1B and 2B). The host RNA polymerase II transcribes the vector (3B), which is translated into the wild-type rhodopsin (4B) that is then transported to the outer segments (5B). AAV, adeno-associated virus.

(Sikkink et al., 2007). As with other retinal dystrophies several inactivating mutations have been identified and there is considerable heterogeneity in disease severity. Foveal schisis (retinal folding) is a characteristic sign and is related to reduced attachment between retinal layers. Ultimately, vitreous hemorrhage and retinal detachment can occur. The protein, retinoschisin, contains discoidin domains seen in cell surface proteins involved in cell adhesion, perhaps explaining the phenotype.

Three mouse models are available (Sikkink et al., 2007). One was generated by replacing exon 3 of the mouse gene with a LacZ/Neo cassette which generated a null model (XLRS1h$^{-/Y}$). Another null model contains a Neo^r cassette in exon/intron 1. Finally a mutation in intron 2 was identified in an ENU-based mutagenesis screen. This creates a new splice site resulting in three transcripts, the normal one and two containing premature stop codons. The phenotype of this mutation might be the result of a haploinsufficiency mechanism.

The feasibility of using a gene delivery approach to treat XLR has been demonstrated using the XLRS1h$^{-/Y}$ and Rs1$^{-/y}$ mouse models (Park et al., 2009; Takada et al., 2008). In one study an AAV8 vector expressing the *Rs-1* gene was delivered by intravitreal injection, resulting in expression in the neural retina and functional restoration as measured by ERG responses. The specific expression in the retina was achieved by the use of a 3.5 Kb fragment of the *Rs-1* promoter because multiple cell types, including photoreceptors, optic nerve, ciliary body, and the peripheral cornea, expressed a marker gene when the CMV promoter was used. It is interesting that successful delivery was achieved using this route when successful delivery in other studies has required subretinal injection to efficiently transduce photoreceptors. This might be due to the disruption of the retina that occurs in the disease. The widespread delivery to multiple cell types could be detrimental for some genes and could lead to unwanted effects if the promoter was widely active.

TARGETING THE COMMON PATHOLOGICAL EVENT

Although retinal degenerative disease is caused by mutations in several genes, in the end, the loss of vision is due to photoreceptor death through apoptosis (Fuchs

& Steller, 2011). Because there is a common final patho-logical event, gene delivery strategies that seek to block various steps in the apoptotic process could be thera-peutic. Apoptosis is also a common pathological event in a number of neuronal degenerative diseases, and a number of gene delivery strategies have been attempted.

Neuronal survival relies on the continued feedback of trophic support through the presence of growth factors. A number of neuronal growth factors, includ-ing brain derived neurotrophic factor (BDNF), ciliary neurotrophic factor (CNTF), glial derived neurotrophic factor (GDNF), nerve growth factor, and FGF2 (fibro-blast growth factor 2, also known as basic fibroblast growth factor; bFGF) have been shown to protect pho-toreceptors (LaVail et al., 1992). Several of these have provided some degree of protection in animal models, and some have reached clinical trials (Musarella & Mac-Donald, 2011). In many of these studies the purified proteins were administered via intravitreal injection, resulting in transient effects due to turnover. Geneti-cally engineered cells expressing neurotrophins have also been tested, but the cells must be placed in a device that is then implanted. Delivery of the gene for a neu-rotrophin with long-term expression avoids some of these problems.

For retinal degeneration, a number of neurotrophins or growth factors including CNTF, BDNF, pigment epi-thelium-derived growth factor, basic fibroblast growth factor (bFGF; FGF2), Lens epithelium-derived growth factor and GDNF have been tested in animal models (LaVail et al., 1992). Many of these studies were short-term, but in those that examined longer term protec-tion, the effects were transient. Although these studies showed that trophic factor delivery preserved retinal anatomy by preventing photoreceptor death, function-ality was not restored in some studies when measured by ERG and behavioral tests.

The viral vectors themselves also have effects. Many viruses express proteins designed to block apoptotic pathways, and viral vectors that still carry these genes can promote cell survival. In addition, even the vectors themselves have been reported to protect photorecep-tors (Takita et al., 2008), but the mechanism has not been determined. This may be due to vector-mediated up-regulation of a growth factor or neurotrophin via an innate signaling pathway. Vectors have also been shown to negatively affect photoreceptor function by suppress-ing ERG responses. Physiologically, an HSV-1 vector lacking a therapeutic gene was shown to suppress both a- and b-wave ERG responses in rats, and this effect could not be completely overcome when a growth factor gene was included in the vector (Spencer et al., 2001).

In addition to using growth factors or neurotrophins to inhibit apoptotic pathways, several steps in apoptosis are valid targets. In photoreceptors where the mutation results in either a dominant negative or toxic gain of function mutation, the buildup of the mutant protein can activate the cell stress or misfolded protein response that in turn can activate apoptosis. Cell death involves the sequential activation of precaspases that then act to cleave and activate effector caspases (Fuchs & Steller, 2011). There are a number of places in the pathway where the process could be blocked. The expression of mutant caspases, or other components of the system with dominant negative effects that inhibit function, might be useful. Cells also contain proteins that are designed to inhibit apoptosis that are collectively referred to as inhibitors of apoptosis (IAPs). The balance between levels of pro-apoptotic and anti-apop-totic proteins is carefully regulated in the cell. Overex-pression of an IAP from a promoter that is not responsive to regulation might be protective, and this has been shown for AAV-mediated delivery of the X-linked inhibi-tor of apoptosis (XIAP) gene in the P23H and S334ter rat models of RP (Leonard et al., 2007). In the P23H model, functional preservation as assessed by ERGs was evident out to 26 weeks, while in the S334ter model the effects were less evident and long-term protection was not achieved. Although it is not clear why the results were different in the two models, it does emphasize that the nature of the mutation and the speed of the retinal degeneration can make an important difference. AAV-mediated delivery of XIAP to transplanted rod precur-sors has been shown to increase the survival of those precursors in the rd9 model of X-linked degeneration, raising the possibility that genetic modification of stem cells by gene delivery might improve stem-cell-mediated therapies (Yao et al., 2011).

Because apoptosis is potentially lethal to cells, the system is regulated at several levels. In addition to IAPs, the amounts of two proteins, Bcl-2 and Bax, are also critical in regulating cell death (Chang, Hao, & Wong, 1993; Fuchs & Steller, 2011). The Bcl-2 protein is anti-apoptotic, and when levels of Bcl-2 are high, apoptosis is inhibited. If Bcl-2 levels drop or if Bax levels increase, the ratio of the two is altered and apoptosis will be trig-gered. Many of the neurotrophins mentioned above affect the levels of Bcl-2 and Bax, and this may be one mechanism for the protective effect. Since Bcl-2 is pro-tective, vector-mediated delivery of the Bcl-2 gene expressed from a constitutively active promoter might be protective.

Apoptosis inducing factor (AIF) is a flavoprotein located in the mitochcondrial intermembrane space and functions as a caspase-independent inducer, perhaps by sensing changes in cellular redox potential. When a signal is received, AIF translocates to the nucleus and causes peripheral chromatin condensation and DNA fragmentation. In the RCS rat model of retinal degeneration, AIF translocates to the nucleus and preventing this event has been shown to be protective. Pigment epithelium-derived factor gene delivery to RPE cells has been shown to prevent retinal degeneration, and inhibition of AIF translocation appears to be a significant component of the mechanism (Murakami et al., 2008).

Although blocking apoptosis is a potentially valuable strategy, the animal studies to date have been somewhat disappointing because very long-term protection has been difficult to achieve. Perhaps combining anti-apoptotic strategies with other approaches will provide improved protection in retinal degenerative diseases.

METHODS FOR RETINAL GENE DELIVERY

Viruses can be defined as a package designed to transfer genomic material from one cell to another. In other words, nature designed a gene delivery system long ago. It is not surprising then that viral gene delivery vectors are the most common gene delivery method. However, other nonviral methods have also been explored. As with any particular system, there are advantages and disadvantages, and the choice of delivery method used will be different depending on a number of specific factors.

VIRAL GENE DELIVERY VECTORS

Several different viral vectors have been used for ocular gene delivery in animal models. These include adeno-associated virus (AAV), adenovirus (Ad) (Sweigard, Cashman, & Kumar-Singh, 2010), lentiviruses (Balaggan & Ali, 2012), herpes simplex virus (HSV) (Liu et al., 1999), and baculoviruses (Bv) (Haeseleer et al., 2001). The most commonly used viral vector systems have been those based on AAV, and AAV vectors are the only ones that have been used in clinical trials to date (den Hollander et al., 2010). To illustrate the advantages and disadvantages of viral-mediated delivery, we will focus on AAV and lentiviral vector systems.

Many factors govern the choice of viral vector that should be used. Perhaps the most important is the ability of a particular vector to transduce the desired cellular target. This is primarily driven by the receptor the virus uses to attach to the cell. Some vectors have a broad host range, and others can be highly selective (Allocca et al., 2007; Coura Rdos & Nardi, 2007). In addition to using the natural receptor affinity, it is possible to exchange viral attachment proteins (pseudotyping) or selectively mutate viral attachment proteins to target specific cellular receptors. For example, retrovirus-based vectors systems commonly use the vesicular stomatitis virus G protein (VSVG) in place of the *env* gene (envelope) resulting in a very broad host range (Bemelmans et al., 2005). Several different serotypes of AAV exist, and they have different receptor specificities that can lead to targeting of specific cell types (Allocca et al., 2007). Although most ocular gene delivery studies to date have not focused on deliberately targeting cells types through receptor specificity, there may be circumstances where this would be desirable.

Another important factor is the carrying capacity of the vector, or how large a transgene can be delivered. Typically, the actual gene is not transduced as most genes that are therapeutically interesting are quite large and have several exons. To reduce the size, the cDNA encoding the protein is cloned into the vector. In addition, a single splice junction is included to generate a spliced mRNA. Even so, some therapeutic genes are too large for some vector systems. The choice of promoter also has an impact on the size of the transgene construct. To date, relatively small promoters have been used, but these may lack specific regulatory sequences that might be important. In the human genome, important regulatory sequences such as enhancers or insulators can be located many kilobases from the actual gene, and inclusion of these elements exceeds the carrying capacity of most vector systems. It is feasible to include such sequences if they are small enough and can function out of the context of the genome, but very little work has been done in this regard. The use of cell type specific promoters can also be used to target expression in a desired target cell. For example, the opsin and rpe65 promoters have been used to drive transgene expression in photoreceptors and retinal pigment epithelial cells, respectively (Bemelmans et al., 2006).

Finally, the vector systems that are most likely to be used are those where it is relatively easy to prepare high titer vector stocks. Delivery to the eye requires the use of small volumes, so the ability to easily prepare stocks of vector with transducing titers of 1×10^8 or greater is an important feature. For studies in animal models small batches of vector are adequate, but for clinical trials and eventual clinical use, the ability to scale up production is an important factor.

ADENO-ASSOCIATED VIRUS VECTORS

AAV is a small, non-enveloped virus that requires a helper virus (usually adenovirus) for replication (Buch,

Bainbridge, & Ali, 2008). A high percentage of the population harbors AAV with no pathological consequences. This a significant advantage because the virus is defective to begin with and does not naturally cause disease. Serotype 2 is the most common, and most people have antibodies to Type 2 AAV. This potentially would affect the efficiency of gene delivery due to neutralization of the vector virus particle, but too date this has not been a problem for delivery to the eye. Although naturally acquired AAV infections generate immune responses, studies have shown that AAV vectors appear to be poor at stimulating the host immune system in mice and humans (Barker et al., 2009). Figure 109.2 shows a schematic diagram of AAV-mediated delivery to a photoreceptor cell.

The genome of AAV is a single-stranded DNA molecule approximately 4.7 Kb long. The genome contains only two open reading frames, and these are deleted in the gene delivery vectors. The ends of the genome contain a 145 base pair inverted repeat that is critical for replication, and the repeats must be retained in the vector. In fact, the only AAV sequences typically retained in the vector are the terminal repeats. This leaves approximately 4.0 to 5.0 Kb of carrying capacity, which is quite small (Buch, Bainbridge, & Ali, 2008).

In order to prepare transducing viral particles, a plasmid carrying the vector DNA is cotransfected into HEK293 cells along with plasmids that express the viral proteins and the necessary helper functions from adenovirus. The vector DNA is the only component of the system that carries a *cis*-acting signal for packaging into virus particles which excludes the helper DNAs. After allowing for the maximal amount of packaging, the cultures are harvested and the vector particles are purified and concentrated. Several methods are available for preparing the final high titer stocks.

One of the distinctive features of AAV is that only a single strand of DNA is packaged. Because of this the genome must be converted into a double-stranded DNA before the transgene can be expressed. This step is carried out by the host cell DNA-dependent DNA polymerase, and transduction is dependent on the efficiency of this step. Cells that are not expressing the necessary enzyme are poorly transduced, and the time it takes the host cell to carry out this step affects the timing of transgene expression. For example, AAV vectors do not efficiently transduce trabecular meshwork cells because the double-stranded DNA is not made (Borrás et al., 2006). Fortunately, a strategy has been devised to overcome this problem (see figure 109.2). It involves constructing the vector genome so that it carries an inverted copy of the transgene separated by a small linker (Kong et al., 2010). When the genome is delivered to the cell, the vector DNA forms a large double-stranded hairpin, effectively turning the single strand into a shorter double-stranded DNA. These self-complementing (scAAV) vectors overcome the problem of second-strand synthesis, but it has the limitation of reducing the carrying capacity of this vector by half.

Initial ocular gene delivery studies used AAV2 as the basis for the vector. There are, however, eight serotypes of AAV with differing receptor specificities and that demonstrate different efficiencies of ocular gene delivery (Allocca et al., 2007; Pang et al., 2010). It was mentioned above that lentiviral vectors can be pseudotyped to change receptor specificity. Pseudotyping is also possible with AAV vectors. This involves replacing the helper plasmid expressing the corresponding capsid protein with one encoding the desired capsid. Such vectors are denoted as rAAVX/Y where X is the type of the genome and Y denotes the capsid protein. For example a vector designated rAAV2/8 would carry an AAV type 2 vector genome packaged in a type 8 capsid. Studies have shown that rAAV2/2, rAAV2/5, and rAAV2/8 tend to transduce photoreceptor cells in dogs but other cells types can be transduced at lower efficiency following subretinal injection. Two pseudotyped vectors, rAAV2/7 and rAAV2/8, appear to be the most efficient at transducing photoreceptors, while; rAAV2/2 vectors can efficiently transduce retinal ganglion cells after intravitreal injection. The type of capsid can also affect the efficiency of gene expression downstream of entry because different capsids vary in their efficiency of uncoating and release of the viral genome. For example, rAAV2/8 vectors have a more rapid onset of transgene expression than rAAV2/5 vectors. Different serotypes of adenovirus are also capable of targeting specific cell types in the retina (Sweigard, Cashman, & Kumar-Singh, 2010).

LENTIVIRAL VECTORS

Lentiviruses are members of the Retroviridae family which have an RNA genome and that rely on a specialized polymerase referred to as reverse transcriptase (RT) for replication. Following entry into the cell, RT uses the RNA genome to copy a single DNA strand. The RT then converts the single-stranded DNA into a double-stranded DNA and at the same time degrades the original RNA genome. The double-stranded DNA then enters the nucleus of the cell, where it integrates somewhat randomly into the host chromosome. The integrated viral DNA (provirus) then can become quiescent, or transcription can be initiated to create more virus particles (see figure 109.3).

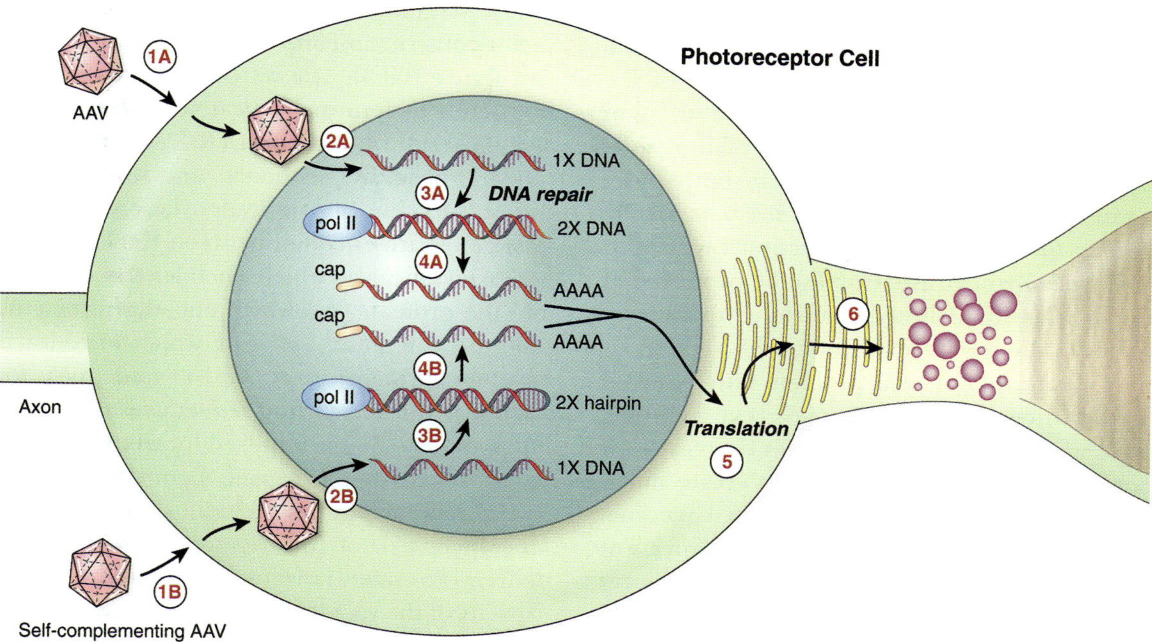

FIGURE 109.2   Gene delivery using adeno-associated virus (AAV) or self-complementing AAV vectors. The vector virus is delivered, attached to cellular receptors, and enters the cell (1A). The single-strand DNA genome is delivered to the nucleus (2A), where host cell DNA repair enzymes complete formation of the double-stranded DNA (3A). Host RNA polymerase II transcribes the mRNA (4A), which is translated into protein (5) that, for rhodopsin, is transported to the outer segments (6). With self-complementary AAV vectors, cellular attachment (1B) and delivery to the nucleus (2B) are identical to standard AAV vectors. The single-stranded genome DNA is converted to a double-stranded hairpin (3B), which is then transcribed by host RNA polymerase II (4B). The mRNA is translated (5), and the protein, rhodopsin, is transported to the outer segments (6).

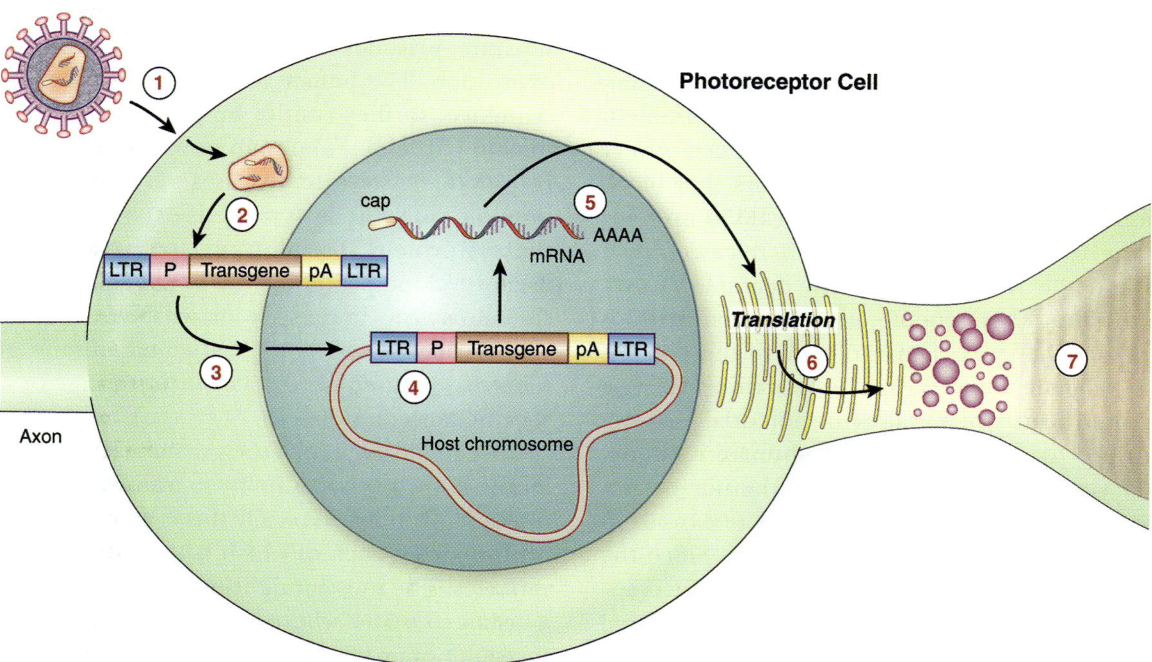

FIGURE 109.3   Retinal gene delivery using lentiviral-based vectors. The vector virus attaches to the cell (1) and enters. The single-stranded genomic RNA is converted to the double-stranded DNA by the viral encoded reverse transcriptase (2). The DNA then translocates to the nucleus (3), where the viral integrase catalyzes the integration of the transgene into the host cell chromosome (4). The host cell RNA pol II then transcribes the messenger RNA (mRNA) (5), which is translated (6), and the protein, rhodopsin, is transported to the outer segments (7). LTR, long terminal repeat; P, promoter driving transgene expression; transgene, the therapeutic gene; pA, polyadenylation signal.

Lentiviruses have a number of advantages for gene delivery, including a carrying capacity of 7 to 8 Kb, the ability to enter the nucleus without disruption of the nuclear envelope, and the ability to persist even in dividing cells due to the integration event (Balaggan & Ali, 2012). In addition, viral particles can be pseudo-typed to either expand the types of cells that can be transduced or to restrict the transduction capacity to specific cell types. Lentiviruses can also be produced at high enough titers to deliver significant numbers of viral particles to the eye in small volumes.

The most significant potential problem with lentiviral vectors is the integration into the host cell genome. Integration, which occurs preferentially into sites of active transcription, has the potential to create muta-tions (insertional mutagenesis) in the host cell genome. However, deleterious consequences of such an event have not been described following ocular gene delivery using these vectors in animal models (Balaggan et al., 2012; Yanez-Munoz et al., 2006). The integration event is catalyzed by the viral integrase protein, which is encoded in the *pol* gene. Integrase also has other essen-tial functions, so the protein must be present in the virus particle to have a functional vector, but specific mutations at positions D64, D116, and E152 (Class I mutations) eliminate the integrase function without inhibiting other essential activities. This information can be used to design nonintegrating vectors that will not cause insertional mutagenesis but still allow trans-gene expression from the episomal vector genome (Philpott & Thrasher, 2007). Another potential problem with lentiviral vectors is related to the presence of very effective species specific restriction factors that block replication (Fadel & Poeschla, 2011). TRIM5α proteins bind to the capsid and interfere with release of the genome and the APOBEC3 protein deaminates cyti-dines in the viral RNA genome, introducing mutations that inactivate the virus. Thus, to achieve the greatest efficiency of gene delivery, species specific lentiviral vectors should be used. This raises the potential problem of patient acceptance of the use of a human immuno-deficiency virus (HIV) based vector, whether or not vector construction and production does not result in the potential generation of infectious virus. Since the restriction factors are known, it is possible that cross-species lentiviral vectors could be engineered that would allow for efficient delivery of transgenes in humans with nonhuman lentiviral vectors.

Several different lentiviruses including those based on HIV, simian immunodeficiency virus (SIV), feline immunodeficiency virus (FIV), and equine infectious anemia virus (EIAV) have been used for ocular gene delivery in various animal models (Balaggan & Ali, 2012). Although different viruses have been used, vector construction and design is similar for all of them (see figure 109.3). The retroviral genome is flanked on both ends by terminal repeated sequences referred to as long terminal repeats (LTRs). The 5' LTR contains the viral promoter sequence, and the 3' LTR contains the termination and polyadenylation sites. All of the viral genes are encoded between the LTRs. The major genes include *gag*, which encodes the capsid proteins and the protease, *pol*, which encodes reverse transcrip-tase and integrase, and *env*, which encodes the viral envelope proteins (e.g., Gp160 and gp41 for HIV). Another critical *cis*-acting sequence is the ψ, or packag-ing, sequence that is required for the vector genome to be packaged in the capsid. Lentiviruses also encode several auxiliary genes including *vif*, *vpr*, *vpu*, *rev*, and *tat*, which, except for *rev*, are not essential for gene delivery. For safety reasons, and to increase the carrying capacity of the vector, all of the genome, with the excep-tion of the LTRs and the ψ sequence, are deleted from the vector construct and are replaced by the desired vector elements. However, in order to produce trans-ducing particles, the *gag*, *pol*, and *rev* genes must be supplied. This is done by providing these genes on separate helper plasmids. Typically, this is done using multiple plasmids to reduce the probability of recom-bination events that could generate an infectious viral genome. Thus, *gag*, *pol*, and *rev* may be cloned into separate plasmids with different promoters to drive expression. The helper vectors also do not have the ψ sequence, so they cannot be packaged into the trans-ducing particles. Finally, an envelope protein must be provided. Because lentiviruses very specifically infect immune cells, the host range is expanded by replacing the *env* protein (pseudotyping). The most commonly used envelope protein is the G protein from vesicular stomatitis virus because it allows for transduction of a wide variety of cell types. Other membrane glycopro-teins may be used to restrict the host range or allow for specific targeting of cells.

Another advantage of the greater carrying capacity of lentiviral vectors is the ability to transduce the genes for multiple proteins. Typically this is accomplished in one of three ways, each of which has advantages and disad-vantages. The most straightforward method is to create a fusion construct where both proteins are expressed in the same open reading frame from a single promoter. One must be careful, however, that the fusion protein retains the essential functions of both proteins. Another strategy takes advantage of the ability of internal ribo-some entry sites (IRES elements) from picornaviruses to promote translation of a downstream orf in a mRNA. A single promoter is used to transcribe a single mRNA with the orfs for both proteins, but an IRES element is included between the orfs allowing for expression of both proteins. In practice, the downstream orf is usually expressed at much lower levels than the upstream 5' orf, and in some vectors, expression of the downstream orf has been extinguished. The third strategy is to construct the vector so that the two genes are expressed from separate promoters.

In order to produce transducing particles, plasmids containing the vector construct, and the helper plasmids (*gag*, *pol*, and VSVG protein), are transfected into cells (usually HEK293 cells). At various times post-transfection, the culture medium is harvested and the virus is purified. The vector titers in unprocessed medium can range from approximately 10^5 to 10^7 transducing particles per milliliter depending on the virus, but this can be increased to 10^8 to 10^9 per milliliter using appropriate procedures. Higher vector titers are critical, because the higher the titer delivered, the greater the numbers of cells that can be transduced, and the smaller the volumes that need to be delivered. For photoreceptor transduction, which requires subretinal injection, small volumes are critical to avoid adverse consequences such as a significant retinal detachment.

OTHER CONSIDERATIONS FOR OCULAR GENE DELIVERY

Route of Delivery

Figure 109.4 shows the various routes that have been used for retinal gene delivery. Gene delivery to photoreceptors is difficult in practice. For the majority of studies in animals to date, and for the human LCA trials, subretinal injection of the vector has been required. Although this method is effective, there are drawbacks. Subretinal injection creates a transitory retinal detachment or bleb. With the volumes that are usually injected, the bleb is small and the fluid is absorbed quickly, but there have been occasions where the detachment does not seal or enlarges. This may be due to the technique or skill of the technician or other factors. Although this does not appear to be a significant concern, the diseased retinas may be more susceptible to damage than is currently realized. Another limitation is that the area of the retina that is transduced is small, so that multiple subretinal injections would be needed to cover the entire human retina. The fovea, which is responsible for color vision and most of our visual acuity, could be targeted, but a serious detachment could result in severe vision loss.

An alternative route for gene delivery is intravitreal injection. In animal studies this route typically results in delivery to retinal pigment epithelial cells (RPEs) or retinal ganglion cells (RGCs). For some diseases, these are the desired target cells. For example, several retinal degenerations are caused by mutations in RPE cells, and glaucoma results in death of RGCs. For some diseases, the retina might be damaged in a way that would allow for access of the vector to photoreceptors. In particular, diseases that result in alteration of the extracellular matrix could allow access to photoreceptors. One interesting possibility is coadministration of matrix-degrading enzymes to transiently allow vector access to photoreceptors.

Injections carry some risk, however, so an alternative delivery method that did not rely on injection would be an advantage. One such method is iontophoresis, where an electrical current is used to transfer naked nucleic acids across the sclera and into the eye. This has been successfully used in animals, but there are potential drawbacks (Souied et al., 2008). The technique requires specialized equipment and a highly trained operator and the parameters must be fine tuned to avoid side effects such as scleral burns. Electrophoresis has also been tested in animal models, but iontophoresis appears to be superior.

Ideally, topical application of the gene delivery vectors would be a significant advantage. One would think that this would only be effective for corneal delivery; however, studies have shown that gene delivery to the retina can be accomplished using this method. Topical application of liposomes can result in delivery of a transgene to RGCs (Masuda et al., 1996). Topical application of a herpes simplex virus vector resulted in delivery to RPE cells with an anterior-to-posterior gradient of transduction with the greatest efficiency nearer the point of application (Spencer et al., 2001). It is not clear how the vectors are accessing the RGC and RPE cells, but this method should be explored further and other vectors should be tested.

Targeting Gene Delivery

Many retinal degenerative diseases involve specific cell types, and the optimal gene therapy strategy would be to deliver and express the transgene only to those cells. There are three potential ways to accomplish this. The first is through the delivery method. Subretinal injection exposes the photoreceptors and RPE cells to the vector, and this has been the preferred method to target these cells because it is the only method that has been shown to result in the efficient transduction of

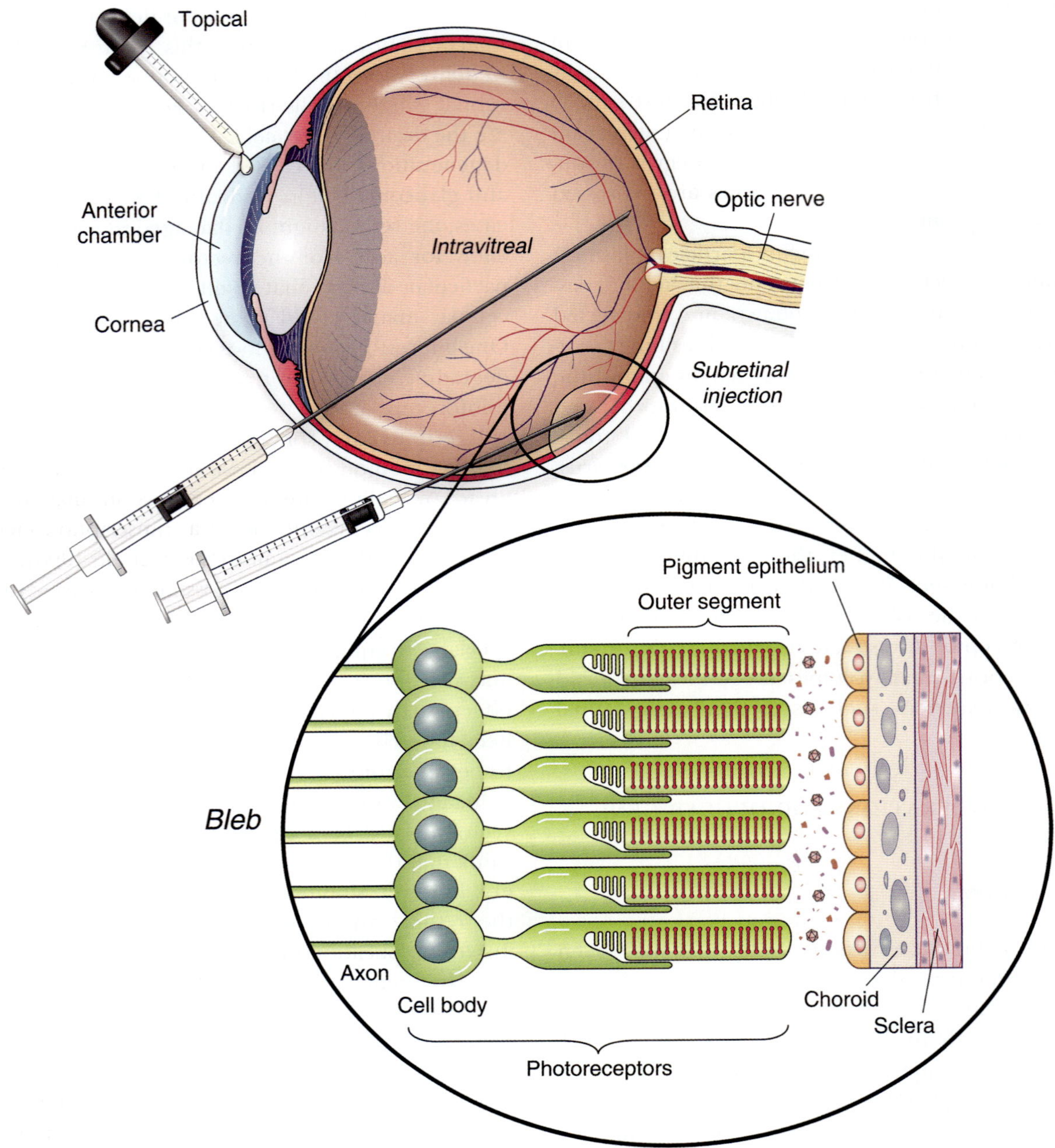

FIGURE 109.4   Routes of administration for retinal gene delivery. The most commonly used route is subretinal injection and is the only way to currently efficiently deliver a transgene to photoreceptor cells. This also results in delivery to the retinal pigment epithelium cells (RPE). This creates a small retinal detachment (bleb) that usually self-seals after a short time. Intravitreal injection is primarily used for retinal ganglion cell delivery although RPE can also be transduced through this route. Topical (eyedrop) application results in transduction of the RPE with HSV-1 vectors, but this route should be tested with other types of vectors. Topical application followed by electrophoresis or iontophoresis has also been used for retinal gene delivery.

photoreceptors. The RPE can be transduced by a number of routes including subretinal injection, intravitreal injection, and even topical instillation of vector on the cornea. The ease of transduction of RPE cells may be related to their enhanced phagocytic capabilities.

A second method involves the use of tissue or cell specific promoters. A number of promoters, some cell specific, have been used to drive transgene expression in photoreceptors, RPE, and RGC. This strategy does not necessarily restrict delivery of the transgene, but it can restrict expression to the desired cell types. For example, the rod opsin and RPE65 promoters result in transgene expression in photoreceptors and RPE cells, respectively (Bemelmans et al., 2006).

Finally, since viruses need receptors to infect cells, specific targeting could be achieved by using the desired viral attachment protein. Most retroviruses normally are very restricted in the cells they can infect. In order to increase the utility of these vectors, the normal viral attachment protein is replaced with another protein. The most commonly used protein for pseudotyping these types of vectors is the attachment protein from vesicular stomatitis virus (G protein), which has a broad host range (Balaggan et al., 2006). Other viral or cellular proteins could be used to pseudotype vectors, but this has not been extensively explored.

It is more difficult to change the binding specificity of viral vectors that are non-enveloped such as AAV or adenovirus. Fortunately, there are several serotypes of these viruses that have different binding specificities and that allow for specific targeting of certain cell types. This was discussed above for AAV vectors. Detailed knowledge of the interaction of these viruses with cellular receptors has resulted in the use of site-directed mutagenesis methods to alter the receptor specificity of AAV and AV vectors although more work is needed to specifically target cells in the eye.

Promoters and Long-Term Transgene Expression

It would be ideal if the transgene could be expressed from its natural promoter, including all of the normal regulatory components, for most applications in the eye. However, this is not feasible for a number of reasons. One is the carrying capacity of the vector. Another is that although there are proximal promoter elements at the 5' end of each gene, other regulatory elements, such as enhancers, insulators, and other regulatory sequences, can be located thousands of base pairs from the genes (Dean, 2011). For these and other reasons, the promoters used to drive gene expression have either been the most proximal elements or abbreviated versions of the proximal promoters. A commonly used promoter is the human cytomegalovirus major immediate early promoter, especially in proof of principle studies. It expresses in several cell types and at high levels but has the disadvantage of not driving long-term expression in many cell types. Long-term expression in photoreceptors has been achieved using photoreceptor specific promoters such as opsin (Bemelmans et al., 2006).

Long-term expression of the transgene is necessary for chronic genetic diseases like many of the retinal degenerations. The shorter the transgene is expressed, the more often the vector would need to be applied, and since current methods involve subretinal injection, this should be avoided. There are a number of reasons why expression could decline after gene delivery. Many of the vectors deliver the transgene as an episomal element that does not replicate, so cell division will gradually reduce the numbers of transduced cells. This is not a major concern in the eye because many of the cells are postmitotic, photoreceptors in particular. This can be avoided by using lentiviral vectors, which integrate naturally, but the capacity for insertional mutagenesis is a potential disadvantages with these vectors (Balaggan & Ali, 2012). Another strategy to avoid vector loss has been to incorporate cellular replication sequences into the episome so that the host cell will copy and segregate the transgene (Kazuki & Oshimura, 2011). Finally, transgene expression may be lost if expression is dependent on a transcription factor that is not present at all times (Loser et al., 1998).

For some diseases, it is possible to express the transgene from an inducible promoter that is active only when a pathological process is occurring. For example, with ocular diseases that involve inflammation, a promoter that is active only when inflammation is activated would restrict expression to those times when it would be needed. Some retinal diseases are the result of the redox status of cells, and by expressing the transgene from a hypoxia or hyperoxia specific promoter, the transgene would only be active when the redox status was altered. Much more work is needed to find additional promoters that will allow us to fine tune transgene expression.

SUMMARY

The recent success of gene therapy trials for LCA and the successes in animal models hold out great hope for significant advances in the therapy of retinal degenerative disorders. A number of creative approaches have been tested for the various classifications of these diseases with promising results in animal models. Continued efforts to address problems associated with gene therapy are likely to result in further advances in gene delivery as an approach to treat retinal degenerative diseases.

ACKNOWLEDGMENTS

The author would like to thank Dr. Monica Sauter, Dr. Nansi Jo Colley, and Inna Larsen for helpful comments and suggestions regarding the chapter and Inna Larsen and Rachel Brummel for administrative assistance. Work in the author's laboratory has been generously supported by the Retina Research Foundation, Houston, Texas, The National Eye Institute Core Grant for Vision Research (P30 EY016665), the National Institute for Allergy and Infectious Diseases, Fight for Sight, Inc., and Research to Prevent Blindness, Inc.

REFERENCES

Acland, G. M., Aguirre, G. D., Ray, J., Zhang, Q., Aleman, T. S., Cideciyan, A. V., et al. (2001). Gene therapy restores vision in a canine model of childhood blindness. *Nature Genetics, 28*, 92–95.

Aguirre, G. K., Komaromy, A. M., Cideciyan, A. V., Brainard, D. H., Aleman, T. S., Roman, A. J., et al. (2007). Canine and human visual cortex intact and responsive despite early retinal blindness from *RPE65* mutation. *PLoS Medicine, 4*, 1117–1128. doi:10.1371/journal.pmed.0040230.

Allocca, M., Mussolino, C., Garcia-Hoyes, M., Sanges, D., Iodice, C., Petrillo, M., et al. (2007). Novel adeno-associated virus serotypes efficiently transducer murine photoreceptors. *Journal of Virology, 81*, 11372–11380. doi:10.1128/JVI.01327-07.

Andrieu-Soler, C., Halhal, M., Boatright, J. H., Padove, S. A., Nickerson, J. M., Stodulkova, E., et al. (2007). Single stranded oligonucleotide-mediated in vivo gene repair in the rd1 retina. *Molecular Vision, 13*, 692–706.

Ashtari, M., Cyckowski, L. L., Monroe, J. F., Marshall, K. A., Chung, D. C., Auricchio, A., et al. (2011). The human visual cortex responds to gene therapy-mediated recovery of retinal function. *Journal of Clinical Investigation, 121*, 2160–2168. doi:10.1172/JCI57377.

Bainbridge, J. W., Smith, A. J., Barker, S. S., Robbie, S., Henderson, R., Balaggan, K., et al. (2008). Effect of gene therapy on visual function in Leber's congenital amaurosis. *New England Journal of Medicine, 358*, 2231–2239. doi:10.1056/NEJMoa0802268.

Balaggan, K. S., Binley, K., Esapa, M., Iqball, S., Askham, Z., Kan, O., et al. (2006). Stable and efficient intraocular gene transfer using pseudotyped EIAV lentiviral vectors. *Journal of Gene Medicine, 8*, 275–285. doi:10.1002/jgm.845.

Balaggan, K. S., & Ali, R. R. (2012). Ocular gene delivery using lentiviral vectors. *Gene Therapy, 19*, 145–153.

Balaggan, K. S., Duran, Y., Georgiadis, A., Thaung, C., Barker, S. E., Buch, P. K., et al. (2012). Absence of ocular malignant transformation after sub-retinal delivery of rAAV2/2 or integrating lentiviral vectors in p53-deficient mice. *Gene Therapy, 19*, 182–188. doi:10.1038/gt.2011.194.

Barker, S. E., Broderick, C. A., Robbie, S. J., Duran, Y., Natkunarajah, M., Buch, P., et al. (2009). Subretinal delivery of adeno-associated virus serotype 2 results in minimal immune responses that allow repeat vector administration in immunocompetent mice. *Journal of Gene Medicine, 11*, 486–497. doi:10.1002/jgm.1327.

Bemelmans, A. P., Bonnel, S., Houhou, L., Dofour, N., Nandrot, E., Helmlinger, D., et al. (2005). Retinal cell type expression specificity of HIV-1-derived gene transfer vectors upon subretinal injection in the adult rat: Influence of pseudotyping and promoter. *Journal of Gene Medicine, 7*, 1367–1374. doi:10.1002/jgm.788.

Bemelmans, A. P., Kostic, C., Crippa, S. V., Hauswirth, W. W., Lem, J., Munier, F. L., et al. (2006). Lentiviral gene transfer

of *Rpe65* rescues survival and function of cones in a mouse model of Leber congenital amaurosis. *PLoS Medicine, 3*, 1892–1903. doi:10.1371/journal.pmed.0030347.

Borrás, T., Xue, W., Choi, V. W., Bartlett, J. S., Li, G., Samulski, R. J., & Chisolm, S. S. (2006). Mechanisms of AAV transduction in glaucoma-associated human trabecular meshwork cells. *Journal of Gene Medicine, 8*, 589–602. doi:10.1002/jgm.886.

Buch, P. K., Bainbridge, J. W., & Ali, R. R. (2008). AAV-mediated gene therapy for retinal disorders: From mouse to man. *Gene Therapy, 15*, 849–857.

Cai, X., Conley, S. M., & Naash, M. I. (2009). RPE65: Role in the visual cycle, human retinal disease, and gene therapy. *Ophthalmic Genetics, 30*, 57–62.

Chadderton, N., Millington-Ward, S., Palfi, A., O'Reilly, M., Tuohy, G., Humphries, M., et al. (2009). Improved retinal function in a mouse model of dominant retinitis pigmentosa following AAV-delivered gene therapy. *Molecular Therapy, 17*, 593–599. doi:10.1038/mt.2008.301.

Chan, F., Hauswirth, W. W., Wensel, T. G., & Wilson, J. H. (2011). Efficient mutagenesis of the rhodopsin gene in rod photoreceptor neurons in mice. *Nucleic Acids Research, 39*, 5955–5966. doi:10.1093/nar/gkr196.

Chang, G.-Q., Hao, Y., & Wong, F. (1993). Apoptosis: Final common pathway of photoreceptor death in rd, rds, and rhodopsin mutant mice. *Neuron, 11*, 595–605.

Cideciyan, A. V., Aleman, T. S., Boye, S. L., Schwartz, S. B., Kaushal, S., Roman, A. J., et al. (2008). Human gene therapy for RPE65 isomerase deficiency activates the retinoid cycle of vision but with slow rod kinetics. *Proceedings of the National Academy of Sciences of the United States of America, 105*, 15112–15117. doi:10.1073/pnas.0807027105.

Cideciyan, A. V., Hauswirth, W. W., Aleman, T. S., Kaushal, S., Schwartz, S. B., Boye, S. L., et al. (2009). Vision 1 year after gene therapy for Leber's congenital amaurosis. *New England Journal of Medicine, 361*, 725–727. doi:10.1056/NEJMc0903652.

Cideciyan, A. V. (2010). Leber congenital amaurosis due to RPE65 mutations and its treatment with gene therapy. *Progress in Retinal and Eye Research, 29*, 398–427. doi:10.1016/j.preteyeres.2010.04.002.

Colella, P., & Auricchio, A. (2012). Gene therapy of inherited retinopathies: A long and successful road from viral vectors to patients. *Human Gene Therapy, 23*, 796–807. doi:10.1089/hum.2012.123.

Coura Rdos, S., & Nardi, N. B. (2007). The state of the art of adeno-associated virus-based vectors in gene therapy. *Virology Journal, 4*, 99. doi:10.1186/1743-422X-4-99.

Dean, A. (2011). In the loop: Long range chromatin interactions and gene regulation. *Briefs in Functional Genomics, 10*, 3–10. doi:10.1093/bfgp/elq033.

den Hollander, A. I., Black, A., Bennett, J., & Cremers, F. P. M. (2010). Lighting a candle in the dark: Advances in genetics and gene therapy of recessive retinal dystrophies. *Journal of Clinical Investigation, 120*, 3042–3053.

Fadel, H. J., & Poeschla, E. M. (2011). Retroviral restriction and dependency factors in primates and carnivores. *Veterinary Immunology and Immunopathology, 143*, 179–189.

Frio, T. R., Wade, N. M., Ransijn, A., Berson, E. L., Bechman, J. S., & Rivolta, C. (2008). Premature termination codons in PRPF31 cause retinitis pigmentosa via haploinsufficiency due to nonsense-mediated mRNA decay. *Journal of Clinical Investigation, 118*, 1519–1531.

Fuchs, Y., & Steller, H. (2011). Programmed cell death in animal development and disease. *Cell, 147,* 742–758.

Gorbatyuk, M., Justilien, V., Liu, J., Hauswirth, W. W., & Lewin, A. S. (2007). Preservation of photoreceptor morphology and function in P23H rats using an allele independent ribozyme. *Experimental Eye Research, 84,* 44–52.

Gorbatyuk, M. S., Knox, T., LaVail, M. M., Gorbatyuk, O. S., Noorwez, S. M., Hauswirth, W. W., et al. (2010). Restoration of visual function in P23H rhodopsin transgenic rats by gene delivery of BiP/Grp78. *Proceedings of the National Academy of Sciences of the United States of America, 207,* 5961–5966. doi:10.1073/pnas.0911991107.

Haeseleer, F., Imanishi, Y., Saperstein, D. A., & Palczewski, K. (2001). Gene transfer mediated by recombinant baculovirus into mouse eye. *Investigative Ophthalmology & Visual Science, 42,* 3294–3300.

Hauswirth, W. W., Aleman, T. S., Kaushal, S., Cideciyan, A. V., Schwartz, S. B., Wang, L., et al. (2008). Treatment of Leber congenital amaurosis due to RPE65 mutations by ocular sub-retinal injection of adeno-associated virus gene vector: Short term results of a phase I trial. *Human Gene Therapy, 19,* 979–990. doi:10.1089/hum.2008.107.

Jiang, L., Zhang, H., Dizhoor, A. M., Boye, S. E., Hauswirth, W. W., Frederick, J. M., et al. (2011). Long-term RNA interference gene therapy in a dominant retinitis pigmentosa mouse model. *Proceedings of the National Academy of Sciences of the United States of America, 108,* 18476–18481. doi:10.1073/pnas.1112758108.

Kazuki, Y., & Oshimura, M. (2011). Human artificial chromosomes for gene delivery and the development of animal models. *Molecular Therapy, 19,* 1591–1601.

Kong, F., Li, W., Li, X., Zheng, Q., Dai, X., Zhou, X., et al. (2010). Self-complementary AAV5 vector facilitates quicker transgene expression in photoreceptor and retinal pigment epithelial cells of normal mouse. *Experimental Eye Research, 90,* 546–554. doi:10.1016/j.exer.2010.01.011.

Kraus, A., Groenendyk, J., Bedard, K., Baldwin, T. A., Krause, K.-H., Dubois-Dauphin, M., et al. (2010). Calnexin deficiency leads to dismyelination. *Journal of Biological Chemistry, 285,* 18928–18938. doi:10.1074/jbc.M110.107201.

LaVail, M. M., Unoki, K., Yasumura, D., & Matthes, M. T., Yancopoulos, G. D., & Steinberg, R. H. (1992). Multiple growth factors, cytokines, and neurotrophins rescue photoreceptors from the damaging effects of constant light. *Proceedings of the National Academy of Sciences of the United States of America, 89,* 11249–11253. doi:10.1073/pnas.89.23.11249.

Leonard, K. C., Petrin, D., Coupland, S. G., Baker, A. N., Leonard, B. C., LaCasse, E. C., et al. (2007). XIAP protection of photoreceptors in animal models of retinitis pigmentosa. *PLoS ONE, 3,* e314. doi:10.1371/journal.pone.0000314.

Liu, X., Brandt, C. R., Gabelt, B. T., Bryar, P. J., Smith, M. E., & Kaufman, P. L. (1999). Herpes simplex virus mediated gene transfer to primate ocular tissues. *Experimental Eye Research, 69,* 385–395. doi:10.1006/exer.1999.0711.

Loser, P., Jennings, G. S., Strauss, M., & Sandig, V. (1998). Reactivation of the previously silenced cytomegalovirus major immediate early promoter in the mouse liver. *Journal of Virology, 72,* 180–190.

Maguire, A. M., High, K. A., Auricchio, A., Wright, F., Pierce, E. A., Testa, F., et al. (2009). Age-dependent effects of RPE65 gene therapy for Leber's congenital amaurosis: A phase 1 dose-escalation trial. *Lancet, 374,* 1597–1605. doi:10.1016/S0140-6736(09)61836-5.

Maguire, A. M., Simonelli, F., Pierce, E. A., Pugh, E. N., Jr., Mingozzi, F., Bennicelli, J., et al. (2008). Safety and efficacy of gene transfer for Leber's congenital amaurosis. *New England Journal of Medicine, 358,* 2240–2248. doi:10.1056/NEJMoa0802315.

Masuda, I., Matsuo, T., Yasuda, T., & Matsuo, N. (1996). Gene transfer with liposomes to the intraocular tissues by different routes of administration. *Investigative Ophthalmology & Visual Science, 37,* 1914–1920.

Millington-Ward, S., Chadderton, N., O'Reilly, M., Palfi, A., Goldmann, T., Kilty, C., et al. (2011). Suppression and replacement gene therapy for autosomal dominant disease in a murine model of dominant retinitis pigmentosa. *Molecular Therapy, 19,* 642–649. doi:10.1038/mt.2010.293.

Murakami, Y., Ikeda, Y., Yonemitsu, Y., Onimaru, M., Nakagawa, K., Hisatomi, R., et al. (2008). Inhibition of nuclear translocation of apoptosis-inducing factor is an essential mechanism of the neuroprotective activity of pigment epithelium-derived factor in a rat model of degeneration. *American Journal of Pathology, 173,* 1326–1338. doi:10.2353/ajpath.2008.080466.

Musarella, M. A., & MacDonald, I. M. (2011). Current concepts in the treatment of retinitis pigmentosa. *Journal of Ophthalmology.* doi:10.1155/2011/753547.

Noorwez, S. M., Sama, R. R. K., & Kaushal, S. (2009). Calnexin improves the folding efficiency of mutant rhodopsin in the presence of pharmacological chaperone 11-cis-retinal. *Journal of Biological Chemistry, 284,* 33333–33342. doi:10.1074/jbc.M109.043364.

O'Reilly, M., Palfi, A., Chadderton, N., Millington-Ward, S., Ader, M., Cronin, T., et al. (2007). RNA interference-mediated suppression and replacement of human rhodopsin in vivo. *American Journal of Human Genetics, 81,* 127–135.

Pang, J., Boye, S. E., Lei, B., Boye, S. L., Everhart, D., Ryals, R., et al. (2010). Self-complementary AAV-mediated gene therapy restores cone function and prevents cone degeneration in two models of Rpe65 deficiency. *Gene Therapy, 17,* 815–826.

Park, T. K., Wu, Z., Kjellstrom, S., Zeng, Y., Bush, P. A., Sieving, P. A., et al. (2009). Intravitreal delivery of AAV8 retinoschisin results in cell type-specific gene expression and retinal rescue in the RS1-KO mouse. *Gene Therapy, 16,* 916–926.

Philpott, N. J., & Thrasher, A. J. (2007). Use of nonintegrating lentiviral vectors for gene therapy. *Human Gene Therapy, 18,* 483–489.

Sikkink, S. K., Biswas, S., Parry, N. R. A., Stanga, P. E., & Trump, D. (2007). X-linked retinoschisis: An update. *Journal of Medical Genetics, 44,* 225–232.

Simonelli, F., Maguire, A. M., Testa, F., Pierce, E. A., Mingozzi, F., Bennicelli, J. L., et al. (2010). Gene therapy for Leber's congenital amaurosis is safe and effective through 1.5 years after vector administration. *Molecular Therapy, 18,* 643–650.

Song, X., Vishnivetskly, S. A., Gross, O. P., Emelianoff, K., Mendez, A., Chen, J., et al. (2009). Enhanced arrestin mutant facilitates photoresponse recovery and protects rhodopsin phosphorylation-deficient rod photoreceptors. *Current Biology, 19,* 700–705.

Souied, E. H., Reid, S. N. M., Piri, N. I., Lerner, L. E., Nusinowitz, S., & Farber, D. B. (2008). Non-invasive gene transfer by iontophoresis for therapy of an inherited retinal degeneration. *Experimental Eye Research, 87,* 168–175.

Spencer, B., Agarwala, S., Gentry, L., & Brandt, C. R. (2001). HSV-1 vector-delivered FGF2 to the retina is neuroprotective but does not preserve functional responses. *Molecular Therapy, 3,* 746–756.

Sweigard, J. H., Cashman, S. M., & Kumar-Singh, R. (2010). Adenovirus vectors targeting distinct cell types in the retina. *Investigative Ophthalmology & Visual Science, 51,* 2219–2228.

Takada, Y., Vijayasarathy, C., Zeng, Y., Kjellstrom, S., Bush, R. A., & Sieving, P. A. (2008). Synaptic pathology in retinoschisis knockout ($Rs1^{-/y}$) mouse retina and modification by rAAV-*Rs-1* gene delivery. *Investigative Ophthalmology & Visual Science, 49,* 3677–3686.

Takita, H., Yoneya, S., Gehlbach, P. L., Wei, L. L., & Mori, K. (2008). An empty E1⁻, E3⁻, E4⁻ adenovirus vector protects photoreceptors from light-induced degeneration. *Journal of Ocular Biology, Diseases, and Informatics, 1,* 30–36. doi:10.1007/s12177-088-9004-4.

Wood, M., Yin, H., & McClorey, G. (2007). Modulating the expression of disease genes with RNA-based therapy. *PLOS Genetics, 3,* 845–854.

Yanez-Munoz, R. J., Balaggan, K. S., MacNeil, A., Howe, S. J., Schmidt, M., Smith, A. J., et al. (2006). Effective gene therapy with non-integrating lentiviral vectors. *Nature Medicine, 12,* 348–353.

Yao, J., Feathers, K. L., Khanna, H., Thompson, D., Tsilfidis, C., Hauswirth, W. W., et al. (2011). XIAP therapy increases the survival of transplanted rod precursors in a degenerating host retina. *Investigative Ophthalmology & Visual Science, 52,* 1567–1572.

110 Retinal Cell Replacement

MANDEEP S. SINGH AND ROBERT E. MACLAREN

Age-related macular degeneration (AMD) and retinitis pigmentosa (RP) are major examples of human outer retinal disease and together are the cause of the majority of cases of irreversible blindness in the Western world (The Eye Diseases Prevalence Research Group, 2004) (see table 110.1). AMD is reported as having an overall prevalence of about 1.6%, with almost half of cases featuring retinal atrophy (Smith et al., 2001); RP is estimated to occur in 1:5000 to 1:3000 people (Hartong, Berson, & Dryja, 2006; Pagon, 1988). It is clear that there are vast numbers of blind patients for whom there is no current cure.

In both these conditions, the loss of visual function occurs due to death of photoreceptor cells in the outer retina (Curcio, Medeiros, & Millican, 1996) (see figure 110.1). In early stages of disease, the therapeutic strategy may in theory aim to arrest disease progression. Treatment options to retard photoreceptor cell death, and hence reduce the rapidity of visual decline, may in the future include specific gene augmentation (Takahashi et al., 1999), knockdown therapy (Jiang et al., 2011), or neuroprotective trophic support (Tao et al., 2002; Faktorovich et al., 1990; Lau et al., 2000). These approaches would demand the presence of sufficient numbers of functional photoreceptors that can be rescued from further degeneration. However, patients often present with vision loss because of relatively advanced disease, when significant photoreceptor death has already occurred. At this stage, rescue interventions would have little effect. The clinical goal of restoring vision therefore shifts to a replacement paradigm, requiring the replacement of retinal photoreceptors as the main thrust of treatment (MacLaren et al., 2007; Singh & MacLaren, 2011; Singh et al., 2013).

A critical question in photoreceptor replacement is whether or not the residual inner retinal neurons would be capable of function after profound photoreceptor loss. Inner retinal remodeling, affecting both neuronal and nonneuronal cells, has been shown to occur in animal models of outer retinal degeneration (Jones & Marc, 2005). It is not fully known to what extent these changes may limit the benefit of photoreceptor cell replacement. However, support for the possibility of functional gain following outer retinal regeneration has come from pioneering work in which electronic microphotodiode array implants were placed subretinally in blind human patients (Zrenner et al., 2011). The restoration of some vision in selected cases implied that downstream neurons such as bipolar cells and ganglion cells, if stimulated appropriately, were capable of relaying meaningful vision to the visual cortex. As these patients received the implant at the stage of almost complete outer retinal loss, the data therefore indicate that useful inner retinal function may remain even after years of photoreceptor degeneration. Furthermore, stimulation of the outer retina has, through this trial, been validated as a strategy to preserve retinotopy. These data support the pursuit of photoreceptor replacement into the outer retina as potential treatment in late stages of disease.

Here we will address the possibility of outer retinal cell replacement as treatment for patients with advanced retinal degeneration. Other aspects of retinal cell therapy—such as those relating to inner retinal replacement, vascular cell replacement, and cell treatment in a trophic rescue paradigm in early disease—are covered in other chapters in this book.

PHOTORECEPTOR REPLACEMENT

RP is prevalent in the young and working-age population. It causes progressive visual loss, classically beginning with night blindness and peripheral visual deficit in the early stages due to rod-specific disease (Hartong, Berson, & Dryja, 2006). Central vision loss is a feature of late disease in classical RP, at the time when the loss of cones occurs following rod death. The theoretical requirements to be fulfilled by replacement of photoreceptors for the repopulation of the outer retina are summarized in table 110.2. The fulfillment of all these criteria would be expected to contribute to the recreation of the physiological functional outer retinal unit (see figure 110.2). The neuronal nature of the cells in question and the requirement for complete differentiation with a light-sensitive outer segment, and an inner synapse, make photoreceptor replacement a challenging task. Nevertheless, it is likely to be considerably easier than aiming, for instance, to replace retinal ganglion cells, which would require long-distance navigation through densely myelinated tissues and reconnection to distant targets in a precise retinotopic map (MacLaren, 1996, 1999; MacLaren & Taylor, 1997).

Foundations of Cell Replacement for Photoreceptor Loss

Early preclinical work featured the use of fetal retinal sheets or fragments of immature neural retina that were transplanted to host animals (Adolph et al., 1994; Chacko et al., 2000; del Cerro et al., 1991; Ghosh, Johansson, & Ehinger, 1999; Gouras et al., 1991; Royo & Quay, 1959; Tansley, 1946). In general, the use of whole retinal sheets or fragments, while demonstrating the possibility of photoreceptor survival after transplantation, faced the problem of the donor inner retinal layers being interposed between donor photoreceptors

TABLE 110.1

Outer retinal disease characteristics

Disease example	Retinitis pigmentosa	Age-related macular degeneration
Primary/main cell type affected	Photoreceptor	Retinal pigment epithelium
Onset	Typically in the young	>50 years of age
Progression	Variable; severe forms progress over a few years to total blindness	Variable; slow progression in the dry form or a more rapid course in the exudative form
Pattern of visual loss	Usually starts in the periphery and progresses to the center of the visual field	Usually occurs in the center of the visual field; the periphery is spared
Inheritance pattern	Autosomal dominant Autosomal recessive X-linked or mitochondrial	Polygenic
Environmental factors	Light exposure	Smoking
Syndromic associations	Well-described	Not known

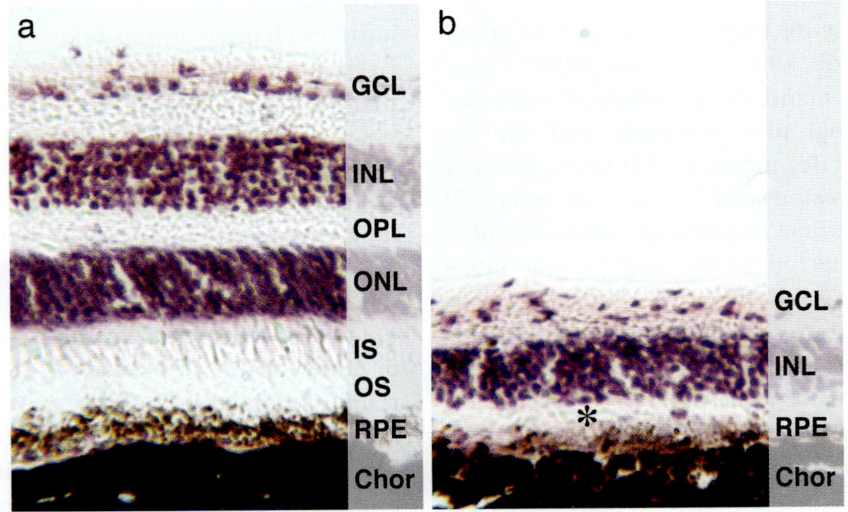

FIGURE 110.1   (a) Normal trilaminar nuclear organization of the retina. In this H&E stained section of the peripheral macaque retina, the nuclei in the neurosensory retina are organized in the ganglion cell layer (GCL), inner nuclear layer (INL) and outer nuclear layer (ONL). The ONL contains the cell bodies of photoreceptors, whose inner segments (IS) and outer segments (OS) are positioned adjacent to the retinal pigment epithelium (RPE). Photoreceptors in the outer retina synapse with inner retinal bipolar neurons whose cell bodies are located in the INL; the synapses are located within the outer plexiform layer (OPL). The outer retina is nourished by the blood supply from vessels in the choroid (Chor). (b) Complete outer retinal degeneration. In this experimental model of outer retinal damage, the retina is devoid of rod and cone photoreceptors and is severely thinned due to complete loss of the ONL. If light sensitivity were to be restored by cell replacement, a new ONL would have to be created by the introduction of new photoreceptors external to the INL—into the same location from which they were lost. The subretinal space (asterisk) is a natural cleavage plane into which cells could be placed surgically.

Target cell	Photoreceptor (PR)	Retinal pigment epithelium (RPE)
Target cell type	Neuronal	Cuboidal epithelium
Embryonic origin	Optic cup; inner layer	Optic cup; outer layer
Layer organization	Yes; cell bodies are organized as the retinal outer nuclear layer	Yes; cells are organized as a monolayer of hexagonal cells
Main pigment	Photosensitive pigments (opsins) localized in PR outer segments	Melanin localized in granules
Polarization	Required: synaptic vs. photosensitive ends	Required: apical vs. basal
Downstream connection	Synapses	None
Basement membrane	None, but require structural support by Müller cells	Bruch's membrane
Tight junctions	None	Zona occludens tight junctions between adjacent RPE cells

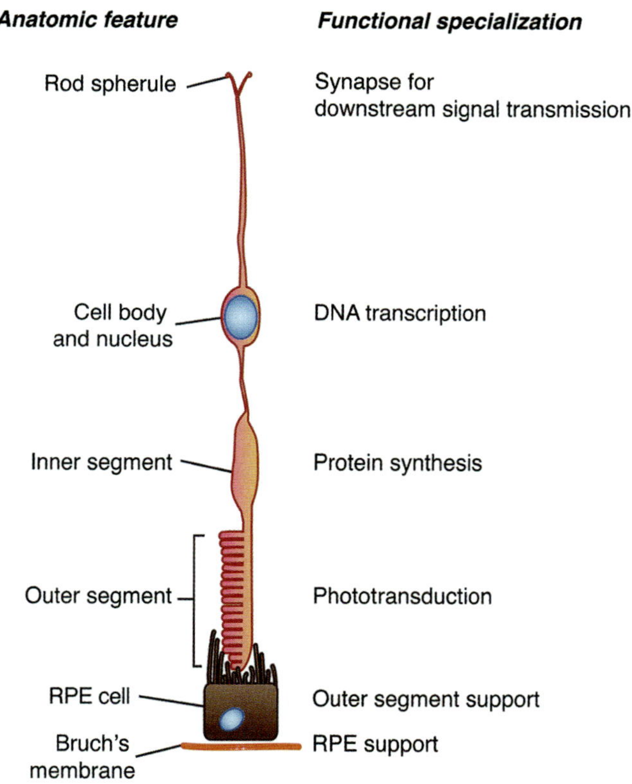

FIGURE 110.2   The functional outer retinal unit. All the anatomic features of the photoreceptor in the outer retina are related to its functions as a light signal detector, transducer and transmitter. In cell-based treatment for retinal degenerations, the aim would be to recreate this unit depicted above. In principle, if only the photoreceptor is lost in a specific degeneration, then that would be the only cell requiring replacement; however if there is concurrent damage to retinal pigment epithelium (RPE) or Bruch's membrane, these would need to be reconstituted in order to support continued photoreceptor function. DNA, deoxyribonucleic acid.

and host retina, thus posing a physical barrier to widespread synaptic integration (Aramant & Seiler, 1995; Ghosh, Johansson, & Ehinger, 1999; Ghosh, Bruun, & Ehinger, 1999; Kwan, Wang, & Lund, 1999; Seiler & Aramant, 1998; Zhang et al., 2004). Mechanical removal of donor inner retina was therefore proposed as a method to create partial retinal sheets (Huang et al., 1998). However, consistent and complete inner retinal removal was found to be challenging, with resulting morphological abnormalities in donor sheets. The transplantation of single-cell suspensions of dissociated retina has been pursued by many groups (Bartsch et al., 2008; Gust & Reh, 2011; Juliusson et al., 1993; Lakowski et al., 2011; MacLaren et al., 2006; Yao et al., 2011a) for the ease of surgical delivery via injection devices. In using cell suspensions, however, techniques would have to be devised to support optimal polarization and layer organization of donor cells.

The technical feasibility and relative safety of fetal neural retinal transplantation was described in humans with RP and AMD (Das et al., 1999; Humayun et al., 2000) although these studies did not demonstrate positive effects on visual function. A rare report of histological data obtained from a patient who underwent fetal neural retinal transplantation 3 years before demise showed the preservation and lamination of grafted tissue in the absence of systemic immune suppression (del Cerro et al., 2000). Going one step further, one group transplanted combined sheets of fetal neural retina together with the underlying retinal pigment epithelium (RPE) layer into RP patients and found graft survival and growth despite human leukocyte antigen (HLA) mismatching and the absence of immune suppression (Radtke et al., 2002). One patient was reported as showing an improvement in visual acuity for up to 1 year (Radtke et al., 2004). From these data it can be seen that the relative immune privilege offered by the subretinal space is a powerful feature that could work in favor of efforts to use exogenous cells for the treatment of disease.

Proof of consistent functional return following photoreceptor replacement was shown in a mouse study in which rod precursors were introduced into the subretinal space of wild-type and rhodopsin knockout mice. The cells integrated into the host outer nuclear layer, and the return of retinal light sensitivity was demonstrated in the degenerate hosts (MacLaren et al., 2006). Specifically, the ontogenetic stage of donor cells was found to be the critical determinant of the extent of successful cell integration. It was shown that survival and integration of transplanted cells were greatest with postmitotic rod precursor cells that were at the *Nrl*-expressing stage, and that the extent of integration decreased with younger donor cells. These data challenged the earlier view that greater donor cell immaturity would result in greater integration and possible functional success. These results have been validated in other studies (Bartsch et al., 2008; Lakowski et al., 2011; West et al., 2008; Yao et al., 2011a), and these data define a goal for the target cell type—the rod photoreceptor precursor, the penultimate cell type before mature rod genesis—that could be used for curative human application. Photoreceptor precursor transplantation has also been shown, in a mouse model of severe blindness, to restore visual function and the tri-laminar retinal structure at the advanced stages of retinal degeneration when many patients would require treatment (Singh et al., 2013). The limitation of this approach is that human donor photoreceptors at this stage are equivalent to those found in second trimester fetuses, and it would not be a clinically relevant possibility to harvest these cells. Hence the donor cells would have to be derived from other sources, as discussed below.

It is known from work in animals that photoreceptor-like cells may be generated from a number of putative sources. Neural stem cells (Akita et al., 2002; Dong et al., 2003), retinal progenitor cells (Klassen et al., 2004, 2007; Qiu et al., 2005; Tomita et al., 2006), and bone marrow mesenchymal stromal cells (Kicic et al., 2003; Tomita et al., 2006) are among the candidates proposed. Only limited data on the transdifferentiation of these cell types into photoreceptors have been shown, and the extent of morphological and functional photoreceptor differentiation remains uncertain. In some studies, immunohistochemical staining revealed the presence of photoreceptor markers in donor cells after transplantation. However, this phenomenon may possibly be attributed to the expression of a variety of proteins by immature or stem-like cells and, thus, may not represent complete and specific photoreceptor differentiation. Functional data from the transplantation of these cell types into retinal disease models are limited, and it should be kept in mind that significant trophic effects may have been the explanation for positive functional outcomes rather than true transdifferentiation and cell replacement. This is especially important to consider as some of these source cells may secrete trophic factors, which may have positive effects on the residual host retina.

Stem Cells in Photoreceptor Regeneration

Considering the application of cell therapy in humans, it is likely that a large number of postmitotic rod precursors would be required to treat a single patient, especially since the rate of incorporation of transplanted cells in animal models is low (about 0.1–1%) (MacLaren et al., 2006; Yao et al., 2011a). In this respect, the need for technology to obtain vast numbers of cells represents a major challenge in the journey toward clinical cell replacement.

The cloning of Dolly the sheep (Campbell et al., 1996) showed that the transfer of a nucleus from an adult somatic cell into a donor oocyte (in this case, from Dolly's genetic older kin) led to the development of an entire adult mammal with every adult cell containing the same original nuclear DNA (excluding mitochondrial DNA, which would be derived from the donor oocyte). Dolly's formed and functional eyes showed that all mature retinal cell types could be derived using this technique. Following from this principle, embryonic stem (ES) cells derived from the inner cell mass of the developing blastocyst (Klimanskaya et al., 2006; Reubinoff et al., 2000; Thomson, 1998) have also been expanded in vitro and driven toward photoreceptor-like and retinal precursor- or progenitor-like cells (Ikeda et al., 2005; Lamba et al., 2006; Osakada et al., 2008; Sugie et al., 2005). These photoreceptor-like cells have been applied in animal models of retinal degeneration and shown to result in some degree of functional restoration (Lamba, Gust, & Reh, 2009). Recently, Eiraku and colleagues successfully generated entire optic cups by culturing mouse ES cells in three-dimensional suspension (Eiraku et al., 2011). In these self-organized cups, layers of primordial neural retinal cells were seen, including *Crx-*, *Rho-*, and *Nrl*-positive outer retinal cells, which are markers specific for photoreceptors. Hence

     MANDEEP S. SINGH AND ROBERT E. MACLAREN

it may be possible in the future to expand a vast quantity of postmitotic rod precursors, at the *Nrl*-expressing stage for potential therapeutic application, from a single stem cell. Furthermore, in theory these cells could be generated in the polarized layer organization of the outer retina, and could be used for outer retinal reconstruction at the stage when no photoreceptors remain.

Somatic cell nuclear transfer and ES cell–based systems pose a significant ethical dilemma to the scientific community as these methods involve harvesting material from other organisms and embryos. As an alternative strategy, in recent years, methods have been devised to reprogram adult skin cells into embryonic-like stages, giving rise to the so-called "induced" pluripotent stem cell (iPS cell) (Takahashi et al., 2007; Takahashi & Yamanaka, 2006; Yu et al., 2007). The manipulation of iPS cells though various protocols can give rise to photoreceptor-like cells (Hirami et al., 2009; Osakada et al., 2009a; Parameswaran et al., 2010). To frame the therapeutic ideal, a single easily accessible adult skin fibroblast may be reprogrammed into millions of photoreceptors that can be reintroduced into the same patient. The additional advantage of adult-derived iPS cell technology is that therapy would be autologous and therefore avoid the need for chronic immune suppression. However, as a result of the reprogramming process, a genomic mutational load is acquired that may result in the risk of malignancy. Efforts have been directed at improving the iPS cell safety profile by avoiding the use of agents associated with higher teratogenicity (Junying et al., 2009; Nakagawa et al., 2008; Osakada et al., 2009b). Of the many such methods under investigation—for example, using nonintegrating vectors (Stadtfeld et al., 2008)—it is unclear for now as to which method may produce the best results. Direct transdifferentiation of adult-derived cells into target cell types, without passing through the iPS cell stage, may represent another way to increase the donor cell safety profile.

One challenge in the possible generation of iPS cell–derived photoreceptors from patients is that iPS cell clones have shown variable and generally lower rates of neuronal differentiation than ES cells (Hu et al., 2010), and it is currently unclear how it may be possible to devise a reliable, reproducible, and highly efficient photoreceptor differentiation protocol. The challenge is compounded in that every cell in the adult patient with retinal degeneration would harbor one or more retinal disease causing mutations. Thus the iPS cell lines expanded from an individual patient may not only respond differently to a standard differentiation protocol but also would require that the gene defect be corrected ex vivo before reintroducing these cells into the same patient.

Toward the Future Ideal: Personalized Cell-Based Regeneration

One recent exciting translational development is that iPS cells were derived from reprogrammed fibroblasts isolated from RP patients, and they could differentiate into rod photoreceptors in vitro (Jin et al., 2011). In the report, the resulting rod cells were proposed as a tool for drug screening or mechanistic dissection of the disease process. These aims are, in general, felt to be the main thrust of iPS cell applications for now, given the many issues that need resolution before considering iPS cells for therapy.

In one other positive development, the single base-pair disease causing mutation in an iPS cell line derived from a patient with gyrate atrophy of the retina was successfully repaired ex vivo (Howden et al., 2011). This study showed not only that it was possible to generate gene-corrected iPS cells from patients but also that the gene correction process did not unduly increase the mutational load in the iPS cell genome—which is a critical consideration if these cells were to be applied for treatment.

Although there are numerous concerns in the development of patient-derived iPS cells for curative purposes—such as the control of malignant potential and infection risk, consistent generation of target cell types, and minimizing immunogenicity—these initial reports may pave the way for the development of adult patient-derived photoreceptors that may be exogenously gene-corrected and then reintroduced to restore the outer retina, as a form of personalized cell-based regeneration.

Cone Photoreceptor Transplantation

Compared to rod cell replacement, relatively little is known about the transplantation of cone photoreceptors. Cone cells are critical to the high-acuity vision provided by the foveal center in humans and also for color vision—indeed, human vision could be said to rely mainly on cone photoreceptors. It is their loss in AMD and also in the late stages of RP that produces the loss of Snellen visual acuity, and so the question of cone replacement merits further investigation. One group has produced some evidence that flow-sorted murine embryonic-stage *Crx*-positive cells (i.e., immature photoreceptors), when transplanted into two mouse models of Leber congenital amaurosis, integrated into the outer nuclear layer of the host and showed characteristics of differentiated cone cells (Lakowski et al., 2010). This thread of research may open the way to the development of cone replacement strategies needed for the treatment of central vision loss found in many retinal degenerations including AMD.

RETINAL PIGMENT EPITHELIUM REPLACEMENT

The clinical need for RPE cell replacement has arisen as the treatment options for RPE degeneration, in diseases such as AMD, are limited. In the dry or atrophic form of AMD, the slow death of RPE cells leads to the gradual loss of foveal photoreceptors in an area of geographic atrophy, leading to central scotomas and the loss of reading vision and face recognition (Tejeria et al., 2002). In the wet or exudative form, the growth of leaky and hemorrhagic new vessels from the choroid that penetrate the diseased Bruch's membrane (BM) results in a relatively rapid loss of central vision (Nowak, 2006). Current pharmacotherapy for exudative AMD (Mitchell et al., 2010) restores only some vision in a subset of patients whereas for the nonexudative or atrophic form of the disease, there is no FDA-approved treatment at present (Kuno & Fujii, 2011).

In work that was described as a breakthrough (Heimann, 1994) for the surgical treatment of AMD, Algvere and coworkers in the 1990s transplanted cultivated human fetal RPE into a number of AMD patients, showing survival and even enlargement of these grafts to cover native RPE defects (Algvere et al., 1994, 1997). Similar pioneering efforts were made by other investigators (Weisz et al., 1999), further energizing the field of RPE transplantation.

Therapeutic Window for Photoreceptor Rescue

In envisioning RPE cell replacement for diseases such as AMD, it is critical to consider the concept of the therapeutic window. Persistent RPE damage results in photoreceptor death and the loss of vision. Hence, to preserve vision, RPE replacement therapy must be undertaken in relatively early disease when the overlying photoreceptors are still relatively healthy. In dry AMD, where the disease process is chronic and visual loss can be mild in early stages, the benefit of early surgical intervention has to be weighed against the significant surgery-related risks to vision at this early stage. Hence any proposed surgical cell replacement treatment method must be shown to be highly safe and effective so as to justify treatment in a

preventive paradigm. If the risk–benefit equation is unfavorable, therapy would have to be delayed until more severe visual loss has occurred. However, by this time, the treatment may not be as effective as early intervention.

Autologous Retinal Pigment Epithelium Translocation

The technique of autologous RPE translocation (MacLaren et al., 2007; Van Meurs & Van Den Biesen, 2003), where a disk of relatively healthy RPE along with the underlying BM and superficial choroid is harvested from the retinal periphery and transferred to the macula, has been reported. This technique has led to visual benefit in some cases. However, the significant drawback is the high rate of complications, including the frequent need for repeat surgery and a relatively high risk of vision loss (Chen et al., 2009; MacLaren et al., 2007). Autologous transplantation of RPE cell suspensions harvested by scraping the peripheral RPE—potentially, a less complex procedure—has also shown only modest functional results. In both approaches, the retinotomies required during surgery carry the risk of intravitreal spread of RPE cells, which gives rise to proliferative vitreoretinopathy (Hilton, Machemer, & Michels, 1983), a process that causes severe retinal detachment requiring difficult corrective surgery. Therefore, RPE replacement, if not carefully targeted to a discrete area and contained effectively in the subretinal space, may lead to severe complications.

The generally poor outcomes of autologous RPE replacement may also reflect the complex requirements of replacement cells (table 110.2). RPE cells must be able to assume the correct apical–basal polarity in a monolayer, with the apices in intimate contact with photoreceptor outer segments, for their protective and supportive role to be carried out. This is probably not to be expected if the cells are injected as a suspension without the aid of a basement membrane substrate. Furthermore, cell adhesion onto the diseased BM, like that which is found in the host macula in AMD, has been shown to be poor (Gullapalli et al., 2005). Natural and artificial substrates for the RPE are being investigated and may, in the future, be used as a surface to support the proper alignment of grafted cells obtained from autologous or heterologous sources. However, even if the cells are aligned well on a replacement substrate, challenges remain with regards to functional transmembrane transport requirements and also the ease of surgical manipulation and placement. This need for a multifunctional basal membrane layer that is critical for RPE survival and function must be considered as an important aspect of RPE cell replacement, and is discussed below.

Embryonic Stem Cell–Derived Retinal Pigment Epithelium in Humans

It is considered a major advance in the field of cell-based treatment for RPE degenerations that early data have been published from the first trial of human ES cell-derived RPE transplantation into patients (Schwartz et al., 2012), 13 years after the discovery of human ES cells. Schwartz and colleagues reported the 4-month outcomes of one patient with AMD and one with Stargardt macular degeneration, both with advanced disease and very low levels of remaining vision, treated with the lowest dose of cells (50,000 cells per injection). In the follow-up period, there was no observable tumor or ectopic tissue formation nor immune rejection, and neither patient suffered visual loss. In the AMD patient, who was reported as being noncompliant with immune suppression, there was no anatomical evidence of donor cell survival. In the other patient, cells were seen to survive and increase in pigmentation. Gains in visual function were reported in the operated eyes of both patients and also in the control eye of the AMD patient. However the authors commented that it was unclear if the functional gains could be attributed to the donor cells.

Based on this encouraging and pioneering work, it may be possible in future to treat patients at an earlier disease stage when there may be a greater chance of photoreceptor preservation or vision rescue. However the study also raises important questions. The optimal cell type for treatment is not fully known (Stargardt disease affects photoreceptors, which usually become dysfunctional before any RPE changes). Moreover, in this work the investigators used a relatively lightly pigmented cell line because it was observed to better survive the preparation and transplantation procedure than (possibly more mature) heavily pigmented cells. Also the ideal timing of treatment and optimal cell dose remain to be clarified.

The work follows from preclinical studies on the application of ES cell-derived RPE (Haruta et al., 2004; Idelson et al., 2009; Klimanskaya et al., 2004; Vugler et al., 2008) in slowing retinal degeneration in animal models (Lamba, Gust, & Reh, 2009; Lu et al., 2009; Lund et al., 2006). While technically showing proof of concept of such an approach, the animal data should be interpreted with caution as the situation in elderly human patients is somewhat different, particularly with the marked degeneration of BM in the latter that may significantly hamper the adhesion and survival of transplanted cells. Furthermore, ES cell-derived RPE donor cells cannot be HLA-matched to the recipient, and hence aggressive immune suppression would have to accompany the transplantation of these cells. A hypothetical alternative approach would be to use autologous cells to create iPS cell–derived RPE, and in theory, this would obviate the requirement for immune suppression. Preclinical data in mouse models have shown the possible application of this approach (Carr et al., 2009).

Although many questions remain, some of which may be answered with the recruitment of more patients observed over a longer follow-up period in clinical trials, the milestone set by Schwartz and colleagues may potentially lay the foundation for other stem cell–based treatments for human blindness. Table 110.3 summarizes the current ES cell trials in progress for RPE regeneration.

STRATEGIES TO MAXIMIZE SURVIVAL AND INTEGRATION OF TRANSPLANTED CELLS

Scaffolds

The use of scaffolds as adjuncts in photoreceptor and RPE replacement has attracted interest in the retinal research community. The use of these materials has been pursued as a strategy to enhance cell survival and also to promote the integration of replacement cells in a manner that more closely approximates the normal, highly ordered layer arrangement of these cells.

In the case of the RPE, the critical role of BM (Spraul & Grossnihlatis, 1997) should be appreciated, not only as an important support structure for the cells in vivo but also as the site of significant pathological changes in AMD (Chong et al., 2005; Hahn, Milam, & Dunaief, 2003; Kamei & Hollyfield, 1999). If healthy and functional BM is not restored in parallel with efforts to reconstitute the overlying RPE monolayer, it is likely that the cells would not survive. Indeed there is in vitro evidence that RPE cell attachment to a substrate prevents apoptosis (Tezel, 1997). Cultured fetal human RPE cells were found to survive poorly when maintained on aged submacular human BM (Gullapalli et al., 2005). Some of these negative outcomes could be avoided by first resurfacing the aged BM with other materials, such as extracellular matrix derived from bovine corneal endothelial cells (Sugino et al., 2011). The improved adhesion of RPE in this context could be in part associated with the up-regulation of integrin expression (Afshari et al., 2010; Gullapalli, Sugino, & Zarbin, 2008).

TABLE 110.3

Clinical trials for retinal pigment epithelium transplantation

Clinicaltrials.gov identifier	NCT01344993	NCT01345006	NCT01469832
Sponsor	Advanced Cell Technology, Inc.		
Condition	Dry age-related macular degeneration	Stargardt's macula dystrophy	Stargardt's macula dystrophy, fundus flavimaculatus, juvenile macular dystrophy
Location(s)	U.S. (California and Pennsylvania)	U.S. (California)	U.K. (London)
Phase	Phase I/II		
Study type	Interventional		
Endpoint classification	Safety study		
Intervention	Subretinal injection of human embryonic stem cell derived retinal pigment epithelial (RPE) cells MA09-hRPE; 50,000 to 200,000 cells transplanted into one eye per patient		
Expected enrollment	12	12	12
Time frame	12 months		
Estimated primary completion date	July 2013	July 2013	December 2013
Status (accessed 29 January 2011 at www.clinicaltrials.gov)	Recruiting		

Other than improving the surface characteristics of native diseased submacular RPE, one other strategy could be to use a naturally available substitute such as lens capsule (Kiilgaard et al., 2002), Descemet membrane (Thumann et al., 1997), amniotic membrane (Capeáns et al., 2003; Ohno-Matsui et al., 2005), retinal inner limiting membrane (Beutel et al., 2007), or donor BM (Del Priore & Tezd, 1998). Collagen polymers (Yeung et al., 2004) could also be used, including those in ultrathin membrane form (Thumann et al., 2009) or as a thermoresponsive polymer that is injectable in the liquid phase containing a cell suspension and solidifies into a gel in situ (Fitzpatrick et al., 2010). The use of synthetic BM substitutes would avoid the potential for disease transfer and donor shortage that could be problematic when using natural material. Combined poly(L-lactic acid) (PLLA) and poly(DL-lactic-co-glycolic acid) (PLGA) has been investigated as thin films (Thomson et al., 1996) and carrier microspheres (Thomson et al., 2010); semipermeable parylene (Lu et al., 2011) and expanded polytetrafluoroethylene (Krishna et al., 2011) may also have potential as artificial RPE scaffolds. The additional advantage of synthetic materials is that their properties may, in theory, be precisely engineered for the desired application—for example, to have a predictable biodegradation time course or a controllable curling/uncurling behavior that could ease surgical insertion. However, it is a significant challenge to design synthetic layers that could simulate the nutrient flow function that is one of the main physiological roles of natural BM sitting between RPE and choriocapillaris.

Somewhat similar considerations exist for photoreceptor replacement, as these cells are normally arranged in a tissue layer. PLLA–PLGA scaffolds (Lavik et al., 2005; Tomita et al., 2005) and a microfabricated polycaprolactone thin film (Sodha et al., 2011) were found to increase the survival of retinal progenitors although it is less clear what further cues may be needed for complete maturation of the progenitor cells. PLGA scaffolds have been designed that deliver active matrix metalloproteinase 2 (MMP2) into the subretinal space along with transplanted progenitor cells in order to increase cellular integration (Tucker et al., 2010), showing how it may be possible to combine biomaterials with compounds that enhance transplantation success. In an effort to more directly address the question of whether ES cell-derived retinal cells could be incorporated into a scaffold to promote the orderly maturation and tissue-level organization of these cells, microchannel PLGA scaffolds were found to support the morphogenesis of human ES cell-derived retinal cells (McUsic, Lamba, & Reh, 2012). Akin to the development of whole optic cups from ES cells (Eiraku et al., 2011) as described above, this may prove to be another way to manufacture organized retinal layers from renewable cell sources for in vitro studies or, conceivably, even for in vivo application for outer retinal reconstruction.

 MANDEEP S. SINGH AND ROBERT E. MACLAREN

Overall, efforts to support the development of cellular grafts into organized layers are important steps toward the potential use of cell-based therapy for human blindness. Bolus suspensions of cells, as used in much of current research, may be sufficient when being delivered as vehicles for paracrine support of existing cells. However the aim to recreate outer retinal photoreceptor or RPE layers when these have degenerated would imply the need to recreate the regular polarization and spatial arrangement of replacement cells. This is one challenge in cell replacement that requires further research in preclinical models.

Immune Modulation

In theory, the relative immune privilege of the central nervous system should extend to both the donor cells and the recipient site when considering retinal and RPE transplantation. The subretinal space is often described as immune privileged (Jiang, Streilein, & McKinney, 1994; Luke Qi, Jorquera, & Streilein, 1993) although it should be appreciated that this privilege is not complete. A number of studies have shown evidence that a significant immune reaction may lead to the destruction of host and donor tissue (Qi Jiang et al., 1995; Zhang & Bok, 1998). This phenomenon is especially relevant in the setting of diseased recipients in which disease-related alterations in the subretinal milieu may impair the immune privilege found in the normal retina.

A specific adaptive T-cell mediated immune response has been suggested to be the cause of the decline in long-term transplanted cell survival in allogeneic rod photoreceptor transplantation. This could be dampened using systemic immune suppression. In one mouse study using wild-type hosts, cyclosporine treatment increased the number of surviving cells at 4 months (West et al., 2010). From these data, it is likely that nonautologous photoreceptor precursor transplantation would need to be augmented by a combination of strategies to increase engraftment and long-term survival of cells.

In a discordant-species xenotransplantation experiment, it was found that fetal rat retinas survived in the subretinal space of non-immune-suppressed rabbits for a longer duration if donor tissue was transplanted as intact sheets rather than as fragmented tissue (Ghosh, Rauer, & Arnér, 2008). From this work it may be seen that methods of donor cell preparation have a significant impact on host response. This concept has not been fully explored experimentally, especially in the setting of using renewable cell sources such as ES cells. The use of autologous adult-derived iPS cells may, in principle, obviate the need for immune suppression.

In human trials, the potential adverse effects of systemic immune suppression with medications in current use (Getts et al., 2011; Smith, Nemeth, & McDonald, 2003) include malignancy, infections, renal toxicity, neurotoxicity, and death. These severe outcomes must be weighed against the potential benefit of the degree of vision restoration offered by any cell-based treatment, and indeed, the considerations may be vastly different than in a patient undergoing solid organ transplantation for liver or heart failure. In the latter situations the diseases in question have a fatal endpoint, which is, in general, not relevant to retinal degenerations. The development of new agents and regimens with a more favorable toxicity profile (Jolly & Watson, 2011; Vincenti et al., 2010) may in the future alter the risk–benefit equation that is relevant to retinal transplantation for vision restoration, especially if the extent of visual recovery is partial at best.

It is notable that the human ES-cell derived RPE transplantation trial currently in progress (Schwartz et al., 2012) has followed a protocol that included short-term systemic administration of tacrolimus and mycophenolate mofetil. The immunogenicity of ES cells is controversial. On the one hand, embryos consisting of 50% nonautologous material are able to survive in the maternal environment, but on the other hand, they are known to express major histocompatibility antigens. Evidence from earlier studies indicated that traditional immune suppression agents did not have a long-term impact on the longevity of intramuscular ES cell grafts in immune competent hosts (Swijnenburg et al., 2008; Toriumi et al., 2009). Also, repeated ES cell transplantation led to more aggressive subsequent rejection (Swijnenburg et al., 2008), and this issue is pertinent when considering that the two eyes in a human patient may require sequential treatment. Further understanding of the nature and duration of immunosuppression required for successful stem cell–based treatments are needed if such strategies will be used to cure blindness.

Whereas local or topical immunosuppression therapy is applicable in corneal transplantation (Panda et al., 2007) for blinding diseases, at present, methods to produce effective local tolerance in the subretinal space are not fully known. One report, however, demonstrated that seeding retinal progenitor cells onto a PLGA scaffold led to the creation of local immune privilege in allogeneic kidney capsules (Tat et al., 2007), and the authors proposed that the synthetic material may have acted as a sink for immunosuppressive factors produced by the progenitors. The development of local immune suppression protocols may be another way of altering the risk–benefit balance for nonautologous retinal cell replacement.

Manipulation of the Local Retinal Environment of the Host

The extracellular matrix of the host may be a structural barrier to the synaptic integration of transplanted photoreceptors. Strategies to overcome this barrier may enhance the functional integration of donor photoreceptors. Some evidence for this was provided by an ex vivo transplantation model where retinal explants from MRL/MpJ mouse, with higher MMP2 levels, were better able to incorporate donor cells in culture compared with explants from wild-type or retinal degeneration (rd1) mice (Tucker et al., 2008). MMP2 delivery has been shown in other studies as having a positive effect on progenitor cell integration into host retinas (Suzuki et al., 2006; Yao et al., 2011b). This approach may be useful if cell therapy is undertaken to restore a partially degenerate retina, in which significant numbers of host photoreceptors and their matrix remain, so that the donor cells would be encouraged to integrate with preexisting outer nuclear layer cells.

Gliosis in the retina may occur either as a consequence of the outer retinal degeneration disease process (Iandiev et al., 2006; Rodrigues, Bardenstein, & Wiggert, 1987) or as a result of the subretinal transplantation procedure (Kinouchi et al., 2003). Retinal cells transplanted into mice deficient in glial fibrillary acid protein and vimentin were shown to integrate robustly and migrated extensively into the host retina after subretinal transplantation (Kinouchi et al., 2003) as a result of the lack of intermediate filaments produced by reactive astrocytes and Müller cells. The specific disruption of the outer limiting membrane, formed by adherens junctions between Müller glia and photoreceptors, has also been shown to produce a relatively high level of rod precursor integration (Pearson et al., 2010; West et al., 2008). Together, these data indicate that interventions directed at host glia and astrocytes may be useful to enhance the synaptic integration of transplanted cells.

Surgical Considerations for Subretinal Cell Delivery

The goal of retinal cell transplantation surgery (Stout & Francis, 2011) would be to safely deliver the cells to the intraocular location that promises the greatest functional integration of transplanted cells. For both photoreceptor and RPE cell replacement, the ideal target location for cell delivery would be the subretinal space. This represents a natural cleavage plane between outer retina and RPE (see figure 110.3). It is from this location that both photoreceptor cells and RPE cells are lost; hence it is logical that this specific location be considered the ideal target to introduce replacement cells.

Intravitreal placement has been described in animal models (Grozdanic et al., 2006; Hill et al., 2009; Meyer et al., 2004); however this approach has not led to successful transretinal migration of replacement cells into the outer retina. Also, since there is no way to contain exogenous material within a specific region or area of the intravitreal space between the lens and the retina, it will not be possible to direct the cells effectively over the disease area which may only be in a specific retinal location—for example, the macula. This is an important consideration given the low rate of survival of transplanted cells currently seen in animal models (MacLaren et al., 2006; West et al., 2010)—to ensure the maximal density of surviving cells by directing the cell load to a discrete and predetermined retinal area.

In human patients as well as animal models in preclinical studies, replacement cells would be optimally delivered to the subretinal location via a direct approach using subretinal cell transplantation (Schwartz et al., 2012). Although this approach is more surgically complex than intravitreal transplantation, the surgeon will be able to deliver the cells directly to the correct anatomical plane and over the correct disease region. In this technique, the potential subretinal space is expanded during the introduction of cell material. As there is no natural access route to the subretinal space, a puncture would have to be made in its inner or outer boundary, that is, the neural retina (inner boundary, internal approach) or the BM–choroid–sclera complex (outer boundary, external approach) (see figure 110.3). In most animal studies, the easiest approach has been to use the transscleral route. However in human intervention, it is likely that a transretinal approach will be utilized as it already has been used in clinical trials for subretinal introduction of viral vectors (Bainbridge et al., 2008; Cideciyan et al., 2008; Maguire et al., 2008). A transscleral approach may also be considered; however subretinal bleeding from choroidal puncture is a potential concern and was reported to occur in 17% of patients undergoing retinal detachment surgery in one randomized clinical trial (Burton et al., 1993). Compression of the eye immediately after choroidal puncture to reduce the risk of bleeding might force any cell suspension back out of the eye (as usually happens with subretinal fluid in detachment surgery). Hence the intraocular pressure would need to be kept low during cell injection to maximize cell delivery; however, this

    MANDEEP S. SINGH AND ROBERT E. MACLAREN

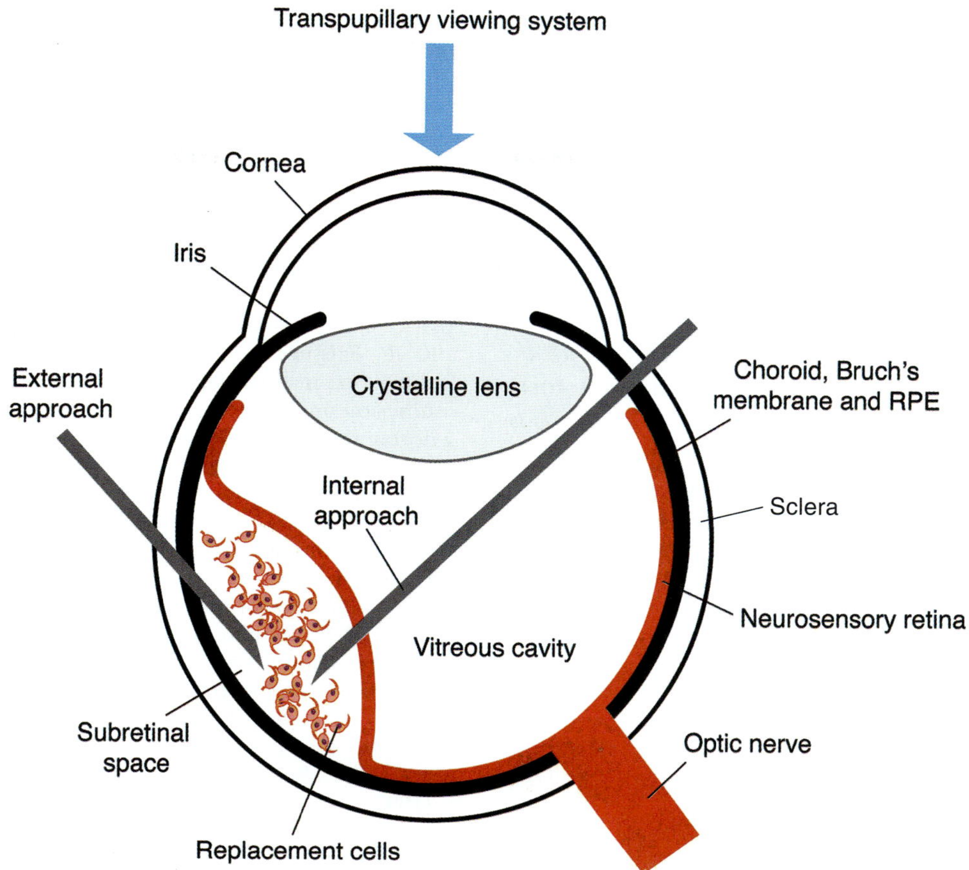

FIGURE 110.3 Surgical anatomy relevant to subretinal cell replacement. The subretinal space is a potential space between the neurosensory retina and retinal pigment epithelium (RPE) and can be accessed via a puncture in either the retina (internal approach) or sclera, choroid, Bruch's membrane, and RPE (external approach). Typically a small-gauge needle or cannula (e.g., 34 gauge) would be used in order to ensure that puncture wounds are self-sealing, so that reflux is minimized. The surgical site is viewed continuously via a transpupillary viewing system used routinely for vitreous and retinal surgery.

would encourage intraocular bleeding. Nevertheless, the external approach would be preferable as it avoids making any hole in the retina, and perhaps further surgical developments may allow the safe delivery of cells through the transscleral external approach.

In subretinal placement via the internal approach, the creation of puncture access creates the potential problem of reflux of transplanted material out of the subretinal space in a retrograde manner though the retinal puncture site. Special maneuvers will have to be designed to minimize or prevent this completely in order to ensure the maximal delivery of cells to the target area. After subretinal transplantation, the retinal bleb created is known to flatten over the following hours due to the absorption of fluid by neighboring cells, and the newly introduced cells then come to lie apposed directly to host RPE and inner retina. It is anticipated that the translation of cell replacement treatments into patients will be rapid, as vitreo–retinal

surgeons are familiar with all of the techniques that may be required (Fine et al., 2007; Fujii et al., 2002).

Combined Replacement as a Future Goal

In general, genetic and acquired insults that initially give rise to dysfunction in one or more specific cell types eventually lead to retinal degeneration involving other classes of cells. This phenomenon is evident in both diseases considered here. In dry or atrophic AMD, primary abnormalities in RPE cells, related to defects in BM, lead eventually to RPE death and then the loss of rods and cones from the outer retina overlying areas of RPE layer atrophy. In order to treat geographic atrophy, multilayered macular reconstruction of BM, RPE, and photoreceptors would provide the ideal cure.

In RP for example, a mutation in rod-specific genes (Wright et al., 2010) initially causes nyctalopia due to absence of light sensitivity in rod photoreceptors but subsequently leads to rod and then cone death later on, resulting in loss of visual acuity. In the advanced stage of this condition, foveal cones would need to be replaced along with peripheral rods in order to approach the normal distribution of photoreceptors. There is evidence to suggest that the presence of rods is critical to support the function of cones (Yang et al., 2009). In advanced RP, pigment epithelial health may also be disrupted, and so replaced photoreceptors would ideally require the transplantation of an RPE monolayer to support their survival.

The combined reconstruction of RPE and photoreceptor laminae represents a therapeutic goal that has received only scant attention, primarily because this is a complex undertaking and significant questions remain with regards to replacement of even a single cell type.

CONCLUSION

Stem cell-derived RPE transplantation, aiming to protect overlying photoreceptors from degeneration and so preserving remaining vision in diseased patients, has recently been launched in a clinical trial. This effort represents the culmination of sustained basic and translational work in retinal cell replacement. However, direct photoreceptor replacement is still in the preclinical stages of development as this task poses greater challenges due to the neuronal nature of the cells and the strict requirement for complex tissue organization.

There are a number of areas for further investigation that may be pursued to optimize cell-based replacement for the outer retina. The optimal dose of cells in relation to the desired spatial density of replacement RPE and photoreceptors, for example, has not been clearly defined in preclinical models. The best way to arrange the cells in layers, perhaps with the use of scaffold adjuncts, also requires further exploration. Mechanistically, the question of donor cell fusion as a potential explanation for the observed beneficial effects in transplantation studies also requires attention.

For translation into patients, questions relating to malignant potential of donor cells, immune suppression, and timing of treatment are among those that remain to be fully answered. However, the vision research community has come a long way since the first description of embryonic ocular transplants in rodents more than half a century ago (Tansley, 1946). The introduction of ES cell derivatives into the human eye is a positive development for this field and will no doubt lead to the proliferation of other trials for outer retinal cell replacement. The launch of human trials signifies the advent of cell-based treatment as a potential reality for the millions of patients worldwide with blindness due to retinal degeneration.

REFERENCES

Adolph, A. R., Zucker, C. L., Ehinger, B., & Bergstrom, A. (1994). Function and structure in retinal transplants. *Journal of Neural Transplantation & Plasticity, 5,* 147–161. doi:10.1155/NP.1994.147.

Afshari, F. T., Kwok, J. C., Andrews, M. R., Blits, B., Martin, K. R., Faissner, A., et al. (2010). Integrin activation or alpha9 expression allows retinal pigmented epithelial cell adhesion on Bruch's membrane in wet age-related macular degeneration. *Brain, 133,* 448–464. doi:10.1093/brain/awp319.

Akita, J., Takahashi, M., Hojo, M., Nishida, A., Haruta, M., & Honda, Y. (2002). Neuronal differentiation of adult rat hippocampus-derived neural stem cells transplanted into embryonic rat explanted retinas with retinoic acid pretreatment. *Brain Research, 954,* 286–293.

Algvere, P. V., Berglin, L., Gouras, P., & Sheng, Y. (1994). Transplantation of fetal retinal pigment epithelium in age-related macular degeneration with subfoveal neovascularization. *Graefe's Archive for Clinical and Experimental Ophthalmology, 232,* 707–716.

Algvere, P. V., Berglin, L., Gouras, P., Sheng, Y., & Kopp, E. D. (1997). Transplantation of RPE in age-related macular degeneration: Observations in disciform lesions and dry RPE atrophy. *Graefe's Archive for Clinical and Experimental Ophthalmology, 235,* 149–158.

Aramant, R. B., & Seiler, M. J. (1995). Fiber and synaptic connections between embryonic retinal transplants and host retina. *Experimental Neurology, 133,* 244–255. doi:10.1006/exnr.1995.1027.

Bainbridge, J. W. B., Smith, A. J., Barker, S. S., Robbie, S., Henderson, R., Balaggan, K., et al. (2008). Effect of gene therapy on visual function in Leber's congenital amaurosis. *New England Journal of Medicine, 358,* 2231–2239. doi:10.1056/NEJMoa0802268.

Bartsch, U., Oriyakhel, W., Kenna, P. F., Linke, S., Richard, G., Petrowitz, B., et al. (2008). Retinal cells integrate into the outer nuclear layer and differentiate into mature photoreceptors after subretinal transplantation into adult mice. *Experimental Eye Research, 86,* 691–700. doi:10.1016/j.exer.2008.01.018.

Beutel, J., Greulich, L., Lüke, M., Ziemssen, F., Szurman, P., Bartz-Schmidt, K. U., & Grisanti, S. (2007). Inner limiting membrane as membranous support in RPE sheet-transplantation. *Graefe's Archive for Clinical and Experimental Ophthalmology, 245,* 1469–1473. doi:10.1007/s00417-007-0566-9.

Burton, R. L., Cairns, J. D., Campbell, W. G., Heriot, W. J., & Heinze, J. B. (1993). Needle drainage of subretinal fluid: A randomized clinical trial. *Retina (Philadelphia, Pa.), 13,* 13–16. doi:10.1097/00006982-199313010-00004.

Campbell, K. H. S., McWhir, J., Ritchie, W. A., & Wilmut, I. (1996). Sheep cloned by nuclear transfer from a cultured cell line. *Nature, 380,* 64–66.

Capeáns, C., Piñeiro Ces, A., Pardo, M., Sueiro-López, C., Blanco, M. J., Domínguez, F., et al. (2003). Amniotic membrane as support for human retinal pigment epithelium

(RPE) cell growth. *Acta Ophthalmologica Scandinavica, 81,* 271–277. doi:10.1034/j.1600-0420.2003.00076.x.

Carr, A. J., Vugler, A. A., Hikita, S. T., Lawrence, J. M., Gias, C., Chen, L. L., et al. (2009). Protective effects of human iPS-derived retinal pigment epithelium cell transplantation in the retinal dystrophic rat. *PLoS ONE, 4,* e8152. doi:10.1371/journal.pone.0008152.

Chacko, D. M., Rogers, J. A., Turner, J. E., & Ahmad, I. (2000). Survival and differentiation of cultured retinal progenitors transplanted in the subretinal space of the rat. *Biochemical and Biophysical Research Communications, 268,* 842–846. doi:10.1006/bbrc.2000.2153.

Chen, F. K., Uppal, G. S., Maclaren, R. E., Coffey, P. J., Rubin, G. S., Tufail, A., et al. (2009). Long-term visual and microperimetry outcomes following autologous retinal pigment epithelium choroid graft for neovascular age-related macular degeneration. *Clinical & Experimental Ophthalmology, 37,* 275–285.

Chong, N. H. V., Keonin, J., Luthert, P. J., Frennesson, C. I., Weingeist, D. M., Wolf, R. L., et al. (2005). Decreased thickness and integrity of the macular elastic layer of Bruch's membrane correspond to the distribution of lesions associated with age-related macular degeneration. *American Journal of Pathology, 166,* 241–251.

Cideciyan, A. V., Aleman, T. S., Boye, S. L., Schwartz, S. B., Kaushal, S., Roman, A. J., et al. (2008). Human gene therapy for RPE65 isomerase deficiency activates the retinoid cycle of vision but with slow rod kinetics. *Proceedings of the National Academy of Sciences of the United States of America, 105,* 15112–15117. doi:10.1073/pnas.0807027105.

Curcio, C. A., Medeiros, N. E., & Millican, C. L. (1996). Photoreceptor loss in age-related macular degeneration. *Investigative Ophthalmology & Visual Science, 37,* 1236–1249.

Das, T., del Cerro, M., Jalali, S., Rao, V. S., Gullapalli, V. K., Little, C., et al. (1999). The transplantation of human fetal neuroretinal cells in advanced retinitis pigmentosa patients: Results of a long-term safety study. *Experimental Neurology, 157,* 58–68.

del Cerro, M., Humayun, M. S., Sadda, S. R., Cao, J., Hayashi, N., Green, W. R., et al. (2000). Histologic correlation of human neural retinal transplantation. *Investigative Ophthalmology & Visual Science, 41,* 3142–3148.

del Cerro, M., Ison, J. R., Bowen, G. P., Lazar, E., & del Cerro, C. (1991). Intraretinal grafting restores visual function in light-blinded rats. *Neuroreport, 2,* 529–532.

Del Priore, L. V., & Tezd, T. H. (1998). Reattachment rate of human retinal pigment epithelium to layers of human Bruch's membrane. *Archives of Ophthalmology, 116,* 335–341.

Dong, X., Pulido, J. S., Qu, T., & Sugaya, K. (2003). Differentiation of human neural stem cells into retinal cells. *Neuroreport, 14,* 143–146.

Eiraku, M., Takata, N., Ishibashi, H., Kawada, M., Sakakura, E., Okuda, S., et al. (2011). Self-organizing optic-cup morphogenesis in three-dimensional culture. *Nature, 472,* 51–56. doi:10.1038/nature09941.

Faktorovich, E. G., Steinberg, R. H., Yasumura, D., Matthes, M. T., & LaVail, M. M. (1990). Photoreceptor degeneration in inherited retinal dystrophy delayed by basic fibroblast growth factor. *Nature, 347,* 83–86.

Fine, H. F., Iranmanesh, R., Iturralde, D., & Spaide, R. F. (2007). Outcomes of 77 consecutive cases of 23-gauge transconjunctival vitrectomy surgery for posterior segment disease. *Ophthalmology, 114,* 1197–1200.

Fitzpatrick, S. D., Jafar Mazumder, M. A., Lasowski, F., Fitzpatrick, L. E., & Sheardown, H. (2010). PNIPAAm-grafted-collagen as an injectable, in situ gelling, bioactive cell delivery scaffold. *Biomacromolecules, 11,* 2261–2267.

Fujii, G. Y., De Juan, E., Jr., Humayun, M. S., Chang, T. S., Pieramici, D. J., Barnes, A., et al. (2002). Initial experience using the transconjunctival sutureless vitrectomy system for vitreoretinal surgery. *Ophthalmology, 109,* 1814–1820. doi:10.1016/S0161-6420(02)01119-3.

Getts, D. R., Shankar, S., Chastain, E. M. L., Martin, A., Getts, M. T., Wood, K., et al. (2011). Current landscape for T-cell targeting in autoimmunity and transplantation. *Immunotherapy, 3,* 853–870.

Ghosh, F., Bruun, A., & Ehinger, B. (1999). Graft-host connections in long-term full-thickness embryonic rabbit retinal transplants. *Investigative Ophthalmology & Visual Science, 40,* 126–132.

Ghosh, F., Johansson, K., & Ehinger, B. (1999). Long-term full-thickness embryonic rabbit retinal transplants. *Investigative Ophthalmology & Visual Science, 40,* 133–142.

Ghosh, F., Rauer, O., & Arnér, K. (2008). Neuroretinal xenotransplantation to immunocompetent hosts in a discordant species combination. *Neuroscience, 152,* 526–533.

Gouras, P., & Du, J. (1991). Gelanze, M., Lopez, R., Kwun, R., Kjeldbye, H., & Krebs, W. Survival and synapse formation of transplanted rat rods. *Journal of Neural Transplantation & Plasticity, 2,* 91–100. doi:10.1155/NP.1991.91.

Grozdanic, S. D., Ast, A. M., Lazic, T., Kwon, Y. H., Kardon, R. H., Sonea, I. M., et al. (2006). Morphological integration and functional assessment of transplanted neural progenitor cells in healthy and acute ischemic rat eyes. *Experimental Eye Research, 82,* 597–607.

Gullapalli, V. K., Sugino, I. K., Van Patten, Y., Shah, S., & Zarbin, M. A. (2005). Impaired RPE survival on aged submacular human Bruch's membrane. *Experimental Eye Research, 80,* 235–248.

Gullapalli, V. K., Sugino, I. K., & Zarbin, M. A. (2008). Culture-induced increase in alpha integrin subunit expression in retinal pigment epithelium is important for improved resurfacing of aged human Bruch's membrane. *Experimental Eye Research, 86,* 189–200.

Gust, J., & Reh, T. A. (2011). Adult donor rod photoreceptors integrate into the mature mouse retina. *Investigative Ophthalmology & Visual Science, 52,* 5266–5272. doi:10.1167/iovs.10-6329.

Hahn, P., Milam, A. H., & Dunaief, J. L. (2003). Maculas affected by age-related macular degeneration contain increased chelatable iron in the retinal pigment epithelium and Bruch's membrane. *Archives of Ophthalmology, 121,* 1099–1105.

Hartong, D. T., Berson, E. L., & Dryja, T. P. (2006). Retinitis pigmentosa. *Lancet, 368,* 1795–1809. doi:10.1016/s0140-6736(06)69740-7.

Haruta, M., Sasai, Y., Kawasaki, H., Amemiya, K., Ooto, S., Kitada, M., et al. (2004). In vitro and in vivo characterization of pigment epithelial cells differentiated from primate embryonic stem cells. *Investigative Ophthalmology & Visual Science, 45,* 1020–1025.

Heimann, K. (1994). A breakthrough in surgery for age-related macular degeneration? *Graefe's Archive for Clinical and Experimental Ophthalmology, 232,* 706.

Hill, A. J., Zwart, I., Tam, H. H., Chan, J., Navarrete, C., Jen, L. S., et al. (2009). Human umbilical cord blood-derived mesenchymal stem cells do not differentiate into neural cell types or integrate into the retina after intravitreal grafting in neonatal rats. *Stem Cells and Development, 18*, 399–409.

Hilton, G., Machemer, R., & Michels, R. (1983). The classification of retinal detachment with proliferative vitreoretinopathy. *Ophthalmology, 90*, 121–125.

Hirami, Y., Osakada, F., Takahashi, K., Okita, K., Yamanaka, S., Ikeda, H., et al. (2009). Generation of retinal cells from mouse and human induced pluripotent stem cells. *Neuroscience Letters, 458*, 126–131. doi:10.1016/j.neulet.2009.04.035.

Howden, S. E., Gore, A., Li, Z., Fung, H.-L., Nisler, B. S., Nie, J., et al. (2011). Genetic correction and analysis of induced pluripotent stem cells from a patient with gyrate atrophy. *Proceedings of the National Academy of Sciences of the United States of America, 108*, 6537–6542.

Hu, B.-Y., Weick, J. P., Yu, J., Ma, L.-X., Zhang, X.-Q., Thomson, J. A., et al. (2010). Neural differentiation of human induced pluripotent stem cells follows developmental principles but with variable potency. *Proceedings of the National Academy of Sciences of the United States of America, 107*, 4335–4340.

Huang, J. C., Ishida, M., Hersh, P., Sugino, I. K., & Zarbin, M. A. (1998). Preparation and transplantation of photoreceptor sheets. *Current Eye Research, 17*, 573–585. doi:10.1076/ceyr.17.6.573.5184.

Humayun, M. S., De Juan, E., Jr., del Cerro, M., Dagnelie, G., Radner, W., Sadda, S. R., et al. (2000). Human neural retinal transplantation. *Investigative Ophthalmology & Visual Science, 41*, 3100–3106.

Iandiev, I., Biedermann, B., Bringmann, A., Reichel, M. B., Reichenbach, A., & Pannicke, T. (2006). Atypical gliosis in Müller cells of the slowly degenerating rds mutant mouse retina. *Experimental Eye Research, 82*, 449–457.

Idelson, M., Alper, R., Obolensky, A., Ben-Shushan, E., Hemo, I., Yachimovich-Cohen, N., et al. (2009). Directed differentiation of human embryonic stem cells into functional retinal pigment epithelium cells. *Cell Stem Cell, 5*, 396–408.

Ikeda, H., Osakada, F., Watanabe, K., Mizuseki, K., Haraguchi, T., Miyoshi, H., et al. (2005). Generation of Rx+/Pax6+ neural retinal precursors from embryonic stem cells. *Proceedings of the National Academy of Sciences of the United States of America, 102*, 11331–11336. doi:10.1073/pnas.0500010102.

Jiang, L. Q., Streilein, J. W., & McKinney, C. (1994). Immune privilege in the eye: An evolutionary adaptation. *Developmental and Comparative Immunology, 18*, 421–431.

Jiang, L., Zhang, H., Dizhoor, A. M., Boye, S. E., Hauswirth, W. W., Frederick, J. M., et al. (2011). Long-term RNA interference gene therapy in a dominant retinitis pigmentosa mouse model. *Proceedings of the National Academy of Sciences of the United States of America, 108*, 18476–18481. doi:10.1073/pnas.1112758108.

Jin, Z. B., Okamoto, S., Osakada, F., Homma, K., Assawachananont, J., Hirami, Y., et al. (2011). Modeling retinal degeneration using patient-specific induced pluripotent stem cells. *PLoS ONE, 6*. doi:10.1371/journal.pone.0017084.

Jolly, E. C., & Watson, C. J. E. (2011). Modern immunosuppression. *Surgery, 29*, 312–318.

Jones, B. W., & Marc, R. E. (2005). Retinal remodeling during retinal degeneration. *Experimental Eye Research, 81*, 123–137. doi:10.1016/j.exer.2005.03.006.

Juliusson, B., Bergstrom, A., Van Veen, T., & Ehinger, B. (1993). Cellular organization in retinal transplants using cell suspensions or fragments of embryonic retinal tissue. *Cell Transplantation, 2*, 411–418.

Junying, Y., Kejin, H., Kim, S. O., Shulan, T., Stewart, R., Slukvin, I. I., et al. (2009). Human induced pluripotent stem cells free of vector and transgene sequences. *Science, 324*, 797–801.

Kamei, M., & Hollyfield, J. G. (1999). TIMP-3 in Bruch's membrane: Changes during aging and in age-related macular degeneration. *Investigative Ophthalmology & Visual Science, 40*, 2367–2375.

Kicic, A., Shen, W. Y., Wilson, A. S., Constable, I. J., Robertson, T., & Rakoczy, P. E. (2003). Differentiation of marrow stromal cells into photoreceptors in the rat eye. *Journal of Neuroscience, 23*, 7742–7749.

Kiilgaard, J. F., Wiencke, A. K., Scherfig, E., Prause, J. U., & La Cour, M. (2002). Transplantation of allogenic anterior lens capsule to the subretinal space in pigs. *Acta Ophthalmologica Scandinavica, 80*, 76–81.

Kinouchi, R., Takeda, M., Yang, L., Wilhelmsson, U., Lundkvist, A., Pekny, M., et al. (2003). Robust neural integration from retinal transplants in mice deficient in GFAP and vimentin. *Nature Neuroscience, 6*, 863–868.

Klassen, H., Kiilgaard, J. F., Zahir, T., Ziaeian, B., Kirov, I., Scherfig, E., et al. (2007). Progenitor cells from the porcine neural retina express photoreceptor markers after transplantation to the subretinal space of allorecipients. *Stem Cells (Dayton, Ohio), 25*, 1222–1230. doi:10.1634/stemcells.2006-0541.

Klassen, H. J., Ng, T. F., Kurimoto, Y., Kirov, I., Shatos, M., Coffey, P., et al. (2004). Multipotent retinal progenitors express developmental markers, differentiate into retinal neurons, and preserve light-mediated behavior. *Investigative Ophthalmology & Visual Science, 45*, 4167–4173.

Klimanskaya, I., Chung, Y., Becker, S., Lu, S. J., & Lanza, R. (2006). Human embryonic stem cell lines derived from single blastomeres. *Nature, 444*, 481–485.

Klimanskaya, I., Hipp, J., Rezai, K. A., West, M., Atala, A., & Lanza, R. (2004). Derivation and comparative assessment of retinal pigment epithelium from human embryonic stem cells using transcriptomics. *Cloning and Stem Cells, 6*, 217–245.

Krishna, Y., Sheridan, C., Kent, D., Kearns, V., Grierson, I., & Williams, R. (2011). Expanded polytetrafluoroethylene as a substrate for retinal pigment epithelial cell growth and transplantation in age-related macular degeneration. *British Journal of Ophthalmology, 95*, 569–573.

Kuno, N., & Fujii, S. (2011). Dry age-related macular degeneration: Recent progress of therapeutic approaches. *Current Molecular Pharmacology, 4*, 196–232.

Kwan, A. S. L., Wang, S., & Lund, R. D. (1999). Photoreceptor layer reconstruction in a rodent model of retinal degeneration. *Experimental Neurology, 159*, 21–33. doi:10.1006/exnr.1999.7157.

Lakowski, J., Baron, M., Bainbridge, J., Barber, A. C., Pearson, R. A., Ali, R. R., et al. (2010). Cone and rod photoreceptor transplantation in models of the childhood retinopathy Leber congenital amaurosis using flow-sorted Crx-positive donor cells. *Human Molecular Genetics, 19*, 4545–4559.

Lakowski, J., Han, Y. T., Pearson, R. A., Gonzalez-Cordero, A., West, E. L., Gualdoni, S., et al. (2011). Effective transplantation of photoreceptor precursor cells selected via cell surface antigen expression. *Stem Cells (Dayton, Ohio), 29*, 1391–1404. doi:10.1002/stem.694.

Lamba, D. A., Gust, J., & Reh, T. A. (2009). Transplantation of human embryonic stem cell-derived photoreceptors restores some visual function in Crx-deficient mice. *Cell Stem Cell, 4*, 73–79.

Lamba, D. A., Karl, M. O., Ware, C. B., & Reh, T. A. (2006). Efficient generation of retinal progenitor cells from human embryonic stem cells. *Proceedings of the National Academy of Sciences of the United States of America, 103*, 12769–12774. doi:10.1073/pnas.0601990103.

Lau, D., McGee, L. H., Zhou, S., Rendahl, K. G., Manning, W. C., Escobedo, J. A., et al. (2000). Retinal degeneration is slowed in transgenic rats by AAV-mediated delivery of FGF-2. *Investigative Ophthalmology & Visual Science, 41*, 3622–3633.

Lavik, E. B., Klassen, H., Warfvinge, K., Langer, R., & Young, M. J. (2005). Fabrication of degradable polymer scaffolds to direct the integration and differentiation of retinal progenitors. *Biomaterials, 26*, 3187–3196.

Lu, B., Liu, Z., Liu, L., Zhu, D., Hinton, D., Thomas, B., et al. (2011). *Semipermeable parylene membrane as an artificial Bruch's membrane*. Paper presented at the 2011 16th International Solid-State Sensors, Actuators and Microsystems Conference, TRANSDUCERS '11, Beijing.

Lu, B., Malcuit, C., Wang, S., Girman, S., Francis, P., Lemieux, L., et al. (2009). Long-term safety and function of RPE from human embryonic stem cells in preclinical models of macular degeneration. *Stem Cells (Dayton, Ohio), 27*, 2126–2135. doi:10.1002/stem.149.

Luke Qi, J., Jorquera, M., & Streilein, J. W. (1993). Subretinal space and vitreous cavity as immunologically privileged sites for retinal allografts. *Investigative Ophthalmology & Visual Science, 34*, 3347–3354.

Lund, R. D., Wang, S., Klimanskaya, I., Holmes, T., Ramos-Kelsey, R., Lu, B., et al. (2006). Human embryonic stem cell-derived cells rescue visual function in dystrophic RCS rats. *Cloning and Stem Cells, 8*, 189–199.

MacLaren, R. E. (1996). Development and role of retinal glia in regeneration of ganglion cells following retinal injury. *British Journal of Ophthalmology, 80*, 458–464.

MacLaren, R. E. (1999). Re-establishment of visual circuitry after optic nerve regeneration. *Eye, 13*, 277–284.

MacLaren, R. E., Pearson, R. A., MacNeil, A., Douglas, R. H., Salt, T. E., Akimoto, M., et al. (2006). Retinal repair by transplantation of photoreceptor precursors. *Nature, 444*, 203–207. doi:10.1038/nature05161.

MacLaren, R. E., & Taylor, J. S. H. (1997). Regeneration in the developing optic nerve: Correlating observations in the opossum to other mammalian systems. *Progress in Neurobiology, 53*, 381–398. doi:10.1016/s0301-0082(97)00041-5.

MacLaren, R. E., Uppal, G. S., Balaggan, K. S., Tufail, A., Munro, P. M. G., Milliken, A. B., et al. (2007). Autologous transplantation of the retinal pigment epithelium and choroid in the treatment of neovascular age-related macular degeneration. *Ophthalmology, 114*, 561–570.

Maguire, A. M., Simonelli, F., Pierce, E. A., Pugh, E. N., Jr., Mingozzi, F., Bennicelli, J., et al. (2008). Safety and efficacy of gene transfer for Leber's congenital amaurosis. *New England Journal of Medicine, 358*, 2240–2248.

McUsic, A. C., Lamba, D. A., & Reh, T. A. (2012). Guiding the morphogenesis of dissociated newborn mouse retinal cells and hES cell-derived retinal cells by soft lithography-patterned microchannel PLGA scaffolds. *Biomaterials, 33*, 1396–1405.

Meyer, J. S., Katz, M. L., Maruniak, J. A., & Kirk, M. D. (2004). Neural differentiation of mouse embryonic stem cells in vitro and after transplantation into eyes of mutant mice with rapid retinal degeneration. *Brain Research, 1014*, 131–144.

Mitchell, P., Korobelnik, J. F., Lanzetta, P., Holz, F. G., Prünte, C., Schmidt-Erfurth, U., et al. (2010). Ranibizumab (*Lucentis*) in neovascular age-related macular degeneration: Evidence from clinical trials. *British Journal of Ophthalmology, 94*, 2–13. doi:10.1136/bjo.2009.159160.

Nakagawa, M., Koyanagi, M., Tanabe, K., Takahashi, K., Ichisaka, T., Aoi, T., et al. (2008). Generation of induced pluripotent stem cells without Myc from mouse and human fibroblasts. *Nature Biotechnology, 26*, 101–106.

Nowak, J. Z. (2006). Age-related macular degeneration (AMD): Pathogenesis and therapy. *Pharmacological Reports, 58*, 353–363.

Ohno-Matsui, K., Ichinose, S., Nakahama, K. I., Yoshida, T., Kojima, A., Mochizuki, M., et al. (2005). The effects of amniotic membrane on retinal pigment epithelial cell differentiation. *Molecular Vision, 11*, 1–10.

Osakada, F., Ikeda, H., Mandai, M., Wataya, T., Watanabe, K., Yoshimura, N., et al. (2008). Toward the generation of rod and cone photoreceptors from mouse, monkey and human embryonic stem cells. *Nature Biotechnology, 26*, 215–224.

Osakada, F., Ikeda, H., Sasai, Y., & Takahashi, M. (2009a). Stepwise differentiation of pluripotent stem cells into retinal cells. *Nature Protocols, 4*, 811–824.

Osakada, F., Jin, Z.-B., Hirami, Y., Ikeda, H., Danjyo, T., Watanabe, K., et al. (2009b). In vitro differentiation of retinal cells from human pluripotent stem cells by small-molecule induction. *Journal of Cell Science, 122*, 3169–3179.

Pagon, R. A. (1988). Retinitis pigmentosa. *Survey of Ophthalmology, 33*, 137–177. doi:10.1016/0039-6257(88)90085-9.

Panda, A., Vanathi, M., Kumar, A., Dash, Y., & Priya, S. (2007). Corneal graft rejection. *Survey of Ophthalmology, 52*, 375–396.

Parameswaran, S., Balasubramanian, S., Babai, N., Qiu, F., Eudy, J. D., Thoreson, W. B., et al. (2010). Induced pluripotent stem cells generate both retinal ganglion cells and photoreceptors: Therapeutic implications in degenerative changes in glaucoma and age-related macular degeneration. *Stem Cells (Dayton, Ohio), 28*, 695–703.

Pearson, R. A., Barber, A. C., West, E. L., MacLaren, R. E., Duran, Y., Bainbridge, J. W., et al. (2010). Targeted disruption of outer limiting membrane junctional proteins (Crb1 and ZO-1) increases integration of transplanted photoreceptor precursors into the adult wild-type and degenerating retina. *Cell Transplantation, 19*, 487–503.

Qi Jiang, L., Jorquera, M., Streilein, J. W., & Ishioka, M. (1995). Unconventional rejection of neural retinal allografts implanted into the immunologically privileged site of the eye. *Transplantation, 59*, 1201–1207.

Qiu, G., Seiler, M. J., Mui, C., Arai, S., Aramant, R. B., de Juan, E., Jr., et al. (2005). Photoreceptor differentiation and integration of retinal progenitor cells transplanted into transgenic rats. *Experimental Eye Research, 80*, 515–525.

Radtke, N. D., Aramant, R. B., Seiler, M. J., Petry, H. M., & Pidwell, D. (2004). Vision change after sheet transplant of fetal retina with retinal pigment epithelium to a patient with retinitis pigmentosa. *Archives of Ophthalmology, 122*, 1159–1165.

Radtke, N. D., Seiler, M. J., Aramant, R. B., Petry, H. M., & Pidwell, D. J. (2002). Transplantation of intact sheets of fetal neural retina with its retinal pigment epithelium in retinitis pigmentosa patients. *American Journal of Ophthalmology, 133*, 544–550.

Reubinoff, B. E., Pera, M. F., Fong, C. Y., Trounson, A., & Bongso, A. (2000). Embryonic stem cell lines from human blastocysts: Somatic differentiation in vitro. *Nature Biotechnology, 18*, 399–404.

Rodrigues, M. M., Bardenstein, D., & Wiggert, B. (1987). Retinitis pigmentosa with segmental massive retinal gliosis. An immunohistochemical, biochemical, and ultrastructural study. *Ophthalmology, 94*, 180–186.

Royo, P. E., & Quay, W. B. (1959). Retinal transplantation from fetal to maternal mammalian eye. *Growth, 23*, 313–336.

Schwartz, S. D., Hubschman, J.-P., Heilwell, G., Franco-Cardenas, V., Pan, C. K., Ostrick, R. M., et al. (2012). Embryonic stem cell trials for macular degeneration: A preliminary report. *Lancet.* doi:10.1016/S0140-6736(12)60028-2.

Seiler, M. J., & Aramant, R. B. (1998). Intact sheets of fetal retina transplanted to restore damaged rat retinas. *Investigative Ophthalmology & Visual Science, 39*, 2121–2131.

Singh, M. S., Charbel Issa, P., Butler, R., Martin, C., Lipinski, D. M., Sekaran, S., et al. (2013). Reversal of end-stage retinal degeneration and restoration of visual function by photoreceptor transplantation. *Proceedings of the National Academy of Sciences of the United States of America 110*, 1101–1106.

Singh, M. S., & MacLaren, R. E. (2011). Stem cells as a therapeutic tool for the blind: Biology and future prospects. *Proceedings of the Royal Society B: Biological Sciences, 278*, 3009–3016. doi:10.1098/rspb.2011.1028.

Smith, J. M., Nemeth, T. L., & McDonald, R. A. (2003). Current immunosuppressive agents: Efficacy, side effects, and utilization. *Pediatric Clinics of North America, 50*, 1283–1300.

Smith, W., Assink, J., Klein, R., Mitchell, P., Klaver, C. C., Klein, B. E., et al. (2001). Risk factors for age-related macular degeneration: Pooled findings from three continents. *Ophthalmology, 108*, 697–704. doi:10.1016/s0161-6420(00)00580-7.

Sodha, S., Wall, K., Redenti, S., Klassen, H., Young, M. J., & Tao, S. L. (2011). Microfabrication of a three-dimensional polycaprolactone thin-film scaffold for retinal progenitor cell encapsulation. *Journal of Biomaterials Science. Polymer Edition, 22*, 443–456.

Spraul, C. W., & Grossnihlatis, H. E. (1997). Characteristics of Drusen and Bruch's membrane in postmortem eyes with age-related macular degeneration. *Archives of Ophthalmology, 115*, 267–273.

Stadtfeld, M., Nagaya, M., Utikal, J., Weir, G., & Hochedlinger, K. (2008). Induced pluripotent stem cells generated without viral integration. *Science, 322*, 945–949.

Stout, J. T., & Francis, P. J. (2011). Surgical approaches to gene and stem cell therapy for retinal disease. *Human Gene Therapy, 22*, 531–535. doi:10.1089/hum.2011.060.

Sugie, Y., Yoshikawa, M., Ouji, Y., Saito, K., Moriya, K., Ishizaka, S., et al. (2005). Photoreceptor cells from mouse ES cells by co-culture with chick embryonic retina. *Biochemical and Biophysical Research Communications, 332*, 241–247.

Sugino, I. K., Gullapalli, V. K., Sun, Q., Wang, J., Nunes, C. F., Cheewatrakoolpong, N., et al. (2011). Cell-deposited matrix improves retinal pigment epithelium survival on aged submacular human Bruch's membrane. *Investigative Ophthalmology & Visual Science, 52*, 1345–1358.

Suzuki, T., Mandai, M., Akimoto, M., Yoshimura, N., & Takahashi, M. (2006). The simultaneous treatment of MMP-2 stimulants in retinal transplantation enhances grafted cell migration into the host retina. *Stem Cells (Dayton, Ohio), 24*, 2406–2411.

Swijnenburg, R. J., Schrepfer, S., Govaert, J. A., Cao, F., Ransohoff, K., Sheikh, A. Y., et al. (2008). Immunosuppressive therapy mitigates immunological rejection of human embryonic stem cell xenografts. *Proceedings of the National Academy of Sciences of the United States of America, 105*, 12991–12996. doi:10.1073/pnas.0805802105.

Takahashi, K., Tanabe, K., Ohnuki, M., Narita, M., Ichisaka, T., Tomoda, K., et al. (2007). Induction of pluripotent stem cells from adult human fibroblasts by defined factors. *Cell, 131*, 861–872.

Takahashi, K., & Yamanaka, S. (2006). Induction of pluripotent stem cells from mouse embryonic and adult fibroblast cultures by defined factors. *Cell, 126*, 663–676.

Takahashi, M., Miyoshi, H., Verma, I. M., & Gage, F. H. (1999). Rescue from photoreceptor degeneration in the RD mouse by human immunodeficiency virus vector-mediated gene transfer. *Journal of Virology, 73*, 7812–7816.

Tansley, K. (1946). The development of the rat eye in graft. *Journal of Experimental Biology, 22*, 221–223.

Tao, W., Wen, R., Goddard, M. B., Sherman, S. D., O'Rourke, P. J., Stabila, P. F., et al. (2002). Encapsulated cell-based delivery of CNTF reduces photoreceptor degeneration in animal models of retinitis pigmentosa. *Investigative Ophthalmology & Visual Science, 43*, 3292–3298.

Tat, F. N., Lavik, E., Keino, H., Taylor, A. W., Langer, R. S., & Young, M. J. (2007). Creating an immune-privileged site using retinal progenitor cells and biodegradable polymers. *Stem Cells (Dayton, Ohio), 25*, 1552–1559.

Tejeria, L., Harper, R. A., Artes, P. H., & Dickinson, C. M. (2002). Face recognition in age related macular degeneration: Perceived disability, measured disability, and performance with a bioptic device. *British Journal of Ophthalmology, 86*, 1019–1026. doi:10.1136/bjo.86.9.1019.

Tezel, T. H. (1997). Reattachment to a substrate prevents apoptosis of human retinal pigment epithelium. *Graefe's Archive for Clinical and Experimental Ophthalmology, 235*, 41–47.

The Eye Diseases Prevalence Research Group. (2004). Causes and prevalence of visual impairment among adults in the United States. *Archives of Ophthalmology, 122*, 477–485. doi:10.1001/archopht.122.4.477.

Thomson, H. A., Treharne, A. J., Backholer, L. S., Cuda, F., Grossel, M. C., & Lotery, A. J. (2010). Biodegradable poly(Œ±-hydroxy ester) blended microspheres as suitable carriers for retinal pigment epithelium cell transplantation. *Journal of Biomedical Materials Research. Part A, 95*, 1233–1243.

Thomson, J. A. (1998). Embryonic stem cell lines derived from human blastocysts. *Science, 282*, 1145–1147.

Thomson, R. C., Giordano, G. G., Collier, J. H., Ishaug, S. L., Mikos, A. G., Lahiri-Munir, D., et al. (1996). Manufacture and characterization of poly(alpha-hydroxy ester) thin films as temporary substrates for retinal pigment epithelium cells. *Biomaterials, 17*, 321–327.

Thumann, G., Schraermeyer, U., Bartz-Schmidt, K. U., & Heimann, K. (1997). Descemet's membrane as membranous support in RPE/IPE transplantation. *Current Eye Research*, *16*, 1236–1238.

Thumann, G., Viethen, A., Gaebler, A., Walter, P., Kaempf, S., Johnen, S., et al. (2009). The in vitro and in vivo behaviour of retinal pigment epithelial cells cultured on ultrathin collagen membranes. *Biomaterials*, *30*, 287–294.

Tomita, M., Lavik, E., Klassen, H., Zahir, T., Langer, R., & Young, M. J. (2005). Biodegradable polymer composite grafts promote the survival and differentiation of retinal progenitor cells. *Stem Cells (Dayton, Ohio)*, *23*, 1579–1588.

Tomita, M., Mori, T., Maruyama, K., Zahir, T., Ward, M., Umezawa, A., et al. (2006). A comparison of neural differentiation and retinal transplantation with bone marrow-derived cells and retinal progenitor cells. *Stem Cells (Dayton, Ohio)*, *24*, 2270–2278. doi:10.1634/stemcells.2005-0507.

Toriumi, H., Yoshikawa, M., Matsuda, R., Nishimura, F., Yamada, S. I., Hirabayashi, H., et al. (2009). Treatment of Parkinson's disease model mice with allogeneic embryonic stem cells: Necessity of immunosuppressive treatment for sustained improvement. *Neurological Research*, *31*, 220–227.

Tucker, B., Klassen, H., Yang, L., Chen, D. F., & Young, M. J. (2008). Elevated MMP expression in the MRL mouse retina creates a permissive environment for retinal regeneration. *Investigative Ophthalmology & Visual Science*, *49*, 1686–1695.

Tucker, B. A., Redenti, S. M., Jiang, C., Swift, J. S., Klassen, H. J., Smith, M. E., et al. (2010). The use of progenitor cell/biodegradable MMP2-PLGA polymer constructs to enhance cellular integration and retinal repopulation. *Biomaterials*, *31*, 9–19.

Van Meurs, J. C., & Van Den Biesen, P. R. (2003). Autologous retinal pigment epithelium and choroid translocation in patients with exudative age-related macular degeneration: Short-term follow-up. *American Journal of Ophthalmology*, *136*, 688–695.

Vincenti, F., Charpentier, B., Vanrenterghem, Y., Rostaing, L., Bresnahan, B., Darji, P., et al. (2010). A phase III study of belatacept-based immunosuppression regimens versus cyclosporine in renal transplant recipients (BENEFIT Study). *American Journal of Transplantation*, *10*, 535–546. doi:10.1111/j.1600-6143.2009.03005.x.

Vugler, A., Carr, A. J., Lawrence, J., Chen, L. L., Burrell, K., Wright, A., et al. (2008). Elucidating the phenomenon of HESC-derived RPE: Anatomy of cell genesis, expansion and retinal transplantation. *Experimental Neurology*, *214*, 347–361.

Weisz, J. M., Humayun, M. S., de Juan, E., Jr., del Cerro, M., Sunness, J. S., Dagnelie, G., et al. (1999). Allogenic fetal retinal pigment epithelial cell transplant in a patient with geographic atrophy. *Retina (Philadelphia, Pa.)*, *19*, 540–545.

West, E. L., Pearson, R. A., Barker, S. E., Luhmann, U. F. O., Maclaren, R. E., Barber, A. C., et al. (2010). Long-term survival of photoreceptors transplanted into the adult murine neural retina requires immune modulation. *Stem Cells (Dayton, Ohio)*, *28*, 1997–2007.

West, E. L., Pearson, R. A., Tschernutter, M., Sowden, J. C., MacLaren, R. E., & Ali, R. R. (2008). Pharmacological disruption of the outer limiting membrane leads to increased retinal integration of transplanted photoreceptor precursors. *Experimental Eye Research*, *86*, 601–611.

Wright, A. F., Chakarova, C. F., Abd El-Aziz, M. M., & Bhattacharya, S. S. (2010). Photoreceptor degeneration: Genetic and mechanistic dissection of a complex trait. *Nature Reviews. Genetics*, *11*, 273–284.

Yang, Y., Mohand-Said, S., Danan, A., Simonutti, M., Fontaine, V., Clerin, E., et al. (2009). Functional cone rescue by RdCVF protein in a dominant model of retinitis pigmentosa. *Molecular Therapy*, *17*, 787–795.

Yao, J., Feathers, K., Khanna, H., Thompson, D., Tsilfidis, C., Hauswirth, W. W., et al. (2011a). XIAP therapy increases survival of transplanted rod precursors in a degenerating host retina. *Investigative Ophthalmology & Visual Science*, *52*, 1567–1572.

Yao, J., Tucker, B. A., Zhang, X., Checa-Casalengua, P., Herrero-Vanrell, R., & Young, M. J. (2011b). Robust cell integration from co-transplantation of biodegradable MMP2-PLGA microspheres with retinal progenitor cells. *Biomaterials*, *32*, 1041–1050.

Yeung, C. K., Chiang, S. W. Y., Chan, K. P., Lam, D. S. C., & Pang, C. P. (2004). The transfer of ocular cells using collagen. *Cell Transplantation*, *13*, 585–594.

Yu, J., Vodyanik, M. A., Smuga-Otto, K., Antosiewicz-Bourget, J., Frane, J. L., Tian, S., et al. (2007). Induced pluripotent stem cell lines derived from human somatic cells. *Science*, *318*, 1917–1920.

Zhang, X., & Bok, D. (1998). Transplantation of retinal pigment epithelial cells and immune response in the subretinal space. *Investigative Ophthalmology & Visual Science*, *39*, 1021–1027.

Zhang, Y., Kardaszewska, A. K., Van Veen, T., Rauch, U., & Perez, M. T. R. (2004). Integration between abutting retinas: Role of glial structures and associated molecules at the interface. *Investigative Ophthalmology & Visual Science*, *45*, 4440–4449. doi:10.1167/iovs.04-0165.

Zrenner, E., Bartz-Schmidt, K. U., Benav, H., Besch, D., Bruckmann, A., Gabel, V. P., et al. (2011). Subretinal electronic chips allow blind patients to read letters and combine them to words. *Proceedings of the Royal Society B: Biological Sciences*, *278*, 1489–1497. doi:10.1098/rspb.2010.1747.

111 Stem Cell Therapies for Visual Disorders

PETER D. WESTENSKOW AND MARTIN FRIEDLANDER

Diseases that lead to catastrophic loss of vision often involve degeneration of the photoreceptors or critical ancillary cells like the retinal pigment epithelium (RPE) or vascular endothelium. Current therapeutic approaches for the treatment of neuro- and vasculodegenerative diseases of the retina such as macular degeneration and diabetic retinopathy usually involve destructive treatments aimed at inhibiting the abnormal growth of new blood vessels that lead to sudden visual loss or death of photoreceptors. With the growing realization that trophic interactions between various cellular components of the retina are critical to maintaining visual homeostasis, it has become fashionable to consider retinal "reconstruction" rather than "destruction" with lasers or molecules that inhibit various cytokines. A key element facilitating the adoption of this approach is the rapidly expanding field of regenerative medicine in which it has become possible to identify, isolate, and propagate a variety of stem cells that can give rise to virtually any cell in the body. These stem cells can be isolated from embryonic or adult tissues and, most recently, can also be generated by exposing adult somatic tissues to key transcription factors resulting in the production of induced pluripotent stem (iPS) cells (Takahashi & Yamanaka, 2006; Yu et al., 2007). Thus, generating various cell types from iPS cells makes it theoretically possible to generate autologous tissues for transplantation under a variety of pathological circumstances, effectively correcting the cellular defect in these diseases.

In this chapter we will discuss the potential use of stem cells for the replacement of defective cell types in a variety of retinal diseases. Unfortunately, it may prove very difficult to "rewire" an adult retina after degenerated photoreceptors are replaced with stem cell-derived rods or cones (Jones et al., 2012; Marc et al., 2008). Given the potential limitations in the plasticity of adult retinotectal projections, it may be more desirable to consider maintenance of the normal adult retina and to facilitate this by preserving, or enhancing, trophic interactions that occur between retinal photoreceptors

and cells that are intimately associated with them such as the RPE, glia, and vasculature. To facilitate this discussion we will first describe the various cell types that participate in these activities and then describe the potential utility of progenitor and stem cells in the treatment of various retinal vasculo- and neurodegenerative diseases.

A DIVERSITY OF RETINAL CELL TYPES; POTENTIAL TARGETS FOR STEM CELL THERAPY

Phototransduction and the ensuing cascade of events culminating in "vision" are extraordinarily biochemically complex and highly energy dependent. A myriad of ocular cell types participate in the visual process, and while the neuronal elements mediate vision, per se, they are exquisitely dependent on a large cast of supporting cell types including glia, vascular endothelium, pericytes, monocytes, and specialized epithelium (the retinal pigmented epithelium). In addition to cells that permanently reside within the retina, there are cells that migrate in and out of the posterior eye under normal and pathological conditions. These migratory cells can be deleterious as well as palliative under a variety of normal and abnormal circumstances. While the detailed structure of the eye is discussed in other chapters of this book, we will briefly review the key cells of the retina that are potential targets for stem cell therapies.

The neurosensory retina is composed of multiple and diverse cell types organized into a highly ordered, multilayered functional architecture. The conversion of light into electrical impulses, phototransduction, is accomplished in the outer cellular layer mainly through the actions of rod and cone photoreceptors and RPE cells. RPE cells occupy the outermost region of the retina and extend long apical processes that wrap around the outer segment tips of photoreceptors. They also partition the retina from the primary blood supply for the outer retinal photoreceptors, the high-flow extraretinal choriocapillaris. Interneurons in the inner nuclear layer are involved in primary information processing. Ganglion cells, whose axons bundle and transmit the integrated output to the visual cortex, establish visual receptive fields and provide the basis of sensory perception. Additionally, glial cells are required for maintaining retinal homeostasis, and iris and ciliary body cells regulate light exposure and ocular pressure. Glia are a diverse group of CNS-specific connective tissue cells and exist as macroglia (e.g., Müller glial cells) and microglia. These cells are critical to normal retinal functioning and are intimately involved in the regulation of vascular and neuronal homeostasis. Primate retinas are highly vascular and, in addition to the choriocapillaris which supplies blood to the outer third of the retina, have three distinct intraretinal plexus layers to support the inner retinal neural networks.

The majority of the neurons in the sensory retina are photoreceptors. Photoreceptors have unique morphologies; their dendritic arbors synapse with interneurons in the inner nuclear layer, and the vitamin-A derived chromophores (11-*cis*-retinal) in photoreceptor outer segments are separated from the cell body by cilia. Photoreceptors are exquisitely sensitive, and single photons of light convert 11-*cis*-retinal to 11-*trans*-retinal. Photoreceptors lack the enzymatic machinery to reisomerize the retinal to its usable form, so retinals are trafficked to neighboring RPE cells that perform the task and transfer 11-*cis*-retinal back to the photoreceptor cells. Deficits in this process can induce retinitis pigmentosa, a group of genetic eye conditions that result in primary photoreceptor cell death and incurable blindness (for review, see Palczewski & Baehr, 2001).

Functional defects in RPE cells or in the choriocapillaris can lead to secondary photoreceptor degeneration. The importance of the RPE in normal retinal functioning cannot be overstated. RPE cells secrete vasculotrophic factors critical to the maintenance of the choriocapillaris integrity (Kurihara et al., 2012; Saint-Geniez et al., 2009). They also regulate adhesion, osmolarity, pH, and water balance in the subretinal space. Melanin pigments in RPE cells absorb stray light to prevent light scatter, and RPE cells couple to form a tight seal that generates the outer blood–retinal barrier. RPE cells are also largely responsible for maintaining the relative immune privilege of the eye and diurnally phagocytose light-damaged membranes and proteins in shed photoreceptor outer segments. RPE and photoreceptors are so codependent that they are considered to be one functional unit (for review, see Strauss, 2005). In fact RPE cell dysfunction or atrophy is characteristic of age-related macular degeneration (AMD). AMD is the leading cause of blindness in industrialized countries (Congdon et al., 2004; Resnikoff et al., 2004), and demographic analyses predict that it will become even more widespread in upcoming years (Friedman et al., 2004).

Another leading cause of blindness is glaucoma (Resnikoff et al., 2004). This is characterized as a multifactorial disease that affects the optic nerve. Loss of vision in glaucoma is induced by ganglion cell atrophy that is frequently, although not always, associated with increased intraocular pressure.

While loss of photoreceptors and RPE lead to vision loss, abnormalities in the retinal and choroidal vasculature are observed in most blinding eye diseases. Mammalian retinal vascular networks develop during late prenatal and early postnatal stages. Oxygen demands increase dramatically in early postnatal life as the neurosensory retina develops and begins functioning. Concomitant with this process is the development and maturation of the retinal vasculature. These vessels provide critical trophic support to the retinal neurons and glia. Once established, the vascular networks become largely quiescent but can become reactivated and proliferate in disease settings. Common vascular changes include (1) abnormalities in vascular permeability leading to retinal edema, (2) intraocular hemorrhage, and (3) neovascularization (NV) that can result in gliosis and tractional retinal detachments (for review, see Stahl et al., 2010). These abnormalities are often observed in retinal vascular diseases such as diabetic retinopathy, retinopathy of prematurity, and the neovascular form of AMD. In neovascular AMD choroid-derived blood vessels penetrate Bruch's membrane and invade the outer retina.

PROSPECTIVE STEM CELL–BASED THERAPIES

Stem cell–based therapies may be employed to treat the shared components of multiple neurodegenerative and neovascular eye diseases. Stem cells are self-renewing unspecialized cells that are classified according to potency. They are sensitive to various signaling cues that control their "stemness" by instructing them either to continue generating new daughter cells or to differentiate into specific cell types (for review, see Weissman, 2000). Pluripotent stem cells have virtually unlimited potential and can form any fetal or adult cell type. Other stem cells (more accurately called "progenitor" cells), including hematopoietic stem cells (HSCs), are multipotent or "committed," and can generate only a limited number of cell types with short- or long-term regeneration potential.

Both embryonic and adult stem cells have been identified. Embryonic stem cells (ESCs) are pluripotent and derived from the inner cell mass of early-stage embryos. Adult stem cell populations are rare and generally have very limited potencies. In the eye, the existence of adult stem cells is controversial, and their clinical uses may be limited by practical considerations; for these reasons they will not be addressed in this review. Adult stem cells isolated from bone marrow or blood, however, may have wide utility in the treatment of many neovascular ocular diseases. These will be addressed in the second section of this chapter.

Since the establishment and regulatory approval of the first ESC lines, researchers have raced to generate their cell type of interest to use as "replacement parts" in diseased tissues. The widespread use of human ESCs (hESCs), however, has been greatly limited due to ethical concerns and limited availability of suitably staged human embryos. These concerns may become moot due to a remarkable, relatively recent observation made by two independent groups. The key observation was that transgenic overexpression of one of two sets of only four transcription factors OSKM: Oct4, Sox2, Klf4, or c-Myc (Takahashi et al., 2007) or OSNL: Oct4, Sox2, Nanog, and Lin28 (Yu et al., 2007) in somatic cells can convert them into induced pluripotent stem cells (iPSCs). The obvious appeal of iPSCs is that they may be used to generate autologous grafts. However, while the use of iPSCs may obviate many of the ethical concerns associated with ESCs, it remains to be determined if iPSCs can serve as adequate or safe alternatives for ESCs. One notable concern is that the virally delivered transcription factors integrate randomly into the genome at multiple loci following retroviral transduction. Malignant transformation could occur if any of these reprogramming transcription factors become reactivated. In fact, iPSC generation and transformation share many molecular mechanisms, and a high incidence of tumorigenesis is observed in iPSC-derived chimeric mice (Aoi et al., 2008; Kim et al., 2004; Markoulaki et al., 2009; Okita, Ichisaka, & Yamanaka, 2007).

Reprogramming processes themselves may induce genetic instability and also increase the risk of iPSC tumor formation. Exhaustive comparative analyses of multiple human iPSC (hiPSC) and hESC lines reveal that although many hiPSC and hESC lines share very similar transcriptomic and epigenetic profiles, others are heterogeneous and exhibit randomly distributed differences that limit or bias the differentiation capacity of the cells (Bock et al., 2011). Furthermore, recent evidence shows that in hiPSCs, selection pressure to obtain rapidly proliferating cell lines may induce chromosomal aneuploidy in nonrandomly distributed loci that may further limit the differentiation capacity and promote tumorigenicity (Hussein et al., 2011; Laurent

et al., 2011; Mayshar et al., 2010), although a recent report using more sensitive techniques disputed these claims (Quinlan et al., 2011). Genetic instability in iPSCs is correlated with high passage numbers, so reprogramming methods that are inefficient and require multiple passages may increase the risk of tumorigenesis (Hussein et al., 2011; Laurent et al., 2011). Furthermore, high cytotoxicity is observed with reprogramming methods that require repeated transfections such as those involving the use of messenger RNA strands (Warren et al., 2010) or recombinant proteins (Zhou et al., 2009).

There is a focused effort, therefore, to reprogram iPSCs using a limited number of viral delivered factors or to use nonintegrating approaches. However, these approaches are inefficient and time-consuming. To generate replacement cells from hiPSCs for treating diseases such as AMD or glaucoma in which the onset occurs in older humans, the dermal cells will be more difficult to reprogram and require even more time to generate. Therefore, to differentiate any tissue of interest from hiPSCs, standardized protocols may need to be adopted to ensure consistent and safe derivation. Many minimally or nonintegrating methods have now been described. Finding the most effective method that works reliably and carries the lowest risk will be imperative for future stem cell–based therapies.

GENERATION OF RETINAL CELLS FOR TRANSPLANTATION FROM STEM CELLS

The eye is an excellent organ target for stem cell–based therapies for several reasons. First, unlike other regions of the CNS it is very accessible, and the effects can be easily and directly observed using an array of noninvasive and sensitive monitoring techniques. Second, the eye is a relatively immune-privileged site, meaning that chronic inflammation is less likely to occur when compared with other organ systems. Third, if any adverse event occurs, the affected cells can be ablated using a molecular "suicide switch," laser coagulation, or, as a last resort, the eye itself can be removed.

Certain blinding diseases including AMD, are also amenable to stem-cell therapies. While AMD spans a broad clinical spectrum of signs and symptoms, there are shared characteristics all leading to photoreceptor and/or RPE atrophy that could be prevented or slowed using cell replacement strategies. AMD is generally classified as a polygenic multifactorial disease that is induced by a confluence of system failures that occurs over decades (the key risk factors for AMD are aging and smoking). Photoreceptor degeneration usually progresses slowly in AMD patients over several months or years after diagnosis. Thus adequate time is available to generate tissue grafts from iPS and to intervene therapeutically after a patient is diagnosed with the disease. Since AMD is polygenic and age related, genetic background will not necessarily lead to recurrence of the same disease within a short period of time. Essentially, RPE transplantation strategies may "reset the clock" for AMD by maintaining photoreceptor homeostasis for a significant number of additional years after transplantation.

Therefore, the transplantation of RPE cells may represent an effective therapy for AMD in many patients, or at least for those in which RPE dysfunction or death is implicated (de Jong, 2006; Zarbin, 2004). Transplanted RPE cells would be expected only to associate at nonspecific regions of Bruch's membrane and with photoreceptor outer segments at their basal and apical surfaces, respectively. Additionally, when compared with other retinal cell types, RPE cells are relatively homogenous, and no subtypes have been identified. Cell-replacement strategies involving retinal neurons, however, are complicated by the fact that different subtypes of photoreceptors or ganglions cells exist and assemble in mosaic patterns to "wire" and complete the retinal circuitry. Furthermore, after the onset of retinal degeneration, the retinal interneurons and glia become migratory and begin to form synapses inappropriately with other neurons in a process termed "remodeling" (Marc, 2005). It is highly doubtful that these remodeling mistakes are reversible. This has not discouraged researchers, however, from generating retinal neurons from stem cells for transplantation. These studies will be briefly addressed in a later section.

RPE TRANSPLANTATION AS A POTENTIAL THERAPY FOR AMD

Prior to the advent of stem cell–based translational medicine, autologous and allogenic approaches provided proof-of-concept validation for RPE transplantation-based AMD therapies. Autologous RPE replacement techniques have been successfully performed in human patients, albeit using highly invasive retinal translocation surgeries (Cahill, Freedman, & Toth, 2003) and peripherally excised RPE grafts (Joussen et al., 2007). Initial attempts involving allogeneic RPE cell transplantation were discouraging since the grafted cells were rejected by the hosts (Algvere, Gouras, & Dafgard Kopp, 1999; Zhang & Bok, 1998). These findings demonstrated that the immune privilege of the eye is not absolute and that patients receiving allografts must be exposed to lifelong immunosuppression for the therapies to be successful.

A growing body of work strongly suggests that RPE can be readily derived from hESCs (Idelson et al., 2009; Klimanskaya et al., 2004; Lund et al., 2006; Meyer et al., 2009; Vugler et al., 2008) and hiPSCs (Buchholz et al., 2009; Carr et al., 2009b; Hirami et al., 2009; Krohne et al., 2012; Meyer et al., 2009; Osakada et al., 2009), and allograft or ideally autograft transplantations represent a much more feasible and effective therapeutic strategy for treating AMD. Pigmented clusters of cells with cobblestone morphologies can be generated from hES and hiPS cultures in 4–7 weeks, but the yield and rate of differentiation can be accelerated by inactivating NODAL, wingless/int-1 (WNT), fibroblast growth factor (FGF), bone morphogenetic protein (BMP) signaling pathways (Meyer et al., 2009; Osakada et al., 2009) or activating transforming growth factor (TGF)-ß pathways (Idelson et al., 2009; Krohne et al., 2012) in vitro to mimic normal steps of embryonic RPE development. Regardless of the differentiation techniques employed, hES– and hiPS–RPE cells express terminal differentiation markers (Buchholz et al., 2009; Carr et al., 2009b; Hirami et al., 2009; Idelson et al., 2009; Klimanskaya et al., 2004; Krohne et al., 2012; Lund et al., 2006; Meyer et al., 2009; Osakada et al., 2009; Vugler et al., 2008) in correct polarized planes (Buchholz et al., 2009; Krohne et al., 2012; Vugler et al., 2008). Once reaching sufficient sizes, the pigmented clusters can be manually removed from the cultures and expanded to generate many more than the required numbers of cells required for grafting.

Multiple gene-profiling experiments have been performed to compare hES–RPE and hiPS–RPE with primary source RPE. From these studies a consensus has formed that both hES–RPE and hiPS–RPE more strongly resemble human fetal (hfRPE) than adult RPE (Buchholz et al., 2009; Carr et al., 2009b; Klimanskaya et al., 2004; Liao et al., 2010; Lu et al., 2009). We arrived at the same conclusion utilizing comparative targeted proteomic analyses. However, we did observe differences in protein expression of some RPE key terminal differentiation markers including Bestrophin, retinal pigment epithelium-specific 65 kDa protein (RPE65), cis-retinaldehyde binding protein (CRALBP), Peropsin, and Connexin 43 in iPS–RPE and hfRPE. Few similarities were observed when iPS–RPE were compared with two aged RPE samples (from 62- or 88-year-old donors), however (Krohne et al., 2012). From these experiments we concluded that our iPS–RPE more closely resemble hfRPE than aged hRPE but had attained a slightly advanced differentiation status compared with our hfRPE cultures.

The finding that iPS–RPE strongly resemble hfRPE using proteomic analyses was significant since findings from a rigorous comparative analysis revealed that hiPS–RPE and hfRPE have disparate transcriptomic profiles (Liao et al., 2010). It is possible that the transcriptomic variations could be due to the lentiviral delivery strategy utilized to generate the iPSCs (Soldner et al., 2009; Yu et al., 2007). We also compared iPS–RPE reprogrammed with different methods with hfRPE and aged hRPE using untargeted high-resolution mass-spectrometry analyses as a measure of functional genomics. The results of this assay revealed that RPE prepared from iPSC using different reprogramming methods differed significantly (Krohne et al., 2012), thus confirming that based on various "omic" profiling experiments the decision of which method to utilize must be carefully considered.

RPE cells have been generated from multiple iPS lines reprogrammed using different methods. Conventional iPS–RPE (OSKM or OSNL; 4F-iPS–RPE) (Buchholz et al., 2009; Carr et al., 2009b; Hirami et al., 2009; Krohne et al., 2012; Liao et al., 2010; Meyer et al., 2009; Osakada et al., 2009), three factor (OSK; 3F-iPS–RPE) (Osakada et al., 2009), two factor and small molecules (OK; 2F-iPS–RPE) (Krohne et al., 2012), and one factor and small molecules (*OCT4* only; 1F-iPS–RPE) (Krohne et al., 2012) have been generated. In addition, nonintegrating episomal DNA plasmid transfections have been utilized to generate episomal derived induced

pluripotent stem cells (EiPSCs). EiPS–RPE have been derived, but these cells have not yet been rigorously characterized (Okita et al., 2011). Interestingly, in our hands the EiPS–RPE differentiate much more rapidly than all other iPS–RPE lines we have tested (see figure 111.1), suggesting again that the integration of transgenes may affect the differentiation capacity of iPSCs. We are currently characterizing these cells and have generated preliminary data that they are capable of forming superb RPE cells.

hES–RPE and hiPS–RPE have also been examined using various other RPE functional assays and both appear to be satisfactorily functional in vitro. Both hES–RPE and hiPS–RPE form fluid filled domes in culture indicating they are tightly coupled and pump ions and water in an apical to basal direction (Mircheff et al., 1990). hiPS–RPE also have functional Na$^+$ and K$^+$ voltage-gated channels and membrane potentials similar to that of hRPE (Kokkinaki, Sahibzada, & Golestaneh, 2011). Lastly, hES–RPE and hiPS–RPE have been shown to phagocytose isolated photoreceptor outer segments in vitro (Buchholz et al., 2009; Carr et al., 2009a, 2009b; Idelson et al., 2009; Krohne et al., 2012; Vugler et al., 2008), and Carr et al. have demonstrated that cultured hES–RPE phagocytosed photoreceptor outer segments from human retinal explants (Carr et al., 2009a). These studies are important because in humans and rodents,

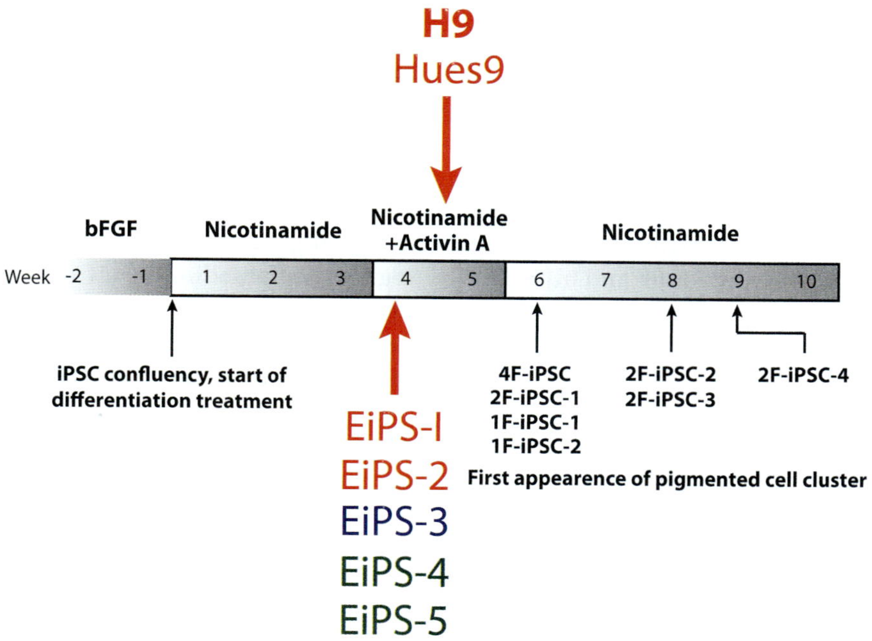

FIGURE 111.1 Rapid differentiation capacity of episomal derived induced pluripotent stem cells (EiPSCs) to form retinal pigmented epithelium. Five EiPS clones have been examined in three independent rounds and compared with embryonic stem cells (H9 and Hues9 lines), iPS derived with all 4- (4F-iPS-RPE), 2- (2F-iPS-RPE), or only 1 reprogramming factor (1F-iPS-RPE). The font color of the clones indicates first (red), second (violet), or current (green) round. The arrows indicate the time when pigmented clones are first observed.

mutations to the MerTK gene induces defects in phagocytosis and retinal degeneration (Charbel Issa et al., 2009; D'Cruz et al., 2000). Since existing methods are not satisfactorily quantitative, we developed a flow cytometry-based method to assess phagocytic activity in cultured RPE and to quantify the receptor density of the key receptors important for RPE phagocytosis. Using this technique we demonstrated that 1F-iPS–RPE express photoreceptor outer segment receptor proteins at similar levels and in similar densities to hfRPE (Westenskow et al., 2012). It will be interesting to compare iPS derived RPE to RPE obtained from normal individuals as well as ones with various retinal degenerative diseases. This can be done by using autopsy specimens to directly derive RPE for analysis or, more interestingly, to isolate recently identified retinal pigment epithelial stem cells and redifferentiate these cells into a variety of mesenchymal derivatives, including RPE itself (Salero et al., 2012). Furthermore, it has been observed that iPS–RPE and hfRPE phagocytose photoreceptor outer segments with similar binding and internalization dynamics as hfRPE do (Westenskow et al., 2012).

The major test of RPE function is to determine if the cells can mediate photoreceptor rescue in dystrophic retinas. Thus far more efforts have been made to treat retinal degenerations with hES–RPE, and a Phase 1/2 clinical trial using hES–RPE that is managed by Advanced Cell Technology Inc. (ACT) is currently underway. The results of this clinical trial will help determine if hES–RPE transplantation is an effective treatment for Stargardt's disease and atrophic AMD. While it is premature to determine if the study will be successful, after four months the transplanted cells appear to have integrated in the subretinal space and are not inducing any obvious adverse side effects (Schwartz et al., 2012). Work done in animal models, however, has provided very encouraging evidence that this technique will be effective. hES–RPE and hiPS–RPE have both been shown to mediate temporary anatomical and functional rescue of photoreceptors in Royal College of Surgeons (RCS) rats that undergo spontaneous RPE-mediated retinal degeneration (Carr et al., 2009b; Idelson et al., 2009; Krohne et al., 2012; Lund et al., 2006; Vugler et al., 2008). It is currently being debated whether RPE grafts function directly with photoreceptors or if they simply provide trophic support to promote their survival.

Some evidence that transplanted RPE cells provide direct support for host photoreceptors comes from examining histological examinations of implanted RPE cells in RCS rat retinas. More rows of photoreceptors cells are observed in regions directly overlying the grafts than in controls (Carr et al., 2009b; Idelson et al., 2009;

Klimanskaya et al., 2004; Krohne et al., 2012; Vugler et al., 2008). Until recently, functional photoreceptor rescue experiments have been less definitive, and involve whole-field electroretinography (ERG) (Idelson et al., 2009) or optokinetic reflex monitoring (Carr et al., 2009b). Unfortunately, these techniques do not provide spatial functional information in RPE-implanted retinas. Utilizing novel technology, we were able to demonstrate that only regions of the retina overlying implanted iPS–RPE cells are functionally active. Using focal ERG, we focused a beam of light with an adjustable diameter over the injected region of the eye that contained several rows of photoreceptors compared with uninjected regions in which few photoreceptors survived. Utilizing this approach, we showed that the uninjected regions of the retina did not respond to light flashes, but that photoreceptors beneath the grafts did (Krohne et al., 2012). Notwithstanding, we cannot rule out the possibility that the grafted RPE cells may have exerted a paracrine trophic rescue function to preserve photoreceptors, similar to that exerted by RPE-produced pigment epithelial derived factor (Houenou et al., 1999) and rod-derived cone viability factor (Yang et al., 2009) (a trophic molecule produced by rods with trophic activity on cones).

To determine whether implanted RPE cells function directly to preserve photoreceptor viability in vivo, we examined the RPE and photoreceptor interface in detail after iPS–RPE implantation. Since the RPE of RCS rats (and humans with MerTK mutations) (Charbel Issa et al., 2009) cannot phagocytose outer segments, a dense layer of cellular debris accumulates in the subretinal space. This debris layer contains materials that are highly autofluorescent (see figure 111.2A; white line). We observed an entirely different distribution of autofluorescent materials in the regions of RCS retinas containing implanted iPS–RPE, however. In these regions the autofluorescent materials are localized in pigmented iPS–RPE (the animals are albino), and the debris layer is far less diffuse (see figure 111.2B, C; white line). Since it is known that much of the fluorescent materials are autofluorescent bisretinoids, we quantified accumulating bisretinoid levels utilizing mass spectrometry. In uninjected eyes, the bisretinoids accumulate in a linear manner at time points consistent with photoreceptor degradation (see figure 111.2D). Direct comparisons of untreated and iPS–RPE implanted eyes showed that less bisretinoids accumulate in RCS retinas implanted with iPS–RPE, suggesting that the implanted cells phagocytose and process them (Krohne et al., 2012). As a final measure, we examined histological preparations of iPS–RPE implanted retinas and discovered that the implanted RPE cells contained large

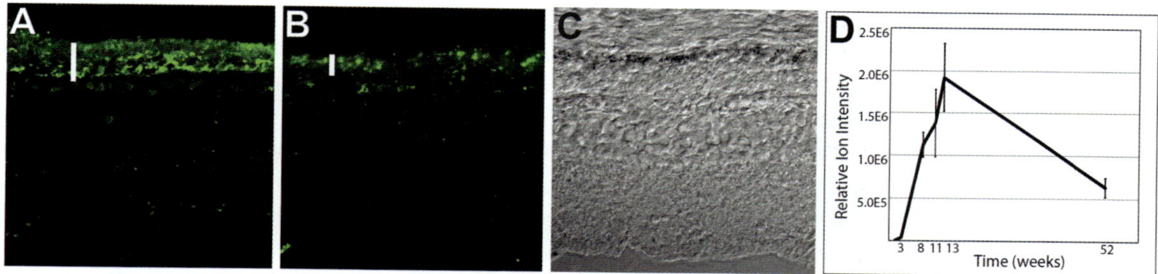

FIGURE 111.2 Autofluorescent bisretinoid accumulation in Royal College of Surgeons (RCS) rat retinas. (A) Autofluorescent bisretinoids accumulate diffusely in the debris layer (white line) of RCS rats. (B) Autofluorescence is observed only in punctate regions of RCS retinas injected with induced pluripotent stem cell derived retinal pigment epithelium (iPS–RPE). (C) Autofluorescence correlates with the presence of implanted iPS–RPE cells (the animal is albino) observed in a bright-field image of the same region. (D) Levels of accumulating bisretinoids levels were measured using mass spectrometry. The amount increases linearly during the period when photoreceptors are degenerating. By 52 weeks the amount decreases steadily, most likely due to macrophage activity.

amounts of autofluorescent materials, indicating that they were actively phagocytosing materials from the debris layer (Krohne et al., 2012). Furthermore, using electron microscopy, we observed outer segment debris in implanted iPS–RPE cells. Cumulatively, we conclude that transplanted RPE cells do function directly in implanted retinas to phagocytose shed photoreceptor outer segments and to promote retinal homeostasis.

DELIVERY OF RPE CELLS TO COMPROMISED RETINAS

Another important issue to resolve is to identify the optimal delivery method for RPE in compromised retinas. Typically, hES– and hiPS–RPE cells have been delivered in suspension with subretinal injections. While overall the results have been positive, there are some concerns regarding the effectiveness of this approach. Cultured RPE cells adhere weakly to aged human Bruch's membranes and may not support RPE grafts (Tezel, Del Priore, & Kaplan, 2004). The results of a recent study performed on several human subjects, however, showed that RPE cells injected in suspension could adhere to compromised Bruch's membranes (Falkner-Radler et al., 2011). Additionally, there are conflicting reports about how long implanted RPE cells remain viable. While some exceptions have been reported, (Carr et al., 2009b; Idelson et al., 2009; Krohne et al., 2012; Lu et al., 2009) transplanted RPE cells generally survive only for a few months after implantation in animal models. Histological analyses reveal that transplanted RPE cells tend either to organize in cellular clumps or to aggregate after implantation. Furthermore, most groups have demonstrated that the RPE cells display ambiguous polarities and form multilayered complexes (Carr et al., 2009b;

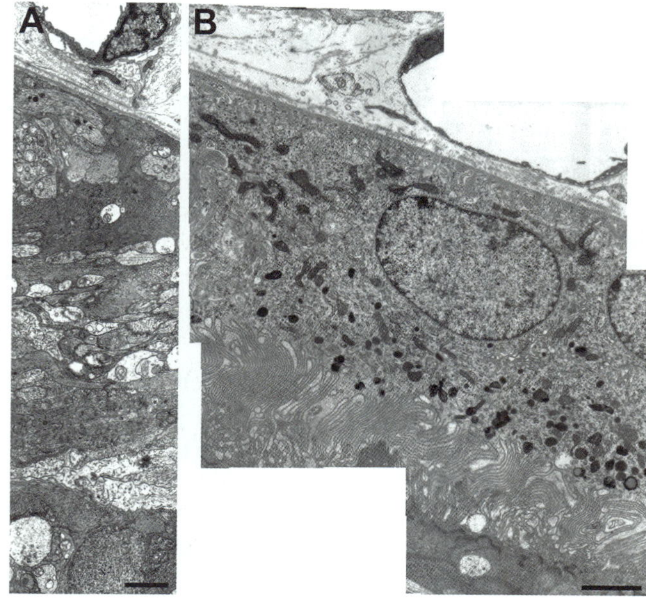

FIGURE 111.3 Ultrastructural analysis of induced pluripotent stem–retinal pigment epithelium (iPS–RPE) 17 months postimplantation. (A) Region of the photoreceptor/RPE interface from an uninjected region of a Royal College of Surgeons rat retina. (B) An image from the same eye is shown with implanted iPS–RPE. Note the presence of both basal infoldings and apical processes.

Idelson et al., 2009; Lu et al., 2009). We observed similar phenomena, but only localized directly around the injection site. At regions flanking the injection site, iPS–RPE clustered in correctly polarized monolayers and could be detected at least 17 months postimplantation (see figure 111.3) (Krohne et al., 2012). In human patients, therefore, an ideal strategy may be to perform the injection close to the optic nerve and inject the cells in the direction of the macula where the RPE cells will

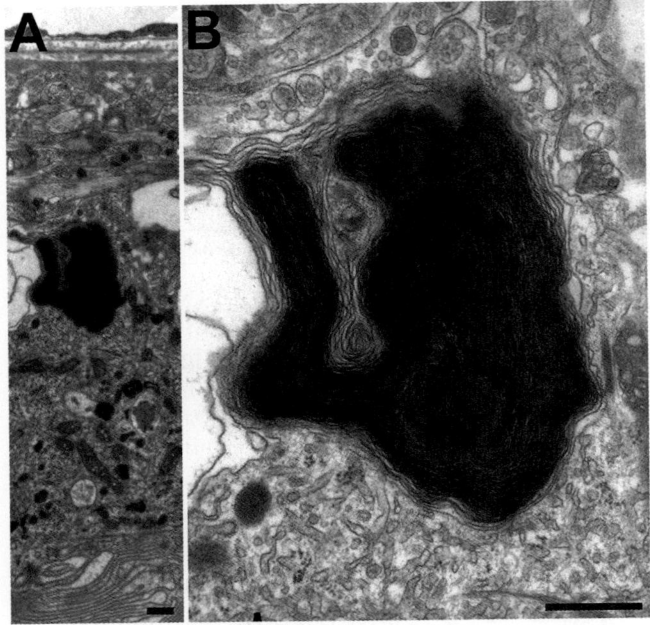

FIGURE 111.4 Evidence of implanted induced pluripotent stem–retinal pigmented epithelium (iPS–RPE) phagocytosis in Royal College of Surgeons (RCS) retinas. (A) An electron micrograph of an iPS–RPE cell (17 months postimplantation) in a RCS retina. Bruch's membrane is visible along the top of the image, and the apical processes can be observed at the bottom of the image. (B) Magnified image of the photoreceptor outer segment is shown.

have the best chance of settling diffusely and forming monolayers.

Some researchers, however, advocate transplantation techniques involving intact RPE and choroid sheets or RPE cells grown over porous substrates engineered to mimic Bruch's membrane (Sheridan, Williams, & Grierson, 2004; Williams et al., 2005). However, the surgical technique to deliver sheets of cells is inherently more complicated and may carry more risks for adverse side effects. Furthermore, it remains to be determined if the scaffolds can support cell growth and are really porous enough (or too porous) to allow RPE cells to function properly. Delivery of RPE cells on scaffolds, however, has clear advantages in retinas with obvious Bruch's membrane and RPE disruptions. In all other cases it may be best to deliver RPE cells in suspension using more simple injection techniques.

FUTURE CHALLENGES OF RPE TRANSPLANTATION APPROACHES

Another question pertinent to iPS technology is whether older source materials can be "rejuvenated" into viable iPS cells utilizing current reprogramming techniques. This critical issue has received very little attention.

There is one report suggesting that premature telomere shortening, DNA chromosomal damage, and increased p21 expression induced cellular senescence in iPS–RPE. Importantly, this was not observed until after the seventh passage of the cultures (Kokkinaki, Sahibzada, & Golestaneh, 2011). Sufficient numbers of iPS–RPE can be readily obtained after the third or fourth passages, however (Buchholz et al., 2009; Carr et al., 2009b; Krohne et al., 2012; Vugler et al., 2008). We have developed proteomic and metabolomics assays used to compare the functional genomics of RPE cells. We are currently working to generate RPE from aged dermal samples and to compare their metabolomic profiles with hfRPE. Hopefully these studies will help to address this key issue.

A practical concern regarding the use of iPS–RPE and personalized medicine in general is time and expense. Certainly the cost of collecting and reprogramming patient dermal samples, deriving autologous grafts of iPS–RPE, generating safety profiles, and performing the surgeries will be high. One alternative approach may be to generate allografts using established human leukocyte antigen (HLA)-homozygous iPSC cell-lines. It has been estimated that approximately 75 and 140 unique donors would be needed to cover ~80% and ~90% of the Japanese population, and roughly 64,000–160,000 individuals would need to be typed to find the donors (Okita, Ichisaka, & Yamanaka, 2007). The clear advantages to this approach are that the waiting time to generate the allografts would be minimized and the safety profiles of the cells could be determined in advance. It is important to also consider that hES–RPE transplantation and lifelong immunosuppression approaches are also expensive. As calculated in the United States in 2011, the cost of the immunosuppressive medications for the major types of organ transplantations range from U.S. $19,300 to $34,600 per year (Karamehic et al., 2011). Furthermore, long-term combined immunosuppressive therapies are not well tolerated by elderly patients (Tezel et al., 2007).

TRANSPLANTATION OF OTHER RETINAL CELL TYPES

Photoreceptor degeneration is observed in hereditary retinopathies including retinitis pigmentosa. While stem cell–derived photoreceptors may theoretically be generated to replace atrophic counterparts, photoreceptor implantation strategies are complicated for several reasons. First, several different types of photoreceptors have been identified in humans that are responsible for establishing trichromatic vision. Besides homogeneous rod photoreceptor populations, short-, middle-, and longwave cones have been identified, and foveal and parafoveal cones have distinct morphologies. It has not yet been determined whether all the different types of photoreceptors can be generated from stem cells in correct stoichiometric ratios. Second, outside of the cone-dense fovea, cones are organized in the retina in distinct mosaic patterns; generating these in vitro would require complicated 3-D culture protocols. Third, rod and cone photoreceptor cells synapse with distinct rod and cone bipolar cells; some of these inter-neurons are components of "on" or "off" pathways. Therefore, to restore color vision, and to provide appropriate processing of the sensory world, layers of photoreceptors in correct mosaic patterns would have to be generated and implanted directly over correct synaptic targets. This is further complicated by the fact that retinal neurons undergo remodeling during degeneration (Marc, 2005), meaning that improper retinal connections will first need to be reversed and correct synapses established to restore proper vision. Fourth, photoreceptor neurons are extremely difficult to maintain in culture.

These challenges may not be insurmountable, however, and stem cell-derived photoreceptors could be utilized to treat some forms of retinitis pigmentosa. Several researchers have shown that cells weakly resembling photoreceptors in culture form outer segments once implanted in degenerating rodent and swine retinas, and some are responsive to flash ERGs (Hirami et al., 2009; Lamba, Gust, & Reh, 2009; Lamba et al., 2010; Meyer et al., 2009; Osakada et al., 2009; Zhou et al., 2011). These groups are quick to argue that restoration of normal vision may not be possible but argue that some vision is better than none. However, until experiments are done in animal models in which vision can actually be tested, it will be impossible to determine if the grafted photoreceptors activate correct pathways. It may actually be more harmful to activate incorrectly wired pathways and transmit incorrect information to the visual cortex.

Similar challenges face stem cell biologists interested in generating ganglion cells from stem cells. Here the challenge may be even greater since at least 18 different types of ganglion cells have been identified in primates with projections that connect the eye and the visual cortex. However, several groups have demonstrated that it may be possible to derive retinal ganglion cells from hESCs (Jagatha et al., 2009) or hiPSCs (Chen et al., 2010; Parameswaran et al., 2010). Notwithstanding, whether or not implanted ganglion cells axons can transverse the extraocular space into the correct compartment of the visual cortex is a serious unanswered question.

LONG-TERM GOAL OF OCULAR STEM CELL BIOLOGY

While extremely premature, one way to generate ocular cells for transplantation that has received a great deal of recent attention may be to generate intact sensory retinas or large patches of ocular tissues from stem cells. Aggregates of mouse and recently human ES cells grown in 3-D cultures self-assembled into structures strongly resembling optic cups (rudimentary sensory retinas) with neural and RPE domains (Eiraku et al., 2011; Nakano et al., 2012). The self-formation of fully stratified 3-D neural retina tissues represents perhaps the next frontier of stem cell biology, the transplantation of artificial retinal tissue sheets rather than simple grafts of homogenous populations of cells.

TROPHIC RESCUE OF RETINAL VESSELS AND NEURONS FOR OCULAR VASCULAR DISEASES

Implantation of a variety of progenitor cell types into retinas in animal models with vascular and/or neuronal degeneration has demonstrated dramatic rescue effects not necessarily related to replacement of the damaged cell types (Marchetti et al., 2010). Rather, a paracrine rescue effect is observed due to production of various trophic molecules by the implanted cells, as well as local target cells themselves. Thus, it may be possible to significantly delay, or even completely prevent, the degeneration of neurons and/or vascular cells subjected to stress and disease that would ordinarily lead to their death and dysfunction. This can be accomplished by implanting bone marrow (Otani et al., 2002, 2004), peripheral blood or cord blood (Marchetti et al., 2011), derived progenitor cells that differentiate into either endothelial cells (Otani et al., 2002) or microglia (Ritter et al., 2006a) that, in turn, either integrate into, or associate with, the retinal vasculature, respectively (see figure 111.5). Furthermore, when retinal vasculature destined to degenerate is rescued using these progenitor cells, there is a profound paracrine rescue effect on photoreceptors that would ordinarily degenerate in these rodent models (see figure 111.6). The vascular rescue can be dissociated from the neuronal stabilization in a different model of outer retinal vascular disease, the *vldlr*$^{-/-}$ mouse. In this model, cone degeneration can be prevented and visual function preserved, even in the face of persistent vascular abnormalities, if a neurotrophic molecule (Neurotrophin-4) is delivered to the sites of vascular abnormalities by gene therapy to a targeted cell population (Dorrell et al., 2009). Similar approaches have been employed in the clinics using encapsulated cell-based therapy to deliver neurotrophic

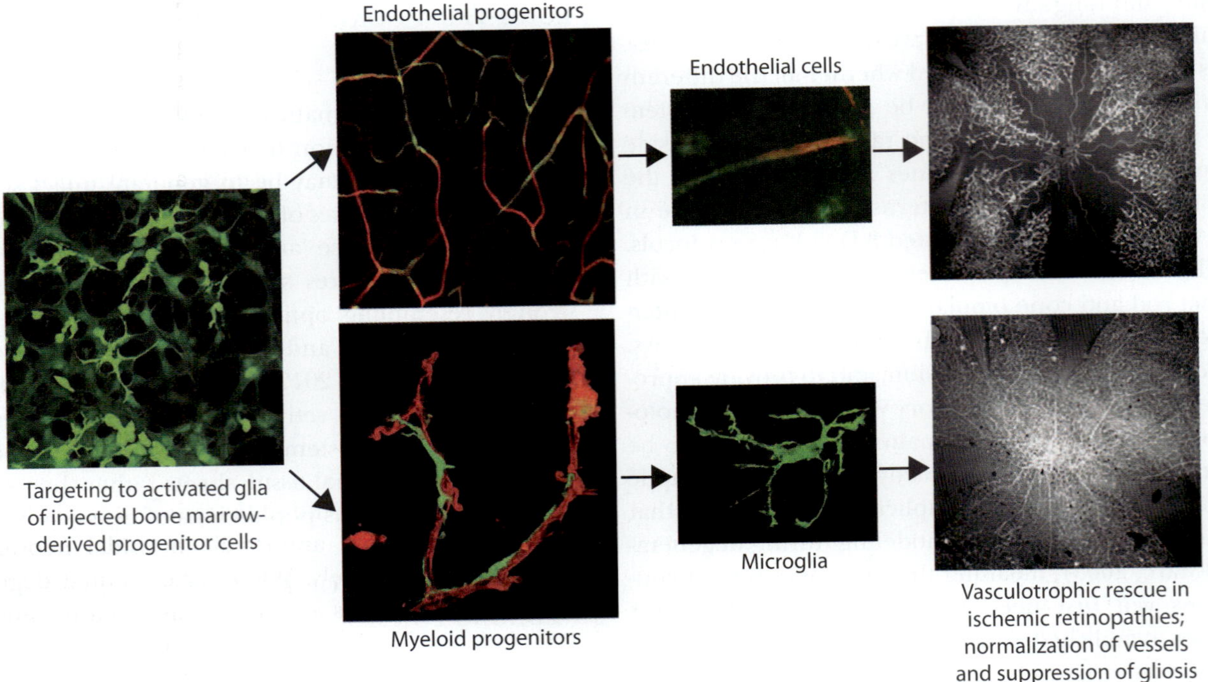

FIGURE 111.5   Cell-based therapy for the treatment of ischemic retinopathies. Bone marrow-derived progenitor cells are injected into the posterior segment of the eye, where the injected cells localize to activated astrocytes. Both endothelial and myeloid progenitor cells localize to form vasculature, where they can differentiate into endothelial cells or microglial cells, each participating in vasculotrophic rescue of ischemic blood vessels. Top far right, control-treated eye; bottom far right, progenitor cell–treated eye. (From Friedlander, 2007.)

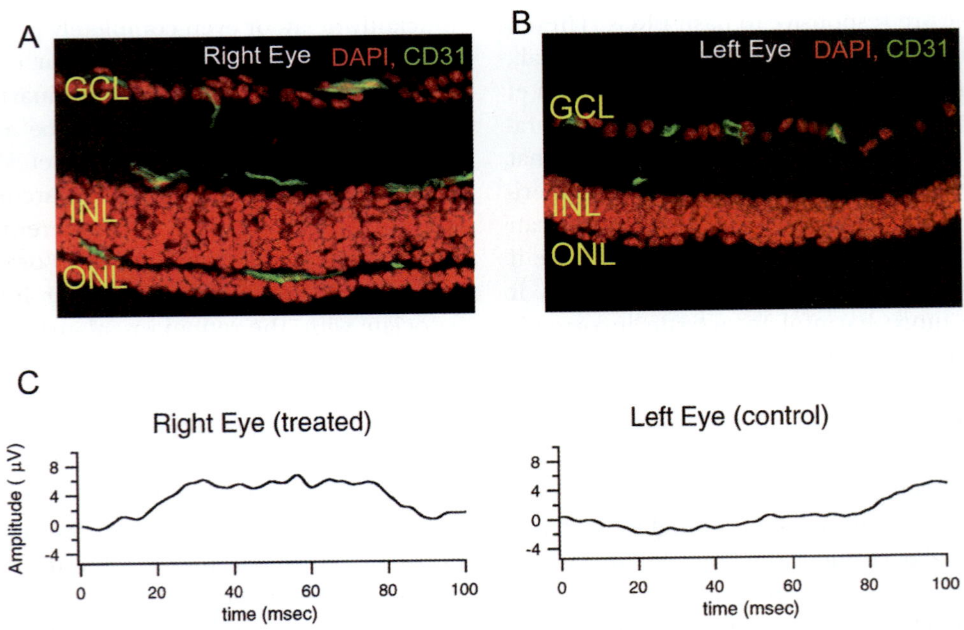

FIGURE 111.6   Endothelial progenitor cells (EPC) injected into retinas of mice with retinal degeneration rescue degenerating blood vessels and retinal function. Electroretinography (ERG) recordings were used to measure the function of EPC-injected or control cell–injected (CD31−) retinas. (A and B) Representative cases of rescued and nonrescued retinas 2 months after injection. Retinal section of Lin- HSC [hematopoietic stem cell]–injected right eye (A) and CD31- control cell–injected left eye (B) of the same animal are shown (green: CD31 stained vasculature; red: DAPI [4',6-diamidino-2-phenylindole] stained nuclei). (C) ERG responses from the same eyes shown in A and B. GCL, ganglion cell layer; INL, inner nuclear layer; ONL, outer nuclear layer. (From Otani et al., 2004.)

molecules to patients with retinal degenerative disease (Sieving et al., 2006). While safety issues do not appear to be a concern, the long-term efficacy of such an approach remains to be determined although preliminary data suggest there may be a positive effect on the maintenance of vision.

One cell type that may be involved in the observed trophic rescue is the myeloid cell, specifically macrophages. There have been several recent reports implicating the involvement of macrophages in choroidal (Csaky et al., 2004; Espinosa-Heidmann et al., 2003; Sengupta et al., 2003) and retinal (Asahara et al., 1997) NV. As discussed below, we have observed that HSC-mediated inhibition of pathological NV in the oxygen induced retinopathy (OIR) model is associated with donor derived F4/80+ cells being present in the inner, intermediate, and outer vascular plexuses. In contrast, in the absence of progenitor cell injection, fewer F4/80 cells are present in the deep layers and are localized to the primary plexus. Recent observations suggest that both endogenous and circulating myeloid/macrophage-type cells are associated with sites of NV and may normalize or exacerbate the angiogenic response to disease and/or injury.

How can this paradoxical observation of macrophages both promoting pathological NV and mediating retinal and vascular rescue be reconciled? It is known that macrophages, depending on the microenvironment in which they reside, will be "polarized" to express differing effector programs. For example, M1 macrophages are strong effectors that target and kill microorganisms and tumor cells and generate considerable amounts of pro-inflammatory cytokines (e.g., TNF, IL-12, ROI, IFNs, and iNOS). In contrast, M2 macrophages dampen and tune inflammatory responses and Th1 T-cells responses, scavenge debris, and promote angiogenesis, tissue remodeling, and repair (i.e., bFGF, EGF, TGF-beta, MMPs, and VEGF). Thus the polarized state of macrophages could regulate maintenance (NOS), proliferation (NOS) and/or development (VEGF, TNF, TGF) of endothelial cells as well as promote inflammation. Given these observations, it becomes necessary to understand the role of donor (intravitreally injected) myeloid cells in the rescue and stabilization of vasculature in retinal degenerative diseases.

In several diseases that can cause irreversible vision loss such as retinal angiomatous proliferation (RAP) (Yannuzzi et al., 2001) and macular telangiectasia (MacTel) (Yannuzzi et al., 2006), abnormal vessels extend from the highly vascularized inner retina into the normally avascular outer retina. The retina of the $vldlr^{-/-}$ mouse exhibits vascular changes very similar to those observed in patients with MacTel and RAP and thus is a good model for studying the general relationship between NV and neurodegeneration. The new vessels observed in both human disease and the $vldlr^{-/-}$ mouse exhibit relatively mild permeability defects accompanied by glial activation and disruption of the RPE. Another feature common to RAP, MacTel, and the $vldlr^{-/-}$ retina is that the nonuniformly distributed, focal vascular lesions are directly associated with neuronal degeneration separated by relatively normal regions. Finally, activation of Müller glial cells and RPE also accompanies the vascular and neuronal changes observed in the VLDLR mouse. Given that the Müller glia span the inner and outer retina, it may be possible to target these cells for expression of trophic and static molecules useful for the stabilization of oxidative-stress sensitive cones and abnormal choroidal neovascular responses, respectively. While other retinal degenerative models, such as $rd1$ or rds mice, have associated vascular abnormalities and neuronal loss, many of these result in uniform degeneration and thus relate more closely to genetic diseases such as retinitis pigmentosa rather than ocular vascular disease. Taken together, these similarities support use of the $vldlr^{-/-}$ mouse as a compelling model to investigate selected human retinal diseases (e.g., MacTel) as well as other, nonocular neurodegenerative disorders with associated NV.

Cell-based therapy represents a newly emerging method by which vascular and neuronal degenerative diseases may be treated. Intravitreal injection of vascular-related progenitor cells prevents vascular regression and protects neurons in mouse models of retinal degeneration (Friedlander, 2006, 2007; Otani et al., 2002, 2004). Injection of myeloid progenitors facilitates vascular repair in models of ischemic retinopathy by accelerating normal revascularization of the superficial and deep retinal vascular plexuses, and decreasing intravitreal vascular pathogenesis (Ritter et al., 2006). These cell based therapies have the potential to correct the underlying vascular abnormalities associated with ischemic retinopathies rather than simply treat the complications resulting from abnormal NV. While we have gained some insight regarding the vasculo- and neurotrophic properties of these cells, the precise mechanism by which they facilitate vascular rescue models of ischemic retinopathy remains unclear. While these studies demonstrate a potential therapeutic role for injected progenitor cells in the treatment of retinal avascular and neurodegenerative diseases, this also raises many interesting questions surrounding the role for endogenous glial and inflammatory cells in retinal response to injuries such as hypoxia or oxidative stress. Microglia, astrocytes, and Müller glia may be critical to maintenance of the normal vascular and neuronal microenvironment in the retina as they are in the brain.

Astrocytes play a critical role during normal inner retinal vascularization (Dorrell, Aguilar, & Friedlander, 2002; Fruttiger, 2002; Penfold et al., 1990; Provis, Sandercoe, & Hendrickson, 2000), and degeneration of retinal astrocytes in ischemic tissues is associated with failure of the blood retinal barrier in oxygen-induced retinopathies (Chan-Ling & Stone, 1991, 1992). We have demonstrated a strong correlation between astrocyte survival, rapid retinal revascularization, and lack of pathological NV in BALB/cByJ retinas when compared with retinas from C57BL/6J mice (Dorrell et al., 2010). We have also described trophic cross talk between astrocytes and retinal microglia and provided evidence that the rescue potential of myeloid progenitor cells may, at least in part, be facilitated by protecting retinal astrocytes from degeneration in the C57BL/6J mouse retina. We have also demonstrated that the injection of astrocytes, astrocyte-conditioned media, and even low levels of VEGF or bFGF also rescues the OIR phenotype by protecting the endogenous astrocytes from hypoxia-related degeneration. In order to further clarify the relationships between glia and regulators of the hypoxic and angiogenic responses, we have created a conditional deletion of the genes encoding VEGF, HIF-1α, HIF-2α, and the negative regulator of HIFα, von Hippel-Lindau protein, in astrocytes.

CONCLUDING REMARKS

Over the past decade, advances in the field of regenerative medicine/therapy/treatment strongly support the potential for exploiting stem and progenitor cells to maintain, and perhaps "fix," abnormal retinal tissue. New technologies that permit the isolation of embryonic, adult and iPSCs provide previously unimaginable ways in which to produce transplantable replacement parts for damaged retinal tissue. The use of autologous adult bone marrow–derived stem cell grafts for the treatment of retinal vascular and degenerative diseases represents a novel conceptual approach that may make it possible to "mature" otherwise immature neovasculature, stabilize existing vasculature to hypoxic damage and/or rescue, replace diseased retinal pigmented epithelial cells, and protect retinal neurons from undergoing apoptosis. This may be possible through the use of autologous bone marrow or cord blood derived HSCs that selectively target sites of NV and gliosis where they provide vasculo- and neurotrophic effects facilitating vascularization of ischemic, and otherwise damaged, retinal tissue. Autologous replacement parts may be differentiated from iPSCs derived from the patient's own skin or from stem/progenitor cells residing in adult tissue "stem cell niches."

Impediments to translating these approaches that have already proven successful in animal models into humans include a variety of mechanical and biochemical considerations. For example, are bone marrow derived progenitor cells in diabetic patients the same as comparable cells from patients without diabetes? Is there a defect in mobilizing such cells for therapeutic applications, or are the cells themselves defective in their vasculo- and neurotrophic activities? Even if autologous bone marrow progenitor cells or iPS derived replacement cells can be generated from patients with the disease(s) to be targeted, will these grafts "take" at sites of ongoing disease? How critical is the diseased retinal microenvironment to graft survival, and will transplanted cells generate their own normalized microenvironment, or will they need a specialized substrate for successful integration into the host? These are all questions that will need to be answered once an appropriate therapeutic cell type is generated and applied to a specific disease. While we have cured many a mouse or rat with retinal vascular and neurodegenerative disease, the utility of such approaches in treating human disease will be determined as these approaches spill into clinical trials, several of which have been initiated.

ACKNOWLEDGMENTS

We are grateful for the collegiality and insightful comments provided by members of the Friedlander laboratory at The Scripps Research Institute and our collaborators from many other institutions. The work discussed in this chapter from our laboratory was supported by generous funding to MF from the National Eye Institute (EY11254; EY017540), the California Institute for Regenerative Medicine (CIRM TR1–01219), the Lowy Medical Research Institute (the MacTel Project), and the Rasmussen Foundation. PW is supported by a Ruth Kirshstein National Service Research Award from the National Eye Institute (EY021416).

REFERENCES

Algvere, P. V., Gouras, P., & Dafgard Kopp, E. (1999). Long-term outcome of RPE allografts in non-immunosuppressed patients with AMD. *European Journal of Ophthalmology, 9,* 217–230.

Aoi, T., Yae, K., Nakagawa, M., Ichisaka, T., Okita, K., Takahashi, K., et al. (2008). Generation of pluripotent stem cells from adult mouse liver and stomach cells. *Science, 321,* 699–702.

Asahara, T., Murohara, T., Sullivan, A., Silver, M., van der Zee, R., Li, T., et al. (1997). Isolation of putative progenitor endothelial cells for angiogenesis. *Science, 275,* 964–967.

Bock, C., Kiskinis, E., Verstappen, G., Gu, H., Boulting, G., Smith, Z. D., et al. (2011). Reference maps of human ES and iPS cell variation enable high-throughput characterization of pluripotent cell lines. *Cell, 144,* 439–452. doi:10.1016/j.cell.2010.12.032.

Buchholz, D. E., Hikita, S. T., Rowland, T. J., Friedrich, A. M., Hinman, C. R., Johnson, L. V., et al. (2009). Derivation of functional retinal pigmented epithelium from induced pluripotent stem cells. *Stem Cells (Dayton, Ohio), 27,* 2427–2434. doi:10.1002/stem.189.

Cahill, M. T., Freedman, S. F., & Toth, C. A. (2003). Macular translocation with 360 degrees peripheral retinectomy for geographic atrophy. *Archives of Ophthalmology, 121,* 132–133.

Carr, A. J., Vugler, A., Lawrence, J., Chen, L. L., Ahmado, A., Chen, F. K., et al. (2009a). Molecular characterization and functional analysis of phagocytosis by human embryonic stem cell-derived RPE cells using a novel human retinal assay. *Molecular Vision, 15,* 283–295.

Carr, A. J., Vugler, A. A., Hikita, S. T., Lawrence, J. M., Gias, C., Chen, L. L., et al. (2009b). Protective effects of human iPS-derived retinal pigment epithelium cell transplantation in the retinal dystrophic rat. *PLoS ONE, 4,* e8152. doi:10.1371/journal.pone.0008152.

Chan-Ling, T., & Stone, J. (1991). Factors determining the migration of astrocytes into the developing retina: Migration does not depend on intact axons or patent vessels. *Journal of Comparative Neurology, 303,* 375–386.

Chan-Ling, T., & Stone, J. (1992). Degeneration of astrocytes in feline retinopathy of prematurity causes failure of the blood-retinal barrier. *Investigative Ophthalmology & Visual Science, 33,* 2148–2159.

Charbel Issa, P., Bolz, H. J., Ebermann, I., Domeier, E., Holz, F. G., & Scholl, H. P. (2009). Characterisation of severe rod–cone dystrophy in a consanguineous family with a splice site mutation in the MERTK gene. *British Journal of Ophthalmology, 93,* 920–925. doi:10.1136/bjo.2008.147397.

Chen, M., Chen, Q., Sun, X., Shen, W., Liu, B., Zhong, X. F., et al. (2010). Generation of retinal ganglion-like cells from reprogrammed mouse fibroblasts. *Investigative Ophthalmology & Visual Science, 51,* 5970–5978. ePub. doi: 10.1167/iovs.09-4504.

Congdon, N., O'Colmain, B., Klaver, C. C., Klein, R., Munoz, B., Friedman, D. S., et al. (2004). Causes and prevalence of visual impairment among adults in the United States. *Archives of Ophthalmology, 122,* 477–485. doi:10.1001/archopht.122.4.477.

Csaky, K. G., Baffi, J. Z., Byrnes, G. A., Wolfe, J. D., Hilmer, S. C., Flippin, J., et al. (2004). Recruitment of marrow-derived endothelial cells to experimental choroidal NV by local expression of vascular endothelial growth factor. *Experimental Eye Research, 78,* 1107–1116.

D'Cruz, P. M., Yasumura, D., Weir, J., Matthes, M. T., Abderrahim, H., LaVail, M. M., et al. (2000). Mutation of the receptor tyrosine kinase gene Mertk in the retinal dystrophic RCS rat. *Human Molecular Genetics, 9,* 645–651.

de Jong, P. T. (2006). Age-related macular degeneration. *New England Journal of Medicine, 355,* 1474–1485. doi:10.1056/NEJMra062326.

Dorrell, M. I., Aguilar, E., & Friedlander, M. (2002). Retinal vascular development is mediated by endothelial filopodia,

a preexisting astrocytic template and specific R-cadherin adhesion. *Investigative Ophthalmology & Visual Science, 43,* 3500–3510.

Dorrell, M. I., Aguilar, E., Jacobson, R., Trauger, S. A., Friedlander, J., Siuzdak, G., et al. (2010). Maintaining retinal astrocytes normalizes revascularization and prevents vascular pathology associated with oxygen-induced retinopathy. *Glia, 58,* 43–54.

Dorrell, M. I., Aguilar, E., Jacobson, R., Yanes, O., Gariano, R., Heckenlively, J., et al. (2009). Antioxidant or neurotrophic factor treatment preserves function in a mouse model of NV-associated oxidative stress. *Journal of Clinical Investigation, 119,* 611–623. doi:10.1172/JCI35977.

Eiraku, M., Takata, N., Ishibashi, H., Kawada, M., Sakakura, E., Okuda, S., et al. (2011). Self-organizing optic-cup morphogenesis in three-dimensional culture. *Nature, 472,* 51–56. doi:10.1038/nature09941.

Espinosa-Heidmann, D. G., Caicedo, A., Hernandez, E. P., Csaky, K. G., & Cousins, S. W. (2003). Bone marrow-derived progenitor cells contribute to experimental choroidal NV. *Investigative Ophthalmology & Visual Science, 44,* 4914–4919.

Falkner-Radler, C. I., Krebs, I., Glittenberg, C., Povazay, B., Drexler, W., Graf, A., & Binder, S. (2011). Human retinal pigment epithelium (RPE) transplantation: Outcome after autologous RPE-choroid sheet and RPE cell-suspension in a randomised clinical study. *British Journal of Ophthalmology, 95,* 370–375. ePub. doi:10.1136/bjo.2009.176305.

Friedlander, M. (2006). Stem cells. In S. J. Ryan (Ed.), *Retina: Vol. 1. Basic science, inherited retinal disease, and tumors* (pp. 23–32). Los Angeles: Elsevier Mosby.

Friedlander, M. (2007). Fibrosis and diseases of the eye. *Journal of Clinical Investigation, 117,* 576–586.

Friedman, D. S., O'Colmain, B. J., Munoz, B., Tomany, S. C., McCarty, C., de Jong, P. T., et al. (2004). Prevalence of age-related macular degeneration in the United States. *Archives of Ophthalmology, 122,* 564–572.

Fruttiger, M. (2002). Development of the mouse retinal vasculature: Angiogenesis versus vasculogenesis. *Investigative Ophthalmology & Visual Science, 43,* 522–527.

Hirami, Y., Osakada, F., Takahashi, K., Okita, K., Yamanaka, S., Ikeda, H., et al. (2009). Generation of retinal cells from mouse and human induced pluripotent stem cells. *Neuroscience Letters, 458,* 126–131. doi:10.1016/j.neulet.2009.04.035.

Houenou, L. J., D'Costa, A. P., Li, L., Turgeon, V. L., Enyadike, C., Alberdi, E., et al. (1999). Pigment epithelium-derived factor promotes the survival and differentiation of developing spinal motor neurons. *Journal of Comparative Neurology, 412,* 506–514.

Hussein, S. M., Batada, N. N., Vuoristo, S., Ching, R. W., Autio, R., Narva, E., et al. (2011). Copy number variation and selection during reprogramming to pluripotency. *Nature, 471,* 58–62. doi:10.1038/nature09871.

Idelson, M., Alper, R., Obolensky, A., Ben-Shushan, E., Hemo, I., Yachimovich-Cohen, N., et al. (2009). Directed differentiation of human embryonic stem cells into functional retinal pigment epithelium cells. *Cell Stem Cell, 5,* 396–408. doi:10.1016/j.stem.2009.07.002.

Jagatha, B., Divya, M. S., Sanalkumar, R., Indulekha, C. L., Vidyanand, S., Divya, T. S., et al. (2009). In vitro differentiation of retinal ganglion-like cells from embryonic stem cell derived neural progenitors. *Biochemical & Biophysics Research Communications, 380,* 230–235. doi:10.1016/j.bbrc.2009.01.038.

Jones, B. W., Kondo, M., Terasaki, H., Lin, Y., McCall, M., & Marc, R. E. (2012). Retinal remodeling. *Japanese Journal of Ophthalmology*. doi:10.1007/s10384-012-0147-2.

Joussen, A. M., Joeres, S., Fawzy, N., Heussen, F. M., Llacer, H., van Meurs, J. C., et al. (2007). Autologous translocation of the choroid and retinal pigment epithelium in patients with geographic atrophy. *Ophthalmology, 114,* 551–560.

Karamehic, J., Ridic, O., Jukic, T., Ridic, G., Slipicevic, O., Coric, J., et al. (2011). Financial aspects of the immunosuppressive therapy. *Medicinski Arhiv, 65,* 357–362.

Kim, S. R., Fishkin, N., Kong, J., Nakanishi, K., Allikmets, R., & Sparrow, J. R. (2004). Rpe65 Leu450Met variant is associated with reduced levels of the retinal pigment epithelium lipofuscin fluorophores A2E and iso-A2E. *Proceedings of the National Academy of Sciences of the United States of America, 101,* 11668–11672. doi:10.1073/pnas.0403499101.

Klimanskaya, I., Hipp, J., Rezai, K. A., West, M., Atala, A., & Lanza, R. (2004). Derivation and comparative assessment of retinal pigment epithelium from human embryonic stem cells using transcriptomics. *Cloning and Stem Cells, 6,* 217–245. doi:10.1089/clo.2004.6.217.

Kokkinaki, M., Sahibzada, N., & Golestaneh, N. (2011). Human induced pluripotent stem-derived retinal pigment epithelium (RPE) cells exhibit ion transport, membrane potential, polarized vascular endothelial growth factor secretion, and gene expression pattern similar to native RPE. *Stem Cells, 29,* 825–835. doi:10.1002/stem.635.

Krohne, T., Westenskow, P. D., Kurihara, T., Friedlander, D., Lehmann, M., Dorsey, A., et al. (2012). Generation of retinal pigment epithelial cells from small molecules and OCT4-reprogrammed human induced pluripotent stem cells. *Stem Cells Translational Medicine, 1,* 96–109. doi:10.5966/sctm.2011-0057.

Kurihara, T., Westenskow, P. D., Bravo, S., Aguilar, E., Friedlander, M. (2012). Targeted deletion of Vegfa in adult mice induces vision loss. *Journal of Clinical Investigation, 122,* 4213–4217.

Lamba, D. A., Gust, J., & Reh, T. A. (2009). Transplantation of human embryonic stem cell-derived photoreceptors restores some visual function in Crx-deficient mice. *Cell Stem Cell, 4,* 73–79.

Lamba, D. A., McUsic, A., Hirata, R. K., Wang, P.-R., Russell, D., & Reh, T. A. (2010). Generation, purification and transplantation of photoreceptors derived from human induced pluripotent stem cells. *PLoS ONE, 5,* e8763. doi:10.1371/journal.pone.0008763.

Laurent, L. C., Ulitsky, I., Slavin, I., Tran, H., Schork, A., Morey, R., et al. (2011). Dynamic changes in the copy number of pluripotency and cell proliferation genes in human ESCs and iPSCs during reprogramming and time in culture. *Cell Stem Cell, 8,* 106–118.

Liao, J.-L., Yu, J., Huang, K., Hu, J., Diemer, T., Ma, Z., et al. (2010). Molecular signature of primary retinal pigment epithelium and stem-cell derived RPE cells. *Human Molecular Genetics.* doi:10.1093/hmg/ddq341.

Lu, B., Malcuit, C., Wang, S., Girman, S., Francis, P., Lemieux, L., et al. (2009). Long-term safety and function of RPE from human embryonic stem cells in preclinical models of macular degeneration. *Stem Cells, 27,* 2126–2135. doi:10.1002/stem.149.

Lund, R. D., Wang, S., Klimanskaya, I., Holmes, T., Ramos-Kelsey, R., Lu, B., et al. (2006). Human embryonic stem cell-derived cells rescue visual function in dystrophic RCS rats. *Cloning and Stem Cells, 8,* 189–199. doi:10.1089/clo.2006.8.189.

Marc, R. (2005). Retinal remodeling. *Journal of Vision, 5,* 5. doi:10.1167/5.12.5.

Marc, R. E., Jones, B. W., Watt, C. B., Vazquez-Chona, F., Vaughan, D. K., & Organisciak, D. T. (2008). Extreme retinal remodeling triggered by light damage: Implications for age related macular degeneration. *Molecular Vision, 14,* 782–806.

Marchetti, V., Krohne, T. U., Friedlander, D. F., & Friedlander, M. (2010). Stemming vision loss with stem cells. *Journal of Clinical Investigation, 120,* 3012–3021. doi:10.1172/JCI42951.

Marchetti, V., Yanes, O., Aguilar, E., Wang, M., Friedlander, D., Moreno, S., et al. (2011). Differential macrophage polarization promotes tissue remodeling and repair in a model of ischemic retinopathy. *Scientific Reports, 1,* 76. doi:10.1038/srep00076.

Markoulaki, S., Hanna, J., Beard, C., Carey, B. W., Cheng, A. W., Lengner, C. J., et al. (2009). Transgenic mice with defined combinations of drug-inducible reprogramming factors. *Nature Biotechnology, 27,* 169–171. doi:10.1038/nbt.1520.

Mayshar, Y., Ben-David, U., Lavon, N., Biancotti, J. C., Yakir, B., Clark, A. T., et al. (2010). Identification and classification of chromosomal aberrations in human induced pluripotent stem cells. *Cell Stem Cell, 7,* 521–531. doi:10.1016/j.stem.2010.07.017.

Meyer, J. S., Shearer, R. L., Capowski, E. E., Wright, L. S., Wallace, K. A., McMillan, E. L., et al. (2009). Modeling early retinal development with human embryonic and induced pluripotent stem cells. *Proceedings of the National Academy of Sciences of the United States of America, 106,* 16698–16703. doi:10.1073/pnas.0905245106.

Mircheff, A. K., Miller, S. S., Farber, D. B., Bradley, M. E., O'Day, W. T., & Bok, D. (1990). Isolation and provisional identification of plasma membrane populations from cultured human retinal pigment epithelium. *Investigative Ophthalmology & Visual Science, 31,* 863–878.

Nakano, T., Ando, S., Takata, N., Kawada, M., Mugurama, K., Sekiguchi, K., Saito, K., Yonemura, S., Eiraku, M., & Sasai, Y. (2012). Self-formation of optic cups and storable stratified neural retina from human ESCs. *Cell Stem Cell, 10,* 771–785.

Okita, K., Ichisaka, T., & Yamanaka, S. (2007). Generation of germline-competent induced pluripotent stem cells. *Nature, 448,* 313–317. doi:10.1038/nature05934.

Okita, K., Matsumura, Y., Sato, Y., Okada, A., Morizane, A., Okamoto, S., et al. (2011). A more efficient method to generate integration-free human iPS cells. *Nature Methods, 8,* 409–412. doi:10.1038/nmeth.1591.

Osakada, F., Jin, Z. B., Hirami, Y., Ikeda, H., Danjyo, T., Watanabe, K., et al. (2009). In vitro differentiation of retinal cells from human pluripotent stem cells by small-molecule induction. *Journal of Cell Science, 122,* 3169–3179. doi:10.1242/jcs.050393.

Otani, A., Dorrell, M. I., Kinder, K., Moreno, S. K., Nusinowitz, S., Banin, E., et al. (2004). Rescue of retinal degeneration by intravitreally injected adult bone marrow-derived lineage-negative hematopoietic stem cells. *Journal of Clinical Investigation, 114,* 765–774. doi:10.1172/JCI21686.

Otani, A., Kinder, K., Ewalt, K., Otero, F. J., Schimmel, P., & Friedlander, M. (2002). Bone marrow-derived stem cells target retinal astrocytes and can promote or inhibit retinal angiogenesis. *Nature Medicine, 8,* 1004–1010. doi:10.1038/nm744.

Palczewski, K., & Baehr, W. (2001). *The retinoid cycle and retinal diseases. eLS.* Wiley.

Parameswaran, S., Balasubramanian, S., Babai, N., Qiu, F., Eudy, J. D., Thoreson, W. B., et al. (2010). Induced pluripotent stem cells generate both retinal ganglion cells and photoreceptors: Therapeutic implications in degenerative changes in glaucoma and age-related macular degeneration. *Stem Cells, 28,* 695–703. doi:10.1002/stem.320.

Penfold, P. L., Provis, J. M., Madigan, M. C., van Driel, D., & Billson, F. A. (1990). Angiogenesis in normal human retinal development: The involvement of astrocytes and macrophages. *Graefe's Archive for Clinical and Experimental Ophthalmology, 228,* 255–263.

Provis, J. M., Sandercoe, T., & Hendrickson, A. E. (2000). Astrocytes and blood vessels define the foveal rim during primate retinal development. *Investigative Ophthalmology & Visual Science, 41,* 2827–2836.

Quinlan, A. R., Boland, M. J., Leibowitz, M. L., Shumilina, S., Pehrson, S. M., Baldwin, K. K., et al. (2011). Genome sequencing of mouse induced pluripotent stem cells reveals retroelement stability and infrequent DNA rearrangement during reprogramming. *Cell Stem Cell, 9,* 366–373.

Resnikoff, S., Pascolini, D., Etya'ale, D., Kocur, I., Pararajasegaram, R., Pokharel, G. P., et al. (2004). Global data on visual impairment in the year 2002. *Bulletin of the World Health Organization, 82,* 844–851.

Ritter, M. R., Banin, E., Moreno, S. K., Aguilar, E., Dorrell, M. I., & Friedlander, M. (2006a). Myeloid progenitors differentiate into microglia and promote vascular repair in a model of ischemic retinopathy. *Journal of Clinical Investigation, 116,* 3266–3276. doi:10.1172/JCI29683.

Saint-Geniez, M., Kurihara, T., Sekiyama, E., Maldonado, A. E., & D'Amore, P. A. (2009). An essential role for RPE-derived soluble VEGF in the maintenance of the choriocapillaris. *Proceedings of the National Academy of Sciences of the United States of America, 106,* 18751–18756.

Salero, E., Blenkinsop, T. A., Corneo, B., Harris, A., Rabin, D., Stern, J. H., et al. (2012). Adult human RPE can be activated into a multipotent stem cell that produces mesenchymal derivatives. *Cell Stem Cell, 10,* 88–95. doi:10.1016/j.stem.2011.11.018.

Schwartz, S. D., Hubschman, J.-P., Heilwell, G., Franco-Cardenas, V., Pan, C. K., Ostrick, R. M., et al. (2012). Embryonic stem cell trials for macular degeneration: A preliminary report. *Lancet, 379,* 713–720.

Sengupta, N., Caballero, S., Mames, R. N., Butler, J. M., Scott, E. W., & Grant, M. B. (2003). The role of adult bone marrow-derived stem cells in choroidal NV. *Investigative Ophthalmology & Visual Science, 44,* 4908–4913.

Sheridan, C., Williams, R., & Grierson, I. (2004). Basement membranes and artificial substrates in cell transplantation. *Graefe's Archive for Clinical and Experimental Ophthalmology, 242,* 68–75. doi:10.1007/s00417-003-0800-z.

Sieving, P. A., Caruso, R. C., Tao, W., Coleman, H. R., Thompson, D. J., Fullmer, K. R., et al. (2006). Ciliary neurotrophic factor (CNTF) for human retinal degeneration: Phase I trial of CNTF delivered by encapsulated cell intraocular implants. *Proceedings of the National Academy of Sciences of the United States of America, 103,* 3896–3901. doi:10.1073/pnas.0600236103.

Soldner, F., Hockemeyer, D., Beard, C., Gao, Q., Bell, G. W., Cook, E. G., et al. (2009). Parkinson's disease patient-derived induced pluripotent stem cells free of viral reprogramming factors. *Cell, 136,* 964–977. doi:10.1016/j.cell.2009.02.013.

Stahl, A., Connor, K. M., Sapieha, P., Chen, J., Dennison, R. J., Krah, N. M., et al. (2010). The mouse retina as an angiogenesis model. *Investigative Ophthalmology & Visual Science, 51,* 2813–2826. doi:10.1167/iovs.10-5176.

Strauss, O. (2005). The retinal pigment epithelium in visual function. *Physiological Reviews, 85,* 845–881.

Takahashi, K., Tanabe, K., Ohnuki, M., Narita, M., Ichisaka, T., Tomoda, K., et al. (2007). Induction of pluripotent stem cells from adult human fibroblasts by defined factors. *Cell, 131,* 861–872. doi:10.1016/j.cell.2007.11.019.

Takahashi, K., & Yamanaka, S. (2006). Induction of pluripotent stem cells from mouse embryonic and adult fibroblast cultures by defined factors. *Cell, 126,* 663–676.

Tezel, T. H., Del Priore, L. V., Berger, A. S., & Kaplan, H. J. (2007). Adult retinal pigment epithelial transplantation in exudative age-related macular degeneration. *American Journal of Ophthalmology, 143,* 584–595. doi:10.1016/j.ajo.2006.12.007.

Tezel, T. H., Del Priore, L. V., & Kaplan, H. J. (2004). Reengineering of aged Bruch's membrane to enhance retinal pigment epithelium repopulation. *Investigative Ophthalmology & Visual Science, 45,* 3337–3348.

Vugler, A., Carr, A. J., Lawrence, J., Chen, L. L., Burrell, K., Wright, A., et al. (2008). Elucidating the phenomenon of HESC-derived RPE: Anatomy of cell genesis, expansion and retinal transplantation. *Experimental Neurology, 214,* 347–361. doi:10.1016/j.expneurol.2008.09.007.

Warren, L., Manos, P. D., Ahfeldt, T., Loh, Y.-H., Li, H., Lau, F., et al. (2010). Highly efficient reprogramming to pluripotency and directed differentiation of human cells with synthetic modified mRNA. *Cell Stem Cell, 7,* 618–630. doi:10.1016/j.stem.2010.08.012.

Weissman, I. L. (2000). Translating stem and progenitor cell biology to the clinic: Barriers and opportunities. *Science, 287,* 1442–1446. doi:10.1126/science.287.5457.1442.

Westenskow, P. D., Moreno, S. K., Krohne, T. U., Zhu, S., Zhang, Z. N., Zhao, T., Xu, Y., Ding, S., & Friedlander, M. (2012). Using flow cytometry to compare the dynamics of photoreceptors outer segment phagocytosis in iPS-derived RPE cells. *Investigative Ophthalmology & Visual Science, 53,* 6282–6290.

Williams, R. L., Krishna, Y., Dixon, S., Haridas, A., Grierson, I., & Sheridan, C. (2005). Polyurethanes as potential substrates for sub-retinal retinal pigment epithelial cell transplantation. *Journal of Materials Science. Materials in Medicine, 16,* 1087–1092. doi:10.1007/s10856-005-4710-y.

Yang, Y., Mohand-Said, S., Danan, A., Simonutti, M., Fontaine, V., Clerin, E., et al. (2009). Functional cone rescue by RdCVF protein in a dominant model of retinitis pigmentosa. *Molecular Therapy, 17,* 787–795. doi:10.1038/mt.2009.28.

Yannuzzi, L. A., Bardal, A. M., Freund, K. B., Chen, K. J., Eandi, C. M., & Blodi, B. (2006). Idiopathic macular telangiectasia. *Archives of Ophthalmology, 124,* 450–460.

Yannuzzi, L. A., Negrao, S., Iida, T., Carvalho, C., Rodriguez-Coleman, H., Slakter, J., et al. (2001). Retinal angiomatous proliferation in age-related macular degeneration. *Retina, 21,* 416–434.

Yu, J., Vodyanik, M. A., Smuga-Otto, K., Antosiewicz-Bourget, J., Frane, J. L., Tian, S., et al. (2007). Induced pluripotent stem cell lines derived from human somatic cells. *Science, 318,* 1917–1920.

Zarbin, M. A. (2004). Current concepts in the pathogenesis of age-related macular degeneration. *Archives of Ophthalmology, 122,* 598–614. doi:10.1001/archopht.122.4.598.

Zhang, X., & Bok, D. (1998). Transplantation of retinal pigment epithelial cells and immune response in the subretinal space. *Investigative Ophthalmology & Visual Science, 39,* 1021–1027.

Zhou, H., Wu, S., Joo, J. Y., Zhu, S., Han, D. W., Lin, T., et al. (2009). Generation of induced pluripotent stem cells using recombinant proteins. *Cell Stem Cell, 4,* 381–384. doi:10.1016/j.stem.2009.04.005.

Zhou, L., Wang, W., Liu, Y., de Castro, J. F., Ezashi, T., Telugu, B. P. V. L., et al. (2011). Differentiation of induced pluripotent stem cells of swine into rod photoreceptors and their integration into the retina. *Stem Cells, 29,* 972–980.

112 Retinal Prostheses

MARK S. HUMAYUN, JAMES D. WEILAND, AND DEVYANI NANDURI

Visual impairment ranging from slight impairment to complete blindness affects millions of people worldwide. In the developed world, one of the major causes of vision loss is disease that involves photoreceptors such as retinitis pigmentosa (RP) and age-related macular degeneration (AMD). Currently, over 15 million, or approximately 1/4,000, people are afflicted with a photoreceptor disease, and these numbers are expected to rise as the population ages (Chader, Weiland, & Humayun, 2009). While both RP and AMD are characterized by the death of photoreceptors, which eventually leads to complete vision loss, the remaining neural cells in the retina are relatively spared, though somewhat disorganized (Jones & Marc, 2005). The prevalence of these various vision disabilities has given rise to a number of therapies that attempt to interface with and activate the remaining visual system.

A variety of visual prosthesis implants have been under development for the past 60 years. These devices can provide artificial vision through the activation of the visual system at different stages such as the cortex, optic nerve, or retina. A retinal prosthesis (one category of visual prosthesis devices) can be placed epiretinally on the surface of the retina, in the subretinal space, or in the suprachoroidal space to provide artificial vision through the electrical activation of retinal cells. The clear advantage of a retinal prosthesis system is that it intervenes at an earlier stage in the visual stream than other types of visual prostheses and is thereby more likely to utilize the processing capabilities of the retina. However, a retinal prosthesis requires an intact optic nerve and cannot be used to treat vision loss due to impairment of the optic nerve.

Proposed therapies other than a visual prosthesis currently under investigation include an optogenetic therapy involving light-sensitive proteins such as Channelrhodopsin-2 (ChR2) and gene replacement therapy. Each therapy has its strengths. Gene replacement technology has been successful clinically in treating one type of RP (i.e., Leber's congenital amaurosis; RPE65 mutation) (Hauswirth et al., 2008). However gene therapy treatment is limited in scope by the specificity to a particular genetic mutation and the presence of residual photoreceptors, requiring intervention during initial stages of photoreceptor degeneration. Optogenetic technology, providing light sensitivity function to remaining neural cells in the retina, has shown success at the preclinical stage. However, a clinically viable therapy will require light stimulation more than 5 orders of magnitude greater than the threshold of cone photoreceptors (Lagali et al., 2008).

Over the past 20 years, retinal prostheses as a viable therapy for vision loss due to retinal degeneration (RD) has made considerable strides. Subretinal prostheses (implanted in the subretinal location of degenerated photoreceptors) have been successful in providing implant recipients with low-level spatial vision. However, groups have yet to implant a fully chronic device in subjects. Epiretinal prosthesis devices, on the other hand, are the only fully long-term solution currently available to partially restore sight to people who are blind from severe RP across multiple RP gene mutations. Starting in 2002, epiretinal prosthesis devices have been implanted in over 40 subjects worldwide by Second Sight Medical Products Inc., a neural prosthesis company located in Southern California. Six subjects were implanted with the first generation of the device (Argus I system) and 32 subjects with the second generation of the device (Argus II) as part of a multisite clinical trial (Ahuja et al., 2011; Mahadevappa et al., 2005). Most recently, the device has also been approved for sale in parts of Europe, and several subjects have been implanted to date with the commercial version. Analogous to a cochlear implant, the system electrically stimulates the inner retinal ganglion cell layer using an implanted microelectrode array.

At the most basic level, when the retina (or any stage along the visual stream) is electrically stimulated, the subject perceives a spot of light called a "phosphene." Conceptually, a visual prosthesis system restores vision by taking an image seen in the visual field (captured in real time by a video camera) and represent each pixel with a phosphene. The final image would look much like a gray-scale digital scoreboard, where each phosphene produced by an electrode can be thought of as a pixel that varies in brightness. As the technology improves, pixel/electrode size would decrease and the resolution of the system would increase. A schematic depicting the concept of how such an epiretinal prosthesis could potentially produce artificial vision is shown in figure 112.1.

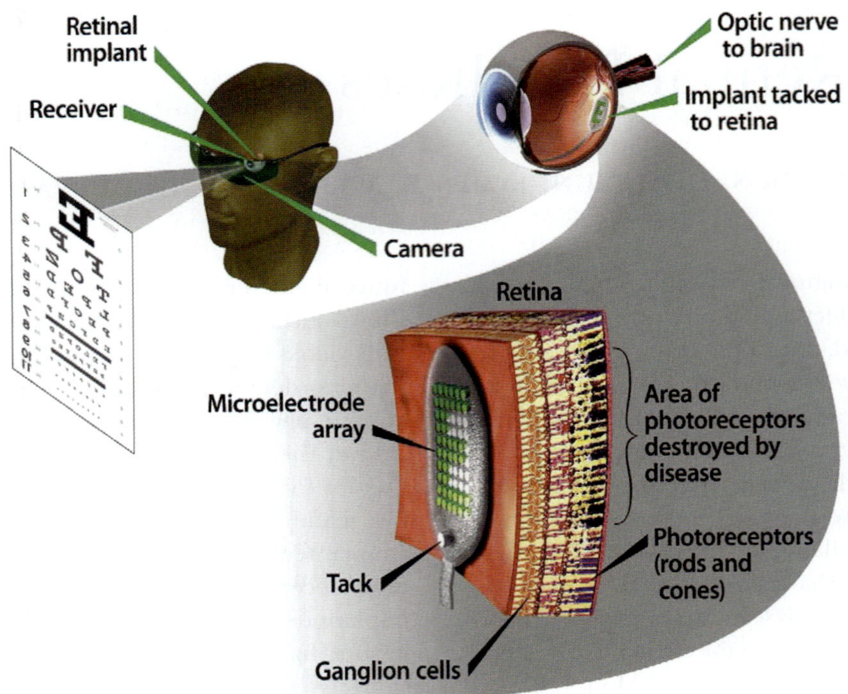

FIGURE 112.1   Concept of an epiretinal prosthesis. The camera captures an image of the world and projects it onto the retina via a microelectrode array. Electrodes stimulate the underlying retinal cells and the subject perceives phosphenes that can be combined into a pixelated view of the world. (Adapted from Chader, Weiland, & Humayun, 2009.)

Presently, the system resolution and capabilities can partially restore visual function, but the technology has not reached the stage where it mimics natural vision. However, the technology continues to improve. This section presents an overview on the development of retinal prosthetic devices. Starting with the history of electrically induced vision and early clinical work with electrical stimulation of the visual system, we first learn about the beginnings of visual prostheses and retinal stimulation. Subsequently, consideration of the effects of RD provides an understanding of the incentive for developing a retinal prosthetic device. A review of the state-of-the-art preclinical retinal electrophysiological research helps to define the technological scope and potential of such a therapy. Lastly, we explore the successful developments of a full-scale system by several groups and published clinical findings in the field from subjects who have been implanted with these remarkable devices.

THE HISTORY OF ELECTRICALLY INDUCED VISION

Development of a Cortical Visual Prosthesis

The first reported use of electrical stimulation to create the sensation of vision was over 2 centuries ago. In 1755, Charles LeRoy, a chemist and physician, briefly induced the first documented phosphenes by discharging a Leyden jar (acting as a high-voltage capacitor) through the head of a blind boy (LeRoy, 1755). One set of wires was wound around the patient's head, likely contacting the skull near the occipital lobes, while the other was connected to his leg, completing the current pathway. When the wire leads were placed in the Leyden jar, the shocks provoked "terrible cries," and the patient reported "flames descending rapidly before his eyes" (Marg, 1991, 1–2). Two hundred years after LeRoy's experiments, in the mid-twentieth century, neurosurgeon Wilder Penfield, while studying the neural origins of epilepsy, discovered that electrical stimulation of the cortex caused subjects to perceive visual percepts described as stars, wheels, disks, spots, streaks, and lines (Penfield & Jasper, 1954). Based on this work, more than a decade later, Drs. Brindley and Lewin implanted the first chronic electrode array in the visual cortex for the purpose of restoring vision. In the procedure, an array of 80 platinum disk electrodes was placed on the surface of the occipital pole of the subject, a 52-year-old woman (Brindley & Lewin, 1968). In 1972, another subject was implanted with an 80-electrode array (Brindley et al., 1972). All these subjects were able to see independent and distinct visual percepts which varied in size, and some spatial mapping and threshold

    MARK S. HUMAYUN, JAMES D. WEILAND, AND DEVYANI NANDURI

experiments were performed. However, experiments attempting to provide functional form vision through combining phosphenes into letters and shapes proved unsuccessful.

In the late 1970s, there was another attempt to develop a chronic cortical prosthesis by the vision scientist William Dobelle. Two blind subjects were implanted with a hexagonal array of 64 platinum disk 1-mm^2 electrodes (Dobelle, Mladejovsky, & Girvin, 1974). Spatial mappings of phosphenes from stimulating various electrodes indicated that stimulation far from the occipital pole led to peripheral phosphenes, and stimulation clustered on the surface of the visual cortex led to clustered phosphenes (Dobelle et al., 1979). Subsequent studies found that simultaneous and interleaved pulse stimuli on pairs of electrodes separated by as much as 1 mm resulted in fused dumbbell-shaped or elongated line percepts (Bak et al., 1990). Another clinical cortical prosthesis study using penetrating microelectrodes in the visual cortex showed that an increase in current causes a decrease in phosphene size (Schmidt et al., 1996). These results suggest that cortical stimulation using multielectrode stimulation could be complicated by electrode–electrode and lateral neural interactions.

One of the advantages of a cortical prosthetic device over other visual prosthetic devices is that it can be used to treat a wider variety of visual impairment diseases (Margalit et al., 2002). At present, work continues to progress in this area with several groups (Normann et al., 2009). Unfortunately, difficulties with safety and surgical techniques may limit the clinical viability of this solution (Kotler, 2002).

Early Work in Retinal Stimulation

Brindley's group also performed experiments with extraocular electrical stimulation. Brindley used a makeshift corneal electrode to apply stimulating pulse trains directly to his own eye and discovered that electrical impulses interfered with his natural vision (Brindley, 1962, 1964). A scientist from Cambridge, Carpenter (1973) also performed extraocular experiments on a human subject and was able to elicit phosphenes by passing alternating currents of 100 Hz. While stimulation was provided extraocularly, the elicited phosphenes were induced by activating the retina with the passing current. These early experiments proved that the visual system could, in fact, be activated by stimulating the retina (in this case, via an extraocular current). The ability of extraocular stimulation to elicit phosphenes clearly demonstrated the potential for a retinal prosthesis to provide artificial vision.

Additional Visual Prosthesis Efforts

LATERAL GENICULATE NUCLEUS STIMULATION The lateral geniculate nucleus (LGN) is located in the thalamus of the brain and serves as a relay center for visual information from the retina through to V1. There are two LGNs for each vertically segmented hemifield. Pezaris and Reid are currently investigating the feasibility of implanting a prosthesis in the LGN (Pezaris & Eskandar, 2009). Recent behavioral research in nonhuman primates has demonstrated that visual or electrical stimulation of LGN specific receptive fields can cause highly localized and repeatable eye movements to the location in space represented by the receptive field of that neuron (Pezaris & Reid, 2007).

Like cortex and retina, retinotopy in the LGN is maintained. Thus, an electrode array implanted in the LGN would be able to produce properly spatially mapped percepts (Schneider, Richter, & Kastner, 2004). Although seemingly highly invasive, surgical implant procedures would be similar to that of Deep Brain Stimulation of the thalamus in human patients for the treatment of Parkinson's disease, which has an established safety profile. A major disadvantage is that a full field prosthesis would require bilateral LGN implants (Pezaris & Eskandar, 2009).

OPTIC NERVE STIMULATION The optic nerve consists of a collection of ganglion cell axons that project information from the retina to higher processing centers in the brain. Based upon the same principles used to develop cuff electrodes for nerve fiber recordings, stimulation of muscle fibers (Fang & Mortimer, 1991) and the vagus nerve (Woodbury & Woodbury, 1991), a visual prosthesis could be developed to stimulate the optic nerve (Hoffer, Loeb, & Pratt, 1981). Using a cuff electrode, one group has shown that optic nerve stimulation can generate visual percepts in a single blind subject (Brelen et al., 2005). Furthermore, variability in phosphene position was reported to be ~5–10° of visual angle (Obeid, Veraart, & Delbeke, 2010). One disadvantage to this approach is that the optic nerve does not maintain retinotopy (Fitzgibbon & Reese, 1996), and could compromise the spatial mapping of the percepts. Although experiments have shown that the location of a percept can be changed with stimulation parameters such as frequency and pulse width, this does not necessarily improve the spatial mapping capabilities of the device (Veraart et al., 2003).

OPTOGENETIC THERAPY Optogenetics is a relatively new subfield of gene therapy that combines genetics and optics to provide cells with a completely new function by rendering them light sensitive. Cells are injected with an adeno-associated virus, which delivers genes to the neural cell. The genes are able to transduce channelrhodopsin, a light-sensitive ion channel into neural cells. These transduced cells can now act similar to photoreceptors by taking in a light stimulus and producing an electrical response that continues through the rest of the visual system. Although a very different approach, optogenetics has some similarities with visual prostheses, namely the attempt to control the timing of signaling between neural cells in the retina. By rendering cells (i.e., bipolar or ganglion cells) further downstream from photoreceptors to be light sensitive, local retinal circuits will be driven (albeit in an unnatural way), leading to complex retinal responses, much like electrical stimulation of the visual system. Likely, a clinical solution would also require an external device (i.e., a pair of glasses with a camera) that will capture the image of the world and translate the image into an amplified blue light signal to drive bipolar or ganglion cells.

Thus far, preclinical development has shown safety and proof-of-concept in both electrophysiological and behavioral experiments (Doroudchi et al., 2011). In vivo experiments that introduce light-sensitive protein channels, such as channelrhodopsin (ChR2), to ON and OFF bipolar or ganglion cells in a degenerate mouse retina show a renewed responsiveness to light stimulus (Bi et al., 2006; Doroudchi et al., 2011; Thyagarajan et al., 2010). Furthermore, optogenetically treated animals have measurable visually evoked potentials (VEPs) and optomotor responses (Tomita et al., 2010) and are able to perform behavioral tasks requiring residual vision (Doroudchi et al., 2011; Thyagarajan et al., 2010). Unfortunately, light stimulation at levels that are 5 orders of magnitude greater than the threshold of cone photoreceptors will be necessary to activate the ChR2 channels (Schnapf, Kraft, & Baylor, 1987), and the operable range of usable light levels could be severely limited (Wang et al., 2007).

Retinal Implants

With the experiences gained from the development of other visual prosthesis devices and neural implants in general, retinal prosthetic devices strive to target neural cells early in the visual pathway prior to neural processing in the LGN and cortex. Presently, there are three approaches for placing the retinal implant: suprachoroidally, epiretinally, and subretinally.

In a suprachoroidal prosthesis, the electrode array is passed through the sclera and sits beneath the choroid, while the electronics are implanted outside of the eye. Surgically, this implant is relatively simple to place, and the electrode to neural tissue distance allows the safe charge level to increase by a factor of 3 compared to direct retinal stimulation (Nakauchi et al., 2007). A number of animal studies have established safety and efficacy of extraocular and suprachoroidal transretinal stimulation (Shivdasani et al., 2010). During preliminary clinical studies, suprachoroidal stimulation was able to induce phosphenes in two subjects with RP (Fujikado et al., 2007). While these findings show promise, with such far distance between stimulating electrode to the neural tissue, the development of a high-resolution extraocular device remains challenging.

Subretinal electrode arrays are implanted in the subretinal space between the pigment epithelial cells and the remaining retina in order to mimic the functionality of the degenerated photoreceptors. The system operates by either directly stimulating the retina with a wire across the skin or with a photodiode microchip that substitutes for photoreceptor function by sensing incoming light and electrically activating the neural retina. There have been several groups working on slightly different subretinal techniques (Chow et al., 2004; Jensen & Rizzo, 2008; Zrenner et al., 1999). In early studies from Chow's group, this light-based pathway had been limited by insufficient power supply to photodiodes. In theory, subretinal stimulation should utilize the computational processing of the retinal circuitry by targeting the retinal bipolar cells, in contrast to an intervention further along in the visual system. However, there is evidence that the degenerated bipolar–ganglion cell synapse and subretinal gliosis of a diseased retina could make it difficult to place the array and stimulate bipolar cells (Marc & Jones, 2003).

Epiretinal implants are implanted on the vitreal, inner surface of the retina in contact with the ganglion cell layer. While the goal has been to stimulate remaining neural circuitry from bipolar to ganglion cells (Greenberg, 1998), electrophysiology studies seem to indicate that ganglion cells are primarily activated (Sekirnjak et al., 2008). There are several groups that have worked on this approach with varying degrees of success (Gerding, Benner, & Taneri, 2007; Humayun et al., 1999a). . Over the last 10 years, Second Sight Medical Products, Inc. (2-Sight) has sponsored two separate 2-Sight clinical trials, implanting more than 30 subjects with epiretinal arrays (Ahuja et al., 2011; Humayun et al., 2003). Based on the success of these trials, the device has been approved for placement in patients with advanced RP in parts of Europe.

RP describes a group of genetic diseases with different causes and biological mechanisms, but with similarities in clinical symptoms and consequences (Daiger, Bowne, & Sullivan, 2007). Retinal degenerative diseases can be inherited through autosomal-recessive, autosomal-dominant, or X-linked genes and other inheritance means. Research into the different triggers of RD from about 50% of all cases has revealed more than 200 different genetic causes (Ayuso & Millan, 2010), with each gene mutating to cause multiple forms of the disease (Daiger, Bowne, & Sullivan, 2007). Correspondingly, many different animal genotype and phenotype mutations of RD exist (Chang et al., 2002). Despite the diversity in the genes and mutations causing photoreceptor cell death, the structural changes in the retina after the onset of RP are remarkably similar across the different genetic causes (Jones & Marc, 2005). Structural modifications in the retina are accompanied by functional changes to neuronal processes (Margolis & Detwiler, 2011). Furthermore, over the last 20 years, the effects of long-term visual deprivation on cortical reorganization have been explored but are still being fiercely debated (Wandell & Smirnakis, 2009).

Structural Changes in Diseased Retina

A great number of animal morphology studies have revealed distinctive stages in the reorganization of the neuronal retina during RD, though the time course of the disease can vary substantially (Jones & Marc, 2005). Effects of remodeling taken from Jones and Marc (2005) are shown in figure 112.2. First, when photoreceptors begin to degenerate, outer segments are shortened (Jones et al., 2003), and rod and cone photoreceptors can bypass connections to horizontal cells (and bipolar cells) and sprout ectopic

neurite projections into the inner nuclear layer (INL), sometimes as far as the ganglion cells (Jones et al., 2003). A small number of rods are also known to migrate into the inner retina (Jones & Marc, 2005). Subsequently, substantial photoreceptor cell death is accompanied by the formation of a subretinal glial seal consisting of Müller cells and the initiation of neuronal cell death (Marc et al., 2003). Lack of photoreceptor glutamatergic input causes significant death and rewiring of the inner nuclear layer that matches human morphology studies (Jones & Marc, 2005). Cone bipolar cell dendrites eventually atrophy, and horizontal cells develop sprouting processes into the inner plexiform layer (IPL) (Gargini et al., 2007). Amacrine cells can sometimes migrate into the ganglion cell layer or toward the outer plexiform layer near the glial seal (Marc et al., 2003), and massive rewiring increases the span of connectivity within the IPL, leading to the formation of recurrent bipolar–bipolar synapses (Jones & Marc, 2005). Despite these many changes, laminar organization of the different layers including presynaptic proteins necessary for synaptic function is preserved (Phillips, Otteson, & Sherry, 2010). While much of the presynaptic connectivity is altered, in rd-1/rd-1 and rd10 mouse models, ganglion cell structure including its dendritic stratification and projections to higher visual centers is maintained (Margolis et al., 2008).

Human morphology studies of degenerate retina have focused largely on quantifying the cell loss within the layers of the retina and eccentricities to the fovea. Within the macular region INL cell loss is 20–60% whereas in the ganglion cell layer (GCL) cell loss ranges between 20 and 80% (Santos et al., 1997). Within the extramacular retina, IPL cell loss is approximately 60% and GCL cell loss ranges from 70 to 80% (Humayun et al., 1999b). The difference between degenerated extramacular and macular cell loss is hardly surprising since RP progresses from the periphery toward the fovea. Although cell death is more substantial for the GCL compared to the INL, the structure of ganglion cells is considerably better preserved than the presynaptic cells (Margolis et al., 2008). While optical coherence tomography (OCT) studies with RP patients have shown a thinning of the retinal nerve fiber layer compared to normal retina, no mention has been made about changes to the axonal pathway patterns (Walia et al., 2007).

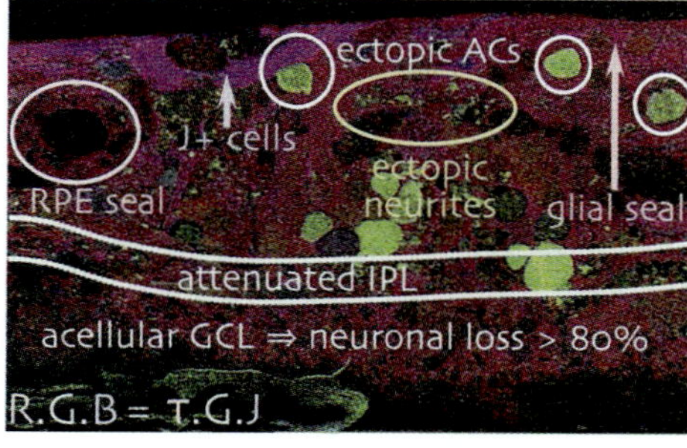

FIGURE 112.2 Retinal remodeling during retinitis pigmentosa (RP). Human retina with RP has significant remodeling and cell loss including cell migration. (Adapted from Jones & Marc, 2005.)

Functional Changes in Diseased Retina

Not surprisingly, such drastic structural changes to the retina are coupled with a number of functional changes to both presynaptic and postsynaptic cells. Lack of photoreceptor input alters ON bipolar cell mGluR6 and iGluR glutamate mediated signaling pathways and causes amacrine and ganglion cells to show spontaneous and strong iGluR mediated responses (Marc et al., 2007; Ye & Goo, 2007), although slow degenerate models may maintain rod bipolar glutamate sensitivity (Barhoum et al., 2008). Electroretinography response studies show a decrease in bipolar cell response that precedes morphological changes (Gargini et al., 2007). Data suggest that ON bipolar cells may even shift behavior to resemble OFF bipolar cells (Stasheff, 2008), and bipolar–bipolar recurrent synapses could compromise signaling pathways in the inner retina (Marc et al., 2003). However, in vivo studies in a degenerate mouse retina that introduce photosensitivity to ON bipolar cells (through genetically encoded neuromodulators) show ganglion cell spiking activity in response to light stimulus (Doroudchi et al., 2011; Lagali et al., 2008). Furthermore, these optogenetically treated animals can perform behavioral tasks requiring functional vision (Doroudchi et al., 2011).

Together these findings imply that the bipolar–ganglion cell synaptic pathway remains functional at some level despite degenerative effects. Though ganglion cell morphology is maintained (Margolis et al., 2008), degeneration-induced spontaneous oscillatory spiking activities in both ON and OFF RGCs have been observed; however intrinsic excitatory and inhibitory response properties are still present (Margolis et al., 2008; Stasheff, 2008). The cause of the spontaneous ganglion cell activity has been linked to presynaptic mechanisms and resonant bipolar–amacrine cell pathways that arise in the absence of photoreceptor input (Margolis & Detwiler, 2011).

Cortical Plasticity

When a blind spot is introduced to the retina via a lesion, the visual system perceptually fills in this "blind spot" to complete the image of the world (Wandell & Smirnakis, 2009). Similarly as RD progresses, the visual system has compensatory mechanisms in order to optimally represent the visual world. It seems plausible that compensatory mechanisms would result in long-term changes to the structure and function of the visual cortex. Recent literature states the conflicting viewpoints on the existence and nature of adult visual cortical plasticity (Wandell & Smirnakis, 2009). Some studies over the last 20 years have supported the concept that the removal of sensory input to V1 results in a change to receptive field size and cortical topography (Gilbert & Wiesel, 1992). In the majority of these experiments a retinal lesion was induced in an animal, and the

response within the cortical lesion projection zone (LPZ) was measured. Cells within the LPZ are initially silenced but eventually begin to respond to visual stimuli placed outside of the retinal lesion. These findings imply a shift in location for the receptive field and possibly a change in the receptive field size (Gilbert, Li, & Piech, 2009; Gilbert & Wiesel, 1992). Response changes within the LPZ may be due to lateral axonal inputs from cells outside of the LPZ and changing dendritic spines (Keck et al., 2008). However, these results mainly relied on electrophysiological recordings from single and populations of cells. On the other hand, functional magnetic resonance imaging (fMRI) studies, which can record long-term cortical activity over a wide field of view, have presented a contrasting viewpoint. Imaging studies in the cortex of primates with retinal lesions show an initial silencing of the LPZ that is not followed by recovery in response (Smirnakis et al., 2005).

Human fMRI studies with AMD and RP patients have both shown that cortex corresponding to the region of RD respond during task-based stimuli, but not passive viewing (Masuda et al., 2010). These findings suggest top-down mechanisms (which are present prior to degeneration) contribute to cortical responses rather than bottom-up sensory processing or newly developed lateral connections. Based on these inconsistent views, the potential for adult cortical plasticity to facilitate the development of vision restoration strategies remains uncertain.

PRECLINICAL STUDIES IN ELECTRICAL STIMULATION OF THE RETINA

Given that visual information is encoded in the retina by the precise temporal firing patterns of ganglion cells, the ability to control a prosthesis subject's perceptual experience will require an understanding of how electrical stimulation interfaces with the different cells of retina and modulates their collective response. Electrophysiological findings can give insight into which cells are activated by different types of stimuli and whether changes to the stimuli, be it the pulse train parameters, the location of stimulus, or the shape and configuration of electrodes, can modulate the final firing output of ganglion cells.

Stimulation Threshold

Threshold quantifies the minimum stimulus necessary to yield the desired response. In psychophysical experiments, threshold is the minimum stimulus required to produce a perceptual change. In visual electrophysiology experiments, threshold refers to the minimum stimulus (charge, voltage, amplitude, or photons) needed to elicit a spike or change in potential from a neural cell. Ganglion cell responses are measured as spikes while the response of bipolar and amacrine cells are measured as a graded change in potential. An equally important factor that limits the stimulus is the maximum allowable charge density for electrical stimulation. Safety requirements set limits to charge density in order to avoid damage to electrode material or neural tissue (Brummer, Robblee, & Hambrecht, 1983; Shannon, 1992). Together, thresholds and charge density limits (in humans) provide the upper and lower bounds for stimulation, respectively, and thus define the dynamic range of possible stimulation intensities.

In retinal electrophysiology experiments, the electrode is positioned either on the epiretinal or subretinal surface. Conventionally, ganglion cell spikes are measured in vitro with epiretinal recording electrodes. Additional techniques to measure cell responses are through whole-cell patch clamp recordings, in vivo recordings of evoked responses in the superior colliculus (Chan et al., 2011), and calcium imaging studies (Behrend et al., 2009). In calcium imaging studies, ganglion cell response is quantified by the degree that a calcium-sensitive molecule fluoresces in response to a stimulus. Intracellular calcium concentration is an indicator of neural activity. With this specialized technique, one is able to observe the response across many ganglion cells simultaneously and gauge the spatial extent of activation.

To characterize response properties of the retina, strength duration curves that provide the relationship between threshold amplitude and pulse duration have been measured across different species. A strength duration curve for a rat retina (values are averaged across 25 cells shown with a solid line) is shown in figure 112.3. The rheobase defines the threshold current for a current pulse of infinite duration and is the asymptote value of the curve. The chronaxie is the stimulus duration time for which the threshold current is at double the amplitude of the rheobase and is also the most energy efficient stimulus. Average threshold currents across different species (i.e., rat, pig, and monkeys) with 6–25 μm electrodes revealed that thresholds for a 0.05-ms pulse were less than 10 μA and decreased with increasing pulse duration. However, the total charge increased with pulse duration (Sekirnjak et al., 2006). Chronaxie values ranged between 100 μs and 400 μs, implying that short pulse durations are the most charge efficient (Fried, Hsueh, & Werblin, 2006). Thresholds also increased by a factor of 2–3 across the different electrode sizes (6–25 μm) and with distance from the

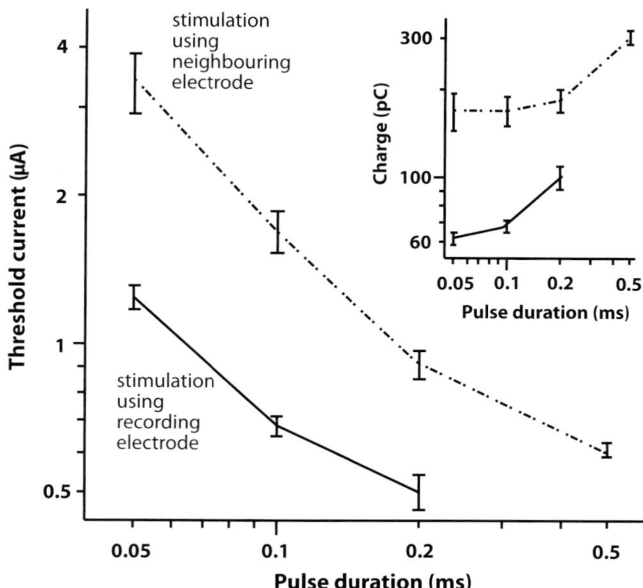

FIGURE 112.3 Strength duration curve. Average change in threshold (large bottom plot) and overall charge (small top plot) as a function of pulse.

stimulating electrode (see figure 112.3, dashed line), demonstrating that current attenuates with distance from the stimulating electrode (Sekirnjak et al., 2006). Comparable to ~10 μA, a separate study with rabbit retina measured thresholds for a 40-μm electrode as ~15 μA (Fried, Hsueh, & Werblin, 2006).

Thus far, threshold studies have shown conflicting results on the effect of RD on threshold. Thresholds for rd1 mice were approximately an order of magnitude higher than wild-type (Jensen & Rizzo, 2008). Similarly, increases in threshold (by nearly a factor of 2) were also noted for in vivo experiments recording evoked responses from the superior colliculus of the RD rat (Chan et al., 2011). However, an in vitro study with very small 10-μm electrodes showed no change in threshold between the P23H rat and normal (Sekirnjak et al., 2009). With a possible increase in thresholds for degenerate retina, short-duration pulses are more likely to have a charge density that is within the safety limits and could provide a greater dynamic amplitude range compared to longer duration pulses.

Temporal Modulation of Retinal Cells

Ideally, retinal stimulation will be able to produce ganglion cell spiking patterns that replicate the temporal and spatial specificity of normal vision. One way to achieve temporal specificity is to activate ganglion cells with a train of short-duration biphasic pulses. With a train of short-duration pulses (cathodic phase < 0.15

ms), a single spike per pulse is generated for frequencies up to 250 Hz (Sekirnjak et al., 2008). On the other hand, with suprathreshold stimulus amplitudes (20 µA–60 µA), electrodes with a diameter of 30 µm generate multiple spikes for pulses < 0.5 ms (Ryu et al., 2011). Furthermore, the number of elicited spikes for each pulse varies among different ganglion cell types (Cai et al., 2011). With high-frequency stimulation > 10 Hz, long latency responses likely attributed to presynaptic mechanisms are suppressed (Sekirnjak et al., 2006). Overall, these findings propose that the frequency of stimulation can control ganglion cell spiking with relatively high temporal specificity. However, adaptive effects causing ganglion cell response to desensitize with successive pulses could limit the operating range of frequencies (Freeman, Rizzo, & Fried, 2011; Jensen & Rizzo, 2007).

RETINAL PROSTHESIS IMPLANTS IN HUMANS

Subretinal Prosthesis

The subretinal implant by Retina Implant AG consists of both intraocular (microphotodiode electrode array) and extraocular (wirelessly operated power control unit) components. The electrode array contains both a microphotodiode array (MPDA) with 1,500 individual light-sensitive elements (see figure 112.4B) and a test field for direct stimulation with 4 × 4 electrodes (see figure 112.4A) for electrical, light-independent stimulation. Each of the MPDA elements includes a light-sensitive photodiode (15×30 µm) that acts independently from other MPDA elements (Zrenner et al., 2011). Each photodiode output controls a differential amplifier which is connected to a titanium nitride (TiN) electrode (50×50 mm) through a contact hole (Zrenner et al., 2011). The test field array consists of quadruple square electrodes of 120×120 µm for which each unit in the quad is a 50×50 µm square electrode. From the intraocular array, a cable from the implanted chip in the eye leads under the temporalis muscle to the exit behind the ear and connects to a wirelessly operated power control unit (see figure 112.4C). The chip covers a visual angle of approximately 11° by 11° (1° approximates 288 µm on the retina) and the distance between two MPDA electrodes is estimated to correspond to a visual angle of 15 min of arc (Zrenner et al., 2011).

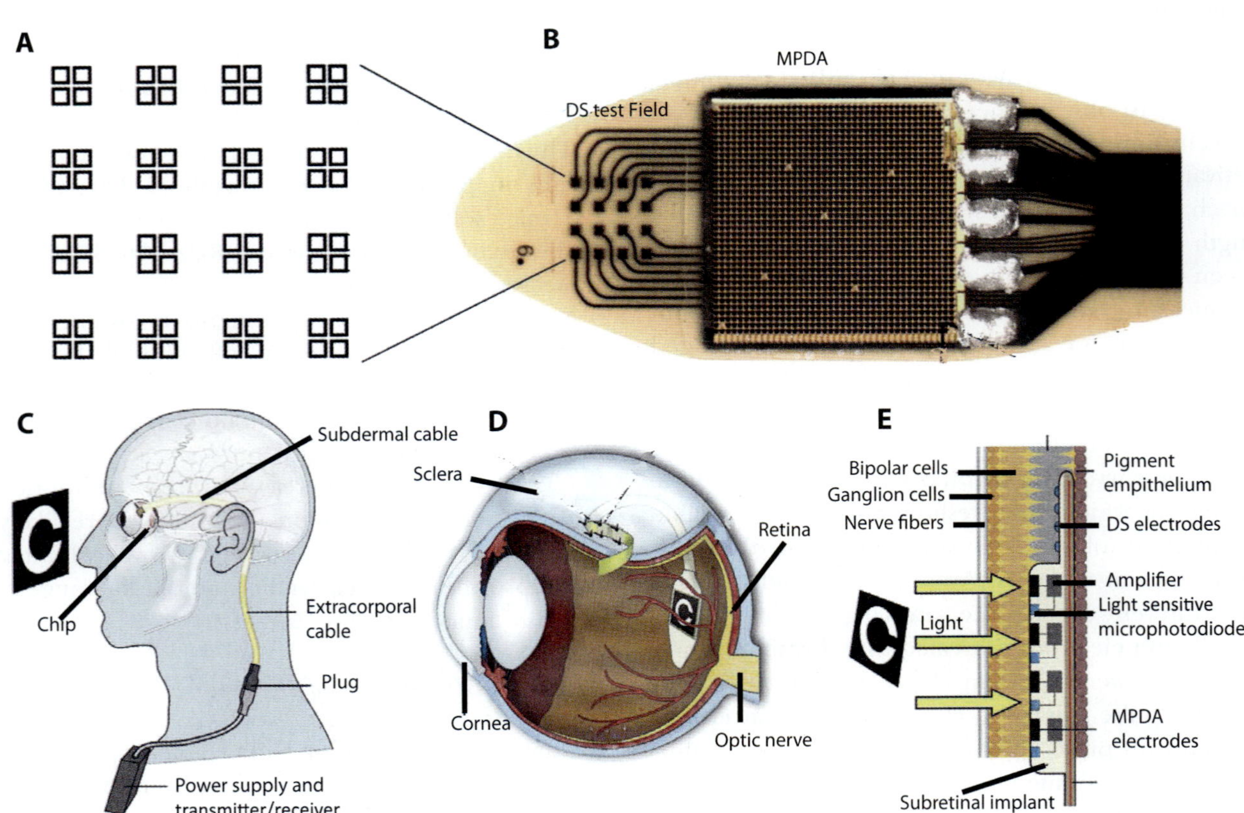

FIGURE 112.4   Schematic of the Retina Implant AG subretinal prosthesis system. (A) Direct stimulation (DS) test field electrode array configuration. (B) Microphotodiode array (MPDA) electrode array. (C and D) Overview of the subretinal implant system. (E) Placement of array in the subretinal space and concept of subretinal stimulation. (Adapted from Zrenner et al., 2011.)

In order to provide artificial vision, an "image" is captured simultaneously by all photodiodes several times per second. The signal from the photodiode is converted to monophasic anodic voltage pulses at each MPDA electrode, and stimulation is delivered via these electrodes to the retinal cells (see figure 112.4E). The amount of current provided to the retina by each electrode is proportional to the light absorbed by each photodiode.

Epiretinal Prosthesis

The Second Sight Medical Products, Inc., epiretinal prosthesis contains both intraocular (electrode array) and extraocular (e.g., glasses, Visual Processing Unit; VPU) components. The first-generation device (Argus I system) has 16 electrodes (see figure 112.5A) while the second-generation device (Argus II system) has 60 electrodes (see figure 112.5B). The intraocular array, implanted epiretinally in the macular region of the retina, consists of 16 platinum disk electrodes (Argus I system) in a 4 × 4 arrangement or 60 electrodes in a 6 × 10 arrangement (Argus II system) and is contained within a clear silicone rubber platform. The array is held in place with a retinal tack. In the Argus I system, the electrodes implanted are 260 or 520 μm in diameter (subtending 0.9° and 1.8° of visual angle, respectively) arranged in an alternating checkerboard pattern spaced 800 μm center-to-center. In the Argus II system, the electrodes implanted are 200 μm in diameter (subtending 0.7° of visual angle) with 525-μm center-to-center spacing.

Custom software on a PC laptop is used to program the external VPU, which in turn sends stimulus commands to the implant. Power and signal information are sent from the VPU through a wire to an external transmitter coil that is attached and aligned magnetically, and coupled inductively, to a secondary coil that is either implanted subdermally in the subject's temporal skull behind the ear (Argus I system—see figure 112.5C) or located along the sclera of the eye (Argus II system—see figure 112.5D). The secondary coil provides power and signal information to an implanted pulse generator (IPG), which decodes the signal and produces the commanded stimulus pulses. The IPG transmits pulses to the array of electrodes via a multi-wire cable that traverses the sclera. Stimulation can be presented using two different protocols: (1) camera mode—real-time video captured by a miniature video camera mounted on the subject's glasses is continuously sampled by the VPU to match the stimulation current amplitude in each electrode to the brightness at the corresponding area of the scene—and (2) direct stimulation mode—the stimulation signal sent to each electrode is independently controlled by the VPU.

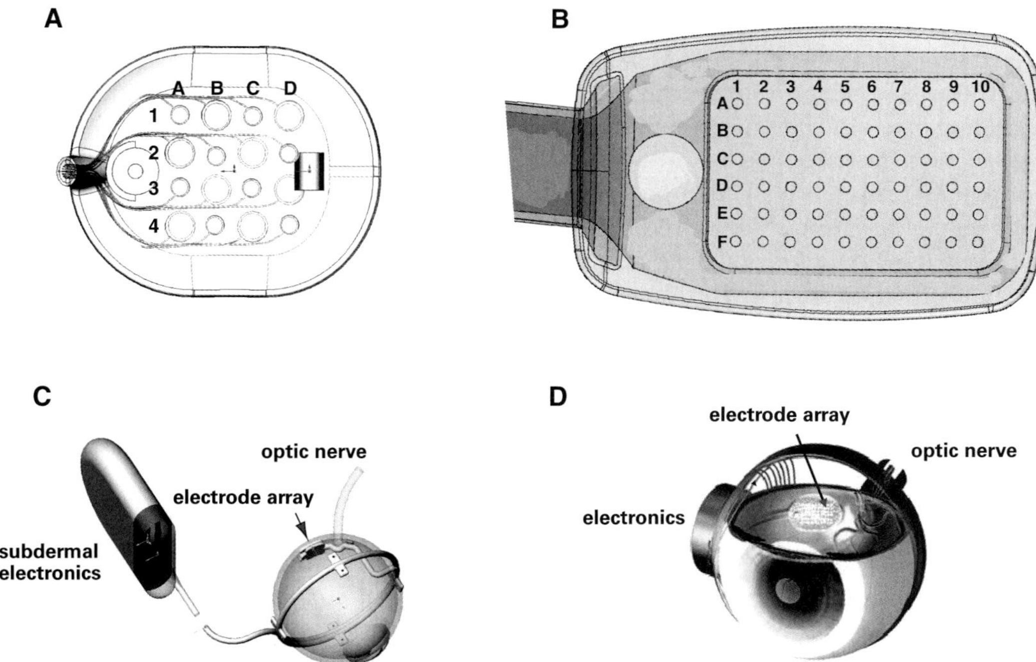

FIGURE 112.5   Schematic of the epiretinal prosthesis first-generation (Argus I) and second-generation (Argus II) arrays and systems. (A) Argus I 4 × 4 electrode array. (B) Argus II 6 ×10 electrode array. (C) Overview of the Argus I implant. (D) Overview of the Argus II implant 32.

Perceptual Threshold

Initial chronic studies with epiretinal prosthesis subjects measured perceptual thresholds for six subjects using a biphasic pulse with a cathode duration of 0.975 ms and electrode–retina distance using OCT to image the retina. Thresholds were mainly below the safety charge density limit and increased with an increase in electrode–retina distance (de Balthasar et al., 2008; Mahadevappa et al., 2005). Though electrode size varied between either 260 or 520 μm, thresholds did not seem to change with electrode size (de Balthasar et al., 2008). In semi-chronic subretinal prosthesis implant subjects, threshold voltage to elicit a percept was assessed in using an up–down staircase procedure. Typical charge transfer of a single electrode at threshold was between 20 and 60 nC per pulse with maximum charge density for direct stimulation electrodes at 600 $\mu C\ cm^{-2}$ (within safe charge limits) (Zrenner et al., 2011).

Properties of Percepts

In epiretinal threshold experiments, subjects anecdotally described percepts as whitish or yellowish round or oval shaped spots (Mahadevappa et al., 2005). Single-electrode experiments show that perceptual threshold and brightness can be manipulated predictably with a variety of parameters such as frequency, pulse duration, and amplitude (Greenwald et al., 2009; Horsager et al., 2009b). Interestingly, threshold is affected by the distribution of charge in time in a way that can be related to the neural integration and adaptive properties of the visual system (Horsager et al., 2009b). Brightness of a single-electrode percept increases as a power function with stimulation intensity, thereby making it possible to code a dynamic range of brightness with different stimulus amplitude levels (Greenwald et al., 2009; Nanduri et al., 2012). At high amplitudes, the brightness of phosphenes tends to saturate, possibly limiting the dynamic range.

When multiple electrodes are stimulated, percept appearance is highly complex (Horsager, Greenberg, & Fine, 2009a). Indeed, even a change in stimulation timing offset between electrodes as far as 1,600 μm can lead to a change in percept appearance (Horsager et al., 2011). These findings suggest that spatiotemporal interactions exist at the neural level and contribute to percept appearance. These results are further supported by findings that the brightness of a single-electrode phosphene can be changed by stimulating an additional electrode (Horsager et al., 2011). Taken together, these results suggest epiretinal stimulation activates a highly complex system of neural cells and that each electrode does not behave independently from the others.

Evidence of Form Vision in Retinal Implant Patients

Successful clinical findings have been reported by Zrenner's subretinal prosthesis group. Presently, clinical studies by Zrenner's group have been reported on acute and semi-chronic implant subjects. (Semi-chronic implants were in place less than 1 year. The current design of this device will not remain functional in the body beyond 1 year, due to failures in the protective coating around the electronics.) Using the implant, these subjects have shown the ability to discriminate bright objects on a dark table, describe patterns, and discriminate letters and short words (Zrenner et al., 2011). When direct stimulation electrodes were used, two subjects reported whitish round dot-like percepts, and one subject reported elongated short whitish or yellowish lines (Wilke et al., 2011; Zrenner et al., 2011). Subjects were also able to distinguish vertically and horizontally oriented electrode stimulations from each other (Wilke et al., 2011). Results are extremely promising and will need to be confirmed in chronically subretinally implanted subjects.

Subjects implanted with chronic epiretinal prostheses have demonstrated functional low-level vision when using the entire system (with a camera) (Yanai et al., 2007). In one experiment, a single subject is able to discriminate oriented visual gratings at four different orientations (Caspi et al., 2009). Furthermore when video input information is scrambled, the subject was unable to perform the task. These findings seem to indicate that the subject had form vision beyond a simple light detector. In another experiment, performed on a larger subject pool, subjects were able to point to a bright square on a dark background monitor (Ahuja et al., 2011). Nearly all subjects showed improvement in the task with device on versus device off. Subjects were also able to discriminate high-contrast letters on a dark background monitor and, using a head scanning technique, discriminate a series of letters to form words and sentences (Humayun et al., 2012). While results show promise of form vision, the ability to create meaningful high-resolution images with future dense arrays is still uncertain.

Summary

Experimental therapies hope to restore sight to blind people with photoreceptor diseases such as AMD and

  MARK S. HUMAYUN, JAMES D. WEILAND, AND DEVYANI NANDURI

RP. Optimally, visual function would be completely restored to its normal functional state by repairing or replacing photoreceptors. An equally ideal outcome would be to develop a therapy that interfaces with the visual system such that the final output response, the spatio–temporal pattern of ganglion cell spikes, are identical to normal vision, though the intermediary stage functioning could be rather different. Unfortunately, current gene therapy solutions are specific to a single gene mutation and thus cannot treat a broad spectrum of retinal disease. Additionally, stem cells, optogenetics, and single-cell electrical stimulation approaches face technical hurdles and are still in the developing stages. Instead, the solution closest to clinical implementation, a retinal prosthesis with relatively large electrodes, is an intermediary solution that offers partial vision restoration. While elicited artificial vision is imperfect, the success of these devices will be determined by their ability to generate useful vision in blind patients. Although intuitively appealing, the visual experience of these subjects is not necessarily like a digital scoreboard. Clinical studies have shown that each electrode does not act as an independent pixel. However, by creating images that contain predictable spatial and contrast information, retinal implant recipients may be able to use this data along with other contextual sensory information to better understand the visual world.

REFERENCES

Ahuja, A. K., Dorn, J. D., Caspi, A., McMahon, M. J., Dagnelie, G., daCruz, L., et al. (2011). Blind subjects implanted with the Argus™ II retinal prosthesis are able to improve performance in a spatial–motor task. *British Journal of Ophthalmology, 95*, 539–543. doi:10.1136/bjo.2010.179622.

Ayuso, C., & Millan, J. M. (2010). Retinitis pigmentosa and allied conditions today: A paradigm of translational research. *Genome Medicine, 2*, 34. doi:10.1186/gm155.

Bak, M., Girvin, J. P., Hambrecht, F. T., Kufta, C. V., Loeb, G. E., & Schmidt, E. M. (1990). Visual sensations produced by intracortical microstimulation of the human occipital cortex. *Medical & Biological Engineering & Computing, 28*, 257–259.

Barhoum, R., Martinez-Navarrete, G., Corrochano, S., Germain, F., Fernandez-Sanchez, L., de la Rosa, E. J., et al. (2008). Functional and structural modifications during retinal degeneration in the rd10 mouse. *Neuroscience, 155*, 698–713. doi:10.1016/j.neuroscience.2008.06.042.

Behrend, M. R., Ahuja, A. K., Humayan, M. S., Weiland, J. D., & Chow, R. H. (2009). Selective labeling of retinal ganglion cells with calcium indicators by retrograde loading in vitro. *Journal of Neuroscience Methods, 179*, 166–172.

Bi, A., Cui, J., Ma, Y. P., Olshevskaya, E., Pu, M., Dizhoor, A. M., et al. (2006). Ectopic expression of a microbial-type rhodopsin restores visual responses in mice with photoreceptor degeneration. *Neuron, 50*, 23–33. doi:10.1016/j.neuron.2006.02.026.

Brelen, M. E., Duret, F., Gerard, B., Delbeke, J., & Veraart, C. (2005). Creating a meaningful visual perception in blind volunteers by optic nerve stimulation. *Journal of Neural Engineering, 2*, S22–S28.

Brindley, G. S. (1962). Beats produced by simultaneous stimulation of the human eye with intermittent light and intermittent or alternating electric current. *Journal of Physiology, 164*, 157–167.

Brindley, G. S. (1964). A new interaction of light and electricity in stimulating the human retina. *Journal of Physiology, 171*, 514–520.

Brindley, G. S., Donaldson, P. E., Falconer, M. A., & Rushton, D. N. (1972). The extent of the region of occipital cortex that when stimulated gives phosphenes fixed in the visual field. *Journal of Physiology, 225*, 57P–58P.

Brindley, G. S., & Lewin, W. S. (1968). The visual sensations produced by electrical stimulation of the medial occipital cortex. *Journal of Physiology, 194*, 54–55P.

Brummer, S. B., Robblee, L. S., & Hambrecht, F. T. (1983). Criteria for selecting electrodes for electrical stimulation: Theoretical and practical considerations. *Annals of the New York Academy of Sciences, 405*, 159–171.

Cai, C., Ren, Q., Desai, N. J., Rizzo, F., III, & Fried, S. I. (2011). Response variability to high rates of electric stimulation in retinal ganglion cells. *Journal of Neurophysiology, 106*, 153–162.

Carpenter, R. H. (1973). Contour-like phosphenes from electrical stimulation of the human eye: Some new observations. *Journal of Physiology, 229*, 767–785.

Caspi, A., Dorn, J. D., McClure, K. H., Humayun, M. S., Greenberg, R. J., & McMahon, M. J. (2009). Feasibility study of a retinal prosthesis: Spatial vision with a 16-electrode implant. *Archives of Ophthalmology, 127*, 398–401.

Chader, G. J., Weiland, J., & Humayun, M. S. (2009). Artificial vision: Needs, functioning, and testing of a retinal electronic prosthesis. *Progress in Brain Research, 175*, 317–332.

Chan, L. L., Lee, E. J., Humayun, M. S., & Weiland, J. D. (2011). Both electrical stimulation thresholds and SMI-32-immunoreactive retinal ganglion cell density correlate with age in S334ter line 3 rat retina. *Journal of Neurophysiology, 105*, 2687–2697.

Chang, B., Hawes, N. L., Hurd, R. E., Davisson, T., Nusinowitz, S., & Heckenlively, J. R. (2002). Retinal degeneration mutants in the mouse. *Vision Research, 42*, 517–525.

Chow, A. Y., Chow, V. Y., Packo, K. H., Pollack, J. S., Peyman, G. A., & Schuchard, R. (2004). The artificial silicon retina microchip for the treatment of vision loss from retinitis pigmentosa. *Archives of Ophthalmology, 122*, 460–469.

Daiger, S. P., Bowne, S. J., & Sullivan, L. S. (2007). Perspective on genes and mutations causing retinitis pigmentosa. *Archives of Ophthalmology, 125*, 151–158.

de Balthasar, C., Patel, S., Roy, A., Freda, R., Greenwald, S., Horsager, A., et al. (2008). Factors affecting perceptual thresholds in epiretinal prostheses. *Investigative Ophthalmology & Visual Science, 49*, 2303–2314. doi:10.1167/iovs.07-0696.

Dobelle, W. H., Mladejovsky, M. G., & Girvin, J. P. (1974). Artifical vision for the blind: Electrical stimulation of visual cortex offers hope for a functional prosthesis. *Science, 183*, 440–444.

Dobelle, W. H., Turkel, J., Henderson, D. C., & Evans, J. R. (1979). Mapping the representation of the visual field by electrical stimulation of human visual cortex. *American Journal of Ophthalmology, 88*, 727–735.

Doroudchi, M. M., Greenberg, K. P., Liu, J., Silka, K. A., Boyden, E. S., Lockridge, J. A., et al. (2011). Virally delivered channelrhodopsin-2 safely and effectively restores visual function in multiple mouse models of blindness. *Molecular Therapy, 19*, 1220–1229. doi:10.1038/mt.2011.69.

Fang, Z. P., & Mortimer, J. T. (1991). Alternate excitation of large and small axons with different stimulation waveforms: An application to muscle activation. *Medical & Biological Engineering & Computing, 29*, 543–547.

Fitzgibbon, T., & Reese, B. E. (1996). Organization of retinal ganglion cell axons in the optic fiber layer and nerve of fetal ferrets. *Visual Neuroscience, 13*, 847–861.

Freeman, D. K., Rizzo, J. F., III, & Fried, S. I. (2011). Encoding visual information in retinal ganglion cells with prosthetic stimulation. *Journal of Neural Engineering, 8*, 035005. doi:10.1088/17412560/8/3/035005.

Fried, S. I., Hsueh, H. A., & Werblin, F. S. (2006). A method for generating precise temporal patterns of retinal spiking using prosthetic stimulation. *Journal of Neurophysiology, 95*, 970–978.

Fujikado, T., Morimoto, T., Kanda, H., Kusaka, S., Nakauchi, K., Ozawa, M., et al. (2007). Evaluation of phosphenes elicited by extraocular stimulation in normals and by suprachoroidal–transretinal stimulation in patients with retinitis pigmentosa. *Graefe's Archive for Clinical and Experimental Ophthalmology, 245*, 1411–1419. doi:10.1007/s00417-007-0563-z.

Gargini, C., Terzibasi, E., Mazzoni, F., & Strettoi, E. (2007). Retinal organization in the retinal degeneration 10 (rd10) mutant mouse: A morphological and ERG study. *Journal of Comparative Neurology, 500*, 222–238.

Gerding, H., Benner, F. P., & Taneri, S. (2007). Experimental implantation of epiretinal retina implants (EPI-RET) with an IOL-type receiver unit. *Journal of Neural Engineering, 4*, S38–S49.

Gilbert, C. D., Li, W., & Piech, V. (2009). Perceptual learning and adult cortical plasticity. *Journal of Physiology, 587*, 2743–2751.

Gilbert, C. D., & Wiesel, T. N. (1992). Receptive field dynamics in adult primary visual cortex. *Nature, 356*, 150–152.

Greenberg, R. J. (1998). *Analysis of electrical stimulation of the vertebrate retina work towards a retinal prosthesis* (Doctoral dissertation). The Johns Hopkins University.

Greenwald, S. H., Horsager, A., Humayun, M. S., Greenberg, R. J., McMahon, M. J., & Fine, I. (2009). Brightness as a function of current amplitude in human retinal electrical stimulation. *Investigative Ophthalmology & Visual Science, 50*, 5017–5025.

Hauswirth, W. W., Aleman, T. S., Kaushal, S., Cideciyan, A. V., Schwartz, S. B., Wang, L., et al. (2008). Treatment of Leber congenital amaurosis due to RPE65 mutations by ocular subretinal injection of adeno-associated virus gene vector: Short-term results of a phase I trial. *Human Gene Therapy, 19*, 979–990. doi:10.1089/hum.2008.107.

Hoffer, J. A., Loeb, G. E., & Pratt, C. A. (1981). Single unit conduction velocities from averaged nerve cuff electrode records in freely moving cats. *Journal of Neuroscience Methods, 4*, 211–225.

Horsager, A., Boynton, G. M., Greenberg, R. J., & Fine, I. (2011). Temporal interactions during paired-electrode stimulation in two retinal prosthesis subjects. *Investigative Ophthalmology & Visual Science, 52*, 549–557.

Horsager, A., Greenberg, R. J., & Fine, I. (2009a). Spatiotemporal interactions in retinal prosthesis subjects. *Investigative Ophthalmology & Visual Science, 51*, 1223–1233. doi:10.1167/iovs.09-3746.

Horsager, A., Greenwald, S. H., Weiland, J. D., Humayun, M. S., Greenberg, R. J., McMahon, M. J., et al. (2009b). Predicting visual sensitivity in retinal prosthesis patients. *Investigative Ophthalmology & Visual Science, 50*, 1483–1491. doi:10.1167/iovs.08-2595.

Humayun, M. S., de Juan, E., Jr., Weiland, J. D., Dagnelie, G., Katona, S., Greenberg, R., et al. (1999a). Pattern electrical stimulation of the human retina. *Vision Research, 39*, 2569–2576. doi:10.1016/S0042-6989(99)00052-8.

Humayun, M. S., Dorn, J. D., da Cruz, L., Dagnelie, G., Sahel, J. A., Stanga, P. E., et al. (2012). Interim results from the International Trial of Second Sight's Visual Prosthesis. *Ophthalmology, 119*, 779–788. doi:10.1016/j.ophtha.2011.09.028.

Humayun, M. S., Prince, M., de Juan, E., Jr., Barron, Y., Moskowitz, M., Klock, I. B., et al. (1999b). Morphometric analysis of the extramacular retina from postmortem eyes with retinitis pigmentosa. *Investigative Ophthalmology & Visual Science, 40*, 143–148.

Humayun, M. S., Weiland, J. D., Fujii, G. Y., Greenberg, R., Williamson, R., Little, J., et al. (2003). Visual perception in a blind subject with a chronic microelectronic retinal prosthesis. *Vision Research, 43*, 2573–2581. doi:10.1016/S0042-6989(03)00457-7.

Jensen, R. J., & Rizzo, J. F., III. (2007). Responses of ganglion cells to repetitive electrical stimulation of the retina. *Journal of Neural Engineering, 4*, S1–S6.

Jensen, R. J., & Rizzo, J. F., III. (2008). Activation of retinal ganglion cells in wild-type and rd1 mice through electrical stimulation of the retinal neural network. *Vision Research, 48*, 1562–1568.

Jones, B. W., & Marc, R. E. (2005). Retinal remodeling during retinal degeneration. *Experimental Eye Research, 81*, 123–137.

Jones, B. W., Watt, C. B., Frederick, J. M., Baehr, W., Chen, C. K., Levine, E. M., et al. (2003). Retinal remodeling triggered by photoreceptor degenerations. *Journal of Comparative Neurology, 464*, 1–16. doi:10.1002/cne.10703.

Keck, T., Mrsic-Flogel, T. D., Vaz Afonso, M., Eysel, U. T., Bonhoeffer, T., & Hubener, M. (2008). Massive restructuring of neuronal circuits during functional reorganization of adult visual cortex. *Nature Neuroscience, 11*, 1162–1167.

Kotler, S. (2002). Vision Quest: A half century of artificial-sight research has succeeded. and now this blind man can see. *Wired, 10*, R1–R3.

Lagali, P. S., Balya, D., Awatramani, G. B., Munch, T. A., Kim, D. S., Busskamp, V., et al. (2008). Light-activated channels targeted to ON bipolar cells restore visual function in retinal degeneration. *Nature Neuroscience, 11*, 667–675. doi:10.1038/nn.2117.

LeRoy, C. (1755). Ou l'on rend compte de quelques tentatives que l'on a faites pour guérir plusieurs maladies par l'électricité. *Histoire de L'Académie Royale des Sciences (Paris)*, Mémoires de Mathématique & de Physique *60*, 81–95.

Mahadevappa, M., Weiland, J. D., Yanai, D., Fine, I., Greenberg, R. J., & Humayun, M. S. (2005). Perceptual thresholds and electrode impedance in three retinal prosthesis subjects. *IEEE Transactions on Neural Systems and Rehabilitation Engineering, 13*, 201–206.

Marc, R. E., & Jones, B. W. (2003). Retinal remodeling in inherited photoreceptor degenerations. *Molecular Neurobiology, 28*, 139–147.

Marc, R. E., Jones, B. W., Anderson, J. R., Kinard, K., Marshak, D. W., Wilson, J. H., et al. (2007). Neural reprogramming in retinal degeneration. *Investigative Ophthalmology & Visual Science, 48*, 3364–3371. doi:10.1167/iovs.07-0032.

Marc, R. E., Jones, B. W., Watt, C. B., & Strettoi, E. (2003). Neural remodeling in retinal degeneration. *Progress in Retinal and Eye Research, 22*, 607–655.

Marg, E. (1991). Magnetostimulation of vision: Direct noninvasive stimulation of the retina and the visual brain. *Optometry and Vision Science, 68*, 427–440.

Margalit, E., Maia, M., Weiland, J. D., Greenberg, R. J., Fujii, G. Y., Torres, G., et al. (2002). Retinal prosthesis for the blind. *Survey of Ophthalmology, 47*, 335–356. doi:10.1016/S0039-6257(02)00311-9.

Margolis, D. J., & Detwiler, P. B. (2011). Cellular origin of spontaneous ganglion cell spike activity in animal models of retinitis pigmentosa. *Journal of Ophthalmology, 1*. doi:10.1155/2011/507037.

Margolis, D. J., Newkirk, G., Euler, T., & Detwiler, P. B. (2008). Functional stability of retinal ganglion cells after degeneration-induced changes in synaptic input. *Journal of Neuroscience, 28*, 6526–6536.

Masuda, Y., Horiguchi, H., Dumoulin, S. O., Furuta, A., Miyauchi, S., Nakadomari, S., et al. (2010). Task-dependent V1 responses in human retinitis pigmentosa. *Investigative Ophthalmology & Visual Science, 51*, 5356–5364. doi:10.1167/iovs.09-4775.

Nakauchi, K., Fujikado, T., Kanda, H., Kusaka, S., Ozawa, M., Sakaguchi, H., et al. (2007). Threshold suprachoroidal–transretinal stimulation current resulting in retinal damage in rabbits. *Journal of Neural Engineering, 4*, S50–S57. doi:10.1088/1741-2560/4/1/S07.

Nanduri, D., Fine, I., Horsager, A., Boynton, G. M., Humayun, M. S., Greenberg, R. J., et al. (2012). Frequency and amplitude modulation have different effects on the percepts elicited by retinal stimulation. *Investigative Ophthalmology & Visual Science, 53*, 205–214. doi:10.1167/iovs.11-8401.

Normann, R. A., Greger, B., House, P., Romero, S. F., Pelayo, F., & Fernandez, E. (2009). Toward the development of a cortically based visual neuroprosthesis. *Journal of Neural Engineering, 6*, 035001.

Obeid, I., Veraart, C., & Delbeke, J. (2010). Estimation of phosphene spatial variability for visual prosthesis applications. *Artificial Organs, 34*, 358–365.

Penfield, W., & Jasper, H. (1954). *Epilepsy and the functional anatomy of the human brain*. Boston: Little, Brown.

Pezaris, J. S., & Eskandar, E. N. (2009). Getting signals into the brain: Visual prosthetics through thalamic microstimulation. *Neurosurgical Focus, 27*, E6.

Pezaris, J. S., & Reid, R. C. (2007). Demonstration of artificial visual percepts generated through thalamic microstimulation. *Proceedings of the National Academy of Sciences of the United States of America, 104*, 7670–7675. doi:10.1073/pnas.0608563104.

Phillips, M. J., Otteson, D. C., & Sherry, D. M. (2010). Progression of neuronal and synaptic remodeling in the rd10 mouse model of retinitis pigmentosa. *Journal of Comparative Neurology, 518*, 2071–2089.

Ryu, S. B., Ye, J. H., Goo, Y. S., Kim, C. H., & Kim, K. H. (2011). Decoding of temporal visual information from electrically evoked retinal ganglion cell activities in photoreceptor-degenerated retinas. *Investigative Ophthalmology & Visual Science, 52*, 6271–6278.

Santos, A., Humayun, M. S., de Juan, E., Jr., Greenberg, R. J., Marsh, M. J., Klock, I. B., et al. (1997). Preservation of the inner retina in retinitis pigmentosa: A morphometric analysis. *Archives of Ophthalmology, 115*, 511–515. doi:10.1001/archopht.1997.01100150513011.

Schmidt, E. M., Bak, M. J., Hambrecht, F. T., Kufta, C. V., O'Rourke, D. K., & Vallabhanath, P. (1996). Feasibility of a visual prosthesis for the blind based on intracortical microstimulation of the visual cortex. *Brain, 119*, 507–522.

Schnapf, J. L., Kraft, T. W., & Baylor, D. A. (1987). Spectral sensitivity of human cone photoreceptors. *Nature, 325*, 439–441.

Schneider, K. A., Richter, M. C., & Kastner, S. (2004). Retinotopic organization and functional subdivisions of the human lateral geniculate nucleus: A high-resolution functional magnetic resonance imaging study. *Journal of Neuroscience, 24*, 8975–8985.

Sekirnjak, C., Hottowy, P., Sher, A., Dabrowski, W., Litke, A. M., & Chichilnisky, E. J. (2006). Electrical stimulation of mammalian retinal ganglion cells with multielectrode arrays. *Journal of Neurophysiology, 95*, 3311–3327.

Sekirnjak, C., Hottowy, P., Sher, A., Dabrowski, W., Litke, A. M., & Chichilnisky, E. J. (2008). High-resolution electrical stimulation of primate retina for epiretinal implant design. *Journal of Neuroscience, 28*, 4446–4456.

Sekirnjak, C., Hulse, C., Jepson, L. H., Hottowy, P., Sher, A., Dabrowski, W., et al. (2009). Loss of responses to visual but not electrical stimulation in ganglion cells of rats with severe photoreceptor degeneration. *Journal of Neurophysiology, 102*, 3260–3269. doi:10.1152/jn.00663.2009.

Shannon, R. V. (1992). A model of safe levels for electrical stimulation. *IEEE Transactions on Bio-Medical Engineering, 39*, 424–426.

Shivdasani, M. N., Luu, C. D., Cicione, R., Fallon, J. B., Allen, P. J., Leuenberger, J., et al. (2010). Evaluation of stimulus parameters and electrode geometry for an effective suprachoroidal retinal prosthesis. *Journal of Neural Engineering, 7*, 036008. doi:10.1088/1741-2560/7/3/036008.

Smirnakis, S. M., Brewer, A. A., Schmid, M. C., Tolias, A. S., Schuz, A., Augath, M., et al. (2005). Lack of long-term cortical reorganization after macaque retinal lesions. *Nature, 435*, 300–307. doi:10.1038/nature03495.

Stasheff, S. F. (2008). Emergence of sustained spontaneous hyperactivity and temporary preservation of OFF responses in ganglion cells of the retinal degeneration (rd1) mouse. *Journal of Neurophysiology, 99*, 1408–1421.

Thyagarajan, S., van Wyk, M., Lehmann, K., Lowel, S., Feng, G., & Wassle, H. (2010). Visual function in mice with photoreceptor degeneration and transgenic expression of channelrhodopsin 2 in ganglion cells. *Journal of Neuroscience, 30*, 8745–8758.

Tomita, H., Sugano, E., Isago, H., Hiroi, T., Wang, Z., Ohta, E., et al. (2010). Channelrhodopsin-2 gene transduced into retinal ganglion cells restores functional vision in genetically blind rats. *Experimental Eye Research, 90,* 429–436. doi:10.1016/j.exer.2009.12.006.

Veraart, C., Wanet-Defalque, M. C., Gerard, B., Vanlierde, A., & Delbeke, J. (2003). Pattern recognition with the optic nerve visual prosthesis. *Artificial Organs, 27,* 996–1004.

Walia, S., Fishman, G. A., Edward, D. P., & Lindeman, M. (2007). Retinal nerve fiber layer defects in RP patients. *Investigative Ophthalmology & Visual Science, 48,* 4748–4752.

Wandell, B. A., & Smirnakis, S. M. (2009). Plasticity and stability of visual field maps in adult primary visual cortex. *Nature Reviews. Neuroscience, 10,* 873–884.

Wang, H., Peca, J., Matsuzaki, M., Matsuzaki, K., Noguchi, J., Qiu, L., et al. (2007). High-speed mapping of synaptic connectivity using photostimulation in Channelrhodopsin-2 transgenic mice. *Proceedings of the National Academy of Sciences of the United States of America, 104,* 8143–8148. doi:10.1073/pnas.0700384104.

Wilke, R., Gabel, V. P., Sachs, H., Bartz Schmidt, K. U., Gekeler, F., Besch, D., et al. (2011). Spatial resolution and perception of patterns mediated by a subretinal 16-electrode array in patients blinded by hereditary retinal dystrophies. *Investigative Ophthalmology & Visual Science, 52,* 5995–6003. doi:10.1167/iovs.10-6946.

Woodbury, J. W., & Woodbury, D. M. (1991). Vagal stimulation reduces the severity of maximal electroshock seizures in intact rats: Use of a cuff electrode for stimulating and recording. *Pacing and Clinical Electrophysiology, 14,* 94–107.

Yanai, D., Weiland, J. D., Mahadevappa, M., Greenberg, R. J., Fine, I., & Humayun, M. S. (2007). Visual performance using a retinal prosthesis in three subjects with retinitis pigmentosa. *American Journal of Ophthalmology, 143,* 820–827.

Ye, J. H., & Goo, Y. S. (2007). The slow wave component of retinal activity in rd/rd mice recorded with a multi-electrode array. *Physiological Measurement, 28,* 1079–1088.

Zrenner, E., Bartz-Schmidt, K. U., Benav, H., Besch, D., Bruckmann, A., Gabel, V. P., et al. (2011). Subretinal electronic chips allow blind patients to read letters and combine them to words. *Proceedings of the Royal Society of London: B Biological Science, 278,* 1489–1497. doi: 10.1098/rspb.2010.1747.

Zrenner, E., Stett, A., Weiss, S., Aramant, R. B., Guenther, E., Kohler, K., et al. (1999). Can subretinal microphotodiodes successfully replace degenerated photoreceptors? *Vision Research, 39,* 2555–2567. doi:10.1016/S0042-6989(98)00312-5.

 MARK S. HUMAYUN, JAMES D. WEILAND, AND DEVYANI NANDURI

CONTRIBUTORS

JOSE MANUEL ALONSO   Department of Biological Sciences, SUNY Optometry, New York, New York

JAYAKRISHNA AMBATI   Departments of Ophthalmology and Visual Sciences and Physiology, University of Kentucky, Lexington, Kentucky

MARK L. ANDERMANN   Division of Endocrinology, Department of Medicine, Beth Israel Deaconess Medical Center, Harvard Medical School, Boston, Massachusetts

BARTON L. ANDERSON   School of Psychology, University of Sydney, Sydney, Australia

JAMES R. ANDERSON   John A. Moran Eye Center, Department of Ophthalmology, University of Utah School of Medicine, Salt Lake City, Utah

SARI ANDONI   Center for Perceptual Systems, Section of Neurobiology, School of Biological Sciences, The University of Texas at Austin, Austin, Texas

ALESSANDRA ANGELUCCI   Department of Ophthalmology and Visual Science, Moran Eye Center, University of Utah, Salt Lake City, Utah

RAJENDRA S. APTE   Departments of Ophthalmology and Visual Science, and Developmental Biology, Washington University School of Medicine, St. Louis, Missouri

WOLFGANG BAEHR   Moran Eye Center, University of Utah, Salt Lake City, Utah

MARK F. BEAR   Howard Hughes Medical Institute, The Picower Institute for Learning and Memory, Department of Brain and Cognitive Sciences, Massachusetts Institute of Technology, Cambridge, Massachusetts

ANDREW H. BELL   MRC Cognition and Brain Sciences Unit, University of Cambridge, Cambridge, United Kingdom

LARRY I. BENOWITZ   Laboratories for Neuroscience Research in Neurosurgery and F. M. Kirby Neurobiology Center, Children's Hospital Boston; Departments of Surgery and Ophthalmology and Program in Neuroscience, Harvard Medical School, Boston, Massachusetts

DAVID M. BERSON   Department of Neuroscience, Brown University, Providence, Rhode Island

MARION BETIZEAU   Stem Cell and Brain Research Institute, INSERM U846, Bron, France; Université de Lyon, Université Lyon I, Lyon, France

ALLISON R. BIALAS   Department of Neurology, F. M. Kirby Neurobiology Center, Children's Hospital Boston; Program in Neuroscience, Harvard Medical School, Boston, Massachusetts

RANDOLPH BLAKE   Department of Psychology, Vanderbilt University, Nashville TN, and Brain and Cognitive Sciences, Seoul National University, Seoul, South Korea

DANIEL BLAND   Queensland Brain Institute and School of Information Technology and Electrical Engineering, and ARC Centre of Excellence in Vision Science, University of Queensland, St. Lucia QLD, Australia

PABLO M. BLAZQUEZ   Department of Otolaryngology, Washington University School of Medicine, Saint Louis, Missouri

GUNNAR BLOHM   Centre for Neuroscience Studies and Canadian Action and Perception Network (CAPnet), Departments of Biomedical and Molecular Sciences, Psychology, Mathematics & Statistics and School of Computing, Queen's University, Kingston, Ontario, Canada

STEWART A. BLOOMFIELD   State University of New York College of Optometry, New York, New York

VINCENT BONIN   Neuro-Electronics Research Flanders, Imec, VIB and KU. Leuven, Leuven, Belgium

ALEXANDER BORST   Department of Systems and Computational Neurobiology, Max-Planck-Institute of Neurobiology, Martinsried, Germany

CONRADO BOSMAN   Center for Neuroscience, University of Amsterdam, Amsterdam, the Netherlands

GEOFFREY M. BOYNTON   Department of Psychology, University of Washington, Seattle, Washington

DAVID H. BRAINARD   Department of Psychology, University of Pennsylvania, Philadelphia, Pennsylvania

CURTIS R. BRANDT   Departments of Ophthalmology and Visual Sciences, and Medical Microbiology and Immunology, and McPherson Eye Research Institute, University of Wisconsin School of Medicine and Public Health, Madison, Wisconsin

FARRAN BRIGGS   Department of Physiology and Neurobiology, Geisel School of Medicine, Dartmouth University, Lebanon, New Hampshire

ANGELA M. BROWN   College of Optometry, Ohio State University, Columbus, Ohio

STEVEN L. BUCK   Department of Psychology, University of Washington, Seattle, Washington

ROBERT W. BURGESS   The Jackson Laboratory, Bar Harbor, Maine

ANDREAS BURKHALTER   Department of Anatomy and Neurobiology, Washington University School of Medicine, St. Louis, Missouri

MARIE E. BURNS   Center for Neuroscience and Departments of Cell Biology and Human Anatomy, and Ophthalmology and Vision Science, University of California, Davis, Davis, California

DAVID BURR   Department of Psychology, University of Florence, Florence, Italy

ANDREW J. CALDER   ARC Center of Excellence in the Study of Cognition and its Disorders, School of Psychology, University of Western Australia, and Medical Research Council, Cognition and Brain Sciences Unit, Cambridge, United Kingdom

EDWARD M. CALLAWAY   Systems Neurobiology Laboratories, The Salk Institute for Biological Studies, La Jolla, California

LEO M. CHALUPA   Office of Vice President for Research, Department of Pharmacology and Physiology, George Washington University, Washington, D.C.

KAREN L. CHENEY   School of Biological Sciences, University of Queensland, St. Lucia QLD, Australia

KEN CHENG   Department of Biological Sciences, Macquarie University, Sydney, Australia

E. J. CHICHILNISKY   The Salk Institute, La Jolla, California

JASON E. COLEMAN   Howard Hughes Medical Institute, The Picower Institute for Learning and Memory, Department of Brain and Cognitive Sciences, Massachusetts Institute of Technology, Cambridge, Massachusetts, and Child Health Research Institute, Department of Pediatrics, University of Florida College of Medicine, Gainesville, Florida

RYAN CONSTANTINE   Moran Eye Center, University of Utah, Salt Lake City, Utah

DYLAN F. COOKE   Center for Neuroscience, Department of Psychology, University of California, Davis, Davis, California

J. DOUGLAS CRAWFORD   Centre for Vision Research, Canadian Action and Perception Network (CAPnet), Departments of Psychology, Biology, and Kinesiology and Health Sciences, and Neuroscience Graduate Diploma Program, York University, Toronto, Ontario, Canada

JOANNA D. CROOK   Department of Biological Structure and The National Primate Research Center, University of Washington, Seattle, Washington

HOLK CRUSE   Biological Cybernetics, Faculty for Biology, University of Bielefeld, Germany

CHRISTINE A. CURCIO   Department of Ophthalmology, University of Alabama School of Medicine, Birmingham, Alabama

DENNIS M. DACEY   Department of Biological Structure and The National Primate Research Center, University of Washington, Seattle, Washington

PIERRE M. DAYE   Laboratory of Sensorimotor Research, National Eye Institute, National Institutes of Health, Bethesda, Maryland

COLETTE DEHAY   Stem Cell and Brain Research Institute, INSERM U846, Bron, France; Université de Lyon, Université Lyon I, Lyon, France

SILMARA de LIMA   Laboratories for Neuroscience Research in Neurosurgery and F. M. Kirby Neurobiology Center, Children's Hospital Boston, and Departments of Surgery and Ophthalmology and Program in Neuroscience, Harvard Medical School, Boston, Massachusetts

JONATHAN B. DEMB   Department of Ophthalmology and Visual Science; and Department of Cellular and Molecular Physiology, Yale University, New Haven, Connecticut

STEVEN H. DeVRIES   Departments of Ophthalmology and Physiology, Northwestern University Feinberg School of Medicine, Chicago, Illinois

DANIEL D. DILKS   Department of Psychology, Emory University, Atlanta, Georgia

CHARLES J. DUFFY   Departments of Neurology, Neurobiology and Anatomy, Ophthalmology, Brain and Cognitive Sciences, and the Center for Visual Science, The University of Rochester Medical Center, Rochester, New York

BASIL EL JUNDI   Department of Biology, Animal Physiology, University of Marburg, Marburg, Germany

MARLA B. FELLER   Department of Molecular and Cell Biology and Helen Wills Neuroscience Institute, University of California, Berkeley, Berkeley, California

DAVID J. FIELD   Department of Psychology, Cornell University, Ithaca, New York

SILVIA C. FINNEMANN   Department of Biological Sciences, Fordham University, Bronx, New York

JEANNE M. FREDERICK   Moran Eye Center, University of Utah, Salt Lake City, Utah

RALPH D. FREEMAN   School of Optometry, University of California, Berkeley, Berkley, California

MARTIN FRIEDLANDER   Department of Cell Biology, The Scripps Research Institute; Division of Ophthalmology, Department of Surgery, Scripps Clinic, La Jolla, California

PASCAL FRIES   Ernst Strüngmann Institute (ESI) for Neuroscience in Cooperation with Max Planck Society, Frankfurt, Germany, and Donders Institute for Brain, Cognition and Behaviour, Nijmegen, the Netherlands

NEERAJ J. GANDHI   Departments of Bioengineering, Otolaryngology, Neuroscience, and Center for the Neural Basis of Cognition (CNBC), University of Pittsburgh, Pittsburgh, Pennsylvania

ENQUAN GAO   Department of Anatomy and Neurobiology, Washington University School of Medicine, St. Louis, Missouri

ANDREW M. GARRETT   The Jackson Laboratory, Bar Harbor, Maine

KARL R. GEGENFURTNER   Department of Psychology, University of Giessen, Giessen, Germany

JOY J. GENG   Department of Psychology, Center for Mind and Brain, University of California, Davis, Davis, California

CHARLES D. GILBERT   Laboratory of Neurobiology, The Rockefeller University, New York, New York

JAMES R. GOLDEN   Department of Psychology, Cornell University, Ithaca, New York

MARK S. GOLDMAN   Department of Neurobiology, Physiology and Behavior, Department of Ophthalmology and Vision Science, University of California, Davis, Davis, California

ADAM GOLDRING   Center for Neuroscience, Department of Psychology, University of California, Davis, Davis, California

MELVYN A. GOODALE   The Brain and Mind Institute, University of Western Ontario, London, ON, Canada

JACQUELINE GOTTLIEB   Department of Neuroscience and the Kavil Institute for Brain Research, Columbia University, New York, New York

R. W. GUILLERY   MRC Anatomical Neuropharmacology Unit, University of Oxford, Oxford, United Kingdom

JASON HABERMAN   Department of Psychology, Harvard University, Cambridge, Massachusetts

AARON M. HAMBY   Department of Molecular and Cell Biology and Helen Wills Neuroscience Institute, University of California, Berkeley, Berkeley, California

JAMES HANDA   Wilmer Ophthalmological Institute, Johns Hopkins Hospital, Baltimore, Maryland

MARY ELIZABETH HARTNETT   Moran Eye Center; Departments of Ophthalmology and Visual Sciences: Neurobiology and Anatomy; and Pediatrics, University of Utah, Salt Lake City, Utah

SAMER HATTAR   Department of Biology, Johns Hopkins University, Baltimore, Maryland

MICHAEL HAWKEN   Center for Neural Science, New York University, New York, New York

ANTHONY HAYES   Division of Psychology, Nanyang Technological University, Singapore

ROBERT F. HESS   McGill Vision Research, Department of Ophthalmology, McGill University, Montreal, Québec, Canada

ARNOLD J. HEYNEN   Howard Hughes Medical Institute, The Picower Institute for Learning and Memory, Department of Brain and Cognitive Sciences, Massachusetts Institute of Technology, Cambridge, Massachusetts

JUDITH A. HIRSCH   Department of Biological Sciences and Neuroscience Graduate Program, University of Southern California, Los Angeles, California

HEIDI J. HOFER   College of Optometry University of Houston, Houston, Texas

UWE HOMBERG   Department of Biology, Animal Physiology, University of Marburg, Marburg, Germany

MARK S. HUMAYUN   Department of Biomedical Engineering and Doheny Eye Institute, University of Southern California, Los Angeles, California

ANDREW T. ISHIDA   Department of Neurobiology, Physiology and Behavior, Department of Ophthalmology and Vision Science, University of California, Davis, Davis, California

GERALD H. JACOBS   Department of Psychological and Brain Sciences, University of California, Santa Barbara, Santa Barbara, California

UDAY K. JAGADISAN   Department of Bioengineering and Center for the Neural Basis of Cognition (CNBC), University of Pittsburgh, Pittsburgh, Pennsylvania

JAN JASTORFF   Laboratorium voor Neuro-en Psychofysiologie, KU Leuven Medical School, Leuven, Belgium

ELIZABETH JOHNSON   Duke Institute for Brain Sciences, Duke University, Durham, North Carolina

ALAN JOHNSTON   Cognitive, Perceptual and Brain Sciences, University College London, London, United Kingdom

BRYAN W. JONES   John A. Moran Eye Center, Department of Ophthalmology, University of Utah School of Medicine, Salt Lake City, Utah

JON H. KAAS   Department of Psychology, Vanderbilt University, Nashville, Tennessee

NANCY KANWISHER   McGovern Institute for Brain Research, Massachusetts Institute of Technology, Cambridge, Massachusetts

EHUD KAPLAN   The Neuroscience Department and the Friedman Brain Institute, The Mount Sinai School of Medicine, New York, New York

SABINE KASTNER   Princeton Neuroscience Institute and Department of Psychology, Princeton University, Princeton, New Jersey

JEREMY N. KAY   Departments of Neurobiology and Ophthalmology, Duke University, Durham, North Carolina

HENRY KENNEDY   Stem Cell and Brain Research Institute, INSERM U846, Bron, France; Université de Lyon, Université Lyon I, Lyon, France

DANIEL KERSTEN   Department of Psychology, University of Minnesota, Minneapolis MN; Department of Brain and Cognitive Engineering, Korea University, Seoul, South Korea

PAUL S. KHAYAT   Cognitive Neurophysiology Laboratory, Department of Physiology, McGill University, Montreal, Québec, Canada

FREDERICK A. A. KINGDOM   Vision Research, McGill University, Montreal, Québec, Canada

LYNNE KIORPES   Center for Neural Science, New York University, New York, New York

DANIEL C. KIPER   Institute of Neuroinformatics, University of Zurich and Swiss Federal Institute of Technology Zurich, Zurich, Switzerland

MARK E. KLEINMAN   Department of Ophthalmology and Visual Sciences, University of Kentucky, Lexington, Kentucky

ELIANA M. KLIER   Department of Neuroscience, Baylor College of Medicine, Houston, Texas

MICHAEL KNIGHT   Queensland Brain Institute and School of Information Technology and Electrical Engineering, and ARC Centre of Excellence in Vision Science, University of Queensland, St. Lucia QLD, Australia

RICHARD H. KRAMER   Department of Molecular and Cell Biology, University of California, Berkeley, California

RICHARD J. KRAUZLIS   Laboratory of Sensorimotor Research, National Eye Institute, National Institutes of Health, Bethesda, Maryland

LEAH KRUBITZER   Center for Neuroscience, Department of Psychology, University of California, Davis, Davis, California

TAKAAKI KUWAJIMA   Department of Pathology and Cell Biology, Columbia University, College of Physicians and Surgeons, New York, New York

MICHAEL F. LAND   School of Life Sciences, University of Sussex, Brighton, United Kingdom

MICHAEL S. LANDY   Department of Psychology and Center for Neural Science, New York University, New York, New York

J. SCOTT LAURITZEN   John A. Moran Eye Center, Department of Ophthalmology, University of Utah School of Medicine, Salt Lake City, Utah

DENNIS M. LEVI   School of Optometry and Helen Wills Neuroscience Institute, University of California, Berkeley, Berkeley, California

MICHAEL S. LEWICKI   Department of Electrical Engineering and Computer Science, Case Western Reserve University, Cleveland, OH

WU LI   State Key Laboratory of Cognitive Neuroscience and Learning, Beijing Normal University, Beijing, China

DELWIN T. LINDSEY   Department of Psychology, Ohio State University, Mansfield, Ohio

GERARD A. LUTTY   Wilmer Ophthalmological Institute, Johns Hopkins Hospital, Baltimore, Maryland

STEPHEN L. MACKNIK   Department of Neurobiology, Barrow Neurological Institute, Phoenix, Arizona

ROBERT E. MacLAREN   Nuffield Laboratory of Ophthalmology, University of Oxford, Oxford, United Kingdom; Oxford Eye Hospital NIHR Biomedical Research Center, Oxford, United Kingdom; Moorfields Eye Hospital NIHR Biomedical Research Center, London, United Kingdom

PATRICIA F. MANESS   Department of Biochemistry and Biophysics, University of North Carolina School of Medicine, Chapel Hill, North Carolina

GEORGE R. MANGUN   Center for Mind and Brain, Departments of Psychology and Neurology, University of California, Davis, Davis, California

ROBERT E. MARC   John A. Moran Eye Center, Department of Ophthalmology, University of Utah School of Medicine, Salt Lake City, Utah

JUSTIN MARSHALL   Queensland Brain Institute, University of Queensland, St. Lucia QLD, Australia

SUSANA MARTINEZ-CONDE   Department of Neurobiology, Barrow Neurological Institute, Phoenix, Arizona

JULIO C. MARTINEZ-TRUJILLO   Cognitive Neurophysiology Laboratory, Department of Physiology, McGill University, Montreal, QC, Canada

CAROL MASON   Departments of Pathology and Cell Biology, Neuroscience, and Ophthalmology, Columbia University, College of Physicians and Surgeons, New York, New York

STEPHEN C. MASSEY   Department of Ophthalmology and Visual Science, University of Texas at Houston, Houston, Texas

KIMBERLEY McALLISTER   Center for Neuroscience, Departments of Neurology and Neurobiology, Physiology & Behavior, University of California, Davis, California

MICHAEL B. McCAMY   Department of Neurobiology, Barrow Neurological Institute, Phoenix, Arizona

D. SCOTT McLEOD   Wilmer Ophthalmological Institute, Johns Hopkins Hospital, Baltimore, Maryland

JOHANNA H. MEIJER   Department of Neurophysiology, Leiden University Medical School, Leiden, the Netherlands

MARKUS MEISTER   Division of Biology, California Institute of Technology, Pasadena, California

NIKOLAOS MELLIOS   Picower Institute for Learning and Memory, Department of Brain and Cognitive Sciences, Massachusetts Institute of Technology, Cambridge, Massachusetts

RANDOLF MENZEL   Institut für Neurobiologie, Freie Universität Berlin, Berlin, Germany

NEIL R. MILLER   Wilmer Eye Institute, Johns Hopkins Hospital, Baltimore, Maryland

STEPHEN L. MILLS   Department of Ophthalmology and Visual Science, University of Texas at Houston, Houston, Texas

RICHARD J. D. MOORE   Queensland Brain Institute and School of Information Technology and Electrical Engineering, and ARC Centre of Excellence in Vision Science, University of Queensland, St. Lucia QLD, Australia

M. CONCETTA MORRONE   Dipartimento di Scienze Fisiologiche Università Degli Studi di Pisa, and IRCCS Fondazione Stella-Mari, Pisa, Italy

J. ANTHONY MOVSHON   Center for Neural Science, New York University, New York, New York

IAN J. MURRAY   Faculty of Life Sciences, University of Manchester, Manchester, United Kingdom

MICHAEL J. MUSTARI   Department of Ophthalmology, Washington National Primate Research Center, University of Washington, Seattle, Washington

IKUE NAGAKURA   Picower Institute for Learning and Memory, Department of Brain and Cognitive Sciences, Massachusetts Institute of Technology, Cambridge, Massachusetts

DEVYANI NANDURI   Department of Biomedical Engineering, University of Southern California, Los Angeles, California

CRISTOPHER M. NIELL   Institute of Neuroscience and Department of Biology, University of Oregon, Eugene, Oregon

AUDE OLIVA   Computer Science and Artificial Intelligence Laboratory (CSAIL), Massachusetts Institute of Technology, Cambridge, Massachusetts

BRUNO A. OLSHAUSEN   Helen Wills Neuroscience Institute, School of Optometry, and Redwood Center for Theoretical Neuroscience, University of California, Berkeley, Berkeley, California

SEIJI ONO   Washington National Primate Research Center, University of Washington, Seattle, Washington

LANCE M. OPTICAN   Laboratory of Sensorimotor Research, National Eye Institute, National Institutes of Health, Bethesda, Maryland

GUY A. ORBAN   Department of Neuroscience, University of Parma, Parma, Italy

ORIN S. PACKER   Department of Biological Structure and The National Primate Research Center, University of Washington, Seattle, Washington

ANDREW J. PARKER   Department Physiology, Anatomy and Genetics, University of Oxford, Oxford, United Kingdom

TATIANA PASTERNAK   Department of Neurobiology and Anatomy, University of Rochester, Rochester, New York

ANGEL M. PASTOR   Departamento de Fisiología Facultad de Biología, Universidad de Sevilla, Sevilla, Spain

NICHOLAS J. PRIEBE   Center for Perceptual Systems, Section of Neurobiology, School of Biological Sciences, The University of Texas at Austin, Austin, Texas

EDWARD N. PUGH, JR.   Center for Neuroscience and Departments of Cell Biology and Human Anatomy, and Physiology and Membrane Biology, University of California, Davis, Davis, California

CHRISTIAN QUAIA   Laboratory of Sensorimotor Research, National Eye Institute, National Institutes of Health, Bethesda, Maryland

ANA RADONJIĆ   Department of Psychology, University of Pennsylvania, Philadelphia, Pennsylvania

RITHWICK RAJAGOPAL   Department of Ophthalmology and Visual Science, Washington University School of Medicine, St. Louis, Missouri

VINOD RANGANATHAN   The Wilmer Eye Institute, Johns Hopkins University School of Medicine, Baltimore, Maryland

GREGG H. RECANZONE   Center for Neuroscience, Department of Neurobiology, Physiology and Behavior, University of California, Davis, Davis, California

GILLIAN RHODES   ARC Center of Excellence in the Study of Cognition and its Disorders, School of Psychology, University of Western Australia, Crawley WA, Australia

FRED RIEKE   Department of Physiology and Biophysics, University of Washington, Seattle, Washington

DARIO L. RINGACH   Departments of Neurobiology and Psychology, David Geffen School of Medicine, University of California, Los Angeles, Los Angeles, California

BOTOND ROSKA   Friedrich Miescher Institute for Biomedical Research, Basel, Switzerland

LINDA RUGGIERO   Department of Biological Sciences, Fordham University, Bronx, New York

ALAPAKKAM P. SAMPATH   Zilkha Neurogenetic Institute, Department of Physiology and Biophysics, University of Southern California, Keck School of Medicine, Los Angeles, California

JOSHUA R. SANES   Center for Brain Science and Department of Molecular and Cellular Biology, Harvard University, Cambridge, Massachusetts

YUKA SASAKI   Department of Cognitive, Linguistic, and Psychological Sciences, Brown University, Providence, Rhode Island

JEFFREY D. SCHALL   Vanderbilt Vision Research Center, Department of Psychology, Vanderbilt University, Nashville, Tennessee

CLIFTON M. SCHOR   Vision Science Group, School of Optometry, University of California at Berkeley, Berkeley, California

AASEF G. SHAIKH   Department of Neurology, Emory School of Medicine, Atlanta Georgia

ROBERT SHAPLEY   Center for Neural Science, New York University, New York, New York

S. MURRAY SHERMAN   Department of Neurobiology, The University of Chicago, Chicago, Illinois

S. SHUSHRUTH   HHMI and Department of Neuroscience, Columbia University, New York, New York

MANDEEP S. SINGH   Nuffield Laboratory of Ophthalmology, University of Oxford, Oxford, United Kingdom

DEAN SOCCOL   Queensland Brain Institute and School of Information Technology and Electrical Engineering, and ARC Centre of Excellence in Vision Science, University of Queensland, St. Lucia QLD, Australia

DAVID C. SOMERS   Department of Psychology and Center for Neuroscience, Boston University, Boston, Massachusetts

FRIEDRICH T. SOMMER   Redwood Center for Theoretical Neuroscience, University of California, Berkeley, Berkeley, California

MARC A. SOMMER   Department of Biomedical Engineering, Center for Cognitive Neuroscience, and Duke Institute for Brain Sciences, Duke University, Durham, North Carolina

CHARLES SPENCE   Crossmodal Research Laboratory, Department of Experimental Psychology, University of Oxford, Oxford, United Kingdom

OLAF SPORNS   Department of Psychological and Brain Sciences, Indiana University, Bloomington, Indiana

MANDYAM V. SRINIVASAN   Queensland Brain Institute and School of Information Technology and Electrical Engineering, and ARC Centre of Excellence in Vision Science, University of Queensland, St. Lucia QLD, Australia

BETH STEVENS   Department of Neurology, F. M. Kirby Neurobiology Center, Children's Hospital Boston; Program in Neuroscience, Harvard Medical School, Boston, Massachusetts

MRIGANKA SUR   Picower Institute for Learning and Memory, Department of Brain and Cognitive Sciences, Massachusetts Institute of Technology, Cambridge, Massachusetts

HARVEY A. SWADLOW   Department of Psychology, University of Connecticut, Storrs, Connecticut

TIMOTHY D. SWEENY   Department of Psychology, University of California, Berkeley, Berkeley, California

ABIGAIL L. D. TADENEV   The Jackson Laboratory, Bar Harbor, Maine

JOSEPH S. TAKAHASHI   Department of Neuroscience, Howard Hughes Medical Institute, University of Texas Southwestern Medical Center, Dallas, Texas

ANDREW TAN   Center for Perceptual Systems, Section of Neurobiology, School of Biological Sciences, The University of Texas at Austin, Austin, Texas

SAUL THURROWGOOD   Queensland Brain Institute and School of Information Technology and Electrical Engineering, and ARC Centre of Excellence in Vision Science, University of Queensland, St. Lucia QLD, Australia

JOHN B. TROY   Department of Biomedical Engineering, Northwestern University, Evanston, Illinois

LESLIE G. UNGERLEIDER   Laboratory of Brain and Cognition, National Institute of Mental Health, Bethesda, Maryland

W. MARTIN USREY   Center for Neuroscience, University of California, Davis, Davis, California

VISHAL S. VAINGANKAR   StumbleUpon, Inc., San Francisco, California

BÉLA VÖLGYI   Department of Ophthalmology, New York University School of Medicine, New York, New York

RÜDIGER VON DER HEYDT   Department of Neuroscience and Krieger Mind/Brain Institute, Johns Hopkins University, Baltimore, Maryland

QING WANG   Departments of Pathology and Cell Biology, and Neuroscience, Columbia University, College of Physicians and Surgeons, New York, New York

QUANXIN WANG   Department of Anatomy and Neurobiology, Washington University School of Medicine, St. Louis, Missouri

XIN WANG   Computational Neurobiology Laboratory, The Salk Institute for Biological Studies, La Jolla, California

YANBIN V. WANG   Department of Ophthalmology and Visual Science; and Department of Cellular and Molecular Physiology, Yale University, New Haven, Connecticut

TAKEO WATANABE   Department of Cognitive, Linguistic and Psychological Sciences, Brown University, Providence, Rhode Island

CARL B. WATT   John A. Moran Eye Center, Department of Ophthalmology, University of Utah School of Medicine, Salt Lake City, Utah

MICHAEL A. WEBSTER   Department of Psychology, University of Nevada, Reno, Reno, Nevada

RÜDIGER WEHNER   Brain Research Institute, University of Zürich, Switzerland

JAMES D. WEILAND   Department of Biomedical Engineering and Doheny Eye Institute, University of Southern California, Los Angeles, California

JOHN S. WERNER   Department of Ophthalmology & Vision Science, and Department of Neurobiology, Physiology and Behavior, University of California, Davis, Davis, California

PETER D. WESTENSKOW   Department of Cell Biology, The Scripps Research Institute, La Jolla, California

DAVID WHITNEY   Department of Psychology, University of California, Berkeley, Berkeley, California

FRANCES WILKINSON   Centre for Vision Research, York University, Toronto, ON, Canada

DAVID R. WILLIAMS   Center for Visual Science, University of Rochester, Rochester, New York

HUGH R. WILSON   Centre for Vision Research, York University, Toronto, ON, Canada

THILO WOMELSDORF   Centre for Vision Research, Department of Biology, York University, Toronto, ON, Canada

ANDRZEJ WRÓBEL   Department of Neurophysiology, Nencki Institute of Experimental Biology, Warsaw, Poland

ROBERT H. WURTZ   Laboratory of Sensorimotor Research, National Eye Institute, National Institutes of Health, Bethesda, Maryland

ALAN YUILLE   Departments of Statistics and Psychology, University of California, Los Angeles, Los Angeles, California; Department of Brain and Cognitive Engineering, Korea University, Seoul, South Korea

DONALD J. ZACK   The Wilmer Eye Institute, Departments of Neuroscience, Molecular Biology and Genetics, and McKusick-Nathans Institute of Genetic Medicine, Johns Hopkins University School of Medicine, Baltimore, Maryland, and Institut de la Vision, Université Pierre et Marie Curie, Paris, France

DAVID S. ZEE   Departments of Neurology, Neuroscience, Ophthalmology and Otolaryngology—Head and Neck Surgery, The Johns Hopkins University School of Medicine, Johns Hopkins Hospital, Baltimore, Maryland

HOUBIN ZHANG   Moran Eye Center, University of Utah, Salt Lake City, Utah

INDEX

bottom-up influences on, 854–855
ecological tasks and, 856
baseball and cricket, 856–857
domestic tasks, 856
driving, 856
magic, 857
during free-viewing, 851f
in invertebrates, 1147–1148
measuring, 851–852, 854f
contact lens methods, 852
electro-oculography (EOG), 852,
854f
optical/feature recognition
methods, 853
measuring three-dimensional,
879–880, 880f
natural, 851
neural and mechanical control
brain stem, 3-D signal, 882–885
cortical, 2-D signals, 882
mechanical factors, 885
two- to three-dimensional
transformation, 885–886
optimal sampling strategy, 859–860
pathological, 860
reflex, 850
top-down influences on, 855
visual consequences of three-
dimensional (head and), 886
accounting for 3-D eye orientation
in vision, memory, and
movement, 889–890
consequences for binocular vision,
887–889
simple geometry effects, 886–887
Eye pursuit movements, motor
learning in, 969, 970f
Eye tracking, future of, 853
Eye tracking methods. *See also* Eye
movements: measuring
capabilities, pitfalls, and advantages
of, 854f
Eyes, 866f
in invertebrates, 1145f, 1150 (*see also*
specific topics)
apposition, 1144–1146, 1145f
compound, 1144–1147
corneal, 1144, 1145f
mirror, 1144, 1145f
single-chambered, 1143–1144,
1145f
in primates, 1235–1236

F2 terminals, 305
Face aftereffects, 535, 539
Face-coding network, 711
coding expression, 716
holistic coding, 716–717
implementation in face network,
719
norm-based coding, 717–719
coding identity, 711
holistic coding, 711–713
implementation in face network,
715–716
norm-based coding, 713–715

Face perception, 734f. *See also* Fusiform
face area (FFA)
perceiving ensembles of faces, 700
Face space, 713–714, 714f
Facial expression, 718, 718f
Facilitation, surround, 433, 434. *See also*
Collinear facilitation
orientation tuning of, 430, 431f
Facilitatory modulations, 426–427. *See*
also Collinear facilitation;
Surround suppression
center and surround stimuli that can
evoke, 426–427, 428f
FAK (focal adhesion kinase), 32, 33
Familiarity, neurons sensitive to, 230
Far surround, 426f, 427
Far-surround facilitation, 428f, 429,
436
Far-surround modulation, 429,
433–435
Far-surround stimulation, 427, 428f,
429, 430f
Far-surround suppression, 427, 429,
430f, 431, 432f, 433, 434, 436,
438–439
Fasciculation, selective, 1297–1298,
1297f
Fastigial oculomotor region (FOR),
973–974
Faulty integration theory, 684, 687, 688
Fear expression on faces, 718, 718f
Feature-based attention (FBA),
1082–1083, 1107
beyond visual cortex, 1116–1117
effects on responses of MT neurons,
1110–1111, 1111f, 1112f, 1113,
1114f, 1115–1116
effects on responses of V4 neurons,
1108–1110, 1109f
with multiple stimuli in receptive
field, 1113, 1115–1116
toward a unified theory of,
1117–1118
Feature binding by synchrony in brain,
1281
Feature binding problem in vision,
1280–1281
Feature detector ganglion cells
(feature detectors), 171–172
direction-selective (DS) cells, 173f,
176–177 (*see also* Direction-
selective (DS) ganglion cells)
diverse circuits using common
components, 177
downstream processing, 178–179
looming detectors, 173f, 175–176
object-motion-sensitive (OMS) cells,
173f, 174–175
pixel sensors vs. feature detectors,
171–172
retinal circuits leading to different,
172–175, 174f, 177
visual features and ecology, 177–178
visual features extracted, 172, 173f,
175, 176
Y cells, 172–174, 173f

Feature extraction, 769
Feature matching (FM), 1107, 1108,
1110, 1111, 1115–1118
Feature processing, clinical tests of,
179
Feature-similarity gain (FSG),
1107–1109, 1117, 1118
effects on responses of MT neurons,
1110–1111, 1112f, 1113, 1114t,
1115–1116
Feature-similarity gain (FSG) model,
1082, 1110, 1117
Feature tracking, 766
Feedback. *See also* Horizontal cell (HC)
feedback transmission
corticogeniculate, 315, 316, 320–321
Feedback connections and surround
modulation, 434–435
Feedback control of saccades, 980,
981f
Feedback mechanisms, negative, 12. *See*
also under Horizontal cell (HC)
feedback transmission
Feedback pathways, modulation of
attentional enhancement by,
343–345, 344f
Feedback (homeostatic) plasticity, 1359
intracellular molecules that mediate,
1360f
mechanisms of, 1362
scaling of intrinsic excitability,
1363
synaptic scaling, 1362–1363
molecules and pathways that link
feedforward plasticity and,
1363–1364
Feedforward and feedback connections
in thalamus, 301, 302f
Feedforward connections and
surround modulation, 432–433
Feedforward model of orientation
selectivity in primary visual
cortex, 368, 368f, 370
Feedforward (Hebbian) plasticity, 1359
intracellular molecules that mediate,
1360f
mechanisms of, 1359–1360
changes in NMDA receptor
subunit, 1360–1361
intracellular signaling, 1361
layer-specific, 1361
modulation by inhibition, 1361–
1362
structural modification, 1362
molecules and pathways that link
feedback plasticity and, 1363–
1364
Feedforward stages. *See* Filter, rectify,
filter
FEFsem (smooth pursuit part of the
frontal eye fields), 898–901,
905
FFA (fusiform face area), 232, 715,
734–735, 737–739, 1083
Fibroblast growth factor 2 (FGF2),
1582

of one type cover retina with a
regular mosaic, 163, 165f
survival of, 1392–1393
Bcl-2 family members, 1393
caspase inhibition, 1393, 1394f
nitric oxide (NO) and reactive
oxygen species (ROS), 1393–
1394
trophic factors, 1393
voltage-gated potassium channels,
1393
Ganglionic eminence (GE), 1409,
1410
Gap junctions. *See* Neuronal gap
junctions in retina
Gaze command centers, 871–873
Gene delivery, 1577
considerations for ocular, 1587–1589
promoters and long-term
transgene expression, 1589
route of delivery, 1587
methods for retinal, 1583–1587
strategies for ocular, 1578–1581
targeting, 1587–1589
targeting the common pathological
event, 1581–1583
Gene delivery vectors, viral, 1583
Gene expression, axon regeneration
and changes in neuronal,
1396–1397, 1396f
Gene therapy for retinal degeneration.
See Gene delivery
Generalized flash suppression (GFS),
827
Geniculocortical afferents, 432. *See also*
Lateral geniculate nucleus
Geniculostriate pathway, 399–400
Geographic atrophy (GA), 1563, 1564,
1570
Germinal zones, cortical, 1408–1410
in primate and mouse, at different
developmental stages, 1411,
1412f
VZ and SVZ, 1407–1408, 1410, 1411,
1414
Gestalt-anchoring models of brightness
and lightness, 503
Gestalt grouping, 1071
Gestalt laws, 665
Gestalt psychology, 665, 1001, 1002
Gibson, E. J., 991
Gibson, J. J., 827, 1191
Glass patterns, 607–608, 609f, 610, 620,
623, 770, 1425–1427, 1426f
Glaucoma, 1612
Glia (of visual system), 1369–1372,
1370f, 1381–1382
distribution in layers of retina,
1370f
how they are generated, 1372–1373
and immune molecules
role in visual experience-
dependent plasticity, 1380–1381
sculpt developing circuits in visual
system, 1377
types of, 1369, 1370f

Glial-derived neurotrophic factor
(GDNF), 1582
Glial fibrillary acidic protein (GFAP),
1370, 1371
Glial palisade, 1373
Global scene-emergent features
representation (scenes), 726
Global visual functions. *See* Higher-
order visual functions
Globs, 521
Glutamate release
presynaptic modulation of, 41–42
properties, 41
rod photovoltage and, 45
Glutamatergic driver, 262f
Glutamatergic inputs, classification of,
267
drivers and modulators, 267–268,
268f
effects of modulation, 268
Glycinergic amacrine cells (GACs),
113–115
Glycosaminoglycans (GAGs), 1537
Glypicans, 1376
Grasp posture, Ponzo illusion and,
1042, 1043f
Grasping. *See* Reaching and grasping
Grating stimuli, stereoscopic, 657–658,
657f
Green-red opponency. *See* Red-green
opponency
Grip aperture, 1041, 1042f
maximum, 1032f
GRK1, 11
GRK7, 11
Group dependencies, modeling,
1254–1255, 1255f
Growth cone behaviors at optic chiasm
midline, 1320
GTP (guanosine triphosphate), 12,
42–44
GTPases, 20, 43
Guanylate cyclase, calcium regulation
of, 13
Guanylate cyclase-activating proteins
(GCAPs), 12–14, 46–47
Guidance molecules that direct retinal
ganglion cell axon divergence,
1320
avoiding the midline, 1320–1321,
1321f
crossing the midline, 1321–1322,
1321f
axon fasciculation and, 1322
Guiding fixations, 856

Half-angle rule, 879, 880f
Halteres, 1199
Hartline, H. K., 3, 63, 218, 367
Hassenstein, B., 1191
Head-centric coordinates, 812, 817,
818, 820, 821
mapping disparity from retinal
angles to linear depth in, 817
Head-centric disparity, 810–812
producing a, 811–812, 812f

Head-centric disparity coding scheme,
812
Head-centric disparity coding vs.
retinal disparity coding, 811,
811f
Hebbian plasticity. *See* Feedforward
(Hebbian) plasticity
Hecht, S., 3, 44
Heeger/Bergen (HB) model, 646
Helmholtz, Hermann von, 1, 401, 403,
935, 950, 1003, 1097, 1247,
1274
Hemineglect, visuospatial, 1087
Hering, Ewald, 478, 499, 512–514, 516,
519–521, 524–526, 809–810
Hering color space, diagram of, 513f
Hering sensations, 525–526
Hering's theory of color vision, 512
Hess, R. F., 631–634
High-density lipoprotein (HDL), 1535,
1535f
High-resolution spatial vision, 1142–
1143
Higher-order inferior parietal lobule
(HM-IPL) region, 783, 784f
Higher-order visual functions, 1424
Holistic coding (face perception),
711–713, 716–717
evidence that it facilitates
identification of faces, 712–713
Holmes, G., 1097
Holoptic vision, 1153–1154
Homeostatic plasticity. *See* Feedback
(homeostatic) plasticity
Homing, 1181–1184, 1182f–1184f
Homosynaptic plasticity. *See* Ocular
dominance (OD) plasticity:
homosynaptic long-term
depression (LTD) and
Honeybees. *See also* Navigation in
honeybees
cognitive maps in, 1179
arguments against, 1186–1187
structure of, 1187–1188
a common frame of spatial memory
in navigation and
communication, 1185–1186
flight tracks back to hive, 1184,
1185f
flying directly home from release
site, 1187
flying into regions they have not
explored, 1187
social communication, 1185
Horizon, used to sense and control
attitude, 1222, 1223, 1224f,
1225
Horizontal cell (HC) feedback
transmission
functional consequences of positive
and negative feedback, 71, 72,
72f
mechanism of negative feedback
synapse, 66–67
ephaptic hypothesis, 68–69
GABA hypothesis, 67–68